AF773784

SNAKES of the WORLD

A Catalogue of Living and Extinct Species

SNAKES of the WORLD

A Catalogue of Living and Extinct Species

Van Wallach
Kenneth L. Williams
Jeff Boundy

CRC Press
Taylor & Francis Group
Boca Raton London New York

CRC Press is an imprint of the
Taylor & Francis Group, an **informa** business

CRC Press
Taylor & Francis Group
6000 Broken Sound Parkway NW, Suite 300
Boca Raton, FL 33487-2742

First issued in paperback 2017

© 2014 by Taylor & Francis Group, LLC
CRC Press is an imprint of Taylor & Francis Group, an Informa business

No claim to original U.S. Government works

ISBN-13: 978-1-4822-0847-4 (hbk)
ISBN-13: 978-1-138-03400-6 (pbk)

This book contains information obtained from authentic and highly regarded sources. Reasonable efforts have been made to publish reliable data and information, but the author and publisher cannot assume responsibility for the validity of all materials or the consequences of their use. The authors and publishers have attempted to trace the copyright holders of all material reproduced in this publication and apologize to copyright holders if permission to publish in this form has not been obtained. If any copyright material has not been acknowledged please write and let us know so we may rectify in any future reprint.

Except as permitted under U.S. Copyright Law, no part of this book may be reprinted, reproduced, transmitted, or utilized in any form by any electronic, mechanical, or other means, now known or hereafter invented, including photocopying, microfilming, and recording, or in any information storage or retrieval system, without written permission from the publishers.

For permission to photocopy or use material electronically from this work, please access www.copyright.com (http://www.copyright.com/) or contact the Copyright Clearance Center, Inc. (CCC), 222 Rosewood Drive, Danvers, MA 01923, 978-750-8400. CCC is a not-for-profit organization that provides licenses and registration for a variety of users. For organizations that have been granted a photocopy license by the CCC, a separate system of payment has been arranged.

Trademark Notice: Product or corporate names may be trademarks or registered trademarks, and are used only for identification and explanation without intent to infringe.

Library of Congress Cataloging-in-Publication Data

Williams, Kenneth L., 1934-
Snakes of the world : a catalogue of living and extinct species / Van Wallach, Kenneth L. Williams, Jeff Boundy.
pages cm
Previous edition by Kenneth L. Williams and V. Wallach.
Includes bibliographical references and index.
ISBN 978-1-4822-0847-4 (hardcover : alk. paper)
1. Snakes. I. Wallach, Van. II. Boundy, Jeff. III. Title.

QL666.O6W455 2014
597.96--dc23
2013047012

Visit the Taylor & Francis Web site at
http://www.taylorandfrancis.com

and the CRC Press Web site at
http://www.crcpress.com

We dedicate this book with warm affection to

Douglas Athon Rossman

a.k.a. "Dag"

Douglas Rossman, Photo Courtesy of the LSU Museum of Natural Science

Emeritus Curator and Professor of Herpetology

Louisiana State University Museum of Natural Science

Baton Rouge, LA (1963–1998)

Doug directed all three of us in our graduate programs
for either Masters (VW) or Doctoral (JB, KLW) degrees,
was an excellent teacher, thoughtful advisor, and supportive friend,
and made each of us better herpetologists
with his encouragement, insightful advice, and constructive criticism
exemplified daily by his dedication to scientific research.

"Mon ami, there are only two kinds of snake:
the good snake (dead) and the bad snake (trop vive!)"
... old Cajun saying from the Bayou

Contents

Introduction

Few works have attempted the challenge of cataloging all of the known species of snakes in the world, the most prominent being the British Museum catalogues of Boulenger (1893a, 1894a, 1896a) and those of F. Werner (1917b, 1921a–b, 1922a, 1923b, 1924c, 1929a) for Recent taxa and Rage (1984b) and Holman (1998a, 2000a) for fossil taxa. Until now no single work has treated all living and extinct snakes. The present work covers all living and fossil snakes described between 1758 and 2012, comprising 3783 species (3509 extant and 274 extinct species) allocated to 651 genera (539 extant and 112 extinct genera). Also included are the names of 54 genera and 302 species that are considered *incertae sedis, nomina dubia, nomina nuda, nomina rejicienda,* and *nomina inedita*, resulting in a grand total of 12,500 primary snake names. A summary of the main historical references follows.

Worldwide works. The starting point of modern binomial nomenclature, Linnaeus' 10th edition (1758), had four genera (*Anguis, Boa, Coluber,* and *Crotalus*) and 101 species of snakes. Laurenti (1768) increased the number of genera to 14 and the species to 104. Merrem (1820) recognized 330 species, placed in 22 genera. Duméril & Bibron (1844) recognized 25 genera and 64 species of scolecophidians and Duméril, Bibron & Duméril (1854) described 139 genera and 484 alethinophidian species, totaling 548 species in 164 genera (see Table I.1). The British Museum catalogues of Gray (1849) and Günther (1858) contained some 544 species and formed the foundation for Boulenger's later, more comprehensive, work. Snakes were placed in 62 genera by Günther (1888). Jan (1863b) listed 789 snake species allocated to 232 genera. Hoffmann (1890) provided descriptions of all snake genera, listing the number of species in each genus and its geographical region. Boulenger's British Museum (Natural History) catalogues (1893–1896) formed the first comprehensive summary of worldwide snakes, and continues to be a critical reference for any systematic work today. Boulenger recognized 1639 species (and 7335 total names) that he placed in 296 genera. Werner updated Boulenger's catalogues in a series of six papers (1917–1929) that resulted in 338 recognized genera and 2242 species. McDiarmid et al. (1999) covered the Scolecophidia, Anilioidea, Henophidia or Booidea, Acrochordoidea, and Viperoidea with subsequent volumes to review the Colubroidea. Table I.1 summarizes the biodiversity of snakes as seen through herpetological publications.

The checklists of Welch (1982, 1983, 1988, 1994a–b), Wilkey (2002a–b), and Brogard (2005) are of little value for any taxonomic work. Along this line, several Australian authors have a habit of universally recognizing every proposed name (Wells & Wellington) or cannibalizing molecular studies and phylogenetic analyses to attach names to every single clade (Hoser). Such works are not only superfluous but also detrimental and a hindrance to taxonomy and nomenclature and they violate clause 4 of the Code of Ethics (ICZN, 1999). They include Wells & Wellington (1984, 1985); Hoser (2000b-c, 2003f, 2004, 2009b-e, 2012b–am); and Wells (2007). The *Australasian Journal of Herpetology*, of which Hoser is the editor, reviewer, publisher, and sole author, is not considered a valid publication under the spirit of the Code and therefore all proposed names are *nomina illegitima* (see Wallach et al., 2009 and Kaiser et al., 2013).

Taxonomic groups. Hahn (1980) compiled the first modern checklist of the Scolecophidia, with synonymies and ranges, followed by McDiarmid et al. (1999). Roux-Estève (1974) treated all African Typhlopidae. Venomous snakes of the world were covered by Golay et al. (1993), David & Ineich (1999), and McDiarmid et al. (1999).

Regional works. North American snakes have been treated by Ernst & Ernst (2003) with an official list of species by Crother (2008). H. Smith & Taylor (1945) produced the only comprehensive work on Mexico, with later summaries by Smith & Smith (1993) and Flores-Villela (1993), and Liner & Casas-Andreu (2008) providing a checklist of species and subspecies. Neotropical snakes of Central and South America were catalogued by Peters & Orejas-Miranda (1970) with Central American species updated by Köhler (2008). Snakes of the West Indies were enumerated by Schwartz & Henderson (1991). Tipton (2005) listed all the genera and species of the New World (North America, Mesoamerica, Caribbean, and South America). Snakes of Europe were listed by Mertens & Wermuth (1960) and Gasc et al. (1997). African snakes have been covered by the following: South Africa (Broadley, 1990), East Africa (Spawls et al., 2002), and West Africa (Chippaux, 2006). The treatises of Bourret (1936), M.A. Smith (1943), and Das (2010) covered Southeast Asia. Mengden (1983) and Cogger (2000) treated the Australian ophiofauna.

Checklists. Recent authors who have provided lists of names of the worldwide serpent fauna include Sokolov (1988), Frank & Ramus (1995), Mattison (1999), and Delhay (2009).

Photo books. There are some books that do not contain any taxonomic or geographical data but are worthy of citation as photo-essay books. These include Seba (1734–1735), Kundert (1974, 1984), Aramata (1990), Lacarrière & Ineich (1992), Lamar (1997b), Moncuit & Daoues (2002), Mocafico (2007), and Laita (2013).

Methods

Due to the length of this catalogue, verbiage has been kept to a minimum and a staccato style has been utilized. Only the most germane topics are briefly mentioned in Remarks but citations to relevant references are provided so that additional information can be traced when desired. The text is organized alphabetically throughout, both by genus and species. Valid genera are listed with author date, and family allocation parenthetically. If the generic name appeared in a work other than the author's, that work is listed on a separate line (preceded by *in*). After each

TABLE I.1

Worldwide Snake Diversity Tabulations from 1758 to 2012

Author	Year	Genera	Species
Linnaeus	1758	4	101
Laurenti	1768	14	104
Merrem	1820	22	330
Duméril, Bibron & Duméril	1844–1854	164	548
Jan	1863	232	789
Hoffmann	1890	402	1760
Boulenger	1893–1896	296	1639
F. Werner	1917–1929	338	2242
Ditmars	1933	333	1653
A.S. Romer (living)	1956	377	—
A.S. Romer (extinct)	1956	26	—
Halliday & Adler	1986	417	2389
Sokolov	1988	433	2620
Rage in Bauchot	1994	428	2600
Frank & Ramus	1995	440	2542
Mattison	1999	466	2969
Delhay	2009	479	2927
McDiarmid in McDiarmid et al.	2012	506	3217
Present work (living)	2012	539	3509
Present work (extinct)	2012	112	274
Present work	Valid	651	3783
Present work	Taxa inquirenda	54	302

genus, the type species and a generalized distribution are presented, with or without sources and taxonomic remarks. Extant and extinct species are numbered and listed in alphabetical order with the following information: original citation, synonyms and subspecies, type or types, type locality, and geographical and geological range. Relevant sources for each taxon (monotypic genus or species) are also provided, along with comments when necessary. When supplemental or revisionary information is presented in later publications we present it and list the original data under Remarks.

Species and genera. Recognition of valid species and genera usually follows the latest revisionary works. In cases where polytypic genera have not been reviewed, the latest literature is utilized. When there is controversy over generic allocation or specific status of a taxon, we provide opposing views under Remarks. For the orthography of a taxon's name, we employ the original spelling unless it was (1) a typographical or accidental error (so noted as *nomen incorrigendum*), (2) incorrectly formed according to the Rules, or (3) a name whose terminus does not agree with gender of the current genus. In the latter two cases the corrected form is used. The original names of valid species include subgeneric names in brackets (subgenera are not included among synonyms). Allocation of genera and species to families follows the classification in Table I.2, with †Ophidia being used for extinct families exhibiting a pre-Serpentes grade of evolution.

Synonyms. We have attempted to list as many nominal taxa as possible. We list under Synonyms mainly primary synonyms, emendations (whether justified or unjustified), and incorrect and corrected original spellings of genera, species, and subspecies (not different combinations, typos, or *lapsus calami*). We do not differentiate between those forms considered junior synonyms and those currently considered valid subspecies. New combinations are cited only when the termination of the specific epithet is different than the primary names, and they are indicated by a dash between name and author. Incorrect secondary spellings are listed in the text but not in the Index (since they are unavailable names). Citations for all these names are included in the Literature Cited. Latin notations following names include these terms: *incertae sedis* (uncertain taxonomic status), *lapsus calami* (slip of the pen), *nomen incorrectum* (incorrect secondary spelling), *nomen corrigendum* (corrected original name), *nomen dubium* (dubious or questionable name), *nomen emendatum* (emendation), *nomen illegitimum* (illegal or invalid name according to the Rules, usually not a binomial or trinomial), *nomen incorrigendum* (incorrect original name), *nomen ineditum* (unpublished name according to the Rules), *nomen nudum* (an unavailable or invalid name, usually lacking a diagnosis), *nomen oblitum* (forgotten name), *nomen praeoccupatum* (occupied name), *nomen protectum* (protected name, applied to a junior synonym in preference to an unused senior synonym or *nomen oblitum*), *nomen rejiciendum* (rejected name by ICZN), *nomen substitutum* (substitute or replacement name), *nomen suppressum* (suppressed name for matters of priority), and *partim* (in part, when specimens belong to more than one taxon). A name may include several of the above categories, in which case only the most relevant is cited. A *nomen emendatum* that is also a *nomen rejiciendum* will be cited as the latter.

For every valid species an abbreviated reference citation is provided along with pagination, figures, and plates referring to the new species. If a figure of a paratype is given (in the absence of a holotype figure) it is so indicated. The original combination of the name is given parenthetically when it is different from that currently employed. If the species was described in more than one publication, the other descriptions are listed in Remarks (listed as supplementary original description when its content is expanded over the initial description). Complete citations of all descriptions are in Literature Cited.

Types. We have tried to determine the location of all primary types by making liberal use of all published type lists, correspondence with various museums regarding types, and visiting certain U.S. museums to examine specimens personally (mainly ANSP, CAS, FMNH, MCZ, SDNHM, USNM). The location of many type specimens remains unknown; types were generally not designated in earlier works and many types have subsequently been lost or destroyed. If the deposition of a type is presumed but not confirmed, the museum acronym is preceded by a "?"

Whenever possible we have listed the status of the type (holotype, lectotype, neotype, or syntype), its location and registry number (including its original museum and number if now different), its total length and gender, and parenthetically the collector and date of collection. In cases where the current catalogue number or museum of deposition is different from that presented in the original description or subsequent publication, the original data are presented in parentheses following the current museum and number. If a third museum is involved, then the original and subsequent numbers are listed parenthetically. Museum acronyms follow, in most cases, Leviton et al. (1985, 1988). We have provided additional acronyms for institutions, private collections, and field series numbers. For simplicity in

TABLE I.2

Number of Living and Extinct Genera and Species by Family

	Living		Extinct	
Family	Genera	Species	Genera	Species
†OPHIDIA				
†Pachyophiidae Nopcsa, 1923	0	0	5	6
†Lapparentophiidae Hoffstetter, 1960	0	0	2	2
†Najashidae Apesteguía & Zaher, 2006	0	0	1	1
SERPENTES *incertae sedis*	0	0	1	1
†Madtsoiidae Hoffstetter, 1961	0	0	10	19
†Dinilysiidae Romer, 1956	0	0	1	1
SCOLECOPHIDIA				
Anomalepididae E.H. Taylor, 1939	4	18	0	0
Gerrhopilidae Vidal et al., 2010	1	15	0	0
Leptotyphlopidae Stejneger, 1891	12	112	0	0
Typhlopidae Gray, 1825	11	260	0	0
Xenotyphlopidae Vidal et al., 2010	1	1	0	0
ALETHINOPHIDIA *incertae sedis*	0	0	5	9
ANILIOIDEA *incertae sedis*	0	0	1	1
Aniliidae Stejneger, 1907	1	1	4	6
UROPELTOIDEA				
Anomochilidae Cundall et al., 1994	1	3	0	0
Cylindrophiidae Fitzinger, 1843	1	10	0	0
Uropeltidae J. Müller, 1831	9	51	0	0
HENOPHIDIA				
PYTHONOIDEA				
Xenopeltidae Bonaparte, 1845	1	2	0	0
Loxocemidae Cope, 1861	1	1	0	0
Pythonidae Fitzinger, 1826	8	42	3	13
BOOIDEA *incertae sedis*	0	0	3	3
Calabariidae Underwood, 1976	1	1	0	0
Boidae Gray, 1825	7	37	18	29
Erycidae Bonaparte, 1831	3	17	11	32
Ungaliophiidae McDowell, 1987	2	3	1	1
BOLYERIOIDEA				
Tropidophiidae Brongersma, 1951	2	34	7	16
Bolyeriidae Hoffstetter, 1946	2	2	0	0
Xenophidiidae Wallach & Günther, 1998	1	2	0	0
CAENOPHIDIA *incertae sedis*	0	0	1	1
ACROCHORDOIDEA				
†Palaeophiidae Lydekker, 1888	0	0	3	24
†Nigerophiidae Rage, 1975b	0	0	5	5
Acrochordidae Bonaparte, 1831	1	3	0	1
VIPEROIDEA				
†Anomalophiidae Auffenberg, 1959	0	0	1	1
†Russellophiidae Rage, 1975a	0	0	2	3
Xenodermatidae Gary, 1849	6	19	0	0
Pareatidae Hoffmann, 1890	3	16	0	0
Viperidae Gray, 1825	48	329	0	13
HOMALOPSOIDEA				
Homalopsidae Jan, 1863	17	55	0	0
ELAPOIDEA *incertae sedis*	2	5	0	0
Elapidae F. Boie, 1827	71	354	1	7
Atractaspididae A.C.L.G. Günther, 1858	13	76	0	0
Lamprophiidae Fitzinger, 1843	13	71	0	0
Prosymnidae Kelly et al., 2009	1	16	0	0

(continued)

TABLE I.2 (CONTINUED)

Number of Living and Extinct Genera and Species by Family

Family	Living		Extinct	
	Genera	Species	Genera	Species
Psammophiidae Bonaparte, 1845	10	61	0	0
Pseudaspididae Cope, 1893	2	2	0	0
Pseudoxyrhophiidae Dowling, 1975	20	83	0	0
COLUBROIDEA *incertae sedis*	4	9	2	2
Pseudoxenodontidae McDowell, 1987	2	11	0	0
Carphophiidae Zaher et al., 2009	4	10	3	7
Dipsadidae Bonaparte, 1838	37	388	1	1
Xenodontidae Bonaparte, 1845	60	325	0	0
Natricidae Boettger, 1883	38	228	4	21
Sibynophiidae Dunn, 1928	2	11	0	0
Calamariidae Bonaparte, 1838	8	88	0	0
Colubridae Oppel, 1811	108	739	16	48

presenting type numbers, we have eliminated all zeros preceding the number (i.e., 00289 = 289), all "R" designations for reptiles (including RENA at RMNH), and for BMNH numbers from the 1800s, we add the prefix "18" to complete the year. When two or more types are catalogued under one number, we refer to them informally with small lower case letters (i.e., 1460a–b) but when a museum has catalogued more than one specimen under a single number we use upper case letters (i.e., 1460A–B). If a type specimen is unknown, unlocated, missing, presumed lost, or lost, we cite just that it is lost; if destroyed it is so noted. If a type is not listed in a type list, we presume it to be lost. A number of museums have been destroyed by fire (CA, IB, MBL, MLS *fide* Almaça & Neves, 1987, Kumar, 2010, and Franco, 2012) or bombing during World War II (BSM, FMSM, ISM, KMJ, LM, LNHM, MSNM, MTD, NMB, NMSW, NMW, SMNS, USTL, ZIB, ZMH, ZMUH, and ZSM *fide* Klauber, 1952, W. Leviton, 1959b, Hellmich *in* Hoge, 1964a, Hahn, 1980, Hoogmoed & Gruber, 1983, Böhme & Bischoff, 1984, W. Ladiges *in* L.D. Wilson, 1990d, M. Podestà *in* Raxworthy & Nussbaum, 1994, Leonardi et al., 1995, Scali, 1995, Schlüter & Hallermann, 1997, Hallermann, 1998, McDiarmid et al., 1999, Fritz, 2002, Cadle, 2007, Franzen & Glaw, 2007, and Köhler & Güsten, 2007) and their types are so noted without further reference. Many types of earlier authors are unknown and it should be pointed out that the following people sent all or part of their collections to the MNHN (C. Peale, 1794–1798; Palisot de Beauvois, 1797–1798; R. Harlan, 1831–1839; C.S. Rafinesque, 1835–1839; and A. Dugès, 1853–1856). The Museum Drottningholmense, from which many Linnaean types originated, is abbreviated as Mus. Drottn.

When a lectotype or neotype has been designated, we cite the authority in question. According to Art. 74.4 of the Code, we designate as lectotypes the descriptions and/or illustrations in older works such as M. Catesby, A. Seba, and P. Russell. Early authors, such as Cope and Hallowell, when describing a new species from a syntypic series, usually only presented data on one individual and did not mention the number of individuals upon which the description was based. It is preferable, therefore, to select the specimen described as the lectotype when it can be determined that more than one specimen was available (Art. 74B).

We have converted all non-metric measurements to the nearest mm when reporting the length of types. Whenever possible we used the conversion rate for that country. However, since the "line," "inch," and "foot" have ranged from 2.0–2.9 mm, 21.3–30.0 mm, and 250–390 mm, respectively, in different countries we employed the standard metric conversion of 2.54 mm, 25.4 mm, and 305 mm, respectively, when unknown. When more than one measurement of length for a type specimen exists in the literature, we present the entire range of values. Usually subsequent measurements of the same individual reveal a shrinkage due to the preservative. If the tail is incomplete, we give the body length and refer to it as SVL or, if only the tip is missing, we list the entire specimen with a "+" after the length. We have included gender and measurements (to the nearest mm) of all available types for which those data have not been published. Since authors in the older literature normally listed measurements for the largest specimen, we cite the reported value as that of longest syntype.

For the collector of the type specimen(s), we do not differentiate between the actual collector, a purchase from native collectors, or specimens in someone's personal collection (but we use the term "coll." after a name to refer to a collection, not to be confused with "collected by"). We merely list the person or collection responsible for procuring the specimens and when a donor separate from a collector is involved, "don." is used to signify that. We refer to all local or indigenous collectors as "natives." When the specific date of collection is unknown, but more inclusive dates are known, such as the dates of a voyage or expedition or the duration of residence of the collector in a certain country/locality, we list the inclusive dates, with a specific Expedition and Voyage abbreviated as Exped. and Voy., respectively. Under each genus the type species is listed as the currently recognized name. See Williams & Wallach (1989) for more complete data on genera type species and designations.

Type locality. The type locality is given verbatim and enclosed in quotes. Whenever it is not presented in English, we translate it in brackets with an "=" sign. In order to maximize the informational content of the type locality, we have included any and all data presented in the original description, including the title. Occasionally more information is presented on the type locality in the introduction or some other part of the work than in the designated "type locality." Following the quoted type locality, in many cases, will be found in brackets the name converted into present day terminology and/or translated into English, more detailed geographical data, the latitude and longitude, and the elevation in meters of the type locality. When a type locality is in the vicinity of a municipality, the coordinates and elevation are included in parentheses after the town. All information within brackets has been determined from other sources, either published literature or personal resources. Elevations and geographic coordinates were determined from several sources: Operational Navigation Charts (ONC) and Tactical Pilotage Charts (TPC) published by the Defense Mapping Agency, travel books and road maps for various countries, the Falling Rain Genomics website Global Gazetteer (www.fallingrain.com), the Atlas of Living Australia website (www.biocache.ala.org.au), Google Earth, and such publications as Crawford-Cabral & Mesquitela (1989) for Angola. It should be noted that these amplified type localities presented in brackets by the authors are intended as aids and not as type locality restrictions. Type locality restrictions are not enclosed in quotes and are given in English (translated, if necessary). Any and all restrictions of type localities, although bearing no official status, are listed with the restricting authority. Valid restrictions are listed under the Type locality field but invalid restrictions are in the Remarks. The geographic coordinates of all possible type localities have been determined. This is obviously impossible for many older names with generalized localities. Whenever a specific type locality was located in a gazetteer, the latitude and longitude are presented in brackets following the name of the country. When the coordinates are preceded by "ca.," it means they are derived from measurements of maps and only approximate, or else if the type locality is a general region the coordinates of the center of that region are listed. If the type locality could not be located but a locality near it was found, the latter's coordinates are given immediately after that particular location parenthetically. When the type locality is near, around, close to, or in the environs of a specific location, we refer to it as the vicinity of that location. When an island is the type locality, the range of latitude and longitude coordinates may be given, preceded by "bet." (= between). When the elevation of the type locality is known, it follows the country name and the latitude and longitude (if known). Whenever additional information on the type locality is presented elsewhere (such as corrections, more precise details, etc.), we list it as emended data, and may paraphrase or translate it in order to fit it into our format of most specific to most general geographic locality, elevation, and coordinates. Corrections (to type localities that are in error) and restrictions of type localities are so noted with the authority and source. For certain older type localities the political boundaries and names have changed and we attempt to provide the modern equivalent. For example, "Bengal" in the early 1800s consisted of the area currently known as Bangladesh and NE India (including Assam, Manipur, Meghalaya, Mizoram, Nagaland, and West Bengal provinces). With fossil species, we consider the geological age and stratigraphic horizon to be part of the type locality, and this information follows the geographical locality of the type specimen, regardless of the order of presentation in the original description. The type locality sequence for an extinct species is thusly: geographical locality; geological period or epoch (geological age, formation or local fauna).

Geological epochs (Cretaceous to Pleistocene) are listed, followed parenthetically with either the North American Land Mammal Ages (NALMA) for North America or the European Land Mammal Ages (ELMA) for the rest of the world and the Neogene Mammal (MN) and Paleogene Mammal (MP) biozones following Gradstein et al., 2004. Fossil times, in millions of years before the present (mya), are listed according to Gradstein et al., 2004, and the International Union of Geological Sciences for 2012. In reference to epochs, we utilize the adjectives "lower, middle and upper," which define formations, in contrast to "early, middle and late," which define ages.

Lists of type specimen are available for the following museums: ADM (E.H. Taylor, 1933; H. M. Smith & Necker, 1943), AMS (Goldmann et al., 1969; Cogger, 1979; Shea & Sadler, 1999), ANSP (Roze, 1958; Malnate, 1971, 1989), BNHS (P. Das et al., 1999b), BYU (W.W. Tanner, 1970), CAS (Slevin & Leviton, 1956), CAS-SU (Leviton, 1953; Leviton & Banta, 1956), CIB (Guo et al., 2000), CM (McCoy & Richmond, 1966; McCoy & Censky, 1982), CMS (Kandamby, 1997), EHT-HMS (E.H. Taylor, 1944), FMNH (Marx, 1958, 1976), FMNH-PR (Bruner, 1991), HLD (J. Köhler & Güsten, 2007), IES (Chamizo-Lara et al., 2000), IMC (Sclater, 1891), IRNSB (Lang, 1990), IZUC (Cekalovic & Artigas, 1981), KU (Duellman & Berg, 1962), KU-VP (Schultze et al., 1985), LSUMZ (Rossman & Good, 1993), MACN (Amalia-Varela, 1999), MB (Bocage, 1896), MCZ (T. Barbour & Loveridge, 1929, 1946; Loveridge, 1961; Ross & Crumley, MS), MDUG (E.H. Taylor, 1933; H.M. Smith & Necker, 1943; Flores-Villela et al., 2010), MHNG (Mahnert, 1976; Schätti & Perret, 1997), MHNM (Olazarri et al., 1970), MHNN (Schätti, 1986), MMC (Clary & Martelli, 1995), MNCN (García-Díez & González-Fernández, 2013), MNHN (Guibé & Roux-Estève, 1972; Roux-Estève, 1979b, 1983), MNHNC (Donoso-Barros & Cardenas, 1965; Ortiz & Nuñez, 1986), MNKNU (Vedmederya et al., 2009), MNRJ (Miranda-Ribeiro, 1955; Soares & Fernandes, 2001), MSNG (Capoccacia, 1961), MSNM (Scali, 1995; Leonardi et al., 1995), MTKD (Schüz, 1929; Obst, 1977; Fritz, 2002), MVZ (Crippen, 1962; Rodriguez-Robles et al., 2003), MZB (Iskandar & Mumpuni, 2003, 2005), MZUT (Elter, 1981), NHRM (Andersson, 1899; Deraniyagala, 1961a), NMBA (Kramer, 1978), NMP (Milkovsky et al., 2011), NMSW (Lampe, 1901, 1911), NMSZ (Herman et al., 1990), NMV (Coventry, 1970), NMW (Grillitsch et al., 1996; Tiedemann & Haupl, 1980; Tiedemann et al., 1994; Tiedemann & Grillitsch, 1999), NSMT (Ota, 1997; Ota & Endo, 1999), NTM (Horner, 1999), OMNH (Hatooka, 1996), QM (Mack & Gunn, 1953; Covacevich, 1971; Ingram & Covacevich, 1981; Covacevich & Couper, 1994), RMNH (Hoogmoed & Gruber, 1983), SAMA (Houston, 1976), SDSNH (Sloan, 1965; Pregill & Berrian, 1984), SMF (Rüppell, 1845; Boettger, 1898; Mertens, 1922, 1952, 1967), SMK (Das & Leh, 2005), SMNS (Schlüter & Hallermann, 1997), SMW (J. Köhler & Güsten, 2007), SMWM (Irish, 1985), THNHM (Chan-ard & Makchai, 2011), UF (Christman in Gilbert, 1974), UIMNH (H.M. Smith et al., 1964; Phillips, 2003 (on internet)), UMMZ (J.A. Peters, 1952; Kluge, 1984), USNM (Cochran, 1961, J.W. Wright et al., 2008), WAM (Anonymous, 1961–1969), ZDKU (Vedmederja et al., 2009), ZFMK (Böhme, 1974, 2010; Böhme & Bischoff, 1984), ZMA (Daan & Hillenius, 1966; van

Tuijl, 1995), ZMB (Bauer, 1998; Das, 1999; Bauer et al., 2002; Hallermann, 2007), ZMH (Meerwarth, 1901; Hallermann, 1998, 2006), ZMUH (L. Müller, 1941), ZMUO (Pethon, 1969), ZMUU (Lönnberg, 1896; Dely, 1961a; Wallen, 1992), ZRC (Das, 2001), ZSI (Das et al., 1998; Das & Gayen, 2004), ZSM (Franzen & Glaw, 2007).

The following sources contain additional information on types and type localities: Andreone & Gavetti, 2007 (M.G. Peracca types), Anonymous, 1958 (southern and East African types), Bauer et al., 1995 (W.C.H. Peters types), Crumley, 1990 (annotated account of types and type locality lists), Dowling et al., 1970 (annotated list of type specimen lists), Dowling & Gilboa, 1974 (annotated list of type specimen lists), Dunn & Stuart, 1951 (critique of type locality restrictions), V. FitzSimons, 1937 (A. Smith types), V. FitzSimons, 1958 (South African types), Flores-Villela et al., 2010 (A. Dugès reptile taxa), Hoogmoed & Gruber, 1983 (Spix & Wagler types), Krecsák, 2007a (A.F.T. Reuss vipers), Liner, 1996a (Nuevo León, México types), Liner, 2000 (H. M. Smith types), J.A. Peters, 1955 (Ecuadorian types), Rodriguez-Schettino, 2000 (Cuban types), Savage, 1974 (Costa Rican types), H.M. Smith & Necker, 1943 (A. Dugès types), H.M. Smith & Taylor, 1950 (Mexican type locality restrictions), E.H. Taylor, 1933 (Alfredo Dugès Museum), E.H. Taylor, 1944a (E.H. Taylor and H.M. Smith types), and Toriba, 1993 (M. Maki types). In addition to the type lists we have made liberal use of numerous biographies, expedition and voyage accounts, online resources (i.e., Wikipedia, Google), and invaluable works such as Adler (1989, 2007, 2012), Bauer (2004), Beolens et al. (2011), Bokermann (1957), C.W. Myers (2000), Papavero (1971), Pietsch & Anderson (1997), Savage (1974), and Vanzolini (1977–1978) for determining localities and specific collection dates (or ranges of dates in the case of voyages, expeditions, or habitations of collectors).

Distribution. We realize that a deficiency of this work is with the geographic ranges of individual species. We have tried to be as complete as possible but time constraints have prevented us from being able to search the entire literature and examine all the material in collections. Being based almost entirely upon literature records, there are undoubtedly errors of omission and commission and we would greatly appreciate being informed of corrections or additions to the listed ranges for a future edition. We have provided primarily a list of all countries from which each species is known, and secondarily a list of all known major administrative regions from which each species has been recorded. These regions include states, provinces, regions, districts, territories, counties, zones, municipalities, prefectures, governates, protectorates, divisions, parishes, and wilayas, in alphabetical order in parenthesis following country. Data has been compiled mostly from the published literature, but some has come from museum specimens examined and also unpublished information provided by colleagues or ourselves.

Consistency of the names of the administrative regions presented a problem for many countries that have changed their names recently. We previously utilized such resources as Foreign Area Studies handbooks of the American University, *Merriam-Webster's New Geographical Dictionary*, U.S. Defense Mapping Agency gazetteers, and Department of Defense maps but recently switched over to Google Earth for current names. Recent name changes employed include Myanmar (Burma), Sulawesi (Celebes), Kalimantan (Borneo), Papua and West Papua (W New Guinea), and Burkina Faso (Upper Volta). Larger geographic regions are denoted as follows: North America = Canada, USA, and Mexico; upper Central America = Guatemala/Belize to Honduras; lower Central America = Nicaragua to Panama; Mesoamerica = Mexico plus Central America; Greater Antilles = Cuba Hispaniola, Jamaica, and Puerto Rico; Lesser Antilles = Windward Islands (Virgin Is. to Dominica) and Leeward Islands (Martinique to Grenada); Amazonia = Amazonian Basin of South America; Guianas = Guyana, Suriname and French Guiana; Latin America = Mexico, Central America, and South America; Eurasia = Europe and Asia; Middle East = Syria, Iraq, Lebanon, Israel, and Jordan; Arabia = Saudi Arabian peninsula; Asia Minor = Turkey, Caucasus, and former SW Russian states; Southwest Asia = Iran to Pakistan; Indochina = Cambodia, Laos, and Vietnam; Southeast Asia = Myanmar, Thailand, and Indochina, whereas Southeastern Asia includes not only Southeast Asia but adjacent regions such as S China, India, West Malaysia, and W Indonesia; East Asia = China, Korea, and Japan; Malay Peninsula = S Myanmar, S Thailand, and West Malaysia; East Indies = Indonesia and Philippines; Borneo = Kalimantan, East Malaysia, Sabah, and Brunei; Greater Sundas = Java, Kalimantan, Sumatra, and Sulawesi; Lesser Sundas = S Indonesian islands; New Guinea = Irian Jaya and Papua New Guinea; Austro-Papua = Australia and New Guinea; Indo-Australia = S or SE Asia, East Indies, New Guinea, and Australia; Sino-Australia = E Asia, East Indies, New Guinea, and Australia; and Australasia = Australia, New Guinea, and Pacific islands. Lastly, F.A.T.A. = Federally Administered Tribal Area in Pakistan.

Questionable records or probable occurrences are discussed where appropriate with references, either under remarks or with a "?" preceding the name of the country or state. Records that are obviously erroneous (based on misidentifications or faulty data) are ignored, or so stated.

For the distribution of a genus we present a general range by geographical area, country, or continent, whereas for the range of a species we list the separate countries from which it is recorded. For marine snakes, we list the bodies of water from which it is known under distribution, but under range we give the countries from whose shores or offshore regions the species has been found.

In listing the geographical distribution, if a species occurs in at least half of the country the country is mentioned without directional adjectives. When a species is known from only a small area of a country, abbreviated prefixes refer to the generalized area (N = northern, E = eastern, S = southern, W = western, NE = northeastern, NW = northwestern, SE = southeastern, SW = southwestern, and cen. = central). Therefore, for example, West Virginia is West Virginia but W Virginia is western Virginia. Additionally, the abbreviation ext. (= extreme) is used when the species' range occupies only a marginal portion of the country in question. In order to save space, only shortened versions of political units are used; such terms as Autonomous Region (in Guangxi Autonomous Region) and Territorio Federal (in Territorio Federal Amazonas) are omitted with the listings as China (Guangxi) and Venezuela (Amazonas). The terms Southern or Western in parentheses refer to provinces such as in Zambia, not to directions within the country, which would precede the country name, i.e., SW Zambia (Western, Southern). Countries are listed, in general, from west to east and north to south. Occasionally, in order to keep continuity from country to country, a different direction is followed.

Following the geographic distribution is the elevational range in meters. We present the minimum and maximum values known to us, mainly from literature reports and Google Earth. In many cases the given range does not reflect the altitudinal range of the species itself, merely the reported ranges that we have found. Ranges are rounded off to the nearest five meters.

NSL (near sea level) refers to values from 0–7 m and BSL (below sea level) is also indicated where appropriate. When one value is far outside the typical range we enclose it parenthetically (before the range when lower and after the range when higher) to indicate that it is either an aberrant figure or possibly was reported in error.

Islands are listed in alphabetical order after the states or provinces. The abbreviation "Is." is omitted from all major islands (i.e., Penang, Java, Sulawesi) and from all islands in Indonesia and Philippines. We do not differentiate between an archipelago, island group, or an island.

Shortened versions of some country names are employed whenever possible such as Brunei Darussalam = Brunei, Democratic People's Republic of Korea = North Korea, Democratic Republic of the Congo (Congo-Brazzaville) = Congo, Lao Democratic People's Republic = Laos, People's Republic of China = China, Republic of Korea = South Korea, Republic of Yemen = Yemen, and Socialist Republic of Vietnam = Vietnam. Sometimes not only the name of the country wherein the type locality is found has been changed but also the state or province. This is a result of changing boundaries, and we have tried to list all locations as they are presently known. The Appendix provides the major references used for each country. Entries under each category are placed in alphabetical order. When an included article covers only a particular region of a country, that region is included in parenthesis after author and date. The disputed territories of Gaza Strip, Golan Heights, and West Bank are included under Israel, and Hong Kong and Tibet (= Xizang) are included under China (but not Taiwan).

When an original description is published more than once, the subsequent description is listed under Remarks as a reprinting of the original. When the original description is brief and a more complete description is then published, we refer to the latter as a supplemental original description.

Living genera and species that are known from fossils have their fossil records listed after the distributions.

Literature cited. We have personally examined originals (or copies thereof) of all generic and species descriptions in addition to all cited references unless noted as [not seen] after the citation. Reference citations for journals follow a standard format thusly: author, date of publication, title of article, journal title, city of publication, (date on volume or title page if different from date of actual publication; in controversial cases the actual publication date is cited in brackets after the citation), (series number or abbreviated name), volume (plus series if represented by a letter), (number, part or section), total pagination, number of figures in text and number of plates. Figures and pages that are numbered are so listed (i.e., figs. 1-3) but unnumbered figures and pages are cited differently (i.e., 3 figs. or 2 pp.) Thus, if a specific number is not provided, figures, plates, and pages are unnumbered. For books the format is: author, date of publication, title, publisher, city, total pagination, number of figures, number of plates, and (list of valid species described therein). An article in a book is presented in either of two formats. If the book is edited, the book title is given first, followed parenthetically by the editors. Otherwise, the author is presented first, followed by the book title. City of publication is given for all foreign journals and obscure domestic journals. If, however, the name of the city is part of the journal title, it is not repeated. When both names of Latin American authors (or those with a double surname) are given, we hyphenate the names. If only the first name is given followed by an initial of the second name, we omit the initial in citations of authors and collectors. When an author distributed advance copies of his paper (and the actual publication date precedes the printed date), we note this in brackets after the citation as a preprint.

We list the titles of books and journal articles exactly as printed, including typographical and grammatical errors, with the sole exception that we have italicized all generic and specific Latin names. Publications in the Romance languages (French, German, Italian, Portuguese, Spanish, etc.) are cited in original form. Publications employing non-Arabic characters (Chinese, Japanese, Russian, Thai, etc.) have their citations transliterated into English with the title placed in quotation marks and the language of the article given parenthetically at the end of the citation.

When more than one edition of a publication exists, the latest edition is cited in full and all previous editions listed in brackets in abbreviated form with only edition, date, pagination, and illustration data. If the work has a different title or authors, it is listed separately. For convenience, subsequent editions are cited in English by their number (second edition, third edition, etc.) rather than by their original designations (deuxième edition, revised, augmented, updated, etc.).

Whereas the journal and book titles are listed verbatim, we have standardized certain notations: figures and plates are listed in Arabic numerals; figures denoted by letters are presented in lower case; numbers of a volume are separated by a dash; numbers, parts, or sections are included parenthetically following the number of the volume (only when both a part and a number are jointly designated do we separate them, in which case the part is listed between the volume and number). In citing figures and plates, when numbers are designated we report them as such (i.e., figs. 1-4); if unnumbered we merely list the total number (i.e., 4 figs.). For unnumbered figures, we count each individual illustration as a separate figure. When "figs." or "illust." are listed without reference to numbers, they were not counted. All line drawings, illustrations, and black and white photographs are listed as figures (plus maps or anything else listed in publication as a figure) unless several different types occur in the same publication. In that case, we list line drawings as figures, photographs as illustrations, and color photographs as colored figures. Plates include black and white or color photographs printed on a separate page as well as designated plates. Journal abbreviations and cities are listed as they were at the time of publication. Hence, certain journals have more than one title or place of publication in their history. Multiple works by an author in a single year are listed chronologically, at least within the same journal.

For all journals the series number is given only for series two and thereafter. If no series is listed then series one is implied. With only a few exceptions authors with multi-worded family names are listed alphabetically under the last name; prefixes such as "de," "do," "la," "van den," "von," etc. are placed after the initials of the personal names (N. de Rooij = Rooij, N. de.). Authors whose prefixes are capitalized and therefore part of their surnames, are listed alphabetically under the prefix, such as De Betta, De Vis, De Waal, Lidth de Jeude, and Van Denburgh. All compound surnames are separated by a hyphen (i.e., Pérez-Higareda, Saint-Girons) and Spanish authors who append only an initial of the second family name are cited using just the first surname.

In the citation of foreign journals or obscure domestic journals, the place of publication follows the abbreviated journal name; the city is omitted from domestic and well-known herpetological journals. Geographic names as nouns are written in their entirety; when used as adjectives they are abbreviated.

Three unpublished manuscripts are worth mentioning as they are cited in the present work: H. Boie's (1823–1825) "Erpétologie de Java" in RMNH, Leiden; J. Wagler's (1825) "Amphibia. Ordo II. Serpentes" in MCZ, Cambridge; and T.E. Cantor's (1834–1837) "Drawings of animals" in RSL, Oxford. Cantor's type specimens, colored sketches, and manuscripts were deposited in RSL (Cantor, 1839a: 31).

Collection Acronyms

ACVS: Arthur C.V. Schott field numbers, Mexico
ADM: (= MDUG)
AG: Academie Groningen, Groningen, the Netherlands (no longer extant) [destroyed by fire in 1906]
AHCCCP: F.N. Chernyshev Central Museum of Geological Research, Leningrad, Russia (also used is CNIGR)
AHU: School of Life Sciences, Anhui University, Hefei, Anhui, China
AIS: Académie Impérial des Sciences, St. Pétersbourg, Russia (no longer extant, now = ZISP)
AK: André Koch field numbers
AM: (= AMS)
AMG: Albany Museum, Grahamstown, Cape of Good Hope, South Africa
AMNH: Department of Ichthyology and Herpetology, American Museum of Natural History, New York, New York, USA
AMS: Australian Museum, Sydney, New South Wales, Australia (includes MMS)
AM-S: Arturo Muñoz field numbers
AMSC: (= FPM)
AMS-FN: Australian Museum field numbers
ANSP: Academy of Natural Sciences of Philadelphia, Philadelphia, Pennsylvania, USA
ASB: Asiatic Society of Bengal, Calcutta, India (no longer extant, now = ASK)
ASFS: Albert Schwartz field numbers
ASIZB: Chinese Academy of Sciences [formerly Academia Sinica], Institute of Zoology [formerly Fan Memorial Institute of Zoology], Beijing, China
ASK: The Asiatic Society, Kolkata, West Bengal, India (includes ASB & IMC)
AWP: Angelo W. Palmisano field numbers
BHSP: Paleontology Collection, Bosnia-Hercegovina Staatsmuseum, Sarajevo, Bosnia and Herzegovina
BK: Bio-Ken Snake Farm, Watamu, Kenya
BM: Bratislava Museum, Bratislava, Slovakia [partially destroyed 9 Sept. 1944 during World War II]
BML: (= MBL)
BMNH: The Natural History Museum [formerly British Museum (Natural History)], London, England, UK
BMUW: Burke Museum of Natural History and Culture, University of Washington, Seattle, Washington, USA
BNHM: (= BNHS)
BNHS: Bombay Natural History Society, Museum, Mumbai [formerly Bombay], India
BPBM: Department of Zoology, Bernice P. Bishop Museum, Honolulu, Hawaii, USA
BRT: (= USTL)
BSCM: (= BSM)
BSM: Department of Agriculture, Bureau of Science, Manila, Luzon, the Philippines (no longer extant) [destroyed January 1945 during World War II]
BSMP: National Institute for Science and Mathematics, University of the Philippines, Manila, Luzon, the Philippines
BSP: (= BSPG)
BSPG: Bayerische Staatssammlung für Paläontologie und Historische Geologie, München, Germany
BSPM: (= BSPG)
BYU: Monte L. Bean Life Science Museum, Brigham Young University, Provo, Utah, USA
CA: Chicago Academy of Sciences, Chicago, Illinois, USA (partially destroyed by fire Oct. 1871)
CAD: Charles A. Domergue private collection
CAS: California Academy of Sciences, Golden Gate Park, San Francisco, California, USA (includes CAS-SU)
CAS-SU: California Academy of Sciences, San Francisco, California, USA (former Stanford University collection, now in CAS)
CBB: Colegio Biffi, Barranquilla, Colombia
CBC: Center for Biodiversity Conservation, Royal University of Phnom Penh, Phnom Penh, Cambodia
CBF: Colección Boliviana de Fauna, Museo Nacional de Historia Natural, Instituto de Ecología, Academia Nacional de Ciencias de Bolivia, La Paz, Bolivia
CC: F.C. Centeno & T.H. Condez field numbers
CG: Chapman Grant private collection
CGM: Cairo Geological Museum, Cairo, Egypt
CHUNB: Universidade Nacional de Brasília, Brasília, Distrito Federal, Brazil
CIB: Chengdu Institute of Biology, Chinese Academy of Sciences [formerly Academia Sinica], Chengdu, Sichuan, China (includes CIS, SBRI & SIBAC)
CIB-FN: Chengdu Institute of Biology field numbers
CIS: (= CIB)
CIT: California Institute of Technology, Los Angeles, California, USA (no longer extant; now = LACM)
CLB: Carlo L. Bonaparte private collection
CM: Carnegie Museum of Natural History, Pittsburgh, Pennsylvania, USA
CMR: Charleston Museum, Raleigh, South Carolina
CMS: National Museum, Colombo, Sri Lanka (includes NMSL)
CNAR: Colección Nacional de Anfibios y Reptiles, Universidad Nacional Autónoma de México, Ciudad México, Distrito Federal, México (includes IBHUNAM, IBM & UNAM)
CNHM: (= FMNH)
CPC: Commonwealth Paleontological Collection, Canberra, Australian Capital Territory, Australia

CRE: Costa Rican Expeditions collection of Jay Savage (to be deposited in LACM)
CRES: Centre for Natural Resources Management and Environmental Studies, Hanoi National University, Hanoi, Vietnam (now = VNUH)
CRUPF: Coleção Zoológica da Répteis da Universidade Federal de Passo Fundo, Passo Fundo, Brazil
CSBR: Caucasian State Biosphere Reserve Collection, Sochi, Russia
CSJ: Museo de Historia Natural, Colégio de San José, Medellin, Antioquia, Colombia
CSJP: Colégio San José de Pamplona, Pamplona, Norte de Santander, Colombia
CSNBR: Scientific Department, Caucasian State Natural Biosphere Reserve, Sochi, Russia
CTR: "Charles T. Ramsden" historical collection (in IES)
CU: Museum of Vertebrates, Cornell University, Ithaca, New York, USA
CUP: Department of Paleontology, Charles University, Prague, Czech Republic
CVRS: Centre Voltaique de la Recherche Scientifique, Ouagadougou, Burkina Faso
CVULA: Colección de Vertebrados de la Universidad de Los Andes, Mérida, Venezuela
CWM: Charles W. Myers field numbers
CWNU: China West Normal University, Nanchong, China
CWW: Wolfgang Weyrauch field numbers
CZACC: Colecciones Zoologicas de la Academia de Ciencias de Cuba, Institute of Ecology and Systematics, La Habana, Habana, Cuba (includes IZAC)
CZGB: Colecão Zoológica Gregório Bondar, Centro de Pesquisas-Comissão Executiva do Plano da Lavoura Cacaueira, Ilhéus, Bahia, Brazil
DGM: Departamento Nacional de Produção Mineral, Setor de Paleontologia, Universidade do Estado de Rio de Janeiro, Rio de Janeiro, Brazil
DHMECN: División de Herpetología del Museo Ecuatoriano de Ciencias Naturales, Quito, Pichincha, Peru
DML: (= WML)
DP-FNSP: (= CUP)
DPL: Dwight P. Lawson field numbers
DRV: David R. Vieites field numbers
DW: Dr. Wilson private collection
DZUSP: Departamento de Zoologia, Universidade de São Paulo, São Paulo, Brazil (= MZUSP)
EBRG: Museo de la Estación Biológia de Rancho Grande, Maracay, Aragua, Venezuela
EBUAP: Laboratorio de Herpetología, Escuela de Biología, Benemérita Universidad Autónoma de Puebla, Puebla, Mexico
EC: Elliott Coues field numbers
EHT: Edward H. Taylor private collection (specimens deposited primarily at CAS, CM, FMNH, KU & UIMNH)
EHT-CC: Edward H. Taylor collection catalogue numbers
EHT-HMS: Edward H. Taylor-Hobart M. Smith private collection (specimens deposited primarily at CAS, CM, FMNH, KU & UIMNH)
EIS: Essex Institute, Salem, Massachusetts (no longer extant; now = Peabody Essex Mus.)
EK: Eugen Kramer private collection
ENS: Eric N. Smith field numbers
EPN: Departamento de Ciencias Biológicas, Escuela Politécnica Nacional, Quito, Pichincha, Ecuador
FAT: Count Francisci Annibalis Turriani private collection, Museo Illustrissimi Turriano, Italy (no longer extant)
FEW: Francis E. W. Venning specimen numbers
FGZC: Frank Glaw field numbers
FHGO: Fundación Herpetológica "Gustavo Orcés," Quito, Pichincha, Ecuador
FHSC: Vertebrate Paleontology Collection, Fort Hays State University, Fort Hays, Kansas, USA
FK: Fred Kraus field numbers
FM: Faizabad Museum, Faizabad, Uttar Pradesh, India
FML: Instituto Herpetológica, Fundación "Miguel Lillo," San Miguel de Tucumán, Tucumán, Argentina
FMNH: Division of Amphibians and Reptiles, Field Museum of Natural History, Chicago, Illinois, USA (includes CNHM)
FMNHG: Field Museum of Natural History, Paleontology Collections, Geology Department, Chicago, Illinois, USA
FMSM: Federated Malay States Museum, Kuala Lumpur, Selangor and Taiping, Perak, West Malaysia (no longer extant, now = NMM) [partially destroyed on 10 March 1945 during World War II]
FPM: Museum of Natural History and Pathological Anatomy, Fort Pitt, Chatham, England, United Kingdom (no longer extant)
FSL: (= UCBL)
FU: Fujian University, Department of Biology, Shanghai, Guangdong, China
FWP: Fred S. Parker field numbers
GECM: Geographical and Exploring Commission of the Republic of México collection of the World Exhibition in New Orleans, LA, 1879–1885 [original collection destroyed by fire 29 Aug. 1884; second collection Oct. 1884–Jan. 1885]
GIH: Geological Institute of Hungary, Budapest, Hungary
GM: Geiseltalmuseum, Martin-Luther-Universität, Halle-Wittenberg, Germany
GMU: Department of Biology, Guangxi Medical University, Nanning, Guangxi Zhuang Autonomous Region, China
GNM: Göteborgs Naturhistoriska Museum [formerly Naturhistoriska Riksmuseet], Göteborg, Sweden (includes NHMG)
GPIBO: Institut für Paläontologie, Rheinische Friedrich-Wilhelm-Universität, Bonn, North Rhine-Westphalia, Germany
GSI/GC: Geologic Survey of India, Nagpur, India
GSI/Pal/CR: Geological Survey of India, Nagpur, India
GSNJ: Geological Survey of New Jersey, Trenton, New Jersey (now = New Jersey Geological and Water Survey)
GU/RSR/VAS: Department of Geology, Vastan Lignite Mine collection of R.S. Rana, H.N.B. Garhwal University, Srinagar, Uttaranchal, India
GVAG: Genevieve V.A. Gee field numbers
HGIM: Hungarian Geological Institute Museum, Budapest, Hungary

H-GSP: Harvard-Geological Survey of Pakistan Project (to be deposited in Islamabad Natural History Museum, Islamabad, Pakistan)
HMG: Hunterian Museum of Geology, University of Glasgow, Glasgow, Scotland, UK
HMS: Hobart M. Smith private collection (specimens deposited primarily at CAS, CM, FMNH, KU & UIMNH)
HNC: Hof-Naturalien-Cabinete, Wien, Austria (no longer extant)
HNHM: Hungarian Natural History Museum [formerly Magyar Természettudományi Múzeum], Budapest, Hungary (includes MNH)
HNU: Animal Museum, Hunan Normal University, Changsha, Hunan, China
HT: Harold Trapido field numbers
HUJ: Department of Zoology, Hebrew University of Jersusalem, Jerusalem, Israel
HUJP: (= HUJ-PAL)
HUJ-PAL: Paleontological Collection, Department of Zoology, Hebrew University of Jersusalem, Jerusalem, Israel
HvM: Hermann von Meyer personal collection
IAV: Instituto de Animales Venenosas "Dr. Jorge W. Abaloz," Santiago del Estero, Argentina
IAvH: Colección Herpetológica, Instituto de Investigación de Recursos Biológicos Alexander von Humboldt, Villa de Leyva, Bogotá, Cundinamarca, Colombia
IB: Herpetological collection Alphonse Richard Hoge, Instituto Butantan, São Paulo, São Paulo, Brazil (destroyed by fire 15 May 2010)
IBHUNAM: (= CNAR)
IBI: (= MZUB)
IBM: Instituto de Biología Herpetological Collections, Universidad Nacional Autónoma de México, Ciudad México, Distrito Federal, México (= CNAR)
ICN: (= ICNMNH)
ICNMHN: Instituto de Ciencias Naturales, Museo de Historia Natural, Universidad Nacional de Colombia, Bogotá, Cundinamarca, Colombia
ID: Indraneil Das field numbers
IEBR: Institute of Ecology and Biological Resources, National Center for Nature, Science and Technology of Vietnam, Hanoi, Vietnam
IES: Instituto de Ecología y Sistemática, Cuba
IFAN: Institut Fondamental d'Afrique Noire [formerly Institut Française d'Afrique Noire], Dakar, Senegal
IGB: Instituto Gabinetlo, Padova, Padua, Italy
IITR: Vertebrate Paleontology Laboratory, Department of Earth Sciences, Indian Institute of Technology, Roorkee, Uttar Pradesh, India
ILS: (= MLS)
IMC: Indian Museum, Calcutta, West Bengal, India (no longer extant, now = ASK)
IMCS: Institut Miquel-Crusafont, Sabadell, Spain
IMRC: (= IRMC)
IMRKL: Institute of Medical Research, Kuala Lumpur, Selangor, West Malaysia
INCDS: Institutul National de Cercetare-Dezvoltare pentru Stirte Biologice, Bucharest, Romania
IND: Laboratoria de Fauna, Division de Fauna Terrestre, Instituto Nacional de Recursos Naturales Renovables y Ambiente, Bogotá, Cundinamarca, Colombia
INHMT: Instituto Nacional de Higiene, Guayaquil, Guayas, Ecuador
INIGM: Instituto Nacional de Investigaciónes Geológico-Mineras, Bogotá, Cundinamarca, Colombia
INIRENA: Instituto de Investigaciones sobre los Recursos Naturales, Universidad Michoacana de San Nicolás de Hidalgo, Morelia, Michoacán, Mexico
INM: Instituto Nacional de Microbiología "Gustav G. Malbrán," Buenos Aires, Argentina
IOAN: P. P. Shirshov Institute of Oceanology, Russian Academy of Sciences, Moscow, Russia
IOC: Instituto Oswaldo Cruz, Rio de Janeiro, Brazil
IPS: Institut de Paleontología "Miguel Crusafont," Sabadell, Cataluña, Spain
IPUB: (= PIUB)
IRD: Institut de Recherche pour le Développement, Dakar, Dakar, Senegal
IRMC: Imperiàle e Reale Museo di Fisica e Storia Naturale della Capitale, Firenze, Italy (no longer extant)
IRSNB: Institut Royal des Sciences Naturelles de Belgique, Bruxelles, Belgium
ISCM: Institut Scientifique Cherifien, Rabat, Moroccos
ISFM: (= ISM)
ISM: Institute of Science Museum, Taipei, Taiwan (no longer extant)
ISMF: Instituto di Sanità Militare di Firenze, Firenze, Italy
ITT: Instituto Technológico de Ciudad Victoria, Ciudad Victoria, Tamaulipas, Mexico
IVO: Serpentario del Instituto Venezolano de Ofidiología, Caracas, Distrito Capital, Venezuela
IVP: Instituto Vital Brazil, Niterói, Rio de Janeiro, Brazil
IVPP: Institute of Vertebrate Paleontology and Palaeoanthropology, Academia Sinica, Beijing, China
IZAC: Instituto de Zoologia, Academia de Ciencias República de Cuba, La Habana, Habana, Cuba (= CZACC)
IZPAN: Institute of Paleozoology, Polska Akademia Nauk [Polish Academy of Sciences], Warsaw, Poland
IZUC: Instituto de Zoología, Universidad de Concepción, Concepción, Chile (no longer extant, now = MZUC)
IZUCS: Collezioni Zoologiche [formerly Istituto di Zoologia], Università de Cagliari, Cagliari, Sardinia, Italy
JA: Museo Jonae Alströmer, Sweden (no longer extant, now = NHR)
JAC: Jonathan A. Campbell field numbers
JD: John Dwyer private collection
JDL: James D. "Skip" Lazell field numbers
JEC: John E. Cadle field numbers
JGF: J. G. Fischer private collection, Hamburg, Germany
JH: Julius Hurter private collection, St. Louis, Missouri, USA
JMR: Juan M. Renjifo field numbers, Colombia
JPO: John P. O'Neill field numbers
JRF: John R. Feick field numbers
JRP: Jeff R. Parmelee field numbers
JSH: Juan J. Silva-Haad field numbers
JU: Department of Geology, University of Jammu, Jammu, India

JWM: Johann W. von Müller private collection
KBIN: (= IRSNB)
KIU: (= KUM)
KIZ: Kunming Institute of Zoology, Chinese Academy of Sciences [formerly Academia Sinica], Kunming, Yunnan, China
KM: (= KMJ)
KMH: Kim M. Howell field numbers
KMJ: Kagoshima Museum, Kagoshima, Japan [partially destroyed during World War II]
KSU: Museum of Natural History, Kharkov State University, Kharkov, Russia
KU: Museum of Natural History, University of Kansas, Lawrence, Kansas, USA
KUM: Kyoto University Museum, Kyoto University [formerly Kyoto Imperial University], Kyoto, Japan
KUVP: Vertebrate Paleontology, University of Kansas Natural History Museum, Lawrence, Kansas, USA
LACM: Natural History Museum of Los Angeles County, Los Angeles, California, USA (includes CIT)
LCB: Loren C. Binford field numbers
LEG: Lawrence E. Griffin field numbers, Philippines
LIV: (= WML)
LM: Naturkundemuseum Leipzig [formerly Leipzig Museum], Leipzig, Germany [partially destroyed during World War II]
LMB: Moravské Zemské Muzeum [Moravian Museum], Brno, Czech Republic
LMK: Laurence M. Klauber private collection, San Diego, California, USA (now in SDSNH)
LMNMO: Herpetological Collection, Landesmuseum für Natur und Mensch, Oldenburg, Germany
LNHM: Lübeck Natural History Museum [formerly Lübeck Museum], Lübeck, Germany [destroyed 28 March 1942 during World War II]
LNK: (= SMNK)
LSNUI: Laboratoire des Sciences Naturelles, Université-Indochinoise, Hanoi, Vietnam (no longer extant, now in MNHN)
LSUMG: Museum of Geology, Louisiana State University, Louisiana State University, Baton Rouge, Louisiana, USA
LSUMZ: Louisiana Museum of Natural History [formerly Museum of Natural Sciences and Museum of Zoology], Louisiana State University, Baton Rouge, Louisiana, USA
MACN: Museo Argentino de Ciencias Naturales "Bernardino Rivadavia" [formerly Museo Nacional de Buenos Aires], Buenos Aires, Buenos Aires, Argentina (formerly MNBA)
MAFR: Museum Adolphi Friderici Regis [also Museum Drottningholmense], Drottningholm, Sweden (no longer extant, now in NHR)
MAGNT: (= NTM)
MAS: Malcolm A. Smith private collection, London, United Kingdom
MB: Museo Blochiano [Marcus E. Bloch private collection], Berlin, Germany (no longer extant; now = ZMB)
MBH: Michael B. Harvey field numbers
MBI: (= MZUB)
MBL: Museu Bocage, Universidade de Lisboa, Lisboa, Portugal (destroyed by fire 18 March 1978)
MBLC: Marine Biological Laboratory, University of Copenhagen, Copenhagen, Denmark
MBLUZ: Museo de Biología de la Universidad del Zulia, Maracaibo, Zulia, Venezuela
MBS: (= NMBA)
MBUCV: Museo de Biología, Universidad Central de Venezuela, Caracas, Distrito Capital, Venezuela
MCG: Museo del Colegio San José de Guanentá, San Gil, Santander, Colombia
MCN: Museu de Ciências Naturales, Fundaçao Zoobotânica do Rio Grande do Sul, Porto Alegre, Rio Grande do Sul, Brazil
MCNA: Museo de Ciências, Naturales de Alava, Vitoria, Spain
MCNC: Museo de Ciências Naturales, Los Caobos (Plaza Morelos), Caracas, Distrito Capital, Venezuela
MCNG: Museo de Ciências Naturales de la UNELLEZ [formerly Museo de Zoología], Guanare, Portuguesa, Venezuela
MCP: Museu de Ciências e Tecnologia da PUCRS [Pontíficia Universidade Católica do Rio Grande do Sul], Porto Alegre, Rio Grande do Sul, Brazil
MCV: Museo Civico di Vicenza, Vicenza, Italy
MCZ: Department of Herpetology, Museum of Comparative Zoology, Harvard University, Cambridge, Massachusetts, USA
MCZ-FS: MCZ field series
MCZ-SC: MCZ slide collection
MCZ-VP: Department of Vertebrate Paleontology, Museum of Comparative Zoology, Harvard University, Cambridge, Massachusetts, USA
MD: Museo Regional do Dundo, Chitato, Luanda-Norte, Angola
MDG: Museum de Geer, Sweden (no longer extant; now in NHR)
MDR: (= MAFR)
MDUG: Museo de Historia Natural "Alfredo Dugès," Universidad de Guanajuato, Guanajuato, México
MFAC: (= MRAC)
MFB: (= CIB)
MG: Museo Laurentii Theodori Gronovii [also Museum Gronovianum], Lugdunum Batavorum (= Leiden) (no longer extant; possibly in RMNH)
MGF: Musée de Gannat, Gannat, France
MGL: Musée Guimet d'Histoire Naturelle, Lyon, France (no longer extant, now = MMC)
MGM: Madras Government Museum, Egmore, Chennai, Tamil Nadu, India
MGPF: (= MGUF)
MGPUF: (= MGUF)
MGUF: Museo di Geologie e Paleontologie, Università di Firenze, Firenze, Italy
MH: Museo Humphrediano, Strasbourg, France (no longer extant) (= MZUS)
MHGI: (= HGIM)
MHNCI: Museu de História Natural Cupão da Imbuia, Curitiba, Paraná, Brazil

MHNG: Muséum d'Histoire Naturelle Genève, Genève, Switzerland
MHNJP: (= MUSM)
MHNL: Muséum d'Histoire Naturelle de Lyon, Lyon, France (no longer extant, now = MMC)
MHNLS: Museo de Historia Natural La Salle, Fundación La Salle de Ciências Naturales [formerly Museo de Ciências Naturales], Caracas, Distrito Capital, Venezuela.
MHNM: Museo Nacional de Historia Natural y Antropología [formerly Museo de Historia Natural de Montevideo], Montevideo, Montevideo, Uruguay
MHNMF: Museum d'Histoire Naturelle de Marseille, Marseille, France
MHNN: Muséum d'Histoire Naturelle de Neuchâtel, Neuchâtel, Switzerland (includes MZN)
MHNPB: Museum d'Histoire Naturelle Pays-Bas, Leiden, the Netherlands (no longer extant, now = RMNH)
MHNR: Muséum d'Histoire Naturelle de La Rochelle, La Rochelle, France
MHNSM: Museo de Historia Natural, Universidad Nacional Mayor de San Marcos, Lima, Lima, Perú
MHNSR: Museo de Historia Natural de San Rafael, San Rafael, Mendoza, Argentina
MHNUC: Museo de Historia Natural Universidad del Cauca, Popoyán, Colombia
MICP: Museo Instituo "Clodomiro Picado," Universidad de Costa Rica, San José, Costa Rica
MISGF: Museum of the Institute of Science of the Government of Formosa, Taipei, Taiwan (no longer extant)
MJU: Museo Jenensis Universitatis, Germany (no longer extant)
MLB: Museo Lugdunensi Batavorum, Leiden, the Netherlands (no longer extant)
MLH: Museum Lampianum, Hannover, Germany (no longer extant, now = ZPW)
MLP: Museo de La Plata, Universidad de La Plata, Buenos Aires, Buenos Aires, Argentina
MLPA: (= MLP)
MLS: Museo de Historia Natural [formerly Museo del Instituto] de La Salle, Bogotá, Cundinamarca, Colombia (destroyed by fire 9–10 April 1948)
MLU: Institut für Zoologie und Zoologisches Sammlung, Martin-Luther-Universität, Halle-Wittenberg, Germany
MM: Museum of Monaco, Monte Carlo, Monaco (no longer extant; now = ZSM)
MMBR: Musée de la Mer, Brest, Finistèrre, France
MMC: Muséum au Musée des Confluences, Lyon, France [to open in 2014] (includes MGL & MNHL)
MMK: Museo de la Misión de Kavanayen, Kavanayen, Bolívar, Venezuela
MML-PV: Museo Municipal de Lamarque Rio Negro, Neuquén, Neuquén, Argentina
MMS: Macleay Museum, University of Sydney, Sydney, Australia (no longer extant; now in AMS)
MMUS: (= MMS)
MN: (= MNRJ)
MNA: Museum of Northern Arizona, Flagstaff, Arizona, USA
MNBA: Museo Nacional de Buenos Aires, Buenos Aires, Argentina (no longer extant, now = MACN)
MNCN: Museo Nacional de Ciencias Naturales, Madrid, Madrid, Spain
MNCR: Museo Nacional de Costa Rica, San José, San José, Costa Rica
MNGL: (= MHGI)
MNH: (= HNHM)
MNHMU: (= MHNM)
MNHN: Muséum National d'Histoire Naturelle, Paris, Ile-de-France, France (includes LSNUI)
MNHNC: Museo Nacional de Historia Natural, Zoología, Santiago, Santiago, Chile
MNHNCU: Museo Nacional de Historia Natural, La Habana, Habana, Cuba
MNHNM: (= MHNM)
MNHNP: Mueo Nacional de Historia National Edificio Patria, Tacuan E/25 de Mayo/Cer. Cora, Asunción, Paraguay
MNHNU: (= MHNM)
MNHSR: Museo de Historia Natural de San Rafael, San Rafael, Argentina
MNK: Museo de Historía Natural "Noel Kempff Mercado," Santa Cruz, Santa Cruz, Bolivia
MNKNU: Museum of Nature, V.N. Karazin Kharkiv National University, Kharkiv, Ukraine
MNKR: (= MNK)
MNRJ: Museu Nacional, Universidade Federal do Rio de Janeiro, Rio de Janeiro, Guanabara, Brazil
MP: Museum Principis, Sweden (no longer extant; now in ZMUU)
MPB: (= MHNPB)
MPCA: Museo Provincial de Cipolletti "Carlos Ameghino," Cipolletti, Río Negro, Argentina
MPEG: Museu Paraense "Emilio Goeldi," Zoologia, Belém, Pará, Brazil
MPI: Museo di Pavia, Lombardy, Pavia, Italy
MPSP: Museu Paulista, São Paulo, São Paulo, Brazil (no longer extant; now = MZUSP)
MRAC: Musée Royal de l'Afrique Centrale [Koninklijk Museum voor Midden Afrika], Tervuren, Flanders, Belgium
MRF: (= IRMC)
MRSN: (= MSNTO)
MSB: Department of Biology, Museum of Southwestern Biology, University of New Mexico, Albuquerque, New Mexico, USA
MSK: Muhammed Sharif Khan private collection, Rabwah, Pakistan
MSNG: Museo Civico di Storia Naturale "Giacomo Doria," Genova, Liguria, Italy
MSNM: Museo Civico di Storia Naturale di Milano, Milano, Lombardia, Italy (includes MZUT) [partially destroyed in 1943 during World War II, including Jan's (1853–1866) collection of 1000 species, types, and catalogues]
MSNTO: Museo Civico di Storia Naturale di Torino [formerly Museo Regionale di Scienze Naturali], Torino, Piemonte, Italy (formerly MRSN)
MSP: Madras Snake Park Trust, Mamallapuram, Tamil Nadu, India
MSUVP: Vertebrate Paleontology Collection, Michigan State University, East Lansing, Michigan, USA

MT: Musée de Toulouse, Toulouse, Haute-Garonne, France
MTD: Museum für Tierkunde, Senckenberg Naturhistorische Sammlungen, Dresden, Saxony, Germany (formerly MTKD) [partially destroyed 13 Feb. 1945 during World War II]
MTKD: (= MTD)
MTR: Miguel T. Rodrigues field numbers
MUSM: Museo de Historia Natural "Javier Prado," Universidad Nacional Major de San Marcos, Lima, Lima, Perú (formerly MHNJP)
MUVP: Vertebrate Paleontology Collection, Midwestern University, Snyder, Texas, USA
MV: Museo Caesareo Vindobonensi, Wien, Austria (no longer extant)
MVP: Museum of Vertebrate Paleontology, University of California, Berkeley, California, USA
MVZ: Museum of Vertebrate Zoology, University of California at Berkeley, Berkeley, California, USA
MW: M. Ward private collection, Australia
MWNH: Museum Wiesbaden, Naturhistorische Landessammlung, Wiesbaden, Hesse, Germany (formerly NMSW and SMW) [partially destroyed Aug. 1940–March 1945 during World War II but some types survived]
MZB: Museum Zoologicum Bogoriense, Juanda 3, Kebun Raya, Bogor, Java, Indonesia
MZFC: Museo de Zoología "Alfonso L. Herrera," Facultad de Ciencias, Universidad Nacional Autónoma de México, México, Distrito Federal, Mexico
MZN: Musée Zoologique de Neuchâtel, Neuchâtel, Switzerland (= MHNN)
MZUB: Museo di Zoologia dell'Università di Bologna, Bologna, Italy (includes IBI)
MZUC: Museo de Zoología de la Universidad de Concepción, Concepción, Concepción, Chile
MZUF: Museo di Storia Naturale di Firenze, Sezione di Zoologia "La Specola," Università degli Studi di Firenze, Firenze, Toscana, Italy (includes NHCL)
MZUN: Museo di Zoologia di Napoli, Università degli Stdui di Napoli Federico II, Naples, Campania, Italy
MZUS: Musée de Zoologique de la Ville de Strasbourg, Université de Strasbourg, Strasbourg, Alsace, France [partially destroyed in 1943–Aug. 1944 during World War II]
MZUSP: Museu de Zoologia da Universidade de São Paulo, São Paulo, Brazil (includes MPSP)
MZUSP-FN: Museu de Zoologia da Universidade de São Paulo field numbers
MZUT: Museo Zoologico, Università di Torino, Torino, Italy (now in MSNM)
NCF: Department of Forest Insects, Northwestern College of Forestry, Xian, Shaanxi, China
NHCL: New Herpetological Collection of Benedetto Lanza, Museo di Storia Naturale di Firenze, Sezione di Zoologia "La Specola," Università degli Studi di Firenze, Firenze, Toscana, Italy (now in MZUF)
NHMB: (= NMBA)
NHMG: (= GNM)
NHMV: (= NMW)
NHMW: (= NMW)
NHR: Department of Vertebrate Zoology, Naturhistoriska Riksmuseet, Stockholm, Stockholm, Sweden (includes MAFR & MDR)
NHRM: (= NHR)
NK: (= MNK)
NLO: Nikolai L. Orlov field numbers
NMB: Naturhistorisches Museum Braunschweig, Brunswick, Germany (no longer extant, now = SNMBR) [destroyed 15 Oct. 1944 during World War II]
NMBA: Naturhistorisches Museum Basel, Basel, Switzerland
NMBE: Naturhistorisches Museum Bern, Bern, Switzerland
NMC: National Museums of Canada, National Museum of Natural Sciences, Ottawa, Ontario, Canada
NMK: National Museum, Nairobi, Central, Kenya
NMM: National Museum, Kuala Lumpur, Selangor, West Malaysia (includes FMSM and SM)
NM/M: Umtali Museum, Umtali, Southern Rhodesia (no longer extant, now = NMZB)
NMP: National Museum, Prague, Czech Republic
NMSL: (= CMS)
NMSR: (= NMZB)
NMSW: Naturhistorischen Museums der Stadt Wiesbaden, Wiesbaden, Hesse, Germany (= MWNH)
NMSZ: National Museum of Scotland [formerly Royal Scottish Museum], Edinburgh, Scotland, UK
NMT: National Museum, Dar es Salaam, Dar es Salaam, Tanzania
NMV: Museum Victoria [formerly National Museum of Victoria], Melbourne, Victoria, Australia
NMW: Naturhistorisches Museum, Wien, Austria
NMWGP: Naturhistorisches Museum, Geologie-Palaontologie, Wien, Austria
NMZB: Natural History Museum, Bulawayo, Zimbabwe (includes NMZB-UM)
NMZB-UM: Natural History Museum, Bulawayo, Zimbabwe (Umtali Museum collection)
NQT: Nguyen Quang Truong field numbers
NRM: (= NHR)
NRS: (= NHR)
NSM: National Science Museum, Bangkok, Thailand
NSMT: National Science Museum, Tokyo, Japan
NTI: Animal Industry Branch, Northern Territory Administration, Alice Springs, Australia
NTM: Northern Territory Museum of Arts and Sciences [formerly Museum and Art Gallery of the Northern Territory], Darwin, Northern Territory, Australia
NTM A/S: Northern Territory Museum of Arts and Sciences, Alice Springs, Australia
NUS: National University of Singapore, Singapore (no longer extant, now = ZRC)
NWNH: Staatliches Museum Wiesbaden, Wiesbaden, Hesse, Germany (= SMW)
NWU: Natural History Museum, Northwestern University, Chicago, Illinois, USA
OEHW: Otto E.H. Wucherer private collection
OF: MHNJP Ophidia collection numbers
OGM: Olduvai Gorge Museum, Olduvai Gorge, Ngorongoro Conservation Area, Tanzania

OMNH: Osaka Museum of Natural History, Osaka, Japan
ORSTOM: Office de Recherche Scientifique et Technologique Outre Mer, Paris, France
OSGP: Olivier S.G. Pauwels field numbers
OUM: Bodleian Library, Oxford University Museum, Oxford, England, UK
PCM: Philadelphia Commercial Museum, Philadelphia, Pennsylvania, USA (no longer extant)
PEM: Port Elizabeth Museum, Port Elizabeth, Cape of Good Hope, South Africa
PEPD: P.E.P. Deraniyagala private collection, Colombo, Sri Lanka
PIAN: Paleontogichesky Institut [Paleontological Institute], Akademia Nauk, Moscow, Russia
PIMUZ: Paläontologischesky Institut und Museum der Universität, Zürich, Switzerland
PIUB: Paläontologisches Institut, Rheinische Friedrich-Wilhelms-Universität Bonn, Bonn, North Rhine-Westphalia, Germany
PK: P.J.R. Kok field numbers
PM: Philadelphia Museum or C.W. Peale Museum, Philadelphia, Pennsylvania, USA (no longer extant, 1846; now = MCZ)
PMNH: (= YPM)
PNM: Philippine National Museum, Manila, Luzon, the Philippines [destroyed January 1945 during World War II]
PT: Proyecto Tupinambis, Argentina field numbers
PTB: Prince Therese von Bayern collection (no longer extant, now = Staatliches Museum für Völkerkunde, Munich, Germany)
PU: Princeton University, Natural History Museum, Princeton, New Jersey, USA
PUL: Purdue University, Lafayette, Indiana, USA (no longer extant)
QCAZ: Pontífica Universidad Católica del Ecuador, Quito, Pichincha, Ecuador
QM: Centre for Biodiversity, Queensland Museum, Brisbane, Queensland, Australia
QMB: (= QM)
QMF: Queensland Museum Fossil Collection, Brisbane, Queensland, Australia
QMP: Quetta Museum, Quetta, Baluchistan, Pakistan (no longer extant)
QSMI: Queen Saovabha Memorial Institute, Thai Red Cross Society, Bangkok, Thailand
RAN: Ronald A. Nussbaum private collection
RBS: Robert B. Stuebing field numbers
RCSM: Museum of Royal College of Surgeons, London, England, UK
RE: Robert E. Elbel field numbers
RGMC: (= MRAC)
Rh-E.F.: Musée de Gannat, Gannat, France
RLB: René L. Bourret field numbers
RM: (= RMBR)
RMAC: (= MRAC)
RMBR: Raffles Museum of Biodiversity Research, Singapore, Singapore
RMNH: Naturalis-Nationaal Natuurhistorisch Museum [formerly Rijksmuseum van Natuurlijke Historie], Leiden, the Netherlands (includes MHNPB & ZMA)
RMNH.RENA: (= RMNH)
ROM: Royal Ontario Museum, Toronto, Ontario, Canada
RSI: Razi State Vaccine and Serum Institute, Tehran, Iran
RSL: Radcliffe Science Library (formerly Bodleian Library), University of Oxford, London, United Kingdom (Cantor sketches in "Drawings of animals," vols. 1–2)
RSM: Royal Scottish Museum, Edinburgh, Scotland, UK
RSR: Rajendra S. Rana private collection (included in GU/VAS)
RT: Richard Thomas private collection, San Juan, Puerto Rico
RUSB: Department of Earth Sciences, Indian Institute of Technology, Roorkee, India
SAM: South African Museum, Cape Town, Cape of Good Hope, South Africa
SAMA: South Australian Museum, Adelaide, South Australia, Australia
SAMA-FN: South Australian Museum field numbers
SBH: S. Blair Hedges field numbers
SBRI: Sichuan Biological Research Institute, Chengdu, Sichuan, China (= CIB)
SCUM: Sichuan University Museum, Chengdu, Sichuan, China
SDSM: Museum of Geology, South Dakota School of Mines, Rapid City, South Dakota, USA
SDSNH: San Diego Society of Natural History, San Diego Natural History Museum, Balboa Park, San Diego, California, USA (includes LMK)
SERI: (= CIB)
SGDB: Geological collection of the opencast mine Bílina [Sbirky geologie, Doly Bílina], Czech Republic
SHNBA: (= MACN)
SIBAC: Southwest Institute of Biology, Academia Sinica, Chengdu, Sichuan, China (= CIB)
SIM: Smithsonian Institution Museum, Washington, D.C., USA (= USNM)
SIPT: State Institute of Pedagogy, Tiraspol, Russia
SJP: (= CSJP)
SM: Selangor Museum, Kuala Lumpur, Selangor, West Malaysia (no longer extant, now = NMM)
SMF: Mertens catalogue, Natur-Museum und Forschungs-Institut Senckenberg, Frankfurt am Main, Germany.
SMF-B: Oskar Boettger catalogue, Museum der Senckenbergischen Naturforschenden Gesellschaft, Frankfurt am Main, Germany (= SMF)
SMF-M: Robert Mertens catalogue, Museum der Senckenbergischen Naturforschenden Gesellschaft, Frankfurt am Main, Germany (= SMF)
SMF-R: Eduard Rüppell catalogue, Museum der Senckenbergischen Naturforschenden Gesellschaft, Frankfurt am Main, Germany (= SMF)
SMK: Sarawak Museum, Kuching, Sarawak, East Malaysia
SMNH: Saskatchewan Museum of Natural History, Regina, Saskatchewan, Canada
SMNHC: Shanghai Museum of Natural History, Shanghai, Jiangxi, China

SMNK: Staatliche Museum für Naturkunde Karlsruhe [formerly Landessammlungen für Naturkunde Karlsruhe], Karlsruhe, Germany
SMNS: Staatliches Museum für Naturkunde in Stuttgart [Stuttgatd State Museum of Natural History], Ludwigsburg, Germany [partially destroyed Sept. 1944 during World War II]
SMP: (= SMPSMU)
SMPSMU: Shuler Museum of Paleontology, Southern Methodist University, Dallas, Texas, USA
SMW: (= MWNH)
SMWN: Staatsmuseum-Windhoek [State Museum], Windhoek, Khomas, Namibia
SNHM: Shanghai Museum of Natural Hisltory, Huanqpu, Shanghai, China
SNM: Singapore National Museum, Singapore (no longer extant, now = ZRC)
SNMBR: Staatliches Naturhistorisches Museum Braunschweig, Brunswick, Germany (includes NMB)
SNP: Sochi National Park, Sochi, Russia
SP: Sabah Parks Zoological Museum, Gunung Kinabalu Park, Sabah, East Malaysia
SPLUE: Paläontologische Sammlung, Friedrich-Alexander Universität [formerly Sammlungen des Palaontologischen Lehrstuhls des Universität Erlangen], Erlangen-Nürnberg, Bavaria, Germany
SRAR: Society for Research of Amur Regiõn, Vladivostok, Primorsky Krai, Russia (no longer extant)
STUM: Santo Tomas University Museum, Rizal, Manila, Luzon, the Philippines
SU: Stanford University, Palo Alto, California, USA (no longer extant, now in CAS as CAS-SU)
SUP: Silliman University, Dumaguete City, Negros Oriental, the Philippines
SYS: Museum of Biology, Sun Yat-Sen University [Zhongshan University], Guangzhou, China
SZE: Sistematik Zooloji Enstitüsü, Bornova-Izmir, Turkey
SZN: Stazione Zoologica "Anton Dohrn," Naples, Campania, Italy (includes UNZM)
TCWC: Texas Cooperative Wildlife Collection, Texas A & M University, College Station, Texas, USA
THNHM: Thailand Natural History Museum, National Science Museum, Bangkok, Thailand
TIU: Science College Museum, Tokyo Imperial University, Tokyo, Japan
TM: (= TMP)
TMP: Transvaal Museum of Natural History, Northern Flagship Institution, Pretoria, Transvaal, South Africa
TMT: Georgian National Museum [formerly Tiflis Museum], Tbilisi, Georgia
TNRC: Thai National Reference Collection, Thailand Institute of Scientific and Technological Research, Bangkok, Krung Thep Mahanakhon, Thailand (= NSM)
TST: T.S. Traill private collection
TUB: Technical University of Berlin, Berlin, Berlin, Germany
UA: Université d'Antananarivo, Antananarivo, Antananarivo, Madagascar
UADZ: Department of Zoology, University of Arequipa, Arequipa, Peru
UANL: Laboratorio de Herpetologia, Universidad Autónoma de Nuevo León, Nuevo León, Mexico
UAZ: Department of Ecology and Evolutionary Biology, University of Arizona, Tucson, Arizona, USA
UBIPRO: Unidad de Biologia, Tecnología y Prototipos, Tlalnepantla, México, Mexico
UCB: (= MVZ)
UCBL: Département des Sciences de la Terre, Université Claude-Bernard, Lyon 1, Rhône-Alps, France
UCBP: (= MVP)
UCG: Universidade Católica de Goiás, Goiânia, Goiás, Brazil
UCM: Museum of Natural History, University of Colorado, Boulder, Colorado, USA
UCMP: University of California, Museum of Paleontology, Los Angeles, California, USA
UCP: Museo de Historia Natural de la Universidad del Cauca, Popoyán, Cauca, Colombia
UCR: Museo de Zoología, Universidad de Costa Rica, San José, San José, Costa Rica
UCV: Universidad Central de Venezuela, Institue of Tropical Medicine, Caracas, Distrito Federal, Venezuela
UF: Florida Museum of Natural History [formerly Florida State Museum], University of Florida, Gainesville, Florida, USA
UFAC-PV: Coleção de Paleovertebrados, Laboratório de Pesquisas Paleontológicas, Universidade Federal do Acre, Rio Blanco, Brazil
UGVP: Vertebrate Paleontology Collection, University of Georgia, Athens, Georgia, USA
UHESM: Hydroelectric Power Plant of Serra da Mesa Collection, Universidade Católica de Goiás, Goiânia, Goiás, Brazil
UIMNH: Museum of Natural History, University of Illinois, Urbana, Illinois, USA
ULABG: Colección de Anfibios y Reptiles del Laboratorio de Biogeografia, Universidad de Los Andes, Mérida, Venezuela
UM: Umtali Museum (no longer extant; now = NMZB)
UMB: Überseemuseum, Breman, Breman, Germany
UMMP: Museum of Paleontology, University of Michigan, Ann Arbor, Michigan, USA
UMMZ: Museum of Zoology, University of Michigan, Ann Arbor, Michigan, USA
UNAH: Departamento de Biología, Universidad Nacional Autónoma de Honduras, Ciudad Universitaria, Tegucigalpa, Francisco Morazán, Honduras
UNAM: Departmento de Zoologia, Universidad Nacional Autónomia de México, Ciudad México, Distrito Federal, Mexico (= CNAR)
UNAM-LT: Estación de Biologia Tropical "Los Tuxtlas," Universidad Nacional Autónomia de México, Catemaco, Veracruz, Mexico
UNAMV: (= UNAM-LT)
UNM: (= MSB)
UNNE-CHC: Coleccíon Herpetológica Corrientes, Universidad Nacional del Nordeste, Corrientes, Corrientes, Argentina
UNS: University of Natural Sciences, Ho Chi Minh City, Ho Chi Minh, Vietnam

UNSM: Nebraska State Museum, University of Nebraska, Lincoln, Nebraska, USA

UNZM: University of Naples Zoological Museum, Naples, Campania, Italy (no longer extant, now = SZN)

UOMNH: University of Oklahoma Museum of Natural History, Norman, Oklahoma, USA

UPNG: University of Papua New Guinea, Port Moresby, Central, Papua New Guinea

UPR: University of Puerto Rico, Mayagüez, Puerto Rico

UPVI: Laboratorie de Paléontologie des Vertébrés, Université Paris VI, Paris, Ile-de-France, France

UR: (= URJ)

URJ: Agriculture and Home Economics Division, University of the Ryukyus, Nishihara, Ryukyu Islands, Japan

US: (= MZUS)

USL: University of Southwestern Louisiana, Lafayette, Louisiana, USA

USNM: National Museum of Natural History [formerly United States National Museum], Smithsonian Institution, Washington, District of Columbia, USA

USNM-FN: National Museum of Natural History field numbers

USTL: Université des Sciences et Techniques du Languedoc, Montpellier II, Languedoc, France [partially destroyed in 1944 during World War II]

UTA: Collection of Vertebrates, Department of Biology, University of Texas at Arlington, Arlington, Texas, USA

UTEP: Laboratory for Environmental Biology, Department of Biological Sciences and Centennial Museum, University of Texas at El Paso, El Paso, Texas, USA

UU: Universiteitmuseum Utrecht, Universiteit Utrecht, Utrecht, the Netherlands

UUZM: Zoological Museum, University of Utah, Salt Lake City, Utah, USA

UV: Departamento de Biología, Universidad del Valle, Cali, Valle, Colombia

UVC: (= UV)

UVG: Uinversidad del Valle de Guatemala, Guatemala City, Guatemala

UZM: (= ZMUC)

UZMK: Universitetets Zoologiske Museum, København, Denmark

VAS: (= GU/RSR/VAS)

VLKE: Vernay-Lang Kalahari Exped., Namibia, Africa, Mar.–Sept. 1930

VM: Vladivostock Museum, Vladivostock, Primorsky Krai, Russia

VNMN: Vietnam National Museum of Nature, Hanoi, Hanoi, Vietnam (includes CRES, VNUH)

VNUH: (= VNMN)

VPI: Vida Preciosa International, Boerne, Texas, USA

VR: J. V. Rueda field numbers

WAM: Western Australian Museum, Perth, Western Australia, Australia

WED: William E. Duellman field numbers

WHC: William Hyde Cabinet, Philadelphia, Pennsylvania

WHM: William Hunter Museum, London, England, UK (now Hunterian Museum and Art Gallery)

WHT: Wildlife Heritage Trust of Sri Lanka, Colombo, Sri Lanka

WML: Clore Natural History Centre, World Museum Liverpool [formerly Derby Museum], Liverpool, England, UK (includes DML)

WRH: W. Ron Heyer field numbers

WS: W. Schultze private collection, Manila, Philippines

YBU: College of Life Science and Food Engineering, Yibin University, Yibin, China

YPM: Peabody Museum of Natural History, Yale University, New Haven, Connecticut, USA (formerly PMNH)

YU: Department of Zoology, Yunnan Normal University, Kunming, Yunnan, China

ZCIKU: (= MNKNU)

ZDEU: Zoology Department, Ege University, Turkey

ZDKU: Museum of Natural History, Kharkiv University, Kharkiv, Ukraine

ZFMK: Zoologisches Forschungsmuseum [formerly Forschunginstitut und Museum] "Alexander Koenig," Bonn, North Rhine-Westphalia, Germany

ZIAS: Zoological Institute, Academia Sinica [Chinese Academy of Sciences], Shanghai, China

ZIB: Zoological Institute, Bonn, North Rhine-Westphalia, German (no longer extant) [destroyed 7–9 March 1945 during World War II]

ZIK: I. I. Schmalhausen Institute of Zoology, National Academy of Sciences of Ukraine, Kiev, Ukraine

ZIKP: Department of Paleozoology, Institute of Zoology, National Academy of Sciences of Ukraine, Kiev, Ukraine

ZIL: (= ZISP)

ZIMG: (= ZMG)

ZIN: (= ZISP)

ZIN PC: (= ZISP)

ZISP: Zoological Institute, Russian Academy of Sciences, St. Petersburg [formerly Leningrad], Russia (includes AIS, ZIL, ZIN & ZIN PC)

ZIUS: Zoologiska Institutionen, Stockholms Universitet, Stockholm, Sweden

ZM: Zoologisches Museum, Kanpur, Uttar Pradesh, India

ZMA: Institut voor Taxonomische Zoölogie [formerly Zoölogisch Museum], Universiteit van Amsterdam, Amsterdam, the Netherlands (now = RMNH)

ZMB: Museum für Naturkunde [formerly Zoologischen Museum], Leibniz-Institut für Evolutions- und Biodiversitätsforschung, Universität-Humboldt zu Berlin, Berlin, Berlin, Germany (includes MB) [partially destroyed 22–23 Nov. 1944 and 3 Feb. 1945 during World War II]

ZMBP: Museum für Naturkunde [formerly Paläontologischen Museum], Leibniz-Institut für Evolutions- und Biodiversitätsforschung, Universität-Humboldt zu Berlin, Berlin, Berlin, Germany

ZMG: Zoologisches Museum, Johann-Friedrich-Blumenbach-Institut für Zoologie und Anthropologie, Göttingen, Göttingen, Germany

ZMH: Zoologisches Museum Hamburg [formerly Zoologisches Institut und Museum], Universität Hamburg, Hamburg, Hamburg, Germany [partially destroyed in July 1943 during World War II, including type collection and catalogues]

ZMK: Zoological Museum, Kristiania, Oslo, Oslo, Norway (no longer extant, now = ZMUO)

ZMMU: Zoological Museum of Mikhail V. Lomonosov, Moscow State University [formerly Imperial Moscow University], Moscow, Russia

ZMT: S. Janashia State Museum of Georgia, Zoological Section, Georgian Academy of Sciences, Tbilisi, Georgia

ZMUC: Vertebrater, Zoologisk Museum, Københavns Universitet, København, København, Denmark

ZMUH: Zoologisches Museum, Universität Halle, Halle, Germany [partially destroyed in 31 March–5 April 1945 during World War II]

ZMUL: Zoologiska Museet, Universitet Lund, Lund, Sweden

ZMUO: Zoologisk Museum, Universitetets i Oslo, Oslo, Norway

ZMUU: Zoologiska Museet, Uppsala Universitet, Uppsala, Uppsala, Sweden (includes MP & MAF)

ZMUZ: (= ZMZ)

ZMZ: Zoologisches Museum, Universität Zürich, Zürich, Switzerland

ZMW: (= NMW)

ZPPAN: (= IZPAN)

ZPUW: (= ZPW)

ZPW: Zakład Paleozoologii, Instytut Zoologiczny, Uniwersytet Wroclawski, Wroclaw, Poland

ZPWM: Zakład Mineralogii, Instytut Geologicznych, Uniwersytet Wroclawski, Wroclaw, Poland

ZRC: Zoological Reference Collection, Department of Life Sciences, National University of Singapore, Singapore (includes FMSM, NUS, RM & SNM)

ZSI: Zoological Survey of India, Kolkata [Calcutta], West Bengal, India

ZSIC: (= ZSI)

ZSM: Zoologisches Staatssammlung München [formerly Zoologisches Sammlung des Bayerischen Staates], München, Bayern, Germany (includes MM) [partially destroyed 24–25 April 1944 and 11 April 1945 [Spix collection] during World War II]

ZSS: Zoologische Sammlung zu Stettin, Szczecin, Poland

ZUEC: Departamento de Zoologia and Museu de História Natural, Universidade Estadual de Campinas "Adão José Cardoso," Campinas, São Paulo, Brazil

ZZS: (= ZZSD)

ZZSD: Zaklad Zoologii Systematycznei i Doswiadczalnej [Institute of Systematic and Experimental Zoology], Polish Academy of Sciences, Cracow, Poland

Acknowledgments

The number of people who have contributed to this effort over the past 30 plus years is so enormous that someone is likely to be forgotten. We thank everyone who has assisted us with literature, information or specimens.

For the use of data from unpublished manuscripts (some now published) we thank: Kraig Adler, Dave Barker, Merel J. Cox, Patrick David, Michel Domínguez, Charles M. Fugler, Hugo von Habsburg und Lothringen, Blair Hedges, Robert W. Henderson, Barry Hughes, Malcolm J. Largen, Hy Marx, Dennis Parmley, Jean-Claude Rage, Jens. B. Rasmussen, Franklin D. Ross, Carlos Pérez-Santos, Alan Resetar, and Zbigniew Szyndlar. For assistance with data on type specimens we thank Aaron Bauer (W. Peters types), Patrick David (MNHN), Esther Dondorp (RMNH), Ned Gilmore (ANSP), Rainer Günther (ZMB), Marinus Hoogmoed (RMNH), Ivan Ineich (MNHN), Rahul Khot (BNHS), Mathias Lang (BMNH, MNHN), Franklin D. Ross (RMNH), and Andrew F. Stimson and E. N. Arnold (BMNH). For unpublished locality data on geographic regions we thank Natasha Carter (Australia), M.J. "Jack" Cox (Thailand), and Jayaditya Purkayastha (Assam, India). For assistance with nomenclatural problems and examination of MNHN types we especially thank Patrick David.

For translation of critical papers we thank: Ingela Chef-Johansson (Swedish, Norwegian, German), Roxane Coombs (French), Marina Dzidziguri (Russian), Katherine Hespe (German), Florence Kil-Bosso (French, German), Mathias Lang (German, Dutch), Herbert Levi (German), Angela Lim (Chinese), Shiu-Ping Osborn (Chinese), Jury Rudyakov (Russian), and Britta Weimann (German).

For providing rare or much needed literature, we thank: Kraig Adler, David Auth, Aaron Bauer, Donald G. Broadley, Jonathan A. Campbell, Roger Conant, Jeanette Covacevich, James R. Dixon, Richard Etheridge, M.J. Fouquette, J. Alan Holman, Marinus Hoogmoed, Barry Hughes, Kathleen Kelly, G.R. McLachlan, K.-Y. Lue, Hidetoshi Ota, Paulo Passos, Jean-Claude Rage, Alan Resetar, Janis A. Roze, Hobart M. Smith, Zbigniew Szyndlar, Paulo E. Vanzolini, John Visser, Harlan D. Walley, and Addison H. Wynn.

For the use of personal libraries, we thank: Richard Estes, Richard Etheridge, Laurence M. Klauber, Ernie Liner, Gregory K. Pregill, Douglas A. Rossman, Karl P. Schmidt, and Ernest E. Williams.

For access to and assistance in institutional libraries, we thank: Academy of Natural Sciences of Philadelphia, California Academy of Sciences, Carnegie Museum, Field Museum of Natural History, Los Angeles County Museum, Louisiana State University, National Museum of Natural History, San Diego Natural History Museum, San Diego State University, University of Florida Department of Herpetology, University of Michigan Museum of Zoology, and University of Texas at Arlington. We are especially indebted to the staff of the Ernst Mayr Library (MCZ) for assistance with literature over the past two decades (Dorothy Bahr, Ronnie Broadfoot, Roxane Coombs, Dana Fisher, Connie Rinaldo, and Mary Sears).

In particular we especially thank the staff of the rare book rooms and collections who have assisted us in our search for literature: Ewell Sale Stewart Library and Wolf Room (ANSP), Rare Book Room (FMNH), Special Collections (MCZ: Roxane Coombs, Dana Fisher), and Laurence M. Klauber Room (SDSNH: Judy Dyer).

KLW acknowledges 1990 and 1995 summer fellowships from Northwestern State University and Fleming Thomas, who enjoyed the challenges given to him in finding rare articles via interlibrary loan.

For opinions and advice on nomenclatural problems we earnestly thank: Patrick David, Jim Dixon, Carl Gans, Donald E. Hahn, Michael Harvey, Edmond V. Malnate, Paula Passos, Franklin D. Ross, Hobart M. Smith, and Larry D. Wilson.

For advice on, and in some cases reading over, specific accounts we wish to thank: Donald Broadley on *Adenorhinos, Atheris, Bitis, Montatheris*, and *Proatheris*; Laurent Chirio on *Prosymna*; Patrick David on *Opisthotropis* and *Amphiesma*; Jim Dixon on *Liophis*; Michael Harvey on *Dipsas*; Travis J. La Duc on *Adelphicos*; Thales de Lema on *Apostolepis* and *Phalotris*; Rafaqat Mansoor on *Spalerosophis*; Zoltán Tamás Nagy & Josef F. Schmidtler on *Eirenis*; Jean-Claude Rage looked at fossil genera starting with "P" and also *Simoliophis*; Dean Ripa on *Lachesis*; Douglas A. Rossman on *Adelophis, Helicops*, Regina, *Seminatrix, Storeria*, and *Thamnophis*; Gernot Vogel on *Boiga* and *Dendrelaphis*; Larry D. Wilson on *Geophis* and *Tantilla*, and all the Central American species accounts, and Wolfgang Wüster on *Naja*.

Last, heartfelt thanks from we three authors go to our academic advisor at LSU, Douglas A. Rossman, who made our graduate study a rewarding, educational, and enjoyable experience.

Valid Genera and Species

A

ACALYPTOPHIS Boulenger, 1896a (*nomen substitutum*) (Elapidae)

Synonyms: *Acalyptus* A.-M.-C. Duméril, 1853 *(nomen praeoccupatum), Acolyptus* – Higgins, 1873 *(nomen incorrectum), Acalyphis* – Fayrer, 1877 *(nomen incorrectum), Acalyphus* – Hoffmann, 1890 *(nomen incorrectum), Acatyptus* – Hoffmann, 1890 *(nomen incorrectum), Pseudodistira* Kinghorn, 1926, *Agalyotophis* – Maass-Berlin, 1933 *(nomen incorrectum), Acalyitephis* – Briceño-Rossi, 1934 *(nomen incorrectum), Acalypthopis* – Rosenfeld, 1963 *(nomen incorrectum),* and *Acalptophis* – Minton & Dunson, 1985 *(nomen incorrectum).*

Type species: *Acalyptus peronii* A.-M.-C. Duméril, 1853.

Distribution: Marine waters of Sino-Australia, including Arafura Sea, Coral Sea, Gulf of Carpentaria, Gulf of Thailand, South China Sea and Timor Sea.

Sources: E.H. Taylor, 1965, McDowell, 1972a, L.A. Smith, 1974, Cogger, 1975, Cogger et al., 1983a, Ehmann, 1992, Golay et al., 1993, O'Shea, 1996, Ineich & Rasmussen, 1997, David & Ineich, 1999, Bauer & Sadlier, 2000, Ineich & Laboute, 2002, Wells, 2007, Xiang & Li, 2009, Zaher et al., 2009, I. Das, 2010, 2012, A.R. Rasmussen et al., 2011 and Hoser, 2012e.

Remarks: A synonym of *Hydrophis fide* Sanders et al., 2013.

1. *Acalyptophis peronii* (A.-M.-C. Duméril, 1853). Mém. Acad. Sci. Paris 23: 522. (*Acalyptus peronii*)

Synonyms: *Acalyptus superciliosus* A.-M.-C. Duméril, Bibron & Duméril, 1854b, *Acalyptus superciliosus peroni* Duméril, Bibron & Duméril, 1854b, and *Pseudodistira horrida* Kinghorn, 1926.

Type: Holotype, MNHN 7711 (F. Péron & C.A. Lesueur, 1800–1804).

Type locality: "Nouvelle-Hollande?" [= Australia].

Distribution: Sino-Australia. Southern China (Guangdong, Hong Kong), Taiwan, Vietnam, Malaysia, E Indonesia (West Papua: Joes Is.), Papua New Guinea (Western), Australia (Northern Territory, Queensland, N Western Australia) and New Caledonia.

ACANTHOPHIS Daudin, 1803c (*nomen susbstitutum*) (Elapidae)

Synonyms: *Acanthurus* Daudin, 1803b *(nomen praeoccupatum), Acantophis* – Fischer von Waldheim, 1808 *(nomen incorrectum), Acanthophis* Leach, 1814 *(nomen praeoccupatum), Ophryas* Merrem, 1820, *Ophyrus* – Bory de Saint-Vincent, 1824 *in* 1822–1831 *(nomen incorrectum), Acantophis* Gray, 1825 *(nomen emendatum), Ophyas* – Gray, 1825 *(nomen incorrectum), Acantophis* Berthold *in* Latreille, 1827 *(nomen emendatum), Ophrias* Cuvier, 1829 *(nomen emendatum), Acantrophis* – Cuvier, 1836 *(nomen incorrectum), Orophias* Oken, 1836 *(nomen emendatum), Acanthopes* – Swainson, 1839 *(nomen incorrectum), Acauthophis* – Steindachner, 1867 *(nomen incorrectum), Acantopis* – Senna, 1886 *(nomen incorrectum), Acanthopsis* Sherborn, 1902 *(nomen emendatum), Acanthiophis* – Nikolsky, 1916 *(nomen incorrectum), Ancanthophis* – Fairley, 1929 *(nomen incorrectum),* and *Acanthopus* – Trethewie, 1956 *(nomen incorrectum).*

Type species: *Acanthophis cerastinus* Daudin, 1803c.

Distribution: Austro-Papua.

Sources: F. Werner, 1923b, Storr, 1981b, Cogger et al., 1983a, Schwaner et al., 1985, Hoser 1989, 1995, 1998b, 2002a, 2012e, Hutchinson, 1990, Ehmann, 1992, Golay et al., 1993, Greer, 1997, Keogh, 1998, David & Ineich, 1999, Cogger, 2000, Fry et al., 2001, 2002, Scanlon & Lee, 2004, Wüster et al., 2005, Sanders et al., 2008 and Zaher et al., 2009.

1. *Acanthophis antarcticus* (G. Shaw & Nodder, 1802 *in* 1789–1813). Nat. Misc. 13: pl. 535, 1 p. (*Boa antarctica*)

Synonyms: *Boa palpebrosa* G. Shaw, 1802 *(nomen oblitum), Acanthophis cerastinus* Daudin, 1803c, *Acanthophis brownii* Leach, 1814, *Boa ambigua* Leach, 1814, *Ophryas acantophis* Merrem, 1820, *Vipera sorda* Salvado, 1851 *(nomen nudum), Boa aculeata* Boulenger, 1896 (*nomen nudum*), *Acanthopus antarcticus* – Trethewie, 1956 *(nomen incorrectum), Acanthophis barkley* – Mollier, Chwetzoff, Frachon & Ménez, 1989 *(nomen incorrectum), Acanthophis schistos* Wells & Wellington, 1985, and *Acanthophis antarcticus cliffrosswellingtoni* Hoser, 2002a.

Type: Holotype, not designated, a 305–380 mm specimen, lost *fide* Cogger et al. (1983: 217).

Type locality: "Australasia."

Distribution: Eastern and S Australia (New South Wales, Queensland, S South Australia, N Victoria, S Western Australia, Boxer, Fraser, Groote Eylandt Hareby, Figure of Eight, Hook, Reevesby and South Twin Peak Is.), NSL–350 m.

Sources: F. McCoy, 1878f, Johnston & Ellins, 1979, Shine, 1980d, F. Parker, 1982, McDowell, 1984, G.R.

Johnston, 1987, Bush et al., 1995, O'Shea, 1996, J.C. Murphy & Schlager, 2003 and Wüster et al., 2005.

Remarks: In accordance with Art. 23.9.2 of the Code (ICZN, 1999), *Boa antarctica* Shaw & Nodder is designated a *nomen protectum* and *Boa palpebrosa* Shaw a *nomen oblitum*.

2. *Acanthophis ceramensis* A.C.L.G. Günther, 1863. Proc. Zool. Soc. London 31(1): 58. (*Acanthophis cerastinus ceramensis*)

Synonyms: *Acanthophis groenveldi* Hoser, 2002a, and *Acanthophis macgregori* Hoser, 2002a.

Types: Syntypes (3), BMNH 1863.2.28.34–36, one male, females, and juveniles (Stevens) 31 Oct.–18 Dec. 1859 or 26 Feb.–4 Apr. 1860.

Type locality: "North Ceram" [= N Seram, Maluku, E Indonesia].

Distribution: Eastern Indonesia (Maluku: Seram, Tanimbar).

3. *Acanthophis hawkei* Wells & Wellington, 1985. Aust. J. Herp. (Suppl. 1): 43.

Synonym: *Acanthophis woolfi* Hoser, 1998b.

Type: Holotype, NTM 3677, an adult specimen (H. van Dyk, 20 April 1977).

Type locality: "1.5 miles south west of Brunette Downs Station Homestead, Barkly Tablelands. Northern Territory, Australia."

Distribution: Northern Australia (Northern Territory), 215 m.

Sources: Hoser, 1998b, Wickramaratna et al., 2003a and Wüster et al., 2005.

4. *Acanthophis laevis* Macleay, 1877. Proc. Linn. Soc. N.S.W. (1877–1878) 2(1): 40–41.

Synonyms: *Acanthophis crotalusei* Hoser, 1998b *(nomen incorrigendum), Acanthophis barnetti* Hoser, 1998b, *Acanthophis crotalusi* Wüster et al., 2001b *(nomen corrigendum),* and *Acanthophis yuwoni* Hoser, 2002a.

Type: Holotype, AMS 31932 (formerly MMS 693), a 398+ mm specimen (W.J. Macleay, July 1875).

Type locality: "Katow, New Guinea" [= Mawatta, S Western Prov., Papua New Guinea, ca. 9°05'S, 142°58'E, elevation 20 m].

Distribution: Papua New Guinea (Central, Eastern Highlands, East Sepik, Enga, Gulf, Madang, Morobe, National Capital District, Oro, Simbu, Southern Highlands, Western, Western Highlands, West Sepik and Karkar Is.), NSL–1800 m.

Sources: Goldman et al., 1969, O'Shea, 1996 and Wüster et al., 2005.

Remarks: Possibly a synonym of *A. praelongus fide* Wüster et al., 2005.

5. *Acanthophis praelongus* E.P. Ramsay, 1877. Proc. Linn. Soc. N.S.W. (1877–1878) 2(1): 72–73.

Type: Holotype, AMS 451, a 419 mm specimen (W. Powell).

Type locality: "Cape York, North Australia" [probably near Somerset, N Queensland].

Distribution: Austro-Papua. Extreme E Indonesia (West Papua, Kai Is.), Papua New Guinea (East Sepik, Western) and N Australia (N Northern Territory, N Queensland, ext. N Western Australia), NSL–600 m.

Sources: Storr, 1981b, O'Shea, 1996 and Wickramaratna et al., 2003a.

6. *Acanthophis pyrrhus* Boulenger, 1898d. Ann. Mag. Nat. Hist. (7) 2(7): 75.

Synonym: *Acanthophis armstrongi* Wells & Wellington, 1985.

Type: Holotype, BMNH 1946.1.18.62, a 530 mm female (E. Gerrard, 1850–1898).

Type locality: "Station Point, Southern Central Australia."

Distribution: Western Australia (S Northern Territory, ext. SW Queensland, W South Australia, Western Australia), NSL– 895 m.

Source: Schulz, 1990a.

7. *Acanthophis rugosus* Loveridge, 1948. Bull. Mus. Comp. Zool. 101(2): 392–393. (*Acanthophis antarcticus rugosus*)

Synonyms: *Acanthophis lancasteri* Wells & Wellington, 1985, *Acanthophis cummingi* Hoser, 1998b *(nomen incorrigendum), Acanthophis lancasteri bottomi* Hoser, 1998, and *Acanthophis cummingae* – Wüster et al., 2001b *(nomen corrigendum).*

Type: Holotype, MCZ 22812, a 595 mm male (P.T.L. Putnam, 1927).

Type locality: "Merauke, southwest Dutch New Guinea" [= Merauke, SE Papua, ext. E Indonesia, 8°30'S, 140°24'E, elevation NSL].

Distribution: Austro-Papua. Eastern Indonesia (SE Papua), possibly S Papua New Guinea (Western) and N Australia (N Northern Territory, NW Queensland, NE Western Australia), NSL–1500 m.

Sources: O'Shea, 1996, Hoser, 1998a, Aplin & Donnellan, 1999, Fry et al., 2001, 2002, Wickramaratna et al., 2003a–b and Wüster et al., 2005.

8. *Acanthophis wellsi* Hoser, 1998b. Monitor 9(2): 37–39, 2 photos. (*nomen corrigendum*)

Synonyms: *Acanthophis wellsei* Hoser, 1998b *(nomen incorrigendum), Acanthophis wellsi* Alpin & Donnellan, 1999 *(nomen corrigendum),* and *Acanthophis wellsei donnellani* Hoser, 2002a.

Type: Holotype, WAM 8886, a 250+ mm juvenile (K.H. Burton, before 26 Nov. 1945).

A

Type locality: "Wittenoom Gorge, WA, Lat: 22°15' Long: 118°23'" [= Blue Asbestos Mine, Western Australia, Australia].

Distribution: Western Australia (NW Western Australia), NSL–800 m.

Sources: Aplin & Donnellan, 1999, Fry et al., 2002, Hoser, 2002a, Wickramaratna et al., 2003a and Wüster et al., 2005.

Remarks: Taxonomic status unclear, possibly a synonym of *A. pyrrhus*.

ACHALINUS W.C.H. Peters, 1869 (Xenodermatidae)

Synonyms: *Ophielaps* Sauvage, 1877, *Cochalinus* Rhumbler, 1910, *Achalinopsis* Steindachner, 1913a, *Ophiolaps* – Maki, 1931 *(nomen incorrectum)*, *Achalimus* – Wang & Wang, 1956 *(nomen incorrectum)*, and *Achalinapsis* – Wang & Wang, 1956 *(nomen incorrectum)*.

Type species: *Achalinus spinalis* W.C.H. Peters, 1869.

Distribution: Southeastern Asia.

Sources: Maki, 1931, Bourret, 1935a, Ota & Toyama, 1989a, Zhao & Adler, 1993 and Zaher et al., 2009.

1. *Achalinus ater* Bourret, 1937b. Bull. Gén. Instr. Publ. Hanoi 17(4): 72. *(nomen substitutum)*

Synonym: *Achalinus niger* Bourret, 1935a *(nomen praeoccupatum)*.

Types: Syntypes (4), MNHN 1935.49 and MNHN 1938.128 [formerly RLB M.454, a 325 mm male, RLB M.455, a 349 mm female, RLB M.793, a 372 mm female, and RLB M.795, a 252 mm male] (R.L. Bourret, 1934), location of other syntypes unknown.

Type locality: "Tonkin (Tam dao alt. 900m)" [Vietnam].

Distribution: Southern China (Gansu, Guangxi, Guizhou) and N Vietnam (Vinh Phuc), 450–1100 m.

Sources: Hu et al., 1973, Zhao & Jiang, 1977, Wen, 1983, Orlov et al., 2000, Zhao, 2006 ,V.S. Nguyen et al., 2009 and I. Das, 2010, 2012.

Remarks: Original description in Bourret (1935a: 103, fig. 2). Types not in MNHN nor listed in MNHN catalogue *fide* David (*in litt*). A synonym of *A. rufescens* Boulenger *fide* M.A. Smith (1943: 127).

2. *Achalinus formosanus* Boulenger, 1908h. Ann. Mag. Nat. Hist. (8) 2(8): 222.

Synonyms: *Achalinopsis sauteri* Steindachner, 1913a, and *Achalinus formosanus chigirai* Ota & Toyama, 1989b.

Type: Holotype, BMNH 1946.1.7.78, an 853–860 mm female (A. Moltrecht, April 1907).

Type locality: "Punkiho, Kagi district, Central Formosa" [= Punkiho, Chiayi Prefecture, cen. Taiwan].

Distribution: Taiwan (Chiayi) and S Japan (Ryukyus: Iriomotejima Is.).

Sources: Maki, 1931, Ota & Toyama, 1989b and Xiang & Li, 2009.

Remarks: Head of holotype illustrated by Ota & Toyama (1989a: fig. 2).

3. *Achalinus hainanus* C. Huang *in* S. Hu, Zhao & Huang, 1975. Acta Zool. Sinica 21(4): 380–381, 384 (English abst.), figs. 3a–c.

Type: Holotype, ASIZB 1076, a 290 mm female (20 Jan. 1964).

Type locality: "Chien Fung Ling, Hainan, altitude 800m" [China].

Distribution: Southern China (Hainan), 800 m. Known only from type locality.

4. *Achalinus jinggangensis* (Zong & Ma, 1983). Acta Herp. Sinica 2(2): 61–62, 63 (English abst.) 3 figs. (*Achalinopsis jinggangensis*)

Type: Holotype, SNHM 82X118, a 460 mm female (native 17 Oct. 1982).

Type locality: "Dajing Village, Jinggangshan, Jiangxi, alt. 940m" [China].

Distribution: Southern China (Jiangxi), 940 m. Known only from type series.

5. *Achalinus meiguensis* S.-Q. Hu & Zhao, 1966. Acta Zootaxon. Sinica 3(2): 162–164, pl. 1, figs. 4–5, pl. 2, figs. 4–6.

Type: Holotype, CIB 639101, an adult female (X.-Y. Tang, 22 May 1963).

Type locality: "Liang Ho Kou, Meigu Hsien, Szechwan, altitude 2,520 m" [= Sichuan Prov., China].

Distribution: Southwestern China (W Sichuan), 1200–2520 m.

Sources: Inger et al., 1990 and P. Guo et al., 1999b.

6. *Achalinus niger* Maki, 1931. Monogr. Snakes Japan: 60–61, figs. 27–28, pl. 19.

Type: Holotype, NSMT H2562 (formerly KIU no. a), a 599–668 mm male (M. Maki, June 1925).

Type locality: "Mt. Arisan, Central Formosa" [= Mt. Ali, ext. S Nantou Co., cen. Taiwan, ca. 23°35'N, 120°49'E, ca. 2000 m].

Distribution: Taiwan (Nantou), 2000 m.

Sources: Huang, 1995, Ota, 1997 and Xiang & Li, 2009.

7. *Achalinus rufescens* Boulenger, 1888c. Ann. Mag. Nat. Hist. (6) 2(7): 43.

Synonyms: *Achalinus meridianus* M.A. Smith, 1923a, and *Stoliczkaia kwangsiensis* Fan, 1931.

Type: Holotype, BMNH 1946.1.12.37, a 290 mm specimen (C. Ford, 1882–1888).

Type locality: "Hongkong," China.

Distribution: Southeastern China (Fujian, Guangdong, Guangxi, Hainan, Hong Kong, Jiangxi, Shaanxi) and N Vietnam (Bac Kan, Cao Bang, Hai Duong, Ha Ting, Lao Cai, Vinh Phuc), 600–1100 m.

Sources: C.H. Pope, 1935, Fang & Wang, 1983, Orlov et al., 2000, Zhao, 2006, V.S. Nguyen et al., 2009 and I. Das, 2010, 2012.

8. *Achalinus spinalis* W.C.H. Peters, 1869. Mber. Königl. Akad. Wiss. Berlin 1869(5): 436–437, pl., figs. 1a–c.

Synonyms: *Ophielaps braconnieri* Sauvage, 1877, and *Achalinus spinalis weigoldi* Mell, 1931a.

Type: Holotype, ZMB 6449, a 360 mm specimen (S. Braconnier).

Type locality: "Japan."

Distribution: Northern Vietnam (Lao Cai, Thai Nguyen, Vinh Phuc), SE China (Fujian, Gansu, Hubei, Jiangsu, Jiangxi, Sichuan, Shaanxi, Zhejiang) and Japan (Ryukyus: Amamioshima, Edatekushima, Ehime, Horshu, Kyushu, Okinawajima and Tokunoshima Is.), 150–1230 m.

Sources: Maki, 1931, C.H. Pope, 1935, Moriguchi & Naito, 1979, Yamamoto, 1981, 1984, M. Mori, 1984, Inger et al., 1990, Ota et al., 1991, Zou & Chen, 1998, Orlov et al., 2000, Zhao, 2006, V.S. Nguyen et al., 2009, I. Das, 2010, 2012 and Yao, 2012.

9. *Achalinus werneri* Van Denburgh, 1912a. New Rept. Amph. Loo Choo Is.: 8.

Synonym: *Achalinus loochooensis* Thompson, 1912.

Type: Holotype, CAS 22064, a 337 mm specimen (J.C. Thompson a.k.a. V. Kühne, 26 April–1 May 1910).

Type locality: "Nase, Amami O shima, Loo Choo Islands, Japan" [= Ryukyu Is.].

Distribution: Extreme S Japan (Ryukyus: Amamioshima and Okinawajima Is.).

Sources: Barbour, 1917a, Maki, 1931, Koba, 1958, Takara, 1962 and M. Mori, 1986.

Remarks: Original description reprinted in Van Denburgh (1912b: 254–255).

ACRANTOPHIS Jan, 1860 *in* Jan & Sordelli, 1860–1866 (Boidae)

Synonyms: *Pelophilus* Duméril & Bibron, 1844 *(nomen praeoccupatum), Acranthophis* – Jan & Sordelli, 1860 *in* 1860–1866 *(nomen incorrectum),* and *Pelophila* Marschall, 1873 *(nomen praeoccupatum).*

Type species: *Acrantophis dumerili* Jan, 1860 *in* Jan & Sordelli, 1860–1866.

Distribution: Madagascar.

Sources: Guibe, 1949a, Stimson, 1969, Kluge, 1991, Glaw & Vences, 1994, 2007, Vences et al., 2001, Glaw, 2003 and Vences & Glaw, 2003.

Remarks: A synonym of *Sanzinia fide* Kluge, 1991. A valid genus *fide* Vences & Glaw, 2003 and Szyndlar & Rage, 2003.

1. *Acrantophis dumerili* Jan, 1860 *in* Jan & Sordelli, 1860–1866. Icon. Gén. Ophid. 1(1): 1, pl. 2, figs. a–b, d, e–g, i–ii, n, p, r, v.

Type: Holotype, formerly MSNM, destroyed in 1943 during World War II.

Type locality: "Amérique Mérid.?" [= ? South America] (in error). Restricted to Amboasary, Madagascar *fide* Henkel & Schmidt (1995: 269).

Distribution: Southern Madagascar (Atananarivo, Fianarantsoa, Toliara), NSL–1325 m.

Source: Matz et al., 1982.

Remarks: Guibé (1949: 101) erroneously listed three MNHN syntypes. Vences & Glaw, 2003 considered Réunion (St. Denis) records erroneous.

2. *Acrantophis madagascariensis* (A.M. C. Duméril & Bibron, 1844). Erpét. Gén. 6: 524. (*Pelophilus madagascariensis*)

Synonym: *Sanzinia madagascariensis volontany* Vences & Glaw, 2003.

Types: Syntypes (3), MNHN 3133, MNHN 7275, and MNHN 8636, longest syntype 2620 mm (C.A.C.J. Bernier, 1831–1834).

Type locality: "île de Madagascar." Restricted to Mahajunga, Madagascar *fide* Henkel & Schmidt (1995: 270).

Distribution: Northern Madagascar (Antsiranana, Mahajanga, N Toamasina, N Toliara), NSL–1075 m.

Sources: Henkel & Schmidt, 1995, McKeown, 1998 and A. Mori et al., 2006.

Remarks: Syntypes are MNHN 43 and MNHN 7329 *fide* Vences & Glaw (2003: 193).

ACROCHORDUS Hornstedt, 1787 (Acrochordidae)

Synonyms: *Acrochordon* Fischer von Waldheim, 1808 *(nomen emendatum), Chersydrus* Cuvier, 1817, *Chersidrus* Oken, 1817 *(nomen emendatum), Verrucator* Merrem *in* Kuhl, 1820 *(nomen nudum), Acrocordus* Gray, 1825 *(nomen emendatum), Chersydreas* Gray, 1825 *(nomen emendatum), Chershydrus* Bonaparte, 1831 *(nomen emendatum), Scroggilurus* Lesson, 1832 *in* Bélanger, 1831–1834, *Verrucator* Schlegel, 1837 *(nomen nudum), Achrocordus* – Swainson, 1839 *(nomen incorrectum), Achrochordus* J.T. Reinhardt, 1843 *(nomen emendatum), Acrocordus* Agassiz, 1844 *(nomen emendatum), Chersydraeas* Gray, 1849a *(nomen emendatum), Potamophis* P. Schmidt, 1852 *(nomen praeoccupatum), Chersydraeus* A.-M.-C. Duméril, Bibron & Duméril,

1854a *(nomen emendatum), Acrochordas* – Fitzinger, 1867 *(nomen incorrectum), Chersydraeus* – Westphal-Castelnau, 1870 *(nomen incorrectum), Chrysydrus* – Boulenger, 1903f *(nomen incorrectum), Acrochodon* – Nutaphand & Tumvipart, 1982 *(nomen incorrectum), Acrocodus* – Nutaphand, 1991b *(nomen incorrectum),* and *Acrocodus* – Nutaphand, 1991b *(nomen incorrectum).*

Type species: *Acrochordus javanicus* Hornstedt, 1787.

Distribution: Indo-Australia.

Fossil records: Lower Miocene, upper Miocene and lower Pliocene of Pakistan, India and Nepal.

Sources: M.A. Smith, 1939, 1943, Bergman, 1958c, Hoffstetter & Gayrard, 1965, Hardaway & Williams, 1976, Groombridge, 1979, McDowell, 1979, 1986, S. Sharma & Nakhasi, 1980, Rieppel, 1988, Heise et al., 1995, McDiarmid et al., 1999, Gravlund, 2001, Kelly et al., 2003, Head et al., 2007 and I. Das, 2010, 2012.

1. *Acrochordus arafurae* McDowell, 1979. J. Herp. 13(1): 75–77, figs. 20b, 21b, 22b, 23b, 26b.

Type: Holotype, AMNH 59887, ca. 840 mm male (skinned out behind head) (A.L. Rand [2nd Archbold-Rand Exped.], Feb. 1936–Jan. 1937).

Type locality: "Papua New Guinea: Western Prov.: Lake Daviumbo" [= vicinity of Kaviananga, Fly River, W Western Prov., SW Papua New Guinea, 7°36'S, 141°18'E, elevation 15 m].

Distribution: Austro-Papua. Eastern Indonesia (Papua), S Papua New Guinea (Central, Western), N Australia (N Northern Territory, N Queensland), NSL–215 m.

Sources: Cogger et al., 1983a, Shine, 1985b, 1986a, S.K. Wilson & Knowles, 1988, Hoser, 1989, D. Houston & Shine, 1993, 1994, O'Shea, 1996, Cogger, 2000 and Madsen & Shine, 2001.

2. †*Acrochordus dehmi* Hoffstetter, 1964. Bull. Soc. Géol. France (7) 6(4): 467–473, fig. 2a.

Type: Holotype, BSPG 1956-II-64, one trunk vertebra (R. Dehm, T. zu Oettingen & H. Vidal).

Type locality: "gisement de Chhoinja (form. Chingi, Miocène supérieur) couches des Siwaliks, au Nord du Pakistan occidental" [= upper Miocene of Punjab, N Pakistan, 33°03'N, 72°30'E].

Distribution: Upper Miocene (3.5–11.1 mya) of Pakistan, India and Nepal.

Sources: Rage, 1984b, West et al., 1991, Rage et al., 2001, Head, 2005 and Head et al., 2007.

3. *Acrochordus granulatus* (J.G. Schneider, 1799). Hist. Amph. 1: 243–244. (*Hydrus granulatus*)

Synonyms: *Anguis granulatus acrochordus* J.G. Schneider, 1801, *Acrochordus fasciatus* G. Shaw, 1802, *Chersydrus annulatus* Gray, 1849a, and *Chersydrus granulatus luzonensis* Loveridge, 1938d.

Type: Holotype, LM, lost *fide* McDowell (1979: 79).

Type locality: Unknown. Restricted to India *fide* M.A. Smith (1943: 134).

Distribution: Indo-Australia. India (Andaman & Nicobars, Andhra Pradesh, Gujarat, Karnataka, Kerala, Maharashtra, Orissa, Tamil Nadu, West Bengal), Sri Lanka (Northern, North-Western, Western), Singapore, Myanmar, S Thailand (Chumphon, Phang Nga, Phetchaburi, Phuket, Songkhla), Cambodia, Vietnam (Bac Ria-Vung Tau, Binh Dinh, Khanh Hoa), China (Hainan), Brunei, East Malaysia (Sabah, Sarawak), Indonesia (Papua, West Papua, Ambon, Bali, Borneo, Flores, Java, Jobi, Komodo, Misool, Schouten, Sulawesi: Central, North, Southeast, Southwest, Sumatra, Ternate, Timor, Weh), Philippines (Bantayan, Cebu, Luzon, Masbate, Palawan, Panay), Papua New Guinea (Central, East Sepik, Gulf, Madang, Milne Bay, Western, d'Entrecasteaux Is., W New Britain, New Ireland, North Solomons), and N Australia (ext. NE Western Australia, ext. N Northern Territory, NE Queensland), NSL–165 m.

Sources: Loveridge, 1938d, E.H. Taylor, 1965, McDowell, 1979, Voris & Glodek, 1980, Cogger et al., 1983a, Tweedie, 1983, S.K. Wilson & Knowles, 1988, Ehmann, 1992, Dutta & Acharjyo, 1993, Zhao & Adler, 1993, I. Das, 1994, Gaulke & Altenbach, 1994, David & Vogel, 1996, O'Shea, 1996, M.J. Cox et al., 1998, Cogger, 2000, Ferner et al., 2001, Whitaker & Captain, 2004, N. Khaire, 2006, V.S. Nguyen et al., 2009 and A. Silva et al., 2010.

4. *Acrochordus javanicus* Hornstedt, 1787. Kon. Svenska Vet. Akad. Handl. 8: 307–308, pl. 12.

Synonyms: *Acrochordus dubius* G. Shaw, 1802, *Acrochordvs iauanicus* – Merrem, 1820 *(nomen incorrectum), Verrucator merremii* Merrem *in* Kuhl, 1820 *(nomen nudum),* and *Potamophis javanica* – P. Schmidt, 1852.

Type: Holotype, not designated, ? NHR or ZIUS, a preserved skin (C.F. Hornstedt, 1702–1785).

Type locality: "Java" [SW Indonesia].

Distribution: Southeast Asia and Greater Sundas. Thailand (Krung Thep Mahanakhon, Nakhon Pathom), Singapore, Vietnam (Binh Phuoc, Can Tho, Dong Nai, Ho Chi Minh City), Brunei, East Malaysia (Sarawak) and Indonesia (Java, Kalimantan, Sumatra), NSL.

Sources: E.H. Taylor, 1965, McDowell, 1979, 1999, Tweedie, 1983, David & Vogel, 1996, M.J. Cox et al., 1998 and V.S. Nguyen et al., 2009.

Remarks: Original description reprinted in Hornstedt (1788: 284–285). Type specimen brought to Europe by Mr. Hornstedt and deposited in Museum of the King of Sweden *fide* Shaw (1802: 574); it was cooked and eaten *fide* Bergman (1958: 145).

ACUTOTYPHLOPS Wallach, 1995a (Typhlopidae)

Synonym: *Martinwellstyphlops* Hoser, 2012am *(nomen illegitimum).*
Type species: *Typhlops subocularis* Waite, 1897b.
Distribution: East Indies.
Sources: Wallach, 1993a, 1995a, O'Shea, 1996, Wallach & Ineich, 1996, McDiarmid et al., 1999, Wallach et al., 2007 and Vidal et al., 2010.

1. *Acutotyphlops banaorum* Wallach, Brown, Diesmos & Gee, 2007. J. Herp. 41(4): 1692–695, figs. 2–4 (paratypes).

Type: Holotype, PNM 9280 (formerly GVAG 219 & FMNH 259604), a 125 mm male (G.V.A. Gee, 28 March 2001).
Type locality: "an irrigation ditch in a muddy area of water pools near Barangay Balbalasang, Municipality of Balbalan, Kalinga Prov., Luzon Island, Philippines (17°29'N, 121°03'E), 900 m above sea level."
Distribution: Northern Phillipines (Luzon), 900–1050 m.

2. *Acutotyphlops infralabialis* (Waite, 1918b). Rec. So. Aust. Mus. 1(1): 35–36, fig. 25. (*Typhlops infralabialis*)

Synonyms: *Typhlops bergi* J.A. Peters, 1948, and *Typhlops adamsi* V.M. Tanner, 1951.
Type: Holotype, AMS 4609, a 305–315 mm specimen (J. Caulfield).
Type locality: "Malaita, Solomon Islands" [= Malaita Island, Solomon Islnads, bet. 8°18'–9°44'S, 160°34'–161°35'E].
Distribution: Eastern Papua New Guinea (North Solomons: Bougainville) and S Solomons (Guadalcanal, Malaita, New Georgia), 15–245 m.
Source: Shea & Sadlier, 1999.

3. *Acutotyphlops kunuaensis* Wallach, 1995a. Asiatic Herpetol. Res. 6: 143–147, figs. 4a–b, 5.

Type: Holotype, MCZ 76964, a 221 mm male (F.S. Parker, 19 Aug. 1963).
Type locality: "Kunua, coastal northwestern Bougainville Island, North Solomons Prov., extreme eastern Papua New Guinea, 5°46'S, 154°43'E, elevation *ca.* 30 m."
Distribution: Extreme E Papua New Guinea (North Solomons: Bougainville), 30 m.

4. *Acutotyphlops solomonis* (H.W. Parker, 1939). Bull. Mus. Roy. Hist. Nat. Belg. 15(60): 4–5, 2 figs. (*Typhlops solomonis*)

Type: Holotype, IRSNB 2029, a 407 mm specimen (J.B. Poncelet, 2 June 1938).
Type locality: "Buin, 10 miles inland on Bougainville, Solomon Islands" [= Buin, Bougainville Island, Solomon Islands, Papua New Guinea, 6°44'S, 155°41'E, elevation 50 m].
Distribution: Eastern Papua New Guinea (Milne Bay, North Solomons: Bougainville), NSL–50 m.

5. *Acutotyphlops subocularis* (Waite, 1897b). Rec. Aust. Mus. 3(3): 69–70, figs. 1–3. (*Typhlops subocularis*)

Synonym: *Typhlops keasti* Kinghorn, 1948.
Type: Holotype, AMS 2202, a 360 mm specimen (G. Brown).
Type locality: "Duke of York Island" [Bismarck Archipelago, Papua New Guinea].
Distribution: Papua New Guinea (Bismarck Arch., Duke of York, New Britain, New Ireland, Umboi) and Solomons (Bougainville, Guadalcanal, Malaita, New Georgia).
Sources: McDowell, 1974a and J.C. Murphy & Schlager, 2003.

ADELOPHIS Dugès *in* Cope, 1879 (Natricidae)

Type species: *Adelophis copei* Dugès *in* Cope, 1879.
Distribution: Southwestern Mexico.
Sources: Rossman & Wallach, 1987 and Zaher et al., 2009.
Remarks: *Adelophis foxi* is nested within *Thamnophis fide* A. de Queiroz et al., 2002, and Zaher et al., 2012.

1. *Adelophis copei* Dugès *in* Cope, 1879. Proc. Amer. Philos. Soc. 18(104): 265–266.

Type: Neotype, MDUG, an unnumbered specimen, designated by H.M. Smith & Necker (1943: 182).
Type locality: "Cupátaro, Mexico" [possibly = Tupátaro, N Michoacán, SW Mexico, 19°30'N, 101°30'W, elevation 2230 m] via neotype selection.
Distribution: Southwestern Mexico (Guanajuato, Jalisco, Michoacán, Morelos), 1215–2580 m.
Source: Dugès, 1887.
Remarks: Original description reprinted in Dugès (1887: 18–20, pl. 3). Holotype lost *fide* Dugès (1887: 20).

2. *Adelophis foxi* Rossman & Blaney, 1968. Occ. Pap. Mus. Zool., Louisiana St. Univ. (35): 1–6, fig. 3 (upper).

Type: Holotype, LSUMZ 14330, a 419 mm female (D.A. Rossman & R.M. Blaney, 16 July 1966).
Type locality: "meadow in pine forest (8,600 feet) 1/4 mi. E Mil Diez (approx. 2 mi. W El Salto), Durango, México."
Distribution: Northwestern Mexico (Durango), 1580–2620 m. Known only from type locality.

A

ADELPHICOS Jan, 1862b (Dipsadidae)

Synonyms: *Rhegnops* Cope, 1866, *Rheynops* – Hoffmann, 1890 *(nomen incorrectum), Adelphisco* – Briceño-Rossi, 1934 *(nomen incorrectum),* and *Adelphicus* – W.W. Tanner & Avery, 1982 *(nomen incorrectum).*
Type species: *Adelphicos quadrivirgatum* Jan, 1862b.
Distribution: Eastern Mexico and upper Central America.
Sources: H.M. Smith, 1942d, J.A. Campbell & Ford, 1982, Cadle, 1984b, LaDuc, 1995, Kohler, 2001 and Zaher et al., 2009.

1. *Adelphicos daryi* J.A. Campbell & Ford, 1982. Occ. Pap. Mus. Nat. Hist. Univ. Kansas (100): 3–5, figs. 2, 4 (upper), 5d.

Type: Holotype, KU 187260, a 487 mm female (J.A. Campbell, 11 July 1980).
Type locality: "San Jorge Muxbal, 5.5 km W San José Pinula, Department of Guatemala, Guatemala, elevation 1844 m."
Distribution: Southeastern Guatemalan (Guatemala), 1300–2135 m.

2. *Adelphicos ibarrorum* J.A. Campbell & Brodie, 1988. Herpetologica 44(4): 416–419, figs. 1–2.

Type: Holotype, UTA 18696, a 521 mm female (J.A. Campbell, 14 June 1987).
Type locality: "3.0 km SSE Chichicastenango, in valley between Paxot and Caminibal, 2100 m, Departamento de Quiché, Guatemala."
Distribution: Southern Guatemala (Quiché), 2000–2100 m.

3. *Adelphicos latifasciatum* Lynch & Smith, 1966. Trans. Kansas Acad. Sci. 69(1): 66–68, fig. 3c. (*Adelphicos veraepacis latifasciatus*)

Type: Holotype, UIMNH 56147, a 333 mm male (T. MacDougall, 6–12 Sept. 1963).
Type locality: "Sierra Madre north of Zanatepec, Oaxaca, Mexico."
Distribution: Southern Mexico (Sierra Madre de Chiapas of Oaxaca), 1000–2000 m.

4. *Adelphicos nigrilatum* H.M. Smith, 1942d. Proc. Rochester Acad. Sci. 8: 182–185. (*Adelphicos veraepacis nigrilatus*)

Type: Holotype, FMNH 100110, a 318 mm female (H.D. Thomas, 1937).
Type locality: "San Cristóbal, Chiapas, Mexico." [= 16°34'N, 93°50'W, elevation 680 m].
Distribution: Southern Mexico (Meseta Central of Chiapas), 680–2900 m.

5. *Adelphicos quadrivirgatum* Jan, 1862b. Arch. Zool. Anat. Fis. 2(1): 19, pl. 7.

Synonyms: *Rhegnops visonbinus* Cope, 1866, *Rhegnops sargii* J.G. Fischer, 1885b, *Adelphicos quadrivirgatus acutirostrum* Bocourt 1883 *in* A.H.A. Duméril, Bocourt & Mocquard, 1870–1909, and *Adelphicos newmanorum* E.H. Taylor, 1950a.
Type: Holotype, formerly MSNM, destroyed in 1943 during World War II.
Type locality: "Giava" [= Java, Indonesia] (in error). Restricted to Jicaltepec, Veracruz, Mexico *fide* H.M. Smith & Taylor (1950a: 348).
Distribution: Mexico and upper Central America. Southeastern Mexico (E Chiapas, Hidalgo, Oaxaca, Puebla, N Querétaro, San Luis Potosí, Tabasco, S Tamaulipas, Veracruz), Belize (Cayo, Orange Walk), Guatemala (Alta Verapaz, Petén), Honduras (Atlántica, Comayagua, ? Colón, Copán, Cortés, Francisco Morazán, Gracias a Dios, Intibucá, Lempira, Ocotepeque, Olancho, Santa Bárbara, Yoro) and Nicaragua (Jinotega), NSL–1900 m.
Sources: L.D. Wilson & Meyer, 1985, J.C. Lee, 1996, 2000, J.A. Campbell, 1998b, Canseco-Marquez et al., 2000, Stafford & Meyer, 2000, Murillo et al., 2003, Mendoza-Quijano et al., 2003, 2005, McCranie et al., 2006, Sunyer et al., 2009 and McCranie, 2011a.
Remarks: Original description reprinted in Jan (1862c: 19, pl. 7). *Adelphicos quadrivirgatum, A. sargii* and *A. visoninus* valid species *fide* H.M. Smith et al., 2001, also supported by Mendoza-Quijano, 2003, 2005.

6. *Adelphicos veraepacis* L.C. Stuart, 1941b. Occ. Pap. Mus. Zool., Univ. Michigan (452): 5–6.

Type: Holotype, UMMZ 89073, a 355 mm male (L.C. Stuart, 21 April 1938).
Type locality: "Cloud forest above Finca Samac, 7 km. west of Cobán, Alta Verapaz, Guatemala. Altitude, about 1500 m."
Distribution: Central Guatemala (Alta Verapaz, Baja Verapaz, Huehuetenango), 1200–2200 m.

AELUROGLENA Boulenger, 1898e (Colubridae)

Synonym: *Aeluroglaena* – Scortecci, 1939a *(nomen incorrectum).*
Type species: *Aeluroglena cucullata* Boulenger, 1898e.
Distribution: Northeast Africa.
Sources: Bogert, 1940, H.W. Parker, 1949, Lanza, 1983a, 1990b, Schätti, 1985a, Largen & Rasmussen, 1993 and Largen & Spawls, 2010.
Remarks: Possibly a synonym of *Meizodon fide* Schätti (1985a: 170).

1. *Aeluroglena cucullata* Boulenger, 1898e. Ann. Mag. Nat. Hist. (7) 2(8): 132–133.

Type: Holotype, BMNH 1946.1.14.87, a 375 mm female (E. Lort-Phillips, 1896–1897).

Type locality: "Goolis Mountains, Somaliland" [= Qar Goilis Mountains, W Togdheer Region, NW Somalia, 9°52'N, 45°55'E].

Distribution: Eastern Ethiopia (Hararge) and NW Somalia (Togdheer, Woqooyi Galbeed), NSL–700 m.

AFRONAJA Wallach, Wüster & Broadley, 2009 (Elapidae)

Synonyms: †*Palaeonaja* Hoffstetter, 1939, †*Paleonaja* – Bogert, 1943 *(nomen incorrectum),* and *Spracklandus* Hoser, 2009c *(nomen illegitimum).*

Type species: *Naja nigricollis* J.T. Reinhardt, 1843.

Distribution: Africa.

Fossil records: Lower Miocene to middle Pleistocene of Europe and Africa. Fossils unassigned to species include: lower Miocene (Orleanian, MN 3–4: 16.0–20.0 mya) of Austria, Saudi Arabia and Spain, upper Miocene (Turolian, MN 13: 4.9–7.3 mya) of Greece, middle Pliocene (Ruscinian, MN 15: 3.2–4.2 mya) of Turkey, upper Pliocene (Villanyian, MN 16: 2.6–3.2 mya) of France, Greece and Spain, and middle Pleistocene (Ionian: 0.13–0.78 mya) of Greece.

Sources: F. Werner, 1923b, Bogert, 1943, Klemmer, 1968, Broadley, 1968c, 1974, Minton et al., 1970, Golay, 1985, Szyndlar & Rage, 1990, Ulber, 1995, Golay et al., 1993, Keogh, 1998, David & Ineich, 1999, Slowinski & Keogh, 2000, Dobiey & Vogel, 2007, Wüster et al., 2007, Hoser, 2009c, 2012a, e, Wallach et al., 2009 and Zaher et al., 2009, 2012.

1. †*Afronaja antiqua* (Rage, 1976). Géol. Méditerr. 3(2): 61–63, figs. 5–6. (†*Naja antiqua*)

Type: Holotype, MNHN BML 903, one incomplete posterior braincase.

Type locality: "Beni Mellal, Maroc, Miocène" [= Beni Mellal, Beni Mellal Prov., cen. Morocco, 32°20'N, 6°21'W, middle Miocene].

Distribution: Middle Miocene (Astaracian, MN 7/8: 11.1–12.8 mya) of Morocco. Known only from type locality.

Source: Rage, 1984b.

Remarks: *Naja incertae sedis fide* Wallach et al. (2009: 32).

2. *Afronaja ashei* (Wüster & Broadley, 2007). Zootaxa (1532): 58–62, figs. 4–5. (*Naja ashei*)

Type: Holotype, NMK S/3993 (formerly BK 100030), a 1507 mm female (R. Taylor, 29 Sept. 2004).

Type locality: "Watamu, Kenya (3°21'S: 40°01'E)."

Distribution: Eastern Africa. Southern Ethiopia (Bale, Gemu G5ofa, Hararge, Shoa, Sidamo), SW Somalia (Gedo, Jubbada Hoose), N Kenya (Coast, North-Eastern) and NE Uganda (Amudat), 300–1750 m.

Sources: Wüster & Broadley, 2007 and Largen & Spawls, 2010.

Remarks: Probably occurs in NE Tanzania and possibly in South Sudan and NE Democratic Republic of the Congo *fide* Wüster & Broadley (2007: 63).

3. *Afronaja crawshayi* (A.C.L.G. Günther, 1894b). Proc. Zool. Soc. London (1893) 62(1): 620. (*Naja nigricollis crawshayi*)

Synonym: *Naja nigricollis crawshawii* – Manjunatha-Kini, Stefansson & Evans, 1987 *(nomen incorrectum).*

Type: Holotype, BMNH, skin of an adult (R.C. Crawshay, 1890–1893, via H.H. Johnston).

Type locality: "Lake Mweru, British Central Africa" [= Lake Meru, N Luapula Prov., ext. N Zambia].

Distribution: Central Africa. Southern Congo (Brazzaville, Plateau, Pool), S Democratic Republic of the Congo (Bandundu, Bas-Congo, Kasai Occidental, Kasai Oriental, Katanga, Kinshasa), Tanzania, Angola (Benguela, Bié, Cuanza Norte, Huíla, Lunda Norte), W and N Zambia Luapula and N Malawi (Northern), 700–1700 m.

Sources: Witte, 1953 and Broadley & Cotterill, 2004.

4. †*Afronaja depereti* (Hoffstetter, 1939). Arch. Mus. Hist. Nat. Lyon 15(3): 66–67, pl. 2, fig. 16. (†*Palaeonaja depereti*)

Type: Holotype, UCBL 92859A, one trunk vertebra.

Type locality: "Pliocène ancien du Roussillon, Fort du Serrat-d'en-Vacquer, près Perpignan" [= Roussillon Basin, eastern Pyrénées-Orientales Department, S France, ca. 42°41'N, 2°53'E; Ruscinian (MN 15), lower Plíocene].

Distribution: Lower Pliocene (Ruscinian, MN 15: 3.2–4.2 mya) of France and Spain.

Fossil records: Upper Miocene (Turolian, MN 13: 4.9–7.3 mya) of Spain and middle Pliocene (Ruscinian, MN 15: 3.2–4.2 mya) of France.

Sources: Rage, 1984b and Szyndlar & Zerova, 1990.

5. †*Afronaja iberica* Szyndlar, 1985a. Estud. Geol. 41(5-6): 455–457, figs. 6–9. (†*Naja iberica*)

Type: Holotype, MNCN Alg. I-31, one basiparasphenoid.

Type locality: "Uppermost Miocene (NM 13) of Algora (Guadalajara, Spain)."

Distribution: Upper Miocene (Turolian, NM 13: 4.9–7.3 mya) of Spain. Known only from type locality.

Remarks: *Naja incertae sedis fide* Wallach et al. (2009: 32).

A

6. *Afronaja katiensis* (Angel, 1922). Bull. Mus. Hist. Nat. Paris 28(1): 40. (*Naja nigricollis katiensis*)

Synonym: *Naja trilepis* Roman, 1968.

Types: Syntypes (6), MNHN 1921.609–10 and MNH 1921.612–15 (Millet-Horsin, 1921).

Type locality: "la région caillouteuse de Bélédougou, à Kati (12 kilomètres au nord de Bamakko), Soudan français" [= Bélédougou region of Mandika people near Kati (12°45'N, 08°04'W, elevation 425 m), S Koulikoro Region, SW Mali].

Distribution: West Africa. Senegal (Diourbel, Kaolack, Kédougou, Tambacounda), Gambia (Western), Guinea (Siguiri), SW Mali (Kayes, Koulikoro, Sikasso), Ivory Coast (Bouna), Burkina Faso (Centre, Centre-Est, Centre-Nord, Centre-Ouest, Est, Hauts-Bassins, Sud-Ouest, Volta-Noire), N Ghana (Northern, Upper East, Upper West), N Togo (Savanes), Niger (Zinder), N Nigeria and N Cameroon (Extreme-Nord, Nord), 150–700 m.

Sources: Villiers, 1950a–b, 1975, Cansdale, 1954, Broadley, 1968a, Roman, 1968, 1973a–b, 1976, 1980, 1989, Hughes & Barry, 1969, Papenfuss, 1969, Hughes, 1983, Schätti, 1986c, Spawls & Branch, 1995, J.-F. Trape & Mané, 2000, 2004, 2006b, Chippaux, 2006, Chirio & LeBreton, 2007 and Segniagbeto et al., 2011.

Remarks: Nima Bélédougou located at 15°04'N, 7°18W, elevation 280 m. Possibly occurs in Benin *fide* Hughes (2013: 153).

7. *Afronaja mossambica* (W.C.H. Peters, 1854). Mber. Königl. Akad. Wiss. Berlin 1854(11): 625. (*Naja mossambica*)

Synonyms: *Naja nigricollis macombique* – Zlotkin, Menashé, Miranda & Lissitzky, 1975 *(nomen incorrectum), Naja massambica massambica* – Grognet, Gatineau, Bougis, Harvey, Coudere, Fromageot & Ménez, 1986 *(nomen incorrectum), Naja mocambique mocambique* – Turell & Middlebrook, 1988 *(nomen incorrectum),* and *Naja mosambica mosambica* – Khole, 1991 *(nomen incorrectum).*

Type: Lectotype ZMB 2811, a 604–1015 mm specimen (W.C.H. Peters, June 1843–Aug. 1847), designated by Golay et al. (1993: 188).

Type locality: "Tette. Sena" [= Tete (16°10'S, 33°36'E, elevation 125 m), Tete Prov., and Vila de Sena (17°26'S, 35°02'E, elevation 50 m), N Sofala Prov., cen. Mozambique].

Distribution: East and S Africa. Southeastern Tanzania (Lindi, Morogoro, Mtwara, Ruvuma, Pemba, Zanzibar Is.), Mozambique (Gaza, Inhambane, Manica, Maputo, Niassa, Sofala, Tete, Zambézia, Mozambique Is.), S Malawi (Southern), SE Democratic Republic of the Congo (S Katanga), SW Angola (Namibe), S Zambia (Central, Eastern, Luapula, Southern, S Western), NE Namibia (Caprivi, Okavango, Oshikoto), E Botswana (Central, Gaborone, Ghanzi, Kgatleng, Kweneng, North East, North West, Southern), Zimbabwe (Bulawayo, Manicaland, Mashonaland Central, Mashonaland East, Mashonaland West, Masvingo, Matabeleland North, Matabeleland South, Midlands), Swaziland and South Africa (Gauteng, KwaZulu-Natal, Limpopo, Mpumalanga, North-West), NSL–1800 m.

Sources: Mertens, 1955, 1971, V.F.M. FitzSimons, 1962, 1974, Witte, 1962, V.J. Wilson, 1965, Broadley, 1971d, 1974, 1983, 1988, Visser, 1972, Stevens, 1974, Broadley & Cock, 1975, Visser & Chapman, 1978, Manaças, 1982, Auerbach, 1985, 1987, Patterson, 1987, Branch, 1988, Broadley & Howell, 1991, Boycott & Culverwell, 1992, Spawls & Branch, 1995, Clauss & Clauss, 2002, Spawls et al., 2002, Wüster & Broadley, 2003, 2007 and Broadley & Blaylock, 2013.

8. *Afronaja nigricincta* (Bogert, 1940). Bull. Amer. Mus. Nat. Hist. 77(1): 89–90, pl. 1, fig. 1. (*Naja nigricollis nigricinctus*)

Synonyms: *Naja nigra* A. Smith, 1838 *(nomen praeoccupatum), Naja nigricollis fasciata* Bocage, 1895 *(nomen praeoccupatum), Merremia haemachates* F. Werner *in* Schultz, 1910 *(lapsus calami),* and *Naja nigricollis woodi* Pringle, 1955.

Type: Holotype, AMNH 51823, a 1100+ mm male (A.S. Vernay, H. Lang & R. Boulton [Vernay Angola Exped.], 6 July 1925).

Type locality: "Munhino, Angola, 101 km. east of Mossamedes via railroad, 14° 55'S., 13° 0'E." [= Munhino, Namibe Prov., SW Angola, 14°57'S, 12°58'E, elevation 435–470 m].

Distribution: Southwestern Africa. Southwestern Angola (Benguela, Cuando Cubango, Cuanza Sul, Cunene, Huambo, Huíla, Luanda, Namibe), Namibia (Erongo, Hardap, Karas, Khomas, Kunene, Okavango, Omaheke, Oshikoto, Otjozondjupa) and W South Africa (Northern Cape, Western Cape), 180–1630 m.

Sources: Mertens, 1955, 1971, Boycott & Haacke, 1979, Haacke, 1979, Buys & Buys, 1983, Branch, 1998, Broadley *in* Bauer & Branch, 2001 and M. Griffin, 2003.

Remarks: Wüster et al. (2007: 445–446) recognized *N. nigricincta* as a distinct species and placed *woodi* as a subspecies.

9. *Afronaja nigricollis* (J.T. Reinhardt, 1843). Kgl. Danske Vidensk. Selsk. Natur. Math. Afh. 10: 269–272, pl. 3, figs. 5–7. (*Naja nigricollis*)

Synonyms: *Naja nigricollis fasciata* Bocage, 1866a *(nomen praeoccupatum), Naja nigricollis melanoleuca* Bocage, 1895 *(nomen praeoccupatum), Naja nigricollis occidentalis* Bocage, 1895, *Naja nigricollis atriceps* Laurent, 1955, *Naja nigrocollis* – Bonta, Vargaftig, Bhargava & Vos, 1970 *(nomen incorrectum),* and *Naja nigricolis* – Teng, Kuo, Lee & Ouyang, 1987 *(nomen incorrectum).*

Type: Lectotype, ZMUC 65204, a 1647 mm specimen (J.R. Chenon, 1826–1836), designated by F.W. Braestrup *in* Hughes & Barry (1969: 1027).

Type locality: "Guinea" [= coast of Ghana *fide* Hughes & Barry, 1969: 1027]. Restricted to vicinity of Accra, Ghana *fide* J.B. Rasmussen & Hughes (1997: 14, 16).

Distribution: Subsaharan Africa. Southwestern Mauritania (Guidimaka, Hodh El Gharbi, Trarza), Senegal (Dakar, Fatick, Kédougou, Kolda, Saint-Louis, Tambacounda, Thiès), Gambia (MacCarthy Island, Western), Guinea–Bissau (Bafatá, Bissau, Quinara, Bolama and Bubaque Is.), Guinea (Boké, Koundara, Kouroussa, Macenta, Nzérékoré, Télimélé), Sierra Leone (Eastern, Northern, Southern, Western), Ivory Coast (Abengourou, Abidjan, Agboville, Bouaflé, Daloa, Lagunes, Toumodi), SW Mali (Kayes, Koulikoro, Ségou, Sikasso), Burkina Faso (Centre, Centre-Nord, Centre-Ouest, Est, Hauts-Bassins, Nord, Sahel, Sud-Ouest, Volta-Noire), Ghana (Accra, Volta, Western), Togo (Centrale, Kara, Maritime, Plateaux), N Benin (Alibori, Atakora, Atlantique, Collines, Cotonou, Zou), S Niger (S Diffa, Dosso, Niamey, S Zinder), Nigeria (Akwa Ibom, Anambra, Bayelsa, Cross River, Delta, Gongola, Kwara, Oyo, Rivers), SW Chad (Chari-Baguirmi, mayo-Kebbi Ouest), Cameroon (Adamaoua, Centre, Extreme-Nord, Est, Littoral, Nord, Nord-Ouest, Ouest, Sud, Sud-Ouest), Central African Republic (Bamingui-Bangoran, Kemo-Gribingui, Lobaye, Ombella-Mpoko, Ouham, Ouham-Pende, Sangha), E Gabon (Haut-Ogooué), N Democratic Republic of the Congo (Maniema, Nord-Kivu, Orientale), SE South Sudan (Eastern Equatoria), W Ethiopia (Gojjam), Somalia (Bay, Hiraan, Jubbada Dhexe, Jubbada Hoose, Shabeellaha Dhexe, Shabeellaha Hoose), Kenya (Central, Coast, Eastern, Nairobi, North-Eastern, Rift Valley), Uganda (Central, Eastern, Nairobi, Northern, Western), Rwanda (Butare, Byumba, Kigali), Burundi (Bujumbura, Gitega, Muramvya, Ruyigi), Tanzania (Arusha, Dodoma, Iringa, Kigoma, Kilimanjaro, Mara, Morogoro, Mtwara, Mwanza, Rukwa, Ruvuma, Singida, Tabora, Kagera), Malawi (Central, Northern, Southern), Angola (Cunene, Huíla, Namibe) and Zambia (Central, Copperbelt, Eastern, Luapula, Lusaka, Northern, North-Western, Southern, Western), NSL–2500 m.

Sources: Andersson, 1903, 1937, Aylmer, 1922, Loveridge, 1936b, 1955b, Monard, 1937, 1951, Bogert, 1940, Uthmöller, 1941b, 1950a, 1952b, 1954a, 1956b, Cansdale, 1954, Mertens, 1955, Laurent, 1956a, 1960, 1964a, 1973a, Vesey-FitzGerald, 1958, 1975, Condamin, 1959, Perret, 1961, Sweeney, 1961, Condamin & Villiers, 1962, V.F.M. FitzSimons, 1962, 1974, Witte, 1962, Doucet, 1963, V.J. Wilson, 1965, Menzies, 1966, Blackwell, 1967, Broadley, 1968c, 1971d, 1974, Leston & Hughes, 1968, M. Stewart, 1968, Hughes & Barry, 1969, Papenfuss, 1969, Roman, 1969, 1973b, 1976, 1980, Hulselmans & Verheyen, 1970, Hulselmans et al., 1970, Visser, 1972, Pitman, 1974, Courtois & Chippaux, 1977, Girardin, 1978, Russelot, 1978, Spawls, 1978, Visser & Chapman, 1978, Hakansson, 1981, Trape, 1981, Manaças, 1982, Buys & Buys, 1983, Hughes, 1983, 2013, Lanza, 1983a, 1990b, MacKay & MacKay, 1985, J.A. Butler & Reid, 1986, Auerbach, 1987, Branch, 1988, Broadley & Howell, 1991, Spawls & Branch, 1995, J.-F. Trape & Roux-Estève, 1995, J.B. Rasmussen & Hughes, 1997, Luiselli & Angelici, 2000, Spawls et al., 2002, Broadley et al., 2003, J.C. Murphy & Schlager, 2003, Greenbaum & Carr, 2005, Chippaux, 2006, Chirio & Ineich, 2006, J.-F. Trape & Mané, 2006b, Chirio & LeBreton, 2007, Wüster & Broadley, 2007, Pauwels & Vande weghe, 2008, Largen & Spawls, 2010, Ullenbruch et al., 2010, Segniagbeto et al., 2011, and Auliya et al., 2012.

Remarks: Probably occurs in SW Botswana *fide* Auerbach (1987: 199).

10. *Afronaja nubiae* (Wüster & Broadley, 2003). J. Zool. London 259(4): 348–351, figs. 1a–d, 3c. (*Naja nubiae*)

Type: Holotype, YPM 3916, a 788 mm male.

Type locality: "Kom Ombo (= Kawn Umû), Aswan Governorate, Egypt (24°28' N, 32°57' E)" [elevation 100 m].

Distribution: Northeastern Africa. Southeastern Egypt (Aswan, Assiut, S El-Bahr El-Ahmar, SE New Valley, Qena, Sohag), Eritrea (Anseba, Gash-Barka, Northern Red Sea, Southern), Sudan (El-Bahr El-Ahmar, Al Khartum, Ash-Shamaliyah, Ash Shaqiyah, Janub Darfur), NE Chad (Ennedi) and cen. Niger (Agadez), 100–2000 m.

Sources: Flower, 1930, Corkill, 1935, Marx, 1968, Spawls & Branch, 1995, J.-F. Trape & Mané, 2006b, and Largen & Spawls, 2010.

Remarks: Chad and Niger populations are isolated *fide* Largen & Spawls (2010: 603) and this species may occur in other mesic habitats in the Sahara *fide* Wüster & Broadley (2003: 354).

11. *Afronaja pallida* (Boulenger, 1896a). Cat. Snakes Brit. Mus. 3: 379. (*Naia nigricollis pallida*)

Synonym: *Naja mosambica pallida* – Khole, 1991 *(nomen incorrectum).*

Type: Lectotype, BMNH 1895.6.14.28, a juvenile (E. Lort-Phillips, Spring 1985), designated by Bogert (1942b: 5).

Type locality: "Inland of Berbera, British Somaliland" [= S of Berbera, Woqooyi Galbeed Region, NW Somalia, ca. 10°25'N, 45°02'E] via lectotype selection.

Distribution: Northeastern Africa. Southern Ethiopia (Bale, Gemu Gofa, Hararge, Kofa, Shoa, Sidamo), Somalia (Awdal, Bakool, Bari, Galguduud, Gedo, Jubbada Dhexe, Jubbada Hoose, Mudug, Nugaal, Sanaag, Shabeellaha Dhexe, Shabeellaha Hoose, Sool, Togdheer, Woqooyi Galbeed), E Kenya (Coast, Eastern, North-Eastern, Rift Valley, South Is.) and ext. N Tanzania (Arusha, Kilimanjaro), NSL–1500 m.

A

Sources: Zavattari, 1930a, Broadley, 1968c, Lanza, 1972, 1983a, Spawls, 1978, Joger, 1984, MacKay & MacKay, 1985, Broadley & Howell, 1991, Spawls & Branch, 1995, Spawls et al., 2002, Wüster & Broadley, 2003 and Largen & Spawls, 2010.

12. †*Afronaja robusta* (Meylan, 1987). Fossil Snakes Laetoli: 80–81, figs. 3.12a–c. (†*Naja robusta*)

Type: Holotype, OGM LAET 3016, five middle trunk vertebrae, presumably from one individual.
Type locality: "*c.* 8 km. w Endulen, northern Tanzania, Laetolil Beds. Locality 10, exact horizon unknown; Upper Pliocene."
Distribution: Upper Pliocene (Gelasian: 1.8–2.6 mya) of Tanzania. Known only from type locality.
Source: Rage & Bailon, 2011.
Remarks: *Naja incertae sedis fide* Wallach et al. (2009: 32).

13. †*Afronaja romani* (Hoffstetter, 1939). Arch. Mus. Hist. Nat. Lyon 15(3): 57–64, pl. 1, figs. 1–16, pl. 2, figs. 1–13. (†*Palaeonaja romani*)

Synonyms: †*Palaeonaja crassa* Hoffstetter, 1939, and †*Naja austriaca* Bachmayer & Szyndlar, 1985.
Type: Holotype, UCBL 92856, a nearly complete skeleton (R. Hoffstetter).
Type locality: "Vindobonien moyen de la Grive-Saint-Alban" [= La Grive-St. Alban, Isère Dept., SE France; Astaracian (NM 7–8), middle Miocene].
Distribution: Lower Miocene (Orleanian, MN 4-5: 13.7–16.9 mya) of France and Germany, middle Miocene (Astaracian, MN 7–8: 11.1–12.8 mya) of France, upper Miocene (Vallesian, MN 9: 9.7–11.1 mya) of Hungary and Ukraine, and upper Miocene (Turolian, MN 11: 7.8–8.7 mya) of Austria.
Sources: Szyndlar & Zerova, 1990, Szyndlar, 1991b, Szyndlar & Schleich, 1993, Szyndlar, 1995, 2005, and Ivanov, 2000.
Remarks: †*Palaeonaja crassa* Hoffstetter, 1939 and †*Naja austriaca* Bachmayer & Szyndlar, 1985 are conspecific *fide* Szyndlar & Rage, 1990.

AFRONATRIX Rossman & Eberle, 1977 (Natricidae)

Type species: *Tropidonotus anoscopus* Cope, 1861e.
Distribution: Tropical W Africa.
Sources: E.H. Taylor & Weyer, 1958, Doucet, 1963, Mertens, 1965c, Menzies, 1966, Leston & Hughes, 1968, Hughes & Barry, 1969, Dunger, 1972, Rossman & Eberle, 1977, Roman, 1980, Hughes, 1983, 2013, J.A. Butler & Reid, 1986, D. Lawson, 1993, J.-F. Trapé & Mané, 1995a, 2006b, Ineich, 2003, Luiselli et al.. 2003, 2005, Chippaux, 2006, Chirio & LeBreton, 2007, Zaher et al., 2008, Ullenburch et al., 2010, W. Böhme et al., 2011, Segniagbeto et al., 2011, and Chirio, 2013.

1. *Afronatrix anoscopus* (Cope, 1861e). Proc. Acad. Nat. Sci. Philadelphia 13(7): 299. (*Tropidonotus anoscopus*)

Synonyms: *Tropidonotus ferox* A.C.L.G. Günther, 1863, *Helicops gendrii* Boulenger, 1910, and *Natrix firestonei* E.H. Taylor & Weyer, 1958.
Type: Holotype, ANSP 6639, a 356 mm specimen (Baker).
Type locality: "Cuba" (in error *fide* Loveridge, 1941a: 118). Corrected to Liberia *fide* Loveridge (1941a: 118).
Distribution: West Africa. Southeastern Senegal (Kédougou), Guinea-Bissau (Southern, Western), Guinea (Boffa, Dalaba, Macenta, Nzérékoré, Télimélé), S Mali (Sikasso), Sierra Leone, Liberia (Bong, Gbarpolu, Loffa, Montserrado, Margibi), S Ivory Coast (Abidjan, Daloa, Dix-Huit Montagnes, Guiglo, Lagunes, Tabou, Boulay Is.), SW Burkina Faso (Hauts-Bassins), S Ghana (Eastern), S Togo (Centrale, Plateaux), Benin (Alibori, Atakora, Borgou, Plateau, Zou), Nigeria (Abia, Adamawa, Akwa Ibom, Anambra, Bauchi, Bayelsa, Benue, Cross River, Delta, Kwara, Lagos, Ogun, Oyo, Plateau, Rivers, Sokoto, Taraba), W Cameroon (Adamaoua, Littoral, Nord-Ouest, Ouest, Sud, Sud-Ouest), and Equatorial Guinea (Bioko Is.), NSL–1200 m.

AFROTYPHLOPS Broadley & Wallach, 2009 (Typhlopidae)

Synonym: *Aspidorhynchus* Fitzinger, 1843 (*nomen praeoccupatum*).
Type species: *Acontias punctatus* Leach, 1819.
Distribution: Subsaharan Africa.
Sources: F. Werner, 1921a, Roux-Esteve, 1974, Wallach, 1993c, McDiarmid et al., 1999 and Broadley & Wallach, 2009.

1. *Afrotyphlops angeli* (Guibé, 1952). Bull. Mus. Nat. Hist. Paris (2) 24(1): 79. (*Typhlops angeli*)

Type: Holotype, MNHN 1951.1, a 365 mm specimen (M. Lamotte, 1943).
Type locality: "Nzo (Mt. Nimba), 500 m." [= Nzao, Nzérékoré Prefecture, Guinea, 07°44'N, 08°45'W, elevation 500 m].
Distribution: Southeastern Guinea (Nzérékoré), 500 m. Known only from type locality.
Sources: Angel et al., 1954 and Ineich, 2003.
Remarks: A valid species *fide* W. Böhme, 2002; previously a synonym of *Typhlops manni*.

2. *Afrotyphlops angolensis* (Bocage, 1866b). J. Sci. Math. Phys. Nat., Lisboa 1(1): 65–66. (*Onychocephalus angolensis*)

Synonyms: *Onychocephalus angolensis* Bocage, 1866a (*nomen nudum*), *Typhlops congicus* Boettger, 1887c, *Typhlops adolfi* Sternfeld, 1910c, *Typhlops dubius* Chabanaud, 1917a, *Typhlops lestradei* Witte, 1933c, *Typhlops ochraceus* Laurent, 1952, *Typhlops angolensis irsaci* Laurent, 1954b (*nomen nudum*), *Typhlops angolensis irsaci* Laurent, 1956a, *Typhlops angolensis polylepis* Laurent, 1954b (*nomen nudum*), *Typhlops angolensis polylepis* Laurent, 1956a, and *Typhlops angolensis symoensi* Laurent, 1960.

Type: Holotype, MBL T79.1134, a 470 mm specimen (F.A.P. Bayão, 1862–1866), destroyed by fire 18 March 1978.

Type locality: "le districto do Duque du Bragança, situado na latitude de Loanda, porém umas 75 leguas para o interior, Portuguesas Africa occidental" [= Duque de Bragança, Malanje District, Angola, 9°08'S, 16°04'E, elevation ca. 1100 m].

Distribution: Central and East Africa. Southern Cameroon (Est, Littoral, Nord-Ouest, Ouest, Sud, Sud-Ouest), S Central African Republic (Baningui-Bangoran, Lobaye, Mbomou), Gabon (Haut-Ogooué, Ogooué-Ivindo, Ogooué-Maritime, Woleu-Ntem), S Congo (Brazzaville, Kouilou, Lekoumou, Pool), Democratic Republic of Congo (Bandundu, Bas-Congo, Equateur, Katanga, Kinshasa, Nord-Kivu, Orientale, Sud-Kivu), Angola (Huambo, Lunda Norte, Malanje), Uganda (Central, Western), W Kenya (Eastern, Nairobi, Western), W Rwanda, NW Burundi, NW Tanzania (Kigomo, Mwanza, Shinyanga) and NE Zambia (Northern), 500–2075 m.

Sources: Laurent, 1956a, 1960, 1964b, Pitman, 1974, J.-F. Trape & Roux-Estève, 1995, Spawls et al., 2002, Broadley et al., 2003, Watkins-Colwell & de Leon, 2003, Branch, 2005, Herrmann et al., 2005, Chirio & Ineich, 2006, Chirio & LeBreton, 2007 and Pauwels, 2008.

3. *Afrotyphlops bibronii* (A. Smith, 1846 *in* 1838–1849). Illust. Zool. So. Africa, Rept.: pl. 51, fig. 2, pl. 54, figs. 5–8, 2 pp. (*Onychocephalus bibronii*)

Types: Syntypes (?), BMNH 1946.1.11.22 (formerly BMNH 1865.5.8.9), longest syntype 309 mm (A. Smith, 12 Aug. 1834–4 Jan. 1836), location of other syntypes unknown.

Type locality: "country to the northward of Latakoo, South Africa" [= Kuruman, N Northern Cape Prov., South Africa *fide* V.F.M. FitzSimons (1962a: 67), 27°27'S, 23°26'E, elevation 1326 m].

Distribution: Southeastern Africa. Eastern Zimbabwe (Manicaland), SE Botswana (South East), Swaziland, Lesotho and South Africa (Eastern Cape, KwaZulu-Natal, Free State, Transvaal), NSL–2000 m.

Sources: V.F.M. FitzSimons, 1974, Broadley, 1983, Branch, 1998, 2001 and Broadley & Blaylock, 2013.

4. *Afrotyphlops blanfordii* (Boulenger, 1889b). Ann. Mag. Nat. Hist. (6) 4(22): 363. (*Typhlops blanfordii*)

Type: Holotype, BMNH 1946.1.11.67 (formerly BMNH 1869.11.4.36), a 320 mm specimen (W.T. Blanford, 1867–1868).

Type locality: "Senafé, Abyssinia" [= Senafe, Southern Prov., SE Eritrea, 14°42'N, 39°25'E, elevation 2450 m].

Distribution: Eritrea (Anseba, Central, Gash-Barka, Northern Red Sea, Southern) and Ethiopia (Bale, Gemu Gofa, Gojjam, Gondar, Hararge, Kefa, Shoa, Sidamo), 1000–2450 m.

Sources: Boulenger, 1893, Laurent, 1956a, 1964b, Largen 1978, 1997, Largen & Rasmussen, 1993 and Largen & Spawls, 2010.

5. *Afrotyphlops congestus* (A.-M.-C. Duméril & Bibron, 1844). Erpét. Gén. 6: 334–336. (*Onychocephalus congestus*)

Synonyms: *Onychophis barrowii* Gray, 1845, *Typhlops crassatus* W.C.H. Peters, 1881, and *Typhlops bocagei* Bethencourt-Ferreira, 1905.

Type: Holotype, not designated, FPM, an 800 mm specimen, lost *fide* A. Grandison *in* Roux-Esteve (1974a: 79).

Type locality: Unknown.

Distribution: Central Africa. Southeastern Nigeria (Akwa Ibom, Cross River), S Cameroon (SW Adamaoua, Est, Centre, Littoral, Sud, Sud-Ouest), Equatorial Guinea (Bioko Is.), Central African Republic (Lobaye, Mbomou, Sangha), Gabon (Moyen-Ogooué, Ngounié, Ogooué-Ivindo, Ogooué-Lolo, Ogooué Maritime, Woleu-Ntem), Congo (Brazzaville, Kouilou, Lekoumou, Sangha), Democratic Republic of Congo (Bandundu, Bas-Congo, Equateur, Kasai Occidental, Kasai Oriental, Nord-Kivu, Oriental, Sud-Kivu), NW Angola (Cabinda), W Uganda (W Central, Western) and E Tanzania (Lindi, Tanga), NSL–1435 m.

Sources: K.P. Schmidt, 1923, Mertens, 1965c, Meirte, 1992, Rödel et al., 1995, J.-F. Trape & Roux-Estève, 1995, Spawls et al., 2002, Branch, 2005, Herrmann et al., 2005, Chirio & Ineich, 2006, Chirio & LeBreton, 2007 and Pauwels, 2008.

6. *Afrotyphlops elegans* (W.C.H. Peters, 1868a). Mber. Königl. Akad. Wiss. Berlin 1868(7): 450, pl. 2, fig. 1. (*Typhlops* [*Ophthalmidion*] *elegans*)

Types: Syntypes (3), ZMB 6066, and ZZSDa–b, 390 mm, 335 mm and 260 mm specimens (H. Dohrn, 1865), latter two lost *fide* Bauer et al. (1995: 79).

Type locality: "Prinzens-insel (Ilha de Príncipe) an der Westküste Africa's" [= Príncipe Island, Gulf of Guinea, West Africa].

Distribution: São Tomé and Príncipe (Príncipe Is.).

Source: Bocage, 1887c.

A

7. *Afrotyphlops fornasinii* (Bianconi, 1849a). Nouvi Ann. Sci. Nat. (2) 10: 107. (*Typhlops fornasinii*) (*nomen corrigendum*)

Synonyms: *Typhlops fornasini* Bianconi, 1849a (*nomen incorrigendum*), *Typhlops fornasinii* – Bianconi, 1850 (*nomen corrigendum*), *Onychocephalus mossambicus* W.C.H. Peters, 1854, *Onychocephalus trilobus* W.C.H. Peters, 1854, *Onychocephalus tettensis* W.C.H. Peters, 1860a, *Typhlops bianconii* Jan, 1860 *in* Jan & Sordelli, 1860–1866, and *Typhlops mossombicus* – F. Werner, 1921a (*nomen incorrectum*).

Type: Holotype, not designated, MBI, a 72 mm specimen (C. Fornasini, 1842–1847).

Type locality: "Mosambico" [= Mozambique].

Distribution: Southeastern Africa. Mozambique (Gaza, Inhambane, Maputo, Bazaruto, Benguerua, Inhaca, Magaruque and Mozambique Is.), SE Zimbabwe (Masvingo) and NE South Africa (ext. NE KwaZulu-Natal), NSL–65 m.

Sources: Laurent, 1964b, V.F.M. FitzSimons, 1974, Broadley, 1983, Branch, 1998 and Broadley & Blaylock, 2013.

Remarks: Original description repeated in Bianconi (1850: 183, pl. 6, figs. 1–1d).

8. *Afrotyphlops gierrai* (Mocquard, 1897b). Bull. Mus. Hist. Nat., Paris 3(4): 122–123. (*Typhlops gierrai*)

Synonym: *Typhlops guirrae* – K.P. Schmidt, 1923 (*nomen incorrectum*).

Type: Holotype, MNHN 1897.24, a 469 mm specimen (A. Gierra, 1895).

Type locality: "Tanga, dans l'Afrique orientale allemande" [= Tanzania].

Distribution: Tanzania (forests in Usambara and Uluguru Mountains), 600–1040 m.

Sources: T. Barbour & Loveridge, 1928, Laurent, 1964b, J. Rasmussen, 1981 and Spawls et al., 2002.

9. *Afrotyphlops kaimosae* (Loveridge, 1935). Bull. Mus. Comp. Zool. 79(1): 5–6. (*Typhlops kaimosae*)

Type: Holotype, MCZ 40060, a 208–215 mm male (A. Loveridge, 7 March 1934)

Type locality: "Kaimosi Forest, near Friends' African Mission, Kakamega district, Nyanza Prov., Kenya Colony."

Distribution: Kenya (Nyanza). Known only from type locality.

10. *Afrotyphlops liberiensis* (Hallowell, 1848). Proc. Acad. Nat. Sci. Philadelphia 4: 59–60, pl., figs. 1–2. (*Onychocephalus liberiensis*)

Synonyms: *Onychocephalus nigro-lineatus* Hallowell, 1848 (*nomen incorrigendum*), *Onychocephalus hallowelli* Jan, 1861b (*nomen nudum*), *Onychocephalus kraussi* Jan, 1861b (*nomen nudum*), *Typhlops hallowelli* Jan, 1863b (*nomen nudum*), *Typhlops kraussi* Jan, 1863b (*nomen nudum*), *Typhlops kraussi* Jan, 1864 *in* Jan & Sordelli, 1860–1866, *Typhlops liberiensis intermedia* Jan, 1863b (*nomen nudum*), *Typhlops hallowelli* Jan, 1864 *in* Jan & Sordelli, 1860–1866, *Typhlops liberiensis intermedia* Jan, 1864 *in* Jan & Sordelli, 1860–1866, *Typhlops punctatus nigrolineatus* – Chabanaud, 1917c (*nomen corrigendum*), and *Typhlops leprosus* E.H. Taylor & Weyer, 1958.

Type: Holotype, ANSP 3243, a 618 mm specimen (T.S. Savage, 1840–1848).

Type locality: "Liberia, Western Coast of Africa."

Distribution: West Africa. Guinea (Dalaba, Dix-Huit Montagnes, Faranah, Kouroussa, Macenta, Nzérékoré, Télimélé, Yomou), Sierra Leone (Eastern), Liberia (Bong, Loffa, Margibi, Montserrado), SW Ivory Coast (Abidjan, Guiglo, Katiola, San Pédro, Tabou) and S Ghana (Accra, Ashanti, S Eastern, S Western), 30–600 m.

Sources: E.H. Taylor & Weyer, 1958, Doucet, 1963, Ineich, 2003, W. Böhme et al., 2011 and Chirio, 2013.

11. *Afrotyphlops lineolatus* (Jan, 1864). Icon. Gén. Ophid.: 24. (*Typhlops* [*Ophthalmidion*] *lineolatus*)

Synonyms: *Typhlops lineolatus* Jan, 1863b (*nomen nudum*), *Typhlops boulengeri* Bocage, 1893c, and *Typhlops tornieri* Sternfeld, 1910c.

Type: Holotype, ZMUU 725 (A. Afzel, late 1700s–early 1800s).

Type locality: "Sierra Leona" [= Sierra Leone].

Distribution: West, cen. and E Africa. Senegal (Dakar, Fatick, Tambacounda, Thiès, Ziguinchor), Gambia (Kombo Saint Mary), Guinea (Dalaba, Dubréka, Télimélé), Sierra Leone (Northern, Southern, Sherbro Is.), Liberia, Ivory Coast (Katiola, Odienné), Ghana (Eastern, Northern), Togo (Plateaux), SW Niger (Dosso), Nigeria (Kaduna), Cameroun (Adamaoua, Est, Nord, Nord-Ouest, Sud, Sud-Ouest), SW Chad (Chari-Baguirmi), Central African Republic (Bamingui-Bangoran, Haut-Mbomou, Haute-Sangha, Mbomou, Ouham), Congo (Brazzaville, Cuvette, Cuvette-Ouest, Kouilou, Pool), Democratic Republic of Congo (Bas-Congo, Katanga, Kinshasa, Maniema, Nord-Kivu, Orientale, Sud-Kivu), Angola (Benguela, Cabinda, Lunda Norte), SW Sudan (Gharb Darfur, Janub Darfur), South Sudan (Central Equatoria, Eastern Equatoria, Western Equatoria), S Ethiopia (Bale, Gemu Gofa, Sidamo), Uganda (Central, Eastern, Northern, Western, Kome Is.), SW Kenya (Central, S Eastern, Nairobi, Nyanza, S Rift Valley, Western), N Tanzania (Arusha, Kigoma, Kilimanjaro, Manyara, Mara, Morogoro, Mwanza, Piwani, Rukwa, Shinyanga, Singida, Tanga, Ukerewe Is.), NSL–2600 m.

Sources: Jan & Sordelli, 1865 *in* 1860–1866, Bocage, 1895c, Laurent, 1956a, 1964b, Pitman, 1974, Largen,

1978, Largen & Rasmussen, 1993, J.-F. Trape & Roux-Estève, 1995, Spawls et al., 2002, J.-F. Trape & Mané, 2006b, Chirio & LeBreton, 2007, Largen & Spawls, 2010, Segniagbeto et al., 2011 and Chirio, 2013.

Remarks: Holotype erroneously reported lost *fide* Roux-Estève (1974a: 80), Hahn (1980: 62) and McDiarmid et al. (1999: 108) but was rediscovered by L. Wallin (*in litt.*, 4 Feb. 1993).

12. *Afrotyphlops manni* (Loveridge, 1941a). Proc. U.S. Natl. Mus. 91(3128): 118. (*Typhlops manni*)

Type: Holotype, USNM 109634, a 343 mm specimen (W.M. Mann, 10 March–17 July 1940).

Type locality: "Harbel, Republic of Liberia" [= Harbel, SW Grand Bassa Co., cen. Liberia, 06°17'N, 10°21'W, elevation 100 m].

Distribution: Central Liberia (Grand Bassa), 100 m. Known only from holotype.

Source: W. Böhme et al., 2011.

13. *Afrotyphlops nanus* Broadley & Wallach, 2009. Zootaxa (2255): 32–33, fig. 5a.

Type: Holotype, BMNH 1897.11.17.16, a 125 mm male (C.S. Belton, 1896).

Type locality: "Samburu, Coast Prov., southeastern Kenya (39°17'S, 3°47'E, elevation 295 m)."

Distribution: Southeastern Kenya (Coast), 295 m. Known only from type locality.

14. *Afrotyphlops nigrocandidus* (Broadley & Wallach, 2000). Afr. J. Herp. 49(2): 166–167. (*Rhinotyphlops nigrocandidus*)

Type: Holotype, ZMUC 52223, a 440 mm male (P. Gravlund & M. Anderson, 17 Jan. 1996).

Type locality: "Kiranzi-Kitungulu Forest Reserve, Udzungwa Mountains, Iringa District, Tanzania (1500 m a.s.l., 08°09'S x 36°05'E)."

Distribution: Southeastern Tanzania (Iringa, Morogoro), 1450–1750 m.

Remarks: Probably inhabits Nguru and Ruhebo Mountains *fide* Broadley & Wallach (2000: 167).

15. *Afrotyphlops punctatus* (Leach *in* Bowdich, 1819). Miss. Cape Coast Castle-Ashantee: 493. (*Acontias punctatus*)

Synonyms: *Typhlops eschrichtii* Schlegel, 1839 *in* 1837–1844, *Onychophis punctata* Gray, 1845, and *Typhlops milleti* Chabanaud, 1920a.

Type: Holotype, BMNH 1946.1.1.47 (T.E. Bowditch, April, 1817–Oct. 1818).

Type locality: "Fantee, Cape Coast Castle to Ashantee" [= Fantee tribal area, between Mansu and Fosu along Amissa (Oki) River, Central Region, S Ghana, elevation 160–180 m].

Distribution: West, Cen. and East Africa. Senegal (Dakar, Kédougou, Tambacounda, Ziguinchor), Gambia (MacCarthy Island, Western), Guinea–Bissau (Bissau, Cacheu, Tombali, Bolama Is.), Guinea (Conakry, Kindia, Kouroussa, Macenta, Nzérékoré, Télimélé), S Mali (Mopti, Sikasso), Sierra Leone (Northern, Southern, Western), Liberia (Margibi, Nimba), Ivory Coast (Bouna, Daloa, Ferkéssédougou, Katiola), Burkina Faso (Centre, Centre-Est, Centre-Nord, Centre-Ouest, Est, Hauts-Bassins, Sud-Ouest, Volta-Noire), S Ghana (Accra, Central, S Eastern, S Western), Togo (Plateaux), Benin (Atakora, Atlantique, Collines, Zou), SW Niger (Tahoua), Nigeria (Benue, Cross River, Delta, Gongola, Plateau), Cameroon (Adamaoua, Centre, Est, Extreme-Nord, Littoral, Nord, Nord-Ouest, Sud, Sud-Ouest), S Chad (Chari-Baguirmi, Mayo-Kebbi), Equatorial Guinea (Bioko Is.), Central African Republic (Bamingui-Bangoran, Gribingui, Haut-Mbomou, Lobaye, Mbomou, Ombella-Mpoko, Ouaka, Ouham, Ouham-Pende, Sangha), N Democratic Republic of Congo (Equateur, Oriental), South Sudan, and Uganda (Central, Eastern, Northern, Western), NSL–1800 m.

Sources: Manaças, 1955, E.H. Taylor & Weyer, 1958, Leston & Hughes, 1968, Pitman, 1974, Roux-Estève, 1974a, Roman, 1980, Hakansson, 1981, J.A. Butler & Reid, 1986, Rödel et al., 1995, W. Böhme et al., 1996, Spawls et al., 2002, Branch, 2005, Herrmann et al., 2005, Chirio & Ineich, 2006, J.-F. Trape & Mané, 2006b, Chirio & LeBreton, 2007, Chirio, 2009, Ullenbruch et al., 2010, Segniagbeto et al., 2011, Auliya et al., 2012, Chirio, 2013 and Hughes, 2013.

16. *Afrotyphlops rondoensis* (Loveridge, 1942). Bull. Mus. Comp. Zool. 91(4): 256–257. (*Typhlops tettensis rondoensis*)

Type: Holotype, MCZ 48066, a 228 mm specimen (A. Loveridge, 5 May 1939).

Type locality: "Nchingidi, 2,700 feet, Rondo Plateau, southeastern Tanganyika Territory." [= near Lindi, Tanzania, 09°36'S, 39°25'E, elevation 825 m].

Distribution: Southeastern Tanzania (SE Lindi, Mtwara), 375–825 m.

Sources: Laurent, 1964b and Spawls et al., 2002.

17. *Afrotyphlops schmidti* (Laurent, 1956a). Ann. Mus. Roy. Congo Belge, Sci. Zool. (48): 71–72, figs. 9–10, pl. 8, fig. 4. (*Typhlops schmidti*)

Synonyms: *Typhlops schmidti wilsoni* Laurent, 1968a (*nomen praeoccupatum*) and *Rhinotyphlops schmidti laurenti* Wallach, 2003 (*nomen substitutum*).

Type: Holotype, MRAC 17996, a 326 mm specimen (IRSAC, 11–14 Dec. 1950).

Type locality: "Nyunza, 700 m., Terr. d'Albertville, Tanganika" [= Nyunza, Katanga Prov., Democratic

Republic of the Congo, 05°57'S, 28°01'E, elevation 700 m].

Distribution: Central Africa. Northeastern Angola (Moxico), SE Democratic Republic of Congo (Katanga) and Zambia (Copperbelt, Eastern, Luapula, Northern, North-Western, Western), 585–1655 m.

Sources: Witte, 1962, Laurent, 1964b, 1968a, Broadley et al., 2003 and Wallach, 2003.

18. *Afrotyphlops steinhausi* (F. Werner, 1909b). Mitt. Naturhist. Mus. Hamburg 26: 209. (*Typhlops steinhausi*)

Synonyms: *Typhlops batesii* Boulenger, 1911b and *Typhlops steinhousi* – Roux-Estève, 1965 *(nomen incorrectum).*

Type: Neotype, MNHN 1964.139, a 420–425 mm male (R. Pujol), designated by Villiers (1968: 1308).

Type locality: "Rép. Centrafricaine: Boukoko" [= Boukoko, Central African Republic, 3°54'N, 17°56'E, elevation 560 m] via neotype selection.

Distribution: Central Africa. Southeastern Nigeria (Cross River), S Cameroon (Centre, Nord-Ouest, Sud, Sud-Ouest), W Central African Republic (Haute-Sangha, Lobaye, Ombella-Mpoko, Ouham), Congo and NE Democratic Republic of Congo (Orientale), NSL–800 m.

Sources: Chirio & Ineich, 2006 and Chirio & LeBreton, 2007.

Remarks: Two ZMH syntypes lost *fide* Hallermann (1998: 216), destroyed July, 1943 during World War II. Illustration of neotype head in Villiers (1963: figs. 1–3).

19. *Afrotyphlops tanganicanus* (Laurent, 1964b). Bull. Mus. Comp. Zool. 130(6): 418–419. (*Typhlops schmidti tanganicanus*)

Type: Holotype, MCZ 57439, a 390 mm female (C.J.P. Ionides, 11 March–26 April 1958).

Type locality: "Liwale, Tanganyika" [= Liwale, Nachingwea District, Southern Prov., Tanzania, 09°,46'S, 37°56'E, elevation 470 m].

Distribution: Southeastern Tanzania (Tanga), 25–470 m.

Source: Spawls et al., 2002.

20. *Afrotyphlops usambaricus* (Laurent, 1964b). Bull. Mus. Comp. Zool. 130(6): 416–417. (*Typhlops boulengeri usambaricus*)

Type: Holotype, MCZ 38699, a 384 mm specimen (R.E. Moreau. June 1955).

Type locality: "Amani, Usambara Mtns" [= Amani, Eastern Usambara Mountains, Tanga Region, Tanzania, 05°06'S, 38°38'E, elevation 750 m].

Distribution: Northeastern Tanzania (E Usambara Mountains of N Tanga), 750 m.

Source: Spawls et al., 2002.

AGKISTRODON Palisot de Beauvois, 1799 (Viperidae)

Synonyms: *Agkishodon* – Palisot de Beauvois, 1799 *(nomen incorrectum), Scytale* Latreille *in* Sonnini & Latreille, 1801a *(nomen praeoccupatum), Cenchris* Daudin, 1803c, *Cenchrus* Link, 1807 *(nomen emendatum), Scytalus* Fischer von Waldheim, 1813 *(nomen emendatum), Tisiphone* Fitzinger, 1826a *(nomen praeoccupatum), Ancistrodon* Wagler, 1830 *(nomen emendatum), Tysiphone* A.-M.-C. Duméril, Bibron & Duméril, 1854b *(nomen emendatum), Agcistrodon* – Bourret, 1935e *(nomen incorrectum), Acontias* Troost, 1836 *(nomen praeoccupatum), Toxicophis* Troost, 1836 *(nomen substitutum), Tisphonus* Swainson, 1839 *(nomen emendatum), Angkistrodon* Gray, 1842b *(nomen emendatum), Agkistodon* – Troschel, 1854 *(nomen incorrectum), Ankistrodon* – Chenu & Desmarest, 1857 *(nomen incorrectum), Triphone* – Westphal-Castelnau, 1870 *(nomen incorrectum), Aucistrodon* – Velasco, 1890a *(nomen incorrectum), Ancistrodoro* – Velasco, 1890c *(nomen incorrectum), Ancistradon* – Velasco, 1897 *(nomen incorrectum), Ancistrodonus* A.L. Herrera, 1899 *(nomen emendatum), Anxistrodon* –Pavloff, 1926 *(nomen incorrectum), Agkiotrodon* – Takahashi, 1930 *(nomen incorrectum), Toricophis* – Haltom, 1931 *(nomen incorrectum), Agpistrodon* – Tchang, 1932 *(nomen incorrectum), Agkisirodon* – Maki, 1933 *(nomen incorrectum), Aghistrodon* – Wettstein, 1938 *(nomen incorrectum), Hykistrodon* – Anon., 1944 *(nomen incorrectum), Agkystrodon* – Matthey, 1949 *(nomen incorrectum), Agkistron* – K.L. Rogers, 1976 *(nomen incorrectum), Agistodon* – Hoge & Romano-Hoge, 1981a *(nomen incorrectum), Seytale* – Cherlin, 1983 *(nomen incorrectum),* and *Aqistrodon* – Nutaphand, 1990 *(nomen incorrectum).*

Type species: *Agkistrodon mokasen* Palisot de Beauvois, 1799.

Distribution: North America and Mesoamerica.

Fossil records: Upper Miocene of USA (Nebraska), lower Pliocene of USA (Nebraska), upper Pliocene of USA (Kansas, Texas), and Pleistocene of USA (Alabama, Florida, Georgia, Kansas, Missouri, Nebraska, Pennsylvania, Texas, Virginia, West Virginia). Fossils unassigned to species include upper Miocene (Clarendonian) of USA (Nebraska), middle Pleistocene (Irvingtonian II) of USA (Nebraska), and upper Pleistocene (Rancholabrean II) of USA (Alabama).

Sources: Rendahl, 1933, Klauber, 1956b, 1958b, Paik et al., 1979, Chen et al., 1984, Chiasson et al., 1989, Maes, 1989, Conant, 1986, 1990, Gloyd, 1990, Gloyd & Conant, 1990, Kardong, 1990, Malnate, 1990, Minton, 1990, 1992, Knight et al., 1992, Golay et al., 1993, Holman, 1995a, 2000a, Parmley & Holman, 1995, Kraus et al., 1996, Cullings et al., 1997, Parkinson et al., 1997, 2000, McDiarmid et al., 1999, J.A. Campbell & Lamar, 2004, Castoe & Parkinson, 2006, Guiher &

Burbrink, 2008, Parmley & Hunter, 2010 and Hoser, 2012d.

1. *Agkistrodon bilineatus* A.C.L.G. Günther, 1863c. Ann. Mag. Nat. Hist. (3) 12(71): 364. (*Ancistrodon bilineatus*)

Synonyms: *Ankistrodon bilineatum* F. Müller, 1877 *(nomen emendatum), Ancistrodonus bilineatus* A.L. Herrera, 1899 *(nomen emendatum), Arkistrodon bilineatus* – Martin del Campo, 1953 *(nomen incorrectum), Trigonocephalus specialis* Recinos, 1954, *Ancistrodon bilinaetus* – Villa, 1962 *(nomren incorrectum), Agkistrodon bilineatus russeolus* Gloyd, 1972a, *Agkistrodon bilineatus howardgloydi* Villa, 1983 (*nomen nudum*), and *Agkistrodon bilineatus howardgloydi* Conant, 1984.

Type: Holotype, BMNH 1864.1.26.396, a 991 mm female (O. Salvin, 1857–1875).

Type locality: "on the Pacific coast of Guatemala."

Distribution: Mexico and upper Central America. Mexico (Campeche, Chiapas, SW Chihuahua, Colima, Guerrero, Jalisco, Michoacán, Morelos, Nayarit, Oaxaca, Quintana Roo, Sinaloa, S Sonora, Yucatán, María Madre Is.), Belize (Belize, Orange Walk), Guatemala, El Salvador (Cuscatlán, La Libertad, Sonsonate), S Honduras (Choluteca, Copán, Valle), Nicaragua (Granada, Masaya) and NW Costa Rica (Guanacaste), NSL–1670 m.

Sources: Villa, 1984, L.D. Wilson & Meyer, 1985, J.A. Campbell & Lamar, 1989, J.C. Lee, 1996, 2000, Blair et al., 1997, Lamar, 1997b, J.A. Campbell, 1998b, G. Köhler, 1999b, H.M. Smith & Chiszar, 2001h, G. Köhler et al., 2005, Lemos-Espinal & Smith, 2006b, Bryson & Mendoza-Quijano, 2007, N. Herrera et al., 2007, Suárez-Velázquez & Luna-Reyes, 2009 and McCranie, 2011a.

2. *Agkistrodon contortrix* (Linnaeus, 1766). Syst. Nat., ed. 12, 1: 373. (*Boa contortrix*)

Synonyms: *Agkistrodon mokasen* Palisot de Beauvois, 1799, *Agkistrodon mokason* – Palisot de Beauvois, 1799 *(nomen incorrectum), Agkishodon mokasen* – Palisot de Beauvois, 1799 *(nomen incorrectum), Cenchris mokeson* Daudin, 1803c, *Scytalus cupreus* Rafinesque-Schmaltz, 1818, *Scytale mockeson* – Say, 1819 *(nomen incorrectum), Cenchris marmorata* F. Boie, 1827, *Acontias atro-fuscus* Troost, 1836 *(nomen incorrigendum), Toxicophis atro-fuscus* Troost, 1836 *(nomen substitutum), Trigonocephalus cenchris* Schlegel, 1837, *Cenchris atrofuscus* – Gray, 1849 *(nomen corrigendum), Trigonocephalus histrionicus* A.-M.-C. Duméril, 1853 (*nomen nudum*), *Ancistrodon contortrix* Baird, 1854 *(nomen emendatum), Agkistrodon contorting* – Abbott, 1869 *(nomen incorrectum), Agkistrodon contortirix* – Keim, 1914 *(nomen incorrectum), Agkistrodon mokasen laticinctus* Gloyd & Conant, 1934, *Agkistrodon mokeson austrinus* Gloyd & Conant, 1943 *(nomen substitutum), Agkistrodon mokeson pictigaster* Gloyd & Conant, 1943, *Agkistrodon contortrix phaeogaster* Gloyd, 1969, and *Agkistrodon cortortrix makasen* – Harding & Welch, 1980 *(nomen incorrectum).*

Type: Holotype, not designated, (A. Garden, 1760–1761), location unknown.

Type locality: "Carolina" [USA]. Restricted to Charleston, South Carolina, USA *fide* K.P. Schmidt (1953a: 224).

Distribution: Southern and E USA (Alabama, Arkansas, Connecticut, Delaware, ext. NW Florida, Georgia, S Illinois, S Indiana, SE Iowa, E Kansas, Kentucky, Louisiana, Maryland, Massachusetts, Mississippi, Missouri, ext. SE Nebraska, N New Jersey, SE New York, North Carolina, S Ohio, Oklahoma, Pennsylvania, South Carolina, Tennessee, Texas, Virginia, West Virginia) and N Mexico (ext. NE Chihuahua, ext. N Coahuila), NSL–1525 m, usually below 500 m.

Fossil records: Upper Miocene (Hemphillian) of USA (Nebraska), lower Pliocene (Hemphillian) of USA (Kansas), middle Pliocene (Blancan II, III) of USA (Kansas, Texas), upper Pliocene (Blancan V) of USA (Kansas, Nebraska), lower/middle Pleistocene (Irvingtonian I) of USA (Kansas), middle/upper Pleistocene (Irvingtonian II) of USA (West Virginia), and upper Pleistocene (Rancholabrean II) of USA (Georgia, Kansas, Pennsylvania, Texas, Virginia, West Virginia).

Sources: Gloyd & Conant, 1938, 1943, Fitch, 1960, Gloyd, 1969, Vial et al., 1977, Ernst & Barbour, 1989, Conant & Collins, 1991, Palmer & Braswell, 1995, Tennant, 1997, Fitch, 1999, Holman, 2000a, Werler & Dixon, 2000, D. Schmidt, 2006b, Guiher & Burbrink, 2008, Douglas et al., 2009 and Ernst & Ernst, 2011.

Remarks: Original description based on Catesby (1743(2): 56, pl. 56). Type not present in ZMUU *fide* Lönnberg, 1896 or NHR *fide* Andersson, 1899. Two western subspecies are distinct species *fide* Douglas et al. (2009: 1164).

3. *Agkistrodon piscivorus* (Lacépède, 1789). Hist. Nat. Serp. 2: 130–131, 424–425. (*Crotalus piscivorus*)

Synonyms: *Crotalus aquaticus* Bonnaterre, 1790, *Scytale piscivora* – Latreille *in* Sonnini & Latreille, 1801, *Coluber tisiphone* G. Shaw, 1802, *Acontias leucostoma* Troost, 1836, *Toxicophis leucostoma* Troost, 1836 *(nomen substitutm), Toxicophis pugnax* Baird & Girard, 1853, *Ancistrodon piscivorus* Cope, 1860 *(nomen emendatum), Agkistrodon piscivorus conanti* Gloyd, 1969, and *Agkistrodon piscivorus laurae* Stewart, 1974 (*nomen nudum*).

Type: Holotype, not designated, location unknown.

Type locality: "Caroline" [USA]. Restricted to vicinity of Charleston, South Carolina, USA *fide* K.P. Schmidt (1953a: 225).

A

Distribution: Southeastern USA (Alabama, Arkansas, Florida, Georgia, ext. S Illinois, S Indiana, ext. SE Kansas, W Kentucky, Louisiana, Mississippi, S and S Missouri, E North Carolina, E Oklahoma, cen. & E South Carolina, W Tennessee, E Texas, SE Virginia), NSL–765 m.

Fossil records: Lower Pleistocene (Irvingtonian I) of USA (Florida), and upper Pleistocene (Rancholabrean II) of USA (Florida, Georgia, Texas).

Sources: Wharton, 1960, 1966, 1969, Burkett, 1966, Mount, 1975, Ernst & Barbour, 1989, Conant & Collins, 1991, Palmer & Braswell, 1995, Tennant, 1997, Holman, 2000a, Werler & Dixon, 2000, J.C. Murphy & Schlager, 2003, Means, 2004c, Guiher & Burbrink, 2008 and Ernst & Ernst, 2011.

Remarks: Official Specific Name *fide* Opinion 1463 (ICZN, 1987). Florida subspecies possibly a valid species *fide* Douglas et al. (2009: 1164).

4. *Agkistrodon taylori* Burger & Robertson, 1951. Univ. Kansas Sci. Bull. 34(5): 213–216, pl. 25, figs. 1–2. (*Agkistrodon bilineatus taylori*)

Synonyms: *Agkistrodon bilineatus taylori* H.M. Smith & Taylor, 1950 *(nomen nudum)*, and *Agkistrodon bilineatus lemosespinali* H.M. Smith & Chiszar, 2001h.

Type: Holotype, UIMNH 10002 (formerly EHT 5514), a 465 mm male (E.H. Taylor, 9 June 1938).

Type locality: "km. 833, 21 kilometers north of Villagrán, Tamaulipas, México."

Distribution: Northeastern Mexico (Hidalgo, Nuevo León, San Luis Potosí, Tamaulipas, N Veracruz), NSL–1500 m.

Sources: Tovar-Tovar & Mendoza-Quijano, 2001, Parkinson et al., 2000, Bryson & Mendoza-Quijano, 2007, Ramírez-Bautista et al., 2010 and Ernst & Ernst, 2011.

AHAETULLA Link, 1807 (Colubridae)

Synonyms: *Aaetula* – Bory de Saint-Vincent, 1816 *in* 1822–1831 *(nomen incorrectum)*, *Dryinus* Merrem, 1820 *(nomen rejiciendum)*, *Aaetula* – Bory de Saint-Vincent, 1822 *in* 1822–1831 *(nomen incorrectum)*, Dryophis Dalman, 1823 *(nomen rejiciendum)*, *Passerita* Gray, 1825 *(nomen rejiciendum)*, *Dryophis* Schlegel, 1826a *(nomen substitutum)*, *Bryopsis* – Schlegel, 1826b *(nomen incorrectum)*, *Tragops* Wagler, 1830, *Fragops* – Bonaparte, 1833 *(nomen incorrectum)*, *Ahaetula* Gray, 1835 *in* Gray & Hardwicke, 1830–1835 *(nomen emendatum)*, *Passirita* Swainson, 1839 *(nomen emendatum)*, *Herpetotragus* Fitzinger, 1843, *Tropidotragops* Fitzinger, 1843, *Driophis* – A.-M.-C. Duméril & Bibron., 1844 *(nomen incorrectum)*, *Dryiphis* – Cantor, 1847 *(nomen incorrectum)*, *Dystyches* Gistel, 1848 *(nomen substitutum)*, *Drijinus* Bleeker, 1856 *(nomen emendatum)*, *Tropidococcyx* A.C.L.G. Günther, 1860, *Tragophis* Fitzinger, 1861, *Ahoetulla* – Westphal-Castelnau, 1870 *(nomen incorrectum)*, *Atraetulla* – F. Müller, 1878b *(nomen incorrectum)*, *Gephyrinus* Cope, 1886, *Dryphis* – Velasco, 1892a *(nomen incorrectum)*, *Trygops* – E. Bartlett, 1896 *(nomen incorrectum)*, *Dryopsis* – Lidth de Jeude, 1898 *(nomen incorrectum)*, *Drophis* – Bourret, 1935 *(nomen incorrectum)*, *Passerina* – Ladiges, 1939 *(nomen incorrectum)*, *Ahetulla* – Angel, 1947 *(nomen incorrectum)*, *Drypohis* – H.W. Parker, 1965 *(nomen incorrectum)*, *Ahaetullua* – Groombridge, 1979 *(nomen incorrectum)*, *Ahaetulia* – Murthy, 1985 *(nomen incorrectum)*, and *Achaetulla* – B.D. Sharma *in* B.D. Sharma & Kumari, 1998a *(nomen incorrectum)*.

Type species: *Ahaetulla mycterizans* Link, 1807.

Distribution: Southeastern Asia and East Indies.

Sources: M.A. Smith, 1926b, 1943, Leviton, 1968a and I. Das, 2010, 2012.

Remarks: Official Generic Name no. 1308 *fide* Opinion 524 (ICZN, 1958a).

1. *Ahaetulla dispar* (A.C.L.G. Günther, 1864a). Rept. Brit. India: 303–304, pl. 23, figs. A, A'. (*Tragops dispar*)

Types: Syntypes (6), BMNH 1946.1.9.37– 42, five females, one juvenile (R.H. Beddome, 1857–1882).

Type locality: "Anamallay Mountains, British India" [= Anamalai Hills, S Western Ghats, E Kerala /W Tamil Nadu States, SW India, ca. 10°22'N, 77°08'E].

Distribution: Southwestern India (Western Ghats of Kerala, Tamil Nadu), 610–1980 m.

Sources: Whitaker & Captain, 2004 and Hutton & David, 2009.

2. *Ahaetulla fasciolata* (J.G. Fischer, 1885a). Arch. Naturg. 51(2): 66–68, pl. 5, fig. 4. (*Tragops fasciolatus*)

Type: Holotype, BMNH 1946.1.8.30, a female (F.J. Grabowsky, Jan. 1881–July 1884).

Type locality: "Südost-Borneo" [= Southeastern Borneo or Kalimantan, Indonesia].

Distribution: Malay Peninsula and Greater Sundas. Southern Thailand (Nakhon Si Thammarat, Ranong), Singapore, West Malaysia (Perak), Brunei, East Malaysia (Sabah, Sarawak) and Indonesia (Kalimantan, Natuna, Sumatra), NSL–985 m.

Sources: Smedley, 1932b, Tweedie, 1983, Stuebing, 1991, Vogel, 1992, David & Vogel, 1996, M.J. Cox et al., 1998, Pauwels & Sumontha, 2007 and Onn *et al.*, 2010.

3. *Ahaetulla fronticincta* (A.C.L.G. Günther, 1858). Cat. Colub. Snakes Brit. Mus.: 158–159. (*Dryiophis fronticincta*)

Types: Syntypes (3), BMNH 1946.1.8.24–25, a male and female (Luna), location of other syntype unknown.
Type locality: "West Indies" (in error).
Distribution: India (West Bengal) and Myanmar (Ayeyarwady, Bago, Mon, Rakhine, Yangon).
Source: Wogan et al., 2008.
Remarks: Endemic to coastal Myanmar with Indian records needing confirmation *fide* Das (2010: 260)..

4. *Ahaetulla mycterizans* (Linnaeus, 1758). Syst. Nat., ed. 10, 1: 226. (*Coluber mycterizans*)

Synonyms: *Natrix flagelliformis* Laurenti, 1768, *Coluber mycterizans* G. Shaw, 1802, *Coluber linkii* Merrem, 1820, *Dryiophis panthoronia* H. Boie *in* Schlegel, 1826a *(nomen nudum), Dryophis xanthozonia* Schlegel, 1826b *(nomen nudum), Dryophis hammatorhynchus* Fitzinger, 1826a *(nomen substitutum), Dryophis parvoninus* Kuhl *in* F. Boie, 1827, *Dryophis xanthozonia* Kuhl *in* F. Boie, 1827, *Herpetotragus boiei* Fitzinger, 1861, *Herpetotragus pavoninus* Fitzinger, 1861, *Herpetotragus russelii* Fitzzinger, 1861, and *Tragops javanicus* Fitzinger *in* Steindachner, 1867.
Type: Holotype, NHR Lin-81 (formerly MAFR), a 730 mm male (Mus. Drottn.).
Type locality: "America" (in error).
Distribution: Malay Peninsula and Greater Sundas. Southern Thailand (Chumphon, Krabbi, Ranong, Trang), West Malaysia (Johor) and W Indonesia (Java, Sumatra), NSL–500 (1290) m.
Sources: Wall, 1905g, E.H. Taylor, 1965, Tweedie, 1983, M.J. Cox, 1991b, Malhotra & Davis, 1991, M.J. Cox et al., 1998 and Miralles & David, 2010.

5. *Ahaetulla nasuta* (Lacépède, 1789). Hist. Nat. Serp. 2: 100, 277–279, pl. 4, fig. 2. (*Coluber nasutus*)

Synonyms: *Ahaetulla mycterizans* Link, 1807 *(nomen praeoccupatum), Coluber nasutus* G. Shaw, 1802, *Dryinus oxyrhynchus* Bell, 1825, *Dryinus russellianus* Bell, 1825, *Dryophis passericki* Schinz, 1833, *Dryophis rostratus* Schinz, 1833, *Dryinus fuscus* A.-M.-C. Duméril, Bibron & Duméril, 1854b, *Drijinus nasutus* Bleeker, 1856 *(nomen emendatum), Dryophis mycterizans anomalus* Annandale, 1906, *Dryophis mycterizans cinereoventer* Wall, 1908j, *Dryophis mycterizans rhodogaster* Wall, 1908j, *Dryophis mycterizans tephrogaster* Wall, 1908g, *Dryophis mycterizans isabellinus* Wall, 1910h, *Dryophis mycterizans lepidorostralis* Wall, 1910h, *Dryophis mycterizans zephrogaster* Wall, 1910h, and *Dryophis mycterizans rhodonotus* Wall, 1921g.
Type: Holotype, not designated, lost *fide* M.J. Cox et al. (2013: 92).
Type locality: "à Malabar, à Ceylan, à Pondichery, au Bengale, à Java, aux îles Philippines et aux Mariannes" [= India, Sri Lanka, Indonesia, Philippines and Mariannas]. Restricted to Sri Lanka *fide* M.A. Smith (1943: 376).
Distribution: Southern and SE Asia. India (Andhra Pradesh, Assam, Bihar, Goa, Gujarat, Karnataka, Kerala, Madhya Pradesh. Maharashtra, Orissa, Tamil Nadu, West Bengal), Sri Lanka (Central, Eastern, North-Central, Northern, North-Western, Sabaragamuwa, Southern, Uva, Western), Bangladesh (Chittagong), Nepal (Bardiya, Dang, Dhanusa, Gorkha, Kailali, Kapilbastu, Makwanpur, Rupandehi, Sunsari, Surkhet), Myanmar (Rangoon, Yangon), Thailand (Chaiyaphum, Chiang Mai, Chon Buri, Kanchanaburi, Krung Thep Mahanakhon, Lopburi, Mae Hong Sam, Nakhon Pathom, Nakhon Ratchasima, Nakhon Sawan, Phra Nakhon, Phuket, Prachuap Kiri Khan, Sara Buri), S Laos (Champasak), Cambodia (Kampong Cham, Kampong Chhnang, Kampong Spoe) and S Vietnam (Ba Ria-Vung Tau, Dak Lak, Dong Nai, Gia Lai, Ho Chi Minh City, Khanh Hoa, Kien Giang, Kom Tum, Ninh Thuan, Tay Ninh, Vinh Long), NSL–1830 m.
Sources: Wall, 1905g, 1921g, Bourret, 1936b, Deraniyagala, 1955, E.H. Taylor & Elbel, 1958, E.H. Taylor, 1965, P. Silva, 1969, 1980a, Campden-Main, 1970a, Deuve, 1970, R. Sharma, 1971, 1976a, Saint-Girons, 1972a, Daniel, 1983, Murthy, 1986, A. Silva, 1990b, 2009, M.J. Cox, 1991b, Malhotra & Davis, 1991, M.J. Cox et al., 1998, Chan-ard et al., 1999, Schleich & Kästle, 2002, Tiwari & Shah, 2004, Whitaker & Captain, 2004, David & Dubois, 2005, N. Khaire, 2006 and V.S. Nguyen et al., 2009.
Remarks: Official Specific Name *fide* Opinion 524 (ICZN, 1958). Among type locality restrictions (Sri Lanka, Guinea and Carolina) of M.A. Smith (1943: 376), only Sri Lanka is valid.

6. *Ahaetulla perroteti* (A.-M.-C. Duméril, Bibron & Duméril, 1854b). Erpét. Gén. 7(2): 899–900. (*Psammophis perroteti*)

Synonyms: *Leptophis canarensis* Jerdon, 1854, and *Dryiophis tropidococcyx* A.C.L.G. Günther, 1858.
Types: Syntypes (5), MNHN 1208, longest syntype 500 mm (G.S. Perrotet, 1822–1832), location of other syntypes unknown.
Type locality: "Indes-Orientales" [= East Indies, including India].
Distribution: Southern India (Western Ghats).
Source: Ganesh & Chandramouli, 2011a.

7. *Ahaetulla prasina* (F. Boie, 1827). Isis von Oken 20(6): 545. (*Dryophis prasina*)

Synonyms: *Dryiophis prasinus* H. Boie *in* Schlegel, 1826a *(nomen nudum), Dryiophis rostratus* H. Boie *in* Schlegel, 1826a *(nomen nudum), Dryophis prasinus* Schlegel, 1826b *(nomen nudum), Dryophis rostratus* Schlegel, 1826b *(nomen nudum), Dryophis rostrata*

A

Reinwardt *in* F. Boie, 1827, *Dryophis prassinus* – Wagler, 1830 *(nomen incorrectum), Tragops prasinus laetus* Cope, 1861c, *Tragops prasinus citrina* F. Müller, 1885 *(nomen nudum), Dryophis prasinus flavescens* Wall, 1910a, *Dryophis griseus* E.H. Taylor, 1922a, *Dryophis preocularis* E.H. Taylor, 1922a, *Dryophis prasinus chinensis* Mell, 1931a, *Dryophis prasinus indicus* Mell, 1931a, and *Ahaetulla prasina suluensis* Gaulke, 1994a, and *Ahaetulla prasina medioxima* Lazell, 2002.

Types: Syntypes (2), RMNH 782 and RMNH 47582 (formerly RMNH 782) (H. Boie & H.C. Macklot, Dec. 1825–Sept. 1827).

Type locality: "Java" [SW Indonesia]

Distribution: Southeastern Asia and East Indies. India (Arunachal Pradesh, Assam, Manipur, Meghalaya, Mizoram, Nagaland, Sikkim, West Bengal), S China (Fujian, Guangdong, Guizhou, Guangxi, Hong Kong, Hunan, Xizang, Yunnan), Bhutan (Mongar), Myanmar, Thailand (Chanthaburi, Chiang Mai, Chumphon, Kanchanaburi, Krung Thep Mahanakhon, Lampang, Loei, Mae Hong Son, Nakhon Ratchasima, Phang Nga, Phetchabun, Phuket, Prachin Buri, Ranong, Roi Et, Surat Thani, Tak, Trang, Trat, Uttarradit), E Cambodia (Kampot, Koh Kong, Mondolkiri), Laos (Champasak), N Vietnam (Bac Giang, Bac Kan, Ba Ria-Vung Tau, Binh Phuoc, Binh Thuan, Ca Mau, Cao Bang, Dak Lak, Da Nang, Dong Nai, Gia Lai, Ha Giang, Hai Duong, Ha Tay, Ha Tinh, Hoa Binh, Ho Chi Minh City, Khanh Hoa, Kien Giang, Kon Tum, Lai Chau, Lang Son, Lao Cai, Nghe An, Ninh Binh, Quang Binh, Quang Nam, Quang Ninh, Quang Tri, Son La, Tay Ninh, Thai Nguyen, Thanh Hua, Thua Thien-Hue, Vinh Phuc, Con Dao Is.), West Malaysia (Kedah, Pahang, Perak, Perlis, Pinang, Sabah, Seribaut Arch.), Singapore, Brunei, East Malaysia (Sabah, Sarawak, Labuan Is.), Indonesia (Bali, Belitung, Java, Kalimantan, Lombok, Riau Arch., Sibau, Sulawesi, Sumbawa, Sumatra) and Philippines (Balabac, Basilan, Bohol, Bongao, Busuanga, Coron, Culion, Dinagat, Jolo, Luzon, Mindanao, Negros, Palawan, Panay, Polillo, Romblon, Samar, Sanga Sanga, Siasi, Sibutu, Tawitawi), 60–2100 m.

Sources: Bergman, 1956a, Deuve, 1962e, E.H. Taylor, 1965, Leviton, 1968a, Tweedie, 1983, Matsui et al., 1984, Lazell & Lu, 1990, M.J. Cox, 1991b, Ross & Lazell, 1991, B.E. Smith, 1993, Zhao & Adler, 1993, Gaulke, 1994a, David & Vogel, 1996, M.J. Cox et al., 1998, Chan-ard et al., 1999, Darevsky, 1999, Hnízdo, 2000, Orlov et al., 2000, Malkmus et al., 2002, Ao et al., 2004, Whitaker & Captain, 2004, McKay, 2006, B.L. Stuart & Emmett, 2006, Zhao, 2006, Teynié & David, 2007, M.F. Ahmed et al., 2009, V.S. Nguyen et al., 2009, Agarwal et al., 2010 and Wangyal, 2011.

Remarks: Original description based on H. Boie's MS (1823–1825) and Russell (1801: pl. 25), with types in MNHN also. A subspecies of *A. mycterizans* (Linnaeus) *fide* Mahendra, 1984. Reinwadrt credited with authorship by Boie. RMNH 779 listed as syntype *fide* Cox et al. (2013: 96).

8. *Ahaetulla pulverulenta* (A.-M.-C. Duméril, Bibron & Duméril, 1854b). Erpét. Gén. 7(2): 812. (*Dryinus pulverulentus*)

Synonyms: *Dryophis pulverulentus* Jan, 1863b (*nec* A.-M.-C. Duméril, Bibron & Duméril, 1854b), *Passerita purpurascens* A.C.L.G. Günther 1864a, *Ahaetulla pulverulenta xanthiscuta* Deraniyagala, 1955, and *Dryophis pulverulenta indica* Deraniyagala, 1955.

Type: Holotype, MNHN 7565, a 954 mm female (T. Leschenault, 1820–1821).

Type locality: "Ceylan" [= Sri Lanka].

Distribution: Sri Lanka and S India (Gujarat, Kerala, Maharashtra, Tamil Nadu).

Sources: P. Silva, 1969, Khaire & Khaire, 1993, Whitaker & Captain, 2004, David & Dubois, 2005, N. Khaire, 2006 and A. Silva, 2009.

Remarks: Holotype collector listed as Reynaud *fide* Duméril, Bibron & Duméril (1854b: 812)

AIPYSURUS Lacépède, 1804 (Elapidae)

Synonyms: *Hydrophus* Fleming, 1822 *(nomen emendatum), Stephanohydra* Tschudi, 1837, *Aspisurus* Gray *in* Grey, 1841b *(nomen praeoccupatum), Aipisurus* Gray, 1846b *(nomen emendatum), Hypotropis* Gray, 1846b, *Aepyurus* Agassiz, 1847 *(nomen emendatum), Tomogaster* Duméril & Bibron *in* Gray, 1849a, *Aipyusurus* – Bavay, 1869 *(nomen incorrectum), Hypothropis* Marschall, 1873 *(nomen emendatum), Aepysurus* F. Müller, 1878b *(nomen emendatum), Pelagophis* W.C.H. Peters & Doria, 1878, *Oepysurus* – Perrier, 1928 *(nomen incorrectum), Aipysuru* – Briceño-Rossi, 1934 *(nomen incorrectum), Aspisurus* – Boquet, 1948 *(nomen incorrectum), Cupysurus* – Boquet, 1948 *(nomen incorrectum), Apysurus* – Barrett, 1950 *(nomen incorrectum), Hipysurus* – Barrett, 1950 *(nomen incorrectum), Aipsurus* – Dowling in Dowling & Duellman, 1978 *(nomen incorrectum), Smithohydrophis* Kharin, 1981, *Aepyrus* – Mengden, 1985 *(nomen incorrectum), Aipysaurus* – K.L. Williams & Wallach, 1989 *(nomen incorrectum), Aepyuurus* – McCarthy *in* Golay, Smith, Broadley, Dixon, McCarthy, Rage, Schätti & Toriba, 1993 *(nomen incorrectum)*, and *Oceanius* Wells, 2007.

Type species: *Aipysurus laevis* Lacépède, 1804.

Distribution: Marine waters of Indo-Australia, including Arafura Sea, Coral Sea, Flores Sea, South China Sea and Timor Sea.

Sources: M.A. Smith, 1926a, Barme, 1968, L.A. Smith, 1974, Kharin, 1981, Cogger et al., 1983a, Schwander et al., 1985, McCarthy, 1986, Ehmann, 1992, Golay et al., 1993, A.R. Rasmussen, 1997, Kelly, 1998, David & Ineich, 1999, Cogger, 2000, Scanlon & Lee, 2004, Kharin & Cheblyukov, 2006, Wells, 2007, Sanders et al., 2008, Zaher et al., 2009 and Hoser, 2012e.

1. *Aipysurus apraefrontalis* M.A. Smith, 1926a. Monogr. Sea Snakes: 24–25, fig. 13.

Type: Holotype, BMNH 1946.1.1.94 (formerly MAS 9032 & BMNH 1926.2.16.5), a 565 mm male (M.A. Smith, 1925).

Type locality: "Ashmore Reefs, Timor Sea" [= Ashmore Reef, Ashmore and Cartier Islands, Timor Sea, NW Australia, ca. 12°16'S, 123°00'E].

Distribution: Northwestern Australia (NW Western Australia: Ashmore and Cartier Is.).

Source: L.A. Smith, 1974.

2. *Aipysurus duboisii* Bavay, 1869. Mém. Soc. Linn. Normandie 15(5): 33–34.

Synonyms: *Aipysurus australis* Sauvage, 1877, *Pelagophis lubricus* W.C.H. Peters & Doria, 1878, and *Aepyrus duboisii* – Ruiter, 1958 *(nomen incorrectum).*

Type: Holotype, formerly MMBR, a 910 mm male (C.F. Dubois or A.J.C. Dubois), lost *fide* McCarthy *in* Golay et al. (1993: 222).

Type locality: "L'ile Lifou, Nouvelle-Calédonie" [= Lifou Is., New Caledonia, 21°00'S, 167°15'E].

Distribution: Australasia. Papua New Guinea (Central: Yule Is.), Australia (Northern Territory, Queensland, Western Australia), New Caledonia and Loyalty Is.

Sources: David & Vogel, 1996, O'Shea, 1996, Ineich & Rasmussen, 1997, Bauer & Sadlier, 2000 and Ineich & Laboute, 2002.

3. *Aipysurus eydouxii* (Bibron *in* Gray, 1849a). Cat. Snakes Brit. Mus.: 59. (*Tomogaster eydouxii*)

Synonyms: *Thalassophis anguillaeformis* P. Schmidt, 1852, *Thalassophis muraenaeformis* P. Schmidt, 1852, *Thalassiophis anguillaeformis* A.-M.-C. Duméril, Bibron & Duméril, 1854b *(nomen emendatum), Aipysurus margaritophorus* Bleeker, 1858c *(nomen nudum), Aipysurus anguilliformis* – Krefft, 1869b *(nomen incorrectum), Aepysurus anguilliformis* – Boettger, 1892b *(nomen incorrectum), Aepyurus eudouxxi* – Ruiter, 1958 *(nomen incorrectum),* and *Aipysurus eidouxii* – Elter, 1981 *(nomen incorrectum).*

Type: Holotype, BMNH 1946.1.6.86 (formerly BMNH III.10.1.a), a juvenile female.

Type locality: "Indian Ocean."

Distribution: Indo-Australia. Thailand (Chanthaburi, Pattani, Phet Buri, Samut Songkhram, Songkhla), S Vietnam (Binh Thuan, Kien Giang), West Malaysia (Johore), Singapore, East Malaysia (Sabah), Indonesia (Java, Kalimantan, Papua), Papua New Guinea (Western), N Australia (Northern Territory, Queensland, Western Australia) and New Caledonia.

Sources: E.H. Taylor, 1965, B.L. Lim & Sawai, 1975, Tweedie, 1983, O'Shea, 1996. Li et al., 2005, V.S. Nguyen et al., 2009, A.R. Rasmussen et al., 2011 and Sanders et al., 2013.

Remarks: Philippine records rejected *fide* E.H. Taylor, 1922a and Sanders et al., 2012.

4. *Aipysurus foliosquama* M.A. Smith, 1926a. Monogr. Sea Snakes: 22–24, fig. 12.

Type: Holotype, BMNH 1946.1.1.96 (formerly MAS 9052 & BMNH 1926.2.16.7), a 545 mm female (M.A. Smith, 1925).

Type locality: "Ashmore Reefs, Timor Sea" [= Ashmore Reef, Ashmore and Cartier Islands, Timor Sea, NW Australia, ca. 12°16'S, 123°00'E].

Distribution: Western Australia (NW Western Australia, Ashmore and Cartier Is., Scott Reef).

Remarks: Sanders et al. (2012: 18) erroneously listed four syntypes (BMNH 1946.1.1.86–88 and BMNH 1946.1.1.96) and one paratype (RMNH 6430).

5. *Aipysurus fuscus* (Tschudi, 1837). Arch. Naturg. 3(1): 335, pl. 8, figs. 1–7. (*Stephanohydra fusca*)

Type: Holotype, ZMB 2824, a 774 mm male (J.L. Schönlein, 1830–1837).

Type locality: "Celebes" [= Sulawesi, cen. Indonesia] (possibly in error *fide* Cogger, 1975: 80).

Distribution: Western Australia (NW Western Australia: Ashmore and Cartier Is.) and ? E Indonesia (Sulawesi).

Remarks: M.A. Smith (1926a: 17) erroneously referred to BMNH 1926.2.16.1 as holotype.

6. *Aipysurus laevis* Lacépède, 1804. Ann. Mus. Hist. Nat. Paris 4: 197, 210, pl. 56, fig. 3.

Synonyms: *Hypotropis jukesii* Gray, 1846, *Aipysurus fuliginosus* A.-M.-C. Duméril, Bibron & Duméril, 1854b, *Aepysurus laevis* – F. Müller, 1878b *(nomen incorrectum), Hipysurus laevis* – Barrett, 1950 *(nomen incorrectum), Aepyurus laevis* – C.J.P. Haas, 1950 *(nomen incorrectum), Oepysurus laevis* – Parsons, 1970 *(nomen incorrectum),* and *Aipysurus laevi* – Khole, 1991 *(nomen incorrectum).*

Type: Neotype, WAM 22384, an adult female, designated by L.A. Smith (1974: 99).

Type locality: "Locker Island, off Onslow (Lat. 21°44'S, Long. 114°46'E), Western Australia" via neotype selection.

Distribution: Australasia. Eastern Indonesia (Kei and Aru Is.), S Papua New Guinea, N Australia (Northern Territory, Queensland, N Western Australia), New Caledonia and Loyalty Is.

Sources: O'Shea, 1996, Ineich & Rasmussen, 1997, Burns & Heatwole, 1998, Bauer & Sadlier, 2000, Burns & Heatwole, 2000, Ineich & Laboute, 2002 and J.C. Murphy & Schlager, 2003,.

A

7. *Aipysurus mosaicus* K.L. Sanders, Rasmussen, Elmberg, Mumpuni, Guinea, Blias, Lee & Fry, 2012. Zootaxa (3431): 7–12, figs. 4a, 5a–b, 6a–c.

Type: Holotype, SAMA 65222, a 549 mm male (B.F. Fry, Oct. 2000).
Type locality: "Gulf of Carpentaria near Weipa, Australia" [= vicinity of Weipa, ext. N Queensland, ext. NE Australia, 12°40'S, 141°50'E, elevation 10 m].
Distribution: Eastern Indonesia (Papua) and N Australia (Northern Territory, Queensland, Swain Reefs and Thursday Is.).
Remarks: Collector of type listed as B.F. Fry *fide* Sanders et al. (2012: 7) but is probably co-author B.G. Fry.

8. *Aipysurus pooleorum* L.A. Smith, 1974. Rec. West. Aust. Mus. 3(2): 97–98. (*Aipysurus laevis pooleorum*)

Type: Holotype, WAM 21366, an adult male (W. Poole & W. Poole, Sept. 1963).
Type locality: "Shark Bay, Western Australia" [= Shark Bay, W West Australia, W Australia, ca. 25°44'S, 113°34'E].
Distribution: Southwestern Australia (Western Australia).
Source: Storr et al., 1986.

9. *Aipysurus tenuis* Lönnberg & Andersson, 1913a. Kon. Svenska Vetensk. Akad. Handl. 52(3): 13–14.

Types: Syntypes (3), NHR 2400a–c, longest syntype 1020 mm (E. Mjöberg, July 1911).
Type locality: "Cape Jaubert, Australia" [= Cape Jaubert, Western Australia, Australia, 18°57'S, 121°33'E].
Distribution: Northwestern Australia (N Western Australia).
Source: Kharin & Cheblyukov, 2006b.

†*ALAMITOPHIS* Albino, 1986 (†Madtsoiidae)

Type species: †*Alamitophis argentinus* Albino, 1986.
Distribution: Upper Cretaceous of Argentina, and lower Eocene of Australia.
Sources: Albino, 1986, 1987b, 2011b and Scanlon, 1993, 2005.

1. †*Alamitophis argentinus* Albino, 1986. Actas IV Congr. Argent. Paleont. Bioestr. 2: 16–18, fig. 1.

Type: Holotype, MACN-RN 27, one middle trunk vertebra.
Type locality: "Ladera sudoeste del Cerro Cuadrado, noreste de la Ea. Los Alamitos, Arroyo Verde, Dpto. Valcheta, Río Negro, Argentina. Sector medio de la Formación Los Alamitos, Campaniano-Maastrichtiano temprano, cretácico tardío."
Distribution: Upper Cretaceous (Campanian-Maastrichtian: 65.5–83.6 mya) of Argentina.
Source: Albino, 2000.

2. †*Alamitophis elongatus* Albino, 1994. Pesquisas 21(1): 59–60, fig. 2.

Type: Holotype, MACN-RN 38, one middle trunk vertebra.
Type locality: "la ladera sudeste del Cerro Cuadrado, Estancia Los Alamitos, Dpto. Valcheta, sudeste de la provincia de Río Negro, Argentina. La Formación Los Alamitos (Campaniano-Maastrichtiano Inferior)."
Distribution: Upper Cretaceous (Campanian-Maastrichtian: 65.5–83.6 mya) of Argentina.

3. †*Alamitophis tingamarra* Scanlon, 2005. Mem. Queensland Mus. 51(1): 224–228, figs. 6c, 7.

Type: Holotype, QMF 19729, one middle-posterior trunk vertebra (H. Godthelp & M. Archer field crews, mid-1980s– early 1990s).
Type locality: "MP1 horizon, Main Quarry,Tingamarra, Boat Mountain, near Murgon, south-east Queensland (26°S, 152°E), Australia; Tingamarra Local Fauna, Early Eocene, 54.6+/-0.05 MYBP."
Distribution: Lower Eocene (Neustrian, MP 8: 54.6 mya) of Australia. Known only from type locality.
Sources: Albino, 1987a, 1993.

†*ALBANERYX* Hoffstetter & Rage, 1972 (Boidae)

Type species: †*Albaneryx depereti* Hoffstetter & Rage, 1972.
Distribution: Middle Miocene of France, and upper Miocene of Ukraine.
Sources: Rage, 1984b, Zerova, 1989 and Augé & Rage, 2000.

1. †*Albaneryx depereti* Hoffstetter & Rage, 1972. Ann. Paleont. (Vert.) 58(1): 105–107, fig. 9, pl. 1, fig. 6.

Type: Holotype, MNHN LGA 1, one posterior caudal vertebra.
Type locality: "La Grìve-Saint-Alban, Tortonien," Isère, France (Astaracian, middle Miocene).
Distribution: Middle Miocene (Astaracian, MN 7/8: 11.1–12.8 mya) of France.

2. †*Albaneryx volynicus* Zerova, 1989. Vestn. Zool. 1989(5): 31–35, figs. 1–12.

Type: Holotype, ZIKP 22-1089, one caudal vertebra.
Type locality: "Middle Sarmatian (late Miocene), Gritsev, Ukraine, USSR."

Distribution: Upper Miocene (Vallesian, MN 9: 9.7–11.1 mya) of Ukraine. Known only from type locality.

ALLUAUDINA Mocquard, 1894b (Pseudoxyrhophiidae)

Synonyms: *Allusdina* – Palacky, 1898 *(nomen incorrectum), Allandina* – Ditmars, 1933 *(nomen incorrectum),* and *Alluadina* – Phelps, 1981 *(nomen incorrectum).*
Type species: *Alluaudina bellyi* Mocquard, 1894b.
Distribution: Madagascar.
Sources: Guibé, 1958, Domergue, 1984c, Lanza, 1990a, Glaw & Vences, 1994, Kelly et al., 2009 and Zaher et al., 2009.

1. *Alluaudina bellyi* Mocquard, 1894b. C.R. Séanc. Soc. Philom. Paris 1894(17): 9–10.

Type: Holotype, MNHN 1893.214, a 312 mm male (C. Alluaud & Belly, 1892–1893).
Type locality: "la montagne d'Ambre, vallée du Saccaranii [= Sakaramy], Madagascar." Emended to Mararaomby, Madagascar *fide* MNHN catalogue.
Distribution: Northern Madagascar (Antsiranana, Mahajanga, N Toamasina, 400–650 m).
Sources: Raxworthy, 1988 and Ziegler, et al., 1997.

2. *Alluaudina mocquardi* Angel, 1939. Bull. Mus. Natl. Hist. Nat. (2) 11(6): 536–538.

Type: Holotype, MNHN 1939.53, a 500 mm male (R. Decary, 23 July 1938).
Type locality: "une grotte totalement obscure de la Mananjeba (district d'Ambilobe), Nord de Madagascar."
Distribution: Northern Madagascar (Antsiranana).
Remarks: Gender of holotype female *fide* Angel (1939: 11); Domergue (1984c: 544) showed it to have hemipenes.

ALSOPHIS Fitzinger, 1843 (Xenodontidae)

Synonyms: *Plesius* Jan, 1857, *Halsophis* Cope, 1894c *(nomen emendatum), Alsophys* – Grant, 1946b *(nomen incorrectum),* and *Asaophis* – Grant, 1946b *(nomen incorrectum).*
Type species: *Psammophis antillensis* Schlegel, 1837.
Distribution: Lesser Antilles.
Sources: J.A. Peters & Orejas-Miranda, 1970, A. Schwartz & Thomas, 1975, Cadle, 1984a, A. Schwartz & R.W. Henderson, 1988, 1991, R.W. Henderson & Sajdak, 1996, Rodríguez-Róbles & Greene, 1996, Vidal et al., 2000, Hedges et al., 2009, R.W. Henderson & Powell, 2009 and Zaher et al., 2009.
Remarks: *Alsophis (sensu lato)* split into several genera *fide* Zaher et al. (2009: 147).

1. *Alsophis antiguae* H.W. Parker, 1933a. Ann. Mag. Nat. Hist. (10) 11(61): 158. (*Alsophis leucomelas antiguae*)

Synonyms: *Herpdodryas punctifer* Martin, 1838, and *Alsophis antiguae sajdaki* R.W. Henderson, 1990.
Types: Syntypes (2): BMNH 1946.1.4.46–47 (formerly BMNH = 1850. 50.4.29.1–2), two juveniles (Gardiner).
Type locality: "Antigua, Lesser Antilles."
Distribution: Lesser Antilles (Antigua).
Sources: Lazell, 1967, R.W. Henderson, 1990, Sajdak & R.W. Henderson, 1991, Henderson et al., 1996a and Daltry et al., 2001.

2. *Alsophis antillensis* (Schlegel, 1837). Essai Phys. Serp. 1: 155, 2: 214. (*Psammophis antillensis*)

Synonym: *Dromicus leucomelas* A.-M.-C. Duméril, Bibron & Duméril, 1854a.
Type: Lectotype, RMNH 768, a 334 mm specimen (F.-X. Donzelot, 1818–1826), designated by Brongersma (1937a: 5, figs. 1c–d).
Type locality: "Guadeloupe," Lesser Antilles via lectotype selection.
Distribution: Lesser Antilles (Guadeloupe, Marie-Galante).
Sources: Brongersma, 1937a, Sajdak & R.W. Henderson, 1990 and Breuil, 2003.

3. *Alsophis manselli* H.W. Parker, 1933a. Ann. Mag. Nat. Hist. (10) 11(61): 157 (*Alsophis leucomelas manselli*)

Types: Syntypes, (13), BMNH 1946.1.4.53 (formerly BMNH 1899.6.29.18), a male (J.W. Gregory), BMNH 1946.1.4.57–62 (formerly BMNH 1931.10.18.179–184) (Gomez), BMNH 1946.1.4.95 (formerly BMNH 1832.4.1.33) (Howes) and BMNH 1946.1.6.71–75) (formerly BMNH 1886.10.16.4–8), males and females (J.S. Hollings).
Type locality: "Montserrat, Lesser Antilles."
Distribution: Lesser Antilles (Montserrat).
Sources: A. Schwartz, 1966 and Sajdak & R.W. Henderson, 1990.

4. *Alsophis rijgersmaei* Cope, 1870. Proc. Amer. Philos. Soc. (1869–1870) 11(81): 154–155.

Synonym: *Alsophis cinereus* S.W. Garman, 1887b.
Types: Syntypes (6), ANSP 5411–16, longest syntype 946 mm (H.E. van Rijgersma, 1863–1870).
Type locality: "small island of St. Martins, in the Spanish West Indies" [= Saint-Martin, Lesser Antilles].
Distribution: Lesser Antilles (Anguilla, Saint-Barthélemy, Saint-Martin).
Sources: Sajdak & R.W. Henderson, 1990, Powell, 1991, Powell et al., 1991, 1992, Howard et al., 2001 and Breuil, 2002, 2003.

A

5. *Alsophis rufiventris* (A.-M.-C. Duméril, Bibron & Duméril, 1854a). Erpét. Gén. 7(1): 668–670. (*Dromicus rufiventris*)

Types: Syntypes (2), MNHN 3559–60 (formerly RMNH), two males, longest syntype 946 mm.
Type locality: Unknown and "Brésil" [= Brazil] (in error).
Distribution: Lesser Antilles (Nevis, Saba, Statia, St. Eustatius, St. Christopher).
Sources: Sajdak & R.W. Henderson, 1990 and Maley et al., 2006.

6. *Alsophis sajdaki* R.W. Henderson, 1990. Carib. J. Sci. 25(3–4): 119–122, figs. 1a, 2 (*Alsophis antiguae sajdaki*)

Type: Holotype, KU 211059 (formerly ASPS V6789), an 857 mm female (R. Thomas, 11 Aug. 1965).
Type locality: "Great Bird Island, St. George Parish, off of the northeastern coast of Antigua, Lesser Antilles."
Distribution: Lesser Antilles (Antigua: Great Bird Is.), NSL–25 m.
Source: Sajdak & R.W. Henderson, 1990.

7. *Alsophis sanctonum* T. Barbour, 1915. Proc. Biol. Soc. Washington 28: 78.

Synonym: *Alsophis leucomelas danforthi* Cochran, 1938.
Type: Holotype, MCZ 10689 (G.K. Noble, summer 1914).
Type locality: "Terre d'en Haut, Iles des Saintes, near Guadeloupe, French West Indies."
Distribution: Lesser Antilles (Iles des Saintes: Ílet-à-Cabrit, Terre-de-Haut).
Sources: Sajdak & R.W. Henderson, 1990 and Breuil, 2003.

9. *Alsophis sibonius* Cope, 1879. Proc. Amer. Philos. Soc. 18: 275–278.

Type: Holotype, USNM 10138, a 640 mm specimen (F. Ober, 1876–1879).
Type locality: "Island of Dominica," Lesser Antilles.
Distribution: Lesser Antilles (Dominica).
Sources: Sajdak & R.W. Henderson, 1990 and Muelleman et al., 2009.

AMASTRIDIUM Cope, 1861a (Dipsadidae)

Synonyms: *Fleischmannia* Boettger, 1898, *Mimometopon* F. Werner, 1903a, *Phrydops* Boulenger, 1905b, *Mimometophon* – Dunn, 1931b *(nomen incorrectum), Phydrops* – Dunn, 1931b *(nomen incorrectum),* and *Amatridium* – Briceño-Rossi, 1934 *(nomen incorrectum).*
Type species: *Amastridium veliferum* Cope, 1861a.
Distribution: Mesoamerica and Colombia.
Sources: Dunn, 1925, 1954, Cadle, 1984b, L.D. Wilson, 1988d, Savage, 2002 and Zaher et al., 2009.

1. *Amastridium sapperi* (F. Werner, 1903a). Abh. Bayer. Akad. Wiss. 22(2): 349–350. (*Mimometopon sapperi*)

Type: Holotype, formerly ZSM, destroyed 24–25 April 1944 during World War II.
Type locality: "Guatemala."
Distribution: Mesoamerica. Mexico (Chiapas, Nuevo León, Oaxaca, Tamaulipas, Veracruz), Belize (Cayo, Stann Creek, Toledo), Guatemala (Petén, Quiché) and Honduras (Cortés), NSL–1600 m.
Sources: , L.D. Wilson & Meyer, 1969, H.M. Smith, 1971, J.C. Lee, 1996, J.A. Campbell, 1998b, 2000, Stafford & Meyer, 2000, Stafford et al., 2010, Townsend & Wilson, 2010, Wilson & Johnson, 2010 and McCranie, 2011a.
Remarks: A valid species *fide* H.M. Smith, 1971 and Savage, 2002.

2. *Amastridium veliferum* Cope, 1861a. Proc. Acad. Nat. Sci. Philadelphia (1860) 12(9): 370–371.

Synonyms: *Fleischmannia obscura* Boettger, 1898, and *Phrydops melas* Boulenger, 1905b.
Type: Holotype, ANSP 3738, ca. 379 mm male (R.W. Mitchell, 1857–1860).
Type locality: "Cocuyas de Veraguas, N. Grenada" [= Cocuyas de Veraguas, a gold mine near the Río Concepción, N Veraguas Prov., cen. Panama, ca. 8°45'N, 81°00'W, elevation 100 m].
Distribution: Lower Central America and Colombia. Nicaragua (Rio San Juan), Costa Rica (Alajuela, Cartago, Guanacaste, Heredia, Limón, Puntarenas), N Panama (Colón, Veraguas) and NW Colombia (Chocó), NSL–1690 m.
Sources: E.H. Taylor, 1951, Villa et al., 1988, Rueda-Almonacid & Rueda-Martínez, 2004 and Solórzano, 2004

AMBLYODIPSAS W.C.H. Peters, 1857a (Atractaspididae)

Synonyms: *Choristodon* A. Smith, 1849 *in* 1838–1849 *(nomen praeoccupatum), Calamelaps* A.C.L.G. Günther, 1866, *Amplyodipsas* – Jan & Sordelli, 1870–1881 *(nomen incorrectum), Calamelas* – Rochebrunne, 1885 *(nomen incorrectum), Rhinocalamus* A.C.L.G. Günther, 1888c, *Clothelaps* Cope, 1895c, *Calumelaps* – Boulenger, 1908k *(nomen incorrectum), Galamelaps* – Bocage, 1897 *(nomen incorrectum), Camelaps* – Calabresi, 1927 *(nomen incorrectum), Calemelaps* – Briceño-Rossi, 1934 *(nomen incorrectum), Rinocalamus* – Briceño-Rossi, 1934 *(nomen incorrectum), Choristocalamus* Witte & Laurent, 1947 *(nomen substitutum), Amblyodipas* – Underwood, 1979 *(nomen* *incorrectum)*, and *Calamelops* – Misonne, 1980 *(nomen incorrectum)*.

Type species: *Calamaria microphthalma* Bianconi, 1852.

Distribution: Subsaharan Africa.

Sources: Witte & Laurent, 1947, Broadley, 1971c, 1983, Underwood & Kochva, 1993, Ulber, 1994, Kelly et al., 2009 and Zaher et al., 2009.

1. *Amblyodipsas concolor* (A. Smith, 1849 *in* 1838–1849). Illust. Zool. So. Africa, Rept. (App.): 18. (*Choristodon concolor*)

Synonym: *Calamelaps mironi* Mocquard, 1905a.

Type: Holotype, RSM, a 190–191 mm specimen (A. Smith, 1831).

Type locality: "Kaffirland to the eastward of the Cape Colony" [= KwaZulu-Natal, South Africa *fide* Loveridge, 1944b: 161].

Distribution: Northeastern South Africa (KwaZulu-Natal, S Limpopo, E Mpumalanga), NSL–780 m.

Sources: V.F.M. FitzSimons, 1937, 1974, Haacke, 1982, Pienaar et al., 1983, Branch, 1988 and Boycott, 1995.

Remarks: Recognized as a subgenus of *Amblyodipsas* (*Choristocalamus*). V.F.M. FitzSimons (1937: 263) discussed type specimen. This species probably occurs in S Mozambique *fide* Boycott (1995: 417).

2. *Amblyodipsas dimidiata* (A.C.L.G. Günther, 1888c). Ann. Mag. Nat. Hist. (6) 1(3): 322–323, pl. 19, fig. c. (*Rhinocalamus dimidiatus*)

Types: Syntypes (3), BMNH 1946.1.2.93–95, longest syntype 394 mm.

Type locality: "Mpwapwa, tropical Africa" [= Mpwapwa, Dodoma Region, cen. Tanzania, 06°21'S, 36°29'E, elevation 1000 m].

Distribution: Central Tanzania (Dodoma), 600–1000 m.

Source: Spawls et al., 2002.

3. *Amblyodipsas katangensis* Witte & Laurent, 1942. Rev. Zool. Bot. Afr. 36(2): 113–114.

Synonym: *Amblyodipsas katangensis ionidesi* Loveridge, 1951a.

Type: Holotype, MRAC 9321, a 300 mm specimen (R.P. Claquin, Nov.–Dec. 1922).

Type locality: "N'Gayu, confluent du Luapula et de la Luombwa (Distr. du Haut Katanga), Congo belge" [= Ngayu, Katanga Prov., Democratic Republic of the Congo].

Distribution: Southeastern Africa. Southeastern Democratic Republic of the Congo (Katanga), N Zambia (Copperbelt) and E Tanzania (Lindi, Ruvuma, Tanga), 40–650 m.

Sources: Broadley, 1971d, Spawls et al., 2002 and Broadley et al., 2003.

Remarks: Possibly occurs in N Mozambique *fide* Spawls et al. (2002: 432).

4. *Amblyodipsas microphthalma* (Bianconi, 1852). Mem. Reale Accad. Sci. Inst. Bologna 1: 167–170, pl. 8, fig. 1. (*Calamaria microphthalma*)

Synonym: *Amblyodipsas microphthalma nigra* Jacobsen, 1986.

Type: Holotype, IBI, a 311 mm specimen (C. Fornasini, 1842–1847).

Type locality: "Inhambane, Mosambico" [= Inhambane, Inhambane Prov., Mozambique, 23°52'S, 35°23'E, elevation 10 m].

Distribution: Southern Mozambique (Gaza, Inhambane, Maputo, Bazaruto and Inhaca Is.) and NE South Africa (NE KwaZulu-Natal, N Limpopo), 10–1415 m.

Sources: V.F.M. FitzSimons, 1974, Pienaar et al., 1983, Branch, 1988, Broadley, 1990b and Marais, 2004.

Remarks: Original description reprinted in Bianconi (1850–1870: 94–108, pl. 12, figs. 1, 1a–d).

5. *Amblyodipsas polylepis* (Bocage, 1873a). J. Sci. Math. Phys. Nat., Lisboa 4(15): 216–217. (*Calamelaps polylepis*)

Synonyms: *Atractaspis hildebrandtii* W.C.H. Peters, 1877d, *Calamelaps miolepis* A.C.L.G. Günther, 1888c, *Calamelaps warreni* Boulenger, 1908k, and *Calamelaps mellandi* Boulenger, 1915a.

Type: Holotype, MBL 1878, a 443 mm female (J.A. d'Anchieta, 1861–1897); destroyed by fire 18 March 1978.

Type locality: "Dondo, intérieur d'Angola" [= Dundo, northeastern Lunda Prov., ext.N Angola, 700 m, 07°22'S, 20°50'E, elevation 700 m].

Distribution: Southeastern Africa. Southern Somalia (Jubbada Dhexe), SE Kenya (Coast, Eastern, Manda Is.), SE Tanzania (Kigoma, Lindi, Manyara, Morogoro, Piwani, Ruvuma, Tanga), Southeastern Democratic Republic of the Congo (Katanga), W Angola (Benguela, Cuanza Norte, Cunene, Luanda), Zambia (Central, Copperbelt, Eastern, Luapula, Lusaka, Northern, North-Western, Southern, Western, Chilubi Is.), S Malawi (Southern, Nchisi Is.), Mozambique (Inhambane, Manica, Maputo, Nampula, Sofala, Tete), Zimbabwe (Bulawayo, Manicaland, Mashonaland Central, Mashonaland East, Mashonaland West, Masvingo, Matebeleland North, Matebeleland South, Midlands), NE Namibia (Caprivi, Okavango), Botswana (Central, Gaborone, Kweneng, North West), Swaziland and NE South Africa (Gauteng, KwaZulu-Natal, Limpopo, Mpumalanga), NSL–1500 m.

Sources: Broadley, 1971d, FitzSimons, 1974, Lanza, 1983a, 1990b, Pienaar et al., 1983, Auerbach, 1987, Branch, 1988, Spawls et al., 2002, Broadley et al., 2003, M. Griffin, 2003 and Broadley & Blaylock, 2013.

A

6. *Amblyodipsas rodhaini* (Witte, 1930a). Rev. Zool. Bot. Afr. 19(1): 1–3, figs. 1–5. (*Rhinocalamus rodhaini*)

Synonym: *Calamelaps ventrimaculatus katangensis* Witte, 1951 (*nomen praeoccupatum*).
Type: Holotype, MRAC 3079, a 439 mm male (J.A.H. Rodhain, 1910–1911).
Type locality: "toute probabilité du Lomami, Congo Belge" [= Lomami, Katanga Prov., Democratic Republic of the Congo, 05°29'S, 25°17'E, elevation 695 m.
Distribution: Southeastern Democratic Republic of the Congo (Katanga), 590–1125 m.
Source: Witte, 1951.

7. *Amblyodipsas teitana* Broadley, 1971c. Occ. Pap. Natl. Mus. Rhodesia 4B(33): 661–662, fig. 12.

Type: Holotype, MCZ 40705, a 434 mm female (A. Loveridge, 27 April 1934).
Type locality: "Mount Mbololo, Teita Hills near Voi, Kenya, at an altitude of c. 1150 metres" [= Mt. Mbololo, Taita Hills, Coast Region, S Kenya, 03°17'S, 38°28'E].
Distribution: Southeastern Kenya (S Coast), 1150 m. Known only from type locality.
Source: Spawls et al., 2002.

8. *Amblyodipsas unicolor* (J.T. Reinhardt, 1843). Kgl. Danske Vidensk. Selsk. Natur. Math. Afh. 10: 236–238, pl. 1, figs. 1–3. (*Calamaria unicolor*)

Synonyms: *Atractaspis metallicus* Rochbrune, 1886, *Calamelaps feae* Boulenger, 1906, and *Calamelaps niangarae* K.P. Schmidt, 1923.
Type: Holotype, ZMUC, a 375 mm specimen, lost *fide* F.W. Braestrup *in* Hughes & Barry (1969: 1024).
Type locality: "Guineakystein" [= Ghana coast or Danish settlement near Accra, Ghana, 05°33'N, 11°47'W, elevation 30 m *fide* J. Rasmussen & Hughes, 1997: 16].
Distribution: West, cen. and E Africa. Senegal (Fatick, Kédougou, Louga, Ziguinchor), Gambia, Guinea-Bissau (Tombali), W Guinea (Télimélé), Sierra Leone, Ivory Coast (Bouaké, Bouna) Burkina Faso (Centre, Centre-Nord, Centre-Ouest, Est, Hauts-Bassins, Sud-Ouest, Volte Noire), Ghana (Accra, Ashanti, Northern), S Togo (Maritime), S Benin (Atakora, Atlantique, Zou), Nigeria (Anambra, Bauchi, Bendel, Benue, Gongola, Lagos, Plateau, Oyo, Western), N Cameroon (Nord), SW Chad (Mayo-Kebbi), Central African Republic (Bamingui-Bangoran, Haut-Mbomou, Mbomou, Ombella-Mpoko, Ouham, Ouham-Pende), SE Congo, Democratic Republic of the Congo (Bandundu, Bas-Congo, Kinshasa, Orientale), SW South Sudan (Western Equatoria), S Uganda (Central, Western) and W Kenya (cen. Rift Valley), NSL–1500 m.
Sources: Witte, 1962, Doucet, 1963, Leston & Hughes, 1968, Hughes & Barry, 1969, Pitman, 1974, Roman, 1980, Hughes, 1983, 2013, Pauwels & Meirte, 1996, J.B. Rasmussen & Hughes, 1997, J.-F. Trape & Mané, 2000, 2006b, Spawls et al., 2002, Chippaux, 2006, Chirio & Ineich, 2006, Chirio & LeBreton, 2007, Ullenbruch et al., 2010, Segniagbeto et al., 2011, Auliya et al., 2012 and Chirio, 2013.

9. *Amblyodipsas ventrimaculata* (Roux, 1907a). Rev. Suisse Zool. 15(1): 78–*79*, *figs. 1–2*. (*Rhinocalamus ventrimaculatus*)

Synonyms: *Calamelaps pellegrini* Angel, 1921a, *Calamelaps ventrimaculatus katangensis* Witte, 1951, and *Calamelaps ventrimaculatus websteri* V.F.M. FitzSimons & Brain, 1958.
Type: Holotype, MHNN 12 (formerly MZN 65), a 425 mm female.
Type locality: "pays des Barotsé, Afrique du Sud" [= Barotseland, W Zambia].
Distribution: Southern Africa. Southeastern Democratic Republic of the Congo (Katanga), W Zambia (Western), W Zimbabwe (Matabeleland North, Matebeleland South), N Namibia (Caprivi, Khomas, Omaheke) and NW Botswana (Ghanzi, Kweneng, North West), 950–1430 m.
Sources: Mertens, 1955, Broadley, 1959, 1971d, V.F.M. FitzSimons, 1974, Broadley & Cock, 1975, Auerbach, 1987, Branch, 1988, Broadley et al., 2003, M. Griffin, 2003 and Broadley & Blaylock, 2013.

†*AMEISEOPHIS* Holman, 1976c (Colubridae)

Type species: †*Ameiseophis robinsoni* Holman, 1976c.
Distribution: Middle Miocene of USA, middle Miocene of USA, and upper Miocene of USA.
Fossil records: Lower to upper Miocene of USA.
Sources: Holman, 1978, 1998b, 2000a, Rage, 1984b and Parmley & Hunter, 2010.

1. †*Ameiseophis robinsoni* Holman, 1976c. Herpetologica 32(4): 421–422, figs. 1a–b.

Type: Holotype, UCM 30222, one trunk vertebra.
Type locality: "Split Rock Formation, Main Micro Locality, Second Bench, (Brachycrus Quarry) NW 1/4, sec. 36, T 29 N, R 90 W, Fremont County, Wyoming, USA; middle Hemingfordian through lower Barstovian (middle Miocene through upper Miocene)." Emended to Late Early to Early Middle Miocene (Middle Hemingfordian to Early Barstovian)-2. Split Rock Formation, Upper Units, Fremont and Natrona Counties, Wyoming *fide* Holman (2000a: 140).
Distribution: Lower Miocene (Hemingfordian: 16.3–20.6 mya) of USA (Delaware, Wyoming), middle Miocene (Barstovian: 13.6–16.3 mya) of USA (Nebraska, South

Dakota, Wyoming), and upper Miocene (Barstovian: 13.6–16.3 mya) of USA (Nebraska).

Fossil records: Lower Miocene (Hemingfordian) of USA (Delaware), lower/middle Miocene (Arikareean/Hemingfordian) of USA (Wyoming), middle Miocene (Barstovian) of USA (Nebraska, South Dakota, Wyoming), and upper Miocene (Clarendonian) of USA (Nebraska).

Remarks: Revised diagnosis in Holman (2000a: 140).

AMNESTEOPHIS C.W. Myers, 2011 (Dipsadidae)

Type species: *Enicognathus melanauchen* Jan, 1863a.

Distribution: Brazil.

Sources: Jan & Sordelli, 1866 *in* 1860–1866, Boulenger, 1894a, Amaral, 1930e–f, J.A. Peters & Orejas-Miranda, 1970, Dixon, 1980 and C.W. Myers, 2011.

1. *Amnesteophis melanauchen* (Jan, 1863a). Arch. Zool. Anat. Fis. 2(2): 267. (*Enicognathus melanauchen*)

Type: Holotype, RMNH 393, a 375 mm male.

Type locality: "Bahia au Brésil" [= Bahia State, NE Brazil, or Salvador, E Bahia State, NE Brazil, 12°58'S, 38°31'W, elevation 30 m].

Distribution: Northeastern Brazil (Bahia). Known only from holotype.

Remarks: Photographs of holotype and illustrations of maxilla and hemipenis in C.W. Myers (2011: figs. 1–5).

AMPHIESMA A.-M.-C. Duméril, Bibron & Duméril, 1854a (Natricidae)

Synonyms: *Tropinotus* Kuhl & Hasselt, 1822a *(partim, nomen nudum), Tropidonotus* H. Boie, 1826 *(partim), Herpetoreas* A.C.L.G. Günther, 1860c, *Tropidonotis* – Velasco, 1890d *(nomen incorrectum), Hebius* J.C. Thompson, 1913a, *Amphieama* – Maki, 1933 *(nomen incorrectum), Amphieasma* – M.S. Khan, 1984a *(nomen incorrectum), Paranatrix* Mahendra, 1984, *Asianatrix* Hoser, 2012ag *(nomen illegitimum), Elliottnatrix* Hoser, 2012ag *(nomen illegitimum), Greernatrix* Hoser, 2012ag *(nomen illegitimum), Sundanatrix* Hoser, 2012ag *(nomen illegitimum), Wellsnatrix* Hoser, 2012ag *(nomen illegitimum),* and *Wellingtonnatrix* Hoser, 2012ag *(nomen illegitimum).*

Type species: *Coluber stolatus* Linnaeus, 1758.

Distribution: Asia and Indonesia.

Sources: Jan, 1865d, Boulenger, 1893a, Rooij, 1917, M.A. Smith, 1943, Malnate, 1960a, 1962, McDowell, 1961, Toriba, 1994b, Wallach, 1998a, Ziegler & Quyet, 2006, David et al., 2007, Zaher et al., 2009 and I. Das, 2010, 2012.

Remarks: Gender of *Amphiesma* is neuter *fide* Toriba, 1994 and David et al., 1999.

1. *Amphiesma andreae* Ziegler & Le, 2006. Zootaxa (1225): 41–47, figs. 1–8.

Type: Holotype, ZFMK 83747, a 609 mm male (T. Ziegler, H.-W. Hermann, N.T.Vu, K.Q. Le, T.H. Nguyen, X.C. Cao, M.T. Luu & H.T. Dinh, 8 Aug. 2004).

Type locality: "adjacent to Phong Nha –Ke Bang National Park, Thuong Hoa commune, Minh Hoa district, Quang Binh Prov., central Vietnam … in a limestone forest valley at an elevation of 450 m."

Distribution: Central Vietnam (Quang Binh), 450 m. Known only from type locality. Probably occurs in Laos.

Sources: Ziegler et al., 2006, David et al., 2007 and V.S. Nguyen et al., 2009.

2. *Amphiesma arquus* David & Vogel, 2010a. Russ. J. Herp. 17(2): 122–124, figs. 1–4.

Type: Holotype, NMW 37943, a 609 mm male (F. Steindachner, 1878).

Type locality: "Borneo," possibly SE Kalimantan *fide* David & Vogel (2010: 124).

Distribution: Indonesia (Borneo: ? SE Kalimantan). Known only from type specimen.

Source: David & Vogel, 2010.

3. *Amphiesma atemporale* (Bourret, 1934f). Bull. Gén. Instr. Pub. Hanoi 13(4): 75, fig. 1. (*Natrix atemporalis*)

Types: Syntypes (2), MNHN 1935.77–78 (formerly RLB M.441, a 388 mm specimen (Ecole d'Agric.) [Tonkin], a 270 mm (svl) specimen (Serv. Geol. coll.), & a 121 mm (svl) specimen.

Type locality: "Tonkin" [Vietnam]. Restricted to Tam Dao, North Vietnam *fide* Malnate & Romer (1969: 3).

Distribution: Southeastern Asia. Southeastern China (Guangdong, Guangxi, Guizhou, Hong Kong, Yunnan), Laos and ext. N Vietnam (Vinh Phuc), 800–915 m.

Sources: Malnate & Romer, 1969, Karsen et al., 1986, Zhao & Adler, 1993, Orlov et al., 2000, V.S. Nguyen et al., 2009 and I Das, 2010.

4. *Amphiesma beddomii* (A.C.L.G. Günther, 1864a). Rept. Brit. India: 269, pl. 22, fig. E. (*Tropidonotus beddomii*) (*nomen substitutum*)

Synonyms: *Spilotes vittatus* Beddome, 1863 *(nomen praeoccupatum),* and *Rhabdophis beddomei* F. Wall, 1923d *(nomen emendatum).*

Type: Lectotype, BMNH 1946.1.15.32, a 208 mm specimen (R.H. Beddome, 1857–1882), designated by Boulenger (1893: 253).

Type locality: "Nilgherries, India" via lectotype selection.

Distribution: Southern India (Kerala, Maharashtra, Tamil Nadu), 310–2100 m.

Sources: Wall, 1919b, Inger et al., 1984, Malhotra & Davis, 1991, Whitaker & Captain, 2004 and N. Khaire, 2006.

5. *Amphiesma bitaeniatum* (Wall, 1925d). J. Bombay Nat. Hist. Soc. (1924–1925) 30(4): 806–807. (*Natrix bitaeniata*)

Type: Holotype, BMNH 1946.1.13.58 (formerly BMNH 1925.4.2.18) (P.M.R. Leonard, 1924).

Type locality: "Kutkai, North Shan States (4,500 feet), Burma" [= Kutkai, N Shan State, NE Myanmar, 23°27'N, 97°56'E, elevation 1350 m].

Distribution: Southeastern Asia. Southern China (Guangdong, Guangxi, Guizhou, W Yunnan), NE Myanmar (Kachin, Shan), N Thailand (Ching Mai), Laos and N Vietnam (Lao Cai, Son La), 1500–2375 m.

Sources: Zhao & Adler, 1993, Xie et al., 1998, V.S. Nguyen et al., 2009 and I. Das, 2010, 2012.

6. *Amphiesma boulengeri* (Gressitt, 1937). Proc. Biol. Soc. Washington 50: 125–127. (*Natrix boulengeri*)

Type: Holotype, MVZ 23623, a 462 mm female (J.L. Gressitt, 5 Aug. 1936).

Type locality: "Tai-yong, alt. 640 meters, E. Kwangtung Prov., southeastern China (lat. 23°34'N., long. 115° 55'E.)."

Distribution: Southeastern Asia. Southeastern China (Fujian, Guangdong, Guangxi, Guizhou, Hainan, Hunan, Jiangxi, Yunnan), E Thailand (Nakhon Ratchasima, Nakhon Si. Thammarat), Cambodia (Kampong Speu, Kampot, Pursat), Laos (Xiangkhouang) and N Vietnam (Bac Kan, Cao Bang, Dak Lak, Guang Binh, Ha Giang, Ha Tay, Kon Tum, Lam Dong, Lao Cai, Nghe An, Quang Tri, Thai Nguyen, Thua Thien-Hue, Vinh Phuc), 540–1000 m.

Sources: Gressitt, 1941b, Zhao & Adler, 1993, Orlov et al., 2000, Zhao, 2006, Grismer et al., 2008, V.S. Nguyen et al., 2009 and I. Das, 2010, 2012.

7. *Amphiesma celebicum* (W.C.H. Peters & Doria, 1878). Ann. Mus. Civ. Stor. Nat. Genova 13: 386–387. (*Styporhynchus celebicus*)

Type: Holotype, MSNG 30160, a 360 mm female (O. Beccari, 1874).

Type locality: "Kandari (Ins. Selebes)" [= Kendari, Sulawesi, E Indonesia, 03°58'S, 122°36'E, elevation 20 m].

Distribution: Eastern Indonesia (Sulawesi, Sula), 20 m.

Sources: Lang & Vogel, 2005 and Koch, 2012.

8. *Amphiesma concelarum* Malnate, 1963. Notulae Nat. (360): 2–3. (*Amphiesma pryeri concelarum*)

Type: Holotype, CAS 21905, a 786 mm male (J.C. Thompson a.k.a. V. Kühne, 23–24 May 1910).

Type locality: "Miyako-shima, Miyako group, Riukiu Islands" [= Miyakojima, Ryukyu Archipelago, S Japan, bet. 24°43'–57'N and 125°14'–28'E].

Distribution: Southern Japan (Miyako Group, Ryukyu Arch.)

Sources: M. Mori, 1986, Ota & Iwanaga, 1997 and Ota & Takahashi, 1997.

9. *Amphiesma craspedogaster* (Boulenger, 1899b). Proc. Zool. Soc. London 67(1): 163, pl. 17, fig. 1. (*Tropidonotus craspedogaster*)

Synonyms: *Tropidonotus gastrotaenia* F. Werner, 1922a, and *Natrix chrysarga chekiangensis* Mell, 1931a.

Types: Syntypes (6), BMNH 1946.1.12.63–68, longest syntype 635 mm (J.D. La Touche, spring 1896 or 1898).

Type locality: "Kuatun, a village about 270 miles from Foochow, in the mountains at the North-west of the Prov. of Fokien, at an altitude of 3000 to 4000 feet or more, China" [= Guadun, ext. N Fujian, SE China, 27°45'N, 117°50'E, elevation 915–1200 m].

Distribution: China (Anhui, Fujian, Gansu, Guangdong, Guangxi, Hubei, Hunan, Jiangsu, Jiangzi, Shaanxi, Shanxi, Sichuan, Zhejiang) and N Vietnam (Vinh Phuc), 915–1290 m.

Sources: Malnate, 1962, Zhao & Adler, 1993, Orlov et al., 2000, Zhao, 2006, V.S. Nguyen et al., 2009 and Yao, 2012.

10. *Amphiesma deschauenseei* (E.H. Taylor, 1934). Proc. Acad. Nat. Sci. Philadelphia 86: 300–301, figs. 2–4, pl. 17, fig. 5. (*Natrix deschauenseei*)

Type: Holotype, ANSP 19927, a 455 mm specimen (R.M. de Schauensee, 1 Jan.–1 March 1933).

Type locality: "Chieng Mai, North Siam" [= Chiang Mai, Chiang Mai Prov., northwestern Thailand, 18°48'N, 98°40'E, elevation 840 m].

Distribution: Northwestern Thailand (Chiang Mai, Uthai Thani), Vietnam (Cao bang, Tuyen Quang), 300–840 m.

Sources: E.H. Taylor, 1965, V.S. Nguyen et al., 2009 and I. Das, 2012.

Remarks: Vietnamese specimens referrable to another species *fide* P. David (*in litt.*) and V.S. Nguyen et al. (2009: 354). Occurs in China *fide* I. Das (2010: 331).

11. *Amphiesma flavifrons* (Boulenger, 1887b). Ann. Mag. Nat. Hist. (5) 20(116): 96. (*Tropidonotus flavifrons*)

Types: Syntypes (4), MNHN 1889.211–13 and MNHN 1889.215, longest syntype 453 mm (J. Whitehead, 1885).

Type locality: "Mount Kina Baloo, North Borneo" [= Mt. Kinabalu, N Sabah, NE East Malaysia, 6°05'N, 116°30'E, elevation 300–1800 m].

Distribution: Borneo. East Malaysia (Sabah, Sarawak), Brunei and Indonesia (Kalimantan), 500–1300 m.

Sources: Malkmus et al., 2002, David & I. Das, 2003 and Auliya, 2006.

Remarks: BMNH 1891.2.25.3, a 330 mm specimen, is possible type *fide* P. Campbell *(in litt.)*.

12. *Amphiesma frenatum* (Dunn, 1923a). J. Malay. Br. Roy. Asiat. Soc. 1(1): 3. (*Natrix frenata*)

Type: Holotype, BMNH 1946.1.15.29, a 260 mm specimen (J.C. Moulton & H.W. Smith, Sept.–Nov. 1920).

Type locality: "Mt. Murud, 2,000 ft., northwestern Sarawak, Borneo."

Distribution: East Malaysia (Sarawak), 610 m. Known only from type locality.

Source: David & I. Das, 2003.

13. *Amphiesma groundwateri* (M.A. Smith, 1922a). J. Nat. Hist. Soc. Siam 4(4): 205–206, pl. 8, fig. 2. (*Natrix groundwateri*)

Type: Holotype, BMNH 1946.1.13.36, a 450 mm male (H.C. Robinson & C.B. Kloss, March 1919).

Type locality: "Tasan, 40 kilometres S.W. of Chumpon, Peninsular Siam" [= Tasan, 40 km SW Chumpon (10°30'N, 99°11'E), S Thailand].

Distribution: Southern Thailand (Chumpon, 300–600 m). Known only from type locality.

Sources: E.H. Taylor, 1965 and David & I. Das, 2003.

14. *Amphiesma inas* (Laidlaw, 1902). Proc. Zool. Soc. London (1901) 71(2): 576, pl. 35, figs. 3–4. (*Tropidonotus inas*)

Type: Holotype, not designated, a 397 mm specimen (Laidlaw & Yapp, 1899–1900).

Type locality: "Gunong Inas, Malay Peninsula" [at or near Sira Rimau, near the base of the mountain Gunong Inas in the northern part of Malaya *fide* E.H. Taylor, 1965: 831].

Distribution: Southern Thailand (Loei, Nakhon Ratchasima, Nakhon Si Thammarat, Prachuap Khiri Khan) and West Malaysia (Pahang, Perak), 950–2000 m.

Sources: E.H. Taylor, 1965, Tweedie, 1983, M.J. Cox, 1991b, M.J. Cox et al., 1998, David & I. Das, 2003, I Das, 2010 and Grismer et al., 2010.

Remarks: Figures of holotype mislabeled as fig. 2a, b *fide* Laidlaw (1902: 576). Holotype listed as BMNH 79 *fide* Cox et al. (2013: 371).

15. *Amphiesma ishigakiense* (Malnate & Munsterman, 1960). Proc. California Acad. Sci. (4) 31(2): 61–63, fig. 2. (*Natrix pryeri ishigakiensis*)

Type: Holotype, CAS 21913, a 932 mm male (J.C. Thompson a.k.a. V. Kühne, 25 May–2 June 1910).

Type locality: "Ishigaki-shima,Yaeyama group, Riukiu Islands" [= Ishigakijima, Ryukyu Arch., S Japan, bet. 24°20'–37'N and 124°04'–21'E].

Distribution: Southern Japan (Yaeyama Group, Ryukyu Arch.).

Sources: M. Mori, 1986 and Ota & Iwanaga, 1997.

16. *Amphiesma johannis* (Boulenger, 1908i). Ann. Mag. Nat. Hist. (8) 2(8): 244. (*Tropidonotus johannis*)

Synonym: *Natrix nigroreticulata* M.L.Y. Chang, 1932.

Types: Syntypes (3), BMNH 1946.1.12.79, male, and BMNH 1946.1.21.88–89, longest syntype 910 mm (J. Graham, 1903–1908).

Type locality: "Yunnan fu" [= Kunming, Yunnan, S China, 25°02'N, 102°43'E, elevation 1950 m].

Distribution: Southern China (Guizhou, Sichuan, Yunnan), 1830–1900 m.

Source: Zhao & Adler, 1993.

17. *Amphiesma kerinciense* David & Das, 2003. Raffles Bull. Zool. 51(2): 414–416, figs. 1–4.

Type: Holotype, ZRC 2.3521, a 516 mm female (D. Yeo & H.H. Tan, 12 June 1996).

Type locality: "Sungai Jalnei Dalam, at base of Gunung Tugu (or Tujuh) [= Mt. Tugo or Tujuh] (01° 42' 59.0" S-101° 21' 43.1" E), Gunung Kerinci [= Mt. Kerinci], Sumatera Barat Prov., Sumatra Island, Indonesia."

Distribution: Western Indonesia (W Sumatra), 1400 m.

18. *Amphiesma khasiense* (Boulenger, 1890a). Fauna Brit. India, Rept. Batr.: 344. (*Tropidonotus khasiensis*)

Synonym: *Natrix gilhodesi* Wall, 1925b.

Types: Syntypes (4), BMNH 1946.1.12.80–82 and BMNH 1946.1.13.45, longest syntype 563 mm (T.C. Jerdon, 1836–1852).

Type locality: "Khási Hills" [Meghalaya, India].

Distribution: Southeastern Asia. Northwestern India (Arunachal Pradesh, Assam, Meghalaya Mizoran, Nagaland, West Bengal), SW China (Guangxi, Xizang, Yunnan), Myanmar (Chin, Kachin), N Thailand (Chiang Mai, Chiang Rai, Loei), Cambodia (Kampot, Koh Kong) and Laos (Phongsali), 400–2000 m.

Sources: Zhao & Adler, 1993, Orlov et al., 2000, Chanhome et al., 2001, Ao et al., 2004, Whitaker & Captain, 2004, B.L. Stuart & Emmett, 2006, V.S. Nguyen et al., 2009, Pauwels et al., 2009b and I. Das, 2010.

Remarks: Vietnam records listed by V.S. Nguyen et al., 2009 are *A. boulengeri fide* David (*in litt.*).

A

19. *Amphiesma leucomystax* David, Bain, Truong, Orlov, Vogel, Thanh & Ziegler, 2007. Zootaxa (1462): 43–46, figs. 1–7.

Type: Holotype, ZFMK 71702, a 595 mm female (T. Ziegler, 12 July 1997).

Type locality: "southeastern border of the Ky Anh-Ke Go lowland forest protected area, surroundings 18°00'N-106°06'E, Cam Xuyen District, Ha Tinh Prov., Vietnam, 125 m asl."

Distribution: Southeast Asia. Northeastern Cambodia (Ratanakiri), Laos (Khammouan, Xekong) and Vietnam (Da Nang, Gia Lai, Ha Tinh, Kon Tum, Nghe An, Quang Binh, Quang Tri, Thua Thien Hue), 100–1300 m.

Sources: B.L. Stuart & Heatwole, 2008, V.S. Nguyen et al., 2009, I. Das, 2010, B.L. Stuart et al., 2010 and Cox et al., 2013.

20. *Amphiesma metusia* Inger, Zhao, Shaffer & Wu, 1990. Fieldiana: Zool. (58): 18–19, fig. 10.

Type: Holotype, CIB (CIB-FN 36982), an 885 mm female (R.F. Inger, E.-M. Zhao, H.B. Shaffer & G.-F. Wu, 29 May 1987).

Type locality: "Wa Shan (29°38"N, 102°56'E), Hongya County, Sichuan Prov., China, alt. 1200 m; in forest clearing."

Distribution: Southern China (Sichuan), 1200–1470 m. Known only from vicinity of type locality,

Source: Zhao & Adler, 1993.

21. *Amphiesma miyajimae* (Maki, 1931). Monogr. Snakes Japan: 49–51, figs. 22a–d, pl. 15. (*Natrix miyajimae*)

Type: Holotype, NSMT H2956 (formerly KIU no. b), a 392–393 mm male (Miyajima, May 1928).

Type locality: "Taihoku, Formosa" [= Taipei City, Taipei Co., N Taiwan, 25°05'N, 121°34'E, elevation 50 m].

Distribution: Northern Taiwan (Taipei), 50–1000 m.

Sources: Toriba, 1993, Zhao & Adler, 1993, Ota, 1997 and Xiang & Li, 2009.

22. *Amphiesma modestum* (A.C.L.G. Günther, 1875). Proc. Zool. Soc. London 43(1): 232. (*Tropidonotus modestus*)

Type: Lectotype, BMNH 1946.1.13.41, a 484 mm male (T.C. Jerdon, 1836–1852), designated by V.S. Nguyen et al. (2009: 356).

Type locality: "Khasi Hills, India" [= Khasi Hills, E Meghalaya State, NE India, 25°35'N, 91°38'E, elevation 1370–1675 m] via lectotype selection.

Distribution: Eastern India (Assam, Meghalaya, Nagaland), SW China (Guizhou, Guangdong, W Yunnan), Myanmar (Kachin, Kayah), N Thailand, Cambodia (Kampot), N Laos (Xiangkhouang) and S Vietnam (Cao Bang, Ha Giang, Ha Tinh, Lam Dong, Lao Cai, Vinh Phuc), 600–1525 m.

Sources: Deuve, 1970, Saint-Girons, 1972, Wen, 1983, Zhao & Adler, 1993, M.J. Cox et al., 1998, Orlov et al., 2000, V.S. Nguyen et al., 2009 and I. Das, 2010.

Remarks: Records of this species from China and Vietnam (excluding above records) represent an undescribed species *fide* David (*in litt.*).

23. *Amphiesma monticola* (Jerdon, 1854). J. Proc. Asiatic Soc. Bengal (1853) 22(6): 530. (*Tropidonotus monticolus*)

Type: Holotype, not designated, a 982+ mm specimen, lost *fide* M.A. Smith (1943: 308).

Type locality: "Wynaad, S. India" [= Wayanad Distr., N Kerala State, SW India].

Distribution: Southern India (S Karnataku, Kerala, Tamil Nadu).

Source: Whitaker & Captain, 2004.

24. *Amphiesma nicobarense* (Sclater, 1891a). J. Asiat. Soc. Bengal 60(3): 241–242, pl. 6, figs. 5a–d. (*Tropidonotus nicobarensis*)

Synonyms: *Tropidonotus nicobaricus* Sclater, 1891a *(nomen incorrigendum)*, and *Tropidonotus nicobariensis* Boulengere, 1893a *(nomen emendatum)*.

Type: Holotype, ZSI 8895, a juvenile (F. A. de Roepstorff, 1875–1883).

Type locality: "Camorta in the Nicobars" [= Camorta Is., Nicobars, Andaman & Nicobars, E India, bet. 8°02'–14'N & 93°27'–33'E, Bay of Bengal].

Distribution: Eastern India (Andaman & Nicobars: Camorta Is.). Known only from holotype.

Sources: Biswas & Sanyal, 1980 and I. Das, 1999.

25. *Amphiesma octolineatum* (Boulenger, 1904a). Ann. Mag. Nat. Hist. (7) 13(74): 132. (*Tropidonotus octolineatus*)

Synonyms: *Tropidonotus pleurotaenia* Boulenger, 1904a, *Tropidonotus quadrilineatus* Boulenger, 1904a, *Tropidonotus parallelus sublaevis* Despax, 1913, and *Natrix septemlineata* K.P. Schmidt, 1925b.

Type: Holotype, BMNH 1946.1.12.60, a 610 mm female (J. Graham, 1904–1904).

Type locality: "Yunnan Fu (altitude about 6000 feet)" [= Kumming, Yunnan, China, 25°02'N, 102°43'E, elevation 1950 m].

Distribution: Southern China (Guangxi, Guizhou, Sichuan, Yunnan), 1830–2100 m.

Sources: He, 1983, Zhao & Adler, 1993 and Zhao, 2006.

26. *Amphiesma optatum* (S.-Q. Hu & Zhao, 1966). Acta Zootaxon. Sinica 3(2): 160–161, 164, pl. 1, figs. 2–3, pl. 2, fig. 3. (*Natrix optata*)

Type: Holotype, CIB 3624, adult male (C.-C. Liu, 9 Aug. 1940).
Type locality: "Liang Feng Kang, Mt. Omei, Szechwan, altitude 700 m, China" [= Liang Feng Gang, Mt. Emei Shan (ca. 29°31'N, 103°20'E), Sichuan Prov., S China, elevation 700 m].
Distribution: Southern China (Guangxi, Guizhou, Hunan, Sichuan) and N Vietnam (Bac Kan, Cao Bang, Lang Son, Thai Nguyen, Vinh Phu), 400–1100 m.
Sources: Zhao & Adler, 1993, David et al., 1998, Xie et al., 1998, Orlov et al., 2000, 2003, V.S. Nguyen et al., 2009 and I. Das, 2010.

27. *Amphiesma parallelum* (Boulenger, 1890a). Fauna Brit. India, Rept. Batr.: 345. (*Tropidonotus parallelus*)

Synonyms: *Tropidonotus dipsas* J. Anderson, 1879, and *Natrix clerki* Wall, 1925b.
Type: Lectotype, BMNH 1946.1.13.53, a 545 mm male (J.D. Hooker, May–Dec. 1849), designated by Kramer (1977: 728).
Type locality: "Sikkim, India" via lectotype selection.
Distribution: Southeastern Asia. Northeastern India (Arunachal Pradesh, Assam, Meghalaya, Mizoran, Nagaland, Sikkim, West Bengal), E Nepal (Ilam, Lalitpur, Sankhuwasabha), SW China (Xizang, Yunnan), Thailand (UMMZ 138282) and N Myanmar (Kachin), 900–3050 m.
Sources: Boulenger, 1913g, Kramer, 1977, Zhao & Adler, 1993, Orlov et al., 2000, Schleich & Kästle, 2002, Ao et al., 2004, Tiwari & Shah, 2004 and I. Das, 2010.
Remarks: Photographs of lectotype in Kramer (1977: figs. 5–8).

28. *Amphiesma pealii* (Sclater, 1891a). J. Asiat. Soc. Bengal 60(3): 241, pl. 6, figs. 4a–c. (*Tropidonotus pealii*)

Types: Syntypes (2), BMNH 1946.1.13.43 and ZSI 4034 (formerly IMC), two males, longest syntype 508 mm (S.E. Peal, 1872–1891).
Type locality: "Sibsagar district of Assam" [= Sibsagar Distr., N Assam State, NE India, ca. 26°59'N, 94°38'E, elevation ca. 100 m].
Distribution: Northeastern India (Assam), ca. 100 m. Known only from type locality.

29. *Amphiesma petersii* (Boulenger, 1893a). Cat. Snakes Brit. Mus. 1: 225, pl. 14, fig. 1. (*Tropidonotus petersii*) (*nomen substitutum*)

Synonym: *Tropidonotus maculatus* W.C.H. Peters, 1871b (*nomen praeoccupatum*), and *Tropidonotus sundanensis* W.C.H. Peters, 1872d (*nomen praeoccupatum*).
Type: Holotype, BMNH 1946.1.13.47, a 370 mm male (A.R. Wallace, 1 Nov. 1854–25 Jan. 1856).
Type locality: "Borneo" [= Sarawak, East Malaysia].
Distribution: Southeast Asia. West Malaysia, Singapore, East Malaysia (Sabah, Sarawak) and W Indonesia (Sumatra).
Sources: Harman, 1961, Tweedie, 1983, Stuebing, 1991, 1994, Lim & Lim, 1992, David & Vogel, 1996 and David & I. Das, 2003, 2010.

30. *Amphiesma platyceps* (Blyth, 1854a). J. Proc. Asiatic Soc. Bengal 23(3): 297–298. (*Tropidonotus platyceps*)

Synonyms: *Tropidonotus dipsas* Blyth, 1854a, *Zamenis himalayanus* Steindachner, 1867, and *Tropidonotus firthi* Wall, 1914c.
Type: Lectotype, ZSI 7482 (formerly IMC 7482), a 692+ mm specimen (W.S. Sherwill, 1843–1854), designated by Malnate (1966: 11).
Type locality: "Darjeeling, Bengal" [=Darjeeling, West Bengal, N India, 27°02'N, 88°16'E, elevation ca. 2100 m] via lectotype selection.
Distribution: Southern Asia. Northern Pakistan (W Khyber Pakhtunkhwa), N India (Kashmir, Uttar Pradesh, West Bengal), Bangladesh, Bhutan, Nepal (Baitadi, Bajhang, Dolakha, Dadeldhura, Dolpa, Doti, Jumla, Kaski, Kavre, Lalitpur, Lamjung, Makwanpur, Manang, Mustang, Nuwakot, Panchthar, Rasuwa, Sankhuwasabha, Sindhupalchok, Solukhumbu, Taplejung) and SW China (Xizang), 1250–3700 m.
Sources: Swan & Leviton, 1962, Malnate, 1966, Kramer, 1977, Nanhoe & Ouboter, 1987, Zhao & Adler, 1993, M.S. Khan, 2002, Tiwari & Shah, 2004 and Whitaker & Captain, 2004.
Remarks: Photographs of lectotype in Kramer (1977: figs. 9–12).

31. *Amphiesma popei* (K.P. Schmidt, 1925a). Amer. Mus. Novit. (157): 3. (*Natrix popei*)

Type: Holotype, AMNH 27763, adult male (C.H. Pope, Dec. 1922–July 1923).
Type locality: "Nodoa, Hainan, China" [= Nada, Zhan County, Hainan Prov., S China].
Distribution: Southern China (Guangdong, Guangxi, Guizhou, Hainan, Hunan, Yunnan) and Vietnam (Vinh Phuc), 600–800 m.
Sources: Malnate, 1962, Zhao & Adler, 1993, Xie et al., 1998 and V.S. Nguyen et al., 2009.

32. *Amphiesma pryeri* (Boulenger, 1887f). Proc. Zool. Soc. London 55(1): 149, pl. 18, fig. 3. (*Tropidonotus pryeri*)

Type: Lectotype, BMNH 1946.1.7.63, a 770 mm male (H. Pryer, 1886), designated by Ota & Iwanaga (1997: 353).

A

Type locality: "Loo Choo Islands" [= Ryukyu Arch., S Japan]. Restricted to Okinawa Island *fide* Malnate & Munsterman (1960: 52). Probably Anamioshima or Okinawajima *fide* Ota & Iwanaga (1997: 353).

Distribution: Extreme S Japan (Ryukyus: Agunijima, Akajima, Amamioshima, Amurojima, Fukajijima, Gishifujima, Gushikawajima, Gusukujima, Hamahigajima, Hatejima, Haterumajima, Hyanzajima, Iejima, Iheyajima, Ikeijima, Ikemajima, Irabujima, Iriomotejima, Ishigakijima, Izenajima, Kakeromajima, Kerumajima, Kikaijima, Kohamajima, Korijima, Kubajima, Kudakajima, Kumejima, Kurimajima, Kuroshima, Maejima, Minnjima, Miyagijima, Miyakojima, Nakajima, Nakanogamijima, Nohojima, Ogamijima, Ohajima, Ojima, Okierabujima, Okinawajima, Sesokojima, Shimojishima, Taketomijima, Taramajima, Tokashikijima, Tokunoshima, Tonakijima, Tsukenjima, Ukejima, Yabuchijima, Yaguchijima, Yakabijima, Yanahajima, Yonagunijima, Yorojima, Yoronjima and Zamamijima Is.).

Sources: Malnate & Munsterman, 1960, Malnate, 1962, 1963, M. Mori, 1982, Toriba, 1990a and Ota et al., 1991.

33. *Amphiesma sanguineum* (Smedley, 1932a). Bull. Raffles Mus (1931) (6): 116. (*Natrix sanguinea*)

Type: Holotype, BMNH 1946.1.8.90, a 475 mm specimen (N. Smedley, 1931).

Type locality: "Tanah Rata, Cameron Highlands, Pahang, Malay Peninsula" [= Tanah Rata, Pahang State, cen. West Malaysia, 4°28'N, 101°23'E, elevation 1500 m].

Distribution: West Malaysia (Johor, Pahang, Selangor), 125–1750 m.

Sources: Tweedie, 1983, Chan-ard et al., 1999, B.L. Lim et al., 2002, David & I. Das, 2003 and Leong & Diesmos, 2004.

34. *Amphiesma sarasinorum* (Boulenger, 1896c). Ann. Mag. Nat. Hist. (6) 17(101): 393–394. (*Tropidonotus sarasinorum*)

Type: Lectotype, NMBA 986, a 530 mm male (P. Sarasin & F. Sarasin, 1893–1896), designated by Kramer (1978: 662).

Type locality: "Loka, Mount Bonthain, 1200 m, Celebes" [= Loka, Mt. Lompobatang, South Sulawesi Prov., SW Sulawesi, cen. Indonesia, 5°19'S, 119°54'E, elevation 1200 m] via lectotype selection.

Distribution: Central Indonesia (SW Sulawesi), 1200–1500 m.

Sources: Lang & Vogel, 2005 and Koch, 2008b.

35. *Amphiesma sarawacensis* (A.C.L.G. Günther, 1872a). Proc. Zool. Soc. London 40(1): 596. (*Tropidonotus sarawacensis*)

Synonyms: *Tropidonotus maculatus* W.C.H. Peters, 1871b, and *Tropidonotus saravacensis* Boulenger, 1893a *(nomen emendatum)*.

Types: Syntypes (3), BMNH 1946.1.13.12–14, two males and a female (A.H. Everett, 1869–1872).

Type locality: "Matang in the district of Sarawak, western Borneo" [= Kampong Matang, First Division, Kidi Distr., W Sarawak, East Malaysia, 1°35'N, 110°14'E, elevation 15 m].

Distribution: Malay Peninsula and Borneo. West Malaysia (Kelantan, Pahang, Perak, Selangor), Brunei, East Malaysia (Sabah, Sarawak) and Indonesia (Kalimantan), 15–1700 m.

Sources: Tweedie, 1983, Stuebing, 1991, Stuebing & Inger, 1999, Malkmus et al., 2002, David & I. Das, 2003, I. Das, 2010, 2012 and Inger & Tan, 2010.

36. *Amphiesma sauteri* (Boulenger, 1909a). Ann. Mag. Nat. Hist. (8) 4(24): 495. (*Tropidonotus sauteri*)

Synonyms: *Natrix copei* Van Denburgh, 1909, *Amphiesma sauteri bourreti* Malnate, 1962, and *Amphiesma sauteri maximus* Malnate, 1962.

Types: Syntypes (2), BMNH 1946.1.13.38–39, a male and female (H. Sauter, 1902–1909).

Type locality: "Kosempo, Formosa" [= Taiwan].

Distribution: Eastern Asia. Southeastern China (Anhui, Fujian, Gansu, Guangdong, Guangxi, Guizhou, Hainan, Hong Kong, Hunan, Jiangxi, Sichuan, Yunnan), Taiwan, Laos and N Vietnam (Bac Kan, Ha Tay, Lang Son, Vinh Phuc), 500–1500 m.

Sources: Malnate, 1962, Kuntz, 1963, Toriba, 1986, Inger et al., 1990, Zhao & Adler, 1993, Xie et al., 1998, Orlov et al., 2000, Zhao, 2006, V.S. Nguyen et al., 2009, Xiang & Li, 2009 and Yao, 2012.

Remarks: V.S. Nguyen et al. (2009: 359) erroneously listed CAS 18004 as holotype.

37. *Amphiesma sieboldii* (A.C.L.G. Günther, 1860c). Proc. Zool Soc. London 28(1): 156–157. (*Herpetoreas sieboldii*)

Type: Holotype, BMNH 1946.1.13.16, a 917 mm male (H.R.A. von Schlagintweit, A. von Schlagintweit, & R. von Schlagintweit, 1854–1858).

Type locality: "Sikkim, Himalaya (7500 feet above the level of the sea)" [= Sikkim State, NE India, elevation 2285 m].

Distribution: Southern Asia. Northeastern Pakistan (Punjab), N India (Arunachal Pradesh, Sikkim), Nepal, Bhutan (Trashiyangtse) and N Myanmar (Shan), 1220–3660 m.

Sources: Malnate, 1966, Nanhoe & Ouboter, 1987, M.S. Khan, 2002, Agarwal et al., 2010, Wangyal, 2011 and I. Das, 2012.

Remarks: A synonym of *A. platyceps* Blyth *fide* Kramer, 1977.

38. *Amphiesma stolatum* (Linnaeus, 1758). Syst. Nat., ed. 10, 1: 219. (*Coluber stolatus*)

Synonyms: *Coluber bilineatus* Lacépède, 1789 (*nomen rejiciendum*), *Coluber cinctum* Lacépède, 1789 (*nomen rejiciendum*), *Coluber malpolon* Lacépède, 1789 (*nomen rejiciendum*), *Elaps bilineatus* Schneider, 1801, *Coluber arboreus* Merrem, 1820, *Tropidonotus ruficeps* W.C.H. Peters, 1869, *Tropidonotus olivaceus* Sclater, 1891, *Tropidonotus stolatus erythrostictus* Wall, 1911, and *Natrix stolata chinensis* Mell, 1931a.
Type: Holotype, NHR Lin-16 (formerly MAFR), a 490 mm male (Mus. Drottn.), lost *fide* Cox et al. (2013: 380).
Type locality: "America" (in error). Corrected to Asia *fide* Linnaeus (1766: 380). Restricted to Dayao Shan, Guangxi, China *fide* Cox et al. (2013: 380).
Distribution: Southern and SE Asia. Pakistan (Khyber Pakhtunkhwa, Punjab, Sindh), India (Andamans & Nicobars: Andaman Is., Andhra Pradesh, Arunachal Pradesh, Assam, Maharashtra, Meghalaya, Nagaland, Orissa, Sikkim, Tripura, Uttar Pradesh, West Bengal), Sri Lanka (Kalutara), Nepal (Banke, Bara, Bardiya, Bhaktapur, Chitwan, Dang, Dadeldhura, Dhankuta, Dhanusa, Dolakha, Doti, Gorkha, Ilam, Jhapa, Kailali, Kanchanapur, Kapilbastu, Kaski, Kathmandu, Lalitpur, Lamjung, Mahottari, Makwanpur, Nuwakot, Parsa, Rasuwa, Rautahat, Rupandehi, Saptari, Sarlahi, Sindhupalchok, Sunsari, Tanahu, Taplejung), Macau, S China (Anhui, Fujian, Guangdong, Guangxi, Guizhou, Hainan, Henan, Hong Kong, Hunan, Jiangxi, Yunnan, Zhejiang, Lanyu Is.), Taiwan, Myanmar (Yangon), Thailand (Chai Nat, Chaiyaphum, Chiang Mai, Loei, Lop Buri, Nakhon Phanom, Nakhon Ratchasima, Nakhon Sawan, Prachuap Khiri Khan, Saraburi, Si Sa Ket, Suphan Buri, Tak, Udon Thani, Uthai Thani), Bangladesh (Chittagong), Cambodia (Kampot, Pursat), Laos and Vietnam (Bac Giang, Bac Kan, Bac Ninh, Binh Duong, Binh Phuoc, Cao Bang, Da Nang, Dong Nai, Gia Lai, Hai Dong, Hai Phong, Hanoi, Ha Tinh, Hoa Binh, Ho Chi Minh City, Lac Cai, Lang Son, Ninh Binh, Nghe An, Phu Tho, Quang Binh, Quang Nam, Quang Ninh, Quang Tri, Son La, Tay Ninh, Thai Nguyen, Thanh Hoa, Thua Thian-Hue, Tuyen Quang, Vinh Phuc, Yen Bai), NSL–2000 m.
Sources: Wall, 1911a, Kuntz, 1963, E.H. Taylor, 1965, P. Silva, 1969, Singh, 1972, Karsen et al., 1986, Nanhoe & Ouboter, 1987, Easton & Leung-Va, 1993, Zhao & Adler, 1993, I. Das, 1994, M.J. Cox et al., 1998, Chan-ard et al., 1999, Lazell, 1999, Orlov et al., 2000, M.S. Khan, 2002, Tiwari & Shah, 2004, Whitaker & Captain, 2004, N. Khaire, 2006, B.L. Stuart & Emmett, 2006, M.F. Ahmed et al., 2009, Mahony et al., 2009, V.S. Nguyen et al., 2009, A Silva, 2009, Xiang & Li, 2009 and Masroor, 2012.

39. *Amphiesma taronense* (M.A. Smith, 1940c). Rec. Indian Mus. 42(3): 482–483. (*Natrix venningi taronensis*)

Type: Holotype, BMNH 1940.6.1.93, a male (R. Kaulback, 1937–1939).
Type locality: "Pangnamdim, Nam Tamai Valley, Upper Burma, 27°42'N, 97°54'E" [= Pangnamdim, N Kachin State, ext. N Myanmar, 27°44'N, 97°53'E, elevation 1000 m].
Distribution: Northern Myanmar (Chin, Kachin), 1000 m.
Sources: I. Das, 2010, 2012.
Remarks: A valid species *fide* I. Das (2010: 335).

40. *Amphiesma venningi* (Wall, 1910j). J. Bombay Nat. Hist. Soc. (1910–19011) 20(2): 345–346. (*Tropidonotus venningi*)

Synonym: *Natrix nigriventer* Wall, 1925b.
Types: Syntypes (3), BMNH 1946.1.21.86, a 724–750 mm male, BMNH 1946.1.13.60, a 686 mm specimen, and BMNH 1946.1.13.49, a juvenile (formerly FEW XXXV, FEW XXXVI & BNHS) (F.E.W. Venning, 26 Sept.–8 Oct. 1908 and 1910).
Type locality: "Haka Chin Hills" [= vicinity of Haka Station, Chin Hills, Chin State, NW Myanmar, elevation 1980 m].
Distribution: Northeastern India (Arunachal Pradesh, Meghalaya), Bangladesh (Rangamati), N Myanmar (Chin, Kachin) and S China (Yunnan), 1040–1980 m.
Sources: Venning, 1910, Kou, 1985, Zhao & Adler, 1993, Captain & Bhatt, 2002, Gayen, 2002, Whitaker & Captain, 2004, Zhao, 2006 and Reza, 2010b.

41. *Amphiesma vibakari* (H. Boie, 1826). Isis von Oken 18(2): 207–208. (*Tropidonotus vibakari*)

Synonyms: *Tropidonotus martensii* Hilgendorf, 1876, *Natrix vibakari ruthveni* Van Denburgh, 1923, *Natrix vibokari* – Pavloff, 1926 *(nomen incorrectum), Natrix vibakari nikolskii* Emelianov, 1929, and *Amphiesma vibakari danjoensis* Toriba, 1986.
Type: Lectotype, RMNH 1068, a 394–400 mm specimen (P.F.B. von Siebold, 1823–1826), designated herein.
Type locality: "Decima, Japan" [= Dejima Is., previously in Nagasaki harbor, now part of Nagasaki, Japan, 32°45'N, 129°52'E, elevation 10 m] (possibly in error *fide* Siebold *in* Temminck & Schlegel, 1838: iii) via lectotype selection.
Distribution: Eastern Asia. Eastern China (Heilongjiang, Jilin, Liaoning), E. Russia, Korea and Japan, NSL–1800 m.
Sources: Maki, 1931, Shannon, 1956, Malnate, 1962, Bannikov et al., 1977, Ota, 1983, M. Mori, 1984, Toriba, 1986, 1990a, Szyndlar & Hung, 1987, Zhao & Adler, 1993, Adnagulov et al., 2000 and San & Lee, 2007.
Remarks: Original description reprinted in H. Boie (1827b: 252–254), in which J.C. Blomhoff is cited as

collector, with syntype illustrated in Temminck & Schlegel (1838: pl. 5, figs. 1–8). RMNH has three paralectotypes (RMNH 47305–07) collected by Siebold in Japan.

42. *Amphiesma viperinum* (Schenkel, 1901). Verh. Naturf. Ges. Basel (1901–1902) 13(1): 155–156. (*Xenochrophis viperinus*)

Type: Holotype, NMBA 1495, a 250 mm male (A. von Mechel, 1895–1901).
Type locality: "Indragiri, Sumatra" [= Indragiri Regency, Sumatra, W Indonesia].
Distribution: Western Indonesia (Sumatra). Known only from type locality.
Sources: Malnate & Minton, 1965, David & Vogel, 1996 and David & I. Das, 2003.

43. *Amphiesma xenura* (Wall, 1907c). J. Bombay Nat. Hist. Soc. (1906–1907) 17(3): 616–617, pl., fig. 2. (*Tropidonotus xenura*)

Type: Holotype, IMC, lost *fide* M.A. Smith (1943: 292).
Type locality: Unknown.
Distribution: India (Arunachal Pradesh, Assam, Meghalaya, Mizoram, Nagaland), Bangladesh and Myanmar (Rakhine, Sagaing).
Sources: Romer, 1945, I. Das, 2004b, Whitaker & Captain, 2004, Wogan et al., 2008, Ali Reza, 2010 and I. Das, 2010.

AMPHIESMOIDES Malnate, 1961 (Natricidae)

Type species: *Tropidonotus ornaticeps* F. Werner, 1924a.
Distribution: Southeastern China and N Vietnam.
Sources: Gressitt, 1941a, Malnate, 1961, Zhao & Adler, 1993, Zhao, 2006, David et al., 2007, Q.T. Nguyen et al., 2010b, I. Das, 2010, 2012 and Shi, 2011.

1. *Amphiesmoides ornaticeps* (F. Werner, 1924a). Sitz. Bayer. Akad. Wiss. Math.-Natur. Kl. Abt. 1 133(1–3): 30–31. (*Tropidonotus ornaticeps*)

Synonym: *Natrix andrewsi* K.P. Schmidt, 1925.
Type: Holotype, NMW 23415, an 840 mm female (A. Owston, 30 May 1903).
Type locality: "Northern Hainan" [= N Hainan Is., Hainan Prov., SE China].
Distribution: Southeastern China (Fujian, E Guangxi, Hainan) and N Vietnam (Bac Giang, Hoa Binh, Nghe An), NSL–720 m.
Remarks: Guangdong, China record rejected *fide* V.S. Nguyen et al. (2010: 50) and Li, 2011.

AMPLORHINUS A. Smith, 1847 *in* 1838–1849 (Pseudoxyrhophiidae)

Synonyms: *Amblorhinus* – A.C.L.G. Günther, 1894 *(nomen incorrectum), Anoplorhinus* – Boulenger, 1912 *(nomen incorrectum),* and *Amplovihirus* – Briceño-Rossi, 1934 *(nomen incorrectum).*
Type species: *Amplorhinus multimaculatus* A. Smith, 1847 *in* 1838–1849.
Distribution: Southeastern Africa.
Sources: Boulenger, 1896a, Bogert, 1940, Broadley, 1959, 1983, V.F.M. FitzSimons, 1962a, Broadley & Cock, 1975, Branch, 1982, 1988, Cadle, 1994, Marais, 2004, Vidal et al., 2008, Kelly et al., 2009, Zaher et al., 2009 and Broadley & Blaylock, 2013.

1. *Amplorhinus multimaculatus* A. Smith, 1847 *in* 1838–1849. Illust. Zool. So. Africa, Rept.: pl. 57, 2 pp.

Synonym: *Dipsas smithi* A.-M.-C. Duméril, Bibron & Duméril, 1854b.
Types: Syntypes (5), BMNH 1946.1.2.71–75, three adult females, an adult male, and a juvenile female (A. Smith, 1828–1837), longest syntype 610 mm.
Type locality: "Southern Africa" [= Cape of Good Hope *fide* Boulenger, 1896a: 125, and Cape Colony, South Africa *fide* V.F.M. FitzSimons, 1962a: 193].
Distribution: Southeastern Africa. Eastern Zimbabwe (Manicaland), Lesotho and SE South Africa (Eastern Cape, Western Cape, KwaZulu-Natal, Limpopo, Mpumalanga), NSL–2590 m.

†*ANILIOIDES* Auffenberg, 1963 (Boidae)

Synonym: *Aniloides* – Estes & Báez, 1985 *(nomen incorrectum).*
Type species: †*Anilioides minuatus* Auffenberg, 1963.
Distribution: Upper Oligocene and lower Miocene of USA.
Sources: Auffenberg, 1963, Holman, 1981b, 2000a and Rage, 1984b.

1. †*Anilioides minuatus* Auffenberg, 1963. Tulane Stud. Zool. 10(3): 164–165, fig. 12.

Type: Holotype, UF 6151, one trunk vertebra (W. Auffenberg, 1954).
Type locality: "Boulder Bar, Thomas Farm, Gilchrist County, Florida; Lower Miocene, Arikareean, Hawthorne formation."
Distribution: Lower Miocene (Hemingfordian: 16.3–20.6 mya) of USA (Florida). Known only from type locality.

2. †*Anilioides nebraskensis* Holman, 1976b. Herpetologica 32(1): 91–92, figs. 3a–d.

Type: Holotype, UNSM 4519, one trunk vertebra.

Type locality: "Loc. I, Lower Gering; University of Nebraska State Museum Locality MO-119 (NW 1/4, SE 1/4, sec. 32, T 20 N, R 52 W, 10.46 km S and 2.01 km W of Bayard, Morrill County, Nebraska, lower Eocene)."

Distribution: Upper Oligocene (Arikareean: 20.6–30.8 mya) of USA (Nebraska, Wyoming) and lower Miocene (Arikareean: 20.6–30.8 mya) of USA (Nebraska).

ANILIOS Gray, 1845 (Typhlopidae)

Synonyms: *Libertadictus* Wells & Wellington, 1984, *Sivadictus* Wells & Wellington, 1985, and *Austrotyphlops* Wallach, 2006.

Type species: *Anilios australis* Gray, 1845.

Distribution: Austro-Papua.

Sources: Krefft, 1869, Waite, 1918, F. Werner, 1921a, Kinghorn, 1929, Cogger et al., 1983a, Wells & Wellington, 1984, S.G. Wilson & Knowles, 1988, Gow, 1989, Shine & Webb, 1990, 1993a, Ehmann, 1992, Greer, 1997, McDiarmid et al., 1999, Shea, 1999, Cogger, 2000, L.D. Wilson & Swan, 2003, Wallach, 2006, Wallach & Glaw, 2009 and Savage & Boundy, 2012.

1. *Anilios affinis* (Boulenger, 1889b). Ann. Mag. Nat. Hist. (6) 4(23): 363. (*Typhlops affinis*)

Synonyms: *Typhlops unguirostris* W.C.H. Peters, 1867d *(partim),* and *Typhlops kenti* Boulenger, 1914c.

Type: Holotype, BMNH 1946.1.10.56 (formerly BMNH 1874.4.29.243), a 170 mm specimen (R.H. Beddome coll.).

Type locality: "Queensland" [Australia].

Distribution: Eastern Australia (ext. N New South Wales, N Northern Territory, E Queensland), 60–235 m.

Sources: Boulenger, 1893, McDowell, 1974, S.G. Wilson & Knowles, 1988, Shea, 1999, Gow, 1989, Weigel, 1990, Covacevich & Couper, 1991, Swan et al., 2004, S.G. Wilson, 2005, Swanson, 2007 and Emmott & Wilson, 2009.

Remarks: *Typhlops kenti* a valid species *fide* Wells & Wellington (1985: 41). Head of type of *Typhlops kenti* illustrated by H.W. Parker (1931: 604). Waite (1894a: 11), Boulenger (1895g: 718), McDowell (1974a: 33), Cogger et al. (1983: 199) and Bauer et al. (2002: 166) discussed *affinis-unguirostris* types.

2. *Anilios ammodytes* (Montague, 1914). Proc. Zool. Soc. London 84(3): 642–643, pl. 1, figs. 8–10. (*Typhlops ammodytes*)

Type: Holotype, BMNH 1946.1.11.23 (formerly BMNH 1913.10.16.1), a 230 mm specimen (P.D. Montague & L. Burns, 29 May–29 Aug. 1912).

Type locality: "Hermite Island, Monte Bello Islands, Western Australia, approximately in lat. 20° 25'S., long. 115° 30' E, 105 miles E.N.E. of North-West Cape, and 40 miles from the mouth of the Fortescue River" [Australia].

Distribution: Western Australia (NW Western Australia: Barrow, Elphick Nob, Legendre, Lowendal and Varanus Is.), NSL–800 m.

Sources: Waite, 1918, Storr, 1981d, Wells & Wellington, 1985, Storr et al., 2002 and S.G. Wilson & Swan, 2003.

Remarks: Resurrected from the synonymy of *Ramphotyphlops (= Anilios) diversus fide* Wells & Wellington (1984: 105).

3. *Anilios aspinus* (Couper, Covacevich & Wilson, 1998). Mem. Queensland Mus. 42(2): 462–464, figs. 3–4. (*Ramphotyphlops aspina*)

Type: Holotype, QM J51541, a 235.5 mm specimen (T. Pulsford, 1990).

Type locality: "Margot Stn, 20km N Barcaldine (23°27'S, 145°16'E), CQ, Australia" [= cen. Queensland, NE Australia, elevation 280 m].

Distribution: Northeastern Australia (cen. Queensland), 280 m. Known only from type series.

Sources: S.G. Wilson & Swan, 2003, Swan, 2005 and Emmott & Wilson, 2009.

4. *Anilios australis* Gray, 1845. Cat. Lizards Brit. Mus.: 135.

Synonyms: *Onychocephalus verticalis* A. Smith, 1846 *in* 1838–1849, *Onychocephalus macrurus* W.C.H. Peters, 1860a, *Typhlops preissi* Jan, 1860 *in* Jan & Sordelli, 1860–1866, and *Typhlops preyssi* – Jan, 1861a *(nomen incorrectum).*

Type: Lectotype, BMNH 1946.1.10.61, a 196 mm specimen (W. Buchanan, 1840–1845), designated by Cogger *in* Cogger, et al. (1983: 195).

Type locality: "W. Australia" [= Western Australia, presumably Perth or adjacent regions *fide* G. Shea *in* Rabosky et al., 2004: 541].

Distribution: Southwestern Australia (SW Western Australia, Bernier, Dirk Hartog, Dorre, East Wallabi and Rottnest Is.), NSL–480 m.

Sources: Waite, 1929, Storr, 1981d, Weigel, 1990, Bush et al., 1995, 2007, Swan, 1995, Shea, 1999, Bauer et al., 2002, Storr et al., 2002, Rabosky et al., 2004 and Swan et al., 2004.

Remarks: Wells & Wellington (1985: 41) recognized *Typhlops preissi* as a distinct species and transferred it to *Sivadictus.*

A

5. *Anilios batillus* (Waite, 1894a). Proc. Linn. Soc. N.S.W. (2) 9(1): 9–10, pl. 1, figs. 1–3. (*Typhlops batillus*)

Synonym: *Typhlina batilla* Hahn, 1980a.
Type: Holotype, AMS 42756 (formerly MM 3 & MM 669), a 320 mm female (W.J. Macleay & G. Masters, Jan. 1874).
Type locality: "Wagga Wagga, New South Wales, Australia" [= 35°07'S, 147°22'E, elevation 145 m]
Distribution: Southeastern Australia (New South Wales), 145 m. Known only from type locality.
Sources: Goldman et al., 1969, Shea, 1999 and Swan et al., 2004.

6. *Anilios bicolor* (P. Schmidt *in* W.C.H. Peters, 1858a). Mber. Königl. Akad. Wiss. Berlin 1858(11): 509. (*Onychocephalus bicolor*)

Type: Lectotype, ZMB 4721, a 275 mm specimen (M.P.F.P. Schmidt), designated by Wells & Wellington (1985: 41).
Type locality: "Adelaide, S.A." [= Adelaide, SE South Australia, S Australia, 34°56'S, 138°36'E, elevation 45 m] via lectotype selection.
Distribution: Southern Australia (W New South Wales, South Australia, NW Victoria, S Western Australia), NSL–635 m.
Sources: Wells & Wellington, 1985, Bauer et al., 2002, Rabosky et al., 2004, Swan & Watharow, 2005 and Bush et al., 2007.
Remarks: See Bauer et al. (2002: 161) for a discussion of publication date. Resurrected from synonymy of *Anilios australis* and placed in *Sivadictus fide* Wells & Wellington (1985: 41).

7. *Anilios bituberculatus* (W.C.H. Peters, 1863a). Mber. Königl. Akad. Wiss. Berlin 1863(4): 233–234. (*Onychocephalus bituberculatus*)

Synonym: *Typhlina bituberculata* Cogger, 1975.
Type: Lectotype, ZMB 4723, a 333 mm specimen (R.H. Schomburgk, 1849–1863), designated by Wells & Wellington (1985:40).
Type locality: "Loos, 4.5km West of Gawler, South Australia" [= vicinity of Buchfelde, SE South Australia, S Australia, 34°36'S, 138°42'E, elevation 50 m] via lectotype selection.
Distribution: Southern Australia (New South Wales, S Northern Territory, SW Queensland, South Australia, NW Victoria, S Western Australia), NSL–920 m.
Sources: Storr, 1981d, Weigel, 1990, Coventry & Robertson, 1991, Bauer et al., 1995, 2002, Swan, 1995, Shea, 1999, Storr et al., 2002, Swan et al., 2004, Swan & Ratharow, 2005, S.G. Wilson, 2005, Bush et al., 2007, Swanson, 2007 and Emmott & Wilson, 2009.
Remarks: ZMB 4723 was longest and first catalogued specimen of four syntypes, both good reasons why it is most likely selected lectotype by Wells & Wellington (1985: 40).

8. *Anilios broomi* (Boulenger, 1898f). Ann. Mag. Nat. Hist. (7) 2(11): 414. (*Typhlops broomi*)

Type: Holotype, BMNH 1946.1.10.64 (formerly BMNH 1898.9.6.2), a 125 mm specimen (R. Broom, 1894–1898).
Type locality: "Muldiva, Queensland" [= Maldiva, near Cairns (16°55'S, 145°46'E), N Queensland, Australia].
Distribution: Northeastern Australia (E Queensland), 30–615 m.
Sources: Covacevich & Couper, 1991, Shea, 1995, 1999 and S.G. Wilson, 2005.

9. *Anilios centralis* (Storr, 1984b). Rec. West. Aust. Mus. 11(3): 313–314, fig. 1. (*Ramphotyphlops centralis*)

Type: Holotype, NTM A/S 344.90 (formerly NTM 317), a male specimen between 165–306 mm (Conser. Comm. Northern Territory, Oct. 1976).
Type locality: "Alice Springs, Northern Territory, in 23°42'S, 133°52'E, central Australia" [Australia, elevation 610 m].
Distribution: Central Australia (S Northern Territory, ext. N South Australia), 425–800 m.

10. *Anilios chamodracaena* (Ingram & Covacevich, 1993). Mem. Queensland Mus. 34(1): 182–184, fig. 3. (*Ramphotyphlops chamodracaena*)

Type: Holotype, QM J40233, a 114–210 mm specimen (G. Ingram & P. Webber, 18 March 1982).
Type locality: "N Camp 'Beagle', ca. 40 km N Aurukun (13°05'S, 141°59'E), Cape York Peninsula" [= N Queensland, NE Australia].
Distribution: Northeastern Australia (ext. N Queensland), NSL–120 m.
Source: Cameron & Cogger, 2000.

11. *Anilios diversus* (Waite, 1894a). Proc. Linn. N.S.W. (2) 9(1): 10–11, pl. 1, figs. 4–6. (*Typhlops diversus*)

Type: Holotype, QM J2943 (formerly D4432), a 212 mm specimen (F.W. Allpuss, 1 June 1887).
Type locality: "Mowen, Central Railway, Queensland, Australia" [= Morven, 26°25'S, 147°07'E, elevation 510 m].
Distribution: Northern Australia (Northern Territory, NW Queensland, N Western Australia, Bathurst, Bumage, Darch, Guluwuru, Mallison, Melville, Vanderlin and Watson Is.), NSL–285 m.
Sources: Storr, 1981d, Swan, 1995, Shea, 1999, Storr et al., 2002, S.G. Wilson, 2005 and Emmott & Wilson, 2009.

12. *Anilios endoterus* (Waite, 1918a). Rec. S. Aust. Mus. 1(1): 32–33, fig. 24. (*Typhlops endoterus*)

Synonyms: *Anilios australis* Gray, 1845 *(partim*, a paralectotype), *Typhlops leonhardii* Sternfeld, 1919, and *Typhlina endotera* McDowell, 1974a.

Type: Holotype, SAMA 88, a 235 mm specimen (F. Scarfe, 22 Jan. 1912).

Type locality: "Hermannsburg, Central Australia" [= Mission Station Hermannsburg, Finke River, S Northern Territory, Australia, 23°56'S, 132°48'E, elevation 595 m].

Distribution: Central Australia (ext. NW New South Wales, Northern Territory, ext. SW Queensland, South Australia, E Western Australia), NSL–870 m.

Sources: Storr, 1981d, Aplin & Donnellan, 1993, Shea, 1999, Swan et al., 2004 and Emmott & Wilson, 2009.

13. *Anilios erycinus* (F. Werner, 1901a). Verh. Zool.-Bot. Ges. Wien 51: 611, 1 fig. (*Typhlops erycinus*)

Synonyms: *Typhlops iridescens* de Jong, 1930b and *Typhlina erycina* McDowell, 1974a.

Type: Holotype, not designated, a 350 mm specimen, possibly ZMB 17665 or ZMB 18963 *fide* Bauer et al. (2002: 170), NMW *fide* Hahn (1980a: 41), or ZMH, destroyed *fide* Iskandar & Colijn (2001: 14).

Type locality: "Deutsch-Neu-Guinea" [= German New Guinea or N Papua New Guinea].

Distribution: New Guinea. Eastern Indonesia (N Papua) and Papua New Guinea (Madang, Morobe, West Sepik, Western), NSL–165 m.

Sources: Wallach, 1993b, O'Shea, 1996, and Kraus & Allison, 2006.

14. *Anilios ganei* (Aplin, 1998). Rec. West Aust. Mus. 19(1): 7–9, figs. 6–8. (*Ramphotyphlops ganei*)

Type: Holotype, WAS 124835, a 258–355 mm (svl) female (B. Bush, 26 Sept. 1995).

Type locality: "Cathedral Gorge, 30 km west of Newman, Western Australia, in 23°17'30"S, 119°28'E" [Australia].

Distribution: Western Australia (NW Western Australia), 200–715 m.

Source: Storr et al., 2007.

Remarks: Collector of holotype L. Gane *fide* Storr et al. (2007: 25).

15. *Anilios grypus* (Waite, 1918a). Rec. So. Aust. Mus. 1(1): 17–18, fig. 7. (*Typhlops grypus*)

Synonyms: *Typhlops gryphus* – Worrell, 1963 *(nomen incorrectum), Typhlina grypa* McDowell, 1974a, and *Typhlina grypha* – Hahn, 1980a *(nomen incorrectum).*

Type: Holotype, NMV D12351 (formerly NMV 7102), a 335 mm specimen.

Type locality: "Australia." Restricted to E Australia *fide* Storr (1981d: 249).

Distribution: Northern Australia (Northern Territory, Queensland, N Western Australia, Barrow, Faure, and West Intercourse Is.), 165–390 m.

Sources: Storr, 1981d, Sadlier, 1990, Covacevich & Couper, 1991, Shea, 1999, Storr et al., 2002, S.G. Wilson, 2005, Bush et al., 2007, and Emmott & Wilson, 2009.

Remarks: Specific locality of holotype unknown *fide* Cogger et al. (1983: 196). Type possibly from E Australia based on its high ventral count *fide* Storr (1981d: 249).

16. *Anilios guentheri* (W.C.H. Peters, 1865). Mber. Königl. Akad. Wiss. Berlin 1865(6): 259–260, pl. figs. 1, 1a–c. (*Typhlops guentheri*) (*nomen corrigendum*)

Synonyms: *Typhlops güntheri* W.C.H. Peters, 1865 *(nomen incorrigendum)* and *Typhlops guentheri* – Boulenger, 1893a *(nomen corrigendum).*

Type: Holotype, BMNH 1946.1.10.88 (formerly BMNH 1857.10.24.53), a 142 mm specimen (J.R. Elsey, Oct. 1855–June 1856).

Type locality: "Nordaustralien" [= North Australia]. Restricted to the major campsite of the North Australian Expedition on the Victoria River, Northern Territory (ca. 15°40'S, 130°20'E) *fide* Shea & Homer (1996: 58, fig. 3).

Distribution: Northern Australia (NW Northern Territory, N Western Australia), NSL–450 m.

Sources: Storr, 1981d, Shea & Horner, 1996, Shea, 1999, Cogger, 2000, and Storr et al., 2002.

17. *Anilios hamatus* (Storr, 1981d). Rec. West. Aust. Mus. 9(3): 251–254, fig. 15. (*Ramphotyphlops hamatus*)

Type: Holotype, WAM 69572, an 85–377 mm (svl) specimen (R.E. Johnstone, 6 May 1980).

Type locality: "garden at Marandoo, Western Australia, in 22°38'S, 118°06'E" [Australia, elevation ca. 730 m].

Distribution: Western Australia (W Western Australia) NSL–440 m.

Sources: Aplin & Donnellan, 1993 and Bush et al., 2007.

Remarks: Aplin & Donnellan (1993: 250, fig. 3) detected three differentiated groups within the species.

18. *Anilios howi* (Storr, 1983). Rec. West. Aust. Mus. 10(4): 315–317, figs. 1–2. (*Ramphotyphlops howi*)

Type: Holotype, WAM 77226, a 210 mm specimen (P. Griffin, 22 April 1982).

Type locality: "Walsh Point, Port Warrender/Mitchell Plateau area, Western Australia, in 14°34' S, 125°51'E" [Australia, elevation NSL].

Distribution: Western Australia (N Western Australia), 35–230 m.

Source: Storr et al., 2007.

A

19. *Anilios kimberleyensis* (Storr, 1981d). Rec. West. Aust. Mus. 9(3): 254–255, fig. 18. (*Ramphotyphlops kimberleyensis*)

Type: Holotype, WAM 41456, a 220–296 mm specimen (J. Van Roon 5 June 1972).

Type locality: "Bigge Island, Western Australia, in 14°32'S, 125°08'E" [Australia, elevation NSL].

Distribution: Northwestern Australia (N Northern Territory, N Western Australia, Bathurst, Bigge, Coronation, Hidden, Irvine, Koolan, Lachlan, Prudhoe, Sir Graham Moore, and South Maret Is.), NSL–455 m.

Source: Storr et al., 2007.

20. *Anilios leptosomus* (Robb, 1972). J. Roy. Soc. West. Aust. 55(2): 39–40, figs. 1a–c. (*Ramphotyphlops leptosoma*)

Type: Holotype, WAM 29623, a 250–282 mm male (R.B. Humphries, 3 Sept. 1967). Two specimens measured 250 and 282 mm, not indicated which was type.

Type locality: "'The Loop', lower Murchison River, Western Australia," [= 'The Loop' or 'Natures Window' in Murchison River Gorge, Kalbarri National Park, Western Australia, Australia, 27°29'S, 114°23'E, elevation ca. 200 m].

Distribution: Western Australia (W Western Australia), NSL–275 m.

Sources: Storr, 1981d and Bush et al., 2007.

21. *Anilios leucoproctus* (Boulenger, 1889b). Ann. Mag. Nat. Hist. (6) 4(23): 361. (*Typhlops leucoproctus*)

Synonyms: *Typhlina leucoprocta* McDowell, 1974a, and *Ramphotyphlops grovesi* Wells & Wellington, 1985 (*nomen nudum*).

Type: Lectotype, BMNH 1946.1.11.84 (formerly BMNH 1885.6.30.19), a 190–220 mm specimen (S. Macfarlane, 28 Aug.–28 Dec. 1875), designated by Wells & Wellington (1985: 40–41).

Type locality: "Fly River, Papua New Guinea," via lectotype selection.

Distribution: Southern Papua New Guinea (Western, Dauan Is.) and ext. NE Australia (ext. N Queensland, Badu, Moa and Murray Is.), 20–240 m.

Sources: Boulenger, 1893a, McDowell, 1974a, Parker, 1982, O'Shea, 1996 and S.G. Wilson, 2005.

Remarks: Type locality restricted to mouth of Fly River *fide* McDowell (1974: 32). A member of *Ramphotyphlops* *fide* McDowell (1974: 31) and Wallach (2006: 14).

22. *Anilios ligatus* (W.C.H. Peters, 1879). Mber. Königl. Akad. Wiss. Berlin 1879(8): 775, pl., figs. 3, 3a–c. (*Typhlops ligatus*)

Synonyms: *Typhlops curtus* Ogilby, 1892, and *Typhlina ligata* McDowell, 1974a.

Type: Holotype, ZMB 6231, a 245 mm specimen (Mus. Godeffroy).

Type locality: "Port Mackay (N. O. Australien)" [= Mackay, Queensland, Australia, 21°09'S, 149°12'E, elevation NSL].

Distribution: Northern and E Australia (N New South Wales, N Northern Territory, E Queensland, N Western Australia, Bathurst, Mooroongga and Raracala Is.), NSL–340 m.

Sources: Storr, 1981d, Swan, 1990, Weigel, 1990, Covacevich & Couper, 1991, Shea, 1999, Storr et al., 2002, Swan et al., 2004, S.G. Wilson, 2005, Swanson, 2007 and Emmott & Wilson, 2009.

Remarks: Wells & Wellington (1985: 41) recognized *Typhlops curtus* as a valid species and allocated it to *Sivadictus*.

23. *Anilios longissimus* (Aplin, 1998). Rec. W. Aust. Mus. 19(1): 4–7, figs. 5a–c. (*Ramphotyphlops longissimus*)

Type: Holotype, WAM 120049, a 268 mm female (L. Whitsed, 22 May 1995).

Type locality: "considerable depth below the ground in a well at Bandicoot Bay, Barrow Island, Western Australia in 20°54'S 115°22'E" [Australia].

Distribution: Northwestern Australia (W Western Australia, Barrow Is.), NSL.

Source: Storr et al., 2002.

24. *Anilios margaretae* (Storr, 1981d). Rec. West. Aust. Mus. 9(3): 259–260, fig. 23. (*Ramphotyphlops margaretae*)

Type: Holotype, WAM 15710, a 306 mm specimen (W.H. Butler, 25 Aug. 1962).

Type locality: "Lake Throssell, Western Australia, in 27° 25'S, 124°18'E" [Australia, elevation ca. 450–500 m].

Distribution: Southwestern Australia (S Western Australia), 450–500 m . Known only from type locality.

25. *Anilios micrommus* (Storr, 1981d). Rec. West. Aust. Mus. 9(3): 260–261, fig. 24. (*Ramphotyphlops micromma*)

Type: Holotype, WAM 1341, a 205 mm specimen (W.R. Richardson, Oct. 1924).

Type locality: "Leopold Downs, Western Australia, in 17° 52'S, 125°26'E" [Australia, elevation 280 m].

Distribution: Northwestern Australia (N Western Australia), 280 m. Known only from type locality.

Source: Storr et al., 2002.

26. *Anilios minimus* (Kinghorn, 1929). Rec. Aust. Mus. 17(4): 190–191, fig. 1. (*Typhlops minimus*)

Synonym: *Typhlina minima* Cogger, 1975.

Type: Holotype, AMS 9692, a 170 mm specimen (H.E. Warren, Dec. 1928).

Type locality: "Groote Eylandt, Gulf of Carpentaria, North Australia, Australia" [= Northern Territory, elevation < 220 m].

Distribution: Northern Australia (Northern Territory, W Queensland, ext. N Western Australia, Elcho and and Groote Eylandt Is.), 25–630 m.

Source: Shea & Horner, 1996.

27. *Anilios nema* (Shea & Horner, 1996). The Beagle 13: 54–55, figs. 1–2. (*Ramphotyphlops nema*)

Type: Holotype, NTM 21665, a 268 mm (svl) specimen (D. Low Choy, 8 Dec. 1995).

Type locality: "Fannie Bay, Darwin, NT, northern Australia" [= Australia, 12°25'S, 130°50'E, elevation NSL].

Distribution: Northern Australia (N Northern Territory), NSL–10 m.

28. *Anilios nigrescens* Gray, 1845. Cat. Spec. Lizards Collect. Brit. Mus. p. 135.

Synonyms: *Typhlops rüppelli* Jan, 1863b (*nomen nudum*), *Typhlops rüppelli* Jan 1864c, *Typhlops temminckii* Jam. 1864 *in* Jan & Sordelli, 1860–1866, *Typhlops reginae* Boulenger, 1889b, *Typhlops ruppelli* – Waite, 1918a (*nomen incorrectum*), and *Typhlops rueppelli* Hahn, 1980a (*nomen emendatum*).

Type: Lectotype, BMNH 1946.1.11.17 (G. Newport, 1840–1845), designated by Cogger et al. (1983: 198).

Type locality: "Parramatta, New South Wales, Australia" [= 33°49'S, 151°00'E, elevation NSL] via lectotype selection.

Distribution: Southeastern Australia (Australian Capital Territory, E New South Wales, SE Queensland, Victoria), NSL–1410 m.

Sources: Boulenger, 1893a, Loveridge, 1934, Rawlinson, 1966, Jenkins & Bartell, 1980, Griffiths, 1987, Hoser, 1989, Swan, 1990, 1995, Weigel, 1990, Covacevich & Couper, 1991, Coventry & Robertson, 1991, Webb & Shine, 1992, 1993b, Shea, 1999, Swan et al., 2004 and S.G. Wilson, 2005.

Remarks: A synonym of *Ramphotyphlops polygrammicus fide* Forcart (1953: 375). *Ramphotyphlops reginae* a valid species *fide* Wells & Wellington (1984: 105).

29. *Anilios nigricaudus* (Boulenger, 1895i). Proc. Zool. Soc. London 63(4): 867, pl. 49, figs. 1, 1a–b. (*Typhlops nigricauda*)

Type: Lectotype, BMNH 1946.1.11.33, a 315 mm specimen (K.T.F. Dahl, 1895), designated by Shea & Homer (1996: 58).

Type locality: "Daly River, North Australia" [= Northern Territory, Australia] via lectotype selection.

Distribution: Northwestern Australia (NW Northern Territory, ext. N Western Australia), NSL–210 m.

Sources: Storr, 1981d, Wells & Wellington 1984, Shea & Homer, 1996 and Cogger, 2000.

Remarks: Shea & Horner (1996: 58) provided data to support full species status.

30. *Anilios nigroterminatus* (H.W. Parker, 1931). Ann. Mag. Nat. Hist. (10) 8(48): 605. (*Typhlops niroterminatus*)

Synonym: *Typhlops kenti* Waite, 1918 (*nomen praeoccupatum*).

Type: Holotype, BMNH 1946.1.11.58 (formerly BMNH 1931.10.10.1), a 270 mm female (B. Grey, 1929–1931).

Type locality: "Roebuck Bay, N. W. Australia" [= Western Australia, 18°04'S, 122°17'E, elevation NSL].

Distribution: Northwestern Australia (N Western Australia), NSL–10 m.

Sources: McDowell, 1974 and Shea, 1999.

Remarks: Resurrected from synonymy of *A. grypus fide* Wallach *in* McDiarmid et al. (1999: 66).

31. *Anilios pilbarensis* (Aplin & Donnellan, 1993). Rec. West. Aust. Mus. 16(2): 244–247, figs. 2a–c. (*Ramphotyphlops pilbarensis*)

Type: Holotype, WAM 90864, a 241 mm male (J. Dell, 26 Sept. 1988).

Type locality: "Site WS2, Woodstock Station, Pilbara region in 21°36'42"S 118°57'20"E, northwestern Western Australia, Australia."

Distribution: Western Australia (NW Western Australia), NSL–685 m.

32. *Anilios pinguis* (Waite, 1897a). Trans. Roy. Soc. So. Aust. (1896–1897) 21(1): 25–26, pl. 3, figs. 1–4. (*Typhlops pinguis*)

Synonym: *Typhlops opisthopachys* F. Werner, 1917a.

Type: Holotype, SAMA 803, a 348 mm specimen (E.C. Stirling, 1890).

Type locality: "South Australia" [Australia].

Distribution: Southwestern Australia (SW Western Australia), NSL–420 m.

Sources: Waite, 1929, Storr, 1981d, Bush et al., 1995, Shea, 1999 and Bush et al., 2007.

Remarks: Distribution restricted to SW Australia *fide* Storr (1981: 262) and Shea (1999: 50).

33. *Anilios polygrammicus* (Schlegel, 1839 *in* 1837–1844). Abbild. Amph.: 40, pl. 32, figs. 35–38. (*Typhlops polygrammicus*)

Synonyms: *Typhlops bipartitus* Sauvage, 1879, *Typhlops torresianus* Boulenger, 1889b, *Typhlops florensis* Boulenger, 1897d, *Typhlops elberti* Roux, 1911, *Typhlops florensis undecimlineatus* Mertens, 1927b,

A

Typhlops florensis brongersmai Mertens, 1929a, *Typhlops soensis* de Jong, 1930a, and *Typhlina polygrammica polygrammica* Hahn, 1980a.

Type: Holotype, RMNH 3712, a 282 mm specimen (S. Müller & H.C. Macklot, 1825–1829).

Type locality: "Timor" [S Indonesia].

Distribution: Austro-Papua. Indonesia (S Papua, Flores, Komodo, Lombok, Moyo, Sumba, Sumbawa, Timor), S Papua New Guinea (Western, Daru and Sturt Is.) and ext. E Australia (E New South Wales, E Queensland, Dunk, Lizard and Murray Is.), NSL–1200 m.

Sources: Mertens, 1927b, 1929a, Brongersma, 1934, Forcart, 1953, Sadlier, 1990, O'Shea, 1996 and Lang, 2011.

Remarks: *Typhlops bipartitus* a valid species *fide* McDiarmid et al. (1999: 93). Java, Indonesia records doubtful *fide* Brongersma (1934:186).

34. *Anilios proximus* (Waite, 1893). Rec. Aust. Mus. 2(5): 60–62, pl. 15, figs. 3–4. (*Typhlops proximus*)

Synonym: *Typhlina proxima* McDowell, 1974a.

Type: Holotype, AMS 131704 (formerly AMS 6411), a 405 mm specimen.

Type locality: "New South Wales, Australia."

Distribution: Southeastern Australia (E New South Wales, ext. SE Queensland, cen. Victoria), NSL–1010 m.

Sources: Hoser, 1989, Swan, 1990, 1995, Weigel, 1990, Covacevich & Couper, 1991, Coventry & Robertson, 1991, Shea, 1999, Swan et al., 2004 and S.G. Wilson, 2005.

Remarks: Type locality listed as "New South Wales" *fide* Cogger et al. (1983: 198), based presumably on data for AMS 6411 as Waite listed both New South Wales and Victoria as localities for the type series of several specimens.

35. *Anilios robertsi* (Couper, Covacevich & Wilson, 1998). Mem. Queensland Mus. 42(2): 460–462, figs. 1–2. (*Ramphotyphlops robertsi*)

Type: Holotype, QM J63736, a 290 mm specimen (L. Roberts, 10 Feb. 1983).

Type locality: "Romeo Ck, via Shipton's Flat (15°50'S, 145°14'E) NEQ" [= near Cooktown, northeastern Queensland, Australia, NSL].

Distribution: Northeastern Australia (ext. NE Queensland), NSL.

Source: S.G. Wilson, 2005

36. *Anilios silvia* (Ingram & Covacevich, 1993). Mem. Queensland Mus. 34(1): 181–182, figs.1–2. (*Typhlops silvia*)

Type: Holotype, QM J27387, a 175 mm specimen (J. Covacevich & P. Filewood, 3–6 February 1976).

Type locality: "Seary's Scrub, Cooloola NP (25°58'S,153°07'E), SEQ" [= Great Sandy Region, Cape York Peninsula, southeastern Queensland, Australia, NSL].

Distribution: Extreme E Australia (ext. SE Queensland, Fraser Is.), NSL–160 m.

Source: S.G. Wilson, 2005.

37. *Anilios splendidus* (Aplin, 1998). Rec. Aust. Mus. 19(1): 1–2, figs. 2, 3a–b, 4a–c. (*Ramphotyphlops splendidus*)

Type: Holotype, WAM 119900, a 512 mm female (T. Heger & N. Heger, 3 April 1995).

Type locality: "Distribution's residence, Milyering Well, Cape Distribution National Park, Western Australia in 22°01'S 113°56'E" [Australia, elevation NSL]

Distribution: Northwestern Australia (ext. Western Australia), NSL. Known only from type locality.

Source: Storr et al., 2002.

38. *Anilios tovelli* (Loveridge, 1945a). Proc. Biol. Soc. Washington 58: 111. (*Typhlops tovelli*)

Type: Holotype, MCZ 48844, a 122 mm specimen (G.T.R. Tovell, 1944).

Type locality: "Koonowarra Sports Ground, about five miles south of Darwin, Northern Territory, Australia" [elevation ca. 30 m].

Distribution: Extreme N Australia (N Territory, Melville Is.), NSL–140 m.

Source: S.G. Wilson & Swan, 2003.

Remarks: Hahn (1980a: 47) erroneously listed holotype as MCZ 48845.

39. *Anilios troglodytes* (Storr, 1981d). Rec. West. Aust. Mus. 9(3): 263, fig. 27. (*Ramphotyphlops troglodytes*)

Type: Holotype, WAM R51043, a 402 mm specimen (B.R. Wilson & S.M. Slack-Smith, 1 Sept. 1975).

Type locality: "Tunnel Cave, Napier Distribution, Western Australia, in 17°37'S, 125°14'E." [= Kimberley region, Australia, ca. 90 m].

Distribution: Northern Australia (NE Queensland, N Western Australia), 10–425 m.

Source: Storr et al., 2002.

40. *Anilios unguirostris* (W.C.H. Peters, 1867d). Mber. Königl. Akad. Wiss. Berlin 1867(11): 708–709, pl., figs. 3, 3a–c. (*Typhlops unguirostris*)

Synonym: *Typhlops curvirostris* W.C.H. Peters, 1879.

Type: Lectotype, ZMB 5884, a 450 mm specimen (A. Dietrich, 1863–1867, via Mus. Godeffroy), designated by McDowell (1974a: 33).

Type locality: "Rockhampton, Queensland" [= Australia, 23°23'S, 150°30'E, elevation NSL] via lectotype selection.

Distribution: Northern Australia (N Northern Territory, N Queensland, N Western Australia, Crocodile, Groote Eylandt, Milingimbi and Yabooma Is.), NSL–700 m.

Sources: Gow, 1977 and Storr, 1981d, Sadlier, 1990, Covacevich & Couper, 1991, Shea, 1999, Bauer et al., 2002, Storr et al., 2002, S.G. Wilson, 2005 and Emmott & Wilson, 2009.

Remarks: Hahn (1980: 48) erroneously reported a ZMB holotype. The second syntype (ZMB 63444, formerly ZMB 5884) was reidentified as *Typhlops (= Anilios) affinis* (Boulenger) *fide* McDowell, (1974a: 33). Types collected by A. Dietrich in NE Queensland for J.C. Godeffroy (Mus. Godeffroy, Hamburg) *fide* Bauer et al. (2002: 166).

41. *Anilios waitii* (Boulenger, 1895g). Proc. Linn. Soc. N.S.W. (1894) (2) 9(4): 718. (*Typhlops waitii*)

Type: Holotype, BMNH 1946.1.11.7 (formerly BMNH 1869.7.27.3), an adult specimen.

Type locality: "N.W. Australia" [= northwest Australia].

Distribution: Western Australia (S Northern Territory, NW South Australia, Western Australia), 245–1040 m.

Sources: Storr, 1981d, Bush et al., 1995, 2007 and Storr et al., 2002.

42. *Anilios wiedii* (W.C.H. Peters, 1867a). Mber. Königl. Akad. Wiss. Berlin 1867(1): 24–25. (*Typhlops wiedii*)

Type: Holotype, ZMB 5845, a 245 mm specimen (A. Dietrich, 1863–1867, via Mus. Godeffroy).

Type locality: "Brisbane (N. O. Australien)" [= Queensland, Australia, 27°30'S, 153°01'E, elevation 25 m].

Distribution: Eastern Australia (E New South Wales, E Queensland), NSL–1005 m.

Sources: Waite, 1894a, Swan, 1990, Weigel, 1990, Covacevich & Couper, 1991, Shea, 1995, 1999, Swan et al., 2004 and S.G. Wilson, 2005.

Remarks: See remarks about collector under *Anilios unguirostris*.

43. *Anilios yampiensis* (Storr, 1981d). Rec. West. Aust. Mus. 9(3): 268–269, fig. 32. (*Typhlops yampiensis*)

Type: Holotype, WAM 26839, a 128 mm specimen (F.C. Van Ingen, March 1966).

Type locality: "Koolan Island, Western Australia, in 16° 08' S, 123°45' E" [= Yampi Sound, NW Kimberley, Australia, elevation NSL].

Distribution: Northwestern Australia (N Western Australia, Koolan Is.), NSL. Known only from type locality.

44. *Anilios yirrikalae* (Kinghorn, 1942). Rec. Aust. Mus. 21(2): 118, fig. 1. (*Typhlops yirrikalae*)

Type: Holotype, AMS 12381, a 182 mm specimen (W.S. Chaseling, 1936–1939).

Type locality: "Yirrikala Mission Station, near Caledon Bay, Arnhem Land, Northern Territory" [= Australia, 12°15'S, 136°53'E, elevation NSL].

Distribution: Northern Australia (N Northern Territory), NSL. Known only from type locality.

ANILIUS Oken, 1816 (Aniliidae)

Synonyms: *Scytale* Gronovius, 1763 *(nomen illegitimum), Tortrix* Oppel, 1811b (*nomen praeoccupatum*), *Helison* Goldfuss, 1820 (*nomen substitutum*), *Elysia* Hemprich, 1820 *(nomen substitutum), Ilysia* Lichtenstein, 1823 *(nomen emendatum), Torquatrix* Haworth, 1825 (*nomen substitutum*), *Ilyssa* Fitzinger, 1826a *(nomen emendatum),* ? *Rouleau* Guérin-Méneville, 1829 *in* 1829–1844, *Illyria* J.P. Müller, 1832 *(nomen incorrectum), Illisia* Schinz, 1833 *(nomen emendatum), Tortryx* – Schinz, 1833 *(nomen incorrectum), Anileus* Agassiz, 1847 *(nomen emendatum), Hysia* Gistel, 1848 *(nomen emendatum), Jlysia* Troschel & Ruthe, 1848 *(nomen emendatum), Hysia* – Palacky, 1898 *(nomen incorrectum), Anilus* – Briceño-Rossi, 1934 *(nomen incorrectum), Inilius* – Beebe, 1946 *(nomen incorrectum),* and *Anileos* – Vanzolini, 1977 *(nomen incorrectum).*

Type species: *Anguis scytale* Linnaeus, 1758.

Distribution: Amazonia.

Sources: Beddard, 1906, Roze, 1958b, 1966a, Lancini, 1986, Chippaux, 1987, Lancini & Kornacker, 1989, Gorzula & Señaris, 1998, Starace, 1998, G. Köhler, 1999, McDiarmid et al., 1999, Boos, 2001, Franklin, 2001, N.J. Silva, 2001, Barrio-Amorgós et al., 2002, Abuys, 2003, J.C. Murphy & Schlager, 2003, Duellman, 2005, Gower et al., 2005, Gómez et al., 2008, Navarrete et al., 2009 and C.J. Cole et al., 2013.

Remarks: Official Generic Name no. 1527 *fide* Opinion 651 (ICZN, 1963b).

1. *Anilius scytale* (Linnaeus, 1758). Syst. Nat., ed. 10, 1: 228. (*Anguis scytale*)

Synonyms: *Anguis annulata* Laurenti, 1768, *Anguis atra* Laurenti, 1768, *Anguis caerulea,* Laurenti, 1768, *Anguis corallina,* Laurenti, 1768, *Anguis fasciata* Laurenti, 1768, *Anguis rostratus* Weigel, 1782, *Coluber anceps* Scopoli, 1788, *Anguis ater* Gmelin, 1789, *Anguis corallinus* Gmelin, 1789, *Anguis rubra* Bonnaterre, 1790, *Anguis ruber* Suckow, 1797, *Anguis fasciatus* J.G. Schneider, 1801, *Anguis americanae* G. Shaw, 1802, *Tortrix coralinus* Oppel, 1811b *(nomen emendatum), Tortrix maculata* Merrem, 1820, *Angueis scytale* – Briceño-Rossi, 1934 *(nomen incorrectum),*

A

Ilisya scytale – Briceño-Rossi, 1934 *(nomen incorrectum)*, and *Anilius scytale phelpsorum* Roze, 1958b.

Type: Lectotype , NHR Lin-13 (formerly MAFR), a 530–630 mm specimen (Mus. Drottn.), designated by McDiarmid et al. (1999: 157).

Type locality: "Indiis" [= India] (in error) via lectotype selection. Designated to Equatorial America *fide* Chippaux (1987: 29).

Distribution: Amazonia. Eastern Colombia (Amazonas, Arauca, Caqueta, Casanare, Cundinamarca, Meta, Putumayo), SE Venezuela (Amazonas, Bolívar, Delta Amacuro, Mérida, Monagas), Guyana (Cuyuni-Mazaruni, Mahica-Berbice, Upper Demerara-Berbice), Surinam (Para), Paramaribo, Sipal), French Guiana (Cayenne, Saint-Laurent-du-Maroni), N Brazil (Amazonas, Ceará, Goías, Mato Grosso, Pará, Rondônia), E Ecuador (? Monagas, Napo) and E Peru (Amazonas, Huanuco, Loreto, Madre de Dios), 30–700 m.

Remarks: Official Specific Name no.1887 *fide* Opinion 651 (ICZN, 1963b).

ANOMALEPIS Jan, 1860 *in* Jan & Sordelli, 1860–1866 (Anomalepididae)

Synonyms: *Anomalolepis* A.C.L.G. Günther, 1893 *in* 1885–1902 *(nomen emendatum)*, *Anaomalepis* – Briceño-Rossi, 1934 *(nomen incorrectum)*, *Anomolepis* – Jan, 1857 *(nomen incorrectum)*, *Anomalepsis* – Kingsley, 1888 *(nomen incorrectum)*, and *Macphieus* Hoser, 2012am *(nomen illegitimum)*.

Type species: *Anomalepis mexicanus* Jan 1860 *in* Jan & Sordelli, 1860–1866.

Distribution: Lower Central America and NW South America.

Sources: Dunn, 1941, Robb & Smith, 1966, J.A. Peters & Orejas-Miranda, 1970, Hahn, 1980a, Kofron, 1988a, Wallach, 1993a, McDiarmid et al., 1999 and Tipton, 2005.

1. *Anomalepis aspinosus* E.H. Taylor, 1939b. Proc. New England Zool. Club 17: 92–93, pl. 5, figs. 4–7.

Type: Holotype, MCZ 14782, a 126 mm specimen (G.K. Noble [Harvard Peru. Exped.], 22 Aug.–16 Sept. 1916).

Type locality: "Perico, Peru." [= Perico, Cajamarca Department, NW Peru, 05°20'S, 78°47'W, elevation 500 m].

Distribution: Northwestern Peru (Cajamarca), 500–1000 m.

Source: Haas, 1968.

2. *Anomalepis colombia* Marx, 1953b. Fieldiana: Zool. 34(17): 197–198.

Type: Holotype, FMNH 54986, a 172 mm specimen (K. von Sneidern, Jan. 1946).

Type locality: "La Selva, Pueblo Rico, Department of Caldas, Colombia. Altitude 1,700 meters" [= Risaralda Department, 05°46'N, 75°46'W].

Distribution: Cordillera Central of W Colombia (Risaralda), 1700 m. Known only from type locality.

Source: Pérez-Santos & Moreno, 1988.

3. *Anomalepis flavapices* J.A. Peters, 1957a. Amer. Mus. Novit. (1851): 3–6.

Type: Holotype, USNM 196349 (formerly JAP 2613), a 146 mm specimen (Gray via G. Orcés).

Type locality: "near Esmeraldas, Esmeraldas Prov., Ecuador." [= 0°59'N, 79°42'W, elevation NSL].

Distribution: Coastal plain of NW Ecuador (Esmeraldas, Manabi), NSL.

Sources: J.A. Peters, 1960 and Pérez-Santos & Moreno, 1991.

4. *Anomalepis mexicanus* Jan 1860 *in* Jan & Sordelli, 1860–1866. Icon. Gén. Ophid. 1(1): pl. 5, fig. 1, pl. 6, figs. 1a–c, g, n, p, r, v, x.

Synonyms: *Anomalepis mexicana* Boulenger, 1893a, and *Anomalepis dentatus* E.H. Taylor, 1939b.

Type: Holotype, formerly MSNM, destroyed in 1943 during World War II.

Type locality: "Messico" [= México] (probably in error *fide* Kofron, 1988: 12).

Distribution: Northeastern Honduras (Colón, Gracias a Dios), Nicaragua, NW Costa Rica (Guanacaste), W Panama (Canal Zone: Barro Colorado Is.) and NW Peru (Amazonas, Cajamarca), NSL–725 m.

Sources: Dunn, 1941, Kofron, 1988a, Villa et al., 1988, G. Köhler, 1999, 2001a–b, 2003, Savage, 2002, Solórzano, 2004, McCranie et al., 2006 and McCranie, 2011a.

Remarks: Expected to occur in Colombia and Ecuador (Dunn, 1944: 47; Kofron, 1988a: 11; Leenders, 2001: 215), elsewhere in Central America and possibly even Mexico (Savage, 2002: 554). Peruvian population possibly a separate species *fide* McCranie (2011: 41).

†*ANOMALOPHIS* Auffenberg, 1959 (†Anomalophiidae)

Synonym: †*Onomalophis* – Porter, 1972 *(nomen incorrectum)*.

Type species: †*Anomalophis bolcensis* Massalongo, 1859.

Distribution: Lower/middle Eocene of Italy.

Sources: Auffenberg, 1959 and Rage, 1984b.

1. †*Anomalophis bolcensis* (Massalongo, 1859). Spec. Photo. Anim. Plant. Foss. Veronensis: 15–16, pls. 3–4. (†*Archaeophis bolcensis*)

Types: Syntypes (3), MCZ-VP 1001–03, portions of the anterior, middle and posterior parts of the vertebral column and their associated ribs, 480 mm, 330 mm and 270 mm specimens (O. di Canossa, 1850, via A. Agassiz).

Type locality: "agri Veronensis;in calcareo eocenio numm. M. Bubulcae" [= La Pesciara (Monte Bolca), Veneto Prov., Italy, 45°36'N, 11°13'E; Ypresian, lower Eocene *fide* Rage, 1984a: 41].

Distribution: Lower Eocene (Ypresian: 48.6–55.8 mya) of Italy (Veneto). Known only from type locality.

Remarks: Age is middle Eocene *fide* Auffenberg (1959: 15) and upper lower Eocene *fide* Tatarinov (1988: 113).

ANOMOCHILUS Berg, 1901 (*nomen substitutum*) (Anomochilidae)

Synonyms: *Anomalochilus* Lidth de Jeude, 1890 (*nomen praeoccupatum*), *Anomachilus* – Terent'ev, 1961 (*nomen incorrectum*), and *Anomolochilus* – Mahendra, 1984 (*nomen incorrectum*).

Type species: *Anomalochilus weberi* Lidth de Jeude *in* Weber, 1890.

Distribution: Malaysia and Indonesia.

Sources: Brongersma & Helle, 1951, McDowell, 1975a, Cundall et al., 1994, McDiarmid et al., 1999, Gower et al., 2005, I. Das et al., 2008a and I. Das, 2010.

1. *Anomochilus leonardi* M.A. Smith, 1940b. Ann. Mag. Nat. Hist. (11) 6(35): 447–448, figs. a–c.

Type: Holotype, BMNH 1946.1.17.4 (formerly BMNH 1940.4.20.1), a 228–271 mm female (G.R. Leonard, 1938–1940).

Type locality: "Sungei Ngeram, near Merapoh, N. Pahang, Malay Peninsula, at about 800 feet altitude" [= West Malaysia, elevation 245 m].

Distribution: West Malaysia (N Pahang, Selangor), 230–245 m.

Sources: B.L. Lim & Sharef bin Kamarudin, 1975, Tweedie, 1983 and Stuebing & Goh, 1993.

Remarks: Type locality listed as "Ngeram R., nr. Merapoh, N Pahang, Malaya, 500 ft." in BMNH catalogue *fide* McDiarmid et al. (1999: 128).

2. *Anomochilus monticola* I. Das, Lakim, Lim & Hui, 2008. J. Herp. 42(3): 585–588, fig. 1.

Type: Holotype, SP 4594, a 520 mm female (B. Benedict, K. Yustic & L. Johhny, 3 March 2004).

Type locality: "Kiau View Trail (06.00N, 116.32E; 1,513 masl; datum: wgs84), Gunung Kinabalu Park, Ranau District, Sabah, Malaysia (northern Borneo)."

Distribution: East Malaysia (Sabah), 1450–1515 m.

Sources: Stuebing & Goh, 1993, Malkmus et al., 2002 and I. Das, 2012.

3. *Anomochilus weberi* (Lidth de Jeude *in* Weber, 1890). Zool. Ergebn. Nieder. Ost.-Ind. 1(2): 181, pl. 15, figs. 1–3. (*Anomalochilus weberi*)

Type: Holotype, RMNH 4329, a 362 mm male (M. Weber, Mar.–Nov. 1888).

Type locality: "Sumatra: Kaju tanam" [= Kayutanam, Sumatera Barat, western Indonesia, 0°33'S, 100°20'E, elevation 122 m].

Distribution: Indonesia (Kalimantan, W Sumatra), 120–145 m.

Sources: Brongersma & Helle, 1951, Cundall & Rossman, 1994 and David & Vogel, 1996.

ANOPLOHYDRUS F. Werner, 1909 (Homalopsidae)

Type species: *Anoplohydrus aemulans* F. Werner, 1909b.

Distribution: Western Indonesia.

Sources: Haas, 1950, David & Vogel, 1996, J.C. Murphy, 2007b and Zaher et al., 2009.

Remarks: Homalopsidae *incertae sedis fide* J.C. Murphy (2007b: 208).

1. *Anoplohydrus aemulans* F. Werner, 1909b. Mitt. Naturhist. Mus. Hamburg 26: 214, fig. 3.

Type: Holotype, ZMH, a 430 mm male (H. Hagedorn, 1908–1909), destroyed in July 1943 during World War II.

Type locality: "Pudang, Distrikt Babongan, Sumatra" [= Ujungpadang, Bakongan District, Aceh Prov., Sumatra, W Indonesia *fide* Hallermann, 1998: 213].

Distribution: Western Indonesia (NW Sumatra). Known only from type specimen.

ANTAIOSERPENS Wells & Wellington, 1985 (Elapidae)

Synonym: *Antairoserpens* Wilson & Swan, 2003 (*nomen incorrectum*).

Type species: *Cacophis warro* De Vis, 1884.

Distribution: Northeastern Australia.

Sources: Mack & Gunn, 1953, Storr, 1979, Wells & Wellington, 1985 and Horner, 1998.

1. *Antaioserpens warro* (De Vis, 1884). Proc. Roy. Soc. Queensland 1(3): 139. (*Cacophis warro*)

Synonyms: *Pseudelaps albiceps* Boulenger, 1898f, *Denisonia rostralis* De Vis, 1911, and *Rhynchelaps fuscicollis* Lönnberg & Andersson, 1913a.

Type: Holotype, QM J188 (F.A. Blackman, 1855–1874)

Type locality: "Warro Station, Port Curtis" [= Warro Station, near Port Curtis (23°53'S, 151°21'E, elevation NSL), E Queensland, E Australia].

A

Distribution: Northeastern Australia (E Queensland), NSL–500 m.

ANTARESIA Wells & Wellington, 1984 (Pythonidae)

Synonym: *Rawlingspython* Hoser, 2009b & 2012b *(nomen illegitimum).*
Type species: *Liasis childreni* Gray, 1842a.
Distribution: Austro-Papua.
Sources: Wells & Wellington, 1984, L.A. Smith, 1985, K.L. Williams, 1987, Kluge, 1993b, D.G. Barker & Barker, 1994, Walls, 1998b, Hoser, 1999a, 2000d, 2003f, 2009b, 2012b, McDiarmid et al., 1999, Cogger, 2000, Torr, 2000, Fyfe & Green, 2003, Rawlings et al., 2008 and Schleip & O'Shea, 2010.

1. *Antaresia childreni* (Gray, 1842a). Zool. Misc. 2(Mar.): 44. (*Liasis childreni*)

Synonym: *Nardoa gilbertii* Gray, 1842a.
Type: Holotype, BMNH 1946.1.16.78 (formerly BMNH IV.4.3a).
Type locality: Unknown. "N.W. Australia" *fide* BMNH catalogue.
Distribution: Northern Australia (N Northern Territory, N Queensland, N Western Australia, Angarbulumardja, Augustus, Barranyi North, Bathurst, Baudin, Berthier, Bickerton, Bigge, Brigadier, Byam Martin, Cape Wessel, Cassini, Centre, Champagny, Coronation, Darcy, Grant, Groote Eylandt, Guluwuru, Hayman, Heywood, Irvine, Koolan, Kingfisher, Lachlan, Long, Marchinbar, Maret, Maria, Melville, Middle Osborn, Milingimbi, Mornington, Muir, North East, North Maret, Saibai, Saint Andrew, Sir Graham Moore, South West, Storr, Sunday, Troughton, Truant, Wessel Marchinbar and Wulalam Is.), NSL–655 m.
Source: Sonnemann, 2000.

2. *Antaresia maculosa* (W.C.H. Peters, 1873c). Mber. Königl. Akad. Wiss. Berlin 1873(10): 608–609. (*Liasis maculosus*)

Synonym: *Antaresia maculosus brentonoloughlini* Hoser, 2003f.
Type: Lectotype, ZMB 5948 (A. Dietrich, 1863–1872, via Mus. Godeffroy), designated by L.A. Smith (1985: 266).
Type locality: "Port Mackay" [= Mackay, east-central Queensland, Australia, 21°09'S, 149°12'E] via lectotype selection.
Distribution: Papua New Guinea (Western) and E Australia (NE New South Wales, E Queensland, Badu, Curtis, Hammond, Hayman, Holbourne, Lindeman, Lizard, Magnetic, Marble and Moa Is.), NSL–700 m.
Sources: Swan, 1990, Covacevich & Couper, 1991, B. Barnett, 1999a and O'Shea et al., 2004.
Remarks: A synonym of *Antaresia childreni* (Gray) *fide* Cogger et al., 1983a. Papua New Guinea population possibly a separate species *fide* O'Shea et al. (2004: 226).

3. *Antaresia perthensis* (Stull, 1932b). Occ. Pap. Boston Soc. Nat. Hist. (8): 26, pl. 2, fig. a. (*Liasis childreni perthensis*)

Type: Holotype, MCZ 24426, a 293–297 mm female (W.S. Brooks, 1927).
Type locality: "Perth, West Australia" [Australia] (possibly in error *fide* L.A. Smith, 1985: 263).
Distribution: Western Australia (W Western Australia, Depuch, Malus, Rosemary and West Lewis Is.), NSL–1040 m.
Sources: F. Mitchell, 1965, Hoser, 1999b, 2009a and J.C. Murphy & Schlager, 2003.
Remarks: Type locality unknown *fide* L.A. Smith (1985: 263).

4. *Antaresia stimsoni* (L.A. Smith, 1985). Rec. West. Aust. Mus. 12(3): 267–269. (*Liasis stimsoni stimsoni*)

Synonyms: *Antaresia saxacola* Wells & Wellington, 1985 *(nomen nudum)*, *Liasis stimsoni orientalis* L.A. Smith, 1985, and *Antaresia saxacola campbelli* Hoser, 2000b.
Type: Holotype, WAM 63108, an 872 mm male (L.A. Smith & R.E. Johnstone, 6 April 1979).
Type locality: "15 km SE of Nullagine, Western Australia, Australia, in 21°58'S, 120°12'E."
Distribution: Australia (NW New South Wales, Northern Territory, W Queensland, South Australia, Western Australia, Barrow, Bernier, Dirk Hartog, Dorre, Hermite and Rosemary Is.), NSL–805 m.
Sources: Bush et al., 1995 and Shea & Sadlier, 1999.

ANTILLOPHIS Maglio, 1970 (Xenodontidae)

Type species: *Dromicus parvifrons* Cope, 1862b.
Distribution: Hispaniola.
Sources: A. Schwartz, 1971, A. Schwartz & Thomas, 1975, A. Schwartz & R.W. Henderson, 1988, 1991, Sosa et al., 1995, Rodríguez-Robles & Greene, 1996, Hedges et al., 2009 and Zaher et al., 2009.
Remarks: A valid genus *fide* Grazziotin et al. (2012: 22) but a synonym of *Hypsirhynchus fide* Hedges et al. (2009: 9).

1. *Antillophis parvifrons* (Cope, 1862b). Proc. Acad. Nat. Sci. Philadelphia 14(1): 79. (*Dromicus parvifrons)*

Synonyms: *Dromicus protenus* Jan, 1863b (*nomen nudum*), *Dromicus protenus* Jan, 1867 *in* Jan & Sordelli, 1866–1870, *Leptophis frenatus* J.G. Fischer,

1883, *Leimadophis alleni* Dunn, 1820a, *Leimadophis parvifrons niger*, Dunn, 1920a, *Leimadophis tortuganus* Dunn, 1920a, *Leimadophis parvifrons lincolni* Cochran, 1931, *Dromicus parvifrons rosamondae* Cochran, Cochran, 1934, *Dromicus parvifrons paraniger* R. Thomas & Schwartz, 1965, and *Dromicus parvifrons stygius* R. Thomas & Schwartz, 1965.

Type: Syntypes (5), MCZ 3344a–b, MCZ 3602a–b, and FMNH 73373 (D.F. Weinland, 1859), longest syntype 673 mm.

Type locality: "Hayti, near Jeremie" [= vicinity of Jérémie (18°39'N, 74°07'W, elevation 15 m), Grand'Anse Dept., SW Haiti].

Distribution: Hispaniola. Haiti (Artibonite, Centre, Grand'Anse, Nord, Nord-Est, Nord-Ouest, Sud, Sud-Est, Grand Cayemite, Grosse Caye, Petite Gonâve, Tortue and Vache Is.) and Dominican Republic Republic (Azua, Baoruco, Barahona, Dajabón, Distrito Nacional, Duarte, Elías Piña, El Seibo, Espillat, Hato Mayor, Independencia, La Altagracia, La Estrelleta, La Romana, La Vega, María Trinidad Sánchez, Monseñor Nouel, Monte Christi, Pedernales, Peravia, Puerto Plata, Salcedo, Samaná, Sánchez Ramirez, San Cristóbal, San Juan, San Pedro de Macorís, Santiago, Santiago Rodriguez, Valverde, Beata and Saona Is.), NSL–15 m.

Sources: Cochran, 1941, R. Thomas & Schwartz, 1965, Maglio, 1970, A. Schwartz & Thomas, 1975, Sajdak & R.W. Henderson, 1982, Schwartz & R.W. Henderson, 1988, 1991, Rodriguez-Robles & Greene, 1996, Hedges et al., 2009, R.W. Henderson & Powell, 2009, Zaher et al., 2009 and Burbrick et al., 2012.

Remarks: Lectotype listed in MCZ *fide* Marx (1958a: 480) but without catalogue number. A member of *Hypsirhynchus fide* Burbrink et al. (2012: 470)

APARALLACTUS A. Smith, 1849 *in* 1838–1849 (Atractaspididae)

Synonyms: *Elapomorphus* A. Smith, 1849 *in* 1838–1849 *(nomen praeoccupatum)*, *Uriechis* W.C.H. Peters, 1854, *Eucritus* Jan, 1857 *(nomen nudum)*, *Elapops* A.C.L.G. Günther, 1859a, *Pariaspis* Cope, 1860e, *Cercocalamus* A.C.L.G. Günther, 1863a, *Metopophis* W.C.H. Peters, 1870b, *Apparalactus* – Palacky, 1898 *(nomen incorrectum)*, *Rouleophis* Chabanaud, 1917a, *Guyomarchia* Angel, 1923b, *Aarallactus* – Briceño-Rossi, 1934 *(nomen incorrectum)*, *Aparallaetus* Witte & Laurent, 1943 *(nomen incorrectum)*, *Periaspis* Loveridge, 1944 *(nomen emendatum)*, *Apparallactus* – Dekeyser & Villiers, 1954 *(nomen incorrectum)*, *Aparrallactus* – Anthony, 1955 *(nomen incorrectum)*, *Aparellactus* – Witte, 1955 *(nomen incorrectum)*, *Pariapsis* – E.H. Taylor & Weyer, 1958 *(nomen incorrectum)*, *Aparalactus* – Guibé, 1970 *(nomen incorrectum)*, and *Plumridgeus* Hoser, 2012ae *(nomen illegitimum)*.

Type species: *Aparallactus capensis* A. Smith, 1849 *in* 1838–1849.

Distribution: Subsaharan Africa.

Sources: F. Werner, 1924c, Loveridge, 1944b, Witte & Laurent, 1947, Cadle, 1983, Wallach, 1991, Underwood & Kochva, 1993, Cadle, 1994, Ulber, 1994, Kelly et al., 2003, 2009, Vidal et al., 2008 and Zaher et al., 2009.

Remarks: Official Generic Name no. 2147 *fide* Opinion 1201 (ICZN, 1982a).

1. *Aparallactus capensis* A. Smith, 1849 *in* 1838–1849. Illust. Zool. So. Africa, Rept. (App.): 16.

Synonyms: *Elalpomorphus capensis* A. Smith, 1849 *in* 1838–1849, *Cercocalamus collaris* A.C.L.G. Günther, 1863a, *Aparallactus bocagii* Boulenger, 1895, *Aparallactus punctatolineatus* Boulenger, 1895, and *Aparallactus lübberti* Sternfeld, 1910a.

Type: Holotype, BMNH 1946.1.8.62, a 166 mm female (A. Smith, 1831).

Type locality: "country (Kaffirland) to the eastward of the Cape Colony, South Africa" [= KwaZulu-Natal Prov., NE South Africa *fide* V.F.M. FitzSimons, 1962a: 154].

Distribution: Southern Africa. Eastern Democratic Republic of the Congo (Katanga, Orientale, Sud-Kivu), S Angola (Benguela, Bié, Cuanza Sul, Huíla, Lunda Norte, Lunda Sul, Namibe), E Burundi, ext. SE Kenya (S Coast), SE Tanzania (Kigoma, Kilimanjaro, Lindi, Manyara, Mtwara, Rukwa, Ruvuma, Tanga, Mafia and Zanzibar Is.), Mozambique (Manica, Maputo, Nampula, Niassa, Sofala, Tete), Malawi (Central, Southern), Zambia (Central, Copperbelt, Eastern, Luapula, Lusaka, Northern, North-Western, Southern, Western), Zimbabwe (Bulawayo, Manicaland, Mashonaland Central, Mashonaland East, Mashonaland West, Masvingo, Matabeleland North, Matabeleland South, Midlands), N Namibia (Erongo, Otjozondjupa), E Botswana (Central, Gaborone, North East, South East, Southern), Swaziland and South Africa (Eastern Cape, Free State, Gauteng, KwaZulu-Natal, Limpopo, Mpumalanga), NSL–1700 m.

Sources: V.F.M. FitzSimons, 1937, 1974, Laurent, 1954a, Mertens, 1955, 1971, Broadley, 1959, 1971d, 1983, Sweeney, 1961, Broadley & Cock, 1975, Pienaar et al., 1983, Auerbach, 1987, Branch, 1988, Branch & McCartney, 1993, Clauss & Clauss, 2002, Spawls et al., 2002, Broadley et al., 2003, M. Griffin, 2003, Marais, 2004, Branch et al., 2005 and Broadley & Blaylock, 2013.

Remarks: Official Specific Name no. 2783 *fide* Opinion 1201 (ICZN, 1982a).

2. *Aparallactus guentheri* Boulenger, 1895d. Ann. Mag. Nat. Hist. (6) 16(92): 172.

Synonym: *Aparallactus uluguruensis* T. Barbour & Loveridge, 1929a.

A

Type: Lectotype, BMNH 1946.1.6.88, adult male (A.A. Simons), designated by Broadley *in* Broadley & Howell (1991: 28).

Type locality: "Lake Nyassa" [= Lake Nyassa, E Malawi] via lectotype selection.

Distribution: Southeastern Africa. Southeastern Kenya (Coast), E Tanzania (Dar es Salaam, Kilimanjaro, Piwani, Ruvuma, Tanga), Malawi (Central, Southern), Mozambique (Manica, Sofala) and E Zimbabwe (Manicaland, Mashonaland Central), 510–1385 m.

Sources: Broadley, 1959, 1983, 1988, Sweeney, 1961, V.F.M. FitzSimons, 1962a, 1974, Broadley & Cock, 1975, J. Rasmussen, 1981, Broadley & Howell, 1991, Spawls et al., 2002 and Broadley & Blaylock, 2013.

Remarks: Boulenger (1896a: 259) listed Lake Nyassa syntype as a female. Broadley & Howell (1991: 28) listed lectotype designation from an uncited Broadley (1968) reference.

3. *Aparallactus jacksonii* (A.C.L.G. Günther, 1888c). Ann. Mag. Nat. Hist. (6) 1(3): 325, pl. 19, fig. e. (*Uriechis jacksonii*)

Synonym: *Aparallactus jacksonii oweni* Loveridge, 1955a.

Type: Holotype, BMNH 1946.1.6.80, a 191 mm female (F.J. Jackson, 1887).

Type locality: "Foot of Kilima-ndjaro, tropical Africa" [= base of Mt. Kilimanjaro, (03°04'S, 37°21'E), Kilimanjaro Region, NE Tanzania].

Distribution: Northeastern Africa. Southeastern South Sudan (Eastern Equatoria), S Ethiopia (Gemu Gofa, Sidamo), S Somalia (Jubbada Dhexe), SE Kenya (S Coast, Eastern, Nairobi, S Rift Valley) and N Tanzania (Arusha, Iringa, Kilamanjaro, Manyara, Tanga), 320–2200 m.

Sources: Pitman, 1974, Lanza, 1983a, 1990b, Largen & Rasmussen, 1993, Spawls et al., 2002 and Largen & Spawls, 2010.

Remarks: *Aparallactus j. oweni* probably a valid species and also occurs in N Uganda *fide* Lanza (1990b: 434).

4. *Aparallactus lineatus* (W.C.H. Peters, 1870b). Mber. Königl. Akad. Wiss. Berlin 1870(8): 643–644, pl. 1, figs. 3, 3a–c. (*Uriechis [Metopophis] lineatus*)

Synonyms: *Uriechis anomala* Boulenger, 1893c, *Elapops heterolepis* Mocquard, 1908, and *Rouleophis chevalieri* Chabanaud, 1917a.

Types: Syntypes (2), ZMB 6913a–b, longest syntype 440 mm female.

Type locality: "Keta (Guinea)" [= Keta, S Volta Region, ext. SE Ghana, 05°55'N, 01°00'E, elevation NSL].

Distribution: West Africa. Southeastern Guinea (Macenta, Nzérékoré), Liberia (Bong, Nimba), Ivory Coast (Aboisso) and SE Ghana (Accra, Volta), NSL–700 m.

Sources: Loveridge, 1938, Doucet, 1963, Leston & Hughes, 1968, Hughes & Barry, 1969, Wallach, 1994, Hughes, 2003, Ineich, 2003, Chippaux, 2006 and W. Böhme et al., 2011.

Remarks: MNHN 1885.698 listed as type *fide* MNHN catalogue. Type locality erroneously listed as Keta, Ivory Coast *fide* Witte & Laurent (1947: 108). Probably occurs in Togo *fide* and Segniagbeto et al. (2011: 353).

5. *Aparallactus lunulatus* (W.C.H. Peters, 1854). Mber. Königl. Akad. Wiss. Berlin 1854(11): 623. (*Uriechis lunulatus*)

Synonyms: *Uriechis concolor* J.G. Fischer, 1884a, *Aparallactus concolor boulengeri* Scortecci, 1931, *Aparallactus liddiardae* H.W. Parker, 1933b, and *Aparallactus lunulatus scortecci* H.W. Parker, 1949.

Type: Holotype, ZMB 4810, a 415 mm female (W.C.H. Peters, June 1843–Aug. 1847).

Type locality: "Tette, Mocambique" [= Tete, Tete Prov., Mozambique, 16°10'S, 33°36'E, elevation 135 m].

Distribution: Subsaharan Africa. Northern Ivory Coast, Burkina Faso (Est, Hauts-Bassins), N Ghana, N Togo (Kara, Savanes), Nigeria (Delta, Kaduna, Plateau), N Cameroon (Adamaoua, Centre, Est, Extreme-Nord, Nord), Central African Republic (Bamingui-Bangoran, Haute-Sangha, Lobaye, Ombella-Mpoko, Ouaka, Ouham, Ouham-Pende), N Democratic Republic of the Congo (Equateur, Orientale), S South Sudan (Central Equatoria), Eritrea (Northern Red Sea, Southern), Ethiopia (Gondar, Hararge, Shoa, Sidamo), Somalia (Jubbada Dhexe, Jubbada Hoose, Mogadishu, Shabeellaha Dhexe, Shabeellaha Hoose, Togdheer, Woqooyi Galbreed), Uganda (N Eastern, Northern, Western), Kenya (Coast, Eastern, North-Eastern, Nyanza, Rift Valley), Tanzania (Arusha, Dodoma, Kilimanjaro, Piwani, Singida, Tabora, Tanga, Lamu Is.), Zambia (Eastern, Southern), Malawi (Central, Southern), Mozambique (Manica, Maputo, Sofala, Tete), Zimbabwe (Bulawayo, Manicaland, Mashonaland Central, Mashonaland East, Mashonaland West, Masvingo, Matabeleland North, Matabeleland South, Midlands), Swaziland and NE South Africa (E Limpopo, E Mpumalanga), NSL–2400 m.

Sources: Broadley, 1959, 1971d, 1983, Sweeney, 1961, V.F.M. FitzSimons, 1962a, 1974, Pitman, 1974, Broadley & Cock, 1975, Hughes, 1983, Lanza, 1983a, 1990b, Pienaar et al., 1983, Roman, 1984, Branch, 1988, Joger, 1990, Broadley & Howell, 1991, Marais, 1992, 2004, Largen & Rasmussen, 1993, Spawls et al., 2002, Broadley et al., 2003, Chippaux, 2006, Chirio & Ineich, 2006, J.-F. Trape & Mané, 2006b, Chirio & LeBreton, 2007, Largen & Spawls, 2010, Segniagbeto et al., 2011 and Broadley & Blaylock, 2013.

Remarks: Original description reprinted in W.C.H. Peters (1855: 52). Official Specific Name no. 2786 *fide* Opinion 1201 (ICZN, 1982a). MNHN 1916.263 and MNH 1916.265 listed as types *fide* MNHN catalogue.

Holotype is female *fide* R. Günther *(in litt.)*. Probably occurs in ext. E Botswana *fide* Auerbach (1987: 172). Possibly occurs in Benin *fide* Hughes (2013: 153).

6. *Aparallactus modestus* (A.C.L.G. Günther, 1859a). Ann. Mag. Nat. Hist. (3) 4(21): 161–162, pl. 4, fig. c. (*Elapops modestus*)

Synonyms: *Pariaspis plumbeatra* Cope, 1860e, *Elapops plumbeater* Cope, 1861c *(nomen emendatum)*, *Elapops petersi* Jan, 1862b, *Aparallactus boulengeri* F. Werner, 1896b, *Aparallactus peraffinis* F. Werner, 1897b, *Aparallactus ubangensis* Boulenger, 1897b, *Aparallactus flavitorques* Boulenger, 1901a, *Aparallactus congicus* F. Werner, 1902, *Aparallactus dolloi* F. Werner, 1902, *Aparallactus batesii* Boulenger, 1907a, *Aparallactus christyi* Boulenger, 1910b, *Aparallactus nigrocollaris* Chabanaud, 1917a, *Aparallactus nigrocollaris roucheti* Chabanaud, 1917a, *Guyomarchia unicolor* Angel, 1923, and *Aparallactus graueri* F. Werner, 1924a.

Type: Holotype, BMNH 1946.1.3.7, a 483 mm female (Rich).

Type locality: "West Africa."

Distribution: West and cen. Africa. Southeastern Guinea (Macenta, Nzérékoré), Sierra Leone, Liberia (Montserrado, Margibi), Ivory Coast (Abidjan, Daloa, Guiglo, Lagunes, Tabou, Toumodi), Burkina Faso, Ghana (S Volta), S Togo (Plateaux), S Benin (Cotonou, Maritime), S Nigeria (Cross River, Lagos), S Cameroon (Adamaoua, Est, Centre, Littoral, Nord-Ouest, Ouest, Sud, Sud-Ouest), Gabon (Estuaire, Haut-Ogooué, Moyen-Ogooué, Ogooué-Ivindo, Ogooué-Maritime, Wolen-Ntem), Congo (Cuvette, Kouilou, Niari, Plateau, Pool, Sangha), SW Central African Republic (Haute-Sangha), Democratic Republic of the Congo (Bandundu, Bas-Congo, Equateur, Kasai Occidental, Katanga, Kinshasa, Maniema, Nord-Kivu, Orientale, Sud-Kivu) and S Uganda (Central, Eastern, Western), NSL–1325 m.

Sources: Boulenger, 1896a, Aylmer, 1922, Loveridge, 1941a, E.H. Taylor & Weyer, 1958, Doucet, 1963, Knoepffler, 1966, Leston & Hughes, 1968, Hughes & Barry, 1969, Roux-Estève, 1969, Pitman, 1974, Hughes, 1983, 2013, J.A. Butler & Reid, 1986, J.-F. Trape & Roux-Estève, 1990, D. Lawson, 1993, J.-F. Trape & Roux-Estève, 1995, Ineich, 2003, Chippaux, 2006 Chirio & Ineich, 2006, Chirio & LeBreton, 2007, Pauwels & Vande weghe, 2008, Ullenbruch et al., 2010, W. Böhme *et al.*, 2011 and Segniagbeto et al., 2011.

Remarks: Boulenger (1896a: 263) listed a specimen from Niger. Possibly occurs in SW Kenya *fide* Spawls et al. (2002: 424).

7. *Aparallactus moeruensis* Witte & Laurent, 1943. Rev. Zool. Bot. Afr. 37(1–2): 178–179.

Type: Holotype, MRAC 240, a 389 mm female (C. Lemaire, 1899).

Type locality: "Pweto (Distr. du Haut-Katanga), Congo Belge" [= Pweto, Katanga Prov., SE Democratic Republic of the Congo, 08°28'S, 28°54'E, elevation 950 m].

Distribution: Southeastern Democratic Republic of the Congo (Katanga), 950 m.

Remarks: A valid species *fide* Resetar (*in litt.*) and Wallach (1991: 475).

8. *Aparallactus niger* Boulenger, 1897a. Ann. Mag. Nat. Hist. (6) 19(110): 154.

Type: Holotype, BMNH 1946.1.6.89, a 335 mm male (W.G. Clements, 1893–1897).

Type locality: "Sierra Leone."

Distribution: West Africa. Guinea (Beyla, Dalaba, Kérouané, Macenta, Nzérékoré, Yomou), W Sierra Leone (Western), N Liberia (Nimba), SW Ivory Coast (Dix-Huit Montagnes) and Ghana, 500–650 m.

Sources: Doucet, 1963, Hughes, 1983, Wallach, 1994, Ineich, 2003, Chippaux, 2006 and W. Böhme et al., 2011.

Remarks: MNHN 1916.236 listed as type *fide* MNHN catalogue.

9. *Aparallactus nigriceps* (W.C.H. Peters, 1854). Mber. Königl. Akad. Wiss. Berlin 1854(11): 623 (*Uriechis nigriceps*)

Synonyms: *Eucritus atrocephalus* Jan, 1857 *(nomen nudum)*, and *Uriechis atriceps* Jan, 1862b *(nomen emendatum)*.

Type: Lectotype, ZMB 4809, a 253–255 mm specimen, designated by Loveridge (1944: 212).

Type locality: "Tette" [= Tete, Tete Prov., Mozambique, 16°10'S, 33°36'E, elevation 135 m] via lectotype selection (in error *fide* Broadley, 1983: 156).

Distribution: Southern Mozambique (Inhambane), NSL–10 m.

Sources: Broadley, 1983, Branch, 1988 and Marais, 1992, 2004.

Remarks: Original description reprinted in W.C.H. Peters (1855: 52).

10. *Aparallactus turneri* Loveridge, 1935. Bull. Mus. Comp. Zool. 79(1): 9–10.

Type: Holotype, MCZ 30117, a 175 mm male (H.J.A. Turner, June 1932).

Type locality: "Sokoki Forest, near Malindi, Coast Prov., Kenya Colony" [= Sokoke Forest, Coast Prov., SE Kenya, ca. 03°32'S, 39°49'E].

Distribution: Southeastern Kenya (E Coast), NSL–20 m.

Sources: Spawls, 1978 and Spawls et al., 2002.

A

11. *Aparallactus werneri* Boulenger, 1895d. Ann. Mag. Nat.Hist. (6) 16(92): 172.

Types: Syntypes (2), BMNH 1946.1.2.97–98, a male and female, longest syntype 390 mm (F. Werner, 1895).
Type locality: "Usambara, German East Africa" [= Usambara Mountains, N Tanga Region, NE Tanzania].
Distribution: Eastern Tanzania (Kilimanjaro, Lindi, Morogoro, Piwani, Tanga), NSL–1600 m.
Sources: Boulenger, 1896a, J. Rasmussen, 1981, Broadley & Howell, 1991 and Spawls et al., 2002.

APLOPELTURA A.-M.-C. Duméril, 1853 (*nomen protectum*) (Pareatidae)

Synonyms: *Aspidocercus* Fitzinger, 1843 *(nomen oblitum), Haplopeltura* Boetger, 1892 *(nomen emendatum), Hoplopeltura* – Palacky, 1898 *(nomen incorrectum), Haplopelturus* – Briceño-Rossi, 1934 *(nomen incorrectum),* and *Aphopeltura* – Nutaphand, 1991b *(nomen incorrectum).*
Type species: *Amblycephalus boa* H. Boie, 1828.
Distribution: Malay Peninsula and East Indies.
Sources: Rooij, 1917, E.H. Taylor, 1922, 1965, David & Vogel, 1996, M.J. Cox et al., 1998, Iskandar & Colijn, 2001, I. Das, 2007b, 2010, 2012, Zaher et al., 2009 and Hoser, 2012q.
Remarks: In accordance with Art. 23.9.2 of the Code (ICZN, 1999), *Aplopeltura* Duméril is designated a *nomen* protectum and *Aspidocercus* Fitzinger a *nomen oblitum*.

1. *Aplopeltura boa* (H. Boie, 1828). Isis von Oken 21(10): 1035. (*Amblycephalus boa*)

Type: Holotype, RMNH 984 (H. Boie, 15-29 July 1827).
Type locality: "Parang, Java" [= Mt. Parang, Magetan Prov., cen. Java, S Indonesia, 7°44'S, 111°20'E, elevation 355 m].
Distribution: Southeast Asia and East Indies. Myanmar, S Thailand (Chumphon, Nakhon Si Thammarat, Pattani, Phang Nga), West Malaysia (Perak), East Malaysia (Sabah, Sarawak), Brunei, Indonesia (Bangka, Java, Kalimantan,Natuna Arch., Nias, Sumatra) and S Philippines (Balabac, Basilan, Mindanao, Palawan), NSL–1500 m.
Remarks: Holotype not in RMNH. Occurs in Myanmar *fide* I. Das (2010: 344).

APODORA Kluge, 1993 (Pythonidae)

Type species: *Liasis papuanus* W.C.H. Peters & Doria, 1878.
Distribution: New Guinea.
Sources: Brongersma, 1956b, Whitaker et al., 1982, F. Parker, 1983, Kluge, 1993b, O'Shea, 1996, McDiarmid et al., 1999, Hoser, 2000b, J.C. Murphy & Schlager, 2003, Rawlings et al., 2008 and Schleip & O'Shea, 2010.

1. *Apodora papuana* (W.C.H. Peters & Doria, 1878). Ann. Mus. Civ. Stor. Nat. Genova 13: 400–401, pl. 3, figs. 1–1b. (*Liasis papuanus*)

Synonyms: *Liasis tornieri* F.Werner, 1897c, *Liasis maximus* F. Werner, 1936, and *Liasi papuano* – F. Werner, 1936c *(nomen incorrectum).*
Type: Holotype, MSNG 29988, a 3450 specimen (L.M. Albertis, 14 May 1872).
Type locality: "Ramoi presso Soron (Costa N.O. della N. Guinea) Nova Guinea austro-occidentali." [= Ramoi, near Sorong, West Papua E Indonesia, ca. 0°53'S, 131°15'E].
Distribution: New Guinea. Eastern Indonesia (Papua, West Papua, Biak, Misool) and Papua New Guinea (Central, East Sepik, Madang, Morobe, Northern, Western, Fergusson Is.).
Comment: A subspecies of *Liasis olivaceus* Gray *fide* Stull, 1935.

APOSTOLEPIS Cope, 1862a (Xenodontidae)

Synonyms: *Elapomojus* Jan, 1862b, *Rhynchonyx* W.C.H. Peters, 1869, *Elapomoius* Senna, 1886 *(nomen incorrectum), Elapomoius* Boulenger, 1896a *(nomen emendatum), Elapohomoeus* Berg, 1898 *(nomen emendatum), Elapomajus* – Palacky, 1898 *(nomen incorrectum), Parapostolepis* Amaral, 1930d, *Parapostelepis* – Briceño-Rossi, 1934 *(nomen incorrectum), Rhynchonys* – A.S. Romer, 1956 *(nomen incorrectum),* and *Apostolepsis* – Fugler, 1983 *(nomen incorrectum).*
Type species: *Elapomorphus flavotorquatus* A.-M.-C. Duméril, Bibron & Duméril, 1854b.
Distribution: South America.
Sources: Strauch, 1884, J.A. Peters & Orejas-Miranda, 1970, Lema, 1978d, 2001, 2002, 2005, Cadle, 1983, Harvey, 1999 and Hofstadler-Deiques & Lema, 2005.
Remarks: Six new species from the Caatinga and Mato Grosso, Mato Grosso do Sul, and São Paulo states, Brazil being described by Lema et al. (pers. comm.).

1. *Apostolepis albicollaris* Lema, 2002a. Comun. Mus. Ciênc. Tecnol. PUCRS (Zool.) 15(2): 228–233, fig. 1 (paratype).

Type: Holotype, MCP 8355, a 433 mm female.
Type locality: "Parque Zoológico do Distrito Federal, Brasilia, Goiás, Brasil."
Distribution: Central Brazil (Distrito Federal, Goiás, Minas Gerais), 1000 m.
Sources: Lema, 2003b and Lema et al., 2004, 2005.

2. *Apostolepis ambinigra* (W.C.H. Peters, 1869). Mber. Königl. Akad. Wiss. Berlin 1869(5): 438–439, pl., fig. 2–2c. (*Rhynchonyx ambiniger*)

Type: Holotype, ZMB 6450, a 395 mm male.
Type locality: "Paraguay."
Distribution: Southern South America. Southwestern Brazil, S Bolivia and Paraguay.
Sources: J.A. Peters & Orejas-Miranda, 1972, Lema, 2001, 2002, 2003b and Lema et al., 2005b.

3. *Apostolepis ammodites* Ferrarezzi, Erritto-Barbo & España-Albuqueque, 2005a. Pap. Avulsos Zool. 45(16): 219–221, figs. 3–4, 7b.

Synonym: *Apostolepis bialbimaculata* Lema & Renner, 2005.
Type: Holotype, IB 65267, an adult female, destroyed by fire 15 May 2010.
Type locality: "Brazil: Tocantins: Palmas (10°02'S, 47°20'W, 230 m altitude)."
Distribution: Southeastern Brazil (Bahia, Districto Federal, Goiás, Mato Grosso, Minas Gerais, São Paulo, Tocantins), 230–860 m.
Source: Lema & Renner, 2007.

4. *Apostolepis arenaria* Rodrigues, 1993a. Mem. Inst. Butantan (1992) 54(2): 54–55, figs. 1–2. (*Apostolepis arenarius*)

Type: Holotype, MZUSP 10027 (MTR 88.6819), a 564 mm female (M. Rodrigues, 5 Sept. 1988).
Type locality: "Brasil: Bahia: Alagoado (9°29'S, 41°21'N)" [= 9°29'S, 41°21'W, elevation 485 m].
Distribution: Northeastern Brazil (cen. Bahia), 485 m.

5. *Apostolepis assimilis* (J.T. Reinhardt, 1861b). Vidensk. Medd. Naturhist. Foren. Kjøbenhavn (1860) 22: 235–236, pl. 4, figs. 1–5. (*Elapomorphus assimilis*)

Synonyms: *Elapomorphus assimilis* J.T. Reinhardt, 1861a *(nomen nudum), Apostolepis tertulianobeui* Lema, 2004., and *Apostolepis parassimilis* Lema & Renner, 2012.
Type: Holotype, ZMUC, a 575 mm specimen (native, 5 May 1852).
Type locality: "Fazenda, Capão dos porcos, der ligger nogle Miil fra den lille Bye Brumado i Campos-Egnene af Provindsen, Minas geraes vest for Serra do espinhaçco" [= Capão dos Porcos Farm, west of Serra do Espinhaço, Minas Gerais State, Brazil *fide* Lema, 2001: 33, ca. 17°30'S, 43°30°W, elevation 650 m].
Distribution: Central Brazil (Bahia, Goiás, Minas Gerais, Paraná, São Paulo), Bolivia, Paraguay and N Argentina (Chaco, Formosa), 650–880 m.
Sources: F. Werner, 1901b, Cei, 1994, Giraudo & Scrocchi, 1998, Freitas, 1999, Valdujo & Nogueira, 2001, Giraudo & Scrocchi, 2002, Lema, 2002, 2004a, 2005, Lema et al., 2004 and Ferrarezzi et al., 2005a.
Remarks: *Apostolepis tertulianobeui* a valid species *fide* Lema (pers. comm.).

6. *Apostolepis borellii* Peracca, 1904a. Boll. Mus. Zool. Anat. Comp. Univ. Torino 19(460): 422–423, 2 figs.

Type: Holotype, MZUT 962, a 197 mm male (A. Borrelli, 1899).
Type locality: "Urucum, Matto Grosso brasiliano" [= Maciço do Urucum, W Mato Grosso do Sul State, Brazil, 19°12'S, 57°36'W, elevation 725 m]
Distribution: Central Bolivia (Santa Cruz) and SW Brazil (W Mato Grosso do Sul) 200–725 m.
Sources: Amaral, 1925b, Parker, 1928, Ferrarezzi, 1993, Harvey, 1999 and Lema & Renner, 2006b
Remarks: A valid species *fide* Lema (pers. comm.). A synonym of *A. nigrolineata fide* Harvey (1999: 399).

7. *Apostolepis breviceps* Harvey, Gonzales & Scrocchi, 2001. Copeia 2001(2): 502–504, fig. 1.

Type: Holotype, MNK 1839, a 319 mm male (G. Sato, 16 Feb. 1999).
Type locality: "in the vicinity of Cerro Cortado (i.e., within 5 km), Cordillera Prov., Santa Cruz Department, Bolivia, 19°32'S, 62°18'W. Elevation at the type locality is approximately 400 m."
Distribution: Eastern Bolivia (SW Santa Cruz), 400 m. Known only from vicinity of type locality.
Source: Lema, 2003b.

8. *Apostolepis cearensis* J. Gomés, 1915. Ann. Paul. Med. Cirurg. 4(6): 122–123, pl. 3, figs. 4–8.

Synonyms: *Apostolepis sanctae-ritae* F. Werner, 1924 *(nomen incorrigendum), Apostolepis amarali* F. Werner, 1925, and *Apostolepis sanctaeritae* – Lema & Fernandes, 1997 *(nomen corrigendum).*
Type: Holotype, IB 882, a 490 mm male (F. Dias da Rocha, 1894–1915), destroyed by fire 15 May 2010.
Type locality: "State of Ceará, Brazil."
Distribution: Northeastern Brazil (Alagoas, Bahia, Ceará, Paraíba, Piauí, Rio Grande do Norte, Sergipe).
Sources: Freitas, 1999, Lema, 2002, 2004a, 2005, Ferrarezzi et al., 2005a and Lema & Renner, 2006b.

9. *Apostolepis cerradoensis* Lema, 2003a. Acta Biol. Leopoldensia 25(1): 124–126, figs. 1–4.

Type: Holotype, UHESM 21800, a 347 mm male (N. J. de Silva, Jr.).
Type locality: "canteiro de obras da barragem da Usina Hidrelétrica Serra da Mesa, Minaçu, Goiás, Brasil."
Distribution: Northeastern Brazil (Goiás). Known only from type locality.

10. *Apostolepis christineae* Lema, 2002c. Facena 18: 41–44, pls. 1–2.

Type: Holotype, MCP 12515, a 352 mm female (C. Strüssmann, 20 Dec. 1987).
Type locality: "Estação Ecológia da Sierra das Araras, Barra dos Bugres, Cáceres, Estado de Mato Grosso, Brasil (15º05'S-57º11'W)" [= elevation 175 m].
Distribution: Central Brazil (Mato Grosso), 175 m. Known only from type locality.
Sources: Lema, 2002d, Lema et al., 2005 and Lema & Renner, 2006.

11. *Apostolepis dimidiata* (Jan, 1862b). Arch. Zool. Anat. Fis. 2(1): 47–48. (*Elapomorphus dimidiatus*)

Synonyms: *Elapomorphus erythronotus* W.C.H. Peters, 1880a, *Apostolepis nigriceps* F. Werner, 1897a, *Apostolepis barrioi* Lema, 1978d, *Apostolepis ventrimaculatus* Lema, 1978d, and *Apostolepis villaricae* Lema, 1978d.
Type: Holotype, formerly MSNM, a 1473 mm female, destroyed in 1943 during World War II.
Type locality: "Brasile" [= Brazil]. Restricted to São Paulo, Brazil *fide* Lema (1986: 74).
Distribution: Southeastern Brazil (Goiás, Mato Grosso do Sul, Paraná, São Paulo), Paraguay (San Pedro) and N Argentina (NE Corrientes, S Misiones), 140 m.
Sources: Lema, 1986, 1993, Cei, 1994, Moura-Leite et al., 1996, Giraudo & Scrocchi, 1998, Lema 2002, 2003a–b, Lema et al., 2004, 2005 and Ferrarezzi et al., 2005a.
Remarks: Harvey, 1999 considered *A. ventrimaculatus* a synonym.

12. *Apostolepis dorbignyi* (Schlegel, 1837). Essai Phys. Serp. 1: 130, 2: 30–31. (*Calamaria d'orbignyi*)

Type: Holotype, MNHN 3664, a 411–433 mm male (A.C.V.M.D. d'Orbigny, 14 April 1830–27 June 1833 [Bolivia] or 2 July 1831–19 Oct. 1831 [Chiquitos] or 19 Dec. 1831–16 March 1832 [Moxos]).
Type locality: "Chilé" [= Chile] (in error). Possibly Mato Grosso, Brazil or E Bolivia (Chiquitos or Moxos) *fide* Koslowsky (1898a: 31) and probably Bolivia *fide* Lema (2001: 35).
Distribution: Southern Bolivia (Chuquisaca, Santa Cruz, Tarija).
Sources: Fugler et al., 1995, Harvey, 1999, Harvey et al., 2001, Lema, 2002, 2004a and Ferrarezzi et al., 2005a.
Remarks: Holotype listed as female *fide* Guibé & Roux-Estève (1972: 130).

13. *Apostolepis flavotorquata* (A.-M.-C. Duméril, Bibron & Duméril, 1854b). Erpét. Gén. 7(2): 836–837. (*Elapomorphus flavotorquatus*) (*nomen corrigendum*)

Synonyms: *Elapomorphus flavo-torquatus* A.-M.-C. Duméril, Bibron & Duméril, 1854b *(nomen incorrigendum),* and *Apostolepis flavotorquata* – Cope, 1862a *(nomen corrigendum).*
Type: Holotype, MNHN 3665, a 500 mm specimen (F.L.N.C.L. Castelnau & E. Deville, June 1843–Aug. 1847), lost *fide* Lema (1978d: 29).
Type locality: "l'intérieur de l'Amerique meridional" [= interior of South America]. Listed as "Bresíl" [= Brazil] *fide* MNHN catalogue.
Distribution: Central Brazil (Bahia, Ceará, Distrito Federal, Goiás, Mato Grosso, Rondônia, Tocantins).
Sources: Freitas, 1999 and Lema, 2002a.

14. *Apostolepis freitasi* Lema, 2004a. Comun. Mus. Cienc. Tecnol. PUCRS (Zool.) 17(1): 4–6, figs. 1a–d.

Type: Holotype, MNRJ 6523, a 181 mm male (M. Locks, July 1991).
Type locality: "Tanque do Aragão, Central municipality of Bahia state, Brazi."
Distribution: Eastern Brazil (Bahia). Known only from type locality.

15. *Apostolepis gaboi* Rodrigues, 1993a. Mem. Inst. Butantan (1992) 54(2): 56–57, fig. 3.

Type: Holotype, MZUSP 10290, a 238 mm female (G. Skuk, 12 Oct. 1990).
Type locality: "Brasil: Bahia: Queimadas (10°23'S, 42°30'W)" [= elevation 400 m].
Distribution: Northeastern Brazil (cen. Bahia), 400 m. Known only from type locality.

16. *Apostolepis goiasensis* Prado, 1943. Mem. Inst. Butantan (1942) 16: 8, pl. 1.

Type: Holotype, IB 10260, a 408 mm female (E. Garbe, 20 Jan. 1942), destroyed by fire 15 May 2010.
Type locality: "Rio Verde, Estado de Goiás, Brasil" [= 17°47'S, 50°55'W, elevation 770 m].
Distribution: Central Brazil (Goiás, Mato Grosso do Sul, Minas Gerais), 370–770 m.
Sources: Harvey et al., 2001, Abes & Ferrarezzi, 2003, Lema, 2003b and Lema et al., 2004, 2005..
Remarks: Illustrations of holotype in Lema (2003b: figs. 1a–b).

17. *Apostolepis intermedia* Koslowsky, 1898a. Rev. Mus. La Plata 8: 30–31, pl. 1, figs. 4–7.

Type: Holotype, MLP, a 228 mm specimen (C. Bach), lost *fide* Lema (2001: 36).
Type locality: "Miranda, Matto-Grosso, Brasil" [= Miranda, Mato Grosso do Sul State, Brazil, 20°15'S, 56°22'W, elevation 130 m].
Distribution: Southern Brazil (Mato Grosso do Sul), 130–155 m.
Sources: Lema, 1993, 2002c and Albuquerque & Lema, 2012.

18. *Apostolepis lineata* Cope, 1887b. Proc. Amer. Philos. Soc. 24(125): 56. (*Apostolepis erythronotus lineatus*)

Types: Syntypes (2), ANSP 11211–12 (H.H. Smith, 1882–1886), with ANSP 11212 deteriorated and discarded *fide* Malnate (1971: 366).

Type locality: "near the village of Chupada, thirty miles north-east of Cuyabá, and near the headwaters of the Xingu, Prov. of Mato Grosso, Brazil" [= Chapada *fide* Malnate (1971: 366) and Chapada dos Guimarães *fide* Lema (2001: 36), 15°27'S, 55°44'W, elevation 775 m].

Distribution: Central Brazil (Mato Grosso), 775 m.

Sources: Lema, 1986, 2002c, Harvey, 1999 and Lema & Renner, 2004.

Remarks: Harvey (1999: 393) reported poor condition of surviving type. Type locality listed as Chapada, a small village of Matto Grosso, Brazil, about twenty miles ENE of Cuyabá, on the plateau. The village itself is about 2,500 feet above sea level, or 1,800 ft. above Cuyabá; but collections are from all surrounding regions as low as 1,500 ft. *fide* H.H. Smith *in* Papavero (1971: 379).

19. *Apostolepis longicaudata* Gomés *in* Amaral, 1921. Ann. Paul. Med. Cirurg. 9(7–8): 110–111, pl. A, figs. 4–7.

Type: Holotype, IB 1684, a 280 mm male (F. de Assis Iglesias, 1916–1918), destroyed by fire 15 May 2010.

Type locality: "Engenheiro Dodt, povoação do municipio de Santa Philomena, Estado de Piauhy, Brasil" [= Engenheiro Dodt, Santa Filomena, Piauí State, Brazil, 9°07'S, 45°55'W, elevation 275 m].

Distribution: Northeastern Brazil (SW Piauí, Tocantins), 250–300 m.

Source: Curcio et al., 2011.

Remarks: The description heading cited Gomés, 1919, referring to the unpublished manuscript from a talk presented in 1918 at the Second South American Hygeine, Microbiology, and Pathology Conference. Brazilian specimens (Espírito Santo and Paraíba, 35–390 m) are listed as possible members of this species *fide* Curcio et al. (2011: 331).

20. *Apostolepis multicincta* Harvey, 1999. Copeia 1999(2): 397–399, fig. 9.

Type: Holotype, MNK 729, a 384 mm male (L. Gonzales & I. Fernández, 1996).

Type locality: "vicinity of Pampagrande, Provincia Florida, Departmento Santa Cruz, Bolivia, 18°5'S, 64°6'W, at an elevation of approximately 1290m."

Distribution: Eastern Bolivia (Santa Cruz), 1240–1800 m.

Sources: Lema, 2002 and Embert & Reichle, 2003.

21. *Apostolepis nelsonjorgei* Lema & Renner, 2004. Biociências 12(2): 140–142, figs. 1–2.

Type: Holotype, UHESM 14414, a 264 mm female.

Type locality: "Minaçu, Goiás, Brazil" [= 13°34'S, 48°13'W, elevation 430 m].

Distribution: Central Brazil (Goiás, Tocantins), 165–560 m.

Sources: Curcio et al., 2011 and Recorder et al., 2011.

22. *Apostolepis niceforoi* Amaral, 1935c. Mem. Inst. Butantan 9: 221, pl., fig. 5.

Type: Holotype, IB 9197, a 365 mm female (H. Nicéforo-María, 1908–1935), destroyed by fire 15 May 2010.

Type locality: "La Pedrera, near the Venezuelan-Colombian boundary, Colombia" [= Bajo Caquetá, Amazonas, SE Colombia, 1°19'S, 63°35'W, elevation 75 m].

Distribution: Southeastern Colombia (Amazonas) and Ecuador, 75–90 m.

Source: Pérez-Santos & Moreno, 1988.

Remarks: Possibly occurs in W Brazil *fide* Lema (2001: 37).

23. *Apostolepis nigrolineata* (W.C.H. Peters, 1869). Mber. Königl. Akad. Wiss. Berlin 1869(5): 439. (*Elapomorphus nigrolineatus*)

Synonyms: *Apostolepis quinquelineata* Boulenger, 1896a, and *Apostolepis rondoni* Amaral, 1925b.

Type: Holotype, ZMB 6447, a 375 mm specimen.

Type locality: "angeblich aus Guinea, wahrscheinlich aber aus Südamerica" [= *lapsus calami* for Guyana *fide* Lema, 2001: 37].

Distribution: Northeastern Brazil (Amapá, W Amazonas, Ceará, Maranhão, W Pará, Rondônia, Roraima), NE Guyana (Demerara-Mahaica, Essequibo Islands-West Demerara), Suriname (Brokopondo), and French Guiana, NSL–165 m.

Sources: Amaral, 1978, Chippaux, 1987, Silva, 1993, Lema & Renner, 1998, Vidal et al., 1988, Harvey, 1999, Lema, 2001, Abuys, 2003, Lema & Albuquerque, 2010, Curio et al., 2011 and C.J. Cole et al., 2013.

Remarks: Possibly occurs in W Brazil (Acre) *fide* Lema (2001: 39). A synonym of *A. flavotorquata fide* Bauer et al. (1995: 71) but valid *fide* Curio et al. (2011: 329). *Apostolepis quinquelineata* and *A. rondoni* are valid species *fide* Lema (pers. comm.).

24. *Apostolepis nigroterminata* Boulenger, 1896a. Cat. Snakes Brit. Mus. 3: 235–236, pl. 10, figs. 2–2c.

Type: Holotype, BMNH 1946.1.9.77 (formerly BMNH 1881.5.13.76), a 217–223 mm male (W. Davis & H.J. Veitch).

Type locality: "Cayaria, north-eastern Peru" [=Pucallpa, Ucayali Department, E Peru, 8°23'S, 74°33'W, elevation 150 m].

A

Distribution: Eastern Peru (Ayacucho, Ucayali) and NE Bolivia (Beni), 150–760 m.
Sources: Harvey, 1999 and Lema & Renner, 2004.
Remarks: Collector of type only W. Davis *fide* Harvey (1999: 400).

25. *Apostolepis phillipsi* Harvey, 1999. Copeia 1999(2): 402–404, fig. 11.

Type: Holotype, UTA 43940, a 432 mm female (B. Phillips; 17 Sept. 1993).
Type locality: "the grounds of Estancia El Refugio, Provincia Velasco, Santa Cruz, Bolivia, 14°45'S; 61°00'W, 0.5–10 m elevation."
Distribution: Southeastern Bolivia (Santa Cruz), <10 m. Known only from type locality.
Sources: Lema, 2002c, 2004b, Michels & Bauer, 2004 and Lema & Renner, 2006.

26. *Apostolepis pymi* Boulenger, 1903b. Ann. Mag. Nat. Hist. (7) 12(69): 353–354

Type: Holotype, BMNH 1946.1.9.82, a 385 mm male (J. Pym).
Type locality: "Brazil." Restricted to Pará State, Brazil *fide* Lema (1997: 194).
Distribution: Northern Brazil (Amazonas, Ceará, W Maranhão, Mato Grosso, E Pará, Rondônia).
Sources: Lema, 1997, Lema & Renner, 1998 and Lema & Albuquerque, 2010.

27. *Apostolepis quirogai* Giraudo & Scrocchi, 1998. Herpetologica 54(4): 470–473, fig. 1.

Type: Holotype, FML 6000, a 228 mm female (R. Martínez, 8 April 1995).
Type locality: "Posal. Das, Misiones, Argentina" [= 27°22'S, 55°54'W, elevation 110 m].
Distribution: Southern Brazil (NE Rio Grande do Sul) and NE Argentina (Entre Ríos, Misiones), 110 m.
Sources: Lema & Cappellari, 2001, 2002, Lema 2003b and Lema et al., 2004, 2005.

28. *Apostolepis serrana* Lema & Renner, 2006b. Ciên. Movimento 8(15): 14–15, 4 figs.

Type: Holotype, BMNH 1972.430, a 337 mm male.
Type locality: "Eastern Serra do Roncador, Rio das Mortes basin, Mato Grosso, Brazil, at 1500 feet high; Cerrado domain, captured near the Indian village Pimentel Barbosa (12°51' S - 51°46' W)."
Distribution: Central Brazil (Mato Grosso), 455 m. Known only from type locality.

29. *Apostolepis striata* Lema, 2004b. Comun. Mus. Ciênc. Tecnol. PUCRS, (Zool.) 17(2): 82–86, fig. 1.

Type: Holotype, CHUNB 12794, a 261 mm male.
Type locality: "Vilhena, estado de Rondônia, Brazil (12°54'5S-60°32'0W)" [= elevation 600 m].
Distribution: Western Brazil (Rondônia), 600 m. Known only from type locality.
Source: Lema & Renner, 2002.

30. *Apostolepis tenuis* Ruthven, 1927. Occ. Pap. Mus. Zool., Univ. Michigan (188): 1–2.

Type: Holotype, UMMZ 64436, a 308 mm specimen (J. Steinbach, Aug. 1906–Nov. 1925).
Type locality: "Buena Vista, Dept. Santa Cruz, Bolivia" [= 17°28'S, 63°40'W, elevation 390 m].
Distribution: Central Bolivia (Beni, Santa Cruz), 390–415 m.
Sources: Hartweg, 1932, J.A. Peters & Orejas-Miranda, 1972, Harvey, 1999 and Lema et al., 2004.

31. *Apostolepis vittata* (Cope, 1887b). Proc. Amer. Philos. Soc. 24(125): 56. (*Rhynchonyx ambiniger vittatus*)

Type: Holotype, ANSP 11293 (H.H. Smith, 1882–1886).
Type locality: "near the village of Chupada, thirty miles north-east of Cuyabá, and near the headwaters of the Xingu, in the Prov. of Mato Grosso, Brazil" [= Chapada *fide* Malnate (1971: 366) and Chapada dos Guimarães *fide* Lema (2001: 36), 15°27'S, 55°44'W, elevation 775 m].
Distribution: Central Brazil (Mato Grosso), Bolivia and Paraguay, 775 m.
Sources: J.A. Peters & Orejas-Miranda, 1972, Harvey, 1999, Harvey et al., 2001, Lema, 2002c and Lema et al., 2005.
Remarks: Type locality listed as "Chapada, a small village of Matto Grosso, Brazil, about twenty miles ENE of Cuyabá, on the plateau. The village itself is about 2,500 feet above sea level, or 1,800 ft. above Cuyabá; but collections are from all surrounding regions as low as 1,500 ft." *fide* H.H. Smith *in* Papavero (1971: 379).

APROSDOKETOPHIS Wallach, Lanza & Nistri, 2010 (Colubridae)

Type species: *Aprosdoketophis andreonei* Wallach, Lanza & Nistri, 2010.
Distribution: Somalia.
Source: Wallach et al., 2010.

1. *Aprosdoketophis andreonei* Wallach, Lanza & Nistri, 2010. Afr. J. Herp. 59(2): 99–102, 3 figs.

Type: Holotype, MZUT 3484, a 392 mm female (Cesarini, 1922–1923)

Type locality: "Addur" [= Xuddor, Oddur], Bakool Region, S Somalia, 04°07'N, 43°53'E, elevation approximately 500 m.

Distribution: Southern Somalia (Bakool), 500 m. Known only from type locality.

---

†*ARCHAEOPHIS* Massalongo, 1859 (†Palaeophiidae)

Synonyms: *Archacophis* – Huene, 1956 *(nomen incorrectum)*, and *Archeophis* – Cadle, 1982 *(nomen incorrectum)*.

Type species: †*Archaeophis proavus* Massalongo, 1859.

Distribution: Lower Eocene of Turkmenistan, and middle/lower Eocene of Italy.

Sources: Rage, 1984b and Scanlon & Lee, 2001.

1. †*Archaeophis proavus* Massalongo, 1859. Spec. Photo. Anim. Plant. Foss. Veron.: 14–15, pls. 1–2.

Type: Holotype, ZMBP MB 1902.1357, a 955–980 mm skeleton with 565 vertebrae and impression of skin (O. Di Canossa, 1850).

Type locality: "agri Veronensis, in calcareo numm. M. Bubulcae" [= la Pesciara (Monte Bolca), Veneto Prov., NE Italy, 45°36'N, 11°13'E; Ypresian, lower Eocene *fide* Rage, 1984a: 36].

Distribution: Lower Eocene (Ypresian: 48.6–55.8 mya) of Italy (Veneto). Known only from type locality.

Sources: Janensch, 1904a–b, 1906b and A.H. Müller, 1968.

Remarks: Age is middle Eocene *fide* Auffenberg (1959: 15) and upper lower Eocene *fide* Tatarinov (1988: 113)..

2. †*Archaeophis turkmenicus* Tatarinov, 1963. Paleont. Zhur. 1963(2): 114, figs. 3–5.

Type: Holotype, PIAN 2030/1, four articulated trunk vertebrae (P.G. Danil'chenko, 1960–1962).

Type locality: "Southwestern Turkmen, Kishlak Uilia; Lower Eocene, Donatins strata" [= Uylya-Kushlyk, Turkmenistan, 38°38'N, 55°48'E; Donatinsk formation, lower Eocene fide Rage, 1984b: 36].

Distribution: Lower Eocene (Donatins: 48.6–55.8 mya) of SW Turkmenistan. Known only from type locality.

Source: Tatarinov, 1988.

Remarks: Rage (1984b: 36) erroneously listed holotype as PIAN 2030/31.

ARCHELAPHE Schulz, Böhme & Tillack, 2011 (Colubridae)

Synonym: *Maculophis* Burbrink & Lawson, 2007 *(nomen nudum)*.

Type species: *Coronella bella* Stanley, 1917.

Distribution: Southeastern Asia.

Sources: Bourret, 1934a, C.H. Pope, 1936, Rendahl, 1937, Hu et al., 1980, Tian et al., 1986, Schulz, 1992a, Zhao & Adler, 1993, Orl 2 ov et al., 2000, Schulz et al., 2000, 2011, Helfenberger 2001, Orlov et al., 2003, 2010, Burbrink & Lawson, 2007, I. Das, 2010, 2012, Orlov et al., 2010 and Ryabov et al., 2011.

1. *Archelaphe bellus* (Stanley, 1917). J. No. China Br. Royal Asiatic Soc. (1916) 47: 84–85, 3 figs., pl. with photo. (*Coronella bella*)

Synonyms: *Coluber leonardi* Wall, 1921b, and *Elaphe leonardi chapaensis* Bourret, 1934a.

Type: Holotype, SMNHC 717-17-1-525, a 950 mm specimen.

Type locality: "Kuatun hills, in north-west Fokien near the borders of Kiangsi Prov., China" [= Guadun, in Wuyishan Mtns., ext. N Fujian Prov., SE China, 27°45'N, 117°50'E, elevation 900–1200 m].

Distribution: Southeastern Asia. China (Fujian, Guangdong, Guangxi, Hunan, Sichuan, W Yunnan), NE India (Nagaland), N Myanmar (Kachin), N Laos and Vietnam (Bac Phan, Bac Thai, NW Ha Tinh, Hoàng Liên Son, Lang Son, Lao Cai, Sa Pa, Vinh Phu), 1000–3000 m.

Remarks: Holotype in FU *fide* Orlov et al. (2010b: 313).

ARGYROGENA F. Werner, 1924a (Colubridae)

Type species: *Coluber fasciolatus* G. Shaw, 1802.

Distribution: Southern Asia.

Sources: Wall, 1914b, M.A. Smith, 1928b, L.D. Wilson, 1967, Singh, 1972, P. Silva, 1980a, Khan, 1993, 2002, Captain & Thakur, 1999, R.C. Sharma, 2003, 2007, M.A.R. Khan, 2004, N. Khaire, 2006 and Whitaker & Captain, 2004.

1. *Argyrogena fasciolata* (G. Shaw, 1802). Gen. Zool., Amph. 3(2): 528. (*Coluber fasciolatus*)

Synonyms: *Coluber hebe* Daudin, 1803c, *Coluber curvirostris* Cantor, 1839, and *Argyrogena rostrata* F. Werner, 1924a.

Type: Lectotype, a 600 mm specimen described and illustrated by P. Russell (1796: 26–27, pl. 21) (native, 1781–1791), designated herein.

Type locality: "Vizagapatam, coast of Coromandel, India" [= Vishakhapatnam, Andhra Pradesh State, SE India, 17°41'N, 83°13'E, elevation 25 m] via lectotype selection.

Distribution: Southeastern Pakistan (Balochistan, Sindh), India (Andhra Pradesh, Bihar, Gujarat, Karnataka, Madhya Pradesh, Maharashtra, Tamil Nadu, Uttar Pradesh, West Bengal), Nepal, Sri Lanka (Northern) and Bangladesh.

Remarks: Original description based on P. Russell (1796: 26–27, pl. 21). Single record from Sri Lanka needs confirmation *fide* E.H. Taylor (1950: 546).

2. *Argyrogena vittacaudata* (Blyth, 1854b). Proc. Asiatic Soc. Bengal 23(7): 740.

Type: Holotype, a 483 mm specimen, lost *fide* I. Das et al. (1998: 157).
Type locality: "vicinity of Darjiling" [= Darjeeling, West Bengal, India, 27°02'N, 88°16'E, elevation 2100 m].
Distribution: Northeastern India (West Bengal).
Remarks: Related to *Coluber* (= *Argyrogena*) *fasciolatus fide* Blyth.

ARIZONA Kennicott *in* Baird, 1859a (Colubridae)

Synonym: *Arizon* – Goode, 1883 *(nomen incorrectum).*
Type species: *Arizona elegans* Kennicott *in* Baird, 1859a.
Distribution: Southwestern USA and N Mexico.
Fossil records: Upper Miocene of USA (Nebraska) and Pleistocene of USA (Arizona, California, New Mexico, Texas).

1. *Arizona elegans* Kennicott *in* Baird, 1859a. U.S.-Mex. Bound. Surv., Rept. (1858) 2(2): 18–19, pl. 13.

Synonyms: *Coluber arizonae* Boulenger, 1894, *Arizona elegans occidentalis* Blanchard, 1924a, *Arizona elegans blanchardi* Klauber, 1946a, *Arizona elegans candida* Klauber, 1946a, *Arizona elegans eburnata* Klauber, 1946a, *Arizona elegans expolita* Klauber, 1946a, *Arizona elegans noctivaga* Klauber, 1946a, *Arizona elegans philipi* Klauber, 1946a, *Arizona elegans arenicola* Dixon, 1960, and *Arizona elegans australis* K.L. Williams, Chrapliwy & Smith, 1961.
Type: Lectotype, USNM 1722 (A.C.V. Schott [U.S.-Mex. Bound. Surv.], 1852–1853), designated by Blanchard (1924: 4).
Type locality: "Lower Rio Grande" [= lower Rio Grande River, SE Texas, S USA] via lectotype selection.
Distribution: Southwestern United States and N Mexico. USA (Arizona, S California, ext. E Colorado, W Kansas, ext. SW Nebraska, ext. S Nevada, New Mexico, Oklahoma, Texas, ext. SE Utah) and Mexico (Aguascalientes, Chihuahua, Coahuila, Durango, Nuevo León, San Luis Potosí, W Sinaloa, W Sonora, S Tamaulipas, Zacatecas), NSL–1830 m.
Fossil records: Upper Pleistocene (Rancholabrean II) of USA (Arizona, California, New Mexico, Texas).
Sources: Klauber, 1946a, Fleet & Dixon, 1971, Dixon & Fleet, 1976, Van Devender & Bradley, 1994, Degenhardt et al., 1996, Keogh, 1996, P.R. Brown, 1997, Rodríguez-Robles & Jesús-Escobar, 1999, Bartlett & Tennant, 2000, Holman, 2000a, Werler & Dixon, 2000, McCranie & Wilson, 2001b, Grismer, 2002 and Pyron & Burbrink, 2009.
Remarks: Official Specific Name no. 2047 *fide* Opinion 717 (ICZN, 1965a). Lectotype closely resembles specimens from Terrel Co., Texas *fide* Dixon & Fleet (1976: 2).

2. *Arizona pacata* Klauber, 1946a. Trans. San Diego Soc. Nat. Hist. 10(17): 379–380 (*Arizona elegans pacata*)

Type: Holotype, SDSNH 17652, a 789 mm male (F.F. Gander, 16 Nov. 1941).
Type locality: "Santo Domingo (lat. 25°30'N), Baja California, Mexico."
Distribution: Northwestern Mexico (ext. SW Baja California Norte, Baja California Sur).
Source: Grismer, 2002.

3. †*Arizona voorhiesi* Parmley & Holman, 1995. J. Vert. Paleont. 15(1): 80–81, figs. 4a–b.

Type: Holotype, MSUVP 1220, one trunk vertebra (UNSM & MSU field crews).
Type locality: "Lemoyne Quarry fossil site (USNM locality Kh-101), is a north-facing road cut on Nebraska State Highway 92 in NE1/4, NW1/4, Sec. 3, 6.4 km W of Lemoyne, Keith County, SW Nebraska; middle Hemphillian mammal age (late Miocene)."
Distribution: Upper Miocene (Hemphillian: 4.9–10.3 mya) of USA (Nebraska).

ARRHYTON A.C.L.G. Günther, 1858 (Xenodontidae)

Synonyms: *Cryptodacus* Gundlach *in* W.C.H. Peters, 1861f, *Arryton* – Cope, 1860e *(nomen incorrectum)*, *Colorhogia* Cope, 1862, and *Arhyton* – Auffenberg, 1958 *(nomen incorrectum).*
Type species: *Arrhyton taeniatum* A.C.L.G. Günther, 1858.
Distribution: Cuba.
Sources: Grant *et. al.*, 1959, A. Schwartz, 1965, Buden, 1966, Maglio, 1970, A. Schwartz & Thomas, 1975, A. Schwartz & Garrido, 1981, A. Schwartz & R.W. Henderson, 1988, 1991, Rodríguez-Robles & Greene, 1996, Vidal et al., 2000, Hedges et al., 2009, R.W. Henderson & Powell, 2009 and Zaher et al., 2009.

1. *Arrhyton ainictum* A. Schwartz & Garrido, 1981. Ann. Carnegie Mus. 50(7): 216–217, fig. 2c, 3c.

Type: Holotype, CZACC 4.4619 (formerly IZAC 4256), a 363 mm male (L. Zayas, 30 July 1974).
Type locality: "Cueva del 18, Francisco, Camagüey Prov., Cuba."
Distribution: Cuba (Camaguey, Las Tunas).
Source: Chamizo-Lara et al., 2005.

2. *Arrhyton dolichura* F. Werner, 1909b. Mitt. Naturhist. Mus. Hamburg 26: 224–225.

Type: Holotype, formerly ZMH, a 410 mm female, destroyed in 1943 during World War II.
Type locality: "Alabama (?)" [USA] (in error). Restricted to La Habana, Habana Prov., Cuba *fide* Grant et al. (1959: 130).
Distribution: Cuba (Habana).
Remarks: For explanation of species spelling see A. Schwartz & Garrido (1981: 220).

3. *Arrhyton procerum* Hedges & Garrido, 1992a. Herpetologica 48(2): 174–175, fig. 3a.

Type: Holotype, MNHNCU 3285, a 371 mm male (E. Alfaro, 12 July 1990).
Type locality: "11.4 km ESE Playa Girón, Matanzas Prov., Cuba, 0 m."
Distribution: Cuba (Matanzas), NSL. Known only from type locality.

4. *Arrhyton redimitum* (Cope, 1862b). Proc. Acad. Nat. Sci. Philadelphia 14(1): 81–82. (*Colorhogia redimita*)

Synonym: *Arrhyton vittatum landoi* A. Schwartz, 1965.
Type: Holotype, USNM 29769 (formerly USNM 5747), a 197 mm specimen (C. Wright, 1856–1862).
Type locality: "Eastern Cuba" [Greater Antilles].
Distribution: Eastern Cuba (Granma, Guantánamo, Santiago de Cuba).
Sources: A. Schwartz, 1965 and Lando & Williams, 1969.

5. *Arrhyton supernum* Hedges & Garrido, 1992a. Herpetologica 48(2): 171, fig. 1b.

Type: Holotype, MNHNCU 2704, a 426 mm female (R. Thomas, 29 July 1989).
Type locality: "the SW slope of El Yunque de Baracoa, Guantánamo Prov., Cuba, 136 m."
Distribution: Extreme E Cuba (Guantánamo), 135 m. Known only from type locality and Monte Libano.

6. *Arrhyton taeniatum* A.C.L.G. Günther, 1858. Cat. Colub. Snakes Brit. Mus.: 244.

Synonym: *Arrhyton fulvum* Cope, 1862b.
Type: Holotype, BMNH 1946.1.21.48, a 229 mm specimen (Zool. Soc. London).
Type locality: "Cuba" [Greater Antilles].
Distribution: Cuba (Cienfuegos, Granma, Guantánamo, Habana, Holguin, Pinar del Río, Santiago de Cuba and Juventud Is.).

7. *Arrhyton tanyplectum* A. Schwartz & Garrido, 1981. Ann. Carnegie Mus. 50(7): 221–222, fig. 2e, 3e.

Type: Holotype, AMNH 77782, a 542 mm male (A. Schwartz, 9 July 1957).
Type locality: "cliffs at San Vicente, Pinar del Río Prov., Cuba."
Distribution: Extreme W Cuba (Pinar del Río).

8. *Arrhyton vittatum* (Gundlach *in* W.C.H. Peters, 1861f). Mber. Königl. Akad. Wiss. Berlin 1861(11): 1003–1004. (*Cryptodacus vittatus*)

Synonym: *Arrhyton bivittatum* Cope, 1862b.
Types: Syntypes (2), ZMB 4096, 283 and 220 mm specimens (J. Gundlach, 1839–1842), other syntype lost *fide* Bauer et al. (1995: 81).
Type locality: "Cuba" [Greater Antilles]. Restricted to Cárdenas, Matanzas Prov., Cuba *fide* & Garrido (1981: 208).
Distribution: A. Schwartz Cuba (Camagüey, Habana, Holguin, Matanzas, Oriente, Pinar del Río, Sancti Spíritus, Via Clara and Pinos Is).

ASPIDELAPS Fitzinger, 1843 (Elapidae)

Synonyms: *Aspidoelaps* A.-M.-C. Duméril & Bibron, 1844 *(nomen emendatum)*, *Cyrtophis* Sundevall *in* A. Smith, 1849 *in* 1838–1849, *Aspidilaps* – Pringle, 1954 *(nomen incorrectum)*, and *Crytophis* – V.F.M. FitzSimons, 1962 *(nomen incorrectum)*.
Type species: *Natrix lubrica* Laurenti, 1768.
Distribution: Southern Africa.
Sources: Mertens, 1955, V.F.M. FitzSimons, 1962a, 1974, Broadley, 1968a, 1983, Buys & Buys, 1983, Branch, 1988, Marais, 1992, 2004, Golay et al., 1993, Spawls & Branch, 1995, Slowinski et al., 1997, Keogh, 1998, Broadley & Baldwin, 2006, Dobiey & Vogel, 2007, Zaher et al., 2009 and Hoser, 2012e.

1. *Aspidelaps lubricus* (Laurenti, 1768). Synop. Rept.: 80. (*Natrix lubrica*)

Synonyms: *Coluber latonia* Daudin, 1803b, *Naia somersetta* A. Smith, 1826, *Aspidelaps lubricus cowlesi* Bogert, 1940, and *Aspidelaps lubricus infuscatus* Mertens, 1954b.
Type: Lectotype, specimen described and illustrated by Seba (1735: pl. 43, fig. 3), lost *fide* Golay et al. (1993: 113).
Type locality: "Africanus, ex Promontorio Bonae Spei" [= Cape of Good Hope, Western Cape Prov., SW South Africa, 34°21'S, 18°28'E, elevation 20 m].
Distribution: Southern Africa. Southwestern Angola (Namibe), Namibia (Erongo, Hardap, Karas, Khomas, Kunene, Omaheke, Oshikoto, Otjozondjupa) and

W South Africa (Eastern Cape, S Free State, Northern Cape, Western Cape), NSL–1415 m.

2. *Aspidelaps scutatus* (A. Smith, 1849 *in* 1838–1849). Illust. Zool. So. Africa, Rept. (App.): 22. (*Cyrtophis scutatus*)

Synonyms: *Naia fula-fula* Bianconi, 1849a *(nomen incorrigendum), Naja fulafula* – W.C.H. Peters, 1882d *(nomen corrigendum), Aspidelaps scutatus bachmani* Mertens, 1954, and *Aspidelaps scutatus intermedius* Broadley, 1968.

Type: Holotype, BMNH 1946.1.17.83 (formerly BMNH 1863.2.21.90), a 190–205 mm specimen (A. Smith, 27 Aug.–10 Sept. 1835).

Type locality: "Kaffirland and the country towards Port Natal, South Africa," (in error *fide* V.F.M. FitzSimons, 1962a: 273, probably from further north, i.e. Mozambique). Restricted to the confluence of the Marico and Crocodile rivers, north-western Transvaal *fide* Broadley (1968d: 1) [= NW Limpopo Prov., NE South Africa, 24°11'S, 26°52'E, elevation NSL].

Distribution: Southern Africa. Southern Mozambique (Gaza, Inhambane, Manica, Maputo, Sofala), SW Zimbabwe (Bulawayo, Masvingo, Matebeleland North, Matebeleland South), Botswana (Central, Gaborone, Ghanzi, Kgaligadi, Kgatleng, Kweneng, North East, North West, Southern), Namibia (Caprivi, Erongo, Hardap, Khomas, Okavango, Omaheke, Oshikoto, Otjozondjupa) and NE South Africa (Gauteng, Limpopo, Mpumalanga), NSL–1670 m.

Sources: Broadley, 1968d, Broadley & Cock, 1975, Pienaar et al., 1978, Auerbach, 1987, Shine et al., 1996, Clauss & Clauss, 2002, M. Griffin, 2003 and Broadley & Blaylock, 2013.

Remarks: Boulenger (1896a: 391) listed a BMNH juvenile from Natal as one of A. Smith's types (confirmed *fide* V.F.M. FitzSimons, 1937: 263), leading Mertens (1955: 108) to restrict type locality to Natal. FitzSimons (1962a: 273) believed type locality of Natal to be in error (as the species does not occur there). Broadley restricted type locality to a site visited by A. Smith in 1835.

ASPIDITES W.C.H. Peters, 1877a (*nomen substitutum*) (Pythonidae)

Synonyms: *Aspidiotes* Krefft, 1864a–b *(nomen praeoccupatum), Aspidiotus* W.C.H Peters, 1876b *(nomen praeoccupatum),* and *Aspiditis* – Briceño-Rossi, 1934 *(nomen incorrectum).*

Type species: *Aspidiotes melanocephalus* Krefft, 1864a.

Distribution: Australia.

Sources: Stimson, 1969, L.A. Smith, 1981a, Storr et al., 1986, S.K. Wilson & Knowles, 1988, Hoser, 1989, 2000b, 2003a, 2012b, Ehmann, 1992, D.G. Barker & Barker, 1994, Wall, 1998b, McDiarmid et al., 1999, Cogger, 2000, Torr, 2000, Rawlings et al., 2008 and Schleip & O'Shea, 2010.

1. *Aspidites melanocephalus* (Krefft, 1864a). Proc. Zool. Soc. London 32(1): 20–22, 2 figs. (*Aspidiotes melanocephalus*)

Synonyms: *Aspidites melanocephalus adelynensis* Hoser, 2000b & 2012b *(nomen illegitimum), Aspidites melanocephalus davieii* Hoser, 2000b *(nomen illegitimum),* and *Aspidites melanocephalus rickjonesii* Hoser, 2009b & 2012b *(nomen illegitimum).*

Type: Holotype, BMNH 1946.1.8.2 (formerly BMNH 1866.2.13.27), a 2388 mm female (G. Krefft, 1861–1864).

Type locality: "Port Denison, N.E. Australia" (in error). Corrected to Bowen, Queensland *fide* Cogger et al. (1983: 202).

Distribution: Northern Australia (Northern Territory, Queensland, N Western Australia, Bathurst, Melville, Montgomery, Quail, Uwins and Valencia Is.), NSL–760 m.

Sources: L.A. Smith, 1981a, Storr et al., 1986, Hoser, 1989, 2000b, 2003a, 2012b, Ehmann, 1992, Wall, 1998b, J.C. Murphy & Schlager, 2003 and Rawlings et al., 2008.

Remarks: Original description reprinted in Krefft (1864b: 225–226, 2 figs.).

2. *Aspidites ramsayi* (Macleay, 1882). Proc. Linn. Soc. N.S.W. (1881–1882) 6(4): 813. (*Aspidiotes ramsayi*)

Synonyms: *Aspidites collaris* Longman, 1913, *Aspidites ramsayi panoptes* Hoser, 2000b & 2012b *(nomen illegitimum), Aspidites ramsayi richardjonesii* Hoser, 2000a *(nomen illegitimum),* and *Aspidites ramsayi neildavieii* Hoser, 2009b & 2012b *(nomen illegitimum).*

Type: Holotype, MMS, a 1905 mm specimen (J. Ramsay via E.P. Ramsay, 1880), lost *fide* Stimson (1969: 2).

Type locality: "near Fort Bourke" [New South Wales, Australia] in error and corrected to Tyndarie *fide* Shea (2012: 149).

Distribution: Central Australia (NW New South Wales, S Northern Territory, SW Queensland, N South Australia, Western Australia), NSL–650 m.

Sources: Waite, 1894b and Krauss, 1994.

ASPIDOMORPHUS Fitzinger, 1843 (Elapidae)

Synonyms: *Pseudelaps* A.-M.-C. Duméril, 1853 *(nomen praeoccupatum), Pseudoelaps* Jan, 1859a *(nomen praeoccupatum), Apisdomorphus* A.-M.-C. Duméril, Birbon & Duméril, 1854b *(nomen emendatum), Psudoelaps* – Hoffmann, 1890 *(nomen incorrectum), Pseudelas* – Briceño-Rossi, 1934 *(nomen incorrectum), Aspidomophus* – Obst, Richter & Jacob, 1984 *(nomen incorrectum),* and *Aspidomorphis* – Schwaner,

Baverstock, Dessauer & Mengden, 1985 *(nomen incorrectum).*

Type species: *Elaps muellerii* Schlegel, 1837.

Distribution: New Guinea and adjacent islands.

Sources: Brongersma, 1934, McDowell, 1967, O'Shea, 1996, Keogh et al., 1998, Scanlon, 2003a, Scanlon & Lee, 2004, Sanders et al., 2008, Metzger et al., 2010, Zaher et al., 2009 and Hoser, 2012e.

Remarks: All three species contain multiple cryptic species *fide* Metzger et al. (2009: 7).

1. *Aspidomorphus lineaticollis* (F. Werner, 1903b). Zool. Anz. (1902) 26(693): 251 (footnote). (*Pseudelaps muelleri lineaticollis*)

Synonyms: *Aspidomorphus mülleri lineatus* Brongersma, 1934 *(nomen incorrigendum), Aspidiomorphus schlegeli* – Brass, 1959 *(nomen incorrectum), Aspidomorphus muelleri lineatus* – Klemmer, 1963 *(nomen corrigendum),* and *Aspidomorphus linegicollis* – McDowell, 1969a *(nomen incorrectum).*

Type: Holotype, NMW 27387, a 475 mm female (B. Geisler & H. Geisler, 1890–1902, via W. Schlüter).

Type locality: "Neuguinea" [= New Guinea].

Distribution: Eastern Papua New Guinea (Madang, Milne Bay, Morobe, Oro, Sudest, Fergusson, Goodenough, Misima, Normanby, Trobriand & Woodlark Is.), NSL–400 m.

2. *Aspidomorphus muellerii* (Schlegel, 1837). Essai Phys. Serp. 1: 182, 2: 452–454, pl. 16, figs. 16–17. (*Elaps muellerii*) *(nomen corrigendum)*

Synonyms: *Elaps mülleri* Schlegel, 1837 *(nomen incorrigendum), Elaps müllerii* Schlegel, 1837 *(nomen incorrigendum), Diemenia muelleri* – A.C.L.G. Günther, 1872b *(nomen corrigendum),* and *Aspidomorphus mülleri interruptus* Brongersma, 1934 *(nomen incorrigendum),* and *Aspidomorphus muelleri interruptus* – Klemmer, 1963 *(nomen corrigendum).*

Types: Syntypes (2), RMNH 1402–03, a male and female, 456 and 405 mm specimens (S. Müller, Aug. 1828).

Type locality: "la côte occidentale de la Nouvelle Guinée" [= Lobo, Triton Bay, West Papua Prov., Indonesia *fide* McDowell, 1967: 521].

Distribution: New Guinea. Eastern Indonesia (West Papua, Mansinam, Misool, Salawati, Seram) and Papua New Guinea (Central, Eastern, East New Britain, East Sepik, Gulf, Milne Bay, New Ireland, Northern, Southern Highlands, Western, Western Highlands, West New Britain, West Sepik, Duke of York Is.), 500–1500 m.

Remarks: Original description listed *E. mülleri* (1: 182 & pl. 16) and *E. müllerii* (2: 452).

3. *Aspidomorphus schlegelii* (A.C.L.G. Günther, 1872b). Ann. Mag. Nat. Hist. (4) 9(49): 35. (*Diemenia schlegelii*)

Synonyms: *Pseudelaps muelleri insulae* T. Barbour, 1908, *Pseudelaps muelleri concolor* F. Werner, 1925, and *Aspidomorphus schlegelli* – Brongersma, 1952 *(nomen incorrectum).*

Types: Syntypes (2), BMNH 1946.1.18.43, a female, and BMNH 1946.1.18.86.

Type locality: "island of Misol" [= Misool, Papua Prov., E Indonesia].

Distribution: New Guinea. Eastern Indonesia (Papua, Batanta, Biak, Jamna, Jobi, Misool, Mysore, Salawati, Waigeu) and ext. NW Papua New Guinea (West Sepik).

ASPIDURA Wagler, 1830 (Natricidae)

Synonyms: *Haplocercus* A.C.L.G. Günther, 1858, *Heptagerrhon* Fitzinger, 1861, *Aspiduraz* – Theobald, 1876 *(nomen incorrectum),* and *Asidura* – Palacky, 1898 *(nomen incorrectum).*

Type species: *Scytale brachyorrhos* F. Boie, 1827.

Distribution: Sri Lanka.

Sources: M.A. Smith, 1943, Deraniyagala, 1955, P. Silva, 1971, 1980a, Gans & Fetcho, 1982, A. Silva, 1990 and Zaher et al., 2009.

Remarks: *Haplocercus* a synonym of *Aspidura fide* Pyron et al. (2013: 975).

1. *Aspidura brachyorrhos* (F. Boie, 1827). Isis von Oken 20(6): 517. (*Scytale brachyorrhos*)

Synonym: *Calamaria scytale* Schlegel, 1837.

Type: Holotype, IRSNB, a 207 mm specimen (P.-A.J. Drapiez, 1819–1827), lost *fide* Gans & Fetcho (1982: 285).

Type locality: "Java" [W Indonesia] (in error). Corrected to Ceylon [= Sri Lanka] *fide* H. Boie (1831: 188). Restricted to Sri Lanka *fide* Gans & Fetcho (1982: 286).

Distribution: Southern Sri Lanka (Central, Western), 350–1220 or 1400 m.

Source: E.H. Taylor, 1953b.

Remarks: Original description based on H. Boie's MS (1823–1825: pl. 22).

2. *Aspidura ceylonensis* (A.C.L.G. Günther, 1858). Cat. Colub. Snakes Brit. Mus.: 15. (*Haplocercus ceylonensis*)

Synonym: *Aspidura carinata* Jan, 1862b.

Types: Syntypes (2), BMNH 1946.1.12.27, a male (R. Templeton, 1839–1851), and BMNH 1946.1.12.6, a 483 mm female (H. Cuming, 1836–1840).

Type locality: "Ceylon" [= Sri Lanka].

A

Distribution: Central Sri Lanka (Central, Sabaragamuwa, Uva), 1070–2140 (2300) m.
Sources: Wall, 1921g, 1923c, M.A. Smith, 1943, E.H. Taylor, 1953b, Deraniyagala, 1955, P. Silva, 1969, 1980a, A. Silva, 1983, 1990b, 2001, 2009 and Somaweera, 2006.
Remarks: Transferred from *Haplocercus* to *Aspidura fide* Pyron et al. (2013: 975).

3. *Aspidura copii* A.C.L.G. Günther, 1864a. Rept. Brit. India: 203, pl. 18, fig. e.

Type: Holotype, BMNH 1946.1.12.9, a 419 mm male.
Type locality: Not stated, but presumed to be Ceylon [= Sri Lanka] *fide* Günther (1864a: 203).
Distribution: Southern Sri Lanka (Central, Uva, Western), 30–2130 m.
Source: Willey, 1906.

4. *Aspidura deraniyagalae* Gans & Fetcho, 1982. Ann. Carnegie Mus. 51(14): 299–301, fig. 6 (left).

Type: Holotype, CM 83455, a 199 mm male (C. Gans, Aug. 1972).
Type locality: "Kanawarella Group Estate (near Namunukula), at 1520 m, Sri Lanka,"[= Uva Prov., SE Sri Lanka].
Distribution: Southeastern Sri Lanka (Uva). Known only from vicinity of type locality, 1500–1520 m.

5. *Aspidura drummondhayi* Boulenger, 1904d. Spolia Zeylan. 2(7): 95–96, pl., 4 figs.

Types: Syntypes (2), BMNH 1946.1.12.45–46, a 202 mm male and 223 mm female (H.M. Drummond-Hay, March–May 1903).
Type locality: "Hopewell estate, Balangoda...the probable elevation being from 3,500 to 4,200 feet above sea level" [= Sabaragamuwa Prov., S Sri Lanka].
Distribution: Southern Sri Lanka (Sabaragamuwa, Southern), 1040–1280 m.

6. *Aspidura guentheri* Ferguson, 1876. Proc. Zool. Soc. London 44(1): 819–820.

Types: Syntypes (12), BMNH 1946.1.12.16–26 and MCZ 28467, males, females and juveniles, longest syntype 157 mm (W. Ferguson, 1839–1877).
Type locality: "Ceylon" [= Sri Lanka].
Distribution: Southern Sri Lanka (Sabaragamuwa, Southern, Western), NSL–500 m.

7. *Aspidura trachyprocta* Cope, 1860d. Proc. Acad. Nat. Sci. Philadelphia 12(2): 75–76.

Synonym: *Heptagerrhon bivittatum* Fitzinger, 1861.
Type: Holotype, ANSP 3308, a 208 mm specimen (H. Cuming, 1836–1840).
Type locality: "Ceylon" [= Sri Lanka].
Distribution: Central Sri Lanka (Central, Uva), 750–2100 m.
Sources: A. Silva, 2009 and Ukuweia & Dawundasekara, 2012.
Remarks: Malnate (1971: 3072) erroneously listed type as ANSP 3309.

ASTHENODIPSAS W.C.H. Peters, 1864a (Pareatidae)

Synonyms: *Amplycephalus* Kuhl & Hasselt, 1822b *(nomen rejiciendum), Amblycephalus* Schlegel, 1826a *(nomen praeoccupatum), Amblycephalus* H. Boie *in* F. Boie, 1827 *(nomen praeoccupatum), Amplyaphalus* – F. Boie, 1827 *(nomen incorrectum), Amblicephalus* Gray *in* Griffith & Pidgeon, 1831 *(nomen emendation), Emblycephalus* – A.-M.-C. Duméril, & bibron, 1844 *(nomen incorrectum), Amblycerhalus* – Kulagin, 1888 *(nomen incorrectum), Astenodipsas* Lidth de Jeude, 1891 *(nomen emendatum),* and *Internatus* Yang & Rao *in* Rao & Yang, 1992.
Type species: *Asthenodipsas malaccana* W.C.H. Peters, 1864a.
Distribution: Southeastern Asia and W Indonesia.
Sources: Rao & Yang, 1992, David & Vogel, 1996, Iskandar & Colijn, 2001, Grossmann & Tillack, 2003, Zaher et al., 2009, I. Das, 2010, 2012 and Hoser, 2012q.

1. *Asthenodipsas laevis* (H. Boie *in* F. Boie, 1827). Isis von Oken 20(6): 520. (*Amblycephalus laevis*)

Synonyms: *Amblycephalus laevis* H. Boie *in* Schlegel, 1826a *(nomen nudum),* and *Amblycephalus laevis* Schlegel, 1826b *(nomen nudum).*
Type: Lectotype, RMNH 986B, a 374 mm female (H. Kuhl & J.C. van Hasselt, Dec. 1820–Sept. 1823), designated by Grossman & Tillack (2003: 180).
Type locality: "Java, Indonesia" via lectotype selection. Restricted to region of Mt. Pangerango and Mt. Salak, West Java, Indonesia *fide* Brongersma (1948b: 14). Emended to Tjihandjawar, at the foot of Mt. Pangerango, W. Java [= Cihanyawar, Nagrak, Jawa Barat, W Java, SW Indonesia, 07°03'S, 108°40'E] *fide* Brongersma (1950: 1499).
Distribution: Malay Peninsula and Greater Sundas. Southern Thailand (Nakhon Si Thammarat, W Phatthalung, E Trang), Laos (Houaphan, Xiangkhouang), West Malaysia (Johore, Pahang, Selangor), East Malaysia (Sabah, Sarawak) and W Indonesia (Bangka, Java, S Kalimantan, Mentawai Arch., Natuna Besar, Siberut, Serasan, Sumatra), 350–1625 m.
Sources: A.C.L.G. Günther, 1864a, Flower, 1896, Rooij, 1917, 1922, M.A. Smith, 1930c, Haas, 1950, B.L. Lim, 1963, 1964, E.H. Taylor, 1965, Tweedie, 1983, Heang, 1987, Dring et al., 1990, M.J. Cox, 1991b, Stuebing, 1991, B.L. Lim et al., 2002, Malkmus et al., 2002 and I. Das, 2010.

Remarks: Original description based on H. Boie's MS (1823–1825). Photograph of lectotype in Grossman & Tillack (2003: fig. 4).

2. *Asthenodipsas malaccanus* W.C.H. Peters, 1864a. Mber. Königl. Akad. Wiss. Berlin 1864(4): 273–274, pl., figs. 3, 3a–3d. (*Asthenodipsas malaccana*)

Synonyms: *Pareas dorsopictus* Edeling, 1870, and *Amblycephalus malaccanus ventrilineatus* Angel, 1941.

Type: Holotype, ZMB 5041, a 215 mm female (W. Baumgarten), lost *fide* R. Günther (*in litt.*).

Type locality: "der Nähe von Malacca" [= Melaka, southwestern West Malaysia]. Corrected to Malay Peninsula *fide* Boulenger (1912c: 210) [= West Malaysia].

Distribution: Malay Peninsula and Greater Sundas. Extreme S Thailand (Yala), West Malaysia (Melaka, Pahang, Selangor), East Malaysia (Sabah, Sarawak), Brunei and W Indonesia (E Kalimantan, Mentawai Arch., Siberuat, Sumatra), 245–1000 m.

Sources: Lidth de Jeude, 1890, Boulenger, 1912c, M.A. Smith, 1930c, Haas, 1950, Batchelor, 1958, E.H. Taylor, 1965, Grandison, 1972, B.L. Lim & Sharef bin Kamarudin, 1975, Tweedie, 1983, Stuebing, 1991, Malkmus et al., 2002 and I. Das, 2007b, 2012.

Remarks: Photograph of holotype in Grossman & Tillack (2003: fig. 3).

3. *Asthenodipsas vertebralis* (Boulenger, 1900a). Ann. Mag. Nat. Hist. (7) 5(27): 307–308. (*Amblycephalus vertebralis*)

Synonym: *Amblycephalus tropidonotus* Lidth de Jeude, 1923.

Type: Holotype, BMNH 1946.1.20.14 (formerly BMNH 1900.7.18.5), a 460 mm specimen (L. Wray, Jr.).

Type locality: "Maxwell's Hill, Larut, 3000 feet elevation, Perak, Malay Peninsula" [= West Malaysia, elevation 915 m].

Distribution: West Malaysia (W Pahang, Perak, Tioman Is.) and W Indonesia (W Sumatra), 915–2010 m.

Sources: Smedley, 1932a, C.P. Haas, 1950, B.L. Lim, 1967a, Grandison, 1972, Sly, 1976, Tweedie, 1983, Stuebing, 1991, David & Vogel, 1996, M.J. Cox et al., 1998, Chan-ard et al., 1999, B.L. Lim et al., 2002, Grossmann & Tillack, 2003, Youmans et al., 2003, Grismer, 2011 and I. Das, 2012.

Remarks: Photograph of type in Grossman & Tillack (2003: fig. 2).

ASTROTIA J.G. Fischer, 1855 (Elapidae)

Synonyms: *Astiotia* – Reid, 1966 *(nomen incorrectum)*, *Astrotea* – Cahill, Heatwole & Goldman, 1973 *(nomen incorrectum)*, and *Aristrotia* – B.D. Sharma *in* B.D. Sharma & Kumari, 1998a *(nomen incorrectum)*.

Type species: *Hydrus stokesii* Gray *in* Stokes, 1846a.

Distribution: Marine waters of Indo-Australia, including Arabian Sea, Arafura Sea, Bay of Bengal, Coral Sea, Flores Sea, Gulf of Thailand, Gulf of Tonkin, Indian Ocean, South China Sea, Strait of Malacca and Timor Sea.

Sources: M.A. Smith, 1926a, E.H. Taylor, 1965, Mao & Chen, 1975, Cogger et al., 1983a, Tweedie, 1983, Gasperetti, 1988, Golay et al., 1993, O'Shea, 1996, A.R. Rasmussen, 1997, Cogger, 2000, M.S. Khan, 2002, 2006, Whitaker & Captain, 2004, Wells, 2007, Kharin & Czeblukov, 2009, V.S. Nguyen et al., 2009, Zaher et al., 2009, Xiang & Li, 2009, I. Das, 2010, 2012, A.R. Rasmussen et al., 2011 and Hoser, 2012e.

Remarks: Original description reprinted in J.G. Fischer, 1856a. Kharin & Czeblokov, 2009 placed this genus with *Distiera* and *Enhydrina* in a new tribe Disteirini. A synonym of *Hydrophis fide* Sanders et al., 2013.

1. *Astrotia stokesii* (Gray *in* Stokes, 1846a). Discov. Australia (App.): 502, pl. 3. (*Hydrus stokesii*)

Synonyms: *Hydrophis schizopholis* P. Schmidt, 1846, *Hydrus annulatus* Gray, 1849, *Hydrophis güntheri* Theobald, 1868b (*nomen praeoccupatum*), *Hydrophis granosa* J. Anderson, 1871a, *Hydrophis guttata* Murray, 1887, *Astrossii stokesii* – W.P. Lowe, 1932 *(nomen incorrectum)*, *Astrokia stokesi* – Ruiter, 1958 *(nomen incorrectum)*, *Astoria stokesi* – Sawai, 1976 *(nomen incorrectum)*, *Astroria stokesi* A. Silva, 1976 *(nomen incorrectum)*, *Astrocia stokessi* – Khole, 1991 *(nomen incorrectum)*, and *Hydrophis guentheri* Cogger et al., 1983a *(nomen corrigendum)*.

Type: Holotype, BMNH 1946.1.17.12 (formerly BMNH III.9.1.d), an adult female (J.L. Stokes, 1837–1843).

Type locality: "Australian Seas."

Distribution Indo-Australia. Pakistan, India (West Bengal), Sri Lanka, Thailand (Chon Buri, Pattani), Vietnam (Gulf of Tonkin to Gulf of Thailand), China, Taiwan, West Malaysia, Singapore, Philippines, E Indonesia (Papua, Aru Is.), S Papua New Guinea (Western) and Australia (New South Wales, Northern Territory, Queensland, Western Australia).

ATHERIS Cope, 1862 (Viperidae)

Synonyms: *Chloroechis* Bonaparte, 1849 *(nomen rejiciendum)*, *Poecilostolus* A.C.L.G. Günther, 1863a, *Antheris* – Giglioli, 1873 *(nomen incorrectum)*, *Adenorhinos* Marx & Rabb, 1965, *Adenorhinus* Dowling *in* Dowling & Duellman, 1978 *(nomen emendatum)*, *Adenorphinus* – Underwood, 1979 *(nomen incorrectum)*, and *Woolfvipera* Hoser, 2012v *(nomen illegitimum)*.

Type species: *Echis squamigera* Hallowell, 1855b.

A

Distribution: West, cen. and East Africa.

Sources: Broadley, 1968a, 1989, 1996b, 1998b, Groombridge, 1980, Ashe & Marx, 1988, Marx *et al.*, 1988, Golay et al., 1993, Herrmann & Joger, 1995, 1997, Spawls & Branch, 1995, Broadley, 1996b, 1998b, J.B. Rasmussen & Howell, 1998, Herrmann et al., 1999, Joger & Courage, 1999, McDiarmid et al., 1999, D. Lawson et al., 2001, Lenk et al., 2001b, Spawls et al., 2002, Mallow et al., 2003, Dobiey & Vogel, 2007, Wüster et al., 2008, Branch & Bayliss, 2009, Phelps, 2010, Menegon *et al.*, 2011 and Hoser, 2012d, v.

Remarks: Official Generic Name *fide* Opinion 1634 (ICZN, 1991). McDiarmid et al. (1999: 353–354) discuss type species of *Atheris* and status of *Chloroechis* Bonaparte.

1. *Atheris acuminata* Broadley, 1998b. Herpetol. J. 8(3): 129–130, figs. 11–12.

Type: Holotype, NMZB 13950, a 440 mm male (Mrs. C. Allen, 3 May 1994).

Type locality: "forest near Nsere Lodge, Kyambura Game Reserve, just south of lake George, Ankole District, western Uganda (00°09'S: 30°08') at an altitude of ca. 950 m.

Distribution: Southwestern Uganda (W Western), 950 m. Known only from holotype.

Source: Spawls et al., 2002.

Remarks: A synonym of *A. hispida fide* Lawson et al., 2001. Taxonomic status problematic *fide* Dobiey & Vogel, 2007.

2. *Atheris anisolepis* Mocquard, 1887c. Bull. Soc. Philom. Paris (1886–1887) (7) 11(1): 89.

Synonym: *Atheris laeviceps* Boettger, 1887a.

Type: Lectotype, MNHN 1886.242, a 650 mm specimen, designated by Broadley (1998b: 128).

Type locality: "Alima [River]- Leketi, Congo" [= Alima River at Leketi, Cuvette Region, Congo, 1°36'S, 14°57'E, elevation 370 m] via lectotype selection.

Distribution: Central Africa. Southern Gabon (Haut-Ogooué), S Congo (Cuvette, Kouilou, Lekoumou, Plateau), W Democratic Republic of the Congo (Bandundu, Bas-Congo), NSL–800 m.

Source: J.-F. Trape & Roux-Estève, 1995.

Remarks: A synonym of *A. squamigera fide* Lawson & Ustach, 2000. *Atheris laeviceps* a valid species *fide* J.-F. Trape & Roux-Estève (1995: 50).

3. *Atheris barbouri* Loveridge, 1930. Proc. New England Zool. Club 11: 107–108.

Type: Holotype, MCZ 29055, a 369 mm male (A. Loveridge, 1 Jan. 1930).

Type locality: "Dabaga, Uzungwe Mountains, southeast of Iringa, Tanganyika Territory, altitude 6000 feet" [= Dabaga, Uzungwa Mountains, Iringa Region, cen. Tanzania, 8°06'S, 35°55'E, elevation 1980 m].

Distribution: Southern Tanzania (Udzungwa and Ukinga Mts. of Dodoma, Iringa, Morogoro), 1700–1980 m.

Source: J.B. Rasmussen & Howell, 1998.

4. *Atheris broadleyi* D. Lawson, 1999. Proc. Biol. Soc. Washington 112(4): 794–797, fig. 1–2.

Type: Holotype, UTA 37801 (formerly DPL 3608), a 660 mm female (D.P. Lawson, 1 June 1994).

Type locality: "vicinity Lipondji village, East Prov., Cameroon (2°23'05"N, 15°25'41"E)" [= elevation 540 m]

Distribution: Central Africa. Southern Cameroon (Adamaoua, Est, Centre, Littoral, Nord-Ouest, Ouest, Sud), SW Central African Republic (Sangha), Gabon (Estuaire, Ngounié, Ogooué-Maritime, Woleu-Ntem) and Congo (Cuvette), 330–985 m.

Sources: D. Lawson et al., 2001 and Chirioi & LeBreton, 2007.

5. *Atheris ceratophora* F. Werner, 1896a. Verh. Zool. Bot. Ges. Wien (1895) 45: 194, pl. 5, figs. 1a–e.

Synonym: *Atheris ceratophorus* – Boulenger, 1896a.

Type: Holotype, BMNH 1946.1.18.23 (formerly BMNH 1895.4.8.5), a 420 mm female (A. Müller. 1893–1895).

Type locality: "Usambara (Deutsch-Ostafrika)" [= Usambara Mountains, Tanga Prov., SE Tanzania].

Distribution: Southeastern Tanzania (Usambara, Udzungwe and Uluguru Mts. of Iringa, Morogoro, Tanga), 700–2000 m

Sources: J. B. Rasmussen & Howell, 1982, W. Böhme, 1987, Emmrich, 1997 and Spawls et al., 2002.

6. *Atheris chlorechis* (Pel, 1851). Ned.Tijdschr. Jagtkunde 1: 172–173. (*Vipera chlorechis*)

Synonyms: *Vipera chloroechis* Schlegel, 1855 *(nomen emendatum), Toxicoa chloroëchis* Cope, 1860 *(nomen emendatum), Atheris polylepis* W.C.H. Peters, 1864c, *Atheris chloroëchis* W.C.H. Peters, 1864c *(nomen emendatum), Atheris proximus* Rochebrune, 1885, and *Atheris chloraechis* Cansdale, 1981 *(nomen emendatum).*

Type: Lectotype, RMNH 1648 (H.S. Pel, 1851), designated by M.S. Hoogmoed *in* Hughes & Barry (1969: 1030).

Type locality: "Boutry" [= Butre, near Fort Batenstein, Western Region, Ghana, 24°49'N, 1°55'W, elevation 35 m] via lectotype selection.

Distribution: Forests of West Africa. Southeastern Guinea (Macenta, Nzérékoré), Sierra Leone (Eastern, Southern), Liberia, S Ivory Coast (Abidjan, Daloa, Guiglo, Lagunes, Tabou, Taï), S Ghana (Eastern, Western) and S Togo (Plateaux), NSL–560 m.

Sources: Doucet, 1963, Leston & Hughes, 1968, Hughes & Barry, 1969, Hughes, 1983, Broadley, 1989, Ineich, 2003, Chippaux, 2006, W. Böhme et al., 2011 and Segniagbeto et al., 2011.

Remarks: Official Specific name *fide* Opinion 1634 (ICZN, 1991). Bogert (1940: 104) considered a subspecies of *A. squamiger* (Hallowell). See Remarks under *A. chloroechis* in McDiarmid et al. (1999: 355) regarding *A. chlorechis* Schlegel, 1855. Occurs in Guinea-Bissau *fide* McDiarmid et al. (1999: 355). Possibly occurs in Benin *fide* Hughes (2013: 151). Luiselli *in* Phelps (2010: 242) indicated that this species probably does not occur in Nigeria.

7. *Atheris desaixi* Ashe, 1968. J. East Africa Nat. Hist. Soc. Natl. Mus. 27(1): 53–55, figs. 1–2, 2 pls.

Type: Holotype, BMNH 1969.338 (formerly NMK 1626), a 645 mm female (F. De Saix, 3 July 1967).

Type locality: "near Chuka, Lat. 0°20'S. Long 37°.35'E... in rain forest at an altitude of c. 1,600 metres, Kenya."

Distribution: Central Kenya (cen. Eastern), 1600–1700 m.

Source: Spawls et al., 2002.

8. *Atheris hirsuta* Ernst & Rödel, 2002. Herpetol. J. 12: 55–57, figs. 2–3.

Type: Holotype, SMNS 11333, a 480 mm male (R. Ernst, 20 Sept. 2000).

Type locality: "about 6 km West of the "Station de Recherche en Ecologie Tropicale" (SRET 5°50'N 7°19'W), Taï National Park, Ivory Coast" [elevation 200 m].

Distribution: Southwestern Ivory Coast (Guiglo), 200 m. Known only from type locality.

9. *Atheris hispida* Laurent, 1955. Rev. Zool. Bot. Afr. 51(1–2): 138–139.

Synonym: *Atheris hispidus* – Meirte, 1992.

Type: Holotype, MRAC 15841, a female (A. Prigogine, 1950).

Type locality: "Lutunguru, Kivu" [= Democratic Republic of the Congo].

Distribution: East Africa. Northeastern Democratic Republic of the Congo (Orientale, Sud-Kivu), ext. SW Uganda (S Western), SW Kenya (Western) and NW Tanzania (Kagera), 800–2400 m.

Sources: Pitman, 1974, Broadley, 1998b and Spawls et al., 2002.

10. *Atheris katangensis* Witte, 1953. Explor. Parc Natl. Upemba, Rept. (6): 301–303, figs. 107a–c, pl. 3, fig. 4, pl. 21, fig. 2.

Type: Holotype, IRSNB 2207, a 397 mm male (G.-F. de Witte, 19 May 1947).

Type locality: "Mubale-Munte (région du confluent), sous-affluent et affluent de la rive droite de la Lufira (alt. 1.480 m)" [= Parc National de l'Upemba, Katanga Prov., Democratic Republic of the Congo].

Distribution: Southeastern Democratic Republic of the Congo (Katanga), 1250–1480 m.

11. *Atheris mabuensis* Branch & Bayliss, 2009. Zootaxa (2113), 44–49, figs. 2–5.

Type: Holotype, PEM 17901, a 303 mm male (native, 20 Oct. 2008).

Type locality: "main forest camp, Mount Mabu (16°17'12"S, 36°24'14"E, 1000m a.s.l., Zambezia Prov., northern Mozambique."

Distribution: Northern Mozambique (Zambézia), 1000–1550 m.

12. *Atheris matildae* Menegon, Davenport & Howell, 2011. Zootaxa (3120): 44–50, fig. 1 (left).

Type: Holotype, MTSN 9344, a 637 mm male (O. Kibure & O. Mwaipungu, Feb. 2009).

Type locality: "forest fragment in Southern Highlands of Tanzania, at about 1995 m."

Distribution: Southwestern Tanzania (SE Mbeya or SW Iringa), 2000 m. Known only from vicinity of type locality.

Remarks: Exact locality not disclosed for conservation reasons.

13. *Atheris nitschei* Tornier, 1902a. Zool. Jahrb. (1901) 15(2): 589–590, 2 figs.

Synonym: *Atheris woosnami* Boulenger, 1906b.

Types: Syntypes (2), ZMB 17669 (Eggel) and ZMB 20480 (H. Ziemann, 1900–1901).

Type locality: "Mpororosumpf, zwischen den Papyrusstauden, Deutsch-Ost-Afrika" [= Mpororo Swamp, Rwanda, near Uganda border *fide* Loveridge (1942: 313) and Broadley (1998: 121)].

Distribution: Eastern Africa. Eastern Democratic Republic of the Congo (ext. NE Katanga, Nord-Kivu, Sud-Kivu), SW Uganda (Kigezi, Toro), W Rwanda, W Burundi, NW Tanzania, 1600–2750 m.

Sources: Loveridge, 1942, Pitman, 1974, D. Lawson et al., 2001 and Spawls et al., 2002.

14. *Atheris rungweensis* Bogert, 1940. Bull. Amer. Mus. Nat. Hist. 77(1): 104–106, fig. 18. (*Atheris nitschei rungweensis*)

Type: Holotype, AMNH 39186, a 450 mm female (R. Boulton & L. Boulton, 1929).

Type locality: "Rungwe Mountains, Tanganyika Territory" [= Rungwe Mtn., Mbeya Region, SW Tanzania, ca. 9°07'S, 33°38'E].

A

Distribution: Eastern Africa. Western Tanzania (Kigoma, Mbeya, Rukwa), NE Zambia (Northern), N Malawi (Northern), 800–2000 m.

Sources: Sweeney, 1961, Broadley, 1998b, Spawls et al., 2002 and Broadley et al., 2003.

Remarks: A valid species *fide* Broadley *in* McDiarmid et al. (1999: 357) and Spawls et al. (2002: 492).

15. *Atheris squamigera* (Hallowell, 1855b). Proc. Acad. Nat. Sci. Philadelphia (1854–1855) 7(5): 193. (*Echis squamigera*)

Synonyms: *Atheris squamatus–* Cope, 1862c *(nomen incorrectum), Toxicoa squamata* – Cope, 1862c *(nomen incorrectum), Poecilostolus burtonii* A.C.L.G. Günther, 1863a, *Atheris lucani* Rochebrune, 1885, *Atheris squamiger* Boulenger, 1896a *(nomen emendatum),* and *Atheris squamigera robusta* Laurent. 1956a ,

Type: Holotype, ANSP 6949, a 546 mm specimen (M.P. Bellonni-Duchaillu).

Type locality: "Near the river Gaboon, Guinea" [= Gabon River, Gabon].

Distribution: Central Africa. Togo, S Nigeria (Akwa Ibom, Bayelsa, Cross River, Rivers), S Cameroon (Adamaoua, Centre, Est, Littoral, Nord-Ouest, Ouest, Sud, Sud-Ouest), Equatorial Guinea (Bioko Is.), Central African Republic (Haut-Mbomou, Lobaye, Mbomou, Ombella-Mpoko, Sangha), Gabon (Estuaire, Haut-Ogooué, Moyen-Ogooué, Ngounié, Ogooué-Ivindo, Ogooué-Lolo, Ogooué-Maritime, Woleu-Ntem), Congo (Brazzaville, Cuvette-Ouest, Lekoumou, Pool, Sangha), Democratic Republic of the Congo (Bandundu, Bas-Congo, Equateur, Kasai Occidental, Kasai Oriental, Kinshasa, Orientale, Sud-Kivu), N Angola (Cabinda, Lunda Norte), S Uganda (Central, Eastern, Western), SW Kenya (Western) and NW Tanzania (Kagera), NSL–1800 m.

Sources: Mertens, 1965c, Leston & Hughes, 1968, Hughes & Barry, 1969, Pitman, 1974, Hughes, 1983, J.A. Butler & Reid, 1986, D.P. Lawson, 1993, 1999, J.-F. Trape & Roux-Estève, 1995, Broadley, 1998b, D.P. Lawson & Ustach, 2000, D.P. Lawson et al., 2000, Spawls et al., 2002, J.C. Murphy & Schlager, 2003, Chippaux, 2006, Chirio & Ineich, 2006, Chirio & LeBreton, 2007, Pauwels & Vande weghe, 2008 and Segniagbeto et al., 2011.

Remarks: Togo record needs confirmation *fide* Segniagbeto et al. (2011: 351). Possibly occurs in Benin *fide* Hughes (2013: 151).

16. *Atheris subocularis* J.G. Fischer, 1888a. Jahrb. Hamburg. Wiss. Anst. 5: 5, pl. 1, figs. 2a–e, pl. 2, fig. 11.

Type: Holotype, BMNH 1946.1.20.80 (formerly BMNH 1899.12.16.139), a 543 mm specimen (J. Voss).

Type locality: "Kamerun" [= Cameroon].

Distribution: Southwestern Cameroon (Sud-Ouest), 240–300 m.

Sources: D. Lawson et al., 2001, Chirio & LeBreton, 2007 and Dobiey & Vogel, 2007.

Remarks: Possibly occurs in SE Nigeria *fide* B. Lawson (2001: 742).

ATRACTASPIS A. Smith, 1849 *in* 1838–1849 (Atractaspididae)

Synonyms: *Brachycranion* Hallowell, 1854c, *Brachychranion* – Cope, 1860b *(nomen incorrectum), Eurystephus* Cope, 1862c, *Actractaspis* – Higgins, 1873 *(nomen incorrectum), Attractaspis* – Palacky, 1898 *(nomen incorrectum), Melanelaps* Wall, 1906b, *Melanoseps* Boulenger, 1920 *(nomen emendatum), Atrataspis* – Hedges, 1983 *(nomen incorrectum),* and *Hoseraspea* Hoser, 2012L *(nomen illegitimum).*

Type species: *Atractaspis bibronii* A. Smith, 1849 *in* 1838–1849.

Distribution: Africa, Middle East and Arabia.

Sources: F. Werner, 1913, Laurent, 1945, 1950a, McDowell, 1961, 1986, Pasqual, 1962, Bourgeois, 1968, Broadley, 1968a, 1991b, Cadle, 1983, 1988, Hughes, 1983, Wallach, 1991, Meirte, 1992, Golay et al., 1993, Underwood & Kochva, 1993, Ulber, 1994, Heise et al., 1995, Spawls & Branch, 1995, David & Ineich, 1999, Duefel & Cundall, 2003, Kelly et al., 2003, 2009, Dobiey & Vogel, 2007 and Zaher et al., 2009.

1. *Atractaspis andersonii* Boulenger, 1905e. Ann. Mag. Nat. Hist. (7) 16(92): 180.

Synonym: *Melanelaps mcphersoni* Wall, 1906b.

Types: Syntypes (6), BMNH 1946.1.18.14–18 and BMNH 1946.1.18.21, longest syntype 550 mm (G.W. Bury, 1901–1903).

Type locality: "El Kubar, Upper Hushabi, South Arabia" [= El Kubar, Ad Dali' Govern., SW Yemen, 13°45'N, 44°45'E, elevation 1350 m].

Distribution: Arabia. Western Saudi Arabia (Jazan, Makkah), Oman (Dhofar) and Yemen (Abyan, Ad Dali', Aden, Ibb), NSL–1460 m.

Sources: Gasperetti, 1988, Y. Werner & Werner, 1990 and David & Ineich, 1999.

Remarks: Five syntypes listed *fide* Boulenger (1905e: 180) and Gasperetti (1988: 427) designated all five as the "holotype" but BMNH catalogue also lists 1946.1.18.21 as a type.

2. *Atractaspis aterrima* A.C.L.G. Günther, 1863c. Ann. Mag. Nat. Hist. (3) 12(71): 363.

Type: Holotype, BMNH 1946.1.20.87 (formerly BMNH 1863.10.5.13), a 361 mm female (Grant).

Type locality: "West Africa."

Distribution: West, cen. and E Africa. Southern Senegal (Kédougou, Tambacounda, Ziguinchor), Gambia (Lower River), Guinea–Bissau (Bolama Is.), Guinea (Boffa, Conakry, Nzérékoré, Télimélé), Sierra Leone (Northern, Southern, Western), Liberia (Grand Cape Mount), Ivory Coast (Aboisso, Abidjan, Adiopodoumé, Bondoukou, Bouna, Daloa, Ferkéssédougou, Lagunes), SW Mali (Sikasso), W Burkina-Faso (Centre, Hauts-Bassins, Sud-Ouest), Ghana (Accra, Ashanti, Central, Volta, Upper West, Western), Togo (Kara, Plateaux), S Benin (Zou), Nigeria (Anambra, Benue, Imo, Kaduna, Lagos, Ondo, Oyo, Plateau, Rivers), Cameroon (Adamaoua, Centre, Ouest, Sud, Sud-Ouest), SW Chad (Chari-Baguirmi), Gabon, Central African Republic (Haute-Sangha, Kemo-Gribingui, Ombella-Mpoko, Ouham), N Democratic Republic of the Congo (N Equateur, N Orientale), N Uganda (W Northern), and SE Tanzania (Iringa, Morogoro), NSL–2000 m.

Sources: Doucet, 1963, Menzies, 1966, Leston & Hughes, 1968, Hughes & Barry, 1969, Pitman, 1974, Hughes, 1983, 2013, J.B. Rasmussen, 2005a, Chippaux, 2006, Chirio & Ineich, 2006, J.-F. Trape & Mané, 2006b, Chirio & LeBreton, 2007, Segniagbeto et al., 2011, Auliya et al., 2012 and Chirio, 2013.

3. *Atractaspis battersbyi* Witte, 1959. Rev. Zool. Bot. Afr. 60(3–4): 350–351.

Type: Holotype, MRAC 20567, a 525 mm male (N'Kele, 1957).

Type locality: "proviennent de Bolobo, (Territ. de Mushie), Congo belge" [= W Equateur Prov., NW Democratic Republic of the Congo, 2°10'S, 16°14'E, elevation 315 m].

Distribution: Western Democratic Republic of the Congo (Equateur, Kinshasa), 315 m.

Remarks: A synonym of *irregularis fide* Hughes, 1983.

4. *Atractaspis bibronii* A. Smith, 1849 *in* 1838–1849. Illust. Zool. So. Africa, Rept.: pl. 71, figs. a–d, 2 pp.

Synonyms: *Atractaspis inornatus* A. Smith, 1849 *in* 1838–1849, *Atractaspis rostrata* A.C.L.G. Günther, 1868, *Atractaspis coarti* Boulenger, 1901a, *Atractaspis katangae* Boulenger, 1901a, and *Atractaspis bitronii* – Falk, 1923 *(nomen incorrectum)*.

Types: Syntypes (2), BMNH 1946.1.18.19, a female, syntypes 591 mm and 457 mm (A. Smith, 1831–1834), location of other syntype unknown.

Type locality: "eastern districts of the Cape Colony, South Africa" [= KwaZulu Natal Prov., E South Africa *fide* Broadley, 1983: 221].

Distribution: Southeastern Africa. Southern Somalia (Shabeellaha Hoose), SE Democratic Republic of the Congo (Katanga), SE Kenya (Coast), Tanzania (Iringa, Kigoma, Lindi, Mbeya, Morogoro, Mtwara, Piwani, Rukwa, Ruvuma, Tanga, Zanzibar Is.), SE Rwanda, E Burundi, Malawi (Southern), Mozambique (= Gaza, Inhambane, Manica, Maputo, Sofala, Bazaruto Is.), NE Angola (Lunda Norte), Zambia (Central, Copperbelt, Eastern, Luapula, Lusaka, Northern, North-Western, Southern, Western), Zimbabwe (Bulawayo, Manicaland, Mashonaland Central, Mashonaland East, Mashonaland West, Masvingo, Matabeleland North, Matabeleland South, Midlands), Botswana (Central, Gaborone, Ghanzi, Kgalagadi, Kgatleng, Kweneng, North West, South East, Southern), Namibia (Erongo, Karas, Khomas, Okavango, Omaheke, Oshikoto, Otjozondjupa), Swaziland and E South Africa (E Eastern Cape, Free State, Gauteng, KwaZulu-Natal, Limpopo, Mpumalanga, North West), NSL–1800 m.

Sources: Bocage, 1867a, Mertens, 1955, 1971, V.F.M. FitzSimons, 1974, Broadley, 1983, 1990b, Lanza, 1983a, 1990b, Pienaar et al., 1983, Auerbach, 1987, 1991b, Branch, 1988, Clauss & Clauss, 2002, Spawls et al., 2002, Broadley et al., 2003, J.C. Murphy & Schlager, 2003 and Broadley & Blaylock, 2013.

5. *Atractaspis boulengeri* Mocquard, 1897a. Bull. Mus. Hist. Nat., Paris 3(1): 54–55.

Synonyms: *Atractaspis matschiensis* F. Werner, 1897b, *Atractaspis schultzei* Sternfeld, 1917, *Atractaspis boulengeri mixta* Laurent, 1945, *Atractaspis boulengeri schmidti* Laurent, 1945, and *Atractaspis boulengeri vanderborghti* Laurent, 1956a.

Type: Holotype, MNHN 1896.562, a 202 mm specimen (E. Haug, 1895–1897).

Type locality: "les environs de Lambaréné, sur le bas Ogooué, Gabon, [= Lambaréné, Moyen-Ogooué Prov., Gabon, 20°42'S, 10°14'E, elevation 25 m].

Distribution: Central Africa. Southwestern Cameroon (Centre, Littoral, Sud, Sud-Ouest), S Congo (Kouilou, Sangha), Gabon (Ogooué-Ivindo, Moyen-Ogooué, Ogooué-Maritime, Woleu-Ntem), SW Central African Republic (Lobaye) and N Democratic Republic of the Congo (Bandundu, Bas-Congo, Equateur, Katanga, Nord-Kivu, Orientale), 10–770 m.

Sources: J.-F. Trape & Roux-Estève, 1995, Chirio & Ineich, 2006, Chirio & LeBreton, 2007 and Pauwels & Vande weghe, 2008.

Remarks: Original description reprinted in Mocquard (1897d: 16–17). Collector of type spelled Huaug *fide* MNHN catalogue.

6. *Atractaspis congica* W.C.H. Peters, 1877d. Mber. Königl. Akad. Wiss. Berlin 1877(10): 616, figs. 2, 2a–2c.

Synonym: *Atractaspis congica orientalis* Laurent, 1945.

Type: Holotype, ZMB 8644.

Type locality: "Chinchoxo (Westafrika)" [= Chinchoxo, Cabinda Prov., NW Angola, 5°33'S, 12°11'E, elevation 20 m].

Distribution: Central Africa. Southwestern Cameroon (Littoral, Sud, Sud-Ouest), S Congo (Brazzaville, Kouilou), S Democratic Republic of the Congo (Bandundu, Bas-Congo, Kasai Occidental, Kasai Oriental, Katanga, Kinshasa), Angola (Lunda Norte, Moxico) and N Zambia (Northern, North-Western), 10–1150 m.

Sources: Laurent, 1964a, Broadley, 1983, Branch, 1988, J.-F. Trape & Roux-Estève, 1995 and M. Griffin, 2003.

Remarks: Probably occurs in N Botswana *fide* Auerbach (1987: 180).

7. *Atractaspis corpulenta* (Hallowell, 1854c). Proc. Acad. Nat. Sci. Philadelphia (1854–55) 7(3): 99. (*Brachycranion corpulentum*)

Synonyms: *Atractaspis leucura* Mocquard, 1886, and *Atractaspis corpulenta kivuensis* Laurent, 1958.

Type: Holotype, ANSP 6902, a 508 mm specimen (H.A. Ford, 1851–1854).

Type locality: "Liberia, W. coast of Africa," (in error *fide* Hughes & Barry, 1969: 1031). Corrected to Gabon *fide* Hallowell (1857b: 70).

Distribution: West and cen. Africa. Sierra Leone, Liberia (Montserrado, Margibi), S Ivory Coast (Guiglo, Tabou), S Ghana, S Nigeria (Cross River, Rivers), SW Cameroon (Centre, Sud, Sud-Ouest), Equatorial Guinea, Gabon (Ogooué-Maritime, Woleu-Ntem), SW Central African Republic (Lobaye), Congo (Plateau, Sangha), N Democratic Republic of the Congo (Equateur, Orientale), 10–965 m.

Sources: E.H. Taylor & Weyer, 1958, Leston & Hughes, 1968, Hughes & Barry, 1969, Courtois, 1979, J.A. Butler & Reid, 1986, D. Lawson, 1993, J.-F. Trape & Roux-Estève, 1995, Chippaux, 2006 and Chirio & Ineich, 2006.

Remarks: Probably occurs in Togo *fide* and Segniagbeto et al. (2011: 353).

8. *Atractaspis dahomeyensis* Bocage, 1887b. J. Sci. Math. Phys. Nat., Lisboa 11(44): 196–197.

Synonyms: *Atractaspis caudalis* Sternfeld, 1908a, and *Attractaspis dohomeyensis* – Khole, 1991 *(nomen incorrectum).*

Type: Holotype, MBL T-108 (formerly MBL 2147), a 490 mm specimen (F. Newton, 1884–1887), destroyed by fire 18 March 1978.

Type locality: "pris à Zomaï, Dahomey" [=Zomai, Benin].

Distribution: West Africa. Southeastern Senegal (Tambacounda), Guinea, S Mali (Sikasso), Ivory Coast (Abidjan, Bouna, Katiola, Lagunes), W Burkina-Faso (Centre, Hauts-Bassins, Sud-Ouest, Volta-Noire), S Ghana (Accra), N Togo (Plateaux, Savanes), N Benin (Alibori, Atakora, Atlantique, Borgou, Zou), S Niger (Maradi), Nigeria, SW Cameroon (Sud-Ouest) and N Central African Republic (Haute-Kotto, Vakaga), 20–200 m.

Sources: Chabanaud, 1916a, Doucet, 1963, Leston & Hughes, 1968, Hughes & Barry, 1969, Stucki-Stirn, 1979, Hughes, 1983, 2013, Chippaux, 2006, J.-F. Trape & Mané, 2006b, Chirio & LeBreton, 2007 and Segniagbeto et al., 2011.

9. *Atractaspis duerdeni* Gough, 1907a. Rec. Albany Mus. 2: 178–179, 1 fig.

Type: Holotype, PEM 7938 (formerly AMG) (S. Blackbeard, 3 May 1904).

Type locality: "Serowe, in the north-eastern Kalahari" [= Serowe, Central District, Botswana, 22°23'S, 26°43'E, elevation 1135 m].

Distribution: Southern Africa. Northern Namibia (Otjozondjupa), SE Botswana (Central, Kweneng, South East, E Southern) and South Africa (Gauteng, Limpopo), 1100–1470 m.

Sources: Auerbach, 1987, Branch, 1988, Broadley, 1991b and Broadley & Blaylock, 2013.

10. *Atractaspis engaddensis* G. Haas, 1950. Copeia 1950(1): 52, fig. 1.

Synonyms: *Atractaspis engaddenis* – Efrati, 1969 *(nomen incorrectum), Attractaspis engaddensis* – Khole, 1991 *(nomen incorrectum),* and *Atractaspis engeddensis* – Kirchberg & Davidson, 1991 *(nomen incorrectum).*

Type: Holotype, HUJ, a 650 mm specimen (H. Mendelssohn, 1940s).

Type locality: "En-Geddi on the western side of the Dead Sea Palestine" [= Ein Gedi, Southern Distr., S Israel, 31°27'N, 35°23'E, elevation 305 m BSL].

Distribution: Middle East, Egypt and Arabia. Southern Israel (SE Northern, Southern, West Bank), W Jordan (Amman, Aqaba, Balqa, Karak, Madaba), NE Egypt (South Sinai) and W Saudi Arabia (Asir, Ha'il, Madinah, Makkah, Tabuk), 300 m BSL–2135 m.

Source: Gasperetti, 1988, Bouskila & Amitai, 2001, Disi et al., 2001, Amr & Disi, 2011 and Bar & Haimovitch, 2011.

11. *Atractaspis engdahli* Lönnberg & Andersson, 1913b. Arkiv Zool. 8(20): 5–6, fig. 1.

Types: Syntypes (2), NHR 2421, a 415 mm specimen (T. Engdahl), location of other syntype unknown.

Type locality: "Kismayu, near Juba river in the northeastern corner of British East Africa" [= Kismayu, Jubbada Hoose Region, S Somalia, elevation NSL].

Distribution: Southern Somalia (Jubbada Dhexe, Jubbada Hose, Shabeellaha Hoose) and NE Kenya (N North-Eastern), NSL–1150 m.

Sources: Lanza, 1983a, 1990b and Spawls et al., 2002.

12. *Atractaspis fallax* W.C.H. Peters, 1867a. Mber. Königl. Akad. Wiss. Berlin 1866(12): 890–891.

Type: Holotype, ZMB 5564 (C. von der Decken, 1864–1865).

Type locality: "Kiriame, Ostafrika" [= Giriama, Coast, Kenya, 3°49'S, 39°35'E, elevation 185 m].

Distribution: Eastern Africa. South Sudan (Eastern Equatoria), SE Ethiopia (Bale, Hararge, Shoa, Sidamo, Wollo), Somalia (Bari, Bay, Gedo, Hiraan, Jubbada Dhexe, Jubbada Hoose, Mogadishu, Mudug, Shabeellaha Dhexe, Shabeellaha Hoose, Sool, Togdheer), SE Kenya (Coast, S Eastern, Rift Valley, Lamu Is.) and ext. NE Tanzania, 185–1200 m.

Sources: Hughes, 1983, Lanza, 1983a, 1990b, Largen & Rasmussen, 1993 and Largen & Spawls, 2010.

13. *Atractaspis irregularis* (J.T. Reinhardt, 1843). Kgl. Danske Vidensk. Selsk. Natur. Math. Afh. 10: 264–266, pl. 3, figs. 1–3. (*Elaps irregularis*)

Synonyms: *Atractaspis bipostocularis* Boulenger, 1905a, *Atractaspis conradsi* Sternfeld, 1908a, *Atractaspis schoutedeni* Witte, 1930b, *Atractaspis babaulti* Angel, 1934a, *Atractaspis irregularis loveridgei* Laurent, 1945, *Atractaspis irregularis parkeri* Laurent, 1945, *Atractaspis irregularis uelensis* Laurent, 1945, *Atractaspis irregularis angeli* Laurent, 1950, and *Atractaspis coalescens* Perret, 1960.

Type: Holotype, ZMUC 6885, a 349 mm specimen (J.R. Chenon, 1826–1836).

Type locality: "Guinea" [= coast of Ghana *fide* Hughes & Barry, 1969: 1027]. Restricted to vicinity of Accra, Ghana *fide* J.B. Rasmussen & Hughes (1997: 14, 16).

Distribution: West, cen. and E Africa. Guinea (Macenta, Nzérékoré, Télimélé), Sierra Leone, Liberia, Ivory Coast, S Ghana, S Togo (Plateaux), S Nigeria (Ondo, Rivers), S Cameroon (Centre, Est, Nord-Ouest, Ouest, Sud, Sud–Ouest), Gabon, Congo (Bouenza, Brazzaville, Kouilou, Plateau, Pool), Central African Republic (Bamingui-Bangoran, Haute-Sangha, Lobaye, Nana-Mambere, Ombella-Mpoko, Ouaka, Sangha), Democratic Republic of the Congo (Equateur, Kinshasa, Maniema, Nord-Kivu, Orientale, Sud-Kivu), N Angola, Uganda (Central, Eastern, Northern, Western), Rwanda, Burundi, cen. Kenya (Central, Nyanza, S Rift Valley, Western), NW Tanzania (Kigoma), S South Sudan, Eritrea (Northern Red Sea) and Ethiopia (Hararge, Kefa, Sidamo, Wollega), NSL–2000 m.

Sources: Bourgeois, 1963b, Aylmer, 1922, Leston & Hughes, 1968, Hughes & Barry, 1969, Pitman, 1974, Hughes, 1983, Largen & Rasmussen, 1993, 1999, J.-F. Trape & Roux-Estève, 1995, J.B. Rasmussen & Hughes, 1997, Spawls et al., 2002, Chippaux, 2006, Chirio & Ineich, 2006, Chirio & LeBreton, 2007, Largen and Spawls, 2010, Segniagbeto et al., 2011 and Chirio, 2013.

Remarks: *Atractaspis coalescens* Perret considered a synonym based on Hughes (pers. comm.) and Wallach (unpubl. data).

14. *Atractaspis leleupi* Laurent, 1950. Rev. Bot. Afr. 43(4): 351–352. (*Atractaspis congica leleupi*)

Type: Holotype, RGMC 18700, a 390 mm female (N. Leleup, Jan. 1950).

Type locality: "Station d'études des l'IRSAC sur le massif des Kundelungu (1.750 m.), Champ de termitières" [= Katanga Prov., SE Democratic Republic of the Congo]. Emended to Station d'études de la Seram, plateau des Kundelungu, Haut Katanga *fide* Laurent (1956a: 310).

Distribution: Southeastern Democratic Republic of the Congo (Katanga), 1750 m.

Sources: Witte, 1953 and Broadley & Cotterill, 2004.

15. *Atractaspis leucomelas* Boulenger, 1895a. Ann. Mus. Civ. Stor. Nat. Genova (2) 15: 16, pl. 4, fig. 2.

Type: Holotype, MSNG 28847, a 575 mm female (V. Bottego, 16-20 Oct. 1892).

Type locality: "Ogaden, Somali" [= Ogaden, Hararge Prov., E Ethiopia, 6°58'N, 44°02'E, elevation 780 m].

Distribution: Northeast Africa. Eastern Ethiopia (Hararge), Djibouti and ext. NW Somalia (Awdal), 285–780 m.

Sources: Lanza, 1983a, 1990b, Largen & Rasmussen, 1993 and Largen & Spawls, 2010.

16. *Atractaspis magrettii* Scortecci, 1928. Atti Soc. Ital. So. Nat. Mus. Civ. Stor. Nat. Milano 67(3–4): 308–309, fig. 6.

Type: Holotype, MSNM 1796, a male (P. Magretti, Jan.–Mar. 1900).

Type locality: "Mandafená, Eritrea."

Distribution: Eastern Sudan (Ash Shaqiyah) and Eritrea (Gash-BarkaNorthern Red Sea, Southern Red Sea), South Sudan (Eastern Equatoria) and W Ethiopia (Illubabor), 400–2400 m.

Sources: Largen & Rasmussen, 1993, Largen, 1997 and Largen & Spawls, 2010.

Remarks: Possibly occurs in N Somalia *fide* Largen & Rasmussen (1993: 333).

17. *Atractaspis microlepidota* A.C.L.G. Günther, 1866. Ann. Mag. Nat. Hist. (3) 18(103): 29, pl. 7, fig. c.

Type: Holotype, BMNH 1946.1.18.6 (formerly BMNH 1866.4.24.2), a 508 female (A.C.L.G. Günther, 1857–1866).

Type locality: "probably...West Africa."

Distribution: Extreme W Africa. Extreme S Mauritania (Guidimaka, Trarza), Senegal (Dakar, Fatick, Kaolack, Thiès), and Gambia, NSL–70 m.

A

Sources: Hughes, 1983, Broadley, 1994, J.-F. Trape & Mané, 2000, 2006b, Spawls et al., 2002, Chippaux, 2006 and J.-F. Trape et al., 2006.

18. *Atractaspis micropholis* A.C.L.G. Günther, 1872b. Ann. Mag. Nat. Hist. (4) 9(49): 36–37, pl. 3, fig. e.

Type: Holotype, BMNH 1946.1.18.7 (formerly BMNH 1865.7.28.5), a 330 mm female (S.G.J. Mivart, 1862–1865).

Type locality: "Africa." Restricted to N Nigeria *fide* J.-F. Trape et al. (2006: 8).

Distribution: West Africa. Senegal (Dakar, Fatick, Kédougou, Louga, Tambacounda, Thiès), NW Burkina Faso (Volta-Noire), S Mali (Koulikoro, Sikasso), S Niger (Maradi, Zinder) and N Nigeria (Bauchi, Borno, Sokoto), NSL–830 m.

Sources: Papenfuss, 1969, Pitman, 1974, Hughes, 1983, Largen & Rasmussen, 1993, Chippaux, 2006, J.-F. Trape & Mané, 2006b and J.-F. Trape et al., 2006.

19. *Atractaspis phillipsi* T. Barbour, 1913b. Proc. Biol. Soc. Washington 26: 148.

Type: Holotype, MCZ 8782 a 312 mm juvenile (J.C. Phillips & G.M. Allen, Feb. 1913).

Type locality: "Singa, Prov. of Sennaar, eastern Anglo-Egyptian, Sudan" [= Singa, Sennar Prov., Sudan, 13°11'N, 33°57'E, elevation 430 m].

Distribution: Southeastern Sudan (Blue Nile, Sennar), 430 m.

Source: Broadley, 1994.

20. *Atractaspis reticulata* Sjöstedt, 1896. Zool Anz. 19: 516–517.

Synonyms: *Atractaspis heterochilus* Boulenger, 1901a, and *Atractaspis reticulata brieni* Laurent, 1956.

Type: Holotype, ZMUU 1796, an 800 mm specimen (native, Feb. 1892).

Type locality: "Kamerun, Ekundu" [= Ekundu, Cameroon] (in error). [= Okundi, Cross River State, Nigeria *fide* Ineich & LeBreton (2007: 636), 6°24'N, 8°48'E, elevation 185 m].

Distribution: Central Africa. Southeastern Nigeria (Cross River), S Cameroon (Centre, Est, Littoral, Nord, Nord-Ouest, Ouest, Sud, Sud-Ouest), SW Central African Republic (Lobaye), Gabon (Ogooué-Ivindo, Ogooué-Maritime), N Congo (Sangha) and Democratic Republic of the Congo (Bandundu, Bas-Congo, Equateur, Kasai Occidental, Kasai Oriental, Kinshasa, Nord-Kivu, Orientale, Sud-Kivu), NSL–1300 m.

Sources: Roux-Estève, 1965, D. Lawson, 1993, Chippaux, 2006 and Chirio & LeBreton, 2007.

Remarks: Supplemental original description in Sjöstedt (1897: 28–29, pl. 1, figs. 1a-d, pl. 3). Occurs in Ghana *fide* Hughes (1983: 342).

21. *Atractaspis scorteccii* H.W. Parker, 1949. Zool. Verh. (6): 109–111.

Type: Holotype, BMNH 1949.2.3.9, an 840 mm male.

Type locality: "Haud, 46°20' E. X 8°15' N., 2100 ft., Somaliland" [= Haud, Togdheer Region, N Somalia, elevation 675 m].

Distribution: Eastern Ethiopia (Hararge) and N Somalia (Bari, Togdheer, Woqooyi Galbeed), 600–1100 m.

Sources: Cherchi, 1958, Lanza, 1983a, 1990b, Largen & Rasmussen, 1993 and Largen & Spawls, 2010.

22. *Atractaspis watsonii* Boulenger, 1908g. Ann. Mag. Nat. Hist. (8) 2(7): 94.

Synonym: *Atractaspis nigra* Pelligrin, 1909.

Type: Holotype, BMNH 1946.1.18.5, a 570 mm female (C.F. Watson, 1906–1908).

Type locality: "Sokoto, Upper Niger, Africa" [= Sokoto, Sokoto State, NW Nigeria, 13°03'N, 5°15'E, elevation 295 m].

Distribution: West Africa. Extreme S Mauritania (Guidimaka, Hodh El Gharbi), N Senegal (Saint-Louis), S Mali (Kayes, Koulikoro, Mopti, Ségou), Burkina Faso (Centre, Centre-Nord, Centre-Sud, Est, Sahel, Volta-Noire), S Niger (S Maradi, S Tahoua), N Nigeria (Gongola, Sokoto), SW Chad (Mayo-Kebbi), N Cameroon (Adamaoua, Extreme-Nord, Nord), N Central Africa Republic (Vakaga) and SW Sudan (Janub Darfur), 100–670 m.

Sources: Chirio & Ineich, 2006, J.-F. Trape & Mané, 2006b, J.-F. Trape et al., 2006 and Chiro & LeBreton, 2007.

Remarks: Previously a synonym of *A. microlepidota*. Possibly occurs in Benin *fide* Hughes (2013: 153).

ATRACTUS Wagler, 1828b (Dipsadidae)

Synonyms: *Brachyura* Kuhl & Hasselt, 1822b (*nomen rejiciendum*), *Urobrachys* Fitzinger, 1843, *Isoscelis* A.C.L.G. Günther, 1858, *Isocelis* – Hoffmann, 1890 (*nomen incorrectum*), *Attractus* – Palacky, 1898 (*nomen incorrectum*), *Atractis* – Tornier, 1904 (*nomen incorrectum*), *Atractopsis* Despax, 1910, and *Atractopisis* – Dunn, 1928b (*nomen incorrectum*).

Type species: *Atractus trilineatus* Wagler, 1828b.

Distribution: Lower Central America and N South America.

Sources: Savage, 1960, Roze, 1961, J.A. Peters & Orejas-Miranda, 1970, Dixon & Soini, 1977, 1986, Hoogmoed, 1980, Cadle, 1984b, 1985, Pérez-Santos & Moreno, 1988, 1991, Martins & Oliveira, 1993, G. Köhler et al., 2001, Schargel & García-Perez, 2002, Schargel & Castoe, 2003, Zaher et al., 2005, 2009, Passos, 2008, Passos et al., 2009d–e, 2010 and Passos & Lynch, 2010.

Remarks: Official Generic Name *fide* Opinion 2210 (ICZN, 2008b).

1. *Atractus acheronius* Passos, Rivas-Fuenmayor & Barrio-Amorgós, 2009b. Amphibia-Reptilia 30(2): 234–236, figs. 1–2.

Type: Holotype, MHNLS 398, a 581623 mm female (C. Alemán, 1950).
Type locality: "upper Río Negro Valley (2200–2400), Sierra de Perijá, municipality of Machiques de Perijá (09°52'N, 72°48'W), state of Zulia, Venezuela."
Distribution: Venezuela (Zulia), 2400 m. Known only from type locality.
Sources: Alemán, 1953 and Passos et al., 2009b.
Remarks: Type erroneously listed as MHNSL 1760 *fide* Alemán (1953: 217). Type locality elevation listed as 2400 m *fide* Alemán (1953: 217).

2. *Atractus albuquerquei* Cunha & Nascimento, 1983b. Bol. Mus. Para. Emílio Goeldi (Zool.) (2) (123): 6–8, figs. 1–3, pl. 1, fig. 1.

Type: Holotype, MPEG 12946, a 418–420 mm male.
Type locality: "Vila Nova, PA-256, próximo ao rio Timboteua, rodovia Tomé-Acu-Parago minas" [= Vila Nova, on the road PA-256, in forested area south of the river Guamá, near the river Timboteua, between Tomé-Açu and Paragominas, Pará State, Brazil, 2°41'S, 47°55'W, elevation 50 m *fide* Zaher et al., 2005: 20].
Distribution: Central Brazil (Acre, Goiás, Mato Grosso, Mato Grosso do Sul, E Pará, Rondônia), 50–735 m.
Source: Zaher et al., 2005.
Remarks: Photograph of type in Zaher (2005: fig. 1).

3. *Atractus alphonsehogei* Cunha & Nascimento, 1983b. Bol. Mus. Para. Emílio Goeldi (Zool.) (2) (123): 25–27, figs. 2–3, pl. 2, fig. 1.

Type: Holotype, MPEG 14928, a 239 mm male.
Type locality: "Bela Vista, 75 km da PA-242, rodovia Braganca-Viseu" [= Pará State, Brazil].
Distribution: Colombia (Amazonas) and Brazil (Maranhas, Pará).
Sources: Cunha & Nascimento, 1984, Martins & Oliveira, 1993 and Prudente & Passos, 2008.

4. *Atractus altagratiae* Passos & Fernandes, 2008. Zootaxa (1849): 60–63, figs. 1–6.

Type: Holotype, MNRJ 7888, a 275 mm male (H. Sick, 1957).
Type locality: "upper Cururú River (07°12'S, 58°04'W, ca. 44 m), tributary of Teles Pires River, Municipality of Itaituba, State of Pará, Brazil."
Distribution: Northeastern Brazil (Pará), 45 m.

5. *Atractus andinus* Prado, 1944. Ciencia 5(4–5): 111, fig. 1.

Type: Holotype, CSJ 516 (formerly CSJ 231), a 287 mm female (R.H. Daniel, 1943).
Type locality: "Andes, Colombia."
Distribution: Central Colombia (Antioquia), 1535 m. Known only from type locality.
Source: Passos et al., 2009c.
Remarks: Original description reprinted in Prado (1946b: 109, fig. 1)

6. *Atractus apophis* Passos & Lynch, 2010. Herpetol. Monogr. (24): 150–152, figs. 1a–c, 2a–b.

Type: Holotype, ICNMNH 10822, a 433 mm male (A. Bello, 30 Oct. 1993).
Type locality: "Parque Arqueológico San Augustín, municipality of San Augustín (01° 53' N, 76° 16' W, ca. 1640 m), department of Huila, Colombia."
Distribution: Colombia (Huila), 1500–1640 m. Known only from type locality.

7. *Atractus arangoi* Prado, 1940b. Mem. Inst. Butantan (1939) 13: 15–16, fig. 1.

Type: Holotype, ILS 136, a 395 mm female (R.P. Fray Miguel, 1938–1939), destroyed by fire 9–10 April 1948.
Type locality: "Colombia."
Distribution: Southern Colombia (Putumayo), 255 m.
Source: Nicéforo-Maria, 1942.

8. *Atractus atratus* Passos & Lynch, 2010. Herpetol. Monogr. (24): 154–156, figs. 5, 6a–b.

Type: Holotype, IAvH 151, a 552 mm female.
Type locality: "Parque Nacional Natural Cueva de Los Guacharos (ca. 1800 m), municipality of Acevedo, department of Huila, Colombia."
Distribution: Colombia (Huila), 1700–2000 m. Known only from vicinity of type locality.

9. *Atractus attenuatus* C.W. Myers & Schargel, 2006. Amer. Mus. Novit. (3532): 2–6, figs. 1–3.

Type: Holotype, AMNH 19998, a 420 mm male (H. Nicéforo-María, 16 May 1921).
Type locality: "Sabanalarga, on Cauca River, [6°51'N, 75°49'W, Department of Antioquia], Colombia. The type locality, Sabanalarga (Nicéforo María, 1942: 87, map), lies on the east bank of the Río Cauca at an elevation of about 1000 m (Paynter, 1997: 372), in the northern end of the Cordillera Central."
Distribution: Colombia (Antioquia), 1000 m. Known only from type locality.
Source: Passos et al., 2009c.

A

10. *Atractus avernus* Passos, Chiesse, Torres-Carvajal & Savage, 2009d. Herpetologica 65(4): 397–398, fig. 8.

Type: Holotype, MLS 2725, a 177 mm female (H. Nicéforo-María).

Type locality: "Paraíso (01° 14' N, 75° 37' W, ca. 480 m), municipality of Florencia, Department of Caquetá, Colombia."

Distribution: Colombia (Caquetá), 480 m.

11. *Atractus badius* (H. Boie *in* F. Boie, 1827). Isis von Oken 20(6): 540. (*Brachyorrhos badius*)

Synonyms: *Brachyorrhos badius* H. Boie *in* Schlegel, 1826a *(nomen nudum), Rabdosoma badium rubinianum* Jan, 1862b, *Rabdosoma badium subbicinctum* Jan, 1862b, *Rabdosoma dubium* Jan, 1862b, and *Atractus micheli* Mocquard, 1904a.

Type: Lectotype, RMNH 120C, a 338 mm female, designated by Hoogmoed (1980: 11).

Type locality: "Java" [W Indonesia] (in error). Corrected to Guyane [= Guianas] *fide* RMNH catalogue and specimen label. Restricted to Cayenne, French Guiana *fide* Savage (1960: 80) and Paramaribo, Suriname *fide* Hoogmoed (1980: 11).

Distribution: Amazonian South America. Colombia (Antioquia, Boyacá, Caldas, Cundinamarca, La Guejira, Magdelena, Norte de Santander, Valle del Cauca), Venezuela (Carabobo, Distrito Federal), Guyana, Suriname (Marowijne, Suriname), French Guiana, Ecuador, Peru (Cusco, Loreto, Madre de Dios) and NE Brazil (Pará), NSL–3005 m.

Sources: Roze, 1966a, Lancini, 1986, Hoogmoed, 1979, 1980, Pérez-Santos & Moreno, 1988, 1991, Cei, 1994, Starace, 1998, Doan & Arriaga, 2000, Kok, 2006 and Passos et al., 2009c.

Remarks: Original description based on H. Boie's MS (1823–1825).

12. *Atractus balzani* Boulenger, 1898k. Ann. Mus. Civ. Stor. Nat. Genova (2) 19(39): 129–130.

Type: Holotype, MSNG CE 28873, a 400 mm specimen (L. Balzan, 1892).

Type locality: "Missiones Mosetenes, Bolivia" [= Campanario de Covendo, La Paz Department, Bolivia, 15°47'S, 66°59'W. elevation 515 m].

Distribution: Bolivia (La Paz), 515 m. Known only from type locality.

13. *Atractus biseriatus* Prado, 1941d. Mem. Inst. Butantan (1940) 14: 26, pl., lower fig.

Type: Holotype, ILS 187, a 237 mm male (N. Díaz-Berrío), destroyed by fire 9–10 April 1948.

Type locality: "Manizales, Colombia" [= 5°04'N, 75°31'W, elevation 2150 m] (in error). Corrected to Villamaría, Colombia *fide* Nicéforo-Maria (1942: 92) and E.R. Dunn *in* Savage (1960: 80) [= Villamaría, Caldas Department, Colombia, 5°03'N, 75°31'W, elevation 1950 m].

Distribution: Central Colombia (Caldas), 1950–2150 m. Known only from type locality.

Sources: Nicéforo-Maria, 1942, Passos & Arredondo, 2009 and Passos et al., 2009c.

14. *Atractus bocki* F. Werner, 1909b. Mitt. Naturhist. Mus. Hamburg 26: 228, fig. 5.

Synonym: *Atractus canedii* Scrocchi & Cei, 1991.

Type: Holotype, formerly ZMH, a 365 mm specimen (C.A. Bock, 1904–1909), destroyed in July 1943 during World War II.

Type locality: "Cochabamba, Bolivia" [= Cochabamba, Cochabamba Department, Bolivia, 17°23'S, 66°10'W, elevation 2575 m].

Distribution: Bolivia (Cochabamba) and NW Argentina (Jujuy, Salta, Tumbaya), 350–2575 m

Sources: Scrocchi & Cei, 1991, Cei, 1994, R. Fernandes, 1995b and Passos et al., 2005, 2007, 2009a.

Remarks: Passos et al., 2009a considered *A. canedii* a synonym.

15. *Atractus bocourti* Boulenger, 1894a. Cat. Snakes Brit. Mus. 2: 306.

Type: Holotype, BMNH 1946.1.2.24, a 390 mm female (H.J. Veitch, 1890–1894).

Type locality: "Acomayo, N. Peru" [= Acomayo, Huanuco Department, Peru, 9°46'S, 76°05'W, elevation 2200 m].

Distribution: Northwestern South America. Colombia, Ecuador and N Peru (Huanuco, Loreto), 90–2200 m.

Sources: Nicéforo-Maria, 1942 and K.P. Schmidt & Walker, 1943c.

16. *Atractus boettgeri* Boulenger, 1896a. Cat. Snakes Brit. Mus. 3: 645.

Type: Holotype, BMNH 1946.1.6.29, a 350 mm female (F.H. Rolle, 1864–1896).

Type locality: "Yungas, Bolivia" [= Sierra de las Yungas, Cochabamba Department, Bolivia].

Distribution: Central Bolivia (Beni, Cochabamba), 195 m.

Sources: C.J. McCoy, 1971, Zaher et al., 2005 and Salazar-Bravo et al., 2010.

Remarks: Type is illustrated in Zaher et al. (2005: 28).

17. *Atractus boulengerii* Peracca, 1896b. Boll. Mus. Zool. Anat. Comp. Univ. Torino 11(252): 1, 1 fig.

Type: Holotype, MSNTO 1832 (formerly MRSN 1727), a male.

Type locality: "America meridionale" [= South America].

Distribution: Columbia (Valle de Cauca), NSL–100 m.

Source: Passos et al., 2009e.

A

18. *Atractus caete* Passos, Fernandes, Bérnils & Moura-Leite, 2010. Zootaxa (2364): 7–9, figs. 1, 2a, 3a–b.

Type: Holotype, MNRJ 16936, a 408 mm female (A.R. Melgarejo, 1986).
Type locality: "municipality of Quebrângulo (09°19'S, 36°28'W, ca. 360 m), state of Alagoas, Brazil."
Distribution: Northeastern Brazil (Alagoas), 300–500 m.

19. *Atractus carrioni* H.W. Parker, 1930a. Ann. Mag. Nat. Hist. (10) 5(26): 208–209, 2 figs.

Type: Holotype, BMNH 1946.1.1.70 (formerly BMNH 1929.10.30.1), a 480 mm female (C. Carrión-Mora, 1929).
Type locality: "Loja, Ecuador (2200 metres)" [= Loja, E Loja Prov., S Ecuador, 3°59'S, 79°12'W, 2200 m].
Distribution: Southern Ecuador (Loja), 1435–2500 m. Known only from vicinity of type locality.
Source: H.W. Parker, 1932a.

20. *Atractus caxiuana* Prudente & Santos-Costa, 2006. Zootaxa (1285): 22–27, figs. 1–3.

Type: Holotype, MPEG 19964, a 249 mm male.
Type locality: "'Estação Científica Ferreira Penna', Flona de Caxiuanã, Pará State, Brazil (1°42'33"S 51°31'45"W), 0 m."
Distribution: Northeastern Brazil (Pará), NSL.
Sources: Passos & Fernandes, 2008 and Prudente & Passos, 2008.
Remarks: Holotype erroneously listed as MPEG 19657 *fide* Prudente & Passos (2010: 402).

21. *Atractus charitoae* Silva-Haad, 2004. Rev. Acad. Colomb. Cienc. 28(108): 417–418, figs. 7–9.

Type: Holotype, ICNMNH 10095 (formerly JSH 1162), a 255 mm male (native, 1989).
Type locality: "Taraira, Colombia, departamento del Vaupés. Región limítrofe entre Colombia y Brasil, en las márgenes del río del mismo nombre (00° 29' 38" N 69° 40' 11" E.)" [elevation 120 m].
Distribution: Colombia (Vaupés), 120 m. Known only from type locality.

22. *Atractus chthonius* Passos & Lynch, 2010. Herpetol. Monogr. (24): 157–159, figs. 7a–c, 8a–b.

Type: Holotype, ICN 5662, a 328 mm male.
Type locality: "Finca Meremburg, Santa Leticia (02° 14' N, 76° 10' W, ca. 2400 m), between municipalities of Popayán in the department of Cauca and La Plata in the department of Huila, Colombia."
Distribution: Colombia (Cauca, Huila), 1500–2400 m.

23. *Atractus clarki* Dunn & Bailey, 1939. Bull. Mus. Comp. Zool. 86(1): 8–9.

Type: Holotype, MCZ 28800, a 313 mm female (native, 1938).
Type locality: "Mine at Santa Cruz de Cana, Darien, Panamá" [= ca. 7°46' N, 77°41' W, 500 m *fide* C.W. Myers, 2003: 10].
Distribution: Eastern Panama (Darién) and NW Colombia (Antioquia, Chaco, Valle de Cauca), 100–1500 m.
Sources: C.W. Myers, 2003, Passos & Arredondo, 2009, and Passos et al., 2009e.

24. *Atractus collaris* Peracca, 1897e. Boll. Mus. Zool. Anat. Comp. Univ. Torino 12(284): 4–5, 1 fig.

Type: Holotype, not designated, location unknown.
Type locality: "a cononacco piccolo villaggio sul Rio Cononaco chef a parte dell'alto bacino idrografico delle Amazzoni. Cononacco si trova a nord ovest di Yquitos e fa parte del Perù, Perù orietale" [= village on Río Cononaco, NW of Iquitos (elevation 200 m), Loreto Department, E Peru].
Distribution: Southeastern Colombia (Amazonas), E Ecuador, and NE Peru (Loreto), 100–200 m.
Sources: Savage, 1955, Dixon & Soini, 1986, Prudente & Passos, 2008, and Vigle, 2008.
Remarks: Type not in MSNTO *fide* Andreone & Gavetti (2007: 87).

25. *Atractus crassicaudatus* (A.-M.-C. Duméril, Bibron & Duméril, 1854a). Erpét. Gén. 7(1): 103–104. (*Rabdosoma crassicaudatum*)

Synonym: *Atractus fuhrmanni* Peracca, 1914.
Types: Syntypes (4), MNHN 601, MNHN 7205, MNHN 1999.7938, and USNM 11138 (formerly USNM 387), longest syntype 420 mm (J.P. Goudot & Riéfer, 1822–1825 or 1837–1842).
Type locality: "la Nouvelle-Grenade...dans les environs de Bogota" [= Bogotá, Cundinamarca Department, Colombia, 4°36'N, 74°05'W, elevation 2630 m].
Distribution: Panama, Colombia (Antioquia, Boyacá, Caldas, Cundinamarca, Santander, Valle del Cauca) and Venezuela, 1000–3500 m.
Source: Auth, 1994.
Remarks: Cochran (1961: 208) listed "New Granada (Panama-Colombia), JPP" as data for the USNM syntype. Nicéforo-María (1942: 93) and Pérez-Santos & Moreno (1988: 72) listed Venezuela as part of Distribution.

26. *Atractus darienensis* C.W. Myers, 2003. Amer. Mus. Novit. (3391): 16–20, figs. 1b, 2b, 3b, 6, 9a.

Type: Holotype, KU 110274 (formerly CWM 6095), a 346 mm female (C.W. Myers, 25 Jan. 1966).
Type locality: "500 m above sea level on the north end of the Serranía de Pirre, Prov. of Darién, eastern Panama.

A

The type locality, a temporary forest camp, is situated roughly at 8°00'N, 77°43'W."

Distribution: Eastern Panama (Darién), 500 m. Known only from type locality.

27. *Atractus depressiocellus* C.W. Myers, 2003. Amer. Mus. Novit. (3391): 20–22, figs. 1c, 2c, 3e.

Type: Holotype, AMNH 119876, a 750 mm female (G. Barratt, Jr., 31 Oct. 1974).

Type locality: "Cerro Azul (Cerro Jefe) region, Prov. of Panamá, Panama. The type locality is in the general region of Cerro Jefe on the Piedras-Pacora Ridge, at about 9°14'N, 79°23'W."

Distribution: Eastern Panama (Panamá), 200–800 m.

Source: C.W. Myers & Schargel, 2006.

28. *Atractus duboisi* (Boulenger, 1880). Bull. Soc. Zool. France 5: 44. (*Rabdosoma duboisi*)

Type: Holotype, IRSNB 2008, a 335 mm male (E. De Ville, 1874).

Type locality: "Andes de l'Équateur" [= Andes of Ecuador].

Distribution: Montane cloud forest of Ecuador (Imbabura, Napo, Morona-Santiago), 1500–2200 m.

Sources: Passos & Arredondo, 2009 and Passos et al., 2009d.

29. *Atractus duidensis* Roze, 1961. Acta Biol. Venez. 3(7): 110–111.

Type: Holotype, AMNH 36609, a 355 mm male (Tyler Exped., 28 Jan. 1929).

Type locality: "Cumbre del Cerro Duida, Territorio Federal Amazonas, Venezuela; unos 2050 metros de altura" [= Cerro Duida Mountain, Amazonas Territory, S Venezuela, ca. 3°21'N, 65°42'W, elevation 2050 m].

Distribution: Southern Venezuela (cen. Amazonas), 2000–2150 m.

Sources: Roze, 1966a and Lancini, 1986.

30. *Atractus dunni* Savage, 1955. Proc. Biol. Soc. Washington 68: 14–15.

Synonym: *Rhabdosoma maculatum* Bocourt, 1883 *in* A.H.A. Duméril, Bocourt & Mocquard, 1870–1909 *(nomen praeoccupatum)*.

Type: Holotype, MNHN 5986 (formerly PM 5986), a 305 mm female.

Type locality: "l'Équateur" [= Ecuador].

Distribution: Northeastern Ecuador (Cotopaxi, Pastaza, Pichincha).

Source: Cisneros-Heredia, 2005a.

Remarks: MNHN 5986 listed as type of *Rabdosoma maculatum* Bocourt *fide* MNHN catalogue. Records from Peru are erroroneous *fide* Cisneros-Heredia (2005a: 93).

31. *Atractus echidna* Passos, Mueses-Cisneros, Lynch & Fernandes, 2009e. Zootaxa (2293): 8–10, fig. 5.

Type: Holotype, UV-C 7718, a 248 mm male.

Type locality: "eastern San Antonio, Flor de la Briza, corregimiento Robles (01°42'N, 78°42'W, ca. 10 m), municipality of Tumaco, department of Nariño, Colombia."

Distribution: Southwestern Colombia (Nariño), 10 m. Known only from type locality.

32. *Atractus ecuadorensis* Savage, 1955. Proc. Biol. Soc. Washington 68: 15–16.

Type: Holotype, FMNH 23529 (formerly CM 23529), a 198 mm male (R.W. Chadwick, March 1936).

Type locality: "'Llangate area', Ecuador (probably refers to the Llanganate Distribution of eastern Tungurahua Prov.)."

Distribution: Central Ecuador (Tungurahua). Known only from type locality.

Source: Passos et al., 2012.

Remarks: Type locality listed as Ecuador: Oriente Prov. (now = Napo, Orellana, Pastaza, Morona-Santiago, Sucumbios and Zamora-Chinchipe) *fide* FMNH catalogue.

33. *Atractus edioi* N.J. Silva, Rodrigues-Silva, Ribeiro, Souza & Souza, 2005. Pap. Avul. Zool. 49(3): 34–36, figs. 2a–b.

Type: Holotype, MZUSP 13371, a 390 mm female (C. do Amaral-Sousa, 29 Jan. 2002).

Type locality: "Brasil: Goiás: Município de Minaçu, UHE Cana Brava (12°25'17"S, 48°09'43"W).

Distribution: Northeastern Brazil (Goiás).

34. *Atractus elaps* (A.C.L.G. Günther, 1858). Cat. Colub. Snakes Brit. Mus.: 241–242. (*Rhabdosoma elaps*)

Synonyms: *Rabdosoma brevifrenum* Jan, 1862b, *Atractus elaps tetrazonus* Amaral, 1931a, and *Geophis diplozeugus* K.P. Schmidt & Walker, 1943a.

Type: Holotype, BMNH 1946.1.6.45, a 381 mm specimen (L. Fraser, 1857–1858).

Type locality: "Guayaquil" [Ecuador]. (probably in error *fide* J.A. Peters (1960a: 506) and Savage (1960: 81).

Distribution: Northern Amazonia. Eastern Colombia (Amazonas, Boyacá, Caqueta, Cauca, Cundinamarca, Meta, Narino, Putumayo, Vaupés), S Venezuela (Amazonas), E Ecuador (Morona-Santiago, Napo, Orellana, Pichincha, Sucumbios, Zamora-Chinchipe), W Brazil (Amazonas, Rondonia) and N Peru (Amazonas, Loreto, Madre de Dios), 100–1640 m.

Sources: Roze, 1966a, Hoogmoed, 1980, Lancini, 1986, Duellman & Salas, 1991, Abuys, 2003, Silva-Haad, 2004, Duellman, 2005, Kok, 2006, Vigle, 2008 and Navarrete et al., 2009.

Remarks: Possibly occurs in Suriname *fide* Hoogmoed (1980: 16) and Guyana *fide* Starace (1998: 125). Kok (2006: 28) noted that the *A. elaps* of Silva-Haad, 2004, are *A. poeppigi.*

35. *Atractus emigdioi* Gonzáles Sponga, 1971. Monogr. Cientif. Inst. Pedagogico (3): 3–4, figs. 2–5.

Type: Holotype, MCNC 5629, a 415 mm female (E. Gonzáles-Sponga, 20 Aug. 1971).
Type locality: "Campamento del M.O.P., a 19 km. de Boconó, en la vía Valera-Trujillo, Estado Trujillo, Venezuela...a 2.100 mts. de altura."
Distribution: Western Venezuela (Lara, Mérida, Trujillo), 2100 m.
Sources: Lancini, 1986 , Barros-Blanco, 2000, Esqueda et al., 2001, Schargel & Castoe, 2003, Esqueda & La Marca, 2005 and Navarrete et al., 2009.

36. *Atractus emmeli* (Boettger, 1888b). Ber. Senck. Naturf. Ges. 1888: 192–195, 3 figs. (*Geophis emmeli*)

Type: Lectotype, SMF 19364 (SMF-B 8311, 1a), a 303 mm male (F. Emmel, 1888), designated by Mertens (1967a: 94).
Type locality: "Mapiri, Nebenfluss des oberen Beni, Bolivien" [= Río Mapiri, tributary of Río Beni, La Paz Department, Bolivia, ca. 15°25'S, 67°49'W, 500 m] via lectotype selection.
Distribution: Northwestern Bolivia (Beni, La Paz) and SE Peru (Arequipa, Cusco, Junin, Loreto, Madre de Dios), 500–2400 m.
Source: K.P. Schmidt & Walker, 1943c.

37. *Atractus eriki* Esqueda, La Marca & Bazo, 2005a. Herpetotropicos 2(2): 88–90, figs. 1–3.

Type: Holotype, ULABG 6693, a 310 mm male (A.R. Angarita Matheus, 10 April 2004).
Type locality: "sector Media Luna, Escuque, 7.38 Km E Valera, aprox. 1000 m de elevación, 9°18'N y 70°40'W, Municipio Escuque, Estado Trujillo, Venezuela."
Distribution: Western Venezuela (Mérida, Táchira, Trujillo), 1000 m.
Source: Navarrete et al., 2009.

38. *Atractus erythromelas* Boulenger, 1903a. Ann. Mag. Nat. Hist. (7) 11(65): 483–484.

Types: Syntypes (8), BMNH 1946.1.7.12–19, longest syntype 480 mm (S. Briceño, 1886–1903).
Type locality: "Merida, Venezuela, at an altitude of 1600 metres" [= Merída, Merída State, W Venezuela, 8°36'N, 71°09'W, elevation 1600 m].
Distribution: Colombia (Antioquia, Norte de Santander) and W Venezuela (Mérida), 1000–1950 m.
Sources: Roze, 1966a, Lancini, 1986, Barros-Blanco, 2000, Esqueda & La Marca, 2005 and Navarrete et al., 2009.

39. *Atractus favae* (Filippi, 1840). Biblio. Ital. 99: 16. (*Calamaria favae*)

Synonyms: *Elaps hypospilus* Wagler, 1830, and *Rabdosoma longicaudatum* A.-M.-C. Duméril, Bibron & Duméril, 1854a.
Type: Holotype, not designated, MPI.
Type locality: Unknown.
Distribution: Northern Guyana (Essequibo Islands-West Demerara, Upper Demerara-Berbice, Upper Demerara-Upper Essequibo) and N Suriname, NSL–20 m.
Sources: Hoogmoed, 1980, Chippaux, 1987, Kok, 2006, Passos et al., 2009b and C.J. Cole et al., 2013.
Remarks: Possibly a distinct genus *fide* Hoogmoed (1980: 19). Possibly occurs in French Guiana *fide* Chippaux (1987: 88).

40. *Atractus flammigerus* (H. Boie *in* F. Boie, 1827). Isis von Oken 20(6): 540. (*Brachyorrhos flammigerus*)

Synonyms: *Brachyorrhos flammigerus* H. Boie *in* Schlegel, 1826a *(nomen nudum),* and *Geophis alasukai* Gasc & Rodrigues, 1979b.
Type: Lectotype, RMNH 118A, a 340 mm female (S.J. Brugmans Cabinet), designated by Hoogmoed (1980: 20).
Type locality: "Guyane" [= French Guiana] via lectotype selection. Restricted to Paramaribo, Suriname *fide* Hoogmoed (1980: 24).
Distribution: Northern South America. Guyana, Suriname (Marowijne, Nickerie, Suriname), French Guiana, Peru (Loreto, Madre de Dios) and Brazil (Amapa, Amazonas, Pará, Rondônia), NSL–500 m.
Sources: Dixon & Soini, 1977, Gasc & Rodriguez, 1979b, Cunha & Nacimento, 1983b, Hoogmoed, 1983, Chippaux, 1987, Duellman & Salas, 1991, Starace, 1998 and Duellman, 2005.
Remarks: Original description based on H. Boie's MS (1823–1825). Some Brazilian records may refer to *A. snethlaeae.*

41. *Atractus francicopaivai* Silva-Haad, 2004. Rev. Acad. Colomb. Cienc. 28(108): 426–428, figs. 24–25.

Type: Holotype, ICN 10100 (formerly JSH 1155), a 180 mm male (J. Silva-Haad, 1989).
Type locality: "Colombia, Puerto Córdoba, en el corregimiento de La Pedrera, en el departamento de Amazonas, a orillas del río Caquetá (01° 09' 38" N, 69° 40' 00" Oeste)" [= elevation 170 m].
Distribution: Southeastern Colombia (NW Amazonas), 170 m.
Source: Kok, 2006.

A

42. *Atractus francoi* Passos, Fernandes, Bérnils & Moura-Leite, 2010. Zootaxa (2364), 12–15, figs. 2b–c, 5.

Type: Holotype, MNRJ 17537 (formerly DZUFRJ 1742), a 473 mm male (M.C. Carlo, 17 July 2006).
Type locality: "Fazenda Recanto, Serra do Piloto (22°50'S, 44°03'W, ca. 600 m).
Distribution: Southeastern Brazil (Rio de Janeiro, São Paulo), 600–1000 m.

43. *Atractus fuliginosus* (Hallowell, 1845a). Proc. Acad. Nat. Sci. Philadelphia (1844–45) 2(9): 243–244. (*Coluber fuliginosus)*

Type: Holotype, ANSP 3333, a 413 mm specimen (S. Ashmead, 1835–1845).
Type locality: "Republic of Columbia, within two hundred miles of Caraccas" (in error). Corrected to Venezuela *fide* Roze (1958d: 2).
Distribution: Northern Venezuela (Distrito Federal, Guárico, Sucre), 1200–2000 m.
Sources: Roze, 1966a, Lancini, 1986 and Navarrete et al., 2009.

44. *Atractus gaigeae* Savage, 1955. Proc. Biol. Soc. Washington 68: 12–13.

Type: Holotype, UMMZ 82887 (formerly UM 82887), a male (C. Altenberg & B.G. Butler, summer 1935).
Type locality: "Morona-Santiago Prov., Ecuador."
Distribution: Ecuador (Morona-Santiago).
Remarks: A subspecies of *A. collaris fide* Dixon & Soini (1977: 34 & 1986: 91) but a valid species *fide* Prudente & Passos (2008: 727).

45. *Atractus gigas* C.W. Myers & Schargel, 2006. Amer. Mus. Novit. (532): 8–12, figs. 5–7.

Type: Holotype, FHGO 194, a 1040 mm female (V. Zak, Jan. 1990).
Type locality: "Bosque Protector Río Guajalito, antigua Hacienda Las Palmeras, old highway between Quito and Santo Domingo, 1900 m, Pichincha, Ecuador (00°14'S, 78°49'W)."
Distribution: Ecuador (Cotopaxi, Pichincha) and Peru (Cajamarca), 600–2300 m.
Source: Passos et al., 2010.

46. *Atractus guentheri* (Wucherer, 1861). Proc. Zool. Soc. London 29(1): 115–116, pl. 19, fig. 1. (*Geophis guentheri*) (*nomen corrigendum*)

Synonyms: *Geophis güntheri* Wucherer, 1861 *(nomen incorrigendum), Rabdosoma univittatum* Jan, 1862b, and *Atractus guentheri* – Amaral, 1930f *(nomen corrigendum).*
Type: Holotype, BMNH 1946.1.1.76 (formerly BMNH 1861.3.23.16), a 327 mm male (O.E.H. Wucherer, 1859–1860).
Type locality: "Caunavieras, which is to the south of the city of Bahia, Prov. of Bahia, Brazil" [= Canavieiras, Bahia State, Brazil, 15°41'S, 38°57'W, near sea level].
Distribution: Amazonian Colombia (Amazonas) and SE Brazil (SE Bahia, Minas Gerais, Rio de Janeiro, Rio Grande do Sul, São Paulo), NSL–300 m.
Sources: R. Fernandes & Puorto, 1994, Lema, 1994, R. Fernandes 1995a, R. Fernandes & Argôlo, 1999, Passos et al., 2010 and Hamdan & Lira-da-Silva, 2012.
Remarks: Nicéforo-María (1942: 93) listed Venezuela.

47. *Atractus guerreroi* C.W. Myers & Donnelly, 2008. Bull. Amer. Mus. Nat. Hist. (308): 114–118, figs. 67–69.

Type: Holotype, EBRG 3403, a 250 mm male (R. Guerrero, 6–8 Feb. 1988).
Type locality: "summit of Auyantepui at 5°46'N, 62°32'W, 2100 m, elevation, Bolívar, Venezuela."
Distribution: Southeastern Venezuela (Bolívar), 2100 m. Known only from type locality.
Source: Navarrete et al., 2009.

48. *Atractus heliobelluomini* Silva-Haad, 2004. Rev. Acad. Colomb. Cienc. 28(108): 428–429, figs. 28–29.

Type: Holotype, ICNMNH 10103 (formerly JSH 639), a 158 mm female (J. Silva-Haad, Nov. 1975).
Type locality: "Colombia, departamento de Amazonas, La Pedrera, en el cerro Jupatí, a 300 m de altitud (01° 19' 42" Sur, 69° 30' 33" Oeste)."
Distribution: Southeastern Colombia (Amazonas), 300 m.
Sources: Kok, 2006 and Passos et al., 2009b.

49. *Atractus hoogmoedi* Prudente & Passos, 2010. Copeia 2010(3): 397–399, figs. 1b, 3.

Type: Holotype, MPEG 13268, a 187 mm juvenile male (F.P. do Nascimento, 1975).
Type locality: "Brazil, state of Pará, municipality of Capitão Poço, 04°45'S, 47°04'W, ca. 48 m elev., at locality of Santa Luzia."
Distribution: Northeastern Brazil (Pará), 50 m.

50. *Atractus hostilitractus* C.W. Myers, 2003. Amer. Mus. Novit. (3391): 22–25, figs. 1d, 2d, 3c, 9b.

Type: Holotype, AMNH 130330, a 374 mm female (T. Quintero, 29 Oct. 1967).
Type locality: "at 'Morti Hydro' [about 100–200 m elev., at 8°52'28"N, 77°54'19"W], Rio Morti, Prov. of Darién, Panama."
Distribution: Eastern Panama (Darién), 100–200 m.

51. *Atractus imperfectus* C.W. Myers, 2003. Amer. Mus. Novit. (3391): 25–28, figs. 3d, 9c, 10.

Type: Holotype, MCZ 50213, head and neck only (Panama Snake Census Surv., 1936).

Type locality: "Piedras-Pacora Ridge, [Prov. of Panamá], Panama." Precise locality is unknown, but a more-or-less central point on this divide is 9°16' N, 79°20' W *fide* C.W. Myers (2003: 25).

Distribution: Central Panama (Panamá).

52. *Atractus indistinctus* Prado, 1940b. Mem. Inst. Butantan (1939) 13: 16, fig. 2.

Type: Holotype, ILS 166, a 444 mm female (C. Hernández Yaruro, 1938–1939), destroyed by fire 9–10 April 1948.

Type locality: "Ocaña, departamento norte de Santander, Colombia" [= Ocaña, Norte de Santander Department, Colombia, 8°14'N, 73°21'W, elevation 1265 m].

Distribution: Northeastern Colombia (Antioquia, Cundinamarca, Norte de Santander), 1200–1265 m.

53. *Atractus insipidus* Roze, 1961. Acta Biol. Venez. 3(7): 106–107.

Type: Holotype, MBUCV 3957, a 235 mm male (Panchenco, 1952).

Type locality: "Poste M-1, cerca río Uraricapará, frontera Venezuela–brasil, unos 400 metrs. de altura."

Distribution: Southeastern Venezuela (Bolívar) and Brazil (Maranhao), 400 m.

Sources: Roze, 1966a, Lancini, 1986, Kok, 2006, Passos & Fernandes, 2008 and Navarrete et al., 2009.

54. *Atractus iridescens* Peracca, 1896b. Boll. Mus. Zool. Anat. Comp. Univ. Torino 11(252): 2–3, 1 fig.

Type: Holotype, MZUT 1830 (formerly MZUT 1726), an adult male (Viale).

Type locality: "America meridionale" [= South America].

Distribution: Northwestern Colombia (Antioquia, Chaco, Nariño), 50–150 m.

Source: Passos et al., 2009e.

55. *Atractus lancinii* Roze, 1961. Acta Biol. Venez. 3(7): 112–114.

Type: Holotype, MCN 574, a 342 mm female (R. Lancini, Aug. 1958).

Type locality: "El Junquito, Distrito Federal, Venezuela; 1.900 metros de altura."

Distribution: Northern Venezuela (Aragua, Carabobo, Distrito Federal, Miranda, Yaracuy), 800–2000 m.

Sources: Roze, 1966a, Lancini, 1986, Barros-Blanco, 2000, Esqueda et al., 2001 and Navarrete et al., 2009.

56. *Atractus lasallei* Amaral, 1931a. Bull. Antivenin Amer. 4(4): 87–88.

Type: Holotype, IB 5315 (formerly ILS 4), a 400 mm male (H. Nicéforo-María, 1908–1931), destroyed by fire 15 May 2010.

Type locality: "Sampedro (north of Medellin, Antioquia), Colombia" [= San Pedro, S of Medellín, Antioquia Department, Colombia, 6°28'N, 75°33'W, elevation 2535 m].

Distribution: Extreme NW South America. Northern Colombia (Antioquia, Cundinamarca) and W Venezuela, 1500–2750 m.

Source: Pérez-Santos & Moreno, 1988.

57. *Atractus latifrons* (A.C.L.G. Günther, 1868). Ann. Mag. Nat. Hist. (4) 1(6): 415, pl. 19, fig. b. (*Geophis latifrons*)

Synonym: *Elaps herthae* Ahl, 1927.

Type: Holotype, BMNH 1946.1.6.52, a 254 mm specimen, (J. Hauxwell,).

Type locality: "Pebas" [= Pébas, Loreto Department, Peru, 3°11'S, 71°47'W, elevation 125 m].

Distribution: Amazonia. Colombia (Amazonas, Caqueta, Putumayo, Vaupés), French Guyana (Cayenne), Suriname (Marowijne), NE Peru (Loreto), Brazil (Acre, Amapá, Amazonas, ? Ceará, Mato Grosso, Pará, Rondônia) and Bolivia (Santa Cruz), 90–500 m.

Sources: Hoogmoed, 1980, Cunha & Nascimento, 1983b, Martins & Oliveira, 1993, Starace, 1998, Silva-Haad, 2004, Embert et al., 2006 and Moravec & Aparcio, 2006.

Remarks: Hoogmoed (1979: 275) listed Amazonian Ecuador.

58. *Atractus lehmanni* Boettger, 1898. Kat. Rept. Senck. Mus. 2: 80.

Type: Lectotype, SMF 19355 (formerly SMF-B 8310a), a 296 mm female (F.C. Lehmann, 1894), designated by Mertens (1967: 91).

Type locality: "Cuenca, Prov. Azanay, Ecuador" [= Cuenca, Azuay Prov., Ecuador, 2°54'S, 78°59'W, elevation 2500 m] via lectotype selection.

Distribution: Colombia (Antioquia, Valle de Cauca) and S Ecuador (Azuay), 2500–2600 m.

Sources: Savage, 1960 and Passos & Arredondo, 2009.

59. *Atractus limitaneus* (Amaral, 1935c). Mem. Inst. Butantan 9: 219–210, figs. 1–3. (*Leptocalamus limitaneus*)

Type: Holotype, IB 9196 (formerly ILS 125), a 175 mm male (H. Nicéforo-María, 1908–1935), destroyed by fire 15 May 2010.

Type locality: "La Pedrera, near the Brazilian-Colombian boundary, Colombia" [= La Pedrera, Río Caquetá,

A

Amazonas Department, SE Colombia, 1°20'S, 69°35'W, elevation 90 m].

Distribution: Southeastern Colombia (E Amazonas), 90 m. Known only from type locality.

Source: Prudente & Passos, 2008.

60. *Atractus longimaculatus* Prado, 1940b. Mem. Inst. Butantan (1939) 13: 17, fig. 3.

Type: Holotype, ILS 170, a 372 mm male (H. Nicéforo-María, July 1938), destroyed by fire 9–10 April 1948.

Type locality: "Região do Quindío, Colombia" [= Nevada Quindío, Caldas Department, Colombia *fide* Savage, 1960: 81] (in error). Corrected to Pacho, a village NE of Zipaquirá, central Colombia *fide* Nicéforo-María (1942: 93) [= Pacho, Cundinamarca Department, cen. Colombia, 5°08'N, 74°10'W, elevation 1830 m].

Distribution: Central Colombia (Cundinamarca), 1830 m.

61. *Atractus loveridgei* Amaral, 1930a. Bull. Antivenin Inst. Amer. 4(2): 28.

Type: Holotype, MCZ 29059, a 360 mm female (H. Nicéforo-María, 1908–1930).

Type locality: "Jerico, central Colombia" [= Jericó, Antioquia Department, Colombia, 5°47'N, 75°47'W, elevation 1850 m].

Distribution: Northern Colombia (Antioquia, Boyacá, Cundinamarca, Santander), 1050–3005 m.

Source: Passos et al., 2009c.

62. *Atractus macondo* Passos, Lynch & Fernandes, 2008. Herpet. J. 18: 180–181, fig. 5.

Type: Holotype, IAvH 15 (formerly IND-R 15), a 383 mm male (G. Toro).

Type locality: "Parque Nacional Natural Isla de Salamanca (10º58'N, 74º30'W, approximately 3m), municipality of Sifionuavo, Department of Magdalena, Colombia."

Distribution: Northern Colombia (Magdalena), NSL.

63. *Atractus maculatus* (A.C.L.G. Günther, 1858). Cat. Colub. Snakes Brit. Mus.: 204–205. *(Isoscelis maculata)*

Type: Holotype, BMNH 1946.1.6.46 (formerly BMNH 1851.3.12.151), a 343 mm male.

Type locality: Unknown. Designated as Brazil *fide* BMNH catalogue.

Distribution: Colombia (Antioquia) and SE Brazil (Alagoas, Bahia, Espírito Santo, Paraná, Pernambuco, Rio de Janeiro, E São Paulo), NSL–500 m.

Sources: R. Fernandes, 1995a, Moura-Leite et al., 1996, R. Fernandes et al., 2000, Marques et al., 2001 and Passos et al., 2010.

Remarks: Nicéforo-María (1942: 93) listed Bolivia and Colombia (Jericó).

64. *Atractus major* Boulenger, 1894a. Cat. Snakes Brit. Mus. 2: 307.

Type: Lectotype, BMNH 1946.9.7.27, a 145 mm male (C. Buckley, 1856–1880), designated by Savage (1960: 50).

Type locality: "Canelos, Napo-Pastaza Prov., Ecuador" [= Canelos, Pastaza Prov., NE Ecuador, 1°35'S, 77°45'W, elevation 500 m] via lectotype selection.

Distribution: Northwestern Amazonia. Eastern Ecuador (Napo), SE Colombia (Amazonas, Cundinamarca, Nariño), S Venezuela (Amazonas, Táchira), Peru (Amazonas, Loreto, Madre de Dios) and W Brazil (Amazonas, Pará), 100–2630 m.

Sources: Roze, 1966a, Lancini, 1986, Duellman & Salas, 1991, Martins & Oliveira, 1993, Schargel & Castoe, 2003, Costa Prudente & Santos Costa, 2004, Silva-Haad, 2004, Duellman, 2005, Esqueda & La Marca, 2005, C.W. Myers & Schargel, 2006, Vigle, 2008 and Navarrete et al., 2009.

65. *Atractus manizalensis* Prado, 1940b. Mem. Inst. Butantan (1939) 13: 17–18, fig. 4.

Type: Holotype, ILS 172, a 370 mm male (N. Díaz-Berrío, 1938–1939), destroyed by fire 9–10 April 1948.

Type locality: "Colombia." Restricted to the región of Manizales, Cordillera Central de las Andes, Colombia *fide* Nicéforo-María (1942: 93) and further restricted to Villamaria, Caldas Department, Colombia *fide* E.R. Dunn *in* Savage (1960: 82) [= 5°03'N, 75°31'W, elevation 1950 m].

Distribution: Colombia (Antioquia, Caldas, Cauca, Quindo), 1950–2150 m.

Sources: R. Fernandes, 1995a, Passos et al., 2009c and Rojas-Morales, 2012b.

66. *Atractus mariselae* Lancini, 1969. Publ. Ocas. Mus. Cienc. Nat. (15): 4–5, figs. 1–3.

Type: Holotype, MCNC 3971, a 325 mm male (M. Urosa-Sambrano, 5 Sept. 1966).

Type locality: "en el jardín posterior de una vieja casa de Boconó, Estado Trujillo, a unos 1.225 metros sobre el nivel del mar, Venezuela" [= Boconó, Trujillo State, W Venezuela, 9°15'N, 70°15'W, elevation 1225 m].

Distribution: Western Venezuela (Táchira, Trujillo), 1225 m.

Sources: Schargel & Castoe, 2003, Esqueda & La Marca, 2005 and Navarrete et al., 2009.

67. *Atractus matthewi* Markezich & Barrio-Amorgós, 2004. Bull. Maryland Herp. Soc. 40(3): 112–117, figs. 1a, 2.

Synonym: *Atractus nororientalis* D. Sánchez, Sousa, Esqueda & Manzanilla, 2005.

Type: Holotype, AMNH 29316, a 365 mm male (G.H.H. Tate, 1921–1929).

Type locality: "Venezuela: Estado Sucre, Carápas, 10°07'56"N, 63°53'00"W."
Distribution: Northeastern Venezuela (Anzoátegui, Sucre), ca. 1400–2130 m.
Sources: Sánchez et al., 2005, Kok et al., 2007 and Navarrete et al., 2009.

68. *Atractus medusa* Passos, Mueses-Cisneros, Lynch & Fernandes, 2009e. Zootaxa (2293): 15–17, figs. 10a–b, 11.

Type: Holotype, IAvH 2981 (formerly VR 1213), a 385 mm male (J.V. Rueda).
Type locality: "Playa Blanca, Gorgona Island (03°00'N, 78°12'W, ca. 0 m), municipality of Guapi, department of Cauca, Colombia."
Distribution: Western Colombia (Cauca: Gorgona Is.), NSL.

69. *Atractus melanogaster* F. Werner, 1916. Zool. Anz. 47(11): 309.

Type: Neotype, ICN 10030, a 283 mm male (J.D. Lynch & J.V. Rueda, 29 May 1981), designated by Passos & Lynch (2010: 160).
Type locality: "Quebrada Perales (2420 m), Vereda La Palma, 11.4 km west of the municipality of Cajamarca (04° 27' N, 75° 26' W), department of Tolima, Colombia" via neotype selection.
Distribution: Central Colombia (Antioquia, Tolima), 1800–2200 m.
Sources: Duellman, 1979 and Passos & Lynch, 2010.
Remarks: LMB holotype lost *fide* M. Berec *in* Passos & Lynch (2010: 160).

70. *Atractus melas* Boulenger, 1908a. Ann. Mag. Nat. Hist. (8) 1(1): 114–115.

Type: Holotype, BMNH 1946.1.6.33 (formerly BMNH 1908.5.29.54), a 235 mm female (M.G. Palmer, 1907–1908).
Type locality: "Los Mangos, S.W. Colombia, altitude 300 m" [= Los Cisneros or Juntas, Valle de Cauca Department, Colombia, 3°46'N, 76°45'W *fide* Savage, 1960: 82 and Passos et al., 2009e: 18].
Distribution: Western Colombia (Antioquia, Chaco, Valles, Valle de Cauca), 80–300 m.
Source: Passos et al., 2009e.

71. *Atractus meridensis* Esqueda & La Marca, 2005. Herpetotropicos 2(1): 10–12, figs. 9–10.

Type: Holotype, ULABG 4341, a 437 mm male (E. La Marca, 22 March 1997).
Type locality: "Venezuela, Estado Mérida, Municipio Pueblo Llano, en patio trasero vivienda, 1705 m.s.n.m., a 1 Km del Puente sobre el río Santo Domingo en la vía principle desde La Mitsús hasta San Domingo, ya 500 m del cruce de acceso a la Población de Las Piedras, 08°52'45"N, 70°39'09"W."
Distribution: Western Venezuela (Barinas, Mérida), 1200–2200 m.
Source: Navarrete et al., 2009.

72. *Atractus michelae* Esqueda & La Marca, 2005. Herpetotropicos 2(10): 14–16, figs. 12–13.

Type: Holotype, ULABG 2672, a 345 mm female (M. Ataroff, 14 June 1989).
Type locality: "Venezuela, Estado Mérida, Municipio Arzobispo Chacón, cerca de la población de Canagua, aprox. 1700 m.s.n.m., coordenadas estimadas 8°10'24"N, 71°28'52"W."
Distribution: Western Venezuela (Mérida, Táchira), 1000–1700 m.
Source: Navarrete et al., 2009.

73. *Atractus microrhynchus* (Cope, 1868b). Proc. Acad. Nat. Sci. Philadelphia 20(2): 102. (*Rhabdosoma microrhynchum*)

Type: Neotype, DHMECN 3008, a 329 mm male (F. Sornoza, March 2005), designated by Passos et al. (2005: 377).
Type locality: "Reserva Biológica Buenaventura (03°38'55"S, 79°45'50"W, ca. 600 m asl), Parroquia piñas, Cantón Piñas, El Oro Province, Ecuador" via neotype designation.
Distribution: Western Ecuador (El Oro, Guayas, Los Ríos) and NW Peru (Tumbes), 50–800 m.
Sources: Dixon & Soini, 1977 and Passos et al., 2009e, 2012b.
Remarks: Holotype erroneously listed as USNM 6693 *fide* J.A. Peters (1960a: 507) and ANSP 6693 *fide fide* Savage (1960: 52), lost *fide* N. Gilmore *in* Passos et al. (2012b: 377).

74. *Atractus mijaresi* Esqueda & La Marca, 2005. Herpetotropicos 2(1): 16–17, figs. 14–15.

Type: Holotype, ULABG 4697, a 369 mm male (N. Jáuregui, 5 May 1997).
Type locality: "Venezuela, Estado Mérida, Distributionl, Mucurubá, parte alta, 2405 m.s.n.m., 08°42'34"N, 70°59'17"W."
Distribution: Western Venezuela (Mérida), 2405 m.
Source: Navarrete et al., 2009.

75. *Atractus modestus* Boulenger, 1894a. Cat. Snakes Brit. Mus. 2: 304, pl. 15, fig. 1.

Type: Holotype, BMNH 1946.1.6.30, a 379–380 mm male (L. Fraser, 1857–1859).
Type locality: "W. Ecuador."
Distribution: Western Ecuador (Azuay, Cotopaxi, Morona-Santiago, Pichincha), 2400–2560 m.

Source: Passos et al., 2007.

76. *Atractus multicinctus* (Jan, 1865 *in* Jan & Sordelli, 1860–1866). Icon. Gén. Ophid. 1(10): 7, pl. 4, figs. 5, a, d–g, n, p. v. *(Rabdosoma badium multicinctum)*

Type: Holotype, formerly MSNM, destroyed in 1943 during World War II.
Type locality: "Lima," Lima Department, Peru," (in error *fide* Savage, 1960: 54, and Passos et al., 2009e: 24).
Distribution: Southwestern Colombia (Cauca, Chocó, Valle del Cauca) and NW Ecuador (Azuay, Cotopaxi, Los Rios, Morona-Santiago), NSL–770 m.
Source: Passos et al., 2009e.

77. *Atractus multidentatus* Passos, Rivas-Fuenmayor & Barrio-Amorgós, 2009b. Amphibia-Reptilia 30(2): 236–238, figs. 3–4.

Type: Holotype, CVULA 7080, a 172 mm female (R. Alarcón-Gallegos, 1979–1982).
Type locality: "quebrada El Paraiso, La Vega (8°32'N, 71°14'W, *ca.* 1000 m), State of Mérida, Venezuela."
Distribution: Western Venezuela (Mérida), 1000 m. Known only from type locality.

78. *Atractus nasutus* Passos, Arredondo, Fernandes & Lynch, 2009c. Copeia 2009(3): 425–428, fig. 1.

Type: Holotype, CSJ 561, a 199 mm male (H. Daniel, 1954).
Type locality: "Colombia, Antioquia department, San Pedro municipality, Vereda La Lana, 06°26'52, 75°36'26"W, ca. 2600 m."
Distribution: Colombia (Antioquia), 2600 m. Known only from vicinity of type locality.

79. *Atractus natans* Hoogmoed & Prudente, 2003. Zool. Meded. 77: 428–437, figs. 1, 3–4.

Synonym: *Atractus emersoni* Silva-Haad, 2004.
Type: Holotype, MPEG 18836, a 308 mm female (M.S. Hogmoed & T.C.S. Ávila-Pires, 1 Aug. 1994).
Type locality: "near flutuarte at confluence of Paraná Apara and Rio Mamirauá, Estação Mamiraua, Municipio de Uarini, Amazonas, Brazil, 3°02'57.2"S 64°50'59.7"W, 50 m."
Distribution: Amazonia. Southeastern Colombia (Amazonas), N Brazil (Amazonas, Para) and ext. NE Peru (Loreto), NSL–100 m.
Sources: Passos & Fernandes, 2008 and Passos et al., 2012b.

80. *Atractus nicefori* Amaral, 1930a. Bull. Antivenin Inst. Amer. 4(2): 28.

Type: Holotype, MCZ 29058, a 420+ mm male (R.H. Apolinar-Enrique & R.H. Daniel).
Type locality: "Jerico, central Colombia" [= Jericó, Antioquia Department, Colombia, 5°47'N, 75°47'W, elevation 1850 m].
Distribution: Central Colombia (Antioquia, Cauca, Cundinamarca), 1570–2600 m.
Source: Passos et al., 2009c.

81. *Atractus nigricaudus* K.P. Schmidt & Walker, 1943c. Field Mus. Nat. Hist., (Zool.) 24(28): 327–328.

Type: Holotype, MCZ 45909, a 367 mm female (W.F. Walker, 18 Dec. 1938).
Type locality: "Huachon, in the upper part of the Paucartambo Valley, east of Cerro de Pasco, Department of Junin, Peru. Altitude 10,000 feet."
Distribution: Central Peru (Junin, Pasco), 1800–3365 m.
Source: Passos & Arredondo, 2009.

82. *Atractus nigriventris* Amaral, 1933c. Mem. Inst. Butantan (1932) 7: 116.

Type: Holotype, IB (formerly ILS 82), a 365 mm female (R.P. Ernesto Catalalina), destroyed by fire 15 May 2010.
Type locality: "Chita (sudeste de San Gil), Colombia" [= Chita, SE of San Gil (6°34'N, 73°08'W, 1170 m), Santander Department, Colombia].
Distribution: Northern Colombia (Boyacá, Santander), 1500–3005 m.

83. *Atractus obesus* Marx, 1960. Fieldiana: Zool. 39(38): 411–413, fig. 71.

Type: Holotype, FMNH 69661, a 762 mm female (P. Hershkovitz, April 1951).
Type locality: "Santa Bárbara, a site at base of Cerro Frontino, upper Río Urrao, a tributary of Río Penserisco, Cordillera Occidental, Antioquia, Colombia... at an altitude of 2700 meters" [= 5°53'N, 75°34'W, elevation 2700 m].
Distribution: Northwestern Colombia (Antioquia, Cundinamarca, Valle del Cauca,), 1400–2700 m.
Sources: C.W. Myers & Schargel, 2006 and Passos et al., 2009c.

84. *Atractus obtusirostris* F. Werner, 1916. Zool. Anz. 47(11): 309–310.

Type: Lectotype, ZMH 4428 (formerly ZMH 33, 181), a 254 mm male (A.H. Fassl, Jan.–March 1910), designated by Passos & Lynch (2010: 163).
Type locality: "Cañon del Tolima, (= Combeima River; west Ibague, 04° 26' N, 75° 14'W, ca. 1100 m), department of Tolima, Colombia" via lectotype selection.
Distribution: Central Colombia (Caldas, Tolima), 1100–2200 m.
Sources: F. Werner, 1924a and Passos & Arredondo, 2009.

A

85. *Atractus occidentalis* Savage, 1955. Proc. Biol. Soc. Washington 68: 16–17.

Type: Holotype, BMNH 1916.5.23.5 (formerly BMNH 1916.5.23.5), a 265 mm male (W. Goodfellow, 1914–1915).
Type locality: "Mindo, Pichincha Prov., Ecuador" [= 2°56'S, 78°47'W, elevation 1280 m].
Distribution: Central Ecuador (Chimborazo, Pichincha), 800–1250 m.
Source: Passos et al., 2012.

86. *Atractus occipitoalbus* (Jan, 1862b). Arch. Zool. Anat. Fis. 2(1): 16–17, pl. 8, fig. 5th (*Rabdosoma occipitoalbum*)

Type: Holotype, ZSM 1395/0, a 508 mm female (F.J.C.M. Wagner, 1857–1860), lost *fide* Franzen & Glaw (2007: 258), probably destroyed 24–25 April 1944 or 11 April 1945 during World War II.
Type locality: "occidentali della Ande nell'Ecuadore all' altezza di 4000 piedi."
Distribution: Southeastern Colombia (Putumayo), E Ecuador (Orellana, Morona-Santiago) and Peru (Cusco, Ucayali), 200–1220 m.
Sources: Henle & Ehrl, 1991, Silva-Haad, 2004, Vigle, 2008 and Passos et al., 2009d.

87. *Atractus ochrosetrus* Esqueda & La Marca, 2005. Herpetotropicos 2(1): 18–20, figs. 16–17.

Type: Holotype, ULABG 4698, a 322 mm male (E. La Marca, 12 Jan. 1999).
Type locality: "Venezuela, Estado Mérida, Municipio Tovar, en la via de Bailadores a "La Y", El Delgadito, approx. 2600 m.s.n.m. , 08°13'06"N, 71°52'26"W."
Distribution: Western Venezuela (Mérida, Táchira), 2600–2705 m.
Source: Navarrete et al., 2009.

88. *Atractus oculotemporalis* Amaral, 1932a. Bull. Antivenin Inst. Amer. 5(3): 67. (*Atractus oculotemporalis*) (*nomen corrigendum*)

Synonyms: *Atractus oculo-temporalis* Amaral, 1932a *(nomen incorrigendum),* and *Atractus oculotemporalis* – J.A. Peters & Orejas-Miranda, 1970 *(nomen corrigendum).*
Type: Holotype, IB 6390 (formerly ILS 46), a 460 mm female (R.H. Apolinar-Enrique & R.H. Daniel), destroyed by fire 15 May 2010.
Type locality: "Jericó, west of Medellín, central Colombia" [= Jericó, Antioquia Department, Colombia, 5°47'N, 75°47'W, elevation 1850 m].
Distribution: Central Colombia (Antioquia), 1850–1970 m. Known only from type locality.
Source: Passos et al., 2009c.

89. *Atractus orcesi* Savage, 1955. Proc. Biol. Soc. Washington 68, 17–18.

Type: Holotype, CAS-SU 15622 (formerly SU 15622), a 307.5 mm male (J. Olalla, Oct. 1952).
Type locality: "Loreto, Napo-Pastaza Prov., Ecuador" [= Loreto, Orellana Prov., NE Ecuador, 00° 41' S, 77° 18' W, 410 m].
Distribution: Southwestern Colombia (Cauca, Huila, Putumayo) and E Ecuador (Morona-Santiago, Orellana, Pastaza, Sucúmbios), 410–3000 m.
Source: Passos et al., 2009d.

90. *Atractus paisa* Passos, Arredondo, Fernandes & Lynch, 2009c. Copeia 2009(3): 428–431, fig. 6.

Type: Holotype, ICNMNH 10698, a 392 mm female (A. Hincapié, 31 March 1996).
Type locality: "Colombia, Antioquia department, Sonsón municipality, 05°43'N, 75°19'W, San Francisco, 2600 m."
Distribution: Northwestern Colombia (Antioquia), 2220–2600 m.

91. *Atractus pamplonensis* Amaral, 1935c. Mem. Inst. Butantan 9: 220, fig. 4.

Type: Holotype, IB 9192, a 315 mm male (R.H. Isidoro-Regis), destroyed by fire 15 May 2010.
Type locality: "Pamplona, (near the Venezuelan-Colombian boundary), Colombia." [= Pamplona, Norte de Santander, Colombia, 7°23'N, 72°39'W, elevation 2400 m].
Distribution: Northeastern Colombia (Norte de Santander) and W Venezuela (Táchira), 1560–2400 m.
Sources: Schargel & García-Pérez, 2002, Esqueda & La Marca, 2005 and Navarrete et al., 2009.
Remarks: Original description reprinted in Amaral (1937b: 1763, fig. 2).

92. *Atractus pantostictus* R. Fernandes & Puorto, 1994. Mem. Inst. Butantan 55(Suppl. 1): 8–12, figs. 1–3, 5.

Type: Holotype, IB 54844, a 273 mm male (M.S. Santos, 30 June 1992), destroyed by fire 15 May 2010.
Type locality: "Franco da Rocha, SP, Brazil,"[= Franco da Rocha, São Paulo State, SE Brazil, 23°19'S, 46°43'W, elevation 770 m].
Distribution: Southeastern Brazil (Distrito Federal, Goiás, Minas Gerais, São Paulo, Tocantins), 200–1200 m.
Sources: R. Fernandes, 1995b, Passos et al., 2005, 2010 and Passos & Fernandes, 2008.

93. *Atractus paraguayensis* F. Werner, 1924a. Sitz. Akad. Wiss. Wien, Math.-Natur. Kl., Abt. 1, 133(1–3): 40.

Type: Holotype, NMW 23443 (formerly NMW 130), a 365 mm female (G. Wieninger, 1900–1904).
Type locality: "Paraguay."
Distribution: Paraguay, NE Argentina (Corrientes, Entre Ríos) and S Brazil (Paraná), 30–1000 m.
Sources: R. Fernandes, 1995b and Passos et al., 2005, 2010.

94. *Atractus paravertebralis* Henle & Ehrl, 1991. Bonn. Zool. Beitr. 42(2): 160–161, figs. 9a–b.

Type: Holotype, ZFMK 39705, a 450 mm specimen (K. Henle & A. Ehrl, 28 Jan. 1983).
Type locality: "Peru; Departamento Madre de Dios; Baja Tambopata; in Sekundärwald" [= Río Tambopata, Madre de Dios Department, SE Peru, ca. 12°53'S, 69°25'W, elevation 225 m].
Distribution: Southeastern Peru (Madre de Dios), 225 m. Known only from type locality.

95. *Atractus paucidens* (Mocquard *in* Despax, 1910). Bull. Mus. Hist. Nat., Paris 16(7): 372. (*Atractopsis paucidens*)

Type: Holotype, MNHN 1906.245 (formerly MNHN 3422A), a 320 mm female (P. Rivet, 1901–1906).
Type locality: "Santo-Domingo de los Colorados, Équateur, 560 mètres" [= Santo Domingo, Pichincha Prov., N Ecuador, 0°15'S, 79°10'W, elevation 550 m].
Distribution: Northwestern Ecuador (Pichincha), 200–600 m.
Source: Passos et al., 2009e.
Remarks: Original description reprinted in Mocquard (1911: 31–32, pl. 2, figs. 2–2c).

96. *Atractus pauciscutatus* K.P. Schmidt & Walker, 1943c. Field Mus. Nat. Hist., (Zool.) 24(28): 326–327.

Type: Holotype, FMNH 36725, a 330 mm female (F. Woytkowski, 14 May 1940).
Type locality: "Carpapata, in the upper Chanchamayo Valley, northeast of Tarma, Department of Junin, Peru. Altitude 10,500 feet."
Distribution: Central Peru (Junin), 3200 m. Known only from type locality.

97. *Atractus peruvianus* (Jan, 1862b). Arch. Zool. Anat. Fis. 2(1): 12–13. (*Rabdosoma peruvianum*)

Type: Holotype, MNHN, a 280 mm specimen.
Type locality: "Peru."
Distribution: Peru. Known only from type specimen.

98. *Atractus poeppigi* (Jan, 1862b). Arch. Zool. Anat. Fis. 2(1): 11–12, pl. 8, fig. 1. (*Rabdosoma pöppigi*)

Type: Holotype, NMW, a 570 mm specimen.
Type locality: "Brasile" [= Brazil].
Distribution: Eastern Colombia (Amazonas, Vaupés), N Peru (Amazonas, Loreto) and W Brazil (Amazonas), 100–250 m.
Sources: Dixon et al., 1976, Martins & Oliveira, 1993, Schargel & Castoe, 2003, Zaher et al., 2005 and Kok, 2006.

99. *Atractus potschi* R. Fernandes, 1995a. J. Herp. 29(3): 417–418, figs. 1–2.

Type: Holotype, IB 48438, a 290 mm male (A. J. Santos, 19 July 1985), destroyed by fire 15 May 2010.
Type locality: "Maceió, State of Alagoas, Brazil" [= 09°40'S, 35°43'W, sea level *fide* Passos et al., 2010: 30].
Distribution: Brazil (Alagoas, Bahia, Sergipe), NSL–1200 m.
Sources: Freitas, 1999, Pires-Lima et al., 2000, Passos et al., 2010 and Angôlo et al., 2011.

100. *Atractus punctiventris* Amaral, 1933c. Mem. Inst. Butantan (1932) 7: 117–118.

Type: Holotype, IB (formerly ILS 102), a 325 mm male (H. Nicéforo-María, 1908–1933), destroyed by fire 15 May 2010.
Type locality: "Villavicencio, Meta (leste de Bogotá), Colombia" [= Villavicencio, Meta Department, Colombia, 4°09'N, 73°38'W, elevation 535 m].
Distribution: Central Colombia (Meta), 465–535 m. Known only from type locality.
Source: Passos & Fernandes, 2008.

101. *Atractus resplendens* F. Werner, 1901a. Verh. Zool. Bot. Ges. Wien 51: 598. (*Atractus torquatus resplendens*)

Type: Holotype, ? NMW or ZMB, a 378 mm specimen (R. Haensch, 1899–1900), location of type unknown *fide* J.A. Peters (1960: 508).
Type locality: "Ecuador."
Distribution: Eastern Ecuador (Chimborazo, Santiago-Zamora, Tungurahua), 1100–1900 m.

102. *Atractus reticulatus* (Boulenger, 1885b). Ann. Mag. Nat. Hist. (5) 16(92): 87. (*Geophis reticulatus*)

Synonym: *Atractus reticulatus scrocchii* Alvarez, Rey & Cei, 1992.
Type: Holotype, BMNH 1946.1.2.7 (formerly BMNH 1885.6.26.10), a 315 mm female (H.F.I. von Ihering, 1880–1885).
Type locality: "S. Lorenzo, on the southern border of the Lagoa dos Patos, Prov. Rio Grande do Sul, Brazil"

A

[= São Lourenço do Sul, Rio Grande do Sul, Brazil, 31°22'S, 51°59'W, elevation 15 m].

Distribution: Southern South America. Paraguay (Rivera), S Brazil (Goiás, Minas Gerais, Paraná, Rio Grande do Sul, Santa Catarina, São Paulo), NE Uruguay (Rivera) and Argentina (Corrientes, Formosa, Tucumán), NSL–1100 m.

Sources: Boulenger, 1894a, Alvarez et al., 1992, Cei, 1994, Lema, 1994, R. Fernandes, 1995b, Giraudo, 2001, Carreira-Vidal et al., 2004, 2005, 2012b and Passos et al., 2005, 2010.

103. *Atractus riveroi* Roze, 1961. Acta Biol. Venez. 3(7): 114–116.

Type: Holotype, AMNH 36615, a 485 mm male (Tyler Duida Exped., 20 April 1929).

Type locality: "Cerro Duida en el subpáramo sabanero de la cumbre, Territorio Federal Amazonas, Venezuela; unos 1800 metros de altura."

Distribution: Southern Venezuela (Amazonas), 1000–2000 m.

Sources: Roze, 1966a, Hoogmoed, 1979, Lancini, 1986 and Navarrete et al., 2009.

104. *Atractus ronnie* Passos, Fernandes & Borges-Nojosa, 2007. Copeia 2007(4): 789–791, figs. 1a–c, 2a–b.

Type: Holotype, MNRJ 14194, a 301 mm female (D. M. Borges-Nojosa, 10 April 1998).

Type locality: "Brazil, Ceará State, Serra de Baturité, municipality of Pacoti, Granja, 04°10'S, 38°55'W, *ca.* 800 m."

Distribution: Northeastern Brazil (Ceará), 500–900 m.

Sources: Passos & Fernandes, 2008, Loebmann et al., 2009 and Passos et al., 2010.

105. *Atractus roulei* Mocquard *in* Despax, 1910. Bull. Mus. Hist. Nat., Paris 16(7): 370–371.

Type: Holotype, MNHN 1906.243, a 450 mm specimen (P. Rivet, 1902–1906).

Type locality: "Alausi, à 2,350 mètres d'altitude, l'Equateur" [= Alausí, Chimborazo Prov., Ecuador, 2°12'S, 78°51'W, elevation 2350 m].

Distribution: Southern Ecuador (Azuay, Bolívar, Chimborazo, El Oro), 1200–2785 m.

Sources: Duellman, 1979 and N.J. Silva, et al., 2005.

Remarks: Original description reprinted in Mocquard *in* Despax (1911: 30–31, pl. 2, figs. 1–1b).

106. *Atractus sanctaemartae* Dunn, 1946a. Occ. Pap. Mus. Zool., Univ. Michigan (493): 2–5.

Synonym: *Atractus nebularis* Bernal-Carlo & Roze, 1997.

Type: Holotype, UMMZ 48298, a 600 mm female (W.W. Brown, 25 July 1899).

Type locality: "San Sebastian, Sierra Nevada de Santa Marta, Colombia, at an altitude of 6500 feet" [= San Sebastian, Sierra Nevada de Santa Marta Mountains, N Magdalena Department, N Colombia, ca. 11°'N, 74°W, elevation 1980 m].

Distribution: Northern Colombia (César, Magdalena), 1200–2500 m.

Sources: Barros-Blanco, 2000, C.W. Myers & Schargel, 2006 and Passos et al., 2008.

Remarks: See Passos et al. (2008: 179–180) for remarks on elevation.

107. *Atractus sanguineus* Prado, 1944. Ciencia 5(4–5): 111, fig. 2.

Type: Holotype, CSJ 517 (formerly CSJ 232), a 424 mm male (R.H. Daniel, 1943).

Type locality: "Yarumal, N. de Medellin, Colombia" [= Yarumal, Antioquia Departament, cen. Colombia, 6°58'N, 75°26'W, elevation 2350 m].

Distribution: Central Colombia (Antioquia), 2300–2350 m. Known only from type locality.

Sources: Passos & Arredondo, 2009 and Passos et al., 2009c.

Remarks: Original description reprinted in Prado (1946b: 110, 1 fig.).

108. *Atractus schach* (H. Boie *in* F. Boie, 1827). Isis von Oken 20(6): 540. (*Brachyorrhos schach*)

Synonym: *Brachyorrhos schach* H. Boie *in* Schlegel, 1826a *(nomen nudum)*.

Type: Lectotype, RMNH 119A, a 192 mm male (RMNH Ancient Cabinet, 1820), designated by Hoogmoed (1980: 34).

Type locality: "Guyane" [= Guyana] via lectotype selection. Restricted to Mamadam, Saramacca River, distr. Brokopando, Suriname *fide* Hoogmoed (1980: 35).

Distribution: Northern Amazonia. Guyana (East Berbice-Correntyne), Suriname (Brokopondo, Commewijne, Nickerie, Saramacca), NW French Guiana (N Saint-Laurent-du-Maroni) and N Brazil (Amazonas, Maranhão, Pará), 20–230 m.

Sources: Hoogmoed, 1980, Cunha & Nascimento, 1983b, Hoogmoed & Avila-Pires, 1991, Martins & Oliveira, 1993, Starace, 199, Abuys, 2003 and C.J. Cole et al., 2013.

109. *Atractus serranus* Amaral, 1930b. Bull. Antivenin Inst. Amer. 4(3): 65.

Type: Holotype, IB 7238, a 700 mm specimen (M. Costilho, 1930), destroyed by fire 15 May 2010.

Type locality: "northermost section of the Serra de Paranapiacaba, State of São Paulo, Brazil" [= near the municipality of São Paulo, E São Paulo State, SE Brazil, 23°33'S, 46°38'W, elevation 800 m *fide* Passos et al., 2010: 43].

A

Distribution: Southeastern Brazil (E Rio de Janeiro, Santa Catarina, São Paulo), 700–1000 m.
Sources: Marques et al., 2001 and Passos et al., 2005, 2010.
Remarks: Amaral (1930b: 65) incorrectly listed the type as IB 5315 *fide* Passos et al. (2010: 40).

110. *Atractus snethlageae* Cunha & Nascimento, 1983b. Bol. Mus. Para. Emilio Goeldi (Zool.) (123): 19–21, pl. 2, figs. 2–3. (*Atractus flammigerus snethlageae*)

Type: Holotype, MPEG 10131, a 218 mm male.
Type locality: "Colônia Nova, rio Gurupi, Pará, rodovia BR-316, 10 km antes do Gurupi" [= Colônia Nova, along Rîo Grupi at BR 316, E Para State, NE Brazil, 1°49'S, 46°10'W, elevation 30 m].
Distribution: Northeastern Brazil (Amapá, E Amazonas, Bahia, Maranhão, Pará, Rondônia), ext. N Bolivia (Pando) and single record for Argentina (Chaco), NSL–500 m.
Sources: Martins & Oliveira, 1993, Frota, 2000a, Silva-Haad, 2004 and Gonzales & Embert, 2008.
Remarks: Unconfirmed reports from Colombia and Peru (Silva-Haad, 2004: 439).

111. *Atractus steyermarki* Roze, 1958c. Acta Biol. Venez. 2(25): 301–302.

Type: Holotype, FMNH 69920 (formerly CNHM 69920), a 270 mm male (C. Griffin, 9 April 1953).
Type locality: "Chimantá Tepui, Estado Bolívar, Venezuela; 1.430 metros de altura" [= ca, 5°18'N, 62°14'W].
Distribution: Southeastern Venezuela (Bolívar) and Guyana (Cuyuni-Mazaruni, Potaro-Siparuni), 1430–2440 m.
Sources: Roze, 1966a, Lancini, 1986, Gorzula & Señaris, 1998, C.W. Myers & Donnelly, 2008, Navarrete et al., 2009 and C.J. Cole et al., 2013.
Remarks: The W Guyana record of Hoogmoed (1979: 275) needs confirmation. Holotype illustrated in C.W. Myers & Donnelly (2008: figs. 70–71).

112. *Atractus surucuru* Prudente & Passos, 2008. J. Herp. 42(4): 724–727, figs. 1–2.

Type: Holotype, MPEG 19146, a 418 mm female (S. Almeida, 19 Nov. 1991).
Type locality: "Serra do Surucucu (02°47'N, 63°40'W, approximately 1000 m), State of Roraima, Brazil."
Distribution: Northern Brazil (Roraima), 1000 m.

113. *Atractus taeniatus* L.E. Griffin, 1916. Mem. Carnegie Mus. 7(3): 173–174, pl. 28, figs. 1–3.

Type: Holotype, CM 117, a 218 mm male (J. Steinbach, Aug. 1906–Nov. 1925).
Type locality: "near Santa Cruz de la Sierra, Bolivia" [= Santa Cruz de la Sierra, Santa Cruz Department, Bolivia, 17°48'S, 63°10'W, elevation 425 m].
Distribution: Southern South America. Southern Brazil (Mato Grosso, Pará, Rio Grande do Sul, Rondônia), E Bolivia (Santa Cruz) and NE Argentina (Corrientes, Entre Ríos, Misiones), 370 m.
Sources: C.J. McCoy, 1971, Cei, 1994, Lema, 1994, R. Fernandes, 1995b, Lema, 1995, Moura-Leite et al., 1996, Rey & Lions, 1997a and Giraudo & Scrocchi, 2002.

114. *Atractus tamaensis* Esqueda & La Marca, 2005. Herpetotropicos 2(1): 20–22, fig. 19.

Type: Holotype, MHNLS 8307, a 283 mm male (A. Agudo, 7 Aug. 1980).
Type locality: "Venezuela, Estado Táchira: Municipia Junin: Betania Macizo de Tamá, apox. 2.100 m.s.n.m., coordenadas estimadas 7°24'21"N, 72°24'29"W."
Distribution: Western Venezuela (Táchira), 2100 m. Known only from type locality.
Source: Navarrete et al., 2009.

115. *Atractus tamessari* Kok, 2006. Zootaxa (1378): 20–26, figs. 1–2a–b, 3, 4a, 4c.

Type: Holotype, IRSNB 2640 (formerly PK 1365), a 397 mm male (P.J.R. Kok, P. Benjamin & G. Seegobin, 23 March 2006).
Type locality: "tributary of Elinkwa River, ESE Kaieteur National Park, ca. 500 m elevation, Potaro-Siparuni District, Guyana (5°08'09"N, 59°25'28"W)."
Distribution: Eastern Venezuela and cen. Guyana (Potaro-Siparuni), 500 m.
Sources: Rivas-Fuenmayor et al., 2012 and C.J. Cole et al., 2013.

116. *Atractus taphorni* Schargel & García-Pérez, 2002. J. Herp. 36(3): 398–400, fig. 1.

Type: Holotype, MCNG 1905, a 423 mm male (A. Lobo, 26 June 1993).
Type locality: "La Carbonera, Cordillera de Mérida, 2200 m, Mérida, Venezuela (8°37'52"N, 71°22'08"W)."
Distribution: Western Venezuela (Mérida), 1700–2200 m.
Sources: Esqueda & La Marca, 2005 and Navarrete et al., 2009.

117. *Atractus thalesdelemai* Passos, Fernandes & Zanella, 2005. Herpetologica 61(2): 210–213, figs. 1–2.

Synonym: *Atractus kangueryensis* Cacciali, Villalba & Yanosky, 2007a.
Type: Holotype, MNRJ 10052, a 303 mm male (N. Zanella, 30 Jan. 2001).

Type locality: "Fazenda da Brigada Militar, Municipality of Passo Fundo (28° 14' 30" S, 52° 21' 27" W), State of Rio Grande do Sul, Brazil."
Distribution: Southern Brazil (Rio Grande do Sul) and S Paraguay (Itapúa), 100–700 m.
Sources: Lema, 1994 and Passos et al., 2010.

118. *Atractus titanicus* Passos, Arredondo, Fernandes & Lynch, 2009c. Copeia 2009(3): 431–433, figs. 7–8.

Type: Holotype, ICNMNH 10697, a 473 mm female (native, 5 March 1996).
Type locality: "Colombia, Antioquia department, Sonsón municipality, 05°43'N, 75°19'W, Quebrada San Andrés, ca. 2400 m."
Distribution: Western Colombia (Antioquia, Caldas, Valle de Cauca), 1800–2400 m.
Source: Rojas-Morales, 2012b.

119. *Atractus torquatus* (A.-M.-C. Duméril, Bibron & Duméril, 1854a). Erpét. Gén. 7(1): 101–103. (*Rabdosoma torquatum*)

Synonyms: *Brachyorrhos torquatus* H. Boie *in* Schlegel, 1826a *(nomen nudum), Brachyorrhos torquatus* F. Boie, 1827 (*nomen nudum*), *Calamaria badia* Schlegel, 1837, *Rabdosoma varium* Jan, 1862b, *Atractus davidhardi* Silva-Haad, 2004, *Atractus janethae* Silva-Haad, 2004 and *Atractus lucilae* Silva-Haad, 2004.
Type: Lectotype, MNHN 437, a male (A.C.V.M.D. d'Orbigny, 11 Nov. 1830–21 June 1831 or 17 Sept.–17 Nov. 1832), designated by Savage (1960: 66).
Type locality: "Santa Cruz, Santa Cruz, Bolivia" [= Santa Cruz de la Sierra, Santa Cruz Department, Bolivia, 17°48'S, 63°10'W, elevation 425 m] via lectotype selection.
Distribution: Amazonia. Colombia (Amazonas), SE Venezuela (Amazonas, Bolívar), Guyana (Upper Demerara-Berbice, Upper Takutu-Upper Essequibo), Suriname (Brokopondo, Nickerie, Para, Saramacca), French Guiana (Cayenne), N Brazil (Amazonas), NE Peru (Loreto) and E Bolivia (Santa Cruz), NSL–600 m.
Sources: Hoogmoed, 1980, Hoogmoed & Avila-Pires, 1991, Martins & Oliveira, 1993, C.L. Barrio-Amorgós et al., 1999, Schargel & Castoe, 2003, Silva-Haad, 2004, Kok, 2006, Navarrete et al., 2009, Passos & Prudente, 2012 and C.J. Cole et al., 2013.
Remarks: Lectotype designation of RMNH 114 and restriction of type locality to Paramaribo, Suriname *fide* Hoogmoed (1980: 36) are invalid. Bolivia (Santa Cruz de La Sierra) record doubtful *fide* Passos & Prudente (2012: 13).

120. *Atractus trihedrurus* Amaral, 1926a. Arch. Mus. Nac. Brazil 26: 105–106, pl. 2, figs. 1–4.

Type: Holotype, IB 3098, a 610 mm male (J. Naderer, 17 Oct. 1924), destroyed by fire 15 May 2010.
Type locality: "S. Bento, Estado de Santa Catharina," Brazil" [= 26°15'S, 49°23'W, ca. 870 m. *fide* Passos et al., 2010: 47].
Distribution: Southern Brazil (Paraná, Rio Grande do Sul, Santa Catarina, São Paulo), 500–1400 m.
Sources: Marques et al., 2001, Lema, 2005 and Passos et al., 2005, 2010.

121. *Atractus trilineatus* Wagler, 1828b. Isis von Oken 21(7): 742, pl. 10, figs. 1–4.

Synonyms: *Coluber brachyurus* Kuhl, 1820 *(nomen praeoccupatum), Brachyorrhos brachyurus* H. Boie *in* Schlegel, 1826a *(nomen nudum), Brachyorrhos kuhli* Schlegel, 1826a *(nomen rejiciendum), Brachyorrhos kuhlii* H. Boie *in* F. Boie, 1827 (*nomen rejiciendum), Rabdosoma lineatum* Duméril, Bibron & Duméril, 1854a, *Rabdosoma punctatovittatum* Jan, 1862b, and *Rabdosoma trivirgatum* Jan, 1862b.
Type: Lectotype, RMNH 48, a 205 mm female, designated by Hoogmoed (1982: 134–135).
Type locality: "Java" [W Indonesia] (in error *fide* Hoogmoed, 1982: 131) via lectotype selection. Restricted to South America *fide* Savage (1960: 83) and British Guiana *fide* Roze (1966a: 88).
Distribution: Northern South America. Northeastern Venezuela (Anzoátegui, Bolívar, Delta Amacuro, Monagas, Sucre), Trinidad & Tobago (Huevos, Little Tobago, Tobago, Trinidad), N Guyana (Essequibo Islands-West Demerara, Upper Demerara-Berbice) and N Brazil (Amazonas, Roraima), NSL–105 m.
Sources: Parker, 1935, Beebe, 1946, Roze, 1966a, Emsley, 1977, Lancini, 1986, Hoogmoed, 1982, 1983, Martins & Oliveira, 1993, J.C. Murphy, 1997, Boos, 2001, Kok, 2006, Navarrete et al., 2009 and C.J. Cole et al., 2013.
Remarks: Official Specific Name *fide* ICZN Opinion 2210 (2008: 239).

122. *Atractus trivittatus* Amaral, 1933c. Mem. Inst. Butantan (1932) 7: 118.

Type: Holotype, IB (formerly ILS 84), a 180 mm male (H. Nicéforo-María, 1908–1932), destroyed by fire 15 May 2010.
Type locality: "Chita, (sudeste de San Gil), Colombia" [= Chita, SE of San Gil (6°34'N, 73°08'W, 1170 m), Santander Department, Colombia].
Distribution: Central Colombia (Boyacá, Casanare, Norte de Santander), 1170–3005 m.

A

123. *Atractus turikensis* Barros-Blanco, 2000. Anartia 11: 3–6. figs. 1, 2a–c (paratype).

Type: Holotype, MBLUZ 301, a 439 mm male (A. Viloria & F. Herrera, 20 March 1991).
Type locality: "Campamento 1800 (fig. 3, ver punto número 1), Mesa Turik, Altitud 1800 msnm, Sierra de Perijá, Estado Zulia, Venezuela."
Distribution: Western Venezuela (Zulia), 1800 m. Known only from vicinity of type locality.
Source: Navarrete et al., 2009.

124. *Atractus typhon* Passos, Mueses-Cisneros, Lynch & Fernandes, 2009e. Zootaxa (2293): 26–29, figs. 17–18.

Type: Holotype, ICNMNH 10901, a 372 mm male (B. Cépeda & J.J. Mueses-Cisneros, 14 July 2006).
Type locality: "Reserva Natural Biotopo Selva Húmeda (01°25'M. 78°17'W, ca. 600 m), vereda Berlin, El Diviso, municipality of Barbacoas, department of Nariño, Colombia."
Distribution: Western Colombia (Nariño), 600 m. Known only from type locality.

125. *Atractus univittatus* (Jan, 1862b). Arch. Zool. Anat. Fis. 2(1): 15. (*Rabdosoma univittata*)

Type: Holotype, ZMH 2851 (formerly ZMH 482), a 300 mm specimen.
Type locality: "Caracas" [= Caracas, Distrito Federal State, Venezuela, 10°29'N, 66°54'W, elevation 925 m].
Distribution: Northern Venezuela (Aragua, Barinas, Cojedes, Distrito Federal, Mérida, Miranda, Portuguesa) and Trinidad & Tobago (Tobago), 100–1100 m.
Sources: Roze, 1966a, Lancini, 1986, J.C. Murphy, 1997, Boos, 2001, Markezich, 2002, Schargel & Castoe, 2003, Esqueda & La Marca, 2005, Rivas-Fuenemayer & Barrio-Amorgós, 2005, Hallermann, 2006 and Navarrete et al., 2009.
Remarks: Possibly a synonym of *A. fuliginosus fide* Rivas-Fuenmayor et al. (2012: 39).

126. *Atractus variegatus* Prado, 1942a. Mem. Inst. Butantan (1941) 15: 379, 1 fig.

Type: Holotype, ILS 201, a 345 mm male (H. Nicéforo-María, 1908–1941), destroyed by fire 9–10 April 1948.
Type locality: "La Uvita, departamento de Boyacá, Colômbia" [= La Uvita, Boyacá Department, Colombia, 6°16'N, 72°33'W, elevation 3050 m].
Distribution: Central Colombia (Boyacá), 3050 m. Known only from vicinity of type locality.
Remarks: Duellman (1979: 455) listed elevation of 4000 m.

127. *Atractus ventrimaculatus* Boulenger, 1905b. Ann. Mag. Nat. Hist. (7) 15(89): 455.

Type: Lectotype, BMNH 1946.1.7.7 (formerly BMNH 1905.5.31.74) (Briceño), designated by Roze (1961b: 108).
Type locality: "Mérida, Venezuela" [= Fuqueros, Estado Mérida, Venezuela, 1630 m] via lectotype selection.
Distribution: Northwestern Venezuela (Mérida, Zulia), 1200–2500 m.
Sources: Roze, 1966a, Lancini, 1986, Barros-Blanco, 2000, Schargel & Castoe, 2003, Esqueda & La Marca, 2005, Rojas-Runjaic et al., 2007 and Navarrete et al., 2009.
Remarks: Syntypes BMNH 1946.1.5.12–15 [Merida, 1630 m] and BMNH 1946.1.7.7 [Fuqueros, 2500 m] *fide* P. Campbell *(in litt.)*.

128. *Atractus vertebrolineatus* Prado, 1941d. Mem. Inst. Butantan (1940) 14: 25–26, 1 fig.

Type: Holotype, ILS 184, a 296 mm male (H. Nicéforo-María, 1908–1940), destroyed by fire 9–10 April 1948.
Type locality: "Ocaña, ao norte de Bucaramanga, Colombia" [= Ocaña, Norte de Santander Department, Colombia, 8°14'N, 73°21'W, elevation 1200 m].
Distribution: Northeastern Colombia (Norte de Santander), 1200 m. Known only from type locality.

129. *Atractus vertebralis* Boulenger, 1904b. Ann. Mag. Nat. Hist. (7) 13(78): 451.

Type: Holotype, BMNH 1946.1.2.6, a 470 mm female (G.R. Ockendon, 1902–1904).
Type locality: "Santo Domingo, Carabaya, Peruvian Andes, altitude 6000 feet" [= Santo Domingo, Cordillera de Carabaya, Puno Department, SE Peru, 1830 m].
Distribution: Southeastern Peru (Cusco, Puno), 1830 m.

130. *Atractus vittatus* Boulenger, 1894a. Cat. Snakes Brit. Mus. 2: 304, pl. 15, fig. 2.

Type: Holotype, BMNH 1946.1.6.47, a 405 mm female.
Type locality: "Caracas, Venezuela" [= Caracas, Distrito Federal State, Venezuela, 10°29'N, 66°54'W, elevation 925 m].
Distribution: Northern Venezuela (Aragua, Distrito Federal), 800–1800 m.
Sources: Duellman, 1979 and Navarrete et al., 2009.
Remarks: A member of *Geophis fide* Savage (1960: 31), an *Atractus fide* Roze (1961: 108–109), and *incertae sedis fide* J.A. Peters & Orejas-Miranda (1970: 327).

131. *Atractus wagleri* Prado, 1945. Ciencia 6(2): 61, fig. 1.

Type: Holotype, ILS 228, a 411 mm female (H. Daniel, 1943), destroyed by fire 9–10 April 1948.

Type locality: "Humbo (Boyacá), Colômbia" [= Humbo, Quíama, Boyacá Department, Colombia, 5°36'N, 74°16'W, 1035 m].
Distribution: Northern Colombia (Boyacá, Santander), 740–1200 m.
Source: Passos & Arredondo, 2009.
Remarks: Original description reprinted in Prado (1946b: 110–111, 1 fig.).

132. *Atractus werneri* Peracca, 1912. Mém. Soc. Neuchâtel Sci. Nat. (1914) 5: 102.

Synonym: *Atractus colombianus* Prado, 1940b.
Type: Holotype, MHNN 91.506 (formerly MZN 75 & MHNN 3), a 245 mm female (O. Fuhrmann, 1910).
Type locality: "Cafetal Argélia, près Viota (C. or Dép. Cundinamarca), alt. 1830 m, Colombie" [= 04° 26' N, 74° 31' W *fide* Passos & Lynch, 2010: 166].
Distribution: Central Colombia (Cundinamarca), 1200–1830 m.
Source: Passos & Lynch, 2010.
Remarks: Schätti (1986a: 98) and Andreone & Gavetti (2007: 65, 88) discussed publication date, which has previously been cited as 1913 or 1914.

133. *Atractus zebrinus* (Jan, 1862b). Arch. Zool. Anat. Phys. 2(1): 11, 15–16, pl. 8, fig. 3rd (*Rabdosoma zebrinum*)

Type: Holotype, formerly MSNM, a 150 mm specimen, destroyed in 1943 during World War II.
Type locality: Unknown.
Distribution: Southeastern Brazil (S Bahia, Espírito Santo, S Minas Gerais, Paraná, Rio de Janeiro, Santa Catarina, São Paulo), 500–1700 m.
Sources: Fernandes et al., 2000 and Passos et al., 2005, 2010.
Remarks: Bahia record needs confirmation *fide* Passos et al. (2010: 52).

134. *Atractus zidoki* Gasc & Rodrigues, 1979a. Bull. Mus. Natl. Hist. Nat. (4) 1A(2): 548–552, fig. 1.

Type: Holotype, MNHN 1977.1601, a 285 mm male (P. Grenand, 1971–1977).
Type locality: "Trois-Sauts, Haut Oyapock, Guyane française" [= village of Trois Sauts, Haut Oyapock basin, Cayenne Department, SE French Guiana, 2°14'N, 52°52'W, elevation 170 m].
Distribution: Northern South America. Eastern Colombia (Amazonas), Suriname (Brokopondo, Marowijne), Guyana (Potaro-Siparuni), French Guiana (Cayenne, Saint-Laurent-du-Maroni) and N Brazil (Amapá, Pará), 150–500 m.
Sources: Hoogmoed, 1980, Cunha & Nascimento, 1983b, 1984, Starace, 1998, Silva-Head, 2004, Means, 2007, Prudente & Passos, 2008 and Prudente & Passos, 2010.

ATRETIUM Cope, 1861e (*nomen substitutum*) (Natricidae)

Synonym: *Tropidophis* Gray, 1849a *(nomen praeoccupatum).*
Type species: *Coluber schistosus* Daudin, 1803d.
Distribution: Southern Asia.
Source: Zaher et al., 2009.

1. *Atretium schistosum* (Daudin, 1803d). Hist. Nat. Rep. 7: 132–133. (*Coluber schistosus*)

Synonyms: *Tropidonotus dimidiatus* H. Boie *in* Schlegel, 1826a, *Tropidonotus dimidiatus* Schlegel, 1826b, *Tropidonotus dimidiatus* F. Boie, 1827, *Tropidonotus moestus* Cantor, 1839, *Tropidonotus surgens* Cantor, 1839, *Coluber flavus* Fitzinger, 1861 (*nomen nudum*), and *Helicops flaviventris* Fitzinger, 1861.
Type: Lectotype, a 483 mm specimen described and illustrated by P. Russell (1801: 5–6, pl. 4) (native, 1781–1791), designated herein.
Type locality: "Bengale" [= Bangladesh and NE India (Assam, Manipur, Meghalaya, Mizoram, Nagaland, or West Bengal)] via lectotype selection.
Distribution: Southern Asia. Eastern India (Andhra Pradesh, Karnataka, Kerala, Orissa, Uttar Pradesh, S West Bengal), Nepal (Dadeldhura, Ilam, Jhapa, Rupandehi, Sankhuwasabha, Surkhet) and Sri Lanka (Central, Eastern, North-Central, North-Western, Northern, Sabaragamuwa, Southern, Uva, Western), 75–1680 m.
Sources: Wall, 1912d, M.A. Smith, 1943, Kramer, 1977, P. Silva, 1969, 1980, Whitaker, 1975, Schleich & Kästle, 2002, Whitaker & Captain, 2004, N. Khaire, 2006, Somaweera, 2006, R.C. Sharma, 2007 and A. Silva, 2009.

2. *Atretium yunnanensis* J. Anderson, 1879. Anat. Zool. West. Yunnan (1878) 1: 822–823. (*Atretium schistosum yunnanensis*)

Synonym: *Helicops schistosus andersonii* Wall, 1909.
Types: Syntypes (3), ZSI 4191–92 (formerly IMC 4191–92), 876 and 629 mm specimens, and ZSI 4196 (formerly IM 4196), a 780 mm male (J. Anderson [First Yunnan Exped.], 13–23 May 1868 [Muangla] & 10–22 Aug. 1868 [Hotha]).
Type locality: "Muangla and Hotha at elevations of 2,000 and 4,500 feet, Western Yunnan" [= Lianghe, Yingjiang County, 24°48'N, 98°18'E, elevation 1050 m, and Husa'achangzuxiang, Longchuan County, 24°28'N, 97°54'E, elevation 1435 m, W Yunnan Prov., SW China].
Distribution: Southern China (ext. W Yunnan), 800–1435 m.
Sources: C.H. Pope, 1935, Zhao & Adler, 1993, I. Das et al., 1998 and Zhao, 2006.

A

ATROPOIDES Werman *in* Campbell & Brodie, 1992 (Viperidae)

Synonyms: *Atropos* Rüppell, 1845 *(nomen praeoccupatum), Atropus* J.W. Müller, 1865 *(nomen emendatum), Atropes* – L.C. Stuart, 1948 *(nomen incorrectum),* and *Adelynhoserserpenae* Hoser, 2012c *(nomen illegitimum).*

Type species: *Trimeresurus nummifer picadoi* Dunn, 1939b.

Distribution: Mesoamerica.

Sources: J.A. Campbell & Lamar,1989, 2004, Werman 1992, Kraus et al., 1996, Werman, 1997, David & Ineich, 1999, McDiarmid et al., 1999, Parkinson, 1999, Castoe et al., 2003, 2005, 2009, Castoe & Parkinson, 2006, Luna-Reyes & Suárez-Velázquez, 2008, E.N. Smith & Ferrari-Castro, 2008, Jardin et al., 2010 and Hoser, 2012d.

1. *Atropoides indomitus* E.N. Smith & Ferrari-Castro, 2008. Zootaxa (1948): 59–66, figs. 1–4.

Type: Holotype, UTA 52952, 573 mm male (J.A. Ferrari-Castro, 17–22 May 2002).

Type locality: "near the edge of Quebrada de Botaderos, Montaña de Botaderos, Departamento de Colón, Honduras, 670 m (15°26'03.4" N 86°08'37.3" W)."

Distribution: Eastern Honduras (Colón, Olancho), 670–1200 m.

Source: McCranie, 2011a.

2. *Atropoides mexicanus* (A.-M.-C. Duméril, Bibron & Duméril, 1854b). Erpet. Gén. 7(2): 1521–1522. (*Atropos mexicanus*)

Synonyms: *Atropos mexicanus* A.-M.-C. Duméril, 1853 (*nomen nudum*), and *Bothriechis nummifera notata* J.G. Fischer, 1880,

Type: Holotype, MNHN 7612.

Type locality: "Coban, capitale de la Prov. de la Vera-Paz, (Republique de Guatemala, Amérique centrale)" [= Cobán, SW Alta Verapaz Dept., cen. Guatemala, 15°28'N, 90°22'W, elevation 1325 m].

Distribution: Mesoamerica. Southern Mexico (Chiapas, S Tabasco), S Belize (Belize, Cayo, Stann Creek, Toledo), Guatemala (Alta Verapaz, Baja Verapaz, Huehuatenango, Izabal, Petén), El Salvador (Chalatenango), NW Honduras (Atlántida, Copán, Cortés, Santa Bárbara), Nicaragua (Matagalpa, Río San Juan), Costa Rica (Alajuela, Cartago, Guanacaste, Limón, Puntarenas, San José) and W Panama (Canal Zone, Chiriquí, Coclé, Colón), NSL–1600 m.

Sources: Mocquard, 1909, K.P. Schmidt, 1941, L.C. Stuart, 1948, 1963, Burger, 1950, E.H. Taylor, 1951, 1954, Mertens, 1952c, Neill & Allen, 1960, R.W. Henderson & Hoevers, 1975, Villa, 1984, L.D. Wilson & Meyer, 1985, Casas-Andreau et al., 1996, J.C. Lee, 1996, 2000, J.A. Campbell, 1998b, Savage, 2002, Solórzano, 2004, Jadin et al., 2010, McCranie, 2011a and Ray, 2011.

3. *Atropoides nummifer* (Rüppell, 1845). Mus. Senck., Amph. 3: 313. (*Atropos nummifer*)

Synonyms: *Bothrops nummifer veraecrucis* Burger, 1950, *Lachesis nummifera* – Boettger, 1898, *Bothriochis mammifera* – Recinos, 1913 *(nomen incorrectum),* and *Adelynhoserserpenae* Hoser, 2012c *(nomen illegitimum).*

Type: Holotype, SMF 21196 (formerly SMF-R III.KK.1a & SMF-B 9544, 1a) (E. Rüppell, 1840).

Type locality: Unknown. Designated as Mexico *fide* Boettger (1898: 138). Restricted to Teapa, Tabasco, Mexico *fide* Burger (1950: 62).

Distribution: Eastern Mexico (Hidalgo, N Oaxaca, E Puebla, ext. SE San Luis Potosí, Veracruz), 670–1800 m.

4. *Atropoides occiduus* (Hoge, 1966a). Mem. Inst. Butantan (1965) 32: 130. (*Bothrops nummifer occiduus*) *(nomen substitutum)*

Synonyms: *Bothrops affinis* Bocourt, 1868 *(nomen praeoccupatum),* and *Bothrops nummifer occidduus* – Hoge & Romano-Hoge, 1981a *(nomen incorrectum).*

Types: Syntypes (3), MNHN 1194–96.

Type locality: "Saint-Augustin (Guatémala), versant occidentale de la Córdillère. 610 mètres d'altitude."

Distribution: Mesoamerica. Southeastern Mexico (SE Chiapas), S Guatemala and W El Salvador (Ahuachapán, Chalatenango, La Libertad, San Vicente, Sonsonate), 610–1700 m.

Sources: Castoe et al., 2003 and G. Köhler et al., 2005.

5. *Atropoides olmec* (Pérez-Higareda, Smith & Juliá-Zertuche, 1985). Bull. Maryland Herp. Soc. 21(3): 97–99, fig. 1. (*Porthidium olmec*)

Type: Holotype, UNAM-LT 1300, a 660 mm female (G. Pérez-Higareda, Sept. 1981).

Type locality: "crest of Cerro Egega, 1100 m, Los Tuxtlas uplands, municipality of Catemaco, southern Veracruz, México" [= 18°25'N, 95°07'W].

Distribution: Southeastern Mexico (Chiapas, SE Oaxaca, S Veracruz) and cen. Guatemala (Alta Verapaz), 530–1100 m.

Source: Castoe et al., 2003.

Remarks: Possibly conspecific with *Atropoides nummifer* *fide* J.A. Campbell & Lamar (1989: 325).

6. *Atropoides picadoi* (Dunn, 1939b). Proc. Biol. Soc. Washington 52: 165. (*Trimeresurus nummifer picadoi*)

Type: Holotype, USNM 37753, a 345 mm female (E. Fernandez).

Type locality: "La Palma, 4500 feet, north of San Jose, Costa Rica" [= La Palma, 1.4 km N San José (9°55'N, 84°05'W), San José Prov., cen. Costa Rica, elevation 1372 m].

Distribution: Costa Rica (Alajuela, Cartago, Guanacaste, Heredia, Limón, Puntarenas, San José) and W Panama (Bocas del Toro, Chiriquí), 15–2000 m.

Sources: E.H. Taylor, 1951, 1954, V. Martinez, 1983, N.J. Scott et al., 1983, Werman, 1984, Savage & Villa, 1986, Solis, 1991, Savage, 2002 and Zaher et al., 2012.

Remarks: The basal species of *Cerrophidion* clade *fide* Zaher et al., 2012, presenting a nomenclatural problem as *Trimeresurus nummifer picadoi* Dunn is type species of *Atropoides* Werman [publ. June 1992], which has priority over *Cerrophidion* Campbell & Lamar [publ. Aug. 1992] either making *Cerrophidion* a junior synonym of *Atropides* or recognizing both genera, but in either case requiring a new generic name for the five "*Atropoides*" species.

†*AUSTRALOPHIS* Gómez, Báez & Rougier, 2008 (Anilioidea *incertae sedis*)

Type species: †*Australophis anilioides* Gómez, Báez & Rougier, 2008.

Distribution: Upper Cretaceous of Argentina.

Sources: Gómez et al., 2008 and Albino, 2011b.

1. †*Australophis anilioides* Gómez, Báez & Rougier, 2008. Cretaceous Res. 29(3): 483–485, fig. 3.

Type: Holotype, MML–PV 181, one trunk vertebra.

Type locality: "Bajo Trapulcó (39°47'02"S, 66°42'17"W), 150 km south-west of Lamarque, Rio Negro Prov., northern Patagonia, Argentina. Lower member of Allen Formation (Malargile Group, southern margin of the Neuquén Basin), Late Campanian-Early Maastrichtian."

Distribution: Upper Cretaceous (Campanian-Maastrichtian: 65.5–83.6 mya) of Argentina.

AUSTRELAPS Worrell, 1963a (Elapidae)

Synonyms: *Austrelops* – Porter, 1972 *(nomen incorrectum)*, and *Australaps* – Carpenter, 1986 *(nomen incorrectum)*.

Type species: *Hoplocephalus superbus* A.C.L.G. Günther, 1858.

Distribution: Southeastern Australia.

Sources; Kinghorn, 1964, S. Phillips, 1980, Shine & Allen, 1980, Lombardi, 1985, Schwaner et al., 1985, Shine, 1987b, S.K. Wilson & Knowles, 1988, Gow, 1989, Hoser, 1989, 2009a, 2012e, Rawlinson, 1991, Ehmann, 1992, Fearn & Munday, 1995, Greer, 1997, David & Ineich, 1999, Cogger, 2000, Sanders et al., 2009 and Zaher et al., 2009.

1. *Austrelaps labialis* (Jan, 1859a). Rev. Mag. Zool. (2) 11: 124, 128. (*Alecto labialis*)

Type: Neotype, SAMA 26414, a 550–554 mm female, designated by Rawlinson (1991: 127).

Type locality: "Islet 477, Pelican Lagoon, Kangaroo Island, South Australia, Australia" via neootype selection.

Distribution: Extreme S Australia (ext. SE South Australia, Kangaroo Is. and islets), NSL–625 m.

Remarks: Supplemental original description in Jan (1859e: pl. c, fig. no. 3).

2. *Austrelaps ramsayi* (Krefft, 1864c). Proc. Zool. Soc. London 32(1): 180–181, 3 figs. (*Hoplocephalus ramsayi*)

Synonyms: *Hoplocephalus bransbyi* Macleay, 1878a, *Hoplocephalus brankysi* – Goldman, Hill & Stanbury, 1969 *(nomen incorrectum)*, *Austrelaps superbus victoria* Saint-Girons & Détrait, 1980 *(nomen nudum)*, and *Austrelaps paulinus* Wells & Wellington, 1985.

Type: Neotype, AMS 31922 (formerly MM MR1362, 541), a 412 mm specimen, designated by Rawlinson (1991: 128).

Type locality: "Moss Vale, New South Wales (34° 33' S, 150° 23' E), Australia, Australia" via lectotype selection.

Distribution: Southeastern Australia (E New South Wales, E Victoria), NSL–2000 m.

Sources: G. Swan, 1990, Hoser, 1991 and Shea & Sadlier, 1985, 1999.

Remarks: Original description reprinted in Krefft (1865a: 67, 3 figs.). Holotype (AMS 6642) lost *fide* Rawlinson (1991: 128).

3. *Austrelaps superbus* (A.C.L.G. Günther, 1858). Cat. Colub. Snakes Brit. Mus.: 217. (*Hoplocephalus superbus*)

Synonyms: *Alecto schmidti* Jan, 1863b (*nomen nudum*), *Alecto schmidtii* Jan, 1873 *in* Jan & Sordelli, 1870–1881, *Austrelaps superbus australia* – Saint-Girons Détrait, 1980 *(nomen incorrectum)*, and *Danisonia superba* – Khole, 1991 *(nomen incorrectum)*.

Type: Lectotype, BMNH 1946.1.20.37, a 725 mm male (R.C. Gunn, July 1832–Jan. 1847), designated by Rawlinson (1991: 126).

Type locality: "Tasmania" [Australia].

A

Distribution: Southeastern Australia (E New South Wales, ext. SE South Australia, S Victoria, Tasmania, Bruny, Cape Barren, Chalky, Clarke, Flinders, Hunter, King, Maria, Outer Sister, Preservation and Three Hummock Is.), NSL–2125 m.

Sources: F. McCoy, 1878b, Littlejohn, 1962 and M.J. Smith, 1975.

AZEMIOPS Boulenger, 1888 (Viperidae)

Synonyms: *Azemops* – Palacky, 1898 *(nomen incorrectum),* and *Azemiophis* – Cope, 1900 *(lapsus calami).*

Type species: *Azemiops feae* Boulenger, 1888h.

Distribution: Eastern Asia.

Sources: C.H. Pope, 1935, Liem et al., 1971, Marx & Rabb, 1972, B. Hu et al., 1980, Zhao & Zhao, 1981, Yuan, 1983a, J.C. Murphy, 1984, Lombard et al., 1986, Tian et al., 1986, Price 1987a, Maes, 1989, Song, 1990, Cadle, 1992, Golay et al., 1993, Knight & Mindell, 1993, Zhao & Adler, 1993, Heise et al., 1995, Orlov, 1995, 1997, David & Ineich, 1999, McDiarmid et al., 1999, Underwood, 1999, Orlov et al., 2000 Kelly et al., 2003, Leviton et al., 2003, Mallow et al., 2003, J.C. Murphy & Schlager, 2003, Zhao, 2006, Wüster et al., 2008, V.S. Nguyen et al., 2009, Phelps, 2010, I. Das, 2012, Hoser, 2012d, and Yao, 2012.

1. *Azemiops feae* Boulenger, 1888h. Ann. Mus. Civ. Stor. Nat. Genova (2) 6: 603, pl. 7.

Type: Holotype, MSNG 30891, a 610 mm specimen (L. Fea, 1886).

Type locality: "Kakhien Hills, north of Tenasserim, Burma" [= Kachin Hills, Tanintharyi State, S Myanmar].

Distribution: Eastern Asia. Southern China (Anhui, Fujian, Gansu, Guangxi, Guizhou, Hunan, Jiangxi, Shaanxi, Sichuan, SE Xizang, Yunnan, Zhejiang), N Myanmar (Kachin), and N Vietnam (Bac Kan, Cao Bang, Vinh Phu), 700–2000 m.

B

BALANOPHIS M.A. Smith, 1938 (Natricidae)

Type species: *Tropidonotus chrysargus ceylonensis* A.C.L.G. Günther, 1858.
Distribution: Sri Lanka.
Sources: Wall, 1921g, M.A. Smith, 1943, Deraniyagala, 1955, McDowell, 1961, P. Silva, 1969, 1980a, Mahendra, 1984, A. Silva, 1990b, 2001, 2009, Somaweera, 2006 and Zaher et al., 2009.
Remarks: Possibly a synonym of *Rhabdophis fide* Young & Kardong (1996: 273).

1. *Balanophis ceylonensis* (A.C.L.G. Günther, 1858). Cat. Colub. Snakes Brit. Mus.: 71. (*Tropidonotus chrysargus ceylonensis*)

Type: Holotype, BMNH 1946.1.15.25, a juvenile male (H. Cuming, 1836–1840).
Type locality: "Ceylon" [= Sri Lanka].
Distribution: Sri Lanka (Central, Sabaragamuwa, Southern, Uva, Western), 915–1220 m.

BAMANOPHIS Schätti & Trape, 2008 (Colubridae)

Type species: *Periops dorri* Lataste, 1888.
Distribution: West Africa.
Sources: Hughes & Barry, 1969, Villiers, 1975, Roman, 1984, 2008, J.-F. Trape, 1997, Hughes, 1983, 2013, Chippaux, 2006, J.-F. Trape & Mané, 2006b, Schätti & Trape, 2008 and Segniagbeto et al., 2011.

1. *Bamanophis dorri* (Lataste, 1888). Naturaliste (2) 10(38): 227. (*Periops dorri*)

Types: Syntypes (2), BMNH 1946.1.13.3 (formerly BMNH 1920.1.20.3742), a 735 mm female, and BMNH 1946.1.14.82 (formerly BMNH 1920.1.20.3785), a 435 mm female (E. Dorr, 3–4 June 1887).
Type locality: "dans le fort et à côté du fort de Bakel, Haut-Senegal" [= Bakel, Senegal, 14°52'N, 12°31'W, elevation 25 m, a French fortress and garrison in the border triangle of Senegal with Mali and Mauritania, below the junction of the Senegal and Falémé Rivers *fide* Schätti & Trape, 2008: 598].
Distribution: West Africa. Southern Mauritania (Assaba, Hodh El Gharbi), S Mali (Kayes, Koulikoro, Mopti, Sikasso), E Senegal (Kédougou, Tambacounda), W Guinea (Kindia, Labé, Mali), S Burkina-Faso (Est, Hauts-Bassins), NE Ghana (N Norther), N Togo (Savanes) and N Benin, 25–600 m.
Remarks: Probably occurs in N Ivory Coast, SW Niger and NW Nigeria *fide* Schätti & Trape (2008: 604).

†*BAVARIOBOA* Szyndlar & Schleich, 1993 (Boidae)

Type species: †*Bavarioboa hermi* Szyndlar & Schleich, 1993.
Distribution: Upper Oligocene to middle Miocene of Germany and Turkey, upper and middle Oligocene of France; lower Miocene of France, Bohemia, Czech Republic, Germany and ? Portugal.
Sources: Szyndlar & Schleich, 1993, 2003, Ivanov, 2002 and Szyndlar & Hosgör, 2012.

1. †*Bavarioboa bachensis* Szyndlar & Rage, 2003. Booidea Oligocene Miocene Europe: 24–25, figs. 5a–e.

Type: Holotype, USTL ESP 601, one trunk vertebra.
Type locality: "Espeyrasse (Phosphorites du Quercy complex, France), middle Oligocene (MP 26)."
Distribution: Middle Oligocene (Arvernian, MP 26: 23.0–28.4 mya) of France.

2. †*Bavarioboa crocheti* Szyndlar & Rage, 2003. Booidea Oligocene Miocene Europe: 28–38, figs. 9a–e.

Type: Holotype, USTL PDS 3102, one trunk vertebra.
Type locality: "Pech-Desse (Phosphorites du Quercy complex, France), late Oligocene (MP 28)."
Distribution: Upper Oligocene (Arvernian, MP 28: 23.0–28.4 mya) of France.

3. †*Bavarioboa hermi* Szyndlar & Schleich, 1993. Stuttgarter Beitr. Naturk. (192): 7–9, fig. 2.

Type: Holotype, BSPG 1976 XXII 5859, one trunk vertebra.
Type locality: "Petersbuch 2 near Eichstätt, Bayern, southern Germany. Lower Miocene (middle Orleanian; MN 4)."
Distribution: Lower Miocene (Orleanian, MN 4: 16.0–16.9 mya) of S Germany and Czech Republic.

B

4. †*Bavarioboa herrlingensis* Szyndlar & Rage, 2003. Booidea Oligocene Miocene Europe: 41–44, figs. 15e–i.

Type: Holotype, SMNS 59441-3, one trunk vertebra.
Type locality: "Herrlingen 11 (Germany), late Oligocene (MP 28/30)."
Distribution: Upper Oligocene (Arvernian, MP 28/30: 23.0–28.4 mya) of Germany.

5. †*Bavarioboa minuta* Szyndlar & Rage, 2003. Booidea Oligocene Miocene Europe: 38–40, figs. 14l–p.

Type: Holotype, SMNS 58196-1, one trunk vertebra.
Type locality: "Herrlingen 8 (Germany), late Oligocene (MP 28)."
Distribution: Upper Oligocene (Arvernian, MP 28: 23.0–28.4 mya) of Germany.

6. †*Bavarioboa ultima* Szyndlar & Rage, 2003. Booidea Oligocene Miocene Europe: 54–56, figs. 24g–k.

Type: Holotype, SMNS 59091-3, one trunk vertebra.
Type locality: "Rothenstein 13 (Germany), middle Miocene (MN 5)."
Distribution: Middle Miocene (Orleanian, MN 5: 13.7–16.0 mya) of Germany.
Source: Szyndlar, 2009.

7. †*Bavarioboa vaylatsae* Szyndlar & Rage, 2003. Booidea Oligocene Miocene Europe: 26–27, figs. 6b–f.

Type: Holotype, USTL MPF 1102, one trunk vertebra.
Type locality: "Mas-de-Pauffié (Phosphorites du Quercy complex, France), middle Oligocene (MP 26)."
Distribution: Middle Oligocene (Arvernian, MP 26: 23.0–28.4 mya) of France.

BITIA Gray, 1842d (Homalopsidae)

Synonyms: *Bitia* Gray, 1840 *(nomen nudum), Hipistes* Gray, 1849a, *Hypistes* – Palacky, 1898 *(nomen incorrectum),* and *Hipistis* – F. Werner, 1929a *(nomen incorrectum).*
Type species: *Bitia hydroides* Gray, 1842d.
Distribution: Tidal rivers and coastal marine waters of SE Asia, including Andaman Sea and Straits of Malacca.
Sources: M.A. Smith, 1943, E.H. Taylor, 1965, Gyi, 1970, Jayne et al., 1995, J.C. Murphy, 2007b, Alfaro et al., 2008 and Zaher et al., 2009.

1. *Bitia hydroides* Gray, 1842d. Zool. Misc. 2(May): 64.

Synonyms: *Homalopsis hydrina* Cantor, 1847, and *Hipistes fasciatus* Gray, 1849a.
Type: Holotype, BMNH 1946.1.2.38 (formerly BMNH III.15.1.a), a dessicated male.
Type locality: Unknown.
Distribution: Southeast Asia. Southern Myanmar (Bago), S Thailand (Bangkok, Phra Nakhon Si Ayutthaya), Singapore and West Malaysia (Kedah, Johor, Pinang, Penang Is.).
Remarks: E.H. Taylor (1965: 927) questioned the Thailand record of Boulenger, 1896a.

BITIS Gray, 1842e (Viperidae)

Synonyms: *Cobra* Laurenti, 1768 *(nomen rejiciendum), Echidna* Merrem, 1820 *(nomen praeoccupatum), Clotho* Gray, 1840 *(nomen nudum), Clotho* Gray, 1842e *(nomen praeoccupatum), Echidne* – A.-M.-C. Duméril & Bibron, 1844 *(nomen incorrectum), Calechidna* Tschudi, 1846, *Echidua* – Reichenow, 1874 *(nomen incorrectum), Echidra* – Kingsley, 1888 *(nomen incorrectum), Bitus* – Hewitt, 1912 *(nomen incorrectum), Hallowellius* A.F.T. Reuss, 1939a, *Hindius* A.F.T. Reuss, 1939a, *Macrocerastes* A.F.T. Reuss, 1939a, and *Keniabitis* Lenk, Herrmann, Joger & Wink, 1999.
Type species: *Echidna arietans* Merrem, 1820 *(nomen protectum).*
Distribution: Africa.
Fossil record: Pliocene and Pleistocene of Tanzania.
Sources: V.F.M. FitzSimons, 1962a, Broadley, 1967a, 1968a, 1983, Haacke, 1975, Derleyn, 1977, Ashe & Marx, 1988, Branch, 1988, Golay et al., 1993, Herrmann & Joger, 1995, Spawls & Branch, 1995, Wallach, 1998a, David & Ineich, 1999, Herrmann et al., 1999, Lenk et al., 1999, McDiarmid et al., 1999, Lenk et al., 2001b, Mallow et al., 2003, Dobiey & Vogel, 2007, Krecsák, 2007, Wüster et al., 2008, Phelps, 2010 and Hoser, 2012d.
Remarks: Pliocene material of Laetoli, Tanzania represents either †*Bitis olduvaiensis* or a new species *fide* Rage & Bailon (2011: 472–473). Lenk et al., 1999 resurrected two subgenera (*Calechidna* and *Macrocerastes*) and described one (*Keniabitis*), which was supported by Wüster et al. (2008: 454), with species as follows: *Bitis (arietans), Calechidna (albanica, armata, atropos, caudalis, cornuta, heraldica, inornata, peringueyi, rubida, schneideri, xeropaga), Keniabitis (worthingtoni),* and *Macrocerastes (gabonica, nasicornis, parviocula, rhinoceros).* Broadley's (pers. comm.) remarks on the Lenk et al., 1999 recognition of *Macrocerastes* as a subgenus was that "this name was not accompanied by a description that states in words, characters that are purported to differentiate the taxon [ICZN 4th Ed., Article 13.1.1]; on the same grounds I rejected *Hindius.*"

1. *Bitis albanica* Hewitt, 1937. Vert. Fauna E Cape Prov. (2): 76–77, pl. 21, fig. 3. (*Bitis cornuta albanica*)

Type: Lectotype, PEM 8279 (formerly AMG 6860), a 193 mm male (W. Pannell), designated by Branch (1999a: 57).

Type locality: "Kleinpoort, near Committees, 25 km ENE of Grahamstown (33°14'S, 26°46'E; 3326BB), Grahamstown District, Eastern Cape Prov., South Africa" via lectotype selection.

Distribution: Southwestern South Africa (Eastern Cape), 50–500 m.

Sources: Branch, 1994, 1997, 1999.

Remarks: Photograph of the lectotype in Branch (1999a: 58). A subspecies of *B. cornuta fide* Mallow et al., 2003.

2. *Bitis arietans* (Merrem, 1820). Syst. Amph.: 152. (*Echidna arietans*)

Synonyms: *Cobra clotho* Laurenti, 1768 *(nomen rejiciendum), Cobra lachesis* Laurenti, 1768 *(nomen rejiciendum), Coluber clotho* Gmelin, 1789 *(nomen rejiciendum), Coluber hebraiecus* Lacépède, 1789 *(nomen rejiciendum), Coluber bitin* Bonnaterre, 1790 *(nomen rejiciendum), Coluber intumescens* Donndorf, 1798 *(nomen rejiciendum), Vipera severa* Latreille *in* Sonnini & Latreille, 1801a *(nomen rejiciendum), Coluber clotho* G. Shaw, 1802 *(nomen rejiciendum), Vipera inflata* Burchell, 1822, *Vipera brachyura* Cuvier, 1829, *Clotho lateristriga* Gray, 1842e, *Bitis lachesis somalica* H.W. Parker, 1949, *Bitis arientans* – Boquet, 1964 *(nomen incorrectum), Bitis arietans peghullae* G.D. Stewart, 1973 *(nomen ineditum), Bitis areitans* – Warrell et al., 1976 *(nomen incorrectum)*, and *Bitis arientas* – Khole, 1991 *(nomen incorrectum)*.

Type: Lectotype, specimen described and illustrated by Seba (1735: 55, pl. 54, fig. 4), designated herein.

Type locality: "Promontorio bonae spei" [= Cape of Good Hope, Western Cape Prov., SW South Africa, 34°21'S, 18°28'E, elevation 20 m] via lectotype selection.

Distribution: Africa and SW Arabia. NW Western Sahara (Saguia El Hamra), SW Morocco (Agadir, Tan Tan, Tiznit), SW Mauritania (Brakna, Guidimaka, Trarza), Senegal (Dakar, Fatick, Kaolack, Kédougou, Louga, Saint-Louis, Tambacounda, Thiès), Gambia (Western), Guinea–Bissau (Bissau, Bolama, Bubaque, Orango and Soga Is.), Guinea (Beyla, Boffa, Boké, Kissidougou, Kouroussa, Macenta, Télimélé), SE Algeria, S Mali (Kayes, Koulikoro, Mopti, Ségou, Sikasso, S Tombouctou), Sierra Leone (Eastern, Northern), Ivory Coast (Bouna, Ferkéssédougou, Toumodi), Ghana (Upper West), Burkina-Faso (Centre, Centre-Est, Centre-Nord, Centre-Ouest, Est, Haute-Bassins, Sud-Ouest, Volta-Noire), Togo (Kara, Maritime, Plateuax, Savanes), N Benin (Alibori, Atakora, Atlantique, Collines, Plateau, Zou), S Niger (Agadez, Maradi, Niamey, Zinder), Nigeria (Gongola, Kaduna), S Chad (Mayo-Kebbi Ouest), SE Sudan, South Sudan, Eritrea (Anseba, Gash-Barka, Northern Red Sea, Southern), Ethiopia (Arussi, Bale, Gemu Gofa, Gojjam, Gondar, Hararge, Illubabor, Kefa, Shoa, Sidamo, Tigre, Wollega, Wollo), Somalia (Awdal, Bakool, Bari, Bay, Galguduud, Gedo, Hiraan, Jubbada Dhexe, Jubbada Hoose, Mogadishu, Mudug, Nugaal, Sanaag, Shabeellaha Dhexe, Shabeellaha Hoose, Sool, Togdheer, Woqooyi Galbeed, Coiama Is.), Cameroon (Adamaoua, Esrt, Centre, Extreme-Nord, Nord, Nord-Ouest), SE Gabon (Haut-Ogooué), Central African Republic (Bamingui-Bangoran, Haut-Mbomou, Haute-Sangha, Haute-Kotto, Mbomou, Nana-Mambere, Ombella-Mpoko, Ouaka, Ouham, Ouham-Pende), Congo (Bouenza, Brazzaville, Cuvette, Cuvette-Ouest, Pool), Democratic Republic of the Congo, Gabon (Bandundu, Bas-Congo, Kasai Occidental, Kasai Oriental, Katanga, Kinshasa, Nord-Kivu, Orientale, Sud-Kivu), Uganda (Central, Eastern, Northern, Western), Kenya (Central, Coast, Eastern, North-Eastern, Nairobi, Nyanza, Rift Valley, Western), Rwanda, Burundi, Tanzania (Arusha, Dar es Salaam, Dodoma, Iringa, Kagera, Kigoma, Kilimanjaro, Lindi, Manyara, Mara, Mbeya, Morogoro, Mtwara, Mwanza, Piwani, Rukwa, Ruvuma, Shinyanga, Singida, Tabora, Tanga), E Angola (Lunda Norte, Moxico), Zambia (Central, Copperbelt, Eastern, Luapula, Lusaka, Northern, North-Western, Southern, Western), Malawi (Central, Northern, Southern), Mozambique (Gaza, Inhambane, Manica, Maputo, Niassa, Sofala, Tete, Zambézia), Zimbabwe (Bulawayo, Manicaland, Mashonaland Central, Mashonaland East, Mashonaland West, Masvingo, Matabeleland North, Matabeleland South, Midlands), Botswana (Central, Gaborone, Ghanzi, Kgalagadi, Kgatleng, Kweneng, North East, North West, South East, Southern), Namibia (Hardap, Karas, Khomas, Okavango, Omaheke, Oshikoto, Otjozondjupa), Swaziland, W Lesotho, South Africa (Eastern Cape, Free State, Gauteng, KwaZulu-Natal, Limpopo, Mpumalanga, Northern Cape, North West, Western Cape), SW Saudi Arabia (Asir, Bahah, Jazan, Makkah), W Oman (Dhofar) and Yemen (Ad Dali', Al Hudaydah, Amanat Al Asimah, Ibb, Sa'dah, Sana'a, Ta'izz), NSL–3500 m.

Sources: Angel, 1933b, H.W. Parker, 1949, Capocaccia, 1955, Mertens, 1955, 1971, Laurent, 1956a, Sweeney, 1961, Laurent, 1964a, Menzies, 1966, J.A. Peters & Broadley, 1967, Leston & Hughes, 1968, Hughes & Barry, 1969, V.F.M. FitzSimons, 1974, Pitman, 1974, Janecek, 1976, Roman, 1980, Broadley, 1983, Buys & Buys, 1983, Hughes, 1983, Pienaar et al., 1983, Auerbach, 1987, Gasperetti, 1988, Schätti & Gasperetti, 1994, Niças, 1995, J.-F. Trape & Roux-Estève, 1995, Schleich et al., 1996, Bons & Geniez, 1997, Herrmann & Joger, 1997, Branch, 1999b, Clauss & Clauss, 2002, Spawls et al., 2002, Geniez et al., 2004, Branch et al., 2005, Chirio & Ineich, 2006, J.-F. Trape & Mané, 2006b, Pauwels & Vande weghe, 2008, Pauwels & Salle, 2009, Largen & Spawls, 2010, Ullenbruch et al., 2010, W. Böhme et al., 2011, Segniagbeto et al., 2011, Auliya et al., 2012, Broadley & Blaylock, 2013 and Chirio, 2013.

Remarks: BMNH 11.2.3a (dry specimen from The Gambia) listed as type *fide* BMNH catalogue. Hybridization discussed by Hughes (1968: 371, *B. gabonica* x *B. nasicornis*) and Broadley & Parker (1976: 81, *B. arietans* x *B. gabonica*).

3. *Bitis armata* (A. Smith, 1826). Edinburgh New Philos. J. 1: 251. (*Vipera armata*)

Synonym: *Vipera atropoides* A. Smith, 1846 *in* 1838–1849.
Type: Neotype, PEM 6796, a 372 mm male (M. Scott, 26 Feb. 1991), designated by Branch (1999a: 56).
Type locality: "Koppie Alleen Road, De Hoop Nature Reserve, Western Cape Prov., South Africa (34°29'S, 20°28'E; 3420AD; reserve grid BB32)" via neotype designation.
Distribution: Extreme SW South Africa (Western Cape), NSL–200 m.
Source: Branch, 1999a.

4. *Bitis atropos* (Linnaeus, 1758). Syst. Nat., ed. 10, 1: 216. (*Coluber atropos*)

Synonyms: *Vipera montana* A. Smith, 1826, *Echidna ocellata* Tschudi, 1845, *Bitis atropos unicolor* V.F.M. FitzSimons, 1959, *Bitis atropus* – G.W. Ferguson, 1977 *(nomen incorrectum),* and *Bitis atropoides* Gans, 1978 *(nomen incorrectum).*
Type: Holotype, NHR Lin-96 [or Lin-97] (formerly MAFR) a 500 mm specimen (Mus. Drottn.).
Type locality: "America" (in error). Probably from Cape of Good Hope [= Western Cape Prov., SW South Africa, 34°21'S, 18°28'E, elevation 20 m] *fide* V.F.M. FitzSimons (1962: 343).
Distribution: Southern Africa. Eastern Zimbabwe (Manicaland), W Mozambique, Lesotho, Swaziland and South Africa (Eastern Cape, Free State, W KwaZulu-Natal, Limpopo, Mpumalanga, Western Cape), NSL–3000 m.
Sources: V.F.M. FitzSimons, 1947, 1962a, Broadley & Cock, 1975, Botha, 1986, Spawls & Branch, 1995, W.R. Branch, 1997, M. Griffin, 2003 and Broadley & Blaylock, 2013.
Remarks: Namibia records doubtful *fide* M. Griffin (2003: 136).

5. *Bitis caudalis* (A. Smith, 1839 *in* 1838–1849). Illust. Zool. So. Africa, Rept.: pl. 7, 2 pp. (*Vipera* [*Cerastes*] *caudalis*) (*nomen substitutum*)

Synonym: *Vipera ocellata* A. Smith, 1838 (*nomen praeoccupatum*).
Types: Syntypes (5), BMNH 1865.5.4.153a–e, 3 males and 2 females (A. Smith, 1828–1837), longest syntype 350 mm, lost *fide* Broadley & Rage *in* Golay et al. (1993: 260).
Type locality: "north of the Cape Colony, South Africa."
Distribution: Southwestern Africa. Southwestern Angola (Namibe), Namibia (Erongo, Hardap, Karas, Khomas, Kunene, Ohangwena, Okavango, Omaheke, Omusati, Oshana, Oshikoto, Otjozondjupa), Botswana (Central, Ghanzi, Kgalagadi, Kweneng, Southern), SW Zimbabwe (Matebeleland North, Matebeleland South) and W South Africa (W Eastern Cape, Gauteng, Limpopo, Northern Cape, North West, Western Cape), NSL–1665 m.
Sources: Mertens, 1955, 1971, V.F.M. FitzSimons, 1962a, Broadley & Cock, 1975, Haacke, 1975, Broadley, 1983, Buys & Buys, 1983, Auerbach, 1987, Branch, 1988, Clauss & Clauss, 2002, M. Griffin, 2003 and Broadley & Blaylock, 2013.
Remarks: V.F.M. FitzSimons (1937: 264) reported the types lost but photographs of syntypes in Auerbach (1987: fig. [p. 203]).

6. *Bitis cornuta* (Daudin, 1803c). Hist. Nat. Rept. 6: 188–190. (*Vipera cornuta*)

Synonyms: *Vipera lophophris* Cuvier, 1829, and *Vipera lophophrys* – Wagler, 1830 *(nomen incorrectum).*
Type: Holotype, not designated, a 300–450 mm specimen (W. Paterson, 31 Aug. 1779), lost *fide* McDiarmid et al. (1999: 362).
Type locality: "between Deepe Kloaf, or Water Val, and Great, or Sea Fountain, country of the Hottentots, and Caffraria" [= Cape of Good Hope, Western Cape Prov., SW South Africa]. Restricted to McDougall's Bay (2916BD), Little Namaqualand, Northern Cape Prov., South Africa *fide* Branch (1999a: 53).
Distribution: Namibia (Hardap, Karas, Kunene, Omaheke) and W South Africa (W Northern Cape, Western Cape), NSL–1000 m.
Sources: Mertens, 1955, 1971, V.F.M. FitzSimons, 1962a, Haacke, 1975 and Branch, 1988, 1989, 1997, 1999a.
Remarks: Holotype illustrated in Paterson (1789 & 1790: pl. 15). BMNH 1905.3.7.162 and TM 57965 from Port Nolloth are topotypes *fide* Branch (1999a: 53).

7. *Bitis gabonica* (A.-M.-C. Duméril, Bibron & Duméril, 1854b). Erpét. Gén. 7(2): 1428–1430, pl. 80 bis, figs. 1–3. (*Echidna gabonica*)

Synonyms: *Bitis javonica* – Suzuki & Iwanaga, 1970 *(nomen incorrectum),* and *Bitis jabonica* – Mohamed et al., 1971 *(nomen incorrectum).*
Type: Lectotype, MNHN 4012 (C.-E. Aubry-Lecomte & Chaval, 1853–1854), designated by Hughes & Barry (1969: 1030).
Type locality: "Gabon" via lectotype selection.
Distribution: Cen, E and S Africa. Southeastern Nigeria (Akwa Ibom, Cross River, Rivers), S Cameroon (Sud Ouest (Adamaoua, Centre, Est, Littoral, Nord-Ouest, Ouest, Sud, Sud-Ouest), Equatorial Guinea, Central African Republic (Bamingui-Bangoran, Basse-Kotto, Haut-Mbomou, Haute-Sangha, Lobaye, Mbomou,

Ombella-Mpoko, Ouaka, Sangha), Gabon (Estuaire, Haut-Ogooué, Ngounié, Ogooué-Ivindo, Ogooué-Lolo, Ogooué-Maritime, Woleu-Ntem), Congo (Brazzaville, Cuvette-Ouest, Kouilou, Likouala, Niari, Sangha), Democratic Republic of the Congo (Bandundu, Bas-Congo, Equateur, Katanga, Maniema, Nord-Kivu, Orientale, Sud-Kivu), South Sudan (Eastern Equatoria), S Uganda (Central, Western), SW Kenya (Western), W Rwanda (Ruhengeri), E Tanzania (Lindi, Mtwara, Piwani, Tanga), E Angola (Lunda Norte, Moxico), N Zambia (Copperbelt, Luapula, Northern, North-Western), N Malawi (Northern), E Zimbabwe (Manicaland), cen. Mozambique (Manica, Sofala) and NE South Africa (NE KwaZulu-Natal), NSL–2300 m.

Sources: V.F.M. FitzSimons, 1962a, Sweeney, 1961, Berry, 1963, Pitman, 1974, Broadley & Cock, 1975, Broadley, 1983, Branch, 1988, Luiselli et al., 1998b, D. Lawson, 1993, J.-F. Trape & Roux-Estève, 1995, Luiselli & Akani, 1998c, 2003, Spawls et al., 2002, J.C. Murphy & Schlager, 2003, Chirio & Ineich, 2006, Chirio & LeBreton, 2007, Pauwels & Vande weghe, 2008 and Broadley & Blaylock, 2013.

Remarks: *Bitis gabonica, B. nasicornis*, and *B. rhinoceros* are at the same taxonomic level *fide* Lenk et al. (1999, 2001).

8. *Bitis heraldica* (Bocage, 1889). J. Sci. Math. Phy. Nat. Lisboa (2) 1(2): 127–128, fig. 1. (*Vipera heraldica*)

Type: Holotype, MBL 2127, a 325 mm specimen (H.C. Capello & R. Ivens, March–April 1884), destroyed by fire 18 March 1978.

Type locality: "sur les bords de la rivière Calae, l'un des affluents du Cunene, entre le 13 et le 14 parallèle à l'est de Caconda, Angola" [= Caluè River, tributary of the Cunene River, E of Caconda, Huíla District, SW Angola, 13°44'S, 15°04'E, elevation 1700 m].

Distribution: Western Angola (Bengo, Bíe, Huambo, Huíla), 1650–1700 m.

Source: Mertens, 1958.

9. *Bitis inornata* (A. Smith, 1838 *in* 1838–1849). Illust. Zool. So. Africa, Rept.: pl. 4, 2 pp. (*Echidna inornata*)

Type: Holotype, BMNH 1946.1.18.13 (formerly BMNH 1865.5.4.166), a 337 mm specimen (J. Leib).

Type locality: "Sneeubergen, or Snow-mountains, which are situated immediately behind the village of Graaff Reynet, South Africa" [= Sneeuwberg, near Graaff-Reinet, W Eastern Cape Prov., South Africa, 32°01'S, 24°13'E, elevation 1600 m].

Distribution: Southern South Africa (Eastern Cape Western Cape), 1600–1800 m.

Sources: V.F.M. FitzSimons, 1962a, Broadley, 1983, 1990 and W.R. Branch, 1988, 1989, 1997, 1999a.

10. *Bitis nasicornis* (G. Shaw & Nodder, 1792 *in* 1789–1813). Nat. Misc. 3: pl. 94, 3 pp. (*Coluber nasicornis*)

Synonyms: *Vipera hexacera* A.-M.-C. Duméril, 1853, and *Bitis nasicorais* – Khole, 1991 *(nomen incorrectum)*.

Type: Holotype, BMNH, an 890 mm specimen (E. Jenkins), lost *fide* Golay et al. (1993: 262).

Type locality: "interior parts of Africa." Restricted to coast of Guinea *fide* J.T. Reinhardt (1843: 273) [= Ghana *fide* Hughes & Barry, 1969: 1030].

Distribution: Central and E Africa. Guinea-Bissau (Bubaque Is.), E Guinea (Dalaba, Kouroussa, Labé, Lélouma, Macenta, Nzérékoré), Liberia (Montserrado, Margibi), S Ivory Coast (Abidjan, Guiglo, Lagunes, Tabou, Toumodi), S Ghana (Accra), ? Benin, SE Nigeria (Anambra, Cross River, Imo, Rivers), S Cameroon (Centre, Est, Littoral, Nord-Ouest, Sud, Sud-Ouest), Equatorial Guinea (Bioko Is.), Gabon (Estuaire, Moyen-Ogooué, Ngounié, Ogooué-Ivindo, Ogooué-Lolo, Ogooué-Maritime, Woleu-Ntem), SW Central African Republic (Lobaye, Ombella-Mpoko, Sangha), S Congo (Kouilou, Lekoumou), Democratic Republic of the Congo (Bandundu, Bas-Congo, Equateur, Orientale, Sud-Kivu, Idjwi Is.), NE Zambia (Luapula, Northern), South Sudan (Eastern Equatoria), S Uganda (Central, Western), SW Kenya (Western) and W Rwanda, NSL–2650 m.

Sources: E.H. Taylor & Weyer, 1958, Doucet, 1963, Mertens, 1965c, Leston & Hughes, 1968, Hughes & Barry, 1969, Pitman, 1974, J.A. Butler & Reid, 1986, D. Lawson, 1993, J.-F. Trape & Roux-Estève, 1995, Lamar, 1997b, Luiselli & Akani, 1998b, 2003, Luiselli et al., 1998c, Spawls et al., 2002, Broadley et al., 2003, Chippaux, 2006, Chirio & Ineich, 2006, Chirio & LeBreton, 2007, Krecsák, 2007, Pauwels & Vande weghe, 2008, Segniagbeto et al., 2011, Auliya et al., 2012 and Hughes, 2013.

Remarks: The holotype was obtained from the master of a Guinea vessel *fide* G. Shaw & Nodder (1792 *in* 1789–1813: 3). Unconfirmed specimens from NE Tanzania (Uluguru and/or Usambara) *fide* Spawls et al. (2002: 481).

11. †*Bitis olduvaiensis* Rage, 1973b. Fossil Vert. Africa 3: 4, 5–6, figs. 1a–d, pl. 1, figs. 5–8.

Type: Holotype, NMT, one left maxilla.

Type locality: "FLK I, Main dig, Zinjanthropus level, Olduvai Gorge, Tanzania" [= Villafranchian Bed I].

Distribution: Lowermost Pleistocene (Villafranchian: 1.0–1.1 mya) of Tanzania. Known only from type locality.

Source: Rage & Bailon, 2011.

B

12. *Bitis parviocula* W. Böhme, 1977. Monit. Zool. Ital. (2, Suppl.) 9(3): 60–67, figs. 1, 2A, 3C.

Type: Holotype, ZFMK 16803, a 752 mm female (G. Nikolaus, 18 Oct. 1975).
Type locality: "Doki River Bridge (8°20'N-35°56'E), bei Yambo (= Yembo), an der Strasse von Metu nach Bedelle, Provinz Illubabor, SW-Äthiopien" [= Ethiopia].
Distribution: Southern Ethiopia (Bale, Illubabor, Kefa), 1700–2800 m.
Sources: Golay et al., 1993, Largen & Rasmussen, 1993, Spawls & Branch, 1995, David & Ineich, 1999, Dobiey & Vogel, 2007 and Largen & Spawls, 2010.
Remarks: A valid species *fide* Golay et al., 1993, McDiarmid et al., 1999 and Mallow et al., 2003.

13. *Bitis peringueyi* (Boulenger, 1888d). Ann. Mag. Nat. Hist. (6) 2(8): 141. (*Vipera peringueyi*)

Type: Holotype, SAM 1852, a 250 mm specimen (L.A. Péringuey, 1884–1889), lost *fide* Stander *in* Golay et al. (1993: 263).
Type locality: "Damaraland, 10 miles east of Walfisch Bay, South Africa." [= 10 mi. E of Walvis Bay, Erongo District, Namibia, ca. 23°01'S, 14°31'E. elevation 5 m].
Distribution: Southern Angola (Cuanza Sul) and W Namibia (W Erongo, W Hardap, W Karas, W Kunene), NSL–1750 m.
Sources: Mertens, 1955, 1971, V.F.M. FitzSimons, 1974, Robinson & Hughes, 1978, Broadley, 1983 and Buys & Buys, 1983.

14. *Bitis rhinoceros* (Schlegel, 1855). Versl. Meded. Akad. Wet. 3: 316–317. (*Vipera rhinoceros*)

Types: Syntypes (5), RMNH 1643–46 (H. Pel, 1850) and RMNH 1647 (H. Pel & Ferrier, 1843).
Type locality: "Goud-kust, fusschen hat Kaap de Drie Punten en hat fort Acre" [= Gold Coast between Cap Three Points and Accra, Ghana]. Designated as Dabocrom (RMNH 1644–47) and Cote d'Or (RMNH 1643) *fide* RMNH catalogue.
Distribution: West Africa. Guinea (Dalaba, Kouroussa, Labé, Lélouma, Macenta, Nzérékoré, Télimélé), Sierra Leone (Eastern, Northern, Southern, Western, Sherbo Is.), Liberia (Montserrado, Margibi), S Ivory Coast (Abidjan, Bouaflé, Daloa, Guiglo, Lagunes, Tabou, Toumodi), S Ghana (Accra), S Togo (Centrale, Plateaux) and Benin, NSL–1500 m.
Sources: E.H. Taylor & Weyer, 1958, Doucet, 1963, Hughes, 1968, Hughes & Barry, 1969, Hughes, 1983, Lenk et al., 1999, 2001, M. Griffin, 2003, Ineich, 2003, Villiers & Condamin, 2005, W. Böhme et al., 2011, Segniagbeto et al., 2011 and Chirio, 2013.
Remarks: A valid species *fide* Lenk et al. (1999: 36). Namibia record from Otjimbingwe in error *fide* M. Griffin (2003: 196).

15. *Bitis rubida* W.R. Branch, 1997. So. Afr. J. Zool. London 32(2): 38–42, fig.1.

Type: Holotype, PEM 12582, a 344 mm male (S.A. Botha, 3 Sept. 1986).
Type locality: "Jeep track above Farm Driehoek, Cedarberg Mountains, Western Cape Prov., 32°25'44"S, 19°12'30"E, alt. 1380 m; 3219AC."
Distribution: Southwestern South Africa (Eastern Cape, S Northern Cape, Western Cape), 300–1380 m.
Source: Branch, 1999aa.

16. *Bitis schneideri* (Boettger, 1886a). Ber. Senck. Naturf. Ges. 13: 8–10, pl. 1, figs. 1a–e. (*Vipera schneideri*)

Synonyms: *Bitis caudalis paucisquamatus* Mertens, 1954b, and *Bitis paucisquamata* – V.F.M. Fitzsimons, 1962.
Type: Holotype, SMF 21019 (formerly SMB-B 9498d), a 282 mm female (O. Schneider, 1886).
Type locality: "Angra Pequenia, Südwest-Afrikas." [= Angra Pequena, Lüderitz Bay, Karas District, Namibia, ca. 26°42'S, 15°22'E].
Distribution: Southwestern Namibia (W Karas) and SW South Africa (NW Northern Cape), NSL–65 m.
Sources: V.F.M. FitzSimons, 1962a, Haacke, 1975, Broadley, 1983, and M. Griffin, 2003.
Remarks: A synonym of *B. caudalis fide* McDiarmid et al., 1999.

17. *Bitis worthingtoni* H.W. Parker, 1932c. J. Linn. Soc. (Zool.) 38(257): 221–222.

Type: Holotype, BMNH 1946.1.20.90 (formerly BMNH 1932.5.2.8), a 275 mm male (E.B. Worthington [Cambridge Exped.], 1930–1931).
Type locality: "shore of Lake Naivasha, Kenya" [=Lake Oleidon, Lake Naivasha National Park, S Rift Valley Prov., S Kenya, 0°46'S, 36°22'E, elevation 1930 m].
Distribution: Southwestern Kenya (S Rift Valley), 1500–2000 m.
Source: Spawls et al., 2002.

18. *Bitis xeropaga* Haacke, 1975. Cimbebasia 4A(5): 116–117, pls. 1–2.

Type: Holotype, TMP 42305, a 395 mm male (A. Maritz, 3 Aug. 1972).
Type locality: "Dreikammberg on north bank of ODistribution River, Lüderitz district, South West Africa (16°52'E, 28°05'S, alt. about 300 m)" [= Karas District].
Distribution: Southern Namibia (Karas) and W South Africa (NW Northern Cape), 300–1000 m.
Sources: Broadley, 1983 and M. Griffin, 2003.

BLYTHIA Theobald, 1868b
(Colubroidea *incertae sedis*)

Synonyms: *Blythea* Theobald, 1876 *(nomen emendatum)*, *Aproaspidops* Annandale, 1912, *Aproaspidelaps* – Phisalix, 1922 *(nomen incorrectum)*, *Asproaspidops* – Phisalix, 1922 *(nomen incorrectum)*, and *Blythis* – B.D. Sharma *in* B.D. Sharma & Kumari, 1998a *(nomen incorrectum)*.

Type species: *Calamaria reticulata* Blyth, 1854a.

Distribution: Southern Asia.

Sources: Boulenger, 1913g, M.A. Smith, 1943, Mahendra, 1984, Zhao & Adler, 1993, Ao et al., 2004, Whitaker & Captain, 2004, Zhao, 2006, R.C. Sharma, 2007, Zaher et al., 2009 and I. Das, 2010, 2012.

1. *Blythia reticulata* (Blyth, 1854a). J. Proc. Asiatic Soc. Bengal 23(3): 287–288. (*Calamaria reticulata*)

Synonym: *Aproaspidops antecursorum* Annandale, 1912.

Types: Syntypes (2), ZSI 7028–29 (formerly IMC), longest syntype 305 mm (W. Robinson, 1841–1854).

Type locality: "Asám" [= Arunachal Pradesh, Assam, Manipur, Meghalaya, Mizoram, Nagaland states, NE India].

Distribution: Southern Asia. Northeastern India (Arunachal Pradesh, Assam, Manipur, Meghalaya, Nagaland), SW China (Xizang) and NW Myanmar (Chin, Kachin), 145–2000 m.

BOA Linnaeus, 1758
(Boidae)

Synonyms: *Constrictor* Laurenti, 1768 *(nomen rejiciendum)*, *Boarius* Rafinesque-Schmaltz, 1815 *(nomen emendatum)*, *Draco* Oken, 1816 *(nomen rejiciendum)*, *Boas* – S.W. Garman, 1884 *(nomen incorrectum)*, *Boaus* A.L. Herrera, 1899 *(nomen emendatum)*, and †*Pseudoepicrates* Auffenberg, 1963.

Type species: *Boa constrictor* Linnaeus, 1758.

Distribution: Latin America and Lesser Antilles.

Fossil records: Lower-middle Miocene of Panama, lower Miocene of USA, upper Pliocene of Argentina, and Pleistocene of Mexico. Possible record from lower Eocene *fide* Albino & Carlini (2008: 85).

Sources: Auffenberg, 1963, Kluge, 1991, 1993a–b, Tolson & R.W. Henderson, 1993, H. Bosch, 1994, Henderson et al., 1995, J.C. Lee, 1996, 2000, Starace, 1998, Wall, 1998a, Vosjoli et al., 1998, McDiarmid et al., 1999, Binder, 2002, G. Köhler et al., 2005, J.R. Berry, 2006, Bonny, 2007, Hes et al., 2007, Russo, 2007, Albino & Carlini, 2008, Navarrete et al., 2009, Ugueto & Rivas, 2010, Albino, 2011a, McCranie, 2011a and Head et al., 2012.

Remarks: Official Generic Name no. 2019 *fide* Opinion 1045 (ICZN, 1976). Vences et al., 2001, recognized *Acrantophis* and *Sanzinia* as distinct genera after Kluge, 1991, synonymized them with *Boa*. *Boa nebulosa* and *B. orophias* are valid species *fide* R.W. Henderson & Powell, 2009.

1. *Boa constrictor* Linnaeus, 1758. Syst. Nat., ed. 10, 1: 215.

Synonyms: *Boa orophias* Linnaeus, 1758, *Constrictor formosissimus* Laurenti, 1768 *(nomen rejiciendum)*, *Constrictor auspex* Laurenti, 1768, *Constrictor diviniloquus* Laurenti, 1768, *Constrictor rex serpentum* Laurenti, 1768 *(nomen rejiciendum)*, *Boa divinatrix* Lacépède, 1789 *(nomen rejiciendum)*, *Coluber glyripeta* Bechstein, 1802b, *Boa imperator* Daudin, 1803b, *Boa eques* Eydoux & Souleyet, 1842, *Boa diviniloquax mexicana* Jan, 1863b, *Boa occidentalis* Phillipi, 1873, *Boa ortonii* Cope, 1878a, *Boa constrictor isthmica* S.W. Garman, 1884, *Epicrates sabogae* Barbour, 1906, *Constrictor constrictor amarali* Stull, 1932b, *Constrictor constritor* – Briceño-Rossi, 1934 *(nomen incorrectum)*, *Constrictor constrictor sigma* H.M. Smith, 1943a, †*Neurodromicus stanolseni* Vanzolini, 1952, †*Neurodromicus barbouri* Vanzolini, 1952, †*Neurodromicus stanolseni* Auffenberg, 1963, *Constrictor constrictor nebulosus* Lazell, 1964, *Boa constrictor melanogaster* Langhammer, 1983, and *Boa constrictor longicauda* Price & Russo, 1991.

Type: Lectotype, NHR Lin-10 (formerly MAFR), an 890 mm specimen (Mus. Drottn.), designated by Hoge (1964d: 51).

Type locality: "Indiis" [= India] (in error *fide* Hoge, 1964a: 51).

Distribution: Latin America. The Lesser Antilles (Dominica, St. Lucia), Mexico (Campeche, Chiapas, ext. SW Chihuahua, Colima, Durango, Guerrero, Hidalgo, Jalisco, México, Michoacán, Morelos, Nayarit, Oaxaca, Puebla, Queretaro, Quintana Roo, San Luis Potosí, Sinaloa, Sonora, Tamaulipas, Veracruz, Yucatán, María Cleofas, María Madre and María Magdalena Is.), Belize (Belize, Cayo, Corazal, Orange Walk, Stann Creek), Guatemala (Alta Verapaz, Izabal, Petén), El Salvador (Ahuachapán, Chalatenango, Cuscatlán, La Libertad, La Paz, Morazán, San Miguel, San Salvador, Santa Bárbara. San Vicente, Sonsonate), Honduras (Atlántida, Choluteca, Colón, Cortés, El Paraíso, Francisco Morazán, Gracias a Dios, Islas de la Bahía, Lempíra, Olancho, Santa Bárbara, Valle, Yoro, Cayo Cochino Grande, Cayo Cochino Pequeña, Guanaja, Roatán and Utila Is.), Nicaragua (Atlántico Norte, Atlántico Sur, Chinandaga, Jinotega, Matagalpa, Rio San Juan, Rivas, Great Corn Is.), Costa Rica (Alajuela, Cartago, Guanacaste, Heredia, Limón, Puntarenas, San José), Panama, Colombia (Amazonas, Antioquia, Arauca, Atlántica, Bolívar, Caldas, Caquita, Casanare, Cauca, Chaco, Cordobá, Cundinamarca, Magdalena, Meta, Sucre, Tolima), Venezuela (Amazonas, Anzoátegui, Apure, Aragua, Barinas, Bolívar, Carabobo, Cojedes,

B

Delta Amacuro, Distrito Federal, Falcón, Guárico, Lara, Mérida, Miranda, Monagas, Nueva Esparta, Portuguesa, Sucre, Táchira, Trujillo, Yaracuy, Vargas, Zulia, Margarita Is.), Trinidad and Tobago (Diego, Monos, Tobago, Trinidad), Guyana (Barimi-Waini, Essequibo Islands-West Demerara, Upper Demerara-Berbice, Upper Takutu-Upper Essequibo), Surinam (Brokopondo, Nickerie, Para, Paramaribo), French Guiana (Cayenne, Saint-Laurent-du-Maroni), Ecuador (Napo), NW Peru (Amazonas, Cajamarca, Huanuco, Junin, La Libertad, Lambayeque, Loreto, Madre de Dios, Pasco, Piura, San Martín, Túmbres), Brazil (Amazonas, Aragoas, Bahia, Goiás, Mato Grosso, Mato Grosso do Sol, Minas Gerais, Pará, Pernambuco, Rio de Janeiro, Rondonia, São Paulo, Sergipe), Bolivia (Beni, Santa Cruz, Tarija), Paraguay (Alto Paraguay, Amambay, Boquerón, Canindeyú, Concepción) and Argentina (Buenos Aires, Chaco, Córdoba, Mendoza, Pasco, Salta, San Luís, Santiago del Estero), NSL–2300 m.

Sources: Quelch, 1898b, Stull, 1932b, E.H. Taylor, 1954, Lazell, 1964, Carrillo de Espinoza, 1966, Cunha, 1968, Stimson, 1969, Matz & Matz, 1970, J.A. Peters & Orejas-Miranda, 1970, Vanzolini et al., 1980, Brygoo, 1983, Langhammer, 1983, L.D. Wilson & Meyer, 1985, Chippaux, 1987, Pérez-Santos & Moreno, 1988, 1991, Price & Russo, 1991, Boos, 1992, Joy, 1992, Pitts et al., 1992, Albino, 1993, Tolson & R.W. Henderson, 1993, H. Bosch, 1994, D.R. Norman, 1994, Casado-B. & Boltan, 1996, J.C. Lee, 1996, 2000, Benitez-Gálvez, 1997, Fogel, 1997, Vosjoli, 1997, J.A. Campbell, 1998b, Gorzula & Señaris, 1998, Starace, 1998, Freitas, 1999, Leynaud & Bucher, 1999, Mijares-Urrutia & Arends, 2000, Vences et al., 2001, Lemos-Espinal et al., 2001, 2002a–b, Markezich, 2002, Savage, 2002, Abuys, 2003, Breuil, 2003, J.C. Murphy & Schlager, 2003, Burbrink, 2005, Duellman, 2005, Boback, 2006, Boback & Carpenter, 2007, N. Herrera et al., 2007, Russo, 2007, Cacciali, 2008, R.W. Henderson & Powell, 2009, Berry, 2010, McCranie & Orellana, 2011, Travers et al., 2011 and C.J. Cole et al., 2013.

Remarks: Official Specific Name no. 2561 *fide* Opinion 1045 (ICZN, 1976).

BOAEDON Duméril, Bibron & Duméril, 1854a (Lamprophiidae)

Synonyms: *Boedon* A.-M.-C. Duméril, 1853 (*nomen nudum*), *Holuropholis* A.-M.-C. Duméril, 1853, *Alopecion* A.-M.-C. Duméril, Bibron & Duméril, 1854a, *Boedon* A.-M.-C. Duméril, Bibron & Duméril, 1854a (*nomen emendatum*), *Eugnathus* A.-M.-C. Duméril, Bibron & Duméril, 1854a, *Boodon* A.C.L.G. Günther, 1858 (*nomen emendatum*), *Catapherodon* Rochebrune, 1885 (*nomen substitutum*), *Eugnatus* – Boacge, 1887c (*nomen incorrectum*), *Boaeodon* – Hoffmann, 1890 (*nomen incorrectum*), *Theleus* Cope, 1893a, *Boaodon* Chabanaud, 1917c (*nomen emendatum*), *Catophaerodon* – V.F.M. FitzSimons, 1962 (*nomen incorrectum*), *Holurophis* – Mertens, 1965c (*nomen incorrectum*), *Bodedon* – Gasperetti, 1974 (*nomen incorrectum*), and *Boadedon* – McCallum, 1986 (*nomen incorrectum*).

Type species: *Lycodon fuliginosus* H. Boie *in* F. Boie, 1827.

Distribution: Africa, Yemen and Seychelles.

Sources: Loveridge, 1957a, V. FitzSimons, 1962a, 1974, Roux-Estève & Guibé, 1965b, Dunger, 1971b, Broadley, 1983, Kelly et al., 2003, 2009, 2011, Marais, 2004, Vital et al., 2008 and Zaher et al., 2009.

Remarks: A valid genus *fide* Kelly et al., 2011.

1. *Boaedon abyssinicus* (Mocquard, 1906a). Bull. Mus. Hist. Nat., Paris 12(5): 249–250. (*Lamprophis abyssinicus*)

Type: Holotype, MNHN 1905.188, a 483 mm female (M. de Rothschild, 1904–1905).

Type locality: "Akaki, en Abyssinie" [= Akaki, Shoa Prov., Ethiopia, 8°57'N, 38°45'E, elevation 2215 m].

Distribution: Southwestern Ethiopia (Gemu Gofa, Illubabor, Kefa, Shoa), 1500–2300 m.

Sources: Largen & Rasmussen, 1993 and Largen & Spawls, 2010.

Remarks: A valid species *fide* Largen & Rasmussen (1993: 345). Photographs of holotype in Largen & Rasmussen (1993: figs. 3a–c).

2. *Boaedon arabicus* H.W. Parker, 1930c. Ann. Mag. Nat. Hist. (10) 6(35): 596–597, 1 fig.

Type: Holotype, BMNH 1903.1.28.14, a 710 mm female (G.W. Bury, 1901–1903).

Type locality: "El Kubar, Amiri Country, S.W. Arabia" [= El Kubar, Ad Dali'' Govern., SW Yemen, 13°48'N, 44°45'E, elevation 1350 m].

Distribution: Western Yemen (Ad Dali', Sana'a, Ta'izz), 500–2200 m.

Sources: Gasperetti, 1988 and Hughes, 1997.

Remarks: A valid species *fide* Hughes (1997: 72).

3. *Boaedon capensis* A.-M.-C. Duméril, Bibron & Duméril, 1854a. Erpét. Gén. 7(2): 364–365. (*Boaedon capense*)

Synonyms: *Boaedon quadrilineatum variegatum* Jan, 1863b (*nomen nudum*), *Boaedon quadrilineatum variegatum* Jan, 1870 *in* 1870–1881, and *Boodon mentalis* A.C.L.G. Günther, 1888c.

Types: Syntypes, MNHN, a 674 mm specimen (A. Smith, 1828–1837), and MNHN, a 567 mm specimen (C.F.F. Krauss, 7 May 1838–22 April 1840).

Type locality: "environs du Cap" [= vicinity of Cape of Good Hope, Western Cape Prov., SW South Africa, ca. 34°21'S, 18°28'E, elevation 20 m].

Distribution: Eastern Africa. Northwestern and SW Somalia (Shabeellaha Dhexe, Shabeellaha Hoose, Togdheer, Woqooyi Galbeed), E Kenya (Coast), E Tanzania (Dar es Salaam, Dodoma, Kilimanjaro, Lindi, Morogoro, Mtwara, Piwani, Tabora, Tanga, Mafia, Pemba and Zanzibar Is.), Malawi (Southern), Mozambique (Gaza, Inhambane, Manica, Maputo, Niassa, Sofala, Tete, Zambezia), Zambia (Central, Eastern, Southern), Namibia (Caprivi, Erongo, Hardap, Karas, Khomas, Kunene, Omaheke, Otjozondjupa), Botswana (Central, Gaborone, Ghanzi, Kgalagadi, Kgatleng, Kweneng, North East, North West, Southern, Tuli), Zimbabwe (Bulawayo, Manicaland, Mashonaland Central, Mashonaland East, Mashonaland West, Masvingo, Matabeleland North, Matabeleland South, Midlands), Swaziland (Lubombo, Manzini, Shiselweni), W Lesotho, South Africa (Eastern Cape, Free State, Gauteng, KwaZulu-Natal, Limpopo, Mpumalanga, Northern Cape, North West, Western Cape), NSL–1350 m.

Sources: Uthmöller, 1937, 1941b, H.W. Parker, 1949, Manaças, 1955, 1959, Vesey-FitzGerald, 1958, Broadley, 1959, Sweeney, 1961, Roux-Estève & Guibé, 1965a–b, Menzies, 1966, Papenfuss, 1969, Pitman, 1974, Broadley & Cock, 1975, Thorpe & McCarthy, 1978, Lanza, 1983, 1990, Pienaar et al., 1983, Auerbach, 1987, Branch, 1988, Gasperetti, 1988, Gras, 1988, Hermann, 1989a, Largen & Rasmussen, 1993, Schätti & Gasperetti, 1994, Hughes, 1997, Clauss & Clauss, 2002, Spawls et al., 2002, Broadley et al., 2003, Branch et al., 2005, Largen & Spawls, 2010 and Broadley & Blaylock, 2013.

Remarks: *Boaedon capensis* a valid species *fide* Hughes, (1997: 73). Taxonomy of *L. fuliginosus* complex *(arabicus-capensis-fuliginosus-lineatus-maculatus)* unresolved and therefore Distributions are only approximate.

4. *Boaedon erlangeri* Sternfeld, 1908a. Sitz. Ges. Naturf. Freude Berlin: 92. (*Boodon erlangeri*)

Type: Holotype, ZMB 27419 (C. von Erlanger & O.R. Neumann, 1900–1901).

Type locality: "Somaliland" [= Somalia] (in error). Corrected to Ethiopia *fide* Largen & Rasmussen (1993: 347).

Distribution: Western Ethiopia (Bale, Gojjam, Illubabor, Kefa, Shoa, Sidamo, Wollega), 820–2800 m.

Sources: Largen & Rasmussen, 1993 and Largen & Spawls, 2010.

Remarks: Photographs of the type in Largen & Rasmussen (1993: figs. 4a–c).

5. *Boaedon fuliginosus* (H. Boie *in* F. Boie, 1827). Isis von Oken 20(6): 551. (*Lycodon fuliginosus*)

Synonyms: *Lycodon fuliginosus* H. Boie in Schlegel, 1826a *(nomen nudum), Lycodon fuliginosus* Schlegel, 1826b *(nomen nudum), Lycodon unicolor* Schlegel, 1837, *Alopecion variegatum* Bocage, 1867b, *Boodon bipraeocularis* A.C.L.G. Günther, 1888c, and *Boodon lineatus lineolata* Bocage, 1895.

Type: Holotype, not designated, formerly RMNH, lost *fide* Brongersma *in* Roux-Estève & Guibé (1965b: 397).

Type locality: "Java" [Indonesia] (in error). Corrected to Africa *fide* V.F.M. FitzSimons (1962a: 119). Restricted to South Africa or possibly Ghana *fide* Brongersma *in* Roux-Estève & Guibé (1965b: 397).

Distribution: Susaharan Africa. Southwestern Morocco (Agadir, Tan Tan), N Western Sahara (Saguia El Hamra), SW Mauritania (Dakhlet Nouadhibou, Guidimaka, Hodh El Gharbi, Inchiri, Trarza), Senegal (Dakar, Fatick, Kaolack, Kédougou, Kolda, Louga, Matam, Saint-Louis, Tambacounda, Ziguinchor), Gambia (MacCarthy Island, Western), Guinea–Bissau (Biombo, Bissau, Tombali, Bubaque Is.), Guinea (Beyla, Boké, Nzérékoré, Télimélé), Sierra Leone (Southern), Liberia (Montserrado), Ivory Coast (Abidjan, Bouaflé, Toumodi), Burkina Faso (Centre, Centre-Nord, Centre-Ouest, Est, Hauts-Bassins, Sud-Ouest, Volta-Noire, Sahel), Ghana (Upper West, Volta), Togo (Centrale, Kara, Maritime, Plateaux, Savanes), N Benin (Alibori, Atakora, Borgou, Collines, Zou), Niger (Dosso, Maradi, Niamey), Nigeria (Bauchi, Gongola, Kaduna, Kwara, Lagos, Oyo, Plateau, Sokoto), SW Chad (Mayo-Kebbi), Cameroon (Adamaoua, Centre, Est, Littoral, Nord, Nord-Ouest, Ouest, Sud, Sud–Ouest), Central African Republic (Bamingui-Bangoran, Bangui, Basse-Kotto, Kemo-Gribingui, Mbomou, Ombella-Mpoko), E Gabon (Haut-Ogooué), S Congo (Bouenza, Brazzaville, Cuvette, Kouilou, Lekoumou, Niari, Plateau, Pool), Democratic Republic of the Congo (Bandundu, Bas-Congo, Equateur, Kasai Occidental, Kasai Oriental, Kinshasa, Katanga, Maniema, Nord-Kivu, Orientale, Sud-Kivu), Angola (Benguela, Bié, Cabinda, Cuanza Norte, Cuanza Sul, Cunene, Huambo, Huíla, Lunda Norte, Malanje, Moxico), South Sudan (Eastern Equatoria, Western Equatoria), Eritrea (Anseba, Central, Gash-Barka, Northern Red Sea, Southern), Ethiopia (Arussi, Bale, Gemu Gofa, Gojjam, Gondar, Hararge, Illubabor, Kefa, Shoa, Sidamo, Tigre, Wollega, Wollo), Somalia (Bari, Bay, Gedo, Hiraan, Jubbada Dhexe, Jubbada Hoose, Mogadishu, Sanaag, Woqooyi Galbeed), W Kenya (Central, N Eastern, Nairobi, N North-Eastern, Nyanza, Rift Valley, Western), NW Uganda (Madi, West Nile), Rwanda (Butare, Byumba, Gisenyi, Kibungo, Kigali), Burundi (Bururi, Muyinga, Ruyigi), W Tanzania (Arusha, Iringa, Kagera, Kigoma, Kilimanjaro, Mara, Mbeya, Mwanza, Rukwa, Ruvuma, Shinyanga, Singida, Tabora, Tanga) and N Zambia (Copperbelt, Luapula, Northern, North-Western, Western), NSL–2150 m.

Sources: Bocage, 1895, K.P. Schmidt, 1923, Andersson, 1937, Uthmöller, 1937, 1941b, H.W. Parker, 1949, Villiers, 1950a, Manaças, 1955, 1959, Mertens, 1955, 1971, Laurent, 1956a, 1964a, Vesey-FitzGerald, 1958, Broadley, 1959, Sweeney, 1961, Johnsen, 1962, Witte,

B

1962, Doucet, 1963, Roussel & Villiers, 1965, Roux-Estève & Guibé, 1965a–b, Menzies, 1966, Papenfuss, 1969, Dunger, 1971b, Pitman, 1974, Broadley & Cock, 1975, Girardin, 1978, Thorpe & McCarthy, 1978, Roman, 1980, Hakansson, 1981, Lanza, 1983, Pienaar et al., 1983, Auerbach, 1987, Branch, 1988, Gras, 1988, Hermann, 1989a, Le Berre, 1989, Largen & Rasmussen, 1993, Schätti & Gasperetti, 1994, J.-F. Trape & Roux-Estève, 1995, Bons & Geniez, 1996, Hughes, 1997, Clauss & Clauss, 2002, Spawls et al., 2002, Broadley et al., 2003, Ineich, 2003, Geneiz et al., 2004, Branch et al., 2005, Chippaux, 2006, Chirio & Ineich, 2006, J.-F. Trape & Mané, 2006b, Chirio & LeBreton, 2007, Largen & Spawls, 2010, Ullenburch et al., 2010 and Segniagbeto et al., 2011, Chirio, 2013 and Hughes, 2013.

Remarks: Original description based on H. Boie's MS (1823–1825). Includes *Lamprophis lineatus* Duméril, Bibron & Duméril *fide* Roux-Estève & Guibé, 1965a–b, and Thorpe & McCarthey, 1978. Type locality restriction to Cape of Good Hope, South Africa *fide* Broadley (1971d: 73) invalid. Taxonomy of *L. fuliginosus* complex *(arabicus-capensis-fuliginosus-lineatus-maculatus)* unresolved and therefore Distributions are only approximate.

6. *Boaedon geometricus* (Schlegel, 1837). Essai Phys. Serp. 1: 142, 2: 111. (*Lycodon geometricus*)

Synonyms: *Coluber geometricus* F. Boie *in* Schlegel, 1826a *(nomen nudum), Coluber geometricus* Schlegel, 1826b *(nomen nudum),* and *Boodon seychellensis* A.C.L.G. Günther, 1888c.

Types: Syntypes (2), MNHN 1888 and MNHN 7292, longest syntype 1040 mm (F. Péron [N.T. Baudin Géographe Voy.], 1800–1804).

Type locality: Unknown. "Australie" [= Australia] *fide* MNHN catalogue (in error). Designated as Seychelles *fide* Nussbaum (1984: 362).

Distribution: Seychelles (Frégate, Mahé, Praslin and Silhouette Is.), NSL–430 m.

Sources: Rendahl, 1939, Honegger, 1966 and Nussbaum, 1984.

Remarks: Original description based on H. Boie's MS (1823–1825).

7. *Boaedon guttatus* (A. Smith, 1843 *in* 1838–1849). Illust. Zool. So. Afr., Rept. : pl. 23, 2 pp. (*Lycodon guttatus*)

Synonym: *Alopecion annulifer* A.-M.-C. Duméril, Bibron & Duméril, 1854a.

Type: ? Holotype, BMNH 1946.1.23.16, a 287 mm specimen (A. Smith, 1834–1836) (Army Med. College, 1908).

Type locality: "beyond Kurrichane, South Africa" [= Gauteng Prov.] (probably in error). Corrected to Cape Province *fide* Broadley (1983: 88) [= Eastern Cape and Western Cape Prov., South Africa].

Distribution: Southern Africa. Southern Namibia (Karas), ext. S Mozambique (Maputo), Swaziland (Lubombo, Manzini, Shiselweni) and South Africa (Eastern Cape, E Free State, Gauteng, KwaZulu-Natal, S Limpopo, Mpumalanga, Northern Cape, Western Cape), NSL–1200 m.

Sources: Broadley, 1983, Branch, 1988 and R. Hermann, 1991.

Remarks: BMNH 1946.1.23.16 not a syntype *fide* V.F.M. FitzSimons (1937: 262).

8. *Boaedon lineatus* A.-M.-C. Duméril, Bibron & Duméril, 1854a. Erpét. Gén. 7(1): 363–364.

Synonyms: *Boaedon quadrivittatum* Hallowell, 1857b, *Boaedon quadrilineatum* A.H.A. Duméril, 1859, *Boodon lineatus angolensis* Bocage, 1895, *Boodon bedriagae* Boulenger, 1906f, and *Boodon lineatus plutonis* F. Werner, 1902.

Types: Syntypes (2), MNHN 6537 and MNHN 6562, two females, longest syntype 785 mm.

Type locality: "Côte-D'or" [= Gold Coast, presumably Ghana *fide* Hughes & Barry, 1969: 1013].

Distribution: West and cen. Africa. Senegal (Dakar, Kaolack, Kédougou, Kolda, Tambacounda, Ziguinchor), Guinea–Bissau (Bissau, Bolama, Cacheu, Gabú, Tombali, Bolama Is.), Guinea (Boffa, Boké, Dubréka, Macenta, Mamou, Nzérékoré, Télimélé), Sierra Leone (Eastern, Northern, Southern, Western), Liberia (Montserrado, Margibi), Ivory Coast (Abidjan), SW Mali (Kayes, Koulikoro, Sikasso), Burkina Faso (Centre, Centre-Ouest, Est, Hauts-Bassins, Sahel), Ghana, Togo (Centrale, Plateaux, Savanes), S Benin (Alibori, Atakora, Atlantique, Collines, Zou), S Niger (Dosso, Maradi), Nigeria (Abia, Adamawa, Benue, Kaduna, Katsina, Lagos, Nasarawa, Niger, Plateau, Sokoto, Taraba), SW Chad (Mayo-Kebbi), São Tome and Príncipe (Príncipe and São Tomé Is.), N Cameroon (Extreme-Nord, Nord, Nord-Ouest, Sud-Ouest), Central African Republic (Bamingui-Bangoran, Bangui, Haut-Kotto, Mbomou, Ouaka, Ouham, Ouham Pende, Sangha), S Gabon (Nyanga), Congo (Kouilou), N Democratic Republic of the Congo (Equateur, Orientale), SE South Sudan (Eastern Equatoria) and Uganda, 100–1700 m.

Sources: Sternfeld, 1908b, Villiers, 1950a, 1951a, Monard, 1951, Manaças, 1957, Taylor & Weyer, 1958, Capoccacia, 1961b, Perret, 1961, Roux-Estève & Guibé, 1965a–b, Dunger, 1971b, Pitman, 1974, Thorpe & McCarthy, 1978, Roman, 1980, Joger, 1990, J.B. Rasmussen, 1991, Largen & Rasmussen, 1993, Hughes, 1997, 2013, Ineich, 2003, Chirio & Ineich, 2006, Kunz, 2006, J.-F. Trape & Mané, 2006b, Chirio & LeBreton, 2007, Pauwels & Vande weghe, 2008, Chirio, 2009, W. Böhme et al., 2011, Segniagbeto et al., 2011, Auliya et al., 2012 and Chirio, 2013.

Remarks: Type locality of MNHN 6537 is Bissao *fide* MNHN catalogue. Taxonomy of *L. fuliginosus* complex *(capensis-fuliginosus-lineatus-maculatus)* unresolved and therefore Distributions are only approximate.

9. *Boaedon maculatus* H.W. Parker, 1932b. Proc. Zool. Soc. London 102(2): 363–364, fig. 3.

Type: Holotype, BMNH 1931.7.20.393, a 250 mm female (R.H.R. Taylor, 10 March 1930).

Type locality: "Bihen (8°25' N., 48°25' E.), 1500 ft., British Somaliland."

Distribution: Northeast Africa. Eastern Ethiopia (Hararge), Djibouti and Somalia (Bakool, Bari, Galguduud, Nugaal, Togdheer), NSL–500 m.

Sources: H.W. Parker, 1949, Lanza, 1983a, Largen & Rasmussen, 1993 and Largen & Spawls, 2010.

Remarks: Taxonomy of *L. fuliginosus* complex *(capensis-fuliginosus-lineatus-maculatus)* unresolved and therefore Distributions are only approximate.

10. *Boaedon olivaceus* (A.H.A. Duméril, 1856). Rev. Mag. Zool. (2) 8: 466. (*Holuropholis olivaceus*)

Synonyms: *Boodon poensis* A.C.L.G. Günther, 1888c, and *Boaedon olivaceus stirnensis* Stucki-Stirn, 1979.

Type: Holotype, MNHN 3408, a 650 mm specimen (C.E. Aubry-Lecomte, 1849–1854).

Type locality: "Gabon."

Distribution: West and cen. Africa. Guinea (Boké, Kindia, Macenta, Nzérékoré), Sierra Leone, S Liberia (Grand Cape Mount, Montserrado), Ivory Coast (Abidjan, Danané, Lagunes), S Ghana (Eastern), S Togo (Plateaux), S Nigeria (Cross River, Lagos, Rivers), S Cameroon (Adamaoua, Centre, Est, Littoral, Nord-Ouest, Ouest, Sud, Sud-Ouest), W Central African Republic (Bangui, Lobaye, Ombella-Mpoko), Equatorial Guinea (Bioko Is.), Gabon (Moyen-Ogooué, Ogooué-Ivindo, Ogooué-Lolo, Ogooué-Maritime, Woleu-Ntem), Congo (Kouilou, Pool, Sangha), Democratic Republic of the Congo (Bandundu, Bas-Congo, Equateur, Kasai Oriental, Kinshasa, Kivu, Orientale), N Angola (Cabinda, Lunda Norte), S Uganda (Central, Eastern, Western) and W Rwanda, NSL–2000 m.

Sources: Bocage, 1895, Sternfeld, 1908a–b, Loveridge, 1941a, Monard, 1951, Laurent, 1954a, Perret & Mertens, 1957a, Witte, 1962, Doucet, 1963, Mertens, 1965c, Knoepffler, 1966, Leston & Hughes, 1968, Hughes & Barry, 1969, Dunger, 1971b, Pitman, 1974, Hughes, 1983, Trape, 1985, J.A. Butler & Reid, 1986, Joger, 1990, D. Lawson, 1993, J.-F. Trape & Roux-Estève, 1995, Spawls et al., 2002, Chippaux, 2006, Chirio & Ineich, 2006, Chirio & LeBreton, 2007, Pauwels & Vande weghe, 2008 and Segniagbeto et al., 2011.

Remarks: Possibly occurs in Benin *fide* Hughes (2013: 151).

11. *Boaedon virgatus* (Hallowell, 1854c). Proc. Acad. Nat. Sci. Philadelphia (1854–1855) 7(3): 98. (*Coelopeltis virgata*)

Synonyms: *Boaedon nigrum* J.G. Fischer, 1856b, *Boaedon quadrivirgatum* Hallowell, 1857b, and *Boodon ventralis* A.C.L.G. Günther, 1888c.

Types: Syntypes (3), ANSP 10260–62, a 710 mm, a 660 mm and a 521 mm specimen (H.A. Ford, 1851–1854).

Type locality: "Liberia, west coast of Africa." Corrected to Gaboon, west coast of Africa [= Gabon] *fide* Hallowell (1857b: 56).

Distribution: West Africa. Gambia (Kombo Saint Mary, Western), SE Guinea (Macenta, Nzérékoré), Sierra Leone (Southern), Liberia (Grand Cape Mount, Montserrado, Margibi, Nimba), Ivory Coast (Abidjan, Daloa, Guiglo, Lagunes, Man, Tabou), S Ghana (Eastern), S Togo (Kara, Plateaux), Benin, S Nigeria (Cross River, Lagos, Ogun, Rivers), S Cameroon (Centre, Est, Littoral, Nord-Ouest, Sud, Sud-Ouest), Equatorial Guinea (Bioko Is.), São Tomé and Príncipe (São Tomé Is.), Gabon (Moyen-Ogooué, Ngounié), Congo (Brazzaville, Kouilou, Lekoumou, Niari) and W Democratic Republic of the Congo (Bandundu, Kinshasa), 10–1000 m.

Sources: Sternfeld, 1908a, 1909b, Loveridge, 1938b, 1941a, Monard, 1951, Perret & Mertens, 1957a, E.H. Taylor & Weyer, 1958, Doucet, 1963, Menzies, 1966, Villiers, 1966, Leston & Hughes, 1968, Hughes & Barry, 1969, Dunger, 1971b, Hakansson, 1981, Trape, 1981, Hughes, 1983, 2013, J.A. Butler & Reid, 1986, D. Lawson, 1993, J.-F. Trape & Roux-Estève, 1995, Ineich, 2003, Chippaux, 2006, Chirio & LeBreton, 2007 and Segniagbeto et al., 2011.

Remarks: Three syntypes of *Boaedon quadrivirgatum* Hallowell (1857b) also syntypes of *Coelopeltis virgata* Hallowell (1854c).

†*BOAVUS* Marsh, 1871 (Tropidophiidae)

Synonyms: †*Protagras* Cope, 1872, and †*Protagaras* – Williston & Gregory, 1925 *(nomen incorrectum)*.

Type species: †*Boavus occidentalis* Marsh, 1871.

Distribution: Lower Eocene of USA, middle Eocene of USA, lower Oligocene of USA, and middle Miocene of USA.

Sources: Marsh, 1877, Gilmore, 1938, Holman, 1979b, 2000a and Rage, 1984b.

1. †*Boavus affinis* Brattstrom, 1955a. J. Paleont. 29(1): 148–149, fig. 1a.

Type: Holotype, LACM 5119 (formerly CIT 5119), one trunk vertebra.

Type locality: "C.I.T. locality 180, upper Eocene, Sespe formation, north side of Simi Valley, Ventura County, California. In a small exposure of red beds in a large

B

canyon on the east side of the first canyon west of Tapo Canyon or 2.9 miles N. 39.5° E. of Santa Susana Bench Mark 961."

Distribution: Middle Miocene (Barstovian: 13.6–16.3 mya) of USA (Texas) and middle Eocene (33.9–46.2 mya) of USA (California).

Remarks: Rage (1984b: 17) noted that assignment to the genus †*Boavus* is doubtful.

2. †*Boavus brevis* Marsh, 1871. Amer. J. Sci. Arts (3) 1(5): 324–325.

Type: Lectotype, YPM 468 (formerly PMNH 468), one trunk vertebra (H.B. Sargent, 3 Sept. 1870), designated by Gilmore (1938: 28).

Type locality: "Grizzley Buttes, Uinta Co., Wyoming. Bridger, Middle Eocene" via lectotype selection.

Distribution: Lower Eocene (Bridgerian: 46.2–50.3 mya) of USA (Wyoming). Known only from type locality.

Remarks: Lectotype illustrated in Gilmore (1938: fig. 8).

3. †*Boavus idelmani* Gilmore, 1938. Geol. Soc. Amer., Spec. Pap. (9): 29–35, figs. 9–10, pls. 1–2.

Type: Holotype, a 960 mm skeleton (M. Idelman via E.S. Weinberg coll., 1938), lost *fide* Bruner (1991: 26).

Type locality: "Fossil Basin, probably near Fossil, Uinta Co., Wyo., Green River formation, Eocene."

Distribution: Middle Eocene (Bridgerian: 46.2–50.3 mya) of USA (Wyoming). Known only from type locality.

Source: Grande, 1980.

Remarks: A cast of the holotype is present in the AMNH 3850.

4. †*Boavus occidentalis* Marsh, 1871. Amer. J. Sci. Arts (3) 1(5): 323–324.

Synonyms: †*Boavus agilis* Marsh, 1871, and †*Protagras lacustris* Cope, 1872.

Type: Lectotype, YPM 511 (formerly PMNH 511), one trunk vertebra (O.C. Marsh, Sept. 1870), designated by Gilmore (1938: 19).

Type locality: "Grizzly Buttes, Uinta County, Wyoming. Middle Eocene, Bridger (horizon B)" via lectotype selection.

Distribution: Lower Eocene (Bridgerian: 46.2–50.3 mya) of USA (Wyoming) and lower Oligocene (Orellan: 33.3–33.9 mya) of USA (South Dakota).

Remarks: Lectotype illustrated in Gilmore (1938: fig. 2).

BOGERTOPHIS Dowling & Price, 1988 (Colubridae)

Synonym: *Bogertophus* – Ernst & Ernst, 2003 *(nomen incorrectum)*.

Type species: *Coluber subocularis* A.E. Brown, 1902b.

Distribution: Southwestern USA and N Mexico.

Fossil records: Upper Pleistocene of SW USA (New Mexico, Texas).

Sources: Dowling, 1957a, Stebbins, 1985, Dowling & Price, 1988, R. Price, 1990a, Staszko & Walls, 1994, Schulz, 1996, Rodríguez-Robles & Jesús-Escobar, 1999, Bartlett & Tennant, 2000, Utiger et al., 2002 and Pyron & Burbrink, 2009a.

1. *Bogertophis rosaliae* (Mocquard, 1899). Nouv. Arch. Mus. Hist. Nat. (4) 1: 321–323, pl. 12, figs. 1–1b. (*Coluber rosaliae*)

Type: Holotype, MNHN 1892.438, an 853 mm female (L. Diguet, 1889–1892).

Type locality: "Santa Rosalia, Basse-Californie" [= Santa Rosalía, Baja California Sur State, NW Mexico, 27°20'N, 112°16'W, elevation 15 m].

Distribution: Extreme SW USA (ext. S California) and NW Mexico (Baja California Norte, Baja California Sur), NSL–1520 m.

Sources: Price, 1990b, P.R. Brown, 1997 and Grismer, 2002.

2. *Bogertophis subocularis* (A.E. Brown, 1902b). Proc. Acad. Nat. Sci. Philadelphia (1901) 53(3): 492–494, fig., pl. 29. (*Coluber subocularis*)

Synonyms: *Elaphe sclerotica* H.M. Smith, 1941L *(nomen rejiciendum)*, and *Elaphe subocularis amplinotus* Webb, 1990.

Type: Holotype, ANSP 13733, a 1590 mm male (E. Meyenberg, 18 June 1901).

Type locality: "Davis Mountains, fifty miles southwest of Pecos, near the head of Toyah Creek, Jeff Davis County, Texas" [USA]. Restricted to vicinity of Madera Canyon, Little Aguja Canyon, and Big Aguja Canyon *fide* D. Rhoads & Salmon (2012: 271).

Distribution: Southwestern USA (S New Mexico, SW Texas) and N Mexico (E Chihuahua, Coahuila, NE Durango, W Nuevo León), 450–1800 m.

Fossil records: Upper Pleistocene (Rancholabrean II) of USA (New Mexico, Texas).

Sources: H.M. Smith, 1953c, Worthington, 1980, Webb & Ferguson, 1986, Parmley, 1990, Webb, 1990, Cranston, 1993a, Degenhardt et al., 1996, Holman, 2000a, Werler & Dixon, 2000, Lemos-Espinal et al., 2004b and D. Rhoads, 2008.

Remarks: Official Specific Name no. 2093 *fide* Opinion 745 (ICZN, 1965c).

BOIGA Fitzinger, 1826a (Colubridae)

Synonyms: *Ibiba* Gray, 1825 *(nomen rejiciendum)*, *Cephalophis* Fitzinger, 1843, *Dipsadomorphus* Fitzinger, 1843, *Eudipsas* Fitzinger, 1843, *Gonyodipsas* Fitzinger, 1843, *Macrocephalus* Fitzinger, 1843 *(nomen* *praeoccupatum), Cephalopholis* – A.-M.-C. Duméril & Bibron, 1844 *(nomen incorrectum), Dipsamorphus* – A.-M.-C. Duméril & Bibron, 1844 *(nomen incorrectum), Goniodipsas* A.-M.-C. Duméril & Bibron, 1844 *(nomen emendatum), Opetiodon* A.-M.-C. Duméril, 1853, *Triglyphodon* A.-M.-C. Duméril, 1853, *Triglijphodon* Bleeker, 1856 *(nomen emendatum), Triglypodon* – Bleeker, 1860 *(nomen incorrectum), Tryglyphodon* A.-H.-A. Duméril, 1859 *(nomen emendatum), Cephalopis* – Westphal-Castelnau, 1870 *(nomen incorrectum), Pappophis* Macleay, 1877, *Trigliphodon* – Rochebrune, 1885 *(nomen incorrectum), Borga* – Hoffmann, 1890 *(nomen incorrectum), Opeliodon* – Hoffmann, 1890 *(nomen incorrectum), Liophallus* Cope, 1895c, *Dipsadoides* Annandale, 1905b, *Dipsadamorphus* – Kinghorn, 1929 *(nomen incorrectum), Opetidon* – Maki, 1933 *(nomen incorrectum), Tryglophodon* – Hughes & Barry, 1969 *(nomen incorrectum), Boigo* – Majupuria, 1981 *(nomen incorrectum),* and *Boigan* – Hedges, 1983 *(nomen incorrectum).*

Type species: *Coluber irregularis* Merrem *in* Bechstein, 1802a.

Distribution: Africa, Asia, East Indies and Pacific islands.

Sources: Stejneger, 1907, F. Werner, 1924c, M.A. Smith, 1943, Haas, 1950, Loveridge, 1957a, Leviton, 1970a, Greene, 1989, I. Das, 1998a, Wallach, 1998a, Gravlund, 2001, Orlov & Ryabov, 2002, Groen, 2008, I. Das, 2010, 2012, Ramadhan et al., 2010, and Hoser, 2012ab

Remarks: Official Generic Name *fide* Opinion 1374 (ICZN, 1986a).

1. *Boiga andamanensis* (Wall, 1909j). Rec. Ind. Mus. 3(1): 153–154. (*Dipsadomorphus andamanensis*)

Types: Syntypes (6), ZSI 7928–30, ZSI 8641, ZSI 15192, and ZSI 15198.

Type locality: "Andamans" [= Andaman & Nicobars, E India, Bay of Bengal].

Distribution: Eastern India (Andaman & Nicobars: Andaman and Long Is.).

Sources: Orlov & Ryabov, 2002 and Whitaker & Captain, 2004.

2. *Boiga angulata* (W.C.H. Peters, 1861c). Mber. Königl. Akad. Wiss. Berlin 1861(4): 688–689. (*Dipsas* [*Dipsadomorphus*] *angulata*)

Synonym: *Dipsas guiraonis* Steindachner, 1867a.

Type: Holotype, ZMB 4000, a 1015 mm male (F. Jagor, 1859–1860).

Type locality: "Insel Leyte, Philippinen" [= Leyte Island, Philippines]. Restricted herein to NE shore of Leyte Is. between Tacloban (11°15'N, 125°00'E) and Dulag (10°57'N, 125°02'E), the 30 km (airline) distance covered by Jagor's sojourn to Leyte.

Distribution: Philippines (Catanduanes, Leyte, Luzon, Mindanao, Negros, Panay, Polillo), NSL–2500 m.

Sources: Leviton, 1970a, Ross & Gonzales, 1992 and Ferner et al., 2001.

Remarks: Probably conspecific with *B. drapiezii* Boie *fide* Leviton (1970a: 297), but a valid species *fide* Gaulke (2011: 240).

3. *Boiga barnesii* (A.C.L.G. Günther, 1869). Proc. Zool. Soc. London 37(1): 506, pl. 40, fig. 2. (*Dipsas barnesii*)

Type: Holotype, BMNH 1946.1.1.33, a 550–584 mm male (R.H. Barnes, 1858–1869).

Type locality: "Ceylon" [= Sri Lanka].

Distribution: Southwestern Sri Lanka (Central, Sabaragamuwa, Western), 220–575 m.

Sources: Boulenger, 1896a, Wall, 1921g, Deraniyagala, 1955, P. De Silva, 1969, 1980a–b, A. Silva & Pradeep, 2005 and Somaweera, 2006.

4. *Boiga beddomei* (Wall, 1909j). Rec. Ind. Mus. 3(2): 152. (*Dipsadomorphus beddomei*)

Synonym: *Boiga ranawanei* Samarawiekrama, Samarawiekrama, Wijesena & Orlov, 2005.

Types: Syntypes (7), BMNH, longest syntype 1095 mm (R.H. Beddome, 1857–1882).

Type locality: "Ceylon; Kandy, Matheran" [=Kandy, Central Prov., Sri Lanka, 7°17'N, 80°38'E, elevation 500 m, and Matheran, Maharashtra State, India, 18°59'N, 73°16'E, elevation 750 m]. Restricted to Sri Lanka *fide* Taylor (1950b: 576).

Distribution: Southern India (Kerala, Maharashtra, Orissa, Tamil Nadu) and Sri Lanka (Central, Eastern), NSL–750 m.

Sources: Wall, 1909g, 1921g, E.H. Taylor, 1950b, Inger et al., 1984, Whitaker & Captain, 2004, Samarawickrama et al., 2005 and N. Khaire, 2006.

Remarks: A synonym of *Boiga ceylonensis* Günther *fide* M.A. Smith, 1943 and Deraniyagala, 1955.

5. *Boiga bengkuluensis* Orlov, Kudryavtzev, Ryabov & Shumakov, 2003. Russ. J. Herp. 10(1): 30–34, col. figs. 1–4.

Type: Holotype, ZMMU 10416, a 1673 mm female (S. Kudryavtsev & S. Mamet, 23 April 1998).

Type locality: "Rejiang (region around Curup and Kapahiang Towns), Rejianglebong Prefecture, Bengkulu Prov., Sumatra, Indonesia; 3°28'S, 102°32'E; 500 m elevation."

Distribution: Western Indonesia (Sumatra) and S Thailand (Pattani), 500 m.

Remarks: The S Thailand record is likely *B. drapiezii fide* Vogel (*in litt.*). Possibly occurs in West Malaysia *fide* I. Das (2010: 261).

B

6. *Boiga bourreti* Tillack, Ziegler & Quyet, 2004. Sauria 26(4): 4–8, figs. 1–10.

Type: Holotype, ZFMK 82921, an 1155 mm female (natives, 25 April 2004)
Type locality: "immergrünen Primärwald nordwestlich des Phong Nha-Ke Bang Nationalparks, Distrikt Minh Hoa, Provinz Quang Binh, Vietnam, 550 m NN."
Distribution: Central Vietnam (Kon Tum, Quang Binh), 550 m.
Sources: Ziegler et al., 2006, 2010 and V.S. Nguyen et al., 2009.

7. *Boiga ceylonensis* (A.C.L.G. Günther, 1858). Cat. Colub. Snakes Brit. Mus.: 176. (*Dipsadomorphus ceylonensis*)

Synonyms: ? *Coluber boyuna* A.-M.-C. Duméril, 1853 (*nomen nudum*), and *Boiga ceylonensis dakhunensis* Deraniyagala, 1955.
Types: Syntypes (7), BMNH 1946.1.1.29, a juvenile male (R. Templeton, 1839–1851), BMNH 1946.1.4.71, an adult female (A. Paul, 1847–1849), BMNH 1946.1.4.75, a juvenile male (R. Templeton, 1839–1851) and BMNH (4), adult and juvenile males (H. Cuming, 1836–1840).
Type locality: "Ceylon" [= Sri Lanka].
Distribution: Southwestern India (Anaimalai and Meghamalai Hills of Maharashtra, Kerala, Tamil Nadu) and Sri Lanka (Central, North-Central, Uva, Western), 310–1525 m.
Sources: Wall, 1909j, 1921g, Deraniyagala, 1955, P. Silva, 1969, 1980, Inger et al., 1984, A. Silva, 1990b, 2001, 2009, N. Khaire, 2006 and Hutton & David, 2009.

8. *Boiga cyanea* (Bibron *in* A.-M.-C. Duméril, Bibron & Duméril, 1854b). Erpét. Gén. 7(2): 1079–1080. (*Triglyphodon cyaneum*)

Synonyms: *Triglyphodon cyaneum* A.-M.-C. Duméril, 1853 (*nomen nudum*), *Dipsas nigromarginata* Blyth, 1854a, *Dipsas hexagonatus* Blyth, 1855b, and *Dipsas bubalina* A.C.L.G. Günther, 1864a.
Type: Holotype, not designated, an 1887 mm specimen (Smith), location unknown.
Type locality: Unknown.
Distribution: Southeastern Asia. Eastern India (Arunachal Pradesh, Assam, Darjeeling, Manipur, Meghalaya, Mizoram, Nagaland, Sikkim, West Bengal, Nicobar Is.), Bangladesh, Nepal (Chitawan, Kathmandu, Makwanpur), Bhutan (Mongar), S China (Yunnan), Myanmar, Thailand (Chachoengsao, Chanthaburi, Chiang Mai, Lampang, Loei, Nakhon Nayok, Nakhon Ratchasima, Phang Nga, Phetchabun, Phuket, Prachin Buri, Roi Et, Si Sa Ket, Surat Thani, Tak, Ubon Ratchathani, Udon Thani, Uttaradit), Cambodia (Kampong Speu, Kampot, Kratie, Mondolkiri, Pursat, Koh Khlap Is.), Laos, Vietnam (Ba Ria-Vung, Binh Phuoc, Dak Lak, Dong Nai, Gia Lai, Kien Giang, Kon Giang, Kon Tum, Lao Cai, Quang Binh, Quang Nam, Tay Ninh), West Malaysia (Perlis) and islands in Gulf of Thailand, 40–2100 m.
Sources: Bourret, 1936b, E.H. Taylor, 1965, Campden-Main, 1970a, C.B. Frith, 1977a, Jiang & Huang, 1984, Ambu & Ng, 1991, B. L. Lim & Ratnam, 1996, Zhao & Adler, 1993, I. Das, 1994, Das & Chandra, 1994, M.J. Cox et al., 1998, Chan-ard et al., 1999, Ao et al., 2004, Tillack et al., 2004, Whitaker & Captain, 2004, Tillack, 2006, B.L. Stuart & Emmett, 2006, B.L. Stuart et al., 2006, M.F. Ahmed et al., 2009, V.S. Nguyen et al., 2009 and S.K. Sharma et al., 2013.

9. *Boiga cynodon* (H. Boie *in* F. Boie, 1827). Isis von Oken 20(6): 549. (*Dipsas cynodon*)

Synonym: *Pareas waandersi* Bleeker, 1860a.
Type: Holotype, RMNH 974, a 2098 mm specimen (H. Kuhl & J.C. van Hasselt, Dec. 1820–Sept. 1823).
Type locality: "Sumatra" [= W Indonesia] (in error). Corrected to Java, Indonesia *fide* RMNH catalogue.
Distribution: Southeastern Asia and East Indies. Eastern India (Andhra Pradesh, Gujarat, Kerala, Madbya Pradesh, Maharashtra, Orissa, Tamil Nadu, Uttar Pradesh, West Bengal), S Thailand (Nakhon Si Thammart, Narathiwat, Phang Nga, Ranong, Surat Thani, Trang, Yala), Cambodia, Myanmar, West Malaysia (Pahang, Pinang, Perlis, Penang and Tioman Is.), Singapore, East Malaysia (Sabah, Sarawak), Brunei, Indonesia (Bali, Bangka, Belitung, Borneo, Java, Kalimantan, Mentawai Arch., Nias, Riau Arch., Sumatra) and Philippines (Basilan, Culion, Dinagat, Leyte, Luzon, Mindanao, Palawan, Panay, Polillo, Romblon, Sibutu, Tawitawi), NSL–1300 m.
Sources: Deuve, 1964, E.H. Taylor, 1965, Leviton, 1968b, Campden-Main, 1970a, Kroon, 1973, Tweedie, 1983, B.E. Smith, 1993, Stuebing, 1994a, David & Vogel, 1996, M.J. Cox et al., 1998, Chan-ard et al., 1999, Ferner et al., 2001, Malkmus et al., 2002, Orlov et al., 2003, Pauwels et al., 2005, I. Das, 2007b, Mohapa et al., 2009 and Grismer, 2011.
Remarks: Original description based on H. Boie's MS (1823–1825), a specimen with 270 V & 145 SC (RMNH 974 has 269 V & 152 SC *fide* M.S. Hoogmoed *in litt.).*

10. *Boiga dendrophila* (F. Boie, 1827). Isis von Oken 20(6): 549. (*Dipsas dendrophila*)

Synonyms: *Coluber peruvianus* G. Shaw, 1802 (*nomen dubium*), *Dipsas dendrophilus* H. Boie *in* Schlegel, 1826a *(nomen nudum), Dipsas dendrophilus* Schlegel, 1826b *(nomen nudum), Dipsas dendrophilus* Fitzinger, 1826a *(nomen nudum), Dipsas dendrophila javanica* Schlegel, 1844, *Triglyphodon gemmicinctum* A.-M.-C. Duméril, 1853 *(nomen nudum), Triglyphodon dendrophilum* – A.-M.-C. Duméril, Bibron & Duméril, 1854b, *Triglyphodon gemmicinctum* A.-M.-C. Duméril, Bibron & Duméril, 1854b, *Triglijphodon gemmicinctum* Bleeker, 1856 *(nomen emendatum), Triglijphodon* *dendrophilum* Bleeker, 1856 *(nomen emendatum), Triglyphodon melanotus* Bleeker *in* Boulenger, 1896a *(nomen ineditum), Dipsadomorphus dendrophilus annectens* Boulenger, 1896a, *Dipsadomorphus dendrophilus latifasciatus* Boulenger, 1896a, *Dipsadomorphus dendrophilus melanotus* Boulenger, 1896a, *Dipsadomorphus dendrophilus multicinctus* Boulenger, 1896a, *Dipsadomorphus dendrophilus regularis* Boulenger, 1896a, *Boiga dendrophila divergens* E.H. Taylor, 1922a, *Naja celebensis* Ahl, 1933b, *Boiga dendrophila occidentalis* Brongersma, 1934, *Boiga dendrophila atra* Kopstein, 1936, and *Boiga dendrophila levitoni* Gauke, Demegillo & Vogel, 2004.

Types: Syntypes (2), MNHN 7626 [Celebes] and RMNH 932 [Java], a female (C.G.C. Reinwardt, April 1816–June 1822).

Type locality: "îles Celebes" [= Sulawesi, Indonesia] and "Java" [= W Indonesia].

Distribution: Southeastern Asia and East Indies. Southern Thailand (Chumphon, Nakhon Si Thammarat, Narathiwat, Pattani, Patthalung, Phang Nga, Ranong, Satun, Trang, Yala), Cambodia (Kampot, Koh Kong), Vietnam (Binh Duong, Gia Lai, Ho Chi Minh City), West Malaysia (Pahang, Pinang, Seribuat Arch.: Besar, Sibu, Tioman), Singapore, East Malaysia (Sabah, Sarawak), Brunei, Indonesia (Babi, Bali, Bangka, Battan, Batu, Belitung, Borneo, Java, Kalimantan, Nias, Pinang, Pulu Ubin, Riau Arch., Sulawesi, Sumatra) and Philippines (Balabac, Catanduanes, Dinagat, Leyte, Luzon, Mindanao, Palawan, Panay, Polillo, Samar), NSL–600 m.

Sources: Mertens, 1961b, Leviton, 1970a, Brongersma, 1934, Tweedie, 1983, G. Vogel, 1990, 2000, Ross & Lazell, 1991, Ross & Gonzales, 1992, B.E. Smith, 1993, I. Das, 1994, David & Vogel, 1996, Lamar, 1997b, M.J. Cox et al., 1998, Chan-ard et al., 1999, Ferner et al., 2001, Malkmus et al., 2002, Orlov et al., 2003, Gauke et al., 2004, Tillack et al., 2004, B.L. Stuart & Emmett, 2006, V.S. Nguyen et al., 2009, Grismer, 2011 and McKay & Lilley, 2012.

11. *Boiga dightoni* (Boulenger, 1894b). J. Bombay Nat. Hist. Soc. (1893–1894) 8(4): 528, 1 pl. (*Dipsas dightoni*)

Type: Holotype, BMNH 1946.1.1.32, an 1100 mm female (S. Dighton, Jan. 1893).

Type locality: "Pirmaad, at an altitude of 3,300 feet" [= Pirmed, Peermade or Azhutha, Travancore Hills, Kerala, SW India, 9°34'N, 76°59'E, elevation 1000 m].

Distribution: Southwestern India (Kerala), 700–1000 m.

Sources: Inger et al., 1984, Murthy, 1985, 1990 and R.C. Sharma, 2007.

12. *Boiga drapiezii* (H. Boie *in* F. Boie, 1827). Isis von Oken 20(6): 549. (*Dipsas drapiezii*)

Synonyms: *Triglijphodon drapiezii* Bleeker, 1856 *(nomen emendatum)*, and *Dipsas drapiezii bancana* W.C.H. Peters, 1867.

Type: Holotype, RMNH 1006 (H. Boie, Dec. 1825–Sept. 1827).

Type locality: "Java" [SW Indonesia] Restricted to region of Mt. Pangerango and Mt. Salak, West Java, Indonesia *fide* Brongersma (1948b: 14). Emended to Tjihandjawar, at the foot of Mt. Pangerango, W. Java [= Cihanyawar, Nagrak, Jawa Barat, W Java, SW Indonesia, 07°03'S, 108°40'E] *fide* Brongersma (1950: 1499).

Distribution: Southeastern Asia and East Indies. Southern Thailand (Krabi, Nakhon Si Thammarat, Pattani, Phang Nga, Ranong, Trang), Vietnam (Gai Lai), West Malaysia (Johor, Perak, Perlis, Selanger, Seribuat Arch.: Tinggi, Tioman), Singapore, East Malaysia (Sabah, Sarawak), Brunei, Indonesia (Borneo, Java, Kalimantan, Mentawai Arch., Natuna Arch., Sumatra) and the Philippines (Ambon, Palawan, Tawitawi), 500–1000 m.

Sources: E.H. Taylor, 1965, Leviton, 1970a, Tweedie, 1983, Matsui et al., 1984, Welch, 1988, Gaulke, 1994b, David & Vogel, 1996, M.J. Cox et al., 1998, Chan-ard et al., 1999, Orlov et al., 2003, Tillack et al., 2004, Chuaynkern & Makchal, 2006, V.S. Nguyen et al., 2009, Onn et al., 2010, Grismer, 2011 and I. Das, 2012.

Remarks: Original description based on H. Boie's MS (1823–1825). Possible syntype is RMNH 1008 (Padang, Sumatra, S. Müller, 1836). S. Müller was in Padang, Sumatra from June 1833–Dec. 1835.

13. *Boiga forsteni* (A.-M.-C. Duméril, Bibron & Duméril, 1854b). Erpét. Gén. 7(2): 1077–1078. (*Triglyphodon forsteni*)

Synonyms: *Triglyphodon forstenii* A.-M.-C. Duméril, 1853 *(nomen nudum)*, *Triglyphodon tesselatum* A.-M.-C. Duméril, 1853 *(nomen nudum)*, *Triglyphodon tessellatum* A.-M.-C. Duméril, Bibron & Duméril, 1854b, *Dipsas forsteni ceylonensis* J. Anderson, 1871b, *Boiga forsteni haematus* Deraniyagala, 1955, *Boiga forsteni marpila* Deraniyagala, 1955, and *Boiga forsteni leucohaematus* Deraniyagala, 1960a.

Type: Holotype, MNHN 7624 (formerly RMNH), a 1472 mm specimen (E.A. Forsten, 1838–1843).

Type locality: Unknown.

Distribution: Southern Asia. India (Andhra Pradesh, Bihar, Gujarat, Karnataka, Kerala, Madhya Pradesh, Maharashtra, Orissa, S Rajasthan, Tamil Nadu, Uttarakhand, Uttar Pradesh, West Bengal), Nepal (Banke, Bardiya, Kailali, Kanchananpur, Rupandehi, Sunsari, Surkhet) and Sri Lanka (Eastern, North-Central, Northern, North-Western, Southern, Uva, Western), 200–2200 m.

Sources: Deraniyagala, 1955, 1960, Daniel, 1962, P. Silva, 1969, Singh, 1972, Hallermann et al., 2001, A. Silva,

2001, 2009, Tiwari & Shah, 2004, Whitaker & Captain, 2004, N. Khaire, 2006 and Mohapatra et al., 2009.

Remarks: MNHN 7624 from Indonesia listed as type *fide* MNHN catalogue. Four color morphs occur in Orissa, India *fide* Mohapatra et al., 2009.

14. *Boiga fusca* (Gray, 1842c). Zool. Misc. 2(Apr.): 54. (*Dendrophis* [*Ahetula*] *fusca*)

Synonyms: *Dipsas boydii* Macleay, 1884b, and *Dipsas ornata* Macleay, 1888.

Type: Holotype, BMNH 1946.1.1.28 (J. Gilbert & J. Gould, 1840–1842).

Type locality: "north coast of New Holland; Port Essington, Australia,"[= ruins at tip of Cobourg Peninsula, Northern Territory, N Australia, ca. 11°09'N, 132°00'E, elevation NSL].

Distribution: Northern and E Australia (E New South Wales, N Northern Territory, E Queensland, N Western Australia, Bathurst, Bigge, Brampton, Byam Martin, Coronation, Cotton, Fletcher, Groote Eylandt, Horn, Indian, Inglis, Jungulu, Koolan, Lindeman, Lizard, Marchinbar, Melville, Middle Osborn, Milingimbi, Molema, Murray, Saint Andrew, Shaw, Tiwi, Wessel Marchinbar and Wigram Is.), NSL–995 m.

Source: Cogger & Lindner, 1974.

Remarks: A valid species *fide* Orlov & Ryabov, 2002.

15. *Boiga gocool* (Gray, 1835 *in* Gray & Hardwicke, 1830–1835). Illust. Indian Zool. 2(19–20): pl. 83, fig. 1. (*Dipsas gocool*)

Synonym: *Boiga gokool* A.C.L.G. Günther, 1864a *(nomen emendatum).*

Type: Holotype, BMNH 1946.1.2.59, an 840 mm female illustrated by Hardwicke in fig. 32 (T. Hardwicke, 1756–1823).

Type locality: "Bengal" [= Bangladesh and NE India (Assam, Manipur, Meghalaya, Mizoram, Nagaland, Sikkim, West Bengal)].

Distribution: South Asia. Northeastern India (Arunachal Pradesh, Assam, Manipur, Meghalaya, Nagaland, North Bengal, Sikkim), Bangladesh and Bhutan, NSL–1000 m.

Sources: Whitaker & Captain, 2004, M.F. Ahmed et al., 2009 and A. Das et al., 2010.

Remarks: M.A. Smith (1943: 529) designated pl. 32 in T. Hardwicke's collection a lectotype. Original specific orthography on plate 83, text and index is *gocool.* Holotype is illustrated in A. Das et al. (2010: 166, fig. 5). Occurs in Myanmar *fide* M.F. Ahmed et al. (2009: 83).

16. *Boiga guangxinensis* Wen, 1998. Sichuan J. Zool. 17(2): 51–52.

Type: Holotype, GMU 82020, adult male (native, 1982).

Type locality: "Longgang Nature Reserve, Longzhou City, Guangxi, Zhuang Autonomous Region, China (22°20'N and 107°4'E)."

Distribution: Southeastern Asia. Southern China (Guangxi), E Laos and N Vietnam (Bac Giang, Cao Bang, Dak Lak, Dong Nai, Gia Lai, Kon Tum, Lam Dong, Lang Son, Lao Cai, Quang Binh, Rinca, Sumba, Tay Ninh, Thai Nguyen, Vinh Phu), 150–1100 m.

Sources: Orlov et al., 2000, 2003, Tillack et al., 2004, Ziegler et al., 2006 and V.S. Nguyen et al., 2009.

17. *Boiga hoeseli* Ramadhan, Iskandar & Subasri, 2010. Asian Herpetol. Res. (2) 1(1): 23–24, figs. 1–2.

Type: Holotype, MZB Oph. 1242, a 1518 mm female (J.M. Vianney, 3 Sept. 1958).

Type locality: "Larantuka, Flores, Nenggara Islands, Indonesia" [= Larantuka, E Flores, Nusa Tenggara Timur Prov., SE Indonesia, 8°21'S, 122°59'E, elevation 25 m].

Distribution: Southern Indonesia (Alor, Flores, Komodo, Lombok, Rinca, Sumba, Sumbawa, Tenggara).

Source: Lang, 2011.

18. *Boiga irregularis* (Merrem *in* Bechstein, 1802a). Lacépède's Natur. Amph. 4: 239–240, pl. 37, fig. 1. (*Coluber irregularis*)

Synonyms: *Hurria pseudoboiga* Daudin, 1803, *Natrix mollii* Wagler, 1825 *(nomen ineditum), Triglyphodon flavescens* A.-M.-C. Duméril, 1853 *(nomen nudum), Triglyphodon flavescens* A.-M.-C. Duméril, Bibron & Duméril, 1854b, *Triglyphodon irregulare* – A.-M.-C. Duméril, Bibron & Duméril, 1854b, *Triglijphodon flavescens* Bleeker, 1856 *(nomen emendatum), Triglijphodon irregulare* Bleeker, 1856 *(nomen emendatum), Dipsas pallida* Jan, 1863, *Pappophis flavigastra* Macleay, 1877, *Pappophis laticeps* Macleay, 1877, and *Dipsas aruanus* A.C.L.G. Günther, 1883.

Type: Holotype, specimen described and illustrated in Bechstein (1802: pl. 37, fig. 1), lost *fide* Iskandar & Colijn (2001: 39).

Type locality: Unknown.

Distribution: East Indies. Indonesia (Papua, Ambon, Aru Is., Batjan, Boano, Buru, Sulawesi, Goram, Halmahera, Kai Is., Manipa, Misool, Morotai, Salawati, Sangihe, Schouten, Seram), Philippines (Sulu Arch., Ternate), Papua New Guinea (Central, East Sepik, Eastern Highlands, Gulf, Milne Bay, Madang, Morobe, Western, Western Highlands, Balaun, Bobo, Boboa, Bougainville, Daru, Duke of York, Fergusson, Goodenough, Karkar, Kiriwina, Manus, New Britain, New Ireland, Normanby and Rabaul Is.) and Guam (introduced), NSL–2286 m.

Sources: Rooij, 1917, Brongersma, 1933e, V.M. Tanner, 1950, McDowell, 1984, Bosch, 1985, Fritts et al., 1987, Fritts, 1988, Savidge, 1991, Ehmann, 1992, O'Shea,

1996, Rodda et al., 1999, J.C. Murphy & Schlager, 2003 and Sajdak, 2010.

Remarks: Official Specific Name *fide* Opinion 1374 (ICZN, 1986a).

19. *Boiga jaspidea* (A.-M.-C. Duméril, Bibron & Duméril, 1854b). Erpét. Gén. 7(2): 1093–1097. (*Triglyphodon jaspideum*)

Synonyms: *Triglyphodon jaspideum* A.-M.-C. Duméril, 1853 (*nomen nudum*), *Dipsas fusca* Motlley & Dillwyn, 1855 (*nomen praeoccupatum*), *Dipsas boops* A.C.L.G. Günther, 1858, and *Dipsadoides decipiens* Annandale, 1905.

Type: Holotype, MNHN 7627 (formerly RMHN) (Dutch Java Exped., 1820–1836).

Type locality: "Java" [SW Indonesia]

Distribution: Southeastern Asia and Greater Sundas. Southern Thailand (Phang Nga, Pattani), Vietnam (Dong Nai, Lam Dong), West Malaysia (Johor, Pinang, Perlis), Singapore, East Malaysia (Sabah, Sarawak, Labuan Is.), Brunei and W Indonesia (Bangka, Borneo, Java, Kalimantan, Mentawai Arch., Nias, Sumatra), NSL–1525 m.

Sources: Westermann, 1942, Brongersma, 1947a, E.H. Taylor, 1965, Tweedie, 1983, Matsui et al., 1984, David & Vogel, 1996, M.J. Cox et al., 1998, Malkmus et al., 2002, Orlov et al., 2003, Tillack et al., 2004, V.S. Nguyen et al., 2009, Ziegler et al., 2010 and I. Das, 2012.

20. *Boiga kraepelini* Stejneger, 1902a. Proc. Biol. Soc. Washington 15: 16–17.

Synonyms: *Boiga sinensis* K.P. Schmidt, 1925, and *Boiga multitemporalis* Bourret, 1935c.

Type: Holotype, ZMH 1565 (O. Warburg, 1888–1889).

Type locality: "Kelung, Formosa" [= Keelung, Keelung Co., Taiwan, 25°07'N, 121°43'E, elevation 90 m].

Distribution: Southeastern Asia. Southeastern China (Anhui, Fujian, Gansu, Guangxi, Guizhou, Hainan, Hunan, Jiangxi, Sichuan, Zhejiang), Taiwan (Hsinchu, Keelung, Kaohsiung, Nantou), Laos (Khammouan) and Vietnam (Bac Giang, Cao Bang, Ha Tinh, Hoa Binh, Lang Son, Nghe An, Quang Tri, Thua Thien-Hue, Vinh Phuc), 90–1100 m.

Sources: Stejneger, 1907, Maki, 1931, C.H. Pope, 1935, Kuntz, 1963, Zhao & Adler, 1993, G. Vogel, 1994, M. Huang et al., 1998, Orlov et al., 2000, 2003, Tillack et al., 2004, Zhao, 2006, Truong et al., 2007, V.S. Nguyen et al., 2009, Xiang & Li, 2009, Ziegler et al., 2010 and Yao, 2012.

21. *Boiga multifasciata* (Blyth, 1860). Proc. Asiatic Soc. Bengal 29(1): 114. (*Dipsas multifasciata*)

Type: Holotype, ZSI 7861 (formerly IMC 7861), a 368 mm specimen (J.C. Browne, 1857–1860).

Type locality: "neighborhood of Subathoo" [= Subhathu, near Simla, Himachal Pradesh State, India, 30°58'N, 77°00'E, elevation 1300 m].

Distribution: Himalayas. Northern India (Arunachal Pradesh, Himachal Pradesh, Sikkim, West Bengal), Nepal (Chitwan, Dolakha, Kaski, Kathmandu, Lalitpur, Sankhuwasabha) and Bhutan (Trashiyangtse), 1300–3100 m.

Sources: Tiwari & Shah, 2004 and Whitaker & Captain, 2004.

22. *Boiga multomaculata* (F. Boie, 1827). Isis von Oken 20(6): 549. (*Dipsas multomaculata*)

Synonyms: *Coluber aldrovandi* Merrem, 1820 (*nomen oblitum*), *Natrix sturmii* Wagler, 1825 (*nomen ineditum*), *Dipsas multimaculatus* H. Boie *in* Schlegel, 1826a (*nomen nudum*), *Dipsas multimaculatus* Schlegel, 1826b (*nomen nudum*), *Dipsas multimaculatus* Duméril, Bibron & Duméril, 1854b (*nomen emendatum*), *Boiga multimaculata hainanensis* Mell, 1931a, *Boiga multimaculata indica* Mell, 1931a, and *Boiga multimaculata sikiangensis* Mell, 1931a.

Types: Syntypes (2), RMNH 978–79 (H. Boie & H.C. Macklot, Dec. 1825–Sept. 1827).

Type locality: "Java" [SW Indonesia]

Distribution: Southeastern Asia and Greater Sundas. Southern China (Fujian, Guangdong, Guangxi, Guanxi, Guizhou, Hainan, Hong Kong, Hunan, Jiangsi, Yunnan, Zhejiang), Macau, E India (Arunachal Pradesh, Assam, Jammu and Kashmir, Nagaland, Sikkim, Uttaranchal, West Bengal), Bangladesh, Myanmar (Yangon), Thailand (Chaiyaphum, Chanthaburi, Chiang Mai, Chon Buri, Kanchanaburi, Khon Kaen, Loei, Lop Buri, Mae Hong San, Nakhon Ratchasima, Phetchabun, Prachin Buri, Rayong, Saraburi, Tak, Udon Thani, Uttardit), Cambodia (Kampong Speu, Kampot), Laos (Luang Prabang), Vietnam (Bac Kan, Binh Dinh, Binh Duong, Cao Bang, Dak Lak, Dak Nang, Da Nang, Dien Bien, Dong Nai, Gia Lai, Ha Tay, Ha Giang, Ha Tinh, Hoa Binh, Ho Chi Minh City, Lang Son, Lao Cai, Nghe An, Ninh Binh, Ninh Thuan, Quang Binh, Quang Tri, Son La, Tay Ninh, Thua Thien-Hue, Vinh Phuc), Malaysia, Singapore, Brunei, and Indonesia (Java, Kalimantan, Sumatra), NSL–2130 m.

Sources: C.H. Pope, 1935, E.H. Taylor, 1965, Karsen et al., 1986, Easton & Leung-Va, 1993, Zhao & Adler, 1993, David & Vogel, 1996, M.J. Cox et al., 1998, Chan-ard et al., 1999, Farkas & Fritz, 1999a, Orlov et al., 2000, Tillack et al., 2004, Whitaker & Captain, 2004, McKay, 2006, B.L. Stuart & Emmett, 2006, V.S. Nguyen et al., 2009 and Ziegler et al., 2010.

Remarks: Original description based on H. Boie's MS (1823–1825) but credit given to Reinwardt. In accordance with Art. 23.9.2 of the Code (ICZN, 1999), *Dipsas multomaculata* F. Boie is designated a *nomen protectum* and *Coluber aldrovandi* Merrem a *nomen* *oblitum*. Sulawesi records doubtful *fide* Land & Vogel (2005: 248).

B

23. *Boiga nigriceps* (A.C.L.G. Günther, 1863c). Ann. Mag. Nat. Hist. (3) 12(71): 359. (*Dipsas nigriceps*)

Synonyms: *Dipsas hoffmanseggii* W.C.H. Peters, 1867, *Dipsadomorphus pallidus* Boulenger, 1903, and *Boiga nigriceps brevicauda* M.A. Smith, 1926b.
Type: Holotype, BMNH 1946.1.2.58, a 1651 mm male (Zool. Soc. London, 1863).
Type locality: Unknown.
Distribution: Southeast Asia and Greater Sundas. Southern China (Guangxi), S Thailand (Nakhon Si Thammarat, Narathiwat, Pattani, Phang Nga, Phatthalung, Ratchaburi, Songkhla, Surat Thoni), West Malaysia (Johor, Selanger, Seribuat Arch.: Aur, Tioman), East Malaysia (Sabah, Sarawak), Brunei and Indonesia (Java, Kalimantan, Mentawai Arch., Nias, Siberut, Simeulue, Sumatra), NSL–1100 m.
Sources: E.H. Taylor, 1965, Tweedie, 1983, Zhao & Adler, 1993, David & Vogel, 1996, M.J. Cox et al., 1998, Chan-ard et al., 1999, Malkmus et al., 2002, Orlov et al., 2003, Wood et al., 2008 and Grismer, 2011.
Remarks: Vogel *(in litt.)* states that *Boiga brevicauda* is a valid species.

24. *Boiga nuchalis* (A.C.L.G. Günther, 1875). Proc. Zool. Soc. London 43(1): 233. (*Dipsas nuchalis*)

Types: Syntypes (5), BMNH 1874.4.29.933–37, four males and a juvenile, longest syntype 1118 mm (R.H. Beddome, 1857–1875).
Type locality: "forests on the western coast of Malabar, Southern India" [= Kerala State, SW India].
Distribution: India (Maharashtra, Kerala, Orissa, Tamil Nadu) and Nepal (Makwanpur), 200–1875 m.
Sources: Wall, 1909j, M.A. Smith, 1943, Inger et al., 1984, Malhotra & Davis, 1991, Schleich & Kästle. 2002, Tiwari & Shah, 2004 and Whitaker & Captain, 2004.
Remarks: A synonym of *B. ceylonensis* Günther *fide* M.A. Smith, 1943.

25. *Boiga ochracea* (Theobald *in* A.C.L.G. Günther, 1868). Ann. Mag. Nat. Hist. (4) 1(6): 425. (*Dipsas ochraceus*)

Synonym: *Dipsadomorphus stoliczkae* Wall, 1909j.
Types: Syntypes (2), BMNH 1946.1.2.60–61, an 889 mm male and 762 mm female (D. Browne).
Type locality: "Pegu" [Myanmar] (in error). Corrected to near Rangoon and at Maulmain *fide* Theobald (1868a: 54), [= Yangon (16°48'N, 96°09'E, elevation 40 m), Yangon Div., and Mawlamyine (16°29'N, 97°38'E, elevation 30 m), Mon Div., S Myanmar].
Distribution: Southern Asia. Northeastern India (Arunachal Pradesh, Assam, Meghalaya, Sikkim, West Bengal, Andaman Is.), Nepal (Chitwan, Gandaki, Ilam, Kaski, Kathmandu, Makwanpur, Sankuwasabha), Bhutan (Mongar), NE Bangledesh (Chittagong, Sylhet) and S Myanmar (Bago, Tanintharyi), 20–2100 m.
Sources: Kramer, 1977, M.J. Cox et al., 1998, Tiwari & Shah, 2004, Whitaker & Captain, 2004, Agarwal et al., 2010, Rahman, 2012a and S.K. Sharma et al., 2013.
Remarks: Günther attributed the name to Theobald. Collector of types W. Theobald *fide* Günther (1868: 425) and Theobald (1876: 196). Günther (1868: 425) lists longest type as 1118 mm. Boulenger (1896a: 66) listed three juveniles from Burma as syntypes. BMNH 1874.4.29.1193–94 and BMNH 1889.3.25.37–39 (from Burma), and BMNH 1872.4.17.119, BMNH 1872.4.17.386, BMNH 1894.12.31.55 and BMNH 1909.3.9.13–15 (from India) listed as syntypes fide BMNH catalogue.

26. *Boiga philippina* (W.C.H. Peters, 1867b). Mber. Königl. Akad. Wiss. Berlin 1867(1): 27. (*Dipsas philippina*)

Type: Holotype, NMW 23401, a 590–690 mm specimen (C. Semper, April 1860–Dec. 1861).
Type locality: "Ylaces im nordwesttheile von Luzon, Philippinen" [= Ilocos Prov., Luzon Is., Philippines *fide* Leviton, 1970a: 312, ca. 16°N, 121°E].
Distribution: Philippines (N Luzon).
Sources: E.H. Taylor, 1922 and Leviton, 1970a.

27. *Boiga quincunciata* (Wall, 1908b). J. Bombay Nat. Hist. Soc. (1907–1908) 18(2): 272–273, pl., figs. 6–10. (*Dipsadomorphus quincunciatus*)

Type: Holotype, ZSI 16131, a 610–914 mm specimen.
Type locality: "almost certainly Tinsukia, Assam" [= Tinsukia, Assam State, NE India, 27°29'N, 95°22'E, elevation 125 m].
Distribution: Northeastern India (Arunachal Pradesh, Assam) and N Myanmar (Kachin), 125 m.
Source: M.A. Smith, 1940b.
Remarks: Type sent to BMNH and lost *fide* M.A. Smith (1943: 353) but BMNH catalogue indicates that it was never received.

28. *Boiga saengsomi* Nutaphand, 1985. Thai Zool. Center Yrbk. 1(1): 4 unnumbered pp., 4 figs.

Synonym: *Boiga mahasomi* Nutaphand, 1985.
Type: Lecotype, NSM 52-3795 (formerly TNRC 52-3795), a 1618–2039 mm female (B. Saengmahasom, 1985), designated by Nutaphand et. al. (1991: 159).
Type locality: "Ban Kanom, Amphoe Khao Phanom, Krabi Prov. (Amphoe Khao Phanom is 50 kilometers northeast of Krabi city at an elevation of 150 meters above sea level)" via lectotype selection.
Distribution: Peninsular Thailand (Krabi, Nakhon Si Thammarat, Surat Thani, Trang), 150–180 m.

Sources: M.J. Cox, 1991b, Nutaphand et. al., 1991, M.J. Cox et al., 1998, Chan-ard et al., 1999 and I. Das, 2010.

Remarks: This species was renamed *B. mahasomi* by Nutaphand in later publications to honor the collector of the type but the action is invalid under the ICZN Code.

29. *Boiga schultzei* E.H. Taylor, 1923. Philippine J. Sci. 22D(5): 552–553, pl. 3, fig. 3.

Type: Holotype, formerly WS, a 1410 mm specimen (C.M. Weber), location unknown.

Type locality: "Palawan, Philippine Islands."

Distribution: Western Philippines (Palawan).

Remarks: Previously a subspecies of *B. drapiezii*.

30. *Boiga siamensis* Nutaphand, 1971. "Poisonous snakes of Thailand": 60–61, fig. p. 64 (upper). (*Boiga* sp. *siamensis* and *Boiga cynodon siamensis*)

Synonym: *Boiga ocellata* Kroon, 1973.

Type: Neotype, FMNH 191997, a 1625 mm male (R.F. Inger, 7 April 1969), designated by Pauwels et al. (2008: 104).

Type locality: "Sakaerat, Nakhon Ratchasima Prov., Thailand" via neotype selection.

Distribution: Southeast Asia. Northeastern India (Arunachal Pradesh, Assam, Meghalaya, Nagaland, West Bengal), Bangladesh (Sylhet), Myanmar (Bago, Kachin), Thailand (Chaiyaphum, Chanthaburi, Chiang Mai, Chumpon, Kamphaeng, Kanchanaburi, Loei, Nakhon Nayok, Nakhon Ratchaburi, Nakhon Ratchasima, Phetchabun, Phetchaburi, Ratchaburi, Sakhon Nakhon Tak, Udon Thani), Cambodia (Kampot, Koh Kong), Laos and S Vietnam (Dong Nai, Lam Dong, Tay Ninh), 50–1780 m.

Sources: Kroon, 1973, Chan-ard et al., 1999, Stuart, 1999, Orlov & Ryabov, 2002, Orlov et al., 2003, Ao et al., 2004, Whitaker & Captain, 2004, Pauwels et al., 2005, B.L. Stuart & Emmett, 2006, V.S. Nguyen et al., 2009, I. Das, 2010 and Rahman, 2012b.

Remarks: FMNH 191997 is also the type of *B. ocellata* Kroon. Tillack et al. (2004: 12) thus incorrectly recognized *B. ocellata*.

31. *Boiga tanahjampeana* Orlov & Ryabov, 2002. Russ. J. Herp. 9(1): 34–37, figs. 1–10.

Type: Holotype, ZISP 21938, a 1605 mm male (B. Soetanto & V.I. Odinchenko, Sept. 1999).

Type locality: "Tanahjampea Island, Sulawesi Selatan Prov., Indonesia."

Distribution: Central Indonesia (Sulawesi). Known only from type locality.

Sources: Lang & Vogel, 2005 and Koch, 2012.

32. *Boiga trigonata* (Schneider *in* Bechstein, 1802a). Lacépède's Naturg. Amph. 4: 256–257, pl. 40, fig. 1. (*Coluber trigonatus*)

Synonyms: *Coluber sagittatus* G. Shaw, 1802, *Coluber catenularis* Daudin, 1803e, and *Dipsadomorphus trigonata melanocephalus* Annandale, 1904.

Type: Lectotype, specimen described and illustrated by P. Russell (1796: 20–21, pl. 15) (P. Russell, 1781–1791), designated herein.

Type locality: "Vizagapatam, coast of Coromandel, India" [= Vishakhapatnam, Andhra Pradesh State, SE India, 17°41'N, 83°13'E, elevation 25 m] via lectotype selection.

Distribution: Southern Asia. Iran (Hormizghan, Kerman, Razavi Khorasan, Sistan va Baluchestan), S Uzbekistan, Turkmenistan, Tajikastan, Afganistan, Pakistan (W Baluchistan, Khyber Pakhtunkhwa, Punjab, Sindh), India (Andhra Pradesh, Bihar, Gujarat, Kashmir, Maharashtra, Manipur, Meghalaya, Nagaland, Orissa, Rajasthan, Sikkim, Tamil Nadu, Uttar Pradesh, Uttaranchal, West Bengal), Bangladesh, Nepal (Bara, Bardiya, Bhojpur, Chitwan, Dadeldhura, Dang, Ilam, Jhapa, Kailali, Kapilbastu, Kaski, Kathmandu, Makwanpur, Parsa, Rupandehi, Saptari, Surkhet, Tanahu), Bhutan and Sri Lanka (Central, Northern, North-Central, North-Western, Uva), NSL–2100 m.

Sources: Wall, 1908e, Deraniyagala, 1955, P. Silva, 1969, 1980, Singh, 1972, Bannikov et al., 1977, Kramer, 1977, Nanhoe & Ouboter, 1987, A. Silva, 1990b, 2001, 2009, Latifi , 1991, Szczerbak, 1994, Hallermann et al., 2001, M.S. Khan, 2002, R.C. Sharma, 2003, 2007, Ananjeva et al., 2004, Tiwari & Shah, 2004, Whitaker & Captain, 2004, N. Khaire, 2006 and Masroor, 2012.

Remarks: Original description based on P. Russell (1796: 20–21, pl. 15). Khan (2002: 97) recognized *Boiga melanocephala* as a distinct species.

33. *Boiga wallachi* I. Das, 1998a. J. So. Asian Nat. Hist. 3(1): 60–64, figs. 1a–b, 2a–c.

Type: Holotype, ZSI 25133 (formerly ID/AN 97), a 1006 mm male (I. Das & S. Bhaskar, 22 March 1994).

Type locality: "Kopen Heat (= Dakoank), 06° 48'N; 93° 41'E, Great Nicobar Island (= Sambelong), Bay of Bengal, India."

Distribution: Eastern India (Andaman & Nicobars: Great Nicobar and Little Nicobar Is.).

Sources: Whitaker & Captain, 2004 and Vijayakumar & David, 2006.

34. *Boiga walli* M.A. Smith, 1943. Fauna Brit. India, Rept. Amph. 3: 349. (*Boiga ochracea walli*)

Type: Holotype, not designated, ? BMNH.

Type locality: "Burma" [= Myanmar].

Distribution: Eastern India (Anadaman & Nicobars) and S Myanmar (Tanintharyi), NSL–300 m.

Source: I Das, 2010.
Remarks: A valid species *fide* I. Das (pers. comm.).

BOIRUNA Zaher, 1996b (Xenodontidae)

Type species: *Oxyrhopus maculatus* Boulenger, 1896a.
Distribution: Southern South America
Fossil records: Lower to middle Pleistocene of Argentina.
Sources: Zaher, 1996b and Zaher et al., 2009.

1. *Boiruna maculata* (Boulenger, 1896a). Cat. Snakes Brit . Mus. 3: 110, pl. 6, figs. 2-2a. (*Oxyrhopus maculatus*)

Synonyms: *Brachyruton occipitoluteum* Duméril, Bibron & Duméril, 1854a, and *Brachyryton plumbeus maculatum* Jan, 1863.
Type: Holotype, BMNH 1946.1.9.33, a 1400 mm female.
Type locality: "Uruguay."
Distribution: Southeast South America. Brazil (Amazonas, Distrito Federal, Goiás, W Mato Grosso do Sul), E Bolivia (Santa Cruz), Paraguay (Alto Paraguay, Asunción, Boquerón, Central, Misiones, Neembucú), Uruguay (Artigas, Cerro Largo, Paysandú, Rocha, Salto, Trienta y Tres) and Argentina (Catamarca, Chaco, Córdoba, Corrientes, Entre Ríos, Formosa, Jujuy, La Pampa, La Rioja, Mendoza, Misiones, Salta, San Luis, Santa Fe, Santiago del Estero, Tucumán), 30–880 m.
Sources: Leynaud & Bucher, 1999, Nugueira, 2001, Cabrera, 2004, Carreira-Vidal et al., 2005, Kacoliris et al., 2006, Cacciali, 2008 and Scanferla et al., 2009.

2. *Boiruna sertaneja* Zaher, 1996. Boll. Mus. Reg. Sci. Nat. Torino 14(2): 297–299, fig. 4a–b.

Type: Holotype, MNRJ 2384, a 1320 mm male.
Type locality: "Barreiras, Bahia, Brazil" [= Barreiras, W Bahia State, Brazil, 12°09'S, 45°00'W, elevation 575 m].
Distribution: Northeastern Brazil (Alagoas, Bahia, Ceará, Minas Gerais, Paraíba, Pernambuco, Piauí, ? Tocantins), NSL–785 m.
Source: Recorder et al., 2011.

BOLYERIA Gray, 1842a (Bolyeriidae)

Synonyms: *Uroleptes* Fitzinger, 1843, *Platygaster* A.-M.-C. Duméril & Bibron, 1844, *Uroleptis* A.-M.-C. Duméril & Bibron, 1844 *(nomen emendatum), Urolepis* – A.-M.-C. Duméril & Bibron, 1844 *(nomen incorrectum), Bolgeria* – Chenu & Desmarest, 1857 *(nomen incorrectum), Eurolepis* Hoffmann, 1890 *(nomen incorrectum), Euroleptes* Hoffmann, 1890 *(nomen incorrectum), Urolepis* Hoffmann, 1890 *(nomen incorrectum), Bolieria* Boulenger, 1893a *(nomen emendatum),* and *Boliera* – Briceño-Rossi, 1934 *(nomen incorrectum).*
Type species: *Eryx multocarinata* H. Boie *in* F. Boie, 1827.
Distribution: Mauritius and nearby islands, Indian Ocean.
Sources: Vinson, 1949, Anthony & Guibé, 1952, Guibé, 1958, Stimson, 1969, Hecht & LaDuke, 1988, Tonge, 1990, Wall, 1998b, McDiarmid et al., 1999 and Winters, 2011.

1. *Bolyeria multocarinata* (H. Boie *in* F. Boie, 1827). Isis von Oken 20(6): 513. (*Eryx multocarinata*)

Synonyms: *Tortrix pseudo-eryx* Schlegel, 1837 *(nomen incorrigendum), Platygaster multicarinatus* A.-M.-C. Duméril & Bibron, 1844 *(nomen emendatum), Bolyeria multicarinata* Gray, 1849 *(nomen emendatum), Tortrix pseudoeryx* – Giebel, 1862 *(nomen corrigendum),* and *Bolieria multicarinata* Boulenger, 1893a *(nomen emendatum).*
Type: Holotype, MNHN 7185, a 540 mm (svl) male (F. Péron & C.A. Lesueur, 1800–1804).
Type locality: "environs de Port Jackson" [Australia] (in error). Corrected to Round Island near Mauritius *fide* MNHN catalogue, Guibé (1958: 204) and Guibé & Roux-Estève (1972: 129).
Distribution: Mauritius and adjacent islands (Flat, Gunner's Quoin, Pass and Round Is.).
Remarks: Original description based on H. Boie's MS (1823–1825). Effectively extinct ca. 1980 *fide* Day (1989: 252).

BORIKENOPHIS Hedges & Vidal *in* Hedges, Couloux & Vidal, 2009 (Xenodontidae)

Type species: *Alsophis portoricensis* J.T. Reinhardt & Lütken, 1862.
Distribution: Puerto Rico and Virgin Islands.
Fossil record: Pleistocene of the West Indies (Puerto Rico).
Sources: Hedges et al., 2009, R.W. Henderson & Powell, 2009 and Zaher et al., 2009.

1. *Borikenophis portoricensis* (J.T. Reinhardt & Lütken, 1862). Vidensk. Medd. Naturhist. Foren. (1862–1863) 24(10–18): 221–123. (*Alsophis portoricensis*)

Synonyms: *Alsophis anegadae* T. Barbour, 1917b, *Dromicus variegatus* K.P. Schmidt, 1926a, *Alsophis nicholsi* Grant, 1937, *Alsophis nicholsi richardi* Grant, 1946b, *Alsophis portoricensis aphantus* A. Schwartz,

1966, and *Alsophis portoricensis prymnus* A. Schwartz, 1966.

Types: Syntypes (3), ZMUC 60460 and NMW 14836a–b, longest syntype 798 mm (A.H. Riise, 1858–1862), latter destroyed 6 Nov. 1944 during World War II.

Type locality: "Portorico" [= Puerto Rico, Greater Antilles].

Distribution: Puerto Rico bank. Puerto Rico (Aguadilla, Arecibo, Bayamón, Carolina, Humacao, Juana Díaz, Ponce, Toa Baja, Utuado, Vegsa Baja, Caya de Muertos, Cayo Santiago, Culebra, Desecheo, Platillo, and Vieques Is.), U.S. Virgin Is. (Buck, Lovango Cay, Mona, Peter, Salt, St. Thomas) and British Virgin Is. (Mosquito Is.), NSL–450 m.

Sources: A. Schwartz, 1966, Philibosian & Yntema, 1976 and Rivero, 1978, 1998.

2. *Borikenophis sancticrucis* (Cope, 1862b). Proc. Acad. Nat. Sci. Philadelphia 14(1): 76. (*Alsophis sancticrucis*)

Synonyms: *Alsophis sanctae-crucis* Boulenger, 1894a (*nomen emendatum*).

Types: Syntypes (2+), ANSP 5404, longest syntype 1270 mm (E.D. Copedon.), other syntypes lost *fide* A. Schwartz & Henderson (1991: 576).

Type locality: "St. Croix Id., W. Indies" [= St. Croix, U.S. Virgin Is., bet. 17°41–47'N and 64°34–54'W].

Distribution: U.S. Virgin Islands (Green Cay, St. Croix).

Sources: A. Schwartz, 1966 and R.W. Henderson & Powell, 1996b.

Remarks: Possibly extinct *fide* Moreno, 1991.

3. *Borikenophis variegatus* (K.P. Schmidt, 1926a). Publ. Field Mus. Nat. Hist. (Zool.) (1917–1929) 12(12): 160–162, fig. 4. (*Dromicus variegatus*)

Type: Holotype, FMNH 266 (W.W. Brown, Jr., 13 Feb. 1892).

Type locality: "Isla Mona, West Indies" [bet. 18°03–07'N and 67°51–57'W].

Distribution: Puerto Rico (Mona Is.).

Remarks: A valid species *fide* Hedges et al. (2009: 8) Possibly occurs on Desecheo Is. *fide* A. Schwartz & Henderson (1991: 574) and R.W. Henderson & Schwartz (2009: 361).

BOTHRIECHIS W.C.H. Peters, 1859b (Viperidae)

Synonyms: *Teleuraspis* Cope, 1860b, *Thamnocenchris* Salvin, 1860, *Thanatos* Posada-Arango, 1889a, *Thanatophis* Posada-Arango, 1889b (*nomen emendatum*), *Botriechis* – Hoffmann, 1890 (*nomen incorrectum*), *Telesuraspis* – Cope, 1891b (*nomen incorrectum*), *Brothriechis* – Brazil, 1911 (*nomen incorrectum*), *Brotriechis* – Brazil, 1911 (*nomen incorrectum*), *Thamnocentris* – Amaral, 1930f (*nomen incorrectum*), *Bothreichis* – L.C. Stuart, 1948 (*nomen incorrectum*), *Teleurasphis* – Burger, 1971 (*nomen incorrectum*), and *Teleuraspia* – Hoge & Romano-Hoge, 1981a (*nomen incorrectum*).

Type species: *Bothriechis nigroviridis* W.C.H. Peters, 1859b.

Distribution: Latin America.

Sources: J.A. Peters & Orejas-Miranda, 1970, Burger, 1971, Hoge & Romano-Hoge, 1971, Villa et al., 1988, J.A. Campbell & Lamar, 1989, 1992, 2004, Gosner, 1989, Maes, 1989, Schätti & Kramer, 1991, 1993, Crother et al., 1992, Golay et al., 1993, Kraus et al., 1996, Vidal et al., 1997, 1999, Werman, 1997, David & Ineich, 1999, McDiarmid et al., 1999, Parkinson, 1999, Salomão et al., 1999, J.A. Campbell & Smith, 2000, Taggart et al., 2001, Castoe & Parkinson, 2006, Luna-Reyes & Suárez-Velázquez, 2008,, Castoe et al., 2009 and Hoser, 2012d.

Remarks: *Bothriechis* and *Bothriopsis* are probably independent lineages *fide* Werman (1997: 32) but LDH enzymes failed to distinguish *Bothriopsis* from *Bothrops*.

1. *Bothriechis aurifer* (Salvin, 1860). Proc. Zool. Soc. London 28(1): 459–460, pl. 32, fig. 1. (*Thamnocenchris aurifer*)

Synonym: *Bothriechis aurifera* – Cope, 1871b.

Type: Holotype, BMNH 1946.1.17.71 (formerly BMNH 1864.1.26.32), an 872 mm specimen (R. Owen, 1859–1860).

Type locality: "Cobán, Vera Paz, Guatemala, 4500 feet" [= Cobán, Alta Verapaz Department, Guatemala, 15°28'N, 90°22'W, elevation 1325 m].

Distribution: Extreme SE Mexico (E Chiapas) and cen Guatemala (Sierra de los Cuchumatanes, Sierra de Chuacus, and Sierra de las Minas in Alta Verapaz, Baja Verapaz, Quiché), 1200–2300 m.

Source: J.C. Murphy & Schlager, 2003.

Remarks: Some specimens have been reported from lower elevations *fide* J.A. Campbell & Lamar (2004: 296).

2. *Bothriechis bicolor* (Bocourt, 1868). Ann. Sci. Nat. (Zool.) (5) 10: 202. (*Bothrops bicolor*)

Synonyms: *Bothrops bernoullii* F. Müller, 1878a, *Bothriechis bernouilii* Cope, 1887c (*nomen emendatum*), *Bothrops bernouillii* Boulenger, 1896a (*nomen emendatum*), and *Bothrops ornatus* Julía-Zertuche & Manuel-Varela, 1978.

Types: Syntypes (4), MNHN 1362, MNHN 6137, and BMNH 1946.1.17.74 (formerly MNHN & BMNH 1895.4.30.38), location other syntype unknown.

Type locality: "des forêts de Saint-Augustin, département de Solola (Guatémala), sur le versant occidental de la Cordillère. 610 mètres d'altitude" [= forests of San

B

Agustin on western slope of cordillera, Sacatepequez Department, Guatemala, 14°33'N, 90°44'W, elevation 610 m.]

Distribution: Southern Mexico (SE Chiapas) and S Guatemala (Chimaltenango, San Marcos, Sololá, Suchitepéquez), 500–2000 m.

Sources: Alvarez del Toro, 1982, J.A. Campbell & Smith, 2000 and Suárez-Velázquez & Luna-Reyes, 2009.

Remarks: MNHN 6137 not listed as a syntype in MHNH catalogue.

3. *Bothriechis lateralis* W.C.H. Peters, 1862c. Mber. Königl. Akad. Wiss. Berlin 1862(12): 674.

Types: Syntypes (4), ZMB 2979–81 and ZMB 8974 (K. Hoffmann, 1853–1859) *fide* Bauer et al. (1995: 80).

Type locality: "Costa Rica vom Vulcan Barbo...aus Veragua" [= Volcán Barba, Heredia Department, Costa Rica, 10°08'N, 84°06'W, and Veragua Prov., cen. Panama].

Distribution: Costa Rica (Cordillera de Tilarán, Cordillera Central, Cordillera de Talamanca in Alajuela, Cartago, Guanacaste, Heredia, Limón, Puntarenas, San José) and W Panama (Chiriquí, Coclé, Lérida, Panamá), 500–2135 m (most abundant between 1200–1800 m *fide* J.A. Campbell & Lamar, 2004: 300).

Sources: E.H. Taylor, 1951, 1954, Villa, 1984, Savage, 2002 and Solórzano, 2004.

Remarks: Records from Nicaragua unconfirmed *fide* J.A. Campbell & Lamar (2004: 300).

4. *Bothriechis marchi* (T. Barbour & Loveridge, 1929b). Bull. Antivenin Inst. Amer. 3(1): 1–3, fig. 1. (*Bothrops nigroviridis marchi*)

Type: Holotype, MCZ 27260, a 560 mm female (D. March, Nov. 1928).

Type locality: "the Gold Mines at Quimistan, Santa Barbara, Honduras Republic" [= El Oro, Municipio de Quimistán, in the Sierra del Espíritu Santo northwest of the town of Quimistán *fide* J.A. Campbell & Lamar, 2004: 301].

Distribution: Northwestern Honduras (Atlántida, Cortés, Santa Bárbara, Yoro) and Nicaragua (Nuevo Segovia), 500–1840 m.

Sources: L.D. Wilson & Meyer, 1985, L.D. Wilson & McCranie, 1992, G. Köhler, 1999b, 2003 and McCranie, 2011a.

Remarks: J.A. Campbell & Smith (2000: 1009) discuss the Santa Bárbara locality.

5. *Bothriechis nigroviridis* W.C.H. Peters, 1859b. Mber. Königl. Akad. Wiss. Berlin 1859(3): 278, pl., fig. 4. (*Bothrops nigroviridis*) (*nomen corrigendum*)

Synonym: *Bothrops nigro-viridis* F. Müller, 1877 *(nomen illegitimum).*

Types: Syntypes (3), ZMB 2986–88, longest syntypes 360 mm and 344 mm (K. Hoffmann, 1853–1859).

Type locality: "Vulcan von Barbo, Costa Rica" [= Volcán Barba, Heredia Department, Costa Rica, 10°08'N, 84°06'W].

Distribution: Costa Rica (Cordillera Central and Cordillera de Talamanca in Cartago, Guanacaste, Heredia, Limón, Puntarenas, San José) and ext. W Panama (Chiriquí), 1150–3000 m.

Sources: E.H. Taylor, 1951, 1954, Villa, 1984, Savage, 2002 and Solórzano, 2004.

6. *Bothriechis rowleyi* (Bogert, 1968b). Amer. Mus. Novit. (2341): 3–7, figs. 1a–c, 2. (*Bothrops rowleyi*)

Synonym: *Bothrops nigroviridis macdougalli* H.M. Smith & Moll, 1969.

Type: Holotype, AMNH 100669, an 875 mm male (J.S. Rowley, 24 April 1967).

Type locality: "a ridge that extends northward from Rancho Vicente, Colonia Rodolfo Figueroa, approximately 5 miles west of Cerro Baúl...approximately 1520 meters, on the headwaters of the Río Grijalva, roughly 30 kilometers to the north and slightly to the east of San Pedro Tapánatepec, in the Distrito de Juchitán, Oaxaca, Mexico."

Distribution: Southern Mexico (NW Chiapas, ext. SE Oaxaca), 1060–1830 m.

Sources: Lynch & Smith, 1965, 1966, H.M. Smith & Moll, 1969 and Gutberlet, 1995.

7. *Bothriechis schlegelii* (Berthold, 1845). Nachr. Georg-Augusts-Univ. Königl. Ges. Wiss. Göttingen (3): 42, pl. 1, figs. 5–6. (*Trigonocephalus schlegelii*)

Synonyms: *Lachesis nitidus* A.C.L.G. Günther, 1859c, *Bothrops nigroadspersus* Steindachner, 1870b, *Teleuraspis nitida* – Cope, 1871b, *Thanatos torvus* Posada-Arango, 1889a, *Thanatophis colgadora* García, 1896, and *Teleuraspis birri* Posada-Arango, 1909.

Type: Holotype, ZFMK 32554 (formerly ZMG 121/261), a 174 mm specimen (C. Degenhardt, 1839–1845).

Type locality: "Provinz Popayan, etwa 2° N.B. und 301° L., Neu-Grenada" [= W Colombia, probably on the Pacific versant *fide* Myers & Böhme, 1996: 17].

Distribution: Latin America. Southern Mexico (N Chiapas), N Guatemala (Izabal region, Petén), S Belize (Cayo, Orange Walk), N Honduras (Atlántida, Colón, Cortés, Gracias a Dios, Olancho, Yoro), E Nicaragua (Jinotega, Matagalpa, Río San Juan), Costa Rica (Alajuela, Cartago, Guanacaste, Heredia, Limón, Puntarenas, San José), Panama (Barro Colorado, Boca del Toro, Colón, Panamá), Colombia (Antioquia, Atlántica, Boyacá, Caldas, Cauca, Norte de Santander, Santander, Valle), W Venezuela (Táchira), N Ecuador (Chimborazo, El Oro, Esmeraldas, Guayas, Manabí, Pichincha) and Peru (Tumbes), NSL–2640 m.

Sources: Pifano et al., 1950, E.H. Taylor, 1951, 1954, Werman, 1984, L.D. Wilson & Meyer, 1985, Schätti & Kramer, 1993, Kuch & Friere, 1995a, J.C. Lee, 1996, 2000, J.A. Campbell, 1998b, Solórzano et al., 1999, G. Köhler, 2000, Koch & Freire, 2001, Savage, 2002, Navarrete et al., 2009, McCranie, 2011a, Rivas-Fuenmayor et al., 2012 and Rojas-Morales, 2012b.

Remarks: Original description reprinted in Berthold (1846: 13, pl. 1, figs. 5–6). See C.W. Myers & Böhme, 1996 for summary of Berthold papers. Type locality designation of near Popayán, Cauca Dept., Colombia, 2°27'N, 76°37'W, elevation 1800 m *fide* Dunn & Stuart (1951: 56) in error *fide* C.W. Myers & Böhme (1996: 17). Records for Amazonas, Venezuela need confirmation *fide* J.A. Campbell & Lamar (2004: 306).

8. *Bothriechis supraciliaris* (E.H. Taylor, 1954). Univ. Kansas Sci. Bull. 36(2): 791–794, fig. 39. (*Bothrops schlegelii supraciliaris*)

Type: Holotype, KU 31977, a 400 mm specimen (native, 1947, via O. Chavarría, 1952).

Type locality: "mountains near San Isidro del General, San José Prov., Costa Rica" [= mountains in vicinity of San Isidro de El General, W slope of Sierra Talamanca, San José Prov., Costa Rica, 9°22'N, 83°42'W, elevation 715 m].

Distribution: Southeastern Costa Rica (Puntarenas, San José) and Panama (Chiriquí), 800–1700 m.

Sources: Solórano et al., 1998, Crother *in* McDiarmid et al. (1999: 247) and De Jesús, 2007.

9. *Bothriechis thalassinus* J.A. Campbell & Smith, 2000. Rev. Biol. Trop. 48(4): 1002–1009, figs. 1, 2a–b.

Type: Holotype, UTA 46526, a 442 mm male (E.N. Smith, 30 Aug. 1997).

Type locality: "Finca la Firmeza, Sierra de Caral, Izabal, Guatemala, 900 m (15°22'29"N, 88°41'44"W)."

Distribution: Eastern Guatemala (Chiquimula, Izabal) and W Honduras (Copán, Ocotepeque, Santa Bárbara), 200–1750 m.

Source: McCranie, 2011a.

BOTHRIOPSIS W.C.H. Peters, 1861a (Viperidae)

Synonyms: *Brothriopsis* – Cope, 1887c *(nomen incorrectum), Bothriops* – Hoffmann, 1890 *(nomen incorrectum), Botriopsis* – Marschall, 1873 *(nomen incorrectum), Bothiriopsis* – Canton, 1895 *(nomen incorrectum),* and *Bothropsis* – Hoge & Romano-Hoge, 1981a *(nomen incorrectum).*

Type species: *Bothrops taeniatus* Wagler *in* Spix, 1824.

Distribution: South America.

Sources: J.A. Campbell & Lamar, 1989, 1992, 2004, Gosner, 1989, Maes, 1989, Werman, 1992, 1997, Golay et al., 1993, David & Ineich, 1999, McDiarmid et al., 1999, Salomão et al., 1999, Vidal et al., 1999, Gutberlet & Campbell, 2001, Castoe & Parkinson, 2006, Fenwick et al., 2009 and Hoser, 2012d.

Remarks: Not a valid genus *fide* Salomão et al., 1999 but valid *fide* McDiarmid et al., 1999 and J.A. Campbell & Lamar, 2004.

1. *Bothriopsis bilineata* (Wied-Neuwied, 1821). Reise Brasil. 2: 339. (*Cophias bilineatus*)

Synonyms: *Coluber arges* Linnaeus, 1758, *Trigonocephalus bilineatus unicolor* Jan 1859b, *(nomen nudum), Trigonocephalus arboreus* Cope, 1870, *Lachesis bilineata* – Boettger, 1898, *Bothrops bilineatus smaragdinus* Hoge, 1966a, and *Bothriopsis bilineatus smaragdina* – J.A. Campbell & Lamar, 1989.

Type: Holotype, not designated, (A.M.P, zu Wied-Neuwied, 2 Jan.–5 Feb. 1816), lost *fide* McDiarmid et al. (1999: 248).

Type locality: "Brasilien" [= Brazil]. Restricted to Villa Vicosa (= ciudad Marobá), on Rio Perhuype, Estado Bahia, Brazil *fide* Hoge & Lancini (1962: 17).

Distribution: Northern South America. Southern Colombia (Amazonas, Vaupés), SE Venezuela (Amazonas, Bolívar), Guyana (Cuyuni-Mazaruni, Essequibo Islands-West Demerara, Potaro-Siparuni), Suriname (Nickerie), French Guiana (Cayenne, Saint-Laurent-du-Maroni), N Brazil (Bahiá, Espírito Santo, Maranhao, Mato Grosso, Pará, Rio de Janeiro, Rondonia), E Ecuador (Morona-Santiago, Napo, Orellana, Pastaza), E Peru (Junin, Loreto, Madre de Dios, Pasco) and N Bolivia (Beni, Cochabamba, La Paz, Pando, Santa Cruz), NSL–1000 m.

Sources: Cunha, 1967, Cunha & Niciemento, 1978, 1993, Morales & McDiarmid, 1996, Starace, 1998, Lehr, 2001, Duellman, 2005, Harvey et al., 2005, Navarrete et al., 2009, Wallach, 2011 and C.J. Cole et al., 2013.

Remarks: Original description reprinted in Wied-Neuwied (1825c: 483).

2. *Bothriopsis chloromelas* (Boulenger, 1912b). Ann. Mag. Nat. Hist. (8) 10(58): 423–424. (*Lachesis chloromelas*)

Synonym: *Bothrops chrysomelas* Hoge & Romano-Hoge, 1981a *(nomen emendatum).*

Types: Syntypes (3), BMNH 1946.1.17.66 (formerly BMNH 1911.12.13.63), and BMNH 1946.1.19.51–52, longest syntype 740 mm (E. Boettger, 1905–1910).

Type locality: "Huancabamba, E. Peru, above 3000 feet" [= Huancabamba, Pasco Dept., Peru, 23°12'S, 75°33'W, elevation 1700 m].

Distribution: Peru (Huánuco, Junín, Loreto, Pasco), 1000–2000+ m.

Sources: Harvey et al., 2003, 2005.

B

3. *Bothriopsis medusa* (Sternfeld, 1920a). Senckenbergiana 2(6): 180–181, figs. 1–2. (*Lachesis medusa*)

Synonym: *Bothrops eneydae* Sandner-Montilla, 1976.

Type: Holotype, SMF 21202 (formerly SMF-M 9540, 1a), a 460 mm male (F. Müller, 1877).

Type locality: "Caracas, Venezuela" [= Caracas, Distrito Federal State, Venezuela, 10°29'N, 66°54'W, elevation 925 m].

Distribution: Cordillera de la Costa of N Venezuela (Aragua, Bolívar, Carabobo, Distrito Federal, Miranda), 475–2000 (2800) m.

Sources: Pifano & Römer, 1949a and Navarrete et al., 2009.

Remarks: The taxonomic status of *B. eneydae* is unclear and the type is lost.

4. *Bothriopsis oligolepis* (F. Werner, 1901c). Abh. Ber. Kön. Zool. Anthro. Ethno. Mus. Dresden (1900/01) 9(2): 12–13. (*Lachesis bilineatus oligolepis*)

Synonyms: *Lachesis peruvianus* Boulenger, 1903b, *Bothrops chrysomelas* – Amaral, 1926b *(nomen incorrectum)*, and *Bothrops peruviana* – Amaral, 1930f.

Type: Holotype, MTD D1714, an 805 mm female (A. Baessler, 1889).

Type locality: "Bolivien" [= Bolivia].

Distribution: Southeastern Peru (Cusco, Junin, Puno) and NW Bolivia (La Paz), 1500–2500 m.

Sources: Schätti & Kramer, 1993, Harvey et al., 2005 and Fenwick et al., 2009.

Remarks: A synonym of *B. pulchra fide* Schätti & Kramer (1993: 244).

5. *Bothriopsis pulchra* (W.C.H. Peters, 1862c). Mber. Königl. Akad. Wiss. Berlin 1862(12): 672–673. (*Trigonocephalus pulcher*)

Synonyms: *Bothrops albocarinata* Shreve, 1934, *Bothrops alticola* H.W. Parker, 1934, *Bothrops albocarinatus* – J.A. Peters & Orejas-Miranda, 1970, and *Bothrops alticolus* – J.A. Peters & Orejas-Miranda, 1970.

Type: Holotype, ZMB 3868, a female (formerly Quito Mus.).

Type locality: Unknown.

Distribution: Northwestern South America. South-central Colombia (Putomayo), Ecuador (Loja, Morona-Santiago, Napo, Pastaza) and N Peru (Amazonas, Cajamarca, Loreto), 300–3000 m.

Sources: J.A. Peters, 1955, Peréz-Santos & Moreno, 1988, 1991, J.A. Campbell & Lamar, 1992 and Kuch, 1997.

Remarks: Official Specific Name *fide* Opinion 1938 (ICZN, 1999b: 220). It was previously believed that the type locality was Quito but J.A. Peters (1955: 347) repeated Peters' (1862: 672) statement that the type was from the museum in Quito. Reports from Amazonian Ecuador and Peru probably refer to another species *fide* J.A. Campbell & Lamar (1989: 221). A synonym of *B. oligolepis fide* Schätti & Kramer, 1993.

6. *Bothriopsis taeniata* (Wagler *in* Spix, 1824). Serp. Brasil. Sp. Nov.: 55, pl. 21, fig. 3. (*Bothrops taeniatus*)

Synonyms: *Atropos castelnautii* A.-M.-C. Duméril, 1853 *(nomen nudum)*, *Bothrops castelnaudi* A.-M.-C. Duméril, Bibron & Duméril, 1854b, *Atropos castelnautii* – A.-M.-C. Duméril, Bibron & Duméril, 1854c *(nomen incorrectum)*, *Teleuraspis castelnaui* – Cope, 1860b *(nomen incorrectum)*, *Bothriopsis quadriscutatus* W.C.H. Peters, 1861, *Bothriopsis castelnavii* – Cope, 1871b *(nomen incorrectum)*, *Bothrops lichenosa* Roze, 1958c, *Bothriopsis quadricarinatus* – Hoge, 1966a *(nomen incorrectum)*, and *Bothrops lichenosus* – Roze, 1966a.

Type: Neotype, MNHN 1582, an 800 mm specimen (F.L.N.C.L. de Castelnau & E. Deville, June 1843–Aug. 1847), designated by Hoogmoed & Gruber (1983: 337).

Type locality: "State of Goiás, Brasil" via neotype selection. Restricted to "stretch of [the Amazon] river between the mouth of the [Río] Tajapuru (01°02'S, 51°02'W) and the mouth of the [Río] Negro (03°08'S, 59°55'W) and the parallel waterways ("paranás" and "furos") normally followed by travelers in small boats" *fide* Vanzolini (1981: xxvi).

Distribution: Amazonia. Southern Colombia (Amazonas, Boyacá, Cundinamarca), SE Venezuela (Bolívar), Guyana (Potaro-Siparuni), Suriname (Nickeria), French Guiana (Cayenne, Saint-Laurent-du-Maroni), E Ecuador (Morona-Santiago, Napo, Orellana, Pastaza, Zamora-Chinchipe), N Brazil (Amazonas, Maranhao, Mato Grosso, Pará, Rondônia, Roraima), E Peru (Huánuco) and ext. N Bolivia (La Paz, Pando), NSL–2135 m.

Sources: Sandner-Montilla, 1978, Cunha & Nascimento, 1972, 1975a, 1982, 1993, Peréz-Santos & Moreno, 1988, 1991, Starace, 1998, I. Freire et al., 2003, Means, 2004a, Harvey et al., 2005, Navarrete et al., 2009 and C.J. Cole et al., 2013.

Remarks: Holotype lost *fide* Hoogmoed & Gruber (1983: 337).

BOTHROCHILUS Fitzinger, 1843 (Pythonidae)

Synonyms: *Nardoa* Gray, 1842a *(nomen praeoccupatum)*, *Botriochilus* – A.-M.-C. Duméril & Bibron, 1844 *(nomen incorrectum)*, *Botriocheilus* – A.-M.-C. Duméril, Bibron & Duméril, 1854b *(nomen incorrectum)*, *Botrochilus* – Hoffmann, 1890 *(nomen incorrectum)*, *Nardoana* Berg, 1901 *(nomen substitutum)*, *Nordoana* – Berg, 1901 *(nomen incorrectum)*, and *Bothrocheilus* – Greene & Burghardt, 1978 *(nomen incorrectum)*.

Type species: *Tortrix boa* Schlegel, 1837.
Distribution: Northeastern Papua New Guinea.
Sources: Jan & Sordelli, 1864 *in* 1860–1866, F. Werner, 1900c, Loveridge, 1948, McDowell, 1975a, Cogger et al., 1983a, Kluge, 1993b, O'Shea, 1996, Lamar, 1997b, McDiarmid et al., 1999, Rawlings et al., 2008 and Schleip & O'Shea, 2010.

1. *Bothrochilus boa* (Schlegel, 1837). Essai Phys. Serp. 1: 129, 2: 22–23. (*Tortrix boa*)

Synonym: *Nardoa schlegelii* Gray, 1842a.
Type: Holotype, MNHN 7172, a 345–390 mm male (R-P. Lesson & P. Garnot [Coquille Voy.], Aug. 1822–March 1825).
Type locality: "Nouvelle Irlande" [= New Ireland, Bismarck Archipelago, eastern Papua New Guinea].
Distribution: Northeastern Papua New Guinea (East New Britain, Morobe, West New Britain, Arawe, Duke of York, Dyaul, Garu, Gasmata, Green, Kavieng, Mioke, Namatanai, New Britain, New Hanover, New Ireland, Tatau and Umboi Is.).
Remarks: Sulawesi and Tokelau Is. records doubtful *fide* McDowell (1975a: 49).

BOTHROCOPHIAS Gutberlet & Campbell, 2001 (Viperidae)

Type species: *Bothrops hyoprora* Amaral, 1935.
Distribution: Northwestern South America.
Sources: J.A. Campbell & Lamar, 1989, 1992, 2004, Golay et al., 1993, David & Ineich, 1999, McDiarmid et al., 1999, Gutberlet & J.A. Campbell, 2001, Castoe & Parkinson, 2006, Fenwick et al., 2009 and Hoser, 2012d.

1. *Bothrocophias campbelli* (Freire-Lascano, 1991). Publ. Trab. Cient. Ecuador, Univ. Técn. Machala 2: [1–2], 3 figs. (*Bothrops campbelli*)

Synonym: *Porthidium almawebi* Schätti & Kramer, 1993 *(nomen substitutum).*
Type: Holotype, INHMT 1956, a 906 mm male (Serpentario Laboratorios Veterinarios, 5 Aug. 1989).
Type locality: "Recinto Huagal-Sacramento, cantón Pallatanga, Provincia de Chimborazo, Ecuador. 1500 – 2000 mts de altura."
Distribution: Western Ecuador (Chimborazo, Cotopaxí, El Oro, W Imbabura, Pichincha), 1300–2000 m.
Sources: J.A. Campbell & Lamar, 1992, Schätti & Kramer, 1993 and Cisneros-Heredia et al., 2006.
Remarks: Official Specific Name *fide* Opinion 1939 (ICZN, 1999a: 220). Photograph of holotype in J.A. Campbell & Lamar (1992: fig. 4 [bottom]).

2. *Bothrocophias colombianus* (Rendahl & Vestergren, 1940). Arkiv. Zool. 33A(1): 15. (*Bothrops microphthalmus colombianus*)

Synonym: *Porthidium colombianum* – Golay, Smith, Broadley, Dixon, McCarthy, Rage, Schätti & Toriba, 1993.
Type: Holotype, NHR 3114, a 1360 mm specimen (K. von Sneidern, 1938–1940).
Type locality: "Colombia, Cauca, La Costa, the pacific coastal slopes of the Western Cordilleras...at the same meridian as El Tambo and at an altitude of about 800 m" [= ca. 2°30'N latitude].
Distribution: Western Colombia (Antioquia, Cauca), 800–2000 m.
Sources: J.A. Campbell & Lamar, 1992, 2004 and Wüster et al., 1999.
Remarks: Transferred to this genus *fide* J.A. Campbell & Lamar, 2004.

3. *Bothrocophias hyoprora* (Amaral, 1935). Mem. Inst. Butantan 9: 222, figs. 7–8. (*Bothrops hyoprora*)

Synonyms: *Bothrops pessoai* Prado, 1939, and *Bothrops hyoprorus* – Klemmer, 1963.
Type: Holotype, IB 9199 (formerly ILS 142), a 240 mm male, destroyed by fire 15 May 2010.
Type locality: "La Pedrera, Colombia, near the Brazilian-Colombian boundary" [= La Pedrera, Amazonas State, SE Colombia, 1°18'S, 69°43'W, elevation 90 m].
Distribution: Northwestern South America. Southeasterm Colombia (Amazonas, Caqueta, Casanare, Putumayo, Vaupés), E Ecuador (Morona-Santiago, Napo, Orellana, Pastazá, Sucumbíos, Zamora-Chinchipo), NE Peru, N Brazil (Acre, Amazonas, Rondônia) and N Bolivia, NSL–1000 m.
Sources: Niceforo-María, 1938, Amaral, 1944d, 1955e, 1977, Hoge, 1949, Dixon & Soini, 1986, Pérez-Santos & Moreno, 1988, 1991, Schätti & Kramer, 1993, Duellman & Mendelson III, 1995, Cisneros-Heredia et al., 2006 and Bernarde et al., 2011.
Remarks: Possibly occurs in S Guianas and an unconfirmed report from SW Venezuela *fide* J.A. Campbell & Lamar (2004: 330).

4. *Bothrocophias microphthalmus* (Cope, 1875b). J. Acad. Nat. Sci. Philadelphia (1876) (2) 8(2): 182–183. (*Bothrops microphthalmus*)

Synonyms: *Lachesis pleuroxanthus* Boulenger, 1912b, and *Bothrops microphthalma* – Amaral, 1930f.
Type: Holotype, ANSP 11515, an 885 mm specimen (J. Orton, autumn 1867–winter 1868).
Type locality: "the Huallaga between Balsa Puerto and Moyabamba, Peru" [= Río Huallaga between Balsa Puerto and Moyobamba (6°02'S, 76°58'W, elevation 880 m), San Martin Dept., N Peru].
Distribution: Northwestern South America. Colombia (Boyacá, Casanare, Cundimarca, Norte de Santander),

B

Ecuador (Chinchipe, Morena-Santiago, Napo, Pastaza, Tungurahua, Zamora), Peru (Amazonas, Cusco, Junin, Loreto, Madre de Dios) and ext. NE Bolivia (Beni, Cochabamba, La Paz), 900–2350 m.

Sources: Nicéforo-Maria, 1964, Kuch & Freire, 1995b, Harvey et al., 2005 and Cisneros-Heredia et al., 2006.

Remarks: Possibly occurs in Brazil and the northern and southern populations may represent separate species *fide* J.A. Campbell & Lamar (2004: 333).

5. *Bothrocophias myersi* Gutberlet & Campbell, 2001. Amer. Mus. Novit. (3316): 7–9, fig. 3 (col.).

Type: Holotype, AMNH 109812, a 689 mm female (C.W. Myers & J.W. Daly, 16–17 Feb. 1973).

Type locality: "Quebrada Guanguí, 0.5 km above the Río Patia, Department of Cauca, Colombia, 100–200 m.

Distribution: Coastal SW Colombia (Cauca, Chocó, Valle de Cauca), 75–200 m.

BOTHROLYCUS A.C.L.G. Günther, 1874b (Lamprophiidae)

Synonym: *Botrolycus* – Hoffmann, 1890 *(nomen incorrectum).*

Type species: *Bothrolycus ater* A.C.L.G. Günther, 1874b.

Distribution: Central Africa.

Sources: Witte, 1962, J.-F. Trape & Roux-Estève, 1995, Chippaux, 2006, Chirio & LeBreton, 2007, Pauwels & Vande weghe, 2008, Kelly et al., 2009 and Zaher et al., 2009.

1. *Bothrolycus ater* A.C.L.G. Günther, 1874b. Proc. Zool. Soc. London 42(1): 444, pl. 57, fig. b.

Synonyms: *Pseudoboodon albopunctatus* Andersson, 1901, and *Pseudoboodon brevicaudatus* Andersson, 1901.

Type: Holotype, BMNH 1946.1.12.42, a 457 mm female (Higgins).

Type locality: "Camaroon Mountain" [= Cameroon].

Distribution: Central Africa. Southwestern Cameroon (Nord-Ouest, Ouest, Sud, Sud-Ouest), Equatorial Guinea, Gabon (Ngounié, Ogooué-Lolo), SW Congo (Kouilou, Lekoumou) and E Democratic Republic of the Congo (Nord-Kivu, Orientale, Sud-Kivu), 10–1500 m.

BOTHROPHTHALMUS W.C.H. Peters, 1863c (Lamprophiidae)

Synonyms: *Bothrophtalmus* – Tornier, 1901 *(nomen incorrectum), Bothopthalmus* – Briceño-Rossi, 1934 *(nomen incorrectum), Bothropthalmus* – Cansdale, 1961 *(nomen incorrectum),* and *Botrophthalmus* – Roux-Estève, 1965 *(nomen incorrectum).*

Type species: *Bothrophthalmus lineatus* W.C.H. Peters, 1863c.

Distribution: West and cen. Africa.

Source: Chippaux, 2006.

1. *Bothrophthalmus brunneus* A.C.L.G. Günther, 1863c. Ann. Mag. Nat. Hist. (3) 12(71): 356, pl. 6, fig. e.

Type: Holotype, BMNH 1946.1.14.1, an 813 mm female.

Type locality: "Fernando Po" [= Equatorial Guinea].

Distribution: Central Africa. Southern Cameroon (Centre, Est, Littoral, Nord-Ouest, Ouest, Sud, Sud-Ouest), Equatorial Guinea (Bioko Is.), Gabon (Ngounié, Ogooué-Lolo, Ogooué-Ivindo) and Congo (Kouilou, Plateau, Sangha), 10–820 m.

Sources: Mertens, 1965c, Stucki-Stirn, 1979, J.-F. Trape & Roux-Estève, 1995, Chirio & LeBreton, 2007 and Pauwels & Vande weghe, 2008.

Remarks: A valid species *fide* Pauwels & David, 2008a.

2. *Bothrophthalmus lineatus* (W.C.H. Peters, 1863c). Mber. Königl. Akad. Wiss. Berlin 1863(6): 287–288. (*Elaphis* [*Bothrophthalmus*] *lineatus*)

Synonyms: *Bothrophthalmus lineatus* Schlegel *in* Lichtenstein & Martens, 1856 *(nomen nudum), Bothrophthalmus melanozostus* Jan, 1863, *Bothrophthalmus lineatus infuscatus* Buchholz & Peters *in* W.C.H. Peters, 1875, and *Bothrophthalmus modestus* J.G. Fischer, 1886.

Type: Holotype, ZMB 1820, a 1045 mm specimen.

Type locality: "Guinea" [= Ghana *fide* Hughes & Barry, 1969: 1013].

Distribution: West and cen. Africa. Southeastern Guinea (Macenta, Nzérékoré), Sierra Leone, Liberia (Montserrado, Margibi, Nimba), Ivory Coast (Abidjan, Daloa, Guiglo, Lagunes, Tabou), S Ghana (Eastern), S Benin (Cotonou, Plateau), Nigeria (Delta), Cameroon (Centre), S Central African Republic (Haut-Mbomou, Lobaye, Sangha), S Congo (Brazzaville), Democratic Republic of the Congo (Bandundu, Bas-Congo, Equateur, Kasai Occidental, Kasai Oriental, Katanga, Kinshasa, Nord-Kivu, Orientale, Sud-Kivu), NE Angola (Lunda Norte) and Uganda (Central, Western), NSL–2300 m.

Sources: Amaral, 1944c, Loveridge, 1957a, E.H. Taylor & Weyer, 1958, Witte, 1962, Doucet, 1963, Thys van den Audenaerde, 1966, Leston & Hughes, 1968, Hughes & Barry, 1969, Pitman, 1974, Hughes, 1983, 2013, Luiselli et al., 1998a, 1999, Spawls et al., 2002, Ineich, 2003, Chippaux, 2006, Chirio & LeBreton, 2007, Kelly et al., 2009, Zaher et al., 2009 and W. Böhme et al., 2011.

Remarks: Probably occurs in Togo *fide* and Segniagbeto et al. (2011: 353).

BOTHROPOIDES Fenwick, Gutberlet, Evans & Parkinson, 2009 (Viperidae)

Type species: *Bothrops neuwiedi* Wagler *in* Spix, 1824.
Distribution: Eastern and S South America.
Sources: Maes, 1989, J.A. Campbell & Lamar, 2004, Fenwick et al., 2009 and Hoser, 2012d.

1. *Bothropoides alcatraz* (Marques, Martins & Sazima, 2002). Herpetologica 58(3): 304–309, figs. 1, 2 (upper), 3 (left). (*Bothrops alcatraz*)

Type: Holotype, IB 62545, a 520 mm male (O.A.V. Marques, M. Martins, M.E. Oliveira, D.N. Pereira & A. Martensen, 15 Dec. 1999), destroyed by fire 15 May 2010.
Type locality: "Alcatrazes Island (24°06'S, 45°42'W), São Sebastião, São Paulo State, Brazil."
Distribution: Brazil (São Paulo: Alcatrazes and Mar Virado Is.), NSL–265 m.
Sources: Wüster et al., 1999, Cicchi et al., 2007 and Barro et al., 2012.

2. *Bothropoides diporus* (Cope, 1862d). Proc. Acad. Nat. Sci. Philadelphia 14(5): 347. (*Bothrops diporus*)

Synonym: *Bothrops neuwiedii meridionalis* Amaral, 1930c.
Type: Holotype, USNM 12380 (formerly ANSP 5401), a 1027 mm specimen (T.J. Page [La Plata Exped.], Sept. 1853–Feb. 1855).
Type locality: "Vermejo River region." Expanded to Vermejo River region (boundary of Paraguay and Argentina)" *fide* Cochran (1961: 161).
Distribution: Southern South America. Southeastern Brazil (Mato Grosso do Sul, Paraná, Rio Grande do Sul, Santa Catarina, São Paulo), Paraguay and Argentina (Catamarca, Chaco, Córdoba, Corrientes, Formosa, Jujuy, La Pampa, La Rioja, Mendoza, Misiones, Neuquén, Río Negro, Salta, San Luis, Santa Fe, Santiago del Estero, Tucumán), NSL–700 m.
Sources: Abalos & Baez, 1963, Yanosky et al., 1996, Martínez & Avila, 2003, V.X. Silva & Rodrigues, 2008 and Minoli et al., 2011.
Remarks: Probably occurs in ext. S Bolivia *fide* J.A. Campbell & Lamar (2004: 414).

3. *Bothropoides erythromelas* (Amaral, 1923). Proc. New England Zool. Club 8: 96–97. (*Bothrops erythromelas*)

Synonym: *Coluber punctulatus* Raddi, 1820 *(nomen oblitum).*
Type: Holotype, IB 3030, a 520–540 mm female (P. Silva, Jan. 1921), destroyed by fire 15 May 2010.
Type locality: "January, near Joazeiro, State of Bahia, Brazil." Corrected to Jaguarary (e Não January, como Sahiu na Nota Previa), proximo de Joazeiro, Estado da Bahía, Brazil *fide* Amaral (1926a: 114).
Distribution: Eastern Brazil (Alagoas, Bahia, Ceará, ext. E Maranhão, Minas Gerais, Paraíba, Pernambuco, Piauí, Rio Grande do Norte, Sergipe), NSL–2000 m.
Sources: Vanzolini et al., 1980 and Freitas, 1999.
Remarks: Supplemental original description in Amaral (1916a: 114–115, pl. 4, figs. 1–3). In accordance with Art. 23.9.2 of the Code (ICZN, 1999), *Bothrops erythromelas* Amaral is designated a *nomen protectum* and *Coluber punctulatus* Raddi a *nomen oblitum.*

4. *Bothropoides insularis* (Amaral, 1922a). Anex. Mem. Inst. Butantan (1921) 1(1): 18–32, pl. 3, figs. 1–5, pl. 4. (*Lachesis insularis*)

Type: Holotype, IB 1996, a 735 mm female (A. do Amaral, Nov. 1920), destroyed by fire 15 May 2010.
Type locality: "Ilha da Queimada Grande, situada no litoral do Estado de S. Paulo, a cêrca de 40 milhas a S.O. da barra de Santos, Brasil."
Distribution: Southeastern Brazil (São Paulo: Queimada Grande Is.), NSL–200 m. Known only from type locality.
Sources: Amaral, 1922b, Hoge et al., 1960, Duarte et al., 1995, Marques et al., 2002, Wüster et al., 2005, Bovo et al., 2012 and Barro et al., 2012.

5. *Bothropoides jararaca* (Wied-Neuwied, 1824a). Isis von Oken 14(10): 1103. (*Cophias jararaca*) (*nomen corrigendum*)

Synonyms: *Vipera maculata* Laurenti, 1768 *(nomen oblitum), Cophias jajaraca* Wied-Neuwied, 1824a *(nomen incorrigendum), Bothrops leucostigma* Wagler *in* Spix, 1824, *Cophias jararaca* – Wied-Neuwied, 1824 *in* 1822–1831 & 1825c *(nomen corrigendum), Cophias jararakka* – Wied-Neuwied, 1825c *(nomen incorrectum),* and *Craspedocephalus brasiliensis* Gray *in* A.C.L.G. Günther, 1858.
Type: Holotype, not designated, a 1400 mm specimen (A.P.M. zu Wied-Neuwied, 5 Feb.–12 April 1816), location unknown.
Type locality: Unknown. Designated as Lagoa d'Arara am Mucuri, Brasil *fide* Wied-Neuwied (1825b: 481) [= Morro d'Arara, 30 km above mouth of Rio Mucuri ext. S. Bahia, SE Brazil, 17°32's, 39°35'W].
Distribution: Southeastern Brazil (S Bahia, Espírito Santo, S Goiás, SE Mato Grosso, S Minas Gerais, Paraná, Rio de Janeiro, N Rio Grande do Sul, São Paulo, Santa Catarina, Anchieta, Búzios, Cananéia, Cardoso, Comprida, Franceses, Porchat, Porcos, Santo Amaro, São Sebatião, São Vicente and Vitória Is.), SE Paraguay (Alto Paraná, Itapúa) and N Argentina (Misiones), NSL–1000 m.

Sources: Amaral, 1925c, Hoge et al., 1978a, Sazima, 1992, Cei, 1994, N. Gomes & Puorto, 1993, Freitas, 1999, Marques et al., 2002, Wüster et al., 2004, Duarte Roche & Van Sluys, 2006, Cicchi et al., 2007, Cacciali, 2008, Monzel, 2009 and Barro et al., 2012.

Remarks: Supplemental original description in Wied-Neuwied (1824b: pl. 3 & 1825c: 470–483). Plate 3 in original description is labelled *Cophias atrox* but corrected in text to *Cophias jararaca*. McDiarmid et al. (1999: 262–263) discussed the type and spelling of specific name. In accordance with Art. 23.9.2 of the Code (ICZN, 1999), *Cophias jararaca* Wied-Neuwied is designated a *nomen protectum* and *Vipera maculata* Laurenti a *nomen oblitum*. Hoogmoed & Gruber (1983: 336) and Frazer & Glaw (2007: 261) discussed taxonomic status of *Bothrops leucostigma* Wagler.

6. *Bothropoides lutzi* (Miranda-Ribeiro, 1915). Arch. Mus. Nac. Rio de Janeiro 17(3): 4, 1 pl., 3 figs. (*Lachesis pictus lutzi*)

Synonyms: *Bothrops iglesiasi* Amaral, 1923 & 1926a, and *Bothrops neuwiedii piauhyensis* Amaral, 1925b.

Type: Holotype, MNRJ AL 5337 (formerly IOC), a 675 mm specimen (A. Lutz, 1912).

Type locality: "Rio S. Francisco, Bahia, Brasil inteiro" [= Río São Francisco, Bahia State, Brazil].

Distribution: Northeastern Brazil (Bahia, Ceará, E Goiás, N Minas Gerais, W Pernambuco, Piauí, Tocantins), NSL–800 m.

Sources: V.X. Silva & Rodrigues, 2008 and Loebmann, 2009c.

Remarks: An unconfirmed record from Americana, São Paulo may represent a disjunct population *fide* J.A. Campbell & Lamar (2004: 415).

7. *Bothropoides mattogrossensis* (Amaral, 1925c). Contr. Harvard Inst. Trop. Biol. Med. 2: 60, pl. 14, fig. 6., pl. 16, fig. 6. (*Bothrops neuwiedii mattogrossensis*)

Synonym: *Bothrops neuwiedii boliviana* Amaral, 1927a.

Type: Holotype, IB 3011, an adult female (D.A. Correia, 25 Feb. 1922), destroyed by fire 15 May 2010.

Type locality: "Miranda, Matto-Grosso" [= Miranda, Mato Grosso do Sul State, SE Brazil, 20°15'S, 56°22'W, elevation 130 m].

Distribution: Southeastern Brazil (S Amazonas [isolated populations], Goiás, Mato Grosso, Mato Grosso do Sul, Rondônia, São Paulo, Tocantins), SE Peru (Puno), Bolivia (Beni, Cochabamba, Santa Cruz, Tarija), Paraguay (Alto Paraguay, Amambay, Boquerón, Presidente Hayes) and Argentina (Salta), NSL–500 m.

Sources: N.J. Silva, 2000, Monteiro et al., 2006, Cacciali, 2008 and V.X. Silva & Rodrigues, 2008.

8. *Bothropoides neuwiedi* (Wagler *in* Spix, 1824). Serp. Brasil. Sp. Nov.: 56–57, pl. 22, fig. 1. (*Bothrops neuwiedi*)

Synonyms: *Crotalus urutú* Liais, 1872 *(nomen incorrigendum), Bothrops urutu* – Lacerda, 1884 *(nomen corrigendum), Bothrops atrox meridionalis* F. Müller, 1885, *Bothrops neuwiedii bahiensis* Amaral, 1925c, *Bothrops neuwiedii goyazensis* Amaral, 1925c, *Bothrops neuwiedii minasensis* Amaral, 1925c, *Bothrops neuwiedii paranaensis* Amaral, 1925c, and *Bothrops neuwiedii fluminensis* Amaral, 1933a.

Type: Lectotype, ZSM 2348/0, a 720–744 mm male (J.B. von Spix & K.F.P. von Martius, Sept. 1818–April 1819), designated by Hoogmoed & Gruber (1983: 338).

Type locality: "provincia Bahiae, Brésil" [= Bahia Prov., NE Brazil].

Distribution: Southeastern Brazil (Bahia, Goiás, Minas Gerais, Paraíba, Paraná, Rio de Janeiro, São Paulo, Santa Catarina) and E Bolivia (Beni, Chuquisaca, Santa Cruz, Tarija), NSL–1000+ m.

Sources: Amaral, 1925c, Fernandez & Abe, 1991, Cei, 1994, D.R. Norman, 1994, Peñaranda et al., 1994, Moro, 1996, Yuki, 1997b, Leynaud & Bucher, 1999, N.J. Silva, 2000, Pérez & Avila, 2000, Pérez et al., 2003, Harvey et al., 2005, V.X. Silva & Rodrigues, 2008, Monzel, 2009 and Rodrigues et al., 2012.

Remarks: ZSM 2348/0 is probably the holotype *fide* McDiarmid et al. (1999: 269) and Franzen & Glaw (2007: 261).

9. *Bothropoides pauloensis* (Amaral, 1925c). Contr. Harvard Inst. Trop. Biol. Med. 2: 59–60, pl. 14, fig. 6, pl. 16, fig. 6. (*Bothrops neuwiedii pauloensis*)

Type: Holotype, IB 3013, adult male (J.S. Queiroz, Sept. 1922), destroyed by fire 15 May 2010.

Type locality: "Leme, São Paulo, Brazil" [= Leme, São Paulo State, SE Brazil, 22°11'S, 47°24'W, elevation 620 m].

Distribution: Southern Brazil (Goiás, Mato Grosso, Mato Grosso do Sul, Minas Gerais, Parana, São Paulo) and E Bolivia (Santa Cruz), NSL–800 m.

Sources: N.J. Silva, 2000, Jansen, 2006, Monteiro et al., 2006 and V.X. Silva & Rosrigues, 2008.

10. *Bothropoides pubescens* (Cope, 1870). Proc. Amer. Philos. Soc. (1869–1870) 11(81): 157–158. (*Trigonocephalus* [*Bothrops*] *pubescens*)

Synonym: *Bothrops neuwiedii riograndensis* Amaral, 1925c.

Type: Neotype, MZUSP 5874, adult male (F. Achaval & C. Prigioni, 1952), designated *fide* X.V. Silva & Rodrigues (2008: 71).

Type locality: "Pozo Hondo, Tambores: Departmento de Tacuarembó: Uruguay" via neotype selection.

Distribution: Southern Brazil (Rio Grande do Sul) and Uruguay (Artigas, Canelones, Cerro Largo, Durango,

Florida, Lavalleja, Maldonado, Rivera, Rocha, Salto, San José, Tacuarembó, Treinta y Tres), NSL–500 m.

Sources: Hoge, 1959d, N.J. Silva, 2000, Carreira-Vidal et al., 2005, M.T. Hartmann et al., 2005 and V.X. Silva & Rodrigues, 2008.

Remarks: Lectotype designation by V.X. Silva & Rodrigues (2008: 71) incorrect since holotype is lost.

BOTHROPS Wagler *in* Spix, 1824 (Viperidae)

Synonyms: *Bathrops* – Gray *in* Griffith & Pidgeon, 1831 *(nomen incorrectum), Botrops* – Senna, 1886 *(nomen incorrectum), Botrops* García, 1896 *(nomen emendatum), Bothropsus* A.L. Herrera, 1899 *(nomen emendatum), Botthrops* – Brazil, 1914 *(nomen incorrectum), Bothropos* – Tomas-Rodas, 1938 *(nomen incorrectum), Bothsrops* – Pifano & Römer, 1949b *(nomen incorrectum), Bothropus* – Tamayo, 1962 *(nomen incorrectum), Daraninus* Hoser, 2012g *(nomen illegitimum),* and *Jackyhoserea* Hoser, 2012g *(nomen illegitimum).*

Type species: *Coluber lanceolatus* Bonnaterre, 1790.

Distribution: Latin America.

Fossil records: Lower/middle Pleistocene (Ensenadan) of Argentina.

Sources: Amaral, 1944c, Hoge, 1966a, J.A. Peters & Orejas-Miranda, 1970, Burger, 1971, Hoge & Romano-Hoge, 1971, Hoge & Romano-Hoge, 1981a, Werman, 1986b, J.A. Campbell & Lamar, 1989, 1992, 2004, Maes, 1989, Schätti et al., 1990, Golay et al., 1993, Schätti & Kramer, 1993, Kraus et al., 1996, Salomão et al., 1997, Vidal et al., 1997, Werman, 1997, 1999, David & Ineich, 1999, McDiarmid et al., 1999, Parkinson, 1999, Salomão et al., 1999, Vidal et al., 1999, Scanferla & Nenda, 2005, Castoe & Parkinson, 2006, Fenwick et al., 2009 and Hoser, 2012d.

Remarks: The following genera and species have been separated from *Bothrops: Atropoides (indomitus, mexicanus, nummifer, occiduus, olmec), Bothriechis (aurifer, bicolor, lateralis, marchi, nigroviridis, rowleyi schlegelii supraciliaris thalassinus), Bothriopsis (bilineata, chloromelas, medusa, oligolepis, pulchra, taeniata), Bothrocophias (campbelli, colombianus, hyoprora, microphthalmus, myersi), Bothropoides (alcatraz, diporus, erythromelas, insularis, jararaca, lutzi, mattogrossensis, neuwiedi, pauloensis, pubescens), Cerrophidion (barbouri, godmani, petlalcalensis, picadoi, tzotzilorum), Porthidium (arcosae, dunni, hespere, lansbergii, nasutum, ophryomegas, porrasi, volcanicum, yucatanicum)* and *Rhinocerophis (alternatus, ammodytoides, cotiara, fonsecai, itapetiningae, jonathani).*

1. *Bothrops andianus* Amaral, 1923. Proc. New England Zool. Club 8: 103–104. (*Bothrops andiana*)

Type: Holotype, MCZ 8832, a 605 mm male (G.F. Eaton, 1912).

Type locality: "Machu Picchu, Department of Cusco, Peru, about 9000–10,000 ft. altitude."

Distribution: Southern Peru (Cusco, Puno) and Bolivia (Cochabamba, La Paz, Santa Cruz), 1800–3300 m.

Source: Harvey et al., 2005.

2. *Bothrops asper* (S.W. Garman, 1884). Mem. Mus. Comp. Zool. (1883) 8(3): 124. (*Trigonocephalus asper*) (*nomen protectum*)

Synonyms: *Bothrops atrox dirus* Jan, 1863 *(nomen nudum), Trigonocephalus xanthogrammus* Cope, 1868 *(nomen oblitum), Trigonocephalus asper lanceolati* Garman, 1884, *Bothrops atrox septentrionalis* F. Müller, 1885, *Bothrops quadriscutatus* Posada-Arango, 1889b, *Bothrops xanthogramma* – Amaral, 1930f, *Bothrops neuwiedii venezuelenzi* Briceõ-Rossi, 1934, *Bothrops xantogrammus* Hoge, 1966a *(nomen emendatum),* and *Bothrops xantogramma* Hoge, 1966a *(nomen emendatum).*

Type: Holotype, MCZ 2718, head only (Maack).

Type locality: "Obispo, on the Isthmus of Darien" [= Panama].

Distribution: Latin America. Southern Mexico (Campeche, Chiapas, Hidalgo, Oaxaca, Quintana Roo, San Luis Potosí, Tabasco, S Tamaulipas, Veracruz, Yucatán), Belize (Belize, Corozal, Orange Walk, Stann Creek, Toledo), Guatemala (Petén), N Honduras (Atlántida, Colón, Cortés, Gracias a Dios, Olancho, Santa Bárbara), E Nicaragua (Atlántico Norte, Atlántico Sur, Jinotega) Costa Rica (Alajuela, Cartago, Guanacaste, Heredia, Limón, Puntarenas, San José, Caño Is.), Panama, Colombia (Cundinamarca, Gorgona Is.), Ecuador (Chimborazo, Pichincha), N Venezuela (Zulia) and Trinidad and Tobago (Trinidad), NSL–2640 m.

Sources: E.H. Taylor, 1951, 1954, Roze, 1958d, Dixon, 1984, J.D. Johnson & Dixon, 1984, L.D. Wilson & Meyer, 1985, Markezich & Taphorn, 1993, J.C. Lee, 1996, 2000, Wüster et al., 1997, J.C. Murphy, 1997, J.A. Campbell, 1998b, Sasa & Barrantes, 1998, Savage, 2002, Townsend et al., 2006, Navarrete et al., 2009, Saldarriaga-Córdoba et al., 2009, Sasa et al., 2009, Suárez-Velázquez & Luna-Reyes, 2009, Wasco, 2009, Ramírez-Bautista et al., 2010 and McCranie, 2011a.

Remarks: A synonym of *Bothrops atrox fide* Schätti & Kramer (1993: 235) and Markezich & Taphorn (1993: 253). Possibly a single variable species *fide* Wüster et al. (1997: 99). J.A. Campbell submitting request to ICZN to conserve the name *B. asper* since both *colombiensis* and *xanthgrammus* have priority.

B

3. *Bothrops atrox* (Linnaeus, 1758). Syst. Nat., ed. 10, 1: 222. (*Coluber atrox*)

Synonyms: *Coluber ambiguus* Gmelin, 1789, *Vipera weigeli* Daudin, 1803e *(nomen substitutum), Cophias holosericeus* Wied-Neuwied, 1821, *Bothrops furia* Wagler *in* Spix, 1824, *Bothrops tessellatus* Wagler *in* Spix, 1824, *Trigonocephalus colombiensis* Hallowell, 1845a, *Bothrops affinis* Gray, 1849a, *Bothrops amarali* Breciño-Rossi, 1934, *Bothrops neuvoiedii venezuelenzi* – Breciño-Rossi, 1934 *(nomen incorrectum), Bothrops lanceolatus aidae* Sandner-Montilla, 1981, *Bothrops atrox aidae* Vanzolini, 1986 *(nomen substitutum),* and *Bothrops lanceolatus nacaritae* Sandner-Montilla, 1990.

Types: Syntypes (2), NHR Lin-100 (formerly MAFR), a 452 mm specimen (Mus. Drottn.), and NHR Lin-101 (formerly MAFR), a 725 mm specimen (Mus. Drottn.).

Type locality: "Asia," (in error *fide* Hoge et al., 1978b: 48). Restricted to Suriname *fide* K.P. Schmidt & Walker (1943c: 295).

Distribution: South America. Southeastern Colombia (Boyacá, Norte de Santander), S Venezuela (Amazonas, Anzoátegui, Apure, Aragua, Barinas, Bolívar, Carabobo, Cojedes, Delta Amacuro, Distrito Federal, Falcón, Guárico, Lara, Mérida, Miranda, Monagas, Portuguesa, Sucre, Táchira, Trujillo, Yaracuy, Vargas, Zulia), Guyana (Barima-Waini, Cuyuni-Mazaruni, Essequibo Islands-West Demerara, Potaro-Siparuni, Upper Demerara-Berbice), Suriname (Marowijne, Nickerie, Saramacca), French Guiana (Cayenne, Saint-Laurent-du-Maroni), Amazonian Brazil (Acre, Amapá, Amazonas, Maranhão, Pará, Rondônia, Roraima), E Ecuador (Chimboraza, Morona-Santiago, Napo, Orellana, Pastaza, Zamora-Chinchipe), E Peru (Amazonas, Loreto, Madre de Dios, Puno) and N Bolivia (Beni, Cochabamba, La Paz, Pando, Santa Cruz), NSL–1585 m.

Sources: Procter, 1918, Amaral, 1925c, Cunha & Nascimento, 1978, 1982a, 1993, Peréz-Santos & Moreno, 1988, 1991, Markezich & Taphorn, 1993, Schätti & Kramer, 1993, Morales & McDiarmid, 1996, Wüster et al., 1996, 1997, 1999, Gorzula & Señaris, 1998, Starace, 1998, Salomão et al., 1999 Mijares-Urrutia & Arends, 2000, Duellman, 2005, Harvey et al., 2005, C.W. Myers & Donnelly, 2008, Navarrete et al., 2009 and C.J. Cole et al., 2013.

Remarks: *Bothrops colombiensis* a valid species *fide* Navarrete et al. (2009: 66).

4. *Bothrops ayerbei* Folleco-Fernández, 2010. Rev. Noved. Colomb. 10(1), 45–48, figs. 7–9.

Type: Holotype, MHNUC SE-237, a 1270 mm male (S. Ayerbe-González, 2 Feb. 2000).

Type locality: "Vereda Corralejas, Municipio de Sotará, Departamento del Cauca (Colombia) (2°16'07"N 76°39'18"W)."

Distribution: Southwestern Colombia (Cauca), 400–1800 m.

5. *Bothrops barnetti* H.W. Parker, 1938b. Ann. Mag. Nat. Hist. (11) 2(11): 447, 450.

Type: Holotype, BMNH 1946.1.18.71 (formerly BMNH 1926.3.24.40), a 573+ mm male (B. Barnett, 1923–1926).

Type locality: "the mouths of the Quebradas Honda and Perines, between Lobitos and Talara, northern Peru" [= mouths of the Quebrada Honda and Quebrada Pariñas, 4°31'S, 81°17'W, elevation NSL, between Lobitos (4° 28'S, 81° 17'W, elevation 25 m) and Talara (4°35'S, 81°16'W, elevation 10 m), Piura Department, NW Peru].

Distribution: Coastal N Peru (La Libertad, Lambayeque, Piura), NSL.

Remarks: Possibly occurs in SW Ecuador *fide* J.A. Campbell & Lamar (2004: 381). The generic placement of this taxon is uncertain.

6. *Bothrops brazili* Hoge, 1954a. Mem. Inst. Butantan (1953) 25(1): 15–16, figs. 1–6, 7b.

Type: Holotype, IB 14721, an 1158 mm female (F. Alves-Soares & H. Renkichi), destroyed by fire 15 May 2010.

Type locality: "Tomé Assú, Acará Mirim River, State of Pará, Brazil."

Distribution: Amazonia. Southern Colombia (Amazonas Guainia, Vaupés), S Venezuela (Amazonas, Apure, Bolívar), Guyana (Cuyuni-Mazaruni, Potaro-Siparuni), Suriname (Nickerie, Saramacca), French Guiana (Cayenne, Saint-Laurent-du-Maroni), E Ecuador (Morona-Santiago, Zamora-Chinchipe), N Brazil (Acre, Amapa, Amazonas, Maranhão, N Mato Grosso, Pará, Rondônia, Roraima) E Peru (Amazonas, Loreto, Madre de Dios) and N Bolivia (Beni), NSL–460 m.

Sources: Cunha & Nascimento, 1975c, 1978, 1993, Peréz-Santos & Moreno, 1988, 1991, Schätti & Kramer, 1993, Morales & McDiarmid, 1996, Starace, 1998, Cisneros-Heredia, 2004a, Navarrete et al., 2009 and C.J. Cole et al., 2013.

7. *Bothrops caribbaeus* (S.W. Garman, 1887b). Proc. Amer. Philos. Soc. 24(126): 285. (*Trigonocephalus caribbaeus*) (*nomen protectum*)

Synonyms: *Bothrops sabinii* Gray, 1842b, *Bothrops subscutatus* Gray, 1842b, and *Bothrops caribbaea* – Schwartz & R.W. Henderson, 1988.

Type: Lectotype, MCZ 4814 (S.W. Garman, 1873–1887), designated by Lazell (1964: 250).

Type locality: "St. Lucia" via lectotype selection. Restricted to Grand'Anse, Saint Lucia *fide* Lazell (1964: 251).

Distribution: Lesser Antilles (coastal Saint Lucia), NSL–200 m.

Sources: Gosner, 1987, Powell & Wittenberg, 1998, Wüster et al., 1999, 2000, Breuil, 2003 and R.W. Henderson & Powell, 2009.
Remarks: Guyana locality data for *B. sabinii* and *B. subscutatus* in error *fide* Underwood (1993: 5).

8. *Bothrops isabelae* Sandner-Montilla, 1979. Mem. Cient. Ofidiol. (4): 3–11, 5 figs.

Type: Holotype, IVO Bi-1, a 1020 mm male (R. Quintero-Serra, 15 June 1979).
Type locality: "Siete kilómetros al SE. de Guanare, Estado Portuguesa, a poco metros de la carretera que va de Guanare a Gualnarito…a 182 mts. sobre el nivel del mar" [= 7 km SE Guanare, Portuguesa State, W Venezuela, ca. 9°01'N, 69°44'W, elevation 182 m].
Distribution: Western Venezuela (Barinas, Mérida, Portuguesa, Yaracuy, Trujillo, Zulia), 180 m.
Sources: Fenwick et al., 2009 and Navarrete et al., 2009.
Remarks: Holotype was alive at time of description. A synonym of *B. atrox fide* Markezich & Taphorn (1993: 253).

9. *Bothrops jararacussu* Lacerda, 1884. Léçons Venin Serpents Brésil: 8–10.

Synonym: *Lachesis jararacuçu* Brazil, 1914 *(nomen emendatum).*
Type: Holotype, MNRJ, a 1500 mm specimen, lost *fide* Vanzolini *in* Golay et al. (1993: 46).
Type locality: "la Prov. do Rio de Janeiro, Brésil" [= Rio de Janeiro State, SE Brazil].
Distribution: Southeastern South America. Southeastern Brazil (SE Bahia, Espírito Santo, SE Mato Grosso, S Minas Gerais, Paraná, Rio de Janeiro, NW Rio Grande do Sul, Santa Catarina, São Paulo, Anchieta, Bom Abrigo, Cannanéia, Cardoso, Comprida, Grande, Porchat, Santo Amaro, São Sebastião, São Vicente and Vitória Is.), ext. SE Bolivia (Santa Cruz), SE Paraguay (Alto Paraná, Amambay, Caaguazú, Canindeyú, Itapúa) and ext. NE Argentina (Misiones), NSL–700 m.
Sources: Amaral, 1925c, Lema & Leitão de Aravjo, 1980, Cei, 1994, Freitas, 1999, Wüster et al., 1999, Duarte-Rocha & Van Sluys, 2006, Cicchi et al., 2007 and Cacciali, 2008.

10. *Bothrops lanceolatus* (Bonnaterre, 1790). Ophiologie: 10–11, pl. 38, fig. 1. (*Coluber lanceolatus*)

Synonyms: *Vipera caerulescens* Laurenti, 1768 *(nomen oblitum), Coluber glaucus* Gmelin, 1789 *(nomen oblitum), Coluber brasiliensis* Lacépède, 1789 *(nomen rejiciendum), Coluber lanceolatus* Lacépède, 1789 *(nomen rejiciendum), Coluber tigrinis* Lacépède, 1789 *(nomen rejiciendum), Coluber hastatus* Suckow, 1797, *Vipera lanceolata* – Latreille *in* Sonnini & Latreille, 1801a, *Vipera brasiliniana* Latreille *in* Sonnini & Latreille, 1801b *(nomen substitutum), Coluber megaera* G. Shaw, 1802, *Vipera brasiliana* – Daudin, 1803c *(nomen corrigendum), Vipera tigrina* – Daudin, 1803c, *Vipera chloris* Gravenhorst, 1807, *Bothrops cinereus* Gray, 1842b, and *Vipera coerulescens* – Hoge & Romano-Hoge, 1981a *(nomen incorrectum).*
Types: Syntypes (2), specimens described and illustrated in Lacépède (1789: 121–131, pl. 5, fig. 1), lost *fide* Guibé *in* Lazell (1964: 255).
Type locality: "Martinique" [= Martinique, Lesser Antilles]. Restricted to Morne Capot, between Ajoupa-bouillon and Lorrain, Martinque *fide* Lazell (1964: 255).
Distribution: Lesser Antilles (Martinque), NSL–1300 m.
Sources: Hoge, 1953e, Lazell, 1964, Gosner, 1987, Wüster et al., 1999, Breuil, 2003 and R.W. Henderson & Powell, 2009.
Remarks: A synonym of *B. atrox/asper fide* J.D. Johnson & Dixon (1984: 329).

11. *Bothrops leucurus* Wagler *in* Spix, 1824. Serp. Brasil. Sp. Nov.: 57–58, pl. 22, fig. 2.

Synonyms: *Bothrops megaera* Wagler *in* Spix, 1824 *(nomen praeoccupatum),* and *Trimeresurus pradoi* Hoge, 1948.
Type: Lectotype, ZSM 2698/0 B, a 282 mm juvenile (J.B. von Spix, Sept. 1818–April 1819), designated by Hoogmoed & Gruber (1983: 335).
Type locality: "provinciae Bahiae, Brésil" [= Bahia State, NE Brazil]. Listed as Bahia, actually Salvador, Brazil *fide* Hoge & Romano-Hoge (1981a: 200) (in error *fide* Vanzolini, 1981: xxiv).
Distribution: Eastern Brazil (Alagoas, Bahia, Ceará, Espírito Santo, Minas Gerais, Pernambuco, Rio de Janeiro, São Paulo, Sergipe), 50–400 [900] m.
Sources: Amaral, 1955d, Wüster et al., 1997, Wüster et al., 1997, 1999, Freitas, 1999, Puorto et al., 2001, Lima Duarte et al., 2003 and Rezende de Carvalho & Barreto Nacimento, 2005.
Remarks: A synonym of *B. neuwiedi* Wagler *in* J.A. Peters & Orejas-Miranda (1970: 49). Closely related to *B. atrox fide* Wüster et al., 1997.

12. *Bothrops lojanus* H.W. Parker, 1930b. Ann. Mag. Nat. Hist. (10) 5(29): 568–569. (*Bothrops lojana*)

Type: Holotype, BMNH 1946.1.19.84 (formerly BMNH 1930.1.30.1), a 488 mm male (C. Carrión-Mora, 1930).
Type locality: "vicinity of Loja City, Ecuador, (2200 metres)" [= near Loja, Loja Prov., S Ecuador, 3°59'S, 79°12'W, elevation 2200 m].
Distribution: Southern Ecuador (E. Loja, Zamora-Chinchipe), 2100–2300 m.
Sources: Peréz-Santos & Moreno, 1991 and Schätti & Kramer, 1993.

B

13. *Bothrops marajoensis* Hoge, 1966a. Mem. Inst. Butantan (1965) 32: 123–125.

Type: Holotype, IB 18143, a 952 mm male (A.R. Hoge, 20 Aug. 1958), destroyed by fire 15 May 2010.
Type locality: "Severino, Island of Marajó, State Pará, Brasil" [= Marajó Is., mouth of Río Amazonas, E Pará State, NE Brazil, ca. 0°53'S, 49°38'W, elevation 15 m].
Distribution: Coastal northern Brazil (Amapá, ? Maranhão, Pará, Marajó Is.), NSL–15 m.
Sources: Wüster et al., 1996, 1997, 1999.
Remarks: A synonym of *B. atrox fide* Cunda & Nascimento (1993: 126).

14. *Bothrops marmoratus* V. Silva & Rodrigues, 2008. Phyllomedusa 7(1): 72–77, figs. 1d, 16a–c, 17a–b.

Type: Holotype, IB 55055, a 585 mm female (F. Rodrigues-Alves, 29 June 1993), destroyed by fire 15 May 2010.
Type locality: "Ipameri (17º43'S, 48º09'W), State of Goiás, Brazil."
Distribution: Central Brazil (Goiás, W Minas Gerais, Tocantins), 785 m.
Remarks: V. Silva & Rodrigues (2008: 72) erroneously listed holotype figures as 12 and 13.

15. *Bothrops moojeni* Hoge, 1966a. Mem. Inst. Butantan (1965) 32: 126–127, pl. 4.

Type: Holotype, IB 23397, a 1468 mm female (Fundação Zoobotânica, 17 April 1962), destroyed by fire 15 May 2010.
Type locality: "Brasília, Distrito Federal, Brasil" [= Brasília, Distrito Federal State, Brazil, 15°47'S, 47°56'W, elevation 1065 m].
Distribution: Eastern Brazil (Bahia, Distrito Federal, Goiás, Maranhão, Mato Grosso, Mato Grosso do Sul, Minas Gerais, Paraná, Piauí, São Paulo, Tocantins), E Bolivia (Santa Cruz), E Paraguay (Alto Paraná, Amambay, Caaguazú, Canindeyú, Concepción, San Pedro) and NE Argentina (Misiones), 250–1500 m.
Sources: Vanzolini, 1961, 1991, Cei, 1994, Freitas, 1999, Wüster et al., 1996, 1997, 1999, Buongermini & Waller, 1999, Harvey et al., 2005, Cacciali, 2008 and Recorder et al., 2011.
Remarks: Possibly a synonym of *B. atrox* (Linnaeus) *fide* J.A. Campbell & Lamar, 1989, 2004, Wüster et al., 1997.

16. *Bothrops muriciensis* Ferrarezzi & Freire, 2001. Bol. Mus. Nac. Rio de Janeiro (440): 2–6, fig. 1.

Type: Holotype, MNRJ 7036, a 619 mm female (S.T. Silva, 20 April 1995).
Type locality: "Brazil: Alagoas: Murici, Fazenda Bananeira (9°14'S, 35°48'W, 640m)"
Distribution: Northeastern Brazil (Alagoas), 400–640 m. Known only from vicinity of type locality.
Source: Freitas et al., 2012.

17. *Bothrops osbornei* Freire-Lascano, 1991. Publ. Trab. Cient. Ecuador, Univ. Técn. Machala 2: [2–4], 3 figs.

Synonym: *Bothriechis mahnerti* Schätti & Kramer, 1991.
Type: Holotype, INHMT 1924, a 396 mm female (Serpentario Laboratorios Veterinarios, 17 June 1989).
Type locality: "Sacramento, Cantón Pallatanga, Provincia del Chimborazo, Ecuador."
Distribution: Eastern Ecuador (Cajamarca, Chimborazo, Cotopaxi, Pichincha) and ext. N Peru (Tumbes), 500–2000 m.
Sources: Schätti & Kramer, 1991, 1993 and Coloma et al., 2000.
Remarks: Type erroneously listed as INHMT 1925 on cover figure (but INHMT 1924 in text and on two other figures).

18. *Bothrops otavioi* Barro, Grazziotin, Sazima, Martins & Sawaya, 2012. Herpetologica 68(3), 421–425, figs. 2–3, 4a.

Type: Holotype, IB 78572 (formerly CC 58), a 456 mm male (F.C. Centeno & T.H. Condez, 11 Aug. 2010).
Type locality: "'Trilha da Vitória' (23º44'28"S, 45º01'16"W; datum = WGS 84: Figs. 1 and 4B), Vitória Island, Ilhabela Archipelago, São Paulo, southeastern Brazil."
Distribution: Southeastern Brazil (São Paulo: Vitória Is.), NSL–200 m. Known only from type locality.

19. *Bothrops pictus* (Tschudi, 1845). Arch. Naturg. 11(1): 166. (*Lachesis picta*)

Type: Lectotype, MHNN 6 (formerly MZN 114) (J.J. von Tschudi, Aug. 1838–Feb. 1841), designated by Schätti (1986a: 100).
Type locality: "Montagnes élevées du Pérou" [= high mountains of Peru] via lectotype selection.
Distribution: Coastal W Peru (Ancash, Arequipa, Ica, Libertad, Lima), 500–2300 m.
Sources: J.A. Campbell & Lamar, 1992 and Yarlequé-Chocas, 2000.
Remarks: Original description reprinted in Tschudi (1846: 61–63, pl. 10). The generic status of this species is uncertain. Werner's (1901) Bolivian record unverified *fide* J.A. Campbell & Lamar (2004: 404).

20. *Bothrops pirajai* Amaral, 1923. Proc. New England Zool. Club 8: 99–100.

Synonym: *Bothrops neglecta* Amaral, 1923 & 1926a.
Type: Holotype, IB 3008, an 1130 mm female (P. da Silva, Dec. 1921), destroyed by fire 15 May 2010.
Type locality: "Ilheos, Bahia, Brazil" [= Ilhéus, SE Bahia State, E Brazil, 14°48'S, 39°02'W, elevation 10 m].
Distribution: Extreme E Brazil (E Bahia, possibly Minas Gerais), NSL–500 m.
Source: Freitas, 1999.

Remarks: Supplemental original description in Amaral (1926a: 120, pl. 4, fig. 8).

21. *Bothrops punctatus* (García, 1896). Ofidios Venen. Cauca: 30–31, pl. 8. (*Lachesis punctata*)

Synonyms: *Thanatos montanus* Posada-Arango, 1889a, *Lachesis monticellii* Peracca, 1910, *Bothrops leptura* Amaral, 1923, and *Bothrops punctatus* – Dunn, 1944.
Type: Holotype, not designated (E. García, 1887–1895), lost *fide* Nicéforo-María (1939: 92).
Type locality: "las montañas del Dagua" [= mtns. of Dagua, Valle del Cauca Department, W Colombia, ca. 3°39'N, 76°42'W].
Distribution: Extreme E Panama (Darién), W Colombia (Antioquia, Cauca, Chocó, Nariño, Valle del Cauca) and NW Ecuador (Esmeraldas) 915–2300 m.
Sources: Freire-Lascano, 1991, Schätti & Kramer, 1991 and Daza, et al., 2005.

22. *Bothrops rhombeatus* García, 1896. Ofidios Venen. Cauca: 20–22, pl. 2. *(nomen corrigendum)*

Synonyms: *Botrops rhomboatus* García, 1896 *(nomen incorrigendum)*, and *Botrops rhombeatus* García, 1896 *(nomen corrigendum)*.
Type: Holotype, not designated (E. García, 1887–1895), location unknown.
Type locality: "las tierras calientes del valle que en las montañas del Cauca, en los bosques espesos, á orillas de los ríos affuentes del Cauca, en el Guachal, el Bolo, el Amaime, y en los cañaverales de las haciendas" [= Río Guachal, Río Bolo and Río Amaime, Valle del Cauca Dept., NW Colombia, bet. 3°36–41'N, 76°25–28'W, elevation 950 m].
Distribution: Western Colombia (Antioquia, Caldas, Cauca, Córdoba, Risaralda, Valle del Cauca), 950–2600 m.
Source: Folleco-Fernández, 2010.
Remarks: Specific name spelled *rhomboatus* (text) and *rhombeatus* (plate).

23. *Bothrops roedingeri* Mertens *in* Titschack, 1942. Beitr. Fauna Perus 11: 284–285.

Type: Holotype, SMF 6017, a 328 mm juvenile (native, 12 June 1936).
Type locality: "Hacienda Huayuri" [= Huayuri, Ica Prov., W Peru, 14°32'S, 75°18'W, elevation 465 m].
Distribution: Coastal W Peru (Ancash, Arequipa, Ica, La Libertad, Lima), NSL–500 m.
Source: Pesantes-Segura, 2000.
Remarks: Type locality spelled Huayri *fide* Vanzolini (1986a: 5) and located near Nazca (14°50'S, 74°56'W, 575 m) *fide* J.C. Campbell & Lamar (1989: 222). Probably a synonym of *B. pictus fide* J.C. Campbell & Lamar (1989: 223).

24. *Bothrops sanctaecrucis* Hoge, 1966a. Mem. Inst. Butantan (1965) 32: 133–134, pl. 9 (middle fig.).

Type: Holotype, IB 24575, a female with 44 mm head, destroyed by fire 15 May 2010.
Type locality: "Oromomo, Rio Secure, upper Beni, Bolivia."
Distribution: Central Bolivia (Beni, Cochabamba, Santa Cruz), NSL–450 m.
Sources: Miranda et al., 1991 and Harvey et al., 2003, 2005.

25. *Bothrops venezuelensis* Sandner-Montilla, 1952. Monogr. Cien. Inst. Terap. Exp. Lab. "Veros" 9(21): 3–4, 6 figs.

Synonyms: *Bothrops venezuelae* Sandner-Montilla, 1961 *(nomen emendatum)*, and *Bothrops pifanoi* Sandner-Montilla & Römer, 1961.
Type: Holotype, UCV BV-1, a specimen ≤ 1200 mm.
Type locality: "Boca de Tigre' en la Serranía de El Avila en el Distrito Federal al Norte de Caracas (Distrito Federal es lo que aquí es Cundinamarca!), Venezuela."
Distribution: Northern Colombia (Boyacá, Norte de Santander) and Venezuela (Anzoátegui, Aragua, Carabobo, Distrito Federal, Falcón, Guárico, Lara, Mérida, Miranda, Monagas, Portuguesa, Sucre, Trujillo, Yaracuy), 590–2800 m.
Sources: J.A. Peters, 1968, Lancini, 1979, Esqueda & LaMarca, 1999, Mijares-Urrutia & Arends, 2000, Markezich, 2002, Navarrete et al., 2009 and Mendoza et al., 2012.

BOULENGERINA Dollo, 1886 (Elapidae)

Synonyms: *Limnonaja* K.P. Schmidt, 1923, *Boulangerina* – Briceño-Rossi, 1934 *(nomen incorrectum)*, *Paranaja* Loveridge, 1944b, *Boulangeria* – Boquet, 1948 *(nomen incorrectum)*, and *Parahaja* – Anon., 1979 *(nomen incorrectum)*.
Type species: *Naja annulata* Buchholz & Peters *in* W.C.H. Peters, 1876a.
Distribution: Subsaharan Africa.
Sources: F. Werner, 1923b, Bogert, 1943, Klemmer, 1968, Broadley, 1968c, 1974, Minton et al., 1970, Golay, 1985, Ulber, 1995, Golay et al., 1993, Keogh, 1998, David & Ineich, 1999, Slowinski & Keogh, 2000, Dobiey & Vogel, 2007, Wüster et al., 2007, Hoser, 2009c, 2012a, e, Wallach et al., 2009 and Zaher et al., 2009, 2012.

1. *Boulengerina annulata* (Buchholz & Peters *in* W.C.H. Peters, 1876a). Mber. Königl. Akad. Wiss. Berlin 1876(2): 119–120. (*Naja annulata*)

Synonyms: *Aspidelaps bocagei* Savage, 1884b, *Boulengerina stormsi* Dollo, 1886, and *Boulengerina dybowskii* Mocquard, 1897d.

B

Type: Holotype, ZMB 8772, a female (R. Buchholz, 1874–1875).

Type locality: "Dorfe Mbusu (Eliva Sonange am Ogowe), Westküste Afrikas" [= Eliva Sonange, near Ogooué River, Ogooué-Maritime Prov., W Gabon, ca. 1°S, 9°E, elevation NSL].

Distribution: Central Africa. Southern Cameroon (Est, Centre, Littoral, Sud, Sud-Ouest), SW Central Africa Republic (Ombella-Mpoko, Sangha), Equatorial Guinea, Gabon (Estuaire, Haut-Ogooué, Moyen-Ogooué, Ngounié, Nyanga, Ogooué-Ivindo, Ogooué-Lolo, Ogooué-Maritime, Woleu-Ntem), S Congo (Brazzaville, Kouilou, Plateau), NW Angola (Cabinda), Democratic Republic of the Congo (Bandundu, Bas-Congo, Equateur, NE Katanga, Kinshasa, Orientale, SE Sud-Kivu), SW Burundi (Bujumbura, Bururi), ext. W Tanzania (Kigoma, Rukwa) and NE Zambia (N Luapula, N Northern), NSL–1065 m.

Sources: Boulenger, 1895i, Witte, 1962, Broadley, 1968a, J.-F. Trape & Roux-Estève, 1995, Spawls et al., 2002, Pauwels & Lavoué, 2004, Chippaux, 2006, Chirio & Ineich, 2006, Chirio & LeBreton, 2007, Wüster et al., 2007 and Pauwels & Vande weghe, 2008.

Remarks: Collector of type is Zenker *fide* R. Günther *(in litt.)*. Lake Mweru, Zambia record requires verification *fide* Broadley et al. (2003: 107).

2. *Boulengerina christyi* Boulenger, 1904c. Ann. Mag. Nat. Hist. (7) 14(79): 14–15.

Type: Holotype, BMNH 1946.1.17.94 (formerly BMNH 1904.6.10.3), a 465 mm specimen (J.E. Dutton, C. Christy & J.L. Todd, Sept. 1903–June 1904).

Type locality: "near Leopoldville, Congo" [= vicinity of Kinshasa, Kinshasa Prov., SW Democratic Republic of the Congo, 4°20'S, 15°19'E, elevation 265 m].

Distribution: Southern Congo (Brazzaville) and SW Democratic Republic of the Congo (Bandundu, Kinshasa), 265–1190 m. Known only from vicinity of type locality.

Sources: Witte, 1962, Broadley, 1968a, J.-F. Trape & Roux-Estève, 1990, J.B. Rasmussen, 1991, J.-F. Trape & Roux-Estève, 1995, Chippaux, 2006 and Wüster et al., 2007.

3. *Boulengerina melanoleuca* (Hallowell, 1857b). Proc. Acad. Nat. Sci. Philadelphia (1857–1858) 9(3): 61–62. (*Naja haje melanoleuca*)

Synonyms: *Aspidelaps bocagei* Sauvage, 1884, *Naja haje leucosticta* J.G. Fischer, 1885b, *Naja melanoleuca subfulva* Laurent, 1955, *Naja melanoleuca aurata* Stucki-Stirn, 1979, and *Naja melanolenca* – Elter, 1981 *(nomen incorrectum)*.

Types: Syntypes (4), ANSP 6875–78, longest syntypes 2070 mm and 1670 mm (H.A. Ford, 1851–1857).

Type locality: "Gaboon country, West Africa" [= Gabon].

Distribution: Subsaharan Africa. Senegal (Dakar, Fatick, Kolda, Tambacounda, Thiès), Gambia (MacCarthy Island, Western), Guinea–Bissau (Bafatá, Oio, Bolama, Bubaque and Orango Is.), Guinea (Dubréka, Kindia, Macenta, Nzérékoré, Télimélé), Sierra Leone (Eastern, Northern, Southern), Liberia (Grand Gedeh, Montserrado, Margibi, Nimba), SW Mali (Kayes, Koulikoro, Sikasso), Ivory Coast (Abidjan, Agnebi, Bouaflé, Bouna, Guiglo, Lagunes, San Pédro, Sassandra, Tabou, Toumodi), W Burkina Faso (Centre-Ouest, Hauts-Bassins, Sud-Ouest, Volta-Noire), Ghana (Accra, Eastern), S Togo (Centrale, Kara, Plateaux), N Benin (Borgou, Zou), Nigeria (Akwa Ibom, Bayelsa, Cross River, Kaduna, Kwara, Oyo, Rivers), Cameroon (Adamaoua, Centre, Est, Extreme-Nord, Littoral, Nord, Nord-Ouest, Ouest, Sud, Sud–Ouest), Equatorial Guinea (Bioko Is.), São Tomé and Príncipe (São Tomé Is.), Gabon (Estuaire, Haut-Ogooué, Moyen-Ogooué, Ngounié, Ogooué-Ivindo, Ogooué-Maritime, Woleu-Ntem), Central African Republic (Bamingui-Bangoran, Haute-Sangha, Haut-Mbomou, Kemo-Gribingui, Lobaye, Ombella-Mpoko, Ouham, Sangha), Congo (Brazzaville, Cuvette, Kouilou, Lekoumou, Platea, Sanghau), Democratic Republic of the Congo (Bandundu, Bas-Congo, Equateur, Kasai Occidental, Kasai Oriental, Katanga, Kinshasa, Maniema, Nord-Kivu, Orientale, Sud-Kivu, Idjwi Is.), South Sudan (Bahr Al Ghazal, Eastern Equatoria), W Ethiopia (Gemu Gofa, Illubabor, Kefa, Shoa), S Somalia (Jubbada Hoose), Kenya (Central, Coast, Eastern, Nyanza, Rift Valley, Western), Uganda (Central, Eastern, W Northern, Western, Busi, Massambwa, Sanga and Sese Is.), Rwanda (Cyangugu, Kibungo), Burundi (Bujumbura, Bururi, Ngozi, Ruyigi), Tanzania (Arusha, Kigoma, Kilimanjaro, Lindi, Mtwara, Mwanza, Shinyanga, Tanga, Kagera, Mafia and Zanzibar Is.), Mozambique (Gaza, Inhambane, Manica, Maputo, Sofala, Tete, Zambézia, Bazaruto, Inhaca and Portuguese Is.), Malawi (Central, Northern, Southern), Zambia (Copperbelt, Central, Eastern, Luapula, Northern, North-Western), Angola (Benguela, Cuanza Norte, Lunda Norte, Malanje), E Zimbabwe (Manicaland) and ext. NE South Africa (E KwaZulu-Natal), NSL–2700 m.

Sources: Bocage, 1895, Angel, 1922, Aylmer, 1922, Flower, 1931, Andersson, 1937, Uthmöller, 1937, 1941a, Bogert, 1940, Monard, 1940, 1951, V.F.M. FitzSimons, 1947, 1962, 1974, H.W. Parker, 1949, Laurent, 1950c, 1956a, 1960, 1964a, 1973a, Villiers, 1950a, 1951b, 1956b, 1966, Loveridge, 1957a, Manaças, 1958, 1982, E.H. Taylor & Weyer, 1958, Capocaccia, 1961b, Sweeney, 1961, P. Johnsen, 1962, Witte, 1962, Doucet, 1963, Mertens, 1965c, Roux-Estève, 1965, 1969a, V.J. Wilson, 1965, Knoepffler, 1966, Menzies, 1966, Blackwell, 1967, Broadley, 1968c, 1971d, 1983, 1990, Leston & Hughes, 1968, Hughes & Barry, 1969, Roman, 1969, 1973b, 1976, 1980, Hulselmans & Verheyen, 1970, Hulsemans et al., 1970, Pitman, 1974, Stevens, 1974, Broadley & Cock, 1975, Courtois & Chippaux,

1977, Girardin, 1978, Rosselot, 1978, Spawls, 1978, Stucki-Stirn, 1979, Trape, 1981, 1985, Derleyn et al., 1983, Lanza, 1983a, MacKay & MacKay, 1985, J.A. Butler & Reid, 1986, Patterson, 1987, Branch, 1988, Broadley & Howell, 1991, Spawls & Branch, 1995, J.-F. Trape & Roux-Estève, 1995, Luiselli & Angelici, 2000, Spawls et al., 2002, Broadley et al., 2003, J.C. Murphy & Schlager, 2003, Chippaux, 2006, Chirio & Ineich, 2006, J.-F. Trape & Mané, 2006b, Chirio & LeBreton, 2007, Pauwels & Vande weghe, 2008, Largen & Spawls, 2010, W. Böhme et al., 2011, Segniagbeto et al., 2011, Broadley & Blaylock, 2013 and Chirio, 2013.

Remarks: *Naja subfulva* a valid species *fide* Chirio & Ineich (2006: 54).

4. *Boulengerina multifasciata* (F. Werner, 1902). Verh. Zool.-Bot. Ges. Wien 52(1): 347. (*Naia multifasciata*)

Synonyms: *Elapechis duttoni* Boulenger, 1904c and *Naja multifasciata anomala* Sternfeld, 1917.

Type: Holotype, IRSNB 2015 (formerly IRSNB 359 & IRSNB 6097), a 525–550 mm male (L. Rom, 16 Jan. 1894).

Type locality: "der oberen Maringa, Congo" [= Upper Moringa River, Tshuapa Distr., SE Equateur Prov., NW Democratic Republic of the Congo, bet. 0°51'N, 20°42'E, elevation 350 m, & 0°17'N, 23°02'E, elevation 525 m].

Distribution: Central Africa. Southeastern Cameroon (Centre, Est, Sud), N Gabon (Ogooué-Ivindo, Woleu-Ntem), N Congo (Sangha), SW Central African Republic (Sangha) and Democratic Republic of the Congo (Bandundu, Bas-Congo, Equateur, Kinshasa, Nord-Kivu, Sufd-Kivu), 300–800 m.

Sources: Loveridge, 1944b, Laurent, 1956b, Perret, 1960 and Trape & Lang, 1990, J.-F. Trape & Roux-Estève, 1990, 1995, Spawls & Branch, 1995, Chippaux, 2006, Chirio & Ineich, 2006, Chirio & LeBreton, 2007 and Pauwels & Vande weghe, 2008.

BRACHYOPHIDIUM Wall, 1921a (Uropeltidae)

Synonyms: *Brahyophlbium* – Murthy, 1986 *(nomen incorrectum), Brachyiophidium* – Murthy, 1990d *(nomen incorrectum),* and *Brahyophlium* – B.D. Sharma *in* B.D. Sharma & Kumari, 1998a *(nomen incorrectum).*

Type species: *Brachyophidium rhodogaster* Wall, 1921a.

Distribution: Southern India.

Sources: Wall, 1922b, 1924f, Procter, 1924, M.A. Smith, 1943, Angel, 1947, Laurent, 1948, Constable, 1949, Gans, 1966, Murthy, 1975, 1981c, 1982b, Rajendran, 1985, McDiarmid et al., 1999, Whitaker & Captain, 2004 and Gower et al., 2005.

1. *Brachyophidium rhodogaster* Wall, 1921a. J. Bombay Nat. Hist. Soc. (1921–1922) 28(1): 41–42, pl., figs. a–d.

Type: Holotype, BMNH 1946.1.15.60 (formerly BMNH 1921.3.4.13), a 178 mm female (St. Joseph's College).

Type locality: Unknown. Designated as Sacred Heart College, Shembaganur, India, ca. 6500 feet [= Sacred Heart College, Shembaganur, 5 km. N Kodaikanal, Palani Hills, S Tamil Nadu State, S India, 10°15'N, 77°29'E, elevation 1980 m] *fide* Wall (1922b: 556).

Distribution: Southern India (Western and Eastern Ghats of Tamil Nadu), 1280–2100 m.

BRACHYOPHIS Mocquard, 1888 (Atractaspididae)

Synonyms: *Brachiophis* Palacky, 1898 *(nomen emendatum),* and *Branchyophis* – Gasperetti, 1974 *(nomen incorrectum).*

Type species: *Brachyophis revoili* Mocquard, 1888.

Distribution: Somalia.

Sources: Witte & Laurent, 1947, Lanza, 1966, 1983a, 1990b, Gasperetti, 1988, Wallach, 1991, Underwood & Kochva, 1993, Vidal et al., 2008, Kelly et al., 2009 and Zaher et al., 2009.

1. *Brachyophis cornii* Scortecci, 1932c. Atti Soc. Ital. Sci. Nat. Mus. Civ. Stor. Nat. Milano 71: 268–269. (*Brachyophis revoili cornii*)

Type: Lectotype, MZUF 2491, a 177 mm female (G. Stefanini & N. Puccioni, 1924), designated by Lanza (1966: 40).

Type locality: "fra Durgale e Magghiole, Somalia settentrionale" [= between Durghale (5°35'N, 48°15'E, elevation 90 m) and Magghiole, E Galguduud Region, cen. Somalia] via lectotype selection.

Distribution: Central Somalia (Galguduud, Mudug), 20–90 m.

Source: Scortecci, 1931.

Remarks: A valid species *fide* Wallach (unpubl. data). Photographs of lectotype in Lanza (1966: pl. 15, fig. 3, pl. 17, figs. 6 [left], 7 [left]).

2. *Brachyophis krameri* Lanza, 1966. Monit. Zool. Ital. 74(1–3): 41–42, fig. 1b, pl. 15, fig. 2, pl. 19, figs. 10–11, pl. 20, fig. 13. (*Brachyophis revoili krameri*)

Type: Holotype, MZUF 3934, a 180 mm female (native, 16 Aug. 1964).

Type locality: "Geriban (Somalia settentrionale, circa 48° 50' E X 7° 15' N)" [= Geriban, Mudug Region, cen. Somalia, ca. 48°50'E, 7°15'N, elevation 200 m].

Distribution: Northern Somalia (NE Mudug), 200 m.

Remarks: Lanza (1966: 41) lists pls. I, V–VII for the new species.

B

3. *Brachyophis revoili* Mocquard, 1888. Mém. Soc. Philom. Paris 1888: 125–128, pl. 11, figs. 3–3d.

Types: Syntypes (3), MNHN 1884.691A, MNHN 1884.692, and BMNH 1946.1.3.86 (formerly MNHN 1884.693), a female, longest syntype 252 mm (G. Révoil, 1884).
Type locality: "Çomalis" [= Somalia].
Distribution: Southern Somalia (Jubbada Hoose, Mogadishu, Shabeellaha Dhexe, Shabeellaha Hoose, Mudug) and W Yemen, NSL–90 m.
Source: Scortecci, 1932a.
BRACHYORRHOS Kuhl

BRACHYORRHOS Kuhl *in* Schlegel, 1826a (Homalopsidae)

Synonyms: *Brachyorrhus* Agassiz, 1847 *(nomen emendatum), Brachyorros* – Doria, 1874 *(nomen incorrectum), Oxyorrhos* J.G. Fischer, 1879, *Atractocephalus* Sauvage, 1879, *Brachyorhos* – Palacky, 1898 *(nomen incorrectum), Brachyorhhus* – Ditmars, 1933 *(nomen incorrectum),* and *Brachyrhos* – Briceño-Rossi, 1934 *(nomen incorrectum).*
Type species: *Coluber albus* Linnaeus, 1758.
Distribution: Indonesia.
Sources: Rooij, 1917, C. Haas, 1950, J.C. Murphy, 2007, 2012b and Zaher et al., 2009.
Remarks: J.C. Murphy (2007: 269) listed genus as Homalopsidae *incertae sedis.* Undetermined *Brachyorrhos* occur on Banda, Batjan, Haruku, Nusa Laut and Saparua *fide* J.C. Murphy et al. (2012: 511).

1. *Brachyorrhos albus* (Linnaeus, 1758). Syst. Nat., ed. 10, 1: 218. (*Coluber albus*)

Synonyms: *Elaps jockischii* Wagler, 1825 (*nomen ineditum*), and ? *Calamaria brachyorrhos* S. Müller, 1857 *(nomen nudum).*
Type: Holotype, NHR Lin-19 (formerly MAFR), a 460 mm specimen (Mus. Drottn.).
Type locality: "Indiis" [= India] (in error). Corrected to Ambon, Indonesia *fide* J.C. Murphy et al. (2012b: 507).
Distribution: Southern Indonesia (Ambon, Aru Is., Banda, Bisa, Haruku, Nusa Laut, Saparua, Seram) and Papua New Guinea (Raja Ampat Is.), NSL–845 m.
Remarks: Iskander & Colijn (2001: 41) list the type as ZIUS from Ambon, Indonesia. *Mainophis robusta* Macleay a synonym *fide* J.C. Murphy (2007: 211). Timor records doubtful *fide* J.C. Murphy et al. (2012b: 505).

2. *Brachyorrhos gastrotaenius* (Bleeker, 1860e). Nat. Tijds. Nederl. Indië (5) 22(2): 42–43. *(Rabdion gastrotaenia)*

Synonym: *Oxyorrhos fusiformis* J.G. Fischer, 1879.
Type: Holotype, BMNH 1863.12.4.25, a 194 mm female (C.G.C.F. Greiner, 1856).
Type locality: "Amboina, Molukschen Archipel" [= Ambon, Maluku, E Indonesia].
Distribution: Eastern Indonesia (Maluku: Ambo, Burun), 775–1450 m.
Remarks: Type locality erroneously designated as Buru, SMNS 1327 listed as lost syntypes, RMNH 3977 as cotype, and BMNH 6312425 as holotype, and publication erroneously listed as Sumatra paper *fide* J.C. Murphy et al. (2012b: 508). Photograph of head of BMNH "holotype" in J.C. Murphy et al. (2012b: fig. 4b).

3. *Brachyorrhos raffrayi* (Sauvage, 1879b). Bull. Soc. Philom. Paris (1878–1879) (7) 3: 62–63 *(Atractocephalus raffrayi)*

Synonym: *Brachyorrhos albus conjunctus* J.G. Fischer, 1880.
Type: Lectotype, MNHN 5174, a 530–580 mm female (M.J.A. Raffray, 1876–1879), designated by J.C. Murphy et al. (2012b: 509).
Type locality: "Ternate, Indonesia" [= Ternate Is.,] via lectotype selection.
Distribution: Eastern Indonesia (Maluku: Batjan [MNHN 5782], Ternate).

4. *Brachyorrhos wallacei* J.C. Murphy, Mumpuni, Lang, Gower & Sanders, 2012b. Raffles Bull. Zool. 60(2): 510, fig. 4d.

Type: Holotype, MZB 3463, a 722 mm female.
Type locality: "Jailolo, Halmahera," Indonesia.
Distribution: Eastern Indonesia (Halmahera).

BRACHYUROPHIS A.C.L.G. Günther, 1863a (Elapidae)

Synonyms: *Rhinelaps* A.C.L.G. Günther, 1872b, *Hornea* Lucus & Frost, 1896, and *Simonelaps* – Mattison, 1986 *(nomen incorrectum).*
Type species: *Brachyurophis semifasciata* A.C.L.G. Günther, 1863a.
Distribution: Australia.
Sources: Thomson, 1934, Kinghorn, 1964, Storr, 1968, 1985, 1986, Cogger et al., 1983a, Shine, 1984a, 1985a, Golay, 1985, Longmore, 1986, Scanlon & Shine, 1988, S.K. Wilson & Knowles, 1988, Gow, 1989, Hoser, 1989, 2012e, Hutchinson, 1990, Cogger, 1992, Clarke & How, 1995, Horner, 1998, Keogh et al., 1998, How & Shine, 1999, Scanlon & Lee, 2004, Sanders et al., 2008 and Zaher et al., 2009.
Remarks: A valid genus *fide* Metzger et al. (2010: 409).

1. *Brachyurophis approximans* (Glauert, 1954). West. Aust. Nat. 4(4): 85, 1 fig. (*Rhynchoelaps approximans*)

Type: Holotype, WAM 10768 (D.W.H. Shilling, June 1953).

Type locality: "Muccan Station, north-western Australia (Lat. 23°30'S., Long. 120°E.)" [= Muccan Station, NW Western Australia, W Australia, 20°40'S, 120°04'E, elevation 95 m].

Distribution: Western Australia (W Western Australia, Barrow and Varanus Is.), 10–1040 m.

Sources: Storr, 1979 and Storr et al., 2002.

Remarks: Type locality coordinates in Glauert (1954: 85) in error *fide* Storr (1968: 89), who cited them as 20°38'.

2. *Brachyurophis australis* (Krefft, 1864c). Proc. Zool. Soc. London 32(1): 180, 4 figs. (*Simotes australis*)

Types: Syntypes (2), formerly BMNH, longest syntype 290–292 mm female (J.C. Cox & J. Francis via G. Krefft), lost *fide* Cogger et al. (1983: 235).

Type locality: "neighbourhood of Port Curtis" and "Clarence River" [= vicinity of Port Curtis, E Queensland, 23°55'S, 151°21'E, elevation NSL, and Clarence River, ext. NE New South Wales, E Australia]. Port Curtis locality emended to Burdekin River in Queensland [= 20°39'S, 146°59'E, elevation 175 m] *fide* Krefft (1869: 52).

Distribution: Southeastern Australia (New South Wales, ext. SE Queensland, SE South Australia, NW Victoria), 20–1045 m.

Sources: A.C.L.G. Günther, 1865, Krefft, 1869, Mackay, 1949, Covacevich & Couper, 1991, Coventry & Robertson, 1991, G. Swan, 1999 and Swan et al., 2004.

Remarks: Original description reprinted in Krefft (1865a: 66). A.C.L.G. Günther (1865: 97) reported that Krefft sent Clarence River specimen BMNH and Boulenger (1896a: 363) listed it but did not call it a type. Krefft (1864c & 1865) reported first specimen from Port Curtis and second from Clarence River, but Krefft (1869: 52) listed Clarence River specimen as first discovered and second specimen from Burdekin River.

3. *Brachyurophis fasciolatus* (A.C.L.G. Günther, 1872b). Ann. Mag. Nat. Hist. (4) 9(49): 34, pl. 5, fig. b. (*Rhinelaps fasciolatus*)

Synonyms: *Vermicella fasciata* Stirling & Zietz, 1893, and *Hornea pulchella* Lucus & Frost, 1896.

Type: Holotype, BMNH 1946.1.19.12, a 335–343 mm female (F.H. DuBoulay, 1862–1872).

Type locality: "Perth, West Australia" [= Perth, SW Western Australia, SW Australia, 31°56'S, 115°50'E, elevation NSL].

Distribution: Southern Australia (NW New South Wales, Northern Territory, ext. SW Queensland, South Australia, Western Australia, Dirk Hartog Is.), NSL–700 m

Sources: Covacevich & Couper, 1991, Bush et al., 1995 and Storr et al., 2002.

4. *Brachyurophis incinctus* (Storr, 1968). J. Roy. Soc. West. Aust. (1967) 50(3): 89–90. (*Vermicella semifasciata incincta*)

Type: Holotype, WAM 26844 (formerly NTM 2301), a 261–292 mm male (K.R. Slater, 16 Oct. 1964).

Type locality: "near Alice Springs, Northern Territory, in 23° 46' S, 133° 53' E, Australia."

Distribution: Central Australia (S Northern Territory, ext. NW Queensland), 50–895 m.

Sources: Covacevich & Couper, 1991 and Horner, 1998.

5. *Brachyurophis morrisi* (Horner, 1998). The Beagle 14: 64–68, fig. 1. (*Simoselaps morrisi*)

Type: Holotype, NTM 22951, a 301 mm male (K. Aland & A. Yumbulul, 10 Oct. 1992).

Type locality: "southern end of Elcho Island, 12°03'S 135°34'E, Northern Territory, Australia."

Distribution: Extreme N Australia (N Northern Territory, Woodah Is.), 10–60 m.

Source: S.K. Wilson & Knowles, 1988.

6. *Brachyurophis roperi* (Kinghorn, 1931). Rec. Aust. Mus. 18(5): 267, fig. 1 (*Rhynchoelaps roperi*)

Synonyms: *Rhynchoelaps campbelli* Kinghorn, 1929, *Rhychoelaps smithii* Kinghorn, 1931, *Rhynchelaps woodjonesii* Thomson, 1934, *Brachyurophis woodjonesii* Worrell, 1963 *(nomen illegitimum),* and *Brachyurophis murrayi* Wells & Wellington, 1985.

Type: Holotype, AMS 9930, a 295 mm specimen (K. Langford-Smith, May 1929).

Type locality: "Roper River, North Australia" [Australia].

Distribution: Northern Australia (Northern Territory, N Western Australia, Bathurst and Melville Is.), NSL–785 m.

Remarks: A valid species *fide* Hutchinson (1990: 404). *Rhynchoelaps campbelli* Kinghorn a valid species *fide* Hutchinson (1990: 404) and Shea & Sadlier (1999: 52). Both taxa prevously synonyms of *S. semifasciatus* (A.C.L.G. Günther).

7. *Brachyurophis semifasciatus* A.C.L.G. Günther, 1863a. Ann. Mag. Nat. Hist. (3) 11(61): 21–22, pl. 3, fig. b. (*Brachyurophis semifasciata*)

Synonyms: *Pseudoelaps rhinostoma* Jan, 1863b, and *Pseudoelaps rhodostomus* Jan, 1873 *in* Jan & Sordelli, 1870–1881.

Type: Lectotype, BMNH 1946.1.19.11, a 300–330 mm male (F.H. DuBoulay, 1862–1863), designated by Wells & Wellington (1985: 44).

Type locality: "Western Australia?" via lectotype selection.

Distribution: Western Australia (Northern Territory, South Australia, S Western Australia, Melville Is.), NSL–805 m.

Sources: Kinghorn, 1929, Hoser, 1989, Bush et al., 1995, Storr et al., 2002, J.C. Murphy & Schlager, 2003 and Edwards, 2009.

Remarks: Type locality originally listed as Baranquilla (New Granada) [= Baranquilla, Colombia] (in error), corrected to West Australia ? *fide* Boulenger (1896a: 364).

†*BRANSATERYX* Hoffstetter & Rage, 1972 (Boidae)

Type species: †*Bransateryx vireti* Hoffstetter & Rage, 1972.

Distribution: Lower Oligocene to lower Miocene of France, Czech Republic, Germany and Kazakhstan.

Sources: Rage, 1984b, Szyndlar, 1994, Szyndlar & Schleich, 1993, Szyndlar & Rage, 1993, Kordikova, 1998 and J. Müller, 1998.

1. †*Bransateryx vireti* Hoffstetter & Rage, 1972. Ann. Paléont. (Vert.) 58(1): 97–102, fig. 7, pl. 2, figs. 1–6.

Synonym: †*Bransateryx septentrionalis* Szyndlar, 1987.

Type: Holotype, MNHN Cod 1, one caudal vertebra.

Type locality: "Coderet (Allier). Oligocène supérieur (Chattien?), France."

Distribution: Upper Oligocene (Arvernian: 23.0–28.4 mya) of France, middle Oligocene (11.1–13.7 mya) of Germany, and lower Miocene (Agenian: 20.0–23.0 mya) of France, Czech Republic and Germany.

BRYGOPHIS Domergue & Bour, 1988 (Pseudoxyrhophiidae) (*nomen substitutum*)

Synonym: *Perinetia* Domergue, 1988 *(nomen praeoccupatum).*

Type species: *Perinetia coulangesi* Domergue, 1988.

Distribution: Madagascar.

Sources: Domergue, 1988, Domergue & Bour, 1988, Glaw & Vences, 1994, Andreone & Raxworthy, 1998, Kelly et al., 2009 and Zaher et al., 2009.

1. *Brygophis coulangesi* (Domergue, 1988). Bull. Mus. Natl. Hist. Nat. (4) 10(1): 135–138, figs. 1a–f. (*Perinetia coulangesi*)

Type: Holotype, MNHN 1977.817 (formerly CAD 801/S), a 1203 mm female (J. Thiel, 8 Nov. 1968).

Type locality: "Forêt de Fiherenana, en zone occidentale de la forêt de l'Est, à une cinquantaine de km, au nord de Périnet, altitude 1000 m environ, Madagascar" [= Fiherenana forest, 50 km N Andasibe (18°56'S, 48°25'E, 915 m), Toamasina Prov., E Madagascar, ca. 18°35'S, 48°15'E, elevation 1000 m].

Distribution: East-central Madagascar (Majunga, Toamasina), 1000–1200 m.

BUHOMA Ziegler, Vences, Glaw & Böhme, 1997 (Elapoidea *incertae sedis*)

Synonym: *Swileserpens* Hoser, 2012t *(nomen illegitimum).*

Type species: *Geodipsas vauerocegae* Tornier, 1902b.

Distribution: Central and East Africa.

Sources: Ziegler et al., 1997, Gravlund, 2001, Spawls et al., 2002, Vidal et al., 2008, Kelly et al., 2009 and Hoser, 2012t.

1. *Buhoma depressiceps* (F. Werner, 1897b). Verh. Zool. Bot. Ges. Wien 47: 402–403. (*Tropidonotus depressiceps*)

Synonyms: *Geodipsas mapanjensis* Andersson, 1901, and *Geodipsas depressiceps marlieri* Laurent, 1956a.

Types: Syntypes (2), ZMB, 331 and 277 mm specimens (P.R. Preuss [Zintgraff Exped.], 1889–1902).

Type locality: "Barombi-Station, Kamerun" [= Barombi Mbo, Sud-Ouest Prov., SW Cameroon, 4°40'N, 9°24'E, elevation 315 m].

Distribution: Central Africa. Southwestern Cameroon (Centre, Littoral, Ouest, Sud, Sud-Ouest), Equatorial Guinea (Bioko Is.), Gabon (Ngounié, Ogooué-Ivindo, Woleu-Ntem), Congo (Kouilou, Lekoumou, Sangha), E Democratic Republic of the Congo (Katanga, Nord-Kivu, Orientale, Sud-Kivu), SW Uganda (Western) and NW Burundi (Bubanza), 315–2200 m.

Sources: Laurent, 1956a, Capoccacia, 1961b, Witte, 1962, Mertens, 1965c, Knoepffler, 1966, Pitman, 1974, Derleyn, 1978a, J.-F. Trape, 1985, J.-F. Trape & Roux-Estève, 1995, Chippaux 2006, Pauwels & Vande weghe, 2006 and Chirio & LeBreton, 2007.

Remarks: *Geodipsas depressiceps marlieri* possibly a valid species *fide* Ziegler et al. (1997: 108).

2. *Buhoma procterae* (Loveridge, 1922). Proc. Zool. Soc. London 92(2): 313–314. (*Geodipsas procterae*)

Type: Holotype, BMNH 1946.1.2.48, a 386 mm male (A. Loveridge, 1 March 1921).

Type locality: "On path in thick bush-country, Uluguru Mts., 3 miles from Morogoro, Tanganyika Territroy, at an altitude of about 3000 ft." [= 3 mi. E Morogoro (6°49'S, 37°40'E), Morogoro Region, E Tanzania, ca. 6°50'S, 37°40'E, elevation 915 m].

Distribution: Southeastern Tanzania (Udzungwa and Uluguru Mts. of Iringa, N Morogoro), 915–2140 m.

Source: J.B. Rasmussen et al., 1995.

3. *Buhoma vauerocegae* (Tornier, 1902b). Zool. Anz. 25(683–684): 703–704. (*Geodipsas vauerocegae*)

Types: Syntypes, ZMB 17557, female (J. Küttner, H. Böhler, & F. Sammler), lost *fide* R. Günther *(in litt.)*
Type locality: "Usambara (Deutsch-Ost-Afrika)" [= Usambara Mountains, Tanga Region, NE Tanzania].
Distribution: Northeastern Tanzania (Usambara and Uluguru Mts. of Tanga, N Morogoro), 1000–1300 m.
Sources: J.B. Rasmussen, 1981 and J.B. Rasmussen et al., 1995.

BUNGARUS Daudin, 1803a (Elapidae)

Synonyms: *Bongarus* Fischer, 1808 *(nomen emendatum), Pseudo-boa* Oppel, 1811b *(nomen praeoccupatum), Aspidoclonion* Wagler, 1828, *Bungaras* – Bonaparte, 1832 *in* 1832–1841 *(nomen incorrectum), Aspidoclonium* Agassiz, 1844 *(nomen emendatum), Megaerophis* Gray, 1849b, *Aspidoclanion* – Mason, 1852 *(nomen incorrectum), Aspidoclonian* – Mason, 1852 *(nomen incorrectum), Bngarus* – Bleeker, 1859b *(nomen incorrectum), Aspidoclanion* – F. Mason, 1860 *(nomen incorrectum), Xenurelaps* A.C.L.G. Günther, 1864a, *Bangurus* – Stoliczka, 1870d *(nomen incorrectum), Henurelaps* – Higgins, 1873 *(nomen incorrectum), Magaerophis* Marschall, 1873 *(nomen emendatum), Pseudoboa* – Boulenger, 1896a *(nomen corrigendum), Aspidoclomion* – Maki, 1933 *(nomen incorrectum), Maegerophis* – Bourret, 1938 *(nomen incorrectum), Bungrus* – Hanvivatvong, Phanuphak, Lowcharoenkul, Benyajati & Sakulramrung, 1987 *(nomen incorrectum),* and *Bungarum* – Ota, 1991 *(nomen incorrectum).*
Type species: *Pseudoboa fasciatus* J.G. Schneider, 1801.
Distribution: Southeastern Asia and East Indies.
Fossil record: Miocene of Pakistan.
Sources: Wall, 1908f, F. Werner, 1923b, Kopstein, 1932a, M.A. Smith, 1943, Klemmer, 1963, Joger, 1984, Toriba, 1987, Golay et al., 1993, Slowinski, 1994, Keogh, 1998, David & Ineich, 1999, Slowinski & Keogh, 2000, Head, 2005, Zaher et al., 2009, I. Das, 2010, 2012 and Hoser, 2012e.
Remarks: Official Generic Name no. 2149 *fide* Opinion 1201 (ICZN, 1982a).

1. *Bungarus andamanensis* Biswas & Sanyal, 1978. J. Bombay Nat. Hist. Soc. 75(1): 179–182, figs. 1–4.

Type: Holotype, ZSI 20895/1, a 715 mm specimen (A. Bayley de Castro, 18 June 1926).
Type locality: "Port Blair, S. Andaman" [= Port Blair, South Andaman Is., Andaman & Nicobars, E India, Bay of Bengal, 11°37'N, 92°44'E, elevation 15 m].
Distribution: Eastern India (Andaman & Nicobars: Long and South Andaman Is.), NSL–15 m.
Source: Whitaker & Captain, 2004.

2. *Bungarus bungaroides* (Cantor, 1839a). Proc. Zool. Soc. London 7(1): 33. (*Elaps bungaroides*)

Synonym: *Alecto bungaroïdes* A.-M.-C. Duméril, Bibron & Duméril, 1854b *(nomen incorrigendum).*
Type: Holotype, BMNH 1946.1.17.91, a juvenile (T.E Cantor, 1834–1839, via British East India Comp.).
Type locality: "Chirra Punji, Indicorum" [= Cherrapunji, East Khasi Hills, S Meghalaya State, NE India, 25°18'N, 91°42'E, elevation 1485 m].
Distribution: Northeastern India (Arunachal Pradesh, Assam, Meghalaya, Nagaland, Sikkim, West Bengal), ext. E Nepal (Ilam), Bhutan, S China (Yunnan), ext. N Myanmar (Kachin) and N Vietnam (Yen Bai), 250–2040 m.
Sources: Shah, 1999, Kizirian et al., 2002, Leviton et al., 2003, Orlov et al., 2003, Tiwari & Shah, 2004, Kuch et al., 2005, Wangyal, 2011 and S.K. Sharma et al., 2013.

3. *Bungarus caeruleus* (J.G. Schneider, 1801). Hist. Amph. 2: 284–285. (*Pseudoboa caerulea*)

Synonyms: *Boa lineata* G. Shaw, 1802, *Boa latotecta* Hermann *in* F. Boie, 1827, *Bungarus arcuatus* A.-M.-C. Duméril, Bibron & Duméril, 1854b, *Bungarus coeruleus* Jan, 1863b *(nomen emendatum), Bungarus coeruleus* – Keegan, 1956 *(nomen incorrectum), Bungarus ceruleus* – Homma, 1974 *(nomen incorrectum),* and *Bungarua caeruleus* – Banerjee & Siddiqui, 1978 *(nomen incorrectum).*
Type: Lectotype, a 737 mm specimen described and illustrated by P. Russell (1796: 1–2, pl. 1), designated by Klemmer (1963: 279).
Type locality: "Vizagapatam (Indien)" [= Vishakhapatnam, Andhra Pradesh State, SE India, 17°41'N, 83°13'E, elevation 25 m] via lectotype selection.
Distribution: Southern Asia. Pakistan (S Balochistan, F.A.T.A., Jammu & Kashmir, Khyber Pakhtunkhwa, Punjab, Sindh), India (Andhra Pradesh, Bihar, Haryana, Himachal Pradesh, Karnataka, Madhya Pradesh, Maharashtra, Orissa, Punjab, Tamil Nadu, Uttar Anchal, Utter Pradesh, West Bengal), Nepal (Banke, Bara, Bardiya, Chitwan, Dang, Dhanusa, Jhapa, Kailali, Kanchanpur, Kapilbastu, Makwanpur, Morang, Nawalparasi, Palpa, Rupandehi, Saptari, Sunsari, Surkhet, Udyapur), E Bangladesh and Sri Lanka (Eastern, North-Central, Northern, North-Western, Sabaragamuwa, Uva, Western), 100–1500 m.
Sources: Wall, 1913d, Bhatnagar, 1960, Kramer, 1977, Khan, 1985a, 2002, Khan & Tasnim, 1986a, Tiwari & Shah, 2004, Whitaker & Captain, 2004, N. Khaire, 2006, Castoe et al., 2007a, Ganesh & Chandramouli, 2011, Masroor, 2012 and S.K. Sharma et al., 2013.
Remarks: Original descrioption based on P. Russell (1796: 1–2, pl. 1). The paralectotype (ZMB 2787) from Museum Blochianum ("India orientali") has been discovered *fide* Bauer (1998: 138). MNHN catalogue lists MNHN 3952 and MNHN 7686–88 as syntypes.

B

4. *Bungarus candidus* (Linnaeus, 1758). Syst. Nat., ed. 10, 1: 223. (*Coluber candidus*)

Synonyms: *Pseudoboa krait* J.G. Schneider, 1801, *Bungarus semifasciatus* F. Boie, 1827, *Bungarus arcuatus* A.-M.-C. Duméril, 1853 (*nomen nudum*), *Bungarus semifasciatus* A.M.C. Duméril, 1853 (*nomen nudum*), *Bungarus javanicus* Kopstein, 1932a, and *Bungaris candidus* – Bruno, 1985 (*nomen incorrectum*).

Type: Holotype, NHR Lin-89 (formerly MAFR), a 925 mm specimen (Mus. Drottn.).

Type locality: "Indiis" [= India] (in error).

Distribution: Southeast Asia and Indonesia. Vietnam (Binh Phuoc, Dak Lak, Dak Nong, Dong Nai, Gia Lai, Ha Tinh, Hoa Binh, Kon Tum, Lam Dong, Ninh Thuan, Phu Yen, Quang Binh, Quang Nam, Quang Tri, Tay Ninh, Thua Thien-Hue), S Cambodia (Kampot, Mondulkiri), Thailand (Chaiyaphum, Chiang Mai, Chon Buri, Kanchanaburi, Khon Kaen, Lampang, Loei, Lop Buri, Nakhon Ratchasima, Pattani, Phang Nga, Phetchabun, Surin, Trang and Koh Kut Is.), West Malaysia (Kedah, Kuala Lumpur, Perak, Perlis), Singapore, and Indonesia (Bawean, Bali, Java, Karimunjawa, Sumatra), NSL–1525 m.

Sources: Wall, 1908f, M.A. Smith, 1914a, Bergman, 1962a, E.H. Taylor, 1965, Tweedie, 1983, Lim & Lim, 1992, Jintakune & Chanhome, 1995, David & Vogel, 1996, M.J. Cox et al., 1998, Kuch et al., 2005, Lang & Vogel, 2005, B.L. Stuart et al., 2006, Kuch & Mebs, 2007, V.S. Nguyen et al., 2009 and I. Das, 2010.

Remarks: Sulawesi record doubtful but possibly occurs on Kalimantan *fide* Lang & Vogel (2005: 259). *Bungarus javanicus* a valid species *fide* Toriba *in* Golay et al. (1993: 120).

5. *Bungarus ceylonicus* A.C.L.G. Günther, 1964a. Rept. Brit. India: 344.

Synonym: *Bungarus ceylonicus karavala* Deraniyagala, 1955.

Type: Lectotype, BMNH 1946.1.19.20, a 664 mm specimen (H. Cuming, 1836–1840), designated by Deraniyagala (1955: 68).

Type locality: "Ceylon" [= Sri Lanka]

Distribution: Southern Sri Lanka (Central, Sabaragamuwa, Southern, Uva, Western), NSL–750 m.

Sources: Deraniyagala, 1955, P. Silva, 1980a and A. Silva, 1990a–b.

6. *Bungarus fasciatus* (J.G. Schneider, 1801). Hist. Amph. 2: 283–284. (*Pseudoboa fasciata*)

Synonyms: *Coluber fasciatus* G. Shaw, 1802, *Coluber platurinus* G. Shaw, 1802, *Bungarus annularis* Daudin, 1803b, *Bungarus semicinctus* Duvernoy, 1832, *Aspidoclonion schneideri* Fitzinger, 1861, *Bungarus fasciatus bifasciatus* Mell, 1930, *Bungarus fasciatus insularis* Mell, 1930, *Bungurus fasciatus* – McDowell, 1969a (*nomen incorrectum*), *Bungaris fasciatus* – H. Fox, 1977 (*nomen incorrectum*), and *Bungrus fasciatus* – Hanvivatvong, Phanuphak, Lowcharoenkul, Benyajati & Sakulramrung, 1987 (*nomen incorrectum*).

Types: Syntypes (2), ZMB 2771–72, longest syntype 1651 mm (P. Russell & natives, 1781–1791).

Type locality: Unknown. "Indien" [= India] *fide* ZMB catalogue. Restricted to "Mansoor Cottah, Bengal, India" [= Ganjam, Orissa State, SE India, 19°23'N, 85°03'E, elevation NSL] *fide* P. Russell (1801: 53).

Distribution: Southeast Asia and Indonesia. Eastern India (N Andhra Pradesh, Arunachal Pradesh, Assam, Bihar, Madhya Pradesh, Maharashtra, Meghalaya, Mizoram, Orissa, Uttar Pradesh, West Bengal), Bangladesh, Nepal (Bardiya, Chitwan, Jhapa, Kailali, Kanchanpur, Kapilbastu, Morang, Parsa, Sunsari), Bhutan, S China (Fujian, Guangdong, Guangxi, Hainan, Hong Kong, Yunnan), Macau, Myanmar (Ayeyarwady, Kachin, Magway, Mandalay, Rakhine, Yangon), Thailand (Ang Thong, Ayutthaya, Chanthaburi, Chiang Mai, Chumphon, Loei, Lop Buri, Mae Hong, Nakhon Pathom, Nakhon Ratchasima, Nakhon Si Thammarat, Nonthaburi, Pattani, Phetchabun, Ranong, Samut Prakan, Saraburi, Sing Buri, Surin, Tak, Uthai Thani), Laos (Attapu, Borikhan, Champasak, Houaphan, Khammouan, Louangphrabang, Phongsali, Salavan, Savannakhet, Vientiane, Xaignabouri, Xekong, Xiangkhouang), Vietnam (Bac Giang, Bac Kan, Bac Ninh, Binh Phuoc, Ca Mau, Cao Bang, Da Nang, Dong Nai, Hai Duong, Hanoi, Ha Tay, Ha Tinh, Hoa Binh, Ho Chi Minh City, Kien Giang, Lang Son, Lao Cai, Nighe An, Ninh Binh, Quang Binh, Quang Nam, Quang Ninh, Quang Tri, Son La, Tay Ninh, Thai Nguyen, Thanh Hoa, Thua Thien-Hue, Tien Giang, Vinh Phuc, Yen Bai), Cambodia, West Malaysia (Pinang), Singapore, East Malaysia (Sabah, Sarawak), Brunei and Indonesia (Java, Kalimantan, Sumatra), NSL–2500 m.

Sources: Wall, 1911c, C.H. Pope, 1935, Leviton, 1955, Bergman, 1962a, E.H. Taylor, 1965, Deuve, 1970, Tweedie, 1983, Karsen et al., 1986, Kuch, 1991, Jintakune & Chanhome, 1995, Zug & Mitchell, 1995, David & Vogel, 1996, Bauer, 1998, M.J. Cox et al., 1998, Xie et al., 1998, Stuebing & Liger, 1999, Zug et al, 1998, Orlov et al., 2000, Malkmus et al., 2002, Schleich & Kästle, 2002, Leviton et al., 2003, Tiwari & Shah, 2004, Whitaker & Captain, 2004, N. Khaire, 2006, Castoe et al., 2007a, M.F. Ahmed et al., 2009, V.S. Nguyen et al., 2009, I. Das, 2012 and S.K. Sharma et al., 2013.

Remarks: Description based on P. Russell (1796: 5, pl. 3). Official Specific Name no. 2785 *fide* Opinion 1201 (ICZN, 1982a). Type locality listed as Indien [= India] *fide* ZMB catalogue.

7. *Bungarus flaviceps* J.T. Reinhardt, 1843. Kgl. Danske Vidensk. Selsk. Natur. Math. Afh. 10: 267–269, pl. 3, fig. 4.

Synonyms: *Megaerophis formosus* Gray, 1849b, *Bungarus flaviceps baluensis* Loveridge, 1938a, and *Maticora intermedia* Westermann, 1942.

Type: Holotype, ZMUC 65301, a 1256 mm specimen (B.W. Westermann, 1801–1817).

Type locality: "Java" [= W Indonesia].

Distribution: Southeast Asia and Greater Sundas. Southern Myanmar (Tanintharyi), S Vietnam (Ba Ria-Vung Tsu, Binh Thuan, Dong Nai, Lam Dong), Cambodia, Thailand (Nakhon Si Thammarat, Phang Nga, Ratchaburi), West Malaysia (Perak, Perlis, Pinang), East Malaysia (Sabah, Sarawak), Brunei and Indonesia (Banka, Belitung, Java, Kalimantan, Nias, Sumatra), 550–1550 m.

Sources: Brongersma, 1933e, 1947a, 1948b, E.H. Taylor, 1965, Tweedie, 1983, Jintakune & Chanhome, 1995, David & Vogel, 1996, J.B. Rasmussen & Hughes, 1997, M.J. Cox et al., 1998, Kuch & Götzke, 2000, Malkmus et al., 2002, Leviton et al., 2003, Orlov et al., 2003, McGuire & Kuch, 2005 and V.S. Nguyen et al., 2009.

8. *Bungarus lividus* Cantor, 1839a. Proc. Zool. Soc. London 7(1): 32.

Type: Lectotype, specimen illustrated in colored sketch of T.E. Cantor (1834–1837: no 1) in RSL, designated by M.A. Smith (1943: 418).

Type locality: "Assam" [= Arunachal Pradesh, Assam, Manipur, Meghalaya, Mizoram or Nagaland State, NE India] via lectotype selection.

Distribution: Northeastern India (Assam, Sikkim, West Bengal), Nepal (Chitwan, Udayapur) and Bangladesh, 340 m.

Sources: Wall, 1911j, Schleich & Kästle, 2002, Tiwari & Shah, 2004 and S.K. Sharma et al., 2013.

9. *Bungarus magnimaculatus* Wall & Evans, 1901. J. Bombay Nat. Hist. Soc. (1900–1901) 13(4): 611–612. (*Bungarus caeruleus magnimaculata*)

Type: Lectotype, BMNH 1908.6.23.90 (formerly IMC), a 1219–1308 mm specimen (J.C. Thompson a.k.a. V. Kühne, Jan.–June 1900), designated by M.A. Smith (1943: 417).

Type locality: "Meiktila, Upper Burma" [= Meiktila, Mandalay State, cen. Myanmar, 20°52'N, 95°51'E, elevation 245 m] via lectotype selection.

Distribution: Central Myanmar (Mandalay, Sagaing, S Shan, Yangon), 10–245 m.

Source: Leviton et al., 2003.

10. *Bungarus multicinctus* Blyth, 1860. Proc. Asiatic Soc. Bengal 29(1): 98.

Synonyms: *Bungarus wanghaotingi* C.H. Pope, 1928b, *Bungarus multicinctum* – Mell *in* Ahl, 1930, *Bungaris multicinctus* – Bertke, Watt & Tu, 1966 *(nomen incorrectum),* and *Bungarus multicinctis* – Bernheimer, Weinstein & Linder, 1986 *(nomen incorrectum).*

Type: Holotype, IMC, a 1220 mm skin without head (R. Swinhoe, 1855–1860 [Amoy] or 7 June–1 July 1858 [Formosa]), lost *fide* M.A. Smith (1943: 416).

Type locality: Unknown. Designated as Amoy [= Xiamen Shi, Fujian Prov., S China] *fide* C.H. Pope (1935: 335), M.A. Smith (1943: 416), Klemmer (1963: 281) and Golay et al. (1993: 120), Formosa [= Taiwan] *fide* Stejneger (1907: 397) and Maki (1931: 153) and both China and Formosa *fide* C.H. Pope (1935: 335).

Distribution: Southeastern Asia. Southeastern China (Anhui, Fujian, Guangdong, Guangxi, Guizhou, Hainan, Hong Kong, Hubei, Hunan, Sichuan, Yunnan, Zhejiang), Macau, Taiwan, Myanmar (Bago, Mandalay, Yangon), N Laos, N Vietnam (Bsc Giang, Bac Kan, Cao Bang, Hai Duong, Hai Phong, Hanoi, Ha Tinh, Hoa Binh, Lang Son, Nghe An, Son La, Thanh Hoa, Thua Thien-Hue, Vinh Phuc), NSL–1500 m.

Sources: Stejneger, 1907, Maki, 1931, Mao, 1961, Kuntz, 1963, Karsen et al., 1986, Easton & Leung-Va, 1993, Xie et al., 1998, Orlov et al., 2000, Kuch et al., 2005b, Kuch & Mebs, 2007, V.S. Nguyen et al., 2009 and Xiang & Li, 2009.

Remarks: The only two snakes reported in Blyth (1860: 98) are represented by dried skins and the other *(Python molurus)* is stated to be from Formosa, suggesting a type locality of Taiwan. Robert Swihoe first lived in Amoy (1855–1860), became British Vice-Consul in Formosa (1861–1873 but earlier collected in Formosa in 1856 and 1858) and then became Consulate in Amoy (1863–1873), Ningpo and Chefoo. Zhao & Adler (1993: 265), David & Ineich (1999: 69), Zhao (2006: 292), Zhang (2009: 127) and Shi (2011:256) are equivocal concerning the type locality.

11. *Bungarus niger* Wall, 1908f. J. Bombay Nat. Hist. Soc. (1907–1908) 18(4): 712–713, 715–716.

Type: Lectotype, BMNH 1946.1.18.63 (formerly BMNH 1860.3.19.1257), a juvenile female (F. Wall, 17 May 1907), designated by Toriba *in* Golay et al. (1993: 121).

Type locality: "Tindharia, eastern Himalayas, India" [= Tindharia, N West Bengal State, NE India, 26°51'N, 88°20'E, elevation 875 m] via lectotype selection.

Distribution: Southern Asia. Northeastern India (Arunachal Pradesh, Assam, Meghalaya, Nagaland, North Bengal, Sikkim, Uttarakand), Nepal (Kaski), Bhutan and N Bangladesh, 100–1800 m.

Sources: Wall, 1911j, Tillack & Grossman, 2001, Schleich & Kästle, 2002, Grosselet et al., 2004, Tiwari & Shah, 2004, Whitaker & Captain, 2004, Theophilus et al.,

B

2008, M.F. Ahmed et al., 2009 and S.K. Sharma et al., 2013.

Remarks: Supplemental original descriptions in wall (1909d: 355 & 1910c: 838–840, pl. figs. 4–7). Type locality restriction to Tindharia, eastern Himalayas [300–1370 m] *fide* M.A. Smith (1943: 417) irrelevant.

12. *Bungarus sindanus* Boulenger, 1897f. J. Bombay Nat. Hist. Soc. (1897–1898) 11(1): 73–74, pl., 4 figs.

Synonym: *Bungarus sindanus razai* Khan, 1985a.

Types: Syntypes (3), BMNH 1946.1.18.54–55, [Sukkur] (L.J. Mountford) and BMNH 1946.1.19.16 [Umarkot] (H.E. Watson), longest syntype 1300 mm.

Type locality: "Umarkot" and "Sukkur, Sindh" [= Umarkot, S Sindh Prov., 25°24'N, 69°43'E, elevation 10 m and Sukkur, 27°41'N, 68°51'E, elevation 60 m, N Sindh Prov., SE Pakistan]. Restricted to Sindh Basin *fide* Boulenger (1897f: 73).

Distribution: Pakistan (N Balochistan, Jammu & Kashmir, Khyber Pakhtunkhwa, Punjab, Sindh) and W India (Gujarat, Madhya Pradesh, Maharashtra, Rajasthan), 10–1180 m.

Sources: Wall, 1913d, Schleich & Kästle, 2002 and Whitaker & Captain, 2004.

Remarks: Resurrected as a valid species *fide* Khan, 1984b, 1985a.

13. *Bungarus slowinskii* Kuch, Kizirian, Truong, Lawson, Donnelly & Krebs, 2005. Copeia 2005(4): 823–828, figs. 3–5.

Type: Holotype, IEBR 1172 (formerly LACM FS 843), a 1350 mm male (Q.T. Nguyen & D. Kizirian, 1 Oct. 2001).

Type locality: "Vietnam, Yen Bai Prov., Van Yen District, from a stream near Na Hau Commune, 21°46'N, 104°32'E, 540 m."

Distribution: Central Laos (Khammouane) and N Vietnam (Lao Cai, Quang Nam, Quang Tri, Thua Thien-Hue, Yen Bai), 140–1300 m. Possibly occurs up Red River Valley in Yunnan, China *fide* Kuch et al. (2005: 828).

Sources: Kien & Truong, 2009, V.S. Nguyen et al., 2009 and Kharin et al., 2011.

14. *Bungarus walli* Wall, 1907a. J. Bombay Nat. Hist. Soc. (1906–1907) 17(3): 608–609, 1 pl., 6 figs.

Type: Lectotype, from BMNH 1946.1.18.51–53 series, a 1511 mm male (F. Wall and natives, 6–29 Aug. 1906), designated by Wall (1924b: 24).

Type locality: "Fyzabad, Oudh" [= Faizabad, Uttar Pradesh, N India, 26°47'N, 82°09'E, elevation 100 m].

Distribution: Northern India (Bihar, Orissa, Uttar Pradesh, West Bengal), SE Nepal and Bangladesh (Chittagong), NSL–155 m.

Sources: Wall, 1924b, 1928, Klemmer, 1963, Whitaker & Captain, 2004, N. Khaire, 2006 and S.K. Sharma et al., 2013.

Remarks: The first new species discovered by F. Wall after 11.5 years of laborious field work, which is why he claimed "sufficient excuse for commemorating the event and attaching my own name to it."

15. *Bungarus wanghaotingi* C.H. Pope, 1928b. Amer. Mus. Novit. (325): 3–4.

Type: Holotype, AMNH 35320, a 484 mm female (W. Granger, Nov. 1926).

Type locality: "Yuan Kiang, southwestern Yunnan Province, China" [= Yuanjiang, SW Yunnan Prov., SW China, 23°36'N, 102°00'E, elevation 400 m].

Distribution: Myanmar (Kachin, Rakhine, Sagaing, Yangon) and SW China (Yunnan), 400–900 m.

Source: A valid species *fide* Leviton et al., 2003.

Remarks: Supplemental original description in C.H. Pope (1929: 469).

C

CAAETEBOIA Zaher, Grazziotin, Cadle, Murphy, Moura-Leitve & Bonatto, 2009 (Xenodontidae)

Type species: *Liophis amarali* Wettstein, 1930.
Distribution: Eastern Brazil.
Sources: Dixon, 1980, Marques et al., 2001, Zaher et al., 2009,Vidal et al., 2010 and Passos et al., 2012a.

1. *Caaeteboia amarali* (Wettstein, 1930). Zool. Anz. 88(1–4): 93–94, fig. 1. (*Liophis amarali*)

Type: Holotype, NMW 23107, a 527 mm specimen (P. Branchhardt don., May 1929).
Type locality: "Bello Horizonte, Staat Minas Geraës, Brasilien" [= Belo Horizonte, cen. Minas Gerais State, E Brazil, 19°55'S, 43°56'W, elevation 865 m].
Distribution: Southeastern Brazil (Bahia, Espírito Santo, Minas Gerais, Parana, Rio de Janeiro, Santa Catarina, São Paulo), 865 m.

CACOPHIS A.C.L.G. Günther, 1863c (Elapidae)

Synonyms: *Petrodymon* Krefft, 1865b, and *Cachophis* – Higgins, 1873 *(nomen incorrectum).*
Type species: *Cacophis krefftii* A.C.L.G. Günther, 1863c.
Distribution: Eastern Australia.
Sources: Krefft, 1869, Shine, 1980b, Cogger et al., 1983a, Mengden, 1983, Shine, 1985a, S.K. Wilson & Knowles, 1988, Hoser, 1989, 2012e, Hutchinson, 1990, Ehmann, 1992, Golay et al., 1993, Greer, 1997, David & Ineich, 1999, Cogger, 2000, Scanlon, 2003a, Scanlon & Lee, 2004 and Sanders et al., 2008.

1. *Cacophis churchilli* Wells & Wellington, 1985. Aust. J. Herp. (Suppl. 1): 45.

Type: Holotype, AMS R74464 (G. Churchill, 1968).
Type locality: "Black Mountain Road, near Kuranda, Queensland, (16 49'S x 14S 38'E), Australia."
Distribution: Northeastern Australia (Queensland, Lindeman Is.), NSL–855 m.
Sources: S.K. Wilson & Knowles, 1988, Covacevich & McDonald, 1993, Shea & Sadlier, 1999 and Scanlon, 2002.

2. *Cacophis harriettae* Krefft, 1869a. Proc. Zool. Soc. London 31(1): 319, fig. 3.

Synonyms: *Cacophis harrietti flavicollis* McDowell, 1967 *(nomen nudum),* and *Cacophis harrietti harrietti* – McDowell, 1967 *(nomen incorrectum).*
Type: Lectotype, AMS 142007 (formerly AMS 6676), a 305 mm specimen (F.A. Blackman, 1855–1869), designated by Cogger (1979: 204).
Type locality: "Warro, Port Curtis, Queensland, Australia" [= 23°25'S, 150°31'E, elevation NSL] via lectotype selection.
Distribution: Eastern Australia (ext. NE New South Wales, E Queensland), NSL–825 m.
Source: Shea & Sadlier, 1999.
Remarks: Supplemental original description in Krefft (1869b: 74, pl. 11, figs. 5–5a.).

3. *Cacophis krefftii* A.C.L.G. Günther, 1863c. Ann. Mag. Nat. Hist. (3) 12(71): 361.

Synonym: *Cacophis fordei* Krefft, 1869b.
Type: Lectotype, BMNH 1946.1.17.60, a 305 mm specimen (J.F. Wilcox, June 1847–Sept. 1850), designated by Cogger et al. (1983: 219).
Type locality: "probably from Port Macquarie" [New South Wales, Australia] (in error). "North of the Clarence River, N.S.W" via lectotype selection. Restricted to north of the Clarence at Ipswich, Queensland *fide* Krefft (1869: 74) [= Ipswich, Queensland, Australia, 27°37'N, 152°46'E, elevation 35 m].
Distribution: Eastern Australia (E New South Wales, ext. SE Queensland), NSL–1005 m.
Remarks: The types were donated by G. Krefft *fide* Günther (1863c: 361).

4. *Cacophis squamulosus* (A.-M.-C. Duméril, Bibron & Duméril, 1854b). Erpét. Gén. 7(2): 1235–1236. (*Pseudelaps squamulosus*)

Synonyms: *Pseudelaps squamulosus* A.-M.-C. Duméril, 1853 (*nomen nudum*), *Diemansia cucullata* A.C.L.G. Günther, 1862b, *Pseudoelaps atropolios* Jan, 1863b (*nomen nudum*), and *Pseudoelaps atropolios* Jan, 1873 *in* Jan & Sordelli, 1870–1881.
Type: Holotype, MNHN 7667 (J. Verreaux, 1844).
Type locality: "Tasmanie" [= Tasmania, Australia]. (in error *fide* Cogger et al., 1983a: 219).
Distribution: Southeastern Australia (E New South Wales, ext. SE Queensland), NSL–1245 m.

C

†*CADURCERYX* Hoffstetter & Rage, 1972 (Boidae)

Type species: †*Cadurceryx filholi* Hoffstetter & Rage, 1972.
Distribution: Middle to upper Eocene of France.
Source: Rage, 1984b.
Remarks: Upper Eocene material believed to represent a second species *fide* Rage (1978: 209).

1. †*Cadurceryx filholi* Hoffstetter & Rage, 1972. Ann. Paléont. (Vert.) 58(1): 91–93, figs. 6a–c, pl. 1, fig. 1.

Type: Holotype, MNHN Qu 301, one anterior cloacal vertebra with both synapophyses and additional processes.
Type locality: "phosphorites du Quercy, anciennes collections, localité non précisée: âge: presque certainement Eocène supérieur, France." Probably Headonian, upper Eocene *fide* Rage (1984b: 24).
Distribution: Middle Eocene (Lutetian: 37.2–48.6 mya) and upper Eocene (Priabonian: 33.9–37.2 mya) of France.

†*CADURCOBOA* Rage, 1978 (Boidae)

Type species: †*Cadurcoboa insolita* Rage, 1978.
Distribution: Middle Eocene of France, upper Eocene of France and United Kingdom.
Sources: Rage, 1978, 1984b, Holman & Harrison, 1998a and Rage & Augé, 2010.
Remarks: An unidentified species known from upper Eocene of United Kingdom *fide* Milner et al. (1982).

1. †*Cadurcoboa insolita* Rage, 1978. Palaeovertebrata 8(2–4): 210–213, figs. 6, 7a–b.

Type: Holotype, UPVI SNB 1000, one trunk vertebra.
Type locality: "Sainte-Néboule, Lot (Phosphorites du Quercy), Eocène supérieur" [France]. Emended to Headonian *fide* Rage (1984b: 31).
Distribution: Middle Eocene (37.2–48.6 mya) and upper Eocene (Priabonian: 33.9–37.2 mya) of France and United Kingdom (England).

CALABARIA Gray, 1858b (Calabariidae)

Synonyms: *Rhoptrura* W.C.H. Peters, 1858, and *Roptrura* Sauvage, 1884 *(nomen emendatum)*.
Type species: *Eryx reinhardtii* Schlegel, 1848.
Distribution: Western and cen. Africa.
Sources: Gray, 1858e, Tornier, 1901, E.H. Taylor & Weyer, 1958, Doucet, 1963, Mertens, 1965c, Leston & Hughes, 1968, Hughes & Barry, 1969, Stimson, 1969, Underwood, 1976, Hughes, 1983, J.A. Butler & Reid, 1986, Kluge, 1993a, 1999a, Heise et al., 1995, J.-F. Trape & Roux-Estève, 1995, Lamar, 1997b, Luiselli & Akani, 1998a, McDiarmid et al., 1999, Luiselli et al., 2002, J.C. Murphy & Schlager, 2003, R. Lawson et al., 2004, Chippaux, 2006, Noonan & Chippindale, 2006, Chirio & LeBreton, 2007, Pauwels & Vande weghe, 2008 and Segniagbeto et al., 2011.
Remarks: A synonym of *Charina fide* Kluge, 1993.

1. *Calabaria reinhardtii* (Schlegel, 1848). Bijdr. Dierk. 1: 2–3, pl. (*Eryx reinhardtii*)

Synonyms: *Calabaria fusca* Gray, 1858b, and *Roptrura petiti* Sauvage, 1884.
Type: Holotype, ZMB 1471, a 457 mm specimen (J.T. Reinhardt, 1845–1847).
Type locality: "Côte d'or" [= Ghana]. Restricted to Aqua Pim (= Akwapim) [= Akwapim, Eastern Prov., S Ghana, 5°58'N, 1°05'W, elevation 120 m] *fide* Hughes & Barry (1969: 1010).
Distribution: West and cen. Africa. Southeastern Guinea (Macenta, Nzérékoré), Liberia (Bong, Gbarpolu, Loffa, Margibi, Montserrado), Ivory Coast (Abidjan), S Ghana (Eastern), Togo (Plateaux), Benin, S Nigeria (Cross River, Delta), S Cameroon (Central, Est, Littoral, Nord-Ouest, Sud, Sud-Ouest), Equatorial Guinea (Bioko Is.), Gabon (Estuaire, Haut-Ogooué, Moyen-Ogooué, Ngounié, Nyanga, Ogooué-Ivindo, Ogooué-Lolo, Ogooué-Maritime), Equatorial Guinea (Bioko Is.), SW Central African Republic (Lobaye, Ombella-Mpoko, Sangha), S Congo (Kouilou, Lekoumou, Niari, Pool) and Democratic Republic of the Congo (Bandundu, Bas-Congo, Equateur, Kasai Occidental, Kasai Oriental, Kinshasa, Kivu, Orientale), NSL–1050 m.
Remarks: Possibly occurs in Benin *fide* Hughes (2013: 151).

†*CALAMAGRAS* Cope, 1873 (Boidae)

Synonyms: †*Aphelophis* Cope, 1873, †*Calamagros* – Hoffmann, 1890 *(nomen incorrectum)*, and †*Aphelopsis* Stromer, 1910 *(nomen emendatum)*.
Type species: †*Calamagras murivorus* Cope, 1873.
Distribution: Lower Eocene of USA, France and Kyrgyzstan, middle Eocene of USA, upper Eocene of Canada, lower Oligocene of USA and Mongolia, upper Oligocene of USA, and lower Miocene of USA.
Sources: Gilmore, 1938, Holman, 1979b, 2000a, and Rage, 1984b.

1. †*Calamagras angulatus* Cope, 1873. Synop. New Vert. Tertiary Colorado: 16.

Type: Holotype, AMNH 1654, one trunk vertebra (E.D. Cope, summer 1873).

Type locality: "Cedar Creek, Colorado (Whitneyan or Orellan, middle or late Oligocene)" [= early Oligocene *fide* Holman, 2000a: 288].
Distribution: Lower Oligocene (Orellan: 33.3–33.9 mya) of USA (Colorado, South Dakota), upper Oligocene (Arikareean: 20.6–30.8 mya) of USA (Nebraska, Wyoming), and lower Miocene (Arikareean: 20.6–30.8 mya) of USA (Nebraska).
Source: Holman, 1976b.
Remarks: Original description reprinted in Cope (1874a: 518).

2. †*Calamagras floridanus* Auffenberg, 1963. Tulane Stud. Zool. 10(3): 162–163, fig. 11.

Type: Holotype, UF 6150, one trunk vertebra (W. Auffenberg, 1954).
Type locality: "Boulder Bar, Thomas Farm, Gilchrist County, Florida, lower Miocene, Arikareean, Hawthorne formation."
Distribution: Lower Miocene (Arikareean: 20.6–30.8 mya) of USA (Florida). Known only from type locality.
Remarks: Listed as middle Miocene *fide* (Holman, 1979b: 211).

3. †*Calamagras gallicus* Rage, 1977b. Herpetologica 33(4): 459–460, figs. 1a–b.

Type: Holotype, MNHN GR 7896, one caudal vertebra.
Type locality: "Grauves (Département de la Marne, France). Lower Eocene = Ypresian (correlated with the Wasatchian of North America)."
Distribution: Lower Eocene (Ypresian: 48.6–55.8 mya) of France. Known only from type locality.

4. †*Calamagras murivorus* Cope, 1873. Synop. New Vert. Tertiary Colorado: 15.

Synonyms: †*Aphelophis talpivorus* Cope, 1873, and †*Calamagras truxalis* Cope, 1873.
Type: Holotype, AMNH 1603, six trunk vertebrae (E.D. Cope, summer 1873).
Type locality: "Cedar Creek, northeastern Colorado" [USA]. Emended to Oligocene, Oreodon Beds *fide* Gilmore (1938: 42) and White River Formation, lower Oligocene (Orellan) of Cedar Creek locality, Logan County, Colorado *fide* Holman (2000a: 63).
Distribution: Lower Oligocene (Orellan: 33.3–33.9 mya) of the USA (Colorado, Nebraska, South Dakota).
Remarks: Original description reprinted in Cope (1874a: 517).

5. †*Calamagras platyspondyla* Holman, 1976b. Herpetologica 32(1): 89, figs. 1a–b.

Type: Holotype, USNM 4512, one trunk vertebra.
Type locality: "Loc. I, Lower Gering formation: University of Nebraska State Museum Locality MO-119 (NW 1/4, SE 1/4, sec. 32, T 20 N, R 52 W, 10.46 km S and 2.01 km W of Bayard, Morrill County, Nebraska), Lower Miocene." Emended to lower Gering formation, late Oligocene (early Arikareean)-6, Durnal locality, Morrill County, Nebraska *fide* Holman (2000a: 64)
Distribution: Lower Oligocene (Arikareean: 20.6–30.8 mya) of USA (Florida), and upper Oligocene (Arikareean: 20.6–30.8 mya) of USA (Nebraska, Wyoming).
Sources: Holman, 1977d and Holman & Harrison, 2001.

6. †*Calamagras primus* M.K. Hecht *in* McGrew, 1959. Bull. Amer. Mus. Nat. Hist. 117(3): 142–143, pl. 54, figs. 1–5.

Type: Holotype, AMNH 3828, one middle trunk vertebra (P.O. McGrew & field crews, 1950–1952).
Type locality: "Locality 5, on the south side of Elk Mountain and Tabernacle Butte area at the northern end of the Green River Basin, about 15 miles from the southwest flank of the Wind River Mountains and 25 miles north of the village of Farson, in southeast corner of Sublette County, Wyoming (T 27-29N, R 103-106 W), [USA]; late Bridgerian, middle Eocene."
Distribution: Middle Eocene (Bridgerian: 46.2–50.3 mya) of USA (Wyoming) and possibly lower Eocene (50.3–55.4 mya) of USA (New Mexico).
Source: Sullivan & Lucus, 1988.

7. †*Calamagras turkestanicus* Danilov & Averianov, 1999. Geodiversitas 21(1): 86–91, figs. 1a–d, 2a–d, 3a–d.

Type: Holotype, ZISP PH 1/2, one anterior/middle trunk vertebra (A.O. Averianov, 1988–1993 and I.G. Danilov & A.O. Averianov, 1995).
Type locality: "Andarak 2, Fergana Valley, Kyrgystan; Alay beds, lower Eocene (late Ypresian)."
Distribution: Lower Eocene (Ypresian: 48.6–55.8 mya) of Kyrgyzstan.

8. †*Calamagras weigeli* Holman, 1972. Canad. J. Earth Sci. 9(12): 1626, figs. 7a–b.

Type: Holotype, SMNH 1437, one posterior trunk vertebra (Saskatchewan Mus. Nat. Hist., 1960–1963).
Type locality: "early Oligocene, Cypress Hills Formation, north branch of Calf Creek in L.S. 4, Sec. 8, Twp. 8, Distribution 22, W 3rd mer., elevation 3600 ft (1100 m), 10 mi (16 km) northwest of Eastend, Saskatchewan." Emended to late Eocene (Chadronian) *fide* Holman (2000a: 67).
Distribution: Upper Eocene (Chadronian: 33.9–38.0 mya) of Canada (Saskatchewan), and lower-middle Miocene (Barstovian, Hemingfordian and Arikareean: 13.6–30.8 mya) of USA (South Dakota, Texas, Wyoming).
Sources: Holman, 1976a, 1976c, 1977b.

C

CALAMARIA H. Boie *in* F. Boie, 1827 (Calamariidae)

Synonyms: *Changulia* Gray, 1835 *in* Gray & Hardwicke, 1830–1835, *Typhlocalamus* A.C.L.G. Günther, 1872a, *Caramaria* – Oshima, 1910 *(nomen incorrectum), Galamaria* – Okada, 1932 *(nomen incorrectum), Keiometopon* E.H. Taylor, 1962, *Calameria* – Witte, 1966 *(nomen incorrectum), Calamalia* – F.-T. Peng & Chang, 1968 *(nomen incorrectum), Crottyreedus* Hoser, 2012aj *(nomen illegitimum), Freudreedus* Hoser, 2012ai *(nomen illegitimum),* and *Oxyreedus* Hoser, 2012ai *(nomen illegitimum).*

Type species: *Calamaria linnaei* H. Boie *in* F. Boie, 1827.

Distribution: Southeastern Asia and East Indies.

Sources: Inger & Marx, 1965, Grismer et al., 2004, Koch et al., 2009, Orlov, 2009, Zaher et al., 2009 and I. Das, 2010, 2012.

Remarks: Official Generic Name *fide* Opinions 18 (ICZN, 1926) & Opinion 2196 (ICZN, 2008).

1. *Calamaria abramovi* Orlov, 2009. Russ. J. Herp. 16(2): 147–151, figs. 1–6.

Type: Holotype, ZISP 25569, a 482 mm female (N.L. Orlov, 15 Sept. 1998).

Type locality: "from above Mang Xang Village, Ngoe Linh Mountain, Dac Glei District, Kon Tum Prov., Vietnam (15°05' N 107°57' E, elevation 1700m)."

Distribution: Southern Vietnam (Kon Tum), 1700 m. Known only from type locality.

Source: Orlov et al., 2010.

2. *Calamaria abstrusa* Inger & Marx, 1965. Fieldiana: Zool. 49: 179–180, fig. 49.

Type: Holotype, NMW 17026, a 170 mm male (J. Schild, 1890–1904).

Type locality: "Padang, Sumatra" [= Padang, Sumatera Barat Prov., W Sumatra, W Indonesia, 0°57'S, 100°21'E, elevation NSL].

Distribution: Western Indonesia (Nias, W Sumatra), NSL.

Source: David & Vogel, 1996.

3. *Calamaria acutirostris* Boulenger, 1896c. Ann. Mag. Nat. Hist. (6) 17(101): 394.

Type: Lectotype, NMBA 1686, a 250 mm male (P. & F. Sarasin, 1893–1896), designated by Inger & Marx (1965: 73).

Type locality: "Celebes: Mount Bonthain, Loka" [= Loka, Mt. Lompobatang, South Sulawesi Prov., Sulawesi, Indonesia, 5°19'S, 119°54'E, elevation 1200 m] via lectotype selection.

Distribution: Central Indonesia (SW Sulawesi), 1065–1200 m.

Sources: Bosch, 1985, Lang & Vogel, 2005 and S.D. Howard & Gillespie, 2007.

Remarks: Lectotype erroneously listed as NMBA 1681 *fide* Inger & Marx (1965: 74).

4. *Calamaria albiventer* (Gray, 1835 *in* Gray & Hardwicke, 1830–1835). Illust. Indian Zool. 2(19–20): pl. 86, figs. 6–9. (*Changulia albiventer*)

Synonyms: *Calamaria indragirica* Schenkel, 1901, and *Calamaria ornata* F. Werner, 1909b.

Types: Syntypes (3), BMNH 1946.1.2.10 (formerly BMNH 1860.3.19.1269a–b) and BMNH 1946.1.2.18 (formerly BMNH 1898.9.22.39), three females (T. Hardwicke, 1756–1823).

Type locality: "Pinang" [= Penang Is., Pinang State, NW Malay Peninsula, 5°22'N, 100°14'E].

Distribution: Malay Peninsula and Sumatra. West Malaysia (Perak, Pinang), Singapore and W Indonesia (E Sumatra), NSL–1370 m.

Sources: Tweedie, 1983, David & Vogel, 1996 and Grismer et al., 2010.

Remarks: M.A. Smith (1943: 530) designated pl. 124 in T. Hardwicke's collection a lectotype. Inger & Marx (1965: 96) listed 3 syntypes with BMNH 1860.3.19.1269 having 2 specimens.

5. *Calamaria alidae* Boulenger, 1920a. Ann. Mag. Nat. Hist. (9) 5(27): 282–283.

Type: Holotype, BMNH 1946.1.3.64 (formerly BMNH 1920.1.16.22), a 220 mm specimen (C.J. Brooks, 1912–1917).

Type locality: "Lebong Tandai in Benkoelen, Sumatra" [= Lebong Tandai, Bengkulu Prov., W Sumatra, W Indonesia, 3°01'S, 101°54'E, elevation 150 m *fide* David & Vogel, 1996: 240].

Distribution: Western Indonesia (W Sumatra), 150 m.

Source: David & Vogel, 1996.

6. *Calamaria apraeocularis* M.A. Smith, 1927. Proc. Zool. Soc. London 97(1): 224–225.

Type: Holotype, BMNH 1946.1.3.48 (formerly MAS 8542 & BMNH 1926.8.20.157), a 285 mm female (M.A. Smith, 15 Jan.–15 Feb. 1924).

Type locality: "Djikoro, on the south-western slopes of Mt. Bonthain, about five kilometers distant from Malakadji, at an altitude of about 1200 metres, South Celebes" [= Cikoro, Mt. Lompobatang, South Sulawesi Prov., Sulawesi, Indonesia, ca. 5°25'S, 119°52'E, elevation 1200 m]

Distribution: Central Indonesia (SW Sulawesi), 1200 m. Known only from type locality.

Sources: Bosch, 1985, Lang & Vogel, 2005 and S.D. Howard & Gillespie, 2007.

7. *Calamaria banggaiensis* Koch, Arida, McGuire, Iskandar & Böhme, 2009. Zootaxa (2196): 21–26, figs. 2, 4, 6, 8.

Type: Holotype, MZB 3230 (formerly AK 182), a 201 mm female (A. Koch & E. Erida, 28 Aug. 2005).
Type locality: "approximately 5 m elevation, southeast of Desa (= village) Banggai (1°37'23"S, 123°32'16E"), Pulau Banggai, Kepulauan Banggai, east of Central Sulawesi (Propinsi Sulawesi Tengah), Indonesia."
Distribution: Central Indonesia (cen. Sulawesi: Banggai and Peleng), NSL.
Source: Koch, 2012.

8. *Calamaria battersbyi* Inger & Marx, 1965. Fieldiana: Zool. 49: 208–209.

Type: Holotype, BMNH 1896.2.17.13, a 92 mm male (F.H. Rolle, 1864–1896).
Type locality: "Tandjong, southeastern Borneo" [= SE Kalimantan].
Distribution: Central Indonesia (SE Kalimantan).

9. *Calamaria bicolor* A.-M.-C. Duméril, Bibron & Duméril, 1854a. Erpét. Gén. 7(1): 78–81.

Synonyms: *Calamaria macrurus* Bleeker, 1860b, *Calamaria hosei* Günther, 1896b, *Calamaria semiannulata* Boettger, 1898, and *Calamaria picteti* Peracca, 1899.
Type: Holotype, RMNH 68, a 281 mm male (S. Müller, spring 1831–May 1832).
Type locality: "Ile de Bornéo" [= Kalimantan, cen. Indonesia].
Distribution: East Malaysia (Sabah, Sarawak) and Indonesia (cen. Java, Kalimantan), 1200 m.
Sources: T. Vogt, 1925, Malkmus et al., 2002 and I. Das, 2012.

10. *Calamaria bitorques* W.C.H. Peters, 1872b. Mber. Königl. Akad. Wiss. Berlin 1872(7): 585–586.

Type: Holotype, ZMB 7444.
Type locality: "Philippinen" [= Philippines]. Listed as Luzon in ZMB catalogue *fide* Inger & Marx (1965: 106).
Distribution: Northern Philippines (Luzón, Panay), 850 m.
Sources: E.H. Taylor, 1922a and Gaulke & Vogel, 2005.

11. *Calamaria boesemani* Inger & Marx, 1965. Fieldiana: Zool. 49: 171–172.

Type: Holotype, GNM 3230 (formerly NHMG), a 116 mm female (W. Kaudern, Dec. 1919).
Type locality: "Pinapuan (?Pinapolan) near Luwuk and Biak, near the tip of the eastern peninsula, Celebes" [= Pinapuan, near Luwuk (0°51'S, 123°03'E, 45 m), E Sulawesi, Indonesia].
Distribution: Central Indonesia (E Sulawesi), 45 m. Known only from holotype.
Sources: Bosch, 1985, Lang & Vogel, 2005 and Koch, 2012.

12. *Calamaria borneensis* Bleeker, 1860b. Nat. Tijds. Nederl. Indië (5) 22(1): 296.

Synonyms: *Calamaria borneensis* Bleeker, 1857a *(nomen nudum), Rabdosoma borneensis* Bleeker, 1857a *(nomen nudum),* and *Rhabdosoma borneensis* Bleeker, 1857a *(nomen nudum).*
Type: Holotype, BMNH 1946.1.3.66, a 220 mm specimen (P. de Bleeker, 1857–1860).
Type locality: "Borneo (Sintang)" [= Pasar Sintang, Kalimantan Barat State, W Kalimantan, Indonesia, 0°05'N, 111°29'E, elevation 25 m].
Distribution: Borneo. East Malaysia (Sarawak) and Indonesia (W Kalimantan), NSL–350 m.
Sources: Lang & Vogel, 2005 and Rooijen & Rooijen, 2007.

13. *Calamaria brongersmai* Inger & Marx, 1965. Fieldiana: Zool. 49: 118–119.

Type: Holotype, GNM 3229 (formerly NHMG), a 225 mm male (W. Kaudern, Dec. 1919).
Type locality: "Penapuan (?Pinapolan) near Luwuk and Biak, near the tip of the eastern peninsula, Celebes" [= Pinapuan, near Luwuk (0°51'S, 123°03'E, 45 m), E Sulawesi, Indonesia].
Distribution: Central Indonesia (cen. and E Sulawesi), 45 m.
Sources: S.D. Howard & Gillespie, 2007 and Koch, 2012.

14. *Calamaria buchi* Marx & Inger, 1955. Fieldiana: Zool. 37: 195–196, fig. 26.

Type: Holotype, FMNH 71697, a 389 mm female (Buch, 1953).
Type locality: "Dalat, Viet Nam, Indo-China. Altitude 1,500 feet" [= Da Lat, Lam Dong Prov., Vietnam, 11°56'N, 108°28'E, elevation 1500 m].
Distribution: Southeast Vietnam (Ha Tinh, Lam Dong), 450–1500 m.
Sources: Ziegler & Le, 2005, Ziegler et al., 2008, V.S. Nguyen et al., 2009 and Orlov et al., 2010.

15. *Calamaria butonensis* S.D. Howard & Gillespie, 2007. J. Herp. 41(2): 237–240, figs. 1–2.

Type: Holotype, MZB 3125, a 176 mm male (G.R. Gillespie, S. Howartd, D. Lockie, M. Scroggie & Boeadi, 21 June 2002).
Type locality: "Kakenauwe Reserve at 5°11'0"S, 122°53'40"E, and 150 m elevation, Sulawesi, Indonesia."
Distribution: Central Indonesia (Sulawesi), 150–430 m.
Sources: G.R. Gillespie et al., 2005 and Koch, 2012.

C

16. *Calamaria ceramensis* Rooij, 1913. Bijdr. Dierk. 19: 26–27.

Synonym: *Calamaria jeudei* Kopstein, 1926.
Types: Syntypes (2), ZMA 10083 (2), 210 and 245 mm specimens (L F. de Beauford, Feb. 1910).
Type locality: "Honitetu, West-Ceram, Archipelago indico, [= Honitétu, W Seram, Maluku, E Indonesia]."
Distribution: Eastern Indonesia (Maluku: Ambon, Saparua, Seram).
Sources: Bosch, 1985, S.D. Howard & Gillespie, 2007 and Koch et al., 2009.
Remarks: Syntype photographs in Koch et al. (2009: figs. 3, 5, 7, 9)

17. *Calamaria concolor* Orlov, Nguyen, Nguyen, Ananjeva & Ho, 2010. Russ. J. Herp. 17(3): 237–239, figs. 1–6.

Type: Holotype, ZISP 30185 (formerly IEBR A.2010.02), a 578 mm male (T.D. Tran, April 2009).
Type locality: "Bach Ma Peak, Bach Ma National Park, on the Eco-tourist trail, Thua Thien-Hue Prov., Vietnam (16°11'16.7" N, 107°50'36.1" E; altitude 1400 m.a.s.l.)."
Distribution: Central Vietnam (Thua Thien-Hue), 1400 m. Known only from type locality.

18. *Calamaria crassa* Lidth de Jeude, 1922. Zool. Meded. 6(4): 248–249.

Type: Holotype, RMNH 4692, a 460 mm male (E. Jacobson, June 1917).
Type locality: "Gunung Talamau (1300 M.) (Ophir districts), Padang Highlands, central-Sumatra" [= Mount Talakmau, Sumatera Barat Prov., W Sumatra, W Indonesia, 0°04'N, 99°59'E, elevation 1300 m *fide* David & Vogel, 1996: 243].
Distribution: Western Indonesia (W Sumatra), 1300 m.
Source: David & Vogel, 1996.

19. *Calamaria curta* Boulenger, 1896d. Ann. Mag. Nat. Hist. (6) 18(103): 62–63.

Type: Holotype, BMNH 1946.1.3.52 (formerly BMNH 1896.4.29.35), a 315 mm female (A.H. Everett, 1896).
Type locality: "S. Celebes, 2000 feet," [= S Sulawesi, E Indonesia, elevation 610 m].
Distribution: Central Indonesia (ext. SW Sulawesi), 610–1800 m.
Sources: Bosch, 1985, Lang & Vogel, 2005 and S.D. Howard & Gillespie, 2007.

20. *Calamaria doederleini* Gough, 1902. Zool. Anz. 25(681): 645–646. (*Calamaria doederleini*) (*nomen corrigendum*)

Synonym: *Calamaria döderleini* Gough, 1902 (*nomen incorrigendum*).
Type: Holotype, MZUS (unnumbered), a 285–288 mm male (C. Koschinsky, 1899), lost *fide* David & Vogel (1996: 49).
Type locality: "East Sumatra, Langkat" [= Langkat Prefecture, Aceh Prov., N Sumatra, W Indonesia, 3°40'N, 98°10'E, elevation 100 m *fide* David & Vogel, 1996: 240].
Distribution: Western Indonesia (N Sumatra), 100 m. Known only from type locality.
Source: David & Vogel, 1996.
Remarks: Possibly a synonym of *C. pavimentata semidoliata* F. Werner *fide* Inger & Marx (1965: 239).

21. *Calamaria eiselti* Inger & Marx, 1965. Fieldiana: Zool. 49: 175–177, fig. 47.

Type: Holotype, NMW 16703.1, a 343 mm male (F. Steindachner, don. 1896).
Type locality: "Padang, Sumatra" [= Padang, Sumatera Barat Prov., Sumatra, W Indonesia, 0°57'S, 100°21'E, elevation NSL].
Distribution: Western Indonesia (W Sumatra), NSL. Known only from type locality.
Source: David & Vogel, 1996.

22. *Calamaria everetti* Boulenger, 1893d. Proc. Zool. Soc. London 61(1): 525.

Type: Holotype, BMNH 1946.1.2 (formerly BMNH 1893.3.6.52), a 100 mm specimen (A.H. Everett, 1872–1876 or 1885–1890).
Type locality: "Sarawak" [= Borneo]. Emended to Senah branch of Sarawak River, First Division *fide* Inger & Marx (1965: 134).
Distribution: Borneo. East Malaysia (Sabah, Sarawak) and Indonesia (N Kalimantan), 1500 m.
Remarks: Does not occur in the Philippines (Palawan) *fide* Gaulke & Vogel (2005: 19).

23. *Calamaria forcarti* Inger & Marx, 1965. Fieldiana: Zool. 49: 184–185.

Type: Holotype, NWM 16710, a 177 mm male (F. Steindachner, 1900).
Type locality: "Deli, Sumatra" [= Medan region (3°35'N, 98°41'E, ca. 25 m), N Sumatra, W Indonesia *fide* David & Vogel, 1996: 49].
Distribution: Western Indonesia (N Sumatra, Nias), 25 m.
Source: David & Vogel, 1996.

24. *Calamaria gervaisii* A.-M.-C. Duméril, Bibron & Duméril, 1854a. Erpét. Gén. 7(1): 76–78.

Synonyms: *Calamaria mindorensis* Boulenger, 1895f, *Calamaria gervaisii iridescens* E.H. Taylor, 1917, *Calamaria tropica* E.H. Taylor, 1922a, *Calamaria hollandi* E.H. Taylor, 1923, and *Calamaria polillensis* E.H. Taylor, 1923.

Types: Syntypes (2), MNHN 7202a–b (formerly MNHN 2314.7202), longest syntype 194 mm (F. Eydoux & F.L.A. Souleyet, 1829–1832), location of other syntype unknown.

Type locality: "Java" [W Indonesia] (in error *fide* Inger & Marx, 1965: 106). Restricted to Luzon, Philippines *fide* Marx & Inger (1955: 182) but rescinded *fide* Inger & Marx (1965: 107).

Distribution: Philippines (Basilan, Catanduanes. Cebu, Corregidor, Luzon, Mindanao, Mindoro, Negros, Panay, Polillo, Romblon, Tablas), NSL–1525 m.

Sources: E.H. Taylor, 1922, Leviton, 1959b and Ross & Gonzales, 1992.

25. *Calamaria gialaiensis* Ziegler, Nguyen & Nguyen, 2009. Curr. Herp. 27(2): 72–75, figs. 1–2.

Type: Holotype, IEBR A.714, a 457 mm male (V.S. Nguyen, April 1999).

Type locality: "Kon Ka Kinh, K Bang District, Gia Lai Province, Vietnam, 1,300 m asl" [= 14°18'N, 108°24'E].

Distribution: Southern Vietnam (Gia Lai), 1300 m. Known only from type locality.

Source: Orlov et al., 2010.

26. *Calamaria gimlettii* Boulenger, 1905b. Ann. Mag. Nat. Hist. (7) 15(89): 456.

Synonyms: *Calamaria doerianense* Brongersma, 1928, and *Calamaria fraseri* E.H. Taylor, 1962.

Type: Holotype, BMNH 1946.1.3.71 (formerly BMNH 1905.2.7.7), a 265 mm specimen (J.D. Gimlette, 1890–1905).

Type locality: "Kelantan, Malay Peninsula" [= Kelantan State, West Malaysia].

Distribution: West Malaysia (Johor, Kelantan, Pahang, Selangor, Negri Sembilan), Singapore (Pawai Is.), Aur Is. and Riau Arch. (Durian, Galang Is.), NSL–1500 m.

Source: I Das, 2010.

Remarks: A valid species *fide* I. Das (2010: 266).

27. *Calamaria grabowskyi* J.G. Fischer, 1885a. Arch. Naturg. 51(1): 50–52, pl. 4, figs. 1a–e.

Synonyms: *Calamaria baluensis* Boulenger, 1893d, and *Calamaria moultoni* Dunn, 1923a.

Types: Syntypes (2), BMNH 1946.1.5.25–26 (formerly BMNH 1886.5.13.14–15), a 457 mm and 428 mm female (F. Grabowsky, 1880–1884).

Type locality: "Distrikt Dusson Timor...bei Telang, ...bei Tameanglaijang, Südost-Borneo" [= Telang and Tamiang Layang, Dusson Timor Distr., SW Kalimantan, cen. Indonesia, ca. 2°06'S, 114°00'E, elevation 25 m].

Distribution: Borneo. East Malaysia (Sabah, Sarawak), Brunei and Indonesia (Kalimantan), 25–1430 m.

Sources: Stuebing, 1994a, Stuebing & Inger, 1999, Malkmus et al., 2002 and I. Das, 2007b.

28. *Calamaria gracillima* A.C.L.G. Günther, 1872a. Proc. Zool. Soc. London 40(1): 594–595, pl. 39, fig. a.

Type: Holotype, BMNH 1946.1.3.62 (formerly BMNH 1872.2.19.77), a 279 mm female (A.H. Everett, 1869–1872).

Type locality: "Matang in the district of Sarawak, Borneo" [= Kampong Matang, First Division, Kidi Distr., W Sarawak, East Malaysia, 1°35'N, 110°14'E, elevation 15 m].

Distribution: East Malaysia (Sarawak), 15–500 m.

29. *Calamaria griswoldi* Loveridge, 1938a. Proc. Biol. Soc. Washington 51: 43–44. (*Calamaria lumbricoidea griswoldi*)

Type: Holotype, MCZ 43580, a 505 mm female (J.A. Griswold, Jr., 14 July 1937).

Type locality: "Luidan River, near Bundutuan, Mount Kinabalu, British North Borneo, at an altitude circa 3,340 feet" [= Luidan River near Bundu Tuhan, Sabah, East Malaysia, 6°00'N, 116°32'E, elevation 1500 m].

Distribution: East Malaysia (Sabah, Sarawak) and Brunei, 1020–1800 m.

Sources: Malkmus, 1987, 1989, Stuebing & Inger, 1999, Malkmus et al., 2002 and I. Das, 2007b.

30. *Calamaria hilleniusi* Inger & Marx, 1965. Fieldiana: Zool. 49: 96–98, fig. 25.

Type: Holotype, ZMA 10078, a 317 mm male (H.A. Lorentz, 24 May 1909).

Type locality: "Samarinda, Indonesian Borneo" [= Samarinda, E Kalimantan, cen. Indonesia, 0°29'S, 117°09'E, elevation NSL].

Distribution: Borneo. East Malaysia (Sabah, Sarawak) and Indonesia (E Kalimantan), NSL.

31. *Calamaria ingeri* Grismer, Kaiser & Yaakob, 2004. Hamadryad 28(1–2): 1–2, figs. 3–4.

Type: Holotype, FMNH 262246, a 177 mm female (G. Grossmann & F. Tillack, 23 July 2002).

Type locality: "98 m above sea level in lowland dipterocarp forest on the west side of Pulau Tioman along the Tekek-Juara trail, 1.9 km east of Kampung Tekek, Pahang, West Malaysia."

Distribution: West Malaysia (Pahang: Tioman Is.), 100 m.

Sources: Grismer & Khang Aun, 2008 and Grismer, 2011.

32. *Calamaria javanica* Boulenger, 1891a. Ann. Mag. Nat. Hist. (6) 7(39): 279–280.

Type: Holotype, BMNH 1946.1.3.41 (formerly BMNH 1873.5.3.21), a 185 mm male (J.C. Ploem).

Type locality: "Java" [SW Indonesia]

Distribution: Western Indonesia (Belitung, Java).

C

33. *Calamaria joloensis* E.H. Taylor, 1922b. Philippine J. Sci. 21D(2): 203–204, pl. 7, figs. 2–3.

Type: Holotype, CAS 60901 (formerly EHT 1855), a 150 mm specimen (E.H. Taylor, 30 Oct. 1920).

Type locality: "central Jolo, Philippine Islands" [= Jolo Island, Sulu Archipelago, SW Philippines].

Distribution: Southwestern Philippines (Sulu Arch.: Jolo). Known only from type locality.

Source: Leviton, 1959b.

34. *Calamaria lateralis* Mocquard, 1890a. Naturaliste (2) 12(80): 154.

Type: Holotype, MNHN 1889.194, a 245 mm male (J. Whitehead, 1885).

Type locality: "Kina Balu, Bornéo" [= Mt. Kinabalu, Sabah, East Malaysia]. Emended to North Borneo, Mt. Kina Balu, between 1,000 and 4,000 or even 6,000 feet *fide* Mocquard (1890b: 118), [= Mt. Kinabalu, N Sabah, NE East Malaysia, 6°05'N, 116°30'E, elevation 300–1800 m].

Distribution: East Malaysia (N Sabah) and W Indonesia (cen. Java).

Source: Malkmus et al., 2002.

Remarks: Supplemental original description in Mocquard (1890b: 136–137, pl. 8, figs. 4a–c). Type listed as MNHN C231789.194 *fide* Inger & Marx (1965: 146).

35. *Calamaria lautensis* Rooij, 1917. Rept. Indo-Aust. Arch. 2: 163–164, fig. 66.

Type: Lectotype, RMNH 4716A, designated by Inger & Marx (1965: 70).

Type locality: "Pulo si Laut, Cocos Islands" via lectotype selection [= Pulau Kokos, 40 km NW of Simeulue, W Indonesia, 2°45'N, 95°50'E].

Distribution: Extreme W Indonesia (Kokos Is.).

Remarks: Lectotype designation of ZMA 10111 (Pulu si Laut, E.R. Jacobson, 21 Aug. 1913) *fide* Daan & Hillenius (1966: 137) appears to be invalid by Inger & Marx's earlier selection. However, Haas (1950: 569) declared the type to be in ZMA.

36. *Calamaria leucogaster* Bleeker, 1860b. Nat. Tijds. Nederl. Indië (5) 21(1): 293.

Synonyms: *Calamaria arcticeps* A.C.L.G. Günther, 1866, *Calamaria beccarii* W.C.H. Peters, 1872d, *Calamaria brookii* Boulenger, 1895, and *Calamaria smithi* Dunn, 1923a.

Types: Syntypes (2), BMNH 1946.1.3.76 (formerly BMNH 1863.12.11.141), a 140 mm specimen, and RMNH 3994, a 225 mm specimen (P. de Bleeker, 1842–1860).

Type locality: "Sumatra (Ampat-lawang)" [= Pendopo or Tanjungraya region, Sumatera Selatan Prov., SE Sumatra, W Indonesia, 2°19'S, 105°12'E, elevation 400 m *fide* David & Vogel, 1996: 237, 245].

Distribution: Borneo. East Malaysia (Sabah, Sarawak), Brunei and Indonesia (? Java, cen. Kalimantan, Sumatra), 100–1000 m.

Sources: David & Vogel, 1996 and Malkmus et al., 2002.

Remarks: MNHN catalogue lists MNHN 1889.194 as a type.

37. *Calamaria linnaei* H. Boie *in* F. Boie, 1827. Isis von Oken 20(6): 539–540.

Synonyms: *Coluber symmetricus* Lacépède, 1789 (*nomen rejiciendum*), *Calamaria maculosa* H. Boie *in* Schlegel, 1826a *(nomen nudum), Calamaria multipunctata* H. Boie *in* Schlegel, 1826a *(nomen nudum), Calamaria reticulata* H. Boie *in* Schlegel, 1826a *(nomen nudum), Calamaria tesselata* H. Boie *in* Schlegel, 1826a *(nomen nudum), Calamaria maculosa* Schlegel, 1826b *(nomen nudum), Calamaria multipunctata* Schlegel, 1826b *(nomen nudum), Calamaria reticulata* Schlegel, 1826b *(nomen nudum), Calamaria tesselata* Schlegel, 1826b *(nomen nudum), Calamaria maculosa* Reinwardt *in* F. Boie, 1827, *Calamaria multipunctata* Reinwardt *in* F. Boie, 1827, *Calamaria reticulata* Reinwardt *in* F. Boie, 1827, *Calamaria versicolor* Ranzani, 1837, *Calamaria hybida* Fitzinger, 1861, *Calamaria linnaei bilineata* Jan, 1862b (*nomen nudum*), *Calamaria linnaei contaminata* Jan, 1862b *(nomen nudum), Calamaria versicolor rhomboidea* Jan, 1862b *(nomen nudum), Calamaria linnaei transversalis* Jan, 1862b *(nomen nudum), Calamaria linnaei tessellata* Jan, 1862b *(nomen nudum), Calamaria rhomboidea interjecta* Jan, 1863b *(nomen nudum), Calamaria linnaei bilineata* Jan & Sordelli, 1865 *in* 1860–66 *Calamaria linaei contaminate* Jan & Sordelli, 1865 *in* 1860–66, *Calamaria versicolor rhomboidea* Jan & Sordelli, 1865 *in* 1860–1866, *Calamaria linnaei transversalis* Jan & Sordelli, 1865 *in* 1860–1866, *Calamaria brevis* Boulenger, 1894a, *Calamaria sondaica* T. Barbour, 1908, *Calamaria linnaei multilineata* F. Werner, 1909c, *Calamaria borneensis ventrimaculata* Holtzinger-Tenerer, 1916, and *Calamaria margaritophora gastropicta* Holtzinger-Tenerer, 1920.

Type: Holotype, RMNH 27, a 243 mm male (H. Kuhl & J.C. van Hasselt, Dec. 1820–Sept. 1823).

Type locality: "Java" [SW Indonesia] Restricted to region of Mt. Pangerango and Mt. Salak, West Java, Indonesia *fide* Brongersma (1948b: 14). Emended to Tjihandjawar, at the foot of Mt. Pangerango, W. Java [= Cihanyawar, Nagrak, Jawa Barat, W Java, SW Indonesia, 07°03'S, 108°40'E] *fide* Brongersma (1950: 1499).

Distribution: Western Indonesia (Bangka, Java), 600–1500 m.

Sources: Duméril, Bibron & Duméril, 1854c, C. Haas, 1941 and Bergman, 1965.s

Remarks: Original description based on H. Boie's MS (1823–1825). Official Specific Name *fide* Opinion 2196 (ICZN, 2008a).

38. *Calamaria longirostris* S.D. Howard & Gillespie, 2007. J. Herp. 41(2): 240–241, figs. 4–5.

Type: Holotype, MZB 3127, an adult female (G.R. Gillespie, S. Howard, D. Lockie, M. Scroggie & Boeadi, June–Sept. 2000–2002).
Type locality: "the Lambusango Reserve at 5°12'59"S, 122°52'10"E, at 400m elevation, Sulawesi, Indonesia."
Distribution: Central Indonesia (Sulawesi), 400 m.
Sources: G.R. Gillespie et al., 2005 and Koch, 2012.

39. *Calamaria lovii* Boulenger, 1887a. Ann. Mag. Nat. Hist. (5) 19(111): 169, 3 figs.

Synonyms: *Calamaria lowi* Mocquard, 1890 *(nomen emendatum), Calamaria ventralis* Cochran, 1923, *Calamaria javanica lineata* Brongersma, 1928, *Calamaria lowi wermuthi* Inger & Marx, 1965, *Calamaria lowi ingermarxi* Darevsky & Orlov, 1992 *(nomen incorrigendum),* and *Calamaria lowi ingermarxorum* Darevsky & Orlov, 1992 *(nomen corrigendum).*
Type: Holotype, BMNH 1946.1.3.63 (formerly BMNH 1887.2.7.18), a 265 mm male (H.B. Low, 1869–1887).
Type locality: "Rejang River, Sarawak, Borneo" [= East Malaysia].
Distribution: Southeast Asia and Indonesia. Southern Thailand, S Vietnam (Gia Lai), West Malaysia, East Malaysia (Sarawak), Brunei and Indonesia (Java, E Kalimantan), NSL–750 m.
Sources: Leong, 2004, Ziegler & Le, 2005, I. Das, 2007b, 2012, Ziegler et al., 2008, V.S.Nguyen et al., 2009, Orlov et al., 2010 and Grismer, 2011.
Remarks: Original spelling as *C. lovii* is evidently (but not explicitly) named after the collector H.B. Low. The use of "v" instead of "w" is due to it's latinization (i.e. "w = v" in Latin).

40. *Calamaria lumbricoidea* H. Boie *in* F. Boie, 1827. Isis von Oken 20(6): 540.

Synonyms: *Calamaria lumbricoidea* H. Boie *in* Schlegel, 1826a *(nomen nudum), Calamaria lumbricoidea* Schlegel, 1826b *(nomen nudum), Calamaria temmincki* A.-M.-C. Duméril, Bibron & Duméril, 1854a, *Calamaria vermiformis* A.-M.-C. Duméril, Bibron & Duméril, 1854a, *Calamaria grayi* A.C.L.G. Günther, 1858, *Calamaria alkeni* Bleeker, 1860b, *Calamaria dimidiata* Bleeker, 1860b, *Calamaria melanorhynchos* Bleeker, 1860b, *Calamaria flaviceps* A.C.L.G. Günther, 1865, *Calamaria philippinica* Steindachner, 1867d, *Calamaria stahlknechtii* Stoliczka, 1873, *Calamaria variabilis* Lidth de Jeude *in* Weber, 1890, *Calamaria vermiformis sumatranus* Lidth de Jeude *in* Weber, 1890, *Calamaria bungaroides* F. Werner, 1901d, *Calamaria göringi* T. Vogt, 1925, and *Calamaria bruegeli* Mertens, 1924.
Type: Lectotype, RMNH 10543 (formerly RMNH 42) (H. Kuhl & J.C. van Hasselt, Dec. 1820–Sept. 1823), designated by Inger & Marx (1965: 77).
Type locality: "Java" [SW Indonesia] Restricted to region of Mt. Pangerango and Mt. Salak, West Java, Indonesia *fide* Brongersma (1948b: 14). Emended to Tjihandjawar, at the foot of Mt. Pangerango, W. Java [= Cihanyawar, Nagrak, Jawa Barat, W Java, SW Indonesia, 07°03'S, 108°40'E] *fide* Brongersma (1950: 1499).
Distribution: Malay Peninsula and East Indies. Southern Thailand (Nakhon Si Thammarat, Narathiwat, Pattani), West Malaysia (Johor, Pahang, Perak, Perlis, Selangor, Tioman Is.), Singapore, East Malaysia (Sabah, Sarawak), Brunei, Indonesia (Great Natuna, W Java, ext. E Kalimantan, Mentawai Arch., Natuna Besar, Nias, Sumatra) and Philippines (Basilan, Bohol, Dinagat, Leyte, Mindanao, Negros, Samar), 200–1675 m.
Sources: T. Vogt, 1925, C. Haas, 1941, Leviton, 1959b, Inger & Marx, 1962, Ross & Lazell, 1991, Stuebing, 1994a, David & Vogel, 1996, M.J. Cox et al., 1998, Malkmus et al., 2002, Orlov et al., 2010, Grismer, 2011 and I. Das, 2012.
Remarks: Original description based on H. Boie's MS (1823–1825).

41. *Calamaria lumholtzii* Andersson, 1923. Nyt Mag. Naturk. 61(7): 123–124.

Synonym: *Calamaria raveni* Cochran, 1923.
Types: Syntypes (2), ZMUO 2010 (formerly ZMK), 209 and 101 mm females (C. Lumholtz, Sept. 1915).
Type locality: "Tumbang Maruwei, Central-Borneo" [= Kalimantan, Indonesia].
Distribution: Central Indonesia (cen. Kalimantan).

42. *Calamaria margaritophora* Bleeker, 1860b. Nat. Tijds. Nederl. Indië (5) 21(1): 285, 94–295.

Synonyms: *Calamaria maculolineata* W.C.H. Peters, 1863d, *Calamaria hoevenii* Edeling, 1870, and *Calamaria margaritifera* Boulenger, 1894a *(nomen emendatum).*
Types: Syntypes (2), BMNH 1946.1.2.9 (formerly BMNH 1863.12.11.146), a 284 mm female, and RMNH 3975A, a 297 mm specimen (H.J. Alken, Nov. 1857).
Type locality: "Java, Sumatra (Ampat-lawang)." Restricted to Ampat Lawang, Sumatra *fide* Inger & Marx (1965: 167) [= Pendopo or Tanjungraya region, Sumatera Selatan Prov., SE Sumatra, W Indonesia, 2°19'S, 105°12'E, elevation 400 m *fide* David & Vogel, 1996: 237, 245].
Distribution: Western Indonesia (SE Sumatra), 945 m.
Source: David & Vogel, 1996.
Remarks: MNHN catalogue lists MNHN 1912.55 as a type. Records from Java incorrect *fide* Inger & Marx (1965: 170). Javan type locality possibly in error as

Bleeker (1860b: 285) only lists *C. margaritophorus* from Sumatra (but in description adds Java as habitat).

43. *Calamaria mecheli* Schenkel, 1901. Verh. Naturf. Ges. Basel (1901–1902) 13(1): 165–166, figs. 4–4b.

Type: Lectotype, NMBA 1699, a 210 mm female (A. von Mechel, 1895–1901), designated by Kramer (1978: 659).
Type locality: "Indragiri, Sumatra" [= Indragiri River region, Riau Prov., E Sumatra, W Indonesia, ca. 0°22'S, 103°26'E *fide* David & Vogel, 1996: 239] via lectotype selection.
Distribution: Western Indonesia (W Sumatra).
Source: David & Vogel, 1996.

44. *Calamaria melanota* Jan, 1865 *in* Jan & Sordelli, 1860–1866. Icon. Gén. Ophid. 1(10): 6, pl. 1, figs. 5, a–b, d, g, p, r, v. (*Calamaria linnaei melanota*)

Synonyms: *Calamaria linnaei gastrogramma* Jan, 1862b (*nomen nudum*), *Calamaria linnaei melanota* Jan, 1862b (*nomen nudum*), *Calamaria benjaminsii* Edeling, 1864a, and *Calamaria electa* T. Barbour, 1927.
Type: Holotype, RMNH 37, a 168–260 mm specimen (C.A.L. Schwaner, 1845).
Type locality: "Borneo" [= Kalimantan, Indonesia].
Distribution: Borneo. Southwestern East Malaysia (Sarawak) and Indonesia (S Kalimantan).

45. *Calamaria modesta* A.-M.-C. Duméril, Bibron & Duméril, 1854a. Erpét. Gén. 7(1): 74–76.

Synonyms: *Calamaria monochrous* Bleeker, 1860b, *Calamaria elegans* Rooij, 1917, *Calamaria Simeulueensis* Rooij, 1817, *Calamaria bogorensis* Lidth de Jeude *in* Weber, 1890, and ? *Calamaria mjoebergi* Lönnberg & Rendahl, 1925.
Type: Holotype, MNHN 3299, a 157 mm female.
Type locality: "ile de Java" [W Indonesia].
Distribution: Greater Sundas. East Malaysia (Sabah) and W Indonesia (W Java, Simeulue, cen. Sumatra), 900–1430 m.
Sources: David & Vogel, 1996 and Malkmus et al., 2002.

46. *Calamaria muelleri* Boulenger, 1896c. Ann. Mag. Nat. Hist. (6) 17(101): 394–395.

Type: Lectotype, NMBA 1690, a 190 mm female (P. Sarasin & F. Sarasin, 1893–1896), designated by Inger & Marx (1965: 99).
Type locality: "Loka, Bonthain Peak, about 3500 feet, Celebes" [= Loka, Mt. Lompobatang, South Sulawesi Prov., Sulawesi, Indonesia, 5°19'S, 119°54'E, elevation 1200 m]
Distribution: Central Indonesia (cen. and ext. SW Sulawesi), 1200–1600 m.
Sources: Bosch, 1985, Lang & Vogel, 2005 and S.D. Howard & Gillespie, 2007.
Remarks: Boulenger's (1896c: 395) type length of 1900 mm is an obvious typo.

47. *Calamaria nuchalis* Boulenger, 1896d. Ann. Mag. Nat. Hist. (6) 18(103): 62.

Type: Holotype, BMNH 1946.1.3.73 (formerly BMNH 1896.4.29.34), a 180 mm male (A.H. Everett, 1896).
Type locality: "S. Celebes" [= Sulawesi, Indonesia].
Distribution: Central Indonesia (cen. and SW Sulawesi).
Sources: Bosch, 1985, Lang & Vogel, 2005 and S.D. Howard & Gillespie, 2007.

48. *Calamaria palavanensis* Inger & Marx, 1965. Fieldiana: Zool. 49: 134–136, fig. 35.

Type: Holotype, CAS 62151, a 289–315 mm male (E.H. Taylor, 1912–1920).
Type locality: "Palawan Island" [Sulu Arch., SW Philippines].
Distribution: Southwestern Philippines (S Palawan).
Source: Leviton, 1959b.

49. *Calamaria pavimentata* A.-M.-C. Duméril, Bibron & Duméril, 1854a. Erpét. Gén. 7(1): 71–73.

Synonyms: *Calamaria quadrimaculata* A.-M.-C. Duméril, Bibron & Duméril, 1854a, *Calamaria siamensis* A.C.L.G. Günther, 1864a, *Calamaria berezowskii* A.C.L.G. Günther, 1896a, *Calamaria pavimentata semidoliata* F. Werner, 1896b, *Calamaria pfefferi* Stejneger, 1901d, *Calamaria pavimentata uniformis* M.A. Smith, 1921a, *Calamaria pavimentata formosana* Maki, 1931, *Calamaria pavimentata banaensis* Bourret, 1934c, *Calamaria pavimentata annamensis* Bourret, 1937a, and *Calamaria pavimentata miyarai* Takara, 1962.
Type: Holotype, MNHN 3298, a 207 mm male.
Type locality: "Java" [W Indonesia] (in error *fide* Inger & Marx, 1965: 212).
Distribution: Southeastern Asia. Southern Myanmar (Bago, Chin, Kachin), Thailand (Lampang, Nakhon Ratchasima, Surat Thani), SE China (Guangdong, Guangxi, Hainan, W Sichuan, Zhejiang, Lanyu Is.), Taiwan, S Japan (Ryukyus: Miyakojima, Okinawajima and Yonagunijima Is.), Laos (Champasak, Xiangkhouang), Vietnam (Cao Bang, Da Nang, Gao Bang, Hai Duong, Ha Tinh, Hoa Binh, Lam Dong, Lao Cai, Nghe An, Quang Binh, Quang Nam, Quang Ninh, Quang Tri, Son Lai, Vinh Phuc, Yen Bai, Cu Lao Ba Mun Is.), Cambodia, and West Malaysia (Pahang, Perak, Pinang, Selangor, Jarak, Pinang and Tioman Is.), 245–2000 m.
Sources: C.H. Pope, 1935, M.A. Smith, 1943, Takara, 1962, E.H. Taylor, 1965, M. Mori, 1984, Ota, 1982, Tweedie, 1983, Gu et al., 1987, Zhao & Adler, 1993, Darevsky, 1999, Lazell, 1999, Orlov et al., 2000, Ziegler & Le, 2005, Zhao, 2006, Wogan et al., 2008, Ziegler et

al., 2008, V.S. Nguyen et al., 2009, Xiang & Li, 2009, Orlov et al., 2010 and Grismer, 2011.

Remarks: MNHN catalogue lists MNHN 1948.92 from Vietnam as holotype. Possibly occurs in Hong Kong, China *fide* Karsen et al. (1998: 162).

50. *Calamaria prakkei* Lidth de Jeude, 1893. Notes Leyden Mus. 15: 252–254.

Type: Lectotype, RMNH 4360, a 172–245 mm male (J.C. Prakke, 1887–1893), designated by Inger & Marx (1965: 121).

Type locality: "North Borneo: Sandakan District: Sandakan Bay" [= Sandakan Bay vicinity E Sabah, East Malaysia, 5°51'N, 118°03'E] via lectotype selection.

Distribution: Singapore and N East Malaysia (N Sabah), 15 m.

51. *Calamaria rebentischi* Bleeker, 1860b. Nat. Tijds. Nederl. Indië (5) 21(1): 293.

Type: Holotype, BMNH 1946.1.4.68 (formerly BMNH 1863.12.4.27), a 270–280 mm male (P. de Bleeker, 1842–1860).

Type locality: "Borneo (Sinkawang)" [= Singkawang, W Kalimantan, Indonesia, 0°55'N, 108°59'E, elevation NSL].

Distribution: Central Indonesia (ext. W Kalimantan). Known only from type locality.

Source: S.D. Howard & Gillespie, 2007.

52. *Calamaria sangi* Q.T. Nguyen, Koch & Ziegler, 2010a. Hamadryad (2009) 34(1): 2–3, figs. 1–6.

Type: Holotype, IEBR 360, a 373 mm female (P.R. Nguyen & A. Kuznetsov, 12 Jan. 2001).

Type locality: "Mang Canh Commune (14°41.950" N, 108°14'642" E), Kon Plong District, Kon Tum Prov., Vietnam, 1.200 m asl."

Distribution: Southern Vietnam (Kon Tum), 1200 m.

Source: Orlov et al., 2010.

53. *Calamaria schlegeli* A.-M.-C. Duméril, Bibron & Duméril, 1854a. Erpét. Gén. 7(1): 81–83.

Synonyms: *Calamaria leucocephala* A.-M.-C. Duméril, Bibron & Duméril, 1854a, *Calamaria agamensis* Bleeker, 1859e *(nomen nudum)*, *Calamaria agamensis* Bleeker, 1860b, *Calamaria dumerili* Bleeker, 1860b, *Calamaria roelandti* Bleeker, 1860b, *Calamaria sinkawangensis* Bleeker, 1860b, *Calamaria cuvieri* Jan, 1862b, *Calamaria nigroalba* A.C.L.G. Günther, 1864a, *Calamaria martapoerensis* Edeling, 1864a *(nomen incorrigendum)*, *Calamaria iris* Boettger, 1873, *Calamaria martapurensis* Edeling, 1865 *(nomen corrigendum)*, and *Keiometopon booliati* E.H. Taylor, 1962.

Type: Holotype, RMNH 10424 (formerly RMNH 68), a 250 mm specimen (S. Müller, 1836–1837).

Type locality: "ile de Bornéo" [= Kalimantan, Indonesia].

Distribution: Peninsular Thailand (Pattani), West Malaysia (Pahang, Perak, Pinang, Selangor, Penang Is.), Singapore, N East Malaysia (Sabah, Sarawak) and Indonesia (Bali, Bangka, Belitung, Java, W and E Kalimantan, Sumatra), NSL–1600 m.

Sources: K.F.L. Lim & Lee, 1989, Stuebing, 1994a, David & Vogel, 1996, S.D. Howard & Gillespie, 2007 and Orlov et al., 2010.

C

54. *Calamaria schmidti* Marx & Inger, 1955. Fieldiana: Zool. 37: 197–198, fig. 27.

Type: Holotype, USNM 130240, a 253 mm female (D.H. Johnson, 1 Aug. 1951).

Type locality: "Bundu Tuhan, Mount Kina Balu, North Borneo, at 4,500 ft" [= Bundu Tuhan, Mt. Kinabalu, Sabah, East Malaysia, 6°00'N, 116°32'E, elevation 1370 m].

Distribution: East Malaysia (Sabah), 1370–1570 m.

Sources: Malkmus, 1989, 1994 and I. Das, 2012.

55. *Calamaria septentrionalis* Boulenger, 1890c. Proc. Zool. Soc. London 58(1): 34.

Types: Syntypes (7), BMNH 1947.3.6.60–61 (formerly BMNH 1891.1.28.1–2) [Hong Kong], 1946.1.3.72 [Hong Kong], 1946.1.7.28–31 [Kiu Kiang], longest syntypes 347 mm and 285 mm females (A.E. Pratt [Kiu Kiang], 1887–1890, and H.M.S. Challenger Voy. [stn. 206], 16 Nov. 1874–6 Jan. 1875 [Hong Kong]).

Type locality: "Kiu Kiang and Hong Kong" [= Wu-suih, about three days' journey or 9 mi. S of the American Central China Mission in Kiu-kiang *fide* Pratt (1892: 3–5) = Wusüeh, SE Hubei Prov., cen. China, 29°50'N, 115°41'E *fide* Pope (1935: 502) and Hong Kong, Hong Kong Prov., SE China *fide* Günther, 1888: 169].

Distribution: Eastern China (Anhui, Fujien, Guangdong, N Hainan, Hong Kong, Hubei, SW Hunan, Jiangsu, Jiangxi, Sichuan, Zhejiang, Chu Shan Is.) and N Vietnam (Cao Bang, Ha Giang, Ha Tay, Ha Tinh, Hoa Binh, Lang Son, Lao Cai, Nghe An, Ninh Binh, Quang Binh, Quang Ninh, Thai Nguyen, Thanh Hoa, Vinh Phuc), 50–1500 m.

Sources: C.H. Pope, 1935, Lazell & Liao, 1986, Zhao & Adler, 1993, Zou & Chen, 1998, Orlov et al., 2000, Ziegler & Le, 2005, Zhao, 2006, Ziegler et al., 2008, V.S. Nguyen et al., 2009 and Orlov et al., 2010.

Remarks: A.C.L.G. Günther (1888: 169) listed four types from Kiu Kiang and one juvenile from Hong Kong. Boulenger (1894a: 349) listed seven types, six from Kiu Kiang (3 males, 3 females) and one juvenile from Hong Kong. Inger & Marx (1965: 218) and V.S. Nguyen et al. (2009: 291) listed two syntypes (BMNH 1947.3.6.60–61) from Kiu Kiang.

C

56. *Calamaria suluensis* E.H. Taylor, 1922a. Snakes Phil. Is.: 189–190.

Synonym: *Calamaria pendleburyi* M.A. Smith, 1931b.
Type: Neotype, FMNH 76294, a 287 mm female (R.F. Inger, 22 May 1956), designated by Inger & Marx (1965: 123).
Type locality: "Deramakot, North Borneo" via lectotype selection [= Sandakan Bay vicinity (5°51'N, 118°03'E), E Sabah, East Malaysia].
Distribution: East Indies. East Malaysia (Sabah, Sarawak), Indonesia (E Kalimantan) and SW Philippines (Cagayan Sulu), 915–1430 m.
Sources: Leviton, 1959b and Malkmus et al., 2002.
Remarks: BSM holotype destroyed January 1945 during World War II; neotype locality is 180 km from Taylor's type locality of Cagayan Sula, Philippines *fide* Inger & Marx (1965: 123).

57. *Calamaria sumatrana* Edeling, 1870. Nat. Tijds. Ned. Indië (7) 31(1): 379–380.

Type: Neotype, ZMA 10237, a 227 mm female (L.P. de Bussy, 1908–1909), designated by Inger & Marx (1965: 181).
Type locality: "Deli, Medan, Sumatra" [= Delitua, Medan, Sumatera Utara Prov., N Sumatra, W Indonesia, 3°30'N, 98°41'E, elevation 20 m *fide* David & Vogel, 1996: 239, 245] via neotype selection.
Distribution: Western Indonesia (cen. and N Sumatra).
Source: David & Vogel, 1996.
Remarks: Syntypes lost *fide* Inger & Marx (1965: 181).

58. *Calamaria thanhi* Ziegler & Le, 2005. Zootaxa (1042): 29–36, figs. 1–6.

Type: Holotype, ZFMK 82920, a 455 mm female (native, 16 June 2003).
Type locality: "adjacent to Phong Nha-Ke Bang National Park, Dan Hoa Commune, Minh Hoa District, Quang Binh Prov., Vietnam."
Distribution: Central Vietnam (Quang Binh).
Sources: Ziegler & Le, 2006, 2007, Ziegler et al., 2008, V.S. Nguyen et al., 2009 and Orlov et al., 2010.

59. *Calamaria ulmeri* Sackett, 1940a. Not. Nat. (41): 2.

Type: Holotype, ANSP 21535, a 307+ mm female (F.A. Ulmer, 11 April 1939).
Type locality: "about 40 kilometers northwest of Blangkedjeren, Atjeh Prov., Sumatra (elevation 6800 ft.)" [= 40 km NW Blangkedjeren, Aceh Prov., Sumatra, W Indonesia, 3°59'N, 97°20'E, elevation 2070 m *fide* David & Vogel, 1996: 238].
Distribution: Western Indonesia (NW Sumatra), 2070–2080 m. Known only from type locality.
Source: David & Vogel, 1996.

60. *Calamaria virgulata* H. Boie *in* F. Boie, 1827. Isis von Oken 20(6): 540.

Synonyms: *Calamaria virgulata* H. Boie *in* Schlegel, 1826a *(nomen nudum)*, *Calamaria virgulata* Schlegel, 1826b *(nomen nudum)*, *Calamaria brachyura* Boulenger, 1895f, *Calamaria anceps* F. Werner, 1896b, *Calamaria gracilis* Boulenger, 1896d, *Calamaria collaris* Boulenger, 1897g, *Calamaria mearnsi* Stejneger, 1907b, *Calamaria albopunctata* T. Barbour, 1908, *Calamaria quinquetaeniata* Despax, 1912, *Calamaria egregia* T. Barbour, 1927, and *Calamaria zamboangensis* Leviton, 1952.
Type: Holotype, RMNH 39 (H. Kuhl & J.C. van Hasselt, Dec. 1820–Sept. 1823).
Type locality: "Java" [SW Indonesia]. Restricted to region of Mt. Pangerango and Mt. Salak, West Java, Indonesia *fide* Brongersma (1948b: 14). Emended to Tjihandjawar, at the foot of Mt. Pangerango, W. Java [= Cihanyawar, Nagrak, Jawa Barat, W Java, SW Indonesia, 07°03'S, 108°40'E] *fide* Brongersma (1950: 1499).
Distribution: East Indies. East Malaysia (Sabah), Indonesia (ext. W Java, N Kalimantan, Nias, Nunukan, Riau Arch., Sulawesi, cen. Sumatra) and S Philippines (Mindanao, Palawan, Sulu Arch.) 30–1800 m.
Sources: Leviton, 1952, 1959b, David & Vogel, 1996, Malkmus et al., 2002, Grismer et al., 2004, Lang & Vogel, 2005 and S.D. Howard & Gillespie, 2007 .
Remarks: Original description based on H. Boie's MS (1823–1825). Photographs of holotype in Inger and Marx (1965: fig. 52).

61. *Calamaria yunnanensis* Chernov, 1962. Zool. Inst. Akad. Sci. SSR Trudy 30: 383, 1 fig.

Type: Holotype, ASIZB, a 245 mm male (A.K. Zagulayev, 31 May 1956).
Type locality: "TSindhun (Ching tung district), Yunnan Prov., China, 1100–1200 m."
Distribution: Southwestern China (Yunnan) and N Laos (Phongsali), 1100–1200 m.
Sources: Zhao & Adler, 1993, Ziegler & Le, 2005, B.L. Stuart & Heatwole, 2008, Ziegler et al., 2008 and Orlov et al., 2010.
Remarks: Possibily a synonym of *C. pavimentata* or *C. septentrionalis fide* Inger & Marx (1965: 239).

CALAMODONTOPHIS Amaral, 1963 (Xenodontidae)

Synonyms: *Calamodon* Amaral, 1935a *(nomen praeoccupatum)* and *Calamadontophis* – Obst, Richter & Jacob, 1984 *(nomen incorrectum)*.
Type species: *Calamodon paucidens* Amaral, 1935a.
Distribution: Southern Brazil.
Sources: Amaral, 1937a, 1977, Bailey, 1966, Pagini & Lema, 1987 and Zaher et al., 2009.

1. *Calamodontophis paucidens* (Amaral, 1935a). Mem. Inst. Butantan 9: 204, fig. 1. (*Calamodon paucidens*)

Type: Holotype, IB 8847, a 360 mm male (O.M. Freitas, 19 Nov. 1934), destroyed by fire 15 May 2010.
Type locality: "S. Simão, Rio Grande do Sul, Brasil."
Distribution: Southern Brazil (Rio Grande do Sul) and Uruguay (Rocha, Treinta y Tres), 40 m.
Sources: Franco et al., 2001, 2006, Carreira-Vidal & Lombardo, 2008 and Carreira-Vidal et al., 2012b.
Remarks: Gender of type female *fide* Amaral (1935a: 204).

2. *Calamodontophis ronaldoi* F.L. Franco, Carvalho Cintra & Lema, 2006. So. Amer. J. Herp. 1(3): 219–221, figs. 1–3.

Synonym: *Calamodontophis dorsolineatus* Lema, 2002 (*nomen nudum*).
Type: Holotype, IB 55914 (formerly MHNCI 1726), a 344 mm female (don. April 1987), destroyed by fire 15 May 2010.
Type locality: "municipality of General Carneiro (26°25'S; 51°18'W–983 m a.s.l.), state of Paraná, Brazil.
Distribution: Southern Brazil (Paraná), 825–985 m.

CALAMOPHIS A.B. Meyer, 1874 (Homalopsidae)

Type species: *Calamophis jobiensis* A.B. Meyer, 1874.
Distribution: New Guinea.
Sources: J.C. Murphy, 2012 and J.C. Murphy et al., 2012b.
Remarks: A valid genus *fide* J.C. Murphy et al. (2012b: 511). Publication date listed as 1875 *fide* McDowell (1975: 77).

1. *Calamophis jobiensis* A.B. Meyer, 1874. Mber. Königl. Akad. Wiss. Berlin 1874(2):135.

Type: Holotype, MTKD 1026 (A.B. Meyer, 1873), destroyed 13–14 Feb. 1945 during World War II.
Type locality: "Jobi, Neu-Guinea" [= Yapen Is., Indonesia]. Restricted to Ansus, SW Yapen Is., West Papua Prov., E Indonesia [= 1°46'S, 135°47'E, elevation NSL] *fide* J.C. Murphy (2012: 517).
Distribution: Eastern Indonesia (West Papua: Yapen Is.), NSL. Known only from type locality.
Source: J.C. Murphy et al., 2012b.

2. *Calamophis katesandersae* J.C. Murphy, 2012. Raffles Bull. Zool. 60(2): 518–520, fig. 3a.

Type: Holotype, MSNG 56343.1, a 223 mm female (A.A. Bruijn, Dec. 1875).
Type locality: "Andai, West Papua, Indonesia (~0°54'58"S, 134°00'25"E)."
Distribution: Eastern Indonesia (West Papua), 35 m. Known only from type locality.

3. *Calamophis ruuddelangi* J.C. Murphy, 2012. Raffles Bull. Zool. 60(2): 520, fig. 3b.

Type: Holotype, MNHN 5175, a 261 mm male (M.J.A. Raffray, 1877).
Type locality: "Ambuaki in the Tamrau Mountains (~0°46'S, 132°57'E) West Papua."
Distribution: Eastern Indonesia (West Papua), 550–770 m.

4. *Calamophis sharonbrooksae* J.C. Murphy, 2012. Raffles Bull. Zool. 60(2): 521.

Type: Holotype, MSNG 30193.1, a 315 mm male (A.A. Bruijn, 1875).
Type locality: "Mount Arfak, West Papua, Indonesia (~1°05'00"S, 133°55'00"E)."
Distribution: Eastern Indonesia (West Papua), 1300–1700 m. Known only from type locality.

CALAMORHABDIUM Boettger, 1898 (Calamariidae)

Type species: *Calamorhabdium kuekenthali* Boettger, 1898.
Distribution: Indonesia.
Sources: Rooij, 1917, Sackett, 1940a, C. Haas, 1950 and Zaher et al., 2009.
Remarks: David & Vogel (1996: 57) noted that ANSP 21951 (*C. "kuenkenthali"*) from Sumatra is probably an undescribed species.

1. *Calamorhabdium acuticeps* Ahl, 1933b. Mitt. Zool. Mus. Berlin 19: 579–580.

Type: Holotype, ZMB 34315, a 114 mm female (G. Heinrich, 21 Nov. 1930).
Type locality: "Nord-Celebes (Ile-Ile, 1700 m)" [= Mt. Ileile, N Sulawesi, Indonesia, 1°0.3'N, 121°50'E, elevation 1700 m].
Distribution: Central Indonesia (N Sulawesi), 1700 m. Known only from holotype.
Sources: Lang & Vogel, 2005 and Koch, 2012.
Remarks: Holotype photographs are in Lang & Vogel (2005: figs. 53–54).

2. *Calamorhabdium kuekenthali* Boettger, 1898. Kat. Rept. Senck. Mus. 2: 82.

Type: Lectotype, SMF 19384 (formerly SMF-B 8330 a), a 205 mm female (W.G. Kükenthals, 1893–1894), designated by Mertens (1967a: 91).

Type locality: "Insel Batjan, Molukken" via lectotype selection, [= Bacan Island, Maluku, E Indonesia, ca. 0°34'N, 127°31'E].

Distribution: Indonesia (Bacan, N Sumatra), 500 m.

Remarks: Sackett's (1940a: 3) specimen (ANSP 21951) agrees in all characters except coloration, but anterior third of body is digested down to the skeleton so head scutellation is unknown.

C

CALLIOPHIS Gray, 1835 *in* Gray & Hardwicke, 1830–1835 (Elapidae)

Synonyms: *Maticora* Gray, 1835 *in* Gray & Hardwicke, 1830–1835, *Gongylocormus* Fitzinger, 1843, *Pseudelaps* Fitzinger, 1843, *Calophis* Agassiz, 1847 *(nomen emendatum), Matirora* – Agassiz, 1844 *in* 1844–1845 *(nomen incorrectum), Doliophis* Girard, 1858a, *Helminthoelaps* Jan, 1858, *Callophis* A.C.L.G. Günther, 1859 *(nomen emendatum), Congylocormus* – Westphal-Castelnau, 1870 *(nomen incorrectum), Adeniophis* W.C.H. Peters, 1871b, *Adenophis* – Sclater, 1891 *(nomen incorrectum), Callophus* – E. Bartlett, 1896a *(nomen incorrectum), Dollophis* – Brazil, 1911 *(nomen incorrectum), Aspis* T. Barbour, 1914 *(nomen praeoccupatum), Calliopis* – Takahashi, 1930 *(nomen incorrectum), Calliophils* – Maki, 1933 *(nomen incorrectum),* and *Daliophis* – Briceño-Rossi, 1934 *(nomen incorrectum).*

Type species: *Calliophis gracilis* Gray, 1835 *in* Gray & Hardwicke, 1830–1835.

Distribution: Southeastern Asia and East Indies.

Sources: M.A. Smith, 1943, Klemmer, 1963, Leviton, 1964b, McDowell, 1986, Welch, 1988, Golay et al., 1993, Keogh, 1988, David & Ineich, 1999, Ota et al., 1999, Slowinski & Keogh, 2000, Slowinski et al., 2001, Castoe et al., 2007a, E.N. Smith et al., 2008, 2012, Zaher et al., 2009, I. Das, 2010, 2012 and Hoser, 2012e.

Remarks: *Maticora* a synonym of *Calliophis fide* Slowinski et al. (2001: 238).

1. *Calliophis beddomei* M.A. Smith, 1943. Fauna Brit. India, Rept. Amph. 3: 423. (*Callophis beddomei*)

Type: Holotype, BMNH 1946.1.17.99, a 565 mm female (R.H. Beddome, 1857–1882).

Type locality: "Shevaroy Hills, British India" [= Shevaroy Hills, Tamil Nadu, S India, ca. 11°50'N, 78°13'E].

Distribution: Southern India (Karnataka, Tamil Nadu), 1380 m.

2. *Calliophis bibroni* (Jan, 1858b). Rev. Mag. Zool. (2) 10: 518, 526–527. (*Elaps bibroni*)

Synonym: *Elaps cerasinus* Beddome, 1864.

Type: Holotype, MNHN 5070, a 489 mm specimen.

Type locality: "Indes orientales" [= Southern Asia and East Indies]. Restricted to Western Ghats *fide* Deepak et al. (2010:1).

Distribution: Western Ghats of SW India (S Karnataka, N Kerala, NW Tamil Nadu), NSL–1220 m.

Source: Deepak et al., 2010.

Remarks: Supplemental original description in Jan (1859e: pl. b, fig. no. 6).

3. *Calliophis bivirgatus* (F. Boie, 1827). Isis von Oken 20(6): 556. (*Elaps bivirgatus*)

Synonyms: *Elaps bivirgatus* H. Boie *in* Schlegel, 1826a *(nomen nudum), Elaps bivirgatus* Schlegel, 1826b *(nomen nudum), Elaps flaviceps* Cantor, 1839, *Elaps bivirgatus javanica* Schlegel, 1844, *Elaps tetrataenia* Bleeker, 1859c, *Elaps bivirgatus quadrivirgata* Jan, 1863b (*nomen nudum*), and *Maticora bivirgata* – Kopstein, 1932b.

Type: Holotype, RMNH 1435, a 595 mm male (H. Kuhl, Dec. 1820–Sept. 1821).

Type locality: "Java" [SW Indonesia] Restricted to region of Mt. Pangerango and Mt. Salak, West Java, Indonesia *fide* Brongersma (1948b: 14). Emended to Tjihandjawar, at the foot of Mt. Pangerango, W. Java [= Cihanyawar, Nagrak, Jawa Barat, W Java, SW Indonesia, 07°03'S, 108°40'E] *fide* Brongersma (1950: 1499).

Distribution: Malay Peninsula and Greater Sundas. Peninsular Thailand (Narathiwat, Pattani, Songkha, Trang, Yala), West Malaysia (Johor, Kedah, Kelantan, Melaka, Pahang, Perak, Perlis, Pinang, Selangor, Penang Is.), Singapore, East Malaysia (Sabah, Sarawak), Brunei and W Indonesia (Bangka, Bantan, Bintan, Java, Kalimantan, Mentawi, Nias, Riau Arch., Sumatra), NSL–1375 m.

Sources: Schlegel & Müller, 1844a, Brongersma, 1948b, Wegner, 1954, E.H. Taylor, 1965, Toriba, 1989c, M.J. Cox et al., 1998, Malkmus et al., 2002 and Auliya, 2006.

Remarks: Original description based on H. Boie's MS (1823–1825) but F. Boie credited authorship to H. Kuhl. Unconfirmed reports from Laos, Cambodia, Myanmar and Vietnam.

4. *Calliophis castoe* E.N. Smith, Ogale, Deepak & Giri, 2012. Zootaxa (3437): 53–62, figs. 3c, 4, 6–7.

Type: Holotype, BNHS 3461, a 536 mm male (H. Ogale, 12 Sept. 2009).

Type locality: "Amboli, Sindhudurg district, Maharashtra, India, [ca. 15.958790° N 73.994686° E)."

Distribution: Southwestern India (Goa, NW Karanataka, S Maharashtra), 10–715 m

5. *Calliophis gracilis* Gray, 1835 *in* Gray & Hardwicke, 1830–1835. Illust. Indian Zool. 2(19–20): pl. 86, figs. 1–3.

Synonym: *Elaps nigromaculatus* Cantor, 1839.

Type: Holotype, BMNH 1946.1.16.17 (T. Hardwicke, 1756–1823).

Type locality: "Penang" [= Penang Is., Pinang State, NW Malay Peninsula, 5°22'N, 100°14'E] via lectotype selection.

Distribution: Malay Peninsula and Sumatra. Peninsular Thailand (Pattani), West Malaysia (Johor, Pinang, Selangor, Penang Is.), Singapore and W Indonesia (Sumatra), NSL–900 m.

Sources: E.H. Taylor, 1965, Tweedie, 1983 and David & Vogel, 1996.

Remarks: M.A. Smith (1943: 530) designated pl. 120 in T. Hardwicke's collection a lectotype.

6. *Calliophis haematoetron* E.N. Smith, K. Manamendra-Arachchi & R. Somaweera, 2008. Zootaxa (1847): 21–26, figs. 1, 2 (upper), 3 (upper).

Type: Holotype, WHT 1621, a 371 mm female (M.M. Bahir & S. Nanayakkara, 7 June 1997).

Type locality: "Wasgomuwa National Park, [Central Prov.], Sri Lanka, ca. 90 m (ca. 7.648056° N 80.93583° E)."

Distribution: Sri Lanka (Central), 90 m.

Remarks: Sister group to all other elapids and therefore a separate genus *fide* Pyron et al. (2013: 976).

7. *Calliophis intestinalis* (Laurenti, 1768). Synop. Rept.: 106. (*Aspis intestinalis*)

Synonyms: *Elaps furcatus* Schneider, 1801, *Coluber intestinalis* G. Shaw, 1802, *Vipera furcata* – Daudin, 1803c, *Elaps gracilis* Wagler, 1825 (*nomen ineditum*), *Maticora lineata* Gray, 1835 *in* Gray & Hardwicke, 1830–1835, *Elaps tri-lineatus* A.-M.-C. Duméril, Bibron & Duméril, 1854b *(nomen incorrigendum)*, *Elaps melanototaenia* Bleeker, 1857a (*nomen nudum*), *Elaps thepassi* Bleeker, 1859c, *Callophis intestinalis bifurcatus* W.C.H. Peters, 1862b *(nomen nudum)*, *Elaps furcatus trilineatus* – Jan, 1863b *(nomen corrigendum)*, *Callophis furcatus nigrotaeniatus* W.C.H. Peters, 1863d, *Callophis intestinalis javanica* A.C.L.G. Günther, 1864a, *Callophis intestinalis malayana* A.C.L.G. Günther, 1864a, *Callophis intestinalis philippina* A.C.L.G. Günther, 1864a, *Callophis bilineatus* W.C.H. Peters, 1881e, *Adeniophis malayanus* – A.B. Meyer, 1886, *Adeniophis philippinus* – A.B. Meyer, 1886, *Elaps sumatranus* Lithe de Jeude *in* Weber, 1890, *Callophis intestinalis suluensis* Steindachner, 1891, *Doliophis intestinalis annectens* Boulenger, 1896a, *Doliophis intestinalis everetti* Boulenger, 1896a, *Doliophis intestinalis vertebralis* F. Werner, 1900d, *Calamaria klossi* M.A. Smith, 1926b, *Maticora intestinalis bilineata* – Loveridge, 1944a, *Maticora intestinalis immaculata* Loveridge, 1944a, *Maticora intestinalis nigrotaeniata* – Haas, 1950, *Maticora intestinalis nigrotaenita* – Wegner, 1954 *(nomen incorrectum)*, and *Elaps melanotaenia* Ulber, 1995 *(nomen nudum)*.

Type: Lectotype, specimen described and illustrated by Seba (1735: 4, pl. 2, fig. 7), designated herein.

Type locality: "Africa," (in error) via lectotype selection. Corrected to Java, Indonesia *fide* Leviton (1964a: 529).

Distribution: Southeast Asia and East Indies. Peninsular Thailand (Nakhon Si Thammarat, Krabi, Pattani, Phang Nga, Phatthalung, Yala), Vietnam (Gai Lai, Lang Son), East Malaysia (Sabah, Sarawak), West Malaysia (Johor, Kelantan, Kuala, Lumpur, Melaka, Negeri, Pahang, Perak, Pinang, Selangor, Bakong, Pinang, Tockus, Terengganua and Tioman Is.), Singapore, Indonesia (Bali, Bangka, Belitung, Java, Kalimantan, Nias, Riau Arch., S Sulawesi, Sumatra), Brunei and Philippines (Balabac, Busuanga, Culion, Dinagat, Jolo, Luzon, Mindanao, Palawan, Samar, S Tagalog), NSL–1525 m.

Sources: E.H. Taylor, 1922a, Wegner, 1954, Ross & Lazell, 1991, B.E. Smith, 1993, Chan-ard et al., 1999, Malkmus et al., 2002, Pauwels et al., 2002, V.S. Nguyen et al., 2009 and McKay & Lilley, 2012.

Remarks: M.A. Smith (1943: 530) designated pl. 122 in T. Hardwicke's collection a lectotype of *Maticora lineata* Gray. Type lost *fide* Toriba *in* Golay et al. (1993: 151). Sulawesi records doubtful *fide* Lang & Vogel (2005: 262).

8. *Calliophis maculiceps* (A.C.L.G. Günther, 1858). Cat. Colub. Snakes Brit. Mus.: 232. (*Elaps maculiceps*)

Synonyms: *Elaps atrofrontalis* Sauvage, 1877, *Callophis maculiceps univirgatus* M.A. Smith, 1915d *(nomen praeoccupatum)*, *Callophis hughi* Cochran, 1927, *Calliophis maculiceps punctulatus* Bourret, 1934b, *Calliophis maculiceps michaelis* Deuve, 1961a, *Calliophis maculiceps smithi* Klemmer, 1963 *(nomen substitutum)*, *Calliophis maculiceps malcolmi* E.H. Taylor, 1965 *(nomen substitutum)*, and *Calliophis maculiceps malcolmi* Nutaphand & Tumvipart, 1982 *(nomen substitutum)*.

Type: Lectotype, BMNH 1858.4.20.6, an adult female (Zool. Soc. London), designated by Leviton et al. (2003: 424).

Type locality: "East Indies" via lectotype selection.

Distribution: Southeast Asia. Southern Myanmar (Mon, Tanintharyi, Yangon), Thailand (Chai Nat, Chanthaburi, Chiang Mai, Chon Buri, Chumphon, Kanchanaburi, Krung Thep Mahanakhon, Lop Buri, Nakhon Ratchasima, Nakhon Sawan, Nakhon Si Thammarat, Nong Khai, Lamphu, Pattani, Phetchabun, Phang Nga, Phuket, Prachuap Khiri Khan, Roi Et, Saraburi, Surat Thani, Trang, Uthai Thani, Tao Is.), Laos (Khammouane, Savannakhet, Sedone, Vientiane), Cambodia (Kampong Chhnang), S Vietnam (Ba

C

Ria-Vung Tau, Binh Duong, Dak Lak, Dong Nai, Gia Lai, Ho Chi Minh City, Kien Giang, Tay Ninh), and West Malaysia (Kedah, Perak, Perlis), NSL–1330 m.

Sources: Cochran, 1927, E.H. Taylor, 1965, Campden-Main, 1969a, Deuve, 1970, Tweedie, 1983, M.J. Cox et al., 1998, Leviton et al., 2003, V.S. Nguyen et al., 2009 and I. Das, 2012.

Remarks: Previously in the genus *Maticora*.

9. *Calliophis melanurus* (G. Shaw, 1802). Gen. Zool., Amph. 3(2): 552–553. (*Coluber melanurus*)

Synonyms: *Vipera trimaculata* Daudin 1803c, and *Calliophis melanurus sinhaleyus* Deraniyagala, 1951.

Type: Holotype, BMNH 1946.1.17.86, a 250–280 mm specimen (P. Russell, June 1788).

Type locality: "near Nerva, Bengal, India" [= Nerva, Uttar Pradesh, NE India].

Distribution: India (Andhra Pradesh, Gujarat, Karnataka, Kerala, Maharashtra, Tamil Nadu, Uttar Pradesh, West Bengal) and Sri Lanka (Central, Eastern, North-Central, North-Western, Sabaragamuwa, Southern, Uva), 500 m.

Sources: E.H. Taylor, 1950, Deraniyagala, 1955, P. Silva, 1980, A. Silva, 1990a–b, Whitaker & Captain, 2004, N. Khaire, 2006 and E.N. Smith et al., 2008.

Remarks: Description based on P. Russell (1796: 12–13, pl. 8).

10. *Calliophis nigrescens* A.C.L.G. Günther, 1862b. Ann. Mag. Nat. Hist. (3) 9(50): 131–132. (*Callophis nigrescens*)

Synonyms: *Elaps malabaricus* Jerdon, 1854, *Callophis concinnus* Beddome, 1863a, *Callophis pentalineatus* Beddome, 1871, *Callophis pectolineatus* – F. Müller, 1887 *(nomen incorrectum)*, and *Callophis nigrescens khandallensis* Wall, 1913g.

Types: Syntypes (2), BMNH 1946.1.17.78 (formerly FPM), a 762 mm female, location of other syntype unknown.

Type locality: "British India" [= India].

Distribution: Southwestern India (Gujarat, Karnataka, Kerala, Maharashtra, Tamil Nadu), 900–2100 m.

Sources: Whitaker & Captain, 2004 and N. Khaire, 2006.

Remarks: Bangladesh record doubtful *fide* Whitaker & Captain (2004: 366).

CALLOSELASMA Cope, 1860b (*nomen substitutum*) (Viperidae)

Synonyms: *Tisiphone* Fitzinger, 1843 *(nomen praeoccupatum), Leiolepis* A.-M.-C. Duméril, 1853 *(nomen praeoccupatum), Liolepis* Agassiz, 1847 *(nomen emendatum), Calloselaema* – Hoffmann, 1890 *(nomen incorrectum), Caloselasma* – Canton, 1895 *(nomen incorrectum)*, and *Leiolephis* – W.L. Burger, 1971 *(nomen incorrectum)*.

Type species: *Trigonocephalus rhodostoma* Kuhl, 1824.

Distribution: Southeast Asia and Indonesia.

Sources: Bergman, 1961b, E.H. Taylor, 1965, Chernov, 1957, Gloyd, 1979, Hoge & Romano-Hoge, 1981a, 1983, Chiasson et al., 1989, Maes, 1989, Gloyd & Conant, 1990, Kardong, 1990, Golay et al., 1993, Jintakune & Chanhome, 1995, Daltry et al., 1996, 1997, Chan-ard et al., 1999, Parkinson et al., 1997, David & Ineich, 1999, McDiarmid et al., 1999, Parkinson, 1999, Malhotra & Thorpe, 2000, Ziegler et al., 2001, Gumprecht et al., 2004, B.L. Stuart & Emmett, 2006, Bain et al., 2007, V.S. Nguyen et al., 2009, Geissler et al., 2011, I. Das, 2012 and Hoser, 2012d.

Remarks: Generic status questionable *fide* J.A. Campbell & Whitmore, 1989. A synonym of *Hypnale fide* Kraus et al., 1996. A valid genus *fide* McDiarmid et al., 1999.

1. *Calloselasma rhodostoma* (Kuhl, 1824). Bull. Sci. Nat. Géol., Paris 2: 80. (*Trigonocephalus rhodostoma*)

Synonyms: *Trigonocephalus orophrias* Oppel *in* Kuhl, 1824 *(nomen nudum), Trigonocephalus rhodostoma* H. Boie *in* Schlegel, 1826a *(nomen nudum), Trigonocephalus rhodostoma* Schlegel, 1826b *(nomen nudum), Trigonocephalus praetextatus* Gravenhorst, 1832, *Calloselasma rhodostomus* – Cope, 1860b, *Ancistrodon annamensis* Angel, 1933a, and *Ankystrodon rhodostoma* – Duong, 1962 *(nomen incorrectum)*.

Type: Holotype, RMNH 1510 (C.G.C. Reinwardt, 1816–1821).

Type locality: "Java" [SW Indonesia]

Distribution: Thailand (Bueng Kan, Chaiyaphum, Chanthaburi, Chiang Mai, Chon Buri, Chumphon, Kalasin, Kamphang Phet, Kanchanaburi, Krabi, Lampang, Lop Buri, Mae Hong Son, Nakhon Phanom, Nakhon Ratchasima, Nakhon Si Thammarat, Nan, Narathiwat, Nong Khai, Pattani, Phang Nga, Phathalung, Phetchabun, Phetchaburi, Phitsanulok, Phrae, Phuket, Prachinburi, Prachuap Khiri Khan, Rayong, Sakon Nakhon, Songkhla, Surat Thani, Tak, Trang, Trat, Udon Thani, Yala), Cambodia (Kam Pot, Koh Khang, Kampong, Koh Kong, Mondolkiri), Laos (Khammouan, Savannakhet, Vientiane, Xekong, Xiangkhouang), S Vietnam (An Giang, Ba Ria-Vung Tau, Binh Duong, Binh Phuoc, Binh Thuan, Dong Nai, Gai Lai, Ninh Thuan, Tay Ninh), West Malaysia (Kedah, Perlis) and Indonesia (Java, Kangean, Karimundjawa), NSL–1900 m.

Remarks: Original description based on H. Boie's MS (1823–1825). Possibly occurs in Myanmar *fide* Dowling & Jenner (1988: 10) but Sumatra unlikely *fide* McDiarmid et al. (1999: 273). Authorship discussed by McDiarmid et al. (1999: 273).

CANDOIA Gray, 1842a (Boidae)

Synonyms: *Cenchris* Gray *in* Griffith & Pidgeon, 1831 *(nomen substitutum), Cenchrus* Swainson, 1839 (*nomen praeoccupatum*), *Tropidoboa* Hombron & Jacquinot *in* Jacquinot & Guichenot, 1842, *Erebophis* A.C.L.G. Günther, 1877, and *Erobophis* – K.L. Williams & Wallach, 1989 *(nomen incorrectum)*.

Type species: *Boa carinata* Schneider, 1801.

Distribution: Australasia.

Sources: Stimson, 1969, McDowell, 1979, M. McCoy, 1980, 2006, Kluge, 1991, Harlow & Shine, 1992, Walls, 1998a, McDiarmid et al., 1999, Austin, 2000 and H.M. Smith et al., 2001c.

1. *Candoia aspera* (A.C.L.G. Günther, 1877). Proc. Zool. Soc. London 45(1): 132, pl. 21. (*Erebophis asper*)

Synonym: *Enygrus asper schmidti* Stull, 1932b.

Type: Holotype, BMNH 1946.1.10.33 (formerly BMNH 1877.2.24.23), a 737 mm specimen (G. Brown, 1875–1877).

Type locality: "Duke-of-York Island" [= New Ireland Prov., Bismarck Archipelago, NE Papua New Guinea].

Distribution: New Guinea. Eastern Indonesia (Papua, Batanta, Biak, Japen, Misool, Salawati, Seleo, Waigeu, Valise) and Papua New Guinea (Central, East Sepik, Enga, Eastern Highlands, Gulf, Madang, Milne Bay, Morobe, Northern, Western, West Sepik and Baluan, Duke of York, Karkar, Manus, New Britain, New Hanover, New Ireland, Los Negros, Lou, Manus, Rambutyo and Umboi Is.), NSL–1000 m.

Sources: Stull, 1932b, O'Shea, 1996 and J.C. Murphy & Schlager, 2003.

Remarks: A single Bougainville record unconfirmed *fide* McDowell (1979: 59).

2. *Candoia bibroni* (A.-M.-C. Duméril & Bibron, 1844). Erpét. Gén. 6: 483–484. (*Enygrus bibroni*)

Synonyms: *Tropidoboa de bibron* Hombron & Jacquinot, 1842 (*nomen illegitimum*), and *Boa australis* Montrousier, 1860.

Types: Syntypes (5), MNHN 61–61A, MNHN 1313, and MNHN 3276–77, longest syntype is 2489 mm (J.B. Hombron & H. Jacquinot, 1837–1840).

Type locality: "l'ile Viti." [= Viti Levu Is., Fiji].

Distribution: Pacific Islands. Solomons (Bellona, Bio, Makira, Olu Malau, Reef Islands, Rennell, San Cristobal, Santa Ana, Santa Cruz, Tapua, Three Sisters, Ugi, Utupua, Vanikoro), Vanuatu (Vanua Lava), New Hebrides (Aoba, Espiritu Santo, Maewo, Malekula, Mau), Loyalty Is. (Lifou, Maré, Ouvéa, Tiga), Fiji Is. (Fulanga, Kandavu, Kia, Koro, Lakemba, Levu, Navandra, Navutuiloma, Ongea, Ovalau, Rotuma, Vanua Belavu, Viwa, Vomo, Yangganga), Wallis and Futuna (Futuna), Western Samoa (Savaii, Upolú) and American Samoa (Ta'u).

Sources: Amerson et al., 1982, Gill, 1995, Crombie & Pregill, 1999, McDiarmid et al., 1999 and Bauer & Sadlier, 2000.

C

3. *Candoia carinata* (Schneider, 1801). Hist. Amph. 2: 261–263. (*Boa carinata*)

Synonyms: *Boa variegata* Thunberg, 1807, *Enygrus ocellatus* Wagler, 1830 *(nomen nudum), Boa ocellata* Wagler, 1830 *(nomen nudum), Cenchris ocellata* Gray *in* Griffith & Pidgeon, 1831 *(nomen nudum), Enygrus carinnatus* – F. Werner, 1898b *(nomen incorrectum), Enygris carinatus* – W.C. Brown & Fehlmann, 1958 *(nomen incorrectum),* and *Candoia carinata tepedeleni* H.M. Smith & Chiszar *in* H.M. Smith, Chiszar, Tepedelen & Breukelen, 2001c.

Type: Lectotype, ZFMK 35503 (formerly ZMG 18a), designated by McDowell (1979: 28).

Type locality: Unknown via lectotype selection. Designated as Amboina [= Ambon Island, S Maluku, SE Indonesia, ca. 3°39'S, 128°10'E] *fide* W. Böhme et al. (1998: 6).

Distribution: Indonesia (Papua, Ambon, Angaur, Ansus, Banda, Batanta, Djamna, Goram, Haruku, Java, Jobi, Karekelong, Laut, Liki, Misool, Salawati, Sangihe, Saparua, Seram, Sulawesi, Tanimbar, Ternate, Timor) and Papua New Guinea (East Sepik, Gulf, Madang, Morobe, Southern Highlands, Western, Aware, Babase, Baluan, Duke of York, Kairiru, Karkar, Lou, Manus, Mioko, New Britain, New Hanover, New Ireland, Manam, Mussau, Rambutyo, Sturt, Tabar and Tench Is.), NSL–1525 m.

Sources: V.M. Tanner, 1950, O'Shea, 1996, Crombie & Pregill, 1999 and Lang & Vogel, 2005.

Remarks: Type rediscovered by W. Böhme et al. (1998: 6). A paralectoatype (ZMB 1496 from "Java, [Indonesia]") exists *fide* Bauer et al. (2002: 167).

4. *Candoia paulsoni* (Stull, 1956). Copeia 1956(3): 185–186. (*Enygrus carinatus paulsoni*)

Synonyms: *Candoia paulsoni mcdowelli* H.M. Smith & Chiszar *in* H.M. Smith, Chiszar, Tepedelen & Breukelen, 2001c, *Candoia paulsoni rosadoi* H.M. Smith & Chiszar *in* H.M. Smith, Chiszar, Tepedelen & Breukelen, 2001c, *Candoia paulsoni sadlieri* H.M. Smith & Chiszar *in* H.M. Smith, Chiszar, Tepedelen & Breukelen, 2001c, *Candoia paulsoni tasmai* H.M. Smith & Tepedelen *in* H.M. Smith, Chiszar, Tepedelen & Breukelen, 2001c, and *Candoia paulsoni vindumi* H.M. Smith & Chiszar *in* H.M. Smith, Chiszar, Tepedelen & Breukelen, 2001c.

Type: Holotype, MCZ 14521, a 567 mm male.

Type locality: "Ugi Island, Solomon Islands" [Papua New Guinea].

C

Distribution: Austro-Papua. Eastern Indonesia (Ambon, Banda, Batanta, Batjan, Goram, Halmahera, Haruku, Misool, Morotai, Salawati, Sangihe, Saparua, NE Sulawesi, Tanimbar, Talaud, Ternate), Papua New Guinea (Central, Madang, Milne Bay, Morobe, National Capitol, Northern, Western, West Sepik, Ambittle, Babase, Bougainville, Buka, Dobu, Fergusson, Goodenough, Kiriwina, Kitava, Misima, Nissan, Normanby, Paneati, Rossel, Samarai, Shortland, Slade, Sudest, Tagula, Tabar, Trobriand, Umboi and Woodlark Is.) and Solomons (Bagga, Bellona, Bio, Choiseul, Fatura, Florida, Ganongga, Gatukai, Guadalcanal, Malaita, Mbanika, Mono, Nggatokae, Ranongga, Rennel, San Cristobal, Santa Ana, Santa Cruz, Santa Isabel, Shortland, Simba, Treasury, Tulagi, Vangunu, Vella Lavella), NSL–1830 m.

Sources: O'Shea, 1996, Kraus & Allison, 2004 and Lang & Vogel, 2005.

5. *Candoia superciliosa* (A.C.L.G. Günther, 1863c). Ann. Mag. Nat. Hist. (3) 12(71): 360–361. (*Enygrus superciliosus*)

Synonym: *Candoia superciliosa crombiei* H.M. Smith & Chiszar *in* H.M. Smith, Chiszar, Tepedelen & Breukelen, 2001c.

Types: Syntypes (2), BMNH 1946.1.16.47 (formerly BMNH 1863.10.2.7) and BMNH 1946.1.16.50 (formerly BMNH 1863.10.2.8), longest syntype 470 mm (G.L. King).

Type locality: "Pelew Islands" [= Palau, ca. 7°31'N, 134°35'E].

Distribution: Palau Islands (Angaur, Babeldaub, Beliliou, Koror, Ngcheangel, Ngeaur, Ngemelachel, Ngerechur, Ngercheu, Ngerduais, Ngerekebesang, Ngerukeuid, Ngeruktabel, Ngetmeduch, Oreor, Pkulaklim).

CANTORIA Girard, 1858a (Homalopsidae)

Synonyms: *Hydrodipsas* W.C.H. Peters, 1859a, *Cantoris* – Oshima, 1944 *(nomen incorrectum)* and *Djokoiskandarus* J.C. Murphy, 2011.

Type species: *Cantoria violacea* Girard, 1858a.

Distribution: Tidal rivers and coastal marine waters of SE Asia and Esat Indies.

Sources: Gyi, 1970, Voris et al., 2002, J.C. Murphy, 2007, 2011, Alfaro et al., 2008 and Zaher et al., 2009.

1. *Cantoria annulata* Jong, 1926a. Zool. Anz. 67(11–12): 304.

Type: Holotype, ZMA 11065, a 625 mm female (W.C. van Heurn, 1920–1921).

Type locality: "Prins Frederik Hendrik Insel Neu-Guinea." [= Yos Sudarso Is., SE Papua Prov., E Indonesia, bet. 7°22'–8°56'S, 137°39'–139°05'E].

Distribution: New Guinea. Eastern Indonesia (Papua) and SW Papua New Guinea (Western, Bobo, Daru Is.), NSL.

Sources: F. Parker, 1982 and O'Shea, 1996.

Remarks: Transferred to *Djokoiskandarus fide* J.C. Murphy (2011: 233).

2. *Cantoria violacea* Girard, 1858a. Proc. Acad. Nat. Sci. Philadelphia (1857) 9(6): 182.

Synonyms: *Hydrodipsas elapiformis* W.C.H. Peters, 1859a, *Cantoria elongata* A.C.L.G. Günther, 1864a *(nomen novum),* and *Cantoria dayana* Stoliczka, 1870a–d.

Type: Holotype, USNM 5523 (C. Wilkes [U.S. Explor. Exped.], 19 Jan.–25 Feb. 1842).

Type locality: "Singapore."

Distribution: Southeastern Asia. India (Andaman & Nicobars: North Andaman, Middle Andaman Is.), Myanmar (Ayeyarwady, Mon), peninsular Thailand (Phuket, Phuket Is.), West Malaysia, Singapore, East Malaysia (Sarawak) and Indonesia (Sumatra, Timor-Leste), NSL.

Sources: Girard, 1858b, C.B. Frith & Boswall, 1978, Tweedie, 1983, Mahendra, 1984, M.J. Cox, 1991b, David & Vogel, 1996, Manthey & Grossmann, 1997, Ghodke & Andrews, 2002 and Whitaker & Captain, 2004.

Remarks: Supplemental original description in Girard (1858b: 156–158, 1858c: pl. 11, figs. 7–10). Sumatra and Timor records doubtful *fide* J.C. Murphy (2007b: 60).

CARAIBA Zaher, Grazziotin, Cadle, Murphy, Moura-Leite & Bonatto, 2009 (Xenodontidae)

Synonym: *Haitiophis* Hedges & Vidal *in* Hedges, Couloux & Vidal, 2009.

Type species: *Liophis andreae* J.T. Reinhardt & Lütken, 1862.

Distribution: Cuba and Hispaniola.

Sources: A. Schwartz & Thomas, 1960, R. Thomas & Garrido, 1967, Hedges et al., 2009, R.W. Henderson & Powell, 2009, Zaher et al., 2009 and Burbrink et al., 2012.

1. *Caraiba andreae* (J.T. Reinhardt & Lütken, 1862). Vidensk. Medd. Naturhist. Foren. Kjøbenhavn (1862–1863) 24(10–18): 214–216. (*Liophis andreae*)

Synonyms: *Dromicus cubensis* S.W. Garman, 1887b, *Leimadophis nebulatus* T. Barbour, 1916a, *Leimadophis andreae orientalis* T. Barbour & Ramsden, 1919, *Dromicus andreae peninsulae* Schwartz & Thomas, 1960, *Dromicus andreae melopyrrha* R. Thomas & Garrido, 1967, and *Antillophis andreae morenoi* Garrido, 1973.

Types: Syntypes (2), ZMUC 60766–67, longest syntype 314 mm (C. Hygom & F. Andréa).
Type locality: "Havanna, Cuba" [= Havana, Habana Proc., Cuba, 23°07'N, 82°23'W, elevation 35 m].
Distribution: Cuba (Camagüey, Cienfuegos, Ciego de Aviula, Granma, Guantánamo, Habana, Holguín, Las Tunas, Matanzas, Pinar del Río, Sancti Spíritus, Santiago de Cuba, Villa Clara, Cantiles, Guajaba, Juventud and Santa María Is.), NSL–100 m.

2. *Caraiba anomala* (W.C.H. Peters, 1863c). Mber. Königl. Akad. Wiss. Berlin 1863(6): 282–283. (*Zamenis anomalus*)

Type: Holotype, ZMB 2269, an 825 mm specimen.
Type locality: Unknown.
Distribution: Haiti (Nord-Ouest, Ouest, Sud-Ouest, Tortue Is.) and W Dominican Republic (Baoruco, Monte Cristi, Beata Is.).

CARPHOPHIS Gervais *in* d'Orbigny, 1843 (Carphophiidae)

Synonyms: *Carphophiops* Gervais *in* d'Orbigny, 1843, *Carpophis* A.-M.-C. Duméril, 1853, *Celuta* Baird & Girard, 1853, *Carphophis* Duméril, Bibron & Duméril, 1854a *(nomen praeoccupatum), Coluta* – Lichtenstein & Martens, 1856 *(nomen incorrectum), Carphoptis* – Cope, 1862f *(nomen incorrectum), Celata* – Westphal-Castelnau, 1870 *(nomen incorrectum), Carphoris* – Hurter, 1893 *(nomen incorrectum), Carphopiops* – McAtee, 1907 *(nomen incorrectum), Carphosis* – Dunn, 1928 *(nomen incorrectum), Caryophis* – Haltom, 1931 *(nomen incorrectum),* and *Carphopis* – A.F. Scott & Snyder, 1968 *(nomen incorrectum).*
Type species: *Coluber amoenus* Say, 1825.
Distribution: Southeastern USA.
Fossil records: Pleistocene of USA.
Sources: Clark, 1970, Rossman, 1973, Cadle, 1984c, Holman, 2000a, Ernst et al., 2003a and Zaher et al., 2009.

1. *Carphophis amoenus* (Say, 1825). J. Acad. Nat. Sci. Philadelphia 4(2): 237–238. (*Coluber amoenus*) (*nomen corrigendum*)

Synonyms: *Coluber amaenus* Say, 1825 *(nomen incorrigendum), Coluber cineritius* Gravenhorst, 1807 *(nomen oblitum), Carphophiops vermiformis* Gervais *in* D'Orbigny, 1843, and *Celuta helenae* Kennicott, 1859a.
Types: Syntypes (5), ANSP (4), 276 mm, 264 mm, 262 mm, 216 mm and 118 mm specimens, and a 244 mm specimen (W.L. Stewart coll.), all lost *fide* Ernst et al. (2003: 774.3).
Type locality: "Pennsylvania" [USA]. Restricted to vicinity of Philadelphia *fide* K.P. Schmidt (1953a: 184).
Distribution: Southeastern USA (Alabama, Connecticut, Delaware, N Georgia, S Illinois, S Indiana, Kentucky, ext. SE Louisiana, Maryland, S Massachusetts, Mississippi, New Jersey, SE NewYork, North Carolina, S Ohio, Pennsylvania, Rhode Island, South Carolina, Tennessee, Virginia, West Virginia), NSL–1310 m.
Fossil records: Lower Pleistocene (Irvingtonian I) of USA (Florida), middle Pleistocene (Irvingtonian) of USA (West Virginia), middle/upper Pleistocene (Irvingtonian II) of USA (Maryland), and upper Pleistocene (Rancholbrean II) of USA (Alabama, Florida, Georgia, Pennsylvania, Tennessee, Virginia, West Virginia).
Sources: Ernst & Barbour, 1989, Conant & Collins, 1991, Palmer & Braswell, 1995 and Ernst et al., 2003b.
Remarks: Var. A of Say (1825: 238) might be considered the holotype. In accordance with Art. 23.9.2 of the Code (ICZN, 1999), *Coluber amoenus* Say is designated a *nomen protectum* and *Coluber cineritius* Gravenhorst a *nomen oblitum.*

2. *Carphophis vermis* (Kennicott, 1859). Proc. Acad. Nat. Sci. Philadelphia 11(3): 99–100. (*Celuta vermis*)

Type: Holotype, USNM 2180 (P.R. Hoy, 1854).
Type locality: "Missouri" [USA]. USNM catalogue entry for type reads Cooper County *fide* Ernst et al. (2003c: 775.1).
Distribution: Central USA (Arkansas, ext. SW Illinois, ext. SE Iowa, E Kansas, ext. N Louisiana, Missouri, ext. SE Nebraska, E Oklahoma, ext. NE Texas, ext. SW Wisconsin), 60–610 m.
Fossil records: Middle/upper Pleistocene (Irvingtonian II) of USA (Arkansas), and upper Pleistocene (Rancholabrean II) of USA (Missouri, Texas).
Sources: Clark, 1968, Fitch, 1999, Werler & Dixon, 2000 and Ernst et al., 2003c.

CASAREA Gray, 1842a (Bolyeriidae)

Synonyms: *Centrophis* Hemprich *in* Fitzinger, 1843, *Caseara* A.-M.-C. Duméril & Bibron, 1844 *(nomen incorrectum), Leptoboa* A.-M.-C. Duméril & Bibron, 1844, *Cascara* – A.-M.-C. Duméril, 1853 *(nomen incorrectum), Leptolon* – Giglioli, 1873 *(nomen incorrectum), Caesarea* – F. Werner, 1890 *(nomen incorrectum),* and *Casaria* – H.W. Parker & Grandison, 1977 *(nomen incorrectum).*
Type species: *Boa dussumieri* Schlegel, 1837.
Distribution: Mauritius, Indian Ocean.
Fossil record: Subrecent of Mauritius.
Sources: Anthony & Guibé, 1952, Guibé, 1958, Stimson, 1969, Vinson, 1975, McAlpine, 1981, Cundall & Irish, 1986, 1989, M.K. Hecht & LaDuke, 1988, McDiarmid et al., 1999, Maiscano & Rieppel, 2007 and Winters, 2011.

C

1. *Casarea dussumieri* (Schlegel, 1837). Essai Phys. Serp. 1: 176, 2: 396–397. (*Boa dussumieri*) (*nomen corrigendum*)

Synonyms: *Boa dussumiri* Schlegel, 1837 *(nomen incorrigendum)* and *Boa dussumieri* Gray, 1842a *(nomen corrigendum)*.

Type: Holotype, MNHN 9, a 420 mm female (J.-J. Dussumier, 1816–1837).

Type locality: "l'île ronde, près de Maurice" [= Round Island, near Mauritius, Indian Ocean, bet. 19°51–52'S and 57°47–48'E].

Distribution: Mascarenes (Flat, Gunner's Quoin, Passe, Round, Mauritius Is.).

Remarks: Incorrect original spelling corrected in Schlegel (1839 *in* 1837–1844: 55, pl. 17, figs. 1–5). Recently extinct on Mauritius *fide* Arnold (1980a: 41) but captive colony on Jersey Island.

CATHETORHINUS Duméril & Bibron, 1844 (Typhlopidae)

Type species: *Cathetorhinus melanocephalus* Duméril & Bibron, 1844.

Distribution: Old World.

Sources: Duméril & Bibron, 1844, Wallach & Pauwels, 2008, Cheke, 2010 and Winters, 2011.

1. *Cathetorhinus melanocephalus* Duméril & Bibron, 1844. Erpét. Gén. 6: 270–271.

Type: Holotype, MNHN 138, a 183 mm male (F. Péron & C.A. Lesueur [N.T. Baudin Géographe Voy.], Oct. 1800–1804).

Type locality: Unknown. Landfalls during the Péron and Lesueur voyage include Mauritius, Australia and Timor *fide* Wallach & Pauwels (2008: 121).

Distribution: Old World.

Remarks: Possibly an extant specimen of fossil *Typhlops cariei* from Mauritius *fide* Cheke (2010: 102).

CAUSUS Wagler, 1830 (Viperidae)

Synonyms: *Distichurus* Hallowell, 1842, *Heterophis* W.C.H. Peters, 1862a, *Dinodipsas* W.C.H. Peters, 1882b, *Cansus* – Mocquard, 1896b *(nomen incorrectum)*, *Caussus* – Lindemann, 1898 *(nomen incorrectum)*, and *Causas* – Briceño-Rossi, 1934 *(nomen incorrectum)*.

Type species: *Sepedon rhombeata* Lichtenstein, 1823.

Distribution: Subsaharan Africa.

Sources: Witte, 1962, Broadley, 1968a, Golay et al., 1993, David & Ineich, 1999, McDiarmid et al., 1999, Underwood, 1999, Lenk et al., 2001b, Mallow et al., 2003, Dobiey & Vogel, 2007, Phelps, 2010 and Hoser, 2012d.

1. *Causus bilineatus* Boulenger, 1905d. Ann. Mag. Nat. Hist. (7) 16(92): 114. (*Causus rhombeatus bilineatus*)

Synonym: *Causus lineatus* Laurent, 1955.

Types: Syntypes (5), BMNH 1905.5.29.36–40, males and females (W.J. Ansorge, June–July 1903, Nov. 1903–Feb. 1904, & end of 1904).

Type locality: "Pungo Andongo, Canhoca, between Benguella and Bihé, Angola" [= between Benguela (12°35'S, 13°25'E) and Bié (12°23'S, 16°57'E), W Angola].

Distribution: Central Africa. Southeastern Democratic Republic of the Congo (Katanga), NE Rwanda (Byumba), Angola (Benguela, Bié, Cuanza Norte, Huambo, Malanje, Moxico, Namibe) and NW Zambia (Luapula, North-Western, Western), 725–1800 m.

Sources: Laurent, 1964a, Manaças, 1982, Spawls & Branch, 1995, Spawls et al., 2002, Broadley et al., 2003 and J.B. Rassmussen, 2005a.

Remarks: Localities listed as Duque de Bragança, Malanje Prov. (09°06'S, 15°57'E), Quissange River, Benguela Prov. (12°26'S, 14°03'E), Caconda, Huíla Prov. (13°44'S, 15°04'E) and Huíla, Huíla Prov. (15°03'S, 13°33'E) *fide* Bocage (1895: 146).

2. *Causus defilippii* (Jan, 1863a). Arch. Zool. Anat. Fis. 2(2): 225–226. (*Heterodon defilippii*) (*nomen corrigendum*)

Synonyms: *Heterodon de filippii* Jan, 1863a *(nomen incorrigendum)*, *Causus rostratus* A.C.L.G. Günther, 1863c, *Causus defilippii* Mocquard, 1892a *(nomen corrigendum)*, *Causus rostratus* Mocquard, 1892b *(nomen praeoccupatum)*, and *Causus dephillippii* – Rose, 1955 *(nomen incorrectum)*.

Type: Holotype, MSNM (formerly MHNG & MZUT 2836), lost *fide* Broadley *in* Golay et al. (1993: 264).

Type locality: "Buenos Ayres" [= Buenos Aires, Argentina] (in error). Corrected to Africa *fide* Broadley (1971d: 102). Restricted to Transvaal [= Limpopo, Mpumalanga and Gauteng Prov., NE South Africa] *fide* Elter (1982: 20, 55).

Distribution: East and S Africa. Eastern Democratic Republic of the Coingo (Sud-Kivu), SE Kenya (Coast), Tanzania (Iringa, Linda, Mbeya, Morogoro, Mtwara, Pwani, Ruvuma, Tanga, Zanzibar), Zambia (Central, Eastern, Northern, Southern), Malawi (Central, Southern), Mozambique (Gaza, Manica, Maputo, Nampula, Niassa, Sofala, Tete, Zambézia), Zimbabwe (Bulawayo, Manicaland, Mashonaland Central, Mashonaland East, Mashonaland West, Masvingo, Matabeleland North, Matabeleland South, Midlands), Swaziland and NE South Africa (KwaZulu-Natal, Limpopo, Mpumalanga), NSL–1800 m.

Sources: A.C.L.G. Günther, 1864b, Mocquard, 1891b, Sweeney, 1961, Broadley, 1971, 1983, FitzSimons, 1974, Broadley & Cock, 1975, Pienaar et al., 1983, Branch, 1988, Spawls & Branch, 1995, Spawls et al., 2002, Branch et al., 2005 and Broadley & Blaylock, 2013.

Remarks: Probably occurs in Namibia *fide* M. Griffin (2003: 130) and E Botswana *fide* Auerbach (1987: 205).

3. *Causus lichtensteinii* (Jan, 1859c). Rev. Mag. Zool. (1858) (2) 11: 511. (*Aspidelaps lichtensteinii*)

Synonym: *Dinodipsas angulifera* W.C.H. Peters, 1882b.

Type: Holotype, formerly NMBA, destroyed in 1943 during World War II.

Type locality: "Côte-d'Or." [= Gold Coast, presumably Ghana].

Distribution: Western and cen. Africa. Southeastern Guinea (Macenta, Nzérékoré), Sierra Leone, Liberia, S Ivory Coast (Guiglo, Tabou), S Ghana (Eastern), S Nigeria, S Cameroon (Adamaoua, Est, Centre, Littoral, Nord-Ouest, Sud, Sud-Ouest), Equatorial Guinea, Gabon (Ogooué-Ivindo, Ogooué-Lolo, Ogooué-Maritime, Woleu-Ntem), Congo (Brazzaville, Kouilou, Pool, Sangha), SW Central Africa Republic (Bamingui-Bangoran, Haut-Mbomou, Haute-Sangha, Lobaye, Mbomou, Ombella-Mpoko, Ouaka, Sangha), SE Sudan, Democratic Republic of the Congo (Bandundu, Bas-Congo, Equateur, Kinshasa, Katanga, Kasai Occidental, Kasai Oriental, Orientale, Sud-Kivu), W Zambia (North-Western), Angola (Benguela, Lunda Norte), S Uganda (Central, Western) and SW Kenya (Western), NSL–2100 m.

Sources: Roux-Estéve, 1965, Leston & Hughes, 1968, Hughes & Barry, 1969, Leston, 1970, Pitman, 1974, Lawson, 1993, J.-F. Trape & Roux-Estève, 1995, Akani et al., 2001, Spawls et al., 2002, Broadley et al., 2003, Chippaux, 2006, Chirio & Ineich, 2006, Chirio & LeBreton, 2007 and Pauwels & Vande weghe, 2008.

Remarks: Probably occurs in Togo *fide* and Segniagbeto et al. (2011: 353). Possibly occurs in Benin *fide* Hughes (2013: 151). See Akani et al. (2001: 192) for discussion of its occurrence in Nigeria.

4. *Causus maculatus* (Hallowell, 1842). J. Acad. Nat. Sci. Philadelphia 8(2): 337–338, pl. 19. (*Distichurus maculatus*)

Type: Holotype, ANSP 6897, a 293 mm specimen (S.M.E. Goheen, 1836–1838).

Type locality: "Liberia, Western Africa."

Distribution: Western and cen. Africa. Southwestern Mauritania (Trarza), W Senegal (Dakar, Fatick, Kaolack, Kédougou, Kolda, Saint-Louis, Tambacounda, Ziguinchor), Gambia (MacCarthy Island), Guinea-Bissau (Bafatá, Cacheu, Tombali), Guinea (Boké, Dalaba, Kouroussa, Macenta, Nzérékoré, Télimélé), Sierra Leone (Eastern, Northern, Southern, Western, Sherbo Is.), Liberia (Bong, Grand Bassa, Grand Cape Mount, Grand Gedeh, Loffa, Montserrado, Margibi, Nimba), S Mali (Kayes, Koulikoro, Mopti, Ségou, Sikasso, S Tombouctou), Ivory Coast (Abidjan, Adzopé, Bouaké, Bouna, Daloa, Dix-Huit Montagnes, Ferkéssédougou, Guiglo, San Pédro, Tabou, Toumodi), Burkina Faso (Centre, Centre-Est, Centre-Nord, Centre-Ouest, Est, Haute-Bassins, Nord, Sahel, Sud-Ouest, Volta-Noire), Ghana (Accra, Ashanti, Brong-Ahafo, Central, Eastern, Northern, Upper East, Upper West, Volta, Western), Togo (Centrale, Kara, Maritime, Plateaux, Savanes), Benin (Alibori, Atakora, Atlantique, Borgou, Zou), S Niger (Dosso, Niamey, S Diffa), Nigeria (Abuja, Akwa Ibom, Anambra, Bayelsa, Bendel, Cross River, Delta, Gongola, Kaduna, Kano, Kwara, Lagos, Oyo, Plateau, Rivers, Sokoto), S Chad (Chari-Baguirmi, Guéra, Mayo-Kebbi, Moyen-Chari), Cameroon (Adamaoua, Centre, Est, Extreme-Nord, Littoral, Nord, Nord-Ouest, Ouest, Sud, Sud-Ouest), Central African Republic (Bamingui-Bangoran, Haute-Sangha, Haut-Mbomou, Kemo-Gribingui, Lobaye, Mbomou, Nana-Mambere, Ombella-Mpoko, Ouaka, Ouham, Ouham-Pende, Sangha, Vakaga), Equatorial Guinea (Bioko Is.), E Gabon (Haut-Ogooué, Ogooué-Ivindo), Congo (Bouenza, Brazzaville, Cuvette, Kouilou, Lekoumou, Likouala, Niari, Pool), Democratic Republic of the Congo (Bandundu, Bas-Congo, Equateur, Kinshasa, Nord-Kivu, Orientale, Sud-Kivu), NE Angola (Cuanza Sul, Lunda Norte, Lunda Sul), E South Sudan, S Ethiopia (Illubabor, Shoa, Wollega), and W Uganda (Western), NSL–1950 m.

Sources: Mertens, 1965c, Hughes, 1978, 2013, Roman, 1980, Hakansson, 1981, J.A. Butler & Reid, 1986, Lawson, 1993, Spawls & Branch, 1995, J.-F. Trape & Roux-Estève, 1995, J.-F. Trape & Mané, 2000, 2006b, Akani et al., 2001, Spawls et al., 2002, Ineich, 2003, Luiselli et al., 2004a, J.B. Rasmussen, 2005a, Chippaux, 2006, Chirio & Ineich, 2006, Chirio & LeBreton, 2007, Pauwels & Vande weghe, 2008, Largen & Spawls, 2010, Ullenbruch et al., 2010, W. Böhme et al., 2011, Segniagbeto et al., 2011 and Chirio, 2013.

5. *Causus resimus* (W.C.H. Peters, 1862a). Mber. Königl. Akad. Wiss. Berlin 1862(1): 277–278, figs. 4–4b. (*Heterophis resimus*)

Synonyms: *Causus jacksonii* A.C.L.G. Günther, 1888c, *Causus nasalis* Stejneger, 1894b, and *Causus resimus angolensis* Bocage, 1895.

Type: Holotype, ZMB 4370, a 295 mm specimen (A. von Barnim & R. Hartman, 1859–1860).

Type locality: "Sennâr, vom Gebel-Ghule" [= Jebol Ghule, Sennar Prov., SE Sudan].

Distribution: Central and E Africa. Ghana (Brong-Ahafo), Nigeria (Abia, Abuja, Borno), S Chad (Chari-Baguirmi, Mayo-Kebbi, Salamat), N Cameroon (Extreme-Nord, Nord), N Central African Republic

(Vakaga), NE Democratic Republic of the Congo (Nord-Kivu, Orientale), SE Sudan (Sennar), E Sudan, South Sudan, S Ethiopia (Bale, Gemu Gofa, Kefa), Somalia (Bay, Gedo, Hiraan, Jubbada Dhexe, Jubbada Hoose, Mogadishu, Shabeellaha Dhexe, Shabeellaha Hoose), Uganda (Central, Eastern, Western), S Kenya (Coast, Nyanza, Rift Valley, Western), Rwanda, N Burundi, Tanzania (Kagera, Mara, Mwanza), and Angola (Benguella, Cuanza Norte, Cuanza Sul, Lunda Norte, Namibe), NSL–1800 m.

Sources: Pitman, 1974, Lanza, 1983a, 1990b, Hughes, 1987, Spawls & Branch, 1995, Akani et al., 2001, Spawls et al., 2002, J.B. Rasmussen, 2005a, Chippaux, 2006, Chirio & Ineich, 2006, Chirio & LeBreton, 2007 and Largen & Spawls, 2010.

Remarks: Distribution consists of numerous disjunct populations.

6. *Causus rhombeatus* (Lichtenstein, 1823). Verzeich. Doubl. Zool. Mus. Univ. Berlin: 106. (*Sepedon rhombeata*)

Synonyms: *Naja v-nigrum* F. Boie, 1827, *Naja rhombeata* – Schlegel, 1837, and *Causus rhombeatus taeniata* Sternfeld, 1912.

Types: Syntypes (2–3), ZMB 2768–69 and ? ZMB 2770, longest syntype 457 mm (G.L.E. Krebs, 1821–1829).

Type locality: "Prom. B. sp." [= Promontorium Bonae Spei or Cape of Good Hope, South Africa]. Restricted to districts of Uitenhage, Kirkwood and Port Elizabeth, Eastern Cape Prov., South Africa *fide* Bauer (2000: 57).

Distribution: Eastern Africa. Northeastern Nigeria (Gongola), Central African Republic (Ouham-Pende), South Sudan, Ethiopia (Bale, Gemu Gofa, Gojjam, Gondar, Hararge, Illubabor, Kefa, Shoa, Sidamo, Tigre), Uganda (Central, Eastern, Northern, Western), SW Kenya (Central, S Coast, Eastern, Nairobi, S Rift Valley, Western), cen. Rwanda, W Burundi, SW Tanzania (S Iringa, Kigoma, Mbeya, Rukwa, Ruvuma), SE Democratic Republic of the Congo (Katanga), NE Angola (Lunda Norte, Moxico), Zambia (Central, Copperbelt, Eastern, Luapula, Lusaka, Northern, North-Western, Southern, Western), Malawi (Central, Northern, Southern), Mozambique (Inhambane, Manica, Maputo, Niassa, Sofala, Tete), Zimbabwe (Bulawayo, Manicaland, Mashonaland Central, Mashonaland East, Mashonaland West, N Masvingo, Matabeleland North, Matabeleland South, Midlands), NE Namibia (Caprivi, E Kavango), N Botswana (North West, North East), Swaziland, Lesotho and E South Africa (Eastern Cape, Free State, Gauteng, KwaZulu-Natal, Limpopo, Mpumalanga, E Western Cape), NSL–2200 m.

Sources: Barros e Cunha, 1935, Sülter, 1962, Witte, 1962, Laurent, 1964a, Hughes & Barry, 1969, Hughes, 1977, FitzSimons, 1974, Pitman, 1974, Broadley & Cock, 1975, Hughes, 1978, Broadley, 1983, Auerbach, 1987, Branch, 1988, Massary & Ineich, 1994, Bauer, 2000, Akani et al., 2001, Spawls et al., 2002, Broadley et al., 2003, M. Griffin, 2003, J.C. Murphy & Schlager, 2003, J.B. Rasmussen, 2005a, Chippaux, 2006, Largen & Spawls, 2010 and Broadley & Blaylock, 2013.

CEMOPHORA Cope, 1860e (Colubridae)

Synonyms: *Stasiotes* Jan, 1862b, *Cemaphora* – Palacky, 1898 *(nomen incorrectum)*, and *Cernophora* – Briceño-Rossi, 1934 *(nomen incorrectum)*.

Type species: *Coluber coccineus* Blumenbach, 1788.

Distribution: Eastern and S USA.

Fossil records: Pleistocene of SE USA.

Sources: K.L. Williams et al., 1966, K.L. Williams & Wilson, 1967, Christman, 1980, Dowling et al., 1983, K.L. Williams, 1985, Ernst & Barbour, 1989, Conant & Collins, 1991, 1998, Trauth, 1993, Holman, 1995a, 2000a, Palmer & Braswell, 1995, Keogh, 1996, Tennant, 1997, Rodríguez-Robles & Jesús-Escobar, 1999, Werler & Dixon, 2000 and Pyron & Burbrink, 2009a.

1. *Cemophora coccinea* (Blumenbach, 1788). Voigt's Mag. Physik Naturg. 5(1): 10–12, pl. 1. (*Coluber coccineus*)

Synonyms: *Coluber doliatus* Linnaeus, 1766 *(nomen rejiciendum)*, *Coluber dumfrisiensis* Sowerby, 1804, *Coluber coccinaenus* Kirtland, 1838 *(nomen emendatum)*, *Cemophora copei* Jan, 1863a, and *Cemophora coccinea lineri* K.L. Williams, Brown & Wilson, 1966.

Type: Holotype, ZMG (A. Garden, 1750–1761), lost *fide* K.L. Williams & Wilson (1967: 112).

Type locality: "Florida und Neu-Spanien" [USA and Mesoamerica]. Restricted to Florida, USA *fide* K.L. Williams & Wilson (1967: 112).

Distribution: Southeastern USA (Alabama, Arkansas, Florida, Georgia, ext. S Illinois, ext. S Indiana, W Kentucky, Louisiana, Maryland, Mississippi, S Missouri, E New Jersey, North Carolina, E Oklahoma, South Carolina, Tennessee, ext. E Texas, Virginia), NSL–600 m.

Fossil records: Lower Pleistocene (Irvingtonian I) of USA (Florida), and upper Pleistocene (Rancholabrean II) of USA (Florida, Virginia).

Remarks: Official Specific Name no. 2187 *fide* Opinion 804 (ICZN, 1967). Collins (1991: 43) suggested that *C. lineri* be recognized as a valid species.

CERASTES Laurenti, 1768 (*nomen protectum*) (Viperidae)

Synonyms: *Aspis* Laurenti, 1768 *(nomen rejiciendum)*, *Cerastes* Rafinesque-Schmaltz, 1815 *(nomen nudum)*, *Haemorrohous* Fitzinger, 1823, *Gonyechis* Fitzinger,

1843, *Gongechis* – A.-M.-C. Duméril & Bibron, 1844 *(nomen incorrectum), Cerates* – K.P. Schmidt, 1941 *(nomen incorrectum),* and *Conyechis* – A.S. Romer, 1956 *(nomen incorrectum).*

Type species: *Coluber cerastes* Linnaeus, 1758.

Distribution: North Africa and SW Asia.

Sources: Villiers, 1950a, Schnurrenberger, 1959, Broadley, 1968a, Papenfuss, 1969, Joger, 1984, Y.L. Werner et al., 1991, Y.L. Werner & Sivan, 1991, 1992, Golay et al., 1993, Herrmann & Joger, 1995, Spawls & Branch, 1995, David & Ineich, 1999, Herrmann et al., 1999, Joger & Courage, 1999, McDiarmid et al., 1999, Lenk et al., 2001b, Mallow et al., 2003, Dobiey & Vogel, 2007, Stümpel & Joger, 2009, Phelps, 2010 and Hoser, 2012d.

Remarks: Official Generic Name no. 1539 *fide* Opinion 661 (ICZN, 1963b).

1. *Cerastes boehmei* Wagner & Wilms, 2010. Bonn Zool. Bull. 17(2): 300–302, figs. 2.1, 3–5.

Type: Holotype, ZFMK 58054, a 219 mm female (T. Holtmann, 1991).

Type locality: "Tunisia, SW Remada, east of Djebei National Park, close to the road midway between Beni Kandeche (Bani Kheddacvhe) and Ksar el Hallouf" [= SW Remada (32°19'N, 10°23'E, 300 m), Tatouine Govern., cen. Tunisia].

Distribution: Central Tunisia (Tatouine), 300 m.

2. *Cerastes cerastes* (Linnaeus, 1758). Syst. Nat., ed. 10, 1: 217. (*Coluber cerastes*)

Synonyms: *Coluber cornutus* Linnaeus *in* Hasselquist, 1762, *Coluber baetaen* Forskal *in* Niebuhr, 1775, *Vipera aegyptiaca* Daudin, 1803c, *Crastes preuini* Fitzinger, 1823, *Cerastes hasselquistii* Gray, 1842e, *Echidna atricauda* A.-M.-C. Duméril, Bibron & Duméril, 1854b *(partim), Vipera avicennae* Jan, 1859a *(part), Cerastes cornutus mutila* Doumergue, 1901, *Cerastes cerates* – Weissenberg et al., 1991 *(nomen incorrectum),* and *Cerastes cerastes hoofieni* Y.L. Werner & Sivan *in* Y.L. Werner, Sivan, Kushnir & Motro, 1999.

Type: Holotype, NHR Lin-98 (formerly MAFR), an anterior body section (F. Hasselquist, 7 July 1750, via. Mus. Drottn.).

Type locality: "Oriente" [= Egypt]. Rashid, Nile delta [= Rasheed, Kafr El-Shaikh Govern., NE Egypt] *fide* Hasselquist (1762: 70–72, 315–322). Restricted to Egypt *fide* Flower (1933: 830) and to southern Judaea *fide* K.P. Schmidt (1939b: 88).

Distribution: North Africa and Arabia. Western Sahara (Rio de Oro, Saguia El Hamra), W Mauritania (Adrar, Assaba, Brakna, Dakhlet Nouadhibou, Gorgol, Inchiri, Tiris Zemmour, Trarza), S Morocco (Boulemane, Figuig, Kaar Es Souk, Ouarzazate, Tan Tan), Algeria (Adrar, El Bayadh, Illizi, Naama, Saida, Tamanrasset), NE Mali (Gao), Niger (Agadez, Diffa, Tahoua, Zinder), Tunisia (Nabeul, Sousse, Zaghouan), Libya (Ajdabiya, Al Butnan, Al Jabal Al Gharbi, Al Jurfah, Al Kurfah, Awbari, Banghazi, Ghat, Nalut, Murzuq, Surt), N Chad (Kanem, Tibetsi), N Sudan (Al Bahr Al Ahmar, Ash Shamaliyah), Egypt (Alexandria, Assiut, Aswan, Beheira, Beni Suef, Cairo, El-Bahr El-Ahmar, Faiyum, Giza, Ismailia, Matrouh, Minya, New Valley, North Sinai, Qena, South Sinai, Sohag, Suez), SW Israel (W Southern), SW Saudi Arabia (Qizan) and Yemen, NSL–1500 m.

Sources: Doumergue, 1901, Kramer & Schnurrenberger, 1963, A. Davidson, 1964, Johann, 1973, Gasperetti, 1988, Y.L. Werner et al., 1991, 1999, Leviton et al., 1992, Sterer, 1992, Y.L. Werner & Sivan, 1992, Schätti & Gasperetti, 1994, Y.L. Werner, 1994, Bons & Geniez, 1996, Schleich et al., 1996, Bogaerts, 1998, Bouskila & Amitai, 2001, J.C. Murphy & Schlager, 2003, Geniez et al., 2004, Chippauz, 2006, J.-F. Trape & Mané, 2006b and Bar & Haimovitch, 2011.

Remarks: Official Specific Name no. 1910 *fide* Opinion 661 (ICZN, 1963c). Photograph of head of "holotype" in Y.L. Werner & Sivan (1992: fig. 1), who listed holotype as lost and NHR Lin-98 as a paratype. Type locality restriction of S Judea *fide* K.P. Schmidt (1939: 88) invalid.

3. *Cerastes gasperettii* Leviton & Anderson, 1967. Proc. California Acad. Sci. (4) 35(9): 183–184, fig. 12 (*Cerastes cerastes gasperettii*)

Synonyms: *Cerastes cerastes karlhartli* Sochurek, 1974 (*nomen nudum*) and *Cerastes gasperettii mendelssohni* Werner & Sivan *in* Y.L. Werner, Sivan, Kushnir & Motro, 1999.

Type: Holotype, CAS 97826, a 580 mm male (J. Gasperetti, 23 Aug. 1964).

Type locality: "Beda Azan [23°41' N., 53°28' E.], Abu Dhabi [Abu Zaby]," United Arab Emirates.

Distribution: Arabia and Middle East. Syria, ext. SE Israel (E Southern), Jordan (Aqaba, Karak, Maan, Tafilah, Zarqa), S Iraq, Saudi Arabia (Asir, Eastern, Jawf, Jazan, Madinah, Makkah, Northern, Qasim, Riyadh, Tabuk), Kuwait, Qatar, United Arab Emirates (Abu Dhabi, Sharjah), Oman (Aden, Al Wusta, Ash Sharqiyah, Dhofar), Yemen (Abyan, Al Jawf, Al Hudaydah, Hadhramaut, Mahwit, Ma'rib, Shabwah) and SW Iran (Khuzestan), NSL–1600 m.

Sources: Y.L. Werner, 1987, 1994, Gasperetti, 1988, Leviton et al., 1992, Schätti & Gasperetti, 1994, Y.L. Werner et al., 1999, Disi et al., 2001, Baker et al., 2004, Amr & Disi, 2011 and Bar & Haimovitch, 2011.

4. *Cerastes vipera* (Linnaeus, 1758). Syst. Nat., ed. 10, 1: 216. (*Coluber vipera*)

Synonyms: *Aspis cleopatrae* Laurenti, 1768, *Coluber aegypticus* Lacépède, 1789 (*nomen rejiciendum*), *Vipera aegyptia* Latreille *in* Sonnini & Latreille,

C

C

1801a, *Vipera aegyptiaca* Daudin, 1803d (*nomen substitutum*), *Aspis cleopatra* Gray, 1842e (*nomen emendatum*), *Cerastes richiei* Gray, 1842e, *Cerastes ritchii* Gray, 1849 (*nomen emendatum*), *Echidna atricauda* A.-M.-C. Duméril, Bibron & Duméril, 1854b (*partim*), *Vipera avicennae* Jan, 1859a (*part*), *Vipera avizennae* – Strauch, 1869 (*nomen incorrectum*), and *Cerastes vipera inornatus* F. Werner, 1929b.

Type: Holotype, NHR Lin-99 (formerly MAFR), a 380 mm male (F. Hasselquist, 7 July 1750, via Mus. Drottn.).

Type locality: "Aegypto" [= Egypt]. Rashid, Nile delta [= Rasheed, Kafr El-Shaikh Govern., NE Egypt] *fide* Hasselquist (1762: 70–72, 315–322).

Distribution: North Africa and Middle East. Western Sahara (Rio de Oro, Saguia El Hamra), Mauritania (Adrar, Brakna, Dahklet Nouadhibou, Inchiri, Tagant, Tiris Zemmour, Trarza), S Morocco (Kaar Es Souk, Ouarzazate, Tan Tan), Algeria (Adrar, Bechar, Illizi, Naama, Tamanrasset), Tunisia (Gafsa, Sousse), N Mali (Gao, Tombouctou), N Niger (Agadez, Zinder), Libya (Ajdabiya, Al Jabar Al Gharbi, Awbari, Banghazi, Surt, Tarabulus), Chad, N Egypt (El-Bahr El-Ahmar, Faiyum, Giza, Ismailia, Matrouh, Minya, N New Valley, North Sinai, South Sinai, Suez), SW Saudi Arabia (Jazan), SW Israel (W Southern) and Jordan, NSL–1500 m.

Sources: Mayet, 1903, Kramer & Schnurrenberger, 1963, Bons & Geniez, 1996, Jooris & Fourmy, 1996, Schleich et al., 1996, Bogaerts, 1998, Bouskila & Amitai, 2001, Geniez et al., 2004, Chippaux, 2006, J.-F. Trape & Mané, 2006b and Bar & Haimovitch, 2011.

Remarks: Record from Lebanon rejected *fide* Joger (1984: 42).

CERATOPHALLUS Cope, 1893a (Natricidae)

Type species: *Coluber vittatus* Linnaeus, 1758.

Distribution: Singapore and Indonesia.

Sources: Bergman, 1950a, 1950b, K.F.L. Lim & Lee, 1989, David & Vogel, 1996 and Manthey & Grossmann, 1997.

Remarks: A valid genus *fide* Dubey et al., 2012, P. Gou et al., 2012, and Zaher et al., 2012.

1. *Ceratophallus vittatus* (Linnaeus, 1758). Syst. Nat., ed. 10, 1: 219. (*Coluber vittatus*)

Type: Holotype, NHR (formerly MAFR), lost *fide* Andersson (1899: 5).

Type locality: "America," (in error).

Distribution: Singapore and Indonesia (Bangka, Java, Sumatra, Weh), NSL–100 m.

Remarks: Introduced into Singapore *fide* K.K.P. Lim & Chou (1990: 54). Sulawesi records doubtful *fide* Lang & Vogel (2005: 258).

†*CERBEROPHIS* Longrich, Bhullar & Gauthier, 2012b (Alethinophidia incertae sedis)

Type species: †*Cerberophis robustus* Longrich, Bhullar & Gauthier, 2012b.

Distribution: Upper Cretaceous of USA.

Source: Longrich et al., 2012a.

1. †*Cerberophis robustus* Longrich, Bhullar & Gauthier, 2012b. Proc. Natl. Acad. Sci. 109(52): 21397–21398, fig. 1-I.

Synonym: Boidae indeterminate Bryant, 1989.

Type: Holotype, UCMP 130696 (formerly V80096), one trunk vertebra.

Type species: "Hell Creek Formation, Garfield/McCone Counties, Montana; Late Cretaceous, ≤1.8 Ma before K-T boundary."

Distribution: Upper Cretaceous (Maastrichtian, 65.5–67.3 mya) of USA (Montana).

Remarks: Illustration of holotype in Bryant (1989: figs. 11a–b).

CERBERUS Cuvier, 1829 (Homalopsidae)

Synonyms: *Hurria* Daudin, 1803b, *Huria* Fischer von Waldheim, 1813 (*nomen incorrectum*), *Hurianus* Rafinesque-Schmaltz, 1815 (*nomen emendatum*), *Hurriah* Rafinesque-Schmaltz, 1815 (*nomen incorrectum*), *Hurrias* Oken, 1817 (*nomen emendatum*), *Strephon* Goldfuss, 1820 (*nomen substitutum*), *Hurrix* – Ritgen, 1828 (*nomen incorrectum*), *Cerbus* – Locard, 1875 (*nomen incorrectum*), *Hyrria* – Hoffmann, 1890 (*nomen incorrectum*), *Cerburus* – Maki, 1931 (*nomen incorrectum*), and *Hurrianus* – Maki, 1931 (*nomen incorrectum*).

Type species: *Hydrus rynchops* Schneider, 1799.

Distribution: Tidal rivers and coastal marine waters of Indo-Australia.

Sources: E.H. Taylor, 1922a, F. Werner, 1923b, Gyi, 1970, Voris et al., 2002, J.C. Murphy, 2007b, Alfaro et al., 2008, Zaher et al., 2009, J.C. Murphy et al., 2012 and J.C. Murphy, 2012a.

1. *Cerberus australis* (Gray, 1842d). Zool. Misc. 2(May): 65. (*Homalopsis australis*)

Synonyms: *Homolopsis australis* Gray, 1842c (*nomen nudum*), *Cerberus rynchops novaeguineae* Loveridge, 1948, and *Cerberus montgomeryi* Wells & Wellington, 1985.

Type: Holotype, BMNH 1946.1.2.40 (J. Gilbert, 1840–1842).

Type locality: "North Coast of New Holland; Port Essington" [= ruins at tip of Cobourg Peninsula, Northern Territory, N Australia, ca. 11°09'N, 132°00'E, elevation NSL].

Distribution: Eastern Indonesia (S Papua, Sumba Is.) and N Australia (N Northern Territory, ext. N Queensland, ext. N Western Australia, Channel, Melville, Mirree and Sir Graham Moore Is.), NSL–250 m.

Sources: O'Shea, 1996 and Karns et al., 2000.

Remarks: J.C. Murphy et al. (2012a: 9) listed type locality as ~12°28'S, 130°54'E.

2. *Cerberus dunsoni* J.C. Murphy, Voris & Karns, 2012a. Zootaxa (3484): 11–12, fig. 5 (paratype).

Type: Holotype, AMNH 116021, an 813+ mm female.

Type locality: "Micronesia, Palau Islands: Arakabesang (formerly Ngerekebesang) Island (~7°29'N, 134°27'E)."

Distribution: Palau (Arakabesang, Babeldaob and Oreor Is.), NSL–80 m.

Source: Crombie & Pregill, 1999.

3. *Cerberus microlepis* Boulenger, 1896a. Cat. Snakes Brit. Mus. 3: 18–19, pl. 2, fig. 2–2a.

Types: Syntypes (2), BMNH 1946.1.7.24–25, two females, longest syntype 660 mm (H. Cuming, 1836–1840).

Type locality: "Philippines." Restricted to vicinity of Lake Buhi, Luzon Island *fide* J.C. Murphy (2007b: 69, 2012: 13).

Distribution: Philippines (Luzon), 60 m.

4. *Cerberus rynchops* (Schneider, 1799). Hist. Amph. 1: 246–247. (*Hydrus rynchops*)

Synonyms: ? *Boa moluroides* Schneider, 1801, *Hydrus cinereus* G. Shaw, 1802, *Coluber cerberus* Daudin, 1803d, *Hurria bilineata* Daudin, 1803d, *Python rhynchops* Merrem, 1820 *(nomen emendatum), Coluber decipiens Oppel* in H. Boie, 1826 *(nomen nudum), Homalopsis rufotaeniatus* Wagler, 1833, *Cerberus grantii* Cantor, 1836b, *Cerberus cinereus* Cantor, 1839, *Cerberus russelii* Fitzinger, 1843, *Homalopsis rhynchops* Cantor, 1847, *Cerberus unicolor* Gray, 1849a, and *Homalopis rhinchops* – F. Mason, 1852 *(nomen incorrectum).*

Type: Lectotype, a 1029 mm specimen described and illustrated by P. Russell (1796: 23, pl. 17) (Snodgrass, July 1788), designated herein.

Type locality: "Ganjam, India" [= Ganjam, Orissa State, SE India, 19°23'N, 85°03'E, elevation NSL] via lectotype selection. Restricted to Ganjam, India (~19°22'N, 85°03'E) *fide* J.C. Murphy et al. (2012a: 14).

Distribution: Southern Asia. India (Andhra Pradesh, Gujarat, Karnataka, Maharashtra, Orissa, Tamil Nadu, West Bengal, Krusadai Is., Andaman & Nicobars: Andaman, Long and Nicobar Is.), Sri Lanka (Eastern, North-Central, Southern, Western), Bangladesh, Myanmar (Ayeyarwady, Mon, Rakhine, Tanintharyi, Yangon) and peninsular Thailand (Phang Nga, Ranong), NSL–10 m.

Sources: Wall, 1918e, M.A. Smith, 1943.H. Taylor, 1965, P. Silva, 1969, Whitaker, 1969, Singh, 1972, Mahendra, 1984, M.J. Cox et al., 1998, A. Silva, 1990b, 2001, 2009, J.C. Murphy & Schlager, 2003, Whitaker & Captain, 2004, N. Khaire, 2006, I. Das, 2010, 2012 and Koch, 2012.

Remarks: Original description based upon Russell (1796: 23, pl. 17). May occur in Pakistan *fide* J.C. Murphy (2007b: 73) based on Murray, 1886.

C

5. *Cerberus schneiderii* (Schlegel, 1837). Essai Phys. Serp. 1: 171, 2: 341, pl. 13, figs. 6–7. (*Homalopsis schneiderii*)

Synonyms: *Elaps boaeformis* Schneider, 1801 (*nomen dubium*), *Hurria schneideriana* Daudin, 1803d (*nomen dubium*), *Python elapiformis* Merrem, 1820 (*nomen dubium*), *Python molurus* Merrem, 1820 (*nomen dubium*), *Homalopsis molurus* H. Boie, 1826, *Coluber obtutsatus* Reinwardt *in* H. Boie, 1826 *(nomen nudum), Homalopsis obtusatus* Fitzinger, 1826a, *rhinchops* Cantor, 1847 *(nomen emendatum),* and *Cerberus acutus* Gray, 1849.

Type: Lectotype, RMNH 1173, an 867 mm female described and illustrated by Schlegel (1837: 341–343, pl. 13, figs. 6–7) (J.T. Reinhardt, 1845–1847), designated by J.C. Murphy et al. (2012a: 17).

Type locality: "Timor" [S Indonesia] via lectotype selection.

Distribution: Southeast Asia and East Indies. Thailand (Nakhon Ratchasima, Pak Phanang, Phuket, Trang), Cambodia, S Vietnam (Ba Ria-Vung Tau, Ca Mau, Ho Chi Minh City, Khanh Hoa, Kien Giang), West Malaysia (Kerala, Pinang, Selangor, Seribuat Arch.: Tioman Is.), Singapore, East Malaysia (Sabah, Sarawak), Brunei, Indonesia (Ambon, Babi, Bali, Bangka, Batjan, Buru, Enggano, Erde, Flores, Fredrick Hendrik, Goram, Halmahera, Java, Kalamantan, Komodo, Lembata, Lombien, Lombok, Mentawai, Nako, Natuna, Nias, Riau Arch., Rinca, Roti Timor, Sanghe, Saparua, Seram, Simeulue, Sula, Sulawesi, Sumatra, Sumbawa, Sumba, Talaud, Ternate, Weh, West Timor, Wetar), Timor-Leste and Philippines (Bantayan, Bohol, Catanduanes, Cebu, Cuyo, Dinagat, Jagoliaou, Jolo, Luzon, Mindanao, Negros, Palawan, Panay, Polillo Romblon), NSL–80 m.

Sources: M.A. Smith, 1943, C. Haas, 1950, R. Bergman, 1955b, E.H. Taylor, 1965, Saint-Girons, 1972a, Tweedie, 1983, Ross & Lazell, 1991, Ehmann, 1992, Ross & Gonzales, 1992, David & Vogel, 1996, O'Shea, 1996, McKay, 2006, V.S. Nguyen et al., 2009 and Kaiser et al., 2011.

Remarks: Photographs of lectotype in J.C. Murphy et al. (2012a: figs. 9a, c).

C

CERCOPHIS Fitzinger, 1843 (Xenodontidae)

Type species: *Dendrophis auratus* Schlegel, 1837.
Distribution: Eastern South America..
Sources: Hoogmoed, 1983, 1997 and Zaher et al., 2009.

1. *Cercophis auratus* (Schlegel, 1837). Essai Phys. Serp. 1: 157, 2: 227–228. (*Dendrophis aurata*)

Type: Holotype, RMNH 813, a 570 mm male (H.H. Dieperink, 1831).
Type locality: "Paramaribo, Suriname" [= Paramaribo, Paramaribo District, Suriname, 5°49'N, 55°10'E, elevation NSL].
Distribution: Suriname (Paramaribo) and SE Brazil (Rio de Janeiro), NSL.

CERROPHIDION J.A. Campbell & Lamar, 1992 (Viperidae)

Type species: *Bothriechis godmani* A.C.L.G. Günther, 1863c.
Distribution: Southern Mexico and Central America.
Sources: J.A. Campbell & Lamar, 1989, 1992, 2004, Minton, 1992, Golay et al., 1993, Kraus et al., 1996, David & Ineich, 1999, López-Luna et al., 1999, McDiarmid et al., 1999, Parkinson, 1999, Castoe et al., 2005, 2009, Castoe & Parkinson, 2006, Jadin, 2010b and Jadin et al., 2011, 2012.

1. *Cerrophidion godmani* (A.C.L.G. Günther, 1863c). Ann. Mag. Nat. Hist. (3) 12(71): 364–365, pl. 6, fig. g. (*Bothriechis godmani*) (*nomen corrigendum*)

Synonyms: *Bothriechis godmanni* A.C.L.G. Günther, 1863c *(nomen incorrigendum), Bothrops brammianus* Bocourt, 1868, *Bothriechis scutigera* J.G. Fischer, 1880, and *Bothriechis trianguligera* J.G. Fischer, 1883.
Type: Lectotype, BMNH 1946.1.18.80 (formerly BMNH 1864.1.26.40), a 446 mm female (F.D. Godman & O. Salvin, 1859–1865), designated by L.C. Stuart (1963: 129).
Type locality: "Duenas and other parts of the tableland of Guatemala" [= San Miguel Dueñas, SW Sacatepéquez Department, S Guatemala, ca. 1450 m, 14°31'N, 90°48'W], via lectotype selection.
Distribution: Southern Mexico (SE Oaxaca, Chiapas) and S Guatemala (Chimaltenango, Guatemala, Huehuetenango, Sacatepéquez, Suchitepéquez, Totonicapán), 1220–3490 m.
Sources: J.A. Campbell & Solórzano, 1992, Sasa, 1997 and G. Köhler, 1999b.

2. *Cerrophidion petlalcalensis* López-Luna, Vogt & Torre-Loranca, 1999. Herpetologica 55(3): 382–387, fig. 1, 3–7, 9.

Type: Holotype, UNAM-LTH 3454, a 395 mm male (M. Angel de la Torre-Loranca; 29 June 1997).
Type locality: "Cerro de Petlalcala, Municipio San Andres Tenejapan, Veracruz, México, 18°47' N and 97°06' W at 2100 m. The locality is 10 km south of the town of Orizaba, Veracruz."
Distribution: Eastern Mexico (W Veracruz), 2100–2300 m. Known only from vicinity of type locality.
Remarks: May occur in SE Puebla and N Oaxaca *fide* López-Luna et al. (1999: 388).

3. *Cerrophidion sasai* Jadin, Townsend, Castoe & Campbell, 2012. Zool. Scripta 41(5): 5–7, figs. 2a–c.

Type: Holotype, UTA 51399, a 713 mm male (M. Sasa, 8 Nov. 2001).
Type locality: "San Ramos de Tres Ríos, Departamento de San José, Costa Rica."
Distribution: Lower Central America. Costa Rica (Alajuela, San José) and ext. W Panama (Chiriquí), 1340–1465 m.
Sources: J.A. Campbell & Solórzano, 1992, Savage, 2002 and Solórzano, 2004.

4. *Cerrophidion tzotzilorum* (J.A. Campbell, 1985). J. Herp. 19(1): 48–50, figs. 1–3. (*Bothrops tzotzilorum*)

Type: Holotype, UTA 9641, a 386 mm male (J.A. Campbell, 8 June 1979).
Type locality: "10.9 km ESE San Cristóbal de Las Casas, Chiapas, Mexico, elevation 2320 m."
Distribution: Extreme SE Mexico (Meseta Central of cen. Chiapas), 2050–2500 m.
Sources: Lazcano-Villarreal-Barrero & Gongora-Arones, 1988, Auth et al., 2000b, G. Köhler, 2001 and Jadin, 2010a.

5. *Cerrophidion wilsoni* Jadin, Townsend, Castoe & Campbell, 2012. Zool. Scripta 41(5): 7–10, figs. 4–5.

Type: Holotype, UTA 52953, a 648 mm female (E.N. Smith, 1 Feb. 2005).
Type locality: "Cerro Azul, Parque Nacional Montaña de Botaderos, Departamaento de Olancho, Honduras, 15.3783a°N/86.14200°W, elevation 1420 m."
Distribution: Upper Central America. El Salvador (Chalatenango, Santa Ana) and Honduras (Cortés, Fraqncisco Morazán, Lempira, Ocotepeque, Olancho, Yoro), 1220–2200 m.
Sources: L.D. Wilson & Myer, 1985, J.A. Campbell & Solórzano, 1992, J.A. Campbell & Lamar, 2004, Wilson & McCranie, 2004 and McCranie, 2011a.

CHAMAELYCUS Boulenger, 1919a
(Lamprophiidae)

Synonym: *Oophilositum* H.W. Parker, 1933b.
Type species: *Chamaelycus christyi* Boulenger, 1919a.
Distribution: West and cen. Africa.
Sources: Witte, 1962, Chifundera, 1990, J.-F. Trape & Roux-Estève, 1995, Kelly et al., 2009 and Zaher et al., 2009.

1. *Chamaelycus christyi* Boulenger, 1919a. Rev. Zool. Afr. 7(1): 22.

Type: Holotype, MRAC, a 370 mm male (C. Christy, 1912–1914).
Type locality: "Madié (district de l'Ituri), Congo Belge" [= Medje, Orientale Prov., NE Democratic Republic of the Congo, 2°49'N, 27°35'E, elevation 735 m].
Distribution: Southern Congo and NE Democratic Republic of the Congo (Nord-Kivu, Orientale), 700–1650 m.
Sources: J.-F. Trape & Roux-Estève, 1990 and Chippaux, 2006.

2. *Chamaelycus fasciatus* (A.C.L.G. Günther, 1858). Cat. Colub. Snakes Brit. Mus.: 196–197. (*Alopecion fasciatum*)

Type: Holotype, BMNH 1946.1.14.29, a 203 mm female (Rich).
Type locality: "West Africa."
Distribution: West and cen. Africa. Southwestern Senegal (Ziguinchor/Kolda), Guinea (Nzérékoré), Sierra Leone, Liberia, Ivory Coast (Dix-Huit Montagnes), S Ghana, S Togo (Plateaux), S Benin, S Nigeria, Cameroon (Est, Centre, Nord-Ouest, Sud, Sud-Ouest), Equatorial Guinea (Littoral, Bioko Is.), Gabon (Moyen-Ogooué, Ngounié, Ogooué-Ivindo, Ogooué-Maritime), SE Central African Republic (Sangha), Congo (Kouilou, Lekoumou, Sangha), Democratic Republic of the Congo (Bandundu, Bas-Congo, Equateur, Kinshasa, Nord-Kivu, Orientale) and ext. W Uganda (W Western), 500–1200 m.
Sources: Doucet, 1963, Mertens, 1965c, Leston & Hughes, 1968, Hughes & Barry, 1969, Hughes, 1983, Spawls et al., 2002, Ineich, 2003, Chippaux, 2006, Chirio & LeBreton, 2007, Pauwels & Vande weghe, 2007 and Segniagbeto et al., 2011.
Remarks: Possibly occurs in Benin *fide* Hughes (2013: 151). *Chamaelycus werneri* a valid species *fide* J.-F. Trape & Roux-Estève (1995: 47).

C

3. *Chamaelycus parkeri* (Angel, 1934b). Bull. Soc. Zool. France 59(5): 417–419. (*Oophilositum parkeri*)

Type: Holotype, MNHN 1934-11, a male (G. Babault, 1912–1920).
Type locality: "Kabulire (dans les forêts situées à l'Ouest du lac Kivu, alt. 1.040 m.)" [= Kabulire, Sud-Kivu Prov., E Democratic Republic of the Congo].
Distribution: Central Africa. Northern Congo (Sangha), N Democratic Republic of the Congo (Equateur, Sud-Kivu, Orientale) and Angola, 400–1040 m.
Sources: J.-F. Trape & Roux-Estéve, 1990, Chippaux, 2006 and Chirio & LeBreton, 2007.
Remarks: Holotype is male *fide* Laurent (1956a: 126). Probably occurs in SE Cameroon *fide* Chirio & LeBreton (2007: 392).

4. *Chamaelycus werneri* (Mocquard, 1902b). Bull. Mus. Hist. Nat., Paris 8(6): 412. (*Lycophidium werneri*)

Type: Holotype, MNHN 1901.551, a 350 mm female (E. Haug, 1895–1901).
Type locality: "à environ 50 kilomètres au Sud-Ouest de Lambaréné, Gabon" [= ca. 50 km SW Lambaréné (0°42'S, 10°14'E, elevation 25 m), SW Moyen-Ogooué Prov., W Gabon].
Distribution: West Africa. Central Cameroon (Adamaoua), Gabon (Moyen-Ogooué) and S Congo (Brazzaville), 1035 m.
Sources: Bogert, 1940, Hughes, 1983 and LeBreton, 1999.

CHAPINOPHIS J.A. Campbell & Smith, 1998
(Dipsadidae)

Type species: *Chapinophis xanthocheilus* J.A. Campbell & Smith, 1998.
Distribution: Guatemala.
Sources: J.A. Campbell & Smith, 1998, G. Köhler, 2001, Zaher et al., 2009 and Mulcahy et al., 2011.

1. *Chapinophis xanthocheilus* J.A. Campbell & Smith, 1998. Herpetologica 54(2): 210–216, figs. 1, 4.

Type: Holotype, UTA 35919, a 518 mm male (R.F. Savage, 8 April 1993).

Type locality: "approximately 5 km ENE Finca Miranda on trail to Aldea Vega Larga, 2300 m, Baja Verapaz, Guatemala...slopes of the Rio Vega Larga watershed at about 15°05'48"N, 90°00'20"W."

Distribution: Central Guatemala (Baja Verapaz), 1830–2300 m.

CHARINA Gray, 1949a (Boidae)

Synonyms: *Wenona* Baird & Girard, 1852c, *Pseudoeryx* Jan, 1862a *(nomen praeoccupatum), Wenonia* – Jan, 1862a *(nomen incorrectum), Chamia* – Briceño-Rossi, 1934 *(nomen incorrectum),* and *Charinae* – Engelmann & Obst, 1981 *(nomen incorrectum).*

Type species: *Tortrix bottae* Blainville, 1835.

Distribution: Western North America.

Fossil records: Lower, middle and upper Miocene of W USA and S-cen. Canada, upper Pliocene of USA, and upper Pleistocene of USA.

Sources: Stimson, 1969, Kluge, 1988a, 1993a, Walls, 1998a, McDiarmid et al., 1999, Holman, 2000a and Lawson et al., 2004.

Remarks: Genus synonymized with *Calabaria fide* Kluge (1993a: 299) but revalidated *fide* Lawson et al. (2004: 292).

1. *Charina bottae* (Blainville, 1835). Nouv. Ann. Mus. Hist. Nat. (3) 4(3): 289–290, pl. 26, figs. 1–1b. (*Tortrix bottae*)

Synonyms: *Wenona isabella* Baird & Girard, 1852c, *Wenona plumbea* Baird & Girard, 1852c, *Charina brachyops* Cope, 1888, and *Charina bottae utahensis* Van Denburgh, 1920c.

Type: Holotype, MNHN 730, a 600 mm specimen (P.E. Botta, 1827–1829).

Type locality: "Californie" [= California, USA]. Restricted to coast Distribution opposite Monterey, California *fide* K.P. Schmidt (1953a: 157).

Distribution: Extreme SW Canada (S British Columbia) and NW USA (N California, Idaho, W Montana, N Nevada, Oregon, Utah, Washington, W Wyoming), 535–2590 m.

Fossil records: Upper Pliocene (Blancan) of USA (Washington), and upper Pleistocene (Rancholabrean II) of USA (California).

Sources: V.M. Tanner & Tanner, 1939, Nussbaum & Hoyer, 1974, Stewart, 1977, 1988, Nussbaum et al., 1983, Stebbins, 1985, H.A. Brown et al., 1995, Bell & Meade, 1996, P.R. Brown, 1997, Bartlett & Tennant, 2000, Rodríguez-Robles et al., 2001 and Parmley & Walker, 2003.

Remarks: Occurs in NW Colorado *fide* Rodríguez-Robles et al. (2001: fig. 1).

2. †*Charina prebottae* Brattstrom, 1958. Bull. So. California Acad. Sci. 57(1): 6–7.0.

Types: Syntypes (2), MVP 45242, two middle trunk vertebrae (R.H. Telford & R.L. Schultz).

Type locality: "Upper Miocene, Barstow formation, Barstow Syncline, San Bernardino County, California" [= Middle Miocene *fide* Holman, 2000a: 69].

Distribution: Lower Miocene (Arikareean-Hemingfordian: 16.3–20.6 mya) of USA (Colorado, Nebraska, South Dakota, Wyoming), middle Miocene (Barstovian: 10.3–16.3 mya) of USA (California, Nebraska, South Dakota, Wyoming), upper Miocene (Hemphillian: 4.9–10.3 mya) of USA (Nebraska, Texas) and W Canada (Saskatchewan), Pliocene (1.8–4.9 mya) of USA (Washington), and Pleistocene (0.01–1.8 mya) of USA (California).

Sources: Holman, 1976a, 1979b, 2003, Parmley, 1988 and Parmley & Holman, 1995.

3. *Charina umbratica* Klauber, 1943d. Trans. San Diego Soc. Nat. Hist. 10(7): 83–84. (*Charina bottae umbratica*)

Type: Holotype, SDSNH 12101, a 264 mm male (C. Searl, 1 July 1929).

Type locality: "Fern Valley, near Idyllwild, San Jacinto Mountains at elevation 5800 ft., Riverside County, California," USA.

Distribution: Southwestern USA (S California), 1560–1765 m.

Sources: Stebbins, 1986 and Rodríguez-Robles et al., 2001.

Remarks: A valid species *fide* Rodríguez-Robles et al. (2001: 234).

†*CHEILOPHIS* Gilmore, 1938 (Boidae)

Type species: †*Cheilophis huerfanoensis* Gilmore, 1938.

Distribution: Lower to middle Eocene of USA.

Sources: H, Miller, 1955, Holman, 1979a, 2000a, Rage, 1984a–b, Sullivan & Lucas, 1988 and Longrich et al., 2012b.

1. †*Cheilophis huerfanoensis* Gilmore, 1938. Geol. Soc. Amer., Spec. Pap. (9): 79–80, figs. 34a–d.

Type: Lectotype, AMNH 5195, one middle trunk vertebra (AMNH Exped., 1918), designated by Rage (1984a: 219).

Type locality: "two miles north of Gardner, Custer County, Colorado, USA. Upper Huerfano, early Bridgerian, early middle Eocene" via lectotype selection.

Distribution: Lower Eocene (Wasatchian: 50.3–55.4 mya) of USA (New Mexico) and middle Eocene (Bridgerian: 46.2–50.3 mya) of USA (Colorado).

C

CHERSODROMUS J.T. Reinhardt, 1861b
(Dipsadidae)

Synonyms: *Opisthiodon* W.C.H. Peters, 1861b and *Schmidtophis* E.H. Taylor, 1949.
Type species: *Chersodromus liebmanni* J.T. Reinhardt, 1861b.
Distribution: Southern Mexico.
Sources: H.M. Smith & Taylor, 1945, Dixon & Ketchersid, 1969, Cadle, 1984b, G. Köhler et al., 2001 and Zaher et al., 2009.

1. *Chersodromus liebmanni* J.T. Reinhardt, 1861b. Vidensk. Medd. Naturhist. Foren. Kjøbenhavn (1860) 22: 243–245, pl. 4, figs. 10–11.

Synonyms: *Chersodromus liebmanni* J.T. Reinhardt, 1861a *(nomen nudum), Chersodromus nigricans* Kröyer *in* J.T. Reinhardt, 1861a, *Opisthiodon torquatus* W.C.H. Peters, 1861b, and *Dirosema collare* F. Werner, 1900a.
Type: Holotype, ZMUC 60561, a 250 mm male (R.W. Liebmann, 1841–1843).
Type locality: "Mexico." Restricted to Cuautlapan, Veracruz, Mexico *fide* H.M. Smith & Taylor (1950: 347) and to Mirador, Veracruz, Mexico *fide* H.M. Smith & Braestrup (1963: 240).
Distribution: Southeastern Mexico (Oaxaca, Veracruz), 1800–1840 m.
Sources: H.M. Smith & Braestrup, 1963, Canseco-Maquez & Flores-Villela, 1995, Casas-Andreu et al., 1996 and Mancilla-Moreno, 1998.

2. *Chersodromus rubriventris* (E.H. Taylor, 1949). Univ. Kansas Sci. Bull. 33(2): 193–194. (*Schmidtophis rubriventris*)

Type: Holotype, LSUMZ 577, a 308 mm specimen (C.R. Shaw, 1946–1947).
Type locality: "near Xilitla (Xilitla Region) San Luis Potosí, México" [= vicinity of Xilitla (21°23'N, 98°59'W, elevation 660 m), San Luis Potosí State, NE Mexico].
Distribution: Central Mexico (E Querétaro, San Luis Potosí), 660 m.

CHILOMENISCUS Cope, 1860f
(Colubridae)

Type species: *Chilomeniscus stramineus* Cope, 1860f.
Distribution: Extreme SW USA and NW Mexico.
Sources: H.M. Smith & Taylor, 1945, Banta & Leviton, 1963, R.W. Murphy & Ottley, 1984, Stebbins, 1985, McPeak, 2000 and Grismer et al., 2002a.

1. *Chilomeniscus savagei* Cliff, 1954. Trans. San Diego Soc. Nat. Hist. 12(5): 71–72, fig. 2.

Type: Holotype, CAS-SU 14034, a 234 mm female (J.M. Savage, 3 April 1952).
Type locality: "southwest coast of Cerralvo Island, Gulf of California, Mexico."
Distribution: Northwestern Mexico (Baja California del Sur: Ceralvo Is.).

2. *Chilomeniscus stramineus* Cope, 1860f. Proc. Acad. Nat. Sci. Philadelphia 12(8): 339.

Synonyms: *Chilomeniscus cinctus* Cope, 1861e, *Chilomeniscus ephippicus* Cope, 1867a, *Chilomeniscus stramineus fasciatus* Cope, 1892a, *Chilomeniscus punctatissimus* Van Denburgh & Slevin, 1921a, and *Chilomeniscus stramineus esterensis* Hoard, 1939.
Types: Syntypes (5), USNM 4674a–b, USNM 6495a–b, and ANSP 3405, longest syntype 229 mm (J. Xantus, Dec. 1859).
Type locality: "Cape St. Lucas, Lower California" [= Cabo San Lucas, Baja California Sur State, NW Mexico, 22°53'N, 109°55'W, elevation 25 m].
Distribution: Southwestern USA (SW Arizona) and NW Mexico (Baja California, W Sonora and Cedros, Monserrate, San José and San Marcos Is.), NSL–915 m.
Sources: Fowlie, 1965, Powers & Banta, 1974b, Wong, 1997 and Bartlett & Tennant, 2000.

CHILORHINOPHIS F. Werner, 1907a
(Atractaspididae)

Synonyms: *Chilorinophis* – Loveridge, 1924c *(nomen incorrectum), Parkerophis* T. Barbour & Amaral, 1927, and *Chilorpinophis* – Isemonger, 1983 *(nomen incorrectum).*
Type species: *Chilorhinophis butleri* F. Werner, 1907a.
Distribution: Eastern Africa.
Sources: Battersby, 1950, Witte & Laurent, 1947, Loveridge, 1958, Wallach, 1991, Meirte, 1992, Underwood & Kochva, 1993, Vidal et al., 2008, Kelly et al., 2009 and Zaher et al., 2009.

1. *Chilorhinophis butleri* F. Werner, 1907a. Kaiserliche Akad. Wiss. Wien 26: 480.

Synonyms: *Parkerophis carpenteri* H.W. Parker, 1927a and *Chilorhinophis carpenteri liwalensis* Loveridge, 1951a.
Type: Holotype, NMW 14729, a 315 mm specimen (F. Werner, 30 March 1905).
Type locality: "Mongalla" [= Mangalla, Central Equatoria Prov., S South Sudan, 5°12'N, 31°46'E, elevation 470 m].
Distribution: Eastern Africa. Southeastern South Sudan (Central Equatoria), SE Tanzania (Lindi, Mtwara) and NE Mozambique (Cabo Delgado), 40–520 m.

Sources: Meirte, 1992 and Spawls et al., 2002.
Remarks: Supplemental original description in F. Werner (1907c: 1881, pl. 3, figs. 8a–d). *Chilorhinophis carpenteri* a synonym *fide* A. Resetar *in* Wallach (1991: 416).

C

2. *Chilorhinophis gerardi* (Boulenger, 1913c). Rev. Zool. Afr. 3(1): 103, 2 figs. (*Apostolepis gerardi*)

Synonym: *Chilorhinophis gerardi tanganyikae* Loveridge, 1951a.
Type: Holotype, MRAC 1205, a 315 mm male (P. Gérard, 1906–1913).
Type locality: "Kikondja, Katanga, Congo Belge" [= Kikondja, cen. Katanga Prov., SE Democratic Republic of the Congo, 8°12'S, 26°25'E, elevation 585 m].
Distribution: Southeastern Africa. Southeastern Democratic Republic of the Congo (Katanga), W Tanzania (Kigoma, Rukwa), Zambia (Central, Copperbelt, Lusaka, Northern, North-Western, Southern) and N Zimbabwe (N Manicaland, Mashonaland Central, Mashonaland East, Mashonaland West, Matabeleland North, Midlands), 585–1300 m.
Sources: FitzSimons, 1974, Broadley, 1983, Branch, 1988, Spawls et al., 2002, Broadley et al., 2003 and Broadley & Blaylock, 2013.

CHIONACTIS Cope, 1860e (*nomen substitutum*) (Colubridae)

Synonym: *Lamprosoma* Hallowell, 1857a *(nomen praeoccupatum).*
Type species: *Rhinostoma occipitale* Hallowell, 1854a.
Distribution: Southwestern USA and NW Mexico.
Fossil records: Pleistocene of SW USA.
Sources: Stickel, 1943, Klauber, 1951, Fowlie, 1965, Stebbins, 1985, Bartlett & Tennant, 2000, Holman, 2000a and Mahrdt et al., 2001

1. *Chionactis occipitalis* (Hallowell, 1854a). Proc. Acad. Nat. Sci. Philadelphia 7(3): 95. (*Rhinostoma occipitale*)

Synonyms: *Lamprosoma annulatum* Baird, 1859a, *Sonora occipitalis klauberi* Stickel, 1941, *Chionactis occipitalis talpina* Klauber, 1951, and *Chionactis saxatilis* Funk, 1967.
Type: Holotype, USNM 8030, a 229 mm specimen (R.S. Williamson [Pacific Railroad Surv.], 1853–1854).
Type locality: "Mohave Desert, California" [USA]. Listed as "Fort Mojave" *fide* USNM catalogue. Restricted to the western Mojave Desert in the vicinity of the Mojave River *fide* Klauber (1951: 154).
Distribution: Southwestern USA (SW Arizona, SE California, S Nevada) and ext. NW Mexico (NE Baja California Norte, ext. NW Sonora), NSL–1500 m.
Fossil records: Upper Pleistocene (Rancholabrean II) of USA (Arizona, California).
Sources: Elvin, 1963, Cross, 1979, P.R. Brown, 1997, Mahrdt et al., 2001b, Grismer, 2002 and D.A. Wood et al., 2008.
Remarks: Holotype based on specimen depicted in Hallowell (1859: pl. 4, figs. 2a–c) *fide* Klauber (1951: 153).

2. *Chionactis palarostris* (Klauber, 1937b). Trans. San Diego Soc. Nat. Hist. 8(27): 363–365. (*Sonora palarostris*)

Synonyms: *Chionactis palarostris organica* Klauber, 1951 and *Chionactis palarostris seri* Blake, 1970 *(nomen ineditum).*
Type: Holotype, SDSNH 26771 (formerly LMK 26771), a 312 mm male (G. Lindsay, April 1937).
Type locality: "5 miles south of Magdalena, Sonora, Mexico." Corrected to six miles south of Hermosillo, Sonora, Mexico *fide* Blake *in* Mahrdt et al. (2001c: 732.1).
Distribution: Southwestern USA (SW Arizona) and NW Mexico (NW Sonora), NSL–760 m.
Fossil records: Upper Pleistocene (Rancholabrean II) of USA (Arizona).
Sources: Mahrdt et al., 2001c and H.M. Smith et al., 2005.

CHIRONIUS Fitzinger, 1826a (Colubridae)

Synonyms: *Erpetodrius* H. Boie *in* F. Boie, 1826, *Erpetodryas* H. Boie *in* Schlegel, 1826a *(nomen emendatum), Erpetodrias* – H. Boie *in* F. Boie, 1827 *(nomen incorrectum), Macrops* Wagler, 1830, *Herpetodrys* Wagler, 1830 *(nomen emendatum),* ? *Coronella* Bonaparte, 1831a *(nomen praeoccupatum), Herpetodryas* Schlegel, 1837 *(nomen emendatum), Hylophis* Fitzinger, 1843, *Phyllosira* Cope, 1862d, *Herpetodryos* – J.W. Müller, 1865 *(nomen incorrectum), Herpetodrias* – Steindachner, 1876 *(nomen incorrectum), Herpedodryas* – F. Müller, 1878b *(nomen incorrectum), Leptastes* Philippi, 1899 *(nomen nudum), Leptastes* Philippi *in* Quijada, 1916, *Herpelodryas* – Guibé, 1958 *(nomen incorrectum),* and *Shironius* – Hoge & Maranhão-Niña, 1964 *(nomen incorrectum).*
Type species: *Coluber carinatus* Linnaeus, 1758.
Distribution: Central America and South America.
Sources: Bailey, 1955, Donoso-Barros, 1969a, J.A. Peters & Orejas-Miranda, 1970, Wiest, 1978, Cunha & Nascimento, 1983c, Dixon & Soini, 1986, Pérez-Santos & Moreno, 1988, Dixon et al., 1995 and Hollis, 2006.
Remarks: Hollis (2006: 444–445) elevated seven subspecies to species: *Chironius carinatus flavopictus* to *C. flavopictus*; *C. carinatus spixii* to *C. spixii*; *C. fuscus leucometapus* to *C. leucometapus*; *C. multiventris cochranae* to *C. cochranae*; *C. m. foveatus* to *C. foveatus*; *C. m. septentrionalis* to *C. septentrionalis*; *C. quadricarinatus maculoventris* to *C. maculoventris.*

1. *Chironius bicarinatus* (Wied-Neuwied, 1820a). Reise Brasil. 1: 181, col. pl. (*Coluber bicarinatus*)

Synonym: *Natrix bicarinata* – Wagler *in* Spix, 1824.
Type: Holotype, ? ZSM 1752/0, an 1125+ mm female (A.M.P. zu Wied-Neuwied, 26 Nov.–19 Dec. 1815.
Type locality: "an dem Sandufer einer Lagoa...unweit Villa do Espirito-Santo, wohin man noch einen Weg von 5 Legoas hat" [= lake near Rio Jacu, within five leagues of Villa do Espírito Santo, Estado Espírito Santo, Brazil, *fide* Wiest, 1978: 79].
Distribution: Southeastern Brazil (E Bahia, Espírito Santo, Goiás, ext. E Minas Gerais, Paraná, Rio de Janeiro, Rio Grande do Sul, Santa Catarina, São Paulo, Sergipe, Búzios, Cananéia, Cardoso, Comprida, Grande, Mar Virado, Porchat, Santo Amaro, São Sebatião and São Vicente Is.), NE Uruguay (Artigas, Cerro Largo, Rio Negro, Salto) and NE Argentina (Chaco, Corrientes, Entre Ríos, Formosa, Missiones, Salta), NSL–1850 m.
Sources: P. Müller, 1969b, Cei, 1994, Carreira-Vidal et al., 2005, Duarte Roche & Van Sluys, 2006, Carreira-Vidal & Lombardo, 2007a, Cicchi et al., 2007 and Morato et al., 2011.

2. *Chironius carinatus* (Linnaeus, 1758). Syst. Nat., ed. 10, 1: 223. (*Coluber carinatus*)

Synonyms: *Coluber subfuscus* Lacépède, 1789 (*nomen rejiciendum*), *Coluber chironius* Donndorff, 1798, *Coluber moschatus* Donndorff, 1798, *Coluber null* Donndorff, 1798, ? *Coluber flaviceps* Gravenhorst, 1807, *Coluber pickeringii* Hallowell, 1845a, *Herpetodryas carinatus carinatissima* Jan, 1863b (*nomen nudum*), *Herpetodryas carinatus decalepis* Jan, 1863b (*nomen nudum*), *Herpetodryas carinatus glabra* Jan, 1863b (*nomen nudum*), *Herpetodryas carinatus macrophthalma* Jan, 1863b (*nomen nudum*), and *Herpetodyas carinatus* – Briceño-Rossi, 1934 (*nomen incorrectum*).
Type: Holotype, NHR Lin-33 (formerly MAFR), a 1500 mm male (Mus. Drottn.).
Type locality: "Indiis" [= India] (in error *fide* Hoge, 1964d: 53). Corrected to Suriname *fide* Hoge (1962c: 53).
Distribution: Northern South America. Venezuela (Amazonas, Bolívar, Delta Amacuro, Distrito Federal, Monagas, Miranda, Sucre), Trinidad & Tobago (Trinidad), Guyana (Cuyuni-Mazaruni, Essequibo Islands-West Demerara, Pomeroon-Supenaam, Upper Takutu-Upper Essequibo), Suriname (Commewijne, Para, Paramaribo), French Guiana (Cayenne, Saint-Laurent-du-Maroni), Peru (Loreto, Madre de Dios) and N Brazil (Alagras, Amazonas, Bahia, Ceara, Maranhao, Pará, Pernambuco, Roraima, Sergipe), NSL–1115 m.
Sources: E.H. Taylor, 1951, Roze, 1966a, Vanzolini et al., 1980, Lancini, 1986, Chippaux, 1987, Duellman & Salas, 1991, Pérez-Santos & Moreno, 1991, Gorzula & Señaris, 1998, Starace, 1998, Markezich, 2002, Abuys, 2003, Solórzano, 2004, Navarrete et al., 2009, Kok, 2010 and C.J. Cole et al., 2013.

3. *Chironius challenger* Kok, 2010. Zootaxa (2611): 32–37, figs. 1, 2a–c, 3a, 4d, 6 (upper and lower left).

Type: Holotype, IRSNB 2659, a 720 female (J.R. Philippe & J.R. Kok, 25 Nov. 2007).
Type locality: "southeastern slope of Maringma Tepui, Cuyuni-Mazaruni District, Guyana (05° 12' N, 060° 35' W, 1500 m elevation)."
Distribution: Venezuela (Bolívar) and Guyana (Cuyuni-Mazaruni, Potaro-Siparuni), 1400–2090 m.
Source: C.J. Cole et al., 2013.

4. *Chironius exoletus* (Linnaeus, 1758). Syst. Nat., ed. 10, 1: 223. (*Coluber exoletus*)

Synonyms: *Coluber pyrrhopogon* Wied-Neuwied, 1824b and *Zaocys tornieri* F. Werner, 1896b.
Type: Holotype, ZMUU 150 (formerly JA), a 1020 mm female, *fide* L. Wallin (1992: 226) and L. Wallin *in* Dixon et al. (1995: 92).
Type locality: "Indiis" [= India] (in error). Partially corrected to South America and India *fide* Gmelin (1789: 1110). Corrected to America *fide* Wagler (1830: 180). Restricted to Brazil *fide* Fitzinger (1843: 60).
Distribution: Lower Central America and South America. Costa Rica (Alajuela, Cartago, Guanacaste, Heredia, Limón, Puntarenas, San José), Panama (Chiriquí, Darién, Panamá), Colombia (Amazonas, Antioquia, Caquetá, Cauca, Huila, Meta, Nariño, Norte de Santander, Vaupés), SE Venezuela (Amazonas, Bolívar, Delta Amacuro, Zulia), Guyana (Cuyuni-Mazaruni, Demerara-Mahaica, Essequibo Islands-West Demerara, Potaro-Siparuni, Upper Demerara-Berbice, Upper Takutu-Upper Essequibo), Suriname (Marowijne, Paramaribo), French Guiana (Cayenne), Brazil (Amazonas, Bahia, Espírito Santo, Goiás, Máranhao, Mato Grosso, Minas Gerais, Pará, E Paraná, Pernambuco, Rio de Janeiro, Santa Catarina, São Paulo, Sergipe, Cananéia, Cardoso, Comprida, Grande, Mar Virado, Porchat, Santo Amaro and São Vicente Is.), Ecuador (Esmeraldas, Guayas, Imbabura, Los Rios, Manabi, Morona-Santiago, Napo, Pastaza, Pichincha), E Peru (Amazonas, Ayacucho, Cusco, Huánuco, Junín, Loreto, Madre de Dios, Puno), N Bolivia (Cochabamba, El Beni, La Paz, Pando, Santa Cruz) and ext. NE Argentina (Misiones), NSL–2400 m.
Sources: P. Müller, 1969b, Hoge et al., 1976–1977, Cunha & Nascimiento, 1982d, 1983, Pérez-Santos & Moreno, 1991, Cei, 1994, Starace, 1998, Frota & Yuki, 2002, Savage, 2002, Duarte-Rocha & Van Sluys, 2006, Cicchi et al., 2007, Navarrete et al., 2009, Kok, 2010 and Morato et al., 2011.
Remarks: Holotype designation of ZMUU 135 *fide* Hoge et al. (1978b: 41–42) in error.

C

5. *Chironius flavolineatus* (Jan, 1863b). Elen. Sist. Ofidi: 80. (*Herpetodryas carinatus flavolineata*)

Type: Lectotype, formerly MSNM, specimen illustrated in Jan & Sordelli (1869 *in* 1866–1870, 2[31]: pl. 2, fig. 5), designated by Dixon et al. (1995: 112), destroyed in 1943 during World War II.

Type locality: "Brazil" via lectotype selection.

Distribution: Eastern South America. Eastern Brazil (Alagoas, Amapá, Amazonas, W Bahia, Ceará, Goiás, E Mato Grosso, Mato Gross do Sul, Minas Gerais, SE Pará, Paraiba, Paraná, W Pernambuco, Rondônia, São Paulo, Tocantins), cen. Bolivia (El Beni, Santa Cruz) and E Paraguay, 20–1380 m.

Sources: Moura-Leite et al., 1996, Colli & Péres, 1997, 1998, Pinto et al., 2010 and Albuquerque & Freitas, 2011.

Remarks: Dixon et al. (1995: 112) claimed figure for type was labeled 3 but it was labeled 5 (there is no fig. 3) and is the third figure from the top.

6. *Chironius flavopictus* (F. Werner, 1909b). Mitt. Naturhist. Mus. Hamburg 26: 220. (*Herpetodryas carinatus flavopicta*)

Type: Lectotype, ZMH 651 (formerly ZMH 3513), a 1530 mm female (F. von Buchwald, 10 June 1902), designated by Dixon et al., (1995: 78).

Type locality: "Ecuador, Guayaquil" via lectotype selection.

Distribution: Lower Central America and NW South America. Southern Costa Rica (Alajuela, Limón, Puntarenas, San José), Panama (Boca del Toro, Canal Zone, Chiriquí, Chocó, Cocle, Colón, Darién, Herrera, Panamá, Veraguas), NW Colombia (Atlántico, Caldas, Cauca, S Chocó, Nariño, Valle, Vichada) and W Ecuador (Esmeraldas, Guayas), NSL–750 m.

Sources: Savage, 2002 and Solórzano, 2004.

Remarks: A valid species *fide* Hollis (2006: 445).

7. *Chironius foveatus* Bailey, 1955. Occ. Pap. Mus. Zool. Michigan (571): 10.

Type: Holotype, MN 1840, a 2281–2267 mm male (V. Rosa).

Type locality: "Rio Fortuna, Ilhéus, Bahia," Brazil.

Distribution: Southeastern Brazil (Bahia, Espírito Santo, Rio de Janeiro, Santa Catarina, São Paulo).

Remarks: A valid species *fide* Hollis (2006: 445).

8. *Chironius fuscus* (Linnaeus, 1758). Syst. Nat., ed. 10, 1: 222–223. (*Coluber fuscus*)

Synonyms: *Coluber saturninus* Linnaeus, 1758, *Natrix cinnamonea* Wagler *in* Spix, 1824, *Natrix dysgena* Wagler, 1825 (*nomen ineditum*), and *Herpetodryas sebastus* Cope, 1861c.

Type: Holotype, NHR Lin-34 (formerly MAFR), a 955–1005 mm specimen (Mus. Drottn.).

Type locality: "Asia" (in error). Corrected to Suriname *fide* Wiest (1978: 159).

Distribution: Northern South America. Eastern Colombia (Amazonas, Caquetá, Meta, Vaupés), S Venezuela (Amazonas, Bolívar, Delta Amacuro), Guyana (Barima-Waini, Cuyuni-Mazaruni, Essequibo Islands-West Demerara, Potaro-Siparuni, Upper Demerara-Berbice, Upper Takutu-Upper Essequibo), Suriname (Brokopondo, Marowijne), French Guiana (Cayenne, St. Laurent-du-Maroni), E Ecuador (Chimborazo, Morona-Santiago, Napo, Pastaza), Brazil (Acre, Amazonas, Bahia, Espírito Santo, Maranhao, Pará, Paraná, Rio de Janeiro, Roraima, São Paulo, Cardoso, Porchat, Santo Amaro and São Vicente Is.), E Peru (Amazonas, Ayacucho, Cusco, Junin, Loreto, Madre de Dios, Puno, Pasco) and N Bolivia (Beni, Cochabamba, La Paz, Santa Cruz), 100–3500 m.

Sources: Pérez-Santos & Morelos, 1991, Starace, 1998, Lehr, 2001, Duellman, 2005, Cicchi et al., 2007, C.W. Myers & Donnelly, 2008, Navarrete et al., 2009, Kok, 2010 and C.J. Cole et al., 2013.

Remarks: Photograph of *C. fuscus* type in Kok (2010: 41, fig. 8) and head of type in Hoge & Maranhao-Niña (1964: figs. 5, 7).

9. *Chironius grandisquamis* (W.C.H. Peters, 1868a). Mber. Königl. Akad. Wiss. Berlin 1868(7): 451. (*Spilotes grandisquamis*)

Synonyms: *Herpetodryas melas* Cope, 1886b, *Herpetodryas schlüteri* F. Werner, 1899a *(nomen incorrigendum), Herpetodryas schlueteri* J.A. Peters & Orejas-Miranda, 1970 *(nomen corrigendum),* and *Herpetodryas vicinus* Boulenger, 1915e, and *Herpetodryas vicimus* – Briceño-Rossi, 1934 *(nomen incorrectum).*

Type: Holotype, ZMB 3940, a 2207 mm male.

Type locality: "Costa Rica."

Distribution: Central America and ext. NW South America. Northern and SE Honduras (Atlántida, Colón, Cortés, El Paraiso, Gracias a Dios, Olancho), E Nicaragua (Atlántico Sur, Chontales, Jinotega, Matagalpa, Rio San Juan, Zelaya), Costa Rica (Alajuela, Cartago, Guanacaste, Heredia, Limón, Puntarenas, San José), Panama (Bocas del Toro, Canal Zone, Chiriquí, Colón, Darién, Panamá, Veraguas), W Colombia (Antioquia, Boyacá, Cauca, Chocó, Cundinamarca, Meta, Nariño, Valle, Gorgona Is.) and NW Ecuador (Cotopaxi, Esmeraldas, Imbabura, Pichincha), NSL–2400 m.

Sources: Amaral, 1944g, E.H. Taylor, 1951, 1954, Pérez-Santos & Moreno, 1988, 1991, G. Köhler, 1999b, Savage, 2002, Solórzano, 2004, McCranie et al., 2006, Mareno-Arias et al., 2008, McCranie, 2011a and Travers et al., 2011.

10. *Chironius laevicollis* (Wied-Neuwied, 1824b). Isis von Oken 17(6): 666. (*Coluber laevicollis*)

Type: Holotype, not designated, an 1858 mm specimen (A.M.P. zu Wied-Neuwied, 26 Nov. 1815), location unknown.

Type locality: Unknown. Designated as "am Ufer des Flusses Itabapuana erlegt, als sie von der Fazenda de Muribeca aus jene Urwälder der Puris durchstreiften" *fide* Wied-Neuwied (1825c: 298–299), [= Lower Rio Itabapoana, on boundary between Espírito Santo and Rio de Janeiro, Brazil *fide* Dixon et al., 1995: 135].

Distribution: Southeastern Brazil (Bahia, Espírito Santo, E Mina Gerais, Paraná, Rio de Janeiro, Santa Catarina, São Paulo, Cardoso, Comprida, São Sebastião and São Vicente Is.), NSL–1000 m.

Sources: P. Müller, 1969b, Marques & Puorto, 1996 and Cicchi et al., 2007.

Remarks: Record from Paraguay doubtful *fide* Dixon et al. (1995: 135).

11. *Chironius laurenti* Dixon, Wiest & Cei, 1995. Mus. Reg. Sci. Nat., Torino (1993) (13): 141–146, fig. 39.

Synonym: *Chironius dixoni* Wiest, 1978 (*nomen ineditum*).

Type: Holotype, AMNH 101815, an 1846 mm male (W.P. Maclean, 8 Sept. 1965).

Type locality: "Bolivia, Department of Beni, Rio Mamoré, ca. 23 km W San Javier" [= ca. 14°36'N, 65°00'W, elevation 150 m].

Distribution: Southeastern Brazil (Mato Grosso) and Bolivia (Beni, Cochabamba, Santa Cruz), 150 m.

12. *Chironius leucometapus* Dixon, Wiest & Cei, 1995. Mus. Reg. Sci. Nat., Torino (1993) (13): 123–126, fig. 31. (*Chironius fuscus leucometapus*)

Synonym: *Chironius fuscus leucometapus* Wiest, 1978 (*nomen ineditum*).

Type: Holotype, AMNH 53317, a 1474 mm male.

Type locality: "Peru, Department of Junin, Chanchamayo."

Distribution: Peru (Huanuco, Junin, San Martin), 500–3500 m.

Remarks: A valid species *fide* Hollis (2006: 445).

13. *Chironius maculoventris* Dixon, Wiest & Cei, 1995. Mus. Reg. Sci. Nat., Torino (1993) (13): 181–184, fig. 55. (*Chironius quadricarinatus maculoventris*)

Synonym: *Chironius multriventris maculoventris* Wiest, 1978 (*nomen ineditum*).

Type: Holotype, UZMK 60816, a 1067 mm male (native, 26 Oct. 1878)

Type locality: "Argentina, Prov. of Corrientes, Plata Staterne."

Distribution: Western Paraguay (Boquerón, Presidente Hayes) and NE Argentina (Chaco, Corrientes, Entre Ríos, Formosa, Salta), 50–275 m.

Remarks: A valid species *fide* Hollis (2006: 445).

14. *Chironius monticola* Roze, 1952a. Acta Biol. Venez. 1(5): 100–102, figs. 1–2.

Type: Holotype, MBUCV 2019, a 1225 mm female (J. Racenis, 15 July 1951).

Type locality: "la carretera Caracas-El Junguito, cerca de El Junquito, Distrito Federal, Venezuela, en una altura de 1.850 metros sobre el nivel del mar."

Distribution: Northwestern South America. Western Colombia (Antioquía, Boyacá, Caldas, Cundinamarca, Nariño), NW Venezuela (Aragua, Distrito Federal, Falcón, Mérida, Miranda, Táchira), Ecuador (Chimborazo, Esmeraldas, Guayas, Loja, Morona-Santiago, Napo, Pastaza, Tungurahua, Zamora-Chinchipe), Peru (Amazonas, Cusco, Huanuco, Junín, Loreto, Madre de Dios, Pasco, Piura, Puno, San Martín) and cen. Bolivia (Cochabamba, Santa Cruz), 90–2930 m.

Sources: Roze, 1966a, Lancini, 1986, Pérez-Santos & Moreno, 1991, Lamar, 1997b, Esqueda & La Marca, 1999, Mijares-Urrutia & Arends, 2000, Navarrete et al., 2009 and Rojas-Morales, 2012b.

15. *Chironius multiventris* K.P. Schmidt & Walker, 1943a. Field Mus. Nat. Hist. (Zool.) 24(26): 282–283.

Type: Holotype, FMNH 38250, a 1226–1250 mm female (Valdivia).

Type locality: "Department of Madre de Dios, Peru." Restricted to Selvas de Rio Madre de Dios *fide* Dixon et al. (1995: 168).

Distribution: Northern South America. Southeastern Colombia (Amazonas, Guainía), S Venezuela (Amazonas), Guyana (Barima-Waini, Cuyuni-Mazaruni, Essequibo Islands-West Demerara, Upper Demerara-Berbice, Upper Takutu-Upper Essequibo), Suriname (Brokopondo, Nickerie, Para), French Guiana (Cayenne), E Ecuador (Chimborazo, Napo, Pastaza), E Peru (Amazonas, Huánuco, Junín, Loreto, Madre de Dios, Pará), Brazil (Acre, Amapa, Amazonas, Maranhão, Rondônia, Roraima, Cananéia, Cardoso, Santo Amaro, São Sebastião and São Vicente Is.) and Bolivia (Beni), NSL–1200 m.

Sources: Chippaux, 1987, Hoogmoed & Avila-Pires, 1991, Pérez-Santos & Moreno, 1991, Moura-Leite et al., 1996, Starace, 1998, Freitas, 1999, Marques et al., 2001, Embert, 2002a, Abuys, 2003, Duellman, 2005, Duarte-Rocha & Van Sluys, 2006, Cicchi et al., 2007, Klaczko et al., 2010, Kok, 2010 and C.J. Cole et al., 2013.

C

16. *Chironius quadricarinatus* (F. Boie, 1827). Isis von Oken 20(6): 548. (*Erpetodryas 4 dricarinatus*)

Synonyms: *Tyria quadricarinata* Fitzinger, 1826a (*nomen nudum*) and *Phyllosira flavescens* Cope, 1862d.

Type: Holotype, RMNH 605 (formerly NMW), a 952 mm male (J. Natterer, 1817–1835).

Type locality: Unknown. Listed as "Bresil" [= Brazil] *fide* RMNH catalogue. Restriction to Asunción, Paraguay *fide* Bailey (1955: 15) invalid.

Distribution: Southeastern South America. Southeastern Brazil (Goiás, Bahia, Mato Grosso, Mato Grosso do Sul, Minas Gerais, Rio de Janeiro, São Paulo, Tocantins), E Bolivia (Santa Cruz, Tarija), Paraguay (Amambay, Boquerón, Central, Concepción, Guairá, Neembucú, Paraguarí, Presidente Hayes, San Pedro) and N Argentina (Chaco, Corrientes), NSL–1100 m.

Sources: Cei, 1994, Argólo, 1998a, Leynaud Bucher, 1999, Embert, 2002b, Cacciali, 2008 and Pinto et al., 2010.

17. *Chironius scurrulus* (Wagler *in* Spix, 1824). Serp. Brasil. Sp. Nov.: 24–26, pl. 8. (*Natrix scurrula*)

Synonyms: *Dendrophis viridis* A.-M.-C. Duméril, Bibron Duméril, 1854a, *Herpetodryas holochlorus* Cope, 1875b, and *Chironius barrioi* Donoso-Barros, 1969a.

Type: Lectotype, ZSM 2628/0, an 1475+ mm male (K.F.P. von Martius, Dec. 1919–Feb. 1820), designated by Hoogmoed & Gruber (1983: 329).

Type locality: "in aquosis fluminis Japura, Brasil" [= Río Japura, Amazonas State, NW Brazil, from 0°24'S, 72°17'W to 3°08'S, 64°66'W *fide* Vanzolini, 1981: xxvi) via lectotype selection.

Distribution: Northwestern South America. Southeastern Colombia (Amazonas, Meta, Vaupés), Venezuela (Amazonas, Bolívar, Delta Amacuro, Monagas, Sucre), Trinidad and Tobago (Trinidad), Guyana (Cuyuni-Mazaruni, Essequibo Islands-West Demerara, Pomeroon-Supenaam, Potaro-Siparuni, Upper Demerara-Berbice, Upper Takutu-Upper Essequibo), Suriname (Brokopondo, Marowijne, Nickerie, Paramaribo), French Guiana (St. Laurent-du-Maroni), Brazil (Amazonas, Maranhao, Pará), E Ecuador (Morona-Santiago, Napo, Pastaza), E Peru (Amazonas, Huánuco, Junín, Loreto, Madre de Dios, Pasco, San Martin) and N Bolivia (El Beni, La Paz, Pando), NSL–840 m.

Sources: Hoge & Naranhão, 1964, Hoge, 1969a, Lancini, 1986, Hoogmoed, 1983, Pérez-Santos & Moreno, 1991, Gorzula & Señaris, 1998, Starace, 1998, Rivas-Fuenmayor & La Marca, 2001a, Duellman, 2005, Navarrete et al., 2009, Kok, 2010 and C.J. Cole et al., 2013.

Remarks: Original description listed plate as pl. 9 in error.

18. *Chironius septentrionalis* Dixon, Wiest & Cei, 1995. Mus. Reg. Sci. Nat., Torino (1993) (13): 173–177, fig. 52. (*Chironius multiventris septentrionalis*)

Synonym: *Chironius multiventris septentrionalis* Wiest, 1978 (*nomen ineditum*).

Type: Holotype, AMNH 98228, a 1651 mm male (W. Beebe, 1942–1946).

Type locality: "Venezuela, Estado Aragua, Rancho Grande, near Maracay" [= Rancho Grande, near Maracay (10°15"N, 67°36'W, 450 m), Aragua State, N Venezuela].

Distribution: Northern Venezuela (Aragua, Carabobo, Distrito Federal, Guárico, Lara, Miranda, Sucre, Trujillo) and Trinidad and Tobago (Trinidad), NSL–1200 m.

Source: Navarrete et al., 2009.

Remarks: A valid species *fide* Hollis (2006: 445).

19. *Chironius spixii* (Hallowell, 1845). Proc. Acad. Nat. Sci. Philadelphia 2(9): 241. (*Coluber spixii*)

Type: Holotype, ANSP 5641, a 1692 mm female (S. Ashmead, 1835–1845).

Type locality: Unknown. Restricted to Venezuela *fide* Roze (1958d: 2).

Distribution: Colombia (Atlántico, Bolívar, Córdoba, Magdalena, Meta, Sucre, Vichada) and Venezuela (Amazonas, Anzoátequi, Apure, Aragua, Barinas, Bolívar, Carabobo, Cojedes, Delta Amacuro, Distrito Federal, Falcón, Guárico, Lara, Miranda, Monagas, Portuguesa, Sucre, Táchira, Yaracuy, Zulia), 50–460 m.

Source: Navarrete et al., 2009.

Remarks: A valid species *fide* Hollis (2006: 445).

20. *Chironius vincenti* (Boulenger, 1891f). Proc. Zool. Soc. London 59(3): 355. (*Herpetodryas carinatus vincenti*)

Type: Lectotype, BMNH 1890.11.25.21A, a 706 mm female (H.H. Smith via F.D. Godman, 1890), designated by Dixon et al. (1995: 195).

Type locality: "St. Vincent, Lesser West Indies" [= Saint Vincent & the Grenadines, Lesser Antilles, 13°15'N, 61°11'W] via lectotype selection.

Distribution: Lesser Antilles (Saint Vincent Is.), 20–600 m.

Sources: Henderson et al., 1988, A. Schwartz & R.W. Henderson, 1988 and R.W. Henderson & Powell, 1996c, 2009.

Remarks: Photograph of the type in Dixon et al. (1995: fig. 59). Possibly endangered *fide* Dixon et al. (1995: 200).

CHITULIA Gray, 1849 (Elapidae)

Synonyms: *Aturia* Gray, 1840 *(nomen nudum), Asturia* Gray, 1842c (*nomen nudum*), *Aturia* Gray, 1842d (*nomen praeoccupatum*), *Dolichodira* Wall, 1921g, *Micromastophis* Wall, 1921g, and *Polypholophis* Wall, 1921g.

Type species: *Chitulia inornata* Gray, 1849a.

Distribution: Marine waters of Indo-Australia, including Andaman Sea, Arabian Sea, Arafura Sea, Bay of Bengal, Bight of Bangkok, Bismarck Sea, Coral Sea, Damman Channel, East China Sea, Gulf of Cambay, Gulf of Carpentaria, Gulf of Martaban, Gulf of Oman, Gulf of Thailand, Indian Ocean, Java Sea, Kimanis Bay, Manila Bay, Marudu Bay, Oquair Bay, Persian Gulf, South China Sea, Strait of Hormuz, Strait of Malacca, SW Pacific Ocean, Visayan Sea, and Yellow Sea.

Sources: Kharin, 2007, 2012, Wells, 2007, I. Das, 2010, 2012, Kharin & Dotsenko, 2012 and Hoser, 2012e.

Remarks: A subgenus of *Hydrophis fide* McDowell (1972a: 226) and A. Rasmussen (1992: 91).

1. *Chitulia belcheri* (Gray, 1849a). Cat. Snakes Brit. Mus.: 46. (*Aturia belcheri*)

Synonym: *Hydrophis beleheri* – L. Rogers, 1903 *(nomen incorrectum).*

Type: Holotype, BMNH 1946.1.1.97 (formerly BMNH III.3.2.a & BMNH 1842.11.22.30), an 805 mm male (E. Belcher, 1843–1846).

Type locality: "New Guinea," (possibly in error *fide* McCarthy & Warrell (1991: 163).

Distribution: Southeast Asia and East Indies. Southern Thailand (Chanthanaburi, Nakhon Si Thammarat, Pattani, Samut Sakhon), S Vietnam, Indonesia (Papua, Java, Karima Djava, S Kalimantan, Sulawesi) and Papua New Guinea (New Britain, Western).

Sources: M.A. Smith, 1935, McCarthy & Warrell, 1991, Kharin, 2005a, A.R. Rasmussen et al., 2007 and V.S. Nguyen et al., 2009.

Remarks: Type reported to be female *fide* Boulenger (1896a: 296).

2. *Chitulia bituberculata* (W.C.H. Peters, 1873a). Mber. Königl. Akad. Wiss. Berlin 1872(12): 855–856, pl. 2, figs. 2–2d. (*Hydrophis bituberculatus*)

Synonyms: *Distura bituberculata* – Oshima, 1944 *(nomen incorrectum)* and *Leioselasma bitubercualata* – Culotta & Pickwell, 1993 *(nomen incorrectum).*

Type: Holotype, ZMB 4647, a 1090–1120 mm female (J. Nietner, 1870).

Type locality: "Colombo (Ceylon)" [= Colombo, Western Prov., Sri Lanka, 6°56'N, 79°52'E].

Distribution: Sri Lanka and S Thailand (Phuket Is.).

Sources: E.H. Taylor, 1950b, P. Silva 1980a, A.R. Rasmussen, 1992 and McCarthy, 1993.

Remarks: Type erroneously listed as ZMB 4677 in original description. A subspecies of *Leioselasma cyanocinctus* (Daudin) *fide* Deraniyagala (1955: 86).

3. *Chitulia inornata* Gray, 1849a. Cat. Snakes Brit. Mus.: 56.

Synonyms: *Chitulia fasciata* Gray, 1849a, *Thalassophis schlegelii* P. Schmidt, 1852, *Hydrophis controversa* Jan, 1857 *(nomen nudum), Hydrophis manillae* Owen, 1859 (*nomen nudum*), *Hydrophis controversus* Jan, 1863b, *Hydrophis longiceps* A.C.L.G. Günther, 1864a (*nomen substitutum*), *Distira cyanosoma* Wall, 1913e, *Hydrophis inormata* – Kinghorn, 1929 *(nomen incorrectum), Hydrophis inornatus ocellus* Kinghorn, 1929, *Hydrophis inormatus* – Beirn, 1979 *(nomen incorrectum),* and *Hydrophis inornataus* – Culotta & Pickwell, 1993 *(nomen incorrectum).*

Type: Holotype, BMNH 1946.1.1.27 (formerly BMNH III.7.1.a), adult female (E. Belcher, 1843–1846).

Type locality: "Indian Ocean" (in error).

Distribution: Northern Australia (Northern Territory, Queensland) and Arabia.

Sources: E.H. Taylor, 1922a, Voris, 1972, A.R. Rasmussen, 1989b, Kharin & Czeblukov, 2005 and Kharin & Dotsenko, 2012.

Remarks: Type reported to be male *fide* Boulenger (1896a: 290). A subspecies of *Hydrophis ornatus* (Gray) *fide* Klemmer (1963: 353) and McDowell (1972a: 232). Specimens from the China and Japan area possibly in error.

4. *Chitulia laboutei* (A. R. Rasmussen & Ineich, 2000). Hamadryad 25(2): 94–97, figs. 1–3. (*Hydrophis laboutei*)

Type: Holotype, MNHN 1999.6574, a 610 mm male (P. Laboute, 22 July 1988).

Type locality: "Chesterfield Reefs, New Caledonia (20°21.98'S, 161°04.87'E), most likely caught at a depth of 62 meters."

Distribution: New Caledonia (Chesterfield Reefs). Known only from the type series.

5. *Chitulia lamberti* (M.A. Smith, 1917c). J. Nat. Hist. Soc. Siam 2(4): 340. (*Hydrophis lamberti*)

Type: Holotype, BMNH 1946.1.9.20 (formerly MAS 1112 & BMNH 1921.2.11.13), an 860 mm female (M.A. Smith, Sept. 1916).

Type locality: "mouth of the Meklong River, Gulf of Siam." Emended to Bight of Bangkok, Thailand *fide* M.A. Smith (1926a: 83).

Distribution: Southeast Asia and Philippines. Thailand (Samut Sakhon, Samut Songkhran), Singapore,

Vietnam (Mekong River) and cen. Philippines (South Gigante).

Sources: A.R. Rasmussen, 1989b, V.S. Nguyen et al., 2009 and A.R. Rasmussen et al., 2011.

C

6. *Chitulia lapemoides* (Gray, 1849a). Cat. Snakes Brit. Mus.: 46. (*Aturia lapemoides*)

Synonyms: *Hydrophis holdsworthii* A.C.L.G. Günther, 1872b, *Hydrophis stewartii* J. Anderson, 1872, *Distira lapemioides* Boulenger, 1890a *(nomen emendatum)*, and *Distira lapimoides* – Phisalix, 1922 *(nomen incorrectum)*.

Type: Lectotype, BMNH 1946.1.7.2 (formerly BMNH III.3.3.a), a 610 mm female (Zool. Soc. London), designated by A.C.L.G. Günther (1864a: 375).

Type locality: "Ceylon" [= Sri Lanka] via lectotype selection.

Distribution: Southern Asia. Eastern Saudi Arabia (Ash Shar Qiyah), Bahrain, United Arab Emirates (Abu Dhabi, Dubai, Sharjah), N Oman, S Iran (Bushehr, Hormozghan), S Pakistan (Balochistan, Sindh), W and E India (Andhra Pradesh, Orissa, Tamil Nadu), Bangladesh, Sri Lanka, S Thailand (Phuket, Samut Sakhon, Phuket Is.), Singapore and West Malaysia (Penang Is.).

Sources: Wall, 1924e, Kennedy, 1937, Volsoe, 1939, Toriba & Sawai, 1981, 1989, Tamiya et al., 1983a, A.R. Rassmussen, 1987, 1989b, 1993, Gasperetti, 1988, Gallagher, 1990, McCarthy & Warrell, 1991, Leviton et al., 1992, M.S. Khan, 2002 and A. Silva, 2009.

Remarks: Type locality redundantly restricted to Ceylon [= Sri Lanka] *fide* E.H. Taylor (1950b: 589).

7. *Chitulia mamillaris* (Daudin, 1803d). Hist. Nat. Rept. 7: 340–342. (*Anguis mamillaris*)

Synonyms: *Hydrophis tesselatus* Murray, 1886 and *Hydrophis mammillaris* Wall, 1906f *(nomen emendatum)*.

Type: Neotype, BMNH 1861.12.30.38, a 650–660 mm male (R.H. Beddome, 1857–1882), designated by M.A. Smith (1926a: 89).

Type locality: "Vizagapatam, coast of Coromandel, India" [= Vishakhapatnam, Andhra Pradesh State, SE India, 17°41'N, 83°13'E].

Distribution: Southern Pakistan (Sindh) and India (Andhra Pradesh, Maharashtra, West Bengal).

Sources: Wall, 1921g, Minton, 1966, Talukdar & Dattagupta, 1980 and M.S. Khan, 2002.

Remarks: Original description based on P. Russell (1796: 49, pl. 44). Holotype lost *fide* M.A. Smith (1926: 89).

8. *Chitulia ornata* (Gray, 1842d). Zool. Misc. 2(May): 61. (*Aturia ornata*)

Synonyms: *Hydrophis ocellata* Gray, 1849, *Hydrophis laevis* Lütken, 1863 *(nomen praeoccupatum)*, *Hydrophis ellioti* A.C.L.G. Günther, 1864a, *Hydrophis godeffroyi* W.C.H. Peters, 1873a, *Distira andamanica* Annandale, 1905a, *Distira mjoebergi* Lönnberg & Andersson, 1913, *Distira godeflovi* – Phisalix, 1922 *(nomen incorrectum)*, *Distira ornate godeffroi* – Takahashi, 1935 *(nomen incorrectum)*, *Hydrophis ornatus maresinensis* Mittleman, 1947, *Hydrophis ornate godeffrayi* – Wang & Wang, 1956 *(nomen incorrectum)*, *Hydrophis maresianus* Klemmer, 1963 *(nomen emendatum)*, *Hydrophis ernata* – Deraniyagala, 1974 *(nomen incorrectum)*, *Hydrophis oranatus* – Tamiya & Puffer, 1974 *(nomen incorrectum)*, and *Hydrophis elliotti* – Culotta & Pickwell, 1993 *(nomen incorrectum)*.

Type: Holotype, BMNH 1946.1.23.72 (formerly BMNH III.3.1.a), a juvenile male.

Type locality: Unknown. Designated as Indian Ocean *fide* A.C.L.G. Günther (1864a: 377).

Distribution: Indo/Sino-Australia. Kuwait, Iran (Hormozghan, Tanb Is.), Saudi Arabia, Bahrain, United Arab Emirates, N Oman, Pakistan (Astola Is.), India (Andaman & Nicobars, Andhra Pradesh, Gujarat, Karnataka, Kerala, Maharashtra, Orissa, Tamil Nadu, West Bengal), Sri Lanka (Western), Myanmar (Rakhine), Thailand (Pattani, Phang Nga, Phuket, Samut Sakhon, Phuket Is.), S Vietnam (Dong Nai, Ho Chí Minh, Thuân Hai), China (Guangdong, Guangxi, Hainan, Shandong), Taiwan, ext. S Japan (Ryukyus: Amamioshima and Ishigakishima Is.), Philippines (Cebu, Gigante, Luzon, Negros, Palawan, Panay, South Gigante), West Malaysia, East Malaysia (Sabah), E Indonesia (Papua: Aru Is., Java), Papua New Guinea (Bismarck Arch., Gulf, West), Australia (New South Wales, Northern Territory, Queensland, Tasmania, Western Australia), New Caledonia and Kiribati.

Sources: Stejneger, 1907a, Wall, 1907h, 1921g, Maki, 1931, Bourret, 1934g, Volsoe, 1939, Mittleman, 1947, Romer, 1958, E.H. Taylor, 1965, Redfield et al., 1978, B.L. Lim, 1982, F. Parker, 1982, M. Mori, 1986, Tian et al., 1986, Gasperetti, 1988, A.R. Rasmussen, 1989b, Stuebing, 1991, Toriba & Nakamoto, 1991, O'Shea, 1996, Bauer & Sadlier, 2000, Ineich & Laboute, 2002, M.S. Khan, 2002, Leviton et al., 2003, Whitaker & Captain, 2008, V.S. Nguyen et al., 2009, Xiang & Li, 2009 and A.R. Rasmussen et al., 2011

9. *Chitulia pachycercos* (J.G. Fischer, 1855). Fam. Seeschlang.: 44–46. (*Hydrophis pachycercos*)

Synonym: *Hydrophis pachycerios* – Jan, 1859b *(nomen incorrectum)*.

Type: Neotype, ZMUC 661230, a 1030 mm female (native, Sept. 2000), designated by A.R. Rasmussen et al. (2007: 290).

Type locality: "fish market at the Phat Thien harbour, ca. 200 km S Nha Trang, Khanh Hoa Prov., S Vietnam" via neotype selection.

Distribution: Southern Vietnam (South China Sea).

Source: A.R. Rasmussen et al., 2011.

Remarks: Holotype (ZMH 390) destroyed in July 1943 during World War II *fide* H.-W. Koepcke *in* A.R. Rasmussen et al. (2007: 288). Previously a synonym of *C. belcheri.*

10. *Chitulia sibauensis* (A.R. Rasmussen, Auliya & Böhme, 2001). Herpetologica 57(1): 24–25, figs. 1, 3. (*Hydrophis sibauensis*)

Type: Holotype, ZFMK 64891, a 735 mm gravid female (M. Auliya, 12 Dec. 1996).
Type locality: "Sibau Kecil (00°52'N, 112°55'E) in Putussibau, Kalimantan, Indonesia.
Distribution: Central Indonesia (Kalimantan).

11. *Chitulia stricticollis* (A.C.L.G. Günther, 1864a). Rept. Brit. India: 376, pl. 25, fig. r. (*Hydrophis stricticollis*)

Synonyms: *Hydrophis neglectus* Wall, 1906f and *Distira neglecta* – Wall, 1909a.
Type: Holotype, BMNH 1946.1.6.90 (formerly BMNH 1860.3.19.1296), a 343 mm specimen (Brit. East India Comp.).
Type locality: "British India" [= India]. Restricted to Bay of Bengal, India *fide* Boulenger (1896a: 284), and M.A. Smith (1943: 459).
Distribution: Southeast Asia. Sri Lanka, E India (Orissa, Tamil Nadu, West Bengal), Bangladesh and Myanmar (Bago, Yangon).
Source: Wall, 1921g.
Remarks: A questionable record from the Sunda Sea (ZMB).

12. *Chitulia torquata* (A.C.L.G. Günther, 1864a). Rept. Brit. India: 369–370, pl. 25, fig. h. (*Hydrophis torquata*)

Synonyms: *Hydrophis diadema* A.C.L.G. Günther, 1864a, *Hydrophis tuberculatus* M.A. Smith, 1915c, *Hydrophis siamensis* M.A. Smith, 1917c, *Hydrophis torquatus aagaardi* M.A. Smith, 1920, *Hydrophis torquatus aagardi* – Tweedie, 1953 *(nomen emendatum), Hydrophys torquatus* – Anthony, 1955 *(nomen incorrectum),* and *Hydrophis torquatus aagaard* – M.W. Lin, 1975 *(nomen incorrectum).*
Types: Syntypes (5), BMNH 1946.1.3.82–86 (formerly BMNH 1925.9.22.1–5), two adult females, one juvenile male, and two juveniles, longest syntype 584 mm (T.E. Cantor, 1836–1846, via British East India Comp.).
Type locality: "coast of Pinang" [= Penang Is., Pinang State, NW Malay Peninsula, 5°22'N, 100°14'E].
Distribution: Southeast Asia and Greater Sundas. Southern China (Guangdong), Myanmar (Bago), W Cambodia (NE Pursat, SW Siem Reap), Thailand (Chachoengsao, Chanthaburi, Samut Sakhon, Samut Songkhram, Pattani), West Malaysia (Johore, Perak, Pinang, Selangor, Penang Is.) and Indonesia ("Borneo," Sumatra).
Sources: E.H. Taylor, 1965, Voris, 1985, Ineich, 1996, V.S. Nguyen et al., 2009 and A.R. Rasmussen et al., 2011.
Remarks: A.C.L.G. Günther (1864a: 370) listed only four syntypes.

C

†*CHOTAOPHIS* Head, 2005 (Colubridae)

Type species: †*Chotaophis padhriensis* Head, 2005.
Distribution: Miocene of Pakistan.
Source: Head, 2005.

1. †*Chotaophis padhriensis* Head, 2005. Palaeont. Electron. 8(1): 18–19, figs. 9.1–9.5.

Type: Holotype, H-GSP 24346, one trunk vertebra (E.H. Linday & W.R. Downs III).
Type locality: "Y-367 (8.95 Ma), Potwar Plateau of north-central Pakistan; Dhok Pathan Formation, Winnewala, Miocene.
Distribution: Middle Miocene (7.3–9.0 mya) of Pakistan.

CHRYSOPELEA H. Boie *in* Schlegel, 1826a (Colubridae)

Synonyms: *Chrysopelea* H. Boie *in* F. Boie, 1826, *Tyria* Fitzinger, 1826a *(nomen praeoccupatum), Chrypopelea* – Schlegel, 1837 *(nomen incorrectum), Coronophis* Fitzinger, 1843, *Dryodynastes* Fitzinger, 1843, *Eremophis* Gistel, 1848 *(nomen substitutum), Oligotropis* A.-M.-C. Duméril, 1853, *Dryiodynastes* – A.-M.-C. Duméril, Bibron & Duméril, 1854b *(nomen incorrectum), Chrysopeleia* – Palacky, 1898 *(nomen incorrectum), Chrysochlora* – Andersson, 1899 *(lapsus calami), Chrysophelea* – Floerikhe, 1912 *(nomen incorrectum), Crysopelea* – E.H. Taylor, 1917 *(nomen incorrectum), Tryria* – Briceño-Rossi, 1934 *(nomen incorrectum),* and *Chrysopeleas* – Saint-Girons, 1972b *(nomen incorrectum).*
Type species: *Coluber ornata* G. Shaw, 1802.
Distribution: Southeastern Asia and East Indies.
Sources: F. Werner, 1924c, Meise & Henning, 1935, M.A. Smith, 1943, C. Haas, 1950, Mertens, 1968b and Mahendra, 1984.

1. *Chrysopelea ornata* (G. Shaw, 1802). Gen. Zool., Amph. 3(2): 477–478. (*Coluber ornatus*)

Synonyms: *Coluber ibiboca* Lacépède, 1789 *(nomen rejiciendum), Coluber ibiboboca* Daudin, 1803c, *Coronella russellii* Blyth, 1854a *(nomen nudum), Chrysopelea ornata ornatissima* F. Werner, 1925, *Chrysopelea ornata lankavae* Deraniyagala, 1945, and *Chrysopelea ornata sinhaleyus* Deraniyagala, 1945.

C

Types: Syntypes (3), specimens described and illustrated by Seba (1734: 148, pl. 94, fig. 7) and Seba (1735: 8, pl. 7, fig. 1 & 60, pl. 61, fig. 2), location unknown.

Type locality: "Ceilonica," "Amboinensis" and "Ceilanica," respectively [= Sri Lanka and Ambon, Indonesia]. Restricted to East Indian islands *fide* Deraniyagala (1945: 106); further restricted to Ceylon [= Sri Lanka] *fide* Leviton (1964c: 137).

Distribution: Southeastern Asia and Indonesia. Eastern India (Arunachal Pradesh, Assam, Bihar, Goa, S Gujarat, Karnataka, Kerala, Madhya Pradesh, Manipur, Mizoram, Nagaland, Orissa, Sikkim, Tripura, Uttar Pradesh, West Bengal), Sri Lanka (Eastern, North-Western, Sabaragamuwa, Southern, Western), Bangladesh, Nepal (Bardiya, Chitwan, Jhapa, Kapilbastu, Makwanpur, Sunsari, Udayapur), Bhutan (Sarpang), Myanmar (Yangon), S China (Fujian, Guangdong, Hainan, Hong Kong, Yunnan), Vietnam (Ba Ria-Vung Tau, Cao Mau, Dak Lak, Da Nang, Dong Nam, Gia Lai, Ho Chi Minh City, Khanh Hoa, Kien Giang, Kon Tum, Nghe An, Ninh Thuan, Quang Binh, Quang Nam, Quang Tri, Son La, Tay Ninh, Thai Nguyen, Tien Giang, Vinh Long, Vinh Phuc, Yen Dai, Hon Thom Is.), Laos (Champasak), Cambodia (Kampong Speu, Kampot), Thailand (Chiang Mai, Lampang, Loei, Lop Buri, Nakhon Phanom, Nakhon Ratchasima, Nakhon Sawan, Nakhon Si Thammarat, Narathiwat, Pattani, Phang Nga, Prachuap Khiri Khan, Ratchaburi, Sakhon Nawan, Trat, Udon Thani), West Malaysia (Kedah, Perlis, Langkawi Is.) and Indonesia (Kalamantan, Java, Sulawesi, Sumatra), NSL–1500 m.

Sources: Wall, 1908a, Leviton, 1964c, E.H. Taylor, 1965, P. Silva, 1969, Campden-Main, 1970a, Saint-Girons, 1972a, G. Sharma & Nakhasi, 1980, Tweedie, 1983, Basu, 1989, A. Silva, 1990b, 2001, 2009, Vyas, 1990, Zhao & Adler, 1993, M.J. Cox et al., 1998, Chan-ard et al., 1999, Darevsky, 1999, Orlov et al., 2000, Schleich & Kästle, 2002, Tiwari & Shah, 2004, Whitaker & Captain, 2004, N. Khaire, 2006, B.L. Stuart & Emmett, 2006, Zhao, 2006, Ziegler et al., 2007, M.F. Ahmed et al., 2009, V.S. Nguyen et al., 2009, I. Das, 2010, 2012 and Wangyal, 2012.

2. *Chrysopelea paradisi* H. Boie *in* F. Boie, 1827. Isis von Oken 20(6): 547.

Synonyms: *Coluber constrictorius* Donndorff, 1798 *(nomen oblitum)*, *Chrysopelea paradisi* H. Boie *in* Schlegel, 1826a *(nomen nudum)*, *Chrysopelia paradisi* Schlegel, 1826b *(nomen nudum)*, *Chrysopelea paradisi celebensis* Mertens, 1968b, and *Chrysopelea paradisi variabilis* Mertens, 1968b.

Type: Neotype, RMNH 885, a female (H. Boie & H.C. Macklot, 1826), designated by Mertens (1968b: 203).

Type locality: "Java" [SW Indonesia].

Distribution: Southeastern Asia and East Indies. Eastern India (Andaman and Nicobars: Andaman Is.), peninsular Myanmar, peninsular Thailand (Phang Nga, Phuket, Satun), West Malaysia (Perak, Pinang, Sibu Tengah Is.), Singapore, East Malaysia (Sabah, Sarawak), Brunei, Indonesia (Bali, Bangka, Belitung, Java, Kalimantan, Mentawai Is., Natuna Is., Nias, Riau Arch., Sulawesi, Sumatra, Weh) and Philippines (Balabac, Bantayan, Banton, Basilan, Bongao, Bubuan, Camiguin, Cebu, Dinagat, Jolo, Kalotkot, Leyte, Luzon, Marongas, Medis, Mindanao, Mindoro, Negros, Palawan, Panay, Polillo, Romblon, Samar, Sibuyan, Siquijor, Sibutu, Sanga Sanga, Tawitawi), NSL–1525 m.

Sources: Leviton, 1964c, E.H. Taylor, 1965, Mertens, 1968b, Tweedie, 1983, Gaulke, 1994b, Gaulke & Altenbach, 1994, Stuebing, 1994a, David & Vogel, 1994, Sison & Gonzales, 1995, M.J. Cox et al., 1998, Chan-ard et al., 1999, Malkmus et al., 2002, Schleich & Kästle, 2002, Whitaker & Captain, 2004, McKay, 2006, I. Das, 2010, 2012 and Grismer, 2011.

Remarks: Original description based on H. Boie's MS (1823–1825). Holotype lost *fide* Guibé *in* Mertens (1968: 203). In accordance with Art. 23.9.2 of the Code (ICZN, 1999), *Chrysopelea paradisi* H. Boie is designated a *nomen protectum* and *Coluber constrictorius* Donndorff a *nomen oblitum*.

3. *Chrysopelea pelias* (Linnaeus, 1758). Syst. Nat., ed. 10, 1: 224. (*Coluber pelias*)

Synonyms: *Chrysopelea erythrochloris* H. Boie *in* Schlegel, 1826a (*nomen nudum*), *Chrysopelia erythrochloris* Schlegel, 1826b *(nomen nudum)*, *Chrysopelea erythrolmelas* F. Boie, 1827 *(nomen nudum)*, *Dendrophis chrysochloros* Schlegel, 1837, *Chrysopelea erythrochloris* Schlegel, 1837 *in* 1837–1844, *Chrysopelea hasseltii* Bleeker, 1860c *(nomen nudum)*, and *Chrysopelea chrysochlora* Boulenger, 1896a *(nomen emendatum)*.

Type: Holotype, NHR Lin-83 (formerly MDG), a 975 mm specimen.

Type locality: "Indiis" [= India].

Distribution: Southeastern Asia and Indonesia. Eastern India (Andaman Is.), Thailand (Chon Buri, Krabi, Nakhon Sawan, Nakhon Si Thammarat, Phetchabun, Prachin Buri, Roi Et, Sakhon Nakhon, Saraburi, Sisaket, Songkhla, Sukhothai, Tak), West Malaysia (Johor, Panang, Perak, Tioman Is.), Singapore, East Malaysia (Sabah, Sarawak), Brunei and Indonesia (Bangka, Java, Kalimantan, Mentawai Is., Natuna Is., Nias, Riau Arch., Sumatra), NSL–600 m.

Sources: Tweedie, 1983, David & Vogel, 1996, M.J. Cox et al., 1998, Chan-ard et al., 1999, Ferner et al., 2001, Malkmus et al., 2002 and I. Das, 2010, 2012.

Remarks: Holotype length listed as 563 mm *fide* Vestergren *in* Mertens (1968b: 210).

4. *Chrysopelea rhodopleuron* F. Boie, 1827. Isis von Oken 20(6): 547.

Synonyms: *Chrysopelea rhodopleuron* H. Boie *in* Schlegel, 1826a *(nomen nudum)*, *Chrysopelia rhodopleuron*

Schlegel, 1826b *(nomen nudum)*, *Chrysopelea vicina* A.C.L.G. Günther, 1872b, and *Chrysopelea viridis* J.G. Fischer, 1880.

Type: Lectotype, RMNH 889, a 1490 mm female (C.G.C. Reinwardt, 26 June–19 July 1821), designated by Mertens (1968: 211).

Type locality: "Ambon, Indonesia" via lectotype selection.

Distribution: Indonesia (Ambon, Aru Is., Banda, Batjan, Buru, Manipa, Misool, Nusa Laut, Sangi, Sulawesi, Sangihe Is., Seram, Tanimbar).

Sources: Lang & Vogel, 2005 and Koch, 2012.

5. *Chrysopelea taprobanica* M.A. Smith, 1943. Fauna Brit. India, Rept. Amph. 3: 254.

Type: Holotype, BMNH 1946.1.9.83 (formerly BMNH 1906.7.21.1), a 960 mm female (Lodge).

Type locality: "Kanthali, Ceylon" [= Kantalai, N Eastern Prov., NE Sri Lanka, 8°21'N, 81°00'E, elevation 50 m].

Distribution: Sri Lanka (Central, Eastern, Northern, North-Central, North-Western), 50 m.

Sources: E.H. Taylor, 1950b and P. Silva, 1969.

Remarks: BMNH 1946.1.9.75 listed as holotype *fide* BMNH catalogue.

†*CHUBUTOPHIS* Albino, 1993 (Boidae)

Type species: †*Chubutophis grandis* Albino, 1993.

Distribution: Lower Eocene of Argentina.

Source: Albino, 1993.

1. †*Chubutophis grandis* Albino, 1993. Hist. Biol. 7: 57–60, fig. 4.

Type: Holotype, MLP 83-III-1-150, one trunk vertebra.

Type locality: "Southern slope of Valle Hermoso, near Boliche Alma Gaucha, southeast of Chubut Prov. Gran Barranca Member, Sarmiento Formation. Casamayoran Land-Mammal Age (Early Eocene)."

Distribution: Lower Eocene (Casamayoran: 48.0–54.0 mya) of Argentina.

CLELIA Fitzinger, 1826a (Xenodontidae)

Synonyms: *Cloelia* Wagler, 1830 *(nomen emendatum)*, *Claelia* Bonaparte, 1831 *(nomen emendatum)*, *Deiropeda* Fitzinger, 1843, *Hydroscopus* Fitzinger, 1843, *Rhinoscytale* Fitzinger, 1843 *(nomen nudum)*, *Chloe* Agassiz, 1847, *Diropeda* Agassiz, 1847 *(nomen emendatum)*, *Brachyrruton* A.-M.-C. Duméril, 1853, *Brachyruton* A.-M.-C. Duméril, Bibron & Duméril, 1854b *(nomen emendatum)*, *Brachyrrhyton* – Lichtenstein & Martens, 1856 *(nomen incorrectum)*, *Brachyrrutron* Marschall, 1873 *(nomen emendatum)*, *Brachyrhyton* – F. Werner, 1890 *(nomen incorrectum)*, *Brachyrhytum* – Boettger, 1898 *(nomen incorrectum)*, and *Barbourina* Amaral, 1924b.

Type species: *Coluber clelia* Daudin, 1803c.

Distribution: Latin America and Lesser Antillies.

Fossil records: Lower/middle Pleistocene (Ensenadan) of Argentina.

Sources: J.A. Peters & Orejas-Miranda, 1970, Achaval, 1973, Cadle, 1984a, Scrocchi & Viñas, 1990, Zaher, 1996, 1999, Vidal et al., 2000, Morato et al., 2003, Reichle & Embert, 2005, Scanferla, 2006, N.J. Scott et al., 2006, Jansen et al., 2009 and Zaher et al., 2009.

Remarks: Zaher et al., 2009 transferred *C. bicolor*, *C. montana* and *C. quimi* to *Mussurana*.

1. *Clelia clelia* (Daudin, 1803c). Hist. Nat. Rept. 6: 330–331, pl. 78. (*Coluber clelia*)

Synonyms: ? *Coluber peruvianus* Gravenhorst, 1807, *Natrix cloelia* – Merrem, 1820, *Clelia daudinii* Fitzinger, 1826a, *Rhinoscytale cloelia* Fitzinger, 1843 *(nomen nudum)*, *Oxiropus cloelia* – Briceño-Rossi, 1934 *(nomen incorrectum)*, *Pseudoba cloelio* – Briceño-Rossi, 1934 *(nomen incorrectum)*, *Clelia clelia immaculata* H.M. Smith, 19421 (*partim*), and *Clelia clelia groomei* Greer, 1965a.

Types: Syntypes (2), formerly MNHN, longest syntype 864 mm (F.M. Daudin & F. Levaillant coll.), lost *fide* H.M. Smith & Taylor (1945: 36).

Type locality: "Suriname en Amérique" [= Suriname, South America].

Distribution: Mesoamerica and South America. Belize (Cayo, Stann Creek, Toledo), N Guatemala (Izabal, Petén), Honduras (Atlántida, Colón, Comayagua, Copán, Cortés, El Paraiso, Gracias a Dios, Olancho, Yoro), Nicaragua (Atlántico Sur, Jinotega, Rivas), Costa Rica (Alajuela, Cartago, Guanacaste, Heredia, Limón, Puntarenas, San José), Panama (Chiriquí, Colón), N and W Colombia (Antioquia, Bolívar, Cauca, Caqueta. César, Chocó, Córdoba, Cundinamarca, La Guajira, Magdalena, Meta, Norte de Santander, Putumayo, Santander, Valle del Cauca, Vaupés, Gorgona Is.), Trinidad and Tobago (Trinidad), Grenada, Dominica, Venezuela (Amazonas, Anzoátegui, Apure, Aragua, Barinas, Bolívar, Carabobo, Cojedes, Delta Amacuro, Distrito Federal, Falcón, Guárico, Lara, Mérida, Miranda, Monagas, Portuguesa, Sucre, Táchira, Trujillo, Yaracuy, Zulia), Guyana (Essequibo Islands-West Demerara), Suriname (Brokopondo, Saramacca, Suriname), French Guiana (Cayenne, Saint-Laurent-du-Maroni), N Brazil (Amapá, Amazonas, Rondônia), Ecuador (Bolívar, Chimborazo, Cotopaxi, Esmeraldas, Guayas, Imbabura, Loja, Los Rios, Manabi, Marona-Santiago, Napo, Orellana, Pastaza, Pichincha, Sucumbios, Tungurahura, Zamora-Chinchipe), W Peru (Amazonas, Cusco, Huánuco, Loreto, Madre de Dios, Pasco, San Martin, Ucayali), Bolivia (Santa Cruz), Paraguay (Alto Paraná, Boquerón, Caaguazú, Caazapá,

C

Central, Itapúa, Misiones, Neembucú, Presidente Hayes, San Pedro) and NE Argentina (E Chaco, Corrientes, E Formosa, E Santa Fe), NSL–2500 m.

Sources: E.H. Taylor, 1951, 1954, Roze, 1966a, Lancini, 1986, Pérez-Santos & Moreno, 1988, Lancini & Kornacker, 1989, Duellman & Salas, 1991, Williams & Francini, 1991, Aquino et al., 1996, Lee, 1996, 2000, Zaher, 1996, 1999, J.C. Murphy, 1997, Tiranti & Avila, 1997, J.A. Campbell, 1998b, Martins & Oliveira, 1998, Starace, 1998, G. Köhler, 1999, 2001, 2003, Mijares-Urrutia & Arends, 2000, Stafford & Meyer, 2000, Rivas-Fuenmayor & La Marca, 2001b, Lehr, 2001, Markezich, 2001, Giraudo, 2002, Savage, 2002, Abuys, 2003, J.C. Murphy & Schlager, 2003, Solórzano, 2004, Duellman, 2005, McCranie et al., 2006, Cisneros-Heredia et al., 2007, Cacciali, 2008, R.W. Henderson & Powell, 2009, Navarrete et al., 2009, McCranie, 2011, Travers et al., 2011 and C.J. Cole et al., 2013.

Remarks: The northernmost extension of the Distribution is unclear and *C. clelia* may occur in S Mexico (Zaher, 1996, Campbell, 1998b). La Pampa, Argentina record of Tiranti & Avila (1997: 109) needs confirmation. Villa et al. (1988: 61) erroneously listed El Salvador.

2. *Clelia equatoriana* (Amaral, 1924b). J. Washington Acad. Sci. 14(9): 201–202. (*Barbourina equatoriana*)

Type: Holotype, USNM 62790, a 300 mm male (F.W. Goding, April 1920).

Type locality: "sent from Guayaquil, Ecuador."

Distribution: Lower Central America and NW South America. Costa Rica (Nuevo Is.), E Panama (Darién), W and N Colombia (Antioquia, Caldas, Cauca, Santander, Valle de Cauca) and NW and Ecuador (Amazonas, Guayas, Napo, Pichincha), 245–1785 m.

Sources: Pérez-Santos & Moreno, 1988, 1991, Zaher, 1996 and Rojas-Morales, 2012b.

Remarks: Costa Rican material is either *C. equatoriana* or *C. scytalina fide* Savage, 2002.

3. *Clelia errabunda* Underwood, 1993. Bull. Nat. Hist. Mus. (Zool.) 59(1): 3–4, fig. 1.

Type: Holotype, BMNH 1889.8.14.25, a 1440 mm male (G.A. Ramage, 1889).

Type locality: "St. Lucia, West Indies" [= Saint Lucia, Lesser Antilles].

Distribution: Lesser Antilles (St. Lucia), now extinct *fide* R.W. Henderson & Powell (2009: 362)

Sources: Underwood, 1995 and R.W. Henderson & Powell, 2009.

Remarks: Record from Dominica undoubtedly in error *fide* Underwood (1995: 5).

4. *Clelia hussami* Morato, Franco & Sanches, 2003. Phyllomedusa 2(2): 94–98, figs. 1a–c.

Type: Holotype, IB 32859, a 602 mm female (L. Miynarczk, 9 Feb. 1972), destroyed by fire 15 May 2010.

Type locality: "município de Cruz Machado (26°01'S, 51°20'W, 940 m), estado do Paraná, Brasil."

Distribution: Southern Brazil (S Paraná, Rio Grande do Sul, N Santa Catarina), 760–1050 m.

Source: Borger-Martins et al., 2005.

5. *Clelia langeri* Reichle & Embert, 2005. J. Herp. 39(3): 380–382, fig. 1.

Type: Holotype, MNK 867, a 1162 mm male (F.A. Langer, 14 April 1994).

Type locality: "the area of Villa Merced (18°6'S, 64°11'W), Prov. Florida, Department of Santa Cruz, Bolivia, about 1300 m altitude in the inter-Andean dry valleys."

Distribution: Bolivia (Chuquisaca, Santa Cruz and probably Tarija), 900–1500 m.

6. *Clelia plumbea* (Wied-Neuwied, 1820a). Reise Brasil. 1: 25. (*Coluber plumbeus*)

Type: Holotype, AMNH 3481, adult male (A.P.M. zu Wied-Neuwied, Sept. 1815).

Type locality: "von São João, etwas nördlich von Cabo Frio, Brasilien" [= between Cabo Frio and Rio São João, Rio de Janeiro, Brazil].

Distribution: Southeastern South America. Southern Brazil (Alagoas, Amazonas, Bahia, Distrito Federal, Espírito Santo, Mato Grosso do Sul, Minas Gerais, Pará, Paraná, Rio de Janeiro, Santa Catarina, São Paulo, Cardoso, Grande, São Sebastião and São Vicente Is.), Paraguay (Alto Paraná, Itapúa) and NE Argentina (Misiones), NSL–1065 m.

Sources: Cunha & Nascimento, 1978, 1993, Pérez-Santos & Moreno, 1988, Lema, 1994, Zaher, 1999, Giraudo, 2002, Giraudo & Scrocchi, 2002, Duarte-Rocha & van Sluys, 2006, N.J. Scott et al., 2006 and Cicchi et al., 2007.

Remarks: Resurrected as a valid species *fide* Zaher (1996: 310).

7. *Clelia scytalina* (Cope, 1867b). Proc. Acad. Nat. Sci. Philadelphia (1866) 18(4): 320. (*Scolecophis scytalinus*)

Synonyms: *Oxyrhopus proximus* Bocourt, 1897, and *Clelia clelia immaculata* H.M. Smith, 1942l.

Type: Holotype, USNM 6581, a 584 mm specimen (C.H. Berendt, 1865–1867).

Type locality: "near Tabasco, Mexico." Emended to San Juan Bautista, Tabasco, Mexico *fide* Cochran (1961: 212).

Distribution: Mesoamerica. Southern Mexico (Chiapas, Colima, Guerrero, Jalisco, Oaxaca, Tabasco, Veracruz), N Belize (Corozal), Guatemala, Costa Rica (Alajuela, Cartago, San José) and Panama, NSL–2000 m.
Sources: García & Ceballos, 1994, Lee, 1996, 2000, Stafford & Meyer, 2000, Savage, 2002, Solórzano, 2004 and Sasa et al., 2010.
Remarks: MNHN catalogue lists MNHN 7871 and MNHN 1899.54–55 as types. Probably occurs in Nicaragua *fide* G. Köhler (1999: 66).

CLONOPHIS Cope, 1889b (Natricidae)

Synonyms: *Cora* Jan, 1863b, *Olonophis* – Surface, 1906 *(nomen incorrectum)*, and *Clanophis* – Hankinson, 1917 *(nomen incorrectum)*.
Type species: *Regina kirtlandii* Kennicott, 1856.
Distribution: Central and NE USA.
Sources: Conant, 1943, Rossman, 1963b, Rossman & Powell, 1985, Ernst & Barbour, 1989, Conant & Collins, 1991, 1998, Harding, 1997, Shupe & Zimmerer, 1998, Alfaro & Arnold, 2001, Zaher et al., 2009 and Frymire & Scott, 2012.

1. *Clonophis kirtlandii* (Kennicott, 1856). Proc. Acad. Nat. Sci. Philadelphia 8(2): 95–96. (*Regina kirtlandii*)

Type: Lectotype, USNM 1514, a 509 mm female (R.W. Kennicott, 1853–1855), designated by Conant (1943: 317).
Type locality: "West Northfield, Illinois" [USA] via lectotype selection.
Distribution: Central USA (N Kentucky, cen. and NE Illinois, Indiana, ext. S Michigan, NE Missouri, Ohio, W Pennsylvania, Tennessee), 90–670 m.
Remarks: USNM 1514 erroneously listed as holotype *fide* Rossman & Powell (1985: 364.1). Pennsylvania records are questionable .

COELOGNATHUS Fitzinger, 1843 (Colubridae)

Synonyms: *Calognathus* – A.-M.-C. Duméril & Bibron, 1844 *(nomen incorrectum)*, *Cynophis* Gray, 1849b, *Compsosoma* A.-M.-C. Duméril, 1853 *(nomen praeoccupatum)*, *Plagiodon* A.-M.-C. Duméril, 1853 *(nomen praeoccupatum)*, *Campsosoma* – Jan, 1857 *Campilodon* Jan, 1861c *(nomen emendatum)*, *Composoma* – E.H. Taylor, 1922 *(nomen incorrectum)*, *Compsoma* – Maki, 1931 *(nomen incorrectum)*, *Plagionon* – Maki, 1931 *(nomen incorrectum)*, *Caelognatus* – Saint-Girons, 1972 *(nomen incorrectum)*, and *Euanedwardsserpens* Hoser, 2012p *(nomen illegitimum)*.
Type species: *Coluber radiatus* F. Boie, 1827.
Distribution: Southeastern Asia and East Indies.
Sources: Schulz, 1996, Helfenberger, 2001, Utiger et al., 2005, Burbrink & Lawson, 2007, I. Das, 2010, 2012 and Hoser, 2012p.

1. *Coelognathus enganensis* (Vinciguerra, 1892). Ann. Mus. Civ. Stor. Nat. Genova (2) 12: 524–525. (*Coluber enganensis*)

Types: Syntypes (2), MSNG 32307–08, two adult females.
Type locality: "Bua Bua, Engano," Indonesia.
Distribution: Southern Indonesia (Lesser Sundas: Engano).
Sources: Auffenberg, 1980, Bosch, 1985 and Lang, 2011.
Remarks: A valid species *fide* Bosch (1985: 22–23).

2. *Coelognathus erythrurus* (A.-M.-C. Duméril, Bibron & Duméril, 1854a). Erpét. Gén. 7(1): 175–177. (*Plagiodon erythrurus*)

Synonyms: *Elaphis melanurus celebensis* Jan, 1863b, *Elaphis melanurus manillensis* Jan, 1863b, *Coluber walli* F. Werner, 1924, and *Elaphe erythrura psephenoura* Leviton, 1979g.
Type: Lectotype, MNHN 7224 (J.R.C. Quoy & J.P. Gaimard [Coquille or Astrolabe Voy.], 1822–1829), designated by Leviton (1979: 103).
Type locality: "Samar Island, Philippine Islands" via lectotype selection.
Distribution: Philippines (Bohol, Camiguin, Catanduanes, Cebu, Dinagat, Guimaras, Jolo, Leyte, Luzón, Masbate, Mindanao, Mindoro, Negros, Pacijan, Palawan, Panay, Polillo, Ponson, Poro, Samar, Siasi) and Indonesia (Butung, Peleng, Sulawesi, Butung, Tawitawi), NSL–1065 m.
Sources: Leviton, 1979, den Bosch, 1985, Schulz, 1988c, 1989f, Ross & Gonzales, 1992, B.E. Smith, 1993 and Gaulke & Altenbach, 1994.

3. *Coelognathus flavolineatus* (Schlegel, 1837). Essai Phys. Serp. 1: 145, 2: 141–143, pl. 5, figs. 11–12. (*Coluber flavolineatus*)

Synonyms: *Coluber flavolineatus* H. Boie *in* Schlegel, 1826a *(nomen nudum)*, *Coluber flavo-lineatus* Schlegel, 1826b *(nomen illegitimum)*, *Coluber melanopsis* F. Boie, 1827, *Coluber melanurus* F. Boie, 1827 *(nomen oblitum)*, *Coluber melanura* – Schlegel, 1837, *Compsosoma melanurus* Bleeker, 1859c *(nomen nudum)*, *Elaphis melanurus bivirgata* Jan, 1863b *(nomen nudum)*, and *Elaphis melanurus bivirgata* Jan, 1867 *in* Jan & Sordelli, 1866–1879.
Types: Syntypes (4), RMNH 418 (H. Boie & H.C. Macklot, 1836) [Java], RMNH 422 (S. Müller, June 1833–Dec. 1835) [Padang, Sumatra], RMNH 424 (H. Boie & H.C. Macklot, Dec. 1825–Sept. 1827) [Java], and RMNH 425 [Sumatra].
Type locality: "îles de Java et de Sumatra" [Indonesia]. Restricted to Java *fide* M.A. Smith (1943: 148).

Further restricted to region of Mt. Pangerango and Mt. Salak, West Java, Indonesia *fide* Brongersma (1948b: 14). Emended to Tjihandjawar, at the foot of Mt. Pangerango, W. Java [= Cihanyawar, Nagrak, Jawa Barat, W Java, SW Indonesia, 07°03'S, 108°40'E] *fide* Brongersma (1950: 1499).

Distribution: Southeast Asia and Indonesia. Eastern India (Andaman & Nicobars: Andaman Is.), ext. S Myanmar (Tanintharyi, Mergui Arch.), Cambodia (Pursat), peninsular Thailand (Nakhon Si Thammarat, Narathiwat, Phang Nga, Phuket, Satun, Surat Thani), Vietnam (Binh Phuac, Gia Lai), West Malaysia (Johor, Kedah, Perak, Pinang, Selangor, Penang and Tioman Is.), Singapore, East Malaysia (Sabah, Sarawak), Brunei, and Indonesia (Bali, Banggai, Bangka, Belitung, Bintan, Galang, Java, Kalimantan, Siberut, Simeulue, Sulawesi, Sumatra, Weh), NSL–1500 m.

Sources: E.H. Taylor, 1965, Tweedie, 1983, Schulz, 1987c, 1988o, Stuebing 1991, David & Vogel, 1996, M.J. Cox et al., 1998, Chan-ard et al., 1999 Malkmus et al., 2002, P. Wood et al., 2003, Whitaker & Captain, 2004, McKay, 2006, J.C. Murphy et al., 2006, Grismer et al., 2008 and V.S. Nguyen et al., 2009.

Remarks: Reinwardt credited as collector by Schlegel (1837: 141). In accordance with Art. 23.9.2 of the Code (ICZN, 1999), *Coluber flavolineatus* Schlegel is designated a *nomen protectum* and *Coluber melanurus* F. Boie a *nomen oblitum.*

4. *Coelognathus helena* (Daudin, 1803c). Hist. Nat. Rept. 6: 277–280, pl. 76, fig. 1. (*Coluber helena*)

Synonyms: *Cynophis bistrigatus* Gray, 1849b, *Herpetodryas malabaricus* Jerdon, 1854, *Cynophis malabariensis carinata* F. Müller, 1878b, and *Elaphe helena monticollaris* Schulz, 1992b.

Type: Lectotype, a 698 mm specimen described and illustrated by P. Russell (1796: 37–38, pl. 32) (P. Russell, April 1781–1791), designated herein.

Type locality: "Vizagapatam, coast of Coromandel, India" [= Vishakhapatnam, Andhra Pradesh State, SE India, 17°41'N, 83°13'E, elevation 25 m] via lectotype selection.

Distribution: Southern Asia. Southeastern Pakistan (S Sindh), India (Andhra Pradesh, Assam, Bihar, Goa, Gujarat, Himachal Pradesh, Jammur and Kashmir, Karnataka, Kerala, Madhya Pradesh, Maharashtra, Nagaland, Oris (Gandaki, Lumbini, Nayayasa, Rajasthan, Tamil Nadu, Utter Pradesh, West Bengal), Nepal (Chitwan, Kaski, Kathmandu, Makwanpur, Palpa, Sunsari, Tanahu), Bangladesh and Sri Lanka (Central, Eastern, North-Central, Northern, Sabaragamuwa, Southern, Uva, Western), 610–1980 m.

Sources: Wall, 1913a, 1921g, E.H. Taylor, 1950b, P. Silva, 1969, Kramer, 1977, Mahendra, 1984, Nanhoe & Ouboter, 1987, Murthy & Chandrasekhar, 1989, A. Silva, 1990b, 2009, Schulz, 1992b, Tiwari & Shah, 2004, Whitaker & Captain, 2004, N. Khaire, 2006, Nande & Deshmukh, 2007, Hutton & David, 2009 and Royans & Knight, 2009.

5. *Coelognathus philippinus* (L.E. Griffin, 1909b). Philippine J. Sci. 4A(6): 597–598. (*Elaphe philippina*)

Type: Lectotype, CAS 62143 (formerly LEG 17 & BSM 291), a 1727 mm female (C.M. Weber, Feb. 1909), designated by Leviton (1979: 114).

Type locality: "Iwabig, Palawan, Philippine Islands" via lectotype selection.

Distribution: Philippines (Balabac, Bongao, Busuanga, Culion, Palawan, Sanga Sanga, Sibutu, Tawitawi).

Sources: W.C. Brown & Alcala, 1970 and Leviton, 1979.

6. *Coelognathus radiatus* (F. Boie, 1827). Isis von Oken 20(6): 536. (*Coluber radiatus*)

Synonyms: *Natrix semiornata* Wagler, 1825 (*nomen ineditum*), *Coluber radiatus* H. Boie *in* Schlegel, 1826a *(nomen nudum)*, *Coluber radiatus* Schlegel, 1826b *(nomen nudum)*, *Coluber quadrifasciatus* Cantor, 1839b, and *Tropidonotus quinque* Cantor, 1839b.

Types: Syntypes (4), RMNH 432, a 1545 mm specimen, RMNH 433, RMNH 47573 (formerly RMNH 432), and RMNH 47574 (formerly RMNH 432) (H. Boie & H.C. Macklot, Dec. 1825–Sept. 1827).

Type locality: "Java, Indonesia" by type locality restriction *fide* C.H. Pope (1935: 261). Further restricted to region of Mt. Pangerango and Mt. Salak, West Java, Indonesia *fide* Brongersma (1948b: 14), ca. 6°43'S, 106°44'E].

Distribution: Southeastern Asia and Indonesia. Northeastern India (Assam, Arunachal Pradesh, S Bihar, Chattisgarh, Madhya Pradesh, Manipur, Meghalaya, Mizoram, Nagaland, Orissa, Sikkim, Uttaranchal, Uttar Pradesh, West Bengal), E Nepal (Dhankuta, Kaski, Kathmandu, Makwanpur, Sankhuwasabha), N Bhutan (Trashigang), Bangladesh, S China (Fujian, Guangdong, Guangxi, Guizhou, Hong Kong, Yunnan, Nan Ao Is.), Vietnam (entire country), Laos (Champasak), Cambodia, Thailand (Loei, Saraburi, Songkhla), Myanmar (Yangon), West Malaysia (Johor, Kedah, Perak, Pinang, Selangor, Terengganu, Mergui Arch.), Singapore, East Malaysia (Sarawak), Brunei and Indonesia (Bangka, Java, Kalimantan, Sumatra), NSL–1480 m.

Sources: Wall, 1914b, Kopstein, 1932a, Leviton et al., 1956, Bergman, 1961a, L. Swan & Leviton, 1962, E.H. Taylor, 1965, R. Bhatnagar, 1969, 1975, Saint Girons, 1972a, Kramer, 1977, Tweedie, 1983, Mahendra, 1984, Karsen et al., 1986, Schulz, 1986d, 1988k, Nanhoe & Ouboter, 1987, Zhao & Adler, 1993, David & Vogel, 1996, M.J. Cox et al., 1998, Chan-ard et al., 1999, Lazell, 1999, Orlov et al., 2000, Schleich & Kästle, 2002, Tiwari & Shah, 2004, Whitaker & Captain, 2004, McKay, 2006, I. Das, 2007b, M.F. Ahmed et al., 2009, V.S. Nguyen et al., 2009 and Wangyal, 2011.

Remarks: Type locality restriction of C.H. Pope (1935: 261) resulted in four Javan syntypes. Kramer (1977: 641) erroneously referred to the series as the holotype. Paralectotypes are RMNH 47573–74, two juveniles.

7. *Coelognathus subradiatus* (Schlegel, 1837). Essai Phys. Serp. 1: 144, 2: 136–137. (*Coluber subradiatus*)

Synonyms: *Elaphis nyctenurus* Jan, 1863b, and *Coluber melanurus timoriensis* Bethencourt-Ferreira, 1897b.

Types: Syntypes (4), RMNH 435, a 1360–1560 mm specimen, RMNH 403, RMNH 47575 (formerly RMNH 403), and RMNH 47576 (formerly RMNH 435) (S. Müller, Oct. 1828– Nov. 1829).

Type locality: "Île de Timor" [= Timor, Lesser Sundas, S Indonesia].

Distribution: Lesser Sundas of S Indonesia (Adonare, Alor, Ende, Flores, Komodo, Lombien, Lombok, Rinca, Roti, Semau, Sumba, Sumbawa, Timor, Wetar) and Timor-Leste, NSL–1185 m.

Sources: Bethencourt-Ferreira, 1898, Boulenger, 1898a, Rooij, 1917, Mertens, 1930, Brongersma, 1934, Kopstein, 1937, C. Haas, 1950, Darevsky, 1964, Auffenburg, 1980, Bosch, 1985, Schulz, 1988c, 1989f, Kaiser et al., 2011 and O'Shea et al., 2012.

Remarks: Most of the RMNH syntypes were exchanged with other institutions *fide* M.S. Hoogmoed (*in litt.*). Paralectotypes now RMNH 403, RMNH 434a–b, 2130 mm and 870 mm specimen, RMNH 47575, RMNH 76, a 560 mm specimen, and BMNH 1946.1.6.16, an adult male.

COLLORHABDIUM Smedley, 1932a (Calamariidae)

Type species: *Collorhabdium williamsoni* Smedley, 1932a.

Distribution: West Malaysia.

Sources: Smedley, 1932a, Tweedie, 1983, Wallach, 1988, Leong & Lim, 2001, B.L. Lim et al., 2002 and Zaher et al., 2009.

1. *Collorhabdium williamsoni* Smedley, 1932a. Bull. Raffles Mus. (1931) (6): 120–121, figs. 3a–b, pl. 2, fig. b.

Types: Syntypes (4), BMNH 1946.1.6.55 (formerly RMBR), ZRC 2.2168–69 (formerly RMBR), longest syntype 236 mm male (K.B. Williamson & native, 1930), location of other type unknown.

Type locality: "Tanah Rata, Cameron Highlands, Malay Peninsula" [= Tanah Rata, Cameron Highlands, Pahang State, cen. West Malaysia, 4°28'N, 101°23'E, elevation 1465 m].

Distribution: West Malaysia (Pahang, Perak) and Singapore, 1465–1860 m.

†*COLOMBOPHIS* Hoffstetter & Rage, 1977 (Aniliidae)

Type species: †*Colombophis portai* Hoffstetter & Rage, 1977.

Distribution: Middle Miocene of Colombia and Venezuela, and upper Miocene of Brazil.

Sources: Hoffstetter & Rage, 1977, Rage, 1984b, Head et al., 2006, Gómez et al., 2008 and Hsiou et al., 2010.

1. †*Colombophis portai* Hoffstetter & Rage, 1977. Ann. Paléont. (Vert.) 63(2): 174–178, fig. 4a, pl. 1, fig. 1.

Type: Holotype, MNHN VIV 6, one middle trunk vertebra.

Type locality: "Los Mangos, près de La Venta, Dépt. Huila, Colombie. Fish Bed de la Formation Villavieja. Miocène moyen (= Friasien)."

Distribution: Middle Miocene (Friasain: 15.5–16.3 mya) of Colombia (Huíla) and Venezuela (Falcón).

Source: Head, 2006.

2. †*Colombophis spinosus* Hsiou, Albino & Ferigolo, 2010. Acta Palaeontol. Pol. 55(3): 371–375, figs. 6. a–e.

Type: Holotype, UFAC-PV 2953, one middle trunk vertebra.

Type locality: "Talismã locality, Punus River, Amazonas State, Brazil. Late Miocene, Solimões Formation."

Distribution: Upper Miocene (Huayquerian: 6.8–9.0 mya) of Brazil (Amazonas).

Remarks: Holotype erroneously listed as UFAC-PV 2359 in Hsiou et al. (2010: fig. 6).

COLUBER Linnaeus, 1758 (Colubridae)

Synonyms: *Caluber* – Sturm, 1802 *(nomen incorrectum)*, *Colubro* – Oken, 1816 *(nomen incorrectum)*, *Scoliophis* Davis, Bigelow & Gray, 1817, *Colaber* – Wyder, 1817 *(nomen incorrectum)*, *Colubur* – Wied-Neuwied, 1825b *(nomen incorrectum)*, *Coluher* – Brüggemann & Wigand, 1838 *(nomen incorrectum)*, *Colubrum* – Gené, 1839 *(nomen incorrectum)*, *Scaliophis* Agassiz, 1845 *(nomen emendatum)*, *Bascanion* Baird & Girard, 1853, *Bascanian* – Slack, 1876 *(nomen incorrectum)*, *Bascaniun* – A.M. Ross, 1878 *(nomen incorrectum)*, *Colubcr* – Rochebrune, 1885 *(nomen incorrectum)*, *Coluteer* – A.W. Butler, 1887 *(nomen incorrectum)*, *Basacnium* – Velasco, 1890a *(nomen incorrectum)*, *Bacanium* – Velasco, 1890 *(nomen incorrectum)*, *Colubus* – E. Bartlett, 1896 *(nomen incorrectum)*, *Colnber* – Seiss, 1903 *(nomen incorrectum)*, *Columber* – C.E. Lord & Scott, 1924 *(nomen incorrectum)*, *Scioliophis* – Gould, 1930 *(nomen incorrectum)*,

C

C

Coluder – Maki, 1931 *(nomen incorrectum), Colober* – Lindberg, 1932 *(nomen incorrectum), Columber* – A.H. Wright & Wright, 1952 *(nomen incorrectum), Coluder* – Bogdanov, 1965 *(nomen incorrectum), Colouber* – Schwab, 1984 *(nomen incorrectum),* and *Colluber* – D.-T. Yang & Su, 1984 *(nomen incorrectum).*

Type species: *Coluber constrictor* Linnaeus, 1758.

Distribution: USA, Mexico, upper Central America.

Fossil records: Upper Miocene to upper Pleistocene of USA. Fossils unassigned to species within the Distribution of *Coluber constrictor* include upper Miocene (Hemphillian) of USA (Nevada), middle Pliocene (Blancan II, III) of USA (Idaho, Nebraska), upper Pliocene (Blancan V) of USA (Idaho), and upper Pleistocene (Rancholabrean II) of USA (Missouri, Tennessee). Possibly some North American *Coluber* fossil records are *Masticophis.*

Sources: Blainville, 1818, Ortenburger, 1928, Inger & Clark, 1943, L.D. Wilson, 1973a, Schätti, 1986b, Schätti & Wilson, 1986, Schätti, 1988a, Holman, 2000a, Nagy et al., 2004 and Utiger et al., 2005.

Remarks: Official Generic Name no. 2148 *fide* Opinion 1201 (ICZN, 1982a). The following genera and species have been separated from *Coluber: Bamanophis (dorri), Dolichophis (caspius, cypriensis, gyarosensis, jugularis, laurenti, schmidti), Hemerophis (socotrae, zebrinus), Hemorrhois (algirus, hippocrepis, nummifera, †pouchetii, ravergieri), Hierophis (andreanus, †caspioides, †freybergi, gemonensis, †robertmertensi, viridiflavus),* and *Platyceps (afarensis, atayeri, brevis, collaris, elegantissimus, florulentus, gracilis, insulanus, karelinii, largeni, messanai, najadum, rhodorachis, rogersi, rubriceps, saharicus, schmidtleri, sinai, somalicus, taylori, thomasi, variabilis, ventromacularus).*

1. *Coluber constrictor* Linnaeus, 1758. Syst. Nat., ed. 10, 1: 216.

Synonyms: *Coluber ovivorus* Linnaeus, 1758, *Coluber ligament* Lacépède, 1789 (*nomen rejiciendum*), *Scoliophis atlanticus* Davis, Bigalow & Gray, 1817, *Coluber relox* Rafinesque-Schmaltz, 1818a, *Coluber flaviventris* Say *in* E. James, 1823, *Coluber mormon* Baird & Girard, 1852a, *Bascanion foxii* Baird & Girard, 1853, *Bascanion fremontii* Baird & Girard, 1853, *Bascanion vetustus* Baird & Girard, 1853, *Bascanium anthicum* Cope, 1862c, *Coryphodon oaxaca* Jan, 1863, *Zamenis conirostris,* Cope, 1895b, *Zamenis stejnegerianus* Cope, 1895b, *Bascaniun constrictor* – A.M. Ross, 1878 *(nomen incorrectum), Zamenis acuminatus* Cope, 1899a, *Coluber ortenburgeri* L.C. Stuart, 1934a, *Coluber constrictor priapus* Dunn & Wood, 1939, *Coluber constrictor haasti* Bell, 1952, *Coluber constrictor paludicola* Auffenberg & Babbitt, 1953, *Coluber constrictor helvigularis* Auffenberg, 1955, *Coluber constrictor etheridgei* L.D.Wilson, 1970, and *Coluber constrictor latrunculus* L.D. Wilson, 1970.

Type: Holotype, ? NHR Lin-456 (formerly MDG), a 1380 mm female (P. Kalm, Sept. 1748–Feb. 1751, via Mus. Drottn.).

Type locality: "Canada," (possibly in error). Probably vicinity of Philadelphia, Pennsylvania, USA *fide* Dunn & Wood (1939: 1).

Distribution: North America. Southern Canada (S British Columbia, SW Ontario, S Saskatchewan), USA (Alabama, Arizona, Arkansas, California, Colorado, Connecticut, Delaware, District of Columbia, Florida, Georgia, Idaho, Illinois, Indiana, Iowa, Kansas, Kentucky, Louisiana, Maine, Maryland, Massachusetts, Michigan, Minnesota, Mississippi, Missouri, Montana, Nebraska, Nevada, New Hampshire, New Jersey, New Mexico, New York, North Carolina, North Dakota, Ohio, Oklahoma, Oregon, Pennsylvania, Rhode Island, South Carolina, South Dakota, Tennessee, Texas, Utah, Vermont, Virginia, Washington, West Virginia, Wisconsin, Wyoming), Mexico (Chiapas, Coahuila, Guerrero, Hidalgo, Nuevo León, Oaxaca, Tamaulipas, Veracruz), N Guatemala (Petén) and SW Belize (Cayo), NSL–2440 m.

Fossil records: Upper Miocene (Hemphillian) of USA (Nebraska, Texas), lower Pliocene (Hemphillian) of USA (Kansas, Nebraska, Oklahoma), middle Pliocene (Blancan II, III) of USA (Kansas, Texas), upper Pliocene (Blancan V) of USA (Kansas, Nebraska), Pleistocene of USA (Illinois), lower/middle Pleistocene (Irvingtonian I) of USA (Florida, Kansas, Pennsylvania), middle Pleistocene (Irvingtonian II) of USA (Nebraska), middle/upper Pleistocene (Irvingtonian II) of USA (Arkansas, Maryland, Pennsylvania, West Virginia), and upper Pleistocene (Rancholabrean I, II) of USA (Alabama, California, Florida, Georgia, Kansas, Nevada, New Mexico, Ohio, Oklahoma, Pennsylvania, Texas, West Virginia).

Sources: Ortenburger, 1928, Shannon, 1950, Auffenberg, 1955, Fitch, 1963, L.D. Wilson, 1970a, 1978, Christman, 1980, P.S. Corn & Bury, 1986, Ernst & Barbour, 1989, Conant & Collins, 1991, H.A. Brown et al., 1995, Palmer & Braswell, 1995, Degenhardt et al., 1996, Lee, 1996, 2000, Harding, 1997, Tennant, 1997, J.A. Campbell, 1998b, Trauth, 1998, Fitch, 1999, Bartlett & Tennant, 2000, Holman, 2000a, D.L Thomas et al., 2000, Werler & Dixon, 2000, Burbrink et al., 2008 and Ramírez-Bautista et al., 2010.

Remarks: Six unnamed clades derived from mDNA analysis *fide* Burbrink et al. (2008: 286). Official Specific Name no. 2784 *fide* Opinion 1201 (ICZN, 1982a).

COLUBROELAPS Orlov, Kharin, Ananjeva, Nguyen & Nguyen, 2009 (Colubridae)

Type species: *Colubroelaps nguyenvansangi* Orlov, Kharin, Ananjeva, Nguyen & Nguyen, 2009.

Distribution: Vietnam.

Sources: Orlov et al., 2009 and Ziegler & Nguyen, 2010.

1. *Colubroelaps nguyenvansangi* Orlov, Kharin, Ananjeva, Nguyen & Nguyen, 2009. Russ. J. Herp. 16(3): 235–236, figs. 1–6, 7d–g, 8a–d, 9a–c, 10–19.

Type: Holotype, ZISP IEBR 25686, a 500 mm female (V.S. Nguyen, May 2003).
Type locality: "Loc Bac Forest Enterprise, Lam Dong Prov., southern Vietnam (elevation 720 m, 11°47'07" N, 107°36'14" E)."
Distribution: Southern Vietnam (Binh Phuoc, Lam Dong), 720 m. Known only from holotype.

COMPSOPHIS Mocquard, 1894b (Pseudoxyrhophiidae)

Synonyms: *Geodipsas* Boulenger, 1896a, *Conspsophie* – Briceño-Rossi, 1934 *(nomen incorrectum)* and *Cempsophis* – Guibé, 1958 *(nomen incorrectum).*
Type species: *Compsophis albiventris* Mocquard, 1894b.
Distribution: Madagascar.
Sources: Guibé, 1958, Cadle, 1996bb, Ziegler et al., 1997, Glaw and Vences, 2007, Glaw et al., 2007b, Kelly et al., 2009 and Zaher et al., 2009.
Remarks: *Geodipsas* synonymized *fide* Ziegler et al. (1997: 110) and Glaw et al. (2007: 60).

1. *Compsophis albiventris* Mocquard, 1894b. C.R. Somm. Seances Soc. Philom. Paris (17): 8–9.

Type: Holotype, MNHN 1893.212, a 167 mm specimen (C. Alluaud & Belly, 1892–1893).
Type locality: "la montagne d'Ambre, Madagascar" [= Ambohitra, N Antsiranana Prov., N Madagascar, 12°30'S, 49°10'E *fide* Cadle, 1996b: 50].
Distribution: Northern Madagascar (N Antsiranana), 860–1250 m. Known only from vicinity of type locality.

2. *Compsophis boulengerii* (Peracca, 1892). Boll. Mus. Zool. Anat. Comp. Univ. Torino 7(112): 3–4, pl. 1, figs. 2, 2a–d. (*Tachymenis boulengerii*)

Synonym: *Geodipsas heimi* Angel, 1936a.
Type: Holotype, MZUT 1874, a 348 mm male (G. Pittarelli).
Type locality: "Madagascar – Valle dell'Umbi (Andrangoloka)" [= Umbi River valley, Andrangolaoka, E Antananarivo Prov., Madagascar, 19°02'S, 47°55'E, elevation 1385 m].
Distribution: Eastern Madagascar (Atananarivo, Fianarantsoa, Toamasina), 600–1400 m.
Source: Cadle, 1996b.

3. *Compsophis fatsibe* (Mercurio & Andreone, 2005). Zootaxa (1903): 62–67, figs. 1–3. (*Geodipsas fatsibe*)

Type: Holotype, MRSN 1922, a 522 mm male (F. Andreone, H. Randriamahazo & J.E. Randrianirina, 28 Jan. 1996).
Type locality: "western slope of Anjanaharibe-Sud Massif, Valley of Analabe River, 14°46.62'S, 49°26.60'E, 1050 m a.s.l., Befandriana Fivondronana, Mahajanga Faritany (Majunga Prov.), Madagascar."
Distribution: Northern Madagascar (Mahajanga), 1050 m.

4. *Compsophis infralineata* (A.C.L.G. Günther, 1882). Ann. Mag. Nat. Hist. (5) 9(52): 265–266, 2 figs (unnumbered). (*Tachymenis infralineatus*)

Type: Holotype, BMNH 1895.10.29.61, a 761 mm female (W.D. Cowan, 1882) *fide* Cadle (1996b: 58).
Type locality: "Eastern Betsileo" [Madagascar], [= N Fianarantsoa Prov., cen. Madagascar].
Distribution: Southeastern Madagascar (Antsiranana, Atananarivo, Fianarantsoa, Toamasina, Toliara), 300–2000 m.
Sources: Boulenger, 1896a and Henkel & Schmidt, 2000.
Remarks: BMNH labeled/catalogued holotype (BMNH 1946.1.7.20 (formerly BMNH 1882.5.8.1), a 787 mm female, in error *fide* Cadle (1996b: 58).

5. *Compsophis laphystius* (Cadle, 1996b). Bull. Mus. Comp. Zool. 155(2): 35–41, fig. 1. (*Geodipsas laphystia*)

Type: Holotype, MCZ 181390 (formerly JEC 13169), a 622 mm male (J.E. Cadle, 2 Jan. 1996).
Type locality: "Talatakely, Ranomafana National Park, 950–1,000 m, Fivondronana Ifanadiana, Fianarantsoa Prov., Madagascar [21°16'S, 47°25'E]."
Distribution: Eastern Madagascar (Antsiranana, Fianarantsoa, Toamasina), 550–1100 m.
Remarks: Some records of *C. infralineata* are very likely this species *fide* Cadle (1996b: 36).

6. *Compsophis vinckei* (Domergue, 1988). Bull. Mus. Natl. Hist. Nat. (4) 10A(1): 140–141, figs. 3a–d. (*Geodipsas vinckei*)

Type: Holotype, MNHN 1977.818 (formerly CAD 973/S), a 495 mm male (M. Vincke, late March 1970).
Type locality: "Station de pisciculture de Périnet (Analamazaotra), Madagascar" [= Lake Alaotra, Toamasina Prov., Madagascar, ca. 17°28'S, 48°32'E, elevation 770 m].
Distribution: Madagascar (Toamasina), 770 m.

C

7. *Compsophis zeny* (Cadle, 1996b). Bull. Mus. Comp. Zool. 155(2): 44–49, figs. 4, 6 (upper). (*Geodipsas zeny*)

Type: Holotype, MCZ 181161, a 281 mm male (J.E. Cadle, 11 Jan. 1993).
Type locality: "Approximately 7 km SW (airline) Midongy du Sud [Midongy Atsimo], near Rianambo ("high waterfall") on Lalampo River, 670 m elev., Fivondronana Midongy du Sud, Prov. Fianarantsoa, Madagascar [23°39'S, 46°57'E]."
Distribution: Southeastern Madagascar (Fianarantsoa), 670–980 m.

†*CONANTOPHIS* Holman & Harrison, 2000 (Booidea *incertae sedis*)

Type species: †*Conantophis alachuaensis* Holman & Harrison, 2000.
Distribution: Lower Oligocene of USA (Florida).
Source: Holman & Harrison, 2000.

1. †*Conantophis alachuaensis* Holman & Harrison, 2000. Acta Zool. Cracov 43(1–2): 128–130, figs. 1a–e.

Type: Holotype, UF 190831, one middle trunk vertebra (UF field crews, 1966–1967).
Type locality: "I-75 Local Fauna (Florida Natural History Museum Locality AL018), 1.5 km WSW Gainesville, Alachua County, Florida; Early Oligocene (Whitneyan)."
Distribution: Lower Oligocene (Whitneyan: 30.8–33.34 mya) of USA (Florida).

CONIOPHANES Hallowell *in* Cope, 1860e (Dipsadidae)

Synonyms: *Scydropus* Jan, 1857 *(nomen nudum)*, *Glaphyrophis* Jan, 1863b, *Hydrocalamus* Cope, 1885a, *Glaphirophis* – Senna, 1886 *(nomen incorrectum)*, *Caniophanes* – Velasco, 1898 *(nomen incorrectum)*, *Conophianes* – Briceño-Rossi, 1934 *(nomen incorrectum)*, *Hidrocalamus* – Briceño-Rossi, 1934 *(nomen incorrectum)*, *Cottonserpens* Hoser, 2012ai *(nomen illegitimum)*, *Daraninserpens* Hoser, 2012ai *(nomen illegitimum)*, *Laidlawserpens* Hoser, 2012ai *(nomen illegitimum)*, and *Smythserpens* Hoser, 2012ai *(nomen illegitimum)*.
Type species: *Coronella fissidens* A.C.L.G. Günther, 1858.
Distribution: Southern USA and Latin America.
Sources: Bailey, 1939a, H.M. Smith & Taylor, 1945, J.A. Peters & Orejas-Miranda, 1970, Cadle, 1984b, 1989, Villa et al., 1988, J.A. Campbell, 1989, Zaher et al., 2009 and Mulcahy et al., 2011.

1. *Coniophanes alvarezi* J.A. Campbell, 1989. Proc. Biol. Soc. Washington 102(4): 1036–1038, figs. 1–3.

Type: Holotype, UTA 12256 (formerly JAC 9389), a 375 mm male (J.A. Campbell, 7 Aug. 1983).
Type locality: "11.3 km ESE Teopisca, 2073 m elevation, Chiapas, Mexico" [= 11.3 KM ESE Teopisca (16°32'N, 92°28'W, elevation 1780 m), Chiapas State, S Mexico].
Distribution: Southern Mexico (Meseta Central of Oaxaca), 2010–2135 m.

2. *Coniophanes bipunctatus* (A.C.L.G. Günther, 1858). Cat. Colub. Snakes Brit. Mus.: 36–37. (*Coronella bipunctata*)

Synonyms: *Glaphyrophis pictus* Jan, 1863a and *Coniophanes bipunctatus biseriatus* H.M. Smith, 1940.
Type: Holotype, BMNH 1946.1.9.58, a 432 mm male.
Type locality: Unknown. Designated as British Honduras [= Belize] *fide* K.P. Schmidt (1941: 504).
Distribution: Mesoamerica. Southeastern Mexico (Campeche, Chiapas, Oaxaca, Quintana Roo, Tabasco, S Veracruz, Yucatán), Belize (Belize, Cayo, Orange Walk, Stann Creek), N Guatemala (Alta Verapaz, Petén), N Honduras (Atlántida, Colón, Cortés, Gracias a Dios, Islas de la Bahía, Roatán), E Nicaragua (Granada, Nueva Segovia), Costa Rica (Heredia, Limón) and NW Panama (Escudo de Veraguas Is.), NSL–1370 m.
Sources: C.W. Myers, 1969a, L.D. Wilson & Hahn, 1973, L.D. Wilson & Meyer, 1985, Lee, 1996, 2000, J.A. Campbell, 1998b, G. Köhler, 1999b, Savage, 2002, Solórzano, 2001, 2004, McCranie et al., 2006 and McCranie, 2011a.

3. *Coniophanes dromiciformis* (W.C.H. Peters, 1863c). Mber. Königl. Akad. Wiss. Berlin 1863(6): 273–275. (*Tachymenis dromiciformis*)

Synonym: *Coniophanes signatus* S.W. Garman, 1892b.
Types: Syntypes (8), ZMB 3729a–b, ZMB 3730a–c and ZMB 4550a–c, longest syntype 390 mm (C. Reiss, 1860)
Type locality: "Umgegend von Guayaquil" [= vicinity of Guayaquil, Guayas Prov., SW Ecuador, 2°12'S, 79°54'W, elevation 10 m].
Distribution: Southwestern Ecuador (Guayas), NSL–100 m.
Sources: C.W. Myers, 1969a and Cadle, 1989.
Remarks: Bauer et al. (1995: 75) listed only five syntypes (ZMB 3729, 4550). Records from Peru (ANSP 3736, MCZ 12426) doubtful *fide* Cadle (1989: 422–423).

4. *Coniophanes fissidens* (A.C.L.G. Günther, 1858). Cat. Colub. Snakes Brit. Mus.: 36. (*Coronella fissidens*)

Synonyms: *Coniophanes proterops* Cope, 1860, *Coniophanes punctigularis* Cope, 1860, *Dromicus chitalonensis* F. Müller, 1878a, *Erythrolamprus violaceus*

Cope, 1887b, *Coniophanes brevifrons* Bailey, 1937a, *Coniophanes fissidens andresensis* Bailey, 1937a, *Coniophanes fissidens dispersus* H.M. Smith, 1941n, *Coniophanes fissidens convergens* Shannon & Smith, 1950, and *Coniophanes fissidens obsoletus* Minton & Smith, 1960.

Type: Lectotype, BMNH 1946.1.8.20, a 489 mm female (A. Sallé, 1846–1856), designated by H.M. Smith (1941n: 104).

Type locality: "Mexico" via lectotype selection. Restricted to San Andres Tuxtla, Veracruz, Mexico by H.M. Smith & Taylor (1950: 350).

Distribution: Mesoamerica and NW South America. Southern Mexico (N Chiapas, Guerrero, Hidalgo, S Michoacán, NE Oaxaca, Puebla, Querétaro, S San Luis Potosí, Tabasco, Tamaulipas, Veracruz), Belize (Belize, Cayo, Stann Creek), Guatemala (Alta Verapaz, Baja Verapaz, Escuintla, Izabal, Petén, Quezaltenango, San Marcos, Santa Rosa, Suchitepéquez, Zacapa), Honduras (Atlántica, Colón, Comayagua, Copán, Cortés, El Paraiso, Francisco Morazán, Gracias a Dios, Lempira, Olancho, Yoro), El Salvador (Ahuachapán, La Libertad, Morazán, San Salvador, Sonsonate), Nicaragua (Matagalpa, Zelaya), Costa Rica (Alajuela, Cartago, Guanacaste, Heredia, Limón, Puntarenas, San José), Panama (Bocas del Toro, Chiriquí, Coclé, Colón, Darién, Panamá), Colombia (Antioquia, Santander, San Andrés Is.), NW Ecuador (Esmeraldas, Pichincha) and ext. NW Peru, NSL–1970 m.

Sources: E.H. Taylor, 1951, 1954, C.B. Fisher, 1968, C.W. Myers, 1969a, L.D. Wilson & Meyer, 1985, J.A. Campbell, 1989, 1998b, Pérez-Santos & Moreno, 1991, Lee, 1996, 2000, G. Köhler, 1999b, Mejenes-López et al., 1999, Savage, 2002, Farr et al., 2004, G. Köhler et al., 2004, McCranie et al., 2006, Ramírez-Bautista et al., 2010 and McCranie, 2011a.

5. *Coniophanes imperialis* (Baird, 1859a). U.S.-Mex. Bound. Surv., Rept. (1858) 2(2): 23, pl. 19, fig. 1. (*Taeniophis imperialis*)

Synonyms: *Glaphyrophis lateralis* Jan, 1863b, *Dromicus clavatus* W.C.H. Peters, 1864b, and *Coniophanes imperialis copei* Hartweg & Oliver, 1938.

Type: Holotype, USNM 2060 (S.L. Van Vliet [U.S.-Mex. Bound. Surv.], 1854–1855).

Type locality: "Brownsville, Texas" [= Cameron Co., ext. SE Texas, USA, 25°54'N, 97°30'W, elevation 10 m] (possibly in error). Corrected to Matamoras, Tamaulipas, Mexico [= Matamoros, ext. NE Mexico, 25°52'N, 97°30'W, elevation 5 m] *fide* Girard (1854: 227), Yarrow (1882: 97) and Cochran (1961: 216).

Distribution: Southern USA to N Central America. USA (S Texas), Mexico (Campeche, Chiapas, Hidalgo, Oaxaca, Quintana Roo, San Luis Potosí, Tabasco, Tamaulipas, Veracruz, Yucatán), Belize (Belize, Cayo, Corozal, Orange Walk, Stann Creek, Toledo), N Guatemala (Alta Verapaz, Huehuetenango, Izabal, Petén) and N Honduras (Atlántida, Colón, Copán, Cortés, Francisco Morazán, Gracias a Dios, Islas de la Bahía, Yoro, Cayo Cochino and Utila Is.), NSL–2200 m.

Sources: C.J. McCoy, 1969, L.D. Wilson & Meyer, 1985, G. Köhler, 1996a, Lee, 1996, 2000, J.A. Campbell, 1998b, Werler & Dixon, 2000, Ramírez-Bautista et al., 2010 and McCranie, 2011a

Remarks: Brownsville and Matamoros face each other across the Rio Grande River [= U.S.-Mexican border].

6. *Coniophanes joanae* C.W. Myers, 1966a. Copeia 1966(4): 665–667, figs. 1–2.

Type: Holotype, KU 93502, a 365–369 mm female (C.W. Myers, 27 May 1965).

Type locality: "on the southeastern slope of Cerro Pirre (also called "Mount Pirri" and "Cerro Cana"), Serranía de Pirre, 1440 m above sea level, Darién, Panama."

Distribution: Eastern Panama (Darién, Panamá), 800–1440 m.

Source: C.W. Myers, 1969a.

7. *Coniophanes lateritius* Cope, 1862a. Proc. Acad. Nat. Sci. Philadelphia (1861) 13(9): 524.

Type: Neotype, BYU 13793, a 193 mm male (W.W. Tanner & W.G. Robinson, Jr., 5 Sept. 1957), designated by W.W. Tanner & Robinson (1960a: 60).

Type locality: "7.5 miles north of Magdalena, northwestern Jalisco, Mexico" via neotype designation.

Distribution: Southwestern Mexico (Colima, ? Guerrero, Jalisco, Michoacán, Morelos, Nayarit, Sinaloa, Sonora), NSL–1850 m.

Sources: H.M. Smith & Grant, 1958a, Wellman, 1959, Ponce-Campos & Smith, 2001 and Castro-Franco & Bustos-Zagal, 2004.

Remarks: Holotype lost *fide* H.M. Smith & Taylor (1945: 41). H.M. Smith & Taylor's (1945: 41) neotype designation of FMNH 100032 (formerly EHT-HMS 5198) for *C. lateritius* invalidated by H.M. Smith & Grant's (1958b: 22) recognition of *C. melanocephalus*.

8. *Coniophanes longinquus* Cadle, 1989. Herpetologica 45(4): 412–417, fig. 2.

Type: Holotype, FMNH 231779, a 336 mm male (J.E. Cadle, 22 June 1987).

Type locality: "1 km (airline) NNW Monte Seco, approximately 80 km ESE Chiclayo, 1300–1340 m elevation, Río Zaña, Department of Cajamarca, Peru."

Distribution: Northwestern Peru (SW Cajamarca), 1200–1430 m.

9. *Coniophanes melanocephalus* (W.C.H. Peters, 1870a). Mber. Königl. Akad. Wiss. Berlin 1869(12): 876–877. (*Tachymenis melanocephala*)

Type: Holotype, ZMB 6656 (Berkenbusch, 1868–1869), lost *fide* Bauer et al. (1995: 75).

C

C

Type locality: "Puebla, Mexicos (Matamoros u.a.O.)." Restricted to "Matamoros (Izúcar), Mexico" [= Izucar de Matamoros, W Puebla State, S Mexico, 18°36'N, 98°28'W, elevation 1285 m] *fide* H.M. Smith & Taylor (1950: 341).
Distribution: Southern Mexico (Michoacán, Morelos, W Puebla), 400–1275 m.
Sources: H.M. Smith & Grant, 1958a, Bauer et al., 1995, Ponce-Campos & Smith, 2001 and Carbajal-Márquez et al., 2011a.
Remarks: A valid species *fide* Ponce-Campos & Smith (2001: 14).

10. *Coniophanes meridanus* K.P. Schmidt & Andrews, 1936. Field Mus. Nat. Hist. (Zool.) 20(18): 179.

Type: Holotype, FMNH 19427, a 313 mm (svl) female (E.R. Blackburn).
Type locality: "Merida, Yucatan" [= Merida, W Yucatán State, SE Mexico, 20°58'N, 89°37'W, elevation 10 m].
Distribution: Yucatán Peninsula of Mexico (N Campeche, Quintana Roo, Yucatán), NSL–270 m.
Sources: C.J. McCoy, 1969 and Lee, 1996, 2000.
Remarks: Type collected by E.W. Andrews *fide* Bailey (1939:43).

11. *Coniophanes michoacanensis* Flores-Villela & Smith, 2009. Herpetologica 65(4): 404–407, figs.1, 2a.

Type: Holotype, MZFC 10398, a 561 mm female (O. Flores-Villela, 26 Feb. 1992).
Type locality: "El Farito, 8 km. NW Caleta de Campos, 17 m elevation, Municipio de Aquila, Michoacán, Mexico (18.1014° N, 102.8173° W)."
Distribution: Southwestern Mexico (Guerrero, Michoacán), NSL–800 m.
Source: Mertz et al., 2012.

12. *Coniophanes piceivittis* Cope, 1870. Proc. Amer. Philos. Soc. (1869–1870) 11(81): 149–150.

Synonyms: *Tachymenis taeniata* W.C.H. Peters, 1870a and *Coniophanes frangivirgatus* J.A. Peters, 1950.
Types: Syntypes (2), USNM 30264–65 (formerly USNM 248), a 542 mm male and 481 mm female (A.-L.-J.-F. Sumichrast, 1855–1869).
Type locality: "Chihuitan, Tehuantepec, Western Mexico" [= Chihuitan, Ciudad Ixtepec, SE Oaxaca State, S Mexico, 16°33'N, 95°05'W, elevation 50 m].
Distribution: Mesoamerica. Southern Mexico (Chiapas, Hidalgo, Oaxaca, Querétaro, Veracruz), Guatemala (Huehuetenango, Quiché, Zacapa), W Honduras (Choluteca, Comayagua, Copán, Francisco Morazán, Yoro), El Salvador (Cuscatlán, La Libertad, San Miguel, San Salvador, Sonsonate), Nicaragua (Granada, Jinotepa, Managua) and Costa Rica (Guanacaste, Puntarenas, San José), NSL–1300 m.
Sources: Hall, 1951, E.H. Taylor, 1951, 1954, Martin, 1955, Neill & Allen, 1960, L.D. Wilson & Meyer, 1985, Harrison, 1992, G. Köhler, 1999b, G. Köhler et al., 2004, N. Herrera et al., 2007 and Ramírez-Bautista et al., 2010.
Remarks: Bailey (1939: 29) indicated both syntypes are 30265.

13. *Coniophanes quinquevittatus* (A.-M.-C. Duméril, Bibron & Duméril, 1854b). Erpét. Gén. 7(2): 975–977. (*Homalopsis quinquevittatus*) (*nomen corrigendum*)

Synonyms: *Homalopsis quinque-vittatus* A.-M.-C. Duméril, Bibron & Duméril, 1854b *(nomen incorrigendum), Scydropus mexicanus* Jan, 1857 (*nomen nudum*), *Calopisma quinquevittatum* – Jan, 1863b *(nomen corrigendum), Calopisma quinquevittatum mexicana* Jan, 1863b (*nomen nudum*), *Coniophanes quinquevittatum mexicana* Jan, 1865c, and *Hydrops lubricus* Cope, 1871b.
Types: Syntypes (2), MNHN 516, a 758 mm specimen (P.M.A. Morelet, 1849), and MNHN 3766.
Type locality: "la Prov. de Peten dans l'Amérique centrale (République de Guatemala)" [= Petén Prov., Guatemala]. Restricted to Flores, Petén, Guatemala *fide* H.M. Smith & Taylor (1950: 318).
Distribution: Southeastern Mexico (Campeche, Chiapas, Tabasco, S Veracruz, Yucatán) and N Guatemala (Petén), NSL–200 m.
Sources: Lee, 1996, 2000 and J.A. Campbell, 1998b.
Remarks: MNHN catalogue lists MNHN 3766 from "Indies occidentals."

14. *Coniophanes sarae* Ponce-Campos & Smith, 2001. Bull. Maryland Herp. Soc. 37(1): 10–11.

Type: Holotype, UNAM 13030, a 125 mm female (P. Ponce-Campos, 1–5 July 1999).
Type locality: "near Tehuantepec, municipality of Chinicuila (18°42'07.7"N, 103°18'22.3"W), 1390 m, Michoacan."
Distribution: Western Mexico (W Michoacán), 800–1500 m.

15. *Coniophanes schmidti* Bailey, 1937. Occ. Pap. Mus. Zool., Univ. Michigan (362): 1–2.

Type: Holotype, UMMZ 73043, a 547 mm male (E. Creaser, 30 June 1932).
Type locality: "Chichen Itzá, Yucatán" [= Chichen-itza, Yucatán State, SE Mexico, 20°41'N, 88°34'W, elevation 35 m].
Distribution: Southeastern Mexico (Campeche, Chiapas, Quintana Roo, Yucatán), Guatemala (cen. Petén) and Belize (Cayo), NSL–600 m.
Sources: C.J. McCoy, 1969, Harrison, 1992, Lee, 1996, 2000 and J.A. Campbell, 1998b.

16. *Coniophanes taylori* Hall, 1951. Univ. Kansas Sci. Bull. 34(1), 208–211, pl. 24, figs. 1–2. (*Coniophanes piceivittis taylori*)

Type: Holotype, FMNH 100868 (formerly EHT-HMS 23523), a 340 mm male (E.H. Taylor, 1940).
Type locality: "Agua del Obispo, Gro." [= Agua del Obispo Hacienda, between Tierra Colorado and Chilpancingo, Guerrero State, W Mexico, 17°18'N, 99°28'W, elevation 915 m *fide* Mendelson et al., 2005: 12].
Distribution: Western Mexico (Guerrero), 915 m.
Remarks: Elevated to specific status by Flores-Villela & Smith, 2009.

†*CONIOPHIS* Marsh, 1892 (Alethinophidia *incertae sedis*)

Synonyms: †*Comophis* – Gowanloch *in* Gowanloch & Brown, 1943 *(nomen incorrectum)* and †*Henophis* J.G. Armstrong, 1976.
Type species: †*Coniophis precedens* Marsh, 1892.
Distribution: Middle Cretaceous of Sudan, upper Cretaceous of USA, Paleocene of Brazil, middle Paleocene of USA, middle Eocene of France and USA, and upper Eocene of France.
Sources: M.K. Hecht, 1959, Fox 1975, Holman, 1979b, 2000, Rage, 1984b, 1998, Bryant, 1989, J.D. Gardner & Cifelli, 1999, Gómez et al., 2008 and Longrich et al., 2012a–b.
Remarks: An undescribed species is known from Utah *fide* J.D. Gardner & Citelli (1999: 91–94). Possibly occurs in Canada and Bolivia. Sister group to all Serpentes *fide* Longrich et al. (2012a: 207).

1. †*Coniophis carinatus* M.K. Hecht *in* McGrew, 1959. Bull. Amer. Mus. Nat. Hist. 117(3): 138–139, pl. 52, figs. 1–5.

Type: Holotype, AMNH 3826, one middle trunk vertebra (P.O. McGrew & field crews, 1950–1952).
Type locality: "Locality 5, on the south side of Elk Mountain and Tabernacle Butte area at the northern end of the Green River Basin, about 15 miles from the southwest flank of the Wind River Mountains and 25 miles north of the village of Farson, in the southeast corner of Sublette County, Wyoming (T 27-29N, R 103-106W), [USA]; late Bridgerian, middle Eocene."
Distribution: Middle Eocene (Bridgerian: 46.2–50.3 mya) of USA (Wyoming, Colorado).

2. †*Coniophis cosgriffi* Armstrong-Ziegler, 1978. J. Paleont. 52(2): 480–482, figs. 1a–e.

Type: Holotype, MNA Pl. 1612, one trunk vertebra (J.G. Armstrong, 1974–1975).
Type locality: "MNA site A (= MNA 107) on the hogback South, 7 1/2' quadrangle, T26N, R16W (108°29'10"W Long., 36°29'10"N Lat.), (thirty miles southwest of Farmington, New Mexico; Fruitland formation, upper Campanian, upper Cretaceous."
Distribution: Upper Cretaceous (Maastrichtian, 65.5–72.1 mya and Campanian, 72.1–83.6 mya) of USA (Montana, New Mexico).
Source: J.D. Gardner & Citelli, 1999.

3. †*Coniophis dabiebus* Rage & Werner, 1999. Paleont. Afr. 35: 94–95, figs. 6–8.

Type: Holotype, TUB Vb-673, one middle trunk vertebra.
Type locality: "Wadi Abu Hashim, Sudan. Wadi Abu Hashim Member of Wadi Milk Formation; Cenomanian."
Distribution: Middle Cretaceous (Cenomanian: 93.9–100.5 mya) of Sudan. Known only from type locality.

4. †*Coniophis platycarinatus* M.K. Hecht *in* McGrew, 1959. Bull. Amer. Mus. Nat. Hist. 117(3): 139, pl. 53, figs. 1–2.

Type: Holotype, AMNH 3827, one middle trunk vertebra (P.O. McGrew and field crews, 1950–1952).
Type locality: "Locality 5, on the south side of Elk Mountain and Tabernacle Butte area at the northern end of the Green River Basin, about 15 miles from the southwest flank of the Wind River Mountains and 25 miles north of the village of Farson, in the southeast corner of Sublette County, Wyoming (T 27-29N, R 103-106W), [USA]; late Bridgerian, middle Eocene."
Distribution: Middle Eocene (Bridgerian: 46.2–50.3 mya) of USA (Wyoming). Known only from type locality.

5. †*Coniophis precedens* Marsh, 1892. Amer. J. Sci. (3) 43(257): 450, figs. 1a–e.

Type: Holotype, USNM 2134, one trunk vertebra (J.B. Hatcher, 1892).
Type locality: "Ceratops beds of Wyoming, Laramie Formation" [Peterson Quarry, Niobrara County, Wyoming (Lance formation, upper Maastrichtian, upper Cretaceous)].
Distribution: Upper Cretaceous (Maastrichtian: 65.5–72.1 mya) of USA (Montana, Wyoming), and possibly Canada (Alberta) and Brazil (Rio de Janeiro).
Sources: Gilmore, 1938, Fox, 1975 and Rage, 2011.
Remarks: Fox (1975: 1558–1560) described *Coniophis* cf. *precedens* from the upper Cretaceous (Campanian) of Alberta, Canada but Rage (1984b: 13) questioned the identification. Rage (2011: 62) listed *Coniophis cf. precedens* from Itaboraí, Rio de Janeiro, Brazil. Sister group to all Serpentes *fide* Longrich et al. (2012a: 207).

CONOPHIS W.C.H. Peters, 1860d (Xenodontidae)

Synonyms: *Eudromus* S.W. Garman, 1884, *Canophis* – Maass-Berlin, 1933 *(nomen incorrectum),* and *Whittonserpens* Hoser, 2012ai *(nomen illegitimum).*

Type species: *Conophis vittatus* W.C.H. Peters, 1860d.
Distribution: Mesoamerica.
Sources: H.M. Smith, 1941a, H.M. Smith & Taylor, 1945, Wellman, 1963, Cadle, 1984c, Villa et al., 1988 Zaher et al., 2009 and Vidal et al., 2010.

C

1. *Conophis lineatus* (A.-M.-C. Duméril, Bibron & Duméril, 1854b). Erpét. Gén. 7(2): 936–938, pl. 73, figs. 1–4. (*Tomodon lineatum*)

Synonyms: *Conophis vittatus* Cope, 1861e (*nec* Peters), *Conophis concolor* Cope, 1867b, *Conophis pulcher* Cope, 1869b, *Conophis pulcher similis* Bocourt, 1886 *in* A.H.A. Duméril, Bibron & Mocquard, 1870–1909, *Conophis lineaticeps* Cope, 1900 (*nomen substitutum*), *Conophis lineatus similis* H.M. Smith, 1941e, *Conophis lineatus dunni* H.M. Smith, 1942m, *Conophis pulcher plagosus* H.M. Smith, 1942m, and *Conophis morai* Pérez-Higareda, López-Luna & H.M. Smith, 2002.
Type: Lectotype, MNHN 3738, a 600 mm specimen (A. Ghiesbreght & F. Schlumberger), designated by Wellman (1963: 267).
Type locality: "Mexique" [= Mexico] via lectotype selection. Restricted to Veracruz, Mexico *fide* H.M. Smith & Taylor (1950: 351).
Distribution: Mesoamerica. Southern Mexico (Campeche, Chiapas, Jalisco, Oaxaca, Querétaro, Quintana Roo, Tabasco, Veracruz, Yucatán), Belize (Belize, Cayo), Guatemala (El Progreso, Huehuetenango, Jutiapa, Petén, Sacatepequez, Santa Rosa, Zacapa), El Salvador (Ahuachapán, Cabañas, Chalatenango, Cuscatlán, La Libertad, La Paz, La Unión, Morazán, San Miguel, San Salvador, Santa Ana, Sonsonate, Usulután), Honduras (Choluteca, Colón, Comayagua, Cortés, El Paraíso, Francisco Marazán, Gracias a Dios, Intibucá, Lempira, Vallé), W Nicaragua (Chinandega, Estelí, León, Managua, Masaya, Matagalpa) and Costa Rica (Alajuela, Guanacaste, Puntarenas, San José), NSL–1500 m.
Sources: Bocourt, 1886 *in* Duméril, Bocourt & Mocquard, 1870–1909, E.H. Taylor, 1951, L.D. Wilson & Meyer, 1985, Lee, 1996, 2000, Auth et al., 1998b, J.A. Campbell, 1998b, G. Köhler, 1999b, Stafford & Meyer, 2000, Savage, 2002, Barragan-Vázquez et al., 2004, G. Köhler et al., 2004, Solórzano, 2004 and McCranie, 2011a.
Remarks: *Conophis concolor* a valid species *fide* Pérez-Higareda et al. (2002: 31). Colombia record of Pérez-Santos & Moreno (1988: 129) in error *fide* Auth et al. (1998: 107).

2. *Conophis morai* Pérez-Higareda, López-Luna & Smith, 2002. Bull. Maryland Herp. Soc. 38(1): 27–31, figs. 1, 3–4.

Type: Holotype, UNAM-LT 3662, a 664 mm male (R. Mora, 12 April 1999).
Type locality: "Ejido Ruiz Cortines on the southeastern slope of San Martin Tuxla Volcano, 1050 m, southern Veracruz, México" [=Ejido Ruiz Cortines, SE Volcan San Martin Tuxtlas, S Veracruz State, SE Mexico, 18°32'N, 95°09'W, elevation 1050 m].
Distribution: Southeastern Mexico (S Veracruz), 200–1050 m.

3. *Conophis vittatus* W.C.H. Peters, 1860d. Mber. Königl. Akad. Wiss. Berlin 1860(10): 519–520, pl., figs. 3–3d.

Synonyms: *Conophis sumichrastii sumichrastii* Cope, 1875a, *Conophis sumichrasiti viduus* Cope, 1875a, and *Conophis vittatus videns* Cope, 1900.
Type: Holotype, ZMB 3819, a 555 mm male (Hamburg dealer).
Type locality: Unknown. Restricted to Acapulco, Guerrero, Mexico *fide* H.M. Smith (1941a: 119) and to Laguna Coyuca, Guerrero, Mexico *fide* H.M. Smith & Taylor (1950: 331).
Distribution: Southwestern Mexico (Chiapas, Colima, ? Durango, Guerrero, Jalisco, México, Michoacán, Morelos, Nayarit, Oaxaca, Puebla) and Guatemala (Huehuetenango), NSL–2000 m.
Sources: H.M. Smith et al., 1993, Benitez-Gálvez, 1997, Matias-Ferrer & Murillo, 2004a, N. Herrera et al., 2007 and Acevedo et al., 2011.
Remarks: Hamburg dealer who sold the type reported the boat originated from "Neu-Orleans im Mississippi." Durango, Mexico record needs confirmation *fide* Webb (2001: 159).

CONOPSIS A.C.L.G. Günther, 1858 (Colubridae)

Synonyms: *Toluca* Kennicott *in* Baird, 1859, *Achirhina* Jan, 1862b, *Epirhina* Jan, 1862b, *Exorhina* Jan, 1862b, *Oxyrhina* Jan, 1862b, *Oxyrrhina* – Troschel, 1863 *(nomen incorrectum), Ogmius* Cope, 1870 *(nomen substitutum), Oxyrrhina* Dugès, 1887 *(nomen emendatum), Ognius* – Ocaranza, 1930 *(nomen incorrectum), Agmius* – Ditmars, 1933 *(nomen incorrectum),* and *Conepsis* – W.W. Tanner & Avery, 1982 *(nomen incorrectum).*
Type species: *Conopsis nasus* A.C.L.G. Günther, 1858.
Distribution: Southern Mexico.
Sources: E.H. Taylor & Smith, 1942b, Bogert & Oliver, 1945, H.M. Smith & Laufe, 1945b, H.M. Smith & Taylor, 1945, Goyenechea, 1995 and Goyenechea & Flores-Villella, 2002, 2006.

1. *Conopsis acuta* (Cope *in* Ferrari-Pérez, 1886). Proc. U.S. Natl. Mus. 9(11): 189. (*Ogmius acutus*)

Type: Holotype, USNM 30552, a 249 mm specimen (A.-L.-J.-F. Sumichrast, 1855–1869).
Type locality: "Tuchitan, on the Pacific side of the isthmus of Tehuantepec" [= Juchitán, Oaxaca State, Mexico *fide* H.M. Smith & Taylor, 1950: 338] (in error

fide Goyenechea & Flores-Villela, 2006: 6). Restricted to 2 miles east of Ixtlán de Juárez, Oaxaca, Mexico *fide* Goyenechea & Flores-Villella (2006: 6), 17°20'N, 96°29'W, elevation 2100 m].

Distribution: Central Mexico (Guerrero, Oaxaca, Puebla, Veracruz), 1800–2660 m.

Sources: García-Vázquez et al., 2008a.

2. *Conopsis amphisticha* (H.M. Smith & Laufe, 1945b). Herpetologica 3(1): 5–10. (*Toluca amphisticha*)

Type: Holotype, USNM 120950, a 262 mm male (W.S. Miller, 15 March 1944).

Type locality: "Xuimygopk, 8,650 feet, about two-thirds of the distance from Ayutla to Cacalotepec, Oaxaca, Mexico."

Distribution: Southern Mexico (Oaxaca, Puebla), 1700–3080 m.

Source: Casas-Andreu et al., 1996.

3. *Conopsis biserialis* E.H. Taylor & Smith, 1942b. Univ. Kansas Sci. Bull. 28(15): 333–325, fig. 2, pl. 31, fig. 1, pl. 35, fig. 9.

Type: Holotype, FMNH 100062 (formerly EHT-HMS 23648 & EHT-CC x-2166), a 293 mm specimen (E.H. Taylor, 1940).

Type locality: "10 mi. west of Villa Victoria, México" [= 10 mi. W Villa Victoria, México State, cen. Mexico, ca. 19°25'N, 100°07'W, elevation 2600 m].

Distribution: Central and S Mexico (Chiapas, Colima, Distrito Federal, Guanajuato, Guerrero, Hidalgo, Jalisco, México, E Michoacán, Morelos, Puebla, Querétaro, Tlaxcala), 1900–2600 m.

Sources: Reyes-Velasco et al., 2009 and Canseco-Márquez & Gutiérrez-Mayén, 2010.

4. *Conopsis lineata* (Kennicott *in* Baird, 1859a). U.S.-Mex. Bound. Surv., Rept. (1858) 2(2): 23, pl. 21, fig. 2. (*Toluca lineata*)

Synonyms: *Stenorrhina defilippii* Jan, 1857 *(nomen nudum), Stenorrhina varians* Jan, 1857 *(nomen nudum), Oxyrhina defilippii* Jan, 1862b, *Oxyrhina varians* Jan, 1862b, and *Chionactis diasii* Cope *in* Ferrari-Pérez, 1886.

Type: Holotype, USNM 2103 (J. Potts, 1848–1855).

Type locality: "Valley of Mexico."

Distribution: Central and S Mexico (District Federal, Guanajuato, ?Guerrero, Hidalgo, Jalisco, México, cen. Michoacán, Morelos, Oaxaca, Puebla, Querétaro, S San Luis Potosí, Tlaxcala, cen. Veracruz, Zacatecas), 1750–3100 m.

Sources: Casas-Andreu et al., 1996, Benitez-Gálvez, 1997 and Carbajal-Márquez et al., 2012.

5. *Conopsis megalodon* (E.H. Taylor & Smith, 1942b). Univ. Kansas Sci. Bull. 28(15): 338–339, fig. 4, pl. 30, fig. 1, pl. 35, fig. 1. (*Toluca megalodon*)

Synonym: *Toluca conica* E.H. Taylor & Smith, 1942b.

Type: Holotype, UIMNH 25077 (formerly EHT-HMS 23640 & EHT-CC x-2162), a 251 mm specimen (E.H. Taylor, 10 July 1940).

Type locality: "summit of Cerro San Felipe, Oaxaca, Oaxaca" [= Cerro San Felipe, Benito Juárez National Park, cen. Oaxaca State, Mexico, 17°10'N, 96°41'W, elevation 3050 m].

Distribution: Southwestern Mexico (cen. Guerrero, Oaxaca), 2200–3050 m.

Sources: Casas-Andreu et al., 1996 and Canseco-Márquez & Gutiérrez-Mayén, 2010 .

6. *Conopsis nasus* A.C.L.G. Günther, 1858. Cat. Colub. Snakes Brit. Mus.: 6–7.

Synonyms: *Oxyrhina maculata* Jan, 1862b, *Epirhina tessellata* Jan, 1862b, *Oxyrrhina maculata anomala* Dugès, 1869, *Conopsis nasus heliae* Cuesta-Terron, 1930, *Conopsis nasus labialis* W.W. Tanner, 1961, and *Gyalopion atavus* Leviton & Banta, 1961.

Type: Neotype, BMNH 1883.4.16.38, a 295 mm male (A. Forrer, 16 April 1883), designated by Goyenechea & Flores-Villela (2000: 286).

Type locality: "Milpas, Durango, Mexico" [= Milpas, Durango State, NW Mexico, 24°00'N, 104°38'W, elevation 1885 m] via neotype selection.

Distribution: Northern Mexico (Aguascalientes, S Chihuahua, Distrito Federal, Durango, Guanajuato, Hidalgo, Jalisco, México, Michoacán, Oaxaco, Puebla, Querétaro, San Luis Potosí, E Sinaloa, Zacatecas), 1515–2950 m.

Sources: W.W. Tanner, 1961, J.D. Johnson, 1978a, Benitez-Gálvez, 1997 and McCranie & Wilson, 2001b.

Remarks: Holotype in BMNH lost *fide* Goyenechea & Flores-Villela (2000: 285). Photograph of neotype in Goyenechea & Flores-Villela (2000: fig. 1).

CONTIA Baird & Girard, 1853 (Carphophiidae)

Synonym: *Lodia* Baird & Girard, 1853.

Type species: *Calamaria tenuis* Baird & Girard, 1852c.

Distribution: Extreme W USA and Canada.

Sources: W.W. Tanner, 1967, Nussbaum et al., 1983, Cadle, 1984c, Stebbins, 1985, H.A. Brown et al., 1995, P.R. Brown, 1997, Leonard & Ovaska, 1998, Bartlett & Tennant, 2000, Feldman & Spicer, 2002, Pinou et al., 2004, Zaher et al., 2009 and Feldman & Hoyer, 2010.

Remarks: A distinct population may exist in the southern Sierra Nevadas of California (Tulare Co.) *fide* Feldman & Hoyer (2010: 262).

C

1. *Contia longicaudae* Feldman & Hoyer, 2010. Copeia 2010(2): 257–261, fig. 4.

Type: Holotype, CAS 231505, a 343 mm female (R.F. Hoyer, 7 July 1998).
Type locality: "California, Mendocino County, 8.6 km E of junction with Highway 1 via State Route 128, 39°10'18"N, 123°39'48"W, 5 m elevation."
Distribution: Northwestern USA (NW California, SW Oregon), NSL–80 m.
Source: Norman et al., 2012.

2. *Contia tenuis* (Baird & Girard, 1852c). Proc. Acad. Nat. Sci. Philadelphia 6(5): 176. (*Calamaria tenuis*)

Synonyms: *Contia mitis* Baird & Girard, 1853, and *Ablabes purpureocauda* A.C.L.G. Günther, 1858.
Type: Holotype, USNM 7289, ca 200 mm specimen (C. Wilkes [U.S. Explor. Exped.], 19–24 June 1841).
Type locality: "Puget Sound" [Washington, USA]. Restricted to vicinity of Gravelly Lake/Chambers Creek, Pierce County, Washington, USA *fide* Leonard & Ovaslea (1998: 4).
Distribution: Extreme SW Canada (SW British Columbia, S Vancouver, South Pender Is.) and W USA (N California, W Oregon, cen. Washington), NSL–2010 m.
Sources: Blaustein et al., 1995, Hoyer, 2001 and Feldman & Spicer, 2002.

†*COPROPHIS* Parris & Holman, 1978 (Alethinophidia *incertae sedis*)

Type species: †*Coprophis dakotaensis* Parris & Holman, 1978.
Distribution: Middle Oligocene of USA.
Sources: Parris & Holman, 1978 and Holman, 2000a.
Remarks: Possibly a member of the Booidea or Anilioidea *fide* Rage (1984b: 58).

1. †*Coprophis dakotaensis* Parris & Holman, 1978. Herpetologica 34(3): 259–260, figs. 1a–b.

Type: Holotype, PU 20732A, one trunk vertebra, (South Dakota School Mines & Tech., before 1969).
Type locality: "Big Badlands of South Dakota, Pennington and Shannon counties, USA; Scenic Member of the Brule Formation, Orellan Age, (approximate Middle Oligcene)."
Distribution: Lower Oligocene (Orellean: 33.3–33.9 mya) of USA (South Dakota). Known only from type locality.

CORALLUS Daudin, 1803b (Boidae)

Synonyms: *Caramus* Rafinesque-Schmaltz, 1815 *(nomen substitutum), Xiphosoma* Wagler *in* Spix, 1824, *Xyphosoma* Gray, 1825 *(nomen emendatum), Xiphophosoma* Schinz, 1833 *(nomen emendatum), Xiphosoma* Gray, 1842a *(nomen praeoccupatum), Xiphosoma* A.-M.-C. Duméril, Bibron & Duméril, 1854a *(nomen praeoccupatum), Chrysenis* Gray, 1860a, *Diphosoma* – A.C.L.G. Günther, 1895 *in* 1885–1902 *(nomen incorrectum), Xenoboa* Hoge, 1954, *Corollus* – D.E. Willard, 1977 *(nomen incorrectum),* and *Corallis* – Pinney, 1981 *(nomen incorrectum).*
Type species: *Boa hortulana* Linnaeus, 1758.
Distribution: Lower Central America, Amazonia and Lesser Antilles.
Sources: Stull, 1935, Carrillo de Espinoza, 1966, Roze, 1966a, Stimson, 1969, J.A. Peters & Orejas-Miranda, 1970, Matz & Matz, 1973, McDowell, 1975a, Chippaux, 1987, Dixon & Soini, 1986, Lancini, 1986, Pérez-Santos & Moreno, 1988, 1991, R.W. Henderson, 1993a, 1997, 2002, Henderson et al., 1995, Stafford & R.W. Henderson, 1996, Walls, 1998a, McDiarmid et al., 1999, R.W. Henderson & Paiers, 2012 and Colston et al., 2013.

1. *Corallus annulatus* (Cope, 1875a). J. Acad. Nat. Sci. Philadelphia (1876) (2) 8(2): 129, pl. 28, figs. 6a–b. (*Xiphosoma annulatum*)

Synonym: *Boa annulata colombiana* Rendahl & Vestergren, 1940.
Type: Holotype, USNM 32480, a 755 mm specimen (W.M. Gabb, 1873–1875).
Type locality: "southern portion of the region of Costa Rica which lies east of the elevated mountain Distribution which traverses that country" [= Limón Prov., E Costa Rica].
Distribution: Central America and NW Colombia. Southeastern Guatemala (Izabal), Honduras (Cortés, Gracias a Dios, Santa Bárbara), SE Nicaragua (Atlántico Norte, Atlántico Sur), E Costa Rica (Alajuela, Cartago, Heredia, Limón), Panama and ext. W Colombia (Antioquia, Chocó, Santander), NSL–400 m.
Sources: Villa et al., 1988, R.W. Henderson, 1993b, 2002, L.D. Wilson & McCranie, 1994, E.N. Smith & Acevedo, 1997, J.A. Campbell, 1998b, Savage, 2002, Solórzano, 2004, Daza, 2005, McCranie et al., 2006 and McCranie, 2011a.

2. *Corallus batesii* (Gray. 1860a). Proc. Zool. Soc. London 28(1): 132. (*Chrysenis batesii*)

Type: Holotype, BMNH 1859.12.28.12, a juvenile (H.W. Bates, May 1850–Dec. 1851, June–Nov. 1852, Sept.–Dec. 1854, June–Sept. 1855 or May 1856–Jan. 1858).
Type locality: "Upper Amazon" [= upper Amazon River, Amazonas State, NW Brazil]. Restricted to the Río Amazonas west of the Rio Negro *fide* Henderson et al. (2009: 575).
Distribution: Amazonia. Colombia (E Amazonas, Antioquia, Boyacá, Caldas, Córdoba, Meta), E

Ecuador (Napo), N Brazil (Acre, Amazonas, Goias, Mato Grosso, Maranhao, Pará, Rondônia, Roraima), E Peru (Amazonas, Cajamarca, Huánuco, Junín, Loreto, Madre de Díos, Pasco) and N Bolivia (Beni, La Paz), 10–1000 m.

Source: Henderson et al., 2009.

3. *Corallus blombergi* (Rendahl & Vestergren, 1941). Arkiv Zool. 33A(5): 1–2, figs. 6–7. (*Boa annulata blombergi*)

Type: Holotype, NRS 3141, a 1320 mm specimen (R. Blomberg, 1937).

Type locality: "Rio Zamora, a tributary of the Rio Santiago, eastern Ecuador" [= Morona-Santiago Prov.] (probably in error *fide* Henderson et al., 2001: 43).

Distribution: Western Ecuador (Esmeraldas, Guayas, Los Ríos), NSL–200 m.

Source: Henderson et al., 2001.

4. *Corallus caninus* (Linnaeus, 1758). Syst. Nat., ed. 10, 1: 215. (*Boa canina*)

Synonyms: *Boa hipnale* Linnaeus, 1758, *Boa aurantiaca* Laurenti, 1768, *Boa exigua* Laurenti, 1768 (*nomen substitutum*), *Boa thalassina* Laurenti, 1768 (*nomen substitutum*), *Boa bojobi* Lacépède, 1789 (*nomen rejiciendum*), *Boa isebequensis* Bonnaterre, 1790, *Boa flavicans* Suckow, 1797, *Boa hipnale* G. Shaw, 1802, *Xiphosoma araramboya* Wagler *in* Spix, 1824, and *Xiphosoma caninum* A.-M.-C. Duméril & Bibron, 1844.

Type: Lectotype, NHR Lin-8 (formerly MAFR), a 1440 mm specimen, designated by Stimson (1969: 10).

Type locality: "America" via lectotype selection.

Distribution: Northern South America. Eastern Venezuela (Amazonas, Bolívar, Monagas), Guyana (Cuyuni-Mazaruni, Essequibo Islands-West Demerara, Potaro-Siparuni, Upper Demerara-Berbice), Suriname (Brokopondo, Para, Saramacca, Sipaliwini, Suriname), French Guiana (Cayenne, Saint-Laurent-du-Maroni) and NE Brazil (Amapá, Amazonas, Pará, Roraima), NSL–1000 m.

Sources: Chippaux, 1987, R.W. Henderson, 1993c, Starace, 1998, Lehr, 2001, Doan & Arizábal, 2002, Abuys, 2003, J.C. Murphy & Schlager, 2003, Kivit & Wiseman, 2005, Vidal et al., 2005, Pizzatto et al., 2007, Beddard, 2008, Henderson et al., 2009, Navarrete et al., 2009, J.D. Sánchez et al., 2009 and C.J. Cole et al., 2013.

Remarks: A complex of multiple, undescribed species *fide* Vidal et al. (2005: 502).

5. *Corallus cookii* Gray, 1842a. Zool. Misc. 2(Mar.): 42.

Synonym: *Corallus hortulanus melanea* Gray, 1849a.

Type: Holotype, BMNH 1946.1.1.50 (formerly BMNH IV.8.1d), an 861 mm male (E. Cook).

Type locality: "America," (in error). Corrected to West Indies *fide* Gray (1849a: 98). Restricted to St. Vincent, Lesser Antilles *fide* Henderson (1997: 207).

Distribution: Lesser Antilles. Saint Vincent and the Grenadines (St Vincent), 70–500 m.

Sources: Tolson & R.W. Henderson, 1993, R.W. Henderson, 2002, Beddard, 2008 and R.W. Henderson & Powell, 2009.

6. *Corallus cropanii* (Hoge, 1954b). Mem. Inst. Butantan (1953) 25(1): 27–28, figs. 1–5, col. pl. (*Xenoboa cropanii*)

Type: Holotype, IB 15200, a 1113–1275 mm male (J. Santos), destroyed by fire 15 May 2010.

Type locality: "Miracatu, State of São Paulo, Brazil" [= Miracatu, Vale do Ribeira, São Paulo State, SE Brazil, 24°17'S, 47°28'W, elevation 50 m].

Distribution: Southeastern Brazil (São Paulo), NSL–60 m.

Sources: Amaral, 1978, Kluge, 1991, R.W. Henderson & Puorto, 1993, Pizzatto et al., 2007 and Machado-Filho et al., 2011.

Remarks: Photograph of the holotype in Machado-Filho et al. (2011: fig. 1).

7. *Corallus grenadensis* (T. Barbour, 1914a). Mem. Mus. Comp. Zool. 44: 327. (*Boa grenadensis*)

Type: Holotype MCZ 7791, an 1170 mm male (G.M. Allen, 20 Aug. 1910).

Type locality: "St. George's, Grenada."

Distribution: Lesser Antilles (Baliceaux, Bequia, Canouan, Carriacou, Grenada, Ile Quatre, Mayreau, Mustique, Union), NSL–520 m.

Sources: R.W. Henderson, 2002 and R.W. Henderson & Powell, 2009.

8. *Corallus hortulanus* (Linnaeus, 1758). Syst. Nat., ed. 10, 1: 215–216. (*Boa hortulana*)

Synonyms: *Boa enydris* Linnaeus, 1758, *Vipera bitis* Laurenti, 1768, *Vipera madarensis* Laurenti, 1768, *Boa ambleocephala* Donndorff, 1798, *Boa merremi* J.G. Schneider, 1801, *Boa annulata* G. Shaw, 1802, *Boa obtusiceps* Bechstein, 1802b, *Boa elegans* Daudin, 1803b, *Corallus obtusirostris* Daudin, 1803b, *Boa stellaris* Gravenhorst, 1807, *Boa broderie* Cuvier, 1816, *Boa elegans Surinamensis* Schneider, 1821, *Xiphosoma dorsuale* Wagler *in* Spix, 1824, *Xiphosoma ornatum* Wagler *in* Spix, 1824, *Xiphosoma hortulanum* Fitzinger, 1826a, *Scytale zigzag* Schinz, 1833, *Boa modesta* A. Reuss, 1834, *Corallus maculatus* Gray, 1842a, *Boa dubia* Guilding *in* Gray, 1849a, and *Boa enydrys albina* Briceño-Rossi, 1934.

C

Type: Holotype, NHR Lin-7 (formerly MDG), a 1275 mm specimen.

Type locality: "America."

Distribution: South America. Colombia (Amazonas, Atlántico, Caquetá, Guaviare, Meta, Putumayo, Vaupés), S Venezuela (Amazonas, Bolívar), Guyana (Barima-Waini, Cuyubi-Mazaruni, Demerara-Mahaica, Essequibo Islands-West Demerara, Mahaica-Berbice, Upper Demerara-Berbice), Suriname (Nickerie, Para, Paramaribo, Sipaliwini), French Guiana (Cayenne, Saint-Laurent-du-Maroni), Brazil (Acre, Amapa, Amazonas, Bahia, Ceara, Espiritu Santo, Goias, Maranhao, Mato Grosso, Minas Gerais, Pará, Paraná, Pernambuco, Rio de Janeiro, Rio Grande do Norte, Rondônia, Roraima, Santa Catarina, São Paulo, Sergipe, Santo Amaro Is.), E Ecuador (Napo), Peru (Cusco, Huánuco, Junín, Loreto, Madre de Dios, Puno, San Martín) and Bolivia (Beni, La Paz, Pando, Santa Cruz), NSL–1000 m.

Sources: Schinz, 1833, Emsley, 1977, Chippaux, 1987, Duellman & Salas, 1991, A. Schwartz & R.W. Henderson, 1991, R.W. Henderson, 1991, 1993d, 1997, 2002, Tolson & R.W. Henderson, 1993a, Puorto & R.W. Henderson, 1994, R.W. Henderson & Hedges, 1995, Henderson et al., 1996b, McDiarmid et al., 1996, Starace, 1998, Freitas, 1999, Savage, 2002, Abuys, 2003, Duellman, 2005, Navarrete et al., 2009, Morato et al., 2011 and C.J. Cole et al., 2013.

9. †*Corallus priscus* Rage, 2001. Palaeovertebrata 30(3–4): 122–124, fig. 6.

Type: Holotype, DGM 1332-R, one middle trunk vertebra.

Type locality: "São José de Itaboraí, State of Rio de Janeiro, Brazil; middle Paleocene."

Distribution: Middle Paleocene (Itaboraian: 57.0–59.0 mya) of Brazil (Rio de Janeiro).

Source: Rage, 2011.

10. *Corallus ruschenbergerii* (Cope, 1875a). J. Acad. Nat. Sci. Philadelphia (1876) (2) 8(2): 129. (*Xiphosoma ruschenbergerii*)

Synonyms: *Xiphosoma ruschenbergii* – Boulenger, 1893a *(nomen incorrectum), Corallus cookei ruschenbergi* – Boettger, 1989 *(nomen incorrectum),* and *Boa salmonidia* Briceño-Rossi, 1934.

Type: Holotype, ANSP 10325, a 1530 mm female (W.S.W. Ruschenberger, 1826–1829 or 1854–1857).

Type locality: "Panama" [= Panama city, Panama].

Distribution: Lower Central America and N South America. Southern Costa Rica (Puntarenas), Panama, Colombia (Antioquia, Bolívar, César, Chocó, Córdoba, Cundinamarca, Magdalena, Sucre, Valle de Cauca), Venezuela (Amacuro, Apure, Barinas, Bolívar, Carabobo, Cojedes, Delta, Distrito Federal, Falcón, Guárico, Miranda, Nueva Esparta, Portuguesa, Sucre, Zulia, Margarita Is.) and Trinidad and Tobago (Tobago, Trinidad), NSL–600 m.

Sources: Mijares-Urrutia & Arends, 2000, R.W. Henderson, 2002 Esqueda & La Marca, 2004, Solórzano, 2004, Rivas-Fuenemayer & Barrio-Amorgós, 2005, J.C. Murphy, 2007, Navarrete et al., 2009, Ugueto & Rivas, 2010 and Sajdak, 2010.

CORONELAPS Lema & Hofstadler-Deiques, 2010 (Dipsadidae)

Type species: *Elapomorphus lepidus* J.T. Reinhardt, 1861b.

Distribution: Southeastern Brazil.

Sources: L. Müller, 1927, Lema & Hofstadler-Deiques, 1995, 2010 and Marques et al., 2001.

1. *Coronelaps lepidus* (J.T. Reinhardt, 1861b). Vidensk. Medd. Natur. Foren. Kjöbenhavn (1860) 22: 239–242, pl. 4, figs. 6–9. (*Elapomorphus lepidus*)

Synonyms: *Elapomorphus lepidus* J.T. Reinhardt, 1861a *(nomen nudum)* and *Elapomorphus coronatus* Sauvage, 1877.

Type: Holotype, ZMUC 63821, a 575 mm female (J.T. Reinhardt, 2 April 1855).

Type locality: "en Fazenda, Feijão cru kaldet, en Miilsvei fra Arrayal de Bicudo, en lille kun faa Aar gammel By ved Rio da casca, altsaa i Provindsen Minas' östlige med Urskov bevoxede Deel" [= Fazenda Feixâo Cru, Minas Gerais, Brazil *fide* L. Müller (1927: 301), ca. 1.5 km from the village of Arraial do Bicudo, near the Casca River, Minas Gerais, Brazil *fide* Lema (1984: 59), or Bicuda, Casca River, Minas Gerais State, Brazil *fide* Lema & Hofstadler-Deiques (1995: 97), ca. 20°13'S, 42°40'W, elevation 375 m].

Distribution: Southeastern Brazil (E Bahia, Espírito Santo, Minas Gerais, Rio de Janeiro, São Paulo), 375 m.

CORONELLA Laurenti, 1768 (Colubridae)

Synonyms: ? *Berus* Wyder, 1817 *(partim), Coronnella* – Fiztinger, 1824 *(nomen incorrectum), Zacholus* Wagler, 1830, *Decagerron* Bonaparte, 1840 *(nomen nudum), Zaccholus* – Schinz, 1840 *(nomen incorrectum), Calonotus* Jan, 1863, *Caronella* – Higgins, 1873 *(nomen incorrectum), Zacholis* – Blanford, 1876 *(nomen incorrectum), Cornella* – Blanchard, 1942 *(nomen incorrectum), Wallophis* F. Werner, 1929a, *Comella* – Maki, 1931 *(nomen incorrectum), Zocholus* – Capocaccia, 1959 *(nomen incorrectum), Coronela* – Hoge & Maranhão-Niña, 1964 *(nomen incorrectum), Coranella* – Taub, 1965 *(nomen incorrectum),* and *Sharonhoserea* Hoser, 2012aa *(nomen illegitimum).*

Type species: *Coronella austriaca* Laurenti, 1768.

Distribution: Europe, NW Africa and SW Asia.

Fossil records: Middle Miocene to upper Pleistocene of Europe. Fossils unassigned to species include lower Pleistocene (Calabrian: 0.8–1.8 mya) of France and upper Pleistocene (Tarantian: 0.01–0.13 mya) of Moldavia.

Sources: Sanchiz, 1980, Holman, 1985, 1998a, 2000b, Szyndlar, 1991a, Capula et al., 1995, Utiger et al., 2002, 2005, Ivanov, 2007 and Hoser, 2012aa.

1. *Coronella austriaca* Laurenti, 1768. Synop. Rept.: 84, pl. 5, fig. 1.

Synonyms: *Coluber levis* Lacépède, 1789 (*nomen rejiciendum*), *Coluber versicolor* Razoumowsky, 1789, *Coluber coronella* Bonnaterre, 1790, *Coluber ferruginosus* Sparrman, 1795, *Coluber oculus cati* Suchow, 1798, *Coluber alpinus* Güldenstedt *in* Georgi, 1801, *Coluber cupreus* Güldenstedt *in* Georgi, 1801, *Coluber ponticus* Güldenstedt *in* Georgi, 1801, *Coluber tetragonus* Latreille *in* Sonnini & Latreille, 1801b, *Coluber thuringicus* Bechstein, 1801, *Coluber gallicus* Hermann, 1804, *Coluber laevis* Hermann, 1804, *Coluber caucasius* Pallas, 1814, *Coluber pustulatus* Rafinesque-Schmaltz, 1814, *Natrix hybrida* Wagler, 1825 (*nomen ineditum*), *Natrix dumfrisiensis* Fleming, 1828, *Coluber nebulosus* Ménétriés, 1832, *Coluber caucasicus* Ménétriés, 1832, *Zacholus fitzingerii* Fitzinger *in* Bonaparte, 1836 (*nomen substitutum*), *Zacholus italicus* Fitzinger *in* Bonaparte, 1836 (*nomen nudum*), *Calamaria coronata* Schlegel, 1837, *Zacholus meridionalis* Fitzinger, 1853 (*nomen nudum*), *Simotes semicinctus* W.C.H. Peters, 1862b, *Coronella austriaca aegyptiaca* Jan, 1863b, *Coronella austriaca caucasica* Jan, 1863b, *Coronella vernalis* Gistel, 1868, *Coronella austriaca marginata* Fatio, 1872, *Coronella austriaca pallida* Fatio, 1872, *Coronella austriaca italica* E. Schrieber, 1875 (*nomen substitutum*), *Coronella laevis leopardina* F. Müller, 1885, *Coronella austriaca fitzingeri conjuncta* Camerano, 1891 (*nomen illegitimum*), *Zacholus italicus* Camerano, 1891 (*nomen nudum*), *Coronella austriaca fasciata* Durigen, 1897, *Coronella austriaca immaculata* Durigen, 1897, *Coronella austriaca concolor* F. Werner, 1897d, *Coronella austriaca lateralis* F. Werner, 1897d, *Coronella austriaca quadritaeniata* F. Werner, 1897d, *Coronella austriaca sparsa* Durigen, 1897, *Coronella austriaca taeniata* Durigen, 1897, *Coronella austriaca scalaris* Sternfeld, 1911, *Coronella austriaca veithi* E. Schrieber, 1912, and *Coronella austriaca acutirostris* Malkmus, 1995a.

Type: Holotype, not designated, location unknown.

Type locality: "circa Viennam" and "America" [= vicinity of Vienna, Austria and America]. Restricted to Austria *fide* Mertens & Müller (1928: 48)

Distribution: Europe and SW Asia. Southern United Kingdom (S England), S Norway, S Sweden (Grinda Is.), ext. S Finland, N Portugal (Braga, Bragansa, Coimbra, Guarda, Porto, Vila Real, Viseu), N Spain (Albacete, Ávila, Barcelona, Burgos, Castellón, Cuenca, Gerona, W Guadalajara, Granada, Guipúzcoa, Logroño, Lugo, N Madrid, Navarra, Orense, Oviedo, Pontevedra, Santander, S Segovia, Soria, Tarragona, Teruel, N Valencia, Vizcaya, Zaragossa), France (Ain, Aisne, Allier, Alpes-Maritimes, Ardèche, Ardennes, Ariège, Aube, Aude, Aveyron, Bas-Rhin, Basses-Alpes, Basses-Pyrénées, Belfort, Calvados, Cantal, Cher, Charente, Charente-Maritime, Corrèze, Côte-d'Or, Côte-du-Nord, Creuse, Deux-Sèvres, Dordogne, Doubs, Drôme, Essonne, Eure, Eure-et-Loir, Finistère, Gard, Gironde, Haut-Rhin, Haute-Garonne, Haute-Loire, Haute-Marne, Haute-Saône, Haute-Savoie, Haute-Vienne, Hautes-Alpes, Hautes-Pyrénées, Hérault, Ille-et-Vilaine, Indre, Indre-et-Loire, Isère, Jura, Loire, Loire-Atlantique, Loiret, Loir-et-Cher, Lozère, Maine-et-Loire, Manche, Marne, Mayenne, Meurthe-Moselle, Meuse, Morbihan, Moselle, Nièvre, Nord, Oise, Orne, Puy-de-Dôme, Pyrénées-Orientales, Rhône, Saône-et-Loire, Sarthe, Savoie, Seine-et-Marne, Seine-Maritime, Somme, Tarn, Val-d'Oise, Var, Vaucluse, Vendée, Vienne, Vosges, Yonne, Yvelines, Corsica Is.), Belgium (Antwerpen, Hainaut, Liège, Limburg, Luxembourg, Namur), Luxembourg, Netherlands (Drente, Friesland, Gelderland, Groningen, Limburg, Noord Brabant, Overijsssel, Utrecht), Germany, Switzerland (Aargau, Appenzell Inner-Rhoden, Appenzell Ausser-Rhoden, Bern, Basel-Landschaft, Basel-Stadt, Fribourg, Genève, Glarus, Graubünden, Jura, Lucerne, Neuchâtel, Nidwalden, Obwalden, St. Gallen, Schaffhausen, Solothurn, Schwyz, Thurgau, Ticino, Uri, Vaud, Valais, Zug, Zürich), Italy (Abruzzi, Basilicata, Calabria, Campania, Emilia-Romagna, Friuli-Venezia Giulia, Lazio, Liguria, Lombardia, Marche, Molise, Piemonte, Puglia, Toscana, Trentino-Alto Adige, Umbria, Valle d'Aosta, Veneto, Sardinia and Sicily Is.), Austria (Burgenland, Carinthia, Lower Austria, Salzburg, Styria, Tyrol, Upper Austria, Vienna, Vorarlberg), Czech Republic (Jihocesky, Jihomoravsky, Praha, Severocesky, Severomoravsky, Stredocesky, Vychodocesky, Zapadocesky), Poland (Dolnoslaskie, Kujawsko-Pomorskie, Lodzkie, Lubelskie, Lubuskie, Malopolskie, Mazowieckie, Opolskie, Podkarpackie, Podlaskie, Pomorskie, Slaskie, Swietokrzyskie, Warminsko-Mazurskie, Wielkopolskie, Zachodniopomorskie), Hungary, Bosnia-Herzegovina, Croatia (Mljet Is.), Macedonia, Romania, Bulgaria (Blagoevgrad, Burgas, Gabrovo, Grand Sofiya, Kardzhali, Lovech, Kyustendil, Montana, Pazardzhik, Pernik, Plovdiv, Rousse, Silistra, Smolyan, Sofiya, Targovishte, Varna, Veliko Tarnovo, Vidin), Albania, N Greece (Samothráki Is.), N Turkey, Moldavia, Ukraine, SW Russia, Georgia, Armenia, Azerbaijan, W Kazakhstan and NW Iran (East Azarbaijan, Central, Ghilan), NSL–2250 m.

Fossil records: Lower Pliocene (Ruscinian, MN 14: 4.2–4.9 mya) of Hungary, middle Pliocene (Ruscinian, MN 15: 3.2–4.2 mya) of Moldavia and Slovakia, upper Pliocene (Villanyian, MN 16: 2.6–3.2 mya) of Austria

C

and Moldavia, lower Pleistocene (Calabrian: 0.8–1.8 mya) of Austria, Czech Republic and Poland, middle Pleistocene (Ionian: 0.13–0.78 mya) Bulgaria, France, Germany, Greece, Romania, Slovakia and United Kingdom, and upper Pleistocene (Tarantian: 0.01–0.13 mya) of Bulgaria, France, Germany, Poland and Serbia.

Sources: Bruno, 1966, 1984, Stewart, 1971, Hopkins, 1974, Bannikov et al., 1977, Bruno & Maugeri, 1977, Spellenberg & Phelps, 1977, Arnold & Burton, 1978, Street, 1979, Goddard, 1980, 1984, Frazer, 1983, Happ, 1985, Engelmann, 1993, Strijbosch & van Gelder, 1993, Venczel, 1994, Malkmus, 1995a, Monney, 1996, Borkin et al., 1997, Mikátová et al., 2001, Petkovski et al., 2000, Cabela et al., 2001, Hofer et al., 2001, F.P. Pérez, 2001, O'Brian, 2002, Spellerberg, 2002, Glowacinski & Rafinski, 2003, J.C. Murphy & Schlager, 2003, Völkl & Käsewieter, 2003, Sindaco et al., 2006, Renner & Vitzthum, 2007, Trapp, 2007, Tillack et al., 2008, Creemers & Delft, 2009, Tuniyev et al., 2009, Valakos et al., 2009, Vacher & Geniez, 2010, Stojanov et al., 2011, Jablonski et al., 2012 and Szyndlar, 2012.

2. *Coronella brachyura* (A.C.L.G. Günther, 1866). Ann. Mag. Nat. Hist. (3) 18(103): 27, pl. 6, figs. a–a′. *(Zamenis brachyurus)*

Type: Holotype, BMNH 1946.1.1.24, a 432 mm female (A. Leith-Adams, 1849–1854).

Type locality: "Poonah (Dekkan)" [= Pune, Deccan Trap Hills, W Maharashtra State, W India, 18°31'N, 73°51'E, elevation 565 m].

Distribution: Western India (Gujarat, Madhya Pradesh, Maharashtra), 10–565 m.

Sources: Abdulali, 1935, M.A. Smith, 1943, Whitaker & Captain, 2004, N. Khaire, 2006, Vyas & Patel, 2007, Nande & Dsehmukh, 2007 and Ingle & Sarsavan, 2011.

3. *Coronella girondica* (Daudin, 1803c). Hist. Nat. Rept. 6: 432–434. (*Coluber girondicus*)

Synonyms: *Coluber meridionalis* Daudin, 1803c, *Coluber riccioli* Metaxa, 1823, *Coluber strigatus* Risso, 1826, *Coluber rubens* Gachet, 1829, *Decagerron girondicum* Hemprich *in* Bonaparte, 1840 *(nomen nudum)*, *Coronella girundica* A.-M.-C. Duméril, Bibron & Duméril, 1854b *(nomen emendatum)*, *Coronella laevis hispanica* Boettger, 1869, and *Rhinechis amaliae* Boettger, 1881b.

Type: Holotype, not designated, a 520 mm specimen (Rodrigues), location unknown.

Type locality: "Bordeaux" [Gironde Department, SW France].

Distribution: Southern Europe and ext. N Africa. Portugal (Beja, Braga, Castelo Branco, Coimbra, Évora, Faro, Guarda, Leiria, Lisboa, Portalegre, Porto, Sertubal, Vila Real), Spain (Álava, Alicante, Almería, Andorra, Aragon, Ávila, Badajoz, Burgos, Cáceres, Cádiz, Castellón, Catalonia, S Ciudad Real, Córdoba, Cuenca, Gerona, Granada, Guadalajara, Guipúzcoa, Huelva, Huesca, Jaén, La Coruña, NE Léon, Lérida, Logroño, Lugo, Madrid, Málaga, Murcia, Navarra, Orense, Oviedo, Pontevedra, Salamanca, Santander, Segovia, W Seville, Soria, Tarragona, Teruel, Toledo, Valencia, Vizcaya, Zamora, Zaragossa), S France (Ain, Alpes-Maritimes, Ardèche, Ariège, Aude, Aveyron, Basses-Alpes, Basses-Pyrenées, Bouches-du-Rhône, Charente Maritime, Corrèze, Dordogne, Drôme, Gard, Gers, Gironde, Haute-Garonne, Hautes-Alpes, Hautes-Pyrénées, Hérault, Isère, Landes, Lot, Lot-et-Garonne, Lozère, Pyrénes-Orientales, Rhône, Tarn, Tarn-et-Garonne, Var, Vauclose), Italy (Abruzzi, Campania, Emilia-Romagna, Friuli-Venezia Giulia, Lazio, Liguria, S Lombardia, Marche, Molise, Piemonte, N Puglia, Toscana, Umbria, Veneto), N Morocco (Al Hoceima, Azilal, Beni Mellal, Chechaouene, Khenitra, Marrakech, Meknes, Nador, Tangier, S Taza, Tetouan), N Algeria (Aïn Témouchent, Annaba, Naama, Oran, Tlemcen) and N Tunisia (Jendouba, Le Kef, Nabeul, Zaghouan), NSL–3000 (3200) m.

Fossil records: Middle Pleistocene (Ionian: 0.13–0.78 mya) of France and Spain, and upper Pleistocene (Tarantian: 0.01–0.13 mya) of Spain.

Sources: Chpakowsky & Chnéour, 1954, Domergue, 1959, Bruno, 1966, 1984, Steward, 1971, Bruno & Maugeri, 1977, Arnold & Burton, 1978, Street, 1979, Castanet & Guyetant, 1989, Dusej, 1993, Agrimi & Luiselli, 1994, Bons & Geniez, 1996, Schleich et al., 1996, Pleguezuelos, 1998, Salvador & Pleguezuelos, 2002, Sindaco et al., 2006, Feriche et al., 2008 and Vacher & Geniez, 2010.

Remarks: Type not in MNHN.

4. †*Coronella miocaenica* Venczel, 1998. Acta Zool. Cracov. 41(1): 6–10, fig. 5a–e, 6a–h.

Type: Holotype, HGIM V20617, one middle trunk vertebra.

Type locality: "Polgárdi 4 "Lower" limestione quarry near the village of Polgárdi (W-Hungary). Uppermost Miocene; Pontian or Upper Turolian (MN 13)."

Distribution: Middle Miocene (Astaracian, MN 7–8: 11.1–12.8 mya) of Romania and upper Miocene (Turolian, MN 13: 4.9–7.3 mya) of Hungary.

Sources: Venczel, 1998, Venczel & Stiuca, 2008 and Szyndlar, 2012.

CRASPEDOCEPHALUS Kuhl & Hasselt, 1822 (Viperidae)

Synonyms: *Atropos* Rafinesque-Schmaltz, 1815 *(nomen nudum)*, *Crasedocephalus* Gray, 1825 *(nomen emendatum)*, *Atropos* Wagler, 1830 *(nomen praeoccupatum)*, *Megaera* Wagler, 1830 *(nomen praeoccupatum)*, *Crassedocephalus* – Gray *in* Griffith & Pidgeon, 1831 *(nomen incorrectum)*, *Craspidocephalus* – Swainson,

1839 *(nomen incorrectum), Megara* – Gray, 1840 *(nomen incorrectum), Megera* – A.-M.-C. Duméril & Bibron, 1844 *(nomen incorrectum), Maegera* – A.-M.-C. Duméril, Bibron & Duméril, 1854b *(nomen incorrectum), Atrops* – Rüppell, 1845 *(nomen incorrectum), Atropus* – J.W. Müller, 1865 *(nomen incorrectum), Atropophis* W.C.H. Peters, 1872d *(nomen substitutum), Atropopos* – C.P.J. Haas, 1950 *(nomen incorrectum), Atropophus* – Leviton, 1968b *(nomen incorrectum), Atropohis* – Hoge & Romano-Hoge, 1975a *(nomen incorrectum),* and *Crasdocephalus* – Hoge & Romano-Hoge, 1981 *(nomen incorrectum).*

Type species: *Craspedocephalus puniceus* Kuhl, 1824.

Distribution: India, SE Asia and Indonesia.

Sources: Maslin, 1942, M.A Smith, 1943, Maslin, 1945, Klemmer, 1963, Leviton, 1964d, W. Burger, 1971, Kramer, 1977, Maes, 1989, Malhotra & Davis, 1991, Golay et al., 1993, Zhao & Adler, 1993, Kraus et al., 1996, David & Vogel, 1998, David & Ineich, 1999, McDiarmid et al., 1999, Parkinson, 1999, Malhotra & Thorpe, 1997, 2000, 2004c, Ziegler et al., 2001, Giannasi et al., 2001a, G. Peng & Fuji, 2001, David et al., 2002, 2009, 2011, Gumprecht et al., 2004, Castoe & Parkinson, 2006, Grismer et al., 2006a, 2008b, Sanders et al., 2006, Dawson et al., 2008, Ganesh et al., 2008, Orlov et al., 2008, 2009, David et al., 2011c, P. Guo & Wang, 2011, Sumontha et al., 2011 and Hoser, 2012d, o.

1. *Craspedocephalus andalasensis* (David, Vogel, Vijayakumar & Vidal, 2006). Zootaxa (1293): 55–64, figs. 23–26. (*Trimeresurus andalasensis*)

Type: Holotype, SMF 22429, a 442 mm male (H.R. Roomaker, 1929).

Type locality: "Nord-Sumatra, Atjeh, Tawar-See, Takengon, 1200 m ü. NN Lake Tawar, Takengon, Aceh Province, Sumatra Island, Indonesia, 1200 msl."

Distribution: Western Indonesia (N Sumatra), 500–1200 m.

Sources: David & Vogel, 1997 and I. Das, 2010, 2012.

2. *Craspedocephalus borneensis* (W.C.H. Peters, 1872d). Ann. Mus. Civ. Stor. Nat. Genova 3: 41–42. (*Atropophis borneensis*)

Synonym: *Bothrops sandakanensis* Lidth de Jeude, 1893.

Types: Syntypes (3), ZMB 7146 (R. Abendroth) and MSNG CE 8383a–b (G. Doria and O. Beccari, 1865–1868).

Type locality: "Sarawack auf Borneo" [= Sarawak, East Malaysia].

Distribution: Borneo. East Malaysia (Sabah, Sarawak), Brunei and cen. Indonesia (Kalimantan), NSL–1200 m.

Sources: Hoge & Romano-Hoge, 1975a, Toriba, 1992b, David & Vogel, 1996, Manthey & Grossmann, 1997, Steubing & Inger, 1999, Ziegler et al., 2001, Malkmus et al., 2002, David et al., 2006 and I. Das, 2007b, 2010, 2012.

Remarks: Previously a synonym of *T. puniceus* but a valid species *fide* David & Gernot (1996: 160).

3. *Craspedocephalus brongersmai* (Hoge, 1969d). Cienc. Cult. (1967) 21(2): 459–460. (*Trimeresurus brongersmai*)

Type: Holotype, RMNH 5654A, a 406 mm male (E. Jacobson, April 1913).

Type locality: "Lugu, Simeulue, Sumatra" [= Lugu, Simeulue Is. off W coast of Sumatra, Indonesia].

Distribution: Western Indonesia (Mentawai Arch.: Siberut and Simeulue).

Sources: Hoge & Romano-Hoge, 1975a, David et al., 2006 and I. Das, 2010, 2012.

Remarks: Possibly occurs on Nias and Batu Is. *fide* David et al. (2006: 68).

4. *Craspedocephalus gramineus* (G. Shaw, 1802). Gen. Zool., Amph. 3(2): 420–421. (*Coluber gramineus*)

Synonyms: *Coluber viridis* Bechstein, 1802, *Vipera viridis* Daudin, 1803c, *Trigonocephalus fario* Jan, 1859b (*nomen nudum*), *Colubar gramineus* – F. Mason, 1860 *(nomen incorrectum), Bothrops viridis fario* Jan, 1863b *(nomen nudum), Bothrops viridis genei* Jan, 1863b (*nomen nudum*), and *Trimeresurus occidentalis* C.H. Pope & Pope, 1933,

Type: Holotype, not designated, a 762 mm specimen described and illustrated in P. Russell (1796: 13–14, pl. 9) (native, Oct. 1788), location unknown.

Type locality: "hills in the vicinity of Vizagapatam, coast of Coromandel, India" [= Vishakhapatnam, Andhra Pradesh State, SE India, 17°41'N, 83°13'E].

Distribution: Southern India (Andhra Pradesh, Goa, Gujarat, Karnataka, Kerala, Madhya Pradesh, Maharashtra, Orissa, Tamil Nadu, ? West Bengal), NSL–400+ m.

Sources: Wall, 1905g, Stejneger, 1927, C.H. Pope & Pope, 1933, Bergman, 1961b, Regenass & Kramer, 1981, David & Vogel, 1998, McDiarmid et al., 1999, Whitaker & Captain, 2004, N. Khaire, 2006 and David et al., 2011c.

Remarks: Type locality in error *fide* Hoge & Romano-Hoge (1981: 257–258), who suggested Java, Indonesia was more accurate.

5. *Craspedocephalus malabaricus* (Jerdon, 1854). J. Asiat. Soc. Bengal (1853) 22(6): 523. (*Trigonocephalus* [*Cophias*] *malabaricus*)

Synonyms: *Trigonocephalus wardii* Jerdon, 1854, *Trimeresurus anamallensis* A.C.L.G. Günther, 1864a, and *Lachesis coorgensis* C. Rao, 1917.

Types: Syntypes (?), longest syntype 600 mm, lost *fide* M.A. Smith (1943: 513).

Type locality: "West Coast, Peninsula of India" [= SW India]. Restricted to Western Ghats, SW India *fide* M.A. Smith (1943: 513).

Distribution: Southwestern India (Western Ghats of Karanataka, Kerala, Maharashtra, Tamil Nadu), 500–2135 m.

Sources: Regenass & Kramer, 1981, Inger et al., 1984, Malhotra & Davis, 1991, David & Vogel, 1998, Whitaker & Captain, 2004, N. Khaire, 2006, Hutton & David, 2009 and Chandramouli & Ganesh, 2010.

Remarks: Types not in BMNH *fide* M. Lang *(in litt.)*. Syntypes of *T. anamallensis* BMNH 1946.1.18.73–74, BMNH 1946.1.19.82, BMNH 1946.1.19.89, BMNH 1946.1.19.93–95, and BMNH 1946.1.20.3.

6. *Craspedocephalus puniceus* Kuhl, 1824. Bull. Sci. Nat. Géol., Paris 2: 80.

Synonyms: *Vipera punicea* H. Boie *in* Schlegel, 1826a *(nomen nudum), Vipera punicea* Schlegel, 1826b *(nomen nudum), Cophias punicea* – F. Boie, 1827, and *Atropos acontia* Gray, 1842b.

Types: Syntypes (3), RMNH 1557 and RMNH 1558a–b, three females (H. Boie & H.C. Macklot, Dec. 1820–Sept. 1823).

Type locality: "Java" [SW Indonesia].

Distribution: Western Indonesia (Java, Mentawai Arch., Simeulue, S Sumatra, Tinjil), 500–1600 m.

Sources: M.A. Smith, 1928a, Hoge & Romano-Hoge, 1976, Tweedie, 1983, M.J. Cox, 1991b, Nutaphand et al., 1991, Toriba, 1992b, Hodges, 1993, David & Vogel, 1996, Inger & Tan Fui Lian, 1996, M.J. Cox et al., 1998, Ziegler et al., 2001, Ryabov et al., 2002, David et al., 2006 and I. Das, 2010, 2012.

Remarks: Photographs of a syntype in David et al. (2006: figs. 5–6). David et al. (2006: 33–37) identified a closely related but possibly undescribed species from cen. Sumatra (including Mentawai, Natuna and Anambas Is.).

7. *Craspedocephalus rubeus* (Malhotra, Thorpe, Mrinalini & Stuart, 2011). Zootaxa (2757): 9–14, fig. 5. (*Cryptelytrops rubeus*)

Type: Holotype, FMNH 262718, a 590 mm female (B.L. Stuart, K. Sok and T. Neang, 1 Nov. 2003).

Type locality: "O Kamen Stream, Seima Biodiversity Conservation Area (formerly part of Samling Logging Concession), O'Rang District, Mondolkiri Province, Cambodia (12.326 N, 107.092 E), 500 m elevation."

Distribution: Eastern Cambodia (Mondolkiri) and S Vietnam (Bien Hoa, Cu Chi, Ho Chi Minh City), NSL–500 m.

8. *Craspedocephalus strigatus* (Gray, 1842b). Zool. Misc. 2(Mar.): 49. (*Trimeresurus strigatus*) *(nomen corrigendum)*

Synonyms: *Trimesurus strigatus* Gray, 1842b *(nomen incorrigendum), Atropos darwinii* A.-M.-C. Duméril, 1853 *(nomen nudum), Atropos darwini* A.-M.-C. Duméril, Bibron & Duméril, 1854b, *Trigonocephalus neelgherriensis* Jerdon, 1854, *Trimesurus neilgherriensis* Beddome, 1862 *(nomen emendatum), Trimeresurus strigolus* – Theobald, 1868b *(nomen incorrectum), Trigonocephalus nilghiriensis* Theobald, 1876 *(nomen emendatum),* and *Trimeresurus strigatus* – Theobald, 1876 *(nomen corrigendum).*

Types: Syntypes (2), BMNH 1946.1.18.78–79 (formerly AMSC & BMNH I.2.3a–b).

Type locality: "Cape of Good Hope?" [= South Africa] (in error) and "Madras?" [= India].

Distribution: Southwestern India (Western Ghats of Kerala, Tamil Nadu), 915–2400 m.

Sources: David & Vogel, 1998 and Whitaker & Captain, 2004.

Remarks: A.F. Stimson doubted that BMNH 1946.1.18.78–79 were types in 1977 note in BMNH catalogue *fide* M. Lang *(in litt.)*, supported by the fact that the collection probably decayed or was destroyed *fide* R. Prys-Jones *in* Alström & Olsson (1999: 549). A member of *Protobothrops fide* Krauss et al. 1996: 769) but a valid *Trimeresurus fide* Herrmann et al. (2004b: 219).

9. *Craspedocephalus trigonocephalus* (Donndorff, 1798). Zool. Beytr. Linn. Natursyst. 3: 203. (*Coluber trigonocephalus*) *(nomen substitutum)*

Synonyms: *Coluber capite-triangulatus* Lacépède, 1789 (*nomen rejiciendum*), *Coluber capitetriangulatus* Bonnaterre, 1790, *Vipera trigonocephala* – Latreille in Sonnini & Latreille, 1801a, *Trigonocephalus nigromarginatus* Kuhl, 1820, *Trigonocephalus sagittiformis* Schinz, 1822 *(nomen substitutum),* and *Megaera olivacea* Gray, 1842b.

Type: Holotype, not designated, a male specimen based upon Lacépède (1789: 132, pl. 5, fig. 2) (King of France's cabinet), location unknown.

Type locality: "Africae regno Whida" [= Kingdom of Whydah, West Africa] (in error). Correctly designated as l'Isle Saint-Eustache [= Sri Lanka] *fide* Lacépède (1789: 132).

Distribution: Sri Lanka (North-Central, North-Western, Sabaragamuwa, Southern, Uva), NSL–1800 m.

Sources: Deraniyagala, 1955, Hoge & Romano-Hoge, 1981a, P.H. Silva, 1980a, A. Silva, 1990a–b, 2001, 2009, Nanayakkara, 1994, Malhotra & Thorpe, 1997, David & Vogel, 1998, McDiarmid et al., 1999 and Somaweera, 2006.

10. *Craspedocephalus wiroti* (Klemmer *in* Trutnau, 1981). Schlangen 2: 188, fig. 43. (*Trimeresurus wiroti*)

Type: Holotype, SMF 69695, a 285 mm male (W. Nutaphand, 20 June 1979).

Type locality: "100–150 km südlich von Surat Thani (Süd-Thailand) an den Hängen des Khao Luang zwischen Amphoe Thung Son und Amphoe Chawang (Provinz Nakhon Si Thammarat)...auf dem Khao Wang Hip in Höhenlagen zwischen 500 und 1196 m bei den aus etwa 10–12 Bambushütten bestehenden Ortschaften Ban Hui Hip und Amphoe Lan Saka" [= vicinity of villages Ban Hui Hip and Amphoe Lan Saka, Khao Wang Hip mtns., Nakhon Si Thammarat Prov., S Thailand, approximately 8°18'N, 99°43'E, elevation 1050 m].

Distribution: Malay Peninsula. Southern Thailand (Krabi, Nakhon Si Thammarat, Narathiwat, Surat Thani, Trang) and West Malaysia (Johore, Pahang, Perak, Selangor), 500–1195 m.

Sources: Tweedie, 1983, Nutaphand et al., 1991, Toriba, 1992, I. Das, 2010, 2012 and Grismer et al., 2010.

Remarks: Nutaphand et al. (1991: 151) listed type as female. A valid species *fide* David et al. (2006: 45).

CRISANTOPHIS Villa, 1971b (Dipsadidae)

Type species: *Conophis nevermanni* Dunn, 1937b.

Distribution: Central America.

Sources: Wellman, 1963, Villa, 1971b, 1988a, L.D. Wilson & Meyer, 1985, Villa et al., 1988, Savage, 2002, Solórzano, 2004, G. Köhler et al., 2005, Zaher et al., 2009, Vidal et al., 2010 and McCranie, 2011a.

1. *Crisantophis nevermanni* (Dunn, 1937b). Copeia 1937(4): 214. (*Conophis nevermanni*)

Type: Holotype, ANSP 22423, a 470 mm female (M. Valerio, 1927–1933).

Type locality: "Rio Poas de Aserri (a few miles south of San José), Costa Rica."

Distribution: Central America. Southwestern Guatemala, El Salvador (San Miguel), ext. S Honduras (Choluteca, Vallé), W Nicaragua (Chinandega, Chontales, Grenada, Jinotega, León, Managua, Matagalpa, Rivas) and NW Costa Rica (Guanacaste, San José), NSL–1395 m.

CROTALUS Linnaeus, 1758 (Viperidae)

Synonyms: *Crotalophorus* Gronovius, 1763 (*nomen illegitimum*), *Crotalophorus* Houttuyn, 1764, *Caudisona* Laurenti, 1768, *Crotalinus* Rafinesque-Schmaltz, 1815 (*nomen emendatum*), *Crotalurus* Rafinesque-Schmaltz, 1820 (*nomen emendatum*), *Crotulurus* – Rafinesque-Schmaltz, 1820 (*nomen incorrectum*), *Uropsophus* Wagler, 1830, *Crotalis* – Duvernoy, 1833 (*nomen incorrectum*), *Urocrotalon* Fitzinger, 1843, *Urocrotalum* Agassiz, 1847 (*nomen emendatum*), *Candisoma* – Chenu & Desmarest, 1857 (*nomen incorrectum*), *Uropscephus* Chenu & Desmarest, 1857 (*nomen incorrectum*), *Campilodon* Jan, 1861c (*nomen emendatum*), *Aploaspis* Cope, 1867a, *Uropsaphus* – Gigliloi, 1873 (*nomen incorrectum*), *Aechmophrys* Coues *in* Wheeler, 1875, *Aploaspsis* – Yarrow, 1875 (*nomen incorrectum*), *Haploaspis* Cope, 1883d (*nomen emendatum*), *Caudisoma* – Hoffmann, 1890 (*nomen emendatum*), *Uropsopus* – Cope, 1894c (*nomen incorrectum*), *Crotallus* – Phisalix, 1922 (*nomen incorrectum*), *Paracrotalus* A.F.T. Reuss, 1930c (*nomen substitutum*), *Candisona* – Haltom, 1931, *Crotalux* – H.W. Parker & Grandison, 1977 (*nomen incorrectum*), *Aploapis* – Hoge & Romano-Hoge, 1981 (*nomen incorrectum*), *Haploapis* – Hoge & Romano-Hoge, 1981 (*nomen incorrectum*), *Uropsophys* – Hoge & Romano-Hoge, 1981 (*nomen incorrectum*), *Crotulus* – Kundert, 1984 (*nomen incorrectum*), *Cottonus* Hoser, 2009d & 2012f (*nomen illegitimum*), *Crutchfieldus* Hoser, 2009d & 2012f (*nomen illegitimum*), *Cummingea* Hoser, 2009d & 2012f (*nomen illegitimum*), *Edwardsus* Hoser, 2009d & 2012f (*nomen illigitimum*), *Hoserea* Hoser, 2009d & 2012f (*nomen illegitimum*), *Matteoea* Hoser, 2009d & 2012f (*nomen illegitimum*), *Mullinsus* Hoser, 2009d & 2012f (*nomen illegitimum*), *Piersonus* Hoser, 2009d & 2012f (*nomen illegitimum*), *Pillotus* Hoser, 2009d & 2012f (*nomen illegitimum*), *Sayersus* Hoser, 2009d & 2012f (*nomen illegitimum*), *Smythus* Hoser, 2009d & 2012f (*nomen illegitimum*), *Piersonius* Hoser, 2012f (*nomen illegitimum*) and *Rattlewellsus* Hoser, 2012f (*nomen illegitimum*).

Type species: *Crotalus horridus* Linnaeus, 1758.

Distribution: North America and Latin America.

Fossil records: Upper Miocene to upper Pleistocene of USA and Pleistocene of Mexico. Fossils unassigned to species include upper Miocene (Clarendonian) of USA (Nebraska), upper Pliocene (Blancan) of USA (Washington), lower/middle Pleistocene (Irvingtonian I) of USA (Texas), and upper Pleistocene (Rancholabrean II) of USA (Utah).

Sources: Klauber, 1936b–c, 1937a, 1938, 1939a, 1940d, 1952, 1956a, 1972, 1997, Gloyd, 1940, A.H. Wright & Wright, 1957, Brattstrom, 1964, Hoge, 1966a, Harris & Simmons, 1977, 1978a–b, B. Armstrong & Murphy, 1979, Hoge & Romano-Hoge, 1981a, Stebbins, 1985, 2003, Stille, 1987, J.A. Campbell & Lamar, 1989, 2004, Maes, 1989, Minton, 1992, Golay et al., 1993, Knight et al., 1993, Ulber, 1993, Holman, 1995a, 2000a, Parmley & Holman, 1995, Kraus et al., 1996, Rubio, 1998, 2010, David & Ineich, 1999, McDiarmid et al., 1999, Bartlett & Tennant, 2000a, Bushar et al., 2001 and R.W. Murphy, 2002, Parmley & Walker, 2003, Castoe & Parkinson, 2006, Valencia-Hernández, 2006, Beaman

& Hayes, 2008, W.K. Hayes et al., 2008, Hoser, 2009d, 2012d, f, Hubbs & O'Connor, 2009, Parmley & Hunter, 2010 and Ernst & Ernst, 2012.

Remarks: Official Generic Name *fide* Opinion 18 (ICZN, 1926).

C

1. *Crotalus abyssus* Klauber, 1930b. Trans. San Diego Soc. Nat. Hist. 6(3): 114–117, pl. 11, fig. 1. (*Crotalus confluentus abyssus*)

Type: Holotype, SDSNH 2216 (formerly LMK 2216), a 905 mm male (E.D. McKee, 15 Sept. 1929).

Type locality: "Tanner Trail 300 ft. below the south rim of the Grand Canyon, Coconino County, Arizona.; altitude approximately 7000 ft." [= USA, elevation 2135 m].

Distribution: Southwestern USA (N Arizona), 1220–2135 m.

Sources: Fowlie, 1965 and Pook et al., 2000.

Remarks: A valid species *fide* Douglas et al. (2008: 38).

2. *Crotalus adamanteus* Palisot de Beauvois, 1799. Trans. Amer. Phil. Soc. 4(42): 368–369, fig. 2.

Synonyms: *Crotalus rhombifer* Latreille *in* Sonnini & Latreille, 1801a, *Crotalus rhombiferus* Brickell, 1805, †*Crotalus adamanteus pleistofloridensis* Brattstrom, 1954, and †*Crotalus giganteus* Brattstrom, 1954.

Type: Holotype, not designated (A.M.F.J. Palisot de Beauvois, summer, 1796), lost *fide* Gillispie (1992: 40).

Type locality: "southern parts of the United States" [= SE USA]. Restricted to Charleston, South Carolina *fide* K.P. Schmidt (1953a: 230).

Distribution: Southeast USA (S Alabama, Florida, Florida Keys, S Georgia, ext. SE Louisiana, ext. S Mississippi, SE North Carolina, E South Carolina), NSL–500 m.

Fossil records: Lower Pleistocene (Irvingtonian I) of USA (Florida) and upper Pleistocene (Rancholabrean II) of USA (Florida).

Sources: Christman, 1980, McCranie, 1980a, Ernst & Barbour, 1989, Conant & Collins, 1991, Palmer & Braswell, 1995, Tennant, 1997, Holman, 2000a and W.H. Martin, 2000.

Remarks: Palisot's entire American collection lost in shipwreck off Halifax, Nova Scotia in 1798 as he returned to Europe (Gillispie, 1992). Extirpated in many parts of Distribution.

3. *Crotalus aquilus* Klauber, 1952. Bull. Zool. Soc. San Diego (26): 24–25, fig. 2. (*Crotalus triseriatus aquilus*)

Synonym: *Crotalus triseriatus quandrangularis* Harris & Simmons, 1978b.

Type: Holotype, MCZ 27843, a 543 mm female (W.W. Brown, 1923).

Type locality: "near Alvarez, San Luis Potosí, Mexico" [= 22°02'N, 100°36'W, elevation 2350 m].

Distribution: Central Mexico (Aguascalientes, Guanajuato, Hidalgo, Jalisco, México, NE Michoacán, Querétaro, S San Luis Potosí, NW Veracruz), 1600–3110 m.

Sources: Dorcas, 1991, 1992, McCranie & Wilson, 2001b, Bryson et al., 2011 and Reyes-Velasco et al., 2012.

4. *Crotalus atrox* Baird & Girard, 1853. Cat. No. Amer. Rept., Serp. 1: 5–6.

Synonyms: *Crotalus cinereous* LeConte *in* Hallowell, 1852a *(nomen rejiciendum)*, *Caudisona atrox sonoraensis* Kennicott, 1861, *Crotalus tortugensis* Van Denburgh & Slevin, 1921b, and *Crotalus tortuguensis* Hoge & Romano-Hoge, 1971 *(nomen emendatum)*.

Type: Holotype, USNM 7761, an 838 mm specimen (J.H. Clark [U.S.-Mex. Bound. Surv.], 1850–1852, via J.D. Graham).

Type locality: "Indianola" [Calhoun County, SE Texas, USA, 28°31'N, 96°19'W, elevation NSL].

Distribution: Southwestern USA (S Arizona, W Arkansas, ext. SE California, S Kansas [possibly introduced], ext. S Nevada, New Mexico, Oklahoma, cen. & W Texas) and N Mexico (ext. NE Baja California Norte, Chihuahua, Coahuila, Durango, ? Guanajuato, Hidalgo, México, Nuevo León, SE Oaxaca, Querétaro, San Luis Potosí, N Sinaloa, Sonora, Tamaulipas, Veracruz, Zacatecas, Tiburón, Tortuga, Turner and San Pedro Martir Is.), NSL–2440 m.

Fossil records: Middle Pliocene (Blancan III) of USA (Texas), middle/upper Pleistocene (Irvingtonian II) of USA (Texas), and upper Pleistocene (Rancholabrean II) of USA (Arizona, California, Nevada, New Mexico, Texas).

Sources: Stejneger, 1940, Klauber, 1927, 1930b, Stebbins, 1985, Van Devender & Bradley, 1994, Degenhardt et al., 1996, P.R. Brown, 1997, Bartlett & Tennant, 2000, Holman, 2000a, Werler & Dixon, 2000, Grismer, 2002, Matlock & Rehmeier, 2002, Gumprecht, 2008, Matias-Ferrer & Murillo, 2004b, Trauth et al., 2004 and Castoe et al., 2007b.

Remarks: Official Specific Name no. 505 *fide* Opinion 365 (ICZN, 1955c).

5. *Crotalus basiliscus* (Cope, 1864). Proc. Acad. Nat. Sci. Philadelphia 16(3): 166. (*Caudisona basilisca*)

Type: Holotype, USNM 53586 (formerly SIM 6118), skin of a 1260 mm specimen (J. Xantus, 1859–1861).

Type locality: "Near Colima, Mexico." Restricted to Colima, Colima, Mexico [= 19°15'N, 103°43'W, elevation 490 m] *fide* H.M. Smith & Taylor (1950: 328).

Distribution: Western Mexico (ext. SW Chihuahua, Colima, Jalisco, NW Michoacán, Nayarit, Oaxaca, Sinaloa, ext. S Sonora, Zacatecas), NSL–2450 m.

Sources: McCranie, 1981, Chiszar & Smith, 2004, Lemos-Espinal et al., 2004a–b and Ahumada-Carrillo et al., 2011.

Remarks: Probably conspecific with *C. molossus* Baird & Girard *fide* McCranie (1981: 283.2).

6. *Crotalus caliginis* Klauber, 1949b. Trans. San Diego Soc. Nat. Hist. 11(6): 90. (*Crotalus viridis caliginis*)

Synonym: *Crotalus viridis caligensis* – Khole, 1991 *(nomen incorrectum).*

Type: Holotype, SDSNH 2800, a mm male (E.H. Quayle, 2 June 1930).

Type locality: "South Coronado Island off the northwest coast of Baja California, Mexico" [= Coronado Sur Is., NW Baja California Norte, ext. NW Mexico, bet. 32°23'–25'N & 117°14'–15'W, elevation NSL–300 m].

Distribution: Northwestern Mexico (N Baja California Norte: Coronado Sur Is.), NSL–300 m. Known only from type locality.

Sources: McPeak, 2000 and Pook et al., 2000.

Remarks: A valid species *fide* Grismer (2001: 20, 2002: 340).

7. *Crotalus catalinensis* Cliff, 1954. Trans. San Diego Soc. Nat. Hist. 12(5): 80–82, pl. 7, fig. 4.

Synonym: *Crotalus catalensis* – Khole, 1991 *(nomen incorrectum).*

Type: Holotype, CAS-SU 15631 (formerly SU 15631), a 701 mm female (B. Firstman, J.C. Briggs & F.S. Cliff, 27 March 1953).

Type locality: "Isla Santa Catalina in the Gulf of California, Mexico." Restricted to northeast end of the Isla Santa Catalina *fide* Leviton & Banta (1956: 218).

Distribution: Northwestern Mexico (Baja California Sur: Santa Catalina Is.).

Sources: R.W. Murphy & Crabtree, 1985, Beaman & Wong, 2001 and Grismer, 2002.

Remarks: Original figure lised as fig. 3 *fide* Cliff (1954: 80).

8. *Crotalus cerastes* Hallowell, 1854a. Proc. Acad. Nat. Sci. Philadelphia (1854–1855) 7(3): 95–96.

Synonyms: *Crotalus cerastes laterorepens* Klauber, 1944, *Crotalus cerastes cercobombus* Savage & Cliff, 1953, and *Cortalus cerastes* – Brode *in* D.F. Williams, Toth, Evans, Brode & Reed, 1991 *(nomen incorrectum).*

Type: Holotype, ANSP 7098 (formerly USNM 352), a 248 mm female (A.L. Heermann [R.S. Williamson, Pacific Railroad Surv.], 1853–1854).

Type locality: "Borders of the Mohave river, and in the desert of the Mohave" [California, USA].

Distribution: Southwestern USA (W Arizona, SE California, S Nevada, ext. SW Utah) and NW Mexico (NE Baja California Norte, NW Sonora, Tiburón Is.), 75 m BSL–1830 m.

Fossil records: Upper Pleistocene (Rancholabrean II) of USA (Arizona).

Sources: Klauber, 1927, 1944, Savage & Cliff, 1953, T.W. Brown & Lillywhite, 1992, Secor, 1994, P.R. Brown, 1997, Bartlett & Tennant, 2000, Holman, 2000a, Grismer, 2002, Douglas et al., 2006 and Sievert, 2008.

Remarks: Elevation usually below 1200 m *fide* J.A. Campbell & Lamar (2004: 538).

9. *Crotalus cerberus* (Coues *in* Wheeler, 1875). Explor. Surv. W of 100th Merid. 5: 606–607. (*Caudisona lucifer cerberus*)

Types: Syntypes (4), ANSP 7085 (formerly EC 509), skin of an 850 mm male, ANSP 7086 (formerly EC 510), ANSP 7087 (formerly EC 672), and ANSP 7088 (formerly EC 511), a head and neck (E. Coues, July 1864).

Type locality: "San Francisco Mountains, Arizona" [= Coconino Co., Arizona].

Distribution: Southwestern USA (cen. Arizona, ext. SW New Mexico), 2440–3660 m.

Sources: Klauber, 1949, Fowlie, 1965, Pook et al., 2000 and Brennan & Holycross, 2006.

Remarks: Two of Coues specimens (ANSP 7086–87) from type locality not considered syntypes *fide* Klauber (1949: 84). A valid species *fide* Douglas et al. (2008: 38).

10. *Crotalus concolor* Woodbury, 1929. Bull. Univ. Utah 20(6): 1–3, figs. 1–2.

Type: Holotype, UUZM 306, a 700 mm male (R.V. Chamberlin, Sept. 1929).

Type locality: "King's Ranch, Garfield Co., at base of the Henry Mtns." [= Utah, USA].

Distribution: Western USA (ext. NW Colorado, NE Utah, ext. SW Wyoming), 1220–2440 m.

Sources: Pook et al., 2000, Lewis, 2011 and Young, 2011.

Remarks: A valid species *fide* Douglas et al. (2008: 38).

11. *Crotalus culminatus* Klauber, 1952. Bull. Zool. Soc. San Diego (26): 65–67. (*Crotalus durissus culminatus*)

Type: Holotype, FMNH 126616 (formerly EHT 5224), a 382 mm female (E.H. Taylor & H.M. Smith, 1936).

Type locality: "El Sabino near Uruapan, Michoacán, Mexico" [= 19°25'N, 102°01'W, elevation 1600 m].

Distribution: Western Mexico (Chiapas, Guerrero, México, Michoacán, Morelos, Oaxaca), 90–1600 m.

Sources: Wüster et al., 2005 and Sanchez de la Vega et al., 2012.

Remarks: A subspecies of *Crotalus simus fide* Campbell and Lamar (2004: 583).

12. *Crotalus durissus* Linnaeus, 1758. Syst. Nat., ed. 10, 1: 214.

Synonyms: *Crotalus dryinus* Linnaeus, 1758, *Caudisona gronovii* Laurenti, 1768, *Caudisona orientalis* Laurenti,

C

1768, *Caudisona terrifica* Laurenti, 1768, *Crotalus exalbidus* Boddaert, 1778, *Crotalus immaculatus* Latreille *in* Sonnini & Latreille, 1801a, *Crotalus strepitans* Daudin, 1803b, *Crotalus cumanensis* Humboldt *in* Humboldt & Bonpland, 1813, *Crotalus loeflingii* Humboldt *in* Humboldt & Bonpland, 1813, *Crotalus cascavella* Wagler *in* Spix, 1824, *Caudisona durissa* – Cope, 1861e, *Crotalus horridus unicolor* Lidth de Jeude, 1887, *Crotalus pulvis* Ditmars, 1905, *Crotalus terrificus collirhombeatus* Amaral, 1927c, *Crotalus terrificus collilineatus* Amaral, 1927c, *Crotalus terrificus crotaminicus* Moura-Gonçalves, 1956 (*nomen nudum*), *Crotalus durissus marajoensis* Hoge, 1966a, *Crotalus durissus ruruima* Hoge, 1966a, *Crotalus durissus trigonicus* Harris & Simmons, 1978a, *Crotalus pifanorum* Sandner-Montilla, 1980, *Crotalus maricela* Rodriguez & Rojas Suárez, 1995 (*nomen nudum*), and *Crotalus maricelae* García-Pérez, 1995 *(nomen nudum)*.

Type: Neotype, RMNH 12857, an adult male, designated by Savage et al. (2005: 370).

Type locality: "2.5 km east of Tibiti, Saramacca District, Suriname (ca. 5°33'N, 55°54'W)" via neotype selection.

Distribution: South America. Colombia (Atlántico, Bolívar, Córdoba, Cundinamarca, Huila, La Guajira, Meta, Tolima), Venezuela (Amazonas, Anzoátegui, Apure, Aragua, Barinas, Bolívar, Carabobo, Cojedes, Delta Amacuro, Distrito Federal, Falcón, Guárico, Lara, Mérida, Miranda, Monagas, Nueva Esparta, Portuguesa, Sucre, Táchira, Trujillo, Yaracuy, Vargas, Zulia, Maragarita and Los Testigos Is.), Netherlands Antilles (Aruba), Guyana (Cuyuni-Mazaruni, Upper Demerara-Berbice, Upper Takutu-Upper Essequibo), NE Suriname (Commewijne), N French Guiana (N Cayenne, N Saint-Laurent-du-Maroni), ext. SE Peru (Puno), Brazil (Amapá, Amazonas, Bahia, Ceara, Goiás, Mato Grosso, Mato Grosso do Sul, Maranhao, Minas Gerais, Pará, Paraiba, Paraná, Pernambuco, Piauí, Rio de Janeiro, Rio Grande Do Norte, Rio Grande do Sul, Rondônia, Roraima, Santa Catarina, São Paulo, Marajo Is.), cen. Bolivia (Beni, Chuquisaca, Cochabamba, La Paz, Santa Cruz), Paraguay (Alto Paraguay, Alto Paraná, Amambay, Asunción, Boquerón, Caaguazú, Caazapá, Canindeyú, Central, Concepción, Cordillera, Guairá, Itapúa, Misiones, Neembucú, Paraguarí, Presidente Hayes, San Pedro), Uruguay (Artigas, Maldonado, Rivera, Treinta y Tres, Tucuarembó), and NE Argentina (Catamarca, Chaco, Córdoba, Corrientes, Entre Ríos, Formosa, Jujuy, La Pampa, La Rioja, Mendoza, Misiones, Salta, San Juan, San Luis, Santa Fe, Santiago del Estero, Tucumán), NSL–1000 [2800] m.

Sources: Gloyd & Kauffeld, 1940, Amaral, 1944e, E.H. Taylor, 1951, 1954, Klauber, 1952, Hoge, 1966a, Lancini, 1967, Harris & Simmons, 1972a–b, 1978a, Vanzolini et al., 1980, Sandner-Montilla, 1983, McCranie, 1984, 1986, 1993a, Villa, 1984, L.D. Wilson & Meyer, 1985, Chippaux, 1987, Cei, 1994, D.R. Norman, 1994, Lee, 1996, Moro, 1996, J.A. Campbell, 1998b, Gorzula & Señaris, 1998, Starace, 1998, Leynaud & Bucher, 1999, Parkinson, 1999, Esqueda et al., 2001, Savage, 2002, Abuys, 2003, Carreira-Vidal et al., 2005, Harvey et al., 2005, Rivas-Fuenemayer & Barrio-Amorgós, 2005, Wüster et al., 2005, Quijada-Mascareñas et al., 2007, Cacciali, 2008, Navarrete et al., 2009, Ugueto & Rivas, 2010, Prigioni et al., 2011 and C.J. Cole et al., 2013.

Remarks: Holotype lost *fide* Klauber (1952: 60). Type locality restriction to Jalapa, Veracruz, Mexico *fide* H.M. Smith & Taylor (1950: 348) rejected *fide* Hoge (1966a: 137). Type locality restriction to Paramaribo, Suriname *fide* J.A. Campbell & Lamar (2004: 540) invalid due to neotype designation. *Crotalus pifanorum* a valid species *fide* Navarrete et al. (2009: 66).

13. *Crotalus enyo* (Cope, 1861e). Proc. Acad. Nat. Sci. Philadelphia 13(7): 293. (*Caudisona enyo*)

Synonyms: *Crotalus enyo cerralvensis* Cliff, 1954 and *Crotalus enyo furvus* Lowe & Norris, 1954.

Type: Lectotype, ANSP 7159 (formerly JX 4663), a 755 mm male, head, skin and tail (J. Xantus, May–June 1859), designated by Klauber (1956a: 34).

Type locality: "Lower California" [Mexico]. Restricted to Cape San Lucas, Baja California, Mexico *fide* Klauber (1956a: 34).

Distribution: Northwestern Mexico (S Baja California Norte, Baja California Sur; Carmen, Cerralvo, Espíritu Santos, Partida Sur, San Francisco, San José, San Marcos Is. in Gulf; Magdalena, Santa Margarita Is. in Pacific).

Sources: Klauber, 1931c, Cliff, 1954, Lowe & Norris, 1954, Beaman & Grismer, 1994, and Grismer, 2002.

14. *Crotalus ericsmithi* J.A. Campbell & Flores-Villela, 2008. Herpetologica 64(2): 247–251, figs. 1–3, 4c.

Type: Holotype, UTA 55372, a 540 mm male (E.N. Smith).

Type locality: Carretera La Laguna–bajitos de la Laguna. Guerrero, Sierra Madre del Sur, Mexico, 1037 m. The coordinates for this locality are 17.55330°N, 100.77472° W."

Distribution: Southwestern Mexico (Guerrero), 1035 m.

15. *Crotalus estebanensis* Klauber, 1949b. Trans. San Diego Soc. Nat. Hist. 11(6): 104–106, pl. 6, fig. 2. (*Crotalus molossus estebanensis*)

Type: Holotype, SDSNH 26792 (formerly LMK 26792), a 737 mm female (G. Allan Hancock Exped., 17 April 1937).

Type locality: "San Estéban Island, in the middle of the Gulf of California, Mexico, Lat. 28° 40' N."

Distribution: Northwestern Mexico (San Estéban Is.).

Remarks: A valid species *fide* Grismer (1999: 462).

16. *Crotalus helleri* Meek, 1905. Field Columb. Mus. Publ. (Zool.) 7(1): 17–18, pl. 2.

Synonyms: *Crotalus hallowelli* Cooper, 1870 *(nomen nudum)* and *Crotalus viridis hellari* – Khole, 1991 *(nomen incorrectum)*.

Type: Holotype, FMNH (formerly CNHM 1272), a 957 mm specimen (E. Heller, May–July 1902).

Type locality: "San Jose, Lower California" [= San José, N Baja California Norte, ext. NW Mexico, 29°32'N, 114°42'W, elevation 535 m].

Distribution: Southwestern USA (SW California, Santa Catalina Is.) and NW Mexico (Baja California Norte, N Baja California Sur), NSL–3350 m.

Fossil records: Upper Pleistocene (Rancholabrean II) of USA (California).

Sources: Klauber, 1949, Holman, 2000a, Pook et al., 2000 and Lemm, 2006.

Remarks: A valid species *fide* Douglas et al. (2008: 38).

17. *Crotalus horridus* Linnaeus, 1758. Syst. Nat., ed. 10, 1: 214.

Synonyms: *Crotalus boiquira* Lacépède, 1789 *(nomen rejiciendum)*, *Crotalus atricaudatus* Latreille *in* Sonnini & Latreille, 1801a, *Crotalus zetazomae* Brickell, 1805, *Crotalinus cyanurus* Rafinesque-Schmaltz, 1818b, *Crotalus catesbaei* Hemprich, 1820, *Crotalus durissus concolor* Jan, 1859b, *Crotalus durissus melanurus* Jan, 1859b, *Crotalus durissus mexicana* Jan, 1863b *(nomen nudum)*, *Crotalus fasciatus* Higgins, 1873, and *Crotalus durissus cincolor* – Notestein, 1905 *(nomen incorrectum)*.

Type: Holotype, NHR Lin-102 (formerly MAFR), a tail with 12 rattle segments (P. Kalm, Sept. 1748–Feb. 1751, via Mus. Drottn.).

Type locality: "America." Restricted to vicinity of New York City, New York, USA *fide* K.P. Schmidt (1953a: 227).

Distribution: Eastern North America. Southeastern Canada (S Ontario) and E USA (Alabama, Arkansas, Connecticut, N Florida, Georgia, Illinois, S Indiana, S and E Iowa, E Kansas, Kentucky, Louisiana, W Maryland, Massachusetts, ext. SE Minnesota, Mississippi, Missouri, ext. SE Nebraska, S New Hampshire, New Jersey, New York, North Carolina, E Oklahoma, S and N Ohio, Pennsylvania, Rhode Island, South Carolina, Tennessee, E Texas, S Vermont, Virginia, West Virginia, SW Wisconsin), NSL–2000 m.

Fossil records: Upper Miocene (Hemphillian) of USA (Nebraska), upper Pliocene (Blancan V) of USA (Nebraska), Pleistocene of USA (Massachusetts), lower Pleistocene (Irvingtonian I [?]) of USA (Pennsylvania), middle/upper Pleistocene (Irvingtonian II) of USA (Arkansas, Maryland, Nebraska, Texas, West Virginia), and upper Pleistocene (Rancholabrean II) of USA (Alabama, Georgia, Indiana, Missouri, Pennsylvania, Tennessee, Virginia, West Virginia).

Sources: Pisani et al., 1972, Odum, 1979, Collins & Knight, 1980, C.W. Brown & Ernst, 1986, W.S. Brown, 1993, Palmer & Braswell, 1995, Harding, 1997, Tennant, 1997, Fitch, 1999, Holman, 2000a, Werler & Dixon, 2000, Clark et al., 2003, J.C. Murphy & Schlager, 2003 and Furman, 2007.

Remarks: Official Specific Name no. *fide* Opinion 92 (ICZN,). Extirpated in many areas of Distribution.

18. *Crotalus intermedius* Troschel *in* J.W. Müller, 1865. Reisen Ver. Staat., Can. Mex. 3: 613–614.

Synonyms: *Crotalus intermedius* Fischer, 1882 *(nomen praeoccupatum)*, *Crotalus omiltemanus* A.C.L.G. Günther, 1895 *in* 1885–1902, *Crotalus triseriatus gloydi* E.H. Taylor, 1942a, *Crotalus gloydi lautus* H.M. Smith *in* H.M. Smith & Laufe, 1945b *(nomen nudum)*, and *Crotalus gloydi lautus* H.M. Smith, 1946.

Type: Holotype, ZIB (formerly JWM), a 419 mm female, destroyed during World War II.

Type locality: "Mexico" [in one of the following states based on Müller's expedition: Distrito Federal, Guerrero, Mexico, Morelos, Oaxaca, Puebla or Veracruz *fide* Klauber, 1952: 10]

Distribution: South-central Mexico (cen. Guerrero, SE Hidalgo, Oaxaca, SE and cen. Puebla, SW Tlaxcala, W Veracruz), 2000–3050 m.

Sources: McCranie, 1991 and Alvarado-Diaz & Campbell, 2004.

Remarks: Michoacán record is *C. transversus* E.H. Taylor (= *C. tancitarensis*) *fide* J.A. Campbell & Lamar (1989: 349).

19. *Crotalus lannomi* W.W. Tanner, 1966b. Herpetologica 22(4): 298, fig. 1b.

Type: Holotype, BYU 23800, a 638 mm female (J.R. Lannom, Jr., 26 June 1966).

Type locality: "1.8 miles west of the pass, Puerto Los Mazos, or 22 miles west by road from the Rio Tuxcacuesco, a branch of the Rio Armeria on Mexican Highway No. 80, Jalisco, Mexico" (in error). Corrected to Arroyo el Tigre, 12 km SW by road from original type locality, at 550 m elevation *fide* J.R. Lannom *in* Reyes-Velasco et al., 2010: 24].

Distribution: Western Mexico (Colima, Jalisco), 550–1150 m.

Sources: W.W. Tanner, 1986, J.A. Campbell & Flores-Villela, 2008 and Reyes-Velasco et al., 2010.

20. *Crotalus lepidus* (Kennicott, 1861). Proc. Acad. Nat. Sci. Philadelphia 13(6): 206. (*Caudisona lepida*)

Synonyms: *Crotalus palmeri* S.W. Garman, 1887c, *Crotalus lepidus klauberi* Gloyd, 1936b, *Crotalus semicornutus* E.H. Taylor, 1944b, *Crotalus lepidus morulus* Klauber, 1952, *Crotalus lepidus maculosus* W.W. Tanner, Dixon & Harris, 1972, *Crotalus lepidus*

C

castaneus Julía-Zertuche & Treviño-Saldaña, 1978a, and *Crotalus lepidus clauberi* – Khole, 1991 *(nomen incorrectum).*

Types: Syntypes (2), two heads (W.H. Emory [U.S.-Mex. Bound. Surv.], 1848–1855), lost *fide* Stejneger (1895: 458).

Type locality: "Presidio del Norte and Eagle Pass" [Texas, USA]. Restricted to Presidio (del Norte), Presidio County, Texas *fide* H.M. Smith & Taylor (1950: 362).

Distribution: Southwestern USA (SE Arizona, S New Mexico, SW Texas) and N and N Mexico (Aguascalientes, Chihuahua, Coahuila, Durango, N Jalisco, E Nayarit, W Nuevo León, W San Luis Potosí, SE Sinaloa, NE Sonora, SW Tamaulipas, S Zacatecas), 600–2700 m.

Fossil records: Upper Pliocene (Blancan V) of USA (Arizona), and lower/middle Pleistocene (Irvingtonian I) of USA (Arizona).

Sources: Gloyd, 1936b, Klauber, 1952, Zweifel, 1954b, Julia-Zertuche & Treviño-Saldaña, 1980, Dorcas, 1992, Van Devender & Bradley, 1994, Degenhardt et al., 1996, Forstner et al., 1997, Schwinford, 1999, Bartlett & Tennant, 2000, Holman, 2000a, Werler & Dixon, 2000, Bryson & Mueller, 2001, Allsteadt et al., 2006, M.S. Price, 2009 and Bryson et al., 2011.

21. *Crotalus lutosus* Klauber, 1930b. Trans. San Diego Soc. Nat. Hist. 6(3): 100–106, pl. 10, fig. 1. (*Crotalus confluentus lutosus*)

Synonym: *Crotalus viridis letuous* – Khole, 1991 *(nomen incorrectum).*

Type: Holotype, SDSNH 1814 (formerly LMK 1814), an 832 mm male (C.B. Perkins, 12 May 1929).

Type locality: "10 miles northwest of Abraham on the road to Joy, Millard County, Utah. Altitude approximately 4650 ft." [= USA, elevation 1415 m].

Distribution: Western USA (NW Arizona, NE California, S Idaho, Nevada, SE Oregon, W Utah), 1220–2745 m.

Fossil records: Upper Pleistocene (Rancholabrean II) of USA (Idaho).

Sources: A.H. Wright & Wright, 1957, Fowlie, 1965, Nussbaum et al., 1983, Holman, 2000a, Pook et al., 2000 and St. John, 2002.

Remarks: A valid species *fide* Douglas et al. (2008: 38).

22. *Crotalus mitchellii* (Cope, 1861e). Proc. Acad. Nat. Sci. Philadelphia 13(7): 293–294. (*Caudisona mitchellii*)

Synonyms: *Caudisona pyrrha* Cope, 1867a, *Crotalus mitcheli* Belding, 1887 *(nomen emendatum), Crotalus goldmani* K.P. Schmidt, 1922, *Crotalus mitchelli aureus* Kallert, 1927 (*nomen nudum*), *Crotalus mitchellii muertensis* Klauber, 1949b, *Crotalus mitchellii angelensis* Klauber, 1963, and *Crotalus mitchellii angelicus* Hoge, 1966 *(nomen emendatum)..*

Type: Holotype, USNM 5291b, an 1118 mm skin (J. Xantus, 1859–1861), lost *fide* Klauber (1956a: 38).

Type locality: "Cape St. Lucas, Lower California" [= Cabo San Lucas, Baja California Sur, Mexico, 22°53'N, 109°55'W, elevation 20 m].

Distribution: Southwestern USA (W Arizona, SE California, ext. SW Nevada, ext. SW Utah) and NW Mexico (ext. NW Sonora, Baja California Norte, Baja California Sur: Angel de la Guarda, Carmen, Cerralvo El Muerto, Espíritu Santo, Monserrate, Partida Sur, Piojo, Smith, Salsipuedes, San José Is. in Gulf; Santa Margarita Is. in Pacific), NSL–2400 m.

Fossil records: Upper Pleistocene (Rancholabrean II) of USA (Nevada).

Sources: Klauber, 1927, 1930a, 1936a, 1952, McCrystal & McCoid, 1986, Strimple, 1992, Grismer, 1993, P.R. Brown, 1997, Bartlett & Tennant, 2000, 1999, 2002, Holman, 2000a, Douglas et al., 2006, 2007 and Meik et al., 2010.

Remarks: The types of *C. enyo* and *C. mitchellii* (four specimens: three *enyo* and one *mitchellii)* were both assigned USNM 5291. Grismer, 1999 & 2002 recognizes *C. mitchellii muertensis* and *C. m. angelensis* as valid species.

23. *Crotalus molossus* Baird & Girard, 1853. Cat. No. Amer. Rept., Serp. 1: 10–11.

Synonyms: *Caudisona molossa* Cope, 1867a *(nomen emendatum), Crotalus molossus nigrescens* Gloyd, 1936a, and *Crotalus basiliscus oaxacus* Gloyd, 1948, and *Crotalus moloosus* – Miranda, 1993 *(nomen incorrectum).*

Type: Holotype, USNM 485, an 861 mm male (J.H. Clark [U.S.-Mex. Bound. Surv.], 1850–1852, via J.D. Graham).

Type locality: "Fort Webster, St. Rita del Cobre, N. Mex." Restricted to Fort Webster Copper Mines, New Mexico *fide* Baird (1859a: 14), [= Santa Rita del Cobre copper mines, Grant County, New Mexico, USA].

Distribution: Southwestern USA (Arizona, SW New Mexico) and Mexico (Aguascalientes, Chihuahua, S Coahuila, Colima, Distrito Federal, Durango, Guanajuato, SW Hidalgo, E Jalisco, SW México, N Michoacán, N Morelos, E Nayarit, SW Nuevo León, W Oaxaca, cen. Puebla, Querétaro, W San Luis Potosí, Sonora, ext. N Sinaloa, ext. SW Tamaulipas, Tlaxcala, W Veracruz, Zacatecas, Tiburón Is.), NSL–2590 (3750) m.

Sources: Stejneger, 1940, Gloyd, 1936a, 1948, Klauber, 1949b, A. Price, 1980, Degenhardt et al., 1996, Grismer, 1999, 2002, Bartlett & Tennant, 2000, Werler & Dixon, 2000, Fernandez et al., 2003 and Lemos-Espinal et al., 2004.

Remarks: Type locality listed as Sonora, Mexico *fide* Baird (1859b: 16) and Yarrow (1882: 78).

24. *Crotalus oreganus* Holbrook, 1840. No. Amer. Herp. 4: 115–117, pl. 24.

Synonyms: *Crotalus oregonus* Holbrook, 1842a *(nomen emendatum), Crotalus lucifer* Baird & Girard, 1852c, *Crotalus hallowelli* Cooper *in* Cronise, 1868 (*nomen nudum*), *Crotalus oreganus niger* Kallert, 1927 (*nomen nudum*), and *Crotalus confluentus decolor* Klauber, 1930b.

Type: Holotype, ANSP 7158 (formerly ANSP 840), a shrivelled 337–381 mm specimen (T. Nuttall, May 1834–Aug. 1835).

Type locality: "west of the Rocky Mountains, on the banks of the Oregon or Columbia river, in the Oregon territory" [= Oregon *fide* ANSP jar labels and Van Denburgh, 1898: 141]. Restricted to The Dalles, Wasco County, Oregon, USA *fide* H.M. Smith & Taylor (1950: 360), to confluence of Columbia and Snake rivers, Washington, USA *fide* K.P. Schmidt (1953: 232) and between Walla Walla, Washington and the Pacific Coast *fide* Klauber (1956a: 53).

Distribution: Extreme SW Canada (S British Columbia) and W USA (NW Arizona, N California, NW Idaho, Oregon, E Washington), NSL–3655 m.

Fossil records: Upper Pleistocene (Rancholabrean II) of USA (California).

Sources: Klauber, 1927, 1949, A.H. Wright & Wright, 1957, Nussbaum et al., 1983, Stebbins, 1985, 2003, Quinn, 1987, M.L. Johnson, 1995, Storm et al., 1995, P.R. Brown, 1997, Holman, 2000a, Pook et al., 2000, K. Ashton & Queiroz, 2001, Douglas et al., 2002 and Brennan & Holycross, 2006

Remarks: Holbrook (1840: pl. 24) incorrectly numbered in text as pl. 29. A valid species *fide* Douglas et al. (2008: 38).

25. *Crotalus ornatus* Hallowell, 1855a. Proc. Acad. Nat. Sci. Philadelphia (1854–1855) 7(5): 192–193.

Type: Holotype, USNM 486, head, tail and connecting skin from an 1134–1162 mm female (A.L. Heermann [J.G. Parke, Pacific Railroad Surv.], 1854).

Type locality: "Near Pecos River, N.W. Texas" [USA].

Distribution: Southwestern USA (New Mexico, W Texas) and N Mexico (N Coahuila), 1220–1980 m.

Remarks: A valid species *fide* C.G. Anderson & Greenbaum, 2012. Photographs of holotype in C.G. Anderson & Greenbaum (2012: fig. 10).

26. *Crotalus polystictus* (Cope, 1865). Proc. Acad. Nat. Sci. Philadelphia 17(4): 191–192. (*Caudisona polysticta*)

Synonyms: *Crotalus lugubris* Jan, 1859b, e (*nomen rejiciendum*), *Crotalus lugubris multimaculata* Jan, 1863b (*nomen oblitum*), *Crotalus lugubris multimaculata* Jan, 1874 *in* Jan & Sordelli, 1870–1881, and *Crotalus jimenezii* Dugès, 1877.

Type: Holotype, formerly USNM, a 603 mm specimen, lost *fide* H.M. Smith & Taylor (1945: 173).

Type locality: "Table Land, Mexico." Restricted to Tupátaro, Guanajuato, Mexico *fide* H.M. Smith & Taylor (1950: 330).

Distribution: South-central Mexico (Aguascalientes, NE Colima, Distrito Federal, S Guanajuato, S Hidalgo, Jalisco, N México, N Michoacán, N Morelos, N Puebla, S Querétaro, ? Tlaxcala, W Veracruz, S Zacatecas), 1450–2740 m.

Sources: McCranie, 1976 and Setzer et al., 2009.

Remarks: May occur in S Durango *fide* J.A. Campbell & Lamar (2004: 571). Official Specific Name no. 506 *fide* Opinion 366 (ICZN, 1955c). In accordance with Art. 23.9.2 of the Code (ICZN, 1999), *Caudisona polysticta* Cope is designated a *nomen protectum* and *Crotalus lugubris multimaculata* Jan a *nomen oblitum*.

27. *Crotalus pricei* Van Denburgh, 1895b. Proc. California Acad. Sci. (1895–1896) (2) 5: 856–857.

Synonym: *Crotalus triseriatus miquihuanus* Gloyd, 1940.

Type: Holotype, CAS-SU 1702 (formerly SU 1702), a 447 mm male (W.W. Price, July–Aug. 1893).

Type locality: "Huachuca Mts., Arizona" [= Cochise Co., SE Arizona, USA].

Distribution: Extreme SW USA (SE Arizona) and N Mexico (Aguascalientes, W Chihuahua, SE Coahuila, Durango, N Nayarit, S Nuevo León, San Luis Potosí, NE Sonora, SW Tamaulipas, SW Zacatecas), 1220–3200 m.

Sources: Gloyd, 1940, Fowlie, 1965, McCranie, 1980b, Bartlett & Tennant, 2000, Alvarado-Diaz & Campbell, 2004 and Reyes-Velasco et al., 2012.

Remarks: Type locality listed in Pima County, Arizona *fide* Leviton (1953b: 132). Probably occurs in E Sinaloa, Jalisco and Zacatecas *fide* McCranie (1980b: 266.2).

28. *Crotalus pusillus* Klauber, 1952. Bull. Zool. Soc. San Diego (26): 34–40.

Type: Holotype, FMNH 39112, a 652 mm male (F.A. Shannon, 25 June–20 July 1941).

Type locality: "Tancítaro, Michoacán, Mexico, altitude 5000 feet."

Distribution: Southwestern Mexico (Guerrero, S Jalisco, W Michoacán), 1525–2380 m.

Sources: Duellman, 1961, Armstrong & Murphy, 1979, McCranie, 1983a, Bryson et al., 2011 and Reyes-Velasco et al., 2012.

Remarks: Distribution as presently known involves three disjunct populations.

29. *Crotalus ravus* Cope, 1865. Proc. Acad. Nat. Sci. Philadelphia 17(4): 191.

Synonyms: *Caudisona rava* – Cope, 1865, *Sistrurus ravus brunneus* Harris & Simmons, 1978b, *Sistrurus*

C

ravus lutescens Harris & Simmons, 1978b, *Sistrurus ravus exiguus* J.A. Campbell & Armstrong, 1979, and *Sistrurus ravus sinaloensis* Juliá-Zertuche, 1982 *(nomen illegitimum)*.

Types: Syntypes (2), USNM 25050–51, a 212 mm and 200 mm specimen (C. Sartorius, 1863).

Type locality: "Table land of Mexico." Emended to south tableland, Veracruz, Mexico *fide* Cochran (1961: 171). Restricted to Totalco, Veracruz, Mexico *fide* H. M. Smith & Taylor (1950: 351).

Distribution: Southern Mexico (Distrito Federal, cen. Guerrero, México, Morelos, ext. N Oaxaca, Puebla, Tlaxcala, W Veracruz), 850–3050 m.

Sources: J.A. Campbell & Armstrong, 1979, McCranie, 1988, Pérez-Higareda & Smith, 1991, Camarillo & Smith, 1992, Benitez-Gálvez, 1997, R.W. Murphy et al., 2002, Campos-Rodriguez & Mendoza-Quijano, 2004 and Bryson et al., 2011.

30. *Crotalus ruber* Cope, 1892a. Proc. U.S. Natl. Mus. (1891–1892) 14(882): 690–691. (*Crotalus adamanteus ruber*)

Synonyms: *Crotalus exsul* S.W. Garman, 1884 *(nomen suppressum), Crotalus lucasensis* Van Denburgh, 1920b, *Crotalus atrox elegans* K.P. Schmidt, 1922, *Crotalus ruber lorenzoensis* Radcliffe & Maslin, 1975, and *Crotalus ruber monserratensis* Harris & Simmons, 1978b.

Type: Holotype, USNM 9209, a 1245 mm male.

Type locality: Unknown. Designated as southern California *fide* Klauber (1956: 41). Restricted to Dulzura, San Diego Co., California, USA *fide* H.M. Smith & Taylor (1950: 356) and to vicinity of San Diego, California, USA *fide* K.P. Schmidt (1953a: 230).

Distribution: Extreme SW USA (ext. SW California) and NW Mexico (Baja Califonia Norte, Baja California Sur, Angel de la Guarda, Danzante, Monserrate, Pond, San José, San Lorenza Sur, San Marcos Is. in Gulf, Cedros, Santa Margarita Is. in Pacific), NSL–1500 m.

Sources: Klauber, 1927, 1930b, 1949a, Radcliffe & Maslin, 1975, Harris & Simmons, 1978b, Grismer et al., 1994, R.W. Murphy et al., 1995, P.R. Brown, 1997, Grismer, 1999, Bartlett & Tennant, 2000, Beaman & Dugan, 2006 and Douglas et al., 2006.

Remarks: Official Specific Name *fide* Opinion 1960 (ICZN, 2000a). The paralectotype (652b) consists of a head and tail only. Grismer (1999: 461) considered *C. ruber lorenzoensis* a distinct species.

31. *Crotalus scutulatus* (Kennicott, 1861). Proc. Acad. Nat. Sci. Philadelphia 13(6): 207–208. (*Caudisona scutulata*)

Synonyms: *Caudisona lepturus* Kennicott, 1861 *(nomen ineditum)*, Crotalus *salvini* A.C.L.G. Günther, 1895 *in* 1885–1902, *Crotalus californicus* Kallert, 1927 *(nomen nudum)*, and *Crotalus confluentus kellyi* Amaral, 1929a.

Type: Holotype, ANSP 7069 (formerly USNM 5027b) (B.J.D. Irwin, 30 Jan. 1861).

Type locality: Unknown. Designated as "Fort Buchanan, (Tucson) Arizona" [= 5 km SW Sonoita in Hog Canyon, Santa Cruz Co., SE Arizona, SW USA, 31°40'N, 110°42'W, elevation 1425 m] *fide* Cope (1900: 1160).

Distribution: Southwestern USA (Arizona, S California, ext. S New Mexico, S Nevada, ext. SW Texas, ext. SW Utah) and N Mexico (Aguascalientes, Chihuahua, Coahuila, Durango, NE Guanajuato, Hidalgo, N and E Jalisco, NE México, W Nuevo León, Puebla, cen. Querétaro, W San Luis Potosí, NE Sonora, ext. SW Tamaulipas, Tlaxcala, N Puebla, W Veracruz, Zacatecas), [NSL] 1800–2500 m.

Fossil records: Upper Pleistocene (Rancholabrean II) of USA (Arizona).

Sources: A. Price, 1982, Stebbins, 1985, Degenhardt et al., 1996, P.R. Brown, 1997, Bartlett & Tennant, 2000, Holman, 2000a, Werler & Dixon, 2000, Matias-Ferrer & Murillo, 2004c and Cardwell et al., 2013.

Remarks: Photographs of the type in Cardwell et al. (2013: figs. 1, 2b–c), who discuss the history and confusion of purported types. Type locality designated as Mojave Desert, California *fide* K.P. Schmidt (1953a: 229). Type locality restriction to Wickenburg, Maricopa Co., Arizona, USA *fide* H.M. Smith & Taylor (1950: 353) invalid.

32. *Crotalus simus* Latreille *in* Sonnini & Latreille, 1801a. Hist. Nat. Rept. 3: 202–203.

Synonym: *Crotalus terrificus copeanus* Amaral, 1937d *(nomen substitutum)*.

Type: Neotype, UTA 52032, an adult male, designated by Savage et al. (2005: 370).

Type locality: "Vicinity of El Arenal, San Vicinte, Departamento de Zacapa, Guatemala, 360 m elevation (14.880176° N, 89.861817° W)" via neotype selection.

Distribution: Mesoamerica. Southern Mexico (ext. SW Campeche, Chiapas, SE Oaxaca, cen. & S Veracruz), Guatemala (Petén), El Salvador (Cabañas, Chalatenango, Cuscatlán, La Libertad, La Unión, Santa Ana, San Miguel, San Vicente, Sonsonate), W Honduras (Choluteca, Comayagua, Cortés, Francisco Morazán, Santa Bárbara, Valle, Yoro), W Nicaragua (Chontales, Matagalpa, Río San Juan) and Costa Rica (Guanacaste, Puntarenas, San José), NSL–2200, usually below 1000 m.

Sources: N. Herrera et al., 2007 and McCranie, 2011a.

Remarks: Holotype lost *fide* Thireau (1991: 4).

33. *Crotalus stejnegeri* Dunn, 1919. Proc. Biol. Soc. Washington 32: 214–216.

Type: Holotype, USNM 46586, a 590 mm male (E.W. Nelson & E.A. Goldman, 19 July 1897).

Type locality: "Plumosas, Sinaloa, Mexico" [= Plomosas, S Sonora, W Mexico, 23°04'N, 105°30'W, elevation 965 m].
Distribution: Northwestern Mexico (W Durango, SE Sinaloa), 480–1200 m.
Sources: W.W. Tanner, 1966, 1976, Collins, 1982, Webb, 1984, Strimple, 1998 and J.A. Campbell & Flores-Villela, 2008.
Remarks: Type erroneously listed as USNM 46486 *fide* Dunn (1919: 215). Probably occurs in N Nayarit *fide* J.A. Campbell & Lamar (1989: 363).

34. *Crotalus stephensi* Klauber, 1930a. Trans. San Diego Soc. Nat. Hist. 6(3): 108–109. (*Crotalus confluentus stephensi*)

Type: Holotype, MVZ 6699, a 591 mm male (J. Grinnell, 8 Oct. 1917).
Type locality: "two miles west of Jackson Springs, Panamint Mts., altitude 6200 ft., Inyo County, California."
Distribution: Southwestern USA (SE California, SW Nevada), 1900 m.
Source: M.E. Douglas et al., 2007.

35. *Crotalus tancitarensis* Alvarado-Díaz & Campbell, 2004. Herpetologica 60(2): 281–283, fig. 1a.

Type: Holotype, INIRENA 309, a 356 mm female (J. Alvarado-Diaz & A. Estrada Virgen, 17 July 2002).
Type locality: "Cerro Tancítaro, Michoacán, Mexico, 3225 m elevation (coordinates 19° 24' 13" N, 102° 19' 45" W)."
Distribution: Western Mexico (Michoacán), 1525–3225 m.
Source: Alvarado-Díaz et al., 2007.

36. *Crotalus tigris* Kennicott *in* Baird, 1859a. U.S.-Mex. Bound. Surv., Rept. (1858) 2(2): 14, pl. 4.

Type: Holotype, USNM 471 (A.C.V. Schott [U.S.-Mex. Bound. Surv.], 1854–1855).
Type locality: "Sierra Verde and Pozo Verde, Gila and Colorado Deserts." [= Sasabe, Sonora, Mexico *fide* Klauber, 1956a: 43, or Arizona, USA *fide* Cochran, 1961: 171].
Distribution: Extreme SW USA (S Arizona, possibly ext. SW New Mexico *fide* Degenhardt et al., 1996: 361) and NW Mexico (Sonora, Tiburón Is.), NSL–2400 m.
Sources: Klauber, 1931c, Bartlett & Tennant, 2000, Grismer 2002 and Douglas et al., 2006.
Remarks: Three syntypes (USNM 471–73) listed *fide* Yarrow (1882: 74), with USNM 473 now = ANSP 7160 *fide* Klauber (1956a: 43). USNM 471 designated the lectotype *fide* Klauber (1956a: 43) and Cochran (1961: 171); however, Kennicott listed USNM 471 as the type and he examined at least two specimens (i.e., dorsal rows 21–23) so his selection represents a holotype.

37. *Crotalus totonacus* Gloyd & Kauffeld, 1940. Bull. Chicago Acad. Sci. 6(2): 12–14, fig. 1.

Synonyms: *Crotalus durissus neoleonensis* Juliá-Zertuche & Trevino-Saldaña *in* Harris & Simmons, 1978b (*nomen nudum*) and *Crotalus durissus tolanacus* – Khole, 1991 (*nomen incorrectum*).
Type: Holotype, CA 4469, a 945–972 mm female (W. Woolems, May 1933)
Type locality: "Panaco Island, about 75 miles south of Tampico, Veracruz, Mexico, 12 miles inland from Cabo Rojo."
Distribution: Northeastern Mexico (Hidalgo, cen. Michoacán, cen. Nuevo León, N Querétaro, E San Luis Potosí, S Tamaulipas, N Veracruz), NSL–1680 m.
Sources: P. Martin 1958, Auth et al., 2000b and Ramírez-Bautista et al., 2010.
Remarks: Photograph of holotype in Gloyd & Conant (1940: pl. 17, fig. 1).

38. *Crotalus transversus* E.H. Taylor, 1944b. Univ. Kansas Sci. Bull. 30(4): 47–49, figs. 1a–e, pl. 6, fig. b.

Type: Holotype, FMNH 100129 (formerly EHT-HMS 30001), a 464 mm female (E. Powell, 14 Aug. 1942).
Type locality: "about 55 km. SW México (city), near Tres Marias (Tres Cumbres), Morelos, México, elevation about 10,000 ft.," (probably in error). Restricted to Laguna Zempoala, Mexico State, Mexico *fide* Davis & Smith (1953: 141) (also in error). Any of six or more lakes known as Lagunas de Zempoala (or Cempoala), NW Morelos, Mexico *fide* J.A. Campbell (1988a: 450.1).
Distribution: South-central Mexico (Distrito Federal, ext. NW Morelos, cen. México), 2870–3430 m.
Sources: J.A. Campbell, 1988a, Camarillo & Campbell 1993, 2002, Alvarado-Diaz & Campbell, 2004 and García-Vázquez et al., 2008b.

39. *Crotalus triseriatus* (Wagler, 1830). Natur. Syst. Amph.: 176. (*Uropsophus triseriatus*)

Synonyms: *Crotalus lugubris* Jan, 1859a (*partim*), *Crotalus triseriata* Cope, 1867a, *Crotalus pallidus* A.C.L.G. Günther, 1895 *in* 1885–1902, *Crotalus triseriatus anahuacus* Gloyd, 1940 (*partim*), and *Crotalus triseriatus armstrongi* J.A. Campbell, 1978.
Types: Syntypes (4), ZMB 2908–11.
Type locality: "Mexico."
Distribution: South-central Mexico (Colima, Distrito Federal, S Hidalgo, cen. Jalisco, México, N Michoacán, N Morelos, cen. Puebla, Tlaxacala, W Veracruz), 2085–4570 m.
Sources: Gloyd, 1940, Harris & Simmons, 1978b, Dorcas, 1992, Reyes-Velasco et al., 2009 and Bryson et al., 2011.

C

Remarks: Official Specific Name no. 507 *fide* Opinion 366 (ICZN, 1955). Type locality restriction to Alvarez, San Luis Potosí, Mexico *fide* H.M. Smith & Taylor (1950a: 342) invalid since *C. triseriatus* does not occur there (Alvarez is the type locality of *C. aquilus*).

40. *Crotalus tzabcan* Klauber, 1952. Bull. Zool. Soc. San Diego (26): 71–73. (*Crotalus durissus tzabcan*)

Type: Holotype, FMNH 36168, a 537 mm male (E.W. Andrews, 28 Oct. 1939).

Type locality: "Kantunil, Yucatán, Mexico" [= Kantunil, cen. Yucatán, SE Mexico, 20°47'N, 89°03'W, elevation 20 m].

Distribution: Yucatan Peninsula. Southeastern Mexico (Campeche, Quintana Roo, Yucatán), NSL–175 m.

Source: Wüster et al., 2005.

41. *Crotalus vegrandis* Klauber, 1941b. Trans. San Diego Soc. Nat. Hist. 9(30): 334–335.

Type: Holotype, CM 17384, a 636 mm male (H.A. Beatty, 1939).

Type locality: "Maturin Savannah, near Uracoa, Sotillo District, State of Monagas, Venezuela" [= Maturín savanna, vicinity of Uracoa, E Monagas State, NE Venezuela, 9°00'N, 62°21'W, elevation 10 m].

Distribution: Northeastern Venezuela (Anzoátegui, E Monagas), 30–300 m.

Sources: Hoge, 1966, Roze, 1966, Lancini, 1967, 1986, Harris & Simmons, 1978, Lancini & Kornacker, 1989, La Marca, 1997, Kornacker, 1999, J.A. Campbell & Lamar, 2004, Navarrete et al., 2009 and Rivas-Fuenmayor et al., 2012.

Remarks: Photographs of holotype in Roze (1966: figs. 74–76) and Lancini (1967: fig. 1). A valid species *fide* Rivas-Fuenmayor et al. (2012: 43).

42. *Crotalus viridis* (Rafinesque-Schmaltz, 1818b). Amer. Month. Mag. Crit. Rev. 4(1): 41. (*Crotalinus viridis*)

Synonyms: *Crotalus confluentus* Say *in* E. James, 1823, *Crotalus confluentis* – Harlan, 1835 *(nomen incorrectum)*, *Crotalus lecontei* Hallowell, 1852a, *Crotalus confluentus pulverulentus* Cope, 1883d, and *Crotalus confluentus nuntius* Klauber, 1935a.

Type: Holotype, not designated, a 610 mm specimen (J. Bradbury, 25 April 1811), location unknown.

Type locality: "the Upper Missouri prairies" [USA]. Restricted to Gross, Boyd County, Nebraska, USA *fide* H.M. Smith & Taylor (1950: 358), and to Pierre, South Dakota, USA *fide* K.P. Schmidt (1953: 231).

Distribution: Extreme SW Canada (SE Alberta, SW Saskatchewan), W USA (NE Arizona, E Colorado, E cen. Idaho, ext. W Iowa, W Kansas, Montana, Nebraska, New Mexico, SW North Dakota, W Oklahoma, W South Dakota, W Texas, ext. SE Utah, Wyoming) and ext. N Mexico (ext. N Chihuahua, ext. NW Coahuila, N Sonora), 1370–3050 m.

Fossil records: Upper Miocene (Hemphillian) of USA (Nebraska), lower Pliocene (Hemphillian) of USA (Oklahoma), middle Pliocene (Blancan II) of USA (Kansas), upper Pliocene (Blancan V) of USA (Kansas), Pleistocene of USA (Iowa), middle/upper Pleistocene (Irvingtonian II) of USA (Colorado, Nebraska), and upper Pleistocene (Rancholabrean I, II) of USA (Kansas).

Sources: Klauber, 1930a, 1935a, 1936c, A.H. Wright & Wright, 1957, Fowlie, 1965, Nussbaum et al., 1983, Stebbins, 1985, Quinn, 1987, Lemos-Espinal et al., 1994, Degenhardt et al., 1996, Conant & Collins, 1998, Hammerson, 1999, Pook et al., 2000, Bartlett & Tennant, 2000, A. Russell & Bauer, 2000, Werler & Dixon, 2000, Holman, 2000a, Ashton & Queiros, 2001, Grismer 2001b, Douglas et al., 2002, Brennan & Holycross, 2006, Lemos-Espinal & Smith, 2007a–b, Holycross et al., 2008, Lewis, 2011 and Young, 2011.

Remarks: The following species separated from *Crotalus viridis (abyssus, caliginis, cerberus, concolor, helleri, lutosus, oreganus) fide* Pook et al., 2000 and Douglas et al. 2008.

43. *Crotalus willardi* Meek, 1906. Field Columb. Mus., Zool. (1905) 7(1): 18–19, pl. 3.

Synonyms: *Crotalus willardi meridionalis* Klauber, 1949, *Crotalus willardi silus* Klauber, 1949, *Crotalus willardi amabilis* J.D. Anderson, 1962, *Crotalus willardi obscurus* Harris & Simmons, 1974 (*nomen nudum*), and *Crotalus willardi obscurus* Harris & Simmons, 1976.

Type: Holotype, FMNH 902, a 415 mm female (F.C. Willard, 1896–1906).

Type locality: "Tombstone, Arizona" [USA] (in error). Corrected to above Hamburg, middle branch of Ramsey Canyon, Huachuca Mtns., Cochise Co., Arizona, USA, elevation ca. 2135 m *fide* Swarth (1921: 83).

Distribution: Extreme SW USA (ext. SE Arizona, ext. SW New Mexico) and NW Mexico (W Chihuahua, W Durango, N Sonora, ext. W Zacatecas), 1460–2750 m.

Sources: Hartman, 1911, Klauber, 1949b, Bogert & Degenhardt, 1961, J.D. Anderson, 1962, Harris & Simmons, 1975, 1976, Stebbins, 1985, Barker, 1992, Degenhardt et al., 1996 and Bartlett & Tennant, 2000.

CROTAPHOPELTIS Fitzinger, 1843 (Colubridae)

Synonyms: *Heterurus* A.-M.-C. Duméril, 1853, ? *Aspidolaeme* Bibron *in* A.-M.-C. Duméril, Bibron & Duméril, 1854, *Crotaphopheltus* – Higgins, 1873 *(nomen incorrectum)*, *Herurus* – Phisalix, 1922 *(nomen incorrectum)*, *Crotophopeltis* – Corkill, 1935 *(nomen incorrectum)*, *Cratophopeltis* – Domergue, 1955 *(nomen incorrectum)*, *Crotapeltis* – Dessauer, 1967

(nomen incorrectum), and *Crottaphopeltis* – Stucki-Stirn, 1979 *(nomen incorrectum)*.
Type species: *Coronella hotamboeia* Laurenti, 1768.
Distribution: Subsaharan Africa.
Sources: Loveridge, 1957a, Broadley, 1968b, Roman, 1974b, J.B. Rassmussen, 1985a, Gravlund, 2001 and Wallach et al., 2010.

1. *Crotaphopeltis barotseensis* Broadley, 1968b. Fieldiana: Zool. 51(10): 135, figs. 1a–b, 2.

Type: Holotype, FMNH 134249, a 544 mm female (R.G. Japp, 24 March 1962).
Type locality: "Kalabo, Barotseland, Zambia" [= Kalabo, NW Western Prov., W Zambia, 5°00'S, 22°41'E, elevation 1025 m].
Distribution: Southern Africa. Western Zambia (Western), ext. NE Namibia (Caprivi) and NW Botswana (W North West), 950–1025 m.
Sources: Auerbach, 1987, Broadley, 1983, Branch, 1988, J.B. Rasmussen, 1997a, Broadley et al., 2003, M. Griffin, 2003 and Broadley & Blaylock, 2013.

2. *Crotaphopeltis braestrupi* J.B. Rassmussen, 1985a. Steenstrupia 11(4): 114, figs. 1–2.

Type: Holotype, CAS 153373, a 480 mm male (L.G. Hoevers & P.M. Johnson, 1 Nov. 1977).
Type locality: "Lower Juba Valley: ca. Mareri, Juba Sugar Project, Somalia" [= N and S of Mareri (0°26'N, 42°43'E, elevation 20 m), Lower Juba Valley, SE Jubbada Dhexe Region, S Somalia].
Distribution: Southern Somalia (Bay, Jubbada Dhexe, Jubbada Hoose, Shabeellaha Dhexe, Shabeellaha Hoose) and SE Kenya (Coast, S North-Eastern), NSL–600 m.
Sources: Lanza, 1983a, 1990b, Largen & Rasmussen, 1993 and Spawls et al., 2002.

3. *Crotaphopeltis degeni* (Boulenger, 1906e). Proc. Zool. Soc. London 76(3–4): 572, figs. 97a–c. (*Leptodira degeni*)

Synonym: *Leptodeira attarensis* F. Werner, 1908.
Types: Syntypes (2), BMNH 1946.1.9.96–97 (formerly BMNH 1906.5.30.7–8), a male and female, longest syntype 450 mm (E.J.E. Degen, 1895–1905).
Type locality: "Entebbe, Uganda" [= Entebbe, Central Prov., S Uganda, 0°03'N, 32°28'E, elevation 1170 m].
Distribution: Central Africa. Northern Cameroon (Extreme-Nord), N Central African Republic (Vakaga), S Sudan (An Nil Al Abyad), E South Sudan (Jonglei), W Ethiopia (Illubabor), S Uganda (Central, Eastern, Western) SW Kenya (Nyanza, S Rift Valley, Western) and SW Tanzania (Iringa-Mbeya, Rukwa), 315–2700 m.
Sources: Pitman, 1974, Largen & Rasmussen, 1993, J.B. Rasmussen, 1997a, J.B. Rasmussen et al., 2000, Spawls et al., 2002, Chippaux, 2006, Chirio & LeBreton, 2007 and Largen & Spawls, 2010.

4. *Crotaphopeltis hippocrepis* (J.T. Reinhardt, 1843). Kgl. Danske Vidensk. Selsk. Natur. Math. Afh. 10: 251–252, pl. 1, figs. 18–20. (*Dipsas hippocrepis*)

Synonym: *Crotaphopeltis acarina* Roman, 1974b.
Type: Holotype, ZMUC 63127, a 240 mm female (J.R. Chenon, 1826–1836).
Type locality: "Guinea" [= coast of Ghana *fide* Hughes & Barry, 1969: 1027]. Restricted to vicinity of Accra, Ghana *fide* J.B. Rasmussen & Hughes (1997: 14, 16).
Distribution: West Africa. Western Guinea (Kindia, Télimélé), Sierra Leone, W Burkina-Faso (Centre-Est, Est, Hauts-Bassins, Sud-Ouest, Volta-Noire), Ivory Coast, Ghana, Togo (Centrale, Plateaux), N Benin (Alibori, Atakora), S Niger (Maradi), N Nigeria (Kaduna, Kano, Plateau), N Cameroon (Adamaoua, N Est, Extreme-Nord, Nord, Nord-Ouest), and Central African Republic (Bamingui-Bangoran, Gribingui, Ouaka, Ouham), 320–1475 m.
Sources: Roman, 1980, J.B. Rasmussen & Hughes, 1997, J.B. Rasmussen et al., 2000, Chippaux, 2006, J.-F. Trape & Mané, 2006b, Chirio & Ineich, 2006, Chirio & LeBreton, 2007, Segniagbeto et al., 2011, Chirio, 2013 and Hughes, 2013.
Remarks: A valid species *fide* J.B. Rassmussen, 1985a.

5. *Crotaphopeltis hotamboeia* (Laurenti, 1768). Synop. Rept.: 85. (*Coronella hotamboeia*)

Synonyms: *Coronella virginica* Laurenti, 1768, *Coluber rufescens* Gmelin, 1789, *Coluber hotamboeja* Bonnaterre, 1790 *(nomen emendatum)*, *Coluber bicolor* Leach *in* Bowdich, 1819, *Ophis albocinctus* Duvernoy, 1832, *Ophis heterurus* Duvernoy, 1832, *Dipsas inornatus* A. Smith, 1849, *Oxyrhopus melanocrotaphus* Cope, 1860e, *Crotaphopeltis hotamboeia kageleri* Uthmöller, 1939, *Tarbophis barnumbrowni* Bogert, 1940, and *Crotaphopeltis hotamboeia ruziziensis* Laurent, 1956a.
Type: Holotype, not designated, location unknown.
Type locality: "India orientali" [= Eastern India] (in error). Corrected to Africa *fide* Loveridge (1957a: 271).
Distribution: Subsaharan Africa. Southwestern Mauritania (Trarza), Senegal (Fatick, Kédougou, Tambacounda), Gambia (MacCarthy Island, Western), Guinea–Bissau (Bafaté, Biombo, Bissau, Cacheu), Guinea (Kissidougou, Kouroussa, Macenta, Télimélé), Sierra Leone (Eastern, Northern, Southern), Liberia, Ivory Coast (Abidjan, Bouaflé, Bouaké, Bouna, Daloa, Guiglo, Lagunes, Tabou, Toumodi), SW Mali (Kayes, Koulikoro, Sikasso, SW Tombouctou), Burkina Faso (Centre, Centre-Est, Centre-Nord, Centre-Ouest, Est, Haute-Bassins, Nord, Sahel, Sud-Ouest, Volta-Noire), N Ghana (Upper East, Upper West), Togo (Kara, Maritime, Plateaux), Benin (Alibori, Atakora, Atlantique, Borgou, Collines), Niger (Tahoua), Nigeria

C

(Cross River, Delta), S Chad (Chari-Baguirmi, Mayo-Kebbi), Cameroon (Adamaoua, Centre, Est, Extreme-Nord, Nord, Nord-Ouest, Ouest, Sud-Ouest), Central African Republic (Bamingui-Bangoran, Kemo-Gribingui, Haut-Kotto, Haut-Mbomou, Haute-Sangha, Mbomou, Ombella Mpoko, Ouaka, Ouham, Ouham-Pende, Sangha, Vakaga), Gabon (Haut-Ogooué, Ngounié, Ogooué-Ivindo, Ogooué-Maritime), Congo (Bouenza, Brazzaville, Cuvette-Ouest, Kouilou, Lekoumou, Pool), Democratic Republic of the Congo (Bandundu, Bas-Congo, Equateur, Kasai Occidental, Kasai Oriental, Katanga, Kinshasa, Maniema, Nord-Kivu, Orientale, Sud-Kivu), NE Angola (Lunda Sul, Moxico), SE Sudan, Eritrea, Ethiopia (Gemu Gofa, Gojjam, Gondar, Hararge, Illubabor, Kefa, Shoa, Sidamo, Wollega), S Somalia (Bakool, Hiraan, Jubbada Hoose, Mogadishu, Shabeellaha Dhexe, Shabeellaha Hoose, Alessandra Is.), Kenya (Coast, S Eastern, North-Eastern, Rift Valley), Angola (Cuando-Cubango), Uganda (Central, Eastern, Northern, Western, Kome and Sese Is.), Rwanda, Burundi, Tanzania (Arusha, Iringa, Kilimanjaro, Lindi, Mara, Mbeya, Morogoro, Mtwara, Mwanza, Pwani, Ruvuma, Tanga, Kagera, Zanzibar Is.), Zambia (Central, Copperbelt, Eastern, Luapula, Lusaka, Northern, North-Western, Southern, Western), Malawi (Central, Northern, Southern), Mozambique (Inhambane, Manica, Maputo, Niassa, Sofala, Tete, Bazaruto and Inhaca Is.), Zimbabwe (Bulawayo, Manicaland, Mashonaland Central, Mashonaland East, Mashonaland West, Masvingo, Matabeleland North, Matabeleland South, Midlands), E Botswana (Gabornoe, Kweneng, North West, South East), N Namibia (Caprivi, Okavango, Otjozondjupa), Swaziland, Lesotho and South Africa (Eastern Cape, Free State, Gauteng, KwaZulu-Natal, Limpopo, Mpumalanga, Northern Cape, North West, Western Cape), NSL–2500 m.

Sources: Mertens, 1955, Doucet, 1963, Menzies, 1966, Leston & Hughes, 1968, Hughes & Barry, 1969, Pitman, 1974, Roman, 1980, Hakansson, 1981, Broadley, 1983, Hughes, 1983, 2013, Lanza, 1983a, 1990b, Pienaar et al., 1983, Auerbach, 1987, Branch, 1988, Branch & McCartney, 1993, Largen & Rasmussen, 1993, J.-F. Trape & Roux-Estève, 1995, Keogh et al., 2000, J.B. Rasmussen et al., 2000, Clauss & Clauss, 2002, Spawls et al., 2002, Broadley et al., 2003, J.-F. Trape & Mané, 2004, 2006b, Branch et al., 2005, Chippaux, 2006, Chirio & Ineich, 2006, Chirio & LeBreton, 2007, Pauwels & Vande weghe, 2008, Largen & Spawls, 2010, Ullenburch et al., 2010, and Segniagbeto et al., 2011, Auliya et al., 2012, Broadley & Blaylock, 2013 and Chirio, 2013.

Remarks: Broadley (1983: 245) suggested that *C. kageleri* Uthmöller, is probably a valid species.

6. *Crotaphopeltis tornieri* (F. Werner, 1908). Sitz. Akad. Wiss. Wien, Math.-Natur. Kl. (1907) 116(1): 1876. (*Leptodira tornieri*)

Type: Holotype, ZMH 652 (formerly ZMH 4576), a juvenile male (A. Müller).

Type locality: "Usambara (Deutsch-Ostafrika)" [= Usambara Mountains, Tanga Prov., NE Tanzania, ca. 4°45'N, 38°30'E].

Distribution: Southeastern Tanzania (Iringa, Lindi, Morogoro, Mtwara, Rukwa, Tanga) and N Malawi (Northern), 610–1220 m.

Sources: Broadley, 1968a, J.B. Rasmussen, 1993a and Spawls et al., 2002.

Remarks: Resurrected by Broadley, 1968b.

CRYOPHIS Bogert & Duellman, 1963 (Dipsadidae)

Type species: *Cryophis hallbergi* Bogert & Duellman, 1963.

Distribution: Southern Mexico.

Sources: Bogert & Duellman, 1963, Cadle, 1984b, Villa et al., 1988, Casas-Andreu et al., 1996, Zaher et al., 2009 and Mulcahy et al., 2011.

1. *Cryophis hallbergi* Bogert & Duellman, 1963. Amer. Mus. Novit. (2162): 4–9, figs. 1, 4–6.

Type: Holotype, KU 70901, a 650 mm male (W.E. Duellman, 27 June 1962).

Type locality: "6 kilometers south of Campamento Vista Hermosa, at an elevation of 1865 meters, at the northern periphery of the Sierra de Juárez, Distrito de Ixtlán, Oaxaca, Mexico...near latitude 17°50' N., longitude 96° 20' W., on the drainage of the Río San Cristobal, to the southwest of Valle Nacional."

Distribution: Southern Mexico (ext. N Oaxaca), 1150–1865 m.

CRYPTOPHIDION Wallach & Jones, 1994 (Colubridae)

Type species: *Cryptophidion annamense* Wallach & Jones, 1994.

Distribution: Central Vietnam.

Sources: Pauwels & Meirte, 1997 and Wallach & Jones, 1997.

1. *Cryptophidion annamense* Wallach & Jones, 1994. Cryptozoology (1992) 11: 5–12, figs. 1–3.

Type: Iconotypes (3) of holotype, MCZ-SC K948–50 (C.E. Snyder, 1968).

Type locality: "West of Da Nang, Quang Nam-Da Nang Prov., central Vietnam, ca. 16°N, 108°E, presumably in the Annam Highlands."
Distribution: Central Vietnam (Quang Nam-Da Nang). Known only from type locality.
Remarks: Holotype lost *fide* Wallach & Jones (1997: 5). A synonym of *Xenopeltis unicolor fide* Pauwels & Meirte (1997: 99–100) and Iskandar & Colijn (2001: 32).

CUBOPHIS Hedges & Vidal *in* Hedges, Couloux & Vidal, 2009 (Xenodontidae)

Type species: *Coluber cantherigerus* Bibron, 1840 *in* Ramón de la Sagra, 1838–1843.
Distribution: Cuba, Bahamas and Cayman Islands.
Sources: Hedges et al., 2009, R.W. Henderson & Powell, 2009 and Zaher et al., 2009.

1. *Cubophis cantherigerus* (Bibron, 1840 *in* Ramón de la Sagra, 1838–1843). Hist. Fís. Polít. Nat. Cuba 2(2): pl. 27, figs. 1–5. (*Coluber cantherigerus*)

Synonyms: *Dromicus angulifer* Bibron, 1843 *in* Ramón de la Sagra, 1838–1843, *Dromicus unicolor* A.-M.-C. Duméril, Bibron & Duméril, 1854a, *Dromicus angulifer adspersus* Gundlach & Peters *in* W.C.H. Peters, 1864b, *Alsophis brooksi* Barbour, 1914a, *Alsophis cantherigerus spielmani* Grant, 1959, *Alsophis cantherigerus pepei* A. Schwartz & Thomas, 1960, and *Alsophis cantherigerus schwartzi* Lando & Williams, 1969.
Types: Syntypes (6), MNHN 3545–46, MNHN 3561–63, and MCZ 2195 (formerly MNHN), longest syntype 1250 mm (D. Ramón de la Sagra, 1824–1835).
Type locality: "Cuba."
Distribution: Cuba (Camagüey, Cienfuegos, Ciego de Aviula, Granma, Guantánamo, Habana, Holguín, Matanzas, Pinar del Río, Sancti Spíritus, Santiago de Cuba, Villa Clara, Juventud, Little Swan, Los Canarreos Arch., Sabana-Camagüey Arch., San Felipe and Swan Is.), NSL–300 m.
Sources: H.M. Smith & Grant, 1958c, A. Schwartz & Thomas, 1960, Lando & Williams, 1969, Novo-Rodriguez & Arazoza-Rodriguez, 1986 and Seidel & Franz, 1994.
Remarks: Supplemental original description in Bibron, 1843 *in* Ramón de la Sagra (1838–1843: 133–134). Original description reprinted in Bibron, 1843 *in* Ramón de la Sagra (1841–1843: 222–224, pl. 27, figs. 1–5).

2. *Cubophis caymanus* (S.W. Garman, 1887a). Proc. Amer. Philos. Soc. 24(126): 276–277. (*Alsophis caymanus*)

Types: Syntypes (5), MCZ 6020a–e, adult, halfgrown and juvenile specimens (W.B. Richardson, 31 Aug. 1886).
Type locality: "Grand Cayman Island" [Cayman Islands, West Indies].
Distribution: Cayman Is. (Cayman Brac, Grand Cayman, Little Cayman Is).
Sources: Grant, 1940c and Hedges et al., 2009.
Remarks: A valid species *fide* Hedges et al. (2009: 9).

C

3. *Cubophis fuscicaudus* (S.W. Garman, 1888). Bull. Essex Inst. 20: 106–107. (*Alsophis fuscicauda*)

Type: Holotype, MCZ 6235, an adult (C.J. Maynard, 1888).
Type locality: "Cayman Brac, Caymans" [West Indies].
Distribution: Cayman Is. (Cayman Brac).
Remarks: A valid species *fide* Hedges et al., 2009.

4. *Cubophis ruttyi* (Grant, 1940c). Bull. Inst. Jamaica (Sci.) (1): 50–51, figs. 5–6. (*Alsophis angulifer ruttyi*)

Type: Holotype, MCZ 44876 (formerly CG 2656), an 840+ mm female (C. Grant, 9 April 1938).
Type locality: "Little Cayman" [Island, Cayman Islands, West Indies].
Distribution: Cayman Is. (Little Cayman Is.).
Remarks: A valid species *fide* Hedges et al. (2009: 9).

5. *Cubophis vudii* (Cope, 1862b). Proc. Acad. Nat. Sci. Philadelphia 14(1): 74. (*Alsophis vudii*)

Synonyms: *Diadophis rubeseans* Cope, 1885b, *Alsophis vudii aterrimus* T. Barbour & Shreve, 1935, *Alsophis vudii raineyi* T. Barbour & Shreve, 1935, *Alsophis vudii utowanae* T. Barbour & Shreve, 1935, and *Alsophis vudii picticeps* Conant, 1937a.
Types: Syntypes (6), ANSP 5567 and ANSP 5569–71 (H.C. Wood, Jr.), and ANSP 5598–99 (H. Bryant, 1859), longest syntype 991 mm.
Type locality: "New Providence Id., Bahamas."
Distribution: Bahama Bank. Bahama Is. (Acklins, Andros, Berry, Binimi, Cat, Crooked, Eleuthera, S Exumas, Grand Bahama, Great Inagua, Little San Salvador, Long, New Providence, S Ragged, Sheep Cay).

CYCLOCORUS A.-M.-C. Duméril, 1853 (Colubroidea *incertae sedis*)

Synonyms: *Cyclochorus* Jan, 1857 *(nomen emendatum)* and *Cyclorus* – Obst, Richter & Jacob, 1984 *(nomen incorrectum)*.
Type species: *Lycodon lineatus* J.T. Reinhardt, 1843.
Distribution: Philippines.
Sources: Leviton, 1967, Alcala, 1986 and Zaher et al., 2009.

C

1. *Cyclocorus lineatus* (J.T. Reinhardt, 1843). Kgl. Danske Vidensk. Selsk. Natur. Math. Afh. 10: 241–244, pl. 1, figs. 7–9. (*Lycodon lineatus*)

Synonyms: *Cyclochorus maculatus* Jan, 1870 *in* Jan & Sordelli, and *Cyclocorus lineatus alcalai* Leviton, 1967.
Type: Holotype, ZMUC 60489, a 318 mm female.
Type locality: "in insula Manillae" [Manila, SW Luzón Is., N Philippines, 14°36'N, 120°59'E, elevation 15 m].
Distribution: Philippines (Catanduanes, Cebu, Guimaras, Inampulugan, Lubang, Luzón, Masbate, Mindoro, Negros, Pacijan, Panay, Polillo, Romblon, Tablas, Zambales), NSL–1525 m.
Sources: Ross & Gonzales, 1992, J.B. Rasmussen & Hughes, 1997 and Ferner et al., 2001.

2. *Cyclocorus nuchalis* E.H. Taylor, 1923. Philippine J. Sci. (1922) 22D(5): 543–545, pl. 3, figs. 1–2.

Synonym: *Cyclocorus nuchalis taylori* Leviton, 1967.
Type: Holotype, CAS 62558 (formerly EHT 1428), a 444 mm male (E.H. Taylor, 30 Sept. 1920).
Type locality: "Pasananka, Zamboanga, Philippine Islands" [= Pasonanca, ext. W Mindanao Is., S Philippines, ca. 6°56'N, 122°04'E, elevation 10 m].
Distribution: Southern Philippines (Basilan, Camiguin, Leyte, Mindanao, Samar, Siargao), NSL–760 m.

CYCLOPHIOPS Boulenger, 1888h (Colubridae)

Synonyms: *Eurypholis* Hallowell, 1861 *(nomen praeoccupatum), Entechinus* Cope, 1895a *Entacanthus* Cope, 1900 *(nomen emendatum), Entracanthus* – Maki, 1931 *(nomen incorrectum),* and *Pheodrys* – Yao, 2012 *(nomen incorrectum).*
Type species: *Cyclophiops doriae* Boulenger, 1888h.
Distribution: Southeastern Asia.
Sources: Bourret, 1936b, M.A. Smith, 1943, Cundall, 1981 and Zhao & Adler, 1993.

1. *Cyclophiops doriae* Boulenger, 1888h. Ann. Mus. Civ. Stor. Nat. Genova (2) 6: 599, pl. 6, figs. 1–1b.

Type: Lectotype, MSNG 30384A (L. Fea, 1886), designated by Capocaccia (1961a: 105).
Type locality: "Monti ad Est di Bhamò: Kakhien Hills (Birmania)" [= mountains to the east of Bhamò (24°15'N, 97°14'E), ext. SE Kachin Division, Myanmar] via lectotype selection.
Distribution: Southern Asia. Northeastern India, N Myanmar (Kachin) and ext. SW China (SE Yunnan), 100–1375 m.
Sources: Wall, 1925b, Murthy, 1985, 1986 and R.C. Sharma, 2003, 2007.

2. *Cyclophiops hamptoni* (Boulenger, 1900d). Ann. Mag. Nat. Hist. (7) 6(34): 409. (*Ablabes hamptoni*)

Type: Holotype, BMNH 1946.1.5.32 (formerly BMNH 1900.9.20.15), a 1050 mm female (H. Hampton, 1898–1900).
Type locality: "Magok, on the Irawaddi, about 12 miles north of Mandalay, Burma" [= Mogok, N Mandalay Div., cen. Myanmar, 22°55'N, 96°32'E, elevation 1125 m]. Restricted to Ruby Mines, Mogok, Myanmar *fide* Wall (1924a: 865).
Distribution: Central Myanmar (Mandalay), 1125 m. Known only from type locality.
Remarks: A synonym of *E. doriae* (Boulenger) *fide* Mahendra (1984: 271).

3. *Cyclophiops herminae* (Boettger, 1895c). Zool. Anz. 18(479): 269–270. (*Ablabes herminae*)

Type: Holotype, SMF 19308 (formerly SMF-B 8277a), a 580 mm male (B. Schmacker, 1895).
Type locality: " Liukiu-Inseln, angeblich von Yaeyama auf Mijakoshima, Südgruppe" [= Miyako-jima Is., Yaeyama group, Okinawa Prefecture, S Ryukyu Archipelago, 24°47'N, 125°20'E].
Distribution: Extreme S Japan (Ryukyus: Aragusukujima, Haterumajima, Hatomajima, Irabujima, Ikemajima, Iriomotejima, Ishigakishima, Kayamajima, Kohamajima, Kurimajima, Kuroshima, Miyakojima, Nakanogamijima, Ogamijima, Shimojijima, Taketomijima, Taramajima and Yonagunijima Is.).
Sources: Stejneger, 1907a, Maki, 1931, Nakamura & Uéno, 1963 and M. Mori, 1986.
Remarks: Original description reprinted in Boettger (1895e: 110–111).

4. *Cyclophiops major* (A.C.L.G. Günther, 1858). Cat. Colub. Snakes Brit. Mus.: 120. (*Cyclophis major*)

Synonyms: *Herpetodryas chloris* Hallowell, 1861, *Coluber delacouri* M.A. Smith, 1930b, and *Liopeltis major bicarinata* Maki, 1931.
Type: Holotype, BMNH 1946.1.1.68, a 775 mm female (H. Cuming, 1836–1840).
Type locality: "Near Ningpo, China" [= vicinity of Ningbo, NE Zhejiang Prov., E China, 29°51'N, 121°33'E, elevation 10 m].
Distribution: Southeastern Asia. Southeastern China (Anhui, Fujian, Gansu, Guangdong, Guizhou, Guangxi, Hainan, Henan, Hong Kong, Hubei, Hunan, Jiangsi, Jiangsu, Shaanxi, Sichuan, Yunnan, Zhejiang, Zhoushan Is.), Taiwan and N Vietnam (Cao Bang, Ha Tay, Lao Cai, Quang Binh, Vinh Phu), 50–1500 m.
Sources: Stejneger, 1907a, Maki, 1931, Bourret, 1934e, C.H. Pope, 1935, Kuntz, 1963, Mao, 1964, Romer, 1979a, Orlov et al., 2000, Ziegler et al., 2006, V.S. Nguyen et al., 2009, Xiang & Li, 2009, I. Das, 2010, 2012 and Yao, 2012.

5. *Cyclophiops multicinctus* (Roux, 1907b). Zool. Anz. 31(24): 762–763. (*Ablabes multicinctus*)

Synonyms: *Ablabes retrofasciatus* Angel, 1920c, *Zamenis moi* M.A. Smith, 1921a, *Ablabes multicinctus bicolor* Angel, 1929, and *Liopeltis sinii* Fan, 1931.
Type: Holotype, NMBA 1 (formerly MHNN 70), a 1060 mm specimen (Klauser).
Type locality: "Tonkin" [= N Vietnam]. Restricted to Red River, N Vietnam *fide* Schätti (1986a:100).
Distribution: Southeast Asia. Southern China (Guangxi, Hainan, Hunan, S Yunnan), N Laos (Xiangkhouang) and Vietnam (Bac Kan, Cao Bang, Da Nang, Hai Duong, Ha Tay, Ha Tinh, Hoa Binh, Koa Tum, Lai Châu, Lao Cai, Ninh Binh, Quang Binh, Quang Nam, Quang Ninh, Quang Tri, Thai Nguyen, Thua Thien-Hue, Vinh Phú), 100–1135 m.
Sources: Bourret, 1934e, C.H. Pope, 1935, M.A. Smith, 1943, Campden-Main, 1970a, Deuve, 1970, Orlov et al., 2000, Zhao, 2006, V.S. Nguyen et al., 2009 and I. Das, 2010, 2012.

6. *Cyclophiops semicarinatus* (Hallowell, 1861). Proc. Acad. Nat. Sci. Philadelphia (1860) 12(11): 493–494. (*Eurypholis semicarinatus*)

Synonyms: *Cyclophis nebulosus* A.C.L.G. Günther, 1868 and *Liopeltis semicarinata fritzei* Maki, 1931.
Type: Lectotype, USNM 7316, a 502 mm specimen (C. Wright, Nov. 1854), designated by Maki (1931a: 112).
Type locality: "Loo Choo, near Napa" [= vicinity of Naha, SW Okinawa-jima, Ryukyu Arch., ext. S Japan] via lectotype selection.
Distribution: Extreme S Japan (Ryukyus: Agunijima, Akajima, Amamioshima, Amurojima, Fukajijima, Gishifujima, Gushikawajima, Gusukujima, Hamahigajima, Hatejima, Haterumajima, Hyanzajima, Iejima, Iheyajima, Ikeijima, Ikemajima, Irabujima, Iriomotejima, Ishigakijima, Izenajima, Kakeromajima, Kerumajima, Kikaijima, Kohamajima, Korijima, Kubajima, Kudakajima, Kumejima, Kurimajima, Kuroshima, Maejima, Minnjima, Miyagijima, Miyakojima, Nakajima, Nakanogamijima, Nohojima, Ogamijima, Ohajima, Ojima, Okierabujima, Okinawajima, Sesokojima, Shimojishima, Taketomijima, Taramajima, Tokashikijima, Tokunoshima, Tonakijima, Tsukenjima, Ukejima, Ukibarujima, Yabuchijima, Yaguchijima, Yakabijima, Yanahajima, Yonagunijima, Yorojima, Yoronjima and Zamamijima Is.).
Sources: Stejneger, 1907a, Maki, 1931, M. Mori, 1982, Toriba, 1991 and Ota et al., 1995.
Remarks: Hallowell's 536 mm length for the type appears to be the mean for both specimens *fide* Stejneger (1907a: 340).

CYCLOTYPHLOPS Bosch & Ineich, 1994 (Typhlopidae)

Type species: *Cyclotyphlops deharvengi* Bosch & Ineich, 1994.
Distribution: Indonesia (Sulawesi).
Sources: Bosch & Ineich, 1994, McDiarmid et al., 1999, Lang & Vogel, 2005 and Koch, 2012.

1. *Cyclotyphlops deharvengi* Bosch & Ineich, 1994. J. Herp. 28(2): 208–209, figs. 4–8.

Type: Holotype, MNHN 1990.4279, a 146 mm specimen (L. Deharveng ['Maros 88' Exped.], 12 July 1988).
Type locality: "Malawa, between Maros and Bone-Watampone, Selatan Prov., southeastern Sulawesi, Indonesia, at an elevation of approximately 500 m."
Distribution: Indonesia (Sulawesi), 500 m. Known only from type locality.

CYLINDROPHIS Wagler, 1828 (Cylindrophiidae)

Synonyms: *Cijlindrophis* Bleeker, 1856 *(nomen emendatum), Cylindrophus* – M.A. Smith, 1914b *(nomen incorrectum), Cilyndrophis* – Briceño-Rossi, 1934 *(nomen incorrectum),* and *Culindrophis* – McDowell, 1975 *(nomen incorrectum).*
Type species: *Anguis ruffa* Laurenti, 1768.
Distribution: Southeastern Asia and Indonesia.
Sources: McDowell, 1975a, Bosch, 1985, Welch, 1988, Cundall et al., 1994, Stuebing, 1994b, L.A. Smith & Sidik, 1998, McDiarmid et al., 1999, Iskandar & Colijn, 2001, Gower et al., 2005 and Gómez et al., 2008.

1. *Cylindrophis aruensis* Boulenger, 1920b. Ann. Mag. Nat. Hist. (9) 6(31): 108.

Types: Syntypes (2), BMNH 1946.1.16.72–73 (formerly BMNH 1900.2.9.16–17), a 169 mm and 167 mm male (F.H. Rolle, 1864–1900).
Type locality: "Aru Island" [E Maluku Prov., E Indonesia].
Distribution: Eastern Indonesia (Maluku: Aru Arch.).
Source: McDowell, 1975a.
Remarks: Record from Aru Islands doubtful and type locality in error *fide* Iskandar & Colijn (2002: 23).

2. *Cylindrophis boulengeri* Roux, 1911. Zool. Jahrb. (Syst.) 30(5): 500.

Type: Holotype, SMF 16996, a 550 mm specimen (J. Elbert, 1909).
Type locality: "Wetar, Iliwaki" [= Iliwaki, Wetar Is., Maluku Prov., SE Indonesia, 7°48'S, 126°18'E].
Distribution: Southeastern Indonesia (Maluku: Baber, Timor, Wetar).

C

C

3. *Cylindrophis engkariensis* Stuebing, 1994b. Raffles Bull. Zool. 42(4): 969–972, figs. 1–2.

Type: Holotype, ZRC (formerly RBS 8821), a 485 mm male (R.B. Stuebing, 12 Aug. 1993).

Type locality: "Nanga Segerak (1°25'N 122°00'E), 245 m asl, headwaters of the Engkari River, Lanjak-Entimau Wildlife Sanctuary, Lubok Antu District, Second Division, Sarawak, Malaysia."

Distribution: East Malaysia (Sarawak), 245 m.

Sources: I. Das, 2010, 2012.

4. *Cylindrophis isolepis* Boulenger, 1896d. Ann. Mag. Nat. Hist. (6) 18(103): 62.

Type: Holotype, BMNH 1946.1.1.47 (formerly BMNH 1896.4.29.45), a 430 mm specimen (A.H. Everett, 1896).

Type locality: "Jampea Island" [= Tanahjampea Is., Selayar Arch., South Sulawesi Prov., S Indonesia, ca. 7°05'S, 120°41'E].

Distribution: Southern Indonesia (S Sulawesi: Tanahjampea Is.).

Sources: Lang & Vogel, 2005 and Koch, 2012.

Remarks: Photographs of the holotype in Lang & Vogel (2005: figs. 128–129).

5. *Cylindrophis lineatus* Blanford, 1881. Proc. Zool. Soc.. London 49(1): 217–218, pl. 20.

Type: Holotype, BMNH 1946.1.16.5 (formerly RMBR & BMNH 1903.6.2.3), a 635 mm specimen (N.B. Dennys, 1878–1881).

Type locality: "island of Singapore" (in error *fide* Stuebing 1991: 327). Corrected to Borneo *fide* Smedley (1932b: 11).

Distribution: East Malaysia (W Sarawak) and Indonesia (Kalimantan)

Sources: C. Haas, 1950, Stuebing, 1991 and I. Das, 2010, 2012.

Remarks: BMNH catalogue lists Singapore as type locality *fide* McDiarmid et al. (1999: 153).

6. *Cylindrophis maculatus* (Linnaeus, 1758). Syst. Nat., ed. 10, 1: 228. (*Anguis maculata*)

Synonyms: *Anguis decussata* Laurenti, 1768, *Anguis hepatica* Laurenti, 1768, *Anguis tessellata* Laurenti, 1768, and *Cylindrophis curticeps* Jerdon, 1854.

Types: Syntypes (3), NHR Lin-14a–c (formerly MAFR), a 420 mm, 325 mm and 270 mm specimen (Mus. Drottn.).

Type locality: "America" (in error). Corrected to Ceylon [= Sri Lanka] *fide* Deraniyagala (1955: 5).

Distribution: Sri Lanka (Central, North-Western, Sabaragamuwa, Southern, Uva, Western), NSL–1200 m.

Sources: Wall, 1921g, M.A. Smith, 1943, Deraniyagala, 1955, P. Silva, 1980 and Bachman, 1985.

Remarks: Considered a subspecies of *C. ruffus* (Laurenti) *fide* Mahendra 1936, 1984.

7. *Cylindrophis melanotus* Wagler, 1828 *in* 1828–1833. Descr. Icon. Amph.: pl. 5, 2 pp. *(nomen corrigendum)*

Synonyms: *Cylindrophe melanoto* Wagler, 1828 *in* 1828–1833 *(nomen incorrigendum), Cylindrophis melanotus* Wagler, 1830 *(nomen corrigendum), Cylindrophis melanota* A.-M.-C. Duméril & Bibron, 1844, *Tortrix rufa celebica* Schlegel, 1844, *Tortrix rufa celebensis* Gray, 1849 *(nomen emendatum), Cylindrophis melanota* Bleeker, 1856 *(nomen emendatum), Cylindrophis celebensis* M.A. Smith, 1927, and *Cylindrophis heinrichi* Ahl, 1933b.

Type: Holotype, RMNH 19 (formerly MLB) (A.J. van Delden, 1825–1828).

Type locality: Unknown. Designated as Célèbes [= Sulawesi, Indonesia] *fide* A.-M.-C. Duméril & Bibron (1844: 595) and RMNH catalogue.

Distribution: Eastern Indonesia (Bacan, Buton, Halmahera, Kepulauan, Sangihe, Sula, Sulawesi, Tabukan), NSL–1200 m.

Sources: Malkmus, 1993a, Lang & Vogel, 2005 and Koch, 2012.

Remarks: Possible types include RMNH 17 (Celebes, E.A. Forsten) and RMNH 18 (Manado, E.A. Forsten) but Forsten collected in East Indies from ca. 1838–1843. A.H.A. Duméril & Bibron (1844: 595) listed collectors of type as J.R.C. Quoy & J.P. Gaimard.

8. *Cylindrophis opisthorhodus* Boulenger, 1897d. Ann. Mag. Nat. Hist. (6) 19(113): 506.

Types: Syntypes (2), BMNH 1946.1.16.48–49 (formerly BMNH 1897.6.21.39–40), longest syntype 230 mm (A.H. Everett, 1895–1896).

Type locality: "Lombok...at an altitude of 1500 feet" [= Lombok Is., Lesser Sundas, S Indonesia, elevation 455 m] .

Distribution: Lesser Sundas of S Indonesia. (Flores, Komodo, Lombok, Sumbawa), NSL–455 m.

Sources: T. Barbour, 1912b, Rooij, 1917, F. Werner, 1922b, Mertens, 1930 and Auffenberg, 1980.

9. *Cylindrophis ruffus* (Laurenti, 1768). Synop. Rept.: 71. (*Anguis ruffa*)

Synonyms: *Anguis rufus* Gmelin, 1789 *(nomen emendatum), Anguis striatus* Gmelin, 1789, *Anguis scytale* P. Russell, 1801 *(nomen praeoccupatum), Anguis rufa* G. Shaw, 1802 *(nomen emendatum), Scytale scheuchzeri* Merrem, 1820, *Eryx binotata* Wagler, 1825 (*nomen ineditum*), *Eryx kuhlii* Wagler, 1825 (*nomen ineditum*), *Eryx shilay* Wagler, 1825 (*nomen ineditum*), *Cylindrophis resplendens* Wagler, 1828a, *Anguis rufa javanica* Gray, 1849a, and *Cylindrophis rufus burmanus* M.A. Smith, 1943.

Type: Holotype, formerly MG, lost *fide* Iskandar & Colijn (2002: 24).

Type locality: "Surinamei" [= Suriname] (in error). Corrected to Java, Indonesia *fide* Schlegel (1837: 11).

Distribution: Southeastern Asia. Southern China (Fujian, Hainan, Hong Kong), S Myanmar (Tanintharyi), Thailand (Chon Buri, Krabi, Nakhon Sawan, Nakhon Si Thammarat, Phang Nga, Phetchabun, Prachin Buri, Roi Et, Sakhon Nakhon, Saraburi, Sisaket, Sukhothai, Tak), Cambodia, Laos (Champasak), Vietnam (An Giang, Ba Ria-Vung Tau, Ca Mau, Can Tho, Da Nang, Dong Nai, Ho Chi Minh City, Kien Giang, Quang Binh, Quang Nam, Quang Tri, Tay Ninh, Thua Thien-Hue, Vinh Phuc), West Malaysia (Pinang), Singapore, East Malaysia (Sabah, Sarawak), Brunei and Indonesia (Bangka, Batjan, Belitung, Buton, Java, Kalimantan, Kepulanan, Riau Arch., Sulawesi, Sangihe Arch., Sumatra), NSL–1675 m.

Sources: M.A. Smith, 1943, Bergman, 1953b, E.H. Taylor, 1965, Tweedie, 1983, Mahendra, 1984, Adler et al., 1992, Toriba, 1992c, David & Vogel, 1996, M.J. Cox et al., 1998, Chan-ard et al., 1999, Orlov et al., 2000, Malkmus et al., 2002, Lang & Vogel, 2005, Ziegler & Phong, 2005, V.S. Nguyen et al., 2009 and I. Das, 2010, 2012.

10. *Cylindrophis yamdena* L.A. Smith & Sidik, 1998. Raffles Bull. Zool. 46(2): 421–423, figs. 2a–b, 3–4.

Type: Holotype, WAM 112252 (to be deposited in MZB), a 628 mm male (R.E. Johnstone, D.J. Kitchener, R.A. Howe; 23 April 1993).

Type locality: "Latdalam, Yamdena Island in 7°59'S, 131°09'E, Tamimbur Archipelago, Indonesia."

Distribution: Southeastern Indonesia (Maluku: Tanimbar Arch.: Yamdena).

D

DABOIA Gray, 1842e (Viperidae)

Synonyms: *Daboia* Gray, 1840 (*nomen nudum*), *Chersophis* Fitzinger, 1843, and *Daboya* – S. Hatori, 1913 *(nomen incorrectum).*

Type species: *Coluber russelii* G. Shaw & Nodder, 1797 *in* 1789–1813.

Distribution: Southern Asia and North Africa.

Fossil records: Middle Pliocene of Spain. Possible records from the lower-middle Miocene of France *fide* Szyndlar & Rage, 1999.

Sources: Szyndlar, 1986, Brodmann, 1987, Golay et al., 1993, McDiarmid et al., 1999, Szyndlar & Rage, 1999, Lenk et al., 2001b, Mallow et al., 2003, Wüster et al., 2008, Stümpel & Joger, 2009, Phelps, 2010 and Hoser, 2012d.

Remarks: Includes *Daboia russelii* and *D, palaestinae* *fide* Groombridge, 1980, 1986, and *D. mauritanica* and *D. deserti fide* Lenk et al., 2001b.

1. *Daboia deserti* (J. Anderson, 1892). Proc. Zool. Soc. London 60(1): 20–23, pl. 1, figs. 6–7. (*Vipera lebetina deserti*)

Type: Holotype, BMNH 1946.1.18.28 (formerly BMNH 1891.5.4.150), a 707 mm female (native, May–June, 1890).

Type locality: "Duirat, on the confines of Tripoli, South Tunisia" [= Douirat, Ghomrassen Governate, Tunisia, 32°59'N, 10°15'E, elevation 400 m].

Distribution: Tunisia (Ariana, Bizerte, Kairouan, Sousse, Tataouine) and NW Libya (Gharyan), NSL–400 m.

Sources: Billing & Schätti, 1984, Nilson & Andrén, 1988b, Herrmann et al., 1992, Spawls & Branch, 1995, Bons & Geniez, 1996, Schleich et al., 1996, Herrmann & Joger, 1997 and Bogaerts, 1998.

Remarks: A valid species *fide* Brodmann, 1987 and Herrmann et al., 1992. Possibly a subspecies of *D. mauritanica fide* U. Joger *in* McDiarmid et al. (1999: 383). Coastal records from Tunisia may belong to this species *fide* Spawls & Branch (1995: 138). Possibly occurs in NE Algeria *fide* Spawls & Branch (1995: 139).

2. †*Daboia maghrebiana* (Rage, 1976). Géol. Méditerr. 3(2): 64–65, figs. 7–8, 1 pl., fig. 14. (†*Vipera maghrebiana*)

Type: Holotype, MNHN BML 905, one right maxilla.

Type locality: "Beni Mellal, Maroc, Miocène" [= Beni Mellal, Morocco, Miocene].

Distribution: Middle Miocene (Astaracian, MN 7/8: 11.1–12.8 mya) of Morocco. Known only from type locality.

Sources: Rage, 1984b, Nilson & Andrén, 1997 and Szyndlar & Rage, 1999.

Remarks: Closely related to *Macrovipera lebetina fide* Rage (1984b: 56).

3. *Daboia mauritanica* (Gray, 1849). Cat. Snakes Brit. Mus.: 27. (*Clotho mauritanica*)

Synonyms: *Echidna mauritanica* A.-M.-C. Duméril & Bibron *in* Guichenot, 1848 *(nomen nudum), Vipera minuta* Eichwald, 1851 (*nomen nudum*), *Vipera confluenta* Cope, 1863a, *Vipera mauretanica* – F. Müller, 1878b *(nomen incorrectum),*

Type: Lectotype, MNHN 4017, a female (A. Guichenot, 1840–1842), designated by J. Anderson (1892: 20).

Type locality: "Algiers" [= Algiers, Algiers Wilaya, NW Algeria, 36°37'N, 3°00'E, elevation 80 m] via lectotype selection.

Distribution; Northwestern Africa. Northern Western Sahara (ext. NE Rio de Oro, Saguia El Hamra), Morocco (Agadir, Azilal, Chechaouene, El Kelaa Des Srarhna, Figuig, Khemisset, Khenitra, Meknes, Nador, Ouarzazate, Oujda, Rabat, Settat, Tangier, Tan Tan, Taounate, Taza, Tetouan, Tiznit), N Algeria (El Oued, Ouargla) and N Tunisia (Ariana, Kairouan, Sousse), 10–2300 m.

Sources: Billing & Schätti, 1984, Brodmann, 1987, Nilson & Andrén, 1988b, Herrmann et al., 1992, Spawls & Branch, 1995, Bons & Geniez, 1996, Schleich et al., 1996, Herrmann & Joger, 1997 and Bogaerts, 1998.

Remarks: David & Ineich (1999: 324–325) discussed authorship.

4. †*Daboia maxima* (Szyndlar, 1988). Acta Zool. Cracov. 31(27): 698–702, fig. 6. (†*Vipera maxima*)

Type: Holotype, MNCN 10032, one trunk vertebra.

Type locality: "Middle Pliocene (MN 15) of Layna (Prov. of Soria, Spain)."

Distribution: Middle Pliocene (Ruscinian, MN 15: 3.2–4.2 mya) Spain (Soria). Known only from type locality.

Remarks: Allocation to *Daboia* due to relationship with *D. russelii fide* Nilson & Andrén (1997: 40).

5. *Daboia palaestinae* (F. Werner, 1938b). Zool. Anz. 122(11–12): 316–317, figs. 3–4. (*Vipera Palaestinae*)

Synonyms: *Vipera palastinae* – Condrea & Vries, 1965 *(nomen incorrectum), Vipera palestina* – J.H. Brown &

Bowles, 1966 *(nomen incorrectum), Vipera palistinae* Minton, 1971 *(nomen emendatum), Daboia palestinae* Esterbauer, 1987 *(nomen emendatum), Vipera palgestinae* – García, Huang & Pérez, 1989 *(nomen incorrectum),* and *Vipera palaestina* – Janssen, Freyvogel & Meier, 1990 *(nomen incorrectum).*

Type: Holotype, ZSM 70/1938, a 975 mm female (E.H. Taylor, 13 July 1936).

Type locality: "Haifa, Palästina" [= Haifa, Hefa District, Israel, 32°50'N, 34°58'E, elevation 20 m].

Distribution: Middle East. Northwestern Jordan (Ajloun, Amman, Balqa, Irbid, Jarash, Karak, Madaba, Tafilah), N Israel (Central, Gaza Strip, Golan Heights, Haifa, Jerusalem, Northern, N Southern, Tel Aviv, West Bank), Lebanon (Beirut, El Jnoub, Liban-Nord, Mont-Liban, Nabatiye) and W Syria (Halab, Homs, Latakia, Tartus), 20–1600 m.

Sources: Mendelssohn, 1963, Hraoui-Bloquet, 1981, Herrmann & Joger, 1997, Bouskila & Amitai, 2001, Disi et al., 2001, Hraoui-Bloquet et al., 2002, Amr & Disi, 2011, Bar & Haimovitch, 2011 and Volynchik, 2011.

Remarks: Photograph of type in Franzen & Glaw (2007: fig. 23). A member of *Daboia fide* Groombridge, 1980, 1986, Lenk et al., 2001b and Mallow et al., 2003.

6. *Daboia russelii* (G. Shaw & Nodder, 1797 *in* 1789–1813). Nat. Misc. 8: pl. 291, 2 pp. (*Coluber russelii*)

Synonyms: *Coluber daboie* Lacépède, 1789 *(nomen rejiciendum), Coluber idolum* Suckow, 1797, *Coluber trinoculus* Schneider *in* Bechstein, 1802, *Vipera daboya* Daudin, 1803c *(nomen substitutum), Vipera elegans* Daudin, 1803c, *Coluber triseriatus* Hermann, 1804, *Vipera daboia* Merrem, 1820 (*nomen substitutum*), *Natrix coillepidota* Wagler, 1825 (*nomen ineditum*), *Natrix semimaculata* Wagler, 1825 (*nomen ineditum*), *Daboia pulchella* Gray, 1842e, *Daboia russellii* Gray, 1849 *(nomen emendatum), Vipera russelli nordicus* Deraniyagala, 1945, *Vipera russellis* – Ouyang & Lin, 1970 *(nomen incorrectum),* and *Vipera ruselli* – Sawai, 1979 *(nomen incorrectum).*

Type: Lectotype, BMNH II.1.1a (P. Russell & natives, 1781–1791), designated by Golay et al. (1993: 268).

Type locality: "coast of Coromandel, India" [= Andhra Pradesh and Tamil Nadu States, SE India] via lectotype selection.

Distribution: Southern Asia. Eastern Pakistan (ext. SE Balochistan, Jammu & Kashmir, Khyber Pakhtunkhwa, Punjab, Sindh), India (Andhra Pradesh, Gujarat, Karnataka, Maharashtra, Nagaland, Madhya Pradesh, Orissa, Rajasthan, Tamil Nadu, Uttar Pradesh, West Bengal), Bangladesh, Nepal (Bardiya, Chitwan, Kanchanpur, Mahottari, Nawalparasi, Parsa, Saptari, Siraha), Bhutan, S China (Hunan, Yunnan), and Sri Lanka (Central, Eastern, North-Central, North-Western, Northern, Sabaragamuwa, Southern, Uva, Western), NSL–2755 m.

Sources: Wall, 1905d, 1907c, 1917, M.A. Smith, 1943, Brongersma, 1958, Bergman, 1961b, P. Silva, 1980a, A. Silva, 1990b, 2009, Wuster et al., 1992, Jintakune & Chanhome, 1995, Belt et al., 1997, Herrmann & Joger, 1997, Nilson & Andrén, 1997, Wüster, 1998b, David & Ineich, 1999, McDiarmid et al., 1999, M.S. Khan, 2002, 2006, J.C. Murphy & Schlager, 2003, Tiwari & Shah, 2004, Whitaker & Captain, 2004, N. Khaire, 2006, Zhao, 2006, Thorpe et al., 2007, Xiang & Li, 2009, I. Das, 2010, 2012, Masroor, 2012 and S.K. Sharma et al., 2013.

Remarks: BMNH catalogue lists BMNH II.1.1a as holotype. Original description based on P. Russell (1796: 10–11, pl. 7), deposited in BMNH and redrawn by Shaw & Nodder (1789 *in* 1789–1813: pl. 291). Orthography of specific epithet discussed by David & Ineich (1999: 313–314) and McDiarmid et al. (1999: 373–375).

7. †*Daboia sarmatica* (Chkhikvadze & Lungu *in* Zerova, Lungu & Chkhikvadze, 1987). Proc. Zool. Inst. Leningrad (1986) 158: 92, fig. 2. (†*Vipera sarmatica*)

Type: Holotype, SIPT 18/72–1, one trunk vertebra.

Type locality: "Kalfa, village in Moldavia State, northern Black Seaside, USSR, late Miocene (middle Sarmatian)."

Distribution: Upper Miocene (Sarmatian: 11.6–12.7 mya) of Russia (Moldavia). Known only from type locality.

Source: Nilson & Andrén, 1997

8. *Daboia siamensis* (M.A. Smith, 1917a). J. Nat. Hist. Soc. Siam 2(3): 223–224, fig. 1. (*Vipera russelli siamensis*)

Synonyms: *Vipera russelli limitis* Mertens, 1927a, *Vipera russeli formosensis* Maki, 1931, *Vipera russellii sublimitis* Kopstein, 1936, *Vipera russellii formosenses* – Sifford & Johnson, 1978 *(nomen incorrectum), Vipera russelli simamensis* – Nakada et al., 1984 *(nomen incorrectum),* and *Vipera russeli burmanus* Maung Maung Aye *in* Gopalakrishnakone & Tan, 1987 *(nomen substitutum).*

Type: Holotype, BMNH 1946.1.18.8 (formerly BMNH 1927.11.24.1), a 565 mm specimen.

Type locality: "Sam Kok, Central Siam, about 60 km. N. of Bangkok" [= Sam Khok, Pathum Thani Prov., Thailand, 14°04"N, 100°31'E, elevation 5 m].

Distribution: Southeast Asia and Lesser Sundas. Southern China (Fujian, Guangdong, Guangxi), Taiwan, Myanmar (Ayeyarwady, Bago, Magway, Mandalay, Sagaing, Shan, Yangon), Thailand (Ayutthaya, Chachoengsao, Chai Nat, Chumphon, Kamphaeng, Lop Buri, Nakhon Nayok, Nakhon Ratchasima, Nakhon Sawan, Prachin Buri, Prachuap Khiri Khan, Samut Prakan, Saraburi, Suphan Buri), Cambodia (Sosophan) and S Indonesia (? Adonara, Ende, Flores, E Java,

Komodo, Lembata, Lombien, Rinca, Solar, Sumbawa), NSL–2755 m.

Sources: Kuntz, 1963, E.H. Taylor, 1965, Auffenberg, 1980, M.J. Cox, 1991b, Wüster et al., 1992, Zhao & Adler, 1993, Belt et al., 1997, M.J. Cox et al., 1998, Chan-ard et al., 1999, Leviton et al., 2003, Thorpe et al., 2007, Lang, 2011, I. Das, 2012 and McKay & Lilley, 2012.

Remarks: Golay et al. (1993: 268) listed two syntypes as BMNH 1920.10.10.1–2. Presence in the Lesser Sundas due to human activities in 18th–19th centuries *fide* Leviton et al., (2003: 440).

D

†*DAKOTAOPHIS* Holman, 1976a (Dipsadidae)

Type species: †*Dakotaophis greeni* Holman, 1976a.

Distribution: Lower to middle Miocene of cen. USA.

Sources: Holman, 1976a, 1979a, 1996a, 2000a and Rage, 1984b.

1. †*Dakotaophis greeni* Holman, 1976a. Herpetologica 32(1): 45–47, figs. 4a–d.

Type: Holotype, SDSM 8380, one trunk vertebra.

Type locality: "Black Bear Quarry II, designated SDSM locality V673, Bennett County, South Dakota, USA, Rosebud Formation, early Hemingfordian, lower middle Miocene."

Distribution: Lower Miocene (Hemingfordian: 16.3–20.6 mya) of USA (Nebraska), lower-middle Miocene (Hemingfordian-Barstovian: 13.6–20.6 mya) of USA (South Dakota), and middle Miocene (Barstovian: 13.6–16.3 mya) of USA (South Dakota, Texas, Wyoming).

Fossil records: Lower Miocene (Hemingfordian) of USA (Nebraska, South Dakota), lower/middle Miocene (Hemingfordian or Barstovian) of USA (Wyoming), and middle Miocene (Barstovian) of USA (Nebraska, South Dakota, Texas).

Remarks: Holman (1979b: 45) erroneously listed the type as SDSM 9380.

DARLINGTONIA Cochran, 1935 (Xenodontidae)

Type species: *Darlingtonia haetiana* Cochran, 1935.

Distribution: Southwestern Hispaniola.

Sources: A. Schwartz, 1970b, A. Schwartz & Thomas, 1965, 1975, A. Schwartz & Rossman, 1976, A. Schwartz, 1986, A. Schwartz & R.W. Henderson, 1988, 1991, H.M. Smith & Wallach, 1993, Powell & R.W. Henderson, 1994a, Rodríguez-Robles & Greene, 1996, Hedges et al., 2009, Henderson and Powell, 2009 and Zaher et al., 2009.

1. *Darlingtonia haetiana* Cochran, 1935. Proc. Boston Soc. Nat. Hist. 40(6): 375.

Synonyms: *Darlingtonia haetiana perfector* A. Schwartz & Thomas, 1965 and *Darlingtonia haetiana vaticinata* A. Schwartz, 1970b.

Type: Holotype, MCZ 38252, a 288 mm female (P.J. Darlington, Oct. 1934).

Type locality: "Roche Croix, northeastern foothills, Massif de la Hotte, Haiti, about 5000 feet altitude."

Distribution: Hispaniola. Southwestern Haiti (Grand'Anse, Ouest, Sud, Sud-Est) and SW Dominican Republic (Perdenales), 300–1700 m.

Remarks: A valid genus *fide* Grazziotin et al. (2012: 22) but a synonym of *Ialtris fide* Hedges et al. (2009: 19).

DASYPELTIS Wagler, 1830 (Colubridae)

Synonyms: *Anodon* A. Smith, 1829 *(nomen rejiciendum), Analcis* Wagler, 1830, *Rachiodon* Jourdan, 1834 *(nomen substitutum), Raphiodon* – Jourdran, 1834 *(nomen incorrectum), Deirodon* Owen, 1841 *in* 1840–1845 *(nomen substitutum), Dirodon* Agassiz, 1847 *(nomen emendatum), Rhachiodon* – Lichtenstein & Martens, 1856 *(nomen incorrectum),* and *Dasipeltis* – Scortecci, 1931 *(nomen incorrectum).*

Type species: *Coluber scaber* Linnaeus, 1758.

Distribution: Subsaharan Africa.

Sources: Gans, 1959, Gravlund, 2001, Kelly, 2003, J.-F. Trape & Mané, 2006a and S. Trape et al., 2012.

Remarks: Official Generic Name no. 979 *fide* Opinion 387 (ICZN, 1956b).

1. *Dasypeltis abyssinica* (A.-M.-C. Duméril, Bibron & Duméril, 1854a). Erpét. Gén. 7(1): 496–497, pl. 81, figs. 1–3. (*Rachiodon abyssinicus*)

Synonym: *Rachiodon abyssinicus* A.-M.-C. Duméril, 1853 (*nomen nudum*).

Type: Lectotype, MNHN 6567, a 755 mm specimen (Cartin-Dillon), designated by J.-F. Trape & Mané (2006a: 53).

Type locality: "Abyssinie" [= Eritrea and Ethiopia] via lectotype designation.

Distribution: Eritrea (Anseba) and Ethiopia.

Sources: Trape & Mané, 2006a.

Remarks: A valid species *fide* J.-F. Trape & Mané (2006a: 53).

2. *Dasypeltis atra* Sternfeld, 1913a. Wiss. Ergebn. Dtsch. Zentr.-Afr.-Exped. (1912) 4: 272. (*Dasypeltis scabra atra*)

Types: Syntypes (3), ZMB, longest syntype 830 mm (R. Grauer, 1909–1911), lost *fide* R. Günther *(in litt.).*

Type locality: "Randbergen am Nordwestufer des Tanganyika-Sees" [= NW shore of Lake Tanganyika, SE Sud-Kivu Prov., E Democratic Republic of the Congo].

Distribution: Eastern Africa. Southeastern South Sudan (Eastern Equatoria), N Ethiopia (Arussi, Gondar, Gojjam, Hararge, Illubabor, Shoa, Wollega), E Democratic Republic of the Congo (Katanga, Nord-Kivu, Orientale, Sud-Kivu, Idjwi Is.), Uganda (Central, Eastern, Northern, Western), W Kenya (Eastern, Nairobi, Rift Valley, Western), N Rwanda, N Burundi, and N Tanzania (Kilimanjaro), (610) 1500–3000 m.

Sources: Gans, 1964, Pitman, 1974, Largen & Rassmussen, 1993, Neças, 1995, Hughes, 1997, Gravlund et al., 1998, Spawls et al., 2002, Broadley & Bates, 2009 and Largen & Spawls, 2010.

3. *Dasypeltis confusa* J.-F. Trape & Mané, 2006a. Bull. Soc. Herp. France (119): 28–31, fig. 1.

Type: Holotype, MNHN 2006.0303 (formerly IRD S-3105), a 696 mm female (native, March 1993–Feb. 1994).

Type locality: "Ibel (12°3'N, 12°23'W), Tambacounda."

Distribution: West Africa. Southern Senegal (Kédougou, Kolda, Tambacounda, Ziguinchor), Gambia (Central River), Guinea-Bissau (Biombo, Bissau, Tombali), Guinea (Boffa, Boké, Faranah, Forécariah, Kindia, Kouroussa, Mandiana, Télimélé), Sierra Leone (Northern, Southern), Togo (Centrale, Kara, Plateaux, Savanes), Benin (Atakora, Collines, Kouffo, Zou), Cameroon (Adamaoua, Centre, Est, Nord-Ouest, Ouest, Sud), Gabon (Haut-Ogooué, Moyen-Ogooué, Ogooué-Maritime), Central African Republic (Bamingui-Bangoran, Haut-Mbomou, Haute-Sangha, Lobaye, Mbomou, Ombella-Mpoko, Ouham, Ouham-Pende, Vakaga), Congo (Brazzaville, Lekoumou, Niari, Pool) and W Democratic Republic of the Congo, 15–1550 m.

Sources: J.-F. Trape & Roux-Estève, 1995, J.-F. Trape & Mané, 2006b, Pauwels & Vande weghe, 2008, Pauwels & Salle, 2009, Segniagbeto et al., 2011, Auliya et al., 2012, Chirio, 2013 and Hughes, 2013.

Remarks: Photographs of holotype in S. Trape et al. (2012: figs. 6a, 7a). Previous records of *D. scabra* from Gabon probably *D. confusa fide* Pauwels & Salle (2009: 23). Possibly a synonym of *D. scabra fide* Hughes (2013: 114).

4. *Dasypeltis fasciata* A. Smith, 1849 *in* 1838–1849. Illust. Zool. So. Africa, Rept.: footnote, p. 2 of text to pl. 73.

Synonyms: *Dipsas carinatus* Hallowell, 1844b (*nomen praeoccupatum*), *Rachiodon scaber subfasciatus* Jan, 1863b (*nomen nudum*), and *Dasypeltis macrops* Boulenger, 1907a.

Type: Holotype, not designated, a 559 mm male (A. Smith, 1829), lost *fide* Gans (1959: 89).

Type locality: "Sierra Leone."

Distribution: Central Africa. Senegal (Fatick, Tambacounda), Gambia (Kombo Saint Mary), Guinea-Bissau, SE Guinea (Macenta, Nzérékoré), S Mali (Sikasso), Liberia, S Ivory Coast (Daloa, Guiglo, Tabou), S Ghana, Togo (Plateaux), S Nigeria (Akwa Ibom, Bayelsa, Rivers), Southern Cameroon (Est, Centre, Littoral, Nord-Ouest, Ouest, Sud, Sud-Ouest), S Chad, Central African Republic (Bamingui-Bangoran, Haute-Sangha, Lobaye, Mbomou, Ombella-Mpoko, Ouham, Ouham-Pende, Sangha), N Gabon (Estuaire, Ogooué-Ivindo, Wolem-Ntem), Congo (Brazzaville, Kouilou, Plateau, Sangha), Democratic Republic of the Congo (Bandundu, Equateur, Kasai Occidental, Kasai Oriental, Kinshasa, Kivu, Orientale) and W Uganda (W Western), NSL–1380 m.

Sources: Doucet, 1963, Pitman, 1974, Roman, 1980, Hakansson, 1981, J.-F. Trape & Roux-Estève, 1995, Spawls et al., 2002, Ineich, 2003, Chippaux, 2006, Chirio & Ineich, 2006, Chirio & LeBreton, 2007 and Segniagbeto et al., 2011.

Remarks: Possibly occurs in Benin *fide* Hughes (2013: 151).

5. *Dasypeltis gansi* J.-F. Trape & Mané, 2006a. Bull. Soc. Herp. France (119): 32–37, fig. 2.

Type: Holotype, IRSNB 2613 (formerly IRD S-6537), a 608 mm male (May 2003–April 2004).

Type locality: "Mahamouda Chérif (12°58'N, 16°30'W), Basse Casamance, Senegal."

Distribution: West Africa. Senegal (Dakar, Kédougou, Kolda, Tambacounda, Saint-Louis, Thiès, Ziguinchor), Gambia, Guinea (Koundara, Siguiri), SW Mali (Kayes, Koulikoro, Ségou, Sikasso), Togo (Kara, Maritime, Plateaux), S Benin (Alibori, Atakora, Atlantique, Borgou, Collines, Kouffo), Nigeria, N Cameroon (Extreme-Nord, Nord), Burkina-Faso (Centre, Centre-Est, Centre-Nord, Centre-Ouest, Est, Haute-Bassins, Nord, Sud-Ouest, Volta-Noire), Niger (Dosso, Maradi, Niamey), SW Chad (Chari-Baguirmi) and South Sudan, NSL–910 m.

Sources: J.-F. Trape & Mané, 2006b, Chiriuo & LeBreton, 2007, Chirio, 2009, Ullenburch et al., 2010, Segniagbeto et al., 2011 and Hughes, 2013.

Remarks: Photographs of holotype in S. Trape et al. (2012: figs. 6d, 7d).

6. *Dasypeltis inornata* A. Smith, 1849 *in* 1838–1849. Illust. Zool. So. Africa, Rept.: pl. 73, figs. a–e, 2 pp.

Synonyms: *Rachiodon immaculatus* A.-M.-C. Duméril, 1853 (*nomen nudum*), *Rachiodon immaculatus* Duméril, Bibron & Duméril, 1854a, and *Rachiodon scaber unicolor* Jan, 1863b (*nomen nudum*).

Type: Holotype, BMNH 1946.1.2.47 (formerly BMNH 1865.5.4.164), a 656–737 mm male (A. Smith, 1825–1837).

D

Type locality: "south-central districts of the Cape Colony, and also in Kaffirland" [= Western Cape, Eastern Cape and KwaZulu Natal provinces, South Africa].

Distribution: Western Swaziland and SE South Africa (Eastern Cape, KwaZulu-Natal, Mpumalunga), NSL–1440 m.

Sources: Broadley, 1983, Branch, 1988 and Cockeran & Haagner, 1993.

Remarks: BMNH catalogue lists BMNH 1946.1.2.47 as "one of the types" (M. Lang, *in litt.*). V.F.M. FitzSimons (1962: 174), Broadley (1990: 261) and Gans (1959: 135) attribute A. Smith (1849: App. p. 20) to this species but it refers to *Crotaphopeltis hotamboeia fide* Branch & Bauer (2005: 8).

D

7. *Dasypeltis latericia* J.-F. Trape & Mané, 2006. Bull. Soc. Herp. France (119): 37–42, fig. 3. (*Dasypeltis gansi latericia*)

Type: Holotype, IRSNB 2615 (formerly IRD S-2088), a 751 mm female (native, March 1992–Feb. 1993).

Type locality: "Boundoukondi (12°31'N, 12°20'W), Tambacounda."

Distribution: West Africa. Southeastern Senegal (Kédougou), Guinea (Koundara, Siguiri, Sikasso) and SW Mali (Bamako, Kayes, Koulikoro), 80–400 m.

Remarks: Photographs of holotype in S. Trape et al. (2012: figs. 6b, 7b). A valid species *fide* S. Trape et al. (2012: 491).

8. *Dasypeltis medici* (Bianconi, 1859). Mem. Accad. Sci. Inst. Bologna 10: 501–502, pl. 26. (*Dipsas medici*)

Synonyms: *Dasypeltis scaber fasciolata* W.C.H. Peters, 1868a, *Dasypeltis elongata* Mocquard, 1888, and *Dasypeltis medica lamuensis* Gans, 1957.

Type: Holotype, IBI, a 531 mm male (Michaeli via C. Fornasini, 1842–1847), lost *fide* Gans (1959: 157).

Type locality: Unknown. Listed as Mosambique [= Mozambique] *fide* Jan & Sordelli (1872: 3 *in* 1870–1881). Probably Inhambane, Mozambique *fide* Broadley (1983: 265).

Distribution: Eastern Africa. Southern Somalia (Jubbada Dhexe, Jubbada Hoose, Alessandra Is.), SE Kenya (Coast, Lamu Is.), E Tanzania (Kilimanjaro, Lindi, Morogoro, Mtwara, Piwani, Tanga, Mafia and Zanzibar Is.), S Zambia (E Southern), Malawi (Northern, Southern), Mozambique (Cabo Delgado, Inhambane, Sofala, Zambézia, Bazaruto Is.) and E Zimbabwe (Manicaland), NSL–1000 m.

Sources: Gans, 1957, Broadley & Cock, 1975, Broadley, 1983, 1990b, Lanza, 1983a, 1990b, Branch, 1988, Spawls et al., 2002, Broadley et al., 2003 and Broadley & Blaylock, 2013.

Remarks: Original description reprinted in Bianconi (1850–1870: 277–278, pl. 14).

9. *Dasypeltis palmarum* (Leach *in* Tuckey, 1818). Exped. River Zaire: 408–409. (*Coluber palmarum*)

Type: Holotype, BMNH 1946.1.2.35, juvenile male (J. Cranch [Tuckey Exped.] via W.E. Leach, 1816).

Type locality: "in palm trees at Embomma" [= Boma, Bas-Congo Prov., SW Democratic Republic of the Congo, 5°51'S, 13°03'E, elevation 15 m].

Distribution: Central Africa. Southern Congo (Kouilou), W Democratic Republic of the Congo (Bandundu, Bas-Congo, Kinshasa) and NW Angola, 15–275 m.

Sources: Laurent, 1956a, J.-F. Trape & Roux-Estève, 1995 and Chippaux, 2006.

10. *Dasypeltis parascabra* S. Trape, Mediannikov & Trape, 2012. C. R. Biol., Paris 335(7): 492–493, figs. 3–5, 6c, 7c.

Type: Holotype, MNHN 2010.438 (formerly IRD 1555-G), a 605 mm gravid female (native, June–Oct. 2005).

Type locality: "Dalakan, Guinea (9°58'N, 9°32'W)" [= Dalakan, Kankan Prov., E Guinea, elevation 400 m].

Distribution: West Africa. Eastern Guinea (Kankan, Kouroussa), Liberia (Nimba), Ivory Coast (Toumodi), Ghana (Eastern), Togo (Plateaux) and S Nigeria (Lagos), NSL–430 m

11. *Dasypeltis sahelensis* J.-F. Trape & Mané, 2006a. Bull. Soc. Herp. France (119): 43–47, fig. 4.

Type: Holotype, MNHN 2006.0315 (formerly IRD S-4804), a 498 mm male (native, June–Oct. 2000).

Type locality: "Tialé (15°14'N, 16°49'W), Senegal" [= Thiès Prov., W Senegal, elevation 25 m].

Distribution: West Africa. Morocco (Agadir, Ouarzazate, Tan Tan), N Western Sahara (W Saguia El Hamra), SW Mauritania (Trarza), N Senegal (Louga, Matam, Saint-Louis, Tambacounda, Thiès), S Mali (Kayes, Koulikoro, Ségou, SW Tombouctou) and S Niger (W Agadez, S Diffa, Dosso, Maradi, Niamey, Tahoua, W Zinder), 20–350 m.

Source: J.-F. Trape & Mané, 2006b.

Remarks: Photographs of holotype in S. Trape et al. (2012: figs. 6e, 7e).

12. *Dasypeltis scabra* (Linnaeus, 1758). Syst. Nat., ed. 10, 1: 223. (*Coluber scaber*)

Synonyms: *Anodon typus* A. Smith, 1828, *Dasypeltis scaber breviceps* W.C.H. Peters, 1864c, *Dasypeltis scaber capensis* W.C.H. Peters, 1864c, *Dasypeltis scaber mossambicus* W.C.H. Peters, 1864c, *Dasypeltis lineolata* W.C.H. Peters, 1878, *Tropidechis dunensis* De Vis, 1911, and *Dasypeltis scabra loveridgei* Mertens, 1954b.

Type: Holotype, NHR (formerly MAFR), a female (Mus. Drottn), lost *fide* Andersson (1899: 5) and Gans (1959: 87).

Type locality: "Indiis" [= India] (in error). Corrected to South Africa *fide* Flower (1933: 818). Restricted to Namibia or Limpopo, South Africa *fide* Gans (1959: 87).
Distribution: Eastern Africa and SW Arabia. NE Egypt (Faiyum), SE Sudan (An Nil Al Azraq, Sennar), Ethiopia (Arussi, Bale, Gemu Gofa, Gojjam, Gondar, Hararge, Shoa, Sidamo, Wollega), Eritrea (Anseba, Northern Red Sea, Southern), Somalia (Bay, Gedo, Jubbada Dhexe, Jubbada Hoose, Mogadishu, Sanaag, Shabeellaha Dhexe, Shabeellaha Hoose, Sool, Togdheer, Woqooyi Galbeed, Alessandra Is.), Kenya (Central, Coast, North-Eastern, Nyanza, S Rift Valley, Western), Uganda (Central, Eastern, Northern, Western), Democratic Republic of the Congo (Bandundu, Equateur, Kasai Occidental, Kasai Oriental, Katanga, Kinshasa, Nord-Kivu, Orientale, Sud-Kivu), E Rwanda, E Burundi, Tanzania (Arusha, Dar es Salaam, Dodoma, Iringa, Kigoma, Kilimanjaro, Lindi, Mara, Mbeya, Morogoro, Mtwara, Mwanza, Piwani, Rukwa, Ruvuma, Shinyanga, Singida, Tabora, Tanga, Kagera, Sese, Ukerewe, and Zanzibar Is.), Mozambique (Gaza, Inhambane, Manica, Maputo, Niassa, Sofala, Tete), Malawi (Central, Northern, Southern), Zambia (Central, Copperbelt, Eastern, Luapula, Lusaka, Northern, North-Western, Southern, Western), NE Angola (Lunda Norte), Namibia (Caprivi, Hardap, Karas, Khomas, Kunene, Ohangwena, Okavango, Omaheke, Omusati, Oshana, Oshikoto, Otjozondjupa), Botswana (Central, Gaborone, Ghanzi, Kagatleng, Kweneng, North East, North West), Zimbabwe (Bulawayo, Manicaland, Mashonaland Central, Mashonaland East, Mashonaland West, Masvingo, Matabeleland North, Matabeleland South, Midlands), Swaziland, W Lesotho and South Africa (Eastern Cape, Free State, Gauteng, KwaZulu-Natal, Limpopo, Mpumalanga, Northern Cape, Western Cape), SW Saudi Arabia (Asir, Jazan) and W Yemen (Ad Dali', Aden, Ibb), NSL–2100 m.
Sources: Jourdan, 1833, Barros e Cunha, 1935, Mertens, 1955, 1971, Sweeney, 1961, Stemmler, 1971b, Pitman, 1974, Broadley & Cock, 1975, Broadley, 1983, Buys & Buys, 1983, Lanza, 1983a, 1990b, Pienaar et al., 1983, Branch, 1988, Gasperetti, 1988, Largen & Rasmussen, 1993, Lopez et al., 1993, Schätti & Gasperetti, 1994, Bons & Geniez, 1996, Schleich et al., 1996, Clauss & Clauss, 2002, Spawls et al., 2002, Broadley et al., 2003, J.C. Murphy & Schlager, 2003, Geniez et al., 2004, Chippaux, 2006, Scheurer, 2007, Largen & Spawls, 2010 and Broadley & Blaylock, 2013.
Remarks: Official Specific Name no. 679 *fide* Opinion 387 (ICZN, 1956b).

†*DAUNOPHIS* Swinton, 1926 (Pythonidae)

Type species: †*Daunophis langi* Swinton, 1926.
Distribution: Pliocene of Myanmar.
Source: Rage, 1984b.

1. †*Daunophis langi* Swinton, 1926. Ann. Mag. Nat. Hist. (9) 17(99): 344–348, pl. 16.

Type: Holotype, HMG V 1230, an 1170 mm nearly complete skeleton (H.F. Lang).
Type locality: "oil-shale exposed in a quarry at Tichara, near Mepale, on the eastern side of the Dawna Hills, South Burma, Pliocene age."
Distribution: Pliocene (1.8–4.9 mya) of Myanmar. Known only from holotype.
Remarks: Genus of doubtful validity *fide* Rage (1984b: 18).

†*DAWSONOPHIS* Holman, 1979b (Boidae)

Type species: †*Dawsonophis wyomingensis* Holman, 1979b.
Distribution: Upper Eocene of USA.
Sources: Rage, 1984b and Holman, 2000a.
Remarks: Possibly an undescribed subfamily of the Boidae *fide* Holman (2000a: 39).

1. †*Dawsonophis wyomingensis* Holman, 1979b. Ann. Carnegie Mus. 48(6): 105–106, fig. 1.

Type: Holotype, CM 14444, one trunk vertebra and fragment of adjacent vertebra (C. Black, M. Dawson & P. Robinson, 1962).
Type locality: "Badwater Creek area, SESE, S. 14, T. 39 N., R. 89 W, Natrona County, Wyoming. Referred to as locality 6. Late Eocene, Uintan Mammal Age. Hendry Ranch Member" [= Middle Eocene *fide* Holman, 2000a: 37].
Distribution: Middle or upper Eocene (Uintan: 42.0–46.2 mya) of USA (Wyoming). Known only from type locality.

DEINAGKISTRODON Gloyd, 1979 (Viperidae)

Synonym: *Dienagkistrodon* – Zhong & Wu, 1981 *(nomen incorrectum).*
Type species: *Halys acutus* A.C.L.G. Günther, 1888b.
Distribution: Southeastern Asia.
Sources: Maki, 1931, C.H. Pope, 1935, Kuntz, 1963, Keegan et al., 1965, Zhao, 1982, 2006, M.-H. Huang & Qu, 1983, Maes, 1989, Gloyd & Conant, 1990, Kardong, 1990, Cadle, 1992, Minton, 1992, Golay et al., 1993, Mao, 1993, Zhao & Adler, 1993, Parkinson et al., 1997, David & Ineich, 1999, P. Guo et al., 1999a, McDiarmid et al., 1999, Parkinson, 1999, Malhotra & Thorpe, 2000, Orlov et al., 2000, Ziegler et al., 2001, J.C. Murphy & Schlager, 2003, Orlov et al., 2003, Gumprecht et al., 2004, V.S. Nguyen et al., 2009, Xiang

& Li, 2009, I. Das, 2010, 2012, Malhotra et al., 2010 and Hoser, 2012d.

1. *Deinagkistrodon acutus* (A.C.L.G. Günther, 1888b). Ann. Mag. Nat. Hist. (6) 1(3): 171–172, pl. 12. (*Halys acutus*)

D

Types: Syntypes (3), BMNH 1946.1.19.56–58 (formerly BMNH 1881.1.30.59–61), three males, longest syntype 1168 mm (A.E. Pratt, 2 April–4 Aug. 1887).

Type locality: "mountains north of Kiu Kiang, on the Yantsze river, China," (in error). Corrected to Wu-suih, about three days' journey or 9 mi. S of the American Central China Mission in Kiu-kiang *fide* Pratt (1892: 3–5), [= Wusüeh, SE Hubei Prov., cen. China, 29°50'N, 115°41'E *fide* C.H. Pope (1935: 502].

Distribution: Southeastern Asia. Central and S China (Anhui, Fujian, Guangdong, Guangxi, Hubei, Hunan, Jiangxi, Zhejiang), Taiwan, N Laos and N Vietnam (Cao Bang, Lai Chau, Lang Son, Lao Cai, Vinh Phuc), NSL–1500 m.

Remarks: Original description reprinted in A.C.L.G. Günther (1892b: 242). Type locality is on the Yangtze River just north of Jiangxi Prov.

DEMANSIA Gray, 1842c (Elapidae)

Synonyms: *Demansia* Gray, 1842b *(nomen nudum), Diemansia* A.C.L.G. Günther, 1858 *(nomen emendatum), Diemensia* – A.C.L.G. Günther, 1862b *(nomen incorrectum), Diemenia* A.C.L.G. Günther, 1863 *(nomen praeoccupatm), Diemennia* A.C.L.G. Günther, 1863 *(nomen substitutum), Diemannia* Steindachner, 1867 *(nomen emendatum), Elapocephalus* Macleay, 1878b *(nomen praeoccupatum), Elapidocephalus* A.C.L.G. Günther *in* Macleay, 1884b *(nomen nudum), Elapocranium* Macleay, 1884b *(nomen praeoccupatum), Demensia* – Hoffmann, 1890 *(nomen incorrectum),* and *Diemena* – Boquet, 1964 *(nomen incorrectum).*

Type species: *Lycodon reticulatus* Gray, 1842c.

Distribution: Austro-Papua.

Sources: F. Werner, 1923b, Storr, 1978, 1985, Cogger et al., 1983a, Golay, 1985, Storr et al., 1986, S.K. Wilson & Knowles, 1988, Hoser, 1989, 2012e, Ehmann, 1992, Golay et al., 1993, Greer, 1997, Keogh, 1998, David & Ineich, 1999, Cogger, 2000, Scanlon, 2003, Shea & Scanlon, 2007, Sanders et al., 2008 and Zaher et al., 2009.

1. *Demansia angusticeps* Macleay, 1888b. Proc. Linn. Soc. N.S.W. (1887–1888) (2) 3(2): 417–418. (*Diemenia angusticeps*)

Type: Holotype, AM 31921 (formerly MMS 712), a 671 mm female (W.W. Froggatt, March–Dec. 1887).

Type locality: "vicinity of King's Sound, North-West Australia." Emended to a few miles round Derby *fide* Macleay (1888a: 1018).

Distribution: Northern Australia (NW Northern Territory, ext. NW Queensland, ext. N Western Australia), 10–415 m.

Sources: Boulenger, 1896a, Shea, 1998 and Shea & Scanlon, 2007.

Remarks: Photograph and illustrations of head of holotype in Shea & Scanlon (2007: figs. 2–3).

2. *Demansia calodera* Storr, 1978. Rec. West. Aust. Mus. 6(3): 294–295, fig. 2. (*Demansia olivacea calodera*)

Type: Holotype, WAM 54992, a 416 mm female (G. Harold & M. Peterson, 29 Aug. 1976).

Type locality: "Tamala, Western Australia, in 26°42' S,113°42' E."

Distribution: Western Australia (W Western Australia, Bernier and Dirk Hartog Is.), NSL–70 m.

Source: Shea & Scanlon, 2007.

Remarks: Photograph and illustrations of head of holotype in Shea & Scanlon (2007: figs. 4–5).

3. *Demansia flagellatio* Wells & Wellington, 1985. Aust. J. Herp. (Suppl. 1): 45.

Type: Holotype, AM 64867, a 535 mm male (D. Stammer).

Type locality: "Mt. Isa district, Queensland, Australia."

Distribution: Northern Australia (NW Queensland, Bustard Is.), NSL–350 m.

Source: Shea & Scanlon, 2007.

Remarks: Illustrations of head of holotype in Shea & Scanlon (2007: fig. 6).

4. *Demansia olivacea* (Gray, 1842c). Zool. Misc. 2(Apr.): 54. (*Lycodon olivaceus*)

Synonyms: *Elapocephalus ornaticeps* Macleay, 1878b, *Demansia elivacea* – McDowell, 1967 *(nomen incorrectum),* and *Demansia olivacia* – Limpus, 1978 *(nomen incorrectum).*

Types: Syntypes (3), from BMNH 1841.10.13.4–45 series, lost *fide* Shea & Scanlon (2007: 125–126).

Type locality: "North Coast of Australia; Port Essington." Restricted to Port Essington, Australia *fide* Cogger & Lindner (1974: 90).

Distribution: Southern Papua New Guinea (Western, Daru Is.) and N Australia (N Northern Territory, N Queensland, N Western Australia, Bathurst, Groote Eylandt, Koolan, Melville and Wargul Wargul Is.), NSL–505 m.

Sources: Cogger & Lindner, 1974, Storr, 1978, Shine, 1980c and Shea & Scanlon, 2007.

Remarks: Boulenger (1896a: 324) erroneously listed two Richardson specimens (BMNH 1946.1.18.40 and BMNH 1946.1.18.95) (formerly BMNH 1844.5.13.7 and

BMNH 1844.9.3) as syntypes *fide* Cogger & Lindner (1974: 90–91).

5. *Demansia papuensis* Macleay, 1877. Proc. Linn. Soc. N.S.W. (1877–1878) 2(1): 40. (*Diemenia papuensis*)

Synonym: *Demansia papuensis melaena* Storr, 1978.
Type: Holotype, AMS 31919 (formerly MMS 713), a 1676 mm specimen (W.J. Macleay, 1875).
Type locality: "New Guinea, 'probably' Hall Sound."
Distribution: Austro-Papua. Eastern Indonesia (SE Papua), S Papua New Guinea (Central, Western) and N Australia (N Northern Territory, N Queensland, ext. N Western Australia, Bremer, Cotton, Elcho, Inglis, Koolan, Milingimbi and Wigram Is.), NSL–590 m.
Source: O'Shea, 1996.
Remarks: Type locality doubtful *fide* Shea (1998: 51–52).

6. *Demansia psammophis* (Schlegel, 1837). Essai Phys. Serp. 1: 182, 2: 455. (*Elaps psammophis*)

Synonym: *Pseudelaps psammophidius* A.-M.-C. Duméril, Bibron & Duméril, 1854b *(nomen substitutum)*.
Types: Syntypes (2), MNHN 7663A, a 353–360 mm male, and MNHN 7666, a 710–720+ mm female lacking a skull (J.R.C. Quoy & J.P. Gaimard [Astrolabe Voy.], July 1827).
Type locality: "Nouvelle Hollande" [= Australia]. Restricted to King George's Sound *fide* A.-M.-C. Duméril, Bibron & Duméril (1854b: 1234).
Distribution: Australia (Australian Capital Territory, New South Wales, Northern Territory, Queensland, South Australia, Victoria, Westwern Australia, Dirk Hartog Is.), NSL–1095 m.
Sources: Guibe & Roux-Esteve, 1972a, Shine, 1980c and Bush et al., 1995.

7. *Demansia quaesitor* Shea *in* Shea & Scanlon, 2007. Rec. Aust. Mus. 59(2): 127–129, figs. 11–12.

Type: Holotype, NTM 16836, a 596 mm male (J. Woinarski, 29 May 1991).
Type locality: "'Hodgson Downs', Mt. Langdon, NT, Australia."
Distribution: Northern Australia (N Northern Territory, ext. NW Queensland, ext. N Western Australia, North Maret Is.), 45–300 m.

8. *Demansia reticulata* (Gray, 1842c). Zool. Misc. 2(Apr.): 54. (*Lycodon reticulatus*)

Synonym: *Demansia reticulata cupreiceps* Storr, 1978.
Type: Lectotype, BMNH 1946.1.19.79 (G. Krefft, 1838), designated by Storr (1978: 297).
Type locality: "Australia."
Distribution: Western Australia (S Northern Territory, South Australia, W Western Australia, Kangaroo Is.), NSL–685 m.

9. *Demansia rimicola* Scanlon *in* Shea & Scanlon, 2007. Rec. Aust. Mus. 59(2): 130–132, figs. 14–15.

Type: Holotype, AMS 62257, an 804 mm male (A. Greer, E. Cameron, H. Cogger, R. Sadlier & P. Webber, 17 July 1977).
Type locality: "61.4 km N Muttaburra via Hughenden Hwy, Qld (22°10'S 144°15'E), Australia."
Distribution: Central Australia (ext. NW New South Wales, N Northern Territory, Queensland, NE South Australia, ext. N Western Australia), 10–375 m.

10. *Demansia rufescens* Storr, 1978. Rec. West. Aust. Mus. 6(3): 292–293. (*Demansia olivacea rufescens*)

Type: Holotype, WAM 52747, a 573 mm female (W.H. Butler, 2 May 1976).
Type locality: "Marandoo minesite, near Mt Bruce, Western Australia, in 22°40' S, 118°10' E" [Australia].
Distribution: Western Australia (W Western Australia, Barrow, Dolphin and Hermite Is.), NSL–1040 m.
Source: Shea & Scanlon, 2007.
Remarks: Illustrations of head of holotype in Shea & Scanlon (2007: fig. 17).

11. *Demansia shinei* Shea *in* Shea & Scanlon, 2007. Rec. Aust. Mus. 59(2): 135–136, figs. 18–19.

Type: Holotype, NTM 6489, a 640 mm male (P. Horner, J. Griffiths & K. O'Brien, 18 March 1979).
Type locality: "Frewena, NT, in 19°25'S 135°24'E, Australia."
Distribution: Northern Australia (N Northern Territory, ext. N Western Australia), 225–420 m.

12. *Demansia simplex* Storr, 1978. Rec. West. Aust. Mus. 6(3): 290–291.

Type: Holotype, WAM 13841 (A.M. Douglas & G.F. Mees, July 1960).
Type locality: "Kalumburu, Western Australia, in 14°17' S, 126°40' E" [Australia].
Distribution: Northern Australia (N Northern Territory, ext. N Western Australia, Melville Is.), NSL–315 m.

13. *Demansia torquata* A.C.L.G. Günther, 1862b. Ann. Mag. Nat. Hist. (3) 9(49): 130, pl. 9, fig. 10. (*Diemansia torquata*)

Type: Holotype, BMNH 1946.1.17.41, a 573–586 mm female (F.M. Rayner, July, 1859).
Type locality: "Percy Islands" [= Queensland, Australia].
Distribution: Northeastern Australia (ext. NW New South Wales, Northern Territory, Queensland, Dent,

Fantome, Hayman, Hook, Long, Palm, Penrith and Magnetic Is.), 40–600 m.

Sources: Shine, 1980c and Shea & Sadlier, 1999.

Remarks: Photograph of holotype in Shea & Scanlon (2007: fig. 21).

D

14. *Demansia vestigiata* (De Vis, 1884b). Brisbane Courier 39(8324): 5. (*Hoplocephalus vestigiatus*)

Synonyms: *Diemenia atra* Macleay, 1884b and *Diemenia maculiceps* Boettger, 1898.

Type: Holotype, QM J206.

Type locality: Unknown.

Distribution: Papua New Guinea (Central, East Sepik, Gulf, Western) and N Australia (N Northern Territory, NE Queensland, ext. N Western Australia, Melville and Mornington Is.), NSL–510 m.

Sources: F. Parker, 1982, Ingram, 1990, Golay et al., 1993, Shea, 1996, 1998 and H.M. Smith and Wallach, 1997.

Remarks: Official Specific Name *fide* Opinion 1940 (ICZN 1999). Original description reprinted in De Vis (1884c: 138–139).

DENDRELAPHIS Boulenger, 1890 (Colubridae)

Synonyms: *Dendrophys* – J. Henle, 1839 *(nomen incorrectum), Dendrophis* Wagler *in* Fitzinger, 1843 *(nomen praeoccupatum), Dendelaphis* – F. Werner, 1901b *(nomen incorrectum), Dendorophis* – Cohn, 1905 *(nomen incorrectum), Dendralaphis* – M.A. Smith, 1915c *(nomen incorrectum), Dendrolaphis* – Luard, 1917 *(nomen incorrectum), Dendrelaphus* – Ditmars, 1933 *(nomen incorrectum), Dendriphis* – Briceño-Rossi, 1934 *(nomen incorrectum), Tachyophis* Mertens, 1934a *(nomen rejiciendum), Dendrellaphis* – Negi, 1992 *(nomen incorrectum), Dendrelophis* – B.D. Sharma *in* B.D. Sharma & Kumari, 1998 *(nomen incorrectum), Charlespiersonserpens* Hoser, 2012ab *(nomen illegitimum), Dorisious* Hoser, 2012ab *(nomen illegitimum), Downieea* Hoser, 2012ab *(nomen illegitimum), Macmillanus* Hoser, 2012ab *(nomen illegitimum),* and *Mulvanyus* Hoser, 2012ab *(nomen illegitimum).*

Type species: *Ahaetulla caudolineata* Gray, 1834 *in* Gray & Hardwicke, 1830–1835.

Distribution: Indo-Australia.

Sources: Rooij, 1917, Meise & Henning, 1932, Mertens, 1934a, M.A. Smith, 1943, Haas, 1950, F. Parker, 1982, Mahendra, 1984, McDowell, 1984, Wallach, 1998a, Ziegler & Vogel, 1999, Rooijen & Vogel, 2008a–b, I. Das, 2010, 2012 and Hoser, 2012ab.

Remarks: Official Generic Name no. 1310 *fide* Opinion 524 (ICZN, 1958a).

1. *Dendrelaphis andamanensis* (J. Anderson), 1871a. Proc. Zool. Soc. London 39(1): 184–185. (*Dendrophis pictus andamanensis*)

Type: Holotype, ZSI 7714 (formerly IM), an 815 mm specimen.

Type locality: "Andamans" [Andaman & Nicobars, E India, Bay of Bengal].

Distribution: Eastern India (Andaman & Nicobars: Little Andaman, Long, Middle Andaman, North Andaman, South Andaman, Great Nicobar Is.).

Sources: Biswas & Sanyal, 1980, R.C. Sharma, 2003, Rooijen & Vogel, 2008a and Vogel & Rooijen, 2011a.

2. *Dendrelaphis ashoki* Vogel & Rooijen, 2011a. J. Herp. 45(1): 106–107, fig. 7.

Type: Holotype, CAS 17222, an 895 mm female (R.H. Beddome, 1857–1882).

Type locality: "Anamalais, India" [= Anaimalai Hills, S Western Ghats, Kerala/Tamil Nadu, S India, ca. 10°22'N, 77°08'E].

Distribution: Southwestern India (Western Ghats of Kerala, Tamil Nadu), 2100 m.

3. *Dendrelaphis bifrenalis* (Boulenger, 1890a). Fauna Brit. India, Rept. Batr.: 338. (*Dendrophis bifrenalis*)

Types: Syntypes (3), BMNH 1946.1.6.4, a 1030 mm female, BMNH 1946.1.10.19, a 991 mm male (A. Paul), and a juvenile.

Type locality: "Ceylon" [= Sri Lanka].

Distribution: Sri Lanka (Central, Eastern, Northern, Sabaragamuwa, Southern, Western), NSL–1150 m.

Sources: Wall, 1921g, E.H. Taylor, 1950b, Deraniyagala, 1955, P. Silva, 1969, A. Silva, 1990b, 2001, 2009, Malhotra & Davis, 1991 and Vogel & Rooijen, 2011c.

Remarks: A subspecies of *D. pictus* (Gmelin) *fide* Meise & Henning, 1932 and Mertens, 1934a.

4. *Dendrelaphis biloreatus* Wall, 1908a. J. Bombay Nat. Hist. Soc. (1907–1908) 18(2): 273–274, pl. figs. 1–5.

Synonym: *Dendrophis gorei* Wall, 1910c.

Type: Holotype, BMNH, a 690–700 mm specimen (native, June, 1907).

Type locality: "Sadiya on the Bramaputra, 70 odd miles north-east of Dibrugarh, Assam" [= Sadiya, Assam State, NE India, 27°50'N, 95°40'E, elevation 125 m].

Distribution: Northeastern India (Arunachal Pradesh, Assam, Nagaland, N West Bengal), SW China (Xizang) and NE Myanmar (Shan, Yangon), NSL–2000 m.

Sources: Wall, 1910c, 1921g, 1923d, Zhao, 2006, I. Das, 2010 and Vogel & Rooijen, 2011b–c.

5. *Dendrelaphis calligastra* (A.C.L.G. Günther, 1867). Ann. Mag. Nat. Hist. (3) 20(115): 53. (*Dendrophis calligastra*)

Synonyms: *Dendrophis aruensis* G. Doria, 1875, *Dendrophis darnleyensis* Macleay, 1877, *Dendrophis katowensis* Macleay, 1877, *Dendrelaphis schlenckeri* Ogilby, 1898, *Dendrophis calligastra keiensis* Mertens, 1926, and *Dendrophis calligaster distinguendus* Meise & Henning, 1932.

Type: Holotype, BMNH 1867.5.6.71, a 914 specimen (E. Dämel, 1867).

Type locality: "Cape York, Australia" [Queensland].

Distribution: Austro-Papua. Eastern Indonesia (Papua, Aru, Babbar, Batanta, Halmahera, Jobi, Kei, Mefoor, Salawati, Schouten, Tanimbar, Ternate, Waigeo, Valise), coastal Papua New Guinea (Central, Eastern Highlands, East New Britain, East Sepik, Gulf, Madang, Milne Bay, Morobe, National Capital District, Western, West New Britain, West Sepik, Bismarck, Daru, d'Entrecasteaux, Ferguson, Louisiade Is.) and ext. NE Australia (NE Queensland, Torres Strait Is.), NSL–1150 m.

Sources: V.M. Tanner, 1950, Cogger et al., 1983a, O'Shea, 1996 and Cogger, 2000.

6. *Dendrelaphis caudolineatus* (Gray, 1834 *in* Gray & Hardwicke, 1830–1835). Illust. Indian Zool. 2(13–14): pl. 81. (*Ahaetulla caudolineata*)

Synonyms: *Leptophis caudalineatus* Cantor, 1847, and *Dendrophis octolineata* A.-M.-C. Duméril, Bibron & Duméril, 1854.

Type: Holotype, BMNH 1946.1.23.21 (formerly BMNH 1858.2.17.19).

Type locality: Unknown. Designated as West Malaysia *fide* Meise & Hennig (1932: 281), but Sri Lanka *fide* BMNH catalogue.

Distribution: Malay Peninsula and Greater Sundas. Southern Myanmar, Mergui Arch., S Thailand (Narathiwat, Pattani, Phang Nga, Phuket, Ranong), West Malaysia (Johor, Pahang, Pinang, Penang and Tioman Is.), Singapore, East Malaysia (Sabah, Sarawak, Labuan Is.), Brunei and Indonesia (Aceh, Babi, Bangka, Batu, Belitung, Besar, Kalimantan, Kangean Arch., Mentawai Arch., Natuna Arch., Nias, Pemanggil, Riau Arch., Seribuat Arch., Tengah, Tinggi, Sumatra), NSL–1525 m.

Sources: Leviton, 1961a, 1970b, P. Silva, 1969, I. Das, 2007b, 2012 and Grismer, 2011.

Remarks: Official Specific Name no. 1572 *fide* Opinion 524 (ICZN, 1958a). Records from Java doubtful *fide* Rooijen & Vogel (2012: 11).

7. *Dendrelaphis caudolineolatus* (A.C.L.G. Günther, 1869). Proc. Zool. Soc. London 37(1): 506, pl. 40, fig. 1. (*Dendrophis caudolineolata*)

Synonyms: *Dendrophis gregorii* Haly, 1888 and *Dendrophis effrenis* F. Werner, 1909b.

Types: Syntypes (2), BMNH 1946.1.23.21, a 610 mm male (R.H. Barnes, 1858–1869) and BMNH 1858.2.17.19 (H.H. Higgins, 1856–1858).

Type locality: "Ceylon" [= Sri Lanka].

Distribution: Sri Lanka (Central, Sabaragamuwa, Southern) and S India (Kerala, Tamil Nadu), NSL–655 m.

Sources: M.A. Smith, 1943, E.H. Taylor, 1965, A. Silva, 1990 and R.C. Sharma, 2003, 2007.

D

8. *Dendrelaphis chairecacos* (F. Boie, 1827). Isis Von Oken 20(6): 541. (*Dendrophis chairecacos*)

Type: Neotype, BMNH 1924.10.13.15, a 485 (svl) mm male (A.F. Abercromby, 1888–1921), designated by Rooijen & Vogel (2009: 197).

Type locality: "Kottayam. Travancore (South India)" [= Kottayam, Kerala State, ext. SW India, 9°36'N, 76°32'E, elevation 5 m] via neotype selection.

Distribution: Southern India (Kerala, Tamil Nadu), NSL.

Sources: Rooijen & Vogel, 2008c, 2009.

Remarks: Skin of holotype in BMNH lost *fide* Bolenger (1894: 88–89). Photograph of neotype in Rooijen & Vogel (2009: fig. 3). Resurrected from synonymy of *D. schokari fide* Rooijen & Vogel, 2009.

9. *Dendrelaphis cyanochloris* (Wall, 1921f). Rec. Ind. Mus. 22(2): 155. (*Dendrophis pictus cyanochloris*)

Type: Lectotype, BMNH 1940.3.4.27 (formerly IM 7734), an 850 mm male, designated by M.A. Smith (1943: 244).

Type locality: "Darjeeling" [West Bengal State, NE India, 27°02'N, 88°16'E, elevation 2100 m] via lectotype selection and restriction *fide* Meise & Henning (1932: 290).

Distribution: Southeastern Asia. Northeastern India (Andaman & Nicobars: Andaman Is., Arunachal Pradesh, Assam, Meghalaya, Nagaland, West Bengal), Bhutan (Mongar), Bangladesh, Myanmar (Kachin, Mandalay), Thailand (Narathiwat, Phang Nga, Prachin Buri, Udon Thani), West Malaysia (Pahang, Pinang, Penang and Tioman Is.) and Singapore, NSL–3000 m.

Sources: E.H. Taylor, 1965, M.J. Cox et al., 1998, Chan-ard et al., 1999, Leong, 2005, Whitaker & Captain, 2004, Vogel & Rooijen, 2007, Rooijen & Vogel, 2008c, M.F. Ahmed et al., 2009, Grismer, 2011, Wangyal, 2011 and Vogel et al., 2012a.

Remarks: Wall's donation of a Burmese specimen from Mergui (Tenasserim Division) that he designated as the type (BMNH 1946.3.4.27) [and type locality restriction of M.A. Smith, 1943: 244] invalid as it was not among the syntypes (only Kachin and Mandalay Divisions). Photographs of the lectotype in Vogel & Rooijen (2007: figs. 7–8).

D

10. *Dendrelaphis flavescens* Gaulke, 1994b. Herpetol. J. 4(4): 138–140, fig. 2. (*Dendrelaphis caudolineatus flavescens*)

Type: Holotype, SMF 74846, a 1260 mm female.
Type locality: "Sanga-Sanga, Sulu Archipelago, Philippines."
Distribution: Southwestern Philippines (Sulu Arch.: Bongao, Bubuan, Sanga-Sanga, Sibutu, Sitanki, Tawitawi).
Sources: E.H. Taylor, 1922a and Rooijen & Vogel, 2012.
Remarks: Photograph of the type in Rooijen & Vogel (2012: fig. 6).

11. *Dendrelaphis formosus* (F. Boie, 1827). Isis von Oken 20(6): 542. (*Dendrophis formosa*)

Synonyms: *Dendrophis formosus* H. Boie *in* Schlegel, 1826a *(nomen nudum)* and *Dendrophis formosus* Schlegel, 1826b *(nomen nudum).*
Type: Holotype, RMNH 877, a 1470 mm specimen (H. Boie & H.C. Macklot, 1825–1827).
Type locality: "Java" [SW Indonesia].
Distribution: Malay Peninsula and Greater Sundas. Southern Thailand (Narathiwat, Phang Nga, Yala), West Malaysia (Kedah, Kelantan, Pahang, Pinang, Perak, Penang Is.), Singapore, East Malaysia (Sabah, Sarawak), Brunei and Indonesia (Bangka, Belitung, Java, Kalimantan, Mentawai Arch., Riau Arch., Sumatra), NSL–1500 m.
Sources: E.H. Taylor, 1965, Tweedie, 1983, Lim & Lim, 1992, David & Vogel, 1996, M.J. Cox et al., 1998, I. Das, 2007b, Vogel & Rooijen, 2007 and Rooijen & Vogel, 2008c.
Remarks: Original description based on H. Boie's MS (1823–1825). Neotype designation of Vogel & Van Rooijen (2007: 30–31) invalid as the specimen appears to be Boie's holotype. Photographs of the type in Vogel & Rooijen (2007: figs. 3–4)

12. *Dendrelaphis fuliginosus* L.E. Griffin, 1909a. Philippine J. Sci. 6D: 55–56.

Type: Neotype, FMNH 67409, a 780 mm male (D.S. Rabor, 1935–1977), designated by Rooijen & Vogel (2012: 12)..
Type locality: "Negros, Philippines" via neotype selection.
Distribution: Central Philippines (Banton, Bantayan, Cagayancillo, Masbate, Mindoro, Negros, Panay, Ticao).
Source: Rooijen & Vogel, 2012.
Remarks: Holotype in BSM destroyed in Jan. 1945 during World War II *fide* Leviton (1970b: 390). Photograph of neotype in Rooijen & Vogel (2012: fig. 7).

13. *Dendrelaphis gastrostictus* (Boulenger, 1894a). Cat. Snakes Brit. Mus. 2: 86–87, pl. 4, figs. 3–3b. (*Dendrophis gastrostictus*)

Synonyms: *Dendrophis meeki* Boulenger, 1895b, *Dendrophis nouhuysii* Lidth de Jeude, 1911b, and *Charlespiersonserpens gastrostictus tyeipperae* Hoser, 2012ab *(nomen illegitimum).*
Type: Holotype, BMNH 1946.1.23.20, an 1120 mm male (M.A. Linden).
Type locality: "N. W. New Guinea" [= West Papua Prov., E Indonesia].
Distribution: New Guinea. Eastern Indonesia (Papua, West Papua) and Papua New Guinea (Bay, Central, Eastern Highlands, Gulf, Milne, Morobe, Simbu, Western, Western Highlands, Normanby and Fergussen Is.), NSL–1475 m.
Source: O'Shea, 1996.

14. *Dendrelaphis girii* Vogel & Van Rooijen, 2011c. Taprobanica 3(2): 80–83, figs. 2–3.

Type: Holotype, BNHS 3494, a 1026 mm male (S. Kehimkar & V. Giri, 30 Sept. 2004).
Type locality: "Castle Rock, District Belgaum, Karanataka, India" [= Castlerock, E Karnataka State, SW India, 15°24'N, 74°20'E, elevation 615 m].
Distribution: Western Ghats of SW India (E Karnataka, Maharashtra, Tamil Nadu), 615 m.

15. *Dendrelaphis grandoculis* (Boulenger, 1890a). Fauna Brit. India, Rept. Batr.: 337. (*Dendrophis grandoculis*)

Type: Lectotype, BMNH 1946.1.23.27, a 1200–1219 mm female (R.H. Beddome, 1857–1882), by locality restriction of Meise & Henning (1932: 286).
Type locality: "Tinnevelly Hills, South India" via type locality restriction.
Distribution: Western Ghats of SW India (S Karnataka, Kerala, Tamil Nadu), 610–1290 m.
Sources: Malhotra & Davis, 1991, Whitaker & Captain, 2004 and Chandramouli & Ganesh, 2010.

16. *Dendrelaphis grismeri* Vogel & Rooijen, 2008. Herpetozoa 21(1–2): 21–27, figs. 15–16.

Type: Holotype, RMNH 40176, an 825 mm female (B. Moens, 1862).
Type locality: "Wahaaij, côte N. O. de Céram (Wahai, northeast coast of Seram)" [= Wahaaij, Seram, Maluku, E Indonesia].
Distribution: Eastern Indonesia (Ambon, Boana, Manipa, Misol, Seram).

17. *Dendrelaphis haasi* Rooijen & Vogel, 2008a. Amphibia-Reptilia 29(1): 108–110, figs. 4, 5a–d.

Type: Holotype, RMNH 5507, a 905 mm male (J.D. Pasteur, 1890).
Type locality: "Nias, Indonesia" [= Nias Is., off NW coast of Sumatra, W Indonesia, ca. 1°08'N, 97°31'E].
Distribution: West Malaysia (Tioman) and W Indonesia (Belitung, Borneo, Java, Mentawai Arch., Nias, Sumatra).
Source: Rooijen & Vogel, 2008a.

18. *Dendrelaphis hollinrakei* Lazell, 2002. Mem. Hong Kong Nat. Hist. Soc. 25: 46–49, figs. 19a, 20.

Type: Holotype, MCZ 173278, a 651 mm female (B. Hollinrake, 1971–1984).
Type locality: "Shek Kwu Chau, 22°12'N, 113°59'E, South China Sea."
Distribution: Southeastern China (Shek Kwu Chau Is.).

19. *Dendrelaphis humayuni* Tiwari & Biswas, 1973. J. Zool. Soc. India 25(1–2): 59–60, fig. 2c.

Type: Holotype, ZSI 22386, an 829 mm specimen (D.P. Sanyal, 1 May 1966).
Type locality: "Campbell Bay, Great Nicobar Island" [Andaman & Nicobars, E India, Bay of Bengal].
Distribution: Eastern India (Andaman & Nicobars: Great Nicobar Is.) and West Malaysia (Malacca, Pinang).
Sources: Biswas & Sanyal, 1977a, 1980, Vijayakumar & David, 2006 and Rooijen & Vogel, 2008c.

20. *Dendrelaphis inornatus* Boulenger, 1897d. Ann. Mag. Nat. Hist. (6) 19(113): 507.

Synonym: *Dendrophis pictus timorensis* M.A. Smith, 1927.
Types: Syntypes (4), BMNH 1946.1.23.23–26, males and females, longest syntype 1200 mm (A.H. Everett, 1895–1896).
Type locality: "Sumba and Savu" [Lesser Sunda Islands, S Indonesia]. Restricted to Sumba *fide* Meise & Hennig (932: 287).
Distribution: Lesser Sundas of S Indonesia (Alor, Flores, Komoto, Lombien, Moyo, Pandar, Rinca, Roti, Sawu, Semau, Sumba, Sumbawa, Timor, Wetar) and Timor-Leste, NSL–300 m.
Sources: Kaiser et al., 2011 and Lang, 2011.

21. *Dendrelaphis kopsteini* Vogel & Rooijen, 2007. Zootaxa (1394): 34–38, figs. 9–12.

Type: Holotype, MNHN 1962.1052, a 1410 mm female (P. Pfeffer).
Type locality: "Endau river, State of Johore, Federation of Malaysia" [= Endau River, Johore State, SE West Malaysia].
Distribution: Malay Peninsula and Sumatra. Peninsular Thailand (Nakhon Si Tammarat, Phang Nga), West Malaysia (Johore, Pahang, Perak, Perlis), Singapore and W Indonesia (Mentawai Arch., Sumatra), NSL–1100 m.
Sources: I. Das, 2007a, 2012 and Rooijen & Vogel, 2008c.

22. *Dendrelaphis levitoni* Rooijen & Vogel, 2012. Zootaxa (3272): 13–15, fig. 9.

Type: Holotype, CAS 15803, a 980 mm female (A. Seale, 6 July 1908).
Type locality: "Puerto Princesa, Palawan, P.I." [= Puerto Princesa, cen. Palawan, SW Philippines, 9°58'N, 118°47'E, elevation 70 m].
Distribution: Southwestern Philippines (Balabac, Busuanga, Candaraman, Palawan), 70 m.

23. *Dendrelaphis lorentzii* (Lidth de Jeude, 1911b). Nova Guinea (Zool.) 9(2): 274, pl. 8, figs. 2a–c. (*Dendrophis lorentzii*)

Synonym: *Charlespiersonserpens jackyhoserae* Hoser, 2012ab *(nomen illegitimum)*.
Type: Holotype, RMNH 4710, an 820 mm specimen (H.A. Lorentz [First Dutch New Guinea Exped.], 18 July 1907).
Type locality: "Sabang, Nouvelle-Guinée" [= Sabang, Merauke, on Lorentz River, S Papua Prov., E Indonesia, 4°46'S, 138°47'E, elevation 30 m].
Distribution: New Guinea. Eastern Indonesia (S Papua) and Papua New Guinea (Morobe, Western), NSL–665 m.
Source: O'Shea, 1996.

24. *Dendrelaphis luzonensis* Leviton, 1961a. Occ. Pap. Nat. Hist. Mus. Stanford Univ. (2): 1–4. (*Dendrelaphis caudolineatus luzonensis*)

Type: Holotype, CAS 61134, a 1264 mm female (E.H. Taylor, April 12–July, 1920).
Type locality: "Los Baños, Laguna Prov., Luzon Island, Philippines" [= 14°10'N, 121°15'E, elevation 30 m].
Distribution: Northern Philippines (Luzon, Camaguin Arch.).
Sources: Rooijen & Vogel, 2012.

25. *Dendrelaphis marenae* Vogel & Rooijen, 2008. Herpetozoa 21(1–2): 13–21, figs. 9–14, 17.

Type: Holotype, MNHN 1994.59, a 960 mm female (Montamo-Ray, 1880).
Type locality: "Prov. Albay, island of Luzon, Philippines."
Distribution: East Indies. Indonesia (Sulawesi: Buton) and Philippines (Balabac, Bantayan, Basilan, Bohol, Bongao, Busuanga, Calauit, Camiguin, Candaraman, Catanduanes, Cebu, Culion, Kalotkot, Leyte, Luzon, Masbate, Mindanao, Mindoro, Negros, Palawan,

Panay, Polillo, Romblon, Samar, Siquijor, Surigao), NSL–1000 m.
Source: Koch, 2012.

26. *Dendrelaphis modestus* Boulenger, 1894a. Cat. Snakes Brit. Mus. 2: 91, pl. 4, fig. 4.

D

Types: Syntypes (2), BMNH 1946.1.6.65–66 (formerly BMNH 1878.1.31.4–5), 1130 mm and 785 (svl) females (H.M.S. Challenger Voy. [Stn. 197], 15–17 Oct. 1874).
Type locality: "Ternate" [Maluku, Indonesia].
Distribution: Eastern Indonesia (Maluku: Bacan, Halmahera, Morotai, Ternate)
Source: Rooijen & Vogel, 2012.

27. *Dendrelaphis ngansonensis* (Bourret, 1935d). Bull. Gén. Instr. Publ. 14(9): 292. (*Dendrophis pictus ngansonensis*)

Type: Holotype, VNUH M.449 (formerly RLB M.449, a 1067 mm specimen (M. Clémençon, 1934).
Type locality: "Ngân-Son, Tonkin (Alt. 700m)" [= Ngân Son, Bac Kan Prov., N Vietnam, 22°26'N, 106°01'E, elevation 700 m].
Distribution: Southeast Asia. Southern China (Hainan, Yunnan), Thailand, N Laos (Xiangkhouang) and N Vietnam (Bac Giang, Bac Kan, Cao Bang, Ha Giang, Hai Duong, Ha Tinh, Lang Son, Ninh Binh, Nghe An, Quang Binh, Quang Nam, Quang Tri, Son La, Thai Nguyen, Thanh Hoa, Thua Thien-Hue, Vinh Phuc), 400–1100 m.
Sources: Ziegler & Vogel, 1999, Orlov et al., 2000, Bain et al., 2007, Nicodemo & Bain, 2007, V.S. Nguyen et al., 2009 and Cox et al., 2013.

28. *Dendrelaphis nigroserratus* Vogel, Rooijen & Hauser, 2012a. Zootaxa (3392): 39–41, figs. 4a, 6.

Type: Holotype QSMI 1282, a 428 mm female (S. Hauser, 13 Oct. 2011).
Type locality: "Umphang Wildlife Sanctuary, Umphang district, province Tak, Thailand. Found on Highway 1090, at 5.1 km north and uphill from the bridge over the Mae Klong Khi, near the Karen village of Mae Klong Khi, c. 1100 m a.s.l."
Distribution: Myanmar (Mergui) and Thailand (Kanchanaburi, Phetchaburi, Tak), 900–1350 m.

29. *Dendrelaphis oliveri* (E.H. Taylor, 1950b). Univ. Kansas Sci. Bull. 33(14): 555–557, pl. 18, fig. 1. (*Ahaetulla oliveri*)

Type: Holotype, formerly EHT-HMS 30388 & EHT-CC 1184, a 756 mm specimen (E.H. Taylor, 1944), location unknown.
Type locality: "12 miles north of Trincomalee, Ceylon" [= near Nilaveli, 18 km N Trincomalee, N Eastern Prov., NE Sri Lanka, ca. 8°43'N, 81°11'E, elevation 3–5 m *fide* Bauer & Silva, 2007: 5].
Distribution: Northeastern Sri Lanka (Eastern), NSL. Known only from type locality.

30. *Dendrelaphis papuensis* Boulenger, 1895e. Ann. Mag. Nat. Hist. (6) 16(95): 409.

Synonym: *Charlespiersonserpens papuensis lizelliottae* Hoser, 2012ab *(nomen illegitimum).*
Types: Syntypes (5), BMNH 1946.1.6.57–61, two males and three females, longest syntype 1050 mm (A.S. Meek, 1894).
Type locality: "Trobriand Islands, British New Guinea" [= Trobriand Is., Milne Bay Prov., E Papua New Guinea].
Distribution: Papua New Guinea (Central, Milne Bay, Morobe, Northern, Western, Fergusson, Rossel, Sudest, Trobriand and Woodlark Is).
Source: O'Shea, 1996.

31. *Dendrelaphis philippinensis* (A.C.L.G. Günther, 1879a). Proc. Zool. Soc. London 47(1): 78, pl. 4. (*Dendrophis philippinensis*)

Synonym: *Dendrelaphis caeruleatus* Griffin, 1909a.
Type: Holotype, BMNH 1946.1.6.69 (formerly BMNH 1877.10.9.62), a 1060 mm female (A.H. Everett, March–May, 1878).
Type locality: "North Mindanao" [Philippines].
Distribution: Eastern and S Philippines (Basilan, Bohol, Camiguin, Cebu, Catanduanes, Dinagat, Kalotkot, Leyte, S Leyte, S Luzon, Mindanao, Polillo, Samar, Siquijor, Surigao).
Source: Rooijen & Vogel, 2012.
Remarks: Photograph of the type in Rooijen & Vogel (2012: fig. 12).

32. *Dendrelaphis pictus* (Gmelin, 1789). Syst. Nat., ed. 13, 1(3): 1116. (*Coluber pictus*)

Synonyms: *Coluber boiga* Lacépède, 1789 (*nomen rejiciendum*), *Coluber decorus* G. Shaw, 1802, *Ahaetulla fasciata* Link, 1807, *Coluber plinii* Merrem, 1820, *Dendrophis polychroa* Reinwardt *in* F. Boie, 1827, *Ahaetulla bellii* Gray *in* Hardwicke & Gray, 1834, *Dendrophis lateralis* Gray, 1835 *in* Gray & Hardwicke, 1830–1835, *Bungarus filum* A.C.L.G. Günther, 1858, *Dendrophis pictus polychroa* Jan, 1863b (*nomen nudum*), and *Ahaetulla boiga intermedia* Mertens, 1927b.
Type: Holotype, NHR Lin-37 (formerly MAFR), a 940 mm specimen (Mus. Drottn.).
Type locality: Unknown. Designated as Waigiu [= Waigeo, West Papua Prov., E Indonesia, 0°12'S, 130°50'E] *fide* F. Boie (1827: 543).
Distribution: Southeastern Asia and East Indies. Southern China (Guangdong, Hainan, Hong Kong, SW Yunnan), Vietnam (An Giang, Bac Giang, Binh Duong, Binh

Thuan, Ca Mau, Dak Nang, Da Nang, Dien Bien, Dong Nai, Gia Lai, Ho Chi Minh City, Khanh Hoa, Kien Giang, Lao Cai, Phu Tho, Phu Yen, Quang Binh, Quang Nam, Quang Ninh, Quang Tri, Soc Trang, Tay Ninh), Thailand (Chaiyaphum, Chanthaburi, Chiang Mai, Chumphon, Khon Kaen, Loei, Lop Buri, Nakhon Pathom, Nakhon Ratchasima, Nakhon Sawan, Nakhon Si Thammarat, Phang Nga, Prachin Buri, Songkhla, Surat Thani, Trang), Cambodia (Preah Vihear), West Malaysia (Perak, Perlis, Aceh, Besar, Mawat, Tioman Is.), Singapore, East Malaysia (Sabah, Sarawak), Brunei, Indonesia (Papua, West Papua, Ambon, Bali, Bangka, Belitung, Belitung, Boano, Borneo, Butung, Flores, Java, Kalimantan, Lombok, Manipa, Mentawei Arch., Misool, Natuna, Nias, Riau Arch., Roti, Sangihe Arch., Seram, Sula Arch., Sulawesi, Sumatra, Ternate, Timor) and Philippines (Bohol, Bongao, Busuanga, Cagayan, Catanduanes, Cebu, Culion, Guimaras, Jolo, Lapac, Luzon, Masbate, Mindanao, Mindoro, Negros, Palawan, Panay, Polillo, Samar, Siargao, Sula, Tawitawi), NSL–1705 m.

Sources: Wall, 1910a, Bergman, 1955d, Leviton, 1970b, Auffenberg, 1980, Karsen et al., 1986, Ross & Gonzales, 1992, Zhao & Adler, 1993, Gaulke, 1994b, Gaulke & Altenbach, 1994, Kovae et al., 1995, David & Vogel, 1996, M.J. Cox et al., 1998, Xie et al., 1998, Chan-ard et al., 1999, Ziegler & Vogel, 1999, Malkmus et al., 2002, Whitaker & Captain, 2004, Rooijen & Vogel, 2008c, Vogel & Rooijen, 2008, 2011a, V.S. Nguyen et al., 2009, Sajdak, 2010, Grismer, 2011 and T. Hartmann et al., 2013.

Remarks: Official Specific Name no. 1574 *fide* Opinion 525 (ICZN, 1958b). Original color sketches of holotypes of *Ahaetulla belli* by Hardwicke (1756–1823) (2: fig. 9) and *Dendrophis lateralis*by Hardwicke (1756–1823) (2: fig. 11) in BMNH. Type locality restrictions of Borneo *fide* Mertens (1930: 188) and Java *fide* Meise & Hennig (1932: 289) are invalid.

33. *Dendrelaphis proarchos* (Wall, 1909d). J. Bombay Nat. Hist. Soc. (1909–1910) 19(2): 347. (*Dendrophis proarchos*)

Type: Holotype, not designated, one specimen (St. Joseph's College, Darjeeling).

Type locality: "Darjeeling" [West Bengal State, NE India, 27°02'N, 88°16'E, elevation ca. 2100 m].

Distribution: Southeastern Asia. Northeastern India (Assam, Sikkim), Myanmar (Kachin, Rakhine, Shan), Laos (Phongsali) and Vietnam (Tonkin), 2100 m.

Source: Vogel & Rooijen, 2011a.

Remarks: Supplemental original description in Wall (1910c: 827–829, figs. a–d), who listed 24 syntypes (of which BMNH 1908.6.23.27–28 are known and BMNH 1908.6.23.27 was described by Vogel & Rooijen, 2011: 108) with type localities of "Dibrugarh" [21 spec.], "Sadiya" [1 spec.], "North Lakhimpur (Dejoo)" [1 spec.], and "Doom Dooma" [1 spec.]. Type locality restriction to Dibrugarh, Assam *fide* M.A. Smith (1943: 242) invalid.

34. *Dendrelaphis punctulatus* (Gray *in* King, 1827). Surv. West. Aust. 2: 432. (*Leptophis punctulatus*)

Synonyms: *Elaps lewisii* Gray *in* Grey, 1841b, *Dendrophis olivacea* Gray, 1842c, *Dendrophis lineolata* Hombron & Jacquinot, 1842, *Dendrophis prasinus* Girard, 1858a, *Dendrophis striolatus* W.C.H. Peters, 1867, *Dendrophis punctulatus astrostriata* A.B. Meyer, 1874, *Dendrophis punctulatus fasciata* A.B. Meyer, 1874, *Dendrophis gracilis* Macleay, 1875, *Dendrophis breviceps* Macleay, 1877, *Dendrophis macrops* A.C.L.G. Günther, 1877, *Dendrophis olivacea* Macleay, 1878b (*nec* Grey), *Dendrophis bilorealis* Macleay, 1884a, *Dendrophis elegans* Ogilby, 1891, and *Dendrophis papuae* Ogilby, 1891.

Type: Holotype, BMNH 1946.1.23.34, a 1219 mm specimen (J. Hunter, 1818–1822).

Type locality: "Careeing Bay, on the north coast, Australia" [= N Western Australia, 15°06'S, 125°00'E *fide* Cogger et al., 1983a: 211].

Distribution: Austro-Papua. Eastern Indonesia (Papua, West Papua), Papua New Guinea (Central, East New Britain, East Sepik, Eastern Highlands, Gulf, Madang, Milne Bay, Northern, Western, West New Britain, West Sepik, New Britain, New Ireland and Normanby Is.) and N and E Australia (E New South Wales, N Northern Territory, E Queensland, ext. N Western Australia, Angustus, Bathurst, Bickerton, Centre, Corneille, Coronation, Cotton, Croker, Groote Eylandt, Hawknest, Hayman, Hidden, King Hall, Koolan, Lachlan, Lindeman, Melville, North, North Goulburn, Vanderlin and Woodah Is.), NSL–1850 m.

Sources: J. White, 1790, Hoser, 1989, O'Shea, 1996, Crombie & Pregill, 1999, Cogger, 2000 and Kraus & Allison, 2004.

35. *Dendrelaphis salomonis* (A.C.L.G. Günther, 1872b). Ann. Mag. Nat. Hist. (4) 9(49): 25. (*Dendrophis salomonis*)

Types: Syntypes (2), BMNH 1946.1.5.97 and BMNH 1946.1.6.11, longest syntype 813 mm (G. Krefft, 1861–1872).

Type locality: "Solomon Islands."

Distribution: Papua New Guinea (Milne Bay, North Solomons, Buka, Buin, Bougainville; Mismia and Sudest Is.) and Solomon Islands (Fauro, Choiseul, Kolombangara, New Georgia, Santa Cruz, Santa Isabel, Malaita, Guadalcanal), 10–20 m.

Source: O'Shea, 1996.

D

36. *Dendrelaphis schokari* (Kuhl, 1820). Zool. Vergl. Anat.: p. 80–82. (*Dipsas schokari*)

Type: Neotype, RMNH 842, a 735 mm female (G.A. Frank, 1853), designated by Rooijen & Vogel (2008b: 39).

Type locality: "Ceylon" [= Sri Lanka] via neotype designation.

Distribution: Sri Lanka (Central, Eastern, Northern, Southern, Western).

Sources: A. Silva, 1990b, 2009, Somaweera, 2006 and Rooijen & Vogel, 2008c, 2009.

Remarks: A valid species *fide* Rooijen & Vogel (2008c: 33). Syntypes in AG destroyed by fire in 1906 *fide* Rooijen & Vogel (2008c: 38). Photographs of neotype in Rooijen & Vogel (2008b: figs. 5–6).

37. *Dendrelaphis striatus* (Cohn, 1905). Zool. Anz. 29(17): 543–544. (*Dendrophis pictus striatus*) (*nomen corrigendum*)

Synonym: *Dendorophis pictus striata* Cohn, 1905 (*nomen incorrigendum*)

Types: Syntypes (2), UMB, a 775 mm female and 640 mm male (P. Cassens-Brake).

Type locality: "Sumatra, Asahau" [= Asahan River, NE Sumatra, W Indonesia, ca. 2°57'N, 99°48'E].

Distribution: Malay Peninsula and Greater Sundas. Peninsular Thailand (Nakhon Si Tammarat, Phang Nga), West Malaysia (Pahang, Perak, Selangor, Tioman Is.) and W Indonesia (Kalimantan, Sumatra), NSL–300 m.

Sources: Tweedie, 1983, G. Vogel, 1990, 1995, David & Vogel, 1996, M.J. Cox et al., 1998, Wood et al., 2003, Rooijen & Vogel, 2008c, Grismer, 2011 and Vogel et al., 2012a.

38. *Dendrelaphis subocularis* (Boulenger, 1888h). Ann. Mus. Civ. Stor. Nat. Genova (2) 6: 600, pl. 6, figs. 2, 2a–2b. (*Dendrophis subocularis*)

Types: Syntypes (2), BMNH 1946.1.6.10, a juvenile male, and MSNG CE 30455, a female, longest syntype 820 mm (M.L. Fea, 1886).

Type locality: "Bhamò, Burma" [= Bhamo, S Kachin Division, N Myanmar, 24°15°N, 97°14'E, elevation 115 m].

Distribution: Southeastern Asia. Northeastern India, S China (Guangdong, Guangxi, Yunnan), Myanmar (Yangon), Thailand (Chiang Mai, Kanchanaburi, Mae Hong Son, Nakhon Ratchasima, Phetchabun, Tak, Udon Thani, Uthai Thani), N Laos, Cambodia and N Vietnam (Hai Duong, Lam Dong, Ninh Thuau, Quang Nam), NSL–1620 m.

Sources: E.H. Taylor, 1965, Campden-Main, 1970a, Saint-Girons, 1972a, Zhao & Adler, 1993, M.J. Cox et al., 1998, Zug et al., 1998, Zhao, 2006 and V.S. Nguyen et al., 2009.

Remarks: An isolated Java population is distinct but unnamed *fide* Rooijen & Vogel (2010: 90).

39. *Dendrelaphis terrificus* (W.C.H. Peters, 1872a). Mber. Königl. Akad. Wiss. Berlin 1872(7): 583–584. (*Dendrophis terrificus*)

Type: Lectotype, ZMB 7416, a 307 mm juvenile (A.B. Meyer, Dec. 1870–Nov. 1871), designated by Leviton (1970b: 389, 391) and Rooijen & Vogel (2012: 20).

Type locality: "Gorontalo, Sulawesi" [Indonesia] via lectotype selection.

Distribution: Central Indonesia (N Sulawesi).

Sources: Lang & Vogel, 2005, Koch, 2012 and Rooijen & Vogel, 2012.

Remarks: Type locality restriction of Borongan, Samar Is., Philippines *fide* Meise & Hennig (1932: 281) invalid.

40. *Dendrelaphis tristis* (Daudin, 1803c). Hist. Nat. Rept. 6: 430–431. (*Coluber tristis*)

Synonyms: *Coluber scandens* Bechstein, 1802a, *Leptophis mancas* Bell, 1825, *Natrix comberi* Wagler, 1825 (*nomen ineditum*), *Natrix diana* Wagler, 1825 (*nomen ineditum*), *Dendrophis maniar* Russell *in* F. Boie, 1827, *Chrysopelea boiei* A. Smith, 1836, *Dendrophis boii* Cantor, 1839, *Coluber cumberi* A.-M.-C. Duméril, 1853 (*nomen nudum*), *Dendrophis pictus vertebralis* Jan, 1863b, *Dendrophis multimaculatus* Higgins, 1873, *Dendrophis helena* F. Werner, 1893, and *Dendrelaphis tristis trapobanensis* Wall, 1921.

Type: Neotype, SMF 58442, a 665 mm male (H. Schetty), designated by Rooijen & Vogel (2008b: 40).

Type locality: "Calcutta, India" [= Kolkata, West Bengal State, NE India, 22°34'N, 88°22'E, elevation 5 m] via neotype designation.

Distribution: Southern Asia. Eastern Pakistan (Punjab), NE India (Andhra Pradesh, Assam, Gujarat, Karnataka, Kerala, Madhya Pradesh, Maharashtra, Orissa, Tamil Nadu, Uttar Pradesh, West Bengal), Sri Lanka (Kalutara), Bangladesh, Nepal (Bardiya, Chitwan, Ilam, Kailali, Mahottari, Makwanpur, Palpa, Panchthar, Rupandehi, Sankhuwasabha, Taplejung), Bhutan (Sarpang) and Myanmar, NSL–2000 m.

Sources: Wall, 1910c, k, P. Silva, 1969, Kramer, 1977, Schleich & Kästle, 2002, Tiwari & Shah, 2004, Whitaker & Captain, 2004, N. Khaire, 2006, Rooijen & Vogel, 2008c, 2009, I. Das, 2010, Masroor, 2012 and Wangyal, 2012.

Remarks: Holotype lost *fide* Rooijen & Vogel (2008b: 37). Photographs of neotype in Rooijen & Vogel (2008b: figs. 9–10). Pakistan record doubtful *fide* M.S. Khan (2002: 206).

41. *Dendrelaphis underwoodi* Rooijen & Vogel, 2008b. Raffles Bull. Zool. 56(1): 191–193, figs. 1–2.

Type: Holotype, RMNH 7447, a 900 mm female (J. Semmelink, 1872–1875).
Type locality: "Radjamandala, West Java, 335 meters" [= Rajamandala, Jawa Barat Prov., Java, W Indonesia, 7°15'S, 108°11'E, elevation 335 m].
Distribution: Western Indonesia (W Java), 335–900 m.

42. *Dendrelaphis walli* Vogel & Rooijen, 2011b. Bonn. Zool. Bull. 60(1): 22–23, fig. 2.

Type: Holotype, CAS 234880, a 710 mm female (A.K. Shein & T. Nyo, 23 Aug. 2005).
Type locality: "Myanmar (Chin State: Mindat township, Mindat District, 21 26 43.5 N, 94 00 24.0 E, 3582 ft)."
Distribution: Southern Myanmar (Chin), 1090 m.

DENDROASPIS Schlegel, 1848 (Elapidae)

Synonyms: *Dinophis* Hallowell, 1852c, *Dendroëchis* J.G. Fischer, 1855 *(nomen incorrigendum)*, *Dendraspis* A.H.A. Duméril, 1856 *(nomen praeoccupatum)*, *Deniraspis* – F. Müller, 1878b *(nomen incorrectum)*, *Denraspis* – Maass-Berlin, 1933 *(nomen incorrectum)*, *Dendrospis* – Briceño-Rossi, 1934 *(nomen incorrectum)*, *Dendroapsis* – Ionides, 1950 *(nomen incorrectum)*, *Dendraspsis* – Anthony, 1955 *(nomen incorrectum)*, and *Dendropsis* – Tanner & Avery, 1982 *(nomen incorrectum)*.
Type species: *Elaps jamesoni* Traill, 1843.
Distribution: Subsaharan Africa.
Sources: Klemmer, 1963, Broadley, 1968a, 1983, Pitman, 1974, Branch, 1979, Saint-Girons & Detrait, 1980, Golay et al., 1993, Slowinskii & Keogh, 2000, Dobiey & Vogel, 2007, Kelly et al., 2009, Zaher et al., 2009 and Hoser, 2012e.
Remarks: Official Generic Name no. 1598 *fide* Opinion 709 (ICZN, 1964b).

1. *Dendroaspis angusticeps* (A. Smith, 1849 *in* 1838–1849). Illust. Zool. So. Africa, Rept.: pl. 70, figs. a–d, 2 pp. (*Naia angusticeps*)

Synonym: *Dendroaspis sjostedti* Lönnberg, 1907.
Type: Holotype, not designated, an 1842 mm specimen (A. Smith, 1831–1834), lost *fide* Broadley *in* Golay et al. (1993: 128).
Type locality: "Natal and in the country to the eastward, towards Delegoa Bay, South Africa" [= KwaZulu Natal, South Africa and Maputo Bay, Mozambique]. Restricted to Natal [= KwaZulu Natal Prov., South Africa] *fide* Loveridge (1942: 306).
Distribution: Eastern South Africa (Eastern Cape, KwaZulu Natal), NSL–210 m.
Sources: FitzSimons, 1946a, 1974, Broadley, 1983 and Sajdak, 2010.
Remarks: Holotype not in RSM or BMNH.

2. *Dendroaspis intermedius* A.C.L.G. Günther, 1865. Ann. Mag. Nat. Hist. (3) 15(86): 98, pl. 3, fig. c. (*Dendraspis intermedius*)

Type: Holotype, BMNH 1946.1.20.48, a 1397 mm male (C.J. Meller, 1861–1864).
Type locality: "Zambesi River" [= Zambezi River, Tete or Sofala/Zambézia Prov., cen. Mozambique].
Distribution: Eastern Africa. Eastern Democratic Republic of the Congo (SE Kasai Oriental, Katanga, Maniema, Orientale), SE Kenya (Coast, Eastern), E Tanzania (Arusha, Coast, Kilimanjaro, Lindi, Mbeya, Morogoro, Mtwara, Piwani, Ruvuma, Tanga, Mafia and Zanzibar Is.), Malawi (Central, Northern, Southern), S Mozambique (Inhambane, Manica, Maputo) and E Zimbabwe (Manicaland), NSL–1700 m.
Sources: Loveridge, 1950, Witte, 1953, Sweeney, 1961, Ionides & Pitman, 1965, Broadley & Cock, 1975, Broadley, 1983, Chifundera, 1990, Spawls et al., 2002 and Broadley & Blaylock, 2013.
Remarks: A valid species *fide* W. Wüster (unpubl. data). Probably occurs in S Somalia *fide* Lanza (1983a: 229). Records from Democratic Republic of the Congo rejected *fide* David & Ineich (1999: 78).

3. *Dendroaspis jamesoni* (Traill, 1843). Edinburgh New Philos. J. 34(67): 54–55, 2 figs. (*Elaps jamesoni*)

Synonyms: *Dendraspis welwitschii* A.C.L.G. Günther, 1865, *Dinophis fasciolatus* J.G. Fischer, 1885b, *Dendraspis neglectus* Bocage, 1888, and *Dendraspis jamesoni kaimosae* Loveridge, 1936a.
Type: Holotype, NMSZ 1869.3.147 (formerly TST & RSM), an 1867 mm specimen (T.S. Traill family).
Type locality: "Demerara, Equinoxial America" [= Guyana, South America] (in error). Designated as West Indies *fide* NMSZ catalogue. Restricted to West Africa *fide* Mertens (1938: 49).
Distribution: West, cen. and E Africa. Guinea-Bissau (Bissau, Bolama Is.), S Ghana (Accra, Volta, Western), S Togo (Plateaux), S Benin (Cotonou), S Nigeria (Akwa Ibom, Bayelsa, Cross River, Rivers), S Cameroon (Adamaoua, Centre, Est, Littoral, Nord-Ouest, Ouest, Sud, Sud-Ouest), Equatorial Guinea (Bioko Is.), São Tomé and Príncipe (São Tomé), Central African Republic (Kemo-Gribingui, Lobaye, Mbomou, Ombella-Mpoko, Ouaka, Sangha), Gabon (Ngounié, Ogooué-Ivindo, Ogooué-Lolo, Ogooué-Maritime, Woleu-Ntem), Congo (Cuvette, Cuvette-Ouest, Kouilou, Pool, Sangha), NE Angola (Lunda Norte), Democratic Republic of the Congo (Bandundu, Bas-Congo, Equateur, Kasai Occidental, Kasai Oriental, Katanga, Kinshasa, Maniema, Nord-Kivu, Orientale, Sud-Kivu, Idjwi Is.), N Zambia (Luapula, Northern,

North-Western), SE Sudan, S South Sudan (SW Eastern Equatoria), S Uganda (Central, Eastern, Western), Rwanda (Butare), SW Burundi (Bururi), SW Kenya (Nyanza, Western) and NW Tanzania (Kagera), NSL–2300 m.

Sources: Loveridge, 1936a, Witte, 1962, Laurent, 1964a, Mertens, 1965c, Leston & Hughes, 1968, Hughes & Barry, 1969, Pitman, 1974, Derleyn, 1977, Hughes, 1983, 2013, Herman et al., 1990, Lawson, 1993, J.-F. Trape & Roux-Estève, 1995, Luiselli et al., 2000, Spawls et al., 2002, Chirio & Ineich, 2006, Chirio & LeBreton, 2007, Pauwels & Vande weghe, 2007, Segniagbeto et al., 2011 and Auliya et al., 2012.

Remarks: Official Specific Name no. 1989 *fide* Opinion 709 (ICZN, 1964b). Original description reprinted in Traill *in* Schlegel (1843: 179, pl. 2, figs. 19–20). Hughes & Barry (1969: 1028), listed two syntypes although Traill clearly stated he had only one specimen at the time of description; he later (*in* Schlegel, 1843: 179) mentioned a headless BMNH specimen as belonging to this taxon. BMNH 1946.1.20.43 listed as a syntype *fide* Hughes & Barry (1969: 1028) and ZMB 2817 reported as a possible syntype. Reports from Benin, Ivory Coast and N Zambia need confirmation.

4. *Dendroaspis polylepis* A.C.L.G. Günther, 1864c. Proc. Zool. Soc. London 32(1): 310. (*Dendraspis polylepis*)

Synonyms: *Dendraspis antinorii* W.C.H. Peters, 1873b, *Dendraspis mamba* Gough, 1907b, *Dendroaspis polyelepis* – Ibrahim & Masr, 1975 *(nomen incorrectum)*, and *Dendroaspis polylepasis* – Khole, 1991 *(nomen incorrectum)*.

Type: Holotype, BMNH 1946.1.20.53, an 1829 mm skin (J. Kirk [Zambezi Exped.], 1858–1864).

Type locality: "regions bordering the Zambezi, including those of the Nyassa Lake." Restricted to Zambesi River, Mozambique *fide* Loveridge (1953: 290).

Distribution: Subsaharan Africa. Senegal (Dakar, Kédougou, Tambacounda, Thiès), Guinea (Kindia, Kouroussa, Télimélé), N Ivory Coast (Ferkéssédougou), SW Burkina-Faso (Hauts-Bassins, Sud-Ouest), Cameroon (Adamaoua), Central African Republic (Bamingui-Bangoran, Haut-Mbomou, Ombella-Mpoko, Ouham, Ouham-Pende), Democratic Republic of the Congo (Equateur, Katanga, Nord-Kivu, Orientale), Angola (Benguela, Huambo, Luanda), SE South Sudan (Eastern Equatoria), South Sudan (Eastern Equatoria), Eritrea (Anseba/Northern Red Sea), Ethiopia (Bale, Gemu Gofa, Gondar, Hararge, Kefa, Shoa, Sidamo), W Somalia (Awdal, Gedo, Jubbada Dhexe, Woqooyi Galbeed), Kenya (Coast, Eastern, Nairobi, N North-Eastern, Nyanza, Rift Valley, Western), N Uganda (Eastern, Northern), E Rwanda, Tanzania (Arusha, Dar es Salaam, Dodoma, Kagera, Kigoma, Kilimanjaro, Lindi, Manyara, Mara, Mtwara, Mwanza, Pwani, Rukwa, Tanga, Manda Is.), Zambia (Central, Copperbelt, Eastern, Luapula, Lusaka, Northern, North-Western, Southern, Western), Malawi (Southern), Mozambique (Gaza, Manica, Maputo, Sofala), Zimbabwe (Bulawayo, Manicaland, Mashonaland Central, Mashonaland East, Mashonaland West, Masvingo, Matabeleland North, Matabeleland South, Midlands), Namibia (Hardap, Khomas, Kunene, Otjozondjupa), Botswana (Central, Gaborone, Ghanzi, Kgalagadi, Kweneng, North West, Southern), Swaziland, Lesotho and NE South Africa (E Eastern Cape, Gauteng, KwaZulu-Natal, Limpopo, Mpumalanga, North West), NSL–1830 [2000] m.

Sources: V.F.M. FitzSimons, 1946a, H.W. Parker, 1949, Loveridge, 1950, Villiers, 1954b, Mertens, 1955, 1971, Sweeney, 1961, Pitman, 1974, Broadley & Cock, 1975, Buys & Buys, 1983, Pienaar et al., 1983, Auerbach, 1987, Clauss & Clauss, 2002, Spawls et al., 2002, Greenbaum et al., 2003, M. Griffin, 2003, J.C. Murphy & Schlager, 2003, Lebreton & Chirio, 2004, Greenbaum & Carr, 2005, J.-F. Trape et al., 2005, Chippaux, 2006, Chirio & Ineich, 2006, J.-F. Trape & Mané, 2006b, Castoe et al., 2007a, Chirio & LeBreton, 2007, Baynham, 2010a, Largen & Spawls, 2010, Maritz & Alexander, 2010, Broadley & Blaylock, 2013 and Chirio, 2013.

Remarks: Probably occurs in Togo *fide* and Segniagbeto et al. (2011: 353).

5. *Dendroaspis viridis* (Hallowell, 1844c). Proc. Acad. Nat. Sci. Philadelphia (1844–1845) 2(6): 172. (*Leptophis viridis*)

Synonyms: *Dinophis hammondii* Hallowell, 1852c, *Dendroëchis reticulata* J.G. Fischer, 1855, and *Dendroaspis viridis hallowelli* Yeomans, 1993 *(nomen ineditum)*.

Type: Lectotype, ANSP 6893, a 572 mm female (S.M.E. Goheen, 1836–1838), designated by Malnate (1971: 3673).

Type locality: "Liberia" via lectotype selection.

Distribution: West Africa. Southwestern Senegal (Ziguinchor), Gambia (Western), Guinea-Bissau (Bolama and Bubaque Is.), Guinea (Boffa, Boké, Conakry, Kouroussa, Macenta, Nzérékoré, Télimélé), Sierra Leone (Northern, Southern, Western, Sherbo Is.), Liberia, Ivory Coast (Abidjan, Bouna, Daloa, Guiglo, Lagunes, Tabou, Toumodi), S Ghana (Eastern, S Western), Togo (Kara, Plateaux), N Benin (Atlantique, Zou) and S Nigeria (Delta), NSL–400 m.

Sources: Doucet, 1963, Menzies, 1966, Leston & Hughes, 1968, Hughes & Barry, 1969, Hakansson, 1981, Greenbaum & Carr, 2005, Chippaux, 2006, J.-F. Trape & Mané, 2006b, Segniagbeto et al., 2011, Auliya et al., 2012, Chirio, 2013 and Hughes, 2013.

Remarks: Two female syntypes (ANSP 6892–93) listed *fide* Hughes & Barry (1969: 1029).

DENDROLYCUS Laurent, 1956a (Lamprophiidae)

Type species: *Lycophidium elapoides* A.C.L.G. Günther, 1874b.
Distribution: West Africa.
Sources: K.P. Schmidt, 1923, Laurent, 1952, Witte, 1962, J.-F. Trape & Roux-Estève, 1995, Chirio & LeBreton, 2007, Gonwouo et al., 2007, Kelly et al., 2009 and Zaher et al., 2009.

1. *Dendrolycus elapoides* (A.C.L.G. Günther, 1874b). Proc. Zool. Soc. London 42(1): 444–445. (*Lycophidium elapoides*)

Synonym: *Lycophidion elapoides angusticinctus* Laurent, 1952.
Type: Holotype, BMNH 1946.1.14.32, a 381–410 mm male (H.H. Higgins, 1856–1974).
Type locality: "Camaroon Mountains" [= Cameroon Mtns., SW Cameroon].
Distribution: Central Africa. Southern Cameroon (Sud, Sud-Ouest), Congo and NE Democratic Republic of the Congo (Nord-Kivu, Orientale, Sud-Kivu), 620–1400 m.

DENDROPHIDION Fitzinger, 1843 (Colubridae)

Synonyms: *Dendrophidium* Agassiz, 1847 *(nomen emendatum)* and *Cacocalyx* Cope, 1895c.
Type species: *Herpetodryas dendrophis* Schlegel, 1837.
Distribution: Latin America.
Sources: J.A. Peters & Orejas-Miranda, 1970, Lieb, 1988, Pérez-Santos & Moreno, 1988,1989a, Villa et al., 1988, Cadle, 2012a–b and Cadle & Savage, 2012.

1. *Dendrophidion apharocybe* Cadle, 2012a. Bull. Mus. Comp. Zool. 160(4): 198–209, figs. 1, 2a, 14c–d.

Type: Holotype, LACM 148593, a 908 mm male (C. Dock, C. Lieb & C. Toft, 9 Dec. 1974).
Type locality: "Finca La Selva, 40 m elevation, Heredia Province, Costa Rica."
Distribution: Lower Central America. Honduras (Gracias a Dios), Nicaragua (Atlántico Norte, Chontales, Matagalpa, Río San Juan), Costa Rica (Alajuela, Guanacaste, Heredia, Limón) and Panama (Bocas del Toro, Coclé, Darién, Panamá, San Blas), 40–1530 m.

2. *Dendrophidion atlantica* Freire, Caramaschi & Gonçalves, 2010. Zootaxa (2719): 63–66, figs. 1–3.

Type: Holotype, MNRJ 17108 (formerly CRE 8598), a 1040 mm male (U. Gonçalves, 5 Dec. 2006).
Type locality: "Mata do Engenbo Coimbra (08°59'S, 35°53'W, 526 m above sea level), Municipality of Ibateguara, State of Alagoas, Northwestern Brazil."
Distribution: Northeastern Brazil (Alagoas), 80–525 m.

3. *Dendrophidion bivittatum* (A.-M.-C. Duméril, Bibron & Duméril, 1854a). Erpét. Gén. 7(1): 540–541. (*Leptophis bivittatus*) (*nomen corrigendum*)

Synonyms: *Leptophis bi-vittatus* A.-M.-C. Duméril, Bibron & Duméril, 1854a *(nomen incorrigendum), Thamnosophis bivittatus* Jan, 1863b *(nomen corrigendum), Tropidonotus subradiatus* Jan, 1865c, and *Herpetodryas tetrataenia* A.C.L.G. Günther, 1872b.
Types: Syntypes (2), MNHN 517 and MNHN 1906.3460, longest syntype 680 mm.
Type locality: "Nouvelle-Grenade (Amérique du Sud)" [= Colombia].
Distribution: Central Colombia (Antioquia, Boyacá, Caldas, Cundinamarca, Santander, Tolima, Valle), 500–2630 m.
Source: Rojas-Morales, 2012b.

4. *Dendrophidion boshelli* Dunn, 1944a. Caldasia 2(10): 475–476.

Type: Holotype, ICNMNH, a 252 mm (svl) female (J. Boshell, Oct. 1943).
Type locality: "Volcanes, 250 m. alt., Municipio de Caparrapí, Cundinamarca, Colombia."
Distribution: Central Colombia (Cundinamarca, Santander), 250–500 m.

5. *Dendrophidion brunneum* (A.C.L.G. Günther, 1858). Cat. Colub. Snakes Brit. Mus.: 116. (*Herpetodryas brunneus*)

Type: Holotype, BMNH 1946.1.12.98, an 867+ mm female.
Type locality: "Guayaquil" [=Guayaquil, Guayas Prov., SW Ecuador, 2°12'S, 79°54'W, elevation 10 m].
Distribution: Central Ecuador (Azuay, Chimborazo, El Oro, Guayas, Imbabura, Loja) and NW Peru (Amazonas, Cajamarca, La Libertad, Piura, Tumbes), 15–2650 m.
Sources: Kuch & Freire, 1993 and Cadle, 2010, 2012b.
Remarks: Photographs of holotype in Cadle (2012b: figs. 28–29). A.C.L.G. Günther (1858: 116) listed length of holotype as 991 mm (see remarks by Cadle, 2012b: 308).

6. *Dendrophidion clarkii* Dunn, 1933a. Occ. Pap. Boston Soc. Nat. Hist. (8): 78.

Type: Holotype, MCZ 34878, a 1067–1078 mm male (E.R. Dunn, August 1932).

Type locality: "El Valle de Anton, Panama" [= Anton Valley, Cocle Prov., cen. Panama, 8°36'N, 80°08'W, elevation 590 m].

Distribution: Lower Central America and NW South America. Costa Rica (Alajuela, Cartago, Limón, Puntarenas, San José), Panama (Bocas del Toro, Coclé, Darién, Panamá, San Blas), W Colombia (Cauca, Chocó, Nariño, Risaralda, Valle del Cauca) and W Ecuador (Carchi, Esmeraldas, Imbabura, Loja, Pichincha), 30–1800 m.

Sources: L.D. Wilson & Meyer, 1985, Lieb, 1991a, Lee, 1996, 2000, J.A. Campbell, 1998b, Stafford & Meyer, 2000, Savage, 2002, Stafford, 2002, Solórzano, 2004, McCranie, 2011a and Cadle & Savage, 2012.

Remarks: Photographs of holotype in Cadle & Savage (2012: figs. 4–6). *Dendrophidion clarkii* separated from *D. nuchale fide* McCranie (2011a: 106–107).

D

7. *Dendrophidion crybelum* Cadle, 2012a. Bull. Mus. Comp. Zool. 160(4): 209–217, figs. 2c, 12, 14b.

Type: Holotype, LACM 148599 (formerly CRE 3182), an 850 mm male (J.E. DeWeese & R.T. Harris, Sept. 1972).

Type locality: "Finca Las Cruces, near San Vito de Java, 4 km S San Vito, 1,200 m elevation, Puntarenas Province, Costa Rica."

Distribution: Southwestern Costa Rica (Puntarenas) and W Panama (Chiriquí), 1100–1330 m.

8. *Dendrophidion dendrophis* (Schlegel, 1837). Essai Phys. Serp. 1: 153, 2: 196–197. (*Herpetodryas dendrophis*)

Synonyms: *Herpetodryas poitei* A.-M.-C. Duméril, Bibron & Duméril, 1854a and *Herpetodryas dendrophis schlegeli* Jan, 1863b.

Type: Lectotype, MNHN 41, an 505–515 (svl) mm female (P.-A. Poiteau, 1818–1822), designated by Lieb (1988: 165).

Type locality: "Cayenne, French Guiana" [= Cayenne, Cayenne Dept., N French Guiana, 4°55'N, 52°20'E, elevation 5 m] via lectotype selection.

Distribution: Northern Amazonia. Southeastern Colombia (Amazonas, Cauca, Cundinamarca) SE Venezuela (Amazonas, Bolívar), Guyana (Mahaica-Berbice, Upper Takutu-Upper Essequibo), Suriname (Paramaribo, Saramacca, Sipaliwini), N French Guiana (N Cayenne, NW Saint-Laurent-du-Maroni), NW Brazil (Alagoas, Amazonas, Pará), E Peru (Amazonas, Junín, Loreto, Madre de Dios) and E Ecuador (Morona-Santiago, Napo, Pastaza), 50–1000 m.

Sources: E.H. Taylor, 1951, Chippaux, 1987, Lieb, 1988, Starace, 1998, Freire, 1999a, Doan & Arizábal, 2002, Abuys, 2003, Natera-Mumaw, 2008, Navarrete et al., 2009, Freire et al., 2010 and C.J. Cole et al., 2013.

Remarks: Photographs of lectotype in Lieb (1988: fig. 2).

9. *Dendrophidion graciliverpa* Cadle, 2012b. Bull. Mus. Comp. Zool. 130(6): 296–305, figs. 20–21, 23, 26a.

Type: Holotype, AMNH 110584 (formerly CWM 12250), a 964 mm male (C.W. Myers & J.W. Daly, 11 Feb. 1974).

Type locality: "3 km E Pasaje, 30 m elevation, El Oro province, Ecuador [03°20'S, 79°49'W]."

Distribution: Western Ecuador (Carchi, Chimborazo, Cotopaxi, El Oro, Esmeraldas, Guayas, Imbabura, Los Ríos, Loja, Pichincha, Santo Domingo de los Tsáchilas), 15–1750 m.

10. *Dendrophidion nuchale* (W.C.H. Peters, 1863c). Mber. Königl. Akad. Wiss. Berlin 1863(6): 285–286. (*Herpetodryas nuchalis*)

Synonym: *Dendrophidion clarkii* Dunn, 1933a.

Types: Syntypes (3), ZMB, longest syntype 1018 mm (J. Gollmer, 1856–1861), lost *fide* G. Peters *in* Lieb (1988: 166–167).

Type locality: ""'Caracas" [= Caracas, Caracas State, N Venezuela, 10°29'N, 66°54'W, elevation 935 m].

Distribution: Northern Venezuela (Aragua, Carabobo, Guárico, Miranda, Yaracuy, Zulia), 100–1270 m.

Sources: Roze, 1966, Lieb, 1988, 1991a, Esqueda et al., 2001, Natera-Mumaw, 2008, Freire et al., 2010 and Cadle & Savage, 2012.

Remarks: Neotype designation of USNM 129579 *fide* Lieb (1988: 167, fig. 4) invalid via Art. 75.2 of the Code *fide* Cadle & Savage (2012: 5). At least seven disjunct populations *fide* Lieb (1988: 167). Separation *D. clarkii* from *D. nuchale* discussed by McCranie (2011a: 106–107).

11. *Dendrophidion paucicarinatum* (Cope, 1894d). Proc. Acad. Nat. Sci. Philadelphia 46: 202–203. (*Drymobius paucicarinatus*)

Type: Holotype, AMNH 17268 (formerly MNCR 116), a 1230 mm specimen (B. Carranza).

Type locality: "La Candelaria, Bruno Carranza, Costa Rica" [= Monte Redondo, Canton de Aserri, San José Prov., Costa Rica [= 9°47'N, 84°08'E, elevation 1170 m] *fide* Savage (1974: 82).

Distribution: Costa Rica (Alajuela, Cartago, Guanacaste, Heredia, Limón, Puntarenas, San José, Bonita Is.) and W Panama (Bocas del Toro, Chiriquí), 1040–1500 m.

Sources: E.H. Taylor, 1951, Lieb, 1988, 1991b, Savage, 2002 and Solórzano, 2004.

12. *Dendrophidion percarinatum* (Cope, 1894a). Proc. Amer. Philos. Soc. (1893) 31: 344–345. (*Drymobius percarinatus*)

Type: Lectotype, AMNH 17366 (formerly MNCR 326), an 815–834+ mm specimen (G.K. Cherrie, 13 Dec. 1891), designated by H.M. Smith (1941h: 73).

Type locality: "Boruca, Costa Rica" [= Boruca, Puntarenas Prov., cen. Costa Rica, 9°00'N, 83°19'W, elevation 545 m] via lectotype selection.
Distribution: Central America and NW South America. Eastern Honduras (Atlántida, Colón, Gracias a Dios, Olancho), E Nicaragua (Atlántico Norte, Boaco, Jinotega), Costa Rica (Alajuela, Cartago, Guanacaste, Heredia, Limón, Puntarenas, San José), Panama (Bocas del Toro, Canal Zone, Colón, Darién, Los Santos, Panamá, San Blas, Veragua), W Colombia (Antioquia, Boyacá, Caldas, Chocó, Risaralda, Valle, Valle del Cauca) and N Venezuela (Aragua, Carabobo, Distrito Federal, Miranda, Zulia), NSL–1200 m.
Sources: E.H. Taylor, 1954, Roze, 1966a, Lancini, 1986, L.D. Wilson & Meyer, 1985, Lieb, 1996, G. Köhler, 1999b, Savage, 2002, McCranie et al., 2006, Rojas-Runjaic & Infante Rivero, 2008, Navarrete et al., 2009, Cadle, 2010, McCranie, 2011a and Cadle, 2012b.
Remarks: Photographs of lectotype in Cadle (2012b: figs. 4–6).

13. *Dendrophidion prolixum* Cadle, 2012b. Bull. Mus. Comp. Zool. 130(6): 282–296, figs. 10–11.

Type: Holotype, AMNH 109721 (formerly CWM 11618), a 754 mm male (C.W. Myers & J.W. Daly, 9 Feb. 1973).
Type locality: "Quebrada Guanguí, 0.5 km above Río Patia (upper Saija drainage), 100–200 m, Cauca department, Colombia [about 02°50'N, 77°25'W]."
Distribution: Western Colombia (Cauca, Chocó, Risaralda, Valle del Cauca) and NW Ecuador (Esmeraldas, Imbabura), 20–1190 m.

14. *Dendrophidion rufiterminorum* Cadle & Savage, 2012. Zootaxa (3513): 23–33, fig. 15.

Type: Holotype, LSUMZ 8901 (formerly AWP 177), a 1459 mm male (S.M. Russell & A.W. Palmisano, 19 July 1963).
Type locality: "1 mi W Baldy Sibun, Cayo district, Belize [17°00'N, 88°46'W]."
Distribution: Central America. Southeastern Guatemala (Alta Verapaz, Izabal), Belize (Cayo, Stann Creek, Toledo), N Honduras (Atlántida, Cortés, Olancho, Yoro), SE Nicaragua (Río San Juan) and Costa Rica (Alajuela, Heredia, Limón, Puntarenas), 30–970 m.
Sources: Savage, 2002, G. Köhler, 2008 and McCranie, 2011a.
Remarks: LSUMZ catalogue lists Russell as sole collector of type *fide* C. Austin *in* Cadle & Savage (2012: 23).

15. *Dendrophidion vinitor* H.M. Smith, 1941h. Proc. Biol. Soc. Washington 54: 74–75.

Type: Holotype, USNM 110662 (formerly HMS 7280), a 510+ mm female (H.M. Smith & R.B. Smith, 21 May 1939).
Type locality: "Piedras Negras, Guatemala" [= Piedras Negras, NW Petén Department, NW Guatemala, 17°11'N, 91°15'W, elevation 195 m].
Distribution: Mesoamerica and ext. NW South America. Southeastern Mexico (Chiapas, Oaxaca, Tabasco, Veracruz, Yucatán), S Belize (Toledo), N Guatemala (Petén, Alta Verapaz), Honduras (Gracias a Dios, Olancho), E Nicaragua (Atlántico Sur, Jinotega, Matagalpa, Zelaya), Costa Rica (Alajuela, Cartago, Guanacaste, Heredia, Limón, Puntarenas, ? San José), Panama (Bocas del Toro, Canal Zone, Darién, Panamá, San Blas) and NW Colombia, NSL–1360 m.
Sources: E.H. Taylor, 1954, Lieb, 1988, 1991c, Casas-Andreu et al., 1996, Lee, 1996, 2000, J.A. Campbell, 1998b, G. Köhler, 1999b, Stafford & Meyer, 2000, Solórzano, 2004, McCranie et al., 2006, McCranie, 2011a and Cadle, 2012a.
Remarks: Photograph of holotype in Cadle (2012a: fig. 1).

D

DENISONIA Krefft, 1869a (Elapidae)

Synonyms: *Desidonia* – Phisalix, 1922 *(nomen incorrectum)*, and *Dewisonia* – Anon., 1944 *(nomen incorrectum)*.
Type species: *Hoplocephalus maculatus* Steindachner, 1867a.
Distribution: Australia.
Sources: Cogger et al., 1983a, Wallach, 1985, S.K. Wilson & Knowles, 1988, Hoser, 1989, 2012e, Hutchinson, 1990, Greer, 1997, David & Ineich, 1999, Cogger, 2000 and Zaher et al., 2009.

1. *Denisonia devisi* Waite & Longman, 1920. Rec. So. Aust. Mus. 1(3): 178–179, fig. 36, pl. 27, fig. 2. (*Denisonia maculata devisi*) (*nomen substitutum*)

Synonym: *Hoplocephalus ornatus* De Vis, 1884a (*nomen praeoccupatum*).
Type: Holotype, QM J199 (F.A. Blackman, 1855–1874).
Type locality: "near Surat, Western Queensland, Australia" [= 27°09'S, 149°04'E, elevation 250 m].
Distribution: Eastern Australia (N New South Wales, S Queensland), 15–375 m.
Source: De Vis, 1884a.

2. *Denisonia maculata* (Steindachner, 1867a). Reise Öster. Novara 1: 81–82, pl. 3, figs. 3–5. (*Hoplocephalus maculatus*)

Synonyms: *Denisonia ornata* Krefft, 1869a and *Hoplocephalus muelleri* J.G. Fischer, 1885b.
Types: Syntypes (2), NMW 19801.1–2, a 318 mm and 241 mm specimen (F. Steindachner [Novara Exped.], 1866).
Type locality: "Neu-Südwales" [= New South Wales, Australia]. Restricted to Rockhampton, Queensland *fide* R. MacKay (1956: 29).

Distribution: Eastern Australia (NE New South Wales, SE Queensland, North West Is.), NSL–260 m.

DIADOPHIS Baird & Girard, 1853 (Carphophiidae)

D

Synonyms: *Diodophis* – A.W. Butler, 1887 *(nomen incorrectum)* and *Diadophus* – Bogert, 1930 *(nomen incorrectum).*

Type species: *Coluber punctatus* Linnaeus, 1766.

Distribution: North America and Mexico.

Fossil records: Upper Miocene to upper Pleistocene of USA.

Sources: Dowling et al., 1983, Holman, 2000a, Pinou et al., 2004 and Zaher et al., 2009.

1. †*Diadophis elinorae* Auffenberg, 1963. Tulane Stud. Zool. 10(3): 170–171, fig. 18 (upper).

Type: Holotype, UF 6413 (W. Auffenberg, early 1950s–1960), one middle trunk vertebra.

Type locality: "Haile VI, Locality A.-(R 17 E, T 9 S, NW 1/4 and SW 1/4, Sec. 24): 0.2 miles east of Route 235, near the town of Haile, Alachua County, Florida; elevation approximately 84 feet; Alachua formation, Middle Pliocene, Hemphillian."

Distribution: Upper Miocene (Hemphillian: 4.9–10.3 mya) of USA (Florida) and lower Pleistocene (Irvingtonian I: 0.9–1.9 mya) of USA (Florida).

Sources: Holman, 1979a, 1981a, Meylan, 1982 and Rage, 1984b.

Remarks: Upper Miocene *fide* Rage (1984b: 46).

2. *Diadophis punctatus* (Linnaeus, 1766). Syst. Nat., ed. 12, 1: 376. (*Coluber punctatus*)

Synonyms: *Coluber torquatus* G. Shaw, 1802, *Coluber edwardsii* Merrem, 1820, *Diadophis amabilis* Baird & Girard, 1853, *Diadophis docilis* Baird & Girard, 1853, *Diadophis regalis* Baird & Girard, 1853, *Diadophis pulchellus* Baird & Girard, 1853, *Ablabes occipitalis* A.C.L.G. Günther, 1858, *Diadophis arnyi* Kennicott, 1859a, *Diadophis dysopes* Cope, 1860e, *Diadophis punctatus pallidus* Cope, 1860e, *Diadophis punctatus stictogenys* Cope, 1860e, *Diadophis texensis* Kennicott, 1860, *Diadophis punctatus laetus* Jan, 1863a, *Diadophis punctatus dugesii* Villada, 1875, *Diadophis punctatus modestus* Bocourt, 1886 *in* A.H.A. Duméril, Bibron & Mocquard, 1870–1909, *Diadophis amabilis occidentalis* Blanchard, 1923b, *Diadophis amabilis similis* Blanchard, 1923b, *Diadophis amabilis vandenburgii* Blanchard, 1923b, *Diadophis anthonyi* Van Denburgh & Slevin, 1923, *Diadophis regalis arizonae* Blanchard, 1923b, *Diadophis regalis blanchardi* K.P. Schmidt & Smith, 1944, and *Diadophis punctatus acricus* Paulson, 1968.

Type: Holotype, not designated (A. Garden, 1760–1761), location unknown.

Type locality: "Carolina" [USA]. Restricted to Charleston, South Carolina *fide* Klauber (1948: 6).

Distribution: North America and Mexico. Southeastern Canada (New Brunswick, S Nova Scotia, S Ontario, S Quebec), USA (Alabama, Arizona, Arkansas, California, Santa Catalina Is., SE Colorado, Connecticut, Delaware, District of Columbia, Florida: Big Pine Key; Georgia, Idaho, ext. W and SE Illinois, Indiana, Iowa, Kansas, Kentucky, Louisiana, Maine, Maryland, Massachusetts, Naushon, Martha's Vineyard and Nantucket Is., Michigan, ext. E Minnesota, Mississippi, Missouri, E Nebraska, NW Nevada, New Hampshire, New Jersey, New Mexico, New York, North Carolina, Ohio, Oklahoma, W Oregon, Pennsylvania, Rhode Island, South Carolina, ext. SE South Dakota, Tennessee, Texas, Utah, Vermont, Virginia, S Washington, West Virginia, Wisconsin) and N Mexico (Aguascalientes, Baja California Norte, Islas Todos Santos Sur and San Martin in Pacific Ocean, NW Chihuahua, Coahuila, Colima, Distrito Federal, Durango, Guanajuato, Hidalgo, Jalisco, México, Michoacán, Nayarit, W Nuevo León, Querétaro, San Luis Potosí, NE Sonora, W Veracruz, Zacatecas), NSL–2400 m.

Fossil records: Middle/upper Pleistocene (Irvingtonian II) of USA (Maryland) and upper Pleistocene (Rancholabrean I, II) of USA (Alabama, Calfornia, Florida, Georgia, Kansas, Missouri, New Mexico, Pennsylvania, Tennessee, Texas, Virginia, West Virginia).

Sources: Blanchard, 1942, K.P. Schmidt & Smith, 1944, H.M. Smith & Taylor, 1945, Logier & Toner, 1961, Gehlbach, 1974, Fitch, 1975, Christman, 1980, Vogt, 1981, Cadle, 1984c, Stebbins, 1985, Ernst & Barbour, 1989, Conant & Collins, 1991, Brown et al., 1995, Palmer & Braswell, 1995, Pinou et al., 1995, Degenhardt et al., 1996, P.R. Brown, 1997, Harding, 1997, Tennant, 1997, Fitch, 1999, R.D. Bartlett & Tennant, 2000, Werler & Dixon, 2000, Grismer, 2001b, McCranie & Wilson, 2001b, 2002 and Fontanella et al., 2008.

Remarks: Probably two distinct species *fide* Pinou et al. (1995: 105) but 14 distinct lineages based on mtDNA *fide* Fontanella et al. (2008: 1066).

DIAPHOROLEPIS Jan, 1863b (Dipsadidae)

Synonyms: *Diaphorelepis* – Dunn, 1944e *(nomen incorrectum), Diapharolepis* – Dunn, 1935 *(nomen incorrectum), Diphorolepis* – Savage, 1966 *(nomen incorrectum), Diaphrolepis* – Holman, 1967 *(nomen incorrectum),* and *Diaphorolepsis* – Phelps, 1981 *(nomen incorrectum).*

Type species: *Diaphorolepis wagneri* Jan, 1863b.

Distribution: Lower Central America and NW South America.

Sources: Bogert, 1964, J.A. Peters & Orejas-Miranda, 1970, Pérez-Santos & Moreno, 1988, 1991 and Zaher et al., 2009.

1. *Diaphorolepis laevis* F. Werner, 1923a. Ann. Naturhist. Mus. Wien 36: 160–161.

Type: Holotype, NMW 14860, a 495 mm specimen (A.H. Fassl, 1910).
Type locality: "Columbien" [= Colombia].
Distribution: Colombia (Amazonas, Cundinamarca, Tolima), 1310–2600 m.

2. *Diaphorolepis wagneri* Jan, 1863b. Elenco Sist. Ofidi: 98.

Type: Holotype, ZSM 2708/0 (formerly MM), a 684 mm female (F.J.C.M. Wagner, 1857–1860).
Type locality: "Ande dell' Ecuadore" [= Andes of Ecuador].
Distribution: Panama and NW South America. Eastern Panama (Darién), Colombia and Ecuador, 100–1460 m.
Sources: F. Werner, 1897, Bogert, 1964, Franzen & Glaw, 2007 and P. Wagner et al., 2012.

DIEUROSTUS Berg, 1901 (*nomen substitutum*) (Homalopsidae)

Synonyms: *Eurostus* Duméril, Bibron & Duméril, 1854b *(nomen praeoccupatum)* and *Eurostrus* – W.C.H. Peters, 1859a *(nomen incorrectum)*.
Type species: *Eurostus dussumierii* A.-M.-C. Duméril, Bibron & Duméril, 1854b.
Distribution: Tidal rivers and coastal marine waters of SW India.
Sources: Gyi, 1970, Mahendra, 1984, J.C. Murphy, 2007, A.B. Kumar & Captain, 2011, Chandrmouli et al., 2012 and A.B. Kumar et al., 2012.
Remarks: A valid genus *fide* A.B. Kumar et al. (2012: 481), previously a synonym of *Enhydris. Hypsirhina malabarica* F. Werner possibly a valid species *fide* A.B. Kumar et al. (2012: 481).

1. *Dieurostus dussumierii* (A.-M.-C. Duméril, Bibron & Duméril, 1854b). Erpét. Gén. 7(2): 953–955, pls. 84, 77, fig. 1. (*Eurostus dussumierii*)

Synonyms: *Eurostus dussumierii* A.-M.-C. Duméril, 1853 (*nomen nudum*) and *Hypsirhina malabarica* F. Werner, 1913.
Type: Lectotype, MNHN 3751, a 640–673 mm female with skull removed, designated by Gyi (1970: 121).
Type locality: "Bengal" [= Bangladesh or NE India] (in error) via lectotype selection. Corrected to Malabar coast *fide* J.C. Murphy (2007b: 109). Restricted to west coast of North Kerala, India *fide* J.C. Murphy (2007b: 109).
Distribution: Southwestern India (Kerala), NSL–45 m.
Remarks: Gyi (1970: 122) measured type as 665 mm. J.C. Murphy (2007b: 109) listed MNHN 3751 and 5751 as paratypes, then (p. 111) as MNHN 3751 and 5753 (3572), and finally (p. 113) as MNHN 3751 (type) and MNHN 5752 (3572). A.B. Kumar et al. (2012: 483) erroneously listed the holotype as MNHN c3458 (which is a collection shelving number). Photograph of lectotype in Chandrmouli et al. (2012: fig. 1a).

D

†*DINILYSIA* Woodward, 1901 (†Dinilysiidae)

Type species: †*Dinilysia patagonica* Woodward, 1901.
Distribution: Upper Cretaceous of Argentina.
Sources: Estes et al., 1970, Frazetta, 1970, M.K. Hecht, 1982, Rage, 1977a, 1984b, Rage & Albino, 1989, Bonaparte, 1991, M. Lee, 1997, Caldwell & Lee, 2001, Caldwell & Albino, 2002, M. Lee & Scanlon, 2002, Scanlon & Lee, 2002, Albino & Caldwell, 2003, Scanlon, 2003b, Albino, 2011a, Filippi & Garrido, 2012 and Zaher & Scanferla, 2012.

1. †*Dinilysia patagonica* Woodward, 1901. Proc. Zool. Soc. London 70(2): 176–179, pl. 20, figs. 1–1c, 2.

Type: Holotype, MLP 26-410, an articulated skull, jaws and most of the vertebral column (S. Roth).
Type locality: "red sandstones of Neuquen, Patagonia, supposed to be of Cretaceous age" [= Boca del Sapo, Neuquén Prov., N Argentina; Neuquén group, Rio Colorado formation, Coniacian, early upper Cretaceous *fide* Bonaparte, 1991: 26].
Distribution: Upper Cretaceous (Santonian: 83.5–85.8 mya, or Coniacian: 85.8–89.3 mya) of Argentina.
Remarks: Photographs of holotype in Zaher & Scanferla (2012: figs. 1, 5, 6b, 20). Estes et al. (1970: 25) reported type locality to be near Río Neuquén, Chubut Territory, but Rage *in* Hecht (1982: 523) corrected it to Neuquén Prov.

DINODON A.-M.-C. Duméril, Bibron & Duméril, 1854a (Colubridae)

Synonyms: *Dinodon* A.-M.-C. Duméril, 1853 *(nomen nudum), Eumesodon* Cope, 1860e, *Lepidocephalus* Hallowell, 1861 *(nomen praeoccupatum), Dianodon* Cope, 1893a *(nomen emendatum), Adiastema* F. Werner, 1925, and *Donodon* – Anon., 1993 *(nomen incorrectum)*.
Type species: *Lycodon rufozonatus* Cantor, 1842a.
Distribution: Asia.

Sources: Stejneger, 1907a, C.H. Pope, 1935, M.A. Smith, 1943, Deuve, 1970, B. Hu et al., 1980, Zhao & Adler, 1993 and I. Das, 2010, 2012.

Remarks: Possibly a synonym of *Lycodon fide* Pyron et al. (2013: 976).

D

1. *Dinodon flavozonatum* C.H. Pope, 1928b. Amer. Mus. Novit. (325): 2–3.

Type: Holotype, AMNH 34371, a 965 mm male (C.H. Pope, April–Sept. 1926).

Type locality: "Kuatun, a village in the mountains of Ch'ungan Hsien, northwest Fukien Prov., China" [= Guadun, in Wuyishan Mtns., ext. N Fujian Prov., SE China, 27°45'N, 117°50'E, elevation 900–1200 m].

Distribution: Southeastern Asia. Southeastern China (Anhui, Fujian, Guangdong, Guizhou, Jiangxi, Sichuan, Yunnan, Zhejiang), N Vietnam (Lai Chau, Vin Phuc) and N Myanmar (E Shan), 400–1100 m.

Sources: C.H. Pope, 1935, Orlov et al., 2000, Zhao, 2006 and I. Das, 2010.

Remarks: Original description reprinted in C.H. Pope (1929: 433–434, figs. 12–13).

2. *Dinodon gammiei* (Blanford, 1878a). Proc. Asiatic Soc. Bengal 1878(6): 141–142. (*Ophites gammiei*)

Type: Holotype, ZSI 8447 (formerly IMC 8447), an 800 mm specimen (J.A. Gammie, 1865–1878).

Type locality: "Cinchona plantations, British Sikkim" [= Sikkim State, NE India]. Emended to Chichona plantation in South-eastern Sikkim, Himalayas *fide* Blanford (1878b: 131), [= Mungpu cinchona plantation, Darjeeling District, N West Bengal, India, ca. 27°00'N, 88°25'E *fide* Mistry et al., 2007: 271].

Distribution: Northeastern India (Arunachal Pradesh, Sikkim, N West Bengal), 1700–2285 m.

Sources: I. Das, 1999a, Mistry et al., 2007 and Agarwal et al., 2010.

Remarks: Supplemental original description in Blanford (1878b: 130–131). Photographs of the type in Mistry et al. (2007: fig.1). Type locality erroneously listed as Cinchona plain, Darjeeling *fide* M.A. Smith (1943: 271).

3. *Dinodon meridionale* Bourret, 1935b. Bull. Gén. Instr. Publ. 14(7): 241–242, fig. 1. (*Dinodon rufozonatum meridionale*)

Type: Holotype, MNHN 1938.129, a 926 mm male (R.L. Bourret, 1934).

Type locality: "Chapa, Indochine française" [= Sa Pa, Sa Pa Distr., W Lao Cai Prov., ext. N Vietnam, 22°20'N, 103°51'E, elevation 1500 m].

Distribution: Southern China (S Guangxi, SE Yunnan), N Laos (Xiangkhouang) and N Vietnam (Bac Giang, Cao Bang, Ha Giang, Hai Duong, Lang Son, Lao Cai, Ninh Binh, Thai Nguyen, Tuyen Guang, Vinh Phuc), 200–1700 m.

Sources: Orlov & Ryabov, 2004 and V.S. Nguyen et al., 2009.

4. *Dinodon orientale* (Hilgendorf, 1880). Sitz. Ges. Naturf. Freude Berlin 1880(8): 115–117, pl., figs. 1–5. (*Ophites orientalis*)

Synonyms: *Ophites japonicus* A.C.L.G. Günther, 1880c, and *Eumesodon carinatus* F. Müller, 1885.

Type: Holotype, ZMB 9419, a 300 mm specimen (F. Hilgendorf, May 1874).

Type locality: "Tokio, Japan" [= Tokyo, Tokyo Prefecture, cen. Japan, 35°41'N, 139°42'E, elevation 35 m].

Distribution: Exteme E Russia (S Kuril Is.) and Japan (Hokkaido, Honshu, Kyushu, Fukuejima, Iwojima, Kanagawa, Shimokoshikijima, Takeshima, Okinoshima, Oshima, Sadogashima, Tanegashima and Yakushima Is.), NSL–45 m.

Sources: Maki, 1931, Bannikov et al., 1977, M. Mori, 1984, Shibata, 1988, Mizukami et al., 1989, Toriba, 1991 and Toda & Okada, 1999.

5. *Dinodon rosozonatum* S.-Q. Hu & Zhao *in* Zhao, 1972 Mater. Herpetol. Res. Chengeu 1: 36

Type: Holotype, CIB 64-III-6089, a 1060 mm male (native, 21 May 1964).

Type locality: "Dali of Diaulo Shan, Hainan Island, Guangdong Prov., China, altitude 200 m."

Distribution: Extreme S China (Hainan) and Vietnam (Quang Binh, Quang Tri), 80–580 m.

Sources: Maslov & Kotlobay, 1998, Zhao et al., 1998, Zhao, 2005, 2006, V.S. Nguyen et al., 2009 and Shi, 2011.

Remarks: Original description reprinted in S.-Q. Hu & Zhao (1975: 379–380, pl. 1, fig. 2).

6. *Dinodon rufozonatum* (Cantor, 1842a). Zool. Chusan: pl. 11. (*Lycodon rufozonatus*) (*nomen corrigendum*)

Synonyms: *Lycodon rufo-zonatus* Cantor, 1842a *(nomen incorrigendum), Dinodon cancellatum* A.-M.-C. Duméril, Bibron & Duméril, 1854a, *Lycodon rufozonatus* A.C.L.G. Günther, 1858 *(nomen corrigendum), Coronella striata* Hallowell, 1856a, *Dinodon rufozonatus formosana* Boettger, 1885b, *Dinodon rufozonatum walli* Stejneger, 1907a, *Adiastema cervinum* F. Werner, 1925, *Dinodon rufozonatum williamsi* K.P. Schmidt, 1925b, and *Dinodon rufozonatum yunnanense* Mell, 1931a.

Type: Holotype, not designated (T.E. Cantor, July 1840–March 1841, via the British East India Comp.), location unknown.

Type locality: "island of Chusan, or Great Chusan, is situated on the East Coast of China, between the 30° and 31° of North latitude and the 122° and 123° of East longitude; separated from the nearest mainland, Keeto

Point, by an arm of the sea, about ten miles across" [= Zhoushan Is., Zhoushan Arch., Zhejiang Prov., E China].

Distribution: East Asia. Southern China (Anhui, Fujian, Gansu, Guangdong, Guangxi, Guizhou, Hainan, Hebei, Heilongjiang, Henan, Hunan, Hubei, Jiangsu, Jiangxi, Jilin, Liaoning, Ningxia, Shaanxi, Shandong, Yunnan, Zhejiang), Taiwan, N Laos, N Vietnam (Ha Tinh, Quang Binh, Quang Tri, Tuyen Quang, Vinc Phuc), Russia (Primorsky), South Korea (Kyonggi-do), North Korea (Hwanghae-pukto, Kaesong-chigu, Kangwon-do, Pyongan-namdo) and S Japan (Ryukyus: Aragusukujima, Haterumajima, Hatomajima, Irabujima, Ikemajima, Iriomotejima, Ishigakishima, Kayamajima, Kohamajima, Kurimajima, Kuroshima, Miyakojima, Nakanogamijima, Nakanoishma, Ogamijima, Okinawajima, Shimojijima, Taketomijima, Taramajima, Tsushima and Yonagunijima Is.), 400–1100 m

Sources: Supplemental original description in Cantor (1842b: 483). M.A. Smith, 1928b, C.H. Pope, 1935, Shannon, 1956, Kuntz, 1963, M. Mori, 1984, 1986, Szyndlar & Hung, 1987, Kudrjavtsev & Mamet, 1989a–b, Toriba, 1990a, Kohno & Ota, 1991, Zou & Chen, 1998, Orlov et al., 2000, Zhao, 2006, San & Lee, 2007, V.S. Nguyen et al., 2009, Xiang & Li, 2009 and Yao, 2012.

7. *Dinodon semicarinatum* (Cope, 1860e). Proc. Acad. Nat. Sci. Philadelphia 12(6): 263. (*Eumesodon semicarinatus*)

Synonym: *Lepidocephalus fasciatus* Hallowell, 1861.

Type: Holotype, ANSP 3479, a 940 mm specimen (W. Heine [No. Pacific Explor. Exped.], 1853).

Type locality: "Loo Choo" [= Ryukyu Islands, S Japan].

Distribution: Extreme S Japan (Ryukyus: Agunijima, Akajima, Amamioshima, Amurojima, Fukajijima, Gishifujima, Gushikawajima, Gusukujima, Hamahigajima, Hatejima, Hyanzajima, Iejima, Iheyajima, Ikeijima, Izenajima, Kakeromajima, Kerumajima, Kikaijima, Korijima, Kubajima, Kudakajima, Kumejima, Maejima, Miyagijima, Nakajima, Nohojima, Ohajima, Ojima, Okierabujima, Okinawajima, Sesokojima, Tokashikijima, Tokunoshima, Tonakijima, Tsukenjima, Ukejima, Ukibarujima, Yabuchijima, Yaguchijima, Yakabijima, Yanahajima, Yorojima, Yoronjima and Zamamijima Is.), NSL–270 m.

Sources: M. Mori, 1986, Toriba, 1990a and A. Mori et al., 1999.

8. *Dinodon septentrionale* (A.C.L.G. Günther, 1875). Proc. Zool. Soc. London 43(1): 233. (*Ophites septentrionalis*)

Synonym: *Dinodon septentrionale chapaensis* Angel & Bourret, 1933.

Type: Holotype, BMNH 1946.1.14.96, a 2108 mm female (T.C. Jerdon, 1836–1852).

Type locality: "northern parts of India." Restricted to the Himalayas or Assam *fide* Blanford (1878a: 131), the Himalayas or Khasi Hills *fide* Boulenger (1893a: 324), and the E. Himalayas or Khasi Hills *fide* M.A. Smith (1943: 270).

Distribution: Northeastern India (Assam, Mizorán, Nagaland, Sikkim, West Bengal), Bhutan (Trashiyangtse), S China (Yunnan), N Myanmar (Kachin, Kayin, Mon), N Thailand (Chiang Mai), N Cambodia (Mondolkiri), N Laos (Xiangkhouang) and N Vietnam (Bac Kan, Cao Bang, Ha Tinh, Lai Cai, Lao Cai, Nghe An, Quang Binh, Thua Thien-Hue, Vinh Phuc), 220–2100 m.

Sources: Kuntz, 1963, E.H. Taylor, 1965, Orlov et al., 2000, 2003, Ao et al., 2004, B.L. Stuart et al., 2006, Zhao, 2006, Yang, 2008, V.S. Nguyen et al., 2009 and Wangyal, 2011.

DIPSADOBOA A.C.L.G. Günther, 1858 (Colubridae)

Synonyms: *Chamaetortus* A.C.L.G. Günther, 1864c, *Anoplodipsas* W.C.H. Peters, 1869, *Dipsadophidium* Lindholm, 1905a, *Dipsoglyphophis* T. Barbour & Amaral, 1927, *Dipsadaboa* – P.S.M. Berry, 1970 *(nomen incorrectum)*, *Dipsoglyphophas* – H.A. Skinner, 1973 *(nomen incorrectum)*, and *Dipsadotoa* – W.W. Tanner & Avery, 1982 *(nomen incorrectum)*.

Type species: *Dipsadoboa unicolor* A.C.L.G. Günther, 1858.

Distribution: Subsaharan Africa.

Sources: Laurent, 1951, Loveridge, 1957a, J.B. Rasmussen, 1979, 1993b, 1996 and Gravlund, 2001.

1. *Dipsadoboa aulica* (A.C.L.G. Günther, 1864c). Proc. Zool. Soc. London 32(1): 310, pl. 26, fig. 2. (*Chamaetortus aulicus*)

Type: Holotype, BMNH 1946.1.8.33, a 330 mm female (J. Kirk [Zambezi Exped.], 1858–1864).

Type locality: "regions bordering the Zambesi, including those of the Nyassa Lake" [= vicinity of Zambezi River (Tete, Sofala & Zambézia Prov.s), Mozambique and Lake Nyassa, S Malawi]. Restricted to Zambezi *fide* Boulenger (1896a: 153) and to Zambezi River *fide* Loveridge (1957a: 272).

Distribution: Eastern Africa. Southern Somalia (Jubbada Dhexe), SE Tanzania (Lindi), S Malawi (Southern), S Mosambique (Gaza, Inhambane, Maputo, Sofala, Tete, Zambézia), SE Zimbabwe (Masvingo, S Matebeleland South, S Midlands), Swaziland and NE South Africa (NE KwaZulu-Natal, Limpopo, E Mpumalanga), NSL–650 m.

Sources: Broadley & Stevens, 1971, Broadley & Cock, 1975, Broadley, 1983, Lanza, 1983a, 1990b, Pienaar et al., 1983, Branch, 1988, J.B. Rasmussen, 1989a, Spawls et al., 2002 and Broadley & Blaylock, 2013.

2. *Dipsadoboa brevirostris* (Sternfeld, 1908b). Mitt. Zool. Mus. Berlin 3(4): 411–412, figs. 5–6. (*Dipsadomorphus brevirostris*)

D

Synonym: *Leptodira guineensis* Chabanaud, 1920b.
Type: Holotype, ZMB, a 580 mm specimen (Riegler), lost *fide* J.B. Rasmussen (1989b: 257).
Type locality: "Jabassi, Kamerun" [= Yabassi, Littoral, Cameroon, 04°30'N, 09°58"E, elevation 125 m].
Distribution: West Africa. Guinea (Boké, Kindia, Koundara, Kouroussa, Macenta, Nzérékoré), Sierra Leone (Northern), Liberia (Montserrado), S Ivory Coast (Toumodi), S Ghana (Accra, Ashanti, Eastern), S Nigeria (Ondo) and SW Cameroon (Littoral), 100–180 m.
Sources: Roux-Estève, 1969a, J.B. Rasmussen, 1989b, Chippaux, 2006, Chirio & LeBreton, 2007, W. Böhme et al., 2011 and Chirio, 2013.
Remarks: Two other syntypes (ZMB 21709–09a) of *Dipsadoboa brevirostris* were redetermined as *D. duchesnii fide* L. Müller (1910: 605) and J.B. Rasmussen (1989b: 250). Probably occurs in Togo *fide* and Segniagbeto et al. (2011: 353).

3. *Dipsadoboa duchesnii* (Boulenger, 1901a). Ann. Mus. Congo (Zool.) 2C(1): 10, pl. 4, figs. 1, 1a–1c. (*Leptodira duchesnii*)

Synonyms: *Dipsadomorphus viridis* Sternfeld, 1908b, *Dipsadomorphus brevirostris* Sternfeld, 1908b *(partim)*, *Leptodira nigeriensis* F. Werner, 1913, *Chamaetortus aulicus ellenbergeri* Chabanaud, 1917a, and *Dipsadomorphus myops* Chabanaud, 1918a.
Type: Lectotype, BMNH 1946.1.8.39 (formerly BMNH 1901.6.19.1), a 780 mm male (C. de la Kethulle & J.M. Gilson, 1893–1896), designated by Hughes & Barry (1969: 1020).
Type locality: "Mandungu, Itimbiri River, Congo-Kinshasa" [= Mandongo, Orientale Prov., NE Democratic Republic of the Congo] via lectotype selection.
Distribution: West and cen. Africa. Southern Nigeria (Akwa Ibom, Bayelsa, Cross River, Edo, Rivers), S Cameroon (Centre, Est, Littoral, Nord-Ouest, Ouest, Sud, Sud-Ouest), SW Central African Republic (Sangha), Gabon (Estuaire, Moyen-Ogooué, Ogooué-Maritime, Woleu-Ntem), Congo (Kouilou, Lekoumou, Sangha), Democratic Republic of the Congo (Bandundu, Bas-Congo, Equateur, Kasai Occidental, Kasai Oriental, Katanga, Kinshasa, Nord-Kivu, Orientale, Sud-Kivu) and W Uganda, NSL–1200 m.
Sources: Leston & Hughes, 1968, Hughes & Barry, 1969, Hughes, 1983, J.A. Butler & Reid, 1986, J.B. Rasmussen, 1989b, 1993, 1994, Lawson, 1993, Luiselli et al., 2001, Chirio & LeBreton, 2007 and Pauwels & Vande weghe, 2008.

4. *Dipsadoboa flavida* (Broadley & Stevens, 1971). Arnoldia, Rhodesia 5(11): 5–10, fig. 1. (*Chamaetortus aulicus flavidus*)

Synonym: *Dipsadoboa flavida broadleyi* J.B. Rasmussen, 1989a.
Type: Holotype, NMZB 23124 (formerly UM), a 655 mm male (R.A. Stevens, Jan. 1969).
Type locality: "a reedbed on Chisambo Tea Estate, Mlanje, southern Malawi, at *c.* 650 metres above sea-level."
Distribution: Eastern Africa. Southern Somalia (Jubbada Dhexe), SE Kenya (Coast, S Eastern, S North-Eastern), E Tanzania (Dar es Salaam, Lindi, Morogoro, Mtwara, Pwani, Ruvuma, Tanga, Zanzibar Is.), S Malawi (Southern) and Mozambique (Maputo, Sofala), NSL–730 m.
Sources: J.B. Rasmussen, 1989a, Broadley, 1990b and Spawls et al., 2002.

5. *Dipsadoboa shrevei* (Loveridge, 1932a). Proc. Biol. Soc. Washington 45: 83–84. (*Crotaphopeltis shrevei*)

Synonym: *Crotaphopeltis hotamboeia kageleri* Üthmöller, 1939.
Type: Holotype, MCZ 32471, a 930 mm male (K.H. Prior, 1931).
Type locality: "Missao de Dondi, Bella Vista, via. Lobito, Angola" [= Missão do Dondi, Huambo Distr., Angola, 12°32'S, 16°15'E, elevation 1750 m].
Distribution: Central Africa. Southern Democratic Republic of the Congo (Bandundu, Bas-Congo, Katanga, Kinshasa, Sud-Kivu), E Tanzania (Kilimanjaro, Lindi), Zambia (Central, Copperbelt, Luapula, Northern, North-Western, Western) and E Angola (Bié, Huambo, Moxico), 780–1750 m.
Sources: Laurent, 1956a, 1964a, Spawls et al., 2002 and Broadley et al., 2003.
Remarks: Photographs of holotype in J.B. Rasmussen (1986: fig. 6).

6. *Dipsadoboa underwoodi* J.B. Rasmussen, 1993b. Steenstrupia 19(4): 169–174, figs. 30–31.

Type: Holotype, ZSM 162/1909/1, a 520 mm male (K.S. Rhode, 1909).
Type locality: "Mukonjefarm, Mundame (04°35'N 09°31'E), Cameroon."
Distribution: West Africa. Southeastern Guinea (Nzérékoré), Sierra Leone, Ivory Coast, S Ghana, Togo, SE Nigeria, S Cameroon (Centre, Est, Nord-Ouest, Ouest, Sud, Sud-Ouest) and Gabon (Moyen-Ogooué, Ogooué-Maritime, Woleu-Ntem), 10–1650 m.
Sources: Ineich, 2003, Chippaux, 2006, Chirio & LeBreton, 2007, Franzen & Glaw, 2007, Pauwels & Vande weghe, 2008 and Segniagbeto et al., 2011.

Remarks: Holotype erroneously listed as ZSM 162E/1909 by Rasmussen (1993b: 169).

7. *Dipsadoboa unicolor* A.C.L.G. Günther, 1858. Cat. Colub. Snakes Brit. Mus.: 183–184.

Synonyms: *Heterurus bicolor* Jan, 1863b (*nomen nudum*), *Heterurus bicolor* Jan, 1871 *in* Jan & Sordelli, 1870–1881, *Dipsadoboa assimilis* Matschie, 1893a, and *Dipsadoboa unicolor viridiventris* Laurent, 1956a.

Type: Holotype, BMNH 1946.1.4.84, a 584 mm female (C. Rich).

Type locality: "West Africa."

Distribution: West and cen. Africa. Guinea-Bissau, Guinea (Boké, Dalaba, Kindia, Kouroussa, Macenta, Nzérékoré, Télimélé), Sierra Leone, Liberia, S Ivory Coast (Abidjan, Dix-Huit Montagnes, Guiglo, Lagunes, Tabou, Toumodi), Ghana, Togo (Plateaux), S Nigeria, Equatorial Guinea (Bioko Is.), SW Cameroon (Centre, Littoral, Nord-Ouest, Ouest, Sud, Sud-Ouest), Gabon, Congo (Cuvette, Kouilou, Lekoumou), Democratic Republic of the Congo (Bandundu, Bas-Congo, Equateur, Kasai Occidental, Kasai Oriental, Kinshasa, Nord-Kivu, Orientale, Sud-Kivu), SW Uganda (Central, Western), W Rwanda, W Burundi, and W Tanzania (Kigoma), 80–3000 m.

Sources: Laurent, 1956a, Doucet, 1963, Mertens, 1965c, Leston & Hughes, 1968, Hughes & Barry, 1969, Pitman, 1974, Witte, 1975, Hughes, 1983, D. Lawson, 1993, Spawls et al., 2002, Ineich, 2003, Chippaux, 2006, Chirio and Ineich, 2006, Chirio & LeBreton, 2007, Segniagbeto et al., 2011 and Chirio, 2013.

Remarks: Probably occus in Central African Republic *fide* Chirio & Ineich (2006: 46).

8. *Dipsadoboa viridis* (W.C.H. Peters, 1869). Monatsb. Akad. Wiss. Berlin 1869(5): 442–443, pl. 1, fig. 4. (*Anoplodipsas viridis*)

Synonyms: *Crotaphopeltis elongata* T. Barbour, 1914c, and *Dipsadoboa elongata gracilis* Laurent, 1956a.

Type: Holotype, ZMB 6451, a 720 mm specimen, lost *fide* J.B. Rasmussen (1993b:174).

Type locality: "Neu-Caledonien" [= New Caledonia] (in error).

Distribution: West and cen. Africa. Southeastern Guinea (Nzérékoré), Liberia, Ivory Coast (Dix-Huit Montagnes), Ghana, Togo (Plateaux), Nigeria, S Cameroon (Est, Centre, Littoral, Sud, Sud-Ouest), Equatorial Guinea (Bioko Is.), SW Central African Republic (Lobaye), Gabon (Estuaire, Moyen-Ogooué, Ngounié, Ogooué-Ivindo, Ogooué-Maritime, Woleu-Ntem), Congo (Kouilou, Sangha), Democratic Republic of the Congo (Bandundu, Bas-Congo, Kasai Oriental, Kinshasa, Kivu, Orientale), Rwanda (Nyansa) and Burundi, 10–2135 m.

Sources: Laurent, 1956a, Doucet, 1963, Mertens, 1965c, Witte, 1975, Courtois & Chippaux, 1977, Bauer et al., 1995, J.-F. Trape & Roux-Estève, 1995, Ineich, 2003, Chippaux, 2006, Chirio & LeBreton, 2007, Pauwels & Vande weghe, 2008 and Segniagbeto et al., 2011.

9. *Dipsadoboa weileri* (Lindholm, 1905a). Jahrb. Nassau Ver. Naturk. Wiesbaden 58: 186–187. (*Dipsadophidium weileri*)

D

Synonym: *Dipsadoboa isolepis* Boulenger, 1907a.

Type: Holotype, MWNH 1238, adult female (J. Weiler, 1905), destroyed Aug. 1940–March 1945 during World War II.

Type locality: "Umgegend von Bibundi (Kamerun)" [= vicinity of Bibundi, Sud-Ouest Prov., SW Cameroon, 4°13,N, 9°01'E, elevation 25 m].

Distribution: West and cen. Africa. Southeastern Guinea (Nzérékoré), Togo, Cameroon (Adamaoua, Est, Centre, Littoral, Nord-Ouest, Ouest, Sud, Sud-Ouest), Gabon (Moyen-Ogooué, Ngounié, Ogooué-Ivindo, Ogooué-Maritime, Woleu-Ntem), Central African Republic (Haute-Sangha, Lobaye, Ouaka, Sangha), Congo, E Democratic Republic of the Congo (Orientale), S South Sudan (SW Eastern Equatoria), Rwanda (Nyansa) and S Uganda (Central, Western), NSL–1000 m.

Sources: L. Müller, 1910, Spawls et al., 2002, Ineich, 2003, Chippaux, 2006, Chirio & LeBreton, 2007, Pauwels & Vande weghe, 2008, W. Böhme et al., 2011 and Segniagbeto et al., 2011.

10. *Dipsadoboa werneri* (Boulenger, 1897b). Ann. Mag. Nat. Hist. (6) 19(111): 281. (*Leptodira werneri*)

Synonym: *Dipsadomorphus reticulatus* F. Werner, 1909d.

Type: Holotype, BMNH 1946.1.8.32 (formerly BMNH 1897.1.29.8), a 370 mm male (A. Müller via F. Werner).

Type locality: "Usambara, German East Africa" [= Usambara Mountains, N Tanga Prov., NE Tanzania].

Distribution: Northeastern Tanzania (Tanga), 880–1200 m.

Sources: J.B. Rasmussen, 1986 and Spawls et al., 2002.

Remarks: Collector of holotype listed as F. Werner *fide* Boulenger (1897b: 281) but Werner sent the specimen to Boulenger. Photographs and head illustration of holotype in J.B. Rasmussen (1986: figs. 1–2).

DIPSAS Laurenti, 1768 (Dipsadidae)

Synonyms: *Bungarus* Oppel, 1811, *Dypsas* Goldfuss, 1820 *(nomen emendatum)*, *Pholidolaemus* Fitzinger, 1843, *Pholidodumus* – A.-M.-C. Duméril & Bibron, 1844 *(nomen incorrectum)*, *Cleone* Gistel, 1848 (*nomen nudum*), *Dipsadomorus* A.-M.-C. Duméril, Bibron & Duméril, 1854a, *Leptognathus* A.-M.-C. Duméril, Bibron & Duméril, 1854a, *Stremmatognathus* A.-M.-C. Duméril, Bibron & Duméril, 1854a, *Dipsamorus* – Chenu & Desmarest, 1857 *(nomen incorrectum)*,

Dipsus – Locard, 1875 *(nomen incorrectum), Leptognatho* – F. Müller, 1878b *(nomen incorrectum), Dipsammodorus* – Lohmeyer, 1882 *(nomen incorrectum), Stemmatognathus* – Hoffmann, 1890 *(nomen incorrectum), Neopareas* A.C.L.G. Günther, 1895 *in* 1885–1902, *Dipsasus* A.L. Herrera, 1899 *(nomen emendatum), Dipsademorus* – Mocquard, 1908 in A.-M.-C. Duméril, Bocourt & Mocquard, 1870–1909 *(nomen incorrectum), Leptognatus* – Phisalix, 1917b *(nomen incorrectum), Heterorhachis* Amaral, 1923, *Strematognathus* – Amaral, 1926 *(nomen incorrectum), Dispas* – Banerji, 1929 *(nomen incorrectum), Leptonathus* – Ocaranza, 1930 *(nomen incorrectum), Hererorhachis* – Ditmars, 1933 *(nomen incorrectum), Leptognatus* – Maass-Berlin, 1933 *(nomen incorrectum), Heterorchachis* – Briceño-Rossi, 1934 *(nomen incorrectum), Heterorhachia* – Briceño-Rossi, 1934 *(nomen incorrectum), Neoparias* – E.H. Taylor, 1951 *(nomen incorrectum),* and *Dipsias* – Deraniyagala, 1955 *(nomen incorrectum).*

Type species: *Dipsas indica* Laurenti, 1768.

Distribution: Latin America.

Sources: J.A. Peters, 1960b, 1965, J.A. Peters & Orejas-Miranda, 1970, Gravlund, 2001, Cadle, 2007, Harvey & Embert, 2009, Harvey et al., 2009, Zaher et al., 2009, Fernandes et al., 2010 and Mulcahy et al., 2011.

1. *Dipsas albifrons* (Sauvage, 1884). Bull. Soc. Philom. Paris (1883–1884) (7) 8: 145. (*Dipsadomorus albifrons*)

Synonym: *Dipsas albifrons cavalheiroi* Hoge, 1950.

Type: Lectotype, IB 17746 (formerly MNHN 6106), a 520 mm female (H. Ferry), designated by Passos et al. (2005: 26) and destroyed by fire 15 May 2010.

Type locality: "Brésil" [= Brazil].

Distribution: Southern Brazil (Espírito Santo, Mato Grosso, Paraná, Rio de Janeiro, Santa Catarina, São Paulo, Alcatrazes, Queimada Grande and São Sebastião Is.), NSL–800 m.

Sources: Hoge, 1950, Marques et al., 2001, Passos et al., 2005 and Cicchi et al., 2007.

Remarks: A listed syntype as MNHN 6106 *fide* J.A. Peters (1960b: 121). Photograph of lectotype in Passos et al. (2005: fig. 5). Distributional gaps discussed in Passos et al. (2005: 30). Paraguay records rejected *fide* Cacciali (2006: 77).

2. *Dipsas alternans* (J.G. Fischer, 1885). Jahrb. Hamburg. Wiss. Anst. 2: 105–107, pl. 4, fig. 8. (*Leptognathus alternans*)

Synonyms: *Sibynomorphus barbouri* Amaral, 1923 & 1926a and *Sibynomorphus garbei* Amaral, 1923 & 1926a.

Type: Neotype, IB 64465, a 730 mm male (F.O. Forenza, 3 Oct. 2001), designated by Passos et al. (2004: 384) and destroyed by fire 15 May 2010.

Type locality: "Sítio Rancho Colina da Vitória, Municipality of Juquitiba (23°55'S, 47°04'W), State of São Paulo, Brazil," by neotype designation.

Distribution: Southeastern Brazil (Espírito Santo, Paraná, Rio de Janeiro, Rio Grande do Sul, Santa Catarina, São Paulo, São Sebastião Is.), NSL–725 m.

Source: Cicchi et al., 2007.

Remarks: Revalidated *fide* Passos et al., 2004. Neotype illustrated by Passos et al. (2004: fig. 1).

3. *Dipsas andiana* (Boulenger, 1896a). Cat. Snakes Brit. Mus. 3: 452, pl. 23, figs. 2–2a. (*Leptognathus andiana*)

Type: Holotype, BMNH 1946.1.20.12.

Type locality: "Quito, Ecuador" [probably a shipping point *fide* Cadle & Myers, 2003: 25].

Distribution: Western Ecuador (Bolívar, Esmeraldas, Los Rios, Pichincha), 1140–1750 m.

Sources: Cadle & Myers, 2003 and Cisneros-Heredia, 2004.

4. *Dipsas articulata* (Cope, 1868c). Proc. Acad. Nat. Sci. Philadelphia 20(2): 135. (*Leptognathus articulata*)

Type: Holotype, ANSP 10113, a 673 mm male (R.W. Mitchell, 1857–1861).

Type locality: "Veraguas, Costa Rica" (in error). Corrected to Cocuyas de Veraguas, New Grenada [= Cocuyas de Veraguas, a gold mine near the Río Concepción, N Veraguas Prov., cen. Panama, ca. 8°45'N, 81°00'W, elevation 100 m] *fide* Cope (1860e: 266).

Distribution: Western Panama (Bocas del Toro, Veraguas), 55–850 m.

Sources: Brattstrom & Howell, 1954, E.H. Taylor, 1954, L.D. Wilson & Meyer, 1985, G. Köhler, 1999b, Savage, 2002, Solórzano, 2004, Lotzkat et al., 2010 and McCranie, 2011a.

5. *Dipsas baliomelas* Harvey, 2008. Herpetologica 64(4): 423–427, figs. 1, 3.

Type: Holotype, UTA 3363, a 393 mm male (J.F. Taulman).

Type locality: "35 km WSW Vista Hermosa (Cañon Sardinata), Serranía de la Macarena, Meta, Colombia, '3700 ft' (=1128 m), between 2°15'–3°15'N latitude and 73°45'–74°10'W longitude."

Distribution: Central Colombia (SW Meta), 1130 m.

6. *Dipsas bicolor* (A.C.L.G. Günther, 1895 *in* 1885–1902). Biol. Cen.-Amer., Rept. Batr.: 178–179, pl. 56, fig. c. (*Neopareas bicolor*)

Synonym: *Neopareas tricolor* Brattstrom & Howell, 1954.

Type: Holotype, BMNH 1894.10.1.39, a 546 mm specimen (R.A. Rix, 1893–1894).

Type locality: "Nicaragua, Chontales Mines" [= Chontales gold and silver mines, vicinity of Santo Domingo, NE

Chontales Dept., Nicaragua, 12°16'N, 85°00'W, elevation 500 m].

Distribution: Central America. Southeastern Honduras (El Paraíso, Gracias a Dios, Olancho), E Nicaragua (Atlántico Norte, Chontales, Jinotega, Nuevo Segovia, Rio San Juan) and Costa Rica (Alajuela, Cartago, Guanacaste, Heredia, Limón, San José), 10–1000 m.

Sources: Belt, 1874, L.D. Wilson & Meyer, 1985, Savage, 2004, Solórzano, 2004, and McCranie, 2011a.

Remarks: A valid species *fide* McCranie (2011a: 279).

7. *Dipsas brevifacies* (Cope, 1866). Proc. Acad. Nat. Sci. Philadelphia 18(2): 127. (*Tropidodipsas brevifacies*)

Synonyms: *Dipsadomorus fasciatus* Bocourt, 1884 and *Leptognathus torquatus* Cope, 1885a.

Type: Holotype, USNM 24886 (formerly USNM 753), a 244 mm male (A.C.V. Schott, 27 Sept. 1865).

Type locality: "Yucatan" [Mexico].

Distribution: Yucatán Peninsula. Southeastern Mexico (Campeche, Quintana Roo, Yucatán, Carmen Is.) and N Belize (Belize, Corozal, Orange Walk), NSL–300 m.

Sources: Kofron, 1982a, Lee, 1996, 2000 and J.A. Campbell, 1998.

8. *Dipsas bucephala* (G. Shaw, 1802). Gen. Zool., Amph. 3(2): 422–423. (*Coluber bucephalus*)

Synonym: *Leptognathus cisticeps* Boettger, 1885a.

Type: Lectotype, specimen described and illustrated by Seba (1834: 71, pl. 43, fig. 4), designated herein.

Type locality: "Ceylon" [= Sri Lanka] (in error) via lectotype selection. Restricted to Brazil *fide* J.A. Peters (1960: 73).

Distribution: Southern South America. Southeastern Brazil (Bahia, Espírito Santo, Rio Grande do Sul, São Paulo), Bolivia (Cochabamba, Santa Cruz), Paraguay (Alto Paraguay, Alto Paraná, Asunción, Canindeyú, Itapúa, Paraguarí) and Argentina (Misiones), 270–2000 m.

Sources: Hoge & Romano-Hoge, 1975b, Cacciali, 2006 and Harvey & Embert, 2009.

9. *Dipsas catesbeji* (Seetzen, 1796b). Myer's Zool. Arch. 2: 66. (*Coluber catesbeji*)

Synonyms: *Dipsas catesbeji* Seetzen, 1795 (*nomen nudum*), *Dipsas catesbaei* F. Boie, 1827 (*nomen emendatum*), *Dipsas catesbyi* Schlegel, 1837 (*nomen emendatum*), and *Sibynomorphus macedoi* Prado & Hoge, 1947.

Types: Syntypes (2), none designated, an 807 mm and 651 mm specimen, lost *fide* J.A. Peters (1960: 56).

Type locality: "Wahrscheinlich Amerika" [= probably America]. Restricted to Suriname and Cayenne [= French Guiana] fide Schlegel (1837: 279).

Distribution: Northern South America. Eastern Colombia (Amazonas, Antioquia, Caquetá, Cundinamarca, Meta, Putumayo, Santander, Vaupés), S Venezuela (Amazonas), N Guyana (Barima-Waini, Cuyuni-Mazaruni, Essequibo Islands-West Demerara, Potaro-Siparuni, Upper Demerara-Berbice), Suriname (Para, S Sipaliwini), French Guiana (Cayenne, Saint-Laurent-du-Maroni) Ecuador (Chimborazo, Morona-Santiago, Napo, Pastaza, Zamora-Chinchipe), E Peru (Amazonas, Huanuco, Junin, Loreto, Madre de Dios, Pasco, Puno, San Martin), N Brazil (Acre, Amazonas, Bahia, Mato Grosso do Sul, Pará, Rondônia) and NE Bolivia (Beni, Cochabamba, La Paz, Pando, Santa Cruz), NSL–940 m.

Sources: J.A. Peters, 1956, Chippaux, 1987, Lancini, 1986, Pérez-Santos & Moreno, 1988, Starace, 1998, Lehr, 2001, Doan & Arizábal, 2002, Abuys, 2003, Duellman, 2005, Harvey & Embert, 2009, Lima & Prudente, 2009, Navarrete et al., 2009 and C.J. Cole et al., 2013.

Remarks: Original orthography of *catesbeji* was presumed a typographical error, but that spelling is consistently used in Seetzen (1795: 320; 1796a: 66) whereas Catesby's name was correctly spelled. Possibly closer to *Sibynomorphus* than other *Dipsas fide* Lima & Prudente (2009: 43–44). Colombian records from Medellín (1538 m) and Bogotá (2630 m) need confirmation *fide* Perez-Santos & Moreno (1988: 142). Paraguay records rejected *fide* Cacciali (2006: 77).

10. *Dipsas chaparensis* Reynolds & Foster, 1992. Herpetol. Monogr. (6): 101–102, fig. 9.

Type: Holotype, USNM 257869, an 830 mm female (M.S. Foster, 14 Nov. 1979).

Type locality: "Parjacti, 83.2 km by road NE Cochabamba on road to Villa Tunari, Chapare Prov., Department of Cochabamba, Bolivia, elevation approximately 2044 m."

Distribution: Bolivia (Cochabamba, Santa Cruz), 1850–2045 m.

Sources: Roze, 1966a, Lancini, 1986, Starace, 1998 and Harvey & Embert, 2009.

11. *Dipsas copei* (A.C.L.G. Günther, 1872b). Ann. Mag. Nat. Hist. (4) 9(49): 30. (*Leptognathus copei*)

Type: Holotype, BMNH 1946.1.21.4 (formerly BMNH 1866.8.14.329), a 650 mm male (T.G. van Lidth de Jeude coll., 1866).

Type locality: "Suriname" [= Suriname].

Distribution: Guianas. Eastern Venezuela (Amazonas), Guyana (Essequibo Islands-West Demeara) and Suriname, NSL–1180 m.

Sources: Abuys, 2003, Passos et al., 2004, Harvey, 2008 and C.J. Cole et al., 2013.

Remarks: A valid species *fide* Harvey, 1999 and Rivas-Fuenmayor et al., 2012. Dipsas *copei* synonymized with *D. incerta* and holotype of *D. copei* (BMNH

1946.1.21.4) designated neotype of *D. incerta fide* Passos et al. (2004: 389); if the two taxa are separate species as Harvey (2008: 436) demonstrated, the neotype designation is invalid. Holotype illustrated by Passos et al. (2004: fig. 5).

D

12. *Dipsas elegans* (Boulenger, 1896a). Cat. Snakes Brit. Mus. 3: 452–453, pl. 23, fig. 3. (*Leptognathus elegans*)

Type: Holotype, BMNH 1946.1.21.77 (formerly BMNH 1871.2.7.36), a 301–310 mm specimen (A.-L.-J.-F. Sumichrast, 1855–1869).

Type locality: "Tehuantepec" [= Oaxaca State, Mexico] (in error).

Distribution: Western Ecuador (Chimborazo, Cotopaxi, Imbabura, Pichincha), 500–1920 m.

Sources: Kofron, 1982 and Cadle, 2005

Remarks: Holotype illustrated by Cadle (2005: fig. 1).

13. *Dipsas ellipsifera* (Boulenger, 1898h). Proc. Zool. Soc. London 66(1): 117–118, pl. 12, fig. 2. (*Leptognathus ellipsifera*)

Type: Lectotype, BMNH 1946.1.21.26 (formerly BMNH 1898.4.28.87), a 465 mm male, designated by Cadle (2005: 77).

Type locality: "Ibarra" [= Ibarra, Imbabura Prov., N Ecuador, 0°22'N, 78°08'W, elevation 2235 m].

Distribution: Northern Ecuador (Imbabura), 570–2600 m.

Source: Cadle, 2005.

Remarks: Photograph of lectotype in Cadle (2005: fig. 2).

14. *Dipsas gaigeae* (Oliver, 1937). Occ. Pap. Mus. Zool., Univ. Michigan (360): 22–23. (*Sibynomorphus gaigeae*)

Type: Holotype, UMMZ 80221, a 287 mm male (J.A. Oliver, 26 July 1935).

Type locality: "Paso del Río, Colima, Mexico." [= Periquillo, Haciendo Paso del Río, Colima State, Mexico *fide* Kofron, 1982b: 289].

Distribution: Western Mexico (Colima, SW Jalisco, Michoacán), NSL–610 m.

Sources: J.A. Peters, 1964, Harris & Simons, 1967, Hale, 1977, Kofron, 1982b and Reyes-Velaso et al., 2008.

15. *Dipsas gracilis* (Boulenger, 1902a). Ann. Mag. Nat. Hist. (7) 9(49): 57. (*Leptognathus gracilis*)

Synonyms: *Leptognathus hammondii* Boulenger, 1920b and *Sibynomorphus macrostomus* Amaral, 1925.

Types: Syntypes (2), BMNH 1946.1.21.24–25, two males, longest syntype 770 mm (W.F.H. Rosenberg, 1896–1901).

Type locality: "St. Javier, 60 feet, North-western Ecuador" [= San Javier, lower Cachabi/Cachavi River, Esmeraldas Prov., NW Ecuador, 1°04'N, 78°47'W, elevation 20 m].

Distribution: Northwestern South America. ? Western Colombia (Chocó), W Ecuador (Esmeraldas, Guayas, Los Rios, Pichincha) and NW Peru (Piura, Tumbes), 20–500 m.

Sources: Pérez-Santos & Moreno, 1988, 1991 and Cadle, 2005.

Remarks: Colombian record doubtful *fide* Cadle (2005: 123): possibly *D. viguieri*. *Dipsas gracilis* and *D. viguieri* possibly conspecific *fide* Harvey (2008: 429).

16. *Dipsas incerta* (Jan, 1863b). Elenco Sist. Ofidi: 101. (*Leptognathus incertus*)

Type: Holotype, formerly MSNM, destroyed in 1943 during World War II.

Type locality: "Guajana francese" [= French Guiana].

Distribution: Northern South America. Southeastern Venezuela (Amazonas, Bolívar), Guyana, Suriname, French Guiana and SW Brazil (Roraima), 525–1020 m.

Sources: Marques et al., 2001, Passos et al., 2004, Harvey, 2008, Navarrete et al., 2009 and Fernandes et al., 2010.

Remarks: *Dipsas copei* synonymized with *D. incerta* and holotype of *D. copei* (BMNH 1946.1.21.4 from Suriname) designated neotype of *D. incerta fide* Passos et al. (2004: 389); if the two taxa are separate species as Harvey (2008: 436) demonstrated, the neotype designation is invalid. Holotype illustrated in Passos et al. (2004: fig. 5).

17. *Dipsas indica* Laurenti, 1768. Synop. Rept.: 90.

Synonyms: *Dipsadomorus indicus* – A.-M.-C. Duméril, Bibron & Duméril, 1854a, *Pholidolaemus gracilis* Fitzinger, 1861, and *Dipsas indica ecuadorensis* J.A. Peters, 1960b.

Type: Lectotype, specimen described and illustrated by Seba (1734: 71, pl. 43, fig. 5), designated and lost by J.A. Peters (1960b: 68).

Type locality: "Zeylona" [= Sri Lanka] (in error). Corrected to Amazonian region of South America [= Amazonia] *fide* J.A. Peters (1960b: 68).

Distribution: South America. Southeastern Colombia (Amazonas, Cundinamarca, Meta, Putumayo, Vaupés), Venezuela (Amazonas, Bolívar), Guyana (Essequibo Islands-West Demerara, Upper Takutu-Upper Essequibo), Suriname (Commewijne), French Guiana (Cayenne, Saint-Laurent-du-Maroni), E Ecuador (Morona-Santiago, Napo, Pastaza), E Peru (Cusco, Loreto, Madre de Dios, Putumayo, Ucayali), Brazil (Acre, Amazonas, Bahía, Ceará, Goiás, Mato Grosso do Sul, Pará, Rondônia), Bolivia (Pando), Paraguay (Alto Paraná, S Boquerón, Itapúa, S Presidente Hayes) and NE Argentina (Misiones), NSL–1000 m.

Sources: J.A. Peters, 1960b, Hoge, 1969b, Hoge & Romano-Hoge, 1975b, Pérez-Santos & Moreno, 1988,

Sazima, 1989, Cei, 1994, Alves & Argôlo, 1998, Starace, 1998, Freitas, 1999, Marques et al., 2001, Argôlo & Alves, 2002, J.C. Murphy & Schlager, 2003, Duellman, 2005, Montechiaro et al., 2006, Cicchi et al., 2007, Harvey & Embert, 2009, Cacciali, 2008, Natera-Mumaw & Battiston, 2008, Navarrete et al., 2009, Sajdak, 2010 and C.J. Cole et al., 2013.

18. *Dipsas maxillaris* (F. Werner, 1909c). Zool. Jahrb. (Abt. Syst. Ökol. Geog. Tiere) 28(3): 279. (*Leptognathus maxillaris*)

Type: Holotype, IRSNB 2026 (formerly IRSNB 120), a 335 mm female (Linden, 17 Nov. 1857).
Type locality: "Tabasco, Mexico" (in error). Corrected to South America *fide* Kofron (1982a: 46).
Distribution: South America. Known only from holotype.
Sources: Laurent, 1949 and Lang, 1990.
Remarks: Type erroneously listed as IRSNB 3042 *fide* Laurent (1949: 15) and J.A. Peters (1960b: 48). Publication date listed as 1909 *fide* cover of separate (nos. 1–5 = 1909, no. 6 and vol. = 1910). F. Werner's bibliography by Wettstein (1941: 36). Possibly a synonym of *D. elegans*.

19. *Dipsas neivai* Amaral, 1926a. Arch. Mus. Nac., Rio de Janeiro 26: 108–109, pl. 2, figs. 9–11.

Type: Holotype, IB 307, a 680 mm male (J.E.A. Coutinho, 30 July 1912), destroyed by fire 15 May 2010.
Type locality: "Caratinga, Estado de Minas Geraes, Brasil" [= Caratinga, E Minas Gerais State, SE Brazil, 19°47'S, 42°08'W, elevation 835 m].
Distribution: Southeastern Brazil (E Bahia, E Minas Gerais), 145–835 m.
Remarks: A valid species *fide* Cicchi et al. (2007: 230).

20. *Dipsas nicholsi* (Dunn, 1933b). Copeia 1933(4): 193–194. (*Sibynomorphus nicholsi*)

Type: Holotype, MCZ 37884, a head and neck (H.C. Clark, early 1933).
Type locality: "Mid-basin of Chagres River and mouth of Pequeni River, Panamá" [= junction of Río Pequeni with Río Boquerón, in upper drainage of Río Chagres, Panamá Prov., Panama, 9°21'N, 79°33'W, elevation 80 m].
Distribution: Central Panama (Canal Zone, Coclé, Colón, Darién, Panamá), NSL–960 m.
Sources: Cadle & Myers, 2003 and C.W. Myers et al., 2007.

21. *Dipsas oreas* (Cope, 1868b). Proc. Acad. Nat. Sci. Philadelphia 20(2): 109. (*Leptognathus oreas*)

Synonyms: *Leptognathus andrei* Sauvage, 1884 and *Leptognathus robusta* L. Müller, 1923.
Type: Holotype, ANSP 10115 (formerly ANSP 6707), a 660–691 mm male (J. Orton, autumn 1867–winter 1968).
Type locality: "the elevated valley of Quito" [Ecuador] (probably in error *fide* J.A. Peters, 1955: 347).
Distribution: Western Ecuador (Chimborazo, Guayas, Loja) and Peru (Cajamarca, Piura), 1400–2900 m.
Sources: Kofron, 1982a, Orcés & Almendáriz, 1987, Cadle & Myers, 2003 and Cadle, 2005.
Remarks: Photograph of holotype in Kofron (1982: fig. 2).

22. *Dipsas pakaraima* MacCulloch & Lathrop, 2004b. Rev. Biol. Trop. 52(1): 240–244, figs. 1–3.

Type: Holotype, ROM 41233, a 625 mm male (R. Edward, 21 Oct. 2000).
Type locality: "northeast plateau of Mount Ayanganna, in the Pakaraima Distribution, western Guyana, 05°24'N, 059°57'W, elevation 1490 m."
Distribution: Guyana (Cuyuni-Mazaruni), 1490 m. Known only from type locality.
Source: C.J. Cole et al., 2013.
Remarks: Similarities between *Dipsas temporalis, D. pakaraima* and *D. vermiculata* separate them from the rest of *Dipsas fide* Harvey (2008: 436).

23. *Dipsas pavonina* Schlegel, 1837. Essai Phys. Serp. 1: 162, 2: 280–281.

Synonym: *Leptognathus pavonimus* – A.-M.-C. Duméril, Bibron & Duméril, 1854b *(nomen incorrectum).*
Types: Syntypes (5), MNHN 435, RMNH 1022a–b, one syntype 620 mm, and RMNH 1024 (formerly MNHN), a 720 mm specimen (T.G. van Lidth de Jeude), location of other syntype unknown.
Type locality: "Guyanes" [= Guianas].
Distribution: Amazonia. Colombia (Amazonas), S Venezuela (Amazonas), Guyana (Demerara-Mahaica, Essequibo Islands-West Demerara), Suriname (Para, Wanica), French Guiana (Cayenne, Saint-Laurent-du-Maroni), N Brazil (Amazonas, Pará, Rondônia), E Ecuador (Morona-Santiago, Napo, Pastaza), NE Peru (Amazonas, Cusco, Loreto, Madre de Dios) and ext. N Bolivia (La Paz, Pando), NSL–200 m.
Sources: Roze, 1966a, Lancini, 1986, Starace, 1998, Abuys, 2003, Harvey & Embert, 2009, Lima & Prudente, 2009, Navarrete et al., 2009 and C.J. Cole et al., 2013.
Remarks: See remarks under *D. catesbyi*. Iquitos, Peru record doubtful *fide* Dixon & Soini (1986: 103). MNHN 435 not listed as a syntype *fide* MNHN catalogue.

24. *Dipsas peruana* (Boettger, 1898). Kat. Rept. Senck. Mus. 2: 128. (*Leptognathus peruana*)

Synonyms: *Leptognathus boettgeri* F. Werner, 1901c, *Leptognathus latifrontalis* Boulenger, 1905c, *Leptognathus boliviana* F. Werner, 1909b, *Leptognathus* *palmeri* Boulenger, 1912b, *Leptognathus polylepis* Boulenger, 1912b, and *Leptognathus latifasciatus* Boulenger, 1913b.

Type: Holotype, SMF 20801 (formerly SMF-B 9448a), a 610 mm female (F. Emmel, 1890).

Type locality: "Sta. Ana in der peruanischen Provinz Cusco" [= Santa Ana, Cusco Department, Peru, 13°31'S, 71°59'W, elevation 3470 m].

Distribution: Western Venezuela (Aragua, Barinas, Lara, Mérida, Miranda), NE Colombia (Boyacá), E Ecuador (Morona-Santiago, Napo, Pastaza), Peru (Amazonas, Cusco, Loreto, Madre de Dios, Puno) and N Bolivia (La Paz, Pando), 500–2400 (3470) m.

Sources: Roze, 1966a, Lancini, 1986, Pérez-Santos & Moreno, 1988, 1991, Esqueda & La Marca, 1999, Fernandes et al., 2002, Cadle, 2005, Harvey & Embert, 2009, Lima & Prudente, 2009 and Rivas-Fuenmayor et al., 2012.

Remarks: *Dipsas latifrontalis* possibly a valid species *fide* Rivas-Fuenmayor et al. (2012: 44).

25. *Dipsas petersi* Hoge & Romano-Hoge, 1975b. Mem. Inst. Butantan 39: 52–56, figs. 2, 4, 9–11. (*Dipsas indica petersi*)

Type: Holotype, IB 23460, a female (V. Rodrigues, 1 Nov. 1963), destroyed by fire 15 May 2010.

Type locality: "Pedro de Toldeo, São Paulo, Brazil" [= Pedro de Toledo, SE São Paulo State, SE Brazil, 24°17'S, 47°14'W, elevation 85 m].

Distribution: Southeastern Brazil (Espirito Santo, Paraná, Rio de Janeiro, E São Paulo, E Paraná, NE Santa Catarina, Cardoso, Santo Amaro, São Sebastião and São Vicente Is.), NSL–85 m.

Source: Cicchi et al., 2007.

Remarks: A valid species *fide* Cicchi et al. (2007: 230).

26. *Dipsas praeornata* (F. Werner, 1909b). Mitt. Natur. Mus. Hamburg 26: 240. (*Leptognathus praeornata*)

Type: Holotype, formerly ZMH, a 733 mm female (R. Rauschenplat), destroyed in July 1943 during World War II.

Type locality: "Venezuela."

Distribution: Cordillera de Costa of N Venezuela (Aragua, District Federal, Libertador, Miranda).

Source: Harvey & Embert, 2008.

Remarks: A valid species *fide* Harvey & Embert (2008: 85) and Harvey (pers. comm.).

27. *Dipsas pratti* (Boulenger, 1897e). Ann. Mag. Nat. Hist. (6) 20(120): 523. (*Leptognathus pratti*)

Synonyms: *Leptognathus triseriatus* Cope, 1899b, *Leptognathus nigriceps* F. Werner, 1916, *Dipsas niceforoi* Prado, 1941b, and *Dipsas tolimensis* Prado, 1941f.

Type: Holotype, BMNH 1946.1.20.52, a 770 mm female (A.E. Pratt, 1896–1897).

Type locality: "Medellin, Republic of Colombia" [= Medellín, Antioquia Department, Colombia, 6°14'N, 75°35'W, elevation 1480 m].

Distribution: Cordillera Central of N Colombia (Antioquia, Cundinamarca, Santander, Tolima) and W Venezuela (Zulia), 1500–2400 m.

Sources: Pérez-Santos & Moreno, 1988, Moreno-Arias et al., 2006, T.R. Barros-Blanco et al., 2012 and Rivas-Fuenmayor et al., 2012.

28. *Dipsas sanctijoannis* (Boulenger, 1911a). Ann. Mag. Nat. Hist. (8) 7(37): 24–25. (*Leptognathus sanctijoannis*) (*nomen corrigendum*)

Synonyms: *Leptognathus sancti-joannis* Boulenger, 1911a *(nomen incorrigendum), Leptognathus sancti-johannis* F. Werner, 1916 *(nomen substitutum), Sibynomorphus caucanus* Rendahl & Vestergren, 1940, and *Dipsas sanctijoannis* J.A. Peters, 1960 *(nomen corrigendum).*

Types: Syntypes (4), BMNH 1946.9.8.56–58 (formerly BMNH 1910.7.11.34–36), a 577 mm female and 600 mm specimen, and MCZ 83209 (formerly BMNH 1946.9.8.55), a 482 mm male (G. Palmer, 1909).

Type locality: "Pueblo Rico, slopes of San Juan River, Colombian Choco, 5200 feet" [= Pueblo Rico, Risaralda Dept., cen. Colombia, 5°11'N, 76°04'W, elevation 2120 m].

Distribution: Cordilleras of N Colombia (Antioquia, Boyacá, Caldas, Cauca, Nariña, Quindio, Risaralda, Valle), 995–2300 m.

Sources: Pérez-Santos & Moreno, 1988 and Rojas-Morales, 2012b.

29. *Dipsas sazimai* Fernandes, Marques & Angólo, 2010. Zootaxa (2691): 58–60, figs. 1–4.

Type: Holotype, MNRJ 15136, a 743 mm male (C.C. Siqueira & W.C. Kiefer, 30 Oct. 2005).

Type locality: "Brazil, state of Rio de Janeiro, municipality of Casimiro de Abreu (22º28'S, 42º12'W, ca. 801 m)."

Distribution: Southeastern Brazil (Bahia, Espírito Santo, Rio de Janeiro, São Paulo), 40–800 m

30. *Dipsas schunkii* (Boulenger, 1908a). Ann. Mag. Nat. Hist. (8) 1(1): 115. (*Leptognathus schunkii*)

Types: Syntypes (3), BMNH 1946.1.21.10–12, longest syntype 920 mm (C. Schunke).

Type locality: "Chanchamayo, E. Peru" [= Chanchamayo Prov., Junin Dept., E Peru].

Distribution: Cordillera Oriental of Peru (Junin, Piura), 1000–1300 m.

31. *Dipsas temporalis* (F. Werner, 1909b). Mitt. Naturhist. Mus. Hamburg 26: 241–242, fig. 13. (*Leptognathus temporalis*)

Synonym: *Leptognathus spurrelli* Boulenger, 1913d.
Type: Holotype, formerly ZMH, a 520 mm specimen (E. Krause, 1900–1909), destroyed in July 1943 during World War II.
Type locality: "Esmeraldas, Ecuador" [= Esmeraldas, Esmeraldas Prov., NW Ecuador, 0°57'N, 79°40'W, elevation 80 m].
Distribution: Panama (Canal Zone, Darién, Panamá, Veraguas), NW Colombia (Chocó, Córdoba, Esmeraldas) and ext. NW Ecuador (Esmeraldas), NSL–1000 m.
Sources: Dunn & Bailey, 1939, Pérez-Santos & Moreno, 1988, 1991, Cadle, 2005, Harvey 2008, Lotzkat et al., 2010 and Carvajal-Cogollo et al., 2011a.
Remarks: Similarities between *Dipsas temporalis, D. pakaraima* and *D. vermiculata* separate them from the rest of *Dipsas fide* Harvey (2008: 436).

32. *Dipsas tenuissima* E.H. Taylor, 1954. Univ. Kansas Sci. Bull. 36(11): 771–774, figs. 28a–c, 29.

Type: Holotype, KU 31961, a 555 mm specimen (E.H. Taylor, 10 July 1952).
Type locality: "approximately 15 km. WSW San Isidro del General, on Dominical Road, in swamp, Costa Rica" [= 15 km WSW San Isidro de General, San José Prov., W Costa Rica, ca. 9°21'N, 83°45'W, elevation 800 m].
Distribution: Southwestern Costa Rica (Puntarenas, San José) and ext. W Panama (Chiriquí), NSL–970 m.
Sources: Savage, 2002 and Solórzano, 2004.

33. *Dipsas trinitatis* H.W. Parker, 1926c. Ann. Mag. Nat. Hist. (9) 18(104): 206–207.

Type: Holotype, BMNH 1926.5.1.1, a 275 mm male (E. Lehner).
Type locality: "Trinity Hill Reserve, Trinidad, British West Indies."
Distribution: Trinidad and Tobago (Trinidad), NSL–60 m.
Sources: J.C. Murphy, 1997, Boos, 2001 and Harvey & Embert, 2009.
Remarks: Previously a subspecies of *D. variegata*.

34. *Dipsas variegata* (A.-M.-C. Duméril, Bibron & Duméril, 1854a). Erpét. Gén. 7(1): 477. (*Leptognathus variegatus*)

Types: Syntypes (2), MNHN 7299 (F. Levaillant, 1765–1780) and RMNH 995 (G. Vrolik, 1800–1830), longest syntype 798 mm.
Type locality: "Suriname."
Distribution: Panama and N South America. Eastern Panama (Canal Zone: Barro Colorado Is.), Colombia (Tolima), Venezuela (Amazonas, Aragua, Barinas, Bolívar, Carabobo, Distrito Federal, Guárico, Mérida, Miranda, Portuguesa, Sucre, Táchira, Yaracuy), Guyana (Barima-Waini, Demerara-Mahaica, Essequibo Islands-West Demerara, Potaro-Siparuni, Upper Demerara-Berbice), Suriname (Marowijne), French Guiana (Cayenne, Saint-Laurent-du-Maroni), Brazil (Alagoas, Amazonas, Ceará, Espírito Santo, Mato Grosso, Pará, Paraná, Santa Catarina, São Paulo, Cananéia and São Sebastião Is.), Ecuador (Chimbaraza, Pichincha, Loja), SE Peru (Madre de Dios) and Bolivia (Cochababa, La Paz), NSL–2154 m.
Sources: Amaral, 1944h, Roze, 1966a, Amaral, 1978, Lancini, 1986, Pérez-Santos & Moreno, 1988, 1991, Porto & Fernandes, 1996, J.B. Rasmussen & Hughes, 1997, Starace, 1998, Freitas, 1999, Markezich, 2002, Calcaño & Barrio, 2003a, Lotzkat et al., 2008, Harvey & Embert, 2009, Navarrete et al., 2009 and C.J. Cole et al., 2013.
Remarks: Type locality listed as Guyane hollandaise [= Suriname] *fide* MNHN catalogue.

35. *Dipsas vermiculata* J.A. Peters, 1960b. Misc. Publ. Mus. Zool., Univ. Michigan (114): 65–67, pl., figs. 2b–c.

Type: Holotype, UMMZ 118063 (formerly EPN 741), a 609 mm male (R. Olalla, 1954).
Type locality: "Chichirota, Lower Bobonaza River, Napo-Pastaza Prov., Ecuador."
Distribution: Eastern Ecuador (Napo, Morona-Santiago, Pastaza) and NW Peru (Amazonas, Loreto), 100–230 m.
Source: Pérez-Santos & Moreno, 1991.
Remarks: Similarities between *Dipsas temporalis, D. pakaraima* and *D. vermiculata* separate them from the rest of *Dipsas fide* Harvey (2008: 436).

36. *Dipsas viguieri* (Bocourt, 1884). Bull. Soc. Philom. Paris (7) 8: 136–137. (*Leptognathus viguieri*)

Type: Holotype, MNHN 6203, a 570 mm male (C. Viguier, Nov. 1876–April 1878).
Type locality: "Isthme de Darien" [= Darién Prov., Panama].
Distribution: Eastern Panama (Canal Zone, Darién, Panamá) and Colombia (? Chocó, Piura), NSL–60 m.
Source: Cadle, 2005.
Remarks: Colombian record of *D. gracilis* may be this species *fide* Cadle (2005: 123).

DIPSINA Jan, 1863a (Psammophiidae)

Synonym: *Dipsida* – F.W. FitzSimons, 1912 *(nomen incorrectum).*
Type species: *Coronella multimaculata* A. Smith, 1847 *in* 1838–1849.

Distribution: Southwestern Africa.

Sources: Broadley, 1983, Auerbach, 1987, Branch, 1988, Chirio & Ineich, 1991, Marais, 2004, Kelly et al., 2008, 2009, Vidal et al., 2008, Zaher et al., 2009 and Broadley & Blaylock, 2013.

1. *Dipsina multimaculata* (A. Smith, 1847 *in* 1838–1849). Illust. Zool. So. Africa, Rept.: pl. 61, figs. a–c, 2 pp. (*Coronella multimaculata*)

Types: Syntypes (2), none designated, longest syntype 330 mm (A. Smith, 1828), location unknown.

Type locality: "in the country of the Bushmen, near to the ODistribution River" [= Bushmanland, NW Northern Cape Prov., South Africa].

Distribution: Southwestern Africa. Namibia (Erongo, Hardap, Karas, Khomas, Kunene, Omaheke, Otjozondjupa), SW Botswana (Kgalagadi) and W South Africa (W Eastern Cape, Northern Cape, North West, Western Cape), NSL–1110 m.

Remarks: Types not present in RSM or BMNH *fide* V.F.M. FitzSimons (1937: 262).

DISPHOLIDUS Duvernoy, 1832 (Colubridae)

Synonyms: *Bucephalus* A. Smith, 1828 *(nomen praeoccupatum), Erymnus* Wagler, 1830, *Dispholides* – Filippi, 1840 *(nomen incorrectum), Bucephala* Gray, 1840 *(nomen emendatum), Dryomedusa* Fitzinger, 1843, *Bucephali* Hopley, 1882, *Dyspholidus* – Peracca, 1909 *(nomen incorrectum), Diopholidus* – R. Kraus & Werner, 1931 *(nomen incorrectum), Dispolidus* – Scortecci, 1943 *(nomen incorrectum),* and *Disphoridus* – Anon., 1978 *(nomen incorrectum).*

Type species: *Bucephalus typus* A. Smith, 1828.

Distribution: Subsaharan Africa.

Sources: Angel, 1933b, Laurent, 1955, 1956a, Mertens, 1955, 1971, Hoesch, 1960, Sweeney, 1961, Broadley, 1962a, 1983, V.F.M. FitzSimons, 1962a, Witte, 1962, Doucet, 1963, Laurent, 1964a, Menzies, 1966, Leston & Hughes, 1968, Hughes & Barry, 1969, Pitman, 1974, Roman, 1980, Branch, 1982, 1988, Buys & Buys, 1983, Hughes, 1983, 2013, Lanza, 1983a, 1990b, Pienaar et al., 1983, Largen & Rasmussen, 1993, David & Ineich, 1999, Broadley & Wallach, 2002, Clauss & Clauss, 2002, Broadley et al., 2003, M. Griffin, 2003, J.C. Murphy & Schlager, 2003, Marais, 2004, Branch & Bauer, 2005, Chippaux, 2006, J.-F. Trape & Mané, 2006b, Chirio & LeBreton, 2007, Largen & Spawls, 2010, Ullenburch et al., 2010, Segniagbeto et al., 2011 and Broadley & Blaylock, 2013.

1. *Dispholidus typus* (A. Smith, 1828). So. Afr. Comm. Adv. 3(144): 2. (*Bucephalus typus*)

Synonyms: *Coluber macrolepidotus* G. Shaw, 1802, *Bucephalus belli* A. Smith, 1828, *Bucephalus gutturalis* A. Smith, 1828, *Bucephalus jardinii* A. Smith, 1828, *Dispholidus lalandii* Duvernoy, 1832, *Dendrophis colubrina* Schlegel, 1837, *Bucephalus viridis* A. Smith, 1838, *Bucephalus capensis* A. Smith, 1841 *in* 1838–1849, *Dendrophis pseudodipsas* Bianconi, 1849a, *Bucephalus typus angolensis* Bocage, 1866a, *Thrasops jacksonii mossambicus* Mertens, 1937, *Dispholidus typus kivuensis* Laurent, 1955, *Dispholidus typus punctatus* Laurent, 1955, and *Dispholidus typus occidentalis* Perret, 1961 (*nomen nudum*).

Type: Holotype, RSM, a 1340 mm male, designated by FitzSimons (1937: 263).

Type locality: "eastern districts of South Africa" via lectotype selection. Restricted to Old Latakoo (= Lattakoo or Lithako), approx. 27°S, 24°E., i.e., btwn. Kuruman and Taung, N. Cape Prov., South Africa *fide* V.F.M. FitzSimons (1962a: 196).

Distribution: Subsaharan Africa. Southern Senegal (Dakar, Tambacounda), Gambia, Guinea–Bissau (Cacheu, Tombali, Bolama Is.), Guinea (Dalaba, Nzérékoré), SW Mali (Koulikoro, Sikasso), Sierra Leone (Northern, Southern), Liberia, Guinea (Dalaba, Kindia, Nzérékoré), Ivory Coast (Bouaké, Bouna, Ferkéssédougou, Guiglo, Tabou), Ghana (Upper West), Togo (Kara, Maritime, Plateaux). S Benin (Alibori, Atakora, Collines, Zou), S Burkina-Faso (Centre-Ouest, Est, Hauts-Bassins, Sud-Ouest, Volta-Noire), Nigeria (Akwa Ibom, Bayelsa, Rivers), Cameroon (Adamaoua, Centre, Est, Nord-Ouest, Ouest), SW Chad (Mayo-Kebbi), Central African Republic (Bamingui-Bangoran, Haute-Sangha, Haut-Mbomou, Kemo-Gribingui, Ombella-Mpoko, Ouaka, Ouham, Ouham-Pende), Gabon, S Congo, Democratic Republic of the Congo (Bandundu, Bas-Congo, Kasai Occidental, Kasai Oriental, Katanga, Kinshasa, Nord-Kivu, Orientale, Sud-Kivu), Eritrea (Northern Red Sea, E Southern, Dahlak Is.), S Ethiopia (Bale, Gemu Gofa, Hararge, Illubabor, Kefa, Shoa, Sidamo), Somalia (Bay, Galguduud, Hiraan, Jubbada Dhexe, Jubbada Hoose, Shabeellaha Dhexe, Shabeellaha Hoose, Woqooyi Galbeed), Kenya (Central, Coast, Eastern, North-Eastern, Nyanza, Western), Uganda (Central, Eastern, Northern, Western) Rwanda, Burundi, Tanzania (Arusha, Dar es Salaam, Dodoma, Iringa, Kigoma, Kilimanjaro, Lindi, Mara, Mbeya, Morogoro, Mtwara, Mwanza, Piwani, Rukwa, Ruvuma, Shinyanga, Singida, Tabora, Tanga, Kagera, Mafia, Pemba and Zanzibar Is.), Mozambique (Gaza, Inhambane, Manica, Sofala, Tete, Bazaruto and Inhaca Is.), Malawi (Central, Northern, Southern), Zambia (Central, Copperbelt, Eastern, Luapula, Lusaka, Northern, North-Western, Southern, Western), Angola (Cuando Cubango, Lunda Norte, Moxico), N Namibia (Caprivi, Erongo, Khomas, Okavango, Omaheke, Oshikoto, Otjozondjupa),

Botswana (Central, Gaborone, Ghanzi, Kgalagadi, Kgatleng, Kweneng, North East, North West, Southern, Tuli), Zimbabwe (Bulawayo, Manicaland, Mashonaland Central, Mashonaland East, Mashonaland West, Masvingo, Matabeleland North, Matabeleland South, Midlands), Swaziland, Lesotho and South Africa (Eastern Cape, NW Free State, Gauteng, KwaZulu-Natal, Limpopo, Mpumalanga, Northern Cape, North West, Western Cape), NSL–2400 m.

Remarks: Original description reprinted in A. Smith (1829: 441–442). Old Latakoo (Lithako) was the principle Kraal of the Batlaping tribe *fide* Loveridge (1957a: 273). Pemba Is., Tanzania population probably a valid species *fide* Hughes *in* Broadley & Wallach, 2003.

DISTEIRA Lacépède, 1804 (Elapidae)

Synonyms: *Disteria* Gray, 1825 *(nomen emendatum)*, *Distera* – Gray in Griffith & Pidgeon, 1831 *(nomen incorrectum)*, *Hydrus* Gray, 1842d *(nomen praeoccupatum)*, *Distira* Agassiz, 1847 *(nomen emendatum)*, *Deisteira* A.-M.-C. Duméril, Bibron & Duméril, 1854b *(nomen emendatum)*, *Distura* – Westphal-Castelnau, 1870 *(nomen incorrectum)*, *Distina* – Williston, 1914 *(nomen incorrectum)*, *Melanomystax* Wall, 1921g, *Disteia* – Maki, 1931 *(nomen incorrectum)*, and *Distera* – Boquet, 1948 *(nomen incorrectum)*.

Type species: *Hydrus major* G. Shaw, 1802.

Distribution: Coastal waters of Australasia, including Arafura Sea, Bay of Bengal, Coral Sea, Gulf of Carpentaria, and Timor Sea.

Sources: Wall, 1909a, M.A. Smith, 1926a, McDowell, 1972a, A.R. Rasmussen, 1994, David & Ineich, 1999, Wells, 2007, Kharin & Czeblukov, 2009, Zaher et al., 2009 and Hoser, 2012e.

Remarks: A synonym of *Hydrophis fide* Sanders et al., 2013.

1. *Disteira kingii* (Boulenger, 1896a). Cat. Snakes Brit. Mus. 3: 276–277. (*Hydrophis kingii*)

Type: Holotype, BMNH 1946.1.10.10 (formerly BMNH II.6.5.a), a 1200 mm male (P.P. King, 1818–1822).

Type locality: "N. Australia" [= northern Australia].

Distribution: Southern Papua New Guinea (Western) and N Australia (Northern Territory, Queensland, Western Australia, Great Barrier Reef and Barrow Is.).

Sources: L.A. Smith, 1974, Cogger, 1975, Cogger et al., 1983a, Storr et al., 1986 and O'Shea, 1996.

Remarks: Holotype is female *fide* Boulenger (1896a: 277).

2. *Disteira major* (G. Shaw, 1802). Gen. Zool., Amph. 3(2): 558–559, pl. 124. (*Hydrus major*)

Synonyms: *Disteira doliata* Lacépède, 1804, *Ophinectes major* Rafinesque-Schmaltz, 1817, *Pelamis shavii* Merrem, 1820 (*nomen substitutum*), *Hydrophis mentalis* Gray, 1842d, *Disteira dumerilii* Jan, 1859b, *Hydrophis lacepedei* Jan, 1859b, *Pelamydoidis major* – Fitzinger, 1861, and *Distira nasalis* De Vis, 1905.

Type: Lectotype, BMNH 1946.1.9.24 (formerly BMNH III.6.8.a), an 1118 mm male, designated by A.C.L.G. Günther (1864a: 364).

Type locality: "Indian Ocean" via lectotype selection.

Distribution: Austro-Papua. Southern Papua New Guinea (Western), Australia (Northern Territory, Queensland, W and N Western Australia, Great Barrier Reef and Melville Is.) and New Caledonia.

Sources: L.A. Smith, 1974, Cogger, 1975, Cogger et al., 1983a, Storr et al., 1986, O'Shea, 1996, Ineich & Rasmussen, 1997, Bauer & Sadlier, 2000 and Ineich & Laboute, 2002.

3. *Disteira nigrocincta* (Daudin, 1803d). Hist. Nat. Rept. 7: 380–382. (*Hydrophis nigrocinctus*)

Synonyms: *Hydrophis nigro-cinctus* – Duvernoy, 1832 *(nomen illegitimum)*, *Hydrophis nigro-cincta* – Viaud-Grand-Marais, 1880b *(nomen illegitimum)*, *Hydrophis nigrocyneta* – Anon., 1902 *(nomen incorrectum)*, *Distira hendersoni* Boulenger, 1903e, and *Distira hendorsoni* – Phisalix, 1922 *(nomen incorrectum)*.

Type: Holotype, BMNH 1946.1.10.13 (formerly BMNH 1896.3.25.5), a 1016 mm female (P. Russell, 1781–1791).

Type locality: "Bengale" [India]. Specified as The Sunderbunds *fide* P. Russell (1801: 57) [= Sundarbans, Ganges River delta, SW Bangladesh and SE West Bengal, India, bet. 21.5–23°N, 88–91.5°E].

Distribution: Southesat Asia. Eastern India, Bangladesh, Myanmar, Thailand, Malaysia and Singapore.

Sources: Wall, 1921g and M.A. Smith, 1943.

Remarks: Original description based upon specimen depicted in P. Russell (1801: 6–7, pl. 6). Type locality is the salt water rivers which intersect that part of Bengal called the Sunderbunds *fide* P. Russell (1801: 7).

4. *Disteira walli* Kharin, 1989. Vestn. Zool. 1989(1): 29–31.

Type: Holotype, BMNH 1864.4.7.6, a 560 mm male (P. de Bleeker, 1842–1860).

Type locality: "Malay Archipelago" [= West Malaysia].

Distribution: West Malaysia. Known only from holotype.

DITAXODON Hoge, 1958a (Xenodontidae)

Type species: *Philodryas taeniatus* Hensel, 1868.

Distribution: Southern Brazil.

Sources: Hoge, 1958a, J.A. Peters & Orejas-Miranda, 1970, Amaral, 1978, Moura-Leite et al., 1996, R.A. Thomas et al., 2006 and Zaher et al., 2009.

1. *Ditaxodon taeniatus* (Hensel, 1868). Arch. Naturg. 34(1): 331–332. (*Philodryas taeniatus*)

Type: Neotype, IB 10545, a 543 mm male, designated by R.A. Thomas et al. (2006: 95), destroyed by fire 15 May 2010.
Type locality: "Palmeire (25°20'S 50°00'W, 860 m), state of Paraná, Brazil" via neotype selection.
Distribution: Southern Brazil (Mato Grosso do Sul, Paraná, Rio Grande do Sul, São Paulo), 800–1800 m.
Remarks: Illustrations of lost holotype (ZMB 5980) mandible, maxilla and scale micro-ornamentation in Hoge (1985a: figs. 5–6, 8c–d). Photograph of destroyed neotype in Thomas (2006: fig. 1). Sea level record doubtful *fide* R.A. Thomas et al. (2006: 94).

DITYPOPHIS A.C.L.G. Günther, 1881 (Pseudoxyrhophiidae)

Synonyms: *Dityophis* – Hoffmann, 1890 *(nomen incorrectum)* and *Didypophis* – Dowling *in* Dowling & Duellman, 1978 *(nomen incorrectum).*
Type species: *Ditypophis vivax* A.C.L.G. Günther, 1881a.
Distribution: Yemen (Socotra Island).
Sources: H.W. Parker, 1949, Corkill & Cochrane, 1965, Showler, 1994, Underwood, 1999, Gravland, 2001, Nagy et al., 2003, Rösler & Wranik, 2004, Lawson et al., 2005, Vidal et al., 2008, Kelly et al., 2009, Zaher et al., 2009 and Razzetti et al., 2011

1. *Ditypophis vivax* A.C.L.G. Günther, 1881a. Proc. Zool. Soc. London 49(4): 462–463, pl. 40.

Type: Holotype, BMNH 1946.1.4.83 (formerly BMNH 1881.3.5.4), a 356 mm female (I.B. Balfour, 1879–1880).
Type locality: "Island of Socotra" [Yemen, Arabian Sea].
Distribution: Yemen (Socotra Is.), NSL–870 m.

DOLICHOPHIS Gistel, 1868 (Colubridae)

Synonym: *Eremiophis* Fitzinger, 1843.
Type species: *Coluber caspius* Gmelin, 1789.
Distribution: Eurasia.
Fossil records: Upper Pliocene to upper Pleistocene of Europe.
Sources: Schätti & Wilson, 1986, Schätti & Utiger, 2001 and Nagy et al., 2004, 2010.

1. *Dolichophis caspius* (Gmelin, 1789). Syst. Nat., ed. 13, 1(3): 1112. (*Coluber caspius*)

Synonyms: *Coluber jaculator* Pallas, 1799, *Coluber bicolor* Georgi, 1800, *Coluber acontistes* Pallas, 1814, *Coluber thermalis* Pallas, 1814, *Coluber trabalis* Pallas, 1814, *Natrix sheltopuszik* Wagler, 1825 (*nomen ineditum*), and *Coluber griseocoeruleus* Dwigubsky, 1832.
Type: Holotype, not designated, location unknown.
Type locality: "ad littora maris Caspii" [= shore of Caspian Sea, SW Asia]. Restricted to Jaikische Steppen, Gebiet des unteren Ural-Flusses [= Yaik steppes of the lower Ural River, Russia and Kazakhstan] *fide* Mertens & Müller (1928: 46).
Distribution: Southeastern Europe and ext. W Asia. Hungary, S Romania (Calarasi, S Caras-Severin, Constanta, Galati, Ialomita, Mehedinti, Tulcea, Vrancea), E Bosnia-Herzegovina, Croatia, Macedonia, S Montenegro, Serbia, Albania, Bulgaria (Blagoevgrad, Burgas, Dobrich, Gabrovo, Haskovo, Kardzhali, Lovech, Kyustendil, Montana, Pazardzhik, Pleven, Plovdiv, Razgrad, Rousse, Shumen, Silistra, Sliven, Stara Zagora, Targovishte, Varna, Veliko Tarnovo, Vidin, Vratsa, Yambol), Macedonia, Greece (Andros, Chios, Corfu, Crete, Euboea, Goekce, Ikaria, Karpathos, Kasos, Kea, Kithnos, Lesvos, Likorima, Limnos, Samos, Samothraki, Serifos, Sifnos, Siros, Spetsai, Tinos, Thasos), W Turkey (Adana, Ankara, Antalya, Balikesir, Bursa, Denizli, Edirne, Gaziantep, Isparta, Istanbul, Izmir, Kastamonu, Kirklareli, Kocaeli, Konya, Manisa, Mugla, Sinop, Tekirday, Imroz Is.), Moldavia, S Ukraine, SW Russia (Dagestan), ext. NW Georgia, ext. NE Turkey and ext. NE Azerbaijan, NSL–915 m.
Fossil records: Upper Pliocene (Villanyian, MN 16–17: 1.8–3.2 mya) of Austria and Hungary, lower Pleistocene (Calabrian: 0.8–1.8 mya) of Austria, Czech Republic, Romania and Ukraine, middle Pleistocene (Ionian: 0.13–0.78 mya) of Austria and Bulgaria, and upper Pleistocene (Tarantian: 0.01–0.13 mya) of Bulgaria and Czech Republic.
Sources: Buresch & Zonkoff, 1934, Fuhn & Vancea, 1961, Baran, 1976a, Basoglu & Baran, 1980, Kotenko et al., 1986, Szczerbak & Böhme, 1993a, Schätti, 1993a, Ataev et al., 1994, Borkin et al., 1997, Holman, 1998a, Szyndlar, 1991a, Petkovski et al., 2000, Baker et al., 2002, Valakos et al., 2004, Ivanov, 2007, Trapp, 2007, Valakos et al., 2008, Kretecki et al., 2009, Tuniyev et al., 2009 and Stojanov et al., 2011.
Remarks: Original description based on Lepechin (1769: 513–514, pl. 21), with type locality near Krassno Jar on the Jaikian Steppes, lower Ural River, Caspian Sea shore *fide* Zinner (1972: 24).

2. *Dolichophis cypriensis* (Schätti, 1985). Rev. Suisse Zool. 92(2): 472–475, figs. 1–2. (*Coluber cypriensis*)

Type: Holotype, MHNG 2206.30, an 869 mm female (native, 14 April 1983).
Type locality: "zwischen Arakapás und Kalokhorio, Zypern" [= between Arakapás and Kalokhorio, Cyprus].
Distribution: Southern Cyprus, NSL–1900 m.

Sources: Utiger & Schätti, 2004, Venchi et al., 2006 and Baier et al., 2009.

Remarks: Illustrations of head of holotype in Baier et al. (2009: figs. 226–227).

3. *Dolichophis gyarosensis* (Mertens, 1968a). Senck. Biol. 49(3–4): 181–184, figs. 1–2, 4. (*Coluber gemonensis gyarosensis*)

Type: Holotype, ZFMK 4332, a 675 mm (svl) male (K. Buchholz, 28–31 Aug. 1963).

Type locality: "Insel Gyaros (= Gioura), Cycladen" [= Gioura Is., Northern Sporades, Greece, 39°23'N, 24°10'E].

Distribution: Southern Greece (N Sporades). Known only from type locality.

Sources: W. Böhme, 1993b and Utiger & Schätti, 2004.

4. *Dolichophis jugularis* (Linnaeus, 1758). Syst. Nat., ed. 10, 1: 225. (*Coluber jugularis*)

Synonyms: *Coluber colloruber* Lacépède, 1789 (*nomen rejiciendum*), *Coluber jugularis* G. Shaw, 1802, *Coluber hannasch* Seetzen, 1855, *Zamenis viridiflavus asiana* Boettger, 1880b, *Zamenis gemonensis tauricus* Venzmer, 1919, and *Dolichophis jugularis zinneri* Cattaneo, 2012.

Type: Holotype, NHR Lin-28 (formerly MAFR), a 1415 mm specimen (F. Hasselquist, 1749–1752).

Type locality: "Aegypto" [= Egypt] (in error *fide* Flower, 1933: 808). Corrected to Palestine or Syria *fide* Flower (1933: 808), Syria *fide* Bodenheimer (1944: 59), and SW Turkey in the surroundings of Selcuk and Ephesus *fide* Zinner (1972: 16).

Distribution: Southwestern Asia. Eastern Greece (Aegean Is., Kálimnos, Kos, Léros, Rhodes, Sími, Tílos), S Turkey (Adana, Antalya, Diyarbakir, Gaziantep, Icel, Isparta, Izmir, Konya, Mugla, Mus, Urfa, Tunceli), Cyprus, Syria (Damascus), W Lebanon (Beirut, Beqaa, El Jnoub, Liban-Nord, Mont-Liban, Nabatiye), Jordan (Ajloun, Amman, Balqa, Irbid, Jarash, Karak, Maan, Madaba, Mafraq, Tafilah), Israel (Central, Gaza Strip, Golan Heights, Haifa, Jerusalem, Northern, N Southern, Tel Aviv, West Bank), ext. NE Egypt (NE North Sinai), N Iraq (As Kuwait, Sulaymaniyah, Al Anbar, Al Basrah), Kuwait and W Iran (Khuzestan), NSL–2000 m.

Sources: Anderson, 1898, Flower, 1933, Basoglu & Baran, 1980, Hraoui-Bloquet, 1981, Schätti, 1987, 1993a, Al-Mohanna et al., 1997, Bouskila & Amitai, 2001, Disi et al., 2001, Hraoui-Bloquet et al., 2002, Valakos et al., 2008, Baier et al., 2009, Amr & Disi, 2011 and Bar & Haimovitch, 2011.

5. *Dolichophis schmidti* (Nikolsky, 1909b). Mitt. Kaula Mus. Tiflis 4: 303–304. (*Coluber schmidti*)

Synonym: *Coluber erythrogaster* Fischer von Waldheim, 1832 (*nomen praeoccupatum*).

Type: Lectotype, MNKNU 14935 (formerly ZCIKU 268), a 390 mm (R.H. Schmidt & A.B. Schelkovnikov, 29 March 1907), designated by Vedmederya et al. (2009: 208).

Type locality: "in salsis Adzi, Mugan centr" [= Mugan' Steppe, Azerbaijan] via lectotype selection.

Distribution: Southwestern Asia. Eastern Turkey (Adana, Amasya, Ankara, Artvin, Bingöl, Gaziantep, Kars, Konya, Urfa, Van), Syria, N Jordan (Amman, Jarash, Mafraq), N Iran (Kordestan), ext. SW Armenia, Azerbaijan, E Georgia, ext. SW Russia and S Turkmenistan, 75–785 m.

Sources: Muskhelishvilli, 1970a, Bosoglu & Baran, 1980, Disi, 1985, Leviton et al., 1992, W. Böhme, 1993, Szczerbak & Böhme, 1993b, Schätti, 1993a, Szczerbak, 1994, Baker et al., 2002 and Amr & Disi, 2011.

Remarks: Photograph of head of holotype in Vedmederya et al. (2009: fig. 8). Original description erroneously cited as vol. 3 (1908) *fide* Schätti (1988: 22). Type locality restriction of Tbilisi, Georgia *fide* Szczerbak & Böhme (1993b: 169) invalid.

DREPANOIDES Dunn, 1928b (*nomen substitutum*) (Xenodontidae)

Synonyms: *Cloelia* Jan, 1863b (*nomen praeoccupatum*), *Drepanodon* Peracca, 1896a (*nomen praeoccupatum*), *Drepranodon* – Amaral, 1930f (*nomen incorrectum*), *Drepanpdon* – Briceño-Rossi, 1934 (*nomen incorrectum*), *Pseudoclelia* Rendahl & Vestergren, 1941, and *Dreprenadon* – Cunha & Nascimento, 1978 (*nomen incorrectum*).

Type species: *Cloelia anomala* Jan, 1863b.

Distribution: Northern South America.

Sources: Peracca, 1896a, J.A. Peters & Orejas-Miranda, 1970, Dixon & Soini, 1986, Vanzolini, 1986a, Pérez-Santos & Moreno, 1988, 1991, Duellman & Salas, 1991, Starace, 1998, Vidal et al., 2000, Duellman, 2005, Zaher et al., 2009, Avila & Kawashita-Ribeiro. 2011 and C.J. Cole et al., 2013.

1. *Drepanoides anomalus* (Jan, 1863b). Elenco Sist. Ofidi: 92. (*Cloelia anomala*)

Synonyms: *Drepanodon astigmaticus* Peracca, 1896a, *Arrhyton quenselii* Andersson, 1901, and *Pseudoclelia guttata* Rendahl & Vestergren, 1941.

Type: Lectotype, MHNN 4 (formerly MZN 100) (J.J. von Tschudi, 1838–1842), designated by Schätti (1986a: 100).

Type locality: "Amérique méridionale" [= South America] via lectotype selection. Restricted to Peru *fide* Peracca (1896a: 5) and Schätti (1986a: 100).

Distribution: Western Amazonia. Southern Colombia (Putumayo), Guyana, E Ecuador (Morona-Santiago, Napo, Pastaza), E Peru (Loreto, Madre de Dios), N Bolivia (Beni, La Paz) and W Brazil (Amazonas, Bahia, Mato Grosso, Pará, Rondônia), 50–500 m.
Remarks: Type locality erroneously listed as Brazil *fide* J.A. Peters (1960a: 518).

D

DROMICODRYAS Boulenger, 1893a (Pseudoxyrhophiidae)

Synonyms: *Lianthera* Cope, 1893a, *Dromycodrias* – Jourdran, 1904 *(nomen incorrectum), Dromicordyas* – Briceño-Rossi, 1934 *(nomen incorrectum),* and *Dromycodryas* – Brygoo, 1983 *(nomen incorrectum).*
Type species: *Herpetodryas bernierii* A.-M.-C. Duméril, Bibron & Duméril, 1854a.
Distribution: Madagascar.
Sources: Jourdran, 1904, Andersson, 1910, Guibé, 1958, Brygoo, 1983, 1987, Gravlund, 2001, A. Mori et al., 2006, Glaw & Vences, 2007, Glaw et al., 2007a, Kelly et al., 2009 and Zaher et al., 2009.

1. *Dromicodryas bernierii* (A.-M.-C. Duméril, Bibron & Duméril, 1854a). Erpét. Gén. 7(1): 211–212, pl. 66, figs. 1–4. (*Herpetodryas bernierii*)

Synonyms: *Herpetodryas bernierii trilineata* Boettger, 1881d *(partim)*, *Herpetodryas bernieri ramavali* Kaudern, 1922, and *Liopholidophis pseudolateralis* Guibé, 1954.
Types: Syntypes (3), MCZ 1966 (formerly MNHN 68), an 843 mm female (C.A.C.J. Bernier, 1844), USNM 6221 (Hérail) and MNHN (R.-P. Lesson & P. Garnot [Coquille Voy.], Aug. 1822–March 1825).
Type locality: "Madagascar" and "l'Ile de France," (latter in error *fide* Brygoo, 1983: 38). Restricted to Madagascar *fide* Bouleneger (1896a: 190).
Distribution: Madagascar (Antsiranana, Atananarivo, Fianarantsoa, Mahajanga, Toamasina, S Toliara), NSL–515 m.
Sources: Henkel & Schmidt, 2000 and A. Mori et al., 2006.

2. *Dromicodryas quadrilineatus* (A.-M.-C. Duméril, Bibron & Duméril, 1854a). Erpét. Gén. 7(1): 212–218. (*Herpetodryas quadrilineatus*)

Synonym: *Herpetodryas bernierii trilineata* Boettger, 1881 *(partim)*.
Type: Holotype, MNHN 7245 (Clouet or C.A.C.J. Bernier, 1847).
Type locality: "Madagascar."
Distribution: Northern Madagascar (Antsiranana, Mahajanga, Toamasina, Nosy Be Is.), NSL–1000 m.
Source: Henkel & Schmidt, 2000.
Remarks: Holotype collected by Mr. Clouet and paratypes by Bernier *fide* Duméril, Bibron & Duméril (1854a: 212) but Bernier is listed as holotype collector *fide* MNHN catalogue.

DROMOPHIS W.C.H. Peters, 1869 (Psammophiidae)

Synonyms: *Philodendros* Fitzinger, 1843 *(nomen rejiciendum)* and *Philodendrus* Agassiz, 1847 *(nomen rejiciendum).*
Type species: *Dendrophis praeornata* Schlegel, 1837.
Distribution: West, cen. and E Africa.
Sources: F. Werner, 1924c, Loveridge, 1940, Brandstätter, 1995, 1996, Hughes, 1999, Gravlund, 2001, Nagy et al., 2003, Lawson et al., 2005, Chippaux, 2006, Kelly et al., 2008, 2009, Vidal et al., 2008, Zaher et al., 2009 and Hoser, 2012u.
Remarks: A valid genus *fide* Hughes (2013: 115) but a synonym of *Psammophis fide* Kelly et al. (2008: 1045).

1. *Dromophis lineatus* (A.-M.-C. Duméril, Bibron & Duméril, 1854b). Erpét. Gén. 7(2): 1124–1126. (*Dryophylax lineatus*)

Synonyms: *Psammophis sibilans tumbensis* Schenkel, 1901 and *Psammophis brevirostris temporalis* F. Werner, 1902.
Types: Syntypes (2), MNHN 7643–44, longest syntype 910 mm (B. d'Arnaud [Second or Third White Nile Exped.], 1840–1850).
Type locality: "Nil blanc" [= White Nile River, bet. Al Khartum (15°35'N, elevatuon 390 m), An Nil Al Abyad Prov., SE Sudan, and Juba (4°42'N, elevation 500 m), Upper Nile, Jonglei or Central Equatoria Prov., South Sudan].
Distribution: West, cen. and E Africa. Senegal (Dakar, Fatick), Gambia, Guinea–Bissau (Bafatá, Bissau, Tombali), Guinea (Boké, Conakry, Dalaba, Kerouane, Kindia, Télimélé, Tumbo Is.), Sierre Leone (Northern, Southern, Western), Ivory Coast (Daloa, Toumodi), SW Mali (Koulikoro), Burkina Faso (Centre, Centre-Est, Centre-Nord, Centre-Ouest, Est, Haute-Bassins, Sahel, Sud-Ouest, Volta-Noire), N Ghana (Upper West), Togo (Kara, Plateaux, Savanes), Benin (Alibori, Atlantique, Collines), NE Nigeria, SW Chad (Mayo-Kebbi), Cameroon (Adamaoua, Centre, Est, Extreme-Nord, Nord, Sud), Central African Republic (Bamingui-Bangoran, Lobaye, Ombella-Mpoko, Ouham, Ouham-Pende, Sangha), Congo (Brazzaville, Plateau), Democratic Republic of the Congo (Bandundu, Bas-Congo, Equateur, Kasai Occidental, Kasai Oriental, Katanga, Kinshasa, Nord-Kivu, Orientale, Sud-Kivu), Angola, SE Sudan (An Nil Al Abyad), E South Sudan (Eastern Equatoria, Jonglei, Upper Nile), W Ethiopia (Illubabor), SW Kenya (Western), Uganda (Central, Northern), W Burundi (Bujumbura), W Tanzania

(S Iringa, W Kigoma, Mbeya), Zambia (Central, Northern, Southern, Western), N Malawi (Northern), ext. W Zimbabwe (W Matebeleland North), ext. NE Nambia (E Caprivi) and ext. NE Botswana (NE North West), 250–1800 m.

Sources: Witte, 1962, Doucet, 1963, Leston & Hughes, 1968, Hughes & Barry, 1969, Pitman, 1974, Broadley & Cock, 1975, Roman, 1980, Broadley, 1983, Hughes, 1983, 2013, Auerbach, 1987, Branch, 1988, Largen & Rasmussen, 1993, J.-F.J.-F. Trape & Roux-Estève, 1995, Broadley et al., 2003, M. Griffin, 2003, Hughes, 2004, Chirio & Ineich, 2006, J.-F. Trape & Mané, 2006b, Chirio & LeBreton, 2007, Largen & Spawls, 2010, Segniagbeto et al., 2011, Broadley & Blaylock, 2013 and Chirio, 2013.

2. *Dromophis praeornatus* (Schlegel, 1837). Essai Phys. Serp. 1: 157–158, 2: 236–237. (*Dendrophis praeornata*)

Synonym: *Dromophis praeornatus gribinguiensis* Angel, 1921.

Type: Holotype, MNHN 7610, a 236–611 mm male (G.S. Perrotet, 1822–1832).

Type locality: "le pays de Walo au Senegal" [= Kingdom of Waalo along lower Senegal River bet. Atlantic Ocean (15°57N, 16°30'W, elevation NSL) and Djeol (16°03'N, 13°24'W, elevation 45 m), N Senegal].

Distribution: West Africa. Senegal (Dakar, Fatick, Kédougou, Matam, Saint-Louis, Thiès), Gambia, Guinea (Kerouane, Kouroussa, Télimélé), S Mali (Kayes, Koulikoro, Mopti, Ségou, Sikasso), Burkina Faso (Centre, Centre-Est, Centre-Nord, Centre-Ouest, Est, Haute-Bassins, Sahel, Sud-Ouest, Volta-Noire), Ivory Coast (Bouna, Daloa), Ghana (Accra, Upper East, Upper West), S Togo (Kara, Plateaux, Savanes), N Benin (Alibori, Atakora), S Niger (Dosso, Maradi, Niamey, Zinder), Nigeria (Plateau), S Chad (Lac), N Cameroon (Extreme-Nord, Nord) and Central African Republic (Ouham, Ouham-Pende, Vakaga), NSL–1210 m.

Sources: Loveridge, 1940, Doucet, 1963, Leston & Hughes, 1968, Hughes & Barry, 1969, J Roman, 1980, Hughes, 1983, 2013, Greenbaum & Carr, 2005, Chippaux, 2006, Chirio & Ineich, 2006, J.-F. Trape & Mané, 2006b, Chirio & LeBreton, 2007, Segniagbeto et al., 2011 and Chirio, 2013.

Remarks: Official Specific Name *fide* Opinion 1384 (ICZN 1986b).

†*DRYINOIDES* Auffenberg, 1958 (Carphophiidae)

Type species: †*Dryinoides oxyrhachis* Auffenberg, 1958.

Distribution: Middle Miocene of N-cen. USA.

Sources: Holman, 1979b, 2000a, Meylan, 1982 and Rage, 1984b.

Remarks: †*Dryinoides* sp. from Pleistocene (Irvingtonian) of USA (Florida) *fide* Meylan (1982: 46).

1. †*Dryinoides oxyrhachis* Auffenberg, 1958. Amer. Mus. Novit. (1874): 2–8, figs. 1 (top row), 2, 3a, 3h, 3m.

Type: Holotype, AMNH 7524, a fragmentary skull and 21 thoracic vertebrae (C.C. Mook & C.S. Williams, 1925).

Type locality: "Madison Valley, Gallatin County, Montana; Lower Clay, Madison Valley formation, upper Miocene (Barstovian)." [= Middle Miocene *fide* Rage (1984b: 46) and Holman, 2000a: 292].

Distribution: Middle Miocene (Barstovian: 13.6–20.6 mya) of USA (Montana). Known only from type locality.

D

DRYMARCHON Fitzinger, 1843 (Colubridae)

Synonyms: *Georgia* Baird & Girard, 1853, *Geoptyas* Steindachner, 1867b, *Morenoa* Dugès, 1905, *Crymarchon* – Briceño-Rossi, 1934 *(nomen incorrectum)*, and *Dryonarchon* – Huene, 1956 *(nomen incorrectum)*.

Type species: *Coluber corais* F. Boie, 1827.

Distribution: Southeastern USA and Latin America.

Fossil records: Pleistocene of USA (Florida), and Holocene of Mexico and USA (Florida).

Sources: Amaral, 1930i and Wüster et al., 2001a.

1. *Drymarchon caudomaculatus* Wüster, Yrausquin, & Mijares-Urrutia, 2001a. Herpetol. J. 11: 159–162, figs. 1a–d.

Type: Holotype, EBRG 3412, a 1562 mm female (W. Wüster & J.L. Yrausquin, 16 July 1997).

Type locality: "On the road to Los Tablones, 1 km from the junction with the new Coro-Churuguara road, Municipio Colima, Estado Falcón, Venezuela. The locality is situated approximately 1 km from the village Las Dos Bocas (11°18'N, 69°24'W), in the eastern foothills of the Sierra de San Luis, at an altitude of approximately 110 m."

Distribution: Northwestern Venezuela (Falcón, Zulia), 110–150 m.

Sources: Mijares-Urrutia & Arends, 2000, Infante-Rivero, 2005 and Navarrete et al., 2009.

2. *Drymarchon corais* (F. Boie, 1827). Isis von Oken 20(6): 537. (*Coluber corais*)

Synonyms: *Coluber reticularis* Daudin, 1803c (*nomen nudum*), *Coluber flavicolis* Hancock, 1830, *Geoptyas collaris* Steindachner, 1867b, *Geoptyas flaviventris* Steindachner, 1867b, *Phrynonax suborbitalis* W.C.H. Peters, 1868, *Spilotes corais xanthurus* A. Brown, 1893,

Phrynonax angulifer F. Werner, 1923a, *Drymanchor corais corais* – Briceño-Rossi, 1934 *(nomen incorrectum)*, and *Drymarcho corais corais* – Briceño-Rossi, 1934 *(nomen incorrectum)*.

Type: Holotype, not designated, lost *fide* Hoge (1964d: 55).

Type locality: "America." Restricted to South America *fide* A.E. Brown (1901: 44) and to Belem, Pará, Brazil *fide* K.P. Schmidt (1953a: 192).

Distribution: South America. Colombia (Guajira), Venezuela (Amazonas, Anzoátegui, Apure, Aragua, Bolívar, Carabobo, Distrito Federal, Falcón, Guárico, Mérida, Miranda, Monagas, Sucre, Zulia), Trinidad and Tobago (Tobago, Trinidad), Guyana (Pomeroon-Supenaam), N Suriname (Para, Paramaribo), N French Guiana (N Cayenne, N Saint-Laurent-du-Maroni), W Ecuador (Loja, Pichincha), ext. NW and E Peru (Cusco, Loreto, Madre a Dios, Piura), N Bolivia (El Beni, La Paz, Pando), Brazil (Alagoas, Amapá, Amazonas, Bahia, Ceará, Maranhão, Mato Grosso, Pará, Pernambuco, Piaui, Rondônia, São Paulo, Tocantins), Paraguay (Alto Paraguay, Boquerón, Central, Concepción, Presidente Hayes, San Pedro) and N Argentina (Chaco), NSL–2750 m.

Sources: Amaral, 1930i, Roze, 1966a, Lancini, 1986, Chippaux, 1987, Pérez-Santos & Moreno, 1988, 1991, Cei, 1994, Starace, 1998, Leynaud & Bucher, 1999, Wüster et al., 2001a, Abuys, 2003, J.C. Murphy & Schlager, 2003, Cacciali, 2008, Mendoza & Lucero, 2009 and Navarrete et al., 2009.

3. *Drymarchon couperi* (Holbrook, 1842a). No. Amer. Herp., ed. 2, 3: 75–77, pl. 16. (*Coluber couperi*)

Type: Holotype, ANSP 3937 (J.H. Couper, 1829–1842)

Type locality: "dry pine hills, south of the Altamaha" [= Altamaha River, Georgia, USA]. Restricted to Wayne Co., Georgia, USA *fide* K.P. Schmidt (1953: 193).

Distribution: Southeastern USA (ext. S Alabama, Florida: Florida Keys; SE Georgia: St. Catherine's Is.; ext. S South Carolina), NSL–150 m.

Fossil records: Lower Pleistocene (Irvingtonian I) of USA (Florida), and upper Pleistocene (Rancholabrean II) of USA (Alabama, Florida).

Sources: Speake et al., 1978, Ashton & Ashton, 1988, Ernst & Barbour, 1989, Collins, 1991, Couper et al., 1996, Crother et al., 2000, Holman, 2000a and Krysko et al., 2011.

4. *Drymarchon margaritae* Roze, 1959b. Nov. Cient. Contr. Ocas. Mus. Hist. Nat. La Salle (25): 1–3, 2 figs.

Type: Holotype, MHNLS 569, a 1300 mm male (J.A. Roze, 18 Dec. 1951).

Type locality: "Cerra de San Francisco de Macanas, Isla de Margarita, Venezuela, 350 m."

Distribution: Northeastern Venezuela (Nueva Esparta: W Margarita Is.), 350–500 m.

Sources: Roze, 1962, Wüster et al., 2001a, Rivas-Fuenmayor et al., 2005 and Ugueto & Rivas, 2010.

Remarks: A valid species *fide* Wüster et al. (2001: 163).

5. *Drymarchon melanurus* (A.-M.-C. Duméril, Bibron & Duméril, 1854a). Erpét. Gén. 7(1): 224–225. (*Spilotes melanurus*)

Synonyms: *Spilotes erebennus* Cope, 1860f, *Morenoa orizabensis* Dugès, 1905, *Drymarchon corais melanocercus* H.M. Smith, 1941m *(nomen rejiciendum)*, *Drymarchon corais rubidus* H.M. Smith, 1941m, *Drymarchon corais unicolor* H.M. Smith, 1941m, and *Drymarchon corais cleofae* Brock, 1942,

Type: Lectotype, MNHN 3185, a 2217 mm specimen, designated by McCranie (1980: 200).

Type locality: "Mexique" [= Mexico] via lectotype selection.

Distribution: Mesoamerica and NW South America, Southeastern USA (S Texas), Mexico (Aguascalientes, Campeche, Chiapas, ext. SW Chihuahua, E Coahuila, Colima, Durango, ? Guanajuato, Guerrero, Hidalgo, Jalisco, Michoacán, México, Morelos, Nayarit, Nuevo León, Oaxaca, Puebla, Querétaro, Quintana Roo, San Luis Potosí, Sinaloa, S Sonora, Tabasco, Tamaulipas, Veracruz, Yucatán, Zacatecas, María Cleofas and María Madre Is.), Belize (Belize, Cayo, Corozal, Orange Walk, Stann Creek, Toledo), Guatemala (Alta Verapaz, Huehuetenango, Petén, Suchitépequez), Honduras (Atlántida, Choluteca, Comayagua, Copán, Cortés, El Paraíso, Francisco Morazán, Gracias a Dios, Islas de la Bahía, Ocotepeque, Olancho, Santa Bárbara, Valle, Yoro, Guanaja, Roatán and Utila Is.), El Salvador (Ahuachapán, Chalatenango, Cuscatlán, La Libertad, Morazán, San Salvador), Nicaragua (Chontales, Jinotega, Managua, Matagalpa, Río San Juan, Zelaya, Ometepec Is.), Costa Rica (Alajuela, Cartago, Guanacaste, Heredia, Limón, Puntarenas, San José), Panama (Canal Zone, Darién, Barro Colorado Is.), Colombia (Amazonas, Antioquia, Atlántico, Bolívar, Caquetá, Cauca, Chocó, Cundinamarca, Distrito Especial, La Guajira, Santander, Tolima, Valle, Vichada), NW Venezuela (Carabobo, Distrito Federal, Mérida, Zulia) and Ecuador, NSL–1600 m.

Sources: Niceforo Maria, 1942, E.H. Taylor, 1951, McCranie, 1980c, Tennant, 1984, 1997, L.D. Wilson & Meyer, 1985, Pérez-Santos & Moreno, 1988, 1991, La Marca et al., 1995, Lee, 1996, J.A. Campbell, 1998, Crother et al., 2000, G. Köhler, 2000, Werler & Dixon, 2000, Wüster et al., 2001a, Savage, 2002, Solórzano, 2004, Duellman, 2005, G. Köhler et al., 2005, McCranie et al., 2005, 2006, N. Herrera et al., 2007, Navarrete et al., 2009, McCranie, 2011a, Muñiz-Martinez & Valdez-Lares, 2011, Orellana & Gutsche, 2011, Travers et al., 2011 and Ahumada-Carrillo & Vázquez-Huizar, 2012b.

Remarks: Official Specific Name no. 2094 *fide* Opinion 746 (ICZN, 1965d). A valid species *fide* Wüster et al. (2001: 163). Lectotype designation of USNM 1416 *fide* H.M. Smith (1963d: 233) invalid *fide* McCranie (1980: 200).

DRYMOBIUS Fitzinger, 1843 (Colubridae)

Synonyms: *Crossanthera* Cope, 1893a and *Drimobius* – Brazil, 1911 *(nomen incorrectum).*

Type species: *Herpetodryas margaritiferus* Schlegel, 1837.

Distribution: Extreme S USA and Latin America.

Fossil record: Holocene of Mexico.

Sources: L.D. Wilson, 1975a and Villa et al., 1988.

1. *Drymobius chloroticus* (Cope, 1886b). Proc. Amer. Philos. Soc. (1885) 23(122): 278. (*Dendrophidium chloroticum*)

Synonym: *Drymobius brunneus* Bocourt, 1890 *in* A.H.A. Duméril, Bocourt & Mocquard, 1870–1909.

Type: Holotype, USNM 6755, a 1048 mm female (H. Hague, 1867–1884).

Type locality: "Guatemala." Restricted to Cobán, Alta Verapaz, Guatemala *fide* Cope (1887b: 69).

Distribution: Southern Mexico (S Chiapas, Guerrero, S Oaxaca, S San Luis Potosí, Veracruz), S Guatemala (Alta Verapaz, Chimaltenango, San Marcos, Santa Rosa, Solola, Suchitepequez), El Salvador (Chalatenango, Santa Ana), Honduras (Atlántida, Comayagua, Copán, Cortés, Francisco Morazán, La Paz, Olancho, Yoro) and N Nicaragua (Estelí, Matagalpa), 500–2200 (2500) m.

Sources: L.D. Wilson, 1970c, 1975b, L.D. Wilson & Meyer, 1985, López-Luna et al., 2001, G. Köhler et al., 2005, McCranie, 2011a and Mendoza-Hernández et al., 2011.

Remarks: *Drymobius chloroticus* and *D. melanotropis* probably conspecific *fide* Savage (2002: 662).

2. *Drymobius margaritiferus* (Schlegel, 1837). Essai Phys. Serp. 1: 151, 2: 184–185. (*Herpetodryas margaritiferus*)

Synonyms: *Coluber chiametla* G. Shaw, 1802 *(nomen rejiciendum), Zamenis tricolor* Hallowell, 1855c, *Drymobius margaritiferus occidentalis* Bocourt, 1890 *in* A.H.A. Duméril, Bocourt & Mocquard, 1870–1909, *Drimobius margaritiferus* – Briceño-Rossi, 1934 *(nomen incorrectum), Drymobius margaritiferus fistulosus* H.M. Smith, 1942j, and *Drymobius margaritiferus maydis* Villa, 1968.

Type: Holotype, MNHN 7309, a 740–766 mm male (J. Barabino, 1831–1836).

Type locality: "Nouvelle Orléans" [= New Orleans, Louisiana, USA] (in error). Restricted to Veracruz, Veracruz, Mexico *fide* H.M. Smith (1942j: 383). Restricted to Cordobá, Veracruz, Mexico *fide* H.M. Smith & Taylor (1950a: 347). Restricted to Brownsville, Texas, USA *fide* K.P. Schmidt (1953a: 192).

Distribution: Exteme S USA (ext. S Texas), Mexico (Campeche, Chiapas, S Chihuahua, E Coahuila, Colima, Guerrero, Hidalgo, Jalisco, México, Michoacán, Morelos, Nayarit, Nuevo León, Oaxaca, ? Querétaro, Quintana Roo, Puebla, San Luis Potosí, Sinaloa, S Sonora, Tabasco, Tamaulipas, Veracruz, Yucatán), Belize (Belize, Cayo, Corozal, Orange Walk, Stann Creek, Toledo), Guatemala (Alta Verapaz, Huehuetenango, Juitapa, Quiché, Petén, Suchitepéquez), El Salvador (Ahuachapán, Cabañas, La Libertad, La Paz, Usulután), Honduras (Atlántida, Choluteca, Colón, Comayagua, Copán, Cortés, El Paraíso, Francisco Morozán, Gracias a Dios, Olancho, Santa Bárbara, Yoro), Nicaragua (Atlántico Sur, Estelí, Granada, León, Managua, Matagalpa, Rio San Juan, Rivas, Zelaya, Great Corn Is.), Costa Rica (Alajuela, Cartago, Guanacaste, Heredia, Limón, Puntarenas, San José), Panama and ext. N Colombia (Bolívar, Córdoba, La Guajira, Magdalena), NSL–1830 m.

Sources: H.M. Smith, 1942j, Villa, 1968, L.D. Wilson, 1974, L.D. Wilson & Meyer, 1985, Pérez-Santos & Moreno, 1988, Lee, 1996, J.A. Campbell, 1998b, G. Köhler, 1999b, Werler & Dixon, 2000, Savage, 2002, Lemos-Espinal et al., 2004a–b, G. Köhler et al., 2005, Carvajal-Cogollo et al., 2011b and McCranie, 2011a.

Remarks: Official Specific Name no. 2850 *fide* Opinion 1246 (ICZN, 1983).

3. *Drymobius melanotropis* (Cope, 1875a). J. Acad. Nat. Sci. Philadelphia (1876) (2) 8(2): 134–135, pl. 26, fig. 1. (*Dendrophidium melanotropis*)

Type: Holotype, USNM 32597, a 1240 mm male (W.M. Gabb, 1873–1875).

Type locality: "southern portion of the region of Costa Rica."

Distribution: Lower Central America. Eastern Honduras (Colón, Gracias a Dios, Olancho), Nicaragua (Atlántico Sur, Estelí, Matagalpa, Zelaya), Costa Rica (Alajuela, Cartago, Guanacaste, Heredia, Limón) and Panama, 60–1200 m.

Sources: Gaige et al., 1937, E.H. Taylor, 1954, L.D.Wilson, 1970c, 1975c, L.D. Wilson & Meyer, 1985, G. Köhler, 1999b, Savage, 2002, Solórzano, 2004, McCranie et al., 2006 and McCranie, 2011a.

Remarks: *Drymobius chloroticus* and *D. melanotropis* probably conspecific *fide* Savage (2002: 662).

4. *Drymobius rhombifer* (A.C.L.G. Günther, 1860e). Proc. Zool. Soc. London 28(1): 236. (*Coryphodon rhombifer*)

Synonym: *Spilotes rhombifer* W.C.H. Peters, 1879.
Type: Holotype, BMNH 1946.1.12.96, an 1100 mm female (L. Fraser, 1857–1859).
Type locality: "Esmeraldas, Ecuador" [= Esmeraldas, Esmeraldas Prov., NW Ecuador, 0°57'N, 79°40'W, elevation 80 m].
Distribution: Lower Central America and NW South America. Eastern Nicaragua (Zelaya), Costa Rica (Alajuela, Cartago, Guanacaste, Heredia, Limón, Puntarenas, San José), Panama (Darién, Panamá), Colombia (Amazonas, Boyacá, Caquetá, Chocó, La Guajira, Meta, Narito, Tolima, Valle), Venezuela (Amazonas, Barinas, Bolívar, Zulia), S Guyana (Upper Takutu-Upper Essequibo), N Brazil (Amazonas, Roraima), Ecuador (Esmeraldas, Napo, Pastaza, Pichincha), E Peru (Loreto, Madre de Dios) and Bolivia (La Paz), NSL–1200 m.
Sources; Roze, 1966a, Lancini, 1986, Pérez-Santos & Moreno, 1988, 1991, O'Shea & Stimson, 1993, Starace, 1998, Passos, 2001, Calcaño & Barrio-Amorgós, 2002, Savage, 2002, Solórzano, 2004, Duellman, 2005, Navarrete et al., 2009 and C.J. Cole et al., 2013.
Remarks: Type erroneously listed as BMNH 1946.1.12.94 *fide* J.A. Peters (1960a: 519).

DRYMOLUBER Amaral, 1930j (Colubridae)

Synonym: *Drimoluber* – Briceño-Rossi, 1934 *(nomen incorrectum).*
Type species: *Herpetodryas dichroa* W.C.H. Peters, 1863c.
Distribution: South America.
Sources: J.A. Peters & Orejas-Miranda, 1970, Amaral, 1978, Lehr et al., 2004 and Caldeira-Costa et al., MS.

1. *Drymoluber apurimacensis* Lehr, Carrillo & Hocking, 2004. Copeia 2004(1): 47–49, fig. 2.

Type: Holotype, MHNSM 20672, a 279 mm female (P. Hocking, 15 Jan. 2001).
Type locality: "Abancay (13°38'38"S, 72°52'33"W, elevation 2500 m.), Provincia de Abancay, Departamento de Apurimác, Peru."
Distribution: Southern Peru (Apurimác), 2500–3300 m.

2. *Drymoluber brazili* (J. Gomés, 1918). Mem. Inst. Butantan 1(1): 81–82, pl. 14, fig. 2. (*Drymobius brazili*)

Synonym: *Drymobius rubriceps* Amaral, 1923 & 1926a.
Type: Holotype, IB 696, a 1590 mm male (Tancredo-Franca, Sept. 1914), destroyed by fire 15 May 2010.
Type locality: "Estação de Engenheiro Lisbôa, perto de Uberaba, Estado de Minas Gerais, Brasil."
Distribution: Southeastern Brazil (Bahía, Ceará, Distrito Federal, Espírito Santo, Goiás, Mato Grosso do Sul, Minas Gerais, Paraiba, Piauí, Rondônia, São Paulo, Tocantins) and Paraguay (Canindeyú), 330–1040 m.
Sources: Amaral, 1935e, Nogueira, 2001, Suzart Argôlo, 2004 and Freitas et al., 2012a.

3. *Drymoluber dichrous* (W.C.H. Peters, 1863c). Mber. Königl. Akad. Wiss. Berlin 1863(6): 284. (*Herpetodryas dichroa*)

Synonyms: *Herpetodryas occipitalis* A.C.L.G. Günther, 1868, and *Spilotes piceus* Cope, 1868b.
Type: Lectotype, ZMB 1661, an 827 mm male (G.W. Freyreiss), designated by Caldeira-Costa et al., MS.
Type locality: "Brasilien" [= Brazil].
Distribution: Northern South America. Colombia (Antioquia, Boyacá, Guaviare, Meta), SE Venezuela (Amazonas, Bolívar), Guyana (Cuyuni-Mazaruni, Essequibo Islands-West Demerara, Upper Demerara-Berbice), Suriname (Brokopondo, Commewjine, Marowjine, Sipaliwini), French Guiana (Cayenne), Ecuador (Azuay, Chimborazo, Morona-Santiago, Napo, Orellano, Pastaza, Sucumbios), E Peru (Amazonas, Cusco, Huánaco, Juinín, Loreto, Madre de Dios, Pasco, San Martin, Ucayali), N Brazil (Acre, Alagoas, Amapá, Amazonas, Bahia, Ceará, Espírito Santo, Maranhão, Mato Grosso, Minas Gerais, Pará, Paraiba, Pernambuco, Rio de Janeiro, Rondônia, Roraima) and Bolivia (Beni, Cochabamba, Santa Cruz), NSL–3500 m.
Sources: Cunha & Nascimento, 1978, Lancini, 1986, Pérez-Santos & Moreno, 1988, 1991, Lancini & Kornacker, 1989, Starace, 1998, Freire, 2000, Passos & Brandão, 2002, Duellman, 2005, Navarrete et al., 2009 and C.J. Cole et al., 2013.

DRYOCALAMUS A.C.L.G. Günther, 1858 (Colubridae)

Synonyms: *Nympha* Fitzinger, 1826a *(nomen praeoccupatum), Odontomus* A.-M.-C. Duméril, 1853 *(nomen praeoccupatum), Hemidipsas* A.C.L.G. Günther, 1858, *Hydrophobus* A.C.L.G. Günther, 1862b, *Nymphophidium* A.C.L.G. Günther, 1864a, *Nymphophidum* Marschall, 1873 *(nomen emendatum), Ulupe* Blanford, 1878a, *Odontonus* – Senna, 1886 *(nomen incorrectum), Dryacalamus* – F. Werner, 1901b *(nomen incorrectum),* and *Hydropholus* – Deraniyagala, 1955 *(nomen incorrectum).*
Type species: *Dryocalamus tristrigatus* A.C.L.G. Günther, 1858.
Distribution: Southern Asia and East Indies.
Sources: M.A. Smith, 1943 and I. Das, 2010, 2012.

1. *Dryocalamus davisoni* (Blanford, 1878a). Proc. Asiatic Soc Bengal 1878(6): 141. (*Ulupe davisoni*)

Synonym: *Odontomus nympha siamensis* Bocourt, 1886.
Type: Holotype, ZSI 3201 (formerly IMC 3201), a 711 mm specimen (Davison), lost *fide* I. Das et al. (1998: 158).
Type locality: "Foot of Nawlabu hill, west of Tavoy" [= Nawlabu Hill, E of Dawei (14°05'N, 98°13'E), N Tanintharyi Div., S Myanmar, elevation 455 m]. Emended to the foot of Nawlabú Hill, east of Tavoy, in evergreen forest, at an elevation of about 1,500 feet above the sea, Tenasserim Burmaniae *fide* Blanford (1878b: 129–130).
Distribution: Southeast Asia. Bhutan, S Myanmar (Tanintharyi), Thailand (Chanthaburi, Chon Buri, Krabi, Nakhon Ratchasima, Nakhon Si Thammarat, Phra Nakhon, Surat Thani), Cambodia (Kampot), Laos and Vietnam (Binh Thuan, Dong Nai, Ho Chi Minh City, Khanh Hoa, Kien Giang, Quang Binh, Quang Tri, Tay Ninh, Thanh Hoa), NSL–1900 m.
Sources: E.H. Taylor, 1965, Saint Girons, 1972a, Chanard et al., 1999, B.L. Stuart & Emmett, 2006, I. Das, 1999a, B.L. Stuart & Emmett, 2006, V.S. Nguyen et al., 2009, Orlov et al., 2011 and Wangyal, 2011.
Remarks: Supplemental original description in Blanford (1878b: 129–130). Mahendra (1984: 211) listed "Southern China."

2. *Dryocalamus gracilis* (A.C.L.G. Günther, 1864a). Rept. Brit. India: 234–235. (*Odontomus gracilis*)

Synonym: *Odontomus fergusonii* Haly, 1888.
Type: Holotype, BMNH 1946.1.13.86, a 533 mm male (R.H. Beddome, 1857–1864).
Type locality: "Anamallay Mountains, Madras Presidency, British India" [= Anamalai Hills, S Western Ghats, E Kerala /W Tamil Nadu States, SW India, ca. 10°22'N, 77°08'E].
Distribution: Southern Asia. Western and Eastern Ghats of India (Andhra Pradesh, Kerala, Orissa, Tamil Nadu), N Sri Lanka (Northern, North-Central, North-Western) and SW Myanmar (Arakan: False Is.), NSL–200 m.
Sources: Wall, 1921g, Jayaram, 1974, Dowling & Jenner, 1988, O'Shea, 2003 and A. Silva & Velarathne, 2005.
Remarks: A synonym of *Dryocalamus nympha* (Daudin) *fide* Mahendra (1984: 211).

3. *Dryocalamus nympha* (Daudin, 1803c). Hist. Nat. Rept. 6: 244–245, pl. 75, fig. 1. (*Coluber nympha*)

Synonyms: *Hydrophobus semifasciatus* A.C.L.G. Günther, 1862b, *Odontomus nympha ceylonensis* F. Müller, 1887, and *Cochliophagus isolepis* Müller, 1924.
Type: Lectotype, BMNH 1946.1.13.69, a 432–440 mm male described and illustrated by P. Russell (1796: 42, pl. 36) (Duffin via P. Russell, 1781–1791), designated by Kucharzewski & Tillack (2008: 47)
Type locality: "Vellore, coast of Coromandel, India" [= Nellore, Andhra Pradesh State, SE India, 14°27'N, 79°59'E, elevation 15 m] via lectotype selection.
Distribution: Southern India (Andhra Pradesh, Karnataka, Orissa, Tamil Nadu) and Sri Lanka (Eastern, North Central, Northern, Sabaragamuwa, Southern), 15 m.
Sources: Wall, 1909c, 1921g, P. Silva, 1969, A. Silva, 1990b, 2001, 2009 and Somaweera, 2006.
Remarks: Description based on P. Russell (1796: 42–43, pls. 36–37).

4. *Dryocalamus philippinus* L.E. Griffin, 1909b. Philippine J. Sci. 4A(6): 596–597.

Type: Neotype, CAS 62174, a 470 mm specimen (L.E. Griffin), designated by A.E. Leviton herein.
Type locality: "Iwahig, Palawan Is., Palawan Prov., S Philippines" via neotype selection.
Distribution: Southern Philippines (Balábac, Palawan).
Remarks: Probably conspecific with *Dryocalamus tristrigatus* Günther *fide* Leviton (1959b: 262).
Remarks: BSM holotype, a 241 mm specimen (W. Schultze), destroyed in Jan. 1945 during World War II. Neotype designation of Leviton (1959b: 263) was unpublished and therefore invalid.

5. *Dryocalamus subannulatus* (A.-M.-C. Duméril, Bibron & Duméril, 1854a). Erpét. Gén. 7(1): 454–455. (*Odontomus subannulatus*) (*nomen corrigendum*)

Synonyms: *Odontomus sub-annulatus* A.-M.-C. Duméril, 1853 *(nomen nudum), Odontomus sub-annulatus* A.-M.-C. Duméril, Bibron & Duméril, 1854a *(nomen incorrigendum), Coronella prosopeion* Bleeker, 1857b *(nomen nudum), Odontomus subannulatus* Bleeker, 1858b *(nomen corrigendum), Hemidipsas ocellata* A.C.L.G. Günther, 1858, *Nymphophidium maculatum* A.C.L.G. Günther, 1864a, and *Dryocalamus davisonii tungsongensis* Nutaphand, 1986.
Type: Holotype, RMNH 985, a 600 mm specimen (S. Müller, 1833–1835).
Type locality: "l'ile de Sumatra" [= W Indonesia]. Restricted to Padang, W. Sumatra *fide* C. Haas (1950: 553).
Distribution: Malay Peninsula and East Indies. Peninsular Thailand (Krabi, Nakhon Si Thammarat, Phang Nga), West Malaysia (Johor, Perak, Pinang, Seribuat Arch.: Sibu, Tinggi, Tioman), Singapore, East Malaysia (Sabah), Brunei, W Indonesia (Mentawai Arch., Siberut Is., Riau Arch., Sumatra) and Philippines (Palawan), 200–350 m.
Sources: Rooij, 1917, Tweedie, 1983, Stuebing, 1991, David & Vogel, 1996, Casper & Burnham, 2001, Wood et al., 2008, Onn et al., 2010 and Grismer, 2011.

6. *Dryocalamus tristrigatus* A.C.L.G. Günther, 1858. Cat. Colub. Snakes Brit. Mus.: 121–122.

Synonym: *Dryocalamus trilineatus* A. Brown, 1902a.
Type: Holotype, BMNH 1946.1.13.78, a 360–368 mm female (Haslar, Mus.).
Type locality: Unknown.
Distribution: Borneo and Philippines. East Malaysia (Sabah, Sarawak), Brunei, N Indonesia (Natuna Arch., Riau Arch.) and S Philippines (Balabac, Palawan), 100–610 m.
Sources: Stuebing, 1991 and Dehling & I. Das, 2006.

D

DRYOPHIOPS Boulenger, 1896a (Colubridae)

Synonyms: *Dryophops* – Phisalix, 1922 *(nomen incorrectum), Driophiops* – Deuve, 1970 *(nomen incorrectum),* and *Dryphiops* – K.K.P. Lim & Chou, 1990 *(nomen incorrectum).*
Type species: *Dipsas rubescens* Gray, 1835 *in* Gray & Hardwicke, 1830–1835.
Distribution: Southeastern Asia and East Indies.
Sources: M.A. Smith, 1930a, C. Haas, 1950 and Leviton, 1964c.

1. *Dryophiops philippina* Boulenger, 1896a. Cat. Snakes Brit. Mus. 3: 195, pl. 9, fig. 2.

Type: Lectotype, BMNH 1946.1.8.7, a male (J. Whitehead Exped., March–July 1895), designated by Leviton (1964c: 142).
Type locality: "Cape Engano, northern Luzon" [= Cape Engaño, Cagayan Prov., N Luzon, Philippines, 18°35'N, 122°08'E, elevation 50 m] via lectotype selection.
Distribution: Philippines (Luzón, Mindanao, Mindoro, Romblon, Sibuyan and Negros), NSL–1000 m.
Sources: E.H. Taylor, 1922a and B.E. Smith, 1993.

2. *Dryophiops rubescens* (Gray, 1835 *in* Gray & Hardwicke, 1830–1835). Illust. Indian Zool. 2(19–20): pl. 84, fig. 2. (*Dipsas rubescens*)

Synonym: *Dendrophis sumatrana* Bleeker, 1857a (*nomen nudum*).
Type: Holotype, BMNH 1946.1.9.62, a female (T. Hardwicke,1756–1823)
Type locality: Unknown. Designated as Bengal, India (in error) *fide* A.C.L.G. Günther (1858: 146). Corrected to Malay Peninsula *fide* Boulenger (1896a: 194).
Distribution: Southeast Asia and East Indies. Peninsular Thailand (Narathiwat, Pattani, Phuket, Surat Thani, Trang), West Malaysia (Kedah, Kelantan, Pahang, Pinang, Perak, Perlis, Selangor, Phuket, Penang and Tioman Is.), Singapore, East Malaysia (Sabah, Sarawak), Brunei, Indonesia (Java, Kalimantan, Mentawai Arch., Natuna Arch., Sumatra) and S Philippines (Corón), NSL–500 m.
Sources: E.H. Taylor, 1925, 1965, Soderberg, 1967a, C.B. Frith, 1977a, Tweedie, 1983, K.F.L. Lim & Lee, 1989, David & Vogel, 1996, M.J. Cox et al., 1998, I. Das, 2007b, 2010, 2012 and Grismer, 2011.
Remarks: Label on jar of type reads "Bengal" but locality probably in Malay Peninsula *fide* E.H. Taylor (1965: 894).

DRYSDALIA Worrell, 1961c (Elapidae)

Type species: *Hoplocephalus coronoides* A.C.L.G. Günther, 1858.
Distribution: Southern Australia.
Sources: Coventry & Rawlinson, 1980, Shine, 1981a, Cogger et al., 1983a, Golay, 1985, S.K. Wilson & Knowles, 1988, Hoser, 1989, 2012e, Hutchinson, 1990, Greer, 1997, David & Ineich, 1999, Cogger, 2000, Sanders et al., 2008 and Zaher.et al., 2009.

1. *Drysdalia coronoides* (A.C.L.G. Günther, 1858). Cat. Colub. Snakes Brit. Mus.: 215–216. (*Hoplocephalus coronoides*)

Synonyms: *Alecto labialis* Jan, 1873 *in* Jan & Sordelli, 1870–1881 *(nomen praeoccupatum), Denisonia nigra* De Vis, 1905, and *Elapognathus orri* Wells & Wellington, 1985.
Type: Lectotype, BMNH 1946.1.17.39 (formerly BMNH 1852.1.15.29), a 423 mm female (R.C. Gunn, July 1832–Jan. 1847), designated by Coventry & Rawlinson (1980: 69).
Type locality: "Van Diemens Land" [= Tasmania, Australia] via lectotype selection.
Distribution: Southeastern Australia (Australian Capital Territory, E New South Wales, ext. SE South Australia, Tasmania, S Victoria, Babel, Badger, Cape Barron, Clarke, Curtis, Deal, Erith, Flinders, Great Dog, Inner Sister, King, Preservation and West Sister Is.), NSL–2125 m.
Sources: F. McCoy, 1878c and Littlejohn, 1962.

2. *Drysdalia mastersii* (Krefft, 1866a). Proc. Zool. Soc. London 34(1): 370. (*Hoplocephalus mastersii*)

Type: Lectotype, BMNH 1946.1.17.42 (formerly BMNH 1866.6.13.2), a 285 mm female (G. Masters, 13 June 1866), designated by Coventry & Rawlinson (1980: 72).
Type locality: "Flinder's Distribution, South Australia" via lectotype selection.
Distribution: Southern Australia (S South Australia, ext. W Victoria, ext. SE Western Australia, Fenelon and St. Francis Is.), NSL–315 m.

3. *Drysdalia rhodogaster* (Jan, 1873 *in* Jan & Sordelli, 1870–1881). Icon. Gén. Ophid. 3(44): 6, pl. 2, figs. 2, a–b, d, f–g, n, p, r, v. (*Alecto rhodogaster*)

Synonyms: *Alecto rhodogaster* Jan, 1863b (*nomen nudum*), *Hoplocephalus collaris* Macleay, 1887b, and *Pseudelaps minutus* D. Fry, 1915.
Type: Holotype, ZMH 747 (formerly ZMH 477), a 314 mm specimen.
Type locality: "Australia."
Distribution: Extreme SE Australia (SE New South Wales), NSL–2000 m.

DUBERRIA Fitzinger, 1826a (Pseudoxyrhophiidae)

Synonyms: *Homalosoma* Wagler, 1830, *Brachyblemma* Wagler, 1833 (*nomen substitutum*), *Duberia* – Gray, 1841b (*nomen incorrectum*), *Omalosoma* – A.-M.-C. Duméril & Bibron, 1844 (*nomen incorrectum*), *Dauberria* – Steindachner, 1867a (*nomen incorrectum*), *Homolosoma* – Sclater, 1891b (*nomen incorrectum*), *Homalosona* – Cope, 1892a (*nomen incorrectum*), and *Homulosoma* – Arnold, 1982 (*nomen incorrectum*).
Type species: *Coluber lutrix* Linnaeus, 1758.
Distribution: Eastern Africa.
Sources: Loveridge, 1944b, Broadley, 1983, Vidal & Hedges, 2002, Lawson et al., 2005, Vidal et al., 2008, Kelly et al., 2009 and Zaher et al., 2009.

1. *Duberria lutrix* (Linnaeus, 1758). Syst. Nat., ed. 10, 1: 216. (*Coluber lutrix*)

Synonyms: *Coluber duberria* Merrem, 1790, *Coluber angustus* Suckow, 1797, *Coluber tetragonus* Latreille *in* Sonnini & Latreille, 1801b, *Coluber arctiventris* Daudin, 1803d, *Coluber eratron* Hermann, 1804, *Cyclophis catenatus* Theobald, 1868b, *Homalosoma abyssinicum* Boulenger, 1894a, *Homalosoma lutrix atriventris* Sternfeld, 1912, *Duberria lutrix currylindahli* Laurent, 1956a, and *Duberria lutrix basilewskyi* Skelton-Bougeois, 1961.
Type: Holotype, not designated, location unknown.
Type locality: "Indiis" [= India] (in error). [= shipment received via the Cape of Good Hope, South Africa *fide* Flower (1933: 818) and Loveridge (1933: 241).
Distribution: Eastern Africa. Ethiopia (Arussi, Bale, Hararge, Shoa, Wollo), SW Kenya (Central, Nairobi, S Rift Valley), E Democratic Republic of the Congo (NE Katanga, Nord-Kivu, E Orientale), NE Tanzania (Kilimanjaro), Mozambique (Manica, Sofala, Tete), Swaziland and E South Africa (Eastern Cape, Free State, Gauteng, KwaZulu-Natal, Limpopo, Mpumalanga, Western Cape), NSL–3250 m.
Sources: Loveridge, 1933, 1942, Laurent, 1956a, Broadley, 1958, Witte, 1962, Pitman, 1974, Branch, 1988, Largen & Rasmussen, 1993, Spawls et al., 2002, J.C. Murphy & Schlager, 2003 and Largen & Spawls, 2009.

2. *Duberria rhodesiana* Broadley, 1958. Occ. Pap. Natl. Mus. So. Rhodesia 3B(22): 215–216. (*Duberria lutrix rhodesiana*)

Type: Holotype, NMZB 916 (formerly NM/M 916), a 275 mm female (D.G. Broadley, 29 Dec. 1956).
Type locality: "Chishawasha, near Salisbury, Southern Rhodesia" [= Chishawasha, Mashonaland East Prov., NE Zimbabwe, 17°46'S, 31°13'E, elevation 1400 m].
Distribution: Northeastern Zimbabwe (Manicaland, Mashonaland Central, Mashonaland East, Mashonaland West), 1100–1550 m.
Sources: Broadley, 1959, Broadley & Cock, 1975 and Broadley & Blaylock, 2013.
Remarks: A valid species *fide* Broadley & Blaylock (2013: 227).

3. *Duberria shirana* (Boulenger, 1894a). Cat. Snakes Brit. Mus. 2: 276, pl. 13, fig. 1. (*Homalosoma shiranum*)

Type: Holotype, BMNH, a 310 mm female (H.H. Johnston, 1890–1894).
Type locality: "Shiré Highlands, Nyasaland" [= Shire Highlands, Southern Region, Malawi, bet. 15°15'–16°00'S, 35°00'–35°15'E].
Distribution: East Africa. Southeastern Democratic Republic of the Congo (SW Katanga, Sud-Kivu), Burundi, SW Uganda (Central, Western), S Tanzania (Iringa, Ruvuma), NE Zambia (Northern) and N Malawi (Northern), 1800–2200 m.
Sources: Loveridge, 1933, 1942, Bogert, 1940, Laurent, 1956a, Vesey-FitzGerald, 1958, Witte, 1962, Broadley & Howell, 1991 and Broadley et al., 2003.
Remarks: A valid species *fide* McDowell (1987: 37).

4. *Duberria variegata* (W.C.H. Peters, 1854). Mber. Königl. Akad. Wiss. Berlin 1854(11): 622. (*Homalosoma variegatum*)

Type: Holotype, ZMB 4802, a 325 mm female (W.C.H. Peters, June 1843–Aug. 1847).
Type locality: "Inhambane" [Inhambane Prov., Mozambique, 23°52'S, 35°23'E, elevation 10 m].
Distribution: Southeastern Africa. Southern Mozambique (Gaza, Inhambane, Maputo, Inhaca Is.) and NE South Africa (KwaZulu-Natal), NSL–50 m.
Sources: Haacke & Burton, 1978, Branch, 1988 and Broadley, 1990a.

†*DUNNOPHIS* M.K. Hecht, 1959 (Tropidophiidae)

Type species: †*Dunnophis microechinis* M.K. Hecht, 1959.

D

Distribution: Middle Eocene-upper Eocene of USA, lower Eocene-lower Oligocene of Belgium, France, Germany and Portugal, and middle Paleocene of USA.

Sources: Rage, 1984b, 1991, McDowell, 1987, Duffaud & Rage, 1997, Holman, 2000a and Szyndlar et al., 2008.

D

1. †*Dunnophis cadurcensis* Rage, 1974. Palaeovertebrata 6(3–4): 297–299, fig. 8.

Type: Holotype, USTL MAL 600, one middle trunk vertebra.

Type locality: "Malpérié, Phosphorites du Quercy, Eocène supérieur" [= Malpérié, Tarn-et-Garonne Prov., France; Phosphorites du Quercy, Headonian, upper Eocene *fide* Rage, 1984b: 29].

Distribution: Upper Eocene (Priabonian, MP 17–20, 33.9–37.2 mya) of France.

2. †*Dunnophis matronensis* Rage, 1973a. C.R. Somm. Soc. Géol. France 3: 76–77, figs. a–e.

Type: Holotype, MNHN GR 7892, one middle trunk vertebra.

Type locality: "Grauves (Marne, France); Cuisien, Éocène moyen" [= lower Rhenanian, lower Eocene *fide* Rage, 1984b: 29].

Distribution: Lower Eocene (Neustrian, MP 8/9: 48.6–55.8 mya, and Rhenanian, MP 11: 48.6 mya) and middle Eocene of France.

Source: Rage & Augé, 2010.

3. †*Dunnophis microechinis* M.K. Hecht, 1959. Bull. Amer. Mus. Nat. Hist. 117(3): 144–145, pl. 56, figs. 1–5.

Type: Holotype, AMNH 3830, one middle trunk vertebra.

Type locality: "Locality 5 (T.27-29N., R. 103–106 W.), vicinity of Tabernacle Butte and Elk Mountain, 25 miles north of the village of Farson, southeast corner of Sublette County, Wyoming, USA."

Distribution: Lower Eocene (Bridgerian: 46.2–50.3 mya) of USA (Wyoming). Known only from type locality.

Source: Holman, 2000a.

E

ECHINANTHERA Cope, 1894c (Xenodontidae)

Synonym: *Caapora* Bailey *in* Di-Bernardo, 1992.
Type species: *Aporophis cyanopleurus* Cope, 1885a.
Distribution: South America.
Sources: Di-Bernardo, 1992, C.W. Myers & Cadle, 1994, Schargel et al., 2005, A.P. Santos et al., 2008, Zaher et al., 2009 and Vidal et al., 2010.

1. *Echinanthera amoena* (Jan, 1863a). Arch. Zool. Anat. Fis. 2(2): 268–269. (*Enicognathus amoenus*)

Type: Holotype, formerly MSNM, a 590 mm specimen, destroyed in 1943 during World War II.
Type locality: Unknown.
Distribution: Southeastern Brazil (S Minas Gerais, Paraná, Rio de Janeiro, São Paulo), 500–1450 m.
Source: Marques et al., 2001.
Remarks: *Liophis incertae sedis fide* Dixon (1980: 5).

2. *Echinanthera cephalomaculata* Di-Bernardo, 1994. Biociências 2(2): 79–81, fig. 1.

Type: Holotype, MCP 5118 (formerly IVB 1478), a 561 mm female (F. Pedro Celestino, Dec. 1992).
Type locality: "Pedra Talhada, município de Quebrângulo, estado de Alagoas, Brasil" [= Serra da Pedra Talhada, Quebrângulo municipality Alagoas State, NE Brazil, (9°19'S, 36°28'W, elevation 380 m].
Distribution: Northeastern Brazil (Alagoas), 380 m. Known only from type locality.

3. *Echinanthera cephalostriata* Di-Bernardo, 1996. The Snake 27(2): 120–123, fig. 2.

Type: Holotype, MCP 5871, a 686 mm male (W. Regis, 18 Jan. 1963).
Type locality: "Serra de Teresópolis, Rio de Janeiro, Brazil" [= 22°25'S, 42°58'W, elevation 895 m].
Distribution: Brazil (Bahia, Espírito Santo, Minas Gerais, Paraná, Rio de Janiero, Santa Catarina, São Paulo, Cardoso, Santo Amaro, São Sebastião and São Vicente Is.), 600–900 m.
Sources: Marques et al., 2001, Cicchi et al., 2007 and Argôlo & Jesus, 2008.

4. *Echinanthera cyanopleura* (Cope, 1885a). Proc. Amer. Philos. Soc. (1884) 22: 191–192. (*Aporophis cyanopleurus*)

Types: Syntypes (2), ANSP 11198–99, longest syntype 805 mm (H.H. Smith [Morgan Exped.], 1870).
Type locality: "Rio Grande do Sul, Brazil, probably São Joao do Monte Negro."
Distribution: Southeastern and S Brazil (Espírito Santo, Minas Gerais, Paraná. Rio de Janeiro, Rio Grande do Sul, Santa Catarina, São Paulo) and NE Argentina (Misiones), 10–1025 m.
Source: Giraudo et al., 1996.

5. *Echinanthera melanostigma* (Wagler *in* Spix, 1824). Serp. Brasil. Sp. Nov.: 17–18, pl. 4, fig. 2. (*Natrix melanostigma*)

Type: Lectotype, ZSM 199/0, a 716+ mm male (J.B. von Spix & K.F.P. von Martius, Sept. 1818–April 1819), designated by Hoogmoed & Gruber (1983: 327)
Type locality: "provinciae Bahiae" [= Bahia State, NE Brazil] via lectotype selection.
Distribution: Southeastern Brazil (SE Bahia, Minas Gerais, Rio de Janeiro, São Paulo, Santo Amaro and São Vicente Is.), 1100 m.
Sources: Lema, 1987, Lima-Silveira et al., 2004a and Cicchi et al., 2007.
Remarks: Dixon (1980: 11) listed as *incertae sedis* under *Liophis*.

6. *Echinanthera undulata* (Wied-Neuwied, 1824b). Isis von Oken 14(6): 667. (*Coluber undulatus*)

Synonym: *Rhadinaea binotata* F. Werner, 1909b.
Type: Holotype, not designated, a 677 mm specimen (Wied-Neuwied, 25 Sept.–23 Nov. 1815), location unknown.
Type locality: "Brasilien" [= Brazil]. Restricted to Parahyba [= Paráiba State], Brazil *fide* Wied-Neuwied (1825c: 332).
Distribution: Southeastern Colombia (Amazonas) and E Brazil (S Minas Gerais, Paraíba, Paraná, Rio de Janeiro, Santa Catarina, São Paulo, Comprida and São Vicente Is.), NSL–1025 m.
Sources: Pérez-Santos, 1986, Marques et al., 2001 and Cicchi et al., 2007.
Remarks: *Incertae sedis fide* C.W. Myers (1974: 22) and Dixon (1980: 17). Possibly occurs in E Ecuador and Guianas *fide* Hoogmoed (1979: 276).

ECHIOPSIS Fitzinger, 1843 (Elapidae)

Synonyms: *Echinopsis* – Tschudi, 1846 *(nomen incorrectum), Alecto* Jan, 1859a *(nomen praeoccupatum), Brachyaspis* Boulenger, 1896a *(nomen praeoccupatum), Brachyaspi* – Brazil, 1911 *(nomen incorrectum),* and *Echropsis* – Wallach, 1985 *(nomen incorrectum).*

Type species: *Naja curta* Schlegel, 1837.

Distribution: Southern Australia.

Sources: Storr, 1980, Shine, 1982, Cogger et al., 1983a, Mengden, 1985a, Storr et al., 1986, Hutchinson, 1990, Bush et al., 1995, Greer, 1997, David & Ineich, 1999, Scanlon & Lee, 2004, Sanders et al., 2008, Zaher et al., 2009 and Hoser, 2012e.

Remarks: A synonym of *Notechis fide* Storr, 1982 and Wallach, 1985.

1. *Echiopsis curta* (Schlegel, 1837). Essai Phys. Serp. 1: 185, 2: 486–487, and (*Naja curta*)

Synonyms: *Hoplocephalus temporalis* A.C.L.G. Günther, 1862b, *Echiopsis carata* – Mushinsky, 1987 *(nomen incorrectum),* and *Echiopsis curta martinekae* Hoser, 2012ac *(nomen illegitimum).*

Type: Holotype, MNHN 953, a 475+ mm male (J.R.C. Quoy & J.P. Gaimard [Astrolabe Voy.], July 1827).

Type locality: "dans les environs du Port du Roi Georges, à la Nouvelle Hollande" [= vicinity of King George's Sound, SW Western Australia, ext. SW Australia].

Distribution: Southern Australia (ext. SW New South Wales, S South Australia, ext. NW Victoria, SW Western Australia), NSL–640 m.

Remarks Original description reprinted in Schlegel (1844 *in* 1837–1844: 140, pl. 48, figs. 18–20).

ECHIS Merrem, 1820 (Viperidae)

Synonyms: *Scythale* I. Geoffroy-Saint-Hilaire, 1827 *(nomen emendatum), Toxicoa* Gray, 1849a, *Enchis* – Haltom, 1931 *(nomen incorrectum), Ecchis* – Rosenfeld, Kelen & Nudel, 1964 *(nomen incorrectum),* and *Turanechis* Cherlin, 1990.

Type species: *Pseudoboa carinata* J.G. Schneider, 1801.

Distribution: Northern Africa, Middle East and S Asia.

Sources: Hughes, 1976a, Arnold, 1980b, Cherlin, 1983, 1990, Joger, 1984, J.S. Ashe & Marx, 1988, Gasperetti, 1988, Cherlin & Borkin, 1990, Auffenberg & Rehman, 1991, Ineich & Tellier, 1992, Golay et al., 1993, Schätti & Gasperetti, 1994, Spawls & Branch, 1995, Herrmann & Joger, 1995, 1997, David & Ineich, 1999, Herrmann et al., 1999, Joger & Courage, 1999, McDiarmid et al., 1999, Lenk et al., 2001b, Mallow et al., 2003, Dobiey & Vogel, 2007, Arnold et al., 2009, Pook et al., 2009, Stümpel & Joger, 2009, Phelps, 2010 and Hoser, 2012d.

Remarks: Genus paraphyletic *fide* Herrmann & Joger (1997: 55).

1. *Echis borkini* Cherlin, 1990. Proc. Zool. Inst., USSR Acad. Sci. 207: 211–212. (*Echis varia borkini*)

Synonym: *Echis varia borkini* Cherlin, 1990b *(nomen nudum).*

Type: Holotype, BMNH 1899.12.5.181, a 555 mm specimen (W.R. Ogilive-Grant & Zorber, 1899).

Type locality: "Arábie, Lahij" [= Lahij, Lahij Governate, Yemen, 13°04'N, 44°53'E, elevation 140 m].

Distribution: Southwestern Saudi Arabia and W Yemen (Aden, Lahij, San'a), 140 m.

Source: Pook et al., 2009.

Remarks: A valid species *fide* Pook et al. (2009: 793).

2. *Echis carinatus* (J.G. Schneider, 1801). Hist. Amph. 2: 285–286. (*Pseudoboa carinata*)

Synonyms: *Boa horatta* G. Shaw, 1802, *Scytale bizonatus* Daudin, 1803b, *Boa horrata* – Gray, 1825 *(nomen incorrectum), Scytale zic zac* Gray, 1825 *(nomen incorrigendum), Echis ziczac* Gary, 1849 *(nomen corrigendum), Vipera noratta* – Jerdon, 1854 *(nomen incorrectum), Coluber gariba* Seetzen, 1855, *Vipera superciliosa* Jan, 1859b, *Echis carinatus nigrocincta* Ingoldby & Procter, 1923 *(nomen nudum), Echis carinatus sinhaleyus* Deraniyagala, 1951, *Echis carinatus sochureki* Stemmler, 1968a, *Echis carinatus astolae* Mertens, 1970, *Echis carinatus sinahleyus* – Hughes, 1976a *(nomen incorrectum), Echis carinas* – Latifi, 1978 *(nomen incorrectum), Echis multisquamatus* Cherlin, 1981, *Echis soshureki soshureki* – Cherlin, 1990 *(nomen incorrectum),* and *Echis carinatus sinhalensis* Auffenberg & Rehman, 1991 *(nomen emendatum).*

Type: Lectotype, a 381 mm specimen described and illustrated by P. Russell (1796: 2–3, pl. 2) (Bonniveaux, 1778), designated herein.

Type locality: "Arni, India" [= Yavatmal, Maharashtra State, W India, 20°23'N, 78°07'E, elevation 450 m] via lectotype selection.

Distribution: Arabia, SW and S Asia. United Arab Emirates (Dubai, Fujairah, Ras al Khaimah, Sharjah), N Oman (Al Batinah, Al Wusta, Ash Sharqiyah, Muscat, Masirah Is.), SE Iraq, Iran (Bushehr, Fars, Hormozghan, Kerman, Khuzestan, Sistan va Baluchestan), Uzbekistan, Turkmenistan, Tadzhikistan, Afghanistan (Helmand), Pakistan (Balochistan, F.A.T.A., Jammu & Kashmir, Khyber Pakhtunkhwa, Punjab, Sindh, Astola Is.), India (Andhra Pradesh, Bihar, Delhi, Goa, Gujarat, Jammu, N Karnataka, Kashmir, Madhya Pradesh, Maharashtra, Orissa, Punjab, Rajasthan, Tamil Nadu), Bangladesh and Sri Lanka (Eastern, Northern, North-Western, Southern, Iranetivu and other Is.) NSL–2000 m.

Sources: Wall, 1908e, 1921g, M.A. Smith, 1943, H.W. Parker, 1949, Deraniyagala, 1951, 1955, Deoras, 1965,

Stemmler, 1965, Singh, 1972, Drewes & Sacherer, 1974, Whitaker, 1978a, P. Silva, 1980a, J. Daniel, 1983, Latifi, 1991, Gasperetti, 1988, A. Silva, 1990b, 2009, Auffenberg & Rehman, 1991, M.S. Khan, 2002, 2006, J.C. Murphy & Schlager, 2003, Whitaker & Captain, 2004, N. Khaire, 2006, Somaweera, 2006, Pook et al., 2009 and Masroor, 2012.

Remarks: Original description based on P. Russell (1796: 2–3, pl. 2). Sokotra record questionable *fide* Golay et al. (1993: 270).

3. *Echis coloratus* A.C.L.G. Günther, 1878. Proc. Zool. Soc. London 46(1): 978. (*Echis colorata*) (*nomen protectum*)

Synonyms: *Echis froenata* A.-M.-C. Duméril, Bibron & Duméril, 1854b (*nomen suppressum*), *Echis coloratus terraesanctae* Babocsay, 2003.

Type: Holotype, BMNH 1946.1.20.84 (formerly BMNH 1878.9.28.6), a 686 mm female (R.F. Burton, 1877).

Type locality: "Jerbel Shárr, at an altitude of 4500 feet above the level of the sea, Midian" [= Jabal Shar Mountain, W Tabuk Govern., ext. NW Saudi Arabia, 27°39'N, 35°45'E, elevation 1370 m].

Distribution: Northeastern Africa, Middle East and Arabia. Eastern Egypt (Al Ahmar, El-Bahr El-Ahmar, Cairo, North Sinai, South Sinai, Suez), NE Sudan, S Israel (SE Northern, Southern, West Bank), W Jordan (Ajloun, Aqaba, Balqa, Irbid, Karak, Maan, Madaba, Tafilah), W Saudi Arabia (Asir, Bahah, N Eastern, Jazan, Madinah, Makkah, Qasim, Riyadh, Tabuk), E United Arab Emirates (Fujairah), Oman (Al Batinah, Ash Sharqiyah, Dhofar, Muscat, Musandam) and Yemen (Abyan, Aden, Hadhramaut, Ma'rib, Sa'dah, Sana'a, Shabwah), NSL–2600 m.

Sources: J. Anderson, 1900, H.W. Parker, 1949, Mendelsson, 1965, Y. Werner, 1973, Arnold & Gallagher, 1977, Joger, 1987, Gasperetti, 1988, Schätti, 1989, Leviton et al., 1992, Wilms & Hulbert, 2000, Bouskila & Amitai, 2001, Disi et al., 2001, Baker et al., 2004, Amr & Disi, 2011 and Bar & Haimovitch, 2011.

Remarks: Official Specific Name no. 2736 *fide* Opinion 1176 (ICZN, 1981b).

4. *Echis hughesi* Cherlin, 1990. Proc. Zool. Inst., USSR Acad. Sci. 207: 212–213, pl. 4, fig. 3.

Synonym: *Echis hughesi* Cherlin, 1990 (*nomen nudum*).

Type: Holotype, BMNH 1931.7.20.405, a 322 mm female (R. Taylor, 1931).

Type locality: "Somalia, 10°02'N, 49°E" [= Within the Sol Haud, about 50 km from the Somalian coast and probably above 3000 ft (864 m) *fide* Hughes, 1976a: 365].

Distribution: Northern Somalia (NE Sanaag), 865 m.

Remarks: Possibly conspecific with *E. pyramidum*.

5. *Echis jogeri* Cherlin, 1990. Proc. Zool. Inst., USSR Acad. Sci. 207: 209, pl. 5, fig. 2.

Synonym: *Echis jogeri* Cherlin, 1990 (*nomen nudum*).

Type: Holotype, MHNH 1993.144, a 283 mm (svl) female.

Type locality: "Mali, 3 km from Tombokto" [= Timbuktu, Tombouctou Regiom, cen. Mali, 16°47'N, 3°01'W, elevation 270 m] (possibly in error *fide* Pook et al., 2009: 801).

Distribution: West Africa. Senegal (Tambacounda), N Guinea and cen. Mali (Timbuktu), 80–270 m.

Remarks: A valid species *fide* Pook et al. (2009: 793). Holotype erroneously listed as MNHN A 144 *fide* Cherlin (1990: 209).

E

6. *Echis khosatzkii* Cherlin, 1990. Proc. Zool. Inst., USSR Acad. Sci. 207: 213–214, pl. 4, fig. 2.

Synonyms: *Coluber hoelleik* Forskal *in* Niebuhr, 1775 (*nomen oblitum*), and *Echis khosatzkii* Cherlin, 1990 (*nomen nudum*).

Type: Holotype, BMNH 1897.3.11.117, a 464 mm male (J.T. Bent, 1893 via J. Anderson).

Type locality: "Arabia, Hadhramaut" [= eastern Yemen].

Distribution: Eastern Yemen (E Aden) and probably S Oman *fide* Arnold et al. (2009: 273).

Sources: Lenk et al., 2001b and Schätti, 2001.

Remarks: A valid species *fide* Pook et al. (2009: 793). In accordance with Art. 23.9.2 of the Code (ICZN, 1999), *Echis khosatzkii* Cherlin is designated a *nomen protectum* and *Coluber hoelleik* Forskal a *nomen oblitum*.

7. *Echis leucogaster* Roman, 1972. Notes Doc. Volta. 5(4): 7–12, figs. 3–8. (*Echis carinatus leucogaster*)

Type: Holotype, CVRS 191, a 700 mm female, lost *fide* Golay et al. (1993: 271).

Type locality: "Boubon, à 20 km au Nord de Niamey, Niger" [= Niamey Dept., ext. SW Niger, 13°36' N, 1°56'E, elevation 200 m].

Distribution: North Africa. Southern Morocco (Tan Tan), S Algeria (Tamanrasset), N Western Sahara (Saguia El Hamra), Mauritania (Adrar, Guidimaka, Hodh Ech Chargui, Hodh El Gharbi, Tagant), Senegal (Dakar, Fatick, Louga, Matam, Saint-Louis, Tambacounda), Gambia, NW Guinea (Koundara), N Mali (Gao, Kayes, Koulikoro, Mopti, Ségou, S Tombouctou), NE Burkina-Faso (Sahel), Niger (Agadez, Diffa, Dosso, Maradi, Niamey, Tahoua, Zinder), N Nigeria and Chad (Kanem), 120–1325 m.

Sources: Roman, 1975, 1976, 1980, Bons & Geniez, 1996, Geniez et al., 2004, Chippaux, 2006 and J.-F. Trape & Mané, 2006b.

Remarks: Cherlin & Borkin (1990: 188) proposed the name *Echis arenicola* for two West African populations whereas the type locality of *E. arenicola* is Egypt. As *E. arenicola* appears to be conspecific with *E. pyramidum*, it is not recognized herein. A subspecies of E. *arenicolus* (= *E. pyramidum*) *fide* Schleich et al. (1996:

541). Possibly a valid species *fide* Pook et al. (2009: 12). Possibly occurs in Tunisia and SW Libya *fide* Cherlin (1990: 217).

8. *Echis megalocephalus* Cherlin, 1990. Proc. Zool. Inst., USSR Acad. Sci. 207: 212, pl. 4, fig. 1.

Synonym: *Echis megalocephalus* Cherlin, 1990 *(nomen nudum).*

Type: Holotype, ZISP 19430.1, a 601 mm male (Russian Navy surgeon).

Type locality: "island in southern Red Sea" [= Nokra Is., Dahlak Archipelago, Red Sea, Ethiopia (= Nokra Is., Northern Red Sea Prov., E Eritrea, 15°42'N, 39°56'E) *fide* Borkin & Cherlin, 1995: 136].

Distribution: Eritrea (Northern Red Sea: Nokra Is.). Known only from type locality.

Sources: Schätti, 2001 and Largen & Spawls, 2010.

Remarks: Probably occurs on Andeber Is. *fide* Schätti (2001: 145). Possibly a synonym of *E. pyramidum.*

9. *Echis ocellatus* Stemmler, 1970. Rev. Suisse Zool. 77(2): 273–274, figs. 1–2. (*Echis carinatus ocellatus*)

Type: Holotype, NMBA 17692, a 400 mm female (M. Lamantellerie, 12 May 1962).

Type locality: "Haute Volta, Garango, 048 N, 033 W" [= Garango, Centre-Est Dept., SE Burkina Faso, 11°48'N, 0°34'W, elevation 300 m].

Distribution: West Africa. Extreme SW Mauritania (Trarza), Senegal (Dakar, Kédougou, Kolda, Matam, Tambacounda), Gambia, E Guinea (Faranah), S Mali (Kayes, Koulikoro, Ségou, Sikasso, S Tombouctou), Ivory Coast (Daloa, Duékoué, Ferkéssédougou, Toumodi), Burkina-Faso (Centre, Centre-Est, Centre-Nord, Centre-Ouest, Est, Haute-Bassins, Sahel, Sud-Ouest, Volta-Noire), Ghana (Accra, Brong-Ahafo, Eastern, Northern, Upper East, Upper West), Togo (Centrale, Kara, Maritime, Plateaux, Savanes), Benin (Alibori, Atakora, Borgou, Collines, Ouémé, Zou), ext. SW Niger (Dosso, Niamey), Nigeria (Anambra, Benue, Gongola, Kaduna, Kwara, Plateau, Sokoto), ext. SW Chad (Chari-Baguirmi, Mayo-Kebbi Est, Tandjile), N Cameroon (Extreme-Nord, Nord) and W Central African Republic (Ombella-Mpoko, Ouham-Pende), NSL–1000 m.

Sources: Doucet, 1963, Hughes, 1976a, 2013, J.-F. Trape & Mané, 2004, 2006b, Chippaux, 2006, Chirio & Ineich, 2006, Chirio & LeBreton, 2007, Ullenbruch et al., 2010 and Segniagbeto et al., 2011.

10. *Echis omanensis* Babocsay, 2004. Syst. Biodiv. 1(4): 506–510, figs. 2–3.

Type: Holotype, BMNH 1973.2113, a 600 mm male (E.N. Arnold, 28 May 1973).

Type locality: "Wadi as Siji, region of Masafi (25°18'N 56°10'E), United Arab Emirates."

Distribution: Northern Oman (Al Dakhiliyah,) and E United Arab Emirates, NSL–1000 m.

Source: Grossmann et al., 2012.

Remarks: A valid species *fide* Pook et al. (2009: 793).

11. *Echis pyramidum* (I. Geoffroy-Saint-Hilaire, 1827 *in* Savigny, 1809–1829). Desc. Rept. Égypt 1(1): 152–154, pl. 8, fig. 1. (*Scytale pyramidum*)

Synonyms: *Echis arenicola* H. Boie *in* F. Boie, 1827, *Echis pavo* A. Reuss, 1834, *Vipera echis* Schlegel, 1837, *Echis carinatus leakeyi* Stemmler & Sochurek, 1969, *Echis pyramidum lucidus* Cherlin, 1990, *Echis varia darevskii* Cherlin, 1990 *(nomen nudum),* and *Echis varia darevskii* Cherlin, 1990.

Type: Lectotype, MNHN 4031, a 457 mm female, designated by Stemmler & Sochurek (1969: 90).

Type locality: "Kaire, Egypte" [= Cairo, Cairo Govern., NE Egypt, 30°03'N, 31°14'E, elevation 30 m] via lectotype selection.

Distribution: North Africa and Arabia. N Algeria, N Tunisia (Medinine), N Libya, Egypt (Assiut, Beni Suef, Cairo, SE El-Bahr El-Ahmar, Faiyum, Matrouh), NE Central African Republic (Vakaga), SE South Sudan (Eastern Equatoria), Eritrea (Anseba, Gash-Barka, Northern Red Sea, Southern Red Sea, Andeber and Nacra Is.), Ethiopia (Gemu Gofa, Hararge, Kefa, Shoa, Sidamo, Tigre, Wollo), Djibiouti, N Somalia (Awdal, Bari, Mudug, Nugaal, Sanaag, Woqooyi Galbeed), N Kenya (N Eastern, N North-Eastern, N Rift Valley), SW Saudi Arabia (Jazan), W Oman (Dhofar) and Yemen (Abyan, Ad Dali', Aden, Al Hudaydah, Hadhramaut), NSL–1700 m.

Sources: Stemmler, 1971a, 1972c, Drewes, 1972, Lanza, 1983a, 1990b, Gasperetti, 1988, Schätti, 1989, 2001, Schleich et al., 1996, Spawls et al., 2002, Chippaux, 2006, Chirio & Ineich, 2006, Pook et al., 2009 and Largen & Spawls, 2010.

Remarks: See discussion of the original description date in McDiarmid et al. (1999: 381).

12. *Echis varius* A. Reuss, 1834. Mus. Senck. 1: 160–162, pl. 7, figs. 2a–b. (*Echis varia*)

Synonym: *Echis carinatus aliaborri* Drewes & Sacherer, 1974.

Type: Holotype, SMF 21058 (formerly SMF-R III.J.J.1.a & SMF-B 9505a), a 565 mm female (E. Rüppell, 1828).

Type locality: "Abyssinien, nördliche Africa" [= Ethiopia, N Africa]. Emended to Nubia *fide* Boettger, 1898: 135 [= SE Egypt or, more probably, the Nubian Desert of NE Sudan].

Distribution: Northeastern Africa. Northeastern Sudan (El-Bahr El-Ahmar), N Somalia (Bari, Mudug, Nugaal, Samaag, Woqooyi Galbeed) and NE Kenya (N North-Eastern), 300 m.

Remarks: Possibly a valid species *fide* Pook et al. (2009: 793). Possibly a subspecies/synonym of *E. pyramidium.*

EIRENIS Jan, 1863a (Colubridae)

Synonyms: *Pediophis* Fitzinger, 1843, *Psilosoma* Jan, 1863b *(nomen praeoccupatum), Eiremias* – Westphal-Castelnau, 1870 *(nomen incorrectum), Pseudocyclophis* Boettger, 1888, *Eireinis* – Taub, 1965 *(nomen incorrectum), Eirinis* – Taub, 1965 *(nomen incorrectum), Eiremis* – J.A. Peters & Orejas-Miranda, 1970 *(nomen incorrectum), Eirensis* – Murthy, 1972b *(nomen incorrectum), Elrensis* – Gasperetti, 1974 *(nomen incorrectum), Collaria* Dotsenko, 1989, and *Eoseirenis* Nagy, Schidtler, Joger & Wink, 2003.

Type species: *Coronella modesta* Martin, 1838.

Distribution: Northeastern Africa and SW Asia.

Sources: Boulenger, 1893a, K.P. Schmidt, 1939b, Chernov, 1948, G. Haas, 1961, R.J. Clark & Clark, 1973, Baran, 1976, Schmidtler & Schmidtler, 1978, Basoglu & Baran, 1980, Dotsenko, 1985, 1989, Latifi, 1991, Nagy et al., 2003, 2004 and Mahlow et al., 2013.

Remarks: Nagy et al., 2003 and Zaher et al., 2012, divided *Eirenis* thusly: *Eirenis (aurolineatus, modestus, persicus), Eoseirenis (decemlineatus),* and *Pediophis (africanus, barani, collaris, coronella, coronelloides, eiselti, levantinus, lineomaculatus, medus, punctatolineatus, rechingeri, rothii, thospitis).* Closely related to *Hierophis fide* Nagy et al. (2004: 231).

1. *Eirenis africanus* (Boulenger, 1914c). Ann. Mag. Nat. Hist. (8) 14(84): 483–484. (*Contia africana*)

Type: Holotype, BMNH 1946.1.5.33, a 400 mm female (W.P. Lowe, [Lynes-Lowe Exped.], 1913–1914).

Type locality: "Erkowit, Red Sea Prov. of the Soudan" [= Ar Kaweit, El-Bahr El-Ahmar Prov., NE Sudan, 18°46'N, 37°06'E, elevation 1100 m].

Distribution: Northeastern Africa. Northeastern Sudan (El-Bahr El-Ahmar), Eritrea (Northern Red Sea) and Djibouti, 1000–2300 m.

Sources: Scortecci, 1930a, Largen & Rasmussen, 1993 and Largen & Spawls, 2010.

2. *Eirenis aurolineatus* (Venzmer, 1919). Arch. Naturg. (1917) 83A(11): 103–104. (*Contia collaris aurolineata*)

Types: Syntypes (2), ZMB, a 224 mm and 206 mm specimen (G. Venzmer, Jan.–Oct. 1916).

Type locality: "Bulghar Dagh aus etwa 1000 m Höhe, im cilicischen Taurus, Süd-Kleinasiens" [= Bolkar Daglari, Taurus Mountains, E Mersin Prov., S Turkey, 37°15'N, 34°20'E, elevation 1000 m].

Distribution: Southern Turkey (Adana, Mersin), 500–1800 m.

Sources: Schmidtler, 1993, 1997.

3. *Eirenis barani* Schmidtler, 1988. Salamandra 24(4): 204–208.

Synonym: *Eirenis barani bischofforum* Schmidtler, 1997.

Type: Holotype, ZSM 293/88, a 306 mm male (J.F. Schmidtler & H. Schmidtler, 27 May 1988).

Type locality: "Akdam, 35 km W Kozan (Prov. Adana), ca. 650 m ü. M., Süden der Türkei."

Distribution: Southern Turkey (Adana), 100–1100 m.

Sources: Schmidtler, 1988, 1997.

4. *Eirenis collaris* (Ménétriés, 1832). Cat. Zool. Voy. Caucase: 67–68. (*Coluber collaris*)

Synonyms: *Coluber reticulatus* Ménétriés, 1832, and *Contia collaris macrospilota* F. Werner, 1903a.

Types: Syntypes (2), ZISP 1546–47 (formerly AIS), longest syntype 279 mm (E. Ménétriés, July 1830).

Type locality: "près du Bèchebermak, non loin de la Côte occidentale de la mer Caspienne, Caucase" [= Beshbarmak, on the W side of the Caspian Sea, in the Caucasus = Mt. Besparmak, Mus Prov., E Turkey, 38°56'N, 41°24'E, elevation 1465 m].

Distribution: Southwestern Asia. Eastern Turkey (Adryaman, Hatay, Kars, Malatya, Mardin, Mus, Sanhurfa, Urfa), ext. SW Armenia, E Georgia, Azerbaijan, SW Russia (Dagestan), ext. NE Iraq and W Iran (Central, East Azarbaijan, Khuzestan, West Azarbaijan, Zanjhan), 500–2500 m.

Sources: Elpatjewsky, 1902, Rai, 1965, R.J. Clark & Clark, 1973, Baran, 1976a, Bannikov et al., 1977, Basoglu & Baran, 1980, Darevsky & Bakradze, 1982, Latifi, 1991, Leviton et al., 1992, Schmidtler & Baran, 1993a, Baran et al., 2004 and Tuniyev et al., 2009.

5. *Eirenis coronella* (Schlegel, 1837). Essai Phys. Serp. 1: 134, 2: 48–49. (*Calamaria coronella*)

Synonyms: *Eirenis fasciatus* Jan, 1863a, *Eirenis coronella fraseri* K.P. Schmidt, 1939b, *Eirenis arabica* G. Haas, 1961, *Eirenis coronella fennelli* Arnold, 1982, and *Eirenis coronella ibrahimi* Sivan & Werner, 2003.

Type: Neotype, HUJ 8410, a 273 mm female (Y.L. Werner, 23 Mar. 1970), designated by Sivan & Werner (2003: 49).

Type locality: "Israel: Negev: near Negev crossroads (Zomet Hanagev), Israeli Grid 13500525" via neotype selection.

Distribution: Southwestern Asia and Egypt. Northeastern Egypt (North Sinai, South Sinai), Syria (Damascus), Lebanon, Israel, Jordan (Ajloun, Amman, Aqabah, Irbid, Jarash, Karak, Maan, Madaba, Mafraq, Tafilah), NW Iraq (Irbil), SW Iran (Khuzestan) and NE and SW Saudi Arabia (Asir, Eastern, Makkah, Northern, Riyadh), 60–2300 m.

Sources: C.A. Reed & Marx, 1959, Baran, 1976, Basoglu & Baran, 1980, Arnold, 1982, Latifi, 1991, Gasperetti, 1988, Leviton et al., 1992, Schätti & Gasperetti, 1994, Disi et al., 2001, Sivan & Werner, 2003, Baker et al.,

2004, Baran et al., 2004, Shwayat et al., 2009 and Amr & Disi, 2011.

Remarks: Holotype lost *fide* Sivan & Werner (2003: 48). Photograph of neotype in Sivan & Werner (2003: fig. 5). Type locality restriction of Syria *fide* K.P. Schmidt (1939: 78) invalidated by neotype selection.

6. *Eirenis coronelloides* (Jan, 1862b). Arch. Zool. Anat. Fis. 2(1): 34. (*Homalosoma coronelloides*)

E

Synonym: *Contia brevicauda* Nikolsky, 1907.

Type: Neotype, HUJ 21223 (formerly ZDEU 122/1975/2), a 224 mm male (native, 11 May 1975), designated by Sivan & Werner (2003: 53).

Type locality: "Turkey: Birecik (Urfa prov.)" via neotype selection.

Distribution: Middle East. Southeastern Turkey (Gaziantep, Sanliurfa, Urfa), Syria (Homs), NW Jordan (Amman, Karak, Mafraq, Zarqa), Israel (Southern, West Bank), ext. NE Egypt (Sinai) N Iraq (Baghdad, Irbil), and W Iran (Kermanshah), 570–750 m.

Sources: Y. Werner, 1988, Nagy et al., 2003, Sivan & Werner, 2003, Amr & Disi, 2011, Avci & Olgun, 2011, Bar & Haimovitch, 2011, and Nilson & Rastegar-Pouyani, 2011.

Remarks: Holotype in MNHN lost *fide* Sivan & Werner (2003: 53). Photograph of neotype in Sivan & Werner (2003: fig. 7).

7. *Eirenis decemlineatus* (A.-M.-C. Duméril, Bibron & Duméril, 1854a). Erpét. Gén. 7(1): 327. (*Ablabes decemlineatus*) (*nomen corrigendum*)

Synonyms: *Ablabes decemlineata* A.-M.-C. Duméril, 1853 (*nomen nudum*), *Ablabes decem-lineatus* A.-M.-C. Duméril, Bibron & Duméril, 1854a (*nomen incorrigendum*), *Eirenis collaris inornata* Jan, 1863a, *Eirenis collaris decemlineata* – Jan, 1863a (*nomen corrigendum*), *Eirenis collaris quadrilineata* Jan, 1863a, and *Contia condoni* Boulenger, 1920h.

Type: Holotype, MNHN 3389, a 760 mm specimen.

Type locality: Unknown.

Distribution: Southeastern Turkey (Adana, Gaziantep, Icel, Mus, Siirt, Tunceli), Syria (Homs), Lebanon (S Beqaa, Mont-Liban), Israel (Central, Golan Heights, Haifa, Jerusalem, Northern, West Bank), NW Jordan (Ajloun, Amman, Balqa, Irbid, Jarash, Madaba, Tafilah, Zarqa), Iraq, and Iran, 915–1350 m.

Sources: R.J. Clark & Clark, 1973, Baran, 1976, Basoglu & Baran, 1980, Hraoui-Bloquet, 1981, Ilani, 1983a, Schmidtler & Eiselt, 1991, Leviton et al., 1992, Bouskila & Amitai, 2001, Disi et al., 2001, Schmidtler et al., 2009, Shwayat et al., 2009, Amr & Disi, 2011, and Bar & Haimovitch, 2011.

Remarks: Has been considered a subspecies of *collaris*. A Cyprus record needs confirmation.

8. *Eirenis eiselti* Schmidtler & Schmidtler, 1978. Ann. Naturhist. Mus. Wien 81: 384–386, pl. 1, fig. 3.

Type: Holotype, CAS 105438, a 298 mm male (R.J. Clark & E. D. Clark, 7 May 1967).

Type locality: "25 km W Viransehir (Prov. Urfa/Türkei), ca. 700 mü NN."

Distribution: Southeastern Turkey (Adana, Adiyaman, Diyarbakir, Gaziantep, Maras, Mardin, Siirt, Urfa), 500–1600 m.

Sources: Baran, 1976 and Schmidtler, 1997.

9. *Eirenis kermanensis* Rajabizadeh, Schmidtler, Orlov & Soleimani, 2012. Russ. J. Herp. 19(4): 311–313, figs. 2a–b, 3a–b. (*Eirenis* [*Pediophis*] *kermanensis*)

Type: Holotype, ICSTZM 7H1062, a 308 mm female (G. Soleimani, May 2008).

Type locality: "Iran, Kerman province, Sarduieh region, Babe Karafs village, (29°16' N 57°15' E), 2800 m a.s.l."

Distribution: Central Iran (Kerman), 2800 m.

10. *Eirenis levantinus* Schmidtler, 1993. Spixiana 16(1): 87–88, figs. 11.

Type: Holotype, ZSM 14/1991, a 334 mm female (J.F. Schmidtler & H. Schmidtler, 23 May 1991).

Type locality: "6 km SW Karaisali (Prov. Adana; Nr. 17 auf der Karte), Süd-Anatolien" [= Karaisali, Adana Prov., S Turkey, 37°15'N, 35°04'E, elevation 240 m].

Distribution: Middle East. Cyprus, S Turkey (Adana, Hatay, Icel), NW Syria, Lebanon (Beqaa, Mont-Liban), and ext. N Israel (N Golan Heights, N Northern), 50–1100 m.

Sources: Ilani, 1983a, Schmidtler, 1997, Sindaco et al., 2000, Hraoui-Bloquet et al., 2002, Franzen & Glaw, 2007, Baier et al., 2009, Schmidtler et al., 2009, and Bar & Haimovitch, 2011.

11. *Eirenis lineomaculatus* K.P. Schmidt, 1939b. Field Mus. Nat. Hist. (Zool.) 24(7): 80–81. (*Eirenis lineomaculata*)

Type: Holotype, FMNH 21909, a 217 mm male (P.Y. Shuwayhat, 1934).

Type locality: "Jordan Valley, Palestine" [= NE Israel].

Distribution: Middle East. Southern Turkey (Antalya, Adana, Hatay, Maras), W Syria, Lebanon (S Beqaa, Liban-Nord, Mont-Liban, Nanatiye), N Israel (Central, Golan Heights, Haifa, Jerusalem, Northern, West Bank), and NW Jordan (Amman, Balqa, Irbid, Karak, Maan, Mafraq, Tafilah, Zarqa), NSL–1000 m.

Sources: Basoglu, 1970, Baran, 1976a, Schmidtler & Schmidtler, 1978, Basoglu & Baran, 1980 and Disi, 1985, Ilani, 1983a, Y.L. Werner, 1995, Bouskila & Amitai, 2001, Disi et al., 2001, Hraoui-Bloquet et al., 2002, Shwayat et al., 2009, Amr & Disi, 2011, and Bar & Haimovitch, 2011.

12. *Eirenis medus* (Chernov *in* Terent'ev & Chernov, 1940). Opred. Presmy. Zem.: 151. (*Contia medus*)

Types: Syntypes (4+), none designated, at least 2 males and 2 females, longest syntype 320 mm, location unknown.
Type locality: "Iran and southern Turkmenia" [= Iran and Turkmenistan].
Distribution: Kopet Mountains of S Turkmenistan and N Iran (Central, Ghilan, Khuzestan, Razavi Khorasan, West Azarbaijan, Turkmenistan, Zanjhan), 20–2000 m.
Sources: Terent'ev & Chernov, 1949, 1965, Bannikov et al., 1977, Latifi, 1991 and Szczerbak, 1994.

13. *Eirenis modestus* (W.C.L. Martin, 1838). Proc. Zool. Soc. London 6(1): 82. (*Coronella modesta*)

Synonyms: *Coluber nigricollis* Dwigubsky, 1832 (*nomen rejiciendum*), *Tyria argonauta* Eichwald, 1839, ? *Psammophis moniliger* Nordmann, 1840 *in* Demidoff, *Tysia argonauta* Carus & Engelmann, 1861 *(nomen incorrectum), Ablabes modestus semimaculata* Boettger, 1876, *Contia modesta werneri* Wettstein, 1937, and *Eirenis modestus cilicius* Schmidtler, 1993.
Types: Syntypes (4), 4BMNH 1850.10.21.21–22, longest syntype 292 mm (K. Abbot [Euphrates Exped.], 1835–1837).
Type locality: "Euphrates" [River] and "Trebizond," [= Trabzon, Trabzon Prov., NE Turkey].
Distribution: Southwestern Asia. Eastern Greece (Chios, Furni, Mytilene, Samos), Cyprus, Turkey (Adana, Adiyaman, Amasya, Ankara, Antalya, Artvin, Aydin, Balikesir, Burdur, Bursa, Cahakkale, Denizli, Hakkari, Hatay, Icel, Istanbul, Izmir, Kars, Kocaeli, Konya, Lesvos, Maras, Mus, Manisa, Mugla, Nigde, Siirt, Trabzon), W Syria, Lebanon (Beirut, Beqaa, Mont-Liban), N Israel, ext. SW Russia (Dagestan), Armenia, S Georgia, Azerbaijan and NW Iran (Central, Zanjhan), NSL–2440 m.
Sources: Elpatjewsky, 1902, G. Haas, 1951, Wettstein, 1953, Bannikov et al., 1977, Baran, 1976, 1986, Basoglu & Baran, 1980, Hraoui-Bloquet, 1981, Schmidtler, 1988, 1993, 1997, Schmidtler & Eiselt, 1991, Obst & Baran, 1993, Schmidtler & Baran, 1993b, Valakos et al., 2004, 2008 and Tuniyev et al., 2009.
Remarks: See Mahlow et al. (2013: 44) for discussion of type locality.

14. *Eirenis persicus* (J. Anderson, 1872). Proc. Zool. Soc.London 40(1): 392–393, fig. 8. (*Cyclophis persicus*)

Synonyms: *Pseudocyclophis walteri* Boettger, 1888d, *Contia angusticeps* Boulenger, 1894a, *Contia transcaspica* Nikolsky, 1903a, *Contia persica nigrofasciata* Nikolsky, 1907, *Contia mcmahoni* Wall, 1911d, and *Contia zebrina* Wall, 1923e.
Type: Holotype, ZSI 4828 (formerly IMC 4828), a 155 mm specimen (Persian Mus. coll.).
Type locality: "Bushire, Persia" [= Bushehr, NW Bushehr Prov., SW Iran, 28°55'N, 50°50'E, elevation 25 m].
Distribution: Southwestern Asia. Southeastern Turkey (Adryaman, Hakkari, Mardin, Sanhurfa, Siirt, Urfa), ext. SE Armenia, S Turkmenistan, N Iraq (At Tamim), Iran (Bakhtaran, Bushehr, Central, East Azarbaijan, Fars, Khuzestan) and Pakistan (Balochistan, Khyber Pakhtunkhwa, Punjab, Sindh), 25–2200 m.
Sources: Wall, 1908d, 1923d, Nikolsky, 1916, Procter, 1921, Corkill, 1932b, M.A. Smith, 1943, C.R. Reed & Marx, 1959, S.C. Anderson, 1963, Rai, 1965, Minton, 1966, G. Haas & Werner, 1969, Eiselt, 1970, Leviton & Anderson, 1970a, Baran, 1976a, 1982, Bannikov et al., 1977, Basoglu & Baran, 1980, M.S. Khan, 1980, 1982, 2002, Dotsenko, 1985, Latifi, 1991, Agasyan, 1987, Szczerbak, 1994, Baran et al., 2004 and Tuniyev et al., 2009.
Remarks: M.A. Smith (1943: 188) erroneously listed the holotype in BMNH. Probably occurs in Afghanistan *fide* Leviton and Anderson (1970a: 197).

15. *Eirenis punctatolineatus* (Boettger, 1893a). Ber. Senck. Naturf. Ges. 1892: 147–148. (*Cyclophis modestus punctatolineata*)

Synonyms: *Zamenis bornmüllerorum* F. Werner, 1903c, *Contia schelkovnikovi* Nikolsky, 1909b, *Contia condoni* Boulenger, 1920f, *Eirenis iranica* K.P. Schmidt, 1939, and *Eirenis punctatolineatus kumerloevei* Eiselt, 1970.
Type: Holotype, SMF 19288 (formerly SMF-B 8261,1a), a female (J. Valentin, 1890).
Type locality: "Russisch-Armenien" [= Karabakh Highlands, E Armenia].
Distribution: Southeastern Turkey (Adryaman, Diyarbakir, Hakkari, Kars, Siirt, Van), Armenia, S Azerbaijan and Iran (Central, East Azarbaijan, Fars, Hamadan, Kerman, Kordestan, Lorestan, West Azarbaijan, Zanjhan), 200–2000 m.
Sources: Eiselt, 1970, 1976, R.J. Clark & Clark, 1973, Baran, 1976a, Bannikov et al., 1977, Basoglu & Baran, 1980, Franzen & Sigg, 1989, Schmidtler & Eiselt, 1991, Baran et al., 2004 and Tuniyev et al., 2009.

16. *Eirenis rechingeri* Eiselt, 1971. Ann. Naturhist. Mus. Wien 75: 375–380, pl. 1, fig. 2, pl. 2, figs. 2, 5.

Type: Holotype, NMW 19688, a 344 mm male (F. Ressl, 16 April 1970).
Type locality: "57 km W Schiras (= 8 km E Dashtarjan, an der alten Strasse nach Schiras), *ca.* 2100 m ü.d.M., Iran."
Distribution: Southwestern Iran (Fars), 2100 m. Known only from type locality.

17. *Eirenis rothii* Jan, 1863a. Arch. Zool. Anat. Fis. 2(2): 259–260.

Type: Holotype, ZSM 75/0, a 787 mm specimen (Roth).
Type locality: "Gerusalemme" [= Jerusalem, E Jerusalem Distr., cen. Israel, 31°46'N, 35°13'E, elevation 775 m].
Distribution: Southwestern Asia. Southern Turkey (Adiyaman, Antalya, Gaziantep, Hatay), Syria (Homs, Latakia), S Lebanon (S Beqaa), N Israel (Central, Gaza Strip, Golan Heights, Haifa, Jerusalem, Northern, Tel Aviv, West Bank), Jordan (Amman, Balqa, Irbid, Jarash, Maan, Madaba, Mafraq, Tafilah) and NW Iraq (Irbil, Kirkuk), 700–2000 m.
Sources: F. Werner, 1939, C.A. Reed & Marx, 1959, Eiselt, 1970, Baran, 1976a, Basoglu & Baran, 1980, Ilani, 1983a, Disi et al., 1988, 2001, Bouskila & Amitai, 2001, Hraoui-Bloquet et al., 2002, Shwayar et al., 2009, Amr & Disi, 2011 and Bar & Haimovitch, 2011.
Remarks: Photograph of the type in Franzen & Glaw (2007: 252, fig. 19).

18. *Eirenis thospitis* Schmidtler & Lanza, 1990. Amphibia-Reptilia 11(4): 363–367, figs. 1–4.

Synonym: *Eirenis hakkariensis* Schmidtler & Eiselt, 1991.
Type: Holotype, MZUF 33663, a 416 mm male (P. Crucitti & A. Campese, 9 Aug. 1988).
Type locality: "2-3 km N-NE of Van, about 2000 m a.s.l. (E Turkey)."
Distribution: Extreme E Turkey (Van), 1500–2000 m.
Source: Schmidtler & Eiselt, 1991.

ELACHISTODON J.T. Reinhardt, 1863 (Colubridae)

Synonyms: *Elastistoda* – Theobald, 1868b *(nomen incorrectum), Elaschistodon* – Palacky, 1898 *(nomen incorrectum),* and *Elanchistodon* – Negi, 1992 *(nomen incorrectum).*
Type species: *Elachistodon westermanni* J.T. Reinhardt, 1863.
Distribution: South-central Asia.
Sources: Wall, 1913c, M.A. Smith, 1943, Gans & Williams, 1954, R. Fleming & Fleming, 1973, Mahendra, 1984, Basu, 1989, Schleich & Kästle, 2002, Tiwari & Shah, 2004, Captain et al., 2005, N. Khaire, 2006, Nande & Deshmukh, 2007 and Dodge, 2008.

1. *Elachistodon westermanni* J.T. Reinhardt, 1863. Overs. Kon. Danske Vidensk. Selsk. Forhand. Kjöbenhavn 1(2): 206–209, figs. 1–5.

Type: Holotype, ZMUC R6401, a 784 mm female (B.W. Westermann, 1801–1817).
Type locality: "Rungpore, et Distrikt i den nordøstlige Deel af Bengalen, som mod Øst begraendses af Floden Berampooter og mod Nord støder op til Bhootan og Cooch Behar" [= district in NE Bengal, bordered in the east by the Brahmaputra River and in the north by Bhutan and Cooch Behar *fide* Neilson *in* Gans (1955: 1) or Rangpur Distr., N Bangladesh, ca. 25°45'N, 89°14'E].
Distribution: Southern Asia. Central Nepal (Chitwan), NE India (NE Bihar, Maharashtra, Uttaranachal, West Bengal) and Bangladesh (Dacca, Rangpur), 250–500 m.

ELAPHE Fitzinger *in* Wagler, 1833 (Colubridae)

Synonyms: *Coluber* F. Boie, 1826 *(nomen praeoccupatum), Elaphis* Bonaparte, 1834 *in* 1832–1841, *Leptophidium* Hallowell, 1861, *Natrix* Cope, 1862c *(nomen praeoccupatum), Phyllophis* A.C.L.G. Günther, 1864a, *Elephas* – F. Werner, 1895 *(nomen incorrectum), Spaniopholis* Mocquard, 1897, *Elaphs* – Fan, 1931 *(nomen incorrectum), Phillophis* Maki, 1931 *(nomen emendatum), Phyliophis* – Gans, 1949 *(nomen incorrectum), Elapho* – Bleakney, 1958 *(nomen incorrectum), Elape* Kuhn, 1963 *(nomen emendatum), Elpahe* – Chiu & Wong, 1974 *(nomen incorrectum),* and *Elahe* – Mahendra, 1984 *(nomen incorrectum).*
Type species: *Coluber quatuorlineatus* Lacépède, 1789.
Distribution: Eurasia.
Fossil records: Lower Pliocene to upper Pleistocene of Europe and Asia. Fossils unassigned to species include lower Miocene (Orleanian, MN 3: 16.9–20.0 mya) of Czech Republic, lower/middle Miocene (Orleanian/Astaracian, MN 5/6: 12.8–16.0 mya) of Germany, middle Miocene (Astaracian, MN 6–8: 11.1–13.7 mya) of France and Hungary, upper Miocene (Villesian, MN 9: 9.7–11.1 mya) of Italy and Ukraine, upper Miocene (Turolian, MN 13: 4.9–7.3 mya) of Italy, middle Pliocene (Ruscinian, MN 15: 3.2–4.2 mya) of France, and Pleistocene of Spain and Ukraine.
Sources: Rooij, 1917, Maki, 1931, C.H. Pope, 1935, Bourret, 1936d, M.A. Smith, 1943, Bhatnagar, 1975, B. Hu et al., 1980, Tian et al., 1986, Szyndlar, 1991a, 2012, Staszko & Walls, 1994, Holman, 1995a, Keogh, 1996, Schulz, 1996, Rodríguez-Robles & Jesús-Escobar, 1999, Helfenberger, 2001, Lenk et al., 2001a, Ivanov, 2002, Utiger et al., 2002, 2005, Burbrink & Lawson, 2007 and Zaher et al., 2012.
Remarks: Official Generic Name no. 1229 *fide* Opinion 490 (ICZN, 1957). Original description of *Elaphis* reprinted in Bonaparte, 1840. The following genera and species have been separated from *Elaphe: Bogertophis (rosaliae, subocularis), Coelognathus (erythurus, enganensis, flavolineatus, helenus, philippinus, radiatus, subradiatus), Euprepiophis (conspicillatus, mandarinus, perlaceus), Gonyosoma (jansenii, oxycephala), Maculophis (bellus), Oocatochus (rufodorsatus), Oreocryptophis (porphyraceus), Orthriophis (cantoris, hodgsonii, moellendorffi, taeniurus), Pantherophis (alleghaniensis, bairdi, †buisi, emoryi,*

gloydi, guttatus, †kansensis, obsoleta, †pliocenicus, slowinskii, spiloides, vulpinus), *Pseudelaphe* (*flavirufa*), *Rhynchophis* (*boulengeri, frenata, prasina*), *Senticolis* (*triaspis*), and *Zamenis* (*†algorensis, hohenackeri, †kohfidischi, †kormosi, lineata, †lobsingensis, longissima, †longivertebrata, †paralongissima, persica, †praelongissima, scalaris, situla, †szyndlari*). Allocation of extinct species based on morphology and geography.

1. *Elaphe anomala* (Boulenger, 1916). Ann. Mag. Nat. Hist. (8) 17(99): 243–244. (*Coluber anomalus*)

Synonyms: *Elaphe schrencki besenbruchi* L. Müller, 1923, and *Elaphe maculata* Ma & Zong, 1984.

Type: Holotype, BMNH 1946.1.13.1 (formerly BMNH 1916.1.15.20), a 1770 mm male (A.L. Hall).

Type locality: "Chihfeng, N. E. Chihli Prov., northern China" [= Chifeng, Nei Mongol Prov., N China, 42°15'N, 118°53'E, elevation 585 m].

Distribution: Northern China (Anhui, Beijing, Gansu, Hebei, Hubei, Jiangsu, Liaoning, Nei Mongol, Shandong, Shaanxi, Tianjin, Zhejiang), North Korea and South Korea, 585 m.

Sources: Ji & Wen, 1997, Xu, 2001, Zhao, 2006, San & Lee, 2007, Treu, 2008, An et al., 2010 and Yao, 2012.

Remarks: Previously a synonym of *E. schrenkii.*

2. *Elaphe bimaculata* K.P. Schmidt, 1925b. Amer. Mus. Novit. (175): 3.

Type: Holotype, AMNH 24640, a 770 mm female (C.H. Pope, Sept.– Oct. 1921).

Type locality: "Ningkwo, Anhwei, China" [= ? Ningbo, Zhejiang Prov., E China, 29°52'N, 121°33'E, elevation 10 m].

Distribution: Eastern China (Anhui, Gansu, Hebei, Hubei, Hunan, Jiangsu, Jiangxi, Shaanxi, Zhejiang), North Korea and South Korea, 1000 m.

Sources: C.H. Pope, 1935, Fang & Wang, 1983, Schulz, 1986b, 1988i, Zhao & Adler, 1993, Zou & Chen, 1998, Zhao, 2006 and Yao, 2012.

Remarks: Supplemental original description in K.P. Schmidt (1927b: 531–532, fig. 17). Previously a synonym of *E. schrenkii.* A subspecies of *E. schrenckii fide* An et al. (2010: 15).

3. *Elaphe carinata* (A.C.L.G. Günther, 1864a). Rept. Brit. India: 295–296, pl. 21, fig. b. (*Phyllophis carinata*)

Synonyms: *Coluber phyllophis* Boulenger, 1891a *(nomen rejiciendum), Spaniopholis souliei* Mocquard, 1897c, *Spaniopholis kreyenbergi* Müller, 1907, *Elaphe osborni* K.P. Schmidt, 1925a, *Coluber camilloschneideri* T. Vogt, 1927, *Elaphe carinata ornithophaga* Bourret, 1936b, *Elaphe carinata yonaguniensis* Takara, 1962, and *Elaphe carinata deqinensis* D. Yang & Hu, 1983.

Type: Holotype, BMNH 1946.1.14.58 (formerly BMNH 1869.9.29.3), a 533 mm specimen.

Type locality: "China." Restricted to Lu Shan Mtns., S of Jiujiang on Lake Poyang, Jiangxi Prov., China, 29°30'N, 116°00'E *fide* Schulz (1992f: 6).

Distribution: Eastern Asia. Southern and E China (Anhui, Beijing, Fujian, S Gansu, Guangdong, Guangxi, Guizhou, Henan, Hubei, Hunan, Jiangsu, Jiangxi, Shaanxi, Shandong, Sichuan, Tianjin, Yunnan, Zhejiang, Lanyu Is.), NW Vietnam (Lai Chau, Lang Son, Lao Cai, Pescadores Is.), Taiwan (Botel-Tobago Is.) and ext. S Japan (Ryukyu Arch.), 450–2900 m.

E

Sources: Stejneger, 1907a, Maki, 1931, Bourret, 1937a, C. Wang & Wang, 1956, C. Wang, 1962, Kuntz, 1963, Kuntz & Ming, 1970, D. Yang & Su, 1984, M. Mori, 1986, W. Gao, 1987, Schulz, 1992c, 1992f, Zhao & Adler, 1993, H.M. Smith et al., 1995, Orlov et al., 2003, Gumprecht, 2004b, Zhao, 2006, V.S. Nguyen et al., 2009, Xiang & Li, 2009 and Yao, 2012.

Remarks: Official Specific Name *fide* Opinion 1867 (ICZN, 1997).

4. *Elaphe climacophora* (H. Boie, 1826). Isis von Oken 18(2): 210–211. (*Coluber climacophorus*)

Synonym: *Coluber virgatus* Schlegel, 1837.

Types: Syntypes (8), RMNH 382, RMNH 386, a 2020 mm specimen, RMNH 390, RMNH 47577 (formerly RMNH 386), RMNH 47578 (formerly RMNH 386), RMNH 47579 (formerly RMNH 386), RMNH 47580 (formerly RMNH 386), RMNH 47581 (formerly RMNH 386) (P.F.B. von Siebold, 1817–1829).

Type locality: "Decima, Japan, [= Dejima Is., previously in Nagasaki harbor, now part of Nagasaki, Japan, 32°45'N, 129°52'E, elevation 10 m].

Distribution: Extreme E Russia (Kuril Arch., Kunashiri Is.) and Japan (Hokkaido, Honshu, Kyushu, Shikoku, Ryukyus: Awajishima, Fukuejima, Hisakeshima, Iwoshima, Kamikoshikijima, Kamishima, Kammurijima, Kuchinoerabujima, Kuchinoshima, Mageshima, Nakadorishima, Nishinoshima, Sadogashima, Shimojima, Shimokoshikijima, Shodoshima, Takeshima, Tanegashima, Tsushima, Ukishima and Yakushima Is.), NSL–1325 m.

Sources: Stejneger, 1907a, Rendahl, 1933, Gans & Oshima, 1952, Koba, 1955, Hadley & Gans, 1972, Bannikov et al., 1977, Goris & Naganuma, 1979, M. Mori, 1982, Ota, 1983, Schulz, 1985b, 1988f, A. Mori, 1994 and Ratnikov, 2004.

Fossil records: Upper Pleistocene of Japan.

Remarks: Original description reprinted in H. Boie (1827b: 257–259). RMNH paralectotypes include RMNH 390 and RMNH 47577–81. Other possible paralectotypes include ZMA 13616a–b and ZMA 13996 *fide* Schulz (1996: 95). Type locality possibly in error *fide* Siebold *in* Temminck & Schlegel (1838: iii), who mistakenly corrected it to East Indies.

5. *Elaphe davidi* (Sauvage, 1884). Bull. Soc. Philom. Paris (7) 8: 144. (*Tropidonotus davidi*)

Synonyms: *Coluber halli* Boulenger, 1914a, and *Elaphe dione coreana* G. Song, 1961.
Type: Holotype, MNHN 6154 (formerly MNHN 1867.86), a 700 mm specimen (J.P.A. David, 1862–1869).
Type locality: "Chine" [= China].
Distribution: Northeastern China (Hebei, Heilongjiang, Nei Monggol, Jilin, Liaoning, Shandong, Shaanxi, Shanxi) and S North Korea (Kaesong, Kangwon, North Hwanghae, Pyongyang, South Hwanghae, South Pyongan), 100–1000 m.
Sources: Szyndlar, 1985b, Szyndlar & Hung, 1987, Jia, 1988, Schulz, 1989a, Zhao & Adler, 1993, Helfenberger & Schätti, 1998, Xu, 2001 and San & Lee, 2007.

E

6. *Elaphe dione* (Pallas, 1773). Reise Russ. Reichs 2(2): 717–718. (*Coluber dione*)

Synonyms: *Coluber tataricus* Güldenstedt *in* Georgi, 1801, *Coluber diana* Latreille, 1802 *in* Sonini & Latreille, *Coluber dione cumana* Pallas, 1814, *Coluber maeota* Pallas, 1814, *Coluber eremita* Eichwald, 1831, *Coluber maeoticus* Rathke, 1837, *Elaphis dione temporalis* Jan, 1863b (*nomen nudum*), *Zamenis pellioti* Mocquard, 1910, *Coluber czerskii* Nikolsky, 1914, *Elaphe dione niger* Golubev, 1923, and *Elaphe dione tenebrosa* Sobolewski, 1929.
Type: Holotype, not designated, location unknown.
Type locality: "in desertis salsis versus mare Caspium, Provinzen Russischen Reich" [= salt steppes near the Caspian Sea, Russia]. Restricted to Gratscheffskoi outpost near Semijarsk, Irtysh area, Semipalatinsk Distr., Kazakhstan *fide* Mertens & Müller (1928: 47).
Distribution: Central Asia. Southern Russia (Dagestan), S Ukraine, ext. SE Georgia, S Armenia, E Azerbaijan (Sara Is.), Kazakhstan, Uzbekistan, Turkmenistan, Kyrgyzstan, Tadzhikistan, N Iran (Mazanderan, Semnan), ext. N Afghanistan (Kunduz), Mongolia (Bayan Khongor, Bayan Ulegei, S Bulagan, Dornod, Dzubkhan, N East Gobi, N Gobi, Khentei, Middle Gobi, Selenga, South Gobi, Tov, Ubsa Nor), N China (Anhui, Beijing, Gansu, Hebei, Heilongjiang, Nei Monggol, Jiangsu, Jilin, Liaoning, Ningxia, Qinghai, Shaanxi, Shanxi, Shandong, Shanxi, Sichuan, Tianjin, Xinjiang), North Korea (Chagang, Kangwon, North Hamgyong, North Hwanghae, North Pyongan, Ryanggang, South Pyongan, South Hwanghae) and South Korea (Kyonggi, Seoul, South Chungchong, and Chejudo Is.), 75–3500 m.
Fossil records: Middle Miocene of Kazakhstan.
Sources: Strauch, 1873, Stejneger, 1907a, Nikolsky, 1916, Emelianov, 1929, Okada, 1935, Shannon, 1956, Webb et al., 1962, Leviton & Anderson, 1970a, Steward, 1971, Bannikov et al., 1977, Schulz, 1986a, 1988h, Szyndlar & Hung, 1987, Borkin et al., 1990a–b, Obst & Szczerbak, 1993, Zhao & Adler, 1993, Szczerbak, 1994, Kordikova, 1998, Kordikova et al., 1998, Adnagulov et al., 2000, Xu, 2001, Ratnikov, 2004, San & Lee, 2007, Tuniyev et al., 2009, True, 2012, and Yao, 2012.

7. *Elaphe quadrivirgata* (H. Boie, 1826). Isis von Oken 18(2): 209–210. (*Coluber quadrivirgatus*)

Synonyms: *Coluber vulneratus* H. Boie, 1826, ? *Leptophidium dorsale* Hallowell, 1861, *Elaphis bilineatus* Hallowell, 1861, *Elaphis quadrivirgatus atra* Jan, 1863b (*nomen nudum*), *Elaphis quadrivirgatum atra* Jan, 1867 *in* Jan and Sordelli, 1866–1870, *Elaphis quadrivirgatus interrupta* Jan, 1863b (*nomen nudum*), *Elaphis quadrivirgatus interrupta* Jan, 1867 *in* Jan & Sordelli, 1866–1870, and *Elapte quadrivirgatus* – Pavloff, 1926 (*nomen incorrectum*).
Types: Syntypes (6), RMNH 396a–f, longest syntype 775 mm (P.F.B. von Siebold, 1823–1829).
Type locality: "Decima, Japan, [= Dejima Is., previously in Nagasaki harbor, now part of Nagasaki, Japan, 32°45'N, 129°52'E, elevation 10 m].
Distribution: Extreme E Russia (Kuril Is.: Kunashiri Is.) and Japan (Hokkaido, Honshu, Kyushu, Ryukyus: Awajishima, Fukuijima, Hisakashima, Kamikoshikijima, Kamishima, Kammurijima, Kozushima, Kuchinoshima, Kushinoerabujima, Mageshima, Mikurajima, Nakadorishima, Niijima, Nishinoshima, Oshima, Sadogashima, Shikinejima, Shikoku, Shodoshima, Shikoko, Shimojima, Shimokoshikijima, Tadanaejima, Takeshima, Tanegashima, Toshima, Yakujima, Yakushima Is.), NSL–2000 m.
Fossil records: Upper Pleistocene of Japan.
Sources: Stejneger, 1907a, Takara, 1962, Nakamura & Uéno, 1963, Bannikov et al., 1977, Ota, 1983, M. Mori, 1984 and Schulz, 1987b, 1988n, 1996.
Remarks: Original description reprinted in H. Boie (1827b: 254–257). Ryukyu record (Okinawa) in error *fide* Maki (1931a: 83). Type locality possibly in error *fide* Siebold *in* Temminck & Schlegel (1838: iii), who mistakenly corrected it to East Indies.

8. *Elaphe quatuorlineata* (Lacépède, 1789). Hist. Nat. Serp. 2(2): 82 & 2(3): 163–164, pl. 7, fig. 1. (*Coluber quatuorlineatus*) (*nomen corrigendum*)

Synonyms: *Coluber quatuor-lineatus* Lacépède, 1789 (*nomen incorrigendum*), *Coluber quadristriatus* Donndorff, 1798, *Coluber nauii* Donndorff, 1798, *Coluber quarter-radiatus* Gmelin, 1799 (*nomen rejiciendum*), *Coluber pictus* Georgi, 1801, *Coluber quadrilineatus* Latreille *in* Sonnini & Latreille, 1801b, *Coluber elaphis* G. Shaw, 1802 (*nomen rejiciendum*), *Elaphis quaterradiatus* A.-M.-C. Duméril, Bibron & Duméril, 1854a, *Coluber quadriradiatus* Erhard, 1858, *Elaphis quaterradiatus* – A.C.L.G. Günther, 1858 (*nomen corrigendum*), *Elaphis dione graeca* Jan, 1863b (*nomen nudum*), *Elaphis cervone* Schreiber, 1875 (*nomen substitutum*), *Coluber quaterradia* – Paolucci,

1880 *(nomen incorrectum)*, *Elaphis sauromates graeca* Bedriaga, 1881 *in* 1881–1882, *Elaphis sauromates muenteri* Bedriaga, 1881 *in* 1881–1882, *Coluber quatuorlineatus* – Boulenger, 1894a *(nomen corrigendum)*, *Elaphe quatuorlineata praematura* F. Werner, 1935, *Elaphe quatuorlineata parensis* Cattaneo, 1999, and *Elaphe quatuorlineata scyrensis* Cattaneo, 1999.

Type: Holotype, not designated, an 1143 mm specimen (King of France's cabinet, Paris), location unknown.

Type locality: Unknown. Designated as Provence, France *fide* Mertens & Müller (1928: 47).

Distribution: Southeastern Europe. Southern Italy (Abruzzi, Basilicata, Calabria, Campania, ? Emilia-Romagna, Lazio, Marche, Molise, Puglia, Toscana, Umbria), Slovenia, Croatia, Bosnia-Herzegovina, Serbia, Montenegro, Albania, Bulgaria, Macedonia and Greece (Sporades and Cyclades: Antimílos, Ias, Iráklia, Kéa, Míkonos, Mílos, Naxos, Páros, Skhoinoúsa, Skíathos, Skíros, Thíra), NSL–2500 m.

Fossil records: Lower Pliocene (Ruscinian, MN 14: 4.2–4.9 mya) of Bosnia & Herzegovina and Hungary, middle Pliocene (Ruscinian, MN 15: 3.2–4.2 mya) of Moldavia and Slovakia, upper Pliocene (Villanyian, MN 16–17: 1.8–3.2 mya) of Austria and Hungary, lower Pleistocene (Calabrian: 0.8–1.8 mya) of Austria, Czech Republic and Hungary, middle Pleistocene (Ionian: 0.13–0.78 mya) of Austria, Bulgaria, Greece and Romania, and upper Pleistocene (Tarantian: 0.01–0.13 mya) of Bulgaria and Croatia.

Sources: Steinheil, 1913a, Buresch & Zonkow, 1934, Wettstein, 1937, 1953b, Q. Buchholz, 1961, Fuhn & Vancea, 1961, Mertens, 1961c, Buchholz & Schultze-Westrum, 1964, Bruno, 1966, 1977, 1984, R.J. Clark, 1967a, 1994, Muskhelishvili, 1970a, Kratzer, 1973, Lotze, 1973, Baran, 1976a, Bannikov et al., 1977, Szyndlar, 1986, Latifi, 1991, Szyndlar, 1991a, 2012, W. Böhme & Szczerbak, 1993, Kubykin, 1994, Szczerbak, 1994, Borkin et al. 1997, Holman, 1998a, Cattaneo, 1999, Petkovski et al., 2000, Ratnikov, 2004, Sindaco et al., 2006, Ivanov, 2007, Trapp, 2007, Valakos et al., 2008 and Stojanov et al., 2011.

Remarks: Official Specific Name no. 1436 *fide* Opinion 490 (ICZN, 1957).

9. *Elaphe rechingeri* F. Werner, 1932. Anz. Akad. Wiss. Wien (Math.-Nat.) 69(20): 232.

Type: Holotype, NMW 16672, an 800 mm (svl) male (K.H. Rechinger, June–July 1932).

Type locality: "Inseln Amorgos, (süd-östliche Cycladen), Cycladenarchipel" [= Amorgos Is., cen. Cyclades, SE Greece, bet. 36°46'–56'N & 25°44'–26°05'E].

Distribution: Southeastern Greece (Amorgos Is.).

Remarks: A valid species *fide* R. Clark (1994: 93).

10. *Elaphe sauromates* (Pallas, 1814). Zool. Rosso-Asiat. 3: 42. (*Coluber sauromates*)

Synonyms: *Coluber alpestris* Pallas, 1814, *Coluber pictus* Pallas, 1814, *Coluber xanthogaster* Andrzejowsky, 1832, *Coluber cereus* Dwigubsky, 1832, *Coluber fulvus* Dwigubsky, 1832, *Coluber poecilopholus* Fischer von Waldheim, 1832, *Coluber taeniothys* Fischer von Waldheim, 1832, *Elaphis parreyssii* Fitzinger, 1833, *Bothriophis erythrogaster* Eichwald, 1837, *Coluber poecilocephalus* Brandt, 1838, and *Sauromates parreysi* – Palacky, 1898 *(nomen incorrectum)*.

Type: Holotype, not designated (P.S. Pallas, 1768–1774), location unknown.

Type locality: "isthmum Perecopensum." Restricted to Perekop Isthmus, Crimea, S Ukraine [= 46°10'N, 33°42'E, elevation NSL] *fide* Mertens & Müller (1928: 47).

Distribution: Southwestern Asia. Albania, S Romania, Bulgaria (Burgas, Dobrich, Gabrovo, Haskovo, Kardzhali, Montana, Pazardzhik, Pleven, Plovdiv, Rousse, Shumen, Silistra, Sliven, Stara Zagora, Varna, Veliko Tarnovo, Yambol), E Greece (Samothrace and Thasos Is.), Turkey (Afyon, Ankara, Antalya, Aydin, Balikesir, Bursa, Diyarbakir, Edirne, Eskisehir, Izmir, Kars, Kayseri, Kocaeli, Konya, Tokat. Trabzon, Usak), ext. W Syria, S Lebanon (Mont-Liban), ext. N Israel (N Golan Heights), Moldavia, S Ukraine, SW Russia (Dagestan), E Georgia, S Armenia, Azerbaijan, SW Kazakhstan, NW Turkmenistan and W Iran (Bakhtaran, East Azarbaijan, Hamadan, West Azarbaijan, Zanjhan), NSL–2000 m.

Sources: Ilani, 1983a, Lenk et al., 2001, Hraoui-Bloquet et al., 2002, Trapp, 2007, Valakos et al., 2008, Tuniyev et al., 2009, Bar & Haimovitch, 2011 and Stojanov et al., 2011.

Remarks: Official Specific Name no. 1437 *fide* Opinion 490 (ICZN, 1957). A valid species *fide* Lenk et al. (2001: 336).

11. *Elaphe schrenckii* (Strauch, 1873). Mém. Acad. Imp. Sci. St. Pétersbourg (7) 21(4): 100–102. (*Elaphis schrenckii*)

Type: Lectotype, ZISP 3706 (formerly AIS 415), a 1950 mm specimen (L. von Schrenck, 1856), designated *fide* Maki (1931a: 101).

Type locality: "Khinggan Military Post, SE Siberia, E Russia" via lectotype selection.

Distribution: Northeastern China (Heilongjiang, Nei Monggol, Jilin), ext. E Mongolia (E Dornov), ext. SE Russia and North Korea (Kangwon), 75–900 m.

Sources: Shannon, 1956, Webb et al., 1962, Bannikov et al., 1977, Ji et al., 1985, Schulz, 1985c, 1988d, 1988g, P. Wang, 1986, Y. Yang et al., 1986, Szyndlar & Hung, 1987, Borkin et al., 1990a–b, Zhao & Adler, 1993, Adnagulov et al., 2000, Ratnikov, 2004, San & Lee, 2007 and Treu, 2008.

E

Remarks: Type locality listed as Chinganski Post *fide* Nikolsky (1916: 141).

12. *Elaphe zoigeensis* S. Huang, Ding, Burbrink, Yang, Huang, Ling, Chen & Zhang, 2012. Asian Herpetol. Res. (2) 3(1): 41–43, figs. 3–4.

Type: Holotype, HUM 20110001, an 880 mm female (native, 26 July 2010).
Type locality: "20 m from mouth of Jiangzha Hot Spring (3200 m elevation, N34°12'18", E102°42'48") in Zoige County, Sichuan, China."
Distribution: Southern China (Sichuan), 3200 m.
Source: Ling et al., 2010.

E

ELAPOGNATHUS Boulenger, 1896a (Elapidae)

Synonyms: *Elicognatus* – Sumichrast, 1880 *(nomen incorrectum),* and *Elapoqnathus* – Anon., 1979 *(nomen incorrectum).*
Type species: *Hoplocephalus minor* A.C.L.G. Günther, 1863c.
Distribution: Extreme SW Australia.
Sources: Cogger et al., 1983a, Storr et al., 1986, Gow, 1989, Hutchinson, 1990, Ehmann, 1992, Golay et al., 1993, Greer, 1997, David & Ineich, 1999, Cogger, 2000, Keogh et al., 2000, Scanlon & Lee, 2004, Sanders et al., 2008 and Zaher et al., 2009.
Remarks: A synonym of *Notechis fide* Storr, 1982; a synonym of *Rhinoplocephalus fide* Wallach, 1985.

1. *Elapognathus coronatus* (Schlegel, 1837). Essai Phys. Serp. 1: 182, 2: 454–455. (*Elaps coronatus*)

Synonyms: *Trimesurus olivaceus* Gray *in* Grey, 1841b, *Elaps melanocephalus* Gray & Neill, 1845 *(nomen praeoccupatum)*, and *Elapognathus resolutus* Wells & Wellington, 1985.
Type: Lectotype, MNHN 321 (formerly MNHN 4696), a 305–309 mm male (J.R.C. Quoy & J.P. Gaimard [Coquille or Astrolabe Voy.], 1822–1829), designated by Coventry & Rawlinson (1980: 67).
Type locality: "King George Sound, Western Australia" via lectotype selection.
Distribution: Extreme SW Australia (ext. SW Western Australia, Mondrain, North Twin Peak and Wilson Is.), NSL–640 m.
Sources: Storr et al., 1986, Bush et al., 1995 and Shea & Sadlier, 1999.

2. *Elapognathus minor* (A.C.L.G. Günther, 1863c). Ann. Mag. Nat. Hist. (3) 12(71): 362–363. (*Hoplocephalus minor*)

Types: Syntypes (5 plus skull), BMNH 1946.1.18.64 [= Boul. a], a female (Swan River, Haslar, 1863), BMNH 1946.1.20.71 [= Boul. b], a male (Swan River), MCZ 77635 (formerly BMNH 1946.1.19.13 [= Boul. c], a 445 mm female, BMNH 1946.1.19.14–15 [= Boul. d–e] (S. Australia), a male and juvenile, and BMNH 1964.962 [= Boul. g] (skull from MCZ 76635, formerly BMNH 1946.1.19.13).
Type locality: "Swan River, S.W. Australia."
Distribution: Extreme SW Australia (ext. SW Western Australia), 10–150 m.
Source: Shine, 1986b.
Remarks: Syntypes denoted above with Boulenger (1896a: 357) catalogue letters. MCZ 76635 has skull removed *fide* Cogger et al. (1983: 225).

ELAPOIDIS F. Boie, 1827 (Natricidae)

Synonyms: *Elapodis* H. Boie *in* F. Boie, 1826 *(nomen nudum), Elapodes* – Schlegel, 1837 *(nomen incorrectum), Elapoides* Jan, 1862b *(nomen emendatum),* and *Elapiodes* – S.W. Garman, 1884 *(nomen incorrectum).*
Type species: *Elapoidis fuscus* F. Boie, 1827.
Distribution: East Malaysia and Indonesia.
Sources: Rooij, 1917, Kopstein, 1938, C. Haas, 1941, 1950, Bergman, 1958a, Capocaccia, 1976, David & Vogel, 1996, Vogel & G. Köhler, 1996 and I. Das, 2010, 2012.

1. *Elapoidis fuscus* F. Boie, 1827. Isis von Oken 20(6): 519. (*Elapoidis fusca*)

Synonyms: *Elapoidis fuscus* H. Boie *in* Schlegel, 1826a *(nomen nudum), Calamaria elapoides* Schlegel, 1837, and *Elaphis sumatranus* Bleeker, 1860b *(nomen nudum).*
Types: Syntypes (2), RMNH 73 and RMNH 10427 (H. Boie & H.C. Macklot, 1825–1832).
Type locality: "Java" [SW Indonesia].
Distribution: East Malaysia (Sabah) and Indonesia (Java, Kalimantan, Sumatra), 1000–1525 m.
Remarks: Original description based on H. Boie's MS (1823–1825). H. Boie sent numerous specimens to RMNH *fide* Schlegel (1837: 44). Collectors probably H. Boie & H.C. Macklot *fide* C. Haas (1950: 559). Boulenger (1893a: 307) listed an exchanged Javan female from RMNH, probably a type.

ELAPOMORPHUS Weigmann *in* Fitzinger, 1843 (Xenodontidae)

Synonyms: *Elapocephalus* A.C.L.G. Günther *in* Gray, 1858g, and *Elapomorpho* Boettger, 1880b *(nomen emendatum), Elapomorphis* – Ditmars, 1933 *(nomen incorrectum), Elapomurphus* – Prado, 1944 *(nomen incorrectum),* and *Elapormorphus* – Bellairs, 1969 *(nomen incorrectum).*
Type species: *Coluber quinquelineatus* Raddi, 1820.

Distribution: Eastern South America.

Sources: Strauch, 1885, J.A. Peters & Orejas-Miranda, 1970, Lema, 1979b, 1984, Ferrarezzi, 1993, Lema & Hofstadler-Deiques, 1995, Hofstadler-Deiques & Lema, 2005, Zaher et al., 2009 and Lema & Hofstadler-Deiques, 2010.

Remarks: *Elapomorphus lepidus* was transferred to *Coronelaps*.

1. *Elapomorphus quinquelineatus* (Raddi, 1820). Mem. Math. Fis. Soc. Ital. Sci. Modena 18(2): 339–340, pl. (*Coluber quinquelineatus*) (*nomen corrigendum*)

Synonyms: *Coluber 5-lineatus* Raddi, 1820 *(nomen incorrigendum), Duberria quinquelineata* – Fitzinger, 1826a *(nomen corrigendum), Calamaria blumii* Schlegel, 1837, *Elapsomorphus blumii* A.-M.-C. Duméril, 1853 *(nomen nudum), Elapocephalus taeniatus* A.C.L.G. Günther, 1858, and *Elapomorphus accedens* Jan, 1862b.

Types: Syntypes (2), formerly IRMC (G. Raddi, 1817–1818), location unknown.

Type locality: "contorni di Rio-Janeiro, Brasile" [= vicinity of Rio de Janeiro, Rio de Janeiro State, SE Brazil, 22°54'S, 43°14'W, elevation NSL].

Distribution: Southeastern Brazil (S Bahia, Espírito Santo, E Minas Gerais, Rio de Janeiro, Rio Grande do Sul, Santa Catarina, São Paulo), NSL–350 m.

Sources: Hoge, 1959e, Lema, 1984, 1992, 1994, Freitas, 1999, Marques et al., 2001 and C.J. Cole et al., 2013.

Remarks: Types not in Modena or MZUF *fide* Hoge (1959e: 267), who reported only a single type (whereas Raddi listed two ventral counts). C.J. Cole et al. (2013: 493) erroneously cited a type as MNHN 3673 (syntype of *Calamaria blumii* Schlegel). Record without specific locality from Guyana or French Guiana (USNM 6180) *fide* C.J. Cole et al. (2013: 493).

2. *Elapomorphus wuchereri* Wucherer *in* A.C.L.G. Günther, 1861b. Ann. Mag. Nat. Hist. (3) 7(41): 415–416, 3 figs.

Type: Lectotype, BMNH 1946.1.2.96, an 1165 mm specimen (O.E.H. Wucherer, 1859–1860).

Type locality: "Ilheus, Bahia" [= Ilhéus, Bahia State, Brazil, 14°48'S, 39°02'W, elevation 10 m].

Distribution: Southeastern Brazil (Bahia, Espírito Santo, ? Rio de Janeiro), 10 m.

Sources: A.C.L.G. Günther, 1861e and Lema & Hofstadier-Deiques, 2010.

Remarks: Original description reprinted in A.C.L.G. Günther (1861e: 15–16, 3 figs.).

ELAPOTINUS Jan, 1862b (Atractaspididae)

Synonym: *Elapotimus* – Briceño-Rossi, 1934 *(nomen incorrectum).*

Type species: *Elapotinus picteti* Jan, 1862b.

Distribution: Unknown.

Sources: Jan & Sordelli, 1860–1866 (13: pl. 3, fig. 1), Witte & Laurent, 1947, Wallach, 1991, Vidal et al., 2008, Kelly et al., 2009 and Zaher et al., 2009.

1. *Elapotinus picteti* Jan, 1862b. Arch. Zool. Anat. Fis. 2(1): 31–32.

Type: Holotype, MHNG 279.37, a 312 mm male (M. Pelletier, Oct. 1840).

Type locality: Unknown. "Tropical Africa ?" *fide* Boulenger (1896a: 245).

Distribution: Unknown. Known only from the holotype and a topotype.

Remarks: According to the MHNG minutes for 1840, Pelletier purchased 17 bottles of exotic reptiles that contained a series of 10 unidentified snakes (MHNG 279.33–279.42). MHNG 279.33–34 are listed as "? Guyane". The locality column for MHNG 279.34–279.42 is blank. Localities for lizards on the same catalogue page include Senegal and Madagascar. As an atractaspidid, the most likely origin is African but Madagascar is also a possibility.

ELAPSOIDEA Bocage, 1866b (Elapidae)

Synonyms: *Elapsoidea* Bocage, 1866a *(nomen nudum), Elapechis* Boulenger, 1896a *(nomen substitutum), Elapoidea* Boulenger, 1896a *(nomen emendatum), Alapsoidea* – Anthony, 1955 *(nomen incorrectum),* and *Elapsoides* – Corkill, 1956 *(nomen incorrectum).*

Type species: *Elapsoidea guentherii* Bocage, 1866b.

Distribution: Subsaharan Africa.

Sources: Loveridge, 1944b, Broadley, 1971b, Golay, 1985, Golay et al., 1993, Jakobsen, 1997, Keogh, 1998, David & Ineich, 1999, Dobiey & Vogel, 2007, Zaher et al., 2009 and Hoser, 2012e.

1. *Elapsoidea boulengeri* Boettger, 1895b. Zool. Anz. 18(468): 62–63.

Type: Holotype, SMF 20555 (formerly SMF-B 9361a), a 170 mm specimen (O. Boettger, 1895).

Type locality: "Sambesi, Ostafrika" [= Zambezi River, East Africa]. Restricted to Boroma, Zambezi River, Zambezia Prov., cen. Mozambique *fide* M. Griffin (2003: 114).

Distribution: Southeastern Africa. Eastern Democratic Republic of the Congo (Sud-Kivu), W and SE Tanzania

(Kigoma, Mbeya, Mtwara, Ruvuma), Zambia (Central, Copperbelt, Eastern, Luapula, Lusaka, Northern, North-Western, Southern, Western), Malawai (Northern, Southern), Mozambique (Gaza, Inhambane, Manica, Nampula, Sofala, Tete, Zambézia), Zimbabwe (Bulawayo, Manicaland, Mashonaland Central, Mashonaland East, Mashonaland West, Masvingo, Matabeleland North, Matabeleland South, Midlands), NE Namibia (Caprivi, Okavango), NE Botswana (Central, North East, North West), Swaziland and N South Africa (N KwaZulu-Natal, W Free State, Limpopo, Mpumalanga, E Northern Cape), NSL–1500 m.

Sources: Sweeney, 1961, Broadley & Cock, 1975, Auerbach, 1987, Broadley, 1998a, M. Griffin, 2003 and Broadley & Blaylock, 2013.

E

2. *Elapsoidea broadleyi* Jakobsen, 1997. Steenstrupia 22: 76–78, figs. 12–13.

Type: Holotype, MZUF 21658, a 770 mm male (G. T. Castellani, 1970).

Type locality: "Gelib (= Jilib, 0°28'N, 42°50'E), Somalia."

Distribution: Southern Somalia (Jubbada Dhexe), 25 m.

3. *Elapsoidea chelazziorum* Lanza, 1979. Monit. Zool. Ital. (2, Suppl.) 12(12): 238–243, figs. 1a–c, 2a–c4a–b. (*nomen corrigendum*)

Synonyms: *Elapsoidea chelazzii* Lanza, 1979 *(nomen incorrigendum),* and *Elapsoidea chelazziorum* Lanza, 1990b *(nomen corrigendum).*

Type: Holotype, MZUF 21964, a 415 mm male (A.M. Simonetta, 27 March 1977).

Type locality: "in the garden of the University Campus (Faculty of Agriculture), about 6 km E of Afgoi (about 02°06'N-45°11'E, SE Somalia)."

Distribution: Southern Somalia (NE Shabeellaha Hoose). Known only from type locality.

Sources: Lanza, 1979, 1983a, 1990b.

4. *Elapsoidea guentherii* Bocage, 1866b. J. Sci. Math. Phys. Nat., Lisboa 1(1): 70–71, pl. 1, figs. 3, 3a–b. (*nomen corrigendum*)

Synonyms: *Elapsoidea güntherii* Bocage, 1866b *(nomen incorrigendum), Elapsoidea hessei* Boettger, 1887a, *Elapsoidea guentheri* – Boettger, 1888e *(nomen corrigendum), Elapsoidea güntherii* – Bocage, 1895 *(nomen incorrectum),* and *Elapsoidea decosteri scalaris* Laurent, 1960.

Type: Lectotype, MBL T-103 (formerly MBL 1935), a 400 mm specimen (J. d'Anchieta, 1865), designated by H.W. Parker (1949: 97), destroyed by fire 18 March 1968.

Type locality: "Cabinda" [= Cabinda, Cabinda Distr., ext. NW Angola, 5°33'S, 12°11'E, elevation 20 m] via lectotype selection.

Distribution: CentralAfrica. Southern Congo (Kouilou), S Democratic Republic of the Congo (Bandundu, Bas-Congo, Katanga, Kinshasa, Oriental, Sud-Kivu), Angola (Benguela, Bié, Cabinda, Cuando Cubango, Cuanza, Huíla, Moxico, Lunda Norte), Zambia (Central, Copperbelt, Northern, North-Western, Southern) and cen. Zimbabwe (Mashonaland Central, Mashonaland East, Mashonaland West, Midlands), NSL–1700 m.

Sources: Laurent, 1956a, Witte, 1962, Leston & Hughes, 1968, Hughes & Barry, 1969, Broadley, 1959, 1983, Fouin, 1969, V.F.M. FitzSimons, 1974, Broadley & Cock, 1975, Branch, 1988, J.-F. Trape & Roux-Estève, 1995, Broadley et al., 2003, Chippaux, 2006 and Broadley & Blaylock, 2013.

5. *Elapsoidea laticincta* (F. Werner, 1919). Denks. Akad. Wiss. Wien Math.-Natur. Kl. 96: 507–508, fig. 8. (*Elapechis laticinctus*)

Type: Holotype, NMW 19068, a 237 mm female (F. Werner, 28 March 1914).

Type locality: "Kadugli, Kordofan, Nuba-Berg-Provinz, Sudan" [= Janub Kurdufan Prov., S. Sudan, 11°01'N, 29°43'E, elevation 525 m].

Distribution: Central Africa. Northern Cameroon (Adamaoua), S Chad, Central African Republic (Mbomou, Ombella-Mpoko, Ouham, Ouham-Pende), N Democratic Republic of the Congo (N Orientale), S Sudan (Janub Kurdufan), S Sudan (Janub Kurdufan), S South Sudan (S Central Equatoria, SW Eastern Equatoria, S Western Equatoria), NW Uganda (W Northern), and S Ethiopia (Sidamo), 270–1000 m.

Sources: Pitman, 1974, Spawls et al., 2002, Chippaux, 2006, Chirio & Ineich, 2006 and Chirio & LeBreton, 2007.

Remarks: Possibily conspecific with *E. semiannulata moebiusi* (Werner) *fide* Broadley, 1971b.

6. *Elapsoidea loveridgei* H.W. Parker, 1949. Zool. Verh. (6): 95–98. (*Elapsoidea sundevallii loveridgei*)

Synonyms: *Elapsoidea decosteri colleti* Laurent, 1956a, and *Elapsoidea decosteri multicincta* Laurent, 1956a.

Type: Holotype, BMNH 1946.1.23.57 (formerly BMNH 1902.5.26.4), a 560 mm female (S.L. Hinde, 1896–1902).

Type locality: "Machakos, Kenya Colony, 5400 ft." [= Machakos, cen. Eastern Prov., cen. Kenya, 1°31'S, 37°16'E, elevation 1645 m].

Distribution: Eastern Africa. Southern Ethiopia (Gemu Gofa, Hararge, Illubabor, Sidamo), S Somalia (Jubbada Dhexe), E Democratic Republic of the Congo (E Nord-Kivu, E Orientale), S Uganda (Central, Eastern, Western), SW Kenya (Central, S Coast, Eastern, Nairobi, Nyanza, S Rift Valley, Western), Rwanda, Burundi (Buyogoma) and NE Tanzania (Arusha, Kilimanjaro, Shinyanga, Tabora), 500–2300 m.

Sources: Laurent, 1956a, Pitman, 1974, Spawls et al., 2002 and Largen & Spawls, 2010.

7. *Elapsoidea nigra* A.C.L.G. Günther, 1888c. Ann. Mag. Nat. Hist. (6) 1(3): 332.

Synonym: *Elapechis niger* – Boulenger, 1896a.
Type: Holotype, BMNH 1946.1.18.95 (formerly BMNH 1879.11.12.3), a 406 mm female (J. Kirk, 1865–1879).
Type locality: "Ushambola" [= Usambara Mtns., N Tanga Prov., NE Tanzania].
Distribution: Northeastern Tanzania (Morogoro, Tanga), 300–1900 m.
Source: Spawls et al., 2002.

8. *Elapsoidea semiannulata* Bocage, 1882a. J. Sci. Math. Phys. Nat., Lisboa 8(32): 303. (*nomen corrigendum*)

Synonyms: *Elapsoidea semi-annulata* Bocage, 1882a *(nomen incorrigendum), Elapsoidea semiannulata* – Bocage, 1895 *(nomen corrigendum), Elapsoidea moebiusi* F. Werner, 1897b, and *Elapsoidea decosteri huilensis* Laurent, 1964a.
Type: Holotype, MBL 1938 (J.A. d'Anchieta, 1882), destroyed by fire 18 March 1978.
Type locality: "Caconda, Angola" [= Caconda, Huila Prov., SW Angola, 13°44'S, 15°04'E, elevation 1635 m].
Distribution: West and cen. Africa. Southern Mauritania (Guidimaka), Senegal (Dakar, Fatick, Kolda, Tambacounda), Gambia (Western), Guinea–Bissau (Bissau, Tombali, Bubaque Is.), Guinea (Boké, Dubréka, Kouroussa, Télimélé), Ivory Coast (Bouna, Toumodi), SW Mali (Kayes, Koulikoro, Sikasso), N Ghana (Northern, Upper East, Upper West, Volta), Togo (Maritime, Plateaux, Savanes), N Benin (Alibori, Atakora, Collines), Burkina Faso (Centre, Centre-Est, Centre-Nord, Centre-Ouest, Est, Haute-Bassins, Sud-Ouest, Volta-Noire), SW Niger (Dosso, Niamey), Nigeria (Bauchi, Benue, Gongola, Kaduna, Kwara, Niger, Plateau), N Cameroon (Extreme-Nord, Nord), S Chad (Mayo-Kebbi, Moyen-Chari), Central African Republic (Bamingui-Bangoran, Kemo-Gribingui, Mbomou, Ombella-Mpoko, Ouaka, Ouham, Ouham-Pende, Vakaga), E Gabon (Haut-Ogooué), Congo (Brazzaville, Cuvette-Ouest, Plateau), Democratic Republic of the Congo (Bandundu, Bas-Congo, Kinshasa, Katanga, Orientale, Sud-Kivu), Angola (Benguela, Bié, Cuanza Norte, Huambo, Huíla, Moxico, Namibe), W Zambia (Western) and NE Namibia (Omaheke, Oshikoto), NSL–2000 m.
Sources: Haacke & Finkeldey, 1967, Roman, 1973b, V.F.M. FitzSimons, 1974, Broadley & Cock, 1975, Pienaar et al., 1983, Branch, 1988, J.-F. Trape & Roux-Estève, 1995, Broadley, 1998a, Spawls et al., 2002, Broadley et al., 2003, Chippaux, 2006, Chirio & Ineich, 2006, J.-F. Trape & Mané, 2006b, Chirio & LeBreton, 2007, Chirio, 2009, Ullenbruch et al., 2010, Segniagbeto et al., 2011, Auliya et al., 2012, Chirio, 2013 and Hughes, 2013.
Remarks: Unconfirmed reports from Ivory Coast, Liberia, Mali, Sierra Leone.

9. *Elapsoidea sundevalli* (A. Smith, 1848 *in* 1838–1849). Illust. Zool. So. Africa, Rept.: pl. 66, figs. a–c, 2 pp. (*Elaps sundevalli*) (*nomen corrigendum*)

Synonyms: *Elaps sundervallii* A. Smith, 1849 *in* 1838–1849 *(nomen incorrigendum), Elaps sunderwallii* A. Smith, 1849 *in* 1838–1849 *(nomen incorrigendum), Elapsoidea sundevallii* – W.C.H. Peters, 1880a *(nomen corrigendum), Elapsoidea decosteri* Boulenger, 1888d, *Elapsoidea sundevallii fitzsimonsi* Loveridge, 1944b, *Elapsoidea sundevallii longicauda* Broadley, 1971b, and *Elapsoidea sundevallii media* Broadley, 1971b.
Type: Holotype, NHR, a 549 mm specimen (C.J. Sunderwall), lost *fide* Hughes, *in litt.*
Type locality: "Southern Africa to the eastward of the Cape Colony" [= KwaZulu-Natal Prov., South Africa].
Distribution: Southern Africa. Southern Mozambique (Gaza, Inhambane, Maputo), ext. S Zimbabwe (S Matabeleland South, S Midlands), W Botswana (Ghanzi, S Kgalagadi, North West), Namibia (Okavango, Omaheke, Otjozondjupa), Swaziland and E South Africa (Free State, Gauteng, KwaZulu-Natal, Limpopo, Mpumalanga, NE Northern Cape), NSL–1600 m.
Sources: Mertens, 1955, 1971, Hughes & Barry, 1969, FitzSimons, 1974, Broadley & Cock, 1975, Broadley, 1983, Pienaar et al., 1983, Branch, 1988 and Broadley & Blaylock, 2013.
Remarks: Specific epithet *sunderwallii* corrected to *sundevalli* in errata sheet (although text, plate and index all use the *–ii* ending).

10. *Elapsoidea trapei* Mané, 1999. Bull. Soc. Herp. France (91): 13–16, fig. 2.

Type: Holotype, MNHN 1999.6505, a 545 mm male (J.-F. Trape, 1 Aug. 1993).
Type locality: "Ndébou, Tambacounda (12°31'N-12°27'W), Senegal" [= Ndébou, S Kédougou Prov., SE Senegal, 12°31'N, 12°27'W, elevation 140 m].
Distribution: Extreme West Africa. Southern Mauritania (Guidimaka), SE Senegal (Dakar, S Kédougou, S Tambacounda), ext. W Mali and ext. N Guinea, NSL–140 m.
Sources: J.-F. Trape & Ineich, 2004, 2006b, Villiers & Condamin, 2005 and J.-F. Trape & Mané, 2006.

EMMOCHLIOPHIS Fritts & Smith, 1969 (Dipsadidae)

Type species: *Emmochliophis fugleri* Fritts & Smith, 1969.
Distribution: Northwestern Ecuador.
Sources: Bogert, 1964, Fritts & Smith, 1969, Hillis, 1990, Sheil & Grant, 2001 and Zaher et al., 2009.

1. *Emmochliophis fugleri* Fritts & Smith, 1969. Trans. Kansas Acad. Sci. 72(1): 60–64, figs. 1a–c, 2a–c.

Type: Holotype, UIMNH 78795, a male (G. Key, 8–10 Feb. 1966).
Type locality: "4 km. E Río Baba bridge, 24 km. S Santo Domingo de los Colorados, Pichincha, Ecuador, ca. 600 m."
Distribution: Ecuador (Pichincha), 600 m. Known only from type locality.

E

2. *Emmochliophis miops* (Boulenger, 1898h). Proc. Zool. Soc. London 66(1): 115, pl. 12, fig. 1. (*Synophis miops*)

Type: Holotype, BMNH 1946.1.12.30 (formerly BMNH 1898.4.28.56), a 385–390 mm female (W.F.H. Rosenberg, Oct. 1897).
Type locality: "Paramba, a farm on the W. bank of the River Mira, at 3500 feet altitude, western Ecuador."
Distribution: Ecuador (Carchi), 775 m. Known only from type locality.
Source: Sheil, 1998.

EMYDOCEPHALUS Krefft, 1869a (Elapidae)

Synonyms: *Emidocephalus* Steindachner, 1913 *(nomen emendatum), Emidocephalus* – Angel, 1950 *(nomen incorrectum), Emydocephalis* – Werler & Keegan, 1963 *(nomen incorrectum),* and *Emydocepharus* – Tamiya, 1987 *(nomen incorrectum).*
Type species: *Emydocephalus annulatus* Krefft, 1869a.
Distribution: Marine waters of Sino-Australia, including Arafura Sea, Bohol Sea, Coral Sea, East China Sea, Great Barrier Reef, Gulf of Carpentaria, SW Pacific Ocean and Timor Sea.
Sources: Boulenger, 1908b, M.A. Smith, 1926a, Voris, 1966, Halstead, 1970, Cogger, 1975, 2000, Cogger et al., 1983a, McCarthy, 1986, Welch, 1988, A.R. Rasmussen, 1997, David & Ineich, 1999, Zaher et al., 2009 and Hoser, 2012e.

1. *Emydocephalus annulatus* Krefft, 1869a. Proc. Zool. Soc. London 37(1): 322.

Synonyms: *Aipysurus chelonicephalus* Bavay, 1869, *Emydocephalus tuberculatus* Krefft, 1869a, and *Emydocephalus annualatus* – M.W. Lin, 1975 *(nomen incorrectum).*
Types: Syntypes (2), AMS 454 and AMS 6633, longest syntype 762 mm.
Type locality: "Probably the Australian seas."
Distribution: Indo-Australia. Vietnam, Philippines (Panglao), N Australia (NW Northern Territory, Queensland, N Western Australia, Ashmore and Cartier Is.), New Caledonia and Loyalty Is.
Sources: Boulenger, 1898l, 1903f, Alcala, 1986, Ehmann, 1992, Ineich & Rasmussen, 1997, Dotsenko, 1999, Bauer & Sadlier, 2000, Cogger, 2000, Ineich & Laboute, 2002, J.C. Murphy & Schlager, 2003 and A.R. Rasmussen et al., 2011.
Remarks: Original description reprinted in Krefft (1869b: 92).

2. *Emydocephalus ijimae* Stejneger, 1898. J. Sci. Coll., Imp. Univ. Tokyo 12(3): 223.

Synonyms: *Emydocephala ijimae* – Mell, 1929 *(nomen incorrectum),* and *Emydocephalis ijimae* – Werler & Keegan, 1963 *(nomen incorrectum).*
Type: Holotype, USNM 36517, an embryo taken from a gravid 890 mm female (Tashiro, 1888).
Type locality: "Riu-Kiu Sea, around Riu-Kiu Archipelago" [= Ryukyus, S Japan].
Distribution: Southern Japan (Ryukyus: Iriomote, Ishigaki, Okinawa, Takara Jima and Yaeyama Is.) and Taiwan (Botel Tobago Is.).
Sources: Stejneger, 1907a, Wall, 1909a, Maki, 1931, Shibata, 1960, Mao & Chen, 1980, M. Mori, 1986 and Xiang & Li, 2009.
Remarks: USNM 36517, although reconfirmed as the holotype *fide* Stejneger (1907a: 413) is listed as a paratype *fide* Cochran (1961: 178). Stejneger (1907a 413), Wall (1909a: 187) and Maki (1931a: 179) erroneously listed original description in vol. 13.

ENHYDRINA Gray, 1849a (Elapidae)

Synonyms: *Entrydrina* – Phipson, 1887 *(nomen incorrectum), Euhydrina* – Phipson, 1888 *(nomen incorrectum), Hydrina* – Perrier, 1928 *(nomen incorrectum), Endydrina* – Schulze, Kükenthal & Hediger, 1929 *in* 1917–1929 *(nomen incorrectum), Enhydria* – Cowan & Hick, 1951 *(nomen incorrectum),* and *Enhydrena* – Chippaux & Goyffon, 1983 *(nomen incorrectum).*
Type species: *Hydrophis schistosus* Daudin, 1803c.
Distribution: Marine waters, estuaries and tidal rivers (up to 600 km upstream in Cambodia) of Indo-Australia, including Andaman Sea, Arabian Sea, Arafura Sea, Banda Sea, Bay of Bengal, Coral Sea, Flores Sea, Gulf of Oman, Gulf of Thailand, Java Sea, Molucca Sea, Persian Gulf, South China Sea, Strait of Malacca and W Indian Ocean.
Sources: M.A. Smith, 1926a, Cogger et al., 1983a, Golay et al., 1993, O'Shea, 1996, A.R. Rasmussen, 1997, David & Ineich, 1999, Wells, 2007, Kharin & Czeblukov, 2009, Zaher et al., 2009, I. Das, 2010, 2012, A. Silva et al., 2011 and Hoser, 2012e.
Remarks: A synonym of *Hydrophis fide* Sanders et al., 2013.

1. *Enhydrina schistosa* (Daudin, 1803c). Hist. Nat. Rept. 7: 386–387. (*Hydrophis schistosus*)

Synonyms: *Enhydris valakadin* Rafinesque-Schmaltz, 1817, *Hydrophis cianura* Rafinesque-Schmaltz, 1817, *Hydrophis hoglin* Rafinesque-Schlmaltz, 1817, Hydrophis *flaviventris* Siebold, 1827, *Hydrus valakadjen* – H. Boie *in* Schlegel, 1826b *(nomen incorrectum), Hydrus valakadyn* – F. Boie *(nomen incorrectum)*, 1827, *Disteira russelii* Fitzinger, 1827, *Polyodontes annulatus* Lesson, 1832 *in* Bélanger, 1831–1834, *Hydrophis bengalensis* Gray, 1842d, *Hydrophis subfasciata* Gray, 1842d, *Thalassophis werneri* P. Schmidt, 1852, *Thalassophis werneria* A.-M.-C. Duméril, Bibron & Duméril, 1854b *(nomen emendatum), Enhydrina valacadyen* – J. Anderson, 1872 *(nomen incorrectum), Entrydrina bengaensis* – Phipsoni, 1887 *(nomen incorrectum), Enhydrina vikadien* – Boettger, 1892a *(nomen incorrectum), Enhydrina valakadien* Boulenger, 1896a *(nomen emendatum), Enhydrina velakadien* – Flower, 1899 *(nomen incorrectum), Enhydrina valacadjen* – Kathariner, 1900 *(nomen incorrectum), Enhydrina schistoza* – Codoceo, 1957 *(nomen incorrectum), Enhydrina schitosa* – Boquet, 1964 *(nomen incorrectum), Enhydrena schistose* – Chippaux & Goyffon, 1983 *(nomen incorrectum), Enhydria schistose* – Murthy, 1986 *(nomen incorrectum)*, and *Enhydrina valacadien* – Culotta & Pickwell, 1993 *(nomen incorrectum).*

Type: Holotype, BMNH 1946.1.10.7 (formerly RCSM & BMNH 1921.7.28.1), a 926 mm male.

Type locality: “Bengale” [= Bangladesh or West Bengal, India]. Restricted to Tranquebar [= Tharangambadi, Tamil Nadu State, SE India, 11°02’N, 79°51’E, elevation 5 m] *fide* M.A. Smith (1926a: 39).

Distribution: Indo-Australia. Iraq, Iran, Oman, United Arab Emirates, Pakistan (S Balochistan, S Sindh), India (Andhra Pradesh, Gujarat, Karnataka, Kerala, Maharashtra, Orissa, Tamil Nadu, West Bengal), Sri Lanka, Bangladesh, Myanmar, Mergui Arch., Thailand (Chumphon, Samut Sakhno, Satun, Songkhla, Surat Thani), Cambodia, Grand Lac, S Vietnam (Binh Thuan to Ba Ria-Vung Tau), West Malaysia (Johore, Kelantan, Negri Sembilan, Pahang, Perak, Perlis, Selangor, Trengganu, Penang Is.), Singapore, East Malaysia (Sabah, Sarawak), Brunei, Indonesia (S Papua, Java, Kalimantan, Moluccas, Sumatra), Seychelles, S Papua New Guinea (Western) and N Australia (Northern Territory, Queensland, Great Barrier Reef).

Sources: Wall, 1909a, 1914g, 1918b, 1919c, Rooij, 1917, Bourret, 1934f, M.A. Smith, 1943, Bergman, 1955c, E.H. Taylor, 1965, Minton, 1966, Chhapgar & Kewalramani, 1967, Halstead, 1970, Saint Girons, 1972, B.L. Lim & Sawai, 1975, Voris et al., 1978, Voris & Jayne, 1979, Voris & Moffett, 1981. Tweedie, 1983, Gasperetti, 1988, Ehmann, 1992, Leviton et al., 1992, Dutta & Acharjyo, 1993, Hallermann et al., 2001, M.S. Khan, 2002, Whitaker & Captain, 2004, N. Khaire, 2006, V.S. Nguyen et al., 2009 and A.R. Rasmussen et al., 2011.

Remarks: Original description based on P. Russell (1801: 11, pl. 10).

2. *Enhydrina zweifeli* Kharin, 1985. Zool. Zhur. 64(5): 786–787, figs. 1–3.

Type: Holotype, AMNH 104340, a 690 mm male (O. Burton, Feb.–March 1966).

Type locality: “New Guinea, District Sepik, mouth of Sepik River” [= East Sepik Distr., Papua New Guinea].

Distribution: Papua New Guinea (East Sepik). Known only from the holotype.

E

ENHYDRIS Latreille *in* Sonnini & Latreille, 1801a (Homalopsidae)

Synonyms: *Enhydra* – Fischer von Waldheim, 1813 *(nomen incorrectum), Enhydrus* Rafinesque-Schmaltz, 1817 *(nomen emendatum), Natrix* Rafinesque-Schmaltz, 1817 *(nomen praeoccupatum), Enydris* Ritgen, 1828 *(nomen emendatum), Hypsirhina* Wagler, 1830, *Hypserhina* Gray, 1834 *in* Gray & Hardwicke, 1830–1835 *(nomen incorrectum), Hypsirina* Gray, 1835 *in* Gray & Hardwicke, 1830–1835 *(nomen emendatum), Potamophis* Cantor, 1836a, *Miralia* Gray, 1842d, *Raclitia* Gray, 1842d, *Hypsiscopus* Fitzinger, 1843, *Pelophis* Fitzinger, 1843, *Enhydra* Agassiz, 1847 *(nomen praeoccupatum), Noterophis* Gistel, 1848, *Phytolopsis* Gray, 1849a, *Hypsirrhina* Lichtenstein & Martens, 1856 *(nomen emendatum), Tachyplotus* J.T. Reinhardt, 1866, *Pythonopsis* W.C.H. Peters, 1871b *(nomen substitutum), Homalophis* W.C.H. Peters, 1871b, *Phytonopsis* – Hoffmann, 1890 *(nomen incorrectum), Hipsirhina* – Palacky, 1898 *(nomen incorrectum), Hypsirhinus* – Perrier, 1928 *(nomen incorrectum), Hypsirhinae* – Maass-Berlin, 1933 *(nomen incorrectum), Hypsirhyna* – Maki, 1931 *(nomen incorrectum), Enhydres* – Briceño-Rossi, 1934 *(nomen incorrectum), Phytonopsis* – C.P.J. Haas, 1950 *(nomen incorrectum), Enhydrous* – Taub, 1966 *(nomen incorrectum)*, and *Hypshirhina* – Tiedemann & Häupl, 1980 *(nomen incorrectum).*

Type species: *Hydrus enhydris* J.G. Schneider, 1799.

Distribution: Tidal rivers, lakes, estuaries, marshes and coastal marine waters of SE Asia and East Indies.

Sources: M.A. Smith, 1943, Gyi, 1970, Welch, 1988, J.C. Murphy & Voris, 1994, Voris et al., 2002, J.C. Murphy, 2007b, Alfaro et al., 2008, Zaher et al., 2009, I. Das, 2010, 2012 and Karns et al., 2010.

Remarks: A polyphyletic genus with *E. bocourti* and *E. punctata* belonging to other genera *fide* Voris et al., 2002 and Karns et al. (2010: 28). J.C. Murphy (2011: 234) placed *polylepis* in the resurrected genus *Pseudoferania* Ogilby. Two undescribed species

(*Enhydris towuti* from cen. Sulawesi and *Enhydris grey* from SE Sulawesi) *fide* Iskandar & Tjan (1996: 43).

1. *Enhydris albomaculata* (A.-M.-C. Duméril, Bibron & Duméril, 1854b). Erpét. Gén. 7(2): 974–975. (*Homalopsis albomaculatus*) (*nomen corrigendum*)

Synonyms: *Homalopsis albo-maculatus* A.-M.-C. Duméril, Bibron & Duméril, 1854b *(nomen incorrigendum),* and *Homalopsis albomaculatus* – A.-M.-C. Duméril, Bibron & Duméril, 1854c *(nomen corrigendum).*

Type: Holotype, MNHN 934, a 370 mm male (O. Kunhardt, 1847).

Type locality: "les environs de Padani, île de Sumatra" [= vicinity of Padang, Sumatera Barat Prov., Sumatra, W Indonesia, 0°57'S, 100°21'E].

Distribution: Western Indonesia (Java, Nias, Simeulue, Sinabang, Sumatra).

Source: David & Vogel, 1996.

Remarks: Java and Sumatra records doubtful *fide* J.C. Murphy (2007b: 85).

2. *Enhydris alternans* (A. Reuss, 1834). Mus. Senck. 1: 155–156, pl. 9, figs. 3a–b. (*Brachyorrhos alternans*)

Synonym: *Homalopsis decussata* Schlegel, 1837.

Type: Lectotype, SMF 19465 (formerly SMF-R III.C.4.a & SMF-B 9004, 1a), a 127 mm specimen (Peitsch, 1827), designated by Mertens (1922: 181).

Type locality: "Java" [Indonesia] via lectotype selection.

Distribution: East Malaysia (Sarawak) and W Indonesia (Bangka, Belitung, Java, Sumatra).

Sources: Bergman, 1960, Iskandar, 1987 and David & Vogel, 1996.

3. *Enhydris bocourti* (Jan, 1865c). Arch. Zool. Anat. Fis. 3(2): 258–259. (*Hypsirhina bocourti*)

Synonyms: *Hypsirhina multilineata* Tirant, 1885b, *Hypsirhina gigantea* F. Werner, 1923a, and *Hypsirhina bocourtii soctDistributionnsis* Bourret 1934d.

Types: Syntypes (2), MNHN 613–613A, longest syntype 490 mm (F. Bocourt, 1862).

Type locality: "Bangkok, R. di Siam" [= Bangkok, Bangkok Prov., Thailand, 13°43'N, 100°29'E, elevation NSL].

Distribution: Southeast Asia. Southern Thailand (Bangkok, Nakhon Sawan, Pattani, Phatthalung, Phra Nakhon Si Ayutthaya, Surat Thani, Trang), Cambodia (Battambang, Kampong Chhang, Kampong Thom, Kampot, Koh Kong, Oddar Meanchey, Pursat, Siem Reap), Laos (Vientiane), S Vietnam (Binh Duong, Ca Mau, Can Tho, Dong Nai, Ho Chí Minh City, Kien Giang, Soc Trang, Tây Ninh) and N West Malaysia (Kedah), 190–200 m.

Sources: M.A. Smith, 1928b, E.H. Taylor, 1965, Campden-Main, 1970a, Saint Girons, 1972a, Tweedie, 1983, M.J. Cox, 1991b, M.J. Cox et al., 1998, B.L. Stuart & Emmett, 2006, V.S. Nguyen et al., 2009, I Das, 2010 and T. Hartmann et al., 2013.

Remarks: India and Myanmar records likely based on *E. sieboldi fide* J.C. Murphy (2007b: 95).

4. *Enhydris chanardi* J.C. Murphy & Voris, 2005. Raffles Bull. Zool. 53(1): 143–147, figs. 1a, 2.

Type: Holotype, USNM 67516, a female (Hugh M. Smith, 1 May 1928).

Type locality: "Bangkok (13°45'N 100°31'E), Thailand."

Distribution: Thailand (Bangkok, Phra Nakhon Si Ayutthaya).

Remarks: Probably endemic to Chao Phraya freshwater swamp forest *fide* J.C. Murphy (2007b: 99).

5. *Enhydris doriae* (W.C.H. Peters, 1871b). Mber. Königl. Akad. Wiss. Berlin 1871(10): 577. (*Homalophis doriae*)

Types: Syntypes (2), MSNG CE 30665 (O. Beccari & M.J. Doria, 1865) and ZMB 7120, an 800 and 780 mm specimen.

Type locality: "Sarawak" [East Malaysia].

Distribution: Borneo. East Malaysia (Sabah, Sarawak), Brunei and Indonesia (W Kalimantan), NSL–500 m.

Sources: J.C. Murphy et al., 2005 and I. Das, 2007b.

Remarks: Original description reprinted in W.C.H. Peters (1872d: 38–39). W.C.H. Peters (1871b: 577) listed two syntypes. Gyi (1970: 130) stated both types are in MSNG but Capoccacia (1961a: 106) listed only one syntype. ZMB catalogue lists three syntypes, ZMS 7120–22.

6. *Enhydris enhydris* (J.G. Schneider, 1799). Hist. Amph. 1: 245–246. (*Hydrus enhydris*)

Synonyms: *Enhydris caerulea* Latreille *in* Sonnini & Latreille, 1801b, *Hydrus atrocaeruleus* G. Shaw, 1802, *Coluber pythonissa* Daudin, 1803d, *Homalopsis aer* Oppel *in* H. Boie *in* Schlegel, 1826a (*nomen nudum*), *Homalopsis aer* Oppel *in* H. Boie, 1826 (*nomen nudum*), *Potamophis lusingtonii* Cantor, 1836b, *Homalopsis olivaceus* Cantor, 1839b, *Hypsirhina bilineata* Gray, 1842d, *Hypsirhina furcata* Gray, 1842d, *Hypsirhina trilineata* Gray, 1842d, *Potamophis lushingtonii* Cantor, 1847 *(nomen nudum), Hypsirhina enhydris albolineata* Jan, 1863b (*nomen incorrectum*), *Hypsirhina enhydris albolineata* Jan, 1865c, *Hypsirhina albolineata* Morice, 1875a *(nomen nudum),* and *Helicops indicus* Annandale, 1905b.

Type: Lectotype, specimen described and illustrated by P. Russell (1796: 35, pl. 30), a 508 mm female (P. Russell, 1781–1791), designated herein.

Type locality: "lake of Ankapilly, coast of Coromandel, India" [= Ankapalle, Andhra Pradesh, E India, 17°40'N, 83°09'E, elevation 20 m] via lectotype selection.

Distribution: Southeastern Asia and Indonesia. India (N Andhra Pradesh, Assam, Bihar, Madhya Pradesh, Meghalaya, Nagaland, Orissa, Uttar Pradesh, West Bengal), E Nepal (Bardiya, Dhanusa, Kailali, Kanchanpur, Sunsari), Bangladesh, Myanmar (Ayeyarwady, Bago, Magwe, Mandalay, Sagaing, Tanintharyi, Yangon), Thailand (Bangkok, Chon Buri, Khon Kaen, Krung Thep Mahanakhon, Lop Buri, Narathiwat, Pattani, Phathalung), Cambodia (Kandal, Kompon Chan, Kampong Chnang, Siem Reap), Laos, Vietnam (Ba Ria-Vung Tau, Ca Mau, Can Tho, Dong Nai, Ho Chi Minh City, Khanh Hoa, Kien Giang, Long An, Phu Yen, Soc Trang, Tay Ninh, Tien Giang), ext. SE China (Guangdong), West Malaysia (Kedah, Pinang, Tioman Is.), Singapore, East Malaysia (Sarawak), Brunei and Indonesia (Papua, Bangka, Java, Kalimantan, Sumatra, Weh), NSL–250 m.

Sources: Wall, 1912d, C.H. Pope, 1935, Bergman, 1955a, Kuntz, 1963, E.H. Taylor, 1965, Campden-Main, 1970a, Saint-Girons, 1972, R. Fleming & Fleming, 1973, G. Sharma & Nakhasi, 1980, Tweedie, 1983, M.J. Cox, 1991b, Stuebing, 1991, David & Vogel, 1996, M.J. Cox et al., 1998, Leong & Tan, 2000, Karns 0et al., 2001, Schleich & Kästle, 2002, Ao et al., 2004, Tiwari & Shah, 2004, Whitaker & Captain, 2004, J.C. Murphy & Voris, 2005, I. Das, 2007b, M.F. Ahmed et al., 2009, V.S. Nguyen et al., 2009 and Grismer, 2011.

Remarks: Unconfirmed records from Hong Kong and Sri Lanka (Boulenger, 1896a: 7). Sulawesi records doubtful, probably *Enhydris matannensis* or *E. plumbea, fide* Lang & Vogel (2005: 251). Cox et al. (2013: 482) listed two syntypes as MHNL 42000343.

7. *Enhydris gyii* J.C. Murphy, Voris & Auliya, 2005. Raffles Bull. Zool. 53(2): 272–273, figs. 1a, 2, 3a–c.

Type: Holotype, ZFMK 65824, a 766 mm female (M. Auliya, 27 Oct. 1996).

Type locality: "Indonesia, Kalimantan Barat, Sungai Kapuas at Putussibau (0°52'N, 112°55'E), where the Sibau River empties into the Kapuas River, Kapuas Hulu Regency, West Kalimantan."

Distribution: Central Indonesia (W Kalimantan), 50 m.

Source: I. Das, 2012.

8. *Enhydris indica* (Gray, 1842d). Zool. Misc. 2(May): 67. (*Raclitia indica*)

Types: Syntypes (2), BMNH 1946.1.2.36, a 324 mm female, and BMNH 1946.1.2.46, a 345 mm male (T. Hardwicke, 1756–1823).

Type locality: "India" (in error). Corrected to Malay Peninsula *fide* Boulenger (1896a: 4).

Distribution: West Malaysia (Perak, Selanger).

Sources: Lim & Kamarudin, 1975 and Tweedie, 1983.

Remarks: BMNH catalogue lists Malay Peninsula as type locality. Probably occurs in Singapore *fide* I.Das (2010: 326).

9. *Enhydris innominata* (Morice, 1875a). Faune Cochinchine Franç.: 58. (*Hypsirhina innominata*)

Type: Holotype, MHNL 42000338, a female (A. Morice, 1870–1875).

Type locality: "Tay-ninh, Cochinchina française" [= Tây Ninh, Tây Ninh Prov., S Vietnam, 11°22'N, 106°07'E, elevation 15 m].

Distribution: Cambodia (Battambang, Kandal, Kampong Cham, Kampong Thom, Pursat, Siem Reap) and S Vietnam (Kien Giang, Tay Ninh).

Sources: M.A. Smith, 1929, E.H. Taylor, 1965, Saint-Girons, 1971a, 1972a–b, M.J. Cox, 1991b, J.C. Murphy & Voris, 2005 and V.S. Nguyen et al., 2009.

10. *Enhydris jagorii* (W.C.H. Peters, 1863b). Mber. Königl. Akad. Wiss. Berlin 1863(6): 245–247. (*Hypsirhina [Eurostus] jagorii*)

Type: Holotype, ZMB 4746, a 470 mm male (F. Jagor, 1861–1863).

Type locality: "Siam" [= Thailand]. Restricted to the vicinity of Bangkok, Thailand *fide* J.C. Murphy (2007b: 131).

Distribution: Southeast Asia. Eastern Thailand (Chanthaburi, Khon Kaen, Nakhon Ratchasima, Nakhon Si Thammarat, Sisaket, Trat, Udon Thani), Cambodia (Kampong Speu), S Laos (Borikhan, Champhon, Khammouan, Savannakhet, Vapikhamthong, Xedon) and Vietnam (Bình Tri Thiên, Hâu Giang).

Sources: E.H. Taylor, 1965, Campden-Main, 1970a, Deuve, 1970, Saint-Girons, 1972a, M.J. Cox, 1991b, M.J. Cox et al., 1998 and J.C. Murphy & Voris, 2005.

11. *Enhydris longicauda* (Bourret, 1934d). Bull. Gén. Instr. Pub. 13(1): 20–22, fig. 4. (*Hypsirhina longicauda*)

Types: Syntypes (3), MNHN 1948.95–96 and MNHN 1938.143 (formerly RLB K.1 [Tonlésap], a 530 mm specimen, RLB M.388 [Siem Réap], a 194 mm juvenile, & RLB M.425 [Tonlé Chma], a 226 mm juvenile) (Chevey, Inst. Océanographique Indochine).

Type locality: "Tonlésap, (Grand Lacs du Cambodge)" [= Tonle Sap Lake, E Bat Dambang, SW Siem Reap, NE Pursat and W Kampong Thom Prov., Cambodia, bet. 12°30'–13°15'N, 103°42'–104°30'E, elevation 5 m], "Siem Réap" [= S Siem Reap Prov., Cambodia, 13°20'N, 103°50'E, elevation 10 m], and "Tonlé Chma" [= Tonle Cham, E Kamphon Cham Prov., Cambodia].

Distribution: Western Cambodia (E Battambang, E Kampong Cham, W Kampong Thom, NE Pursat, S Siem Reap), NSL–40 m.

Sources: M.A. Smith, 1943 and Saint-Girons, 1971.

E

12. *Enhydris maculosa* (Blanford, 1881). Proc. Zool. Soc. London 49(1): 226. (*Hypsirhina maculosa*) (*nomen substitutum*)

Synonyms: *Hypsirhina maculata* Blanford, 1879a–b (*nomen praeoccupatum*), and *Hypsirhina blanfordii* Boulenger, 1890a (*nomen substitutum*).

Type: Neotype, BMNH 1913.6.12.1, a 308 mm female, designated by J.C. Murphy (2007a: 282).

Type locality: Unknown via neotype selection. Designated as Bassein in Pegu *fide* Blanford (1879a: 217) [= Pathein, Ayeyarwady Div., SW Myanmar, 16°46'N, 94°44'E, elevation 20 m].

Distribution: Southern Myanmar (Ayeyarwady, Bago), NSL–20 m.

Source: I. Das, 1999a.

Remarks: Original description based on Blanford (1879a: 216–217) with supplemental original description in Blanford (1879b: 130–131). Holotype, ZSI 8207 (formerly IMC 8207), a 300 mm female (Baker), lost *fide* Sclater (1891: 245) and I. Das et al. (1998: 157). Record from Nias, Indonesia unconfirmed *fide* Iskandar & Colijn (2001: 91).

13. *Enhydris matannensis* (Boulenger, 1897g). Proc. Zool. Soc. London 65(1): 225–226, pl. 15, fig. 1. (*Hypsirhina matannensis*)

Type: Holotype, NMBA 1735, a 240 mm male (P. Sarasin & F. Sarasin, 1893–1896).

Type locality: "Lake Matanna, South-eastern Celebes" [= Matano Lake, Central Sulawesi, bet. 2°25–34'S, 121°13–29'E, elevation 400 m].

Distribution: Central Indonesia. Sulawesi (Buton, Muna), 400–415 m.

Sources: Iskandar, 1979 and Koch, 2012.

14. *Enhydris pahangensis* Tweedie, 1946. Ann. Mag. Nat. Hist. (11) 13(98): 142–144, 1 fig.

Type: Holotype, BMNH 1947.1.1.70, a 220 mm male (M.W.F. Tweedie, April 1940).

Type locality: "Kuala Tahan, River Tembeling, Pahang, between 500 and 1,000 feet altitude, and over 70 miles from the east coast of the Malay Peninsula." Emended to King George V National Park (*fide* Gyi, 1970: 117–118) or the Taman Negara Game Reserve (*fide* Tweedie, 1983: 102) [= N Pahang State, West Malaysia 4°17'N, 102°25'E, elevation 150–300 m].

Distribution: West Malaysia (N Pahang, Terengganu), 150–300 m.

Sources: Tweedie, 1983, Onn, 2009 and I. Das, 2012.

15. *Enhydris pakistanica* Mertens, 1959b. Senck. Biol. 40(3–4): 117–119, pl. 10, figs. 1a–b.

Type: Holotype, SMF 56340, a 708 mm male (M.G. Konieczny, 5 June 1958).

Type locality: "Jati, unteres Indus-Gebiet, Sindh, West-Pakistan" [= Jati, Induis River delta, Sindh Prov., Pakistan, 24°21'N, 68°16'E, elevation 10 m].

Distribution: Southern Pakistan (S Sindh), NSL–10 m.

Sources: Minton, 1966 and M.S. Khan, 2002.

16. *Enhydris plumbea* (H. Boie *in* F. Boie, 1827). Isis von Oken 20(6): 550. (*Homalopsis plumbea*)

Synonyms: *Homalopsis plumbea* H. Boie *in* Schlegel, 1826a (*nomen nudum*), *Homalopsis plumbea* Schlegel, 1826b (*nomen nudum*), *Hypsirhina hardwickii* Gray, 1834 *in* Gray & Hardwicke, 1830–1835, and *Helicops wettsteini* Amaral, 1929b.

Type: Lectotype, RMNH 1163 (H. Kuhl & J.C. van Hasselt, Dec. 1820–Sept. 1823), designated by Gyi (1970: 82).

Type locality: "Java, Indonesia" via lectotype selection. Restricted to region of Mt. Pangerango and Mt. Salak, West Java, Indonesia *fide* Brongersma (1948b: 14). Emended to Tjihandjawar, at the foot of Mt. Pangerango, W. Java [= Cihanyawar, Nagrak, Jawa Barat, W Java, SW Indonesia, 07°03'S, 108°40'E] *fide* Brongersma (1950: 1499).

Distribution: Southeastern Asia and Indonesia. Eastern India (Andaman & Nicobars: Andaman and Great Nicobar Is.), S China (Fujian, Guangdong, Guangxi, Hainan, Hong Kong, Jiangsu, Jiangxi, Yunnan, Zhejiang), Taiwan, Myanmar (Mandalay), Thailand (Bangkok, Chanthaburi, Chiang Mai, Chon Buri, Kanchanaburi, Khon Kaen, Lampang, Nakhon Ratchasima, Nakhon Sawan, Pattani, Phang Nga, Phetchaburi, Phrae, Prachuap Khiri Khan, Sakon Nakon, Saraburi, Trang), Laos (Attapu, Bolikhamsai, Champasak, Houaphan, Khammouan, Louangphrabang, Phongsali, Savannakhet, Vientiane, Xaignabouli, Xekong, Xiangkhouang), Vietnam (An Giang, Bac Giang, Bac Kan, Bac Lieu, Bac Ninh, Ba Ria-Vung Tau, Ben Tre, Binh Dinh, Binh Duong, Binh Phuoc, Binh Thuan, Ca Mau, Can Tho, Cao Bang, Dak Lak, Dak Nong, Da Nang, Dien Bien, Dong Nai, Dong Thap, Gia Lai, Ha Giang, Hai Duong, Hai Phong, Ha Nam, Hanoi, Ha Tay, Ha Tinh, Hau Giang, Hoa Binh, Ho Chi Minh City, Hung Yen, Khanh Hoa, Kien Giang, Kon Tum, Lai Chau, Lam Dong, Lang Son, Lao Cai, Long An, Nam Dinh, Nghe An, Ninh Binh, Ninh Thuan, Phu Tho, Phu Yen, Quang Binh, Quang Nam, Quang Ngai, Quang Ninh, Quang Tri, Soc Trang, Son La, Tai Ninh, Thai Binh, Thai Nguyen, Thanh Hoa, Thua Thien-Hue, Tien Giang, Tra Vinh, Tuyen Quang, Vinh Phuc, Vinh Long, Yen Bai), Cambodia (Kampong Chan, Kampong Cham, Kampot, Pursat, Siem Reap), West Malaysia (Johore, Kedah, Kelantan, Perlis, Pinang, Terengganu, Tioman Is.), East Malaysia (Sabah), Brunei and Indonesia (Bali, Bangka, Belitung, Java, Kalimantan, Sulawesi, Sumatra), NSL–1500 m.

Sources: Stejneger, 1907a, Maki, 1931, C.H. Pope, 1935, Bergman, 1960, Kuntz, 1963, E.H. Taylor, 1965, Rossman & Scott, 1968, Deuve, 1970, Saint-Girons, 1972a, Romer, 1979a, Tweedie, 1983, Karsen et al., 1986, M.J.Cox, 1991b, Zhao & Adler, 1993, David & Vogel, 1996, M.J. Cox et al., 1998, Chan-ard et al., 1999, Farkas & Fritz, 1999b, Orlov et al., 2000, Ghodke & Andrews, 2002, Whitaker & Captain, 2004, I. Das, 2007b, Grismer et al., 2008, V.S. Nguyen et al., 2009, Xiang & Li, 2009, Grismer, 2011 and Koch, 2012.

Remarks: Original description based on H. Boie's MS (1823–1825). Page column in original description erroneously numbered "560."

17. *Enhydris punctata* (Gray, 1849a). Cat. Snakes Brit. Mus.: 68. (*Phytolopsis punctata*)

Synonyms: *Eurostus heteraspis* Bleeker, 1857a *(nomen nudum), Tachyplotus hedemanni* J.T. Reinhardt, 1866, *Pythonopsis borneensis* W.C.H. Peters, 1871b, and *Hypsirhina hageni* Lidth de Jeude, 1890.

Type: Holotype, BMNH 1946.1.2.37, a 370 mm female (Warwick).

Type locality: "India," (in error *fide* J.C. Murphy (2007b: 160).

Distribution: Malaysia and Greater Sundas. West Malaysia (Johore, Selangor, Terengganu), East Malaysia (Sarawak), Brunei and Indonesia (Bangka, Belitung, Java, W Kalimantan, Sumatra), NSL.

Sources: Tweedie, 1983, Mahendra, 1984, David & Vogel, 1996, J.C. Murphy et al., 2005 and I. Das, 2007b.

Remarks: Probably a collection of several species that are isolated in peat swamp forests and associated streams *fide* J.C. Murphy (2007b: 160). Indian record discounted *fide* J.C. Murphy (2007b: 161).

18. *Enhydris smithii* (Boulenger, 1914e). J. Nat. Hist. Soc. Siam 1(2): 69–70, 3 figs. (*Hypsirhina smithii*)

Type: Holotype, BMNH 1948.1.2.27, a 400 mm female (native boy via Miller and M.A. Smith).

Type locality: "Messrs. L.T. Leonowens, Ltd., river Menam at Bangkok" [= Louis T. Leonowens Company on Menam River at Bangkok, Bangkok Prov., cen. Thailand, 13°44'N, 100°31'E, elevation NSL].

Distribution: Central Thailand (Bangkok, Krung Thep Mahanakhon, Phra Nakhon Si Ayutthaya, Prachuap Khiri Khan), NSL.

Sources: E.H. Taylor, 1965, M.J. Cox, 1991b, M.J. Cox et al., 1998, J.C. Murphy & Voris, 2005 and I. Das, 2010, 2012.

Remarks: A valid species *fide* M.J. Cox (1991b: 191) and I Das (2010: 328).

19. *Enhydris subtaeniata* (Bourret, 1934d). Bull. Gén. Instr. Pub. 14(1): 19–20, fig. 3. (*Hypsirhina enhydris subtaeniata*)

Type: Lectotype, MNHN 1958.0474 (formerly RLB M.388), a 202–211 mm specimen, designated by I. Ineich *in* J.C. Murphy (2007b: 166).

Type locality: "Kampong Speu, Cambodia" [= Kampong Speu, Kampong Speu Prov., S Cambodia, 11°28'N, 104°30'E, elevation 40 m] via lectotype selection.

Distribution: Indochina. Eastern Thailand (Khon Kaen), Laos, Cambodia (Kampong Speu) and Vietnam (Ca Mau, Ho Chi Minh City, Kien Giang, Kon Tum, Soc Trang, Tay Ninh), NSL–40 m.

Sources: E.H. Taylor, 1965, Deuve, 1970a, V.S. Nguyen et al., 2009 and Voris et al., 2012.

20. *Enhydris vorisi* J.C. Murphy, 2007a. Hamadryad 31(2): 283–285, figs. 1a, 1c. 1e.

Type: Holotype, KU 92465, a 586 mm female (K.K. Gyi).

Type locality: "Maubin, Ayeyarwady Division, Myanmar (about 16.73°N 95.65°E)."

Distribution: Myanmar (Ayeyarwady), NSL. Known only from type locality.

E

ENULIOPHIS McCranie & Villa, 1993 (Dipsadidae)

Type species: *Leptocalamus sclateri* Boulenger, 1894a.

Distribution: Central America and N South America.

Sources: Netting, 1936b, E.H. Taylor, 1954, H. Greene, 1977, Pérez-Santos & Moreno, 1988, Lips, 1993a, McCranie & Villa, 1993, Doan & Arizábal, 2002, Savage, 2002, G. Köhler, 2003, McCranie et al., 2003, G. Köhler et al., 2004, McCranie, 2004b, Solórzano, 2004, Zaher et al., 2009, McCranie, 2011a and Travers et al., 2011.

Remarks: A synonym of *Enulius fide* Savage, 2002.

1. *Enuliophis sclateri* (Boulenger, 1894a). Cat. Snakes Brit. Mus. 2: 251, pl. 12, fig. 1. (*Leptocalamus sclateri*)

Type: Holotype, BMNH 1946.1.21.72, a 380 mm female (P.L. Sclater).

Type locality: "S. America" [= South America] (in error).

Distribution: Lower Central America and NW South America. Northeastern Honduras (Gracias a Dios), Nicaragua (Jinotega, Matagalpa), Costa Rica (Alajuela, Cartago, Heredia, Limón, Puntarenas, San José), Panama (Canal Zone, Herrera, Barro Colorado Is.), W Colombia (Antioquia, Santander, Tolima) and Peru (Madre de Dios), NSL–1285 m.

ENULIUS Cope, 1871a (Dipsadidae)

Synonyms: *Leptocalamus* A.C.L.G. Günther, 1872b, and *Enulis* – L.D. Wilson & Meyer, 1985 *(nomen incorrectum).*
Type species: *Liophis flavitorques* Cope, 1869b.
Distribution: Latin America.
Sources: Boulenger, 1894a, Dunn, 1938a, E.H. Taylor, 1951, H.M. Smith, Arndt & Sherbrook, 1967, J.A. Peters & Orejas-Miranda, 1970, McCranie & Villa, 1993, McCranie & G. Köhler, 1999 and Zaher et al., 2009.

E

1. *Enulius bifoveatus* McCranie & G. Köhler, 1999. Carib. J. Sci. 35(1–2): 15–17, figs. 1–3b.

Type: Holotype, SMF 78514, a 321 mm male (G. Köhler & E. G. Köhler, 30 Aug. 1997).
Type locality: "between Savannah Bight and East End (16°29.19'N, 85°50.30'W), less than 10 m elevation, Isla de Guanaja, Islas de la Bahia, Honduras."
Distribution: Honduras (Islas de la Bahía: N Guanaja), NSL.
Sources: McCranie et al., 2005 and McCranie, 2011a.

2. *Enulius flavitorques* (Cope, 1869b). Proc. Acad. Nat. Sci. Philadelphia (1868) 20: 307–308. (*Liophis flavitorques*)

Synonyms: *Enulius murinus* Cope, 1871a, *Leptocalamus torquatus* A.C.L.G. Günther, 1872b, *Geophis unicolor* J.G. Fischer, 1880, *Enulius sumichrasti* Bocourt, 1883 *in* A.H.A. Duméril, Bocourt & Mocquard, 1870–1909, and *Geagras longicaudatus* Cope, 1885a.
Type: Holotype, ANSP 3695, a 445 mm specimen (S. Buckow).
Type locality: "the Magdalena river, New Grenada" [= Colombia].
Distribution: Mesoamerica. Southern Mexico (Chiapas, Colima, Guerrero, S Jalisco, México, Michoacán, Morelos, Oaxaca, Puebla), S Guatemala (Jutiapa), El Salvador (Cuscatlán, La Libertad, San Miguel, San Salvador, Santa Ana), Honduras (Choluteca, Colón, Comayagua, Cortés, Francisco Morazán, Gracias a Dios, Intibucá, Islas de la Bahía, Yoro, Utila Is.), W Nicaragua (Chinandega, Granada, Managua, Matagalpa), W Costa Rica (Guanacaste, Puntarenas, San José), Panama (Chiriquí, Canal Zone, Darién, Los Santos, Barro Colorado, Flamenco and Pedro Gonzales Is.) ext. N Colombia (Atlántico, César, La Guajira, Sucre) and ext. NW Venezuela (Zulia), NSL–1800 m.
Sources: Booth, 1959, L.D. Wilson & Meyer, 1985, Pérez-Santos & Moreno, 1988, G. Köhler, 1999b, Savage, 2002, Gutsche, 2003, Solórzano, 2004, Canseco-Márquez et al., 2005, McCranie et al., 2005, Navarrete et al., 2009, McCranie, 2011a and Orellana, 2011.

3. *Enulius oligostichus* H.M. Smith, Arndt & Sherbrooke, 1967. Nat. Hist. Misc. (186): 1–2.

Type: Holotype, UIMNH 62740, a 275 mm male (R.G. Arndt & W.C. Sherbrooke, 8 July 1963).
Type locality: "on Mexico Highway 54 between San Blas, Nayarit, and Mexico Highway 15 (5-15 miles NE San Blas), Mexico."
Distribution: Western Mexico (Jalisco, Nayarit, S Sinaloa), 1190 m.
Sources: McDiarmid & Bezy, 1971 and Ponce-Campos, 2003.

4. *Enulius roatanensis* McCranie & G. Köhler, 1999. Carib. J. Sci. 35(1–2): 17–21, figs. 5–6a–b.

Type: Holotype, SMF 78515, a 346 mm female (G. Köhler & E. G. Köhler, 2 May 1998).
Type locality: "near Mud Hole Bay (16°20.88'N, 86°32.05W), less than 10 m elevation, Isla de Roatán, Islas de la Bahía, Honduras."
Distribution: Honduras (Islas de la Bahía: Roatán), NSL.
Sources: McCranie et al., 2005 and McCranie, 2011a.

†*EOANILIUS* Rage, 1974 (Aniliidae)

Synonym: †*Eonilius* – Rage, 1974 *(nomen incorrectum).*
Type species: †*Eoanilius europae* Rage, 1974.
Distribution: Upper Eocene of France and United Kingdom, and upper Oligocene of Germany.
Sources: Milner et al., 1982, Rage, 1984b, Cundall et al., 1993, Szyndlar & Schleich, 1993, Duffaud & Rage, 1997 and Gómez et al., 2008

1. †*Eoanilius europae* Rage, 1974. Palaeovertebrata 6(3–4): 278–281, figs. 1a–b, 2a–b.

Type: Holotype, USTL MAL 550, one middle trunk vertebra.
Type locality: "Malpérié; Eocène supérieur, Phosphorites du Quercy" [= Tarn-et Garonne Dept., France; Headonian *fide* Rage, 1984b: 13].
Distribution: Upper Eocene (Priabonian, MP 17–20: 33.9–37.2 mya) of France and United Kingdom.

2. †*Eoanilius oligocenicus* Szyndlar, 1994. J. Vert. Paleont. 14(1): 24–26, figs. 1f–j.

Type: Holotype, SMNS 58196/2, one middle trunk vertebra.
Type locality: "Herrlingen 8 (upper Oligocene, middle Arvernian, MP 28), near Ulm in Baden-Württemberg (southern Germany)."
Distribution: Upper Oligocene (Chattian, MP 28: 23.0–28.4 mya) of Germany.

EPACROPHIS Hedges, Adalsteinsson & Branch *in* Adalsteinsson, Branch, Trape, Vitt & Hedges, 2009 (Leptotyphlopidae)

Type species: *Glauconia boulengeri* Boettger *in* Voeltzkow, 1913.
Distribution: East Africa.
Sources: Hahn, 1980a, Wallach, 1998b and McDiarmid et al., 1999.

1. *Epacrophis boulengeri* (Boettger *in* Voeltzkow, 1913). Reise Ostafrika: 354, pl. 25, fig. 1. (*Glauconia boulengeri*)

Type: Holotype, SMF 16700 (formerly SMF-B 7066, a), a 110 mm specimen (A. Voeltzkow, 10–14 Feb. 1903).
Type locality: "Insel Manda, Br. Ostafrika" [= Manda Island, Coast Region, E Kenya, 2°16'S, 40°57'E, elevation < 10 m].
Distribution: Eastern Kenya (N Coast: Manda & Lamu Is.), NSL–10 m.
Sources: Boulenger, 1893a, 1915c, Loveridge, 1936c, 1957a, Hallermann & Rödel, 1995, Broadley & Wallach, 1996, Wallach, 1996c, Spawls et al., 2002 and Broadley & Wallach, 2007a.
Comment: Mertens (1922: 178 & 1967: 89) lists Voeltzkow collection date as 1905.

2. *Epacrophis drewesi* (Wallach, 1996c). J. Afr. Zool. 110(6): 426–428, fig. 1. (*Leptotyphlops drewesi*)

Type: Holotype, CAS 85756, a 143 mm female (R.E. Leach & E.S. Ross, 26 Nov. 1957).
Type locality: "Base of north slope of Mt. Kenya, 10 km south of and uphill from Isiolo (0°21'N, 37°35'E, 1250 m), south-central Eastern Province, central Kenya (about 1400 m)."
Distribution: Central Kenya (cen. Eastern), 1400 m. Known only from type locality.
Sources: Spawls et al., 2002, 2006 and Broadley & Wallach, 2007a.

3. *Epacrophis reticulatus* (Boulenger, 1906d). Ann. Mag. Nat. Hist. (7) 18(108): 441. (*Glauconia reticulata*)

Type: Holotype, BMNH 1946.1.16.53 (formerly BMNH 1905.11.7.42), a 190 mm specimen (G.W. Bury, 1905).
Type locality: "Wagga, Goolis Mountains, near Berbera, altitude 3000–4000 feet, Somaliland" [= Wagar, W Togdheer Region, NW Somalia, 10°02'N, 45°35'E, elevation 900–1200 m].
Distribution: Northwestern Somalia (Togdheer, Woqooyi Galbeed), 900–1315 m.
Sources: Boulenger, 1915d, Parker, 1949, Hahn, 1978, Lanza, 1983a, 1990b, Largen & Rasmussen, 1993, Hallermann & Rödel, 1995, Wallach, 1996 and Broadley & Wallach, 2007a.
Comment: Possibly occurs in Ethiopia *fide* Largen & Rasmussen (1993: 324).

EPHALOPHIS M.A. Smith, 1931a (Elapidae)

Type species: *Ephalophis greyae* M.A. Smith, 1931a.
Distribution: Marine waters of W Australia.
Sources: McDowell, 1969b, 1974b, Burger & Natsuno, 1974, L. Smith, 1974, Cogger et al., 1983a, McCarthy, 1986, Gow, 1989, Golay et al., 1993, O'Shea, 1996, A.R. Rasmussen, 1997, David & Ineich, 1999, Cogger, 2000, Zaher et al., 2009, Bush & Maryan, 2011 and Hoser, 2012e.

1. *Ephalophis greyae* M.A. Smith, 1931a. Proc. Zool Soc. London 101(2): 397–398, pl. 1. (*nomen corrigendum*)

Synonyms: *Ephalophis greyi* M.A. Smith, 1931a *(nomen incorrigendum),* and *Ephalophis greyae* – Shea, 1996 *(nomen corrigendum).*
Type: Holotype, BMNH 1946.1.6.92 (formerly BMNH 1930.11.17.3), a 410 mm female (B. Grey, March 1930).
Type locality: "Cape Boileau, north-west coast of Australia" [= N of Broome, N Western Australia].
Distribution: Western Australia (NW Western Australia).

EPICRATES Wagler, 1830 (Boidae)

Synonyms: *Chilabothrus* A.-M.-C. Duméril & Bibron, 1844, *Chilobothrus* A.-M.-C. Duméril & Bibron, 1844 *(nomen emendatum), Chilobotrus* A.-M.-C. Duméril & Bibron, 1844 *(nomen incorrectum), Cliftia* Gray, 1849a, *Epicarsius* J.G. Fischer, 1856b, *Homalochilus* J.G. Fischer, 1856b, *Chilabotrus* Marschall, 1873 *(nomen emendatum), Piesigaster* Seoane, 1881, *Epicrafes* – Kempff-Mercado, 1975 *(nomen incorrectum), Epicratis* – Willard, 1977 *(nomen incorrectum),* and *Boella* H.M. Smith & Chiszar, 1992b.
Type species: *Boa cenchria* Linnaeus, 1758.
Distribution: Lower Central America, West Indies and South America.
Fossil records: Pleistocene of the West Indies (Bahamas, Cuba, Puerto Rico).
Sources: Amaral, 1955f, Stimson, 1969, J.A. Peters & Orcjas-Miranda, 1970, Matz & Matz, 1973, Sheplan & Schwartz, 1974, Tolson, 1987, Kluge, 1988a–b, 1989, Tolson & R.W. Henderson, 1993, Henderson et al., 1995, Rodriguez-Robles & Greene, 1996, Walls, 1998a, McDiarmid et al., 1999, Noonan & Chippindale, 2006, Passos & Fernandes, 2008, R.W. Henderson & Powell, 2009 and Rivera et al., 2011.

E

Remarks: Rivera et al., 2011 supported conclusions of Passos & Fernandes, 2008.

1. *Epicrates alvarezi* Abalos, Báez & Nader, 1964. Acta Zool. Lilloana 20: 218–223, fig. 3. (*Epicrates cenchria alvarezi*)

Type: Holotype, IAV, an 1100 mm female (M. Mendieta, 1 July 1959).

Type locality: "Forres, Robles Department, Santiago del Estero, Argentina."

Distribution: Southeastern Bolivia (Santa Cruz, Tarija), W Paraguay (Alto Paraguay, Boquerón) and N Argentina (Chaco, Córdoba, Corrientes, Jujui, Santiago del Estero, San Miguel, Tucumán), 50–715 m.

Sources: Cacciali, 2008 and Bellosa & Bisplinghof, 2012.

E

2. *Epicrates angulifer* Bibron, 1840 *in* Ramón de la Sagra, 1838–1843. Hist. Fís. Polít. Nat. Cuba 2(2): 129–130, pl. 25, figs. 1–3.

Synonym: *Boa brachyura* Filippi, 1848.

Type: Lectotype, MNHN 3292, probably a 1990 mm specimen, designated by Stimson (1969: 11).

Type locality: "Cuba" via lectotype selection.

Distribution: West Indies. Cuba (Camaguey, Ciego de Avila, Cienfuegos, Guantánamo: Juventud Is.; Arch. de los Canarreos: Cayo Cantiles; Arch. Sabana-Camagüey: Cayo Guajaba, Cayo Santa María; Habana, Pinar del Río, Villa Clara, Sancti Spíritus, Santiago de Cuba; Matanzas), NSL–325 m.

Sources: F. Werner, 1901b, Lando & Williams, 1969, A. Schwartz & R.W. Henderson, 1991 and R.W. Henderson & Arias B., 2001.

Remarks: Original description reprinted in Sagra (1839–1843: 215–217, pl. 25). Authorship has been attributed to Cocteau & Bibron, but H.M. Smith & Grant (1958: 216) demonstrated that the snake section was written solely by Bibron (as per footnote, p. 120). Spanish edition listed as vol. 8 *fide* A. Schwartz & Thomas (1975: 183) et al.

3. *Epicrates assisi* Machado, 1944. Bol. Inst. Vital Brazil 4(27): 61, pl. 1 (middle fig.). (*Epicrates cenchria assisi*)

Synonym: *Epicrates cenchria xerophilus* Amaral, 1955f.

Type: Holotype, IVB, lost *fide* Passos & Fernandes (2009: 19).

Type locality: "Campina Grande, Paraíba, Brasil" [= Campina Grande, E Paríba State, NE Brazil, 7°14'S, 35°53'W, elevation 465 m].

Distribution: Northeastern Brazil (Bahia, Ceará, Paraíba, Pernambuco, Piauí, Rio Grande do Norte, Sergipe), NSL–465 m.

Source: Bellosa & Bisplinghof, 2012.

Remarks: Original description date (1945) and volume (5) incorrectly cited *fide* Passos & Fernandes (2009: 19).

4. *Epicrates cenchria* (Linnaeus, 1758). Syst. Nat., ed. 10, 1: 215. (*Boa cenchria*)

Synonyms: *Boa flavescens* Boddaert, 1783, *Coluber tamachia* Scopoli, 1788, *Boa cenchris* – Gmelin, 1789, *Coluber dubius* Gmelin, 1789, *Coluber bitin* Bonnaterre, 1790, *Boa aboma* Daudin, 1803b, *Boa annulifer* Daudin, 1803b, *Boa ternatea* Daudin, 1803b, *Python liberiensis* Hallowell, 1845, *Epicrates cenchria gaigei* Stull, 1938, and *Epicrates cenchria hygrophilus* Amaral, 1955f.

Type: Holotype, NHR Lin-6 (formerly MAFR), a 475 mm specimen (Mus. Drottn.).

Type locality: "Surinamei" [= Surinam].

Distribution: Amazonia. Colombia (Amazonas, Boyacá, Casanare, Coqueta, Huila, Meta, Putumayo, Vaupé), E Venezuela (Amazonas, Bolívar, Monagas, Táchira), Guyana (Cuyuni-Mazaruni, Potaro-Siparuni, Upper Demerara-Berbice), Suriname (Saramacca, S Sipaliwini), French Guiana (Cayenne, Saint-Laurent-du-Maroni), Ecuador (Morona, Napo, Pastaza), Brazil (Amazonas, Amapá, Bahia, Ceará, Espírito Santo, Goiás, Mato Grosso, Mato Grosso do Sul, Minas Gerais, Pará, Paraiba, Paraná, Pernambuco, Piauí, Rio Grande do Norte, Rondônia, São Paulo, Sergipe), E Peru (Amazonas, Apurímac, Cusco, Huánuco, Junín, Loreto, Madre de Dios, San Martin) and E Bolivia (Beni, Cochabamba, La Paz, Pando, Santa Cruz, Tarija), NSL–2750 m.

Sources: Stull, 1938, Amaral, 1944f, 1955f, Carrillo de Espinoza, 1966, Roze, 1966a, Cunha & Nascimento, 1978, Vanzolini et al., 1980, Chippaux, 1987, Pérez-Santos & Moreno, 1988, Abuys, 1989, Duellman & Salas, 1991, Fugler & Riva, 1990, J.D. Williams & Francini, 1991, Cei, 1994, Marca & Soriano, 1995, Starace, 1998, Waller & Buongermini-P., 1998, Leynaud & Bucher, 1999, H.E.A. Boos, 2001, Markezich, 2002, Passos & Franco, 2002, Savage, 2002, Abuys, 2003, Solórzano, 2004, Duellman, 2005, Pizzatto et al., 2007, Barrio-Amorgós & Diaz de Pascual, 2009, Navarrete et al., 2009, Passos & Fernandes, 2009, Bellosa & Bisplinghof, 2012 and C.J. Cole et al., 2013.

5. *Epicrates chrysogaster* (Cope, 1871a). Proc. Amer. Philos. Soc. (1870) 11: 557–558. (*Homalochilus chrysogaster*)

Synonyms: *Dendrophilus tortilis* J.T. Reinhardt & Lütken, 1862 *(nomen ineditum)*, *Epicrates relicquus* T. Barbour & Shreve, 1935, and *Epicrates chrysogaster schwartzi* Buden, 1975.

Type: Holotype, ANSP 10322, a 665 mm specimen (probably female) (A.J. Ebell).

Type locality: "Turk's Island" [Grand Turk, Turks & Caicos Is., West Indies].

Distribution: Southern Bahamas (Acklins, Crooked, Great Inagua, Sheep Cay) and Turks and Caicos (Big Ambergris Cay, Caicos, Grand Turk, Long Cay, Middle Caicos, North Caicos), NSL–10 m.

Sources: Buden, 1975, R.W. Henderson & Powell, 2009 and Reynolds, 2012.

6. *Epicrates crassus* Cope, 1862d. Proc. Acad. Nat. Sci. Philadelphia 14(5): 349.

Synonym: *Epicrates cenchria polylepis* Amaral, 1935e.

Type: Holotype, USNM 12413 (formerly SIM 5409), a 920 mm specimen (T.J. Page [Paraguay Exped.], Sept. 1853–Jan. 1856).

Type locality: "Cadosa, Parana River, Paraguay." Corrected to Gardosa, Paraná River, Paraguay *fide* J.D. Williams & Francini (1991: 60).

Distribution: Southern South America. Southeastern Brazil (Goiás, Mato Grosso, Mato Grosso do Sul, Minas Gerais, Pará, Paraná, Rondônia, São Paulo, Tocantins), Bolivia (Santa Cruz), Paraguay (Alto Paraná, Amambay, Boquerón) and ext. NE Argentina (Missiones), 170–490 m.

Sources: J.D. Williams & Francini, 1991, D.R. Norman, 1994, Lema, 2002, Passos & Fernandes, 2009 and Bellosa & Bisplinghof, 2012.

7. *Epicrates exsul* Netting & Goin, 1944. Ann. Carnegie Mus. 30: 71–73, figs. 1a–c.

Type: Holotype, CM 21408, a 555 mm male (A.C. Twomey, 6 Feb. 1942).

Type locality: "Near Blackrock (approximately 26° 49' N. lat. and 77° 25' 30" W. long.) on the east coast of Great Abaco in the Bahama Islands."

Distribution: Northern Bahamas (Elbow Cay, Grand Bahama, Great Abaco, Little Abaco) and Little Bahama Bank, NSL–10 m.

Source: Sheplan & Schwartz, 1974.

Remarks: Holotype erroneously listed as CM 21048 in original description.

8. *Epicrates fordii* (A.C.L.G. Günther, 1861f). Proc. Zool. Soc. London 29(1): 142, pl. 23. (*Pelophilus fordii*)

Synonyms: *Chilabothrus maculatus* J.G. Fischer, 1888c, *Epicrates fordii agametus* Sheplan & Schwartz, 1974, and *Epicrates fordi manototus* A. Schwartz, 1979b.

Type: Holotype, BMNH 1946.1.1.55 (formerly BMNH 1862.3.10.4), a 678 mm female (A. Sallé, 1846–1856).

Type locality: "Western Africa" (in error). Restricted to the vicinity of Port-au-Prince, Dépt. de l'Ouest, Haiti *fide* Sheplan & Schwartz (1974: 104). Corrected to República Dominicana *fide* Wetherbee (1987: 63).

Distribution: Hispaniola. Haiti (Artibonite, Nord, Nord-Ouest, Ouest, Cabrit and Gonâve Is.) and ext. NW and S República Dominicana (Azua, Barahona, Independencia, Monte Cristi, Peravia, Catalina and Saona Is.), NSL–305 m.

Sources: A.C.L.G. Günther, 1861d, J.G. Fischer, 1888c, Boulenger, 1893a, Cochran, 1941, Sheplan & Schwartz, 1974, A. Schwartz, 1979b, Obst et al., 1984 and R.W. Henderson & Powell, 2004.

9. *Epicrates gracilis* (J.G. Fischer, 1888c). Jahrb. Hamburg. Wiss. Anst. (1887) 5: 35–36, pl. 3, figs. 8a–b. (*Chilabothrus gracilis*)

Synonym: *Epicrates gracilis hapalus* Sheplan & Schwartz, 1974.

Types: Syntypes (2), NMW 21369, an 895 mm specimen (F.H. Rolle via F. Steindachner don.), and formerly ZMH, an 834 mm specimen (F.H. Rolle), destroyed in July 1943 during World War II.

Type locality: "Cap Hayti" [= Cap Haitien, Nord Dept., N Haiti, 19°45'N, 72°12'W, elevation 5 m].

Distribution: Northern and SW Hispaniola. Southern Haiti (Ouest, Sud) and N and SW República Dominicana (Barahona, Dajabón, El Seibo, Espaillat, La Altagracia, Monte Cristi, Puerto Plata, Samaná, Santiago, Santiago Rodriguez), NSL–305 m.

Sources: Cochran, 1941 and R.W. Henderson & Powell, 2002.

10. *Epicrates inornatus* (J.T. Reinhardt, 1843). Kgl. Danske Vidensk. Selsk. Natur. Math. Afh.(4) 10: 253–257, pl. 1, figs. 21–23. (*Boa inornata*)

Synonyms: *Piesigaster boettgeri* Seoane, 1881, and *Boella tenella* H.M. Smith & Chiszar, 1992b.

Types: Syntypes (3), ZMUC 5597–98 and ZMUC 55101, longest syntypes 1857 mm and 1800 mm (P.S. Ravn, 1816–1839).

Type locality: "insula Porto-Rico, dansk-vestindiske" [= Puerto Rico, Danish West Indies].

Distribution: Puerto Rico (Bayamón, Caguas, Humacao, Las Marías, Luguillo, Mayagüez, Naguabo, Río Grande), NSL–1050 m.

Sources: Rivero, 1978, 1998, Reagan, 1984, Wallach & Smith, 1992, J.B. Rasmussen & Hughes, 1997 and Pizzatto et al., 2007.

11. *Epicrates maurus* Gray, 1849a. Cat. Snakes Brit. Mus.: 96.

Synonyms: *Cliftia fusca* Gray, 1849a, *Epicarsius cupreus* J.G. Fischer, 1856b, *Epicrates cupreus concolor* Jan, 1863b (*nomen nudum*), *Epicrates cupreus concolor* Jan, 1864 *in* Jan & Sordelli, 1860–1866, *Epicrates cenchria barbouri* Stull, 1938, *Epicrates maurus colombianus* Matz, 2004, and *Epicrates maurus guayanensis* Matz, 2004.

Type: Holotype, BMNH 1946.1.10.40 (formerly BMNH VI.6.3a & BMNH 1846.7.23.2a), a 1209 mm male (D. Dyson, 1846).

Type locality: "Venezuela."

Distribution: Lower Central America and N South America. Nicaragua, Costa Rica (Guanacaste, Limón, Puntarenas, possibly ext. W San José), Panama (Canal

Zone, Coclé, Darién, Barro Colorado Is.), NW Colombia (Arauca, Atlántico, Antioquia, Bolívar, Caldas, Cansanare, César, Chocó, Córdoba, Cundinamarca, Magdalena, Norte de Santander, Sucre), Venezuela (Amazonas, Anzoátegui, Apure, Aragua, Barinas, Bolívar, Carabobo, Cojedes, Delta Amacuro, Distrito Federal, Falcón, Guárico, Lara, Mérida, Miranda, Monagas, Nueva Esparta, Portuguesa, Sucre, Táchira, Trujillo, Yaracuy, Vargas, Zulia, Margarita Is.), Trinidad & Tobago (Tobago, Trinidad), N Guyana (Barima-Waini, Demerara-Mahaica, Essequibo Islands-West Demerara), Suriname (Coronie, Nickerie, Sipaliwini), N French Guiana (Cayenne, Saint-Laurent-du-Maroni) and N Brazil (Amapá, Pará, Roraima), NSL–2630 m.

Sources: Roze, 1966a, Emsley, 1977, Matz et al., 1982, Chippaux, 1987, Pérez-Santos & Moreno, 1988, Gorzula & Señaris, 1998, Starace, 1998, Mijares-Urrutia & Arends, 2000, Abuys, 2003, Solórzano, 2004, Rivas-Fuenemayer & Barrio-Amorgós, 2005, Navarrete et al., 2009, Passos & Fernandes, 2009, Ugueto & Rivas, 2010, Bellosa & Bisplinghof, 2012 and C.J. Cole et al., 2013.

Remarks: A valid species *fide* Chippaux (1987: 37) and Passos & Fernandes (2008: 14).

12. *Epicrates monensis* Zenneck, 1898. Zeit. Wiss. Zool. 64(1–2): 64–66, pl. 3, figs. 58–62.

Synonym: *Epicrates inornatus granti* Stull, 1933.

Types: Syntypes (4), ZMH 2034a–d, a 1010 mm (svl) specimen, an 810 mm (svl) female, a 500 mm female and a 350 mm female (C. Beck, 1894), destroyed in July 1943 during World War II.

Type locality: "Insel Mona bei Porto Rico" [= Mona Is. near Puerto Rico]

Distribution: Greater Antilles. Puerto Rico (Cayo Diablo and Mona Is.), U.S. Virgin Is. (St. Thomas) and British Virgin Is. (Great Camanoe, Guana, Necker, St. Thomas, Tortola, Virgin Gorda), NSL–50 m.

Sources: Meerwarth, 1901, K.P. Schmidt, 1926a, Stull, 1933, Rivero, 1978, 1998, Rivero et al., 1982, Nellis et al., 1983, Tolson, 1988 and Tolson et al., 2007.

Remarks: McDiarmid et al. (1999: 198) listed 5 syntypes. *Epicrates granti* a valid species *fide* R.W. Henderson & Powell (2009: 250).

13. *Epicrates striatus* (J.G. Fischer, 1856b). Abh. Ges. Naturwiss. Ver. Hamburg 3: 102–106, pl. 2, figs. 2a–b. (*Homalochilus striatus*)

Synonyms: *Homalochilus multisectus* Cope, 1862b, *Homalochilus strigilatus* Cope, 1862b, *Epicrates versicolor* Steindachner, 1864, *Epicrates striatus fosteri* T. Barbour, 1941, *Epicrates striatus ailurus* Sheplan & Schwartz, 1974, *Epicrates striatus exagistus* Sheplan & Schwartz, 1974, *Epicrates striatus fowleri* Sheplan & Schwartz, 1974, *Epicrates striatus mccraniei* Sheplan & Schwartz, 1974, and *Epicrates striatus warreni* Sheplan & Schwartz, 1974.

Types: Syntypes (2), ZMH 53–54, a 1902 mm and 1223 mm specimen (A. Sallé, 1849), destroyed in July 1943 during World War II.

Type locality: "St. Thomas," and "St. Domingo" [= St. Thomas, U.S. Virgin Islands and Santo Domingo, Dominican Republic]. Restricted to the vicinity of Santo Domingo, Distrito Nacional, Dominican Republic *fide* Sheplan & Schwartz (1974: 66).

Distribution: Bahamas and Hispaniola. Bahamas (Alligator Cay, Andros, Berry, Booby Cay, Cat, Chub Cay, Compass Cay, East Bimini, Easter Cay, Great Eleuthera, Exuma, Great Exuma, Great Harbour Cay, Little Ragged, Long, Mangrove Cay, Margaret Cay, New Providence, North Andros, North Bimini, Rose, South Bimini), Haiti (Nord, Nord-Ouest, Ouest, Sud, Gonáve, Tortue and Vache Is.) and Dominican Republic (SW Azua, Barahona, Dajabón, Distrito Nacional, Duarte, El Seibo, Espaillat, Independencia, La Vega, Monte Cristi, Puerto Plata, Samaná, La Altagracia, Pedernales, Sánchez Ramírez, San Cristóbal, San Juan, San Pedro de Macorís, Valverde, Saona Is.), NSL–1220 m.

Source: Meerwarth, 1901.

Remarks: Probably occurs on South Andros Is. and possibly Green Cay *fide* Sheplan & Schwartz, 1974.

14. *Epicrates subflavus* Stejneger, 1901a. Proc. U.S. Natl. Mus. 23(1218): 469–470.

Type: Holotype, USNM 14507 (Zool. Soc. Philadelphia, 25 Nov. 1885).

Type locality: "Jamaica" [Greater Antilles].

Distribution: Jamaica (S Clarendon, Hanover, S St. Ann, S St. Catherine, St. Elizabeth, St. Thomas, Trelawny; Westmoreland, Goat Is.), NSL–40 m.

Sources: Oliver, 1982, 1986.

EPICTIA Gray, 1845 (Leptotyphlopidae)

Synonyms: *Stenostoma* Wagler *in* Spix, 1824 *(nomen praeoccupatum), Stenostona* – Bory de Saint-Vincent, 1829 *in* 1822–1831 *(nomen incorrectum), Sabrina* Girard, 1858a, *Stenostome* – Westphal-Castelnau, 1870 *(nomen incorrectum), Stenostomophis* Rochebrune, 1884 *(nomen substitutum), Stenostomus* – Hoffmann, 1890 *(nomen incorrectum), Sternostoma* – Freiberg, 1939 *(nomen incorrectum), Etenostoma* – Roux-Estève, 1965 *(nomen incorrectum),* and *Crishagenus* Hoser, 2012am *(nomen illegitimum).*

Type species: *Typhlops undecimstriatus* Schlegel, 1839 *in* 1837–1844.

Distribution: North America and Latin America.

Sources: Hahn, 1980, Wallach, 1998b and McDiarmid et al., 1999.

1. *Epictia albifrons* (Wagler *in* Spix, 1824). Serp. Brasil. Sp. Nov.: 68–69, pl. 25, fig. 3. (*Stenostoma albifrons*)

Type: Holotype, ZSM 1348/0, a 183 mm specimen (J.B. von Spix & K.F.P. von Martius, 25 July–21 Aug. 1819), destroyed 11 April 1945 during World War II.
Type locality: "in adjacentibus urbis Para [Santa Maria de Belém do Grão Pará]" [= near Belém, Pará, Brazil, 01°27'S, 48°30'W, elevation NSL]. Possibly in error *fide* Cunha & Nascimento (1978: 31 and 1993: 11) and Avila-Pires et al. (2010: 70).
Distribution: Amazonia. Colombia (Antioquía, Cundinamarca), SE Venezuela (Amazonas, Bolívar, Cojedes, Falcón, Lara, Monagas, Nueva Esparta, Portuguesa, Margarita Is.), Netherlands Antilles (Bonaire), Trinidad & Tobago (Trinidad & Patos Is.), Guyana (Barima-Waini, Cuyuni-Mazaruni, Essequibo Islands-West Demerara, Potaro-Siparuni, Upper Demerara-Berbice), Suriname (Brokopondo, Commewijne, Marowijne, Nickerie, Saramacca, Suriname), French Guiana (Cayenne), Brazil (Amapa, Amazonas, Bahia, Mato Grosso, Pará, Rondônia, Roraima), E Peru (Amazonas, Loreto, Pasco) and E Bolivia (Beni, La Paz), NSL–2000 m.
Sources: Nicéforo María, 1942, Bailey & Carvalho, 1946, H.M. Smith & List, 1958, Abalos et al., 1964, R. Thomas, 1965a, 1975, Orejas-Miranda, 1967, L.D. Wilson & Hahn, 1973, Emsley, 1977, Cunha & Nascimento, 1978, Vanzolini et al., 1980, Hoogmoed & Gruber, 1983, Cei, 1986, 1994, Fugler & Riva, 1990, Carrillo de Espinoza & Icochea, 1995, Vanzolini, 1996, Gorzula & Señaris, 1998, Starace, 1998, Kornacker, 1999, J.C. Murphy, 1999, Lehr, 2000, Boos, 2001, Markezich, 2002, Buurt, 2005, Rivas-Fuenemayer & Barrio-Amorgós, 2005, Börschig, 2007, Franzen & Glaw, 2007, Avila-Pires et al., 2010 and C.J. Cole et al., 2013.
Remarks: A synonym of *Leptotyphlops tenellus fide* Franco & Pinto (2009: 239) and others. *Leptotyphlops tenellus* recognized as distinct from *L. albifrons* by Klauber (1939b: 59), Orejas-Miranda (1967: 439) and Wallach *in* McDiarmid et al. (1999: 20). Populations on Bonaire, Margarita and Trinidad of uncertain affinity (could be *Leptotyphlops albifrons, L. goudotii* or *L. magnamaculatus*). References to this species in Ecuador are likely in error *fide* Cisneros-Heredia (2008: 179). The records from W Peru (Carrillo de Espinoza & Icochea, 1995; Lehr, 2002: 196) are likely in error. Specimens reported from Paraguay, Uruguay and NW Argentina are probably in error (Cei, 1994). This species complex is in need of revision. Klauber (1939b: 59) stated that "*L. albifrons*, as usually defined, is a complex of a number of species and subspecies," K.P. Schmidt & Walker (1943b: 305) asserted that there is "no doubt that the *Leptotyphlops albifrons* of authors is an *omnium gatherum* of species of varying degrees of distinctness," and R. Thomas (1965a: 6) stated that "there is obviously more than one species [of *Leptotyphlops albibrons]* involved in the material that I have seen."

2. *Epictia albipuncta* (Burmeister, 1861). Reise La Plata-Staaten 2: 527–528. (*Stenostoma albipuncta*)

Synonyms: *Stenostoma albipunctum* Jan, 1861, *Stenostoma flavifrons* Weyenbergh *in* Napp, 1876, and *Leptotyphlops abipunctus* Hahn, 1980a.
Type: Holotype, MLU IZH 461 (H. Burmeister, 1857–1860).
Type locality: "Tucumán" [= San Miguel de Tucumán, Tucumán Province, N Argentina, 26°48'S, 65°13'W, elevation 470 m].
Distribution: Southern South America. Southern Peru, SE Bolivia (Beni, Chuquisaca, La Paz, Tarija), Paraguay (Alto Paraguay, Boquerón, Central, Neembucú, Presidente Hayes) and N Argentina (Buenos Aires, Catamarca, Chaco, Córdoba, Corrientes, Entre Ríos, Formosa, Jujuy, Mendoza, Misiones, Neuquén, Salta, Santa Fe, Santiago del Estero, Tucumán), 30–1250 m.
Sources: Vanzolini, 1966, Freiberg, 1982, Fugler, 1982, Laurent, 1984, Fabrezi et al., 1985, Fugler & Riva, 1990, Cei, 1994, Aquino et al., 1996, Leynaud & Bucher, 1999, Kretzschmar, 2006, Börschig, 2007 and Cacciali, 2008.
Remarks: See McDiarmid et al. (1999: 20) and Kretzschmar (2006: 48) for remarks about the original description, which previously was attributed to Jan, 1861. Type locality possibly in error *fide* Laurent (1984: 29).

3. *Epictia alfredschmidti* (Lehr, Wallach, G. Köhler & Aguilar, 2002). Copeia 2002(2): 131–134, figs. 1–2. (*Leptotyphlops alfredschmidti*)

Type: Holotype, MHNSM 20068, a 255 mm male (E. Lehr & C. Aguilar, 4 Feb. 1997).
Type locality: "Malvas (09°55'36"S, 77°39'00"W, elevation 2940 m), Provincia de Huarmey, Departamento de Ancash, Peru."
Distribution: Western Peru (Ancash), 2940–3090 m.
Source: Lehr, 2002.

4. *Epictia ater* (E.H. Taylor, 1940d). Univ. Kansas Sci. Bull. (1939) 26(15): 536–538, fig. 4. (*Leptotyphlops ater*)

Synonym: *Leptotyphlops nasalis* Taylor, 1940d.
Type: Holotype, USNM 79947, a 185 mm specimen (H.C. Kellers, 1928–1929).
Type locality: "Managua, Nicaragua" [12°08'N, 86°15'W, elevation 100 m].
Distribution: Central America. El Salvador (Cuscatlán, La Libertad, Morazán, San Miguel, San Salvador, Santa Ana, San Vicente, Sonsonate), W Honduras (Choluteca, Comayagua, Copán, Cortés, Francisco Morazán, Gracias a Dios, Intibucá, Lempira, Santa Barbara,

Valle, Yoro), W Nicaragua (Carazo, Chinandega, Estelí, Granada, Managua, Rivas) and NW Costa Rica (Alajuela, Guanacaste, N Puntarenas, Murciélagos Is.), NSL–1100 m.

Sources: Dunn & Saxe, 1950, Villa, 1983, 1990b, L.D. Wilson & Meyer, 1985, Sasa & Solórzano, 1995, G. Köhler, 1999b, Léon-Soler & Solórzano, 2000, Marineros, 2000, 2001a–b, 2003, Savage, 2002, Solórzano, 2004, G. Köhler et al., 2006 and Porras, 2011.

Remarks: Type erroneously listed as USNM 79957 *fide* E.H. Taylor (1940d: 536 & 1944a: 156).

5. *Epictia australis* (Freiberg & Orejas-Miranda, 1968). Physis 28(76): 145–147, figs. 1–2. (*Leptotyphlops australis*)

Type: Holotype, MACN 12525, a 245 mm female (M. Birabén & M.I. Hylton Scott, Jan. 1953).

Type locality: "Valcheta, prov. Rio Negro, Argentina" [= 40°41'S, 66°10'W, elevation 185 m].

Distribution: Northern Argentina (Buenos Aires, Catamarca, Córdoba, La Pampa, La Rioja, Mendoza, Neuquén, Río Negro, San Luis, Tucumán), Paraguay (Boquerón, Central, Concepción, Presidente Hayes), S Brazil (W Rio Grande do Sul) and Uruguay, 75–1000 m.

Sources: Orejas-Miranda & Achaval, 1969, Lema & Fabián-Beurmann, 1977, Miranda & Tio-Vallejo, 1985, Cei, 1986, 1994, Lema, 1987, Scrocchi, 1990a, J.D. Williams & Francini, 1991, Aquino et al., 1996, Cabrera, 2004, Scolaro, 2006 and Börschig, 2007.

6. *Epictia bakewelli* (Oliver, 1937). Occ. Pap. Mus. Zool. Univ. Michigan (360): 16–17, fig. 1a. (*Leptotyphlops bakewelli*)

Synonym: *Leptotyphlops gadowi* Duellman, 1956.

Type: Holotype, UMMZ 80228 (J.A. Oliver & A. Bakewell, 26 July 1935).

Type locality: "Paso del Río, Colima, Mexico" [18°54'N, 103°53'W, elevation 30 m].

Distribution: Southwestern Mexico (Colima, Guerrero, Jalisco, Michoacán, Oaxaca), 30–850 (1560) m.

Sources: H.M. Smith, 1943a, H.M. Smith & Taylor, 1945 and Duellman, 1961, 1965.

Remarks: Previously a subspecies of *Leptotyphlops goudotii.*

7. *Epictia borapeliotes* (Vanzolini, 1996). Pap. Avulsos Zool. 39(15): 282–284. (*Leptotyphlops borapeliotes*)

Type: Holotype, MZUSP 9595 (formerly MR 87.6701) (M. Rodrigues, 1 Oct. 1987).

Type locality: "Santo Inácio, Bahia, Brazil" [= 11°07'S, 42°43'W, elevation 525 m].

Distribution: Northeastern Brazil (Alagoas, Bahia, Ceará, Paraíba, Pernambuco, Rio Grande do Norte, Sergipe), NSL–845 m.

Sources: Cordeiro & Hoge, 1974, Vanzolini et al., 1980, Ulloa, 2006 and Roberto & Veiga, 2009.

8. *Epictia clinorostris* Arredondo & Zaher, 2010. So. Amer. J. Herp. 5(3): 190–194, figs. 1a–b, 2a–b.

Type: Holotype, MZUSP 17480, a 206 mm female (A.P. Soares, 5 Sept. 2007).

Type locality: "Toricoejo locality, bank of Rio das Mortes, (15°14'21.6"S – 53°14'36.6"W), Municipality of Barra do Garças, State of Mato Grosso, Brazil" [elevation ca. 500 m].

Distribution: Central Brazil (W Goiás, E Mato Grosso), 500 m.

9. *Epictia collaris* (Hoogmoed, 1977). Zool. Meded. 51(7): 100–103, figs. 1a–c, pl. 1, pl. 2, figs. c–d. (*Leptotyphlops collaris*)

Type: Holotype, RMNH 13468, a 102 mm specimen Suriname Exped., 12 Feb. 1949).

Type locality: "Base Camp Nassau Mountains, distr. Marowijne, Suriname" [= ca. 04°48'N, 54°36'W, elevation ca. 500 m].

Distribution: Northern Suriname (Brokospondo, Marowijne) and N French Guiana (N Cayenne), NSL–475 m.

Sources: Hoogmoed, 1977, Gasc & Rodrigues, 1980, Chippaux, 1987, Starace, 1998 and Abuys, 2003.

10. *Epictia columbi* (Klauber, 1939b). Trans. San Diego Soc. Nat. Hist. 9(14): 62–64, figs. 3a–b. (*Leptotyphlops columbi*)

Type: Holotype, CM 1364, a 183 mm specimen (W.W. Worthington, 1909).

Type locality: "Watling (or San Salvador) Island, Bahama Islands" [= 24°03'N, 74°29'W, elevation NSL].

Distribution: Bahamas (San Salvador Is. and satellites), NSL.

Sources: Legler, 1959b, R. Thomas, 1965a, A. Schwartz & R.W. Henderson, 1991 and R.W. Henderson & Powell, 2009.

11. *Epictia diaplocia* (Orejas-Miranda, 1969). Comun. Zool. Mus. Hist. Nat. Montevideo 10(124): 5–8, pl. 1, fig. 2. (*Leptotyphlops diaplocius*)

Type: Holotype, AMNH 55654, a 170 mm specimen.

Type locality: "Requena, Montecarmelo, Peru" [= Loreto Dept., NE Peru, 04°59'S, 73°59'W, elevation 110 m].

Distribution: Eastern Peru (Loreto, Madre de Dios, San Martín) and NW Brazil (Amazonas, Rondônia), 45–1500 m.

Sources: Nascimento et al., 1988, Duellman & Salas, 1991, Carrillo de Espinoza & Icochea, 1995, Martins & Oliveira, 1999 and Duellman, 2005.

12. *Epictia goudotii* (A.-M.-C. Duméril & Bibron, 1844). Erpét. Gén. 6: 330. (*Stenostoma goudotii*)

Synonyms: *Stenostoma goudottii* – Cope, 1875a *(nomen incorrectum),* and *Stenostoma fallax* W.C.H. Peters, 1857b.

Type: Holotype, MNHN 1068, a 148–151 mm male (J.P. Goudot, 1822–1826 or 1837–1842).

Type locality: "la vallée de la Magdeleine, à la Nouvelle-Grenade" [= valley of the Río Magdalena, cen. Colombia].

Distribution: Panamá (Canal Zone), Colombia (Atlántica, Bolivar, Cesar, Cundinamarca, Magdalena, Meta, Santander, Tolima, Salamanca Is.) and Venezuela (Amazonas, Anzoátegui, Bolívar, Distrito Federal, Falcón, Monagas, Nueva Esparta, Sucre, Yaracuy), NSL–1535 m.

Sources: Nicéforo María, 1942, Roze, 1966a, Emsley, 1977, Pérez-Santos & Moreno, 1988, Lancini & Kornacker, 1989, Kornacker, 1999, Mijares-Urrutia & Arends, 2000, Navarrete et al., 2009, Pinto et al., 2010 and Ugueto & Rivas, 2010.

13. *Epictia magnamaculata* (E.H. Taylor, 1940d). Univ. Kansas Sci. Bull. (1939) 26(15): 532–533, fig. 1. (*Leptotyphlops magnamaculata*)

Synonym: *Leptotyphlops albifrons margaritae* Roze, 1952b.

Type: Holotype, USNM 54760, a 167 mm female (F.J. Dyer, 9 April 1916).

Type locality: "Utilla Id., Honduras" [= Útila Island, Islas de la Bahía, Honduras, 16°06'N, 86°55'W].

Distribution: Caribbean islands of SE Mexico (Quintana Roo: Cozumel Is.), N Honduras (Gracias a Dios, Islas de la Bahía, Bahia, Barbareta, Cisne, Cayo Cochino Grande, Cayo Cochino Pequeña, Grande, Guanaja, Roatán, Swan and Útila Is.), Colombia (Providencia, San Andrés) and Venezuela (Nueva Esparta: Margarita Is.), NSL–25 m.

Sources: Dunn & Saxe, 1950, L.D. Wilson & Hahn, 1973, Lee, 1996, L.D. Wilson & Meyer, 1985, G. Köhler, 2001, 2003, McCranie et al., 2005, Pinto et al., 2010 and McCranie, 2011.

Remarks: Possibly also includes the populations on Suma Is., Bonaire, Margarita and Trinidad (McCranie et al., 2005: 127).

14. *Epictia melanoterma* (Cope, 1862d). Proc. Acad. Nat. Sci. Philadelphia 14(5): 350. (*Stenostoma melanoterma*)

Synonym: *Leptotyphlops melanotermus* Orejas-Miranda, 1964.

Types: Syntypes (2), ANSP 3297–98 (T.J. Page [Paraguay Exped.], Sept. 1853–Feb. 1855).

Type locality: "Corrientes," Argentina.

Distribution: Southern South America. Southern Peru, SE Bolivia, Paraguay and N Argentina (Córdoba, Santa Fe), 20–1295 m.

Sources: Laurent, 1984, Cei, 1993, Leynaud & Bucher, 1999 and Börschig, 2007.

Remarks: A synonym of *Epictia albipuncta fide* Kretzchmar, 2006.

E

15. *Epictia melanura* (K.P. Schmidt & Walker, 1943b). Field Mus. Nat. Hist. (Zool.) 24(27): 303–304. (*Leptotyphlops melanurus*)

Type: Holotype, FMNH 34269, a 135 mm specimen (C.L. Hoyle, Nov. 1939).

Type locality: "Chiclin, Libertad, Peru" [= Chiclin, La Libertad Dept., NW Peru, 07°50'S, 79°10'W, elevation 115 m].

Distribution: Northwestern Peru (La Libertad), 115 m. Known only from type locality.

Sources: K.P. Schmidt & Walker, 1943b and Carrillo de Espinoza & Icochea, 1995.

16. *Epictia munoai* (Orejas-Miranda, 1961). Acta Biol. Venez. 3(5): 85–89, figs. 1a–c. (*Leptotyphlops munoai*) (*nomen corrigendum*)

Synonyms: *Leptotyphlops muñoai* Orejas-Miranda, 1961 *(nomen incorrigendum), Leptotyphlops munoai* – J.A. Peters & Orejas-Miranda, 1970 *(nomen corrigendum),* and *Leptotyphlops munaoi* – Hahn, 1980a *(nomen incorrectum).*

Type: Holotype, MBUCV 4547 (formerly MHNM 68), a 140 mm specimen (M.A. Klappenbach & P.R. San Martin, 12 Oct. 1956).

Type locality: "Pozo Hondo, Tambores, Departamento de Tacuarembó, República Oriental del Uruguay" [= 31°57'S, 56°15"W, elevation 215 m].

Distribution: Southeastern South America. Southern Brazil (Rio Grande do Sul, Santa Catarina), Paraguay (Central, Concepción, Neembucú, Presidente Hayes), Uruguay (Artigas, Canelones, Cerro Largo, Colonia, Duranzo, Flores, Florida, Lavelleja, Maldonado, Paysandú, Río Negro, Rivera, Rocha, Salto, San José, Soriano, Tacuarembó, Treinta y Tres) and NE Argentina (Buenos Aires, Corrientes, La Pampa, Misiones), 50–290 m.

Sources: Orejas-Miranda, 1962, Miranda et al., 1982, Miranda & Tio-Vallejo, 1985, J.D. Williams & Francini, 1991, Cei, 1994, Lema, 1994, Aquino et al., 1996, Giraudo, 2001, Carreira-Vidal, 2002, Achaval & Olmos, 2003, Carreira-Vidal et al., 2005, Cacciali, 2008 and Francisco et al., 2012.

E

17. *Epictia peruviana* (Orejas-Miranda, 1969). Comun. Zool. Mus. Hist. Nat. Montevideo 10(124): 9–10, pl. 2, fig. 1. (*Leptotyphlops peruvianus*)

Type: Holotype, AMNH 52437, a 94 mm specimen.
Type locality: "Chanchamayo, Departamento Junín, Peru" [= Chanchamayo, near San Ramon, Chanchamayo District, N Departamento de Junín, cen. Peru, 11°08'S, 75°21'W, elevation 900 m].
Distribution: Central Peru (Junín), 900 m. Known only from type locality.
Sources: Rodrígues et al., 1993 and Carrillo de Espinoza & Icochea, 1995 and Young & León, 1999.

18. *Epictia phenops* (Cope, 1875a). J. Acad. Nat. Sci. Philadelphia (1876) (2) 8(2): 128. (*Stenostoma phenops*)

Types: Syntypes (8), USNM 12444 and USNM 30289–95, longest syntype 156 mm (A.-L.-J.-F. Sumichrast, 1855–1869).
Type locality: "Tehuantepec" [= Tehuantepec, Oaxaca, México, 16°21'N, 95°16'W, elevation 60 m] via type locality restriction of H.M. Smith & Taylor (1945: 24 & 1950: 340).
Distribution: Southeastern México (N Chiapas, Hidalgo, Oaxaca, N Querétaro, Quintana Roo, Tamaulipas, Veracruz, Yucatán, Cozumel Is.), Guatemala (Petén) and Belize (Corozal), NSL–1330 m.
Sources: H.M. Smith, 1943a, H.M. Smith & Taylor, 1945, Cochran, 1961, Peréz-Higareda & Smith, 1991, Villela et al., 1991, Lee, 1996, G. Köhler et al., 2006, Flores-Benabib & Flores-Villela, 2008, Dixon & Lemos-Espinal, 2010, Márquez & Mayén, 2010, Ramírez-Bautista et al., 2010 and McCranie, 2011.

19. *Epictia rubrolineata* (F. Werner, 1901c). Abh. Ber. Kön. Zool. Anthro.-Ethno. Mus. Dresden (1900–1901) 9(2): 6. (*Glauconia albifrons rubrolineata*)

Synonym: *Leptotyphlops rubrolineatus* Hahn, 1980a.
Type: Holotype, MTKD D-1752 (A. Baessler, 1899), destroyed 13 Feb. 1945 during World War II.
Type locality: "Lima, Peru" [= Lima, Lima Dept., W Peru, 12°03'S, 77°02'W, elevation 150 m].
Distribution: Western Peru (W Lima), 150 m. Known only from type locality.
Sources: Schüz, 1929, Obst, 1977 and Carrillo de Espinoza & Icochea, 1995.

20. *Epictia rufidorsa* (E.H. Taylor, 1940d). Univ. Kansas Sci. Bull. (1939) 26(15): 533–535, fig. 2. (*Typhlops rufidorsum*)

Synonym: *Leptotyphlops rufidorsus* K.P. Schmidt & Walker, 1943b.
Type: Holotype, USNM 49993, a 265 mm specimen (C.H. Townsend, 16 April 1913).
Type locality: "Lima, Peru" [= Lima, Lima Dept., W Peru, 12°03'S, 77°02'W, elevation 150 m]. Emended to Rimac Valley, 5300 ft. *fide* Cochran (1961: 194).
Distribution: Western Peru (La Libertad, Lima), 1615 m.
Sources: K.P. Schmidt & Walker, 1943b and Carrillo de Espinoza & Icochea, 1995.

21. *Epictia signata* (Jan, 1861b). Arch. Zool. Anat. Fis. 1(2): 188. (*Stenostoma signatum*)

Synonyms: *Glauconia signata* Boulenger, 1893a, *Leptotyphlops amazonicus* Orejas-Miranda, 1969, and *Leptotyphlops signatus* Hahn, 1980a
Type: Holotype, MNHN 3235, an 87 mm specimen (G. Jan, 1858).
Type locality: Unknown. Restricted to northern Amazonian region of South America *fide* Hahn (1979: 59).
Distribution: Northwestern Amazonia. Colombia and SE Venezuela (S Amazonas, Bolívar) and E Ecuador (Morona-Santiago), 125–2020 m.
Sources: Jan & Sordelli, 1861 *in* 1860–1866, Jan, 1864, Boulenger, 1893a, Hoogmoed, 1977, Hahn, 1979a, Miyata, 1982, Lancini & Kornacker, 1989, Kornacker, 1999, Pinto, 2010 and Pinto et al., 2010.
Remarks: Holotype length erroneously reported by Jan (1864: 36) to be 130 mm, which is length of Sordelli's figure in Jan & Sordelli [1861 *in* 1860-1866 (livr. 2): pl. 5, fig. 3]. Orton's ANSP 3290, which Orejas-Miranda (1969: 4) suggested was from Amazonian Ecuador, likely in error for Venezuela *fide* Cisneros-Heredia (2008: 179).

22. *Epictia striatula* (H.M. Smith & Laufe, 1945a). Proc. Biol. Soc. Washington 58: 29–31, text figs. a–b, pl. 5, figs. a–b. (*Leptotyphlops striatula*)

Synonym: *Leptotyphlops striatulus* McDiarmid et al., 1999.
Type: Holotype, USNM 98889, a 241 mm specimen (M. Cárdenas, 1921–1945).
Type locality: "Yamachi, in the Southern Yungas, Bolivia" [= Yanacachi, La Paz Dept., W Bolivia, 16°22'S, 67°45'W, elevation 1480 m].
Distribution: Western Bolivia (El Beni, Chuquisaca, La Paz) and Argentina (Salta), 155–2510 m.
Sources: Laurent, 1984, Fugler & Riva, 1990, Cei, 1994, Börschig, 2007 and Pinto, 2010.

23. *Epictia subcrotilla* (Klauber, 1939b). Trans. San Diego Soc. Nat. Hist. 9(14): 61–62, figs. 2a–b. (*Leptotyphlops subcrotilla*)

Synonym: *Leptotyphlops subcrotillus* K.P. Schmidt & Walker, 1943b.
Type: Holotype, CAS 14554, a 188 mm specimen (G. Baer, 1902).

Type locality: "Grau Tombes, Northern Peru" (in error). Corrected to Grau, Tumbez *fide* Schmidt & Walker, 1943b: 303), [= Miguel Grau, Tumbes Dept., ext. N Peru, 03°34'S, 80°27'W, elevation 15 m].

Distribution: Western Peru (La Libertad, Piura, Tumbes) and SW Ecuador (El Oro, Esmeraldas, Guayas, Manabí, Santa Elena), 10–95 m.

Sources: K.P. Schmidt & Walker, 1943b, Miyata, 1982, Carrillo de Espinoza & Icochea, 1995, Purtschert, 2007 and Cisnero-Heredia, 2008.

Remarks: Ecuadorian specimens from El Oro, Esmeraldas and Manabí provinces may belong to this species *fide* Cisneros-Heredia (2008: 179).

24. *Epictia teaguei* (Orejas-Miranda, 1964). Comun. Zool. Mus. Hist. Nat. Montevideo 8(103): 4–6, pl. 2, figs. a–c, pl. 3, figs. a–c. (*Leptotyphlops teaguei*)

Type: Holotype, MNHNU CH 910 (formerly CWW 12526A), a 139 mm specimen (W. Weyrauch, 24 June 1956).

Type locality: "la margen izquierda del Río Chotano (Altitud: 2.350 metros) entre Chota y Cutervo, norte de Peru, vertiente oriental de la Cordillera Occidental; en savana, bajo piedras" [= Río Chotano, between Chota (06°34'S, 78°39'W) and Cutervo (06°23'S, 78°49'W), Cajamarca Dept., NW Peru, elevation 2350 m].

Distribution: Northwestern Peru (Cajamarca), 2200–2700 m. Known only from vicinity of type locality.

Sources: Laurent, 1984, Carrillo de Espinoza & Icochea, 1995 and Lehr et al., 2002.

25. *Epictia tenella* (Klauber, 1939). Trans. San Diego Soc. Nat. Hist. 9(14): 59–61, figs. 1a–b. (*Leptotyphlops tenella*)

Type: Holotype, AMNH 14269, a 177 mm specimen (W. Beebe, 1919).

Type locality: "Kartabo, British Guiana" [= Kartabo, Essequibo Islands-West Demerara Region, Guyana, ca. 06°35'N, 58°35'W, elevation NSL].

Distribution: Northern South America. Southeastern Venezuela (Bolivar), Trinidad (St. George, St. Joseph), Guyana (Essequibo Islands-West Demerara, Potaro-Siparuni) Suriname (Brokopondo, Marowijne, Nickerie, Para, Saramacca, Suriname), French Guiana (Cayenne) and N Brazil (Amazonas, Mato Grosso, Pará), NSL–495 m.

Sources: Orejas-Miranda & Zug, 1974, Hoogmoed, 1977, Chippaux, 1987, J.C. Murphy, 1997, Starace, 1998, Boos, 2001 and Abuys, 2003.

Remarks: Considered a separate species (Klauber, 1939a, Orejas-Miranda, 1967; Orejas-Miranda *in* J.A. Peters & Orejas-Miranda, 1970; Wallach *in* McDiarmid et al., 1999), a junior synonym (Hoogmoed & Gruber, 1983), or senior synonym (Franco & Pinto, 2009) of *Leptotyphlops albifrons.*

26. *Epictia tesselata* (Tschudi, 1845). Arch. Naturg. 11(1): 162–163. (*Typhlops tesselatum*)

Synonyms: *Typhlps tessellatus* Jan, 1861b *(nomen emendatum),* and *Stenostoma albifrons tessellata* Jan & Sordelli, 1861 *in* 1860–1866 *(nomen emendatum).*

Type: Lectotype, MHNN 15 (formerly MZN 60) (J.J. von Tschudi, Aug. 1838–Feb. 1841), designated by Schätti (1986a: 102).

Type locality: "Lima" [= Lima, Lima Dept., W Peru, 12°03'S, 77°02'W, elevation 155 m] via lectotype selection.

Distribution: Western Peru (Lima), 155–990 m. Known only from vicinity of type locality.

Sources: Jan & Sordelli, 1862 *in* 1860–1866, K.P. Schmidt & Walker, 1943b, Carrillo de Espinoza, 1970, Carrillo de Espinoza & Icochea, 1995 and Lehr et al., 2002.

Remarks: Original description reprinted in Tschudi (1846: 46–47).

27. *Epictia tricolor* (Orejas-Miranda & Zug, 1974). Proc. Biol. Soc. Washington 87(16): 167–171, figs. 1a–f. (*Leptotyphlops tricolor*)

Type: Holotype, MHNJP 3487 (formerly MHNJP 669), a 295 mm male (A. Ramos, 24 Feb. 1966).

Type locality: "Peru: Ancash Department, Huaylas Province, Huaylas District: Yunca Pampa, (2700 m)."

Distribution: Western Peru (Ancash, Cajamarca, Lima), 2700–3250 m.

Sources: Zug, 1977, Carrillo de Espinoza & Icochea, 1995, Lehr et al., 2002 and Aguilar et al., 2007.

28. *Epictia undecimstriata* (Schlegel, 1839 *in* 1837–1844). Abbild. Amph.: 36–37. (*Typhlops undecimstriatus*)

Type: Holotype, MNHN, a 236 mm specimen (A.C.V.M.D. d'Orbigny, Nov. 1830), lost *fide* Hahn (1980a: 28).

Type locality: "St. Crux de la Sierra, Amerika" [= Santa Cruz de la Sierra, Santa Cruz Dept., E Bolivia, 17°48'S, 63°10'W, elevation 425 m].

Distribution: Eastern Bolivia (Santa Cruz), 420 m. Known only from type locality.

Sources: Gray, 1845, J.A. Peters & Orejas-Miranda, 1970, Fugler & Riva, 1990 and Börschig, 2007.

29. *Epictia unicolor* (F. Werner, 1913). Mitt. Naturhist. Mus. Hamburg (1912) 30: 21. (*Glauconia bilineata unicolor*)

Type: Holotype, ZMH 8401 (formerly ZMH 4823), a 165–172 mm specimen (A. Brückmann, 9 Nov. 1910).

Type locality: "wahrscheinlich von Brasilien" [= Brazil].

Distribution: Brazil. Known only from type locality.

Source: Boundy & Wallach, 2008.

E

30. *Epictia vellardi* (Laurent, 1984). Acta Zool. Lilloana 38(1): 30–31, figs. 1a–c. (*Leptotyphlops vellardi*)

Type: Holotype, FML 110, a 179 mm female (A. Vellard, Dec. 1944).
Type locality: "Ciudad de Formosa. prov. de Formosa" [= Ciudad de Formosa, SE Formosa Prov., NE Argentina, 26°11'S, 58°11'W, elevation 65 m].
Distribution: Brazil (Mato Grosso do Sul) and ext. N Argentina (Chaco, Formosa), 55–185 m.
Sources: Laurent, 1984, Miranda & Tio-Vallejo, 1985, J.D. Williams & Francini, 1991, Cei, 1994, Leynaud & Bucher, 1999 and Francisco et al., 2012.

31. *Epictia weyrauchi* (Orejas-Miranda, 1964). Comun. Zool. Mus. Hist. Nat. Montevideo 8(103): 1–4, pl. 1. (*Leptotyphlops weyrauchi*)

Type: Holotype, MNHNU CH 911 (R. Golbach, 14 June 1962).
Type locality: "la Ciudad de Tucumán (Altitud: 436 metros), Prov. Tucumán, Rep. Argentina."
Distribution: Argentina (Chaco, Córdoba, Formosa, Salta, Santiago del Estero, Tucumán), 435 m.
Sources: Fabrezi et al., 1985, Cei, 1993, Vanzolini, 1996 and Leynaud & Bucher, 1999.
Remarks: A synonym of *L. albipunctus fide* Kretzschmar, 2006.

ERISTICOPHIS Alcock & Finn, 1897 (Viperidae)

Synonym: *Eristocophis* Wall, 1906d *(nomen emendatum).*
Type species: *Eristicophis macmahonii* Alcock & Finn, 1897.
Distribution: South Asia.
Sources: Blanford, 1897, Wall, 1906d, M.A. Smith, 1943, Krishna & Dave, 1956, Marx & Rabb, 1965, Minton, 1966, Joger, 1984, Latifi, 1985, J.S., Werman, 1986a, Ashe & Marx, 1988, Department of Defense, 1991, Golay et al., 1993, David & Ineich, 1999, Herrmann et al., 1999, McDiarmid et al., 1999, Lenk et al., 2001b, M.S. Khan, 2002, Mallow et al., 2003, Wüster et al., 2008, Phelps, 2010 and Hoser, 2012d.

1. *Eristicophis macmahonii* Alcock & Finn, 1897. J. Asiat. Soc. Bengal (1896) 65(4): 564–565, pl. 15, figs. 1–1a.

Synonyms: *Eristocophis mcmahoni* F. Wall, 1925a *(nomen emendatum), Pseudocerastes latirostris* Guibé, 1957, and *Eristophis macmohoni* – Khole, 1991 *(nomen incorrectum).*
Types: Syntypes (8), BMNH 1946.9.8.11 (formerly BMNH 1896.12.22.14), head and part of body, BMNH 1946.9.8.61 (formerly BMNH 1896.12.12.11), and ZSI 14179–84 (formerly IMC), longest syntype 648 mm (A.H. MacMahon & F.P. Maynard [Afghan-Baluch Bound. Comm.], 30 March–May 1896).
Type locality: "Amirchah, 3300 feet, Zeh, 2500 feet, Drana Koh, Robat I., 4300 feet, Afghan-Baluch boundary" [= Robat I (29°33'N, 63°38'E, 1310 m), S Afghanistan and Drana Koh (29°14'N, 61°57'E, 900 m), Zeh (29°15'N, 62°11'E, 760 m) and Amir Cháh (29°15'N, 62°30'E, 1000 m), W Pakistan]. W. Baluchistan *fide* BMNH catalogue or the desert south of the Helmand River, Baluchistan *fide* M.A. Smith (1943: 493).
Distribution: Southwest Asia. Southeastern Iran (Kerman, Sistan va Baluchestan), SW Afghanistan (Helmand, Heart, Nimroze) and NW Pakistan (SW Balochistan), 760–1310 m.
Remarks: BMNH 1946.9.8.61 recorded as holotype in BMNH catalogue *fide* I. Das *in* McDiarmid et al. (1999: 383). Alcock referred to as sole author of new genus and species *fide* Blanford (1897: 295). Unconfirmed record from Rajasthan, NW India *fide* Krishna & Dave, 1956. At least one of the type localities (Robat I) appears to be in Helmand Prov., Afghanistan. Two of the type localities are spelled differently on their map (pl. 11), Daran Koh and Amir Cháh.

ERPETON Lacépède, 1801 (Homalopsidae)

Synonyms: *Rhinopirus* Merrem, 1820 *(nomen substitutum), Herpeton* Oken, 1816 *(nomen rejiciendum), Herpeton* Wagler, 1830 *(nomen emendatum), Rhinopyrus* Bonaparte, 1831 *(nomen emendatum), Herpetum* Agassiz, 1847 *(nomen emendatum),* and *Herpedon* – Maass-Berlin, 1933 *(nomen incorrectum).*
Type species: *Erpeton tentaculatus* Lacépède, 1801.
Distribution: Tidal rivers, lakes, ponds and coastal marine waters of Indochina, including the Gulf of Thailand.
Sources: Morice, 1875b, M.A. Smith, 1914c–d, Bourret, 1936b, E.H. Taylor, 1965, Soderberg, 1966b, Campden-Main, 1970a, Cornelissen, 1970, Gyi, 1970, Saint-Girons, 1972a, Winokur, 1977, Martinez & Behler, 1988, K.K.P. Lim & Lee, 1989, M.J. Cox, 1991b, M.J. Cox et al., 1998, Voris et al., 2002, J.C. Murphy & Schlager, 2003, J.C. Murphy, 2007b, Alfaro et al., 2008, V.S. Nguyen et al., 2009, Zaher et al., 2009 and I. Das, 2010, 2012.

1. *Erpeton tentaculatum* Lacépède, 1801. Bull. Soc. Philom. Paris 2(46): 169.

Synonyms: *Erpeton tentaculatus* – Latreille *in* Sonnini & Latreille, 1801b, *Rhinopirus erpeton* Merrem, 1820, and *Homalopsis herpeton* Schlegel, 1837.
Type: Holotype, MNHN 22, a 500+ mm specimen (old Dutch coll.).
Type locality: Unknown.

Distribution: Indochina. Southern Thailand (Bangkok, Krung Thep Mahanakhon, Nakhon Pathom, Nakhon Si Thammarat, Phattalung, Phra Nakhon Si Ayutthaya, Prachuap Khiri Khan), S Cambodia (Kampot, Kandal, Kornpong Chhnang) and S Vietnam (An Gian, Binh Duong, Kien Giang, Long An, Tay Ninh), NSL–25 m.

Remarks: Original description reprinted in Lacépède (1803: 284, pl. 50).

ERYTHROLAMPRUS F. Boie, 1826 (Xenodontidae)

Synonyms: *Erythrolampus* – Gray *in* Griffith & Pidgeon, 1831 *(nomen incorrectum), Erythrophis* Fitzinger, 1843, *Rhinaspis* Fitzinger, 1843 *(nomen rejiciendum), Erythrolampas* – F. Müller, 1878b *(nomen incorrectum), Erythrlomprus* – Amaral, 1930f *(nomen incorrectum), Erithrolamprus* – Briceño-Rossi, 1934 *(nomen incorrectum), Erythroglamprus* – Machado, 1945 *(nomen incorrectum), Erythtrolamprus* – Villa, 1962 *(nomen incorrectum), Erythrolomprus* – Engelmann & Obst, 1981 *(nomen incorrectum),* and *Erythrocamprus* – Phelps, 1981 *(nomen incorrectum).*

Type species: *Coluber aesculapii* Linnaeus, 1758.

Distribution: Central America and South America.

Sources: J.A. Peters & Orejas-Miranda, 1970 and Vidal et al., 2000, 2010.

1. *Erythrolamprus aesculapii* (Linnaeus, 1758). Syst. Nat., ed. 10, 1: 220. (*Coluber aesculapii*)

Synonyms: *Coluber agilis* Linnaeus, 1758, *Coluber scutata* Laurenti, 1768, *Coluber nigrofasciatus* Lacépède, 1789 (*nomen rejiciendum*), *Coluber elegantissimus* G. Shaw, 1802, *Coluber linnaei* G. Shaw, 1802 (*nomen substitutum*), *Coluber atrocinctus* Daudin, 1803c, *Coluber venustissimus* Wied-Neuweid, 1822, *Coluber binatus* Lichtenstein, 1823, *Elaps biannulatus* Wagler, 1825 (*nomen ineditum*), *Lycodon maximiliani* F. Boie, 1827, *Erythrolamprus bauperthuisii* A.-M.-C. Duméril, Bibron & Duméril, 1854b, *Erythrolamprus intricatus* A.-M.-C. Duméril, Bibron & Duméril, 1854b, *Erythrolamprus milberti* A.-M.-C. Duméril, Bibron & Duméril, 1854b, *Erythrolamprus albostolatus* Cope, 1860e, *Erythrolamprus aesculapii confluentus* Jan, 1863a, *Erythrolamprus aesculapii dicranta* Jan, 1863a, *Erythrolamprus aesculapii monozona* Jan, 1863a, *Erythrolamprus aesculapii tetrazona* Jan, 1863a, and *Erythrolamprus baileyi* Roze, 1959a.

Types: Syntypes (3), NHR Lin-84 (formerly MAFR), a 610 mm specimen (Mus. Drottn.), NHR Lin-85 (formerly MAFR), 700 mm specimen (Mus. Drottn.), and ZMUU 92 (C. Grill, 1747).

Type locality: "Indiis" [= India] (in error *fide* Hoge, 1964d: 56). Corrected to Brazil *fide* Amaral (1977: 176). Restricted to Popayan, Bahia, Brazil *fide* Gasc & Rodrigues (1980: 575).

Distribution: South America. Colombia (Atlántico, Boyacá, Caldas, Cauca, Magdalena, Meta, Guaviare, Nariño, Norte de Santander, Santander, Tolima, Valle, Vaupés), Trinidad and Tobago (Trinidad), Venezuela (Amazonas, Bolívar, Delta Amacuro, Monagas, Sucre), Guyana (Essequibo Islands-West Demerara, Pomeroon-Supenaam, Upper Demerara-Berbice), Suriname (Brokopondo, Suriname), French Guiana (Cayenne, Saint-Laurent-du-Maroni), Brazil (Amapá, Amazonas, Bahia, Brazilia, Goiás, Mato Grosso, Minas Gerais, Pará, Paraná, Rio de Janeiro, Rondônia, Santa Catarina, São Paulo, Barnabé, Cananéia, São Sebastião and São Vicente Is.), E Ecuador (Napo, Zamora-Chinchipe), Bolivia (El Beni, Cochabamba, La Paz, Pando, Santa Cruz, Sergipe), Peru (Huancabamba, Junin, Loreto, Madre de Dios, Pasco, San Martin, Ucayali), Paraguay (Amambay, Canindeyú, Central, Concepción, Itapúa, Misiones, Neembucú, Paraguarí) and ext. NE Argentina (Misiones), NSL–2300 m.

Sources: Roze, 1959a, 1966a, Amaral, 1978, Cunha & Nascimento, 1978, 1980, Abuys, 1983d, Dixon & Sonini, 1986, Lancini, 1986, Chippaux, 1987, Pérez-Santos & Moreno, 1988, Campbell & Lamar, 1989, Cei, 1994, J. Hardy & Boos, 1995, Marques, 1996b, Starace, 1998, H.E.A. Boos, 2001, Marques et al., 2001, Doan & Arazábal, 2002, Cicchi et al., 2007, Cacciali, 2008, Navarrete et al., 2009, Morato et al., 2011 and C.J. Cole et al., 2013.

Remarks: NHR Lin-84 and Lin-85 lengths erroneously listed as 350 and 390 mm, respectively, *fide* Andersson (1899: 16). The lectotype designation of J.A. Peters (1960a: 520) is invalid as NHR possesses two syntypes and Peters did not provide a catalogue number.

2. *Erythrolamprus bizonus* Jan, 1863a. Arch. Zool. Anat. Fis. 2(2): 314. (*Erythrolamprus aesculapii bizona*)

Types: Syntypes (3), MHNG 464.30, MSNM and NMW.

Type locality: "Bahia, Messico, Popayan, Cayenne, Brasile, Montevideo, Colombia" [= Mexico; Popayan, Colombia; Cayenne, French Guiana; Bahia, Brazil; Montevideo, Uruguay]. Restricted to Colombia *fide* Dunn & Bailey (1939: 12).

Distribution: Lower Central America and ext. NW South America. Costa Rica (Alajuela, Cartago, Guanacaste, Puntarenas, San José), Panama (Canal Zone, Darién, Barro Colorado Is.), Colombia (Antioquia, Atlántico, Bolívar, Caldas, Cauca, César, Cundinamarca, Guaviare, La Guajira, Magdalena, Meta, Nariño, Norte de Santander, Santander, Valle), Trinidad and Tobago (Trinidad) and N Venezuela (Apure, Aragua, Barinas, Carabobo, Cojedes, Distrito Federal, Falcón, Guárico, Lara, Miranda, Portuguesa, Táchira, Trujillo, Zulia), NSL–2630 m.

Sources: E.H. Taylor, 1951, 1954, Roze, 1966a, Lancini, 1986, Pérez-Santos & Moreno, 1988, J. Hardy & Boos, 1995, G. Köhler, 1999b, Mijares-Urrutia & Arends, 2000, H.E.A. Boos, 2001, Markezich, 2002, Savage, 2002, Solórzano, 2004, Rivas-Fuenemayer & Barrio-Amorgós, 2005, Navarrete et al., 2009 and Rojas-Morales, 2012b.

Remarks: Type locality restriction of H.M. Smith & Taylor (1950: 364) rejected *fide* Dunn & Stuart (1951: 56). Possibly occurs in SE Honduras and Nicaragua *fide* G. Köhler (1999: 72).

E

3. *Erythrolamprus guentheri* S.W. Garman, 1884. Mem. Mus. Comp. Zool. (1883) 8(3): 63, 154.

Synonym: *Erythrolamprus albertguentheri* Grazziotin, Zaher, Murphy, Scrocchi, Benavides, Zhang & Bonatto, 2012 *(nomen substitutum).*

Type: Holotype, BMNH 1946.19.57.

Type locality: "Mexico (?)" (in error).

Distribution: Eastern Ecuador (? Azuay, Morona-Santiago, Napo, Orellana, Pastaza) and Peru (Amazonas), 600–950 m.

Sources: Fugler & Walls, 1978 and Perez-Santos & Moreno, 1991.

Remarks: Original description on p. 63 as *E. venustissimus* var. D; the name *guentheri* on p. 154. Günther (1858: 48) and Boulenger (1896a: 203) list only a single juvenile specimen that should be the holotype. Three syntypes listed as BMNH 1946.1.8.5–6 and BMNH 1946.1.8.38.

4. *Erythrolamprus mimus* (Cope, 1869b). Proc. Acad. Nat. Sci. Philadelphia (1868) 20: 307. (*Opheomorphus mimus*)

Synonyms: *Erythrolamprus aesculapii impar* K.P. Schmidt, 1936d, and *Erythrolamprus mimus micrurus* Dunn & Bailey, 1939.

Types: Syntypes (?), ANSP 3689, longest syntype 342 mm (E. Day), location of other syntypes unknown.

Type locality: "a mining district in the higher regions of Equador or New Grenada" [= Ecuador or Colombia].

Distribution: Central America and NW South America. El Salvador, Honduras (Comayagua, Cortés, El Paraíso, Gracias a Dios, Olancho, Yoro), Nicaragua (Jinotega, Matagalpa, Rio San Juan, Zelaya), Costa Rica (Alajuela, Cartago, Guanacaste, Heredia, Limón, Puntarenas), Panama (Canal Zone, Darién), W Colombia (Atlántico, Antioquia, Caquetá, Cauca, Chocó, Cundinamarca, Putumayo, Santander, Valle), E and NW Ecuador (Bolívar, Esmeraldas, Los Rios, Pichincha), Brazil (Rondônia) and E Peru, NSL–2000 m.

Sources: L.D. Wilson & Meyer, 1985, Pérez-Santos & Moreno, 1988, 1991, Campbell & Lamar, 1989, Savage, 2002, Cisneros-Heredia, 2004c, Solórzano, 2004 and McCranie, 2011a.

Remarks: Malnate (1971: 369) lists ANSP 3689 as the holotype but Cope (1869b: 307) mentioned "specimens."

5. *Erythrolamprus ocellatus* W.C.H. Peters, 1868b. Mber. Königl. Akad. Wiss. Berlin 1868((12): 642.

Type: Holotype, ZMB 5059.

Type locality: Unknown. "Süd Amerika" [= South America] *fide* ZMB catalogue. Restricted to Tobago *fide* Emsley (1966a: 129).

Distribution: Trinidad and Tobago (Tobago), 45–140 m.

Sources: Emsley, 1966a–b, 1977, J. Hardy, 1982, J. Hardy & Boos, 1995, J.C. Murphy, 1997 and H.E.A. Boos, 2001.

Remarks: Taxonomic status questionable *fide* J.C. Murphy (1997: 174).

6. *Erythrolamprus pseudocorallus* Roze, 1959a. Acta Biol. Venez. 2(35): 530–533.

Type: Holotype, MBUCV 3789, a 720 mm male (native, 1958).

Type locality: "Cerca de Maracaibo, Estado Zulia,Venezuela, en la región montañosa" [= in the mountains near Maracaibo (10°39'N, 71°38'W), Zulia State, N Venezuela].

Distribution: Northern South America. Colombia (Antioquia, Cundinamarca), W Venezuela (Distrito Federal, Mérida, Trujillo, Zulia) and NW Brazil (Amazonas), 100–1130 m, usually above 800 m.

Sources: Roze, 1966a, Lancini, 1986, Pérez-Santos & Moreno, 1988, J.C. Murphy, 1997, Fuentes & Barrio-Amorgós, 1999a, Curcio et al., 2009b and Navarrete et al., 2009.

Remarks: Probably occurs in ext. NE Colombia (La Guajira) *fide* Pérez-Santos & Moreno (1988: 164).

ERYX Daudin, 1803d (Boidae)

Synonyms: *Clothonia* Daudin, 1803d, *Erix* A.-M.-C. Duméril, 1806 *(nomen emendatum), Haemorrohous* Aldrovandi *in* Fitzinger, 1823, *Gongylophis* Wagler, 1830, *Gonylophis* – Swainson, 1839 *(nomen incorrectum), Cusoria* Gray, 1849a, *Cursoria* A.C.L.G. Günther, 1864a *(nomen emendatum), Congylophis* – Westphal-Castelnau, 1870 *(nomen incorrectum), Eryse* – Angel, 1938 *(nomen incorrectum),* †*Crythiosaurus* Gilmore, 1943, *Gongliophis* – Mishima & Yamazato, 1970 *(nomen incorrectum), Erys* – Tiwari & Sharma, 1971 *(nomen incorrectum), Ery* – Witte, 1975 *(nomen incorrectum), Gonglyophis* – Branch, 1980 *(nomen incorrectum), Neogongylophis* Tokar, 1989a, and *Pseudogongylophis* Tokar 1989a.

Type species: *Anguis jaculus* Linnaeus, 1758.

Distribution: Southeastern Europe, N and E Africa and SW and cen. Asia.

Fossil records: Lower Miocene of Czech Republic, France, Germany and Saudi Arabia, middle Miocene of France and Spain, upper Miocene of Mongolia, Spain and Ukraine, lower Pliocene of Greece and Ukraine, middle Pliocene of Greece and Ukraine, upper Pliocene of France, Greece and Turkey, Pliocene of Greece, and Pleistocene of Greece.

Sources: Jan, 1865b, Tzarevsky, 1916, Stull, 1935, Stimson, 1969, Rage, 1972, Matz, 1974, Latifi, 1985, Stafford, 1986, Tokar, 1986, 1989a, 1995a–b, Sorensen, 1988, I. Das, 1991, Szyndlar & Schleich, 1994, Walls, 1998a and McDiarmid et al., 1999.

Remarks: Fossils referrable to the Boidae are known from the Miocene of Italy and Morocco, and the Pliocene of France and Italy. *Gongylophis* a synonym of *Eryx fide* Noonan & Chippindale (2006: 352) but valid *fide* Tokar, 1995a–b and McDiarmid et al., 1999 (including *G. colubrinus, G. conicus, G. muelleri* and *G. whitakeri*). †*Eryx moldaviensis* Redkozubov, 2003 *(nomen nudum)* listed from MN 14–15 (2.9–4.5 mya) of Moldava.

1. *Eryx borrii* Lanza & Nistri, 2005. Trop. Zool. 18(1): 95–99, figs. 15–17.

Type: Holotype, BMNH 1900.11.28.4, a 390 mm female (A. Donaldson-Smith, 28 Nov. 1900).

Type locality: "Biji (400 m a.s.l.), Somalia" [= Biji, Woqooyi Galbeed Region, NW Somalia, ca. 10°10'N, 44°06'E, elevation 400 m].

Distribution: Northwestern Somalia (Woqooyi Galbeed), 400 m. Known only from holotype.

Sources: Parker, 1949 and Largen & Rasmussen, 1993.

2. *Eryx colubrinus* (Linnaeus, 1758). Syst. Nat., ed. 10, 1: 228. (*Anguis colubrina*)

Synonyms: *Eryx thebaicus* Reuss, 1834, *Eryx scutata* Gray, 1842a, *Eryx jaculus sennariensis* Jan, 1863b *(nomen nudum)*, *Eryx jaculus sennariensis* Jan, 1864 *in* Jan & Sordelli, 1860–1866, *Eryx thebaicus loveridgei* Stull, 1932b, and *Eryx rufescens* Ahl, 1933a.

Type: Holotype, not designated, lost *fide* Flower (1933: 805).

Type locality: "Aegypto" [= Egypt]. Restricted to Thebes, Qena Prov., Upper Egypt *fide* Flower (1933: 804).

Distribution: Northern Africa and Yemen. Central Niger (Agadez), Chad (Borkou, Chari-Baguirmi, Mayo-Kebbi Est), Egypt (Assiut, Beni Suef, S El-Bahr El-Ahmar, Giza, Minya, Qena, Sohag), E Sudan (El-Bahr El-Ahmar, Al Jazirah, Ash Sharqiyah, Sennar), Eritrea, E Ethiopia (Bale, Gemu Gofa, Hararge, Shoa, Sidamo, Tigre, Wollo), Somalia (Awdal, Bakool, Bari, Bay, Galguduud, Gedo, Jubbada Dhexe, Jubbada Hoose, Mogadishu, Mudug, Shabeellaha Dhexe, Shabeellaha Hoose, Togdheer, Woqooyi Galbeed), Kenya (Coast, N Eastern, North-Eastern, N Rift Valley), NE Tanzania (Arusha, Kilimanjaro, Morogoro), and W Yemen, 20–800 m.

Sources: J. Anderson, 1898, Stull, 1932b, Flower, 1933, Villiers, 1950a, 1975, Papenfuss, 1969, Lanza, 1983a, 1990b, Gasperetti, 1988, Le Berre, 1989, Largen & Rasmussen, 1993, Spawls et al., 2002, J.C. Murphy & Schlager, 2003, Lanza & Nistri, 2005, Baha El Din, 2006, Chippaux, 2006, J.-F. Trape & Mané, 2006b and Largen & Spawls, 2010.

3. *Eryx conicus* (Schneider, 1801). Hist. Amph. 2: 268. (*Boa conica*)

Synonyms: *Boa viperina* G. Shaw, 1802, *Boa ornata* Daudin, 1803b *(nomen substitutum)*, *Erix bengalensis* Guérin-Méneville, 1830 *in* 1829–1844, *Eryx conicus laevis* W.C.H. Peters, 1869, *Eryx conicus brevis* Deraniyagala, 1951, and *Eryx conicus gansi* Rajendran, 1986.

Types: Syntypes (3), ZMB 1470 (C.S. John via M.E. Bloch coll.), Museo Barbyensi, and specimen described and illustrated by P. Russell (1796: 5–6, pl. 4).

Type locality: "India orientali" [= eastern India] and "Madras" [= Chennai, Tamil Nadu State, SE India, 13°04'N, 80°15'E, elevation 10 m]. Restricted to Tronquebar [= Tharangambadi, Tamil Nadu State, SE India, 11°02'N, 79°51'E, elevation 5 m] *fide* Stimson (1969: 17).

Distribution: Southern Asia. Southern Pakistan (Balochistan, S Khyber Pakhtunkhwa, Punjab, Sindh), India (Andaman & Nicobars, Andhra Pradesh, Bihar, Goa, Gujarat, Jammu and Kashmir, Karnataka, Kerala, Madhya Pradesh, Maharashtra, Orissa, Rajasthan, Tamil Nadu, Uttar Pradesh, West Bengal), N Sri Lanka (E Eastern, Northern), Bangladesh and Nepal (Bardiya, Mugu, Parsa, Surkhet, Udayapur), 100–2680 m.

Sources: Wall, 1911f, M.A. Smith, 1943, Deraniyagala, 1951, 1963, Minton, 1966, Singh, 1972, Rieppel, 1978, P. Silva, 1980a, Daniel, 1983, Khaire & Khaire, 1986, Tokar, 1986a, M.A.R. Khan, 1988, Hallermann et al., 2001, M.S. Khan, 2002, Schleich & Kästle, 2002, Whitaker & Captain, 2004 and N. Khaire, 2006.

Remarks: Original description based upon P. Russell (1796: 5–6, pl. 4) from Madras. Jar label for ZMB 1470 reads Tronquebar *fide* Bauer et al. (2002: 167). Daudin (1803b: 213) stated Barby. Kluge (1993a: 297) listed type locality as Tranquebar, Tanjore District, SE Madras.

4. *Eryx elegans* (Gray, 1849a). Cat. Snakes Brit. Mus.: 107. (*Cusoria elegans*)

Synonym: *Eryx jaculus czarewskii* Nikolsky, 1916.

Type: Holotype, BMNH 1843.7.21.70 (formerly BMNH IV.19.1a), adult female (East India Company).

Type locality: "Afghanistan."

Distribution: Southwestern Asia. Southern Turkmenistan, N Iran (East Azarbaijan, Razavi Khorasan) and N Afghanistan (Badghis, Kabul), 1800–2440 m.

Sources: Boulenger, 1893a, Nikolsky, 1916, Terent'ev & Chernov, 1949, 1965, Fuhn & Vancea, 1961, S.C. Anderson & Leviton, 1969, Bannikov et al., 1977, Tokar, 1989b and Szczerbak, 1994.

5. *Eryx jaculus* (Linnaeus, 1758). Syst. Nat., ed. 10, 1: 228. (*Anguis jaculus*)

Synonyms: *Anguis cerastes* Linnaeus, 1758, *Boa turcica* G. Olivier, 1801, ? *Tortrix gracilis* Merrem, 1820, *Eryx familiaris* Eichwald, 1831, *Boa charontis* Seetzen, 1855, *Eryx jaculus teherana* Jan, 1865 *in* De Filippi, *Eryx persicus* Nikolsky, 1907, *Eryx montanus* Nikolsky, 1909b, *Eryx jaculus proprius* Tzarevsky, 1916, and *Eryx jaculus urmianus* Rostombekov, 1928.

Type: Lectotype, NHR Lin-12 (formerly MAFR), a 530 mm specimen, designated by J. Anderson (1898: 244), lost *fide* Stimson (1969: 18).

Type locality: "Aegypto" [= Egypt].

Distribution: Southeastern Europe, SW Asia and N Africa. Macedonia, SE Romania, Bulgaria (Blagoevgrad, Haskovo, Kardzhali, Plovdiv, Pazardzhik, Pleven, Sliven, Stara Zagora, Veliko Tarnovo, Vratsa), SW Albania, Macedonia, Greece (Amorgós, Antikáros, Attica, Ios, Iráklia, Kálimnos, Káros, Kas, Kérkira, Kímolos, Límnos, Lesvos, Mílos, Náxos, Páros, Peloponnese, Políaigos, Síkinos, Spétsai, Tínos), Turkey (Adana, Ankara, Antalya, Aydin, Balikesir, Bitlis, Hatay, Icel, Istanbul, Izmir, Kars, Kayseri, Mugla, Siirt, Urfa, Usak), Israel (Central, Gaza Strip, Golan Heights, Haifa, Jerusalem, Northern, Southern, Tel Aviv, West Bank), Syria (Damascus), Lebanon (Beqaa, Mont-Liban), Jordan (Ajloun, Amman, Balqa, Irbid, Jarash, Karak, Maan, Madaba, Mafraq, Tafilah), NE Saudi Arabia, ext. SW Russia (Dagestan), SE Georgia, S Azerbaijan, Armenia, S Turkmenistan, Iraq, NW Iran (Central, East Azerbaijan, Fars, Kermanshahan, Khuzestan, Kordestan, Mazandaran, Razavi Khorasan, Sistan va Baluchestan, West Azerbaijan), NE Morocco (Figuig, Nador, Oujda), N Algeria (Batna, M'Sila, Ouahran), Tunisia (Beja, Gafsa, Gabès, Kairouan, Kasserine, Le Kef, Medenine, Sfax, Siliana, Sousse, Tebessa, Tunis, Zaghouan, Kerkennah Is.), N Libya (Benghazi, Jabal al-Akhdar, Tripoli) and N Egypt (Alexandria, Beheira, Cairo, Daqahlia, Giza, Kafr El-Shaikh, Matrouh, North Sinai, Sharkeya), NSL–1500 m.

Sources: Strauch, 1873, J. Anderson, 1898, Mayet, 1903, Nikolsky, 1916, F. Werner, 1921b, Wall, 1923f, Chernov, 1939, F. Werner, 1939, Terent'ev & Chernov, 1949, 1965, G. Haas, 1951, Chpakowsky & Chnéour, 1954, Wettstein, 1953b, Karaman, 1955, Khalaf, 1959, Marx, 1968, R.J. Clark, 1967b, 1989, Lotze, 1973, Bannikov et al., 1977, Hraoui-Bloquet, 1981, Wütschert, 1984, Blanc & Nouira, 1988, Le Berre, 1989, Petrusev et al., 1990, Tokar, 1991, Tokar & Obst, 1993, Szyndlar & Schleich, 1994, Bons & Geniez, 1996, Schleich et al., 1996, Petkovski et al., 2000, Bouskila & Amitai, 2001, Disi et al., 2001, Joger, 2003, Valakos et al., 2004, Trapp, 2007, Tuniyev et al., 2009, Amr & Disi, 2011, Bar & Haimovitch, 2011 and Stojanov et al., 2011.

Remarks: Illustration of lectotype in J. Anderson (1898: pl. 33A, fig. 1). Type locality restriction to Misir, Turkey *fide* Baran (1976a: 15) invalid. Probably occurs in Sinai, Egypt *fide* Y. Werner, 1982.

6. *Eryx jayakari* Boulenger, 1888e. Ann. Mag. Nat. Hist. (6) 2(12): 508–509.

Synonym: *Eryx fodiens* Annandale, 1913.

Type: Holotype, BMNH 1946.1.7.99 (formerly BMNH 1888.12.29.4), a 400 mm female (A.S.G. Jayakar, 1878–1888).

Type locality: "Muscat, Arabia" [= Muscat, Muscat Govern., NE Oman, 23°37'N, 58°32'E, elevation 45 m].

Distribution: Arabia and Iran. Southern Saudi Arabia (Eastern, Jazan, Najran, Qasim, Riyadh), Kuwait (Fao Is.), Bahrain, United Arab Emirates (Abu Dhabi, Fujairah, Ras al Khaimah, Sharjah), Oman (Al Wusta, Ash Sharqiyah, Dhofar, Muscat), Yemen (Al Hudaydah, Hadhramaut, Shabwah) and SW Iran (Bushehr, Khuzestan), NSL–1100 m.

Sources: Leviton & Anderson, 1967, Gasperetti, 1988, Latifi, 1991, Leviton et al., 1992 and Schätti & Gasperetti, 1994.

7. *Eryx johnii* (P. Russell, 1802). Indian Serp. 2: 18–20, pls. 16–17 (left fig.). (*Boa johnii*)

Synonyms: *Boa anguiformis* Schneider, 1801, *Eryx amica* Wagler, 1825 (*nomen ineditum*), *Eryx indicus* Schlegel, 1837, *Tortrix eryx* Schlegel, 1837, and *Eryx maculatus* Hallowell, 1849.

Type: Lectotype, a 762 mm specimen described and illustrated in P. Russell (1801: 18–19, pl. 16) (C.S. John), designated by M.A. Smith (1943: 113–114), lost *fide* Stimson (1969: 19).

Type locality: "Tranquebar, coast of Coromandel, India" [= Tharangambadi, Tamil Nadu State, SE India, 11°02'N, 79°51'E, elevation 5 m]

Distribution: Southern Asia. Eastern Iran (Baluchistan, N Sistan va Baluchestan), SE Afghanistan (Kandahar), Pakistan (Balochistan, Khyber Pakhtunkhwa, Punjab, Sindh) and India (Andhra Pradesh, Gujarat, Himachal Pradesh, Jammu and Kashmir, Karnataka, Kerala, Maharashtra, Orissa, Punjab, Rajasthan, Tamil Nadu, Uttar Pradesh, West Bengal) and cen. and Nepal (Banke, Bardiya, Mahottari, Parsa), NSL–960 m.

Sources: Wall, 1911f, 1923f, M.A. Smith, 1943, Minton, 1966, Stemmler, 1969a, Bhati & Wadhawan, 1974, R. Sharma, 1974, 1982, Mahendra, 1984, Khaire & Khaire, 1986, M.S. Khan & Tasnim, 1986b, 1987, M.S. Khan, 2002, Tiwari & Shah, 2004, Whitaker & Captain, 2004 and N. Khaire, 2006.

8. *Eryx miliaris* (Pallas, 1773). Reise Russ. Reichs. 2(2): 718. (*Anguis miliaris*)

Synonyms: *Anguis helluo* Pallas, 1814, *Eryx miliaris koslowi* Bedriaga, 1907, *Eryx miliaris roborowskii* Bedriaga, 1907, *Eryx miliaris nogaiorum* Nikolsky, 1910b, *Eryx miliaris incerta* Tzarevsky, 1916, *Eryx miliaris rarus* Tzarevsky, 1916, *Eryx miliaris tritus* Tzarevsky, 1916, *Eryx tataricus bogdanovi* Tzarevsky, 1916, *Eryx nogajorum* –Terent'ev, 1926 *(nomen incorrectum),* and *Eryx rickmersi* F. Werner, 1930b.

Type: Holotype, not designated, a 350 mm specimen, lost *fide* Stimson (1969: 19).

Type locality: "versus mare Caspium, Provinzen Russischen Reichs" [= near Caspian Sea, Russia]. Restricted to N shore of the Caspian Sea between the Volga and Ural rivers, Kazakhstan *fide* Darevsky (1993: 60).

Distribution: Western Asia. Southwestern Russia, Kazakhstan, Uzbekistan, Turkmenistan, Tadzhikistan, Kyrgyzstan, Iran (Esfahan, Hamadan, Khuzestan, Mazandaran, Razavi Khorasan, Sistan va Baluchestan, West Azarbaijan) W and NW Afghanistan (Badghis), NW China (Gansu, Nei Monggol, Xinjiang) and S Mongolia, 150–955 m.

Sources: Boulenger, 1889d, Nikolsky, 1916, F. Werner, 1921b, 1930b, C.H. Pope, 1935, Terent'ev & Chernov, 1949, 1965, Steward, 1971, Bannikov et al., 1977, Latifi, 1985, Li & Wang, 1989, Darevsky, 1993, Zhao & Adler, 1993, Scherbak, 1994, Shaposhninov, 2001, Xu, 2001, Tuniyev et al., 2009 and Yao, 2012.

9. †*Eryx mongoliensis* (Gilmore, 1943). Bull. Amer. Mus. Nat. Hist. 81(4): 377–379, fig. 16. (†*Crythiosaurus mongoliensis*)

Type: Holotype, AMNH 6629, an imperfect skull, posterior portions of both rami and two cervical vertebrae (AMNH Cen. Asiatic Exped., 1925).

Type locality: "Grand Canyon, north of Tsagan Nor, Inner Mongolia; Hsanda Gol, Lower Oligocene."

Distribution: Lower Oligocene (28.4–33.9 mya) of Mongolia. Known only from type locality.

10. *Eryx muelleri* (Boulenger, 1892a). Ann. Mag. Nat. Hist. (6) 9(49): 74–75. (*Gongylophis muelleri*)

Synonym: *Eryx muelleri subniger* Angel, 1938.

Type: Holotype, BMNH 1891.11.20.2, a 370 mm male (E. Marno, 1872–1876).

Type locality: "Sennar, Nubia, [= Sennar Prov., SE Sudan, ca. 13°N, 34°E].

Distribution: West Africa. Southern Mauritania (Assaba, Brakna, Fatick, Kaolack, Hodh Ech Chargui, Trarza), Senegal (Dakar, Matam, Saint-Louis, Tambacounda, Thiès, Ziguinchor), Gambia, Sierra Leone, S Mali (S Gao, Kayes, Koulikoro, Mopti, Ségou, S Tombouctou), N Burkina Faso (Centre, Centre-Est, Centre-Nord, Centre-Ouest, Est, Sahel, Sud-Ouest, Volta-Noire), N Ivory Coast (Daloa), N Ghana (Upper East, Upper West), N Togo (Savanes), N Benin (Alibori), Niger (Agadez, Diffa, Dosso, Maradi, Niamey, Zinder), SW Chad (Chari-Baguirmi, Kanem), N Nigeria (Borno, Sokoto), N Cameroon (Extreme-Nord), SW Chad, N Central African Republic (Vakaga) and SE Sudan (Sennar), 300–470 m.

Sources: Boulenger, 1893a, Angel, 1938, Doucet, 1963, Leston & Hughes, 1968, Hughes & Barry, 1969, Roman, 1980, Hughes, 1983, 2013, Le Berre, 1989, Li & Wang, 1989, Lanza & Nistri, 2005, Villiers & Condamin, 2005, Chippaux, 2006, J.-F. Trape & Mané, 2006b, Chirio & LeBreton, 2007, Chirio, 2009 and Segniagbeto et al., 2011.

11. †*Eryx primitivus* Szyndlar & Schleich, 1994. Amphibia-Reptilia 15(3): 241–243, fig. 3j–l.

Type: Holotype, IPS 9147, one posterior caudal vertebra.

Type locality: "Gorafe 5 (Granada, Spain); middle Pliocene (MN 15)."

Distribution: Middle Pliocene (Ruscinian, MN 15: 3.2–4.2 mya) of Spain. Known only from type series.

12. *Eryx somalicus* Scortecci, 1939b. Ann. Mus. Civ. Stor. Nat. "Giacomo Doria," Genova (1935–1939) 58: 269–270.

Type: Lectotype, MSNM 581 (formerly MSNM 2118), a 225 mm specimen (G. Scortecci, 1931), designated by Lanza & Nistri (2005: 88).

Type locality: "Dintorni di Mogadiscio" [= vicinity of Mogadishu, Mogadishu Region, S Somalia, 2°02'N, 45°21'E, elevation NSL] via lectotype selection.

Distribution: Northeast Africa. Extreme NE Central African Republic, Somalia (Bari, Galguduud, Mogadishu, Shabeellaha Dhexe, Sool, Togdheer) and E Ethiopia (E Hararge), NSL–1065 m.

Sources: H.W. Parker, 1949, Gans & Laurent, 1965, Lanza, 1983, 1990b, Largen & Rasmussen, 1993, Lanza & Nistri, 2005, Chippaux, 2006 and Largen & Spawls, 2010.

Remarks: Holotype illustrated in Lanza & Nistri (2005: figs. 8–9, 11–12).

13. *Eryx tataricus* (Lichtenstein *in* Eversmann, 1823). Reise Orenburg Buchara: 146–147. (*Boa tatarica*)

Synonym: *Eryx speciosus* Tzarevsky, 1916.

Type: Lectotype, ZMB 1461, a 420 mm specimen (E.F. Eversmann, 1820–1822), designated by Bauer et al. (2002: 167).

Type locality: "Tataria" [= Tatarstan, SW Russia] via lectotype selection. Restricted to Aral Sea [SW Kazakhstan and NW Uzbekistan] *fide* M.S. Khan (2002: 72) invalid.

Distribution: Central Asia. Western Tadzhikistan, Turkmenistan, SE Uzbekistan, ext. S Mongolia (S Bayan Hongor, S Gobi Altay, S Ömönö Gobi), ext. N China (Gansu, Nei Monggol, Ningxia, Xinjiang), W Iran (Central, East Azarbaijan, Khuzestan, West Azarbaijan, Zanjhan), 380–2650 m.

Sources: Chernov, 1959, Terentev & Chernov, 1965, Minton, 1966, Bannikov et al., 1977, Borkin et al., 1990a–b, Zhao & Adler, 1993 and M.S. Khan, 2002.

Remarks: Original description reprinted in Lichtenstein (1823: 104).

14. *Eryx vittatus* Chernov, 1959. Trudy Inst. Zool. Parazit. 98: 129–130, fig. 10. (*Eryx tataricus vittatus*)

Type: Holotype, ZISP 14009, a male (S.A. Chernov & A.V. Givozdev, 1934).

Type locality: "Gissar Valley, 20 Km. southwest of Dushanbe, Tadzhik."

Distribution: West-central Asia. Eastern Kazakhstan, NW Kyrgyzstan, E Uzbekistan, ext. W China, E Iran, N Afghanistan (Badakhstan, Faryab, Kabul) and W Pakistan (NW Balochistan), 700–1800 m.

Sources: S.C. Anderson & Leviton, 1969, Leviton & Anderson, 1970a and Tokar, 1989a.

15. *Eryx whitakeri* I. Das, 1991. J. Bombay Nat. Hist. Soc. 88(1): 93–96, figs. 1–3.

Type: Holotype, ZSI 24810, a 550 mm female (Mangalore Snake Park, April 1990).

Type locality: "Mangalore, Karnataka State, India" [= Mangalore, W Karnataka State, SW India, 12°55'N, 74°51'E, elevation 30 m].

Distribution: Southwestern India (Goa, W Karnataka, W Kerala, S Maharashtra), NSL–30 m.

Sources: Whitaker & Captain, 2004 and N. Khaire, 2006.

ETHERIDGEUM Wallach, 1988 (*nomen substitutum*) (Calamaridae)

Synonym: *Padangia* F. Werner, 1924a *(nomen praeoccupatum).*

Type species: *Padangia pulchra* F. Werner, 1924a.

Distribution: Western Indonesia.

Sources: Wallach, 1988, David & Vogel, 1996, Zaher et al., 2009 and I. Das, 2010, 2012.

1. *Etheridgeum pulchrum* (F. Werner, 1924a). Anz. Akad. Wiss. Wien (4): 3. (*Padangia pulchra*)

Type: Holotype, NMW 23449, a 143–145 mm (svl) female (J. Schild, 1890–1904).

Type locality: "Padang, Sumatra" [= Padang, Sumatera Barat Prov., W Sumatra, W Indonesia, 0°57'S, 100°21'E, elevation NSL].

Distribution: Western Indonesia (W Sumatra), NSL. Known only from the type specimen.

Remarks: Supplemental original description in F. Werner (1924b: 54–55, fig. 8).

EUNECTES Wagler, 1830 (Boidae)

Synonyms: *Cenchris* Gronovius, 1763 *(nomen illegitimum), Cenchrus* Link, 1807 *(nomen emendatum), Cenchrias* – Oken, 1816 *(nomen incorrectum), Eunectus* – Swainson, 1839 *(nomen incorrectum), Cenchrina* – A.-M.-C. Duméril, Bibron & Duméril, 1854b *(nomen incorrectum), Eunectis* – Plateau, 1866 *(nomen incorrectum), Euneces* – Lydekker, 1901 *(nomen incorrectum), Eunetes* – Brazil, 1914 *(nomen incorrectum), Enectes* – Briceño-Rossi, 1934 *(nomen incorrectum), Enectus* – Briceño-Rossi, 1934 *(nomen incorrectum), Eunictes* – W.W. Tanner & Avery, 1982 *(nomen incorrectum),* and *Maxhoserboa* Hoser, 2012w *(nomen illegitimum).*

Type species: *Boa murina* Linnaeus, 1758.

Distribution: South America.

Fossil records: Middle Miocene of Colombia and upper Miocene of Brazil.

Sources: Belluomini et al., 1960, Stimson, 1969, J.A. Peters & Orejas-Miranda, 1970, Matz, 1981, Petzold, 1983, Henderson et al., 1995, Wall, 1998a, Wallach, 1998a, McDiarmid et al., 1999, Dirksen & Böhme, 2005, Noonan & Chippindale, 2006 and Bisplinghof & Bellosa, 2007, Hsiou & Albino, 2009 and Hoser, 2012w.

1. *Eunectes beniensis* Dirksen, 2002. Anakondas: 169–174, figs. 115, 117–121.

Type: Holotype, AMNH 101924, a 1700 mm male (W.P. McLean, 30 Aug. 1964).

Type locality: "Trinidad, Beni, Bolivia" [= Trinidad, Beni Dept., NE Bolivia, 14°50'N, 64°54'W, elevation 160 m].

Distribution: Northeastern Bolivia (Beni), 160 m.

Source: Dirksen & Bohme, 2005.

Remarks: Type locality erroneously listed as 11°08'N, 66°10'W *fide* Dirksen & Böhme (2005: 225.).

2. *Eunectes deschauenseei* Dunn & Conant, 1936. Proc. Acad. Nat. Sci. Philadelphia 88: 505–506, pl. 14, fig. 2.

Type: Holotype, ANSP 20891, a 2477+ mm female (?) (Philadelphia Zoo via R.M. de Schauensee, 6 Oct. 1924).
Type locality: "island of Marajo" [mouth of Río Amazonas, E Pará State, N Brazil, ca. 0°53'S, 49°38'W, elevation 15 m].
Distribution: Guianas. Guyana, Suriname (Marowijne), French Guiana (Cayenne, Saint-Laurent-du-Maroni) and NE Brazil (Amapá, Pará, Marajó Is.), NSL–15 m.
Sources: Amaral, 1948, 1977, P. Müller, 1970, Chippaux, 1987, Starace, 1998, Colares-Canto, 2000, Dirksen & R.W. Henderson, 2002, Abuys, 2003 and Pizzatto et al., 2004.
Remarks: Possibly occurs in Suriname *fide* Dirksen & Henderson (2002: 755.2).

3. *Eunectes murinus* (Linnaeus, 1758). Syst. Nat., ed. 10, 1: 215. (*Boa murina*)

Synonyms: *Boa scytale* Linnaeus, 1758, *Boa glauca* Boddaert, 1783, *Coluber raninus* Bonnaterre, 1790, *Boa gigas* Latreille *in* Sonnini & Latreille, 1801a, *Boa anacondo* Daudin, 1803b, *Boa aquatica* Wied-Neuwied, 1824b, and *Eunectes barbouri* Dunn & Conant, 1936.
Type: Holotype, NHR Lin-9 (formerly MAFR), a 930 mm specimen (Mus. Drottn.).
Type locality: "America." Restricted to South America *fide* Dirksen (2002: 31).
Distribution: Amazonia. Southeastern Colombia (Amazonas, Caquetá, Meta, Putumayo, Vaupés), Venezuela (Amazonas, Anzoátegui, Apure, Barinas, Bolívar, Delta Amacuro, Guárico, Monagas, Portuguesa, Sucre, Zulia), Trinidad and Tobago (Trinidad), Guyana (Cuyuni-Mazaruni, Essequibo Islands-West Demerara, Mahaica-Berbice, Potaro-Siparuni, Upper Demerara-Berbice, Upper Takutu-Upper Essequibo), Suriname (Brokopondo, Nickerie, Para, Paramaribo, Saramacca, Wanica), N French Guiana (Cayenne, Saint-Laurent-du-Maroni), Brazil (Amazonas, Amapá, Bahia, Ceará, Goias, Maranhão, Mato Grosso, Mato Grosso do Sul, Minas Gerais, Pará, Piauí, Rondônia, Roraima, Santa Catarina, São Paulo, Sergipe), E Peru (Amazonas, Cusco, Loreto, Madre de Dios), NE Bolivia (Beni) and E Paraguay (Amambay, Canindeyú), NSL–750 m.
Sources: Carrillo de Espinoza, 1966, Roze, 1966a, Emsley, 1977, Bellluomini et al., 1978, Holstrom & Behler, 1981, Chippaux, 1987, Lancini, 1986, Duellman & Salas, 1991, Strimple, 1993,1994, J.C. Murphy & R.W. Henderson, 1997, Dirksen & Böhme, 1998a–b, 2005, Starace, 1998, Abuys, 2003, J.C. Murphy & Schlager, 2003, Duellman, 2005, Rivas et al., 2007, Cacciali, 2008, Infante-Rivero et al., 2008, Navarrete et al., 2009, Silva-Leite et al., 2010, Morato et al., 2011 and C.J. Cole et al., 2013.

4. *Eunectes notaeus* Cope, 1862b. Proc. Acad. Nat. Sci. Philadelphia 14(1): 70.

Synonym: *Epicrates wieningeri* Steindachner, 1903b.
Types: Syntypes (2), ANSP and USNM 4707, a 2845 mm specimen (T.J. Page [Paraguay Exped.], Sept. 1853–Jan. 1856), lost *fide* J.A. Peters *in* Stimson (1969: 22).
Type locality: "Paraguay River and confluents" [Paraguay-Uruguay]. Restricted to 50 km SW Fuerte Coimbra on Río Paraguay at intersection of Bolivia, Brazil and Paraguay [= 20°10'S, 58°10'W, elevation 85 m] *fide* Dirksen (2002: 41).
Distribution: Southern South America. Southwestern Brazil (Mato Grosso, Mato Grosso do Sul, Paraná, Rio Grande do Sul, São Paulo), Bolivia, Paraguay (Alto Paraguay, Boquerón, Central, Misiones, Neembucú, Presidente Hayes), Uruguay (Artigas, Salto) and NE Argentina (Chaco, Corrientes, Entre Ríos, Formosa, Misiones, Salta, Santa Fe), 85–100 m.
Sources: Cope, 1899b, Devincenzi, 1925, Lema & Fabián-Beurmann, 1977, Amaral, 1978, J.D. Williams & Francini, 1991, Cei, 1994, D.R. Norman, 1994, Moura-Leite et al., 1996, Leynaud & Bucher, 1999, Carreira-Vidal et al., 2005 and Cacciali, 2008.

5. †*Eunectes stirtoni* Hoffstetter & Rage, 1977. Ann. Paleont. (Vert.) 63(2): 180–183, figs. 5b, 6a, pl. 1, figs. 2–3.

Type: Holotype, MNHN VIV 7, one right prootic (R. Hoffstetter, J. de Porta, J. Perico & F. Etayo, 1966).
Type locality: "Los Mangos près La Venta, Dépt. Huila, Colombia Fish Bed de la Formation Villavieja, Miocene Moyen (Friasien)" [= Los Mangos (= Les Manguiers) près La Venta, (3°19' lat. N, 75°8' long. W, 440 m d'alt.), situé à 5 km au nord-est de Villavieja, immédiatement au nord de la Quebrada La Venta se situé dans la valleé supérieure du Rio Magdalena, département de Huila, Colombie. Fish Bed de la Formation Villavieja. Miocène moyen (Friasian)].
Distribution: Middle Miocene (Friasien: 15.5–16.3 mya) of Colombia. Known only from type locality.
Source: Hsiou & Albino, 2009.

†*EUPODOPHIS* Rage & Escuillié, 2002 (†Pachyophiidae)

Synonym: †*Podophis* Rage & Escuillie, 2000 (*nomen praeoccupatum*).
Type species: †*Podophis descouensi* Rage & Escuillié, 2000.
Distribution: Upper Cretaceous of Lebanon

Sources: Rage & Escuillié, 2000, 2002, 2003a–b, Rieppel et al., 2003 and Bardet et al., 2008.

Remarks: Replacement name for †*Podophis* Rage & Escuillié, 2000.

1. †*Eupodophis descouensi* (Rage & Escuillié, 2000). C.R. Acad. Sci. Paris (Sci. Terre) 330: 513–520, figs. 1, 1a–c, 2a–d. (†*Podophis descouensi*)

E

Type: Holotype, MGF Rh-E.F. 9001-03, a nearly complete skeleton.

Type locality: "Al Nammoura, Lebanon. Age: Cenomanian."

Distribution: Middle Cretaceous (Cenomanian: 93.9–100.5 mya) of Lebanon.

Sources: Rage & Escuillié, 2000 and Houssaye et al., 2011.

EUPREPIOPHIS Fitzinger, 1843 (Colubridae)

Synonyms: *Proterodon* Hallowell, 1861, *Proterdon* – Maki, 1931 *(nomen incorrectum)*, and *Sinoelaphe* Hoser, 2012ad *(nomen illegitimum)*.

Type species: *Coluber conspicillatus* H. Boie, 1826.

Distribution: East-central Asia.

Sources: Schulz, 1996, Helfenberger, 2001, Utiger et al., 2002, 2005 and Burbrink & Lawson, 2007.

1. *Euprepriophis conspicillatus* (H. Boie, 1826). Isis von Oken 18(2): 211–212. (*Coluber conspicillatus*)

Synonyms: *Proterodon tessellatus* Hallowell, 1861, *Coluber conspicillatus nagasakai* Namiye, 1903, and *Elaphe japonica* Maki, 1931.

Types: Syntypes (7), RMNH 399, an adult, RMNH 400a–b, two juveniles, RMNH 401a–b, two adults, and BMNH (2), a male and juvenile, longest syntype a 381 mm female (J.C. Blomhoff, 1817–1824).

Type locality: "Decima, Japan, [= Dejima Is., previously in Nagasaki harbor, now part of Nagasaki, Japan, 32°45'N, 129°52'E, elevation 10 m].

Distribution: Extreme E Russia (Kunashiri Is.) and Japan (Hokkaido, Honshu, Kyushu, Ryukyus: Fukuejima, Hisakeshima, Kagoshima, Kamishima, Mageshima, Nakadorishima, Nishinoshima, Okinawajima, Takeshima, Tanegashima, Shimojima, Yakushima Is.), 100–3000 m.

Fossil records: Upper Pleistocene of Japan.

Sources: Stejneger, 1907a, Maki, 1931, M. Mori, 1984, Schulz, 1988a, 1988d and Toriba, 1990a.

Remarks: Original description reprinted in H. Boie (1827b: 260–263). Possible types listed in Schulz (1996: 99) and Boulenger (1896a: 52). Type locality possibly in error *fide* Siebold *in* Temminck & Schlegel (1838: iii), who mistakenly corrected it to East Indies.

2. *Euprepiophis mandarinus* (Cantor, 1842a). Zool. Chusan: pl. 12. (*Coluber mandarinus*)

Synonyms: *Ablabes pavo* Annandale, 1912, *Elaphe takasago* Takahashi, 1930, *Holarchus roulei* Angel & Bourret, 1933, and *Elaphe mondarina* – Yao, 2012 *(nomen incorrectum)*.

Type: Holotype, BMNH 1946.1.12.97 (formerly BMNH 1860.3.19.6241), a 700 mm female (T.E. Cantor, July 1840–March 1841, via British East India Comp.).

Type locality: "island of Chusan, or Great Chusan, is situated on the East Coast of China, between the 30° and 31° of North latitude and the 122° and 123° of East longitude; separated from the nearest mainland, Keeto Point, by an arm of the sea, about ten miles across" [= Zhoushan Is., Zhoushan Arch., Zhejiang Prov., E China].

Distribution: Eastern Asia. China (Anhui, Fujian, Gansu, Guangdong, Guangxi, Guizhou, Hebei, Henan, Hubei, Hunan, Jiangsu, Jiangxi, Liaoning, Shaanxi, Sichuan, Tianjin, Xizang, Yunnan, Zhejiang), Taiwan, NE India (E Arunchal Pradesh), N Myanmar (Kachin, Sagaing) and N Vietnam (Dak Lak, Ha Tinh, Lai Chai, Lang Son, Lao Cai, Son La, Tuyen Quang, Vinh Phuc), 450–3000 m.

Sources: H.-F. Li et al., 1986, T. Smith, 1989, Schulz, 1990c, Schulz & Münzenmaier, 1990, Zhao, 1990a, Wei et al., 1992, Zhao & Adler, 1993, Orlov et al., 2000, Gumprecht, 2004c, Quyet & Ziegler, 2004, Whitaker & Captain, 2004, Zhao, 2006, Bain et al., 2007, V.S. Nguyen et al., 2009, Xiang & Li, 2009, I. Das, 2010, 2012 and Yao, 2012.

Remarks: Supplemental original description in Cantor (1842b: 483). Possibly occurs in Laos *fide* I. Das (2010: 279).

3. *Euprepiophis perlaceus* (Stejneger, 1929). Proc. Biol. Soc. Washington 42: 129–130. (*Elaphe perlacea*)

Type: Holotype, USNM 76257, an 1150 mm male (D.C. Graham, June 1928), lost *fide* Schulz (1996: 183).

Type locality: "Yachow prefecture, Szechwan, China" [= Ya'an, Sichuan Prov., W China, ca. 30°N, 103°E].

Distribution: Western China (Sichuan).

Sources: C.H. Pope, 1935, Deng et al., 1983, Schulz, 1989b, Zhao, 1990a and Zhao & Adler, 1993.

Remarks: A synonym of *E. mandarinus* (or a very closely related species) *fide* Utiger et al., 2002.

EXALLODONTOPHIS Cadle, 1999 (Pseudoxyrhophiidae)

Type species: *Pararhadinaea albignaci* Domergue, 1984b.

Distribution: Madagascar.

Sources: Cadle, 1999, Glaw et al., 2005b and Zaher et al., 2009.

Remarks: Previously in *Pararhadinaea*.

1. *Exallodontophis albignaci* (Domergue, 1984b). Bull. Mus. Natl. Hist. Nat. (4) 6A(1): 153–155, figs. 2a–c, 3a. (*Pararhadinaea albignaci*)

Type: Holotype, MNHN 1982.1221 (formerly CAD 970/S), a 237 mm male (J. Thiel, 5 Jan. 1970).

Type locality: "Forêt tropicale d'Analamazaotra (Périnet); altitude 900 m" [=Analamazaotra forest, Andasibe, S Toamasina Prov., E Madagascar, 18°56'S, 48°25'E, elevation 900 m].

Distribution: East-central Madagascar (S Toamasina), 300–900 m.

EXILIBOA Bogert, 1968c (Ungaliophiidae)

Type species: *Exiliboa placata* Bogert, 1968c.

Distribution: Southern Mexico.

Sources: Stimson, 1969, L. Hardy, 1989, Zaher, 1994, Casas-Andreu et al., 1996, Walls, 1998a, McDiarmid et al., 1999 and Wilcox et al., 2002.

1. *Exiliboa placata* Bogert, 1968c. Amer. Mus. Novit. (2354): 6–13, figs. 1b, 2–4, 5a–b, 6a–b, 7, 8b.

Type: Holotype, AMNH 100000, a 412 mm female (J.S. Rowley & C.M. Bogert, 10 Aug. 1967).

Type locality: "near latitude 17° 37' N. and longitude 96° 25' W., at an elevation of approximately 2300 meters on the headwaters of the Río Valle Nacional on the northern slopes of the Sierra de Juárez, in the state of Oaxaca, Mexico,"

Distribution: Sierra de Juárez of S Mexico (N Oaxaca), 2300–2500 m.

F

F

†*FALSERYX* Szyndlar & Rage, 2003 (Tropidophiidae)

Type species: †*Falseryx petersbuchi* Szyndlar & Rage, 2003.
Distribution: Lower Oligocene of Belgium, and lower Miocene of Czech Republic and Germany.
Sources: Szyndlar & Rage, 2003 and Szyndlar et al., 2008.

1. †*Falseryx neervelpensis* Szyndlar, Smith & Rage, 2008. Zool. J. Linn. Soc. 152: 395–399, fig. 1.

Type: Holotype, IRSNB 240, one trunk vertebra.
Type locality: "Boutersem TGV (Belgium). Earliest Oligocene (MP 21)."
Distribution: Lower Oligocene (Rupelian, MP 21: 28.4–33.9 mya) of Belgium. Known only from type locality.

2. †*Falseryx petersbuchi* Szyndlar & Rage, 2003. Non-erycine Booidea Europe: 60–62, figs. 26k–o.

Type: Holotype, BSPG 1976 XXII 6124, one trunk vertebra.
Type locality: "Petersbuch 2 (Germany), lower Miocene (MN 4)."
Distribution: Lower Miocene (Orleanian, MN 4: 16.0–16.9 mya) of Germany and Czech Republic.

FARANCIA Gray, 1842d (Carphophiidae)

Synonyms: *Abastor* Gray, 1849a, and *Faranica* – F. Werner, 1929a *(nomen incorrectum).*
Type species: *Coluber abacurus* Holbrook, 1836.
Distribution: Southeastern USA.
Fossil records: Pleistocene of SE USA.
Sources: H.M. Smith, 1938, J.C. Mitchell, 1982a, Cadle, 1984c, Cundall & Rossman, 1984, Dundee & Rossman, 1989, Ernst & Barbour, 1989, Conant & Collins, 1991, 1998, Palmer & Braswell, 1995, Tennant, 1997, Holman, 2000a, Gravlund, 2001 and Pinou et al., 2004.

1. *Farancia abacura* (Holbrook, 1836). No. Amer. Herp. 1: 119–120, pl. 23. (*Coluber abacurus*)

Synonyms: *Homalopsis reinwardtii* Schlegel, 1837, *Farancia drummondii* Gray, 1842d, *Farancia fasciata* Gray, 1849a, and *Homalopsis crassa* Blyth, 1854a.
Type: Holotype, ANSP 5146, a 1346 mm female (J.E. Holbrook, 1822–1836).
Type locality: "South Carolina" [= USA]. Restricted to Charleston, South Carolina *fide* K.P. Schmidt (1953a: 185).
Distribution: Southeastern USA (Alabama, Arkansas, Florida, S Georgia, ext. S Illinois, ext. SW Indiana, W Kentucky, Louisiana, Mississippi, ext. SE Missouri, E North Carolina, ext. SE Oklahoma, E South Carolina, W Tennessee, E Texas and ext. SE Virginia), NSL–150 m.
Fossil records: Lower Pleistocene (Irvingtonian I) of USA (Florida), and upper Pleistocene (Rancholabrean II) of USA (Florida).
Sources: Karges & McDaniel, 1982, McDaniel & Karges, 1983, Dundee, 1992, Robinette & Trauth, 1992, Palmer & Braswell, 1995 and Werler & Dixon, 2000.

2. *Farancia erytrogramma* (Palisot de Beauvois *in* Sonnini & Latreille, 1801b). Hist. Nat. Rept. 4: 141–142. (*Coluber erytrogrammus*)

Synonyms: *Coluber erythrogrammus* Daudin, 1803c *(nomen emendatum), Duberria ancoralis* Berthold, 1842, *Homalopsis parviceps* Blyth, 1854a, and *Farancia erytrogramma seminola* Neill, 1964.
Type: Holotype, not designated (A.M.F.J. Palisot de Beauvois, spring–summer, 1796), lost *fide* Gillispie (1992: 40).
Type locality: "Amérique septentrionale" [= North America]. Restricted to lower Cooper River, in the vicinity of Charleston, South Carolina *fide* Harper (1940: 714) and Neill (1964: 261).
Distribution: Coastal plain of SE USA (cen. and S Alabama, N and cen. Florida, S Georgia, SE Louisiana, S Maryland, S Mississippi, E North Carolina, E South Carolina, E Virginia), NSL–150 m.
Sources: Gordon, 1957, Neill, 1964, Gibbons et al., 1977, J.C. Mitchell, 1982b and R. Miller & Zyla, 1992.
Remarks: Palisot's entire American collection lost in shipwreck off Halifax, Nova Scotia in 1798 as he returned to Europe (Gillispie, 1992). Restriction of the type locality to Alachua County, Florida *fide* K.P. Schmidt (1953a: 185) rejected *fide* Neill (1957: 141 and 1964: 261).

FERANIA Gray, 1842d (Homalopsidae)

Synonyms: *Pythomorphus* Fitzinger, 1843, *Phytomorphus* Gray, 1849 *(nomen emendatum), Trigonurus* A.-M.-C. Duméril, 1853 *(nomen praeoccupatum), Phytonomorphus* – A.-M.-C. Duméril, Bibron & Duméril, 1854b *(nomen incorrectum), Feranioides*

Carlleyle, 1869, *Feronia* – Phipson, 1888 *(nomen incorrectum), Feranoides* Boulenger, 1896a *(nomen emendatum),* and *Feraniodes* – Gyi, 1970 *(nomen incorrectum).*

Type species: *Homalopsis sieboldii* Schlegel, 1837.

Distribution: Rivers of South Asia.

Sources: Gyi, 1970 and A.B. Kumar et al., 2012.

Remarks: A valid genus *fide* A.B. Kumar et al. (2012: 483), previously a synonym of *Enhydris.*

1. *Ferania sieboldii* (Schlegel, 1837). Essai Phys. Serp. 1: 172, 2: 349–350, pl. 13, figs. 4–5. (*Homalopsis sieboldii*)

Synonym: *Feranioides jamnaeticus* Carlleyle, 1869.

Type: Holotype, RMNH 1168, a 760 mm specimen.

Type locality: "Bengale" [= Bangladesh and NE India (Assam, Manipur, Meghalaya, Mizoram, Nagaland, West Bengal)].

Distribution: Southern Asia. Northern India (Bihar, Karnataka, Madhya Pradesh, Maharashtra, Nagaland, Uttaranchal, Uttar Pradesh, West Bengal), Bangladesh, Nepal (Bardiya, Chitwan, Kanchanpur) and Myanmar (Bago), NSL–200 m.

Sources: Wall, 1898, 1908k, M.A.R. Khan, 1988, Hallermann et al., 2001, Schleich & Kästle, 2002, Tiwari & Shah, 2004, Whitaker & Captain, 2004, Thakur & Watve, 2008 and S.K. Sharma et al., 2013.

Remarks: Penang record doubtful *fide* Wall (1923c: 350 and 1924a: 867). Possibly occurs in Myanmar *fide* I. Das (2012: 150).

FICIMIA Gray, 1849a (Colubridae)

Synonyms: *Ficinia* – Gray, 1849a *(nomen incorrectum), Amblymetopon* A.C.L.G. Günther, 1858, *Amblymepon* – Marschall, 1873 *(nomen incorrectum), Ficinea* – Velasco, 1896 *(nomen incorrectum),* and *Ficinea* – Velasco, 1890a *(nomen incorrectum).*

Type species: *Ficimia olivacea* Gray, 1849a.

Distribution: South-central USA, Mexico and upper Central America.

Fossil record: Pleistocene of USA (New Mexico).

Sources: H.M. Smith & Taylor, 1941, L.M. Hardy, 1975b–c, 1990 and Mendoza-Quijano & Smith, 1993.

1. *Ficimia hardyi* Mendoza-Quijano & Smith, 1993. J. Herp. 27(4): 407–409, fig. 1.

Type: Holotype, MZFC 4877, a 358 mm male (F. Mendoza-Quijano, 13 Jan. 1985).

Type locality: "3.8 km NW Zoquizoquipán (10 km W Zacualtipán), municipality of Metztitlán, Hidalgo, Mexico, 1480 m."

Distribution: Central Mexico (Hidalgo, San Luis Potosí), 1200–2280 m.

Sources: Hernández-Ibarra et al., 1999 and Ramirez-Bautista et al., 1999.

2. *Ficimia olivacea* Gray, 1849a. Cat. Snakes Brit. Mus.: 80.

Synonym: *Ficimia elaiacroma* Jan, 1862b.

Types: Syntypes (2), BMNH 1946.1.5.44–45, a 451 and 361 mm male (H. Finck).

Type locality: "Mexico." Restricted to Orizaba, Veracruz, Mexico *fide* H.M. Smith & Taylor (1950a: 350).

Distribution: East-central Mexico (Chiapas, Distrito Federal, Hidalgo, Morelos, Oaxaca, S Puebla, Querétaro, San Luis Potosí, S Tamaulipas and Veracruz), 100–2300 m.

Sources: L.M. Hardy, 1978, Casas-Andreu et al., 1996 and Castro-Franco & Bustos Zagal, 2004.

3. *Ficimia publia* Cope, 1866. Proc. Acad. Nat. Sci. Philadelphia 18(2): 126.

Synonyms: *Ficimia ornata* Bocourt, 1883 *in* A.H.A. Duméril, Bocourt & Mocquard, 1870–1909, *Ficimia publia taylori* H.M. Smith, 1947, and *Ficimia publia wolffsohni* Neill, 1965.

Type: Lectotype, USNM 16428 (formerly SIM 625 or 726), a 233 mm male (A.C.V. Schott, 28 July 1865), designated by L.M. Hardy (1975b: 135).

Type locality: "Yucatan" [= Yucatán State, Mexico]. Restricted to Chichén Itzá, Yucatán, Mexico *fide* H.M. Smith & Taylor (1950a: 352).

Distribution: Mesoamerica. Southeastern Mexico (Campeche, Chiapas, Guerrero, Hidalgo, Jalisco, Morelos, Oaxaca, Puebla, Quintana Roo, Tabasco, Veracruz, Yucatán), Belize (Cayo, Corozal, Orange Walk), Guatemala (Escuintla, Petén) and NW Honduras (Atlántida, Cortés, Yoro), NSL–1620 m.

Sources: L.M. Hardy, 1980b, L.D. Wilson & Meyer, 1985, García & Valtierra-Azotla, 1996, Lee, 1996, J.A. Campbell, 1998b, Canseco-Márquez & Gutiérrez-Mayén, 2010 and McCranie, 2011a.

4. *Ficimia ramirezi* H.M. Smith & Langebartel, 1950. J. Washington Acad. Sci. (1949) 39(12): 411–412, fig. 1 (right).

Type: Holotype, UIMNH 3767, a 313–325 mm male (J. Ramirez, 6 March 1949).

Type locality: "1 league north of Niltepec, Oaxaca, Isthmus of Tehuantepec, México" [= 5 km N Santiago Niltepec, Oaxaca, S Mexico, ca. 16°36'N, 94°37'W, elevation 80 m].

Distribution: Southern Mexico (Oaxaca), 80 m. Known only from type locality.

Source: L.M. Hardy, 1979.

F

5. *Ficimia ruspator* H.M. Smith & Taylor, 1941. J. Washington Acad. Sci. 31(8): 364–365, figs. 5, 10, 12.

Type: Holotype, UIMNH 25064 (formerly EHT-X 1523 & EHT-HMS 23646), a 142–164 mm female (E.H. Taylor & H.M. Smith, 29 July 1940).

Type locality: "3 miles east of Tixtla (about 10 miles east of Chilpancingo), Guerrero," Mexico.

Distribution: Southern Mexico (cen. Guerrero). Known only from type locality.

Source: L.M. Hardy, 1980a.

Remarks: Possibly occurs in Morelos *fide* L.M. Hardy (1980a: 243.1).

6. *Ficimia streckeri* E.H. Taylor, 1931. Copeia 1931(1): 5–6.

Type: Holotype, KU 9140, a 285–298 mm female (E.H. Taylor, 13 July 1930).

Type locality: "three miles east of Rio Grande City, Texas." [= Starr County].

Distribution: Extreme S USA (S Texas) and NE Mexico (ext. NE Hidalgo, E Nuevo León, Puebla, E San Luis Potosí, Tamaulipas, Veracruz), NSL–1500 m.

Sources: Hardy, 1976a, Canseco-Marquez et al., 2000 and Werler & Dixon, 2000.

Remarks: Holotype erroneously listed as KU 4140 *fide* E.H. Taylor (1931: 5) and L.M. Hardy (1975b: 158 and 1990: 471.1).

7. *Ficimia variegata* (A.C.L.G. Günther, 1858). Cat. Colub. Snakes Brit. Mus.: 7–8. (*Amblymetopon variegatum*)

Types: Syntypes (2), BMNH 1946.1.5.49, a 177 mm female (A. Sallé, 1846–1856), and BMNH 1946.1.6.78, a 385–394 mm female (H. Cuming, 1828–1830).

Type locality: "Mexico." Restricted to Guichicovi, Oaxaca, Mexico *fide* H.M. Smith & Taylor (1950a: 338).

Distribution: Southern Mexico (Hidalgo, NE Oaxaca, S Veracruz), 50–2140 m.

Sources: L.M. Hardy, 1975b, 1980c.

FIMBRIOS M.A. Smith, 1921a (Xenodermatidae)

Type species: *Fimbrios klossi* M.A. Smith, 1921a.

Distribution: Indochina.

Sources: M.A. Smith, 1921a, 1943, Campden-Main, 1970a, Ziegler et al., 2008, V.S. Nguyen et al., 2009, Zaher et al., 2009 and I. Das, 2010, 2012.

1. *Fimbrios klossi* M.A. Smith, 1921a. Proc. Zool. Soc. London 91(2): 425, pl. 1, fig. 1.

Types: Syntypes (3), BMNH 1946.1.15.87–88 (formerly MAS 2143 and MAS 2145), a 345 mm and 310 mm male, and BMNH 1965.1 (formerly MAS 2144), a 395 mm female (M.A. Smith & C. B. Kloss, March–May, 1917).

Type locality: "at Dalat and Camly at 1500 metres, on the Langbian Plateau, Southern Annam, (Indo-China)" [= Da Lat (11°56'N, 108°28'E, 1485 m) and Cam Ly (12°01'N, 108°30'E, 1500 m), Lam Dong Prov., S Vietnam]. Restricted to Da Lat, S Vietnam *fide* Orlov et al. (2003: 220).

Distribution: Indochina. Cambodia, Laos (Champasak) and S Vietnam (Da Lac, Gia Lai, Kon Tum, Lam Dong, Quang Tri, Thua Thien-Hue), 690–1800 m.

Sources: Orlov et al., 2003b and B.L. Stuart & Heatwole, 2008.

2. *Fimbrios smithi* Ziegler, David, Miralles, Kien & Truong, 2008. Zootaxa (1729): 39–41, figs. 2–8.

Type: Holotype, IEBR 3157, a 440 mm male (V.K. Doan & V.H. Nguyen, 15 May 2005).

Type locality: "Phong Nha–Ke Bang National Park, Quang Binh Prov., central Vietnam, in the Cha Noi region at an altitude of ca. 350 m above sea level."

Distribution: Central Vietnam (Quang Binh), 350 m. Known only from type locality.

†*FLORIDAOPHIS* Holman, 2000c (Colubridae)

Type species: †*Floridaophis auffenbergi* Holman, 2000c.

Distribution: Lower Oligocene of USA.

Source: Holman, 1999.

1. †*Floridaophis auffenbergi* Holman, 2000c. Acta Zool. Cracov. (1999) 42(3): 450–451, fig. 2.

Type: Holotype, UF 190884, one trunk vertebra (UF field crews, 1966–1967).

Type locality: "I-75 Local Fauna (Florida Natural History Museum Locality AL018), 1.5 km WSW Gainesville, on the W side of I-75, Micanopy Quadrangle, NE1/4, Sec. 4, T10S, R19E, Alachua County, Florida; Early Oligocene (Whitneyan)."

Distribution: Lower Oligocene (Whitneyan: 30.8–33.3 mya) of the USA (Florida). Known only from type locality.

FORDONIA Gray, 1842d (Homalopsidae)

Synonyms: *Hydropsis* Fitzinger, 1843, *Hemiodontus* A.-M.-C. Duméril, 1853 *(nomen substitutum),* and *Ferdonia* – Phipson, 1888 *(nomen incorrectum).*

Type species: *Homalopsis leucobalia* Schlegel, 1837.

Distribution: Tidal rivers, brackish estuaries and coastal marine waters of Indo-Australia.

Sources: Schlegel & Müller, 1844a, Kopstein, 1931, C. Haas, 1950, Bergman, 1960, Gyi, 1970, C.B. Frith & MacIver, 1978, Cogger et al., 1983a, F. Parker,

1982, Tweedie, 1983, Storr et al., 1986, S.K. Wilson & Knowles, 1988, Gow, 1989, David & Vogel, 1996, O'Shea, 1996, M.J. Cox et al., 1998, Cogger, 2000, Voris et al., 2002, Whitaker & Captain, 2004, J.C. Murphy, 2007b, Alfaro et al., 2008, V.S. Nguyen et al., 2009, Zaher et al., 2009 and I. Das, 2010, 2012.

1. *Fordonia leucobalia* (Schlegel, 1837). Essai Phys. Serp. 1: 171, 2: 345–346, pl. 13, figs. 8–9. (*Homalopsis leucobalia*)

Synonyms: *Fordonia unicolor* Gray, 1849a, *Hemiodontus chalybaeus* Jan, 1863b, *Fordonia bicolor* Theobald, 1868a, *Fordonia papuensis* Macleay, 1877, and *Fordonia variabilis* Macleay, 1878b.
Type: Lectotype, RMNH 1161, a 690 mm female (H.C. Macklot & S. Müller, 1825–1832), designated by Iskandar & Colijn (2001: 92).
Type locality: "Timor, Indonesia" via lectotype selection.
Distribution: Indo-Australia. Eastern India (West Bengal, Andamans and Nicobars: Nicobar Is.), Bangladesh, S Myanmar (Yangon), ext. S Thailand (Phuket, Satun, Phuket Is.), Cambodia, ext. S Vietnam (Ca Mau, Soc Trang), West Malaysia (Penang), Singapore, East Malaysia (Sarawak), Philippines (Luzon, Mindanao), Indonesia (S Papua, Ambon, Java, Kalimantan, Seram, Sumatra, Timor, Weh, Yos Sudarso), Papua New Guinea (Central, Gulf, Western, Daru, Bobo, Bristow and Yule Is.) and N Australia (N Northern Territory, N Queensland, N Western Australia, Bentick, Elcho, Groote Eylandt, Marv, Melville, Milingimbi, Mornington and Yam Is.), NSL–195 m.
Remarks: J.C. Murphy erroneously listed the collector as unknown.

FURINA A.-M.-C. Duméril, 1853 (Elapidae)

Synonyms: *Brachysoma* Fitzinger, 1843 *(nomen praeoccupatum)*, *Glyphodon* A.C.L.G. Günther, 1858, *Mainophis* Macleay, 1877, and *Lunelaps* Worrell, 1961.
Type species: *Calamaria diadema* Schlegel, 1837.
Distribution: Austro-Papua.
Sources: Storr, 1981a, Shine, 1981b, Cogger et al., 1983a, Shine, 1985a, S.K. Wilson & Knowles, 1988, Gow, 1989, Hoser, 1989, 2012e, Hutchinson, 1990, Greer, 1997, Cogger, 2000, Scanlon, 2003a, Scanlon & Lee, 2004, Sanders et al., 2008 and Zaher et al., 2009.

1. *Furina barnardi* (Kinghorn, 1939). Rec. Aust. Mus 20(4): 258–259, figs. 1–2. (*Glyphodon barnardi*)

Type: Holotype, CAS 77798, a 480 mm specimen (H.G. Barnard, 1936).
Type locality: "Fifteen miles south from Duaringa, Queensland" [Australia].
Distribution: Northeastern Australia (Queensland), 20–400 m.
Source: D. Ferguson et al., 2012.

2. *Furina diadema* (Schlegel, 1837). Essai Phys. Serp. 1: 131, 2: 32–33. (*Calamaria diadema*)

Synonyms: *Rabdion occipitale* Girard, 1858a, *Cacophis blackmanii* Krefft, 1869b, and *Denisonia bancrofti* De Vis, 1911.
Types: Syntypes (2), MNHN 3941, a 205 mm male, and MNHN 7668, a 260–320 mm male (J.R.C. Quoy & J.P. Gaimard [Coquille or Astrolabe Voy.], 1822–1829).
Type locality: "Nouvelle Hollande" [= Australia]. Restricted to Port Jackson, New South Wales, Australia *fide* Duméril, Bibron & Duméril (1854b: 1239).
Distribution: Eastern Australia (Australian Capital Territory, New South Wales, S Northern Territory, E Queensland, E South Australia, Bentick Is.), NSL–1115 m.

F

3. *Furina dunmalli* (Worrell, 1955). Proc. Roy. Zool. Soc. N.S.W. (1953–1954): 41–42, 3 figs., 1 pl. (*Glyphodon dunmalli*)

Type: Holotype, AMS 14809 (formerly JD), a 750 mm female (W. Dunmall, Jan. 1954).
Type locality: "Glenmorgan, Queenland" [Australia, 27°15'S, 149°41'E, elevation 290 m].
Distribution: Eastern Australia (SE Queensland), 90–290 m.
Remarks: Collector of type listed as Worrell *fide* Cogger (1979: 205).

4. *Furina ornata* (Gray, 1842c). Zool. Misc. 2(Apr.): 55. (*Elaps ornatus*)

Synonyms: *Brachysoma simile* Macleay, 1878b, *Denisonia bancrofti* De Vis, 1911, and *Pseudelaps christieanus* Fry, 1915.
Type: Holotype, BMNH 1946.1.21.76, a male (W. Buchanan, 1840–1842).
Type locality: "Australia, (Swan River?)."
Distribution: Australia (Northern Territory, Queensland, N South Australia, Western Australia, Barrow, Bathurst, Cotton, Groote Eylandt, Guluwuru, Horn, Koolan, Melville, North West, Prince of Wales, Thursday and Wessel Marchinbar Is.), NSL–1015 m.
Source: Worrell, 1945.

5. *Furina tristis* (A.C.L.G. Günther, 1858). Cat. Col. Snakes Brit.Mus.: 211. (*Glyphodon tristis*)

Synonyms: *Brachysoma triste* – A.C.L.G. Günther, 1863a, *Mainophis robusta* Macleay, 1877, *Denisonia fenestrata* De Vis, 1905, and *Furina somarei* Wells & Wellington, 1984.
Type: Holotype, BMNH 1946.1.18.99, an 813 mm male (J. MacGillivray, 1842–1850).
Type locality: "N.E. Coast of Australia."
Distribution: Southern Papua New Guinea (Central, Western, Daru Is.) and ext. N Australia (ext. NE Northern Territory, ext. N Queensland, Torres Strait and Lizard Is.), NSL–90 m.
Sources: Whitaker et al., 1982, F. Parker, 1982 and O'Shea, 1996.

G

†*GAIMANOPHIS* Albino, 1996a (Boidae)

Type species: †*Gaimanophis tenuis* Albino, 1996a.
Distribution: Lower Miocene of Argentina.
Source: Albino, 1996a.

1. †*Gaimanophis tenuis* Albino, 1996a. Neues Jb. Geol. Paläont. Abh. 199(3): 421–424, fig. 2.

Type: Holotype, MACN-CH 1004-1, one middle trunk vertebra.
Type locality: "near Gaiman, southern cliff of the lower Chubut river valley, north-east Chubut Prov., Argentina; Trelew Member, Sarmiento Formation. Colhuehuapian land-mammal age (early Miocene)."
Distribution: Lower Miocene (Colhuehuapian: 17.5–21.0 mya) of Argentina.

†*GANSOPHIS* Head, 2005 (Colubridae)

Type species: †*Gansophis potwarensis* Head, 2005.
Distribution: Upper Miocene of Pakistan.
Source: Head, 2005.

1. †*Gansophis potwarensis* Head, 2005. Palaeont. Electron. 8(1): 14–15, fig. 7b.

Type: Holotype, H-GSP 49900, one trunk vertebra (E.H. Linday & W.R. Downs III).
Type locality: "Y-908 (6.78 Ma), Potwar Plateau of north-central Pakistan, Dhok Pathan Formation, Winnewala, Miocene."
Distribution: Middle Miocene (6.8 mya) of Pakistan. Known only from holotype.

GARTHIUS Malhotra & Thorpe, 2004c (Viperidae)

Type species: *Trimeresurus chaseni* M.A. Smith, 1931b.
Distribution: East Malaysia.
Sources: Malhotra & Thorpe, 2004c, I. Das, 2010, 2012 and Malhotra et al., 2010.

1. *Garthius chaseni* (M.A. Smith, 1931b). Bull. Raffles Mus. (5): 29, pl. 2, fig. 1. (*Trimeresurus chaseni*)

Type: Holotype, BMNH 1946.1.18.84 (formerly BMNH 1929.12.22.115), a 645 mm male (F.N. Chasen & H.M. Pendlebury, April–May 1929).
Type locality: "Kiau, a Dusun village at the foot of the mountain, Mt. Kinabalu, British North Borneo, 3,000 ft. (The lower mountain zone)," [= Kampong Kiau, Mt. Kinabalu, N Sabah, NE East Malaysia, 6°02'N, 116°31'E, elevation 915 m].
Distribution: Northeastern East Malaysia (N Sabah), 915–1550 m. Known only from vicinity of type locality.

GEAGRAS Cope, 1875a (Colubridae)

Synonym: *Sphenocalamus* J.G. Fischer, 1883.
Type species: *Geagras redimitus* Cope, 1875a.
Distribution: Southern Mexico.
Sources: Hartweg & Oliver, 1940, H.M. Smith, 1943a, L.D. Wilson, 1987d and Casas-Andreu et al., 1996.

1. *Geagras redimitus* Cope, 1875a. J. Acad. Nat. Sci. Philadelphia (1876) (2) 8(2): 141–142.

Synonyms: *Sphenocalamus lineolatus* J.G. Fischer, 1883, and *Tantilla depressa* Dunn, 1928a.
Type: Holotype, USNM 30115 (formerly SIM 8), a 162–166 mm male (A.-L.-J.-F. Sumichrast, 1855–1869).
Type locality: "west side of the State of Tehuantepec, Mexico" [= Pacific versant of Isthmus of Tehuantepec, Oaxaca State, Mexico].
Distribution: Southern Mexico (SE Oaxaca), 100–1000 m.

GEOPHIS Wagler, 1830 (*nomen substitutum*) (Dipsadidae)

Synonyms: *Catostoma* Wagler, 1830 *(nomen oblitum), Rabdosoma* A.-M.-C. Duméril, 1853, *Rhabdosoma* Lichtenstein & Martens, 1856 *(nomen praeoccupatum), Colobognathus* W.C.H. Peters, 1859b, *Geophidium* W.C.H. Peters, 1861e, *Colophrys* Cope, 1868c, *Parageophis* Bocourt, 1883 *in* A.H.A. Duméril, Bocourt & Mosquard, 1870–1909, *Catastoma* – Hoffmann, 1890 *(nomen incorrectum), Dirosema* Boulenger, 1894a, *Seophis* – Tornier, 1904 *(nomen incorrectum), Colophris* – Briceño-Rossi, 1934 *(nomen incorrectum),* and *Geophidrium* – Briceño-Rossi, 1934 *(nomen incorrectum).*
Type species: *Catostoma chalybeum* Wagler, 1830.

Distribution: Latin America.

Sources: H.M. Smith, 1941c, H.M. Smith & Taylor, 1945, Downs, 1967, Cadle, 1984b, Villa et al., 1988, Lips & Savage, 1994, L.D. Wilson et al., 1998, Townsend, 2006a, 2009, L.D. Wilson & Townsend, 2007, Savage & Watling, 2008, Zaher et al., 2009 and Pavón-Vázquez et al., 2011.

Remarks: In accordance with Art. 23.9.2 of the Code (ICZN, 1999), *Geophis* Wagler is designated a *nomen protectum* and *Catostoma* Wagler a *nomen oblitum*.

1. *Geophis anocularis* Dunn, 1920b. Proc. Biol. Soc. Washington 33: 127.

Type: Holotype, USNM 46556, a 287 mm male (E.W. Nelson & E.A. Goldman, 26 July 1894).

Type locality: "Totontepec, Oaxaca, Mexico" [= Totontopec Villa de Morelos, Oaxaca State, S Mexico, 17°15'N, 96°02'W, elevation 1850 m].

Distribution: Southern Mexico (Oaxaca), 1850 m. Known only from the vicinity of type locality.

Sources: J.A. Campbell et al., 1983 and Casas-Andreu et al., 1996.

Remarks: Resurrected from synonymy of *G. dubius fide* J.A. Campbell et al., 1983.

2. *Geophis bellus* C.W. Myers, 2003. Amer. Mus. Novit. (3391): 30–33, figs. 11–14, 15c.

Type: Holotype, KU 110703, a 201 mm male (C.W. Myers, 13 Dec. 1964).

Type locality: "at 700 m above sea level near community of Altos de Pacora (east of Cerro Jefe), Prov. of Panamá, central Panama. The type locality is a few km northeastward of the summit of Cerro Jefe, upper Rio Pacora drainage, at roughly 9°15'N, 79°22'W."

Distribution: Eastern Panama (Darién), 700 m.

Source: Savage & Watling, 2008.

3. *Geophis betaniensis* Restrepo & Wright, 1987. J. Herp. 21(3): 191–193, figs. 1–3.

Type: Holotype, UV 7360, a 296 mm female (J.H. Restrepo, 18 Feb. 1984).

Type locality: "Eastern slope of the Cordillera Occidental, at Corregimiento de Betania, 76°21'W and 4°23'N, 1680 m elev., Municipio de Bolivar, Departamento del Valle, Colombia."

Distribution: Western Colombia (Valle), 1680 m. Known only from type locality.

Sources: Pérez-Santos & Moreno, 1989, Lips & Savage, 1994, C.W. Myers, 2003 and Savage & Watling, 2008.

4. *Geophis bicolor* A.C.L.G. Günther, 1868. Ann. Mag. Nat. Hist. (4) 1(6): 416.

Type: Lectotype, BMNH 1946.1.1.64, a 372 mm male (H. Doorman), designated by Downs (1967: 55).

Type locality: "neighbourhood of Mexico City, Mexico" via lectotype selection.

Distribution: Mexico (Distrito Federal, W Jalisco, México, Michoacán, Veracruz), 1800–2600 m.

Sources: Lara-Góngora & Flores-Villela, 1978, 1980 and H.M. Smith & Flores-Villela, 1993.

5. *Geophis blanchardi* E.H. Taylor & Smith, 1939. Univ. Kansas Sci. Bull. (1938) 25(13): 245–247, figs. 2a–d.

Type: Holotype, FMNH 100037 (formerly EHT-HMS 5194 & EHT-CC 2194), a 388 mm female (E.H. Taylor, 14 Aug. 1936).

Type locality: "about two miles southwest of Acultzingo, Veracruz, Mexico."

Distribution: Cloud forest of S Mexico (Oaxaca, SE Puebla, cen. Veracruz), 2240–2510 m.

Sources: Canseco-Márquez et al., 2004 and Canseco-Márquez & Austin, 2005.

6. *Geophis brachycephalus* (Cope, 1871b). Proc. Acad. Nat. Sci. Philadelphia 23(2): 211–212. (*Colobognathus brachycephalus*)

Synonyms: *Colobognathus dolichocephalus* Cope, 1871b, *Geophis moestus* A.C.L.G. Günther, 1872b, *Geophis chalybaea quadrangularis* A.C.L.G. Günther, 1893 *in* 1885–1902, and *Geophis bakeri* E.H. Taylor, 1954

Type: Holotype, ANSP 3337, a 305–355 mm male (C.H. Van Patten, 1865–1871).

Type locality: "near San José, Costa Rica" [= near San José, San José Prov., Costa Rica, 9°56'N, 84°05'W, elevation 1160 m].

Distribution: Costa Rica (Alajuela, Cartago, Heredia, Limón, Puntarenas, San José) and W Panama (Bocas del Toro, Chiriquí), 15–2115 m.

Sources: E.H. Taylor, 1951, 1954, Pérez-Santos & Moreno, 1988, Sasa, 1993, Savage, 2002, C.W. Myers, 2003 and Savage & Watling, 2008.

Remarks: South American records of this species are tentatively assigned to *G. nigroalbus fide* C.W. Myers (2003: 38).

7. *Geophis cancellatus* H.M. Smith, 1941c. Smithson. Misc. Coll. 99(19): 1–2.

Type: Holotypc, USNM 46440, a 410 mm female (E.W. Nelson & E.A. Goldman, Feb. 1896).

Type locality: "Chicharras, Chiapas, México, ca. 1035 m elevation."

Distribution: Southeastern Mexico (Chiapas), 1035–1230 m.

Sources: Hartweg, 1959 and Landy et al., 1966.

G

8. *Geophis carinosus* L.C. Stuart, 1941b. Occ. Pap. Mus. Zool., Univ. Michigan (452): 3–4.

Type: Holotype, UMMZ 89082, a 240 mm male (L.C. Stuart, 2 Aug. 1940).
Type locality: "Finca San Francisco, 27 km. northeast of Nebaj, El Quiché, Guatemala. Altitude, about 1175 m."
Distribution: Southeastern Mexico (Chiapas, Oaxaca, S Veracruz) and W Guatemala (Sierra de los Cuchumatanes of El Quiché and Huehuetenango), 800–1500 m.
Sources: Pérez-Higareda & Smith, 1991, Casas-Andreu et al., 1996, Lee, 1996 and Nieto-Montes de Oca, 2003.

9. *Geophis chalybeus* (Wagler, 1830). Natur. Syst. Amph.: 194. (*Catostoma chalybeum*)

Synonym: *Rhabdosoma guttulatum* Cope, 1885b.
Type: Holotype, ? ZSM 1940/0 (H.M. von Leuchtenberg), lost *fide* Franzen & Glaw (2007: 249), or ? MSNM, destroyed in 1943 during the World War II.
Type locality: "Mexico."
Distribution: Eastern Mexico (cen. Veracruz), 1150–1800 m.
Sources: Pérez-Higareda & Smith, 1991 and Torre-Loranca et al., 2000.
Remarks: Holotype possibly extant *fide* Downs (1967: 37).

10. *Geophis championi* Boulenger, 1894a. Cat. Snakes Brit. Mus. 2: 321–322, pl. 16, fig. 3.

Type: Holotype, BMNH 1946.1.1.77, a 250 mm male (G.C. Champion & F.D. Godman, 1879–1883).
Type locality: "Chiriqui, Panama" [= Chiriqui Prov., W Panama].
Distribution: Werstern Panama (Chiriquí), 1370 m.
Sources: Dunn, 1942 and C.W. Myers, 2003.

11. *Geophis damiani* L.D. Wilson, McCranie & Williams, 1998. Proc. Biol. Soc. Washington 111(2): 410–413, figs. 1–2.

Type: Holotype, USNM 498356, a 327 mm male (D. Almendarez, J.R. McCranie, L.D. Wilson & K.L. Williams, 26 July 1995).
Type locality: "2.5 airline km NNE La Fortuna (15°26'N, 87°18'W), 1750 m elev., Departmento de Yoro, Honduras."
Distribution: Honduras (Yoro), 1550–1750 m.
Sources: McCranie & Castañeda, 2004a, Townsend, 2009 and McCranie, 2011a.

12. *Geophis downsi* Savage, 1981. Copeia 1981(3): 549–551, fig. 1.

Type: Holotype, LACM 130254, an adult male (N.J. Scott & OTS field crew, 14–16 July 1972).
Type locality: "Costa Rica: Provincia de Puntarenas; Canton Corredores; Las Cruces Field Station and Botanical Garden, 4 km S San Vito de Jaba, 1,200 m."
Distribution: Southwestern Costa Rica (Cordillera Costeña of SE Puntarenas), 1100–1200 m.
Sources: Savage, 1981, 2002 and Solórzano, 2004.

13. *Geophis dubius* (W.C.H. Peters, 1861e). Mber. Königl. Akad. Wiss. Berlin 1861(10): 923–924. (*Geophidium dubium*)

Synonym: *Geophis fuscus* J.G. Fischer, 1886.
Type: Holotype, ZMB 4064, a 260 mm female.
Type locality: Unknown.
Distribution: Southern Mexico (cen. and S Oaxaca, Puebla, cen. Veracruz), 1255–2800 m.
Sources: Bogert & Porter, 1966b, J.A. Campbell et al., 1983, Pérez-Higareda & H.M. Smith, 1988, 1991, H.M. Smith & Pérez-Higareda, 1991, Casas-Andreu et al., 1996 and Nieto-Montes de Oca, 2003.

14. *Geophis duellmani* H.M. Smith & Holland, 1969. Trans. Kansas Acad. Sci. 72(1): 47–49, figs. 1, 2a, 3b.

Type: Holotype, KU 87447, adult female (R.L. Holland, 19 June 1964).
Type locality: "2 mi. S Vista Hermosa, 1750 m., Oaxaca, Mexico."
Distribution: Southern Mexico (N Oaxaca), 1570–1830 m.
Sources: J.A. Campbell et al., 1983 and H.M. Smith & Flores-Villela, 1993.

15. *Geophis dugesii* Bocourt, 1883 *in* A.H.A. Duméril, Bocourt & Mocquard, 1870–1909. Miss. Sci. Mex. Amér. Cen., Rept. 3(9): 573–574, pl. 37, figs. 1, 1a–e.

Synonym: *Geophis aquilonaris* Legler, 1959c.
Type: Holotype, MNHN 1883.278, a 220 mm male (O. Navarro via A.A.D. Dugès).
Type locality: "Tangancicuaro (Mexique)" [= Tangancicuaro de Arista, Michoacán State, SW Mexico, 19°54'N, 102°08'W, elevation 1700 m].
Distribution: Western Mexico (Aguascalientes, SW Chihuahua, Durango, Guanajuato, Jalisco, Michoacán, E Sinaloa, Sonora, Zacatecas), 1400–2500 m.
Sources: Dugès, 1884, H.M. Smith, 1939, Hartweg, 1959, Legler, 1959c, Webb, 1977, W.W. Tanner, 1985, Rodriguez-Torres & Vázquez-Díaz, 1990, Vázquez-Díaz & Quintero-Díaz, 1999, Enderson & Bezy, 2007, Recchio et al., 2007, Arenas-Monroy et al., 2012, García-Balderas & Quintero-Díaz, 2012 and R.A. Villa et al., 2012b.

16. *Geophis dunni* K.P. Schmidt, 1932b. Copeia 1932(1): 8.

Type: Holotype, MCZ 31870, a 310–367 mm female (W.B. Richardson, 1891–1927).
Type locality: "Matagalpa, Nicaragua" [= Matagalpa, Matagalpa Dept., cen. Nicaragua, 12°55'N, 85°55'W, elevation 735 m].
Distribution: Central Nicaragua (Matagalpa), 735 m. Known only from type locality.
Sources: G. Köhler, 1999b, Townsend, 2006, 2009 and Savage & Watling, 2008.
Remarks: Holotype removed from the stomach of a *Micrurus nigrocinctus.*

17. *Geophis fulvoguttatus* Mertens, 1952b. Zool. Anz. 149(5–6): 134–135.

Type: Holotype, SMF 43248, a 150 mm specimen (A. Zilch, 26–27 Aug. 1951).
Type locality: "Hacienda Monte Cristo, 2200 m. H., Gebirge von Metapan, Dept. Santa Ana, El Salvador."
Distribution: Western El Salvador (Santa Ana) and SW Honduras (Copán, Ocotepeque), 1680–2200 m.
Sources: Mertens, 1952c, J.A. Campbell et al., 1983, McCranie & Wilson, 1991, G. Köhler, 1996b, G. Köhler et al., 2005, Townsend & Wilson, 2006b, Townsend, 2009 and McCranie, 2011a.
Remarks: Probably occurs in highlands of S Guatemala *fide* Townsend and Wilson (2006b: 158).

18. *Geophis godmani* Boulenger, 1894a. Cat. Snakes Brit. Mus. 2: 322, pl. 16, fig. 4.

Type: Lectotype, BMNH 1946.1.6.41, a 400–401 mm female (O. Salvin & F.D. Godman, 1861–1875), designated by Downs (1967: 72).
Type locality: "Irazu, Costa Rica" [= Volcan Irazú, Cartago Prov., Costa Rica, 9°59'N, 83°51'W] via lectotype selection.
Distribution: Central Costa Rica (Alajuela, SE Cartago, Puntarenas, San José) and W Panama (Chiriquí, Panamá), NSL–2200 m.
Sources: E.H. Taylor, 1951, 1954, Savage, 1981, 2002, Lips & Savage, 1994, C.W. Myers, 2003 and Solórzano, 2004.

19. *Geophis hoffmanni* (W.C.H. Peters, 1859b). Mber. Königl. Akad. Wiss. Berlin 1859(3): 276, pl., figs. 2, 2a–c. (*Colobognathus hoffmanni*)

Synonyms: *Geophis acutirostris* E.H. Taylor, 1954, and *Geophis bartholomewi* Brattstrom & Howell, 1954.
Type: Lectotype, ZMB 1870, a 254 mm female (K. Hoffmann, 1853–1859), designated by Downs (1967: 156).
Type locality: "Costa Rica" via lectotype selection.
Distribution: Lower Central America and Colombia. Eastern Honduras (Colón, El Paraíso, Gracias a Dios, Olancho), Nicaragua (Nueva Segovia, Matagalpa, Zelaya), Costa Rica (Alajuela, Cartago, Guanacaste, Heredia, Limón, Puntarenas, San José), Panama (Chiriquí, Canal Zone, Cocle) and N Colombia, 20–1670 m.
Sources: E.H. Taylor, 1951, 1954, H.M. Smith & Smith, 1964, L.D. Wilson & Meyer, 1985, G. Köhler, 1999b, Savage, 2002, C.W. Myers, 2003, Solórzano, 2004, Townsend, 2006, 2009, Savage & Watling, 2008 and McCranie, 2011a.
Remarks: Lectotype listed as ZMB 4003 *fide* Savage & Watling (2008: 588).

G

20. *Geophis immaculatus* Downs, 1967. Misc. Publ. Mus. Zool., Univ. Michigan (131): 90–91.

Type: Holotype, UMMZ 107297, a 305 mm female (L.C. Stuart, 18 April 1949).
Type locality: "Guatemala, Quetzaltenango, Finca Lorena, Pacific versant of Guatemala; about 1700 meters above sea level."
Distribution: Southern Mexico (Chiapas) and SW Guatemala (Quetzaltenango), 1700 m.
Source: Espinosa et al., 1999.

21. *Geophis incomptus* Duellman, 1959. Occ. Pap. Mus. Zool., Univ. Michigan (605): 3–5, fig. 2, pl. 1 (lower, a paratype).

Type: Holotype, UMMZ 118840 (formerly WED 12437), a 275 mm male (W.E. Duellman & J. Wellman, 18 June 1958).
Type locality: "Dos Aguas, Michoacán, México (2100 meters)."
Distribution: Southwestern Mexico (Sierra de Coalcomán of Michoacán), 2100 m. Known only from type locality.
Sources: Bogert & Porter, 1966a and J.A. Campbell & Murphy, 1977a.

22. *Geophis isthmicus* (Boulenger, 1894a). Cat. Snakes Brit. Mus. 2: 307–308. (*Atractus isthmicus*) (*nomen substitutum*)

Synonym: *Rhabdosoma zebrinum* Bocourt, 1883 *in* A.H.A. Duméril, Bocourt & Mocquard, 1870–1909 *(nomen praeoccupatum).*
Type: Holotype, MNHN 1984, a 155 mm female.
Type locality: "Tehuantepec" [= Tehuantepec, Oaxaca, Mexico].
Distribution: Southern Mexico (Oaxaca). Known only from the holotype.
Sources: Bocourt, 1883 *in* A.H.A. Duméril, Bocourt & Mocquard, 1870–1909, Dunn, 1928 and Casas-Andreu et al., 1996.

23. *Geophis juarezi* Nieto-Montes de Oca, 2003. Herpetologica 59(4): 574–578, figs. 1a–c.

Type: Holotype, MZFC 2236, a 275+ mm female (P. García & natives, April 1986)
Type locality: "vicinity of Metates, municipality of Santiago Comaltepec, Sierra de Juárez, Oaxaca, Mexico, 17°39' 03"N, 96°21' 26"W, 900 m elevation."
Distribution: Southern Mexico (N Oaxaca), 800–1200 m. Known only from vicinity of the type locality.

24. *Geophis juliai* Pérez-Higareda, Smith & López-Luna, 2001. Bull. Maryland Herp. Soc. 37(2): 42–44, fig. 1.

G

Type: Holotype, UNAM-LT 2775, a 314 mm male (native, 10 Aug. 1984).
Type locality: "Estación de Biología Tropical Los Tuxtlas (EBTLT), 150m, southern Veracruz, Mexico."
Distribution: Eastern Mexico (S Veracruz), 150–600 m.

25. *Geophis laticinctus* H.M. Smith & Williams, 1963. Herpetologica 19(1): 24–27, fig. 1.

Synonym: *Geophis laticinctus albiventris* H.M. Smith & Holland, 1969.
Type: Holotype, UIMNH 51352, a 384 mm female (T. MacDougall, 23 Dec. 1961).
Type locality: "±1800 m., near Colonia Francisco I. Madero, municipality of Jitotol, Chiapas, México" [= vicinity of Francisco I. Madero (16°48'N, 93°46'W), W Chiapas State, SE Mexico, elevation ca. 1800 m].
Distribution: Southeastern Mexico (W Chiapas, Oaxaca), 600–2000 m.
Sources: J.D. Johnson, 1976, 1979.
Remarks: Type locality elevation probably in error as the highest mountains within 10 km are 1100 m.

26. *Geophis laticollaris* H.M. Smith, Lynch & Altig, 1965. Nat. Hist. Misc., Acad. Chicago Nat. Sci. (180): 2–4, figs. 1–2.

Type: Holotype, UIMNH 57170, a 139 mm female (R. Altig, 29 June 1964).
Type locality: "3 miles south of Putla, Oaxaca, Mexico" [= ca. 16°59'N, 97°56'W, elevation 750 m].
Distribution: Southern Mexico (Oaxaca), 750 m. Known only from type locality.
Sources: H.M. Smith & Chiszar, 1992a and Pavón-Vázquez et al., 2011.
Remarks: Previously a synonym of *G. sallei*. Possible relationship to *G. russatus*.

27. *Geophis latifrontalis* S.W. Garman, 1884. Mem. Mus. Comp. Zool. (1883) 8(3): 103.

Synonym: *Geophis semiannulatus* H.M. Smith, 1941g.
Type: Holotype, MCZ 4538, a 387 mm female (E. Palmer, Aug. 1879).
Type locality: "mts. of Alvarez, fifty miles south of San Luis Potosi, Mexico."
Distribution: Central Mexico (SE Hidalgo, Querétaro, S San Luis Potosí, SW Tamaulipas), 800–2600 m.
Sources: H.M. Smith, 1941 and Dixon & Thomas, 1974.

28. *Geophis maculiferus* E.H. Taylor, 1942a. Univ. Kansas Sci. Bull. (1941) 27(7): 119–121, fig. 1.

Type: Holotype, UIMNH 25078 (formerly EHT-HMS 23552 & EHT-CC X-1648), a 152 mm male (E.H. Taylor, 14 Aug. 1940).
Type locality: "on the `Huetamo' road, about a kilometer north of the village of Cício, Michoacán, 17 Km. south of the Mexico-Guadalajara highway, Mexico" [= 1 km N Tzitzio, Michoacán, Mexico, 19°36'N, 100°55'W, elevation 1630 m *fide* Duellman, 1961: 97].
Distribution: Southwestern Mexico (NE Michoacán), 1630 m. Known only from type locality.
Sources: E.H. Taylor, 1941, Duellman, 1961 and Bogert & Porter, 1966a.
Remarks: Duellman (1961: 97) erroneously listed the type as female.

29. *Geophis mutitorques* (Cope, 1885b). Proc. Amer. Philos. Soc. 22: 384. (*Rhabdosoma mutitorques*)

Synonym: *Rhabdosoma longiceps* Cope *in* Ferrari-Pérez, 1886.
Type: Lectotype, ANSP 14762, a 337 mm female (S. Bernad, 1885), designated by Downs (1967: 105).
Type locality: "Zacualtipan, Mexico" [Hidalgo State, elevation 1975 m] via lectotype selection. Emended to highlands [= 2000–2200 m] above Zacualtipan, Mexico *fide* Malnate (1971: 369).
Distribution: Mexico (Hidalgo, Puebla, Querétaro, San Luis Potosí, Veracruz), usually 1500–2525 m.
Sources: H.M. Smith, 1941 and Bogert & Porter, 1966.

30. *Geophis nasalis* (Cope, 1868c). Proc. Acad. Nat. Sci. Philadelphia 20(2): 131, 3 figs. (*Catostoma nasale*)

Type: Lectotype, ANSP 3320, a 293 mm female (C.H. Van Patten, 1865–1868), designated by Downs (1967: 160).
Type locality: "near the city of Guatemala" [= near Guatemala city (14°38'N, 90°32'W, elevation 1520 m), Guatemala Dept., S Guatemala].
Distribution: Southeastern Mexico (SE Chiapas) and S Guatemala (Chimaltenango, Guatemala, Sacatepéquez, San Marcos, Santa Rosa, Sololá, Suchitepéquez), 600–1830 m.

31. *Geophis nephodrymus* Townsend & Wilson, 2006a. Proc. Biol. Soc. Washington 119: 151–155, figs. 1–2.

Type: Holotype, UF 142577, a 253 mm female (S.M. Hughes & J.H. Townsend, 11 July 2004).
Type locality: "Sendero Las Minas" (15°29.525'N, 88°12.705'W), 1570 m elevation, Parque Nacional El Cusuco, Sierra de Omoa, Departamento de Cortés, Honduras."
Distribution: Honduras (Sierra de Omoa in Cortés), 1540–1780 m.
Sources: Townsend 2006a, 2009, Townsend et al., 2006 and McCranie, 2011a.

32. *Geophis nigroalbus* Boulenger, 1908j. Ann. Mag. Nat. Hist. (8) 2(12): 522. (*nomen corrigendum*)

Synonyms: *Geophis nigro-albus* Boulenger, 1908j *(nomen incorrigendum), Catastoma nigroalbum* – Amaral, 1930f *(nomen corrigendum),* and *Geophis negroalbus* – C.W. Myers, 2003 *(nomen incorrectum).*
Type: Holotype, BMNH 1946.1.6.50 (formerly BMNH 1909.4.30.71), a 134–135 mm male (N.G. Palmer).
Type locality: "Pavas" [= La Cumbre, Valle del Cauca Dept., Colombia, 3°41'N, 76°35'W, elevation 1400 m].
Distribution: Panama (Panamá) and Colombia (Antioquia, Santander, Valle del Cauca), 1350–1680 m.
Sources: Restrepo & Wright, 1987, Pérez-Santos & Moreno, 1989, Lips & Savage, 1994, Ibánez D., et al., 2001, C.W. Myers, 2003 and Savage & Watling, 2008.
Remarks: Photographs of the holotype are in C.W. Myers (2003: figs. 15a, 16).

33. *Geophis nigrocinctus* Duellman, 1959. Occ. Pap. Mus. Zool., Univ. Michigan (605): 1–3, fig. 1, pl. 1 (upper).

Type: Holotype, UMMZ 118841 (formerly WED 12433), a 368 mm male (J. Wellman, 18 June 1958).
Type locality: "Dos Aguas, Michoacán, México (2100 meters)" [= Dos Aguas, SW Michoacán State, SW Mexico, 18°48'N, 102°56'W, elevation 2215 m].
Distribution: Southwestern Mexico (Jalisco, SW Michoacán), 1900–2100 m.
Sources: J.A. Campbell & Murphy, 1977a and Flores-Villela et al., 1995.

34. *Geophis occabus* Pavón-Vázquez, García-Vázquez, Blancas-Hernández & Nieto-Montes de Oca, 2011. Herpetologica 67(3): 334–339, fig. 1.

Type: Holotype, MZFC 25530, a 178 mm female (J.C. Blancas-Hernández, summer, 2006).
Type locality: "El Molote, municipality of Atoyac de Álvarez, Guerrero, Mexico (17°25'14.4"N, 100°10'15.7"W, datum = WGS84), elevation 1787 m."
Distribution: Southwestern Mexico (Sierra Madre del Sur of S Guerrero), 1785–2015 m.

35. *Geophis omiltemanus* A.C.L.G. Günther, 1893 *in* 1885–1902. Biol. Cen.-Amer., Rept. Batr.: 92, pl. 33, fig. a. (*Geophis omiltemana*)

Type: Lectotype, BMNH 1946.1.6.37, a 292 mm male (H.H. Smith, 1887–1889), designated by Downs (1967: 124).
Type locality: "Omilteme, Guerrero, Mexico" [= Omilteme, cen. Guerreo, SW Mexico, 17°33'N, 99°41'W, elevation 2440 m] via lectotype selection.
Distribution: Southwestern Mexico (Guerrero, W Oaxaca), 2440 m.
Sources: L.M. Hardy 1976 and J.A. Campbell, 1988.

G

36. *Geophis petersii* Boulenger, 1894a. Cat. Snakes Brit. Mus. 2: 321, pl. 16, fig. 2.

Type: Lectotype, BMNH 1946.1.6.31, a 200–208 mm male (H. Doorman), designated by Downs (1967: 166).
Type locality: "City of Mexico" [= Mexico city, Distrito Federal, cen. Mexico, 19°26'N, 99°08'W, elevation 2245 m] via lectotype selection. Restricted to Pátzcuaro, Michoacán, Mexico *fide* H.M. Smith & Taylor (1950a: 335) but rejected *fide* Downs (1967: 168).
Distribution: Southern Mexico (Distrito Federal, N Michoacán), 1800–2500 m.
Sources: Hartweg, 1959 and Duellman, 1961.
Remarks: Type locality dubious *fide* Duellman (1961: 98) and Downs (1967: 166).

37. *Geophis pyburni* J.A. Campbell & Murphy, 1977a. J. Herp. 11(4): 397–399, figs. 1–2.

Type: Holotype, UTA 4404, a 299 mm female (J.A. Campbell, 8 July 1974).
Type locality: "Rancho La Pastilla, ca. 2.5 airline kms. W. Dos Aguas, Sierra de Coalcomán, Michoacán, Mexico, 2164 m" [= 18°49'N, 102°58'W].
Distribution: Southwestern Mexico (SW Michoacán), 2000–2180 m. Known only from type locality.

38. *Geophis rhodogaster* (Cope, 1868c). Proc. Acad. Nat. Sci. Philadelphia 20(2): 130–131, 2 figs. (*Colophrys rhodogaster*)

Type: Lectotype, ANSP 3317, a 305 mm female (C.H. Van Patten, 1865–1868), designated by Downs (1967: 93).
Type locality: "the elevated country in the neighborhood of the city of Guatemala" [= highlands near Guatemala city (14°38'N, 90°32'W, elevation 1520 m), Guatemala Dept., S Guatemala] via lectotype selection.
Distribution: Upper Central America. Southeastern Mexico (SW Chiapas), S Guatemala (Baja Verapaz, Chimaltenango), Guatemala (San Marcos, Sololá), W El Salvador (Chalatenango, Santa Ana) and Honduras (Ocotepeque), 1500–2745 m.
Sources: E.N. Smith, 1995, Nieto Montes de Oca, 2003, G. Köhler et al., 2005, Townsend, 2006, 2009, McCranie & Castañeda, 2007 and McCranie, 2011a.

39. *Geophis rostralis* (Jan, 1865 *in* Jan & Sordelli, 1860–1866). Icon. Gén. Ophid. 1(12): 8, pl. 2, figs. 2, a–b, f–g, n–p, r, v. (*Elapoides rostralis*)

Type: Holotype, ZMB 6407.
Type locality: "Mexique" [= Mexico].
Distribution: Southern Mexico (Sierra Madre del Sur of Oaxaca), 1680 m.
Sources: H.M. Smith, 1959, Bogert & Porter, 1966, Downs, 1967 and J.A. Campbell et al., 1983.
Remarks: Holotype listed as MNHN *fide* original description (p. 8). Possibly a synonym of *G. dubius fide* H.M. Smith et al. (1959: 267). See remarks in L.D. Wilson & Townsend (2007: 17).

G

40. *Geophis russatus* H.M. Smith & Williams, 1966a. J. Ohio Herp. Soc. 5(3): 90–91, fig. 1. (*Geophis sallaei russatus*)

Type: Holotype, UIMNH 61409, a 323 mm male (T. MacDougall, 25 May 1965).
Type locality: "La Concepción, near Putla, Oaxaca, México."
Distribution: Southern Mexico (Oaxaca), 1000 m. Known only from type locality.
Sources: Casas-Andreu et al., 1996 and Pavón-Vázquez et al., 2011.

41. *Geophis ruthveni* F. Werner, 1925. Sitz. Akad. Wiss. Wien Math.-Natur. Kl. Abt. I, 134(1–2): 60.

Type: Holotype, NMW 16508, a 245 mm male (native, 1899).
Type locality: "Sarapigui, Brasilien" [= Sarapigui, Brazil] (in error). Corrected to Sarapiqui, Costa Rica [= Puerto Viejo de Sarapiqui, Heredia Prov., N Costa Rica, 10°27'N, 84°00'W, elevation 40 m] *fide* Dunn *in* Savage (1960f: 31) and Downs (1967: 75).
Distribution: Northwestern Costa Rica (Alajuela, Guanacaste, Heredia, Limón), 100–1600 m.
Sources: Savage, 1981, 2002 and Solórzano, 2004.

42. *Geophis sallaei* Boulenger, 1894a. Cat. Snakes Brit. Mus. 2: 318, pl. 16, fig. 1.

Type: Lectotype, BMNH 1946.1.6.27, a 232 mm female (A. Sallé, 1846–1856), designated by Downs (1967: 168).
Type locality: "Mexico." Restricted to Pluma Hidalgo, Oaxaca, Mexico [= Pluma Hidalgo, S Oaxaca State, S Mexico, 15°56'N, 96°25'W, elevation 1240 m] *fide* H.M. Smith & Taylor (1950a: 338).
Distribution: Southern Mexico (Sierra Madre del Sur of Oaxaca), 1000–2000 m.
Sources: Hartweg, 1959, H.M. Smith & Chiszar, 1992a, Casas-Andreu et al., 1996 and Pavón-Vázquez et al., 2011.

43. *Geophis semidoliatus* (A.-M.-C. Duméril, Bibron & Duméril, 1854a). Erpét. Gén. 7(1): 93–95. (*Rabdosoma semidoliatum*) (*nomen corrigendum*)

Synonyms: *Rabdosoma semi-doliatum* A.-M.-C. Duméril, Bibron & Duméril, 1854a *(nomen incorrigendum), Catastoma semidoliatum* – Cope, 1860b *(nomen corrigendum),* and *Geophis semidoliata* – A.C.L.G. Günther, 1893 *in* 1885–1902.
Type: Lectotype, MNHN 4522, a 310 mm female (Parduracki), designated by Downs (1967: 133).
Type locality: "Mexique" [= Mexico]. Restricted to Córdoba, Veracruz, Mexico *fide* H.M. Smith & Taylor (1950a: 347).
Distribution: Eastern Mexico (Hidalgo, E Puebla, cen. and S Veracruz), 500–1400 m.
Sources: Hartweg, 1959, Pérez-Higareda & Smith, 1991, Canseco-Marquez et al., 2000, Pérez-Higareda et al., 2001 and Ramírez-Bautista et al., 2010.

44. *Geophis sieboldi* (Jan, 1862b). Arch. Zool. Anat. Fis. 2(1): 21–22, pl. 17. (*Elapoides sieboldi*)

Types: Syntypes (4), MSNM (2), a 345 mm female and 155 mm male, NMW, a 360 mm male, and MM, a 340 mm female, probably all destroyed in 1943 during World War II.
Type locality: "Messico" [= Mexico] and "Guadalupa" [= ? Guadalupe, Oaxaca, Mexico].
Distribution: Southwestern Mexico (México, S Michoacán).
Sources: Lara-Góngora & Flores-Villela, 1980 and Savage and Watling, 2008.
Remarks: Distribution uncertain *fide* Downs (1967: 171). Guerrero record in error *fide* Pavón-Vázquez et al. (2011: 339).

45. *Geophis talamancae* Lips & Savage, 1994. Proc. Biol. Soc. Washington 107(2): 410–413, fig. 1.

Type: Holotype, LACM 147196 (formerly CRE 5343), a 212–218 mm female (K.R. Lips, 1 Sept. 1992).
Type locality: "Costa Rica: Puntarenas Prov.: Cantón Coto Brus: Zona Protectora Las Tablas, Finca Jaguar, 1800 m elevation."
Distribution: Southwestern Costa Rica (Puntarenas) and W Panama (Chiriquí), 1200–1800 m.
Sources: Savage, 2002, C.W. Myers, 2003, Solórzano, 2004 and Savage and Watling, 2008.
Remarks: Holotype is a juvenile (not an adult) *fide* C.W. Myers (2003: 35).

46. *Geophis tarascae* Hartweg, 1959. Occ. Pap. Mus. Zool., Univ. Michigan (601): 1–4, fig. 1, pl. 1.

Type: Holotype, UMMZ 99151, a 302 mm male (N. Hartweg, 18 July 1947).

Type locality: "at 5500 feet above sea level in the Parque Nacional on the outskirts of Uruapan, Michoacán" [Mexico].
Distribution: Southwestern Mexico (Cordillera Volcánica of Jalisco and Michoacán), 1400–1965 m.
Sources: Dixon, 1968 and Medica et al., 1975.

47. *Geophis tectus* Savage & Watling, 2008. Zool. J. Linn. Soc. 153: 585–586.

Type: Holotype, MCZ 19326, a 253 mm male (E.R. Dunn & C.B. Duryea, July–Aug. 1923).
Type locality: "La Loma (= Buena Vista), Distrito de Chiriquí Grande, Provincia de Bocas del Toro, Panama; c. 8°50'N, 82°13'W (300 m)."
Distribution: Panama (Bocas del Toro, Panamá), 40–1700 m.

48. *Geophis zeledoni* E.H. Taylor, 1954. Univ. Kansas Sci. Bull. 36(11): 693–695, figs. 4a–c.

Type: Holotype, KU 31992, a 397 mm female (E.H. Taylor, 24 July 1952).
Type locality: "Finca Zeledón, between Volcán Barba and Volcán Poás (elev. circa 6000 ft.), Costa Rica."
Distribution: Costa Rica (probably Alajuela, Heredia, San José), 1830–2100 m.
Sources: Savage, 2002, C.W. Myers, 2003, Solórzano, 2004 and Savage & Watling, 2008.

GERARDA Gray, 1849a (Homalopsidae)

Synonyms: *Campylodon* A.-M.-C. Duméril, 1853 *(nomen praeoccupatum), Campilodon* Jan, 1861c *(nomen emendatum), Campitodon* Jan, 1862b *(nomen incorrectum), Heleophis* F. Müller, 1884, *Heliophis* – Cope, 1887c *(nomen incorrectum),* and *Gerardia* Boulenger, 1890 *(nomen praeoccupatum).*
Type species: *Coluber prevostianus* Eydoux & Gervais, 1837a.
Distribution: Tidal rivers and coastal marine waters of SE Asia and East Indies.
Sources: M.A. Smith, 1943, B.L. Lim, 1963b, E.H. Taylor, 1965, Gyi, 1970, Singh, 1972, P. Silva, 1980a, Tweedie, 1983, M.A.R. Khan, 1988, Auth et al., 1990, M.J. Cox, 1991b, M.J. Cox et al., 1998, Porej, 2001, Voris et al., 2002, Whitaker & Captain, 2004, N. Khaire, 2006, J.C. Murphy, 2007b, Alfaro et al., 2008, Zaher et al., 2009 and Das et al., 2013.

1. *Gerarda prevostiana* (Eydoux & Gervais, 1837a). Mag. Zool., Paris 7(3): 5–7, pl. 16, figs. 4–6. (*Coluber* [*Homalopsis*] *prevostianus*)

Synonyms: *Gerarda bicolor* Gray, 1849a, and *Heleophis flavescens* F. Müller, 1884.
Types: Syntypes (2), MNHN 3758 and MNHN 7593, longest syntype 540 mm (J.F.T. Eydoux & F.L.A. Souleyet, Feb. 1836–Nov. 1837).
Type locality: "Manille" [= Manila, Manila Prov., SW Luzon Is., N Philippines, 14°36'N, 120°59'E, elevation 10 m].
Distribution: Southeastern Asia and East Indies. India (Andhra Pradesh, Gujarat, Karnataka, Kerala, Maharashtra, Orissa, N Tamil Nadu, West Bengal), SW Sri Lanka (Western), Bangladesh, Myanmar (Ayeyarwady, Bago), S Thailand (Chon Buri), W West Malaysia (Perak, Selangor), East Malaysia (W Sarawak) and SW Philippines (Palawan), NSL–90 m.
Remarks: Original description reprinted in Eydoux & Gervais (1837b: 70–72, pl. 30, figs. 4–6). Plates incorrectly labeled 15 in text (1837a: 5) and 29 in text (1837b: 70).

†*GERINGOPHIS* Holman, 1976b (Boidae)

Type species: †*Geringophis depressus* Holman, 1976b.
Distribution: Lower and upper Oligocene of USA, and lower, middle and upper Miocene of USA.
Sources: Holman, 1979b, 1982d, 1995b, 2000a, Rage, 1984b and Holman & Harrison, 2001.

1. †*Geringophis depressus* Holman, 1976b. Herpetologica 32(1): 90, figs. 2a–c.

Type: Holotype, UNSM 4517, one trunk vertebra.
Type locality: "Loc. III. Apparently from center, S edge, SE 1/4, sec. 27, T 31 N, R 52 W, 4 miles S, 1/4 mile W of Crawford, Dawes County, Nebraska; Gering formation, Arikareean, lower Miocene."
Distribution: Lower Oligocene (Orellan: 33.3–33.9 mya) of USA (South Dakota), upper Oligocene (Arikareean: 20.6–30.8 mya) of USA (Wyoming), lower Miocene (Arikareean: 20.6–30.8 mya) of USA (Nebraska), middle Miocene (Barstovian: 13.6–16.3 mya) of USA (Nebraska), and upper Miocene (Barstovian: 13.6–16.3 mya) of USA (Nebraska).
Sources: Holman, 1977c–e, 1981b and Parmley & Holman, 2009.

2. †*Geringophis robustus* Holman & Harrison, 2001. Acta Zool. Cracov. 44(1): 30–33, figs. 5–7.

Type: Holotype, UF 190837, one trunk vertebra (UF field crews, 1966–1967)
Type locality: "I–75 Local Fauna (Florida Natural History Museum Locality AL018), 1.5 km WSW Gainesville, Alachua County, Florida. Early Oligocene (Whitneyan)."
Distribution: Lower Oligocene (Whitneyan: 30.8–33.3 mya) of USA (Florida).

3. †*Geringophis vetus* Holman, 1982d. Herpetologica 38(4): 490–491, fig. 1.

Type: Holotype, KUVP 49126, one trunk vertebra (KU field crews, 1970's).

Type locality: "Site KU-NEBR-22, Brule Formation, Late Orellan Land Mammal Age of the middle Oligocene, in the vicinity of Toadstool Park area near Orella, Sioux County, Nebraska, 42°53'N; 103°35'W" [USA].

Distribution: Lower Oligocene (Orellan: 33.3–33.9 mya) of USA (Colorado, Nebraska and South Dakota).

4. †*Geringophis yatkolai* Holman, 1977c. Herpetologica 33(4): 443–444, figs. 1a–e. (*nomen corrigendum*)

G

Synonyms: †*Geringophis yatkolae* Holman, 1977c *(nomen incorrigendum),* and †*Geringophis yatkolai* – Holman, 2000 *(nomen corrigendum).*

Type: Holotype, UNSM 46514, one trunk vertebra (D.A. Yatkola).

Type locality: "Harrison Formation, Locality 2, Lizard Quarry, Sioux County, Nebraska, late early Miocene" [USA].

Distribution: Lower Miocene (Arikareean: 20.6–30.8 mya) of USA (Nebraska). Known only from type locality.

GERRHOPILUS Fitzinger, 1843 (Gerrhopilidae)

Type species: *Typhlops ater* Schlegel, 1839 *in* 1837–1844.

Distribution: Asia.

Sources: Hahn, 1980a, Wallach, 1998b, McDiarmid et al., 1999 and Vidal et al. 2010.

1. *Gerrhopilus andamanensis* (Stoliczka, 1871). J. Asiat. Soc. Bengal 40(2): 428–429, pl. 25, figs. 9–12. (*Typhlops andamanensis*)

Type: Holotype, NMW 15427 (formerly IMC), a 165 mm specimen (F. Stoliczka, Feb. 1874).

Type locality: "Andaman islands" [= Andaman & Nicobars, E India, Bay of Bengal, ca. 12'30'N, 92°45'E].

Distribution: Eastern India (Andaman & Nicobars: Andaman Is.). Known only from type locality.

Sources: Wall, 1923c, M.A. Smith, 1943, Biswas & Sanyal, 1980, R.C. Sharma, 2003, 2007 and Kraus, 2005.

2. *Gerrhopilus ater* (Schlegel, 1839 *in* 1837–1844). Abbild. Amph.: 39–40, pl. 32, figs. 29–31. (*Typhlops ater*)

Synonym: *Typhlops ater suturalis* Brongersma, 1934.

Type: Holotype, RMNH 3714, a 130 mm specimen (S. Müller, 1825–1832).

Type locality: "Inner von Java" [= interior of Java, W Indonesia].

Distribution: Eastern Indonesia (West Papua, Bali, Buton, Halmahera, Salawati, Java, N Sulawesi, Ternate, Waigeo), NSL–1000 m.

Sources: Boulenger, 1893a, M.A. Smith, 1943, McDowell, 1974a, Inger et al., 1984, Murthy, 1990, Wallach, 1996d, 2009, R.C. Sharma, 2003, 2007 and Whitaker & Captain, 2004 and Koch, 2012.

3. *Gerrhopilus beddomii* (Boulenger, 1890a). Fauna Brit. India, Rept. Batr.: 237. (*Typhlops beddomii*)

Synonym: *Typhlops beddomei* Wall, 1923c *(nomen emendatum).*

Types: Syntypes (13), BMNH 1946.1.10.69–72 [Anaimalai], 1946.1.11.27–30 (formerly BMNH 1884.5.8.39–42) [Vizagapatnam], BMNH 1946.1.11.93–95 [Travancore], BMNH 1946.1.10.48 (formerly BMNH 1874.4.29.247) [Travancore], and MCZ 22372 (formerly BMNH) [Travancore], longest syntype 127 mm (R.H. Beddome, 1857–1882).

Type locality: "Kimedy Hills (Vizagapatam district) and in the Anaimalai and Travancore Hills, between 2000 and 5000 feet" [= Kimedy Hills, NE Andhra Pradesh State, E India, and Anamalai Hills, E Kerala/W Tamil Nadu States, and Travancore Hills, S Kerala State, S Western Ghats, SW India, 600–1500 m].

Distribution: Southern India (Andhra Pradesh, Kerala, Maharashtra, Tamil Nadu), 10–1525 m.

Sources: Boulenger, 1993a, M.A. Smith, 1943, McDowell, 1974a, Inger et al., 1984, Murthy, 1990, Wallach, 1996d, 2009, R.C. Sharma, 2003, 2007 and Whitaker & Captain, 2007.

4. *Gerrhopilus bisubocularis* (Boettger, 1893d). Zool. Anz. 16(427): 336–337. (*Typhlops bisubocularis*)

Type: Holotype, SMF 16683, a 131 mm specimen (H. Fruhstorfer, 1893).

Type locality: "West-Java" [Indonesia].

Distribution: Indonesia (Jawa Barat).

Sources: Mertens, 1967a, Wallach, 1996d and Kraus, 2005.

Remarks: Possibly occurs on Sumatra *fide* Iskandar & Colijn (2001: 17).

5. *Gerrhopilus ceylonicus* (M.A. Smith, 1943). Fauna Brit. India, Rept. Amph.: 55–56. (*Typhlops ceylonicus*)

Type: Holotype, IMC, a 140 mm specimen.
Type locality: "Peradeniya, Ceylon" [= Peradiniya, Central Prov., Sri Lanka, 7°16'N, 80°36'E, elevation 480 m].
Distribution: Sri Lanka (Central), 480 m. Known only from type locality.
Sources: Deraniyagala, 1955, P. Silva, 1980 and A. Silva, 1990.
Remarks: A subspecies of *G. mirus fide* Deraniyagala (1955: 18). A valid species *fide* Hahn (1980: 54), McDiarmid et al. (1999: 95) and Somaweera (2006: 206).

6. *Gerrhopilus depressiceps* (Sternfeld, 1913b). Sitz. Ges. Naturf. Freunde Berlin 1913(8): 384. (*Typhlops depressiceps*)

Synonyms: *Typhlops monochrus* T. Vogt, 1932b, and *Typhlops monochrous* Loveridge, 1948 *(nomen emendatum).*
Type: Holotype, ZMB 23986, a 328 mm specimen (? Sammler).
Type locality: "Neuguinea" [= Papua New Guinea].
Distribution: Papua New Guinea (East Sepik, Madang, Milne Bay, Kiriwina, Morobe and Oro Is.).
Sources: McDowell, 1974a, O'Shea, 1996, Wallach, 1996d, Fougopoulos, 2001, Bauer et al., 2002 and Kraus, 2005.
Remarks: Type description clearly based on single specimen *fide* Wallach & Günther (2002: 163).

7. *Gerrhopilus floweri* (Boulenger *in* Flower, 1899). Proc. Zool. Soc. London 67(3): 654, pl. 37, fig. 2. (*Typhlops floweri*)

Type: Holotype, BMNH 1946.1.10.55 (formerly BMNH 1898.11.8.26), a 210 mm specimen (S.S. Flower, Nov. 1896–Sept. 1898).
Type locality: "Siam" [= Thailand].
Distribution: Southern Thailand (Bangkok, Chanthaburi, Phra Nakhon Si Ayutthaya, Rayong) and S Vietnam (Hué), 110–205 m.
Sources: E.H. Taylor, 1965, Wallach, 1996d, 2001, 2004, Niyomwan, 1999, Ineich et al., 2002, Kraus, 2005 and I. Das, 2010.

8. *Gerrhopilus fredparkeri* (Wallach *in* O'Shea, 1996). Snakes Papua New Guinea: 60. (*Typhlops fredparkeri*)

Type: Holotype, MCZ 142651 (formerly MCZ-F 14163), a 149 mm female (F. Parker, 24 Oct. 1973).
Type locality: "Korobosea (9°29'S, 147°11'E), 9 km east of Port Moresby, Central Province, Papua New Guinea, 40–60 m."
Distribution: Southeastern Papua New Guinea (Central), 40–60 m. Known only from type locality.
Source: Kraus, 2005.
Remarks: Supplemental original description in Wallach (1996d: 111–113, figs. 2a–c).

9. *Gerrhopilus hades* (Kraus, 2005). J. Herp. 39(4): 592–593, figs. 1a–b. (*Typhlops hades*)

Type: Holotype, BPBM 20819 (formerly FK 10449), a 127 mm male (F. Krause, 14 May 2004).
Type locality: "forest along Rupu River at Bibikea, 11.33537°S, 154.22470°E, 280 m, Rossel Island, Milne Bay Province, Papua New Guinea."
Distribution: Southeastern Papua New Guinea (Milne Bay: Rossel Is.), NSL–280 m.

10. *Gerrhopilus hedraeus* (Savage, 1950). Proc. California Zool. Club 1(10): 49–50, figs. 2a–c. (*Typhlops hedraea*)

Synonym: *Gerrhopilus carollinehoserae* Hoser, 2012am *(nomen illegitimum).*
Type: Holotype, CAS-SU 12346 (formerly SU 12346), a 130 mm specimen (A.W. Herre, 9 Aug. 1940).
Type locality: "a coco palm bole about 1500 feet above Luzuriaga, approximately six miles southwest of Dumaguete, Oriental Negros, Philippine Islands" [= 10 km SW Dumaguete, Negros Is., cen. Philippines, ca. 09°16N, 123°13'E, elevation 450 m].
Distribution: Philippines (Bohol, Camotes, Luzon, Marinduque, Mindanao, Mindoro, Negros), 15–450 m.
Sources: McDowell, 1974b, A.H. Wynn & Leviton, 1993 and Wallach, 1996d.
Remarks: Illustrations of holotype mislabeled as figs. 1a–c in Savage (1950: 49) (but corrected in reprint).

11. *Gerrhopilus inornatus* (Boulenger, 1888b). Ann. Mag. Nat. Hist. (6) 1(5): 344. (*Typhlops inornatus*)

Type: Holotype, BMNH 1946.1.11.80 (formerly BMNH 1888.3.21.6), a 170 mm specimen (H.O. Forbes, 1885–1886).
Type locality: "Camp of Sogere, in interior, 1750 feet above sea" [= Sogeri, Central District, Papua New Guinea, 09°25'S, 147°26'E, elevation 475–535 m].
Distribution: Papua New Guinea (Central, Milne Bay, Morobe, Northern, Western, Western Highlands), 275–2730 m.
Sources: Boulenger, 1893a, McDowell, 1974b, O'Shea, 1996, Wallach, 1996d, Kraus & Allison, 2004 and Kraus, 2005.

12. *Gerrhopilus mcdowelli* (Wallach *in* O'Shea, 1996). Snakes Papua New Guinea: 61. (*Typhlops mcdowelli*)

Type: Holotype, UPNG 7502, a 199 mm male.
Type locality: "Hombron's Bluff (9° 23' S, 147° 20' E), Central Province, Papua New Guinea, ca. 600 m."
Distribution: Southeastern Papua New Guinea (Central), 50–600 m.
Source: Shea & Wallach, 2000.
Remarks: Supplemental originbal description in Wallach (1996d: 108–111, figs. 1a–c).

G

13. *Gerrhopilus mirus* (Jan, 1860 *in* Jan & Sordelli, 1860–1866). Icon. Gén. Ophid. 1(1): 1, pl. 5, fig. 7, pl. 6, figs. 7a–c, f–g, n, p, r, v, x. (*Typhlops mirus*)

Type: Holotype, RMNH 3721, a 135 mm specimen.
Type locality: "Ceylan" [= Sri Lanka].
Distribution: Sri Lanka (Central, Uva, Western), 10–1825 m.
Sources: Wall, 1921, E.H. Taylor, 1947, Deraniyagala, 1955, P. Silva, 1980a, A. Silva, 1990, I. Das, 1996, Wallach, 1996d, R. Sharma, 2003, 2007 and Somaweera, 2006.

14. *Gerrhopilus oligolepis* (Wall, 1909d). J. Bombay Nat. Hist. Soc. (1909–1910) 19(2): 339–340, 1 fig. (*Typhlops oligolepis*)

Type: Holotype, BMNH 1946.1.11.85 (formerly BMNH 1924.5.19.1), a 140 mm specimen (Seal via Darjeeling Mus.).
Type locality: "on a road in the Nagri Valley below Darjeeling at an altitude of about 5,000 feet" [= West Bengal State, NE India, ca. 27°00'N, 88°12'E, elevation 1525 m].
Distribution: Northeast India (West Bengal), 1525 m.
Sources: G.E. Shaw et al., 1938, M.A. Smith, 1943 and Schleich & Kästle, 2002.
Remarks: Probably occurs in E Nepal *fide* Schleich & Kästle (2002: 1004).

15 *Gerrhopilus tindalli* (M.A. Smith, 1943). Fauna Brit. India, Rept. Amph. 3: 53–54. (*Typhlops tindalli*)

Types: Syntypes (3), BMNH 1946.1.8.92–93, (formerly BMNH 1891.7.4.1–2) and BMNH 1946.1.8.95 (formerly BMNH 1893.11.2.1), longest syntype 175 mm.
Type locality: "Nilambur, Malabar district" [= Nilambur, Kerala, S India, 11°16'N, 76°13'E, elevation 40 m].
Distribution: Southern India (Kerala, Tamil Nadu), 40–750 m.
Sources: M.A. Smith, 1943 and R. Sharma, 2003, 2007.

†*GIGANTOPHIS* C.W. Andrews, 1901 (†Madtsoiidae)

Type species: †*Gigantophis garstini* C.W. Andrews, 1901.
Distribution: Upper Eocene of Egypt and Libya.
Sources: C.W. Andrews, 1906, Hoffstetter, 1961b and Rage, 1984b.

1. †*Gigantophis garstini* C.W. Andrews, 1901. Geol. Mag. (4) 8(10): 437–438, figs. 1a–c.

Types: Syntypes, CGM C.10022, ca. 20 vertebrae and fragments of two ribs.
Type locality: "Egypt." Emended to Qasr-el-Sagha beds (Middle Eocene): north of Birket-el-Qurun, the Fayûm, Egypt *fide* C.W. Andrews (1906:307).
Distribution: Upper Eocene (Bartonian: 37.2–40.4 mya) of Egypt and Libya.

GLOYDIUS Hoge & Romano-Hoge, 1981a (*nomen substitutum*) (Viperidae)

Synonyms: *Trigonocephalus* Kuhl & Hasselt, 1822b (*nomen praeoccupatum*), *Halys* Gray, 1849a *(nomen praeoccupatum)*, *Trigonocephalophis* – Bleeker, 1860e *(nomen incorrectum)*, *Trigonocephalo* – Higgins, 1873 *(nomen incorrectum)*, *Halyx* – Palacky, 1898 *(nomen incorrectum)*, *Trigonophalus* – Palacky, 1898 *(nomen incorrectum)*, *Trigonscephalus* – Brazil, 1911 *(nomen incorrectum)*, *Habys* – Steindachner, 1913b *(nomen incorrectum)*, *Gloydins* – Hoge & Romano-Hoge, 1983 *(nomen incorrectum)*, and *Cloydius* – Yao, 2012 *(nomen incorrectum)*.
Type species: *Coluber halys* Pallas, 1776.
Distribution: Eurasia.
Sources: Hoge & Romano-Hoge, 1981a, Y.-C. Chen et al., 1984, Maes, 1989, Gloyd & Conant, 1990, Kardong, 1990, A. Knight et al., 1992, Minton, 1992, Golay et al., 1993, Zhao & Adler, 1993, Kraus et al., 1996, Cullings et al., 1997, Parkinson et al., 1997, Vidal et al., 1997, David & Ineich, 1999, McDiarmid et al., 1999, Orlov & Barabanov, 1999, Parkinson, 1999, Shen et al., 1999, Gumprecht et al., 2004, Castoe & Parkinson, 2006, Malhotra et al., 2010, Hoser, 2012d and Xu et al., 2012.

1. *Gloydius blomhoffi* (H. Boie, 1826). Isis von Oken 18(2): 214–215. (*Trigonocephalus blomhoffi*)

Synonyms: *Trigonocephalus affinis* Gray, 1849a, *Trigonocephalus blomhoffii megaspilus* Cope, 1860b, and *Agkistrodon blomhoffii dubitatus* Gloyd, 1977.
Type: Holotype, ZMA 15179 (J.C. Blomhoff, 1817–1824).
Type locality: "Decima, Japan, [= Dejima Is., previously in Nagasaki harbor, now part of Nagasaki, Japan, 32°45'N, 129°52'E, elevation 10 m]
Distribution: Extreme E Russia (Kunashir and Sakhalin Is.) and Japan (Hokkaido, Honshu, Kyushu, Shikoko, Amakusa, Awaji, Chichijima, Goto, Hachijojima, Iki, Ishigakijima, Izuoshima, Mageshima, Mishima, Oki, Oshima, Sado, Tanegashima, Teuri, Tsushima, Yagishiri, Yakushima Is.), NSL–1500 m.

Sources: Maki, 1931, Takara, 1962, Gloyd, 1972b, 1977, Bannikov et al., 1977, M. Mori, 1982, Szyndlar & Hung, 1987, Toriba, 1988, 1989b, Ji et al., 1989, Paik et al., 1993, Isogawa et al., 1994 and Wang & Zhao, 2007.

Remarks: Original description reprinted in H. Boie (1827b: 266–268) with illustration of type in Siebold (1838: 6, figs. 1–10). *Gloydiud ussuriensis* a subspecies *fide* Gloyd & Conant (1990: 309) and Ji et al. (1989: 189). Conspecific with *G. halys fide* Ji et al. (1989: 185). Records from Kuril Arch. rejected *fide* Orlov & Barabanov (1999: 169). Type locality possibly in error *fide* Siebold *in* Temminck & Schlegel (1838: iii), who mistakenly corrected it to East Indies. Pescadores record suspect *fide* Gloyd & Conant (1990: 281).

2. *Gloydius brevicaudus* (Stejneger, 1907a). Bull. U.S. Natl. Mus. (58): 463–464. (*Agkistrodon blomhoffii brevicaudus*)

Synonyms: *Agkistrodon blomhoffii brevicaudatus* Mori, 1928 *(nomen emendatum), Ankistrodon halys brevicaudus* – Pavloff, 1926 *(nomen incorrectum), Ancistrodon blomhoffii brevicaudus coloratus* Pavloff, 1932 (*nomen illegitimum*), *Ancistrodon blomhoffii brevicaudus rubrum* Pavloff, 1932 (*nomen illegitimum*), and *Agkistrodon blomhoffii siniticus* Gloyd, 1977.

Type: Holotype, USNM 17507, a 595 mm female (P.L. Jouy, 1885).

Type locality: "Fusan, Korea" [= Busan, Busan Region, SE South Korea (Kwangwon, Kyonggi, North Chungchong, North Kyongsang, South Chungchong, South Kyongsang,), 35°11'N, 129°05'E, elevation 70 m].

Distribution: Northern China (Gansu, Heliongjiang, Liaoning, Nei Monggol), North Korea and South Korea, 70–1100 m.

Sources: Toriba, 1988, Paik et al., 1998, P. Guo et al., 1999a, Orlov & Barabanov, 1999, 2000, Xu, 2001, Watkins-Colwell & Leenders, 2002, San & Lee, 2007, Wang & Zhao, 2007 and Yao, 2012.

Remarks: Original description lists holotype as a male. A valid species *fide* Paik et al. (1998: 101) and P. Guo et al., (1999a.: 38). Previously a subspecies of *G. blomhoffii*.

3. *Gloydius halys* (Pallas, 1776). Reise Russ. Reichs 3(2): 703. (*Coluber halys*)

Synonyms: *Echidna aspis pallasii* Merrem, 1820, *Trigonocephalus caraganus* Eichwald, 1831, *Ancistrodon halys caucasicus* Nikolsky, 1916, *Ancistrodon halys paramonovi* Nikolsky, 1931, *Ancistrodon halys persicus* Rendahl, 1933, *Ancistrodon halys stejnegeri* Rendahl, 1933, *Agkistrodon halys cognatus* Gloyd, 1977, *Agkistrodon halys boehmei* Nilson, 1983, *Agkistrodon halys karaguana* – Jucker, 1987 *(nomen incorrectum),* and *Agkistrodon halys mogoi* Bour, 1993.

Type: Neotype, ZISP 14784, a 714 mm male (P.S. Mikhno, 20 July 1930), designated by Orlov & Barabanov (1999: 175).

Type locality: "Borgaiskaya steppe, 84 km W from Kyakhta town, Burin-Khan mountain, approximate 50°25' N, 105°15' E," via neotype selection.

Distribution: Central Asia. Southeastern Azerbaijan, N Iran (Gilan, Mazandaran, N Razavi Khorasan), Afghanistan, Tadzhikistan, Uzbekistan, Turkmenistan, Kazakhstan, Kyrgystan, Mongolia (Arhangay, Bayanhongor, Bayanolgiy, Bulgan, Dornod, Dornogovi, Dundgovi, Dzavhan, Govi-Altay, Hentiy, Hovsgol, Hovd, Omnogovi, Ovorhangay, Selenge, Suhbaatar, Tov, Uvs), Russia (S Siberia) and China (Gansu, Hebei, Jiangsu, Nei Monggol, Qinghai, Shanxi), 75–4000 m.

Sources: Rendahl, 1933, Koba, 1938, Shannon, 1956, Goris 1967, Bannikov et al., 1977, Conant, 1982, Nilson, 1983, Ji et al., 1989, 1991, Toriba, 1989b, Borkin et al., 1990a–b, Cadle, 1992, Bour, 1993, Paik et al., 1993, Szczerbak, 1994, Kubykin & Brushko, 1998, Paik et al., 1998, Zou & Chen, 1998, Orlov & Barabanov, 1999, 2000 and Tuniyev et al., 2009.

Remarks: Syntypes lost *fide* Orlov & Barabanov (1999: 171). *Trigonocephalus caraganus* possibly a synonym of *G. saxatilis fide* Paik et al. (1998: 101).

G

4. *Gloydius himalayanus* (A.C.L.G. Günther, 1864a). Rept. Brit. India: 393–394, pl. 24, figs. a–a'. (*Halys himalayanus*)

Synonym: *Agkistrodon himalayana* – Underwood, 1979.

Types: Syntypes (2), BMNH 1946.1.18.75 (formerly BMNH 1860.3.19.1189), (T.E. Cantor, 1834–1855, via British East India Comp.), and BMNH 1946.1.19.64 (formerly BMNH 1860.3.19.1358) (H.R.A. von Schlagintwelt, 1854–1858), two males, longest syntype 635 mm.

Type locality: "Garhval, Himalayas (altitude 9000 feet),"[= Garhwal, Himalaya Mtns., Uttar Pradesh State, N India, elevation 2745 m].

Distribution: Himalayas of N Pakistan (Jammu, Khyber Pakhtunkhwa, N Punjab), N India (N Haryana, Himachal Pradesh, Kashmir, N Punjab, Uttaranchal, Uttar Pradesh) and Nepal (Baglung, Dolpa, Doti, Ilam, Jumla, Kalikot, Kaski, Khotang, Manang, Mugu, Mustang, Myagdi, Parbat), 1525–3050 (4800) m.

Sources: Wall, 1910g, Hubrecht, 1882, Kramer, 1977, Mahendra, 1984, M.S. Khan & Tasnim,1986c, Nanhoe & Ouboter, 1987, Hallermann et al., 2001, M.S. Khan, 2002, Tiwari & Shah, 2004, Whitaker & Captain, 2004 and S.K. Sharma et al., 2013.

Remarks: Type locality erroneously listed as N West Pakistan *fide* Leviton (1968b: 559). Records from Afghanistan, Bangladesh and Sikkim, India need confirmation *fide* Whitaker & Captain (2007: 414).

5. *Gloydius intermedius* (Strauch, 1868). Trudy. Russ. Estest. 1: 295–296. (*Trigonocephalus intermedius*)

Synonyms: *Trigonocephalus vittatus* Maack, 1859 (*nomen nudum*), *Ankistrodon halys intermedius* – Pavloff, 1926 (*nomen incorrectum*), *Ancistrodon halys intermedius viridis* Pavloff, 1932 (*nomen illegitimum*), and *Agkistrodon shedaoensis continentalis* Zhao, 1980.

Type: Lectotype, ZISP 2221, a 614 mm female (Schrenck, 1855), designated by Orlov & Barabanov (1999: 186).

Type locality: "Cape Tyr, Amur River, Amurskaya Region, Russian Far East, approximate 49°11' N, 130°54' E" via lectotype selection.

Distribution: Eastern Asia. Eastern Russia, NW China (Beijing, Gansu, Heilonjiang, Liaoning, Nei Monggol, Ninxia, Qinghai, Shanxi, Xinjiang), North Korea and South Korea, 500–3000 m.

Sources: Schmidt, 1926b, Joger, 1984, M.-P. Huang, 1984, 1989, P. Guo et al., 1999a, Orlov & Barabanov, 1999, 2000, Adnagulov et al., 2000, Xu, 2001, Zhao, 2006 and Yao, 2012.

6. *Gloydius lijianlii* F. Jiang & Zhao, 2009. Acta Zool. Sinica 34(3), 642–645, fig. 2.

Type: Holotype, CIB 95886, a male (X.-H. Wang & F. Jiang, 27 Sept. 2008).

Type locality: "Daheishan Island (37°58'N, 121°35'E; alt. 174 m), Shandong, China."

Distribution: Eastern China (Shandong: Daheishan Is.), 175 m.

7. *Gloydius monticola* (F. Werner, 1922a). Anz. Akad. Wiss. Wien 59(24–25): 222. (*Ancistrodon blomhoffii monticola*)

Synonym: *Agkistrodon strauchi popei* Mell, 1931a.

Types: Syntypes (2), NMW 17089.1–2, two juveniles (H. Handel-Mazzetti, June 1915).

Type locality: "Yao-Schan bei Lidjiang, NW-Yünnan, 3600 m, Süd-China." [= Yaoshan, near Lijiang (26°54'N, 100°14'E), Yunnan Prov., SW China, elevation 3600 m].

Distribution: Southwestern China (NW Yunnan), 3600–4000 m.

Source: C.H. Pope, 1935.

Remarks: Original description reprinted in F. Werner (1924b: 49).

8. *Gloydius saxatilis* (Emelianov, 1937). Bull. Far East. Br. Acad. Sci. USSR 1937(24): 26–31, 39–40, figs. 1a–c, 2. (*Ancistrodon saxatilis*)

Synonym: *Agkistrodon shedaoensis continentalis* Zhao, 1980, and *Gloydius saxatilis changdaoensis* J.-L. Li, 1999.

Types: Syntypes (39), ZIAS 9456, 13291–92 [Vladivostok] and ZIAS 13308 [Suchan], location of other syntypes unknown.

Type locality: "near Vladivostok, Voroshilov-Ussuriski (formerly Nikolsk-Ussurisk) and river Sutchan" [= Vladivostok, 43°10,N, 131°56'E, elevation 65 m, Ussuriysk, 43°48'N, 131°57'E, elevation 20 m, S Primorsky Krai, Siberia, ext. E Russia].

Distribution: East Asia. Extreme E Russia (ext. SE Siberia), NE China (Heilongjiang, Jilin, Liaoning, Nei Monggol, Shanxi), North Korea and South Korea.

Sources: Paik & Yang, 1989, Ji et al., 1991, Paik et al., 1993, 1998, P. Guo et al., 1999a, Xu, 2001, Zhao, 2006 and San & Lee, 2007.

Remarks: Only two syntypes *fide* McDiarmid et al. (1999: 307): ZIAS #7, Zoolog. Cab. GDU, vic. Vladivostok (A.A. Emelianov, 30 Sept. 1927), a 220 mm juvenile, and ZIAS #17, D.V.U. Suchan (Gasovskii, 8 Aug. 1927), a 625 mm male. A subspecies of *G. intermedius fide* Gloyd & Conant (1990: 357).

9. *Gloydius shedaoensis* (Zhao, 1979b). Acta Herp. Sinica 1(1): 4–6, figs. 1–2. (*Agkistrodon shedaoensis*)

Synonym: *Gloydius shedaoensis qianshanensis* J.-L. Li, 1999.

Type: Holotype, CIB 79-I-0005, adult male (native, 19 June 1979).

Type locality: "Shedao (Snake Island), situated about 24 nautical miles from and to the north-west of Lüshun (Port Arther), Liaoning Province, altitude below 215 m." [= 38°57'N, 120°59'E, NE China].

Distribution: Northeastern China (Liaoning: Shedao Is.), NSL–215 m. Known only from type locality.

Sources: Zhao, 1980, J.-L. Li, 1995, Orlov & Barab, anov, 1999 and Shine et al., 2002.

10. *Gloydius strauchi* (Bedriaga, 1912). Przewalski Cen.-Asien Reisen, Amph. Rept. 3(1): 728–732, pl. 10, figs. 1, 1a–1d. (*Ancistrodon strauchi*)

Synonyms: *Agkistrodon tibetanus* T. Barbour, 1912a, *Agkistrodon halys qinlingensis* M.-T. Song & Chen, 1985, and *Agkistrodon halys liupanensis* N. Liu, Song & Luo, 1989.

Type: Lectotype, ZISP 8534 (Dy-tschu, 1884), designated by Orlov & Barabanov (1999: 169).

Type locality: "Daudsen-Lu (= Ta-Tsian-lu), Sichuan Province, China" via lectotype selection.

Distribution: Western China (Gansu, Ningxia, Qinghai, S Shaanxi, Sichuan, E Xizang), 3150–4600 m.

Sources: C.H. Pope, 1935, P. Guo et al., 1999a and Orlov & Barabanov, 2000, J.C. Murphy & Schlager, 2003 and Yao, 2012.

Remarks: *Gloydius qinlingensis* and *G. liupanensis* valid species *fide* Xu et al., 2012.

11. *Gloydius tsushimaensis* (Isogawa, Moriya & Mitsui, 1994). Jap. J. Herp. 15(3): 102–106, fig. 2. (*Agkistrodon tsushimaensis*)

Type: Holotype, OMNH R3934, a 563 mm male (Yomeishu Seizo Research team, 28 July 1987).
Type locality: "Yora-Naiin, Izuhara-cho, Shimoagata-gun, Nagasaki Prefecture, Japan."
Distribution: Japan (Kyushu, Tsushima Is.), NSL–20 m.

12. *Gloydius ussuriensis* (Emelianov, 1929). Snakes Far East. Distr.: 123–130, figs. 38–41, pl. 1, fig. 49, pl. 5. (*Ancistrodon blomhoffii ussuriensis*)

Synonym: *Agkistrodon caliginosus* Gloyd, 1972b.
Type: Lectotype, SRAR 125/46, a 645 mm male (P. Pravdin, 1927), designated by Chernov (1934: 356).
Type locality: "the river Suchan" [= Suchan River, Primorsky Krai, Siberia, ext. E Russia] via lectotype selection.
Distribution: East Asia. Extreme E Russia (SE Siberia), NE China (Heilongjiang, Jilin, Liaoning, Nei Monggol), North Korea, and South Korea (Quelpart Is.).
Sources: Stejneger, 1907a, Chernov, 1934, Toriba, 1988, Golay et al., 1993, Paik et al., 1993, 1998, P. Guo et al., 1999a, Orlov & Barabanov, 1999, 2000, Adnagulov et al., 2000, Xu, 2001 and San & Lee, 2007.
Remarks: Lectotype designation of ZISP 13327 *fide* Orlov & Barabanov (1999: 188) invalid. Toriba, 1988 considered *caliginosis* a synonym.

†*GOINOPHIS* Holman, 1976b (Serpentes *incertae sedis*)

Type species: †*Goinophis minusculus* Holman, 1976b.
Distribution: Lower Miocene of USA.
Sources: Rage, 1984b and Holman, 2000a.

1. †*Goinophis minusculus* Holman, 1976b. Herpetologica 32(1): 92–93, figs. 4a–d.

Type: Holotype, UNSM 4521, one trunk vertebra.
Type locality: "Loc. I. (Durnal Locality). Lower Gering (earliest part of the lower Miocene): University of Nebraska State Museum Locality MO-119 (NW 1/4, SE 1/4, sec. 32, T 20 N, R 52 W, 10.46 km S and 2.01 km W of Bayard, Morrill County, Nebraska" [USA].
Distribution: Lower Miocene (Arikareean: 20.6–30.8 mya) of USA (Nebraska. and Wyoming).

GOMESOPHIS Hoge & Mertens, 1959 (Xenodontidae)

Type species: *Tachymenis brasiliensis* Gomés, 1918.
Distribution: Southeastern Brazil.
Sources: J.A. Peters & Orejas-Miranda, 1970, Amaral, 1978, Costa Prudente & Brandão, 1998, Marques et al., 2002 and Zaher et al., 2009.

1. *Gomesophis brasiliensis* (Gomés, 1918). Mem. Inst. Butantan 1(1): 78–80, pl. 14, fig. 1. (*Tachymenis brasiliensis*)

Type: Holotype, IB 1316, a 466 mm female (Ribeiro & Irmaos, May 1917).
Type locality: "Pindamonhangaba, Estado de São Paulo, à margem do rio Paraíba do Sul, Brasil" [= Pindamonhangaba, São Paulo State, SE Brazil, 22°55'S, 45°28'W, elevation 555 m].
Distribution: Southeastern Brazil (Distrito Federal, Minas Gerais, Paraná, Rio Grande do Sul, São Paulo), 555 m.

G

GONGYLOSOMA Fitzinger, 1843 (Colubridae)

Synonym: *Gonglyosoma* – F. Werner, 1929a *(nomen incorrectum)*.
Type species: *Coronella baliodeira* F. Boie, 1827.
Distribution: Southeastern Asia and Greater Sundas.
Sources: Leviton, 1964a, Grismer et al., 2003 and I. Das, 2010, 2012.

1. *Gongylosoma baliodeira* (F. Boie, 1827). Isis von Oken 20(6): 539. (*Coronella baliodeira*)

Synonyms: *Coronella baliodeira* H. Boie *in* Schlegel, 1826a *(nomen nudum)*, *Coronella baliodeira* Schlegel, 1826b *(nomen nudum)*, *Ablabes baliodeirus* – A.-M.-C. Duméril, Bibron & Duméril, 1854a, *Ablabes baliode-rus* – A.-M.-C. Duméril, Bibron & Duméril, 1854a *(nomen incorrectum)*, *Ablabes baliodeiroides* Bleeker, 1857a *(nomen nudum)*, *Ablabes baliodirus cinctus* J.G. Fischer, 1885b, and *Liopeltis baliodeirus cochranae* E.H. Taylor, 1962.
Types: Syntypes (2), RMNH 4035 and BMNH (formerly RMNH), one specimen male (? H. Kuhl, Dec. 1820–Sept. 1821).
Type locality: "Java" [Indonesia].
Distribution: Malay Peninsula and Greater Sundas. Southern Thailand (Chanthaburi), West Malaysia (Pahang, Perak, Pinang, Selangor), Singapore, East Malaysia (Sabah, Sarawak), Brunei and W Indonesia (Bali, Java, Kalimantan, Nias Arch., Riau Arch., Sumatra, Weh), NSL–1525 m.
Sources: Rooij, 1917, Sworder, 1922, C. Haas, 1941, E.H. Taylor, 1962, 1965, B.L. Lim, 1963, Bergman, 1963, Leviton, 1964a, Stuebing, 1991, David & Vogel, 1996, Malkmus et al., 2002, McKay, 2006 and I. Das, 2007b, 2010.
Remarks: Original description based on H. Boie's MS (1823–1825).

2. *Gongylosoma longicaudum* (W.C.H. Peters, 1871b). Mber. Königl. Akad. Wiss. Berlin 1871(10): 574–575. (*Ablabes longicaudus*)

Synonyms: *Ablabes quinquestriatus* F. Müller, 1878b, and *Diadophis bipunctatus* Lidth de Jeude, 1890.
Types: Syntypes (2), MSNG CE 30385 and ZMB 7142 (G. Doria & O. Beccari, 1865), longest syntype 385 mm.
Type locality: "Sarawak (Borneo)" [= East Malaysia].
Distribution: Malay Peninsula and Greater Sundas. East Malaysia (Sabah, Sarawak), West Malaysia (Pahang, Pinang, Selangor, Penang Is.) and W Indonesia (Java, Sumatra), NSL–610 m.
Sources: B.L. Lim, 1967a, Grandison, 1972, Tweedie, 1983 and David & Vogel, 1996.
Remarks: Original description reprinted in W.C.H. Peters (1872d: 35–36, pl. 5, figs 1–1c). Doria listed as sole collector *fide* W.C.H. Peters (1871b: 575).

G

3. *Gongylosoma mukutense* Grismer, Das & Leong, 2003. Herpetologica 59(4): 566–567, figs. 2–3.

Type: Holotype, ZRC 2.5143, a 429 mm male (L.L. Grismer, C.A. Ledbetter, S. Guerrero, S.C. Newbold & R.E. Diaz, 10 July 2001).
Type locality: "on the coast 0.5 km south of Kampung Mukut on the trail to Sungai Raya, Pulau Tioman (02°43-54'N; 104°07-3.5'E), Pahang, West Malaysia, at 10 m elevation."
Distribution: West Malaysia (Pahang: Tioman Is.), 10 m. Known only from vicinity of type locality.
Sources: Grismer et al., 2003b, 2008a.

4. *Gongylosoma nicobariensis* (Stoliczka, 1870a). Proc. Asiatic Soc. Bengal 1870(4): 106. (*Ablabes nicobariensis*)

Synonym: *Abalbes nikobariensis* – Theobald, 1876 *(nomen incorrectum).*
Type: Holotype, ZSI 7201 (formerly IMC 7201), a 440–445 mm female (F. Stoliczka, 1870).
Type locality: "Nicobars" [= Nicobar Is., Andaman & Nicobars, E India, Bay of Bengal]. Emended to the Nancowry haven on Camorta (Nicobars) *fide* Stoliczka (1870d: 185).
Distribution: India (Nicobar Is.: Camorta). Known only from type locality.
Sources: M.A. Smith, 1943 and I. Das, 1999a.
Remarks: Original description reprinted in Stoliczka (1870b: 107) and supplemental original description in Stoliczka (1870d: 184–185, pl. 11, figs. 1, 1a–b).

5. *Gongylosoma scriptum* (Theobald, 1868a). J. Linn. Soc. London 10(41): 42. (*Ablabes scriptus*)

Type: Holotype, ZSI 7207 (formerly IMC 7207) (T.M. Berdmore, 1855–1856).
Type locality: "Martaban, British Birma" [= Mottama, Myanmar, 16°34'N, 97°37'E, elevation 10 m].
Distribution: Southern Myanmar (Mon) and S Thailand (Chon Buri, Nakhon Si Thammarat, Phet Buri, Phuket, Panjang Is.), 10 m.
Sources: M.A. Smith, 1930a, 1943 and E.H. Taylor, 1965.
Remarks: Theobald (1868b: 49) is a repeat of the original description.

GONIONOTOPHIS Boulenger, 1893a (*nomen substitutum*) (Lamprophiidae)

Synonyms: *Gonionotus* Mocquard, 1889 *(nomen praeoccupatum), Gonyonotus* Boulenger, 1891b *(nomen emendatum), Cephalosimus* Chabanaud, 1917a, *Gonionothophis* – Briceño-Rossi, 1934 *(nomen incorrectum), Goniotophis* – Monard, 1951 *(nomen incorrectum), Gonisonotophis* – Dunger, 1971a *(nomen incorrectum),* and *Gonionotohis* – Stucki-Stirn, 1979 *(nomen incorrectum).*
Type species: *Gonionotus brussauxi* Mocquard, 1889.
Distribution: West and cen. Africa.
Sources: Mocquard, 1887a, Boulenger, 1891d, 1893a, Sternfeld, 1908b, Angel, 1933b, Loveridge, 1939, 1957a, McDowell, 1961, 1986, Doucet, 1963, Villiers, 1975, Kelly et al., 2003, Villiers & Condamin, 2005, Vidal et al., 2008 and Zaher et al., 2012.
Remarks: *Mehelya* synonymized with *Gonionotophis fide* Vidal et al., 2008, but valid *fide* Zaher et al., 2012.

1. *Gonionotophis brussauxi* (Mocquard, 1889). Bull. Soc. Philom. Paris (1888–1889) (8) 1(12): 146–148, pl. 2, figs. 1–4. (*Gonionotus brussauxi*)

Synonyms: *Gonionotus vossi* Boettger, 1892b, *Simocephalus insignis* Chabanaud, 1917a, and *Gonionotophis brussauxi prigoginei* Laurent, 1956a.
Type: Holotype, MNHN 1890.54, a 440 mm female (E. Brussaux, 1886–1889).
Type locality: "Loudinia-Niari, sur le fleuve Niari, entre Loango sur le littoral, et Brazzaville sur le fleuve Congo" [= between Ludinia and Niari Rivers, French Congo *fide* Loveridge (1939: 150).
Distribution: Central Africa. Southern Cameroon, Gabon (Moyen-Ogooué, Ogooué-Ivindo, Ogooué-Lolo, Ogooué-Maritime, Woleu-Ntem), Central African Republic (Ngotto Forest), Congo (Kouilou, Lekoumou, Sangha), N Democratic Republic of the Congo (Equateur, Nord-Kivu, Orientale), W Uganda (Western) and NE Angola (Lunda Norte), 10–1000 m.
Sources: Perret, 1961, Laurent, 1954a, 1956a, Witte, 1962, D. Lawson, 1993, J.-F. Trape & Roux-Estève, 1995, Spawls et al., 2002, Ineich, 2003, Chippaux, 2006, Chirio & LeBreton, 2007 and Ineich et al., 2007.
Remarks: MNHN catalogue entry for type reads Monts du Mayumbe à 80 km. de Loango entre Loango et Loudima (Rep. du Congo) *fide* M. Lang (*in litt.*). Togo record questionable *fide* Chippaux (2006: 61).

2. *Gonionotophis grantii* (A.C.L.G. Günther, 1863c). Ann. Mag. Nat. Hist. (3) 12(71): 360, pl. 5, fig. f. (*Simocephalus grantii*)

Synonym: *Simocephalus rostralis* Sternfeld, 1910b.

Types: Syntypes (2), BMNH 1946.1.12.47–48 (formerly BMNH 1863.10.5.3a–b), two males, longest syntype 410 mm (Grant).

Type locality: "West Africa."

Distribution: Western Africa. Western Senegal, Guinea-Bissau (Bissau), Guinea (Nzérékoré), Ivory Coast (Bouna, Toumodi), S Burkina Faso (Centre, Centre-Est, Est, Hauts-Bassins, Sud-Ouest, Volta-Noire), Ghana (Upper East), Togo (Kara, Plateaux), SW Niger (Niamey), Nigeria (Kwara, Plateau, Taraba), Cameroon (Adamaoua, Est, Extreme-Nord), SW Chad and Central African Republic (Ombella-Mpoko, Ouham), 200–1170 m.

Sources: Boulenger, 1893a, F. Werner, 1901b, Perret, 1961, *fide* Hughes & Barry (1969: 1011), Roman, 1969, 1980, Roux-Estève, 1969a, Dunger, 1971a, Hughes, 1983, 2013, Chippaux, 2006, Chirio & Ineich, 2006, J.-F. Trape & Mané, 2006b, Chirio & LeBreton, 2007 and Segniagbeto et al., 2011.

Remarks: Possibly occurs in Benin *fide* Hughes (2013: 152).

3. *Gonionotophis klingi* Matschie, 1893a. Sitz. Ges. Natur. Freunde Berlin 1893: 172–173.

Type: Holotype, ZMB 13951, a 420 mm specimen (O.A.R. Büttner, 1890–1891).

Type locality: "Bismarckburg, Togo, West-Afrika" [= near Sotouboua, Centrale Region, Togo, 8°34'N, 0°59'E, elevation 380 m].

Distribution: West Africa. Southeastern Guinea (Nzérékoré), Liberia (Sinoe), Ivory Coast (Guiglo, Tabou, Toumodi), S Ghana (SE Eastern), S Togo (Centrale, Plateaux) and SE Nigeria (Lagos), 380 m.

Sources: Dunger, 1971a, Leston & Hughes, 1968, Hughes & Barry, 1969, Roux-Estève, 1969a, Dunger, 1971a, Huselmans et al., 1971, Hughes, 1983, Chippindale, 1989, Ineich, 2003, Chippaux, 2006 and Segniagbeto et al., 2011.

Remarks: Hughes & Barry (1969: 1011) pointed out erroneous type locality in Gans, 1959 and Loveridge, 1957a. Possibly occurs in Benin *fide* Hughes (2013: 151).

GONYOPHIS Boulenger, 1891c (Colubridae)

Synonym: *Goniophis* – Palacky, 1898 *(nomen incorrectum).*

Type species: *Gonyosoma margaritatum* W.C.H. Peters, 1871b.

Distribution: Malaysia.

Sources: Rooij, 1917, C. Haas, 1950, Tweedie, 1983, Stuebing, 1991, Malkmus et al., 2002 and I. Das, 2010, 2012.

1. *Gonyophis margaritatus* (W.C.H. Peters, 1871b). Mber. Königl. Akad. Wiss. Berlin 1871(10): 578. (*Gonyosoma margaritatum*)

Type: Holotype, MSNG 30613, a 1530 mm specimen (G. Doria & O. Beccari, 1865).

Type locality: "Sarawak (Borneo)" [= East Malaysia].

Distribution: West Malaysia (Johore, Kelantan, Pahang, Perak, Selangor), Singapore and East Malaysia (Sabah, Sarawak), NSL–670 (2000) m.

Remarks: Original description reprinted in W.C.H. Peters (1872d: figs. 3, 3a–c). Doria listed as collector *fide* W.C.H. Peters (1871b: 578).

GONYOSOMA Wagler, 1828a (Colubridae)

Synonyns: *Gonyasoma* Wagler, 1828 *(nomen emendatum), Goniosoma* A.-M.-C. Duméril & Bibron, 1844 *(nomen emendatum), Gonysoma* Agassiz, 1847 *(nomen emendatum), Alopecophis* Gray, 1849b, *Gonijosoma* Bleeker, 1856 *(nomen emendatum), Aepidea* Hallowell, 1861, *Aepidia* Marschall, 1873 *(nomen emendatum), Allophis* W.C.H. Peters, 1872c., and *Gynyosoma* – Murthy, 1986 *(nomen incorrectum).*

Type species: *Coluber oxycephalus* F. Boie, 1827.

Distribution: Southeastern Asia, Indonesia and Philippines.

Sources: M.A. Smith, 1943, Dowling, 1958, Lopez & Maxson, 1995, Schulz, 1996, Gravlund, 2001, Helfenberger, 2001 and Burbrink & Lawson, 2007.

Remarks: Official Generic Name no. 1230 *fide* Opinion 490 (ICZN, 1957). Burbrink & Lawson (2007: 186) transferred *Elaphe frenata* and *E. prasina* to *Rhadinophis* but they form a monophyletic clade with *Rhynchophis fide* Zaher et al., 2012.

1. *Gonyosoma jansenii* Bleeker, 1858e. Nat. Tijds. Nederl. Indië (4) 16(2): 242.

Synonym: *Allophis nigricaudus* W.C.H. Peters, 1872c.

Type: Holotype, BMNH 1946.1.6.20 (formerly BMNH 1863.12.4.134), an adult male (A.J.F. Jansen, 1848–1858, via P. de Bleeker).

Type locality: "Manado, Celebes" [= Manado, ext. NE Sulawesi, 1°30'N, 124°50'E, elevation NSL].

Distribution: Eastern Indonesia (Sulawesi).

Sources: Boulenger, 1894a, P. Müller, 1895, Rooij, 1917, C. Haas, 1950, E.H. Taylor, 1965 and Bosch, 1985, Gumprecxht, 2004e and Koch, 2012.

Remarks: P. Bleeker listed as source of type *fide* Boulenger (1894a: 58).

G

2. *Gonyosoma oxycephalum* (F. Boie, 1827). Isis von Oken 20(6): 537. (*Coluber oxycephalus*)

Synonyms: *Coluber oxycephalus* H. Boie *in* Schlegel, 1826a *(nomen nudum), Coluber oxycephalus* Schlegel, 1826b *(nomen nudum), Gonyosoma viride* Wagler, 1828a, *Alopecophis chalybeus* Gray, 1849b, *Gonijosoma oxijcephalum* Bleeker, 1856 *(nomen emendatum), Aepidea robusta* Hallowell, 1861, *Coluber deroyi* F. Werner, 1923, *Coluber floweri* F. Werner, 1925, and *Composoma elegans* F. Werner, 1926.

Type: Holotype, RMNH 677 (C.G.C. Reinwardt, 1816–1821).

Type locality: "Java" [SW Indonesia].

G

Distribution: Southeastern Asia and Indonesia. Eastern India (Andaman & Nicobars: Andaman Is.), S Myanmar (Rangoon, Tanintharyi), Thailand (Chiang Mai, Chiang Rai, Kanchanaburi, Krabi, Mae Hong Son, Nakhon Ratchasima, Narathiwat, Pattani, Phang Nga, Phuket, Ratchaburi, Satun, Tak, Trang, Trat, Uthai Thani), Laos, Cambodia (Kampot, Koh Kong, Siem Reap), S Vietnam (Dong Nai, Lâm Dong, Phú Khánh, Tây Ninh), West Malaysia (Kedah, Johore, Kelantan, Melaka, Pahang, Pinang, Perak, Penang and Tioman Is.), Singapore, East Malaysia (Sabah, Sarawak, Labuan Is.), Brunei, Indonesia (Bali, Banka, Belitung, Borneo, Java, Kalimantan, Karimata, Lombok, Mentawai Is.: South Pagai, Natuna Is., Nias, Riau Arch., Sumatra) and Philippines (Balabac, Bohol, Bongao, Dinagat, Lubang, Luzon, Mindoro, Negros, Palawan, Panay), 20–1300 m.

Sources: Flower, 1899, T. Barbour, 1912b, Robinson & Kloss, 1916, Rooij, 1917, M.A. Smith, 1928b, Mertens, 1930, Bourret, 1936b, Leviton, 1959b, E.H. Taylor, 1965, Campden-Main, 1970a, Deuve, 1970, Saint Girons, 1972a, H. Müller, 1991, Gaulke, 1994b, Staszko & Walls, 1994, David & Vogel, 1996, M.J. Cox et al.,1998, Chan-ard et al., 1999, Ferner et al., 2001, Malkmus et al., 2002, Gumprecht, 2004e, Whitaker & Captain, 2004, McKay, 2006, B.L. Stuart & Emmett, 2006 and I. Das, 2007b, 2010, 2012.

Remarks: Original description based on H. Boie's MS (1823–1825). Official Specific Name no. 1435 *fide* Opinion 490 (ICZN, 1957). Eastern Himalayas *fide* Flower (1899: 668) and S China *fide* Leviton (1959b: 301). Sulawesi records doubtful, probably *Gonyosoma jansenii, fide* Lang & Vogel (2005: 248). RMNH 675 and RMNH 677b collected by S. Müller in Borneo *fide* RMNH catalogue.

GRAYIA A.C.L.G. Günther, 1858 (Natricidae)

Synonyms: *Heteronotus* Hallowell, 1857b *(nomen praeoccupatum), Graya* A.-H.-A. Duméril, 1859 *(nomen incorrectum), Glaniolestes* Cope *in* Slack, 1862 *(nomen substitutum), Lejonotus* Jan, 1863b, *Xenurophis* A.C.L.G. Günther, 1863c, *Macrophis* Bocage, 1866a, and *Gryia* – Stucki-Stirn, 1979 *Graya* A.-M.-C. Duméril, 1859 *(nomen incorrectum).*

Type species: *Coluber smythii* Leach, 1818.

Distribution: West cen. and E Africa.

Sources: Mocquard, 1897d, Boulenger, 1909b, 1915a, Witte, 1953, 1962, Laurent, 1956a, Perret, 1961, Dunger, 1972, J.-F. Trape & Roux-Estève, 1995, Kelly et al., 2003, Chippaux, 2006 and Chirio & Ineich, 2006.

1. *Grayia caesar* (A.C.L.G. Günther, 1863c). Ann. Mag. Nat. Hist. (3) 12(71): 357, pl. 6, fig. c. (*Xenurophis caesar*)

Synonym: *Grayia longicaudata* Mocquard, 1891.

Type: Holotype, BMNH 1946.1.6.38 (formerly BMNH 1843.1.10.7), a 470 mm (svl) specimen (W. Raddon).

Type locality: "Bioko" [= Bioko Is., Equatorial Guinea, 3°13–3°48'N, 8°25–8°58'E].

Distribution: Central Africa. Southern Cameroon (Est, Littoral, Nord-Ouest, Sud, Sud-Ouest), Equatorial Guinea (Bioko Is.), Gabon (Ogooué-Maritime, Woleu-Ntem), Congo (Kouilou, Plateau, Sangha), SW Central African Republic (Sangha, Haute-Sangha), Congo (Kouilou, Sangha) and Democratic Republic of the Congo (Bandundu, Bas-Congo, Equateur, Katanga, Nord-Kivu, Orientale, Sud-Kivu), 265–1500 m.

Sources: Mertens, 1965c, Roux-Estève, 1965, D. Lawson, 1993, Chirio & LeBreton, 2007 and Pauwels & Vande weghe, 2008.

2. *Grayia ornata* (Bocage, 1866a). J. Sci. Math. Phys. Nat., Lisboa 1(1): 47. (*Macrophis ornatus*)

Synonyms: *Grayia furcata* Mocquard, 1887c, and *Grayia striata* Sternfeld, 1910c.

Types: Syntypes (3), ZMB 7772 (formerly MBL), two adults and a juvenile, longest syntype 1640 mm (F.A.P. Bayão, 1864).

Type locality: "l'interior d'Angola (Duque de Braganca)" [= Duque de Bragança, cen. Malanje Distr., Angola, 09°06'S, 15°57'E, elevation 1110 m].

Distribution: Central Africa. Cameroon (Adamaoua, Centre, Est, Littoral, Nord-Ouest, Ouest, Sud), Equatorial Guinea, Gabon (Estuaire, Moyen-Ogooué, Ngounié, Nyanga, Ogooué-Ivindo, Ogooué-Lolo, Ogooué-Maritime, Woleu-Ntem), Congo (Brazzaville, Kouilou, Plateau, Sangha), W Central African Republic (Bamingui-Bangoran, Haute-Sangha, Lobaye, Ombella-Mboko, Sangha), Democratic Republic of the Congo (Bandundu, Bas-Congo, Equateur, Kasai Occidental, Kasai Oriental, Katanga, Kinshasa, Maniema, Nord-Kivu, Orientale, Sud-Kivu), NE Angola (Lunda Norte, Malanje) and W Zambia (North-Western, Western), NSL–1140 m.

Sources: Laurent, 1964a, Knoepffler, 1966, J.B. Rasmussen, 1991, Broadley et al., 2003 and Chirio & LeBreton, 2007.

Remarks: Original description reprinted in Bocage (1866b: 67–68, pl. 1, figs. 2–2b).

3. *Grayia smythii* (Leach *in* Tuckey, 1818). Exped. River Zaire: 409. (*Coluber smythii*)

Synonyms: *Coluber laevis* Hallowell, 1844b *(nomen praeoccupatum), Coronella triangularis* Hallowell, 1854c, *Grayia silurophaga* A.C.L.G. Günther, 1858, *Lejonotus schlegeli* Jan, 1863b, and *Grayia smithi* Laurent, 1956a *(nomen emendatum).*

Type: Lectotype, BMNH 1946.1.1.67 or BMNH 1946.1.5.16, a female specimen with 147 V + 93 SC (J. Cranch [Tuckey Exped.] via W.E. Leach, 1816), designated by Boulenger (1894a: 287).

Type locality: "near Embomma, Congo" [= vicinity of Boma, Bas-Congo Prov., SW Democratic Republic of the Congo, 5°51'S, 13°03'E, elevation 15 m].

Distribution: West, cen. and E Africa. Southeastern Senegal (Kédougou, Kolda, Tambacounda, Ziguinchor), Gambia (Western), Guinea–Bissau (Bissau, Gabú), Guinea (Boké, Dabola, Kindia, Kouroussa, Macenta, Nzérékobé, Télimélé), Sierra Leone (Eastern, Northern, Southern, Western), Liberia (Grand Cape Mount, Montserrado, Nimba), Ivory Coast (Abidjan, Bouaflé, Bouaké, Bouna, Daloa, Ferkéssédougou, Guiglo, Katiola, Lagunes, Tabou, Toumodi), S Burkina Faso (Centre, Hauts-Bassins, Volta-Noire), Ghana (Ashanti, Eastern), Togo (Plateaux), Benin (Atlantique, Borgou, Plateau), SW Niger (Dosso), Nigeria (Akwa Ibom, Bauchi, Cross River, Delta, Lagos, Oyo, Plateau, Rivers, Sokoto), Cameroon (Adamaoua, Centre, Littoral, Nord, Nord-Ouest, Sud, Sud-Ouest), Equatorial Guinea (Bioko Is.), SW Chad (Mayo-Kebbi), Gabon, Congo (Brazzaville, Kouilou, Likouala, Sangha), Central African Republic (Bamingui-Bangoran, Haut-Kotto, Kemo-Gribingui, Lobaye, Mbomou, Ombella-Mpoko, Ouaka, Ouham, Ouham-Pende), Democratic Republic of the Congo (Bandundu, Bas-Congo, Equateur, Katanga, Kinshasa, Maniema, Nord-Kivu, Orientale, Sud-Kivu), Angola (Bié, Lunda Norte), South Sudan, Uganda (Central, Eastern, Northern, Western, Kome Is.), SW Kenya (Nyanza, Western) and NW Tanzania (Mara, Mwanza, Kagera), 15–1385 m.

Sources: Sternfeld, 1908c, Manaças, 1957, Laurent, 1956a, 1964a, Doucet, 1963, Mertens, 1965c, Roux-Estève, 1965, Leston & Hughes, 1968, Hughes & Barry, 1969, Roman, 1969, Dunger, 1972, Pitman, 1974, Roman, 1980, Joger, 1981, Hughes, 1983, 2013, J.A. Butler & Reid, 1986, D. Lawson, 1993, Akani & Luiselli, 2001b, Spawls et al., 2002, Ineich, 2003, Greenbaum & Carr, 2005, Luiselli et al., 2005, J.F. Trape & Mané, 2006b, Chirio & LeBreton, 2007, Segniagbeto et al., 2011, Auliya et al., 2012 and Chirio, 2013.

Remarks: Laurent (1956a: 135), Meirtre (1992: 113) and Hughes (2013: 116) argue for correcting *smythii* to *smithii*. However, there is no indication that the species was named after the Dr. Smith of the expedition and P. David pointed out that the name Smith can be transliterated into the Latin as *Smythius*, the genetive form being *Smythii*. Also, Art. 32.5 of the Code prohibits changing an incorrect original spelling unless there is clear evidence of an inadvertent error or else a correction in an errata sheet at the time of publication.

4. *Grayia tholloni* Mocquard, 1897d. Bull. Soc. Philom. Paris (8) 9(1): 11–12.

Synonym: *Grayia fasciata* Boulenger, 1901a.

Types: Syntypes (4), MNHNa–b (F.-R. Thollon, 1875–1882), and MTa–b, longest syntype 810 mm, shortest syntype 282 mm.

Type locality: "Brazzaville," and "Congo francais" [= Brazzaville, Brazzaville Region, S Congo, 4°16'S, 15°17'E, elevation 285 m, and French Congo = Congo].

Distribution: West and cen. Africa. Southern Senegal (Dakar, Fatick, Kaolack, Kédougou), W Gambia (Banjul), N Nigeria (Adamawa, Benue, Kebbi, Plateau), SW Chad (Mayo-Kebbi), Cameroon (Adamaoua, Centre, Nord-Ouest, Ouest), Central African Republic (Bamingui-Bangoran, Ouham, Ouham-Pende), Congo (Brazzaville, Plateau, Pool), Democratic Republic of the Congo (Bandundu, Bas-Congo, Kasai Occidental, Kasai Oriental, Katanga, Kinshasa, Orientale), NE Angola (Lunda), NW Zambia (Luapula, North-Western), South Sudan, SE Ethiopia (Gemu Gofa, Illubabor), SW Kenya (Nyanza, Western), Uganda (Central, Eastern, Northern, Western), E Rwanda (Byumba, Kibungo), E Burundi (Bubanza, Bujumbura, Ruyigi) and W Tanzania (Kigoma, Mara, Mwanza, Rukwa, Kagera, Ukewere Is.), 265–1600 m.

Sources: Boulenger, 1901a, Laurent, 1954b, Dunger, 1972, Pitman, 1974, Joger, 1990, Broadley & Howell, 1991, Largen & Rasmussen, 1993, J.-F. Trape & Mané, 1995b, 2006, Pauwels, 1996, Spawls et al., 2002, Broadley et al., 2003, Chirio & LeBreton, 2007 and Largen & Spawls, 2010.

Remarks: Possibly occurs in Benin *fide* Hughes (2013: 152).

GRYPOTYPHLOPS W.C.H. Peters, 1881b (Typhlopidae)

Synonyms: *Gryptotyphlops* Boulenger, 1893a *(nomen emendatum),* and *Cryptotyphlops* – Nikolsky, 1916 *(nomen incorrectum).*

Type species: *Onychocephalus acutus* A.-M.-C. Duméril & Bibron, 1844.

Distribution: India.

Sources: Hahn, 1908a, Murthy & Rao, 1989, Wallach, 1994b, 1998b, 2003, McDiarmid et al., 1999, Whitaker & Captain, 2004 and N. Khaire, 2006.

G

1. *Grypotyphlops acutus* (A.-M.-C. Duméril & Bibron, 1844). Erpét. Gén. 6: 333–334. (*Onychocephalus acutus*)

Synonyms: *Onychocephalus unilineatus* A.-M.-C. Duméril & Bibron, 1844, *Typhlops russellii* Gray, 1845 (*non* Schlegel), *Onychocephalus westermanni* Lütken, 1863, *Typhlops excipiens* Jan, 1864, *Onychocephalus malabaricus* Beddome *in* A.C.L.G. Günther, 1875 (*nomen nudum*), *Typhlops psittacus* F. Werner, 1903b, *Typhlops acuta* Constable, 1949, and *Typhlops acuts* – Rajendran, 1986 (*nomen incorrectum*).

Type: Neotype, UF 19902, a 381 mm specimen (W. Auffenberg, 23 June 1964), designated by Wallach (1994b: 214).

Type locality: "one of the Kanheri Caves at the base of the Kanheri Hills, Kanheri National Park, 5 mi. E Borivli and *ca.* 20 mi. NNE Bombay, west-central India, elevation ca. 180 m, 19°14'N, 72°51'E" via neotype selection.

Distribution: Peninsular India (Andhra Pradesh, Bihar, Goa, Gujarat, Karnataka, Kerala, Madhya Pradesh, Maharashtra, Orissa, Pondicherry, Rajasthan, Tamil Nadu, West Bengal), NSL–300 m.

Sources: Wall, 1918a, M.A. Smith, 1943 and R. Sharma, 2003, 2007.

Remarks: Holotype in MNHN lost *fide* Hahn (1980a: 49).

GYALOPION Cope, 1860e (Colubridae)

Synonym: *Gyalopium* Cope, 1863c (*nomen emendatum*).

Type species: *Gyalopion canum* Cope, 1860e.

Distribution: Southwestern USA and N Mexico.

Fossil records: Upper Pleistocene of SW USA.

Sources: H.M. Smith & Taylor, 1941, L.M. Hardy, 1975a, 1976b, Holman, 1995a and R.D. Bartlett & Tennant, 2000.

1. *Gyalopion canum* Cope, 1860e. Proc. Acad. Nat. Sci. Philadelphia 12(6): 243.

Type: Holotype, USNM 5284 (formerly SIM 4675), a 188–190 mm male (B.J.D. Irwin, 1857–1860).

Type locality: "near Ft. Buchanan, Arizona" [= 5 km SW Sonoita in Hog Canyon, Santa Cruz Co., SE Arizona, SW USA, 31°40'N, 110°42'W, elevation 1425 m].

Distribution: Southwestern USA (ext. SE Arizona, S New Mexico, SW Texas) and N Mexico (Chihuahua, Coahuila, E Durango, Jalisco, Michoacán, S Nuevo León, Querétaro, NW San Luis Potosí, N Sonora, W Zacatecas), 300–2600 m.

Fossil records: Upper Pleistocene (Rancholabrean II) of USA (Arizona, New Mexico, Texas).

Sources: Stebbins, 1985, T. Van Devender & Bradley, 1994, Degenhardt et al., 1996, Ponce-Campos & Huerta-Ortega, 1999, Holman, 2000a, Werler & Dixon, 2000 and Ponce-Campos et al., 2001.

Remarks: H.M. Smith & Taylor (1945: 71) erroneously listed the types as USNM 16427–28 (= *Ficimia publia fide* L.M. Hardy, 1975a: 117).

2. *Gyalopion quadrangulare* (A.C.L.G. Günther, 1893 *in* 1885–1902). Biol. Cen.-Amer., Rept. Batr.: 99, pl. 35, fig. a. (*Ficimia quadrangularis*)

Synonym: *Ficimia desertorum* E.H. Taylor, 1936a.

Type: Holotype, BMNH 1946.1.1.75, a 305–319 mm specimen (A. Forrer, 1883).

Type locality: "Mexico, Presidio de Mazatlan in Sinaloa" [= Mazatlan, S Sinaloa State, NW Mexico, 23°14'N, 106°25'W, elevation 10 m].

Distribution: Extreme SW USA (ext. S-cen Arizona) and NW Mexico (S Chihuahua, Nayarit, Sinaloa, E Sonora), 15–1250 m.

Sources: H.M. Smith et al., 2005a–b and H.M. Smith & Lemos-Espinal, 2006b.

H

†*HAASIOPHIS* Tchernov, Rieppel, Zaher, Polcyn & Jacobs, 2000 (†Pachyophiidae)

Type species: †*Haasiophis terrasanctus* Tchernov, Rieppel, Zaher, Polcyn & Jacobs, 2000.
Distribution: Upper Cretaceous of Israel.
Sources: Tchernov et al., 2000, Lee & Scanlon, 2002, Rage & Escuillié, 2003, Rieppel et al., 2003 and Bardet et al., 2008.
Remarks: Possibly a synonym of *Simoliophis fide* Rage & Escuillié (2003: 7).

1. †*Haasiophis terrasanctus* Tchernov, Rieppel, Zaher, Polcyn & Jacobs, 2000. Science 287(5460): 2011–2012, figs. 1–2.

Type: Holotype, HUJ-PAL EJ 695, cranium, 155 trunk vertebrae and 12 caudal vertebrae from a 717 mm (svl) specimen.
Type locality: "Limestone quarries of 'Ein Yabrud, Judean hills, 20 km north of Jerusalem. Aminadav Formation or the slightly younger Bet-Meir Formation, middle part of the Judea Group, early to middle Cenomanian, basal Upper Cretaceous" [Israel].
Distribution: Middle Cretaceous (Cenomanian: 93.9–100.5 mya) of Israel.

HAPSIDOPHRYS J.G. Fischer, 1856b (Colubridae)

Synonyms: *Hapsidrophis* – Jan, 1857 *(nomen incorrectum), Gastropyxis* Cope, 1861c, *Hapsidophis* Reichenow, 1874 *(nomen emendatum), Hapidophrys* – Sauvage, 1877 *(nomen incorrectum), Tropidophidion* F. Werner, 1902, *Hapsidophris* – Sternfeld, 1908b *(nomen incorrectum), Gastrophyxis* – Boulenger, 1919 *(nomen incorrectum), Gastrohis* – Briceño-Rossi, 1934 *(nomen incorrectum), Gastropyrix* – Roux-Estève, 1965 *(nomen incorrectum), Hapsidiphrys* – Knoepffler, 1968 *(nomen incorrectum),* and *Gastrophixis* – Stucki-Stirn, 1979 *(nomen incorrectum).*
Type species: *Hapsidophrys lineatus* J.G. Fischer, 1856b.
Distribution: West and cen. Africa.
Sources: Sternfeld, 1917, Bogert, 1940, Loveridge, 1942, 1957a, Laurent, 1956a, Witte, 1962, Doucet, 1963, Roux-Estève, 1965, Thys van den Audenaerde, 1965, Broadley, 1966a, Dunger, 1973, Pitman, 1974, Villiers, 1975, J.-F. Trape & Roux-Estève, 1995, Villiers & Condamin, 2005, Chirio & Ineich, 2006, Jesus et al., 2009 and Pauwels & Vande weghe, 2008.

1. *Hapsidophrys lineatus* J.G. Fischer, 1856b. Abh. Naturwiss. Ver. Hamburg 3: 111, pl. 2, figs. 5a–b.

Synonyms: *Dendrophis nigrolineatus* Lichtenstein, 1856, *Hapsidophrys lineatus coeruleus* J.G. Fischer, 1856b, and *Gastropyxis orientalis* F. Werner, 1909d.
Types: Syntypes (3), ZMH 4399–4400 (formerly ZMH 102) and ZMH 39, a 958 mm, 932 mm and 781 mm specimen (C. Weiss, 1848), latter syntype lost.
Type locality: "Elmine (West-Afrika)" [= Elmina, Central Region, Ghana, 5°05'N, 1°21'W, elevation NSL].
Distribution: West and cen. Africa. Southeastern Guinea, Sierra Leone, SE Guinea (Macenta, Nzérékoré), Sierra Leone (Southern), Liberia, Ivory Coast (Albidjan, Daloa, Lagunes, San Pédro), S Ghana, S Togo (Plateaux), S Benin (Maritime), S Nigeria (Abia, Anambra, Bayelsa, Delta, Lagos, Rivers), Cameroon (Adamaoua, Centre, Est, Littoral, Nord-Ouest, Sud, Sud-Ouest), Central African Republic (Haut-Mbomou, Haute-Sangha, Lobaye, Mbomou, Ombella-Mpoko, Sangha), Gabon (Haut-Ogooué, Moyen-Ogooué, Ogooué-Ivindo, Ogooué-Lolo, Ogooué-Maritime), Equatorial Guinea (Bioko Is.), Congo (Brazzaville, Cuvette, Kouilou, Lekoumou, Pool, Sangha), Democratic Republic of the Congo (Bandundu, Bas-Congo, Equateur, Kasai Occidental, Kasai Oriental, Katanga, Kinshasa, Nord-Kivu, Orientale, Sud-Kivu), SW Uganda (Central, Western), SW Kenya (Nyanza, cen. Rift Valley, Western), NSL–1830 m.
Sources: Doucet, 1963, Mertens, 1965c, Knoepffler, 1966, Menzies, 1966, Luiselli et al., 2001, Spawls et al., 2002, Chippaux, 2006, Chirio & LeBreton, 2007, Segniagbeto et al., 2011 and Hughes, 2013.
Remerks: Possibly occurs in Rwanda *fide* Spawls et al. (2002: 364).

2. *Hapsidophrys principis* (Boulenger, 1906f). Ann. Mus. Civ. Stor. Nat. Genova (3) 2(42): 213, fig. 8. (*Gastropyxis principis*)

Types: Syntypes (3), MSNG CE 28144A–B and BMNH 1946.1.5.99, longest syntype 1150 mm.
Type locality: "Prince's Island, Gulf of Guinea, West Africa" [= Príncipe Is., São Tomé & Príncipe, 1°32'–1°43'N, 7°20'–7°28'E].
Distribution: São Tomé and Príncipe (Príncipe Is.), 100–300 m.
Sources: Manaças, 1958 and Capoccacia, 1961b.

Remarks: A valid species *fide* Jesus et al., 2009.

3. *Hapsidophrys smaragdinus* (Schlegel, 1837). Essai Phys. Serp. 1: 158, 2: 237–238. (*Dendrophis smaragdina*)

Synonyms: *Chrysopelea smaragdina* H. Boie *in* Schlegel, 1826a *(nomen nudum), Chrysopelia smaragdina* Schlegel, 1826b *(nomen nudum), Leptophis gracilis* Hallowell, 1844a, *Hapsidophrys coeruleus* J.G. Fischer, 1856b, and *Tropidophidion steini* F. Werner, 1902.

Type: Lectotype, RMNH 909 (D.F. Eschricht, 1830–1837), designated by Hoogmoed *in* Hughes & Barry (1969: 1017), longest syntype 750 mm.

Type locality: "Côte de Guinée" [= coastal Ghana *fide* Hughes & Barry, 1969: 1017] via lectotype selection

H

Distribution: West and cen. Africa. Gambia (Western), Guinea–Bissau (Bolama Is.), SE Guinea (Macenta, Nzérékoré), Sierra Leone (Southern), Liberia (Bong, Grand Gedah, Lofa, Bushrod Is.), Ivory Coast (Abidjan, Guiglo, Lagunes, Tabou, Toumodi), Ghana (Brong-Ahafo, Eastern, Volta), Togo (Maritime, Plateaux), Benin (Atlantique, Borgou, Cotonou, Maritime, Zou), S Nigeria (Akwa Ibom, Anambra, Bayelsa, Cross River, Delta, Lagos, Rivers), S Cameroon (Centre, Est, Littoral, Nord-Ouest, Ouest, Sud, Sud-Ouest), Equatorial Guinea (Littoral, Bioko Is.), Gabon (Estuaire, Haut-Ogooué, Moyen-Ogooué, Ngounié, Nyanga, Ogooué-Ivindo, Ogooué-Lolo, Ogooué-Maritime, Woleu-Ntem), Central African Republic (Haut-Mbomou, Haute-Sangha, Kemo-Gribingui, Lobaye, Mbomou, Ombella-Mpoko, Sangha), Congo (Brazzaville, Cuvette, Cuvette-Ouest, Kouilou, Lekoumou, Niari, Plateau, Pool, Sangha), E Democratic Republic of the Congo (Bandundu, Bas-Congo, Equateur, Kasai Occidental, Kasai Oriental, Kinshasa, Nord-Kivu, Orientale, Sud-Kivu), N Angola (Cabinda, Lunda Norte), S Uganda (Central, Western, Bunjako and Sese Is.) and NW Tanzania (Arusha, Kagera, Rubondo Is.), 600–2200 m.

Sources: Johnsen, 1962, Laurent, 1964a, Doucet, 1963, Mertens, 1965c, Thys van den Audenaerde, 1966, Menzies, 1966, Villiers, 1966, Leston & Hughes, 1968, Hughes & Barry, 1969, Leston, 1970, Dunger, 1973, Hakansson, 1981, Hughes, 1983, 2013, J.A. Butler & Reid, 1986, J.B. Rasmussen, 1991, D. Lawson, 1993, Gravlund, 2000a, Spawls et al., 2002, Leaché, 2005, Villiers & Condamin, 2005, Chippaux, 2006, Chirio & LeBreton, 2007, Segniagbeto et al., 2011 and Auliya et al., 2012.

Remarks: Possibly occurs in Rwanda and Burundi *fide* Spawls et al. (2002: 365).

†*HEADONOPHIS* Holman, 1993 (Colubroidea *incertae sedis*)

Type species: †*Headonophis harrisoni* Holman, 1993.

Distribution: Upper Eocene of United Kingdom.

Source: Holman, 1993.

1. †*Headonophis harrisoni* Holman, 1993. Tertiary Res. 14(4): 152–153, figs. 1a–e.

Type: Holotype, MSUVP 1342, one trunk vertebra (D.L. Harrison, late 1980s–1992).

Type locality: "Hatherwood Point, SW Headon Hill Locality, Isle of Wight, England. Totland Bay Member of the Headon Hill Formation (Upper Eocene)."

Distribution: Upper Eocene (Chattian: 33.9–33.9 mya) of United Kingdom (England). Known only from type locality.

†*HECHTOPHIS* Rage, 2001 (Boidae)

Type species: †*Hechtophis austrinus* Rage, 2001.

Distribution: Middle Paleocene of Brazil.

Sources: Rage 2001, 2011.

1. †*Hechtophis austrinus* Rage, 2001. Palaeovertebrata 30(3–4): 116–118, fig. 2.

Type: Holotype, DGM 1326-R, one middle trunk vertebra (J.S. Carvalho, 1949).

Type locality: "São José de Itaboraí, State of Rio de Janeiro, Brazil; middle Paleocene."

Distribution: Middle Paleocene (Itaboraian: 57.0–59.0 mya) of Brazil.

†*HELAGRAS* Cope, 1883a (Boidae)

Type species: †*Helagras prisciformis* Cope, 1883a.

Distribution: Lower and middle Paleocene of USA, and middle Oligocene of USA.

Sources: Holman, 2000a and Longrich et al., 2012b.

Remarks: Hoffstetter & Rage (1972: 111) and Holman (1979a: 209) considered it an Erycine boid.

1. †*Helagras orellanensis* Holman, 1983. J. Herp. 17(4): 417–419, fig. 1.

Type: Holotype, KUVP 49127, one trunk vertebra (KU field crews, 1970s).

Type locality: "University of Kansas Museum of Natural History Site KU-NEBR-22, Brule Formation, late Orellan Land Mammal Age of the middle Oligocene, Toadstool Park Area near Orella, Sioux County, Nebraska, 42°53'N, 103° 35'W."

Distribution: Middle Oligocene (Orellan: 33.3–33.9 mya) of USA (Nebraska). Known only from type locality.

Remarks: Lower Oligocene *fide* Holman (2000a: 75).

2. †*Helagras prisciformis* Cope, 1883a. Proc. Amer. Philos. Soc. 20: 545–546.

Type: Holotype, AMNH 1628, two posterior trunk vertebrae (D. Baldwin, 1881).
Type locality: "Puerco Eocene." Emended to San Juan Basin, northwestern New Mexico, USA *fide* Gilmore (1938: 81) and Rage (1984b: 25).
Distribution: Lower Paleocene (Puercan: 63.3–65.5 mya) and middle Paleocene (Torrejonian: 60.2–63.3 mya) of USA (New Mexico).
Sources: Gilmore, 1938, Holman, 1979a and Rage, 1984b.
Remarks: Original description reprinted in Cope (1883b: 191).

HELICOPS Wagler, 1828a (Xenodontidae)

Synonyms: *Tachynectes* Fitzinger, 1843, *Uranops* Fitzinger, 1843, *Stranops* – A.-M.-C. Duméril & Bibron, 1844 *(nomen incorrectum), Tachynectes* Gray, 1849, and *Uranops* Gray, 1849.
Type species: *Coluber carinicaudus* Wied-Neuwied, 1825c.
Distribution: South America.
Sources: Rossman *in* J.A. Peters & Orejas-Miranda, 1970, Rossman, 1974, Cadle, 1984a, Dixon & Soini, 1986, Pérez-Santos & Moreno, 1988, 1991, Vidal et al., 2000 and Zaher et al., 2009.

1. *Helicops angulatus* (Linnaeus, 1758). Syst. Nat., ed. 10, 1: 217. (*Coluber angulatus*)

Synonyms: *Coluber alidras* Linnaeus, 1758, *Coluber Surinamensis* G. Shaw, 1802, *Natrix aspera* Wagler *in* Spix, 1824, *Natrix lampronota* Wagler, 1825 (*nomen ineditum*), *Helicops cyclops* Cope, 1869b, *Helicops fumigatus* Cope, 1869b, and *Helicops caralina* Briceño-Rossi, 1934.
Type: Holotype, NHR Lin-17 (formerly MAFR), a 600 mm female (Mus. Drottn.).
Type locality: "Asia" (in error *fide* Hoge, 1964d: 56).
Distribution: Northern South America. Colombia (Amazonas, Atlántico, Antioquia, Boyacá, Caquetá, Casanare, Guaviare, Meta, Vaupés), Venezuela (Amazonas, Anzoátegui, Apure, Barinas, Bolívar, Cojedes, Delta Amacuro, Guárico, Mérida, Monagas, Portuguesa, Sucre, Zulia), Trinidad and Tobago (Trinidad), Guyana (Barima-Waini, Cuyuni-Mazaruni, Demerara-Mahaica, Essequibo Islands-West Demerara, Mahaica-Berbice, Pomeroon-Supenaam, Upper Demerara-Berbice), Suriname (Commewijne, Coronie, Marowijne, Nickerie, Saramanca, Suriname), French Guiana (Cayenne, Saint-Laurent-du-Maroni), E Ecuador (Napo, Pastaza), E Peru (Loreto, Madre de Dios, Pasco, San Martin), E Bolivia (Beni, Santa Cruz) and N Brazil (Acre, Amazonas, Bahia, Ceará, Distrito Federal, Goiás, Maranhão, ? Mato Grosso, Pará, ? Pernambuco, Rondônia, Roraima, Sergipe, Tocantins), NSL–2410 m.
Sources: Hoge, 1960d, Roze, 1966a, Emsley, 1977, Cunha & Nascimento, 1978, Abuys, 1983d, Chippaux, 1987, Dixon & Soini, 1986, Lancini, 1986, Duellman, 1990, 2005, Duellman & Salas, 1991, Henle & Ehrl, 1991, Perez-Santos & Moreno, 1991, Gorzula & Señaris, 1998, Starace, 1998, Freitas, 1999, Frota. 2005, Rivas-Fuenemayer & Barrio-Amorgós, 2005, Infante-Rivero et al., 2008, Navarrete et al., 2009, Roberto et al., 2009, Morato et al., 2011 and C.J. Cole et al., 2013.
Remarks: ZMUU has one Linnaean specimen *fide* Lönnberg (1896: 7).

2. *Helicops carinicaudus* (Wied-Neuwied, 1825c). Beitr. Natur. Brasil. 1: 300–304. (*Coluber carinicaudus*)

Synonyms: *Homalopsis carinicauda* – Schlegel, 1837, *Helicops trivittatus* Cope, 1878b, and *Helicops carinicauda triserialis* Lindholm *in* Lampe & Lindholm, 1902.
Type: Holotype, AMNH 3365, a 909 mm specimen (A.M.P. zu Wied–Neuwied, 26 Nov.–19 Dec. 1815).
Type locality: "Urwäldern am Flusse Itapémirim (Itapémiri), Brasilien" [= Brazil, Estado do Espírito Santo, Río Itapémirim, from label in bottle containing holotype].
Distribution: Southeastern Brazil (Espírito Santo, Paraná, Rio de Janeiro, Rio Grande do Sul, Santa Caterina, São Paulo, Cannanéia, Cardoso, Comprida, Porchat, Santo Amaro and São Vicente Is.), NSL–925 m.
Sources: Bertoni, 1918, Lema, 1958, 1961a, Vaz-Ferreira & Sierra de Soriano, 1960, Lema & Fabían-Beurmann, 1977, J.D. Williams & Francini, 1991, Hofstadler-Dieques & Cechin, 1992, Marques et al., 2001, Yuki & Lema, 2006 and Cicchi et al., 2007.
Remarks: Colombian record (Magdalena Valley) needs confirmation *fide* Pérez-Santos & Moreno (1989: 174).

3. *Helicops danieli* Amaral, 1938b. Mem. Inst. Butantan (1937) 11: 232–233.

Type: Holotype, IB 9872, a 235 mm female ("Hno. Daniel" J.G. Patiño, 1928–1938), destroyed by fire on 15 May 2010.
Type locality: "Carare (Santander), Colombia" [= Carare, Puerto Parra, W Santander Dept., NW Colombia, 6°47'N, 74°06'E, elevation 90 m].
Distribution: Northwestern Colombia (Antioquia, Atlántico, Bolívar, Chocó, Córdoba, Santander, Sucre), NSL–2050 m.
Sources: Nicéforo-Maria, 1942 and Rossman, 2002b.
Remarks: Usually occurs below 500 m *fide* Rossman (2002b: 592).

4. *Helicops gomesi* Amaral, 1921. Anex. Mem. Inst. Butantan 1(1): 7–10, pl. 1, figs. 1–4.

Type: Holotype, IB 1843, a 695 mm female (A.V. d'Alkmim, 4 Dec. 1919), destroyed by fire on 15 May 2010.
Type locality: "Estação Costa Pinto, na linha Sorocabana, Estado de S. Paulo, Brasil."
Distribution: Southeastern Brazil (São Paulo), 750 m.
Source: Valdujo & Nogueira, 2001.

5. *Helicops hagmanni* Roux, 1910a. Zool. Anz. 36(25): 439–440.

Synonym: ? *Uranops lepricurii* Fitzinger, 1867.
Type: Holotype, NMBA 6281, a 630–634 mm male (G.A. Hagmann, 1894–1910).
Type locality: "in der Nähe von Santarem, N. Brasilien" [= vicinity of Santarem, Pará State, N Brazil, 2°26'S, 54°42'W, elevation 35 m].
Distribution: Northern South America. Southern Colombia (Amazonas, Caquetá, Vaupés [UTA R3782]), S Venezuela (Amazonas) and N Brazil (Amapá, Amazonas, Pará, Rondônia), 35–70 m.
Sources: Hoge, 1964d, Hoge et al., 1973, Rossman, 1975, Cunha & Nascimento, 1978, Hoogmoed, 1979 and Navarrete et al., 2009.
Remarks: Type listed as female *fide* Roux (1910: 439). Type locality erroneously listed as near to Suriname *fide* Hoge (1967: 220) and Amazonas Dept. *fide* Rossman (1975: 415). Probably occurs in Peru *fide* Rossman (1975: 415).

H

6. *Helicops infrataeniatus* Jan, 1865c. Arch. Zool. Anat. Fis. 3(2): 253.

Synonyms: *Helicops infrataeniatus* Jan, 1863b (*nomen nudum*), *Helicops carinicaudus gastrosticta* Jan, 1863b (*nomen nudum*), *Helicops carinicaudus gastrosticta* Jan, 1865c, *Helicops trivittatus* Cope, 1877 (*non* Gray), *Calopisma septemvittatum* J.G. Fischer, 1879, *Helicops baliogaster* Cope, 1885a, and *Helicops pictiventris* F. Werner, 1897a.
Type: Lectotype, ZMH 4331 (formerly ZMH B88 da), a 515 mm male, designated by Rossman (2000: 126).
Type locality: "Suriname" [= Suriname] (in error *fide* Rossman, 2000: 126) via lectotype selection.
Distribution: Southern Brazil (Mato Grosso do Sul, Paraná, Rio Grande do Sul, Santa Catarina, São Paulo), E Paraguay (Alto Paraná, Amambaty, Canindeyú, Itapúa, San Pedro), Uruguay (Artigas, Canelones, Cerro Largo, Rivera, Rocha, Salto, Tacuarenbo, Treinta y Tres) and N Argentina (Buenos Aires, Chaco, Corrientes, Entre Ríos, Formosa, Misiones, Santa Fe), NSL–925 m.
Sources: Cei, 1994, Lema, 1995, Rossman, 2000, Carreira-Vidal et al., 2005, Yuki & Lema, 2005, J. Köhler & Güsten, 2007, Cacciali, 2008 and Filho & Aoki, 2011.
Remarks: Photographs of the lectotype in Rossman (2000: fig. 2).

7. *Helicops leopardinus* (Schlegel, 1837). Essai Phys. Serp. 1: 173, 2: 358–359. (*Homalopsis leopardina*)

Synonyms: *Helicops prieurii* A.-M.-C. Duméril, 1853 (*nomen nudum*), *Helicops leprieurii* A.-M.-C. Duméril, Bibron & Duméril, 1854a, and *Helicops prieurii* A.-M.-C. Duméril, Bibron & Duméril, 1854c (*nomen emendatum*).
Types: Syntypes (2), RMNH 1158, a 480 mm male, and RMNH 1159, a 600 mm female (N.C. de Fremery).
Type locality: Unknown.
Distribution: South America. Southern Colombia (Amazonas), Guyana, Suriname (Saramacca), N French Guiana (Cayenne, Saint-Laurent du Moroni), Brazil (Amapá, Bahia, Ceara, Distrito Federal, Goias, Mato Grosso, Pará, Pernambuco, Piauí, Tocantins), E Peru (Loreto, Madre de Dios), Bolivia (Beni, Cochabamba, La Paz, Santa Cruz), Paraguay (Alto Paraguay, Alto Paraná, Amambay, Asunción, Caaguazú, Caazapá, Canindeyú, Central, Concepción, Cordillera, Guairá, Itapúa, Misiones, Neembucú, Paraguarí, Presidente Hayes, San Pedro, Kururu, Modesto, Talavera and Yacyreta Is.) and N Argentina (Buenos Aires, Chaco, Corrientes, Entre Ríos, Formosa, Jujuy, Misiones, Salta, Santa Fe), NSL–340 m.
Sources: Vanzolini et al., 1980, Abuys, 1983d, Chippaux, 1987, F.D. Williams & Francini, 1991, Cei, 1994, Starace, 1998, Freitas, 1999, Leynaud & Bucher, 1999, Campos-Nogueira, 2001, Doan & Arizábal, 2002, Frota, 2005, Cacciali, 2008 and Recorder et al., 2011.

8. *Helicops modestus* A.C.L.G. Günther, 1861c. Ann. Mag. Nat. Hist. (3) 7(42): 425–426, 2 figs.

Synonyms: *Helicops leprieurii moestus* Jan, 1863b (*nomen nudum*), *Helicops leprieurii moesta* Jan, 1865c, and *Helicops assimilis* J.T. Reinhardt, 1866.
Type: Holotype, BMNH 1946.1.14.42 (formerly RMNH), a 525 mm female.
Type locality: "North America" (in error). Corrected to Tropical America *fide* A.C.L.G. Günther (1861c: 426).
Distribution: Central and S Brazil (Bahia, Distrito Federal, Goiás, Minas Gerais, Rio de Janeiro, São Paulo), 970–1010 m.
Sources: Nogueira, 2001 and Freitas & Santos, 2011.

9. *Helicops pastazae* Shreve, 1934. Occ. Pap. Boston Soc. Nat. Hist. (8): 129–130.

Type: Holotype, MCZ 36993, a 322 mm female (C. Spencer, Jan. 1931–Aug. 1932).
Type locality: "along the Pastaza River, from Canelos to the Marañon River, Ecuador" [= Río Pastaza bet. Canelos (1°35'S, 77°45'W, 500 m), Pastaza Prov., Ecuador and Río Marañon junction (4°55'S, 76°25'W, 130 m), Loreto Dept., Peru].
Distribution: Northwestern South America. Venezuela (Barinas, Bolívar, Táchira, Zulia), Colombia (Amazonas, Antioquia, Caquetá, Meta, Norte de Santander), E Ecuador (Morona-Santiago, Napo,

Pastaza, Zamora-Chinchipe) and N Peru (Amazonas, Loreto), NSL–1140 m.
Sources: Rossman, 1976, Markezich & Rossman, 1992, Markezich, 2002 and Navarrete et al., 2009.
Remarks: Type erroneously listed as MCZ 36963 *fide* Shreve (1934: 130). Collection date listed as 1932 *fide* Rossman (1976: 4).

10. *Helicops petersi* Rossman, 1976. Occ. Pap. Mus. Zool., Louisiana St. Univ. (50): 2–4, fig. 1 (paratype).

Type: Holotype, USNM 196360 (formerly JAP 2758), a 573 mm male (J.A. Peters, 31 Oct.–3 Nov. 1958).
Type locality: "from the east bank of the Mishualli River, 1 mile NE Tena, Napo-Pastaza Prov., Ecuador" [= 1 mi. NE Tena along Río Misahuallí, Napo Prov., NE Peru, ca. 0°59'S, 77°48'W, elevation 535 m].
Distribution: Northeastern Ecuador (Napo, Orellana, Sucumbios), 300–570 m.

11. *Helicops polylepis* A.C.L.G. Günther, 1861c. Ann. Mag. Nat. Hist. (3) 7(42): 426–427.

Synonyms: *Tachynectes chrysostictus* Cope, 1862b, and *Helicops spixii* Jan, 1863b.
Type: Holotype, BMNH 1946.1.13.25, a 426 mm male.
Type locality: "Upper Amazon" [= Río Amazonas, Amazonas State, NW Brazil].
Distribution: Northern South America. Southern Colombia (Amazonas, Meta, Putumayo), Ecuador (Oriente, [FMNH 23484]), E Peru (Loreto, Madre de Dios), N Bolivia (Beni, Cochabamba, La Paz) and N Brazil (? Amapá, Amazonas, Maranhão [UOMNH 31518], Mato Grosso, Pará), 200–1000 m.
Sources: F. Werner, 1901b, Cunha & Nascimento, 1978, 1993, Duellman, 1990, 2005, Duellman & Salas, 1991 and Frota, 2005.

12. *Helicops scalaris* Jan, 1863b. Elenco Sist. Ofidi: 76.

Synonym: *Helicops hogei* Lancini, 1964.
Type: Holotype, MZUT, a 410 mm male, lost *fide* O. Elter *in* Rossman (2002a: 3).
Type locality: "Venezuela."
Distribution: Extreme NW South America. Northern Colombia (Norte de Santander) and Venezuela (Amazonas, Delta Amacuro, Mérida, Táchira, Trujillo, Zulia), NSL–500 m.
Sources: Jan, 1868 *in* Jan & Sordelli, 1866–1870, Lancini-V, 1964, 1986, Roze, 1966a, Perez-Bravo, 1978, Hoogmoed, 1979, Lancini, 1986, Rossman, 2002a and Navarrete et al., 2009.
Remarks: Photograph of the holotype of *H. hogei* in Rossman (2002a: fig. 11).

13. *Helicops tapajonicus* Frota, 2005. Phyllomedusa 4(1): 62–66, figs. 1–3.

Type: Holotype, MCP 15380, a 582 mm female (J. Lúcio, 20 May 2001).
Type locality: "Comunidade de Parauá (02°50'38" S, 55°10'54" W), margem direita do rio Tapajós, municipio e Santarém, estado do Pará, Brasil."
Distribution: Northeastern Brazil (Pará), 15 m.

14. *Helicops trivittatus* (Gray, 1849a). Cat. Snakes Brit. Mus.: 70–71. (*Myron trivittatus*)

Type: Holotype, BMNH 1946.1.13.99, an adult female (Zool. Soc. London).
Type locality: Unknown.
Distribution: Northeastern Brazil (Amapá, NE Mato Grosso, Pará, Tocatins, Marajó Is.), NSL–300 m.
Sources: Boulenger, 1893a, Cunha & Nascimento, 1993 and Rossman, 2010.
Remarks: Type specimen discussed by Rossman (2010: 272). A second specimen was labeled "India?" (in error).

15. *Helicops yacu* Rossman & Dixon, 1975. Herpetologica 31(4): 412–413.

Type: Holotype, TCWC 39093, a 352 mm (svl) female (P. Soini, May 1972).
Type locality: "Moropon, Departamento de Loreto, Peru."
Distribution: Eastern Peru (Loreto) and W Brazil (Acre), 100 m.
Remarks: Probably a subspecies of *H. pastazae* Shreve *fide* Rossman & Abe (1979: 9).

HELMINTHOPHIS W.C.H. Peters, 1860d (Anomalepididae)

Synonyms: *Idiotyphlops* Jan, 1860 *in* Jan & Sordelli, 1860–1866, and *Helmintophis* – Hoffmann, 1890 *(nomen incorrectum)*.
Type species: *Typhlops frontalis* W.C.H. Peters, 1860d.
Distribution: Lower Central America and NW South America.
Sources: F. Werner, 1921a, Dunn, 1944, Robb & Smith, 1966, J.A. Peters & Orejas-Miranda, 1970, Hahn, 1980a, Pérez-Santos & Moreno, 1988, Wallach, 1993a, McDiarmid et al., 1999 and Tipton, 2005.
Remarks: An undescribed species present on Trinidad *fide* Wynn & Wallach (unpubl. data).

1. *Helminthophis flavoterminatus* (W.C.H. Peters, 1857b). Mber. Königl. Akad. Wiss. Berlin 1857(8): 402. (*Typhlops flavoterminatus*)

Synonym: *Typhlops multilatus* F. Werner, 1900a.

Types: Syntypes (6), ZMB 1426–30, and ZMH, longest syntype 240 mm (C. Moritz), the latter destroyed in July 1943 during World War II.

Type locality: "Carácas" [= Caracas, Distrito Federal State, N Venezuela, 10°29'N, 66°54'W, elevation 925 m].

Distribution: Northwestern Venezuela (Amazonas, Aragua, Distrito Federal, Miranda, Zulia) and W Colombia (Chocó, Norte de Santander), NSL–1800 m. Introduced into Mauritius.

Sources: A.C.L.G. Günther, 1870, Roze, 1966a, Lancini & Kornacker, 1989, Wallach & Günther, 1997, Mijares-Urrutia & Arends, 2000 and Navarrete et al., 2009.

Remarks: ZMB catalogue records Moritz as collector, not Gollmer and Moritz as stated in original description *fide* Bauer et al. (2002: 158).

H

2. *Helminthophis frontalis* (W.C.H. Peters, 1860d). Mber. Königl. Akad. Wiss. Berlin 1860(10): 517–518, pl., figs. 1, 1a–1c. (*Typhlops frontalis*)

Types: Lectotype, ZMB 3925, possibly a 145 mm female (K. Hoffmann, 1853–1859), designated by Hahn (1980a: 2).

Type locality: "Costa Rica" via lectotype selection.

Distribution: Costa Rica (Alajuela, Heredia, San José) and W Panama (Chiriquí), 95–1435 m.

Sources: Bocourt, 1882 *in* A.H.A. Duméril, Bocourt & Mocquard, 1870–1909, E.H. Taylor, 1951, Savage, 2002 and Solórzano, 2004.

3. *Helminthophis praeocularis* Amaral, 1924a. Proc. New England Zool. Club 9: 28.

Type: Holotype, MCZ 17960, a 214 mm specimen (H. Nicéforo-Maria, March 1924).

Type locality: "Honda, Colombia" [= Honda, upper Río Magdalena Valley, Tolima Dept., cen. Colombia, 05°12'N, 74°45'W, elevation 210 m].

Distribution: Central Colombia (Antioquia, Meta, Norte de Santander, Santander, Tolima), 200–1280 m.

Source: Medem, 1965.

HELOPHIS Witte & Laurent, 1942 (*nomen substitutum*) (Natricidae)

Synonym: *Pelophis* Witte, 1922 *(nomen praeoccupatum).*

Type species: *Pelophis schoutedeni* Witte, 1922.

Distribution: Democratic Republic of the Congo.

Sources: Witte, 1933a, 1962, Witte & Laurent, 1942 and Laurent, 1956b.

1. *Helophis schoutedeni* (Witte, 1922). Rev. Zool. Bot. Afr. 10(3): 318–319. (*Pelophis schoutedeni*)

Types: Syntypes (2), MRAC 2468, a 446 mm specimen (H. Schouteden, 24 Feb. 1924), and MRAC 2469, a 540 mm specimen (H. Schouteden, 23 May 1921).

Type locality: "Kwamouth (district du Moyen-Congo), Congo" [= W Bandundu Region, Democratic Republic of the Congo, 3°20'S, 16°10'E] and "Tondu, sur le Lac Tumba (district de l'Équateur), Congo" [= W Equateur Prov., Democratic Republic of the Congo, 0°50'S, 18°10'E].

Distribution: Western Democratic Republic of the Congo (Bandundu, Bas-Congo, Equateur, Kinshasa), 285–450 m.

HEMACHATUS J. Fleming, 1822 (Elapidae)

Synonyms: *Sepedon* Merrem, 1820 *(nomen praeoccupatum), Haemacathus* Gray *in* Griffith & Pidgeon, 1831 *(nomen emendatum), Haemachates* – Duvernoy, 1833b *(nomen incorrectum), Spedon* – Schinz, 1833 *(nomen incorrectum), Sepidon* Swainson, 1839 *(nomen emendatum), Hamachates* – A.-M.C., Duméril, Bibron & Duméril, 1854b *(nomen incorrectum), Sipedon* – Lockington, 1886 *(nomen incorrectum), Merremia* Berg, 1901 *(nomen substitutum), Sepedonophis* H.M. Smith, 1936 *(nomen substitutum), Hemachtus* – J.H. Welsh, 1966 *(nomen incorrectum), Haemachatus* – H.A. Skinner, 1973 *(nomen incorrectum), Haemachates* – Golay, Smith, Broadley, Dixon, McCarthy, Rage, Schätti & Toriba, 1993 *(nomen incorrectum),* and *Hamachates* – Golay, Smith, Broadley, Dixon, McCarthy, Rage, Schätti & Toriba, 1993 *(nomen incorrectum).*

Type species: *Coluber haemachates* Bonnaterre, 1790.

Distribution: Southeastern Africa.

Sources: V.F.M. FitzSimons, 1962a, 1974, Broadley, 1962c, 1983, Broadley & Cock, 1975, Keogh, 1998, Marais, 2004, Dobiey & Vogel, 2007, Zaher et al., 2009, Hoser, 2012e and Broadley & Blaylock, 2013.

1. *Hemachatus haemachates* (Bonnaterre, 1790). Tabl. Ency. Méth., Ophiol.: 31–32, pl. 37, fig. 2. (*Coluber haemachates*)

Synonyms: *Coronella catus* Laurenti, 1768 *(nomen oblitum), Vipera haemachate* Lacépède, 1789 *(nomen rejiciendum), Coluber hikkanella* Bonnaterre, 1790, *Hemachatus vulgaris* Fleming, 1822, *Natrix diphorolepidota* Wagler, 1825 *(nomen ineditum), Naia capensis* A. Smith, 1826, *Sepedon hemachata* – Flower, 1929, and *Hemachatus haemachatus* – Moroz, Grotto, Goldblum & Vries, 1967 *(nomen incorrectum).*

Type: Holotype, formerly MNHN, a 423 mm specimen (King of France's cabinet, Paris), lost *fide* Broadley *in* Golay et al. (1993: 138).

Type locality: "Le Japon, la Perse" [= Japan and Iran] (in error). Corrected to South Africa *fide* Bogert (1940: 87) and V.F.M. FitzSimons (1962a: 287).

Distribution: Southeastern Africa. Eastern Zimbabwe (Manicaland), Swaziland, Lesotho and South Africa

(Eastern Cape, Free State, Gauteng, KwaZulu-Natal, Mpumalanga Western Cape), NSL–2500 m.

Remarks: NHR Lin-1632 (MAFR, Mus. Drottn.) is a Linnaean specimen. In accordance with Art. 23.9.2 of the Code (ICZN, 1999), *Coluber haemachates* Bonnaterre is designated a *nomen protectum* and *Coronella catus* Laurenti a *nomen oblitum.*

HEMEROPHIS Schätti & Utiger, 2001 (Colubridae)

Type species: *Zamenis socotrae* A.C.L.G. Günther, 1881a.

Distribution: Yemen (Socotra Is.).

Sources: Showler, 1994, Schätti & Utiger, 2001, Nagy et al., 2004 and Utiger et al., 2005.

1. *Hemerophis socotrae* (A.C.L.G. Günther, 1881a). Proc. Zool. Soc. London 49(2): 463, pl. 41. (*Zamenis socotrae*)

Types: Syntypes (3), BMNH 1946.1.14.97–99, one male and two females, longest syntype 826–860 mm (I.B. Balfour, 1879–1880).

Type locality: "Island of Socotra" [= Socotra Is., Yemen, Arabian Sea].

Distribution: Yemen (Socotra Is.), NSL– 900 m.

Source: Razzetti et al., 2011.

2. *Hemerophis zebrinus* (Broadley & Schätti, 2000). Madoqua (1997) 19(2): 172–174, figs. 2–3, 5. (*Coluber zebrinus*)

Type: Holotype, SMWN 8046, a 385 mm male (M. Lindeque & M. Griffin, 10 April 1991).

Type locality: "near the Canene River, at Ruacana, western Owamboland, Namibia (17°25'S, 14°10'E)."

Distribution: Namibia (Damaraland, Kaokoland, Owambo), 780 m.

Source: Schätti, 1986b.

HEMIASPIS Fitzinger, 1861 (Elapidae)

Synonyms: *Drepanodontis* Worrell, 1961a, and *Hemiaspus* – Vitt, 1983 *(nomen incorrectum).*

Type species: *Alecto signata* Jan, 1859a.

Distribution: Eastern Australia.

Sources: Cogger et al., 1983a, Shine, 1987a, Gow, 1989, Hutchinson, 1990, Greer, 1997, Cogger, 2000, Scanlon & Lee, 2004, Sanders et al., 2008, Zaher et al., 2009 and Hoser, 2012e.

1. *Hemiaspis damelii* (A.C.L.G. Günther, 1876). J. Mus. Godeffroy 5(12): 46. (*Hoplocephalus damelii*)

Synonyms: *Hoplocephalus dämelii* – F. Müller, 1878b *(nomen incorrigendum), Hoplocephalus suboccipitalis* Ogilby, 1892, and *Denisonia daemelii* Boulenger, 1896a *(nomen emendatum).*

Types: Syntypes (3+), BMNH 1946.1.18.89–90 [Peak Downs], and ZMH 4369 (formerly ZMH 824) [Queensland], longest syntype 381 mm (Dämel), location of other syntypes unknown *fide* Cogger et al. (1983: 226).

Type locality: "Rockhampton and Peak Downs, Australia" [= Rockhampton, 23°23'S, 150°31'E, elevation 15 m, and Peak Downs, 23°04'S, 148°10'E, elevation 220 m, SE Queeensland, Australia].

Distribution: Eastern Australia (SE Queensland, NE New South Wales), NSL–340 m.

2. *Hemiaspis signata* (Jan, 1859a). Rev. Mag. Zool. (2) 11: 124, 128. (*Alecto signata*)

Synonyms: *Hoplocephalus signatus* – De Vis, 1884, and *Denisonia vagrans* S.W. Garman, 1901.

Type: Holotype, SMF-B 9338,2a, a 460 mm specimen (W. Kirchner & A. becker, 1847–1859), lost *fide* Cogger et al. (1983: 226).

Type locality: "Nouvelle-Hollande" [= Australia]. Restricted to Sydney, New South Wales, Australia *fide* Boettger (1898: 117).

Distribution: Eastern Australia (ext. E Queensland, ext. E New South Wales, S Victoria), NSL–1320 m.

Remarks: Supplemental original description in Jan (1859e: pl. C, fig. 4). Hallerman (1998: 206) listed holotype as ZMH 4433 (formerly ZMH 775) from Sydney (Kraft, 1861).

HEMIBUNGARUS W.C.H. Peters, 1862b (Elapidae)

Synonyms: *Brachyrhynchus* Fitzinger, 1843 *(nomen praeoccupatum), Hemibungars* – Noguchi, 1909 *(nomen incorrectum), Hemibungaros* – Brazil, 1914 *(nomen incorrectum), Hemibungaus* – Takahashi, 1930 *(nomen incorrectum),* and *Hemibngarus* – Maik, 1933 *(nomen incorrectum).*

Type species: *Elaps calligaster* Wiegmann *in* Meyen, 1834.

Distribution: Philippines.

Sources: Golay et al., 1993, David & Ineich, 1999, Ota et al., 1999, Slowinski et al., 2001, Castoe et al., 2007, Zaher et al., 2009 and Hoser, 2012e.

Remarks: A valid genus *fide* Castoe et al. (2007a: 824).

1. *Hemibungarus calligaster* (Wiegmann *in* Meyen, 1834). Nova Acta Caesar.-Leop. Carol. 17(1): 253–254, pl. 20, fig. 2. (*Elaps calligaster*)

Synonyms: *Elaps caligaster* – Jan, 1858, *(nomen incorrectum), Naja fasciata* W.C.H. Peters, 1861c *(nomen* *praeoccupatum), Hemibungarus gemianulis* W.C.H. Peters, 1872b, *Hemibungarus gemmiannulis* – Boettger, 1886c *(nomen incorrectum),* and *Hemibungarus mcclungi* E.H. Taylor, 1922a.

Type: Holotype, ZMB 2742, a 376–377 mm specimen (F.J.F. Meyen, 1830–1832), lost *fide* R. Günther *(in litt.)*

Type locality: "Manila" [= Manila, Manila Prov., SW Luzon Is., N Philippines, 14°36'N, 120°59'E, elevation 10 m].

Distribution: Philippines (Catanduanes, Cebu, Luzon, Masbate, Mindoro, Negros, Panay, Polillo), 10–800 m.

Sources: Leviton, 1964b, McDowell, 1987, Ross & Gonzales, 1992, Gaulke & Altenbach, 1994, Kucharzewski & Tillack, 2007 and Siler et al., 2011.

Remarks: Records for Balabac, Paraqua, Mindanao and Samar need confirmation *fide* Leviton (1964b: 544) and Kucharzewski & Tillack (2007: 54).

H

HEMIRHAGERRHIS Boettger, 1893c (Psammophiidae)

Synonyms: *Rhamphophis* – Boulenger, 1892d *(lapsus calami), Hemirhagerchis* – Palacky, 1898 *(nomen incorrectum), Hemirhagerchis* – Palacky, 1898 *(nomen incorrectum), Hemirhagerhis* – Palacky, 1898 *(nomen incorrectum), Hemirhaggeris* – Scortecci, 1931 *(nomen incorrectum), Hemihagerrhis* – Briceño-Rossi, 1934 *(nomen incorrectum), Hemiphagerris* – Ionides, 1950 *(nomen incorrectum), Hemirhaggerhis* – Laurent, 1964a *(nomen incorrectum), Hemirhaggerrhis* – H.A. Skinner, 1973 *(nomen incorrectum), Hemirrhagerris* – H.A. Skinner, 1973 *(nomen incorrectum), Hemirhagerris* – Dowling *in* Dowling & Duellman, 1978 *(nomen incorrectum),* and *Hermirhagerrhis* – Spawls, 1978 *(nomen incorrectum).*

Type species: *Hemirhagerrhis kelleri* Boettger, 1893c.

Distribution: Central and East Africa.

Sources: Bogert, 1940, Loveridge, 1957a, V.F.M. FitzSimons, 1962a, Lanza, 1983a, 1990b, Largen & Rasmussen, 1993, Brandstätter, 1995, Broadley & Hughes, 2000, Spawls et al., 2002, Vidal et al., 2008, Kelly et al., 2009 and Zaher et al., 2009.

1. *Hemirhagerrhis hildebrandtii* (W.C.H. Peters, 1878). Mber. Königl. Akad. Wiss. Berlin 1878(3): 205–206, pl. 2, fig. 6. (*Ablabes hildebrandtii*)

Type: Holotype, ZMB 9282.

Type locality: "Kitui (Ukamba)" [= Kitui Catholic Church, Ukamba Distr., S Eastern Prov., Kenya, 1°31'S, 37°14'E, elevation 1615 m].

Distribution: Northeastern Africa. South Sudan, SE Ethiopia (Bale, Hararge, Sidamo), SW Somalia (Bakool, Gedo), N Kenya (N North-Eastern, N Rift Valley) and NE Tanzania (Arusha, Kilimanjaro, Tanga), 150–1615 m.

Source: Largen & Spawls, 2010.

Remarks: Resurrected as a valid species *fide* Broadley & Hughes (2000: 9).

2. *Hemirhagerrhis kelleri* Boettger, 1893c. Zool. Anz. 16(417): 129.

Synonym: *Amplorhinus taeniatus* Sternfeld, 1908e.

Type: Lectotype, SMF 19906 (formerly SMF-B 9119, 1a), possibly a 350 mm adult female (C. Keller, 1893), designated by Mertens (1922:181).

Type locality: "Abdallah, südlich vom Webiflusse, Somaliland" [= Abdula, S of Webi River, Hararge Prov., E Ethiopia, 9°27'N, 42°16'E, elevation 2485 m] via lectotype selection.

Distribution: Northeastern Africa. South Sudan, SE Ethiopia (Bale, Hararge, Sidamo), Somalia (Bay, Gedo, Hiraan, Jubbada Dhexe, Jubbada Hoose, Mogadishu, Shabeellaha Dhexe, Shabeellaha Hoose, Togdheer), Kenya (Coast, North-Eastern, Lamu Is.) and N Tanzania (Arush), 200–2485 m.

Source: Largen & Spawls, 2010.

Remarks: Three syntypes listed in original description but one specimen listed by Boettger (1898: 99). Lectotype listed as holotype *fide* Broadley & Hughes (2000: 6–7).

3. *Hemirhagerrhis nototaenia* (A.C.L.G. Günther, 1864c). Proc. Zool. Soc. London 32(1): 309, pl. 26, fig. 1 (*Coronella nototaenia*)

Synonyms: *Amplorhinus güntheri* Mocquard, 1906 *(nomen incorrigendum).*

Type: Holotype, BMNH 1946.1.8.51, a 368 mm male (J. Kirk, 1858–1864).

Type locality: "regions bordering the Zambesi" [= vicinity of Zambezi River, Mozambique]. Emended to Rios de Sena, Zambezi *fide* Boulenger, 1896a: 126 [= Villa de Senna, S bank of Zambezi River, Sofala Prov., Mozambique, 17°27'S, 35°02'E, elevation 75 m].

Distribution: West, cen. and E Africa. Southeastern Burkina Faso (Est), N Togo (Kara), N Benin (Alibori), S Niger (Maradi), SW Chad (Chari-Baguirmi), Central African Republic (Ombella-Mpoko, Ouham, Vakaga), S Sudan (Janub Kurdufan) W South Sudan (Bahr El Ghazal, Western Equatoria), Somalia (Bay, Galguduud, Gedo, Jubbada Dhexe, Jubbada Hoose, Shabeellaha Hoose), E Kenya (Eastern, North-Eastern, N Rift Valley), E Tanzania (Kilimanjaro, Lindi, Morogoro, Mwanza, Mtwara, Piwani, Rukwa, Tanga), E Democratic Republic of the Congo (Katanga, Orientale), SW Angola (Huambo, Huíla), Zambia (Central, Copperbelt, Eastern, Luapula, Lusaka, Northern, North-Western, Southern, Western), Malawi (Northern), S Mozambique (Gaza, Inhambane, Manica, Niassa, Sofala, Tete), Zimbabwe (Manicaland, Mashonaland Central, Mashonaland East, Mashonaland West, Masvingo, Matabeleland North, Matabeleland South, Midlands), N Botswana (Central, North West), N Namibia (Caprivi,), Swaziland, and NE South Africa

(KwaZulu-Natal, Limpopo, Mpumalanga), NSL–1600 m.

Sources: Mertens, 1955, 1971, Sweeney, 1961, Witte, 1962, Pitman, 1974, Broadley & Cock, 1975, Broadley, 1983, Roman, 1980, 1984, Branch, 1988, Haagner, 1990, Chirio & Ineich, 1993, Broadley et al., 2003, Marais, 2004, Branch et al., 2005, Chirio & Ineich, 2006, Chippaux, 2006, J.-F. Trape & Mané, 2006b, Chirio & LeBreton, 2007, Segniagbeto et al., 2011, Broadley & Blaylock, 2013 and Hughes, 2013.

Remarks: Probably occus in Cameroon *fide* Chirio & LeBreton (2007: 448) and possibly occurs in Uganda *fide* Pitman (1974: 146). Record from Bas Congo, Democratic Republic of the Congo doubtful *fide* Thys van den Audenaerde (1965: 380).

4. *Hemirhagerrhis viperina* (Bocage, 1873a). J. Sci. Math. Phys. Nat. Lisbon (1)1: 222–223. (*Psammophylax viperinus*)

Type: Holotype, MBL 1715, destroyed by fire 18 March 1978.

Type locality: "Dombe (Benguella)" [= Dombe Grande, Benguela Distr., W Angola, 12°57'S, 13°06'E, elevation 45 m].

Distribution: Southwestern Angola (Benguela, Huambo, Huíla, Namibe) and N Namibia (Erongo, Kunene, Omaheke, W Otjozondjupa), 45–1400 m.

Sources: Mertens, 1955, 1971 and Elzen, 1980.

Remarks: A valid species *fide* Broadley & Hughes (2000: 7).

HEMORRHOIS F. Boie, 1826 (Colubridae)

Synonyms: *Haemorrhois* F. Boie, 1827 *(nomen emendatum),* and *Periops* Wagler, 1830.

Type species: *Coluber hippocrepis* Linnaeus, 1758.

Distribution: Southern Eurasia, Middle East and North Africa.

Fossil records: Upper Pliocene to upper Pleistocene of Morocco and Spain.

Sources: Welch, 1980, 1983a, Schätti & Wilson, 1986, Schátti & Utiger, 2001 and Nagy et al., 2004.

1. *Hemorrhois algirus* (Jan, 1863b). Elenco Sist. Ofidi: 60. (*Periops algira*)

Synonyms: *Periops hippocrepis algira* Jan, 1857 (*nomen nudum*), *Coluber algirus intermedius* F. Werner, 1929b, and *Coluber algirus villiersi* Bons, 1962.

Types: Syntypes (4), MNHN 3575, MNHN 7464, and formerly MSNM, destroyed in 1943 during World War II.

Type locality: "Algeri, and Sphax d' Algeria" [= Tunisia].

Distribution: Northwestern Africa. Northern Mauritania (Adrar, Dakhlet Nouadhibou, Tiris Zemmour), Western Sahara (Rio de Oro, Saguia El Hamra), Southern Morocco (Agadir, Kaar Es Souk, Ouarzazate, Tan Tan, Tata, Tiznit), S Algeria (El Bayadh, Oran, Tamanrasset), S Tunisia (Gabès, Gafsa, Kairouan, Kasserine, Kebili, Medenine, Sfax, Sidi Bou Zid, Sousse, Tataounie, Tunis, Tozeur, Djerba and Kerkennah Is.), Malta, N Libya (Ajbabiya, Al Jabar Al Akhdar, Al Murqub, Banghazi, Darnah, Gharyan, Misratah, Surt) and NW Egypt (NW Matrouh), NSL–1600 m.

Sources: Doumergue, 1901, Mayet, 1903, Chpakowsky & Chnéour, 1954, Kramer & Schnurrenberger, 1963, Pozuelo, 1974, Bruno, 1984, Schätti, 1986d, 1993a, Lang, 1993, Bons & Geniez, 1996, Schleich et al., 1996, Bogaerts, 1998, Geniez et al., 2004, Baha El Disi, 2006, Chippaux, 2006, Carranza et al., 2006 and J.-F. Trape & Mané, 2006b.

Remarks: MNHN catalogue lists MNHN 1961.315 from Mauritania as holotype. A subspecies of *Platyceps florulentus* Geoffroy-Saint-Hilaire *fide* Kramer & Schnurrenberger, 1963.

2. *Hemorrhois hippocrepis* (Linnaeus, 1758). Syst. Nat. ed., 10, 1: 226. (*Coluber hippocrepis*)

Synonyms: *Coluber domesticus* Linnaeus, 1766, *Natrix bahiensis* Wagler *in* Spix, 1824, *Coluber diadema* Gené, 1834, and *Coluber hippocrepis nigrescens* Cattaneo, 1985.

Type: Holotype, NHR Lin-29 (formerly MAFR), a 340 mm specimen (P. Loeflingius, 1751, via Mus. Drottn.).

Type locality: "America" (in error). Corrected to Spain, Iberian Peninsula *fide* Bruno (1968a: 176).

Distribution: Iberian Peninsula and ext. NW Africa. Southern Portugal (Beja, Braga, Bragansa, Castelo Branco, Coimbra, Évora, Faro, Guarda, Leiria, Lisboa, Portalegre, W Santarém, Setubal, Vila Real, Viseu), S Spain (S Albacete, Alicante, Almería, Ávila, Badajoz, Barcelona, Cáceres, Cádiz, Castellón, Ciudad Real, Córdoba, Gerona, Granada, Huelva, Huesca, Jaén, Lérida, Madrid, Málaga, Murcia, Salamanca, Seville, Tarragona, W Toledo, Valencia, Zamora), W Italy (S Sardinia), NW Morocco (Agadir, Al Hoceima, Al Jadida, Azilal, Beni Mellel, Boulemane, Casablanca, Chechaouene, El Kelaa des Srarhna, Essaouira, Fes, Kenitra, Khemisset, Khenitra, Khouribga, Marrakech, Meknes, Nador, Oujda, Rabat, Safi, Settat, Tangier, Tan Tan, Taonate, Taza, Tetouan, Tiznit), N Algeria (El Bayadh, Naama, Tlemcen) and Tunisia (Bizerte, Gabès, Jendouba, Nabeul, Tunis, Zembra Is.), NSL–2660 m.

Fossil records: Upper Pliocene of Morocco, middle Pleistocene (Ionian: 0.13–0.78 mya) of Spain, and upper Pleistocene (Tarantian: 0.01–0.13 mya) of Spain.

Sources: Doumergue, 1901, Mayet, 1903, Pozuelo, 1974, Bruno & Mangeri, 1977, Salvador, 1985, 1998, Schätti, 1986d, 1993b, Blanc, 1988, Bons & Geniez, 1996, Schleich et al., 1996, Bogaerts, 1998, Holman, 1998a, Pleguezuelos & Feriche, 1999, Bailon, 2000, Corti et al., 2000, Salvador & Pleguezuelos, 2002, Carranza et al., 2006, Sindaco et al., 2006 and Feriche et al., 2008.

3. *Hemorrhois nummifera* (A. Reuss, 1834). Mus. Senck. 1: 135–137. (*Coluber nummifer*)

Synonyms: *Coluber tyria* Linnaeus, 1758 (*nomen rejiciendum*), *Tyria linnaei* Fitzinger, 1826a (*nomen oblitum*), *Coluber barbarus* H. Boie *in* Schlegel, 1826a (*nomen nudum*), *Coluber barbarus* Schlegel, 1826b (*nomen nudum*), *Coluber barbarus* F. Boie, 1827 (*nomen oblitum*), *Zamenis caudaelineatus* A.C.L.G. Günther, 1858 (*partim*), *Periops neglectus* Jan, 1863b, *Zamenis glazunowi* Nikolski, 1896 (*partim*), *Zamenis ater* Venzmer, 1919, *Elaphe dione plumbea* Chernov, 1926, *Zamenis venzmeri* F. Werner, 1929 (*partim*), *Coluber nummifer pallaryi* Angel, 1936c, and *Elaphe quatuorlineatus sauromates bahtiyari* Bodenheimer, 1944 (*nomen illegitimum*).

H

Type: Holotype, SMF 18236 (formerly SMF-R III.O.4.a & SMF-B 7407 a), a 914 mm female (E. Rüppel, 1832).

Type locality: "Aegypten" [= Egypt].

Distribution: Southwestern Asia. Greece (Kalymnos, Kastellorizo, Kos, Leipsi, Leros, Syme, Rhodes), Cyprus, S Turkey, W Syria (Homs, Latakia), Lebanon (Beirut, Mont-Liban, Nabatiye), Israel (Central, Gaza Strip, Golan Heights, Haifa, Jerusalem, Northern, N Southern, Tel Aviv, West Bank), Jordan (Ajloun, Amman, Balqa, Irbid, Jarash, Karak, Maan, Tafilah), NE Egypt (Cairo, South Sinai), S Armenia, N Iraq and NW Iran, 100–2000 m.

Sources: F. Werner, 1939, Schätti & Agarian, 1985, Szczcerbak, 1994, Y.L. Werner, 1995, Bouskila & Amitai, 2001, Disi et al., 2001, Hraoui-Bloquet et al., 2002, Baier et al., 2009, Schätti et al., 2009, Tuniyev et al., 2009, Amr & Disi, 2011 and Bar & Haimovitch, 2011.

Remarks: Official Specific Name *fide* Opinion 2282 (ICZN, 2011). In accordance with Art. 23.9.2 of the Code (ICZN, 1999), *Tyria linnaei* Fitzinger and *Coluber barbarus* F. Boie are designated *nomina oblita*.

4. *Hemorrhois ravergieri* (Ménétriés, 1832). Cat. Zool. Voy. Caucase: 69–70. (*Coluber ravergieri*)

Synonyms: *Coluber maculatus* Dwigubsky, 1832, *Coluber cliffordi* Martin, 1838, *Zamenis caudaelineatus* A.C.L.G. Günther, 1858 (*partim*), *Periops caudilineatus neva* Jan *in* DeFilippi, 1865, *Zamenis fedtschenkboi* Strauch, 1873 (*partim*), *Zamenis kaufmanni* Grum-Grzimailo, 1885, *Zamenis glazunowi* Nikolsky (1896 (*partim*), and *Coluber ravergieri cernovi* Mertens, 1952d..

Types: Syntypes (2), ZISP 8681–82 (formerly AIS), longest syntype 280–305 mm (Ravergier, 1828–1832).

Type locality: "Géorgie," "Prés de Bakou," and "côtes occidentales de la mer Caspienne, Caucase." [= Georgia; Baku, Azerbaijan; W coast of Caspian Sea in Caucasus Mtns., E Azerbaian]. Restricted to Georgia *fide* Flower (1933: 812) and Schätti & Agasian (1985: 110).

Distribution: Middle East and SW Asia. Lebanon (Mont-Liban), N Jordan (Mafraq), Israel (Golan Heights), NE Turkey, SE Georgia, Armenia, Azerbaijan, E Kazakhstan, Uzbekistan, Turkmenistan, Kyrgyzstan, Tadzhikistan, N Iran, N Afghanistan, cen. Pakistan (N Balochistan, Khyber Pakhtunkhwa), W China (Zingjiang) and W Mongolia, 30–4000 m.

Fossil records: Upper Miocene/lower Plioicene of Kazakhstan.

Sources: Minton, 1966, Leviton & Anderson, 1970a, Baran, 1976a, Bannikov et al., 1977, Hraoui-Bloquet, 1981, Schätti & Agasian, 1985, W. Böhme, 1993c, Disi, 1993, Zhao & Adler, 1993, Szczcerbak, 1994, M.S. Khan, 1997, 2002, Kordikova et al., 1998, Disi et al., 2001, Hraoui-Bloquet et al., 2002, Tuniyev et al., 2009 and Amr & Disi, 2011.

†*HERENSUGEA* Rage, 1996 (†Madtsoiidae)

Type species: †*Herensugea caristiorum* Rage, 1996.

Distribution: Upper Cretaceous of Spain.

Sources: Rage, 1996, 1999.

1. †*Herensugea caristiorum* Rage, 1996. C. R. Acad. Sci. Paris (2A) 322: 606–607, fig. 2.

Type: Holotype, MCNA 5387, one middle trunk vertebra.

Type locality: "le Crétacé terminal de Laño (Pays Basque espagnol)" [= upper Cretaceous of Laño, Condado de Treviño Prov., Basque Country, N Spain, 42°40'N, 2°37'W]. Clarified as a sand quarry near the village of Laño, about 20 km SE of the city of Vitoria, Basque Country, Spain; Upper Cretaceous (Early Maastrichian) *fide* Astibia et al. (1990: 462).

Distribution: Upper Cretaceous (Maastrichian: 65.5–72.1 mya) of Spain. Known only from type locality.

HETERODON Latreille *in* Sonnini & Latreille, 1801b (Carphophiidae)

Synonyms: *Hetorodon* – Latreille *in* Sonnini & Latreille, 1801 *(nomen incorrectum), Heterodus* – Goode, 1883 *(nomen incorrectum), Heterodoa* – Adcock, 1922 *(nomen incorrectum),* and *Heteroden* – F. Werner, 1922c *(nomen incorrectum).*

Type species: *Heterodon platirhinos* Latreille *in* Sonnini & Latreille, 1801b.

Distribution: Extreme S Canada, cen. and E USA and N Mexico.

Fossil records: Upper Miocene to upper Pleistocene of E USA. Fossils unassigned to species include middle/upper Pleistocene (Irvingtonian II) of USA (Arkansas), and lower Pleistocene (Irvingtonian I) of USA (South Dakota).

Sources: Edgren, 1952, Bennett, 1979, Meylan, 1982, D.R. Platt, 1983, Cadle, 1984c, Ernst & Barbour, 1989, Conant & Collins, 1991, 1998, Holman, 1995a, 2000a, Pinou et al., 2004 and J.R. Berry, 2012.

1. †*Heterodon brevis* Auffenberg, 1963. Tulane Stud. Zool. 10(3): 174, fig. 22.

Type: Holotype, UF 6153, a middle trunk vertebra (W. Auffenberg, early 1950s–1960).

Type locality: "Haile VI, Locality A. 0.2 miles east of route 235, near the town of Haile, Alachua County, Florida, elevation approximately 84 feet. Middle Pliocene, Hemphillian, Alachua formation."

Distribution: Upper Miocene (Hemphillian: 4.9–10.3 mya) of USA (Florida). Known only from type locality.

Source: Rage, 1984b.

2. *Heterodon kennerlyi* Kennicott, 1860. Proc. Acad. Nat. Sci. Philadelphia 12(8): 336–337.

Type: Lectotype, USNM 1282 (C.B.R. Kennerly & D.N. Couch [A.W. Whipple, U.S.-Mex. Bound. Surv.], 1853–1854), designated by Cope (1900: 774).

Type locality: "Matamoros, Mexico" (in error) via lectotype selection. Restricted to lower Rio Grande River, Texas and Matamoros, Mexico (latter in error) *fide* Coues & Yarrow (1878: 271). Sonora corrected to south of the Gila River in Santa Cruz or Cochise Co., SE Arizona, USA *fide* Bogert (1939a: 198–199), Stejneger (1940: 204) and Stickel (1943: 120).

Distribution: Southern USA (ext. SE Arizona, S New Mexico, SW Texas) and NE Mexico (Aguascalientes, Chihuahua, Coahuila, NE Durango, Nuevo León, NE Sonora, N San Luis Potosí, Tamaulipas, N Zacatecas), NSL–2440 m.

Sources: Ernst & Ernst, 2003 and J.R. Berry, 2012.

Remarks: Kennicott (1860: 337) listed Kennerly as collector and Cochran (1961: 185) listed Couch. A valid species *fide* H.M. Smith et al. (2003: 19). Lectotype designation of USNM 1282 by Cope (1900: 774) possibly invalid. If so, three syntypes are USNM 7290 (A.C.V. Schott, 1852–1858) [Rio Grande], and USNM 1282a–b (C.B.R. Kennerly & D.N. Couch [Whipple, U.S.-Mex. Bound. Surv.], 1853–1854 [Sonora]).

3. *Heterodon nasicus* Baird & Girard, 1852a. Proc. Acad. Nat. Sci. Philadelphia 6(1): 70. (*Heterodon nasicum*)

Synonyms: *Heterodon cognatus* Baird & Girard, 1853, and *Heterodon nasicus gloydi* Edgren, 1952.

Type: Neotype, USNM 1249, a 659 mm female (A.C.V. Schott [U.S.-Mex. Bound. Surv.], 1852–1853), designated by Webb & Eckerman (1998: 104).

Type locality: "Eagle Pass, Texas, USA" via neotype selection. Restricted to the Río Grande approximately 4.3 air miles (7 km) southwest El Indio or approximately 19 air miles (30.6 km) downstream from Eagle Pass, Maverick County, Texas *fide* Webb & Eckerman (1998: 102).

Distribution: North America. Southern Canada (SE Alberta, S Manitoba, S Saskatchewan) and cen. USA (E and NW Colorado, W Illinois, NW and SE Iowa, Kansas, W and SE Minnesota, NW and SE Missouri, E Montana, Nebraska, E New Mexico, North Dakota, Oklahoma, South Dakota, Texas, E Wyoming), NSL–2440 m.

Fossil records: Upper Miocene (Hemphillian) of USA (Arizona), upper Pliocene (Blancan V) of USA (Kansas, Nebraska), lower/middle Pleistocene (Irvingtonian I) of USA (Florida, Kansas), middle Pleistocene (Irvingtonian II) of USA (Nebraska), middle/upper Pleistocene (Irvingtonian II) of USA (Colorado), and upper Pleistocene (Rancholabrean II) of USA (Kansas, New Mexico, Oklahoma, Texas).

Sources: Logier & Toner, 1961, D.R. Platt, 1969, Stebbins, 1985, Degenhardt et al., 1996, Vazquez-Dias & Quintero-Diaz, 1998, Webb & Eckerman, 1998, Hammerson, 1999, Walley & Eckerman, 1999, R.D. Bartlett & Tennant, 2000, Werler & Dixon, 2000, K.J. Irwin, 2001, Parmley & Peck, 2002, H.M. Smith et al., 2003, 2005, T. Schmidt, 2009 and J. Berry, 2012.

Remarks: Holotype (SIM 1272) lost *fide* R.G. Webb & Eckerman (1998: 100). USNM 1285 erroneously listed as the holotype *fide* Yarrow (1883: 141). Type locality restriction to Amarillo, Texas *fide* Edgren (1952a: 203), cited by K.P. Schmidt (1953a: 179), invalid due to neotype selection.

4. *Heterodon platirhinos* Latreille *in* Sonnini & Latreille, 1801b. Hist. Nat. Rept. 4: 32–37, pl. 4, figs. 1–3.

Synonyms: *Boa porcaria* Lacépède, 1789 (*nomen rejiciendum*), *Coluber cacodaemon* G. Shaw, 1802, *Coluber heterodon* Daudin, 1803b, *Scytale niger* Daudin, 1803d *(nomen rejiciendum), Coluber thraso* Harlan, 1835, *Heterodon annulatus* Troost, 1836, *Heterodon niger* Troost, 1836, *Heterodon tigrinus* Troost, 1836, *Heterodon platyrhinos* Schlegel, 1837 *(nomen emendatum), Heterodon atmodes* Baird & Girard, 1853, *Heterodon platyrhinus* – A.-M.-C. Duméril, Bibron & Duméril, 1854a, and *Heterodon browni* Stejneger, 1903.

Type: Holotype, not designated, location unknown.

Type locality: "les environs de Philadelphie" [= vicinity of Philadelphia, Pennsylvania, USA].

Distribution: Extreme S Canada (SE Ontario) and E USA (Alabama, Arkansas, Connecticut, Delaware, Florida, Georgia, Illinois, Indiana, S Iowa, Kansas, Kentucky, Louisiana, Maryland, Massachusetts, E Michigan, SE Minnesota, Mississippi, Missouri, NE and S Nebraska, S New Hampshire, New Jersey, SE and W New York, North Carolina, Ohio, Oklahoma, Pennsylvania, Rhode Island, South Carolina, ext. SE

H

South Dakota, Tennessee, E Texas, Virginia, West Virginia, Wisconsin), NSL–1000 m.

Fossil records: Upper Miocene (Hemphillian) of USA (Nebraska), middle Pliocene (Blancan II, III) of USA (Kansas, Nebraska, Texas), upper Pliocene (Blancan V) of USA (Kansas, Nebraska), lower Pleistocene (Irvingtonian I) of USA (Florida, Pennsylvania), lower/middle Pleistocene (Irvingtonian I) of USA (Kansas), middle Pleistocene (Irvingtonian II) of USA (Nebraska), middle/upper Pleistocene (Irvingtonian II) of USA (Maryland, West Virginia), and upper Pleistocene (Rancholabrean II) of USA (Alabama, Florida, Georgia, Kansas, Missouri, Oklahoma, Pennsylvania, Tennessee, Texas, Virginia, West Virginia).

Sources: D. Davis, 1946, Munyer, 1967, D.R. Platt, 1969, Blem, 1981, Palmer & Braswell, 1995, J. Harding, 1997, Tennant, 1997, Werler & Dixon, 2000 and J.C. Murphy & Schlager, 2003.

Remarks: Official Specific Name no. 686 *fide* Opinion 390 (ICZN, 1956).

H

5. †*Heterodon plionasicus* J.A. Peters, 1953. J. Paleont. 27(3): 328–331, figs. 3, 5–6.

Type: Holotype, UMMP 28113, one left maxilla (UMMZ Paleont. field crews, 1947 & 1951).

Type locality: "Upper Pliocene, Rexroad formation, Rexroad fauna. Locality UM-K1-47, in Fox Canyon, in sec. 35, T. 34 S., R 30 W., XI Ranch, Meade County, Kansas."

Distribution: Lower Pliocene (Hemphillian) of USA (Oklahoma), and middle Pliocene (Blancan II, III) of USA (Kansas, Texas).

Source: Rage, 1984b.

6. *Heterodon simus* (Linnaeus, 1766). Syst. Nat., ed. 12, 1: 375. (*Coluber simus*)

Synonym: *Heterodon catesbyi* A.C.L.G. Günther, 1858.

Type: Holotype, not designated, (A. Garden, 1760–1761), lost *fide* Edgren (1953: 640).

Type locality: "Carolina" [USA]. Restricted to the vicinity of Charleston, South Carolina *fide* Klauber (1948: 6).

Distribution: Southeastern USA (Alabama, Georgia, N Florida, S Mississippi, SE North Carolina, South Carolina), NSL–150 m.

Fossil records: Lower Pleistocene (Irvingtonian I) of USA (Florida), and upper Pleistocene (Rancholabrean II) of USA (Florida).

Sources: Meylan, 1985, Ashton & Ashton, 1988, Palmer & Braswell, 1995, Tennant, 1997 and Tuberville et al., 2000.

Remarks: Official Specific Name no. 685 *fide* Opinion 390 (ICZN, 1956).

HETEROLIODON Boettger, 1913 (Pseudoxyrhophiidae)

Synonym: *Heteroloidon* – Phelps, 1981 *(nomen incorrectum)*.

Type species: *Pseudoxyrhopus occipitalis* Boulenger, 1896a.

Distribution: Madagascar.

Sources: Guibé, 1958, Glaw & Vences, 1994, Raxworthy & Nussbaum, 1994, Cadle, 1999, Nussbaum & Raxworthy, 2000, Glaw et al., 2005b, Nagy et al., 2005, Kelly et al., 2009 and Zaher et al., 2009.

1. *Heteroliodon fohy* Glaw, Vences & Nussbaum, 2005. Herpetologica 61(3): 276–278, figs. 1–2.

Type: Holotype, ZSM 548/2000, a 280 mm male (M. Landsberger, M. Vences, F. Glaw, & K. Schmidt, 13 March 2000).

Type locality: "Montagne des Français (12° 19' 17" S, 49° 20' 13" E, ca. 170 m elevation), Antsiranana Prov., northern Madagascar."

Distribution: Extreme N Madagascar (Antsiranana), NSL–170 m.

Source: Megson et al., 2009.

2. *Heteroliodon lava* Nussbaum & Raxworthy, 2000. Herpetologica 56(4): 494–497, figs. 3–4.

Type: Holotype, UMMZ 221787 (formerly RAN 55066), a 440 mm female (J.-B. Ramanamanjato, A. Razafiminantsoa & A. Razafiminantsoa, 25 April 1996).

Type locality: "Lac Vert, Ankarana Special Reserve, 12°52'S, 49°06'E, 100 m elevation, Ambilobe Fivondronana, Antsiranana Prov., Madagascar."

Distribution: Western Madagascar (Antsiranana, Mahajanga), 100 m.

3. *Heteroliodon occipitalis* (Boulenger, 1896a). Cat. Snakes Brit. Mus. 3: 613–614. (*Pseudoxyrhopus occipitalis*)

Synonym: *Heteroliodon torquatus* Boettger, 1913.

Type: Holotype, BMNH 1946.1.12.28, a 310 mm specimen (J.T. Last, 1890–1895).

Type locality: "S.W. Madagascar."

Distribution: Southwestern Madagascar (Toliara), NSL–280 m.

HEURNIA Jong, 1926a (Homalopsidae)

Type species: *Heurnia ventromaculata* Jong, 1926a.

Distribution: Eastern Indonesia.

Sources: Jong, 1927, Gyi, 1970, O'Shea, 1996, J.C. Murphy, 2007b and Zaher et al., 2009.

1. *Heurnia ventromaculata* Jong, 1926a. Zool. Anz. 67(11–12): 303–304.

Type: Holotype, ZMA 11066, a 702–730 mm male (W.C. van Heurn, 12 July 1920).
Type locality: "Pionierbivak am Mamberamofluss, Nord-Neu-Guinea" [= Pionierbivak, Mamberano River, N Papua Prov., E Indonesia, 2°04'S, 137°50'E, elevation 35 m].
Distribution: Eastern Indonesia (N Papua). Known only from type locality.
Remarks: Type erroneously listed as 30 mm in length *fide* Jong (1927: 305). Photographs of the type in J.C. Murphy (2007b: pl. 34, figs. a–f).

HIEROPHIS Fitzinger *in* Bonaparte, 1834 (Colubridae)

Synonyms: †*Tamnophis* Rochebrune, 1880, and †*Sansanosaurus* Kuhn, 1939a *(nomen substitutum).*
Type species: *Coluber viridiflavus* Lacépède, 1789.
Distribution: Southern Eurasia.
Fossil records: Lower Oligocene to upper Pleistocene of Europe. Fossils unassigned to species include lower Miocene (Orleanian, MN 4, 5: 13.7–16.9 mya) of Portugal and Spain, middle Miocene (Astaracian, MN 6–8: 11.1–13.7 mya) of Hungary, and middle Pliocene (Ruscinian, MN 15: 3.2–4.2 mya) of Spain.
Sources: Schätti & Wilson, 1986, Schätti, 1988a, Gruber, 1989, Szyndlar, 1991a, Schätti & Utiger, 2001, Nagy et al., 2004 and Szyndlar, 2009.
Remarks: Genus resurrected for species formerly in *Coluber fide* Schätti, 1988a.

1. *Hierophis andreanus* (F. Werner, 1917c). Verh. Zool. Bot. Ges. Wien 67(5–6): 207–208. (*Zamenis andreana*)

Synonym: *Zamenis hotsoni* Boulenger, 1920.
Types: Syntypes (2), ZMG 160, a 660 mm male, and ZMG 164, a 750 mm female (F.C. Andreas, 1877–1878 & 1905).
Type locality: "Persien (Provinz Fars)" [= Fars Province, SW Iran].
Distribution: Iran (Bushehr, Fars, Kerman, Kermanshah, Lorestan), 1000 m.
Sources: Schätti, 2001b and Rajabizadeh & Rastegar-Pouyani, 2009.

2. †*Hierophis arcuatus* (H. Meyer, 1845). Fauna der Vorwelt: 42, pl. 2, fig. 5. (†*Coluber arcuatus*)

Type: Holotype, SMNK TE/57, a section of axial skeleton.
Type locality: "Molasse-Mergel Steinbrüche von Öningen" [= Oeningen, Baden-Württemberg State, SW Germany; Astaracian, middle Miocene].
Distribution: Middle Miocene (Astaracian, MN 7–8: 11.1–112.8 mya) of Germany. Known only from holotype.
Sources: Rage, 1984b and Szyndlar, 2012.
Remarks: Generic allocation uncertain.

3. †*Hierophis cadurci* (Rage, 1974). Palaeovertebrata 6(3–4): 295–297, fig. 7. (†*Coluber cadurci*)

Type: Holotype, USTL MGT 3505, one trunk vertebra.
Type locality: "Mas-de-Got, Iocus A, Oligocène moyen, Phosphorites du Quercy" [= Mas-de-Got Dept., France; Upper Suevian, Middle Oligocene *fide* Rage, 1984b: 44].
Distribution: Lower Oligocene (Rupelian, MP 21–23: 28.4–33.9 mya) of France, upper Oligocene (Chattian, MP 25: 23.0–28.4 mya) of France, and lower Miocene (Agenian–Orleanian, MN 1–4: 16.0–23.0 mya) of France, Germany and Spain.
Sources: Rage, 1984b, 1988, Rage et al., 1992, Augé & Rage, 1995, Szyndlar & Alférez, 2005 and Szyndlar, 2012.
Remarks: Generic allocation uncertain.

H

4. †*Hierophis caspioides* (Szyndlar & Schleich, 1993). Stuttgarter Beitr. Naturk. (192): 23–26, figs. 6a–j. (†*Coluber caspioides*)

Type: Holotype, BSPG 1976 XXII 7096, one trunk vertebra.
Type species: .
Type locality: "Peterrsbuch 2 near Eichstätt, Bayern, southern Germany. Lower Miocene (middle Orleanian; MN 4)."
Distribution: Lower Miocene (Orleanian, MN 3–5: 13.7–20 mya) of Austria, Czech Republic, France and Germany.
Sources: Ivanov, 2002, Ivanov & Böhme, 2011 and Szyndlar, 2012.

5. †*Hierophis dolnicensis* (Szyndlar, 1987). J. Vert. Paleont. 7(1): 65, figs. 8b–f.

Type: Holotype, CUP 4558, one trunk vertebra.
Type locality: "MN4 locality of Dolnice (West Czechoslovakia) near to the town of Cheb in western-most Bohemia, lower Miocene (Early Orleanian or Ottnangian)."
Distribution: Lower Miocene (Orleanian, MN 3–4: 16.0–20.0 mya) of Czech Republic, and middle Miocene (Astaracian, MN 7–8: 11.1–12.8 mya) of France.
Source: Ivanov, 2002 and Szyndlar, 2012.

6. †*Hierophis freybergi* (G. Brunner, 1954). Neues Jb. Geol. Paläont., Abh. 100(1): 108, figs. 18.1–2. (†*Coluber freybergi*)

Types: Syntypes (5), BSPG 1982 X, one maxilla, one palatine, one dentary and two fragmentary compound bones.
Type locality: "Fuchsloch, bei Siegmannsbrunn (Oberfr.). (Eine mediterrane Riss-Würm-Fauna)." [= Fuchsloch, Bayern, Germany; middle Pleistocene, Riss-Würm].
Distribution: Pleistocene (Riss-Würm Interglacial, 0.116–0.128 mya) of Germany. Known only from type series.
Sources: Rage, 1984b and Holman, 1998a.
Remarks: Possibly conspecific with *H. viridiflavus* (Lacépède) *fide* Rage (1984b: 44) and Holman (1998a: 104).

H

7. *Hierophis gemonensis* (Laurenti, 1768). Synop. Rept.: 76. (*Natrix gemonensis*)

Synonyms: *Coluber pethola* Georgi, 1801 *(nomen praeoccupatum), Zamenis atrovirens dalmatina* F. Müller, 1878b, *Zamenis gemonensis laurenti* Bedriaga, 1882, and *Zamenis atrovirens jaculator* Sordelli, 1886.
Type: Neotype, MHNG 1357.70, a 916 mm male (W. Zinniker, May 1962), designated by Opinion 1686 (ICZN, 1992).
Type locality: "Cegar (Zadar, Yugoslavia)" [= Zadar, Zadar Prov., Croatia, 44°07'N, 15°15'E, elevation 25 m] via neotype selection.
Distribution: Southeastern Europe. Northeastern Italy (Venezia), W Slovenia, W Croatia (Ada, Brac, Cres, Hvar, Korcula, Krk, Lastovo, Losinj, Mljet, Rab, Silba, Sipan, Solta, Veli Palagruza and Vis Is.), SW Bosnia-Herzegovina, Montenegro, Albania and Greece (Macedonia: Agria Gramvousa, Angístri, Antikithira, Aspronisi, Corfù, Dia, Dragoáda, Effira, Egina, Eúboia, Gavdopoúla, Gávdos, Giura, Gramvoúsa, Gyáros, Ipdili, Itháke, Janusada, Kefallinía, Kérkyra, Kíthira, Kriti, Lefkáda, Lerkas, Paxos, Platía, Póros, Skíathos, Skópelos, Spétses, Spinalónga, Theocloros, Tolon, Zákinthos and Zogriá Is.), NSL–2450 m.
Fossil records: Lower Pliocene (Ruscinian, MN 14: 4.2–4.9 mya) of Bosnia & Herzegovina, middle Pliocene (Ruscinian, MN 15: 3.2–4.2 mya) of Moldavia, upper Pliocene (Villanyian, MN 16: 2.6–3.2 MYA) of Austria, lower Pleistocene (Calabrian: 0.8–1.8 mya) of Austria and Czech Republic, middle Pleistocene (Ionian: 0.13–0.78 mya) of Austria, Bulgaria and Italy, and upper Pleistocene (Tarantian: 0.01–0.13 mya) of Bulgaria, Greece and Italy.
Sources: Mertens, 1968a, Bruno, 1977, 1980, 1984, Kock, 1979, Schätti, 1983a, Bruno & Maugeri, 1990, Henle, 1993, Holman, 1998a, Szyndlar, 1991a, 2012, Ivanov, 2007, Trapp, 2007, Valakos et al., 2008 and Jablonski et al., 2012.
Remarks: Holotype lost *fide* Schätti et al. (1991: 50). Official Specific Name *fide* Opinion 1686 (ICZN,1992). *Hierophis laurenti* a valid species *fide* Ferri (1994: 73).

8. †*Hierophis hungaricus* (Bolkay, 1913). Mitt. Jahrb. Kgl. Ungar. 2: 223–224, pl. 12, fig. 3. (†*Zamenis hungaricus*)

Type: Holotype, HGIM Ob-4464/vt. 76, one right quadrate.
Type locality: "Polgárdi" [= quarries of Somló-Hill and Kószár-Hill near the village of Polgárdi, W Hungary; late Miocene (Pontian or Upper Turolian *fide* Venczel, 1998: 1].
Distribution: Lower Miocene (Orleanian, MN 5: 13.7–16.0 mya) of Germany, middle Miocene (Astaracian, MN 7–8: 11.1–12.8 mya) of Romania, upper Miocene (Villesian, MN 9: 9.7–11.1 mya) of Hungary, upper Miocene (Turolian, MN 13: 4.9–7.3 mya) of Hungary, and lower Pliocene (Ruscinian, MN 14: 4.2–4.9 mya) of Hungary.
Sources: Szyndlar, 1991a, 2012, Vogel, 1994b, Venczel, 1998, Venczel & Stiuca, 2008 and Szyndlar, 1991a, 2000, 2005, 2012, Ivanov, 2007 and Ivanov & Böhme, 2011.
Remarks: Generic allocation uncertain but similar to *H. gemonensis* and *H. caspioides fide* Venczel (1998: 6).

9. †*Hierophis pouchetii* (Rochebrune, 1880). Nouv. Arch. Mus. Hist. Nat. (2) 3: 281, pl. 12, figs. 9a–d. (†*Tamnophis pouchetii*)

Types: Syntypes (7), MNHN SA 9872, six trunk vertebrae and one rib.
Type locality: "Colline de Sansan" [= Sansan, Gers Prov., France; Astaracian, middle Miocene *fide* Rage, 1984b: 44].
Distribution: Lower Miocene (Orleanian, MN 4, 5: 13.7–16.9 mya) of France and Germany, middle Miocene (Astaracian, MN 6: 12.8–13.7 mya) of France, and upper Miocene (Villesian, MN 9: 9.7–11.1 mya) of Hungary.
Sources: Rage, 1981, Augé & Rage, 2000 and Szyndlar, 2012.
Remarks: Generic allocation uncertain.

10. †*Hierophis robertmertensi* (Mlynarski, 1964). Senck. Biol. 45(3–5): 331–332, figs. 9–12. (†*Coluber robertmertensi*)

Types: Syntypes (2), ZZSD T.R.F. 1/63, one quadrate and one compound bone (Steinbruches, 1962).
Type locality: "Rebielice Królewskie bei Klobuck im NW-Teil dei von Krakau-Wiekau, Polen; Jungpliozán" [= Rebielice Królewskie, N Slaskie Prov., Poland]
Distribution: Lower Pliocene (Ruscinian, MP 14–15, 3.2–4.9 mya) of Moldova and Poland.
Remarks: Generic allocation uncertain but similar to *H. viridiflavus fide* Mlynarski (1964: 332).

11. †*Hierophis suevicus* (Fraas, 1870) Jahresh. Ver. Vaterland. Natuuk. Wurttemb. (1869) 26: 291. (†*Naja suevica*)

Type: Lectotype, SMNS 51026, one middle trunk vertebra (Bauer), designated by Rage (1984b: 45).
Type locality: "Steinheim, Baden-Württemberg, Western Germany; Astaracian, Middle Miocene" via lectotype selection.
Distribution: Lower Miocene (Orleanian, MN 3: 16.9–20.0 mya) of Czech Republic, and middle Miocene (Astaracian, MN 7–8: 11.1–12.8 mya) of Germany.
Sources: Ivanov, 2002 and Szyndlar, 2012.
Remarks: Generic allocation uncertain.

12. *Hierophis viridiflavus* (Lacépède, 1789). Hist. Nat. Serp. 2: 137–146, pl. 6, fig. 1. (*Coluber viridiflavus*)

Synonyms: *Coluber luteostriatus* Gmelin, 1789, *Coluber vulgaris* Bonnaterre, 1790, *Coluber communis* Donndorff, 1798, *Coluber franciae* Suckow, 1798, *Coluber sardus* Suckow, 1798, *Coluber petolarius* Georgi, 1800, *Coluber atroviriens* G. Shaw, 1802, *Coluber personatus* Daudin, 1803c, *Coluber plutonius* Daudin, 1803c, ? *Coluber xanthurus* Rafinesque-Schmaltz, 1810, *Coluber melanepis* Rafinesque-Schmaltz, 1814 *(nomen rejiciendum)*, *Coluber pustulatus* Rafinesque-Schmaltz, 1814, *Coluber uccellator* Rafinesque-Schmaltz, 1814, *Coluber glaucoides* Millet de la Turtandière, 1828, ? *Coluber profulax* Costa, 1828, *Coluber viridiflavus carbonarius* Bonaparte, 1833 *in* 1832–1841, †*Coluber fossilis* Pomel, 1853, †*Coluber gervaisii* Pomel, 1853, *Zamenis caspicus* Jan, 1867 *in* Jan & Sordelli, 1866–1870, *Zamenis aristotelis* Gistel, 1868, *Zamenis viridiflavus ocellata* Betta. 1874, *Zamenis atrovirens caudaelineata* F. Müller, 1878b (*nomen nudum*), †*Coluber etruriae* Portis, 1890, *Coluber viridiflavus kratzeri* Kramer, 1971, and *Coluber viridiflavus antoniimanujeli* Capolongo, 1984.
Type: Holotype, not designated, location unknown.
Type locality: "plusieurs Provincés de France, & sur-tout dans les méridionales" [= numerous provinces throughout the south of France].
Distribution: South-central Europe. Extreme NE Spain (Andorra, Gerona, Huesca, Lérida, Navarra), France (Ain, Allier, Alpes-Maritimes, Ardèche, Ardennes, Ariège, Aube, Aude, Aveyron, Bas-Rhin, Basses-Alpes, Basses-Pyrénées, Cantal, Cher, Charente, Charente-Maritime, Corrèze, Côte-d'Or, Creuse, Deux-Sèvres, Dordogne, Doubs, Drôme, Essonne, Eure-et-Loir, Gard, Gers, Gironde, Haut-Rhin, Haute-Garonne, Haute-Loire, Haute-Marne, Haute-Saône, Haute-Savoie, Haute-Vienne, Hautes-Alpes, Hautes-Pyrénées, Hérault, Indre, Indre-et-Loire, Isère, Jura, Landes, Loire, Loire-Atlantique, Lot, Lot-et-Garonne, Lozère, Maine-et-Loire, Marne, Meurthe-Moselle, Meuse, Morbihan, Moselle, Nièvre, Puy-de-Dôme, Pyrénées-Orientales, Rhône, Saöne-et-Loire, Savoie, Tarn, Tarn-et-Garonne, Val-d'Oise, Var, Vaucluse, Vendée, Vienne, Vosges, Yonne, Yvelines, Corsica Is.), Corsica, S Switzerland (Genève, Graubunden, Ticino, Valais, Vaud), W Slovenia, NW Croatia, Greece (Gyaros Is.), Italy (Sardinia, Santo Antioco, Asinava, Budelli, Caprera, Corcelli, Maddalena, Molara, Razzoli, Santa Maria, San Pietro, Spargi, Tavolara, Vacca, Capraia, Elba, Giannurti, Gíglio, Gorgona, Montecristo, Palmarola, Pianosa, Ponza, Santo Stefano, Ventoténe, Capri, Ischia, Cirella, Krk, San Domino, San Nicola, Pelagosa, Veli Brijun, Sant'Andrea, Sicilia, Alicudi, Basiluzzo, Filicudi, Lipari, Panarea, Salina, Stromboli, Vulcano, Ustica, Favignana, Leranzo, Marettimo, Isola Grande, Cavallo, Farina, Forana, Giroglia, Gargalu, Lavezzi, Luigi, Giafferi, Mezzomare, Piana and Gozo Is.) and Malta, NSL–2100 m.
Fossil records: Lower Pliocene (Ruscinian, MN 14: 4.2–4.9 mya) of Bosnia & Herzegovina, middle Pliocene (Ruscinian, MN 15: 3.2–4.2 mya) of Moldavia, Poland, and Slovakia, upper Pliocene (Villanyian, MN 16: 2.6–3.2 mya) of Austria, Hungary, Moldavia, Poland, Romania and Ukraine, lower Pleistocene (Calabrian: 0.8–1.8 mya) of Austria and Czech Republic, middle Pleistocene (Ionian: 0.13–0.78 mya) of Austria, Bulgaria, France, Germany, Italy and Spain, and upper Pleistocene (Tarantian: 0.01–0.13 mya) of France and Spain.
Sources: Bruno, 1975, 1977, 1984, Arnold & Burton, 1978, Rage, 1984b, Schätti & Vanni, 1986, Szyndlar, 1986, 1991a, 2012, Gruber, 1989, Heimes, 1993, Schätti, 1993a, Holman, 1998a, Hofer et al., 2001, Nagy et al., 2002, Deltino & Szyndlar, 2003, Ivanov, 2007, Renner & Vitzthum, 2007, Rato et al., 2008, Valakos et al., 2008 and Vacher & Geniez, 2010.
Remarks: Official Specific Name *fide* Opinion 1686 (ICZN, 1992).

HIMALAYOPHIS Malhotra & Thorpe, 2004c (Viperidae)

Type species: *Trimeresurus tibetanus* Z.-Y. Huang, 1982.
Distribution: China and Nepal.
Sources: David & Tong, 1997 and Tiwari & Shah, 1999, Schleich & Kästle, 2002, Tiwari & Shah, 2004 and S.K. Sharma et al., 2013.

1. *Himalayophis tibetanus* (Z.-Y. Huang, 1982). Acta Fudan Univ. (Nat. Sci.) 21(1): 116–118, figs. 1–5. (*Trimeresurus tibetanus*)

Synonym: *Trimeresurus karanshahi* Orlov & Helfenberger, 1997.
Type: Holotype, Fudan Univ. 80001, a 640 mm female (J.R. Sog Lang, 5 May 1979).

Type locality: "Naylam, Chokesumo, altitude 3200m, Tibetan" [= Nayalam, S Xizang Prov., SW China, 28°09'N, 85°59'E, elevation 3200 m].
Distribution: Southwestern China (Xizang) and N Nepal (Lalitpur, Sindhupalchok), 1400–3200 m.

†*HISPANOPHIS* Szyndlar, 1985a (Colubridae)

Type species: †*Hispanophis coronelloideus* Szyndlar, 1985a.
Distribution: Upper Miocene of Spain.
Source: Szyndlar, 1985a, 2012.

H

1. †*Hispanophis coronelloideus* Szyndlar, 1985a. Estud. Geol. 41(5–6): 453–455, fig. 5.

Type: Holotype, MNCN Alg. I-27, one basiparasphenoid.
Type locality: "Uppermost Miocene (Turolian, NM 13) of Algora (Guadalajara, Spain)."
Distribution: Upper Miocene (NM 13, 4.9–7.3 mya) of Spain. Known only from type locality.

†*HOFFSTETTERELLA* Rage, 1998 (Aniliidae)

Type species: †*Hoffstetterella brasiliensis* Rage, 1998.
Distribution: Middle Paleocene of Brazil.
Sources: Rage, 1998, 2011, Gómez et al., 2008 and Albino, 2011b.

1. †*Hoffstetterella brasiliensis* Rage, 1998. Palaeovertebrata 27(3–4): 132–137, fig. 16.

Type: Holotype, DGM 1323-R, one middle trunk vertebra (L.I. Price & D. de Campos, 1968).
Type locality: "Itaboraí, state of Rio de Janeiro, Brazil; middle Palaeocene."
Distribution: Middle Paleocene (Itaboraian: 57.0–59.0 mya) of Brazil. Known only from type locality.

HOLOGERRHUM A.C.L.G. Günther, 1858 (Natricidae)

Synonym: *Holorgermi* – Briceño-Rossi, 1934 *(nomen incorrectum).*
Type species: *Hologerrhum philippinum* A.C.L.G. Günther, 1858.
Distribution: Philippines.
Sources: Leviton, 1959b, Ross & Gonzales, 1992, R.M. Brown et al., 2001 and Zaher et al., 2009.

1. *Hologerrhum dermali* R.M. Brown, Leviton, Ferner & Sison, 2001. Asiatic Herpetol. Res. 9: 17–19.

Type: Holotype, PNM 2711, a 288 mm female (R.M. Brown, 9 April 1992).
Type locality: "at 1510 m above sea level in the area known locally as "Hanggud Tubig" ("Big Water"), on the western face of Mt. Madja-as, Barangay Alojipan, Municipality of Culasi, Antique Prov., Panay Island, Philippines (11°23'N, 122°09'E)."
Distribution: Central Philippines (Panay Is.), 450–1510 m.
Source: Ferner et al., 2001.

2. *Hologerrhum philippinum* A.C.L.G. Günther, 1858. Cat. Colub. Snakes Brit. Mus.: 186–187.

Type: Holotype, BMNH 1946.1.2.41, a 292 mm female (H. Cuming, 1836–1840).
Type locality: "Philippine Islands." Restricted to Luzon Island *fide* Leviton (1959b: 311).
Distribution: Northern Philippines (Catanduanes, Luzon, Marinduque, Polillo).

HOMALOPSIS Kuhl & Hasselt, 1822 (Homalopsidae)

Synonyms: *Homolopsis* – Gray *in* Griffith & Pidgeon, 1831 *(nomen incorrectum), Homolopsis* Bonaparte, 1832 *(nomen emendatum), Homatopsis* – J. Müller, 1832 *(nomen incorrectum), Homalpepsis* – Cantor, 1847 *(nomen incorrectum), Homalopis* – Mason, 1852 *(nomen incorrectum), Pythonia* Blyth, 1859, *Pythonella* Carlleyle, 1869 *(lapsus calami),* and *Honalopsis* – Deuve, 1970 *(nomen incorrectum).*
Type species: *Coluber buccatus* Linnaeus, 1758.
Distribution: Tidal rivers, canals, lakes and ponds and coastal marine waters of SE Asia and Indonesia.
Sources: Boulenger, 1890a, Rooij, 1917, F. Werner, 1923b, M.A. Smith, 1943, B.L. Lim,1964, E.H. Taylor, 1965, Deuve, 1970, Gyi, 1970, Saint-Girons, 1972a, Stuebing, 1991, David & Vogel, 1996, M.J. Cox et al., 1998, Chan-ard et al., 1999, Schleich & Kästle, 2002, Voris et al., 2002, B.L. Stuart et al., 2006, J.C. Murphy, 2007b, Alfaro et al., 2008, Zaher et al., 2009 and J.C. Murphy et al., 2012.

1. *Homalopsis buccata* (Linnaeus, 1758). Syst. Nat., ed. 10, 1: 217. (*Coluber buccatus*)

Synonyms: *Coluber monilis* Linnaeus, 1758, *Coluber subalbidus* Gmelin, 1789, *Coluber margaritaceus* Donndorff, 1798, *Coluber horridus* Daudin, 1803d, and *Homalopsis albomaculata* A.-M.-C. Duméril, 1853 (*nomen nudum*).
Type: Neotype, FMNH 250114, a 496 mm female (K. Lim, 1 May 1992), designated by J.C. Murphy et al. (2012: 10) .

Type locality: "Singapore" via neotype selection.

Distribution: West Malaysia and Indonesia. West Malaysia (Johore, Melaka, Perak, Selangor), East Malaysia (Sabah, Sarawak) and Indonesia (Bangka, Java, Kalimantan, Sumatra), NSL–410 m.

Sources: V.S. Nguyen et al., 2009, I. Das, 2010, 2012 and J.C. Murphy et al., 2012.

Remarks: Holotype lost *fide* Andersson (1899: 16). Sulawesi record doubtful *fide* Lang & Vogel (2005: 253).

2. *Homalopsis hardwickii* Gray, 1842d. Zool. Misc. 2(May): 65.

Type: Holotype, BMNH 1946.1.7.26, a 683 mm male.

Type locality: Unknown. Designated as NE India *fide* J.C. Murphy et al. (2012: 12).

Distribution: Probably NE India and cen. Nepal (Chitwan), 200 m.

Sources: Mitchell & Zug, 1986, Tiwari & Shah, 2004 and J.C. Murphy et al., 2012.

3. *Homalopsis mereljcoxi* J.C. Murphy, Voris, Murthy, Traub & Cumberbatch, 2012. Zootaxa (3208): 13–14, figs. 1c, 7b, 8.

Type: Holotype, FMNH 263756, a 457 mm female (D.R. Karns & J.C. Murphy, 16 June 2004).

Type locality: "Thailand, Nakhon Ratchasima Prov., Wang Nam Khieo, Udom Sap (subdistrict), Ban Badan Reservoir (14°31'04" N, 101°58'25" E)."

Distribution: Thailand (Bangkok, Krung Thep, Mahanakhon, Nakhon Ratchasima, Nakhon Si Thammarat, Pattani, Phang Nga, Phattalung, Phuket, Ranong, Songkhla, Trang, Uttaradit), Cambodia (Chhnang, Siem Reap) and Vietnam (Can Tho, Vinh Thuan), NSL–235 m.

4. *Homalopsis nigroventralis* Deuve, 1970. Mém. Off. Rech. Sci. Tech. Outre-Mer, Paris (39): 183–185, pl. 21, figs. 1–4, pl. 22, figs. 1–3, 7. (*Homalopsis buccata nigroventralis*)

Type: Holotype, not designated, lost *fide* David (2009: 23), destroyed by fire *fide* J.C. Murphy et al. (2012: 16).

Type locality: "Nam Ngum River Valley, Laos."

Distribution: Mekong River valley of E Thailand (Sakon Nakhon),p Cambodia (Kratie, Ratanakiri, Siem Pang, Stung Treng, Koh Kapeung and Koh Khlap Is.) and Laos (Champasak, Louang Phrabang, Vientiane), 150–285 m.

Sources: E.H. Taylor & Elbel, 1958, B.L. Stuart et al., 2006, J.C. Murphy, 2007b, I. Das, 2010, 2012 and J.C. Murphy et al., 2012.

Remarks: A resurrected species *fide* B.L. Stuart et al. (2006: 149–150).

5. *Homalopsis semizonata* Blyth, 1855a. Proc. Asiatic Soc. Bengal 24(2): 187–188. (*Homolopsis semizonata*)

Type: Holotype, ZSI 8137, a 728 mm male (T.M. Berdmore, 1855).

Type locality: "Martaban (= Mottama, southern Myanmar)" *fide* I. Das et al. (1998: 149) [= Mottama, Mon Div., S Myanmar, 16°32'N, 97°36'E, elevation NSL].

Distribution: Myanmar (Mon, Yangon), NSL.

Sources: Firth, 1977 and J.C. Murphy et al., 2012.

Remarks: Type locality seems to be "Scheu Gyen, Pegu" *fide* Blyth (1855: 187). Possibly occurs in Phuket, Thailand *fide* Firth, 1977.

HOMOROSELAPS Jan, 1858b (Atractaspididae)

Synonyms: *Poecilophis* A.C.L.G. Günther, 1859b, *Homeroselaps* – Cope, 1862c *(nomen incorrectum)*, and *Homorelaps* Boulenger, 1896a *(nomen emendatum)*.

Type species: *Coluber lacteus* Linnaeus, 1758.

Distribution: Southern Africa.

Sources: McDowell, 1968, Kochva & Wollberg, 1970, V.F.M. FitzSimons, 1974, Branch, 1981, 1982, Broadley, 1983, Cadle, 1983, McCarthy, 1985, Lombard et al., 1986, McDowell, 1986, Branch, 1988, Wallach, 1991, Golay et al., 1993, Vidal & Hedges, 2002, Kelly et al., 2003, 2009, Slowinski & Keogh, 2004, Lawson et al., 2005, Nagy et al., 2005, Slowinski & Lawson, 2005, Dobiey & Vogel, 2007, Zaher et al., 2009 and Hoser, 2012e.

Remarks: Official Generic Name no. 2145 *fide* Opinion 1201 (ICZN, 1982a).

1. *Homoroselaps dorsalis* (A. Smith, 1849 *in* 1838–1849). Illust. Zool. So. Africa, Rept. (App.): 21. (*Elaps dorsalis*)

Synonym: *Elaps dorsalis* A. Smith, 1849 *in* 1838–1849 *(nomen praeoccupatum)*.

Type: Holotype, not designated, lost *fide* Broadley *in* Golay et al. (1993: 143).

Type locality: "Kaffirland, and the country towards Port Natal, southern Africa" [= KwaZulu-Natal, South Africa *fide* Loveridge, 1944b: 161]. Restricted to the vicinity of Durban, Natal Province, South Africa *fide* David & Ineich (1999: 100).

Distribution: South Africa (Free State, Gauteng, KwaZulu-Natal, Limpopo, Mpumalanga), and Swaziland, 525–1535 m.

Remarks: Holotype not present in RSM or BMNH *fide* V.F.M. FitzSimons (1937: 264). See Broadley (1983: 215) for discussion of nomenclatural history.

2. *Homoroselaps lacteus* (Linnaeus, 1758). Syst. Nat., ed. 10, 1: 220. (*Coluber lacteus*)

Synonyms: *Coluber domicella* Linnaeus, 1758, *Coluber domicellarum* Lacépède, 1789 (*nomen rejiciendum*), *Coluber guineensis* Bonnaterre, 1790, *Coluber picatus* Shaw & Nodder, 1798 *in* 1789–1813, *Coluber siamensis* Donndorff, 1798, *Vipera lactea* – Latreille in Sonnini & Latreille, 1801, Coluber *hygeiae* G. Shaw, 1802, *Coluber iphisa* Daudin, 1803c, *Coluber melanogaster* Hermann, 1804, ? *Elaps fuscus* Merrem, 1820, *Elaps punctatus* A. Smith, 1826, *Elaps domicella* Wagler, 1830, *Elaps bipunctiger* A.-M.-C. Duméril, Bibron & Duméril, 1854b, *Poecilophis hygiae* A.C.L.G. Günther, 1859b *(nomen emendatum)*, *Elaps hygiae chrysopeleoides* F. Müller, 1887, and *Micropechis crucifer* De Vis, 1905.

Type: Holotype, NHR Lin-92 (formerly MAFR), a 480 mm male (Mus. Drottn.), lost *fide* Broadley *in* Golay et al. (1993: 144).

Type locality: "Indiis" [= India] (in error). Corrected to South Africa *fide* V.F.M. FitzSimons (1962a: 283).

Distribution: South Africa (Eastern Cape, Free State, Gauteng, KwaZulu-Natal, Limpopo, Mpumalanga, W Northern Cape, Western Cape), Lesotho and Swaziland, NSL–1200 m.

Remarks: Official Specific Name no. 2781 *fide* Opinion 1201 (ICZN, 1982a).

H

HOPLOCEPHALUS Wagler, 1830 (Elapidae)

Synonyms: *Elapocormus* Fitzinger, 1843, *Hoplicephalus* – Lydekker, 1901 *(nomen incorrectum)*, *Hipocephalus* – Brazil, 1914 *(nomen incorrectum)*, *Hoplosephalus* – Brazil, 1914 *(nomen incorrectum)*, *Hoplocephalis* – Boquet, 1948 *(nomen incorrectum)*, and *Hyplocephalus* – J.H. Brown, 1973 *(nomen incorrectum)*.

Type species: *Naja bungaroides* Schlegel, 1837.

Distribution: Eastern Australia.

Sources: F. Werner, 1923b, Kinghorn, 1964, Cogger et al., 1983a, Shine, 1983a, Longmore, 1986, S.K. Wilson & Knowles, 1988, Gow, 1989, Hoser, 1989, 2012e, Hutchinson, 1990, G. Swan, 1990, Covacevich & Couper, 1991, Greer, 1997, Cogger, 2000, Keogh et al., 2003, Sanders et al., 2008 and Zaher et al., 2009.

1. *Hoplocephalus bitorquatus* (Jan, 1859a). Rev. Mag. Zool. (2) 11: 124, 128. (*Alecto bitorquata*)

Synonyms: *Hoplocephalus sulcans* De Vis, 1884c, *Hoplocephalus waitii* Ogilby, 1894, *Denisonia angulata* De Vis, 1905, and *Denisonia revelata* De Vis, 1911.

Type: Holotype, SMF 20533 (formerly SMF-B 9349a), a 330 mm specimen (W. Kirchner, 1848).

Type locality: "Nouvelle-Hollande" [= Australia].

Distribution: Eastern Australia (NE New South Wales, E Queensland), NSL–1040 m.

Remarks: Supplemental original description in Jan (1859e: pl. c, fig. 5). ZMB catalogue entry reads Sydney, Neusüdwales (in error *fide* Cogger et al., 1983a: 227, probably because Kirchner was the German consul in Sydney).

2. *Hoplocephalus bungaroides* (Schlegel, 1837). Essai Phys. Serp. 1: 184, 2: 477–478. (*Naja bungaroides*)

Synonyms: *Naja bungaroides* H. Boie *in* Schlegel, 1828 *(nomen nudum)*, *Naja bungaroides* Schlegel, 1837, *Alecto variegata* A.-M.-C. Duméril, 1853 (*nomen nudum*), *Alecto variegata* A.-M.-C. Duméril, Bibron & Duméril, 1854b, and *Hoplocephalus variegatus* – A.C.L.G. Günther, 1858.

Types: Syntypes (2), MNHN 7678, a 340 mm male lacking a skull (J.R.C. Quoy & J.P. Gaimard [Coquille or Astrolabe Voy.], 1822–1829) and MNHN 7679, a 620–640 mm male (P.F. Keraudren, 1800–1804).

Type locality: "environs de Port-Jackson" [New South Wales, Australia] and "Indes," (latter in error), respectively. Corrected to Nouvelle-Hollande [= Australia] by A.-M.-C. Duméril, Bibron & Duméril (1854b: 1254).

Distribution: Southeastern Australia (E New South Wales), NSL–740 m.

Sources: Guibé & Roux-Estévè, 1972 and Hosmer, 1952.

3. *Hoplocephalus stephensii* Krefft, 1869b. Snakes Aust.: 58–59, pl. 6, fig. 7.

Synonyms: *Hoplocaphalus stephensi* – McDowell, 1967 *(nomen incorrectum)*, and *Haplocephalus stephensii* – Khole, 1991 *(nomen incorrectum)*.

Type: Holotype, AMS 37361 (formerly AMS 6588, exch. JH [1897] & USNM 55999 *fide* Cogger, 1979: 206), a 762 mm specimen.

Type locality: "Port Macquarie, on the Hastings River" [= Port Macquarie, E New South Wales, SE Australia, 31°26'S, 152°54'E, elevation 10 m].

Distribution: Eastern Australia (ext. SE Queensland, NE New South Wales), NSL–1025 m.

Remarks: ZMB 8547 labelled a type but probably in error *fide* Cogger et al. (1983: 228).

†*HORDLEOPHIS* Holman, 1996b (Boidae)

Type species: †*Hordleophis balconae* Holman, 1996b.

Distribution: Upper Eocene of United Kingdom.

Source: Holman, 1996b.

1. †*Hordleophis balconae* Holman, 1996b. Tertiary Res. 17(1–2): 11–13, figs. 1a–e.

Type: Holotype, MSUVP 1361, one trunk vertebra.

Type locality: "Hordle Cliff, Hampshire, England. Rodent Bed upper part of the Totland Bay Member of the Headon Hill Formation (Upper Eocene)."
Distribution: Upper Eocene (33.9–37.2 mya) of United Kingdom (England).

HORMONOTUS Hallowell, 1857b (Lamprophiidae)

Synonyms: *Homonotus* – Gray, 1858b *(nomen incorrectum),* and *Hormonothus* – Stucki-Stirn, 1979 *(nomen incorrectum).*
Type species: *Lamprophis modestus* A.-M.-C. Duméril, Bibron & Duméril, 1854a.
Distribution: West and cen. Africa.
Sources: K.P. Schmidt, 1923, Angel, 1933b, Bogert, 1940, Witte, 1962, Doucet, 1963, Mertens, 1965c, Roux-Estève, 1965, Leston & Hughes, 1968, Hughes & Barry, 1969, Pitman, 1974, Villiers, 1975, Hughes, 1983, 2013, D. Lawson, 1993, J.-F. Trape & Roux-Estève, 1995, Spawls et al., 2002, Villiers & Condamin, 2005, Chippaux, 2006, Chirio & Ineich, 2006, Chirio & LeBreton, 2007, Pauwels & Vande weghe, 2008, Kelly et al., 2009, Zaher et al., 2009, Segniagbeto et al., 2011 and Chirio, 2013.

1. *Hormonotus modestus* (A.-M.-C. Duméril, Bibron & Duméril, 1854a). Erpét. Gén. 7(1): 429–430. (*Lamprophis modestus*)

Synonyms: *Hormonotus audax* Hallowell, 1857b, *Heterolepis glaber* Jan, 1863b *(nomen nudum), Heterolepis glaber* Jan, 1870 *in* Jan & Sordelli, 1870–1881, and *Boodon vossii* J.G. Fischer, 1888a.
Type: Holotype, RMNH 304, a 581 mm female (H.S. Pel, 1840–1854).
Type locality: "côte de Guinée" [= Ghana]. Restricted to Dabocrom, Ghana *fide* Hughes & Barry (1969: 1015).
Distribution: West and cen. Africa. Guinea (Nzérékoré, Télimélé), Sierra Leone, S Ivory Coast (Abidjan, Danané, Dix-Huit Montagnes, Guiglo, Lagunes, Tabou), S Ghana, S Togo (Plateaux), S Benin (Maritime, Plateau), S Nigeria, S Cameroon (Centre, Est, Littoral, Sud-Ouest), S Central African Republic (Lobaye, Mbomou, Ombella-Mpoko), Equatorial Guinea (Bioko Is.), Gabon (Haut-Ogooué, Moyen-Ogooué, Ogooué-Maritime), Congo (Kouilou, Sangha), Democratic Republic of the Congo (Bandundu, Bas-Congo, Equateur, Kinshasa, Orientale, Sud-Kivu), W Uganda (Central, Western), ext. SW Kenya (Western) and Angola, 10–1300 m.

†*HUBEROPHIS* Holman, 1977 (Boidae)

Type species: †*Huberophis georgiensis* Holman, 1977a.
Distribution: Upper Eocene of USA.
Sources: Holman, 1977, 1979b, 2000a and Rage, 1984b, 2011.

1. †*Huberophis georgiensis* Holman, 1977a. J. Herp. 11(2): 143–144, figs. 2a–c.

Type: Holotype, UGVP 51, one trunk vertebra (M. Voorhies).
Type locality: "red sandy clay 50 ft above the base of the Twiggs Clay, in an abandoned Kaolin (Clay) Pit (Pit no. 22, J.M. Huber Corporation) from approximately 9 mi SSE of Macon and 4 mi ENE of Huber, Twiggs County, Georgia; Upper Eocene (Twiggs Clay)."
Distribution: Upper Eocene (Duchesnean: 38.0–42.0 mya) of USA (Georgia). Known only from type locality.

HYDRABLABES Boulenger, 1891b (Natricidae)

Synonyms: *Hydrables* – Shelford, 1901, and *Hydrablabis* – F. Werner, 1925 *(nomen incorrectum).*
Type species: *Ablabes periops* A.C.L.G. Günther, 1872a.
Distribution: Borneo.
Sources: Rooij, 1917c, Haas, 1950, Haile, 1959, Stuebing, 1991, Stuebing & Inger, 2002, Zaher et al., 2009 and I. Das, 2010, 2012.
Remarks: Both taxa are conspecific *fide* Stuebing & Inger, 2002.

1. *Hydrablabes periops* (A.C.L.G. Günther, 1872a). Proc. Zool. Soc. London 40(1): 595–596, fig. 3. (*Ablabes periops*)

Types: Syntypes (2), BMNH 1946.1.14.16–17, a 398 mm male and juvenile (A.H. Everett, 1869–1872).
Type locality: "Matang in the district of Sarawak, Borneo" [= Kampong Matang, First Division, Kidi Distr., W Sarawak, East Malaysia, 1°35'N, 110°14'E, elevation 15 m].
Distribution: Borneo. East Malaysia (Sabah, Sarawak), Brunei and cen. Indonesia (W Kalimantan), 15–600 m.
Sources: Das, 2007, 2012.

2. *Hydrablabes praefrontalis* (Mocquard, 1890a). Naturaliste (2) 12(80): 154. (*Ablabes periops praefrontalis*) (*nomen corrigendum*)

Synonyms: *Ablabes periops prefrontalis* Mocquard, 1890a *(nomen incorrigendum),* and *Hydrablabes praefrontalis* – Boulenger, 1891b *(nomen corrigendum).*
Types: Syntypes (2), MNHN 1889.202–03, longest syntype 436 mm (J. Whitehead, 1885).
Type locality: "Kina Balu, Bornéo," Emended to North Borneo, Mt. Kina Balu, between 1,000 and 4,000 or even 6,000 feet *fide* Mocquard (1890b: 118), [=

Mt. Kinabalu, N Sabah, NE East Malaysia, 6°05'N, 116°30'E, elevation 300–1800 m].

Distribution: East Malaysia (N Sabah), 300–1800 m.

Remarks: Original description reprinted in Mocquard (1890b: 137–138, pl. 9, figs. 1, 1a–c).

HYDRAETHIOPS A.C.L.G. Günther, 1872b (Natricidae)

H

Synonym: *Hydraethips* – Briceño-Rossi, 1934 *(nomen incorrectum).*

Type species: *Hydraethiops melanogaster* A.C.L.G. Günther, 1872b.

Distribution: Central Africa.

Sources: Boulenger, 1920c, Perret, 1961, Chippaux, 2006, Chirio & LeBreton, 2007, Pauwels & Vande weghe, 2008 and Zaher et al., 2009.

1. *Hydraethiops laevis* Boulenger, 1904b. Ann. Mag. Nat. Hist. (7) 13(78): 450–451.

Types: Syntypes (2), BMNH 1946.1.15.35–36, longest syntype 570 mm (G.L. Bates, 1901–1904).

Type locality: "Efulen, S. Cameroon" [= Efulen, W Sud Prov., SW Cameroon, 2°46'N, 10°43'E, elevation 420 m].

Distribution: Southern Cameroon (Sud) and Gabon (Ngounié), 420–670 m.

Sources: Girardin, 1978.

Remarks: Elevation of type locality 670 m *fide* Pauwels & Vande weghe (2008: 224).

2. *Hydraethiops melanogaster* A.C.L.G. Günther, 1872b. Ann. Mag. Nat. Hist. (4) 9(49): 28, pl. 3, fig. g.

Synonyms: *Helicops marginatus* J.G. Fischer, 1883, and *Helicops lineofasciatus* Sauvage, 1884.

Types: Syntypes (4), BMNH 1946.1.7.72–75, one male and three females, longest syntype 610 mm.

Type locality: "Gaboon, West Africa" [= Gabon].

Distribution: Central Africa. Cameroon (Adamaoua, Centre, Est, Littoral, Ouest, Nord-Ouest, Sud, Sud-Ouest), Central African Republic (Haute-Sangha, Lobaye, Mbomou, Ouaka, Sangha), Gabon (Estuaire, Moyen-Ogooué, Ngounié, Nyanga, Ogooué-Ivindo, Ogooué-Lolo, Ogooué-Maritime), Congo (Kouilou, Lekoumou, Sangha) and N Democratic Republic of the Congo (Bandundu, Bas-Congo, Equateur, Kinshasa, Nord-Kivu, Orientale, Sud-Kivu), NSL–1250 m.

Sources: Witte, 1962, Roux-Estève, 1965, Knoepffler, 1966, J.-F. Trape, 1981, 1985, Joger, 1990, J.B. Rasmussen, 1991, J.-F. Trape & Roux-Estève, 1995 and Chirio & Ineich, 2006.

Remarks: Probably occurs in Equatorial Guinea.

HYDRELAPS Boulenger, 1896a (Elapidae)

Synonym: *Hidrelaps* – Briceño-Rossi, 1934 *(nomen incorrectum).*

Type species: *Hydrelaps darwiniensis* Boulenger, 1896a.

Distribution: Mangrove creeks, estuaries and shallow marine waters of Austro-Papua, including Arafura Sea and Gulf of Carpentaria.

Sources: M.A. Smith, 1926a, Brongersma, 1956c, McDowell, 1969b, 1972a, L.A. Smith, 1974, Cogger, 1975, 1992, Whitaker et al., 1982, Cogger et al., 1983a, F. Parker, 1982, Storr et al., 1986, Gow, 1989, O'Shea, 1996, A.R. Rasmussen, 1997, Keogh, 1998, David & Ineich, 1999, Wells, 2007, Kharin, 2008, Zaher et al., 2009 and Hoser, 2012e.

Remarks: Kharin (2008: 103) proposed the subfamily Hydrelapinae. Wells (2007: 20) placed in the family Hydrelapidae.

1. *Hydrelaps darwiniensis* Boulenger, 1896a. Cat. Snakes Brit. Mus. 3: 270, pl. 12, figs. 1–la.

Type: Lectotype, BMNH 1946.1.1.91 (formerly BMNH 1884.9.13.24), an adult female, longest syntype 435 mm (R.G.S. Buckland, 1877–1884), designated by Kharin (2008: 106).

Type locality: "Port Darwin, North Australia" [= Northern Territory, Australia].

Distribution: Austro-Papua. Eastern Indonesia (S Papua), SW Papua New Guinea (Gulf, Western, Daru Is.) and N Australia (N Western Australia, Northern Territory, NW Queensland, Melville Is.), NSL–10 m.

Remarks: Photographs of the lectotype in Kharin (2008: figs. 4–5).

HYDRODYNASTES Fitzinger, 1843 (Xenodontidae)

Synonyms: *Lejosophis* Jan, 1863a *(nomen emendatum), Leiosophis* – Jan, 1863a *(nomen incorrectum), Cyclagras* Cope, 1885a *(nomen substitutum), Ciclagras* – Brazil, 1911 *(nomen incorrectum), Xyclagras* – Freiberg, 1939 *(nomen incorrectum), Dugandia* Dunn, 1944d, *Hydrodinastes* – Cunha & Nascimento, 1978 *(nomen incorrectum), Leitosphis* – Cunha & Nascimento, 1978 *(nomen incorrectum), Cyglagras* – Engelmann & Obst, 1981 *(nomen incorrectum),* and *Azdronestes* – Fugler, 1983 *(nomen incorrectum).*

Type species: *Coluber bicinctus* Hermann, 1804.

Distribution: South America.

Sources: Hoge, 1958b, J.A. Peters & Orejas-Miranda, 1970, Dowling & Gibson, 1970a, Cadle, 1984a, Chippaux, 1987, Starace, 1998, Franco et al., 2007 and Zaher et al., 2009.

1. *Hydrodynastes bicinctus* (Hermann, 1804). Observ. Zool.: 276. (*Coluber bicinctus*)

Synonyms: *Coronella anguiformis* Laurenti, 1768 (*partim, nomen oblitum*), *Elaps annulatus* Schneider, 1801 (*partim, nomen oblitum*), *Coluber thalia* Daudin, 1803d (*partim, nomen oblitum*), *Elaps schrankii* Wagler *in* Spix, 1824, *Coluber cuvieri* Kaup, 1827, and *Hydrodynastes bicinctus schultzi* Hoge, 1966b.
Type: Holotype, not designated, location unknown.
Type locality: Unknown.
Distribution: Amazonia. Southeastern Colombia (Amazonas, Meta, Vaupéss), S Venezuela (Amazonas, Bolívar), Guyana (Upper Takutu-Upper Essequibo), S Suriname (S Sipaliwini), N French Guiana (Cayenne, Saint-Laurent-du-Maroni) and N Brazil (Amazonas, Distrito Federal, Goiás, Pará, Paraná, Rondônia, São Paulo), NSL–750 m.
Sources: Roze, 1966a, P. Muller, 1975, Cunha & Nascimento, 1978, Abuys, 1983d, Lancini, 1986, Pérez-Santos, 1986, Chippaux, 1987, Pérez-Santos & Moreno, 1988, Kuch, 1990, N.J. Silva & Sousa, 2002, Abuys, 2003, Navarrete et al., 2009 and C.J. Cole et al., 2013.
Remarks: In accordance with Art. 23.9.2 of the Code (ICZN, 1999), *Coluber bicinctus* Hermann is designated a *nomen protectum* and *Coronella anguiformis* Laurenti, *Elaps annulatus* Schneider, and *Coluber thalia* Daudin are *nomina oblita*.

2. *Hydrodynastes gigas* (A.-M.-C. Duméril, Bibron & Duméril, 1854a). Erpét. Gén. 7(1): 761–763. (*Xenodon gigas*)

Synonym: *Xenodon gigas* A.-M.-C. Duméril, 1853 (*nomen nudum*).
Types: Syntypes (3), MNHN 2493a–c, longest syntype 2270 mm (A.C.V.M.D. d'Orbigny, 22 April 1827–20 April 1828).
Type locality: "la Prov. de Corrientes (Confédér. du Rio de la Plata), l'Amérique du Sud" [= Corrientes Prov., ext. NE Argentina].
Distribution: South America. Suriname, N French Guiana (Cayenne, Saint-Laurent-du-Maroni, Cayenne), Brazil (Amapá, Distrito Federal, Goiás, Maranhão, Mato Grosso, Mato Grosso do Sul, Pará, Paraíba, Paraná, Piauí, Rio Grande do Sul, Rondônia, São Paulo, Tocantins, Marajo Is.), E Bolivia (Beni), Paraguay (Alto Paraguay, Alto Paraná, Asunción, Caaguazú, Caazapá, Canindeyú, Central, Concepción, Cordillera, Guairá, Itapúa, Misiones, Neembucú, Paraguarí, Presidente Hayes, San Pedro) and N Argentina (Chaco, Corrientes, Entre Ríos, Formosa, Misiones, Santa Fe), NSL–100 m.
Sources: P. Müller, 1975, Amaral, 1978, Hoge et al., 1978b, Hoogmoed, 1979, J.A. Campbell & Murphy, 1984, Bels, 1987, Chippaux, 1987, F.E. Hayes, 1991, Strüssmann & Sazima, 1991, J.D. Williams & Francini, 1991, Cei, 1994, D.R. Norman, 1994, Fugler et al., 1995, Moura-Leite et al., 1996, Bernarde & Moura-Leite, 1999, Leynaud & Bucher, 1999, Abuys, 2003, Pereira-Filho & Montingelle, 2006, Cacciali, 2008 and Silva-Leite et al., 2009.

3. *Hydrodynastes melanogigas* Franco, Fernandes & Bentim, 2007. Zootaxa (1613): 58–61, 4 figs.

Type: Holotype, IB 65796, a 1795 mm male (M.M.T. da Rocha, M.E.V. Calleffo & S.R. Travaglia-Cardoso, 21–30 Jan. 2002), destroyed by fire 15 May 2010.
Type locality: "Municipality of Palmas (10°12'46"S 48°21'37"W, 230 m above sea level), State of Tocantins, Brazil."
Distribution: Central Brazil (Tocantins), 200–230 m.

HYDROMORPHUS W.C.H. Peters, 1859b (Dipsadidae)

Type species: *Hydromorphus concolor* W.C.H. Peters, 1859b.
Distribution: Central America.
Sources: Nelson, 1966, Villa, 1970b, 1990a, Savage & Donnelly, 1988 and Cunha & Nascimento, 1993.

1. *Hydromorphus concolor* W.C.H. Peters, 1859b. Mber. Königl. Akad. Wiss. Berlin 1859(3): 277, pl. figs. 3, 3a–3c.

Synonym: *Hydromorphus clarki* Dunn, 1942.
Type: Holotype, ZMB 2493, an 850 mm female (K. Hoffmann, 1853–1859).
Type locality: "Costa Rica."
Distribution: Central America. Eastern Guatemala (Izabal), NW Honduras (Atlántida, Cortés, Gracias a Dios, Olancho, Santa Bárbara, Yoro), Nicaragua (Jinotega, S Matagalpa), Costa Rica (Alajuela, Cartago, Guanacaste, Heredia, Limón, Puntarenas, San José) and Panama (Bocas del Toro, Panamá, Bastimentas Is.), NSL–1500 m.
Sources: E.H. Taylor, 1951, 1954, L.D. Wilson & Meyer, 1985, Crother, 1989, Solórzano et al., 1989, G. Köhler, 1999b, Savage, 2002, Solórzano, 2004, McCranie et al., 2006 and McCranie, 2011a.

2. *Hydromorphus dunni* Slevin, 1942. Proc. California Acad. Sci. (4) 23(32): 474, pl. 39, figs. 2–3.

Type: Holotype, CAS 78939, adult male (J.R. Slevin, 30 July 1939).
Type locality: "Vicinity North of Boquete, Chiriqui Prov., Panama."
Distribution: Western Panama (W Chiriquí), 1150–1250 m. Known only from type locality.

HYDROPHIS Latreille *in* Sonnini & Latreille, 1801b (Elapidae)

Synonyms: *Hydrophys* Bory de St. Vincent, 1825 *(nomen emendatum), Hydrus* Wagler, 1830, *Nauticophis* Lesson, 1832 *in* Bélanger, 1831–1834 *(nomen substitutum), Polyodontes* Lesson, 1832 *in* Bélanger, 1831–1834 *(nomen praeoccupatum), Hydrophes* Agassiz, 1846 *(nomen emendatum), Noterophis* Gistel, 1848, *Polydontes* – Gray, 1849a *(nomen incorrectum), Colubrinus* A.-M.-C. Duméril, Bibron & Duméril, 1854b, *Hijdrophis* Bleeker, 1856 *(nomen emendatum), Hydrophtis* – Gervais & Beneden, 1859 *(nomen incorrectum), Hydrophia* – Cope, 1860b *(nomen incorrectum), Hidrophis* – Kulagin, 1888 *(nomen incorrectum), Polyodontus* Boulenger, 1896a *(nomen emendatum), Porrecticollis* Wall, 1921g, *Enhydrophis* – Maki, 1933 *(nomen incorrectum), Hygrophis* – Brygoo, 1983 *(nomen incorrectum),* and *Porreticollis* – Culotta & Pickwell, 1993 *(nomen incorrectum).*

Type species: *Hydrus fasciatus* J.G. Schneider, 1799.

Distribution: Marine waters of Indo-Australia, including Arafura Sea, Arabian Sea, Bay of Bengal, Chilka Lake, Coral Sea, Gulf of Carpentaria, Gulf of Thailand, Java Sea, Indian Ocean, South China Sea, Strait of Madura, Strait of Malacca and Timor Sea.

Sources: Wall, 1909a, 1914g, M.A. Smith, 1926a, 1943, C. Haas, 1950, Klemmer, 1963, Barme, 1968, Halstead, 1970, McDowell, 1972a, Cogger, 1975, 1992, Minton, 1975, Voris, 1977, Cogger et al., 1983a, Burger & Natsuno, 1984, Schwaner et al., 1985, Ehmann, 1992, A.R. Rasmussen, 1994, 1997, David & Ineich, 1999, A.R. Rasmussen et al., 2001, Kelly et al., 2003, Kharin, 2004b, Scanlon & Lee, 2004, Kharin & Cheblukov, 2007, A.R. Rasmussen et al., 2007, Wells, 2007, Sanders et al., 2008, Zaher et al., 2009, I. Das, 2010, 2012, Hoser, 2012e and Ukuwela et al., 2012.

Remarks: Official Generic Name no. 2146 *fide* Opinion 1201 (ICZN, 1982a). The following genera and species have been separated from *Hydrophis: Chitulia (belcheri, bituberculatus, inornatus, laboutei, lamberti, lapemoides, mamillaris, ornata, sibauensis, sticticollis, torquatus), Disteira (kingii, major, nigrocincta, walli), Leioselasma (coggeri, cynocincta, czeblukovi, elegans, melanocephala, pacifica, semperi, spiralis), Mediohydrophis (klossi, melanosoma),* and *Polyodontognathus (caerulescens).*

1. *Hydrophis atriceps* A.C.L.G. Günther, 1864a. Rept. Brit. India: 371, pl. 25, figs. i–i'.

Synonyms: *Aturia lindsayi* Gray, 1842d, *Hydrophis alcocki* Wall, 1906g, *Disteira cincinnatii* Van Denburgh & Thompson, 1908, and *Distira cincinatii* – Phisalix, 1922 *(nomen incorrectum).*

Types: Syntypes (2), BMNH 1946.1.2.62 (formerly BMNH 1862.11.1.255) and BMNH 1863.9.29.5, two males, longest syntype 787 mm.

Type locality: "Siam" [= Thailand].

Distribution: Indo-Australia. Thailand (Krung Thep Maharakhom, Surat Thani, Songkhla), Cambodia (Kaoh Kong), Vietnam (Dong Nai), China (Fujian, Guangdong, Guangxi, Hainan, Liaoning), E West Malaysia (Kelantan), East Malaysia (Sabah, Sarawak), Indonesia (Papua, Aru Is., Java, Kalimantan, Sumatra), Philippines (Luzon), S Papua New Guinea (Gulf) and N Australia (Northern Australia, Queensland).

Sources: Bourret, 1934g, M.A. Smith, 1935, Bergman, 1962b, E.H. Taylor, 1965, Tian et al., 1986, Stuebing, 1991, O'Shea, 1996 and V.S. Nguyen et al., 2009.

Remarks: A valid species *fide* Cogger et al., 1983a, Gow, 1989 and Cogger, 2000. Previously a subspecies of *Hydrophis fasciatus* (Schrieber). Possibly occurs in West Malaysia *fide* I. Das (2012: 149).

2. *Hydrophis brookii* A.C.L.G. Günther, 1872a. Proc. Zool. Soc. London 40(1): 597, fig. 5.

Synonyms: *Hydrophis consobrinus* M.A. Smith, 1917c, *Hydrophis brookei* Rooij, 1917 *(nomen emendatum),* and *Disteira consobrina* Oshima, 1944.

Type: Holotype, BMNH 1946.1.1.57 (formerly BMNH 1872.2.16.58), a 914 mm female (A.H. Everett, 1872–1876 or 1885–1890).

Type locality: "Sarawak, Borneo" [East Malaysia].

Distribution: Southeast Asia and Greater Sundas. Southern Thailand (Krung Thep Mahanakhon, Songkhla, Trang), ext. S Vietnam (Dong Nai, Ke Ga Cape, Phan Ri, Phan Thiet, Vung Tau), West Malaysia (Johore, Perak, Pinang), Singapore, N East Malaysia (Sarawak) and W Indonesia (N Java).

Sources: E.H. Taylor, 1965, B.L. Lim & Sawai, 1975 and V.S. Nguyen et al., 2009.

3. *Hydrophis donaldi* Ukuwela, Sanders & Fry, 2012. Zootaxa (3201): 48–51, fig. 2.

Type: Holotype, QM J90700, an 860 mm male (B.G. Fry, 15 Oct. 2000).

Type locality: "Gulf of Carpentaria, Weipa, Queensland, Australia (12°35'10.88"S, 141°57'47.21"E)."

Distribution: Northern Australia (N Queensland).

Remarks: Generic allocation uncertain.

4. *Hydrophis fasciatus* (J.G. Schneider, 1799). Hist. Amph. 1: 240–241. (*Hydrus fasciatus*)

Synonyms: *Hydrophis laticauda* Latreille *in* Sonnini & Latreille, 1801b, *Anguis xiphura* Hermann, 1804, *Hydrophis lanceolatus* Oken, 1817, *Hydrus chlorus* Merrem, 1820, *Hydrus chloris* Gray *in* Griffith & Pidgeon, 1831 *(nomen emendatum), Pelamis lindsayi* Gray *in* Griffith & Pidgeon, 1831, *Colubrinus* *hydrus* A.-M.-C. Duméril, Bibron & Duméril, 1854b, *Hydrophis leptodira* Boulenger, 1896a, *Hydrophis rhombifer* Boulenger, 1900a, *Hydrophis lindsays* – Mocquard, 1904b *(nomen incorrectum), Hydrophis laindsaya* – Wall, 1921g *(nomen incorrectum), Hydrophis fasciatus faciatus* – M.W. Lin, 1975 *(nomen incorrectum),* and *Pelamis fasciataus* – Culotta & Pickwell, 1993 *(nomen incorrectum).*

Type: Lectotype, ZMB 2836, a juvenile female (M.E. Block, 1760), designated by Iskandar & Colijn (2001: 134).

Type locality: Unknown. "East Indies" via lectotype selection.

Distribution: Southern Asia and Greater Sundas. Pakistan, India (Andhra Pradesh, Karnataka, Kerala, Orissa, Tamil Nadu, West Bengal), Bangladesh, Myanmar (Rakhine), China (Fujian, Guangdong, Guangxi, Hainan), Vietnam (Con Dao Is.), West Malaysia (Perak, Perlis, Phuket, Selangor, Angsa, Langkawi and Phuket Is.), Singapore and Indonesia (N Java, Sumatra).

Sources: Wall, 1921g, Singh, 1972, B.L. Lim & Sawai, 1975, M.S. Khan, 2002, V.S. Nguyen et al., 2009 and A. Silva et al., 2011.

Remarks: Official Specific Name no. 2782 *fide* Opinion 1201 (ICZN, 1982a). ZMB catalogue lists type locality as "ostindien" and ZMB 2837 as holotype but "Sudamerika" *fide* Bauer (1998: 137).

5. *Hydrophis macdowelli* Kharin, 1983. Zool. Zhur. 62(11): 1751–1753, figs. 1–4.

Type: Holotype, ZISP 19678, a 708 mm male (U.P. Shuntov, 1969).

Type locality: "North Australian shelf."

Distribution: Australasia. Papua New Guinea (Western), N Australia (Northern Territory, Queensland, N Western Australia) and New Caledonia.

Sources: Gow, 1989, O'Shea, 1996, Ineich & Rasmussen, 1997, Bauer & Sadlier, 2000 and Ineich & Laboute, 2002.

6. *Hydrophis obscurus* Daudin, 1803d. Hist. Nat. Rept. 7: 375–376.

Synonyms: *Hydrophis cloris* Daudin, 1803d, *Hydrophis shootur* Rafinesque-Schmaltz, 1817, *Pelamis chloris* Merrem, 1820 *(nomen emendatum), Leioselasma obscura* – Fitzinger, 1827, *Hydrophis subcinctus* Gray, 1842d, *Hydrophis coronata* A.C.L.G. Günther, 1864a, *Hydrophis latifasciata* A.C.L.G. Günther, 1864a, *Hydrophis subcincta* – A.C.L.G. Günther, 1864a, and *Porreticollis obscurus* – Culotta & Pickwell, 1993 *(nomen incorrectum).*

Type: Holotype, BMNH 1946.1.9.27 (formerly BMNH 1896.3.25.4), a 914 mm male (P. Russell, 1781–1791).

Type locality: "Coromandel" [India] (in error). Specified as The Sunderbunds *fide* P. Russell (1801: 57) [= Sundarbans, Ganges River delta, SW Bangladesh and SE West Bengal, India, bet. 21.5–23°N, 88–91.5°E]

Distribution: Southern Asia. Eastern India (Kerala, Orissa, Tamil Nadu, West Bengal, Barkuda Is.), Bangladesh (Chittagong) and Myanmar (Irrawaddy, Mergui Arch., Tanintharyi).

Sources: Wall, 1921g, F. Scott et al., 1977, Ahmed, 1978 and M.A.R. Khan, 1988.

Remarks: Original description based on P. Russell (1801: 9, pl. 8).

7. *Hydrophis parviceps* M.A. Smith, 1935. Dana Report (8): 5, fig. 2.

Type: Holotype, ZMUC 66182 (formerly MBLC St. 3692), an 890 mm male (R.S. Dana crew, 11 April 1929).

Type locality: "coast of Cochin China (9°59' N 107°23.5' E)" [= south of Vung Tau, Dong Nai Prov., S Vietnam].

Distribution: Southern Vietnam (Bihn Thuan, Dong Nai, Han Tam).

Sources: E.H. Taylor, 1963, V.S. Nguyen et al., 2009 and A.R. Rasmussen et al., 2011, 2012.

Remarks: Photograph and head illustrations of holotype in A.R. Rasmussen et al. (2012: figs. 1–2).

8. *Hydrophis vorisi* Kharin, 1984a. Zool. Zh. 63(4): 630–631, figs. 1–4.

Type: Holotype, AMNH 58869, a 635 mm female (Second Archibold Exped., 1937).

Type locality: "East Bank of Fly River opposite Stuart Island, Western District, Papua New Guinea."

Distribution: Southern Papua New Guinea (Western) and N Australia (Torres Strait).

Sources: Cogger, 1975 and O'Shea, 1996.

Remarks: India and Myanmar erroneously included as part of Distribution *fide* Gow (1989: 159).

HYDROPS Wagler, 1830 (Xenodontidae)

Synonyms: *Higina* Gray, 1842d, *Hygina* Gray, 1849 *(nomen emendatum),* and *Hydrophs* – Orcés & Matheus, 1988 *(nomen incorrectum).*

Type species: *Elaps martii* Wagler *in* Spix, 1824.

Distribution: South America.

Sources: Roze, 1957b, 1957c, J.A. Peters & Orejas-Miranda, 1970, Cunha & Nascimento, 1978, Dixon & Soini, 1986, Pérez-Santos, 1986, Pérez-Santos & Moreno, 1988, Vidal et al., 2000, Albuquerque, 2002, Scrocchi et al., 2005, Albuquerque & Lema, 2008 and Zaher et al., 2009.

1. *Hydrops caesurus* Scrocchi, Ferreira, Giraudo, Avila, & Motte, 2005. Herpetologica 61(4): 470–472, fig. 2.

Type: Holotype, MNHNP 6698 (formerly MCZ-FS 2-11689), a 656 mm female (Consultora FORACRO, 4 Aug. 1994).

Type locality: "Departamento Itapúa; Isla Paloma, Canal de los Jesuitas, Paraguay" [= Los Jesuitas canal, Paloma Is., S Itapúa Dept., ext. SE Paraguay, 27°28'S, 56°25'W, elevation 75 m].

Distribution: Brazil (Mato Grosso, Mato Grosso do Sul), S Paraguay (Alto Paraná, Canindeyú, Central, Itapúa, Misiones, Neembucú, Presidente Hayes, Paloma and Yacyretá Is.) and Argentina (Corrientes, San Miguel), NSL–185 m.

Source: Cacciali, 2008.

H

2. *Hydrops martii* (Wagler *in* Spix, 1824). Serp. Brasil. Sp. Nov.: 3–4, pl. 2, fig. 2. (*Elaps martii*)

Synonyms: *Coronella anguiformis* Laurenti, 1768 (*partim, nomen oblitum*), *Elaps annulatus* Schneider, 1801 (*partim, nomen oblitum*), *Coluber thalia* Daudin, 1803d (*partim, nomen oblitum*), *Hydrops martii ornatissimus* Jan, 1857 (*nomen nudum*), and *Hydrops callostictus* A.C.L.G. Günther, 1868.

Type: Holotype, ZSM 1844/0, a 596 mm specimen (J.B. von Spix & K.F.P. von Martius, May–July 1819).

Type locality: "provincia Maranhao ad flumen Itapicurú" [= between Caxias (4°50'S, 43°21'W) and Arrarial (2°37'S, 44°41'W), Río Itapicuru, Maranhão Prov., Brazil *fide* Franzen & Glaw, 2007: 253].

Distribution: Amazonia. Southeastern Colombia (E Amazonas, Vaupés), S Venezuela (SE Amazonas), N Brazil (Amazonas, Maranhão, Pará, Roraîma, Marajo Is.) and NE Peru (Loreto), NSL–250 m.

Sources: Albuquerque, 2000 and Rivas-Fuenmayor et al., 2012.

Remarks: Plate figure of original description erroneously listed in text as fig. 1. Photograph of holotype in Albuquerque (2000: fig. 1) and Franzen & Glaw (2007: fig. 20). In accordance with Art. 23.9.2 of the Code (ICZN, 1999), *Elaps martii* Wagler is designated a *nomen protectum* and *Coronella anguiformis* Laurenti, *Elaps annulatus* Schneider, and *Coluber thalia* Daudin are *nomina oblita*.

3. *Hydrops triangularis* (Wagler *in* Spix, 1824). Serp. Brasil. Sp. Nov.: 5–6, pl. 2a, fig. [1] (right). (*Elaps triangularis*) (*nomen protectum*)

Synonyms: *Coronella anguiformis* Laurent, 1768 (*partim, nomen oblitum*), *Elaps annulatus* Schneider, 1801 (*partim, nomen oblitum*), *Coluber thalia* Daudin, 1803d (*partim, nomen oblitum*), *Coluber nicandri* Merrem, 1820, *Higina fasciata* Gray, 1842d, *Pseudoeryx inagnitus* Bocourt, 1894, *Hydrops triangularis bassleri* Roze, 1957c, *Hydrops triangularis bolivianus* Roze, 1957c, *Hydrops triangularis neglectus* Roze, 1957c, and *Hydrops triangularis venezuelensis* Roze, 1957c.

Type: Holotype, ZSM 1846/0, a 611 mm male (J.B. von Spix, Nov.–Dec. 1819).

Type locality: "prope Ega ad lacum Teffé cum flumine Solimöens confluentem" [= Téfé, confluence of Lago Tefé with Río Amazonas, Amazonas State, NW Brazil, 3°19'S, 64°43'W, elevation 50 m].

Distribution: South America. Southeastern Colombia (Amazonas, Meta, Vaupés), S Venezuela (Amazonas, Apure, Barinas, Bolívar, Cojedes, Delta Amacuro, Portuguesa), Trinidad and Tobago (Trinidad), Guyana (Cuyuni-Mazaruni, Essequibo Islands-West Demerara), Suriname (Brokopondo, Para, Sipaliwini), French Guiana (Cayenne, Saint-Laurent-du-Maroni), Ecuador (Napo), N Brazil (Amazonas, Amapá, Maranhao, Pará, Piauí, Rondônia, Tocantins), E Peru (Loreto), NE Bolivia (Beni, Santa Cruz) and Paraguay (Presidente Hayes), NSL–240 m.

Sources: Roze, 1966a, Nicéforo-María, 1970, J.D. Williams & Couturier, 1984, Lancini, 1986, Chippaux, 1987, Cei, 1994, Yuki, 1997a, Buongerminip & Waller, 1998, Starace, 1998, Rivas-Fuenmayor & Fuentes, 2000, Albuquerque, 2001, Markezich & Dixon, 2001, Guedes et al., 2002 and Franzen & Glaw, 2007, Albuquerque & Lema, 2008, Navarrete et al., 2009 and C.J. Cole et al., 2013.

Remarks: Original description listed figure as pl. 2a, fig. 1 but both figures are unnumbered and *E. triangularis* is on the right. In accordance with Art. 23.9.2 of the Code (ICZN, 1999), *Elaps triangularis* Wagler is designated a *nomen protectum* and *Coronella anguiformis* Laurenti, *Elaps annulatus* Schneider, and *Coluber thalia* Daudin are *nomina oblita*.

HYPNALE Fitzinger, 1843 (Viperidae)

Synonyms: *Hynale* – Chenu & Desmarest, 1857 (*nomen incorrectum*), *Hypuale* – Canton, 1895 (*nomen incorrectum*), and *Hypinale* – Hoge & Romano-Hoge, 1983 (*nomen incorrectum*).

Type species: *Cophias hypnale* Merrem, 1820.

Distribution: Southern India and Sri Lanka.

Sources: Wall, 1921g, M.A. Smith, 1943, Deraniyagala, 1955, Gloyd, 1977, Hoge & Romano-Hoge, 1981a, Chiasson et al., 1989, Maes, 1989, Gloyd & Conant, 1990, Kardong, 1990, Kraus et al., 1996, Parkinson et al., 1997, David & Ineich, 1999, McDiarmid et al., 1999, Parkinson, 1999, Gumprecht et al., 2004 and Maduwage et al., 2009.

Remarks: An undescribed species occurs in Galle, Sri Lanka *fide* Maduwage et al. (2009: 20–23, fig. 9).

1. *Hypnale hypnale* (Merrem, 1820). Tent. Syst. Amph.: 155. (*Cophias hypnale*)

Synonyms: *Trimesurus ceylonensis* Gray, 1842b, *Trigonocephalus hypnalis* – Blyth *in* Kelaart, 1852, *Hypnale affinis* J. Anderson, 1871b, and *Ancistrodon millardi* Wall, 1908b.

Type: Neotype, BNHS 2531.759, a 350 (svl) mm female (H.S. Wise, 31 Dec. 1907), designated by Maduwage et al. (2009: 4).

Type locality: "Castle Rock in Karnataka State, India, 15°23'52"N, 74°19'56"E, alt. 620 m" via neotype selection.

Distribution: Southwestern India (Karnataka, Kerala, Tamil Nadu) and Sri Lanka (Central, Eastern, North-Central, Northern, North-Western, Sabaragamuwa. Southern, Uva, Western), NSL–1525 m.

Sources: P. Silva, 1980a, Nanayakkara, 1989, A. Silva, 1990a–b, 2001, 2009, Whitaker & Captain, 2004 and N. Khaire, 2006.

Remarks: Original description based on Russell (1801: 26–27, pl. 22). Type specimen(s) lost *fide* Maduwage et al. (2009: 3). Photograph of neotype in Maduwage et al. (2009: 9, fig. 3a).

2. *Hypnale nepa* (Laurenti, 1768). Synop. Rept.: 97. (*Coluber nepa*)

Synonym: *Hypnale walli* Gloyd, 1977.

Type: Neotype, WHT 6515, a 367 (svl) mm male (S.V. Nanayakkara, 4 March 2006), designated by Maduwage et al. (2009: 10).

Type locality: "Agra Arboretum, Agarapatana, 06°50'N, 80°40'E, alt. 1665 m" via neotype selection.

Distribution: Southern Sri Lanka (Central, Sabaragamuwa, Uva, Western), 1250–1850 m.

Source: J.C. Murphy & Schlager, 2003.

Remarks: Original description based on Seba (1735: 55–56, pl. 55, figs. 1, 2, 4). Type specimen(s) lost *fide* Maduwage et al. (2009: 3). Photographs of neotype in Maduwage et al. (2009: figs. 4a–e, 5b). Maduwage et al., 2009 considered *H. walli* a synonym of *nepa*.

3. *Hypnale zara* (Gray, 1849a). Cat. Snakes British Mus.: 15. (*Trigonocephalus zara*)

Type: Holotype, BMNH 1946.1.19.96, a 303 (svl) male (Argent coll.)

Type locality: "Singapore," (in error *fide* Maduwage et al., 2009: 25).

Distribution: Sri Lanka (Central, Sabaragamuwa, Southern), 60–1310 m.

Remarks: Photograph of holotype in Maduwage et al. (2009: fig. 7a).

HYPOPTOPHIS Boulenger, 1908g (Atractaspididae)

Synonyms: *Michellia* L. Müller, 1911, and *Hypostrophis* – Phisalix, 1922 *(nomen incorrectum).*

Type species: *Hypoptophis wilsonii* Boulenger, 1908g.

Distribution: Central Africa.

Sources: Witte & Laurent, 1947, Laurent, 1950c, Witte, 1953, 1962, Broadley, 1971d, Underwood & Kochva, 1993, Kelly et al., 2009 and Zaher et al., 2009.

1. *Hypoptophis wilsonii* Boulenger, 1908g. Ann. Mag. Nat. Hist. (8) 2(7): 93–94.

Synonym: *Michellia katangae* L. Müller, 1911.

Type: Holotype, BMNH 1946.1.2.99 (formerly BMNH 1908.6.16.5), a 560 mm female (H. Wilson, June 1904–Dec. 1907).

Type locality: "Inkongo, on the Sankuru River, in the Kassai Prov. of the Congo" [= Inkongo, Kasai Oriental Prov., S Democratic Republic of the Congo, 4°55'S, 23°15'E, elevation 520 m].

Distribution: Southeastern Democratic Republic of the Congo (Kasai Occidental, Kasai Oriental, Katanga) and NW Zambia (North-Western, Western), 520–1095 m.

H

HYPSIGLENA Cope, 1860e (Dipsadidae)

Synonyms: *Pseudodipsas* W.C.H. Peters, 1860d, *Comastes* Jan, 1863b, *Hypseglena* Scudder, 1882 *(nomen emendatum), Pseudodipas* – W.W. Tanner, 1944 *(nomen incorrectum), Hypseglena* – Hoge & Garcia, 1949 *(nomen incorrectum), Eridiphas* Leviton & Tanner, 1960, and *Hypsoglena* – Obst, Richter & Jacob, 1984 *(nomen incorrectum).*

Type species: *Hypsiglena ochrorhyncha* Cope, 1860e.

Distribution: North America.

Fossil records: Upper Pleistocene of SW USA.

Sources: W.W. Tanner, 1939, Mulcahy, 2008, Zaher et al., 2009 and Mulcahy et al., 2011.

1. *Hypsiglena affinis* Boulenger, 1894a. Cat. Snakes Brit. Mus. 2: 210–211, pl. 8, fig. 1.

Types: Syntypes (3), BMNH 1946.1.1.21 (J.M. Cameron), BMNH 1946.1.1.22 (F.D. Godman, 1885–1889), and BMNH 1946.1.1.23 (A.G. Butler), longest syntype 310 mm.

Type locality: "Zacatecas" [= Zacatecas State, Mexico], "Mezquital del Oro, Zacatecas" [= Mezquital del Oro, S Zacatecas State, Mexico, 21°13'N, 103°22'W, elevation 1255 m], and "Jalisco," = Jalisco State, Mexico].

Distribution: Western Mexico (Jalisco, ext. S Zacatecas), 1000–2100 m.

2. *Hypsiglena chlorophaea* Cope, 1860e. Proc. Acad. Nat. Sci. Philadelphia 12(6): 247.

Synonyms: *Hypsiglena ochrorhynchus deserticola* W.W. Tanner, 1946, *Hypsiglena ochrorhynchus lorealus* W.W. Tanner, 1946, *Hypsiglena torquata catalinae* W.W. Tanner, 1966a, and *Hypsiglena torquata tiburonensis* W.W. Tanner, 1981.
Types: Syntypes (2), USNM 4676, longest syntype 380 mm (B.J.D. Irwin, 1857–1860), location of other syntype unknown.
Type locality: "Fort Buchanan, Arizona" [= 5 km SW Sonoita in Hog Canyon, Santa Cruz Co., SE Arizona, SW USA, 31°40'N, 110°42'W, elevation 1425 m].
Distribution: North America. Canada (ext. S British Columbia), W USA (Arizona, E California, W Colorado, S Idaho, Nevada, E Oregon, Utah, E Washington) and W Mexico (ext. SW Chihuahua, Sonora), NSL–1545 m.
Fossil records: Upper Pleistocene (Rancholabrean II) of USA (Arizona, California, Nevada).
Sources: P.R. Brown, 1997, Grismer, 1999, 2001b, 2002 and Holman, 2000a.

H

3. *Hypsiglena jani* (Dugès, 1865). Mem. Acad. Sci. Lett. Montpellier 6: 32–33. (*Liophis janii*)

Synonyms: *Hypsiglena texana* Stejneger, 1893b, and *Hypsiglena torquata dunklei* E.H. Taylor, 1939a
Type: Neotype, USNM 9889 (native, 1877), designated by H.M. Smith & Necker (1943: 204).
Type locality: "Guanájuato, Guan., Mexico" [= Guanajuato, Guanajuato State, cen. Mexico, 21°01'N, 101°15'W, elevation 2010 m].
Distribution: USA (E Arizona, SE Colorado, New Mexico, W Oklahoma, W and S Texas) and Mexico (Aguascalientes, N and E Chihuahua, Coahuila, E Durango, Guanajuato, Hidalgo, Jalisco, Nuevo León, Querétaro, San Luis Potosí, Tamaulipas, Zacatecas), NSL–2450 m.
Fossil records: Upper Pleistocene (Rancholabrean II) of USA (New Mexico, Texas).
Sources: E.H. Taylor, 1939a, W.W. Tanner, 1943, 1946, 1966a, 1981, 1985, 1987, Dixon, 1965, R.W. Murphy & Ottley, 1984, Stebbins, 1985, Dixon & Dean, 1986, Van Devender & Bradley, 1994, H.A. Brown et al., 1995, Degenhardt et al., 1996, Macey et al., 1996, R.S. Bartlett & Tennant, 2000, Holman, 2000a, McAfee & Gilardi, 2000, Werler & Dixon, 2000, Lemos-Espinal et al., 2004b and Reyes-Velasco et al., 2012.

4. *Hypsiglena ochrorhyncha* Cope, 1860e. Proc. Acad. Nat. Sci. Philadelphia 12(6): 246–247.

Synonyms: *Hypsiglena venusta* Mocquard, 1899, *Hypsiglena nuchalatus* W.W. Tanner, 1943, *Hypsiglena ochrorhynchus klauberi* W.W. Tanner, 1946, *Hypsiglena ochrorhynchus tortugaensis* W.W. Tanner, 1946, *Hypsiglena ochrorhynchus unaocularus* W.W. Tanner, 1946, *Hypsiglena torquata gularis* W.W. Tanner, 1954, *Hypsiglena torquata baueri* Zweifel, 1958, and *Hypsiglena torquata martinensis* W.W. Tanner & Banta, 1962a,
Types: Syntypes (11), USNM 5283a–c, USNM 9216, USNM 69538–40, USNM 131733, ANSP 3537–38, and MCZ 9503 (formerly USNM 5283d), longest syntype 295 mm (J. Xantus, Dec. 1859)
Type locality: "Cape St. Lucas, California" [= Cabo San Lucas, S Baja California Sur, NW Mexico, 22°53'N, 109°55'W, elevation 25 m].
Distribution: Western USA (S California), W Mexico (Baja California Norte, Baja California Sur, Salsipuedes, San Estaban, San Lorenzo Sur, Angel de la Guarda, Carmen, Cedros, Ceralvo, Danzante, Montserrat, Partida Norte, Partida Sur, San Esteban, San José, San Lorenzo, San Marcos, Santa Catalina, Tiburon and Tortuga Is.), NSL–2745 m.
Fossil records*:* Upper Pleistocene (Rancholabrean II) of USA (California).
Sources: Grismer, 1999, 2002 and Holman, 2000a.

5. *Hypsiglena slevini* W.W. Tanner, 1943. Great Basin Nat. 4(1–2): 53 –54.

Synonym: *Eridiphas slevini marcosensis* Ottley & Tanner, 1978.
Type: Holotype, CAS 53631, a 213 mm male (J.R. Slevin, 14 June 1921).
Type locality: "Puerto Escondido, Lower California" [= a few miles south of Loreto (25°59'N, 111°21'W, elevation NSL), Baja California Sur, NW Mexico].
Distribution: Western Mexico (S Baja California Norte, Baja California Sur, Coronados, Cerralvo, Danzante, La Peña and San Marcos Is.), NSL–1600 m.
Sources: W.W. Tanner, 1946, Ottley & Tanner, 1978, Cadle, 1984b, R. Murphy & Ottley, 1984, Castro Franco & Gaviño, 1990, Mellink, 1996, Grismer, 1999, 2002, Youmans et al., 2002 and Mulcahy & Archibald, 2003.
Remarks: Grismer (1999: 458) considered *E. sleveni marcosensis* a distinct species.

6. *Hypsiglena tanzeri* Dixon & Lieb, 1972. Contr. Sci. Nat. Hist. Mus. Los Angeles Co. (222): 2–5, figs. 1a–b, 2 (right).

Type: Holotype, TCWC 34079, a 328 mm male (F. Guyer, 13 April 1971).
Type locality: "5 km E. Jalpan, Querétaro, Mexico, 762 m, 99° 27' W., 21° 13 N."
Distribution: Central Mexico (Querétaro), 760–1070 m.
Source: Dixon & Dean, 1986.
Remarks: A subspecies of *H. torquata fide* W.W. Tanner (1981: 141).

7. *Hypsiglena torquata* (A.C.L.G. Günther, 1860j). Ann. Mag. Nat. Hist. (3) 5(27): 170–171, pl. 10, fig. a. (*Leptodeira torquata*)

Synonyms: *Pseudodipsas fallax* W.C.H. Peters, 1860d, and *Comastes quincunciatus* Jan, 1863b.

Types: Syntypes (2), BMNH 1946.1.1.15 (formerly BMNH 1861.12.30.97), a female, and WML, 451 and 417 mm specimens (Derby Mus.).

Type locality: "Nicaragua; Island of Laguna," (in error).

Distribution: Western Mexico (Colima, Guerrero, Jalisco, Michoacán, Morelos, Nayarit, Oaxaca, ext. S Puebla, S Sinaloa), NSL–800 m.

Sources: Dixon and Dean, 1986, Boundy et al. *in* Crother, 2000, H.M. Smith et al., 2005b and Canseco-Márquez & Gutiérrez-Mayén, 2010.

HYPSIRHYNCHUS A.C.L.G. Günther, 1858 (Xenodontidae)

Synonym: *Hipsirhynchus* – Briceño-Rossi, 1934 *(nomen incorrectum).*

Type species: *Hypsirhynchus ferox* A.C.L.G. Günther, 1858.

Distribution: Hispaniola.

Sources: A. Schwartz, 1971, A. Schwartz & Thomas, 1975, A. Schwartz & R.W. Henderson, 1988, 1991, Sosa et al., 1995, Rodríguez-Robles & Greene, 1996, Hedges et al., 2009 and Zaher et al., 2009.

1. *Hypsirhynchus ferox* A.C.L.G. Günther, 1858. Cat. Colub. Snakes Brit. Mus.: 49.

Synonyms: *Hypsirhynchus ferox exedrus* A. Schwartz, 1971, and *Hypsirhynchus ferox paracrousis* A. Schwartz, 1971.

Type: Holotype, BMNH 1946.1.4.96, a 756 mm female (A. Sallé via H. Cuming, 1859).

Type locality: "Barbadoes" [= Barbados Is.] (in error). Corrected to Santo Domingo *fide* Boulenger (1894a: 118). Restricted incorrectly to the vicinity of Port-au-Prince, Departement de l'Ouest, Haiti *fide* A. Schwartz (1971: 74). Corrected restriction to Republica Dominica *fide* Wetherbee (1987: 63).

Distribution: Hispaniola. Haiti (Artibonite, Nord-Ouest, Nord, Ouest, Sud-Est, Gonave Is.) and República Dominicana (Azua, Barahona, Buoruco, Dajabón, Distrito Nacional, Independencia, Monte Cristi, Pedernales, Peravia, San Juan, Valverde, Saona Is.), NSL–1700 m.

Remarks: The source of H. Cuming's Hispaniolan reptile collection was Auguste Sallé (whose expedition he sponsored) *fide* Wetherbee (1987: 63).

2. *Hypsirhynchus scalaris* Cope, 1862b. Proc. Acad. Nat. Sci. Philadelphia 14(1): 72.

Type: Holotype, MCZ 3611 (formerly MCZ 1517), a 483 mm specimebn (D.F. Weinland, 1859).

Type locality: "Hayti, (near Jeremie)" [= vicinity of Jérémie (18°39'N, 74°07'W, elevation 15 m), Grand'Anse Dept., SW Haiti].

Distribution: Southwestern Haiti (Grand'Anse, Ouest, Sud, Sud-Est), NSL–305 m.

Source: R.W. Henderson & Powell, 2009.

Remarks: Collector of type listed as A.F. Weinland *fide* Cope (1862b: 72).

I

IALTRIS Cope, 1862b (Xenodontidae)

Synonym: *Jaltris* Cope, 1862e *(nomen emendatum).*
Type species: *Philodryas dorsalis* A.C.L.G. Günther, 1858.
Distribution: Hispaniola.
Sources: A. Schwartz & Thomas, 1975, A. Schwartz & Rossman, 1976, A. Schwartz, 1986, A. Schwartz & R.W. Henderson, 1988, 1991, H.M. Smith & Wallach, 1993, Powell & R.W. Henderson, 1994a, Rodríguez-Robles & Greene, 1996, Hedges et al., 2009, Henderson and Powell, 2009 and Zaher et al., 2009.

1. *Ialtris agyrtes* A. Schwartz & Rossman, 1976. Stud. Fauna Curaçao Carib. Is. 50(165): 90–91.

Type: Holotype, LSUMZ 28564, a 455 mm (svl) female (A. Schwartz, 25 July 1969).
Type locality: "Barreras, Azua Province, República Dominicana" [= Barrea, S Azua Prov., SW Dominican Republic, 18°19'N, 70°54'W, elevation 220 m].
Distribution: Hispaniola. Southwestern Domincan Republic (Azua, Pedernales), 220 m.
Source: Powell & R.W. Henderson, 1994b.

2. *Ialtris dorsalis* (A.C.L.G. Günther, 1858). Cat. Colub. Snakes Brit. Mus.: 126. (*Philodryas dorsalis*)

Synonyms: *Dromicus mentalis* A.C.L.G. Günther, 1862b, *Ialtris vultuosa* Cope, 1862b, and *Dromicus w-nigrum* F. Werner, 1909b.
Type: Holotype, BMNH 1946.1.2.77, a 1090–1118 mm male (A. Sallé, 1849).
Type locality: "St. Domingo" [= Santo Domingo, Distrito Nacional Prov., Dominican Republic, 18°30'N, 69°59'W, elevation 50 m].
Distribution: Hispaniola. Haiti (Grand'Anse, Nord, Ouest, Sud, Sud-Est, Gônave, Tortuga and Vache Is.) and República Dominicana (Barahona, Duarte, La Altagracia, Monseñor Nouel, Pedernales, Samaná), NSL–1000 m.
Sources: F. Werner, 1924a, Cochran, 1941 and Zippel et al., 1994.
Remarks: Restriction to vicinity of Port-au-Prince, Ouest Dept., Haiti *fide* A. Schwartz & Rossman (1976: 80), in the belief that A.C.L.G. Günther's (1858: 126) Santo Domingo referred to Hispaniola, is invalid because Santo Domingo is an older name for the Dominican Republic as pointed out by Wetherbee (1986a: 116).

3. *Ialtris parishi* Cochran, 1932. Proc. Biol. Soc. Washington 45: 189–190.

Type: Holotype, USNM 80773, a 945+ mm male (L.H. Parish & W.M. Perrygo, 7 April 1930).
Type locality: "ten miles east of Baradères, southwestern peninsula of Haiti" [= Grand'Anse Dept.].
Distribution: Hispaniola. Extreme NW and SW Haiti (Grand'Anse, Tortue Is.), NSL–140 m.
Sources: Cochran, 1941 and Powell & R.W. Henderson, 1994c.
Remarks: W.M. Perrygo is collector of type *fide* original data tag.

IGUANOGNATHUS Boulenger, 1898c (Colubridae)

Synonym: *Iguanagnathus* – E.H. Taylor, 1922 *(nomen incorrectum).*
Type species: *Iguanognathus werneri* Boulenger, 1898c.
Distribution: Western Indonesia.
Sources: Rooij, 1917, Underwood, 1967, David & Vogel, 1996 and K. Jackson et al., 1999.

1. *Iguanognathus werneri* Boulenger, 1898c. Ann. Mag. Nat. Hist. (7) 2(7): 74, 4 figs.

Type: Holotype, BMNH 1946.1.6.34, a 350 mm female G. Schneider, Jr., 1897–1898, via (F. Werner).
Type locality: "Sumatra" [W Indonesia].
Distribution: Western Indonesia (Sumatra). Known only from holotype.

IMANTODES A.-M.-C. Duméril, 1853 (Dipsadidae)

Synonyms: *Bungarus* Oppel, 1811a *(nomen praeoccupatum), Himantodes* Cope, 1860e *(nomen emendatum), Himantodas* – Briceño-Rossi, 1934 *(nomen incorrectum), Himantodos* – Briceño-Rossi, 1934 *(nomen incorrectum), Imantodos* – Briceño-Rossi, 1934 *(nomen incorrectum), Imantrodus* A.L. Herrera, 1899 *(nomen emendatum), Maclachlanus* Hoser, 2012x *(nomen nudum), Maconchieus* Hoser, 2012x *(nomen illegitimum),* and *Neilsimpsonus* Hoser, 2012x *(nomen illegitimum).*
Type species: *Coluber cenchoa* Linnaeus, 1758.
Distribution: Latin America.

Sources: Cope, 1894b, F. Werner, 1924c, J.A. Peters & Orejas-Miranda, 1970, C.W. Myers, 1982, Cadle, 1984b, Pérez-Santos, 1986, Pérez-Santos & Moreno, 1988, 1991, Villa et al., 1988, Daza et al., 2009, Zaher et al., 2009, Mulcahy et al., 2011 and Hoser, 2012x.

1. *Imantodes cenchoa* (Linnaeus, 1758). Syst. Nat., ed. 10, 1: 226. (*Coluber cenchoa*)

Synonyms: *Coluber virgatus* Thunberg, 1787, *Dipsas weigelii* Fitzinger, 1826a (*nomen substitutum*), *Himantodes leucomelas* Cope, 1861e, *Dipsas cenchoa rhombeata* F. Müller, 1880 (*nomen nudum*), *Himantodes anisolepis* Cope, 1894b, *Himantodes semifasciatus* Cope, 1894b and *Himantodes platycephalus* Cope, 1899b.

Types: Syntypes (3), NHR Lin-66 (formerly MAFR) (Mus. Drottn.), and ZMUU 46a–b (formerly MP), a male with everted hemipenes.

Type locality: "America." Restricted to Suriname *fide* C. Myers (1982: 14). Restricted to vicinity of Bogotá, Colombia *fide* J.D. Williams & Francini (1991: 67).

Distribution: Latin America. Mexico (Campeche, Chiapas, Oaxaca, Puebla, Quintana Roo, San Luis Potosí, Tabasco, Tamaulipas, Veracruz), Belize (Belize, Corazal, Orange Walk, Stann Creek, Toledo), Guatemala (Alta Verapaz, Baja Verapaz, Escuintla, Huehuetenango, Izabal, Petén, Quezaltenango, San Marcos, Sololá, Zacapa), N Honduras (Atlántida, Colón, Comayagua, Copán, Cortés, El Paraíso, Gracias a Dios, Islas de la Bahía, Olancho, Yoro, Utila Is.), E Nicaragua (Atlántico Sur, Jinotega, Matagalpa, Zelaya), Costa Rica (Alajuela, Cartago, Colón, Guanacaste, Heredia, Limón, San José, Puntarenas), Panama (Bocas del Toro, Canal Zone, Chiriquí, Coclé, Darién, Los Santos, Panamá, San Blas, Barro Colorado and San Miguel Is.), Colombia (Amazonas, Antioquia, Atlántico, Bolívar, Boyacá, Caldas, Caquetá, Cauca, Chocó, Córdoba, Cundinamarca, La Guajira, Magdalena, Meta, Norte de Santander, Putumayo, Risaralda, Santander, Sucre, Tolima, Valle, Vaupés), NW and S Venezuela (Amazonas, Aragua, Bolívar, Carabobo, Distrito Federal, Falcón, Guárico, Lara, Mérida, Miranda, Portuguesa, Zulia), Trinidad and Tobago (Tobago, Trinidad), Guyana (Essequibo Islands-West Demerara, Potaro-Siparuni, Upper Demerara-Berbice), Suriname (Para), French Guiana (Cayenne, Saint-Laurent-du-Maroni), Brazil (Acre, Amapá, Amazonas, Bahia, Ceará, Espírito Santo, Goiás, N Mato Grosso, Minas Gerais, Rio de Janeiro, Pará, Paraná, Rondônia, Santa Catarina, São Paulo, Cardoso and Grande Is.), E Ecuador (Carchi, Cotopaxi, El Oro, Esmeraldas, Guayas, Morona-Santiago, Napo, Orellana, Pastaza, Pichincha, Sucumbíos), E Peru (Loreto, Madre de Dios), Bolivia (La Paz, Santa Cruz), Paraguay and N Argentina (Chaco, Formosa), NSL–2200 m.

Sources: L.C. Stuart, 1963, Roze, 1966a, Cunha & Nascimento, 1978, Savage & Scott, 1985, L.D. Wilson & Meyer, 1985, Lancini, 1986, Chippaux, 1987, Nascimento & Lema-Verde, 1989, Cei, 1994, H.M. Smith et al., 1995, Lee, 1996, Marques, 1996c, J.A. Campbell, 1998b, Gorzula & Señaris, 1998, G. Köhler, 1998a, Starace, 1998, Esqueda & La Marca, 1999, Freitas, 1999, Bérnils et al., 2000, Boos, 2001, Campos-Nogueira, 2001, Q. Marques et al., 2001, Markezich, 2002, Savage, 2002, Abuys, 2003, Duellman, 2005, Cicchi et al., 2007, García-Vázquez et al., 2009, Navarrete et al., 2009, H.C. Costa et al., 2010c, McCranie, 2011a, Rojas-Morales, 2012b and C.J. Cole et al., 2013.

2. *Imantodes chocoensis* Torres-Carvajal, Yánez-Muñoz, Quirola, Smith & Almendáriz, 2012. ZooKeys (244), 95–102, figs. 1–2, 3a.

Type: Holotype, QCAZ 7984, a 97 mm male (E. Carrillo-Ponce, I.G. Tapia & E.E. Tapia, 24 April 2007).

Type locality: "4 km N Durango, 1.0283°N; -78.5950°W (DD), 253 m, Provincia Esmeraldas, Ecuador."

Distribution: Northwestern Ecuador (Carchi, Esmeraldas), 115–260 m.

3. *Imantodes gemmistratus* Cope, 1861e. Proc. Acad. Nat. Sci. Philadelphia 13(7): 296–297. (*Himantodes gemmistratus*)

Synonyms: *Himantodes cenchoa elegans* Jan, 1863b (*nomen nudum*), *Himantodes cenchoa elegans* Jan, 1871 *in* Jan & Sordelli, 1870–1881, *Dipsas cenchoa reticulata* F. Müller, 1880, *Dipsas cenchoa reticulata* Müller, 1882 (*nomen nudum*), *Leptognathus stratissima* Cope, 1886b, *Dipsas gemmistrata latistrata* Cope, 1887c, *Dipsas gracillima* A.C.L.G. Günther, 1895 *in* 1885–1902, *Dipsas splendida* A.C.L.G. Günther, 1895 *in* 1885–1902, *Himantodes hemigenius* Cope, 1899b, *Imantodes luciodorsus* Oliver, 1942b, and *Imantodes splendidus oliveri* H.M. Smith, 1942k.

Type: Holotype, ANSP 156, a 762 mm specimen (J.M. Dow, 1860–1861), lost *fide* E.V. Malnate *in* P. Yingling (1972: 23).

Type locality: "San Salvador, Centr. America," Near Isalco, San Salvador [= vicinity of Volcán Izalco, N Sonsonate Dept., W El Salvador, 13°49'N, 89°38'W] *fide* Cope (1860e: 265).

Distribution: Latin America. Southern Mexico (Campeche, Chiapas, SW Chihuahua, Colima, Guerrero, Jalisco, México, Michoacán, Morelos, Nayarit, Oaxaca, Puebla, Querétaro, Quintana Roo, Sinaloa, S Sonora, Tamaulipas, Veracruz, Yucatán, María Madre Is.), SW Guatemala (Escuintla, Petén, Retalhuleu, San Marcos, Santa Rosa, Sololá, Suchitepéquez), W Honduras (Choluteca, Cortés, Francisco Morazán, Valle), W El Salvador (Ahuachapán, Cabañas, La Libertad, La Unión, San Ana, San Salvador, Sonsonate), SW Nicaragua (Granada, Jinotega, Managua, Rivas), Costa Rica (Alajuela, Cartago, Guanacaste, ? Heredia, Limón, Puntarenas, San José), Panama (Canal Zone, Chiriquí,

Coclé, Darién, Herrera, Panamá) and W Colombia (Tolima), NSL–2500 m.

Sources: H.M. Smith, 1942k, Zweifel, 1959c, L.C. Stuart, 1963, Yingling, 1972, Savage & Scott, 1985, L.D. Wilson & Meyer, 1985, H.M. Smith et al., 1995, Lee, 1996, 2000, J.A. Campbell, 1998b, Canseco-Marquez et al., 2000, Savage, 2002, G. Köhler et al., 2005, H.M. Smith & Lemos-Espinal, 2006a, N. Herrera et al., 2007 and McCranie, 2011a.

Remarks: ANSP 4133 (near Volcano Isalco, San Salvador Dept., El Salvador, J.M. Dow donor) possible type *fide* ANSP catalogue.

4. *Imantodes inornatus* Boulenger, 1896a. Cat. Snakes Brit. Mus. 3: 88, pl. 5, fig. 1. (*Himantodes inornatus*)

Synonym: *Himantodes unonatus* – Briceño-Rossi, 1934 *(nomen incorrectum).*

Types: Syntypes (2), BMNH 1946.1.2.63a, an 820 mm female, and BMNH 1946.1.2.63b, head and neck of adult (E. Rothschuh, 1893–1894).

Type locality: "Hacienda Rosa de Jericho, 3250 ft., Nicaragua" [= Jericó, Matagalpa Dept., Nicaragua].

Distribution: Central America and NW South America. Extreme NW Honduras (Gracias a Dios, Olancho), Nicaragua (Jinotega, Matagalpa), Costa Rica (Alajuela, Cartago, Heredia, Limón, Puntarenas, San José), Panamá (Bocas del Toro, Colón, Darién, San Blas), W Colombia (Chocó, Valle) and NW Ecuador (Esmeraldas, Manabí), NSL–1600 m.

Sources: Savage & Scott, 1985, L.A. Wilson et al., 1981a, L.D. Wilson & Meyer, 1985, Savage, 2002, Solórzano, 2004, Ortega-Andrade et al., 2010, Sajdak, 2010, McCranie, 2011a and Travers et al., 2011.

Remarks: Record from Honduras (Cortés) doubtful *fide* McCranie (2011a: 320).

5. *Imantodes lentiferus* Cope, 1894b. Amer. Nat. 28(331): 613–614. (*Himantodes lentiferus*)

Synonym: *Himantodes fulviventer* Poeppig *in* Obst, 1977.

Type: Lectotype, ANSP 11459, a male (J. Hauxwell or J. Orton, 1867–1877), designated by C.W. Myers (1982: 40).

Type locality: "upper Amazonian drainage, either Pebas, Peru, or eastern Ecuador" via lectotype selection.

Distribution: Amazonia. Southern Colombia (Amazonas, Putumayo, Vaupés), Venezuela (Amazonas, Bolívar), NW Guyana (Barima-Waini), Suriname (Saramacca), French Guiana (SE Cayenne, Saint-Laurent-du-Maroni), N Brazil (Acre, Amazonas, Mato Grosso, Pará, Rondônia), E Ecuador (Morona-Santiago, Napo, Orellana, Pastaza, Pichincha, Sucumbíos, Zamora-Chinchipe), E Peru (Amazonas, Huanuco, Loreto, Madre de Dios, San Martín) and Bolivia (Cochabamba, La Paz), 90–1150 m.

Sources: Cunha & Nascimento, 1978, 1993, Chippaux, 1987, Donnelly & Myers, 1991, Fugler et al., 1995, Starace, 1998, Duellman, 2005, Navarrete et al., 2009, Avila & Kawashita-Ribeiro, 2011, Sampaio & Lima-Maciel, 2012 and C.J. Cole et al., 2013.

Remarks: Holotype of *Himantodes fulviventer* discovered by Obst (1977: 169).

6. *Imantodes phantasma* C.W. Myers, 1982. Amer. Mus. Novit. (2738): 3–7, figs. 1–3, 4d, 4e.

Type: Holotype, KU 110217, a 1053 mm male (C.W. Myers & T. Quintero, 24 Jan. 1966).

Type locality: "at 1030 m. elevation on the southeastern slope of Cerro Cituro, a peak on the northern end of the Serranía de Pirre, Province of Darién, Republic of Panama."

Distribution: Extreme E Panama (E Darién), 1030–1100 m. Known only from vicinity of type locality.

7. *Imantodes tenuissimus* Cope, 1867b. Proc. Acad. Nat. Sci. Philadelphia (1866) 18(4): 317–318. (*Himantodes tenuissimus*)

Type: Holotype, USNM 6563 (formerly ACVS 903), an 838 mm specimen (A.C.V. Schott, 7 Jan. 1866).

Type locality: "Yucatan, Mexico." Restricted to Chichen Itza, Yucatán State, SE Mexico *fide* H.M. Smith & Taylor (1950a: 352).

Distribution: Yucatán Peninsula. Southeastern Mexico (Campeche, Quintana Roo, Yucatán), NSL–100 m.

Sources: J.C. Lee, 1980, 1996, 2000.

Remarks: Chiapas, Mexico record doubtful *fide* J.C. Lee (1996: 328).

†*INCONGRUELAPS* Scanlon, Lee & Archer, 2003 (Elapidae)

Type species: †*Incongruelaps iteratus* Scanlon, Lee & Archer, 2003.

Distribution: Upper Miocene of Australia.

Source: Scanlon et al., 2003.

1. †*Incongruelaps iteratus* Scanlon, Lee & Archer, 2003. Géobios 36: 580–588, fig. 3c.

Type: Holotype, QM F42691, one middle trunk vertebra.

Type locality: "Encore Site, Gregory River, Riversleigh World Heritage Fossil Property, north-west Queensland, Australia (19°01'S, 138°40'E); Late Miocene, approximately 10 MYA."

Distribution: Upper Miocene (10.0 mya) of Australia. Known only from type locality.

†*INDOPHIS* Rage & Prasad, 1992 (†Nigerophiidae)

Type species: †*Indophis sahnii* Rage & Prasad, 1992.
Distribution: Upper Cretaceous of India.
Sources: Rage & Prasad, 1992 and LaDuke et al., 2010.

1. †*Indophis sahnii* Rage & Prasad, 1992. Neues Jb. Geol. Paläont. Abh. 187(1): 84–89, figs. 1–5.

Type: Holotype, JU 500, one trunk vertebra.
Type locality: "intertrappean bed outcropping 2 km northeast of Naskal, Pargi Taluq, Andhra Pradesh (about 70 km west of Hyderabad), India; late Cretaceous (Maastrichtian)."
Distribution: Upper Cretaceous (Maastrichtian: 65.5–72.1 mya) of India. Known only from type series.
INYOKA Branch & Kelly

INYOKA Branch & Kelly *in* Kelly, Branch, Broadley, Barker & Villet, 2011 (Lamprophiidae)

Type species: *Lamprophis swazicus* N. Schaefer, 1970.
Distribution: South Africa and Swaziland.
Sources: Visser, 1978, 1979a, Branch, 1988 and R. Hermann, 1993.
Remarks: Not congeneric with *Lamprophis fide* Visser, 1978, 1979a, a view supported by Vidal et al. (2008: 58).

1. *Inyoka swazicus* (N. Schaefer, 1970). Ann. Cape Prov. Mus. (Nat. Hist.) 8(14): 205–207, fig. 1. (*Lamprophis swazicus*)

Type: Holotype, PEM 1514/81, a 568 mm specimen (J. Culverwell, Oct. 1968?).
Type locality: "Forbes Reef, Swaziland (26° 42' S, 31° 05' E)" [= Manzini Distr., W Swaziland, elevation 930 m].
Distribution: Northeastern South Africa (N KwaZulu-Natal, Limpopo, Mpumalanga) and W Swaziland (Hhohho, Manzini), 930–1900 m.

†*ITABORAIOPHIS* Rage, 2011 (Booidea *incertae sedis*)

Type species: †*Itaboraiophis depressus* Rage, 2011.
Distribution: Middle Palaeocene of Brazil.
Source: Rage, 2011.

1. †*Itaboraiophis depressus* Rage, 2011. Palaeovertebrata (2008) 36(1–4): 46–49, fig. 3.

Type: Holotype, DGM 1357-R, one middle trunk vertebra (J.S. Carvalho & O.S. Santos, 1953).
Type locality: "Itabouraí, State of Río de Janeiro, Brazil; middle Palaeocene."
Distribution: Middle Paleocene (Itaboraian: 57.0–59.0 mya) of Brazil.

ITHYCYPHUS A.C.L.G. Günther, 1873a (Pseudoxyrhophiidae)

Synonyms: *Periscopus* Fitzinger, 1843, *Ichthycyphus* – Cope, 1886c *(nomen incorrectum), Itycyphus* – Boettger, 1898 *(nomen incorrectum), Ithyciphus* – Phisalix, 1922 *(nomen incorrectum),* and *Pseudoithycyphus* Domergue, 1988.
Type species: *Herpetodryas goudotii* Schlegel, 1837.
Distribution: Madagascar.
Sources: Domergue, 1986, Glaw & Vences, 1994, Nagy et al., 2003, Lawson et al., 2005, Kelly et al., 2009 and Zaher et al., 2009
Remarks: Domergue, 1988, erected a new subgenus (*Pseudoithycyphus*).

I

1. *Ithycyphus blanci* Domergue, 1988. Bull. Mus. Natl. Hist. Nat. (4) 10A(1): 138–139, figs. 2a–f.

Type: Holotype, MNHN 1977.819 (formerly CAD 1112/S), a 610 mm male (C.P. Blanc, 15 Dec. 1972).
Type locality: "Massif montagneux de Marojezy, vers 300 m d'altitude, la région malgache" [= Antsiranana Province, Madagascar].
Distribution: Extreme N Madagascar (E Antsiranana), 300 m. Known only from type locality.

2. *Ithycyphus goudotii* (Schlegel, 1837). Essai Phys. Serp. 1: 152, 2: 187–188. (*Herpetodryas goudotii*)

Synonyms: *Ithycyphus caudolineatus* A.C.L.G. Günther, 1873a, and *Stenophis longicaudus* Boettger, 1913.
Type: Holotype, RMNH 690, a 650 mm specimen (J.P. Goudot [Barberie Voy.], 1827–1836).
Type locality: "Madagascar." Emended to the vicinity of Tamatave, Toamasina Prov., Madagascar *fide* Domergue (1986: 415).
Distribution: Eastern Madagascar (Fianarantsoa, Toamasina, Sainte Marie Is.), 10–270 m.
Sources: Guibé, 1958 and Henkel & Schmidt, 2000.
Remarks: MNHN 3855 is possibly a type *fide* Domergue (1986: 412).

3. *Ithycyphus miniatus* (Schlegel, 1837). Essai Phys. Serp. 1: 146–147, 2: 148–149. (*Coluber miniatus*)

Type: Holotype, MNHN 3852, a 1325–1340+ mm male (R.-P. Lesson & P. Garnot [Coquille Voy.], Aug. 1822–March 1825).
Type locality: "'île de France" [= Mauritius] (in error). Corrected to Madagascar *fide* Duméril, Bibron & Duméril (1854b: 1122).

Distribution: Western and NW Madagascar (W Antsiranana, NW Mahajanga, NW Toliara, Nosy Be Is.), 20–700 m.

Sources: Jourdran, 1904, Guibé, 1958 and A. Mori et al., 2006.

Remarks: Record from Mayotte, Comoro Is. rejected *fide* Blanc (1972: 516) and Domergue (1986: 419).

4. *Ithycyphus oursi* Domergue, 1986. Bull. Mus. Natl. Hist. Nat. (4) 8A(2): 424–425, figs. 6a–c, 7a–b.

Type: Holotype, MNHN 1982.1225 (formerly CAD 143/S), a 1535 mm female (J. de Saint-Ours, 1 Feb. 1961).

Type locality: "Entrée est de la piste de Soalara, un peu au sud du village d'Ambatry (Betioky) (province de Tuléar), région malgache; altitude 300 m" [= SW Toliara Prov., SW Madagascar].

Distribution: Southern Madagascar (Toliara), NSL–680 m.

5. *Ithycyphus perineti* Domergue, 1986. Bull. Mus. Natl. Hist. Nat. (4) 8A(2): 419–421, figs. 4, 5a–c.

Type: Holotype, MNHN 1982.1226 (formerly CAD 454/S), a 1200 mm female (J. Thiel, Jan. 1964).

Type locality: "Analamazotra (Périnet), aux environs des bassins de pisciculture. Altitude 900 m."

Distribution: Northern and E Madagascar (Antsiranana, Toamasina, Nosy Be Is.), 20–900 m.

Source: Henkel & Schmidt, 2000.

K

†*KELYOPHIS* LaDuke, Krause, Scanlon & Kley, 2010 (†Nigerophiidae)

Type species: †*Kelyophis hechti* LaDuke, Krause, Scanlon & Kley, 2010.
Distribution: Upper Cretaceous of Madagascar.
Source: LaDuke et al., 2010.

1. †*Kelyophis hechti* LaDuke, Krause, Scanlon & Kley, 2010. J. Vert. Paleont. 30(1): 126–128, fig. 10.

Type: Holotype, UA 9682, one middle trunk vertebra (Mahajanga Basin field party, 1993–2007).
Type locality: "MAD93-01, Berivotra Study Area, Mahajanga Basin, northwestern Madagascar; Upper Cretaceous (Maastrichtian) Maevarano Formation."
Distribution: Upper Cretaceous (Maastrichtian: 65.5–72.1 mya) of Madagascar.

KERILIA Gray, 1849a (Elapidae)

Type species: *Kerilia jerdonii* Gray, 1849a.
Distribution: Marine waters of SE Asia and Indonesia, including Andaman Sea, Bay of Bengal, Gulf of Thailand, Pattani Bay, Singgora Sea, South China Sea and Strait of Malacca.
Sources: Wall, 1921g, M.A. Smith, 1926a, 1943, Bourret, 1934g, 1936b, Deraniyagala, 1955, E.H. Taylor, 1965, Voris, 1977, B.L. Lim, 1982, A.R. Rasmussen & Anderson, 1990, Stuebing, 1991, Zhao & Adler, 993, A.R. Rasmussen, 1997, David & Ineich, 1999, Whitaker & Captain, 2004, I. Das, 2007b, 2010, 2012, V.S. Nguyen et al., 2009, Xiang & Li, 2009, Zaher et al., 2009, A.R. Rasmussen et al., 2011 and Hoser, 2012e.
Remarks: A synonym of *Hydrophis fide* Sanders et al., 2013.

1. *Kerilia jerdonii* Gray, 1849a. Cat. Snakes Brit. Mus.: 57.

Synonyms: *Eryx shiddil* Wagler, 1825 (*nomen ineditum*), *Hydrus schiddil* F. Boie, 1827 (*nomen nudum*), *Hydrus cantori* Jerdon, 1854, *Kerilia jerdoni siamensis* M.A. Smith, 1926a, and *Hydrophis jerdone* – Culotta & Pickwell, 1993 *(nomen incorrectum)*.
Type: Holotype, BMNH 1946.1.10.11 (formerly BMNH III.8.1.a), an adult male (T.C. Jerdon, 1836–1852).
Type locality: "Madras" [= Chennai, NE Tamil Nadu State, SE India, 13°04'N, 80°15'E, elevation 10 m].
Distribution: Southeastern Asia and Greater Sundas. Southern India (Andhra Pradesh, Gujarat, Kerala, Maharashtra, Orissa, Tamil Nadu, West Bengal), Sri Lanka (Eastern, Western), Myanmar (Mergui Arch.), S Thailand (Pattani, Phuket, Songkhla, Phuket Is.), S Vietnam (Bac Lieu, Ba Ria-Vung Tau, Ben Tre, Binh Thuan, Ca Mau, Ho Chi Minh City, Soc Trang, Tien Giang, Tra Vinh), S China, Taiwan, Singapore, West Malaysia (Penang Is.), East Malaysia (Sabah), Brunei and W Indonesia (Kalimantan, Sumatra).

KOLPOPHIS M.A. Smith, 1926a (Elapidae)

Synonym: *Kolphophis* – Dowling *in* Dowling & Duellman, 1978 *(nomen incorrectum)*.
Type species: *Distira annandalei* Laidlaw, 1902.
Distribution: Marine waters of Southeast Asia and Indonesia, including Gulf of Thailand, Java Sea, Pattani Bay, Singgora Sea and South China Sea.
Sources: Rooij, 1917, M.A. Smith, 1926a, 1943, Bourret, 1934g, 1936b, E.H. Taylor, 1965, McDowell, 1972a, Voris, 1977, B.L. Lim, 1982, I. Das, 1993, A.R. Rasmussen, 1997, David & Ineich, 1999, I. Das, 2007b, 2010, 2012, V.S. Nguyen et al., 2009, A.R. Rasmussen et al., 2011 and Hoser, 2012e.
Remarks: A synonym of *Lapemis fide* McDowell *in* Golay et al. (1993: 243).

1. *Kolpophis annandalei* (Laidlaw, 1902). Proc. Zool. Soc. London (1901) 71(2): 579–580, pl. 35, fig. 1. (*Distira annandalei*)

Synonyms: *Kolpophis annandali* – J.H. Brown, 1973 *(nomen incorrectum)*, *Kolpopis annandalei* – G. Haas, 1973 *(nomen incorrectum)*, and *Kolpophis annanadalei* – Tu, 1974 *(nomen incorrectum)*.
Type: Holotype, BMNH 1946.1.17.56 (formerly BMNH 1926.10.18.1), a 500 mm female (N. Annandale, 1899–1900).
Type locality: "Patani, Malay Peninsula" [= Pattani, Pattani Prov., S Thailand, 6°54'N, 101°15'E]. Specified as Patani Bay *fide* M.A. Smith (1926a: 107).
Distribution: Southeast Asia and Greater Sundas. Southern Thailand (Songkhla, Pattani), Cambodia, S Vietnam (from Binh Thuan to Ba Ria-Vung Tau), West

Malaysia, Singapore, Brunei and W Indonesia (Java, Kalimantan).

Remarks: Possibly occurs in Sumatra *fide* I. Das (2012: 149).

†*KREBSOPHIS* Rage & Werner, 1999 (†Russellophiidae)

Type species: †*Krebsophis thobanus* Rage & Werner, 1999.

Distribution: Middle Cretaceous of Sudan.

Source: Rage & Werner, 1999.

1. †*Krebsophis thobanus* Rage & Werner, 1999. Palaeont. Afr. 35: 100–102, figs. 16, 17 (paratype).

Type: Holotype, TUB-SFB69 Vb-681, one middle trunk vertebra.

Type locality: "Wadi Abu Hashim, N Sudan; Wadi Milk Formation, Mid-Cretaceous (Cenomanian)" [= Ash Shamaliyah Prov., N Sudan, ca. 17°09'N, 31°01'E].

Distribution: Middle Cretaceous (Cenomanian: 93.9–100.5 mya) of Sudan. Known only from type locality.

L

LACHESIS Daudin, 1803b (Viperidae)

Synonyms: *Trigonocephalus* Oppel, 1811a, *Trigalus* Rafinesque-Schmaltz, 1815 *(nomen substitutum), Cophias* Merrem, 1820, *Lachensis* – Gray, 1825 *(nomen incorrectum), Lestes* Gistel, 1848 *(nomen nudum), Lachoesis* – Liais, 1872 *(nomen incorrectum), Trigonophalus* – Palacky, 1898 *(nomen incorrectum), Lanchesis* – F. Werner, 1909b *(nomen incorrectum), Lachess* – Amaral, 1930f *(nomen incorrectum), Lachesus* – L. Ribeiro, 1939 *(nomen incorrectum), Lahcesis* – Anon., 1979 *(nomen incorrectum),* and *Lachesi* – Hoge & Romano-Hoge, 1981b *(nomen incorrectum).*

Type species: *Crotalus mutus* Linnaeus, 1766.

Distribution: Lower Central America and N South America.

Sources: Hoge, 1966a, J.A. Peters & Orejas-Miranda, 1970, Abuys, 1988a, Campbell & Lamar, 1989, 2004, Maes, 1989, Cullings et al., 1997, David & Ineich, 1999, McDiarmid et al., 1999, Parkinson, 1999, Ripa, 1999, Castoe & Parkinson, 2006 and Hoser, 2012d.

1. *Lachesis acrochorda* (García, 1896). Ofidios Venen. Cauca: 23–25, pl. 4. (*Bothrops acrochordus*) (*nomen corrigendum*)

Synonyms: *Botrops arochordus* García, 1896 *(nomen incorrigendum), Bothrops verrucosus* Posada-Arango, 1909, and *Lachesis calcaratus* Posada-Arango, 1909.

Type: Holotype, not designated, location unknown.

Type locality: "las selvas del Chocó, á orillas de los ríos Atrato, San Juan, Dagua y Telembí y … el camino de Buenaventura" [= Chocó forests, Atrato, San Juan, Dagua and Telembí riverbanks, and the Buenaventura highway, Chocó, Valle del Cauca, Cauca and Nariño Dept., W Colombia].

Distribution: Panama and NW South America. Eastern Panama (Darién), Colombia (Antioquia, Atlántico, Bolívar, Caldas, Cauca, Chocó, Córdoba, Cundinamarca, Nariño, Santander, Sucre, Tolima, Valle de Cauca) and Ecuador (Esmeraldas, Manabi), NSL–820 m.

Sources: Nicéforo-Maria, 1929, Pérez-Santos, 1986, Pérez-Santos & Moreno, 1988, 1991, Ripa, 2004 and Jaramillo et al., 2010.

2. *Lachesis melanocephala* Solórzano & Cedras, 1986. J. Herp. 20(3): 463–464, fig. 1. (*Lachesis muta melanocephala*)

Type: Holotype, MICP 301, a 2150 mm male (A. Solórzano & L. Cerdas, 31 May 1984).

Type locality: "9 km northern of Ciudad Neily in south-eastern Provincia de Puntarenas, Costa Rica."

Distribution: Southwestern Costa Rica (Puntarenas, San José) and W Panama (Chiriquí), NSL–1600 m.

Sources: E.H. Taylor, 1951, Ripa, 2001, Savage, 2002, J.C. Murphy & Schlager, 2003, Solórzano, 2004, Dwyer & Perez, 2009 and Jaramillo et al., 2010.

3. *Lachesis muta* (Linnaeus, 1766). Syst. Nat., ed. 12, 1: 373. (*Crotalus mutus*)

Synonyms: *Coluber crotalinus* Gmelin, 1789, *Scytale ammodytes* Latreille *in* Sonnini & Latreille, 1801a, *Scytale catenatus* Latreille *in* Sonnini & Latreille, 1801a, *Coluber alecto* G. Shaw, 1802, *Lachesis ater* Daudin, 1803b, *Scytale catenata* – Schinz, 1822, *Bothrops surucucu* Wagler *in* Spix, 1824, *Lachesis rhombeata* Wied-Neuwied, 1824a, *Trigonocephalus brasiliensis* Liais, 1872, *Trigonocephalus rhumbeatus* Liais, 1872 *(nomen emendatum), Lachoesis rhumbeata* Liais, 1872 *(nomen emendatum),* and *Lachesis muta noctivaga* Hoge, 1966a.

Type: Holotype, not designated, location unknown.

Type locality: "Surinamei" [= Suriname, or more likely the Guianas].

Distribution: Northern South America. Colombia (Amazonas, Casanare, Meta, Vaupés), E Venezuela (Amazonas, Anzoátegui, Bolívar, Delta Amacuro, Monagas, Sucre, Jotacuay Is.), Trinidad and Tobago (Trinidad), Guyana (Cuyuni-Mazaruni, Essequibo Islands-West Demerara, Potaro-Siparuni, Upper Demerara-Berbice, Upper Takutu-Upper Essequibo), Suriname (Brokopondo, Para, Saramacca), French Guiana (Cayenne, Saint-Laurent-du-Maroni), Brazil (Acre, Amapa, Amazonas, Bahia, Ceará, Espírito Santo, Goias, Maranhão, Mato Grosso, Minas Gerais, Pará, Paraíba, Pernambuco, Rio de Janeiro, Rio Grande do Norte, Rondônia, São Paulo), Ecuador (Orellana, Morona-Santiago, Napo, Pichincha), E Peru (Loreto, Madre de Dios) and NE Bolivia (Beni, Cochabamba, La Paz, Pando, Santa Cruz), 100–1000 m.

Sources: Beebe, 1946, Roze, 1966a, Cunha & Nascimento, 1975c, 1978, 1993, Emsley, 1977, Dixon & Soini, 1986, Chippaux, 1987, Ripa, 1994, Visinoni, 1994, Gorzula & Señaris, 1998, Starace, 1998, Borges-Nojosa &

Lima-Verde, 1999, Freitas, 1999, D.S. Fernandes et al., 2004, Duellman, 2005, Harvey et al., 2005, Navarrete et al., 2009 and C.J. Cole et al., 2013.

Remarks: Holotype possibly in NHR, a head in rather bad condition which does not belong to any *Crotalus* and possibly it is the head of a *Lachesis mutus fide* Andersson (1899: 27).

4. *Lachesis stenophrys* Cope, 1875a. J. Acad. Nat. Sci. Philadelphia (1876) (2) 8(2): 152.

Type: Holotype, USNM 32479, a 495 mm male (W.M. Gabb, 1873–1875).

Type locality: "Sipurio, Costa Rica" [= Sipurio, SE Limón Prov., SE Costa Rica, 9°32'N, 82°55'W, elevation 200 m].

Distribution: Lower Central America. Nicaragua (Chontales), N Costa Rica (Alajuela, Heredia, Limón) and N Panama (Bocas del Toro, Canal Zone, Chiriquí, Colón, Panamá, Veraguas, Barro Colorado Is.), NSL–1000 m.

Sources: Vial & Jiménez-Porras, 1967, Martínez & Bolaños, 1982, Villa, 1984, Solórzano, 1887 et al., Villa et al., 1988, Ripa, 1994, Zamudio & Greene, 1997 and Savage, 2002.

L

LAMPROPELTIS Fitzinger, 1843 (Colubridae)

Synonyms: *Sphenophis* Fitzinger, 1843, *Ophibolus* Baird & Girard, 1853, *Osceola* Baird & Girard, 1853, *Ophibulus* – Westphal-Castelnau, 1870 *(nomen incorrectum), Bellophis* Lockington, 1877, *Stilosoma* A. Brown, 1890, *Opzibolus* – Velasco, 1890a *(nomen incorrectum), Ophilobus* – Velasco, 1890a *(nomen incorrectum), Ophylobus* – Velasco, 1891 *(nomen incorrectum), Stylosoma* Cope, 1893a *(nomen emendatum), Oreophis* Dugès, 1897a, *Stilosema*– Palacky, 1898 *(nomen incorrectum), Stylophis* Berg, 1901 *(nomen substitutum), Ophobolus* – Speck, 1923 *(nomen incorrectum), Triaenopholis* F. Werner, 1924a, *Styliphis* – Briceño-Rossi, 1934 *(nomen incorrectum), Lamtropeltis* – Conant, 1937b *(nomen incorrectum), Ophiobolus* – Nicoli, 1963 *(nomen incorrectum), Lampropelitis* – T.H. Bullock & Barrett, 1968 *(nomen incorrectum), Lamoropeltis* – Kashyap & Sohoni, 1973 *(nomen incorrectum), Lampropeltus* – McDermid, Board & Agar, 1977 *(nomen incorrectum), Lamplopeltis* – Anon., 1978 *(nomen incorrectum), Campropeltis* – Gruber & Pace, 1978 *Campilodon* Jan, 1861c *(nomen emendatum),* and *Eksteinus* Hoser, 2012z *(nomen illegitimum).*

Type species: *Coluber getulus* Linnaeus, 1766.

Distribution: North America and Latin America.

Fossil records: Upper Miocene to upper Pleistocene of USA and S-cen. Canada. Fossils unassigned to species include middle Pliocene (Blancan V) of USA (Idaho), and upper Pleistocene (Rancholabrean II) of USA (Utah).

Sources: Blanchard, 1920b, 1921, Blaney, 1973, Dowling et al., 1983, Stebbins, 1985, Markel, 1990, Conant & Collins, 1991, 1998, Keogh, 1996, Rodríguez-Robles & Jesús-Escobar, 1999, Holman, 2000a, Bryson et al., 2005, 2007, Applegate, 2007, Pyron & Burbrink, 2009a and Hoser, 2012z.

Remarks: *Stilosoma* congeneric with *Lampropeltis fide* Rodríguez-Robles & Jesús-Escobar, 1999.

1. *Lampropeltis alterna* (A.E. Brown, 1902c). Proc. Acad. Nat. Sci. Philadelphia (1901) 53(1): 612–613, pl. 34. (*Ophibolus alternus*)

Synonym: *Lampropeltis blairi* Flury, 1950.

Type: Holotype, ANSP 14977, a 710 mm female (E. Meyenberg, 1901–1902).

Type locality: "Davis Mountains, Jeff Davis county, Texas" [USA]. Restricted to vicinity of Madera Canyon, Little Aguja Canyon, and Big Aguja Canyon *fide* D. Rhoads & Salmon (2012: 271).

Distribution: Extreme S USA (SE New Mexico, SW Texas) and N Mexico (ext. E Chihuahua, S Coahuila, NE Durango, ext. W Nuevo León), 400–2220 m.

Fossil records: Upper Pleistocene (Rancholabrean II) of USA (Texas).

Sources: Webb, 1961, Gehlbach & Baker, 1962, Gehlbach & McCoy, 1965, Gehlbach, 1967, Tanzer, 1970, Worthington and Arvizo, 1974, Miller, 1978, Garstka, 1982, Cranston, 1991, Degenhardt et al.,1996, Hilken & Schlepper, 1998, Bartlett & Tennant, 2000, Dixon & Werler, 2000, Salmon et al., 2004, Merker & Merker, 2005 and D. Rhoads & Salmon, 2012.

Remarks: Probably a subspecies of *L. mexicana* S.W. Garman *fide* Bryson et al. (2007: 682).

2. *Lampropeltis californiae* (Blainville, 1835). Nouv. Ann. Mus. Hist. Nat. Paris (3) 4(3): 292, pl. 27, figs. 1, 1a–b. (*Coluber [Ophis]californiae*)

Synonyms: *Coronella balteata* Hallowell, 1853a, *Ophibolus boylii* Baird & Girard, 1853, *Lampropeltis boylii conjuncta* Cope, 1861e, *Coronella pseudogetulus* Jan, 1863a, *Liophis poecilogyrus californica* Jan, 1863b, *Ophibolus getulus eiseni* Yarrow, 1882, *Lampropeltis nitida* Van Denburgh, 1895a, *Lampropeltis getulus yumensis* Blanchard, 1919, *Lampropeltis catalinensis* Van Denburgh & Slevin, 1921b, and, *Lampropeltis getulus nigritus* Zweifel & Norris, 1955.

Type: Holotype, not designated, a 660 mm specimen (P.E. Botta [Le Héros Exped.], 1827–1829), location unknown.

Type locality: "Californie" [= Alta California or California, USA]. Restricted to vicinity of Fresno, California *fide* K.P. Schmidt (1953: 207).

Distribution: Southwestern USA (W Arizona, California, S Colorado, E and S Nevada, S Oregon, ext. S Utah) and

NW Mexico (Baja California Norte, Baja California Sur, Angel de la Catalina, Cerralvo, Montserrat, Salsipuedes, San Pedro Martir, San Lorenzo Norte, Santa Catalina and Tortuga Is.], N Sinaloa, Sonora), NSL–1830 m.

Fossil records: Upper Miocene (Hemphillian) of USA (Arizona), and upper Pleistocene (Rancholabrean II) of USA (Arizona, California, Nevada).

Sources: Klauber, 1939e, Brattstrom, 1955b–c, Blaney, 1977, 1979a, Zweifel, 1980, 1981a–b, Zweifel & Dessauer, 1983, R.W. Murphy & Ottley, 1984, Briggs & McKeown, 1994, Parmley, 1994, H.A. Brown et al., 1995, Cranston & Loza, 1995, P.R. Brown, 1997, Grismer, 1999, 2002, Grismer & Hollingsworth, 2000, Holman, 2000a, Parmley & Peck, 2002, Hubbs, 2009 and Pyron & Burbrink, 2009b–c.

Remarks: *Lampropeltis catalinensis* Van Denburgh & Slevin a valid species *fide* Grismer (1999: 459).

3. *Lampropeltis calligaster* (Harlan, 1827). J. Acad. Nat. Sci. Philadelphia (1826–1827) 5(2): 359. (*Coluber calligaster*)

Synonyms: *Coluber rhombo-maculatus* Holbrook, 1840 *(nomen incorrigendum), Coluber rhombomaculatus* – Holbrook, 1840 *(nomen corrigendum), Ophibolus evansii* Kennicott *in* Baird, 1859, ? *Coronella tigrina* Jan, 1863b, and *Lampropeltis calligaster occipitolineata* Price, 1987b.

Types: Syntypes (?), not designated, PM, longest syntype 1220 mm, and a stuffed skin (T. Say, June 1819– Oct. 1820), lost *fide* Blanchard (1921: 116).

Type locality: "Missouri" [USA]. Restricted to vicinity of St. Louis, Missouri, USA *fide* K.P. Schmidt (1953a: 205).

Distribution: Eastern USA (Alabama, Arkansas, Florida, Georgia, Illinois, W Indiana, S Iowa, Kansas, W Kentucky, Louisiana, S Maryland, Mississippi, Missouri, ext. SE Nebraska, North Carolina, Oklahoma, South Carolina, Tennessee, E Texas, Virginia), NSL–915 m.

Fossil records: Upper Pliocene (Blancan V) of USA (Nebraska), lower/middle Pleistocene (Irvingtonian I) of USA (Kansas), middle/upper Pleistocene (Irvingtonian II) of USA (Arkansas), and upper Pleistocene (Rancholabrean I, II) of USA (Florida, Georgia, Kansas, Pennsylvania, Texas, Virginia).

Sources: P.W. Smith, 1961, Mount, 1975, Blaney, 1979b, Ashton & Ashton, 1988, T.R. Johnson, 1987, Ernst & Barbour, 1989, Palmer & Braswell, 1995, Tennant, 1997, Fitch, 1999 and Werler & Dixon, 2000.

Remarks: One syntype had eight entire subcaudals *fide* Harlan (1835: 122).

4. *Lampropeltis elapsoides* (Holbrook, 1838). No. Amer. Herp. 2: 123–125, pl. 28. (*Coluber elapsoides*) (*nomen corrigendum*)

Synonyms: *Coluber elapsoïdes* Holbrook, 1838 *(nomen incorrigendum), Coluber elapsoides* – Holbrook, 1838 *(nomen corrigendum), Ophibolus elapsoideus* S.W. Garman, 1894 *(nomen emendatum),* and *Lampropeltis elapsoides virginiana* Blanchard, 1920a.

Type: Holotype, not designated, location unknown.

Type locality: "South Carolina and Georgia" [SE USA]. Restricted to vicinity of Charleston, South Carolina, USA *fide* K.P. Schmidt (1953: 208).

Distribution: Southeastern USA (Alabama, Florida, Georgia, E Kentucky, SE Louisiana, Mississippi, North Carolina, South Carolina, Tennessee), NSL–450 m.

Fossil records: Lower Pleistocene (Irvingtonian I) of USA (Florida), and upper Pleistocene (Rancholabrean II) of USA (Florida, Georgia).

Sources: Blanchard, 1920a, K.L. Williams, 1974, 1988, 1994, T. Mills, 1995 and Palmer & Braswell, 1995.

Remarks: Previously a subspecies of *L. triangulum,* recognized as a valid species *fide* Pyron & Burbrink, 2009a.

5. *Lampropeltis extenuata* (A. Brown, 1890). Proc. Acad. Nat. Sci. Philadelphia 42: 199. (*Stilosoma extenuata*)

Synonyms: *Stilosoma extenuatum arenicola* Highton, 1956, and *Stilosoma extenuatum multistictum* Highton, 1956.

Type: Holotype, ANSP 3351, a 532 mm male (N.P. Fry, summer).

Type locality: "Lake Kerr, Florida" [Marion County, USA].

Distribution: Extreme SE USA (cen. and N Florida), NSL–30 m.

Fossil records: Upper Pleistocene (Rancholabrean II) of USA (Florida).

Sources: Highton, 1956, 1976, Ashton & Ashton, 1988, Ernst & Barbour, 1989, Dowling & Maxson, 1990, Conant & Collins, 1991, Rossi & Rossi, 1993, Keogh, 1996, Tennant, 1997 and Holman, 2000b.

Remarks: Head illustration of holotype in Highton (1956: fig. 2).

6. *Lampropeltis getula* (Linnaeus, 1766). Syst. Nat., ed. 12, 1: 382. (*Coluber getulus*)

Synonyms: *Coluber catena* Lacépède, 1789 *(nomen rejiciendum), Coluber labiatus* Kuhl, 1820, *Lampropeltis getulus brooksi* T. Barbour, 1919, *Lampropeltis getulus floridana* Blanchard, 1919, *Triaenopholis arenarius* F. Werner, 1924a, *Lampropeltis getulus sticticeps* Barbour & Engels, 1942, *Lampropeltis getulus goini* Neill & Allen, 1949, and *Lampropeltis getula meansi* Krysko & Judd, 2006.

Type: Holotype, not designated (A. Garden, 1760–1761), location unknown.

Type locality: "Carolina" [USA]. Restricted to vicinity of Charleston, South Carolina *fide* Klauber (1948: 6).

Distribution: Southeastern USA (SE Alabama, Delaware, Florida, Georgia, Maryland, S New Jersey, North

Carolina: Outer Banks; ext. SE Pennsylvania, South Carolina, Virginia, West Virginia), NSL–760 m.

Fossil records: Lower Pleistocene (Irvingtonian I) of USA (Florida), and upper Pleistocene (Rancholabrean II) of USA (Florida, Georgia, Virginia).

Sources: Lazell & Musick, 1973, Blaney, 1977, 1979a, Holman, 1979a, Christman, 1980, Ernst & Barbour, 1989, Cranston, 1994a, Palmer & Braswell, 1995, Tennant, 1997, D. Schmidt, 2004c, Krysko & Judd, 2006, Hubbs, 2009, Pyron & Burbrink, 2009b–c and Parmley & Hunter, 2010.

7. *Lampropeltis holbrooki* Stejneger, 1902b. Proc. U.S. Natl. Mus. 25(1282): 152. (*nomen substitutum*)

Synonym: *Coronella sayi* Holbrook, 1842a (*nomen praeoccupatum*).

Type: Lectotype, a 1200 mm specimen described and illustrated by Holbrook (1842a: 99–101, pl. 22), designated herein.

Type locality: "valley of the Mississippi…from Louisiana, high up Red river, from Missouri, from Arkansas, and…from Alabama, United States" *fide* Holbrook (1842a: 101) via lectotype selection. Restricted to central or northeastern Arkansas *fide* Blaney (1977: 78).

Distribution: Central USA (Arkansas, ext. SE Colorado, S Iowa, Kansas, Louisiana [W of Mississippi River], Missouri, SE Nebraska, E New Mexico, Oklahoma, E Texas), NSL–1200 m..

Fossil records: Upper Miocene (Clarendonian) of USA (Nebraska), upper Miocene (Hemphillian) of USA (Nebraska), middle Pliocene (Blancan II) of USA (Kansas), upper Pliocene (Blancan V) of USA (Kansas), middle/upper Pleistocene (Irvingtonian II) of USA (Arkansas), and upper Pleistocene (Rancholabrean I, II) of USA (Kansas, Missouri, Oklahoma).

Sources: Blaney, 1977, 1979a, Ernst & Barbour, 1989, Degenhardt et al., 1996, Fitch, 1999, Hammerson, 1999, Werler & Dixon, 2000, Hubbs, 2009, Pyron & Burbrink, 2009b–c and Parmley & Hunter, 2010.

Remarks: Original description in Holbrook (1842a: 99–101, pl. 22).

8. *Lampropeltis knoblochi* E.H. Taylor, 1940e. Copeia 1940(4): 253–255, figs. 1–2 (paratype).

Synonym: *Lampropeltis pyromelana woodini* W.W. Tanner, 1953.

Type: Holotype, FMNH 100013 (formerly EHT-HMS 23017 & EHT-CC x-1687), a 717 mm specimen (I. Knobloch, 1939)

Type locality: "Mojárachic, Chihuahua, Mexico."

Distribution: Southwestern USA (ext. SE Arizona, SW New Mexico) and NW Mexico (W Chihuahua, E Sonora), 840–2730 m.

Sources: W.W. Tanner, 1083, Hubbs, 2004, Lemos-Espinal et al., 2004a–b and Burbrink et al., 2011.

Remarks: Possibly occurs in N Durango *fide* W.W. Tanner (1985: 635) and Hubbs, 2004. Resurrected *fide* Burbrink et al., 2011.

9. *Lampropeltis mexicana* (S.W. Garman, 1884). Mem. Mus. Comp. Zool. (1883) 8(3): 66. (*Ophibolus triangulus mexicanus*)

Synonyms: *Coronella leonis* A.C.L.G. Günther, 1893 *in* 1885–1909, *Oreophis boulengeri* Dugès, 1897a, *Lampropeltis thayeri* Loveridge,1924a, and *Lampropeltis greeri* Webb, 1961.

Types: Syntypes (2), MCZ 4652, a 435 mm female, and MCZ 4653, a 774 mm female (E. Palmer, Aug. 1879).

Type locality: "Mexico, near San Luis Potosi."

Distribution: Northern Mexico (Aguascalientes, ext. SE Coahuila, S Durango, Guanajuato, Jalisco, Nayarit, S Nuevo León, San Luis Potosí, W Tamaulipas, S Zacatecas), 1035–2500 m.

Sources: Webb, 1961, Gehlbach & Baker, 1962, Gehlbach & McCoy, 1965, Gehlbach, 1967, Garstka, 1982, Parmley, 1990,Van Devender & Bradley, 1994, Hilken & Schlepper, 1998, Bryson et al., 2001, McCranie & Wilson, 2001b, Quintero-Díaz et al., 2001, Salmon et al., 2001, Hansen & Bryson, 2009 and Hansen et al., 2011.

10. *Lampropeltis nigra* (Yarrow, 1882). Proc. U.S. Natl. Mus. 5(299): 438. (*Ophibolus getulus niger*)

Types: Syntypes (2), USNM 12149a–b (R. Ridgway, 1881).

Type locality: "Wheatland, Indiana" [USA].

Distribution: Eastern USA (Alabama, Georgia, S Illinois, S Indiana, Kentucky, E Louisiana, Mississippi, S Ohio, Tennessee, W Virginia, West Virginia), NSL–750 m.

Fossil records: Upper Pleistocene (Rancholabrean II) of USA (Alabama, Pennsylvania, Tennessee).

Sources: P.W. Smith, 1961, Blaney, 1977, 1979a, Hubbs, 2009, Pyron & Burbrink, 2009b–c and Burbrink et al., 2011.

11. *Lampropeltis pyromelana* (Cope, 1867a). Proc. Acad. Nat. Sci. Philadelphia (1866) 18(4): 305. (*Ophibolus pyromelanus*)

Synonym: *Lampropeltis pyromelana infralabialis* W.W. Tanner, 1953.

Type: Lectotype, USNM 11421, a 632–775 mm male (E. Coues, 1865), designated by Cope (1900: 908).

Type locality: "Fort Whipple, Arizona" via lectotype selection.

Distribution: Southwestern USA (Arizona, ext. E Nevada, SW New Mexico, W Utah), 850–2375 m.

Fossil records: Upper Pleistocene (Rancholabrean II) of USA (Arizona, Nevada, New Mexico).

Sources: Van Denburgh, 1922, Fowlie, 1965, W.W. Tanner, 1953, Cranston, 1993b, Degenhardt et al., 1996,

Bartlett & Tennant, 2000, Hubbs, 2004 and Burbrink et al., 2011.

Remarks: Cochran (1961: 204) listed USNM 7845 as a syntype.

12. *Lampropeltis ruthveni* Blanchard, 1920a. Occ. Pap. Mus. Zool., Univ. Michigan (81): 8–10, pl. 1, fig. 2.

Type: Holotype, USNM 46558, skin and head of 745 mm specimen (E.W. Nelson, 2 Aug. 1892).

Type locality: "Patzcuaro, Michoacan, Mexico" (in error). Corrected to "Potrenaro, Michoacán, Mexico *fide* Garstka (1982: 30) and USNM catalogue entry.

Distribution: West-central Mexico (Hidalgo, Jalisco, Morelos, N Michoacán, Querétaro), 1515–2600 m.

Sources: Garstka, 1982, K.L. Williams, 1978, Hilken & Schlepper, 1998, Hubbs, 2004 and Roth-Monzón et al., 2011.

13. †*Lampropeltis similis* Holman, 1964. Copeia 1964(4): 635, fig. 3.

Type: Holotype, UNSM 61035, one trunk vertebra (J.A. Tihen & C.J. Chantell, 1962–1963).

Type locality: "Near the Norden-Johnstown Road, about 300 yards south of the bridge across the Niobrara River; on the Ed McCall Ranch, SE 1/4, SW 1/4 of Sect. 33, T 33 N, R 23 W, Brown County, Nebraska. Valentinian; Valentine formation, Norden Bridge Quarry." [= Barstovian].

Distribution: Middle Miocene (Barstovian: 13.6–16.3 mya) of Canada (Saskatchewan) and USA (Nebraska, South Dakota), middle/upper Miocene (Clarendonian: 10.3–13.6 mya) of USA (Kansas, South Dakota), and upper Miocene (Hemphillian: 4.9–10.3 mya) of USA (Texas).

Sources: Holman, 1970, 1973c, 1975, 1979a, M. Green & Holman, 1977, Rage, 1984b, Parmley, 1987 and Parmley & Hunter, 2010.

14. *Lampropeltis splendida* (Baird & Girard, 1853). Cat. No. Amer. Rept., Serp. 1: 83–84. (*Ophibolus splendidus*)

Synonym: †*Lampropeltis intermedius* Brattstrom, 1955b.

Type: Holotype, USNM 1726, head, tail and connecting skin of a 915 mm specimen (J.H. Clark [U.S.-Mex. Bound. Surv.], 1851, via J.D. Graham).

Type locality: "Sonora, Mexico" (in error). Corrected to south of the Gila River in Santa Cruz or Cochise Co., SE Arizona, USA *fide* Bogert (1939a: 198–199), Stejneger (1940: 204) and Stickel (1943: 120).

Distribution: Southwestern USA (E Arizona, S New Mexico, ext. SW Texas) and N Mexico (Chihuahua, Coahuila, Durango, San Luis Potosí, Zacatacas)., NSL–2135 m.

Fossil records: Middle Pliocene (Blancan III) of USA (Texas), lower/middle Pleistocene (Irvingtonian I) of USA (Arizona), middle/upper Pleistocene (Irvingtonian II) of USA (Texas), and upper Pleistocene (Rancholabrean I, II) of USA (New Mexico, Texas).

Sources: Stejneger, 1940, Blaney, 1977, 1979a, Zweifel, 1980, Zweifel & Dessauer, 1983, Degenhardt et al., 1996, Hubbs, 2009 and Pyron & Burbrink, 2009b–c.

Remarks: Yarrow (1882: 93) and Cope (1900: 919) listed Graham's Sonora type specimen as USNM 1709 (and USNM 1726 from Pecos River, Texas) but Cochran (1961: 204) listed the type as USNM 1726 (skin).

15. *Lampropeltis triangulum* (Lacépède, 1789). Hist. Nat. Serp. 2: 86, 331. (*Coluber triangulum*)

Synonyms: *Coluber americanus* Suckow, 1797, *Coluber eximius* Harlan, 1827, *Pseudoelaps Y* Berthold, 1842, *Ophibolus clericus* Baird & Girard, 1853, *Ophibolus gentilis* Baird & Girard, 1853, *Lampropeltis amaura* Cope, 1860e, *Lampropeltis micropholis* Cope, 1860e, *Lampropeltis polyzona* Cope, 1860e, *Lampropeltis annulata* Kennicott *in* Cope, 1860e, *Lampropeltis multistriata* Kennicott, 1860, *Coronella doliata conjuncta* Jan, 1863b, *Coronella doliata formosa* Jan, 1863b, *Coronella formosa abnorma* Bocourt, 1886 *in* A.H.A. Duméril, Bocourt & Mocquard, 1870–1909, *Coronella formosa anomala* Bocourt, 1886 *in* A.H.A. Duméril, Bocourt & Mocquard, 1870–1909, *Coronella formosa oligozona* Bocourt, 1886 *in* A.H.A. Duméril, Mocquard & Bocourt, 1870–1909, *Ophibolus doliatus collaris* Cope, 1889a, *Ophibolus doliatus occipitalis* Cope, 1889a, *Ophibolus doliatus parallelus* Cope, 1889a, *Ophibolus doliatus syspilus* Cope, 1889a, *Ophibolus doliatus temporalis* Cope, 1893b, *Lampropeltis micropholis arcifera* F. Werner, 1903a, *Lampropeltis pyrrhomelaena celaenops* Stejneger, 1902b, *Lampropeltis doliatus coccineus* Wright & Bishop, 1916, *Lampropeltis triangulum nelsoni* Blanchard, 1920a, *Lampropeltis polyzona blanchardi* L.C. Stuart, 1935c, *Lampropeltis triangulum schmidti* L.C. Stuart, 1935a, *Lampropeltis triangulum gaigae* [*sic*] Dunn, 1937a, *Lampropeltis doliata taylori* W.W. Tanner & Loomis, 1957, *Lampropeltis triangulum andesiana* K.L. Williams, 1978, *Lampropeltis triangulum conanti* K.L. Williams, 1978, *Lampropeltis triangulum hondurensis* K.L. Williams, 1978, *Lampropeltis triangulum sinaloae* K.L. Williams, 1978, *Lampropeltis triangulum smithi* K.L. Williams, 1978, *Lampropeltis triangulum stuarti* K.L. Williams, 1978, *Lampropeltis triangulum campbelli* Quinn, 1983, and *Lampropeltis triangulum dixoni* Quinn, 1983.

Type: Holotype, not designated, a 792 mm specimen (King of France's cabinet, Paris), lost *fide* K.L. Williams (1978: 75).

Type locality: "Amérique" [= America]. Restricted to vicinity of New York City, New York, USA *fide* K.P. Schmidt (1953a: 208).

Distribution: North America, Mesaoamerica and NW South America. Extreme SE Canada (ext. SE Ontario and ext. SW Quebec) USA (N Alabama, cen. and E Arizona, Arkansas, Colorado, Connecticut, Delaware, District of Columbia, ext. N Georgia, Illinois, Indiana, Iowa, Kansas, Kentucky, cen. & N Louisiana, SE Maine, Maryland, Massachusetts, S Michigan, SE Minnesota, N Mississippi, Missouri, SE Montana, Nebraska, NE Nevada, New Hampshire, New Jersey, New Mexico, New York, Ohio, Oklahoma, Pennsylvania, Rhode Island, South Dakota, Tennessee, Texas, Utah, Vermont, Virginia, West Virginia, S Wisconsin, Wyoming), Mexico (Campeche, Chiapas, ext. SW Chihuahua, NE Coahuila, Colima, Guanajuato, Guerrero, Hidalgo, Jalisco, México, Michoacán, Morelos, Nayarit, Nuevo León, Oaxaca, Puebla, Querétaro, Quintana Roo, San Luis Potosí, Sinaloa, ext. S Sonora, Tabasco, Tamaulipas, Veracruz, Yucatán, María Madre Is.), Guatemala (Alta Verapaz, Escuintla, Jutiapa, Retalhuleu, Petén), Belize (Cayo, Orange Walk, Stann Creek), El Salvador (La Libertad, La Paz, Morazán, San Miguel, San Salvador, Santa Ana), Honduras (Atlántida, Choluteca, Colón, Copán, Cortés, El Paraíso, Francisco Morazán, Gracias a Dios, Olancho, Santa Bárbara, Yoro), Nicaragua (Estelí, Granada, Managua, Matagalpa, Río San Juan, Zelaya), Costa Rica (Alajuela, Cartago, Guanacaste, Heredia, Limón, Puntarenas, San José), Panama (Canal Zone, Chiriquí, Darién, Panamá), W Colombia (Antioquia, Atlántico, Caldas, Caqueta, Cauca, Cesar, Chocó, Cundinamarca, Huila, Magdalena, Tolima, Valle), W Venezuela (Distrito Federal, Mérida, Táchira, Trujillo, Zulia) and W Ecuador, NSL–3330 m.

L

Fossil records: Upper Miocene (Hemphillian) of USA (Nebraska), lower Pliocene (Hemphillian) of USA (Oklahoma), middle Pliocene (Blancan II, III) of USA (Kansas, Nebraska, Texas), upper Pliocene (Blancan V) of USA (Kansas, Nebraska), lower Pleistocene (Irvingtonian I [?]) of USA (Pennsylvania), lower/middle Pleistocene (Irvingtonian I) of USA (Kansas), middle Pleistocene (Irvingtonian II) of USA (Nebraska), middle/upper Pleistocene (Irvingtonian II) of USA (Arkansas, Maryland, Texas, West Virginia), and upper Pleistocene (Rancholabrean I, II) of USA (Alabama, Kansas, Missouri, Nevada, Ohio, Oklahoma, Pennsylvania, Tennessee, Texas, Virginia, West Virginia, Wisconsin).

Sources: L.C. Stuart, 1935a, c, H.M. Smith, 1942e, W.W. Tanner & Loomis, 1957, P.W. Smith, 1961, K.L. Williams, 1978, 1988, 1994, Christman, 1980, J.C. Lee, 1980, 1996, 2000, Nussabaum et al., 1983, Quinn, 1983, L.D. Wilson & Meyer, 1985, Pérez-Santos, 1986, Ernst & Barbour, 1989, Parmley, 1994, Van Devender & Bradley, 1994, & Braswell, 1995, Degenhardt et al., 1996, Kardon & Holmback, 1996, Porras, 1996, Palmer J. Harding, 1997, Tennant, 1997, J.A. Campbell, 1998b, Barrio-Amorgós & Navarrete, 1999, Fitch, 1999, Klarsfeld, 1999, H.M. Smith et al., 1999, Werler & Dixon, 2000, M. Armstrong et al., 2001, Savage, 2002, Gumprecht, 2006a, 2007, J.C. Murphy & Schlager, 2003, Solórzano, 2004, Navarrete et al., 2009, McCranie, 2011a, Hamilton et al., 2012 and Rojas-Morales, 2012b.

Remarks: Official Specific Name no. 2186 *fide* Opinion 804 (ICZN, 1967). It is likely that populations of this wide-ranging species will be recognized as multiple species *fide* Bryson et al., 2007.

16. †*Lampropeltis vetustum* (Auffenberg, 1963). Tulane Stud. Zool. 10(3): 186–187, fig. 35 (lower). (†*Stilosoma vetustum*)

Type: Holotype, UF 6467, one trunk vertebra (W. Auffenberg, June 1954).

Type locality: "Haile VI, locality A (R 17 E, T 9 S, NW1/4 and SW1/4, sec. 24), 0.2 miles east of Route 235, near the town of Haile, Alachua County, Florida; elevation approximately 84 feet. Middle Pliocene, Hemphilian, Alachua fm."

Distribution: Upper Miocene (Hemphilian: 4.9–10.3 mya) of USA (Florida).

Sources: Holman, 1979a, 2000b and Rage, 1984b.

17. *Lampropeltis webbi* Bryson, Dixon & Lazcano-Villarreal, 2005. J. Herp. 39(2): 208–210, figs. 1–2.

Type: Holotype, UANL 5684, a 266 mm female (R. Bryson, D. Hartman & J. Banda, 30 June 2000).

Type locality: "4.0 km west of El Palmito on Hwy. 40, Municipio Concordia, Sinaloa, Mexico (23° 33' 14.2" N, 105° 50' 47.2" W), 2000 m elevation."

Distribution: Western Mexico (W Durango, S Sinaloa), 2000 m.

Remarks: Photographs of this species in Applegate (2007: 73, 75), who suggested that it possibly occurs in Nayarit.

18. *Lampropeltis zonata* (Blainville, 1835). Nouv. Ann. Mus. Hist. Nat. Paris (3) 4(3): 293–294. (*Coluber zonatus*)

Synonyms: *Bellophis zonatus* Lockington, 1877, *Ophibolus getulus multicinctus* Yarrow, 1882, *Coronella multifasciata* Bocourt, 1886 *in* A.H.A. Duméril, Bocourt & Mocquard, 1870–1909, *Lampropeltis agalma* Van Denburgh & Slevin, 1923, *Lampropeltis herrerae* Van Denburgh & Slevin, 1923, *Lampropeltis zonata parvirubra* Zweifel, 1952, and *Lampropeltis zonata pulchra* Zweifel, 1952.

Type: Holotype, MNHN, a 415 mm specimen (P.E. Botta, 1827–1829), lost *fide* Bocourt *in* Stejneger (1902b: 153).

Type locality: "Californie" [= Alta California or California, USA]. Restricted to northern California, USA *fide* K.P. Schmidt (1953a: 210).

Distribution: Western USA (California, SW Oregon, S Washington) and ext. NW Mexico (cen. Baja California Norte, Todos Santos Sur Is.), NSL–2750 m.
Fossil records: Upper Pleistocene (Rancholabrean II) of USA (Nevada).
Sources: Stejneger, 1902b, Blanchard, 1921a, Linsdale, 1932, Burt, 1936, J.L. Peters, 1938, and Klauber, 1943c, Zweifel, 1952, 1974, M.P. Hayes, 1975, McGurty, 1988, Cranston, 1994b, H.A. Brown et al., 1995, P.R Brown, 1997, Rodríguez-Robles et al., 1999, Bartlett & Tennant, 2000, Grismer, 2002, Hubbs, 2004, Mulks, 2004 and Bryson et al., 2007.
Remarks: *Lampropeltis herrerae* a valid species *fide* Grismer, 1999.

LAMPROPHIS Fitzinger, 1843 (Lamprophiidae)

Synonyms: *Iridophis* Fitzinger, 1861, and *Pachyophis* F. Werner, 1924a *(nomen praeoccupatum)*, and *Trachyophis* F. Werner, 1925 *(nomen substitutum)*.
Type species: *Coluber aurora* Linnaeus, 1758.
Distribution: Southern Africa.
Sources: Loveridge, 1957a, V.F.M. FitzSimons, 1962a, 1974, Roux-Estève & Guibé, 1965b, Dunger, 1971b, Broadley, 1983, Scheidt, 1996, Kelly et al., 2003, 2009, Marais, 2004, Vital et al., 2008 and Zaher et al., 2009.

1. *Lamprophis aurora* (Linnaeus, 1758). Syst. Nat., ed. 10, 1: 219. (*Coluber aurora*)

Type: Holotype, NHR Lin-20 (formerly MAFR), a 660 mm female (Mus. Drottn.).
Type locality: "America" (in error). Corrected to Africa *fide* V.F.M. FitzSimons (1962a: 112).
Distribution: Southern Africa. Eastern Botswana (Central), Lesotho, Swaziland and South Africa (Eastern Cape, KwaZulu-Natal, Free State, Gauteng, Limpopo, Mpumalanga, North-West, Western Cape), NSL–1350 m.
Sources: V.F.M. FitzSimons, 1974, Branch, 1988, Hermann, 1989b and Broadley & Blaylock, 2013.

2. *Lamprophis fiski* Boulenger, 1887i. Proc. Zool. Soc. London 55(3): 398, pl. 34.

Type: Holotype, BMNH 1946.1.12.41, a 315 mm specimen (W.G. Atherstone via G.H.R. Fisk).
Type locality: "Touw's River, S.-Africa" [= Touws River, Western Cape Prov., SW South Africa, 33°20'S, 20°02'E, elevation 810 m].
Distribution: Southwestern South Africa (Eastern Cape, Northern Cape, Western Cape), NSL–1275 m.
Sources: Visser, 1978, 1979a, Branch, 1988, Branch & Haagner, 1991 and Bartys et al., 2012..
Remarks: Possibly occurs in S Namibia *fide* M. Griffin (2003: 128).

3. *Lamprophis fuscus* Boulenger, 1893a. Cat. Snakes Brit. Mus. 1: 322, pl. 20, fig. 4.

Synonym: *Coluber simplex* Gravenhorst, 1807 *(nomen oblitum)*.
Type: Holotype, BMNH 1946.1.12.43, a 560 mm male.
Type locality: "Cape of Good Hope" [= Cape of Good Hope, Western Cape Prov., SW South Africa, 34°21'S, 18°28'E, elevation 20 m].
Distribution: Southeastern South Africa (Eastern Cape, Free State, Gauteng, S KwaZulu-Natal, S Mpumalanga, Western Cape), NSL–900 m.
Sources: Broadley, 1983, Branch, 1988 and R. Hermann, 1991.
Remarks: In accordance with Art. 23.9.2 of the Code (ICZN, 1999), *Lamprophis fuscus* Boulenger is designated a *nomen protectum* and *Coluber simplex* Gravenhorst a *nomen oblitum*.

LANGAHA Bonnaterre, 1790 (Pseudoxyrhophiidae)

Synonyms: *Langaha* Lacepède, 1789 *(nomen rejiciendum)*, *Langaya* G. Shaw, 1802 *(nomen emendatum)*, *Langaia* Cuvier, 1817 *(nomen emendatum)*, *Amphistrate* Goldfuss, 1820, *Langhaha* Ritgen, 1828 *(nomen emendatum)*, *Xiphorhynchus* Wagler, 1830, *Xiphorhina* Fitzinger, 1843, *Xiphorrhynchus* Hoffmann, 1890 *(nomen incorrectum)*, *Xyphorhynchus* Boulenger, 1896a *(nomen emendatum)*, *Xiphoryncus* – Brongersma, 1972 *(nomen incorrectum)*, and *Lanagaha* – Anon., 1976 *(nomen incorrectum)*.
Type species: *Langaha madagascariensis* Bonnaterre, 1790.
Distribution: Madagascar.
Sources: F. Werner, 1924c, Guibé, 1948, 1949b, 1958, Domergue, 1988, Nagy et al., 2003, Glaw & Vences, 2007, Kelly et al., 2009 and Zaher et al., 2009.

1. *Langaha alluaudi* Mocquard, 1901. Bull. Mus. Hist. Nat., Paris 7(6): 253.

Type: Holotype, MNHN 1901.193, a female (C. Alluaud, 1892–1893).
Type locality: "Andrahomana, sud de Madagascar" [= Tanjon'Andavaka, SE Toliara Prov., S Madagascar, 25°12'S, 46°38'E, elevation NSL].
Distribution: Southern Madagascar (Toliara), NSL–135 m.
Source: Brongersma, 1972.
Remarks: Original description reprinted in Mocquard (1902a: 12–13, pl. 1, fig. 4).

2. *Langaha madagascariensis* Bonnaterre, 1790. Tabl. Ency. Méth., Ophiol.: 71, pl. 35, fig. 4.

Synonyms: *Langaya nasuta* Shaw & Nodder, 1790 *in* 1789–1813, *Amphisbaena langaha* J.G. Schneider,

L

1801, *Langaha cristagalli* A.-M.-C. Duméril, Bibron & Duméril, 1854b, *Langaha ensifera* A.-M.-C. Duméril, Bibron & Duméril, 1854b, and *Langaha intermedia* Boulenger, 1888a.

Type: Lectotype, an 867 mm specimen described and illustrated in Brugnière (1784: 132–133, pl. 2), designated herein.

Type locality: "ile de Madagascar" via lectotype selection.

Distribution: Madagascar (Antsiranana, Mahajanga, Toamasina, Toliara), NSL–540 m.

Sources: Brongersma, 1972, Lamar, 1977, Love, 1993 and Henkel & Schmidt, 2000.

Remarks: Original description based on Brugnière (1784: 132–133, pl. 2). Official Specific Name *fide* Opinion 1463 (ICZN, 1987).

3. *Langaha pseudoalluaudi* Domergue, 1988. Bull. Mus. Natl. Hist. Nat. (4) 10A(1): 142–143, figs. 4a–e.

Type: Holotype, MNHN 1986.1352 (formerly CAD 664/S), a 1290 mm female (P. Randrianarivelo, 10 Oct. 1966).

Type locality: "Région d'Ambilobe, nord de Madagascar" [= vicinity of Ambilobe (13°12'S, 49°03'E, elevation 35 m), N Antsiranana Prov., N Madagascar].

Distribution: Madagascar (N Antsiranana, Mahajanga, Toliara), 35–170 m.

L

LAPEMIS Gray, 1835 *in* Gray & Hardwicke, 1830–1835 (Elapidae)

Synonyms: *Enhydris* Merrem, 1820 *(nomen praeoccupatum), Lapemys* Agassiz, 1847 *(nomen emendatum), Pelamydoidis* Fitzinger, 1861, *Lampemis* – Westphal-Castelnau, 1870 *(nomen incorrectum), Lepemis* – Angel, 1950 *(nomen incorrectum),* and *Pelamydoides* Iskandar & Colijn, 2001 *(nomen nudum).*

Type species: *Hydrus curtus* G. Shaw, 1802.

Distribution: Marine waters of Indo-Australia, including Arabian Sea, Arafura Sea, Bay of Bengal, Coral Sea, Coromandel Coast, Gulf of Carpentaria, Gulf of Oman, Gulf of Papua, Gulf of Thailand, Indian Ocean, Java Sea, Kimanis Bay, Malabar Coast, Manila Bay, Marudu Bay, Persian Gulf, South China Sea, Strait of Hormuz, Strait of Madura, Strait of Malacca, Taiwan Strait, Taraba Bay, Timor Sea, Torres Strait and Yellow Sea.

Sources: Siebold, 1833, Stejneger, 1907a, Wall, 1909a, 1918e, 1921g, Rooij, 1917, E.H. Taylor, 1922a, 1965, M.A. Smith, 1926a, 1943, Maki, 1931, Bourret, 1934g, Kennedy, 1937, Volsoe, 1939, Bergman, 1949c, C. Haas, 1950, Minton, 1966, 1975, B.L. Lim & Balasingam, 1969, McDowell, 1972a, B.L. Lim & Sawai, 1975, Capocaccia, 1976, J. Romer, 1979a, Glodek & Voris, 1982, B.L. Lim, 1982, Whitaker et al., 1982, F. Parker, 1982, Joger, 1984, Storr et al., 1986, Tian et al., 1986, M.A.R. Khan, 1988, M.S. Khan, 2002, Gow, 1989, Hoser, 1989, 2012e, Gallagher, 1990, Gritis & Voris, 1990, Han et al., 1991, Ehmann, 1992, Leviton et al., 1992, Golay et al., 1993, Sanyal et al., 1993, Zhao & Adler, 1993, O'Shea, 1996, A.R. Rasmussen, 1997, David & Ineich, 1999, Ferner et al., 2001, Whitaker & Captain, 2004, N. Khaire, 2006, I. Das, 2007b, 2010, 2012, A.R. Rasmussen et al., 2007a, 2011, Wells, 2007, Xiang & Li, 2009, Zaher et al., 2009, and A. Silva et al., 2011.

Remarks: A synonym of *Hydrophis fide* Sanders et al., 2013.

1. *Lapemis curtus* (G. Shaw, 1802). Gen. Zool., Amph. 3(2): 562. (*Hydrus curtus*)

Synonyms: *Hydrophis flaviventris* Siebold, 1827, *Lapemis hardwickii* Gray, 1835 *in* Gray & Hardwicke, 1830–1835, *Hydrophis pelamidoides* Schlegel, 1837, *Lapemis loreatus* Gray, 1843, *Hydrus pelamoides* Hallowell, 1845 *(nomen emendatum), Hydrophis pelamiodes* – A.-M.-C. Duméril, 1853 *(nomen incorrectum), Hijdrophis pelamidioides* – Bleeker, 1856 *(nomen emendatum), Hydrophis pelamidoides annulata* J.G. Fischer, 1856a, *Hydrophis problematicus* Jan, 1859b, *Hydrophis propinquus* Jan, 1859b, *Hydrophis abbreviatus* Jan, 1863b, *Hydrophis brevis* Jan, 1863b, *Hydrophis loreata* – A.C.L.G. Günther, 1864a, *Hydrophis fayreriana* J. Anderson, 1871b, *Hydrophis fayeri* Fayer, 1871 *(nomen emendatum), Hydrophis pelamidoides unimaculatus* W.C.H. Peters *in* Martens, 1876, *Enhydris hardtwickei* – Pagenstecher, 1881 *(nomen incorrectum), Hydrophis hardwickei* Boettger, 1888c *(nomen emendatum), Hydrophis abreviatus* – Casto de Elera, 1895 *(nomen incorrectum), Lepemis hardwickii* – Angel, 1950 *(nomen incorrectum), Lapemis curtis hardwickei* – Hattori et al., 1974 *(nomen incorrectum), Lapemis hardwiskii* – Arseniev et al., 1981 *(nomen incorrectum), Hydrophis pelmoides* – M.S. Khan, 1982 *(nomen incorrectum),* and *Hydrophis abbreviataus* – Culotta & Pickwell, 1993 *(nomen incorrectum).*

Type: Holotype, BMNH 1946.1.17.59 (formerly BMNH III.2.1.a), a 300 mm female.

Type locality: "East-India." Emended to India *fide* Boulenger (1896a: 300).

Distribution: Indo-Australia. Bahrain, Iran (Hormozghan, Sistan va Baluchestan, Qeshm Is.), United Arab Emirates, Oman, Pakistan (Balochistan, Sindh), India (Andaman & Nicobars, Andhra Pradesh, Karnataka, Kerala, Maharashtra, Orissa, Tamil Nadu), Sri Lanka, Bangladesh, S Myanmar (Bassein, Mergui Arch.), China (Fujian, Guangdong, Guangxi, Hainan, Hong Kong, Shandong), Vietnam (Dong Nai, Thuân Hai), Cambodia (Koh Kong), Thailand (Chanthaburi, Chumphon, Nakhon Si Thammarat, Phet Buri, Samut Sakhon, Songkhla, Surat Thani, Trang), West Malaysia (Johore, Kelantan, Melaka, Pahang, Perak, Selangor, Terengganu, Langkawi Is.), Singapore, East Malaysia

(Sabah, Sarawak), Brunei, Indonesia (Papua, Flores, Java, Lingga Arch., Maluku, Sulawesi, Sumatra, Sumba), Papua New Guinea (Central, East Sepik, Gulf, Morobe, Western), Philippines (Luzon, Negros, Panay) and N Australia (Northern Territory, Queensland, Western Australia, Ashmore and Cartier Is.).

Remarks: Presence in Japan doubtful *fide* Stejneger (1907a: 436) and Taiwan doubtful *fide* Mao & Chen (1980: 51). *Lapemis hardwickii* a valid species *fide* Leviton et al. (2003: 436).

†*LAPPARENTOPHIS* Hoffstetter, 1960a (†Lapparentophiidae)

Type species: †*Lapparentophis defrennei* Hoffstetter, 1960a.

Distribution: Lower Cretaceous of Algeria.

Sources: Rage, 1984b and Cuny et al., 1990.

1. †*Lapparentophis defrennei* Hoffstetter, 1960a. Bull. Soc. Géol. France (1959) (7) 1(9): 899–900, fig. 2.

Type: Holotype, MNHN 1959.11A, one trunk vertebra (P. Defrenne, Dec. 1947).

Type locality: "le 'continental intercalaire' (Crétacé inférieur) des buttes d'In Akhamil (Sahara), 30 km au SSE d'Ohanet (N du Sahara central, bordure du Tinrhert) et environ 120 km au NW d'Edjeleh" [= lower Cretaceous of Ouargla Wilaya, Algeria].

Distribution: Lower Cretaceous (Albian: 100.5–113 mya) of Algeria. Known only from type locality.

LATICAUDA Laurenti, 1768 (Elapidae)

Synonyms: *Platurus* Latreille *in* Sonnini & Latreille, 1801a, *Platyurus* Ritgen, 1828 *(nomen emendatum),* and *Platura* Gray, 1840 *(nomen emendatum), Laticaude* – Chenu & Desmarest, 1857 *(nomen incorrectum), Laticaudatus* – Higgins, 1873 *(nomen incorrectum), Laticaudata* – Oshima, 1910 *(nomen incorrectum), Lalicauda* – Steindachner, 1913 *(nomen incorrectum), Colubrinus* Cyrén, 1934 *(lapsus calami), Laticouda* – Gasperetti, 1974 *(nomen incorrectum),* and *Latacauda* – Dowling *in* Dowling & Duellman, 1978 *(nomen incorrectum).*

Type species: *Coluber laticaudatus* Linnaeus, 1758.

Distribution: Marine waters and rocky shores of Sino-Australia, including Andaman Sea, Arafura Sea, Banda Sea, Bay of Bengal, Bismarck Sea, Bohol Straits, Coral Sea, East China Sea, Flores Sea, Gulf of Thailand, Java Sea, Lake Te-Nggano, Manila Bay, Maruda Bay, Melanesia, Moluccan Sea, Philippine Sea, Polynesia, Prony Bay, Seram Sea, South China Sea, Solomon Sea, South China Sea, South Pacific Ocean, Strait of Malacca, Sulawesi Sea, Sulu Sea, Taiwan Strait, Tasman Sea, Timor Sea, and West Pacific Ocean.

Sources: Stejneger, 1907a, Rooij, 1917, F. Werner, 1924c, Maki, 1931, C. Haas, 1950, Werler & Keegan, 1963, Halstead, 1970, Voris, 1977, Mao & Chen, 1980, Gorman, 1981, Kharin, 1984c, McCarthy, 1986, R.W. Murphy, 1988, Slowinski, 1989, Ehmann, 1992, Golay et al., 1993, Zhao & Adler, 1993, Greer, 1997, A.R. Rasmussen, 1997, Keogh, 1998, Keogh et al., 1998, David & Ineich, 1999, Scanlon & Lee, 2004, Kharin & Cheblukov, 2006a, Wells, 2007, Sanders et al., 2008, Zaher et al., 2009, Heatwole, 2010, Kharin et al., 2010 and Hoser, 2012e.

Remarks: The species *L. schistorhynchus* and *L. semifasciatus* have been transferred to *Pseudolaticauda*.

1. *Laticauda colubrina* (J.G. Schneider, 1799). Hist. Amph. 1: 238–240. (*Hydrus colubrinus*)

Synonyms: *Coluber platycaudatus* Oken, 1836 *(nomen substitutum),* and *Laticauda celubrina* – Deraniyagala, 1977 *(nomen incorrectum).*

Type: Holotype, ZMB 9078 (formerly MB), a juvenile male (Lampe coll.).

Type locality: Unknown. "Ostindisches Meer" [= East Indian Ocean] *fide* ZMB catalogue (Bauer, 1998: 139).

Distribution: Southeast Asia and Indo-Australia. Eastern India (Andaman & Nicobars: Andaman and Nicobar Is., ? West Bengal), Bangladesh, Myanmar (Tanintharyi, Ramri Is.), Thailand (Chumphon, Phang Nga, Phuket, Raja Is.), Vietnam, West Malaysia (Negri Sembilan, Pinang, Selangor), Singapore (Biola and Salo Is.), East Malaysia (Sabah, Sarawak, Kalampunian, Keraman, Layang-Layang, Layang-Layangan and Matunggung, Is.), Brunei, Indonesia (Papua, Ambon, Aru Is., Babi, Bali, Banda, Batu, Belitung, Enggano, Flores, Halmahera, Java. Jobil, Kai Is., Moluccas, Riau Arch., Roti, Simeulue, Sulawati, Sulawesi, Sumatra, Ternate, Timor, Weh, Wetar), Timor-Leste, Philippines (Bantayan, Bohol, Bubuan, Cebu, Gato, Jolo, Luzon, Negros, Palawan, Panay, Romblon, Samar, Tulian), Papua New Guinea (Central, Madang, Milne Bay, Morobe, Admiralty Is., Bismarck Arch., Bougainville, Karkar, Lion, Louisiade Arch., Manubada, Manus, New Britain, New Hanover, Pigeon, Saidor, Schouten and Woodlark Is.), Solomon Is. (New Georgia, Rennell, Ugi), Vanuatu (Aneitum, Efate, Tana), New Caledonia (Belep, Nouméa, Ouen, Pines, Poum), Loyalty Is. (Lifou, Maré, Ouvéa), E Australia (New South Wales, Queensland), New Zealand (Bay of Islands), S Japan (Ryukyus: Miyako and Yaeyaema Is.), S Taiwan (Pingtung, Lan Yu Is.), Palau, Carolines (Ponape), Samoa, Fiji (Aiwa, Fulanga, Kandavu, Kia, Komo, Lomaloma, Maldeula, Taveuni, Vanua Levu, Viti Levu, Vomo) and Tonga (Ha'apai, Kelefesia, Niuafo'ou, Tongatapu, Vava'u), NSL–10 m.

Sources: Rooij, 1922, E.H. Taylor, 1922a, 1965, Maki, 1931, Herre, 1942, M.A. Smith, 1943, Brongersma,

1956c, Volsøe, 1956, 1958, Gail & Rageau, 1958, Keegan, 1960, Kuntz, 1963, Saint-Girons, 1964, 1990, McCann, 1966, Cogger, 1975, B.L. Lim & Sawai, 1975, Capocaccia, 1976, Pernetta, 1977, M. McCoy, 1980, Gorman et al., 1981, Heatwole, 1981, Tweedie, 1983, Bosch, 1985, Ota et al., 1985, 1986, M. Mori, 1986, M.A.R. Khan, 1988, Bauer & Vindum, 1990, Engkamet et al., 1991, Stuebing, 1991, McCarthy, 1993, O'Shea, 1996, Shetty & Prasad, 1996, M.J. Cox et al., 1998, David & Ineich, 1999, Bauer & Sadlier, 2000, Ferner et al., 2001, Ineich & Laboute, 2002, Shine et al., 2002, J.C. Murphy & Schlager, 2003, Kraus and Allison, 2004, Whitaker & Captain, 2004, Heatwole et al., 2005, Kharin, 2006, I. Das, 2007b, 2010, 2012, Xiang & Li, 2009 and C. Sanchez et al., 2012.

Remarks: Photographs of holotype in Kharin & Czeblukov (2006: figs. 11a–b). Photograph of Linnaeus' type of *L. colubrina* (NHR 88) in Heatwole et al. (2005: pl. 1 [upper]) and Kharin & Czeblukov (2006: fig 9), the latter erroneously designating it as lectotype of *L. laticaudata*. Unconfirmed reports from Mexico, El Salvador and Nicaragua doubtful *fide* McCarthy (1986: 134).

2. *Laticauda crockeri* Slevin, 1934. Proc. California Acad. Sci. (4) 21(15): 186.

L

Synonym: *Laticauda laticaudata wolffi* Vølsoe, 1958.

Type: Holotype, CAS 72001, a 479 mm male (T. Crocker [Templeton Crocker Exped.], 8 June 1933).

Type locality: "Lake Tungano, situated about two miles inland from the east point of Rennell Island, Solomon Islands" [= Te-Nggano Lake, Rennell Is., S Solomons, 160°30'E, 11°45'S, elevation 20 m].

Distribution: Southern Solomons (Rennell Is.). Known only from type locality.

Sources: Volsøe, 1958a–b, M. McCoy, 1980 and Cogger et al., 1987.

3. *Laticauda frontalis* (De Vis, 1905). Ann. Queensland Mus. (6): 48. (*Platurus frontalis*)

Type: Holotype, QM J202, a 327 mm specimen.

Type locality: "New Guinea," (possibly in error *fide* Cogger & Heatwole, 2006: 248).

Distribution: Southwest Pacific. Eastern Papua New Guinea (New Britain), New Caledonia, Loyalty Is. and Vanuatu (Efate, Elephant, Espiritu Santo).

Sources: Shine et al., 2002, Heatwole et al., 2005 and Kharin et al., 2010.

Remarks: Photograph of holo type in Cogger & Heatwole (2006: fig. 1). A valid species *fide* Cogger & Heatwole (2006: 247).

4. *Laticauda guineai* Heatwole, Busack & Cogger, 2005. Herpetol. Monogr. (19): 97–98, pl. 1 (middle).

Type: Holotype, AMS 69514, a 910 mm male (F. Parker, 1 Feb.–22 Mar. 1977).

Type locality: "Bava Island, near Port Moresby, Papua New-Guinea."

Distribution: Southwest Pacific. Southeastern Papua New Guinea (Central, National Capital Province, Milne Bay, Bava, Lion, Manu, Manubada, Motupore and New Britain Is.), Loyalty Is. and Vanuatu.

5. *Laticauda laticaudata* (Linnaeus, 1758). Syst. Nat., ed. 10, 1: 222. (*Coluber laticaudatus*)

Synonyms: *Laticauda scutata* Laurenti, 1768 *(nomen substitutum), Platurus fasciatus* Latreille *in* Sonnini & Latreille, 1801b, *Platurus laurenti* Rafinesque-Schmaltz, 1817 (*nomen praeoccupatum*), *Aspisurus laticaudatus* – Gray *in* Grey, 1841b *(nomen incorrectum), Platurus fasciatus* Jan, 1859b *(nomen praeoccupatum), Platurus fischeri* Jan, 1859b, *Platurus affinis* J. Anderson, 1871a, *Platurus fisheri* – Tirant, 1884 *(nomen incorrectum), Platurus muelleri* Boulenger, 1896a, *Laticaudata laticaudata* – Oshima, 1910 *(nomen incorrectum), Laticauda laticauda* – Brehm, 1913 *(nomen incorrectum),* and *Laticauda laticoudata* – Khole, 1991 *(nomen incorrectum).*

Type: Lectotype, NHR Lin-87 (formerly MAFR), a 690 mm specimen (Mus. Drottn.), designated by Stejneger (1907a: 402).

Type locality: "Indiis" [= India] via lectotype selection.

Distribution: Southeast Asia and Indo-Australia. Eastern India (Andaman & Nicobars: Andaman Is., West Bengal), Sri Lanka, Bangladesh, Myanmar (Tanintharyi: Ramri Is.), Thailand (Chanthaburi), West Malaysia (Pinang, Selangor), Singapore, East Malaysia (Sabah, Kalampuian Damit and Si Amil Is.), S Japan (Ryukyus: Okinawa and Ishigakishima Is.), China (Fujian), Taiwan, Philippines (Bantayan, Gato, Luzon, Mindanao, Mindoro, Samar, Sulu Arch.: Jolo), Indonesia (Ambon, Aru Is.: Babi, Poeloe, Jobi; Babi, Buru, Flores, Halmahera, Java, Kai Is., Mansalar, Nias, Simeulue, N Sulawesi, Sumatra, Ternate, Timor, Weh), Papua New Guinea (Central, Milne Bay, Duke of York, Schouten, New Ireland, New Britain and Samarui Is.), E Australia (New South Wales, Queensland, Tasmania, Great Barrier Reef), Vanuatu, Etate Is., New Caladonia (Anse Vata, Goro, Grand Taenia, Nounéa, Petit Taenia, Signal), Fiji Is., Tonga Is. and Society Is.

Sources: W.C.H. Peters, 1877a, Rooij, 1922, E.H. Taylor, 1922a, 1965, M.A. Smith, 1943, Saint-Girons, 1964, 1990, Klemmer, 1967, Cogger, 1975, Bosch, 1985, Tian et al., 1986, Toriba & Yosida, 1986, M.A.R. Khan, 1988, Bauer & Vindum, 1990, Stuebing, 1991, Ehmann, 1992, O'Shea, 1996, Bauer & Sadlier, 2000, Ineich & Laboute, 2002, Xiang & Li, 2009, I. Das, 2010, 2012 and A.R. Rasmussen et al., 2011.

Remarks: Kharin & Czeblukov (2006a: 233) erroneously designated NHR 88 (via figs. 9a–b = *L. colubrina fide* Heatwole et al., 2005: pl. 1, upper fig.) as lectotype.

6. *Laticauda saintgironsi* Cogger & Heatwole, 2006. Rec. Aust. Mus. 58(2): 249–254, figs. 3a–b.

Type: Holotype, AMS 162999 (formerly AMS-FN 34602), an 815 mm male (H.G. Cogger & H.F. Heatwole, 14 Sept. 1993).
Type locality: "Porc-épic Island, Le Lagon, near Noumea, New Caledonia, 22°19'45.6"S 166°34'08.5E."
Distribution: Southwest Pacific. New Caledonia (Alter, Atire, Bayonnaise, Belep, Lange, Maître, Moro, Mouac, Nou, Noumea, Pins, Porc-Epic, Redika, Signal and Uatio Is.), Loyalty Is. (Lifou, Maré, Ouvéa), Caroline Is. and New Zealand.
Remarks: Original description reprinted in Heatwole et al. (2005: 6, 23, 98, 136, figs. 19, 23).

LEIOHETERODON A.-M.-C. Duméril, Bibron & Duméril, 1854a (Pseudoxyrhophiidae)

Synonyms: *Anomalodon* Jan, 1863a, *Lioheterodon* Boulenger, 1893a *(nomen emendatum), Licheterodon* – Maass-Berlin, 1933 *(nomen incorrectum),* and *Michaelnicholsus* Hoser, 2012t *(nomen illegitimum).*
Type species: *Heterodon madagascariensis* A.-M.-C. Duméril, Bibron & Duméril, 1854a.
Distribution: Madagascar.
Sources: Methuen & Hewitt, 1913, Angel, 1931, G.S. Myers, 1949, Guibé, 1958, Brygoo, 1983, Glaw & Vences, 1994, Nagy et al., 2003, Lawson et al., 2005, Kelly et al., 2009 and Hoser, 2012t.

1. *Leioheterodon geayi* Mocquard, 1905c. Bull. Mus. Hist. Nat., Paris 11(5): 288–289. (*Lioheterodon geayi*)

Synonym: *Lioheterodon voeltzkowi* Boettger, 1913.
Type: Holotype, MNHN 1905.137, a 930 mm male (F. Geay, 1904–1905).
Type locality: "de la plaine de Fiherena" [Toliara Prov., S Madagascar].
Distribution: Southwestern Madagascar (Toliara), NSL–290 m.

2. *Leioheterodon madagascariensis* (A.-M.-C. Duméril, Bibron & Duméril, 1854a). Erpét. Gén. 7(1): 776–777, pl. 69. (*Heterodon madagascariensis*)

Types: Syntypes (6), BMNH 1946.1.15.19, MCZ 2202 (formerly MNHN 87), a 1644 mm male, MNHN 7501, MNHN 7558–59, and MNHN 1991.1629, longest MNHN syntype 1570 mm specimen (C.A.C.J. Bernier, 1831–1846 & L.-H. Boivin, 1847–1852).
Type locality: "Madagascar."
Distribution: Madagascar (Antsiranana, Fíanarantsoa, Mahajanga, Toamasina, Toliara, Nosy Be and Sainte Marie Is.) and Comoro Is. (Grande Comore), 100–1200 m.
Sources: Desquez, 1884, Conant, 1938, Arnoult & Fourmanoir, 1967, J.A. Campbell & Murphy, 1977b, Raxworthy, 1988, Wallach, 1985, Meirte, 1993 and Henkel & Schmidt, 2000.

3. *Leioheterodon modestus* (A.C.L.G. Günther, 1863c). Ann. Mag. Nat. Hist. (3) 12(71): 356–357. (*Heterodon modestus*)

Type: Holotype, BMNH 1946.1.7.79 (formerly RMNH), a 940 mm specimen (1863).
Type locality: "Madagascar."
Distribution: Western Madagascar (Antsiranana, Atananarivo, Mahajanga, Toliara), 20–650 m.
Sources: Henkel & Schmidt, 2000 and A. Mori et al., 2006.

LEIOPYTHON Hubrecht, 1879a (Pythonidae)

Synonym: *Liopython* O'Shaughnessy, 1880 *(nomen emendatum).*
Type species: *Liasis albertisii* W.C.H. Peters & Doria, 1878.
Distribution: Austro-Papua.
Sources: Brongersma, 1953, 1956b, McDowell, 1975a, Whitaker et al., 1982, F. Parker, 1982, Kluge, 1993b, D.G. Barker & Barker, 1994, O'Shea, 1996, Walls, 1998b, McDiarmid et al., 1999, Cogger, 2000, Schleip, 2008, Schleip & O"Shea, 2010 and Hoser, 2012b.

1. *Leiopython albertisii* (W.C.H. Peters & Doria, 1878). Ann. Mus. Civ. Stor. Nat. Genova 13: 401–403, pl. 3, figs. 2. (*Liasis albertisii*)

Synonyms: *Leiopython gracilis* Hubrecht, 1879a, *Leiopython albertisii barkeri* Hoser, 2000b *(nomen incorrigendum), Leiopython barkerorum* – Wüster et al., 2001b *(nomen corrigendum),* and *Leiopython albertisii barkerorum* Hoser, 2009b & 2012b *(nomen illegitimum).*
Types: Syntypes (2), MSNG 29989 [Andai] (O. Beccari, 11 June 1875) and MSNG 29990 [Kapaor], a 2230 mm specimen (L.M. d'Albertis, April 1872).
Type locality: "prope Andai presso Dorei" and "Kapaor fra i Papua Onin in Nova Guinea boreali occidentali" [= Andai, near Dorei, N West Papua Prov., 0°55'S, 134°00'E, and Kapaor, Fak Fak District, Onin Peninsula, West Papua Prov., E Indonesia].
Distribution: New Guinea. Eastern Indonesia (West Papua, Biak, Salawati) and Papua New Guinea (Eastern Highlands, East Sepik, Enga, Madang, Morobe, Northern, Northern Highlands, Western Highlands, West Sepik, Bismarck Arch., d'Entrecasteaux Arch., Mussau and Normanby Is.), NSL–1600 m.

L

2. *Leiopython bennettorum* Hoser, 2000b. Ophidia Rev. (1): 19. (*Leiopython albertisii bennettorum*) (*nomen corrigendum*)

Synonyms: *Leiopython albertisii bennetti* Hoser, 2000b *(nomen incorrigendum),* and *Leiopython albertisii bennettorum* – Wüster *et al.*, 2001b *(nomen corrigendum).*
Type: Holotype, BPBM 5452, a 635 mm female (A.C. Ziegler, 13 June 1967).
Type locality: "near Wau, PNG, Lat: 7° 20' Long: 146° 40'" [= vicinity of Wau (7°20'S, 146°43'E, elevation 1140 m), Morobe Prov., Papua New Guinea].
Distribution: Papua New Guinea (Morobe), 1050–1400 m.
Remarks: A valid species *fide* by Schleip (2008: 658).

3. *Leiopython biakensis* Schleip, 2008. J. Herp. 42(4): 661–662.

Type: Holotype, RMNH 10193, a 1670 mm female (Fleet Air Arm Royal Netherlands Navy, 1952–1953).
Type locality: "Biak, Schouten Islands, Indonesia" [= Biak, Supiori Is., Papua Prov., E Indonesia, 1°11'S, 136°06'E, elevation 10 m].
Distribution: Eastern Indonesia (Biak), 10 m.

L

4. *Leiopython fredparkeri* Schleip, 2008. J. Herp. 42(4): 659–660, figs. 6e, g.

Type: Holotype, CAS 118906, a 1620 mm male (F. Parker, 7 July 1967).
Type locality: "near Wau, PNG" [= vicinity of Wau (7°20'S, 146°43'E, elevation 1140 m), Morobe Prov., Papua New Guinea].
Distribution: Papua New Guinea (Morobe), 1100–1500 m.

5. *Leiopython hoserae* Hoser, 2000b. Ophidia Rev. (1): 19.

Type: Holotype, AMNH 107150, a 2413 mm male (F. Parker, Aug. 1969).
Type locality: "Wipim (= Wipam), Western District, PNG, Lat. 8°40' Long. 142°55'."
Distribution: Austro-Papua. Eastern Indonesia (Papua, Aru Arch.), Papua New Guinea (Central, Chimbu, Eastern Highlands, Gulf, Southern Highlands, Western) and N Australia (Queensland: Torres Strait Is.), NSL–80 m.
Remarks: Redescribed by Schleip (2008: 656–657) who supported its recognition.

6. *Leiopython huonensis* Schleip, 2008. J. Herp. 42(4): 660–661, figs. 6i–j.

Type: Holotype, AMNH 95535, a 776 mm female (L. van Royen, 1964)
Type locality: "about 16 km west of Lae" [= 16 km W Lae (6°43'S, 146°59'E, elevation 35 m), Morobe Prov., Papua New Guinea].
Distribution: Eastern Papua New Guinea (Morobe), 35 m.

LEIOSELASMA Lacépède, 1804 (Elapidae)

Synonyms: *Lejoselasmus* Fischer von Waldheim, 1808 *(nomen emendatum), Enhydris* Wagler, 1830 *(nomen praeoccupatum), Leioselama* Bonaparte, 1831 *(nomen emendatum), Leyoselasma* – Cuvier & Voigt, 1832 *(nomen incorrectum), Lielasma* Agassiz, 1847 *(nomen emendatum), Noterophis* Gistel, 1848 *(nomen substitutum),* and *Lioselasma* Wall, 1921g *(nomen emendatum).*
Type species: *Hydrophis cyanocinctus* Daudin, 1803d.
Distribution: Marine waters and mangrove swamps including Arabian Sea, Arafura Sea, Astrolabé Bay, Bay of Bengal, Bight of Bangkok, Bismarck Sea, Coral Sea, Coromandel Coast, East China Sea, Geographe Bay, Gulf of Carpentaria, Gulf of Martaban, Gulf of Thailand, Hainan Strait, Half Moon Bay, Idzu Sea, N Indian Ocean, Java Sea, Karachi Sea, Kimanis Bay, King Sound, Malabar Coast, Manila Bay, Marudu Bay, Okinawa Sea, Persian Gulf, Philippine Sea, Safety Bay, Shark Bay, South China Sea, Strait of Hormuz, Strait of Madura, Strait of Malacca, Sulawesi Sea, Taiwan Strait, Taraba Bay, Timor Sea, Torres Strait and Yellow Sea.
Sources: Wall, 1909a, Rooij, 1917, M.A. Smith, 1926a, 1943, Halstead, 1970, McDowell, 1972a, Minton, 1975, Voris, 1977, Kharin, 1984b, 2005b, Golay et al., 1993, David & Ineich, 1999, Wells, 2007, I. Das, 2010, 2012 and Hoser, 2012e.
Remarks: Formerly in the genus *Hydrophis* and still considered so by some.

1. *Leioselasma coggeri* Kharin, 1984b. Zool. Zh. 63(10): 1538–1541, fig. 1b (upper right).

Type: Holotype, ZISP 19681, an 845 mm female (V.Z. Boldyrev, March 1975).
Type locality: "Port Suva, Fiji Islands."
Distribution: Indonesia and SW Pacific. Indonesia (N Sulawesi, Timor), Fiji Is., New Caledonia and Vanuatu.
Sources: Ineich & Rasmussen, 1997, Bauer & Sadlier, 2000, Ineich & Laboute, 2002 and Kharin, 2005a.

2. *Leioselasma cyanocincta* (Daudin, 1803d). Hist. Nat. Rept. 7: 383–385. (*Hydrophis cyanocinctus*)

Synonyms: *Leioselasma striata* Lacépède, 1804, *Hydrophis chittul* Rafinesque-Schmaltz, 1817, *Hydrophis sublaevis* Gray 1842d *(partim), Hydrophis aspera* Gray, 1849a, *Hydrophis subannulata* Gray, 1849a, *Hydrophis westermani* Jan, 1859b, *Hydrophis westermnani* Jan, 1863b *(nomen emendatum),*

Hydrophis trachyceps Theobald, 1868b, *Hydrophis crassicollis* J. Anderson, 1871b, *Hydrophis tuberculata* J. Anderson, 1871b, *Hydrophis dayanus* Stoliczka, 1872b, *Hydrophis tenuicollis* W.C.H. Peters, 1872e, *Hydrophis asperrimus* Murray, 1886, *Hydrophis phipsoni* Murray, 1887, *Hydrophis taprobanica* Haly, 1887, *Distira saravacensis* Boulenger, 1900f, *Distira sarawacensis* Shelford, 1901 *(nomen emendatum), Hydrophis cyanocyneta* – Anon., 1902 *(nomen incorrectum), Distira longissima* Rosén, 1905, *Distina cyanocincta* – Aiyar, 1906 *(nomen incorrectum), Distra cyanocincta* – Castellani & Chalmers, 1913 *(nomen incorrectum), Hydrophis asperriums* – G.S. Myers, 1947 *(nomen incorrectum), Hydrophys cyanocinctus* – Anthony, 1955 *(nomen incorrectum), Disteira saravaciensis* – Culotta & Pickwell, 1993 *(nomen incorrectum),* and *Hydrophis asperimus* – Culotta & Pickwell, 1993.

Type: Holotype, BMNH 1946.1.9.23 (formerly BMNH 1896.3.25.6), a 1524 mm female (P. Russell, 1781–1791).

Type locality: "Bengale" [= Bengal, India]. The Sunderbunds *fide* P. Russell (1801: 57) [= Sundarbans, Ganges River delta, SW Bangladesh and SE West Bengal, India, bet. 21.5–23°N, 88–91.5°E]

Distribution: South, SE and E Asia. Kuwait, Saudi Arabia, Bahrain, United Arab Emirates (Abu Dhabi, Das Is.), Oman, Iran (Bushehr, Hormozghan, Basuida and Qeshm Is.), Pakistan (Balochistan, Sindh), India (Andhra Pradesh, Kerala, Maharashtra, Orissa, Tamil Nadu, West Bengal), Sri Lanka, Bangladesh, Myanmar (Ayeyarwady. Tanintharyi), Thailand (Chanthaburi, Chumphon, Krung Thep Mahanakhon, Nakhon Si Thammarat, Pattani, Samut Sakhon, Singgora, Songkhla, Trang), Vietnam (Gulf of Tonkin to central coast) West Malaysia (Johor, Kelantan, Melaka, Perak, Pinang, Selangor), Singapore, East Malaysia (Sabah, Sarawak), Brunei, Indonesia (Java, W Kalimantan, Aru Is.), Papua New Guinea (Madang), Vietnam (Dong Nai, Minh Hai, Thuân Hai), Philippines (Cebu, Luzon, Sulu Arch.), China (Fujian, Guangdong, Guangxi, Hainan, Jiangsu, Liaoning, Shandong, Zhejiang), Hong Kong, W Taiwan (P'eng Hu Is.) and S Japan (Kyushu: Kumamoto, Ryukyus: Amamioshima and Misaki Is.).

Sources: P. Russell, 1801–1810, Stejneger, 1907a, Wall, 1907g, 1918d, 1919a, c, 1921g, Maki, 1931, Bourret, 1934g, C.H. Pope, 1935, C. Haas, 1950, Romer, 1954, Deraniyagala, 1955, Wang & Wang, 1956, Kuntz, 1963, E.H. Taylor, 1965, Minton, 1966, B. L. Lim & Balasingam, 1969, Singh, 1972, Voris, 1972, Mao & Chen, 1980, M. Mori, 1984, Latifi, 1991, Tian et al., 1986, Gasperetti, 1988, M.A.R. Khan, 1988, Stuebing, 1991, Leviton et al., 1992, I. Das, 2007b, San & Lee, 2007, V.S. Nguyen et al., 2009, Xiang & Li, 2009, A. Silva et al., 2011 and Rezaie-Atagholipour et al., 2012.

Remarks: Original description based on P. Russell (1801: 10, pl. 9).

3. *Leioselasma czeblukovi* Kharin, 1984b. Zool. Zhur. 63(10): 1542–1543, fig. 1b (lower left).

Synonym: *Hydrophis geometricus* L.A. Smith, 1986.

Type: Holotype, ZISP 19800, a 970 mm female (V.P. Shuntov, Feb. 1968).

Type locality: "Arafura Sea" [= SE Indonesia/N Australia].

Distribution: Papua New Guinea (Sepik, Lesson Is.) and NW Australia (Western Australia).

Sources: L.A. Smith, 1986, A.R. Rasmussen & Smith, 1997, Kharin, 2005b and Kharin & Hallermann, 2010.

4. *Leioselasma elegans* (Gray, 1842d). Zool. Misc. 2(May): 61. (*Aturia elegans*)

Synonyms: *Asturia elegans* Gray, 1842c *(nomen nudum), Distira grandis* Boulenger, 1896a, and *Hydrophys elegans* – Anthony, 1955 *(nomen incorrectum).*

Type: Holotype, BMNH 1946.1.3.89 (formerly BMNH III.6.5.b), a juvenile female (J. Gilbert, 1840–1842).

Type locality: "North Coast of Australia; Port Essington" [= Port Essington, ruins at tip of Cobourg Peninsula, Northern Territory, N Australia, ca. 11°09'N, 132°00'E, elevation NSL].

Distribution: Austro-Papua. Eastern Indonesia (Papua, Aru Arch.), Papua New Guinea (East Sepik, Gulf, Morobe, New Britian, Western) and Australia (N New South Wales, Northern Territory, Queensland, Western Australia, Ashmore and Great Barrier Reef, Cartier, Dirk Hartog and Murray Is.).

Sources: McDowell, 1972a, L.A. Smith, 1974, Cogger, 1975, Limpus, 1975, Redfield et al., 1978, F. Parker, 1982, Whitaker et al., 1982, Cogger et al., 1983a, Storr et al., 1986, Gow, 1989, Ehmann, 1992 and O'Shea, 1996.

5. *Leioselasma melanocephala* (Gray, 1849a). Cat. Snakes Brit. Mus.: 53. (*Hydrophis sublaevis melanocephala*)

Synonym: *Disteira orientalis* Stejneger, 1901d.

Type: Lectotype, BMNH 1946.1.9.22 (formerly BMNH 1847.3.4.68), adult female (E. Belcher [Samarang Exped.], 1847), designated by Stejneger (1907a: 426).,

Type locality: "Indian Ocean," (probably in error *fide* Stejneger (1907a: 425) and M.A. Smith (1926a: 65). "China Sea or even…the Ryukyus" via lectotype selection.

Distribution: Sino-Australia. Southern Japan (Ryukyus: Iriomotejima, Ishigakishima, Miyakoshima, Okinawa and Yaeyamashima Is.), China (Fujian, Guangdong, Zhejiang), Vietnam, Taiwan (T'ai-nan, P'eng Hu Is.) and NW Australia (Ashmore and Cartier Is. and Scott Reef).

Sources: Stejneger, 1907a, Mao & Chen, 1980, Cogger et al., 1983a, M. Mori, 1984, Storr et al., 1986, Tian et al., 1986, Ehmann, 1992, V.S. Nguyen et al., 2009 and Xiang & Li, 2009.

Remarks: Collection date of lectotype erroneously listed as 1897 *fide* Stejneger (1907a: 426). A subspecies of *L. spiralis* (Shaw) *fide* Klemmer (1963: 354) and Gasperetti (1988: 316). A synonym of *L. cyanocinctus fide* Cheh et al. (1987: 143).

6. *Leioselasma pacifica* (Boulenger, 1896a). Cat. Snakes Brit. Mus. 3: 278–279, pl. 12, fig. 2. (*Hydrophis pacificus*)

Synonym: *Distira macfarlani* Boulenger, 1896a.
Type: Holotype, BMNH 1946.1.10.14 (formerly BMNH 1883.3.17.15), an 840–940 mm male (Mus. Godeffroy).
Type locality: "New Britain" [= West or East New Britain Prov., Bismarck Archipelago, E Papua New Guinea].
Distribution: Australasia. Eastern Indonesia (Papua), Papua New Guinea (Gulf, New Britian, Western [MCZ 142380]) and Gilbert Is.
Sources: Whitaker et al., 1982, O'Shea, 1996 and Kharin, 2005a.
Remarks: A synonym of *L. cyanocincta* (Daudin) *fide* Wall (1909a: 219); a synonym of *Hydrophis belcheri* (Gray) *fide* M.A. Smith (1926a: 52); a synonym of *L. elegans* (Gray) *fide* McDowell (1972a: 217).

L

7. *Leioselasma semperi* (S.W. Garman, 1881). Bull. Mus. Comp. Zool. 8(3): 85. (*Hydrophis semperi*)

Type: Holotype, MCZ 4352, a 323 mm female (C.G. Semper, 1877).
Type locality: "Lake Taal, Luzon Island, Philippines" [Batangas Prov., SW Luzon].
Distribution: Northern Philippines (SW Luzon). Known only from type locality.
Remarks: A synonym of *L. melanocephala* (Gray) *fide* McDowell (1972a: 216).

8. *Leioselasma spiralis* (G. Shaw, 1802). Gen. Zool., Amph. 3(2): 564–565, pl. 125. (*Hydrus spiralis*)

Synonyms: *Hydrus brugmannii* H. Boie *in* Schlegel, 1826a *(nomen nudum), Hydrophis brugmansii* F. Boie, 1827 *(nomen emendatum), Hydrophis melanurus* Wagler, 1828a, *Hydrophis sublaevis* Gray 1842d *(partim), Hydrophis rappii* Jan, 1863b (*nomen nudum*), *Hydrophis robusta* A.C.L.G. Günther, 1864a, *Hydrophis robustus* – Theobald, 1868b, *Hydrophis rappii* Jan, 1872 *in* Jan & Sordelli, 1870–1881, *Hydrophis temporalis* Blanford, 1881, *Hydrophis bishopii* Murray, 1884, *Hydrophis aurifasciata* Murray, 1886, *Hydrophis melanocinctus* Wall, 1906g, *Distira spiralis typica* Raj, 1926, and *Hydrophis rapii* – Culotta & Pickwell, 1993 *(lapsus calami).*
Type: Holotype, BMNH 1946.1.6.94 (formerly BMNH III.6.10.c), a 381 mm specimen.
Type locality: Unknown. Designated as Indian Ocean *fide* A.C.L.G. Günther (1864a: 366).
Distribution: Southern Asia and East Indies. Iran (Bushehr, Sistan va Baluchestan), E Saudi Arabia, United Arab Emirates, Oman, Bahrain, SE Pakistan (Sindh), India (Andhra Pradesh, Gujarat, Kerala, Maharashtra, Orissa, Tamil Nadu), Sri Lanka, Bangladesh, Myanmar (Ayeyarwady, Bago, Mon, Tanintharyi, Mergui Arch.), ext. S Thailand (Phang Nga, Phuket Is.), Vietnam, West Malaysia (Kedah, Melaka, Perak, Perlis, Pinang, Selangor, Penang Is.), Singapore, East Malaysia (Sabah, Sarawak), Philippines (Tablas), Indonesia (Java, Sumatra, Sumba, Sulawesi: South; Waigeo) and New Caledonia.
Sources: Wall, 1907h, 1911b, 1919a, Minton, 1966, Singh, 1972, Voris, 1972, B.L. Lim & Sawai, 1975, C.B. Frith, 1977b, B.L. Lim, 1982, Gasperetti, 1988, Stuebing, 1991, Leviton et al., 1992, Ineich & Rasmussen, 1997, Bauer & Sadlier, 2000, Ineich & Laboute, 2002 and A.R. Rasmussen et al., 2011.
Remarks: Holotype erroneously listed by J.E. Gray as BMNH III.6.10.a *fide* M.A. Smith (1926a: 50). Luzon, Philippine record doubtful *fide* E.H. Taylor (1922a: 239). ZMB record of Mergusi, Philippines (M.A. Smith, 1926a: 50) probably a *lapsus* for Mergui Arch., Myanmar.

LEPTODEIRA Fitzinger, 1843 (Dipsadidae)

Synonyms: *Leptodira* Agassiz, 1847 *(nomen emendatum), Megalops* Hallowell, 1861, *Leptogeira* – F. Müller, 1878b *(nomen incorrectum), Crossmanus* Hoser, 2012x *(nomen illegitimum), Ginafabaserpenae* Hoser, 2012x *(nomen illegitimum),* and *Lukefabaserpens* Hoser, 2012x *(nomen illegitimum).*
Type species: *Coluber annulatus* Linnaeus, 1758.
Distribution: Extreme S USA and Latin America.
Sources: F. Werner, 1924c, Duellman, 1958a, 1966b, J.A. Peters & Orejas-Miranda, 1970, Willard, 1972, Cadle, 1984b, Villa et al., 1988, Daza et al., 2009, Zaher et al., 2009, Mulcahy et al., 2011 and Hoser, 2012x.

1. *Leptodeira annulata* (Linnaeus, 1758). Syst. Nat., ed. 10,1: 224. (*Coluber annulatus*)

Synonyms: *Coronella taeniata* Laurenti, 1768, *Coluber albofuscus* Lacépède, 1789 (*nomen rejiciendum*), *Coluber candidulus* Lacépède, 1789 (*nomen rejiciendum*), *Coluber epidaurius* Hermann, 1804, *Coluber ashmeadii* Hallowell, 1845a, *Megalops maculatus* Hallowell, 1861, *Eteirodipsas annulata rhomboidalis* Jan, 1863b (*nomen nudum*), *Leptodeira personata* Cope, 1869b, *Dipsas approximans* A.C.L.G. Günther, 1872b, *Leptodira rhombifera* A.C.L.G. Günther, 1872b, *Eteirodipsas wieneri* Sauvage, 1884, *Leptodira ocellata* A.C.L.G. Günther, 1895 *in* 1885–1902, *Leptodira nycthemera* F. Werner, 1901a, *Tarbophis dipsadomorphoides* Ahl, 1925, *Urotheca aureorostris* Briceño-Rossi,

1934, *Leptodeira smithi* E.H. Taylor, 1938, *Leptodeira rhombifera kugleri* Shreve, 1947, *Leptodeira annulata cussiliris* Duellman, 1958a, and *Leptodeira annulata pulchriceps* Duellman, 1958a.

Type: Lectotype, ZMUU 9 (formerly MP), a male, designated by Duellman (1958a: 47).

Type locality: "America" via lectotype selection. Restricted to the lower Amazon River, Pará State, Brazil *fide* Duellman (1958a: 48).

Distribution: Latin America. Mexico (Chiapas, Colima, Guerrero, Hidalgo, México, Michoacán, Oaxaca, Puebla, San Luis Potosí, Sinaloa, S Tamaulipas, Veracruz), SE Guatemala (Baja Verapaz, Chiquimula, El Progresso, Escuintla, Guatemala, Izabal, Jutiapa, Sacatepéquez, Santa Rosa, Sololá, Zacapa), El Salvador (Ahuachapán, Cabañas, Cuscatlán, La Libertad, La Unión, Morazán, San Ana, San Miguel, San Vicente, San Salvador, Sonsonate), Honduras (Choluteca, Comayagua, Colón, Copán, Cortés, El Paraíso, Francisco Morazán, Gracias a Dios, Intibucá, La Paz, Lempira, Ocotepeque, Olancho, Santa Bárbara, Valle, Yoro), Nicaragua (Boaco, Chinandega, Chontales, Estelí, Granada, Jinotega, León, Managua, Matagalpa, Nueva Segovia, Río San Juan, Rivas, Zelaya), Costa Rica (Alajuela, Cartago, Guanacaste, Heredia, Limón, Puntarenas, San José), Panama (Canal Zone: Colorado Is.: Chiriqui, Coclé, Colón, Darién, Herrera, Los Santos, Panamá: Gorgona Is. Perlas Arch.: San José), Colombia (Amazonas, Antioquia, Arauca, Atlántico, Bolívar, Boyaca, Caquetá, Casanare, Cauca, César, Chocó, Córdoba, Cundinamarca, La Guajira, Magdalena, Meta, Norte de Santander, Putumayo, Santander, Sucre, Tolima, Valle, Vaupés, Vichada, Gorgona Is.), Venezuela (Amazonas, Anzoátegui, Apure, Aragua, Barinas, Bolívar, Carabobo, Cojedes, Delta Amacuro, Distrito Federal, Falcón, Guárico, Lara, Mérida, Miranda, Monagas, Nueva Esparta, Portuguesa, Sucre, Táchira, Trujillo, Yaracuy, Vargas, Zulia, Margarita Is.), Trinidad and Tobago (Huevos, Tobago, Trinidad), Guyana (Demerara-Mahaica, Essequibo Islands-West Demerara, Upper Demerara-Berbice, Upper Takutu-Upper Essequibo), Suriname (Suriname), French Guiana (Cayenne, Saint-Laurent-du-Maroni), Brazil (Acre, Amapá, Amazonas, Bahia, Distrito Federal, Espiríto Santo, Goiás, Maranhao, Mato Grosso, Mato Grosso do Sul, Pará, Paraiba, Pernambuco, Rio de Janeiro, Rondônia, Roraima, São Paulo, Tocantins, Marajo Is.), E Ecuador (Chimborazo, Morona-Santiago, Napo, Pastaza, Zamora-Chinchipe), E Peru (Amazonas, Huanuco, Junín, Loreto, Madre de Dios, San Martín, Pasco, Cedros Is.), Bolivia (Beni, Chuquisaca, Cochchamba, La Paz, Pando, Santa Cruz, Tarija), Paraguay (Alto Paraguay, Alto Paraná, Amambay, Asunción, Boquerón, Caaguazú, Caazapá, Canindeyú, Central, Concepción, Cordillera, Guairá, Itapúa, Misiones, Neembucú, Paraguarí, Presidente Hayes, San Pedro) and Argentina (Chaco, Corrientes, Entre Ríos, Formosa, Salta, Santa Fe, Tucumán), NSL–2300 m.

Sources: L.E. Griffin, 1917, Dunn, 1936, Beebe, 1946, E.H. Taylor, 1951, 1954, Mertens, 1952c, Hoge, 1964d, Roze, 1966a, Emsley, 1977, Cunha & Nascimento, 1978, Medem, 1979, L.D. Wilson & Meyer, 1985, Pérez-Santos, 1986, Pérez-Santos & Moreno, 1988, 1991, J.D. Williams & Francini, 1991, Cei, 1994, D.R. Norman, 1994, Fugler et al., 1995, Vitt, 1996, Benitez-Gálvez, 1997, Starace, 1998, Freitas, 1999, Leynaud & Bucher, 1999, Lehr, 2001, Valdujo & Nogueira, 2001, Markezich, 2002, Savage, 2002, Duellman, 2005, Rivas-Fuenemayer & Barrio-Amorgós, 2005, G. Köhler et al., 2006, Cacciali, 2008, C.W. Myers & Donnelly, 2008, Navarrete et al., 2009, Sajdak, 2010, Ugueto & Rivas, 2010, McCranie, 2011a and C.J. Cole et al., 2013.

Remarks: Lectotype designation of NHR Lin-69 *fide* Hoge (1964d: 59) invalid. Type locality listed as Amazon Basin *fide* Pérez-Santos (1986: 442) and as Amazon *fide* Chippaux (1987: 62). This taxon is a complex of multiple species *fide* Daza et al. (2009: 663). McCranie (2011a: 325) transferred the Honduran group into *L. rhombifera*.

2. *Leptodeira bakeri* Ruthven, 1936. Occ. Pap. Mus. Zool., Univ. Michigan (330): 1–2.

Type: Holotype, UMMZ 57130, a 591 mm female (H.B. Baker, 6 Aug. 1922).

Type locality: "Bubali, Aruba" [= Bubali, Noord, Aruba, Netherlands Antilles, 12°33'N, 70°03'W, elevation NSL].

Distribution: Netherlands Antilles (Aruba Is.) and N Venezuela (Falcón).

Sources: Brongersma, 1940, 1959, Mijares-Urrutia et al., 1995 and Navarrete et al., 2009.

Remarks: Suriname record doubtful *fide* Duellman (1958a: 114). A subspecies of *L. annulata fide* Brongersma (1940: 128).

3. *Leptodeira frenata* (Cope *in* Ferrari-Pérez, 1886). Proc. U.S. Natl. Mus. 9(11): 184. (*Sibon frenatum*)

Synonyms: *Sibon annulata yucatanensis* Cope, 1887c, *Leptodira dunckeri* F. Werner, 1913, and *Leptodeira yucatanensis malleisi* Dunn & Stuart, 1935.

Type: Holotype, GECM 298, a 305 mm female (F. Ferrari-Pérez, Oct. 1884–Jan. 1885), lost *fide* Duellman (1958a: 61).

Type locality: "Republic of Mexico, State of Vera Cruz, Jalapa" [= Jalapa, Veracruz State, E Mexico, 19°33'N, 96°55'W, elevation 1370 m].

Distribution: Mexico and upper central America. Southeastern Mexico (Campeche, Chiapas, Quintana Roo, Tabasco, Veracruz, Yucatán, Cozumel Is.), Belize (Belize, Cayo, Corozal, Orange Walk, Stann Creek, Toledo) and N Guatemala (N Petén), NSL–450 m.

Sources: Duellman & Werler, 1955, R.W. Henderson & Hoevers, 1975, J.C. Lee, 1980, 1996, 2000, J.A.

Campbell, 1998b, Stafford & Meyer, 2000 and Reyes-Velasco & Mulcahy, 2010.

4. *Leptodeira nigrofasciata* A.C.L.G. Günther, 1868. Ann. Mag. Nat. Hist. (4) 1(6): 425. (*Leptodira nigrofasciata*)

Synonym: *Leptodira mystacina* Cope, 1870.

Type: Holotype, BMNH 1946.1.8.37, a 356 mm male (B.C. Seemann, 1866–1867).

Type locality: "Nicaragua."

Distribution: Mexico and upper central America. Southern Mexico (Chiapas, S Guerrero, Oaxaca), E Guatemala (Chiquimula), El Salvador (Cuscatlán, La Libertad, San Salvador), Honduras (Choluteca, Copán, Cortés, El Paraíso, Francisco Morazán, Valle, Yoro), W Nicaragua (Chinandega, Granada, Managua, Matagalpa, Nueva Segovia) and Costa Rica (Alajueta, Cartago, Guanacaste, Heredia, Limón, Puntarenas, San José), NSL–1300 m.

Sources: E.H. Taylor, 1951, 1954, Hidalgo, 1980a, Alvarez del Toro, 1982, Golder, 1984, L.D. Wilson & Meyer, 1985, Dowling & Jenner, 1987, G. Köhler, 1999b, Savage, 2002, G. Köhler et al., 2006 and McCranie, 2011a.

L

5. *Leptodeira ornata* (Bocourt, 1884). Bull. Soc. Philom. Paris (1883–1884) (7) 8(3): 141–142. (*Comastes ornatus*)

Synonyms: *Coluber ignobilis* Laurenti, 1768, *Leptodira affinis* A.C.L.G. Günther, 1895 *in* 1885–1902, *Leptodeira albofusca* Boulenger, 1896a *(partim)*, ? *Leptodeira dunckeri* Werner, 1913, and *Leptodeira larcorum* K.P. Schmidt & Walker, 1943b.

Types: Syntypes (2), MNHN 6201a–b, longest syntype 347 mm (C. Viguier, Nov. 1876–April 1878).

Type locality: "Isthme de Darien, Amérique inter-tropicale" [= Darién Prov., Panama].

Distribution: Lower Central America and NW South America. Costa Rica (Alajueda, Cartago, Guanacaste, Heredia, Limón, Puntarenas, San José), Panama (Bocas del Toro, Canal Zone, Chiriquí, Coclé, Darién, Panamá, San Blas, Barro Colorado Is.), W Colombia (Atlántico, Antioquia, Bolívar, Boyaca, Caldas, Cauca, Chocó, Cordoba, Cundinamarca, La Guajira, Magdalena, Meta, Putumayo, Santander, Norte del Santander, Tolima, Valle), W Venzuela (? Falcón, Zulia), Ecuador (Bolívar, El Oro, Esmeraldas, Guayas, Imbabura, Los Rios, Pichincha) and NW Peru (Cajamarca, La Libertad), NSL–2000 m.

Source: Navarrete et al., 2009.

Remarks: A valid species *fide* Daza et al. (2009: 663) and Daza (2010: 33).

6. *Leptodeira polysticta* A.C.L.G. Günther, 1895 *in* 1885–1902. Biol. Cen.-Amer., Rept. Batr.: 172, pl. 55, fig. a. (*Leptodira polysticta*)

Synonyms: *Leptodeira albofusca* Boulenger, 1896a *(partim)*, and *Leptodeira annulata taylori* H.M. Smith, 1941k.

Type: Lectotype, BMNH 1946.1.8.45, a female (F.D. Godman, 1861–1869), designated by H.M. Smith & Taylor (1950a: 316).

Type locality: "Belize, British Honduras" [= Belize city, Belize, 17°30'N, 88°11'W, elevation NSL] via type locality restriction *fide* Smith & Taylor (1950: 316).

Distribution: Mesoamerica. Mexico (Campeche, Chiapas, Colima, Guerrero, Jalisco, México, Michoacán, Nayarit, Oaxaca, Puebla, Quintana Roo, Sinaloa, S Sonora, Tabasco, Veracruz, Yucatán), Belize (Belize, Cayo, Corozal, Orange Walk, Stann Creek, Toledo), Guatemala (Alta Verapaz, Chimaltenango, Guatemala, Izabal, Petén, Quezaltenango, Retalhuleu, San Marcos, Santa Rosa, Sololá), El Salvador (Ahuachapán, Chalatenango, La Libertad, Morazán, San Salvador, Santa Ana, San Vicente), N Honduras (Atlántida, Colón, Copán, Cortés, Gracias a Dios, Intabucá, Islas de la Bahía, Lempira, Ocotopeque, Olancho, Santa Bárbara, Yoro, Cayo Cochino Pequeño Is.), E Nicaragua (Atlántico Norte, Chinandega, Chontales, Estelí, Granada, Jinotega, Managua, Matagalpa, Río San Juan, Zelaya) and Costa Rica (Alajueda, Cartago, Guanacaste, Heredia, Limón, Puntarenas, San José), 75–2200 m.

Remarks: A valid species *fide* J.A. Campbell, 1982, 1998, Daza et al. (2009: 663) and Daza (2010: 33).

7. *Leptodeira punctata* (W.C.H. Peters, 1866). Mber. Königl. Akad. Wiss. Berlin 1866(2): 93–94. (*Crotaphopeltis punctata*)

Synonym: *Leptodira pacifica* Cope, 1869b.

Type: Holotype, ZMB 5378, a 493 mm male (Mus. Godeffroy).

Type locality: "Südafrica" [= South Africa] (in error). Restricted to Mazatlán, Sinaloa, Mexico *fide* H.M. Smith & Taylor (1950a: 343).

Distribution: Western Mexico (Jalisco, Nayarit, Sinaloa), NSL–1500 m.

Sources: Davis & Dixon, 1957, Shannon & Humphrey, 1964 and Greer, 1965b.

Remarks: Date of publication erroneously cited as 1867 *fide* Duellman (1958a: 94). Type collected in Mexico, transported to South Africa, probably from Mazatlán seaport, where purchased by Mr. Godeffroy *fide* Duellman (1958a: 94).

8. *Leptodeira rubricata* (Cope, 1894a). Proc. Amer. Philos. Soc. (1893) 31: 347. (*Sibon septentrionale rubricatum*)

Type: Holotype, AMNH 17367 (formerly MNCR 346), a 660 mm specimen (G.K. Cherrie, 1894–1897).
Type locality: "Boca Mala, Pacific slope of Costa Rica" [Puntarenas Prov.].
Distribution: Southwestern Costa Rica (Puntarenas) and Panama (USNM 52493), NSL.
Sources: E.H. Taylor, 1951, Savage, 2002 and Solórzano, 2004.

9. *Leptodeira septentrionalis* (Kennicott *in* Baird 1859a). U.S.-Mex. Bound. Surv., Rept. 2(2): 16, pl. 8, fig. 1. (*Dipsas septentrionalis*)

Type: Lectotype, USNM 131739 (formerly SIM 2288), a 607 mm female (S. Van Vliet, 1853–1859), designated by H.M. Smith & Taylor (1950a: 361).
Type locality: "Brownsville, Cameron Co., Texas" [= 25°54'N, 97°30'W, elevation 5 m] via lectotype selection.
Distribution: USA and E Mexico. Extreme S USA (ext. S Texas) and E Mexico (Coahuila, Guanajuato, Hidalgo, Jalisco, México, Nuevo León, Puebla, Querétaro, San Luis Potosí, Tamaulipas, Veracruz), NSL–1990 m.
Sources: R.W. Henderson & Hoevers, 1975, J.C. Lee, 1980, 1996, 2000, N.J. Scott et al., 1983, L.D. Wilson & Meyer, 1985, Pérez-Santos, 1986, Pérez-Santos & Moreno, 1988, G. Köhler, 1999b, Mijares-Urrutia & Arends, 2000, Stafford & Meyer, 2000, Werler & Dixon, 2000, J.C. Murphy & Schlager, 2003, Campos-Rodríguez et al., 2004, G. Köhler et al., 2006, McCranie et al., 2005, 2006, McCranie, 2011a and Travers et al., 2011.
Remarks: *Leptodeira polysticta* a valid species *fide* J.A. Campbell (1998b: 221–222), Daza et al. (2009: 680) and Daza (2010: 33).

10. *Leptodeira splendida* A.C.L.G. Günther, 1895 *in* 1885–1902. Biol. Cen.-Amer., Rept. Batr.: 171, pl. 53, fig. b. (*Leptodira splendida*)

Synonyms: *Leptodeira bressoni* E.H. Taylor, 1938, *Leptodeira ephippiata* H.M. Smith & Tanner, 1944, and *Leptodeira duellmani*, J.A. Peters, 1954.
Types: Syntypes (2), BMNH 1946.1.9.94–95, a 660 mm female and juvenile (A. Boucard, 1854–1870).
Type locality: "Mexico, Izucar" [= Izucar de Matamoros, Puebla State, cen. Mexico, 18°36'N, 98°28'W, elevation 1275 m].
Distribution: Western Mexico (ext. SW Chihuahua, Colima, Durango, Guerrero, Jalisco, México, Michoacán, Morelos, SW Puebla, Sinaloa, S Sonora, Zacatecas), 500–1700 m.
Sources: Benitez-Gálvez, 1997, Lemos-Espinal et al., 2004b, Enderson & Bezy, 2007, Ahumada-Carrillo et al., 2011 and Valdez-Lares & Muñiz-Martinez, 2011a.

11. *Leptodeira uribei* (Ramírez-Bautista & Smith, 1992). Bull. Maryland Herp. Soc. 28(3): 83–86, figs. 1–5, 9. (*Pseudoleptodeira uribei*)

Type: Holotype, CNAR 7528 (formerly IBHUNAM), a 481 mm male (A. Rodríguez-Palafox, 27 Nov. 1991).
Type locality: "Estación de Biología 'Chamela' (EBCH, less than 5 km from the Pacific coast, at approximately 19°30'N, 105°03'W, el. 10-584 m), Mexico" [= Jalisco].
Distribution: Western Mexico (Colima, Guerrero, W Jalisco, Michoacán, S Oaxaca), 10–585 m.
Sources: Mijangos et al., 2006, Reyes-Velasco et al., 2009, 2010 and Mertz et al., 2011.

LEPTODRYMUS Amaral, 1927b (Colubridae)

Type species: *Masticophis pulcherrimus* Cope, 1874b.
Distribution: Central America.
Sources: Bogert, 1939a, 1947, E.H. Taylor, 1951, Mertens, 1952c, J.A. Peters & Orejas-Miranda, 1970, L.D. Wilson & Meyer, 1985, Villa et al., 1988, Savage, 2002, Solórzano, 2004 and McCranie, 2011a.

1. *Leptodrymus pulcherrimus* (Cope, 1874b). Proc. Acad. Nat. Sci. Philadelphia 26(1): 65. (*Masticophis pulcherrimus*)

Synonyms: *Zamenis bitaeniatus* Boettger, 1898, and *Leptodrymus clarki* Amaral, 1927b.
Type: Lectotype, ANSP 14688a, an 875 mm specimen (G. Davidson, 1873–1874), designated by Bogert (1939a: 182).
Type locality: "western side of Central America" via lectotype selection.
Distribution: Central America. Southwestern Guatemala (Retalhuleu), SW El Salvador (Sonsonate), Honduras (Atlántida, Choluteca, Cortés, Francisco Morazán, Intibucá, Olancho, Yoro), W Nicaragua (Carazo, Managua, Matagalpa, Rivas: Ometepe Is.) and NW Costa Rica (Guanacaste, San José), NSL–1300 m.
Remarks: ANSP 14688 consists of two syntypes *fide* Malnate (1971: 370).

LEPTOPHIS Bell, 1825 (Colubridae)

Synonyms: *Ahaetulla* Gray, 1825 *(nomen rejiciendum)*, *Ahoetula* Gray, 1831 *(nomen substitutum)*, *Dendrophis* H. Boie *in* Fitzinger, 1826a *(nomen rejiciendum)*, *Dendrophis* H. Boie *in* F. Boie, 1827 *(nomen rejiciendum)*, *Dendrophis* H. Boie *in* Schlegel, 1826a *(nomen rejiciendum)*, *Leptopthis* – Kelaart, 1853 *(nomen incorrectum)*, *Diplotropis* A.C.L.G. Günther, 1872b, *Lepthophis* – García-Cubas, 1893 *(nomen incorrectum)*, *Leptophius* A.L. Herrera, 1899 *(nomen emendatum)*,

L

Thalerophis Oliver, 1948 *(nomen substitutum)*, and *Talerophis* – Beltrán, 1953 *(nomen incorrectum)*.

Type species: *Coluber ahaetulla* Linnaeus, 1758.

Distribution: Latin America.

Sources: Oliver, 1942a, 1948, J.A. Peters & Orejas-Miranda, 1970, Mertens, 1973, Pérez-Santos & Moreno, 1988, 1991 and Villa et al., 1988.

Remarks: Official Generic Name no. 1309 *fide* Opinion 524 (ICZN, 1958a). Original description reprinted in Bell (1826: 104–105 & 1830: 1035–1038).

1. *Leptophis ahaetulla* (Linnaeus, 1758). Syst. Nat., ed. 10, 1: 225. (*Coluber ahaetulla*)

Synonyms: *Coluber filiformis* Lacépède, 1789 (*nomen rejiciendum*), *Coluber boiguatrara* Gravenhorst, 1807, *Coluber caninana* Raddi, 1820, *Coluber cepedii* Merrem, 1820, *Coluber viridis* Raddi, 1820, *Coluber richardii* Bory de St. Vincent, 1823 *in* 1822–1831, *Coluber liocercus* Wied-Neuwied, 1824a, *Ahaetula linnei* Gray, 1831, *Ahaetulla occidentalis* A.C.L.G. Günther, 1859c, *Thrasops marginatus* Cope, 1862d, *Ahaetulla nigromarginata* A.C.L.G. Günther, 1866, *Thrasops praestans* Cope, 1869b, *Herpetodryas affinis* Steindachner, 1870b, *Ahaetulla urosticta* W.C.H. Peters, 1873c, *Leptophis ortonii* Cope, 1875b, *Thrasops sargii* J.G. Fischer, 1881, *Leptophis ultramarinus* Cope, 1894d, *Leptophis bocourti* Boulenger, 1898h, *Leptophis flagellum* Andersson, 1901, *Leptophis rostralis* Lönnberg, 1902, *Leptophis argentinus* F. Werner, 1903a, *Leptophis occidentalis insularis* T. Barbour, 1905, *Leptophis vertebralis* F. Werner, 1909b, *Leptophis maximus* Weller, 1930, and *Leptophis ahuetulla* – Briceño-Rossi, 1934 *(nomen incorrectum)*. *Leptophis ahaetulla bolivianus* Oliver, 1942a, *Leptophis coeruleodorsus* Oliver, 1942a, *Leptophis copei* Oliver, 1942a, *Leptophis occidentalis chocoensis* Oliver, 1942a, and *Leptophis santamartensis* Bernal-Carlo & Roze, 1994.

Type: Lectotype, ZMUU 2 (C. Gyllenborg, 1744), designated by Savage & Oliver (1956: 149).

Type locality: "America" via first revisor selection of Andersson (1899: 22). Restricted to South America *fide* Savage & Oliver (1956: 148) and to Suriname *fide* Hoge (1964d: 60).

Distribution: Latin America. Southeastern Mexico (Campeche, Chiapas, Oaxaca, Puebla, Quintana Roo, Tabasco, Veracruz, Yucatán), Belize (Belize, Cayo, Orange Walk, Stann Creek, Toledo), N Guatemala (Alta Verapaz, Petén), Honduras (Atlántida, Colón, Cortés, El Paraíso, Francisco Morazán, Gracias a Dios, Olancho, Yoro), Nicaragua (Chontales, Matagalpa, Río San Juan, Zelaya), Costa Rica (Alajuela, Cartago, Guanacaste, Heredia, Limón, San José, Puntarenas), Panama (Bocas del Toro, Canal Zone: Barro Colorado Is.; Chiriquí, Darién, San Miguel Is.; Los Santos), Colombia (Amazonas, Antioquia, Arauca, Atlántico, Bolívar, Boyacá, Caquetá, Cauca, Chocó, Córdoba, Cundinamarca, Magdalena, Meta, Nariño, Norte de Santander, Putumayo, Risaralda, Sucre, Tolima, Vaupés, Valle, Gorgona Is.), Venezuela (Amazonas, Anzoátegui, Apure, Aragua, Bolívar, Carabobo, Cojedes, Delta Amacuro, Distrito Federal, Falcón, Guárico, Lara, Miranda, Monagas, Nueva Esparta, Portuguesa, Táchira, Trujillo, Yaracuy, Zulia, Margarita Is.), Trinidad and Tobago (Tobago, Trinidad), Guyana (Baramita-Waini, Essequibo Islands-West Demerara, Mahaica-Berbice, Pomeroon-Supenaam, Potaro-Siparuni, Upper Demerara-Berbice, Upper Takutu-Upper Essequibo), Suriname (N Nickerie, Marowijne, Para, Suriname), French Guinea (Cayenne), Brazil (Amapá, Amazonas, Bahia, Distrito Federal, Espírito Santo, Goías, Maranhão, Mato Grosso, Minas Gerais, Pará, Pernambuco, Rio de Janeiro, Rio Grande do Norte, São Paulo, Baraja and Marajo Is.), W Ecuador (Esmeraldas, Guayas, Imbabura, Napo), E Peru (Amazonas, Junín, Loreto, Madre de Dios, San Martín), E Bolivia (Beni, Santa Cruz, Tarija), Paraguay (Alto Paraguay, Alto Paraná, Amambay, Asunción, Boquerón, Caaguazú, Caazapá, Canindeyú, Central, Concepción, Cordillera, Guairá, Itapúa, Misiones, Neembucú, Paraguarí, Presidente Hayes, San Pedro), Uruguay (Artigas) and NE Argentina (Chaco, Corrientes, Entre Ríos, Formosa, Misiones), NSL–2750 m.

Sources: Beebe, 1946, E.H. Taylor, 1951, 1954, Roze, 1966a, N.J. Scott, 1969, R.W. Henderson & Hoevers, 1975, Emsley, 1977, Medem, 1979, J.C. Lee, 1980, 1996, 2000, Abuys, 1983a, L.D. Wilson & Meyer, 1985, Lancini, 1986, Fugler & Riva, 1990, J.D. Williams & Francini, 1991, Cei, 1994, D.R. Norman, 1994, Yanosky et al., 1996, Meneghel & Achaval, 1997, J.A. Campbell, 1998b, Starace, 1998, Freitas, 1999, Leynaud & Bucher, 1999, Canseco-Marquez et al., 2000, Stafford & Meyer, 2000, Savage, 2002, Souza & Silva, 2002a., Duellman, 2005, Rivas-Fuenemayer & Barrio-Amorgós, 2005, McCranie et al., 2006, Cacciali, 2008, Albuquerque, 2009, Navarrete et al., 2009, Ugueto & Rivas, 2010, McCranie, 2011a, Albuquerque et al., 2012 and C.J. Cole et al., 2013.

Remarks: Official Specific Name no. 1571 *fide* Opinion 524 (ICZN, 1958a). Paralectotypes include NHR Lin-37, NHR Lin-40–41, and ZMUU 90. Lectotype designation of NHR Lin-40 *fide* Hoge (1964d: 60) invalid (antedated by Savage & Oliver, 1956). L.D. Wilson & Meyer (1985: 65) inadvertently listed Chile as part of Distribution. *Leptophis coeruleodorsus* and *L. occidentalis* valid species *fide* Navarrete et al. (2009: 61).

2. *Leptophis cupreus* (Cope, 1868b). Proc. Acad. Nat. Sci. Philadelphia 20(2): 106. (*Thrasops cupreus*)

Type: Holotype, ANSP 5202 (formerly USNM 6666), a 521 mm female (J. Orton, fall 1867–winter 1868).

Type locality: "River Napo and Maranon, Ecuador" [= Napo/Orellana Prov., NE Ecuador, or Loreto Dept., NE Peru].

L

Distribution: Southeastern Colombia (Cauca, Guaviare, Meta, Gorgona Is.), Venezuela (Bolívar), E Ecuador (Napo, Orellana, Pastaza) and NE Peru (Loreto), 250–1880 m.

Sources: J.A. Peters & Orcés-V., 1960, Dixon & Soini, 1986, Pérez-Santos, 1986 and Albuquerque & McDiarmid, 2010.

Remarks: A valid species *fide* J.A. Peters & Orcés V. (1960: 139), who erroneously reported the type lost. Photographs of the holotype in Albuquerque & McDiarmid (2010: 326–327, figs. 1–2).

3. *Leptophis depressirostris* (Cope, 1861c). Proc. Acad. Nat. Sci. Philadelphia (1860) 12(12): 557. (*Philothamnus depressirostris*)

Synonyms: *Diplotropis bilineata* A.C.L.G. Günther, 1872, *Leptophis aeruginosus* Cope, 1875a, *Leptophis saturatus* Cope, 1875a, *Leptophis bileneatus* – Briceño-Rossi, 1934 *(nomen incorrectum).* and *Leptophis saturus* – Briceño-Rossi, 1934 *(nomen incorrectum).*

Type: Holotype, ANSP 5207, an 1155 mm female (R.W. Mitchell, 1857–1860).

Type locality: "Cocuyas de Veraguas, New Granada" [= Cocuyas de Veraguas, a gold mine near the Río Concepción, N Veraguas Prov., cen. Panama, ca. 8°45'N, 81°00'W, elevation 100 m]

Distribution: Central America and NW South America. Western Honduras (Gracias a Dios), E Nicaragua (Chontales, Jinotega, Matagalpa, Río San Juan, Zelaya), Costa Rica (Alajuela, Cartago, SE Guanacaste, Heredia, Limón, Puntarenas, San José), Panama (Colón, Veraguas), W Colombia (Cauca, Chocó, Nariño, Valle, Gorgona Is.), NW Ecuador (Carachi, Esmeraldas) and Peru (Lambayeque, Loreto), NSL–1300 m.

Sources: E.H. Taylor, 1951, Pérez-Santos, 1986, G. Köhler, 1999b, Savage, 2002, Venegas, 2005, G. Köhler & Ferrari, 2009 and McCranie, 2011a.

Remarks: Type locality listed as Colombia *fide* E.H. Taylor (1951: 85). Record from Junín, Peru doubtful *fide* Oliver (1948: 207).

4. *Leptophis diplotropis* (A.C.L.G. Günther, 1872b). Ann. Mag. Nat. Hist. (4) 9(49): 25–26, pl. 6, fig. a. (*Ahaetulla diplotropis*)

Synonyms: *Leptophis diplotropis aeneus* Bocourt, 1897 *in* A.H.A. Duméril, Bocourt & Mocquard, 1870–1909, and *Leptophis diplotropis forreri* H.M. Smith, 1943a.

Types: Syntypes (3), BMNH 1946.1.4.1–3, one male and two females, longest syntype 838 mm (A.-L.-J.-F. Sumichrast, 1855–1869).

Type locality: "Tehuantepec" [= vicinity of Tehuantepec (16°19'N, 95°14'W, 40 m), S Oaxaca State, S Mexico].

Distribution: Western Mexico (Chiapas, ext. SW Chihuahua, Colima, Durango, Guanajuato, Guerrero, Jalisco, México, Michoacán, Morelos, Nayarit, Oaxaca, Puebla, Sinaloa, ext. S Sonora, María Magdalena and María Madre Is.), NSL–2575 m.

Sources: H.M. Smith, 1943a, W.B. Davis & Smith, 1953, Alvarez del Toro & Smith, 1956, H.M. Smith & Grant, 1958a, Zweifel, 1960, Telford, 1964, McDiarmid et al., 1976, Pérez-Higareda & Smith, 1989, Benitez-Gálvez, 1997, Auth et al., 2000b, Sánchez-Luna et al., 2009, Canseco-Máquez & Guitiérrez-Mayén, 2010 and Valdez-Lares & Muñiz-Martinez, 2011b.

Remarks: Collector listed as Boucard *fide* A.C.L.G. Günther (1872b: 14) but F. Sumichrast *fide* BMNH catalogue. If the latter, then the type locality is the western side of the Isthmus of Tehuantepec *fide* Sumichrast (1873: 246 & 1880: 183). Boulenger (1894a: 110) listed all types as female; at least one male is present *fide* M. Lang (*in litt.*).

5. *Leptophis mexicanus* A.-M.-C. Duméril, Bibron & Duméril, 1854a. Erpét. Gén. 7(1): 536–537.

Synonyms: *Leptophis mexicanus yucatanensis* Oliver, 1942a, *Leptophis mexicanus septentrionalis* Mertens, 1972a, and *Leptophis mexicanus hoeversi* R.W. Henderson, 1976.

Type: Lectotype, MNHN 3455, a 965 mm specimen, designated by Mertens (1973: 146).

Type locality: "Mexique" [= Mexico] via lectotype selection. Restricted to Potrero Viejo, Veracruz, Mexico *fide* H.M. Smith & Taylor (1950a: 350).

Distribution: Mesoamerica. Eastern Mexico (Campeche, Chiapas, Hidalgo, Michoacán, Nuevo León, Querétaro, Quintana Roo, E Oaxaca, Puebla, San Luis Potosí, Tabasco, S Tamaulipas, Veracruz, Yucatán), Belize (Belize, Cayo, Corozal, Orange Walk, Stann Creek, Toledo), N Guatemala (Alta Verapaz, Baja Verapaz, Izabal, Péten), El Salvador (Cabañas, Chalatenango, Cuscatlán, La Libertad, La Paz, San Salvador), Honduras (Atlántida, Colón, Comayagua, Copán, Cortés, Francisco Morazán, Gracias a Dios, Islas de la Bahía, Lempira, Olancho, Yoro, Cayos Cochino and Utila Is.), Nicaragua (Atlanta Sur, Granada, Managua, Matagalpa, Río San Juan, Zelaya, Great Corn and Ometepe Is.) and Costa Rica (Alajueta, Guanacaste, Heredia, Limón, Puntarenas, San José), NSL–1600 m.

Sources: E.H. Taylor, 1951, 1954, Mertens, 1972a, Villa, 1972, L.D. Wilson & Hahn, 1973, R.W. Henderson & Hoevers, 1975, R.W. Henderson, 1976, J.C. Lee, 1980, 1996, 2000, L.D. Wilson & Meyer, 1985, Carmarillo-R., 1995, Mancilla-Moreno & Ballardo, 1995, J.A. Campbell, 1998b, G. Köhler, 1999b, Auth et al., 2000b, Stafford & Meyer, 2000, Savage, 2002, Solórzano, 2004, G. Köhler et al., 2006, Frazier et al., 2007, N. Herrera et al., 2007, García-Vázquez et al., 2009 and McCranie, 2011a.

6. *Leptophis modestus* (A.C.L.G. Günther, 1872b). Ann. Mag. Nat. Hist. (4) 9(49): 26, pl. 6, fig. c. (*Ahaetulla modesta*)

Type: Holotype, BMNH 1946.1.21.50, a 1300–1321 mm male (O. Salvin, 1857–1875).
Type locality: "banks of the Rio Chisoy, below the town of Cubulco" [= Río Chixoy or Negro, Baja Verapaz Dept., Guatemala].
Distribution: Southern Mexico and upper Central America. Extreme SE Mexico (NE Chiapas), cen. Guatemala (Baja Verapaz), NW El Salvador (Santa Ana) and SW Honduras (Comayagua, Cortés, Intibucá, Lempira, Ocotepeque), 1335–2590 m.
Sources: A.C.L.G. Günther, 1895 *in* 1885–1902, Hoyt, 1964, K.L. Williams & Smith, 1966, L.D. Wilson et al., 1978, 1986, L.D.Wilson & Meyer, 1985, 1986, McCranie & Wilson, 1993, P. Holm, 1995 and McCranie, 2011a.

7. *Leptophis nebulosus* Oliver, 1942a. Occ. Pap. Mus. Zool., Univ. Michigan (462): 12–14.

Type: Holotype, MCZ 15287 (formerly MNCR), a 798 mm male.
Type locality: "Cariblanca, Costa Rica" [= Cariblanco, Heredia Prov., Costa Rica, 10°16'N, 84°11'W, elevation 880 m].
Distribution: Central America. Northeastern Honduras (Colón, Gracias a Dios), E Nicaragua (Matagalpa, Río San Juan) and Costa Rica (Cartago [UMMZ 117655], Guanacaste, Heredia, Limón, Puntarenas) and Panama (Coclé, Panamá), NSL–1600 m.
Sources: E.H. Taylor, 1951, 1954, L.D. Wilson & Meyer, 1985, Savage, 2002, Jaramillo & Ibáñez, 2003, McCranie, 2011a and Ray, 2011.

L

8. *Leptophis riveti* Despax, 1910. Bull. Mus. Hist. Nat., Paris 16(7): 368–369.

Synonym: *Leptophis brevior* Boulenger, 1914b.
Type: Holotype, MNHN 1906.259, a 576 mm specimen (P. Rivet, 1901–1906).
Type locality: "Gualaquiza, Équateur" [= Gualaquiza, S Morona-Santiago Prov., E Ecuador, 3°24'S, 78°34'W, elevation 730 m].
Distribution: Southwestern Costa Rica (S Puntarenas), Panama (Canal Zone), NW Colombia (Antioquia, Chocó), Ecuador (Cañar, Esmeraldas, Guayas, Loja, Los Ríos, Pichincha [MCZ 154613]) and E Peru (Amazonas), NSL–1525 m.
Sources: Hardy, 1995 and Savage, 2002.
Remarks: Original description reprinted in Despax (1911: 26–28, pl. 1, figs. 1, 1a–1b).

9. *Leptophis stimsoni* K. Harding, 1995. Trop. Zool. 8(2): 222–223, figs. 1–2.

Type: Holotype, BMNH 1940.3.11.84, a 593 mm male (I.T. Sanderson, 1937).
Type locality: "Mount Aripo, Northern Distribution, Trinidad, altitude about 914 m."
Distribution: Trinidad and Tobago (Trinidad), 135–915 m
Sources: J.C. Murphy, 1997 and Boos, 2001.

LEPTOTYPHLOPS Fitzinger, 1843 (Leptotyphlopidae)

Synonyms: *Stenostoma* A.-M.-C. Duméril & Bibron, 1844, *Glauconia* Gray, 1845, *Stenostomophis* Rochebrune, 1885, *Rhamphostoma* Boulenger, 1893a *(nomen emendatum), Leptotyphlus* – Devincenzi, 1925 *(nomen incorrectum), Leptatyphlops* – Brongersma, 1933 *(nomen incorrectum), Leptotyphlos* – Briceño-Rossi, 1934 *(nomen incorrectum), Leptotypheops* – Sundersingh, 1960 *(nomen incorrectum), Leptophlops* – Cunha & Nascimento, 1978 *(nomen incorrectum), Glaukonia* – Tiedemann & Häupl, 1980 *(nomen incorrectum), Leptothyphlops* – Pérez-Santos, 1986 *(nomen incorrectum), Karimdaouesus* Hoser, 2012am *(nomen illegitimum), Ottobreus* Hoser, 2012am *(nomen illegitimum),* and *Teesleptotyphlops* Hoser, 2012am *(nomen illegitimum).*
Type species: *Typhlops nigricans* Schlegel, 1839 *in* 1837–1844.
Distribution: Africa.
Sources: Hahn, 1980a, Wallach, 1998b, Broadley & Broadley, 1999, McDiarmid et al., 1999 and Broadley & Wallach, 2007a.
Remarks: Official Generic Name no. 2155 *fide* Opinion 1207 (ICZN, 1982b). The following genera and species have been separated from *Leptotyphlops: Epacrophis (boulengeri, drewesi, reticulatus), Epictia (albifrons, albipunctata, alfredschmidi, ater, australis, borapeliotes, collaris, columbi, diaplocia, goudotii, magnamaculata, melanurus, munoai, nasalis, peruviana, rubrolineata, rufidorsa, signata, striatula, subcrotilla, teaguei, tenella, tesselata, tricolor, undecimstriata, vellardi), Leptotyphlops (aethiopicus, conjunctus, distanti, emini, howelli, incognitus, jacobseni, kafubi, keniensis, latirostris, macrops, mbanjensis, merkeri, monticolus, nigricans, nigroterminus, pembae, pitmani, pungwensis, scutifrons, sylvicolus, telloi), Mitophis (asbolepis, calypso, leptepileptus, pyrites), Myriopholis (adleri, albiventer, algeriensis, blanfordi, boueti, braccianii, burii, cairi, erythraeus, filiformis, ionidesi, longicauda, macrorhyncha, macrura, nasirostris, natatrix, nursii, perreti, phillipsi, rouxestevae, tanae, wilsoni, yemenica), Namibiana (labialis, latifrons, occidentalis, rostrata, Rena (affinis, boettgeri, bressoni, dissecta, dulcis, humilis, maxima, myopica, unguirostris), Rhinoleptus (koniagui, parkeri), Siagonodon (acutirostris, borrichianus, cupinensis, septemstriatus), Tetracheilostoma (bilineatum, breuili, carlae), Tricheilostoma (bicolor, broadleyi, dissimilis, greenwelli, sundewalli),* and *Trilepida (anthracina, brasiliensis, brevissima, dimidiata, dugandi, fulfinosa,*

guayaquilensis, jani, joshuai, koppesi, macrolepis, nicefori, salgueiroi).

1. *Leptotyphlops aethiopicus* Broadley & Wallach, 2007a. Zootaxa (1408): 38–40, pl. 10, fig. 4.

Type: Holotype, BMNH 1911.12.13.17, a 145 mm specimen (G. Kristensen, 1910–1911).

Type locality: "near Harrar [=Harar], Hararge Region, Ethiopia (09°18'N, 42°08'E, elevation ca. 2000 m)."

Distribution: Northeastern Africa. Southern Ethiopia (Hararge, Sidamo), S Somalia (Gedo) and N Kenya (N Eastern), 1365–2100 m.

Sources: Lanza, 1983a, 1990b and Largen & Spawls, 2010.

2. *Leptotyphlops conjunctus* (Jan, 1861b). Arch. Zool. Anat. Fis. 1(2): 189–190. (*Stenostoma conjunctum*)

Synonym: *Glauconia conjuncta* Boulenger, 1890d, and *Stenostoma groutii* Cope, 1875a.

Type: Holotype, SMNS 2519 A, a 111 mm specimen (C.F.F. Krauss, 1860).

Type locality: Unknown. Designated as Cap de Bonne Espérance [= Cape of Good Hope, South Africa] *fide* Jan & Sordelli (1861 *in* 1860–1866: 2). Restricted to Eastern Cape Province, South Africa *fide* Broadley & Watson (1976: 483 & 493).

Distribution: Southeastern Africa. Lesotho, W Swaziland and E South Africa (Eastern Cape, Free State, Gauteng, KwaZulu-Natal, Mpumalanga, S Transvaal), 100–1720 m.

Sources: Jan & Sordelli, 1861 *in* 1860–1866, Boulenger, 1893a, 1915c, FitzSimons, 1956, 1962a, Broadley & Watson, 1976, Broadley, 1983, Bates, 1996, Branch, 1998, Broadley & Broadley, 1999 and Marais, 2004.

Remarks: Krauss collected in South Africa from May 1838–April 1840.

3. *Leptotyphlops distanti* (Boulenger, 1892c). Naturalist Transvaal: 175–176, 3 figs. (*Glauconia distanti*)

Type: Holotype, BMNH 1946.1.10.82 (formerly BMNH 1890.12.16.10), a 130 mm specimen.

Type locality: "Pretoria, Transvaal" [= Pretoria, Gauteng Prov., NE South Africa, 25°44'S, 28°13'E, elevation 1300 m].

Distribution: Southern Mozambique (Gaza, Maputo) and NE South Africa (Eastern Cape, Free State, Gauteng, Kwazulu-Natal, Limpopo, Mpumalanga, North West), NSL–1500 m.

Sources: Boulenger, 1893a, 1915c, Mertens, 1955, V.F.M. FitzSimons, 1962a, 1974, Broadley, 1983, Pienaar et al., 1983, Branch, 1988, 1998, 2001 and Broadley & Wallach, 1997a–b and Marais, 2004.

4. *Leptotyphlops emini* (Boulenger, 1890d). Ann. Mag. Nat. Hist. (6) 6(31): 91. (*Glauconia emini*)

Type: Lectotype, BMNH 1946.1.10.87 (formerly BMNH 1890.5.14.1) (E. Pasha, 1890), designated by Broadley & Wallach (2007a: 34–35).

Type locality: "Karagwe, Victoria Nyanza, Tanzania (ca. 01°S, 31°E, elevation above 1200 m" via lectotype selection [= Karagwe, Karagwe District, Kiwa Magharibi Region, ext. NW Tanzania, 1°30'S, 30°59'E, elevation 1400 m].

Distribution: East Africa. Eastern Democratic Republic of the Congo (Kivu), SE South Sudan (Eastern Equatoria), Uganda (S Central, Western), Kenya (Central, Coast, Eastern, Rift Valley), Rwanda (Gisenyi), Burundi, Tanzania (Lindi, Morogoro, Mtwara, Mwanza, Rukwa, Kagera) and NE Zambia (N Northern), 650–1490 m.

Sources: Boulenger, 1893a, 1915c–d, Witte, 1941, Loveridge, 1942, 1957a, Laurent, 1956, Pitman, 1974, Broadley & Wallach, 1996, 2007a, Wallach, 1996c and Spawls et al., 2002.

Remarks: Type locality listed as Bukoba, Karagwe district, Tanganyika Territory *fide* Loveridge (1957a: 247).

L

5. *Leptotyphlops howelli* Broadley & Wallach, 2007a. Zootaxa (1408): 30–32.

Type: Holotype, NMZB 10455 (formerly KMH 7041), a 158 mm male (K.M. Howell, 5 Sept. 1990).

Type locality: "Mchungu Forest Reserve, Rufiji District, Piwani Region, Tanzania (07°44'S, 39°17'E, elevation 15 m)."

Distribution: Eastern Tanzania (SE Piwani) and SW Kenya (N Nyanza), 15–1180 m.

6. *Leptotyphlops incognitus* Broadley & Watson, 1976. Occ. Pap. Natl. Mus. Rhodesia 5B(8): 494–496, fig. 14. (*Leptotyphlops conjunctus incognitus*)

Type: Holotype, NMZB-UM 6121, a 166 mm specimen (R.A. Barton, Nov. 1963).

Type locality: "Umtali, Rhodesia" [= Mutare, Manicaland, E Zimbabwe, 18°58'S, 32°39'E, elevation 1100 m].

Distribution: Southeastern Africa. Southern Zambia (Central, Eastern, Southern), S Malawi (Southern), Zimbabwe (Manicaland, Mashonaland Central, Mashonaland East, Matebeleland North, Matabeleland South), S Mozambique (Inhambane, Manica, Maputo, Sofala, Bazaruto, Benguerra and Santa Carolina Is.), E Swaziland, Lesotho and NE South Africa (Gauteng, KwaZulu-Natal, Limpopo, Mpumalanga), 45–1550 m.

Sources: De Waal, 1978, Broadley, 1988, 1992a, Broadley & Wallach, 1997b, Branch, 1998, Broadley & Broadley, 1999, Broadley et al. 2003, M. Griffin, 2003, Marais, 2004 and Broadley & Blaylock, 2013.

7. *Leptotyphlops jacobseni* Broadley & Broadley, 1999. Syntarsus (5): 19–20.

Type: Holotype, TM 57919, a 151 mm male (C.R. Owen, 5 Aug. 1984).

Type locality: "Buffelsvley, 388KT Lydenburg District, Mpumalanga Province, South Africa (24°52'S: 30°19'E)" [= elevation ca. 1000 m].

Distribution: Eastern South Africa (NE Eastern Cape, Limpopo, N Mpumalanga), 750–1700 m.

Source: Marais, 2004.

8. *Leptotyphlops kafubi* (Boulenger, 1919b). Rev. Zool. Afr. (1919–1920) 7(2): 186–187. (*Glauconia kafubi*)

Types: Syntypes (3), BMNH 1946.1.11.7–8 (formerly BMNH 1919.8.26.13–14) and MRAC 2085, longest syntype 105 mm (L. Stappers, 1911–1919).

Type locality: "la rivière Kafubo, à Élisabethville" [= Kafubu River, Lubumbashi, SE Katanga Prov., SE Democratic Republic of the Congo, 11°40'S, 27°29'E, elevation 1250 m].

Distribution: Southeastern Africa. Eastern Angola, SE Democratic Republic of the Congo (Katanga), Zambia (Central, Copperbelt, Northern, North-Western), Malawi (Central) and N Zimbabwe (Mashonaland West), 585–1500 m.

Sources: Witte, 1953, Broadley, 1971, Broadley & Watson, 1976, Broadley, 1991, Broadley & Broadley, 1999 and Broadley et al., 2003.

9. *Leptotyphlops keniensis* Broadley & Wallach, 2007a. Zootaxa (1408): 38, pl. 10, fig. 3.

Type: Holotype, NMK/O 2903, a 189 mm female (F.J. McCartney, 24 Feb. 1983).

Type locality: "Nyeri, Central Province, Kenya (00°25' S, 36°56' E, just below 2000 m)"

Distribution: Southern Kenya (Central, Eastern) and NE Tanzania (Kilimanjaro), 1250–2000 m.

10. *Leptotyphlops latirostris* (Sternfeld, 1913a). Wiss. Deutsch. Zentr.-Afrika Exp. (1912) 4(2): 264.

Type: Holotype, ZMB 22376, a 155 mm specimen (R. Grauer, 1907–1908).

Type locality: "Nordwestufer des Tanganyika" [= NW shore of Lake Tanganyika, elevation above 775 m, E Kivu Region, Democratic Republic of the Congo].

Distribution: Eastern Democratic Republic of the Congo (Sud-Kivu) and SW Burundi (Bujumbura, Bururi), 700–1565 m.

Sources: Witte, 1933, 1962, Laurent, 1956 and Broadley & Wallach, 2007a.

11. *Leptotyphlops macrops* Broadley & Wallach, 1996. Copeia 1996(1): 162–165, figs. 1a–c.

Type: Holotype, NMZB 11352, a 198 mm male (A. Cockle, 10 Sept. 1991).

Type locality: "Mkwaja forest, Pangani District, Tanga Region, Tanzania (5°52'S, 38°47'E), altitude below 100 m."

Distribution: Southeastern Kenya (S Coast) and NE Tanzania (Piwani, Tanga), NSL–100 m.

Sources: Wallach, 1996c, Spawls et al., 2002 and Broadley & Wallach, 2007a.

Remarks: Problematic records from W Kenya (Central, Western) at 1600–2285 m *fide* Spawls et al. (2002: 303).

12. *Leptotyphlops mbanjensis* Broadley & Wallach, 2007a. Zootaxa (1408): 37, pl. 10, fig. 2.

Type: Holotype, MCZ 48040, a 132–135 mm female (A. Loveridge, 26 April 1939)

Type locality: "Mbanja, ca. 6 km N Lindi, Lindi District, Lindi Region, Southern Province, Tanzania (09°24'S, 39°45'E, 130 m)."

Distribution: Southeastern Tanzania (Lindi), 90–130 m.

Remarks: Holotype erroneously listed as MCZ 45040 *fide* Broadley & Wallach (2007a: 37).

13. *Leptotyphlops monticolus* (Chabanaud, 1917a). Bull. Mus. Hist. Nat., Paris (1916) 22(7): 366–367, figs. 7–9. (*Glauconia monticola*)

Type: Holotype, MNHN 1916.213, a 138 mm specimen (E. Gromier & P. Le Petit, 1911).

Type locality: "Congo belge: volcans de Kivori (altitude 1,500 mètres)" [= volcanoes of the Virunga Mountains, E. Kivu Region, Democratic Republic of the Congo, elevation 1500 m].

Distribution: Eastern Democratic Republic of the Congo (Nord-Kivu, Sud-Kivu) and W Rwanda (Western), 900–1965 m.

Sources: Witte, 1941, Laurent, 1956, and Broadley & Wallach, 2007a.

Remarks: Possibly occurs in W Burundi *fide* Broadley & Wallach (2007a: fig. 5).

14. *Leptotyphlops nigricans* (Schlegel, 1839 *in* 1837–1844). Abbild. Amph.: 38, pl. 32, figs. 21–24. (*Typhlops nigricans*)

Types: Syntypes (3), MNHN 3232, a 125 mm specimen, RMNH 3705, and ZMB 5244, one of latter syntypes 105 mm (P.A. Delalande, 1818–1820).

Type locality: "auf dem Vorgebirge der guten Hoffnung" [= Cape of Good Hope, Western Cape Prov., SW South Africa, 34°21'S, 18°28'E, elevation 20 m].

Distribution: Southern South Africa (Eastern Cape, Western Cape), NSL–1450 m.

Sources: V.F.M. FitzSimons, 1956, 1962a, 1974, Guibé & Roux-Estève, 1972, Broadley & Watson, 1976,

Broadley, 1983, Branch, 1988, 1998, Marais, 1992, 2004, Broadley & Wallach, 1996, Broadley & Broadley, 1999, Bauer et al., 2002 and Marais, 2004.

Remarks: Official Specific Name no. 2797 *fide* Opinion 1207 (ICZN, 1982b).

15. *Leptotyphlops nigroterminus* Broadley & Wallach, 2007a. Zootaxa (1408): 40–42, pl. 12, fig. 1.

Type: Holotype, MCZ 54813, a 159 mm male (C.J.P. Ionides, 7 July 1956).

Type locality: "Karema, eastern shore of Lake Tanganyika, Mpanda District, Rukwa Region, Tanzania (06°50'S, 30°50'E, elevation 950 m)."

Distribution: Southwestern Kenya (S Rift Valley) and W Tanzania (Rukwa, Singida, Tabora), 775–1585 m.

16. *Leptotyphlops nursii* (J. Anderson *in* Boulenger, 1896a). Cat. Snakes Brit. Mus. 3: 591. (*Glauconia nursii*)

Type: Lectotype, BMNH, 1946.1.16.91 (formerly BMNH 1895.11.27.1), a 242 mm specimen (C.G. Nurse, ca. 1890), designated by Hahn (1978: 487).

Type locality: "Aden" [= Yemen, 12°46'N, 45°01'E, elevation NSL] via lectotype selection.

Distribution: Northeastern Africa and SW Arabia. Southeastern Egypt (S El-Bahr El-Ahmar), N Sudan (Ash Shamaliyah), E Ethiopia (NE Hararge), NW Somalia (Woqooyi Galbeed), SW Saudi Arabia (Makkah, Najran, Riyadh,) and Yemen (Al Bayda, Al Hudaydah, Hadhramaut, Ibb, Lahij, Sadah, Ta'izz), NSL–1825 m.

Sources: Hahn, 1978, Gasperetti, 1988, Schätti & Gasperetti, 1994, Baha El Din, 2006, Broadley & Wallach, 2007a and Largen & Spawls, 2010.

17. *Leptotyphlops pembae* Loveridge, 1941. Proc. Biol. Soc. Washington 54: 177–178. (*Leptotyphlops emini pembae*)

Type: Holotype, MCZ 46116, a 201 mm female (R.H.W. Pakenham, 4 Oct. 1940).

Type locality: "Wingwi Pwana, Pemba Island, East Africa" [= Wingwi Pwana, NE Pemba Island, Tanzania Territory, NE Tanzania, 05°01' S, 39°50' E, elevation NSL].

Distribution: Northeastern Tanzania (Mafia and Pemba Is.), NSL–50 m.

Sources: Loveridge, 1957a, Pakenham, 1983, Broadley & Howell, 1991, Spawls et al., 2002 and Broadley & Wallach, 2007a.

18. *Leptotyphlops pungwensis* Broadley & Wallach, 1997b. Arnoldia, Zimbabwe 10(11): 116–117, fig. 1.

Type: Holotype, NMZB-UM 7251 (formerly UM 7251), a 90 mm female (D.G. Broadley, 8 Nov. 1963).

Type locality: "between Muda and Lamego railroad sidings, Pungwe Flats, Mozambique (19°22'S: 34°22'E), altitude ca. 50 metres" [= Sofala Prov., cen. Mozambique, elevation 25 m].

Distribution: Central Mozambique (Manica, Sofala), 25–50 m.

Sources: Broadley & Watson, 1976, Branch, 1998, Broadley & Broadley, 1999 and Marais, 2004.

19. *Leptotyphlops scutifrons* (W.C.H. Peters, 1854). Verh. Preuss. Akad. Wiss. Berlin 1854(11): 621. (*Stenostoma scutifrons*)

Synonyms: *Glauconia merkeri* F. Werner, 1909d, *Glauconia okahandjana* Ahl, 1924b, and *Leptotyphlops scutifrons pitmani* Broadley & Wallach, 2007a.

Type: Holotype, ZMB 4826, an 83 mm specimen (W.C.H. Peters, 21 Dec. 1845).

Type locality: "Sena" [= Vila da Sena, Zambezi River, Sofala Prov., cen. Mozambique, 17°27'S, 35°02'E, elevation 65 m]. Emended to "dem Zambeze bei Sena" *fide* Peters (1882d: 104).

Distribution: Southern Africa. Angola (Benguela, Bié, Cuanza Norte, Cuanza Sul, Huambo, Huíla, Malanje, Namibe), E Zambia (Eastern), S Malawi (Southern), SE Tanzania (Lindi, Mtwara), Mozambique (Inhambane, Manica, Sofala, Tete, Zambézia), Namibia (Caprivi, Hardap, Khomas, Okavango, Omaheke, Oshikoto, Otjozondjupa), Botswana (Central, Ghanzi, Kgalagadi, Kweneng, North East, North West, South East), Zimbabwe (Eastern, Manicaland, Mashonaland Central, Mashonaland East, Mashonaland West, Masvingo, Matabeleland North, Matabeleland South, Midlands), Swaziland and E South Africa (Eastern Cape, Free State, Gauteng, KwaZulu-Natal, Limpopo, Mpumalanga, North West, Northern Cape), 10–2025 m.

Sources: W.C.H. Peters, 1882d, Boulenger, 1893a, 1915c, Sternfeld, 1908f, V.F.M. FitzSimons, 1935b, 1956, 1962a, 1974, Loveridge, 1936c, 1942, 1957a, Monard, 1937, Mertens, 1955, 1971, Broadley, 1959, 1983, Broadley & Watson, 1976, De Waal, 1978, Pienaar et al., 1983, Auerbach, 1987, Patterson, 1987, Branch, 1988, 1998, Bates, 1996, Wallach, 1996c, Broadley & Wallach, 1997a–b, 2007a, Branch, 1998, 2001, 2005, Broadley & Broadley, 1999, Bauer et al., 2002, Razzetti & Msuya, 2002, Spawls et al., 2002, Broadley et al., 2003, M. Griffin, 2003, J.C. Murphy & Schlager, 2003, Marais, 2004 and Broadley & Blaylock, 2013.

Remarks: Original description reprinted in W.C.H. Peters (1855: 51).

20. *Leptotyphlops sylvicolus* Broadley & Wallach, 1997a. Durban Mus. Novit. (22): 37–40, fig. 1.

Type: Holotype, NMZB-UM 11903 (formerly UM 11903), a 90 mm male (E. Moll, Sept. 1966).

Type locality: "Hlogwene Forest, Kwazulu-Natal, South Africa" [= ca. 29°15'S, 31°27'E, elevation 25–100 m].
Distribution: Southeastern South Africa (Eastern Cape, N KwaZulu-Natal), 20–960 m.
Sources: Branch, 1998, Hahn & Wallach, 1998, Broadley & Broadley, 1999 and Marais, 2004.

21. *Leptotyphlops tanae* Broadley & Wallach, 2007a. Zootaxa (1408): 28–29, pl. 8, fig. 4.

Type: Holotype, MCZ 40099, a 99 mm male (A. Loveridge, 14–21 June 1934).
Type locality: "old village of Ngatana, ca. one mile northwest from new village of Wema, in the lower Tana River, Coast Region, Kenya (02°30'S, 40°15'E, elevation 50 m)."
Distribution: Northeastern Africa. Southern Ethiopia (Gemu Gofa), S Somalia (Mogadishu) and SE Kenya (NE Coast), NSL–460 m.
Sources: Loveridge, 1936c, Lanza, 1983, 1990, Largen & Rasmussen, 1993 and Largen & Spawls, 2010.

22. *Leptotyphlops telloi* Broadley & Watson, 1976. Occ. Pap. Natl. Mus. Rhodesia 5B(8): 500–501, fig. 16.

Type: Holotype, NMZB-UM 30635 (formerly UM 30635), a 143–152 mm specimen (J.L.P.L. Tello, 8 Feb. 1974).
Type locality: "Estatuane, Moçambique" [= Estatuane, SW Maputo Prov., S Mozambique, 26°30'S, 32°08'E, elevation 200 m].
Distribution: Eastern Swaziland and S Mozambique (SW Maputo), 200–530 m.
Sources: Broadley, 1983, Branch, 1988, 1998 and Broadley & Wallach, 1997b, Broadley & Broadley, 1999 and Marais, 2004.

LEPTUROPHIS Boulenger, 1900f (Colubridae)

Synonym: *Sphecodes* A.-M.-C. Duméril, 1853 *(nomen praeoccupatum).*
Type species: *Sphecodes albofuscus* A.-M.-C. Duméril, Bibron & Duméril, 1854a.
Distribution: Malay Peninsula and Indonesia.
Sources: Rooij, 1917, C. Haas, 1950, Grandison, 1972, B.L. Lim & Sharef bin Kamarudin, 1975, McDowell, 1987, Stuebing, 1991, David & Vogel, 1996, Manthey & Grossman, 1997, M. J. Cox et al., 1998, Chan-ard et al., 1999, Stuebing & Inger, 1999, Malkmus et al., 2002, Cota et al., 2011 and Grismer, 2011.
Remarks: A synonym of *Lycodon fide* McDowell (1987: 42) and I. Das (2012: 44).

1. *Lepturophis albofuscus* (A.-M.-C. Duméril, Bibron & Duméril, 1854a). Erpét. Gén. 7(1): 394–397. (*Sphecodes albofuscus*) (*nomen corrigendum*)

Synonyms: *Sphecodes albofuscus* A.-M.-C. Duméril, 1853 (*nomen nudum*), *Sphecodes albo-fuscus* A.-M.-C. Duméril, Bibron & Duméril, 1854a *(nomen incorrigendum), Ophites albofuscus* – A.C.L.G. Günther, 1858 *(nomen corrigendum),* and *Lepturophis borneensis* Boulenger, 1900f.
Type: Holotype, formerly MNHN, a 632 mm specimen, lost *fide* Iskandar & Colijn, 2001: 65.
Type locality: "Sumatra" [W Indonesia].
Distribution: Malay Peninsula and Greater Sundas. Southern Thailand (Nakhon Si Thammarat, Phatthalung, Songkhla, Yala), West Malaysia (Selangor, Pahang, Johore, Tioman Is.), Singapore, East Malaysia (Sabah, Sarawak) and Indonesia (Kalimantan, Nias, Sumatra), 215–1100 m.
Remarks: Type listed in SMK *fide* Cox e al. (2013: 224).

LETHEOBIA Cope, 1869c (Typhlopidae)

Synonyms: *Cottontyphlops* Hoser, 2012am *(nomen illegitimum), Gleesontyphlops* Hoser, 2012am *(nomen illegitimum), Laidlawtyphlops* Hoser, 2012am *(nomen illegitimum), Pillotttyphlops* Hoser, 2012am *(nomen illegitimum), Trioanotyphlops* Hoser, 2012am *(nomen illegitimum), Whybrowtyphlops* Hoser, 2012am *(nomen illegitimum),* and *Wilsontyphlops* Hoser, 2012am *(nomen illegitimum).*
Type species: *Letheobia pallida* Cope, 1869c.
Distribution: Central and E Africa and the Middle East.
Sources: F. Werner, 1921a, Witte, 1962, Pitman, 1974, Roux-Estève, 1974, Hahn, 1977, 1980a, Broadley, 1983, Wallach, 1994, Wallach & Ineich, 1996, McDiarmid et al., 1999, Franzen & Wallach, 2002 and Broadley & Wallach, 2007, 2009.

1. *Letheobia acutirostrata* (Andersson, 1916). Meddel. Göteborgs Mus. Zool. Afd. (9) 17(5): 23–24, fig. 3. (*Typhlops acutirostratus*)

Synonyms: *Typhlops avakubae* K.P. Schmidt, 1923, and *Typhlops caecus pitmani* Witte, 1961.
Type: Holotype, NHMG 1375, a 360 mm specimen (G. Göransson, 1910).
Type locality: "Irebu, Belgian Congo" [= Irebu, Bandundu Prov., NE Democratic Republic of the Congo, 00°37'S, 17°45'E, elevation 300 m].
Distribution: Northern Democratic Republic of the Congo (Bandundu, Bas-Congo, Equateur, Kinshasa, Oriental), 150–400 m.
Sources: Witte, 1966 and Young & Wallach, 1998.
Remarks: Previously a synonym of *Typhlops* (= *Rhinotyphlops*) *graueri fide* Witte (1962: 43).

Resurrected as a species *fide* Young & Wallach (1998: 252).

2. *Letheobia ataeniata* (Boulenger, 1912). Ann. Mus. Civ. Stor. Nat. Genova (3) 5: 331. (*Typhlops unitaeniatus ataeniatus*)

Synonym: *Typhlops ataenitus* – Gans & Laurent *in* Gans, Laurent & Pandit, 1965 *(nomen incorrectum)*.
Types: Syntypes (3), BMNH 1946.1.10.78 (formerly BMNH 1912.6.6.12) and MSNG (2) (C. Citerni, Mar.–July 1911).
Type locality: "Dolo, la frontiera Italo-Etiopica" [= Ethiopia-Somalia border, 04°11'N, 42°05'E, elevation 200 m].
Distribution: Northeast Africa. Southeastern Ethiopia (Bale, Hararge, Sidamo), Somalia (Bay, Gedo, Jubbada Dhexe, Jubbada Hoose, Mogadishu, Mudug, Shabeellaha Dhexe, Shabeellaha Hoose, Togdheer) and ext. NE Kenya (N North-Eastern), 25–1000 m.
Sources: Gans & Laurent, 1965, Lanza, 1983a, 1990b, Largen & Rasmussen, 1993, Wallach, 1994, Spawls et al., 2002 and Largen & Spawls, 2010.
Remarks: Broadley & Wallach (2007b: 56) discussed type locality.

3. *Letheobia caeca* (A.H.A. Duméril, 1856). Rev. Mag. Zool. (2) 8(10): 462–463, pl. 21, figs. 4, 4a–4c. (*Onychocephalus caecus*)

Synonym: *Typhlops coecus* Jan, 1863b *(nomen emendatum)*.
Types: Syntypes (2), MNHN 1063 and MNHN 1994.1166 (formerly MNH 1063A), longest syntype 380 mm (C.-E. Aubry-Lecomte, 1853–1856).
Type locality: "côte du Gabon, l'Afrique occidentale" [= Gabon, West Africa].
Distribution: Central Africa. Southern Cameroon (Centre, Sud, Sud-Ouest), Equatorial Guinea (Rio Muni), Gabon (Moyen-Ogooué) and Congo (Kouilou), 65–1300 m.
Sources: J.-F. Trape & Roux-Estève, 1990, 1995, Chirio & LeBreton, 2007 and Pauwels & Vande weghe, 2008.

4. *Letheobia crossii* (Boulenger, 1893a). Cat. Snakes Brit. Mus. 1: 52, pl. 3, figs. 5a–c. (*Typhlops crossii*)

Type: Holotype, BMNH 1946.1.11.55 (formerly BMNH 1892.6.23.3), a 270 mm specimen (W.H. Crosse, 1890–1893).
Type locality: "Asaba, 180 miles up the Niger, West Africa" [= Nigeria, 06°11'N, 06°45'E, elevation 55 m].
Distribution: Southern Nigeria (Anambra, Bayelsa, Cross River) and cen. Togo (Centrale, Kara), 55–545 m.
Sources: J.A. Butler & Reid, 1986, Trape & Segniagbeto, 2008 and Segniagbeto et al., 2011.

5. *Letheobia debilis* (Joger *in* G. Peters & Hutterer, 1990). Vert. Tropics: 93–94, figs. 3–5. (*Rhinotyphlops debilis*)

Type: Holotype, ZFMK 33684, a 380 mm specimen (N. Degallier & J.-P. Hervé).
Type locality: "near Bangui, Central African Republic" [= 04°23'N, 18°33'E, elevation 450 m].
Distribution: Central African Republic (Haut-Mbomou, Ombella-Mpoko), 450–800 m.
Source: Chirio & Ineich, 2006.

6. *Letheobia decorosa* (Buchholz & Peters *in* W.C.H. Peters, 1875b). Mber. Königl. Akad. Wiss. Berlin 1875(3): 197–198. (*Typhlops decorosus*)

Synonym: *Typhlops buchholzi* W.C.H. Peters, 1881b.
Type: Holotype, ZMB 8322, a 330 mm specimen (R. Buchholz, 1875).
Type locality: "Cameruns" [= Cameroon].
Distribution: Cameroon (Adamaoua, Est, Littoral, Nord, Nord-Ouest, Ouest, Sud) and W Central African Republic (Haute-Sangha, Sangha), 10–1775 m.
Sources: Chirio & Ineich, 2006 and Chirio & LeBreton, 2007.

7. *Letheobia episcopus* (Franzen & Wallach, 2002). J. Herp. 36(2): 177–180, figs. 1–2. (*Rhinotyphlops episcopus*)

Type: Holotype, ZFMK 74224, a 318 mm female (W. Bischoff, 8 April 2000).
Type locality: "3 km north of Halfeti (37°16'N, 37°51'E) on road to Savasan Köyü, upper slopes of Euphrates River Valley, Sanliurfa Province, southeastern Turkey, about 500 m above sea level."
Distribution: Southeastern Turkey (Sanliurfa), 500 m. Known only from vicinity of type locality.
Source: Franzen, 2000.
Remarks: Specific epithet is a noun in apposition and thus requires a "*-us*" ending *fide* Franzen & Wallach (2007: 261).

8. *Letheobia erythraea* (Scortecci, 1928). Atti Soc. Ital. Sci. Nat. Mus. Civ. Stor. Nat. Milano 67(3–4): 291–292, fig. 1. (*Typhlops erythraeus*) (*nomen corrigendum*)

Synonym: *Typhlos erythraeus* Scortecci, 1928 (*nomen incorrigendum*).
Type: Holotype, MSNM 3353 (formerly MSNM 2015), a 205 mm specimen (L. Fossati).
Type locality: "Sanganeiti" [= Segheneyti, Southern Prov., SE Eritrea, 15°03'N, 39°11'E, elevation 2180 m].
Distribution: Southeastern Eritrea (Northern Red Sea, Southern), 1800–2180 m.
Sources: Largen & Rasmussen, 1993, Largen, 1997, Wallach & Lanza, 2004 and Largen & Spawls, 2010.

9. *Letheobia feae* (Boulenger, 1906f). Ann. Mus. Civ. Stor. Nat. Genova (3) 2(42): 209, fig. 5. (*Typhlops feae*)

Synonym: *Typhlops principis* Boulenger, 1906.
Type: Holotype, MSNG 28871, a 205 mm specimen (L. Fea, 1901).
Type locality: "Vista Alegre, San Thomé, 200–300 m" [= São Tomé Island, São Tomé and Príncipe, Gulf of Guinea].
Distribution: São Tomé and Príncipe (São Tomé Is.), 200–300 m.

10. *Letheobia gracilis* (Sternfeld, 1910c). Mitt. Zool. Mus. Berlin 5(1): 70. (*Typhlops gracilis*)

Synonyms: *Typhlops katangensis* Witte, 1933, and *Typhlops leptosoma* Witte, 1933.
Type: Holotype, ZMB 22030, a 450 mm specimen (H. Fromm, 1910).
Type locality: "Kitungulu in Urungu, Deutsch-Ostafrika" [= Sumbawanga District, Tanzania, ca. 04°32'S, 34°13'E, elevation 1370 m].
Distribution: Southeastern Africa. Southeastern Democratic Republic of the Congo (Katanga), S Tanzania (Morogoro, Rukwa) and NE Zambia (Luapula, Northern), 700–1400 m.
Source: Spawls et al., 2002.

L

11. *Letheobia graueri* (Sternfeld, 1913a). Wiss. Deutsch. Zentr.-Afrika Exp. (1912) 4(2): 264. (*Typhlops graueri*)

Synonyms: *Typhlops leptosoma polli* Laurent, 1956, and *Typhlops graveri* – Robb, 1967 *(nomen incorrectum).*
Type: Holotype, ZMB 27161, a 355 mm specimen (R. Grauer, 1907–1908).
Type locality: "Urwald hinter den Randbergen am Nordwestufer des Tanganyika" [= rainforest behind boundary mountains on northwest shore of Lake Tanganyika, Democratic Republic of the Congo].
Distribution: East Africa. Southeastern Democratic Republic of the Congo (Bandundu, Bas-Congo, Equateur, Kasai Occidental, Kasai Orientale, Katanga, Orientale, Sud-Kivu), W Uganda (Western), SW Burundi (Bururi) and W Tanzania (Kigoma), 750–2000 m.
Source: Spawls et al., 2002.

12. *Letheobia jubana* Broadley & Wallach, 2007b. Zootaxa (1515): 44–45, fig. 3c.

Type: Holotype, CAS 151201, a 347 mm female (P.M. Johnson, 25 Jan. 1979).
Type locality: "vicinity of Mareri, Juba Sugar Project, west bank of Juba River, Middle Juba region, southern Somalia (00°30'N, 42°46'E, elevation ca. 20 m)."
Distribution: Southern Somalia (Jubbada Dhexe), 20 m. Known only from type locality.

13. *Letheobia kibarae* (Witte, 1953). Explor. Parc Natl. Upemba 6: 150–151, figs. 35a–c. (*Typhlops kibarae*)

Type: Holotype, IRSNB 2201, a 525 mm specimen (G.-F. de Witte, 31 Jan. 1949).
Type locality: "Kaswabilenga (région du cours inférieur de la Lupiala, affluent de la rive droit de la Lufira), Parc National de l'Upemba" [= Katanga, SE Democratic Republic of the Congo, 08°48'S, 26°41'E, elevation 700 m].
Distribution: Southeastern Demoncratic Republic of the Congo (Katanga), 700–1250 m.

14. *Letheobia largeni* Broadley & Wallach, 2007b. Zootaxa (1515): 48, fig. 5b.

Type: Holotype, BMNH 1974.5162, a 270–274 mm female (E. McConnell, 19 April 1974).
Type locality: "Gambela, Wollega District, Ethiopia (08°15'N, 34°35'E, elevation 515 m)." Emended to vicinity of Gambela, on the Baro River *fide* Largen & Spawls (2010: 444).
Distribution: Southwestern Ethiopia (Illubabor), 515 m. Known only from type locality.
Source: Largen & Spawls, 2010.

15. *Letheobia leucosticta* (Boulenger, 1898b). Ann. Mag. Nat. Hist. (7) 1(2): 124. (*Typhlops leucostictus*)

Type: Holotype, BMNH 1946.1.11.24 (formerly BMNH 1897.5.8.16), a 225 mm specimen.
Type locality: "Liberia."
Distribution: Liberia. Known only from type locality.
Source: Wallach, 1993b.

16. *Letheobia lumbriciformis* (W.C.H. Peters, 1874b). Mber. Königl. Akad. Wiss. Berlin 1874(6): 377. (*Onychocephalus lumbriciformis*)

Synonym: *Typhlops kleebergi* F. Werner, 1904.
Types: Syntypes (2), ZMB 8045, a 445 mm specimen, and ZMB 63440 (formerly ZMB 8045A), a 160 mm specimen (J.M. Hildebrandt, 1872–1874).
Type locality: "Küste Zanzibar" [= Zanzibar coast, Tanzania].
Distribution: Extreme SE Kenya (S Coast) and ext. NE Tanzania (Tanga, Zanzibar Is.), NSL–55 m.
Source: Spawls et al., 2002.
Remarks: Possibly occurs on Pemba Is. *fide* Spawls et al. (2002: 295).

17. *Letheobia newtoni* (Bocage, 1890). J. Sci. Math. Phys. Nat., Lisboa (2) 2(5): 61–62. (*Typhlops newtoni*)

Synonym: *Typhlops naveli* Angel, 1920.

Type: Holotype, MBL 1890, a 40 mm specimen (F. Newton, 1890), destroyed by fire 18 March 1978.
Type locality: "l'Ilheo das Rolas, ilot situé au sud de l'île St Thomé" [= Rolas Island, south of São Tomé Is., São Tomé and Príncipe, Gulf of Guinea].
Distribution: São Tomé and Príncipe (Rolas and Príncipe Is.).

18. *Letheobia obtusa* (W.C.H. Peters, 1865). Mber. Königl. Akad. Wiss. Berlin 1865(6): 260–261, pl. fig. 2. (*Typhlops obtusus*)

Synonym: *Typhlops obtusus algravei* Laurent, 1968.
Types: Syntypes (2), BMNH 1946.1.11.31–32 (formerly BMNH 1864.10.29.15–16), a 303 mm and 296 mm specimen.
Type locality: "Thale des Shireflusses (Mossambique)" [= Shire River, Malawi]. Emended to vallée du Schire, Nyassaland *fide* Roux-Estève (1974: 69) and Shire Valley *fide* Broadley (1983: 39).
Distribution: Southeastern Africa. Southern Malawi (Southern), N Mozambique (Zambézia) and NE Zimbabwe (Manicaland), 600–1415 m.
Sources: Laurent, 1964b, 1968a, Broadley, 1983, Branch, 1998 and Broadley & Blaylock, 2013.

19. *Letheobia pallida* Cope, 1869c. Proc. Acad. Sci. Philadelphia (1868) 20(5): 322.

Synonym: *Typhlops pallidus* Boulenger, 1893a.
Type: Lectotype, MCZ 5723 (formerly EIS 504), a 196 mm female (C. Cooke, 1860–1865), designated by Broadley & Wallach (2007: 45).
Type locality: "Zanzibar" [= Zanzibar Is., Tanzania].
Distribution: Northeastern Tanzania (Tanga, Pemba and Zanzibar Is.) and SE Kenya (Coast), NSL–50 m.
Sources: Pitman, 1974 and Pakenham, 1983.
Remarks: See Broadley & Wallach (2007: 45) for a discussion of the type specimens. In the 1860's when Cooke collected the types (plus numerous other specimens now deposited in the MCZ all with the label "Zanzibar"), Zanzibar referred to 1) a large region including Tanzania, Kenya and S Somalia, 2) the island of Zanzibar, and 3) the capital city of Zanzibar (06°10'S, 39°12'E, elevation NSL). Even though Pakenham (1983: 21) cited the type locality as "Zanzibar Island," Cooke's locality may have been the city, island, or mainland area.

20. *Letheobia pauwelsi* Wallach, 2005. Afr. J. Herp. 54(1): 86–89, fig. 1.

Type: Holotype, IRSNB 2580 (formerly OSGP 806), a 310 mm male (O.S.G. Pauwels, 18 Nov. 2001).
Type locality: "Eastern base of Mont Iboundji, Dpartment de Offoué-Onoy, Ogooué-Lolo Province, central Gabon, ca. 1° 08' S, 11° 52' E, 580 m."
Distribution: Central Gabon (Ogooué-Lolo), 580 m. Known only from type locality.
Source: Pauwels & Vande weghe, 2008.

21. *Letheobia pembana* Broadley & Wallach, 2007b. Zootaxa (1515): 45.

Type: Holotype, SMF 16688, a 158 mm male (A. Voeltzkow, 1905).
Type locality: "Pemba Island, Tanzania (05°15'S, 39°50'E, elevation NSL–100 m)."
Distribution: Northeastern Tanzania (Pemba Is.), NSL–100 m. Known only from type locality.

22. *Letheobia praeocularis* (Stejneger, 1894a). Proc. U.S. Natl. Mus. (1893) 16(970): 709–710. (*Typhlops praeocularis*)

Synonym: *Typhlops praeocularis lundensis* Laurent, 1964.
Type: Holotype, USNM 20799, a 340 mm specimen (J.H. Camp, 1887–1893).
Type locality: "Leopoldville, or Stanley Pool, Congo Free State, Africa" [= Kinshasa, Democratic Republic of the Congo, 04°20'S, 15°19'E, elevation 240 m].
Distribution: Central Africa. Southern Congo (Brazzaville), SW Democratic Republic of the Congo (Bandundu, Bas-Congo, Kinshasa) and NE Angola (Lunda Norte), 240–700 m.
Sources: J.-F. Trape & Roux-Estève, 1995, J.B. Rasmussen, 1997b and Chirio & LeBreton, 2007.
Remarks: Probably occurs in Nigeria and Cameroon *fide* Chirio & LeBreton (2007: 328).

L

23. *Letheobia rufescens* (Chabanaud, 1917a). Bull. Mus. Hist. Nat., Paris (1916) 22(7): 365–366, figs. 4–6. (*Typhlops rufescens*)

Synonyms: *Typhlops vrydaghi* Laurent, 1956, and *Typhlops vrijdaghi* Witte, 1966 *(nomen emendatum)*.
Type: Holotype, MNHN 1916.262, a 422 mm specimen (J. du Rouchet de Chazotte, 1908).
Type locality: "Congo français" [= Haut-Oubangui, Central African Republic *fide* Roux-Estève, 1974: 230].
Distribution: Central African Republic (Bamingui-Bangoran, Haut-Mbomou, Mbomou, Ombella-Mpoko, Ouham) and N Democratic Republic of Congo (Orientale), 450–650 m.
Sources: Joger, 1990 and Chirio & Ineich, 2006.

24. *Letheobia scorteccii* (Gans & Laurent, 1965). Ann. Mus. Roy. Congo Belge Tervuron (8) (134): 56–60, fig. 12, pl. 6. (*Typhlops scorteccii*)

Type: Holotype, MCZ 74455, a 221 mm specimen (C. Gans, 25 July 1961).
Type locality: "10 km south of Afgoí (40 km south of Mogadiscio), on the sand dune road to Merca, Benadir

Province, Somali Republic" [= 10 km S Afgoye on road to Merca, NE Shabeellaha Hoose Region, S Somalia, 2°04'N, 45°03'E, elevation 80 m].

Distribution: Southern Somalia (Benadir, Mogadishu, Shabeellaha Hoose), NSL–100 m.

25. *Letheobia simoni* (Boettger, 1879). Ber. Senck. Naturf. Ges. 1879a: 58–60. (*Onychocephalus simoni*)

Type: Lectotype, SMF 16673 (formerly SMF-B 7051,1a), a 196 mm specimen (H. Simon, 1879), designated by Mertens (1922: 178).

Type locality: "Haiffa, Syrien" [= Haifa, Palestine Distr., Israel, 32°49'N, 35°02'E, elevation 10 m] via lectotype selection.

Distribution: Middle East. Southwestern Turkey (Urfa), Israel (Central, Gaza Strip, Golan Heights, Haifa, Jerusalem, Northern, Southern, Tel Aviv, West Bank), SW Syria, and NW Jordan (Ajloun, Balqa, Irbid, Jarash, Madaba), 10–585 m.

Sources: Disi, 1985, Disi et al., 1988, Wallach, 1994, Franzen, 2000, Bouskila & Amitai, 2001, Disi et al., 2001, Amr & Disi, 2011 and Bar & Haimovitch, 2011.

L

26. *Letheobia somalica* (Boulenger, 1895h). Proc. Zool. Soc. London 63(3): 536, pl. 30, fig. 1. (*Typhlops somalicus*)

Synonyms: *Typhlops acutirostris* Mocquard, 1905, and *Typhlops somalicu* – K.P. Schmidt, 1923 *(nomen incorrectum).*

Type: Holotype, BMNH 1946.1.11.54 (formerly BMNH 1895.6.11.5), a 450 mm specimen (A. Donaldson-Smith, 19 Dec. 1894).

Type locality: "Beearso, Galla or western Somali-land" [= Bio Asse River, E Ethiopia, 06°53'N, 42°18'E, elevation 600 m].

Distribution: Ethiopia (Bale, Gojjam, Gondar, Hararge, Kefa, Shoa, Wollega), 600–2200 m.

Sources: Largen, 1978, Largen & Rasmussen, 1993 and Largen & Spawls, 2010.

27. *Letheobia stejnegeri* (Loveridge, 1931). Copeia 1931(3): 92–93, figs. 1–2. (*Typhlops stejnegeri*)

Synonym: *Typhlops vanderysti* Witte, 1933.

Type: Holotype, USNM 23979, a 418 mm specimen (D.W. Snyder, 1897).

Type locality: "Luebo District, Belgian Congo" [= Luebo Distr., Kasai Occidental Prov., S Democratic Republic of the Congo].

Distribution: Democratic Republic of the Congo (Bandundu, Kasai Occidental, Kasai Oriental, Kinshasa), 300–475 m.

Source: Witte, 1933.

28. *Letheobia sudanensis* (K.P. Schmidt, 1923). Bull. Amer. Mus. Nat. Hist. 49(1): 51–52, fig. 2. (*Typhlops sudanensis*)

Type: Holotype, AMNH 11677, a 469 mm specimen (H. Lang & J.P. Chapin, Nov. 1911).

Type locality: "Faradje, Sudan (ca. 3° 40' N., 29° 40' E.)" [= Ituri, Orientale Prov., NE Democratic Republic of the Congo, 03°43'N, 29°42'E, elevation 820 m].

Distribution: Northeastern Democratic Republic of the Congo (Orientale), 700–1300 m.

Source: Pitman, 1974.

Remarks: Type locality erroneously listed as Garamba (04°10'N, 29°40'E) *fide* K.P. Schmidt (1923: 4).

29. *Letheobia swahilica* Broadley & Wallach, 2007b. Zootaxa (1515): 46–47, fig. 4b.

Type: Holotype, MCZ 40076, a 166 mm male (A. Loveridge, 14 June 1934).

Type locality: "old village of Ngatana, ca. one mile north-west from new village of Wema, in the lower Tana River, Coast Region, Kenya (02°30'S, 40°15'E, elevation 50 m)."

Distribution: Southeastern Kenya (Coast) and NE Tanzania (Tanga, Zanzibar Is.), NSL–50 m.

30, *Letheobia toritensis* Broadley & Wallach, 2007b. Zootaxa (1515): 47–48, fig. 4c.

Type: Holotype, MCZ 53324, a 270 mm female (J.S. Owen, 19 June 1950).

Type locality: "Torit, Equatoria Province, Sudan (04°27'N, 32°31'E, elevation 625 m)" [= Torit, Eastern Equatoria Prov., SE South Sudan, 4°25'N, 32°35'E, elevation 620 m].

Distribution: Southeastern South Sudan (Eastern Equatoria), 620–1200 m.

31. *Letheobia uluguruensis* (T. Barbour & Loveridge, 1928). Mem. Mus. Comp. Zool. 50(2): 104–105. (*Typhlops uluguruensis*)

Type: Holotype, MCZ 23080, a 240 mm male (A. Loveridge, 6 Oct. 1926).

Type locality: "Nyange, Uluguru Mountains, Tanganyika Territory" [= Tanzania, 06°52'S, 37°46'E, elevation 760–850 m].

Distribution: Eastern Tanzania (Uluguru Mtns. of Morogoro), 760–1000 m.

Sources: Spawls et al., 2002 and Gower et al., 2004.

32. *Letheobia unitaeniata* (W.C.H. Peters, 1878b). Mber. Königl. Akad. Wiss. Berlin 1878(3): 205, pl. 2, fig. 5. (*Typhlops unitaeniatus*)

Type: Holotype, ZMB 9246, a 157 mm specimen (J.M. Hildebrandt, 1872–1878).

Type locality: "Districten von Taita, Kenia" [= Taita Hills, Kenya, ca. 03°25'S, 38°25'E].
Distribution: Eastern Africa. Southern Somalia (Bay, Gedo, Jubbada Dhexe, Jubbada Hoose, Mogadishu, Shabeellaha Hoose, Alessandra Is.), E Kenya (Coast, S Eastern) and ext. NE Tanzania (Kilamanjaro, Tanga), 20–1600 m.
Sources: Largen, 1978, Lanza, 1983a, 1990b, Largen & Rasmussen, 1993, Wallach, 1994 and Spawls et al., 2002.
Remarks: Uganda (Taru) record of Gans & Laurent (1965: 60) is from Taru, Kenya.

33. *Letheobia wittei* (Roux-Estève, 1974a). Mem. Mus. Natl. Hist. Nat. (2) 87A: 239–240, fig. 179a–c. (*Rhinotyphlops wittei*)

Type: Holotype, IRSNB 2511, a 285 mm specimen (R. Cremer & M. Neumann [Congo Exped.], 23 Aug. 1948).
Type locality: "Vandia-Poshe, rég. riv. Liki-Bembe, terr. Libenge, Distr. Ubangui, Prov. Équateur, Zaire,"[= Vandia-Poshe, NW Equateur Prov., NW Democratic Republic of the Congo, ca. 04°30'N, 19°00'E, elevation 520 m].
Distribution: Northwestern Democratic Republic of the Congo (NW Equateur), 520 m.
Sources: Joger, 1990 and Chirio & Ineich, 2006.
Remarks: Possibly occurs in Central African Republic (Ombella-Mpoko) *fide* Joger (1990: 95).

LIASIS Gray, 1842a (Pythonidae)

Synonyms: *Liasis* Gray, 1840 *(nomen nudum), Lisalia* Gray, 1849a, *Simalia* Gray, 1849a *(partim), Liases* – Bleeker, 1860g *(nomen incorrectum),* and *Katrinus* Hoser, 2000b.
Type species: *Liasis mackloti* A.-M.-C. Duméril & Bibron, 1844.
Distribution: Indo-Australia.
Fossil Record: Lower Pliocene of Australia.
Sources: Kluge, 1993b, D.G. Barker & Barker, 1994, Walls, 1998b, McDiarmid et al., 1999, Rawlings et al., 2008, Schleip & O'Shea, 2010 and Hoser, 2012b.
Remarks: Official Generic Name *fide* Opinion 1514 (ICZN, 1988).

1. †*Liasis dubudingala* Scanlon & Mackness, 2002. Alcheringa 25(3): 428–432, figs. 1, 2e.

Type: Holotype, QMF 9132, one middle trunk vertebra (M. Archer, 1975).
Type locality: "Main Quarry (Lat. 19°43'S, Long. 145°36'E), Allingham Formation, Bluff Downs Station, northeastern Queensland, Australia. Early Pliocene."
Distribution: Lower Pliocene (4.0–4.5 mya) of Australia. Known only from type locality.
Source: Archer & Wade, 1976.

2. *Liasis fuscus* W.C.H. Peters. 1873c. Mber. Kønigl. Akad. Wiss. Berlin 1873(10): 607–608.

Synonyms: *Liasis cornwallisius* A.C.L.G. Günther, 1879b, *Nardoa crassa* Macleay, 1885b, *Lisalia fusca* Wells & Wellington, 1985, and *Katrinus fuscus jackyae* Hoser, 2003a.
Type: Holotype, ZMB 7840, a 955 mm specimen (A. Dietrich, 1863–1872, via Mus. Godeffroy), lost *fide* Cogger et al. (1983: 203).
Type locality: "Port Bowen" [= Port Bowen, E Queensland, Australia, 20°01'S, 148°15'E, elevation NSL].
Distribution: Papua New Guinea (Western) and N Australia (N Northern Territory, N Queensland, ext. N Western Australia, Groote Eylandt, Maria, McClure, Melville and Mornington Is.), NSL–425 m.
Sources: O'Shea, 1996, C. Hay, 1998 and Torr, 2000.
Remarks: A synonym of *L. mackloti fide* Kluge (1993: 6).

3. *Liasis mackloti* A.-M.-C. Duméril & Bibron, 1844. Erpét. Gén. 6: 440–442.

Synonyms: *Python timorensis* S. Müller & Schlegel, 1844 *(nomen rejiciendum), Python timoriensis* S. Müller, 1857 *(nomen rejiciendum), Liasis timoriensis* W.C.H. Peters, 1876b *(nomen rejiciendum), Liasis macloti* – T. Barbour, 1912b *(nomen incorrectum), Liasis mackloti dunni* Stull, 1932b, and *Liasis mackloti savuensis* Brongersma, 1956b.
Type: Lectotype, MNHN 1625, longest type specimen 457 mm (S. Müller & H.C. Macklot, Sept. 1829), designated by Brongersma (1968: 57).
Type locality: "Timor" [Lesser Sundas, SE Indonesia] via lectotype selection.
Distribution: Southeastern Indonesia (Alor Roti, Savu, Semau, Timor, Wetar) and Timor-Leste, NSL–10 m.
Sources: Stull, 1932b, C. Haas, 1950, L.A. Smith, 1981b, O'Shea, 1996, Hoser, 2000b, C.K. Carmichael, 2007 and O'Shea et al., 2012.
Remarks: Official Specific Name no. 2426 *fide* Opinion 939 (ICZN, 1970).

4. *Liasis olivaceus* Gray, 1842a. Zool. Misc. 2(Mar.): 45. (*Liasis olivacea*)

Synonyms: *Liasi olivaceo* – F. Werner, 1936 *(nomen incorrectum),* and *Liasis olivaceus barroni* L.A. Smith, 1981.
Type: Holotype, BMNH 1946.1.1.56 (formerly BMNH IV.4.4a), a 965 mm specimen (J. Gilbert, 1840–1842).
Type locality: "North Australia; Port Essington" [= ruins at tip of Cobourg Peninsula, N Northern Territory, N Australia, ca. 11°09'N, 132°00'E, elevation NSL].
Distribution: Northern Australia (N Northern Territory, NW Queensland, N Western Australia, Augustus, Bickerton, Groote Eylandt, Heywood, Inglis, Melville,

North, St. Andrew, Vanderlin and Warnawi Is.), NSL–650 m.

Sources: L.A. Smith, 1981b, Cogger, 2000 and Torr, 2000.

LICHANURA Cope, 1861e (Boidae)

Synonym: *Lichanotus* – Wallace, 1876 *(lapsus calami).*

Type species: *Lichanura trivirgata* Cope, 1861e.

Distribution: Southwestern USA and NW Mexico.

Fossil records: Upper Miocene and upper Pleistocene of SW USA. Fossils tentatively assigned to *Lichanura* include middle Miocene (Clarendonian) of USA (California), upper Miocene (Clarendonian) of USA (California), and upper Miocene (Hemphillian) of USA (Nebraska).

Sources: Holman, 2000a, Wood et al., 2008 and Limburg et al., 2011.

1. *Lichanura orcutti* Stejneger, 1889. West. Amer. Sci. 6(46): 83.

Synonyms: *Lichanura simplex* Stejneger, 1890a, and *Lichanura roseofusca gracia* Klauber, 1931a.

Type: Holotype, USNM 15503, an 870 mm specimen (C.R. Orcutt, April 1889).

Type locality: "Colorado Desert, San Diego County, California."

Distribution: Southwestern USA (S California, W Arizona, S Nevada), 100–915 m.

Fossil records: Upper Pleistocene (Rancholabrean II) of USA (California).

Source: Stejneger, 1890a.

Remarks: Original description reprinted in Stejneger (1890a: 96–97, fig. 1). A valid species *fide* Wood et al. (2008: 496).

2. *Lichanura trivirgata* Cope, 1861e. Proc. Acad. Nat. Sci. Philadelphia 13(7): 304–305.

Synonyms: *Lichanura myriolepis* Cope, 1868a, *Lichanura roseofusca* Cope, 1868a, *Lichanura trivirgata bostici* Ottley, 1978, *Lichanura trivirgata saslowi* Mehrtens 1987 (*nomen nudum*), *Lichanura trivirgata saslowi* Spiteri, 1988, *Lichanura trivirgata myriolepsis* – McDiarmid, Campbell & Touré, 1999 *(nomen incorrectum),* and *Lichanura trivirgata arizonia* C. Scott, 1996.

Type: Lectotype, USNM 15502 (formerly SIM 5023), a 291 mm male (J. Xantus, Sept. 1859), designated by Cochran (1961: 195).

Type locality: "the southern region of Lower California" [= S Baja California, Mexico]. Emended to Cape San Lucas, Lower California [= Cabo San Lucas, S Baja California Sur State, NW Mexico, 22°53'N, 109°55'W, elevation 25 m] *fide* Stejneger (1891: 515).

Distribution: Southwestern USA (SW Arizona, ext. S California) and NW Mexico (Baja California Norte, Baja California Sur, Sonora, Angel de la Guarda, Cedros, Cerralro, Mejia, Natividad, San Marcos and Tiburon, Is.), NSL–1370 m.

Sources: Stejneger, 1891, Klauber, 1933, Fowlie, 1965, Gorman, 1965, Kurfess, 1967, Bostic, 1971, Ottley, 1978, Ottley et al., 1980, Yingling, 1982, R.W. Murphy & Ottley, 1984, Stebbins, 1985, Spiteri, 1988, 1991, 1993, 1994, P.R. Brown, 1997, Bartlett & Tennant, 2000, Grismer, 2002 and J.C. Murphy & Schlager, 2003.

Remarks: USNM 5023 is also the type of *Thamnophis cyrtopsis cyclides* Cope. Gorman (1965: 283) suggested that *L. roseofusca* Cope is a valid species.

LIMNOPHIS A.C.L.G. Günther, 1865 (Natricidae)

Synonym: *Lymnophis* – Cope, 1887c *(nomen emendatum).*

Type species: *Limnophis bicolor* A.C.L.G. Günther, 1865.

Distribution: Southern Africa.

Remarks: *Limnophis* Günther not preoccupied by *Limnophis* Blanchard, 1845 (Diptera) *contra* K.L. Williams & Wallach (1989: 86) as it is a misspelling for *Limnobia* Meigen.

1. *Limnophis bangweolicus* (Mertens, 1936). Zool. Anz. 114(9–10): 284–285. (*Helicops bangweolicus*)

Type: Holotype, SMF 22172, a 520 mm female (F. Haas, 26 Oct. 1931).

Type locality: "Nsombo, Nordende das Bangweolo-Sees, Nord-Rhodesien" [= Nsombo, Lake Bangweulu, Northern Prov., Zambia, 10°48'S, 30°00'E, elevation 1200 m].

Distribution: Southwestern Africa. Eastern Angola (Lunda Sul, Moxico), SE Democratic Republic of the Congo (Katanga), Zambia (Luapula, Northern, Southern, Western), ext. W Zimbabwe (NW Matebeleland North), NE Namibia (Caprivi) and N Botswana (North West), 950–1690 m.

Sources: Broadley et al., 2003 and Broadley & Blaylock, 2013.

Remarks: A valid species *fide* Griffin (2003: 96).

2. *Limnophis bicolor* A.C.L.G. Günther, 1865. Ann. Mag. Nat. Hist. (3) 15(86): 96–97, pl. 2, fig. c.

Types: Syntypes (2), BMNH, 660 mm and 530 mm females (F.A.P. Bayão, 1864).

Type locality: "the province Duque de Bragance (Angola)" [= Duque de Bragança, cen. Malanje Prov., N Angola, 9°06'S, 15°57'E, elevation 1110 m].

Distribution: Southern Africa. Southern Democratic Republic of the Congo (Bandundu, Bas-Congo, Katanga), Angola (Benguela, Huíla, Cabinda, Luanda,

Lunda Norte, Malanje), W Zambia (NW North-Western) and ext. W Zimbabwe (W Matebeleland North), 100–1725 m.

Sources: Witte, 1953, 1962, Mertens, 1963, Laurent, 1964a, Broadley & Cock, 1975, Broadley & Blake, 1979, Broadley, 1983, Auerbach, 1987, Branch, 1988, H.M. Smith et al., 1992, M. Griffin, 2003 and Zaher et al., 2009.

Remarks: Bocage listed as collector *fide* A.C.L.G. Günther (1865: 97) and Boulenger (1893a: 275) but Bocage (1866a: 47, 1866b: 68, and 1895a: 77) sent the specimens to BMNH and listed Bayão as collector. Democratic Republic of the Congo (Bas-Congo) record doubtful *fide* Thys van der Audenaerde (1965: 370).

LIOHETEROPHIS Amaral, 1934b (Xenodontidae)

Type species: *Lioheterophis iheringi* Amaral, 1934b.
Distribution: Brazil.
Sources: J.A. Peters & Orejas-Miranda, 1970 and Amaral, 1978.

1. *Lioheterophis iheringi* Amaral, 1934b. Mem. Inst. Butantan (1933–1934) 8: 187–188.

Type: Holotype, IB 8362, a 360 mm female (R.T.G.W. von Ihering, 1933), destroyed by fire 15 May 2010.
Type locality: "Campina Grande, Estado da Parahyba, nordeste do Brasil" [= Campina Grande, E Paraiba State, NE Brazil, 7°14'S, 35°53'W, elevation 465 m].
Distribution: Extreme NE Brazil (Paraíba), 465 m. Known only from holotype.

LIOPELTIS Fitzinger, 1843 (Colubridae)

Synonyms: *Phragmitophus* A.C.L.G. Günther, 1862b, *Leopeltis* – J.F. James, 1887 *(nomen incorrectum)*, *Lopeltis* – Takahashi, 1930 *(nomen incorrectum)*, *Lipopeltis* – Gharpurey, 1962 *(nomen incorrectum)*, and *Liopelitis* – Kuntz, 1970 *(nomen incorrectum)*.
Type species: *Herpetodryas tricolor* Schlegel, 1837.
Distribution: Southern and SE Asia and East Indies.
Sources: Boulenger, 1894a, M.A. Smith, 1943 and Welch, 1988.

1. *Liopeltis calamarius* (A.C.L.G. Günther, 1858). Cat. Colub. Snakes Brit. Mus.: 250–251. (*Cyclophis calamaria*)

Synonyms: *Homalosoma baliolum* Jan, 1862b, *Cyclophis nasalis* A.C.L.G. Günther, 1864a, and *Cyclophia calamaria* – Phipson, 1888 *(nomen incorrectum)*.
Types: Syntypes (2), BMNH 1946.1.21.64, two males, longest syntype 343 mm (Zool. Soc. London).
Type locality: "Ceylon" [= Sri Lanka].
Distribution: India (Andhra Pradesh, Bihar, Karnataka, Kerala, Maharashtra, Tamil Nadu, Uttar Pradesh) and Sri Lanka (North-Central, North-Western, Southern, Uva), 1190–1690 m.
Sources: Ferguson, 1902, Wall, 1921d, 1924b, Deraniyagala, 1955, P. Silva, 1969, Murthy, 1986, A. Silva, 1990b, 2009, Malhotra & Davis, 1991, N. Khaire, 2006 and Srinivasulu & I. Das, 2008.

2. *Liopeltis frenatus* (A.C.L.G. Günther, 1858). Cat. Colub. Snakes Brit. Mus.: 120–121. (*Cyclophis frenatus*)

Type: Holotype, BMNH 1946.1.1.72, a 686 mm male (W. Griffith, June 1937).
Type locality: "Mountains near Affghanistan" [= Afghanistan] (in error). Corrected to Khasi Hills, India *fide* Boulenger (1894a: 280).
Distribution: Southeast Asia. Northeastern India (Arunachal Pradesh, Assam, Meghalaya, Nagaland), SW China (Xizang Zizhigu, Yunnan), N Myanmar (Kachin), N Laos (Xiangkhoang) and cen. Vietnam (Da Nang, Dong Nai, Gia Lai, Lao Cai, Quang Binh), 600–1830 m.
Sources: Bourret, 1936b, Deuve, 1970, Zhao & Adler, 1993, Ao et al., 2004, Whitaker & Captain, 2004, Ziegler et al., 2007, V.S. Nguyen et al., 2009 Southwest Pacific.
Remarks: Griffith collected in Afghanistan from Nov. 1838–June 1841. Pakistan (Sindh) record of Murray (1884) rejected *fide* Wall (1924a: 865).

3. *Liopeltis philippinus* (Boettger, 1897). Zool. Anz. 20(531): 164–165. (*Ablabes philippinus*)

Type: Lectotype, SMF 19318 (formerly SMF-B 8281,1a), a 628 mm male (O.F. von Moellendorff & O. Koch, 1897), designated by Mertens (1922: 180).
Type locality: "Inseln Culion (Calamianes), Philippien" [= Culion Island, Calamianes group, SW Philippines] via lectotype selection and type locality restriction *fide* Leviton (1964a: 371).
Distribution: Southwestern Philippines (Busuanga, Culion, N Palawan).
Sources: Boettger, 1898, E.H. Taylor, 1922a and Leviton, 1964a.
Remarks: Boettger's (1897: 165) Philippine record for Samar Island doubtful *fide* Leviton (1964a: 370).

4. *Liopeltis rappii* (A.C.L.G. Günther, 1860c). Proc. Zool. Soc. London 28(1): 154–155, pl. 26, fig. b. (*Ablabes rappii*)

Synonym: *Ablabes owenii* A.C.L.G. Günther, 1860.
Type: Holotype, BMNH 1946.1.5.61, a 419 mm female (H.R.A. von Schlagintweit, A. von Schlagintweit, & R. von Schlagintweit, 1854–1858).

L

Type locality: "Sikkim (5340 feet above the level of the sea)" [= Sikkim State, NE India].

Distribution: Southern Asia. Himalayas of Nepal (Kaski) and N India (Himachal Pradesh, Sikkim, N West Bengal), 340–2285 m.

Sources: Kramer, 1977, Murthy, 1986 and Schleich & Kästle, 2002 and Tiwari & Shah, 2004.

Remarks: Holotype listed as male *fide* Kramer (1977: 741). Record from Simla, W Himalayas doubtful *fide* M.A. Smith (1943: 186).

5. *Liopeltis stoliczkae* (Sclater, 1891a). J. Asiat. Soc. Bengal 60(3): 234, pl. 6, figs. 1a–b. (*Ablabes stoliczkae*)

Types: Syntypes (2), ZSI 3955 (formerly IMC 3955) (J. Butler, 1869–1876) [Assam], and IMC 8725 (F. Stoliczka, 1862–1874) [unknown], latter lost *fide* I. Das et al. (1998: 145).

Type locality: "Samagooting in the Naga Hills of Assam" [= Naga Hills, Nagaland, NE India] and unknown.

Distribution: Southeast Asia. Northeastern India (Nagaland, Sikkim, N West Bengal), cen. Myanmar (Kayah), cen. Laos (Bolikhamsai) and E Cambodia (Mondulkiri), 200–700 m.

L

Sources: Ao et al., 2004, B.L. Stuart et al., 2006, B.L. Stuart & Heatwole, 2008 and I. Das, 2010.

Remarks: Sclater (1891: 18) erroneously listed IMC 8725 as the Assam type collected by Butler and IMC 3955 as the no data specimen collected by Stoliczka.

6. *Liopeltis tricolor* (Schlegel, 1837). Essai Phys. Serp. 1: 151–152, 2: 187, pl. 6, figs. 16–18. (*Herpetodryas tricolor*)

Synonyms: *Coluber tricolor* H. Boie *in* Schlegel, 1826a *(nomen nudum),* and *Coluber tricolor* Schlegel, 1826b *(nomen nudum).*

Types: Syntypes (3), RMNH 492, RMNH 671 (H. Kuhl & J.C. van Hasselt, Dec. 1820–Sept. 1823), and RMNH 679 (H. Boie & H.C. Macklot, Dec. 1825–1832), longest syntype 560 mm.

Type locality: "Java" [SW Indonesia] Restricted to region of Mt. Pangerango and Mt. Salak, West Java, Indonesia *fide* Brongersma (1948b: 14). Emended to Tjihandjawar, at the foot of Mt. Pangerango, W. Java [= Cihanyawar, Nagrak, Jawa Barat, W Java, SW Indonesia, 07°03'S, 108°40'E] *fide* Brongersma (1950: 1499).

Distribution: Southeast Asia and East Indies. Southern Thailand, S Vietnam (Lam Dong), West Malaysia (Pinang), Singapore, East Malaysia (Sarawak), Brunei, Indonesia (Bangka, Bintan, Java, W Kalimantan, Mentawai Arch., Nias, Riau Arch., Sumatra, Weh) and SW Philippines (Bubuan, Palawan), 300–1200 m.

Sources: Rooij, 1917, C. Haas, 1950, Leviton, 1964a, Tweedie, 1983, Stuebing, 1991, M.J. Cox et al., 1998, V.S. Nguyen et al., 2009 and I Das, 2010.

LIOPHIDIUM Boulenger, 1896a (Pseudoxyrhophiidae)

Synonyms: *Idiophis* Mocquard, 1901, *Dakaria* F. Werner, 1925 *(nomen praeoccupatum), Wernerodakaria* Strand, 1928 *(nomen substitutum), Parasibynophis* Leviton & Munsterman, 1956, *Dekaria* – Tiedemann & Häupl, 1980 *(nomen incorrectum),* and *Liophidum* – Obst, Richter & Jacob, 1984 *(nomen incorrectum).*

Type species: *Liophidium trilineatum* Boulenger, 1896a.

Distribution: Madagascar, Mascarenes and Comoros.

Sources: Leviton & Munsterman, 1956, Guibé, 1958, E. Morgan, 1973, Domergue, 1984a, Cadle, 1999, Glaw & Vences, 2007, Franzen et al., 2009, Zaher et al., 2009 and Vieites et al., 2010.

Remarks: Morgan (1973: 9) erroneously listed type species as *Coronella torquata* Boulenger, 1888.

1. *Liophidium apperti* Domergue, 1984a. Bull. Mus. Natl. Hist. Nat. (1983) (4) 5A(4): 1111–1113, figs. 1a–c, 2.

Type: Holotype, MNHN 1982.442 (formerly CAD 873/S), a 238 mm male (C.A. Domergue, 17 Nov. 1968).

Type locality: "7 km au nord de la petite ville de Befandriana-sud, 2 km, à l'est de la route nationale no 9, Tuléar-Morombe, la région Malgache" [= Toliara Province, SW Madagascar].

Distribution: Southern Madagascar (S Toliara), 200–615 m.

Source: Ziegler et al., 1996.

2. *Liophidium chabaudi* Domergue, 1984a. Bull. Mus. Natl. Hist. Nat. (1983) (4) 5A(4): 1117–1119, figs. 6a–c.

Type: Holotype, MNHN 1978.2789 (formerly CAD 730/S), a 446 mm female (C.A. Domergue, 1970–1978).

Type locality: "Village d'Ankorongo, non loin de l'aérodrome de Tuléar, de la région Malgache" [= Toliara Province, SW Madagascar].

Distribution: Southwestern Madagascar (SW Toliara), 10–200 m.

3. *Liophidium maintikibo* Franzen, Jones, Raselimanana, Nagy, Cruze, Glaw & Vences, 2009. Amphibia-Reptilia 30(2): 174–177, figs. 1a–b, 2–3.

Type: Holotype, ZSM 2052/2007, a 255 mm specimen (J. Jones, 20 Nov. 2004).

Type locality: "near Kirindy research station (20°03'S, 44°39'E), elevation below 100 m, about 60 km north of Morondava (district of Morondava, region of Menabe, Toliara Province), western Madagascar."

Distribution: Western Madagascar (Toliara), < 100 m. Known only from type locality.

4. *Liophidium pattoni* Vieites, Ratsoavina, Randrianianina, Nagy, Glaw & Vences, 2010. Salamandra 46(1): 3–7, figs. 1a–c, a–d2.

Type: Holotype, ZSM 186/2009 (formerly DRV 5948), a 417 mm male (M. Vences, D.R. Vieites, F.M. Ratsoavina & R.D. Randrianiaina, 28 June 2009).
Type locality: "Angozongahy at the western side of the Makira plateau, within the newly created reserve 'Makira Natural Park', 15°26'13.3"S, 49°07'07.0"E, 1009 m above sea level, district of Mandritsara, region of Sofia, province of Mahajanga, northeastern Madagascar."
Distribution: Northeastern Madagascar (Mahajanga), NSL–1100 m.

5. *Liophidium rhodogaster* (Schlegel, 1837). Essai Phys. Serp. 1: 152–153, 2: 193–194. (*Herpetodryas rhodogaster*)

Type: Holotype, MNHN 1956, a 530+ mm female lacking a skull (J.R.C. Quoy & J.P. Gaimard, 1817–1829).
Type locality: "'île de Madagascar."
Distribution: Eastern Madagascar (Atananarivo, Antsiranana, Fianarantsoa, Toamasina, Nosy Be Is.) 20–1300 m.
Source: Henkel & Schmidt, 2000.

6. *Liophidium therezieni* Domergue, 1984a. Bull. Mus. Natl. Hist. Nat. (1983) (4) 5(4): 1114–1116, figs. 4a–c, 5a.

Type: Holotype, MNHN 1982.444 (formerly CAD 679/S), a 726 mm male (Y. Therezien, 14 Feb. 1966).
Type locality: "Forêt d'Anatelo, en bordure du massif calcaire de l'Ankarana (sous-préfecture d'Ambilobe) de la région malgache" [= Anatelo forest, Ambilobe Prefect., NW Antsiranana Province, N Madagascar].
Distribution: Extreme N Madagascar (N Antsiranana), 20–230 m.
Source: Franzen et al., 2009.

7. *Liophidium torquatum* (Boulenger, 1888a). Ann. Mag. Nat. Hist. (6) 1(2): 103–104, pl. 5, fig. 3. (*Coronella torquata*)

Type: Holotype, BMNH 1946.1.3.38, a 410 mm female (R. Baron, 1875–1888).
Type locality: "Madagascar."
Distribution: Madagascar (Antsiranana, Fianarantsoa, Mahajanga, Toamasina, S Toliara), 100–950 m.
Sources: Ziegler et al., 1996, Henkel & Schmidt, 2000 and A. Mori et al., 2006.

8. *Liophidium trilineatum* Boulenger, 1896a. Cat. Snakes Brit. Mus. 3: 599.

Type: Holotype, BMNH 1946.1.15.66, a 330 mm male (J.T. Last, 1890–1896).
Type locality: "S.W. Madagascar."
Distribution: Southwestern Madagascar (SW Toliara), 50–100 m.
Source: Ziegler et al., 1996.

9. *Liophidium vaillanti* (Mocquard, 1901). Bull. Mus. Hist. Nat., Paris 7(6): 253. (*Idiophis vaillanti*)

Synonyms: *Idiophis vaillanti extensa* Boettger, 1913, and *Dakaria subpunctata* F. Werner, 1925.
Type: Lectotype, MNHN 1901.186, a 608 mm female (C. Alluaud, 1892–1893), designated herein by E.C. Morgan.
Type locality: "Isaka, le sud de Madagascar" [= Isaka-Ivondro, 30 km NW Taolagnaro, SE Toliara Prov., S Madagascar] via lectotype selection.
Distribution: Madagascar (Mahajanga, Toliara) and the Mascarenes (Reunion), NSL–800 m.
Sources: Mocquard, 1902a, M.A. Smith, 1928b and A. Mori et al., 2006.
Remarks: Lectotype designation of MNHN 1901.186 *fide* Morgan (1973: 134) invalid.

LIOPHIS Wagler, 1830 (Xenodontidae)

Synonyms: *Iaculatrix* Boddaert, 1783 *(nomen nudum)*, *Dromicus* Bibron, 1843 *in* Ramón de la Sagra, 1838–1857, *Calophis* Fitzinger, 1843, *Leimadophis* Fitzinger, 1843, *Opheomorphus* Fitzinger, 1843, *Pariopeltis* Fitzinger, 1843, *Pseudophis* Fitzinger, 1843, *Caelophis* – A.-M.-C. Duméril & Bibron, 1844 *(nomen incorrectum)*, *Calliophis* Agassiz, 1847 *(nomen emendatum)*, *Limadophis* Agassiz, 1847 *(nomen emendatum)*, *Dromiscus* – Kingsley, 1888 *(nomen incorrectum)*, *Lhiophis* – Peracca, 1896b *(nomen incorrectum)*, *Adromicus* – Cope, 1900 *(nomen incorrectum)*, *Liopis* – Rosén, 1911 *(nomen incorrectum)*, *Lemiadophis* T. Barbour, 1916c *(nomen incorrectum)*, *Leophis* – Briceño-Rossi, 1934 *(nomen incorrectum)*, and *Oreomorphus* – Cochran, 1961 *(nomen incorrectum)*.
Type species: *Coluber cobella* Linnaeus, 1758.
Distribution: Central America, West Indies and South America.
Fossil record: Pleistocene of West Indies (Barbados).
Sources: Roze, 1964, J.A. Peters & Orejas-Miranda, 1970, Cunha, 1976b, Dixon 1980, 1989, Pérez-Santos & Moreno, 1988, Vidal et al., 2000, Curcio et al., 2009a, Hedges et al., 2009, Zaher et al., 2009 and Vidal et al., 2010.
Remarks: *Liophis amarali* Wettstein, 1930 has been placed in a new genus (*Caaeteboia*) by Zaher et al., 2009. *Liophis* is paraphyletic *fide* Curcio et al. (2009a:

66–67) and Zaher et al. (2009: 146) and considered a synonym of *Erythrolamprus fide* Grazziotin et al. (2012: 14).

1. *Liophis almadensis* (Wagler *in* Spix, 1824). Serp. Brasil. Sp. Nov.: 30–31, pl. 10, fig. 3. (*Natrix almadensis*) (*nomen corrigendum*)

Synonyms: *Natrix almada* Wagler *in* Spix, 1824 *(nomen incorrigendum), Liophis almadensis* Wagler *in* Spix, 1824 *(nomen corrigendum)*, ? *Natrix bahiensis* Wagler *in* Spix, 1824, *Liophis wagleri* Jan, 1857, *Liophis conirostris* A.C.L.G. Günther, 1858, *Liophis y-graecum* W.C.H. Peters, 1882, and *Trigonocephalus scolecomorphus* Bacqué, 1906.

Type: Lectotype, ZSM 2688/0, a 170 mm specimen (J.B. von Spix & K.F.P. von Martius, Oct.–Nov. 1819), designated by Hoogmoed & Gruber (1983: 331).

Type locality: "Almada, Bahia, Brazil" [= Castelo Novo, Bahia State, NE Brazil, ca. 14°39'S, 39°10'W] via lectotype selection.

Distribution: Southern South America. Southern and E Brazil (Bahia, Distrito Federal, Goiás, Maranhão, Mato Grosso, Mato Grosso do Sul, Minas Gerais, S Pará, Paraíba, Parana, Pernambuco, Rio de Janeiro, Rio Grande do Norte, Rio Grande do Sul, Santa Catarina, São Paulo, Marajo Is.), E Bolivia (Beni, Santa Cruz), Paraguay (Alto Paraguay, Alto Paraná, Amambay, Caaguazú, Caazapá, Canindeyú, Central, Concepción, Cordillera, Guairá, Itapúa, Misiones, Neembucú, Paraguarí), NE Uruguay (Rivera, Tacuarembó, Trienta y Tres) and N Argentina (Chaco, Corrientes, Entre Ríos, Formosa, Misiones, Santa Fe), 25–130 m.

Sources: Serie, 1921, 1936, Hoge, 1953, Barth, 1957, Cordeiro & Hoge, 1974, Abalos & Mischis, 1975, Amaral, 1978, Lema & Fabián-Beurmann, 1977, Lema, 1987, Dixon, 1991, Cei, 1994, Dixon et al., 1993, Rufino, 1999, Leynaud & Bucher, 1999, Giraudo & Scrocchi, 2002, Carreira-Vidal et al., 2005, Cacciali, 2008 and Prigioni et al., 2011.

Remarks: Original orthography *Natrix almada* in text, *Natrix almadensis* on plate. See Hoogmoed & Gruber (1983: 331) for discussion.

2. *Liophis andinus* Dixon, 1983b. Ann. Carnegie Mus. 52(6):129–132, figs. 5–6d.

Type: Holotype, CM 2808, a 485 mm male (J. Steinbach, Oct. 1921).

Type locality: "Incachaca, 2500 m, Cochabamba, Bolivia."

Distribution: Central Bolivia (Cochabamba), 2500 m. Known only from type locality.

3. *Liophis atraventer* Dixon & Thomas, 1985. Herpetologica 41(3): 260–261, fig. 1.

Type: Holotype, MZUSP 5066, a 505 mm male (W.R. Heyer, 28 Oct. 1968).

Type locality: "Boracéia (Estação Biológica), 28°38'S, 45°50'W, São Paulo, Brazil."

Distribution: Southeastern Brazil (Rio de Janeiro, São Paulo), 950 m.

Source: Passos & Fernandes, 2001.

Remarks: Latitude of type locality in error: probably 23°38'S.

4. *Liophis breviceps* Cope, 1860e. Proc. Acad. Nat. Sci. Philadelphia 12(6): 252.

Synonyms: *Liophis longiventris* Amaral, 1925b, and *Liophis canaima* Roze, 1957a.

Type: Holotype, ANSP 3697, a 442 mm female (C.J. Hering, 1860).

Type locality: "Suriname" [= Suriname].

Distribution: Amazonia. Southeastern Colombia (Amazonas, Meta, Vaupés), E Venezuela (Amazonas, Bolívar, Delta Amacuro, Monagas), Guyana (Barima-Waini, Cuyuni-Mazaruni, Demerara-Mahaica, Essequibo Islands-West Demerara, Potaro-Siparuni, Upper Demerara-Berbice), Suriname (Brokopondo, Suriname), French Guiana (Cayenne, Saint-Laurent-du-Maroni), E Ecuador (Napo, Pastaza), NE Peru (Loreto), N Brazil (Amazonas, Mato Grosso, ? Pará, Roraima) and Bolivia (Beni), NSL–500 m.

Sources: Beebe, 1946, Roze, 1966a, Dixon, 1983a, Pérez-Santos, 1986, Gorzula & Señaris, 1998, Starace, 1998, Frota, 2001a, D.S. Fernandes et al., 2002, Rivas-Fuenmayor & Molina, 2003, Navarrete et al., 2009 and C.J. Cole et al., 2013.

Remarks: Publication date erroneously listed as 1861 *fide* J.A. Peters & Orejas-Miranda (1970: 177).

5. *Liophis carajasensis* Cunha, Nascimento & Ávila-Pires, 1985. Publ. Mus. Avulsas Para. Emílio Goeldi (40): 53–54, pls. 4–5.

Type: Holotype, MPEG 16.611, a 429 mm male (Mus. Goeldi, 18 Nov. 1983).

Type locality: "campo rupestre do platô N1, Serra Norte, da área de Carajas, Pará, Brasil, 600–800 m."

Distribution: Northern Brazil (SE Pará), 600–800 m. Known only from type locality.

6. *Liophis ceii* Dixon, 1991. Texas J. Sci. 43(3): 230–233.

Type: Holotype, TCWC 53409, a 524 mm male (J.R. Dixon, 18 May 1977).

Type locality: "near Tucumán, Tucumán, Argentina" [= vicinity of Tucumán (26°49'S, 65°13'W, elevation 465 m), Tucumán Prov., N Argentina].

Distribution: Southeastern Bolivia (Santa Cruz, Tarija) and NW Argentina (Catamarca, Jujuy, Salta, Santiago del Estero, Tucumán), 1000–2000 m.

Source: Cei, 1994.

7. *Liophis cobellus* (Linnaeus, 1758). Syst. Nat., ed. 10, 1: 218–219. (*Coluber cobella*)

Synonyms: *Coluber venosa* Laurenti, 1768, *Coluber cobella dimidiata* Weigel, 1783, *Coluber serpentinus* Daudin, 1803d, *Coluber cenchrus* Daudin, 1803d, *Coluber cancellatus* Gravenhorst, 1807, *Coluber melanogaster* Gravenhorst, 1807, and *Rhadinae cobella* – Briceño-Rossi, 1934 *(nomen incorrectum).*

Types: Syntypes (5): ZMUU 4A–B (C. Gyllenborg, 1744), ZMUU 91 (MAF, 1745), and NHRa–b, a 650 mm and 590 mm specimen (MAFR).

Type locality: "America." Restricted to Suriname *fide* J.R. Dixon (*in litt.*).

Distribution: Guianas. Eastern Venezuela (Amazonas, Bolívar, Delta Amacuro, Monagas, Sucre), Trinidad and Tobago (Trinidad), Guyana (Barima-Waini, Cuyuni-Mazaruni, Demerara-Mahaica, East Berbice-Corentyne, Essequibo Islands-West Demerara, Mahaica-Berbice, Potaro-Siparuni), N Suriname (Brokopono, Commewijne, Marowijne, Nickerie, Saramacca, Suriname) and N French Guiana (Cayenne, Saint-Laurent-du-Maroni), NSL–2630 m.

Sources: Emsley, 1977, Dixon, 1983a, J.C. Murphy, 1997, Starace, 1998, Freitas, 1999 and Frota, 2001b, Fernandes et al., 2002, Navarrete et al., 2009 and C.J. Cole et al., 2013.

Remarks: Lectotype restricted to ZMUU without number *fide* Hoge (1964d: 59). *Liophis trebbaui* Roze, *L. ingeri* Roze and *L. torrenicola* Donnelly & Myers have been considered synonyms.

8. *Liophis cursor* (Bonnaterre, 1790). Tabl. Ency. Méth., Ophiol.: 27–28, pl. 42, fig. 3. (*Coluber cursor*)

Synonyms: *Coluber cursor* Lacépède, 1789 *(nomen rejiciendum), Coluber fugitivus* Donndorff, 1798, *Liophis putnami* Cope, 1862b, and *Liophis cursor andreoides* F. Werner, 1924a.

Types: Syntypes (2), ANSP 5580, a 654 mm specimen (Jardin de Plantes), and MNHN, a 624 mm specimen (M. de Chanvalon, King of France's cabinet, Paris).

Type locality: "Martinique" [Lesser Antilles].

Distribution: Lesser Antilles (W Martinique, Rocher de Diamant Is.).

Sources: Lazell, 1967, A. Schwartz & Thomas, 1975, Dixon, 1981, A. Schwartz & R.W. Henderson, 1991, Breuil, 2003 and R.W. Henderson & Powell, 2009.

9. *Liophis epinephelus* Cope, 1862b. Proc. Acad. Nat. Sci. Philadelphia 14(1): 78. (*Liophis epinephalus*)

Synonyms: *Liophis reginae albiventris* Jan, 1863a, *Liophis reginae quadrilineata* Jan, 1863a, *Opheomorphus alticolus* Cope, 1868b, *Zamenis ater* A.C.L.G. Günther, 1872b, *Liophis fraseri* Boulenger, 1894a, *Liophis bimaculatus* Cope, 1899b, *Liophis bipraeocularis* Boulenger, 1903b, *Liophis opisthotaenius* Boulenger, 1908a, *Liophis pseudocobellus* Peracca, 1914, *Liophis cobellus alticolus* Amaral, 1931, *Liopihis bipraeoculares* – Briceño-Rossi, 1934 *(nomen incorrectum), Leimadophis taeniurus juvenalis* Dunn, 1937b, *Leimadophis bimaculatus lamonae* Dunn, 1944b, *Leimadophis epinephalus ecuadorensis* Laurent, 1949, and *Liophis epinephelus kogiorum* Bernal-Carlo, 1994.

Type: Holotype, ANSP 3688 (formerly SIM 4305), a 527 mm specimen (N. Michler Exped., 1857–1858).

Type locality: "Truando, New Granada" [= Truandó, N Chocó Dept., W Colombia, 7°22'N, 77°06'W, elevation NSL].

Distribution: Lower Central America and NW South America. Costa Rica (Alajuela, Cartago, Guanacaste, Heredia, Limón, Puntarenas, San José, Bonita Is.), Panama (Bocas del Toro, Canal Zone, Chiriquí, Coclé, Colón, Darién, Herrera, Los Santos, Panamá, Veraguas, Barro Colorado Is.), W Colombia (Antioquia, Bolívar, Boyaca, Caldas, Caquetá, Cauca, Cesar, Chocó, Córdoba, Cundinamarca, Huila, La Guajira, Magdalena, Nariño, Norte de Santander, Risaralda, Santander, Tolima,Valle), W Venezuela (Aragua, Falcón, Mérida, Táchira, Trujillo, Zulia), W Ecuador (Azuay, Bolívar, Cañar, Carchi, El Oro, Esmeraldas, Guayas, Imbabura, Loja, Los Ríos, Manabí, Morona-Santiago, Pichincha, Tungurahua) and W Peru (Cajamarca, Huancavelica, Piura, La Libertad, Lima), NSL–3400 m.

Sources: Dixon, 1983e, Pérez-Santos, 1986, Bernal-Carlo, 1994, Rivas-Fuenmayor & Fuentes, 2002, Savage, 2002, Rojas-Runjaic et al., 2007, Navarrete et al., 2009 and Rojas-Morales, 2012b.

Remarks: Original orthography corrected in errata and addenda sheet *fide* Cope (1862e: 594).

10. *Liophis festae* (Peracca, 1897f). Boll. Mus. Zool. Anat. Comp. Univ. Torino 12(300): 16–17. (*Rhadinaea festae*)

Type: Holotype, MZUT 227 (formerly MZUT 2178), an adult male (E. Festa, Sept. 1895–July 1897).

Type locality: "Valle del Rio Santiago, Ecuador" [= Río Santiago, W Loreto, E Peru, ca. 3°49'S, 77°45'W, elevation 215 m].

Distribution: Eastern Ecuador (Chinchipe-Zamora, Morona-Santiago, Napo, Pastaza) and N Peru (Amazonas, Cajamarca), 200–1300 m.

Sources: Dixon & Markezich, 1979 and Frota & Yuki, 2001.

Remarks: Photograph of the holotype in Andreone & Gavetti (2007: fig. 11). Occurs in S Colombia *fide*

Dixon (1989: 13) but not listed *fide* Pérez-Santos & Moreno, 1988.

11. *Liophis frenatus* (F. Werner, 1909b). Mitt. Naturhist. Mus. Hamburg 26: 224. (*Rhadinaea frenata*)

Synonyms: *Rhadinaea brazili* Amaral, 1923, and *Liophis brazili* Amaral, 1926a.

Type: Holotype, ZMH, a 430 mm male (Rolle), destroyed in July 1943 during World War II.

Type locality: "Paraguay."

Distribution: Southeastern South America. Southeastern Brazil (Mato Grosso do Sul, S Minas Gerais, Paraná, São Paulo), Paraguay (Alto Paraná, Boquerón, Itapúa, Presedente Hayes, San Pedro) and Argentina (Corrientes, Misiones), 760–880 m.

Sources: Hoge et al., 1976, Dixon, 1983a, Freitas, 1999, Giraudo, 1999, Valdujo & Nogueira, 2001, Arzamendia & Giraudo, 2002 and D.S. Fernandes et al., 2002 and Cacciali, 2008.

12. *Liophis guentheri* Peracca, 1897b. Boll. Mus. Zool. Anat. Comp. Univ. Torino 12(274): 11–12.

Synonym: *Erythrolamprus albertguentheri* Grazziotin, Zaher, Murphy, Scrocchi, Benavides, Zhang & Bonatto, 2012 *(nomen substitutum)*.

Types: Syntypes (6), MZUT 1510.1–4 (formerly MZUT 1964), longest syntype 890 mm (A. Borelli, 1895–1896), BMNH 1946.1.5.69, a female (H. Druce), and BMNH unnumbered.

Type locality: "Caiza (Chaco boliviano)" [= Caiza, ext. S Tarija Dept., S Bolivia, 22°02'S, 63°41'W, elevation 625 m].

Distribution: Southern South America. Southern Bolivia (Tarija), Paraguay (Chaco, Presidente Hayes) and N Argentina (Chaco, Formosa, Salta, Santiago del Estero), 130–625 m.

Sources: Dixon, 1987, Cei, 1994 and Leynaud & Bucher, 1999.

13. *Liophis ingeri* Roze, 1958c. Acta Biol. Venez. 2(25): 303–305, fig. 2. (*Liophis cobella ingeri*)

Type: Holotype, FMNH 74038, a 630 mm male (J.A. Steyermark & J. Wurdack, 12 Feb. 1955).

Type locality: "Chimantá Tepui, Estado Bolívar, Venezuela. Colectado entre rocas a log largo del río Tirica, en la sección central del campo en la cumber; 1,900 metros de altura."

Distribution: Southeastern Venezuela (Bolívar), 1900 m.

Sources: D.S. Fernandes et al., 2002 and C.W. Myers & Donnelly, 2008.

Remarks: Previously a subspecies of *L. cobella*.

14. *Liophis jaegeri* (A.C.L.G. Günther, 1858). Cat. Colub. Snakes Brit. Mus.: 37. (*Coronella jaegeri*)

Synonyms: *Liophis dorsalis* W.C.H. Peters, 1863c, *Aporophis coralliventris* Boulenger, 1894c, and *Rhadinaea lineata* Jensen, 1900 (*nomen praeoccupatum*).

Types: Syntypes (2), BMNH 1946.1.5.78, a 584 mm female (G. Gardiner, 1836–1841) and BMNH 1946.1.9.12, a juvenile female (M. Clausen).

Type locality: "Brazil."

Distribution: Southeastern South America. Southeastern Brazil (Bahia, Espírito Santo, Mato Grosso do Sul, Minas Gerais, Paraná, Rio de Janeiro, Rio Grande do Sul, Santa Catarina, São Paulo), Paraguay (Alto Paraná, Amambay, Caaguazú, Caazapá, Canindeyú, Central, Concepción, Guairá, Itapúa, Misiones, Neembucú, Paraguarí, San Pedro, Concepción Is.), S Uruguay (Artigas, Canelones, Montevideo, Rocha, Salto) and NE Argentina (Buenos Aires, Chaco, Corrientes, Entre Ríos, Formosa, Misiones, Santa Fe, Treinta y Tres), NSL–850 m.

Sources: Boulenger, 1894a, Miranda et al., 1983, Dixon, 1985b, 1987a, Gudynas, 1989, J.D. Williams & Francini, 1991, Cei, 1994, Freitas, 1999, Leynaud & Bucher, 1999, Carreira-Vidal et al., 2005, Cacciali, 2008 and Prigioni et al., 2011.

15. *Liophis janaleeae* Dixon, 2000a. Copeia 2000 (2): 487–488.

Type: Holotype, BMNH 1874.8.4.62, a 422 mm male (A.H. Roff, 1874).

Type locality: "Moyobamba, Peru, 854 m,"[= Moyobamba, N San Martín Prov., Peru, 6°02'S, 76°58'W, elevation 850 m].

Distribution: Northern Peru (Huánuco, San Martín), 850–2700 m.

16. *Liophis juliae* (Cope, 1879). Proc. Amer. Philos. Soc. 18: 274–275. (*Aporophis juliae*)

Synonyms: *Leimadophis mariae* T. Barbour, 1914a, and *Dromicus juliae copeae* H.W. Parker, 1936b.

Type: Holotype, USNM 10152, a 640–668 mm female (F.A. Ober, 1877).

Type locality: "Island of Dominica" [= Dominica, Lesser Antilles].

Distribution: Lesser Antilles (Dominica, Guadeloupe: Basse-Terre and Grande-Terre, Marie-Galante, Terre-de-Bas des Saintes).

Sources: A. Schwartz & Thomas, 1975, Dixon, 1981, A. Schwartz & R.W. Henderson, 1991, Malhotra & Thorpe, 1999, Breuil, 2002, 2003 and R.W. Henderson & Powell, 2009.

17. *Liophis maryellenae* Dixon, 1985b. Proc. Biol. Soc. Washington 98(2): 295–297, fig. 2b.

Type: Holotype, AMNH 62202, a 325 mm male (R. M. Gilmore, 7 March 1936).

Type locality: "Annapolis (16°20'S-48°58'W), Goias, Brazil."

Distribution: Southeastern Brazil (W Bahia, Distrito Federal, S Goiás, Minas Gerais, Tocantins), 600–1040 m.

Sources: Dixon, 1987, Argôlo, 1999a, Freitas, 1999, Valdujo & Nogueira, 2001 and França, 2004.

18. *Liophis melanotus* (G. Shaw, 1802). Gen. Zool., Amph. 3(2): 534. (*Coluber melanotus*)

Synonyms: *Coluber raninus* Merrem, 1820, *Coluber vittatus* Hallowell, 1845a, *Liophis melanonotus* Cope, 1860e *(nomen emendatum), Aporophis crucifer* Ahl, 1925, *Leiphis melanotus* – Briceño-Rossi, 1934 *(nomen incorrectum), Liophis melanotus lamari* Dixon & Michaud, 1992, and *Liophis melanotus nesos* Dixon & Michaud, 1992.

Type: Lectotype, a 457–610 mm specimen described and illustrated in Seba (1735: 62, pl. 62, fig. 3), designated herein.

Type locality: "Africana, Promontorii Bonae Spei" [= Cape of Good Hope, South Africa] (in error *fide* Dixon, 1989: 15) via lectotype selection.

Distribution: Northern South America. Colombia (Antioquia, Arauca, Atlántico, Bolívar, Boyaca, Cesar, Córdoba, Cundinamarca, La Guajira, Magdalena, Meta, Santander del Norte, Sucre, Tolima, Valle del Cauca), Venezuela (Amazonas, Anzoátegui, Apure, Aragua, Barinas, Bolívar, Carabobo, Cojedes, Delta Amacuro, Distrito Federal, Falcón, Guárico, Lara, Mérida, Miranda, Monagas, Portuguesa, Sucre, Táchira, Trujillo, Yaracuy, Vargas, Zulia, Cubagua and Margarita Is.) and Trinidad and Tobago (Tobago, Trinidad), NSL–2630 m.

Sources: Roze, 1966a, Emsley, 1977, Lancini, 1986, Pérez-Santos, 1986, A. Schwartz & R.W. Henderson, 1988, 1991, Dixon & Michaud, 1992, Gorzula & Señaris, 1998, Renjifo & Lundberg, 1999, Markezich, 2002, Rivas-Fuenemayer & Barrio-Amorgós, 2005, Navarrete et al., 2009 and Ugueto & Rivas, 2010.

Remarks: Grenada record rejected *fide* Emsley (1977: 222).

19. *Liophis miliaris* (Linnaeus, 1758). Syst. Nat., ed. 10, 1: 220. (*Coluber miliaris*)

Synonyms: *Coluber merremii* Wied-Neuwied, 1821, *Coluber dictyodes* Wied-Neuwied, 1824b, *Coluber bicolor* A. Reuss, 1834, *Ablabes purpurans* A.-M.-C. Duméril, Bibron & Duméril, 1854a, *Coronella australis* A.C.L.G. Günther, 1858, *Ophiomorphus merremmii* – Cope, 1862b *(nomen incorrectum), Liophis collaris* Jan, 1863a, *Coronella orientalis* A.C.L.G. Günther, 1864a, *Rhadinaea chrysostoma* Cope, 1868b, *Coronella poecilolaemus* A.C.L.G. Günther, 1872b, *Rhadinaea orina* L.E. Griffin, 1916, *Dromicus amazonicus* Dunn, 1922, *Rhadinaea merremii natricoides* F. Werner, 1927, and *Liophis mossoroensis* Hoge & Lima-Verde, 1973.

Type: Holotype, NHR Lin-56 (formerly MAFR), a 600 mm specimen (Dahlberg via Mus. Drottn.), designated by Gans (1964b: 39).

Type locality: "Indiis" [= India] (in error *fide* Dixon, 1989: 15). Corrected to Suriname *fide* Dixon (1983d: 799).

Distribution: South America. Southeastern Colombia (Amazonas, Cundinamarca, Meta, Tolima, Vaupés), Venezuela (Amazonas, Bolívar), E Ecuador (Morona-Santiago, Napo, Pastaza), E Peru (Loreto, Madre de Dios), N Guyana (Cuyuni-Mazaruni, East Berbice-Corentyne, Potaro-Siparuni, Upper Demerara-Berbice), N Suriname (Nickerie, Para, N Sipaliwini, Suriname), N French Guiana (Cayenne, Saint-Laurent-du-Maroni), Brazil (Alagoas, Amapá, Amazonas, Bahia, Ceará, Espírito Santo, Goiás, Maranhão, Mato Grosso, Mato Grosso do Sul, Minas Gerais, Pará, Paraná, Pernambuco, Piaui, Rio de Janeiro, Rio Grande do Norte, Rio Grande Do Sul, Rondônia, Santa Catarina, São Paulo, Sergipe, Anchieta, Bom Abrigo, Cananéia, Cardoso, Comprida, Couves, Grande, Porchat, Santo Amaro, São Sebatião, São Vicente and Vitória Is.) and E Bolivia (Beni), 100–1960 m.

Sources: Gans, 1964b, Hoge & Gans, 1965, Cunha & Nascimento, 1970, Dixon, 1983d, Pérez-Santos, 1986, Chippaux, 1987, Nascimento et al., 1987, Fugler & Riva, 1990, Henle & Ehrl, 1991, J.D. Williams & Francini, 1991, Cei, 1994, Marques & Castro-Souza, 1993, Lema, 1994, Starace, 1998, Freitas, 1999, Frota & Santos, 2002, Fuentes-R., 2003, Carreira-Vidal et al., 2005, Duarte-Rocha & Van Sluys, 2006, Cicchi et al., 2007, C.W. Myers & Donnelly, 2008, Navarrete et al., 2009 and C.J. Cole et al., 2013.

Remarks: Photographs of the holotype in Gans (1964b: figs. 23a–e). Another syntype (ZMUU 28) lost long ago *fide* Lönnberg (1896: 37). Type locality restriction to the port of Santos, São Paulo State, Brazil *fide* Gans (1964b: 35) rejected by Dixon (1983d: 796).

20. *Liophis oligolepis* Boulenger, 1905b. Ann. Mag. Nat. Hist. (7) 15(89): 455.

Type: Holotype, BMNH 1946.1.4.66, a juvenile male (M.A. Robert, 1902–1905).

Type locality: "Igapé-Assu, Pará, Brazil." Emended to Igarapé-Açu, Pará, Brazil *fide* Cunha & Nascimento (1993: 73).

Distribution: Colombia, Venezuela (Amazonia), Brazil (Acre, Maranhão, Pará, Rondônia), Peru (Iquitos region) and N Bolivia, 40–100 m.

Sources: Cunha et al., 1985, Dixon & Soini, 1986 and Cunha & Nascimento, 1993.

L

21. *Liophis ornatus* (S.W. Garman, 1887b). Proc. Amer. Philos. Soc. 24(126): 281. (*Dromicus ornatus*)

Synonyms: *Dromicus giganteus* Jan, 1863b (*nomen nudum*), and *Leimadophis boulengeri* T. Barbour, 1914a.

Types: Syntypes (6), MCZ 6135, a 1058 mm female (S.W. Garman, 1 April 1879), MCZ 6136, a 756 mm male (S.W. Garman, 14 Feb. 1879), MCZ 6137, an 875 mm female (S.W. Garman, 14 Feb. 1879), MCZ 176765, an 857 mm female (S.W. Garman, 14 Feb. 1879), MCZ 176766, a 550 mm (svl) male (S.W. Garman, 14 Feb. 1879), and BMNH 1946.1.9.18 (formerly MCZ), a female (S.W. Garman, 14 Feb. 1879).

Type locality: "St. Lucia, West Indies." Emended to Point Castries, St. Lucia *fide* MCZ catalogue.

Distribution: Lesser Antilles (Maria Major, St. Lucia).

Sources: Dixon, 1981, A. Schwartz & R.W. Henderson, 1988, 1991 and R.W. Henderson & Powell, 2009.

Remarks: Only MCZ 6135 has locality data of St. Lucia; all others from Point Castries. Extirpated from St. Lucia *fide* R.W. Henderson & Powell (2009: 369).

22. *Liophis perfuscus* Cope, 1862b. Proc. Acad. Nat. Sci. Philadelphia 14(1): 77.

L

Synonym: *Liophis rufus* Jan, 1863a.

Types: Syntypes (2), USNM 6044a–b, a 423–432 mm male and 301 mm female (T.N. Gill, 1859–1862).

Type locality: "Barbadoes" [= Barbados, West Indies].

Distribution: Lesser Antilles (Barbados).

Sources: Emsley, 1963b, Dixon, 1981, A. Schwartz & R.W. Henderson, 1991 and R.W. Henderson & Powell, 2009.

Remarks: USNM 6044 erroneously listed as holotype *fide* A. Schwartz & Thomas (1975: 183). Last seen in 1963 and possibly extinct *fide* R.W. Henderson & Powell (2009: 369).

23. *Liophis poecilogyrus* (Wied-Neuwied, 1824a). Isis von Oken 14(6): 669. (*Coluber poecilogyrus*)

Synonyms: *Coluber m-nigrum* Raddi, 1820 (*nomen rejiciendum*), *Coluber alternans* Lichtenstein, 1823 (*nomen rejiciendum*), *Natrix forsteri* Wagler *in* Spix, 1824 (*nomen rejiciendum*), *Coluber doliatus* Wied-Neuwied, 1825c (*nomen rejiciendum*), *Xenodon schottii* Schlegel, 1837, *Liophis merremii sublineatus* Cope, 1860e, *Liophis subfasciatus* Cope, 1862b, *Opheomorphus doliatus caesius* Cope, 1863d, *Liophis reginae ornatissima* Jan, 1863a, *Liophis typhlus gastrosticta* Jan, 1863a, *Liophis verecundus* Jan, 1863a, *Liophis cobella flaviventris* Jan, 1866 *in* Jan & Sordelli, 1866–1870, *Liophis reginae viridicyanea* Jan, 1866 *in* Jan & Sordelli, 1866–1870, *Opheomorphus meleagris* Cope *in* Boulenger, 1894a (*lapsus calami*), *Rhadinaea dichroa* F. Werner, 1899a, *Rhadinaea praeornata* F. Werner, 1909d, *Leimadophis poecilogyrus reticulatus* H.W. Parker, 1931, *Leimadophis poecilogyrus albadspersus* Amaral, 1944j, *Leimadophis poecilogyrus amazonicus* Amaral, 1944j, *Leimadophis poecilogyrus franciscanus* Amaral, 1944j, *Leimadophis poecilogyrus intermedius* Amaral, 1944j, *Leimadophis poecilogyrus montanus* Amaral, 1944j, *Leimadophis poecilogyrus pictostriatus* Amaral, 1944j, *Leimadophis poecilogyrus pineticola* Amaral, 1944j, *Leimadophis poecilogyrus platensis* Amaral, 1944j, *Leimadophis poecilogyrus xerophilus* Amaral, 1944j, and *Dromicus poecilogyrus lancinii* Hoge, Romano & Cordeiro, 1978 (*nomen substitutum*).

Types: Syntypes (5), AMNH 3593, a 517 mm female, and AMNH 3594, a 378 mm female, and ZMH 4429–31 (formerly ZMH 787) (A.P.M. zu Wied-Neuwied, 26 Nov.–19 Dec. 1815).

Type locality: "von Barra de Jucú, unweit des Flusses Espírito Santo, Brasilien" [= Barra do Jucu, near Río Espírito Santo, E Espírito Santo State, E Brazil, ca. 20°20'S., 40°20'W, elevation NSL].

Distribution: South America. Venezuela (Amazonas, Anzoátegui, Bolívar, Guárico), S Guyana (Potaro-Siparuni, Upper Takutu-Upper Essequibo), Suriname (Commewijne, Nickerie, Suriname), Brazil (Alagoas, E Amazonas, Bahia, Ceará, Distrito Federal, Espírito Santo, Goiás, Maranhão, Mato Grosso, Mato Grosso do Sul, Minas Gerais, Pará, Paraiba, Paraná, Pernambuco, Piauí, Rio de Janeiro, Rio Grande do Norte, Rio Grande do Sul, S Rondônia, Santa Catarina, São Paulo, Sergipe, Cardoso, Grande, Marajo, Santo Amaro and São Vicente Is.), E Bolivia (Beni, Cochabamba, La Paz, Santa Cruz, Tarija), Paraguay (Alto Paraguay, Alto Paraná, Amambay, Asunción, Boquerón, Caaguazú, Caazapá, Canindeyú, Central, Concepción, Cordillera, Guairá, Itapúa, Misiones, Neembucú, Paraguarí, Presidente Hayes, San Pedro), Uruguay (Artigas, Cerro Largo, Colonia, Flores, Florida, Lavalleja, Maldonado, Montevideo, Paysandú, Rio Negro, Rivera, Rocha, Salto, San José, Soriano, Tacuarembó, Treinta y Tres) and Argentina (Buenos Aires, Chaco, Córdoba, Corrientes, Entre Ríos, Formosa, Jujuy, La Pampa, La Rioja, Misiones, Salta, San Luis, Santa Fe, Santiago del Estero, Tucumán), NSL–300 m.

Sources: Gomes, 1918, H.W. Parker, 1931c, Amaral, 1944j, Hoge, 1964e, Markezich, 1976, Vanzolini et al., 1980, Halloy, 1985b, Cei, 1986, 1994, J.D. Williams & Francini, 1991, Dixon & Markezich, 1992, D.R. Norman, 1994, Smith et al., 1994, Rufino, 1998a, Freitas, 1999, Fuentes & Barrio-Amorgós, 1999b, Leynaud & Bucher, 1999, Rivas-Fuenmayor et al., 2002, Carreira-Vidal et al., 2005, Duarte-Roche & Van Sluys, 2006, Cicchi et al., 2007, Natera-Mumaw et al., 2007, Cacciali, 2008, Navarrete et al., 2009 and C.J. Cole et al., 2013.

Remarks: Original description reprinted in Wied-Neuwied (1824b: 1103 & 1824c: pl. 44). Official Specific Name *fide* Opinion 1832 (ICZN, 1996a). Records from Ecuador (J.A. Peters, 1960a: 523 & J.A. Peters and Orejas-Miranda, 1970: 145) rejected *fide* Dixon (1989: 20).

24. *Liophis problematicus* C.W. Myers, 1986. Amer. Mus. Novit. (2853): 2–5, figs. 1–6.

Type: Holotype, FMNH 64733, a 275 mm male (H.H. Heller, 22 Nov.–20 Dec. 1950).

Type locality: "at 1520 m above sea level at San Juan, Tambopata [Río], Sandia [Province of], Department of Puno, Peru. (= San Juan del Oro, about 30 km ENE of the town of Sandia, on the west bank of the Río Tambopata—at 14°13' S, 69°10' W.)"

Distribution: Extreme SE Peru (NE Puno), 1520 m. Known only from type locality.

25. *Liophis reginae* (Linnaeus, 1758). Syst. Nat., ed. 10, 1: 219. (*Coluber reginae*)

Synonyms: ? *Coluber crassicaudus* Gmelin, 1789, *Coluber violaceus* Lacépède, 1789 (*nomen rejiciendum*), *Coluber graphicus* G. Shaw, 1802, *Natrix semilineata* Wagler *in* Spix, 1824, *Liophis reginae maculata* Steindachner, 1867b, *Leimadophis reginae macrosoma* Amaral, 1935e, *Leimadophis reginae maculicauda* Hoge, 1953f, *Liophis miliaris intermedius* Henle & Ehrl, 1991, and *Liophis dorsocorallinus* Esqueda, Natera, La Marca & Ilija-Fistar, 2005b.

Type: Lectotype, NHR Lin-45 (formerly MAFR), a 200–250 mm specimen (Mus. Drottn.), designated by Hoge (1964d: 57–58) and Dixon (1983b: 114).

Type locality: "Indiis" [= India] (in error *fide* Hoge, 1964d: 57). Restricted to Suriname by Hoge (1964d: 57) and Dixon (1983b: 114).

Distribution: South America. Southeastern Colombia (Amazonas, Arauca, Boyacá, Caqueta, Casanare, Meta, Putumayo, Vaupés), N Venezuela (Amazonas, Anzoátegui, Aragua, Barinas, Bolívar, Carabobo, Delta Amacuro, Distrito Federal, Falcón, Guárico, Lara, Mérida, Miranda, Monagas, Portuguesa, Sucre, Trujillo, Yaracuy), Trinidad and Tobago (Tobago, Trinidad), Guyana (Barimi-Waini, Cuyuni-Mazaruni, East Berbice-Corentyne, Essequibo Islands-West Demerara, Potaro-Siparuni, Upper Takutu-Upper Essequibo), Suriname (Brokopondo, Commewijne, Coronie, Marowijne, Nickerie, Saramacca, Suriname), French Guiana (Cayenne, Saint-Laurent-du-Maroni), Brazil (Acre, Alagoas, Amazonas, Bahia, Ceará, Espírito Santo, Goiás, Marahhao, Mato Grosso, Mato Grosso do Sul, Minas Gerais, Pará: Marajo Is.; Pernambuco, Rio de Janeiro, Rio Grande do Sul, Rondônia, Santa Catarina, São Paulo), E Ecuador (Loja, Los Ríos, Morona-Santiago, Napo, Pastaza, Tungurahua, Zamora-Chinchipe), E Peru (Ayacucho, Cusco, Huanuco, Junin, Loreto, Madre de Dios, Pasco, San Martín), Bolivia (Beni, Chuquisaca, Cochabamba, La Paz, Pando, Santa Cruz, Tarija), Paraguay (Amambay, Central, Presidente Hayes, San Pedro) and ext. N Argentina (Formosa, Misiones, Salta), NSL–3000 m.

Sources: Roze, 1966a, Emsley, 1977, Cunha & Nascimento, 1978, 1993, Dixon, 1983b, Dixon & Soini, 1986, Lancini, 1986, Chippaux, 1987, Fugler & Riva, 1990, J.D. Williams & Francini, 1991, Cei, 1994, Starace, 1998, Esqueda & La Marca, 1999, Freitas, 1999, Leynaud & Bucher, 1999, Mijares-Urrutia & Arends, 1999, Lehr, 2001, Dixon & Tipton, 2003, Rivas-Fuenmayor & Molina, 2003, Duellman, 2005, Navarrete et al., 2009 and C.J. Cole et al., 2013.

Remarks: *Liophis dorsocorallinus* is a synonym *fide* J.R. Dixon (*in litt.*).

26. *Liophis sagittifer* (Jan, 1863b). Elenco Sist. Ofidi: 82. (*Liopeltis sagittifer*)

Synonyms: *Chlorosoma sagittifer* Jan *in* Burmeister, 1861 *(nomen nudum)*, *Liophis pulcher* Steindachner, 1867b, *Rhadinaea modesta* Koslowsky, 1895, *Liophis trifasciatus* F. Werner, 1899a, and *Zamenis argentinus* Bréthès, 1917.

Type: Holotype, MLU IV 102, an adult male, destroyed during World War II.

Type locality: "Mendoza" [= Mendoza, Mendoza Prov., W Argentina, 32°53'S, 68°51'W, elevation 780 m].

Distribution: Southern South America. Southern Bolivia (Chuquisaca, Santa Cruz, Tarija), NW Paraguay (Chaco, Nueva Asunción), Uruguay, ext. S Brazil (W Rio Grande do Sul) and Argentina (Buenos Aires, Catamarca, Chaco, Chubut, Córdoba, Formosa, Jujuy, La Pampa, La Rioja, Mendoza, Neuquén, Rio Negro, Salta, San Juan, San Luis, Santa Fe, Santiago del Estero, Tucumán), 100–1000 m.

Sources: Jan & Sordelli, 1866–1870, Abalos et al., 1964, Abalos & Mischis, 1975, Dixon & Thomas, 1982, Halloy, 1985a, Cei, 1986, 1994, Vanzolini, 1986b, Lema, 1987, J.D. Williams & Francini, 1991, Arzamendia & Giraudo, 1999, Leynaud & Bucher, 1999, Scolaro, 2005, 2006 and Avila, 2009b.

27. *Liophis semiaureus* (Cope, 1862d). Proc. Acad. Nat. Sci. Philadelphia 14(5): 348. (*Opheomorphus merremii semiaureus*) (*nomen corrigendum*)

Synonyms: *Opheomorphus meremmii semiaureus* Cope, 1862d *(nomen incorrigendum)*, and *Liophis reginae ornata* Jan, 1863b.

Type: Holotype, USNM 4655, a female (T.J. Page [Paraguay Exped.], Sept. 1853–Jan. 1856).

Type locality: "Paraguay." Probably the lower Paraguay River *fide* Gans (1964: 39).

Distribution: Southern South America. Southern Brazil (S Rio Grande do Sul), Paraguay (Alto Paraná, Canindeyú, Central, Itapúa), Uruguay (Artigas, Canelones, Cerro Largo, Colonia, Durazno, Florida, Lavalleja, Maldonado, Montevideo, Paysandú, Rio Negro, Rivera, Rocha, Salto, San José, Treinta y Tres) and N Argentina (Buenos Aires, Córdoba, Corrientes, Entre Ríos, Formosa, Misiones, ? Salta, Santa Fe, Treinta y Tres).

Sources: Gans, 1964, J.D. Williams & Francini, 1991, Cei, 1994, Giraudo, 2002, Giraudo et al., 2006, Cacciali, 2008 and Prigioni et al., 2011.

28. *Liophis taeniogaster* Jan, 1863a. Arch. Zool. Anat. Fis. 2(2): 292–293.

Synonym: *Liophis cobella dyticus* Dixon, 1983.
Types: Syntypes (4), ZSM, an 880 mm specimen [Brazil], MHNN, an 870 mm specimen [Brazil], NMW, a 620 mm specimen [So. Amer.], and MHNG, a 500 mm specimen [So. Amer.].
Type locality: "Brasile," and "Amer. Merid" [= Brazil and South America].
Distribution: Central South America. Colombia, NE Brazil (Amapá, Bahia, Ceará, Mato Grosso, Pará, Bantanal Is.), Ecuador, Peru (Loreto, Madre de Dios) and Bolivia.
Sources: D.S. Fernandes et al., 2002 and Duellman, 2005.
Remarks: A valid species *fide* D.S. Fernandes et al. (2002: 9) and Frota & Santos (2003: 388).

29. *Liophis taeniurus* Tschudi, 1845. Arch. Naturg. 11(1): 164.

L

Type: Lectotype, MHNN 9 (formerly MZN 79), a 737+ mm female (J.J. von Tschudi, Aug. 1838–Feb. 1841), designated by Schätti (1986a: 101).
Type locality: "Dans les régions boisées chaudes (du Pérou)" [= hot rainforest regions of Peru] via lectotype selection.
Distribution: Northwestern South America. Colombia (Nariño), Ecuador (Loja, Pichincha, Tungurahua), E Peru (Ayacucho, Cajamarca, Cusco, Huanuco, Junín, Loreto, Pasco, Puno) and Bolivia (La Paz), 1400–3700 m.
Sources: Dixon & Markezich, 1979, Pérez-Santos, 1986, Pérez-Santos & Moreno, 1988 and Dixon, 2000a.
Remarks: Original description reprinted in Tschudi (1846: 51, pl. 5). Page 164 of original description mis-numbered "166."

30. *Liophis torrenicolus* Donnelly & Myers, 1991. Amer. Mus. Novit. (3017): 41–45, figs. 26, 29. (*Liophis torrenicola*)

Type: Holotype, AMNH 136211 (formerly CWM 19983), a 457 mm male (C.W. Myers & M.A. Donnelly, 26 Feb. 1990).
Type locality: "Summit Camp 1, 1030 m elevation, north side of Cerro Guaiquinina, Bolívar, Venezuela."
Distribution: Southeastern Venezuela (Amazonas, Bolívar), 1030–2000 m.
Sources: Roze, 1958c, D.S. Fernandes et al., 2002, Barrio-Amorgós & Brewer-Carias, 2008, C.W. Myers & Donnelly, 2009 and Navarrete et al., 2009.
Remarks: Possibly a synonym of *L. cobellus*.

31. *Liophis trebbaui* Roze, 1958c. Acta Biol. Venez. 2(22): 243–270, fig. 11.

Type: Holotype, AMNH 61019, a 583 mm female (W.H. Phelps, 2 April 1938), lost *fide* C.W. Myers & Donnelly (2008: 126).
Type locality: "Auyantepui, Estado Bolívar, Venezuela."
Distribution: Eastern Venezuela (Bolívar), 1020 m.
Sources: D.S. Fernandes et al., 2002 and C.W. Myers & Donnelly, 2008.

32. *Liophis triscalis* (Linnaeus, 1758). Syst. Nat., ed. 10, 1: 224. (*Coluber triscalis*)

Synonym: *Coluber corallinus* Linnaeus, 1758.
Type: Lectotype, NHR Lin-146 (formerly MAFR), a 450 mm specimen described and illustrated in Seba (1735: 18–19, pl. 17, fig. 1) (L. Ulrika, 1752), designated herein.
Type locality: "Indiis" [= India] (in error *fide* Dixon, 1981: 302) via lectotype selection. Corrected to Curaçao *fide* Roze (1966a: 163).
Distribution: Netherlands Antilles (Curaçao).
Sources: Brongersma, 1940, 1959, Roze, 1966a, Dixon, 1981 and Buurt, 2001, 2005.
Remarks: Holotype of *Coluber triscalis* Linnaeus lost *fide* Dixon (1981: 302). NHR Lin-146 is holotype of *Coluber corallinus* Linnaeus.

33. *Liophis typhlus* (Linnaeus, 1758). Syst. Nat., ed. 10, 1: 218. (*Coluber typhlus*)

Synonyms: ? *Coluber albiventris* Reuss, 1834, *Liophis typhlus prasina* Jan, 1863b (*nomen nudum*), *Liophis typhlus olivacea* Jan, 1863a, *Xenodon isolepis* Cope, 1870, *Opheomorphus brachyurus* Cope, 1887b, *Liophis elaeoides* L.E. Griffin, 1916, *Liophis macrops* F. Werner, 1925, and *Liophis elacoides* – Briceño-Rossi, 1934 *(nomen incorrectum)*.
Type: Holotype, NHR (formerly MDG), a 630 mm specimen.
Type locality: "Indiis" [= India] (in error *fide* Hoge, 1964d: 58).
Distribution: South America. Eastern Colombia (Amazonas, Boyacá, Caqueta, Casanare, Cundinamarca, Meta, Putumayo, Vaupés), Venezuela (Amazonas, Bolívar, Monagas), Guyana (Barima-Waini, Cuyuni-Mazaruni, Demerara-Mahaica, East Berbice-Corentyne, Essequibo Islands-West Demerara, Potaro-Siparuni, Upper Demerara-Berbice), Suriname (Commewijne, Nickerie, Paramaribo, Saramacca), French Guiana (Cayenne, Saint-Laurent-du-Maroni), Brazil (Acre, Amapá, Amazonas, Bahia, Distrito Federal, Goiás, Maranhão, Mato Grosso, Mato Grosso do Sul, Minas Gerais, Pará, Pernambuco, Rio de Janeiro, Rondônia, São Paulo), E Ecuador (Morona-Santiago, Napo, Pastaza), E Peru (Amazonas, Huanuco, Junín, Loreto, Madre de Dios, Pasco, San Martín), SE Bolivia (Beni, Cochabamba, La Paz, Santa Cruz) and N Paraguay, 100–1640 m.

Sources: Amaral, 1926b, Beebe, 1946, Roze, 1966a, Cunha & Nascimento, 1978, Miranda & Couturier, 1984, Lancini, 1986, Chippaux, 1987, Dixon, 1987, Starace, 1998, Freitas, 1999, Leynaud & Bucher, 1999, Duellman, 2005, Navarrete et al., 2009 and C.J. Cole et al., 2013.

34. *Liophis viridis* A.C.L.G. Günther, 1862a. Ann. Mag. Nat. Hist. (3) 9(50): 58–59, pl. 9, fig. 2.

Synonym: *Liophis typhlus prasina* Jan, 1863b *(nomen nudum)*, and *Liophis typhlus prasina* Jan, 1866 *in* Jan & Sordelli, 1866–1870.
Type: Holotype, BMNH 1946.1.5.69, a 559 mm female (native, 1858–1861).
Type locality: "Pernambuco" [= Pernambuco State, Brazil].
Distribution: Northeastern Brazil (Alagoas, Bahia, Ceará, Maranhão, Minas Gerais, Rio Grande do Norte, Paraíba, Pernambuco, Piauí, Sergipe), NSL–510 m.
Sources: K.P. Schmidt & Inger, 1951, Cordeiro & Hoge, 1974, Vanzolini et al., 1980, Vitt, 1983, Dixon, 1987 and Freitas, 1999.
Remarks: Two types reported *fide* A.C.L.G. Günther (1862a: 59) but BMNH catalogue lists only one. If second type is found, BMNH 1946.1.5.69 will become lectotype via designation by Dixon (1987a: 180).

35. *Liophis vitti* Dixon, 2000a. Copeia 2000(2): 485–488, figs. 3a–b.

Type: Holotype, KU 179506, a 496 mm male (J.D. Lynch, 31 May 1977).
Type locality: "Maldonado, 1410 m, Carchi, Ecuador (0°54'N; 78°06'W)."
Distribution: Extreme N Ecuador (Carchi), 1070–1650 m. Known only from type locality.

36. *Liophis williamsi* (Roze, 1958a). Breviora (88): 1. (*Urotheca williamsi*)

Type: Holotype, MCZ 51329, a 420 mm female (G. Marcuzzi, Oct. 1949).
Type locality: "El Junquito, Distrito Federal, Venezuela, 1900 meters."
Distribution: Cordillera de la Costa of N Venezuela (Aragua, Distrito Federal), 1200–2000 m.
Sources: Roze, 1964, Test et al., 1966, Dixon, 1983b, C.W. Myers, 1986 and Navarrete et al., 2009.

37. *Liophis zweifeli* (Roze, 1959c). Amer. Mus. Novit. (1934): 4–7, fig. 1. (*Leimadophis zweifeli*)

Type: Holotype, MBUCV 95, a 645 mm female (J. Racenis, 6 Sept. 1949).
Type locality: "Rancho Grande in the state of Aragua, Venezuela, at an elevation of 1100 meters."
Distribution: Northern Venezuela (Amazonas, Aragua, Carabobo, Distrito Federal, Miranda), 675–1800 m.
Sources: Roze, 1966, Dixon, 1983b and Rivas-Fuenmayor et al., 2012.
Remarks: A valid species *fide* Rivas-Fuenmayor et al. (2012: 46). Status of Amazonas population uncertain *fide* Rivas-Fuenmayor et al. (2013: 46).

LIOPHOLIDOPHIS Mocquard, 1904 (Pseudoxyrhophiidae)

Synonym: *Liopholiodophis* – Maass-Berlin, 1933 *(nomen incorrectum)*.
Type species: *Liopholidophis grandidieri* Mocquard, 1904a.
Distribution: Madagascar.
Sources: Jourdran, 1904, Guibé, 1958, Domergue, 1973, Cadle, 1996a, Underwood, 1999, Nagy et al., 2003, Glaw & Vences, 2007, Glaw et al., 2007a, Kelly et al., 2009 and Zaher et al., 2009.
Remarks: Closely related to *Liophidium fide* Glaw et al. (2007: 299).

1. *Liopholidophis dimorphus* Glaw, Nagy, Franzen & Vences, 2007a. Zool. Scripta 34(4): 294–297, figs. 3–4.

Type: Holotype, ZSM 252/2004 (formerly FGZC 491), a 1072 mm male (F. Glaw, M. Puente, R.D. Randrianiaina & A. Razafimanantsoa, 21 Feb. 2004).
Type locality: "Montagne d'Ambre, northern Madagascar, 12°30'S, 49°10E, c. 800-1000 m above sea level."
Distribution: Northern Madagascar (Antsiranana), 800–1000 m. Known only from type locality.

2. *Liopholidophis dolicocercus* (Peracca, 1892). Bol. Mus. Zool. Anat. Comp. Univ. Torino 7(112): 1–3, pl., figs. 1, 1a–d. (*Dromicus dolicocercus*)

Type: Holotype, MZUT 857, a 427 mm male.
Type locality: "Madagascar – Valle dell'Umbi (Andrangoloka)" [= Umbi River Valley, Andrangolaoka, Toamasina Prov., E Madagascar, 19°02'S, 47°55'E, elevation 1400 m].
Distribution: Eastern Madagascar (Antsiranana, Fianarantsoa, Toamasina), 800–1400 m.
Remarks: Holotype listed as MZUT 796 *fide* Cadle (1996a: 460).

3. *Liopholidophis grandidieri* Mocquard, 1904a. Bull. Mus. Hist. Nat., Paris 10(6): 304–305.

Type: Holotype, MNHN 1902.103, a 1636 mm male (A. Grandidier, 1867–1891).
Type locality: "l'embouchure du Saint-Augustin, Madagascar" [= mouth of Saint-Augustin River, Madagascar] (in error *fide* Cadle, 1996a: 460).

Corrected to town of Ianantsony, mouth of Onilahy River, SW Toliara Prov., SW Madagascar [ca. 23°33'S, 43°46'E, elevation NSL] *fide* Cadle (1996a: 460).

Distribution: Southern Madagascar (Fianarantsoa, SW Toliara, Toamasina), NSL–1375 m.

4. *Liopholidophis rhadinaea* Cadle, 1996a. Bull. Mus. Comp. Zool. 154(5): 373–380, figs. 1–2.

Type: Holotype, MCZ 180395, a 720 mm male (J.E. Cadle, 20–26 Dec. 1991).

Type locality: Talatakely, Ranomafana National Park, 950–1,000 m, Fivondronana Ifanadiana, Fianarantsoa Province, Madagascar [21°16'S, 47°25'E].

Distribution: Eastern Madagascar (Fianarantsoa, Toamasina), 235–1000 m.

5. *Liopholidophis sexlineatus* (A.C.L.G. Günther, 1882). Ann. Mag. Nat. Hist. (5) 9(52): 264–265, 2 figs. (*Dromicus sexlineatus*)

Synonym: *Dromicus macrocercus* A.C.L.G. Günther, 1882.

Types: Syntypes (3), BMNH 1946.1.13.17–19 (formerly BMNH 1882.5.8.2–4), three females, longest syntype 610 mm (W.D. Cowan, 1874–1881).

Type locality: "Eastern Betsileo, Madagascar" [= N Fianarantsoa Prov., cen. Madagascar].

Distribution: Eastern Madagascar (Antananarivo, Fianarantsoa, Toamasina), 670–2450 m.

Source: Langrand & Gordman, 1996.

6. *Liopholidophis varius* (J.G. Fischer, 1884b). Jahrb. Hamburg. Wiss. Anst. 1: 36–38, pl. 3, fig. 8. (*Leptophis varius*)

Synonyms: *Liopholidophis pinguis* H.W. Parker, 1925b, and *Philodryas pallidus* F. Werner, 1927.

Type: Lectotype, ZMH 4319 (formerly ZMH 1175), a 617–623 mm female, designated by Cadle (1998: 434).

Type locality: "Madagascar" via lectotype selection.

Distribution: Northeastern Madagascar (Andasibe, Antananarivo, cen. Toamasina) 435–1065 m.

Sources: H.W. Parker, 1925b, Cadle, 1998 and R.A. Thomas et al., 2001.

Remarks: Photographs of the lectotype in Cadle (1998: fig. 1).

L

LIOTYPHLOPS W.C.H. Peters, 1881b (*nomen substitutum*) (Anomalepididae)

Synonyms: *Rhinotyphlops* W.C.H. Peters, 1857b (*nomen praeoccupatum*), *Liotyphiops* – W.W. Tanner & Avery, 1982 *(nomen incorrectum), Hawkeswoodus* Hoser, 2012am *(nomen illegitimum),* and *Kraussus* Hoser, 2012am *(nomen illegitimum).*

Type species: *Rhinotyphlops albirostris* W.C.H. Peters, 1857b.

Distribution: Central America and South America.

Sources: Robb & Smith, 1966, J.A. Peters & Orejas-Miranda, 1970, Hahn, 1980a, Dixon & Kofron, 1984, McDiarmid et al., 1999, Tipton, 2005, Freire et al., 2007, Silva Haad et al., 2008 and Centeno et al., 2010.

Remarks: An undescribed species from Barra do Tapirapes, Mato Grosso, Brazil *fide* Dixon & Kofron (1984: 259).

1. *Liotyphlops albirostris* (W.C.H. Peters, 1857b). Mber. Königl. Akad. Wiss. Berlin 1857(8): 402. (*Rhinotyphlops albirostris*)

Synonyms: *Typhlops emunctus* S.W. Garman, 1884, *Helminthophis petersii* Boulenger, 1889b, *Helminthophis canellei* Mocquard, 1903, *Helminthophis bondensis* L.E. Griffin, 1916, *Liotyphlops cucutae* Dunn, 1944c, *Liotyphlops caracasensis* Roze, 1952b, and *Liotyphlops rowani* H.M. Smith & Grant, 1958b.

Types: Syntypes (2), ZMB 9529 and ZMB 63438, a 190 mm and 158 mm specimen (J. Warszewicz, 1844–1850).

Type locality: "Veragua" [Veraguas, Veraguas Prov., Panama, 8°52'N, 80°54'W, elevation 35 m].

Distribution: Lower Central America and NW South America. Southwestern Costa Rica, Panama (Coclé, Canal Zone, Chiriquí, Colón, Panamá, Veraguas, Barro Colorado and San José Is.), NW Colombia (Antioquia, Atlántico, Bolívar, Córdoba, Cundinamarca, Magdalena, Meta, Santander, Santander del Norte, Sucre, Tolima), N Ecuador (Guayas), NW Venezuela (Aragua, Carabobo, Cojedes, Distrito Federal, Falcón, Mérida, Miranda, Portuguesa, Sucre, Zulia, Toas Is.) and Netherlands Antilles (Bonaire, Curaçao), NSL–1640 m.

Sources: Bocourt, 1882 *in* Duméril, Bocourt & Mocquard, 1870–1909, Brongersma, 1940, 1959, E.H. Taylor, 1951, Roze, 1952b, 1966a, Valdivieso & Tamsitt, 1963, Savage, 1980a, 2002, Pérez-Santos, 1986, Lancini & Kornacker, 1989, Mijares-Urrutia & Arends, 2000, Markezich, 2002, La Marca & Soriano, 2004, Rojas-Runjaic & Rivero, 2004, Buurt, 2005 and Navarrete et al., 2009.

2. *Liotyphlops anops* (Cope, 1899b). Philadelphia Comm. Mus. Sci. Bull. (1): 10–11, pl. 4, figs. 1a–f. (*Helminthophis anops*)

Synonym: *Liotyphlops metae* Dunn, 1944c.
Types: Syntypes (2), PCM 14a–b, longest syntype 372 mm, lost *fide* Dunn (1944c: 48).
Type locality: Probably "in the neighborhood of Bogota, department of New Grenada, Colombia." [= near Bogotá (4°36'N, 74°05'W, elevation 2620 m), Cundinamarca Dept., Colombia]. Probably within a 50 mile radius of Bogotá *fide* Dunn & Stuart (1951: 56).
Distribution: Central Colombia (Cundinamarca, Meta, Santander), 250–1040 m.
Sources: Dunn, 1944c and Pérez-Santos, 1986.
Remarks: Syntypes exhibited in the New Granada Dept. of 1898 Chicago World Exposition. Some Colombian (Cundinamarca & Santander) records might be *L. argaleus fide* Dixon & Kofron (1984: 260).

3. *Liotyphlops argaleus* Dixon & Kofron, 1984. Amphibia-Reptilia (1983) 4(2–4): 261–262, fig. 6 (paratype).

Type: Holotype, MCZ 67933 (formerly MLS 50), a 266 mm specimen (H. Nicéforo-María, 1922–1983).
Type locality: "La Selva, Cundinamarca, Colombia" (in error). Corrected to Hacienda "La Selva," south of Cúcuta, Norte de Santander Dept., Colombia, ca. 7°54'N, 72°30'W, elevation 300 m *fide* MCZ catalogue and data label.
Distribution: Northern Colombia (Cundinamarca, Norte de Santander, Santander), 160–1600 m.
Source: Pérez-Santos & Moreno, 1988.
Remarks: Discussion of type locality *fide* Dixon & Kofron (1984: 262) and Pérez-Santos & Moreno (1988: 414) irrelevant as the two specimens sent by Nicéforo-Maria (MLS 50–51, now MCZ 67933–67934) had the specific locality of Hacienda La Selva.

4. *Liotyphlops beui* (Amaral, 1924a). Proc. New England Zool. Club 9: 29. (*Helminthophis beui*)

Type: Holotype, IB 1806, a 290 mm specimen (T. Beu, 1915–1924), destroyed by fire 15 May 2010.
Type locality: "Butantan, S. Paulo, Brazil" [= Bairro Butantan, São Paulo, São Paulo State, SE Brazil, 23°43'S, 46°43'W, elevation 750 m].
Distribution: Southeastern Brazil (Espírito Santo, Goias, Mato Grosso, Mato Grosso do Sul, Paraná, Rio Grande do Sul, Rio de Janeiro, São Paulo), SE Paraguay (Alto Paraná) and ext. NE Argentina (Misiones), 155–915 m.
Sources: Vanzolini, 1986, Cei, 1994, Giraudo, 1994, Lema, 1994 and Marques et al., 2001, 2004, 2005.

5. *Liotyphlops caissara* Centeno, Sawaya & Germano, 2010. Herpetologica 66(1): 88–89, figs. 2–3.

Type: Holotype, IB 76774, an 195 mm female (F.C. Centeno & K.R. Zamudio, 17 Jan. 2007), destroyed by fire 15 May 2010.
Type locality: "Trilha da Água Branca in Ilha de São Sebastião (23° 50' 45" S, 45° 21' 12" W), municipality of Ilhabela, São Paulo state, southeastern Brazil, 300 m above sea level."
Distribution: Southeastern Brazil (São Paulo), 300 m. Known only from type locality.

6. *Liotyphlops haadi* Silva-Haad, Franco & Maldonado, 2008. Biota Colomb. 9(2): 296–297, figs. 1–2.

Type: Holotype, IAvH 5434, a 180 mm specimen (V. Tanimuka, Feb. 2007).
Type locality: "southern Colombian Amazon, in the Amazonas Department, middle region of the Caquetá River, Vereda de los Engleses, La Pedrera district (01°19'42"S, 69°30'33"W)."
Distribution: Southeastern Colombia (Amazonas), 65–85 m.

L

7. *Liotyphlops schubarti* Vanzolini, 1948. Rev. Bras. Biol. 8(3): 379–381, figs. 1–2.

Type: Holotype, MZUSP 2665 (formerly DZUSP 425), a 191 mm male (O. Schubart, 1936–1948).
Type locality: "Cachõeira de Emas, no Município de Pirassununga, Estado de São Paulo" [= Brazil, 21°55'S, 47°21'W, elevation 585 m].
Distribution: Southeastern Brazil (Minas Gerais, São Paulo), 585–700 m.

8. *Liotyphlops ternetzii* (Boulenger, 1896a). Cat. Snakes Brit. Mus. 3: 584. (*Helminthophis ternetzii*)

Synonyms: *Helminthophis incertus* Amaral, 1924a, and *Helminthophis collenettei* H.W. Parker, 1928.
Type: Holotype, BMNH 1946.1.11.77 (formerly BMNH 1895.5.17.282), a 335 mm specimen (C. Ternetz, 1895–1896).
Type locality: "Paraguay."
Distribution: Southeastern South America. Guyane, Brazil (E Amazonia, Brazilia, Ceará, Distrito Federal, Goiás, Mato Grosso, Minas Geraís, Pará, Paraná, Rio de Janeiro, São Paulo), S Paraguay (Amambay, Caazapá, Canendiyu, Itapúa, Presidente Hayes), Uruaguay (Rio Negro, Salto) and N Argentina (Corrientes, Entre Ríos, Formosa, Jujuy, Misiones, Salta), 25–915 m.
Sources: Amaral, 1955a, Roig & Cei, 1961, Cunha & Nascimento, 1975b, 1978, 1993, Melgarejo & Meneghel, 1984, Chippaux, 1987, Giraudo, 1994, Kretzschmar, 1998, Starace, 1998, Carreira-Vidal, 2004, Carreira-Vidal et al., 2005, Cacciali, 2008 and Loebmann, 2009b.

Remarks: Possibly occurs in French Guiana *fide* Starace (1998: 71). Type locality of Suriname for *Helminthophis incertus* Amaral rejected *fide* Dixon & Kofron (1984: 256).

9. *Liotyphlops trefauti* Freire, Caramaschi & Argôlo, 2007. Zootaxa (1393): 20–23, figs. 1a–e (paratype).

Type: Holotype, MZUSP 12178, a 362 mm male (E.M.X. Freire, 4 Nov. 1994).
Type locality: "Fazenda Bananeira, Municipality of Murici (9°14'S, 35°48'W, 640 m), State of Alagoas, Brazil."
Distribution: Northeastern Brazil (Alagoas, Bahia), 50–640 m.
Source: Freire et al., 2007.

10. *Liotyphlops wilderi* (S.W. Garman, 1883). Sci. Observ. 4(5–6): 48. (*Typhlops wilderi*)

Synonym: *Helminthophis guentheri* Boulenger, 1889b.
Types: Syntypes (3), MCZ 5126, a 169 mm female, MCZ 19585 (formerly CU), a 185–187 mm female, FMNH 73387 (formerly CU & MCZ 18138), a 170 mm specimen (J.G. Branner, 1878–1879).
Type locality: "São Cyriaco, Brazil" [= Cipriano, Minas Gerais State, SE Brazil, 19°45'S, 43°57'W, elevation 850 m]. Emended to near Serra Providencia, Minas Gerais, Brazil *fide* Hammar (1908: 335).
Distribution: Southeastern Brazil (Minas Gerais, Rio de Janeiro, Grande Is.) and Paraguay.
Sources: Boulenger, 1908e, Hammar, 1908 and Amaral, 1978.
Remarks: MCZ 5126 was recorded as the holotype in 1883 by S.W. Garman *fide* MCZ catalogue and specimen data label. Two CU specimens were exchanged with MCZ in May 1924 (MCZ 18138, 19585s). The former specimen was subsequently exchanged with FMNH in 1954 (FMNH 73387). Hammar (1908: 335) mentioned two CU specimens as the types and that three specimens were from Cyriaco, near Serra Providencia, Minas Gerais, Brazil. Hahn (1980a: 5) suggested the two CU syntypes were lost, and Dixon & Kofron (1984: 253) mentioned one being lost.

†*LITHOPHIS* Marsh, 1871 (Boidae)

Type species: †*Lithophis sargenti* Marsh, 1871.
Distribution: Middle Eocene of USA.
Sources: Gilmore, 1938, Holman, 1979b and Rage, 1984a.

1. †*Lithophis sargenti* Marsh, 1871. Amer. J. Sci. Arts (3) 1(5): 325.

Type: Holotype, YPM 2719, one fragmentary trunk vertebra (H.B. Sargent, 1870).
Type locality: "in the Eocene 'Mauvaises Terres' beds at Grizzly Buttes, near Fort Bridger, Wyoming Territory" [Uinta Co., Wyoming, USA, Bridger formation, middle Eocene].
Distribution: Middle Eocene (Bridgerian: 46.2–50.3 mya) of USA (Wyoming). Known only from type locality.

LOVERIDGELAPS McDowell, 1970 (Elapidae)

Type species: *Hoplocephalus elapoides* Boulenger, 1890b.
Distribution: Solomon Islands.
Sources: McDowell, 1970, M. McCoy, 1980, Golay et al., 1993, O'Shea, 1996, Keogh, 1998, David & Ineich, 1999, Zaher et al., 2009 and Hoser, 2012e.

1. *Loveridgelaps elapoides* (Boulenger, 1890b). Proc. Zool. Soc. London 58(1): 30, pl. 2, fig. 3. (*Hoplocephalus elapoides*)

Type: Holotype, BMNH 1946.1.18.98 (formerly BMNH 1889.3.29.19), a 750 mm male (C.M. Woodford, 1886–1889).
Type locality: "Florida Island, or Gela, north of Guadalcanal, Solomon Islands" [= Nggela Sule Is., South Solomons, bet. 9°00–12'S & 160°03–25'E].
Distribution: South Solomons (Florida, Gizo, Guadalcanal, Malaita, Nggela and Santa Isabel Is.).

LOXOCEMUS Cope, 1861d (Loxocemidae)

Synonyms: *Plastoseryx* Jan, 1862a, *Platoseryx* Marschall, 1873 *(nomen emendatum), Loxonemus* – Boucard, 1884 *(nomen incorrectum),* and *Loxocenus* – Velasco, 1890a *(nomen incorrectum).*
Type species: *Loxocemus bicolor* Cope, 1861d.
Distribution: Mesoamerica.
Sources: Bocourt, 1882, *in* A.H.A. Duméril, Bocourt & Mocquard, 1870–1909, E.H. Taylor, 1940c, 1954, Mertens, 1951, G. Haas, 1955, Nelson & Meyer, 1967, McDowell, 1975a, L.D. Wilson & Meyer, 1985, Mora, 1987, Pitts et al., 1992, Cundall et al., 1994, García & Ceballos, 1994, McDiarmid et al., 1999, Savage, 2002, Wilcox et al., 2002, J.C. Murphy & Schlager, 2003, G. Köhler et al., 2005, N. Herrera et al., 2007 and M. Lee et al., 2007.

1. *Loxocemus bicolor* Cope, 1861d. Proc. Acad. Nat. Sci. Philadelphia 13(3): 77.

Synonyms: *Xenopeltis alveolarius* Wagler, 1833 (*nomen oblitum*), *Plastoseryx bronni* Jan, 1862a, *Loxocemus sumichrasti* Bocourt, 1876, and *Loxocemus unicolor* Herter, 1960 (*nomen nudum*).

Type: Holotype, USNM 4948 (J.M. Dow, 1860–1861), lost *fide* H.M. Smith & Taylor (1945: 27).

Type locality: "La Union, San Salvador" [= La Unión, La Unión Dept., ext. E El Salvador, 13°29'N, 87°54'W, elevation 65 m].

Distribution: Mesoamerica. Southwestern Mexico (Chiapas, Colima, Guerrero, Jalisco, Michoacán, Morelos, Nayarit, Oaxaca: Laguna Superior Is.), Central America: SE Guatemala (Juitapa), W Honduras (Atlántida, Choluteca, Comayagua, Cortés, Valle, Yoro), ext. E El Salvador (La Libertad, La Unión, San Miguel, Santa Ana, San Vincinte, Sonsonate, Usulután), W Nicaragua (Carazo, Chinandega, León, Managua, Masaya, Matagalpa) and NW Costa Rica (Guanacaste), NSL–600 m.

Remarks: In accordance with Art. 23.9.2 of the Code (ICZN, 1999), *Loxocemus bicolor* Cope is designated a *nomen protectum* and *Xenopeltis alveolarius* Wagler a *nomen oblitum*.

LYCODON H. Boie *in* Fitzinger, 1826a (Colubridae)

Synonyms: *Lycodon* H. Boie *in* F. Boie, 1826, *Cercaspis* Wagler, 1830,

Ophites Wagler, 1830, *Lycodontos* – Wagler, 1833 *(nomen incorrectum),* Sphecodes A.-M.-C. Duméril, 1853, *Lijcodon* Bleeker, 1856 *(nomen emendatum), Leptorhytaon* A.C.L.G. Günther, 1858, *Leptorhyton* A.C.L.G. Günther, 1858 *(nomen incorrectum), Tetragonosoma* A.C.L.G. Günther, 1858, *Euprotodon* Cope, 1860c, *Tytleria* Theobald, 1868b, *Ophitis* – Edeling, 1870 *(nomen incorrectum), Lycondon* – W.F. Sinclair, 1889 *(nomen incorrectum), Ceraspis* – Hoffmann, 1890 *(nomen incorrectum), Tetragosoma* – Hoffmann, 1890 *(nomen incorrectum), Anoplophallus* Cope, 1893a *(nomen substitutum), Haplonodon* L.E. Griffin, 1910, *Tyleria* – Wall, 1921g *(nomen incorrectum), Lycon* – Maki, 1931 *(nomen incorrectum), Lyocdon* – Donoso-Barros, 1969b *(nomen incorrectum),* and *Lgcodon* – Biswas & Sanyal, 1980 *(nomen incorrectum).*

Type species: *Coluber aulicus* Linnaeus, 1758.

Distribution: Southern and SE Asia and East Indies.

Sources: Boulenger, 1893a, Wall, 1909b, Rooij, 1917, E.H. Taylor, 922a, 1965, M.A. Smith, 1943, C. Haas, 1950, Leviton, 1965c, Welch, 1988, Zhao & Adler, 1993, Ota & Ross, 1994, Adler & Zhao, 1995, I. Das, 2010, 2012 and Zhang et al., 2011.

Remarks: Zhao & Adler (1993: 246) replaced *Lycodon* with the name *Ophites* Wagler, incorrectly *fide* Adler & Zhao (1995: 75). *Cercasis* Wagler a synonym of *Lycodon fide* Pyron et al. (2013: 976).

1. *Lycodon alcalai* Ota & Ross, 1994. Copeia 1994(1): 159–161, figs. 2a–b, 3a–c.

Type: Holotype, PNM 990, a 787 mm male (R.S. Kennedy & field crew, 30 May 1985).

Type locality: "Batan Island (20°25'N, 121°58'E), Batan Island Group; northern Philippines; approximately 2.5 km ENE of Basco on the W slope of Mt. Iraya, alt. 150 m."

Distribution: Extreme N Philippines (Batan), 130–320 m.

2. *Lycodon aulicus* (Linnaeus, 1758). Syst. Nat., ed. 10, 1: 220–221. (*Coluber aulicus*)

Synonyms: *Lycodon aulica* – Laurenti, 1768, ? *Coluber scalaris* Gravenhorst, 1807, *Lycodon subfuscus* Cantor, 1839b, *Lycodon aulicum* – A.-M.-C. Duméril, Bibron & Duméril, 1854a, *Lijcodon aulicum* Bleeker, 1856 *(nomen emendatum), Lycodon anamallensis* A.C.L.G. Günther, 1864a, *Lycodon aulicus oligozonatus* Wall, 1909b, and *Lycodon osmanhilli* E.H. Taylor, 1950b.

Type: Holotype, NHR ? Lin-21 (formerly MAFR), a 250 mm specimen (Mus. Drottn.).

Type locality: "America" (in error). Corrected to India *fide* Kramer (1977: 742).

Distribution: Southern Asia. Eastern Pakistan (Punjab, Sindh), India (Andhra Pradesh, Assam, Bihar, Goa, Gujarat, Jammu, Karnataka, Kashmir, Kerala, Maharashtra, Madhya Pradesh, Manipur, Mizoram, Nagaland, Orissa, Rajasthan, Sikkim, Tamil Nadu, Uttar Pradesh, West Bengal), Sri Lanka (Central, Eastern, North-Central, Northern, Northwestern, Saharagamuwa, Southern, Uva, Western), Nepal (Bardiya, Bhojpur, Chitwan, Jhapa, Kanchanpur, Kaski, Kathmandu, Latlipur, Myagdi, Palpa Parsa, Rupandehi, Sindhupalchok), Bangladesh and N Myanmar (Shan, Yangon), NSL–2130 m.

Sources: Laidlaw, 1902, Wall, 1909b, 1921g, Herklots, 1935, Rendahl, 1937, E.H. Taylor, 1953b, Deraniyagala, 1955, Acharji, 1958, P. Silva, 1969, 1990b, 2001, 2009, Singh, 1972, R.C. Sharma, 1974, M.A.R. Khan, 1988, M.S. Khan, 1977, 2002, Kramer, 1977, J. Daniel, 1983, Nanhoe & Ouboter, 1987, Dutta & Acharjyo, 1990, Zhao & Adler, 1993, Henkel & Schmidt, 2000, Hallermann et al., 2001, Tiwari & Shah, 2004, Whitaker & Captain, 2004, N. Khaire, 2006, Somaweera, 2006, Mukherjee & Bhupathy, 2007, M.F. Ahmed et al., 2009 and Ganesh & Chandramouli, 2011.

Remarks: Andersson (1899: 16) stated the type represented variety D (= *capucinus*) of Boulenger (1893a: 353) but pl. 12, fig. 2, of Linnaeus, 1754, clearly depicts variety A (= *aulicus*). *Lycodon osmanhilli* Taylor a valid species *fide* A. Silva (2009: 82, pl. 35).

3. *Lycodon bibonius* Ota & Ross, 1994. Copeia 1994(1): 162–164, figs. 6a–b, 7a–c (all paratypes).

Type: Holotype, PNM 2044, a 511 mm female (R.I. Crombie & H. Ota, 21 May 1989).

Type locality: "Camiguin Island (18°55'N, 121°55'E), Babuyan Island Group, northern Philippines; ca. 1.5 km E of Mambit, alt. 70 m."

Distribution: Extreme N Philippines (Camiguin), NSL–70 m.

4. *Lycodon butleri* Boulenger, 1900c. J. Bombay Nat. Hist. Soc. (1900–1902) 13(2): 336.

Type: Holotype, BMNH (1946.1.14.90 formerly SM), a 540 mm female (A.L. Butler, 1898).

Type locality: "Perak hills, at an altitude of 5,000 feet" [= Larut Hills, Perak State, W West Malaysia *fide* Boulenger, 1900e: 193].

Distribution: Malay Peninsula. Southern Thailand (Krabi, Trang) and West Malaysia (Pahang, Perak, Selangor), 1220–2030 m.

Sources: B.L. Lim, 1956, 1967b, Tweedie, 1983, M.J. Cox, 1991a, Chan-ard et al., 1999 and Daltry & Wüster, 2002.

Remarks: Original description reprinted in Boulenger (1900e: 193).

5. *Lycodon capucinus* H. Boie *in* F. Boie, 1827. Isis von Oken 20(6): 551.

L

Synonyms: *Natrix fenestrata* Wagler, 1825 (*nomen ineditum*), *Natrix plebeja* Wagler, 1825 (*nomen ineditum*), *Lycodon capucinus* H. Boie *in* Schlegel, 1926a *(nomen nudum), Lycodon capucinus* Schlegel, 1926b *(nomen nudum), Lycodon unicolor* F. Boie, 1827, *Lycodon hebe* Schlegel, 1837, *Lycodon atropurpureus* Cantor, 1839b, *Lycodon aulicum cochinchinensis* Jan, 1863b (*nomen nudum*), *Lycodon aulicum picturatum* Jan, 1863b (*nomen nudum*), *Lycodon aulicum reticulatum* Jan, 1863b (*nomen nudum*), and *Tytleria hypsirhinoides* Theobald, 1868b.

Type: Lectotype, a 483 mm specimen described and illustrated in P. Russell (1801: 41, pl. 37), designated herein.

Type locality: "Java" [W Indonesia] via lectotype selection. Restricted to region of Mt. Pangerango and Mt. Salak, West Java, Indonesia *fide* Brongersma (1948b: 14). Emended to Tjihandjawar, at the foot of Mt. Pangerango, W. Java [= Cihanyawar, Nagrak, Jawa Barat, W Java, SW Indonesia, 07°03'S, 108°40'E] *fide* Brongersma (1950: 1499).

Distribution: Southeast Asia and East Indies. Eastern India (Andaman & Nicobars: Andaman, Long and Nicobar Is.), S China (Fujian, Gansu, Guangdong, Hong Kong, Hunan, Yunnan), Laos, Vietnam (Ba Ria-Vung Ta, Bing Duong, Da Nang, Ho Chi Minh City, Khanh Hoa, Kien Giang, Tay Ninh, Cu Lao Panjang Is.), Cambodia (Kampong Chhnang), Thailand (Bangkok, Chaiyaphum, Chiang Mai, Chon Buri, Krung Thep Mahanakhon, Loei, Lop Buri, Nakhon Ratchasina, Phang Nga, Phuket, Prachuap Khiri Khan, Ratchaburi), S Myanmar (Tanintharyi, Yangon), Maldives, West Malaysia (Kedah, Pinang), Singapore, Indonesia (Adonara, Alor, Babar, Bali, Buton, Ende, Flores, Java, Kalao, Komodo, Lomblen, Moyo, Padar, Rinca, Roti, Salajar, Savu, Sawu, Semau, Solar, Sulawesi, Sumatra, Sumba, Sumbawa, Timor, Wetar), Timor-Leste, Philippines (Bantayan, Cebu, Cuyo, Leyte, Luzon, Masbate, Mindanao, Mindoro, Negros, Palawan, Panay, Romblon, Samar) and ? Mauritius, NSL–700 m.

Sources: Flower, 1899, E.H. Taylor, 1922a, 1965, Bourret, 1936b, W. Phillips, 1958, Campden-Main, 1970a, Deuve, 1970, Saint-Girons, 1972a, Romer, 1979a, Auffenberg, 1980, Biswas & Sanyal, 1980, Bosch, 1985, Tikader & I. Das, 1985, Karsen et al., 1986, Fritts, 1993, Gaulke & Altenbach, 1994, Ota & Ross, 1994, David & Vogel, 1996, M.J. Cox et al., 1998, Chan-ard et al., 1999, Darevsky, 1999, Daltry & Wüster, 2002, Whitaker & Captain, 2004, Zhao, 2006, Mukherjee & Bhupathy, 2007, V.S. Nguyen et al., 2009, Kaiser et al., 2011, Lang, 2011 and Harikrishnan et al., 2012.

Remarks: Original description based on H. Boie's MS (1823–1825) and P. Russell (1801: 41, pl. 37). Type locality erroneously listed as Trivandrum, India *fide* E.H. Taylor & Elbel (1958: 1136). Introduced to Christmas Island *fide* L.A. Smith (1988: 251).

6. *Lycodon cardamomensis* Daltry & Wüster, 2002. Herpetologica 58(4): 499–502, figs. 1–2, 4.

Type: Holotype, BMNH 2000.70, a 316 mm male (J.C. Daltry, 9 Feb. 2000).

Type locality: "from 500 m above sea level at 12° 25' N ; 103° 02' E in the Phnom Samkos Wildlife Sanctuary, Pursat Province, Cardamom Mountains, Southwest Kampuchea."

Distribution: Northeastern Thailand (Chanthaburi) and Cambodia (Pursat), 500 m.

Sources: Pauwels et al., 2005, Vogel & Brachtel, 2008, Vogel & Luo, 2011 and I. Das, 2012.

7. *Lycodon carinatus* (Kuhl, 1820). Beiträge Zool. Vergl. Anat.: 95–96. (*Hurria carinata*)

Types: Syntypes (2), MNHN and RMNH (formerly MHNPB), longest syntype 343 mm.

Type locality: Unknown. Designated as Insel Ceylon [= Sri Lanka] *fide* F. Boie (1837: 517).

Distribution: Sri Lanka (Central, Northern, Sabaragamuwa, Southern, Western), 150–1200 m.

Sources: Schlegel, 1837, Wall, 1921g, M.A. Smith, 1943, Deraniyagala, 1955, P. Silva, 1969, P. H. Silva, 1980a, A. Silva, 1990b, 2001, 2009, Somaweera, 2006 and Zaher et al., 2009.

Remarks: Transferred from *Cercaspis* to *Lycodon fide* Pyron et al. (2013: 976).

8. *Lycodon chrysoprateros* Ota & Ross, 1994. Copeia 1994(1): 165–167, figs. 9a–c. (*Lycodon chrysoprateros*) (*nomen corrigendum*)

Synonym: *Lycondon chrysoprateros* Ota & Ross, 1994 (*nomen incorrigendum*).
Type: Holotype, PNM 2045, a 727 mm female (R.I. Crombie, 18 March 1990).
Type locality: "east side of Dalupiri Island (19°05'N, 121°15'E), Babuyan Island Group, northern Philippines."
Distribution: Extreme N Philippines (Dalupiri).

9. *Lycodon davidi* Vogel, Nguyen, Kingsada & Ziegler, 2012. NW J. Zool. 8(2): 345–347, figs. 1–3.

Type: Holotype, IEBR A.2011.7 (formerly NQT 2010.39), a 390 mm male (T.Q. Nguyen & P. Kingsada, 6 May 2010).
Type locality: "Nam Cave (18°54.597'N, 102°25.998'E, 268 m above sea level), near Ban Muang Song, Vang Vieng District, Vientiane Province, Laos."
Distribution: Northern Laos (Vientiane), 270 m.

10. *Lycodon dumerilii* (Boulenger, 1893a). Cat. Snakes Brit. Mus. 1: 368–369. (*Stegonotus dumerilii*)

Synonym: *Dryocalamus mccroryi* E.H. Taylor, 1922b.
Type: Lectotype, BMNH 1946.1.15.6 (formerly BMNH 1877.10.9.67), a female (A.H. Everett, May–July, 1877 or March–May, 1878), designated *fide* Leviton (1965c: 123).
Type locality: "Surigao, Surigao Province, Mindanao Island, Philippines" via lectotype selection.
Distribution: Southern Philippines (Basilan, SW Mindanao).
Sources: Leviton, 1965c and Ota & Ross, 1994.

11. *Lycodon effraenis* Cantor, 1847a. J. Asiat. Soc. Bengal 16(182): 917–918, pl. 40, fig. 2.

Synonyms: *Tetragonosoma effrene* A.C.L.G. Günther, 1858 *(nomen emendatum), Lycodon ophiteoides* Bleeker, 1859b, *Lycodon adspersus* Jan, 1863b, *Lycodon effrenis* Boulenger, 1893a *(nomen emendatum),* and *Lycodon ophitevides* – E. Bartlett, 1895a *(nomen incorrectum).*
Type: Holotype, BMNH 1946.1.13.84, a 318 mm specimen (W. Norris, 1844–1847).
Type locality: "Pinang" [= Penang Is., Pinang State, NW West Malaysia, 5°25'N.L. 100°19'E.].
Distribution: Southeastern Asia and Greater Sundas. Thailand (Krabbi, Nakhon Si Thammarat), West Malaysia (Kelantan, Pahang, Pinang, Selangor, Penang Is.), East Malaysia (Sarawak), Brunei and Indonesia (Kalimantan, Sumatra), 30–700 m.
Sources: Boulenger, 1912c, B.L. Lim, 1967a, Tweedie, 1983, K.F.L. Lim & Lee, 1989, Stuebing, 1991, David & Vogel, 1996, M.J. Cox et al., 1998, Chan-ard et al., 1999, Daltry & Wüster, 2002 and I. Das, 2007b, 2012.
Remarks: Original description reprinted in Cantor (1886: 176–178). Boulenger (1893a: 356) and subsequent authors erroneously cite p. 70 as original description.

12. *Lycodon fasciatus* (J. Anderson, 1879). Anat. Zool. Res. W. Yunnan: 827, pl. 78, fig. 2. (*Ophites fasciatus*)

Synonyms: *Ophites subminiatus* Morice, 1875a, and *Dinodon yunnanensis* F. Werner, 1922a.
Type: Neotype, BMNH 1901.4.26.1, an 815 mm male (H. Hampton, 1898–1901), designated by Vogel & Luo (2011: 32).
Type locality: "'Mogok', Mandalay Division, Myanmar" [= 22°55'N, 96°30'E, elevation 1120 m] via neotype selection.
Distribution: Southeastern Asia. Southern China (Fujian, Gansu, Guangdong, Guangxi, Guizhou, Hubei, Shaanxi, Sichuan, Xizang, Yunnan), NE India (Arunachal Pradesh, Assam, Meghalaya, Nagaland, N West Bengal), Bhutan, Bangladesh, N Myanmar (Chin, Kachin, Shan), N Thailand (Chiang Mai, Loei, Nakhon Ratchasima, Tak), N Laos and N Vietnam (Cao Bang, Hai Duong, Ha Tinh, Kon Tum, Lai Chau, Lao Cai, Quang Binh, Son Lai, Vinh Phuc), 500–2300 m.
Sources: Wall, 1911c, 1923a, C.H. Pope, 1935, Bourret, 1936b, Dueve, 1962b, E.H. Taylor, 1965, Tian et al., 1986, Y. Zhang & Tang, 1990, Zhao & Adler, 1993, Chan-ard et al., 1999, Orlov et al., 2000, Daltry & Wüster, 2002, Ao et al., 2004, V.S. Nguyen et al., 2009, Vogel et al., 2009, Vogel & David, 2010, Vogel & Luo, 2011, Zhang et al., 2011, Wangyal, 2012 and Yao, 2012.
Remarks: Holotype lost during World War II *fide* Das *in* Vogel & Luo (2011: 29). Photographs of neotype in Vogel & Luo (2011: figs. 1a–b, 2).

L

13. *Lycodon fausti* Gaulke, 2002. Spixiana 25(1): 87–88, figs. 2–3.

Type: Holotype, PNM 7271, a 337 mm juvenile (E. Sanchez, 2001).
Type locality: "Baranguay Guia, Municipality Pandan, Antique Province, W Panay Peninsula, Philippines."
Distribution: Central Philippines (Panay).

14. *Lycodon ferroni* Lanza, 1999. Trop. Zool. 12(1): 90–94, figs. 1–2.

Type: Holotype, MZUF 36690, a 382 mm male (C. Ferron, 28 April 2001).
Type locality: "Lungib Ginbagsangan, 12°00'00"N-124°55'40"E, about 32 km by air NNE of Catbalogan, 40 m a.s.l. [Barrio Kag-Toto-Og; Provincia Samar Occidental (= Western Samar Province), Samar Island, central-eastern Philippines."
Distribution: Central Philippines (Samar), 40 m.

15. *Lycodon flavicollis* Mukherjee & Bhupathy, 2007. Russ. J. Herp. 14(1): 22–25, figs. 2–4.

Type: Holotype, ZSI 25641, a 543 mm male (D. Mukherjee, 7 Sept. 2003).
Type locality: "Mangarai village located on the periphery of the Aneikatti Reserve Forest (600 m a.s.l. 11°05' N, 76°47' E), Western Ghats, Coimbatore District, Tamil Nadu State, southwestern India."
Distribution: Southwestern India (S Tamil Nadu), 600 m.

16. *Lycodon flavomaculatus* Wall, 1907b. J. Bombay Nat. Hist. Soc. (1906–1907) 17(3): 612–614, fig. 1, pl., 2 figs.

Types: Syntypes (6), BNHSa–e, all specimens ca. 350 mm (C. Hudson & Millard) and BMNH (formerly FM).
Type locality: "Sangli (S. Maratha District), Kirkee, Poona, Nasik, and Dharwar, all places on the western Ghats inside the Bombay Presidency" [= W Maharashtra and NW Karnataka States, India]. Restricted to Kirkee [= Khadki, Pune, W Maharashtra State, W India, 18°34'N, 73°52'E, elevation 560 m] *fide* M.A. Smith (1943: 262).
Distribution: Western Ghats of W India (Karnataka, Kerala, Maharashtra), 550–660 m.
Sources: Murthy, 1985, 1991, Captain, 1999, Whitaker & Captain, 2004, N. Khaire, 2006 and Mukherjee & Bhupathy, 2007.
Remarks: Partial type locality restriction of Oudi *fide* M.A. Smith (1943: 262) invalid as not one of original localities. A synonym of *L. striatus fide* Captain, 1999.

17. *Lycodon futsingensis* (C.H. Pope, 1928a). Amer. Mus. Novit. (320): 5–6. (*Dinodon futsingensis*)

Type: Holotype, AMNH 34106, a 796 mm female (C.H. Pope, April–Sept. 1926).
Type locality: "Ch'ungan Hsien, northwest Fukien Province, China" [= Fujing Xian Co., 25°44"N, 119°22'E, Fujian Prov., China].
Distribution: Southern China (Fujian, Hong Kong) and Vietnam (Bắc Kan, Lam Dong, Lào Cai, Hà Tinh, Quang Binh, 15–625 m.
Sources: Zhao & Adler, 1993, Ziegler, 2002, Vogel et al., 2009, 2012b and Vogel & Luo, 2011.
Remarks: Previously a synonym of *Dinodon septentrionale*.

18. *Lycodon gongshan* Vogel & Luo, 2011. Zootaxa (2807), 33–35, figs. 3–4.

Type: Holotype, KIZ 730034, a 963 mm male.
Type locality: "'Bapo, Gongshan, China', today Dulongjiang Township, Gongshan County, Yunnan Province, People's Republic of China."
Distribution: Southern China (Hengduan Mts. of Yunnan)

19. *Lycodon jara* (G. Shaw, 1802). Gen. Zool., Amph. 3(2): 525. (*Coluber jara*)

Synonyms: *Coluber bipunctatus* Cantor, 1839b, and *Lycophidion bipunctatum* W.C.H. Peters, 1863d.
Type: Lectotype, a 381 mm specimen described and illustrated in P. Russell (1796: 19–20, pl. 14) (Snodgrass), designation herein.
Type locality: "Ganjam, India" [= Ganjam, Orissa State, E India, 19°23'N, 85°03'E, elevation 5 m].
Distribution: Southern Asia. Northeastern India (Arunachal Pradesh, Assam, Haryana, Orissa, Sikkim, Uttarkhand, Uttar Pradesh, West Bengal), Bangladesh, Nepal (Bardiya, Chitwan, Kanchanpur, Udayapur), Bhutan (Sarpang) and N Myanmar (Kachin), 150–340 m.
Sources: Wall, 1907b, M.A.R. Khan, 1988, Basu, 1989, Schleich & Kästle, 2002, Tiwari & Shah, 2004, Whitaker & Captain, 2004, Mukherjee & Bhupathy, 2007, I. Das, 2010 and Wangyal, 2012.
Remarks: G. Shaw (1802: 525) erroneously cited Russell's plate as "44."

20. *Lycodon kundui* M.A. Smith, 1943. Fauna Brit. India, Rept. Amph. 3: 260.

Type: Holotype, BMNH 1946.1.13.68, a 225 mm juvenile (Kundu).
Type locality: "Gyobyu, Taikkyi Township, Pegu district" [= Yangon Div., S Myanmar, 17°22'N, 96°03'E, elevation 100 m].
Distribution: Southern Myanmar (Yangon), 100 m. Known only from type locality.

21. *Lycodon laoensis* A.C.L.G. Günther, 1864a. Rept. Brit. India: 317–318.

Types: Syntypes (3), BMNH 1946.1.13.64–65, a 419 mm male and female (H. Mouhot, 1858–1860) [Laos], and BMNH 1946.1.13.76, a female (Price) [Thailand]. [one female 470 mm].
Type locality: "Laos Mountains" and "probably Siam" [= Laos Mtns. and Thailand]. Restricted to Laos, French Indo-China [= Laos Mtns., Laos] *fide* M.A. Smith (1943: 259).
Distribution: Southeastern Asia. Northeastern India (Arunachal Pradesh), S China (Yunnan), Laos (Borikhan, Kamphaeng Na Khon, Vientiane), Cambodia (Kampot, Siem Reap), Thailand (Chachoengsao, Chiang Mai, Chon Buri, Krabi, Loei, Nakhon Ratchasima, Nakhon Si Thammarat, Narathiwat, Pattani, Phang Nga, Phitsanulok, Saraburi, Si Saket, Songkhla, Trang), Vietnam (Bac Giang, Binh Phuoc, Dong Nai, Ho Chi Minh City, Lam Dong, Quang Nam, Tay Ninh) and N West Malaysia (Kedah, Kelantan, Perlis), NSL–1820 m.
Sources: Boulenger, 1893a, M.A. Smith, 1915d, Bourret, 1936b, B.L. Lim, 1958, E.H. Taylor, 1965, Campden-Main, 1970a, Deuve, 1970, Saint-Girons, 1972a,

L

Tweedie, 1983, Zhao & Adler, 1993, M.J. Cox et al., 1998, Chan-ard et al., 1999, Daltry & Wüster, 2002, Whitaker & Captain, 2004, Bain et al., 2007a and V.S. Nguyen et al., 2009.
Remarks: Laos Mountains may be located ca. 15°N, 105°E in S Laos or SE Thailand.

22. *Lycodon liuchengchaoi* Zhang, Jiang, Vogel & Rao, 2011. Zootaxa (2982): 60–64, figs. 1–4.

Type: Holotype, CWNU 867001, a 747 mm female (Q. Deng, July 1986).
Type locality: "Tangialhe National Nature Reserve (32.5439°N, 104.8322°E, elevation 1360m), Qingchuan County, Sichuan Province, P. R. China."
Distribution: Southern China (Sichuan), 1230–1360 m.

23. *Lycodon mackinnoni* Wall, 1906c. J. Bombay Nat. Hist. Soc. (1906–1907) 17(1): 29–30, 3 figs.

Type: Holotype, BMNH 1946.1.13.81 (P.W. Mackinnon, 1895–1906).
Type locality: "neighbourhood of Mussoorie, Himalayas, at an altitude of 6,100 feet" [= Mussoorie, Uttar Pradesh, N India, 30°27'N, 78°04'E, elevation 1900 m].
Distribution: Western Himalayas of N India (Uttar Pradesh), 1860–1900 m.
Sources: Wall, 1923d and Lamba, 1987.

24. *Lycodon muelleri* A.-M.-C. Duméril, Bibron & Duméril, 1854a. Erpét. Gén. 7(1): 382–384.

Synonym: *Haplonodon philippinensis* L.E. Griffin, 1910.
Type: Holotype, MNHN 1320, a 558 mm specimen.
Type locality: "Java" [Indonesia] (in error). Corrected to Philippines *fide* Leviton (1965c: 126). Restricted to Luzon Island, Philippines *fide* Leviton (1965c: 126).
Distribution: Northern Philippines (Batan, Catanduanes, Luzon, Mindoro, Polillo).
Sources: Leviton, 1965c, Ross & Gonzales, 1992 and Ota & Ross, 1994.

25. *Lycodon multifasciatus* (Maki, 1931). Monogr. Snakes Japan: 127–129, figs. 83a–e. (*Dinodon septentrionale multifasciatum*)

Type: Holotype, KIU A, a 695 mm female (S. Nischiishigaki, April 1928).
Type locality: "Ishigaki-shima, Loo Choo Islands" [= Ishigakijima, Okinawa Prefecture, Ryukyu Archipelago, Japan].
Distribution: Extreme S Japan (Ryukyus: Iriomatejima, Ishigakijima and Miyakojima Is.).
Sources: M. Mori, 1984, 1986, Ota, 1988, Vogel & Brachtel, 2008 and Vogel et al., 2009, 2012b.
Remarks: Colored plate of original description in Maki (1931–1933: pl. 47).

26. *Lycodon ophiophagus* Vogel, David, Pauwels, Sumontha, Norval, Hendrix, Vu & Ziegler, 2009. Trop. Zool. 22(2): 157–160, figs. 11–13, 15.

Type: Holotype, QSMI 596, a 909 mm female (M. Sumontha, 30 Dec. 2002).
Type locality: "Lamru Waterfall, Khao Lak-Lamru National Park, Phang-Nga Province, southern Thailand."
Distribution: Southern Thailand (Chumphon, Phang Nga, Ranong).
Source: Vogel et al., 2012b.

27. *Lycodon paucifasciatus* Rendahl *in* M.A. Smith, 1943. Fauna Brit. India, Rept. Amph. 3: 267.

Type: Holotype, NHR 3095, a 763 mm specimen (B. Björkegren, July–Aug. 1938 or Feb.–April 1939).
Type locality: "Thua Lun, Annam, 50 km. south of Hué" [= Thua Luu, Thua Thien-Hue Prov., cen. Vietnam, 16°15'N, 107°56'E, elevation 150 m].
Distribution: Central Vietnam (Quang Binh, Thua Thien-Hue), NSL–255 m.
Sources: V.S. Nguyen et al., 2009 and Vogel et al., 2009, 2012b.
Remarks: Type locality spelling corrected to Thua Luu *fide* Campden-Main (1970a: 35).

L

28. *Lycodon ruhstrati* (J.G. Fischer, 1886). Abh. Naturw. Ver. Hamburg 9(1): 64–65, pl. 2, figs. 6a–d. (*Ophites ruhstrati*)

Synonym: *Lycodon ruhstrati abditus* Vogel, David, Pauwels, Sumontha, Norval, Hendrix, Vu & Ziegler, 2009.
Types: Syntypes (2), LMNMO 918 (formerly LMNMO 37c & 73), a 583+ mm male, and LMNMO 919 (formerly LMNMO 37b & 74), a 248–250 mm specimen (E.K.A. Ruhstrat, 1884–1886).
Type locality: "Süd-Formosa" [= S Taiwan].
Distribution: Southeastern Asia. Southern China (Anhui, Fujian, S Gansu, Guangdong, Guangxi, Guizhou, Hong Kong, Jiangsu, Jiangxi, Shaanxi, Sichuan, Yunnan, Zhejiang), Taiwan (Nantou), N Laos (Xiangkhouang) and N Vietnam (Bac Giang, Cao Bang, Da Nang, Ha Tinh, Hoa Binh, Lao Cai, Quang Binh, Son La, Thanh Hoa, Vinh Phù), NSL–1500 m.
Sources: C.H. Pope, 1935, Romer, 1966, 1979a, Karsen et al., 1986, Zhao & Adler, 1993, Orlov et al., 2000, Zhao, 2006, Ziegler et al., 2007, Vogel & Brachtal, 2008, Kucharzewski et al., 2009, V.S. Nguyen et al., 2009, Vogel et al., 2009, 2012b, Xiang & Li, 2009, Vogel & Luo, 2011 and Yao, 2012.
Remarks: Original description pagination was erroneously cited as page 16 *fide* Boulenger (1893a: 363). Photographs of syntypes in Kucharzewski et al. (2009: figs. 1–3).

29. *Lycodon solivagus* Ota & Ross, 1994. Copeia 1994(1): 168–169, figs. 10 a–b, 11a–c.

Type: Holotype, PNM 2046, a 946 mm female (H. Ota, 28 April 1989).

Type locality: "Cabatacan Barrio, Lasam, Cagayan Province, Luzon Island, northern Philippines (18°00'N, 121°35'E)."

Distribution: Extreme N Philippines (N Luzon), NSL.

30. *Lycodon stormi* Boettger, 1892b. Zool. Anz. 15(405): 418–419.

Type: Holotype, LNHM, a 597 mm specimen (H. Storm, 1887–1892), destroyed 28 March 1942 during World War II.

Type locality: "Celebes" [= Sulawesi, E Indonesia].

Distribution: Eastern Indonesia (Sulawesi).

Sources: Bosch, 1985, Lang & Vogel, 2005 and Koch, 2012.

31. *Lycodon striatus* (G. Shaw, 1802). Gen. Zool., Amph. 3(2): 527. (*Coluber striatus*)

L

Synonyms: ? *Coluber malignus* Daudin, 1803e, *Coluber galathea* Daudin, 1803e, *Lycodon napei* A.-M.-C. Duméril, 1853 (*nomen nudum*), *Lycodon napei* A.-M.-C. Duméril, Bibron & Duméril, 1854a, *Contia bicolor* Nikolsky, 1903b, *Lycodon striatus nordicus* Deraniyagala, 1955, and *Lycodon striatus sinhaleyus* Deraniyagala, 1955.

Type: Lectotype, a 356 mm specimen described and illustrated in P. Russell (1796: 22, pl. 16), designation herein.

Type locality: "Vizagapatam, coast of Coromandel, India" [= Vishakhapatnam, Andhra Pradesh State, SE India, 17°41'N, 83°13'E, elevation 25 m].

Distribution: Southern Asia. Extreme SW Tadzhikistan, ext. S Turkmenia, Turkmenistan, S Uzbekistan, E Iran (Sistan va Baluchestan), Pakistan (Balochistan, Khyber Pakhtunkhwa, Punjab, W Sindh), India (Andhra Pradesh, Bihar, Himachal Pradesh, Jammu, Jharkhand, Karnataka, Kashmir, Madhya Pradesh, Maharashtra, Orissa, Tamil Nadu, Uttar Pradesh, West Bengal), Nepal (Bardiya, Ilam) and Sri Lanka (Central, Uva, Western), NSL–1830 m.

Sources: Wall, 1909b, Chernov, 1935, Minton, 1966, P. Silva, 1969, 1990b, 2009, K.K. Tiwari & Sharma, 1971, Mahajan & Agrawal, 1976, Bannikov et al., 1977, M.S. Khan, 1977, 1982b, 2002, Murthy et al., 1979, J. Daniel, 1983, Murthy, 1985, Latifi, 1991, Sanyal et al., 1993, Szczerbak, 1994, Captain, 1999, Tiwari & Shah, 2004, Whitaker & Captain, 2004, N. Khaire, 2006, Somaweera, 2006 and Mukherjee & Bhupathy, 2007.

Remarks: Original description based on P. Russell (1796: 22, pl 16).

32. *Lycodon subcinctus* F. Boie, 1827. Isis von Oken 20(6): 551.

Synonyms: *Coluber platurinus* G. Shaw, 1802 (*nomen nudum*), *Coluber platyrhinus* Merrem, 1820 (*nomen substitutum*), *Natrix semifasciata* Wagler, 1825 (*nomen ineditum*), *Lycodon subcinctus* H. Boie *in* Schlegel, 1826a (*nomen nudum*), *Lycodon subcinctus* Schlegel, 1826b (*nomen nudum*), *Lycodon platurinus* Cantor, 1847, *Lycodon succinctus* – A.-M.-C. Duméril, Bibron & Duméril, 1854a (*nomen incorrectum*), *Elapoides annulatus* Sauvage, 1884, *Anoplophallus maculatus* Cope, 1895a, and *Lycodon subcinctus sealei* Leviton, 1955.

Type: Lectotype, a 953 mm specimen described and illustrated in P. Russell (1801: 44, pl. 41), designated herein.

Type locality: "Java" [W Indonesia] via lectotype selection.

Distribution: Southeastern Asia and East Indies. Myanmar (Kachin, Tanintharyi, S China (Fujian, Guangdong, Guangxi, Hainan, Hong Kong), Laos (Borikhan), Vietnam (Bac Kan, Ha Tay, Lai Chau, Son La, Tay Ninh, Thanh Hoa, Vinh Phuc), Cambodia, S Thailand (Nakhon Ratchasima, Nakhon Si Thammarat, Narathiwat, Pattani, Phang Nga, Phetchabun, Satun), West Malaysia (Pahang, Pinang, Perak, Penang Is.), Singapore, East Malaysia (Sabah, Sarawak), Brunei, Indonesia (Alor, Bali, Flores, Java, Kalimantan, Lombok, Mentawi, Nias, Simeulue, Sumatra, Sumba, Sumbawa), Timor-Leste and ext. SW Philippines (Palawan), NSL–1770 m.

Sources: Flower, 1899, K.P. Schmidt, 1927a, Kopstein, 1930, Mertens, 1930, Herklots, 1935, C.H. Pope, 1935, Bourret, 1936b, Leviton, 1955, 1965c, E.H. Taylor, 1965, Deuve, 1970, Saint Girons, 1972, Romer, 1979a, Tweedie, 1983, Karsen et al., 1986, Tian et al., 1986, Stuebing, 1991, Zhao & Adler, 1993, Ota & Ross, 1994, David & Vogel, 1996, M.J. Cox et al., 1998, Chan-ard et al., 1999, Orlov et al., 2000, Daltry & Wüster, 2002, Malkmus et al., 2002, Zhao, 2006, V.S. Nguyen et al., 2009, Kaiser et al., 2011 and Lang, 2011.

Remarks: Original description based on H. Boie's MS (1823–1825), Seba (1734: 173, pl. 109, fig. 7) and P. Russell (1801). Mindanao, Philippine record doubtful *fide* Leviton (1965c: 129).

33. *Lycodon synaptor* Vogel & David, 2010. Bonn Zool. Bull. 57(2): 290–293, fig. 1–3, 6a.

Type: Holotype, BMNH 1905.1.30.63, a 463 mm female (J. Graham [Yunnan Exped.], Mar.–Nov. 1868 or Jan.–March 1875).

Type locality: "'Tongchuan, Yunnan', today Dougechuan, 100 km north of Kumming, Yunnan Province, People's Republic of China."

Distribution: Southern China (Yunnan), 1400 m. Known only from vicinity of type locality.

Source: Vogel & Luo, 2011.

34. *Lycodon tessellatus* Jan, 1863b. Elenco Sist. Ofidi: 96. (*Lycodon tessellatum*)

Type: Holotype, NMW 21708, a 344 mm male (native, 24 April 1848).
Type locality: "Manilla" [=Manila, Manila Prov., SW Luzon Is., N Philippines, 14°36'N, 120°59'E, elevation 10 m].
Distribution: Northern Philippines (S Luzon). Known only from holotype and dataless specimen.
Sources: Jan & Sordelli, 1870–1881, Leviton, 1965c, Ota & Ross, 1994 and Ota, 2000.

35. *Lycodon tiwarii* Biswas & Sanyal, 1965. Proc. Zool. Soc. Calcutta 18(2): 137–139, pl. 5.

Type: Holotype, ZSI 20849, a 768 mm specimen (K.K. Tiwari, 5 Feb. 1959).
Type locality: "Mayabundar, North Andaman, Andaman Islands."
Distribution: Eastern India (Andaman & Nicobars: North Andaman Is.).
Source: Vijayankumar & David, 2006.

36. *Lycodon travancoricus* (Beddome, 1870). Madras Month. J. Med. Sci. 2: 169. (*Cercaspis travancoricus*)

Types: Syntypes (2), BMNH 1946.1.13.75, a 381 mm female (R.H. Beddome, 1857–1870), location of other syntype unknown.
Type locality: "Travancore hills, Attraymallay, 5,000 feet elevation, Madras Presidency" [= Attraymallay, Travancore Hills, Kerala State, SW India, 1525 m].
Distribution: Southern India (Andhra Pradesh, Karnataka, Kerala, Laccadives: Androth; Madhya Pradesh, Maharashtra, Orissa, Tamil Nadu), 145–1800 m.
Sources: Boulenger, 1893a, Wall, 1905c, Adiyodi, 1963, Minton, 1966, M.S. Khan, 1982, 2002, J. Daniel, 1983, Inger et al., 1984, Murthy, 1985, Whitaker & Captain, 2004, N. Khaire, 2006, Mukherjee & Bhopathy, 2007, Hutton & David, 2009 and Chandramouli & Ganesh, 2010.
Remarks: Boulenger (1893a: 355) listed one type (either holotype or lectotype). Record from Pakistan (Sindh) doubtful *fide* J. A. Peters *in* Minton (1966: 133) and Whitaker & Captain (2007: 192).

37. *Lycodon zawi* Slowinski, Pawar, Win, Thin, Gyi, Oo & Tun, 2001. Proc. California Acad. Sci. (4) 52(20): 398–403, figs. 1, 2a, 3a.

Type: Holotype, CAS 210323, a 480 mm male (H. Win, T. Thin, K. Wiseman, J. Lovette, & J.V. Vindum, 17 July 1999).
Type locality: "Alaungdaw Katapha National Park (AKNP) (22°19'N, 94°29'E), Sagaing Division, Myanmar."
Distribution: Southeast Asia. Bangladesh, NE India (Assam, Meghalaya, Mizoram) and W Myanmar (Rakhine, Sagaing), 145–500 m.
Sources: Whitaker & Captain, 2004, Ali Reza, 2010 and I. Das, 2012.

LYCODONOMORPHUS Fitzinger, 1843 (Lamprophiidae)

Synonyms: *Lycodontomorphus* Agassiz, 1847 *(nomen emendatum), Ablabes* A.-M.-C. Duméril, Bibron & Duméril, 1854a, *Ablates* – Beddome, 1862 *(nomen incorrectum), Neusterophis* A.C.L.G. Günther, 1863d, *Ablables* – Senna, 1886 *(nomen incorrectum), Ablabophis* Boulenger, 1893a, *Glypholycus* A.C.L.G. Günther, 1893a, *Ablabs* – Palacky, 1898 *(nomen incorrectum), Glypholicus* – Palacky, 1898 *(nomen incorrectum), Nerophidion* F. Werner, 1924a, *Glupholycus* – Briceño-Rossi, 1934 *(nomen incorrectum), Nerophidon* – Neave, 1940a *(nomen incorrectum), Ablades* – Guibé, 1958 *(nomen incorrectum), Albabophis* – Duguy, 1970 *(nomen incorrectum), Abladophis* – Fukada, 1975 *(nomen incorrectum),* and *Abladophis* – Barnard, 1996 *(nomen incorrectum).*
Type species: *Coluber rufulus* Lichtenstein, 1823.
Distribution: Central, SE and S Africa.
Sources: Loveridge, 1958, Broadley, 1967b, Branch, 1988, Kelly et al., 2003, 2009, Vidal et al., 2008 and Zaher et al., 2009.

1. *Lycodonomorphus bicolor* (A.C.L.G. Günther, 1894c). Proc. Zool. Soc. London (1893) 62(1): 629, fig. 1. (*Glypholycus bicolor*)

Synonym: *Nerophidion hypsirhinoides* F. Werner, 1924a.
Types: Syntypes (6), BMNH 1946.1.7.81–86, five males and one female, two longest syntypes 660 mm and 530 mm (E. Coode-Hore, 1889).
Type locality: "shores of Lake Tanganyika" [= W Kigoma or W Rukwa Prov., W Tanzania].
Distribution: Lake Tanganyika (East Africa). Southeastern Democratic Republic of the Congo (E Katanga, Sud-Kivu), SW Burundi (W Bururi), W Tanzania (W Kigoma, W Rukwa) and NE Zambia (N Northern, Mutondwe and Nkumbula Is.), 780 m.
Sources: M.A. Smith, 1928b, Witte, 1952, 1962, Vesey-FitzGerald, 1958, Broadley, 1971d and Madsen & Osterkamp, 1982.
Remarks: Type locality erroneously listed as Lake Victoria *fide* Vesey-FitzGerald (1958: 36).

2. *Lycodonomorphus inornatus* (A.-M.-C. Duméril, Bibron & Duméril, 1854a). Erpét. Gén. 7(1): 434–435. (*Lamprophis inornatus*)

Synonyms: *Lamprophis inornatus* A.-M.-C. Duméril, 1853 (*nomen nudum*), *Boodon infernalis* A.C.L.G.

Günther, 1858, and *Pachyophis temporalis* F. Werner, 1924a.

Type: Holotype, not designated, a 711 mm specimen, location unknown.

Type locality: "des environs du Cap de Bonne-Espérance" [= vicinity of Cape of Good Hope, Western Cape Prov., SW South Africa, 34°21'S, 18°28'E, elevation 20 m].

Distribution: Eastern South Africa (Eastern Cape, E Free State, Gauteng, KwaZulu-Natal, Limpopo, Mpumalanga, Western Cape), NSL–1780 m.

Sources: M.A. Smith, 1928b, Branch, 1988 and R. Hermann, 1993.

Remarks: Type not in MNHN. Original description erroneously cited on p. 464 *fide* V.F.M. FitzSimons (1962a: 114) and Broadley (1983: 84). Namibia records rejected *fide* M. Griffin (2003: 134).

3. *Lycodonomorphus laevissimus* (A.C.L.G. Günther, 1862b). Ann. Mag. Nat. Hist. (3) 9(49): 124–125, pl. 9, fig. 4. (*Natrix laevissima*)

Synonyms: *Grayia lubrica* Sclater, 1898, *Hydrablabes melanogaster* F. Werner, 1923a, *Lycodonomorphus laevissimus fitzsimonsi* Raw, 1973, and *Lycodonomorphus laevissimus natalensis* Raw, 1973.

L

Type: Holotype, BMNH 1946.1.12.78 (formerly FPM R9 & BMNH 1861.11.7.23), an 850–864 mm male.

Type locality: "probably from the East Indies," (in error *fide* V.F.M. FitzSimons, 1962: 104). Corrected to South Africa *fide* V. FitzSimmons (1962a: 104).

Distribution: Swaziland and E South Africa (Eastern Cape, N Free State, KwaZulu-Natal, Mpumalanga), NSL–1715 m.

Sources: Raw, 1973, Broadley, 1983, Branch, 1988 and Haagner & Branch, 1994.

Remarks: Holotype label has "Mediterranean" crossed out, replaced by "Cape" *fide* Haagner & Branch (1994: 238).

4. *Lycodonomorphus leleupi* (Laurent, 1950b). Rev. Zool. Bot. Afr. 43(4): 351. (*Ablabophis whytei leleupi*)

Type: Holotype, MRAC 17637, a 582 mm female (N. Leleup, April 1950).

Type locality: "Station d'études [de l'IRSAC] des Kundelungu (1.750 m.)" [IRSAC Biol. Stn., Kundelungu National Park, Haut-Katanga Prov., SE Democratic Republic of the Congo, ca. 10°38'S, 27°50'E].

Distribution: Southeastern Democratic Republic of the Congo (Katanga), 1750 m.

Sources: Witte, 1953, 1962, Laurent, 1956a, Sweeney, 1961, Broadley & Cock, 1975 and Broadley, 1983, 1988.

Remarks: Loveridge (1958: 14) erroneously listed type locality as being located in "northwest Lake Tanganyika."

5. *Lycodonomorphus mlanjensis* Loveridge, 1953. Bull. Mus. Comp. Zool. 110(3): 253–255. (*Lycodonomorphus rufulus mlanjensis*)

Type: Holotype, MCZ 51050, a 735 mm male (A. Loveridge, 1 April 1949).

Type locality: "Ruo River, Mlanje Mountain, Nyasaland" [= Ruo River, S of Mulanje Mtns., Southern Region, S Malawi, 16°06'S, 35°39'E, elevation 715 m].

Distribution: Southern Malawi (Southern), E Zimbabwe (Manicaland, Mashonaland East) and S Mozambique, 715–1500 m.

Sources: Broadley & Cotterill, 2004 and Broadley & Blaylock, 2013.

6. *Lycodonomorphus obscuriventris* V.F.M. FitzSimons, 1964. Kodoe (7): 26–28, 1 fig. (*Lycodonomorphus whytii obscuriventris*)

Type: Holotype, TMP 26665, a 483 mm female (U. de V. Pienaar, 26 Feb. 1959).

Type locality: "Ngirivane Windmill — approximately 24° 20' S Lat. and 31° 40'E Long. — Kruger National Park, E. Transvaal, South Africa" [= NE Mpumalanga Prov., NE South Africa, elevation 350 m].

Distribution: Southern Mozambique (Gaza, Inhambane, Sofala, Zambézia), ext. SE Zimbabwe (S Manicaland, SE Masvingo), E Swaziland and NE South Africa (E Limpopo, E Mpumalanga), 350 m.

Sources: Broadley, 1995c and Broadley & Blaylock, 2013.

7. *Lycodonomorphus rufulus* (Lichtenstein, 1823). Verzeich. Doubl. Zool. Mus. Univ. Berlin: 105. (*Coluber rufulus*)

Synonym: *Coronella leucopilus* A. Smith, 1831.

Type: Holotype, ZMB 1759, a 610 mm specimen (K.H. Bergius, 1815–1817), lost *fide* R. Günther (*in litt.*)

Type locality: "Afr. austr." [= South Africa]. Restricted to vicinity of Cape Town, south-western part of the Western Cape Province of South Africa (between 33°30'–34°15'S and 18°15'–19°00'E) *fide* Bauer (2000: 55).

Distribution: Southen Africa. Eastern Zimbabwe (Manicaland, Mashonaland East, Midlands), Mozambique (Manica, Niassa, Sofala, Inhaca Is.), Swaziland, Lesotho and South Africa (Eastern Cape, Free State, Gauteng, KwaZulu-Natal, Limpopo, Mpumalanga, Western Cape), 365–2135 m.

Sources: Barros e Cunha, 1935, Broadley & Cock, 1975, Pienaar et al., 1983, Auerbach, 1987, Branch, 1988, Broadley, 1990a, Haagner & Cockeran, 1993, Bauer, 2000 and Broadley & Blaylock, 2013.

Remarks: Restriction of type locality to Swartkops River, Uitenhage area *fide* ffolliott & Liversidge (1971: 46) invalid and their designation of G.L.E. Krebs as collector of type also incorrect *fide* ZMB catalogue. Probably occurs in SE Botswana *fide* Auerbach (1987: 155). Namibia record doubtful *fide* M. Griffin (2003: 134).

8. *Lycodonomorphus subtaeniatus* Laurent, 1954a. Publ. Cult. Comp. Diamantes Angola (23): 38–40, figs. 1–4. (*Lycodonomorphus subtaeniatus subtaeniatus*)

Type: Holotype, MRAC 14864 (formerly RGMC 14864), a 589 mm female (N'Kele, 1941).

Type locality: "Keseki, près de Kwamouth" [= Keseki, near Kwamouth (3°10'S, 16°13'E, elevation 305 m), SW Equateur Prov., W Democratic Republic of the Congo].

Distribution: Western Democratic Republic of Congo (Bandundu, Bas-Congo, S Equateur, Kasai Occidental, Kasai Oriental, Kinshasa, Katanga) and NE Angola (Lunda Norte), 265–725 m.

Source: Witte, 1962.

Remarks: Laurent (1954a: 39) listed 504 mm as trunk length and 85 mm as tail length, implying an overall length of 589 mm but the 17% tail length ratio indicates a 504 mm length.

9. *Lycodonomorphus upembae* Laurent, 1954a. Publ. Cult. Comp. Diamantes Angola (23): 41–43, figs. 5–8. (*Lycodonomorphus subtaeniatus upembae*)

Type: Holotype, MRAC 5201 (formerly RGMC 5201), a 729 mm female (A. Bayet, 1925).

Type locality: "Nyonga" [= Nyonga, Katanga Prov., SE Democratic Republic of the Congo, 8°35'S, 26°18'E, elevation 565 m].

Distribution: Southeastern Democratic Republic of the Congo (Bandundu, Bas-Congo, Katanga, Sud-Kivu), 565–895 m.

Source: Witte, 1953, 1962 and Broadley & Cotterill, 2004.

10. *Lycodonomorphus whytii* (Boulenger, 1897h). Proc. Zool. Soc. London 65(3): 802, pl. 46, fig. 2. (*Glypholycus whytii*)

Type: Holotype, BMNH 1946.1.13.93, a 565 mm female (A. Whyte, 1896).

Type locality: "Fort Hill, Masuku district, 4000 ft., Northern Nyasaland" [= Chitipa, Chitipa District, ext. Northern Region, N Malawi, 9°42'S, 33°16'E, elevation 1280 m].

Distribution: Southeastern Africa. Southwestern Tanzania (S Iringa, Kigoma, Morogoro, W Ruvuma), N Malawi (Northern) and SE Zimbabwe (ext. S Manicaland, SE Masvingo), 270–1525 m.

Sources: Cott, 1935, Broadley & Cock, 1975, Broadley, 1983, Pienaar et al., 1983, Branch, 1988, Spawls et al., 2002 and J.B. Rasmussen, 2004.

Remarks: Holotype erroneously listed as BMNH 1946.1.13.96 and type locality listed as Chipata [= E Zambia, 13°38'S, 32°39'E, elevation 1100 m] *fide* J.B. Rasmuseum (2004: 157 & 160).

LYCODRYAS A.C.L.G. Günther, 1879c (Pseudoxyrhophiidae)

Synonym: *Lycodrias* – Phisalix, 1922 *(nomen incorrectum)*.

Type species: *Dipsadoboa maculata* A.C.L.G. Günther, 1858.

Distribution: Comoros.

Sources: Guibé, 1958, Brygoo, 1983, 1987, Domergue, 1994, Glaw & Vences, 1994, Zaher et al., 2009 and Hawlitschek et al., 2012.

1. *Lycodryas cococola* Hawlitschek, Nagy & Glaw, 2012. PLoS One 7(8): 11–12, fig. 4g.

Synonym: *Lycodryas cococola innocens* Hawlitschek, Nagy & Glaw, 2012.

Type: Holotype, ZSM 41/2020 (formerly FGZC 1512), a 672 mm male (O. Hawlitschek, J. Berger & B. Brückmann, 13 Feb. 2010).

Type locality: "Comoros Archipelago, Comoros, Grand Comoro, Lac Salé; volcanic rocks around lake (11.37375°S; 43.37306°E, 72 m elevation)."

Distribution: Comoros (Grande Comoro Is.), 10–600 m.

L

2. *Lycodryas maculatus* (A.C.L.G. Günther, 1858). Cat. Colub. Snakes Brit. Mus.: 183. (*Dipsadoboa maculata*)

Synonyms: *Dipsas gaimardii comorensis* W.C.H. Peters, 1874a, *Lycodryas sanctjohannis* A.C.L.G. Günther, 1879c, and *Lycodryas sanctijohannis mayottensis* Boettger, 1913.

Type: Holotype, BMNH 1946.1.21.47, a 655–710 mm male (M. Parzudaki).

Type locality: "Central America" (in error). Corrected to Comoro Archipelago, [Indian Ocean] *fide* Domergue (1970: 451).

Distribution: Comoros (Anjouan, Grande Comore, Moheli, Mayotte), NSL–100 m.

Sources: Boulenger, 1896a, Domergue, 1970 and Blanc, 1971, 1972.

LYCOGNATHOPHIS Boulenger, 1893a (Natricidae)

Synonyms: *Scopelophis* Fitzinger, 1843 *(nomen rejiciendum)*, and *Lycognathopsis* – Palacky, 1898 *(nomen incorrectum)*.

Type species: *Psammophis seychellensis* Schlegel, 1837.

Distribution: Seychelles.

Sources: Rendahl, 1939, Nussbaum, 1984, Dowling, 1990, Vidal et al., 2008 and Zaher et al., 2009.

Remarks: Official Generic Name *fide* Opinion 1856 (ICZN, 1996b).

1. *Lycognathophis seychellensis* (Schlegel, 1837). Essai Phys. Serp. 1: 155, 2: 212–213. (*Psammophis seychellensis*)

Synonym: *Thrasops citrinus* Cope, 1869c.
Types: Syntypes (2), MNHN 3509, a 280 mm male (F. Péron), and MNHN 7408, a 660–670 mm specimen (J.-J. Dussumier, 1816–1837).
Type locality: "îles Seychelles" [= Seychelles Arch., Indian Ocean, bet. 4°12–49'S and 55°12–57'E].
Distribution: Seychelles (Frégate, Mahé, Praslin, Silhouette Is.), NSL–915 m.
Remarks: Official Specific Name *fide* Opinion 1856 (ICZN, 1996b).

LYCOPHIDION Fitzinger, 1843 (Lamprophiidae)

Synonyms: *Lycophydion* Agassiz, 1844 *(nomen emendatum), Lycophidium* Agassiz, 1847 *(nomen emendatum), Lissophis* Hallowell, 1857b, *Metoporhina* A.C.L.G. Günther, 1858, *Metophorhina* – V.F.M. FitzSimons, 1962 *(nomen incorrectum), Cryptolycus* Broadley, 1968e, *Lycophodium* – Anthony, 1955 *(nomen incorrectum), Lycophydium* – Villiers, 1956 *(nomen incorrectum), Licophidion* – Lanza, 1972 *(nomen incorrectum),* and *Jacobclarkus* Hoser, 2012ae *(nomen illegitimum).*
Type species: *Lycodon capense* A. Smith, 1831.
Distribution: Southwestern Arabia and Subsaharan Africa.
Sources: Witte, 1962, Laurent, 1968b, Guibé & Roux-Estève, 1972b, Hughes, 1983, Broadley, 1991a, 1992b, 1996a, Broadley & Hughes, 1993, Vidal & Hedges, 2002, Lawson et al., 2005, Nagy et al., 2005, Vidal et al., 2008, Kelly et al., 2009 and Zaher et al., 2009.

1. *Lycophidion acutirostre* A.C.L.G. Günther, 1868. Ann. Mag. Nat. Hist. (4) 1(6): 427, pl. 19, fig. d. (*Lycophidium acutirostre*)

Types: Syntypes (5), BMNH 1946.1.14.34–38 (formerly BMNH 1868.2.29.16), five 203–280 mm females (J. Kirk, 1858–1864).
Type locality: "Zanzibar," (in error *fide* Laurent, 1968: 471). Corrected to Mozambique *fide* Broadley (1996a: 12).
Distribution: Southeastern Africa. Southern Tanzania (Lindi), N Mozambique (Nampula, Sofala, Mozambique Is.) and S Malawi (Southern), NSL–615 m.
Source: Spawls et al., 2002.

2. *Lycophidion albomaculatum* Steindachner, 1870a. Sber. Akad. Wiss. Wien 62(3): 334. (*Lycophidion horstockii albomaculata*)

Type: Holotype, NMW 23464 (F. Steindachner, Oct.–Dec. 1868).
Type locality: "Umgebung von Dakar, Senegambien" [= vicinity of Dakar (14°42'N, 17°27'W, elevation 10 m), Dakar Region, W Senegal].
Distribution: West Africa. Southern Senegal (Dakar, Kédougou, Kolda, Tambacounda, Thiès, Ziguinchor), Gambia (Kombo Saint Mary, Western), Guinea-Bissau (Biombo, Bissau, Bolama Is.), Guinea (Boffa, Boké, Télimélé) and SW Mali (Kayes, Koulikoro, Sikasso), NSL–105 m.
Sources: Manaças, 1955, Villers, 1975, Hakansson, 1981, J.-F. Trape & Mané, 2004, 2006b, Chippaux, 2006, Auliya et al., 2012 and Chirio, 2013.
Remarks: Original but unnamed description based on two specimens from Gambia (A. Smith coll.), described and illustrated in Günther (1866: 29, pl. 7, fig. a). Holotype donated to NMW in Feb. 1869 by Steindachner *fide* Tiedemann et al. (1994: 70).

3. *Lycophidion capense* (A. Smith, 1831). So. Afr. Quart. J. 1(5): 18. (*Lycodon capensis*)

Synonyms: *Lycodon horstokii* Schlegel, 1837, *Lycophidium abyssinicum* Boulenger, 1893a, *Lycophidium jacksoni* Boulenger, 1893a, *Lycophidion capense loveridgei* Laurent, 1968b, *Lycophidion capense vermiculatum* Laurent, 1968b, and *Lycophidion gambensis* Garcia, 1991 *(nomen ineditum).*
Types: Syntypes (2), none designated, RSM, longest syntype 432 mm male, lost *fide* V.F.M. FitzSimons (1937: 262).
Type locality: "near Cape Town," and "close to Port Elizabeth, South Africa." Restricted to Port Elizabeth, Eastern Cape Prov., South Africa *fide* De Waal (1978: 91).
Distribution: Africa and ext. SW Arabia. Northeastern Egypt (Faiyum), SE Sudan (Al Jazirah, Sennar), S South Sudan (Central Equatoria, Eastern Equatoria, Western Equatoria), Eritrea, Ethiopia (Bale, Gondar, Gojjam, Hararge, Illubabor, Shoa, Sidamo, Wollega), W Somalia (Hiraan, ? Jubbada Hoose, Mogadishu, Shabeellaha Dhexe, Shabeellaha Hoose, Togdheer, Woqooyi Galbeed), S Kenya (Central, Coast, Nairobi, North-Eastern), Uganda (Central, Eastern, Northern, Western), NE Democratic Republic of the Congo (Nord-Kivu, Orientale, Sud-Kivu), Rwanda (Gikongoro), Burundi (Ruyigi), Tanzania (Coast, Dar es Salaam, Kigoma, Kilimanjaro, Lindi, Morogoro, Mtwara, Mwanza, Rukwa, Tanga, Ukewere and Zanzibar Is.), S Zambia (Central, Copperbelt, Eastern, Lusaka, Southern), Malawi (Central, Northern, Southern), Mozambique (Gaza, Inhambane, Manica, Maputo, Sofala, Zambézia), Zimbabwe (Bulawayo, Manicaland, Mashonaland Central, Mashonaland East, Mashonaland West, Masvingo, Matabeleland North, Matabeleland South, Midlands), Namibia (Erongo, Hardap, Karas, Khomas, Kunene, Okavango, Omaheke, Otjozondjupa), Botswana (Central, Gaborone, Kgalagadi, Kweneng, North West, South East), E Swaziland, E South Africa

(Eastern Cape, Free State, Gauteng, KwaZulu-Natal, Limpopo, Mpumalanga, E Northern Cape, North West, Western Cape) and W Yemen, NSL–2500 m.

Sources: Sternfeld, 1910e, 1912, Scortecci, 1932a, Flower, 1933, Monard, 1951, Mertens, 1955, 1964a, V.F.M. FitzSimons, 1962a, Thys van den Audenaerde, 1965, Broadley, 1971d, 1983, 1991a, Pitman, 1974, Branch, 1976, 1988, De Waal, 1978, Broadley & Blake, 1979, Trape, 1981, Derleyn et al., 1983, Lanza, 1983a, Pienaar et al., 1983, Gasperetti, 1988, Joger, 1990, Broadley & Hughes, 1993, Largen & Rasmussen, 1993, Clauss & Clauss, 2002, Spawls et al., 2002, Broadley et al., 2003, Marais, 2004, Chippaux, 2006, Chirio & Ineich, 2006 and Broadley & Blaylock, 2013.

Remarks: Syntypes not in BMNH *fide* FitzSimons (1937: 262). Type locality erroneously listed as Kurrichane, Rustenberg Distr., Gauteng Prov., South Africa *fide* Laurent (1968b: 473). Baha El Din (2006: 243) indicated that the Egyptian population is likely extinct. Probably occurs in E Botswana *fide* Broadley & Blake (1979: 10).

4. *Lycophidion depressirostre* Laurent, 1968b. Bull. Mus. Comp. Zool. 136(12): 472–473, fig. 7.

Type: Holotype, MCZ 53348, a 467–469 mm female (J.S. Owen, 20 May 1950).

Type locality: "Torit, Sudan" [= Torit, Eastern Equatoria Prov., SE South Sudan, 4°25'N, 32°35'E, elevation 620 m].

Distribution: Eastern Africa. Eastern Central African Republic (Bamingui-Bangoran, Mbomou, Ouaka, Ouham), SW Sudan (Shamal Darfur), SE South Sudan (Eastern Equatoria), S Ethiopia (Hararge, Kefa, Shoa, Sidamo), Somalia (Bakool, Bay, Hiraan, Jubbada Dhexe, Jubbada Hoose, Mogadishu, Shabeellaha Dhexe, Shabeellaha Hoose, Togdheer), Kenya (Coast, Eastern, North-Eastern, Rift Valley, Lamu Is.), S Uganda (Central) and E Tanzania (Dar es Salaam, Iringa, Lindi, Morogoro, Mtwara, Piwani, Ruvuma, Tanga), 620–1200 (2000) m.

Sources: Lanza, 1983a, Broadley & Hughes, 1993, Largen & Rasmussen, 1993, Spawls et al., 2002, Chippaux, 2006, Chirio & Ineich, 2006 and Largen & Spawls, 2010.

5. *Lycophidion hellmichi* Laurent, 1964a. Publ. Cult. Comp. Diamantes Angola (67): 95–97, fig. 29.

Type: Holotype, MD 3824, a 471 mm female (H. Baumann, 20 March 1954).

Type locality: "Kapolopopo, désert de Moçâmedes, ±15.55 S 12.42 E, ± 450 m" [= Capolopopo, Mossamedes desert, Namibe Distr., ext. SW Angola, 15°53'S, 12°48'E, elevation 515 m].

Distribution: Southwestern Angola (Benguela, Huambo, Huíla, Namibe) and NW Namibia (Erongo, Kunene, Otjozondjupa), NSL–1215 m.

Sources: Mertens, 1955, 1971, Broadley, 1983, 1991b, Branch, 1988 and M. Griffin, 2003.

Remarks: Questionable record from "Kongostaten" *fie* Broadley (1996a: 18).

6. *Lycophidion irroratum* (Leach *in* Bowdich, 1819). Miss. Cape Coast Castle-Ashantee: 494. (*Coluber irroratus*)

Synonyms: *Hypsirhina maura* Gray, 1842d, and *Lycophidion guttatus* Jan, 1863b.

Type: Holotype, BMNH 1946.1.14.30, a juvenile (T.E. Bowdich, April 1817–Oct. 1818).

Type locality: "Fantee, Cape Coast castle to Ashantee" [= Fantee tribal area, between Mansu and Fosu along Amissa (Oki) River, Central Region, S Ghana, elevation 160–180 m]. Restricted to vicinity of Cape Coast, Central Region, SW Ghana *fide* Leston (1970: 142).

Distribution: West and cen. Africa. Southwestern Senegal (Fatick/Ziguinchor), Guinea–Bissau (Bissau, Tombali), Guinea (Boffa, Kindia, Nzérékoré, Télimélé), Sierra Leone (Southern), Liberia (Montserrado), Ivory Coast (Abidjan, Bouna, Lagunes, Toumodi), S Mali (Sikasso), S Burkina Faso (Centre-Est), Ghana (Ashanti, Central, Eastern, Western), S Togo (Plateaux), S Benin (Collines), S Nigeria (Oyo), Cameroon (Centre, Nord, Nord-Ouest), Central African Republic (Bamingui-Bangoran, Haut-Mbomou, Ombella-Mpoko), N Democratic Republic of the Congo (Bandundu, Bas-Congo, Orientale) and SE South Sudan (Eastern Equatoria), 500–1950 m.

Sources: Bowdich, 1819, Villiers, 1950a, 1965, 1975, Manaças, 1955, Perret, 1961, Doucet, 1963, Blackwell, 1967, Leston & Hughes, 1968, Hughes & Barry, 1969, Hughes, 1983, 2013, Broadley & Hughes, 1993, Ineich, 2003, Villiers & Condamin, 2005, Chippaux, 2006, Chirio & Ineich, 2006, J.-F. Trape & Mané, 2006b, Chirio & LeBreton, 2007, Segniagbeto et al., 2011, Auliya et al., 2012 and Chirio, 2013.

Remarks: Type locality restriction to coastal region west of Takoradi, Western Region, SW Ghana *fide* Hughes & Barry (1969: 1008) invalid.

7. *Lycophidion laterale* Hallowell, 1857b. Proc. Acad. Nat. Sci. Philadelphia (1857–1858) 9(3): 58–59.

Synonyms: *Lycophidion laterale ocellata* Sternfeld, 1908b, and *Lycophidion polylepis* Boulenger, 1919b.

Type: Holotype, ANSP 10268, a 367 mm female (H.A. Ford, 1851–1857).

Type locality: "Gaboon country, West Africa" [= Gabon].

Distribution: West and cen. Africa. Southern Ivory Coast, S Ghana (Accra, Brong-Ahato, Eastern), S Togo (Plateaux), S Nigeria, S Cameroon (Centre, Est, Littoral, Nord-Ouest, Sud, Sud-Ouest), Equatorial Guinea (Bioko Is.), W Gabon (Moyen-Ogooué, Ngounié, Nyanga, Ogooué-Maritime), S Central African Republic (Haut-Mbomou, Lobaye, Mbomou, Ombella-Mpoko, Sangha), Congo (Brazzaville, Kouilou),

Democratic Republic of the Congo (Bandundu, Bas-Congo, Equateur, Kasai, Katanga, Kinshasa [LSUMZ 40710], Nord-Kivu, Orientale, Sud-Kivu) NW Angola (Cabinda) and W Uganda (ext. W Western), 25–1880 m.

Sources: Bocage, 1895, Sternfeld, 1908b, Villiers, 1953, 1975, Doucet, 1963, Mertens, 1965c, Thys van den Audenaurde, 1965, Leston & Hughes, 1968, Hughes & Barry, 1969, Guibé & Roux-Estève, 1972, Hughes, 1983, D. Lawson, 1993, J.-F. Trape & Roux-Estève, 1995, Spawls et al., 2002, Villiers & Condamin, 2005, Chippaux, 2006, Chirio & Ineich, 2006, Chirio & LeBreton, 2007 and Segniagbeto et al., 2011.

Remarks: Possibly occurs in Benin *fide* Hughes (2013: 151).

8. *Lycophidion meleagre* Boulenger, 1893a. Cat. Snakes Brit. Mus. 1: 337, pl. 21, figs. 2, 2a–2b. (*Lycophidium meleagris*)

Type: Lectotype, BMNH 1946.1.14.40, a 260 mm female, designated by Broadley (1996: 9).

Type locality: "Ambrizette, Angola" [= Ambrizette, W Zaire Prov., ext. NW Angola, 07°14'S, 12°52'E, elevation NSL] via lectotype selection.

Distribution: Central Africa. Northwestern Angola (Cabinda, Zaire), Democratic Republic of the Congo (Bandundu, Bas-Congo, Equateur, Kasai Occidental, Kasai Oriental, Katanga, Orientale), SE Kenya (S Coast) and NE Tanzania (Morogoro, Tanga), NSL–1600 m.

Sources: Barbour & Loveridge, 1928, Loveridge, 1957a, J.B. Rasmussen, 1981 and Spawls et al., 2002.

9. *Lycophidion multimaculatum* Boettger, 1888e. Ber. Senck. Naturf. Ges. 1888: 67–68. (*Lycophidium capense multimaculata*)

Type: Lectotype, SMF 17973 (formerly SMF-B 7340, a), a female (P. Hesse, 5 Oct. 1886), designated by Mertens (1922: 179).

Type locality: "Povo Netonna, unterer Kongo" [= vicinity of Banana, lower Congo, or mouth of Congo River, Bas-Congo Prov., W Democratic Republic of the Congo, 6°01'S, 12°25'E, elevation NSL] via lectotype selection.

Distribution: Central Africa. Cameroon (Adamaoua, Est, Centre, Nord, Ouest), E Central African Republic (Bamingui-Bangoran, Haut-Mbomou, Haute-Sangha, Ombella-Mpoko, Ouaka, Ouham, Vakaga), S Gabon, S Congo (Bouenza, Brazzaville, Kouilou), Democratic Republic of the Congo (Bandundu, Bas-Congo, Equateur, Kasai Occidental, Kasai Oriental, Katanga, Kinshasa, Orientale, Sud-Kivu), W Tanzania (Rukwa), Angola (Benguela, Bié, Cabinda, Cuando Cubango, Cuanza Norte, Huambo, Huíla, Lunda Norte, Malanje, Moxico, Namibe), W Zambia (Copperbelt, Northern, North-Western, Western), and NE Namibia (Caprivi, Okavango), NSL–2450 m.

Sources: Boettger, 1888, Mertens, 1955, 1971, Laurent, 1964a, Broadley et al., 2003, M. Griffin, 2003, Chippaux, 2006 and Chirio & LeBreton, 2007.

Remarks: Mertens (1922: 179 and 1967: 96) listed lectotype collection date as April 1887, probably date received in Germany.

10. *Lycophidion namibianum* Broadley, 1991a. Ann. Transvaal Mus. 35(14): 210–212, figs. 1, 3 (paratype).

Type: Holotype, TMP 52730, a 475 mm female (H. Bachran, 1966).

Type locality: "west of Usakos, Karibib District, Namibia (2115 C4)."

Distribution: Northwestern Namibia (Erongo, Khomas, Kunene, Otjozondjupa), 350–1665 m.

Source: M. Griffin, 2003.

11. *Lycophidion nanum* (Broadley, 1968e). Arnoldia, Rhodesia 4(10): 2–3, figs. 1–3. (*Cryptolycus nanus*)

Type: Holotype, NMZB-UM 6988, a 219 mm male (L.A. Balarin & D.G. Broadley, 3 Dec. 1963).

Type locality: "Xiluvo, Manica e Sofala District, Mozambique at ca. 250 metres" [= Xiluvo, Sofala Prov., Mozambique, 19°15'S, 34°04'E, elevation 250 m].

Distribution: Eastern Zimbabwe (E Manicaland) and cen. Mozambique (Manica, Sofala), 35–300 m.

Sources: Broadley & Cock, 1975, Branch, 1994, Marais, 2004 and Broadley & Blaylock, 2013.

12. *Lycophidion nigromaculatum* (Schlegel *in* W.C.H. Peters, 1863c). Mber. Königl. Akad. Wiss. Berlin 1863(6): 288. (*Alopecion [Lycodon] nigromaculatus*)

Synonym: *Lycophidion melanogaster* Hoogmoed *in* Bauer, Günther & Klipfel, 1995 *(lapsus calami).*

Type: Holotype, RMNH 88 (H.S. Pel, 1840–1855).

Type locality: "Guinea" [= coast of Ghana]. Restricted to Dabocrom, Ghana *fide* Hughes & Barry (1969: 1015).

Distribution: West Africa. Southeastern Guinea (Nzérékoré), Liberia, Ivory Coast (Tabou), S Ghana (Central, Eastern, Western) and S Togo (Plateaux), 15–280 m.

Sources: Jan, 1870 *in* Jan & Sordelli, 1870–1881, Leston & Hughes, 1968, Hughes & Barry, 1969, Villiers, 1975, Hughes, 1983, J.-F. Trape & Roux-Estève, 1995, Ineich, 2003, Villiers & Condamin, 2005, Chippaux, 2006 and Segniagbeto et al., 2011.

Remarks: Type locality is listed as Dabo Crom *fide* RMNH catalogue. *Lycophidion melanogaster* Hoogmoed *in* Bauer et al. (1995: 69) is a MS name from the old RMNH catalogue. Schlegel is the author following Art. 50.1.1 as the description makes it clear that Peters did not examine the specimen (the description is brief and there are no segmental counts or measurements listed) and he attributed the name to Schlegel. Possibly occurs in Benin *fide* Hughes (2013: 152).

13. *Lycophidion ornatum* H.W. Parker, 1936d. Novit. Zool. (1936–1937) 40(1): 122–123.

Type: Holotype, BMNH 1946.1.14.28 (formerly BMNH 1936.8.1.690), a 344 mm female (K. Jordan, April 1934).

Type locality: "Congulu, Angola" [= Congulo, Cuanza Sul Distr., N Angola, 10°52'S, 14°17'E, elevation 570 m].

Distribution: Central and East Africa. Nigeria, ext. E Cameroon (Adamaoua, Est, Nord), Congo (Lekoumou), S Democratic Republic of the Congo (Kasai Occidental, Kasai Oriental, Katanga, Kivu, Idjwi Is.), N Angola (Cuanza Sul), SE South Sudan (Eastern Equatoria), S Uganda (Central, Eastern, Western), Kenya (Eastern, Nairobi, Western), W Rwanda (Butare, Cyangugu, Gisenyi), W Burundi (Bujumbura, Bururi, Ruyigi) and W Tanzania (Arusha, Kigoma), 570–2900 m.

Sources: Witte, 1953, Laurent, 1956a, Pitman, 1974, Broadley & Hughes, 1993, Derleyn et al., 1983, J.-F. Trape & Roux-Estève, 1995, Spawls et al., 2002, Chippaux, 2006 and Chirio & LeBreton, 2007.

Remarks: Loveridge (1957a: 252) listed Ghana [MCZ 49606].

14. *Lycophidion pembanum* Laurent, 1968b. Bull. Mus. Comp. Zool. 136(12): 478, fig. 10. (*Lycophidion capense pembanum*)

Type: Holotype, BMNH 1940.1.18.74, a 267 mm male.

Type locality: "Pemba Island, Tanzania." Restricted to Mtambile (5°23'S, 39°42'E, elevation 30 m), Pemba Is., Zanzibar Arch., Zanzibar Region, NE Tanzania *fide* Broadley (1996a: 18).

Distribution: Northeastern Tanzania (Pemba Is.), 20–60 m.

Source: Spawls et al., 2002.

15. *Lycophidion pygmaeum* Broadley, 1996a. Syntarsus (3): 13–14, fig. 9.

Type: Holotype, TMP 56602, a 288 mm female (J.D. Visser, Feb. 1982).

Type locality: "St Lucia, Hlabisa District, KwaZulu, South Africa" [= St. Lucia, KwaZulu Natal Prov., E South Africa, 28°23'S, 32°25'E, elevation 25 m].

Distribution: Eastern South Africa (KwaZulu-Natal), 20–630 m. Known only from type series.

16. *Lycophidion semiannule* W.C.H. Peters, 1854. Mber. Königl. Akad. Wiss. Berlin 1854(11): 622. (*Lycophidium semiannulis*)

Type: Holotype, ZMB 4813, a 256–260 mm male (W.C.H. Peters, July–Oct. 1846).

Type locality: "Tette" [Mozambique] (in error). Corrected to Inhambane, Mozambique *fide* Broadley *in* Bauer et al. (1995: 72).

Distribution: Mozambique (Inhambane, Benguerua Is.), NSL–125 m.

Sources: Laurent, 1964a, Broadley, 1983 and Branch, 1988.

Remarks: Original description reprinted in W. Peters (1855: 51).

17. *Lycophidion semicinctum* A.-M.-C. Duméril, Bibron & Duméril, 1854a. Erpét. Gén. 7(1): 414–415.

Synonyms: *Lycophidion semicinctum* A.-M.-C. Duméril, 1853 (*nomen nudum*), and *Lycophidion gambiense* Rochebrune, 1885.

Type: Holotype, RMNH 307, a 283 mm specimen (D.F. Eschricht, 1830–1854).

Type locality: "Cap de Bonne-Espérance, Afrique australe" [= Cape of Good Hope, South Africa] (in error). Corrected to Ghana *fide* Guibé & Roux-Estève (1972b: 394).

Distribution: West Africa. Southeastern Guinea (Nzérékoré), SW Mali (Koulikoro, Sikasso), Sierra Leone (Northern, Southern), Ivory Coast (Bouaké, Toumodi), Burkina Faso (Centre, Centre-Est, Centre-Nord, Centre-Ouest, Est, Hauts-Bassins, Nord, Sud-Ouest, Volta-Noire), Ghana (Accra, Northern, Upper East, Upper West, Volta), Togo (Plateaux), Benin (Alibori, Atakora, Collines, Zou), SW Niger (Dosso), Nigeria, N Cameroon (Adamaoua, Extreme-Nord, Nord), SW Chad (Mayo-Krebbi) and Central African Republic (Bamingui-Bangoran, Kemo-Gribingui, Mbomou, Ombella-Mpoko, Ouaka, Sangha, Vakaga), 250–1125 m.

Sources: Jan 1870 *in* Jan & Sordelli, 1870–1881, Sternfeld, 1908b, 1910a, Angel, 1933b, Manaças, 1957, Doucet, 1963, Leston & Hughes, 1968, Hughes & Barry, 1969, Papenfuss, 1969, Roux-Estève, 1969, Villiers, 1975, Roman, 1980, Hakansson, 1981, Hughes, 1983, 2013, Joger, 1990, Ineich, 2003, Villiers & Condamin, 2005, Chippaux, 2006, Chirio & Ineich, 2006, J.-F. Trape & Mané, 2006b, Chirio & LeBreton, 2007, Ullenburch et al., 2010 and Segniagbeto et al., 2011.

18. *Lycophidion taylori* Broadley & Hughes, 1993. Herpetol. J. 3(1): 12–13, fig. 2.

Type: Holotype, BMNH 1949.2.1.76, a 237 mm male (R.H.R. Taylor, 1932–1933).

Type locality: "Borama District (43°E: 10°N) at ca. 1375 metres, along the boundary between northern (formerly British) Somalia and Ethiopia" [= Woqooyi Galbeed Region, NW Somalia, elevation 1375 m].

Distribution: West and E Africa. Senegal (Kaolack), SE Chad (Ouaddai), Central African Republic (Vikaga), E Ethiopia (Hararge, Kefa, Shoa), Djibouti, S Somalia (Jubbada Hoose, Mogadishu, Shabeellaha Dhexe), NW Kenya (N Rift Valley) and NE Tanzania (Kilimanjaro), 540–1525 m.

Sources: H.W. Parker, 1949, Lanza, 1972, Broadley & Hughes, 1993, Largen & Rasmussen, 1993, Gravlund, 2000b, Spawls et al., 2002, Chippaux, 2006, Chirio, 2009 and Largen & Spawls, 2010.

Remarks: Photograph of holotype in Broadley & Hughes (1993: fig. 2).

19. *Lycophidion uzungwense* Loveridge, 1932c. Bull. Mus. Comp. Zool. 72(10): 375. (*Lycophidion capense uzungwensis*)

Type: Holotype, MCZ 30117, a 219 mm male (A. Loveridge, 1 Jan. 1930).

Type locality: "Dabaga, Uzungwe Mountains, Tanganyika Territory" [= SW Kigoma Region, W Tanzania, ca. 6°S, 30°E].

Distribution: Southern Tanzania (Uzungwe Mtns. of Iringa, Kigoma), 1000–1940 m.

Sources: Loveridge, 1933 and Spawls et al., 2002.

20. *Lycophidion variegatum* Broadley, 1969. Arnoldia, Rhodesia 4(27): 1–4, fig. 1, pls. 1–2.

Type: Holotype, NMZB-UM 10666, a 367 mm male (D.G. Broadley, 20 March 1966).

Type locality: "Chitora River, 40 km. south of Umtali, Rhodesia at ca. 920 metres" [= E Manicaland Prov., E Zimbabwe].

Distribution: Southeastern Africa. Southern Zambia (Southern), Zimbabwe (Manicaland, Mashonaland West, Masvingo, Matabeleland North, Matabeleland South, Midlands) and NE South Africa (N KwaZulu-Natal, Limpopo, Mpumalanga), 400–990 m.

Sources: Branch, 1976, 1988, Broadley & Cock, 1975, Broadley, 1983, Pienaar et al., 1983, Broadley et al., 2003 and Broadley & Blaylock, 2013.

L

LYGOPHIS Fitzinger, 1843 (Xenodontidae)

Synonyms: *Lycophis* Agassiz, 1844 *(nomen emendatum)*, *Aporophis* Cope, 1878a *(nomen substitutum)*, *Digophis* – Briceño-Rossi, 1934 *(nomen incorrectum)*, *Ligofis* – Briceño-Rossi, 1934 *(nomen incorrectum)*, and *Ligophis* – Briceño-Rossi, 1934 *(nomen incorrectum)*.

Type species: *Coluber lineatus* Linnaeus, 1758.

Distribution: South America.

Sources: J.A. Peters & Orejas-Miranda, 1970 and Curcio et al., 2009a.

Remarks: Resurrected from synonymy of *Liophis fide* Zaher et al. (2009: 147).

1. *Lygophis anomalus* (A.C.L.G. Günther, 1858). Cat. Colub. Snakes Brit. Mus.: 37–38. (*Coronella anomala*)

Synonyms: *Lygophis rutilus* Cope, 1862b, and *Coronella pulchella* Jan, 1863a.

Type: Holotype, BMNH 1946.1.9.13, a 474 mm (svl) male (Haslar, Mus.).

Type locality: "Banks of the Parana" [= Río Paraná, Argentina]. Probably from near the city of Santa Fe, Santa Fe Prov., Argentina *fide* Dixon (1985a: 570).

Distribution: Southeastern South America. Southern Brazil (Minas Gerais, Rio Grande do Sul, Santa Catarina, São Paulo), Uruguay (Artigas, Canelones, Cerro Largo, Durazno, Flores, Florida, Maldonado, Montevideo, Paysandú, Rio Negro, Rivera, Rocha, Salto, San José, Soriano, Tacuarembó, Treinta y Tres) and NE Argentina (Buenos Aires, Chaco, Córdoba, Corrientes, Entre Ríos, Formosa, La Rioja, Santa Fe, Salta, Treinta y Tres), 300–1600 m.

Sources: Dixon, 1985a, Cei, 1986, 1994, Leynaud & Bucher, 1999, Freitas, 1999, Carreira-Vidal et al., 2005 and Prigioni et al., 2011.

2. *Lygophis dilepis* Cope, 1862b. Proc. Acad. Nat. Sci. Philadelphia 14(1): 81.

Type: Holotype, USNM 5861, a 648 mm specimen (T.J. Page [Paraguay Exped.], 1853–1858).

Type locality: "Paraguay."

Distribution: Southeastern South America. Eastern Brazil (Bahia, Ceará, Mato Grosso do Sul, Minas Gerais, Paraíba, Pernambuco, Piauí, Rio Grande do Norte), SE Bolivia (S Santa Cruz), Paraguay (Alto Paraguay, Alto Paraná, Amambay, Asunción, Boquerón, Caaguazú, Caazapá, Canindeyú, Central, Concepción, Cordillera, Guairá, Itapúa, Misiones, Neembucú, Paraguarí, Presidente Hayes, San Pedro) and NE Argentina (Chaco, Corrientes, Formosa, Santa Fe), 10–650 m.

Sources: Hoge, 1953g, Lema, 1985a, 1989b, Michaud & Dixon, 1987, Yanosky et al., 1993, Cei, 1994, Leynaud & Bucher, 1999 and Cacciali, 2008.

Remarks: Holotype not listed in Cochran, 1961 for USNM.

3. *Lygophis elegantissimus* (Koslowsky, 1896a). Rev. Mus. La Plata (1895) 7: 154, pl. 1. (*Rhadinaea elegantissima*)

Type: Holotype, MLP, a 630 mm female (J. Koslowsky & N. Alboff, 1890), lost *fide* J.D. Williams *in* Dixon (1985a: 571).

Type locality: "Sierra de la Ventana, Provincia de Buenos Aires" [= Sierra de la Ventana, S Buenos Aires Prov., E Argentina, 38°14'S, 61°40'W, elevation 300 m].

Distribution: Eastern Argentina (S Buenos Aires), 300 m. Known only from vicinity of type locality.

Sources: Cranwell, 1942, Dixon, 1985a and Cei, 1994.

4. *Lygophis flavifrenatus* Cope, 1862b. Proc. Acad. Nat. Sci. Philadelphia 14(1): 80–81.

Synonyms: *Dromicus amabilis* Jan 1863b (*nomen nudum*), *Dromicus amabilis* Jan, 1867 *in* Jan & Sordelli,

1866–1870, *Zamenis parallelus* T. Vogt, 1927 *(nomen praeoccupatum),* and *Zamenis vogti* F. Werner, 1929b *(nomen substitutum).*

Types: Syntypes (2), USNM 5397, a 656–666 mm female, and ANSP 4604 (formerly USNM 5398) (T.J. Page [Paraguay Exped.], 22 May–5 July 1854).

Type locality: "Rio Vermejo region" [= Río Bermejo, 122 mile stretch between 26°10'S, 59°39'W and 26°52'S, 58°28'W, Formosa Prov., N Argentina *fide* Page, 1859: 110, 256].

Distribution: Southeastern South America. Southeastern Brazil (Mato Grosso do Sul, Paraná, Rio Grande do Sul, São Paulo), Paraguay (Guaivá, Presidente Hayes), NE Argentina (Chaco, Corrientes, Formosa, Misiones) and N Uruguay (Artigas, Paysandú, Salto, Tacuarembó, Treinta y Tres), NSL–725 m.

Sources: Hoge, 1953g, Meneghel & Achaval, 1983, Michaud & Dixon, 1987, J.D. Williams & Francini, 1991, Cei, 1994, Kucharzewski & Tillack, 2006 and Carreira-Vidal & Lombardo, 2007b.

Remarks: Type locality is "Paraguay" *fide* USNM catalogue and specimen data tag, with "Buenos Ayres" later added in USNM catalogue and systematic cards. Cochran (1961: 196) listed "Buenos Aires, Argentina" as type locality (point of departure of expedition). The type locality is probably in Argentina as the Río Bermejo flows into the Río Paraguay at the Argentina-Paraguay border and Puerto Bermejo is in Argentina.

5. *Lygophis lineatus* (Linnaeus, 1758). Syst. Nat., ed. 10, 1: 221. (*Coluber lineatus*)

Synonyms: *Coluber minervae* Linnaeus, 1758, *Coluber jaculatrix* Linnaeus, 1766, *Coluber iaculatrix* Boddaert, 1783, *Iaculatrix surinamensis* Boddaert, 1783 *(nomen nudum)*, *Coluber terlineatus* Lacépède, 1789 *(nomen rejiciendum)*, and *Natrix sentzenii* Merrem, 1820.

Types: Syntypes (3), NHR Lin-49–51 (formerly MAFR), 750 mm, 580 mm and 555 mm specimens (Mus. Drottn.).

Type locality: "Asia," (in error *fide* Hoge et al., 1978b: 46).

Distribution: Panama and N South America. Eastern Panama (Coclé, Herrera, Los Santos, Veraguas), Colombia (Antioquia, Atlántico, Bolívar, Caquetá, Córdoba, Cundinamarca, Guaviare, La Guajira, Magdalena, Meta, Nariño, Santander, Sucre, Tolima, Vaupés, Vichada), Venezuela (Amazonas, Anzoátegui, Apure, Aragua, Barinas, Bolívar, Carabobo, Cojedes, Delta Amacuro, Distrito Federal, Falcón, Guárico, Lara, Mérida, Miranda, Monagas, Nueva Esparta, Portuguesa, Sucre, Táchira, Trujillo, Yaracuy, Vargas, Zulia), Guyana (East Berbice-Corentyne, Essequibo Islands-West Demerara, Potaro-Siparuni, Upper Takutu-Upper Essequibo), Suriname (Marowijne, Paramaribo, Saramacca, Suriname), French Guiana (Cayenne, Saint-Laurent-du-Maroni) and N Brazil (Amapá, Amazonas, ? Bahia, Pará, Roraima, Marajo Is.), NSL–900 m

Sources: Hoge, 1953f, Roze, 1966a, Vanzolini et al., 1980, Lancini, 1986, Pérez-Santos, 1986, Michaud & Dixon, 1987, Lema, 1989b, Starace, 1998, Esqueda & La Marca, 1999, Freitas, 1999, Mijares-Urrutia & Arends, 2000, Markezich, 2002, Rivas-Fuenemayer & Barrio-Amorgós, 2005, Navarrete et al., 2009 and C.J. Cole et al., 2013.

Remarks: Costa Rican (Limón) and Ecuadorian (Esmeraldas, Manabí) records doubtful *fide* Michaud & Dixon (1987: 3), latter specimens probably transported from Panama.

6. *Lygophis meridionalis* (Schenkel, 1901). Verh. Naturf. Ges. Basel (1901–1902) 13(1): 160–166. (*Aporophis lineatus meridionalis*)

Synonym: *Aporophis lineatus lativittatus* Müller, 1928.

Type: Lectotype, NMBA 1454, a juvenile male (C. Ternetz, 1895–1896), designated by Kramer (1978: 658).

Type locality: "Paraguay: Bemalcue; Mte. Sociedad," (partially in error). [= Cerrito, 28 km N of Asunción, SE Presidente Hayes Dept., Paraguay, 24°58'S, 57°33'W, elevation 75 m *fide* Cacciali & Wüest, 2009: 383].

Distribution: Central and S South America. Eastern Bolivia (Beni, Santa Cruz), cen. and SE Brazil (Distrito Federal, Espírito Santo, Goiás, Mato Grosso, Mato Grosso do Sul, Minas Gerais, ext. S Pará, Paraná, São Paulo, Tres Estados Is.), Paraguay (Alto Paraguay, Alao Paraná, Amambay, Caaguazú, Caazapá, Canindeyú, Central, Concepción, Cordillera, Itapúa, Misiones, Neembucú, Presidente Hayes, San Pedro) and ext. NE Argentina (Corrientes, Formosa, Misiones), NSL–400 m.

Sources: Hoge, 1953f, Michaud & Dixon, 1987, Cei, 1994, Leynaud & Bucher, 1999, Valdujo & Nogueiro, 2001, Padua Alneida et al., 2006, Cacciali, 2008 and Cacciali & Wüest, 2009.

Remarks: Cacciali & Wüest (2009: 383) have shown that Bermalcue (= Bernal Cué) and Monte Sociedad (= Cerrito) are two different localities. *Aporophis lineatus lativittatus,* whose type was destroyed during World War II, possibly a synonym of *L. dilepis* as both *L. meridionalis* and *L. dilepis* occur in Bolivia (Dixon, *in litt.*).

7. *Lygophis paucidens* Hoge, 1953b. Mem. Inst. Butantan (1952) 24(2): 189–190, pl. 4–5.

Type: Holotype, IB 12016 (formerly MNRJ), a 484 mm female (H. Sick, 24 Sept. 1948), destroyed by fire 15 May 2010.

Type locality: "Mato Verde, Estado de Goiás, Brasil" (in error). [= Mato Verde, on Río Araguaia, ext. NE Mato Grosso State, cen. Brazil, 11°11'S, 50°51'W, elevation 185 m].

Distribution: Central South America. Eastern Brazil (Bahia, Distrito Federal, S Goiás, NE Mato Grosso, Minas Gerais, Paraíba, Pernambuco, Piauí), 80–780 m.

Sources: Hoge, 1953f, Lema, 1985a, 1989b, Michaud & Dixon, 1987, Freitas, 1999, Nogueira, 2001 and Valdujo & Nogueira, 2001.

Remarks: Mato Verde is near Mato Grosso-Goiás border but clearly in Mato Grosso *fide* Hoge (1953b: 191, map).

8. *Lygophis vanzolinii* (Dixon, 1985a). Copeia 1985(3): 567–569.

Type: Holotype, AMNH 65263, a 748 mm male (P. Horen, 1943).

Type locality: "Achiras, Córdoba, Argentina" [= Achiras, W Córdoba Prov., cen. Argentina, 33°10'S, 65°00'W, elevation 850 m].

Distribution: Central Argentina (Córdoba, San Luis), 850 m.

Source: Cei, 1994.

LYSTROPHIS Cope, 1885a (Xenodontidae)

Synonyms: *Rhinostoma* Fitzinger, 1826a *(nomen rejiciendum), Rhinostoma* Wagler, 1830 *(nomen praeoccupatum), Rhinosiphon* Fitzinger, 1843 *(nomen rejiciendum), Rhinosyphon* A.-M.-C. Duméril, Bibron & Duméril, 1854a *(nomen emendatum), Lystrophys* – E. Santos, 1943 *(nomen incorrectum),* and *Lysorhophis* – H.M. Smith & White, 1955 *(nomen incorrectum).*

Type species: *Heterodon dorbignyi* Bibron *in* A.-M.-C. Duméril, Bibron & Duméril, 1854a.

Distribution: Southern South America.

Fossil records: Middle-Upper Pleistocene of Argentina (Buenos Aires).

Sources: A. Ferreira, 1949, Orejas-Miranda, 1966a, J.A. Peters & Orejas-Miranda, 1970, Hoge et al., 1976a, Amaral, 1978, J.D. Williams & Francini, 1991, Scrocchi & Cruz, 1993, Scanferla et al., 2005 and Zaher et al., 2009.

Remarks: A synonym of *Xenodon fide* Zaher et al., 2009.

L

1. *Lystrophis dorbignyi* (Bibron *in* A.-M.-C. Duméril, Bibron & Duméril, 1854a). Erpét. Gén. 7(1): 772–774. (*Heterodon dorbignyi*)

Synonyms: *Rhinostoma proboscidea* Fitzinger, 1826a *(nomen nudum), Vipera nasua* Oppel *in* Wagler, 1830 *(nomen rejiciendum), Rhinostoma nasua* Wagler, 1830, *Rhinostoma nasuum* A.-M.-C. Duméril, Bibron & Duméril, 1854b *(nomen emendatum), Lystrophis dorbignyi chacoensis* Lema, 1994, *Lystrophis dorbignyi orientalis* Lema, 1994, and *Lystrophis dorbignyi uruguayensis* Lema, 1994.

Types: Syntypes (5), MNHN 3633, MNHN 7554–55 (A.C.V.M.D. d'Orbigny, Oct. 1826–Nov. 1828), USNM 420, a 388 mm (svl) stuffed skin (A.F.C.P. de Saint-Hilaire or Dupré, Jan.–May 1820), and IB (formerly MNHN 7556) (C. Gaudichaud-Beaupré, Dec. 1817–Feb. 1818, 1831–1833 or Feb.–July 1836), longest syntype 480 mm, the latter destroyed by fire 15 May 2010.

Type locality: "Amérique du Sud: Buenos-Ayres" [= Buenos Aires, Argentina, 20 Jan. 1827–8 Nov. 1828], "Montevideo" [= Montevideo, Uruguay, 29 Oct. 1826–20 Jan. 1827], "Sainte-Catherine" [= Santa Catarina, Brazil], and "Brésil" [= Brazil].

Distribution: Southern South America. Southern Brazil (Rio Grande do Sul), S Paraguay (Alto Paraná, Amambay, Asunción, Caaguazú, Caazapá, Canindeyú, Central, Concepción, Cordillera, Guairá, Itapúa, Misiones, Neembucú, Paraguarí, San Pedro), Uruguay (Artigas, Canelones, Cerro Largo, Colonia, Durazno, Florida, Lavalleja, Maldonado, Montevideo, Paysandú, Río Negro, Rivera, Rocha, Salto, San José, Soriano, Tacuarembó, Treinta y Tres) and Argentina (Buenos Aires, Chaco, Córdoba, Corrientes, Entre Ríos, La Pampa, Río Negro, Salta, Santa Fe, Santiago del Estero).

Sources: Lema, 1961b, 1994, Gudynas, 1979, Cei, 1994, Reati, 1996, Leynaud & Bucher, 1999. Carreira-Vidal et al., 2005, Nenda & Cacivio, 2007 and Cacciali, 2008.

Remarks: Official Specific Name no. 1966 *fide* Opinion 698 (ICZN, 1964). Contrary to Opinion 698, Bibron must be the author according to Art. 50a because Duméril, Bibron & Duméril (1854a: 774, 776) explicitly stated that the name was proposed by Bibron, who studied the material from the d'Orbigny expedition.

2. *Lystrophis histricus* (Jan, 1863a). Arch. Zool. Anat. Fis. 2(2): 224. (*Heterodon histricus*)

Type: Holotype, formerly MSNM, a 350 mm specimen, destroyed in 1943 during World War II.

Type locality: Unknown. Designated as South America *fide* Jan & Sordelli [1865 *in* 1860–1866, 1(livr. 11): pl. 4, fig. 2].

Distribution: Southern South America. Southern Brazil (Mato Grosso, Paraná, Rio Grande do Sul), Paraguay, E Uruguay (Cerro Largo, Lavalleja, Maldonado, Treinta y Tres) and NE Argentina (Chaco, Corrientes, Formosa, La Pampa, Misiones).

Sources: Hoge et al., 1976, Viñas and Olmedo, 1988, Cei, 1994, Carreira-Vidal et al., 2005 and Prigioni et al., 2012.

Remarks: Original description reprinted in Jan (1863b: 44).

3. *Lystrophis matogrossensis* Scrocchi & Cruz, 1993. Pap. Avul. Zool. 38(10): 181–183, pls. 3–4.

Type: Holotype, IB 53544, a 441 mm female, destroyed by fire 15 May 2010.

Type locality: "Miranda, Mato Grosso do Sul, Brasil" [= Miranda, W Mato Grosso do Sul, SW Brazil, 20°15'S, 56°22'W, elevation 130 m].

Distribution: Brazil (Mato Grosso, Mato Grosso do Sul) and Bolivia (Santa Cruz), 535 m.
Sources: Giraudo, 1997 and Jansen, 2008.

4. *Lystrophis nattereri* (Steindachner, 1867a). Reise Öster. Novara 1: 90. (*Heterodon nattereri*)

Type: Holotype, NMW 23378, a 355 mm specimen (J. Natterer, 1817–1835).
Type locality: "Brasilien" [= Brazil]. Emended to interior of Brazil *fide* Steindachner (1864: 233).
Distribution: Eastern Brazil (Bahia, Distrito Federal, Goías, S Mato Grosso do Sul, Minas Gerais, Paraná, S Piauí, São Paulo), 720–1200 m.
Sources: Hoge et al., 1976, Argôlo, 2002a and Silveira et al., 2004c.

5. *Lystrophis pulcher* (Jan, 1863a). Arch. Zool. Anat. Fis. 2(2): 222. (*Heterodon pulcher*)

Type: Holotype, formerly MSNM, destroyed in 1943 during World War II.
Type locality: "Bolivia."
Distribution: Southern South America. Eastern Bolivia (Santa Cruz, Tarija), W Paraguay (Alto Paraguay, Boquerón, Central, Neembucú, Presidente Hayes, San Pedro) and Argentina (Catamarca, Chaco, Córdoba, Corrientes, Entre Dios, Formosa, Jujuy, Junín, La Rioja, Salta, San Juan, San Luis, Santa Fe, Santiago del Estero, Tucumán), 100–400 m.
Sources: Scrocchi & Cruz, 1993, Leynaud & Bucher, 1999, Nenda & Cacivio, 2007, Cacciali, 2008 and Jansen, 2008.
Remarks: Revalidated by Scrocchi and Cruz, 1993.

6. *Lystrophis semicinctus* (Bibron *in* A.-M.-C. Duméril, Bibron & Duméril, 1854a). Erpét. Gén. 7(1): 774–776. (*Heterodon semicinctus*) (*nomen corrigendum*)

Synonyms: *Heterodon semi-cinctus* Bibron *in* A.-M.-C. Duméril, Bibron & Duméril, 1854a *(nomen incorrigendum), Heterodon semicinctus* Jan, 1863a *(nomen corrigendum),* and *Lystrophis semicinctus weiseri* Müller, 1928.
Types: Syntypes (3), MNHN 1209, MNHN 1521, and MNHN 3634, a skin (A.C.V.M.D. d'Orbigny, 20 Jan. 1827–8 Nov. 1828 [Buenos Aires] and 11 Nov. 1830–21 June 1831 and 17 Sept.–17 Nov. 1832 [Santa Cruz]), longest syntype 420 mm.
Type locality: "Buenos-Ayres et à Santa-Cruz" [= Buenos Aires, Argentina, 34°36'S, 58°23'W, elevation 175 m, and Santa Cruz de la Sierra, Santa Cruz Dept., Bolivia, 17°48'S, 63°10'W, elevation 425 m].
Distribution: Southern South America. Southern Brazil (Rio Grande do Sul), S Bolivia (Cochabamba, Santa Cruz, Tarija), Paraguay and Argentina (Buenos Aires, Catamarca, Chubut, Córdoba, Corrientes, Entre Ríos, La Pampa, La Rioja, Mendoza, Neuquén, Rio Negro, San Luis, Tucumán), 150–4000 m.
Sources: Koslowsky, 1895, P. Muller, 1969a, Scrocchi & Cruz, 1993, Cei, 1994, D.R. Norman, 1994 and Giambelluca, 1999.

LYTORHYNCHUS W.C.H. Peters, 1862a (Colubridae)

Synonyms: *Chatachlein* Jan, 1863a, *Acontiophis* A.C.L.G. Günther, 1875, *Catachlaena* Blanford, 1881 *(nomen emendatum), Lythorhynchus* – Bocage, 1882a *(nomen incorrectum), Lythorynchus* – Olivier, 1894 *(nomen incorrectum), Lytorynchus* – Olivier, 1899 *(nomen incorrectum), Lithorhynchus* Doumergue, 1901 *(nomen praeoccupatum), Lytorhyncus* – Andersson, 1903 *(nomen incorrectum), Acanthiophis* – Nikolsky, 1916 *(nomen incorrectum), Lytovinchus* – Zavattari, 1929 *(nomen incorrectum), Aconitophis* – Mertens, 1969a *(nomen incorrectum), Acantiophis* – Leviton & Anderson, 1970 *(nomen incorrectum),* and *Lytorhynebus* – Gasperetti, 1974 *(nomen incorrectum).*
Type species: *Heterodon diadema* A.-M.-C. Duméril, Bibron & Duméril, 1854a.
Distribution: Southwestern Asia and N Africa.
Sources: Minton, 1966, Leviton & Anderson, 1970b, Leviton, 1977 and Gasperetti, 1988.

1. *Lytorhynchus diadema* (A.-M.-C. Duméril, Bibron & Duméril, 1854a). Erpét. Gén. 7(1): 779–780. (*Heterodon diadema*)

Synonyms: *Lithorhynchus diadema hirouxii* Doumergue, 1901, *Lytorhynchus gaddi* Nikolsky, 1907, *Lithorhynchus diadema hoggarense* Angel, 1944, *Lytorhynchus diadema arabicus* G. Haas, 1952a, and *Lytorhynchus diadema mesopotamicus* G. Haas, 1952a.
Type: Holotype, MNHN 7560, a 380 mm female (P.K.A. Schousboe, 1850).
Type locality: "Algérie et du désert de l'ouest de l'Afrique Septentrionale" [= Algeria and desert of W North Africa].
Distribution: North Africa and SW Asia. Mauritania (Adrar, Dakhlet Nouadhibou, Inchiri, Tagant, W Tiris Zemmour, Trarza), Western Sahara (Rio de Oro, Saguia El Hamra), S Morocco (Kaar Es Souk, Ouarzazate, Oujda, Tan Tan, Tata), Western Sahara (Rio de Oro, Saguia El Hamra), Algeria (Béchar, El Bayadh, Illizi, Naama, Tamanrasset), S Tunisia (Gabès, Gafsa, Kairouan, Kasserine, Tozeur), N Niger (Agadez), N Libya (Ajdabiya, Al Jabar Al Akhdar, Al Jurfah, Banghazi, Gharyan, Darnah, Tarabulus, Tripoli), NE Sudan, Egypt (Cairo, El-Bahr El-Ahmar, Faiyum, Giza, Ismailia, Matrouh, New Valley, North Sinai, South Sinai, Suez), Saudi Arabia (Asir, Eastern, Ha'il, Jawf, Jazan, Madinah, Makkah, Qasim, Riyadh), Kuwait, Bahrain, United Arab Emirates (Abu Dhabi,

Ras al Khaimah, Sharjah), Oman (Al Wusta, Dhofar, Muscat), SW Yemen (Lahij), S Israel (Central, Gaza Strip, Southern, Tel Aviv), Syria (Homs), Jordan (Maan, Waddi Ramm), Iraq (Al Basrah, Baghdad, Ninawa) and SW Iran (Bushehr, Khuzestan), NSL–2300 m.

Sources: Mayet, 1903, Angel & Lhote, 1938, G. Haas, 1952a, Chpakowsky & Chnéour, 1954, Arnold 1980b, Le Berre, 1989, Latifi, 1991, Leviton et al., 1992, Martens, 1993, Schätti & Gasperetti, 1994, Bons & Geniez, 1996, Schleich et al., 1996, Bouskila & Amitai, 2001, Disi et al., 2001, Geniez et al., 2004, Chippaux, 2006, J.-F. Trape & Mané, 2006b, Smid, 2010 and Bar & Haimovitch, 2011.

2. *Lytorhynchus gasperetti* Leviton, 1977. J. Saudi Arabia Nat. Hist. Soc. (19): 17–19, figs. 1a–c.

Type: Holotype, CAS 134150, a 380 mm female (J. Gasperetti, 9 Jan. 1972).

Type locality: "Khasawiyah (16°56'N, 42°37'E), Saudi Arabia" [= Jazan Prov., SW Saudi Arabia].

Distribution: Extreme SW Saudi Arabia (S Asir, Jazan), NSL–2000 m.

Source: Farang & Banaja, 1980 and Gasperetti, 1988.

L

3. *Lytorhynchus kennedyi* K.P. Schmidt, 1939b. Field Mus. Nat. Hist. (Zool.) 24(7): 75.

Type: Holotype, FMNH 19586, a 380 mm male (H. Field & R.A. Martin, 21 May 1934).

Type locality: "between Homs and Palmyra, Syria" [= bet. Homs (34°44'N, 36°43'E, elevation 500 m), and Palmyra (34°33'N, 38°17'E, elevation 400 m), Homs Distr., Syria].

Distribution: Syria (Homs) and Jordan (Aqabah, Maan, Mafraq, Tafilah), 400–500 m.

Sources: Leviton et al., 1992, Martens, 1993 and Amr & Disi, 2011.

Remarks: Probably occus in SW Iraq fide Leviton et al. (1992: 98). A subspecies of *L. diadema fide* Schleich et al. (1996: 496).

4. *Lytorhynchus maynardi* Alcock & Finn, 1897. J. Asiat. Soc. Bengal (1896) 65(4): 562–563, pl. 14, figs. 1, 1a–1c.

Types: Syntypes (4), BMNH 1946.1.14.79, a male, and ZSI 14223–25 (formerly IMC), longest syntype 381–400 mm (A.H. MacMahon & F.P. Maynard [Afghan-Baluch Bound. Comm.], 30 March–May 1896).

Type locality: "Near Robat I, just south of the Koh Malik-do-khand, Afghan-Baluch Boundary, 4500 feet" [= ext. S Helmand Prov., S Afghanistan, 29°33'N, 63°38'E, elevation 1310 m].

Distribution: Southwest Asia. Eastern Iran (Razavi Khorasan), ext. SW Afghanistan and W Pakistan (ext. N Balochistan), 790–1680 m.

Sources: F. Werner, 1929a, M.A. Smith, 1943, Baloutch, 1976 and M.S. Khan, 2002.

Remarks: Wall (1923d: 619) implied only two types and M.A. Smith (1943: 192) listed only three types.

5. *Lytorhynchus paradoxus* (A.C.L.G. Günther, 1875). Proc. Zool. Soc. London 43(1): 232–233, fig. 5. (*Acontiophis paradoxa*)

Synonym: *Lytorhynchus monticornis* F. Werner, 1927.

Type: Holotype, BMNH 1946.1.14.75, a 305 mm male (T.C. Jerdon, 1836–1852).

Type locality: "probably within the Himalayan region or in Khassya" [= Himalayan region or Khassia, Bulandshahar Distr., Uttar Pradesh, N India]. Restricted to N India *fide* M.A. Smith (1943: 191) and to Zangipur, N Sindh *fide* Minton (1966: 130)..

Distribution: Eastern Pakistan (Punjab, Sindh) and NW India (Rajasthan), NSL–150 m.

Sources: M.A. Smith, 1943, M.S. Khan, 1977, 1982a, 2002, Whitaker & Captain, 2004 and Baig et al., 2008.

Remarks: Type locality restriction of Zangipur, N Sindh, Pakistan *fide* M.S. Khan (2002: 111) invalid.

5. *Lytorhynchus ridgewayi* Boulenger, 1887e. Ann. Mag. Nat. Hist. (5) 20(120): 413–414.

Synonyms: *Lythorhynchus ridgewaji roseni* Elpatjewsky & Sabanejew, 1906, and *Lytorhynchus gabrielis* F. Werner, 1938a.

Types: Syntypes (2), BMNH 1946.1.14.77–78, two males, longest syntype 425 mm (J.E.T. Aitchison, 27–29 May 1885).

Type locality: "Chin Kilak, Afghanistan" [= Chinkalok, 38 km NW of Herat, Herat Prov., NW Afghanistan, ca. 34°32'N, 61°53'E, elevation 600 m].

Distribution: Southwestern Asia. Extreme SW Turkmenistan, E Iran (Central, Fars, Kerman, Khuzestan, Mazandaran, Sistan va Baluchestan), W Afghanistan (Helmand, Herat) and W Pakistan (Balochistan, Punjab), 600–2440 m.

Sources: Boulenger, 1889d, Leviton & Anderson, 1970b, Bannikov et al., 1977, Latifi, 1991, Szczerbak, 1994 and M.S. Khan, 2002.

M

MACRELAPS Boulenger, 1896a (Atractaspididae)

Synonym: *Macrolaps* – Maass-Berlin, 1933 *(nomen incorrectum).*
Type species: *Uriechis microlepidotus* A.C.L.G. Günther, 1860i.
Distribution: Southern Africa.
Sources: Witte & Laurent, 1947, V.F.M. FitzSimons, 1962a, 1974, Branch, 1981, 1982, 1988, Broadley, 1983, Wallach, 1991, Underwood & Kochva, 1993, Nagy et al., 2005, Kelly et al., 2009 and Zaher et al., 2009.

1. *Macrelaps microlepidotus* (A.C.L.G. Günther, 1860i). Ann. Mag. Nat. Hist. (3) 5(27): 168–169, pl. 9. (*Uriechis microlepidotus*)

Synonym: *Atractaspis natalensis* W.C.H. Peters, 1877c.
Types: Syntypes (2), BMNH 1946.1.3.9, a 940 mm female (T. Ayres), and BMNH 1946.1.8.67, a juvenile (J.H. Gurney, 1858–1860).
Type locality: "Port Elizabeth, in Algoa Bay, South Africa" (in error). Corrected to D'Urban Port Natal *fide* Gurney (1860: 342) [= Durban, Natal Province, NE South Africa].
Distribution: Eastern South Africa (Eastern Cape, KwaZulu-Natal), NSL–525 m.

MACROCALAMUS A.C.L.G. Günther, 1864a (Calamariidae)

Type species: *Macrocalamus lateralis* A.C.L.G. Günther, 1864a.
Distribution: Malay Peninsula.
Sources: Grandison, 1972, Tweedie, 1983, Vogel & David, 1999, Norsham & Lim, 2003, David & Pauwels, 2004, Zaher et al., 2009 and I. Das, 2010, 2012.

1. *Macrocalamus chanardi* David & Pauwels, 2004. Raffles Bull. Zool. 52(2): 637–639, fig. 4.

Type: Holotype, BMNH 1900.6.14.17, a 264 mm female (A.L. Butler).
Type locality: "Larut Hills, Perak, 3500-4500 ft." [= Bukit Larut, Perak State, West Malaysia *fide* David & Pauwels, 2004: 637].
Distribution: Central West Malaysia (Pahang, Perak), 1065–1500 m.

2. *Macrocalamus gentingensis* Norsham & Lim, 2003. Hamadryad (2002) 27(1): 83–87, figs. 1–3.

Type: Holotype, ZRC 2.5062, a 378 mm female (O.K. Sang, June 2001).
Type locality: "Genting Highlands, about 32 km north-east of Kuala Lumpur, in the vicinity of the water pump at 1181 m in elevation, Pahang, Malay Peninsula (03°26'N, 101°47'E)."
Distribution: Central West Malaysia (Pahang), 1180–1690 m.

3. *Macrocalamus jasoni* Grandison, 1972. Bull. Brit. Mus. Nat. Hist. (Zool.) 23(4): 88–90, fig. 4.

Type: Holotype, BMNH 1967.2283, a 760 mm female (J.A. Bullock, 9 April 1967).
Type locality: "track below camp 5 at about 5800 ft, Gunong Benom, C. Pahang, W. Malaysia" [= Mt. Benom, Pahang State, cen. West Malaysia, 1765 m].
Distribution: Central West Malaysia (Pahang), 1765–1980 m. Known only from type locality.
Source: Vogel & David, 1999.

4. *Macrocalamus lateralis* A.C.L.G. Günther, 1864a. Rept. Brit. India: 199, pl. 18, fig. d.

Synonym: *Macrocalamus lateralis* A.C.L.G. Günther, 1863c *(nomen nudum)*
Type: Holotype, BMNH 1946.1.7.23, a 305 mm male (T. Hardwicke, 1756–1823).
Type locality: "East Indies, probably from the continent." Restricted to Malay Peninsula *fide* Flower (1899: 674) and to mainland opposite Penang Is., Pinang State, West Malayasia *fide* David & Pauwels (2004: 640).
Distribution: Peninsular Thailand (Narathiwat) and West Malaysia (Pinang), 400 m.
Sources: Flower, 1899, Boulenger, 1912, M.A. Smith, 1922, B.L. Lim, 1963, 1967a, Tweedie, 1983, M.J. Cox et al., 1998, Chan-ard et al., 1999, Vogel & David, 1999 and B.L. Lim et al., 2002.
Remarks: Type locality designation of India *fide* A.C.L.G. Günther (1863c: 350) in error. Type locality restriction to Cameron Highlands, Pahang State, West Malaysia *fide* B.L. Lim (1963a: 101) in error.

5. *Macrocalamus schulzi* Vogel & David, 1999. Raffles Bull. Zool. 47(2): 318–322, figs. 1–3 (paratype), pls. 1, 4–5, 9–12.

Type: Holotype, ZFMK 51159, a 383 mm male (K.-D. Schulz, June–July 1989).

Type locality: "Tanah Rata (ca 4°29›N, 101°23›E), Cameron Highlands, Pahang, West Malaysia, Malaysia" [= Tanah Rata, Pahang State, cen. West Malaysia, 4°29'N, 101°23'E, elevation 1430 m].

Distribution: Central West Malaysia (Pahang), 1000–1800 m.

6. *Macrocalamus tweediei* B.L. Lim, 1963a. Bull. Natl. Mus. Singapore (32): 100, fig. 1, pl. 2.

Type: Holotype, ZRC 2.2174 (formerly IMRKL 57656), a 500 mm specimen (B.L. Lim & H.E. McClure, 2 Oct. 1959).

Type locality: "Gunong Brinchang, Cameron Highlands, Pahang at an elevation of 6,000 ft., Malaya" [= Mt. Batu Berinchang, Pahang State, cen. West Malaysia, elevation 1830 m].

Distribution: Central West Malaysia (Pahang, Selangor), 1500–1970 m.

Sources: M.J. Cox et al., 1998, Vogel & David, 1999 and Chan-ard et al., 1999.

M

7. *Macrocalamus vogeli* David & Pauwels, 2004. Raffles Bull. Zool. 52(2): 642–643, figs. 7–9.

Type: Holotype, BMNH 1968.764, a 192 mm male (FMSM Meteorol. Surv., Sept. 1921).

Type locality: "'Camp Padang, Gunong Tahan, Pahang, Malaya, 5.400–5.700 ft.,' Gunung Tahan, Pahang, West Malaysia, Malaysia."

Distribution: Central West Malaysia (Pahang), 1650–1750 m.

MACROPISTHODON Boulenger, 1893a (Natricidae)

Synonyms: *Macropisthistodon* – Steindachner, 1913b *(nomen incorrectum), Macropistodon* – M.A. Smith, 1915c *(nomen incorrectum), Macropisthon* – Maki, 1933 *(nomen incorrectum),* and *Maciophistodon* – Briceño-Rossi, 1934 *(nomen incorrectum).*

Type species: *Amphiesma flaviceps* A.-M.-C. Duméril, Bibron & Duméril, 1854a.

Distribution: Southern and SE Asia and W Indonesia.

Sources: Boulenger, 1893a, C. Haas, 1950, McDowell, 1961, E.H. Taylor, 1965, Tweedie, 1983, Stuebing, 1991, Kelly et al., 2003, Zaher et al., 2009, I. Das, 2010, 2012 and Cadle, 2011.

1. *Macropisthodon flaviceps* (A.-M.-C. Duméril, Bibron & Duméril, 1854a). Erpét. Gén. 7(1): 738–739. (*Amphiesma flaviceps*)

Synonyms: *Amphiesma flaviceps semifasciata* Jan, 1863b *(nomen nudum), Amphiesma flaviceps semifasciata* Jan, 1865c, *Tropidonotus leucomelas* A.C.L.G. Günther, 1864a, *Amphiesma rufo-torquatum* Edeling, 1865 *(nomen incorrigendum),* and *Amphiesma rufotorquatum* – Boulenger, 1893a *(nomen corrigendum).*

Type: Holotype, MNHN 3601 (formerly RMNH), a 710 mm specimen.

Type locality: "Bornéo" [= East Malaysia or Kalimantan, Indonesia].

Distribution: Malay Peninsula and Greater Sundas. Extreme S Thailand (Surat Thani), West Malaysia (Pahang, Perak, Pinang), East Malaysia (Sarawak) and Indonesia (Bangka, Kalimantan, Nias, Sumatra), NSL–1300 m.

Sources: M.A. Smith, 1922b, 1930a, David & Vogel, 1996 and I. Das, 2012.

Remarks: M.A. Smith (1943: 529) designated pl. 35 in Hardwicke (1756–1823) (2: fig. 35) the lectotype of *Tropidonotus leucomelas* A.C.L.G. Günther.

2. *Macropisthodon plumbicolor* (Cantor, 1839b). Proc. Zool. Soc. London 7(1): 54. (*Tropidonotus plumbicolor*)

Synonyms: *Xenodon viridis* A.-M.-C. Duméril, 1853 *(nomen nudum), Trigonocephalus elliotti* Jerdon, 1854, *Xenodon viridis* A.-M.-C. Duméril, Bibron & Duméril, 1854a, *Amphiesma brachyura* Jan, 1863b *(nomen nudum), Amphiesma brachyura* Jan, 1865c, and *Macropisthodon plumbicolor palabariya* Deraniyagala, 1955.

Type: Lectotype, specimen illustrated in colored sketch of T.E. Cantor (1834–1837) in RSL, designated by M.A. Smith (1943: 314).

Type locality: "Malwa (Saugor), India" [= Sagar, Madhya Pradesh State, cen. India, 23°50'N, 78°45'E, elevation 535 m] via lectotype selection.

Distribution: Southern Asia. India (Andhra Pradesh, Goa, Gujarat, Haryana, Karnataka, Kerala, Madhya Pradesh, Maharashtra, Orissa, Punjab, Rajasthan, Tamil Nadu, West Bengal), Bangladesh, Myanmar and Sri Lanka (Central, North-Central, Northern, Uva), 600–2135 m.

Sources: Wall, 1905a, 1906a, 1921g, 1923d, M.A. Smith, 1943, Deraniyagala, 1955, P. Silva, 1969, R. Sharma, 1971, 1974, 1976a, Daniel, 1983, M.S. Khan, 1983, M.A.R. Khan, 1988, A. Silva, 1990b, 2001, 2009, Whitaker & Captain, 2004, N. Khaire, 2006, I. Das, 2010 and Vyas, 2011.

Remarks: Single Pakistan (Sindh) record of Murray (1884: 380) doubtful.

3. *Macropisthodon rhodomelas* (H. Boie *in* F. Boie, 1827). Isis von Oken 20(6): 535. (*Tropidonotus rhodomelas*)

Synonym: *Tropidonotus rhodomelas* H. Boie in Schlegel, 1826a *(nomen nudum), Tropidonotus rhodomelas* Schlegel, 1826b *(nomen nudum),* and *Tropidonotus mortoni* Theobald, 1868b.

Type: Lectotype, RMNH 1078, designated by R.F. Inger herein.

Type locality: "Java" [W Indonesia] via lectotype selection. Restricted to region of Mt. Pangerango and Mt. Salak, West Java, Indonesia *fide* Brongersma (1948b: 14). Emended to Tjihandjawar, at the foot of Mt. Pangerango, W. Java [= Cihanyawar, Nagrak, Jawa Barat, W Java, SW Indonesia, 07°03'S, 108°40'E] *fide* Brongersma (1950: 1499).

Distribution: Malay Peninsula and Greater Sundas. Southern Thailand (Nakhon Si Thammarat, Trang), West Malaysia (Pahang, Perlis, Pinang), Singapore, East Malaysia (Sabah, Sarawak) and W Indonesia (Bangka, Java, Sumatra), 150–1330 m.

Sources: E.H. Taylor, 1965, K.F.Lim & Lee, 1989, David & Vogel, 1996, M.J. Cox et al., 1998 and Malkmus et al., 2002.

Remarks: Original description based on H. Boie's MS (1823–1825). Single Sulawesi record of Rooij (1917: 92) doubtful *fide* Bosch (1985: 26).

MACROPROTODON Guichenot, 1850 (Colubridae)

Synonyms: *Macroprodon* – Zavattari, 1930 *(nomen incorrectum), Macropodon* – Maass-Berlin, 1933 *(nomen incorrectum),* and *Macroptodon* – Buresch & Zonkow, 1934 *(nomen incorrectum).*

Type species: *Macroprotodon mauritanicus* Guichenot, 1850.

Distribution: Southwest Europe, N Africa and ext. SW Asia.

Fossil records: Fossils unassigned to species include middle Pliocene (Ruscinian, MN 15: 3.2–4.2 mya) of Spain and upper Pliocene (MN 16/17) of Morocco.

Sources: J. Anderson, 1898, Mosauer & Wallis, 1927, Roux, 1939, Chpakowsky & Chnéour, 1954, Lanza & Bruzzone, 1960, Steward, 1971, Bruno, 1977, 1984, Andrada, 1980, J.B. Rasmussen, 1985b, Gisbert & García-Perea, 1986, Wade, 1988, 2001, Busack & McCoy, 1990, Pleguezuelos et al., 1994, Bons & Geniez, 1996, Schleich et al., 1996, Bailon, 2000, Corti & Luiselli, 2001, Carranza et al., 2004 and Szyndlar, 2012.

Remarks: Molecular study of Carranza et al., 2004, suggested five or six species *(M. abubakeri, brevis, cucullatus, ibericus, mauritanicus, textilis)* but results are inconclusive.

1. *Macroprotodon abubakeri* Wade, 2001. Bull. Nat. Hist. Mus. London (Zool.) 67(1): 95–100, figs. 2, 5a, 6f.

Type: Holotype, BMNH 1913.7.3.14, a 464 mm male (W. Rothschild & E. Hartet, Feb. 1908–June 1912).

Type locality: "Oran, Algeria" [= Oran, Oran Wilaya, N Algeria, 35°42'N, 0°38'W, elevation 120 m].

Distribution: Northwest Africa. Northern Morocco (Al Hoceima, Nador, Oujda), S Spain (Melilla) and NW Algeria (Ain Defla, Ain Témouchent, Algiers, Annaba, El Bayadh, Laghouat, Mascara, Mostaganem, Oran, Setif, Tlemcen, Habibas Is.), NSL–1000 m.

2. *Macroprotodon brevis* (A.C.L.G. Günther, 1862a). Ann. Mag. Nat. Hist. (3) 9(49): 58. (*Coronella brevis*)

Synonym: *Macroprotodon cucullatus ibericus* Busack & McCoy, 1990.

Type: Holotype, BMNH 1946.1.9.87 (R.T. Lowe, 1858–1861).

Type locality: "small island (without name) off the coast of Mogador" [= Essaouira islet, W of Mogador Is., SW of Essaouira, Morocco, 31°30'N, 9°47'W, elevation NSL].

Distribution: Iberian Peninsula and N Morocco. Northern Morocco, Spain (Almería, Avila, Badajoz, Centa, Mallorca, Melilla, Menorca, Barcelona, Cáceres, Cadiz, Castellón, Ciudad Real, Córdoba, Huelva, Jaén, Madrid, Málaga, Murcia, Salamanca, Sevilla, Toledo, Valencia, Zamora, Zaragoza, Baleares Is.) and Portugal (Algarve, Alentejo, Alto Alentejo, Baixo, Beira Litoral, Estremadura, Minho, Trás-os-Montes), NSL–600 m.

Source: Feriche et al., 2008.

Remarks: Supplemental original description in A.C.L.G. Günther (1862c: 48–52).

3. *Macroprotodon cucullatus* (I. Geoffroy-Saint-Hilaire, 1827 *in* Savigny, 1809–1829). Descr. Rept. Égypte 1(1): 148–151, pl. 8, figs. 3–3'. (*Coluber cucullatus*)

Synonyms: *Lycognathus taeniatus* A.-M.-C. Duméril, Bibron & Dumeril, 1854a, *Lycognathus textilis* A.-M.-C. Duméril, Bibron & Duméril, 1854a, *Oxyrhopus scolopax* Gasco, 1876, *Macroprotodon maroccanus* W.C.H. Peters, 1882b, and *Macroprotodon cucullatus melanocephala* Mosauer & Wallis, 1927.

Type: Holotype, not designated, a 356 mm specimen (F. Schoushoé), lost *fide* Busack & McCoy (1990: 268).

Type locality: "Egypte" [= Egypt]. Restricted to Lower Egypt *fide* Mertens & Müller (1928: 50).

Distribution: Northern Africa and Middle East. Western Sahara (Rio de Oro, Saguia El Hamra), Morocco (Al Hoceima, Boulemane, Figuig, Nador, Oujda), N Algeria (Al Butnan, Al Jazair, Annaba, Batna, Béchar, Djelfa, El Bayadh, Médéa, Mostaganem, Oran, Ouahran, Qacentina, Sétif, Saida, Tamanrasset, Tlemcen, Habibas Is.), Tunisia (Gabès, Nabeul, Tozeur,

Tunis, Zembra Is.), S Italy (Lampedusa Is.), N Libya (Ajdabiya, Al Akhdar, Al Jabal, Az Zawiyah, Al Khaliji, Banghazi, Butnan, Darnah, Gharyan, Misratah, Surt, Tarabulus), N Egypt (Alexandria, Beheira, Beni Suef, Cairo, Faiyum, Gharbia, Giza, Ismailia, Matrouh, North Sinai), SW Israel (Gaza Strip, W Southern) and Syria, NSL–2500 m.

Sources: Doumergue, 1901, Chpakowsky & Chnéour, 1954, Kramer & Schnurrenberger, 1963, Blanc, 1988, Bouskila & Amitai, 2001, Geniez et al., 2004, Sindaco et al., 2006 and Bar & Haimovitch, 2011.

Remarks: Type locality restriction to El Aricha region (34°13'N, 1°01'W), Naâma Wilaya, Algeria *fide* Doumergue (1901: 283) and Wade (2001: 91) invalid. See Carranza et al. (2004: 530) on the status of this species.

4. *Macroprotodon mauritanicus* Guichenot, 1850. Explor. Sci. Algérie: 22, pl. 2, fig. 2.

Synonym: *Macroprotodon mauritanicus* Duméril & Bibron *in* Guichenot, 1850 *(nomen ineditum)*.

Types: Syntypes (4), MNHN 2171, a 469 mm male, MNHN 1994.2339, a 452 mm female, MNHN 1994.2340, a 487 mm female, and MNHN 1994.2341, a 557 mm male (A. Guichenot, 1840–1842).

Type locality: "Algérie" [= Algeria].

Distribution: Spain (Mallorca), N Algeria (Algiers, Annaba, Batna, Béjaïa, Biskra, Bordj Bou Arreridj, Constantine, Guelma, Medea, Setif), N Libya and N Tunisia (Ben Arous, Bizerte, Jendouba, Nabeul, Sfax, Tunis, Zaghouan, Galita Is.).

Sources: Wade, 2001 and Carranza et al., 2004.

MACROVIPERA A.F.T. Reuss, 1927a (Viperidae)

Synonym: *Laidlawus* Hoser, 2012k *(nomen illegitimum)* and *Maxhoservipera* Hoser, 2012k *(nomen illegitimum)*.

Type species: *Coluber lebetinus* Linnaeus, 1758.

Distribution: Greece, Middle East, W Asia and N Africa.

Fossil records: Lower Miocene to upper Pliocene of Europe and extreme NW Africa, and middle Pleistocene of Greece. Fossils unassigned to species include lower Miocene (Orleanian, MN 4: 16.0–16.9 mya) of Spain, middle Miocene (Astaracian, MN 7/8: 11.1–12.8 mya) of Romania, upper Miocene (Turiolian, MN 11–12: 7.3–8.7 mya) of Hungary, upper Pliocene of France and Morocco. *Macrovipera* is included with a group, which as unassigned fossil material, are referred to as the "Oriental vipers," and may include any of the following records: lower Miocene (Orleanian, MN 4–5: 13.7–16.9 mya) of France, Germany, Portugal and Russia, middle Miocene (Astaracian, MN 6-8: 11.1–13.7 mya) of France and Germany, upper Miocene (Vallesian, MN 9–10: 8.7–11.1 mya) of Hungary, Turkey and Ukraine, upper Miocene (Turolian, MN 13: 4.9–7.3 mya) of Greece, Hungary, Italy and Ukraine, lower Pliocene (Ruscinian, MN 14: 4.2–4.9 mya) of Spain and Ukraine, upper Pliocene (Villanyian, MN 16: 2.6–3.2 mya) of Greece, upper Pliocene/lower Pleistocene of Morocco, and middle Pleistocene (Ionian: 0.13–0.78 mya) of Greece.

Sources: Brodmann, 1987, Herrmann et al., 1992, Golay et al., 1993, Schleich et al., 1996, Herrmann & Joger, 1997, Nilson & Andrén, 1997, David & Ineich, 1999, McDiarmid et al., 1999, Bailon, 2000, Lenk et al., 2001b, Szyndlar & Rage, 2002, Mallow et al., 2003, Szyndlar & Rage, 2007, Dobiey & Vogel, 2007, Venczel & Stiuca, 2008, Stümpel & Joger, 2009 and Phelps, 2010.

1. †*Macrovipera burgenlandica* (Bachmayer & Szyndlar, 1987). Ann. Naturhist. Mus. Wien 88A: 33–37, figs. 5.1–5.3, pl. 1, figs. 5–6. (†*Vipera burgenlandica*)

Type: Holotype, NMWGP 1986/3, one basiparasphenoid.

Type locality: "Kohfidisch, province of Burgenland, southeastern Austria. Late Miocene (Mammalian stage – Lower Turolian; Mammalian European continental biozone MN 11; Pontian in Paratethyan terms)."

Distribution: Upper Miocene (Turolian, MN 11: 7.8–8.7 mya) of Austria.

2. †*Macroviopera gedulyi* (Bolkay, 1913). Mitt. Jahrb. Kgl. Ungar. Geol. Reichsanst. 21(7): 225–226, pl. 12. figs. 9–12. (†*Vipera gedulyi*)

Types: Syntypes, HGIM ob.–4467/vt. 74, 97 skull bones, 19 maxillae and 206 fangs.

Type locality: "Polgárdi, Hungary" [= Turolian, upper Miocene *fide* Rage, 1984b: 56].

Distribution: Upper Miocene (Turolian, MN 13: 4.9–7.3 mya) of Hungary. Known only from type locality.

Sources: Bolkay, 1920, Rage, 1984b, Szyndlar, 1991a, Venczel, 1994, 1998 and Nilson & Andrén, 1997.

Remarks: Closely related to *Vipera ammodytes fide* Rage (1984b: 56).

3. †*Macrovipera kuchurganica* (Zerova *in* Zerova, Lungu & Chkhikvadze, 1987). Proc. Zool. Inst., Acad. Sci. Leningrad (1986) 158: 95–96, fig. 3. (†*Vipera kuchurganica*)

Type: Holotype, ZIK 37-2536, one trunk vertebra.

Type locality: "Kuchurgan, near Odessa, Ukraine, USSR, Middle Pliocene."

Distribution: Middle Pliocene (Ruscinian, MN 14–15: 3.2–4.9 mya) of Moldova and Ukraine.

Sources: Zerova, 1993, Nilson & Andrén, 1997 and Redkozubov, 2003.

Remarks: Transferred to *Macrovipera fide* Zerova, 1993.

4. *Macrovipera lebetina* (Linnaeus, 1758). Syst. Nat., ed. 10, 1: 218. (*Coluber lebetinus*)

Synonyms: *Vipera obtusa* Dwigubsky, 1832, *Vipera euphratica* W.C.L. Martin, 1838, *Vipera libethina* Filippi, 1865 *(nomen emendatum), Vipera peilei* Murray, 1892, *Vipera lietina* – Wall, 1908i *(nomen incorrectum), Macrovipera lebetina cypriensis* A.F.T. Reuss, 1933c, *Vipera lebetina turanica* Chernov *in* Terent'ev & Chernov, 1940, *Vipera lebentina* – Khalil, 1963 *(nomen incorrectum), Vipera libertina* – Roitman, 1967 *(nomen incorrectum), Vipera lebtina obtuse* – Minton, 1968 *(nomen incorrectum), Vipera lebethina* – Elter, 1981 *(nomen incorrectum), Vipera lebentina lebentina* – Billings & Schätti, 1984 *(nomen incorrectum),* and *Vipera lebetina transmediterranea* Nilson & Andrén, 1988b.

Type: Holotype, not designated (formerly MAFR), lost *fide* Andersson (1899: 29).

Type locality: "Oriente" [= Eastern Asia]. Restricted to Cyprus *fide* Mertens & Müller (1928: 52).

Distribution: North Africa, Middle East and SW Asia. Northern Algeria (Aïn Témouchent, Naama, Oran, Tlemcen), N Tunisia (Ben Arous, Gafsa, Jendouba, Kairouan, Kasserine, Medinine, Nabeul, Zaghouan), E Turkey, Cyprus, Syria (Halab), Lebanon (Mont-Liban), Jordan (Karak, Tafilah), S ext. SW Russia (Dagestan), Armenia, SE Georgia, Azerbaijan, Turkmenistan, Uzbekistan, Tadzhikistan, Iraq, Iran (Bakhtaran, Central, East Azarbaijan, Esfahan, Fars, Ghilan, Hamadan, Khuzestan, Kordestan, Lorestan, Mazandaran, Razavi Khorasan, Semnan, West Azarbaijan, Zanjhan), Afghanistan, Pakistan (Balochistan) and NW India (Kashmir), NSL–2500 m.

Fossil records: Middle Pleistocene (Ionian: 0.13–0.78 mya) of Greece.

Sources: Steindachner, 1897, Doumergue, 1901, Mayet, 1903, F. Werner, 1939, Chpakowsky & Chnéour, 1954, Balozet, 1957, Bannikov et al., 1977, Billing & Schätti, 1984, Nilson & Andrén, 1988b, Chikin, 1991, Szyndlar, 1991a, Esterbauer, 1992, Herrmann et al., 1992, Leviton et al., 1992, Szczerbak, 1994, Spawls & Branch, 1995, Bons & Geniez, 1996, Schleich et al., 1996, Herrmann & Joger, 1997, Nilson & Andrén, 1997, David et al., 1999, Szyndlar & Rage, 1999, Disi et al., 2001, Hraoui-Bloquet et al., 2002, M.S. Khan, 2002, J.C. Murphy & Schlager, 2003, Geniez et al., 2004, Whitaker & Captain, 2004, Arikau et al., 2005, Göçmen et al., 2006, 2007a, Smedt, 2006, Baier et al., 2009, Tuniyev et al., 2009 and Amr & Disi, 2011.

Remarks: Type locality restriction of Mertens & Müller invalid *fide* David & Ineich (1999: 322).

5. *Macrovipera schweizeri* (F. Werner, 1935). Sitz. Akad. Wiss. Wien 144(1–2): 117, figs. 18–20. (*Vipera lebetina schweizeri*)

Synonyms: *Vipera lebetina siphnensis* Wettstein, 1953a, *Vipera lebetina schweitzeri* – González, 1991 *(nomen incorrectum),* and *Vipera lebetina cernovi* Chikin & Szczerbak, 1992.

Type: Lectotype, MCZ 37008, a 680–690 mm female (H. Schweizer, July 1932), designated by Mertens (1951b: 208).

Type locality: "Milos" [= Milos Is., SW Cyclades, S Greece, bet. 36°39–46'N, 24°19–33'E] via lectotype selection.

Distribution: Southern Greece (Cyclades Arch.: Kimolos, Milos, Polyaigos, Sifnos).

Sources: Nilson & Andrén, 1988b, Herrmann & Joger, 1997, Nilson et al., 1999, Mallow et al., 2003, Smedt, 2006, Kreiner, 2007 and Valakos et al., 2008.

Remarks: Golay et al. (1993: 276) incorrectly listed the type as MCZ 37000. Previously a subspecies of *M. lebetina*.

6. †*Macrovipera ukrainica* (Zerova, 1992). Neues Jb. Geol. Paläont. Abh. 184(2): 236–243, fig. 1. (†*Vipera ukrainica*)

Type: Holotype, ZIKP 22-1786, one left maxilla.

Type locality: "limestone quarry on the right bank of the river Khomora, 3 miles westward from the village Gritsev, Shepetovsky district of the Khmelnitsky region, Ukraine; Lower (Novomoskovsky) horizon of the Middle Sarmatian (Upper Miocene, Vallesian, MN9a)."

Distribution; Upper Miocene (Vallesian, MN 9: 9.7–11.1 mya) of Ukraine. Known only from type series.

Sources: Nilson & Andrén, 1997 and Zerova, 1993.

Remarks: Transferred to *Macrovipera fide* Zerova, 1993.

M

MADAGASCAROPHIS Mertens, 1952e (Pseudoxyrhophiidae)

Synonyms: *Eteirodipsas* Jan, 1863b, *Heterodipsas* – Palacky, 1898 *(nomen incorrectum), Hetaerodipsas* Berg, 1901 *(nomen emendatum), Eteirodispsas* – Jourdran, 1903 *(nomen incorrectum), Madagascariophis* – J.A. Campbell & Murphy, 1977b *(nomen incorrectum),* and *Madagacarophis* – Dowling, Highton, Maha & Maxson, 1983 *(nomen incorrectum).*

Type species: *Dipsas colubrina* Schlegel, 1837.

Distribution: Madagascar.

Sources: Jourdran, 1904, Guibé, 1958, Brygoo, 1977, Domergue, 1987, Nagy et al., 2003, 2007, Glaw & Vences, 2007, Kelly et al., 2009 and Zaher et al., 2009.

Remarks: An undescribed species exists in ext. N Madagascar *fide* Nagy et al. (2007: 241).

1. *Madagascarophis colubrinus* (Schlegel, 1837). Essai Phys. Serp. 1: 161, 2: 273–274. (*Dipsas colubrina*)

Synonyms: *Eteirodipsas colubrina citrina* Boettger, 1877, *Madagascarophis colubrinus insularis* Domergue, 1987, *Madagascarophis colubrinus occidentalis*

Domergue, 1987, *Madagascarophis colubrinus pastoriensis* Domergue, 1987, and *Madagascarophis colubrinus septentrionalis* Domergue, 1987.

Type: Lectotype, MNHN 3891, a 375–408 mm male (J.P. Goudot & Milius [Barberie Voy.], 1827–1836), designated by Domergue (1987: 460, 472).

Type locality: "Madagascar" via lectotype selection. Restricted to vicinity of Tamatave (Toamasina), Madagascar *fide* Domergue (1987: 460).

Distribution: Madagascar (Antananarivo, Antsiranana, N Fianarantsoa, Mahajanga, Toamasina, N Toliara, Nosy Be and Nosy Komba Is.), NSL–1400 m.

Source: Henkel & Schmidt, 2000.

Remarks: Lectotype listed as female *fide* Guibé & Roux-Estève (1972: 132).

2. *Madagascarophis meridionalis* Domergue, 1987. Bull. Mus. Natl. Hist. Nat. (4) 9A(2): 482–483, fig. 3 (left).

Type: Holotype, MNHN 1985.378 (formerly CAD 202/S), an 870 mm male (native, 23 June 1961).

Type locality: "village de Mahavelo, plateau Karimbola, une trentaine de kilomètres à l'ouest de Beloha (Androy), altitude 150m, la région malgache" [= ext. S Toliara Province, ext. SW Madagascar].

Distribution: Southern Madagascar (Fianarantsoa, Toliara), NSL–1070 m.

M

3. *Madagascarophis ocellatus* Domergue, 1987. Bull. Mus. Natl. Hist. Nat. (4) 9A(2): 484, fig. 3 (right).

Type: Holotype, MNHN 1977.1056 (formerly CAD 190/S), an 826 mm male (native, 20 June 1961).

Type locality: "hameau de Zampongotra, plateau Karimbola (pays Antandroy), vers la bifurcation de la piste Beloha-Marolinta-Ampotaka, altitude 156m, la région malgache" [= ext. S Toliara Province, ext. S Madagascar].

Distribution: Southwestern Madagascar (SW Toliara), NSL–350 m.

Source: Henkel & Schmidt, 2000.

†*MADTSOIA* Simpson, 1933 (†Madtsoiidae)

Type species: †*Madtsoia bai* Simpson, 1933.

Distribution: Upper Cretaceous of Argentina, Madagascar, Niger and Spain, lower Eocene of Argentina, and Paleocene of Argentina, Australia, Brazil and Morocco.

Sources: Simpson, 1933, Hoffstetter, 1959, 1961a, Broin et al., 1974, Rage, 1984b, Astiba et al., 1990, Le Loeuff, 1991, Albino, 1993, Scanlon, 2005, LaDuke et al., 2010, Albino, 2011b and Mohabey et al., 2011.

1. †*Madtsoia bai* Simpson, 1933. Bull. Amer. Mus. Nat. Hist. 67(1): 10–12, figs. 1–3, 5–6.

Type: Holotype, AMNH 3154, 45 posterior trunk vertebrae and ribs (G.G. Simpson, 17 Jan. 1931).

Type locality: "*Notostylops* Beds, Cañadón Vaca, Chubut, Patagonia, Argentina, Eocene" [= "Cow Canyon," cen. Patagonia, S Chubut Prov., S Argentina].

Distribution: Upper Cretaceous (Campanian-Maastrichtian: 65.5–83.6 mya), upper Paleocene (54.0–57.0 mya), and lower Eocene (Casamayorian: 48.0–54.0) of Argentina, and middle Paleocene (Itaboraian: 57.0–59.0) of Brazil.

2. †*Madtsoia camposi* Rage, 1998. Palaeovertebrata 27(3–4): 116–121, fig. 2.

Type: Holotype, DGM 1311-R (field crew, 1949), one middle trunk vertebra.

Type locality: "Itaboraí, state of Rio de Janeiro, Brazil; middle Palaeocene."

Distribution: Middle Paleocene (Itaboraian: 57.0–59.0) of Brazil. Known only from type series.

3. †*Madtsoia madagascariensis* Hoffstetter, 1961a. Bull. Mus. Natl. Hist. Nat. (2) 33(2): 152–159, figs. 2–3.

Type: Holotype, MNHN MAJ 5, one trunk vertebra (R. Lavocat, 1955).

Type locality: "provient du N de Berivotra, localité située à 35 km au SE de Majunga et à 25 km au NNW de Marovoay, Madagascar, peut-être santonien, plus probablement campanien, le Crétacé supérieur." Emended to 'Gîte du Guide', north of Berivotra, Mahajanga Prov., Madagascar, upper Cretaceous (Santonian or Campanian) *fide* Rage (1984b: 30).

Distribution: Upper Cretaceous (Santonian, 83.6–86.3 mya, or Campanian: 72.1–83.6 mya) of Madagascar and Niger, and Paleocene of Morocco.

Sources: Rage, 1999 and LaDuke et al., 2010.

4. †*Madtsoia pisdurensis* Mohabey, Head & Wilson, 2011. J. Vert. Paleont. 31(3): 580–592, figs. 2a–e, 3.

Type: Holotype 225/GSI/PAL/CR/10, two posterior trunk vertebrae.

Type locality: "PIS I, less than 1 km from the village of Pisdura (20°21'N, 79°02'E), Maharashtra State, central India, Lameta Formation, Maastrichtian age, upper Cretaceous."

Distribution: Upper Cretaceous (Maastrichtian: 65.5–72.1 mya) of India. Known only from type locality.

MAGLIOPHIS Zaher, Grazziotin, Cadle, Murphy, Moura-Leite & Bonatto 2009 (Xenodontidae)

Type species: *Dromicus exiguus* Cope, 1862b.

Distribution: Puerto Rico and Virgin Islands.
Sources: A. Schwartz, 1967, Rivero, 1978, 1998, T. Campbell & Campbell, 1999, R.W. Henderson & Powell, 2009 and Zaher et al., 2009.

1. *Magliophis exiguum* (Cope, 1862b). Proc. Acad. Nat. Sci. Philadelphia 14(1): 79. (*Dromicus exiguus*)

Synonym: *Dromicus exiguus subspadix* A. Schwartz, 1967.
Types: Syntypes (5) (formerly ANSP & USNM), longest syntype 434 mm (Rüse, 1855–1862), lost *fide* A. Schwartz & Henderson (1991: 585).
Type locality: "St. Johns and St. Thomas, W.I." [= U.S.Virgin Islands]. Restricted to St. Thomas, U.S. Virgin Islands *fide* A. Schwartz (1967: 3).
Distribution: Puerto Rico Bank. Southern Puerto Rico (Arroyo, Guánica, Guayanilla, Culebra and adj. islands), U.S. Virgin Is. (Hassel, St. Thomas) and British Virgin Is. (Anegada, Guana, Peter, Tortola and Virgin Gordo), NSL–1800 m.
Remarks: *Magliophis e. subspadix* possibly a valid species *fide* R.W. Henderson & Powell (2009: 370).

2. *Magliophis stahli* (Stejneger, 1904). Ann. Rep. U.S. Natl. Mus. 1902: 695–697, fig. 161. (*Leimadophis stahli*)

Type: Holotype, USNM 27323, an adult female (A. Stahl, 1900).
Type locality: "Bayamon, Puerto Rico" [= Bayamón, Bayamón Municip., N Puerto Rico, 18°22'N, 66°10'W, elevation 20 m].
Distribution: Northern Puerto Rico (Aguada, Aguadilla, Arecibo, Bayamón, Isabella, Lares, Los Rábanos, Manatí, Maricao, Maunabo, Mayaguez, Morovis, Naguabo, Quebradillas, Río Grande, San Juan, Vega Baja, Verde Is.), NSL–1800 m.
Source: Zaher et al., 2009.
Remarks: A valid species *fide* Zaher et al., 2009.

MALPOLON Fitzinger, 1826a (Psammophiidae)

Synonyms: *Coelopeltis* Wagler, 1830, *Caelopeltis* Bonaparte, 1831 *(nomen emendatum)*, *Rhabdodon* Fleischmann, 1831 (*nomen rejiciendum*), *Bothriophis* Eichwald, 1837, *Malpodon* – Schinz, 1840 *(nomen incorrectum)*, *Rabdolon* – A.-M.-C. Duméril, Bibron & Duméril, 1854b *(nomen incorrectum)*, *Malpolum* – Cope, 1896b *(nomen incorrectum)*, *Coclopeltis* – Tristram, 1860 *(nomen incorrectum)*, *Colopeltis* – Garcia, 1896 *(nomen incorrectum)*, *Coelopetis* – Nikolsky, 1916 *(nomen incorrectum)*, *Coelopeltes* – Caziot, 1922 *(nomen incorrectum)*, *Coelopaltis* – Flower, 1926 *(nomen incorrectum)*, *Calopeltis* – Dowsett, 1928 *(nomen incorrectum)*, and *Molpolon* – Szyndlar, 1981 *(nomen incorrectum)*.
Type species: *Coluber monspessulanus* Hermann, 1804.
Distribution: Southern Europe, SW Asia and N Africa.
Fossil records: Middle Pliocene to upper Pleistocene of S Europe and extreme NW Africa. Fossils unassigned to species include upper Pliocene (Villanyian, MN 16: 2.6–3.2 mya) of France, Greece and Morocco, and upper Pleistocene (Tarantian: 0.01–0.13 mya) of Morocco.
Sources: J. Anderson, 1898, Flower, 1933, K.P. Schmidt, 1939b, Lanza & Bruzzone, 1960, Pasteur & Bons, 1960, Kramer & Schnurrenberger, 1963, Bons, 1967, Marx, 1968, Y. Werner, 1971, Woerkom, 1982, Osman & Sir, 1988, Bailon, 1989, 2000, Gruber, 1989, Le Berre, 1989, Latifi, 1991, Szyndlar, 1991a, Leviton et al., 1992, Brandstätter, 1995, Gravlund, 2001, Bailon & Aouraghe, 2002, Nagy et al., 2003, Lawson et al., 2005, Kelly et al., 2009 and Zaher et al., 2009.
Remarks: †*Palaeomalpolon borealis* Szyndlar possibly a synonym *fide* Holman (1998a: 116–117). The species *moilensis* has been transferred to the resurrected genus *Rhagerhis.*

1. *Malpolon fuscus* (Fleischmann, 1831). Dalmatiae Nova Serp. Gen.: 26–30, pl. 2, figs. a–e. (*Rhabdodon fuscus*)

Synonyms: *Coluber flexuosus* G. Fischer, 1832 and *Coluber monspessulana neumayeri* Bonaparte, 1838 *in* 1832–1841.
Type: Holotype, not designated, a 1270 mm specimen, location unknown.
Type locality: "Dalmatia et Istria prope ad hortes et urbes" [= gardens and urban Croatia and Istrian Peninsula].
Distribution: Southeastern Europe and SW Asia. Northern Italy (Liguria, Trentino-Alto Adige, Veneto), Croatia (Cres, Korcula, Krk and Mljet Is.), Bosnia-Herzegovina, Montenegro, N Macedonia, Albania, S Bulgaria (Blagoevgrad, Burgas, Haskovo, Kardzhali, Plovdiv, Yambol), Cyprus, Turkey (Adana, Ankara, Balikesir, Diyarbakir, Hatay, Istanbul, Izmir, Kars, Konya, Mardin, Siirt, Urfa), Greece (Aitolia and Akarnania, Alonissos, Arkadhia, Akhaia, Argolis, Arta, Attica, Drama, Evritania, Éuboea, Evros, Évvola, Florina, Fokis, Fthiotis, Grevena, Hydra, Idhra, Ilia, Imathia, Ioannina, Kardhitsa, Kastoria, Kefallinía, Kérkira, Khalkidhiki, Khíos, Kilkis, Korinthia, Kos, Kozani, Lakonia, Larisa, Lesvos, Levkas, Magnisia, Messinia, Páros, Pella, Pieria, Preveza, Rodhopi, Salamis, Sámos, Samothráki, Serrai, Skiathos, Skópelos, Thásos, Thesprotia, Thessaloniki, Trikala, Voiotia, Xanthi, Zakinthos), Armenia, Azerbaijan, ext. SW Russia, N Iraq (Dohuk, Khordestan) and Iran (Bakhtaran, Central, East Azerbaijan, Hamadan, Hormozghan, Khuzestan, Mazandaran, Razavi Khorasan, West Azerbaijan, Zanjhan), NSL–700 m.

Sources: Gasc, 1997, De Haan, 1999, Valakos et al., 2004, Carranza et al., 2006, Tuniyev et al., 2009, Afrasiab & Mohamad, 2011 and Jablonski et al., 2012.

Remarks: A valid species *fide* Carranza et al. (2006: 538).

2. *Malpolon insignitus* (I. Geoffroy-Saint-Hilaire, 1827 *in* Savigny, 1809–1829). Descr. Rept. Égypte 1(1): 147–148, pl. 7, fig. 6. (*Coluber insignitus*)

Synonyms: *Coluber neumayeri* Fitzinger, 1826a (*nomen nudum*), *Coluber vermiculatus* Ménétriés, 1832, *Coluber virens* Dwigubsky, 1832, and *Bothriophis distinctus* Eichwald, 1837.

Type: Holotype, not designated, location unknown.

Type locality: "Égypte" [= Egypt]. Restricted to lower Egypt *fide* Mertens & Müller (1928: 51).

Distribution: North Africa and Middle East. Eastern Morocco, Algeria, Tunisia (Gafsa, Lampedusa Is.), N Libya (Al Jabal, Al Murqub, Al Nuquat Al Khams, Banghazi, Tarabulus, Tripoli), N Sudan, N Egypt (Alexandria, Beheira, El-Bahr El-Ahmar, Damietta, Fayum, Kafr El-Shaikh, Matrouh, North Sinai, Sharkeya), Israel (Central, Gaza Strip, Golan Heights, Haifa, Jerusalem, Northern, Tel Aviv, West Bank), Jordan (Al Karak, Al Mafraq, Amman, Balqa Irbid, Ma'an), Lebanon (Beirut, Beqaa, El Jnoub, Liban-Nord, Mont-Liban, Nabatiye), (Amman, Balqa, Irbid, Jarash, Karak, Maan, Mafraq) and S Syria (Homs), NSL–1400 m.

Sources: Kramer & Schnurrenberger, 1963, Baran, 1976a, Bannikov et al., 1977, Hraoui-Bloquet, 1981, Ilani, 1983a, Bouskila & Amitai, 2001, Disi et al., 2001, Hraoui-Bloquet et al., 2002, Carranza et al., 2006 and Baier et al., 2009, Amr & Disi, 2011, Bar & Haimovitch, 2011 and Stojanov et al., 2011.

Remarks: A valid species *fide* Carranza et al. (2006: 538).

3. †*Malpolon laurenti* (Depéret, 1897). Mém. Soc. Géol. France (Paléont.) 7(3): 170–172, pl. 18, figs. 4–9. (†*Coelopeltis laurenti*)

Types: Syntypes, UCBL FSL 92857, several vertebrae.

Type locality: "fort du Serrat d'en Vacquer, Roussillon, vers le milieu du Pliocène" [= Pyrénées-Orientales, France; Ruscinian, Lower Pliocene *fide* Rage, 1984a: 48].

Distribution: Lower or middle Pliocene (Ruscinian, MN 15: 3.2–4.2 mya) of France. Known only from type locality.

Sources: Pusch, 1837, 1842, Hoffstetter, 1939, Rage, 1984b, Szyndlar, 1988, 2012 and Bailon, 1991.

4. †*Malpolon mlynarskii* Szyndlar, 1988. Acta Zool. Cracov. 31(27): 689–696, figs. 1–2.

Type: Holotype, MNCN 10000, one posterior braincase.

Type locality: "Layna (also known as Cerro Pelado) is situated in the Spanish province of Soria, north-east of Madrid, Spain; Middle Pliocene (MN 15)."

Distribution: Middle Pliocene (MN 15: 3.2–4.2 mya) of Spain. Known only from type locality.

Source: Szyndlar, 2012.

5. *Malpolon monspessulanus* (Hermann, 1804). Observ. Zool. 1: 283. (*Coluber monspessulanus*)

Synonyms: *Coluber hannasch asuaed* Forskal *in* Niebuhr, 1775 (*nomen oblitum*), *Natrix lacertina* Wagler *in* Spix, 1824, *Coluber rupestris* Risso, 1826, *Coluber esculapii* Dugès, 1827, *Coluber vermiculatus* Ménétriés, 1832, *Coelopeltis wagleri* Ranzani, 1836, *Coluber genetta* Eichwald, 1841 (*nomen nudum*), *Coluber temporalis* Eichwald, 1841 (*nomen nudum*), *Coluber monspeliensis* Gervais, 1848, ? *Coluber atratus* Costa, 1871 (*nomen nudum*), *Coelopeltis monspessulana occidentalis* F. Werner, 1909c, *Coelopeltis turkmena* Moritz, 1917, and *Malpolon monspessulanus saharatlanticus* Geniez, Cluchier & de Haan, 2006.

Type: Holotype, not designated, MZUS, a 762 mm specimen (D. Roubioux, 1798).

Type locality: "Monspelio" [= ? Alsace, France]. Restricted to Montpellier, S France *fide* Mertens & Müller (1928: 51).

Distribution: Southwestern Europe and NW Africa. Portugal (Aveiro, Beja, Braga, Bragança, Castelo Branco, Coimbra, Évora, Faro, Guarda, Leiria, Lisboa, Portalegre, Porto Santarém, Setúbal, Viana do Castelo, Vila Real, Viseu), Spain (S Albacete, Alicante, Almería, W Badajoz, Barcelona Gerona, Burgos, Cáceres, Cádiz, Córdoba, Granada, Huelva, Huesca, E Jaén, León, Lérida, Lugo, W Madrid, Málaga, Murcia, Orense, Pamploma, Pontevedra, W Salamanca, Sevilla, Tarragona, E Valencia, Valladolid, W Zamora, Balearic Is.), SE France (Alpes-Maritimes, Ardèche, Aude, Basses-Alpes, Bouchés-du-Rhône, Drôme, Gard, Hérault, Lozère, Pyrénées-Orientales, Vaucluse, Var), NW Italy (Liguria), NW Algeria (El Bayadh, Naama, Oran, Sidi Bel Abbes, Tlemcen), Tunisia (Gabès, Gafsa, Kairouan, Nabeul, Sidi Bou Zid, Siliana, Sousse, Jendouba, Kebili, Sfax, Tataouine, Tozeur, Tunis, Kerkennah and Zembra is.), Morocco (Agadir, Al Hoceima, Al Jadida, Azilal, Beni Mellel, Boulemane, Casablanca, Chechaouene, El Kelaa des Srarhna, Essaouira, Fes, Figuig, Kaar es Souk, Kenitra, Khemisset, Khenitra, Khouribga, Marrakech, Meknes, Nador, Oujda, Ouarzazate, Rabat, Safi, Settat, Tangier, Tan Tan, Taonate, Tata, Taza, Tetouan, Tiznit) and NW Western Sahara (W Rio de Oro, Saguia El Hamra), NSL–2100 (3040) m.

Fossil records: Middle Pliocene (Ruscinian, MN 15: 3.2–4.2 mya) of France, middle Pleistocene (Ionian: 0.13–0.78 mya) of France and Spain, and upper Pleistocene (Tarantian: 0.01–0.13 mya) of France.

M

Sources: Camerano, 1891, Doumergue, 1901, Mertens & Wermuth, 1960, Bruno, 1977, 1984, Arnold & Burton, 1978, Salvador, 1985, Brodmann, 1987, Blanc, 1988, Blanc & Nouira, 1988, Pleguezuelos & Moreno, 1989, Bruno & Maugeri, 1990, Bons & Geniez, 1996, Schleich et al., 1996, Bogaerts, 1998, Holman, 1998a, Petkovski et al., 2000, Ibrahim, 2001, Geniez et al., 2004, 2006, Carramza et al., 2006, Sindaco et al., 2006, Trapp, 2007, Feriche et al., 2008, Valakos et al., 2008, Vacher & Geniez, 2010 and Szyndlar, 2012.

Remarks: In accordance with Art. 23.9.2 of the Code (ICZN, 1999), *Coluber monspessulanus* Hermann is designated a *nomen protectum* and *Coluber hannasch asuaed* Forskal a *nomen oblitum*.

MANOLEPIS Cope, 1885a (Xenodontidae)

Synonyms: *Monolepis* – A.C.L.G. Günther, 1895 *in* 1885–1902 *(nomen incorrectum), Manelopis* – Phisalix, 1922 *(nomen incorrectum),* and *Manolopis* – Maass-Berlin, 1933 *(nomen incorrectum).*

Type species: *Dromicus putnami* Jan, 1863b.

Distribution: Southwestern Mexico.

Sources: Jan & Sordelli, 1867 *in* 1866–1870, Hartweg & Oliver, 1940, H.M. Smith & Taylor, 1945, Werler & Smith, 1952, J.D. Johnson, 1978b, Villa et al., 1988, Zaher et al., 2009 and Vidal et al., 2010.

1. *Manolepis putnami* (Jan, 1863b). Elenco Sist. Ofidi: 67. (*Dromicus putnami*)

Synonym: *Tomodon nasutus* Cope, 1864.

Type: Holotype, MCZ 824, a 546 mm male (E.S.C. Cammar, 1 May 1861).

Type locality: "St. Blas Jalisco, Messico" [= San Blas, Nayarit, Mexico] (possibly in error, but could be the vicinity of San Blas). Corrected to west coast of Mexico *fide* MCZ catalogue.

Distribution: Southwestern Mexico (Chiapas, Colima, Guerrero, Jalisco, Michoacán, Nayarit, Oaxaca), NSL–1900 m.

Remarks: Type was found on a ship that sailed from the port of San Blas, Mexico.

MASTICOPHIS Baird & Girard, 1853 (Colubridae)

Synonyms: *Mastigophis* – A.C.L.G. Günther, 1865 *(nomen incorrectum), Masticophiss* – Velasco, 1893 *(nomen incorrectum), Nasticophis* – Boulenger, 1896a *(nomen incorrectum), Mesticophis* – Velasco, 1896 *(nomen incorrectum), Maticophis* – Briceño-Rossi, 1934 *(nomen incorrectum),* and *Masticophus* – A.H. Wright & Wright, 1952 *(nomen incorrectum).*

Type species: *Leptophis taeniata* Hallowell, 1852a.

Distribution: Southwestern USA, Mesoamerica and NW Latin America.

Fossil records: Middle Pliocene to upper Pleistocene of S USA. Fossils unassigned to species include upper Pleistocene (Rancholabrean II) of USA (Arizona, Texas). Due to difficulty in distinguishing this genus from *Coluber* (Auffenberg, 1963, Holman, 1981a, Schätti, 1987), it may also be known from the Pleistocene of Georgia, Kansas, Missouri and/or Tennessee.

Sources: Ortenburger, 1928, Schätti, 1986b, Nagy et al., 2004, Utiger et al., 2005 and Collins & Taggart, 2008.

Remarks: A synonym of *Coluber fide* Lopez & Maxson, 1995, 1996 and Utiger et al., 2005. A valid genus *fide* Nagy et al., 2004 and Pyron & Burbrink *in* Crother, 2012.

1. *Masticophis anthonyi* (Stejneger, 1901b). Proc. U.S. Natl. Mus. 23(1231): 715–717. (*Bascanion anthonyi*)

Type: Holotype, USNM 24390, a 1450 mm male (A.W. Anthony, spring 1897).

Type locality: "Clarion Island, Revilla Gigedo group, west coast of Mexico, about 400 miles southwest from Cape St. Lucas, Lower California" [= Clarion Is., Revillagigedo Arch., Colima, W Mexico, 18°22'N,114°44°W].

Distribution: Western Mexico (Colima). Known only from type locality.

M

2. *Masticophis aurigulus* (Cope, 1861e). Proc. Acad. Nat. Sci. Philadelphia 13(7): 301. (*Drymobius aurigulus*)

Type: Holotype, USNM 5793, anterior 520 mm of mangled body (J. Xantus, 1859–1861).

Type locality: "Cape St. Lucas, Lower California, Mexico" [= Cabo San Lucas, S Baja California Sur State, NW Mexico, 22°53'N, 109°55'W, elevation 25 m].

Distribution: Western Mexico (ext. S Baja California Sur), NSL–1600 m.

Sources: Grismer, 1990a, 1990b, 1999, 2002.

Remarks: A valid species *fide* Grismer (1999: 459).

3. *Masticophis barbouri* (Van Denburgh & Slevin, 1921a). Proc. California Acad. Sci. (4) 11(6): 96. (*Coluber barbouri*)

Type: Holotype, CAS 49157, an adult female (J.R. Slevin, 30 May 1921).

Type locality: "Isla Partida, Espiritu Santo Island, Gulf of California, Mexico."

Distribution: Mexico (Baja California: Espíritu Santo, Partida Sur), NSL–575 m.

Source: Grismer, 1990a–b, 1999, 2002.

Remarks: A valid species *fide* Grismer (1999: 459). Previously a subspecies of *M. aurigulus.*

4. *Masticophis bilineatus* Jan, 1867 *in* Jan & Sordelli, 1866–1870. Icon. Gén. Ophid. 2(22): 3, pl. 6, figs. 2, 2a–b, e–g, n, r, v.

Synonyms: *Masticophis bilineatus* Jan, 1863b (*nomen nudum*), *Bascanium semilineatum* Cope, 1892a, and *Masticophis bilineatus lineolatus* Hensley, 1950.

Type: Holotype, MTD 15523, a 467 mm (svl) female.

Type locality: "Mexique" [= Mexico]. "Messico occid. ?" [= W Mexico] *fide* Jan (1863: 65). Restricted to Guaymas, Sonora, Mexico *fide* H.M. Smith & Taylor (1950a: 344) and Casas Grandes, Chihuahua, Mexico *fide* K.P. Schmidt (1953a: 190).

Distribution: Extreme SW USA (S Arizona, ext. SW New Mexico) and W Mexico (Aguascalientes, ext. SW Chihuahua, Colima, W Durango, Jalisco, Nayarit, Oaxaca, Sinaloa, E Sonora, Zacatecas, Tiburón Is.), NSL–2500 m.

Sources: Jan & Sordelli, 1867 *in* 1861–1870, H.M. Smith, 1941j, Fowlie, 1965, Stebbins, 1985, Camper & Dixon, 1994, Camper, 1996a, Degenhardt et al., 1996, Grismer, 1999, 2002, R.D. Bartlett & Tennant, 2000, McCranie & Wilson, 2001b and Lemos-Espinal et al., 2004a–b.

Remarks: Jan's 1863b description does not adequately separate *Masticophis bilineatus* from *M. flagellum*. Grismer (1999: 459) considered *M. bilineatus slevini* a distinct species.

M

5. *Masticophis flagellum* (G. Shaw, 1802). Gen. Zool., Amph. 3(2): 475. (Coluber flagellum)

Synonyms: *Coluber flagelliformis* Daudin, 1803c, ? *Coluber americanus* Gravenhorst, 1807, *Coluber testaceus* Say *in* E. James, 1823, *Herpetodryas psammophis* Schlegel, 1837, *Psammophis flavigularis* Hallowell, 1852a, *Bascanium flagelliforme piceum* Cope, 1875a, *Bascanium flagelliforme bicinctum* Yarrow, 1883a, *Bascanion flagellum frenatum* Stejneger, 1893b, *Masticophis flagellum lineatulus* H.M. Smith, 1941i, *Masticophis flagellum ruddocki* Brattstrom & Warren, 1953, and *Masticophis flagellum cingulum* Lowe & Woodin, 1954.

Type: Lectotype, specimen described and illustrated in Catesby (1743: 54, pl. 54).

Type locality: "Carolina and Virginia" [USA]. Restricted to Charleston, South Carolina, USA *fide* K.P. Schmidt (1953a: 188).

Distribution: Southern USA (Alabama, Arizona, Arkansas, California, E Colorado, Florida, Georgia, ext. SW Illinois, Kansas, S cen. Kentucky, Louisiana, Mississippi, Missouri, ext. SW Nebraska, W Nevada, New Mexico, SE North Carolina, Oklahoma, South Carolina, Tennessee, Texas, ext. SW Utah) and N Mexico (Aguascalientes, NE Baja California Norte, Chihuahua, Coahuila, E Durango, Nuevo León, Querétaro, San Luis Potosí, N Sinaloa, Sonora, Tamaulipas, N Veracruz, N Zacatecas), NSL–2500 m.

Fossil records: Middle Pliocene (Blancan III) of USA (Texas), upper Pliocene (Blancan V) of USA (Nebraska), lower Pleistocene (Irvingtonian I) of USA (Florida), middle/upper Pleistocene (Irvingtonian II) of USA (Arkansas), and upper Pleistocene (Rancholabrean II) of USA (California, Florida, Nevada, New Mexico).

Sources: L.D. Wilson, 1970b, 1973b, Christman, 1980, R.W. Murphy & Ottley, 1984, Ernst & Barbour, 1989, Van Devender & Bradley, 1994, Palmer & Braswell, 1995, Degenhardt et al., 1996, P.R. Brown, 1997, Tennant, 1997, Holman, 2000a, Werler & Dixon, 2000 and Quintero-Diaz et al., 2007.

6. *Masticophis fuliginosus* (Cope, 1895b). Amer. Nat. 29(343): 679–680. (*Zamenis lateralis fuliginosus*)

Types: Syntypes (2), USNM 15135–36, an 815+ mm and 665 mm specimen.

Type locality: "St. Margarita Island, Lower California" [= Baja California Sur, Mexico].

Distribution: Southwestern USA (ext. SW California) and Mexico (Baja California Norte, Baja California Sur, Coronados, Carmen, Cerralvo, Danzante, Espiritu Santo, Magdalena, Montserrate, Partida Sur, San Ildefonso, San José, San Marcos, Santa Margarita and Tiburón Is.), BSL–2515 m.

Source: Grismer, 2002.

Remarks: A valid species *fide* Grosmer (2002: 290). Previously a subspecies of *M. flagellum*.

7. *Masticophis lateralis* (Hallowell, 1853a). Proc. Acad. Nat. Sci. Philadelphia 6(7): 237. (*Leptophis lateralis*)

Synonym: *Masticophis lateralis euryxanthus* Riemer, 1954.

Type: Holotype, ANSP 5365, a 461 mm male (A.L. Heermann [Pacific Railroad Exped.], 1853).

Type locality: "California" [USA]. Probably Sierra Nevada foothills, N California *fide* Jennings (1983a: 343.1).

Distribution: Extreme W USA (California) and ext. NW Mexico (NW Baja California Norte, E Baja California Sur), NSL–2250 m.

Fossil records: Upper Pleistocene (Rancholabrean II) of USA (California).

Sources: Jennings, 1983a, Grismer, 1990a, 2002, Grismer & Mahrdt, 1996, P.R. Brown, 1997, R.D. Bartlett & Tennant, 2000 and Holman, 2000a.

Remarks: Type locality restriction to San Diego, California, USA *fide* K.P. Schmidt (1953: 189) in error *fide* Jennings (1983a: 343.1).

8. *Masticophis mentovarius* (A.-M.-C. Duméril, Bibron & Duméril, 1854a). Erpét. Gén. 7(1): 187. (*Coryphodon mentovarius*) (*nomen corrigendum*)

Synonyms: *Coryphodon mento-varius* A.-M.-C. Duméril, Bibron & Duméril, 1854a (*nomen incorrigendum*), *Liophis wagleri varia* Jan, 1863b, *Bascanion* *suboculare* Cope, 1867b, *Spilotes corais suborbitalis* W.C.H. Peters, 1868b, *Bascanium mentovarium* – Cope, 1879 *(nomen corrigendum), Bascanion lineatus* Bocourt, 1890 *in* A.H.A. Duméril, Bocourt & Mocquard, 1870–1909, *Coluber striolatus* Mertens, 1934b, *Masticophis flagellum variolosus* H.M. Smith, 1943a, and *Coluber mentovarius centralis* Roze, 1953b.

Types: Syntypes (2), MNHN 3199 and MNHN 3331, two females.

Type locality: "Mexique" [= Mexico]. Restricted to Tehuantepec, Oaxaca, Mexico *fide* H.M. Smith & Taylor (1950a: 340).

Distribution: Latin America. Mexico (Aguascalientes, Campeche, Chiapas, SW Chihuahua, Colima, S Durango, Guanajuato, Guerrero, Jalisco, S México, Michoacán, Morelos, Nayarit, La Peña and San Juanito Is.; Oaxaca, S Puebla, Querétaro, ext. N Quintana Roo, S San Luis Potosí, Sinaloa, S Sonora, Tabasco, N Veracruz, Yucatán, S Zacatecas, María Madre and María Magdalena Is.), Belize (Belize, Cayo, Corozal, Orange Walk), Guatemala (Alta Verapaz, Jutiapa, Petén, Santa Rosa), E El Salvador (Ahuachapán, Cabañas, Chalatenango, Cuscatlán, La Libertad, La Paz, Morazán, San Miguel, San Salvador, Usulutan), Honduras (Comayagua, Cortés, Francisco Morazán, Gracias a Dios, Olancho, Valle), Nicaragua (Atlántico Norte, Boaco, Chontales, Granada, Madriz, Managua, Masaya, Matagalpa, Rivas: Ometepe Is.), Costa Rica (Cartago, Guanacaste, Puntarenas, San José), cen. Panama (Coclé), ext. N Colombia (La Guajira, Magdalena) and N Venezuela (Anzoátegui, Apure, Aragua, Bolívar, Carabobo, Cojedes, Distito Federal, Falcón, Guárico, Lara, Mérida, Monagas, Nueva Esparta, Sucre, Zulia, Margarita Is.), NSL–2170 m.

Sources: H.M. Smith, 1942c, E.H. Taylor, 1951, Roze, 1953b, 1966a, Lancini., 1966, 1986, R.W. Henderson & Hoevers, 1975, J.D. Johnson, 1977, 1982, J.C. Lee, 1980, 1996, L.D. Wilson & Meyer, 1985, Villa et al., 1988, Castro-Franco & Gaviño de la Torre, 1990, Villa, 1990b, 1993, Hellmich, 1991, Auth et al., 1997a, 1998a, Benitez-Gálvez, 1997, J.A. Campbell, 1998b, Esqueda & Le Marca, 1999, Franklin & Franklin, 1999, Mijares-Urrutia & Arends, 2000, Lemos-Espinal et al., 2004b, G. Köhler et al., 2005, Navarrete et al., 2009 and Ugueto & Rivas, 2010.

9. *Masticophis schotti* Baird & Girard, 1853. Cat. No. Amer. Rept., Serp. 1: 160–161.

Synonym: *Masticophis ruthveni* Ortenburger, 1923.

Type: Holotype, USNM 1972, a 943 (svl) mm male (A.C.V. Schott [U.S.-Mex. Bound. Comm.], 1852–1853).

Type locality: "Eagle Pass, Tex." [Maverick Co., Texas, USA].

Distribution: Southern USA (S Texas) and N cen. Mexico (E Coahuila, Guanajuato, Hidalgo, N Michoacán, Nuevo León, Querétaro, San Luis Potosí, Tamaulipas, cen. Veracruz), NSL–2395 m.

Sources: Camper & Dixon, 1994, Camper, 1996b, Werler & Dixon, 2000 and Auth et al., 2000.

10 *Masticophis slevini* Lowe & Norris, 1955. Herpetologica 11(2): 93–95. (*Masticophis bilineatus slevini*)

Type: Holotype, SDSNH 3826, a 1206 mm female (G. Bancroft, 18 April 1930).

Type locality: "San Esteban Island, Gulf of California, Sonora, Mexico."

Distribution: Western Mexico (Sonora: San Esteban Is.), NSL–430 m.

Source: Grismer, 2002.

Remarks: A valid species *fide* Grismer (1999: 459).

11. *Masticophis taeniatus* (Hallowell, 1852a). Proc. Acad. Nat. Sci. Philadelphia 6(5): 181. (*Leptophis taenita*)

Synonyms: *Masticophis ornatus* Baird & Girard, 1853, *Coluber taeniatus girardi* Stejneger & Barbour, 1917, and *Masticophis taeniatus australis* H.M. Smith, 1941j.

Type: Holotype, USNM 2110, a 741–807 mm male (S.W. Woodhouse, 1849–1850).

Type locality: "New Mexico" [USA]. Emended to west of the Rio Grande, New Mexico, USA *fide* Cochran (1961: 193). Restricted to Shiprock, San Juan County, New Mexico, USA *fide* H.M. Smith & Taylor (1950a: 359) and specimen label on holotype.

Distribution: Western USA (Arizona, E California, W Colorado, S Idaho, Nevada, New Mexico, E Oregon, SW Texas, Utah, S Washington) and NE Mexico (Aguascalientes, Chihuahua, W Coahuila, Durango, NE Jalisco, N Zacatecas), NSL–3075 m.

Fossil records: Upper Pleistocene (Rancholabrean II) of USA (Nevada).

Sources: Van Devender & Lowe, 1977, W.S. Parker & Brown, 1980, W.S. Parker, 1982, Camper & Dixon, 1990, 1994, H.A. Brown et al., 1995, Stumpel, 1995, Camper, 1996c, Degenhardt et al., 1996, P.R. Brown, 1997, R.D. Bartlett & Tennant, 2000, Holman, 2000a, Werler & Dixon, 2000, McCranie & Wilson, 2001 and Sigala-Rodríguez et al., 2008.

Remarks: Spelling of epithet corrected to *taeniata* by Hallowell (1854f: 133).

MASTIGODRYAS Amaral, 1934a (Colubridae)

Synonyms: *Eudryas* Fitzinger, 1843 *(nomen praeoccupatum), Coryphodon* A.-M.-C. Duméril, Bibron & Duméril, 1854a *(nomen praeoccupatum), Endryas* – A.-M.-C. Duméril & Bibron, 1844 *(nomen incorrectum), Coryphedon* – Lohmeyer, 1882 *(nomen incorrectum), Goryphodon* – Bocourt, 1884 *(nomen incorrectum), Coriphodon* – Velasco, 1896 *(nomen incorrectum),*

M

Coryphdom – Briceño-Rossi, 1934 *(nomen incorrectum)*, and *Dryadophis* L.C. Stuart, 1939 *(nomen substitutum)*.

Type species: *Mastigodryas danieli* Amaral, 1934a.

Distribution: Latin America.

Sources: L.C. Stuart, 1941a, H.M. Smith, 1963b, J.A. Peters & Orejas-Miranda, 1970, L.D. Wilson & Meyer, 1985, Pérez-Santos, 1986, Pérez-Santos & Moreno, 1988, Hernández & Rada de M., 1992 and Montingelli, 2009.

Remarks: Official Generic Name no. 2008 *fide* Opinion 1035 (ICZN, 1975). Revision of genus in progress *fide* Montingelli, 2009.

1. *Mastigodryas alternatus* (Bocourt, 1884). Bull. Soc. Philom. Paris (1883–1884) (7) 8(3): 133–134. (*Coryphodon alternatus*)

Synonym: *Dryadophis sanguiventris* E.H. Taylor, 1954.

Type: Holotype, MNHN 6202, a 380 mm specimen (C. Viguier, Nov. 1876–April 1878).

Type locality: "Isthme de Darien" [= Daríen Prov., Panama].

Distribution: Honduras (Colón, Gracias a Dios, Olancho), Nicaragua (Atlántico Norte, Atlántico Sur, Chontales, Estelí, Masaya, Matagalpa, Monagua, Rio San Juan, Zelaya), Costa Rica (Cartago, Guanacaste, Limón, Puntarenas, San José), Panama (Bocas del Toro, Canal Zone, Chiriquí, Saboga Is.) and Ecuador (Manabi: Plata Is.), NSL–300 m.

Sources: L.D. Wilson & Meyer, 1985 and McCranie, 2011a.

Remarks: Resurrected from synonymy of *M. melanolomus fide* McCranie (2011a: 157).

2. *Mastigodryas amarali* (L.C. Stuart, 1938). Copeia 1938(1): 7–8. (*Eudryas amarali*)

Type: Holotype, USNM 22534, an 897 mm female (W. Robinson, 13 July 1895).

Type locality: "Marguerita Island, Venezuela" [= Isla Margarita, Nueva Sparta State, NE Venezuela].

Distribution: Northeastern Venezuela (Monagas, Sucre, Nueva Esparta, Margarita Is.), NSL– 200 m.

Sources: Roze, 1966a, Lancini, 1986, Navarrete et al., 2009 and Ugueto & Rivas, 2010.

Remarks: Single Tobago record rejected *fide* J.C. Murphy (1997: 186).

3. *Mastigodryas bifossatus* (Raddi, 1820) Mem. Math. Fis. Soc. Ital. Sci. 18(2): 333–334. (*Coluber bifossatus*)

Synonyms: *Coluber capistratus* Lichtenstein, 1823, *Coluber lichtensteinii* Wied-Neuweid, 1825a, *Coluber pantherinus* Schlegel, 1837, *Drymobius bifossatus striatus* Amaral, 1931a, *Drymobius bifossatus triseriatus* Amaral, 1931a, *Dryadophis bifossatus villelai* Hoge, 1953b, and *Mastigodryas bifossatus lacerdai* Cunha & Nacimento, 1978.

Type: Holotype, ? MZUF (formerly IRMC), a 1575 mm specimen (G. Raddi, 1817–1818).

Type locality: "Neri di Rio-Janeiro, Brasile" [= vicinity of Rio de Janeiro (22°54'S, 43°14'W, elevation NSL), S Rio de Janeiro State, SE Brazil].

Distribution: South America. Eastern Colombia (Meta), Venezuela (Amazonas, Apure, Aragua, Barinas, Bolívar, Cojedes, Guárico, Monagas, Portuguesa), Guyana, N French Guiana (Cayenne, Saint-Laurent-du-Maroni), E Peru, E Bolivia (La Paz, Santa Cruz), Brazil (Bahia, Ceará, Goiás, Maranhão, Mato Grosso, Minas Gerais, Pará, Pernambuco, Rio de Janeiro, Rio Grande do Sul, São Paulo, Sergipe, Bananal Is.), S Paraguay (Alto Paraná, Amambay, Asunción, Caaguazú, Caazapá, Canindeyú, Central, Concepción, Cordillera, Guairá, Itapúa, Misiones, Neembucú, Paraguarí, San Pedro) and N Argentina (Chaco, Corrientes, Entre Ríos, Formosa, Jujuy, Misiones, Salta, Santa Fe, Tucumán), 50–760 m.

Sources: Roze, 1966a, Cunha & Nascimento, 1978, 1993, Lancini, 1986, Hernández & Rada de M., 1992, Starace, 1998, Markezich, 2002, Cacciali, 2008, Navarrete et al., 2009 and Montingelli & Zaher, 2011.

4. *Mastigodryas boddaerti* (Seetzen, 1796) Meyer's Zool. Arch. 2: 59. (*Coluber boddaerti*)

Synonyms: *Coluber fuscus* Hallowell, 1845a, *Herpetodryas rappii* A.C.L.G. Günther, 1858, *Ablabes frenatus* Berthold, 1859, *Ablabes tesellatus* Jan *in* Berthold, 1859, *Herpetodryas bilineatus* Jan, 1863b, *Herpetodryas reticulata* W.C.H. Peters, 1863c, *Drymobius boddaerti californiensis* Bocourt, 1890 *in* A.H.A. Duméril, Bocourt & Mocquard, 1870–1909, *Eudryas dunni* L.C. Stuart, 1933, and *Eudryas ruthveni* L.C. Stuart, 1933.

Type: Holotype, not designated, a 635 mm specimen, lost *fide* Hoge (1964d: 55).

Type locality: Unknown. Restricted to Suriname, South America *fide* Merrem (1820: 110).

Distribution: Northern South America. Colombia (Atlántico, Antioquia, Bolívar, Boyacá, Caldas, Cauca, Chocó, Córdoba, Cundinamarca, La Guajira, Magdalena, Meta, Nariño, Santander, Sucre, Tolima, Valle, Gorgona Is.), Venezuela (Amazonas, Anzoátegui, Apure, Aragua, Barinas, Bolívar, Carabobo, Cojedes, Delta Amacuro, Distrito Federal, Falcón, Guárico, Lara, Mérida, Miranda, Monagas, Portuguesa, Sucre, Táchira, Trujillo, Yaracuy, Vargas, Zulia), Guyana (Baramita-Waini, East Berbice-Corentyne, Essequibo Islands-West Demerara, Upper Demerara-Berbice, Upper Takutu-Upper Essequibo, Santa Rosa Is.), Trinidad and Tobago (Tobago, Trinidad), Suriname (Saramacca, Suriname), French Guiana (Cayenne, Saint-Laurent-du-Maroni), W Ecuador (Esmeraldas, Guayas, Loja, Manabi, Pichincha), Peru (Ancash,

Casco, Junín, Loreto), N Bolivia (Beni, La Paz, Pando, Santa Cruz) and N Brazil (Amazonas, Bahia, Ceará, Mato Grosso, Pará: Marajó Is.; Rondônia), NSL–2200 m.

Sources: Hoge, 1964c, Roze, 1966a, Lancini, 1986, Nascimento & Lima-Verde, 1989, Hernández & Rada de M., 1992, Starace, 1998, C.W. Myers & Donnelly, 2008, Navarrete et al., 2009, Montingelli & Zaher, 2011 and C.J. Cole et al., 2013.

5. *Mastigodryas bruesi* (T. Barbour, 1914a). Mem. Mus. Comp. Zool. 44(2): 337–338. (*Alsophis bruesi*)

Type: Holotype, MCZ 7792, a 1010 mm specimen (G.M. Allen & C.T. Brues, 17 Aug.–25 Sept. 1910).

Type locality: "near St. George's, Grenada" [St. George Parish].

Distribution: Leeward Islands. St. Vincent & the Grenadines (Baliceaux, Bequia, Caille, Carriacou, Mustique, Quatre, Union), Grenada and Barbados (introduced).

Sources: A. Schwartz & R.W. Henderson, 1985, 1988, 1991, B.T. Greene et al., 2003 and R.W. Henderson & Powell, 2009.

Remarks: T. Barbour & Loveridge (1929: 208) erroneously listed five syntypes with Allen & Barbour as collectors. A population in Barbados is introduced *fide* B.T. Greene et al. (2003: 777).

6. *Mastigodryas cliftoni* (L. Hardy, 1964). Copeia 1964(4): 714. (*Dryadophis cliftoni*)

Synonym: *Dryadophis fasciatus* L. Hardy, 1963 (*nomen praeoccupatum*)

Type: Holotype, KU 73489, a 1532 mm male (P.L. Clifton, 29 Aug. 1962).

Type locality: "Plumosas, 22 kilometers east of Matatán, elevation 770 meters, Sinaloa, México."

Distribution: Western Mexico (SW Chihuahua, Jalisco, Nayarit, S Sinaloa, Zacatecas), 100–1700 m.

Sources: L. Hardy, 1963, Ponce-Campos & Huerta-Ortega, 1998, Lemos-Espinal et al., 2004b and R.A. Villa et al., 2011.

Remarks: Original description in L. Hardy (1963: 669–671, figs. 1a–b).

7. *Mastigodryas danieli* Amaral, 1934a. Mem. Inst. Butantan (1933–1934) 8: 158–159.

Type: Holotype, IB 8694, an 895 mm female (I. Daniel), destroyed by fire 15 May 2010.

Type locality: "Medellin, Colombia," = [Medellín, Antioquia Dept., NW Colombia, 6°14'N, 75°35'W, elevation 1480 m].

Distribution: Northwestern Colombia (Antioquia, Magdalena, Tolima, Valle), 250–2000 m.

Remarks: Official Specific Name no. 2545 *fide* Opinion 1035 (ICZN, 1975).

8. *Mastigodryas dorsalis* (Bocourt, 1890) *in* A.H.A. Duméril, Bocourt & Mocquard, 1870–1909. Miss. Sci. Mex. Amér.-Cen., Rept. 3(12): 724–725, pl. 51, figs. 2, 2a–2d. (*Drymobius [Eudryas] dorsalis*)

Types: Syntypes (2), MNHN 7391 and MNHN 91.257, longest syntype 1140 mm (Comm. Scient., 1865–1866).

Type locality: "le plateau de Guatemala" [= highlands of Guatemala].

Distribution: Upper Central America. Southern Guatemala (Baja Verapaz, Guatemala, Izabal, Sololá, San Marcos), El Salvador (Ahuachapán, Chalatenango, Cuscatlán, La Libertad, La Unión, Morazán, San Vicente, Santa Ana, Usulután), Honduras (Copán, Cortés, El Paraíso, Francisco Morazán, Intibucá, La Paz, Lempira, Ocotepeque, Olancho, Santa Bárbara, Yoro) and N Nicaragua (Estelí, Jinotega, Matagalpa), 450–2290 m.

Sources: Villa, 1971a, L.D. Wilson & Meyer, 1985, G. Köhler et al., 2005 and McCranie, 2011a.

9. *Mastigodryas heathii* (Cope, 1875b). J. Acad. Nat. Sci. Philadelphia (1876) (2) 8(2): 179–180. (*Drymobius heathii*)

Type: Holotype, ANSP 11544, a 780 mm specimen (E.R. Heath).

Type locality: "in the valley of Jequetepeque, which extends from the Cordillera of Caxamarca to near the coast at Pacasmayo, Peru" [= Jequetepeque Valley, from Cajamarca (7°20'S, 78°22'W, elevation 3400 m), Cajamarca Dept. to Pacasmayo (7°24'S, 79°33'W, elevation 35 m), N La Libertad Dept., W Peru].

Distribution: Coastal Peru (Ancash, Cajamarca, La Libertad, Lima, Puno) and Ecuador (Esmeraldas), 800–2600 m.

Sources: L.C. Stuart, 1941a, J.A. Peters, 1960a, Ortega-Andrade et al., 2010 and Cadle, 2012a.

Remarks: Photograph of holotype in Cadle (2012: fig. 1).

10. *Mastigodryas melanolomus* (Cope, 1868c). Proc. Acad. Nat. Sci. Philadelphia 20(2): 134–135. (*Masticophis melanolomus*)

Synonyms: *Herpetodryas laevis* J.G. Fischer, 1881, *Dromicus coeruleus* J.G. Fischer, 1885b, *Drymobius boddaerti modesta* F. Werner, 1903a, *Eudryas boddaertii gaigeae* L.C. Stuart, 1933, *Eudryas boddaertii mexicanus* L.C. Stuart, 1933, *Eudryas slevini* L.C. Stuart, 1933, *Dryadophis melanolomus veraecrucis (nomen substitutum)*, L.C. Stuart, 1941a, *Dryadophis melanolomus stuarti* H.M. Smith, 1943a, and *Dryadophis melanolomus tehuanae* H.M. Smith, 1943a.

Type: Holotype, USNM 24985, an 1124 mm specimen (A.C.V. Schott, 1865–1866).

Type locality: "Yucatan" [Mexico]. Restricted to Chichen Itzá, Yucatán, Mexico *fide* H.M. Smith & Taylor (1950a: 352).

Distribution: Latin America. Southern Mexico (Campeche, Chiapas, Colima, Guerrero, Hidalgo, Jalisco, Michoacán, Nayarit, Oaxaca, Puebla, Querétaro, Quintana Roo, San Luis Potosí, Sinaloa, Tabasco, Tamaulipas, Veracruz, Yucatán, María Cleofas and María Magdalena Is.), Belize (Belize, Cayo, Corozal, Orange Walk, Stann Creek, Toledo), N Guatamala (Alta Verapaz, Izabal, Petén), El Salvador (Ahuachapán, San Salvador, Santa Ana), N Honduras (Atlántida, Comayagua, Copán, Cortés, Francisco Morazán, Islas de la Bahía, Santa Bárbara, Yoro, Cochino Grande and Utila Is.) and Nicaragua (Jinotega), 1180–1345 m.

Sources: L.C. Stuart, 1933, H.M. Smith 1943a, E.H. Taylor, 1951, Hidalgo, 1981a, L.D. Wilson & Meyer, 1985, Lee, 1996, J.A. Campbell, 1998b, G. Köhler, 1999b, Canseco-Marquez et al., 2000, Torres-Carvajal, 2004, G. Köhler et al., 2005, McCranie et al., 2005, 2006, McCranie, 2011a, Montingelli & Zaher, 2011 and Travers et al., 2011.

11. *Mastigodryas moratoi* Montingelli & Zaher, 2011. J. Herp. 45(1): 112–113, fig. 4a.

Type: Holotype, MZUSP 5371, an 1151 mm male (Exped. Permanente Amazonia, 4 Nov. 1972).

Type locality: "approximately 50 m of altitude, on the north bank of the Negro River in Tapera, State of Amazonas, Brazil (00°25'S, 64°35'W)."

Distribution: Northern Brazil (Amazonas, Roraima) and N Guyana, 50–60 m.

12. *Mastigodryas pleii* (A.-M.-C. Duméril, Bibron & Duméril, 1854a). Erpét. Gén. 7(1): 661–663. (*Dromicus pleii*)

Synonyms: *Dromicus pleii tristriata* Jan, 1863b (*nomen nudum*), *Dromicus pleii cicatricosa* Jan, 1863b (*nomen nudum*), *Herpetodryas quinquelineatus* Steindachner, 1870b, *Dromicus maculivittis* W.C.H. Peters, 1877b, and *Alsophis pulcher* S.W. Garman, 1887b.

Type: Lectotype, MNH 3549, a 690 mm specimen (L.D. Bauperthuis, 1841), designated by type locality restriction of Hoge & Lancini (1960a: 61).

Type locality: "la province de Venézuela dans la Colombie" [= Zulia State, NW Venezuela] via lectotype selection.

Distribution: Lower Central America and NW South America. Eastern Panama (Cocle, Los Santos), N Colombia (Antioquia, Atlántico, Bolívar, Cundinamarca, La Guajira, Magdalena, Norte de Santander, Santander, Sucre, Tolima, Vallee) and Venezuela (Amazonas, Anzoátegui, Apure, Aragua, Barinas, Bolívar, Carabobo, Cojedes, Delta Amacuro, Distrito Federal, Falcón, Guárico, Lara, Mérida, Miranda, Monagas, Nueva Esparta, Portuguesa, Sucre, Táchira, Trujillo, Yaracuy, Vargas, Zulia, Margarita and Testigos Is.), 230–1970 m.

Sources: Valdivieso & Tamsitt, 1963, Roze, 1966a, Lancini, 1986, Pérez-Santos & Moreno, 1988, Hernández & Rada de M., 1992, Esqueda & La Marca, 1999, Mijares-Urrutia & Arends, 2000, Navarrete et al., 2009, Ugueto & Rivas, 2010 and Montingelli & Zaher, 2011.

Remarks: Type locality restriction of Hoge & Lancini (1960a: 61) does not select a lectotype in absence of a museum or catalogue number.

13. *Mastigodryas pulchriceps* (Cope, 1868b). Proc. Acad. Nat. Sci. Philadelphia 20(2): 105. (*Masticophis pulchriceps*)

Synonyms: *Herpetodryas reissii* W.C.H. Peters, 1868 and *Coluber fasciatus* Rosén, 1905.

Type: Holotype, ANSP 5710 (formerly USNM 6704), a 533 mm specimen (J. Orton, autumn 1867–winter 1868), lost *fide* L.C. Stuart (1941: 50).

Type locality: "the plateau valley of Quito" [Ecuador] (in error *fide* J.A. Peters, 1955: 347).

Distribution: Southwestern Colombia (Tolima) and W Ecuador (El Oro, Esmeraldas, Guayas, Pichincha), 750–1285 m.

Sources: Pérez-Santos & Moreno, 1988, 1991, Agostini et al., 2003 and Ortega-Andrade, 2010.

MEDIOHYDROPHIS Kharin, 2004b (Elapidae)

Type species: *Hydrophis klossi* Boulenger, 1912c.

Distribution: Marine waters of Indo-Australia, including Bay of Pattani, Bight of Bangkok, Brunei Bay, Coral Sea, Gulf of Thailand, Orie Bay, Strait of Madura and Strait of Malacca.

Source: Kharin, 2004b.

1. *Mediohydrophis klossi* (Boulenger, 1912c). Vert. Fauna Malay Penin., Rept. Batr.: 190. (*Hydrophis klossi*)

Type: Holotype, BMNH 1946.1.10.8 (formerly SM & BMNH 1920.6.3.7), a 900 mm female.

Type locality: "Thuala, Selangor, Malay Peninsula" (in error). Corrected to Kuala Selangor [= Kuala Selangor Distr., Selangor State, W West Malaysia, 3°20'N, 101°15'E, elevation NSL] *fide* Stuebing (1991: 355).

Distribution: Malay Peninsula and Greater Sundas. Southern Thailand (Chanthaburi, Pattani, Phuket, Singgora, Songkhla, Trang), West Malaysia (Perak, Selangor), Singapore, East Malaysia (Sabah) and W Indonesia (Sumatra).

Sources: E.H. Taylor, 1965 and Tweedie, 1983.

2. *Mediohydrophis melanosoma* (A.C.L.G. Günther, 1864a). Rept. Brit. India: 367–368, pl. 25, fig. e. (*Hydrophis melanosoma*)

Synonyms: *Hydrophis floweri* Boulenger, 1898g, *Distira wrayi* Boulenger, 1900a, *Hydrophis flower* – Phisalix, 1922 *(nomen incorrectum), Hydrophis melanosorana* – Tu, 1974 *(nomen incorrectum),* and *Distira melanaosoma* – Culotta & Pickwell, 1993 *(nomen incorrectum).*

Type: Holotype, BMNH 1946.1.10.6 (formerly RCSM & BMNH 1896.1.25.1), a 1245 mm female.

Type locality: Unknown.

Distribution: Indo-Australia. West Malaysia (Johore, Perak), Singapore, East Malaysia (Sabah), Indonesia (Java, Sulawesi: South; Sumatra), Papua New Guinea (Milne Bay) and N Australia (Queensland, Western Australia, Great Barrier Reef).

Sources: O'Shea, 1996, Stuebing & Inger, 1999 and Cogger, 2000.

MEGATYPHLOPS Broadley & Wallach, 2009 (Typhlopidae)

Type species: *Onychocephalus mucruso* W.C.H. Peters, 1854.

Distribution: East and S Africa.

Source: Broadley & Wallach, 2009.

1. *Megatyphlops anomalus* (Bocage, 1873b). J. Sci. Math. Phys. Nat., Lisboa 4(15): 248–249, pl. 1, fig. 3. (*Onychocephalus anomalus*)

Synonyms: *Typhlops anchietae* Bocage, 1886 and *Typhlops anchiate* – Rose, 1950 *(nomen incorrectum).*

Type: Syntypes (4), MBL 1177-D, longest syntype 188 mm specimen, destroyed by fire 18 March 1978.

Type locality: "Huilla, dans l'intérieur de Mossamedes" [= Huíla, Huíla Province, Angola, 15°04'S, 13°33'E, elevation 1695 m].

Distribution: Southwestern Angola (Benguela, Bié, Cuanza Sul, Huambo, Huíla, Namibe), 910–1735 m.

2. *Megatyphlops brevis* (Scortecci, 1929). Atti Soc. Ital. Sci. Nat. Mus. Civ. Stor. Nat. Milano 68(3–4): 267–269, 3 figs. (*Typhlops brevis*)

Type: Holotype, MSNM 2033, a 182 mm female (L. Luppi, 25 Jan. 1926).

Type locality: "Chisimajo" [= Kismayu, Jubbada Hoose Region, S Somalia, 00°23'S, 42°28'E, elevation NSL].

Distribution: Northeastern Africa. Southeastern South Sudan (Eastern Equatoria), N Uganda (Northern), Somalia (Bari, Jubbada Hoose, Shabeellaha Hoose, Shabeellaha Dhexe, Togdheer), S Ethiopia (Bale, Hararge, Kefa, Sidamo, Wollo) and SW Kenya (Coast, Eastern, Nyanza, Rift Valley, Western), NSL–2000 m.

Sources: Lanza, 1983a, 1990b, Spawls et al., 2002, Wallach & Lanza, 2004 and Largen & Spawls, 2010.

3. *Megatyphlops mucruso* (W.C.H. Peters, 1854). Verhandl. Königl.-Preuss. Akad. Wiss. Berlin 1854(11): 621. (*Onychocephalus mucruso*)

Synonyms: *Onychocephalus dinga* W.C.H. Peters, 1854, *Onychocephalus varius* W.C.H. Peters, 1860a, *Typhlops riparius* W.C.H. Peters, 1881a, *Typhlops mandensis* Stejneger, 1894a, *Typhlops latirostris* Sternfeld, 1910c, *Typhlops viridiflavus* Peracca, 1912a, *Typhlops excentricus* Procter, 1922, and *Typhlops schlegeli mucroso* H.W. Parker, 1936d *(nomen emendatum).*

Type: Lectotype, ZMB 3963, a 245–395 mm specimen (W.C.H. Peters, Jan.–June, 1845), designated by Loveridge (1933: 216).

Type locality: "Macanga" [= Makanga, Tete Prov., Mozambique, 15°27'S, 33°31'E, elevation 450 m] via type locality restriction of Loveridge (1933: 216).

Distribution: Eastern Africa. Northeastern Angola (Lunda Norte), S Democratic Republic of the Congo (Bas-Congo, Kasai Occidental, Kasai Oriental, Katanga, Kinshasa), SE Kenya (Coast), Tanzania (Arusha, Dar es Salaam, Dodoma, Iringa, Lindi, Mbeya, Morogoro, Mtwara, Mwanza, Pwani, Rukwa, Tabora, Tanga, Manda and Zanzibar Is.), Zambia (Central, Copperbelt, Eastern, Luapula, Lusaka, Northern, North-Western, Southern, Western), Malawi (Central, Northern, Southern), N Mozambique (Manica, Nampula, Sofala, Tete, Zambesia), Zimbabwe (Bulawayo, Manicaland, Mashonaland Central, Mashonaland East, Mashonaland West, Masvingo, Matabeleland North, Matabeleland South, Midlands) and ext. E Botswana (North East), NSL–1740 m.

Sources: Spawls et al., 2002, Branch, 2005 and Broadley & Blaylock, 2013.

Remarks: Complex history of types discussed by McDiarmid et al. (1999: 86).

4. *Megatyphlops schlegelii* (Bianconi, 1849a). Nouvi Ann. Sci. Nat., Bologna (2) 10: 106–107. (*Typhlops schlegelii*)

Synonyms: *Onychocephalus petersii* Bocage, 1873b, *Typhlops humbo* Bocage, 1886, and *Typhlops hottentotus* Bocage, 1893c.

Type: Holotype, IBI, a 522 mm specimen (C. Fornasini, 1842–1847).

Type locality: "Mozambico" [= Mozambique]. Restricted to Inhambane, Mozambique *fide* Roux-Estève (1974a: 165).

Distribution: Southern Africa. Southern Mozambique (Gaza, Inhambane, Maputo), S Zimbabwe (S Masvingo), SW Angola (Benguela, Cuando Cubango, Cunene), N Namibia (Caprivi, Khomas, Kunene, Ohangwena, Okavango, Omaheke, Omusati, Oshikoto, Otjozondjupa), Botswana (Central, Gaborone, Kgatleng,

North-East, North-West), Swaziland and South Africa (Gauteng, KwaZulu-Natal, Limpopo, Mpumalanga), NSL–1775 m.

Sources: Mertens, 1955, 1971, FitzSimons, 1974, Broadley & Cock, 1975, Largen, 1978, Broadley, 1983, Pienaar et al., 1983, Auerbach, 1987, Largen & Rasmussen, 1993, Branch, 1998, 2001, M. Griffin, 2003, J.C. Murphy & Schlager, 2003 and Broadley & Blaylock, 2013.

Remarks: Original description reprinted in Bianconi (1850: 183–184, pl. 6, figs. 2, 2a–d & 1850–1870: 13–14, pl. 3, figs. 2, 2a–d.).

MEHELYA Csiki, 1903 (*nomen substitutum*) (Lamprophiidae)

Synonyms: *Heterolepis* A. Smith, 1847 *in* 1838–1849 *(nomen praeoccupatum), Simocephalus* Gray *in* A.C.L.G. Günther, 1858 (*nomen oblitum*), *Grobbenia* Poche, 1903 *(nomen praeoccupatum), Siebenrockia* Poche, 1904 *(nomen substitutum), Simocaphalus* – F.W. FitzSimons, 1912 *(nomen incorrectum), Symocephalus* – Calabresi, 1927 *(nomen incorrectum), Mehelia* – Scortecci, 1929 *(nomen incorrectum), Mehelys* – Laurent, 1956b *(nomen incorrectum), Mehely* – Stucki-Stirn, 1979 *(nomen incorrectum),* and *Mehylya* – W.W. Tanner & Avery, 1982 *(nomen incorrectum).*

Type species: *Heterolepis capensis* A. Smith, 1847 *in* 1838–1849.

Distribution: Subsaharan Africa.

Sources: Mocquard, 1887a, Boulenger, 1891d, 1893a, Csiki, 1904, Sternfeld, 1908b, Angel, 1933b, Loveridge, 1939, 1957a, McDowell, 1961, 1986, Doucet, 1963, Dunger, 1971a, Villiers, 1975, Kelly et al., 2003, Villiers & Condamin, 2005 and Vidal et al., 2008.

Remarks: A valid genus *fide* Zaher et al., 2012. In accordance with Art. 23.9.2 of the Code (ICZN, 1999), *Mehelya* Csiki is designated a *nomen protectum* and *Simocephalus* Gray a *nomen oblitum.*

1. *Mehelya capensis* (A. Smith, 1847 *in* 1838–1849). Illust. Zool. So. Africa, Rept.: pl. 55, 2 pp. (*Heterolepis capensis*)

Synonyms: *Heterolepis gueinzii* W.C.H. Peters, 1874b, and *Mehelya grayi* Petzold, 1975.

Type: Neotype, BMNH 1891.9.15.9 (formerly SAM), a male [head and anterior extremity of body and tail], designated by Lanza & Broadley, 1913.

Type locality: "Delagoa Bay" [= Maputo Bay, Maputo Prov., S Mozambique, ca. 26°S, 32.5°E, elevation NSL] via neotype selection.

Distribution: Eastern Africa. Southeastern Democratic Republic of the Congo (Katanga), E Angola (Cabinda, Cuando Cubango, Huambo, Moxica, Namibe), E Tanzania (Dar es Salaam, Lindi, Ruvuma, Tanga, Zanzabar Is.), SE Kenya (Coast, Eastern), Zambia (Central, Copperbelt, Eastern, Luapula, Lusaka, Northern, North-Western, Southern, Western), Malawi (Central, Northern, Southern), Mozambique (Gaza, Inhambane, Manica, Maputo, Sofala), Zimbabwe (Bulawayo, Manicaland, Mashonaland Central, Mashonaland East, Mashonaland West, Matabeleland North, Matebeleland South, Midlands), E Botswana (Gaborone, Kgatleng, Kweneng, North West), Swaziland and E South Africa (Gauteng, KwaZulu-Natal, Limpopo, Mpumalanga), NSL–1000 m.

Sources: Witte, 1953, 1962, 1966, 1975, Laurent, 1956a, 1960, Vesey-FitzGerald, 1958, Sweeney, 1961, Pitman, 1974, Derleyn, 1978b, Spawls, 1978, Haacke, 1981, Trape, 1981, Hoevers & Johnson, 1982, Broadley, 1983, Derleyn et al., 1983, Lanza, 1983a, Pienaar et al., 1983, Auerbach, 1987, Patterson, 1987, Branch, 1988, Joger, 1990, Branch & McCartney, 1993, Largen & Rasmussen, 1993, Shine et al., 1996, Spawls et al., 2002, Broadley et al., 2003, J.C. Murphy & Schlager, 2003, Chippaux, 2006, Chirio & Ineich, 2006, Chirio & LeBreton, 2007, Chirio, 2009, Broadley & Blaylock, 2013 and Lanza & Broadley, 2013.

Remarks: Holotype lost *fide* V.F.M. FitzSimons (1937: 262). Records from Cameroon (Chirio & LeBreton, 2007), and Centrale African Republic and Gabon (Chippaux, 2006) need confirmation.

2. *Mehelya chanleri* (Stejneger, 1894b). Proc. U.S. Natl. Mus. (1893) 16(970): 726–727. (*Simocephalus chanleri*)

Synonyms: *Simocephalus unicolor* Boulenger, 1910b, *Mehelya somaliensis* Lönnberg & Andersson, 1913b, and *Mehelia fiechteri* Scortecci, 1929.

Type: Holotype, USNM 20126, a mm specimen (G. Denhardt via. W.A. Chanler, 1892).

Type locality: "Wange, Island Manda, north of Lamu, East Africa," (partly in error). Corrected to Wange (not Manda Is.), Kenya [= Wange, Lamu Prov., E Kenya, 1°30'S, 40°37'E, elevation 60 m] *fide* Lanza & Broadley, 1913.

Distribution: East Africa. Eastern Democratic Republic of the Congo (Sud-Kivu), Rwanda (Biuma, Butare, Byumba, Eastern, Northern), Burundi (Bubanza, Bujumbura, Bururi, Ruyigi), NW Tanzania (Arusha, Kigoma, Rukwa, Tanga), ext. SW Uganda (S Western), Kenya (Central, Coast, Eastern, Lamu and Manda Is.) Somalia (Hiraan, Jubbada Dhexe, Jubbada Hoose, Mogadishu, Shabeellaha Dhexe, Shabeellaha Hoose, Woqooyi Galbeed, Alessandra Is.), E Ethiopia (Hararge) and Eritrea, NSL–2200 m.

Sources: Scortecci, 1929, Loverdige, 1939, Lanza, 1983a, 1990b, Broadley & Howell, 1991, Largen, 1997 and Lanza & Broadley, 2013.

Remarks: A valid species *fide* Largen & Spawls, 2010 and Lanza & Broadley, 2013.

M

3. *Mehelya crossii* (Boulenger, 1895c). Ann. Mag. Nat. Hist. (6) 16(91): 33. (*Simocephalus crossii*)

Synonym: *Simocephalus riggenbachi* Sternfeld, 1910b.

Type: Holotype, BMNH 1946.1.14.49, a 410 mm female (W.H. Crosse, 1890–1895).

Type locality: "near Asaba, about 150 miles up the Niger" [= vicinity of Asaba, Niger River, SE Bendel State, Nigeria].

Distribution: West Africa. Senegal (Dakar, Ziguinchor/Kolda), Guinea, S Mali (Kayes, Koulikoro, Ségou, Sikasso), Ivory Coast, W Burkina Faso (Centre, Est, Hauts-Bassins, Sud-Ouest, Volta-Noire), Ghana (Eastern), Togo (Centrale, Kara, Maritime, Plateaux, Savanes), Benin (Alibori, Atakora, Zou), S Chad, SW Niger (Dosso), Nigeria (Anambra, Bendel, Benue, Borno, Cross River, Delta, Imo, Kaduna, Lagos, Plateau, Rivers), N Cameroon (Extreme-Nord, Nord), and S Central African Republic (Ombella-Mpoko, Haut-Mbomou, Haut-Sangha), 220–1175 m.

Sources: Doucet, 1963, Villiers, 1963, 1975, Leston & Hughes, 1968, Hughes & Barry, 1969, Papenfuss, 1969, Roman, 1969, 1980, Hulsemans & Verheyen, 1970, Dunger, 1971a, Courtois & Chippaux, 1977, Roman, 1980, Hughes, 1983, 2013, J.A. Butler & Reid, 1986, Joger, 1990, Villiers & Condamin, 2005, Chippaux, 2006, Chirio & Ineich, 2006, J.-F. Trape & Mané, 2006b, Chirio & LeBreton, 2007, Chirio, 2009 and Segniagbeto et al., 2011.

4. *Mehelya egbensis* Dunger, 1966. Amer. Mus. Novit. (2268): 2–7, figs. 1–4.

Type: Holotype, AMNH 96100, a 235 mm male (D. Warren, Oct. 1964).

Type locality: "Egbe, Nigeria (latitude 8°14'N., longitude 5°31'E.)" [= Egbe, W Kogi State, SW Nigeria, 8°13'N, 5°31'W, 350 m].

Distribution: Southwestern Nigeria (Oyo, Kogi), 350 m.

Sources: Dunger, 1971a and Chippaux, 2006.

5. *Mehelya gabouensis* (J.-F. Trape & Mané, 2005). Bull. Soc. Herp. France (115): 24–30, figs. 1–2.

Type: Holotype, MNHN 2004.493 (formerly IRD S-3425), a 364–370 mm male (native, Nov. 1998).

Type locality: "Haute-Casamance (lieu de provenance exact incertain, probablement environs de Vélingara, 13°09'N-14°06'W), Sénégal" [= vicinity of Vélingara (13°09'N, 14°06'W, elevation 50 m), E Kolda Prov., SE Senegal].

Distribution: Southeastern Senegal (Kolda), 30–50 m.

Source: Trape & Mané, 2006b.

Remarks: Color photographs of holotype in J.-F. Trape & Mané (2006: 131).

6. *Mehelya guirali* (Mocquard, 1887a). Bull. Soc. Philom. Paris (1886–1887) (7) 11(1): 23–27, pl. 2, figs. 3, 3a–3c. (*Heterolepis guirali*)

Synonym: *Simocephalus baumanni* Sternfeld, 1908a.

Types: Syntypes (2), MNHN 1885.580, an 1130 mm (svl) specimen (A. Lanchier, Dec. 1880–March 1883) [Niger] and MNHN, a 1060 mm specimen (L. Guiral, Dec. 1880–March 1883) [San-Benito].

Type locality: "Niger" and "San-Benito, l'Ouest africain" [= Niger and Rio de São Benito or Mbini, Gabon].

Distribution: West Africa. Southeastern Guinea (Macenta, Nzérékoré), Sierra Leone (Eastern, Southern), S Liberia (Nimba), S Ivory Coast (Abidjan, Lagunes), S Togo (Plateaux), Benin (Cotonou, Zou), S Nigeria (Cross River, Delta, Lagos, Ogun, Rivers), SW Chad (Chari-Baguirmi), SW Cameroon (Littoral, Nord-Ouest, Sud, Sud-Ouest), Gabon (Moyen-Ogooué, Ogooué-Maritime), W Gabon (Moyen-Ogooué, Ogooué-Maritime), S Congo (Kouilou, Pool) and W Democratic Republic of the Congo (Bandundu), NSL–1200 m.

Sources: Villiers, 1956b, 1975, Doucet, 1963, Menzies, 1966, Leston & Hughes, 1968, Hughes & Barry, 1969, Roux-Estève, 1969a, Dunger, 1971a, Hughes, 1983, 2013, J.A. Butler & Reid, 1986, J.B. Rasmussen, 1991, D. Lawson, 1993, Villiers & Condamin, 2005, Chippaux, 2006, Chirio & LeBreton, 2007, Pauwels & Vande weghe, 2008, W. Böhme et al., 2011 and Segniagbeto et al., 2011.

Remarks: Figure 3 in original plate 2 mislabelled as Fig. 1. Type locality may be Benito Point, Bendel State, S Nigeria, 5°26'N, 5°25'E.

7. *Mehelya laurenti* Witte, 1959. Rev. Zool. Bot. Afr. 60 (3–4): 348–349.

Type: Holotype, MRAC 21033, a 465 mm male (R.P. Lootens, 1957).

Type locality: "d'Ikela (Territ. d'Ikela), Congo belge" [= Ikela, SE Equateur Region, NW Democratic Republic of the Congo, 00°37'S, 20°01'E, elevation 350 m].

Distribution: Central Democratic Republic of the Congo (SE Equateur), 350 m. Known only from holotype.

Remarks: Possible synonym of *M. stenophthalmus* (Mocquard) *fide* Thys van den Audenaerde (1965: 373).

8. *Mehelya nyassae* (A.C.L.G. Günther, 1888c). Ann. Mag. Nat. Hist. (6) 1(3): 328. (*Simocephalus nyassae*)

Synonym: *Gonionotophis degrijsi* F. Werner, 1906.

Type: Holotype, BMNH 1946.1.14.52, a 432 mm specimen.

Type locality: "Lake Nyassa, Tropical Africa" [= Lake Malawi, E Malawi, SW Tanzania or NW Mozambique].

Distribution: Eastern Africa. Southern Somalia (Jubbada Dhexe), E Kenya (Coast, North-Eastern), E Tanzania (Lindi, Mtwara, Ruvuma, Tanga, Zanzibar Is.), Rwanda, Burundi, SE Democratic Republic of the Congo (Katanga, Sud-Kivu), Zambia (Central, Copperbelt,

Eastern, Luapula, Lusaka, Northern, North-Western, Southern, Western), Malawi (Central, Northern, Southern), Mozambique (Inhambane, Maputa, Nampula), Zimbabwe (Manicaland, Mashonaland Central, Mashonaland West, Masvingo, Matebeleland North), Botswana (Central, Gaborone, Kgalagadi, North West), N Namibia (Caprivi, Khomas), NE Swaziland and E South Africa (Gauteng, E KwaZulu-Natal, Limpopo, Mpumalanga), NSL–1200 m.

Sources: Loveridge, 1936b, 1955b, 1957a, Witte, 1953, Sweeney, 1961, Broadley, 1971d, 1983, Spawls, 1978, Haacke, 1981, Lanza, 1983a, 1990b, Pienaar et al., 1983, Auerbach, 1987, Branch, 1988, Shine et al., 1996, Spawls et al., 2002, Broadley et al., 2003, Marais, 2004, Broadley & Blaylock, 2013 and Lanza & Broadley, 2013.

9. *Mehelya poensis* (A. Smith, 1847 *in* 1838–1849). Illust. Zool. So. Africa, Rept.: p. 2 of text to pl. 55. (*Heterolepis poensis*)

Synonyms: *Heterolepis bicarinatus* A.-M.-C. Duméril, Bibron & Duméril, 1854a and ? *Heterolepis mabilli* Rochebrune, 1886.

Type: Holotype, BMNH 1946.1.14.20, a 1041 mm specimen (T. Vogel & J. Ansell [Niger Exped.], 1841–1842).

Type locality: "Bioko" [= Bioko Is., Equatorial Guinea, 3°13–3°48'N, 8°25–8°58'E].

Distribution: West and cen. Africa. Guinea-Bissau, SE Guinea (Macenta, Nzérékoré), Sierra Leone (Northern, Southern), Liberia (Montserrado, Margibi, Nimba), S Ivory Coast (Abengourou, Abidjan, Agnebi, Daloa, Guiglo, Lagunes, Moyen-Comoe, Tabou), S Ghana (Eastern), S Togo (Plateaux), Benin (Cotonou), S Nigeria (Benue, Cross River, Lagos, Oyo, Rivers), S Cameroon (Centre, Est, Littoral, Nord-Ouest, Ouest, Sud, Sud-Ouest), Equatorial Guinea (Bioko Is.), Gabon (Estuaire, Ogooué-Ivindo, Ogooué-Maritime), S Central African Republic (Haute-Sangha, Lobaye, Mbomou, Ombella-Mpoko, Sangha), Congo (Brazzaville, Kouilou, Plateau, Pool), N Angola (Cabinda, Cuanza Norte, Lunda), Democratic Republic of the Congo (Bandundu, Bas-Congo, Equateur, Kasai Occidental, Kasai Oriental, Katanga, Kinshasa, Orientale, Sud-Kivu), SW South Sudan (Western Equatoria), S Uganda (Central, Eastern, Western), N Rwanda and Burundi (Bujumbura), NSL–2195 m.

Sources: Bogert, 1940, Villiers, 1950a, 1958, 1966, Witte, 1953, 1962, E.H. Taylor & Weyer, 1958, Johnsen, 1962, Doucet, 1963, Mertens, 1965c, Roux-Estève, 1965, Knoepffler, 1966, Menzies, 1966, Blackwell,1967, Leston & Hughes, 1968, Hughes & Barry, 1969, Dunger, 1971a, Pitman, 1974, Derleyn, 1978b, Hughes, 1983, 2013, J.A. Butler & Reid, 1986, J.-F. Trape, 1985, Joger, 1990, J.-F. Trape & Roux-Estève, 1995, Spawls et al., 2002, Villiers & Condamin, 2005, Chippaux, 2006, Chirio & Ineich, 2006, Chirio & LeBreton, 2007, Pauwels & Vande weghe, 2008, Segniagbeto et al., 2011 and Auliya et al., 2012.

10. *Mehelya savorgnani* (Mocquard, 1887a). Bull. Soc. Philom. Paris (1886–1887) (7) 11(1): 27–31, pl. 2, figs. 4, 4a–b. (*Heterolepis savorgnani*)

Synonyms: *Simocephalus phyllopholis* F. Werner, 1901d, *Simocephalus butleri* Boulenger, 1907c, and *Mehelya lamani* Lönnberg, 1910.

Type: Holotype, MNHN 1886.227, a 1130 mm specimen (P.P.F.C. Savorgnan de Brazza [Brazza Mission Scientifique du Congo], Dec. 1879–June 1882).

Type locality: "Ogôoué, l'Ouest africain" [= vicinity of Ogooué River, Ogooué-Maritime, Moyen-Ogooué, Ogooué-Ivindo, Ogooué-Lolo or Haut-Ogooué Prov., W, cen. or E Gabon].

Distribution: Central Africa. Southern Cameroon (Adamaoua, Centre, Est, Nord-Ouest, Sud, Sud-Ouest), Central African Republic (Bamingui-Bangoran, Basse-Kotto, Kemo-Gribingui, Lobaye, Mbomou, Ombella-Mpoko, Ouaka, Ouham), Gabon (Haut-Ogooué, Moyen-Ogooué), Congo (Brazzaville, Kouilou, Pool, Sangha), Democratic Republic of the Congo (Bandundu, Bas-Congo, Equateur, Kasai Occidental, Kasai Oriental, Katanga, Kinshasa, Nord-Kivu, Orientale, Sud-Kivu), N Angola, S Sudan (Al-Qadariff), South Sudan (Bahr Al Ghazal, Eastern Equatoria), W Ethiopia (Ilubabar), Uganda (Central, Eastern, W Northern, Western) and SW Kenya (Nyanza, S Rift Valley, Western), NSL–1530 m.

Sources: J.-F. Trape & Roux-Estève, 1995, Chirio & Ineich, 2006, Chirio & LeBreton, 2007, Pauwels & Van de Weghe, 2008 and Lanza & Broadley, 2013.

Remarks: Figure 4 in original plate 2 mislabelled as fig. 2. A valid species *fide* Chirio & LeBreton (2007: 484), Largen & Spawls (2010: 486) and Broadly *(in litt.)*.

11. *Mehelya stenophthalmus* (Mocquard, 1887a). Bull. Soc. Philom. Paris (1886–1887) (7) 11(1): 16–19, pl. 1, figs. 1, 1a–b. (*Heterolepis stenophthalmus*)

Synonym: *Gonionotophis microps* Boulenger, 1911b.

Types: Syntypes (3), MNHN 1885.615–1b (M.A. Chaper, 1882–1884) [Assinie] and MNHN 1885.715 (Boisguilleaume) [Lopez], longest syntype 700 mm.

Type locality: "Assinie, l'Ouest africain" [= Assini, Aboisso Dept., ext. SE Ghana, 5°07'N, 3°17'W] and "cap Lopez, l'Ouest africain" [= Cape Lopez, N Ogooué-Maritime Prov., W Gabon, 0°38'S, 8°42'E, elevation NSL].

Distribution: West and cen. Africa. Guinea–Bissau, SE Guinea (Macenta, Nzérékoré), Ivory Coast, Ghana (Western), S Togo (Centrale, Plateaux), S Benin (Atlantique), Nigeria (Lagos, Oyo), Cameroon (Centre, Est, Littoral, Nord-Ouest, Sud, Sud-Ouest), Gabon (Ogooué-Ivindo, Ogooué-Maritime), S Central African Republic (Bamingui-Bangoran, Haute-Sangha,

Haut-Kotto, Kemo-Gribingui, Lobaye, Mbomou, Ombella-Mpoko, Ouham), S Congo (Brazzaville, Cuvette, Pool, Sangha), Democratic Republic of the Congo (Bandundu, Bas-Congo, Equateur, Katanga, Kinshasa, Sud-Kivu) and SW Uganda (Western), NSL–2600 m.

Sources: Villiers, 1954a, Laurent, 1956a, Roux-Estève, 1965, Knoepffler, 1966, Dunger, 1971a, Pitman, 1974, Witte, 1975, Courtois & Chippaux, 1977, D. Lawson, 1993, J.-F. Trape & Roux-Estève, 1995, Spawls et al., 2002, Villiers & Condamin, 2005, Chippaux, 2006, Chirio & Ineich, 2006, Chirio & LeBreton, 2007, Pauwels & Vande weghe, 2008, Segniagbeto et al., 2011 and Hughes, 2013.

12. *Mehelya vernayi* Bogert, 1940. Bull. Amer. Mus. Nat. Hist. 77(1): 28–30, figs. 1a–e, 2.

Type: Holotype, AMNH 51795, a 1035 mm female (A.S. Vernay, H. Lang & R. Boulton, 13–19 May 1925).

Type locality: "Hanha, Angola" [= Hanha do Norte, Benguela Distr., Angola, 12°16'S, 13°42'E, elevation 35 m].

Distribution: Southwestern Angola (Benguela) and NW Namibia (Erongo, Kunene, Oshikoto, Otjozondjupa), 35–100 m.

Sources: Mertens, 1955, 1971, Elzen, 1980, Haacke, 1981, Broadley, 1983 and Branch, 1988.

MEIZODON J.G. Fischer, 1856b (Colubridae)

Synonym: *Mizodon* W.C.H. Peters, 1861a *(nomen emendatum).*

Type species: *Meizodon regularis* J.G. Fischer, 1856b.

Distribution: Subsaharan Africa.

Sources: Loveridge, 1957a, Roux-Estève, 1969b, Schätti, 1985a and Spawls et al., 2002.

1. *Meizodon coronatus* (Schlegel, 1837). Essai Phys. Serp. 1: 133–134, 2: 46–47. (*Calamaria coronata*)

Synonyms: *Coryphodon margaritiferus* A.C.L.G. Günther, 1858 and *Coronella bitorquata* A.C.L.G. Günther, 1860h.

Type: Lectotype, RMNH 90, a 186 mm female (D.F. Eschricht, 1830–1837), designated by Hughes & Barry (1969: 1018).

Type locality: "Gold Coast" [= presumably Ghana *fide* Hughes & Barry, 1969: 1018] via lectotype selection.

Distribution: West Africa. Southern Senegal (Dakar, Fatick, Kaolack, Kédougou, Tambacounda), Guinea (Dalaba, Koundara, Kouroussa, Macenta, Nzérékoré, Télimélé), SW Mali (Kayes, Koulikoro, Mopti, Sikasso), Sierra Leone, Liberia, Ivory Coast (Bouaké, Toumodi), S Burkina Faso (Centre, Centre-Ouest, Est, Hauts-Bassins, Sud-Ouest, Volta-Noire), Ghana, Togo (Plateaux), N Benin (Atakora, Borgou), S Nigeria (Akwa Ibom, Anambra, Cross River, Edo, Rivers), SW Chad (Mayo-Kebbi), SW Niger (Dosso), Cameroon (Adamaoua, Centre, Nord, Nord-Ouest, Ouest) and Central African Republic (Bamingui-Bangoran, Haut-Mbomou) and N Democratic Republic of the Congo (Equateur, Hait-Zaire), NSL–1300 m.

Sources: A.C.L.G. Günther, 1861a, Witte, 1962, Doucet, 1963, Roman, 1980, J.A. Butler & Reid, 1986, Joger, 1990, Luiselli et al., 2001, J.-F. Trape & Mané, 2004, Chippaux, 2006, Chirio & Ineich, 2006, J.-F. Trape & Mané, 2006b, Chirio & LeBreton, 2007, W. Böhme et al., 2011, Segniagbeto et al., 2011, Chirio, 2013 and Hughes, 2013.

2. *Meizodon krameri* Schätti, 1985a. Monit. Zool. Ital. (2) 20(8): 155–157, figs. 2a–b.

Type: Holotype, MCZ 40323, a 494–497 mm female (A. Loveridge, 5 June 1934).

Type locality: "Kau, Tana River, Kenya" [= Tana River delta, E Coast Prov., SE Kenya, 2°29'S, 40°28'E].

Distribution: Extreme SE Kenya (Coast). Known only from type locality.

3. *Meizodon plumbiceps* (Boettger, 1893b). Zool. Anz. 16(416): 117–118. (*Coronella plumbiceps*)

M

Synonyms: *Coronella somalica* Scortecci, 1932 and *Meizodon loveridgei* Bogert, 1940.

Type: Holotype, SMF 62453 (formerly ZMUZ), a 576 mm male (C. Keller [Ruspoli Exped.], 1892).

Type locality: "Ogadeen, Somaliland" [= Ogaden, Somali Prov., E Ethiopia, 6°58'N, 44°02'E, elevation 790 m].

Distribution: Northeast Africa. Eastern Ethiopia (Bale, Hararge, Sidamo), S Somalia (Bay, Jubbada Dhexe, Jubbada Hoose, Shabeellaha Dhexe, Shabeellaha Hoose) and ext. NE Kenya (N North-Eastern), 20–1500 m.

Sources: Hoevers & Johnson, 1982, Lanza, 1983a, 1990b, Largen & Rasmussen, 1993 and Largen & Spawls, 2010.

Remarks: Yemen record (Scortecci, 1932) erroneous, originating from Somalia *fide* Schätti (1985a: 165).

4. *Meizodon regularis* J.G. Fischer, 1856b. Abh. Ges. Naturwiss. Verh. Hamburg 3: 112–114, pl. 3, figs. 3a–c.

Synonyms: *Coronella elegans* Jan, 1863a, *Coronella regularis praeornata* Angel, 1933b, and *Meizodon coronata camerunensis* Monard, 1951.

Type: Holotype, ZMH 2638 (formerly ZMH 77), a 706 mm male (L. Wolf, Nov. 1847–Jan. 1851).

Type locality: "Peki" [= Peki, S Volta Region, SE Ghana, 6°32'N, 0°15'E, elevation 400 m].

Distribution: West and cen. Africa. Guinea (Beyla, Dalaba, Dubréka, Kérouané, Labé, Macenta,

Nzérékoré), Liberia (Nimba), Ivory Coast (Bouaké, Toumodi), Ghana, Togo (Kara, Plateaux), S Benin (Atlantique), Benin, S Chad (Moyen-Chari), cen. Cameroon (Adamaoua, Centre, Est, Nord, Nord-Ouest, Ouest), Central African Republic (Bamingui-Bangoran, Ouaka, Ouham, Ouham-Pende), Democratic Republic of the Congo (Orientale), S Sudan, SE South Sudan (Eastern Equatoria), Ethiopia (Gojjam, Gondar, Hararge, Illubabor), N Uganda, (N Eastern, Northern Western) and SW Kenya (Nyanza), 150–2000 m.

Sources: A.C.L.G. Günther, 1861a, Pitman, 1974, Largen & Rasmussen, 1993, Spawls et al., 2002, Chippaux, 2006, Chirio & Ineich, 2006, Chirio & LeBreton, 2007, Largen & Spawls, 2010, Ullenburch et al., 2010, Segniagbeto et al., 2011 and Hughes, 2013.

5. *Meizodon semiornatus* (W.C.H. Peters, 1854). Mber. Königl. Akad. Wiss. Berlin 1854(11): 622. (*Coronella semiornata*)

Synonyms: *Zamenis fischeri* W.C.H. Peters, 1879, *Coronella inornata* Fischer, 1884a, *Coronella scheffleri* Sternfeld, 1908a, *Zamenis citernii* Boulenger, 1912a, *Zamenis tchadensis* Chabanaud, 1918a, and *Coronella semiornata mossambicae* Cott, 1935.

Types: Syntypes (3), ZMB 4804 and ZMB 4805a–b, a 530 mm, 440 mm and 260 mm specimen (W.C.H. Peters, Dec. 1844–Sept. 1845).

Type locality: "Tette" [= Tete, Tete Prov., W Mozambique, 16°10'S, 33°36'E, elevation 125 m].

Distribution: North and E Africa. Southwestern Chad (Chari-Baguirmi), N Cameroon (Extreme-Nord), SE Sudan (Al Jazirah, Nahr An Nil), SE South Sudan (Eastern Equatoria), ext. S Ethiopia (SW Bale, Sidamo), S Somalia (Gedo, Jubbada Dhexe, Jubbada Hoose, Mogadishu, Shabeellaha Dhexe, Shabeellaha Hoose), Kenya (Coast, Eastern, Nairobi, Rift Valley), E Tanzania (Arusha, Dar es Salaam, Lindi, Mtwara, Mwanza, Rukwa, Ruvuma, Singida, Tanga), Zambia (Central, Copperbelt, Eastern, Luapula, Lusaka, Northern, North-Western, Southern, Western), S Malawi (Southern), S Mozambique (Gaza, Maputo, Sofala, Tete), Zimbabwe (Mashonaland Central, Mashonaland West, Masvingo, Matabeleland North, Midlands), ext. N Botswana (N North West), E Swaziland and ext. NE South Africa (ext. N KwaZulu-Natal), NSL–2200 m.

Sources: Broadley, 1971d, 1983, Pitman, 1974, Broadley & Cock, 1975, Lanza, 1983a, 1990b, Auerbach, 1987, Branch, 1988, Gasperetti, 1988, Largen & Rasmussen, 1993, Spawls et al., 2002, Broadley et al., 2003, Chippaux, 2006, Chirio & LeBreton, 2007, Largen & Spawls, 2010 and Broadley & Blaylock, 2013.

Remarks: Original description reprinted in W.C.H. Peters (1855: 51–52). Possibly occurs in Namibia *fide M.* Griffin (2003: 129).

M

MELANOPHIDIUM A.C.L.G. Günther, 1864a (Uropeltidae)

Synonyms: *Mellanophidium* – Higgins, 1873 *(nomen incorrectum), Melanophium* – Cardew, 1897 *(nomen incorrectum),* and *Melanophidim* – Briceño-Rossi, 1934 *(nomen incorrectum).*

Type species: *Plectrurus wynaudensis* Beddome, 1863a.

Distribution: Southern India.

Sources: M.A. Smith, 1943, Gans, 1966, Murthy, 1973, 1981c, 1982b, McDiarmid et al., 1999 and Gower et al., 2005.

1. *Melanophidium bilineatum* Beddome, 1870. Madras Month. J. Med. Sci. (2) 2: 169.

Types: Syntypes (2), BMNH 1946.1.15.75 (formerly BMNH 1872.1.2.5) and MNHN 1895.111, longest syntype 370 mm (R.H. Beddome, 1857–1882).

Type locality: "near the summit of the Peria peak in the Wynaad, elevation 5,000 feet, and also at a similar elevation on the Tirrhioot peak, Madras Presidency." Emended to west of Manantoddy [= NE Kerala State, SW India, 11°48'N, 76°01'E, elevation 1525 m] *fide* M.A. Smith (1943: 67).

Distribution: Southwestern India (Western Ghats of W Karnataka, NE Kerala), 1200–1525 m.

Source: A.C.L.G. Günther, 1875.

2. *Melanophidium punctatum* Beddome, 1871. Madras Month. J. Med. Sci. (2) 4: 401–402.

Types: Syntypes (4), BMNH 1946.1.4.73 (formerly BMNH 1872.1.2.6), BMNH 1946.1.15.48–49 (formerly BMNH 1874.4.29.121–22), and MNHN 1895.116, longest syntype 457 mm (R.H. Beddome, 1857–1882).

Type locality: "Travancore... in the Muti-kuli vayal, a little valley on the Asamboo Distribution (4,500 feet elevation)" and "Peermede, Madras Presidency" [= Ashambu Hills, elevation 1370 m, and Azhutha, elevation 915 m, Kerala State, SW India].

Distribution: Southwestern India (Western Ghats of Goa, W Karnataka, E Kerala, W Maharashtra, SW Tamil Nadu), 1200–1500 m.

Sources: A.C.L.G. Günther, 1875, Wall, 1923i, Rajendran, 1985, Ghate & Yazdani, 1991, Whitaker & Captain, 2004 and N. Khaire, 2006.

Remarks: Beddome mentioned only 3 specimens but Gans (1966a: 10) listed the MNHN specimen as a type. Possible lectotype designated by Welch (1988: 16) by restriction (? inadvertently) of type locality to Muti-Kuli Vayal.

3. *Melanophidium wynaudense* (Beddome, 1863a). Madras Month. J. Med. Sci. 6: 48, pl. 2, fig. 9. (*Plectrurus wynaudensis*)

Synonyms: *Melanophidium wynandense* A.C.L.G. Günther, 1863 *(nomen emendatum), Plectrurus wynadensis* Beddome, 1867 *(nomen emendatum), Melanophidium wynaadense* Theobald, 1876 *(nomen emendatum), Plectrurus wynaadensis* Theobald, 1876 *(nomen emendatum),* and *Melanophidium wynadense* Boulenger, 1893a *(nomen emendatum).*
Types: Syntypes (4), BMNH 1946.1.15.46 (formerly BMNH 1864.29.84), BMNH 1946.1.15.74 (formerly BMNH 1864.3.9.6), and MNHN 1895.108–09 (R.H. Beddome, 1857–1882).
Type locality: "Cherambady in the Wynaud, Madras Presidency." Emended to Wynand, elevation 3500 feet, Southern India [= Cherambadi, N Kerala State, SW India, 11°32'N, 76°17'E, elevation 1065 m] *fide* Beddome (1863b: 228), M.A. Smith (1943: 67) and Constable (1949: 115).
Distribution: Southwestern India (Western Ghats of W Karnataka, NE Kerala, NW Tamil Nadu), 585–2120 m.
Sources: Wall, 1923i and Ganesh et al., 2012b.
Remarks: Original description reprinted in Beddome (1863b: 228). Discussion of specific epithet in McDiarmid et al. (1999: 131).

†*MENARANA* LaDuke, Krause, Scanlon & Kley, 2010 (†Madtsoiidae)

Type species: †*Menarana nosymena* LaDuke, Krause, Scanlon & Kley, 2010.
Distribution: Upper Cretaceous of Madagascar and Spain.
Sources: Rage, 1996, 1999 and LaDuke et al., 2010.

1. †*Menarana laurasiae* (Rage, 1996). C.R. Acad. Sci. Paris (2) 322: 605–606, fig. 1. (†*Madtsoia laurasiae*)

Type: Holotype, MCNA 5337, one middle trunk vertebra.
Type locality: "Crétacé supérieur de Laño (Pays Basque espagnol)" [= upper Cretaceous of Laño, Condado de Treviño Prov., Basque Country, N Spain, 42°40'N, 2°37'W]. Clarified as a sand quarry near the village of Laño, about 20 km SE of the city of Vitoria, Basque Country, Spain; Upper Cretaceous (Early Maastrichian) *fide* Astibia et al. (1990: 462).
Distribution: Upper Cretaceous (Maastrichian: 65.5–72.1 mya) of Spain. Known only from type locality.
Source: Rage, 1999.

2. †*Menarana nosymena* LaDuke, Krause, Scanlon & Kley, 2010. J. Vert. Paleont. 30(1): 119–125, figs. 5–9.

Type: Holotype, UA 9684.1–6, numerous vertebrae, fragmentary ribs and a partial braincase (Mahajanga Basin Project field crew, 1993–2007).
Type locality: "MAD93–14, Berivotra Study Area, Mahajanga Basin, northwestern Madagascar; Upper Cretaceous (Maastrichtian) Maevarano Formation."
Distribution: Upper Cretaceous (Maastrichtian: 65.5–72.1 mya) of Madagascar.

†*MESSELOPHIS* Baszio, 2004 (Tropidophiidae)

Type species: †*Messelophis variatus* Baszio, 2004.
Distribution: Middle Eocene of Germany.
Sources: Szyndlar et al., 2008 and Rage, 2011.
Remarks: An "enigmatic booid" *fide* Rage (2011: 41).

1. †*Messelophis ermannorum* Schaal & Baszio, 2004. Cour. Forsch.-Inst. Senck. (252): 68–74, figs. 1a–d.

Type: Holotype, SMF ME 1812, one trunk vertebra.
Type locality: "UNESCO Weltnaturerbe Fossilienfundstätte Grube Messel bei Darmstadt, Hessen, Deutschland; Mittlere Messel-Formation (Ölschiefer), Mittel-Eozän, Unteres Lutetium, MP 11, Fossilienfundstätte Grube Messel."
Distribution: Middle Eocene (Lutetian, MP 11: 48.6 mya) of Germany.

2. †*Messelophis variatus* Baszio, 2004. Cour. Forsch.-Inst. Senck. (252): 49–60, figs. 1, 3, 5–6, 8–9.

Type: Holotype, SMF ME 1828, one nearly complete articulated skeleton and four isolated vertebrae from same specimen.
Type locality: "World Natural Heritage Messel Pit Fossil Site near Darmstadt; Middle Eocene (MP 10/11), Messel formation."
Distribution: Lower-middle Eocene (Lutetian, MP 10–11: 48.6–55.8 mya) of Germany.

†*MICHAUXOPHIS* Bailon, 1988 (Aniliidae)

Type species: †*Michauxophis occitanus* Bailon, 1988.
Distribution: Upper Pliocene of France.
Sources: Bailon, 1988, 1989 and Gómez et al., 2008.

1. †*Michauxophis occitanus* Bailon, 1988. C.R. Acad. Sci. Paris (2) 306: 1256–1258, figs. a–e.

Type: Holotype, USTL BAC II 911, one trunk vertebra.
Type locality: "Balaruc II, Hérault, France, Pliocène supérieur (Villanyien), zone 16."
Distribution: Upper Pliocene (Villanyian, MN 16: 2.6–3.2 mya) of France. Known only from type locality.

MICRELAPS Boettger, 1880b (Atractaspididae)

Synonyms: *Elaposchema* Mocquard, 1888 and *Miculaps* – Briceño-Rossi, 1934 *(nomen incorrectum).*
Type species: *Micrelaps muelleri* Boettger, 1880b.
Distribution: East Africa and ext. SW Asia.
Sources: Witte & Laurent, 1947, Pitman, 1974, Underwood & Kochva, 1993, Vidal et al., 2008, Kelly et al., 2009 and Zaher et al., 2009.
Remarks: Vidal et al., 2008, studies do not support allocation to the Atractaspidinae (or Atractaspididae) *contra* Kelly et al., 2009 and Zaher et al., 2009.

1. *Micrelaps bicoloratus* Sternfeld, 1908a. Sitz. Ges. Natur. Freunde Berlin 1908(4): 93.

Synonym: *Micrelaps bicoloratus moyeri* J.B. Rasmussen, 2002.
Type: Holotype, ZMB 20725, a 240 mm female (G. Scheffler, 1906–1908).
Type locality: "Kibwezi (Brit. Ostafrika)" [= Kibwezi, S Eastern Prov., S Kenya, 2°25'S, 37°58'E, elevation 900 m].
Distribution: Southeastern Kenya (Central, Coast, Eastern, Nairobi, Rift Valley, Lamu Is.) and NE Tanzania (Arusha, Kilimanjaro, Manyara), NSL–2000 m.
Sources: J.B. Rasmussen, 2002 and Spawls et al., 2002.
Remarks: A valid species *fide* Spawls et al. (2002: 435).

M

2. *Micrelaps muelleri* Boettger, 1880b. Ber. Senck. Naturf. Ges. (1879–1880): 137–138, pl. 3, figs. 2a–d. (*nomen corrigendum*)

Synonyms: *Micrelaps mülleri* Boettger, 1880b *(nomen incorrigendum)* and *Micrelaps muelleri* Y.L. Werner, 1965 *(nomen corrigendum).*
Type: Lectotype, SMF 20349 (formerly SMF-B 9249, a) (H. Simon, 1879), designated by Mertens (1922: 181), longest syntype 335 mm.
Type locality: "Jerusalem" [= Jerusalem, E Jerusalem Distr., cen. Israel, 31°46'N, 35°13'E, elevation 775 m] via lectotype selection.
Distribution: Middle East. Syria (Lattakia), NW Lebanon (Liban-Nord), Israel (Central, Golan Heights, Haifa, Jerusalem, Northern, Tel Aviv, West Bank) and NW Jordan (Balqa, Irbid, Jarash), BSL–1800 m.
Sources: Lortet, 1883, G. Haas, 1951, 1952b, Battersby, 1953, Hoofien, 1972, Hraoui-Bloquet, 1981, Bischoff & Schmidtler, 1997, Bouskila & Amitai, 2001, Disi et al., 2001, Hraoui-Bloquet et al., 2002, Y.L. Werner et al., 2006, Amr & Disi, 2011 and Bar & Haimovitch, 2011.
Remarks: Boettger (1898: 111) list collection date of types as 1879–1884. Lectotype erroneously listed as SMF 4249a *fide* Y.L. Werner et al. (2006:36).

3. *Micrelaps tchernovi* Y.L. Werner, Babocsay, Carmely & Thuna, 2006. Zool. Middle East 38: 40–42, figs. 2b, 5c, 6, 9c, 10–11.

Type: Holotype, HUJ 16864, a 558 mm female (E. Tchernov, 5 July 1988).
Type locality: "Ubeidiyya Site, Jordan Valley, Israel [IG 20292328]."
Distribution: Northeastern Israel (SE Northern, NE West Bank) and W Jordan (Balqa, Irbid), 300 m BSL–80 m.
Sources: Amr & Disi, 2011 and Bar & Haimovitch, 2011.

4. *Micrelaps vaillanti* (Mocquard, 1888). Mém. Cent. Soc. Philom. Paris 1888: 123–124, pl. 12, figs. 1a–c. (*Elaposchema vaillanti*)

Synonyms: *Micrelaps boettgeri* Boulenger, 1896g, *Micrelaps nigriceps* Sternfeld, 1910c, and *Rhinocalamus meleagris* Sternfeld, 1908e.
Type: Holotype, MNHN 1884.696, a 282 mm male (G.E.J. Révoil, 1877–1888).
Type locality: "pays Çomalis" [= Somalia].
Distribution: Northeast Africa. Southeastern Sudan (An Nil Al Abyad, Janub Kurdufan), SE South Sudan (Eastern Equatoria), SE Ethiopia (Bale, Hararge, Shoa, Sidamo), Somalia (Gedo, Jubbada Dhexe, Jubbada Hoose, Mogadishu, Shabeellaha Dhexe, Togdheer, Woqooyi Galbeed), Kenya (Coast, Eastern, E Nairobi, North-Eastern, S Rift Valley), N Uganda (E Northern) and NE Tanzania (Arusha, Kilimanjaro [MCZ 20948]), NSL–1830 m.
Sources: Pitman, 1974, Spawls, 1978, Largen & Rasmussen, 1993, J.B. Rasmussen, 2002, Spawls et al., 2002 and Largen & Spawls, 2010.

MICROCEPHALOPHIS Lesson, 1832 *in* Bélanger, 1831–1834 (Elapidae)

Synonyms: *Microcephalus* Schlegel, 1837 *(nomen praeoccupatum), Liopala* Gray, 1840, *Liopola* – Gray, 1842d *(nomen incorrectum), Microcephalolepis* – Hoffmann, 1890 *(nomen incorrectum), Microicephaloides* – Barrett, 1950 *(nomen incorrectum), Michocephalophis* – Haile, 1959 *(nomen incorrectum), Microcephalopis* – Sharma, 1982 *(nomen incorrectum),* and *Michocephalaphis* – Culotta & Pickwell, 1993 *(nomen incorrectum).*
Type species: *Hydrus gracilis* G. Shaw, 1802.
Distribution: Marine waters of Indo-Australia, including Arabian Sea, Bay of Bengal, Coromandel Coast, Coral Sea, Great Barrier Reef, Gulf of Carpenteria, Gulf of Mandura, Gulf of Oman, Gulf of Papua, Gulf of Thailand, Malabar Coast, Palk Strait, Persian Gulf, Strait of Hormuz, Strait of Hainan, Strait of Malacca and Torres Strait.
Sources: Wall, 1921g, 1926b, M.A. Smith, 1926a, 1943, Bourret, 1936b, E.H. Taylor, 1950b, Klemmer, 1963,

Minton, 1966, Halstead, 1970, McDowell, 1972a, Voris, 1977, Joger, 1984, Murthy, 1985, M.A.R. Khan, 1988, David & Ineich, 1999, M.S. Khan, 2002 and Wells, 2007.

Remarks: Considered a synonym of *Hydrophis* (or as a subgenus) by Golay et al., 1993 as well as others.

1. *Microcephalophis cantoris* (A.C.L.G. Günther, 1864a). Rept. Brit. India: 374–375, pl. 25, fig. u. (*Hydrophis cantoris*)

Synonyms: *Distira gillespiae* Boulenger, 1899d and *Microcephalophis cantori* – M.W. Lin, 1975 *(nomen incorrectum).*

Type: Holotype, BMNH 1946.1.18.30 (formerly BMNH 1860.3.19.1302), a 1067–1099 mm male (T.H. Cantor, 1834–1847, via British East India Comp.).

Type locality: "Pinang" [= Penang Is., Pinang Prov., NW Malay Peninsula, 5°22'N, 100°14'E] (possibly in error *fide* Minton, 1966: 151).

Distribution: Southern Asia. Pakastan (Sindh), E India (Andaman & Nicobars: Andaman Is., Andhra Pradesh, Gujarat, Karnataka, N Kerala, Maharashtra, Orissa, Tamil Nadu, West Bengal), Bangladesh (Chittagong, Khulna, Patuakhali), Myanmar and W West Malaysia (Pinang, Penang Is.).

Source: I. Das & Chandra, 1994.

Remarks: A.C.L.G. Günther, 1864a mentioned an adult and juvenile specimen but based description on Cantor's adult holotype.

2. *Microcephalophis gracilis* (G. Shaw, 1802). Gen. Zool., Amph. 3(2): 560–561. (*Hydrus gracilis*)

Synonyms: *Hydrophis kadell-nagam* F. Boie, 1827 *(nomen incorrigendum), Hydrus gracilus* F. Mason, 1852 *(nomen emendatum), Thalassophis microcephala* P. Schmidt, 1852, *Hydrophis leprogaster* A.-M.-C. Duméril, 1853 *(nomen nudum), Hydrophis leprogaster* A.-M.-C. Duméril & Bibron *in* J.G. Fischer, 1856a, *Hydrophis guentheri* Theobald, 1868b, *Hydrophis guntherii* Murray, 1884 *(nomen emendatum), Hydrophis rostralis* M.A. Smith, 1917c, *Microcephaloides gracilis* – Barrett, 1950 *(nomen incorrectum), Distevia gracilis* – Okada, 1953 *(nomen incorrectum), Microcephalus graculis* – Saint-Girons, 1967 *(nomen incorrectum), Hydrophis spiralis spiralis* Leviton & Aldich, 1984 *(lapsus calami), Microcephalophus gracilis* – Vitt, 1987 *(nomen incorrectum), Microcephalophis gracilis microcepalophis* – Culotta & Pickwell, 1993 *(nomen incorrectum),* and *Microcephalopis gracilis* – Culotta & Pickwell, 1993 *(nomen incorrectum).*

Type: Holotype, BMNH 1946.1.17.37 (formerly BMNH III.4.1.a), a 610 mm male.

Type locality: Unknown.

Distribution: Indo-Australia. Southeastern Iraq (Al Basrah), Iran (Bushehr, Hormozghan, Khuzestan, Sistan va Baluchestan), N Oman (Muscat), Pakistan (Sindh), India (Andhra Pradesh, Gujarat, Karnataka, Maharashtra, Tamil Nadu), N Sri Lanka (N Eastern, Northern), Bangladesh, Myanmar (Mergui Arch., Tanintharyi), Thailand (Pattani, Phuket, Singgora, Songkhla), Vietnam, S China (Fujian, Guangdong, Guangxi, Hainan, Hong Kong), Macao, West Malaysia (Perak, Pinang, Selanger), Singapore, East Malaysia (Sarawak), W Indonesia (Bangka, N Java, Madura, Sumatra), S Papua New Guinea (Gulf) and N Australia (Queensland).

Sources: Bleeker, 1857b, Rooij, 1917, C.H. Pope, 1935, Kennedy, 1937, Vølsoe, 1939, C. Haas, 1950, Suvatti, 1950, Deraniyagala, 1955, Haile, 1959, E.H. Taylor, 1965, Tu & Tu, 1970, Romer, 1972, Singh, 1972, Tu, 1974, Cogger, 1975, 1992, P. Silva, 1980a, 1990, B.L. Lim, 1982, Whitaker et al., 1982, Cogger et al., 1983a, Tweedie, 1983, Voris & Voris, 1983, Tian et al., 1986, Dowling & Jenner, 1988, Gasperetti, 1988, Gow, 1989, K.K.P. Lim & Chou, 1990, M.J, Cox, 1991b, Stuebing, 1991, Leviton et al., 1992, V.S. Nguyen et al., 2009 and A. Silva et al., 2011.

†*MICRONATRIX* Parmley & Hunter, 2010 (Natricidae)

Type species: †*Micronatrix juliescottae* Parmley & Hunter, 2010.

Distribution: Upper Miocene of USA.

Source: Parmley & Hunter, 2010.

1. †*Micronatrix juliescottae* Parmley & Hunter, 2010. J. Herp. 44(4): 534–536, fig. 5.

Type: Holotype, USNM 100260, one anterior trunk vertebra.

Type locality: Pratt Slide fossil locality: University of Nebraska State Museum locality Bw123; 6.5 km north of Johnstown, Brown County, north-central Nebraska; NE1/4, SW1/4 section 30. T. 3/N, R.23W; Merrit Dam Member of the Ash Hollow Formation; Age is late Clarendonian North American Land Mammal Age (late Miocene)."

Distribution: Upper Miocene (Clarendonian: 10.3–13.6 mya) of USA (Nebraska).

MICROPECHIS Boulenger, 1896a (Elapidae)

Synonyms: *Elaposoma* Fitzinger, 1843, *Mikropechis* – Maass-Berlin, 1933 *(nomen incorrectum)* and *Microphechis* – Bulmer, Menzies & Parker, 1975 *(nomen incorrectum).*

Type species: *Coluber ikaheca* Lesson, 1829 *in* Duperrey, 1826–1832.

Distribution: New Guinea.

Sources: Loveridge, 1948, Klemmer, 1963, McDowell, 1970, 1984, Talukdar & Dattagupta, 1980, F. Parker, 1982, Golay et al., 1993, O'Shea, 1996, Greer, 1997, Keogh, 1998, David & Ineich, 1999, Sanders et al., 2008, Zaher et al., 2009 and Hoser, 2012e.

1. *Micropechis ikaheca* (Lesson, 1829 *in* Duperrey, 1826–1832). Voy. Coquille (Zool.), Atlas: pl. 5. (*Coluber ikaheca*)

Synonyms: *Naja elaps* Schlegel, 1837, *Trimeresurus ophiophagus* A.-M.-C. Duméril, Bibron & Duméril, 1854b *(partim), Trimeresurus ikaheca* Jan, 1863b *(nomen emendatum), Coluber ikaheka* W.C.H. Peters & Doria, 1878 *(nomen emendatum),* and *Ophiophagus ikaheka fasciatus* J.G. Fischer, 1884c.

Type: Holotype, MNHN 7669, a 1422–1448 mm specimen (R.-P. Lesson & P. Garnot [Coquille Voy.], 1824).

Type locality: "Nouvelle-Guinée, dans les forêts épaisses qui bordent le havre de Doréry" [= Manokwari, NW West Papua Prov., E Indonesia, 0°52'S, 134°05'E, elevation 70 m].

Distribution: New Guinea. Eastern Indonesia (Papua, West Papua, Aru Arch., Batanta, Misool, Num, Sarera Bay Is., Yapen) and Papua New Guinea (Eastern Highlands, East Sepik, Morobe, Milne Bay, Western, Daru and Karkar Is.), NSL–1200 m.

Remarks: Supplemental original description in Lesson (1830 *in* Duperrey, 1826–1832: 54–55). Original orthography of species in plate (1829), text (1830) and index (1831) was *Coluber ikaheca*.

M

MICROPISTHODON Mocquard, 1894b (Pseudoxyrhophiidae)

Type species: *Micropisthodon ochraceus* Mocquard, 1894b.

Distribution: Madagascar.

Sources: Guibé, 1958, Domergue, 1969, Brygoo, 1983, 1987, Glaw & Vences, 1994, Nagy et al., 2003, Kelly et al., 2009 and Zaher et al., 2009.

1. *Micropisthodon ochraceus* Mocquard, 1894b. C.R. Somm. Séances Soc. Philom. Paris (17): 8.

Type: Holotype, MNHN 1886.173, a 687 mm female (H. Deyrolle).

Type locality: "Nossi-Bé, Madagascar" [= Nosy Be Is., W Antsiranana Prov., ext. NW Madagascar, bet. 13°12'–25'S and 48°10'–22'E, elevation < 225 m].

Distribution: Northeastern Madagascar (Antsiranana, Toamasina, Nosy Be Is.), 25–800 m.

MICRUROIDES K.P. Schmidt, 1928c (Elapidae)

Synonyms: *Micruvides* – Hess, 1980 *(nomen incorrectum)* and *Micruriroides* – J.D. Williams, 1983 *(nomen incorrectum).*

Type species: *Elaps euryxanthus* Kennicott, 1860.

Distribution: Southwestern USA and NW Mexico.

Sources: Ditmars, 1939, K.P. Schmidt & Davis, 1941, Bogert & Oliver, 1945, H.M. Smith & Taylor, 1945, Zweifel & Norris, 1955, Fowlie, 1965, Roze, 1967, 1974, 1983, 1996, Hoge & Romano-Hoge, 1971c, Cadle & Sarich, 1981, Cadle, 1983, McCarthy, 1985, J.A. Campbell & Lamar, 1989, 2004, Slowinski, 1991, 1995, Golay et al., 1993, Degenhardt et al., 1996, Goldberg, 1997, Keogh, 1998, David & Ineich, 1999, R.D. Bartlett & Tennant, 2000, Slowinski & Keogh, 2000, Grismer, 2002, Lemos-Espinal et al., 2004, Castoe et al., 2007, Guardado & Gatica-Colima, 2009, Zaher et al., 2009 and Ernst & Ernst, 2011.

1. *Micruroides euryxanthus* (Kennicott, 1860). Proc. Acad. Nat. Sci. Philadelphia 12(8): 337–338. (*Elaps euryxanthus*)

Synonyms: *Micruroides euryxanthus australis* Zweifel & Norris, 1955, *Micruroides euryxanthus neglectus* Roze, 1967, *Micruroides euryxanthinus* – Gans & Maderson, 1970 *(nomen incorrectum), Micruroides euryzathes* – Simonb, 1979 *(nomen incorrectum), Micruroides euroxanthus* – J.T. Henderson & Bieber, 1986 *(nomen incorrectum),* and *Micruroides euryzanthus* – Lombard, Marx & Rabb, 1986 *(nomen incorrectum).*

Type: Holotype, USNM 1122, a 441 mm male (T.H. Webb).

Type locality: "North America." Emended to Sonora, Mexico *fide* Yarrow (1882: 82).

Distribution: Extreme SW USA (S Arizona, SW New Mexico) and NW Mexico (ext. W Chihuahua, S Sinaloa, Sonora, Tiburón Is.), NSL–1800 m.

Remarks: Type locality restriction of Guaymas, Sonora, Mexico *fide* K.P. Schmidt (1953a: 222) rejected by Roze (1974: 163.3). West Texas (El Paso) record [MCZ 22645] requires confirmation *fide* Roze (1974: 163.3).

MICRURUS Wagler *in* Spix, 1824 (Elapidae)

Synonyms: *Elaps* J.G. Schneider, 1801 *(nomen suppressum), Elops* – Bleeker, 1857b *(nomen incorrectum), Elapsus* A.L. Herrera, 1899 *(nomen emendatum), Leptomicrurus* K.P. Schmidt, 1937, *Michurus* – Barrett, 1950 *(nomen incorrectum), Micruris* – Wacha & Christiansen, 1974 *(nomen incorrectum), Micurus* – J.W. Simpson, 1975 *(nomen incorrectum), Micrupus* – Federsoni, 1981 *(nomen incorrectum),*

Erythrokomophis Slowinski *in* Roze, 1994 *(nomen nudum)*, *Monadophis* Slowinski *in* Roze, 1994 *(nomen nudum)*, *Binghamus* Hoser, 2012f *(nomen illegitimum)*, *Hoserelapidea* Hoser, 2012f *(nomen illegitimum)*, and *Troianous* Hoser, 2012f *(nomen illegitimum)*.

Type species: *Micrurus spixii* Wagler *in* Spix, 1824.

Distribution: Southeastern USA and Latin America.

Fossil records: Lower to middle Miocene of W Europe and middle Miocene to upper Pleistocene of SE USA. Fossils unassigned to species include middle Miocene (Barstovian) of USA (Nebraska) and upper Miocene (Hemphillian) of USA (Florida).

Sources: K.P. Schmidt, 1937, Klemmer, 1963, Hoge & Romano-Hoge, 1966, 1971, 1972, Roze, 1967, 1983, 1994, 1996, J.A. Peters & Orejas-Miranda, 1970, Hoge & Romano-Hoge, 1971, Cadle & Sarich, 1981, McCarthy, 1985, Sandner-Montilla, 1985, Roze & Bernal-Carlo, 1988, J.A. Campbell & Lamar, 1989, 2004, Grantsou, 1991, Slowinski, 1991, 1994, 1995, Savage & Slowinski, 1992, Scrocchi, 1992a, Golay et al., 1993, Ulber, 1995, Keogh, 1998, David & Ineich, 1999, N.J. Silva & Sites, 1999, 2001, Slowinski & Keogh, 2000, Lamar, 2003, Rage & Bailon, 2005, Castoe et al., 2007a, Zaher et al., 2009, Ivanov & Böhme, 2011, Hoser, 2012e, f and Marques et al., 2013.

Remarks: Official Generic Name no. 2144 *fide* Opinion 1201 (ICZN, 1982a). *Leptomicrurus* a valid genus *fide* J.A. Campbell & Lamar, 2004.

1. *Micrurus albicinctus* Amaral, 1925b. Comm. Lin. Telegr. Matto Grosso Amazonas (84): 26–27, pl., figs. 7–10.

Synonym: *Micrurus waehnerorum* Meise, 1938.

Type: Holotype, MNRJ 376 (formerly MNRJ 88), a 492 mm female (C.M. da Silva Rondon [Comiaaão Rondon], 1908–1914).

Type locality: "regiões central e septentrional de Matto Grosso, Brasil" [= cen. or N Mato Grosso state, cen. Brazil].

Distribution: Southeastern Colombia (Amazonas) and W Brazil (Acre, Amazonas, Mato Grosso, Rondônia), 100–250 m.

Sources: Cunha & Nascimento, 1991 and N.J. Silva, 1993.

Remarks: Probably occurs in Bolivia *fide* Campbell & Lamar (1989: 123).

2. *Micrurus alleni* K.P. Schmidt, 1936c. Field Mus. Nat. Hist. (Zool.) 20(20): 209–211, fig. 25. (*Micrurus nigrocinctus alleni*)

Synonyms: *Micrurus nigrocinctus yatesi* Dunn, 1942, *Micrurus alleni richardi* E.H. Taylor, 1951, and *Micrurus alleni vatesi* – Hoge & Romano-Hoge, 1971 *(nomen incorrectum)*.

Type: Holotype, UMMZ 79794 (formerly UMMZ 307), a 1068 mm female (M.J. Allen, 13 July 1935).

Type locality: "Rio Mico, seven miles above Rama, Siquia District, Nicaragua," (partially in error). Corrected to 16.1 km above Recero [*sic* = El Recreo, 12°10'N, 84°21'W, elevation 30 m], Río Mico, Siquía District, Atlántico Sur Dept., SE Nicaragua *fide* Kluge (1984: 65).

Distribution: Lower Central America. Eastern Honduras (Gracias a Dios), Nicaragua (Atlántico Norte, Atlántico Sur, Granada, Matagalpa, Río San Juan), Costa Rica (Alajuela, Cartago, Heredia, Limón, Puntarenas, San José) and Panama (Bocas del Toro, Chiriquí, Darién), NSL–1400 (2000) m.

Sources: E.H. Taylor, 1951, 1954, Savage & Vial, 1974, R.T. Taylor et al., 1974, N.J. Scott et al., 1983, Villa, 1984, Villa et al., 1988, McCranie, 1993b, G. Köhler, 1999b, Savage, 2002 and McCranie, 2011a.

Remarks: Cadle (2012: 234) listed type locality as 12°10'N, 84°19'W, elevation 50 m. Granada, Nicaragua record doubtful *fide* Villa (1984: 9). The allopatric Pacific population of *M. alleni* possibly a distinct species *fide* J.A. Campbell & Lamar (2004: 147–148).

3. *Micrurus altirostris* (Cope, 1860b). Proc. Acad. Nat. Sci. Philadelphia (1859) 11(12): 345. (*Elaps altirostris*)

Synonyms: *Elaps heterochilus* Mocquard, 1877b, and *Micrurus lemniscatus multicinctus* Amaral, 1944k.

Type: Holotype, ANSP 6857, a 674–749 mm male.

Type locality: "South America."

Distribution: Southeastern South America. Southern Brazil (Paraná, Rio Grande do Sul, Santa Catalina, ext. S São Paulo), Uruguay (Artigas, Canelones, Cerro Largo, Colonia, Durazno, Flores, Florida, Maldonado, Paysandú, Rivera, Rio Negro, Salto, San José, Soriano, Tacuarembó, Treinta y Tres), SE Paraguay (Alto Paraná, Itapuá, Misiones) and Argentina (Corrientes, Entre Ríos, Misiones), NSL–400 m.

Sources: Devincenzi, 1925, Abalos & Mischis, 1975, Achával, 1976, Achával et al., 1976, 1978, N.J. Silva & Sites, 1999, 2001, Giraudo & Scrocchi, 2002, Carreira-Vidal et al., 2005, Machado et al., 2005, Di-Bernardo et al., 2007 and Cacciali, 2008.

Remarks: Illustrations of the holotype in N.J. Silva & Sites (1999: fig. 8). Previously a synonym of *M. frontalis*.

4. *Micrurus ancoralis* (Jan, 1872 *in* Jan & Sordelli, 1870–1881). Icon. Gén. Ophid. 3(42): 5, pl. 4, figs. 2, 2a–b, d–e, g, n, p, r, v. (*Elaps marcgravii ancoralis*)

Synonyms: *Elaps marcgravi ancoralis* Jan, 1863b (*nomen nudum*), *Elaps rosenbergi* Boulenger, 1898h, and *Micrurus ancoralis jani* K.P. Schmidt, 1936b.

Type: Holotype, ZSM 210/0, a 780 mm male (F.J.C.M. Wagner, 1857–1860), destroyed 24–25 April 1944 during World War II.

Type locality: "Ecuadore" [= Ecuador].

Distribution: Panama and ext. NW South America. Eastern Panama (Darién), W Colombia (Antioquia, Cauca, Chocó, ? Meta, Nariño, Risaralda, Valle) and NW Ecuador (Esmeraldas, El Oro, Guayas, Pichincha), NSL–1500 (2300) m.

Sources: Boulenger, 1896a, K.P. Schmidt, 1936b, H.M. Smith, 1958, Miyata, 1980, Pérez-Santos, 1986 and Almendáriz, 1991.

5. *Micrurus annellatus* (W.C.H. Peters, 1871a). Mber. Königl. Akad. Wiss. Berlin 1871(8): 402. (*Elaps annellatus*)

Synonyms: *Elaps balzani* Boulenger, 1898b, *Elaps regularis* Boulenger, 1902d, *Micrurus annellatus montanus* K.P. Schmidt, 1954, and *Micrurus annellatus bolivianus* Roze, 1967.

Type: Holotype, ZMB 7185, a female (R. Abendroth, 1870–1871).

Type locality: "Pozuzu, in dem Hochlande vom Peru." [= Pozuzo, Pasco Dept., cen. Perú, 10°04'S, 75°33'W, elevation 745 m].

Distribution: Northwestern South America. Southern Ecuador, E Peru (Cusco, Huánuco, Junín, Loreto, Madre de Dios, Puno), W Brazil (Acre, Amazonas) and Bolivia (Beni, Chuquisaca, Cochabamba, La Paz, Pando), 200–2000 m.

M

Sources: K.P. Schmidt, 1954, Hoge & Romano-Hoge, 1969b, 1973, 1981b, Duellman & Salas, 1991, Harvey et al., 2003 and Duellman, 2005.

Remarks: Colombian record doubtful *fide* J.A. Campbell & Lamar (2004: 149).

6. *Micrurus averyi* K.P. Schmidt, 1939a. Field Mus. Nat. Hist. (Zool.) 24(6): 45–47, fig. 5.

Type: Holotype, FMNH 30956, a 670 mm female (E.R. Blake [Sewell Avery Exped.], 22 Sept. 1938).

Type locality: "Boundary Commission's 'Boundary Camp', at about 2,000 feet above sea level, at head of Itabu Creek. This creek enters the New River, an affluent of the Courantyne, at New River Depot, Courantyne District, near the Brazilian border, at Latitude 1°40' N. and Longitude 58° W., British Guiana" [= ext. S East Berbice-Corentyne Region, ext. S Guyana, 1°34'N, 58°17'W, elevation 600 m].

Distribution: Northern South America. Southern Guyana (S East Berbice-Corentyne), S Suriname (Nickerie) and N Brazil (NE Amazonas, Mato Grosso, Pará), 100–700 m.

Sources: Vanzolini, 1985, Velloso-Calleffo, 1997, Passos, 2003 and C.J. Cole et al., 2013.

Remarks: Occurence in Suriname *fide* Brongersma (1967: 73) due to type locality being claimed by The Netherlands at that time *fide* Hoogmoed *in* Vanzolini (1985: 77).

7. *Micrurus baliocoryphus* (Cope, 1860b). Proc. Acad. Nat. Sci. Philadelphia (1859) 11(12): 346. (*Elaps baliocoryphus*)

Synonym: *Micrurus frontalis mesopotamicus* Barrio & Miranda-Ribeiro, 1967.

Type: Holotype, ANSP 6842, a 457 mm male (Kennedy).

Type locality: "Buenos Ayres" [= Buenos Aires, Argentina]. Corrected and restricted to Villa Federal, Entre Ríos, Argentina *fide* Roze (1983: 324).

Distribution: Southern South America. Southern Brazil (Mato Grosso do Sul, Paraná), Paraguay (Amambay, Boquerón, Caaguazú, Central, Neembucú, Presidente Hayes) and N Argentina (Chaco, Corrientes, Entre Ríos, Formosa, S Misiones), NSL–200 m.

Sources: Abalos & Mischis, 1975, N.J. Silva & Sites, 1999, 2001, Di-Bernardo et al., 2007 and Cacciali, 2008.

Remarks: Illustrations of the holotype in N.J. Silva & Sites (1999: fig. 15). Previously a subspecies of *M. frontalis*.

8. *Micrurus bernadi* (Cope, 1887c). Bull. U.S. Natl. Mus. (32): 87. (*Elaps bernadi*)

Synonym: *Elaps bernadi* Cope *in* Ferrari-Pérez, 1886 (*nomen nudum*).

Type: Holotype, ANSP 14767, an 807 mm female (S. Bernad, 1885–1887).

Type locality: "Zacualtipan, Hidalgo, Mexico" [= Zacualtipan, NE Hidalgo State, E Mexico, 20°39'N, 98°39'W, elevation 1990 m].

Distribution: Mexico (E Hidalgo, N Puebla, N Veracruz), (50) 1400–2000 m.

Sources: K.P. Schmidt, 1933c, 1958, Roze, 1983, 1996 and J.A. Campbell, 2000.

9. *Micrurus bocourti* (Jan, 1872 *in* Jan & Sordelli, 1870–1881). Icon. Gén. Ophid. 3(42): 5, pl. 6, figs. 2, 2a–b, d, g, n, p, r, v, y. (*Elaps bocourti*)

Synonym: *Micrurus ecuadorianus* K.P. Schmidt, 1936b.

Type: Holotype, MNHN 869, a 283 mm male (E.T. Liautaud [Danaïde Exped.], 1843).

Type locality: Unknown. "Indes Occidentales" [= West Indies] *fide* MNHN catalogue (in error). Corrected and restricted to Río Daule, Guayas Province, SW Ecuador *fide* Roze (1967: 8).

Distribution: Western Ecuador (Guayas, Loja, Los Rios, Pichincha, Zamore-Chinchipe) and ext. NW Peru (? Tumbes), NSL–1455 m.

Source: Roze, 1989.

Remarks: MNHN 869 is a syntype of *Elaps circinalis*.

10. *Micrurus bogerti* Roze, 1967. Amer. Mus. Novit. (2287): 9–11, fig. 3.

Type: Holotype, AMNH 96952, a 344 mm male (W. Beebe, 1937).

Type locality: "Tangola-Tangola (Tangolunda), east of Puerto Angel, Oaxaca, Mexico" [= Tangolunda, S Oaxaca, S Mexico, 15°46'N, 96°06'W, elevation 10 m].
Distribution: Southeastern Mexico (ext. SE Oaxaca, probably ext. SW Chiapas), NSL–400 m.
Sources: Villa et al., 1988 and J.D. Johnson, 1989.

11. *Micrurus brasiliensis* Roze, 1967. Amer. Mus. Novit. (2287): 25–26, fig. 9. (*Micrurus frontalis brasiliensis*)

Type: Holotype, UMMZ 108880, a 865–885 mm male (J.R. Bailey, 1942).
Type locality: "Barreiras, Bahia, Brazil" [= Barreiras, W Bahia State, Brazil, 12°09'S, 45°00'W, elevation 575 m].
Distribution: Central Brazil (Bahia, Goiás, Minas Gerais, Tocantins), NSL–600 m.
Sources: Cordeiro & Hoge, 1974, Freitas, 1999 and N.J. Silva & Sites, 1999, 2001.
Remarks: Illustrations of the holotype in N.J. Silva & Sites (1999: fig. 20).

12. *Micrurus browni* K.P. Schmidt & Smith, 1943. Field Mus. Nat. Hist. (Zool.) 29(2): 29–30.

Synonyms: *Micrurus nuchalis taylori* K.P. Schmidt & Smith, 1943 and *Micrurus browni importunus* Roze, 1967.
Type: Holotype, FMNH 38494, a 689 mm male (W.W. Brown, 1932–1943).
Type locality: "Chilpancingo, Guerrero, Mexico" [= Chilpancingo de los Bravos, cen. Guerrero State, SW Mexico, 17°33'N, 99°30'W, elevation 1270 m].
Distribution: Southern Mexico (Sierra Madre del Sur of Chiapas, Guerrero, México, Michoacán, Oaxaca, S Puebla), W Guatemala (Sacatepéquez, Santa Rosa) and Honduras (Itibucá, La Paz, Ocotepeque), NSL–2000 m.
Sources: Blaney & Blaney, 1978, Casas-Andreu & López-Forment, 1978, Camarillo & Smith, 1992, J.A. Campbell, 2000, Suárez-Velázquez & Luna-Reyes, 2009, McCranie, 2011a and Reyes-Velasco et al., 2012.
Remarks: J.A. Campbell & Lamar (2004: 154) list erroneous records of this species.

13. *Micrurus camilae* Renjifo & Lundberg, 2003. Rev. Acad. Colomb. Cien. Exact. Fis. Nat. 27(102): 142–144, fig. la.

Type: Holotype, ICNMNH 8506, a 265 mm male (M. Lundberg, 9 Nov. 1999).
Type locality: "Columbia, departamento de Córdoba, municipio de Tierra Alta, campamento de Funcionarios, represa Hidroeléctrica de Urrá (8°1'3.4"N, 76°11'8.1"W) a 90 metros sobre el nivel del mar."
Distribution: Northwestern Colombia (Córdoba), 90 m. Known only from type locality.

14. *Micrurus carvalhoi* Roze, 1967. Amer. Mus. Novit. (2287): 33–34, fig. 11. (*Micrurus lemniscatus carvalhoi*)

Type; Holotype, USNM 76341, a 923 mm male.
Type locality: "Catanduva, São Paulo, Brazil" [= Catanduva, N São Paulo State, SE Brazil, 21°08'S, 48°58'W, elevation 500 m].
Distribution: Southeastwern Brazil (Bahia, Distrito Federal, Goias, Mato Grosso, Mato Grosso do Sur, Minas Gerais, Paraná, Pernambuco, Rio de Janeiro, Rio Grande do Norte, São Paulo) and N Argentina (N Corrientes, Misiones), 500 m.
Sources: N.J. Silva & Sites, 2001 and Giraudo & Scrocchi, 2002.
Remarks: Probably occurs in Paraguay *fide* Roze (1996: 190).

15. *Micrurus catamayensis* Roze, 1989. Amer. Mus. Novit. (2932): 3–4, fig. 1

Type: Holotype, BMNH 1935.11.3.103, a 495 mm male (C. Carrión-Mora, 1935).
Type locality: "2 km west of Loja, Catamayo Valley, Ecuador, 1568 m" [= E Loja Prov., S Ecuador, 3°59'S, 79°13'W].
Distribution: Ecuador (Loja), 1000–1800 [2360] m.

M

16. *Micrurus circinalis* (A.-M.-C. Duméril, Bibron & Duméril, 1854b). Erpet. Gén. 7(2): 1210–1211. (*Elaps circinalis*)

Synonym: *Elaps riisei* Jan, 1858b & 1859e.
Type: Lectotype, MNHN 3912, a 280 mm female (Mus. Geoffroy), designated by Roze (1989: 14).
Type locality: Unknown. Designated as Martinique (in error) *fide* Roze (1955: 472). Restricted to Trinidad, Trinidad & Tobago, West Indies *fide* Roze (1989: 14).
Distribution: Northeastern Venezuela (Anzoátegui, Monagas, Sucre) and Trinidad and Tabago (Trinidad), NSL–400 m.
Sources: Roze, 1989, 1994, J.C. Murphy, 1997, Mijares-Urrutia & Arends, 2000 and Navarrete et al., 2009.
Remarks: Possibly a synonym of *M. psyches*.

17. *Micrurus clarki* K.P. Schmidt, 1936c. Field Mus. Nat. Hist. (Zool.) 20(20): 211–212.

Type: Holotype, MCZ 38390, a 267–274 mm male (H.C. Clark, 1934).
Type locality: "Yavisa, Darien, Panama" [= Yaviza, cen. Darién Prov., ext. E Panama, 8°09'N, 77°42'W, elevation 15 m].
Distribution: Lower Central America and Colombia. Southeastern Costa Rica (Puntarenas), SW Panama (Chiriquí, Coclé, Colón, Darién. San Blas) and NW Colombia (Chocó), NSL–900 m.
Sources: E.H. Taylor, 1951, H.M. Smith, 1958, Savage & Vial, 1974, Solórzano & Cerdas, 1985, Pérez-Santos,

1986, Savage & Villa, 1986, Villa et al., 1988, Ryan, 2002 and Savage, 2002.

18. *Micrurus collaris* (Schlegel, 1837). Essai Phys. Serp. 1: 181, 2: 448–449. (*Elaps collaris*)

Synonyms: *Elaps collaris* H. Boie *in* Schlegel, 1826a *(nomen nudum), Elaps collaris* Schlegel, 1826b *(nomen nudum), Elaps collaris* Gray, 1831 *(nomen nudum), Elaps gastrodelus* A.-M.-C. Duméril, 1853 *(nomen nudum), Elaps gastrodelus* A.-M.-C. Duméril, Bibron & Duméril, 1854b, and *Leptomicrurus collaris breviventris* Roze & Bernal-Carlo, 1988.

Type: Lectotype, RMNH 1444, a 404 mm male (S.J. Brugmans, 1815–1819), designated by Brongersma (1966: 250).

Type locality: Unknown. Designated as the Guianas *fide* Hoge & Romano-Hoge (1966: 4).

Distribution: Northern South America. Southeastern Venezuela (Bolívar), Guyana (Cuyuni-Mazaruni, East Berbice-Corentyne, Potaro-Siparuni), Suriname (Brokopondo, Marswijne), French Guiana (N Cayenne) and ext. NE Brazil (Pará), NSL–700 m.

Sources: J.C. Thompson, 1913a, K.P. Schmidt, 1937, Brongersma, 1966, Roze, 1958e, 1966a, 1967, Lancini, 1979, Abuys, 1987a, Chippaux, 1987, Starace, 1998, Navarrete et al., 2009 and C.J. Cole et al., 2013.

Remarks: Original description based on H. Boie's MS (1823–1825).

M

19. *Micrurus corallinus* (Merrem, 1820). Tent. Syst. Amph.: 144–145. (*Elaps corallinus*)

Synonyms: *Elaps corallinus* Wied-Neuwied, 1820b *(nomen praeoccupatum)* and *Coluber corallinus* Raddi, 1820 *(nomen praeoccupatrum).*

Type: Lectotype, MNHN 3911, a 279 mm female (A.P.M. zu Wied-Neuwied, Sept. 1815–May 1817), designated by Roze (1966b: 370).

Type locality: "Rio de Janeiro, Cabo Frio, Campos de Coaytacases, Parahyba, Brazil" [= all in vicinity of Rio de Janeiro (22°54'S, 43°13'W, elevation 15 m), Rio de Janeiro State, SE Brazil] via lectotype selection.

Distribution: Southeastern South America. Southeastern Brazil (Bahia, Espírito Santo, ? Goiás, Mato Grosso do Sul, Minas Gerais, Paraíba, Paraná, Rio de Janeiro, Rio Grande do Norte, Rio Grande do Sul, Santa Catarina, São Paulo, Alcatrazes, Anchieta, Barnabé, Búzios, Cananéia, Cardoso, Comprida, Grande, Porchat, Santo Amaro, São Sebatião and São Vicente Is.), E Paraguay (Alto Paraná, Itapúa), ext. NE Argentina (Misiones) and ext. NE Uruguay, NSL–600 m.

Sources: Wied-Neuwied, 1820b, 1825c, Bertoni, 1915, Serie, 1915b, 1921, 1936, Amaral, 1925a, 1926?, Vaz-Ferreira & Sierra de Soriano, 1960, Azevedo, 1963, Roze, 1966b, Lema & Fabián-Beurmann, 1977, Lema, 1987, J.D. Williams & Francini, 1991, Cei, 1994, Casais e Silva Nunes, 1996, Marques, 1996a, Freitas, 1999, Freire, 2001, N.J. Silva & Sites, 2001, Giraudo & Scrocchi, 2002, Duarte-Rocha & Van Sluys, 2006 and Cacciali, 2008.

Remarks: Tucumán, Argentina record needs confirmation.

20. *Micrurus decoratus* (Jan, 1858b). Rev. Mag. Zool. (2) 10: 518, 525–526. (*Elaps decoratus*)

Synonyms: *Elaps decoratus* A.C.L.G. Günther, 1859 *(nomen praeoccupatum), Elaps fischeri* Amaral, 1921a, and *Elaps ezequieli* Lutz & Mello, 1922 & 1923.

Types: Syntypes (3), formerly MSNM, a 450 mm specimen, destroyed in 1943 during World War II, MZUT, a 420 mm specimen, lost *fide* Elter (1981: 67), and ZFMK, a 390 mm specimen, lost *fide* W. Böhme (2010: 106).

Type locality: "Mexique" [= Mexico] (in error *fide* Roze 1967: 14). Corrected and restricted to Serra da Bocaina, São Paulo, Brazil, [ca. 23°00'S, 46°33'W] *fide* Hoge & Romano (1973: 122).

Distribution: Southeastern Brazil (ext. S Minas Gerais, Paraná, Rio de Janeiro, NE Rio Grande do Sul, Santa Catarina, São Paulo), NSL–1500 m.

Sources: Jan, 1872 *in* Jan & Sordelli, 1870–1881, Amaral, 1922a, 1977, Barth, 1957, Lema & Azevedo, 1969, Lema, 1987, N.J. Silva & Sites, 2001, Marques, 2002 and Marques et al., 2002.

Remarks: Supplemental original description in Jan (1859e: pl. b, fig. no. 5). Jan (1858: 526) described three syntypes with the Milan specimen first, but did not designate a holotype, and Jan & Sordelli [1872 *in* 1870–1881, 3(livr. 42): pl. 6, fig. 4] illustrated the type from MSNM. A.C.L.G. Günther (1859: 85) inadvertently designated a lectotype by describing Jan's Milan specimen and Roze's (1996: 151) designation of the Milan specimen as the holotype is a lectotype designation, and that specimen is lost.

21. *Micrurus diana* Roze, 1983. Mem. Inst. Butantan (1982) 46: 324–325. (*Micrurus frontalis diana*)

Type: Holotype, FMNH 159889, a 998–1009 mm male (R.F. Steinbach, 7–20 April 1973).

Type locality: "vicinity of Santiago, Provincia Chiquiticos, Departamento Santa Cruz, Bolivia, 700 meters" [= Serranía de Santiago, Santa Cruz Dept., Bolivia (18°16'S, 60°00'W, elevation 400 m) *fide* N.J. Silva & Sites (199: 170).

Distribution: Eastern Bolivia (Santa Cruz), 400–700 m.

Sources: Roze, 1994, N.J. Silva & Sites, 1999 and Harvey et al., 2003.

Remarks: Illustrations of the holotype in N.J. Silva & Sites (1999: fig. 24).

22. *Micrurus diastema* (A.-M.-C. Duméril, Bibron & Duméril, 1854b). Erpét. Gén. 7(2): 1222. (*Elaps diastema*)

Synonyms: *Elaps epistema* A.-M.-C. Duméril, Bibron & Duméril, 1854b, *Elaps affinis* Jan, 1858b & 1859e, *Elaps apiatus* Jan, 1858b & 1859, *Elaps aglaeope* Cope, 1860b, *Elaps corallinus crebripunctatus* W.C.H. Peters, 1870a, *Elaps alienus* F. Werner, 1903b, *Elaps fulvius sapperi* F. Werner, 1903a, *Elaps guatemalensis* Ahl, 1927, *Micrurus affinis mayensis* K.P. Schmidt, 1933c, *Micrurus affinis stantoni* K.P. Schmidt, 1933c, and *Micrurus diastema macdougalli* Roze, 1967.

Type: Lectotype, MNHN 7657, a 607 mm female (Ducommun, 1838), designated by Roux-Estève (1983: 82).

Type locality: "Mexique" [= Mexico]. Restricted to Potrero Viejo, Veracruz, Mexico [18°53'N, 96°51'W, elevation 620 m] *fide* Roze (1983: 318).

Distribution: Mesoamerica. Southeastern Mexico (Campeche, N Chiapas, Hidalgo, Oaxaca, Puebla, Quintana Roo, Tabasco, Veracruz, Yucatán), Belize (Belize, Cayo, Corozal, Orange Walk, Stann Creek), N Guatemala (Alta Verapaz, Huehuetenango, Izabal, Petén) and NW Honduras (Comayagua, Copán, Cortés, Santa Bárbara), NSL–1680 m.

Sources: K.P. Schmidt, 1933c, L.C. Stuart, 1948, Zweifel, 1959a, Fraser, 1970, 1973, R.W. Henderson & Hoevers, 1975, Blaney & Blaney, 1979a, J.C. Lee, 1980, 1996, 2000, Roux-Estève, 1983, L.D. Wilson & Meyer, 1985, Villa et al., 1988, Pérez-Higareda & Smith, 1990, 1991, Benitez-Gálvez, 1997, J.A. Campbell, 1998b, 2000, Stafford, 2000a and McCranie, 2011a.

Remarks: Type locality restriction of Colima, Mexico *fide* K.P. Schmidt (1933c: 39) invalid as Roux-Estève (1983: 82) has shown that MNHN 4620 (now = MNHN 7656) was not a syntype, having been collected in 1859.

23. *Micrurus dissoleucus* (Cope, 1860b). Proc. Acad. Nat. Sci. Philadelphia (1859) 11(12): 345. (*Elaps dissoleucus*)

Synonyms: *Elaps gravenhorstii* Jan, 1858b & 1859e (*nomen oblitum*), *Elaps melanogenys* Cope, 1860c, *Elaps hollandi* L.E. Griffin, 1916, *Micrurus dunni* T. Barbour, 1923, and *Micrurus dissoleucus nigrirostris* K.P. Schmidt, 1955a.

Type: Holotype, ANSP 6781, a 571 mm female (C.D. Meigs, 1850–1859).

Type locality: "Venezuela." Restricted to Maracaibo, Zulia, Venezuela [= 10°39'N, 71°38'W, elevation 25 m] *fide* Roze (1955: 479).

Distribution: Eastern Panama (Canal Zone, Coclé), N Colombia (Antioquia, Atlántico, Bolívar, Cesar, ? Chocó, La Guajira, Magdalena, Norte de Santander, Valle) and N Venezuela (Apure, Aragua, Barinas, Bolívar, Carabobo, Cojedes, Delta Amacuro, Distrito Federal, Falcón, Guárico, Mérida, Miranda, Portuguesa, Sucre, Trujillo, Zulia), NSL–550 (1000) m.

Sources: Amaral, 1931b, K.P. Schmidt, 1936b, Evans, 1947, Roze, 1955, 1966a, H.M. Smith, 1958, Pérez-Santos, 1986, Villa et al., 1988, Gorzula & Señaris, 1998 and Navarrete et al., 2009.

Remarks: In accordance with Art. 23.9.2 of the Code (ICZN, 1999), *Elaps dissoleucus* Cope is designated a *nomen protectum* and *Elaps marcgravi gravenhorstii* Jan a *nomen oblitum.*

24. *Micrurus distans* (Kennicott, 1860). Proc. Acad. Nat. Sci. Philadelphia 12(8): 338. (*Elaps distans*)

Synonyms: *Elaps fulvius hypostema* Jan, 1863b (*nomen nudum*), *Elaps diastema michoacanensis* Dugès, 1891, *Micrurus distans oliveri* Roze, 1967, and *Micrurus distans zweifeli* Roze, 1967.

Type: Holotype, USNM 1144, a 537 mm male (J. Potts).

Type locality: "Batosegachie, Chihuahua" [= Batosegachi, Guazapares, SW Chihuahua, NW Mexico, 27°24'N, 108°19'W, elevation 1430 m].

Distribution: Western Mexico (Aguascalientes, ext. SW Chihuahua, Colima, NW Guerrero, Jalisco, S Michoacán, Nayarit, Sinaloa, S Sonora, Zacatecas), NSL–1525 m.

Sources: K.P. Schmidt & Smith, 1943, Zweifel, 1959b, J.A. Campbell, 2000, Quintero-Diaz et al., 2000, Ernst & Ernst, 2011 and Ahumada-Carrillo & Vázquez-Huizar, 2012a.

Remarks: Collector listed as J. Potts *fide* USNM catalogue and original description but listed as A. Schott on data tags with specimen *fide* Cochran (1961: 177): one data tag lists Chihuahua, another tag Batosegachie, Chihuahua.

25. *Micrurus dumerilii* (Jan, 1858b). Rev. Mag. Zool. (2) 10: 517, 522–523. (*Elaps dumerilii*)

Synonyms: *Elaps marcgravii* A.-M.-C. Duméril, Bibron & Duméril, 1854b (*nomen praeoccupatum*), *Elaps colombianus* L.E. Griffin, 1916, *Micrurus antioquiensis* K.P. Schmidt, 1936b, *Micrurus carinicauda* K.P. Schmidt, 1936b, *Micrurus transandinus* K.P. Schmidt, 1936b, *Micrurus dumerilii venezuelensis* Roze, 1989, and *Micrurus c. dumarilii* – Khole, 1991 (*nomen incorrectum*),

Type: Holotype, MNHN 3923, a 660 mm male (Barot).

Type locality: "Colombie" [= Colombia]. Emended to Carthagène, Colombie [= Cartagena, ext. N Bolívar Dept., Colombia, 10°25'N, 75°32'W, elevation 10 m] *fide* Jan (1859c: 509).

Distribution: Panama and NW South America. Eastern Panama (Darién), W Colombia (Antioquia, Atlántico, Bolívar, Boyaca, Cauca, Chocó, Cundinamarca, Magdalena, Nariño, Norte de Santander, Risaralda, Santander, Valle, Tolima), Venezuela (Apure, Aragua, Barinas, Carabobo, Distrito Federal, Falcón, Lara, Mérida, Miranda, Táchira, Trujillo, Yaracuy, Zulia) and NW Ecuador (Esmeraldas, Pichincha), NSL– 2135 m.

Sources: Jan & Sordelli, 1872 *in* 1870–1881, Roze, 1955, 1966a, 1989, Miyata, 1980, Roux-Estève, 1983, Pérez-Santos, 1986, Barrio-Amorgós, 2001 and Navarrete et al., 2009.

Remarks: Supplemental original description in Jan (1859e: pl. a, fig. no. 15).

26. *Micrurus elegans* (Jan, 1858b). Rev. Mag. Zool. (2)10: 517, 524. (*Elaps elegans*)

Synonyms: *Micrurus elegans verae-pacis* K.P. Schmidt, 1933c *(nomen incorrigendum)* and *Micrurus elegans veraepacis* – L.C. Stuart, 1950 *(nomen corrigendum)*.

Type: Lectotype, formerly MSNM, a 470 mm female described and illustrated by Jan (1858: 524 & 1859a: pl. B, no. 1) and Jan & Sordelli (1872 *in* 1870–1881, 3(42): 5, pl. 5, fig. 2), designated by K.P. Schmidt (1958: 208), destroyed in 1943 during World War II.

Type locality: "Mexique" [= Mexico]. Restricted to Jalapa, Veracruz, Mexico *fide* H.M. Smith & Taylor (1950a: 348).

Distribution: Southeastern Mexico (N Chiapas, NE Oaxaca, S Tabasco, S Veracruz) and cen. Guatemala (Alta Verapaz, Huehuetenango), 800–1830 m.

Sources: L.C. Stuart, 1948, 1963, H.M. Smith & Smith, 1976, Alvarez del Toro, 1982, Villa et al., 1988 and Pérez-Higareda & Smith, 1991.

M

Remarks: Supplemental original description in Jan (1859e: pl. b, fig. no. 1). All 3 syntypes (MZUT, a 495 mm male, ZFMK, a 490 mm female, and MSNM, a 470 mm female) are lost or were destroyed in 1943 during World War II *fide* Elter (1981: 67), Golay et al. (1996: 165) and W. Böhme (2010: 106). Lectotype illustrations reproduced in K.P. Schmidt (1958: fig. 34), Roze (1996: fig. 56) and J.A. Campbell & Lamar (2004: fig. 54).

27. *Micrurus ephippifer* (Cope, 1886b). Proc. Amer. Philos. Soc. (1885) 23(122): 281. (*Elaps ephippifer*)

Synonym: *Micrurus ephippifer zapotecus* Roze, 1989.

Type: Holotype, USNM 3085, a female (A.-L.-J.-F. Sumichrast, 1855–1869).

Type locality: "Pacific side of the Isthmus of Tehuantepec" [= SE Oaxaca State, S Mexico].

Distribution: Southern Mexico (E Oaxaca), NSL–2300 m.

Sources: Roze, 1989, Camarillo, 1995 and J.A. Campbell, 2000.

28. *Micrurus filiformis* (A.C.L.G. Günther, 1859b). Proc. Zool. Soc. London 27(1): 86–87, pl. 18, figs. b–b′. (*Elaps filiformis*)

Synonym: *Micrurus filiformis subtilis* Roze, 1967.

Type: Holotype, BMNH 1946.1.20.13, a 390–432 mm male (H.H. Higgins, 1856–1859).

Type locality: "Para, Tropical America" [=Pará State, N Brazil]. Probably vicinity of Belém [= ca. 1°27'S, 48°30'W] *fide* Cunha & Nascimento (1973: 282).

Distribution: Amazonian South America. Southeastern and E Colombia (Amazonas, Caqueta, Casanare, Cundinamarca, Meta, Putumayo, Vaupés), E Ecuador, NE Peru (Loreto) and N Brazil (Amazonas, Maranhão, Pará), NSL–500 m.

Sources: Hoge & Romano-Hoge, 1971, Cunha & Nascimento, 1973, 1978, 1993, Dixon & Soini, 1986, Pérez-Santos, 1986, Alméndariz, 1991, Henle & Ehrl, 1991, Silva Haad, 1994 and Feitosa et al., 2007.

Remarks: Holotype redescribed by Feitosa et al. (2007: 152). J.A. Campbell & Lamar (2004: 215) indicated probably in S Venezuela and also S Guyana and French Guiana.

29. *Micrurus frontalis* (A.-M.-C. Duméril, Bibron & Duméril, 1854b). Erpét. Gén. 7(2): 1223–1224. (*Elaps frontalis*)

Type: Lectotype, MNHN 854, a 501–529 mm male (P. Claussen, 1844), designated by Roze (1996: 172).

Type locality: "Brazil" via lectotype selection. Restricted to Lagoa Santa, Minas Gerais, Brazil *fide* Vanzolini *in* N.J. Silva & Sites (1999: 149).

Distribution: Southeastern Brazil (Distrito Federal, Espírito Santo, Goiás, Mato Grosso, Mato Grosso do Sul, Minas Gerais, Rio de Janeiro, São Paulo, Bananal Is.), Bolivia (Santa Cruz) and Paraguay (Amambay, Central, Concepción, Cordillera, Paraguarí, San Pedro), NSL–700 m.

Sources: Serié, 1915b, 1936, Amaral, 1944k, Shreve, 1953, Vaz-Ferreira & Sierra de Soriano, 1960, Abalos & Nader, 1962, Abalos et al., 1964, Lema & Fabián-Beurmann, 1977, Hoge & Romano-Hoge, 1981b, Roux-Estève, 1983, Fugler & Riva, 1990, Scrocchi, 1990b, Cei, 1994, Lema, 1994, D.R. Norman, 1994, Roze, 1994, N.J. Silva & Sites, 1999, 2001, Di-Bernardo et al., 2007, Cacciali, 2008 and Jansen, 2008.

Remarks: Illustrations of the lectotype in N.J. Silva & Sites (1999: fig. 3). Type locality restriction of Corrientes and Misiones, Argentina *fide* J.A. Peters & Orejas-Miranda (1970: 208) invalid. Shreve (1953: 4) suggested, but did not designate, Lagoa Santa, Minas Gerais, Brazil as the type locality.

30. *Micrurus frontifasciatus* (F. Werner, 1927). Sitz. Akad. Wiss. Wien, Math.-Natur. Kl. (1926) 135(7–8): 250–251. (*Elaps frontifasciatus*)

Type: Holotype, NMW 18298, a 1040 mm male (O. Staudinger, 28 May 1898).

Type locality: "Bolivia."

Distribution: Southeastern Peru (Madre de Dios) N Bolivia (Beni, Chuquisaca, Cochabamba, La Paz, Pando, Santa Cruz) and N Brazil (Pará), NSL–2500 m.

Sources: Fugler & Riva, 1990 and Guimarâes-Guedes, 2000.

31. *Micrurus fulvius* (Linnaeus, 1766). Syst. Nat., ed. 12, 1: 381. (*Coluber fulvius*)

Synonyms: *Coluber nigrorufus* Lacépède, 1789 *(nomen rejiciendum)*, *Vipera fulvia* – Harlan, 1827, *Elaps flavius* – Gray, *in* Griffith & Pidgeon, 1831 *(nomen incorrectum)*, *Micrurus fulvius barbouri* K.P. Schmidt, 1928c, *Micrulus fulvius* – Boquet, 1964 *(nomen incorrectum)*, *Mucrurus fulvins* – Garza, 1984 *(nomen incorrectum)*, *Micrurus fluvius* – Dart, O'Brien & McNally, 1990 *(nomen incorrectum)*, and *Micrurus fulvis* – Savage & Slowinski, 1990 *(nomen incorrectum)*.

Type: Holotype, a female (A. Garden, 1760–1761), lost *fide* Roze & Tilger (1983: 316.1).

Type locality: "Carolina" [USA]. Restricted to vicinity of Charleston, South Carolina, USA *fide* Klauber (1948: 6).

Distribution: Southeastern USA (S Alabama, Florida, S Georgia, ext. SE Louisiana, S Mississippi, SE North Carolina, E South Carolina), NSL–400 m.

Fossil records: Lower Pleistocene (Irvingtonian I) of USA (Florida), and upper Pleistocene (Rancholabrean II) of USA (Florida).

Sources: A.H. Wright & Wright, 1957, Graham, 1977, Jackson & Franz, 1981, Roze & Tilger, 1983, Ashton & Ashton, 1988, Ernst & Barbour, 1989, Conant & Collins, 1991, Palmer & Braswell, 1995, Tennant, 1997, Holman, 2000a and Ernst & Ernst, 2011.

32. †*Micrurus gallicus* Rage & Holman, 1984. Géobios 17(1): 97–98, figs. 6a–e.

Type: Holotype, UCBL 285013, one trunk vertebra.

Type locality: "La-Grive-Saint-Alban (Isère), France; zone MN 7, Astaracien, Miocène français."

Distribution: Lower Miocene (Orleanian, MN 4–5: 13.7–16.9 mya) of France and Germany, and middle Miocene (Astaracian, MN 6: 12.8–13.7 mya) of France.

Sources: Szyndlar & Schleich, 1993, Augé & Rage, 2000, Ivanov, 2000, Rage & Bailon, 2005 and Ivanov & Böhme, 2011.

33. *Micrurus hemprichii* (Jan, 1858b). Rev. Mag. Zool. (2)10: 517, 523. (*Elaps hemprichii*)

Synonyms: *Micrurus hemprichi ortoni* K.P. Schmidt, 1953b and *Micrurus rondonianus* Roze & Silva, 1990.

Types: Syntypes (2), NMW, a 350 mm specimen, and formerly MSNM, a 450 mm specimen, destroyed in 1943 during World War II.

Type locality: "Colombie" [= Colombia].

Distribution: Northern South America. Southeastern Colombia (Amazonas, Caquetá), S Venezuela (Amazonas, Bolívar), Guyana (Barima-Waini, Cuyuni-Mazaruni, East Berbice-Corentyne, Potaro-Siparuni), N Suriname (Brokopondo, Marowijne, Nickarie, Para, Suriname), N French Guiana (Cayenne, Saint-Laurent-du-Maroni), E Ecuador (Morona Santiago, Napo), E Peru (Amazonas, Loreto, Madre de Dios), NW Brazil (N Amazonas, Maranhao, Pará, Rondônia, Marajo Is.) and Bolivia (Beni, Cochabamba, La Paz), NSL–1200 m.

Sources: K.P. Schmidt, 1953b, J.A. Peters, 1960a, Roze, 1966a, 1994, Hoge & Romano-Hoge, 1972, Cunha & Nascimento, 1978, 1993, Miyata, 1982, Dixon & Soini, 1986, Pérez-Santos, 1986, Abuys, 1987a, Pérez-Santos & Moreno, 1988, 1991, N. Jorge de Silva, 1993, Silva Haad, 1994, Gorzula & Señaris, 1998, Starace, 1998, N.J. Silva & Sites, 2001, Harvey et al., 2003 and Navarrete et al., 2009.

Remarks: Supplemental original description in Jan (1859e: pl. a, fig. no. 7). Type locality restrictions to vicinity of Bartica, Guyana *fide* K.P. Schmidt (1953b: 166) and Venezuela *fide* Hoge & Lancini (1962: 12) invalid. Roze (1996: 181) and J.A. Campbell & Lamar (2004: 217) listed the MSNM specimen as holotype.

34. *Micrurus hippocrepis* (W.C.H. Peters, 1861e). Mber. Königl. Akad. Wiss. Berlin 1861(10): 925. (*Elaps hippocrepis*)

Type: Holotype, ZMB 4065, a 680 mm female (J. Sivers & Schaum), lost *fide* R. Günther *(in litt.)*.

Type locality: "Santo Tomas de Guatemala" [= Puerto Matías de Galvez, Guatemala *fide* L.C. Stuart (1963: 126) or Puerto Matías de Castilla, E Izabal Dept., E Guatemala, 15°42'N, 88°37'W, elevation NSL].

Distribution: Central Belize (Cayo, Stann Creek, Toledo) and NE Guatemala (Izabal), NSL–1000 m.

Sources: R.W. Henderson & Hoevers, 1975, Villa et al., 1988, Lee, 1996, J.A. Campbell, 1998b, Stafford, 2000a and Stafford & Meyer, 2000.

Remarks: Possibly a color morph of *M. diastema fide* J.A. Campbell & Lamar (2004:172).

35. *Micrurus ibiboboca* (Merrem, 1820). Tent. Syst. Amph.: 142–143. (*Elaps ibiboboca*)

Synonyms: *Elaps marcgravii* Wied-Neuwied, 1820b, *Elaps ibibiboca* – Boulenger, 1896a *(nomen incorrectum)*, and *Micrurus ibiboca* – Klemmer, 1963 *(nomen incorrectum)*.

Type: Holotype, AMNH 3937, a 318 mm male (A.P.M. zu Wied-Neuwied, 25 July–17 Aug. 1816).

Type locality: "Brasilia" [= Brazil]. Emended to mouth of Río Belmonte, Bahia Prov., Brazil *fide* Wied-Neuwied (1825c: 420). Restricted to E Brazil, south of the Amazon *fide* MacCulloch & Lathrop (2004a: 143).

Distribution: Eastern Brazil (Alagoas, Bahia, Ceará, Espírito Santo, Maranhão, Paraíba, Pernambuco, Piauí, Rio de Janeiro, Rio Grande do Norte, Sergipe), NSL–1200 m.

Sources: Roze, 1966b, Vanzolini et al., 1980, Freitas, 1999 and N.J. Silva & Sites, 2001.

Remarks: Guyana, Suriname and French Guiana populations are *M. lemniscatus fide* J.A. Campbell & Lamar (1989: 121) and MacCulloch & Lathrop (2004a: 143).

36. *Micrurus isozonus* (Cope, 1860c). Proc. Acad. Nat. Sci. Philadelphia 12(2): 73. (*Elaps isozonus*)

Synonym: *Elaps omissus* Boulenger, 1920b.

Types: Syntypes (2), ANSP 6804–05, a male and female (Wilson).

Type locality: "South America." Restricted to Caracas, Venezuela [= Caracas, Distrito Federal State, N Venezuela, 10°29'N, 66°54'W, elevation 925 m] *fide* Roze (1955: 486).

Distribution: Northeastern Colombia (Arauca, Casanare, Meta), Venezuela (Amazonas, Anzoátegui, Apure, Aragua, Barinas, Bolívar, Carabobo, Cojedes, Delta Amacuro, Distrito Federal, Falcón, Guárico, Lara, Miranda, Monagas, Nueva Esparta, Portuguesa, Sucre, Trujillo, Yaracuy, Vargas, Zulia, Margarita Is.), S Guyana (Upper Takutu-Upper Essequibo) and N Brazil (Roraima), NSL–1400 m.

Sources: K.P. Schmidt, 1936b, Roze, 1966a, Medem, 1969, Gorzula & Señaris, 1998, Markezich, 2002, Navarrete et al., 2009, Ugueto & Rivas, 2010 and C.J. Cole et al., 2013.

Remarks: ANSP 6836 is possibly a third syntype.

M

37. *Micrurus langsdorffi* (Wagler *in* Spix, 1824). Serp. Brasil. Sp. Nov.: 10–12, pl. 2, fig. 1. (*Elaps langsdorffi*)

Synonyms: *Elaps langsdorfii* – Jan, 1858b *(nomen incorrectum), Elaps batesii* A.C.L.G. Günther, 1868, *Elaps imperator* Cope, 1868b, and *Micrurus mimosus* Amaral, 1935c.

Type: Lectotype, ZSM 2250/0, a 713 mm female (K.F.P. von Martius, Dec. 1819–Feb. 1820), designated by Hoogmoed & Gruber (1983: 326).

Type locality: "Río Japurá, Brazil" [= bet. mouth of Río Japurá, Brazil (3°08'S, 64°66'W) and Araracuara, Colombia (0°24'S, 72°17'W) *fide* Vanzolini, 1981: xxvi] via lectotype selection.

Distribution: Western Amazonia. Southern Colombia (Amazonas, Caquetá, Putumayo, Vaupés), NW Brazil (Amazonas), E Ecuador (Napo) and NE Peru (Loreto), 80–450 m.

Sources: Soini, 1974, Cunha & Nascimento, 1982c, Silva Haad, 1994, Lamar, 1997 and Franzen & Glaw, 2007.

Remarks: Figure in original description listed as fig. 2 in text. Possibly occurs in S Venezuela *fide* J.A. Campbell & Lamar (1989: 123). Discussion of taxonomic problems found in J.A. Campbell & Lamar (2004: 174).

38. *Micrurus laticollaris* (W.C.H. Peters, 1870a). Mber. Königl. Akad. Wiss. Berlin 1869(12): 877–878. (*Elaps marcgravii laticollaris*)

Synonym: *Micrurus laticollaris maculirostris* Roze, 1967.

Type: Lectotype, ZMB 6659, a male (Berkenbusch), designated by K.P. Schmidt (1958: 211).

Type locality: "Puebla (probably Matamoros), Mexico" via lectotype selection. Restricted to Izúcar de Matamoros, Puebla, Mexico *fide* H.M. Smith & Taylor (1950a: 341).

Distribution: Southwestern Mexico (Colima, N Guerrero, Jalisco, Michoacán, Morelos, NW Oaxaca, Puebla), 300–1915 m.

Sources: K.P. Schmidt, 1936b, Roze, 1989, Benitez-Gálvez, 1997 and Castro-Franco & Bustos Zagal, 2004.

39. *Micrurus latifasciatus* K.P. Schmidt, 1933c. Field Mus. Nat. Hist. (Zool.) 20(324): 35.

Synonym: *Micrurus nuchalis* K.P. Schmidt, 1933c.

Type: Holotype, MCZ 22135, an 887 mm female (A.W. Anthony, 1925).

Type locality: "Finca El Cipres, Volcan Zunil, Suchetepequez, Guatemala."

Distribution: Extreme SE Mexico (S Chiapas, SE Oaxaca) and SW Guatemala (Sachitepequez), NSL–1350 m.

Sources: K.P. Schmidt & Smith, 1943 and Villa et al., 1988.

40. *Micrurus lemniscatus* (Linnaeus, 1758). Syst. Nat., ed. 10, 1: 224. (*Coluber lemniscatus*)

Synonyms: *Coluber javanus* Laurenti, 1768, *Natrix lemniscata* – Laurenti, 1768, *Micrurus helleri* K.P. Schmidt & Schmidt, 1925, *Micrurus lemniscatus diutius* Burger, 1955, and *Micrulus lemniscatus* – Boquet, 1964 *(nomen incorrectum).*

Type: Lectotype, NHR Lin-93 (formerly MAFR), a 675–680 mm male (Mus. Drottn.), designated by Roze (1989: 14).

Type locality: "Asia," (in error *fide* Roze, 1967: 32) via lectotype selection.

Distribution: South America. Colombia (Amazonas, Caqueta, Meta, Putumayo), E Venezuela (Amazonas, Barinas, Bolívar, Carabobo, Cojedes, Delta Amacuro, Monagas), Trinidad and Tobago (Trinidad), Guyana (Barima-Waini, Cuyuni-Mazaruni, Essequibo Islands-West Demerara, Mahaica-Berbice, Potaro-Siparuni, Upper Takutu-Upper Essequibo), Suriname (Para, Paramaribo, Saramacca), French Guiana (Cayenne, Saint-Laurent-du-Maroni), N Brazil (Acre, Amapá, Amazonas, Ceará, Goiás, Maranhão, Mato Grosso, Pará, Rio Grande do Norte, Rondônia, Roraima), E Ecuador (Morona-Santiago, Napo, Pastaza), E Peru (Amazonas, Huánuco, Junin, Loreto, Madre de Dios), N Bolivia (Beni, Cochabamba, La Paz, Pando, Santa Cruz) and Paraguay, NSL–445 m.

Sources: Amaral, 1944k, Beebe, 1946, Emsley, 1977, Cunha & Nascimento, 1978, 1982a, 1993, Pérez-Santos,

1986, Abuys, 1987a, Chippaux, 1987, N.J. Silva, 1993, 1996, Casais e Silva, 1996, Starace, 1998, Di-Bernardo et al., 2001, N.J. Silva & Sites, 2001, Guimarães Guedes, 2003, Harvey et al., 2003, Duellman, 2005, Natera-Mumaw et al., 2007, Navarrete et al., 2009 and C.J. Cole et al., 2013.

Remarks: Official Specific Name no. 2779 *fide* Opinion 1201 (ICZN, 1982a). Roze (1989: 14) confused the NHR Linnean catalogue numbers: smaller specimen (500 mm) is Lin-93, larger specimen (680 mm) is Lin-95. Roze intended to designate the larger specimen (Lin-95) as the lectotype but instead selected the smaller one (Lin-93). Type locality restriction of Belém, Pará, Brazil *fide* K.P. Schmidt & Walker (1943a: 294) invalid *fide* Roze (1967: 32).

41. *Micrurus limbatus* Fraser, 1964. Copeia 1964. (3): 570–571, fig. 1 (middle).

Synonym: *Micrurus limbatus spilosomus* Pérez-Higareda & Smith, 1990.

Type: Holotype, UMMZ 123858, a 585 mm female (D.C. Robinson, 11 July 1959).

Type locality: "on the southern slope of Vocán San Martin, 7 airline miles north of San Andrés Tuxtla, Veracruz, Mexico, at 1,050 m" [= Volcán San Martín Sierra Tuxtlas, S Veracruz State, S Mexico, 18°32'N, 95°12'W, elevation 1050 m].

Distribution: Southern Mexico (Sierra de Los Tuxtlas of S Veracruz), 10–1500 m.

Sources: Pérez-Higareda & Smith, 1990, 1991 and B. Greene et al., 1998.

42. *Micrurus margaritiferus* Roze, 1967. Amer. Mus. Novit. (2287): 35–36, fig. 12.

Type: Holotype, AMNH 53362, a 580 mm female (H. Bassler, Aug. 1929).

Type locality: "Boca Río Santiago-Río Marañón, Peru."

Distribution: Northern Peru (Amazonas), 150–600 m. Possibly E Ecuador *fide* J.A. Campbell & Lamar (1989: 130).

43. *Micrurus medemi* Roze, 1967. Amer. Mus. Novit. (2287): 41–42, fig. 14. (*Micrurus psyches medemi*)

Type: Holotype, AMNH 96998, a 639 mm male (H. Nicéforo-María, 1950).

Type locality: "Villavicencio, Meta, Colombia" [= Villavicencio, Meta Dept., cen. Colombia, 4°09'N, 73°38'W, elevation 535 m].

Distribution: Central Colombia (Meta), 250–600 m. Known only from vicinity of the type locality.

Source: Roze, 1994.

44. *Micrurus meridensis* Roze, 1989. Amer. Mus. Novit. (2932): 5–7, figs. 2–3. (*Micrurus dissoleucus meridensis*)

Type: Holotype, USNM 217256, a 390 mm male (native, 27 March 1966).

Type locality: "1 km NE of Lagunilla, Mérida, Venezuela, 915 m."

Distribution: Western Venezuela (Mérida), 915 m.

Sources: Roze, 1994 and Navarrete et al., 2009.

45. *Micrurus mertensi* K.P. Schmidt, 1936b. Field Mus. Nat. Hist. (Zool.) 20(19): 192–193.

Type: Holotype, SMF 20714 (formerly SMF-B 9420 a), an 880 mm male (M. Bamberger, 1887).

Type locality: "Pacasmayo, Peru" [= Pacasmayo, NW La Libertad Dept., NW Peru, 7°24'S, 79°33'W, elevation 25 m].

Distribution: Extreme S Ecuador (Loja) and NW Peru (Ancash, La Libertad, Lambayeque, Piura), NSL–1700 m.

Remarks: Holotype listed as SMF-B 9420b *fide* Mertens (1967: 100) but Boettger (1898: 125) recorded it as SMF 9420 a, and it was erroneously listed as SMF 94206 *fide* K.P. Schmidt (1936b: 192). Taxonomic problems discussed by J.A. Campbell & Lamar (2004: 177).

46. *Micrurus mipartitus* (A.-M.-C. Duméril, Bibron & Duméril, 1854b). Erpét. Gén. 7(2): 1220–1221. (*Elaps mipartitus*)

Synonyms: *Elaps decussatus* A.-M.-C. Duméril, Bibron & Duméril, 1854b, *Elaps semipartitus* A.-M.-C. Duméril, Bibron & Duméril, 1854b, *Elaps multifasciatus* Jan, 1858b, *Elaps anomalus* Boulenger, 1896a, *Elaps fraseri* Boulenger, 1896a, *Elaps hertwigii* F. Werner, 1896c, *Elaps mentalis* Boulenger, 1896a, *Elaps calamus* Boulenger, 1902a, *Elaps aequicinctus* F. Werner, 1903b, *Elaps microps* Boulenger, 1913d, *Elaps decipiens* F. Werner, 1927, *Micrurus mipartitus popayanensis* Ayerbe, Tidwell & Tidwell, 1990, and *Micrurus mipartitus rozei* Golay, Chiszar, Smith & van Breukelen, 1999.

Type: Holotype, MNHN 3915, a 762 mm female (J.P. Goudot, 1822–1825 or 1837–1842), lost *fide* J.A. Peters (1960a: 531).

Type locality: "probablement de la Nouvelle Grenada" [probably Colombia]. Restricted to Rio Sucio, Colombia [= Ríosucio, Chocó Dept., NW Colombia, 7°27'N, 77°07'W, elevation 30 m] *fide* Klemmer (1963: 314) and Pérez-Santos (1986: 779).

Distribution: Lower Central America and NW South America. Southern Nicaragua (Chontales, Matagalpa, Rio San Juan, Zelaya Sur), Costa Rica (Alajuela, Cartago, Guanacaste, Heredia, Limón, Puntarenas), Panama (Bocas del Rio, Canal Zone: Barro Colorado Is.; Darién, Panamá, Veraguas), Colombia (Antioquia, Atlántico, Boyacá, Caldas, Casanare, Cauca, Chocó,

Cundinamaraca, Huila, La Guajira, Magdalena, Meta, Nariño, Norte de Santander, Putumayo, Quindío, Risaralda, Santander, Tolima, Valle, Gorgona Is.), NW Venezuela (Aragua, Barinas, Carabobo, Distrito Federal, Falcón, Mérida, Lara, Mérida, Miranda, Portuguesa, Táchira, Trujillo, Yaracuy, Zulia) and W Ecuador (Esmeraldas, Guayas, Loja, Manabi, Pichincha), NSL–2750 m.

Sources: Dunn, 1944, E.H. Taylor, 1951, Roze, 1966a, Test et al., 1966, Savage & Vial, 1974, Miyata, 1980, Roux-Estève, 1983, Villa, 1984, Mijares-Urrutia & Arends, 1993, Villa et al., 1988, N.J. Silva, 1993, Lamar, 1997, Markezich, 2002, Savage, 2002, Navarrete et al., 2009 and Rojas-Morales, 2012b.

Remarks: Supplemental original description in Jan (1859: pl. a, fig. no. 1). Confusion exists about the type locality. The type was collected by Goudot, who lived in Bogotá, Colombia. David & Ineich (1996: 146) listed the type locality as Río Sucio (or Río Senio), Chocó Dept., Colombia. In discussing the coloration of the type, Duméril et al. (1854b: 1221) used the terms *"rio-sucio ou senio,"* which appear to refer to the snake's morphology. *Micrurus multifasciatus* a valid species *fide* Roze, 1996, David & Ineich, 1996, and J.A. Campbell & Lamar, 2004 but a synonym *fide* Savage (2002: 713).

M

47. *Micrurus multiscutatus* Rendahl & Vestergren, 1940. Arkiv Zool. 33A(1): 9, figs. 3b, 4.

Type: Holotype, NHR 3131A, an 842 mm female (K. von Sneidern, 1937).

Type locality: "Colombia, Cauca, El Tambo. A village situated 20 km west of Popayan on the eastern slopes of the western Cordilleras at an altitude of 1745 m and on the watershed between the Rio Cauca and the Rio Patia" [= 2°27'N, 76°48'W].

Distribution: Western Colombia (Cordillera Occidental of Cauca), 150–1745 m.

Remarks: A synonym of *M. mipartitus fide* Savage (2002: 713).

48. *Micrurus narduccii* (Jan, 1863a). Arch. Zool. Anat. Fis. 2(2): 222. (*Elaps narduccii*)

Synonym: *Elaps melanotus* W.C.H. Peters, 1881a.

Type: Holotype, formerly MSNM, a 440 mm male, destroyed in 1943 during World War II.

Type locality: "Bolivia." Restricted to vicinity of Buenavista, Santa Cruz Prov., Bolivia *fide* Roze & Bernal-Carlo (1988: 590).

Distribution: Northwestern South America. Southeastern Colombia (Putumayo), E Ecuador (Morona-Santiago, Napo, Pastaza), NE Peru (Amazonas, Loreto, Madre de Dios, San Martín), E Bolivia (Cochabamba, Santa Cruz) and NW Brazil (Acre, ? Amazonas), 130–1300 m.

Sources: Duellman, 1978, Fugler & Riva, 1990 and Visinoni, 1996.

Remarks: Jan (1863b) erroneously cited as original description *fide* Roze & Bernal-Carlo (1988: 590, 607). Colombian records discussed by Campbell & Lamar (1989: 134).

49. *Micrurus nattereri* K.P. Schmidt, 1952. Fieldiana: Zool. 34(4): 27–29. (*Micrurus surinamensis nattereri*)

Type: Holotype, SMF 20708, a 460 mm female (G. Hübner, 1895).

Type locality: "between Guaramoca and San Fernando, Venezuela" (in error). Corrected to between Guaramaco and San Fernando de Atabapo in upper Orinoco River, Amazonas Terr. and Bolívar State, Venezuela *fide* Hoge & Lancini (1962: 12) [latter locality = W Amazonas Terr., S Venezuela, 4°03'N, 67°42'W, elevation 90 m].

Distribution: Western Amazonia. Eastern Colombia (Cundinamarca), SW Venezuela (Amazonas, Apure, Bolívar) and NW Brazil (Amapá, Maranhão), 50–250 m.

Sources: Hoge & Lancini, 1960b, Passos & Fernandes, 2005 and Navarrete et al., 2009.

Remarks: Previously a subspecies of *M. surinamensis.*

50. *Micrurus nebularis* Roze, 1989. Amer. Mus. Novit. (2932): 9–10, fig. 5.

Type: Holotype, AMNH 103118, a 557 mm male (B. Hallberg, 13 July 1968).

Type locality: "Vivero Rancho Teja, 3 km east of Ixtlán de Juárez, Oaxaca, Mexico, 2370 m," (partially in error). Corrected to elevation of 2255 m *fide* Roze (1996: 201).

Distribution: Southern Mexico (Sierra de Juárez in cen. Oaxaca), 1585–2255 m.

Sources: J.A. Campbell, 2000 and Canseco-Márquez & Campbell, 2003.

51. *Micrurus nigrocinctus* (Girard, 1854). Proc. Acad. Nat. Sci. Philadelphia (1854–1855) 7(6): 226. (*Elaps nigrocinctus*)

Synonyms: *Elaps divaricatus* Hallowell, 1854a, *Elaps melanocephalus* Hallowell, 1861 *(nomen praeoccupatum), Micrurus nigrocinctus zunilensis* K.P. Schmidt, 1932a, *Micrurus nigrocinctus mosquitensis* K.P. Schmidt, 1933c, *Micrurus nigrocinctus coibensis* K.P. Schmidt, 1936c, *Micrurus nigrocinctus wagneri* Mertens, 1941b, *Micrurus nigrocinctus ovandoensis* K.P. Schmidt & Smith, 1943, *Micrurus pachecoi* E.H. Taylor, 1951, *Micrurus nigrocinctus ovadoensis* – Alvarez del Toro, 1960 *(nomen incorrectum),* and *Micrurus nigrocinctus babaspul* Roze, 1967.

Type: Lectotype, USNM 7347, a 737 mm female (J.M. Gilliss, 1842–1853), designated by Roze (1996: 202).

Type locality: "Taboga Island, Panama" [= Taboga Is., Bay of Panama, Panama Prov., Panama, bet. 8°47–48'N,

79°32–34'W] via lectotype selection. Emended to Taboga, on the bay of Panama, Central America *fide* Girard (1855: 211),

Distribution: Southeastern Mexico, Central America and ext. NW South America. Southeastern Mexico (S Chiapas, ext. SE Oaxaca), S Belize (Belize, Cayo, Orange Walk, Stann Creek), S and cen. Guatemala (Escuintla, Jutiapa, Retalhuleu, San Marcos, Santa Rosa, Suchitepequez), Honduras (Atlántida, Choluteca, Colón, Comayagua, Copán, Cortés, El Paraíso, Francisco Morazán, Gracias a Dios, Olancho, Valle, Yoro), S El Salvador (Ahuachapán, Cabañas, Cuscatlán, La Libertad, Morazán, Santa Ana, San Salvador, Sonsonate), Nicaragua (Atlántico Norte, Atlántico Sur, Boaco, Chinandega, Chontales, Granada, Jinotega, León, Managua, Masaya, Matagalpa, Nuevo Segovia, Río San Juan, Rivas, Zelaya, Corn, Great Corn and Ometepe Is.), Costa Rica (Alajuela, Cartago, Guanacaste, Heredia, Limón, San José, Puntarenas), Panama (Bocas del Toro, Canal Zone, Chiriquí, Coclé, Colón, Darién, Herrera, Panamá, San Jose, Veraguas, Barro Colorado, Coiba, Gigante, San Miguel and Taboga Is.) and ext. NW Colombia (Antioquia, Magdalena), NSL–1600 (2000) m.

Sources: Girard, 1855, K.P. Schmidt, 1936c, E.H. Taylor, 1951, 1954, Landy et al., 1966, R.W. Henderson & Hoevers, 1975, Villa, 1984, L.D. Wilson & Meyer, 1985, Villa et al., 1988, Lee, 1998, 2000, J.A. Campbell, 1998b, G. Köhler, 1999b, Stafford, 2000a, Savage, 2002, G. Köhler et al., 2005, McCranie et al., 2006 and McCranie, 2011a.

Remarks: Supplemental description and original illustration in Girard (1855: 210–211, pl. 25, figs. 1–6). Colombian records of Providencia Is. (Dunn, 1945: 365) and Chocó Dept. (Pérez-Santos, 1986: 783) doubtful *fide* J.A. Campbell & Lamar (1989: 135).

52. *Micrurus obscurus* (Jan, 1872 *in* Jan & Sordelli, 1870–1881). Icon. Gén. Ophid. 3(41): 5, pl. 6, figs. 3, 3a–b, d–e, g, n, p, r, v. (*Elaps corallinus obscura*)

Synonyms: *Elaps heterozonus* W.C.H. Peters, 1881 and *Elaps princeps* Boulenger, 1905b.

Type: Holotype, formerly MSNM, destroyed in 1943 during World War II.

Type locality: "Lima" [Peru] (in error). Corrected to E Peru *fide* K.P. Schmidt & Walker (1943a: 294). Restricted to Iquitos, Peru [= E Loreto Dept., NE Peru, 3°45'S, 73°15'W, elevation 90 m] *fide* KP. Schmidt (1953c: 175).

Distribution: Southeastern Colombia (Amazonas, Caquetá, Meta, Putumayo, Vaupés, Vichada), Ecuador (Morona-Santiago, Napo, Pastaza), E Peru (Huanuco, Loreto, Madre de Dios, Pasco), NE Brazil (Amazonas) and Bolivia (Beni, La Paz, Pando, Santa Cruz), 100–600 m.

Sources: Silva Haad, 1994, Harvey et al., 2003 and Duellman, 2005.

53. *Micrurus oligoanellatus* Ayerbe-González & López-López, 2005. Noved. Colomb. 8(1), 42–43, fig. 1.

Synonym: *Micrurus oligoanellatus* Ayerbe & López, 2002 *(nomen ineditum).*

Type: Holotype, MHNUC HE-357-Se (formerly CSA 349), a 625 mm female (C.E. González-Orozco, 6 Oct. 2000).

Type locality: "la trocha entre El Cocal y la Reserva Natural Tambito, Municipio de El Tambo, Departamento del Cauca, Colombia a 1300 metros de altura sobre el nivel del mar en el costado Oeste de la Cordillera Occidental, 2° 35' N; 77° 03' W" [Colombia].

Distribution: Northwestern Colombia (Cauca), 1000–1500 m. Known only from type locality.

Remarks: Original description in Ayerbe & López, 2002. Mem. II Symp. Ivest. Cien. Biol., Dept. Biol., Univ. Cauca, Asoc. Colombia Cien. Biol., Capitulo Popayán (12): 2 pp., 1 fig.

54. *Micrurus ornatissimus* (Jan, 1858b). Rev. Mag. Zool. (2)10: 516, 521–522. (*Elaps ornatissimus*)

Synonym: *Elaps buckleyi* Boulenger, 1896a.

Type: Holotype, formerly MSNM, a 550 mm male, destroyed in 1943 during World War II.

Type locality: "Amérique centrale" [= Central America] (in error). Emended to ? Mexico *fide* Jan & Sordelli (1872 *in* 1870–1881: 5) (also in error *fide* K.P. Schmidt, 1936b: 191). Restricted to Amazon Basin *fide* K.P. Schmidt (1936b: 191), Río Putumayo, Colombia *fide* K.P. Schmidt (1955a: 345) and Ecuador or Peru *fide* Cunha & Nascimento (1982c: 8).

Distribution: Northwestern Amazonia. Southeastern Colombia (Amazonas, Vaupés), E Ecuador (Morona-Santiago, Napo, Pastaza, Zamora-Chinchipe) and N Peru (Loreto, ? Madre de Dios), 500–1200 m.

Sources: K.P. Schmidt, 1955a, J.A. Peters, 1960a, Dixon & Soini, 1978, 1986, Fugler & Walls, 1978, Cunha & Nascimento, 1982c and Silva Haad, 1994.

Remarks: Supplemental original description in Jan (1859e: pl. a, fig. no. 3). A subspecies of *M. langsdorffi* Wagler *fide* Roze (1983: 328), J.A. Campbell & Lamar (1989: 125) and Golay et al. (1993: 172).

55. *Micrurus pacaraimae* Carvalho, 2002. Pap. Avul. Zool. 42(8): 184–185, figs. 1–4.

Type: Holotype, MZUSP 8565, a 355 mm male (C.M. Carvalho, 10 Jan. 1985).

Type locality: "Brasil: Vila Pacaraima: Rodovia BR-174 na fronteira com a Venezuela (04°31'N, 61°09'W)."

Distribution: Northwestern Brazil (Roraima), 30–40 m.

56. *Micrurus pachecogili* J.A. Campbell, 2000. Proc. Biol. Soc. Washington 113(1): 292–293, figs. 1–2.

Type: Holotype, UTA 12546 (formerly JAC 9752), a 639 mm male (E. Pacheco-Gil, Nov. 1983).

Type locality: "5.6 km SSW Zapotitlán, Salinas, 1494 m, Puebla, Mexico" [= 18°18'N, 97°31'W].

Distribution: Southern Mexico (N Oaxaca and SE Puebla), 620–1540 m.

Source: Canseco-Marquez & Campbell, 2003.

57. *Micrurus paraensis* Cunha & Nascimento, 1973. Publ. Avul. Mus. Para. Emílio Goeldi 20: 276–278, fig. 1. (*Micrurus psyches paraensis*) (*nomen corrigendum*)

Synonyms: *Micrurus psiches paraensis* Cunha & Nascimento, 1973 (*nomen incorrigendum*) *Micrurus donosoi* Hoge, Cordeiro & Romano-Hoge, 1979 and *Micrurus paraensis debruini* Abuys, 1987b.

Type: Holotype, MPEG 851, a 460 mm male (O.R. da Cunha & F.P. do Nascimento, 1971).

Type locality: "Icoaraci, Belém, Pará, Amazonia" [= Icorací, Belém, Pará State, N Brazil, 1°18'S, 48°28'W, elevation 15 m].

Distribution: Northern Brazil (W Maranhão, SW Mato Grosso, Pará, Rondônia) and S Suriname (Nickerie), NSL–400 m.

Sources: Hoge et al., 1978c, Hoge & Romano-Hoge, 1981b, Cunha & Nascimento, 1982, 1993, Nascimento et al., 1987, 1988, N.J. Silva, 1993, Roze, 1994, Strussmann & André de Carvalho, 1998 and Feitosa et al., 2007.

M

58. *Micrurus peruvianus* K.P. Schmidt, 1936b. Field Mus. Nat. Hist. (Zool.) 20(19): 193.

Type: Holotype, MCZ 17385, a 412–415 mm male (G.K. Noble [Harvard peru. Exped.], 22 Aug.–16 Sept. 1916).

Type locality: "Perico, Department of Cajamarca, Peru" [= Perico, Cajamarca Department, NW Peru, 05°20'S, 78°47'W, elevation 500 m].

Distribution: Northern Peru (Amazonas, Cajamarca) and S Ecuador, 450–1500 m.

Sources: Duellman, 1979 and Kuch et al., 2002.

59. *Micrurus petersi* Roze, 1967. Amer. Mus. Novit. (2287): 45–46, fig. 16. (*Micrurus steindachneri petersi*)

Type: Holotype, USNM 158295, a 667 mm female (J.A. Peters, 1 Aug. 1962).

Type locality: "One mile south of Plan de Milagro on the trail to Pan de Azúcar, Morona-Santiago, Province, Ecuador, 5600 feet."

Distribution: Andes of SE Ecuador (Morona-Santiago), 1000–1705 (2800) m.

Remarks: Possibly occurs in Colombia and Peru *fide* J.A. Campbell & Lamar (2004: 187).

60. *Micrurus proximans* H.M. Smith & Chrapliwy, 1958. Herpetologica 13(4): 270–271. (*Micrurus diastema proximans*)

Type: Holotype, UIMNH 40369, a 420 mm female (W.H. Davis, W.Z. Lidicker & J.R. Winklemann, 16 July 1956).

Type locality: "4 mi. NE San Blas, Nayarit, Mexico."

Distribution: Western Mexico (Colima, Nayarit), NSL–1900 m.

Sources: Zweifel, 1959b, Nordgren & Villa, 2011 and Reyes-Velasco et al., 2012.

61. *Micrurus psyches* (Daudin, 1803e). Hist. Nat. Rept. 8: 320–321, pl. 100, fig. 1. (*Vipera psyches*)

Type: Holotype, not designated, a 244 mm male (F. Levaillant, 1765–1780, via Dufresne coll.), location unknown.

Type locality: "intérieur de la colonie de Suriname" [interior of Suriname].

Distribution: Northern South America. Eastern Venezuela (Bolívar, Monagas, Sucre), Trinidad and Tobago (Trinidad: Gaspar Grande Is.), Guyana (Essequibo Islands-West Demerara, Potaro-Siparuni), Suriname (Brokopondo, Marowijne, Nickerre, Para, Saramacca), French Guiana (Cayenne, Saint-Laurent-du-Maroni) and N Brazil (Amapá), NSL–500 m. Probably Roraima, Brazil *fide* J.A. Campbell & Lamar (1989: 139).

Sources: Beebe, 1946, Roze, 1966a, 1994, Emsley, 1977, Boos, 1983a–b, Roux-Estève, 1983, Pérez-Santos, 1986, Abuys, 1987c, Chippaux, 1987, Starace, 1998, C. Molina et al., 2002, Navarrete et al., 2009 and C.J. Cole et al., 2013.

Remarks: MNHN 7654, a 490 mm female, long considered the type, is not Daudin's specimen *fide* Roux-Estève (1983: 91–92); MNHN file card has "type" crossed out and notation "seen by Duméril, not Daudin's type."

62. *Micrurus putumayensis* Lancini, 1962b. Publ. Ocas. Mus. Cien. Nat. Caracas (Zool.) (3): 1. (*nomen substitutum*)

Synonym: *Micrurus schmidti* Lancini, 1962a (*nomen praeoccupatum*).

Type: Holotype, AMNH 110058 (formerly MCNC 1117), a 624 mm female (J. Cáceres, March 1959).

Type locality: "Puerto Socorro (a 270 Km. NE. de Iquitos), Rio Putumayo, Departamento de Loreto, República del Perú."

Distribution: Northwestern Amazonia. Southeastern Colombia (S Amazonas, Putumayo), NE Peru (N Loreto) and NW Brazil (W Amazonas), NSL–300 m.

Sources: Lancini, 1962a, Lema, 1972 and Dixon & Soini, 1986.

Remarks: Original description in Lancini (1962a: 1–3, 1 fig.). Photograph of holotype in Roze (1996: fig. 60). Roze (1983: 334) cited date as 1963. Roze (1996: 210) and J.A. Campbell & Lamar (2004: 190) erroneously listed holotype as AMNH 1100058.

63. *Micrurus pyrrhocryptus* (Cope, 1862d). Proc. Acad. Nat. Sci. Philadelphia 14(5): 347. (*Elaps pyrrhocryptus*)

Synonyms: *Elaps simonsii* Boulenger, 1902c and *Micrurus tricolor* Hoge, 1957.

Type: Holotype, ANSP 5395 (T.J. Page [Paraguay Exped.], 1853–1856), lost *fide* N.J. Silva & Sites (1999: 174).

Type locality: "Parana, Paraguay, Vermejo and Uruguay Rivers, Paraguay." Restricted to Vermejo River, Argentine Chaco *fide* K.P. Schmidt (1936b: 199). Emended to Río Bermejo *fide* Scrocchi (1990b: 358).

Distribution: Southern South America. Southeastern Brazil (Mato Grosso), Bolivia (Chuquisaca, Santa Cruz, Tarija), N Paraguay (Alto Paraguay, Boquerón) and Argentina (Catamarca, Chaco, Córdoba, Corrientes, Entre Ríos, Formosa, Jujuy, La Pampa, La Rioja, Mendoza, Neuquén, Rio Negro, Salta, San Juan, San Luis, Santa Fé, Santiago del Estero, Tucumán), 100–600 m.

Sources: Page, 1859, Scrocchi, 1990b, Cei, 1994, Roze,1994, N.J. Silva & Sites, 1999, 2001, Leynaud & Bucher, 1999, Giraudo & Scrocchi, 2002, Di-Bernardo et al., 2007 and Cacciali, 2008.

Remarks: Illustrations of the holotype of *M. tricolor* in N.J. Silva & Sites (1999: fig. 31). *Micrurus tricolor* a synonym *fide* Harvey et al. (2003: 28).

64. *Micrurus remotus* Roze, 1987. Rev. Fr. Aquar. 14(3): 110, fig. 1. (*Micrurus psyches remotus*)

Type: Holotype, USNM 266100, a 491 mm male (R.W. McDiarmid, 6 Feb. 1985).

Type locality: "Base Camp of Cerro de la Neblina, Territorio Federal Amazonas, Venezuela, 0°55'N, 66°10'W, 90 m."

Distribution: Western Amazonia. Eastern Colombia (Guaínia, Vaupés), S Venezuela (Amazonas) and N Brazil (N Amazonas), 90–1700 m.

Sources: Roze, 1994 and Navarrete et al., 2009.

65. *Micrurus renjifoi* (Lamar, 2003). Rev. Biol. Trop. 51(3–4): 806–808, 2 col. figs. (*Leptomicrurus renjifoi*)

Type: Holotype, UTA 3490, a 427 mm gravid female (F. Barney, Aug. 1969).

Type locality: "the south side of the Río Tomo, near its juncture with the Río Orinoco, 97 km SSW of Puerto Carreño, Commisaría del Vichada, Colombia, 115 m."

Distribution: Colombia (Vichada), 115 m. Known only from vicinity of type locality.

66. *Micrurus ruatanus* (A.C.L.G. Günther, 1895 *in* 1885–1902). Biol. Cen.-Amer., Rept. Batr.: 185, pl. 57, fig. b. (*Elaps ruatanus*)

Type: Lectotype, BMNH 1946.1.21.20 (G.F. Gaumer, 1886, via O. Salvin & F.D. Godman), designated by Roze (1996: 213).

Type locality: "Roatán Island, Honduras" [16°23'N, 86°30'W] via lectotype selection.

Distribution: Northern Honduras (Islas de la Bahía: Roatán Is.).

Sources: L.D. Wilson & Hahn, 1973, L.D. Wilson, 1984, Luykx et al., 1992, L.D. Wilson et al., 1992, McCranie et al., 2005 and McCranie, 2011a.

67. *Micrurus sangilensis* Nicéforo-María, 1942. Rev. Acad. Colomb. Cien. Exact. Fis. Nat. 5(17): 98–99, pl. 3, fig. 10. (*Micrurus ecuadorianus sangilensis*)

Type: Holotype, MCG 2A, a 545 mm male (R.H. Silvano-Jorge, Aug. 1937).

Type locality: "San Gil, Departamento de Santander, Colombia."

Distribution: Northeast Colombia (Boyacá, Cundinamarca, Santander), 1020–1735 m.

Sources: K.P. Schmidt, 1955a, Pérez-Santos, 1986 and Pérez-Santos & Moreno, 1988.

M

68. *Micrurus scutiventris* (Cope, 1870). Proc. Amer. Philos. Soc. (1869–1870) 11(81): 156–157. (*Elaps scutiventris*)

Synonyms: *Leptomicrus schmidti* Hoge & Romano-Hoge, 1966 (*nomen incorrigendum*), *Leptomicrurus schmidti* Hoge & Romano-Hoge, 1966 (*nomen praeoccupatum*), and *Micrurus karlschmidti* Romano-Hoge, 1972 (*nomen substitutum*).

Type: Holotype, ANSP 6801, a 445 female (J. Orton, 1867–1870).

Type locality: "Pebas on the Amazon in Equador" [= Pebas, Loreto Dept., NE Peru, 3°19'S, 71°52'W, elevation 100s m].

Distribution: Northwestern South America. Southeastern Colombia (Amazonas, Vaupés, Vichada), E Ecuador (Pastaza), NE Peru (Amazonas, Loreto) and NW Brazil (Amazonas), 80–1200 m.

Source: Roze & Bernal-Carlo, 1988.

69. *Micrurus serranus* Harvey, Aparicio & Gonzalez, 2003. Ann. Carnegie Mus. 72(1): 34–41, figs. 6a, 6c, 13.

Type: Holotype, UTA 34561 (formerly MBH 1960), a 514 mm male (M.B. Harvey, 14 Feb. 1992).

Type locality: "3 km N of Samaipata on the road to Mairana, Florida Province, Santa Cruz, Bolivia."

Distribution: Bolivia (Chuquisaca, Cochabamba, Potosí, Santa Cruz, Tarija), 1200–2750 m.

Source: Muñoz-Saravia et al., 2009.

70. *Micrurus silviae* Di-Bernardo, Borges-Martins & Silva, 2007. Zootaxa (1447): 1–17, figs. 1–2, 7.

Type: Holotype, CRUPF 1488, a 1113 mm male (V.L. Mello de Paulo, 21 March 2005).

Type locality: "Passo Fundo Municipality, Miranda Stream (within the drainage basin of the Companhia Riograndense de Saneamento—CORSAN dam), in the W-NW of Rio Grande do Sul, southern Brazil."

Distribution: Southern Brazil (Rio Grande do Sul) and Paraguay (Itapúa), 120–740 m.

Source: Cacciali et al., 2011.

71. *Micrurus spixii* Wagler in Spix, 1824. Serp. Brasil. Sp. Nov.: 48–49, pl. 18.

Synonyms: *Elaps ehrhardti* L. Müller, 1926 and *Micrurus spixii martiusi* K.P. Schmidt, 1953c.

Type: Holotype, ZSM 209/0, an 1142 mm male (J.B. von Spix, Nov. 1819–Feb. 1820).

Type locality: "ad flumen Solimöens" [= Río Solimões or upper Amazon River, bet. mouth of Río Negro and Peruvian border, Amazonas State, NW Brazil].

Distribution: Amazonia. Southern Venezuela (Amazonas), Brazil (Amazonas, Bahia, N Goiás, W Maranhão, Mato Grosso, Pará, Rondônia, Marajo Is.), cen. Bolivia (Beni, Santa Cruz) and N Paraguay, 50–1200 m.

Sources: J.C. Thompson, 1913c, Roze, 1966a, Cunha & Nascimento, 1978, 1993, Hoogmoed & Gruber, 1983, Pérez-Santos, 1986, Pérez-Santos & Moreno, 1988, Fugler & Riva, 1990, Henle & Ehrl, 1991, N.J. Silva, 1993, Morales & McDiarmid, 1996, N.J. Silva & Sites, 2001, Lehr, 2001 and Navarrete et al., 2009.

Remarks: Official Specific Name no. 2780 *fide* Opinion 1201 (ICZN, 1982a). Photograph of the type of *Elaps ehrhardti* in Franzen & Glaw (2007: fig. 21). Possibly occurs in W Guyana, S Suriname, SE French Guiana and Acre, Brazil *fide* J.A. Campbell & Lamar (2004: 228).

72. *Micrurus spurrelli* (Boulenger, 1914b). Proc. Zool. Soc. London 82 (2): 817, pl. 2, figs. 3–3a. (*Elaps spurrelli*)

Synonyms: *Micrurus nicefori* K.P. Schmidt, 1955a, and *Elaps spurelli* – Rize, 1983 *(nomen incorrectum).*

Type: Holotype, BMNH 1946.1.17.19, a 230 mm female (H.G.F. Spurrell, 1913–1914).

Type locality: "at or near Peña Lisa, Condoto, Choco, Colombia" [= Peña Lisa, Río Condoto, S Chocó, W Colombia, ca. 5°06'N, 76°42'W, elevation 100 m].

Distribution: Extreme W Colombia (S Chocó), NSL–400 m. Known only from vicinity of type locality.

Source: Pérez-Santos & Moreno, 1988.

73. *Micrurus steindachneri* (F. Werner, 1901a). Verh. Zool.-Bot. Ges. Wien 51: 599. (*Elaps steindachneri*)

Synonyms: *Elaps fasslii* Steindachner *in* F. Werner, 1927 and *Micrurus steindachneri orcesi* Roze, 1967.

Type: Holotype, NMW 15750, an 800 mm male (R. Haensch, 1899–1900).

Type locality: "Ecuador."

Distribution: Central Ecuador (W Morona-Santiago, W Napo, W Pastaza, Tungurahua) and Peru (Amazonas), 600–2000 m.

Sources: Roze, 1970, 1983, Duellman, 1979 and Tiedemann & Häupl, 1980.

74. *Micrurus stewarti* T. Barbour & Amaral, 1928. Bull. Antivenin Inst. Amer. 1(4): 100.

Synonym: *Micrurus schmidti* Dunn, 1940.

Type: Holotype, MCZ 24924, a 380–382 mm male (T.H. Stewart, Jr., 1926).

Type locality: "inland from Nombre de Dios, Serrania de la Bruja, Republic of Panama, approximate altitude 1200 meters" [= S of Nombre de Díos (9°34'N, 79°28'W, elevation 20 m)in the Serrania de la Bruja (which rises to 800 m), E Colón Province, cen. Panama].

Distribution: Central Panama (Boca del Toro, ext. E Coclé, E Colón. ext. W Panamá), 100–1200 m.

Source: Villa et al., 1988.

Remarks: Holotype erroneously reported as a female *fide* Barbour & Amaral (1928: 100).

75. *Micrurus stuarti* Roze, 1967. Amer. Mus. Novit. (2287): 47–48, fig. 17.

Type: Holotype, UMMZ 106708, a 593 mm male (L.C. Stuart, 15 May–30 June 1952).

Type locality: "Finca La Paz, San Marcos, Guatemala, 1345 meters."

Distribution: Southwestern Guatemala (San Marcos, Suchitepéquez), 800–1600 m.

Sources: Villa et al., 1988 and J.A. Campbell & Vannini, 1989.

Remarks: Possibly occurs in Mexico (Chiapas) and Guatemala (Escuintla) *fide* J.A. Campbell & Lamar (1989: 150).

76. *Micrurus surinamensis* (Cuvier, 1817). Règne Anim. 2: 84. (*Elaps surinamensis*)

Synonyms: *Coluber cinctus* Lacépède, 1789 (*nomen rejiciendum*), *Coluber melaleucus* Gravenhorst, 1807 (*nomen oblitum*), *Vipera anguiformis* Kuhl, 1820, *Elaps grandis* Wagler, 1825 (*nomen ineditum*), *Elaps scutelliger* Wagler, 1825 (*nomen ineditum*), and *Elaps zenega* Wagler, 1825 (*nomen ineditum*).

Type: Lectotype, MNHN 3926, an 862 mm male (F. Levaillant, 1765–1780), designated by K.P. Schmidt (1952: 26).

Type locality: "Suriname," via lectotype selection.

Distribution: Northern South America. Southeastern Colombia (Amazonas, Caquetá, Meta, Putumayo, Vaupés), Guyana (Cuyuni-Mazaruni, Portaro-Siparuni, Upper Demerara-Berbice, Upper Demerara-Upper Essequibo), Suriname (Marowijne, Para, Paramaribo, Saramacca), French Guiana (Caquita, Cayenne, Putumayo, St. Laurent du Moroni), NW Brazil (Acre, Amazonas, Goias, Mato Grosso, Pará, Rondônia, Roraima), E Ecuador (Morona-Santiago, Napo, Pastaza), E Peru (Amazonas, Loreto, Madre de Dios, Pasco) and N Bolivia (Beni, Cochabamba, Pando, Santa Cruz), 30–575 m.

Sources: K.P. Schmidt, 1952, Roze, 1955, Hoge, 1958d, Cunha & Nascimento, 1978, Parent, 1982, Roux-Estève, 1983, Dixon & Soini, 1986, Pérez-Santos, 1986, Abuys, 1987a, Chippaux, 1987, Pérez-Santos & Moreno, 1988, 1991, Fugler & Riva, 1990, Rodriguez & Cadle, 1990, N.J. Silva, 1993, Silva Haad, 1994, Starace, 1998, N.J. Silva & Sites, 2001, Harvey et al., 2003, Duellman, 2005, Passos & Fernandes, 2005, Morata et al., 2011 and C.J. Cole et al., 2013.

Remarks: Original description based on Seba (1735: 91, pl. 86, fig. 1). In accordance with Art. 23.9.2 of the Code (ICZN, 1999), *Elaps surinamensis* Cuvier is designated a *nomen protectum* and *Coluber melaleucus* Gravenhorst a *nomen oblitum*. A record from Chacó, Colombia requires confirmation.

77. *Micrurus tamaulipensis* Lavin-Murcio & Dixon, 2004. Phyllomedusa 3(1): 4–5, fig. 1.

Type: Holotype, ITT 751, a 602–632 mm male (P.A. Lavin-Murcio, April 2002).

Type locality: "Sierra de Tamaulipas, Rancho La Sauceda, ca. 50 km N Gonzales, 750–1000 m elevation, (23°06'47"N, 98°20'19"W), Tamaulipas, Mexico."

Distribution: Northeastern Mexico (Tamaulipas), 750–1000 m.

78. *Micrurus tener* (Baird & Girard, 1853). Cat. No. Amer. Rept., Serp. 1: 22–23. (*Elaps tenere*)

Synonyms: *Elaps tristis* Baird & Girard, 1853, *Elaps fitzingeri* Jan, 1858b & 1859, *Micrurus fitzingeri microgalbineus* B.C. Brown & Smith, 1942, *Micrurus fulvius maculatus* Roze, 1967, *Micrurus fulvius microgalbineos* – Possani, Varela-Julía, Alagon, Julía-Zertuche & Fletcher, 1978 *(nomen incorrectum)*, and *Micrurus fulvius macrogalbineus* – Harding & Welch, 1980 *(nomen incorrectum)*.

Type: Lectotype, USNM 1119, an adult female (J.H. Clark [U.S. Mex. Bound. Surv.], 1850–1852 via J.D. Graham), designated by Roze & Tilger (1983: 316.2).

Type locality: "San Pedro of Rio Grande" [= San Pedro on Rio Grande River, Cameron Co., ext. S Texas, S USA] via lectotype selection.

Distribution: Southern USA (SW Arkansas, W Louisiana, E Texas) and Mexico (NE Coahuila, Colima, N Guanajuato, Hidalgo, Morelos, NE Nuevo León, Pueblo, Querétaro, San Luis Potosí, Tamaulipas), NSL–2000 m.

Fossil records: Middle/upper Pleistocene (Irvingtonian II) of USA (Texas), and upper Pleistocene (Rancholabrean II) of USA (Texas).

Sources: Tennant, 1984, 1895, Benitez-Gálvez, 1997, Conant & Collins, 1998, Holman, 2000a, Werler & Dixon, 2000, Reyes-Velasco et al., 2009, Ramírez-Bautista et al., 2010 and Ernst & Ernst, 2011.

Remarks: Type locality restriction to New Braunfels, Comal County, Texas, USA *fide* H.M. Smith & Taylor (1950a: 361) invalid. Possibly occurs in N Veracruz *fide* Pérez-Higareda & Smith (1991: 74). Records from Distrito Federal, México and Michoacán, Mexico are in error *fide* J.A. Campbell & Lamar (2004: 196).

79. *Micrurus tschudii* (Jan, 1858b). Rev. Mag. Zool. (2)10: 517, 524–525. (*Elaps tschudii*)

Synonym: *Micrurus olssoni* K.P. Schmidt & Schmidt, 1925.

Type: Holotype, NMW 18284, a 310 mm specimen (I. Pfeiffer, 1856).

Type locality: "Pérou" [= Peru]. Restricted to the Rimac Valley, Department of Lima, Peru *fide* K.P. Schmidt & Walker (1943: 320).

Distribution: Extreme SW Ecuador (Loja) and ext. W Peru (Ancash, Apurimac, Cajamarca, La Libertad, Lambeyeque, Lima, Piura, Tumbes), NSL–1455 m.

Sources: K.P. Schmidt, 1925c, 1936b and Pérez-Santos & Moreno, 1991.

Remarks: Supplemental original description in Jan (1859e: pl. b, fig. no. 2). Possibly occurs in NW Bolivia *fide* J.A. Campbell & Lamar (1989: 153). Roze (1996: 224) and Campbell & Lamar (2004: 232) erroneously reported the type to be unknown and apparently lost.

MIMOPHIS A.C.L.G. Günther, 1868 (Psammophiidae)

Synonym: *Mimorphis* – Maass-Berlin, 1933 *(nomen incorrectum)*.

Type species: *Psammophis mahfalensis* Grandidier, 1867.

Distribution: Madagascar.

Sources: A.C.L.G. Günther, 1868, Boulenger, 1896a, Jourdran, 1904, Mocquard, 1909, Guibé, 1958, Brygoo, 1983, Glaw & Vences, 1994, Brandstätter, 1995, Henkel & Schmidt, 2000, Vidal & Hedges, 2002, Aprea et al., 2003, Nagy et al., 2003, Vidal et al., 2008, Kelly et al., 2009 and Zaher et al., 2009.

1. *Mimophis mahfalensis* (Grandidier, 1867). Rev. Mag. Zool. (2)19: 234. (*Psammophis mahfalensis*)

Synonyms: *Mimophis madagascariensis* A.C.L.G. Günther, 1868 and *Mimophis mahafalensis albiceps* Jourdran, 1904.

Types: Syntypes (2), MNHN 1459 and MNHN 1716 (A. Grandidier, 1866).

Type locality: "Machikova et Anhoulabé, la côte sud-ouest de Madagascar" [= Machikova and Anhoulabé, Toliara Prov., SW coast of Madagascar].

Distribution: Madagascar (Antsiranana, Atananarivo, Fianarantsoa, Mahajanga, Toamasina, Toliara), NSL–600 m.

†*MIOCOLUBER* Parmley, 1988 (Colubridae)

Type species: †*Miocoluber dalquesti* Parmley, 1988.

Distribution: Upper Miocene of USA.

Sources: Parmley, 1988 and Holman, 2000a.

1. †*Miocoluber dalquesti* Parmley, 1988. J. Vert. Paleont. 8(3): 324–325, figs. 3a–e.

M

Type: Holotype, MSUVP 1209, one trunk vertebra (W.W. DalOuest).

Type locality: "Higgins Quarry B, locality 24-B of Reed and Longnecker (1932); center, S1/2, SE1/4, Sec. 176, Bld. 43 of Houston and Texas Central Railroad Survey; approximately 3 km south and 1.6 km east of Higgins, Lipscomb County, Texas."

Distribution: Upper Miocene (Hemphillian: 4.9–10.3 mya) of USA (Texas). Known only from type locality.

†*MIONATRIX* Sun, 1961 (Natricidae)

Type species: †*Mionatrix diatomus* Sun, 1961.

Distribution: Middle Miocene of China.

Source: Rage, 1984b.

Remarks: Similar to *Natrix fide* Rage, 1884b.

1. †*Mionatrix diatomus* Sun, 1961. Vert. Palasiat. 1961(4): 309, 312, figs. 1–2, pls. 1–2.

Types: Syntypes (2), IVPP V.993, a 650–700 mm imprint of a complete skeleton, and IVPP V.994, a 1000 mm imprint of axial skeleton.

Type locality: "Miocene diatomaceous beds of Shanwang, Shantung, China."

Distribution: Middle Miocene (16.0–18.0 mya) of China. Known only from type specimens.

MITOPHIS Hedges, Adalsteinsson & Branch *in* Adalsteinsson, Branch, Trape, Vitte & Hedges, 2009 (Leptotyphlopidae)

Type species: *Leptotyphlops pyrites* R. Thomas, 1965a.

Distribution: Hispaniola.

Sources: Hahn, 1980a, A. Schwartz & R.W. Henderson, 1991, Wallach, 1998b, McDiarmid et al., 1999 and R.W. Henderson & Powell, 2009.

1. *Mitophis asbolepis* (R. Thomas, McDiarmid & Thompson, 1985). Proc. Biol. Soc. Washington 98(1): 209–210, fig. 4. (*Leptotyphlops asbolepis*)

Type: Holotype, UF 54802, a 156 mm female (F.G. Thompson, 29 Jan. 1976).

Type locality: "west slope of Loma del Aguacate, up the mountainside from La Salina (Puerto Alejandro) on the east side of the Bahia de Neiba, 350 m, Sierra Martín García, Provincia de Barahona, Dominican Republic."

Distribution: Southwestern Dominican Republic (SW Barahona), 350–400 m. Known only from vicinity of type locality.

2. *Mitophis calypso* (R. Thomas, McDiarmid & Thompson, 1985). Proc. Biol. Soc. Washington 98(1): 206–209, figs. 2a, 3 (*Leptotyphlops calypso*)

Type: Holotype, USNM 236659, a 166 mm male (R.W. McDiarmid, 22 Feb. 1975).

Type locality: "6.5 km S Las Galeras, Provincia de Samaná, Dominican Republic" [= ca. 19°15'N, 69°12'W, elevation 65 m].

Distribution: Northeastern Dominican Republic (Samaná), 25–65 m. Known only from vicinity of type locality.

3 *Mitophis leptepileptus* (R. Thomas, McDiarmid & Thompson, 1985). Proc. Biol. Soc. Washington 98(1): 211–213, figs. 2b, 5. (*Leptotyphlops leptepileptus*)

Type: Holotype, USNM 236661, a 184 mm female (R. Thomas, 19 July 1978).

Type locality: "valley of the Rivière Soliette, on the north slopes of the Massif de la Selle, 5 km airline NW Fond Verettes, 366 m, Département de l'Ouest, Haiti."

Distribution: Southern Haiti (E Ouest), 365 m. Known only from vicinity of type locality.

4 *Mitophis pyrites* (R. Thomas, 1965a). Breviora (222): 2–4, figs. 1, 2a. (*Leptotyphlops pyrites*)

Type: Holotype, MCZ 77239, a 133 mm specimen (R. Thomas, 3 July 1964).

Type locality: "southern outskirts of the town of Pedernales, approx. 1 km from the center of town,

Pedernales Province, Dominican Republic, West Indies" [= 18°02'N, 71°44'W, elevation 10 m].
Distribution: Hispaniola. Southeastern Haiti (SE Sud-Est) and SW Dominican Republic (Independencia, Pedemales), 10–185 m.
Source: R. Thomas et al., 1985.

MIXCOATLUS Jadin, Smith & Campbell, 2011 (Viperidae)

Type species: *Agkistrodon browni* Shreve, 1938.
Distribution: Southern Mexico.
Sources: J.A. Campbell & Lamar, 2004 and Jadin et al., 2011, 2012.
Remarks: Previously in the genera *Cerrophidion* and *Ophryacus*.

1. *Mixcoatlus barbouri* (Dunn, 1919). Proc. Biol. Soc. Washington 32: 213–214. (*Lachesis barbouri*)

Type: Holotype, USNM 46347, a 370 mm specimen (E.W. Nelson & E.A. Goldman, 20 May 1903).
Type locality: "Omilteme, Guerrero, Mexico" [= Omilteme, cen. Guerrero, S Mexico, 17°30'N, 99° 40'W, elevation 2135 m].
Distribution: Southwestern Mexico (Sierra Madre del Sur of cen. Guerrero), 2135–3295 m.
Sources: Shreve, 1938, H.M. Smith, 1941i and J.A. Campbell, 1977, 1988b.

2. *Mixcoatlus browni* (Shreve, 1938). Copeia 1938(1): 9. (*Agkistrodon browni*)

Type: Holotype, MCZ 42678, a 445 mm male (W.W. Brown, 1936).
Type locality: "Omilteme, Guerrero, Mexico" [= Omilteme, cen. Guerrero, S Mexico, 17°30'N, 99° 40'W, elevation 2135 m].
Distribution: Southwestern Mexico (Guerrero), 2135 m.

3. *Mixcoatlus melanurus* (L. Müller, 1924). Mitt. Zool. Mus. Berlin 11(1): 92–93. (*Trimeresurus melanurus*)

Synonyms: *Bothrops melanura* – Amaral, 1930f, *Trimeresurus garciai* H.M. Smith, 1940, *Trimeresurus garciae* – E.H. Taylor, 1944a *(nomen incorrectum)*, and *Porthidium melanurum* – J.A. Campbell & Lamar, 1989.
Type: Holotype, ZMB 26403, a 423 mm female (C.A. Purpus, 1897–1923).
Type locality: "Mexico." Restricted to Cacaloapam [= Cacaloapan], Puebla, Mexico *fide* H.M. Smith & Taylor (1950a: 341).
Distribution: Southern Mexico (NW Oaxaca, SE Puebla), 1600–2400 m.
Sources: Casas-Andreu et al., 1996 and Gutberlet, 1998.

MONTASPIS Bourquin, 1991 (Lamprophiidae)

Type species: *Montaspis gilvomaculata* Bourquin, 1991.
Distribution: Southern Africa.
Sources: Bourquin, 1991, Marais, 1992, Branch et al., 1993, Lambiris, 1997, Kelly et al., 2009 and Zaher et al., 2009.

1. *Montaspis gilvomaculata* Bourquin, 1991. Ann. Transvaal Mus. 35(12): 200–202, figs. 1a–c, 2–4.

Type: Holotype, TMP 68088, a 395–412 mm female (D. Tomlinson, Jan. 1980).
Type locality: "Organ Pipes Pass, near the 'The Camel', Cathedral Peak Forest Reserve, Natal, South Africa, 28°00'35" S, 29°11'11" E; at *c.* 2860 m above sea level."
Distribution: Eastern South Africa (Drakensberg of SW KwaZulu-Natal), 1870–2865 m.

MONTATHERIS Broadley, 1996b (Viperidae)

Synonym: *Hindius* A.F.T. Reuss, 1939a (*nomen nudum*).
Type species: *Vipera hindii* Boulenger, 1910b.
Distribution: Kenya.
Sources: Kramer, 1961a, Pitman, 1965, Broadley, 1996b, Herrmann & Joger, 1995, Spawls & Branch, 1995, David & Ineich, 1999, Herrmann et al., 1999, McDiarmid et al., 1999, Spawls et al., 2002, Mallow et al., 2003, Dobiey & Vogel, 2007, Krecsák, 2007 and Phelps, 2010.

1. *Montatheris hindii* (Boulenger, 1910b). Ann. Mag. Nat. Hist. (8) 5(30): 513. (*Vipera hindii*)

Synonym: *Hindius hindii* A.F.T. Reuss, 1939a *(nomen nudum)*.
Type: Holotype, BMNH 1946.1.18.10 (formerly BMNH 1908.9.17.7), a 300 mm male (S.L. Hinde, 1902–1908).
Type locality: "Fort Hall, Kenya District, 4000 ft" [= Muranga, Central Prov., cen. Kenya, 0°43'S, 37°10'E, elevation 1285 m] (in error). Corrected to Aberdare Mtns., N Central Prov., cen. Kenya, bet. 0°10–50'S, 36°30–50'E, elevation 2200–3700 m *fide* Loveridge, 1936b: 44].
Distribution: Central Kenya (Central, cen. Eastern), 2700–3800 m.
Remarks: Holotype probably sent from Fort Hall by Hinde when stationed there *fide* Loveridge (1957: 301) and collected elsewhere.

MONTIVIPERA Nilson, Tuniyev, Andrén, Orlov, Joger & Herrmann, 1999a (Viperidae)

Type species: *Daboia xanthina* Gray, 1849a.

Distribution: Middle East.

Fossil records: Lower Miocene and upper Pliocene of Europe. Fossils unassigned to species include upper Pliocene (Villanyian, MN 16–17: 1.8–3.2) of France and Spain. *Montivipera* is included with a group, which as unassigned fossil material, are referred to as the "Oriental vipers", and may include any of the following records: lower Miocene (Orleanian, MN 4–5: 13.7–16.9 mya) of France, Germany, Portugal and Russia, middle Miocene (Astaracian, MN 6-8: 11.1–13.7 mya) of France and Germany, upper Miocene (Vallesian, MN 9–10: 8.7–11.1 mya) of Hungary, Turkey and Ukraine, upper Miocene (Turolian, MN 13: 4.9–7.3 mya) of Greece, Hungary, Italy and Ukraine, lower Pliocene (Ruscinian, MN 14: 4.2–4.9 mya) of Spain and Ukraine, upper Pliocene (Villanyian, MN 16: 2.6–3.2 mya) of Greece, upper Pliocene lower/Pleistocene of Morocco, and middle Pleistocene (Ionian: 0.13–0.78 mya) of Greece.

Sources: Golay et al., 1993, Nilson et al., 1999, Lenk et al., 2001b, Szyndlar & Rage, 2002, Mallow et al., 2003, Krecsák, 2007, Stümpel & Joger, 2009 and Phelps, 2010.

M

1. *Montivipera albicornuta* (Nilson & Andrén, 1985a). Amphibia-Reptilia 6(2): 207–210, fig. 1. (*Vipera albicornuta*)

Type: Holotype, RSI 3098, a 573 mm male (M. Latifi, 1971).

Type locality: "Abhar in the Zanjan valley, between Tabriz and Teheran, NW Iran" [= Abhar, Zanjan Prov., N Iran, 36°08'N, 49°13'E, elevation 1545 m].

Distribution: North-central Iran (Zanjan), 1545 m.

Sources: Tuniyev et al., 2009 and Rajabizadeh et al., 2011.

Remarks: Possibly conspecific with *M. raddei.*

2. *Montivipera albizona* (Nilson, Andrén & Flärdh, 1990). Amphibia-Reptilia 11(3): 287–289, figs. 1–3. (*Vipera albizona*)

Type: Holotype, GNM 5022, a 266 mm female (G. Nilson, 2 June 1989).

Type locality: "from the regions of the Kulmac Daglari mountain Distribution, central Turkey, close to the Anatolian 'Diagonal'."

Distribution: Central Turkey.

Sources: Schätti et al., 1991 and Mulder, 1994.

3. *Montivipera bornmuelleri* (F. Werner, 1898a). Zool. Anz. 21(555): 218–220. (*Vipera bornmuelleri*) (*nomen corrigendum*)

Synonym: *Lachesis libanotica* Hemprich, 1827 *(nomen nudum), Vipera bornmülleri* F. Werner, 1898a *(nomen incorrigendum), Vipera bornmuelleri* – Mertens, 1967 *(nomen corrigendum),* and *Vipera bonnmulleri* – Khole, 1991 *(nomen incorrectum).*

Type: Lectotype, ZSM 1653, (J.B. Bornmüller), designated by F. Werner (1922: 210, fig. 2a), destroyed 24–25 April 1944 during World War II.

Type locality: "Libanon, 1800 m" [= Lebanon, 1800 m] via lectotype selection. Restricted to Bcherré, Lebanon *fide* Mertens (1967b: 154).

Distribution: Middle East. Southern Lebanon (Mont-Liban), SW Syria (Ma'an) and ext. N Israel (N Golan Heights), 450–2500 m.

Sources: Mertens, 1967b, Hraoui-Bloquet, 1981, Nilson & Andrén, 1985b, Ilani, 1983a, Gruber, 1989, Nilson et al., 1990, Herrmann & Joger, 1997, Hraoui-Bloquet et al., 2002 and Bar & Haimovitch, 2011.

Remarks: Type locality restriction of Kar Boghaz, Cilicischer Taurus, Kleinasien *fide* Mertens & Müller (1928: 52) invalid. A valid species *fide* David & Ineich (1999: 334). Previously a synonym or subspecies of *V. xanthina.*

4. *Montivipera bulgardaghica* (Nilson & Andren, 1985b). J. Herp. 19(2): 277–280, fig. 1. (*Vipera bulgardaghica*)

Type: Holotype, GNM 1618, a 481 mm male (M. Holtz, 1897).

Type locality: "Cilician Taurus (Kar Boghaz, Bulgar Dagh, 2500m), Province Nigde (= Karbogaz, Bo lkar dagi Ulukisla), south central Anatolia, Turkey."

Distribution: South-central Turkey (Bolkar Daghlari Mts. of Anatolia), 2500 m.

Sources: Nilson & Andrén, 1985b, Gruber, 1989, Nilson et al., 1990 and Mulder, 1994

Remarks: Holotype is also a paratype of *V. bornmuelleri.* A valid species *fide* David & Ineich (1999: 334). A synonym of *M. albizona fide* Mulder (1994: 82).

5. *Montivipera kuhrangica* Rajabizadeh, Nilson & Kami, 2011. Russ. J. Herp. 18(3): 235–238, figs. 1–2.

Type: Holotype, ZMGU 2203, a 667 mm male (M. Sadeghi, May 2008).

Type locality: "Tulip valley (Darreye Lake or Dashte Lake) (32°36' N 50°11' E), 8 km north east of Chelgerd village, Kuhrang region, Chaharmahal and Bakhtiari province (the province is located in highest part of central Zagros mountains), Iran. Altitude 2490 m."

Distribution: Iran (Zagros Mts. of Bakhtiari, Chaharmahal), 2490 m.

6. ***Montivipera latifi*** **(Mertens, Darevsky & Klemmer, 1967). Senck. Biol. 48(3): 161–166, figs. 1–5. (*Vipera latifi*)**

Type: Holotype, SMF 62585, a 542 mm male (M. Latifi, 2 Oct. 1966).
Type locality: "Hochtal von Lar (2180–2900 m Höhe), süwestlich des Demavend-Gipfels in Elburs-Gebirge, nordöstlich von Teheran, Iran" [= SW Damavand Mtn. summit, upper Lar Valley, Elburz Mtns., S Mazandaran Prov., N Iran, ca. 35°54'N, 52°01'E, 2180–2900 m].
Distribution: Northern Iran (S Mazandran), 2180–2900 m.
Sources: Nilson & Andrén, 1985a–b, Schätti et al., 1991 and Rajabizadeh et al., 2011.
Remarks: Probably a synonym of *M. raddei fide* Schätti et al., 1991.

7. **†*Montivipera platyspondyla*** **(Szyndlar, 1987). J. Vert. Paleont. 7(1): 67–68, figs. 10c–g. (†*Vipera platyspondyla*)**

Type: Holotype, CUP 940, one trunk vertebra.
Type locality: "Lower Miocene (MN 4) locality of Dolnice near to the town of Cheb in westernmost Bohemia (Czechoslovakia)."
Distribution: Lower Miocene (Orleanian, MN 4: 16.0–16.9 mya) of Czech Republic and Germany.
Sources: Szyndlar & Schleich, 1993, Nilson & Andrén, 1997 and Szyndlar & Rage, 2002.

8. ***Montivipera raddei*** **(Boettger, 1890a). Zool. Anz. 13(327): 62–63. (*Vipera raddei*)**

Synonyms: *Vipera raddii* Boulenger, 1896a *(nomen emendatum)* and *Vipera raddei kurdistancia* Nilson & Andrén, 1986.
Types: Syntypes (2), ZMT (Christoph), lost *fide* Nilson & Andrén (1986: 5).
Type locality: "Kasikoperan in Armenien" [= Kozkoparan, Tuzluca Distr., Igdir Prov., NE Turkey, 40°01'N, 43°26'E, elevation 1810 m].
Distribution: Southwestern Asia. Eastern Turkey (Kars), S Armenia, NW Iran (West Azerbaijan) and possibly ext. NE Iraq, (200) 1100–3000 m.
Sources: Groombridge, 1980, Nilson & Andrén, 1988a, Schätti et al., 1991, Leviton et al., 1992, Herrmann & Joger, 1997, Tuniyev et al., 2009 and Rajabizadeh et al., 2011.
Remarks: BMNH 6.7.96 is a topotype *fide* Golay et al. (1993: 288). A valid species *fide* Groombridge (1980: 69). Previously a subspecies of *V. xanthina* (Gray).

9. ***Montivipera wagneri*** **(Nilson & Andrén, 1984a). Bonn. Zool. Beitr. 35(1–3): 178–181, figs. 1–3. (*Vipera wagneri*)**

Type: Holotype, ZFMK 23495, a 291 mm female (F.J.C.M. Wagner, 1846).
Type locality: "vicinity of lake Urmia ("Armenisch-Persische Grenze"), province Azarbaijan, N.W. Iran," (probably in error *fide* Nilson & Andrén (1984: 181).
Distribution: Mountains of NE Turkey (Kars) and adjacent NW Iran (West Azerbaijan), 1600–1900 m.
Sources: Joger et al., 1988, Nilson et al., 1990, Mulder, 1994, Herrmann & Joger, 1997 and Tuniyev et al., 2009.
Remarks: Holotype is erroneously labeled ZFMK 32495 in Nilson & Andrén (1984: fig. 1). Photographs of holotype in P. Wagner et al. (2012: fig. 4). Status uncertain; possibly a synonym of *V. xanthina*.

10. ***Montivipera xanthina*** **(Gray, 1849a). Cat. Snakes Coll. Brit. Mus.: 24. (*Daboia xanthina*)**

Synonym: *Vipera xanthina santhina* – González, 1991 *(nomen incorrectum)*.
Type: Lectotype, BMNH 1946.1.18.11, adult female (C. Fellows, April 1838), designated by Nilson & Andrén (1986: 63).
Type locality: "Xanthus (Kinik), province Mugla, S. W. Turkish Anatolia" via lectotype selection.
Distribution: Northeastern Greece (Macedonia, Leros, Lesvos, Chios, Oinousses, Thrace) and W Turkey (W Anatolia), NSL–2700 m.
Sources: Eiselt & Baran, 1970, Bannikov et al., 1977, Gärdenfors, 1980, Nilson & Andrén, 1985a, 1985b, 1986, 1992, van Wingerde, 1986, Dimitropoulos, 1987, 1990, Schätti & Baran, 1988, Schätti et al., 1991, Mulder, 1994, Herrmann & Joger, 1997, Nilson et al., 1999, Szyndlar & Rage, 1999, Smedt, 2006, Valakos et al., 2004, 2008 and Trapp, 2007.
Remarks: Schatti et al. (1991: 305) included *M. albizona*, *M. bornmuelleri*, and *M. bulgardaghica* as synonyms.

MORELIA Gray, 1842a (Pythonidae)

Synonyms: *Spilotos* Lacépède, 1804 *(nomen oblitum)*, *Simalia* Gray, 1849a, *Aspidopython* A.B. Meyer, 1874, *Chondropython* A.B. Meyer, 1874, *Condropython* – Sauvage, 1879a *(nomen incorrectum)*, *Hypaspistes* Ogilby, 1891 *(nomen praeoccupatum)*, *Australiasis* Wells & Wellington, 1984, *Chondrophython* – M.J. Smith & Plane, 1985 *(nomen incorrectum)*, *Montypythonoides* M.J. Smith & Plane, 1985, *Nyctophilopython* Wells & Wellington, 1985, *Lenhoserus* Hoser, 2000b, and *Jackypython* Hoser, 2009a & 2012b *(nomen illegitimum)*.
Type species: *Coluber spilotus* Lacépède, 1804.
Distribution: Australasia.
Fossil Records: Upper Oligocene and middle-upper Miocene of Australia.
Sources: Rooij, 1917, C. Haas, 1950, Stimson, 1969, Matz & Matz, 1971, McDowell, 1975a, Matz, 1980, M.J. Smith, 1981a, 1985, Cogger et al., 1983a, S.K. Wilson & Knowles, 1988, Gow, 1989, Hoser, 1989, 2009b,

2012b, Underwood & Stimson, 1990, Kluge, 1993b, D.G. Barker & Barker, 1994, McDiarmid et al., 1999, Cogger, 2000, Kortlang & Green, 2001, Mense, 2006, Swan, 2007, Rawlings et al., 2008, Flagle & Stoops, 2009, and Schleip & O'Shea, 2010.

1. *Morelia amethistina* (J.G. Schneider, 1801). Hist. Amph. 2: 254–259. (*Boa amethistina*)

Synonyms: *Coluber flavocaeruleus* Lacépède, 1789 (*nomen rejiciendum*), *Python amethystinus* Daudin, 1803 *(nomen emendatum), Natrix oularjawa* Wagler, 1825 (*nomen ineditum*), *Boa amethystina* – Gray *in* Griffith & Pidgeon, 1831, *Aspidopython jakati* A.B. Meyer, 1874, *Liasis duceboracensis* A.C.L.G. Günther, 1879b, *Hypaspistes dipsadides* Ogilby, 1891, *Liasis clarki* T. Barbour, 1914e, *Australiasis funki* Hoser, 2012b *(nomen illegitimum),* and *Morelia wellsi* Hoser, 2012b *(nomen illegitimum).*

Type: Lectotype, ZMB 1485 (M.E. Bloch), designated by Stimson (1969: 23), lost *fide* Kluge (1993b: 5).

Type locality: Unknown.

Distribution: Austro-Papua. Eastern Indonesia (Papua, Aru, Biak, Gag, Kai, Misool, Salawati, Yapen), Papua New Guinea (Central, Chimbu, Milne Bay, Eastern Highlands, East Sepik, Gulf, Madang, Morobe, New Ireland, Northern, Western, Western Highlands, West Sepik, Daru, Duke of York, Fergusson, Normanby; Rossel, Trobriand and Umboi Is.) and N Australia (N Queensland, Cornwallis and Murray Is.), NSL–1600 m.

Sources: Whitaker et al., 1982, F. Parker, 1982, O'Shea, 1996, B. Oliver, 1998, Harvey et al., 2000, Hoser, 2000b, Torr, 2000 and Kraus & Allison, 2004.

Remarks: Records for the Philippines (Mindanao), Solomons, Timor and Samao erroneous *fide* McDowell (1975a: 60).

2. *Morelia azurea* (A.B. Meyer, 1874). Mber. Königl. Akad. Wiss. Berlin 1874(2): 134. (*Chondropython azureus*)

Synonym: *Chondropython pulcher* Sauvage, 1878.

Types: Syntypes (3), MTKD 638–39 (A.B. Meyer, 1873), destroyed 13 Feb. 1945 during World War II, and ZMB 8832, a male (A.B. Meyer, 1873), lost *fide* Stimson (1969: 9).

Type locality: "Mysore" [= Biak Is., Schouten Arch., N West Papua Prov., E Indonesia]. Emended to Kodor, Mysore *fide* Obst (1977: 179).

Distribution: New Guinea. Eastern Indonesia (N Papua, West Papua, Biak, Salawati, Sorong, Yapen) and N Papua New Guinea (East Sepik, Madang, Morobe, Northern, West Sepik), NSL–1900 m.

Sources: Rawings & Donnellan, 2003, Maxwell, 2005 and Schleip & O'Shea, 2010.

Remarks: Schleip & O'Shea (2010: 45) urged the designation of a neotype.

3. *Morelia boeleni* (Brongersma, 1953). Proc. Kon. Ned. Akad. Weten. 56C(3): 317–318, pl. 1, figs. 1–5. (*Liasis boeleni*)

Synonym: *Liasis taronga* Worrell, 1958.

Type: Holotype, RMNH 9651, a 2190 mm female (K.W.J. Boelen, 25 Dec. 1952).

Type locality: "Dimija (3° 56' S, 136° 18' E), Wissel Lakes, Dutch New Guinea, about 1750 m (5700 feet) above sea level" [= W Papua Prov., E Indonesia].

Distribution: New Guinea. Eastern Indonesia (W Papua) and Papua New Guinea (Central, Eastern Highlands, Morobe, Western, Western Highlands), 1000–2000 m.

Sources: Brongersma, 1956b, 1969, F. Parker, 1982, Whitaker et al., 1982, O'Shea, 1996 and Flagle & Stoops, 2009.

4. *Morelia bredli* (Gow, 1981). Aust. J. Herp. 1(1): 29–30, pl. 1, fig. 4a. (*Python bredli*)

Type: Holotype, NTM 4535, a 2049 mm male (K. Roth, 12 Jan. 1977).

Type locality: "Pitchie Ritchie Park, Alice Springs, Northern Territory, central Australia (23°42', 133°51')."

Distribution: Central Australia (S Northern Territory), 430–775 m.

Sources: Barker & Barker, 1994, Fyfe, 1994, Cogger, 2000 and Mense, 2006.

5. *Morelia carinata* (L.A. Smith, 1981a). Rec. West. Aust. Mus. 9(2): 220–222, figs. 5–6. (*Python carinatus*)

Type: Holotype, WAM 45352, a 1975 mm specimen (L.A. Smith, R.E. Johnstone & J.A. Smith, 14 Jan. 1973).

Type locality: "Mitchell River Falls, Admiralty Gulf on north-west coast of Kimberley, Western Australia, Australia (14°50'S, 125°42'E)."

Distribution: Northwestern Australia (ext. N Western Australia), 10–205 m.

Sources: Weigel & Russell, 1993, Torr, 2000, Mense, 2006 and Hoser, 2009a.

6. *Morelia clastolepis* Harvey, Barker, Ammerman & Chippindale, 2000. Herpetol. Monogr. (14): 158–161, figs. 2a, 3c–d.

Type: Holotype, UTA 44486 (formerly VPI 68), a 1476–1588 mm female (natives, 1993).

Type locality: "island of Ambon, Maluku, Indonesia."

Distribution: Indonesia (S Maluku: Ambon, Seram).

Remarks: Probably occurs on Haruku and Saparua *fide* Harvey et al. (2000: 161).

M

7. *Morelia kinghorni* (Stull, 1933). Occ. Pap. Mus. Zool., Univ. Michigan (267): 3–4. (*Liasis amethistinus kinghorni*)

Type: Holotype, MCZ 35022, a male (W.E. Schevill, 1931–1932).
Type locality: "Lake Barrine, Queensland, Australia."
Distribution: Extreme NE Australia (ext. NE Queensland), NSL–755 m.
Remarks: Resurrected *fide* Wells & Wellington (1985: 41).

8. *Morelia nauta* Harvey, Barker, Ammerman & Chippindale, 2000. Herpetol. Monogr. (14): 163–166, figs. 1a, 3a–b, 9.

Type: Holotype, UTA 44482 (formerly VPI 62), an 1816 mm female (natives, 1996).
Type locality: "Yamdena Island of the Tanimbar group, Maluku, Indonesia."
Distribution: Eastern Indonesia (Tanimbars: Yamdena).

9. *Morelia oenpelliensis* (Gow, 1977a). Aust. Zool. 19(2): 133–136, figs. 1–3. (*Python oenpelliensis*)

Type: Holotype, NTM 840, a 3560 mm male (B. Jukes, 17 June 1975).
Type locality: "6.5 km S. W. of Oenpelli, Northern Territory, Australia (12°21'S., 133°01'E.)."
Distribution: Extreme N Australia (ext. N Northern Territory), NSL–290 m.
Sources: Begg & Martin, 1980, Sadlier, 1990 and Torr, 2000.

10. †*Morelia riversleighensis* (L.A. Smith & Plane, 1985). BMR J. Aust. Geol. Geophy. 9: 191–192, fig. 2. (†*Montypythonoides riversleighensis*)

Synonym: †*Morelia antiquus* M.J. Smith & Plane, 1985.
Type: Holotype, QMF 12926, one right maxilla fragment with 9 teeth.
Type locality: "Henk's Hollow, approximately 3.6 km southwest of Tedford's (1967) Site B, Riversleigh, Queensland, Australia; Riversleigh local fauna, middle Miocene."
Distribution: Middle Miocene (10.4–16.3 mya) of Australia.
Source: Scanlon & Mackness, 2001.

11. *Morelia spilota* (Lacépède, 1804). Ann. Mus. Hist. Nat. 4: 209. (*Coluber spilotus*)

Synonyms: *Python punctatus* Merrem, 1820, *Vipera spilotes* Merrem, 1820 *(nomen emendatum)*, *Python peronii* Wagler, 1828a, *Morelia punctata* Gray, 1842a, *Morelia variegata* Gray, 1842a, *Morelia argus* A.-M.-C. Duméril & Bibron, 1844, *Morelia spilotes* Gray, 1849 *(nomen emendatum)*, *Morelia argus fasciolata* Jan, 1863b (*nomen nudum*), *Morelia argus fasciolata* Jan & Sordelli 1864 *in* 1860–1866, *Python spilotes macrospila* F. Werner, 1909e, *Python spilotus imbricatus* L.A. Smith, 1981a, *Morelia cheynei* Wells & Wellington, 1984, *Morelia mcdowelli* Wells & Wellington, 1984, *Morelia metcalfei* Wells & Wellington, 1985, *Morelia harrisonii* Hoser, 2000b, *Morelia macburniei* Hoser, 2003a, and *Morelia mippughae* Hoser, 2003a *(nomen dubium)*.
Type: Holotype, MNHN 3272, a 2000+ mm specimen (Boudin), lost *fide* Stimson (1969: 28).
Type locality: "Nouvelle-Hollande" [= Australia].
Distribution: Austro-Papua. Eastern Indonesia (E Papua), S Papua New Guinea (Western, Central, Yule Is.) and Australia (N and SW Western Australia, N Northern Territory, Queensland, New South Wales, N Victoria, S South Australia, Abrolhos, Bathurst, Garden, Groote Eylandt, Melville, Milingimbi, Mondrain, St. Francis, Thursday and West Wallabi Is.), NSL–1125 m.
Sources: J. White, 1790, F. McCoy, 1878g, Mackay, 1950, Storr, 1965, G. Swan, 1990, Covacevich & Couper, 1991, Bush et al., 1995, Slip & Shine, 1998a–e, Shea & Sadlier, 1999, Torr, 2000, Mense, 2006, Schleip & O'Shea, 2010 and Mutton & Julander, 2011.

12 *Morelia tracyae* Harvey, Barker, Ammerman & Chippindale, 2000. Herpetol. Monogr. (14): 166–169, figs. 1b, 3e–f.

Type: Holotype, UTA 44473, a 2934 mm male (skin dealer, Feb. 1995).
Type locality: "near Tobelo, Halmahera Island, Maluku, Indonesia" [= Tobelo, NE Halmahera Is., Maluku, Indionesia, 1°44'N, 128°00'E, elevation 30 m].
Distribution: Central Indonesia (Batjan, NW Halmahera), 30 m.
Source: J.C. Murphy & Schlager, 2003.

13 *Morelia viridis* (Schlegel, 1872). De Dierentuin Nat. Artis Mag., Rept. (4): 54. (*Python viridis*)

Synonyms: *Chondropython viridis shireenae* Hoser, 2003b and *Chondropython viridis adelynhoserae* 2009a & 2012b.
Type: Lectotype, RMNH 4672 (formerly Amsterdam Zoo) (Rosenberg, 1866), designated by Stimson (1969: 8).
Type locality: "Aru Islands" [E Indonesia] via lectotype selection.
Distribution: Austro-Papua. Eastern Indonesia (Papua, Aru Arch.), S Papua New Guinea (Central, Chimbu, Eastern Highlands, Enga, Gulf, Milne Bay, Northern, Western, Western Highlands, Normanby Is.) and ext. NE Australia (ext. N Queensland), NSL–2000 m.
Sources: McDowell, 1975a, O'Shea, 1996, Weier & Vitt, 1999, Hoser, 2000b, 2003a, Torr, 2000, J.C. Murphy & Schlager, 2003, Rawlings & Donnellan, 2003, Kivit & Wiseman, 2005, Maxwell, 2005 and Arth & Baus, 2006.
Remarks: Solomons records rejected *fide* McDowell (1975a: 72).

MUSSURANA Zaher, Grazziotin, Cadle, Murphy, Moura-Leite & Bonatto, 2009 (Xenodontidae)

Type species: *Oxyrhopus bicolor* Peracca, 1904b.
Distribution: Southern South America.
Source: Zaher et al., 2009.

1. *Mussurana bicolor* (Peracca, 1904b). Rev. Suisse Zool. 12(3–4): 667–668. (*Oxyrhopus bicolor*)

Type: Holotype, MHNG 677.47, a 657 mm male (Dupuis, 1903).
Type locality: "Nord de Santa Fé, Argentine" [= N of Santa Fe (31°38'S, 60°42'W, elevation 20 m), Santa Fe Prov., NE Argentina].
Distribution: Southern South America. Southeastern Brazil (S Mato Grosso, Mato Grosso do Sul, Paraná, Rio Grande do Sul, São Paulo), Peru (Huánuco, Loreto), Bolivia (Beni, Santa Cruz), Paraguay (Alto Paraguay, Asunción, Central, Cordillera, Itapúa, Neembucú, Presidente Hayes) and N Argentina (Chaco, Corrientes, Formosa, Jujuy, La Rioja, Misiones, San Luis, Santa Fe, Tucumán), 125–465 m.
Sources: Vanzolini, 1986a, Cei, 1994, Aquino et al., 1996, Couturier & Faivovich, 1996, Cacivio, 1999, Leynaud & Bucher, 1999, N.J. Scott et al., 2006, Cacciali, 2008 and Jansen, 2008.

M

2. *Mussurana montana* (Franco, Marques & Puorto, 1997). J. Herp. 31(4): 486–487, fig. 2. (*Clelia montana*)

Type: Holotype, IB 22505, a 946 mm male (J. Morteira, 27 Nov. 1962), destroyed by fire 15 May 2010.
Type locality: "city of Guaratinguetá (22°49'S; 45°93'W), Sodré mill, state of São Paulo, Brazil."
Distribution: Southeastern Brazil (São Paulo).

3. *Mussurana quimi* (Franco, Marques & Puorto, 1997). J. Herp. 31(4): 484–485, fig. 3. (*Clelia quimi*)

Type: Holotype, IB 53743, a 1090 mm male (L.M. Gelpe, 27 March 1990), destroyed by fire 15 May 2010.
Type locality: "Santo António farm, city of Itu (23°16'S 47°19'W), state of São Paulo (SP), Brazil."
Distribution: Southern South America. Brazil (Bahia, Espírito Santo, Federal District, Goiás, Minas Gerais, Paraná, Santa Catarina, São Paulo), Paraguay (Itapúa) and N Argentina (Misiones), 180 m.
Sources: Giraudo, 1999, Lima-Silveira & Agostini-Cotta, 2006 and N.J. Scott et al., 2006.

MYERSOPHIS E.H. Taylor, 1963 (Colubridae)

Type species: *Myersophis alpestris* E.H. Taylor, 1963.
Distribution: Philippines.
Sources: E.H. Taylor, 1963 and Leviton, 1983.

1. *Myersophis alpestris* E.H. Taylor, 1963. Copeia 1963(2): 430–431.

Type: Holotype, EHT-HMS 3109 (formerly EHT-CC 4820), a 664 mm female (E.H. Taylor, 23 May 1961), lost *fide* Leviton (1983: 212).
Type locality: "elevation circa 1,980 m in mountains near Banaue, Mountain Province, Luzon, Philippines."
Distribution: Northern Philippines (Luzon), 1980 m. Known only from type locality.
Remarks: The paratype, KU 203013, a 176 mm male (E.H. Taylor, 23 May 1961) is extant.

MYRIOPHOLIS Hedges, Adalsteinsson & Branch *in* Adalsteinsson, Branch, Trape, Vitt & Hedges, 2009 (Leptotyphlopidae)

Synonyms: *Ramphostoma* Jan & Sordelli, 1860 *in* 1860–1866 *(nomen praeoccupatum), Bobbottomus* Hoser, 2012am *(nomen illegitimum), Longinidis* Hoser, 2012am *(nomen illegitimum),* and *Scanlonus* Hoser, 2012am *(nomen illegitimum).*
Type species: *Stenostoma longicaudum* W.C.H. Peters, 1854.
Distribution: Africa, Arabia and SW Asia.
Sources: Hahn, 1980a, Wallach, 1998b, McDiarmid et al., 1999 and Hedges et al., 2009.

1. *Myriopholis adleri* (Hahn & Wallach, 1998). Hamadryad 23(1): 51, fig. 1. (*Leptotyphlops adleri*)

Type: Holotype, MNHN 1978.1910, a 135 mm male (Evard, 1974).
Type locality: "Bongor (10°17'N, 15°22'E), Mayo Kebbi Prefecture, Chad" [elevation 330 m].
Distribution: West Africa. Southern Senegal (Kolda, Tambacounda), W Burkina Faso (Volta-Noire), SW Niger (Dosso), N Benin (Alibori), SW Chad (Mayo Kebbi), N Cameroon (Extrême-Nord) and N Central African Republic (Vakaga), 110–900 m.
Sources: J.F Trape, 2002, 2005, J.-F. Trape & Mané, 2004, 2006b, Chirio & Ineich, 2006, Chirio & LeBreton, 2007, Chirio, 2009 and Hughes, 2013.

2. *Myriopholis albiventer* (Hallermann & Rödel, 1995). Stuttgart. Beitr. Naturk. A (532): 3–5, figs. 1a–c, 2a–b. (*Leptotyphlops albiventer*)

Type: Holotype, SMNS 8454.5, a 148 mm specimen (M.O. Rödel, 4 Sept. 1994).

Type locality: "Ivory Coast, Comoé National Park, 8°45'N, 3°47'W" [= Bouna Dept., NE Ivory Coast, elevation 250 m].

Distribution: West Africa. Guinea Bissau, S Mali (Sikasso), and N Ivory Coast (Bouna, Ferkéssédougou), 250–360 m.

Source: J.-F. Trape & Mané, 2006b.

3. *Myriopholis algeriensis* (Jacquet, 1898). Bibliogr. Anat. 4(2): 79–81, figs. 1–7. (*Glauconia algeriensis*)

Synonym: *Glauconia natatrix* Andersson, 1937.

Type: Holotype, not designated, ? INCDS, a 280 mm specimen (native, March 1894).

Type locality: "environs du Hammam Salahin, bain situé à six kilometers au nord de Biskra, Algérie" [= Hammam Salhin, 6 km. N Biskra, Biskra, NE Algeria, ca. 34°54'N, 05°45'E, elevation 165 m].

Distribution: Northwestern Africa. Southern Morocco (Agadir, Kaar es Souk, Ouarzazate), Algeria (Béchar, Biskr, Illizi, Tindouf), SW Libya (Awbari), S. Tunisia (Tataouine), NE Western Sahara (Saguia El Hamra), NW Mauritania (Adrar, Tagant), N Mali (Gao, Tombouctou), N Burkina Faso (Volta-Noire), N Niger (Agadez), and NW Nigeria (Bauchi, Sokoto), 10–1300 m.

Sources: Boulenger, 1920d, Villiers, 1956, Papenfuss, 1969, Roman, 1980, Le Berre, 1989, Bons & Geniez, 1996, Schleich et al., 1996, Hahn & Wallach, 1998, Joger & Lambert, 2002, Trape, 2002, Geniez et al., 2004, J.-F. Trape & Mané, 2006b and D.J. Harris et al., 2010.

4. *Myriopholis blanfordii* (Boulenger, 1890a). Fauna Brit. India, Rept. Batr.: 243, figs. 72a–b. (*Glauconia blanfordii*)

Synonyms: *Glauconia laticeps* Nikolsky, 1907 and *Glauconia carltoni* T. Barbour, 1908.

Type: Lectotype, BMNH 1946.1.16.85 (orig. BMNH 1869.8.28.58), a 237 mm specimen (A. Leith-Adams, 1861–1868), designated by Hahn (1978: 485).

Type locality: "Sindh" [= Sindh Prov., SE Pakistan].

Distribution: Southwestern Asia. Southern Iran (Bushehr, Fars, Lorestan, Sistan va Baluchestan), E Afghanistan (Nangarhar, Nimruz), Pakistan (Balochistan, Khyber Pakhtunkhwa, Punjab, Sindh) and NW India (Gujarat, Haryana, Punjab, Rajasthan), 20–1525 m.

Sources: Alcock & Finn, 1897, Wall, 1923c, M.A. Smith, 1943, R.J. Clark, 1990, M.S. Khan, 2002, R. Sharma, 2003, 2007 and Broadley & Wallach, 2007a.

Remarks: Head of lectotype illustrated by Hahn (1978: 486, fig. 7).

5. *Myriopholis boueti* (Chabanaud, 1917c). Bull. Mus. Hist. Nat., Paris 23(1): 9–10, figs. 1–3. (*Glauconia boueti*)

Synonym: *Glauconia debilis* Chabanaud, 1918b.

Type: Holotype, MNHN 1917.24, a 166 mm specimen (G. Bouet, 1911).

Type locality: "Soudan: Djenné" [= Djenné, Mopti Region, S Mali, 13°54'N, 04°33'W, elevation 275 m].

Distribution: West Africa. Southern Mauritania (Brakna, Hodh El Gharbi), Senegal (Dakar, Fatick, Saint-Louis, Tambacounda, Goré Is.), Gambia, Guinea–Bissau, S Mali (Koulikoro, Mopti), Burkina Faso (Centre, Centre-Ouest, Est, Sud-Ouest, Volta-Noire), SW Niger (Dosso) and S Chad (Chari-Baguirmi, Guéra), NSL–350 m.

Source: Boulenger, 1920c, Villiers, 1950a–b, 1951b, 1953, 1954a, 1955, 1956a, Roman, 1980, 1984, Hahn & Wallach, 1998, J.-F. Trape & Mané, 2000, 2006b, J.-F. Trape, 2005, Villiers & Condamin, 2005 and Chirio, 2009.

Remarks: A subspecies of *Leptotyphlops narirostris fide* Villiers (1950: 58). Type locality listed as Sudan (error for French Sudan = Mali) *fide* Chabanaud (1917: 10). Possibly occurs in N Benin *fide* Hughes (2013: 153).

6. *Myriopholis braccianii* (Scortecci, 1928). Atti. Soc. Ital. Sci. Nat. Mus. Civ. Stor. Nat. Milano 67(3–4): 294–295, fig. 3. (*Glauconia braccianii*)

Synonyms: *Glauconia variabilis* Scortecci, 1928 and *Glauconia fiechteri* Scortecci, 1929.

Type: Holotype, MSNM 3351 (formerly MSNM 1917), a 116–119 mm specimen (G.F. Turati).

Type locality: "Adi Ugri, Eritrea" [= Mendefera, Southern Red Sea Prov., Eritrea, 14°53'N, 38°49'E, elevation 1900 m].

Distribution: Northeastern Africa. Western Sudan (Janub Darfur, Shamal Darfur), South Sudan (Bahr-el-Ghazel, Western Equatoria), Eritrea (Anseba, Gash-Barka, Southern), S Ethiopia (Illubabor, Sidamo), Somalia (Bari, Hiraan, Jubbada Hoose, Mogadishu, Shabeellaha Dhexe, Shabeellaha Hoose) and NE Kenya (N Coast), 100–1900 m.

Sources: Loveridge, 1957a, Lanza, 1983a, 1990b, Largen & Rasmussen, 1993, Wallach & Lanza, 2004, Broadley & Wallach, 2007a and Largen & Spawls, 2010.

7. *Myriopholis burii* (Boulenger, 1905e). Ann. Mag. Nat. Hist. (7) 16(92): 178. (*Glauconia burii*)

Type: Holotype, BMNH 1946.1.16.56 (formerly BMNH 1903.3.6.48), a 210 mm specimen (G.W. Bury, 1900–1903).

Type locality: "El Kubar, Upper Hushabi, near Yemen frontier, South Arabia" [= Khawbar, upper Haushabi, SW Yemen, 13°48'N, 44°45'E, elevation 1350 m].

Distribution: Southwestern Yemen (Lahij), 1350–1500 m.

Sources: Corkill & Cochrane, 1966, Hahn, 1978, Gasperetti, 1988, Hahn & Wallach, 1998, Broadley & Wallach, 2007a and Egan, 2007.

8. *Myriolepis cairi* (A.-M.-C. Duméril & Bibron, 1844). Erpét. Gén. 6: 323–325. (*Stenostoma cairi*)

Synonyms: *Stenostoma fitzingeri* Jan, 1861b, *Glauconia variabilis* Scortecci, 1928, and *Leptotyphlops macrorhynchus bilmaensis* Angel, 1936b.

Type: Holotype, MNHN 3231, a 235–241 mm specimen (Birr).

Type locality: "environs du Caire" [= vicinity of Cairo, Cairo Govern, NE Egypt, 30°03'N, 31°15'E, elevation 200 m].

Distribution: Northeastern Africa. Eastern Niger (E Agadez), N Chad (Borkou), E Egypt (Assiut, Beheira, Beni Suef, Cairo, Gharbia, Giza, Menufia, Minya, New Valley, Qalyubia, Qena, Sohag, South Sinai, Suez; Elephantine, Kitchener & Rhodes Is.), cen. Sudan (An Ni Al Abyadl, Al Khartum, Senner, Shamal Kurdufan), E Eritrea (Northern Red Sea: Dahlak and Entedebir Is.) and W Somalia (Jubbada Dhexe, ? Sanaag, Woqooyi Galbeed), NSL–1300 m.

Sources: Boulenger, 1893a, 1915d, 1920d, Anderson, 1898, Parker, 1932, Flower, 1933, Kramer & Schnurrenberger, 1963, Marx, 1968, Pitman, 1974, Lanza, 1983a, 1990b, Abdeen et al., 1991a–b, Largen & Rasmussen, 1993, Hallermann & Rödel, 1995, Largen, 1997, Hahn & Wallach, 1998, Le Berre, 1999, J.-F. Trape, 2002, Baha El Din, 2006, J.-F. Trape & Mané, 2006b and Broadley & Wallach, 2007a.

Remarks: A record for Mauritania needs confirmation *fide* McDiarmid et al. (1999: 25).

M

9. *Myriopholis erythraeus* (Scortecci, 1928). Atti. Soc. Ital. Sci. Nat. 67(3–4): 293–294, fig. 2. (*Glauconia erythraea*)

Type: Holotype, MSNM 3349 (formerly MSNM 1916), a 157–158 mm specimen (F. Fatigati).

Type locality: "Massaua" [= Massawa, Northern Red Sea Prov., Eritrea, 15°37'N, 39°28'E, elevation NSL].

Distribution: Northeastern Africa. Eastern Eritrea (Anseba, Northern Red Sea, Southern, Southern Red Sea, Entedebir Is.), Ethiopia (Illubabor, Shoa) and N Somalia (Bari, Sanaag), NSL–1970 m.

Sources: H.W. Parker, 1949, Leviton, 1986, Wallach & Lanza, 2004, Broadley & Wallach, 2007a and Largen & Spawls, 2010.

10. *Myriopholis filiformis* (Boulenger, 1899a). Bull. Liverpool Mus. 2: 7–8, pl. 9, figs. 2, 2a–b. (*Glauconia filiformis*)

Type: Lectotype, BMNH 1946.1.11.1 (orig. BMNH 1899.12.15.109), a 153–155 mm specimen (H.O. Forbes & W.R. Ogilvie-Grant, 30 Dec. 1898–16 Jan. 1899), designated by Hahn (1978: 480–481).

Type locality: "Sokotra (Jenna-agahan, 1200–2500 feet)" [= Jenna-agahan, E Sokotra Is., Yemen, 365–760 m] via lectotype selection.

Distribution: Yemen (Socotra Is.), 100–775 m.

Sources: Boulenger, 1903g, 1915d, Corkill & Cochrane, 1966, Hahn, 1978, Schätti & Desvoignes, 1999, Rösler &Wranik, 2004, 2006 and Razzetti et al., 2011.

11. *Myriopholis hamulirostris* (Nikolsky, 1907). Ann. Mus. Zool. Acad. Imp. Sci. St.- Pétersbourg 10(3–4): 286–288, figs. 4–5. (*Leptotyphlops hamulirostris*)

Type: Lectotype, MHNG 1326.72 (formerly ZIL 10299-305 & EK 3986), a 257 mm specimen (N.A. Zarudny, 28 March 1904), designated by Kramer & Schnurrenberger (1963: 478, fig. 2).

Type locality: "Aguljaschker in Arabistano" [= Aguliaschxer, Khuzestan Prov., SW Iran, 31°40'N, 49°40'E, elevation 835 m] via lectotype selection.

Distribution: Arabia and SW Asia. Saudi Arabia (Makkah, Najran, Riyadh), Oman (Dhofar, Muscat), United Arab Emirates (Abu Dhabi, Sharjah), Iraq (Baghdad), Iran (Busher, Fars, Ilam, S Kerman, Khuzestan, Sistan va Baluchestan), Afghanistan, Pakistan (Balochistan, Punjab, Sindh) and W India (Rajasthan), 230–1550 m.

Sources: Wettstein, 1951, Kramer & Schnurrenberger, 1963, Corkill & Cochrane, 1966, Hahn, 1978, Gallagher & Arnold, 1988, Gasperetti, 1988, Latifi, 1991, Schätti & Gasperetti, 1994, McDiarmid et al., 1999, Egan, 2007 and R. Sharma, 2007.

12. *Myriopholis ionidesi* (Broadley & Wallach, 2007a). Zootaxa (1408): 27–28, pl. 8, fig. 3. (*Leptotyphlops ionidesi*)

Type: Holotype, MCZ 57440 (formerly MCZ X-8333), a 145 mm female (C.J.P. Ionides, 27 April 1958).

Type locality: "Liwale, Southern Province, Tanzania (09°47'S, 38°00'E, elevation 600 m)" [= Liwale, Lindi, SE Tanzania, 09°48'S, 37°56'E, elevation 520 m].

Distribution: Southeastern Tanzania (Dar es Salaam, Lindi, Mtwara), N Malawi (Northern) and N Mozambique (Nampula), NSL–700 m.

13. *Myriopholis longicauda* (W.C.H. Peters, 1854). Verh. Preuss. Akad. Wiss. Berlin 1854(11): 621. (*Stenostoma longicaudum*)

Synonyms: *Glauconia brevirostralis* V.F.M. FitzSimons, 1930 and *Leptotyphlops longicaudus* – Broadley & Watson, 1976.

Type: Holotype, ZMB 4827, a 208 mm specimen (W.C.H. Peters, May, 1845).

Type locality: "Tette" [= Tete, Zambezi River, Tete Prov., W Mozambique, 16°10'S, 33°36'E, elevation 130 m].

Emended to Tette on the Zambezi River *fide* W.C.H. Peters (1882d: 103).

Distribution: Southeastern Africa. Southeastern Zambia (Central, Southern), Malawi (Northern, Southern), Mozambique (Gaza, Inhambane, Manica, Nampula, Sofala, Tete, Mozambique Is.), E Botswana (E Central, E North West), Zimbabwe (Manicaland, Mashonaland Central, Mashonaland East, Mashonaland West, Masvingo, Matebeleland North, Matebeleland South), Swaziland and NE South Africa (Gauteng, Limpopo, Mpumalanga), NSL–1550 m.

Sources: W.C.H. Peters, 1882d, Boulenger, 1893a, 1915c, Loveridge, 1957a, Broadley, 1959, Sweeney, 1961, V.F.M. FitzSimons, 1962a, 1974, Broadley & Watson, 1976, Broadley, 1983, 1988, Lanza, 1983a, 1990b, Pienaar et al., 1983, Auerbach, 1987, Branch, 1988, Hallermann & Rödel, 1995, Broadley & Wallach, 1996, 1997b, Branch, 1998, 2001, 2005, Spawls et al., 2002, Broadley et al., 2003, Marais, 2004 and Broadley & Blaylock, 2013.

Remarks: Original description reprinted in Peters (1855: 50–51). Probably occurs in Namibia *fide* M. Griffin (2003: 128).

14. *Myriopholis macrorhyncha* (Jan, 1860 *in* Jan & Sordelli, 1860–1866). Icon. Gén. Ophid. 1(1): pl. 5, fig. 12, pl. 6, figs. 12, 12a–b, e, g, n, p, r, v, x. (*Stenostoma macrorhynchum*)

Synonyms: *Glauconia macrorhynchus* Boulenger, 1890d and *Glauconia fiechteri* Scortecci, 1928.

Type: Holotype, formerly MSNM, a 230 mm specimen, destroyed in 1943 during World War II.

Type locality: "Sennaar" [= Sennar, Sennar Prov., SE Sudan, 13°34'N, 33°34'E, elevation 430 m].

Distribution: Northeastern Africa. Northern Egypt (Cairo, Faiyum, Giza, Matrouh, South Sinai), SE Sudan (Sennar), S Ethiopia (Illubabor), NE Somalia (NW Bari, NE Sanaag) and N Kenya (N Eastern, North-Eastern, Rift Valley), NSL–900 m.

Sources: Jan, 1861b, 1864d, Boulenger, 1893a, 1915d, 1920d, Villiers, 1952a, Kramer & Schnurrenberger, 1965, Marx, 1968, Hahn, 1978, Lanza, 1983a, 1990b, Le Berre, 1989, Largen & Rasmussen, 1993, Schleich et al., 1996, Saleh, 1997, Hahn & Wallach, 1998, Spawls et al., 2002, Wallach, 2003, Baha El Din, 2006, Broadley & Wallach, 2007a, Yildiz et al., 2009, 2010, Largen & Spawls, 2010 and Masroor, 2012.

Remarks: In need of revision as *M. macrorhyncha* probably occurs only in NE Africa. M.A. Smith (1943: 60) mentioned that the Asian populations called "*macrorhynchus*" may not be true *M. macrorhyncha*. Hahn (1978: 483) suggested that "several distinct subspecies" could be recognized. Recently the Distribution was believed to include West Africa (from Senegal to Nigeria), North Africa (from Morocco and Mauritania to Egypt), Turkey, the Middle East (Syria to Israel), Arabian Peninsula and SW Asia (from Iraq to W India), a highly unlikely scenario considering the size, dispersal ability and gene flow possibilities in these tiny snakes. A similar case involves *Leptotyphlops nigricans (sensu lato),* which was thought to extend from Ethiopia to the Cape of South Africa but which in fact is a series of isolated populations including, from north to south, *L. aethiopicus* (Ethiopia), *L. keniensis* (Kenya), *L. howelli* (Kenya-Tanzania), *L. macrops* (Kenya-Tanzania), *L. pembae* (Tanzania), *L. mbanjensis* (Tanzania), *L. emini* (Sudan-Tanzania), *L. monticolus* (DRC-Rwanda), *L. kafubi* (DRC-Zambia), *L. jacobseni* (NE South Africa) and *L. nigricans* (SW South Africa) (Broadley & Wallach, 2007a).

15. *Myriopholis macrura* (Boulenger, 1903g). Bull. Liverpool Mus. 2: 89, pl. 11, figs. 3, 3a–b. (*Glauconia macrura*)

Synonym: *Glauconia longicauda* Boulenger, 1899a *(nomen praeoccupatum).*

Type: Lectotype, BMNH 1946.1.11.75 (formerly BMNH 1899.12.5.111), a 165–170 mm specimen (H.O. Forbes & W.R. Ogilvie-Grant, 30 Dec. 1898–16 Jan. 1899), designated by Hahn (1978: 480).

Type locality: "Sokotra (Jena-agahan, 1200–2500 feet)" [= Jenna-agahan, E Sokotra Is., Yemen, 365–760 m] via lectotype selection.

Distribution: Yemen (Socotra Is.), NSL–995 m.

Sources: Boulenger, 1903g, 1915d, Parker, 1949, Corkill & Cochrane, 1966, Hahn, 1978, Schätti & Desvoignes, 1999, Rösler &Wranik, 2004, 2006 and Razzetti et al., 2011.

M

16. *Myriopholis narirostris* (W.C.H. Peters, 1867d). Mber. Königl. Akad. Wiss. Berlin 1867(11): 708, pl., figs. 2, 2a–c. (*Stenostoma narirostre*)

Type: Holotype, ZMB 5915 (Mus. Godeffroy).

Type locality: "Lagos (West-afrika)" [= Lagos, Lagos State, SW Nigeria, 06°27'N, 03°24'E, elevation NSL].

Distribution: West Africa. Guinea-Bissau (Biombo, Bissau), W Guinea (Télimélé), Ivory Coast (Ferkéssédougou), Ghana (Accra, Eastern, Upper West), Burkina Faso (Centre, Est, Sud-Ouest, Volta-Noire), S Togo (Plateaux), Nigeria (Lagos, Plateau), S Cameroon (Ouest, Sud) and Central African Republic (Bamingui-Bangoran), NSL–1250 m.

Sources: Boulenger, 1893a, 1920c, Villiers, 1950d, 1951a, 1952b, 1953, Monard, 1951, Perret, 1961, Doucet, 1963, Hughes & Barry, 1969, W. Böhme, 1978, Stucki-Stirn, 1979, Roman, 1980, Hughes, 1983, 1988, Roman, 1984, Hallermann & Rödel, 1995, Wallach & Hahn, 1997, Hahn & Wallach, 1998, LeBreton, 1999, Chirio & LeBreton, 2007, Chirio, 2012, Segniagbeto et al., 2011, Auliya et al., 2012 and Chirio, 2013.

Remarks: Possibly occurs in Benin *fide* Hughes (2013: 152).

17. *Myriopholis parkeri* (Broadley, 1999a). Arnoldia, Zimbabwe 10(14): 142–143, fig. 1. (*Leptotyphlops parkeri*)

Type: Holotype, BMNH 1949.2.3.47, a 160 mm female (R.H.R. Taylor; 1932–1933).
Type locality: "Degeh Bar, Harerge Region, Ethiopia (8°13'N: 43°30'E) at 1067m" [= Degeh Bur, Hararge Region, SE Ethiopia, 08°13'N, 43°34'E, elevation 1065 m].
Distribution: Southeastern Ethiopia (Hararge), 1065 m. Known only from type locality.
Sources: Parker, 1949, Largen & Rasmussen, 1993, Broadley & Wallach, 2007a, Largen & Spawls, 2010 and Boundy, 2013.

18. *Myriopholis perreti* (Roux-Estève, 1979a). Rev. Suisse Zool. 86(2): 464–466, fig. 1. (*Leptotyphlops perreti*)

Type: Holotype, MHNG 1453.11, a 233 mm male (J.-L. Perret, 1967).
Type locality: "Ebolowa, Cameroun" [= Ebolowa, W Sud Province, SW Cameroon, 02°55'N, 11°09'E, elevation 600 m].
Distribution: Southeastern Cameroon (Est, Sud) and SE Gabon (Haut-Ogooué), 410–585 m.
Sources: Hallermann & Rödel, 1995, LeBreton, 1999, Chirio & LeBreton, 2007, Pauwels & David, 2008b and Pauwels & Vande weghe, 2008.
Remarks: Elevation of Ebolowa 800 m *fide* Chirio & LeBreton (2007: 358).

M

19. *Myriopholis phillipsi* (T. Barbour, 1914). Proc. New England Zool. Club 5: 87–88, 2 figs. (*Leptotyphlops phillipsi*)

Type: Holotype, MCZ 9650, a 243 mm female (J.C. Phillips & W.M. Mann, 1914).
Type locality: "Petra, Arabia" [= Petra, Ma'an Govern., SW Jordan, 30°20'N, 35°27'E, elevation 915 m].
Distribution: Southeastern Turkey (Adiyaman, Gaziantep, Hatay, Mardin, Sanliurfa, Siirt, Urfa), Syria, Israel (Jerusalem, Southern, West Bank) and Jordan (Aqabah, Balqa, Maan, Madaba, Mafraq), 475–1000 m.
Sources: Corkill & Cochrane, 1966, Werner & Drook, 1967, Hoofien, 1972, Baran, 1976, 1978, 1982, Hahn, 1978, Y.L. Werner, 1987, 1995, Bouskila & Amitai, 2001, Disi et al., 2001, Ugurtas et al., 2006, Yildiz et al., 2009, Amr & Disi, 2011 and Bar & Haimovitch, 2011.

20. *Myriopholis rouxestevae* (J.-F. Trape & Mané, 2004). Bull. Soc. Herp. France (109): 8–9, fig. 3. (*Leptotyphlops rouxestevae*)

Type: Holotype, MNHN 2000.2515 (formerly ORSTOM S-5673), a 133 mm specimen.
Type locality: "Ndébou, Tambacounda (12°31'N–12°27'W)" [= Ndébou, Kédougou Dept., SE Senegal, 12°31'N, 12°20'W, elevation 165 m].
Distribution: Southern Senegal (Kédougou, Kolda) and SW Mali (Koulikoro), 145–165 m.
Source: J.-F. Trape & Mané, 2006b.

21. *Myriopholis wilsoni* (Hahn, 1978). J. Herp. 12(4): 478–479, fig. 1. (*Leptotyphlops wilsoni*)

Type: Holotype, BMNH 1957.1.10.28, a 100 mm specimen (H.D. Gwynne, 1956).
Type locality: "Kirschon, Socotra Island; elevation 634 meters" [= Wadi Qishn, Socotra Is., Al Hudaydah Govern., Yemen, 12°34'N, 54°02'E, elevation 1000 m].
Distribution: Yemen (Socotra Is.), 170–1065 m.
Sources: Corkill & Cochrane, 1966, Hahn, 1978, Hallermann & Rödel, 1995, Schätti & Desvoignes, 1999, Rösler &Wranik, 2004, 2006 and Razzetti et al., 2011.
Remarks: Type locality spelled "Kischen" *fide* Corkill & Cochrane (1966: 496).

22. *Myriopholis yemenica* (Scortecci, 1933). Atti. Soc. Ital. Sci. Nat. Mus. Civ. Stor. Nat. Milano 72(2): 165–166, 3 figs. (*Leptotyphlops yemenicus*)

Type: Holotype, MSNM 3354, a 91–92 mm specimen (G. Franchini, 1933).
Type locality: "Yemen."
Distribution: Yemen. Known only from type locality.
Source: Wallach & Lanza, 2004.

MYRON Gray, 1849a (Homalopsidae)

Synonym: *Neospades* De Vis, 1889.
Type species: *Myron richardsonii* Gray, 1849a.
Distribution: Tidal rivers and coastal marine waters of Australasia.
Sources: Gyi, 1970, Whitaker et al., 1982, Cogger et al., 1983a, F. Parker, 1982, Storr et al., 1986, S.K. Wilson & Knowles, 1988, Gow, 1989, O'Shea, 1996, Cogger, 2000, J.C. Murphy, 2007b, 2011, Alfaro et al., 2008 and Zaher et al., 2009.

1. *Myron karnsi* J.C. Murphy, 2011. Raffles Bull. Zool. 59(2): 231–232, figs. 2a–d.

Type: Holotype, SMF 19569, a 397 mm female.
Type locality: Indonesia Aru, Kobroor, Selrutti (about 5°46'S and 134°31'E)."
Distribution: Eastern Indonesia (Aru Arch.). Known only from type locality.

2. *Myron resetari* J.C. Murphy, 2011. Raffles Bull. Zool. 59(2): 232, figs. 3a–d (paratype).

Type: Holotype, QM J52861, a 395 mm male..
Type locality: Broome, Western Australia (about 17°58'S and 122°14'E)."
Distribution: Australia (Western Australia), 10 m. Known only from type locality.

3. *Myron richardsonii* Gray, 1849a. Cat. Snakes Brit. Mus.: 70.

Synonym: *Neospades kentii* De Vis, 1889.
Type: Holotype, BMNH 1946.1.2.43, a female without skull (J. Richardson).
Type locality: "N.-West Australia." Restricted to Buffalo Creek, Northern Territory, Australia, ca. 12°40'S, 131°29'E *fide* J.C. Murphy (2011: 232).
Distribution: Southern Papua New Guinea (Western, Gulf, Daru Is.) and N Australia (ext. N Western Australia, ext. N Northern Territory, ext. N Queensland), NSL–40 m.

MYRROPHIS A.B. Kumar, Sanders, George & Murphy, 2012 (Homalopsidae)

Type species: *Hypsirhina chinensis* Gray, 1842d.
Distribution: Tidal rivers and coastal marine waters of Eastern Asia.
Sources: Gyi, 1970, I. Das, 2010, 2012 and A.B. Kumar et al., 2012.
Remarks: A valid genus *fide* A.B. Kumar et al. (2012: 484). Previously a synonym of *Enhydris*.

1. *Myrrophis bennettii* (Gray, 1842d). Zool. Misc. 2(May): 67. (*Hypsirhina bennettii*)

Synonym: *Hypsirhina maculata* A.-M.-C. Duméril, Bibron & Duméril, 1854b.
Type: Holotype, BMNH 1946.1.2.49, male (G. Bennett).
Type locality: "China."
Distribution: Southeastern Asia. Southeastern China (Fujian, Guangdong, Guangxi, Hainan, Hong Kong), Macau and Vietnam (Quang Ninh).
Sources: Stejneger, 1907a, C.H. Pope, 1935, Romer, 1979a, Karsen et al., 1986, Tian et al., 1986, Iskandar, 1987, Easton & Leung-Va, 1993, Zhao & Adler, 1993, Zhao, 2006, V.S. Nguyen et al., 2009 and A.B. Kumar et al., 2012.
Remarks: Java record in error *fide* J.C. Murphy (2007b: 91).

2. *Myrrophis chinensis* (Gray, 1842d). Zool. Misc. 2(May): 66. (*Hypsirhina chinensis*)

Synonym: *Hypsirhina sinensis* Stanley, 1915 *(nomen emendatum)*.
Type: Holotype, BMNH 1946.1.2.42, a 419 mm male (J.R. Reeves).
Type locality: "China."
Distribution: Southeastern Asia. Southeastern China (Anhui, Fujian, Guangdong, Guangxi, Hainan, Hong Kong, Hubei, Jiangsu, Jiangxi, Zhejiang), Taiwan and N Vietnam (Bac Giang, Dak Lak, Da Nong, Hai Duong, Hanoi, Ha Tinh, Hoa Binh, Nghe An, Son La, Thai Nguyen, Thanh Hoa, Vinh Phuc), NSL–200 (1100) m.
Sources: Maki, 1931, C.H. Pope, 1935, Romer, 1979a, Karsen et al., 1986, Zhao & Adler, 1993, Orlov et al., 2000, V.S. Nguyen et al., 2009, Xiang & Li, 2009, I. Das, 2010 and A.B. Kumar et al., 2012.

N

NAJA Laurenti, 1768 (Elapidae)

Synonyms: *Naia* Merrem, 1820 *(nomen emendatum)*, *Aspis* Wagler, 1830 *(nomen praeoccupatum)*, *Tomyris* Eichwald, 1831, *Najia* – Owen, 1841a *(nomen incorrectum)*, *Naje* – C.C. Vogt, 1851 *(nomen incorrectum)*, *Naya* – Kelaart, 1853 *(nomen incorrectum)*, *Haja* – Higgins, 1873 *(nomen incorrectum)*, *Naga* Nicholson, 1874 *(nomen emendatum)*, and *Noja* – Oshima, 1944 *(nomen incorrectum)*.

Type species: *Coluber naja* Linnaeus, 1758.

Distribution: Asia and East Indies.

Sources: F. Werner, 1923b, Bogert, 1943, Klemmer, 1968, Minton et al., 1970, Golay, 1985, Szyndlar & Rage, 1990, Jintakune & Chanhome, 1995, Ulber, 1995, Golay et al., 1993, Keogh, 1998, David & Ineich, 1999, Slowinski & Keogh, 2000, Dobiey & Vogel, 2007, Wüster et al., 2007, Hoser, 2009c, 2012a, e, Wallach et al., 2009, I. Das, 2010, 2012, Groen, 2010a–b and Zaher et al., 2009, 2012.

Remarks: *Boulengerina* (*B. annulata* and *B. christyi*) and *Paranaja* (*P. multifasciata*) congeneric with *Naja fide* Wüster et al. (2007: 445). Wallach et al., 2009 allocated cobras (*Naja*) to four subgenera, here considered genera based on Wüster et al., 2007 and Zaher et al., 2012: *Afronaja (ashei, katiensis, mossambica, nigricollis, nigricincta, nubiae, pallida), Boulengerina (annulata, christyi, melanoleuca, multifasciata), Naja (atra, kaouthia, mandalayensis, naja, oxiana, philippinensis,†N. romani, sagittifera, samarensis, siamensis, sputatrix, sumatrana),* and *Uraeus (anchietae, annulifera, arabica, haje, nivea, senegalensis)*. Five extinct species (†*N. antiqua*, †*N. depereti*, †*N. iberica*, †*N. robusta*, †*N. romani*) are placed in *Afronaja* based on dentition *fide* Bogert, 1943 and Szyndlar & Rage, 1990.

1. *Naja atra* Cantor, 1842a. Zool. Chusan: pl. 10.

Synonyms: *Naja tripudians nigra* Gray, 1834, *Naja tripudians scopinucha* Cope, 1860b, *Naja lutescens larvata* Steindachner, 1867 *(nomen praeoccupatum)*, *Naga tripudians monocellata* Nicholson, 1874, *Naja tripudians unicolor* W.C.H. Peters *in* Martens, 1876 *(nomen substitutum)*, and *Naja naja atrox* Sarkar & Devi, 1968 *(lapsus calami)*.

Type: Holotype, not designated (T.E. Cantor, July 1840–March 1841), lost *fide* Golay et al. (1993: 185).

Type locality: "island of Chusan, or Great Chusan, is situated on the East Coast of China, between the 30° and 31° of North latitude and the 122° and 123° of East longitude; separated from the nearest mainland, Keeto Point, by an arm of the sea, about ten miles across" [= Zhoushan Is., Zhoushan Arch., Zhejiang Prov., E China].

Distribution: Southeastern Asia. Southeastern China (Anhui, Fujian, Guangdong, Guangxi, Guizhou, Hong Kong, Hubei, Hunan, Macau, Sichuan, Yunnan, Zhejiang, Hainan, Lammock and Zhoushan Is.), Taiwan (T'ai-pei, T'ai-nan), N Laos (Bolikhamsai), and Vietnam (Bac Giang, Bac Kan, Cao Bang, Hai Duong, Hai Phong, Hanoi, Ha Tay, Ha Tinh, Hoa Binh, Hoang Lien Son, Lang Son, Lao Cai, Nghe An, Ninh Binh, Phu Tho, Quang Binh, Quang Ninh, Quang Tri, Son La, Thai Binh, Thai Nguyen, Thanh Hoa, Thua Thien-Hue, Tuyen Quang, Vinh Phu, Cu Lao Ba Mun Is.), NSL–1600 m.

Sources: Stejneger, 1907a, Stanley, 1915, Maki, 1931, Bourret, 1934f, 1936a–b, 1939?, C.H. Pope, 1935, Herklots, 1938, Kuntz, 1963, Mao, 1963, E.H. Taylor, 1965, Deuve, 1970, Romer, 1972, 1979a, Tian et al., 1986, Lue, 1987, 1990, M.J. Cox, 1991b, Wüster & Thorpe, 1992b, Easton & Leung-Va, 1993, Zhao & Adler, 1993, Wüster et al., 1995, Wüster, 1996, Xie et al., 1998, Darevsky, 1999, Orlov et al., 2000, Slowinski & Wüster, 2000, Leviton et al., 2003, V.S. Nguyen et al., 2009 and Xiang & Li, 2009.

Remarks: Supplemental original description in Cantor (1842b: 482–483).

2. *Naja kaouthia* Lesson, 1831. Bull. Sci. Nat. Géol., Paris 25(4): 122.

Synonyms: *Naja tripudians fasciata* Gray, 1834, *Naja larvata* Cantor, 1839a, *Naja tripudians viridis* Wall, 1913b, *Naja kaothia* – Wall, 1913i *(nomen incorrectum)*, *Naja naja kauothia* – Kabara & Fischer, 1972 *(nomen incorrectum)*, *Naja kaouthia suphanensis* Nutaphand, 1986b, *Naja kauthia* – Khole, 1991 *(nomen incorrectum)*, and *Naja naja combodia* Khole, 1991 *(lapsus calami)*.

Type: Holotype, formerly MNHN, a 2475 mm specimen illustrated in Lesson, 1832 *in* Bélanger (1831–1834: pl. 2) (R.-P. Lesson [Astrolabe Voy.], 1826–1829), lost *fide* Kramer (1977: 752).

Type locality: "Inde continentale" [= continental India]. Restricted to Bengale [= NE India or Bangladesh] *fide* Lesson (1834: 313).

Distribution: Southeastern Asia. India (Arunachal Pradesh, Assam, Bihar, Delhi, Haryana, Manipur, Meghalaya, Mizoram, Nagaland, Orissa, Sikkim,

Uttar Pradesh, West Bengal), Nepal (Darchula, Jhapa, Kanchanpur, Nawalparasi, Rupandehi, Solukhumbu, Saptari, Siraha, Sunsari), Bhutan, Bangladesh, Myanmar (Ayeyarwady, Bago, Chin, Kachin, Magway, Mandalay, Rakhnie, Sagaing, Shan, Yangon), SW China (Xizang, Yunnan), Thailand (Ang Thong, Chachoengsao, Chanthaburi, Chiang Mai, Chon Buri, Kanchanaburi, Krabi, Krung Thep Mahanakhon, Loei, Lop Buri, Mae Hong Son, Nakhon Ratchasima, Nakhon Sawan, Nan, Nakhon Pathom, Nakhon Si Thammarat, Narathiwat, Nonthaburi, Pathum Thani, Pattani, Phang Nga, Phra Nakhon Si Ayutthaya, Phatthalung, Phitsanulok, Phuket, Samut Prakan, Saraburi, Satun, Sing Buri, Songkhla, Suphan Buri, Tak, Trang), West Malaysia (Kedah, Kelantan, Perak, Perlis, Pinang), Cambodia (Chhnang, Kampong, Kampot, Siem Reap), S Laos (Balikhansai) and S Vietnam (Dak Lak, Ho Chi Minh City, Kien Giang, Lam Dong, Quang Tri), NSL–3200 m.

Sources: Stejneger, 1907a, Bourret, 1934d, 1934f, C.H. Pope, 1935, M.A. Smith, 1943, Shaw et al., 1946b, Deraniyagala, 1960b, 1961b, Acharji, 1964, E.H. Taylor, 1965, Soderberg, 1965, 1967a, 1973, Campden-Main, 1970a, Saint-Girons, 1972a, Singh, 1972, Biswas & Sanyal, 1980, Sahi & Duda, 1986, Tikader & I. Das, 1985, Tiwari, 1985, Murthy, 1986, Tian et al., 1986, Duckett, 1987, K.F.L. Lim & Lee, 1989, Wüster & Thorpe, 1989, 1991, 1992a–b, 1994, M.J. Cox 1991b, Zhao & Adler, 1993, Jintakune & Chanhome, 1995, Wüster et al., 1995, Wüster, 1996, M.J. Cox et al., 1998, Wüster et al., 1998a, Chan-ard et al., 1999, Slowinski & Wüster, 2000, Nutaphand, 2001, Leviton et al., 2003, Ao et al., 2004, Tiwari & Shah, 2004, Whitaker & Captain, 2004, Zhao, 2006, M.F. Ahmed et al., 2009, V.S. Nguyen et al., 2009 and S.K. Sharma et al., 2013.

Remarks: Supplemental original description in Lesson, 1832, *in* Bélanger (1831–1834: 312–314, pl. 2).

3. *Naja mandalayensis* Slowinski & Wüster, 2000. Herpetologica 56(2): 260–264, fig. 1.

Type: Holotype, CAS 207097, an 828 mm male (J.B. Slowinski, 6 May 1998).

Type locality: "near Monywa (N 22°13'; E 95°20'), Sagaing Division, Myanmar."

Distribution: Central Myanmar (Magwe, Mandalay, Sagaing, Shan), 70–90 m.

Source: Leviton et al., 2003.

Remarks: Probably occurs in NE India (Bihar) *fide* Slowinski & Wüster (2000: 264).

4. *Naja naja* (Linnaeus, 1758). Syst. Nat., ed. 10, 1: 221. (*Coluber naja*)

Synonyms: *Naja brasiliensis* Laurenti, 1768, *Naja fasciata* Laurenti, 1768, *Naja lutescens* Laurenti, 1768, *Naja maculata* Laurenti, 1768, *Naja non naja* Laurenti, 1768, *Coluber brasiliae* Lacépède, 1789 (*nomen rejiciendum*), *Coluber caecus* Gmelin, 1789, *Coluber rufus* Gmelin, 1789, *Coluber non naja* Lacépède, 1789 (*nomen rejiciendum*), *Coluber peruvii* Lacépède, 1789 (*nomen rejiciendum*), *Vipera pileata* Lacépède, 1789 (*nomen rejiciendum* and *nomen nudum*), *Coluber naja brasiliensis* Bonnaterre, 1790, *Coluber naja peruvianus* Bonnaterre, 1790, *Coluber rufoalbus* Suckow, 1797, *Naja tripudians* Merrem, 1820 *(nomen substitutum)*, *Naja nigra* Gray, 1830 *in* Gray & Hardwicke, 1830–1835, *Naja goerou* Lesson, 1832 *in* Bélanger, 1831–1834, *Naga tripudians binocellata* Nicholson, 1874, *Naia naia polyocellata* Deraniyagala, 1939, *Naja naja gangetica* Deraniyagala, 1945, *Naja naja madrasiensis* Deraniyagala, 1945, *Naja naja indusi* Deraniyagala, 1960b, *Naja naja bombaya* Deraniyagala, 1961b, *Naja naja karachiensis* Deraniyagala, 1961b, *Naja maja* – F. Miranda, Rochat & Lissitzky, 1964 *(nomen incorrectum)*, *Notechis tripudians* Chapman, 1968 *(lapsus calami)*, *Naja naja ceylonicus* Chatnam & Dimari, 1974, and *Naja naja Indian* – Ménez, 1985 *(nomen incorrectum)*, and *Naja tripudis* – Khole, 1991 *(nomen incorrectum)*.

Type: Lectotype, NHR Lin-90 (formerly MAFR), an 1140 mm male (Mus. Drottn.), designated herein.

Type locality: "India orientali" [= eastern India, including Sri Lanka] via lectotype selection. Restricted to India *fide* M.A. Smith (1943: 427).

Distribution: Southern Asia. Pakistan (Balochistan, F.A.T.A., Jammu & Kashmir, Khyber Pakhtunkhwa, Punjab, Sindh), Nepal (Banke, Bardiya, Chitwan, Dadeldhura, Dang, Jhapa, Kailali, Kanchanpur, Kathmandu, Makwanpur, Parsa, Rupandehi, Sankhuwasabha, Sunsari, Saptari, Sindhuli, Surkhet, Tehrathum), Bhutan (Sarpang), India (Andhra Pradesh, Bihar, Delhi, Goa, Gujarat, Haryana, Himachal Pradesh, Jammu and Kashmir, Karnataka, Kerala, Madhya Pradesh, Maharashtra, Orissa, Punjab, Rajasthan, Tamil Nadu, Uttar Pradesh, West Bengal), W Bangladesh and Sri Lanka (Central, N Eastern, North-Central, Northern, North-Western, Sabaragamuwa, Uva, Western), NSL–2000 m.

Sources: Bannerman & Pocha, 1905, Bannerman, 1907, Wall, 1913b, f, 1921g, 1922a, Deraniyagala, 1939, 1945, 1960b, 1961b, M.A. Smith, 1943, Deoras, 1965, Minton, 1966, P. Silva, 1980a, Tiwari & Sharma, 1970, R. Sharma, 1971, 1976a, Kramer, 1977, Singh, 1972, Fleming & Fleming, 1973, M.A.R. Khan, 1988, M.S. Khan, 1983, 2002, Joger, 1984, Wüster & Thorpe, 1987, 1989, 1992a–b, A. Silva, 1990a–b, 2001, 2009, Jintakune & Chanhome, 1995, Wüster, 1996, 1998a, Das & Palden, 2000, Slowinski & Wüster, 2000, Tiwari & Shah, 2004, Whitaker & Captain, 2004, N. Khaire, 2006, Castoe et al., 2007, Einfalt, 2012, Masroor, 2012, Wangyal, 2012 and S.K. Sharma et al., 2013.

Remarks: Original description based on Seba (1734: 71, pl. 44, fig. 1; 1735: 90–91, pl. 85, fig. 1; 1735: 95, pl. 89, figs. 1–4; 1735: 96, pl. 90, figs. 1–2; 1735: 99, pl. 94, fig. 1; 1735: 102–103, pl. 97, figs. 1–4). Original color sketches of holotype of *Naja tripudians* by Hardwicke

(1756–1823) (1: figs. 169–170) in BMNH. Type locality restriction to Sri Lanka *fide* Deraniyagala (1945: 108) invalid. Holotype listed in ZMUU *fide* Deraniyagala (1955: 74). Neotype designation of Sri Lanka specimen in BMNH *fide* Deraniyagala (1949: 109) invalid, based on *Naja naja polyocellata* Deraniyagala (1939: 235, pl. 211, figs. a–d). One syntype (460 mm specimen described and figured in Linnaeus 1754: 30, pl. 21, fig. 3) is lost *fide* Deraniyagala (1960b: 61), Wallin (1992: 226) and Golay et al. (1993: 189). Another syntype exists in NHR *fide* Andersson (1899: 17) and Deraniyagala (1960: 49). Possibly occurs in E Afghanistan *fide* Wüster & Thorpe (1992: 75).

5. *Naja oxiana* (Eichwald, 1831). Zool. Spec. Rossiae 3: 171. (*Tomyris oxiana*)

Synonym: *Naja noxiana* Munjal & Elliott, 1972 *(nomen incorrectum).*

Type: Holotype, ZISP 8728, a 762 mm specimen (K.E.I. Eichwald, 1825–1826), lost *fide* Golay et al. (1993: 190).

Type locality: "caspii maris littore orientali, prope Oxum amnem ad pristinum hujus ostium in sinu balchanensi" [= eastern shore of Caspian Sea near the former mouth of the Amu Darya (= Oxus) River, NW Turkmenistan or SW Kazakhstan].

Distribution: Southwestern Asia. Southwestern Tadzhikistan, Turkmenistan, Uzbekistan, NE Iran (Mazandaran, Razavi Khorasan, Semnan), N and E Afghanistan (Kabul, Nangarhar), N Pakistan (N Balochistan, F.A.T.A., Gilgit-Baltistan, Khyber Pakhtunkhwa, Punjab) and ext. NW India (Himachal Pradesh, Jammu and Kashmir, Khyber Pakhtunkhwa, N Punjab), 245–2100 m.

Sources: Strauch, 1873, Nikolsky, 1899b, 1905, 1916, M.A. Smith, 1943, Minton, 1966, S.C. Anderson & Leviton, 1969, Leviton & Anderson, 1970a, R. Sharma, 1974, Mahajan & Agrawal, 1976, M.S. Khan, 1983, 2002, Sahi & Duda, 1986, Latifi, 1991, Kudrjavtsev & Mamet, 1989a, Wüster & Thorpe, 1991, 1992a–b, Szczerbak, 1994, Wüster, 1996, 1998a, Slowinski & Wüster, 2000, Ananjeva et al., 2004, Whitaker & Captain, 2004 and Masroor, 2012.

Remarks: Supplemental original description in Eichwald (1841–1842: 104–105, pl. 20, figs. 1–2). Amu Darya River currently enters the Aral Sea in N Uzbekistan but in the past has also entered the Caspian Sea (and at times both seas).

6. *Naja philippinensis* E.H. Taylor, 1922a. Snakes Phil. Is.: 265–268. (*Naja naja philippinensis*)

Synonyms: *Naja philippinesis* – Hati, Mandal & Hati, 1990 *(nomen incorrectum)* and *Naja philippeninsis* – Khole, 1991 *(nomen incorrectum).*

Type: Holotype, BSM 7, a 1000 mm female (C. Canonizado), destroyed in January 1945 during World War II.

Type locality: "Manila, Philippine Islands" [= Manila, Manila Prov., SW Luzon Is., N Philippines, 14°36'N, 120°59'E, elevation 10 m].

Distribution: Northern Philippines (Luzon, Mindoro, Marinduque, Masbate).

Sources: Leviton, 1965b, Wüster & Thorpe, 1990, 1992b, Gaulke & Altenbach, 1994, R.M. Brown et al., 1996, Wüster, 1996 and Slowinski & Wüster, 2000.

Remarks: Probably occurs on Polillo and Catanduanes Is., Philippines *fide* Wüster & Thorpe (1990: 341).

7. *Naja sagittifera* Wall, 1913b. J. Bombay Nat. Hist. Soc. (1913–1914) 22(2): 247–248. (*Naja tripudians sagittifera*)

Type: Holotype, ZSI 4011 (formerly IMC) (N. Annandale, 1900–1913).

Type locality: "the Andamans" [= Andaman & Nicobars, E India, Bay of Bengal].

Distribution: Eastern India (Andaman & Nicobars: South Andaman Is.).

Sources: Biswas & Sanyal, 1980, Wüster & Thorpe, 1994, Wüster et al., 1995, Wüster, 1996 and Slowinski & Wüster, 2000.

Remarks: Resurrected as a valid species *fide* Wüster et al. (1995: 504).

8. *Naja samarensis* W.C.H. Peters, 1861c. Mber. Königl. Akad. Wiss. Berlin 1861(4): 690. (*Naja tripudians samarensis*)

Synonyms: *Naja naja saramensis* – Bourret, 1936 *(nomen incorrectum)* and *Naja naja samarenis* – Trethewie, 1971 *(nomen incorrectum).*

Type: Holotype, ZMB 3955 (F. Jagor, July 1860).

Type locality: "Loquilocun, Insel Samar" [= Loquilocon, Paranas Pueblo, W Samar, cen. Philippines, 11°48'N, 125°06'E].

Distribution: Southeastern Philippines (Bohol, Camiguin [Sur], Dinagat, Leyte, Mindanao, Samar), NSL–1000 m.

Sources: E.H. Taylor, 1922a, Leviton, 1965b, Wüster & Thorpe, 1990, 1992b, Ross & Lazell, 1991, B.E. Smith, 1993, Wüster, 1996, 1998a and Slowinski & Wüster, 2000.

Remarks: Probably occurs on Siargao and Basilan Is., Philippines *fide* Wüster & Thorpe (1990: 341).

9. *Naja siamensis* Laurenti, 1768. Synop. Rept.: 91.

Synonyms: *Naja naja siamenesis* – Kamel, 1974 *(nomen incorrectum)* and *Naja naja isanensis* Nutaphand *in* Nutaphand & Tumvipart, 1982.

Type: Neotype, BMNH 1987.678, a 936 mm female, designated by Wüster et al. (1997: 776).

Type locality: "Ban Phai district hospital" [= Bao Phai Distr., Khon Kaen Prov., NE Thailand] via neotype selection.

N

Distribution: Southeast Asia. Thailand (Buriram, Chainat, Chanthaburi, Chilang Mai, Chiang Saen, Chon Buri, Kamphaeng Phet, Kanchanaburi, Khon Kaen, Lampang, Mae Hong Son, Nakhon Phanom, Nakhon Ratchasima, Nakhon Sawan, Phetchabun, Phitsanulok, Phrae, Sing Buri, Si Saket, Suphan Buri, Trat, Udon Thani, Uthai Thani, Uttaradit, Yasothon), Cambodia (Trapeang Chan), Laos (Champasak, Savannakhet) and S Vietnam (Binh Phuoc, Dong Nai, Gia Lai, Ho Chi Minh City, Kien Giang, Kon Tum, Tay Ninh), NSL–260 m.

Sources: Viravan et al., 1991, Wüster & Thorpe, 1994, Jintakune & Chanhome, 1995, Wüster et al., 1995, M.J. Cox et al., 1998, Chan-ard et al., 1999, 2000, Slowinski & Wüster, 2000, Teyné & David, 2007 and V.S. Nguyen et al., 2009.

Remarks: Type specimen(s) lost *fide* Wüster et al. (1997: 772). Probably occurs in Myanmar adjacent to Thailand *fide* Wüster et al. (1995: 503).

10. *Naja sputatrix* F. Boie, 1827. Isis von Oken 20(6): 557.

Synonyms: *Naja sputatrix* H. Boie *in* Schlegel, 1826a *(nomen nudum), Naja sputatrix* Schlegel, 1826b *(nomen nudum),* Coluber *castaneus* Oppel *in* F. Boie, 1827 *(nomen nudum), Elaps fuscus* Merrem, 1820, *Naja leptocoryphaea* Berthold, 1842, and *Naja tripudians sondaica* Schlegel, 1844 *in* 1837–1844 & Schlegel *in* S. Müller & Schlegel, 1844b *(partim).*

Types: Syntypes (2), RMNH (C.G.C. Reinwardt, 1816–1821), and specimen described and illustrated in P. Russell (1801: 41, pl. 36).

Type locality: "Java" [SW Indonesia]

Distribution: Lesser Sundas of S Indonesia (Alor, Bali, Flores, Java, Komodo, Lomblen, Lombok, Rinca, Sumbawa), NSL–700 m.

Sources: Rooij, 1917, Dunn, 1927, Mertens, 1930, Bergman, 1962a, Capoccacia, 1976, Auffenberg, 1980, Wüster & Thorpe, 1989, 1990, Wüster, 1996, McKay, 2006 and Lang, 2011.

Remarks: Original description based on H. Boie's MS (1823–1825) and P. Russell (1801: 41, pl. 36). Sulawesi record doubtful *fide* Lang & Vogel (2005: 260) and Timor record requires confirmation *fide* I. Das (2010: 317).

11. *Naja sumatrana* F. Müller, 1887. Verh. Naturf. Ges. Basel 8(2): 277–278. (*Naja tripudians sumatrana*)

Synonyms: *Naja tripudians sondaica* Schlegel, 1844 *in* 1837–1844 & Schlegel *in* S. Müller & Schlegel, 1844b *(partim), Naia tripudians leucodira* Boulenger, 1896a, *Naia tripudians miolepis* Boulenger, 1896a, *Naja tripudians paucisquamis* Peracca, 1899, *Naja sputatrix malayae* Deraniyagala, 1960b, and *Naja naja moilepis* – Trethewie, 1971 *(nomen incorrectum).*

Type: Holotype, NMBA 2244, a female (S. Müller, Dec. 1825–Sept. 1837).

Type locality: "Solok (Sumatra)" [= Solok, W Sumatra, W Indonesia, 00°48'S, 100°39'E, elevation 400 m].

Distribution: Malay Peninsula and East Indies. Southern Thailand (Nakhon Si Thammarat, Narathiwat, Pattani, Phang Nga, Phatthalung, Surat Thani, Trang), West Malaysia (Johor, Kedah, Melaka, Negeri Sembilon, Perak, Perlis, Pinang, Selangor), Singapore, East Malaysia (Sabah, Sarawak), Brunei, Indonesia (Bangka, Belitung, Bintan, Bulan, Galang, Kalimantan, Karimun, Lingga Arch., Penuba, Riau Arch., Sumatra) and ext. SW Philippines (Culion, Palawan), NSL–1370 m.

Sources: E.H. Taylor, 1922a, C. Haas, 1950, Leviton, 1965b, 1968b, Wüster, 1988, 1996, Wüster & Thorpe, 1989, 1990, 1992b, M.J. Cox, 1991b, Stuebing, 1991, David & Vogel, 1996, M.J. Cox et al., 1998, Slowinski & Wüster, 2000, Malkmus et al., 2002 and I. Das, 2007b.

Remarks: Probably occurs on Balabac and Calamianes Is., Philippines *fide* Wüster & Thorpe (1990: 341).

†*NAJASH* Apesteguía & Zaher, 2006 (†Najashidae)

Type species: †*Najash rionegrina* Apesteguía & Zaher, 2006.

Distribution: Upper Cretaceous of Argentina.

Sources: Apesteguía & Zaher, 2006, J. Gauthier et al., 2012 and Palci et al., 2013.

Remarks: Palci et al. (2013: 132, 136) listed the holotype as MPCA 389, MPCA 391–397 and MPCA 400 and showed the "type locality" to be sites from 5–30 km distant from one another.

1. †*Najash rionegrina* Apesteguía & Zaher, 2006. Nature 440(7087): 1037–1039, figs. 1–2.

Type: Holotype, MPCA 390–98 and MPCA 400, one dentary, one splenial, 122 vertebrae, and a pelvis with two femurs, one fibula, and a right tibia.

Type locality: "Upper section of the Candeleros Formation (Cenomanian-Turonian) at 'La Buitrera', Rio Negro Province, north Patagonia, Argentina." Emended to 'La Buitrera' fossil locality between the villages of El Chocón and Cerro Policía, about 80 km SW Cipolletti close to the S shore of Lake Ezequiel Ramos-Mexía *fide* Makovicky et al. (2005: 1007).

Distribution: Middle Cretaceous (Cenomanian: 97.0–100.5 mya) of Argentina. Known only from type locality.

NAMIBIANA Hedges, Adalsteinsson & Branch *in* Adalsteinsson, Branch, Trape, Vitt & Hedges, 2009 (Leptotyphlopidae)

Type species: *Leptotyphlops occidentalis* V.F.M. FitzSimons, 1962b.
Distribution: Southwestern Africa.
Sources: Hahn, 1980a, Wallach, 1998b, Broadley & Broadley, 1999 and McDiarmid et al., 1999.

1 *Namibiana gracilior* (Boulenger, 1910a). Ann. So. Afr. Mus. 5(9): 524. (*Glauconia gracilior*)

Types: Syntypes (11), BMNH 1946.1.16.51 (formerly BMNH 1910.8.5.2), BMNH 1946.1.16.93 (formerly BMNH 1910.8.5.3), SAM 2398, SAM 2418–19, SAM 3359, SAM 9182, SAM 9184, SAM 8068, SAM 43241 and SAM 43542, longest syntype 240 mm.
Type locality: "Cape Colony, South Africa. Van Rynsdorp (A.W. Rogers); Matjesfontein (Purcell); Clanwilliam (Leipoldt)" [= Vanrhynsdorp, N Western Cape, South Africa, 31°36'S, 18°44'E, elevation 120 m; Matjesfontein, S Eastern Cape, South Africa, 33°33'S, 26°20'E, elevation 150 m; Clanwilliam, N Western Cape, South Africa, 32°11'S, 18°53'E, elevation 100 m].
Distribution: Southern Namibia (Lüderitz, Rehoboth) and SW South Africa (S Eastern Cape, W Northern Cape, Western Cape), 100–190 m.
Sources: V.F.M. FitzSimons, 1962a, 1974, Elzen, 1980, Broadley, 1983, Branch, 1988, 1998, Marais, 1992, 2004, Branch & Bauer, 1995, Broadley & Wallach, 1997a and Broadley & Broadley, 1999.
Remarks: Broadley & Broadley (1999: 13) erroneously list Robertson (SAM 1858) as a type locality.

2. *Namibiana labialis* (Sternfeld, 1908a). Sitz. Ges. Naturf. Freunde Berlin 1908(4): 92. (*Glauconia labialis*)

Type: Holotype, ZMB 21691, a 170 mm male (O. Dempwolff, 1899–1901).
Type locality: "Outgo (D. S. W. Africa)" [= Outjo, Outjo Prov., Namibia, 20°07'S, 16°09'E, elevation 1265 m].
Distribution: Southwestern Angola (Cunene) and NW Namibia (Erongo, Khomas, Kunene, Oshikoto, Otjozondjupa), 240–2220 m.
Sources: Mertens, 1955, 1971, V.F.M. FitzSimons, 1962a, 1974, Broadley & Watson, 1976, Broadley, 1983, Branch, 1988, 1998, Bauer et al., 1993, Broadley & Broadley, 1999, Bauer et al., 2002 and Marais, 2004.

3. *Namibiana latifrons* (Sternfeld, 1908a). Sitzber. Ges. Naturf. Freunde Berlin 1908(4): 94–95. (*Glauconia latifrons*)

Synonyms: *Stenostoma scutatum* Peters, 1865 and *Glauconia lepezi* Boulenger, 1901a.
Type: Lectotype, BMNH 1864.6.14.5, a 175 mm specimen (J.J. Monteiro, 1861–1864), designated by Broadley & Watson (1976: 483).
Type locality: "Benguela, Angola" [= Benguela, Benguela Prov., SW Angola, 12°35'S, 13°24'E, elevation 10 m] via lectotype selection.
Distribution: Southwestern Democratic Republic of the Congo (Bas-Congo) and W Angola (W Benguela, Cuanza Sul), NSL–30 m.
Sources: Broadley & Watson, 1976 and Broadley & Broadley, 1999.

4. *Namibiana occidentalis* (V.F.M. FitzSimons, 1962b). Ann. Transvaal Mus. 24(2–3): 239–240, fig. 1. (*Leptotyphlops occidentalis*)

Type: Holotype, TMP 5017, a 206 mm specimen (G. van Kerkhoff, 31 March 1912).
Type locality: "Keetmanshoop, Great Namaqualand, South-West Africa" [= Keetmanshoop, Keetmanshoop Prov., S Namibia, 26°34'S, 18°08'E, elevation 1000 m].
Distribution: Namibia (Erongo, Gobabis, Hardap, Karas, Khomas, Kunene, Omaheke) and W South Africa (Northern Cape), 20–1425 m.
Sources: V.F.M. FitzSimons, 1962a, 1974, Broadley & Watson, 1976, Broadley, 1983, Branch, 1988, 1998, 2001 Marais, 1992, 2004, Bauer et al., 1993, Broadley & Broadley, 1999 and Bauer & Branch, 2001.

5. *Namibiana rostrata* (Bocage, 1886). J. Sci. Math. Phys. Nat., Lisboa 11(43): 173–174. (*Stenostoma rostratum*)

Synonym: *Leptotyphlops rostratus* Hahn, 1980a.
Type: Holotype, MBL, a 192 mm specimen (J.A. d'Anchieta, 1861–1866), destroyed by fire 18 March 1978.
Type locality: "Humbe, sur les bords du Cunene" [= Humbe, border of Cunene River, Cunene Prov., SW Angola, 16°41'S, 14°54'E, elevation 1100 m].
Distribution: Western Angola (Cunene, Luanda, Malanje, Namibe), 1060–1325 m.
Sources: Boulenger, 1893a, Bocage, 1895, Hahn, 1980 and Broadley & Broadley, 1999.
Remarks: The Kibwezi, Kenya records (ZMB) cited by Wallach *in* McDiarmid et al. (1999: 41) are *Leptotyphlops scutifrons merkeri*.

†*NANOWANA* Scanlon, 1997 (†Madtsoiidae)

Type species: †*Nanowana godthelpi* Scanlon, 1997.
Distribution: Lower Miocene of Australia.
Source: Scanlon, 1997.

1. †*Nanowana godthelpi* Scanlon, 1997. Mem. Queensland Mus. 41(2): 396–399, figs. 1a–f, 2a–i.

Type: Holotype, QMF 31379, partial to complete two maxillae, two palatines and two pterygoids of single individual.
Type locality: "early Miocene (System B) Camel Sputum Site, Godthelp Hill, Riversleigh, NW Queensland, Australia."
Distribution: Lower Miocene (16.3–24.0 mya) of Australia. Known only from type locality.

2. †*Nanowana schrenki* Scanlon, 1997. Mem. Queensland Mus. 41(2): 403–407, figs. 9a–d.

Type: Holotype, QMF 31395, one right palatine.
Type locality: "early Miocene Upper Site, Godthelp Hill, Riversleigh, NW Queensland, Australia."
Distribution: Lower Miocene (16.3–24.0 mya) of Australia. Known only from type locality.

NATRICITERES Loveridge, 1953 (Natricidae)

Synonyms: *Neusterophis* W.C.H. Peters, 1877c, *Natriceteres* – Barbault, 1971 *(nomen incorrectum)*, and *Natriceres* – Wallach, 1980 *(nomen incorrectum)*.
Type species: *Coronella olivacea* W.C.H. Peters, 1854.
Distribution: Subsaharan Africa.
Sources: Sternfeld, 1908b–c, Aylmer, 1922, Loveridge, 1953, 1958, Broadley, 1966b, Hughes & Barry, 1969, Dunger, 1972, Villiers, 1975, D. Lawson, 1993, Akani & Luiselli, 2001a, Villiers & Condamin, 2005, Chippaux, 2006 and Zaher et al., 2009.

1. *Natriciteres bipostocularis* Broadley, 1962b. Occ. Pap. Natl. Mus. So. Rhodesia 3B(26): 785–786. (*Natriciteres olivacea bipostocularis*)

Type: Holotype, NMSR/M 3524, a 282 mm female (L.D.E.F. Vesey-FitzGerald, 1 Oct. 1959).
Type locality: "Chisansa, Abercorn, Northern Rhodesia at 5300 ft." [= Mbala, N Northern Prov., NE Zambia, 8°51'S, 31°22'E, elevation 1615 m].
Distribution: Southern Africa. Southeastern Democratic Republic of the Congo (Katanga), Zambia (Northern, North-Western) and Angola (Huíla, Malanje), 1320–1615 m.
Sources: Broadley & Pitman, 1960, Broadley et al., 2003 and Broadley & Cotterill, 2004.

2. *Natriciteres fuliginoides* (A.C.L.G. Günther, 1858). Cat. Colub. Snakes Brit. Mus.: 39. (*Coronella fuliginoides*)

Synonym: *Mizodon longicauda* A.C.L.G. Günther, 1863.
Types: Syntypes (3), BMNH 1946.1.12.77, an adult male (Haslar Mus.), BMNH 1956.2.23.51, juvenile male (Rich coll.) [both West Africa], and juvenile male (W. Logan) [Old Calabar].
Type locality: "West Africa" and "Old Calabar" [= Nigeria].
Distribution: West and cen. Africa. Eastern Guinea (Beyla), Sierra Leone, Ivory Coast, Ghana, SW Burkina Faso (SW Centre-Ouest, Hauts-Bassins), S Togo (Plateaux), S Nigeria (Anambra, Bayelsa, Benue, Cross River, Delta, Lagos, Rivers), S Cameroon (Centre, Est, Littoral, Nord-Ouest, Sud, Sud-Ouest), Equatorial Guinea (Bioko Is.), Gabon (Estuaire, Haut-Ogooué, Moyen-Ogooué, Ngounié, Nyanga, Ogooué-Ivindo, Ogooué-Lolo, Ogooué-Maritime, Woleu-Ntem), S Central African Republic (Lobaye, Mbomou, Ombella-Mpoko, Sangha), Congo (Cuvette, Kouilou, Lekoumou, Niari, Plateau) and Democratic Republic of the Congo (Bas-Congo, Bandundu, Equateur, Katanga, Kinshasa, Maniema, Nord-Kivu, Orientale, Sud-Kivu), NSL–800 m.
Sources: Angel, 1933b, Bogert, 1940, Monard, 1951, Witte, 1953, 1962, Capocaccia, 1961b, Mertens, 1965c, Roux-Estève, 1965, Knoepffler, 1966, Leston & Hughes, 1968, Hughes & Barry, 1969, Dunger, 1972, Roman, 1980, Hughes, 1983, J.A. Butler & Reid, 1986, Joger, 1990, J.-F. Trape & Roux-Estève, 1995, Chippaux, 2006, Chirio & Ineich, 2006, Chirio & LeBreton, 2007, Pauwels & Vande weghe, 2008 and Segniagbeto et al., 2011.
Remarks: Original description listed three syntypes but Boulenger (1893a: 217) and Hughes & Barry (1969: 1011) only cite Günther's first two specimens from West Africa as types. Possibly occurs in Benin *fide* Hughes (2013: 152) but occurs in Benin and Ivory Coast *fide* Chirio & LeBreton (2007: 494).

N

3. *Natriciteres olivacea* (W.C.H. Peters, 1854). Mber. Königl. Akad. Wiss. Berlin 1854(11): 622–623. (*Coronella olivacea*)

Synonyms: *Coronella dumerilii* A.C.L.G. Günther, 1860b, *Enicognathus punctatostriatus* Jan, 1863a, *Neusterophis atratus* W.C.H. Peters, 1877c, and *Grayia giardi* Dollo, 1886.
Type: Holotype, ZMB 4803, a 385 mm specimen (W.C.H. Peters, June 1843–Aug. 1847).
Type locality: "Tette" [= Tete, Tete Prov., W Mozambique, 16°10'S, 33°36'E, elevation 125 m].
Distribution: Subsaharan Africa. Southwestern Senegal (Fatick/Kaolack), Guinea-Bissau, Guinea (Beyla), Sierra Leone, Liberia, S Mali (Gao, Sikasso, Tombouctou), Ivory Coast (Bouaflé, Toumodi), S Burkina Faso (Centre-Ouest), S Ghana (Volta), S Togo (Centrale, Kara, Plateaux), S Benin (Atlantique, Collines), Nigeria (Benue, Borno, Cross River, Kaduna, Kwara, Lagos, Oyo, Plateau, Sokoto), Cameroon (Adamaoua, Centre, Nord, Nord-Ouest, Sud-Ouest), Central African

Republic (Bamingui-Bangoran, Haut-Mbomou, Kemo-Gribingui, Nana Mambere, Ombella-Mpoko, Ouham), Gabon (Haut-Ogooué, Moyen-Ogooué, Ogooué-Maritime), Congo (Brazzaville, Kouilou, Likouala, Pool), Democratic Republic of the Congo (Bandundu, Bas-Congo, Equateur, Kasai Occidental, Kasai Oriental, Kinshasa, Katanga, Nord-Kivu, Orientale, Sud-Kivu), Sudan (Al Khartum, An Nil Al Abyad), South Sudan (Bahr al Ghazal, Eastern Equatoria), W Ethiopia (Gemu Gofa, Gojjam, Illubabor, Shoa, Sidamo), S Somalia (Mogadishu), S Kenya (Coast, Nyanza, S Rift Valley, Western), Uganda (Central, Eastern, Northern, Western, Buvuma, Damba, Kome and Sese Is.), Rwanda (Butare, Byumba, Cyangugu, Gisenyi, Kibungo, Kigali, Ruhengeri), Burundi (Bubanza, Bururi, Gitega, Ruyigi), Tanzania (Arusha, Dar es Salaam, Kigoma, Kilimanjaro, Morogoro, Mugharibi, Mwanza, Rukwa, Tanga, Kagera, Mafia, Ukerewe and Zanzibar Is.), Mozambique (Faza, Maputo, Nampula, Niassa, Tete, Zambézia), Malawi (Central, Northern, Southern), Zambia (Central, Copperbelt, Eastern, Luapula, Lusaka, Northern, North-Western, Southern, Western), Zimbabwe (Bulawayo, Manicaland, Mashonaland Central, Mashonaland East, Mashonaland West, Matabeleland North, Matebeleland South, Midlands), N Botswana (North West), N Angola (Cabinda, Cuanza Norte, Cuanza Sul, Luanda, Lunda Norte, Malanje) and NE Namibia (Caprivi), NSL–2300 m.

Sources: K.P. Schmidt, 1923, T. Barbour & Loveridge, 1928, Loveridge, 1929, 1936b, Bogert, 1940, Witte, 1941, 1953, 1962, 1966, 1975, Villiers, 1950a, 1966, Monard, 1951, Laurent, 1954a, 1956a, 1960, Vesey-FitzGerald, 1958, Broadley, 1959, 1983, Sweeney, 1961, Doucet, 1963, Blackwell, 1967, Papenfuss, 1969, Roux-Estève, 1969a, Dunger, 1972, Pitman, 1974, Broadley & Cock, 1975, Girardin, 1978, Spawls, 1978, J.B. Rasmussen, 1981, Derleyn et al., 1983, Hughes, 1983, 2013, Lanza, 1983a, 1990b, Auerbach, 1987, Branch, 1988, Broadley & Howell, 1991, Largen & Rasmussen, 1993, J.-F. Trape & Roux-Estève, 1995, Spawls et al., 2002, Broadley et al., 2003, Chirio & Ineich, 2006, J.-F. Trape & Mané, 2006b, Chirio & LeBreton, 2007, Pauwels & Vande weghe, 2008, Largen & Spawls, 2010, Ullenburch et al., 2010, Segniagbeto et al., 2011, Auliya et al., 2012 and Broadley & Blaylock, 2013.

Remarks: Original description reprinted in W.C.H. Peters (1855: 52).

N

4. *Natriciteres pembana* (Loveridge, 1935). Bull. Mus. Comp. Zool 79(1): 8. (*Natrix olivacea pembana*)

Type: Holotype, MCZ 19112, a 288 mm female (S. bin Asmani, 6 Oct. 1923).

Type locality: "Chake Lake, Pemba Island, Tanganyika Territory" [= Chake Lake, Chake, Pemba Is., Tanga Prov., NE Tanzania, 5°15'S, 39°46'E, elevation NSL].

Distribution: Northeastern Tanzania (Pemba Is.), NSL.

Sources: Loveridge, 1925, Broadley, 1966 and Spawls et al., 2002.

5. *Natriciteres sylvatica* Broadley, 1966b. Arnoldia, Rhodesia 2(35): 5–7. (*Natriciteres variegata sylvatica*)

Type: Holotype, NMZB-UM 8354 (formerly UM 8354), a 332 mm male (H.C. Garbett, March 1965)

Type locality: "Inyangani Tea Estates, Inyangani District, Rhodesia, at 2,500 feet."

Distribution: Southeastern Africa. Southern Tanzania (Iringa, Lindi, Mbeya, Morogoro, Ruvuma), Mozambique (Gaza, Maputo, Niassa, Sofala, Zambézia), Malawi (Central, Northern, Southern), E Zimbabwe (Manicaland) and ext. NE South Africa (KwaZulu-Natal), NSL–2000 m.

Sources: Spawls et al., 2002, Branch et al., 2005 and Broadley & Blaylock, 2013.

6. *Natriciteres variegata* (W.C.H. Peters, 1861a). Mber. Königl. Akad. Wiss. Berlin 61(3): 358. (*Mizodon variegatus*)

Synonyms: *Natrix olivacea uluguruensis* Loveridge, 1935.

Type: Holotype, ZMB 2057, a 328 mm male (H.S. Pel, 1840–1855), lost *fide* Hughes & Barry (1969: 1011).

Type locality: "Goldküste, Westafrika" [= Ghana *fide* Hughes & Barry, 1969: 1011].

Distribution: West and cen. Africa. Guinea (Beyla, Boffa, Kérouané, Kouroussa, Macenta, Nzérékoré, Télimélé), Sierra Leone (S Northern, Southern, Western), Liberia (Grand Cape Mount, Margibi, Montserrado), S Ivory Coast (Abidjan, Danané, Dix-Huit Montagnes, Guiglo, Lagunes, San Pédro, Tabou, Toumodi), S Ghana (Accra, Ashanti, Eastern, S Volta, Western), S Togo (Plateaux), S Benin (Atlantique, Cotonou, Maritime), Burkina Faso (Est, Hauts-Bassins), S Nigeria (Akwa Ibom, Bayelsa, Bendel, Benue, Cross River, Delta, Edo, Lagos, Oyo, Rivers), Cameroon (Centre, Est, Littoral, Nord, Nord-Ouest, Ouest, Sud, Sud-Ouest), Central African Republic (Haute-Sangha, Ouaka, Ouham), Tanzania (Iringa, Lindi, Mbeya, Ruvuma, Tanga: Pemba Is.), S Malawi (Southern), Mozambique and E Zimbabwe (E Manicaland), 45–1830 m.

Sources: Monard, 1937, Bogert, 1940, Loveridge, 1941a, Pakenham, 1947, Villiers, 1950a, 1954a, Condamin, 1958, 1959, E.H. Taylor & Weyer, 1958, Broadley, 1959, 1971d, Manaças, 1959, Doucet, 1963, Menzies, 1966, Blackwell, 1967, Leston & Hughes, 1968, Roux-Estève, 1969a, Hulselmans & Verheyen, 1970, Hulselmans et al., 1971, Dunger, 1972, Stevens, 1974, Broadley & Cock, 1975, Girardin, 1978, J.A. Butler & Reid, 1986, Broadley & Howell, 1991, Chirio & Ineich, 2006, J.-F. Trape & Mané, 2006b, Chirio & LeBreton, 2007, Ullenburch et al., 2010, Chirio, 2013 and Hughes, 2013.

NATRIX Laurenti, 1768 (Natricidae)

Synonyms: *Hydrus* J.G. Schneider, 1799, *Tropidonotus* Kuhl & Hasselt, 1822a *(nomen nudum), Tropidonotus* Kuhl *in* F. Boie, 1827, *Tropedinotus* – Holbrook, 1842b *(nomen incorrectum), Elaphrodytes* Gistel, 1848 *(nomen substitutum), Xyleus* Gistel, 1848 *(nomen nudum), Hydrophilophis* P. Schmidt, 1852 *(nomen substitutum), Tropidophorus* Jan, 1863b, *Tropidonotes* – Jan *in* Filippi, 1865 *(nomen incorrectum), Tropidonatus* – Higgins, 1873 *(nomen incorrectum),* †*Pylmophis* Rochebrune, 1880, *Tropidenotus* – Bedriaga, 1883 *(nomen incorrectum),* †*Pilemophis* Lydekker, 1888 *(nomen emendatum), Tropidonotis* – Velasco, 1890b *(nomen incorrectum), Tropiaonotus* – Stebbing, 1908 *(nomen incorrectum), Tropidnodotus* – Phisalix, 1917c *(nomen incorrectum), Tropitonotus* – Mertens, 1921 *(nomen incorrectum),* †*Pylemophis* – Williston & Gregory, 1925 *(nomen incorrectum), Tropedonotus* – Roddy, 1928 *(nomen incorrectum), Natryx* – Buresch & Zonkow, 1934 *(nomen incorrectum), Natirix* – Pantuwatana, Imlarp & Marshall, 1969 *(nomen incorrectum), Tropidontus* – Tiedemann & Häupl, 1980 *(nomen incorrectum),* †*Plylmophis* – Rage, 1981 *(nomen incorrectum),* †*Palaeomalpolon* Szyndlar, 1981, *Guystebbinsus* Hoser, 2012aa *(nomen illegitimum),* and *Jackyhosernatrix* Hoser, 2012aa *(nomen illegitimum).*

Type species: *Coluber natrix* Linnaeus, 1758.

Distribution: Europe, W Asia and ext. N Africa.

Fossil records: Lower Oligocene to upper Pleistocene of Europe. Fossils unassigned to species include upper Miocene (Vallesian, MN 9–13: 4.9–11.1 mya) of Italy, Moldavia and Spain, lower Pliocene (Ruscinian, MN 14: 4.2–4.9 mya) of Bulgaria, middle Pliocene (Ruscinian, MN 15: 3.2–4.2 mya) of Greece and Slovakia, upper Pliocene (Villanyian, MN 16: 2.6–3.2 mya) of Austria and Ukraine, lower Pleistocene (Calabrian: 0.8–1.8 mya) of Greece and Spain, middle Pleistocene (Ionian: 0.13–0.78 mya) of Bulgaria, Italy and Spain, and upper Pleistocene (Tarantian: 0.01–0.13 mya) of Italy.

Sources: Fatio, 1872, Camerano, 1891, Schreiber, 1912, G. Hecht, 1930a–b, Bodenheimer, 1944, Angel, 1946, Mertens & Wermuth, 1960, McDowell, 1961, Hellmich, 1962, Stemmler, 1967, Steward, 1971, Fretey, 1975, Bruno, 1977, Arnold & Burton, 1978, Street, 1979, Hotz & Broggi, 1982, Lanza, 1983a, Matz, 1983a, Engelmann et al., 1985, Lanka & Vit, 1985, Gruber, 1989, 2009, Bruno & Maugeri, 1990, Szyndlar, 1991a, 1999, 2012, R. Günther, 1996, Holman, 1998a, 2000b, Aragón-Rebollo et al., 2006, Guicking et al., 2006, Ivanov, 2007, Zaher et al., 2009, Cadle, 2011 and Hoser, 2012aa.

1. †*Natrix borealis* (Szyndlar, 1981). Acta Geol. Polon. 31(1–2): 89–90, fig. 12. (†*Palaeomalpolon borealis*)

Type: Holotype, ZZSD RF/290/80, one parabasisphenoid (M. Mlynarski, 1974).

Type locality: "Kozi Grzbiet Hill near Checiny in the Holy Cross Mts, Central Poland, Early Pleistocene (Cromerian)."

Distribution: Lower Pleistocene (Cromerian: 0.465–0.85 mya) of Poland.

Remarks: Rage (1984b: 50) corrected the holotype name from basisphenoid to parabasisphenoid. A species of *Natrix fide* Szyndlar *in* Rage (1984b: 50) but *Palaeomalpolon* possibly a valid genus.

2. †*Natrix longivertebrata* Szyndlar, 1984a. Acta Zool. Cracov. 28(1): 71–73, figs. 29.6–10.

Type: Holotype, ZZSD RK I-10000, one middle trunk vertebra.

Type locality: "Rebielice Królewskie I (RK I) near Klobuck, Kraków-Wielun Upland, central Southern Poland, Upper Pliocene (MN 16)."

Distribution: Lower Miocene (Orleanian, MN 4: 16.0–16.9 mya) of France, middle Miocene (Astaracian, MN 6–8, 11.1–13.7 mya) of France, upper Miocene (Vallesian-Turolian, MN 9–13: 4.9–11.1 mya) of Austria, Hungary and Ukraine, lower Pliocene (Ruscinian, MN 14: 4.2–4.9 mya) of Hungary, middle Pliocene (Ruscinian, MN 15: 3.2–4.2 mya) of Moldavia and Poland, and upper Pliocene (Villanyian, MN 16: 2.6–3.2 mya) of Moldavia and Poland.

Sources: Bachmayer & Szyndlar, 1985, 1987, Mlynarski et al., 1985, Szyndlar, 1991b, 2012, Szyndlar & Zerova, 1992, Venczel, 1994, 1998, 2001, Augé & Rage, 2000, Redkozubov, 2003 and Szyndler & Rage, 2005.

3. *Natrix maura* (Linnaeus, 1758). Syst. Nat., ed. 10, 1: 219. (*Coluber maurus*)

Synonyms: *Coluber viperinus* Latreille *in* Sonnini & Latreille, 1801b, *Coluber natricula* Hermann, 1804, *Coluber pseudoechidna* Hermann, 1804, *Vipera vissena* Rafinesque-Schmaltz, 1814, *Natrix cherseoides* Wagler *in* Spix, 1824, *Natrix ocellata* Wagler *in* Spix, 1824, ? *Coluber guttatus* Risso, 1826, ? *Coluber palustris* Risso, 1826, *Tropidonotus oppelii* Duméril *in* F. Boie, 1827, *Coluber viperinus aurolineatus* Gervais, 1836, *Natrix viperina vittata* Gené, 1839, *Natrix viperina bilineata* Bonaparte, 1840 *(nomen nudum), Natrix viperina nigrotorquata* Bonaparte, 1840 *(nomen nudum), Coluber terstriatus* Duméril *in* Bonaparte, 1840, *Natrix viperina bilineata* Bonaparte, 1840, *Tropidonotus bonnellii* Fitzinger, 1853 *(nomen nudum), Tropidonotus chersoïdes* A.-M.-C. Duméril, Bibron & Duméril, 1854a *(nomen illegitimum), Tropidonotus viperinus bilineata* Jan, 1865c *(nomen substitutum), Coluber viperinus ichthyocharis* Gistel,

N

1868, *Tropidonotus viperinus incerta* Fatio, 1872, and *Tropidonotus viperinus nigra* Mosauer & Wallis, 1927.

Type: Lectotype, NHR Lin-214 (formerly MAFR) (Ulriksdals Castle via Mus. Drottn.), designated herein.

Type locality: "Algiriae" [= Algeria] via lectotype selection.

Distribution: Southwestern Europe and ext. N Africa. France (Ain, Allier, Alpes-Maritimes, Ardèche, Ariège, Aude, Aveyron, Basses-Alpes, Basses-Pyrénées, Bouchés-du-Rhône, Cantal, Cher, Charente, Charente-Maritime, Corrèze, Corsica, Côte-d'Or, Creuse, Deux-Sèvres, Dordogne, Drôme, Essonne, Eure-et-Loir, Gard, Gers, Gironde, Haut-Rhin, Haute-Garonne, Haute-Loire, Haute-Savoie, Haute-Vienne, Hautes-Alpes, Hautes-Pyrénées, Hérault, Ille-et-Vilaine, Indre, Indre-et-Loire, Isère, Jura, Landes, Loire, Loire-Atlantique, Loiret, Loir-et-Cher, Lot, Lot-et-Garonne, Lozère, Maine-et-Loire, Mayenne, Nièvre, Puy-de-Dôme, Pyrénées-Orientales, Rhône, Saône-et-Loire, Sarthe, Savoie, Seine-et-Marne, Tarn, Tarn-et-Garonne, Val-d'Oise, Var, Vaucluse, Vendée, Vienne, Yonne, Yvelines, Hyères Is.), ext. SW Switzerland (Genève, Valais, Vaud), W Italy (W Emilia-Romagna, Liguria, Lombardia, Piemonte, Asinara, Caprera, Maddalena, San Pietro, Sardinia and Sant'Antioco Is.), Spain (Alava, Albacete, Alicante, S Almería, Avilo, Badajoz, Barcelona, Burgos, Caceres, Cádiz, S Castellón, Ciudad Real, Cordoba, Gerona, Granada, Guadelajara, Guipúzcoa, Huelva, Huesca, SE Jaén, La Coruña, Lérida, Lugo, Madrid, Majorca, Málaga, Minorca, Murcia, Navarra, Orense, Oviedo, Palencia, Pontevedra, Salamanca, Santander, Segovia, W Seville, Soria, Tarragona, SW Teruel, Toledo, Valencia, Valladolid, Vizcaya, W Zamora, E Zaragoza, Baleares Is.), Portugal (Aveiro, Beja, Braga, Bragança, Castelo Branco, Coimbra, Evora, Faro, Guarda, Leiria, Lisboa, Portalegre, Porto, Santarém, Setubal, Viana de Castelo, Vila Real, Viseu), Morocco (Agadir, Beni Mellal, Casablanca, Al Jadida, Fès, Kenitra, Kaar es Souk, Marrakech, Meknès, Nador, Ouarzazate, Oujda, Rabat, Safi, Settat, Tangier, Taza, Tetonan), Algeria (Béchar, Biskra), N Tunisia (Gabès, Gafsa, Medenine, Nabeul, Sousse, Tozeur, Tunis, Djerba and Galita Is.) and N Libya (Az Zawiyah, Misratah, Tarabulus), NSL–2600 m.

Fossil records: Middle Pliocene (Ruscinian, MN 15: 3.2–4.2 mya) of France and upper Pleistocene (Tarantian: 0.01–0.13 mya) of France.

Sources: Chpakowsky & Chnéour, 1954, Kramer & Schnurrenberger, 1963, Bons, 1967, Salvador, 1974, 1985, Schätti, 1982, Bruno, 1984, Le Berre, 1989, Fons et al., 1991, Duguy & Saint-Girons, 1993, Bons & Geniez, 1996, Schleich et al., 1996, Holman, 1998a, F.P. Pérez, 2001, Sindaco et al., 2006, Kreiner, 2007, Feriche et al., 2008, Klesius, 2009 and Szyndlar, 2012.

Remarks: Records from Turkey (Maras) and Italy (Sicily) need confirmation. This species composed of 4 lineages *fide* Guicking et al. (2006: 135).

N

4. *Natrix megalocephala* Orlov & Tuniyev, 1986a. Proc. Zool. Inst., USSR Acad. Sci. 158: 117, figs. 1–2, 3 (left), 5 (left), 6 (upper).

Type: Holotype, ZISP 11846, a 940 mm female (K. Satunin, 1909).

Type locality: "Pitsunda, West Caucasus, Georgia, USSR" [= ext. NW Georgia, 43°09'N, 40°21'E]. Emended to Pitsunda, Abkhazia, Georgia, W Caucasus *fide* Orlov & Tuniyev (1992: 44).

Distribution: Northwestern Azerbaijan, W Georgia, ext. SW Russia and NE Turkey, NSL–1500 m.

Sources: Gruber, 1989, Orlov & Tuniyev, 1992, Ilgaz et al., 2005 and Tuniyev et al., 2009.

5. †*Natrix merkurensis* Ivanov, 2002. Acta Palaeont. Polon. 47(3): 525–526, figs. 8d (1–2).

Type: Holotype, SGDB Ah-315, one fragmentary right compound bone.

Type locality: "Merkur-North locality, in the opencast brown coal mine near Tusimice, in the area of the former village Ahníkov, close to Chomutov (northwestern Bohemia), Czech Republic. Orleanian (MN 3a). Early Miocene."

Distribution: Lower Miocene (Orleanian, MN 3–4: 16.0–20.0 mya) of Czech Republic and France.

Sources: Ivanov, 2002, Rage & Bailon, 2005 and Szyndlar, 2012.

6. †*Natrix mlynarskii* Rage, 1988a. Acta Zool. Cracov. 31(13): 464, fig. 1.1.

Type: Holotype, MNHN QU 17181, one middle trunk vertebra.

Type locality: "Phosphorites du Quercy, France; probably the Oligocene of Mas-de-Got."

Distribution: Lower Oligocene (MP 22: 28.4–33.9 mya) of France. Known only from type specimens.

Source: Szyndlar, 2012.

7. †*Natrix natricoides* (Augé & Rage *in* Ginsburg, 2000). Mém. Mus. Natl. Hist. Nat. 183: 293–294, fig. 24. (†*Neonatrix natricoides*)

Type: Holotype, MNHN Sa 23683, one trunk vertebra.

Type locality: "Sansan, Gers, Southwestern France; Astaracien (MN 6), Miocène moyen."

Distribution: Lower to middle Miocene (Orleanian-Astaracian, MN 4–6: 12.8–13.7 mya) of France.

Sources: Rage & Bailon, 2005 and Szyndlar, 2012.

Remarks: Transferred to *Natrix fide* Szyndlar (2005: 41).

8. *Natrix natrix* (Linnaeus, 1758). Syst. Nat., ed. 10, 1: 220. (*Coluber natrix*)

Synonyms: *Natrix gronoviana* Laurenti, 1768, *Natrix vulgaris* Laurenti, 1768, *Coluber scutatus* Pallas, 1773,

Coluber bipedalis Scopoli, 1788, *Coluber arabicus* Gmelin, 1789, *Coluber bipes* Gmelin, 1789, *Coluber tyrolensis* Gmelin, 1789, *Coluber capitetriangulatus* Lacépède, 1789 (*nomen rejiciendum*), *Coluber helveticus* Lacépède, 1789 (*nomen protectum*), *Coluber semimonile* Lacépède, 1789 (*nomen rejiciendum*), *Coluber torquatus* Lacépède, 1789 (*nomen rejiciendum*), *Coluber azureus* Donndorff, 1798, *Coluber helvetus* Donndorff, 1798, *Coluber aesculapii* Sturm, 1799, *Coluber gronovius* Bechstein, 1802, *Coluber scopolianus* Daudin, 1803e, *Coluber decorus* Gravenhorst, 1807, *Coluber discinctus* Gravenhorst, 1807, *Coluber irroratus* Gravenhorst, 1807, *Coluber minutus* Pallas, 1814, *Coluber persa* Pallas, 1814, *Vipera vissena* Rafinesque-Schmaltz, 1814, *Coluber viperinus* Rafinesque-Schmaltz, 1814, *Natrix hybridus* Merrem, 1820, *Coluber bicephalus* Hufeland & Osann, 1825, *Coluber natrix minax* Fitzinger, 1826a (*nomen nudum*), *Coluber natrix murorum* Fitzinger, 1826a (*nomen nudum*), *Coluber scopolii* Risso, 1826, *Coluber siculus* Cuvier, 1829, *Tropidonotus ater* Eichwald, 1831, *Tropidonotus persicus* Eichwald, 1831, *Coluber bilineatus* Bibron & Bory de Saint-Vincent, 1832, *Coluber niger* Dwigubsky, 1832, *Coluber ponticus* Ménétriés, 1832, *Coluber bilineatus* Birbron & Bory, 1833, *Coluber natrix dalmatina* Schinz, 1833, *Natrix torquata minax* Bonaparte, 1834 *in* 1832–1841, *Natrix torquata murorum* Bonaparte, 1834 *in* 1832–1841, *Natrix cetti* Gené, 1839, *Natrix rubetaria* Bonaparte, 1840, *Tropidonotus natrix colchica* Nordmann *in* Demidoff, 1840, *Tropidonotus natrix nigra* Nordmann *in* Demidoff, 1840, ? *Coluber jenisonii* Gistel *in* Gistel & Bromme, 1850, *Natrix torquata nigrescens* Betta, 1853, *Tropidonotus tessellatus nigrescens* Betta, 1857, *Natrix torquata senescens* Miná Palumbo, 1863, *Tropidonotus natrix bilineata* Jan, 1863b (*nomen nudum*), *Tropidonotus natrix nigra* Jan, 1863b (*nomen nudum*), *Tropidonotus natrix picturata* Jan, 1863b (*nomen nudum*), *Tropidonotus natrix subbilineata* Jan, 1863b (*nudum nomen*), *Tropidonotus natrix bilineatus* Jan, 1865c, *Tropidonotus natrix nigra* Jan, 1865c, *Tropidonotus natrix picturata* Jan, 1865c, *Tropidonotus natrix subbilineata* Jan, 1865c, *Tropidonotus fallax* Fatio, 1872, *Tropidonotus sparsus* Schreiber, 1875, *Tropidonotus natrix concolor* Ninni, 1880, *Tropidonotus natrix lineata* Ninni, 1880, *Tropidonotus natrix nigrescens* Ninni, 1880 (*nomen substitutum*), *Tropidonotus natrix nigri-torquata* Ninni, 1880 (*nomen incorrigendum*), *Tropidonotus natrix moreoticus* Bedriaga, 1881–1882, *Tropidonotus natrix bulsanensis* Gredler, 1882, *Tropidonotus natrix astreptophorus* Seoane, 1884, *Tropidonotus natrix natrix albo-torquata* Camerano, 1891 (*nomen illegitimum*), *Tropidonotus natrix albiventris* Dürigen, 1897, *Tropidonotus natrix fasciatus* Dürigen, 1897, *Natrix vibakari continentalis* Nikolsky, 1925, *Tropidonotus natrix albotorquata* – G. Hecht, 1930a (*nomen corrigendum*), *Tropidonotus natrix algericus* G. Hecht, 1930a, *Tropidonotus natrix bithynius* G. Hecht, 1930a, *Tropidonotus natrix britannicus* G. Hecht, 1930a, *Tropidonotus natrix bucharensis* G. Hecht, 1930a, *Tropidonotus natrix cephallonicus* G. Hecht, 1930a, *Tropidonotus natrix corsus* G. Hecht, 1930a, *Tropidonotus natrix cypriacus* G. Hecht, 1930a, *Tropidonotus natrix dystiensis* G. Hecht, 1930a, *Tropidonotus natrix nigrotorquata* – G. Hecht, 1930a (*nomen corrigendum*), *Tropidonotus natrix schirvanae* G. Hecht, 1930a, *Tropidonotus natrix syrae* G. Hecht, 1930a, *Tropidonotus natrix syriacus* G. Hecht, 1930a, *Natrix natrix schweizeri* L. Müller, 1932b, *Natrix natrix lanzai* Kramer, 1971, *Natrix natrix gotlandica* Nilson & Andrén, 1981a, and *Natrix natrix fusca* Cattaneo, 1990.

Type: Holotype, ZMUU 3 (C. Gyllenborg, 1744).

Type locality: "Europa" [= Europe]. Restricted to Sweden *fide* Mertens & Müller (1928: 49).

Distribution: Eurasia and ext. N Africa. Southern Norway (Akershus, Aust-Agder, Bergen, Buskerud, Hedmark, Hordaland, More og Ramsdal, Oppland, Oslo, Ostfold, Rogaland, Sogn og Fjordane, Sør Trøndelag, Telemark, Vest-Agder, Vestfold), Sweden (Alvsborg, Blekinge, Gävleborg, Göteborg and Bohus, Gotland, Halland, SW Jämtland, Jönköping, Kalmar, Kranoberg, Kristianstad, Malmöhus, Östergötland, Skaraborg, Södermanland, Stockholm, Uppsala, Värmland, Västmanland, E Västerbotten, E Västernorrland, Angesön, Farö, Gotska Gräso, Holmön, Möja, Öland and Sandön Is.), S Finland (Häme, Keski-Suomi, W Kuopio, Kymi, SW Lappi, Mikkeli, SW Oulu, Uusimaa Vaasa Turku-Pori, Jädön and Nämdön Is.), United Kingdom (England, Wales, Channel Is. and Isle of Wight), Portugal (Almería, Aveiro, Beja, E Braga, N Bragança, N Castelo Branco, Coimbra, Faro, E Guarda, S Leiria, Lisboa, Portalegre, Santarém, Setubal, E Viana de Castelo, Vila Real, Viseu), Spain (Álava, Alicante, Andorra, Aragon, Ávila, N Badajoz, Burgos, Cáceres, Cádiz, Castellón, Catalonia, Ciudad Real, E Córdoba, Cuenca, Gerona, Granada, Guadalajara, Guipúzcoa, Huelva, Huesca, NW Jaén, La Coruña, Léon, Lérida, Logroño, Lugo, Madrid, Málaga, Murcia, Navarra, Orense, Oviedo, Pontevedra, Salamanca, Santander, Segovia, Soria, Tarragona, Teruel, Toledo, E Valencia, Valladolid, Vizcaya, Zamora, Zaragossa), Andorra, France (Ain, Aisne, Allier, Alpes-Maritimes, Ardèche, Ardennes, Ariège, Aube, Aude, and Aveyron, Bas-Rhin, Basses-Alpes, Basses-Pyrénées, Belfort, Bouchés-du-Rhône, Calvados, Cantal, Cher, Charente, Charente-Maritime, Corrèze, Côte-d'Or, Côte-du-Nord, Creuse, Deux-Sèvres, Dordogne, Doubs, Drôme, Essonne, Eure, Eure-et-Loir, Finistère, Gard, Gers, Gironde, Haut-Rhin, Haute-Garonne, Haute-Loire, Haute-Marne, Haute-Saône, Haute-Savoie, Haute-Vienne, Hautes-Alpes, Hautes-Pyrénées, Hauts-de-Seine, Hérault, Ille-et-Vilaine, Indre, Indre-et-Loire, Isère, Jura, Landes, Loir-et-Cher, Loire, Loire-Atlantique, Loiret, Lot, Lot-et-Garonne, Lozère, Maine-et-Loire, Manche, Marne, Mayenne, Meurthe-Moselle, Meuse, Morbihan, Moselle, Nièvre, Nord, Oise, Orne, Paris,

Pas-de-Calais, Puy-de-Dôme, Pyrénées-Orientales, Rhône, Saône-et-Loire, Sarthe, Savoie, Seine-et-Marne, Seine-Maritime, Seine St.-Denis, Somme, Tarn, Tarn-et-Garonne, Val-d'Oise, Val-de-Marne, Var, Vaucluse, Vendée, Vienne, Vosges, Yonne, Yvelines, Corsica, Hyères and Oléron Is.), Belgium (Antwerpen, Flemish Brabant, Liège, Luxembourg, Namur, Walloon Brabant), Luxembourg, Netherlands (Drente, Friesland, Gelderland, Groningen, Limburg, Noord Holland, Overijsssel, Utrecht, Zuid Holland), Germany (Baden-Württemberg, Bavaria, Berlin, Brandenburg, Bremen, Hamburg, Hesse, Lower Saxony, Mecklenburg-Vorpommern, North Rhine-Westphalia, Rhineland-Palatinate, Saarland, Saxony, Saxony-Anhalt, Schleswig-Holstein, Thuringia, Borkum and Juist Is.), Denmark (Aarhus. Bornholm, Fredericksborg, Fyn: Langeland Is.; København, S Nordjylland, Ribe, Ringkøbing, Roskilde, Sønderjylland, Storstrøm, Vejle, Vestsjaelland, Viborg), Switzerland (Aargau, Appenzell Inner-Rhoden, Appenzell Ausser-Rhoden, Bern, Basel-Landschaft, Basel-Stadt, Fribourg, Genève, Glarus, Graubünden, Jura, Lucerne, Neuchâtel, Nidwalden, Obwalden, St. Gallen, Schaffhausen, Solothurn, Schwyz, Thurgau, Ticino, Uri, Vaud, Valais, Zug, Zürich), Austria (Burgenland, Carinthia, Lower Austria, Salzburg, Styria, Tyrol, Upper Austria, Vienna, Vorarlberg), Italy (Abruzzi, Basilicata, Calabria, Campania, Emilia-Romagna, Friuli-Venezia Giulia, Lazio, Liguria, Lombardia, Marche, Molise, Piemonte, Puglia, Toscana, Trentino-Alto Adige, Umbria, Valle d'Aosta, Veneto, Elba, Sardinia and Sicily Is.), Poland (Dolnoslaskie, Kujawsko-Pomorskie, Lodzkie, Lubelskie, Lubuskie, Malopolskie, Mazowieckie, Opolskie, Podkarpackie, Podlaskie, Pomorskie, Slaskie, Swietokrzyskie, Warminsko-Mazurskie, Wielkopolskie, Zachodniopomorskie), Czech Republic (Jihocesky, Jihomoravsky, Praha, Severocesky, Severomoravsky, Stredocesky, Vychodocesky, Zapadocesky), Hungary (Bacs-Kiskun, Baranya, Bekes, Borsod-Abauj-Zemplen, Budapest, Csongrad, Fejer, Gyor Moson Sopron, Hajdu-Bihar, Heves, Jasz-Nagykun-Szolnok, Komarom Esziergom, Nograd, Pest, Somogy, Szabolcs-Szatmar Bereg, Tolna, Vas, Veszprem, Zala), Romania (Alba, Arad, Arges, Bacau, Bihor, Bistrita-Nasaud, Botosani, Braila, Brasov, Bucharest, Buzau, Calarasi, Caras-Severin, Cluj, Constanta, Covasna, Dambovita, Dorj, Galati, Giurgiu, Gorg, Harghita, Hunedoara, Ialomita, Iasi, Maramures, Mehedinti, Mures, Neamt, Olt, Prahova, Salaj, Satu Mare, Sibiu, Suceava, Teleorman, Timis, Tulcea, Vaslui, Valcea, Vrancea), Bosnia (Cres, Croatia, Krk, Macedonia, Rab, Serbia, Slovenia, Vojvodina, Pag and Korcula Is.), Bulgaria (Blagoevgrad, Burgas, Dobrich, Gabrovo, Grand Sofiya, Haskovo, Kardzhali, Lovech, Kyustendil, Montana, Pazardzhik, Pernik, Pleven, Plovdiv, Razgrad, Rousse, Shumen, Silistra, Sliven, Sofiya, Stara Zagora, Targovishte, Varna, Veliko Tarnovo, Vidin, Vratsa, Yambol), Albania, Greece (Aitolía and Akarnanía, Akhaïa, Andros, Antíparos, Argolís, Arkadhía, Arta, Attikí, Dhodhekánisos, Dráma, Evros, Evvoia, Florina, Fokís, Ilía, Imathía, Ioánnina, Kárpathos, Kavála, Kéa, Kefallinía, Kérkira, Khalkidhikí, Khardhítsa, Khíos, Kikládhes, Kímolos, Korinthía, Kos, Kozáni, Lakonía, Lárisa, Levkás, Magnisía, Messinía, Míkonos, Mílos, Náxas, Páros, Piería, Préveza, Ródhos, Sámos, Serrai, Síros, Thesprotía, Thessaloníki, Tínos, Tríkala, Voiotía, Zákinthos, Xanthi, Léros, Lésvos, Límnos, Políaigos, Samothráki, Skíras and Thásos Is.), Turkey (Adana, Afyon, Amasya, Ankara, Antalya, Artvin, Aydin, Balikesir: Bolu, Burdur, Bursa, Denizli, Edirne, Gaziantep, Hatay, Icel, Isparta, Istanbul, Izmir, Kars, Kayseri, Kirklareli, Kocaeli, Konya, Mugla, Samsan, E Skisehir, Trabzon, Tunceli, Usak, Van, Marmara and Imroz Is.), Cyprus, N Syria (Homs), Lebanon (Bayrut), Belorussia, Estonia (Hiiumaa and Saaremaa Is.), Latvia, Lithuania, Moldavia, Ukraine, SW Russia, Georgia, Armenia, Azerbaijan, N Kazakhstan, W Turkmenistan, ext. N Mongolia (Hovsgol, N Selenge, N Hentiy, NW Dornod), ext. NW China (Xinjiang), ext. N Iran (Central, East Azarbaijan, Ghilan, Mazandaran), N Morocco (Beni Mellal, Chechaouene, Fès, Gharb-Chrarda-Beni Hssen, Kenitra, Meknès, Tetouan), ext. N Algeria (Algiers, Blida, Tiaret) and ext. N Tunisia (Jendouba, Nabeul), NSL–2500 m.

Fossil records: Upper Miocene (Turolian, MN 13: 4.9–7.3 mya) of Hungary, lower Pliocene (Ruscinian, MN 14: 4.2–4.9 mya) of Russia, upper Pliocene (Villanyian, MN 16–17: 1.8–3.2 mya) of Hungary and Moldavia, Pleistocene of Ukraine, lower Pleistocene (Calabrian: 0.8–1.8 mya) of Austria, Czech Republic, Poland and Romania, middle Pleistocene (Ionian: 0.13–0.78 mya) of Austria, France, Germany, Netherlands, Poland, Romania and United Kingdom, and upper Pleistocene (Tarantian: 0.01–0.13 mya) of Bulgaria, Croatia, Germany, Poland, Serbia and United Kingdom.

Sources: Strauch, 1873, Boulenger, 1891h, Doumergue, 1901, Bruner, 1908, Steinheil, 1913a, 1913c, 1914b, Nikolsky, 1916, Grodzinski, 1928, Bird, 1936, Werner, 1939, Fejérváry-Lángh, 1943, Hvass, 1945, Mertens, 1947b, Terent'ev & Chernov, 1949, 1965, Chpakowsky & Chnéour, 1954, M.A. Smith, 1954, Domergue, 1959, Gislén & Kauri, 1959, Fuhn & Vancea, 1961, Davidson, 1964, Salvador, 1974, Brelih & Dzukic, 1975, 1985, Baran, 1976a, Bannikov et al., 1977, Dely, 1978, Thorpe, 1979, 1984a–b, 1989a–b, Basoglu & Baran, 1980, Bruno, 1980, 1984, 1989, Nilson & Andrén, 1981a, Sparreboom, 1981, Frazer, 1983, Cabela & Tiedemann, 1985, Beshkov, 1986, Tian et al., 1986, Tonge, 1986, Frazer, 1989, Borkin et al., 1990a–b, Cattaneo, 1990, Latifi, 1991, Szyndlar, 1991c, Veith, 1991a–b, Eckstein, 1993, Zhao & Adler, 1993, Szczerbak, 1994, Venczel & Sen, 1994, Bons & Geniez, 1996, R. Günther, 1996, Schleich et al., 1996, Borkin et al., 1997, Holman, 1998a, Cabela et al., 2001, Hofer et al., 2001, Mikátová et al., 2001, Paracuellos, 2001, F.P. Pérez, 2001, Spellerberg, 2002, Glowacinski & Rafinski, 2003, Sindaco et al.,

2006, Ivanov, 2007, Renner & Vitzthum, 2007, Trapp, 2007, Baier et al., 2009, Creemers & Delft, 2009, Tuniyev et al., 2009, Vacher & Geniez, 2010, Stojanov et al., 2011, Jablonski et al., 2012 and Szyndlar, 2012.

Remarks: This species composed of five lineages *fide* Guicking et al., (2006: 136).

9 †*Natrix parva* Szyndlar, 1984a. Acta Zool. Cracov. 28(1): 47–49, fig. 17.

Type: Holotype, ZPW IZ-6/R/4, one trunk vertebra.

Type locality: "Mala Cave (MA) at Zelce Hill near Dzialoszyn (Northern Kraków-Wielun Upland in central Southern Poland), Upper Miocene (Turolian/ Pontian), layer 7+8."

Distribution: Upper Miocene (MN 7–8) of Poland. Known only from type locality.

Source: Szyndlar, 1991b.

Remarks: Doubt about species validity based upon type locality to actually be upper Pleistocene; if so, would be a synonym of *Natrix natrix fide* Szyndlar (1991b: 241).

10. †*Natrix rudabanyaensis* Szyndlar, 2005. Paleontogr. Ital. (2004) 90: 38–40, fig. 5c.

Type: Holotrype, GIH V.22174, one trunk vertebra.

Type locality: "locality 2, Rudabánya, NE Hungary; Late Miocene, early Vallesian (biozone MN 9)."

Distribution: Middle Miocene (Astaracian, MN 7–8: 11.1–12.8 mya) of Romania and upper Miocene (Vallesian, MN 9: 9.7–11.1 mya) of Hungary.

Sources: Venczel & Stiuca, 2008 and Szyndlar, 2012.

11. †*Natrix sansaniensis* (Lartet, 1851). Notice Colline Sansan: 40. (†*Coluber sansaniensis*)

Type: Lectotype, MNHN SA 9873, one trunk vertebra, designated by Rage (1981: 538).

Type locality: "Sansan, Gers, France" via lectotype selection.

Distribution: Lower Miocene (Orleanian, MN 3–5: 13.7–20.0 mya) of Czech Republic, France and Germany, and middle Miocene (Astaracien, MN 6–8: 11.1–13.7 mya) of France and Hungary.

Sources: Rochebrune, 1880, Szyndlar, 1984a, Rage, 1981, Szyndlar & Schleich, 1993, Augé & Rage, 2000, Ivanov, 2000, 2002, Redkozubov, 2003 and Szyndlar, 2005, 2012.

12. *Natrix tessellata* (Laurenti, 1768). Synop. Rept.: 87. (*Coronella tessellata*)

Synonyms: *Coluber hydrus* Pallas, 1771, *Tropidonotus bipes* Gmelin, 1789, *Coluber hydrophilus* Lindaker, 1791, ? *Coluber ponticus* Güldenstaedt, 1791, *Coluber caspius* Latreille, 1801b *(nomen praeoccupatum)*, *Coluber ponticus* Georgi, 1801, *Hydrus caspius* G. Shaw, 1802, *Coluber pseudoechidna* Hermann, 1804, *Enhydris caspia* Oken, 1816, ? *Pelamis fuscatus* Rafinesque-Schmaltz, 1817, *Coluber gabinus* Metaxá, 1823, *Coluber palustris* Risso, 1826, *Tropidonotus gracilis* Eichwald, 1831, *Tropidonotus tantalus* Eichwald, 1831, *Coluber gabina* Bonaparte, 1832 *in* 1832–1841, *Coluber griseus* Dwigubsky, 1832, *Coluber muravievii* Dwigubsky, 1832, *Coluber reticulatus* Ménétriés, 1832, *Natrix tessellata albo-lineolata* Bonaparte, 1834 *in* 1832–1841 *(nomen incorrigendum)*, *Coluber elaphoides* Brandt, 1838, *Natrix hydrus immaculata* Fitzinger *in* Bonaparte, 1840 *in* 1832–1841, *Natrix tessellata flavescens* Massalongo, 1854, *Tropidonotus tessellatus concolor* Jan, 1863b *(nomen nudum)*, *Tropidonotus tessellates concolor* Jan, 1865c, *Tropidonotus tessellatus decipiens* Betta, 1865, *Coluber tessellatus nigerrima* Gistel, 1868 *(nomen nudum)*, *Tropidonotus tessellatus laurenti* Bedriaga, 1881, *Trepidonotus tesselatus* – Lortet, 1883 *(nomen incorrectum)*, *Coluber sabinus* – Camerano, 1891 *(nomen incorrectum)*, *Tropidonotus elephoides* – Camerano, 1891 *(nomen incorrectum)*, *Tropidonotus tessellatus albolineata* – Camerano, 1891 *(nomen corrigendum)*, *Tropidonotus tessellatus flavescens* F. Werner, 1891, *Tropidonotus tessellatus sparsus* Dürigen, 1897, *Tropidonotus tessellatus rubromaculosus* Dürigen, 1897, *Tropidonotus tessellatus viperinoides* Dürigen, 1897 *(nomen substitutum)*, *Tropidonotus tessellatus hagenbecki* F. Werner, 1897c, *Tropidonotus tessellatus lineaticollis* F. Werner, 1897c, *Tropidonotus tessellatus nigerrima* F. Werner, 1897c, *Tropidonotus tessellatus vosseleri* F. Werner, 1903c, *Tropidonotus tessellatus heinrothi* G. Hecht, 1930a, *Natrix dubbioisi* Scortecci, 1932a, *Natrix tessellata cyreni* Sochurek, 1956, *Natrix tessellata gaydai* Sochurek, 1956, and *Natrix dunni* Malnate, 1968.

Type: Holotype, not designated, formerly FAT, location unknown.

Type locality: "Japidia, vulgo Cars" [= Karst country or Kras Plateau, S Carniola, W Slovenia, 45°48'N, 14° 00'E *fide* Flower, 1933: 807, and Street, 1979: 140].

Distribution: Southeastern Europe, Egypt, Arabia and Asia. Southwestern Germany (N Rhineland-Palatinate), S Switzerland (Genève, Graubunden, Schwyz, St. Gallen, Nidwalden, Obwalden, Ticino, Uri, Vaud, Zürich), Italy (Abruzzi, Basilicata, N Calabria, Campania, Emilia-Romagna, Friuli-Venezia Giulia, Lazio, Liguria, Lombardia, Marche, Molise, Piemonte, Puglia, Toscana, Trentino-Alto Adige, Umbria, Veneto), E Austria (Burgenland, Carinthia, Lower Austria, E Styria, E Upper Austria, Vienna), Czech Republic (Jihocesky, Jihomoravsky, Praha, Severocesky, Stredocesky, Zapadocesky), Hungary (Budapest, Fejér, Somogy, Vesprém), Bosnia-Herzegovina, Ada, Cres and Krk Is., Croatia, Kosovo, Macedonia, Montenegro, Serbia, Slovenia, Romania (Arges, Bacau, Bihor, Brasov, Bucharest, Calarasi, Caras-Severin, Cluj, Constanta, Galati, Gorg, Hunedoara, Mehedinti, Timis), Bulgaria (Blagoevgrad, Burgas, Dobrich, Gabrovo, Grand Sofiya, Haskovo, Kardzhali, Lovech, Kyustendil, Montana, Pazardzhik, Pernik, Pleven, Plovdiv, Razgrad, Rousse, Shumen, Silistra, Sliven,

Smolyan, Sofiya, Stara Zagora, Targovishte, Varna, Veliko Tarnovo, Vidin, Vratsa, Yambol), N Albania, Greece (Aitolía and Akaranía, Akhaïa, Arta, Attikí, Dhodhekánisos, Evritanía, Evros, Evvoia, Florina, Fokís, Fthíotis, Grevená, Ilía, Imathía, Ioánnina, Kardhítsa, Kastoria, Kérkíra, Kilkís, Kikládhes, Kriti, Korinthía, Lakonía, Lésvos, Levkás, Magnisía, Messinía, Piería, Préveza, Sámos, Serrai, Voiotía, Corfu, Kythira, Lefkada, Ródhos, Sérifos, Thesprotía and Tínos Is.), Turkey (Adana, Ankara, Antalya, Artvin, Bitlis, Burdur, Bursa, Canakkale, Diyarbakir, Edirne, Eskisehir, Gaziantep, Icel, Istanbul, Izmir, Kars, Kayseri, Kocaeli, Konya, Manisa, Maras, Mardin, Mugla, Mus, Nigde, Rize, Siirt, Trabzon, Tunceli, Urfa, Van), Cyprus, Syria (Damascus, Halab, Homs), Lebanon (Beqaa, Beirut, El Jnoub, Liban-Nord, Mont-Liban, Nabatiye), Israel (Central, Gaza Strip, Golan Heights, Haifa, Jerusalem, Northern, Tel Aviv, West Bank), NW Jordan (Ajloun, Amman, Balqa, Irbid, Jarash, Karak, Maan, Madaba, Zarqa), NE Egypt (Cairo, Damietta, Daqahlia, Faiyum, Gharbia, Giza, Kafr El-Shaikh, Menufia, Sharkeya, Suez), Yemen, Iraq (Al Anbar, Al-Basrah, An Najaf, Babil, Baghdad, Diyala, Dhi Qar, Ninawa, Salah ad-Din), Iran (Central, East Azarbaijan, Esfahan, Fars, Hamadan, Khuzestan, Kordestan, Lorestan, Razavi Khorasan, West Azarbaijaln, Zanjhan), Moldavia, S Ukraine (Serpilar and Zmeinyi Is.), SW Russia, Georgia, Armenia, Azerbaijan, S Kazakhstan, Uzbekistan (Vozrozhdeniya Is.), Turkmenistan, Kyrgystan, Tadzhikistan, N Afghanistan (Balkh), N Pakistan (Gilgit-Baltistan, Khyber Pakhtunkhwa) and ext. NW China (Xinjiang), NSL–2800 m.

Fossil records: Upper Miocene (Turolian, MN 13: 4.9–7.3 mya) of Hungary, upper Pliocene (Villanyian, MN 16–17: 1.8–3.2 mya) of Hungary, Pleistocene of Ukraine, lower Pleistocene (Calabrian: 0.8–1.8 mya) of Austria, Czech Republic, Hungary, Poland and Romania, middle Pleistocene (Ionian: 0.13–0.78 mya) of Greece and Romania, and upper Pleistocene (Tarantian: 0.01–0.13 mya) of Croatia, Germany and Greece.

Sources: Lortet, 1883, J. Anderson, 1898, Bedriaga, 1912, Fejérváry, 1922, Wall, 1923c, Corkill, 1932b, Scortecci, 1932a, Flower, 1933, Stepánek, 1934, Angel, 1936c, F. Werner,1938a, 1939, M.A. Smith, 1940b, 1943, Fejérváry-Langh, 1943, Terent'ev & Chernov, 1949, 1965, G. Haas, 1951, Mertens, 1957c, 1969b, Fuhn & Vancea, 1961, Beshkov & Beron, 1964, Pozzi, 1966, Marx, 1968, G. Haas & Werner, 1969, Korneva,1969, Král, 1969, Leviton & Anderson, 1970, Y. Werner, 1971, Brelih & Dzukic, 1974, Baran, 1976a, Obst, 1976, Bannikov et al., 1977, Leviton, 1977, Dely, 1978, Gruschwitz, 1978, 1986, Krapp & Böhme, 1978, Lanka, 1978, Sklenár & Rocek, 1979, Basoglu & Baran, 1980, Hraoui-Bloquet, 1981, Bruno, 1984, Cabela & Tiedemann, 1985, Gasperetti, 1988, Chondropoulos, 1989, Latifi, 1991, Szyndlar, 1991a, Leviton et al., 1992, Zhao & Adler, 1993, Szczerbak, 1994, R. Günther, 1996, Lada & Nedosekin, 1997, Holman, 1998a, Bouskila & Amitai, 2001, Cabela et al., 2001, Disi et al., 2001, Hofer et al., 2001, Mikátová et al., 2001, Hraoui-Bloquet et al., 2002, M.S. Khan, 2002, Sindaco et al., 2006, Ivanov, 2007, Trapp, 2007, Tuniyev et al., 2009, Vacher & Geniez, 2010, Amr & Disi, 2011, Bar & Haimovitch, 2011, Stojanov et al., 2011, Jablonski et al., 2012 and Masroor & Mebert, 2012.

Remarks: Holotype not in MRSN or NMW. Leviton et al. (1992: 101) cited the type locality as an alpine meadow probably in Italy. Record for Italy (Sicily) needs confirmation. Relict population on Cyprus probably or nearly extinct *fide* Baier et al. (2009: 260). This species composed of three lineages *fide* Guicking et al., (2006: 135).

†*NEBRASKOPHIS* Holman, 1973a (Colubridae)

Type species: †*Nebraskophis skinneri* Holman, 1973a.

Distribution: Upper Eocene to lower Oligocene and middle to upper Miocene of USA. Fossils unassigned to species include upper Eocene (Chadronian) of USA (Georgia).

Sources: Holman, 1973a, 1977e, 1979a, 1982a, 2000a, Rage, 1984b, Parmley & Holman, 1995 and Parmley & Holman, 2003.

1. †*Nebraskophis oligocenicus* Holman, 1999. Acta Zool. Cracov. 42(3): 452–454, fig. 3.

Type: Holotype, UF 190833, one trunk vertebra (UF field crews, 1966–1967).

Type locality: "I-75 Local Fauna (Florida Natural History Museum Locality AL018), 1.5 km WSW Gainesville, on the W side of I-75; Micanopy Quadrangle, NE1/4, Sec. 4, T10S, R19E, Alachua County, Florida; Early Oligocene (Whitneyan)."

Distribution: Lower Oligocene (Whitneyan: 30.8–33.3 mya) of USA (Florida). Known only from type locality.

2. †*Nebraskophis skinneri* Holman, 1973a. Contr. Mus. Paleont., Univ. Michigan 24(12): 131–132, figs. 4a–b.

Type: Holotype, MSUVP 708, one trunk vertebra (M.F. Skinner, C.W. Hibbard, J.A. Holman & MSU field crews, 1967–1971).

Type locality: "one mile north of the Niobrara River on the Ranch of Franklin H. Egelhoff in the SE corner of the NE1/4 SW1/4 sec. 29, T 33 N, R 23 W, Keya Paha County, Nebraska. Upper Miocene, lower Valentine Formation, Egelhoff local fauna, [= middle Miocene].

Distribution: Middle Miocene (Barstovian 13.6–16.3 mya) of USA (Nebraska) and upper Miocene (Hemphillian: 4.8–9.0 mya) of USA (Nebraska).

Remarks: Original description modified in Holman (2000a: 175–176, fig. 111).

†*NEONATRIX* Holman, 1973a (Natricidae)

Type species: †*Neonatrix elongata* Holman, 1973a.
Distribution: Fossils unassigned to species include lower Miocene (Orleanian, MN 4–5: 13.7–16.9 mya) of Austria and Germany, middle Miocene (Astaracian, MN 6–8: 11.1–13.7 mya) of Slovakia, upper Miocene (Vallesian, MN 9: 9.7–11.1 mya) of Hungary, and lower Pliocene (Ruscinian, MN 14: 4.2–4.9 mya) of Hungary.
Sources: Holman, 1973a, 1979a, 1982a, Rage, 1984b, 2000a, Szyndlar, 1987, 1991a and Szyndlar & Schleich, 1993.
Remarks: †*Neonatrix* should be restricted to North American species with European species in a different genus *fide* Szyndlar & Schleich (1993: 23).

1. †*Neonatrix crassa* Rage & Holman, 1984. Géobios 17(1): 96–97, figs. 5a–e.

Type: Holotype, UCBL FSL 285 012, one middle trunk vertebra.
Type locality: "la fissure M de La-Grive-Saint-Alban (Isère), France; zone MN 7, Astaracien, le Miocène français."
Distribution: Middle Miocene (Astaracian, MN 7–8: 11.1–12.8 mya) of France. Known only from type series.
Source: Szyndlar, 2012.

2. †*Neonatrix elongata* Holman, 1973a. Contr. Mus. Paleont., Univ. Michigan 24(12): 130–131, figs. 3a–b.

Type: Holotype, MSUVP 710, one posterior trunk vertebra (M.F. Skinner, C.W. Hibbard, J.A. Holman & MSU field crews, 1967–1971).
Type locality: "one mile north of the Niobrara River on the Ranch of Franklin H. Egelhoff in the SE corner of the NE 1/4 SW1/4 sec. 29, T 33 N, R 23 W, Keya Paha County, Nebraska. Upper Miocene, lower Valentine formation, Egelhoff local fauna."
Distribution: Lower Miocene (Hemingfordian: 16.3–20.6 mya) of USA (South Dakota, Texas), middle Miocene (Barstovian: 13.6–16.3 mya) of USA (Nebraska, South Dakota, Texas), and upper Miocene (Hemphillian: 4.8–9.0 mya) of USA (Nebraska).
Sources: Holman, 1976a, 1977b, 1977e, 1978, 1979a, 1996a, 2000a and Parmley & Holman, 1995.
Remarks: Original description modified by Holman (2000: 196–197, fig. 125).

3. †*Neonatrix europaea* Rage & Holman, 1984. Géobios 17(1): 94–96, figs. 4a–h.

Type: Holotype, UCBL 285 008, one middle trunk vertebra.
Type locality: "la fissure M de La-Grive-Saint-Alban (Isère), France; zone MN 7, Astaracien, le Miocène français."
Distribution: Lower-middle Miocene (Orleanian-Astaracian, MN 4–8: 11.1–16.9 mya) of France.
Sources: Augé & Rage, 2000, Ivanov, 2000, Rage & Bailon, 2005 and Szyndlar, 2005, 2012.

4. †*Neonatrix infera* Holman, 1996a. Tertiary Res. 17(1–2): 7–8, figs. 1a–e.

Type: Holotype, MSUVP 1423, one trunk vertebra (J.A. Holman, R.D. Weigel, D. Gillette, B.H. Slaughter & MSU field crew, 1967).
Type locality: "Trinity River Local Fauna, at longitude 95 degrees 03', latitude 30 degrees 41', 11.3 km northeast of Coldspring, San Jacinto County, Texas, USA; Flemming Formation, Early Miocene, Barstovian Land Mammal Age."
Distribution: Middle Miocene (Barstovian: 13.6–16.3 mya) of USA (Texas). Known only from type locality.
Remarks: Original description modified by Holman (2000a: 198, fig. 126). Possible record from upper Miocene of Nebraska *fide* Parmley & Hunter (2010: 538).

5. †*Neonatrix magna* Holman, 1982a. Trans. Nebraska Acad. Sci. 10: 35–36, figs. 6a–d.

Type: Holotype, MSUVP 943, one trunk vertebra (A. Griggs, J.A. Holman, M. Podell & C. Wellstead, July 1976).
Type locality: "Norden Bridge Quarry, near the Norden-Johnstown Road, about 274 m south of the Norden Bridge across the Niobrara River, at an elevation of 661 m in SE 1/4 SW 1/4 Sec. 33, T. 33 N., R. 23 W., Brown County, Nebraska; lowermost part of Valentine Formation, (Miocene: Late Barstovian)."
Distribution: Middle Miocene (Barstovian: 13.6–16.3 mya) of USA (Nebraska). Known only from type locality.
Remarks: Original description modified by Holman (1987a: 116 and 2000a: 199–200, fig. 127).

6. †*Neonatrix nova* Szyndlar, 1987. J. Vert. Paleont. 7(1): 61–62, figs. 6b–f.

Type: Holotype, CUP 5197, one trunk vertebra.
Type locality: "Dolnice near to the town of Cheb in westernmost Bohemia (Czechoslovakia); Lower Miocene (Early Orleanian or Ottnangian; European Land Mammalian Biozone MN 4 = Upper Hemingfordian in North America)."
Distribution: Lower Miocene (Orleanian, MN 4: 16.0–16.9 mya) of Czech Republic. Known only from type locality.

NERODIA Baird & Girard, 1853 (Natricidae)

Synonym: *Funkus* Hoser, 2012h *(nomen illegitimum).*
Type species: *Coluber sipedon* Linnaeus, 1758.
Distribution: North America and Cuba.
Fossil records: Middle/upper Miocene to upper Pleistocene of USA. Fossils unassigned to species include lower Pliocene (Hemphillian) of USA (Kansas), lower/middle Pleistocene (Irvingtonian I) of USA (Arizona), middle/upper Pleistocene (Irvingtonian II) of USA (Kansas, Pennsylvania), and upper Pleistocene (Rancholabrean I) of USA (Oklahoma).
Sources: Clay, 1938b, Cliburn, 1960, McDowell, 1961, Conant, 1975, Rossman & Eberle, 1977, N.J. Rossman et al., 1982, R. Lawson, 1985, 1987, Dundee & Rossman, 1989, Ernst & Barbour, 1989, Chiasson & Lowe, 1989, Conant & Collins, 1991, Holman, 1995a, 2000a, Dwyer & Kaiser, 1997, Wallach, 1998, Alfaro & Arnold, 2001, Gibbons & Dorcas, 2004, Zaher et al., 2008 and Cadle, 2011.

1. *Nerodia clarkii* (Baird & Girard, 1853). Cat. No. Amer. Rept., Serp. 1: 48. (*Regina clarkii*)

Synonyms: *Tropidonotus medusa* A.C.L.G. Günther, 1858, *Nerodia compressicauda* Kennicott, 1860, *Tropidonotus ustus* Cope, 1860b, *Tropidonotus compsolaemus* Cope, 1861a, *Tropidonotus cubanus* Gundlach *in* W.C.H. Peters, 1861f, *Tropidonotus ornatus* Jan, 1863b (*nomen nudum*), *Tropidonotus ornatus* Jan, 1865c, *Tropidonotus compressicaudus walkeri* Yarrow, 1883a, *Natrix compressicauda bivittata* Cope, 1889a, *Natrix compressicauda obscura* Lönnberg, 1894, and *Natrix compressicauda taeniata* Cope, 1895b.
Type: Holotype, USNM 2264, an 806 mm specimen (J.H. Clark [U.S.-Mex. Bound. Surv.], 1850–1852, via J.D. Graham).
Type locality: "Indianola, Texas" [USA].
Distribution: Coastal marshes of SE USA (Alabama, Florida, Florida Keys, Louisiana, Mississippi, Texas) and Cuba (Camagüey, Lattebana, Villa Clara), NSL–25 m.
Sources: Stejneger, 1940, Carr, 1940, F. Cook, 1962, Mount, 1975, Dunson, 1979, Linzey, 1979, Ashton & Ashton, 1981, Myer, 1988, R. Lawson et al., 1991, Rodrígues-Robles & Greene, 1996, Tennant, 1997 and Werler & Dixon, 2000.

2. *Nerodia cyclopion* (A.-M.-C. Duméril, Bibron & Duméril, 1854a). Erpét. Gén. 7(1): 576–577. (*Tropidonotus cyclopion*)

Types: Syntypes (2), MNHN 3481a–b, longest syntype 1130 mm.
Type locality: "Nouvelle-Orléans" [= New Orleans, Louisiana, USA] and not stated. Restricted to New Orleans, Louisiana, USA *fide* Stejneger & Barbour (1917: 95).
Distribution: South-central USA (ext. SW Alabama, E and S Arkansas, ext. NW Florida, ext. SW Illinois, ext. W Kentucky, Louisiana, W and S Mississippi, ext. SE Missouri, ext. W Tennessee, SE Texas), NSL–150 m.
Fossil records: Middle Pliocene (Blancan II) of USA (Kansas).
Sources: P. Smith, 1961, Garton et al., 1970, R.W. Barbour, 1971, Mount, 1975, Sanderson, 1983, 1993, T.R. Johnson, 1987, Meade, 1991, Tennant, 1997 and Werler & Dixon, 2000.

3. *Nerodia erythrogaster* (Forster *in* Bossu, 1771). Travels Louisiana 1: 364. (*Coluber erythrogaster*)

Synonyms: *Tropidonotus transversus* Hallowell, 1852a, *Nerodia woodhousii* Baird & Girard, 1853, *Nerodia couchii* Kennicott, 1860, *Tropidonotus bisectus* Cope, 1887d, *Tropidonotus roulei* Chabanaud, 1917d, *Natrix erythrogaster flavigaster* Conant, 1949, *Natrix erythrogaster neglecta* Conant, 1949, *Natrix erythrogaster bogerti* Conant, 1953, and *Natrix erythrogaster alta* Conant, 1963c.
Type: Neotype, USNM 126890, a 1318 mm female (T.M. Beckett via E.B. Chamberlain, 19 March 1948), designated by Conant (1949: 10).
Type locality: "near Parker's Ferry, Edisto River Swamp, Charleston County, South Carolina" [USA] via neotype selection.
Distribution: Southern USA (Alabama, Arkansas, S Delaware, N Florida, Georgia, Illinois, Indiana, ext. SE Iowa, SE and S Kansas, W Kentucky, Louisiana, SE Maryland, S Michigan, Mississippi, Missouri, ext. SE New Mexico, E North Carolina, W Ohio, Oklahoma, South Carolina, W Tennessee, Texas, ext. SE Virginia, ext. S Wisconsin) and NE Mexico (N Coahuila, NE Durango, Nuevo León, ext. NW Tamaulipas, cen. Zacatecas), NSL–2040 m.
Fossil records: Upper Miocene/lower Pliocene (Hemphillian) of USA (Indiana), middle Pliocene (Blancan III) of USA (Texas), lower Pleistocene (Irvingtonian I) of USA (Florida), lower/middle Pleistocene (Irvingtonian I) of USA (Texas), and upper Pleistocene (Rancholabrean I, II) of USA (Florida, Indiana, Texas, Virginia).
Sources: Conant, 1949, 1951, 1969, Webb, 1970, Minton, 1972, Martof et al., 1980, T.R. Johnson, 1987, McCranie, 1990, Meade, 1991, Palmer & Braswell, 1995, Degenhardt et al., 1996, Harding, 1997, Tennant, 1997, Werler & Dixon, 2000, Christiansen & Leclere, 2002 and Makowsky et al., 2010.
Remarks: Original description based on Catesby (1843: 46, pl. 46). Holotype presumed lost *fide* Conant (1949: 10). Holotype of *Tropidonotus roulei* examined and found to be *Nerodia erythogaster fide* P. David (pers. comm.).

4. *Nerodia fasciata* (Linnaeus, 1766). Syst. Nat., ed. 12, 1: 378. (*Coluber fasciatus*)

Synonyms: *Coluber vampum* Lacépède, 1789 (*nomen rejiciendum*), *Coluber porcatus* Latreille, 1801b, *Natrix fasciata pictiventris* Cope, 1895b, *Natrix fasciata confluens* Blanchard, 1923a, and *Natrix sipedon engelsi* T. Barbour, 1943.

Type: Holotype, not designated (A. Garden, 1760–1761), location unknown.

Type locality: "Carolina" [USA]. Restricted to vicinity of Charleston, South Carolina, USA *fide* Klauber (1948: 6).

Distribution: Southeastern USA (S Alabama, Arkansas, Florida, S Georgia, ext. S Illinois, ext. W Kentucky, Louisiana, W and S Mississippi, ext. SE Missouri, E North Carolina, Shackleford Banks, SE Oklahoma, E South Carolina, ext. W Tennessee, E Texas), NSL–200 m.

Fossil records: Lower Pleistocene (Irvingtonian I) of USA (Florida), and upper Pleistocene (Rancholabrean II) of USA (Florida, Georgia).

Sources: Conant, 1963a, Schwaner & Mount, 1976, Blaney & Blaney, 1979b, Myer, 1988, R. Lawson et al., 1991, Palmer & Braswell, 1995, Tennant, 1997, Werler & Dixon, 2000, D. Schmidt, 2004b and Mebert, 2008.

5. *Nerodia floridana* (Goff, 1936). Occ. Pap. Mus. Zool., Univ. Michigan (327): 1–3. (*Natrix cyclopion floridana*)

Type: Holotype, UMMZ 77229, a 1054 mm female (E.R. Allen, March 1935).

Type locality: "near Leesburg, Florida, at the Alsa–brook Prairie between Lake Griffin and Lake Yale" [USA].

Distribution: Southeastern USA (ext. S Alabama, Florida, S Georgia, S South Carolina), NSL–100 m.

Fossil records: Lower Pleistocene (Irvingtonian I) of USA (Florida) and upper Pleistocene (Rancholabrean II) of USA (Florida).

Sources: Betz, 1963, Sanderson, 1983, 1993, Tennant, 1997 and J.S. Thompson & Crother, 1998.

6. *Nerodia harteri* (Trapido, 1941). Amer. Midl. Nat. 25(3): 673–675, fig. 1. (*Natrix harteri*)

Type: Holotype, USNM 110927 (formerly HT 1302), a 694 mm male (P. Harter, early April 1940).

Type locality: "rocky stretches along the Brazos River north of Palo Pinto, Palo Pinto County, Texas, 830 to 840 feet in altitude."

Distribution: South-central USA (Brazos River of cen. Texas), 250–550 m.

Sources: Mecham, 1983, F. Rose & Selcer, 1989, N.J. Scott et al., 1989, Densmore et al., 1992 and Werler & Dixon, 2000.

7. †*Nerodia hibbardi* (Holman, 1968). Copeia 1968(1): 153–154, fig. 1. (†*Natrix hibbardi*)

Type: Holotype, UMMP V53444, one left mandible with coalesced articular, surangular and prearticular from a 900 mm specimen.

Type locality: "Hagerman Quadrangle, NE 1/4, Sec. 17, T 7 S, R 13 E, outcrop on the Snake River from the town of Hagerman downstream to Homedale, Twin Falls County, Idaho; Glenns Ferry Formation, upper Pliocene."

Distribution: Middle-upper Pliocene (Blancan: 1.9–3.6 mya) of USA (Idaho and Texas).

Sources: K.L. Rogers, 1976 and Holman, 1979a, 2000.

8. †*Nerodia hillmani* (R.L. Wilson, 1968). Contr. Mus. Paleont., Univ. Michigan 2(2): 100–102, figs. 10a–c. (†*Natrix hillmani*)

Type: Holotype, UMMP V55701, one trunk vertebra (UMMP field crews, 1941–1966).

Type locality: "WaKeeney local fauna site, University of Michigan locality UM-K6-59, northeast of Ogallala, on the Lowell Hillman Ranch, 2350-2550 feet south and 75 feet east of the NW corner Sec. 22, R22W, T11S, Trego County, Kansas. Elevation is 2555 feet; Ogallala Formation, Lower Pliocene."

Distribution: Upper Miocene (Clarendonian: 10.3–13.6 mya) of USA (Kansas, Nebraska).

Sources: Rage, 1984b, Holman, 2000 and Parmley & Hunter, 2010.

9. *Nerodia paucimaculata* (Tinkle & Conant, 1961). SW Nat. 6(1): 34–36, figs. 1–2. (*Natrix harteri paucimaculata*)

Type: Holotype, AMNH 85542, a 652 mm male (D.W. Tinkle & D. L. McGregor, 8 Oct. 1960).

Type locality: "The Colorado River south of the city limits of Robert Lee, Coke County, Texas" [USA].

Distribution: South-central USA (Colorado and Concho Rivers of cen. Texas), 300–350 m.

Sources: Mecham, 1983, Rose & Selcer, 1989, N.J. Scott et al., 1989 and Densmore et al., 1992

10. *Nerodia rhombifer* (Hallowell, 1852a). Proc. Acad. Nat. Sci. Philadelphia 6(5): 177. (*Tropidonotus rhombifer*)

Synonyms: *Nerodia holbrookii* Baird & Girard, 1853, *Tropidonotus pogonias* A.-M.-C. Duméril, Bibron & Duméril, 1854a, *Natrix rhombifera blanchardi* Clay, 1938a, and *Natrix rhombifera werleri* Conant, 1953.

Type: Lectotype, ANSP 5047, a 734–813 mm male (S.W. Woodhouse, 1849–1850), designated by Conant (1969: 55).

Type locality: "Arkansas River between Keystone and Tulsa, Tulsa Co., Oklahoma" [USA] via lectotype selection.

Distribution: Central and SE USA (cen. Alabama, Arkansas, ext. W Florida, W Illinois, ext. SW Indiana, ext. SE Iowa, SE and S Kansas, ext. W Kentucky, Louisiana, Mississippi, Missouri, Oklahoma, W Tennessee, Texas) and E Mexico (ext. SW Campeche, N Chiapas, Coahuila, N Hidalgo, Nuevo León, Puebla, E San Luis Potosí, Tabasco, Tamaulipas, Veracruz), NSL–2235 m.

Fossil records: Upper Pliocene (Blancan V) of USA (Nebraska).

Sources: Conant, 1969, McAllister, 1985, Villa et al., 1988, Lee, 1996, Werler & Dixon, 2000, Pérez-Higareda et al., 2002, Brandley et al., 2010 and Meza-Parral et al., 2010.

Remarks: Type locality restriction of Fort Smith, Arkansas, USA *fide* K.P. Schmidt (1953a: 161) in error. A genetic difference exists between populations in the USA east and west of the Mississippi River *fide* Brandley et al. (2010: 558).

11. *Nerodia sipedon* (Linnaeus, 1758). Syst. Nat., ed. 10, 1: 219. (*Coluber sipedon*)

Synonyms: ? *Coluber ichthyoplaga* Rafinesque-Schmaltz, 1818a (*nomen dubium*), *Coluber fulviventer* Rafinesque-Schmaltz, 1820, *Coluber atrifuscus* Harlan, 1827 *(nomen praeoccupatum), Coluber caudaschistosus* Harlan, 1827 *(nomen incorrigendum), Coluber poecilogaster* Wied-Neuwied, 1839 *in* 1838–1841, *Coluber sipeolon* – Holbrook, 1840 *(nomen incorrectum), Tropidonotus niger* Holbrook, 1842b, *Nerodia gassizii* Baird & Girard, 1853, *Tropidonotus obliquus* Hallowell, 1856c, *Natrix fasciata pleuralis* Cope, 1892a, *Natrix sipedon insularum* Conant & Clay, 1937, and *Natrix sipedon williamengelsi* Conant & Lazell, 1973.

Type: Holotype, not designated (P. Kalm, Sept. 1748–Feb. 1751), location unknown.

Type locality: "America septentrionali" [= North America]. Restricted to vicinity of New York City, New York, USA *fide* K.P. Schmidt (1953a: 161).

Distribution: Extreme SE Canada (ext. SE Ontario, ext. S Quebec, Pelée Is.) and cen. and E USA (Alabama, Arkansas, E Colorado, Connecticut, Delaware, W Florida, N Georgia, Illinois, Indiana, Iowa, Kansas, Kentucky, E Louisiana, S Maine, Maryland, Massachusetts, Michigan, SE Minnesota, Mississippi, Missouri, Nebraska, S New Hampshire, New Jersey, New York, North Carolina, Outer Banks; Ohio, Put-in-Bay Arch.; E Oklahoma, Pennsylvania, Rhode Island, W South Carolina, ext. SE South Dakota, Tennessee, ext. NE Texas, Vermont, Virginia, West Virginia, Wisconsin), NSL–1675 m.

Fossil records: Middle Pliocene (Blancan II, III) of USA (Kansas, Nebraska), upper Pliocene (Blancan V) of USA (Kansas), lower Pleistocene (Irvingtonian I) of USA (Pennsylvania, South Dakota), lower/middle Pleistocene (Irvingtonian I) of USA (Kansas), middle Pleistocene (Irvingtonian II) of USA (Nebraska), middle/upper Pleistocene (Irvingtonian II) of USA (West Virginia), and upper Pleistocene (Rancholabrean I, II) of USA (Alabama, Georgia, Kansas, Missouri, Ohio, Pennsylvania, Tennessee, Virginia, West Virginia).

Sources: Conant, 1963a, Conant & Clay, 1963, Conant & Lazell, 1973, Schwaner & Mount, 1976, Blaney & Blaney, 1979b, Frost & Collins, 1989, Palmer & Braswell, 1995, Hardy, 1997, Tennant, 1997, Weatherford et al., 1998, Fitch, 1999, Hammerson, 1999, Werler & Dixon, 2000, Mebert, 2008 and Walley et al., 2012.

Remarks: Probably occurs in S New Brunswick, Canada *fide* Gorham (1970b: 16).

12. *Nerodia taxispilota* (Holbrook, 1838). No. Amer. Herp. 2: 113–114, pl. 25. (*Coluber taxispilotus*)

Synonym: *Tropidonotus taxispilotus brocki* Heilprin, 1887.

Types: Syntypes (2), MCZ 16, a 50 mm head, and MCZ 21, a 635 mm specimen (J.E. Holbrook, 1822–1836).

Type locality: "sea-board of South Carolina" and "neighbourhood of the Altamaha river in Georgia, United States."

Distribution: Southeastern USA (E Alabama, Florida, Georgia, E North Carolina, South Carolina, SE Virginia), NSL–300 m.

Fossil records: Upper Pleistocene (Rancholabrean II) of USA (Florida).

Sources: Mount & Schwaner, 1970, McCranie, 1983b, Palmer & Braswell, 1995 and Tennant, 1997.

Remarks: Original description reprinted in Holbrook (1842b: 35–36, pl. 8). T. Barbour & Loveridge (1929a: 353) suggested MCZ 16 and/or MCZ 21 were the types, given to L. Agassiz by Holbrook and catalogued from Charleston, SC. Longest syntype 889 mm *fide* Holbrook (1838: 114).

†*NIGEROPHIS* Rage, 1975b (†Nigerophiidae)

Synonyms: *Nigerophoides* – Barrie, 1990 *(nomen incorrectum)* and *Nigrophis* – Nietzke & Kornacker, 2002 *(nomen incorrectum).*

Type species: †*Nigerophis mirus* Rage, 1975b.

Distribution: Lower Paleocene of Mali or Niger and upper Paleocene of Kazakhstan.

Sources: Rage, 1975b, 1983b, 1984b, 1987, Kordikova, 1998 and LaDuke et al., 2010.

1. †*Nigerophis mirus* Rage, 1975b. C.R. Acad. Sci. Paris 281D(9): 515–516, figs. 1a–e.

Type: Holotype, USTL SES 100, one trunk vertebra.

Type locality: "Krebb de Sessao, 'Terme V'; République du Niger. Paléocène" [= Kahal Séssao, ext. E Gao

Region, E Mali, 16°55'N, 4°12'E, or ext. W Tahoua Dept., W Niger].

Distribution: Lower Paleocene (55.8–65.5 mya) of Mali or Niger. Known only from type locality.

Remarks: The Sessao Mesa is just west of the Mali-Niger border.

NINIA Baird & Girard, 1853 (Dipsadidae)

Synonyms: *Streptophorus* A.-M.-C. Duméril, Bibron & Duméril, 1854a, *Sterptophorus* – Chenu & Desmarest, 1857 *(nomen incorrectum),* and *Streptophurus* – Briceño-Rossi, 1934 *(nomen incorrectum).*

Type species: *Ninia diademata* Baird & Girard, 1853.

Distribution: Latin America.

Sources: Dunn, 1935, E.H. Taylor, 1951, Burger & Werler, 1954, J.A. Peters & Orejas-Miranda, 1970, Cadle, 1984b, Villa et al.,1988, Savage & Lahanas, 1991, G. Köhler et al., 2005 and Zaher et al., 2009.

1. *Ninia atrata* (Hallowell, 1845a). Proc. Acad. Nat. Sci. Philadelphia (1844–1845) 2(9): 245–246. (*Coluber atratus*)

Synonyms: *Coluber atratus* Gmelin, 1789 *(nomen suppressum),* ? *Coluber subatrata* Daudin, 1803d *(nomen oblitum), Streptophorus drozii* A.-M.-C. Duméril, Bibron & Duméril, 1854a, *Streptophorus lansbergi* A.-M.-C. Duméril, Bibron & Duméril, 1854a, *Streptophorus langsbergii* A.-M.-C. Duméril, Bibron & Duméril, 1854c *(nomen emendatum), Streptophorus sebae schmidti* Jan, 1862b, *Ninia spilogaster* W.C.H. Peters, 1881a, and *Streptophurus atractus* – Briceño-Rossi, 1934 *(nomen incorrectum).*

Types: Syntypes (3), ANSP 3410–12, 318–356 mm specimens (S. Ashmead, 1835–1845).

Type locality: "Republic of Columbia, within two hundred miles of Caraccas" (in error). Corrected to Venezuela *fide* Roze (1958d: 2) [= within 320 km of Caracas, Venezuela].

Distribution: Panama and N South America. Panama (Canal Zone: Barro Colorado Is.; Darién, Panamá), W Colombia (Amazonas, Antioquia, Boyacá, Caldas, Cauca, Chocó, Cundinamarca, Guaviare, Magdalena, Meta, Norte de Santander, Santander, Tolima, Valle, Vaupés), Venezuela (Aragua, Barinas, Bolívar, Carabobo, Distrito Federal, Falcón, Guárico, Lara, Mérida, Miranda, Monagas, Portuguesa, Sucre, Táchira, Trujillo, Yaracuy), Trinidad & Tobago (Tobago, Trinidad) and N Ecuador (Carchi, Esmeraldas, Pastaza, Pichincha), NSL–1900 m.

Sources: E.H. Taylor, 1951, 1954, Roze, 1966a, J.R. Meyer & Wilson, 1971, Emsley, 1977, Miyata, 1980, Hidalgo, 1981b, L.D. Wilson & Meyer, 1985, Lancini, 1986, Pérez-Santos, 1986, Savage & Villa, 1986, Pérez-Santos & Moreno, 1988, 1991, McCranie & Wilson, 1995, J.C. Murphy, 1997, Markezich, 2002, La Marca et al., 2004, Angarita-Sierra, 2009 and Navarrete et al., 2009.

Remarks: Official Specific Name no. 1859 *fide* Opinion 644 (ICZN, 1963). Hallowell mentioned four syntypes but only three are extant (Burger & Werler, 1954: 649, Roze, 1958d: 3 and Malnate, 1971: 363). The type locality is essentially Venezuela as all states except Táchira are within 320 air-miles of Caracas. In accordance with Art. 23.9.2 of the Code (ICZN, 1999), *Coluber atratus* Hallowell is designated a *nomen protectum* and *Coluber subatrata* Daudin a *nomen oblitum.*

2. *Ninia celata* McCranie & Wilson, 1995. J. Herp. 29(2): 225–228, fig. 1a (paratype).

Type: Holotype, KU 31971, 241 mm female (E.H. Taylor, 19 July 1952).

Type locality: "Cinchona (Isla Bonita), Atlantic side of Volcán Poás, 1200 m elevation (see Savage, 1974: 79), Provincia de Alajuela, Costa Rica."

Distribution: Central Costa Rica (Alajuela, W Cartago, S Heredia, Limón, N San José) and W Panama (Boca del Toro), 650–1360 m.

Sources: Savage, 2002 and Solórzano, 2004.

3. *Ninia diademata* Baird & Girard, 1853. Cat. No. Amer. Rept., Serp. 1: 49–50.

Synonyms: *Streptophorus bifasciatus* A.-M.-C. Duméril, Bibron & Duméril, 1854a, *Streptophorus labiosus* Bocourt, 1883 *in* A.H.A. Duméril, Bocourt & Mocquard, 1870–1909, *Ninia diademata plorator* H.M. Smith, 1942f, and *Ninia diademata nietoi* Burger & Werler, 1954.

Type: Holotype, USNM 12122, a 219 mm specimen (J. Fairie).

Type locality: "Orizaba, Mex." [= Orizaba, Veracruz State, Mexico; 18°51'N, 97°06'E, elevation 1240 m].

Distribution: Southeastern Mexico and upper Central America. Southeastern Mexico (Chiapas, Hidalgo, Oaxaca, Puebla, Querétaro, E San Luis Potosí, Tabasco, Veracruz), S Belize (Toledo), Guatemala (Alta Verapaz, Chimatenango, Petén, San Marcos, Sololá, Suchitepequez) and NW Honduras (Atlántida, Comayagua, Copán, Cortés, Ocotepeque, Santa Bárbara, Yoro), NSL–1830 m.

Sources: L.C. Stuart, 1948, R.W. Henderson & Hoevers, 1975, J.C. Lee, 1980, 1996, L.D. Wilson & Meyer, 1985, Pérez-Higareda & Smith, 1991, Benitez-Gálvez, 1997, J.A. Campbell, 1998b and McCranie, 2011a.

Remarks: F. Sumichrast listed collector as *fide* Cochran (1961: 201).

4. *Ninia espinali* McCranie & Wilson, 1995. J. Herp. 29(2): 228–230, fig. 1b (paratype).

Type: Holotype, USNM 333030, a 377 mm male (J.R. McCranie, K.L. Williams & L.D. Wilson, 9 Aug. 1987).

Type locality: "El Portillo de Ocotepeque (14°28'N, 89°04'W), 1910 m elevation, Departamento de Ocotepeque, Honduras."

Distribution: Northern El Salvador (Chalatenango, Comayagua, Copán, Cortés, La Paz, Lempira, Ocotepeque, Santa Bárbara) and W Honduras (Comayagua, Cortés, Ocotepeque), 1040–2240 m.

Sources: Greenbaum et al., 2002, Luque-Montes & Townsend, 2009 and McCranie, 2011a.

5. *Ninia hudsoni* H.W. Parker, 1940. Ann. Mag. Nat. Hist. (11) 5(27): 270–271.

Type: Holotype, BMNH 1946.1.15.38 (formerly BMNH 1939.1.1.9), a 436 mm male (C.A. Hudson, 1938–1939).

Type locality: "New River, British Guiana" [= New River, S East Berbice-Corentyne District, SE Guyana, elevation 230 m]. Emended to S British Guiana, altitude of 750 feet *fide* Burger & Werler (1954: 649).

Distribution: Northern and NW South America. Southern Guyana (S East Berbice-Corentyne), W Brazil (Acre, Rondônia), E Ecuador (Napo, Pastaza) and Peru (Madre de Dios, Pasco), 230–1500 m.

Sources: Pérez-Santos & Moreno, 1991, N.J. Silva, 1993, Lehr & Fernandez, 2000a and C.J. Cole et al., 2013.

Remarks: Probably occurs in S Colombia (Putamayo) *fide* Pérez-Santos (1986: 564).

N

6. *Ninia maculata* (W.C.H. Peters, 1861e). Mber. Königl. Akad. Wiss. Berlin 1861(10): 924–925. (*Streptophorus maculatus*)

Synonyms: *Ninia sebae tessellatus* Cope, 1875a and *Streptophorus subtessellatus* F. Werner, 1909b.

Types: Syntypes (5), ZMB 1872, ZMB 1873a–b, and ZMB 1874a–b, longest syntype 335 mm female (K. Hoffmann & von Frantzius, 1853–1859).

Type locality: "Costa Rica."

Distribution: Central America. Honduras (Colón, Olancho), E Nicaragua (Atlántico Norte, Atlántico Sur, Chontales, Jinotega, Zelaya), Costa Rica (Alajuela, Cartago, Guanacaste, Heredia, Limón, Puntarenas, San José) and Panama (Bocas del Toro, Canal Zone, Chiriquí, Darién, Barro Colorado Is.), NSL–1800 m.

Sources: L.C. Stuart, 1948, E.H. Taylor, 1954, Savage & Villa, 1986, B.E. Smith & Campbell, 1996, G. Köhler, 1999b, McCranie et al., 2001, 2006, Savage, 2002 and McCranie, 2011a.

Remarks: Possibly occurs in W Colombia (Chocó) *fide* Pérez-Santos (1986: 564).

7. *Ninia pavimentata* (Bocourt, 1883 *in* A.H.A. Duméril, Bocourt & Mocquard, 1870–1909). Miss. Sci. Mex. Amér. Cen., Rept. 3(9): 549–550, pl. 32, figs. 8, 8a–d, pl. 33, fig. 2. (*Streptophorus maculatus pavimentatus*)

Types: Syntypes (4), MNHN 1152a–d, one male and 3 females, longest syntype 179 mm (Comm. Sci. Mexico Cen. Amer., 1865–1867).

Type locality: "Haute Vera Paz (République du Guatemala)" [= Alta Verapaz Dept., N Guatemala].

Distribution: Guatemala (Alta Verapaz, Baja Verapaz, Huehuetenango, Izabal, San Marcos) and Honduras (Cortés, Yoro), 1120–1825 m.

Sources: B.E. Smith & Campbell, 1996, Townsend et al., 2005a, 2008 and McCranie, 2011a.

8. *Ninia psephota* (Cope, 1875a). J. Acad. Nat. Sci. Philadelphia (1876) (2) 8(2): 146. (*Catostoma psephotum*)

Synonyms: *Streptophorus oxynotus* F. Werner, 1909b and *Ninia cerroensis* E.H. Taylor, 1954.

Type: Holotype, USNM 61971, a 480 mm specimen (W.M. Gabb, 1873–1875).

Type locality: "higher points on the Pico Blanco, chiefly in the rainy zone, at from 5000 to 7000 feet, southern Costa Rica" [= Cerro Pico Blanco, Limón Prov., E Costa Rica, 9°17'N, 83°02'W, elevation 1500–2100 m].

Distribution: Costa Rica (Alajuela, Cartago, Heredia, Limón, Puntarenas, San José) and W Panama (Chiriquí), 1090–2285 m.

Source: Savage, 2002.

Remarks: Probably occurs in Guanacaste, Costa Rica *fide* Savage & Lahanas (1991: 53).

9. *Ninia sebae* (A.-M.-C. Duméril, Bibron & Duméril, 1854a). Erpét. Gén. 7(1): 515–517. (*Streptophorus sebae*)

Synonyms: *Elapoidis fasciatus* Hallowell, 1855c, *Streptophorus sebae collaris* Jan, 1862b, *Streptophorus sebae dorsalis* Bocourt, 1883 *in* A.H.A. Duméril, Bocourt & Mocquard, 1870–1909, *Streptophorus sebae punctulatus* Bocourt, 1883 *in* A.H.A. Duméril, Bocourt & Mocquard, 1870–1909, *Ninia sebae morleyi* K.P. Schmidt & Andrews, 1936, and *Ninia sebae immaculata* K.P. Schmidt & Rand, 1957.

Type: Lectotype, MNHN 3443, a 309 mm specimen, designated herein.

Type locality: "Mexico" via lectotype selection. Restricted to Veracruz State, Mexico *fide* K.P. Schmidt & Andrews (1936: 171). Further restricted to Veracruz, Veracruz State, Mexico *fide* H.M. Smith & Taylor (1950a: 351).

Distribution: Mesoamerica. Southern Mexico (Campeche, Chiapas, Oaxaca, Puebla, Quintana Roo, Veracruz, Yucatán), Belize (Belize, Cayo, Corozal, Orange Walk, Stann Creek, Toledo), Guatemala (Chimaltenango, Petén, Quezaltenango, San Marcos, Sololá,

Suchitepéquez), El Salvador (Ahuachapán, Cabañas, Chalatenango, Cuscatlán, La Libertad, Morazán, San Salvador, San Vicente, Usulután), Honduras (Atlántida, Choluteca, Colón, Comayagua, Copán, Cortés, El Paraíso, Francisco Morazán, Gracias a Dios, Intibucá, Ocotepeque, Olancho, Santa Bárbara), Nicaragua (Atlántico Norte, Atlántico Sur, Chontales, Granada, Jinotega, Managua, Matagalpa, Río San Juan), Costa Rica (Alajuela, Cartago, Guanacaste, Heredia, Limón, Puntarenas, San José) and Panama (Bocas del Toro, ? Chiriquí), NSL–2000 m.

Sources: K.P. Schmidt, 1941, E.H. Taylor, 1951, 1954, Burger & Werler, 1954, Mertens, 1952a, K.P. Schmidt & Rand, 1957, Duellman, 1963, R.W. Henderson & Hoevers, 1975, J.C. Lee, 1980, 1996, 2000, L.D. Wilson & Meyer, 1985, Savage & Villa, 1986, Pérez-Higareda & Smith, 1991, Benitez-Gálvez, 1997, J.A. Campbell, 1998b, G. Köhler, 1999b, Stafford & Meyer, 2000, Savage, 2002, Solórzano, 2004, Kohler et al., 2005, N. Herrera et al., 2007, Ponce A. et al., 2008 and McCranie, 2011a.

Remarks: Lectotype designation of MNHN 3778 *fide* Burger & Werler (1954: 655) invalid (MNHN 3778 = *Pseudoboa neuwiedi fide* P. David, *in litt.)*. Lectotype designation of MNHN 7303 *fide* L.D. Wilson & Meyer (1982: 81 & 1985: 73) also invalid as the specimen is lost (seen by A. Hoge in 1978 so possibly lost in Butantan fire of 2010) *fide* P. David *(in litt.)*. MNHN 3443 represents only extant syntype.

NOTECHIS Boulenger, 1896a (Elapidae)

Synonym: *Oplocephalus* Gray *in* Griffith & Pidgeon, 1831 *(nomen oblitum), Ophicephalus* – Swainson, 1839 *(nomen incorrectum), Hoplocephalus* Agassiz, 1844 *(nomen emendatum),* and *Nothechis* – Briceño-Rossi, 1934 *(nomen incorrectum).*

Type species: *Naja scutata* W.C.H. Peters, 1861c.

Distribution: Southern Australia.

Fossil record: Pleistocene of Australia.

Sources: F. McCoy, 1878c, F. Werner, 1923b, Kinghorn, 1956, Worrell, 1963, Kinghorn, 1964, Rawlinson, 1974, 1991, Shine, 1977a–b, 1987c, Storr, 1982, 1985, Schwaner et al., 1985, Longmore, 1986, Storr et al., 1986, S.K. Wilson & Knowles, 1988, Hoser, 1989, 2012e, Hutchinson, 1990, Schwaner & Sarre, 1990, Ehmann, 1992, Golay et al., 1993, Greer, 1997, Keogh et al., 1998, 2005, Cogger, 2000, Scanlon & Lee, 2004 and Sanders et al., 2008.

Remarks: In accordance with Art. 23.9.2 of the Code (ICZN, 1999), *Notechis* Boulenger is designated a *nomen protectum* and *Oplocephalus* Gray a *nomen oblitum.*

1. *Notechis ater* (Krefft, 1866). Proc. Zool. Soc. London 34(1): 370. (*Hoplocephalus ater*)

Synonyms: *Notechis scutatus niger* Kinghorn, 1921, *Notechis scutatus occidentalis* Glauert, 1948, *Notechis ater humphreysi* Worrell, 1963b, *Notechis ater serventyi* Worrell, 1963b, *Notechis nigra* – Trethewie, 1971, *Notechis ater humphrysi* – Mengden, 1985a *(nomen incorrectum), Notechis atra niger* – DuBourdieu, Kawaguchi & Shier, 1987 *(nomen incorrectum),* and *Notechis ater kangaroo* – Mollier, Chwetzoff, Frachon & Ménez, 1989 *(lapsus calami).*

Type: Holotype, AMS 131696 (formerly AMS 6577) (G. Masters, 1860–1866).

Type locality: "Flinder's Distribution, South Australia," (in error). Corrected to Pt. Lincoln, [= Port Lincoln, S South Australia, S Australia, 34°40'S, 135°53'E, elevation NSL] *fide* AMS catalogue, Cogger (1979: 205) and Shea & Sadlier (1999: 50).

Distribution: Southern Australia (S South Australia, Tasmania, SW Western Australia, Babel, Cat, Chappell, Christmas, East Franklin, Flinders, Forsyth, Goat, Grindal, Hopkins, Kangaroo, King, Louth, Lusby, Marum, New Year, Partney, Reevesby, Roxby, Spilsby, St. Francis, St. Peter, West Franklin and Williams Is.), NSL–1195 m.

Sources: Rawlinson, 1967, 1991, Fearn, 1993, Shea & Sadlier, 1999 and Means, 2004b.

2. *Notechis scutatus* (W.C.H. Peters, 1861c). Mber. Königl. Akad. Wiss. Berlin 1861(4): 690. (*Naja scutata*)

N

Synonyms: ? *Boa laevis* Lacépède, 1804 *(nomen oblitum), Oplocephalus cuvieri* Gray *in* Griffith & Pidgeon, 1831 *(nomen oblitum), Alecto fasciolata* Jan, 1863b *(nomen nudum), Hoplocephalus fuscus* Steindachner, 1867a, *Alecto fasciolata* Jan, 1873 *in* Jan & Sordelli, 1870–1881, *Notechis scutatis* – Trethewie & Khaled, 1973 *(nomen incorrectum), Notechis edwardsi* Wells & Wellington, 1985, *Notechis longmorei* Wells & Wellington, 1985, and *Notechis schwaneri* Wells & Wellington, 1985.

Type: Neotype, NMV D47618, a 1018 mm male (P.A. Rawlinson, A.J. Coventry & P.B. Mather, 27 Jan. 1976), designated by Rawlinson (1991: 130).

Type locality: "'The Brothers', 10 km NE of Benambra, Victoria (36° 56' S, 147° 45' E), Australia" via neotype selection.

Distribution: Southeastern Australia (Australian Capital Territory, New South Wales, ext. SE Queensland, ext. SE South Australia, Victoria, Badger, Blyth, Carnac, East Franklin, Flinders, Garden, Goat, Gundal, Hareby, Hopkins, King, Lusby, Reevesby, Roxby, St. Francis, St. Peter, West Franklin, Williams and Winceby Is.), NSL–1470 m.

Sources: Kellaway & Thomson, 1932, Worrell, 1963b, Rawlinson, 1967, M.J. Smith, 1975, Jenkins & Bartell, 1980, Gow, 1983, Schwaner, 1985b, 1989, G.

Swan, 1990, Covacevich & Couper, 1991, Mirtschin & Davis, 1992a–b, Bush et al., 1995, & Shea & Sadlier, 1999, Cogger, 2000, Storr et al., 2002, J.C. Murphy & Schlager, 2003, Swan et al., 2004 and Swanson, 2008.

Remarks: Holotype (ZMB 2815) lost *fide* Cogger et al. (1983: 229) and Rawlinson (1991: 129). In accordance with Art. 23.9.2 of the Code (ICZN, 1999), *Naja scutata* W.C.H. Peters is designated a *nomen protectum* and *Boa laevis* Lacépède and *Oplocephalus cuvieri* Gray a *nomina oblita*.

NOTHOPSIS Cope, 1871b (Dipsadidae)

Synonyms: *Notopsis* – Hoffmann, 1890 *(nomen incorrectum),* and *Nothopis* – Briceño-Rossi, 1934 *(nomen incorrectum).*

Type species: *Nothopsis rugosus* Cope, 1871b.

Distribution: Central America and NW South America.

Sources: Dunn & Dowling, 1957, J.A. Peters & Orejas-Miranda, 1970, Miyata, 1980, Pérez-Santos, 1986, Pérez-Santos & Moreno, 1988, 1991, Villa et al., 1988, G. Köhler, 1999b, Savage, 2002, Solórzano, 2004, McCranie et al., 2006, Zaher et al., 2009 and McCranie, 2011a.

N

1. *Nothopsis rugosus* Cope, 1871b. Proc. Acad. Nat. Sci. Philadelphia 23(2): 201–203, pl. 17, figs. 1–7.

Synonyms: *Nothopsis affinis* Boulenger, 1905b and *Nothopsis torresi* E.H. Taylor, 1951.

Type: Holotype, USNM 12427, a 381 mm female (T.O. Selfridge, 1869–1871).

Type locality: "Isthmus of Darien" [Panama]. Emended to Caribbean Coast of the Isthmus of Darién, Panama *fide* Dunn & Dowling (1957: 255).

Distribution: Central America and NW South America. Honduras (Colòn, Gracias a Dios), Ext. S Nicaragua (Río San Juan), Costa Rica (Alajuela, Heredia, Limón, Puntarenas, San José), Panama (Bocas del Toro, Chiriquí, Darién, Colón, Panamá, Veraguas), W Colombia (Chocó, Valle) and NW Ecuador (Esmeraldas, Pichincha), NSL–1000 m.

†*NUBIANOPHIS* Rage & Werner, 1999 (†Nigerophiidae)

Type species: †*Nubianophis afaahus* Rage & Werner, 1999.

Distribution: Middle Cretaceous of Sudan.

Sources: Rage & Werner, 1999 and LaDuke et al., 2010.

1. †*Nubianophis afaahus* Rage & Werner, 1999. Palaeont. Afr. 35: 97–100, figs. 13–14.

Type: Holotype, TUB-SFB69 Vb-1041, one middle trunk vertebra.

Type locality: "Wadi Abu Hashim, Sudan. Wadi Abu Hashim Member of Wadi Milk Formation; Cenomanian."

Distribution: Middle Cretaceous (Cenomanian: 93.9–100.5 mya) of Sudan. Known only from type locality.

O

OCYOPHIS Cope, 1886c (Xenodontidae)

Synonym: *Plesius* Jan, 1857 *(nomen nudum).*
Type species: *Natrix atra* Gosse, 1851.
Distribution: Jamaica.
Sources: Hedges et al., 2009, R.W. Henderson & Powell, 2009 and Zaher et al., 2009.
Remarks: A valid genus *fide* Grazziotin et al. (2012: 22).

1. *Ocyophis ater* (Gosse, 1851). Nat. Jamaica: 228–232. (*Natrix atra*)

Synonyms: *Natrix capistrata* Gosse, 1851, *Plesius inconstans* Jan, 1857 (*nomen nudum*), *Dromicus ater inconstans* Jan, 1863b (*nomen nudum*), *Dromicus ater transversus* Jan, 1863b (*nomen nudum*), and *Dromicus ater transverses* Jan, 1867 *in* Jan & Sordelli, 1866–1870.
Types: Syntypes (5), BMNH 1946.1.4.65 (formerly BMNH 1845.8.17.8) and BMNH 1946.1.5.6 (formerly BMNH 1845.8.17.45), BMNH 1946.1.4.94 and BMNH 1946.1.23.81–82, two females (P.H. Gosse, 6 Dec. 1844–11 July 1846).
Type locality: "Jamaica." Emended to Bluefields, Jamaica [= Westmoreland Parish, W Jamaica, 18°10'N, 78°02'W, elevation 35 m] *fide* A.C.L.G. Günther (1858: 127).
Distribution: Jamaica (Kingston & St. Andrews, St. Ann, Trelawny, Westmoreland), very rare *fide* Grant (1940c: 123) or extinct *fide* A. Schwartz & Thomas (1975: 170) and R.W. Henderson & Powell (2009: 365).
Sources: Grant, 1940b and R.W. Henderson & Powell, 1996a.
Remarks: Gosse (1851: 228) listed a single specimen with 180 V + 137 SC. A.C.L.G. Günther (1858: 127) stated that "Var. A. no. e" from Bluefields is the holotype (half-grown specimen, Gosse coll.). Boulenger (1894a: 122) listed two female syntypes from Jamaica with Var. A. no. a–b (175 V + 160 SC and 181 V + ? SC). Boulenger's only Bluefields specimen is a male "Var. A. no. d" (178 V +? SC), without collector listed. Günther's designation of a type may have precedence over Boulenger's.

2. *Ocyophis melanichnus* (Cope, 1862b). Proc. Acad. Nat. Sci. Philadelphia 14(1): 76–77. (*Alsophis melanichnus*)

Type: Holotype, MCZ 1522, a 914 mm specimen (D.F. Weinland, 1859), lost *fide* A. Schwartz & Henderson (1988: 205).
Type locality: "Hayti, near Jeremie" [= vicinity of Jérémie (18°39'N, 74°07'W, elevation 15 m), Grand'Anse Dept., SW Haiti].
Distribution: Hispaniola. Haiti (Grand'Anse) and Dominican Republic (La Vega), 15–100 m.
Source: Powell & R.W. Henderson, 1998b.
Remarks: Very rare *fide* A. Schwartz & Thomas (1975: 172) and A. Schwartz (1980; 97) or extinct *fide* A. Schwartz & Henderson (1991: 572).

OGMODON W.C.H. Peters, 1864a (Elapidae)

Synonyms: *Labionaris* Brocchi, 1876 and *Agmodon* – Boquet, 1948 *(nomen incorrectum).*
Type species: *Ogmodon vitianus* W.C.H. Peters, 1864a.
Distribution: Fiji Islands.
Sources: McDowell, 1969a, 1970, Gorham, 1970a, Golay et al., 1993, Zug & Ineich, 1993, Keogh, 1998, Scanlon & Lee, 2004, Zaher et al., 2009 and Hoser, 2012e.

1. *Ogmodon vitianus* W.C.H. Peters, 1864a. Mber. Königl. Akad. Wiss. Berlin 1864(4): 275–276, pl., figs. 4, 4a–4f.

Synonym: *Labionaris filholii* Brocchi, 1876.
Type: Lectotype, ZMB 5118, a 355–359 mm specimen (Gräffe & Godeffroy), designated by Zug & Ineich (1993: 18).
Type locality: "Viti-Levu, Fiji Islands" [bet. 17°20'–18°17'S and 177°15'–178°40'E] via lectotype selection.
Distribution: Fiji Islands (SE Viti Levu), 10–150 m.
Remarks: Original description reprinted in W. Peters (1864d: 8). McDowell (1970: 179) erroneously cited publication date as 1865.

†*OGMOPHIS* Cope, 1884b (Boidae)

Synonym: †*Ogmophus* – Douglass, 1903 *(nomen incorrectum).*
Type species: †*Ogmophis oregonensis* Cope, 1884b.
Distribution: Middle Eocene of USA, upper Eocene of Canada, lower and middle Oligocene of USA, Oligocene of USA, lower, middle and upper Miocene of USA, upper Miocene of Poland, and Pliocene of USA.
Sources: Gilmore, 1938, Holman, 1979b, 2000a and Rage, 1984b.

1. †*Ogmophis arenarum* Douglass, 1903. Ann. Carnegie Mus. (1903–1904) 2(2): 171. (*nomen corrigendum*)

Synonym: *Ogmophus arenarum* Douglass, 1903 *(nomen incorrigendum),* and *Ogmophis arenarum* – Lambe, 1908 *(nomen corrigendum).*

Type: Lectotype, CM 744, the vertebra illustrated in Gilmore (1938: figs. 12a–d), one trunk vertebra (E. Douglass, 1899), designated herein.

Type locality: "Flint creek beds, within two or three miles of the village of New Chicago in Granite County, western Montana; Miocene."

Distribution: Middle Miocene (Barstovian: 13.6–16.3 mya) of USA (Montana). Known only from the type series.

2. †*Ogmophis compactus* Lambe, 1908. Contr. Canad. Paleont. 3(4): 20–21, pl. 1, figs. 26–30.

Type: Lectotype, NMC 6237, one trunk vertebra (L.M. Lambe, 1904), designated by Gilmore (1938: 39).

Type locality: "4 miles above east end of Post Office, Cypress Hills, Saskatchewan Canada; Oligocene."

Distribution: Upper Eocene (Chadronian: 33.9–38.0 mya) of Canada (Saskatchewan) and lower Oligocene (Orellan: 33.3–33.9 mya) of USA (Nebraska).

Sources: Holman, 1972, 2000a.

O

3. †*Ogmophis europaeus* Szyndlar *in* Mlynarski, Szyndlar, Estes & Sanchez, 1982. Estud. Geol. 38: 112–113, fig. 9.

Type: Holotype, ZPW OP 86/2, one trunk vertebra (Ryziewicz, 1950–1958).

Type locality: "Opole, Silesia. Southern Poland. Late Miocene (Sarmatian, MN 7)." Emended to a suburb of the town of Opole, in its district called Nowa Wies Królewska, Opole Voivodship, SW Poland *fide* Mlynarski (1984: 130).

Distribution: Upper Miocene (Sarmatian, MN 7: 11.1–12.8 mya) of Poland. Known only from type series.

Source: Mlynarski, 1984.

4. †*Ogmophis miocompactus* Holman, 1976a. Herpetologica 32(1): 41–42, fig. 1.

Type: Holotype, SDSM 8372, one trunk vertebra.

Type locality: "Black Bear Quarry II, designated SDSM locality V673, of Bennett County, South Dakota, USA; early Rosebud Formation, Hemingfordian (middle Miocene)."

Distribution: Lower Miocene (Arikareean: 20.6–30.8 mya) of USA (Nebraska, Wyoming), middle Miocene (Hemingfordian: 16.3–20.6 mya) of USA (South Dakota), and middle Miocene (Barstovian: 13.6–16.3 mya) of USA (South Dakota, Texas).

Sources: Holman, 1976c, 1977b, 1978, 1981b, 1996a, 2000a.

Remarks: Type locality is lower Miocene *fide* Holman (2000a: 84).

5. †*Ogmophis oregonensis* Cope, 1884b. Vert. Tertiary Form. West 3(2): 783, pl. 58a, figs. 10, 11–11b, 12–12a.

Types: Syntypes (5), AMNH, four middle trunk and one anterior trunk vertebrae (J.L. Wortman), lost *fide* Gilmore (1938: 36).

Type locality: "the John Day beds near the John Day River, Oregon; John Day epoch, upper Oligocene."

Distribution: Upper Oligocene (Arikareean: 20.6–30.8 mya) of USA (Oregon). Known only from type series.

6. †*Ogmophis pliocompactus* Holman, 1975. Univ. Michigan Paps. Paleont. (12): 61–62, fig. 2g.

Type: Holotype, MSUVP 784, one trunk vertebra (J.A. Holman, 1969–1973).

Type locality: "locality UM-K6-59 on the Lowell Hillman Ranch 2350-2550 ft S and 75 ft E of the NW corner, Sec. 22, R. 22 W, T. 11 S, elevation 2255 ft., Trego County, Kansas; lower Pliocene (Clarendonian), Ogallala Formation."

Distribution: Upper Miocene (Clarendonian: 10.3–13.6 mya) of USA (Kansas, Texas).

Remarks: Holotype erroneously listed as MSUVP 772 *fide* Rage (1984b: 27).

7. †*Ogmophis voorhiesi* Holman, 1977a. J. Herp. 11(2): 144, figs. 3a–b.

Type: Holotype, UGVP 47, one trunk vertebra (M. Voorhies).

Type locality: "an abandoned Kaolin (Clay) Pit (Pit No. 22, J.M. Huber Corporation) from approximately 9 mi SSE of Macon and 4 mi ENE of Huber, Twiggs County, Georgia; Upper Eocene of the Twiggs clay, Duchesnean."

Distribution: Middle Eocene (Duchesnean: 38.0–42.0 mya) of USA (Georgia, Wyoming).

OLIGODON H. Boie *in* Fitzinger, 1826a (Colubridae)

Synonyms: *Calamaria* Fitzinger, 1826a *(nomen nudum), Calamaria* F. Boie, 1826 *(nomen nudum), Simotes* A.-M.-C. Duméril, 1853 *(nomen praeoccupatum),* ? *Conorhina* Bibron *in* A.-M.-C. Duméril, Bibron & Duméril, 1854a, *Limotes* – Lohmeyer, 1882 *(nomen incorrectum), Tripeltis* Cope, 1886c, *Holarchus* Cope, 1886c, *Oligodurus* – Kingsley, 1888 *(nomen* *incorrectum), Simodes* – Palacky, 1898 *(nomen incorrectum), Dicraulax* Cope, 1893a, *Sinotes* – Angel, 1929 *(nomen incorrectum), Tripelis* – Maki, 1933 *(nomen incorrectum), Oligodom* – Terent'ev & Chernov, 1949 *(nomen incorrectum), Oligidon* – G. Haas, 1952b *(nomen incorrectum), Simoctes* – Deraniyagala, 1955 *(nomen incorrectum), Simoetes* – Deraniyagala, 1955 *(nomen incorrectum), Oligondon* – Cochran, 1961 *(nomen incorrectum), Oligodan* – Acharji & Mukherjee, 1966 *(nomen incorrectum), Cligodon* – B.L. Lim & Sharef bin Kanarudin, 1975 *(nomen incorrectum), Cottonkukri* Hoser, 2012af *(nomen illegitimum), Crottykukrius* Hoser, 2012af *(nomen illegitimum), Dannyelfakharikukri* Hoser, 2012af *(nomen illegitimum), Daviekukri* Hoser, 2012af *(nomen illegitimum), Funkikukri* Hoser, 2012af *(nomen illegitimum), Geddykukrius* Hoser, 2012af *(nomen illegitimum), Harrigankukriae* Hoser, 2012af *(nomen illegitimum), Hoserkukriae* Hoser, 2012af *(nomen illegitimum), Hugheskukri* Hoser, 2012af *(nomen illegitimum), Moseselfakharikukri* Hoser, 2012af *(nomen illegitimum), Ninkukri* Hoser, 2012af *(nomen illegitimum), Oxykukrius* Hoser, 2012af *(nomen illegitimum), Sammykukriae* Hoser, 2012af *(nomen illegitimum),* and *Smythkukri* Hoser, 2012af *(nomen illegitimum).*

Type species: *Coluber bitorquatus* F. Boie, 1827.

Distribution: Asia and East Indies.

Sources: Boulenger, 1894a, Rooij, 1917, E.H. Taylor, 1922c, Wall, 1923g, Bourret, 1936b, M.A. Smith, 1943, C. Haas, 1950, Leviton, 1963a, Pauwels et al., 2002, Tillack, 2008, Tillack & Günther, 2009, I. Das, 2010, 2012, Green, 2010, Green et al., 2010, David & Vogel, 2012a and David et al., 2012.

1. *Oligodon affinis* A.C.L.G. Günther, 1862a. Ann. Mag. Nat. Hist. (3) 9(50): 58.

Type: Holotype, BMNH 1946.1.5.52, a 267 mm female (R.H. Beddome, 1857–1861).

Type locality: "Anamallay Hills" [= Anamalai Hills, S Western Ghats, E Kerala /W Tamil Nadu States, SW India, ca. 10°22'N, 77°08'E].

Distribution: Southwestern India (Kerala, Tamil Nadu), 100–1200 m.

Sources: Inger et al., 1984 and Whitaker & Captain, 2004.

Remarks: Probably occurs in S Karnataka.

2. *Oligodon albocinctus* (Cantor, 1839b). Proc. Zool. Soc. London 7(1): 50. (*Coronella albocincta*)

Synonyms: *Coronella puncticulatus* Gray, 1853 and *Simotes amabilis* A.C.L.G. Günther, 1868.

Type: Lectotype, specimen illustrated in colored sketch of T.E. Cantor (1834–1837) in RSL, designated by M.A. Smith (1943: 211).

Type locality: "Cherrapungi, Assam" [= Cherrapunji, East Khasi Hills, Meghalaya State, NE India, 25°18'N, 91°42'E, elevation 1485 m] via lectotype selection.

Distribution: Southern Asia. Nepal (Ilam, Kathmandu, Makwanpur), NE India (Arunachal Pradesh, Assam, Manipur, Meghalaya, Nagaland, Sikkim, E Uttar Pradesh, N West Bengal), Bhutan, SW China (Yunnan, Xizang), Bangladesh (Rangpur, Chittagong Hills), N Vietnam (Vinh Phuc) and NW Myanmar (Chin, Kachin, Mandalay, Rakhine), 60–1980 m.

Sources: A.C.L.G. Günther, 1868, Venning, 1910, Wall, 1911e, 1914a, 1923g, 1925d, Acharji & Kripalani, 1951, Kramer, 1977, Zhao & Adler, 1993, Orlov et al., 2000, Tiwari & Shah, 2004, Whitaker & Captain, 2004, M.F. Ahmed et al., 2009, V.S. Nguyen et al., 2009 and Wangyal, 2011.

Remarks: Holotype lost *fide* Iskandar & Colijn (2001: 70). Wagner (1975: 65) synonymized *O. juglandifer* (Wall) with *O. albocinctus.*

3. *Oligodon ancorus* (Girard, 1858a). Proc. Acad. Nat. Sci. Philadelphia (1857) 9(6): 182. (*Xenodon ancorus*)

Synonyms: *Simotes aphanospilus* Cope, 1860e, *Simotes phaenochalinus* Cope, 1860e, *Simotes ancoralis* Jan, 1863a, *Holarchus burksi* E.H. Taylor, 1918b, and *Oligodon rhombifer* F. Werner, 1924a .

Type: Holotype, USNM 5521 (formerly USNM 20) (C. Wilkes [U.S. Explor. Exped.], 12–21 Jan. 1842).

Type locality: "exotic to North America." Emended to Manilla (*sic*) *fide* Girard (1858b: 168), [= Manila, Manila Prov., SW Luzon Is., N Philippines, 14°36'N, 120°59'E, elevation 10 m].

Distribution: Northern Philippines (Luzon, Mindoro), 10–275 m.

Sources: E.H. Taylor, 1922a, Leviton, 1959b, 1963a, Tillack, 2008 and Tillack & Günther, 2009.

Remarks: Supplemental original description in Girard (1858b: 167–168). Philippine (Samar Is.) record needs confirmation.

4. *Oligodon annamensis* Leviton, 1953a. J. Washington Acad. Sci. 43(12): 423–424, fig. 1.

Type: Holotype, USNM 90408, a 249 mm female (E. Poilane, 11 March 1933).

Type locality: "Blao, Haut Donai, Annam, French Indo-China" [= Bao Loc, Lam Dong Prov., S Vietnam, 11°32'N, 107°48'E, elevation 800 m].

Distribution: Southern Vietnam (Lam Dong), 800 m. Known only from vicinity of type locality.

Sources: Leviton, 1960, Campden-Main, 1970a and V.S. Nguyen et al., 2009.

5. *Oligodon annulifer* (Boulenger, 1893d). Proc. Zool. Soc. London 61(1): 524. (*Simotes annulifer*)

Type: Holotype, BMNH 1946.1.3.34, a 160 mm specimen (A.H. Everett, 1869–1885).

Type locality: "North Borneo" [= Sabah, East Malaysia].

Distribution: Borneo. East Malaysia (Sabah), Brunei and Indonesia (W Kalimantan), NSL–480 m.

Sources: Stuebing, 1991, David & Vogel, 1996, Stuebing & Inger, 1999, Iskandar, 2004 and Tillack & Günther, 2009.

Remarks: Holotype illustrated in Tillack & Günther (2009: figs. 8a–d). Kalimantan record doubtful.

6. *Oligodon arnensis* (G. Shaw, 1802). Gen. Zool., Amph. 3(2): 526. (*Coluber arnensis*)

Synonyms: *Coluber russelius* Daudin, 1803c, *Coluber monticolus* Cantor, 1839b, *Simotes albiventer* A.C.L.G. Günther, 1864a, and *Simotes russellii ceylonensis* F. Müller, 1887.

Type: Lectotype, a 457 mm specimen described and illustrated by P. Russell (1796: 43–44, pl. 38) (Bonniveaux, Oct. 1788), designated herein.

Type locality: "Arnee in the East Indies" [= Arni, N Tamil Nadu State, SE India, 12°40'N, 79°17'E, elevation 150 m] via lectotype selection.

Distribution: Southern Asia. Eastern Afghanistan (Nangarhar), Pakistan (Balochistan, Khyber Pakhtunkhwa, Sindh), cen. Nepal (Bara, Chitwan, Dolakha, Kaski, Kathmandu, Parsa, Sindhupalchok), India (Andhra Pradesh, Bihar, Goa, Himachal Pradesh, Jammu and Kashmir, Karnataka, Kerala, Manipur, Maharashtra, Orissa, Tamil Nadu, Uttar Pradesh, West Bengal), Sri Lanka (Central, Eastern, Northern, North-Western, Southern, Western) and Bangladesh (Rangpur), 100–2000 m.

O

Sources: Wall, 1914a, 1921g, Deraniyagala, 1936, 1955, Acharji, 1946, E.H. Taylor, 1950b, A. Bhatnagar, 1959, L. Swan & Leviton, 1962, Acharji & Mukherjee, 1966, Minton, 1966, Král, 1969, P. Silva, 1969, Tiwari & Sharma, 1971, Murthy, 1972a, 1986, Mahajan & Agrawal, 1976, R.C. Sharma, 1976a, Kramer, 1977, Murthy et al., 1979, M.S. Khan, 1982a–b, 2002, Daniel, 1983, M.A.R. Khan, 1988, A. Silva, 1990b, 2009, Sanyal et al., 1993, Tiwari & Shah, 2004, Whitaker & Captain, 2004 and N. Khaire, 2006.

Remarks: Original description based on P. Russell (1796: 43–44, pl. 38).

7. *Oligodon barroni* (M.A, Smith, 1916). J. Nat. Hist. Soc. Siam 2(1): 46–47, pl., figs. 4–4a. (*Simotes barroni*)

Synonym: *Holarchus taeniatus caudaensis* Bourret, 1934c.

Types: Syntypes (3), BMNH 1946.1.3.21–22 and BMNH 1946.1.3.26 (formerly BMNH 1917.5.14.16a–c), three males, longest syntype 380 mm (P.A.R. Barron).

Type locality: "Hup Bon, E. of Sriracha, S. E. Siam" [= Hup Bon, Si Racha, Chon Buri Prov., SE Thailand, ca. 13°06'N, 101°06'E, elevation 135 m].

Distribution: Southeast Asia. Southern Thailand (Bangkok, Champasak, Chanthaburi, Chon Buri, Phra Nakhon Si Ayutthaya, Rayong, Saraburi, Koh Lam Is.), Cambodia (Kaoh Kong, Pursat), S Laos (Champasak) and S Vietnam (Binh Duong, Cao Bang, Dong Nai, Gia Lai, Hai Duong, Khanh Hoa, Thua Thien-Hue), 60–780 m.

Sources: Bourret, 1934c, E.H. Taylor, 1965, Campden-Main, 1970a, Deuve, 1970, Saint-Girons, 1972a, M.J. Cox et al., 1998, Teynié & David, 2007, David et al., 2008b and V.S. Nguyen et al., 2009.

Remarks: Two topotypes incorrectly listed in BMNH catalogue as syntypes *fide* David et al. (2008: 36).

8. *Oligodon bitorquatus* (F. Boie, 1827). Isis von Oken 20(6): 519. (*Coluber bitorquatus*)

Synonyms: *Oligodon bitorquatus* H. Boie *in* Schlegel, 1826a *(nomen nudum),* Calamaria *oligodon* Schlegel, 1837, *Oligodon subquadratum* A.-M.-C. Duméril, Bibron & Duméril, 1854a, and *Rhabdosoma amboinense* Bleeker, 1860e.

Type: Lectotype, RMNH 10429 (formerly RMNH 53) (S. Müller, 1825–1827), designated herein by R.F. Inger.

Type locality: Unknown. Corrected to Java *fide* H. Boie (1826a: 236), Fitzinger (1826a: 56) and RMNH catalogue.

Distribution: Southern Indonesia (Java, Sumbawa, Timor), 330–1525 m.

Sources: Mertens, 1930, Kopstein, 1938, Tillack & Günther, 2009 and Lang, 2011.

Remarks: Original description based on H. Boie's MS (1823–1825). Records from Sumatra, Ambon and Kalimantan doubtful *fide* T. Barbour (1912b: 24), Mertens (1930: 308), C. Haas (1950: 555) and (Lang, 2011: 111).

9. *Oligodon booliati* Leong & Grismer, 2004. Asiatic Herpetol. Res. 10: 12–14, figs. 1–3.

Type: Holotype, ZRC 2.5153, a 510 mm female (T.M. Leong & K.M. Crane, 16 July 2001).

Type locality: "primary forest along the Tekek-Juara trail, ca. 150 m asl., Pulau Tioman (Pahang, Malay Peninsula)."

Distribution: West Malaysia (Pahang: Tioman Is.), 150–300 m.

Sources: Grismer & Khang Aun, 2008 and Grismer, 2011.

Remarks: Probably a "colour pattern class" and not a valid species *fide* Green (2010: 93).

10. *Oligodon brevicaudus* A.C.L.G. Günther, 1862a. Ann. Mag. Nat. Hist. (3) 9(50): 58. (*Oligodon brevicauda*)

Type: Holotype, BMNH 1946.1.3.35, a 381 mm female (R.H. Beddome, 1857–1861).

Type locality: "Anamallay Hills" [= Anamalai Hills, S Western Ghats, E Kerala /W Tamil Nadu States, SW India, ca. 10°22'N, 77°08'E].

Distribution: Southwestern India (Western Ghats of Kerala, Tamil Nadu), 915–1005 m.
Sources: A.C.L.G. Günther, 1864a, Cope, 1886c, Campden-Main, 1970b and Pauwels et al., 2002.
Remarks: Possibly occurs in SW Karnataka and W Tamil Nadu, India.

11. *Oligodon calamarius* (Linnaeus, 1758). Syst. Nat., ed. 10, 1: 216. (*Coluber calamarius*)

Synonym: *Oligodon templetoni* A.C.L.G. Günther, 1862a.
Types: Syntypes (2), NHR Lin-62 (formerly MAFR), a 210 mm specimen, and ZMUU 143 (formerly JA).
Type locality: "America" (in error). Corrected to Sri Lanka *fide* M.A. Smith (1943: 229).
Distribution: Southwestern Sri Lanka (Central, Sabaragamuwa, Southern, Western), 300–1280 m.
Sources: A.C.L.G. Günther, 1864a, Wall, 1921g, Deraniyagala, 1955, Brongersma et al., 1966, P. Silva, 1969, 1980a and A. Silva, 1990.
Remarks: M.A. Smith (1943: 228) restricted the type to Stockholm but without specific number..

12. *Oligodon catenatus* (Blyth, 1854a). J. Asiat. Soc. Bengal 23(3): 287. (*Calamaria catenata*)

Synonym: *Oligodon herberti* Boulenger, 1905f.
Type: Holotype, not designated, IMC, a 432 mm specimen (Robinson), lost *fide* Sclater (1891a: 231).
Type locality: "Asám from the Khásya hills, or other upland territory" [= vicinity of Khasi Hills, E Meghalaya State, NE India, 25°35'N, 91°38'E, elevation 1370–1675 m].
Distribution: Southeastern Asia. Northeastern India (Meghalaya), S China (Fujian, Guangdong, Guangxi), N Myanmar (Kachin, Sagaing, N Shan) and N Laos (Phongsali, Xiangkhouang), 400–1825 m.
Sources: Wall, 1925b, C.H. Pope, 1935, Bourret, 1939a–d, M.A. Smith, 1940c, Deuve, 1970, B.Q. Hu et al., 1980, Tian et al., 1986, Pauwels et al., 2002, Zhao, 2006 and V.S. Nguyen et al., 2009.
Remarks: Cambodia record doubtful *fide* Saint-Girons (1972a: 61).

13. *Oligodon chinensis* (A.C.L.G. Günther, 1888b). Ann. Mag. Nat. Hist. (6) 1(3): 169–170. (*Simotes chinensis*)

Type: Holotype, BMNH 1946.1.3.28, a 470 mm female (A.E. Pratt, 1887–1890).
Type locality: "mountains north of Kiu Kiang, on the Yantsze river, China," (in error). Corrected to Wu-suih, about three days' journey or 9 mi. S of the American Central China Mission in Kiu-kiang *fide* Pratt (1892: 3–5), [= Wusüeh, SE Hubei Prov., cen. China, 29°50'N, 115°41'E *fide* C.H. Pope (1935: 502].
Distribution: Southeastern China (Anhui, Fujian, Guangdong, Guangxi, Guizhow, Hainan, Henan, Hunan, Jiangsu, Jiangxi, Sichuan, Yunnan, Zhejiang) and Vietnam (Bac Kan, Cao Bang, Gai Lai, Hai Duong, Hai Phong, Ha Tinh, Lang Son, Lao Cai, Nghe An, Quang Binh, Quang Ninh, Vinh Phuc), 400–1830 m.
Sources: K.P. Schmidt, 1927a, C.H. Pope, 1929, 1935, Chang & Fang, 1931, Bourret, 1939d, Zhao & Adler, 1993, Orlov et al., 2000, Zhao, 2006, David et al., 2008a, V.S. Nguyen et al., 2009 and David et al., 2012.
Remarks: Original description reprinted in A.C.L.G. Günther (1892b: 239). M.A. Smith (1943: 206) listed the type locality as Lushan (= Mt. Lu Shan, 29°30'N, 115°55'E), which is south of Jiujiang on the Yangtze; Günther possibly misstated the direction of the type locality from Jiujiang as there are no mountains to the north of the Yangtze River.

14. *Oligodon cinereus* (A.C.L.G. Günther, 1864a). Rept. Brit. India: 215. (*Simotes cinereus*)

Synonyms: *Simotes swinhonis* Swinhoe, 1863 (*nomen nudum*), *Simotes swinhonis* A.C.L.G. Günther, 1864a, *Simotes multifasciatus* Jan, 1865 *in* Jan & Sordelli,1860–1866, *Simotes semifasciatus* J. Anderson, 1871b, *Holarchus dolleyanus* Cope, 1895a, *Simotes violaceus fasciatus* Wall & Evans, 1901, *Simotes violaceus pallidocinctus* Bourret, 1934b, *Holarchus violaceus tamdaoensis* Bourret, 1935c, *Holarchus lightfooti* Rendahl, 1937, *Holarchus violaceus poilani* Bourret, 1939c, and *Holarchus violaceus plurimaculatus* Bourret, 1941b.
Type: Holotype, BMNH 1946.1.1.25 (formerly BMNH 1861.8.12.8), a 432 mm (svl) male (M. Mouhot, 1858–1860).
Type locality: "Gamboja" [= Cambodia].
Distribution: Southeast Asia. Northeastern India (Arunachal Pradesh, Assam, Bihar, Meghalaya, Nagaland, Uttar Pradesh), Bangladesh (Chittagong Hills), Bhutan, S China (Fujian, Guangdong, Guangxi, Guizhou, Hainan, Hong Kong, Hunan, Yunnan), Taiwan, Myanmar (Bago, Kachin, Kayah, Mandalay, Rakhire, Sagaing, Shan, Tanintharyi, Yangon), Thailand (Chiang Mai, Chiang Rai, Kanchanaburi, Lampang, Loei, Nakhon Ratchasima, Phetchabun, Phet Buri, Phitsanulok, Phrae, Prachuap Khiri Khan, Surat Thani, Udon Thani, Ko Chang and Ko Tao Is.), Cambodia (Kampong Chhnang, Mondolkiri), Laos (Attapu, Champasak, Louangphrabang, Phongsali, Xekong, Xiangkhouang) and Vietnam (Bac Kan, Binh Thuan, Gai Lai, Hai Duong, Hai Phong, Ha Tinh, Lang Son, Lao Cai, Nghe An, Quang Binh, Quang Ninh, Vinh Phac), 100–2450 m.
Sources: E.H. Taylor & Elbel, 1958, E.H. Taylor, 1965, Campden-Main, 1970a, Deuve, 1970, Saint-Girons, 1972a, Wagner, 1975, Romer, 1979a, B.Q. Hu et al., 1980, Karsen et al., 1986, Tian et al., 1986, M.A.R. Khan, 1988, Stuebing, 1991, Zhao & Adler, 1993, B.L. Stuart et al., 2006, Zhao, 2006, Teynié & David, 2007,

V.S. Nguyen et al., 2009, Agarwal et al., 2010, David et al., 2011b, Geissler et al., 2011, Wangyal, 2011 and David et al., 2012.

Remarks: Occurs in Sabah, East Malaysia *fide* Stuebing (1991: 334).

15. *Oligodon cruentatus* (A.C.L.G. Günther, 1868). Ann. Mag. Nat. Hist. (4) 1(6): 417. (*Simotes cruentatus*)

Types: Syntypes (5), BMNH 1946.1.4.15–19, four males, longest syntypes 381 mm and 330 mm gravid female (W. Theobald, Jr., 1862–1873).

Type locality: "Pegu" [= Bago, S Bago Div., S Myanmar, 17°20'N, 96°30'E, elevation 15 m].

Distribution: Northern India [MCZ 2279] and Myanmar (Bago, Kachin, Mandalay, Yangon), 15 m.

16. *Oligodon cyclurus* (Cantor, 1839b). Proc. Zool. Soc. London 7(1): 50. (*Coronella cyclura*)

Synonyms: *Coronella violacea* Cantor, 1839b and *Simotes bicatenatus* A.C.L.G. Günther, 1864a.

Type: Neotype, BMNH 1940.3.4.41, a 578 mm male, designated by David et al. (2011: 2).

Type locality: "Subdistrict of Kaligunj, Lalmonirhat District, near Rangpur, Bangladesh" via neotype selection.

Distribution: Northeastern India (Assam, Arunachal Pradesh, Bihar, Himachal Pradesh, Nagaland, Meghalaya, Uttar Pradesh, West Bengal), Bangladesh (Dinajpur, Rangpur, Sylhet) and NW Myanmar (Ayeyarwady, Bago, Kachin, Yangon), 30–1525 m.

Sources: J. Anderson, 1871a, Wall, 1925d, Wagner, 1975, Zhao & Adler, 1993, Chan-ard et al., 1999, Orlov et al., 2000, Schleich & Kästle, 2002, Das, 2004a, David et al., 2008a, 2011b, 2012 and Rahman, 2012c.

Remarks: Colored sketch of T.E. Cantor (1834–1837: no. 15) in RSL designated lectotype *fide* M.A. Smith (1943: 202). Holotype lost *fide* E.H. Taylor (1965: 765). Neotype designations of Campden-Main (1970b: 764, USNM 72067 from Bangkok, Thailand) and Wagner (1975: 62, BMNH 1940.3.4.41 from Kalinganz, Bangladesh) invalid *fide* David et al. (2011: 3–4).

17. *Oligodon deuvei* David, Vogel & Rooijen, 2008. Zootaxa (1965): 28–35, figs. 19–22.

Type: Holotype, MNHN 1974.1366, a 289 mm male (S. Poilane, 1932).

Type locality: "'Arboretum de Trang Bôm', now Arboretum of Bien Hoa, Dong Nai Province, Vietnam."

Distribution: Southeast Asia. Laos (Vientiane), Cambodia (Pursat) and S Vietnam (Dong Nai, Ho Chi Minh City), 25–175 m.

Remarks: This species has been confused with *O. mouhoti* and probably occurs in NE Thailand *fide* David et al. (2008: 35). Photograph of this species as *Oligodon* sp. in Grismer et al. (2008: fig. 19).

18. *Oligodon dorsalis* (Gray, 1835 *in* Gray & Hardwicke, 1830–1835). Illust. Indian Zool. 2(19–20): pl. 85, fig. 1. (*Elaps dorsalis*)

Type: Holotype, BMNH 1946.1.1.60, a 406 mm male (T. Hardwicke, 1756–1823).

Type locality: Unknown. Designated as Chittagong [= Chittagong Hills, SE Bangladesh] *fide* M.A. Smith (1943: 530).

Distribution: Southeastern Asia. Northeastern India (Assam, Manipur, Meghalaya, Mizoram, Nagaland, West Bengal), Bangladesh (Chittagong), Bhutan, NW Myanmar (Chin, Kachin, Sagaing) and N Thailand (Loei), 1300–1980 m.

Sources: J. Anderson, 1871a, Venning, 1910, M.A. Smith, 1940c, M.A.R. Khan, 1988, M.J. Cox et al., 1998, Das & Palden, 2000 and Ao et al., 2004.

Remarks: M.A. Smith (1943: 530) designated pl. 123 in T. Hardwicke's collection a lectotype. A.C.L.G. Günther (1864a: 211) erroneously suggested the type locality as probably Afghanistan. Probably occurs in Laos *fide* M.J. Cox (1991b: 227).

19. *Oligodon durheimi* Baumann, 1913. Zool. Jb. (Syst., Ökol. Geog.) (1912–1913) 34(3): 269–270, fig. C [= 3], figs. a–e.

Type: Holotype, NMBE, a 368 mm specimen (E. Durheim, 1907).

Type locality: "Battak Gebirge von West-Sumatra zum Geschenke, die alle aus einer Höhe von 800–1000 m" [= Toba Massif, Battak Highlands, near Tapanuli (2°51'N, 98°16'E), W Sumatra, elevation 800–1000 m].

Distribution: Western Indonesia (W Sumatra), 800–10000 m. Known only from holotype.

Sources: David & Vogel, 1996 and I Das, 2010.

20. *Oligodon eberhardti* Pellegrin, 1910. Bull. Soc. Zool. France 35(2): 30–32. (*Oligodon herberti eberhardti*)

Synonym: *Macrocalamus wongii* Fan, 1931.

Type: Holotype, MNHN 1908.13, a 438 mm specimen (P.A. Eberhardt, 1906–1910).

Type locality: "Tam Dao, au nord-ouest de Hanoï (Tonkin)" [= Tam Dao, Vinh Phuc Prov., N Vietnam, 21°27'N, 105°39'E, elevation 910 m].

Distribution: Vietnam (Cao Bang, Dac Kan, Ha Tay, Lai Chau, Lao Cai, Son La, Tuyen Quang, Vinh Phuc) and S China (Guangxi), 515–1200 m.

Sources: Boulenger, 1905f, Zhao & Adler, 1993, Orlov et al., 2000, V.S. Nguyen et al., 2009 and I. Das, 2010.

Remarks: This species confused with *O. catenatus* Blyth and also occurs in Myanmar, Laos and Cambodia *fide* V.S. Nguyen et al. (2009: 326).

O

21. *Oligodon erythrogaster* Boulenger, 1907d. Rec. Ind. Mus. 1(3): 216.

Type: Holotype, ZSI 15850 (formerly IMC 15850), a 510 mm specimen (J.M. Smith).

Type locality: "Nagarkote, Nepal, altitude 6,000 feet" [= Nagarkot, Kabhre Distr., Bagmati Zone, cen. Nepal, 27°42'N, 85°31'E, elevation 2000 m].

Distribution: Himalayas of Nepal (Bhaktapur, Dolpa, Gorkha, Kalikot, Kaski, Lalitpur, Rasuwa) and NE India (Sikkim, N West Bengal), 850–2600 m.

Sources: Wall, 1910f, 1913h, 1923g, G.E. Shaw & Shebbeare, 1929d, G.E. Shaw et al., 1940b, M.A. Smith, 1951, M.A. Smith & Battersby, 1953, L. Swan & Leviton, 1962, Fleming & Fleming, 1973, Kramer, 1977 and Tiwari & Shah, 2004.

Remarks: Holotype listed as IMC 16108 *fide* Wall (1923g: 321). I. Das et al. (1998: 149) referred to ZSI 15850 as a syntype.

22. *Oligodon erythrorhachis* Wall, 1910e. J. Bombay Nat. Hist. Soc. (1909–1910) 19(4): 923–924, pl., 8 figs.

Type: Holotype, BMNH 1946.1.5.51, a 375 mm female (S.G.C. Gore, 1899–1910).

Type locality: "Namsang, Jaipur, Assam." Emended to Namsang near Jaipur, at the foot of the Naga Hills *fide* Wall (1923b: 309), [= NE Assam Prov., NE India, 27°15'N, 95°23'E, elevation 150 m].

Distribution: Northeastern India (NE Assam), 150 m. Known only from holotype.

23. *Oligodon everetti* Boulenger, 1893d. Proc. Zool. Soc. London 61(1): 524.

Type: Holotype, BMNH 1946.1.3.14, a 370 mm female (A.H. Everett, 1869–1885).

Type locality: "Mt. Kina Balu, Borneo" [= Mt. Kinabalu, N Sabah, East Malaysia, 6°05'N, 116°33'E].

Distribution: Borneo. East Malaysia (Sabah) and Indonesia (S Kalimantan), 600–1000 m.

Sources: Stuebing, 1991 and Malkmus et al., 2002.

24. *Oligodon fasciolatus* (A.C.L.G. Günther, 1864a). Rept. Brit. India: 218–219, pl. 20, fig. b. (*Simotes fasciolatus*)

Synonyms: *Simotes cochinchinensis* A.C.L.G. Günther, 1864a, *Simotes albocinctus dorsolateralis* Wall, 1910d, *Simotes smithi* F. Werner, 1925, and *Oligodon cyclurus superfluens* E.H. Taylor, 1965.

Types: Syntypes (2), BMNH 1946.1.4.6–7 (H. Mouhot, 1858–1860), longest syntype 483 mm.

Type locality: "Pachebone" [= Petchabun or, Phetchabun Prov. Thailand, 16°25'N, 101°10'E, elevation 165 m].

Distribution: Southeast Asia. Northeastern India, E Myanmar (Kachin, Kayah, Mon, Shan), Thailand (Bangkok, Chaiyaphum, Chanthaburi, Chiang Mai, Chiang Rai, Chon Buri, Chuchoengsao, Chumphon, Kanchanaburi, Krung Thep Mahanakhon, Loei, Mae Hong Son, Nakhon Phanom, Nakhom Ratchasima, Nakhon Si Thammarat, Phetchabun, Phetchaburi, Phichit, Phrae, Prachuap Khiri Khan, Ratchaburi, Ranong, Rayong, Sakon Nakhon, Samut Prakan, Saraburi, Sisaket, Tak, Trang, Ubon Ratchathani, Uthai Thani, Ko Chang Is.), ext. S China (Yunnan), N Laos (Louangphrabang, Phongsali, Savannakhet, Xiangkhouang), Cambodia (Kampot, Koh Kong, SW Pursat, Siem Riep) and S Vietnam (An Giang, Bac Ria-Vung-Tau, Ca Mau, Cao Bang, Dak Lak, Da Nang, Dong Nai, Gia Lai, Hai Duong, Ha Tay, Ha Tinh, Ho Chi Minh City, Kien Giang, Lai Chou, Lam Dong, Nghe An, Phu Khanh, Quang Tri, Son La, Vinh Phuc, Con Son Is.), NSL–900 m.

Sources: M.A. Smith, 1928b, Bourret, 1936b, E.H. Taylor, 1965, Campden-Main, 1970a, Deuve, 1970, Saint-Girons, 1972a, M.J. Cox et al., 1998, B.L. Stuart & Emmett, 2006, David et al., 2008a, V.S. Nguyen et al., 2009, I. Das, 2010 and Green, 2010.

Remarks: E.H. Taylor (1965: 765) suggested type locality possibly Phet Buri, SW Thailand. Con Son Is., Vietnam specimens problematical *fide* Wagner (1975: 69).

25. *Oligodon forbesi* (Boulenger, 1883). Proc. Zool. Soc. London 51(1): 387–388, pl. 42. (*Simotes forbesi*)

Types: Syntypes (2), BMNH 1946.1.3.98–99, two females (H.O. Forbes), longest syntype 305 mm.

Type locality: "Timor Laut Islands" [= Pulau Yamdena, Tanimbar Islands, SE Maluku, SE Indonesia, 7°36'S, 131°25'E].

Distribution: Eastern Indonesia (SE Maluku: Babar, Damar, Yamdena).

Sources: Rooij, 1917 and Brongersma, 1933g.

26. *Oligodon formosanus* (A.C.L.G. Günther, 1872b). Ann. Mag. Nat. Hist. (4) 9(49): 20. (*Simotes formosanus*)

Synonyms: *Simotes swinhonis* Swinhoe, 1863 *(nomen nudum)*, Simotes *bicatenatus* F. Müller, 1878b, *Simotes hainanensis* Boettger, 1894, *Holarchus nesiotis* T. Barbour, 1908, *Holarchus formosanus brunnea* Mell, 1930, and *Holarchus formosanus violaceoides* Mell, 1930.

Type: Holotype, BMNH 1946.1.4.42, a 559 mm male (R. Swinhoe, June–July 1858).

Type locality: "Takou, Formosa" [= Kaohsiung, Kaohsiung Prov. S Taiwan, 23°01'N, 120°40'E, elevation 325 m].

Distribution: Southeastern East Asia. Southern China (Fujian, Guangdong, Guangxi, Guizhou, Hainan, Hong Kong, Hunan, Jiangsu, Jiangxi, Macau, Yunnan, Zhejiang, Lanyu and Nan Ao Is.), Taiwan (Kaohsiung, Lanyu, Miaoli, Pingtung, Tainan, Taipei, Lutao and Botel-Tobago Is.) and N Vietnam (Cao Bang, Ha Giang, Hanoi, Thanh Hoa, Vinh Phuc), 100–900 m.

Sources: Stejneger, 1907a, Oshima, 1910, K.P. Schmidt, 1927a, Mell, 1930, Takahashi, 1930, Fan, 1931, Maki, 1931, Bourret, 1934c, 1934f, C.H. Pope, 1935, Herklots, 1935, Wang & Wang, 1956, B.Q. Hu et al., 1959, 1980, Wang, 1962, Kuntz, 1963, Tian et al., 1986, Lue, 1987, 1990, Lue et al., 1989, Huang et al., 1990, Ota, 1991, Easton & Leung-Va, 1993, Zou & Chen, 1998, Lazell, 1999, Orlov et al., 2000, David et al., 2008a, 2012, V.S. Nguyen et al., 2009 and Xiang & Li, 2009.

27. *Oligodon hamptoni* Boulenger, 1918. Proc. Zool. Soc. London 89(1): 9–10, fig. 1.

Type: Holotype, BMNH 1946.1.1.71, a 540 male (H. Hampton, 1898–1905).
Type locality: "Mogok, Upper Burma" [= Mogok, N Mandalay Div., N Myanmar, 22°55'N, 96°30'E, elevation 300 m].
Distribution: Northern Myanmar (Kachin, Mandalay), 300–1500 m.
Sources: Wall, 1925d, Wallach & Bauer, 1996 and I. Das, 2010.

28. *Oligodon inornatus* (Boulenger, 1914e). J. Nat. Hist. Soc. Siam 1(2): 68–70, pl., 3 figs (upper). (*Simotes inornatus*)

Type: Holotype, BMNH 1946.1.4.25, a 580 mm male (P.A.R. Barron).
Type locality: "Nong Kai Ploi, E. of Sriracha, Siam" [= Nong Kai Ploi, east of Si Racha (13°10'N, 100°56'E, elevation 40 m), Chon Buri Prov., S Thailand].
Distribution: Central Thailand (Chiang Mai, Chon Buri, Lampang, Lamphun, Loei, Tak, Trat, Uthai Thani, Ko Kut Is.), NW Cambodia (Kampot, Koh Kong) and Laos (Champasak), 50–400 m.
Sources: M.A. Smith & Kloss, 1915, Bourret, 1936b, E.H. Taylor, 1965, Saint-Girons, 1972a, M.J. Cox, 1991b, M.J. Cox et al., 1998, B.L. Stuart & Emmett, 2006, Teynié & David, 2007 and David et al., 2012.

29. *Oligodon jintakunei* Pauwels, Wallach, David & Chanhome, 2002. Nat. Hist. J. Chulalongkorn Univ. 2(2): 8–11, figs. 1–5.

Type: Holotype, QSMI 385, a 448 mm male (P. Jintakune, 1990).
Type locality: "Krabi Province, Thailand."
Distribution: Southern Thailand (Krabi). Known only from holotype.

30. *Oligodon joynsoni* (M.A. Smith, 1917b). J. Nat. Hist. Soc. Siam 2(4): 276–277. (*Simotes longicauda joynsoni*)

Synonym: *Simotes longicauda* Boulenger, 1903b.
Type: Holotype, BMNH 1946.1.4.23 (formerly MAS 1116), a 760 mm male (H.W. Joynson, June 1917).
Type locality: "valley of the Maa Yome, Muang Ngow, N. Siam" [= Maa Yome Valley, Muang Ngow, Lampang Prov., N Thailand, 18°47'N, 99°58'E, elevation 300 m].
Distribution: Southeastern Asia. Northwestern Thailand (Chiang Mai, Kamphaeng, Lampang, Loei, Nakhon Ratchasima), S China (S Yunnan) and Laos, 300–850 m.
Sources: M.A. Smith, 1922a, E.H. Taylor, 1965, Wagner, 1975, M.J. Cox, 1991b, M.J.Cox et al., 1998, Zhao, 2006, David et al., 2011b and Jiang et al., 2012.
Remarks: Occurs in Vietnam *fide* M.J. Cox et al. (1998: 58). Guangxi, China record identified as *O. nagao fide* Jiang et al. (2012: 319).

31. *Oligodon juglandifer* (Wall, 1909d). J. Bombay Nat. Hist. Soc. (1909–1910) 19(2): 349. (*Simotes albocinctus juglandifer*)

Type: Holotype, BNHS 210, a 717 mm specimen (A.W. Wright).
Type locality: "neighbourhood of Darjeeling, about the Railway between the foot hills and Tindharia, *i.e.,* at an altitude between about 500 and 2,800 feet" [= Tindharia, Darjeeling Distr., ext. N West Bengal State, NE India, 26°51'N, 88°20'E, elevation 150–850 m].
Distribution: Northeastern India (Sikkim, West Bengal), 850 m.
Sources: M. Smith, 1943, Mahendra, 1984 and Green, 2010.
Remarks: M.A. Smith (1943: 207) erroneously reported the type (which he believed was in BMNH) lost. Wall (1923d: 630) listed Nepal. A valid species *fide* Green (2010: 112).

32. *Oligodon kampucheaensis* Neang, Grismer & Daltry, 2012. Zootaxa (3388): 44–45, figs. 2–3, 4a–d.

Type: Holotype, CBC 1464, a 428 mm male (T. Neang, S. Hun & M. Meang, 27 Jan. 2011).
Type locality: "Chruk Prul stream at N 12°11'55.1", E 103°03'35.3", 330 m above sea level in the Phnom Samkos Wildlife Sanctuary, Cardamom Mountains, Pursat Province, southwest Cambodia."
Distribution: Southwestern Cambodia (W Pursat), 330 m.

33. *Oligodon kheriensis* Acharji & Ray, 1936. Rec. Ind. Mus. 38(4): 519–520.

Type: Holotype, ZSI 20391, an adult female.
Type locality: "North Kheri Division, Eastern Circle, Kheri-Lakimpur, in the United Provinces" [= vicinity of Kheri (27°54'N, 80°48'E, elevation 150 m), Uttar Pradesh State, N India].
Distribution: Northern India (Uttar Pradesh) and SW Nepal (Kanchanpur), 195 m.
Sources: Schleich & Kästle, 2002, Tiwari & Shah, 2004 and David et al., 2011b.
Remarks: Previously a synonym of *O. cyclurus*.

34. *Oligodon kunmingensis* Kou & Wu, 1993. Acta Zootax. Sinica 18(3): 379–381, figs. 1–4.

Type: Holotype, YU 906452, a 309 mm male (native, 22 June 1990).
Type locality: "western hill of Kunming city (25.0°N, 102.7°E), Yunnan, altitude 1900m" [= hills to the W of Kunming, Yunnan Prov., SW China, 25°02'N, 102°43'E, elevation 1900 m].
Distribution: Southwestern China (Yunnan), 1900 m. Known only from type locality.

35. *Oligodon lacroixi* Angel & Bourret, 1933. Bull. Soc. Zool. France 58(3–4): 138–139.

Type: Lectotype, MNHN 1772.2 (formerly MNHN 1933.1), a 574 mm specimen (R.L. Bourret, 1930–1931), designated by Orlov et al. (2010b: 310).
Type locality: "Chapa, Tong-king" [= Sa Pa, Sa Pa Distr., W Lao Cai Prov., ext. N Vietnam, 22°20'N, 103°51'E, elevation 1500 m] via lectotype selection.
Distribution: Southern China (Sichuan, Yunnan) and N Vietnam (Lao Cai), 500–1600 m.
Sources: Bourret, 1934a, 1935b, 1936a–b, 1939d, Tian et al., 1986, Zhao & Adler, 1993, Pauwels et al., 2002, V.S. Nguyen et al., 2009 and Orlov et al., 2010b.
Remarks: Photograph of the lectotype in Orlov et al. (2010b: fig. 2).

36. *Oligodon lungshenensis* Zheng & Huang, 1978 *in* Huang, Zheng & Fang. J. Fujian Normal Univ. (Nat. Sci.) 1978(2): 92–93, fig. 2.

Synonym: *Oligodon guizhouensis* Li *in* Wu, Li, Xu, Dong & Liu, 1979.
Type: Holotype, ? FNU 64099, a 185–233 mm male (native, 1994).
Type locality: "Longsheng Co., Guangxi Zhuang Autonomous Region, China."
Distribution: Southern China (Guangxi, Guizhou), 1210 m.
Sources: Tian et al., 1986, D.J. Li, 1989 and Zhao & Adler, 1993.

37. *Oligodon macrurus* (Angel, 1927). Bull. Mus. Hist. Nat., Paris 33(6): 497–498. (*Simotes violaceus macrurus*)

Type: Holotype, MNHN, a 447 mm specimen (P. Chevy, 17 May 1926), lost *fide* V.S. Nguyen et al. (2009: 328).
Type locality: "Annam: Pointe Lagan (sur les dunes)" [= Cape Lagan or mui LaGan, NE Binh Thuan Prov., S Vietnam, 11°10'N, 108°42'E, elevation NSL].
Distribution: Southern Vietnam (Binh Thuan, Ho Chi Minh City, Khanh Hoa, Ninh Thuan), NSL.
Sources: Geissler et al., 2011a and David et al., 2012.
Remarks: Holotype apparently not in MNHN. Type locality listed within Ninh Thuan Prov., Vietnam *fide* Geissler et al. (2011a: 321).

38. *Oligodon maculatus* (E.H. Taylor, 1918b). Philippine J. Sci. 13D(6): 364–365, pl. 1. (*Holarchus maculatus*)

Type: Holotype, CM 2571 (formerly EHT 40), a 299 mm specimen (E.H. Taylor, Aug. 1912).
Type locality: "Bunawan, Agusan, P.I." [= Bunawan, Agusan del Sur Prov., NE Mindanao Is., SE Philippines, 8°10'N, 126°01'E, elevation 30 m].
Distribution: Southern Philippines (N Mindanao), 30–850 m.
Sources: E.H. Taylor, 1925 and Leviton, 1963a.

39. *Oligodon mcdougalli* Wall, 1905b. J. Bombay Nat. Hist. Soc. (1904–1905) 16(2): 251–252, 3 figs.

Type: Holotype, BNHS 963, a 337–349 mm male (E. McDougall).
Type locality: "Sandarang, Burma" (in error). Corrected to Sandoway [= Thandwe, S Rakhine State, SW Myanmar, ca.18°28'N, 94°22'E, elevation NSL] *fide* Wall (1923d: 626).
Distribution: Southwestern Myanmar (S Rakhine), NSL. Known only from the holotype.
Source: Dowling & Jenner, 1988.

40. *Oligodon melaneus* Wall, 1909d. J. Bombay Nat. Hist. Soc. (1909–1910) 19(2): 349–350, 3 text figs., pl., figs. 4–6.

Types: Syntypes (2), BMNH 1946.1.3.11 and BNHS 958, a 330 mm male and 300–333 mm gravid female (A.W. Wright, July 1908).
Type locality: "neighbourhood of Darjeeling, about the Railway between the foot hills and Tindharia, *i.e.*, at an altitude between about 500 and 2,800 feet." Emended to Sukna, Darjiling District, 540 feet *fide* Wall (1923g: 316), [= Sukna, Darjeeling Distr., ext. N West Bengal State, NE India, *ca.* 26°47'N, 88°22'E, elevation 165 m].
Distribution: Northeastern India (N West Bengal), 165 m.
Remarks: Wall (1909d: 350) reported the length of the female to be 333 mm, but M.A. Smith (1943: 229) listed the male to be 330 mm and the female 300 mm.

41. *Oligodon melanozonatus* Wall, 1922d. Rec. Ind. Mus. 24(1): 29–30, figs. 1a–d.

Synonym: *Ologodon melazonotus* M.A. Smith, 1943 *(nomen emendatum).*
Types: Syntypes (2), BMNH 1946.1.5.48 (formerly IMC 16798) and ZSI 16799, a 510–520 mm and 300 mm female (M. de Courey & 32nd Sikh Pioneers, Dec. 1911).
Type locality: "Upper Rotung Valley, Abor Hills, Assam frontier, at about 2000 feet elevation" [= upper Rotung Valley, Abor Hills, NE Arunachal Pradesh Terr., NE India, 28°08'N, 95°12'E, elevation ca. 600 m.

Distribution: Northeastern India (Arunachal Pradesh) and SW China (Xizang), 600–610 m.
Sources: Annandale, 1912 and Zhao & Adler, 1993.

42. *Oligodon meyerinkii* (Steindachner, 1891). Sitz. Akad. Wiss. Wien, Math.-Natur. Kl. 100(1): 292–293. (*Simotes meyerinkii*)

Types: Syntypes (2), NMW 25828a–b.
Type locality: "Sulu-Inseln" [= Sulu Arch., ext. S Philippines].
Distribution: East Malaysia (Sabah) and ext. S Philippines (Sulu Arch.: Bongao, Jolo, Papahang, Sibutu, Tawitawi).
Sources: T. Barbour, 1912b, E.H. Taylor, 1918b, 1922a, Leviton, 1963a and Gaulke, 1993b, 1994b.

43. *Oligodon modestus* A.C.L.G. Günther, 1864a. Rept. British India: 210.

Type: Holotype, BMNH 1946.1.5.54, a 330 mm male (H. Cuming, 1836–1840).
Type locality: "Philippine Islands." Restricted to Negros Oriental Prov., Negros Island, Philippines *fide* Leviton (1963a: 474).
Distribution: Southern Philippines (Cebu, Negros, Panay, Romblon, Tablas), NSL–950 m.
Sources: A.C.L.G. Günther, 1879a, E.H.Taylor, 1922a, Leviton, 1963a and Sison & Gonzales, 1995.
Remarks: Records from Luzon and Mindanao doubtful *fide* Leviton (1963a: 474).

O

44. *Oligodon moricei* David, Vogel & Rooijen, 2008. Zootaxa (1965): 40–45, figs. 29–34.

Type: Holotype, MNHN 1919.137, a 1443 mm female (M. Krempf).
Type locality: "'Nha-Trang', now Nha Trang, Khanh Hoa Province, Vietnam" [= Nha Trang, E Khanh Hoa Prov., S Vietnam, 12°14'N, 109°12'E, elevation 10 m].
Distribution: Vietnam (Khanh Hoa), 10 m. Known only from the holotype.

45. *Oligodon mouhoti* (Boulenger, 1914e). J. Nat. Hist. Soc. Siam 1(2): 70. (*Simotes taeniatus mouhoti*)

Synonym: *Simotes taeniatus* A.C.L.G. Günther, 1861g (*partim*).
Type: Holotype, BMNH 1946.1.3.32, a male (H. Mouhot, 1858–1860).
Type locality: "Gamboja" [= Cambodia].
Distribution: Southern Thailand (Bangkok, Krung Thep Mahanakhon, Nakhon Ratchasima, Nakhon Sawan, Phetchaburi, Phra Nakhon Si Ayutthaya, Prachuap Khiri Khan, Trat, Uthai Thanis) and Cambodia, NSL–10 m.
Sources: A.C.L.G. Günther, 1864a, Cochran, 1930, Bourret, 1936a, Deuve, 1963e, 1970, E.H. Taylor, 1965, Campden-Main, 1969b, 1970a, Saint-Girons, 1972a, M.J. Cox, 1991b, M.J. Cox et al., 1998, David et al., 2008b and V.S. Nguyen et al., 2009.
Remarks: Records from Vietnam (Binh Duong, Dak Nang) doubtful *fide* David et al. (2008b: 28).

46. *Oligodon multizonatum* Zhao & Jiang, 1981. Acta Herp. Sinica 5(7): 54, figs. 1.1–1.3, 2, col. pl.

Synonym: *Oligodon multizonatus* – Zhao & Adler, 1993.
Type: Holotype, CIB 80-II-289, a 530 mm male (native, 17 Aug. 1980).
Type locality: "Luding, Sichuan, alt. 1,400 m., China" [= Luding, cen. Sichuan Prov., S China, 29°54'N, 102°15'E, elevation 1400 m].
Distribution: Central China (Gansu, Sichuan), 1350–1400 m.
Sources: Yao & Cheng, 1981, Tian et al., 1986 and Zhao & Adler, 1993.

47. *Oligodon nagao* David, Nguyen, Nguyen, Jiang, Chen, Teynié & Ziegler, 2012. Zootaxa (3498): 47–51, figs. 1a–b.

Type: Holotype, VNMN A.2012.1, a 653 mm male (T.T. Nguyen, 30 June 2009).
Type locality: "Huu Lien forest, Huu Lung District, Lang Son Province, Vietnam, at elevation of about 300 m a.s.l."
Distribution: Indo-China. Southern China (Guangxi), N Vietnam (Cao Bang, Lang Son) and cen. Laos (Khammouane), 170–475 m.
Source: Jiang et al., 2012.

48. *Oligodon nikhili* Whitaker & Dattatri, 1982. J. Bombay Nat. Hist. Soc. 79(3): 630–631, pl. 1.

Type: Holotype, BMNH 1982.1358 (formerly MSP), a 423 mm male (R. Whitaker, Z. Whitaker & S. Dattatri, 20 Sept. 1981).
Type locality: "Tiger shola, 1500 mts., a few kilometers below Shembaganur on the Kodaikanal ghat road, Palni Hills, Tamilnadu, South India."
Distribution: Extreme S India (SW Tamil Nadu), 1500 m. Known only from type locality.
Sources: Whitaker, 1982 and Whitaker & Captain, 2004.

49. *Oligodon ningshaanensis* Yuan, 1983b. Acta Herp. Sinica (2) 2(2): 65–67, 4 figs.

Type: Holotype, NCF 810218, a 620 mm male (native, 30 July 1982).
Type locality: "Huoditang, Ningshan County, Shaanxi, China, alt. 1650 m" [= Huoditang, Ningshan Co., S Shaanxi Prov., cen. China, 33°29'N, 108°27'E, elevation 1650 m].
Distribution: Central China (S Shaanxi), 1650 m. Known only from vicinity of type locality.
Sources: Tian et al., 1986 and Zhao & Adler, 1993.

50. *Oligodon notospilus* A.C.L.G. Günther, 1873b. Proc. Zool. Soc. London 41(1): 169, pl. 18, fig. a.

Synonyms: *Oligodon schadenbergi* Boettger, 1895d and *Oligodon iwahigensis* L.E. Griffin, 1909b.
Type: Holotype, BMNH 1946.1.3.23, a 267 mm specimen (A.B. Meyer, 1870–1873).
Type locality: "Mindanao" [= Mindanao Is., S Philippines].
Distribution: Southern Philippines (Balabac, Busuanga, Calamian, Mindanao, Palawan).
Sources: E.H. Taylor, 1922a, Leviton, 1963a, Gaulke, 1999 and David et al., 2012.
Remarks: Previously a subspecies of *O. vertebralis*.

51. *Oligodon ocellatus* (Morice, 1875a). Faune Cochin. Franç.: 57. (*Simotes ocellatus*)

Synonyms: *Simotes brevicauda* Steindachner, 1867a *(nomen praeoccupatum)* and *Oligodon analepticos* Campden-Main, 1970b *(nomen substitutum)*.
Types: Syntypes (4), MMC 42000354a–b (formerly MHNL 1569a–b), a male and female, MHNL 42000347 (formerly MHNL 1571), a male, and MHNL 42000359 (formerly MHNL 1572), a male, syntypes 539–663 mm (A. Morice).
Type locality: "Tay-ninh, Cochinchine Française" [= Tay Ninh, Tay Ninh Prov., S Vietnam, 11°18'N, 106°06'E, elevation 10 m].
Distribution: Indo-China. Eastern Cambodia (Kampong Cham, Ratanakiri), Laos (Champasak) and S Vietnam (Binh Dinh, Dac Lac, Dong Nai, Lam Dong, Quang Nam, Quang Ngai, Tay Ninh), NSL–800 m.
Sources: Campden-Main, 1970a, Saint-Girons, 1972a, Wagner, 1975, Teynié & David, 2007, David et al., 2008a, 2012 and V.S. Nguyen et al., 2009.
Remarks: Photographs of type of *Simotes brevicauda* in David et al. (2008a: figs. 12–14).

52. *Oligodon octolineatus* (J.G. Schneider, 1801). Hist. Amph. 2: 299. (*Elaps octolineatus*)

Type: Neotype, ZMB 2099, a 412 mm male, designated by Tillack & Günther (2009: 269, figs. 1a–d).
Type locality: "Java" [W Indonesia] via neotype selection.
Distribution: Malaysia and Greater Sundas. West Malaysia, Singapore, East Malaysia (Sabah, Sarawak), Brunei and Indonesia (Bali, Bangka, Belitung, Bintan, Java, Kalimantan, Mentawai Arch., Nias, Riau Arch., Sumatra), NSL–1000 m.
Sources: T. Barbour, 1912b, Dammerman, 1926, Westermann, 1942, Mertens, 1969c, Tweedie, 1983, K.F.L. Lim & Lee, 1989, K.K.P. Lim & Chou, 1990, Stuebing, 1991, David & Vogel, 1996, Malkmus et al., 2002, Lang & Vogel, 2005, McKay, 2006 and Tillack & Günther, 2009.
Remarks: Holotype (ZMB 2094) lost *fide* Tillack & Günther (2009: 268–269). Mertens (1969c: 340) demonstrated that ZMB 10612 could not be Schneider's type. Stuebing (1991: 334) erroneously listed the type locality as Borneo via lectotype designation. Sulawesi and Philippine records doubtful *fide* Lang & Vogel (2005: 250) and Tillack & Günther (2009: 267–268).

53. *Oligodon ornatus* Van Denburgh, 1909. Proc. California Acad. Sci. (4) 3(3): 53–54.

Synonyms: *Holarchus torquatus konishii* Oshima, 1910, and *Simotes musyi* Roux, 1919a.
Type: Holotype, CAS 18005, a 343 mm male (V. Kühne, 7 March 1909).
Type locality: "Shinchiku, Formosa" [= Hsinchu, Hsinchu Co., NW Taiwan, 24°48'N, 120°58'E, elevation 25 m].
Distribution: Southeastern China (Anhui, Fujian, Guangdong, Guangxi, Hunan, Jiangxi, Sichuan, Zhejiang) and Taiwan (Chiayi, Hsinchu, Keelung, Nantou, New Taipei, Taipei).
Sources: C.H. Pope, 1929, 1935, Takahashi, 1930, Maki, 1931, Wang & Wang, 1956, Kuntz, 1963, Lue, 1987, 1990, Lue et al., 1989, Zhao & Adler, 1993, Zhao, 2006 and Xiang & Li, 2009.

54. *Oligodon perkinsi* (E.H. Taylor, 1925). Philippine J. Sci. 26D(1): 108–109. (*Holarchus perkinsi*)

Type: Holotype, MCZ 25725 (formerly EHT 1164), a 672 mm female (G.A. Perkins, Sept. 1923).
Type locality: "Culion, Culion Island" [Calamianes Group, Palawan Prov., W Philippines, bet. 11°46–59'N and 119°52'–120°05'E].
Distribution: West-central Philippines (Palawan: Culion Is.).
Source: Leviton, 1963a.

O

55. *Oligodon petronellae* Roux *in* Rooij, 1917. Rept. Indo-Aust. Arch. 2: 131.

Synonyms: *Oligodon ornatus* Roux, 1914 *(nomen praeoccupatum), Oligodon rouxi* Ruthven, 1921 *(nomen substitutum)*, and *Oligodon annulifer confluens* F. Werner, 1924a.
Type: Holotype, MHNG 767.2, a 470+ mm male.
Type locality: "Sumatra" [W Indonesia].
Distribution: Western Indonesia (Sumatra), NSL.
Sources: Roux, 1914, David & Vogel, 1996 and Tillack & Günther, 2009
Remarks: Holotype redescribed and illustrated in Tillack & Günther (2009: 271, figs. 2a–d)

56. *Oligodon planiceps* (Boulenger, 1888h). Ann. Mus. Civ. Stor. Nat. Genova (2) 6: 597–598, pl. 5, fig. 2. (*Simotes planiceps*)

Type: Holotype, MSNG CE 30257, a 130 mm specimen (G.B. Comotto via L. Fea, 1887).
Type locality: "Minhlà, Burma, North of Tenasserim" [= Minhlà, Thayet Distr., S Magwe Div., cen. Myanmar, 19°58'N, 95°01'E, elevation 70 m].

Distribution: Southwestern Myanmar (Bago, Irrawaddy, S Magwe, Yangon), NSL–150 m.
Sources: Wall, 1923g and C.H. Pope, 1935.
Remarks: Northern Myanmar or S China record doubtful *fide* Wall (1923g: 308).

57. *Oligodon praefrontalis* F. Werner, 1913. Mitt. Naturhist. Mus. Hamburg (1912) 30: 25.

Type: Holotype, ZMH, a 241 mm specimen (W. Schwinghammer, 17 May 1910), lost *fide* Hallermann *in* Tillack & Günther (2009: 284).
Type locality: "Sabang auf Pulo Weh an der Nordwestküste von Sumatra" [= Sabang, E Weh Is., off NW coast of N Sumatra, Aceh Prov., ext. NW Indonesia, 5°53'N, 95°19'E, elevation NSL].
Distribution: Extreme NW Indonesia (Aceh: Weh). Known only from holotype.
Sources: Pauwels et al., 2002 and Tillack & Günther, 2009.
Remarks: Holotype not in NMW *fide* H. Grillitsch & D. Siderits *in* Tillack & Günther (2009: 285). *Incertae sedis fide* Tillack & Günther (2009: 265).

58. *Oligodon propinquus* Jan, 1862a. Arch. Zool. Anat. Fis. 2(1): 38.

Type: Holotype, ZMH 4432 (formerly ZMH 132 or 192), a 285 mm specimen.
Type locality: "Giava" [= Java, W Indonesia].
Distribution: Western Indonesia (Java). Known only from the holotype.
Source: Jan & Sordelli, 1876 *in* 1870–1881.
Remarks: Holotype illustrated in Jan (1862b: pl. e, no. 4) and Jan & Sordelli (1876 *in* 1870–1881 [48]: pl. 1, fig. 1).

O

59. *Oligodon pseudotaeniatus* David, Vogel & Rooijen, 2008. Zootaxa (1965): 18–23, figs. 10–13.

Type: Holotype, BMNH 1938.8.7.37, a 297 mm male (M.A. Smith, 1916).
Type locality: "'Khorat', Khorat Plateau, Nakhon Ratchasima Province, Thailand" [= ca. 14°55'N, 101°58E, elevation 210–225 m].
Distribution: Southern Thailand (Bangkok, Nakhon Ratchasima, Phra Nakhon Si Ayutthaya, Saraburi), NSL–210 m.

60. *Oligodon pulcherrimus* F. Werner, 1909b. Mitt. Naturhist. Mus. Hamburg 26: 227, fig. 4.

Synonym: *Simotes annulifer bipartita* Despax, 1912.
Type: Neotype, ZMB 70756, a 275 mm male (A. Michalczyk, 18 Feb. 1995, via P.e.V. Leipzig), designated by Tillack & Günther (2009: 274, figs. 3a–d).
Type locality: "settlement Balinka, approx. 7 km southwest of Bukittinggi, West Sumatra, Indonesia, ca. 1200m a.s.l. [ca. 0°21' S, 100°19' E]" via neotype selection.
Distribution: Western Indonesia (Sumatra), 800–1000 m.
Sources: David & Vogel, 1996, Pauwels et al., 2002 and Tillack & Günther, 2009.
Remarks: Holotype in ZMH destroyed during World War II *fide* Hallermann (1998: 214).

61. *Oligodon purpurascens* (Schlegel, 1837). Essai Phys. Serp. 1: 139, 2: 90–91, pl. 3, figs. 13–14. (*Xenodon purpurascens*)

Synonyms: *Simotes albocinctus* A.-M.-C. Duméril, 1853 (*nomen nudum*). *Simotes albocinctus* A.-M.-C. Duméril, Bibron & Duméril, 1854a (*nomen praeoccupatum*), *Simotes tri-notatus* A.-M.-C. Duméril, Bibron & Duméril, 1854a (*nomen incorrigendum*), *Simotes labuanensis* A.C.L.G. Günther, 1864a, *Simotes trinotatus* – A.C.L.G. Günther, 1864a (*nomen corrigendum*), *Simotes crassus* Theobald, 1868b, *Simotes obscurus* Theobald, 1868b, *Simotes catenifer* Stoliczka, 1873, *Simotes dennysi* Blanford, 1881, *Simotes affinis* J.G. Fischer, 1885b, and *Oligodon purpurascens kangeanicus* Mertens, 1959a.
Types: Syntypes (4), RMNH 242a–c (H. Boie & H.C. Macklot, Dec. 1820–Sept. 1823), and UU (Klinkenberg), longest syntype 640 mm.
Type locality: "au pied du mont Parang à l'île de Java" [= at base of Mt. Parang, Jawa Barat Prov., W Java, SW Indonesia, ca. 7°32'S, 108°20'E].
Distribution: Malay Peninsula and Indonesia. Southern Thailand (Nakhon Si Thammarat, Pattani, Phang Nga, Surat Thani), West Malaysia (Johor, Kelantan, Pahang, Perak, Perlis, Pinang, Selangor, Terengganu, Perhentian Is.), Singapore, East Malaysia (Sabah, Sarawak), Brunei and Indonesia (Bangka, Galang, Java, Kalimantan, Kangean, Karimata, Mentawai, Nias, Riau, Serutu, Sipura, Sumatra, Timur, Weh), NSL–1840 m.
Sources: M.A. Smith, 1930a, Deuve, 1963a–b, E.H. Taylor, 1965, Tweedie, 1983, K.F.L. Lim & Lee, 1989, M.J. Cox, 1991b, Stuebing, 1991, David & Vogel, 1996, M.J. Cox et al., 1998, Chan-ard et al., 1999, Malkmus et al., 2002, Pauwels et al., 2002, I. Das, 2007b, Tillack & Günther, 2009, Grismer, 2011, Rooijen et al., 2011 and David et al., 2012.

62. *Oligodon saintgironsi* David, Vogel & Pauwels, 2008. Zootaxa (1939): 21–27, figs. 1–7.

Type: Holotype, MNHN 1974.1264, a 538 mm male (Poilane, 1930–1940).
Type locality: "'Arboretum de Trang Bôm', now Arboretum of Bien Hao, Dong Nai Province, southern Vietnam."
Distribution: Southern Cambodia and S Vietnam (Dong Nai, Ho Chi Minh City), NSL–10 m.

63. *Oligodon signatus* (A.C.L.G. Günther, 1864a). Rept. Brit. India: 215, pl. 20, figs. f–f'. (*Simotes signatus*)

Synonyms: *Simotes subcarinatus* A.C.L.G. Günther, 1872a and *Simotes annulifer annulata* Lidth de Jeude, 1922.

Type: Lectotype, BMNH 1946.1.3.20, a 533 mm female illustrated by Hardwicke (1756–1823, 2: fig. 52) (T. Hardwicke, 1756–1823), designated by M.A. Smith (1943: 529).

Type locality: "Singapore" [= Singapore Island *fide* David & Vogel, 1996: 103] via lectotype selection.

Distribution: Southern West Malaysia (Melaka), Singapore, East Malaysia (Sabah, Sarawak) and W Indonesia (Sumatra), NSL–350 m.

Sources: Tweedie, 1983, David & Vogel, 1996 and Tillack & Günther, 2009.

Remarks: Java record needs confirmation *fide* C. Haas (1950: 557). *Simotes subcarinatus* Günther a valid species *fide* I. Das (2010: 293).

64. *Oligodon splendidus* (A.C.L.G. Günther, 1875). Proc. Zool. Soc. London 43(1): 231, pl. 33. (*Simotes splendidus*)

Type: Holotype, BMNH 1946.1.4.8, a 508 mm female (R.H. Beddome, 1857–1875).

Type locality: "the Wynad" [S India] (in error). Corrected to Myanmar *fide* Wall (1908g: 782 & 1923d: 631).

Distribution: Northern Myanmar (Magwe, Mandalay, Sagaing, Shan, Yangon), 100–300 m.

Sources: Evans, 1905, Wall, 1908g, 1908l, 1923g, 1925d, Venning, 1914 and Zug et al., 1998.

65. *Oligodon sublineatus* A.-M.-C. Duméril, Bibron & Duméril, 1854a. Erpét. Gén. 7(1): 57–58. (*Oligodon sublineatum*)

Type: Holotype, MNHN 3239, a 180 mm specimen (Leschenault).

Type locality: "Ceylan" [= Sri Lanka].

Distribution: Southwestern Sri Lanka (Central, Sabaragamuwa, Southern, Western), NSL–1220 m.

Sources: Wall, 1921g, 1923d, 1923g, E.H. Taylor, 1950b, Deraniyagala, 1955, P. Silva, 1969, 1980a, A. Silva, 1990b, 2001, 2009 and Somaweera, 2006.

Remarks: Nicobar Is. record rejected *fide* Wall (1923d: 627 & 1923g: 315).

66. *Oligodon taeniatus* (A.C.L.G. Günther, 1861g). Proc. Zool. Soc. London 29(1): 189. (*Simotes taeniatus*)

Synonym: *Simotes quadrilineatus* Jan, 1865 *in* Jan & Sordelli, 1860–1866.

Type: Lectotype, BMNH 1946.1.3.27, a male (H. Mouhot, 1858–1860), designated by David et al. (2008b: 8, 10).

Type locality: "Cambodia" via lectotype selection.

Distribution: Southeast Asia. Southern China (? Yunnan), Thailand (Bangkok, Chaiyaphum, Chon Buri, Khan Kaen, Krung Thep Mahanakhon, Nakhon Ratchasima, Nakhon Si Thammarat, Nonthaburi, Pattani, Petchaburi, Prachin Buri, Prachuap Khiri Khan, Saraburi, Songkhla), Cambodia (Battambang, Kandal, Koh Kong, Kampong Chhnang, Phnom Penh, Preah Vihear, Siem Reap), Laos (Champasak) and S Vietnam (Bai Ria-Vung Tau, Cao Bang, Dong Nai, Ha Giang, Hai Duong, Ho Chi Minh City, Khanh Hoa, Kien Giang, Lam Dong, Nghe An, Ninh Binh, Quang Binh, Soc Trang, Tay Ninh, Thai Nguyen, Tien Giang, Vinh Phuc), 50–1800 m.

Sources: Cochran, 1930, E.H. Taylor & Elbel, 1958, Deuve, 1963a, E.H. Taylor, 1965, Campden-Main, 1969b, 1970a, Saint-Girons, 1972a, Zhao & Adler, 1993, M.J. Cox et al., 1998, Zug et al., 1998, Orlov et al., 2000, Teynié & David, 2007, David et al., 2008b, V.S. Nguyen et al., 2009 and T. Hartmann et al., 2013.

Remarks: *Simotes quadrilineatus* Jan a valid species *fide* I. Das, 2010.

67. *Oligodon taeniolatus* (Jerdon, 1854). J. Asiat. Soc. Bengal (1853) 22(6): 528. (*Coronella taeniolata*)

Synonyms: *Coluber taeniolatus* Daudin, 1803b *(nomen oblitum)*, *Natrix scalaris* Wagler, 1825 (*nomen ineditum*), *Xenodon dubium* Jerdon, 1854, *Oligodon subgriseum* A.-M.-C. Duméril, Bibron & Duméril, 1854a *(nomen incorrigendum)*, *Oligodon subgriseus* – Gray, 1858 *(nomen corrigendum)*, *Oligodon elliotti* A.C.L.G. Günther, 1864a, *Oligodon fasciatus* A.C.L.G. Günther, 1864a, *Oligodon spilonotus* A.C.L.G. Günther, 1864a, *Oligodon subgriseus alternans* Bethencourt-Ferreira, 1897a, and *Oligodon taeniolatus ceylonicus* Wall, 1921g.

Type: Holotype, not designated (T.C. Jerdon, 1836–1852), location unknown.

Type locality: "Madras, India" [= Chennai, NE Tamil Nadu State, SE India, 13°04'N, 80°15'E, elevation 10 m].

Distribution: Southern Turkmenistan, ext. NE Iran (Mazandaran), Afghanistan (Herat), Pakistan (E Balochistan, Khyber Pakhtunkhwa, Punjab, Sindh), India (Andhra Pradesh, Bihar, Goa, Karnataka, Kerala, Maharashtra, Orissa, Tamil Nadu, Uttar Pradesh), Bangladesh and Sri Lanka (Eastern, North Central, North-Western, Western), 520–610 m.

Sources: Wall, 1921g, 1923g, E.H. Taylor, 1950b, Deraniyagala, 1955, Minton, 1966, Brück, 1968, Sharma, 1971, 1976a, K. Tiwari & Sharma, 1971, M.S. Khan, 1977, 1982a–b, 2002, 2006, P. Silva, 1980a, Daniel, 1983, M.A.R. Khan, 1988, Dutta & Acharjyo, 1990, A. Silva, 1990b, 2009, Latifi, 1991, Szczerbak, 1994, Whitaker & Captain, 2004 and N. Khaire, 2006.

Remarks: Original description based on P. Russell (1796: 24–25, pl. 19). *Coluber taeniolatus* Daudin requires conservation.

68. *Oligodon theobaldi* (A.C.L.G. Günther, 1868). Ann. Mag. Nat. Hist. (4) 1(6): 417. (*Simotes theobaldi*)

Synonym: *Simotes beddomii* Boulenger, 1890a.
Type: Holotype, BMNH 1946.1.4.9, a 406 mm female (W. Theobald, Jr., 1862–1873).
Type locality: "Pegu" [= Bago, S Bago Div., S Myanmar, 17°20'N, 96°30'E, elevation 15 m].
Distribution: Northeastern India (W Meghalaya, Nagaland) and Myanmar (Bago, Kachin, Magwe, Mandalay, Shan, Tanintharyi, Yangon), 15–250 m.
Sources: Wall, 1925d and Ao et al., 2004.

69. *Oligodon torquatus* (Boulenger, 1888h). Ann. Mus. Civ. Stor. Nat. Genova (2) 6: 597, pl. 5, figs. 1–1a. (*Simotes torquatus*)

Types: Syntypes (8), BMNH 1946.1.4.21–22, a female and juvenile, and MSNG CE 30258a–f, longest syntype 290 mm (L. Fea, 1887).
Type locality: "Bhamò, Burma" [= Bhamò, S Kachin State, NE Myanmar, 24°15'N, 97°14'E, elevation 115 m].
Distribution: Northern Myanmar (Ayeyarwady, S Kachin), 115–140 m.
Source: Wall, 1923g.
Remarks: Lectotype restrictions of M.A. Smith (1943: 219, for BMNH) and Iskandar & Colijn (2001: 76, for MSNG 30258) invalid.

O

70. *Oligodon travancoricus* Beddome, 1877b. Proc. Zool. Soc. London 45(1): 685. (*Oligodon travancoricum*)

Type: Holotype, BMNH 1946.1.5.53, a juvenile male (R.H. Beddome, 1857–1877).
Type locality: "South-Travancore mountains, Madras Presidency, 3000 feet elevation" [= S India]. Emended to Tinnevelly hills [= ext. S Tamil Nadu, S India] *fide* Boulenger (1894a: 236).
Distribution: Southern India (Western Ghats of Andhra Pradesh, S Kerala, S Tamil Nadu), 815–985 m.
Sources: Ferguson, 1895, Sanyal et al., 1993, Ganesh et al., 2009 and Chandramouli & Ganesh, 2010.

71. *Oligodon trilineatus* (A.-M.-C. Duméril, Bibron & Duméril, 1854a). Erpét. Gén. 7(1): 636–637. (*Simotes trilineatus*)

Synonym: *Simotes trilineatus* A.-M.-C. Duméril, 1853 (*nomen nudum*).
Type: Holotype, MNHN 3541 (P.-M. Diard, 1829–1848).
Type locality: "grandes Indes" [= Greater Sundas, W Indonesia].
Distribution: Western Indonesia (Mentawai Arch., Nako, Nias, Sumatra), NSL–500 m.
Sources: J.G. Fischer, 1885b, Lidth de Jeude, 1890, David & Vogel, 1996 and Tillack & Günther, 2009.
Remarks: Photograph of the holotype in Tillack & Günther (2009: fig. 7).

72. *Oligodon unicolor* (Kopstein, 1926). Zool. Meded. 9(2–3): 107. (*Simotes unicolor*)

Type: Holotype, RMNH, a 300 mm specimen, (F. Kopstein, May 1924).
Type locality: "Samlakki, Süd-Tenimber" [= Yamdena, Tanimbar Islands, SE Indonesia, 7°36'S, 131°25'E].
Distribution: Southeastern Indonesia (Tanimbars: Yamdena).

73. *Oligodon venustus* (Jerdon, 1854). J. Asiat. Soc. Bengal (1853) 22(6): 528. (*Xenodon venustum*)

Synonyms: *Simotes binotatus* A.-M.-C. Duméril, 1853 (*nomen nudum*) and *Simotes binotatus* A.-M.-C. Duméril, Bibron & Duméril, 1854a.
Type: Holotype, not designated, a 305 mm specimen, lost *fide* M.A. Smith (1943: 222).
Type locality: "West Coast of S India." Restricted to North Canara district [= Uttara Kannada Distr., coastal Karnataka State, SW India] *fide* M.A. Smith (1943: 222).
Distribution: Southern India (Anaimalai Hills of Karnataka, Kerala, S Tamil Nadu), 610–2135 m.
Sources: Malhotra & Davis, 1991 and Hutton & David, 2009.

74. *Oligodon vertebralis* (A.C.L.G. Günther, 1865). Ann. Mag. Nat. Hist. (3) 15(86): 91–92. (*Simotes vertebralis*)

Type: Holotype, BMNH 1946.1.3.33, a 345–356 mm male (L.L. Dillwyn, 1855–1865).
Type locality: "Bandjermassing" [= Banjarmasin, SW Kalimantan Selantan, S Kalimantan, Indonesia, 3°20'S, 114°35'E, elevation NSL].
Distribution: Borneo and S Philippines. East Malaysia (N Sabah), cen. Indonesia (Kalimantan), NSL–70 m.
Sources: Stuebing, 1991 and Malkmus et al., 2002.

75. *Oligodon waandersi* (Bleeker, 1860f). Nat. Tijds. Nederl. Indië (5) 22(2): 83. (*Rabdion waandersi*)

Synonym: *Rabdion cruciatum* Bleeker, 1860e.
Type: Holotype, BMNH 1946.1.2.52, a 209 mm male (H.L. van Bloemen-Waanders, 1841–1845).
Type locality: "Boni" [= N Sulawesi, cen. Indonesia, 0°27'S, 121°48'E].
Distribution: Central Indonesia (Buton, Sula, Sulawesi), NSL–1200 m.
Sources: A.C.L.G. Günther, 1865, M.A. Smith, 1927, Bosch, 1985, Lang & Vogel, 2005 and Koch, 2012.

76. *Oligodon wagneri* David & Vogel, 2012a. Zootaxa (3201): 59–62, figs. 1–3.

Synonym: *Oligodon pseudocolor* F. Wagner, 1976. *(nomen ineditum)*
Type: Holotype, MNHN 1891.233, a 235 mm male (J. Claine, 23 July 1891).
Type locality: "'Poulo Nias', *i.e.* Pulau Nias (Nias Island), Nias Archipelago, Sumatera Utara Province, off the northern part of the west coast of Sumatra, Indonesia."
Distribution: Western Indonesia (Nias). Known only from the holotype.

77. *Oligodon woodmasoni* (Sclater, 1891a). J. Asiat. Soc. Bengal 60(3): 235–236, pl. 6, fig. 2. (*Simotes woodmasoni*) (nomen corrigendum)

Synonyms: *Simotes wood-masoni* Sclater, 1891a *(nomen incorrigendum)* and *Simotes woodmasoni* – Sclateri, 1891 *(nomen corrigendum).*
Types: Syntypes (2), ZSI 8459 (formerly IMC 8459) (J. Wood-Mason, 1872) [Andaman] and ZSI 12547 (formerly IMC 12547) (F.A. de Roepstorff) [Nicobar].
Type locality: "Andamans" and "Nicobars" [= Anadaman & Nicobars, E India, Bay of Bengal].
Distribution: Eastern India (Andaman & Nicobars).
Sources: Biswas & Sanyal, 1980, Tikader & I. Das, 1985 and Vijayakunar & David, 2006.

OMOADIPHAS G. Köhler, Wilson & McCranie, 2001 (Colubridae)

Type species: *Omoadiphas aurula* G. Köhler, Wilson & McCranie, 2001.
Distribution: Honduras.
Sources: G. Köhler et al., 2001 and McCranie & Cruz-Díaz, 2010.

1. *Omoadiphas aurula* G. Köhler, Wilson & McCranie, 2001. Senck. Biol. 81(1–2): 271–273, figs. 1–3.

Type: Holotype, SMF 78865, a 210 mm female (Fund. Hector Rodrigo Pastor Fasquelle).
Type locality: "a coffee finca near Buenos Aires, 1250 elevation, Depto. Cortés, Honduras. This locality lies on the southern slopes of the Sierra de Omoa at about 15°30.03' N, 88°11.13' W."
Distribution: Northwestern Honduras (Cortés), 1250–1900 m.
Sources: Townsend, 2006, Townsend et al., 2006 and McCranie, 2011a.

2. *Omoadiphas cannula* McCranie & Cruz-Díaz. 2010. Zootaxa (2690): 54–57, figs. 1–2.

Type: Holotype, UNAH 1740, a 326 mm male (G.A. Cruz, 13 Sept. 1989).
Type locality: "Montaña de Peña Blanca, Sierra de Agalta, 4 km N of Catacamas, 14°53'26"N, 85°53'46"W, 1250 m elev., departamento de Olancho, Honduras."
Distribution: Eastern Honduras (Olancho), 1250 m. Known only from type locality.
Source: McCranie, 2011a.

3. *Omoadiphas texiguatensis* McCranie & Castañeda, 2004b. Proc. Biol. Soc. Washington 117(3): 312–315, figs. 1–3.

Type: Holotype, USNM 559599, a 169 mm female (F.E. Castaneda & J.R. McCranie, 3 Sept. 2003).
Type locality: "approximately 2.5 airline km NNE of La Fortuna, 15°25'49"N, 87°18'32"W 1690 m elev., Cerro Texíguat Wildlife Refuge, Departamento de Yoro, Honduras."
Distribution: Northwestern Honduras (Yoro), 1690 m.
Source: McCranie, 2011a.

OOCATOCHUS Helfenberger, 2001 (Colubridae)

Type species: *Tropidonotus rufodorsatus* Cantor, 1842a.
Distribution: Eastern Asia.
Sources: Stejneger, 1907a, Emelianov, 1929, Shannon, 1956, Stewart, 1970, Bannikov et al., 1977, Sura, 1981a, Wen, 1983, Schulz, 1985a, 1988e, 1996, H.-F. Li et al., 1986, Szyndlar & Hung, 1987, Steehouder, 1992a, Zhao & Adler, 1993, Adnagulov et al., 2000, Helfenberger, 2001, Xu, 2001, Utiger et al., 2002, 2005, Ratnikov, 2004, Burbrink & Lawson, 2007 and San & Lee, 2007.
Remarks: A synonym of *Coronella fide* Zaher et al., 2012.

1. *Oocatochus rufodorsatus* (Cantor, 1842a). Zool. Chusan: pl. 13. (*Tropidonotus rufodorsatus*)

Synonyms: *Ablabes sex-lineatus* H.M.C. Duméril, Bibron & Duméril, 1854a *(nomen incorrigendum), Coronella sexlineata* Jan, 1863a *(nomen corrigendum), Simotes conradi* W.C.H. Peters, 1874c, *Simotes herzi* Boettger, 1886b, and *Elaphe rufodorsata coreensis* G.D. Stewart, 1970 *(nomen nudum).*
Types: Syntypes (2), BMNH 1946.1.7.32–33 (formerly BMNH 1843.7.21.34), two adult females (T.E. Cantor, July 1840–March 1841, via British East India Comp.).
Type locality: "island of Chusan, or Great Chusan, is situated on the East Coast of China, between the 30° and 31° of North latitude and the 122° and 123° of East longitude; separated from the nearest mainland, Keeto Point, by an arm of the sea, about ten miles across"

O

[= Zhoushan Is., Zhoushan Arch., Zhejiang Prov., E China].

Distribution: Eastern Asia. Extreme E Russia, E China (Anhui, Beijing, Fujian, Guangxi, Guangdong, Heilongjiang, Hainan, Hebei, Henan, Hubei, Hunan, Nei Monggol, Jiangsu, Jiangxi, Jilin, Liaoning, Shandong, Shanxi, Tianjin, Zhejiang), ext. SE Russia, North Korea (Kangwon, North Hamgyong, North Pyongan, Pyongyang, South Pyongan) and South Korea (Kyonggi, Seoul), 75–215 m.

Remarks: Supplemental original description reprinted in Cantor (1842b: 483). Record from Taiwan unconfirmed.

OPHEODRYS Fitzinger, 1843 (Colubridae)

Synonyms: *Opheodris* A.-M.-C. Duméril & Bibron, 1844 *(nomen emendatum), Cyclophis* A.C.L.G. Günther, 1858, *Cyclophes* – Higgins, 1873 *(nomen incorrectum), Opheadrys* – W.H. Smith, 1882 *(nomen incorrectum), Phyllophilophis* S.W. Garman, 1884, *Tropidodryas* Cope, 1885a, *Phylophilophis* – O.P. Hay, 1887 *(nomen incorrectum), Cyclophide* – Boettger, 1888d *(nomen incorrectum), Philophyllophis* S.W. Garman, 1892 *(nomen emendatum), Cylophis* – Boulenger, 1896a *(nomen incorrectum), Cyclophys* – K.P. Schmidt & Necker, 1936 *(nomen incorrectum), Ophodryos* – J.A. Burton, 1991 *(nomen incorrectum),* and *Liochlorophis* Oldham & Smith, 1991.

O

Type species: *Coluber aestivus* Linnaeus, 1758.

Distribution: Southern Canada, E USA and ext. N Mexico.

Fossil records: Pleistocene of SE Canada and USA. Fossils unassigned to species include upper Miocene (Clarendonian) of USA (Nebraska), and upper Pleistocene (Rancholabrean II) of USA (Nebraska, Tennessee).

Sources: Cundall, 1981, 1986, Plummer, 1981, 1984, Morris, 1982, Ernst & Barbour, 1989, Conant & Collins, 1991, Holman, 1995a, Palmer & Braswell, 1995, and Tennant, 1997, Holman, 2000a, T. Schmidt, 2002, Ernst & Ernst, 2003 and Parmley & Hunter, 2010.

1. *Opheodrys aestivus* (Linnaeus, 1766). Syst. Nat., ed. 12, 1: 387. (*Coluber aestivus*)

Synonyms: *Coluber subviridis* Lacépède, 1789 (*nomen rejiciendum*), *Leptophis majalis* Baird & Girard, 1853, *Opheodrys aestivus carinatus* Grobman, 1984, and *Opheodrys aestivus conanti* Grobman, 1984.

Type: Neotype, USNM 92473, a male (C.W. Burn, 19 July 1933), designated by Grobman (1984: 161).

Type locality: "one mile west of Parksville, McCormick Co., SC" [South Carolina, USA] via neotype selection.

Distribution: Eastern USA (Alabama, Arkansas, Delaware, Florida, Georgia, S Illinois, S Indiana, SE Kansas, Kentucky, Louisiana, Maryland, Mississippi, Missouri, SE Nebraska, S New Jersey, North Carolina, ext. S Ohio, E Oklahoma, ext. S Pennsylvania, South Carolina, Tennessee, E Texas, Virginia: Assateague, Parramore, Revel and Smith Is., W West Virginia) and NE Mexico (ext. N Coahuila, E Nuevo León, Tamaulipas), NSL–1525 m.

Fossil records: Upper Pleistocene (Rancholabrean II) of USA (Alabama, Florida, Georgia, Pennsylvania, Texas, Virginia).

Sources: Burger, 1947, Christman, 1980, Cundall, 1981, 1986, Plummer, 1981, 1984, Morris, 1982, Grobman, 1984, Ernst & Barbour, 1989, Conant & Collins, 1991, Holman, 1995a, Palmer & Braswell, 1995, Tennant, 1997, Walley & Plummer, 2000, Walley & Janis, 2002 and Klesius & Klesius, 2010.

Remarks: Holotype presumed lost *fide* Grobman (1984: 161). Type locality restriction to Charleston, South Carolina, USA *fide* Schmidt (1953: 191) invalid due to neotype selection.

2. *Opheodrys vernalis* (Harlan, 1827). J. Acad. Nat. Sci. Philadelphia (1826–1827) 5(2): 361. (*Coluber vernalis*)

Synonyms: *Coluber viridis* Bonnaterre, 1790 (*nomen oblitum*), *Coluber cyaneus* G. Shaw, 1802 (*nomen oblitum*), *Coluber viridicaeruleus* Oken, 1836, *Opheodrys vernalis blanchardi* Grobman, 1941, and *Opheodrys vernalis borealis* Grobman, 1992.

Type: Holotype, none designated, ANSP, lost *fide* Grobman (1941: 10).

Type locality: "Pennsylvania and New Jersey" [USA]. Restricted to vicinity of Philadelphia, Pennsylvania, USA *fide* K.P. Schmidt (1953a: 191).

Distribution: Southern Canada (S Manitoba, New Brunswick, Nova Scotia, SE Ontario, Prince Edward Island, ext. S Quebec, ext. SE Saskatchewan, Cape Breton Is.), cen. and NE USA (Colorado, Connecticut, ext. SE Idaho, Illinois, NW Indiana, Iowa, Maine: Applecore and Orr's Is.; W Maryland, Massachusetts, Michigan, Minnesota, N Missouri, possibly ext. E Montana, Nebraska, New Hampshire, N New Jersey, New Mexico, New York, North Dakota, Ohio, Pennsylvania, Rhode Island: Dutch Is.; ext. NE, SE and ext. SW South Dakota, ext. SE Texas, E Utah, Vermont, W Virginia, West Virginia, Wisconsin, E Wyoming) and ext. N Mexico (NW Chihuahua), NSL–2745 m.

Fossil records: Lower Pleistocene (Irvingtonian I) of USA (Florida, Pennsylvania), middle/upper Pleistocene (Irvingtonian II) of USA (Maryland), and upper Pleistocene (Rancholabrean II) of Canada (Ontario) and USA (Indiana, Ohio, Virginia, West Virginia).

Sources: Blanchard, 1933, Grobman, 1941, 1991, 1992, Logier & Toner, 1961, Cook, 1964, 1967, 1977a, Van Devender & Lowe, 1977, Holman, 1981c, Stebbins,

1985, Oldman & Smith, 1991, H.M. Smith et al., 1991, Harding, 1997, Degenhardt et al., 1996, Werler & Dixon, 2000, Hammerson, 1999 and Walley, 2003.

Remarks: In accordance with Art. 23.9.2 of the Code (ICZN, 1999), *Coluber vernalis* Harlan is designated a *nomen protectum* and *Coluber viridis* Bonnaterre and *Coluber cyaneus* G. Shaw a *nomina oblita.*

OPHIOPHAGUS A.C.L.G. Günther, 1864a (Elapidae)

Synonyms: *Hamadryas* Cantor, 1836a *(nomen praeoccupatum), Dendraspis* Fitzinger, 1843 *(nomen rejiciendum), Hamadryus* – Higgins, 1873 *(nomen incorrectum), Ophiofagus* – Canton, 1895 *(nomen incorrectum), Ophiphagus* – Noguchi, 1909 *(nomen incorrectum), Ophiophagous* – Wall, 1925a *(nomen incorrectum),* and *Ophiophaqus* – Murthy, 1986 *(nomen incorrectum).*

Type species: *Hamadryas hannah* Cantor, 1836a.

Distribution: Southern and SE Asia and East Indies.

Sources: Cantor, 1838, Müller & Schlegel, 1844b, Hasselt, 1857, Fayrer, 1872, Beddard, 1904, Rooij, 1917, E.H. Taylor, 1922a, 1965, Wall, 1924c, 1925a, C.H. Pope, 1935, Bourret, 1936a–b, Kopstein, 1936, Herklots, 1938, Bogert, 1943, M.A. Smith, 1943, G.E. Shaw et al., 1946a, Oliver, 1956, Hoesel, 1959, Deraniyagala, 1960b, 1961b, Leviton, 1965b, Campden-Main, 1970a, Deuve, 1970, Saint-Girons, 1972a, Fleming & Fleming, 1973, Soderberg, 1966c, 1973, Biswas, 1975, Biswas et al., 1976, Acharjyo et al., 1978, Romer, 1979a, Biswas & Sanyal, 1980, B.Q. Hu et al., 1980, Acharjyo & Murthy, 1983, Daniel, 1983, Tweedie, 1983, Murthy, 1985, 1986, Tian et al., 1986, Zhao et al., 1986, Murthy & Acharjyo, 1987, M.A.R. Khan, 1988, Zhong & Zhou, 1988, Narayan & Rosalind, 1990, Zhong, 1990, Ross & Lazell, 1991, Stuebing, 1991, Golay et al., 1993, Zhao & Adler, 1993, Jintakune & Chanhome, 1995, Das & Whitaker, 1996, David & Vogel, 1996, M.J. Cox et al., 1998, Keogh, 1998, David & Ineich, 1999, Slowinski & Keogh, 2000, Orlov et al., 2000, Leviton et al., 2003, J.C. Murphy & Schlager, 2003, Tiwari & Shah, 2004, Whitaker & Captain, 2004, N. Khaire, 2006, McKay, 2006, Zhao, 2006, Castoe et al., 2007, M.F. Ahmed et al., 2009, V.S. Nguyen et al., 2009, Zaher et al., 2009, I. Das, 2010, 2012, Hoser, 2012e, Koch, 2012, Subba-Rao, 2012 and S.K. Sharma et al., 2013.

Remarks: Official Generic Name no. 1599 *fide* Opinion 709 (ICZN, 1964b).

1. *Ophiophagus hannah* (Cantor, 1836a). Asiatic Res. 19(1): 87–91, pl. 10, figs. a–b, pl. 11, figs. a–d, pl. 12, figs. a–f. (*Hamadryas hannah*)

Synonyms: *Naja bungaroides* H. Boie *in* Schlegel, 1828 *(nomen oblitum), Hamadryas elaps* Lesson, 1830 *in* Duperrey, 1826–1832, *Naja bungarus* Schlegel, 1837, *Hamadryas ophiophagus* Cantor, 1838, *Naja vittata* Elliott, 1840, *Dimeresurus boiei* Bleeker, 1857b *(nomen nudum), Naja fasciata* W.C.H. Peters, 1861c, *Naja ingens* Hassett, 1882b, *Naja hamadryad* Purananda, 1956 (*in* Buckley & Porges), *Dendraspis hannah borneensis* Deraniyagala, 1960b, *Dendraspis hannah sinensis* Deraniyagala, 1960b, *Dendraspis hannah brunnea* Deraniyagala, 1961b, *Dendraspis hannah nordicus* Deraniyagala, 1961b, *Ophiophagus hanna* – R.L. Irwin, Oliver, Mohamed & Haast, 1970 *(nomen incorrectum), Ophiaphagus hannah* – J. White, 1981 *(nomen incorrectum), Ophiophagua hannah* – Jadhav & Kapre, 1991 *(nomen incorrectum),* and *Ophiphagus hannah* – Khole, 1991 *(nomen incorrectum).*

Type: Holotype, BMNH 1996.451 (formerly RSL), a 2433–2990 mm dry specimen (? skin) (T.E. Cantor, 1834–1836, via British East India Comp.).

Type locality: "in the Sunderbuns" and "not far from Calcutta" [= Sundarbans, Ganges River delta, SW Bangladesh and SE West Bengal State, NE India, bet. 21.5–23°N and 88–91.5°E and Kolkata, S West Bengal State, NE India, 22°34'N, 88°22'E, elevation 10 m]. Emended to Bengal *fide* Cantor (1838: 73).

Distribution: Northern Pakistan (NE Punjab), India (Andaman & Nicobars: Andaman Is., Andhra Pradesh, Arunachal Pradesh, Assam, Bihar, Gujarat, Karnataka, Kerala, Maharasktra, Manipur, Meghalaya, Mizoram, Nagaland, Orissa, Sikkim, Tamil Nadu, Uttarakhand, Uttar Pradesh, West Bengal), Nepal (Bardiya, Chitwan, Dolakha, Jhapa, Kaski, Kathmandu, Lamjung, Morang, Parsa, Rautahat, Saptari, Sunsari,), E Bhutan, Bangladesh, Myanmar (Ayeyarwady, Bago, Chin, Kachin, Karen, Magwe, Mandalay, Mon, Sagaing, Shan, Tanintharyi, Yangon), S China (Fujian, Guangdong, Guangxi, Guizhou, Hainan, Hong Kong, Hunan, Jiangxi, Sichuan, SE Xizang, Yunnan, Zhejiang), Vietnam (most of the country), Laos (Champasak, Khammouan, Salavan, Vientiane, Xiangkhouang), Cambodia (Kampong Cham, Siem Reap), Thailand (Bangkok, Chachoengsao, Chaiyaphum, Chanthaburi, Chiang Mai, Chon Buri, Kanchanaburi, Krung Thep Mahanakhon, Loei, Lop Buri, Nakhon Phathom, Nakhon Ratchasima, Nakhon Si Thammarat, Nartahiwat, Phang Nga, Phatthalung, Phetchabun, Prachuab Khiri Khan, Puttani, Ratchaburi, Rayong, Samut Prakan, Samut Sakhon, Saraburi, Songkhla, Surin, Tak, Trang, Ubon Ratchasima, Uthai Thani, Yala, Koh Lak Is.), West Malaysia (Johor, Kedah, Kelantan, Melaka, Negeri Sembilan, Perak, Perlis, Pinang, Tioman Is.), Singapore, Brunei, East Malaysia (Sabah, Sarawak), Indonesia (Bali, Bangka, Belitung, Galang, Java, Kalimantan, Nias, Peleng, Simeulue, Sulawesi, Sumatra) and Philippines (Balabac, Dinagat, Jolo, Luzon, Lubang, Mindanao, Mindoro, Negros, Palawan), NSL–2180 m.

Remarks: Official Specific Name no. 990 *fide* Opinion 709 (ICZN, 1964). Original description reprinted in Cantor (1838: 73–75). Holotype, colored sketches and MS of Cantor sent to RSL in 1839. In accordance with

Art. 23.9.2 of the Code (ICZN, 1999), *Hamadryas hannah* Cantor is designated a *nomen protectum* and *Naja bungaroides* H. Boie a *nomen oblitum*.

OPHRYACUS Cope, 1887c (Viperidae)

Synonym: *Atropus* A.C.L.G. Günther, 1895 *in* 1885–1902 *(nomen emendatum)*.
Type species: *Trigonocephalus undulatus* Jan, 1859b.
Distribution: Southern Mexico.
Sources: Ferrari-Pérez, 1886, A.C.L.G. Günther, 1895 *in* 1885–1902, Mocquard, 1909, Martin del Campo, 1935, H.M. Smith, 1941i, H.M. Smith & Taylor, 1945, Burger, 1971, J.A. Campbell & Lamar, 1989, 2004, Pérez-Higareda & Smith, 1991, Werman, 1992, Crother et al., 1992, David & Ineich, 1999, McDiarmid et al., 1999, Parkinson, 1999, Taggart et al., 2001, Castoe & Parkinson, 2006 and Hoser, 2012d.

1. *Ophryacus undulatus* (Jan, 1859b). Rev. Mag. Zool. (2) 11: 155, 157. (*Trigonocephalus* [*Atropos*] *undulatus*)

Synonyms: *Bothrops undulata* – Amaral, 1930f and *Bothrops sphenophrys* H.M. Smith, 1960.
Types: Syntypes (2), formerly MSNM, longest syntype 560 mm, destroyed in 1943 during World War II.
Type locality: "Mexique" [= Mexico]. Restricted (by inference) to Orizaba, Veracruz, Mexico *fide* H.M. Smith & Taylor (1950a: 350).
Distribution: Southern Mexico (cen. Guerrero, Hidalgo, Oaxaca, Puebla, W Veracruz), 1490–2800 m.
Sources: Canseco-Marquez et al., 2000 and Mendoza-Paz et al., 2006.
Remarks: Supplemental original description in Jan (1859e: pl. e, fig. no. 6).

OPISTHOTROPIS A.C.L.G. Günther, 1872b (Natricidae)

Synonyms: *Opisthotropos* – E. Bartlett, 1875 *(nomen incorrectum)*, *Calamohydrus* Boulenger, 1888c, *Helicopsoides* Mocquard, 1890a–b, *Lepidognathus* Lidth de Jeude, 1891, *Trimerodytes* Cope, 1895a, *Opithotropis* – Palacky, 1898 *(nomen incorrectum)*, *Tapinophis* Boulenger, 1899b, *Liparophis* Peracca, 1904b, *Cantonophis* F. Werner, 1909d, *Opisthotrophis* – Bellairs, 1969 *(nomen incorrectum)*, and *Opistotrophis* – Engelmann & Obst, 1981 *(nomen incorrectum)*.
Type species: *Opisthotropis ater* A.C.L.G. Günther, 1872b.
Distribution: Eastern and SE Asia and East Indies.
Sources: Boulenger, 1891b, C.H. Pope, 1935, Bourret, 1936a–b, M.A. Smith, 1943, C. Haas, 1950, W.C. Brown & Leviton, 1961, E.H. Taylor, 1965, B.Q. Hu et al., 1980, Tian et al., 1986, Ji, 1992, Zhao & Adler, 1993, Orlov et al., 1998, Zhao, 2006, J.C. Murphy et al., 2008, Ziegler et al., 2008, Zaher et al., 2009, David et al., 2011a and Yang et al., 2011.
Remarks: J.C. Murphy et al. (2008: 177) transferred *O. praemaxillaris* back to *Paratapinophis*.

1. *Opisthotropis alcalai* W.C. Brown & Leviton, 1961. Occ. Pap. Nat. Hist. Mus. Stanford Univ. (8): 2–4, fig. 1.

Type: Holotype, CAS-SU 22250 (formerly SU 22250), a 532 mm male (A.C. Alcala & field crew, 28 April 1959).
Type locality: "Cugat Creek, on the west side of Dapitan Peak, Mt. Malindang, Zamboanga del Norte Province, Mindanao Island, at an altitude of 2500 feet, in the Philippine Islands" [= ext. N Zambonga del Norte Province, W Mindanao, SE Philippines, 8°13'N, 123°38'E].
Distribution: Southern Philippines (W Mindanao), 760 m. Known only from type locality.
Source: Alcala, 1986.

2. *Opisthotropis andersonii* (Boulenger, 1888c). Ann. Mag. Nat. Hist. (6) 2(7): 44. (*Calamohydrus andersonii*)

Type: Holotype, BMNH 1946.1.5.26, a 240–245 mm male (C. Ford).
Type locality: "Hongkong" [China].
Distribution: China (Hong Kong) and Vietnam (Hai Phong, Quang Ninh, Baitsilong Arch.: Cu Lao Ba Mun Is.), NSL–300 m.
Sources: Boulenger, 1893a, Wall, 1903, Romer, 1952, 1979a, Karsen et al., 1986, Darevsky, 1999, V.S. Nguyen et al., 2009 and I. Das, 2010, 2012.

3. *Opisthotropis atra* A.C.L.G. Günther, 1872b. Ann. Mag. Nat. Hist. (4) 9(49): 16, pl. 3, fig. b. (*Opisthotropis ater*)

Type: Holotype, BMNH 1946.1.15.33, a 339–340 mm male (H.H. Higgins, 1856–1872).
Type locality: "West Africa," (in error *fide* Boulenger, 1914c: 483, and C.H. Pope, 1935: 164).
Distribution: Unknown. Known only from holotype.
Source: Boulenger, 1893a.

4. *Opisthotropis balteata* (Cope, 1895a). Proc. Acad. Nat. Sci. Philadelphia (1894) 46: 426, pl. 10, figs. 2a–d, 2f. (*Trimerodytes balteatus*)

Synonyms: *Liparophis bedoti* Peracca, 1904b and *Opisthotropis multicinctata* Fan, 1931.
Type: Holotype, ANSP 11435, a 377 mm male (F.P. Gilman, 1886–1894).
Type locality: "Hainan" [= Hainan Is., Hainan Prov., ext. SE China].

O

Distribution: Southern China (Guangdong, Guangxi, Hainan, Hong Kong) and N Vietnam (Hai Phong), NSL–350 m.
Sources: Peracca, 1904b, Steindachner, 1906, K.P. Schmidt, 1927a, Fan, 1931, Romer, 1979a, Karsen et al., 1986, 1998, V.S. Nguyen et al., 2009 and I. Das, 2010, 2012.
Remarks: Malnate (1971: 363) listed the collector of type as F. Gillman and original plate as pl. 10, fig. 1. M.A. Smith (1943: 331) and W.C. Brown & Leviton (1961: 4) erroneously listed the type in MCZ. Cambodia record rejected *fide* Saint-Girons (1972a: 73).

5. *Opisthotropis cheni* Zhao, 1999. Sichuan J. Zool. 18(3): 96.

Type: Holotype, CIB 80029 (formerly MFB 9339), a 455 mm male (Y.-H. Chen, July 1993).
Type locality: "near Workers and Staffs Hospital, Mt. Mang, Yizhang Co., Hunan Province, China; 480 meters" [= 25°00'N, 112°50'E].
Distribution: Southern China (N Guangdong, S Hunan), 480–1300 m.
Sources: Zhao, 2004, 2006 and Li et al., 2010.
Remarks: Supplemental original description in Zhao (2004: 342).

6. *Opisthotropis cucae* David, Pham, Nguyen & Ziegler, 2011. Zootaxa (2758): 44–47, figs. 1–2, 3a–d.

Type: Holotype, IEBR A.924, a 464 mm female (T.C. Ho & T.C. Pham, 22 May 2009).
Type locality: "Chu Mom Ray National Park, Sa Thay District, Kon Tum Province (14°26.227' N, 107°43.268'E), 740 m asl, Vietnam."
Distribution: Central Vietnam (Kon Tum), 740 m. Known only from type locality.

7. *Opisthotropis daovantieni* Orlov, Darevsky & Murphy, 1998. Russ. J. Herp. 5(1): 61–62, figs. 1–2.

Type: Holotype, FMNH 252009, a 510 mm female (I.S. Darevsky & N.L. Orlov, 25 June 1995).
Type locality: "Buoenloy, Ankhe District, Gia Lai Province, southern Vietnam (14°20'N, 108°36'E), 750m."
Distribution: Southern Vietnam (Gia Lai, Kon Tum), 750 m.
Sources: V.S. Nguyen et al., 2009 and I. Das, 2010, 2012.

8. *Opisthotropis guangxiensis* Zhao, Jiang & Huang, 1978. Mater. Herpetol. Res. 4: 21.

Synonym: *Opisthotropis yaoshanensis* Huang & Zheng *in* Huang, Zheng, & Fang, 1978.
Type: Holotype, CIB 602488, a 455 mm male (native, 19 June 1960).
Type locality: "Mt. Yao, Guangxi Zhuang Autonomous Region, China."
Distribution: Southern China (Guangxi, Hunan), 950–1220 m.
Sources: Zhao et al., 1979, P. Guo et al., 1999b and Zhao, 2006.
Remarks: Supplemental original description in Zhao et al. (1979: 313, fig. 1).

9. *Opisthotropis jacobi* Angel & Bourret, 1933. Bull. Soc. Zool. France 58(3–4): 130–131.

Types: Syntypes (2), MNHN 1933.10, a 475 mm male (R.L. Bourret, 1930–1931), location of other syntype unknown.
Type locality: "Chapa, station de villégiature située au Tonkin, dans la province de Lao Kay, près de la frontière du Yunnan, à une altitude moyenne de 1.500 mètres" [= Sa Pa, Sa Pa Distr., W Lao Cai Prov., ext. N Vietnam, 22°20'N, 103°51'E, elevation 1500 m].
Distribution: Southern China (Yunnan) and N Vietnam (Cao Bang, Bac Kan, Ha Tay, Lang So, Lao Cai, Tuyen Quang, Vinh Phu), 900–1600 m.
Sources: Tian et al., 1986, V.S. Nguyen et al., 2009 and I. Das, 2010.

10. *Opisthotropis kikuzatoi* (Okada & Takara, 1958). Bull. Biogeogr. Soc. Japan 20(3): 1–3, figs. 1a–c. (*Liopeltis kikuzatoi*)

Type: Holotype, URJ 1, a 533 mm male (K. Kikuzato, 15 Sept. 1956).
Type locality: "Kume-jima of Okinawa-group, Ryukyu Islands" [Japan]. Emended to the upper reaches of Shirasegawa River, at a point 1,000 meters from the summit of Mt. Uegusuku-dake (309 m), Kume-jima Island, Ryukyu Archipelago *fide* Toyama (1983: 34).
Distribution: Extreme S Japan (Ryukyus: Kumejima Is.), 10–1000 m.
Sources: Takara, 1962, Toyama, 1983, M. Mori, 1984, Ota & Mori, 1985b and Ota, 1993.

11. *Opisthotropis kuatunensis* C.H. Pope, 1928a. Amer. Mus. Novit. (320): 5.

Type: Holotype, AMNH 34437, a 666 mm female (C.H. Pope [Third AMNH Asiat. Exped.], April–Sept. 1926).
Type locality: "Kuatun, a village in the mountains of Ch'ungan Hsien, northwest Fukien Province, China" [= Guadun, in Wuyishan Mtns., ext. N Fujian Prov., SE China, 27°45'N, 117°50'E, elevation 900–1200 m].
Distribution: Southeastern China (Fujian, Guangxi, Hong Kong, Jiangxi, Zhejiang), 200–1200 m.
Sources: C.H. Pope, 1929, Reitinger & Lee, 1978, Romer, 1978, 1979a, Wen, 1983, Zhong, 1984, Karsen et al., 1986, Tian et al., 1986 and Zhao, 2006 .
Remarks: Supplemental original description in C.H. Pope (1929: 423–425, fig. 10).

12. *Opisthotropis lateralis* Boulenger, 1903b. Ann. Mag. Nat. Hist. (7) 12(69): 350–351.

Synonym: *Tapinophis shini* Mell, 1930.
Type: Holotype, BMNH 1946.1.13.21, a 360 mm male (H. Fruhstorfer, June 1990–April 1903).
Type locality: "the Man-Son Mountains, Tonkin, altitude 3000 to 4000 feet" [= Mau Son Mtns., SE Lang Son Prov., ext. NE Vietnam, 21°51'N, 106°58'E, elevation 900–1200 m].
Distribution: Southern China (Guangxi, Guizhou, Hong Kong, Hunan) and N Vietnam (Bac Giang, Cao Bang, Hoa Binh, Lang Son, Quang Ninh, Vinh Phuc), 100–1200 m.
Sources: Bourret, 1939d, Reitinger et al., 1978, Romer, 1979a, Karsen et al., 1986, Zhao, 2006, V.S. Nguyen et al., 2009 and I. Das, 2010, 2012.

13. *Opisthotropis latouchii* (Boulenger, 1899b). Proc. Zool. Soc. London 67(1): 164, pl. 18, figs. 1, 1a–1c. (*Tapinophis latouchii*)

Synonym: *Cantonophis praefrontalis* F. Werner, 1909g.
Type: Holotype, BMNH 1946.1.15.37, a 455 mm female (J.D. La Touche, spring 1896 or 1898).
Type locality: "Kuatun, a village about 270 miles from Foochow, in the mountains at the North-west of the Province of Fokien, at an altitude of 3000 to 4000 feet or more, China" [= Guadun, ext. N Fujian, SE China, 27°45'N, 117°50'E, elevation 900–1200 m].
Distribution: Southeastern China (Anhui, Fujian, Guangdong, Guangxi, Guizhou, Hunan, Jiangxi, Sichuan, Zhejiang), 100–1425 m.
Sources: F. Werner, 1909d, C.H. Pope, 1929, Gressitt, 1941a, Tian et al., 1986 and B.H. Li, 1987.

14. *Opisthotropis maculosa* B.L. Stuart & Chuaynkern, 2007. Curr. Herp. 26(1): 37–39, figs. 3–5. (*Opisthotropis maculosus*)

Type: Holotype, THNHM 5343 (formerly FMNH 265798), a 520 mm male (Y. Chuaynkern, C. Chuechat, S. Makchai & B.L. Stuart, 7 Sept. 2004).
Type locality: "Thailand, Nong Khai Province, Boong Klar District, Phu Wua Wildlife Sanctuary, at marker 027 on footpath to Lad Phleuy Waterfall, 18°14'31.0"N 103°57'44.5"E, 190 m asl."
Distribution: Northeastern Thailand (Nong Khai) and ext. S China (Guangdong, Guangxi), 190–1150 m.
Sources: I. Das, 2010, 2012 and Yang et al., 2011.
Remarks: Probably occurs in N Laos and N Vietnam *fide* Yang et al. (2011: 67).

15. *Opisthotropis maxwelli* Boulenger, 1914c. Ann. Mag. Nat. Hist. (8) 14(84): 483.

Type: Holotype, BMNH 1946.1.13.22, a 305 mm female (J.P. Maxwell, 1899–1914).
Type locality: "South Fokien, China" [= S Fujian Prov., SE China].
Distribution: Southeastern China (Fujian, Guangdong, Guangxi, Jiangxi), 550–900 m.
Sources: Gressitt, 1941a and Y. Zhang & Tang, 1990.

16. *Opisthotropis rugosa* (Lidth de Jeude, 1890). Zool. Ergebn. Nieder. Ost-Indien 1(2): 187–192, pl. 16, figs. 1–5. (*Lepidognathus rugosus*)

Type: Holotype, RMNH 4330, a 473 mm specimen (M. Weber, Mar.–Nov. 1888).
Type locality: "Sumatra: Kaju tanam" [= Kayu Tanam, Sumatra, W Indonesia, 0°32'S, 100°20'E, elevation 200 m].
Distribution: Western Indonesia (Sumatra), 200 m. Known only from the holotype.
Sources: Rooij, 1917, C. Haas, 1950 and David & Vogel, 1996.

17. *Opisthotropis spenceri* M.A. Smith, 1918. J. Nat. Hist. Soc. Siam 3(1): 13, pl. 2.

Type: Holotype, BMNH 1946.1.15.31 (formerly MAS 1178), a 600 mm female (F.A. Spencer, Sept. 1917).
Type locality: "Muang Ngow, in a hill stream at about 300 metres elevation, N. Siam" [= Muang Ngao, Lampang Prov., NW Thailand, 18°46'N, 99°59'E, elevation 300 m].
Distribution: Northwestern Thailand (Lampang), 300 m. Known only from type locality.
Source: E.H. Taylor, 1965.

18. *Opisthotropis tamdaoensis* Ziegler, David & Vu, 2008. Zoosyst. Evol. 84(2): 198–203, figs. 1a–d, 2.

Type: Holotype, VNUH 10606, a 555 mm male (native, June 2006).
Type locality: "Bac (Silver) stream, Tam Dao, Vinh Phuc Province, northern Vietnam, ca. 750 m above sea level."
Distribution: Northern Vietnam (Vinh Phuc), 750 m. Known only from type locality.

19. *Opisthotropis typica* (Mocquard, 1890a). Naturaliste (2) 12(80): 154. (*Helicopsoides typicus*)

Type: Holotype, MNHN 1889.216, a 390 mm female (J. Whitehead, 1885).
Type locality: "Kina Balu, Bornéo." Emended to North Borneo, Mt. Kina Balu, between 1,000 and 4,000 or even 6,000 feet *fide* Mocquard (1890b: 118), [= Mt. Kinabalu, N Sabah, NE East Malaysia, 6°05'N, 116°30'E, elevation 300–1800 m].
Distribution: Northeastern East Malaysia (N Sabah, Sarawak), Brunei and ext. SW Philippines (Palawan), 75–900 m.

Sources: Mocquard, 1890b, Alcala, 1986, Stuebing, 1991, Mori, 1993, Stuebing & Inger, 1998, Malkmus et al., 2002 and I. Das, 2007b, 2010, 2012.

Remarks: Supplemental original description in Mocquard (1890b: 141–142, pl. 9, figs. 3, 3a–3c).

OREOCALAMUS Boulenger, 1899 (Calamariidae)

Type species: *Oreocalamus hanitschi* Boulenger, 1899c.

Distribution: Malaysia.

Sources: M.A. Smith, 1931b, C. Haas, 1950, Tweedie, 1983, Stuebing, 1991 and Malkmus et al., 200f2.

1. *Oreocalamus hanitschi* Boulenger, 1899c. Ann. Mag. Nat. Hist. (7) 4(24): 453.

Type: Holotype, BMNH 1946.1.7.8 (formerly RM), a 375 mm male (R. Hanitsch, March 1899).

Type locality: "Mount Kina Balu, North Borneo, 4200 feet" [= Mt. Kinabalu, N Sabah, NE East Malaysia, 6°05'N, 116°30'E, 1280 m].

Distribution: Central West Malaysia (NW Pahang) and East Malaysia (Sabah, Sarawak), 1000–1800 m.

Remarks: Original description repeated in Boulenger *in* Hanitsch (1900: 73).

OREOCRYPTOPHIS Utiger, Schätti, & Helfenberger, 2005 (Colubridae)

Synonym: *Oreophis* Utiger, Helfenberger, Schätti, Schmidt, Ruf & Ziswiler, 2002 (*nomen praeoccupatum*)

Type species: *Coluber porphyraceus* Cantor, 1839b.

Distribution: Southeast Asia and ext. W Indonesia.

Sources: Hubrecht, 1886, K.P. Schmidt, 1927, Rendahl, 1937, Kuntz, 1963, E.H. Taylor, 1965, Deuve, 1970, Tweedie, 1983, Mahendra, 1984, Schulz, 1989c, 1996, Zhao & Adler, 1993, David & Vogel, 1996, M.J. Cox et al., 1998, Schulz & Helfenberger, 1998, Chan-ard et al., 1999, Orlov et al., 2000, Lenk et al., 2001a, Utiger et al., 2002, 2005, Tiwari & Shah, 2004, Whitaker & Captain, 2004, Zhao, 2006, Burbrink & Lawson, 2007, Teynié & David, 2007, V.S. Nguyen et al., 2009, Xiang & Li, 2009, I. Das, 2010, 2012, Wangyal, 2011 and Yao, 2012.

1. *Oreocryptophis porphyraceus* (Cantor, 1839b). Proc. Zool. Soc. London 7(1): 51. (*Coluber porphyraceus*)

Synonyms: *Psammophis nigrofasciatus* Cantor, 1839b, *Coronella callicephalus* Gray, 1853, *Simotes vaillanti* Sauvage, 1877, *Liopeltis kawakamii* Oshima, 1910, *Elaphe porphyracea pulchra* K.P. Schmidt, 1925b, *Elaphe porphyracea hainana* Mell, 1931a, *Elaphe porphyracea sikiangensis* Mell, 1931a, *Elaphe porphyracea longilineata* Bourret, 1934f, *Elaphe porphyracea coxi* Schulz & Helfenberger, 1998, and *Elaphe porphyracea laticincta* Schulz & Helfenberger, 1998.

Type: Holotype, BMNH 1946.1.6.22 (T.E. Cantor, 1834–1839, via Brit. East India Comp.).

Type locality: "Míshmee Hills, Asám" [= Mishmi Hills, NE Arunachal Pradesh Prov., NE India, ca. 28°22'N, 95°48'E].

Distribution: Southeastern Asia. Northeastern India (Arunachal Pradesh, Assam, Manipur, Meghalaya, Nagaland, Sikkim, N West Bengal), Nepal (Ilam, Kaski), Bhutan, China (Anhui, Fujian, S Gansu, Guangdong, Guangxi, Guizhou, Hainan, Henan, Hong Kong, Hubei, Hunan, Jiangsu, Jiangxi, Shaanxi, Sichuan, Xizang, Yunnan, Zhejiang), Taiwan, N Myanmar (Chin, Kachin, Karem, Mandalay, Sagaing, Shan), NW Thailand (Chang Mai, Chiang Rai, Loei, Mae Hong Son), N Laos (Bolikhamsai, Champasak, Houaphan, Xiangkhouang), N Vietnam (Bac Kan, Cao Bang, Lang Son, Lao Cai, Quang Binh, Son La, Thai Nguyen, Vinh Phú), West Malaysia (Pahang, Perak), Singapore and ext. W Indonesia (Sumatra), 115–2600 m

Remarks: BMNH 1946.1.14.69 from Singapore listed as a type on website. Record from the Andaman Islands doubtful *fide* Schulz (1996: 190).

ORIENTOCOLUBER Kharin, 2011 (Colubridae)

Synonym: *Yeomansus* Hoser, 2012ak (*nomen illegitimum*).

Type species: *Masticophis spinalis* W.C.H. Peters, 1866a.

Distribution: Eastern Asia.

Sources: Stejneger, 1907, Emelianov, 1929, C.H. Pope, 1929, 1935, Maki, 1931, Shannon, 1956, Bannikov et al., 1977, Szyndlar & Hung, 1987, Kudrjavtsev & Mamet ,1989a, Borkin et al., 1990a–b, Zhao & Adler, 1993, Zou & Chen, 1998, Xu, 2001, Zhao, 2006, San & Lee, 200,7 Kharin, 2011 and Yao, 2012.

1. *Orientocoluber spinalis* (W.C.H. Peters, 1866a). Mber. Königl. Akad. Wiss. Berlin 1866(2): 91–92. (*Masticophis spinalis*)

Synonyms: *Zamenis bocourti* Jan, 1866 *in* Jan & Sordelli, 1860–1866, and *Zamenis cataphoronotus* Jan, 1867 *in* Jan & Sordelli, 1866–1870.

Type: Holotype, ZMB 5367, an 835 mm specimen (Wessel).

Type locality: "Mexico" (in error). Corrected to eastern Asia *fide* Stejneger (1907: 351).

Distribution: Eastern Asia. Eastern Kazakhstan, S Mongolia (Bayanhongor, S Dornogaovi, Dundgovi, S Govi-Altay, Omnogovi, Ovorhangay), N China (Anhui, Gansu, Hebei, Heilongjiang, Henan, Nei

Monggol, Ningxia, Jiangsu, Jilin, Liaoning, Shaanxi, Shandong, Shanxi, Xinjiang), North Korea (Kangwon, North Hamgyong, North Hwanghae, Pyongyang, South Hwanghae) and South Korea (Seoul, South Chungchong, Cheju Is.), 1245–1900 m.

Remarks: Photograph of holotype in Kharin (2011: fig. 2).

ORTHRIOPHIS Utiger, Helfenberger, Schätti, Schmidt, Ruf & Ziswiler, 2002 (Colubridae)

Synonym: *Martinekea* Hoser, 2012m *(nomen illegitimum)*.

Type species: *Elaphe taeniurus* Cope, 1861c.

Distribution: Southeastern Asia and Greater Sundas.

Sources: Zhao & Adler, 1993, Schulz, 1996, Utiger et al., 2002, 2005 and Burbrink & Lawson, 2007.

1. *Orthriophis cantoris* (Boulenger, 1894a). Cat. Snakes Brit. Mus. 2: 35. (*Coluber cantoris*)

Synonym: *Coluber reticularis* Cantor, 1839b.

Type: Lectotype, a 1200 mm specimen illustrated in colored sketch of T.E. Cantor (1834–1837) in RSL, designated by M.A. Smith (1943: 152).

Type locality: "Chirra Púnji, India" [= Cherrapunji, East Khasi Hills, Meghalaya State, NE India, 25°18'N, 91°42'E, elevation 1485 m] via lectotype selection.

Distribution: Southern Asia. Northeastern India (Arunachal Pradesh, Meghalaya, Nagaland, Sikkim, N West Bengal), E Nepal (Ilam), Bhutan and N Myanmar (Kachin), 700–2450 m.

Sources: A.C.L.G. Günther, 1864a, M.A. Smith, 1943, L. Swan & Leviton,1962, Schulz, 1990b, Helfenberger et al., 2000, Schleich & Kästle, 2002, Tiwari & Shah, 2004, Whitaker & Captain, 2004, Agarwal et al., 2010, I. Das, 2010, 2012 and Wangyal, 2011.

2. *Orthriophis hodgsonii* (A.C.L.G. Günther, 1860c). Proc. Zool. Soc. London 28(1): 156, pl. 27. (*Spilotes hodgsonii*)

Type: Lectotype, BMNH 1946.1.6.15, a 1295 mm female (H.R.A. von Schlagintweit, A. von Schlagintweit, & R. von Schlagintweit, 1854–1858), designated by Kramer (1977: 739).

Type locality: "Tsomoriri, Ladak, Tibet" [= Lake Moriri, Ladak, ext. NW India, 32°53'N, 78°21'E, elevation 4635 m] via lectotype selection.

Distribution: Southern Asia. Southwestern China (Sichuan, Xizang), ext. N India (N Bihar, Himachal Pradesh, Jammu Kashmir, Meghalaya, N Punjab, Sikkim, Uttaranchal, N Uttar Pradesh, N West Bengal) and Nepal (Baitadi, Bajhang, Dadeldhura, Dolakha, Dolpa, Jhapa, Ilam, Kaski, Kathmandu, Makwanpur, Palpa, Panchthar, Rasuwa, Solukhumbu), 1000–5000 m.

Sources: M.A. Smith & Battersby, 1953, L. Swan & Leviton, 1962, Kramer, 1977, S.Q. Li, 1983, Zhao & Li, 1985, Schulz, 1986c, 1988j, Nanhoe & Ouboter, 1987, Zhao & Adler, 1993, Schleich & Kästle, 2002, Tiwari & Shah, 2004, Whitaker & Captain, 2004 and S.K. Sharma et al., 2013.

Remarks: Type locality restricted to Nepal *fide* M.A. Smith (1943: 152) but lectotype designation via type locality restriction invalid because two syntypes exist.

3. *Orthriophis moellendorffi* (Boettger, 1886b). Zool. Anz. 9(231): 520–521. (*Cynophis moellendorffi*)

Synonyms: *Elaphis rhombifer* Jan, 1863b and *Elaphe moellendorffii tonkinensis* Bourret, 1934f.

Type: Lectotype, SMF 18476 (formerly SMF-B 8056, 1a), a 1655 mm male (O. Herz, 1886), designated by Mertens (1922: 180).

Type locality: "Nan-ning am Ju-tshang, Prov. Guanghsi" [= Jutshang River near Nanning, S Guangxi Prov., SW China, 22°49'N, 108°22'E, elevation 85 m] via lectotype selection.

Distribution: Southern China (Guangxi, Guangdong, Hainan) and N Vietnam (Bac Kan, Ha Tinh, Hoa Binh, Lang Son, Nghe An, Ninh Binh, Quang Binh, Quang Tai, Son La, Thai Nguyen, Tuyen Quang, Vinh Phuc), 50–500 m.

Sources: Boettger, 1886b, Mell, 1922, 1931a, Bourret, 1936b, Romer, 1953a, Schulz, 1988b, 1989e, Zhao & Adler, 1993, Orlov et al., 2000, Gumprecht, 2004a, V.S. Nguyen et al., 2009 and I. Das, 2010, 2012.

Remarks: Supplemental original description in Boettger (1888c: 72–74, pl. 1, figs. 1a–e).

4. *Orthriophis taeniurus* (Cope, 1861c). Proc. Acad. Nat. Sci. Philadelphia (1860) 12(12): 565–566. (*Elaphe taeniurus*)

Synonyms: *Coluber nuthalli* Theobald, 1868b, *Elaphis yunnanensis* J. Anderson, 1879, *Elaphis grabowskyi* J.G. Fischer, 1885a, *Coluber schmackeri* Boettger, 1895c, *Coluber taeniurus ridleyi* A. Butler, 1899b, *Coluber vaillanti* Mocquard, 1905a, *Coluber friesi* F. Werner, 1926, *Coluber taeniurus anderssoni* Rendahl, 1937, *Coluber taeniurus pallidus* Rendahl, 1937, *Elaphe formosa* Bellairs, 1969 (*nomen nudum*), and *Elaphe taeniura mocquardi* Schulz, 1996.

Type: Lectotype, ANSP 3840, a 1626 mm male (D.B. McCartee, 1844–1860), designated by Stejneger (1907a: 319).

Type locality: "Ningpo, China" [= Ningbo, NE Zhejiang, E China, 29°52'N, 121°32'E, elevation NSL] via lectotype selection.

Distribution: Eastern Asia and Indonesia. Extreme SE Russia, China (Anhui, Beijing, Fujian, Gansu, Guangdong, Guangxi, Guizhou, Hainan, Hebei, Henan, Hubei, Hunan, Jiangsu, Jiangxi, Liaoning, Shaanxi, Shanxi, Sichuan, Tianjin, Yunnan, Xizang,

Zhejiang), ext. S Japan (Ryukyus: Haterumajima, Ikemajima, Irabujima, Ishigakijima, Iriomotejima, Kohamajima, Kurimajima, Kuroshima, Miyakojima, Nakanogamijima, Shimojishima, Taketomijima, Taramajima and Yonagunijima Is.), NE India (Arunachal Pradesh, Sikkim, West Bengal), Bhutan, Myanmar (Chin, Kachin, Tanintharyi), Thailand (Chanthaburi, Chiang Mai, Loei, Pattani, Phetchaburi, Prachin Buri, Prachuap Khiri Khan, Ranong, Trang, Yala), N Laos, N Vietnam (Bac Kan, Cao Bang, Gia Lai, Hai Duong, Ha Tinh, Kon Tum, Lao Cai, Quang Binh, Son La, Thai Nguyen, Thua Thien-Hue, Vinh Phuc), West Malaysia (Johor, Kedah, Pahang, Perak, Perlis, Selangor, Seribuat Arch.: Palau Tioman), Singapore, East Malaysia (Sabah, Sarawak), Brunei and Indonesia (Kalimantan, Sumatra), NSL–3100 m.

Sources: A. Butler, 1899a–b, Mocquard, 1905d, Stejneger, 1907a, Kuntz, 1963, E.H. Taylor, 1965, Bannikov et al., 1977, Tweedie, 1983, M. Mori, 1984, Schulz, 1987a, 1988L–m, Szyndlar & Hung, 1987, Oxtoby, 1988, Stuebing 1991, R. Wang et al., 1991, Zhao & Adler, 1993, David & Vogel, 1996, M.J. Cox et al., 1998, T. Smith, 1998, Chan-ard et al., 1999, Orlov et al., 2000, Ratnikov, 2000, Malkmus et al., 2002, Gumprecht, 2004d, Pauwels et al., 2004, Ratnikov, 2004, Whitaker & Captain, 2004, I. Das, 2007b, 2010, 2012, V.S. Nguyen et al., 2009, Xiang & Li, 2009, Grismer, 2011, Wangyal, 2011 and Yao, 2012.

Remarks: Angel (1929: 79) reported two specimens from Laos but they are not mentioned by Saint-Girons, 1972a. Korean and Hong Kong records doubtful *fide* C.H. Pope (1935: 275), Szyndlar & Hung (1987: 38) and Karsen et al., 1998.

OVOPHIS Burger *in* Hoge & Romano-Hoge, 1981a (Viperidae)

Synonyms: *Ovophis* Burger, 1971 (*nomen nudum*), *Oxyus* Hoser, 2012j (*nomen illegitimum*), *Oviophis* – Yao, 2012 (*nomen incorrectum*), and *Ovophs* – Yao, 2012 (*nomen incorrectum*).

Type species: *Trimeresurus monticola* A.C.L.G. Günther, 1864a.

Distribution: Southeastern Asia.

Sources: Maslin, 1942, C. Haas, 1950, Burger, 1971, Hoge & Romano-Hoge, 1981a, H.M. Smith & Chiszar, 1988, Maes, 1989, Stuebing, 1991, Kraus et al., 1996, David & Ineich, 1999, McDiarmid et al., 1999, Toda et al., 1999, Malhotra & Thorpe, 2000, 2004c, Ziegler et al., 2001, Gumprecht et al., 2004, Castoe & Parkinson, 2006 and Malhotra et al., 2011a.

1. *Ovophis convictus* (Stoliczka, 1870a). Proc. Asiatic Soc. Bengal 1870(2): 108. (*Trimeresurus convictus*)

Type: Holotype, ZSI 3082 (F. Stoliczka, 1866–1870).

Type locality: "Penang" [Malaysia]. Emended to top of Western hill on Pennang, at an elevation of about 2400 feet [= Western Hill, Penang Is., Pinang State, NW West Malaysia, 5°26'N, 100°15'E, elevation 730 m] *fide* Stoliczka (1870d: 225).

Distribution: West Malaysia (Pahang, Pinang, Perak, Penang Is.) and Singapore, 610–1735 m.

Sources: J. Anderson, 1871b, Tweedie, 1941b–c, 1983, E.H. Taylor, 1965, B.L. Lim, 1967, 1991, K.K.P. Lim & Chou, 1990 and I. Das, 1999a, 2010, 2012.

Remarks: Original description reprinted in Stoliczka (1870b: 108) and supplemental original description in Stoliczka (1870d: 224–226, pl. 12, figs. 1–1b). Previously a synonym of *O. monticola*.

2. *Ovophis gracilis* (Ôshima, 1920). Ann. Rep. Inst. Sci. Govern. Formosa 8(2): 10–11, 78–79, pl. 12, figs. 5–6. (*Trimeresurus gracilis*)

Type: Holotype, ISM, a 478 mm specimen (Y. Kikuchi, Oct. 1918), destroyed 31 May 1945 during World War II.

Type locality: "Mt. Noko, Nanto, Central Formosa" [= Nenggao Mt., Nantou Prov., cen. Taiwan, 24°00'N, 121°16'E, elevation 3220 m].

Distribution: Central Taiwan (Chiayi, Nantou), 2000–3535 m.

Sources: Maki, 1931, C. Wang & Wang, 1956, Keegan & Matsui, 1964, Zhao & Adler, 1993, Zhao, 2006, Malhotra et al., 2009 and Xiang & Li, 2009.

Remarks: Generic status uncertain but related to *Ovophis okinavensis fide* Malhotra & Thorpe, 2000, P. Guo et al., 2009b and Malhotra et al., 2009.

3. *Ovophis makazayazaya* (Takahashi, 1922). Jap. Venom. Snakes: [11], pl. 4, 4 figs. (*Trimeresurus makazayazaya*)

Synonyms: *Trimeresurus monticola formosensis* Mell, 1931a and *Trimeresurus orientalis* K.P. Schmidt, 1925b.

Type: Holotype, not designated, a 724 mm female (S. Takahashi, 15 June 1922), location unknown.

Type locality: "Makazayazaya, Formosa" [= vicinity of Beiyè village, Pingtung Co., S Taiwan, 22°43'N, 120°467'E, elevation 1500 m].

Distribution: Eastern Asia. Southeastern China (Anhui, Fujian, Gansu, Guangdong, Guangxi, Guizhou, Hong Kong, Hunan, Shaanxi, Sichuan, SE Xizang, Yunnan, Zhejiang), Taiwan and N Vietnam, 500–2200 m.

Sources: Takahashi, 1930, C.H. Pope, 1929, 1935, Maki, 1931, Hu et al., 1980, Lue et al., 1989, Zhang & Zhao, 1990, Zhao, 1995, 2006, Xiang & Li, 2009 and Malhotra et al., 2011a.

Remarks: A valid species *fide* Malhotra et al., 2011a.

4. *Ovophis monticola* (A.C.L.G. Günther, 1864a). Rept. Brit. India: 388, pl. 24, fig. b. (*Trimeresurus monticola*)

Synonyms: *Parias maculata* Gray, 1853 and *Trimeresurus monticola meridionalis* Bourret, 1935d.

Type: Lectotype, BMNH 1946.1.19.91, a 250 mm female (J.D. Hooker, May–Dec. 1849), designated by Iskandar & Colijn (2001: 155).

Type locality: "Sikkim, India" [= Sikkim State, NE India] via lectotype selection.

Distribution: Southeastern Asia and Sumatra. Northeastern India (Arunachal Pradesh, Assam, Manipur, Meghalaya, Nagaland, North Bengal, Sikkim, Uttaranchal), W China (Gansu, Sichuan, Yunnan), Nepal (Bajhang, Bajura, Bhaktapur, Bhojpur, Dadeldhura, Dolakha, Doti, Ilam, Jumla, Kaski, Kavre, Khotang, Lalitpur, Lamjung, Makwanpur, Manang, Mugu, Mustang, Nuwakot, Panchthar, Rasuwa, Sankhuwasabha, Sindhupalchok, Surkhet, Udayapur), Bhutan, Myanmar (Bago, Chin, Kachin, Sagaing, Shan, Tanintharyi), N Thailand (Chumporn), S Laos (Champasak, Xe Kong), ext. NE or SW Cambodia, Vietnam (Binh Tri Thien, Hoang Lien Son, Lam Dong, Quang Nam-Da Nang, Vinh Phu), 500–3000 m.

Sources: Stoliczka, 1871, Steindachner, 1906, Wall, 1911k, 1925a, 1926b, 1928, Bourret, 1936b, Tweedie, 1941b–c, 1983, M.A. Smith, 1943, E.H. Taylor, 1965, Campden-Main, 1970a, Deuve, 1970, Saint-Girons, 1972a, Fleming & Fleming, 1978, Kramer, 1977, B.Q. Feng, 1986, Iskandar, 1986, Tien et al., 1986, Lue, 1987, 1990, Nanhoe & Ouboter, 1987, Inger et al., 1990, Zhang & Tang, 1990, M.J. Cox, 1991b, Lim & Chou, 1990, David & Vogel, 1996, David & Tong, 1997, Chan-ard et al., 1999, Leviton et al., 2003, Ao et al., 2004, Tiwari & Shah, 2004, Whitaker & Captain, 2004, P. Guo et al., 2006a, B.L. Stuart & Heatwole, 2008, M.F. Ahmed et al., 2009, I. Das, 2010, 2012, Malhotra et al., 2011a, Yao, 2012 and S.K. Sharma et al., 2013.

Remarks: Original description based on Gray (1853: 392). Type locality restriction of Nepal *fide* M.A. Smith (1943: 508) invalid, as well as Kramer's (1977: 757) designation of BMNH 1946.1.18.76 (a 533 mm female from Nepal) as syntype *fide* McDiarmid et al. (1999: 318).

5. *Ovophis okinavensis* (Boulenger, 1892b). Ann. Mag. Nat. Hist. (6) 10(58): 302. (*Trimeresurus okinavensis*)

Types: Syntypes (2), BMNH 1946.1.18.69–70 (formerly BMNH 1892.9.3.17–18), two females, longest syntype 350 mm (P.A. Holst, 1889–1892).

Type locality: "Okinawa or Great Loo Choo, the Loo Choo Islands" [= Okinawashima, Ryukyu Islands, S Japan].

Distribution: Extreme S Japan (Ryukyus: Akajima, Amamioshima, Amurojima, Fukajijima, Gishifujima, Gushikawajima, Gusukujima, Hatejima, Iejima, Iheyajima, Izenajima, Kakeromajima, Kerumajima, Kubajima, Kumejima, Kuroshima, Maejima, Nakajima, Nohojima, Okinawajima, Takashikijima, Tonakijima, Tokunoshima, Ukejima, Yakabijima, Yaguchijima, Yanahajima, Yorojima and Zamamijima Is.).

Sources: Wall, 1905f, Stejneger, 1907, Takahashi, 1922, 1930, Maki, 1931, Keegan, 1958, Koba, 1958, 960, 1961, 1963, 1968, Nakamura & Uéno, 1962, Takara, 1962, Koba et al., 1970, Koba & Tanaka, 1973, M. Moriu, 1982 and Malhotra et al., 2009, 2011a.

Remarks: Taxonomic status uncertain *fide* Malhotra et al. (2009: 117, 118). New genus description for this species and *Trimeresurus gracilis* in progress *fide* Malhotra et al., 2010.

6. *Ovophis tonkinensis* (Bourret, 1934a). Bull. Gen. Inst. Pub. Hanoi 13(7): 138. (*Trimeresurus tonkinensis*)

Type: Holotype, M.261, a 183 mm specimen (R. Bourret, 1930–1931).

Type locality: "Chapa., province de Laokay (Tonkin), l'Indochine française, à l'altitude moyenne de 1.600m" [= Sa Pa, Sa Pa Distr., W Lao Cai Prov., ext. N Vietnam, 22°20'N, 103°51'E, elevation 1600 m].

Distribution: Southern China (Guangdong, Guangxi, Hainan) and N Vietnam (Lao Cai, Gia Lai), 900–1600 m.

Sources: David, 1995, Orlov, 1997, Orlov et al., 2000, Gumprecht et al., 2004 and I. Das, 2010, 2012.

Remarks: Supplemental original description in Bourret (1934b: 154–155, fig. 2). A valid species *fide* Hoge & Romano-Hoge, 1981a, Golay et al., 1993 and Gumprecht et al., 2004. A synonym of *O. monticola fide* McDiarmid et al. (1999: 319) and a subspecies of *O. monticola fide* Orlov (1997: 69). Hoge & Romano-Hoge (1981: 264) restricted type locality to Chapa, Tonkin.

7. *Ovophis zayuensis* (Jiang *in* Zhao & Jiang, 1977). Acta Zool. Sinica 23(1): 67, 69, 71, figs. 1–3. (*Trimeresurus monticola zayüensis*)

Synonyms: *Ovophis monticola zhaokentangi* Zhao, 1995 and *Trimeresurus zaynensis* – Orlov & Helfenberger, 1997 *(nomen incorrectum)*.

Type: Holotype, CIB 73-I-5024 (formerly SBRI), adult male (Sichuan Biol. Res. Inst., 22 July 1973).

Type locality: "Zayü Xian, Xizang Autonomous Region, China, 1,800 m."

Distribution: Northeastern India, Myanmar and SW China (Xizang, Yunnan), 1400–2070 m.

Sources: Zhao, 1995 and P. Guo et al., 2006a.

Remarks: A valid species *fide* Zhao (1995: 37) and Malhotra et al., 2011a.

OXYBELIS Wagler, 1830 (Colubridae)

Synonyms: *Dryinus* Wagler *in* Spix, 1824, *Oxibelis* A.-M.-C. Duméril & Bibron, 1844 *(nomen emendatum), Plastor* Gistel, 1848 *(nomen substitutum), Oxibelis* – Velasco, 1890a *(nomen incorrectum), Oxibilis* – Velasco, 1892b *(nomen incorrectum), Oxybelus* A.L. Herrera, 1899 *(nomen emendatum), Oxybaelis* – Nopcsa, 1923 *(nomen incorrectum), Oxybelius* – Kempff-Mercado, 1975 *(nomen incorrectum), Oxybellis* – J.D. Johnson, Ely & Webb, 1976 *(nomen incorrectum),* and *Oxibelys* – Medem, 1979 *(nomen incorrectum).*

Type species: *Dryinus aeneus* Wagler *in* Spix, 1824.

Distribution: Extreme SW USA and Latin America.

Sources: E.H. Taylor, 1951, 1954, Keiser *in* J.A. Peters & Orejas-Miranda, 1970, Pérez-Santos, 1986, Pérez-Santos & Moreno, 1988, 1991, Keiser, 1991 and Vidal et al., 2000.

1. *Oxybelis aeneus* (Wagler *in* Spix, 1824). Serp. Brasil. Sp. Nov.: 12–14, pl. 3. (*Dryinus aeneus*)

Synonyms: *Coluber acuminatus* Wied-Neuwied, 1824b, *Dryinus auratus* T. Bell, 1825, *Dryophis vittatus* Girard, 1854, *Oxybelis microphthalmus* T. Barbour & Amaral, 1926, and *Oxybelis potosiensis* E.H. Taylor, 1942a.

Type: Lectotype, ZSM 2645/0, a 1214 mm female (J.B. von Spix, Nov.–Dec. 1819), designated by Hoogmoed & Gruber (1983: 326).

Type locality: "sylvis adjacentibus flumini Solimöens, prope Ega" [= forest bordering the Solimöens or Río Amazonas near Tefé (3°19'S, 64°43'W, elevation 50 m), cen. Amazonas State, NW Brazil].

Distribution: Latin America. Extreme SW USA (ext. S Arizona), Mexico (Aguascalientes, Campeche, Chiapas, SW Chihuahua, Colima, W Durango, Guerrero, Hidalgo, Jalisco, México, Michoacán, Morelos, Nayarit, S Nuevo León, Oaxaca, Puebla, Querétaro, Quintana Roo, SE San Luis Potosí, Sinaloa, Sonora, Tabasco, S Tamaulipas, Veracruz, Yucatán, S Zacatecas, Blanca, María Cleofas, María Madre and María Magdalena Is.), Belize (Belize, Cayo, Orange Walk, Stann Creek, Toledo), Guatemala (Alta Verapaz, Petén), Honduras (Atlántida, Choluteca. Colón, Comayagua, Copán, Cortés, El Paraíso, Francisco Morazán, Gracios a Dios, Intibucá, Islas de la Bahía, Lempira, Olancho, Santa Bárbara, Valle, Yoro, Barbareta, Elena, Guanaja, Roatán and Utila Is.), El Salvador (Ahuachapán, Chalatenango, Cuscatlán, La Libertad, La Paz, Morazán, San Salvador, Usulután, Los Blancos Is.), Nicaragua (Atlántico Sur, Boaco, Chinandega, Grenada, Managua, Matagalpa, Río San Juan, Rivas, Zelaya, Corn, Great Corn and Ometepe Is.), Costa Rica (Alajueta, Cartago, Guanacaste, Heredia, Limón, Puntarenas, San José), Panama (Canal Zone, Coclé, Colón, Darién, Herrera, Los Santos; Panamá, Barro Colorado, Pedro González, Naranjas, Saboga, San Miguel and Taboga Is.), Colombia (Antioquia, Atlántico, Bolívar, Boyacá, Cauca, Cesar, Chocó, Cundinamarca, La Guajira, Magdalena, Risaralda, Santander, Tolima, Valle, Vaupés, Gorgona Is.), Venezuela (Amazonas, Anzoátegui, Apure, Aragua, Barinas, Bolívar, Carabobo, Cojedes, Delta Amacuro, Distrito Federal, Falcón, Guárico, Lara, Mérida, Miranda, Monagas, Nueva Esparta, Portuguesa, Sucre, Táchira, Trujillo, Yaracuy, Vargas, Zulia, Margarita and Los Testigos Is.), Trinidad and Tobago (Huevos, Tobago, Trinidad), Netherland Antilles (Aruba), Guyana (Cuyuni-Mazaruni, Essequibo Islands-West Demerara, Potaro-Siparuni, Upper Takutu-Upper Essequibo), Suriname (Para, Paramaribo, Suriname), French Guiana (Cayenne, Saint-Laurent-du-Maroni), Brazil (Acre, Alagoas, Amazonas, Amapá, Bahia, Ceará, Distrito Federal, Espírito Santo, Goiás, Maranhao, Mato Grosso, Minas Gerais, Pará, Pernambuco, Rio de Janeiro, Rio Grande do Norte, Rondônia, Roraima, Tocantins), Ecuador (Azuay, Esmeraldas, Guayas, Imbabura, Loja, Manabí: Plata Is.), E Peru (Ancash, Loreto, Madre de Dios) and E Bolivia (Beni, Santa Cruz), NSL–1915 [2750] m.

Sources: Beebe, 1919, 1946, Mole, 1924, H.W. Parker, 1935, Brongersma, 1940, E.H. Taylor, 1942a, Bogert & Oliver, 1945, H.M. Smith & Taylor, 1945, Cochran, 1946, Mertens, 1952c, Hoge, 1953b, 1967, Duellman, 1961, L.C. Stuart, 1963, Fowlie, 1965, Busack, 1966, Roze, 1966a, Boos & Quesnel, 1968, L.M. Hardy & McDiarmid, 1969, Keiser, 1974, 1982, R.W. Henderson & Hoevers, 1975, Emsley, 1977, Cunha & Nascimento, 1978, 1980, 1993, Hoge et al., 1978, J.C. Lee, 1980, 1996, Vanzolini et al., 1980, Abuys, 1985a, L.D. Wilson & Meyer, 1985, Vanzolini, 1986a, Chippaux, 1987, Duellman & Salas, 1991, Flores-Villela et al., 1991, Pérez-Higareda & Smith, 1991, N.J. Silva, 1993, Van Devender et al., 1994, J.A. Campbell, 1998b, Starace, 1998, Vazquez-Diaz et al., 1998, G. Köhler, 1999b, Lehr & Aguilar, 2000a, Nogueira, 2001, Savage, 2002, Hernandez, 2004, Duellman, 2005, G. Köhler et al., 2005, McCranie et al., 2006, Cisneros-Heredia & Touzet, 2007, Navarrete et al., 2009, Ugueto & Rivas, 2010, McCranie, 2011a and C.J. Cole et al., 2013.

Remarks: Holotype erroneously listed as specimen from ZMB 2376–2384 *fide* Wermuth *in* L.C. Stuart (1963: 108). Keiser (1974: 9) listed ZSMH 2645/0 as the holotype. Trinidad and Tobago record (Huevos Is.) needs confirmation *fide* Boos (1983b: 21). See Franzen & Glaw (2007: 250) for type data and a figure of head.

2. *Oxybelis brevirostris* (Cope, 1861c). Proc. Acad. Nat. Sci. Philadelphia (1860) 12(12): 555–556. (*Dryophis brevirostris*)

Synonym: *Oxybelis coerulescens* Jan, 1863b.

Type: Holotype, USNM 31349 (formerly ANSP), a 1028 mm specimen (R.W. Mitchell, 1857–1860).

Type locality: "Veraguas, New Grenada" [= Cocuyos de Veraguas, a gold mine near the Río Concepción, N Veraguas Province, cen. Panama, ca. 8°45'N, 81°00'W, elevation 100 m].

Distribution: Lower Central America and ext. NW South America. Eastern Honduras (Colón, Gracias a Dios, Olancho), E Nicaragua (Atlántico Norte, Chontales, Jinotega, Río San Juan), Costa Rica (Alajuela, Cartago, Guanacaste, Heredia, Limón, Puntarenas, ? San José), Panama (Veraguas), W Colombia (Cauca, Chocó, Nariño, Risaralda, Valle, Gorgona Is.) and W Ecuador (Azuay, El Oro, Esmeraldas, Guayas, Los Ríos, Pichincha), NSL–2500 m.

Sources: E.H. Taylor, 1951, 1954, Medem, 1979, Miyata, 1980, N.J. Scott et al., 1983, L.D. Wilson & Meyer, 1985, Savage & Villa, 1986, G. Köhler, 1999b, Savage, 2002, McCranie et al., 2006, Cisneros-Heredia & Touzet, 2007, McCranie, 2011a and Travers et al., 2011.

3. *Oxybelis fulgidus* (Daudin, 1803c). Hist. Nat. Rept. 6: 352–354, pl. 80. (*Coluber fulgidus*)

Synonym: *Dryiophis catesbyi* Schlegel, 1837.

Type: Holotype, MNHN, a 1524 mm specimen (H. Ruiz-López, June 1784–Aug. 1786 via Velasques via Polony).

Type locality: "Saint-Domingue, dans les environs du Port-au-Prince," (in error *fide* Roze, 1966a: 192). Corrected to Río Santo Domingo, headwaters of Río Urubamba at old Muña, near Cusco, Cusco Dept., SE Peru, 13°36'S, 71°50'W, elevation 3150 m *fide* Wetherbee (1986b: 7). Suggested as Suriname *fide* K.P.Schmidt (1941: 506). Restricted to Chichén Itzá, Yucatán, Mexico *fide* H.M. Smith & Taylor (1950a: 352).

Distribution: Southern Mexico, Central America and N South America. Southeastern Mexico (Campeche, Oaxaca, Quintana Roo, Yucatán), Belize (Belize, Cayo, Corozal, Orange Walk, Stann Creek, Toledo), N Guatemala (Alta Verapaz, Petén), El Salvador (Cuscatlán, La Libertad, San Miguel, San Salvador, Santa Ana, San Vicente), Honduras (Choluteca, Colón, Cortés, Francisco Morazán, Gracias a Dios, Islas de la Bahía, Olancho, Yoro, Elena and Utila is.), W Nicaragua (Chinandega, Chontales, Granada, León, Managua, Masaya, Río San Juan, Ometepe Is.), Costa Rica (Alajuela, Cartago, Guanacaste, Heredia, Limón, Puntarenas), Panama (Canal Zone, Coclé, Colón, Herrera, Panamá, Barro Colorado Is.), S Colombia (Amazonas, Caquetá), E Venezuela (Amazonas, Anzoátegui, Bolívar, Delta Amacuro, Monagas, Sucre, Zulia, Patos Is.), Guyana (Cuyuni-Mazaruni, Essequibo Islands-West Demerara, Potaro-Siparuni, Upper Demerara-Berbice, Upper Takutu-Upper Essequiba), Suriname (Nickerie, Para), French Guiana (Cayenne Saint-Laurent-du-Maroni), Ecuador (Los Ríos, Morona-Santiago, Napo, Pastaza), E Peru (Cusco, Loreto, Madre a Dios), N Brazil (Amapá, Amazonas, Maranhão, Mato Grosso, Pará, Rondônia) and NE Bolivia (Beni), NSL–1600 (3150) m.

Sources: H.W. Parker, 1935, Beebe, 1946, L.C. Stuart, 1948, 1963, Roze, 1966a, Hoge, 1967, R.W. Henderson & Hoevers, 1975, Cunha & Nascimento, 1978, Hoge et al., 1978, J.C. Lee, 1980, Boos, 1983b, Abuys, 1985a, L.D. Wilson & Meyer, 1985, Dixon & Soini, 1986, Lancini, 1986, Vanzolini, 1986b, Chippaux, 1987, Duellman & Salas, 1991, J.A. Campbell, 1998b, Starace, 1998, Strussmann & André de Carvalho, 1998, G. Köhler, 1999b, T. Barros-Blanco & Pirela, 2000, Mijares-Urrutia & Arends, 2000, Stafford & Meyer, 2000, Savage, 2002, Solórzano, 2004, G. Köhler et al., 2005, Cisneros-Heredia & Touzet, 2007, Navarrete et al., 2009, McCranie, 2011a and C.J. Cole et al., 2013.

Remarks: Wetherbee (1986b: 15) has shown that the type was probably sent to Velasques in Mexico via Dr. Polony of Santo Domingo to Daudin in France.

4. *Oxybelis wilsoni* Villa & McCranie, 1995. Rev. Biol. Trop. 43(1–3): 297–303, figs. 1a–c.

Type: Holotype, USNM 318010, a 1982 mm male (J. R. McCranie, 20 Nov. 1989), lost *fide* McCranie (1999: 11).

Type locality: "Rocky Point, Santa Elena on the eastern end of Isla de Roatán, Departamento de Islas de la Bahía, Honduras."

Distribution: Northern Honduras (Islas de Bahía: Roatán), NSL–90 m.

Sources: Keiser, 1969, Groves, 1995, McCranie, 1999, 2011, Savage, 2002 and McCranie et al., 2005.

Remarks: McCranie (1999: 11–12) indicated that the types (holotype and paratypes) are presently lost.

OXYRHABDIUM Boulenger, 1893a (*nomen substitutum*) (Xenodermatidae)

Synonyms: *Stenognathus* A.-M.-C. Duméril, Bibron & Duméril, 1854a *(nomen praeoccupatum)*, *Stegognathus* – Hoffmann, 1890 *(nomen incorrectum)*, and *Oxyrhabdinium* – W.W. Tanner & Avery, 1982 *(nomen incorrectum)*.

Type species: *Stenognathus modestus* A.-M.-C. Duméril, Bibron & Duméril, 1854a.

Distribution: Philippines.

Sources: Boulenger, 1893a, E.H. Taylor, 1922a, Leviton, 1957, 1965a, Vidal et al., 2008 and Kelly et al., 2009.

1. *Oxyrhabdium leporinum* (A.C.L.G. Günther, 1858). Cat. Colub. Snakes Brit. Mus.: 12. (*Rhabdosoma leporinum*)

Synonyms: *Stenognathus brevirostris* W.C.H. Peters, 1872b and *Oxyrhabdium leporinum visayanum* Leviton, 1957.

Type: Holotype, BMNH 1946.1.13.98, a 711 mm female (H. Cuming, 1836–1840).

Type locality: "Philippine Islands."
Distribution: Philippines (Cebu, Luzon, Negros), 450–2000 m.
Remarks: A.C.L.G. Günther's type series was composite; one of the two syntypes was *Oxyrhabdium modestum*, so BMNH 1946.1.13.98 is the holotype. Probably occurs on Panay *fide* Leviton (1957: 289). *Oxyrhabdium visayanum* a valid species *fide* R. Crombie (pers. comm.).

2. *Oxyrhabdium modestum* (A.-M.-C. Duméril, Bibron & Duméril, 1854a). Erpét. Gén. 7(1): 504–506. (*Stenognathus modestus*)

Synonym: *Geophis schadenbergi* J.G. Fischer, 1885b.
Types: Syntypes (2), MNHN 7301a–b, longest syntype 433 mm.
Type locality: "Java" [W Indonesia] (in error *fide* W.C.H. Peters, 1872b: 587, A.C.L.G. Günther, 1873b: 169, and E.H. Taylor, 1922a: 100). Designated as Mindanao Island, southern Philippines *fide* Leviton (1957: 291).
Distribution: Southern Philippines (Basilan, Bohol, Dinagat, Leyte, Mindanao, Negros, Samar), NSL–1920 m.
Remarks: Negros, Luzon and Calamines records doubtful *fide* Leviton (1965a: 411).

OXYRHOPUS Wagler, 1830 (Xenodontidae)

Synonyms: *Oxyrrhopus* Fitzinger, 1843 *(nomen emendatum), Sphenocephalus* Wiegmann *in* Fitzinger, 1843, *Oxyropus* Gervais *in* d'Orbigny, 1847 *(nomen emendatum), Otyrrhopus* – Sumichrast, 1880 *(nomen incorrectum), Oxyrophus* – Hoffmann, 1890 *(nomen incorrectum), Oxirhopus* – Brazil, 1911 *(nomen incorrectum), Erythroxyrhopus* J.C. Thompson, 1913b, *Erythroxirhopus* – Briceño-Rossi, 1934 *(nomen incorrectum), Oxiropus* – Briceño-Rossi, 1934 *(nomen incorrectum), Oxyrhophus* – Neill & Allen, 1959 *(nomen incorrectum),* and *Oxirhopus* – J.D. Williams, 1983 *(nomen incorrectum).*
Type species: *Coluber petolarius* Linnaeus, 1758.
Distribution: Latin America.
Sources: F. Werner, 1924c, Bailey *in* J.A. Peters & Orejas-Miranda, 1970, Cadle, 1984a, Zaher, 1996, Vidal et al., 2000 and Zaher et al., 2009.
Remarks: Several undescribed species exist *fide* Bailey *in* J.A. Peters & Orejas-Miranda (1970: 229) and Henle & Ehrl (1991: 166).

1. *Oxyrhopus clathratus* A.-M.-C. Duméril, Bibron & Duméril, 1854b. Erpét. Gén. 7(2): 1026–1028.

Synonyms: *Oxyrhopus doliatus viperina* F. Werner, 1903b and *Clelia clathrata pulcherrima* L. Müller, 1923.
Type: Lectotype, MNHN 3791, a 944–945 mm specimen (F.L.N.C.L.de Castelnau, June 1843–Aug. 1847), designated by Bernardo et al. (2012: 139).
Type locality: "Brazil" via lectotype selection.
Distribution: Southeastern Brazil (Bahia, Espírito Santo, S Minas Gerais, Rio de Janeiro, Rio Grande do Sul, Santa Catarina, São Paulo, Cananéia, Cardoso, Grande, São Sebastião and São Vicente Is.) and ext. NE Argentina (Misiones), 60–930 m.
Sources: J.D. Williams & Francini, 1991, Cei, 1994, Giraudo, 1999, Argôlo, 2001a, Duarte-Rocha & Van Sluys, 2006, Cicchi et al., 2007, Medeiros et al., 2010 and Bernardo et al., 2012.
Remarks: Photographs of lectotype in Bernardo et al. (2012: figs. 4a–e).

2. *Oxyrhopus doliatus* A.-M.-C. Duméril, Bibron & Duméril, 1854b. Erpét. Gén. 7(2): 1020–1022.

Synonyms: *Oxyrrhopus doliatus* A.-M.-C. Duméril, 1853 *(nomen nudum)* and *Oxyrhopus venezuelanus* Shreve, 1947.
Type: Holotype, MNHN 325, a 566 mm female (Joly).
Type locality: "trés-probablement originaire du Brésil" [= very probably Brazil] (in error *fide* J.A. Peters & Orejas-Miranda, 1970: 231).
Distribution: Northern Venezuela (Aragua, Carabobo, Distrito Federal, Falcón, Lara, Mérida, Miranda, Trujillo, Yaracuy), NSL–300 [1220] m.
Sources: Zaher & Caramaschi, 1993, Reynolds et al., 1997, Rivas-Fuenmayor, 2002 and Navarrete et al., 2009.
Remarks: J.A. Peters & Orejas-Miranda (1970: 231) reported it known only from Cusco, Peru.

O

3. *Oxyrhopus erdisii* (T. Barbour, 1913a). Proc. Acad. Sci. Philadelphia 65(3): 506, pl. 17, figs. 3–4. (*Drepanodon erdisii*)

Type: Holotype, MCZ 8829, a 497 mm female (G.F. Eaton & E.C. Erdis, 1912).
Type locality: "Machu Pichu, 9,000–10,000 feet altitude, Department of Cusco, Southern Peru." Elevation emended to 9,500 feet *fide* Barbour & Loveridge (1929: 254).
Distribution: Eastern Peru (Cusco, Madre de Dios), 2750–3050 m.
Source: Zaher & Caramaschi, 1993.

4. *Oxyrhopus fitzingeri* (Tschudi, 1845). Arch. Naturg. 11(1): 165. (*Siphlophis fitzingeri*)

Synonym: *Oxyrhopus fitzingeri frizzelli* K.P. Schmidt & Walker, 1943b.
Type: Lectotype, MHNN 13 (formerly MZN 99), a 760 mm female (J.J. von Tschudi, Aug. 1838–Feb. 1841), designated by Schätti (1986a: 102).

Type locality: "Champs de cannes ("Rohrfelder") bordent les fleuves des régions côtières [du Pérou]" [= marshes bordering the rivers of coastal Peru] via lectotype selection.

Distribution: Southwestern Ecuador (Guayas, Loja) and NW Peru (N Ancash, La Libertad, Lambayeque, Lima, Piura), 300–1830 m.

Sources: Tschudi, 1846, Jan & Sordelli, 1870–1881, K.P.Schmidt & Walker, 1943b, J.A. Peters, 1960a, Miyata, 1982 and Pérez-Santos & Moreno, 1991.

Remarks: Original description reprinted in Tschudi (1846: 56–57, pl. 8).

5. *Oxyrhopus formosus* (Wied-Neuwied, 1820a). Reise Brasil. 1: 381–386. (*Coluber formosus*)

Synonyms: *Oxyrhopus leucocephalus* A.-M.-C. Duméril, Bibron & Duméril, 1854b, *Oxyrhopus labialis* Jan, 1863b, *Oxyrhopus submarginatus* W.C.H. Peters, 1871a, *Clelia peruviana* L.E. Griffin, 1916, and *Oxyrhopus iridescens* F. Werner, 1927.

Type: Holotype, not designated, a 380 mm specimen (A.M.P. zu Wied-Neuwied, 2 Jan.–12 Apr. 1815), location unknown.

Type locality: "den grossen Waldungen fand, die der Fluss Mucuri durchströmt" [= dense woodland bordering the Mucuri River]. Emended to "Diese unvergleichlich schöne Natter wurde an der Ufern des Mucuri von den Indianern, welche diesen Fluss beschifften, getödtet, als ich mich zu Morro d'Arara an der Lagoa dieses Namens aufhielt" *fide* Wied-Neuwied (1825c: 385) [= Morro d'Arara, 30 km above mouth of Río Mucuri, ext. S Bahia, SE Brazil, 17°32'S, 39°35'W].

Distribution: Northern South America. Colombia (Córdoba, Santander), SE Venezuela (Amazonas, Bolívar), Brazil (Acre, Amazonas, Bahia, Maranhão, Pará, Rondônia), E Ecuador (Napo, Pastaza), Peru (Huanuco, Junín, Loreto [LSUMZ 14598, 26846], Madre de Dios, Piura) and Bolivia, 200–1000 m.

Sources: Wied-Neuwied, 1822 *in* 1822–1831, L.E. Griffin, 1916, Cunha & Nascimento, 1978, 1983a, 1993, Duellman, 1978, Gasc & Rodrigues, 1980, Abuys, 1985a, Pérez-Santos, 1986, Chippaux, 1987, Pérez-Santos & Moreno, 1988, 1991, Duellman & Salas, 1991, N.J. Silva, 1993, Starace, 1998, Mattei & Barrio-Amorgós, 1999, Duellman, 2005, Lynch, 2009, MacCulloch et al., 2009, Navarrete et al., 2009, Carvajal-Cogollo et al., 2011c and Uetz et al., 2011.

Remarks: Wied-Neuwied was staying at Morro d'Arara, a village near the Mucuri River in S Bahia, when he collected the type. Lake Arara, in Amazonas State (3°26'S, 61°22'W), is not the type locality.

O

6. *Oxyrhopus guibei* Hoge & Romano-Hoge, 1978. Mem. Inst. Butantan (1977) 40–41: 58, figs. 3, 5, 7. (*Oxyrhopus trigeminus guibei*)

Type: Holotype, IB 27382, an 1175 mm female (native, 3 July 1967), destroyed by fire 15 May 2010.

Type locality: "Londrina, state of Paraná, Brazil" [= Londrina, N Paraná State, SE Brazil, 23°18'S, 51°09'W].

Distribution: Southeastern South America. Eastern Bolivia (Beni, Cochabamba, Santa Cruz), S Brazil (Alagoas, Bahia, S Goiás, S Mato Grosso, Mato Grosso do Sul, Minas Gerais, Paraná. Rio de Janeiro, São Paulo, Tocantins), S Paraguay (Alto Paraguay, Alto Paraná, Amambay, Canindeyú, Central, Itapúa, Misiones, Neembucú, Paraguarí) and NE Argentina (Chaco, Corrientes, Formosa, Misiones), NSL–1000 m.

Sources: Zaher & Caramaschi, 1993, Yanosky et al., 1996, Freire, 1999b, Giraudo & Scrocchi, 2002, Cacciali, 2008 and Lima-Silveira, 2008.

Remarks: Possibly occurs in NW Brazil (Acre, Amazonas) and E Peru (Loreto) *fide* Zaher & Caramaschi (1993: 825).

7. *Oxyrhopus leucomelas* (F. Werner, 1916). Zool. Anz. 47(11): 310. (*Tropidodipsas leucomelas*)

Type: Holotype, LMB, a 240 mm female (A.H. Fassl, Jan.–March 1910).

Type locality: "Coñon del Tolima, 1700 m (Central-cordilleren), der Anden von Kolumbien" [= Tolima Canyon, 1700 m, above Ibagué (4°27'N, 75°14'W) in Cordillera Central of Andes, Tolima Dept., cen. Colombia].

Distribution: Northwestern South America. Colombia (Antioquia, Boyacá, Cauca, Huila, Magdalena, Meta, Norte de Santander, Nariño, Risaralda, Tolima, Valle del Cauca), Venezuela (Tochira), E Ecuador (Napo, Tungurahua) and Peru (Ayachuco [LSUMZ 26873], Huánuco), 1000–3050 m.

Sources: Downs, 1961, J.A. Peters, 1960a, Medem, 1965, Pérez-Santos, 1986, Pérez-Santos & Moreno, 1988, 1991, Henle & Ehrl, 1991, Lynch, 2009 and Calcaño & Barrio-Amorós, 2010.

8. *Oxyrhopus marcapatae* (Boulenger, 1902d). Ann. Mag. Nat. Hist. (7) 10(59): 401–402. (*Homalocranium marcapatae*)

Synonym: *Drepanodon eatoni* Ruthven *in* T. Barbour, 1913a.

Type: Holotype, BMNH 1946.1.8.72, a 390 mm female (G.R. Ockenden, 1902).

Type locality: "Marcapata Valley, Andes of E. Peru" [= Cusco Dept., SE Peru].

Distribution: Southeastern Peru (Cusco), 1830–2745 m.

Source: T. Barbour, 1913a.

9. *Oxyrhopus melanogenys* (Tschudi, 1845). Arch. Naturg. 11(1): 163. (*Sphenocephalus melanogenys*)

Synonyms: *Tachymenis bitorquata* A.C.L.G. Günther, 1872b and *Oxyrhopus melanogenys orientalis* Cunha & Nascimento, 1983a.

Type: Lectotype, MHNN 14 (formerly MZN 99) [longest syntype 755 mm], (J.J. von Tschudi, Aug. 1838–Feb. 1841), designated by Schätti (1986a: 102),

Type locality: "Republica Peruana" [= Peru]. Emended to "den Urwäldern des Chanchamayogebietes" *fide* Tschudi (1846: 49). Further emended to "dans les forêts vierges de la région de Chanchamayo (Pérou)" [= virgin forests of the Chanchamayo region, Peru] *fide* Schätti (1986a: 102).

Distribution: Amazonia. Southeastern Venezuela (Amazonas, Bolívar), E Ecuador (Morona-Santiago, Napo, Pastaza), Guyana (Cuyuni-Mazaruni, Essequibo Islands-West Demerara, Potaro-Siparuni, Upper Demerara-Berbice), French Guiana (Cayenne, Saint-Laurent-du-Maroni), E Peru (Amazonas, Cajamarca, Huánuco, Junín, Loreto, Madre de Dios, San Martin), N Brazil (Acre, Amazonas, Bahia, Ceará, Maranhão, Mato Grosso, Pará, Rondônia, Roraima) and N Bolivia (Beni, La Paz), NSL–1200 m.

Sources: Tschudi, 1846, Amaral, 1937c, J.A. Peters, 1960a, Duellman, 1978, Dixon & Soini, 1986, Cunha & Nascimento, 1978, 1983a, 1993, Pérez-Santos, 1986, Nascimento et al., 1987, 1988, Pérez-Santos & Moreno, 1988, 1991, Henle & Ehrl, 1991, N.J. Silva, 1993, Zaher & Caramaschi, 1993, Starace, 1998, Frota, 2000b, Duellman, 2005, Means & Kalamandeen, 2007, Loebmann & Roberto, 2009, Lynch, 2009, MacCulloch et al., 2009, Navarrete et al., 2009, Rivas-Fuenmayor et al., 2012 and C.J. Cole et al., 2013.

Remarks: Supplemental original description in Tschudi (1845: 49, pl. 4). Pérez-Santos & Moreno (1988: 262) cited a specimen from Santander, Colombia.

10. *Oxyrhopus occipitalis* (Wagler *in* Spix, 1824). Serp. Brasil. Sp. Nov.: 21–23, pl. 6, fig. 2. (*Natrix occipitalis*)

Synonym: *Oxyrhopus occipitalis* Jan, 1863b *(nomen praeoccupatum).*

Type: Lectotype, ZSM 2053/0, a 746 mm male (J.B. von Spix, Dec. 1819–Jan. 1820), designated by Hoogmoed & Gruber (1983: 328).

Type locality: "ad flumen Solimöens" [= Río Solimões or upper Amazon River, bet. mouth of Río Negro and Peruvian border, Amazonas State, NW Brazil].

Distribution: Amazonia. Eastern Colombia (Amazonas, Boyacá, Meta, Nariño, Norte de Santander, Putumayo, Santander, Vaupés), S Venezuela (Amazonas, Bolívar), Guyana (Demerara-Mahaica, Potaro-Siparuni, Upper Demerara-Berbice), Suriname, French Guiana (Cayenne, Saint-Laurent-du Maroni), W Brazil (Amapá, Amazonas, Rondônia), E Ecuador, and NE Peru, 100–1050 m.

Sources: Hoge et al., 1973, Vanzolini, 1986b, N.J. Silva, 1993, Franzen & Glaw, 2007, Lynch, 2009, MacCulloch et al., 2009 and C.J. Cole et al., 2013.

11. *Oxyrhopus petolarius* (Linnaeus, 1758). Syst. Nat., ed. 10, 1: 225. (*Coluber petolarius*)

Synonyms: *Coluber petola* Linnaeus, 1758, *Coluber pethola* Linnaeus, 1766, *Coluber petalarius* Linnaeus, 1766, *Coronella cerastoides* Laurenti, 1768, *Coronella fasciata* Laurenti, 1768, *Coluber zebra* Bonnaterre, 1790, ? *Coluber aurantiacus* Gravenhorst, 1807, *Coluber digitalis* A. Reuss, 1834, *Lycodon semifasciatus* Tschudi, 1845, *Colubrinus hydrus* A.-M.-C. Duméril, Bibron & Duméril, 1854b *(lapsus calami), Oxyrhopus bi-prae-ocularis* A.-M.-C. Duméril, Bibron & Duméril, 1854b *(nomen incorrigendum), Oxyrhopus immaculatus* A.-M.-C. Duméril, Bibron & Duméril, 1854b, *Oxyrhopus multifasciatus* A.-M.-C. Duméril, Bibron & Duméril, 1854b, *Oxyrhopus sebae* A.-M.-C. Duméril, Bibron & Duméril, 1854b, *Oxyrhopus spadiceus* A.-M.-C. Duméril, Bibron & Duméril, 1854b, *Oxyrrhopus doliatus semicinctus* Cope, 1887c, *Oxyrhopus bipraeocularis* Boulenger, 1896a *(nomen corrigendum), Oxyrrhopus intermedius* F. Werner, 1899c, *Oxyrhopus doliatus aequifasciata* F. Werner, 1909b, ? *Tarbophis splendidus* Ahl, 1924a, *Clelia cornelii* L. Müller, 1928, *Drepanpdon alternatus* – Briceño-Rossi, 1934 *(nomen incorrectum),* and *Clelia baileyi* H.M. Smith, 1942.

Types: Syntypes (8), ZMUU 8 (C. Gyllenborg, 1744), ZMUU 45 (MP, 1745), ZMUU 151 (JA, 1749), NHR Lin-73 (formerly MAFR), an 845 mm specimen (Mus. Drottn.), NHR Lin-74 (formerly MAFR), a 490 mm specimen (Mus. Drottn.), and NHR Lin-75 (formerly MAFR), 740 mm, 450 mm and 410 mm specimens (Mus. Drottn.).

Type locality: "Africa," (in error *fide* Roze, 1966a: 195).

Distribution: Latin America. Southeastern Mexico (Chiapas, Oaxaca, Puebla, S Veracruz), SW Belize (Belize, Cayo), N Guatemala (Alta Verapaz, Petén), N Honduras (Atlántida, Colón, Cortés, Gracias a Dios, Olancho), W El Salvador (Santa Ana), E Nicaragua (Atlántico Norte, Atlántico Sur, Chontales, Jinotega, Matagalpa, Nueva Segovia, Rio San Juan), Costa Rica (Alajuela, Cartago, Heredia, Limón, Puntarenas, San José), Panama (Canal Zone, Coclé, Colón, Darién, Herrera, Panamá, Veraguas, Barro Colorado Is.), Colombia (Amazonas, Antioquia, Atlántico, Bolívar, Boyacá, Caldas, Caquetá, Cauca, Chocó, Córdoba, Cundinamarca, Huila, Magdalena, Meta, Nariño, Norte de Santander, Putumayo, Risaralda, Santander, Tolima, Valle, Vaupés), Venezuela (Amazonas, Anzoátegui, Apure, Aragua, Barinas, Bolívar, Carabobo, Cojedes, Delta Amacuro, Distrito Federal, Falcón, Guárico, Lara, Mérida, Miranda, Monagas, Portuguesa, Sucre, Táchira, Trujillo, Yaracuy, Vargas, Zulia), Trinidad and Tobago (Trinidad), Guyana (Barima-Waini, Essequibo

O

Islands-West Demerara, Upper Demerara-Berbice), Suriname (Nickerie, Para, Paramanca, Paramaribo), French Guiana (Cayenne, Saint-Laurent-du-Maroni), Ecuador (Carchi, Guayas, Imbabura, Los Ríos, Manabí, Morona-Santiago, Napo, Pastaza, Pichincha, Zamora-Chinchipe), E Peru (Amazonas, Loreto, Madre de Dios, Pasco), Brazil (Amapá, Amazonas, Bahia, Ceará, Distrito Federal, Goiás, Mato Grosso, Mato Grosso do Sul, Pará, Rio de Janeiro, Rondônia, São Paulo, Sergipe), E Bolivia (La Paz, Santa Cruz) and Argentina (Córdoba, Formosa, Misiones), NSL–2750 m.

Sources: J. Gomes, 1918, Mertens, 1942, H.M. Smith, 1942m, K.P. Schmidt & Walker, 1943b, H.M. Smith & Taylor, 1945, Beebe, 1946, E.H. Taylor, 1951, 1954, Roze, 1957a, 1966a, L.C. Stuart, 1963, Hoge, 1967, R.W. Henderson & Hoevers, 1975, H.M. Smith & Smith, 1976, Amaral, 1978, Emsley, 1977, Cunha & Nascimento, 1978, 1983a, 1993, Duellman, 1978, 2005, Fugler & Walls, 1978, Gasc & Rodrigues, 1980, J.C. Lee, 1980, 1996, Fugler, 1984, Abuys, 1985a, L.D. Wilson & Meyer, 1985, Bailey, 1986, Dixon & Soini, 1986, Lancini, 1986, H.M. Smith et al., 1986, Vanzolini, 1986b, Chippaux, 1987, Pérez-Santos & Moreno, 1988, 1991, Rodriguez & Cadle, 1990, Flores-Villela et al., 1991, Pérez-Higareda & Smith, 1991, N.J. Silva, 1993, Fugler et al., 1995, Duellman & Mendelson III, 1995, Reati, 1996, Yanosky et al., 1996, J.A. Campbell, 1998b, Gorzula & Señaris, 1998, Starace, 1998, Esqueda & La Marca, 1999, Giraudo, 1999, G. Köhler, 1999b, Stafford & Meyer, 2000, Lehr, 2001, Markezich, 2002, Savage, 2002, Rivas-Fuenemayer & Barrio-Amorgós, 2005, García-Vázquez et al., 2009, Lynch, 2009, MacCulloch et al., 2009, Navarrete et al., 2009, Lotzkat et al., 2010, McCranie, 2011a and C.J. Cole et al., 2013.

Remarks: See MacCulloch et al. 2009 for remarks on confusion between this species and *O. melanogenys,* especially in the Guiana Shield area. There are likely two species involved split into *petola* and *petolarius* by some; however problem involves nomenclatural and other problems that need solving in a detailed study. Thus we retain what is likely two species under this name. See Savage (2011: 223–224) for a details on application of the name *Coluber petolarius* and synonyms.

O

12. *Oxyrhopus rhombifer* A.-M.-C. Duméril, Bibron & Duméril, 1854b. Erpét. Gén. 7(2): 1018–1019.

Synonyms: *Oxyrrhopus rhombifer* A.-M.-C. Duméril, 1853 (*nomen nudum*), *Oxyrrhopus subpunctatus* A.-M.-C. Duméril, 1853 (*nomen nudum*), *Oxyrhopus dorbignyi* A.-M.-C. Duméril, Bibron & Duméril, 1854b, *Oxyrhopus subpunctatus* Duméril, Bibron & Duméril, 1854b, *Coronella bachmanni* Weyenberg, 1876a, *Oxyrhopus rhombifer inaequifasciata* F. Werner, 1909b, *Leptodira weiseri* L. Müller, 1923, *Oxyrhopus rhombifer septentrionalis* Vellard, 1943, and *Pseudoboa ornata* Hoge & Mertens, 1955.

Type: Holotype, MNHN 1705, a 712 mm specimen (A.C.V.M.D. d'Orbigny, 22 April 1827–20 April 1828).

Type locality: "province de los Corrientes" [= Corrientes Prov., ext. NE Argentina].

Distribution: Southern South America. Southern Brazil (Bahia, Ceara, Distrito Federal, Goiás, Mato Grosso, Mato Grosso do Sul, Minas Gerais, Pará, Paraná, Pernambuco, Rio de Janeiro, Rio Grande do Sul, Rondônia, Santa Catarina, São Paulo), SE Bolivia (Santa Cruz, Tarija), Paraguay (Alto Paraguay, Boquerón, Itapúa, Misiones, Neembucú), Uruguay (Artigas, Canelones, Cerro Largo, Colonia, Durazno, Flores, Lavalleja, Moldonado, Paysandú, Rivera, Salto, Sariano, Trienta y Tres) and N Argentina (Buenos Aires, Catamarca, Chaco, Chubut, Córdoba, Corrientes, Entre Ríos, Formosa, Jujuy, La Pampa, La Rioja, Mendoza, Rio Negro, Salta, San Juan, San Luis, Santa Fe, Santiago de Estero, Tucumán), 100–1500 m.

Sources: Serie, 1921, 1936, Cordeiro & Hoge, 1974, Amaral, 1978, Lema & Fabián-Beurmann, 1977, Lema, 1980, J.D. Williams & Francini, 1991, Cei, 1994, N.J. Silva, 1993, Rey & Lions, 1997b, Avila & Morando, 1999, Freitas, 1999, Leynaud & Bucher, 1999, Argôlo & Freitas, 2000, Carreira-Vidal et al., 2005, Cacciali 2008, Carrera & Javier-Avila, 2008 and Hamdan & Lira-Da-Silva, 2008.

13. *Oxyrhopus trigeminus* A.-M.-C. Duméril, Bibron & Duméril, 1854b. Erpét. Gén. 7(2): 1013–1015.

Synonyms: *Oxyrrhopus trigeminus* A.-M.-C. Duméril, 1853 (*nomen nudum*) and *Erythroxirhopus trigeminus* – Briceño-Rossi, 1934 *(nomen incorrectum).*

Type: Lectotype, MNHN 7599, designated by Zaher & Caramaschi (1993: 818).

Type locality: "l'état de Bahia, Brésil" [=Bahia State, E Brazil] via lectotype selection.

Distribution: Central South America. Southeastern Venezuela (Amazonas, Bolívar), E Brazil (Alagoas, Amapá, Bahia, Ceará, Distrito Federal, S Goiás, Mato Grosso do Sul, Minas Gerais, Pará, Paraíba, Pernambuco, Rio de Janeiro, Rio Grande do Norte, Rondônia, São Paulo, Marajo Is.), Bolivia (La Paz) and Peru (Loreto), NSL–300 m.

Sources: Amaral, 1935e, Hoge et al., 1973, 1975, 1978, Cordeiro & Hoge, 1974, Vanzolini et al., 1980, Cascimento et al., 1991, N.J. Silva, 1993, Zaher & Caramaschi, 1993, Freitas, 1999, Franklin, 2001 and Navarrete et al., 2009.

Remarks: Probably occurs in Maranhão, Piauí and Sergipe, Brazil *fide* Zaher and Caramaschi (1993: 820).

14. *Oxyrhopus vanidicus* Lynch, 2009. Pap. Avul. Zool. 49(25): 333–335.

Type: Holotype, ICNMNH 10683 (formerly JMR 4106), an adult male (J.M. Renjifo, 13 May 2003).

Type locality: "Km 10, Leticia-Tarapacá road, municipality of Leticia, Amazonas, Colombia" [= Km. 10, Leticia (4°05'S, 69°34'W)–Tarapacá (2°53'S, 69°45'W) road, SE Amazonas Prov., ext. SE Colombia].

Distribution: Southeastern Colombia (Amazonas, Meta, Putumayo, Vaupés), Brazil (Amazonas, Pará), NE Ecuador (Sucumbios) and NE Peru (Loreto), 200–470 m.

Remarks: Specimens of this species have been misidentified as *O. melanogenys, O. petola semifasciatus* and *O. trigeminus fide* Lynch (2009: 336). Possibly a synonym of *O. melanogenys fide* Rivas-Fuenmayor et al. (2012: 47).

OXYURANUS Kinghorn, 1923 (Elapidae)

Synonyms: *Parademansia* Kinghorn, 1955, *Oxyuramus* – Battersby & Swinton, 1960 *(nomen incorrectum),* and *Oxyurans* – Anon., 1978 *(nomen incorrectum).*

Type species: *Pseudechis scutellatus* W.C.H. Peters, 1867d.

Distribution: Austro-Papua.

Sources: Kinghorn, 1929, Worrell, 1969, Covacevich et al., 1981, Cogger et al., 1983a, Shine & Covacevich, 1983, Longmore, 1986, S.K. Wilson & Knowles, 1988, Gow, 1989, Hoser, 1989, 2002b, 2012e, Covacevich & Couper, 1991, Ehmann, 1992, Golay et al., 1993, Masci & Kendall, 1995, Greer, 1997, David & Ineich, 1999, Cogger, 2000, Scanlon & Lee, 2004, Doughty et al., 2007, Sanders et al., 2008 and Zaher et al., 2009.

1. *Oxyuranus microlepidotus* (F. McCoy, 1879b). Prod. Zool. Victoria 3: 12–13, pl. 23, figs. 2–2a, 3–3a. (*Diemenia microlepidota*)

Synonyms: *Diemenia ferox* Macleay, 1882 and *Parademansia microlepidtus* – Khole, 1991 *(nomen incorrectum).*

Type: Lectotype, NMV D12354 (formerly NMV 493 & NMV 12871), specimen with prepared skull (W. Blandowski, 1849–1858), designated by Coventry (1970: 121).

Type locality: "Junction of Murray and Darling Rivers" [NW Victoria, SE Australia] via lectotype selection.

Distribution: Central Australia (W New South Wales, ext. SE Northern Territory, ext. SW Queensland, N South Australia), 10–130 m.

Sources: Kinghorn, 1923, Covacevich & Wombey, 1976, Broad et al., 1979 and Mirtschin, 1981, 1982a–b.

2. *Oxyuranus scutellatus* (W.C.H. Peters, 1867d). Mber. Königl. Akad. Wiss. Berlin 1867(11): 710–711. (*Pseudechis scutellatus*)

Synonyms: *Pseudechis wilesmithii* De Vis, 1911, *Oxyuranus maclennani* Kinghorn, 1923, *Oxyuranus scutellatus canni* K.R. Slater, 1956, *Oxyuranus maclennoni* – Slater, 1956 *(nomen incorrectum), Oxuranus scutallatus* – Manjunatha-Kini & Iwanaga, 1986 *(nomen incorrectum), Oxyuranus scutulatus* – Khole, 1991 *(nomen incorrectum), Oxyuranus scutellatus barringeri* Hoser, 2002b, *Oxyuranus scutellatus adelynhoserae* Hoser, 2009c & 2012i *(nomen illegitimum),* and *Oxyuranus scutellatus andrewwilsoni* Hoser, 2009c & 2012i *(nomen illegitimum).*

Type: Holotype, ZMB 5883, a 2230 mm specimen (Mus. Godeffroy).

Type locality: "Rockhampton" [E Queensland, E Australia, 23°23'S, 150°30'E, elevation 10 m].

Distribution: Southern Papua New Guinea (Central, Gulf, Milne Bay, Western, Daru Is.) and N Australia (ext. NE New South Wales, N Northern Territory, N and E Queensland, ext. N Western Australia, Bathurst, Koolan, Melville and Thursday Is.), NSL–800 m.

Sources: Thomson, 1933, 1950, MacKay, 1952, K.R. Slater 1956, 1968, C.R. Johnson, 1975a, F. Parker, 1982, Whitaker et al., 1982, Storr et al., 1986, O'Shea, 1996, B. Barnett, 1999b, J.C. Murphy & Schlager, 2003 and Wüster et al., 2005.

3. *Oxyuranus temporalis* Doughty, Maryan, Donnellan & Hutchinson, 2007. Zootaxa (1422): 52–54, figs. 3a–c, 4a–b, 5a.

Type: Holotype, WAM 166250 (formerly SAMA-FN BS-9075), a 970 mm female (M.N. Hutchinson, 22 Sept. 2006).

Type locality: "east of the Walter James Distribution, Western Australia (24°40'06"S, 128°45'52"E), Australia."

Distribution: Central Australia (E Western Australia), 415–515 m.

Source: Brennan et al., 2012.

P

†*PACHYOPHIS* Nopcsa, 1923 (†Pachyophiidae)

Synonym: *Mesophis* Bolkay, 1925.
Type species: †*Pachyophis woodwardi* Nopcsa, 1923.
Distribution: Upper Cretaceous of Bosnia (marine in life).
Sources: Bolkay, 1925, A.M. Müller, 1968, McDowell, 1987, M.S,Y. Lee et al., 1999, Caldwell & Albino, 2001, Scanlon & Lee, 2002, Rage & Escuillié, 2003a–b, Bardet et al., 2008 and Houssaye, 2010.
Remarks: Holotype illustrated in Lee et al. (1999: figs. 1–23). A lizard *fide* Rage (1984b: 62). †*Mesophis* a valid genus *fide* M.S.Y. Lee et al. (1999: 517).

1. †*Pachyophis woodwardi* Nopcsa, 1923. Palaeontographica 55(4): 118–122.

Synonym: *Mesophis nopcsai* Bolkay, 1925.
Type: Holotype, NMW A3919, skull fragments and 101 vertebrae and ribs, ca. 400 mm in length.
Type locality: "die Beschaffenheit der Plattenkalke von Bilek...über die fossilen Kohlen Bosniens und der Herzegovina, aus dem Neoc om." Emended to "Östlich von Bilek, in dem Vorort Selista (ost-Hercegovina)" [= Seliste, near Bileca, Bosnia; early upper Cretaceous, Cenomanian *fide* Rage, 1984b: 62].
Distribution: Upper Cretaceous (Cenomanian: 93.9–100.5 mya) of Bosnia. Known only from type locality.

P

†*PACHYRHACHIS* G. Haas, 1979 (†Pachyophiidae)

Synonyms: *Ophiomorphus* G. Haas, 1980 *(nomen praeoccupatum)* and *Estesius* Wallach, 1984 *(nomen substitutum)*.
Type species: †*Pachyrhachis problematicus* G. Haas, 1979.
Distribution: Upper Cretaceous of Israel (marine in life).
Sources: G. Haas, 1979, 1980a, Rage, 1984b, McDowell, 1987, Caldwell & Lee, 1997, M.S.Y. Lee & Caldwell, 1998, M.S.Y. Lee et al., 1998, Zaher, 1998, Scanlon et al., 1999, Zaher & Rieppel, 1999, Caldwell, 2000, Rage & Escuillié, 2000, 2003a–b, Rieppel & Zaher, 2000, Tchernov et al., 2000, Caldwell & Albino, 2001, M.S.Y. Lee & Scranlon, 2002, Scanlon & Lee, 2002, Rieppel et al., 2003, Polcyn et al., 2005 and Bardet et al., 2008.
Remarks: A lizard *fide* Rage, 1984b. A snake *fide* Caldwell & Lee (1997: 705).

1. †*Pachyrhachis problematicus* G. Haas, 1979. Bull. Mus. Natl. Hist. Nat. (4) 1C(1): 61–64, figs. 1–6.

Synonym: †*Ophiomorphus colberti* G. Haas, 1980b.
Type: Holotype, HUJ-PAL 3659, nearly complete skeleton of a 1700 mm specimen.
Type locality: "Lower Cenomanian banked limestone of Ein Jabrud, near Ramallah, 20 km North of Jerusalem on the El-Bira – Jericho road" [= Ein Jabrud, West Bank, cen. Israel; upper Cretaceous].
Distribution: Upper Cretaceous (Cenomanian: 93.9–100.5 mya) of Israel.

†*PALAEONATRIX* Szyndlar *in* Mlynarski, Szyndlar, Estes & Sanchiz, 1982 (Natricidae)

Synonym: †*Dolniceophis* Rage & Rocek, 1983.
Type species: †*Palaeonatrix silesiaca* Szyndlar *in* Mlynarski, Szyndlar, Estes & Sanchiz, 1982.
Distribution: Lower Miocene of Czech Republic and Germany, and middle Miocene of France and Poland. Fossils unassigned to species include lower Miocene (Orleanian, MN 4: 16.0–16.9 mya) of Austria.
Sources: Rage, 1984b and Szyndlar, 1991b, 1998.

1. †*Palaeonatrix lehmani* (Rage & Rocek, 1983). Casopis Min. Geol. 28(1): 18–19, pls. 1, pl. 2, fig 1. (†*Dolniceophis lehmani*)

Type: Holotype, CUP 3930, one trunk vertebra.
Type locality: "Dolnice locality near the town of Cheb, West Bohemia, Czechoslovakia, 50°04'N, 12°22'E; Ottnangian, lower fossiliferous level Dolnice 1 (fine calcareous silts), Lower Miocene."
Distribution: Lower Miocene (Orleanian, MN 4: 16.0–16.9 mya) of Czech Republic, France and Germany, and middle Miocene (Astaracian, MN 6: 12.8–13.7 mya) of France.
Sources: Szyndlar, 1987, Szyndlar & Schleich, 1993, Augé and Rage, 2000, Rage & Bailon, 2005 and Szyndlar, 2005.

2. †*Palaeonatrix silesiaca* Szyndlar *in* Mlynarski, Szyndlar, Estes & Sanchiz, 1982. Estud. Geol. 38: 114–115, fig. 10.

Type: Holotype, ZPW OP 86/24, one trunk vertebra (Z. Ryziewicz, 1950–1958).

Type locality: "Nowa Wies Królewska II (NWK II), Opole, Silesia, southern Poland; Late Miocene (MN 7)" [= Sarmatian (Astaracian), middle Miocene *fide* Rage, 1984b: 50].
Distribution: Middle Miocene (Astaracian, MN 7–8: 11.1–12.8 mya) of Poland. Known only from type series.
Source: Szyndlar, 1984a.

†*PALAEOPHIS* Owen, 1840 (†Palaeophiidae)

Synonyms: †*Dinophis* Marsh, 1869 *(nomen praeoccupatum)*, †*Paloeophis* – Beneden, 1871 *(nomen incorrectum)*, †*Titanophis* Marsh, 1877 *(nomen substitutum)*, †*Paleophis* – Palacky, 1898 *(nomen incorrectum)*, †*Paleophis* – Lynn, 1934 *(nomen emendatum)*, †*Palelophis* – Huene, 1956 *(nomen incorrectum)*, †*Vialovophis* Nessov & Odovitschenko, 1984, and †*Nessovophis* Averianov, 1997.
Type species: †*Palaeophis toliapicus* Owen, 1840.
Distribution: Upper Cretaceous and Paleocene of Morocco, upper Paleocene of Kazakhstan, upper Paleocene or lower Eocene of USA, lower Eocene of Belgium, Denmark, France, Germany, India, Ivory Coast, Kyrgystan, Mali, Morocco, United Kingdom and USA, middle Eocene of Belgium, Italy, Mali, Nigeria, Ukraine, United Kingdom, Uzbekistan, and USA, and upper Eocene of Kazakhstan and Ukraine.
Sources: Marsh, 1877, Gilmore, 1938, Kuhn, 1939a, 1963, Holman, 1979b, 1982c, 2000a, Rage & Wouters, 1979, Rage, 1984b, Rage & Holman, 1984 and Rage et al., 2003, 2008.
Remarks: Upper Eocene record from United Kingdom doubtful *fide* Rage (1983b: 227).

1. †*Palaeophis africanus* Andrews, 1924. Bull. Geol. Surv. Nigeria 7: 43, 39–41, figs. 1a–c.

Type: Holotype, BMNH 4964, one trunk vertebra (J.D. Falcóner).
Type locality: "Nigeria, the exact locality being Oshosun, Abeokuta Province on the railway 27 1/4 miles north of Lagos" [= Ososun, Ogun State, SW Western Region, ext. SW Nigeria; Oshoshun formation, middle Eocene *fide* Rage, 1984b: 34].
Distribution: Middle Eocene (37.2–48.6 mya) of Nigeria.

2. †*Palaeophis casei* Holman, 1982c. J. Vert. Paleont. 2(2): 163–166, fig. 1.

Type: Holotype, PU 23488, one trunk vertebra (G.R. Case and P. Borodin, Nov.–Dec. 1980).
Type locality: "Case Estuary Number 1 Site at Meridian, Lauderdale County, Mississippi, at latitude 32°25'N, 89°27'W; the Bashi Marl, an upper unit of the Wilcox Group, early Eocene," (partially in error). Corrected to Tuscahoma formation, at Sec. 6, T6N, R16E *fide* Parmley and Case (1988: 334).
Distribution: Upper Paleocene (Clarkfordian: 55.4–56.8 mya) or lower Eocene (Wasatchian: 50.3–55.4 mya) of USA (Mississippi). Known only from type locality.
Sources: Parmley & Case, 1988, Holman et al., 1991 and Weems, 1999.

3. †*Palaeophis colossaeus* Rage, 1983a. C.R. Acad. Sci. Paris (3) 296(22): 1741–1743, figs. 1a–e.

Type: Holotype, MNHN TGE 614, one middle trunk vertebra from a 9000 mm specimen.
Type locality: "butte de Tamaguilelt, Mali; Lutétien (Eocène moyen)" [= Tamaguélelt hills, S Gao Region, E Mali].
Distribution: Middle Eocene (Lutetian: 37.2–48.6 mya) of Mali. Known only from type series.

4. †*Palaeophis ferganicus* Averianov, 1997. Russ. J. Herp. 4(2): 132–135, fig. 4.

Type: Holotype, ZISP PC 1/35, one middle trunk vertebra (A.O. Averianov, 1988).
Type locality: "Andarak 2, Lyailyak District, Osh Region, Kyrgystan. Early Eocene (early Ypresian)."
Distribution: Lower Eocene (Ypresian, MP 7: 48.6–55.8 mya) of Kyrgystan.

5. †*Palaeophis grandis* (Marsh, 1869). Amer. J. Sci. Arts (2) 48(144): 397–400. (†*Dinophis grandis*)

Type: Holotype, YPM 2762, one trunk vertebra (W.S. Kimball).
Type locality: "the Eocene greensand, near Shark River, Monmouth county, New Jersey." Emended to Shark River formation *fide* Gilmore (1938: 51).
Distribution: Lower-middle Eocene (38.0–55.4 mya) of USA (Maryland, New Jersey).
Remarks: Possibly a synonym of †*Palaeophis littoralis* Cope *fide* Gilmore (1938: 49) or †*P. typhaeus* Owen *fide* Rage (1984b: 34).

6. †*Palaeophis littoralis* Cope, 1869a. Proc. Acad. Nat. Sci. Philadelphia (1868) 20(4): 234.

Synonym: †*Palaeophis littoralis* Cope, 1868d *(nomen nudum)*.
Types: Syntypes (3), AMNH 2392 (formerly GSNJ), three fragmentary vertebrae (Knieskern), lost *fide* H. Miller (1955: 2).
Type locality: "green sand of Squankum, Monmouth Co., N. J" [= New Jersey, USA]. Emended to Eocene green sand bed of Shark River *fide* Cope (1869a: 234), Manasquam Marl, Eocene *fide* Gilmore (1938: 48), and Ypresian, lower Eocene *fide* Rage (1984b: 34).
Distribution: Upper Paleocene (Clarkforkian: 55.4–56.8 mya) of USA (Mississippi) and lower to middle Eocene

P

(38.0–55.4 mya) of USA (Alabama, Mississippi, New Jersey).

Sources: Cope, 1869a, H. Miller, 1955, Parmley and Case, 1988 and Holman et al., 1990.

Remarks: Possibly a synonym of †*Palaeophis toliapicus* Owen *fide* Rage (1984b: 34).

7. †*Palaeophis maghrebianus* Arambourg, 1952. Notes Mém. Serv. Géol. Maroc (92): 291–294, fig. 58, pl. 41, fig. 10.

Type: Holotype, MNHN APH 5, one trunk vertebra.

Type locality: "Djemaïa; les niveaux ypéslens des ouled Abdoundu Maroc" [= Douar Jemaa, Fez Prov., N Morocco; Ypresian, lower Eocene].

Distribution: Lower Eocene (Ypresian: 48.6–55.8 mya) of Morocco.

Source: Bardet et al., 2010.

8. †*Palaeophis nessovi* Averianov, 1997. Russ. J. Herp. 4(2): 135, figs. 3f–i.

Type: Holotype, ZISP PC 1/36, one middle trunk vertebra.

Type locality: "Kachar, iron ore quarry, 52 km W of Kustanay city, Kustanay Province, northern Kazakhstan, Upper Eocene (Priabonian)."

Distribution; Upper Eocene (Priabonian: 33.9–37.2 mya) of Kazakhstan. Known only from type locality.

Source: Kordikova, 1998.

9. †*Palaeophis tamdy* (Averianov, 1997). Russ. J. Herp. 4(2): 131–132, figs. 2a–j. (†*Nessovophis tamdy*)

Type: Holotype, ZISP PC 7/34, one anterior trunk vertebra.

Type locality: "Dzheroi 2, white quartz sands in a large hollow near the eastern Muruntau-Tamdy road, Central Kyzylkum Desert, Bukhara Viloyat, Uzbekistan. Middle Eocene (Bartonian)."

Distribution: Middle Eocene (Bartonian: 37.2–40.4 mya) of Uzbekistan. Known only from type locality.

10. †*Palaeophis toliapicus* Owen, 1840. Proc. Geol. Soc. London (1839–1840) 3(66): 165–166.

Types: Syntypes (2), BMNH 38998 and BMNH 39447 (formerly RCSM), two nodules with several vertebrae (J. Hunter & J.S. Bowerbank).

Type locality: "London clay from the Isle of Sheppey" [= Kent, England, United Kingdom; lower Eocene: upper Cuisian *fide* Rage (1983b: 222) or Ypresian *fide* Rage, 1984b: 34].

Distribution: Lower Eocene (Ypresian or Cuisian: 48.6–55.8 mya) of Belgium, Denmark, France, United Kingdom and USA (Mississippi).

Sources: Owen, 1841a, Lydekker, 1888, Rage, 1983b, Augé et al., 1997 and Weems, 1999.

Remarks: Original description reprinted in Owen (1841b: 209–210, pl. 22).

11. †*Palaeophis typhaeus* Owen, 1850. Palaeont. Soc. Monogr. 1(3): 56–61, pl. 13, figs. 5–8, pl. 14, figs. 1–3, 7–9, 16–17, 26–28.

Synonyms: †*Palaeophis porcatus* Owen, 1850 and †*Palaeophis oweni* Zigno, 1882.

Type: Holotype, BMNH, one trunk vertebra illustrated in pl. 13, figs. 5–8 (F. Dixon, J.S. Bowerbank & G.A. Combe), lost *fide* Rage (1984b: 34).

Type locality: "Eocene clay at Bracklesham, Sussex" [= Bracklesham, Sussex, England, United Kingdom; Bracklesham group, Lutetian or uppermost Ypresian *fide* Rage (1984b: 34); 50°45'N, 0°54'E].

Distribution: Lower Eocene (Ypresian: 48.6–55.8 mya) of United Kingdom (England) and middle Eocene (Cuisian and Lutetian: 40.4–48.6 mya) of Belgium, Italy, United Kingdom and ? France.

Sources: Lydekker, 1888, Rage, 1983b and Augé et al., 1997.

Remarks: Lydekker (1888: 259) listed 4 syntypes (BMNH 25600, 33207, 33209–10).

12. †*Palaeophis udovichenkoi* Averianov, 1997. Russ. J. Herp. 4(2): 135–137, figs. 5a–e.

Type: Holotype, ZISP PC 8/37, one middle trunk vertebra.

Type locality: "Prolom, limestone quarry 8 km NE of Belogorsk city in the Crimea, Ukraine, Middle-?Upper Eocene (Bartonian-?Priabonian)."

Distribution: Middle or upper Eocene (Bartonian: 37.2–40.4 mya or Priabonian: 33.9–37.2 mya) of Ukraine. Known only from type locality.

13. †*Palaeophis vastaniensis* Bajpai & Head, 2007. Gondwana Geol. Mag. 22(2): 86–88, figs. 2a–c.

Type: Holotype, IITR/SB/VLM 914, three articulated precloacal vertebrae.

Type locality: "Vastan Lignite Mine, Taluka Mangrol, District Surat, Gujarat, western India; early Eocene Cambay Shale, Shallow Benthic Zone 8 *(N. globulus* Zone)."

Distribution: Lower Eocene (Ypresian: 48.6–55.8 mya) of India.

Source: Rage et al., 2008.

14. †*Palaeophis virginianus* Lynn, 1934. John Hopkins Univ. Stud. Geol. 11: 245–249, pl. 17, figs. 1–6. (†*Paleophis virginianus*)

Types: Syntypes (2), USNM 13640–41, two trunk vertebrae (W.G. Lynn, 2 Dec. 1933 and 27 March 1934).

Type locality: "Belvedere Beach, Virginia, in Eocene deposits of the Aquia Creek stage" [= Ypresian (?Cuisian), lower Eocene *fide* Rage, 1984b: 34].

Distribution: Lower Eocene (Wasatchian: 50.3–55.4 mya) of USA (Alabama, Georgia, Mississippi, New Jersey, Virginia).

P

Sources: Holman, 1977a, Holman and Case, 1988, Parmley and Case, 1988, Holman et al., 1991 and Weems, 1999.

15. †*Palaeophis zhylan* (Nessov and Udovitschenko, 1984). Paleont. Sbornik 21: 71–72, figs. 18–20. (†*Vialovophis zhylan*)

Type: Holotype, AHCCCP 18/12173, one middle trunk vertebra (L.A. Nessov, 1977).
Type locality: "Kazakh Republic, Chimkent Region, Quarry Dzilga [= Zhylga], Paleogene."
Distribution: Upper Paleocene and Pleistocene (Gelasian: 1.8–2.6 mya) of Kazakhstan.
Source: Kordikova, 1998.

16. †*Palaeophis zhylga* (Averianov, 1997). Russ. J. Herp. 4(2): 132, figs. 3a–e. (†*Nessovophis zhylga*)

Type: Holotype, ZISP PC 45/32, one anterior trunk vertebra (A.O. Averianov, I.G. Danilov, L.A. Nessov, O.R. Potapova, or A.S. Rezvyi, 1989, 1990 or 1996).
Type locality: "Zhylga 1b, the white sand quarry near Zhylga 1a, the white quartz sands quarry, 10 km NW to Zhylga railway station, Chimkent Province, Kazakhstan. Lower Eocene (Ypresian)."
Distribution: Lower Eocene (Ypresian: 48.6–55.8 mya) and upper Paleocne of Kazakhstan.
Source: Kordikova, 1998.
Remarks: Possibly a member of the †Nigerophiidae *fide* Rage et al. (2003: 700).

†*PALAEOPYTHON* Rochebrune, 1880 (Pythonidae)

Synonyms: †*Paleopython* Palacky, 1898 *(nomen emendatum)*, †*Palaeophyton* – Nopcsa, 1926 *(nomen incorrectum)*, †*Palaeopithon* – Ginsburg, 1970 *(nomen incorrectum)*, and †*Palaepython* – Rage, 1974 *(nomen incorrectum)*.
Type species: †*Python cadurcensis* Filhol, 1877.
Distribution: Middle Eocene of France and Germany, upper Eocene of France and Switzerland, and ? lower Oligocene of France.
Sources: Rochebrune, 1880, 1884, Lydekker, 1888, Kuhn, 1939a–b, 1963, Rage, 1974, 1984b and Rage & Augé, 2010.

1. †*Palaeopython cadurcensis* (Filhol, 1877). Ann. Sci. Nat. Géol., Paris 8: 490–493, pls. 52–53, figs. 435–444. (†*Python cadurcensis*)

Types: Syntypes (8), MNHN QU 16317, one left dentary [figs. 437–438], MNHN QU 16318, five articulated middle trunk vertebrae [figs. 441–442], MNHN QU 16319, one middle trunk vertebra [figs. 439–440, 443–444], and MNHN QU 16321, one right maxilla [figs. 435–436] (H. Filhol).
Type locality: "divers gisements des les Phosphorites du Quercy, près de Caylux, des départements du Lot, du Tarn-et-Garonne, et de l'Aveyron, l'éocène supérieur dans cette partie du sud-ouest de la France" [probably Headonian, upper Eocene *fide* Rage, 1984b: 19].
Distribution: Middle Eocene (Bartonian: 37.2–40.4 mya) and upper Eocene (Priabonian: 33.9–37.2 mya) of France and ? Switzerland.

2. †*Palaeopython ceciliensis* Barnes, 1927. Jahrb. Halle. Verband. Erforsch. Mitteldeutsch. Bodensch. (2) 6: 9–10, pl. 1, fig. 7.

Type: Holotype, GM 2978, one partial trunk vertebra (J. Walther, 1925).
Type locality: "Grube Cecilie, der Braunkohle des Geiseltals; obere mittel Kohle, subherzynen Braunkohlen formation, Eozäne" [=SW of Halle, Germany; middle Rhenanian (= Lutetian), Middle Eocene *fide* Rage, 1984b: 19].
Distribution: Middle Eocene (Lutetian: 40.4–48.6 mya) of Germany.
Remarks: Erroneously allocated to †*Paleryx* by Kuhn (1939b: 127 and 1963: 14) *fide* Rage (1984b: 20). The incomplete vertebra (holotype) is unsatisfactory for accurate comparison *fide* Szyndlar & Böhme (1993: 392).

3. †*Palaeopython filholii* Rochebrune, 1880. Nouv. Arch. Mus. Hist. Nat. (2) 3: 277–278, pl. 12, figs. 5a–c.

Type: Holotype, MNHN QU 16322 (formerly QU 322), one posterior trunk vertebra.
Type locality: "Eocène.—Phosphorites du Quercy" [France; probably Headonian, upper Eocene *fide* Rage, 1984b: 19].
Distribution: Upper Eocene (Priabonian: 33.9–37.2 mya) of France.
Remarks: Lower Eocene *fide* Rage (1974: 287).

4. †*Palaeopython fischeri* Schaal, 2004. Cour. Forsch.-Inst. Senck. (252): 36–44, figs. 1–7.

Type: Holotype, SMF ME 929, one trunk vertebra.
Type locality: "Grube Messel bei Darmstadt, Hessen, Deutschland. Mittleres Eozän, Unteres Lutetium, MP 11, Messel-Ölschiefer-Formation."
Distribution: Middle Eocene (Lutetian, MP 11: 48.6 mya) of Germany.

5. †*Palaeopython neglectus* Rochebrune, 1884. Mém. Soc. Sci. Nat. Saóne et Loire 5: 154–155, pl. 2, figs. 5a–b.

Type: Lectotype, MNHN QU 16326, one trunk vertebra, designated by Rage (1984b: 20).

P

Type locality: "Phosphorites du Quercy, France; upper Eocene or Oligocene" via lectotype selection.
Distribution: Upper Eocene (Priabonian: 33.9–37.2 mya) or Oligocene (Rupelian/Chattian: 23.0–33.9 mya) of France. Known only from lectotype.
Remarks: Generic allocation dubious *fide* Rage (1984b: 20).

†*PALEOFARANCIA* Auffenberg, 1963 (Carphophiidae)

Type species: †*Paleofarancia brevispinosus* Auffenberg, 1963.
Distribution: Upper Miocene of USA.
Sources: Holman, 1979b, 2000a and Rage, 1984b.

1. †*Paleofarancia brevispinosus* Auffenberg, 1963. Tulane Stud. Zool. 10(3): 169, fig. 16.

Type: Holotype, UF 5566, one middle trunk vertebra (S. Telford, Sr., 1952).
Type locality: "bottom of the stream formed by Lithia Springs, close to its head near the Alafia River, Hillsborough County, Florida (R21E, T30S, Sec. 16); Alachua formation, Middle Pliocene (?), Hemphillian."
Distribution: Upper Miocene (Hemphillian: 4.9–10.3 mya) of USA (Florida). Known only from type locality.
Remarks: Modified description provided in Holman (2000a: 121–122). Age questionable *fide* Auffenberg (1963: 135) and Rage (1984b: 50).

P

†*PALEOHETERODON* Holman, 1964 (Xenodontidae)

Type species: †*Paleoheterodon tiheni* Holman, 1964.
Distribution: Middle Miocene of USA and France, upper Miocene of USA, and upper Miocene–lower Pliocene of USA. Fossils unassigned to species include middle/upper Miocene (Clarendonian) of USA (Kansas).
Sources: Holman, 1979b, Rage, 1984b and Parmley & Hunter, 2010.

1. †*Paleoheterodon arcuatus* Rage and Holman, 1984. Géobios 17(1): 93–94, figs. 3a–e.

Type: Holotype, UCBL 285 001, one trunk vertebra.
Type locality: "fissure M de La-Grive-Saint-Alban (Isère), France; zone MN 7, Astaracien, the French Miocene."
Distribution: Middle Miocene (Astaracian, MN 6–8: 11.6–13.7 mya) of France.
Sources: Augé & Rage, 2000 and Szyndlar, 2012.

2. †*Paleoheterodon tiheni* Holman, 1964. Copeia 1964(4): 633, fig. 1.

Type: Holotype, UNSM 61031, one trunk vertebra (J.A. Tihen & C.A. Chantell, summer 1962 or 1963).
Type locality: "Near the Norden-Johnstown Road, about 300 yards south of the bridge across the Niobrara River; on the Ed McCall Ranch, SE 1/4, SW 1/4 of Sect. 33, T 33 N, Rm 23 W; Brown County, Nebraska; Valentinian; Valentine formation; Norden Bridge Quarry" [= Barstovian, middle Miocene *fide* Rage, 1984b: 50].
Distribution: Middle Miocene (Barstovian: 13.6–16.3 mya) of USA (Nebraska, South Dakota), upper Miocene (Clarendonian: 10.3–13.6 mya and Hemphillian: 16.3–20.8 mya) of USA (Nebraska, Texas), and upper Miocene–upper Pliocene (Hemphillian-Blancan: 1.8–10.3 mya) of USA (Indiana).
Sources: R.L. Wilson, 1968, Holman, 1973a, 1975, 1977e, 1978, 1987a, 1995b, 2000a, Voorhies et al., 1987 Parmley, 1988 and Parmley & Holman, 1995.

†*PALERYX* Owen, 1850 (Pythonidae)

Synonyms: †*Palaeryx* – Massalongo, 1859 *(nomen incorrectum)*, and †*Paleoeryx* – Palacky, 1898 *(nomen incorrectum)*.
Type species: †*Paleryx rhombifer* Owen, 1850.
Distribution: Middle Eocene of Germany, and upper Eocene of United Kingdom.
Source: Rage, 1984b.
Remarks: *Paleryx cayluxi* Stefano is *incertae sedis fide* Rage (1984: 22).

1. †*Paleryx rhombifer* Owen, 1850. Palaeont. Soc. Monogr. 1(3): 67, pl. 13, figs. 29–32.

Synonym: †*Paleryx depressus* Owen, 1850.
Type: Holotype, BMNH 25259, one anterior trunk vertebra (A.P. Falcóner).
Type locality: "the Eocene sand at Hordwell" [= Hordwell, Hampshire, England; Headonian, upper Eocene *fide* Rage, 1984b: 20].
Distribution: Upper Eocene (Priabonian, MP 17–20: 33.9–37.2 mya) of United Kingdom (England).
Sources: Lydekker, 1888 and Rage & Ford, 1980.

2. †*Paleryx spinifer* Barnes, 1927. Jahrb. Halle. Verband. Erforsch. Mitteldeutsch. Bodensch. (2) 6: 7–9, pl. 1, figs. 1–6.

Types: Syntypes (5), GM 2979, articulated vertebrae, GM 5822, isolated vertebrae, GM 5823, one parietal, GM 5824, one pterygoid, and GM 5826, one maxilla (J. Walther, 1925).
Type locality: "Grube Cecilie, der Braunkohle des Geiseltals; obere mittel Kohle, Subherzynen

Fraunkohlen formation, Eozäne" [= SW of Halle, Germany; middle Rhenanian (= Lutetian), middle Eocene *fide* Rage, 1984b: 20].

Distribution: Middle Eocene (Lutetian: 40.4–48.6 mya) of Germany.

Remarks: Generic allocation doubtful *fide* Rage (1984b: 20) and Szyndlar & Böhme (1993: 392). Kuhn's (1939b: 125) assignment of skeleton to this species invalid *fide* Rage (1984b: 20).

PANTHEROPHIS Fitzinger, 1843 (Colubridae)

Synonyms: *Scotophis* Baird & Girard, 1853, *Scotophisus* A.L. Herrera, 1899 *(nomen emendatum)*, †*Palaeoelaphe* Gilmore, 1938, and *Mintonius* Collins & Taggart, 2008.

Type species: *Coluber guttatus* Linnaeus, 1758.

Distribution: Southern Canada, E United States and NE Mexico.

Fossil records: Middle Miocene to upper Pleistocene of S-cen. Canada and USA. Fossils unassigned to species include middle/upper Miocene (Clarendonian) of USA (Kansas), upper Miocene/lower Pliocene (Hemphillian) of USA (Indiana), and upper Pleistocene (Rancholabrean II) of USA (South Carolina).

Sources: Dowling et al., 1983, Schulz, 1996, Holman, 2000a, Utiger et al., 2002, 2005, Burbrink & Lawson, 2007, Collins & Taggart, 2008 and Pyron & Burbrink, 2009a.

Remarks: A synonym of *Pituophis fide* Burbrink & Lawson (2007: 186–187). Collins & Taggart (2008: 16) proposed *Mintonius* for *vulpinus* and *gloydi*, resurrected *Scotophis* for *bairdi, obsoletus, alleghaniensis*, and *spiloides* and retained *emoryi, guttatus*, and *slowenskii* in *Pantherophis*. *Pantherophis* should include *Mintonius* and *Scotophis* and exclude *Pituophis fide* Pyron & Burbrink (2009a: 528).

1. *Pantherophis alleghaniensis* (Holbrook, 1836). No. Amer. Herp. 1: 111–112, pl. 20. (*Coluber alleghaniensis*)

Synonyms: *Coluber quadrivittata* Holbrook, 1836, *Scotophis confinis* Baird & Girard, 1853, *Elaphis holbrooki* A.-M.-C. Duméril, 1853 (*nomen nudum*), *Elaphis holbrookii* Duméril, Bibron & Dumril, 1854a, *Scotophis obsoletus lemniscatus* Cope, 1889a, *Elaphe quadrivittata deckerti* Brady, 1932, *Elaphe williamsi* T. Barbour & Carr, 1940, *Elaphe quadrivitatta parallela* T. Barbour & Engels, 1942, and *Elaphe obsoleta rossalleni* Neill, 1949.

Types: Syntypes (2), ANSP 16792 (Wilkins), location of second syntype unknown.

Type locality: "Highlands of the Hudson" [New York] and "summit of the Blue Ridge in Virginia."

Distribution: Eastern USA (Connecticut, Florida, Delaware, Georgia, Maryland, Massachusetts, New Jersey, New York, E and cen. North Carolina, E Pennsylvania, Rhode Island, South Carolina, W Vermont, E Virginia, E West Virginia)

Fossil records: Lower Pleistocene (Irvingtonian I) of USA (Florida) and upper Pleistocene (Rancholabrean II) of USA (Florida, Georgia, Pennsylvania, Virginia).

Sources: Burbrink et al., 2000, Burbrink, 2001 and Gibbs et al., 2006.

Remarks: Malnate (1971: 362) corrected collector of type from Wilkens to Wilkins. Hybrids exist between *P. alleghaniensis* and *P. spiloides fide* Gibbs et al., 2006.

2. *Pantherophis bairdi* (Yarrow *in* Cope, 1880). Bull. U.S. Natl. Mus. (17): 41. (*Coluber bairdi*)

Type: Holotype, USNM 10403 (W.F. von Manteuffel, 1878).

Type locality: "near Fort Davis, Tex. This post is in the Apache Mountains, 50 miles from the Mexican border or Rio Grande, Northwest of Presidio del Norte" [= Jeff Davis Co., Texas].

Distribution: South-central USA (W Texas) and NE Mexico (N Coahuila, Nuevo León, W Tamaulipas), 900–1800 (2900) m.

Sources: Olson, 1977, R. Lawson & Lieb, 1990, Schulz & Philippes, 1991, Cranston & Cranston, 1996, Werler & Dixon, 2000, Burbrink et al., 2000, Burbrink, 2001 and Farr et al., 2009.

3. †*Pantherophis buisi* (Holman, 1973b). Copeia 1973(3): 575–578, figs. 1–2. (†*Elaphe buisi*)

Type: Holotype, UMMP V-57403, a partial skeleton (C. Hibbard & W. Akersten, Map, 1968).

Type locality: "Ogallala Formation, Upper Middle Pliocene. Buis Ranch local fauna, Beaver County, Oklahoma, NE ¼ SW ¼ Sec. 5 T 5 N, R 26 E. C. M. 36°45' N, 100°30' W."

Distribution: Lower Pliocene (Hemphillian: 3.6–4.9 mya) of USA (Oklahoma). Known only from type locality.

Remarks: Characters fit between *Pantherophis obsoletus* and *Bogertophis subocularis fide* Holman (2000a: 160).

4. *Pantherophis emoryi* (Baird & Girard, 1853). Cat. No. Amer. Rept., Serp. 1: 157–158. (*Scotophis emoryi*)

Synonyms: *Coluber rhinomegas* Cope, 1860c, *Elaphe laeta intermontanus* Woodbury & Woodbury, 1942, *Elaphe quivira* Burt, 1946 (*nomen nudum*), and *Elaphe guttata meahllmorum* H.M. Smith, Chiszar, Staley & Tepedelen, 1994.

Type: Holotype, USNM 2257, a 1043–1048 mm specimen (J.H. Clark [U.S.-Mex. Bound. Surv.], 1848–1852, via J.D. Graham), lost *fide* Dowling (1951: 43).

Type locality: "Howard Springs, Tex." [= 20 miles SW Ozona, Crockett Co., Texas, USA *fide* Dowling, 1951: 43].

P

Distribution: South-central and W USA (Arkansas, W and SE Colorado, SW Illinois, cen. N and W Kansas, S Missouri, SE Nebraska, New Mexico, Oklahoma, cen. and W Texas, E Utah) and NE Mexico (E Chihuahua, Coahuila, NE Durango, N Hidalgo, Nuevo León, ext. N Querétaro, San Luis Potosí, Tamaulipas, ext. N Veracruz, NE Zacatecas), NSL–1900 m.

Fossil records: Upper Miocene (Hemphillian) of USA (Nebraska), middle Pliocene (Blancan III) of USA (Texas), upper Pliocene (Plancan V) of USA (Nebraska), middle/upper Pleistocene (Irvingtonian II) of USA (Arkansas, Texas), and upper Pleistocene (Rancholabrean II) of USA (New Mexico, Texas).

Sources: Baird, 1859, Dowling, 1951, Fitch, 1999, Hammerson, 1999, Werler & Dixon, 2000, Burbrink, 2002, Lemos-Espinal et al., 2004c and H.M. Smith, 2005b.

Remarks: Holotype illustrated in Baird (1859: pl. 12). Yarrow (1882: 101) erroneously listed type locality as Fort Smith, Arkansas with Whipple as collector. Elevated to specific status *fide* Burbrink (2001: 473).

5. *Pantherophis guttatus* (Linnaeus, 1766). Syst. Nat., ed. 12, 1: 385–386. (*Coluber guttatus*)

Synonyms: *Coluber maculatus* Lacépède, 1789 (*nomen rejiciendum*), *Coluber maculatus* Bonnaterre, 1790, *Coluber ludovicianus* Suckow, 1797, *Coluber compressus* Donndorff, 1798, *Coluber moniligera* Latreille *in* Sonnini & Latreille, 1801b, *Coluber carolinianus* G. Shaw, 1802, *Coluber molossus* Daudin, 1803c, *Coluber pantherinus* Daudin, 1803c, ? *Coluber zeae* Gravenhorst, 1807, ? *Coluber zonatus* Gravenhorst, 1807, *Coluber floridanus* Harlan, 1827, *Coluber guttatus sellatus* Cope, 1889a, and *Coluber rosaceus* Cope, 1889a.

P

Type: Holotype, ZMUU 147 (formerly JA 1749) (A. Garden, 1760–1761).

Type locality: "Carolina" [USA]. Restricted to vicinity of Charleston, South Carolina, USA *fide* Klauber (1948: 6).

Distribution: Southeastern USA (Alabama, S Delaware, Florida: Bahia, Boca Grande, Honda, Indian, Key West, Lower Matecumbe, Vaca, Little Pine, Little Torch, Little Pine, Big Pine, Middle Torch, Ramrod, Stock, Summerland and Vaca Is.; Georgia, cen. Kentucky, E Louisiana, Maryland, Mississippi, S New Jersey, S North Carolina, South Carolina, S Tennessee, Virginia, ext. W West Virginia) and introduced into Mexico (Hidalgo), Grand Cayman, Martinique, Turks and Caicos Is. (Grande Turk) and U.S. Virgin Is. (St. Thomas), NSL–1830 m.

Fossil records: Lower Pleistocene (Irvingtonian I) of USA (Florida), and upper Pleistocene (Rancholabrean II) of USA (Alabama, Florida, Georgia, Pennsylvania, Tennessee, Virginia).

Sources: J.C. Mitchell, 1977, Christman, 1980, Ashton & Ashton, 1988, Ernst & Barbour, 1989, Conant & Collins, 1991, H.M. Smith et al., 1994, Palmer & Braswell, 1995, Vaughan et al., 1996, Tennant, 1997, Quijano et al., 1998, W. Schmidt, 1999, D. Bartlett & Tennant, 2000, D. Schmidt, 2000, Burbrink, 2002, Kunz, 2004, G. Köhler & Berg, 2005, R.W. Henderson & Powell, 2009, Pritzel, 2011 and Reynolds & Riggs, 2011.

6. †*Pantherophis kansensis* (Gilmore, 1938). Geol. Soc. Amer., Spec. Pap. (9): 63–64, figs. 24a–b, 25a–d. (†*Palaeoelaphe kansensis*)

Synonym: †*Elaphe nebraskensis* Holman, 1964.

Type: Holotype, USNM 13500, several disarticulated vertebrae, including one cervical vertebra [fig. 24] and one trunk vertebra [fig. 25] (J.B. Hatcher, Oct. 1884).

Type locality: "'Quarry E', Long Island, Phillipps Co., Kans. Republican River formation, Lower Pliocene."

Distribution: Middle Miocene (Clarendonian and Barstovian: 10.3–16.3 mya) of USA (Nebraska, South Dakota, Texas) and upper Miocene (Hemphillian: 4.9–10.3 mya) of Canada (Saskatchewan) and USA (Kansas, Nebraska, Texas).

Fossil records: Middle Miocene (Barstovian) of Canada (Saskatchewan) and USA (Nebraska, South Dakota, Texas), upper Miocene (Clarendonian) of USA (Nebraska), upper Miocene (Clarendonian/Hemphillian) of USA (Kansas), and upper Miocene (Hemphillian) of USA (Nebraska, Texas).

Sources: Auffenberg, 1963, Holman, 1964, 1970, 1973a, 1977b–c, 1979a, 1987a, 1995a, Rage, 1984b, Parmley & Holman, 1995 and Parmley & Hunter, 2010.

Remarks: Type vertebrae may originate from more than one individual *fide* Gilmore (1938: 64). Revised diagnosis in Holman (2000a: 162). Closer to *Pantherophis vulpinus* than *P. obsoletus fide* Holman (2000a: 162–163).

7. *Pantherophis obsoletus* (Say *in* E. James, 1823). Exped. Pittsburgh–Rocky Mts. 1: 140. (*Coluber obsoletus*)

Synonyms: *Coluber reticulatus* Lacépède, 1789 (*nomen rejiciendum*), *Scotophis laetus* Baird & Girard, 1853, and *Scotophis lindheimeri* Baird & Girard, 1853.

Types: Syntypes (3), ANSP (T. Say, 29 Aug.–26 Sept., 1819), longest syntype 1515 mm, lost *fide* Dowling (1952b: 4).

Type locality: "the Missouri from the vicinity of Isle au Vache to Council Bluff" [= Missouri River between Cow Island (39°27'N, 94°58'W, elevation 300 m), W Platte Co., NW Missouri and Council Bluffs (41°15'N, 95°52'W, elevation 240 m), SW Pottawattamie Co., SE Iowa, USA]. Restricted to Council Bluffs, Iowa, USA *fide* K.P. Schmidt (1953a: 197) and vicinity of Cow Island, Leavenworth Co., Kansas, USA *fide* Dundee (1996b: 78).

Distribution: Central USA (Arkansas, S, E Iowa, E Kansas, Louisiana, ext. SE Minnesota, Missouri, SE Nebraska, Oklahoma, cen. and E Texas), NSL–1200 m.

Fossil records: Upper Miocene (Hemphillian) of USA (Nebraska), lower Pliocene (Hemphillian) of USA (Kansas, Oklahoma), middle Pliocene (Blancan II, III) of USA (Kansas), upper Pliocene (Blancan V) of USA (Kansas), lower/middle Pleistocene (Irvingtonian I) of USA (Kansas), and upper Pleistocene (Rancholabrean II) of USA (Kansas, Texas).

Sources: Neill, 1949a, A.H. Wright & Wright, 1957, Burbrink et al., 2000, D. Schmidt, 2000, 2004a, Werler & Dixon, 2000 and Burbrink, 2001.

8. †*Pantherophis pliocenicus* (Holman, 1968). Copeia 1968(1): 155–156, fig. 3. (†*Elaphe pliocenica*)

Type: Holotype, MSUVP V-53170, one trunk vertebra from a 1070 mm specimen.

Type locality: "Glenns Ferry Formation, upper Pliocene, Hagerman Quadrangle, NW 1/4, Sec. 5, T 8 S, R 13 E, Twin Falls County, Idaho."

Distribution: Middle Pliocene (Blancan: 3.6 mya) of USA (Idaho) and upper Pliocene (Blancan) of USA (Washington).

Sources: Holman, 1979a, 2000, Rage, 1984a and Parmley & Walker, 2003.

9. *Pantherophis ramspotti* Crother, White, Savage, Eckstut, Graham & Gardner, 2011. Intern. Scholar. Res. Network Zool. 2011(11), 10–11.

Type: Holotype, USNM 578514 (formerly JRP 1395), an 880 mm female (J.R. Parmelee, 21 June 2002).

Type locality: "Warren Co. (R24W, T76N, Sec. 30), Iowa" [USA].

Distribution: Northern USA (Iowa, S Minnesota, N Missouri, E Nebraska, SE South Dakota), 230–500 m.

Sources: Conant, 1940, A.H. Wright & Wright, 1957, R.C. Vogt, 1981, Holman, 1982b, 1982e, 1984c, 1987b, Ernst & Barbour, 1989, Powell, 1990, and Conant & Collins, 1991, 1998, Burbrink et al., 2000 and Crother et al., 2011b.

Remarks: Probably occurs in Wisconsin and N Illinois.

10. *Pantherophis slowinskii* (Burbrink, 2002). Mol. Phylogen. Evol. 25: 472–476. (*Elaphe slowinskii*)

Type: Holotype, LSUMZ 55922, an 1122 mm male (J. Boundy and J.B. Slowinski, 6 May 1993).

Type locality: "Longleaf Pine Highway, 0.5 miles west of Hwy 119, Natchitoches Parish, Louisiana," USA.

Distribution: Southern USA (W Louisiana and ext. E Texas), NSL–70 m.

11. *Pantherophis spiloides* (A.-M.-C. Duméril, Bibron & Duméril, 1854a). Erpét. Gén. 7(1): 269–270. (*Elaphis spiloides*)

Synonym: *Elaphis spiloides* A.-M.-C. Duméril, 1853 (*nomen nudum*).

Type: Holotype, MNHN 827, an 1176 mm specimen.

Type locality: "Amérique du Nord de La Nouvelle-Orléans" [= New Orleans, Louisiana, USA, 29°57'N, 90°04'W, elevation NSL].

Distribution: Southern Canada (ext. S Ontario) and E and SE USA (Alabama, NE Arkansas, Florida (W panhandle), ? NW Georgia, Illinois, Indiana, Kentucky, SE Louisiana, S Michigan, Mississippi, ext. NW North Carolina, Ohio, cen. and W Pennsylvania, Tennessee, ext. W Virginia, West Virginia and ext. SW Wisconsin), NSL–600 m.

Fossil records: Upper Pleistocene (Rancholabrean II) of USA (Tennessee).

Sources: Burbrink et al., 2000, Burbrink 2001 and Gibbs et al., 2006.

Remarks: Type locality in error *fide* Blaney (1971: 424–425) but J. Boundy examined holotype and it is a typical S Louisiana snake. Hybrids exist between *P. alleghaniensis* and *P. spiloides fide* Gibbs et al., 2006.

12. *Pantherophis vulpinus* (Baird & Girard, 1853). Cat. No. Amer. Rept., Serp. 1: 75–76. (*Scotophis vulpinus*)

Synonyms: *Elaphis rubriceps* A.-M.-C. Duméril, 1853 (*nomen nudum*), *Elaphis rubriceps* A.-M.-C. Duméril, Bibron & Duméril, 1854a, and *Elaphe vulpina gloydi* Conant, 1940.

Type: Neotype, USNM 9969, a 1242 mm male (P.R. Hoy), designated by Conant (1940: 10).

Type locality: "Racine, Wisconsin" [= Racine, E Racine Co, E Wisconsin, USA, 42°44'N, 87°47'W, elevation 185 m] via neotype selection.

Distribution: Extreme S Canada (S Ontario) and N USA (Illinois, NW Indiana, Iowa, E Michigan, SE Minnesota, NE Missouri, NE Nebraska, N Ohio, SE South Dakota, Wisconsin), 135–450 m.

Fossil records: Upper Miocene (Hemphillian) of USA (Nebraska), middle Pliocene (Blancan II, III) of USA (Idaho, Nebraska, Texas), upper Pliocene (Blancan V) of USA (Kansas, Nebraska, Washington), lower Pleistocene (Irvingtonian I) of USA (Pennsylvania, South Dakota), middle Pleistocene (Irvingtonian II) of USA (Nebraska), middle/upper Pleistocene (Irvingtonian II) of USA (Arkansas, Maryland, West Virginia), and upper Pleistocene (Rancholabrean I, II) of USA (Alabama, Georgia, Kansas, Missouri, Ohio, Oklahoma, West Virginia, Wisconsin).

Sources: Conant, 1940, A.H. Wright & Wright, 1957, R.C. Vogt, 1981, Holman, 1982b, 1982e, 1984c, 1987b, Ernst & Barbour, 1989, Powell, 1990, and Conant & Collins, 1991, 1998, Collins, 1991, Harding, 1997, Parmley &

Walker, 1998, Burbrink et al., 2000, Parmley & Walker, 2003 and Crother et al., 2011b.

Remarks: Syntypes (USNM 1570, Gross Ile, Michigan, C. Fox, and USNM 7269, Racine, Wisconsin, P.R. Hoy) lost *fide* Conant (1940: 10).

†*PARACOLUBER* Holman, 1970 (Colubridae)

Type species: †*Paracoluber storeri* Holman, 1970.

Distribution: Upper Miocene–lower Pliocene of USA, lower Miocene of USA, middle Miocene of Canada, and upper Miocene of USA.

Sources: Holman, 1976c, 1979a, 2000a, Parmley, 1984, 1988, Rage, 1984b, Parmley & Holman, 1995 and Parmley & Hunter, 2010.

Remarks: Possibly lower Miocene of USA (California).

1. †*Paracoluber storeri* Holman, 1970. Canad. J. Earth Sci. 7(5): 1321, figs. 2a–b.

Type: Holotype, ROM 7707, one posterior trunk vertebra (L.S. Russell & J. Storer, 1967).

Type locality: "Kleinfelder Farm locality, near Rockglen, Saskatchewan, Lsd. 16, Sec. 31, Tp. 2, Rge. 30, W. of 2nd Meridian; Barstovian (Upper Miocene) Wood Mountain Formation" [= middle Miocene].

Distribution: Lower Miocene (Hemingfordian: 16.3–20.6 mya) of USA (Wyoming), middle Miocene (Barstovian: 13.6–16.3 mya) of Canada (Saskatchewan), upper Miocene (Hemphillian: 4.9–10.3 mya) of USA (Nebraska, Texas), and upper Miocene/lower Pliocene (Hemphillian) of USA (Indiana).

P

†*PARAEPICRATES* M.K. Hecht *in* McGrew, 1959 (Boidae)

Type species: †*Paraepicrates brevispondylus* M.K. Hecht *in* McGrew, 1959.

Distribution: Middle Eocene of USA.

Sources: M.K. Hecht, 1959, Kluge, 1988a and Holman, 2000a.

Remarks: A valid genus *fide* Rage (2001: 121).

1. †*Paraepicrates brevispondylus* M.K. Hecht *in* McGrew, 1959. Bull. Amer. Mus. Nat. Hist. 117(3): 143–144, pl. 55, figs. 1–5.

Type: Holotype, AMNH 3829, one middle trunk vertebra (P.O. McGrew & field crews, 1950–1952).

Type locality: "Locality 5 an the south side of Elk Mountain and Tabernacle Butte area at the northern end of the Green River Basin, about 15 miles from the southwest flank of the Wind River Mountains and 25 miles north of the villaage of Farson, in southeast corner of Sublette County, Wyoming (T 27–29N, R 103–106 W), [USA]; late Bridgerian, middle Eocene.

Distribution: Middle Eocene (Bridgerian: 46.2–50.3 mya) of USA (Wyoming)

Remarks: A species of *Lichanura fide* Kluge (1988a: 229).

PARAHELICOPS Bourret, 1934c (Natricidae)

Type species: *Parahelicops annamensis* Bourret, 1934c.

Distribution: Indo-China.

Sources: J.C. Murphy et al., 2008, Ziegler et al., 2008, Zaher et al., 2009, I. Das, 2010, 2012 and Yang et al., 2011.

Remarks: A synonym of *Opisthotropis* Günther *fide* M.A. Smith (1943: 334).

1. *Parahelicops annamensis* Bourret, 1934c. Bull. Gén. Instr. Publ. Hanoi 14(9): 170–171, fig. 2.

Type: Holotype, MNHN 1938.117 (formerly RLB M.327), a 460 mm specimen (R.L. Bourret, 1930–1934).

Type locality: "Bana, station d'altitude située à 1.500m d'altitude au bord de la Baie de Tourane, Annam" [= Ba Na, Da Nang Prov., cen. Vietnam, 15°59'N, 107°59'E, elevation 1500 m].

Distribution: Southern Laos (Xe Kong) and cen. Vietnam (Da Nang, Kon Tum), 1280–1500 m.

Sources: Bourret, 1939d, M.A. Smith, 1943, Orlov et al., 1998, Orlov et al., 2003 and B.L. Stuart, 2006.

2. *Parahelicops boonsongi* E.H. Taylor & Elbel, 1958. Univ. Kansas Sci. Bull. 38(13): 1156–1159, fig. 31.

Type: Holotype, FMNH 135328 (formerly RE 5500 & EHT-HMS 31707), a 621+ mm male (R.E. Elbel, 2 June 1955).

Type locality: "Phu Nam Lang (mt.), 1780 m., Ban Khok (subvillage), Na Phung (village), Dan Sai (district), Loei (province), Thailand" [= ca. 17°23'N, 101°06'E].

Distribution: North-central Thailand (SW Loei), 700–1780 m.

Sources: E.H. Taylor, 1965, M.J. Cox, 1991b and M.J. Cox et al., 1998.

Remarks: Possibly occurs in S China, Laos and Vietnam *fide* E.H. Taylor (1965: 841) and M.J. Cox (1991b: 241).

PARAHYDROPHIS Burger & Natsuno, 1974 (Elapidae)

Synonyms: *Parahydrophis* Voris, 1969 *(nomen ineditum)* and *Parhydrophis* – Minton, 1975 *(nomen incorrectum).*

Type species: *Distira mertoni* Roux, 1910b.

Distribution: Tidal rivers and marine waters of Austro-Papua, including Arafura Sea and Gulf of Carpentaria.

Sources: M.A. Smith, 1926a, McDowell, 1969b, 1974b, Burger & Natsuno, 1974, Gow, 1989, Golay et al., 1993, A.R. Rasmussen, 1997, David & Ineich, 1999, Cogger, 2000, Wells, 2007, Zaher et al., 2009 and Hoser, 2012e.
Remarks: Previously a member of *Ephalophis*.

1. *Parahydrophis mertoni* (Roux, 1910b). Abh. Senck. Naturf. Ges. (1910–1911) 33: 222, pl. 13, figs. 4–4a. (*Distira mertoni*)

Type: Holotype, NMBA 6246, a 380 mm male (H. Merton, 5 May 1908).
Type locality: "Sungi Waskai, Wokam, Aru-Inseln, südöstlichen Molukken" [= Sungi Waskai, Wokam Is., Aru Arch., SE Maluku, ext. SE Indonesia, 5°37'S, 134°30'E].
Distribution: Extreme SE Indonesia (Aru Is.) and ext. N Australia (ext. N Northern Territory, NW Queensland), NSL.

†*PARAOXYBELIS* Auffenberg, 1963 (Colubridae)

Type species: †*Paraoxybelis floridanus* Auffenberg, 1963.
Distribution: Lower Miocene of USA.
Sources: Holman, 1979b, 2000a and Rage, 1984b.

1. †*Paraoxybelis floridanus* Auffenberg, 1963. Tulane Stud. Zool. 10(3): 189–190, fig. 39 (left).

Type: Holotype, UF 5134, one trunk vertebra (W. Auffenberg, March 1954).
Type locality: "Boulder Bar, Thomas Farm, Gilchrist County, Florida (R15E, T7S, Sec. 20); Hawthorne fm., Lower Miocene (zone B) (Arikareean)" [= early Hemingfordian *fide* Woodburne, 1987: 201].
Distribution: Lower Miocene (Hemingfordian: 16.3–20.6 mya) of USA (Florida). Known only from type locality.
Remarks: Holotype erroneously listed as UF 5143 *fide* Auffenberg (1963: 190). Taxonomic status doubtful *fide* Rage (1984b: 50): possibly an *Oxybelis, Leptophis* or *Uromacer*.

PARAPHIMOPHIS Grazziotin, Zaher, Murphy, Scrocchi, Benavides, Zhang & Bonatto, 2012 (Xenodontidae)

Type species: *Oxyrrhopus rusticus* Cope, 1878b.
Distribution: Southern South America.
Sources: Achaval, 1973, Halloy & Laurent, 1984, Cei, 1986, Scrocchi & Viñas, 1990, Marques & Lema, 1992, Cei, 1994, Giraudo & Arazamendia, 1997a, Leynaud & Bucher, 1999, Cabrera, 2004, Carriera et al., 2005 and Scolaro, 2005, 2006.

1. *Paraphimophis rusticus* (Cope, 1878b). Proc. Amer. Philos. Soc. (1877) 17(101): 92–93. (*Oxyrrhopus rusticus*)

Type: Holotype, ANSP 11326 (formerly SIM).
Type locality: "supposed to be the Argentine Confederation" [= Argentina].
Distribution: Southern South America. Southeastern Brazil (S Minas Gerais, Paraná, Rio de Janeiro, Rio Grande do Sul, Santa Catarina, São Paulo), Paraguay, Uruguay (Artigas, Canelones, Colonia, Maldonado, Montevideo, Rio Negro, Salto, San José) and Argentina (Buenos Aires, Chubut, Córdoba, Corrientes, Entre Ríos, Formosa, Jujuy, La Pampa, Mendoza, Misiones, Nequén, Rio Negro, Salta, San Luis, Santa Fe, Santiago del Estero, Tucumán).
Remarks: Previously in the genus *Clelia*.

PARAPISTOCALAMUS Roux, 1934 (Elapidae)

Type species: *Parapistocalamus hedigeri* Roux, 1934.
Distribution: Eastern Papua New Guinea.
Sources: E.E. Williams & Parker, 1964, McDowell, 1969a, 1987, M. McCoy, 1980, Golay et al., 1993, O'Shea, 1996, David & Ineich, 1999 and Zaher et al., 2009.

1. *Parapistocalamus hedigeri* Roux, 1934. Verh. Naturf. Ges. Basel 45: 78–80, figs. 1–2.

Synonym: *Denisonia furva* Brazenor, 1947.
Type: Holotype, NMBA 11967, a 305 mm female (H. Hediger, 1931–1934).
Type locality: "Distr. Buin, S. de l'île Bougainville (l'Archipel des Salomon)" [= Buin Distr., S Bougainville Is., N Solomon Arch., E Papua New Guinea, 6°50'S, 155°44'E].
Distribution: Papua New Guinea (N Solomons: SE Bougainville), NSL–610 m.

P

†*PARAPLATYSPONDYLIA* Holman & Harrison, 1998b (Boidae)

Type species: †*Paraplatyspondylia batesi* Holman & Harrison, 1998b.
Distribution: Upper Eocene of United Kingdom.
Sources: Holman & Harrison, 1998b and Rage, 2011.

1. †*Paraplatyspondylia batesi* Holman & Harrison, 1998b. Acta Zool. Cracov. 41(1): 24–27, fig. 1a–e.

Type: Holotype, MSUVP 1434, one middle trunk vertebra (D.L. Harrison field crew, 1995).

Type locality: "Hordle Cliff, Hampshire, England. Rodent Bed in the upper part of the Totland Bay Member, MP17, of the Headon Hill Formation (Upper Eocene).
Distribution: Upper Eocene (Priabonian, MP 17: 33.9–37.2 mya) of United Kingdom (England). Known only from type locality.

PARAPOSTOLEPIS Amaral, 1930d (Xenodontidae)

Type species: *Apostolepis polylepis* Amaral, 1922a.
Distribution: Northeastern Brazil.
Sources: J.A. Peters & Orejas-Miranda, 1970, Amaral, 1978, Lema, 2003b, and Hofstadler-Deiques & Lema, 2005.
Remarks: A valid genus *fide* Lema *(pers. comm.)*.

1. *Parapostolepis polylepis* (Amaral, 1922a). Anex. Mem. Inst. Butantan (1921) 1(1): 13–14, pl. 1, figs. 5–8. *(Apostolepis polylepis)*

Type: Holotype, IB 1681, a 620 mm female (F. de Assis-Iglesias, 1916–1918), destroyed by fire 15 May 2010.
Type locality: "Povoação Eng.ro Dodt, município de Santa Filomena, Estado do Piauí, Brasil" [= Engenheiro Dodt, Santa Filomena, Piauí State, Brazil, 8°49'S, 45°56'W, elevation 270 m].
Distribution: Northeastern Brazil (S Piauí), 270 m.

PARARHABDOPHIS Bourret, 1934a (Natricidae)

Type species: *Pararhabdophis chapaensis* Bourret, 1934a.
Distribution: Northern Vietnam.
Sources: Bourret, 1936b, M.A. Smith, 1943, Orlov et al., 2003, V.S. Nguyen et al., 2009, Zaher et al., 2009 and I. Das, 2010, 2012.

1. *Pararhabdophis chapaensis* Bourret, 1934a. Bull. Gén. Instr. Publ. Hanoi 14(7): 131–132, fig. 2.

Type: Holotype, MNHN 1938.125 (formerly RLB M.270), a 690+ mm male (R.L. Bourret, 1930–1931).
Type locality: "Chapa, province de Laokay (Tonkin), à l'altitude moyenne de 1.600 m., Indochine française" [= Sa Pa, NW Lao Cai Prov., ext. N Vietnam, 22°21'N, 103°50'E, elevation 1500–1600 m].
Distribution: Northern Vietnam (NW Lao Cai), 1500–1600 m. Known only from the holotype.

PARARHADINAEA Boettger, 1898 (Pseudoxyrhophiidae)

Synonyms: *Pseudorhadinaea* Boulenger, 1899e *(nomen substitutum), Rhabdotophis* F. Werner, 1909d, and *Pararhadinea* – Boettger, 1913 *(nomen incorrectum).*
Type species: *Pararhadinaea melanogaster* Boettger, 1898.
Distribution: Madagascar.
Sources: Guibé, 1958, Domergue, 1984b, Glaw & Vences, 1994, Raxworthy & Nussbaum, 1994, Cadle, 1999, Kelly et al., 2009 and Zaher et al., 2009.
Remarks: *Pararhadinaea albignaci* Domergue has been transferred to *Exallodontophis fide* Cadle, 1999.

1. *Pararhadinaea melanogaster* Boettger, 1898. Kat. Rept. Senck. Mus. 2: 33–34.

Synonyms: *Rhabdotophis subcaudalis* F. Werner, 1909d and *Pararhadinaea melanogaster marojejyensis* Domergue, 1984b.
Type: Holotype, SMF 17885 (formerly SMF-B 7313,1a), a female (A. Voeltzkow, 1897).
Type locality: "Insel Nossibé" [Madagascar], [= Nosy Be Is., W Antsiranana Prov., ext. NW Madagascar, bet. 13°12'–25'S and 48°10'–22'E, elevation < 225 m].
Distribution: Northern Madagascar (N Antsiranana, Nosy Be Is.), 30–500 m.
Sources: F. Werner, 1909d, 1929a, Boettger, 1913, Mertens, 1967a, Domergue, 1984b, Glaw et al., 2005b and Franzen et al., 2009.

PARASUTA Worrell, 1961c (Elapidae)

Type species: *Elaps gouldii* Gray, 1841a.
Distribution: Australia.
Sources: Worrell, 1961c, Cogger et al., 1983a, Wallach, 1985, Longmore, 1986, Storr et al., 1986, Gow, 1989, Hutchison, 1990, Greer, 1997, Keogh et al., 1998, Cogger, 2000, Sanders et al., 2008 and Hoser, 2012e.

1. *Parasuta dwyeri* (Worrell, 1956). Aust. Zool. 12(3): 202–203, figs. 1a–c, 3. *(Denisonia dwyeri)*

Synonym: *Parasuta robertsoni* Wells and Wellington, 1985.
Type: Holotype, AMS 14808, a 309 mm specimen (W. Dunmall).
Type locality: "Glenmorgan area, Queensland" [= vicinity of Glenmorgan, SE Queensland, E Australia, 27°15'S, 149°41'E, elevation 290 m].
Distribution: Southeastern Australia (Australian Capital Territory, New South Wales, ext. SE Queensland, NE Victoria), NSL–1395 m.
Sources: Coventry, 1971 and Shea & Sadlier, 1999.
Remarks: A subspecies of *Suta spectabilis fide* Cogger (1992: 692).

P

2. *Parasuta flagellum* (F. McCoy, 1878d). Prod. Zool. Victoria 2: 7–8, pl. 11, figs. 1, 1a–d. (*Hoplocephalus flagellum*)

Synonyms: *Hoplocephalus flagellum* McCoy, 1867 (*nomen nudum*) and *Parasuta harveyi* Wells & Wellington, 1985.
Type: Lectotype, NMV D4259, a 311 mm specimen, designated by Coventry (1970: 122).
Type locality: "Boroondara, Victoria" [=Boroondara (37°50'S, 145°03'E, elevation 80 m) (or Kew East (37°48'S, 145°03'E, elevation 50 m *fide* Coventry, 1970: 122), S Victoria, ext. SE Australia] via lectotype selection.
Distribution: Southeastern Australia (Australian Capital Territory, SE New South Wales, ext. S South Australia, Victoria, Kangaroo Is.), 10–590 m.

3. *Parasuta gouldii* (Gray, 1841a). Ann. Mag. Nat. Hist. 7(42): 91. (*Elaps gouldii*)

Type: Holotype, BMNH 1946.1.21.73, a female (J. Gould, 1837–1839).
Type locality: "Western Australia."
Distribution: Southwestern Australia (SW Western Australia), NSL–445 m.
Sources: Storr, 1981c and Bush et al., 1995.
Remarks: A member of *Rhinoplocephalus fide* Bush et al., 1995.

4. *Parasuta monachus* (Storr, 1964). West. Aust. Nat. 9(4): 89–90. (*Denisonia monachus*)

Type: Holotype, WAM 20606 (J. Douglas, 26 June 1963).
Type locality: "Kalgoorlie, Western Australia; Lat. 30° 43' S, Long. 121° 27' E" [= Kalgoorlie, S Western Australia, SW Australia, 30°45'S, 121°28'E, elevation 385 m].
Distribution: Southwestern Australia (SW Northern Territory, South Australia, S Western Australia), 105–835 m.
Sources: Storr, 1981c and Hoser, 1989.

5. *Parasuta nigriceps* (A.C.L.G. Günther, 1863c). Ann. Mag. Nat. Hist. (3) 12(71): 362. (*Hoplocephalus nigriceps*)

Synonym: *Denisonia nigrostriata brevicauda* F.J. Mitchell, 1951.
Type: Holotype, BMNH 1946.1.20.70, a 406 mm male.
Type locality: "probably from Australia."
Distribution: Southern Australia (SW New South Wales, S South Australia, NW Victoria, S Western Australia, Flinders Is.), NSL–630 m.
Sources: Coventry, 1971, Storr, 1981c and Bush et al., 1995.

6. *Parasuta spectabilis* (Krefft, 1869b). Snakes Aust.: 61, pl. 12, fig. 4. (*Hoplocephalus spectabilis*)

Synonyms: *Denisonia spectabilis nullarbor* Storr, 1981e and *Rhinoplocephalus spectabilis bushi* Storr, 1988.
Types: Syntypes (6), AMS 131133–35 (formerly AMS 6593–95), AMS 131137 (formerly AMS 6597), AMS 131139 (formerly AMS 6599), and BMNH 1866.6.13.1, longest syntype 305 mm (G. Masters).
Type locality: "Port Lincoln, South Australia" [= Port Lincoln, S South Australia, S Australia, 34°44'S, 135°52'E, elevation NSL].
Distribution: Southern Australia (New South Wales, South Australia, W Victoria, SE Western Australia), NSL–1095 m.
Sources: Storr, 1981c, Hoser, 1989 and Shea & Sadlier, 1999.

PARATAPINOPHIS Angel, 1929 (Natricidae)

Type species: *Paratapinophis praemaxillaris* Angel, 1929.
Distribution: Northern Indochina.
Sources: Deuve, 1970, J.B. Rasmussen, 1982a, Mo et al., 1984, Zhao, 2006, J.C. Murphy et al., 2008, I. Das, 2010, 2012, David et al., 2011 and Yang et al., 2011.

1. *Paratapinophis praemaxillaris* Angel, 1929. Bull. Mus. Hist. Nat., Paris (2) 1(1): 78, figs. a–d.

Types: Syntypes (2), MNHN 1928.63–64, a 214 mm and 210 mm specimen (J.T. Delacour [Second Delacour-Lowe Mission], 15 Dec. 1925–13 Jan. 1926).
Type locality: "Xieng-Khouang, Haut-Laos" [= Xieng Khoang, Xiangkhouang Prov., N Laos, 19°17'N, 103°25'E].
Distribution: Northern Indochina. Southern China (Yunnan), N Thailand (Nan, Chiang Mai) and N Laos (Xiangkhouang), 475–1400 m.

†*PARAUNGALIOPHIS* Rage, 2011 (Ungaliophiidae)

Type species: †*Paraungaliophis pricei* Rage, 2011.
Distribution: Middle Palaeocene of Brazil.
Source: Rage, 2011.

1. †*Paraungaliophis pricei* Rage, 2011. Palaeovertebrata (2008) 36(1–4): 41–44, fig. 1.

Type: Holotype, DMG 1365-R, one middle trunk vertebra.
Type locality: "Itaboraí, State of Rio de Janeiro, Brazil; middle Palaeocene."
Distribution: Middle Paleocene (Itaboraian, 57.0–59.0 mya) of Brazil.

P

PAREAS Wagler, 1830 (Pareatidae)

Synonyms: *Pareos* – A.-M.-C. Duméril & Bibron, 1844 *(nomen incorrectum), Eberhardtia* Angel, 1920c, *Eberhardia* – Neave, 1939 *(nomen incorrectum), Parea* – Nutaphand, 1991b *(nomen incorrectum), Dannyleeus* Hoser, 2012q *(nomen illegitimum),* and *Katrinahoserserpenea* Hoser, 2012q *(nomen illegitimum).*

Type species: *Pareas carinatus* Wagler, 1830.

Distribution: Eastern and SE Asia and W Indonesia.

Sources: Boulenger, 1914c, Wall, 1922c, Maki, 1931, C.H. Pope, 1935, Bourret, 1936b, Deuve, 1970, Rao & Yang, 1992, Zhao & Adler, 1993, K. Guo & Deng, 2009, Zaher et al., 2009, I. Das, 2010, 2012, Y. Guo et al., 2011 and Hoser, 2012q.

Remarks: Consists of two clades, with *P. carinatus* and *P. nuchalis* possibly a separate genus *fide* Y. Guo et al. (2011: 61) and Pyron et al. (2011: 334).

1. *Pareas boulengeri* (Angel, 1920a). Bull. Mus. Hist. Nat., Paris 26(2): 113–114. (*Amblycephalus boulengeri*)

Types: Syntypes (3), MNHN 1912.349–51, a 470 mm, 460 mm and 182 mm specimen (R.P. Cavalerie).

Type locality: "Koeï Tchéou (Chine)" [= Guizhou Prov., SE China, ca. 27°N, 107°E].

Distribution: China (Anhui, Fujian, Gansu, Guangdong, Guangxi, Guizhou, Henan, Hunan, Jiangsu, Jiangxi, Shaanxi, Sichuan, Yunnan, Zhejiang), 315–1360 m.

Sources: C.H. Pope, 1929, Chang, 1936, Wen, 1983, Zhao, 2006 and Yao, 2012.

Remarks: A subspecies of *P. monticolus fide* Bourret (1936b: 426).

P

2. *Pareas carinatus* Wagler, 1830. Nat. Syst. Amph.: 181.

Synonyms: *Amblycephalus carinatus* H. Boie *in* Schlegel, 1826a (*nomen nudum*), *Amblycephalus carinatus* Schlegel, 1826b (*nomen nudum*), *Amblycephalus carinatus* H. Boie, 1828 (*nomen nudum*), *Pareas berdmorei* Theobald, 1868b, and *Amblycephalus carinatus unicolor* Bourret, 1934d.

Types: Syntypes (4), RMNH 954a–b and BMNH, a male (formerly RMNH) (C.G.C. Reinwardt, 1816–1821).

Type locality: "Asia." Emended to Java *fide* RMNH catalogue and Schlegel (1837: 286).

Distribution: Southeast Asia and Indonesia. Southern China (Yunnan), S Myanmar (Tanintharyi), Thailand (Chiang Mai, Chiang Rai, Chumphon, Kanchanaburi, Lamphum, Loei, Nakhon Ratchasima, Nakhon Si Tammarat, Narathiwat, Pattani, Phetchabun, Saraburi, Songkhla, Tak, Uthai Thani), N Laos (Bolikhamsai), S Cambodia (Koh Kong, Kampong Chhnang, Kampong Speu, Kampot, Mondulkiri), Vietnam (Dok Lak, Dong Nai, Lang Dong, Quang Binh, Quang Nam, Quang Tri, Son La, Tay Ninh, Vinh Phu), N West Malaysia (Terengganu), East Malaysia (Sabah), Brunei and Indonesia (Bali, Java, Kalimantan, Lombok, Sumatra), 20–1780 m.

Sources: Schlegel, 1837, Rooij, 1917, M.A. Smith, 1923a, 1930a, 1943, Wall, 1925a, Brongersma, 1929, Cochran, 1930, Kopstein, 1930, 1938, 1941, Mertens, 1930, Bourret, 1935c, 1936b, C. Haas, 1950, E.H. Taylor & Elbel, 1958, E.H. Taylor, 1965, Campden-Main, 1970a, Saint-Girons, 1972a, Capocaccia, 1976, Tweedie, 1983, M.J. Cox, 1991b, Stuebing, 1991, David & Vogel, 1996, M.J. Cox et al., 1998, Chan-ard et al., 1999, Orlov et al., 2000, McKay, 2006, B.L. Stuart & Emmett, 2006, Ziegler et al., 2006, Bain et al., 2007, I. Das, 2007b, and V.S. Nguyen et al., 2009.

Remarks: Official Specific Name no. 2452 *fide* Opinion 963 (ICZN, 1971).

3. *Pareas chinensis* (T. Barbour, 1912a). Mem. Mus. Comp. Zool. 40(4): 132, pl. 2, fig. 1. (*Amblycephalus chinensis*)

Synonyms: *Amblycephalus kuangtungensis* T. Vogt, 1922, *Amblycephalus yunnanensis* T. Vogt, 1922, and *Amblycephalus niger* C.H. Pope, 1928b.

Type: Holotype, MCZ 7326, a 463 mm female (W.R. Zappey & J.E. Thayer [Arnold Arboretum Exped.], 1909).

Type locality: "Luluping, western Szechwan, China." [= E and S of Washan at ca. 29°44'N, 103°15'E *fide* C.H. Pope, 1935: 371]. Restated as Lu-lu-ping *fide* T. Barbour & Amaral (1929: 210).

Distribution: Southeastern China (Anhui, Fujian, Guangdong, Guangxi, Guizhou, Hong Kong, Jiangsu, Jiangxi, Sichuan, Yunnan, Zhejiang) and N Vietnam (Hoang Lien Son), 500–2200 m.

Sources: C.H. Pope, 1929, Bourret, 1934a, Chang, 1936, B.Q. Hu et al., 1959, 1980, Tian et al., 1986, Wu et al., 1979, Huang et al., 1990, Ota et al., 1997 and Zhao, 2006.

Remarks: Official Specific Name no. 2092 *fide* Opinion 744 (ICZN, 1965b). A subspecies of *P. monticola fide* Bourret (1936b: 426). Possibly a synonym of *P. formosus fide* Jiang, 2004.

4. *Pareas formosensis* (Van Denburgh, 1909). Proc. California Acad. Sci. (4) 3(3): 55. (*Amblycephalus formosensis*)

Synonyms: *Psammodynastes compressus* Oshima, 1910 and *Amblycephalus komaii* Maki, 1931.

Type: Holotype, CAS 18006, a 274 mm specimen (V. Kühne, 27 March 1909).

Type locality: "Kanshirei, Formosa" [= Kuan-tzu Ling, Baihe Distr., Tainan Prov., SW Taiwan, 23°21'N, 120°30'E, elevation 365 m].

Distribution: Taiwan (Chiayi, Kaohsiung, Nantou, Pingtung, Tainan, Taipei), 315–500 m.

Sources: C.S. Wang & Wang, 1956, Kuntz, 1963, Lue et al., 1989, Lue, 1990, Lue & Lai, 1992, Ota et al., 1997 and Xiang & Li, 2009.

5. *Pareas hamptoni* (Boulenger, 1905f). J. Bombay Nat. Hist. Soc. (1904–1905) 16(2): 236, pl., fig. 2. (*Amblycephalus hamptoni*)

Synonyms: *Eberhardtia tonkinensis* Angel, 1920c, and *Amblycephalus carinatus hainanus* M.A. Smith, 1923a.

Type: Holotype, BMNH 1946.1.20.16, a 555 mm male (H. Hampton, 1898–1905).

Type locality: "the neighbourhood of Mogok, Upper Burma" [= vicinity of Mogok (22°55'N, 96°30'E, elevation 1135 m), SE Sagaing Div., N Mandalay State, N Myanmar].

Distribution: Southeastern Asia. Southern China (Guizhou, Guangxi, Hainan, Yunnan), N Myanmar (Kachin, N Mandalay, Shan), NE Thailand (Narathiwat, Tak), N Laos (Champasak, Xaignbouli) and Vietnam (Bac Kan, Bac Thai, Binh Tri Thien, Cao Bang, Dak Lak, Dong Nai, Ha Giang, Hai Duong, Ha Tinh, Kon Tum, Lam Dong, Lao Cai, Quang Binh, Quang Nam, Thai Nguyen, Vinh Phuc), 500–1980 m.

Sources: Wall, 1905e, 1922c, Bourret, 1934c, 1935b–c, 1937b, 1939a, 1939d, M.A. Smith, 1943, E.H. Taylor, 1965, Romer, 1966, 1979a, Deuve, 1970, Campden-Main, 1970a, B.Q. Hu et al., 1980, Tian et al., 1986, M.J. Cox, 1991b, Ota et al., 1997, Orlov et al., 2000, Zhao, 2006, Ziegler et al., 2007, V.S. Nguyen et al., 2009, I. Das, 2010 and Vogel, 2010.

Remarks: Possibly occurs in Thailand (Nan) *fide* M.J. Cox (1991b: 259).

6. *Pareas iwasakii* (M. Maki, 1937). Trans. Nat. Hist. Soc. Formosa 27(169): 217–218, 3 figs. (*Amblycephalus formosensis iwasakii*)

Type: Holotype, ? KUM, a 220 mm male (Iwasaki, 26 Jan. 1937), lost *fide* Toriba (1993: 72).

Type locality: "Ishigaka-jima, Riukyu Archipelagoes" [= Ishigakishima Is., Yaeyama Group, S Ryukyu Arch., ext. S Japan, 24°24'N, 124°12'E].

Distribution: Extreme S Japan (Ryukyus: Iriomotejima and Ishigakijima Is.).

Sources: Takara, 1962, Nakamura & Uéno, 1963, M. Mori, 1984, 1986 and Ota et al., 1997.

7. *Pareas macularius* Theobald, 1868a. J. Linn. Soc. London (Zool.) 10(41): 54–55.

Synonyms: *Pareas modestus* Theobald, 1868a, *Pareas andersonii* Boulenger, 1888h, and *Amblycephalus tamdaoensis* Bourret, 1935c.

Types: Syntypes (3), BMNH 1925.12.22.1, BMNH 1926.3.17.9, and BMNH 1946.1.20.8 (formerly IMC 8024–26), longest syntype 610 mm (T.M. Berdmore, 1855–1856 via W. Theobald, Jr.).

Type locality: "Tenasserim" [= Tanintharyi Div., S Myanmar]. Emended to Martaban, Tenasserim [= Mottama, Mon Div., S Myanmar, 16°34'N, 97°37'E, elevation 10 m] *fide* Sclater (1891: 67).

Distribution: Southeastern Asia. Northeastern India (Sikkim, West Bengal), Bangladesh, S China (Guangdong, Guangxi, Guizhou, Yunnan), Myanmar (Kachin, Mandalay, Mon, Shan, Tanintharyi, Yangon), N Thailand (Chiang Mai), N Laos (Xiangkhouang) and N Vietnam (Bac Kan, Cao Bang, Hai Duong, Hoa Binh, Lai Chau, Nghe An, Quang Binh, Vinh Phuc), NSL–1300 m.

Sources: Wall, 1926b, Rendahl, 1937, M.A. Smith, 1943, E.H. Taylor & Elbel, 1958, E.H. Taylor, 1965, B.Q. Hu et al., 1980, Murthy, 1986, Tian et al., 1986, M.A.R. Khan, 1988, M.J. Cox, 1991b, Huang, 2004, Whitaker & Captain, 2004, Ziegler et al., 2007 and V.S. Nguyen et al., 2009.

Remarks: Theobald (1868a: 55) listed two syntypes from Tenasserim in ASB but Sclater (1891: 67) listed three (IMC 8024–26) as syntypes from Martaban, Tenasserim (all collected by Berdmore).

8. *Pareas margaritophorus* (Jan *in* Bocourt, 1866). Nouv. Arch. Mus. Hist. Nat. 2(1): 8. (*Leptognathus margaritophorus*)

Synonyms: *Pareas macularius* Blyth, 1859 (*nomen nudum*), *Pareas modestus* Theobald, 1868, *Pareas moellendorffi* Boettger, 1885b, *Pareas andersonii* Boulenger, 1888, and *Amblycephalus tamdaoensis* Bourett, 1935.

Type: Holotype, MNHN 599, a 250 mm specimen (F. Bocourt, 1861).

Type locality: "des provinces méridionales du Royaume de Siam" [= southern provinces of Thailand]. Probably the vicinity of Bangkok, Thailand *fide* Milne-Edwards (1866: 1).

Distribution: Southeastern Asia. Southern China (Guangdong, Guangxi, Guizhou, Hainan, Hong Kong, Lantan, Stonecutter's Is.), S Myanmar (Tanintharyi), Thailand (Champasak, Chanthaburi, Chiang Mai, Krung Thep Mahanakhon, Nakhon Ratchasima, Narathiwat, Pattani), Cambodia (Kampon Speau, Kampot, Koh Kong, Oddar Meanchey, Pursat, Stung Treng), Laos (Bolikhamsai, Vientiane, Xiangkhouang), Vietnam (Gia Lai, Hai Duong, Ha Tay, Hoa Binh, Ho Chi Minh City, Kien Giang, Lam Dong, Ninh Binh, Quang Tri, Tay Ninh, Vinh Phuc), N West Malaysia (Johor, Kedah, Kelantan, Perak, Perlis) and Singapore, NSL–1640 m.

Sources: Flower, 1899, K.P. Schmidt, 1927b, Cochran, 1930, Fan, 1931, Bourret, 1934b, 1939b, M.A. Smith, 1943, Deuve, 1962d, 1970, E.H. Taylor, 1965,

Campden-Main, 1970a, Saint-Girons, 1972a, Romer, 1979a, B.-Q. Hu et al., 1980, Tweedie, 1983, Tian et al., 1986, M.J. Cox, 1991b, M.J. Cox et al., 1998, Chanard et al., 1999, Orlov et al., 2000, B. Lee and Koh, 2005, B.L. Stuart et al., 2006, Grismer et al., 2008a, V.S. Nguyen et al., 2009 and T. Hartmann et al., 2013.

Remarks: Occurs in Brunei *fide* I. Das (2007: 20).

9. *Pareas monticola* (Cantor, 1839b). Proc. Zool. Soc. London 7(1): 53–54. (*Dipsas monticola*)

Synonym: *Pareas nuchalis* A.C.L.G. Günther, 1863c (*nomen nudum*).

Type: Holotype, BMNH 1946.1.20.5 (formerly OUM), a male (T.E. Cantor, 1834–1839, via British East India Comp.).

Type locality: "Naga Hills (Asám)" [= Nagaland State, NE India].

Distribution: Southern Asia. Northeastern India (Arunachal Pradesh, Assam, Meghalaya, Mizoram, Nagaland, North Bengal, Sikkim), Bangladesh, SW China (Yunnan, Xizang), N Myanmar (Chin) and Vietnam (Lao Cai, Vinh Phu), 100–2500 m.

Sources: Wall, 1908c , 1909d , 1910c, 1922c, Annandale, 1912, M.A. Smith, 1943, Swan & Leviton, 1962, Murthy, 1986, Tian et al., 1986, M.A.R. Khan, 1988, Orlov et al., 2000, 2003, Ahmed et al., 2009, V.S. Nguyen et al., 2009 and I Das, 2010.

Remarks: Nicobar Is. record rejected *fide* Wall (1922c: 23) and (1925a: 245).

10. *Pareas nigriceps* K. Guo & Deng, 2009. Zootaxa (2008): 54–58, figs. 1a–d, 2.

Type: Holotype, HNU 200505004, a 521 mm male (H.-M. Yan and K.-J. Guo, 28 May 2005).

Type locality: "Xiaohei Hill, Gaoligong Mountains National Nature Reserve (24.83671°N 98.76185°E, 2067 m a.s.l.), Longling County, Baoshan District, Yunnan province, Perople's [sic] Republic of China."

Distribution: Southern China (W Yunnan), 2010–2070 m.

Remarks: Holotype length listed as 396 mm *fide* K. Guo & Deng (2009: fig. 2).

11. *Pareas nuchalis* (Boulenger, 1900f). Proc. Zool. Soc. London 69(2): 185, pl. 17, figs. 1, 1a–b. (*Amblycephalus nuchalis*)

Type: Holotype, SMK, a 490 mm specimen (A.H. Everett, 1869–1872).

Type locality: "Matang, Borneo" [= Kampong Matang, First Division, Kidi Distr., W Sarawak, East Malaysia, 1°35'N, 110°14'E, elevation 15 m].

Distribution: Borneo. East Malaysia (Sabah, Sarawak) and cen. Indonesia (Kalimantan), 15–850 m.

Sources: C. Haas, 1950, Stuebing, 1991 and Malkmus et al.,2002.

Remarks: C. Haas (1950: 529) erroneously listed the type as being in BMNH.

12. *Pareas stanleyi* (Boulenger, 1914c). Ann. Mag. Nat. Hist. (8) 14(84): 484. (*Amblycephalus stanleyi*)

Synonym: *Amblycephalus sinensis* Stanley, 1917.

Type: Holotype, BMNH 1946.1.20.4, a 440 mm male (A. Stanley, 1898–1914).

Type locality: "N. W. Fokien, China" [= NW Fujian Prov., SE China]. Emended to Kuatun, village in the mountains of Ch'ungan Hsien, NW Fukien, China *fide* C.H. Pope (1929: 462), [= Guadun, in Wuyishan Mtns., ext. N Fujian Prov., SE China, 27°45'N, 117°50'E, elevation 900–1200 m].

Distribution: Southeastern China (Fujian, Guizhou, Hunan, Jiangxi, Sichuan, Zhejiang), 365–1200 m.

Source: Zhao, 2006.

PARIAS Gray, 1849a (Viperidae)

Synonyms: *Rerias* – Kingsley, 1888 *(nomen incorrectum)* and *Pareas* – Hoffmann, 1890 *(nomen incorrectum).*

Type species: *Megaera flavomaculatus* Gray, 1842b.

Distribution: Malay Peninsula and East Indies.

Sources: Maslin, 1942, M.A Smith, 1943, Maslin, 1945, Klemmer, 1963, Leviton, 1964d, W. Burger, 1971, Hoge & Romano-Hoge, 1981a, Malhotra & Davis, 1991, Golay et al., 1993, Kraus et al., 1996, David & Vogel, 1998, David & Ineich, 1999, McDiarmid et al., 1999, Parkinson, 1999, Malhotra & Thorpe, 1997, 2000, 2004c, Ziegler et al., 2001, Giannasi et al., 2001a, G. Peng & Fuji, 2001, David et al., 2002, 2009, 2011, Gumprecht et al., 2004c, Castoe & Parkinson, 2006, Grismer et al., 2006a, 2008b, Sanders et al., 2006, Dawson et al., 2008, Ganesh et al., 2008, Orlov et al., 2008, 2009, I. Das, 2010, 2012, David et al., 2011c, P. Guo & Wang, 2011, Sumontha et al., 2011 and Hoser, 2012d, o.

1. *Parias flavomaculatus* (Gray, 1842b). Zool. Misc. 2(Mar.): 49. (*Megaera flavomaculatus*)

Synonyms: *Megaera ornata* Gray, 1842b, *Megaera variegata* Gray, 1842b, *Trimeresurus schadenbergi* J.G. Fischer, 1885b, *Trimerisurus halieus* L.E. Griffin, 1910, and *Trimeresurus mcgregori* E.H. Taylor, 1919.

Type: Lectotype, BMNH 1946.1.19.34 (formerly BMNH I.3.la), an adult male (H. Cuming, 1836–1840), designated by Iskandar & Colijn (2001: 158).

Type locality: "Philippine Islands" via lectotype selection. Restricted to Luzon Is., Philippines *fide* Leviton (1964b: 260).

Distribution: Philippines (Agutayan, Batan, Bohol, Camiguin, Catanduanes, Dinagat, Jolo, Leyte, Luzon,

Mindanao, Mindoro, Negros, Penay, Polillo), 400–800 m.

Sources: Leviton, 1964b, 1964d, Ross & Lazell, 1991, Ross & Gonzales, 1992, Lamar, 1997b, Sajdak, 2010 and Siler et al., 2011.

2. *Parias hageni* (Lidth de Jeude, 1886). Notes Leyden Mus. 8: 52–54, pl. 2, fig. 6. (*Bothrops hageni*)

Type: Lectotype, RMNH 819 (B. Hagen, 1881–1886), designated by Brongersma (1933c: 7).

Type locality: "Deli, Sumatra" [= Deli, North Sumatra Prov., Sumatra, W Indonesia, 3°37'N, 98°44'E, elevation 20 m] via lectotype selection.

Distribution: Malay Peninsula and Sumatra. Peninsular Thailand (Pattani, Phang Nga, Songkhla, Surat Thani, Trang), Singapore, West Malaysia (Johor, Pahang, Perak, Selangor) and Indonesia (Batu, Bangka, Mentawai Arch., Nias, Simeulue, Sumatra), 20–1000 m.

Sources: Brongersma, 1933c, Grandison, 1972, Tweedie, 1983, M.J. Cox, 1991b, B.L. Lim, 1991, B.L. Lim et al., 1995, David & Vogel, 1996, M.J. Cox et al., 1998, David & Vogel, 1998, Chan-ard et al., 1999, Ryabov et al., 2002, Sanders et al., 2002 and I. Das, 2010.

3. *Parias malcomi* (Loveridge, 1938a). Proc. Biol. Soc. Washington 51: 45. (*Trimeresurus sumatranus malcolmi*)

Type: Holotype, MCZ 43604, a 1060 mm male (J.A. Griswold, Jr. [Harvard Primate Exped.], 20 July 1937).

Type locality: "Sungii River, near Bundutuan, Mount Kinabalu, British North Borneo, at an altitude *circa* 3,000 feet" [= Sungii River, vicinity of Bundu Tuhan, Mt. Kinabalu, Sabah, East Malaysia, elevation 915 m].

Distribution: East Malaysia (Sabah) and Brunei, 915–1700 m.

Sources: Stuebing & Inger, 1998, Keller, 2008 and I. Das, 2012.

Remarks: A valid species *fide* Stuebing & Inger (1998: 227).

4. *Parias mcgregori* (E.H. Taylor, 1919). Philippine J. Sci. 14D(1): 110–111, figs. 1a–c. (*Trimeresurus flavomaculatus mcgregori*)

Type: Neotype, CAS 60525, a female, designated by Leviton (1964d: 262).

Type locality: "Batan Island, Philippines" [= Batan Is., Batanes Prov., ext. N Philippines] via neotype selection.

Distribution: Extreme N Philippines (Batan, Sabtang).

Source: Gumprecht, 2002.

Remarks: Holotype (BSM 748, an 865 mm R.C. McGregor, 12 June 1907) destroyed January 1945 during World War II.

5. *Parias schultzei* (L.E. Griffin, 1909b). Philippine J. Sci. 4A(1): 601. (*Trimeresurus schultzei*)

Type: Holotype, PNM 315 *fide* E.H. Taylor (1922a: 294), a 330 mm specimen (W. Schultze), destroyed Jan. 1945 during World War II.

Type locality: "Iwahig, Palawan" [= Iwahig, cen. Palawan Is., SW Philippines, ca. 9°44'N, 118°40'E, elevation NSL].

Distribution: Philippines (Balabac, Palawan).

Sources: Leviton, 1964d and Alcala, 1986.

6. *Parias sumatranus* (Raffles, 1822). Trans. Linn. Soc. London 13(2): 334. (*Coluber sumatranus*)

Synonyms: *Coluber sumatrensis* H. Boie *in* Schlegel, 1826a *(nomen emendatum)* and *Trigonocephalus formosus* Müller & Schlegel, 1844a *in* Temminck, 1842–1845.

Type: Holotype, not designated, a 1220–1525 mm specimen (T.S.B. Raffles, Mar.–Dec. 1818), location unknown.

Type locality: "Island of Sumatra" [W Indonesia]. Restricted herein to vicinity of Bengkulu city, SW Sumatra, W Indonesia.

Distribution: Malay Peninsula and Greater Sundas. Peninsular Thailand (Narathiwat, Pattani), West Malaysia (Johor, ? Kedah, ? Kelantan, ? Perak, Perlis), Singapore, East Malaysia (Sabah, Sarawak) and Indonesia (Bangka, Belitung, N Kalimantan, Mentawai Arch., Nias, Simeulue, Sumatra), NSL–1000 m.

Sources: M.A. Smith, 1928a, E.H. Taylor, 1965, Regenass & Kramer, 1981, M.J. Cox, 1991b, David & Vogel, 1996, Ryabov et al., 2002, Sanders et al., 2002 and I. Das, 2010.

Remarks: Bengkulu was a trading center established by the East India Company and Raffles was stationed there from March 1818 until the end of the year. Possibly occurs on Palawan, Philippines *fide* Das (2010: 306). Holotype formerly in BMNH *fide* Cox et al. (2013: 762).

PAROPLOCEPHALUS Keogh, Scott & Scanlon, 2000 (Elapidae)

Type species: *Brachyaspis atriceps* Storr, 1980.

Distribution: Southwestern Australia.

Source: Keogh et al., 2000.

1. *Paroplocephalus atriceps* (Storr, 1980). Rec. West. Aust. Mus. 8(3): 397–399, fig. 1. (*Brachyaspis atriceps*)

Type: Holotype, WAM 67330, a 379 mm (svl) specimen (P. Griffin and G. Baron, 6 Oct. 1979).

Type locality: "Lake Cronin in the southern interior of Western Australia, in 32°23'S, 119°45'E."
Distribution: Southwestern Australia (S Western Australia), 270–440 m.

†*PATAGONIOPHIS* Albino, 1986 (†Madtsoiidae)

Type species: †*Patagoniophis parvus* Albino, 1986.
Distribution: Upper Cretaceous of Argentina and lower Eocene of Australia.
Sources: Albino, 1987b, Scanlon, 2005 and Albino, 2011b.

1. †*Patagoniophis australiensis* Scanlon, 2005. Mem. Queensland Mus. 51(1): 217–220, fig. 1e.

Type: Holotype, QMF 19710, one middle trunk vertebra (H. Godthelp & M. Archer field crews, mid-1980s–early 1990s).
Type locality: "MP1 horizon, Main Quarry, Tingamarra, Boat Mountain, near Murgon, south-east Queensland (26°S, 152°E), Australia; Tingamarra Local fauna, Early Eocene, 54.6+/-0.05 MYBP."
Distribution: Lower Eocene (54.6 mya) of Australia. Known only from type locality.

2. †*Patagoniophis parvus* Albino, 1986. Actas IV Congr. Argent. Paleont. Bioestr. 2: 18, fig. 2.

Type: Holotype, MACN RN 33, one trunk vertebra.
Type locality: "Ladera sudoeste del Cerro Cuadrado, noreste de la Ea. Los. Alamitos, Arroyo Verde, Dpto. Valcheta, Río Negro, Argentina. Sector medio de la Formación Los Alamitos, Campaniano-Maastrichtiano temprano, Cretácio tardío" [= SW slope of Cerro Cuadrado, NE of Los Alamitos, 41°57'S, 66°21'W, proximal zone of Arroyo Verde, 41°55'S, 66°10'W, Valcheta Dept., SE Río Negro Prov., S Argentina; middle section of Los Alamitos formation, early Campanian-Maastrichtian, upper Cretaceous].
Distribution: Upper Cretaceous (Campanian-Maastrichtian: 65.5–83.6 mya) of Argentina. Known only from type locality.

P

†*PAULACOUTOPHIS* Rage, 2011 (Booidea *incertae sedis*)

Type species: †*Paulacoutophis perplexus* Rage, 2011.
Distribution: Middle Palaeocene of Brazil.
Source: Rage, 2011.

1. †*Paulacoutophis perplexus* Rage, 2011. Palaeovertebrata (2008) 36(1–4): 52–55, fig. 8.

Type: Holotype, DGM 1367-R, one middle trunk vertebra (1968).
Type locality: "Itaboraí, State of Río de Janeiro, Brazil; middle Palaeocene."
Distribution: Middle Paleocene (Itaboraian: 57.0–59.0 mya) of Brazil.

PELAMIS Daudin, 1803d (Elapidae)

Synonyms: *Ophinectes* Rafinesque-Schmaltz, 1817 *(nomen emendatum), Pelamys* Wagler, 1830 *(nomen praeoccupatum), Pelamides* Cuvier & Voigt, 1832 *(nomen emendatum), Elaphrodytes* Gistel, 1848 *(nomen substitutum), Pelamijs* Bleeker, 1856 *(nomen emendatum), Hydrus* Boulenger, 1890a *(nomen praeoccupatum), Pelamydrus* Stejneger, 1910 *(nomen susbstitutum), Belamydrus* – Steindachner, 1913b *(nomen incorrectum), Pelamydus* – Briceño-Rossi, 1934 *(nomen incorrectum), Pelamiris* – Daniel, 1949 *(nomen incorrectum), Pelamytrus* – Imazumi, Kuroda & Okada, 1950 *(nomen incorrectum), Pelamus* – J.H. Brown, 1973 *(nomen incorrectum),* and *Pelemis* – Dowling *in* Dowling & Duellman, 1978 *(nomen incorrectum).*
Type species: *Anguis platura* Linnaeus, 1766.
Distribution: Marine waters of Indian and Pacific Oceans, including Andaman Sea, Arabian Sea, Arafura Sea, Banda Sea, Banderas Bay, Bass Strait, Bay of Bengal, Bismarck Sea, Blanca Bay, Celebes Sea, Coral Sea, Culebra Bay, East China Sea, Formosa Strait, Gulf of Aden, Gulf of California, Gulf of Cambay, Gulf of Carpentaria, Gulf of Dulce, Gulf of Fonseca, Gulf of Guayaquil, Gulf of Kutch, Gulf of Mannar, Gulf of Martaban, Gulf of Oman, Gulf of Panamá, Gulf of Thailand, Gulf of Tonkin, Honda Bay, N Indian Ocean, Java Sea, Kanggava Bay, Laccadive Sea, Los Angeles Bay, Magdalena Bay, Manila Bay, Molucca Sea, Noumea Lagoon, Persian Gulf, Possiet Bay, Sagami Bay, Savu Sea, Sea of Japan, Solomon Sea, South China Sea, Strait of Hormuz, Strait of Malacca, Sulu Sea, Suruga Bay, Tasman Bay, Timor Sea, Tomini Sea and Yellow Sea.
Sources: Strauch, 1873, Murray, 1884, F. Werner 1900c, Stejneger, 1907a, Wall, 1909a, 1919c, 1921g, Rooij, 1917, E.H. Taylor, 1922a, 1951, 1954, 1965, Slevin, 1926, M.A. Smith, 1926a, 1943, Kopstein, 1930, Maki, 1931, Burt & Burt, 1932, Tchang, 1932, Bourret, 1934g, 1936b, H.M. Smith & Taylor, 1945, C. Haas, 1950, V.M. Tanner, 1951, Oliver & Shaw, 1953, Lanza, 1954, Romer, 1954, 1979, Deraniyagala, 1955, Shannon, 1956, Loveridge, 1957a, 1959, Gail & Rageau, 1958, Guibé, 1958, Zweifel, 1960, Duellman, 1961, V.F.M. FitzSimons, 1962, Shuntov, 1962, 1963, Villa, 1962, 1984, C.-S. Wang, 1962, Keegan & Werler, 1963, Kuntz, 1963, Klawe, 1964, Goris, 1965, McCann, 1966, Minton, 1966, 1975, Visser, 1967, Barme, 1968, Halstead, 1970, Tu & Tu, 1970, Gallagher, 1971, 1990, Graham, et al., 1971, Kropach, 1972, 1975, McDowell, 1972a, G. Pickwell, 1972, Hecht et al., 1974, Cogger,

1975, 1992, Tu, 1976, Bannikov et al., 1977, F. Scott et al., 1977, Dunson & Minton, 1978, L.D. Wilson et al., 1978, Medem, 1979, Hidalgo, 1980b, Mao & Chen, 1980, M. McCoy, 1980, G.V. Pickwell & Culotta, 1980, Robb, 1980, Nutaphand & Tumvipart, 1982, Brygoo, 1983, 1987, J.C. Daniel, 1983, Lanza, 1983a, 1990b, G.V. Pickwell et al., 1983, Tweedie, 1983, Joger, 1984, Bosch, 1985, L.D. Wilson & Meyer, 1985, Alcala, 1986, Ineich, 1986, 1988, Karsen et al., 1986, M. Mori, 1986, Pérez-Santos, 1986, Storr et al., 1986, Patterson, 1987, Szyndlar & Hung, 1987, Branch, 1988, Gasperetti, 1988, M.A.R. Khan, 1988, J.A. Campbell & Lamar, 1989, 2004, Gow, 1989, Hoser, 1989, 2012e, Bauer & Vindum, 1990, Huang et al., 1990, A. Silva, 1990, B.L. Lim, 1991, Pérez-Santos & Moreno, 1991, Stuebing, 1991, Leviton et al., 1992, Golay et al., 1993, Zhao & Adler, 1993, Bush et al., 1995, Spawls & Branch, 1995, O'Shea, 1996, A.R. Rasmussen, 1997, M.J. Cox et al., 1998, Switak, 1998, I. Das, 1999, 2007b, 2010, 2012, Bauer & Sadlier, 2000, Ineich & Laboute, 2002, Spawls et al., 2002, J.C. Murphy & Schlager, 2003, Whitaker & Captain, 2004, Kharin, 2007, San & Lee, 2007, Wells, 2007, V.S. Nguyen et al., 2009, Xiang & Li, 2009, Zaher et al., 2009, Ernst & Ernst, 2011, McCranie, 2011a and A.R. Rasmussen et al., 2011.

Remarks: Gender of *Pelamis* is feminine *fide* Lanza & Boscherini (2000: 328). A synonym of *Hydrophis fide* Sanders et al., 2013.

1. *Pelamis platura* (Linnaeus, 1766). Syst. Nat., ed. 12, 1: 391. (*Anguis platura*)

Synonyms: *Anguis platuros* – Gmelin, 1789 *(nomen incorrectum), Hydrus bicolor* J.G. Schneider, 1799, *Natrix dorsalis* Rafinesque-Schmaltz, 1817, *Ophinectes lutens* Rafinesque-Schmaltz, 1817, *Pelamis schneideri* Rafinesque-Schmaltz, 1817 (*nomen substitutum*), *Hydrophis pelamis* Schlegel, 1837, *Hydrophis variegata* Siebold, 1838, *Pelamis ornata* Gray, 1842d, *Pelamis bicolor maculata* A.-M.-C. Duméril, Bibron & Duméril, 1854b *(nomen nudum), Pelamis bicolor sinuata* A.-M.-C. Duméril, Bibron & Duméril, 1854b, *Pelamis bicolor variegata* A.-M.-C. Duméril, Bibron & Duméril, 1854b, *Pelamijs bicolor* – Bleeker, 1856 *(nomen emendatum), Hydrophis bicolor alternans* J.G. Fischer, 1856a (*nomen substitutum*), *Hydrophis bicolor chinensis* Jan,, 1859a, *Pelamis tricolor* Bennett, 1862, *Pelamides platurus* – Blyth, 1863 *(nomen incorrectum), Hydrophis bicolor maculata* Jan, 1863b (*nomen nudum*), *Hydrophis bicolor maculata* Jan, 1872 *in* Jan & Sordelli, 1870–1881, *Pelamis platurus* – Stoliczka, 1872, *Hydrus platyurus* – Lydekker, 1901 *(nomen incorrectum), Hydrus platurus linneaus* – Castellani & Chalmers, 1910 *(lapsus calami), Hydrus platurus pallidus* Wall, 1921g, *Hydrus platurus subobscurus* Wall, 1921g, *Pelamydrus platalus* – Nagai, 1928 *(nomen incorrectum), Pelamas platurus* – Dean, 1938 *(nomen incorrectum), Pelamiris platurus* – Daniel, 1949 *(nomen incorrectum), Pelamis platura brunnea* Deraniyagala, 1955, *Pelamis platura fasciata* Deraniyagala, 1955, *Pel amis platura leucostriata* Deraniyagala, 1955, *Pelamis platura neuricatenata* Deraniyagala, 1955, *Pelamis platura neurileucura* Deraniyagala, 1955, *Pelamis platura tricolor* Deraniyagala, 1955, *Pelamis platyura* – Deraniyagala, 1977 *(nomen incorrectum), Palemis platurus* – Felger & Moser, 1985 *(nomen incorrectum),* and *Pelamydrus platarus* – Culotta & Pickwell, 1993 *(nomen incorrectum).*

Type: Holotype, not designated (F. Ziervogel, Naturalien-Cabinett), lost *fide* L.C. Stuart (1963: 125).

Type locality: Unknown. Restricted to "ad littus insulae maris pacifici, Pine-isle" *fide* Gmelin (1789: 1122), [= shore of Pacific Ocean island, Pine Isle = Île des Pins, ext. SE New Caledonia, ca. 22°37'S, 167°30'E].

Distribution: Indian and Pacific Oceans from Africa to the Americas. Central Somalia (E Mudug), E Kenya (E Coast), E Tanzania (E Lindi), W Namibia (W Erongo), E South Africa (Eastern Cape, KwaZulu-Natal, Western Cape), Madagascar (Toamasina, Toliara), Seychelles, Mauritius, Kuwait, E Saudi Arabia, Bahrain, United Arab Emirates (Ajman, Dubai, Ras al Khaimah, Sharjah), N Oman (Masirah Is.), S Yemen, S Iran (Hormozghan; Hormuz and Larak Is.), Pakistan, India (Andaman & Nicobars: Andaman and Nicobar Is., Kerala, Maharashtra, Orissa, Tamil Nadu, West Bengal), Laccadives, Maldives, Sri Lanka, Bangladesh, Myanmar, Thailand (Songkhla), Vietnam (entire coast), ext. SE Russia (Kunashir Is.), Japan (W Hokkaido, Honshu, Kyushu, Shikoku, Amamioshima and Tsushima Is.), S Korea, SE China (Fujian, Guangdong, Hainan, Hong Kong, Zhejiang), Taiwan, Pescadores and Botel-Tobago Is., Philippines (Gigante, Luzon), E West Malaysia, Singapore, East Malaysia (Sabah, Sarawak), Brunei, Indonesia (Papua, Ambon, Bangka, S Flores, N S Java, Kalimantan, Mansinam, Num, Obi, Sulawesi, Sumatra, SW Sumba, Ternate, Weh), Papua New Guinea (New Britian), Australia (New South Wales, Northern Territory, Queensland, Tasmania, Victoria, Western, Chesterfield Reefs, Cocos and Christmas Is.), N New Zealand (North and Kermadec Is.), Palau (Angaur Is.), Solomons (New Georgia, Rennell), New Hebrides, New Caledonia, Pine Is., Fiji Is., Samoa, Tonga, Cook Is., Marquesas, Tahiti, Tuamotu Arch., Manihiki Atoll, W USA (Hawaii: Oahu; ext. SW California: San Clemente Is.), W Mexico (Baja California Norte, Baja California Sur, Chiapas, Colima, Guerrero, Jalisco, Michoacán, Nayarit, Sinaloa, Sonora, María Magdalena, Santa Margarita and Espiritu Santo Is.), S Guatemala, El Salvador (La Libertad, La Paz, La Union, San Miguel, Sonsonate), S Honduras (Choluteca, Valle), W Nicaragua (León), W Costa Rica (Guanacaste, Puntarenas), S Panama (Chiriquí, Darién, Panamá, Veraguas, Culebra, Flamenco, Naos and Venado Is.), W Colombia (Cauca, Chóco, Gorgona Is.), Ecuador (Esmeraldas, Guayas, Manabí, Galapagos Is.: Caamaño, Genovesa, Salvador, Santa Cruz, South Plaza), N Peru and Chile (Easter Is.).

Remarks: Holotype from the Cabinet of Natural Wonders of F. Ziervogel, pharmacist to the Royal Court in Stockholm *fide* E. Thorsell *(in litt.)*. Holotype in NHR *fide* Wall (1921g: 421). Type locality erroneously listed as Suriname *fide* J.A. Peters (1960a: 535). Undocumented reports exist for the Red Sea and Caribbean Ocean.

PELIAS Merrem, 1820 (Viperidae)

Synonyms: *Berus* Oken, 1816 *(nomen rejiciendum), Berus* Swainson, 1839, *Pelius* Bell, 1839 *(nomen emendatum), Polygerrus* Bonaparte, 1840 *(nomen nudum), Echidnoides* Mauduyt, 1844, *Acridophaga* A.F.T. Reuss, 1927a, *Mesocoronis* A.F.T. Reuss, 1927a, *Mesohorinis* – A.F.T. Reuss, 1927d *(nomen incorrectum), Mesohoronis* – A.F.T. Reuss, 1927d *(nomen incorrectum), Tzarevscya* A.F.T. Reuss, 1929a, *Tcarevscya* – A.F.T. Reuss, 1930c *(nomen incorrectum),* and *Mesocoponis* – A.F.T. Reuss, 1935a *(nomen incorrectum).*

Type species: *Vipera berus* Linnaeus, 1758.

Distribution: Eurasia.

Sources: Schwarz, 1936, F.A. Reuss, 1939b, Wittmann, 1954, Groombridge, 1980, Jirousek, 1983, Obst, 1983, W. Böhme & Joger, 1984, Bozhansky & Kudryavcev, 1986, Nilson & Andrén, 1986, 1988a–b, Brodmann, 1987, Marx & Nörstrom, 1988, Ashe & Marx, 1988, Golay et al., 1993, Bergman, 1994, Leenders, 1994, Venczel & Sen, 1994, David & Ineich, 1999, Herrmann, et al., 1999, McDiarmid et al., 1999, Lenk et al., 2001, Mallow et al., 2003, Ananjeva et al., 2004, Garrigues et al., 2005, Dobiey & Vogel, 2007, Stümpel & Joger, 2009, Tuniyev et al., 2009, Phelps, 2010 and Hoser, 2012d.

Remarks: A valid genus *fide* Ananjeva et al., 2004 and Tuniyev et al., 2009. Previously in the genus *Vipera.*

P

1. *Pelias altaica* (Tuniyev, Nilson & Andrén, 2010). Russ. J. Herp. 17(2): 110–112, figs. 1–4. (*Vipera [Pelias] altaica*)

Type: Holotype, GNM Re.ex. 6639, a 363 mm female (G. Nilson, 1 June 1997).

Type locality: "Altay, eastern Kazakhstan, 4 km N. Village Chernyaevka, at river Kalgyr (Kalgir), E. Kazakhstan (48°41'59" N 85°2'24.6"E), 218–327 m above sea level."

Distribution: Eastern Kazakhstan, 200–1200 m.

2. *Pelias anatolica* (Eiselt & Baran, 1970). Ann. Naturhist. Mus. Wien 74: 357–360, pl. 1, figs. 3–4, pl. 2, figs. 1–2, pl. 3, fig. 3. (*Vipera ursinii anatolica*)

Type: Holotype, SZE 170 (formerly 246), a 315 mm female (A. Budak, 13 July 1969).

Type locality: "Ciglikara Ormani, Elmali, Vilayet Antalya, SW-Türkei" [= Ciglikara Ormani, Kuhu Dagh Mtns., Antalya Prov., SW Turkey, elevation 2500 m].

Distribution: Southwestern Turkey (Antalya), 1650–2500 m.

Sources: Joger et al., 1992 and Mallow et al., 2003.

3. *Pelias barani* (W. Böhme & Joger, 1984). Amphibia-Reptilia (1983) 4(2–4): 267–268, figs. 1–2. (*Vipera barani*)

Type: Holotype, ZFMK 35444, a 540 mm female (K. Gutsche, May 1981).

Type locality: "ca. 60 km N Adapazari, Türkei, 400 m N.N." [= ca. 60 km N Adapazari (40°50'N, 30°25'E), Sakarya Prov., NW Turkey, elevation 400 m].

Distribution: Northwestern Turkey (Sakarya), 400 m. Known only from type locality.

Sources: Billing et al., 1990, Herrmann & Joger, 1997, Franzen & Heckes, 2000, Baran et al., 2001, Kutrup, 2003, Kalyabina-Hauf et al., 2004 and Tuniyev et al., 2009.

4. *Pelias berus* (Linnaeus, 1758). Syst. Nat., ed. 10, 1: 217. (*Coluber berus*)

Synonyms: *Coluber chersea* Linnaeus, 1758, *Coluber prester* Linnaeus, 1761, *Coluber vipera anglorum* Laurenti, 1768, *Coluber melanis* Pallas, 1771, *Coluber scytha* Pallas, 1773, *Coluber niger* Lacépède, 1789 *(nomen rejiciendum), Coluber schytha* Bonnaterre, 1790 *(nomen emendatum), Vipera schyta* Latreille, 1801a *(nomen emendatum), Coluber caeruleus* Sheppard, 1804, *Vipera communis* Leach, 1817 *(nomen substitutum), Vipera melaenis* Pallas, 1814 *(nomen emendatum), Coluber chersea marasso* Pollini, 1818 (*nomen nudum*), *Vipera marasso* Sette, 1821, *Vipera limnaea* Bendiscioli, 1826, *Vipera trilamina* Millet, 1828 *(nomen substitutum), Coluber scitta* Glückselig, 1832 *(nomen emendatum), Vipera torva* Lenz, 1832 *(nomen substitutum), Vipera marasso* Sette *in* Bonaparte, 1835 *(nomen nudum), Polygerrus paradoxus* Bonaparte, 1840 *(nomen nudum), Vipera prester gagatina* Freyer, 1842, *Pelias dorsalis* Gray, 1842e, *Echidnoides trilamina* Mauduyt, 1844a, *Vipera pelias* Soubeiran, 1855, *Pelias aspis niger* Cope, 1860b (*nomen nudum*), *Vipera berus concolor* Jan, 1863b (*nomen nudum*), *Vipera berus lymnaea* Jan, 1863b *(nomen nudum), Vipera berus lymnaea* Jan, 1873 *in* Jan & Soprdelli, 1870–1881 *(nomen emendatum), Vipera berus typica montana* Méhély, 1893 (*nomen illegitimum*), *Vipera berus nigra* Bethencourt-Ferreira, 1897a, *Vipera berus bosniensis* Boettger *in* Mojsisovics, 1889, *Vipera berus montana* Méhely, 1894, *Pelias berus lugubris* Kashchenko, 1902, *Vipera berus pseudaspis* Schreiber, 1912, *Vipera berus brunneomarcata* A.F.T. Reuss, 1923, *Vipera berus luteoalba* A.F.T. Reuss, 1923, *Vipera berus luteachersea* Reuss, 1923, *Vipera berus ochracea* *asymmetrica* A.F.T. Reuss, 1923 *(nomen illegitimum), Vipera berus rudolphi-marchica* A.F.T. Reuss, 1924a *(nomen incorrigendum), Vipera berus bilineata* A.F.T. Reuss, 1924b, *Vipera berus zobrysi* A.F.T. Reuss, 1924c *(nomen nudum), Vipera berus chersea splendens* A.F.T. Reuss, 1925a *(nomen illegitimum), Vipera berus coronis leopardina* A.F.T. Reuss, 1925b *(nomen illegitimum), Vipera berus coronis zamenoides* A.F.T. Reuss, 1925b *(nomen illegitimum), Vipera berus ochracea splendens* A.F.T. Reuss, 1925a *(nomen illegitimum), Vipera berus punctata* F. Reuss, 1925a, *Vipera berus rutila* A.F.T. Reuss, 1925a, *Vipera coronis* A.F.T. Reuss, 1925b, *Vipera coronis leopardina* A.F.T. Reuss, 1925b, *Vipera coronis nigroides* A.F.T. Reuss, 1925b, *Vipera coronis zamenoides* A.F.T. Reuss, 1925b, *Vipera coronis leopardina chersoides* A.F.T. Reuss, 1925c *(nomen illegitimum), Vipera coronis leopardina coerulea* A.F.T. Reuss, 1925c *(nomen illegitimum), Vipera coronis leopardina grisescens* A.F.T. Reuss, 1925c *(nomen illegitimum), Vipera coronis leopardina ochracea* A.F.T. Reuss, 1925c *(nomen illegitimum), Vipera coronis leopardina rosacea* A.F.T. Reuss, 1925c *(nomen illegitimum), Vipera coronis aspoides* A.F.T. Reuss, 1925c, *Vipera coronis rudolphoides* A.F.T. Reuss, 1925c, *Vipera berus rudolphitrans* A.F.T. Reuss, 1926 *(nomen nudum), Vipera berus sudetica* Sure, 1926 *(nomen nudum), Vipera berus sudetica coronoides* Sure, 1926 *(nomen nudum), Vipera coronis beroides* Sure, 1926 *(nomen nudum), Vipera coronis eimeri* Sure, 1926 *(nomen nudum), Vipera coronis flava* Sure, 1926 *(nomen nudum), Vipera rudolphi marchici* Sure, 1926 *(nomen nudum), Pelias rudolphi marchici* A.F.T. Reuss, 1927a *(nomen nudum), Pelias sudetica* A.F.T. Reuss, 1927a *(nomen nudum), Vipera aspoides annulata* A.F.T. Reuss, 1927b, *Vipera berus marchici* A.F.T. Reuss, 1927b *(nomen incorrigendum), Vipera berus rudolphi* A.F.T. Reuss, 1927b *(nomen nudum), Vipera coronis beroides* A.F.T. Reuss, 1927b, *Vipera eimeri* A.F.T. Reuss, 1927b, *Vipera rudolphoides* A.F.T. Reuss, 1927b, *Vipera aspoides* A.F.T. Reuss, 1927c *(nomen nudum), Vipera coronis leopardina* A.F.T. Reuss, 1927c *(nomen nudum), Vipera coronis nigroides* A.F.T. Reuss, 1927c *(nomen nudum), Vipera coronis zamenoides* A.F.T. Reuss, 1927c *(nomen nudum), Vipera eimeri* A.F.T. Reuss, 1927c *(nomen nudum), Vipera kochi* A.F.T. Reuss, 1927c *(nomen nudum), Vipera rudolphoides* A.F.T. Reuss, 1927c *(nomen nudum), Mesocoronis heinrothi* A.F.T. Reuss, 1929b *(nomen nudum), Pelias elberfeldi* A.F.T. Reuss, 1929b *(nomen nudum), Pelias marchici* A.F.T. Reuss, 1929b *(nomen nudum), Pelias meridionalis* A.F.T. Reuss, 1930a *(nomen nudum), Pelias meridionalis* A.F.T. Reuss, 1930b, *Pelias meridionalis schreiberi* A.F.T. Reuss, 1930b, *Mesocoronis bosniensis atrobosniensis* A.F.T. Reuss, 1930c, *Mesocoronis coronis nigroides flavocincta* A.F.T. Reuss, 1930c *(nomen illegitimum), Mesocoronis coronis ornata* A.F.T. Reuss, 1930c *(nomen nudum), Pelias flavescens* A.F.T. Reuss, 1930c *(nomen nudum), Pelias schöttleri* A.F.T. Reuss, 1930b *(nomen nudum), Pelias marchici* A.F.T. Reuss, 1930c, *Pelias schöttleri* A.F.T. Reuss, 1930c *(nomen incorrigendum), Pelias subalpina* A.F.T. Reuss, 1930c *(nomen nudum), Pelias tyrolensis* A.F.T. Reuss, 1930c, *Pelias flavescens* Sure, 1930 *(nomen nudum), Vipera bosnensis* – F. Werner, 1930c *(nomen incorrectum), Pelias neglecta* A.F.T. Reuss, 1932, *Vipera berus sphagnosa* Krassawzeff, 1932, *Pelias occidentalis* A.F.T. Reuss, 1933b, *Pelias occidentalis oldesloensis* A.F.T. Reuss, 1933b *(nomen nudum), Pelias occidentalis orbensis* A.F.T. Reuss, 1933b *(nomen nudum), Pelias sudetica steinii* A.F.T. Reuss, 1935b *(nomen nudum), Mesocoronis pseudaspoides* A.F.T. Reuss, 1936 *(nomen nudum), Mesocoronis heinrothi* A.F.T. Reuss, 1936 *(nomen nudum), Pelias sudetica steini* A.F.T. Reuss, 1937, *Pelias sudetica steini emarcata* A.F.T. Reuss, 1937 *(nomen illegitimum), Mesocoronis pseudaspoides* A.F.T. Reuss, 1937, *Mesocoronis parabosniensis* A.F.T. Reuss, 1937, *Pelias flavescens* A.F.T. Reuss, 1938c, *Vipera mesoccoronis* – Boquet, 1964 *(nomen incorrectum),* and *Coluber coeruleus* – Rage & Schätti *in* Golay et al., 1993 *(nomen incorrectum).*

Type: Neotype, ZMUU 7454, a 455 mm male (2 June 1934), designated by Krecsák & Wahlgren (2008: 2357).

Type locality: "Berthaga, Uppsala (59°51'N 17°34'E), Uppland, Sweden" via neotype selection.

Distribution: Eurasia. United Kingdom (England, Scotland, Wales), Sweden, Norway, Finland, Netherlands (Drente, Friesland, Gelderland, Groningen, Limburg, Noord Brabant, Overijssel, Utrecht), Belgium (Antwerpen, Hainaut, Luxembourg, Namur), France (Ain, Aisne, Allier, Ardèche, Ardennes, Aube, Aveyron, Calvados, Cantal, Corrèze, Côte-d'Or, Côtes-du-Nord, Creuse, Doubs, Essonne, Eure, Eure-et-Loir, Finistère, Haut-Rhin, Haute-Loire, Haute-Marne, Haute-Savoie, Haute-Vienne, Ille-et-Vilaine, Indre, Indre-et-Loire, Jura, Loir-et-Cher, Loire, Loire-Atlántique, Loiret, Maine-et-Loire, Lozère, Manche, Marne, Mayenne, Morbihan, Oise, Orne, Pas-de-Calais, Puy-de-Dôme, Saône-et-Loire, Seine-et-Marne, Seine-Maritime, Somme, Yvelines), Germany (Baden-Württemberg, Bavaria, Brandenburg, Bremen, Hamburg, E Hesse, Lower Saxony, Mecklenburg-Vorpommern, North Rhine-Westphalia, Saxony, Saxony-Anhalt, Schleswig-Holstein, Thuringia), Austria (Carinthia, Lower Austria, Salzburg, Styria, Tyrol, Upper Austria, Vorarlberg), Poland (Dolnoslaskie, Kujawsko-Pomorskie, Lodzkie, Lubelskie, Lubuskie, Malopolskie, Mazowieckie, Opolskie, Podkarpackie, Podlaskie, Pomorskie, Slaskie, Swietokrzyskie, Warminsko-Mazurskie, Wielkopolskie, Zachodniopomorskie), Hungary, Switzerland (Bern, Fribourg, Genève, Glarus, Graubünden, Lucerne, Neuchâtel, Obwalden, St. Gallen, Schaffhausen, Schwyz, Ticino, Uri, Vaud, Valais, Zug, Zürich), N Italy (Emilia-Romagna, Friuli-Venezia Giulia, Lombardia, Piemonte, Trentino-Alto Adige), Czech Republic (Jihocesky, Jihomoravsky, Praha, Severocesky, Severomoravsky, Stredocesky,

Vychodocesky, Zapadocesky), Estonia, Latvia, Bosnia-Herzegovina, Croatia, Slovenia, Albania, Macedonia, Bulgaria (Blagoevgrad, Gabrovo, Kyustendil, Lovech, W Montana, Pazadzhik, Smolyan, S Shumen, S Sofiya, S Targovishte, Varna, Veliko Tarnovo, S Vidin), N Greece (Dráma, Évros, Thesprotía, Thessaloníki), Ukraine, Belarus, Russia (Moldavia), N Kazakhstan, ext. N Mongolia (Bayanolgiy, Hovsgol, Selenge) and N China (Jilin, Xinjiang), NSL–3000 m.

Fossil records: Lower Pliocene (Ruscinian, MN 14: 4.2–4.9 mya) of Hungary, lower Pleistocene (Calabrian: 0.8–1.8 mya) of Austria, Czech Republic and Poland, middle Pleistocene (Ionian: 0.13–0.78 mya) of Austria, Germany, Romania and United Kingdom, and upper Pleistocene (Tarantian: 0.01–0.13 mya) of Germany, Poland, Serbia and United Kingdom.

Sources: Frommhold, 1964, Prestt, 1971, Bannikov et al., 1977, Bruno, 1977, 1984, Saint-Girons, 1980b, Frazer, 1983, Vedmederya et al., 1986, Naulleau, 1987b, Szyndlar, 1987, Borkin et al., 1990a–b, Gillette, 1990, Szyndlar, 1991a, Zhao & Adler, 1993, Nilson et al., 1994, 1995, Höggren, 1995, Lindell, 1995, Schiemenz, 1995, R. Günther, 1996, Andrén et al., 1997, Borkin et al., 1997, Forsman, 1997, Herrmann & Joger, 1997, Nilson & Andrén, 1997, Holman, 1998a, Adnagulov et al., 2000, Cabela et al., 2001, Hofer et al., 2001, F.P. Pérez, 2001, Spellerberg, 2002, Völkl & Thièsmeier, 2002, Carlsson, 2003, Glowacinski & Rafinski, 2003, J.C. Murphy & Schlager, 2003, Joger & Wollsen, 2004, Kalyabina-Hauf et al., 2004, Sindaco et al., 2006, Smedt, 2006, Ivanov, 2007, San & Lee, 2007, Trapp, 2007, Valakos et al., 2008, Creemers & Delft, 2009, Vacher & Geniez, 2010, Zinenko et al., 2010, Stojanov et al., 2011 and Jablonski et al., 2012.

Remarks: Photograph of neotype in Krecsák & Wahlgren (2008: fig. 2). *Pelias bosniensis* either closely related species or synonym of *P. barani fide* Herrmann & Joger, 1997.

P

5. *Pelias darevskii* (Vedmederya, Orlov & Tuniyev, 1986). Proc. Zool. Inst., USSR Acad. Sci. 157: 58–59. (*Vipera darevskii*)

Synonym: *Vipera kaznakovi darevskii* Vedmederya, 1984 (*nomen nudum*).

Type: Holotype, ZISP 19934, a 421 mm female (I.S. Darevsky, June 1980).

Type locality: "south-eastern part of Dzavachet mountain ridge in the border of Armenia." Restricted to Legli Mtn., the Mokrye (Wet) Mountains, Lesser Caucasus, Gukasynsky region, Armenia [ca. 41°10'N, 43°59'E, elevation 3000 m] *fide* Orlov & Tuniyev (1990: 21).

Distribution: Lesser Caucasus of NW Armenia (Gegharkunik, Kotayk, Lori, Shirak, Tavush) and ext. NE Turkey (Artvin), 2500–3000 m.

Sources: Billing et al., 1990, Orlov & Tuniyev, 1986b, 1990, Höggren et al., 1993, Tuniyev et al., 2009, 2012 and Avci et al., 2010.

Remarks: Photographs of holotype in Orlov & Tuniyev (1990: figs. 10a–c). Occurrence in S Georgia doubtful *fide* Joger and Nilson *in* David & Ineich (1999: 335).

6. *Pelias dinniki* (Nikolsky, 1913). Herpetol. Caucas.: 176–179. (*Vipera berus dinniki*)

Synonyms: *Vipera berus dinnicki* Nikolsky, 1916 *(nomen emendatum), Vipera tigrina* Tzarevsky, 1917, and *Vipera kaznakowi orientalis* Vedmederya, 1984 *(nomen nudum).*

Type: Lectotype, ZDKU 26044, an adult female (N.Y. Dinnik, 1908), designated by Vedmederya et al. (1986b: 58).

Type locality: "upper reaches of the Malaya (Small) Laba River, Northern Caucasus and Svanetla, Georgia, elevation 2440 m" via lectotype selection.

Distribution: Greater Caucasus of NW Azerbaijan, N Georgia and ext. SW Russia, 1500–3000 m.

Sources: Nikolsky, 1916, Vedmederya et al., 1986b, 1990, Billing et al., 1990, Orlov & Tuniyev, 1990, Höggren et al., 1993, Nilson et al., 1994, 1995, Herrmann & Joger, 1997, Smedt, 2006 and Tuniyev et al., 2009.

Remarks: Photograph of lectotype in Orlov & Tuniyev (1990: fig. 5). Previously a synonym of *P. kaznakovi.*

7. *Pelias ebneri* (Knoepffler & Sochurek, 1955). Bergenländ. Heimat. 17(4): 185–186. (*Vipera ursinii ebneri*)

Type: Holotype, NMW 14889, a 330 mm male (R. Ebner, 1936)

Type locality: "Elbursgebirge zwischen Rhema und Demawend in 2700 m. Nord-Persien" [= between Rhema and Damavand volcano (35°57'N, 52°07'E), Elburz Mtns., Mazandaran Prov., N Iran, elevation 2700 m].

Distribution: Northern Iran (Mazandaran) and ext. SE Azerbaijan (Lerik), 2000–2700 m.

Sources: Tuniyev et al., 2009, 2010.

Remarks: A synonym of *P. ursinii fide* McDiarmid et al. (1999: 410).

8. *Pelias eriwanensis* (A.F.T. Reuss, 1933c). Nachr. Aquar. Terrar.-Ver. (26): 373. (*Acridophaga renardi eriwanensis*)

Type: Holotype, ZMB 55160, a 425 mm male (Ramme & Richter, 2 Sept. 1929).

Type locality: "Lake Sevan above Shordzha at 2000 m a.s.l. [Yerevan], Armenia," *fide* Krecsák (2007: 658).

Distribution: Asia Minor. Northeastern Turkey (Kars), Armenia and W Azerbaijan, 2000 m.

Sources: Reuss, 1935a, Höggren et al., 1993, Nilson et al., 1994, 1995, 1999, Krecsák, 2007 and Tuniyev et al., 2009, 2010.

Remarks: Neotype designation *fide* Nilson et al. (1999: 105) invalid. A valid species *fide* David & Ineich (1999:

336) but a synonym of *P. ursinii fide* McDiarmid et al. (1999: 410). Photographs of holotype in Krecsák (2007: 657, fig. 1).

9. *Pelias kaznakovi* (Nikolsky, 1909a). Mitt. Kaukas. Mus. 4(3): 174. (*Vipera kaznakovi*)

Synonyms: *Coluber kaznakowi* Nikolsky, 1916 *(nomen emendatum), Mesocoronis kaxnakovi* – A.F.T. Reuss, 1927a *(nomen incorrectum), Vipera berus ornata* Basoglu, 1947 *(nomen praeoccupatum),* and *Vipera kaznokovi* – Bogert, 1961 *(nomen incorrectum).*
Type: Lectotype, ZMT 4408 (Y.V. Voronov), designated by Kramer (1961b: 713).
Type locality: "USSR, Georgien, Kreis Suchum, Cebelda, an der Ostküste des Schwarzen Meeres" [= Tsebelda or Tsebel'dinskoye, Sukhumi Distr., Abkhazia Prov., NW Georgia, 43°01'N, 41°16'E, elevation 440 m] via lectotype restriction.
Distribution: Asia Minor. Extreme NE Turkey (Artvin), SW Georgia (Abkhazia) and Armenia, 440–2400 (3000) m.
Sources: Darevsky, 1956, Kramer, 1961b, Kretz, 1971, Bannikov et al., 1977, Saint-Girons, 1980b, Orlov & Tuniyev, 1990, Höggren et al., 1993, Nilson et al., 1994, 1995, Billing et al., 1990, Herrmann & Joger, 1997, Tuniyev & Ostrovskikh, 2001, Smedt, 2006, Afsar & Afsar, 2009 and Tuniyev et al., 2009.
Remarks: Lectotype designation of Orlov & Tuniyev (1990: 6). The species complex of *Pelias kaznakovi, P. orlovi,* and *P. magnifica* possibly conspecific. Original description reprinted in Nikolsky (1910a: 81–84).

10. *Pelias lotievi* (Nilson, Tuniyev, Orlov, Höggren & Andrén, 1995). Asiatic Herpetol. Res. 6: 21–23, fig. 22. (*Vipera lotievi*)

Type: Holotype, ZISP 20309, a 422 mm female (K. Lotiev, 20–23 July 1986).
Type locality: "Armkhi, Checheno-Ingushetia, Russia, below Mt. Stolovaya, 2000 m. altitude."
Distribution: Greater Caucasus of N Georgia, N Azerbaijan and SW Russia (Dagestan), 1200–2700 m.
Sources: Tuniyev et al., 2009, 2010.

11. *Pelias magnifica* (Tuniyev & Ostrovskikh, 2001). Russ. J. Herp. 8(2): 123–126, fig. 9. (*Vipera magnifica*)

Type: Holotype, CSNBR 541, a 436 mm male (B.S. Tuniyev, 24 June 1998).
Type locality: "Shakhghireevskoye Gorge, the Malaya Laba River, Krasnodarskii region, Russia."
Distribution: Greater Caucasus in SW Russia (Krasnodar Krai) and NW Georgia, 700–1000 m.
Source: Tuniyev et al., 2009.
Remarks: The species complex of *Pelias kaznakovi, P. orlovi,* and *P. magnifica* possibly conspecific. A synonym of *P. dinniki fide* Smedt (2006: 149).

12. *Pelias nikolskii* (Vedmederya, Grubant & Rudayeva, 1986a). Vesnik Kharkov Univ. (288): 84. (*Vipera nikolskii*)

Type: Holotype, ZDKU 14703, a 680 mm female (K. Pengo, 1867).
Type locality: "near the Uda River in the vicinity of Kharkov," Russia.
Distribution: Southwestern Russia.
Remarks: A valid species *fide* Ananjeva et al. (2004: 205).

13. *Pelias olguni* Tuniyev, Avci, Tuniyev, Agasian & Agasian, 2012. Russ. J. Herp. 19(4): 323–325, fig. 4.

Type: Holotype, SNP 866, a 496 mm female (B. Tuniyev, S. Tuniyev, A. Avci, 21 July 2011).
Type locality: "Turkey, Ardahan Province, Town Posof, Mt. Ilgar-Dag, (2020 m a.s.l.)."
Distribution: Lesser Caucasus of NE Turkey (Ardahan, Artvin), 2020–2100 m.

14. *Pelias orlovi* (Tuniyev & Ostrovskikh, 2001). Russ. J. Herp. 8(2): 117–123, figs. 1–4. (*Vipera orlovi*)

Type: Holotype, CSNBR 528, a 447 mm male (S.V. Ostrovskikh, 12 May 1997).
Type locality: "Papai Mountain, Krasnodarskii region, Russia."
Distribution: Southwestern Russia (Krasnodar Krai), 820–1100 m.
Source: Tuniyev et al., 2009.
Remarks: A synonym of *P. dinniki fide* Mallow et al., 2003 and Smedt, 2006.

15. *Pelias pontica* (Billing, Nilson & Sattler, 1990). Zool. Scripta 19(2): 228–229, figs. 1–3. (*Vipera pontica*)

Type: Holotype, GNM 5021 (formerly GNM Re.ex. 5021), a 277 mm male (U. Sattler, June 1989).
Type locality: "Coruh valley, province Artvin, northeastern Turkey."
Distribution: Extreme NE Turkey (Artvin) and cen. Georgia.

16. *Pelias renardi* Christoph, 1861. Bull. Soc. Impér. Nat. Moscou 34(4): 600–606.

Synonyms: *Acridophaga uralensis* A.F.T. Reuss, 1929a, *Vipera renardi parursinii* Nilson & Andrén, 2001, *Vipera renardi tienshanica* Nilson & Andrén, 2001 and *Vipera renardi bashkirovi* Garanin, Pavlov & Bakiev, 2004.

Types: Syntypes (2), none designated, longest female 533 mm, longest male 477 mm (S. von Renard, 1840–1860), lost *fide* Golay et al. (1993: 291).

Type locality: "Sarepta allenthalben in der Steppe, südlichen Russland" [= Krasnoarmeysky Rayon, Volgogradskaya Distr., SW Russia, 48°31'N, 44°31'E, elevation 10 m].

Distribution: Central Asia. Ukraine, SW Russia, N Azerbaijan, Uzbekistan, Kazakhstan, Kyrgyzstan, Tadzhikistan and NW China (Xinjiang), 10 m.

Sources: Boulenger, 1893f, Nilson et al., 1994, 1995, Ostrovskikh, 1997, Nilson & Andrén, 2001, Bakleu et al., 2004, Tuniyev et. al., 2009, 2010 and Gvozdík et al., 2012.

Remarks: A valid species *fide* David & Ineich (1999: 340) but a synonym of *P. ursinii fide* McDiarmid et al. (1999: 409).

17. *Pelias sachalinensis* (Tzarevsky, 1917). Ann. Mus. Zool. Acad. Sci. Petrograd (1916) 21: 37. (*Vipera berus sachalinensis*)

Synonym: *Coluber sachalinensis continentalis* Nikolsky, 1927b.

Type: Holotype, ZISP 7960 *fide* I.S. Darevsky *in* McDiarmid et al. (1999: 399).

Type locality: "Sachalin" [= Sakhalin Island, E Russia].

Distribution: North Korea, N Mongolia and E Russia (SE Siberia, Sakhalin & Santarskie Is.).

Source: Szyndlar & Hung, 1987.

Remarks: A valid species *fide* Adnagulov et al. (2000: 150) and Joger et al., 2007. Syntypes ZISP 11992–93 *fide* Golay et al. (1993: 284).

18. *Pelias seoanei* (Lataste, 1879). Bull. Soc. Zool. France 4: 132. (*Vipera berus seoanei*)

Synonyms: *Vipera seoanei cantabrica* Braña & Bas, 1983 and *Vipera seoannei* – Bon, 1987 *(nomen incorrectum)*.

Types: Syntypes (2), BMNH 1946.1.21.44–45 (formerly BMNH 1920.1.20.786).

Type locality: "montibus Gallaecorum et Cantabrorum, d'Espagne" [= mountains of Galicia and Cantabrici, Spain]. Restricted to Cabanas, La Coruna Prov., Spain *fide* Mertens & Müller (1928: 52).

Distribution: Extreme SW France (Basses-Pyrènées), N Portugal (Braga, Porto, Viana de Castelo, Vila Real) and NW Spain (Álava, N Burgos, Vila Guipúzcoa, La Coruña, Léon, Lugo, Navarra, Orense, Oviedo, N Palencia, Pontevedra, Santander, Vizcaya), NSL–1500 (1900) m.

Sources: Saint-Girons & Duguy, 1976, Salvador, 1985, Castanet & Guyetant, 1989, Herrmann & Joger, 1997, Pleguezuelos, 1997, Salvador & Pleguezuelos, 2002, Malkmus, 2005, Smedt, 2006 and Vacher & Geniez, 2010.

19. *Pelias ursinii* Bonaparte, 1835 *in* 1832–1841. Icon. Fauna Ital. 2(12): 60 [p. 1 of *Pelias chersea* account], pl. 78, fig. 2 [*Pelias berus* pl.].

Synonyms: *Coluber dorsolaevibrunneo* Boddaert, 1783 *(nomen oblitum)*, *Coluber orientalis* Gmelin, 1789 *(nomen oblitum)*, *Coluber foetidus* Güldenstedt *in* Georgi, 1801 *(nomen suppressum)*, *Vipera berus rákosiensis* Méhely, 1894 *(nomen incorrigendum)*, *Vipera ursinii rakosiensis* – Méhely, 1894 *(nomen corrigendum)*, *Vipera macrops* Méhely, 1911, *Vipera ursinii rudolphi* A.F.T. Reuss, 1924a, *Coluber ursinoides* Nikolsky, 1927 *(nomen illegitimum)*, *Pelias berus uralensis* R. Reuss, 1929a *(nomen nudum)*, *Acridophaga uralensis* R. Reuss, 1929a, *Vipera berus ralosiensis* – F. Werner, 1929c *(nomen incorrectum)*, *Vipera orsinii* – Ghidini, 1938 *(nomen incorrectum)*, *Vipera ursinii wettsteini* Knoepffler & Sochurek, 1955, *Vipera orsinii* Ghidini, 1958 *(nomen emendatum)*, *Vipera ursinii graeca* Nilson & Andrén, 1988a, *Vipera ursinii rakoniensis* – González, 1991 *(nomen incorrectum)*, and *Vipera ursinii moldavica* Nilson, Andrén & Joger, 1993.

Type: Holotype, ANSP 6915, a juvenile (formerly CLB 674 & DW 6914-15) (C.-L.-J.-L. Bonaparte, 1814–1835, via Wilson).

Type locality: "monti dell'Abruzzo prossimi alla provincia d'Ascoli, Italica" [= hills of Abruzzo near the province of Ascoli, Italy].

Distribution: Southeastern Europe. Southeastern France (Alpes-Maritimes, Basses-Alpes, Vaucluse, Var), cen. Italy (Abruzzi, Marche, Umbria), E Austria (N Burgenland, Lower Austria, Vienna), Romania, Bulgaria (Shumen, Sofiya), Bosnia-Herzegovina, Croatia, Krk Is., N Macedonia, Montenegro, Moldavia, NE Albania, Greece (Ioannina) and Armenia, 250–3000 m.

Fossil records: Lower Pleistocene (Calabrian: 0.8–1.8 mya) of Czech Republic and middle Pleistocene (Ionian: 0.13–0.78 mya) of France.

Sources: Boulenger, 1893e, A.F.T. Reuss, 1935b, Stugren, 1955, Lanza, 1958, Kramer, 1961b, Bruno, 1967, 1977, 1984, Eiselt & Baran, 1970, Eiselt, 1976, Bannikov et al., 1977, Saint-Girons, 1980b, Dely, 1983, Dely & Stohl, 1984, Nilson & Andrén, 1988a, Billing et al., 1990, Borkin et al., 1990a–b, 1997, Agrimi & Luiselli, 1992, Kammel, 1992a-b, Höggren et al., 1993, Nilson et al., 1993, Zhao & Adler, 1993, Nilson et al., 1994, Herrmann & Joger, 1997, Holman, 1998a, Újvári et al., 2000, Cabela et al., 2001, Krecsák & Zamfirescu, 2001, Krecsák et al., 2003, De Smedt, 2006, Sindaco et al., 2006, Kreiner, 2007, Trapp, 2007, Valakos et al., 2008, Vacher & Geniez, 2010, Stojanov et al., 2011 and Gvozdík et al., 2012.

Remarks: Lectotype designation of ANSP 6915 *fide* Kramer (1961b: 713) invalid *fide* McDiarmid et al. (1999: 411). In accordance with Art. 23.9.2 of the Code (ICZN, 1999), *Pelias ursinii* Bonaparte is designated a *nomen protectum* and *Coluber dorsolaevibrunneo* Boddaert and *Coluber orientalis* Gmelin a *nomina oblita*.

PELTOPELOR A.C.L.G. Günther, 1864a (Viperidae)

Synonym: *Peltocolor* – Hoffmann, 1890 *(nomen incorrectum).*
Type species: *Trimesurus macrolepis* Beddome, 1862.
Distribution: Southern India.
Source: M.A. Smith, 1943, Gumprecht et al., 2004, Vogel, 2006 and Hutton & David, 2009.

1. *Peltopelor macrolepis* (Beddome, 1862). Madras Quart. J. Med. Sci. 5: 2, pl. 2, fig. 6. (*Trimesurus macrolepis*)

Type: Lectotype, BMNH 1946.1.18.72 (formerly BMNH 1861.12.30.80), a male (R.H. Beddome, 1857–1862), designated by Toriba *in* Golay et al. (1993: 101).
Type locality: "Anamallay Mountains; at 6,000 feet elevation" [= Anaimalai Hills, Western Ghats, E Tamil Nadu/W Kerala States, SW India, ca. 10°22'N, 77°08'E] via lectotype selection."
Distribution: Southern India (Western Ghats of Kerala, Tamil Nadu), 610–1980 m.

PHALOTRIS Cope, 1862a (Xenodontidae)

Synonyms: *Elapsomorphus* A.-M.-C. Duméril, 1853 *(nomen nudum)* and *Phalotus* Scudder, 1884 *(nomen emendatum).*
Type species: *Elapomorphus tricolor* A.-M.-C. Duméril, Bibron & Duméril, 1854b.
Distribution: Southern South America.
Sources: Ferrarezzi, 1993, 1994, Puorto & Ferrarezzi, 1994, Hofstadler-Deiques & Lema, 2005, Cacciali et al., 2007 and Zaher et al., 2009.

1. *Phalotris bilineatus* (A.-M.-C. Duméril, Bibron & Duméril, 1854b). Erpét. Gén. 7(2): 839–840. (*Elapomorphus bilineatus*)

Synonyms: *Elapomorphus spegazzinii* Boulenger, 1913f, *Elapomorphus suspectus* Amaral, 1924b, and *Elapomorphus bollei* Mertens, 1954a.
Type: Holotype, MNHN 3667, a 348 mm specimen (A.C.V.M.D. d'Obigny, 1826–1833).
Type locality: "province de los Corrientes" [= Corrientes Prov., ext. NE Argentina].
Distribution: Argentina (Buenos Aires, Chaco, Chubut, Córdoba, Corrientes, Entre Ríos, Formosa, Jujuy, La Pampa, Mendoza, Misiones, Río Negro, Salta, San Luis, Santa Fe, Santiago del Estero, Tucumán), NSL–500 m.
Sources: Lema, 1978b, Scolaro & Cei, 1979, Cei, 1986, 1994, J.D. Williams & Francini, 1991, Leynaud & Bucher, 1999, Avila et al., 2001 and Scolaro, 2005, 2006.
Remarks: *Phalotris spegazzinii* and *P. suspectus* valid species *fide* Lema (pers. comm.).

2. *Phalotris concolor* Ferrarezzi, 1994. Mem. Inst. Butantan (1993) 55(Suppl. 1): 26–28, figs. 2a–c.

Type: Holotype, IB 55018, a 550 mm female (R.N. Feio & G. Cotta, 1990), destroyed by fire 15 May 2010.
Type locality: "Brasil: Minas Gerais: Municipio de Cristália (17S, 43W, próximo a Grão Mogol), margem do rio Tacambiruçu, afluente da margem direita do rio Jequitinhonha."
Distribution: Brazil (Minas Gerais). Known only from type locality.
Source: Lema, 2002c.

3. *Phalotris cuyanus* (Cei, 1984). Bol. Mus. Cienc. Nat. Anthro. Juan Cornelio Moyano (4): 48–49, figs. a–d. (*Elapomorphus cuyanus*)

Type: Holotype, MHNSR 300, a 442 mm male (H.A. Lagiglia, 20 April 1975).
Type locality: "Los Reyunos, San Rafael, Mendoza, 900 m s. m., área de Sierra Pintada, Argentina."
Distribution: West-central Argentina (Mendoza, San Juan), 900–1000 m.
Sources: Cei, 1986, Lema et al., 2005a and Leynaud et al., 2005.
Remarks: Type locality elevation listed as 900–1000 m *fide* Cei (1984: 48) but corrected to 900 m *fide* Cei (1986: 416).

P

4. *Phalotris labiomaculatus* Lema, 2002b. Comm. Mus. Ciênc. Tecnol. PUCRS (Zool.) 15(2): 203–207, figs. 1–2.

Type: Holotype, MCP 10693, a 583 mm male (R.N. Yuke, Sept. 1983).
Type locality: "within Esperança Farm, at Buriti Grande, Porto Franco Municipality, State of Maranhão, Brazil (6.20°5'S-47.24°W)."
Distribution: Northeastern Brazil (Maranhão, Tocantins), 200 m.
Source: França et al., 2005.

5. *Phalotris lativittatus* Ferrarezzi, 1994. Mem. Inst. Butantan (1993) 55(Suppl. 1): 28–30, figs. 3a–c.

Type: Holotype, IB 26188, a 775 mm male (J.M. Guerra, 25 May 1966), destroyed by fire 15 May 2010.
Type locality: "Brasil: São Paulo: municipio de Matão (22S, 48W)."
Distribution: Southeastern Brazil (Santa Catarina, São Paulo), 325 m.
Sources: Lema, 2002b and Silveira Vasconcelos & Gomes dos Santos, 2009.

6. *Phalotris lemniscatus* (A.-M.-C. Duméril, Bibron & Duméril, 1854b). Erpét. Gén. 7(2): 840–841. (*Elapormorphus lemniscatus*)

Synonyms: *Elapsomorphus lemniscatus* A.-M.-C. Duméril, 1853 *(nomen nudum), Elapomorphus iheringi* Strauch, 1884, *Phalotris melanopleurus* Cope, 1885a, *Elapomorphus trilineatus* Boulenger, 1889a, and *Elapomorphus lemniscatus divittatus* Lema, 1984.
Type: Holotype, MNHN 3668, a 245–249 mm specimen (C.R. Darwin, 24 Dec. 1832–4 July 1835 [Chile] or 26 July 1832–4 Dec. 1833 [Uruguay]).
Type locality: "Amérique du Sud" [= South America]. Emended to Chili *fide* MNHN catalogue. Restricted to Montevideo, Uruguay *fide* Lema (1984: 67).
Distribution: Southern South America. Southeastern Brazil (Rio Grande do Sul, Santa Catarina, São Paulo), S Bolivia, E Paraguay, Uruguay (Artigas, Canalones, Cerro Largo, Durazano, Lavalleja, Maldonado, Montevideo, Paysandú, Rivera, Rocha, San José, Soriano) and N Argentina (Corrientes, Entre Ríos, Treinta y Tres), 500 m.
Sources: Prado, 1941a, 1941c, Lema, 1970, 1977b, 1978a, 1979a–b, 1985b, 1994, J.D. Williams & Francini, 1991, Cei, 1994, E.F. Schaeffer, 1998, Giraudo, 1999, Carreira-Vidal et al., 2005 and Prigioni et al., 2011.
Remarks: *Phalotris divittatus, P. iheringi* and *P. trilineatus* valid species *fide* Lema (pers.comm.).

7. *Phalotris matogrossensis* Lema, Agostini & Cappellari, 2005. Iheringia 95(1): 71–75, figs. 12–21.

Type: Holotype, MCP 12801, a 507 mm male (C. Strüssmann).
Type locality: "Cuiabá, Mato Grosso, bairro Cophema, 15°37'S 56°05'O," Brazil."
Distribution: Southeastern Brazil (S Goiás, S Matto Grosso, Mato Grosso do Sul, Minas Gerais, Paraná, Rio Grande do Sul, São Paulo) and SE Paraguay (? Caaguazú).
Source: Jansen & G. Köhler, 2008.

8. *Phalotris mertensi* (Hoge, 1955). Senck. Biol. 36(5–6): 301–304, pl. 27, fig. 1, pl. 29, fig. 3. (*Elapomorphus mertensi*)

Type: Holotype, IB 16425, a 930 mm female (F. Wenceslau-Garcia, 13 April 1955), destroyed by fire 15 May 2010.
Type locality: "Serra Azul, São Paulo, Brasilien" [= Serra Azul, NE São Paulo State, SE Brazil, 21°19'S, 47°34'W, elevation 615 m].
Distribution: Southeastern Brazil (Goiás, Mato Grosso, S Minas Gerais, N Paraná, São Paulo), 510–615 m.
Sources: Lema et al., 2005a and Forlani et al., 2010.

9. *Phalotris multipunctatus* Puorto & Ferrarezzi, 1994. Mem. Inst. Butantan (1993) 55(Suppl. 1): 40–42, figs. 1a–c.

Type: Holotype, IB 43635, a 380 mm male (C.W. Marchesan, Jr., 24 Oct. 1980), destroyed by fire 15 May 2010.
Type locality: "Brasil: Estado de São Paulo: município de Brotas (22 17S, 48 07W): Fazenda Elba."
Distribution: Southern Brazil (Mato Grosso, São Paulo).

10. *Phalotris nasutus* (Gomés, 1915). Ann. Paul. Med. Med. Cirurg. 4(6): 121, pl. 3, figs. 1–3. (*Elapomorphus nasutus*)

Type: Holotype, IB 873, a 800 female (A. Alonso, March 1915), destroyed by fire 15 May 2010.
Type locality: "Paineiras, perto de Uberaba, Estado de Minas Geraes, Brasil" [= Paineiras, cen. Minas Gerais State, SE Brazil, 18°54'S, 45°32'W, elevation 625 m].
Distribution: Central and SE Brazil (Distrito Federal, Goiás, Mato Grosso, Mato Grosso do Sul, Minas Gerais, São Paulo) and Bolivia (Santa Cruz), 500–1000 m.
Sources: Hoge & Garcia, 1949, Ferrarezzi, 1994, Lema, 1999, 2002b, Valdujo & Nogueira, 2001 and Jansen, 2008.
Remarks: Type locality listed as Peiropolis, Minas Gerais, Brazil *fide* Vanzolini (1986a: 7).

11. *Phalotris nigrilatus* Ferrarezzi, 1994. Mem. Inst. Butantan (1993) 55(Suppl. 1): 30–31, figs. 4a–c.

Type: Holotype, FML 709, a 715 mm female (R. Golbach, 1973).
Type locality: "Paraguai: Dept. San Pedro: Carumbé" [= Estancia Carumbé, San Pedro, Dept., cen. Paraguay, 23°54'S, 56°20'W, elevation 250 m].
Distribution: South-central Paraguay (San Pedro), 250 m.
Source: Cacciali et al., 2007b.

12. *Phalotris punctatus* (Lema, 1979b). Rev. Bras. Biol. 39(4): 836–851, figs. 1–6. (*Elapomorphus punctatus*)

Type: Holotype, MLPA 151, a 393 mm male (C. Piorno).
Type locality: "Rosário de la Frontera, Província de Salta, Argentina" [= Rosario de la Frontera, Salta Prov., N Argentina, 25°48'S, 64°58'W, elevation 790 m].
Distribution: Northern Argentina (Catamarca, Chaco, Córdoba, Corrientes, Formosa, Misiones, Salta, Santa Fe), 425–1270 m.
Sources: J.D. Williams & Francini, 1991, Leynaud & Bucher, 1999 and Leynaud et al., 2005.

13. *Phalotris reticulatus* (W.C.H. Peters, 1860d). Mber. Kønigl. Akad. Wiss. Berlin 1860(10): 518–519. (*Elapomorphus reticulatus*)

Type: Holotype, ZMB 3811, a 318 mm specimen.
Type locality: "Brasilien" [= Brazil].
Distribution: Brazil (Minas Gerais, São Paulo), 805 m.
Sources: Giraudo, 1999 and Ferrarezzi et al., 2005b.

14. *Phalotris sansebastiani* Jansen & G. Köhler, 2008. Senck. Biol. 88(1): 103–110, figs. 1–2.

Type: Holotype, MNK 4345, a 524 mm male (native, Jan. 2007).
Type locality: "Hacienda San Sebastian, S 16°21.676', W 62° 0.017', 550 m, Provincia Ñuflo de Chávez, Departament of Santa Cruz, Bolivia."
Distribution: Eastern Bolivia (Santa Cruz), 550 m.

15. *Phalotris tricolor* (A.-M.-C. Duméril, Bibron & Duméril, 1854b). Erpét. Gén. 7(2): 837–839. (*Elapomorphus tricolor*)

Synonym: *Elapomorphus punctatus* Lema, 1979b.
Type: Holotype, MNHN 520, a 938 mm skin (A.C.V.M.D. d'Orbigny, 11 Nov. 1830–21 June 1831 or 17 Sept.–17 Nov. 1832).
Type locality: "Santa-Cruz" [Bolivia]. Restricted to Santa Cruz de la Sierra à Chuquisaca, Santa Cruz, Bolivia *fide* Vanzolini (1948: 383).
Distribution: Southern South America. Eastern Bolivia (Santa Cruz), S Brazil (Mato Grosso, W São Paulo), Paraguay (Alto Paraguay, Amambay, Asunción, Boquerón, Caaguazú, Caazapá, Central, Concepción, Cordillera, Guairá, Misiones, Neembucú, Paraguarí, Presidente Hayes, San Pedro) and Uruguay, 100–370 m.
Sources: Vanzolini, 1948, Hoge, 1955, J.D. Cei, 1994, Leynaud & Bucher, 1999, Leynaud et al., 2005, Cacciali, 2008 and Jansen & Köhler, 2008.
Remarks: Specimens previously identified as *P. tricolor* are *P. punctatus fide* Giraudo & Scrocchi (2002: 29).

PHILODRYAS Wagler, 1830 (Xenodontidae)

Synonyms: *Chlorosoma* Wagler, 1830, *Orophis* Fitzinger, 1843, *Phylodryas* Fitzinger, 1843 *(nomen emendatum), Taeniophis* Girard, 1854, *Taenophis* – Girard, 1854 *(nomen incorrectum), Callirhinus* Girard, 1858a, *Euophrys* A.C.L.G. Günther, 1858, *Philodrias* – Senna, 1886 *(nomen incorrectum), Atomophis* Cope, 1887c *(nomen substitutum), Dirrhox* Cope, 1887a *(nomen nudum), Agratomus* Cope, 1887c, *Dirrhox* Cope, 1887c *(nomen substitutum), Monobothris* Cope, 1894c, *Pseudablabes* Boulenger, 1896a, *Pachyurus* Philippi, 1899 (*nomen nudum*), *Stenodeiros* Philippi, 1899 *(nomen nudum),* Rhinodryas F. Werner, 1903a, *Pachyurus* Philippi *in* Quijada, 1916, *Stenodeiros* Philippi *in* Quijada, 1916, *Pseudoblabes* – Phisalix, 1922 *(nomen incorrectum), Platyinion* Amaral, 1923, *Pseuduromacer* F. Werner, 1924a, *Phylodrias* – Amaral, 1926d *(nomen incorrectum), Chloroma* – Briceño-Rossi, 1934 *(nomen incorrectum), Cholrosoma* – Freiberg, 1939 *(nomen incorrectum), Estherophis* Donoso-Barros, 1962, *Phylodryas* – Abalos, Baez & Nader, 1964 *(nomen incorrectum), Philodryass* – Donoso-Barros, 1966 *(nomen incorrectum), Platynion* J.A. Peters & Orejas-Miranda, 1970 *(nomen emendatum), Philodry* – Cunha & Nascimento, 1978 *(nomen incorrectum), Incaspis* Donoso-Barros, 1974, *Pseadablabes* – Vaz-Ferreira, Covelo de Zollesi & Achaval, 1976 *(nomen incorrectum), Platinion* – Dixon & Thomas, 1985 *(nomen incorrectum),* and *Xenoxybelis* Rodrigues-Machado, 1993.
Type species: *Coluber olfersii* Lichtenstein, 1823.
Distribution: South America.
Sources: F. Werner, 1924c, Amaral, 1930e, J.A. Peters & Orejas-Miranda, 1970, R.A. Thomas, 1976a, Duellman, 1978, Cadle, 1984a, David & Ineich, 1999, Zaher, 1999, Vidal et al., 2000, Zaher et al., 2009 and Navarro-Cornejo, 2012.
Remarks: Gender of genus is feminine *fide* Donnelly and Myers (1991: 46). *Pseudablabes* and *Xenoxybelis* are synonyms *fide* Zaher et al., 2009.

1. *Philodryas aestiva* (A.-M.-C. Duméril, Bibron & Duméril, 1854b). Erpét. Gén. 7(2): 1111–1115. (*Dryophylax aestivus*)

Synonyms: *Philodryas carinatus* Hensel, 1868, *Philodryas campicola* Jensen, 1900, *Philodryas subcarinatus* Boulenger, 1902b, *Pseuduromacer lugubris* F. Werner, 1924a, *Philodryas aestivus manegarzoni* Orejas-Miranda, 1959a and *Philodryas aestivus levisquamus* R.A. Thomas, 1976 (*nomen ineditum*).
Types: Syntypes (3), MNHN 3842, an adult female (Gaudichaud) [Sainte-Catherine], MNHN (Clossen) [Brésil], and MNHN [méridional du Nouveau-Monde ou des Antilles], longest syntype 840 mm.
Type locality: "l'ile Sainte-Catherine" [= Santa Catarina Is., Santa Catarina State, S Brazil, 27°36'S, 48°30'W], "Brésil" [= Brazil], and "continent méridional du Nouveau-Monde ou des Antilles" [= South America or Antilles].
Distribution: Southern South America. Central Bolivia (Cochabamba, Tarija), SE Brazil (Bahia, Distrito Federal, Goiás, Mato Grosso do Sul, Minas Gerais, Paraná, Rio de Janeiro, Rio Grande do Sul, Santa Catarina, São Paulo), Uruguay (Artigas, Canelones, Cerro Largo, Colonia, Flores, Florida, Kocha, Lavalleja, Maldonado, Montevideo, Paysandú, Rio Negro, Rivera, Rocha, Salto, San José, Sariano, Tacuarembó, Treinta y Tres), S Paraguay (Alto Paraná, Caaguazú, Central, Guairá, Itapúa, Misiones, Neembucú, Paraguarí) and N Argentina (Buenos Aires, Chaco, Córdoba, Corrientes,

P

Entre Ríos, Formosa, Jujuy, Santiago del Estero, Misiones, Salta, Santa Fe, Treinta y Tres, Tucumán), NSL–2500 m.

Sources: Amaral, 1925b, 1977, M.A. Smith, 1929b, Serie, 1936, Orejas-Miranda, 1959b, Abalos et al., 1964, A. Barrio et al., 1977, Lema & Fabián-Beurmann, 1977, R.A. Thomas, 1977b, R.A. Thomas et al., 1977, Lema, 1987, J.D. Williams & Francini, 1991, Avila, 1996, Agostini, 1998, Freitas, 1999, Leynaud & Bucher, 1999, Argôlo, 2002b, Carreira-Vidal et al., 2005, Cacciali, 2008, Celsi et al., 2008 and Prigioni et al., 2011.

Remarks: Lectotype designation of MNHN 3842 *fide* R.A. Thomas (1976a: 86) invalid.

2. *Philodryas agassizii* (Jan, 1863a). Arch. Zool. Anat. Fis. 2(2): 260. (*Eirenis agassizii*)

Synonyms: *Philodryas paucisquamis* W.C.H. Peters, 1863 and *Liopeltis brevicauda* Jan, 1863b.

Types: Syntypes (2), MCZ 870, a 154 mm specimen, and MCZ 177823, a 347 mm male (J. Wyman, 1861).

Type locality: "Uruguay" (in error). Corrected to Paraguay *fide* original data tag and MCZ Catalogue.

Distribution: Southern South America. Southern Brazil (Distrito Federal, Goiás, Minas Gerais, Parana, Rio Grande do Sul, São Paulo), Paraguay, Uruguay (Artigas, Cerro Largo, Lavalleja, Maldonado, Montevideo, Río Negro) and NE Argentina (Buenos Aires, Chaco, Córdoba, Entre Ríos, La Pampa, Misiones, Rio Negro, San Luis, Santa Fe, Santiago del Estero), 150–1200 m.

Sources: Jan & Sordelli, 1860–1866, Boulenger, 1885a, 1886a, Devincenzi, 1925, Amaral, 1930f, 1977, Vaz-Ferreira & Sierra de Soriano, 1960, J.A. Peters & Orejas-Miranda, 1970, Lema & Fabián-Beurmann, 1977, Vinas-Dameri and Guida, 1989, J.D. Williams & Francini, 1991, Reati, 1996, Kiefer, 1998, Giraudo, 1999, Leynaud & Bucher, 1999, Nogueira, 2001 and Perez et al., 2012.

Remarks: A member of *Philodryas* (previously in the genus *Pseudablabes) fide* Grazziotin et al. (2012: 12). Records for Chaco and Santa Fe, Argentina need confirmation *fide* J.D. Williams & Francini (1991: 77).

3. *Philodryas argentea* (Daudin, 1803c). Hist. Nat. Rept. 6: 336–338. (*Coluber argenteus*)

Synonym: *Coluber argentatus* Merrem, 1820 *(nomen emendatum).*

Type: Holotype, not designated, an 1104 mm specimen, location unknown.

Type locality: Unknown.

Distribution: South America. Southeastern Colombia (Meta, Caquetá), S Venezuela (Amazonas), S Guyana (Upper Takutu-Upper Essequibo), Suriname (Saramacca, Brokopondo), French Guiana (Saint-Laurent de Maroni, Cayenne), E Ecuador (Napo, Pastaza, Zamora-Chinchipe), E Peru (Loreto), N Brazil (Acre, Amapá, Amazonas, Bahia, Goiás, Mato Grosso, Pará, Rondônia), Bolivia (Beni, Cochabamba) and Paraguay, NSL–600 m.

Sources: Roze, 1966a, Hoge, 1967, Hoogmoed, 1979, Vanzolini et al., 1980, Abuys, 1985a, Dixon & Soini, 1986, Lancini V., 1986, Vanzolini, 1986b, Chippaux, 1987, Keiser, 1989, Rodrigues-Machado, 1993, Fugler & Riva, 1998. Gorzula & Señaris, 1998, Starace, 1998, Freitas, 1999, Doan & Arriaga, 2000, Cisneros-Heredia & Touzet, 2007, Navarrete et al., 2009 and C.J. Cole et al., 2013.

Remarks: The type is not in MNHN *fide* M. Lang *(in litt.).* A member of *Philodryas* (previously in the genus *Xenoxybelis) fide* Grazziotin et al. (2012: 12).

4. *Philodryas arnaldoi* (Amaral, 1933b). Mem. Inst. Butantan (1932) 7: 100–101, figs. 2–3, 4.2. (*Chlorosoma arnaldoi*)

Type: Holotype, IB 7807, an 896+ mm female, destroyed by fire on 15 May 2010..

Type locality: "S. Bento, Sta. Catharina, Brasil" [= São Bento, Santa Catarina State, ext. S Brazil].

Distribution: Southern Brazil (Paraná, Rio Grande do Sul, Santa Catarina, São Paulo), NSL–785 m.

Source: Lema, 1962b.

5. *Philodryas baroni* Berg, 1895. An. Mus. Nac. Buenos Aires 4: 189–191, figs. 1.1–1.4.

Synonyms: *Rhinodryas königi* F. Werner, 1903a and *Philodryas baroni fuscoflavescens* Serie, 1915a.

Types: Syntypes (3), MACN, a 1430 mm female (A. Sohncke) [Chaco], and a 1280 mm female and 1180 mm male (M. Barón-Morlat) [Tucumán].

Type locality: "Tucuman et Chaco australis (Riacho de Oro)" [= Tucumán and Riacho de Oro, S Chaco, Argentina].

Distribution: Bolivia, NW Paraguay (W Alto Paraguay, Boquerón, Caaguazú) and N Argentina (Catamarca, Chaco, Córdoba, Formosa, Salta, Santa Fe, Santiago del Estero, Tucumán), 150–300 m.

Sources: Koslowsky, 1898b, Abalos et al., 1964, Cei, 1986, J.D. Williams & Francini, 1991, Arzamendia, 1999, Leynaud & Bucher, 1999 and Cacciali, 2008.

6. *Philodryas boliviana* Boulenger, 1896a. Cat. Snakes Brit. Mus. 3: 132, pl. 9, fig. 1. (*Philodryas bolivianus*)

Synonym: *Liophis bolivianus* F. Werner, 1909b.

Type: Lectotype, BMNH 1946.1.2.86, a 690–692 mm female (M.R. Rolle), designated by R.A. Thomas herein.

Type locality: "Charobamba, Bolivia" [= Coroico, La Paz Dept., cen. Bolivia, 16°12'S, 67°44'W, elevation 1690 m] via lectotype selection.

Distribution: Bolivia (? Chuquisaca, La Paz, Santa Cruz, Tarija), 1690–2600 m.

P

Sources: F. Werner, 1909c, Amaral, 1944a and Fugler et al., 1995.

Remarks: A valid species *fide* R.A. Thomas (1976a: 111–117, fig. 15). Lectotype designation of BMNH 1946.1.2.86 *fide* R.A. Thomas (1976a: 111) invalid. Boulenger (1896a: 132) erroneously listed the two adult syntypes as males. Type locality erroneously listed as Cochabamba *fide* Amaral (1926d: 31).

7. *Philodryas chamissonis* (Wiegmann *in* Meyen, 1834). Nova Acta Acad. Caesar. Leop.-Carol. 17(1): 246–250, pl. 19. (*Coronella chamissonis*)

Synonyms: ? *Coluber aesculapii* Molina, 1782, *Coluber chamissonis* Hemprich *in* Wiegmann, 1834 *(nomen nudum), Psammophis temminckii* Schlegel, 1837, *Taeniophis tantillus* Girard, 1854, *Coluber regius* Philippi, 1899 *(nomen nudum), Dromicus chamissonis subtetragona* Philippi, 1899, *Liophis gracilis* Philippi, 1899 (*nomen nudum*), *Liophis insularis* Philippi, 1899 (*nomen nudum*), *Liophis luctuosa* Philippi, 1899 (*nomen nudum*), *Taeniophis multilineata* Philippi, 1899 *(nomen nudum), Coronella plumbea* Philippi *in* Quijada, 1916 *(nomen nudum), Dromicus albiventris* Philippi *in* Quijada, 1916 (*nomen nudum*), *Dromicus cinereus* Philippi *in* Quijada, 1916 *(nomen nudum), Dromicus concolor* Philippi *in* Quijada, 1916 *(nomen nudum), Dromicus leucurus* Philippi *in* Quijada, 1916 *(nomen nudum), Dromicus liguanus* Philippi *in* Quijada, 1916 *(nomen nudum), Dromicus maculatus* Philippi *in* Quijada, 1916 *(nomen nudum), Dromicus marmorata* Philippi *in* Quijada, 1916 *(nomen nudum), Dromicus tenuis* Philippi *in* Quijada, 1916 *(nomen nudum), Dromicus variegatus* Philippi *in* Quijada, 1916 *(nomen nudum), Liophis gracilis* Philippi *in* Quijada, 1916 (*nomen nudum*), *Liophis insularis* Philippi *in* Quijada, 1916 *(nomen nudum), Liophis lutea* Philippi *in* Quijada, 1916 *(nomen nudum), Liophis marmorata* Philippi *in* Quijada, 1916 *(nomen nudum), Pachyurus mucronatus* Philippi *in* Quijada, 1916 *(nomen nudum)*, and *Taeniophis albigularis* Philippi *in* Quijada, 1916 *(nomen nudum)*.

Type: Holotype, ZMB, a 762 mm specimen (F.J.F. Meyen, 1830–1832), lost *fide* R. Günther (*in litt.*)

Type locality: "Chile, bei Tollo" [= Tollo, Santiago Prov., cen. Chile, 33°24'S, 70°23'W, elevation 1380 m].

Distribution: Central Chile (Antofagasta, Atacama, Concepción, Coquimbo, Malleco, Santiago, Valdavía, Valparaíso), NSL–1900 m.

Sources: Philippi, 1899, Quijada, 1916, Donoso-Barros & Candiani, 1950, Donoso-Barros, 1954, 1966, 1970, Donoso-Barros & Cardenas, 1964 and Webb & Greer, 1969.

8. *Philodryas cordata* Donnelly & Myers, 1991. Amer. Mus. Novit. (3017): 46–49, figs. 32–33.

Type: Holotype, AMNH 136212, a 740+ mm male (C.W. Myers, 24 Feb. 1990).

Type locality: "FUDECI Summit Camp 1, 1030 m elevation, north side of Cerro Guaiquinima, Bolívar, eastern Venezuela, *ca.* 5°56'N, 63°42'W."

Distribution: Eastern Venezuela (SW Bolívar), 1030–1520 m. Known only from type locality.

Sources: Mägdefrau et al., 1991 and Navarrete et al., 2009.

9. *Philodryas georgeboulengeri* Grazziotin, Zaher, Murphy, Scrocchi, Benavides, Zhang & Bonatto, 2012. Cladistics 28(5): 456–457. (*nomen substitutum*)

Synonym: *Oxybelis boulengeri* Procter, 1923 *(nomen praeoccupatum).*

Type: Holotype, BMNH 1946.1.9.86, a juvenile male (A.C. McDougall, 1922).

Type locality: "Trinidad, River Mamore, Bolivia" [= Trinidad, Beni Dept., E Bolivia, 14°50'S, 64°54'W, elevation 160 m].

Distribution: Eastern Peru (Cuzco) and E Bolivia (Beni), 160 m.

Sources: Keiser, 1989, Duellman & Salas, 1991, Fugler et al., 1995, Lehr, 2002 and Costa-Prudente et al., 2008.

Remarks: Original description in Procter (1923: 1062–1063, figs. 1a–c). A valid species fide Keiser, 1989. A member of *Philodryas* (previously in *Xenoxybelis*) *fide* Pyron et al. 2011.

10. *Philodryas laticeps* F. Werner, 1900a. Zool. Anz. 23(612): 198.

Synonyms: *Philodryas oligolepis* Gomes *in* Amaral, 1921 and *Philodryas affinis* L. Müller, 1928.

Type: Holotype, ZMB 15704, a 1045–1060 mm female (F. Weichberger).

Type locality: "Sta. Catarina, Brasilien" [= Santa Catarina State, S Brazil].

Distribution: South and SE Brazil (Espírito Santo, Minas Gerais, Santa Catarina) and Bolivia (Cochabamba, Santa Cruz), 190–420 m.

Source: Zaher et al., 2008.

Remarks: Neotype designation of FMNH 168003, a 1081 mm male from Río Chaparé at Río Mamoré (15°56'S, 64°41'W), Cochabamba Dept., Bolivia (W.P. McLean, 30 July 1965) *fide* R.A. Thomas (1976a: 235) invalid. Photograph of holotype in Zaher et al. (2008: figs. 2 a–f).

11. *Philodryas livida* (Amaral, 1923). Proc. New England Zool. Club 8: 91–93. (*Platyinion lividum*)

Type: Holotype, IB 3000, a 730 mm male (E. Guarini, 18 Oct. 1920), destroyed by fire 15 May 2010.

Type locality: "Dorizon, State of Paraná, Brazil" (in error). Corrected to Guaycurús, Estado de Matto

Grosso, Brasil [= Guaicurus, W Mato Grosso do Sul State, cen. Brazil, 20°07'S, 56°48'W, elevation 145 m] *fide* Amaral (1925b: 5).

Distribution: Southern Brazil (Mato Grosso do Sul, Goiás, ? Paraná, São Paulo), 145–880 m.

Sources: R.A. Thomas & Fernandes, 1996 and Valdujo & Nogueira, 1999.

Remarks: Supplemental original description in Amaral (1926a: 113, pl. 3, figs. 10–12). Amaral (1977: 164) erroneously reverted to Dorizon, Paraná as the type locality.

12. *Philodryas mattogrossensis* Koslowsky, 1898a. Rev. Mus. La Plata 8: 29–30, pl. 1, figs. 1–3.

Synonyms: *Philodryas ternetzii* Schenkel, 1901, *Philodryas erlandi* Lönnberg, 1902, and *Philodryas boulengeri* F. Werner, 1909b.

Type: Holotype, MLP, an 1195 mm female (C. Bach).

Type locality: "Miranda (Matto-Grosso), Brasil" [= Miranda, N Mato Grosso State, SW Brazil, 8°23'S, 57°57'W, elevation 225 m].

Distribution: Southern South America. Southern Brazil (Mato Grosso, Mato Grosso do Sul, Minas Gerais, São Paulo), S Bolivia (Chaco, Santa Cruz, Tarija), N Paraguay (Alto Paraguay, Boquerón, Concepción, Presidente Hayes) and ext. N Argentina (Chaco, Formosa, Salta), 150–500 m.

Sources: Lönnberg, 1902, L. Müller, 1928, J.D. Williams & Francini, 1991, Nobuo-Yuki, 1992, Lobo & Scrocchi, 1994, D.R. Norman, 1994, Leynaud & Bucher, 1999 and Cacciali, 2008.

P

13. *Philodryas nattereri* Steindachner, 1870b. Sitz. Akad. Wiss. Wien Math.-Natur. Kl., Abt. I, 62: 345–346, pl. 7, figs. 1–3.

Synonym: *Psammophis molochina* Berthold, 1846 *(nomen rejiciendum)*.

Type: Holotype, NMW 23400, an adult female (J. Natterer, 1822–1823), lost *fide* R.A. Thomas (1976a: 145).

Type locality: "Matogrosso" [= Mato Grosso or Mato Grosso do Sul State, Brazil].

Distribution: Eastern and S Brazil (Bahia, Ceará, Distrito Federal, S Goiás, S Mato Grosso, Mato Grosso do Sul, Minas Gerais, Paraiba, Pernambuco, Rio Grande do Norte, Sergipe, Tocantins), NSL–760 m.

Sources: Cordeiro & Hoge, 1974, R.A. Thomas, 1977c, Vanzolini et al., 1980, Freitas, 1999, Morato et al., 2011 and Recorder et al., 2011.

Remarks: Official Specific Name no. 2802 *fide* Opinion 1211 (ICZN, 1982d).

14. *Philodryas olfersii* (Lichtenstein, 1823). Verzeich. Doubl. Zool. Mus. Univ. Berlin: 104–105. (*Coluber olfersii*)

Synonyms: *Coluber herbeus* Wied-Neuwied, 1825c, *Coluber pileatus* Wied-Neuwied, 1825c, *Philodryas latirostris* Cope, 1862b, *Philodryas reinhardtii* A.C.L.G. Günther, 1862b, *Philodryas argentinus* L. Müller, 1924, and *Philodryas carbonelli* Roze, 1957a.

Type: Holotype, ZMB, lost *fide* R.A.Thomas (1976a: 153).

Type locality: "Brasil."

Distribution: South America. Central Colombia (Meta), Venezuela (Amazonas, Bolívar, Carabobo, Guárico, Miranda), S Guyana (Upper Demerara-Berbice, Upper Takutu-Upper Essqequibo), Suriname (Commewijne, Marowijne Nickerie, Saramacca), French Guiana (Cayenne), cen. Peru (Cusco, Huánuco, Junín, Loreto, Madre de Dios, Pasco, Puno, San Martín), Brazil (Alagoas, Bahia, Ceará, Distrito Federal, Espírito Santo, Goiás, Maranhao, Paraiba, Mato Grosso, Minas Gerais, Pará, Paraná, Pernambuco, Piauí, Rio de Janeiro, Rio Grande do Norte, Rio Grande do Sul, Rondônia, Santa Catarina, São Paulo, Sergipe), N Bolivia (Beni, Cochabamba, La Paz, Santa Cruz), Paraguay (Alto Paraguay, Alto Paraná, Amambay, Asunción, Caaguazú, Caazapá, Canindeyu, Central, Concepción, Cordillera, Guairá, Itapúa, Misiones, Neembucú, Paraguarí, Presidente Hayes, San Pedro), Uruguay (Artigas, ? Montevideo, Rivera, Salto) and N Argentina (Chaco, Corrientes, Entre Ríos, Formosa, Jujuy, Misiones, Salta, Santa Fe), NSL–500 m.

Sources: Orejas-Miranda, 1959b, Roze, 1966a, Hoge et al., 1975, R.A. Thomas, 1975, Lema & Fabián-Beurmann, 1977, Amaral, 1978, Vanzolini et al., 1980, Abuys, 1985b, Pérez-Santos, 1986, Vanzolini, 1986b, Chippaux, 1987, Pérez-Santos & Moreno, 1988, J.D. Williams & Francini, 1991, Fugler & Riva, 1990, Quaini & Arzamendia, 1998, Starace, 1998, Freitas, 1999, Leynaud & Bucher, 1999, Carreira-Vidal et al., 2005, 2012b, Cacciali, 2008, Navarrete et al., 2009 and C.J. Cole et al., 2013.

15. *Philodryas patagoniensis* (Girard, 1858a). Proc. Acad. Nat. Sci. Philadelphia (1857) 9(6): 182. (*Callirhinus patagoniensis*)

Synonyms: *Euophrys modestus* A.C.L.G. Günther, 1858 and *Liophis poecilostictus* Jan, 1863a.

Types: Syntypes (2), USNM 5536, a 927 mm female, and ANSP 3806 (formerly USNM), a 584 mm specimen (C. Wilkes [U.S. Explor. Exped.], 1838–1842), latter lost *fide* E. Malnate *in* R.A. Thomas (1976a: 172).

Type locality: "Coast of Patagonia" [S Argentina]. Emended to the coast of Patagonia, about the mouth of the Rio Negro *fide* Girard (1858b: 141), [= near mouth of Río Negro on border between provinces of Buenos Aires and Río Negro, S Argentina, 41°02'S, 62°47'W, elevation NSL].

Distribution: Southern South America. Southern Brazil (Bahia, Distrito Federal, Goiás, Mato Grosso, Mato Grosso do Sul, Minas Gerais, Pará, Paraná, Rio de Janeiro, Rio Grande do Norte, Rio Grande do Sul, São Paulo), Bolivia (Beni, Cochabamba, Santa Cruz), Paraguay (Alto Paraguay, Alto Paraná, Amambay, Asunción, Boquerón, Caaguazú, Caazapá, Canindeyu, Central, Concepción, Cordillera, Guairá, Itapúa, Misiones, Neembucú, Paraguarí, Presidente Hayes, San Pedro), Uruguay (Artigas, Canelones, Colonia, La Valleja, Maldonado, Montevideo, Rocha, Soriano, Tacuarembó, Treinte y Tres), cen. Chile (Santiago) and Argentina (Buenos Aires, Catamarca, Chaco, Chubut, Córdoba, Corrientes, Entre Ríos, Formosa, Jujuy, La Pampa, La Rioja, Misiones, Mendoza, Nequén, Río Negro, Salta, San Juan, San Luis, Santa Fe, Santiago del Estero, Treinta y Tres, Tucumán), NSL–1500 m.

Sources: Girard, 1858b, Lema & Fabián-Beurmann, 1977, R.A. Thomas et al., 1977, J.D. Williams & Francini, 1991, Lema, 1994, Lobo & Scrocchi, 1994, D.R. Norman, 1994, Fugler et al., 1995, R.A. Thomas & Fernandes, 1996, Freitas, 1999, Leynaud & Bucher, 1999, Avila et al., 2001, Nogueira, 2001, Passos & Fernandes, 2002a, Scolaro, 2005, 2006, Cacciali, 2008, López & Giraudo, 2008 and Prigioni et al., 2011.

Remarks: Supplemental original description in Girard (1858b: 139–142, pl. 12, figs. 1–6). Lectotype designation of USNM 5536, a 927 mm female (U.S. Explor. Exped., 1838–1842), *fide* R.A. Thomas (1976a: 172) invalid.

16. *Philodryas psammophidea* A.C.L.G. Günther, 1872b. Ann. Mag. Nat. Hist. (4) 9(49): 23–24, pl. 4, fig. a.

Synonyms: *Dirrhox lativittatus* Cope, 1887b, *Liophis trifasciatus* F. Werner, 1899a, *Philodryas lineatus* F. Werner, 1909b, *Philodryas werneri* L. Müller, 1926, *Philodryas pallidus* F. Werner, 1927, and *Philodryas psammophideus andensis* R.A. Thomas, 1976 (*nomen ineditum*).

Type: Holotype, BMNH 1946.1.2.79, a 686 mm female.

Type locality: "Tucuman" [N Argentina].

Distribution: Southern South America. Southwestern Brazil (Distrito Federal, NE Mato Grosso), SE Bolivia (Chuquisaca, Cochabamba, Santa Cruz, Tarija), Uruguay (Artigas, Canelones, Cerro Largo, Durazno, Flores, Florida, Maldonado, Montevideo, Paysandú, Rio Negro, Rivera, Rocha, Salto, San José, Soriano, Tacuarembó, Trienta y Tres), Paraguay (Boquerón) and Argentina (Buenos Aires, Catamarca, Chaco, Córdoba, Corrientes, Formosa, Jujuy, La Pampa, La Rioja, Mendoza, Nequén, Río Negro, Salta, San Juan, San Luis, Santa Fe, Santiago del Estero, Tucumán), 150–2750 m.

Sources: Cei, 1974, 1986, Cei et al., 1978, Bonino, 1987, J.D. Williams & Francini, 1991, Leynaud & Bucher, 1999, R.A. Thomas et al., 2001, Franco, 2003, Carreira-Vidal et al., 2005, Scolaro, 2005, 2006 and Franzen & Glaw, 2007.

17. *Philodryas simonsii* Boulenger, 1900b. Ann. Mag. Nat. Hist. (7) 6(32): 185–186.

Synonyms: *Dromicus angustilineatus* K.P. Schmidt & Walker, 1943b, *Dromicus inca* K.P. Schmidt & Walker, 1943c, and *Incaspis cercostropha* Donoso-Barros, 1974.

Type: Holotype, BMNH 1946.1.4.98 (formerly BMNH 1900.3.30.18), a 780 mm male (P.O. Simons, 1899–1901).

Type locality: "Cajamarca, 9000 ft., Peru" [= Cajamarca, S Cajamarca Dept., NW Perú, 7°10'S, 78°31'W, elevation 2745 m].

Distribution: Southern South America. Extreme S Ecuador (Loja), W Peru (Ancash, Arequipa, Ayacucho, Cajamarca, Lima, Tacna) and ext. N Chile (Tarapacá), NSL–3050 m.

Sources: K.P. Schmidt & Walker, 1943b, Donoso-Barros, 1966, R.A. Thomas, 1977a and Franco, 2003.

18. *Philodryas tachymenoides* (K.P. Schmidt & Walker, 1943b). Field Mus. Nat. Hist. (Zool.) 24(27): 309–310. (*Dromicus tachymenoides*)

Type: Holotype, FMNH 34261, a 1055 mm female (C.C. Sanborn, Oct. 1939).

Type locality: "Chucurapi, a sugar hacienda near Mollendo, in the Department of Arequipa, Peru" [Chucurapi, near Mollendo (17°01'S, 72°01'W, elevation 95 m), S Arequipa Dept., S Peru].

Distribution: Southern Peru (Ancash, Arequipa, Lima, Puno, Tacna) and ext. N Chile (Tarapacá), NSL–3400 m.

Sources: Donoso-Barros, 1962, 1966, 1970 and Aguilar et al., 2007.

19. *Philodryas trilineata* (Burmeister, 1861). Reise La Plata-Staaten: 529. (*Herpetodryas trilineatus*)

Synonyms: *Dryophylax burmeisteri* Jan *in* Burmeister, 1861, *Dromicus chilensis* Steindachner, 1867b and *Philodryas arenarius* Andersson, 1898.

Type: Lectotype, ZMUH IV.108, an 1156 mm male (K.H.K. Burmeister, 1857–1860), designated by L. Müller (1941: 198), destroyed 31 March–5 April 1945 during World War II.

Type locality: "Mendoza." Emended to Mendoza, Argentina *fide* Jan and Sordelli (1870–1881: 10).

Distribution: Bolivia (Santa Cruz) and W Argentina (Chubut, Córdoba, Catamarca, La Pampa, La Rioja, Mendoza, Neuquén, Río Negro, Salta, San Juan, San Luis, Tucumán), 1000–2000 m.

Sources: F. Werner, 1896c, Andersson, 1898, Berg, 1898, Cei et al., 1978, Cei, 1986,1994, J.D. Williams & Francini, 1991, Lobo & Scrocchi, 1994 and Scolaro, 2005, 2006.

Remarks: Lectotype illustrated in Jan & Sordelli (1870–1881, 3[49]: pl. 5, fig. 1).

20. *Philodryas varia* (Jan, 1863a). Arch. Zool. Anat. Fis. 2(2): 297. (*Liophis wagleri varia*)

Synonyms: *Philodryas borellii* Peracca, 1897b and *Philodryas patagoniensis haywardi* Laurent, 1973b.

Type: Holotype, MNHN 7502, a 380–387 mm male (A.C.V.M.D. d'Orbidny, 1834).

Type locality: "Veracruz" [Mexico] (in error). Corrected to Santa Cruz de la Sierra, Santa Cruz Dept., Bolivia [17°48'S, 63°10'W, elevation 425 m] *fide* R.A. Thomas and Johnson (1984: 80).

Distribution: Southern South America. Southern Bolivia (Cochabamba, W Santa Cruz, Tarija) and NW Argentina (Catamarca, Jujuy, Salta, Tucumán), 500–2000+ m.

Sources: R.A. Thomas et al., 1977, J.D. Johnson, 1982, R.A. Thomas & Johnson, 1984, J.D. Williams & Francini, 1991 and Fugler et al., 1995.

Remarks: Type locality restriction of Rodriguez Clara, Veracruz, Mexico *fide* H.M. Smith and Taylor (1950a: 350) invalid. Buenos Aires, Argentina records need confirmation *fide* J.D. Williams & Francini (1991: 74). A synonym of *Masticophis mentovarius fide* H.M. Smith & Taylor (1945: 97), based solely on the incorrect type locality, but removed *fide* J.D. Johnson (1982: 259.2).

21. *Philodryas viridissima* (Linnaeus, 1758). Syst. Nat., ed. 10, 1: 226. (*Coluber viridissimus*)

Synonyms: ? *Coluber purpurescens* Gmelin, 1789, ? *Coluber subcyaneus* Lacépède, 1789 (*nomen rejiciendum*), *Coluber janthinus* Daudin, 1803c, ? *Natrix caerulescens* Merrem, 1820, and *Philodryas crassifrons* Cope, 1861c.

Type: Holotype, ZMUU 149 (formerly JA).

Type locality: "Surinamei" [= Suriname, or Guianas].

Distribution: Northern South America. Southeastern Colombia (Amazonas, Meta), E Venezuela (Amazonas, Bolívar, Monagas), Guyana (Essequibo Islands-West Demerara, Potaro-Siparuni, Upper Demerara-Berbice, Upper Takutu-Upper Essequibo), Suriname (Marowijne, Nickerie, Paramaribo, Saramacca), French Guiana (Cayenne, Saint-Laurent-du-Maroni), Brazil (Acre, Amazonas, Bahia, Maranhão, Mato Grosso, ? Minas Gerais, Pará, Rondônia, Marajo Is.), E Ecuador (Morona-Santiago, Napo, Pastaza), E Peru (Loreto, Madre de Dios, Pasco, San Martin) and N Bolivia (Beni, Cochabamba, La Paz, Santa Cruz), NSL–985 m.

Sources: Beebe, 1946, Roze, 1966a, Cunha & Nascimento, 1978, 1993, Abuys, 1985b, Dixon & Soini, 1986, Lancini-V., 1986, Pérez-Santos, 1986, Vanzolini, 1986b, Chippaux, 1987, Pérez-Santos & Moreno, 1988, 1991, Nascimento et al., 1991, Marques & Franco, 1998, Starace, 1998, Duellman, 2005, Zaher et al., 2008, Navarrete et al., 2009, Freitas et al., 2012b and C.J. Cole et al., 2013.

Remarks: Syntype NHR Lin-38 (formerly MAFR) not present in NHR *fide* R.A. Thomas (1976a: 226).

PHILOTHAMNUS A. Smith, 1847 *in* 1838–1849 (Colubridae)

Synonyms: *Chlorophis* Hallowell, 1857b, *Herpetaethiops* A.C.L.G. Günther, 1866, *Chlorophis* Theobald, 1868b, *Philotamnus* – Bocage, 1872 *(nomen incorrectum), Herpeaethiops* – Giglioli, 1873 *(nomen incorrectum), Philothalmus* – Sauvage, 1884b *(nomen incorrectum), Phylothamnus* – Bedriaga, 1892 *(nomen incorrectum), Chloropsis* – Bocage, 1895 *(nomen incorrectum), Oligolepis* Boulenger, 1895d, *Clorophis* – Odhner, 1908 *(nomen incorrectum), Chlorophi* – Lampe, 1911 *(nomen incorrectum), Pithothamnus* – F.W. FitzSimons, 1912 *(nomen incorrectum), Philotamnus* – Briceño-Rossi, 1934 *(nomen incorrectum), Philothamnua* – Gasperetti, 1974 *(nomen incorrectum),* and *Philothamus* – Gasperetti, 1974 *(nomen incorrectum).*

Type species: *Dendrophis semivariegatus* A. Smith, 1847 *in* 1838–1849.

Distribution: Subsaharan Africa.

Sources: A.C.L.G. Günther, 1863b, Bocage, 1882b, Loveridge, 1951b, 1957a, 1958, Hughes, 1985 and Wallach, 1998a.

1. *Philothamnus angolensis* Bocage, 1882b. J. Sci. Math. Phys. Nat., Lisboa 9(33): 7.

Synonyms: *Dendrophis albovariata* A. Smith, 1840 *in* 1839–1849, *Ahaetulla emini* A.C.L.G. Günther, 1888a, *Ahaetulla shirana* A.C.L.G. Günther, 1888c, *Philothamnus guentheri* Pfeffer, 1893, and *Chlorophis vernayi* FitzSimons, 1932.

Types: Syntypes (2), MBL, a 980 mm specimen (J.A. d'Anchieta, 1867) [Capangombe] and a juvenile (J. Horta, 1873) [Angola], destroyed by fire 18 March 1978.

Type locality: "Capangombe (Angola)," and "Angola." Restricted to Capangombe, Angola *fide* Loveridge (1951b: 9), [= Capangombe, W Namibe Prov., SW Angola, 15°06'S, 13°09'E, elevation 525 m].

Distribution: Central and E Africa. Cameroon (Adamaoua, Centre, Est, Nord, Nord-Ouest, Ouest, Sud, Sud-Ouest), W Central African Republic (Bamingui-Bangoran, Gribingui, Haute-Sangha, Ombella-Mpoko, Sangha), Congo (Brazzaville, Kouilou, Niari), Democratic Republic of the Congo (Bandundu, Bas-Congo, Katanga, Kinshasa, Kivu, Orientale, Idjwi Is.), Sudan (Al Khartum, An Nil Al Aybad), E South Sudan (Eastern Equatoria, Junglei), Uganda (Central, Western, Kome and Nkose Is.), Rwanda (Butare, Byumba, Cyangugu, Gisenyi, Kibungo, Kigali), Burundi (Bujumbura, Bururi, Gitega), Tanzania (Iringa, Kigoma, Mbeya, S Morogoro, Mwanza, Ruvuma, Ziwi Magharibi), Zambia (Northern, North-Western, Western), Malawi

P

(Central, Northern, Southern), Mozambique (Gaza, Manica, Maputo, Sofala, Zambézia), Zimbabwe (Manicaland, Mashonaland Central, Mashonaland East, Mashonaland West, Masvingo, Matebeleland North, Midlands), Angola (Benguela, Cabinda, Cuanza Norte, Cuanza Sul, Cunene, Huambo, Huíla, Lunda Norte, Malanje, E Moxico, Namibe, Zaire), cen. Namibia (Caprivi, Erongo, Otjozondjupa), N Botswana (North West) and ext. NE South Africa (NE KwaZulu-Natal), 65–2600 m.

Sources: Bogert, 1940, Mertens, 1955, 1971, Laurent, 1956a, Sweeney, 1961, Witte, 1962, Broadley, 1971d, 1983, Branch, 1988, Joger, 1990, J.-F. Trape & Roux-Estève, 1995, Clauss & Clauss, 2002, Spawls et al., 2002, Broadley et al., 2003, Chippaux, 2006, Chirio & LeBreton, 2007, Largen and Spawls, 2010 and Broadley & Blaylock, 2013.

2. *Philothamnus battersbyi* Loveridge, 1951a. Bull. Mus. Comp. Zool. 106(4): 190–191. (*Philothamnus irregularis battersbyi*)

Type: Holotype, MCZ 40155, a 1045 mm female (A. Loveridge, 12–23 Dec. 1933).

Type locality: "Sipi Forest at 6,000 feet, Mount Elgon, eastern Uganda" [= Bugisu Distr., ext. E Uganda, 1°07'N, 34°35'E, elevation 1830 m].

Distribution: Northeastern Africa. Southeastern South Sudan (Eastern Equatoria), cen. Ethiopia (Gondar, Gemu Gofa, Gojjam, Kefa, Shoa, Wollo), S Somalia (Bay, Hiraan, Jubbada Dhexe, Jubbada Hoose, Shabeellaha Hoose), Kenya (Central, Coast, Eastern, Nairobi, Nyanza, Rift Valley, Western), Uganda (Central, Eastern, S Western), NE Democratic Republic of the Congo (E Kivu, NE Orientale), Rwanda, Burundi and N Tanzania (Arusha, Kigoma, Kilimanjaro, Mara, Tanga, Kagera, Ukerewe Is.), NSL–2500 m.

Sources: Witte, 1962, 1975, Pitman, 1974, Spawls, 1978, J.B. Rasmussen, 1981, Lanza, 1983a, 1990b, Broadley & Howell, 1991, Largen & Rasmussen, 1993, Spawls et al., 2002 and Largen & Spawls, 2009.

3. *Philothamnus bequaerti* (K.P. Schmidt, 1923). Bull. Amer. Mus. Nat. Hist. 49(1): 75–76, fig. 3. (*Chlorophis bequaerti*)

Type: Holotype, AMNH 12080, a 652 mm male (H. Lang & J. P. Chapin [Lang-Chapin Belgian Congo Exped.], Nov. 1910).

Type locality: "Niangara, Belgian Congo, ca. 3°40'N, 27°50'E" [= Niangara, ext. N Orientale Prov., NE Democratic Republic of the Congo, 3°42'N, 27°53'E, elevation 675 m].

Distribution: Central Africa. Central Cameroon (Adamaoua, Centre), S Chad (Moyen-Chari), Central African Republic (Bamingui-Bangoran, Haute-Sangha, Kemo-Gribingui, Ombella-Mpuko, Ouaka, Ouham, Ouham-Pende, Vakaga), S Congo (Pool), Democratic Republic of the Congo (? S Bandundu, Bas-Congo, N Equateur, N Orientale), Uganda (Central, Eastern, Northern, Western, Kome Is.), E South Sudan (Eastern Equatoria, Junglei) and W Ethiopia (Gondar, Illubabor, Wollega), 100–1200 m.

Sources: Hughes, 1985, Joger, 1990, Largen & Rasmussen, 1993, Spawls et al., 2002, Chippaux, 2006, Chirio & Ineich, 2006, Chirio & LeBreton, 2007 and Largen & Spawls, 2010.

4. *Philothamnus carinatus* (Andersson, 1901). Bih. Kon. Sven. Vet.-Akad. Handl. (4) 27(5): 9–11. (*Chlorophis carinatus*)

Synonyms: *Herpetaethiops bellii* A.C.L.G. Günther, 1866 *(nomen oblitum)* and *Philothamnus nigrofasciatus* Buchholz & Peters *in* W.C.H. Peters, 1875 *(nomen rejiciendum)*.

Types: Syntypes (5), NHR 1972, a 700 mm female (Knöppel, 1886) [Cameroon], NHR 1973–76, an 815 mm female, 730+ mm female, 595 mm male, and 300 mm male (K.W. Knutson and G. Valdau, 1885) [Mapanja].

Type locality: "Cameroon" and "Mapanja, Cameroon." Restricted to Mapanja [= Mapanja, Sud-Ouest Prov., SW Cameroon, 4°07'N, 9°07'E, elevation 1600 m] *fide* Mertens (1964a: 227).

Distribution: Central Africa. Southeastern Guinea (Macenta, Nzérékoré), W Ivory Coast (Danané, Dix-Huit Montagnes), S Ghana (Accra, Ahafo, Ashanti, Central, Eastern, Volta, Western), Togo (Plateaux), SW Nigeria, S Cameroon (Adamaoua, Est, Centre, Littoral, Nord-Ouest, Sud, Sud-Ouest), S Central African Republic (Haute-Sangha, Lobaye, Mbomou, Ombella-Mpoko, Sangha), Equatorial Guinea (Littoral, Bioko Is.), Gabon (Estuaire, Haut-Ogooué, Moyen-Ogooué, Ngounié, Ogooué-Ivindo, Ogooué-Lolo, Ogooué-Maritime, Woleu-Ntem), Congo (Kouilou, Lekoumou, Sangha), NW Angola (Cabinda), Democratic Republic of the Congo (Bandundu, Bas-Congo, Equateur, Kasai Occidental, Kasai Oriental, Katanga, Kinshasa, Nord-Kivu, Orientale, Sud-Kivu), SW Uganda (Western), SW Kenya (Western), W Rwanda, W Burundi and W Tanzania (Kigoma), NSL–2300 m.

Sources: Witte, 1953, 1962, Angel et al., 1954, Doucet, 1963, Mertens, 1965c, Knoepffler, 1966, Villiers, 1966, Pitman, 1974, Trape, 1981, 1985, Joger, 1990, J.B. Rasmussen, 1991, J.-F. Trape & Roux-Estève, 1995, Spawls et al., 2002, Chippaux, 2006, Chirio & Ineich, 2006, Chirio & LeBreton, 2007, Pauwels & Vande weghe, 2008 and Segniagbeto et al., 2011.

Remarks: Official Specific Name no. *fide* Opinion 1214 (ICZN, 1982e). NHR catalogue lists NHR 1973–76 syntypes under NHR 1973 (4). In accordance with Art. 23.9.2 of the Code (ICZN, 1999), *Chlorophis carinatus* Andersson is designated a *nomen protectum* and *Herpetaethiops bellii* Günther a *nomen oblitum*.

5. *Philothamnus dorsalis* (Bocage, 1866b). J. Sci. Math. Phys. Nat., Lisboa 1(1): 48. (*Leptophis dorsalis*)

Type: Lectotype, MBL (J.A. d'Anchieta, 1864), designated by Bocage (1882b: 10), destroyed by fire 18 March 1978.
Type locality: "Molembo, Afrique occidentale,"[= Molembo, West Africa]. Emended to Molembo de la Côte de Loango au nord du Zaire *fide* Bocage (1882b: 10), [= Malembo, Cabinda Prov., ext. NW Angola, 05°20'S, 12°11'E, elevation 110 m].
Distribution: Central Africa. Southern Cameroon (Est, Ouest, Sud), W Gabon (Moyen-Ogooué, Ogooué-Maritime), S Congo (Bouenza, Brazzaville, Kouilou, Plateau, Pool), Democratic Republic of the Congo (Bandundu, Bas-Congo, Equateur, Katanga, Kinshasa, Nord-Kivu, Orientale) and N Angola (Benquela, Cabinda, Huíla, Lunda Norte, Zaire), 10–1700 m.
Sources: J.B. Rasmussen, 1991, J.-F. Trape & Roux-Estève, 1995, Chippaux, 2006, Chirio & LeBreton, 2007 and Pauwels & Vande weghe, 2008.

6. *Philothamnus girardi* Bocage, 1893a. J. Sci. Math. Phy. Nat. (2) 3(9): 46.

Type: Lectotype, BMNH 1946.1.6.3, a 736 mm female (F. Newton, 1892), designated by Hughes (1985: 525).
Type locality: "Annobon Island" [= Pagalu Is., Equatorial Guinea, 1°26'S, 5°38'E] via lectotype selection.
Distribution: Equatorial Guinea (Pagalu Is.).
Sources: Mertens, 1934b and Jesus et al., 2009.
Remarks: Supplemental original description reprinted in Bocage (1893b: 47–48). Boulenger (1894a: 102) listed the lectotype as male and Hughes (1985: 525) listed it as a neotype. Possibly occurs in Congo (Kouilou) *fide* J.-F. Trape & Roux-Estève (1995: 39).

P

7. *Philothamnus heterodermus* (Hallowell, 1857b). Proc. Acad. Nat. Sci. Philadelphia (1857–1858) 9(3): 54. (*Chlorophis heterodermus*)

Synonym: *Chlorophis cyaneus* G. Hecht, 1929b.
Type: Holotype, ANSP 5219, a 540 mm specimen (H.A. Ford).
Type locality: "Gaboon country, West Africa" [= Gabon].
Distribution: West and cen. Africa. Guinea–Bissau (Bafatá, Gabú, Tombali, Orango Is.), Guinea (Gaoual), S Sierra Leone (S Northern, Southern, Western), Liberia (Maryland, Montserrado, Margibi, Nimba), S Ivory Coast (Abidjan, Bas-Sassandra, Bouna, Danané, Dix-Huit Montagnes, Guiglo, Lagunes, Sassandra, Tabou, Toumodi), S Ghana (Accra, Ashanti, Central, Eastern, Western), S Togo (Plateaux), S Benin (Cotonou), S Nigeria (Bendel, Benue, Cross River, Delta, Lagos, Ogun, Oyo, Rivers), Cameroon (Adamaoua, Centre, Est, Littoral, Nord, Nord-Ouest, Ouest, Sud, Sud-Ouest), ext. S Chad (Legone Oriental), Central African Republic (Haute-Sangha, Kemo-Gribingui, Lobaye, Mbomou, Ombella-Mpoko, Sangha), Gabon (Moyen-Ogooué, Ogooué-Ivindo, Woleu-Ntem), Congo (Brazzaville, Kouilou, Lekoumou, Niari, Pool, Sangha), Democratic Republic of the Congo (Bandundu, Bas-Congo, Équateur, Katanga, Kinshasa, Orientale, Sud- Kivu), SW Uganda (Western), N Rwanda, Burundi, Tanzania (Mbeya), and W Angola (Cuanza Sul), NSL–1900 m.
Sources: Villiers, 1950b, 1966, E.H. Taylor & Weyer, 1958, Witte, 1962, Doucet, 1963, Knoepffler, 1966, Menzies, 1966, Blackwell, 1967, Leston & Hughes, 1968, Hughes & Barry, 1969, Roux-Estève, 1969a, Dunger, 1973, Trape, 1981, 1985, Hughes, 1983, 2013, J.A. Butler & Reid, 1986, Joger, 1990, J.-F. Trape & Roux-Estève, 1995, Spawls et al., 2002, M. Griffin, 2003, Chirio & Ineich, 2006, Chirio & LeBreton, 2007, Pauwels & Vande weghe, 2008, Segniagbeto et al., 2011 and Auliya et al., 2012.

8. *Philothamnus heterolepidotus* (A.C.L.G. Günther, 1863b). Ann. Mag. Nat. Hist. (3) 11(64): 286. (*Ahaetulla heterolepidota*)

Synonyms: *Ahaetulla gracillima* A.C.L.G. Günther, 1888c, *Chlorophis gracilis* Sternfeld, 1910b, *Chlorophis schubotzi* Sternfeld, 1912, and *Philothamnus heterolepidotus chlorophis* Stucki-Stirn, 1979.
Type: Holotype, BMNH 1946.1.10.21, a 657–662 mm female.
Type locality: "Africa." Restricted to West Africa *fide* Boulenger (1894a: 96).
Distribution: West and cen. Africa. Guinea-Bissau (Biombo), Sierra Leone (Southern, Sherbo Is.), Ivory Coast (Toumodi), S Ghana (Accra, S Central, Volta, S Western), S Benin (Atlantique, Plateau), SW Nigeria (Lagos, Ogun), Cameroon (Adamaoua, Centre, Sud), Central African Republic (Ombella-Mpoko, Ouham, Ouham-Pende, Vakaga), S Congo (Pool, Sangha), Democratic Republic of the Congo (Bandundu, Bas-Congo, Equateur, Kasai Occidental, Kasai Oriental, Kinshasa, Nord-Kivu, Orientale, Sud-Kivu), Sudan (Al Khartum, An Nil Al Abyad), South Sudan (Bahr al Ghazal, Eastern Equatoria, Junglei), Uganda (Central, Eastern, Northern, Western, Kome Is.), W Ethiopia (Illubabor), SW Kenya (Western), NW Tanzania (Kigoma, Mtwara, Mwanza, Kagera), Rwanda (Butare, Kibungo), Burundi (Bururi, Rutana), Zambia (Luapula, Northern, North-Western, Western) and Angola (Benguela, Bié, Cabinda, Cuanza Sul, Huambo, Huíla, Lunda Norte, Malanje, Moxico, Zaire), NSL–2000 m.
Sources: Boulenger, 1894a, Laurent, 1954a, 1956a, 1964a, Menzies, 1966, Leston & Hughes, 1968, Hughes & Barry, 1969, Broadley, 1971d, Dunger, 1973, Pitman, 1974, Spawls, 1978, Trape, 1981, Hughes, 1983, 2013, Largen & Rasmussen, 1993, J.-F. Trape & Roux-Estève, 1995, Spawls et al., 2002, Broadley et al., 2003, Chippaux, 2006, Chirio & Ineich, 2006, Chirio & LeBreton, 2007 and Largen & Spawls, 2010.
Remarks: BMNH catalogue lists BMNH 1946.1.10.25 from Congo as a type. Probably occurs in Togo *fide* Segniagbeto et al. (2011: 353). Mozambique record

doubtful *fide* Loveridge (1958: 105); Dar es Salaam, Lake Rukwa and Lamu Is., Tanzania records doubtful *fide* Spawls et al. (2002: 356).

9. *Philothamnus hoplogaster* (A.C.L.G. Günther, 1863b). Ann. Mag. Nat. Hist. (3) 11(64): 286. (*Ahaetulla hoplogaster*)

Synonyms: *Philothamnus neglectus* W.C.H. Peters, 1867a and *Chlorophis oldhami* Theobald, 1868b.

Types: Syntypes (3), BMNH 1946.1.6.1, BMNH 1946.1.6.6, and BMNH 1946.1.6.14, longest syntype 650–660 mm (T. Ayers).

Type locality: "Port Natal" [= Durban, E KwaZulu-Natal Prov., E South Africa, 29°52'S, 31°02'E, elevation NSL].

Distribution: Eastern Africa. Southeastern South Sudan (Eastern Equatoria), S Democratic Republic of the Congo (Bas-Congo, Kasai Occidental, Kasai Oriental, Katanga, Kinshasa), S Uganda (Central), SE Kenya (Coast), Burundi (Bururi), Tanzania (Das es Salaam, Iringa, Kigoma, Kilimanjaro, Lindi, Mbeya, Morogoro, Mtwara, Piwani, Rukwa, Ruvuma, Tanga, Mafia Is.), Angola (Benguela, Bié, Lunda Norte, Namibe), Zambia (Central, Copperbelt, Eastern, Luapula, Lusaka, Northern, North-Western, Southern, Western), Malawi (Central, Northern, Southern), Mozambique (Cabo Delgado, Gaza, Inhambane, Manica, Maputo, Niassa, Sofala, Tete, Zambézia), Zimbabwe (Bulawayo, Manicaland, Mashonaland Central, Mashonaland East, Mashonaland West, Masvingo, Matabeleland North, Matabeleland South, Midlands), ext. NE Botswana (North East, North West), cen. Namibia (Damaraland), Swaziland and South Africa (E Cape, KwaZulu-Natal, Transvaal), NSL–1800 m.

Sources: Uthmöller, 1937, 1941a, Laurent, 1956a, 1964a, FitzSimons, 1974, Broadley & Cock, 1975, Spawls, 1978, Broadley, 1983, Pienaar et al., 1983, Auerbach, 1987, Branch, 1988, Boycott & Culverwell, 1992, Spawls et al., 2002, Broadley et al., 2003 and Broadley & Blaylock, 2013.

Remarks: Namibia record rejected *fide* FitzSimons (1962: 148) but accepted *fide* Hughes (1985: 517). Probably occurs in Central African Republic *fide* Chirio & Ineich (2006: 51).

10. *Philothamnus hughesi* J.-F. Trape and Roux-Estève, 1990. J. Afr. Zool. 104(5): 376–378, fig. 1.

Type: Holotype, MNHN 1989.2543, an 810 mm female (D. Schwarz, 22 Dec. 1982).

Type locality: "Gangalingolo (10 Km au sud-ouest de Brazzaville), Congo" [= Gangalingolo, S Brazzaville Region, S Congo, 4°18'S, 15°13'E, elevation 300 m].

Distribution: Central Africa. Cameroon (Adamaoua, Centre, Ouest, Nord-Ouest), S Chad, Central African Republic (Haute-Sangha, Lobaye, Mbomou, Ombella-Mpoko, Ouaka, Ouham), Gabon, Congo (Brazzaville), N Democratic Republic of the Congo (Equateur, Orientale), SE South Sudan (Eastern Equatoria) and S Uganda (S Central), 300–2100 m.

Sources: Roux-Estève, 1965, Joger, 1990, J.-F. Trape & Roux-Estève, 1995, Spawls et al., 2002, Chippaux, 2006, Chirio & Ineich, 2006 and Chirio & LeBreton, 2007.

11. *Philothamnus irregularis* (Leach *in* Bowdich, 1819). Miss. Cape Coast Castle-Ashantee: 494. (*Coluber irregularis*)

Synonyms: *Coluber azureus* Lacépède, 1789 (*nomen rejiciendum*), *Coluber caeruleus* Lacépède, 1789 (*nomen rejiciendum*), *Coluber caerulescens* Daudin, 1803d, *Coluber caesius* Cloquet, 1818 *in* Levrault, 1816–1830, *Dendrophis chenonii* J.T. Reinhardt, 1843, and *Ahaehulla fraenata* F. Müller, 1885.

Type: Holotype, BMNH 1965.641 (T.E. Bowdich, April 1817–Oct. 1818).

Type locality: "Fantee, Cape Coast Castle to Ashantee" [= Fantee tribal area, between Mansu and Fosu along Amissa (Oki) River, Central Region, S Ghana, elevation 160–180 m].

Distribution: West Africa. Senegal (Dakar, Fatick, Kaolack, Kédougou, Louga, Tambacounda), Gambia (Kombo Saint Mary, MacCarthy Island, Western), Guinea–Bissau (Biomo, Bissau, Bolama, Cacheu, Gabú, Bolama Is.), Guinea (Beyla, Boffa, Boké, Kérouané, Nzérékoré, Télimélé), Sierra Leone (Eastern, Northern, Southern, Western, Sherbro Is.), Liberia (Bong, Grand Cape Mount, Montserrado, Nimba), S Mali (S Gao, Koulikoro, Mopti, Ségou, Sikasso), Ivory Coast (Abidjan, Bouaké, Bouna, Danané, Ferkéssédougou, Lagunes, Toumodi), Burkina Faso (Centre, Centre Est, S Centre-Nord, Centre Ouest, Est, Hauts-Bassins, Sud-Ouest, Volta-Noire), Ghana (Accra, S Central, Eastern, Northern, Upper East, Upper West, Volta, S Western), S Togo (Kara, Maritime, Plateaux), S Benin (Alibori, Atakora, Atlantique, Maritime, Zou), SW Niger (Dosso, Niamey), Nigeria (Abia, Bauchi, Bayelsa, Benue, Borno, Delta, Imo, Jigawa, Kaduna, Kogi, Kwara, Lagos, Niger, Oyo, Plateau, Rivers, Soko to, Taraba), N Cameroon (Adamaoua, Centre, Littoral, Extreme-Nord, Nord, Nord-Ouest, Sud, Sud-Ouest), SW Chad (Mayo-Kebbi), Central African Republic (Kemo-Gribingui), Democratic Republic of the Congo (Katanga, Kinshasa, Orientale), NSL–1400 m.

Sources: Bowdich, 1819, Villiers, 1953, 1966, 1975, E.H. Taylor & Weyer, 1958, Doucet, 1963, Menzies, 1966, Leston & Hughes, 1968, Hughes & Barry, 1969, Papenfuss, 1969, Dunger, 1973, Miles et al., 1978, Roman, 1980, Hughes, 1983, 1984, 2013, Greenbaum & Carr, 2005, Villiers & Condamin, 2005, Chippaux, 2006, J.-F. Trape & Mané, 2006b, Chirio & LeBreton, 2007, Chirio, 2009, Ullenburch et al., 2010, Segniagbeto et al., 2011, Auliya et al., 2012 and Chirio, 2013.

Remarks: Official Specific Name no. 234 *fide* Opinion 328 (ICZN, 1955a).

12. *Philothamnus macrops* (Boulenger, 1895d). Ann. Mag. Nat. Hist. (6) 16(92): 171. (*Oligolepis macrops*)

Type: Holotype, BMNH 1946.1.10.32, a 215 mm specimen.

Type locality: "Usambara, German East Africa" [= Usambara Mtns., Tanga Prov., ext. NW Tanganyika, ca. 4°45'S, 38°30'E].

Distribution: Eastern Tanzania (Iringa, Lindi, Mbeya, Mtwara, Tanga, Zanzibar Is.), NSL–2000 m.

Sources: Loveridge, 1928, J.B. Rasmussen, 1981 and Spawls et al., 2002.

Remarks: Possibly occurs in N Mozambique *fide* Spawls et al. (2002: 358).

13. *Philothamnus natalensis* (A. Smith, 1848 *in* 1838–1849). Illust. Zool. So. Africa, Rept.: pl. 64, figs. 1, 1ab, 2, 2a–b, 3, 3a–b, 2 pp. (*Dendrophis* [*Philothamnus*] *natalensis*)

Synonyms: *Dendrophis subcarinatus* Jan, 1863b (*nomen nudum*), *Dendrophis subcarinatus* Jan, 1869 *in* Jan & Sordelli, 1866–1870, and *Philothamnus natalensis occidentalis* Broadley, 1966a.

Type: Holotype, not designated, an 820 mm specimen (A. Smith, 1831), location unknown.

Type locality: "neighbourhood of Port Natal" [= vicinity of Durban (29°52'S, 31°02'E, elevation NSL), E KwaZulu-Natal Prov., E South Africa].

Distribution: Southeastern Africa. Eastern Zimbabwe (Manicaland, Masvingo, Midlands), S Mozambique (Gaza, Inhaca, Inhambane, Manica, Maputo, Sofala, Bazaruto and Magarngue Is.), NE Swaziland and SE South Africa (Eastern Cape, Free State, Gauteng, KwaZulu-Natal, Limpopo, Mpumalanga, North West, Western Cape), NSL–1570 m.

Sources: FitzSimons, 1937, 1974, Broadley & Cock, 1975, De Waal, 1978, Broadley, 1983, 1990, Pienaar et al., 1983, Branch, 1988 and Broadley & Blaylock, 2013.

Remarks: BMNH catalogue lists 7 paratypes (BMNH 1862.9.2.27, 1865.7.27.5, 1876.11.9.4, 1877.8.8.6, 1887.12.6.9, 1878.10.12.23, 1953.1.1.4) from South Africa whereas FitzSimons (1937: 262) reported no types in BMNH. Namibia records need confirmation.

P

14. *Philothamnus nitidus* (A.C.L.G. Günther, 1863b). Ann. Mag. Nat. Hist. (3) 11(64): 286–287. (*Ahaetulla nitida*)

Synonyms: *Ahaetulla lagoensis* A.C.L.G. Günther, 1872b, *Philothamnus irregularis longifrenatus* Buchholz and Peters *in* W.C.H. Peters, 1875, and *Philothamnus nitidus loveridgei* Laurent, 1960.

Types: Syntypes (2), BMNH 1946.1.10.28, an 813 mm male [Demerara, i.e. Lagos], and BMNH 1946.1.6.2, a female [Nigeria].

Type locality: "Demerara, South America" [= Demerara, Guyana] (in error) and not stated. Corrected to Lagos, Nigeria [= Lagos, Nigeria, 6°27'N, 3°24'E, elevation NSL] *fide* Boulenger (1894a: 101).

Distribution: West, cen. and E Africa. Southeastern Guinea, Sierra Leone (Southern), S Ivory Coast (Abidjan, Guiglo, Lagunes, Moyen-Cavally, Tabou, Tai), S Ghana (Accra, Central, Eastern, Volta), S Togo (Plateaux), S Benin (Maritime, Zou), SW Nigeria (Lagos), S Cameroon (Adamaoua, Centre, Est, Littoral, Nord-Ouest, Sud, Sud-Ouest), SW Central African Republic (Haute-Sangha, Lobaye), Equatorial Guinea (Bioko Is.), Gabon (Haut-Ogooué, Moyen-Ogooué, Nyanga), Congo (Brazzaville, Kouilou, Pool), Congo (Cuvette, Kouilou, Lekoumou, Sangha), Democratic Republic of the Congo (Bandundu, Bas-Congo, Equateur, N Katanga, Kinshasa, Nord-Kivu, Orientale, Sud-Kivu), NE Angola (Lunda Norte), S Uganda (Central, Western, Kome Is.), SW Kenya (S Rift Valley, Western), Rwanda (Kigali), Burundi (Rutana) and SW Tanzania (Rukwa), NSL–1200 m.

Sources: Boulenger, 1894a, Witte, 1962, Doucet, 1963, Laurent, 1964a, Mertens, 1965c, Roux-Estève, 1965, Leston & Hughes, 1968, Hughes & Barry, 1969, Dunger, 1973, Hughes, 1983, 2013, Joger, 1990, J.-F. Trape & Roux-Estève, 1990, 1995, Boycott & Culverwell, 1992, Spawls et al., 2002, Chippaux, 2006, Chirio & Ineich, 2006, Chirio & LeBreton, 2007 and Segniagbeto et al., 2011.

Remarks: BMNH 1946.1.10.28 is type of *Ahaetulla lagoensis*, but BMNH catalogue reads type of *A. nitidus fide* M. Lang (pers. comm.). Hughes (1985: 524) listed BMNH 1946.1.10.24 from "Demerara, Guyana" as one of Günther's types of *P. nitidus*.

15. *Philothamnus ornatus* Bocage, 1872. J. Sci. Math. Phys. Nat., Lisboa 4(13): 80–81.

Types: Syntypes (3), MBL (2), a male and female (J.A. d'Anchieta, 1871) [Huilla], and MBL, a female (Hopffer, 1871) [Cacheu], longest syntype 710 mm, destroyed by fire 18 March 1978.

Type locality: "Cacheu, sur la côté de Guinée, Afrique occidentale" [= Cacheu, Guinea–bissau] (in error *fide* Loveridge, 1958: 84) and "Huilla" [Angola]. Restricted to Huilla, Mossamedes Distr., Angola *fide* Bogert (1940: 52).

Distribution: Central Africa. Southern Cameroon (Centre, Est, Littoral, Nord-Ouest), NE Democratic Republic of the Congo (NE Orientale), SW Tanzania (Rukwa), W Angola (Benguela, Bié, Cunene, Huambo, Huíla), Zambia (Eastern, Northern, North West), NE Namibia (Caprivi, Okavango), N Botswana (North West) and NE Zimbabwe (Manicaland, Mashonaland East, Mashonaland West), 300–1500 m.

Sources: Bocage, 1895, 1896, Bogert, 1940, FitzSimons, 1974, Broadley, 1971, 1983, Broadley & Cock, 1975, Auerbach, 1987, Branch, 1988, Spawls et al., 2002, M. Griffin, 2003, Chippaux, 2006, Chirio & LeBreton, 2007 and Broadley & Blaylock, 2013.

16. *Philothamnus pobeguini* (Chabanaud, 1917a). Bull. Mus. Hist. Nat., Paris (1916) 22(7): 371–372, fig. 12. (*Chlorophis heterodermus pobeguini*)

Type: Holotype, MNHN 1916.206, a 780 mm male (H. Pobéguin, 1908).

Type locality: "Guinée française" [= Guinea].

Distribution: Guinea (Boké, Forécariah, Macenta, Nzérékoré, Télimélé), NSL–200 m.

Remarks: A valid species *fide* J.-F. Trape *in* Chirio (2013: 85).

17. *Philothamnus punctatus* W.C.H. Peters, 1867a. Mber. Königl. Akad. Wiss. Berlin 1866(12): 889–890.

Synonyms: *Dendrophis melanostigma* Jan, 1863b (*nomen nudum*), *Dendrophis melanostigma* Jan, 1869 *in* Jan & Sordelli, 1866–1870, *Ahaetulla kirkii* A.C.L.G. Günther, 1868, and *Philothamnus punctatus sansibaricus* Pfeffer, 1893.

Types: Syntypes (7), ZMB 1919a–b, ZMB 1920a–b, ZMB 5571a–c, one male and two females, sex of others unknown (C. von der Decken).

Type locality: "Zanzibarküste, Ostafrika" [= Zanzibar coast, East Africa or coastal Kenya and Tanzania].

Distribution: East Africa. Eastern Ethiopia (Bale, Hararge, Sidamo), Somalia (Awdal, Bay, Gedo, Hiraan, Jubbada Dhexe, Jubbada Hoose, Mogadishu, Shabeellaha Dhexe, Shabeellaha Hoose, Togdheer, Woqooyi Galbeed), E Kenya (E Coast, N Eastern, N North-Eastern), E Tanzania (Dar es Salaam, Iringa, Kilimanjaro, Lindi, Mtwara, Piwani, Ruvuma, Tanga, Pemba and Zanzibar Is.), NE Malawi (Northern) and N Mozambique (Cabo Delgado, Niassa, Sofala, Zambézia), 200–1500 m.

Sources: Lanza, 1983a, Largen & Rasmussen, 1993, Spawls et al., 2002, Branch et al., 2005 and Largen & Spawls, 2010.

18. *Philothamnus ruandae* Loveridge, 1951b. Bull. Inst. Roy. Sci. Nat. Belgique 27(37): 2–3. (*Philothamnus heterodermus ruandae*)

Type: Holotype, IRSNB 2053 (formerly IRSNB I.G. 15860), a 922 mm female (F.L. Hendrickx, 1947).

Type locality: "Mulungu, Lake Kivu, Congo Belge" [= Mulungu, Sud-Kivu Prov., E Democratic Republic of the Congo, 2°19'S, 28°48'E, elevation 1800 m].

Distribution: East Africa. Eastern Democratic Republic of the Congo (Katanga, Nord-Kivu, Sud-Kivu), SW Uganda (S Western), W Rwanda and NW Burundi, 700–2350 m.

Sources: Laurent, 1956a, Witte, 1962, 1975, Pitman, 1974 and Spawls et al., 2002.

19. *Philothamnus semivariegatus* (A. Smith, 1847 *in* 1838–1849). Illust. Zool. So. Africa, Rept.: pls. 59–60, 2 pp. (*Dendrophis semivariegata*)

Synonyms: *Philothamnus smithii* Bocage, 1882b and *Ahaetulla bocagii* A.C.L.G. Günther, 1888c.

Type: Lectotype, NMSZ 1859.13.1257a (formerly RSM), a 1067 mm specimen with skull and viscera removed (A. Smith, July 1828), designated by Bogert (1940: 56).

Type locality: "Busman Flat, South Africa" [= Bushman Flat, W Northern Cape Prov., W South Africa, ca. 29°40'S, 17°53'E, elevation 935 m] via lectotype selection.

Distribution: Subsaharan Africa. Southern Senegal (Dakar, Fatick, Kédougou, Kolda, Thiès, Ziguinchor), Gambia (MacCarthy Island), Guinea–Bissau (Bissau, Quinara), Guinea (Beyla, Dubréka, Macenta, Télimélé), Sierra Leone (Eastern, Northern, Southern), Liberia (Bong), Ivory Coast (Abidjan, Bouaké, Ferkéssédongou, Toumodi), S Mali (S Gao, Koulikoro, Mopti, Sikasso), Ghana (Accra, E Brong-Ahafo, Central, Northern, Upper East, Upper West, Volta), N Togo (Centrale, Kara, Maritime, Plateaux), Burkina Faso (Centre-Est, Centre-Ouest, Est, Sud-Ouest, Volta-Noire), Benin (Alibori, Atlantique, Collines, Cotonou, Maritime, Zou), SW Niger (Dosso, Niamey), Nigeria (Bauchi, Benue, Delta, Kwara, Lagos, Plateau), Cameroon (Adamaoua, Centre, Extreme-Nord, Littoral, Nord, Nord-Ouest, Ouest, Sud-Ouest), Equatorial Guinea (Bioko Is.), Congo (Kouilou), Central African Republic (Bamingui-Bangoran, Haute-Kotto, Haut-Mbomou, Kemo-Gribingui, Lobaye, Ouaka, Ouham, Ombella-Mpoko, Ouham-Pende, Vakaga), Democratic Republic of the Congo (Bandundu, Bas-Congo, Equateur, Kasai Occidental, Kasai Oriental, Katanga, Kinshasa, Maniema, Nord-Kivu, Orientale, Sud-Kivu), Angola (Benguela, Bié, Cuanza Sul, Cunene, Huíla, Lunda Norte, Moxico, Namibe), S Sudan (Al Khartum, Nahr An Nil), SE South Sudan (Eastern Equatoria), S Eritrea (Southern, Southern Red Sea), Ethiopia (Gemu Gofa, Illubabor, Sidamo), Somalia (Bay, Gedo, Hiraan, Jubbada Dhexe, Jubbada Hoose, Mogadishu, Shabeellaha Dhexe, Shabeellaha Hoose, Sool, Togdheer, Woqooyi Galbeed), Uganda (Central, Eastern, Northern, Western, Kome and Sese Is.), Kenya (Coast, S Eastern, North-Eastern, Nyanza, Rift Valley, Western, Lamu and Manda Is.), Tanzania (Arusha, Dar es Salaam, Dodoma, Iringa, Kigoma, Kilimanjaro, Lindi, Mara, Mbeya, Morogoro, Mtwara, Mwanza, Piwani, Rukwa, Ruvuma, Shinyanga, Singida, Tabora, Tanga, Kagera, Jambiani, Mafia, Pemba, Ukwere and Zanzibar Is.), Rwanda, Burundi (Bururi), Zambia (Central, Copperbelt, Eastern, Luapula, Lusaka, Northern, North-Western, Southern, Western), Malawi (Central, Northern, Southern), Mozambique (Gaza, Inhambane, Manica, Maputo, Nampula, Tete, Sofala, Zambézia, Bazaruto and Querimba Is.), Zimbabwe (Bulawayo, Manicaland, Mashonaland Central, Mashonaland East, Mashonaland West, Masvingo,

Matebeleland North, Matebeleland South, Midlands), Botswana (Central Gaborone, Ghanzi, Kgalagali, Kweneng, North East, North West, South East, Southern, Tuli), N Namibia (Caprivi, Khomas, Kunene, Omaheke, Omusati, Otjozondjupa), Swaziland, E South Africa (Eastern Cape, NW Free State, Gauteng, KwaZulu-Natal, Limpopo, Mpumalanga, E Northern, North West), NSL–2000 m.

Sources: Sternfeld, 1908f, Scortecci, 1932, Corkill, 1935, Orton & Morrison, 1946, Villiers, 1948, 1950a, 1956b, 1975, H.W. Parker, 1949, Laurent, 1950c, 1954a, 1956a, 1960, 1964a, Witte, 1953, 1962, Mertens, 1955, 1971, Sweeney, 1961, Doucet, 1963, Menzies, 1966, Leston & Hughes, 1968, Hughes & Barry, 1969, Papenfuss, 1969, Roux-Estève, 1969a, Broadley, 1971d, 1990, Dunger, 1973, V.F.M. FitzSimons, 1974, Pitman, 1974, W. Böhme, 1978, Spawls, 1978, J.B. Rasmussen, 1981, Hughes, 1983, 2013, Lanza, 1983a, 1990b, Pienaar et al., 1983, McLachlan, 1984, Haacke, 1985, Auerbach, 1987, Branch, 1988, Gasperetti, 1988, Boycott & Culverwell, 1992, Largen & Rasmussen, 1993, D. Lawson, 1993, J.-F. Trape & Roux-Estève, 1995, Clauss & Clauss, 2002, Spawls et al., 2002, Broadley et al., 2003, Villiers & Condamin, 2005, Chippaux, 2006, Chirio & Ineich, 2006, Chirio & LeBreton, 2007, Largen & Spawls, 2010, Ullenburch et al., 2010, Segniagbeto et al., 2011, Auliya et al., 2012, Broadley & Blaylock, 2013 and Chirio, 2013.

Remarks: Official Specific Name no. 235 *fide* Opinion 328 (ICZN, 1955a). Yemen record rejected *fide* Gasperetti (1988: 265) and Spawls et al. (2002: 363).

P

20. *Philothamnus thomensis* Bocage, 1882a. J. Sci. Math. Phys. Nat., Lisboa 8(32): 302.

Type: Lectotype, BMNH 1946.1.21.60, a 900 mm specimen, designated by Hughes (1985: 524).

Type locality: "ile de S. Thomé, Angola" [= São Tomé Is., São Tomé and Príncipe, Gulf of Guinea] via lectotype selection.

Distribution: São Tomé and Príncipe (São Tomé Is.).

Source: Jesus et al., 2009.

PHIMOPHIS Cope, 1860d (*nomen substitutum*) (Xenodontidae)

Synonyms: *Rhinosimus* A.-M.-C. Duméril, 1853 *(nomen nudum), Rhinosimus* A.-M.-C. Duméril, Bibron & Duméril, 1854b *(nomen rejiciendum), Pilmophis* – Rochebrune, 1884 *(nomen rejiciendum), Phymophis* – Aleman, 1953 *(nomen incorrectum),* and *Rhinosomus* – Hoge, 1953c *(nomen incorrectum).*

Type species: *Rhinosimus guerini* A.-M.-C. Duméril, Bibron & Duméril, 1854b.

Distribution: Lower Central America and South America.

Sources: J.A. Peters & Orejas-Miranda, 1970, Cadle, 1984a, Abdala, 1990, Rodrigues, 1993b and Zaher et al., 2009.

Remarks: Official Generic Name no. 1585 *fide* Opinion 698 (ICZN, 1955a).

1. *Phimophis guerini* (A.-M.-C. Duméril, Bibron & Duméril, 1854b). Erpét. Gén. 7(2): 991–992, pl. 72, figs. 1–3. (*Rhinosimus guerinii*)

Synonyms: *Rhinosimus guerinii* A.-M.-C. Duméril, 1853 *(nomen nudum), Rhinostoma scytaloides* F. Werner, 1913, and *Rhinosimus amarali* Mello, 1926.

Type: Holotype, MNHN 7594, a 758 mm specimen.

Type locality: Unknown. Restricted to Bahia, Brazil [= Salvador, E Bahia State, E Brazil, 12°58'S, 38°32'W, elevation NSL] *fide* Serié (1936: 47).

Distribution: Eastern South America. Suriname or French Guiana, E Brazil (Alagoas, Bahia, Goiás, Mato Grosso do Sul, Minas Gerais, S Piauí, São Paulo) and N Argentina (Chaco, Córdoba, Corrientes, Entre Ríos, Formosa, Misiones, Santa Fe, Tucumán), 20–300 m.

Sources: Hoge, 1953c, Hoogmoed, 1982, Abdala, 1990, J.D. Williams & Francini, 1991, Cei, 1994, Leynaud & Chiaraviglio, 1996, Yanosky et al., 1996, Freitas, 1999, Giraudo & Scrocchi, 2002 and Silva Lisboa et al., 2011.

Remarks: Official Specific Name no. 1968 *fide* Opinion 698 (ICZN, 1964b). A second specimen from E. Deville mentioned, but not described, by Duméril, Bibron & Duméril (1854b: 992) originated from Bahia, Brazil.

2. *Phimophis guianensis* (Troschel *in* Schomburgk, 1848). Versuch Fauna Flora Brit.-Guiana: 653. (*Heterodon guianensis*)

Synonyms: *Coluber contaminatus* Gravenhorst, 1807 *(nomen oblitum), Rhinostoma guntheri* Cope, 1860e, *Rhinostoma guianense* – Briceño-Rossi, 1934 *(nomen incorrectum),* and *Rhinostoma guiamnense* – Briceño-Rossi, 1934 *(nomen incorrectum).*

Type: Holotype, not designated, a 762 mm specimen (R.H. Schomburgk, 1834–1836 or 1840–1844), location unknown.

Type locality: "in der unmittelbaren Umgebung von Pirara, Britisch-Guiana" [= vicinity of Pirara (3°37'N, 59°40'W, elevation 80 m), W Rupununi Region, SW Guyana].

Distribution: Panamá and N South America. Eastern Panama (Coclé, Darien, Los Santos, Panamá), Colombia (Atlántico, Bolívar, Caldas, Chocó, Córdoba, Cundinamarca, Huila, La Guajira, Magdalena, Meta, Sucre, Vaupés), N Venezuela (Amazonas, Anzoátegui, Apure, Aragua, Barinas, Bolívar, Carabobo, Cojedes, Delta Amacuro, Distrito Federal, Falcón, Guárico, Lara, Mérida, Miranda, Monagas, Nueva Esparta, Portuguesa, Sucre, Táchira, Trujillo, Yaracuy, Vargas, Zulia, Cubagua and Margarita Is.), Guyana (Demerara-Mahaica, Upper Demerara-Upper Essequibo),

Suriname (Para) and French Guiana (Cayenne, Saint-Laurent-du-Maroni), NSL–900 m.

Sources: Serie, 1915, Dunn, 1949, Roze, 1966a, Amaral, 1978, Gasc & Rodrigues, 1980, Abuys, 1985b, Lancini-V., 1986, Pérez-Santos, 1986, Chippaux, 1987, Pérez-Santos & Moreno, 1988, Gorzula & Señaris, 1998, Starace, 1998, Mijares-Urrutia & Arends, 2000, Giraudo & Scrocchi, 2002, Navarrete et al., 2009, Ugueto & Rivas, 2010 and C.J. Cole et al., 2013.

Remarks: Official Specific Name no. 1967 *fide* Opinion 698 (ICZN, 1964). In accordance with Art. 23.9.2 of the Code (ICZN, 1999), *Heterodon guianensis* Troschel is designated a *nomen protectum* and *Coluber contaminatus* Gravenhorst a *nomen oblitum*. Probably occurs in NE Brazil.

3. *Phimophis vittatus* (Boulenger, 1896a). Cat. Snakes Brit. Mus. 3: 115, pl. 5, fig. 3. (*Rhinostoma vittatum*)

Type: Holotype, BMNH 1946.1.9.29, a 620+ mm female (E.W. White).

Type locality: "Buenos Ayres, Argentina," (in error *fide* J.A. Peters and Orejas-Miranda, 1970: 246 & Cei, 1994: 651).

Distribution: Southern South America. Southern Bolivia (Tarija), Paraguay and N Argentina (Catamarca, Chaco, Córdoba, Formosa, La Rioja, Salta, San Juan, San Luis, Santa Fe, Santiago del Estero, Tucumán, Aragonés Is.), 200–800 m.

Sources: Peracca, 1897b, Bertoni, 1915, Abalos et al., 1964, Abdala, 1990, J.D. Williams & Francini, 1991, Cei, 1994, Leynaud & Bucher, 1999 and Giraudi & Scrocchi, 2002.

Remarks: Argentine records of Entre Ríos and Santa Fe need confirmation *fide* J.D. Williams & Francini (1991: 76).

PHYLLORHYNCHUS Stejneger, 1890b (Colubridae)

Synonym: *Phyllorhyncus* – Pinney, 1981 *(nomen incorrectum).*

Type species: *Phyllorhynchus browni* Stejneger, 1890b.

Distribution: Southwestern USA and NW Mexico.

Fossil record: Pleistocene of SW USA.

Sources: Klauber, 1935b, Bogert & Oliver, 1945, Fowlie, 1965, Hardy & McDiarmid, 1969, McDiarmid & McCleary, 1993, Goldberg, 1996, Bartlett & Tennant, 2000 and Cadle, 2011.

1. *Phyllorhynchus browni* Stejneger, 1890b. Proc. U.S. Natl. Mus. 13(802): 152–155, 3 figs.

Synonyms: *Phyllorhynchus browni lucidus* Klauber, 1940c, *Phyllorhynchus browni fortitus* Bogert and Oliver, 1945, and *Phyllorhynchus browni klauberi* Shannon & Humphrey, 1959a.

Type: Holotype, USNM 15719, a 325 mm specimen (H. Brown).

Type locality: "Tucson, Arizona" [Pima Co., USA].

Distribution: Southwestern USA (S Arizona) and NW Mexico (Sinaloa, Sonora), 300–915 m.

Source: Stebbins, 1985.

Remarks: Possibly two valid species *(P. browni* and *P. fortitus) fide* H.M. Smith et al. (2005: 68) and Cadle (2011: 4).

2. *Phyllorhynchus decurtatus* (Cope, 1869b). Proc. Acad. Nat. Sci. Philadelphia (1868) 20(5): 310–311. (*Phimothyra decurtata*)

Synonyms: *Phyllorhynchus decurtatus perkinsi* Klauber, 1935b, *Phyllorhynchus decurtatus nubilus* Klauber, 1940c, *Phyllorhynchus decurtatus norrisi* H.M. Smith & Langebartel, 1951, *Phyllorhynchus arenicola* Savage & Cliff, 1954, and *Phyllorhynchus decurtatus porelli* Powers & Banta, 1974.

Type: Holotype, ANSP 5489, a 350 mm specimen (W.M. Gabb, 1867).

Type locality: "upper part of Lower California" [= Baja California Norte State, Mexico].

Distribution: Southwestern USA (SW Arizona, S California, S Nevada, ext. SW Utah) and NW Mexico (Baja California Norte, Baja California Sur, W Sonora, Angel de la Guarda, San Marcos, Montserrate, San José and Cerralvo Is.), NSL–910 m.

Fossil records: Upper Pleistocene (Rancholabrean II) of USA (Arizona).

Sources: Stejneger, 1890b, Atsatt, 1923, R.W. Murphy & Ottley, 1980, 1984, McCleary & McDiarmid, 1993, P.R. Brown, 1997, Grismer, 1999, 2000, Holman, 2000a and S.A. Gardner & Mendelson, III, 2004.

PITUOPHIS Holbrook, 1842b (Colubridae)

Synonyms: *Churchillia* Baird & Girard, 1852a, *Pityophis* Hallowell, 1852 *(nomen emendatum), Pythyophis* – Voigtländer, 1860 *(nomen incorrectum), Petuophis* – J.K. Lord, 1866 *(nomen incorrectum), Curchillia* Marschall, 1873 *(nomen emendatum), Pitiofis* – Dugès, 1887 *(nomen incorrectum), Pituoplus* – Cragin, 1881 *(nomen incorrectum), Pithyophis* A.L. Herrera, 1889 *(nomen emendatum), Pityophisus* A.L. Herrera, 1889 *(nomen emendatum), Pitiophis* – Herrera, 1890 *(nomen incorrectum), Epiglottophis* Cope, 1891a, *Pityopis* – Velasco, 1895 *(nomen incorrectum), Pitnophis* – F.C. Baker & Woodruff, 1897 *(nomen incorrectum), Pityophisus* Herrera, 1899 *(nomen emendatum), Pityphis* – R.H. Johnson, 1902 *(nomen incorrectum), Churchilla* – E.B. Branson, 1904 *(nomen emendatum), Piturophis* – Cyrén, 1934 *(nomen incorrectum), Pituophia* – Stanek, 1962 *(nomen incorrectum),* and *Pituophrys* – Halpin, 1983 *(nomen incorrectum),*

Type species: *Coluber melanoleucus* Daudin, 1803c.
Distribution: Southwestern Canada, USA, Mexico and Guatemala.
Fossil records: Upper Miocene to upper Pleistocene of USA. Fossils unassigned to species include upper Miocene (Clarendonian) of USA (Nebraska), and upper Pliocene (Blancan V) of USA (Idaho).
Sources: Van Denburgh & Slevin, 1919, Stull, 1932a, 1940, H.M. Smith & Taylor, 1945, H.M. Smith & Kennedy, 1951, Himmelstein, 1973b, Dowling et al., 1983, Sweet, 1984, J.L. Knight, 1986, Holman, 1995a, 2000a, Keogh, 1996, Rodrígues-Robles & Jesús-Escobar, 1999, Parmley & Walker, 2003, Burbrink & Lawson, 2007, Collins & Taggart, 2008, Pyron & Burbrink, 2009a, Parmley & Hunter, 2010 and Hirschkorn & Skubowius, 2011.

1. *Pituophis catenifer* (Blainville, 1835). Nouv. Ann. Mus. Hist. Nat. (3) 4(3): 290–291, pl. 26, figs. 2, 2a–b. (*Coluber catenifer*)

Synonyms: *Coluber sayi* Schlegel, 1837, *Pityophis affinis* Hallowell, 1852a, *Churchillia bellona* Baird & Girard, 1852a, *Pituophis annectens* Baird & Girard, 1853, *Pituophis mcclellanii* Baird & Girard, 1953, *Pituophis wilkesii* Baird & Girard, 1853, *Pityophis heermanni* Hallowell, 1853a, *Elaphis reticulatus* A.-M.-C. Duméril, 1853 (*nomen nudum*), *Rhinechis mexicanus* A.-M.-C. Duméril, 1853 (*nomen nudum*), *Elaphis reticulatus* A.-M.-C. Duméril, Bibron & Duméril, 1854a, *Pituophis mexicanus* A.-M.-C. Duméril, Bibron & Dumeril, 1854a, *Pituophis catenifer deserticola* Stejneger, 1893b, *Pituophis catenifer stejnegeri* Van Denburgh, 1920a, *Pituophis catenifer rutilus* Van Denburgh, 1920a, *Pituophis catenifer coronalis* Klauber, 1946b, *Pituophis catenifer fuliginatus* Klauber, 1946b, and *Pituophis catenifer pumilus* Klauber, 1946c.
Type: Holotype, MNHN 57, a 1080 mm specimen (P-E. Botta, March–Oct. 1827).
Type locality: "Californie" [= Alta Californie or California, W USA]. Restricted to vicinity of San Francisco, California, USA *fide* K.P. Schmidt (1953a: 202), but probably southern California *fide* Klauber (1947: 13).
Distribution: Extreme SW Canada (SE Alberta, S British Columbia, SW Saskatchewan), cen. and W USA (Arizona, California: Santa Cruz, Santa Rosa and San Miguel Is.; Colorado, Idaho, Illinois, W Indiana, Iowa, Kansas, S Minnesota, Missouri, Montana, Nebraska, Nevada, New Mexico, SW North Dakota, Oklahoma, Oregon, South Dakota, Texas, Utah, Washington, SW Wisconsin, Wyoming) and N Mexico (N Baja California Norte, Chihuahua, Coahuila, N Durango, Nuevo León, San Luis Potosí, Sinaloa, Sonora, Tamaulipas, Veracruz, Zacatecas, San Martin, South Coronado and Tiburon Is.), NSL–2895 m.
Fossil records: Upper Miocene (Hemphillian) of USA (Arizona, Nebraska), middle Pliocene (Blancan II, III) of USA (Kansas, Nebraska, Texas), upper Pliocene (Blancan V) of USA (Kansas, Washington), Pleistocene of USA (Illinois), middle/upper Pleistocene (Irvingtonian II) of USA (Arkansas, Colorado), and upper Pleistocene (Rancholabrean I, II) of USA (Arizona, California, Kansas, Missouri, Nevada, New Mexico, Oklahoma, Texas, Utah).
Sources: Klauber, 1946b, 1947, Zweifel, 1954b, Logier & Toner, 1961, K.L. Williams et al., 1961, Conant, 1965, Bullock & Tanner, 1966, Froom, 1972, W.S. Parker and Brown, 1980, Pérez-Higareda, 1981, R.W. Murphy & Ottley, 1984, Gutzke et al., 1985, Stebbins, 1985, Sweet & Parker, 1990, Conant & Collins, 1991, T. Van Devender & Bradley, 1994, H.A. Brown et al., 1995, Reichling, 1995, Degenhardt et al., 1996, P.R. Brown, 1997, Grismer, 1997, 2002, Harding, 1997, Moriarty, 1998, Fitch, 1999, Rodríguez-Robles & Jesús-Escobar, 2000, Werler & Dixon, 2000, Loeffler, 2001 and Parmley & Peck, 2002.
Remarks: *Pituophis fulginatus* a valid species *fide* McPeak (2000: 78).

2. *Pituophis deppei* (A.-M.-C. Duméril, Bibron & Duméril, 1854a). Erpét. Gén. 7(1): 268. (*Elaphis deppei*)

Synonyms: *Elaphis deppei* A.-M.-C. Duméril, 1853 (*nomen nudum*), *Elaphis pleurostictus* A.-M.-C. Duméril, Bibron & Duméril, 1854a, *Arizona jani* Cope, 1861a, and *Pituophis deppei pholidostictus* Jan, 1863b.
Type: Holotype, ? RMNH, a 1658 mm specimen (F. Deppe, Dec. 1824–Jan. 1827).
Type locality: "Mexique" [= Mexico]. Restricted to San Juan Teotihuacán, México State, Mexico *fide* H.M. Smith and Taylor (1950a: 334).
Distribution: Mexico (Aguascalientes, S Chihuahua, S Coahuila, Colima, Distrito Federal, Durango, Guanajuato, Hidalgo, Jalisco, México, Michoacán, S Nuevo León, Oaxaca, Puebla, Querétaro, SW San Luis Potosí, Sonora, Tamaulipas, Veracruz, Zacatecas), 1490–2990 m.
Sources: Zweifel, 1954b, Duellman, 1960, Conant, 1965, Pérez-Higareda & Smith, 1991, Kardon, 1995, Casas-Andreu et al., 1996, McCranie & Wilson, 2001b, Lemos-Espinal et al., 2004c, H.M. Smith et al., 2005a–b and Ramírez-Bautista et al., 2011.
Remarks: RMNH 355 (or RMNH 379a) from Mexico, originally catalogued as *Lycodon pholidostictus*, was exchanged from ZMB in 1843.

3. *Pituophis insulanus* Klauber, 1946. Trans. San Diegpo Soc. Nat. Hist. 11(1): 11–14, pl. 1, fig. 1. (*Pituophis catenifer insulanus*)

Type: Holotype, CAS 56353, a 1210 mm female (J.R. Slevin 7 Aug. 1922).

P

Type locality: "Cedros (Cerros) Island off the west coast of Baja California, Mexico."
Distribution: Extreme NW Mexico (Baja California Norte: Cedros Is.).
Source: Grismer, 2002.
Remarks: A valid species *fide* Grismer, 1994, 2001a, 2002.

4. *Pituophis lineaticollis* (Cope, 1861e). Proc. Acad. Nat. Sci. Philadelphia 13(7): 300–301. (*Arizona lineaticollis*)

Synonyms: *Pituophis deppei brevilineata* K.P. Schmidt & Shannon, 1947, and *Pituophis deppei gibsoni* L.C. Stuart, 1954a.
Type: Neotype, UMMZ 114668, adult female (W.E. Duellman, 29 Feb. 1956), designated by Duellman (1960: 607).
Type locality: "24 kilometers northwest of Ciudad Oaxaca, Oaxaca, Mexico" via neotype selection. Emended to 1676 m elevation *fide* Kluge (1984: 49).
Distribution: Southern Mexico (Chiapas, Colima, Guerrero, Jalisco, México, Michoacán, Morelos, Oaxaca, Puebla) and SW Guatemala (Chimaltenango, Quiché, Sacatépequez), 1000–3000 m.
Sources: H.M. Smith & Taylor, 1945, Duellman, 1960, Casas-Andreu et al., 1996, Benitez-Gálvez, 1997, Ponce-Campos & Ustach, 2004 and Reyes-Velasco et al., 2012.
Remarks: Holotype lost *fide* Duellman (1960: 607).

5. *Pituophis melanoleucus* (Daudin, 1803c). Hist. Nat. Rept. 6: 409–410. (*Coluber melanoleucus*)

Synonyms: *Pituophis melanoleucus mugitus* T. Barbour, 1921b, and *Pituophis lodingi* Blanchard, 1924b.
Type: Holotype, not designated (W. Bartram, July 1765–Oct. 1777), lost *fide* Adler (2004: 290).
Type locality: "Floride" [= Florida, USA] and "Pensilvanie, Caroline méridionale et de la Floríde, Amérique septentrionale" [= Pennsylvania, southern Carolina and Florida, USA]. Restricted to Carolina *fide* Barbour (1921b: 117) and to Charleston, South Carolina, USA *fide* K.P. Schmidt (1953a: 201).
Distribution: Southeastern USA (SE New Jersey, S Kentucky, S West Virginia, W Virginia, Tennessee, S North Carolina, South Carolina, ext. E Louisiana, SE Mississippi, Alabama, Georgia, Florida), NSL–600 m.
Fossil records: Lower Pleistocene (Irvingtonian I) of USA (Florida), and upper Pleistocene (Rancholabrean II) of USA (Florida, Georgia, Pennsylvania).
Sources: Moore, 1893, Conant, 1956, Himmelstein, 1973a, Zappalorti et al., 1983, Ernst & Barbour, 1989, Sweet & Parker, 1990, Conant & Collins, 1991, Palmer & Braswell, 1995, Reichling, 1995 and Rodrigues-Robles & Jesús-Escobar, 2000.
Remarks: Original description based on Bartram (1791: 276) [Florida] with mention of Palisot de Beauvois [Pennsylvania, Carolina and Florida].

6. *Pituophis ruthveni* Stull, 1929. Occ. Pap. Mus. Zool., Univ. Michigan (205): 1–2. (*Pituophis melanoleucus ruthveni*)

Type: Holotype, USNM 76278, a 1520 mm male (W.D. Harris, 24 March 1927).
Type locality: "Longleaf, Louisiana" [Rapides Parish, Louisiana, USA].
Distribution: Southern USA (N Louisiana, ext. E Texas), 30–150 m.
Sources: Dundee & Rossman, 1989, Reichling, 1995, Rodrígues-Robles & Jesús-Escobar, 2000, Werler & Dixon, 2000 and Grismer, 2001.

7. *Pituophis vertebralis* (Blainville, 1835). Nouv. Ann. Mus. Hist. Nat. (3) 4(3): 293, pl. 27, figs. 2, 2a–b. (*Coluber vertebralis*)

Synonyms: *Pityophis haematois* Cope, 1860f, *Pituophis catenifer bimaris* Klauber, 1946b, and *Pituophis catenifer insulanus* Klauber, 1946b.
Type: Holotype, MNHN 63, a 530 mm specimen (P.-E. Botta, March–Oct. 1827).
Type locality: "mont Liban, la Californie" [= ? Mount Lebanon, Alta Californie, or California, W USA] (possibly in error). Restricted to Cabo San Lucas, Baja California Sur, Mexico *fide* H.M. Smith and Taylor (1950a: 322).
Distribution: Extreme NW Mexico (Baja California Norte, Baja California Sur, Cedros, Magdalena, San Jose and Santa Margarita Is.), NSL–500 m.
Sources: Klauber, 1946b, Sweet & Parker, 1990, Grismer, 1994, 1997, 2001 and Rodrigues-Robles & Jesús-Escobar, 2000.
Remarks: A valid species *fide* Pyron & Burbrink (2009: 527). A synonym of *P. catenifer fide* Rodrigues-Robles & Jesús-Escobar, 2000.

P

PLAGIOPHOLIS Boulenger, 1893a (Pseudoxenodontidae)

Synonyms: *Trirhinopholis* Boulenger, 1893a, *Trichinopholis* – Palacky, 1898 *(nomen incorrectum), Trirhinopolis* – M.A. Smith, 1915a *(nomen incorrectum), Plaglopholis* – Briceño-Rossi, 1934 *(nomen incorrectum),* and *Trirhinopholis* – Deuve, 1970 *(nomen incorrectum).*
Type species: *Plagiopholis blakewayi* Boulenger, 1893a.
Distribution: Southeastern Asia.
Sources: Bourret, 1936b, M.A. Smith, 1943, Zhao & Adler, 1993, Zhao, 2006 and Zaher et al., 2009.

1. *Plagiopholis blakewayi* Boulenger, 1893a. Cat. Snakes Brit. Mus. 1: 301–302, pl. 19, figs. 3, 3a–b.

Synonym: *Oligodon kunmimgensis* Kou & Wu, 1993.
Type: Holotype, BMNH 1946.1.12.35, a 245 mm male (Blakeway).
Type locality: "Toungyi, Shan States, mountains of Burma, 5000 feet" [= Taunggyi, SW Shan State, E Myanmar, elevation 1525 m].
Distribution: Northern Myanmar (Kachin, Shan), NW Thailand (Chiang Mai) and SW China (Guizhou, Sichuan, S Yunnan), 1300–2200 m.
Sources: Wall, 1921g, 1923i, 1925d, Anonymous, 1973, Tian et al., 1986, Zhao et al., 1986, Tillack et al., 2006 and I. Das, 2010.

2. *Plagiopholis delacouri* Angel, 1929. Bull. Mus. Hist. Nat., Paris (2) 1(1): 78–79.

Types: Syntypes (2), MNHN 1928.65–66, longest syntype 189 mm (J.T. Delacour [Second Delacour-Lowe Mission], 15 Dec. 1925–13 Jan. 1926).
Type locality: "Xieng-Khouang, Haut-Laos" [= Xiang Khoang, Xiangkhouang Prov., N Laos, 19°17'N, 103°25'E].
Distribution: Northern Laos (Xiangkhouang) and ext. N Vietnam (Lao Cai, Vinh Phuc), 600–2500 m.
Sources: Deuve, 1970, Orlov et al., 2003, V.S. Nguyen et al., 2009 and I. Das, 2010.

3. *Plagiopholis nuchalis* (Boulenger, 1893a). Cat. Snakes Brit. Mus. 1: 419–420, pl. 28, fig. 1. (*Trirhinopholis nuchalis*)

P

Synonym: *Oligodon evansi* Wall, 1913e.
Type: Holotype, BMNH 1946.1.13.94, a 340 mm female (E.W. Oates, 1867–1893).
Type locality: "Toungyi, S. Shan States, mountains of Burma" [= Taunggyi, SW Shan State, cen. Myanmar, 20°47'N, 97°02'E, elevation 1415 m].
Distribution: Southeastern Asia. Southern China (Yunnan) E Myanmar (Kachin, Karen, Sagaing, Shan), NW Thailand (Chiang Mai, Kanchanaburi, Lampang, Lamphun, Loei, Mae Hong Son) and NW Vietnam (Hoa Binh, Lao Cai, Lai Chau, Nghe An), 600–1620 m.
Sources: Wall, 1919d, E.H. Taylor, 1965, Deuve, 1970, Cheke, 1973, B.-Q. Hu et al., 1980, Tian et al., 1986, M.J. Cox, 1991b, M.J. Cox et al., 1998, Orlov et al., 2003, Tillack et al., 2006, V.S. Nguyen et al., 2009 and I. Das, 2010, 2012.
Remarks: Possibly occurs in Laos *fide* Deuve (1970: 116). Specimens from high elevations (1400–2500 m) in Vietnam possibly *P. delacouri* or *P. styani fide* N. Orlov *in* Tillack et al. (2006: 185).

4. *Plagiopholis styani* (Boulenger, 1899b). Proc. Zool. Soc. London 67(1): 164–165, pl. 18, figs. 2–2a. (*Trirhinopholis styani*)

Types: Syntypes (2), BMNH 1946.1.14.11–12, a 350 mm male and a juvenile (J.D. La Touche, spring 1896/1898).
Type locality: "Kuatun, a village about 270 miles from Foochow, in the mountains at the North-west of the Province of Fokien, at an altitude of 3000 to 4000 feet or more, China" [= Guadun, in Wuyishan Mtns., ext. N Fujian Prov., SE China, 27°45'N, 117°50'E, elevation 900–1200 m].
Distribution: Southern China (Anhui, Fujian, S Gansu, Guangxi, Hunan, Jiangxi, Sichuan, Zhejiang) and ext. N Vietnam (Lang Son, Lao Cai, Vinh Phuc), 670–1500 m.
Sources: C.H. Pope, 1935, Bourret, 1939d, 1941b, B.-Q. Hu et al., 1980, Tian et al., 1986, B.H. Li, 1987, Inger et al., 1990, Huang et al., 1999, Orlov et al., 2003, Zhao, 2006, V.S. Nguyen et al., 2009, Xiang & Li, 2009 and I. Das, 2010, 2012.

5. *Plagiopholis unipostocularis* Zhao, Jiang & Huang, 1978. Mater. Herpetol. Res., Chengdu 4: 21.

Type: Holotype, CIB 505121, a 300 mm female.
Type locality: "Yunnan Prov., China."
Distribution: Southwestern China (Yunnan). Known only from holotype.
Remarks: Supplemental original description in Zhao et al. (1979: 313, fig. 2).

PLATYCEPS Blyth, 1860 (Colubridae)

Synonyms: *Tyria* Bonaparte, 1840 *(nomen praeoccupatum), Dendrophilus* Fitzinger, 1843 *(nomen praeoccupatum), Pseudoperiops* Jan, 1857 *(nomen nudum), Etairejus* Jan, 1857 *(nomen oblitum), Dendrophlius* – Bedriaga, 1882 *in* 1881–1882 *(nomen incorrectum), Acanthocalyx* Cope, 1895c, and *Tylanthera* Cope, 1895c.
Type species: *Coluber ventromaculatus* Gray, 1834 *in* Gray & Hardwicke, 1830–1835.
Distribution: Southwestern Asia and N Africa.
Fossil records: Upper Miocene of Europe.
Sources: Schätti & Wilson, 1986, Schätti & Utiger, 2001 and Nagy et al., 2004.
Remarks: In accordance with Art. 23.9.2 of the Code (ICZN, 1999), *Platyceps* Blyth is designated a *nomen protectum* and *Pseudoperiops* Jan and *Etairejus* Jan are designated *nomina oblita*.

1. *Platyceps afarensis* Schätti & Ineich, 2004. Rev. Suisse Zool. 111(4): 686–687, fig. 1.

Type: Holotype, MNHN 2001.650, a 1035 mm male (CECAP personnel, 1999–2000).

Type locality: "Arta (11°31'N 42°51'E, approx. 705 ma.s.l.), Djibouti."
Distribution: Djibouti, 705 m. Known only from type locality.

2. *Platyceps atayevi* (Tuniyev & Shammakov, 1993). Asiatic Herpetol. Res. 5: 2–5, fig. 1. (*Coluber atayevi*)

Type: Holotype, CSBR 420, a 672 mm male (B.S. Tuniyev, 12 May 1991).
Type locality: "environs of Saivan Village, Saivan-Nokhur Plateau, western Kopet-Dag, Bakharden Region, Turkmenistan."
Distribution: Turkmenistan (Ashgabad, Bakharden), 1200–1300 m.
Remarks: A valid species *fide* Nagy et al., 2004. Previously a synonym of *P. najadum*.

3. *Platyceps bholanathi* (R. Sharma, 1976b). Comp. Physiol. Ecol. 1(3): 106, fig. 1. (*Coluber bholanathi*)

Type: Holotype, ZSI 21337, a 992–1190 mm male (B. Nath, 9 Sept. 1962).
Type locality: "Nagarjana Hill (16º31'N, 79º14'E), Guntur District, Andhra Pradesh (India)."
Distribution: Southern India (Andhra Pradesh 530–610 m).
Sources: Schätti, 1986a, Sanyal et al., 1993 and Seetharamaraju & Srinivasulu, 2013.

4. *Platyceps brevis* (Boulenger, 1895a). Ann. Mus. Civ. Stor. Nat. Genova (2) 15(35): 13–14, pl. 3, figs. 3, 3a–b. (*Zamenis brevis*)

Synonyms: *Zamenis tripraeocularis* F. Werner, 1925, *Zamenis boschisi* Scortecci, 1930a, and *Coronella semiornata fuscorosea* Loveridge, 1935.
Type: Holotype, MSNG 28848, a 200 mm male (V. Bottego, 16 Oct.–12 Dec. 1892).
Type locality: "Ogaden" [= Hararge Prov., E Ethiopia].
Distribution: Northeastern Africa. Somalia (Bari, Bay, Gedo, Mudug, Nugaal, Togdheer, Woqooyi Galbeed), SE Ethiopia (Bale, Gemu Gofa, Hararge, Sidamo), Kenya and NE Tanzania, 200–1400 m.
Sources: Lanza, 1963, 1983a, 1990b, Largen & Rasmussen, 1993, Spawls et al., 2002, Schätti & Charvet, 2003 and Largen & Spawls, 2010.

5. *Platyceps collaris* (F. Müller, 1878b). Verh. Naturf. Ges. Basel 6(4): 599. (*Zamenis dahlii collaris*)

Synonyms: *Zamenis dahlii rubriceps* Venzmer, 1919, and *Coluber rubriceps thracius* Rehák, 1985.
Type: Lectotype, NMBA 1166, adult male, designated by Schätti et al. (2001: 15).
Type locality: "Beirut" [= Beirut, Beirut Distr., Lebanon] via lectotype selection.
Distribution: Middle East. Southeastern Bulgaria (Burgas), Turkey (Marden), Syria, Lebanon (Beirut, Beqaa, El Jnoub, Mont-Liban, Nabatiye), Israel (Central, Gaza Strip, Golan Heights, Haifa, Jerusalem, Northern, N Southern, Tel Aviv, West Bank) and W Jordan (Ajloun, Amman, Balqa, Irbid, Jarash, Jawf, Karak, Maan, Tafilah), NSL–1900 m.
Sources: Baran, 1976a, Hraoui-Bloquet, 1981, Obst, 1981, Rehák, 1985, Ilani, 1986, Bousklia & Amitai, 2001, Disi et al., 1988, 2001, Schätti, 1993a, El-Oran, 1994, Schätti et al., 2001, Hraoui-Bloquet et al., 2002, Amr & Disi, 2011, Bar & Haimovitch, 2011 and Stojanov et al., 2011.
Remarks: *Platyceps rubriceps* a valid species *fide* Nagy et al., 2004 and Amr, 2009a.

6. *Platyceps elegantissimus* (A.C.L.G. Günther, 1878). Proc. Zool. Soc. London 46(1): 977–978, pl. 62. (*Zamenis elegantissimus*)

Type: Holotype, BMNH 1946.1.15.7, a 660 mm female (R.F. Burton, 1851–1853).
Type locality: "mountains east of El Muwayláh, Midian" [= mtns. E of Almuwaylih (27°41'N, 35°29'E), Tabuk Prov., NW Saudi Arabia].
Distribution: Middle East and Arabia. Southern Israel (E Southern), SW Jordan (Aqaba, Maan, Madaba, Tafilah) and N Saudi Arabia (Ha'il, Jazan, Madinah, Makkah, Qasim, Riyadh, Tabuk), NSL–1650 m.
Sources: Gasperetti, 1976b, 1988, Y. Werner & Sivan, 1991, Schätti & Gasperetti, 1994, Bouskila & Amitai, 2001, Disi et al., 2001, Baker et al., 2004, Egan, 2007, Amr & Disi, 2011 and Bar & Haimovitch, 2011.
Remarks: Possibly occurs in Sinai *fide* Marx, 1968.

7. *Platyceps florulentus* (I. Geoffroy-Saint-Hilaire, 1827 *in* Savigny, 1809–1829). Descr. Rept. Égypte 1(1): 146, 151, pl. 8, figs. 2–2'. (*Coluber florulentus*)

Synonyms: *Coluber ruber* Gmelin, 1789 (*nomen oblitum*), *Coluber keniensis* H.W. Parker, 1932c, and *Coluber florulentus perreti* Schätti, 1988b.
Type: Lectotype, a 760 mm specimen illustrated in pl. 8, fig. 2 (Expéd. Armée Franç., 1798–1801), lost *fide* Kramer & Schnurrenberger (1963: 486).
Type locality: "Égypte" [= Egypt].
Distribution: Northeastern Africa. Northern Cameroon (Extreme-Nord, Nord), NE Nigeria (Borno, Plateau, Taraba), E Egypt (Alexandria, Assiut, Aswan, Beheira, Beni Suef, Cairo, Damietta, Daqahlia, El-Bahr El-Ahmar, Faiyum, Gharbia, Giza, Ismailia, Matrouh, Menufia, Minya, E New Valley, North Sinai, Qalyubia, Qena, Sharkeya, Sohag, Suez), E Sudan (El-Bahr El-Ahmar, Al Khartum, Al Nil Al Abyad, An Nil Al Azraq, Ash Shamaliyah, Nahr An Nil, Sennar, Shamal Kurdufan), Eritrea (Anseba, Gash-Barka, Northern Red Sea), Ethiopia (Arussi, Gojjam, Gondar, Hararge, Kefa, Shoa), N Somalia (Bari, Woqooyi Galbeed), W Kenya (cen. Eastern) and N Uganda (W Northern), 160–2400 m.

Sources: Pitman, 1974, Pozuelo, 1974, Lanza, 1983a, 1990b, Schätti, 1988b, Largen & Rasmussen, 1993, Bons & Geniez, 1996, Broadley & Schatti, 1997, Schätti, 2001, Spawls et al., 2002, Chirio & LeBreton, 2007 and Largen & Spawls, 2010.

Remarks: In accordance with Art. 23.9.2 of the Code (ICZN, 1999), *Coluber florulentus* Geoffroy-Saint-Hilaire is designated a *nomen protectum* and *Coluber ruber* Gmelin a *nomen oblitum.*

8. *Platyceps gracilis* (A.C.L.G. Günther, 1862b). Ann. Mag. Nat. Hist. (3) 9(49): 125–126. (*Zamenis gracilis*)

Synonym: *Coluber cinereus* Linnaeus, 1758.

Types: Syntypes (6), BMNH 1946.1.15.1–4, two females and two juveniles (W. Masters), BMNH 1946.1.13.82, a female (Sykes), and BMNH 1946.1.13.85 (formerly FPM), a female.

Type locality: "western India (Kurrachee)," "India" and "unknown," respectively.

Distribution: Central India (Madhya Pradesh, Maharashtra).

Sources: M.A. Smith, 1943, Schätti, 1986b, Whitaker & Captain, 2004 and N. Khaire, 2006.

Remarks: Five syntypes in BMNH *fide* A.C.L.G. Günther (1862b: 125) but six *fide* BMNH catalogue (M. Lang, *in litt.*).

9. *Platyceps insulanus* (Mertens, 1965a). Senck. Biol. 46(1): 5–9, figs. 1–2. (*Coluber insulanus*)

Type: Holotype, SMF 60027, a 910+ mm male (W. Kost, 26 Nov. 1964).

Type locality: "Insel Sarso, Farsan-Archipel, Rotes Meer, SW-Arabien" [= Saso Is., Farsan Arch., Jazan Govern., SW Saudi Aribia, Red Sea, 16°51'N, 41°36'E].

Distribution: Southwestern Saudi Arabia, (Sarso Is.), NSL–30 m.

Source: Largen & Rasmussen, 1993.

P

10. *Platyceps karelinii* (Brandt, 1838). Bull. Acad. Imp. Sci. St. Pétersbourg 3(16): 243. (*Coluber* [*Tyria*] *karelinii*)

Synonyms: *Coluber chesneii* W.C.L. Martin, 1838, *Choristodon brachycephalus* Severzow, 1873, and *Zamenis rogersi* J. Anderson, 1893.

Types: Syntypes (6), ZISP 1695–1700 (formerly AIS 1695–1700), longest syntype 807 mm (G.S. Karelin, 1828–1836).

Type locality: "sur la côte orientale de la mer Caspienne, dans la Turcomanie et dans le nord de la Perse, notamment dans les environs d'Astrabad" [= eastern shore of Caspian Sea, within Turkey and northern Iran, notably the vicinity of Gorgan, Golestan Prov., N Iran]. Restricted to eastern shore of Caspian Sea *fide* Nikolsky (1916: 98).

Distribution: Northeastern Africa, Middle East and SW Asia. Northeastern Libya (Al Fatih, Butnan), N Egypt (Cairo, Matrouh, North Sinai, South Sinai, Suez), S Israel (Southern, West Bank), Jordan (Ajloun, Amman, Aqaba, Balqa, Irbid, Karak, Maan, Madaba, Mafraq, Tafilah, Zarqa), Lebanon, Syria (Damascus, Daría, Dayr az Zawr, Halab, Homs), NE Saudi Arabia (Eastern, Tarut Is.), Kuwait, Bahrain, United Arab Emirates (Sir Bani Yas Is.), SW Kazakhstan, Kyrgyzstan, Tadzhikistan, Turkmenistan, Uzbekistan, E Iraq (Al Anbar, Baghdad, Basrah, Dhi Qar, Diyala, Maysan, Ninawa, Salah Al-Din, Sulaymaniyah, Wasit), Iran (Bushehr, Central, Fars, Golestan, Kermanshah, Khorasan Razavi, Khuzestan, Markazi, Semnan, Sistan va Baluchestan, South Khorasan, Yazd), Afghanistan (Balkh, Helmand, Herat, Kandahar, Nimruz) and SW Pakistan (N Balochistan), NSL–2200 m.

Sources: Strauch, 1873, Minton, 1966, Mertens, 1969a, Bannikov et al., 1977, Latifi, 1991, Gasperetti, 1988, Szczerbak et al., 1993, Schleich et al., 1996, M.S. Khan, 1997, 2002, Disi et al., 2001, Baha El Din, 2006, Amr & Disi, 2011, Bar & Haimovitch, 2011 and Schätti et al., 2012.

Remarks: *Zamenis rogersi* J. Anderson a synonym *fide* Schätti et al., 2012.

11. *Platyceps largeni* (Schätti, 2001a). Russ. J. Herp. 8(2): 142–144, fig. 2. (*Coluber largeni*)

Type: Holotype, BMNH 1973.3211, ca. 600 mm female (M.J. Largen, 7 Jan. 1970).

Type locality: "Sarad Island, Dahlak, Ethiopia" [= Sarad Is., Dahlak Arch., Northern Red Sea Prov., E Eritrea, 15°50'N, 39°55'E].

Distribution: Eastern Eritrea (Southern Red Sea: Andeber, Nocra and Sarad Is.).

Source: Largen & Spawls, 2010.

Remarks: Possibly a synonym of *P. taylori.*

12. *Platyceps manseri* (Leviton, 1986). Fauna Saudi Arabia 8: 438, figs. 1a–d. (*Coluber manseri*)

Type: Holotype, CAS 140514, a 375 mm female (P. Manser, July–Sept. 1975).

Type locality: "vicinity of Hakimah (17°01'N 42°50'E) (elev. approx. 80 m), southern Tihama, Saudi Arabia" [= Hakimat Abu Arish, Jazan Prov., SW Saudia Arabia].

Distribution: Southwestern Saudi Arabia (Jazan), 80–200 m.

Sources: Gasperetti, 1988 and Egan, 2007.

Remarks: Previously a subspecies or synonym of *P. varaiabliis.*

13. *Platyceps messanai* (Schätti & Lanza, 1989). Boll. Mus. Reg. Sci. Nat. Torino 7(2): 414–418, figs. 1–6. (*Coluber messanai*)

Type: Holotype, MZUF 5468, a 542 mm male (E. Granchi, B. Lanza & A. Simonetta, 9 Aug. 1969).

Type locality: "Wadi Run, Central Nogal Valley, Somalia (about 8°48'20" N-48°53'30" E)."
Distribution: Northeastern Somalia (Nugaal), 500 m. Known only from type locality.
Source: Schätti & Charvet, 2003.

14. *Platyceps mintonorum* (Mertens, 1969a). Stutt. Beitr. Naturk. (197): 56–60, fig. 17. (*Coluber karelini mintonorum*)

Type: Holotype, SMF 62942, a 1550 mm male (M.G. Konieczny, 12 April 1962).
Type locality: "Zangi-Nawar, 27 km südwestlich Nushki, Distr. Chagai, West-Pakistan" [= Balochistan, SW Pakistan, 29°26'N, 65°47'E, 900 m].
Distribution: Southwestern Asia. Southeastern Iran, S Afghanistan (Helmand, Kandahar) and SW Pakistan (Balochistan), 700–1060 m.
Source: Schätti & Stutz, 2005 and Schätti et al., 2012.

15. *Platyceps najadum* (Eichwald, 1831). Zool. Spec. Rossiae 3: 174. (*Tyria najadum*)

Synonyms: *Tyria dahlii* Fitzinger, 1826a (*nomen nudum*), *Tyria ocellata* Eichwald, 1831, *Coluber olivaceus* Dwigubsky, 1832, *Coluber fascicularis* Ménétriés, 1832, *Coluber dahlii* Schinz, 1833, *Zamenis dahlii immaculata* Schreiber, 1912, *Coluber najadum kalymnensis* B. Schneider, 1980, and *Coluber najadum albitemporalis* Darevsky & Orlov, 1994.
Type: Holotype, not designated, a 760 mm specimen (K.E.I. Eichwald, 1825–1826), location unknown.
Type locality: "urbis Bacuáe, ad mare caspium" [= Baku, on the Caspian Sea, E Azerbaijan, 40°25'N, 49°52'E, elevation 20 m].
Distribution: Southeastern Europe amd SW Asia. Croatia (Krk and Pag Is.), S Bosnia-Herzegovina, Montenegro, S Bulgaria, (Blagoevgrad, Haskovo, Kardzhali, Pazardzhik, Plovdiv, Sliven, Yambol), Albania, Macedonia, Greece (Aitolia and Akarnania, Arkadhia, Akhaia, Argolis, Arta, Attica, Drama, Evritania, Evros, Evvoia, Florina, Fokis, Fthiotis, Grevena, Ilia, Imathia, Ioannina, Kardhitsa, Kastoria, Kefallinia, Kerkira, Khalkidhiki, Kilkis, Korinthia, Kozani, Lakonia, Larisa, Lesvos, Levkas, Magnisia, Messinia, Pella, Pieria, Preveza, Rodhopi, Serrai, Thesprotia, Thessaloniki, Trikala, Voiotia, Xanthi, Zakinthos, Chios, Corfu, Kavala, Lesbos, Limnos, Rhodos, Samos, Samothraki and Thasos Is.), Cyprus, Turkey, Syria (Damascus), Lebanon (Liban-Nord, Mont-Liban), Armenia, SW Russia (Dagestan), Georgia, Armenia, Azerbaijan, Iraq and N Iran, NSL–2200 m.
Sources: F. Werner, 1939, Baran, 1976a, Bruno, 1984, Latifi, 1991, Darevsky & Szczerbak, 1993, Schätti, 1993a, 2004, Tuniyev & Shammakov, 1993, Darvevsky and Orlov, 1994, Szczerbak, 1994, Tokar et al., 1995, Tuniyev et al., 1999, Petkovski et al., 2000, Hraoui-Bloquet et al., 2002, Valakos et al., 2004, Schätti et al., 2005, Trapp, 2007, Valakos et al., 2008, Baier et al., 2009, Tuniyev et al., 2009, Stojanov et al., 2011, Jablonski et al., 2012 and Paphiles & Valakos, 2012.
Remarks: Presence on Cyprus needs confirmation *fide* Baier et al. (2009: 261–262).

16. †*Platyceps planicarinatus* (Bachmayer & Szyndlar, 1985). Ann. Naturhist. Mus. Wien 87A: 84–86, figs. 1.17–1.21, pl. 1, figs. 4–6. (†*Nanus planicarinatus*)

Type: Holotype, NMWGP 1984/97, one trunk vertebra.
Type locality: "Kohfidisch fissure deposits of Burgenland, Austria; Late Pannonian (= Pontian, MN 11) in the Paratethyian terms or Early Pliocene."
Distribution: Upper Miocene (Turolian, MN 11–13: 4.9–8.7 mya) of Austria and Italy.
Source: Bachmayer & Szyndlar, 1987, Delfino, 2002 and Szyndlar, 2012.
Remarks: Possibly related to *P. najadum fide* Bachmayer & Szyndlar (1987: 30).

17. *Platyceps rhodorachis* (Jan *in* Filippi, 1865). Viaggio Persia 1: 356. (*Zamenis rhodorachis*)

Synonyms: ? *Natrix scheuchzeri* Merrem, 1820 (*nomen dubium*), ? *Coluber schneideri* Merrem, 1820 (*nomen dubium*), *Etairejus rhodorachis* Jan, 1857 (*nomen nudum*), *Zamenis ladacensis* J. Anderson, 1871b, *Gonyosoma dorsale* J. Anderson, 1872, *Zamenis rhodorhachis* Boulenger, 1891g (*nomen emendatum*), *Zamenis ladacensis subnigra* Boettger, 1893b, *Zamenis rhodorhachis tessellata* F. Werner, 1909c, *Coluber karelini mintonorum* Mertens, 1969a, and *Coluber rhodorachis kashmirensis* M.S. Khan & Khan, 2000.
Type: Lectotype, MSNG CE 30312 (formerly MSNG 3320), a 585 mm female (Ratchy via G. Jan, 1864), designated by Kramer & Schnurrenberger (1963: 501).
Type locality: "Umgebung von Schiras Zentral-persien" [= vicinity of Shiraz (29°37'N, 52°32'E, elevation 1545 m), Fars Prov., SW Iran] via lectotype selection.
Distribution: Northeastern Africa, Arabia and SW Asia. Eritrea (Northern Red Sea, Southern Red Sea, Difnein Is.), E Ethiopia (Hararge), Djibouti (Maskali and Musha Is.), N Somalia (Awdal, Mudug, Nugaal, Sanaag, Sool, Togdheer, Woqooyi Galbeed), Saudi Arabia (Bahah, Asir, Riyadh, Ha'il, Jazan, Madinah, Makkah, Najran, Qasim, Riyadh, Tabuk), United Arab Emirates (Fujairah, Ras al Khaimah, Sharjah), Oman (Al Batinah, Al Dakhiliyah, Al Wusta, Ash Sharqiyah, Dhofar, Musandam, Muscat, Masirah Is.), Yemen (Abyan, Ad Dali', Aden, Al Bayda, Al Hudaydah, Al Mahrah, Amanat Al Asimah, Hadhramaut, Sana'a, Ta'izz), N Israel (Haifa, Northern, West Bank), Jordan (Amman, Aqaba, Balqa, Irbid, Karak, Maan, Mafraq, Tafilah, Zarqa), Syria, N Iraq (Sulaymaniyah), Iran (Central, Esfahan, Fars, Khuzestan, Kordestan, Razavi Khorasan, Saheli, Semnan, Sistan va Baluchestan), W Tadzhikistan, Kazakhstan, S Turkmenistan & S Uzbekistan, Afghanistan, Pakistan (Balochistan,

F.A.T.A., Gilgit-Baltistan, Jammu & Kashmir, Khyber Pakhtunkhwa, Punjab, Sindh), and N India, NSL–2800 m.

Sources: J. Anderson, 1898, Kramer & Schnurrenberger, 1963, Minton, 1966, Mertens, 1969a, Leviton & Anderson, 1970a, Lanza, 1983a, 1990b, Bannikov et al., 1977, Latifi, 1991, Gasperetti, 1988, Schätti, 1989, 1993a, 2006a–b, Leviton et al., 1992, Largen & Rasmussen, 1993, Ataev et al., 1994, Schätti & Gasperetti, 1994, Szczerbak, 1994, Schleich et al., 1996, M.S. Khan, 1997, 2002, 2006, M.S. Khan & Khan, 2000, Bouskila & Amitai, 2001, Disi et al., 2001, Baker et al., 2004, Egan, 2007, Largen & Spawls, 2010, Afrasiab & Mohamad, 2001, Amr & Disi, 2011, Bar & Haimovitch, 2011, Grossmann et al., 2012 and Masroor, 2012.

Remarks: This taxon is a complex of species yet to be resolved *fide* Schätti (2006: 77).

18. *Platyceps saharicus* Schätti & McCarthy, 2004. Rev. Suisse Zool. 111(4): 693–700, fig. 1.

Type: Holotype, FMNH 72108, a 1270 mm female (H. Hoogstraal, 14 May 1953).

Type locality: "Egypt: St. Catherine's Monastery area, Wadi el Sheikh" [= Wadi el Sheikh, vicinity of St. Catherine's Monastery (28°33'N, 33°59'E, elevation 1585 m), South Sinai Govern., NE Egypt].

Distribution: North Africa and Middle East. Southeastern Algeria (Illizi, Tamanrasset), SE Libya (Al Kufrah), N Chad (Borkou-Ennedi-Tibesti), NE Sudan, E Egypt (Alexandria, El-Bahr El-Ahmar, ext. SW New Valley, North Sinai, Sharkeya, South Sinai, Suez, Shadwan Is.), Israel (Jerusalem, Southern, West Bank), W Jordan (Irbid, Ma'an) and W Saudi Arabia (Jazan, Tabuk), 550–1700 m.

Sources: Geniez & Gaulthier, 2008 and Bar & Haimovitch, 2011.

Remarks: Occurs south to Somalia and east to Afghanistan *fide* Baha El Din (2006: 258).

19. *Platyceps schmidtleri* (Schätti & McCarthy, 2001). Herpetozoa 14(1–2): 82–85, figs. 1–2. (*Coluber schmidtleri*)

Synonym: *Zamenis ventrimaculatus semifasciatus* Werner, 1917 *(nomen praeoccupatum).*

Type: Holotype, FMNH 170927, a 655 mm male (E.D. Womochel, 14 Oct. 1968).

Type locality: "'Fars, 5 mi. N. E. Yasoodi' (Yasuj, Boyerahmad-va-Kogiluye, 30°40'N, 51°36'E)" [= 3 km NE Yasuj, Fars Prov., Iran].

Distribution: Southwestern Iran (Zagros Mtns. of Fars, Khuzestan), 1500–2000 m.

Remarks: Related to *P. collaris* and *P. najadum.*

20. *Platyceps scortecci* (Lanza, 1963). Atti Soc. Ital. Sci. Nat. Mus. Civ. Stor. Nat. Milano 102(4): 379–381, figs. 1a–1d. (*Coluber scortecci*)

Type: Holotype, MSNM 2146, a 293 mm female (G. Scortecci, 1931).

Type locality: "Dusa Mareb (Somalia centrale)" [= Dhuusamareeb, Galguduud Region, cen. Somalia, 5°32N, 46°23'E, elevation 250 m].

Distribution: Central Somalia (Nugaal, Galguduud), 250 m.

Source: Schätti, 1986b.

21. *Platyceps sinai* (K.P. Schmidt & Marx, 1956). Fieldiana: Zool. 39(4): 30–32, fig. 4. (*Lytorhynchus sinai*)

Type: Holotype, USNM 134989, a female (R.E. Kuntz).

Type locality: "Wadi Feiran, Sinai Peninsula" [Egypt].

Distribution: Middle East and N Africa. Southeastern Israel (Southern, West Bank), W Jordan (Aqaba, Maan, Madaba, Tafilah) and NE Egypt (South Sinai), 1500–1650 m.

Sources: Marx, 1968, Leviton, 1986, Sindaco et al., 1995, Y.L. Werner, 1995, Bouskila & Amitai, 2001, Disi et al., 2001, Baker et al., 2004, Amr & Disi, 2011 and Bar & Haimovitch, 2011.

22. *Platyceps smithi* (Boulenger, 1895h). Proc. Zool. Soc. London 63(3): 536–537, pl. 30, figs. 2–2a. (*Zamenis smithi*)

Synonym: *Coronella semiornata fuscorosea* Loveridge, 1935.

Type: Lectotype, BMNH 1946.1.13.79, a male (A. Donaldson-Smith, 23 Aug. 1894), designated by Schätti (1988b: 113), longest syntype 560 mm.

Type locality: "Shebeli" [= Webi Shebeli, Dacata River, SE Hamarro Hadad, Hararge Prov., E Ethiopia, 7°20'N, 42°17'E, elevation 600 m] via lectotype selection.

Distribution: Northeastern Africa. Southern Somalia (Bay, Gedo, Hiraan, Jubbada Dhexe, Jubbada Hoose, Shabeellaha Dhexe), S Ethiopia (Bale, Gemu Gofa, Hararge, Sidamo) and Kenya (Coast, Eastern, ext. N North-Eastern, N Rift Valley), 100–1300 m.

Sources: Lanza, 1983a, 1990b, Largen & Rasmussen, 1993 and Spawls et al., 2002.

Remarks: Possibly a synonym of *P. brevis.*

23. *Platyceps somalicus* (Boulenger, 1896g). Ann. Mus. Civ. Stor. Nat. Genova (2) 17(37): 11. (*Zamenis somalicus*)

Type: Holotype, MSNG 29059, a 440 mm female (E. Ruspoli, 1893).

Type locality: "Audo Mts., between Webi Shebeli and Web, Gallaland" [= Audo Mtns., Bale Prov., SE Ethiopia, ca. 6°30'N, 41°30'E, elevation 1500 m].

Distribution: Southern Ethiopia (Bale), 1500 m.

P

Sources: Parker, 1949, Schätti, 1986b, Largen & Rasmussen, 1993, Schätti & Charvet, 2003 and Largen & Spawls, 2010.

24. *Platyceps taylori* (H.W. Parker, 1949). Zool. Verh. (6): 40–44, fig. 7. (*Coluber taylori*)

Type: Holotype, BMNH 1949.2.1.56, a 500 mm female (native, 4 March 1933).

Type locality: "north western part of the mountainous OGO, Borama district, British Somaliland (42° 45' E. X 10° 20' N.), 4000 ft." [= Ogo Mtns., N Hararge Prov., E Ethiopia, 10°20'N, 42°45'E, elevation 1220 m].

Distribution: Northeast Africa. Eritrea (Northern Red Sea, Southern Red Sea), E Ethiopia (Hararge, Shoa), Djibouti and N Somalia (Bari, Woqooyi Galbeed), NSL–1220 m..

Sources: Lanza, 1972, 1983a, 1990b, Schatti, 1988a, 2001, 2007, Largen & Rasmussen, 1993 and Largen & Spawls, 2010.

25. *Platyceps thomasi* (H.W. Parker, 1931a). Ann. Mag. Nat. Hist. (10) 8(47): 516–517. (*Coluber thomasi*)

Type: Holotype, BMNH 1946.1.14.73 (formerly BMNH 1931.7.16.68), a 213–217 mm female (B. Thomas, 7 Nov. 1930).

Type locality: "Ain, 1500 ft., between Salalah and Rub'al Khali, S. E. Arabia" [= Ain, between Salalah (17°00'N, 54°06'E) and Rub-al-Khail (18°40'N, 53°12'E), W Dhofar Region, W Oman, ca. 17°43'N, 53°34'E, elevation 450 m].

Distribution: Western Oman (Dhofar) and E Yemen, NSL–460 m.

Sources: Arnold, 1980b, Leviton, 1986, Schätti, 1986b, Gasperetti, 1988, Egan, 2007 and Smid, 2010.

Remarks: Sympatric with *P. variabilis* in SE Oman *fide* Smid (2010: 330).

26. *Platyceps variabilis* (Boulenger, 1905e). Ann. Mag. Nat. Hist. (7) 16(92): 178–179. (*Zamenis variabilis*)

Type: Lectotype, BMNH 1946.1.14.70 (formerly BMNH 1903.1.28.12), a female, (G.W. Bury, 1900–1903), designated by Gasperetti (1988: 409), longest syntype 380 mm.

Type locality: "Schaf Ravine, South Arabia" [= Yemen] via lectotype selection.

Distribution: Western Oman (Dhofar) and SW Yemen (Ad Dali', Al Hudaydah), (25) 355–1525 m.

Sources: Leviton, 1986, Gasperetti, 1988, Schätti, 1989, Schätti & Gasperetti, 1994, Egan, 2007 and Smid, 2010.

Remarks: Boulenger (1905e:179) reported 10 syntypes but listed 13 specimens.

27. *Platyceps ventromaculatus* (Gray, 1834 *in* Gray & Hardwicke, 1830–1835). Illust. Indian Zool. 2(13–14): pl. 80, fig. 1. (*Coluber ventromaculatus*)

Synonyms: *Pseudoperiops persicus* Jan, 1857 *(nomen nudum), Zamenis ventrimaculatus* A.C.L.G. Günther, 1858 *(nomen emendatum), Platyceps semifasciatus* Blyth, 1860, *Platyceps fasciatus* Stoliczka, 1870c *(nomen emendatum), Platyceps subfasciatus* Loveridge, 1957 *(nomen emendatum),* Coluber *ventromaculatus bengalensis* M.S. Khan & Khan, 2000, *Coluber ventromaculatus indusai* M.S. Khan & Khan, 2000, and *Coluber ventromaculatus khanorum* Barabanov, 2002 (*nomen substitutum*).

Type: Holotype, BMNH 1946.1.11.42, a gravid female (T. Hardwicke, 1756–1823).

Type locality: Unknown. Designated as Bengal *fide* K.P. Schmidt (1939b: 74).

Distribution: Southern Asia. Pakistan (ext. SE Balochistan, Jammu & Kashmir, Khyber Pakhtunkhwa, Punjab, Sindh) and NW India (Delhi, Gujarat, Maharashtra, Punjab, Rajasthan, Uttar Pradesh), NSL–500 (1000) m.

Sources: Wall, 1914b, Minton, 1966, Mertens, 1969a, Gasperetti, 1988, Leviton et al., 1992, Disi, 1993, M.S. Khan, 1997, 2002, 2006, M.S. Khan & Khan, 2000, Schätti, 2006b, Schätti & Schmitz, 2006, Egan, 2007, Baig et al., 2008, Soorae et al., 2009, Vyas, 2011 and Masroor, 2012.

Remarks: Records from Himachal Pradesh and Uttaranchal, India require confirmation *fide* Schätti & Schmitz (2006: 757).

PLATYPLECTRURUS A.C.L.G. Günther, 1868 (Uropeltidae)

P

Synonyms: *Platyplecturus* – Phisalix, 1917b *(nomen incorrectum), Wallia* F. Werner, 1925, *Platyplactrurus* – W.W. Tanner & Avery, 1982 *(nomen incorrectum),* and *Platyplecturus* – Skoczylas, 1978 *(nomen incorrectum).*

Type species: *Plectrurus trilineatus* Beddome, 1867.

Distribution: Southern India and Sri Lanka.

Sources: Beddome, 1886, M.A. Smith, 1943, Gans, 1966, Murthy, 1981c, 1982b, Rajendran, 1985, McDiarmid et al., 1999 and Aengals, 2009.

1. *Platyplectrurus madurensis* Beddome, 1877a. Proc. Zool. Soc. London 45(1): 167.

Synonyms: *Wallia inexpectata* F. Werner, 1925, *Platyplectrurus ruhunae* Deraniyagala, 1954, and *Platyplectrurus madurensis ruhanae* Gans, 1966 *(nomen emendatum).*

Types: Syntypes (8), BMNH 1883.1.12.56–58, BMNH 1946.1.15.78–80, and MNHN 1895.115a–b, four longest syntypes 356 mm, 319 mm, 280 mm and 270 mm.

Type locality: "About Kodiukarnal on the Pulney Mountains (Madura district), Southern India, 600 feet elevation" [= Kodaikanal, S Tamil Nadu, ext. S India, 10°14'N, 77°29'E, elevation 2100 m].

Distribution: Southern India (Western Ghats of SE Kerala, SW Tamil Nadu) and S Sri Lanka (Southern), 1200–2200 m.

Sources: Wall, 1923a, M.A. Smith, 1928b, Deraniyagala, 1955, Murthy, 1974, 1983b and A. Silva, 1990.

2. *Platyplectrurus trilineatus* (Beddome, 1867). Madras Quart. J. Med. Sci. 11: 14, pl. 1, fig. 1. (*Plectrurus trilineatus*)

Synonym: *Platyplectrurus bilineatus* Beddome, 1886.

Types: Syntypes (3), BMNH 1946.1.15.72 (formerly BMNH 1866.12.15.10), MNHN 1895.112, and 1895.113.

Type locality: "Anamallay forests, Madras Presidency; elevation 4,000 feet" [=W Tamil Nadu State, S India]. Emended to Anamallay forests above Ponachi [= Pollachi, 10°40'N, 77°01'E] at an elevation of 4000 feet *fide* Beddome (1886: 32).

Distribution: Southern India (Western Ghats of SE Kerala, SW Tamil Nadu), 1300–1860 m.

Sources: Murthy, 1983b and Ganesh, 2011.

†*PLATYSPONDYLIA* Rage, 1974 (Tropidophiidae)

Type species: †*Platyspondylia lepta* Rage, 1974.

Distribution: Upper Eocene of France and middle-upper Oligocene of France and Germany.

Sources: Underwood, 1976, Rage, 1984b, 2011, McDowell, 1987, Holman & Harrison, 1998b, Szyndlar & Rage, 2003 and Szyndlar et al., 2008.

P

1. †*Platyspondylia germanica* Szyndlar & Rage, 2003. Booidea Oligocene Miocene Europe: 81–83, figs. 39a–e.

Type: Holotype, SMNS 87146-2, one trunk vertebra.

Type locality: "Herlingen 11 (Germany), late Oligocene (MP 28/30)."

Distribution: Upper Oligocene (Arvernian, MP 28–30: 23.0–28.4 mya) of Germany.

2. †*Platyspondylia lepta* Rage, 1974. Palaeovertebrata 6(3–4): 288–293, fig. 4a.

Type: Holotype, MNHN PFR 6350, one middle trunk vertebra.

Type locality: "Phosphorites du Quercy, Pech-du-Fraysse, Oligocène Supérieur" [= Pech-du-Fraysse, Tarn-et-Garrone Dept., SW France; upper Oligocene, Arvernian].

Distribution: Upper Oligocene (Arvernian, MP 28–30: 23.0–28.4 mya) of France.

3. †*Platyspondylia sudrei* Rage, 1988c. Palaeontographica 205A: 22–24, figs. 20–21.

Type: Holotype, USTL BRT 1345, one middle trunk vertebra (Pernotet).

Type locality: "Le Bretou, Phosphorites du Quercy, France; niveau-repère du Bretou, Headonien inférieur (= Marinésien; Bartonien `anté-Ludien'), Eocène supérieur" [= Bretou, Tarn-et-Garonne Dept., SW France; upper Eocene, early Headonian].

Distribution: Upper Eocene (Priabonian, MP 17–18: 33.9–37.2 mya) of France.

Source: Rage, 2006.

PLECTRURUS A.-M.-C. Duméril & Duméril, 1851 (Uropeltidae)

Synonyms: *Maudia* Gray, 1858c, *Plecturus* – W.C.H. Peters, 1861h *(nomen incorrectum), Mandia* – Hoffmann, 1890 *(nomen incorrectum),* and *Pelctrurus* – Mahendra, 1984 *(nomen incorrectum).*

Type species: *Plectrurus perrotetii* A.-M.-C. Duméril & Bibron *in* A.-M.-C. Duméril & Duméril, 1851.

Distribution: Southern India.

Sources: Jan, 1865b, Beddome, 1886, M.A. Smith, 1943, Gans, 1966, Murthy, 1981c, 1982b, Rajendran, 1985, McDiarmid et al., 1999 and Aengals, 2009.

1. *Plectrurus aureus* Beddome, 1880. Proc. Zool. Soc. London 48(2): 182.

Types: Syntypes (2), BMNH 1946.1.1.54 (formerly BMNH 1883.1.12.53) and MNHN 1895.107 (R.H. Beddome, 1857–1880).

Type locality: "Chambra mountain in Wynad, near Kalpatty, Malabar, at 4,500 and 6000 feet elevation" [= Chembra Peak, Wyanad hills of Western Ghats, N Kerala State, SW India, ca. 11°30'N, 76°05'E, elevation 1370–1830 m].

Distribution: Southwestern India (Western Ghats of W Karnataka), 1370–2050 m.

2. *Plectrurus guentheri* Beddome, 1863a. Madras Quart. J. Med. Sci. 6: 48, pl. 2, fig. 10. (*nomen corrigendum*)

Synonyms: *Plectrurus güntheri* Beddome, 1863a *(nomen incorrigendum)* and *Plectrurus guentheri* – Beddome, 1863b *(nomen corrigendum).*

Type: Holotype, BMNH 1946.1.16.32 (formerly BMNH 1883.1.12.54), a 330–356 mm specimen (R.H. Beddome, 1857–1863).

Type locality: "Walaghat, in the Wynaud, on the Western slopes of the Neilgherries, Madras Presidency" [= Walaghat, Nilgiri Hills, cen. Westrn Ghats, Kerala State, SW India]. Emended ro Walahghat, on the western slopes of the Nilgherries (3500 feet elevation) *fide* Beddome (1863b: 228).

Distribution: Southwestern India (Western Ghats of E Kerala, SW Tamil Nadu), 1065–2250 m.

Source: Murthy, 1983b.

Remarks: Supplemental original description in Beddome (1863b: 228, pl. 27).

3. *Plectrurus perrotetii* A.-M.-C. Duméril & Bibron *in* A.-M.-C. Duméril & Duméril, 1851. Cat. Méth. Rept. Mus. Paris: 224.

Synonyms: *Plectrurus perrotetti* Gray, 1858c *(nomen emendatum), Plectrurus brevis* A.C.L.G. Günther, 1875 *(nomen nudum), Plectrurus perrotteti* A.C.L.G. Günther, 1875 *(nomen emendatum), Plectrurus davidsoni* Beddome, 1886, and *Plectrurus davisonii* Boulenger, 1893a *(nomen emendatum).*

Types: Syntypes (15–20), BMNH 1946.1.1.41, MNHN 170, MNHN 190, MNHN 6996, and ZMB 4036, longest syntype 248 mm (G.S. Perrotet, 1822–1831 & A. Delessert), location of other syntypes unknown.

Type locality: "Monts Nilgerrhy (Indes orient.)" [= Nilghiri hills, Western Ghats, W Tamil Nadu State, S India, ca. 11°24'N, 76°45'E].

Distribution: Southwestern India (Western Ghats of W Karnataka, NE Kerala, SW Tamil Nadu), 1065–2250 m.

Sources: Murthy, 1983b, 1990, J.C. Murphy & Schlager, 2003, Whitaker & Captain, 2004 and Hutton & David, 2009.

Remarks: Supplemental original description in Duméril, Bibron & Duméril (1854a: 167–168, pl. 59, figs. 4–4a).

PLESIODIPSAS Harvey, Rivas-Fuenmayor,

PLESIODIPSAS Harvey, Rivas-Fuenmayor, Portilla & Ruede-Almonacid, 2009 (Dipsadidae)

Type species: *Tropidodipsas perijanensis* Aleman, 1953.

Distribution: Colombia and Venezuela.

Sources: Harvey et al., 2009 and Zaher et al., 2009.

1. *Plesiodipsas perijanensis* (Aleman, 1953). Mem. Soc. Cienc. Nat. La Salle 13(35): 217–218, 1 fig. (*Tropidodipsas perijanensis*)

Type: Holotype, MHNLS 4478, a 910 mm female (R. Urbano, 9 March 1952).

Type locality: "Jamayaujaina, Sierra de Perijá, Estado Zulia, a 1.700 metros de altura, Venezuela."

Distribution: Colombia (César, Santander) and Venezuela (Zulia), 1350–2100 m.

Sources: Roze, 1966a, Lancini, 1986 and Navarrete et al., 2009.

Remarks: Previously in the genus *Dipsas.*

†*PLESIOTORTRIX* Rochebrune, 1884 (Boidae)

Type species: †*Plesiotortrix edwardsi* Rochebrune, 1884.

Distribution: Upper Eocene or Oligocene of France.

Source: Rage, 1984b.

1. †*Plesiotortrix edwardsi* Rochebrune, 1884. Mém. Soc. Sci. Nat. Saone-et-Loire 5: 156, pl. 2, figs. 6–6a.

Type: Holotype, MNHN QU 16332, three articulated trunk vertebrae.

Type locality: "Phosphorites du Quercy" [Tarn-et-Garonne Prov., SW France, upper Eocene or Oligocene].

Distribution: Upper Eocene (33.9–37.2 mya) or Oligocene (23.0–33.9 mya) of France. Known only from type series.

PLIOCERCUS Cope, 1860e (Dipsadidae)

Synonyms: *Elapochrus* W.C.H. Peters, 1860c, *Pleiocercus* Salvin, 1861 *(nomen emendatum), Pleiokerkos* Cope, 1862b *(nomen emendatum), Cosmiosophis* Jan, 1863a, *Plioceras* – K.P. Schmidt, 1936a *(nomen incorrectum),* and *Elapochrous* – H.M. Smith & Taylor, 1945 *(nomen incorrectum).*

Type species: *Pliocercus elapoides* Cope, 1860e.

Distribution: Latin America.

Sources: H.M. Smith & Taylor, 1945, J.A. Peters & Orejas-Miranda, 1970, H.W. Greene & McDiarmid, 1981, Cadle, 1984b, L.D. Wilson & Meyer, 1985, Savage & Crother, 1989, H.M. Smith et al., 1995, 1996, H.M. Smith & Chiszar, 1996, L.D. Wilson et al., 1996, L.D. Wilson & McCranie, 1997, H.M. Smith & Chiszar, 2000b and Zaher et al., 2009.

Remarks: A synonym of *Urotheca* fide Savage and Crother (1989: 341). H.M. Smith & Chiszar, 2001b–g recognized *P. andrewsi, bicolor, dimidiatus, elapoides,* and *euryzonus* as valid species. Possibly only a single species *fide* L.D. Wilson et al. (1996: 62), but *P. euryzonus* and *P. elapoides* sympatric in Honduran Mosquitia *fide* L.D. Wilson et al., 2003.

1. *Pliocercus elapoides* Cope, 1860e. Proc. Acad. Nat. Sci. Philadelphia 12(6): 253–254.

Synonyms: *Elapochrus deppei* W.C.H. Peters, 1860c, *Pleiocercus aequalis* Salvin, 1861, *Liophis tricinctus* Jan, 1863a, *Pliocercus salvinii* F. Müller, 1878b, *Pliocercus sargii* J.G. Fischer, 1881, *Liophis elapoides diastema* Bocourt, 1886 *in* Duméril, Bocourt & Mocquard, 1870–1909, *Pliocercus bicolor* H.M. Smith, 1941b, *Pliocercus elapoides laticollaris* H.M. Smith, 1941b, *Pliocercus elapoides semicinctus* K.P. Schmidt, 1941, *Pliocercus andrewsi* H.M. Smith,

1942h, *Pliocercus elapoides schmidti* H.M. Smith, 1942h, *Pliocercus elapoides celatus* H.M. Smith, 1943b, *Pliocercus elapoides salvadorensis* Mertens, 1952a, *Pliocercus andrewsi pacificus* H.M. Smith & Chrapliwy, 1957, *Pliocercus elapoides hobartsmithi* Liner, 1960, *Pliocercus elapoides occidentalis* H.M. Smith & Landy, 1965, *Pliocercus psychoides* H.M. Smith & Chiszar, 1996, and *Pliocercus wilmarai* Smith, Pérez-Higareda & Chiszar *in* H.M. Smith & Chiszar, 1996.

Types: Syntypes (4), ANSP 3810–13, longest syntype 502 mm (R.M. de Oca).

Type locality: "near Jalapa, Mexico" [= vicinity of Jalapa, cen. Veracruz State, E Mexico, 19°33'N, 96°55'W, elevation 1455 m].

Distribution: Southeastern Mexico and upper Central America. Southeastern Mexico (Campeche, Chiapas, Hidalgo, Oaxaca, Puebla, Quintana Roo, San Luis Potosí, Tabasco, S Tamaulipas, Veracruz, Yucatán), Belize (Cayo, Stann Creek, Toledo), Guatemala (Alta Verapaz, Baja Verapaz, Chimaltenango, Escuintla, Huehuetenango, Petén, Quetzaltenango, Retalhuleu, Santa Marcos, Santa Rosa, Suchitepequez), W Honduras (Colón, Comayagua, Copán, Cortés, Francisco Morazán, La Paz, Ocotepeque, Olancho, Yoro) and El Salvador (Chalatenango, La Libertad, Morazán, San Vicente, Sonsonate, Usulután), NSL–1980 m.

Sources: H.M. Smith, 1941b, 1942h, Mertens, 1952a, L.C. Stuart, 1963, H.W. Greene, 1969, R.W. Henderson & Hoevers, 1975, J.C. Lee, 1980, 1996, H.M. Smith et al., 1989, J.A. Campbell, 1998b, H.M. Smith & Chiszar, 2001a, 2001c–d, 2001f, G. Köhler et al., 2005, McCranie et al., 2006, Ramírez-Bautista et al., 2010 and McCranie, 2011a.

Remarks: *Pliocercus aequalis* and *P. bicolor* valid species *fide* Pérez-Higareda & Smith, 1986. H.M. Smith et al. (1995: 208) discussed the date of publication of *Elapochrus deppei*.

2. *Pliocercus euryzonus* Cope, 1862b. Proc. Acad. Nat. Sci. Philadelphia 14(1): 72–73.

Synonyms: *Liophis splendens* Jan, 1863a, *Pliocercus dimidiatus* Cope, 1865, *Pliocerus rurizona* – Briceño-Rossi, 1934 *(nomen incorrectum), Urotheca aureorostris* Briceño-Rossi, 1934 *(nomen nudum), Pliocercus annellatus* E.H. Taylor, 1951, *Pliocercus arubricus* E.H. Taylor, 1954, and *Pliocercus euryzonus burghardti* H.M. Smith & Chiszar, 1996.

Type: Holotype, USNM 4303 (formerly SIM 4303), a 603 mm specimen (N. Michler [Darien Surv. Exped.], 1857–1860), lost *fide* J.A. Peters (1960a: 535).

Type locality: "Region of the Truando, New Grenada" [= vicinity of Río Truandó, N Chocó Dept., W Colombia].

Distribution: Lower Central America and NW South America. Northeastern Honduras (Gracias a Dios), E Nicaragua (Atlántico Norte, Atlántico Sur, Matagalpa, Río San Juan), Costa Rica (Alajuela, Cartago, Guanacaste, Heredia: Bonita Is.; Limón, Puntarenas, San José), Panama (Bocas del Toro, Canal Zone, Chiriquí, Coclé, Colón, Darién, San Blas, Veraguas, Barro Colorado Is.), W Colombia (Antioquia, Boyaca, Caldas, Cauca, Chocó, Cundinamarca, Huila, Meta, Nariño, Risaralda, Santander, Valle del Cauca) and NW Ecuador (Carchi, Cotopaxi, Esmeraldas, Imbabura, Los Rios, Pichincha), NSL–2750 m.

Sources: E.H. Taylor, 1954, N.J. Scott, 1969, Pérez-Santos, 1986, Pérez-Santos & Moreno, 1988, 1991, H.M. Smith & Chiszar, 2001e, 2001g, Savage, 2002, McCranie et al., 2006 and McCranie, 2011a.

Remarks: Record from Brazil rejected *fide* Savage & Crother, 1989.

POECILOPHOLIS Boulenger, 1903b (Atractaspididae)

Synonyms: *Poicilopholis* – Hughes, 1983 *(nomen incorrectum)* and *Poecilipholis* – Knoepffler, 1968 *(nomen incorrectum).*

Type species: *Poecilopholis cameronensis* Boulenger, 1903b.

Distribution: West Africa.

Sources: Perret, 1961, Girardin, 1978, Wallach, 1991, Underwood & Kochva, 1993, Chippaux, 2006, Chirio & LeBreton, 2007 and Vidal et al., 2008.

1. *Poecilopholis cameronensis* Boulenger, 1903b. Ann. Mag. Nat. Hist. (7) 12(69): 352–353.

Type: Holotype, BMNH 1946.1.1.69, a 520 mm specimen (G.L. Bates).

Type locality: "Efulen, S. Cameroon" [= Efoulan, W Sud Prov., ext. SW Cameroon, 2°56'N, 11°05'E, elevation 660 m].

Distribution: Southwestern Cameroon (Sud), 420–1060 m.

POLEMON Jan, 1858b (Atractaspididae)

Synonyms: *Microsoma* Jan, 1858, *Miodon* A.H.A. Duméril, 1859, *Urobelus* Reinhardt, 1861b, *Mijodon* – Hoffmann, 1890 *(nomen incorrectum), Myodon* – Hoffmann, 1890 *(nomen incorrectum), Palaemon* – Bocage, 1895 *(nomen incorrectum), Microsona* – Canton, 1895 *(nomen incorrectum), Cynodontophis* F. Werner, 1902, *Mioden* – Sternfeld, 1908e *(nomen incorrectum), Elapocalamus* Boulenger, 1911b, *Melanocalamus* Witte, 1941, *Myodon* – Lema, 1978 *(nomen incorrectum),* and *Cynodantophis* – Phelps, 1981 *(nomen incorrectum).*

Type species: *Polemon barthii* Jan, 1858b.

Distribution: West, cen. and E Africa.

Sources: Witte & Laurent, 1943, 1947, Loveridge, 1944b, Witte, 1962, Bourgeois, 1968, Hughes, 1983, Wallach,

1991, Underwood and Kochva, 1993, Nagy et al., 2005, Vidal et al., 2008, Kelly et al., 2009 and Zaher et al., 2009.

1. *Polemon acanthias* (Reinhardt, 186lb). Vidensk. Medd. Naturhist. Foren. Kjøbenhavn (1860) 2: 229–238, pl. 3, figs. 1–4. (*Urobelus acanthias*)

Synonym: *Calamaria acanthias* Kröyer, 1853 *(nomen nudum).*
Type: Lectotype, ZMUC 63128, a 315 mm specimen (Schönning, 1850), designated by F.W. Braestrup *in* Hughes & Barry (1969: 1024).
Type locality: "Guinea" [= coast of Ghana *fide* Hughes & Barry, 1969: 1024] via lectotype selection.
Distribution: West Africa. Southern Guinea (Macenta, Nzérékoré), Sierra Leone, Liberia (Bong, Montserrado), S Ivory Coast (Abidjan, Daloa, Danane, Dix-Huit Montagnes, Guiglo, Man, San Pédro, Tabou), S Ghana (Central, Eastern), S Togo (Plateaux) and S Nigeria (Ondo), NSL–1200 m.
Sources: Aylmer, 1922, Angel, 1933b, Villiers, 1950a, 1954a, Doucet, 1963, L.R. Cole, 1967, Condamin, 1959, Doucet, 1963, Leston & Hughes, 1968, Hughes & Barry, 1969, Leston, 1970, Hughes, 1983, Roman, 1984, Chippaux, 2006 and Segniagbeto et al., 2011.
Remarks: Reinhardt attributed authorship to Kröyer, recognized by A.C.L.G. Günther (1888c: 323).

2. *Polemon barthii* Jan, 1858b. Rev. Mag. Zool. (2) 10: 516, 520.

Type: Holotype, ZSM 2167/0, a 490–510 mm female (H. Barth, 1857–1859), destroyed 24–25 April 1944 during World War II.
Type locality: "Guinée" [= coast of Ghana *fide* Hughes & Barry, 1969: 1024].
Distribution: West Africa. Liberia (Bong), Ivory Coast (Guiglo, Tabou), S Ghana (Eastern) and SW Cameroon.
Sources: L. Müller, 1910, Angel, 1933b, Doucet, 1963, Leston & Hughes, 1968, Hughes & Barry, 1969, Villiers, 1975, Hughes, 1983, 2003, Roman, 1984, Villiers & Condamin, 2005 and Chippaux, 2006.
Remarks: Supplemental original description in Jan (1859c: pl. 5, figs. 1–8). Photograph of the holotype in Franzen & Glaw (2007: fig. 15). Probably occurs in Togo *fide* and Segniagbeto et al. (2011: 353).

3. *Polemon bocourti* Mocquard, 1897a. Bull. Mus. Hist. Nat., Paris 3(1): 54.

Synonym: *Aparallactus hagmanni* Gough, 1902.
Types: Syntypes (2), MNHN 1896.553–54, 540 mm and 350 mm specimens (E. Haug, 1895–1897).
Type locality: "les environs de Lambaréné, sur le bas Ogooué, Gabon" [= vicinity of Lambaréné, Moyen-Ogooúe Region, W Gabon, 0°42'S, 10°13'E, elevation 25 m].
Distribution: Central Africa. Southeastern Guinea, S Cameroon (Centre, Nord-Ouest, Sud, Sud-Ouest), Equatorial Guinea, Gabon (Moyen-Ogooué), Congo (Brazzaville, Cuvette) and W Democratic Republic of the Congo (Bas-Congo, Bandundu, Equateur, Kasai Occidental, Kasai-Oriental, Kinshasa), 25–1200 m.
Sources: L. Müller, 1910, Witte, 1933a, 1962, Witte & Laurent, 1943, Doucet, 1963, Thys van den Audenaerde, 1965, Villiers, 1966, 1975, J.-F. Trape & Roux-Estève, 1995, Hughes, 2003, Villiers & Condamin, 2005, Chippaux, 2006, Chirio & LeBreton, 2007 and Pauwels & Vande weghe, 2008.
Remarks: Original description reprinted in Bocourt (1897d: 13–14). Photograph of syntype in Pauwels & Vande weghe (2008: fig. 284).

4. *Polemon christyi* (Boulenger, 1903b). Ann. Mag. Nat. Hist. (7) 12(69): 354. (*Miodon christyi*)

Synonyms: *Miodon unicolor* K.P. Schmidt, 1923 and *Melanocalamus leopoldi* Witte, 1941.
Type: Holotype, BMNH 1946.1.8.88, a 430 mm female (C. Christy).
Type locality: "Uganda."
Distribution: Eastern Africa. Southeastern South Sudan (Eastern Equatoria), E Democratic Republic of the Congo (Katanga, Nord-Kivu, Orientale, Sud-Kivu), Uganda (Central, Eastern, Western), SW Kenya (Western), Rwanda (Bwmba, Kibungo, Ruhengeri), Burundi (Bujumbura, Ngozi), SW Tanzania (Rukwa) and N Zambia (Copperbelt, N Northern, North-Western), NSL–1700 m.
Sources: Witte, 1941, 1953, 1962, 1975, Broadley, 1971d, Pitman, 1974, Spawls, 1978, Joger, 1990, Broadley & Howell, 1991, Spawls et al., 2002 and Chirio, 2009.
Remarks: Central African Republic record doubtful *fide* Chirio & Ineich (2006: 58).

5. *Polemon collaris* (W.C.H. Peters, 1881e). Sitz. Ges. Naturf. Freunde Berlin 1881(9): 148–149. (*Microsoma collare*)

Synonyms: *Miodon collaris brevior* Witte & Laurent, 1947 and *Miodon collaris longior* Witte & Laurent, 1947.
Type: Holotype, ZMB 10045 (M.-F.W.A. von Mechow, 1873–1881).
Type locality: "Macange, Cuango, West-Afrika" [= Malanje, Malanje Prov., N Angola, 9°33'S, 16°20'E, elevation 1125 m].
Distribution: Central Africa. Southeastern Nigeria (Cross River), S Cameroon (Adamoua, Centre, Est, Littoral, Nord-Ouest, Sud, Sud-Ouest), Equatorial Guinea (Bioko Is.), Gabon (Haut-Ogooué, Moyen-Ogooué, Ogooué-Ivindo), SW Central African Republic (Haute Sangha, Lobaye, Ombella-Mpoko, Ouaka, Sangha), S Congo (Brazzaville, Kouilou, Lekoumou, Plateau), Democratic Republic of the Congo (Bandundu,

Bas-Congo, Equateur, Kasai Occidental, Kasai Oriental, Katanga, Kinshasa, Nord-Kivu, Orientale, Sud-Kivu), SW Uganda (W Western) and N Angola (Benguela, Bié, Cuanza, Huambo, Malanje, Uigé), NSL–2000 m.

Sources: Laurent, 1956a, Perret & Mertens, 1957**b**, Witte, 1962, 1975, Bourgeois, 1963c, Mertens, 1965c, Knoepffler, 1966, Roux-Estéve, 1965, Thys van den Audenaerde, 1965, Villiers, 1966, Pitman, 1974, J.A. Butler & Reid, 1986, Joger, 1990, J.-F. Trape & Roux-Estève, 1995, Spawls et al., 2002, Chippaux, 2006, Chirio & Ineich, 2006, Chirio & LeBreton, 2007 and Pauwels & Vande weghe, 2008.

Remarks: Type locality erroneously listed as Kwango, French Equatorial Africa [= Congo] *fide* Loveridge (1944b: 175) and Mertens (1964a: 231).

6. *Polemon fulvicollis* (Mocquard, 1887c). Bull. Soc. Philom. Paris (1886–1887) (7) 11(1): 65–66. (*Microsoma fulvicollis*)

Synonyms: *Miodon gracilis* Witte & Laurent, 1943 (*nomen praeoccupatum*) and *Polemon fulvicollis laurenti* Resetar & Marx, 1981 *(nomen substitutum)*.

Type: Holotype, MNHN 1886.211, a 200 mm specimen (P.P.F.C. Savorgnan de Brazza [Brazza Mission Scientifique du Congo], Nov. 1875–June 1882).

Type locality: "Franceville, Congo" [= Franceville, Haut-Ogooué Region, E Gabon, 1°38'S, 13°35'E, elevation 400 m].

Distribution: Central Africa. Southern Cameroon (Centre, Est, Sud), Congo (Lekoumou, Sangha), E Gabon (Haut-Ogooué, Moyen-Ogooué), N Democratic Republic of the Congo (Equateur, Kivu, Orientale) and W Uganda, 350–860 m.

Sources: Villiers, 1966, J.-F. Trape & Roux-Estève, 1990, 1995, Spawls et al., 2002, Chippaux, 2006 Chirio & LeBreton, 2007 and Pauwels & Vande weghe, 2008.

7. *Polemon gabonensis* (A.H.A. Duméril, 1856). Rev. Mag. Zool. (2) 8: 468–470. (*Elapocalamus gabonensis*)

Synonyms: *Elapomorphus caecutiens* A.C.L.G. Günther, 1888c, *Miodon gabonensis schmidti* Witte & Laurent, 1947, and *Miodon gabonensis brachyurus* Laurent, 1960.

Type: Holotype, MNHN 3674, a 550 mm specimen (C.E. Aubry-Lecomte, 1853–1856).

Type locality: "côte du Gabon, l'Afrique occidentale" [= coast of Gabon, West Africa] (in error *fide* Loveridge (1944a: 173). Corrected to Congo *fide* Loveridge (1944a: 173).

Distribution: West and cen. Africa. Southeastern Ghana (Accra), S Togo (Maritime), SE Nigeria (Cross River), S Cameroon (Adamaoua, Centre, Est, Littoral, Nord-Ouest, Ouest, Sud, Sud-Ouest), SW Central African Republic (Haute-Sangha, Lobaye, Ombella-Mpoko), Equatorial Guinea, Gabon, Congo, Democratic Republic of the Congo (Bandundu, Bas-Congo, Equateur, Nord-Kivu, Orientale, Sud-Kivu, Idjwi Is.) and W Uganda (W Western), 50–1450 m.

Sources: A.H.A. Duméril, 1859, L. Müller, 1910, Angel, 1933b, Laurent, 1953, 1956a, Witte, 1962, 1975, Doucet, 1963, Roux-Estève, 1965, Thys van den Audenaerde, 1965, Hughes, 1983, 2013, Joger, 1990, Spawls et al., 2002, Chippaux, 2006 and Chirio & LeBreton, 2007.

Remarks: Possibly occurs in N Rwanda *fide* Spawls et al. (2002: 429).

8. *Polemon gracilis* (Boulenger, 1911b). Ann. Mag. Nat. Hist. (8) 8(45): 371. (*Elapocalamus gracilis*)

Type: Holotype, BMNH 1946.1.8.86, a 285 mm female (G.L.Bates, 1895–1911).

Type locality: "Bitye, South Cameroon" [= Bityé, E Sud Dept., S Cameroon, 3°01'N, 12°21'E, elevation 650 m].

Distribution: Central Africa. Nigeria, SW Cameroon (Sud, Sud-Ouest), Gabon and Congo, 500–800 m.

Sources: Perret, 1960, Resetar & Marx, 1981 and Chirio & LeBreton, 2007.

9. *Polemon graueri* (Sternfeld, 1908a). Sitz. Ges. Natur. Freunde Berlin 1908: 94. (*Miodon graueri*)

Type: Holotype, ZMB, a 270 mm male (R. Grauer, 1905–1908).

Type locality: "Entetbe (Uganda)" [= Entebbe, Wasiki Distr., West Mengo Prov., cen. Uganda, 00°03'N, 32°28'E, elevation 1195 m].

Distribution: Central Africa. Eastern Democratic Republic of Congo (Nord-Kivu, Orientale, Sud-Kivu, Idjwi Is.), Uganda (Central, Eastern, Western) and W Rwanda, 800–1830 m.

Sources: Pittman, 1974, Witte, 1975 and Spawls et al., 2002.

Remarks: Previously a subspecies of *P. fulvicollis*.

10. *Polemon griseiceps* (Laurent, 1947). Bull. Mus. Roy. Hist. Nat. Belg. 23(16): 11–12. (*Miodon griseiceps*)

Type: Holotype, BMNH 1946.1.21.90 (formerly BMNH 1913.10.29.11), a 502 mm male (G.L. Bates, 1895–1931).

Type locality: "Bitye" [= Bityé, E Sud Dept., S Cameroon, 3°01'N, 12°21'E, elevation 650 m].

Distribution: West Africa. Southern Cameroon (Centre, Sud), SW Central African Republic and Congo (Brazzaville, Cuvette-Ouest, Sangha), 500–1240 m.

Sources: Villiers, 1966, Joger, 1990, J.B. Rasmussen, 1991, J.-F. Trape & Roux-Estève, 1995, Chippaux, 2006 and Chirio & LeBreton, 2007.

Remarks: Welch (1982: 143) listed Central African Republic. Chirio & Ineich (2006: 58) considered the Central African Republic record unlikely.

11. *Polemon neuwiedi* (Jan, 1858b). Rev. Mag. Zool. (2) 10: 516, 519–520. (*Microsoma neuwiedi*)

Synonym: *Cynodontophis werneri* L. Müller, 1910.
Type: Holotype, USTL, a 172 mm specimen (A.M.P. zu Wied-Neuwied via A. Westphal-Castelnau), destroyed in 1944 during World War II.
Type locality: "côtes de Guinée (Christiansbourg), " [= Christian Borg, S Accra Region, SE Ghana, 5°33'N, 0°11'W, elevation 10 m].
Distribution: West Africa. Southern Mali (Sikasso), SW Burkina Faso (Hauts-Bassins), NE Ivory Coast (Bouna), S Ghana (Accra, Western), S Togo (Maritime), S Benin (Collines) and Nigeria, 10–200 m.
Sources: Angel, 1933, Doucet, 1963, Leston & Hughes, 1968, Hughes & Barry, 1969, Hughes, 1983, 2013, Roman, 1984, Rödel et al., 1995, Chippaux, 2006, J.-F. Trape & Mané, 2006b and Segniagbeto et al., 2011.
Remarks: Supplemental original description in Jan (1859c: pl. 4, figs. 1–10).

12. *Polemon notatus* (W.C.H. Peters, 1882c). Sitz. Ges. Naturf. Freunde Berlin 1882(8): 127–128. (*Microsoma notatum*)

Synonym: *Cynodontophis aemulans* F. Werner, 1902.
Type: Holotype, ZMB 10271.
Type locality: Unknown. Designated as Cameroon *fide* Chirio & LeBreton (2007: 654).
Distribution: West Africa. Southern Cameroon (Centre, Est, Littoral, Sud, Sud-Ouest), W Gabon (Ogooué-Maritime), S Congo (Brazzaville, Kouilou), SW Central African Republic (Haute-Sangha) and W Democratic Republic of the Congo (Bandundu, Bas-Congo, Equateur, Kinshasa), 10–1300 m.
Sources: L. Müller, 1910, Perret & Mertens, 1957b, Witte, 1962, Villiers, 1966, J.-F. Trape & Roux-Estève, 1995, Chippaux, 2006, Chirio & Ineich, 2006, Chirio & LeBreton, 2007 and Pauwels & Vande weghe, 2008.

13. *Polemon robustus* (Witte & Laurent, 1943). Rev. Zool. Bot. Afr. 37(1–2): 167. (*Miodon robustus*)

Type: Holotype, MRAC 9829, a 645 mm female (H. Schouteden, 1934).
Type locality: "Bolobo (Congo belge: Distr. du Lac Léopold II), Congo belge" [= Bolobo, ext. NW Bandundu Region, ext. W Democratic Republic of the Congo, 2°10'S, 16°14'E, elevation 315 m].
Distribution: Southwestern Central African Republic (Haute-Sangha, Lobaye) and W Democratic Republic of the Congo (Bandundu, Bas-Congo, Equateur, Kinshasa), 315–450 m.
Sources: Witte, 1962, Bourgeois, 1963c, Thys van den Audenaerde, 1965, Chippaux, 2006, Chirio & Ineich, 2006 and Chirio & LeBreton, 2007.
Remarks: Probably occurs in E Cameroon *fide* Chirio & LeBreton (2007: 656).

†*POLLACKOPHIS* Holman, 1998b (Colubridae)

Type species: †*Pollackophis depressus* Holman, 1998b.
Distribution: Lower Miocene of USA.
Sources: Holman, 1998b, 2000a.

1. †*Pollackophis depressus* Holman, 1998b. Delaware Geol. Surv., Spec. Publ. (21): 145, figs. 2a–e.

Type: Holotype, USNM 483397, one trunk vertebra.
Type locality: "From the Pollack Farm Site, Delaware Geological Survey Site Id11-a (39°14'08"N, 75°34'36"W), near Cheswold, Kent County, Delaware. Creswold sands of the Calvert Formation; lower Mioclene, Hemingfordian Age."
Distribution: Lower Miocene (Hemingfordian: 16.3–20.6 mya) of USA (Delaware).

POLYODONTOGNATHUS Wall, 1921g (Elapidae)

Synonyms: *Polyodontognatus* – Bourret, 1934g *(nomen incorrectum)* and *Polydontognathus* – Schulze, Kükenthal, Hediger & Hesse, 1935 *in* 1932–1935 *(nomen incorrectum).*
Type species: *Hydrus caerulescens* G. Shaw, 1802.
Distribution: Marine waters of Indo-Australia, including Arabian Sea, Bay of Bengal, Bight of Bangkok, Gulf of Carpentaria, Gulf of Martaban, Gulf of Thailand, N Indian Ocean, Java Sea, Morrison Bay, Sea of Singgora, Strait of Malacca and Yellow Sea.
Sources: Kharin, 2004a, N. Khaire, 2006, Zhao, 2006, Wells, 2007, V.S. Nguyen et al., 2009 and Hoser, 2012e.

1. *Polyodontognathus caerulescens* (G. Shaw, 1802). Gen. Zool., Amph. 3(2): 561–562. (*Hydrus caerulescens*)

Synonyms: *Hydrophis hybrida* Schlegel, 1844 *in* 1837–1844, ? *Hydrophis colubrinus* Jerdon, 1854, *Hydrophis protervus* Jan, 1859b, *Hydrophis wertmani* Jan, 1859a, *Hydrophis frontalis* Jan, 1863b, *Hydrophis polydonta* Jan, 1863b (*nomen nudum*), *Hydrophis polydontus* Jan, 1872 *in* Jan & Sordelli, 1870–1881, *Hydrophis caerulescens thai* M.A. Smith, 1920, *Hydrophis coerulescens* – Boquet, 1964 *(nomen incorrectum), Hydrophis caerulscens* – Deraniyagala, 1977 *(nomen incorrectum),* and *Hydrophis coerelescens* – M.W. Lin, 1975 *(nomen incorrectum).*
Type: Holotype, BMNH 1946.1.3.90 (formerly BMNH III.6.13.a), a 610 mm male (P. Russell, 1781–1791).
Type locality: "East-India." Emended to Vizagapatam [= Vishakhapatnam, Andhra Pradesh State, SE India, 17°41'N, 83°13'E, elevation 25 m] *fide* BMNH label, Wall (1921g: 379) and Cogger et al. (1983: 250).

P

Distribution: Indo-Australia. Pakistan (SW Sindh), India (Andhra Pradesh, Karnataka, Kerala, Maharashtra, Orissa, Tamil Nadu, West Bengal), Sri Lanka, Bangladesh, Myanmar (Mergui Arch., Tanintharyi), Thailand (Chanthaburi, Pattani, Phet Buri, Samut Sakhon, Samut Songkhram, Phuket Is.), Cambodia (Koh Kong), Vietnam (coastal regions of central and S), China (Guangdong, Shandong), Taiwan, West Malaysia (Johore, Kelantan, Pahang, Pinang, Perak, Selangor, Terengganu, Angsa Is.), Singapore, East Malaysia (Sabah, Sarawak), Indonesia (N Java) and N Australia (Northern Territory, Queensland).

Sources: Wall, 1921g, Bourret, 1934g, 1936b, E.H. Taylor, 1965, Minton, 1966, Tian et al., 1986, M.A.R. Khan, 1988, Stuebing, 1991 and M.S. Khan, 2000.

POPEIA Malhotra & Thorpe, 2004c (Viperidae)

Type species: *Trimeresurus popeorum* M.A. Smith, 1937.

Distribution: Southeast Asia and Indonesia.

Sources: Maslin, 1942, M.A Smith, 1943, Maslin, 1945, Klemmer, 1963, Leviton, 1964d, W. Burger, 1971, Kramer, 1977, Hoge & Romano-Hoge, 1981a, Malhotra & Davis, 1991, Golay et al., 1993, Kraus et al., 1996, David & Vogel, 1998, David & Ineich, 1999, McDiarmid et al., 1999, Parkinson, 1999, Malhotra & Thorpe, 1997, 2000, 2004c, Ziegler et al., 2001, Giannasi et al., 2001a, G. Peng & Fuji, 2001, David et al., 2002, 2009, 2011, Gumprecht et al., 2004c, Castoe & Parkinson, 2006, Grismer et al., 2006a, 2008b, Sanders et al., 2006, Dawson et al., 2008, Ganesh et al., 2008, Orlov et al., 2008, 2009, I. Das, 2010, 2012, David et al., 2011c, P. Guo & Wang, 2011, Sumontha et al., 2011 and Hoser, 2012d, o.

P

1. *Popeia barati* (Regenass & Kramer, 1981). Rev. Suisse Zool. 88(1): 189–190, figs. 5–8. (*Trimeresurus popeorum barati*)

Type: Holotype, NMBA 2587, a 505 mm male (M. Müller, 1885).

Type locality: "Solok, Sumatra" [= Solok Regency, Sumatera Barat Prov., W-cen. Sumatra, W Indonesia].

Distribution: Western Indonesia (Sumatra, Mentawai Arch.: Pagai), 390–1600 m.

Sources: David & Gernot, 1996, Vogel et al., 2004 and David et al., 2009.

2. *Popeia buniana* Grismer, Grismer & McGuire, 2006a. Zootaxa (1305): 9–13, fig. 2.

Type: Holotype, ZRC 2.6176, a 699 mm male (J.A. McGuire & J.L. Grismer, 21 July 2002).

Type locality: "Tekek-Juara Trail at 2° 49' 18.0' N x 104° 10' 28.5"E at 295 m, Pulau Tioman, Pahang State, West Malaysia" [= Tekek-Juara trail, Tioman Is., E Pahang State, SE West Malaysia, 2°49'N, 104°10'E, elevation 295 m].

Distribution: Southeastern West Malaysia (Pahang: Tioman Is.), 240–810 m.

Sources: Grismer, 2005, 2011 and Grismer and Khang Aun, 2008.

3. *Popeia fucatus* (Vogel, David & Pauwels, 2004). Zootaxa (727): 24–30, figs. 6–9, 13. (*Trimeresurus fucatus*)

Type: Holotype, MNHN 1990.4283, a 704 mm male.

Type locality: "Province of Nakhon Si Thammarat, Thailand."

Distribution: Myanmar (Taninthayi, Kanmaw Is.), Thailand (Chumphon, Krabi, Nakhon Si Thammarat, Phang Nga, Prachuap Khiri Khan, Surat Thani, Trang) and West Malaysia (Johor, Kedah, Pahang, Perak, Pinang, Selangor, Terengganu, Langkawi and Penang Is.), 400–1280 m.

Sources: I. Das, 2012, 2012.

4. *Popeia nebularis* (Vogel, David & Pauwels, 2004). Zootaxa (727): 38–45, figs. 14–17. (*Trimeresurus nebularis*)

Synonym: *Popeia inornata* Sanders, Malhotra, Gumprecht, Thorpe & Kuch, 2004.

Type: Holotype, USNM 142425, an 881 mm female (H. Baker, 17 Oct. 1959).

Type locality: "Gunung Brinchang [now Gunung Batu Berinchang], Cameron Highlands, State of Pahang, West Malaysia."

Distribution: West Malaysia (Pahang), 1370–1765 m.

Sources: Sanders et al., 2004 and David et al., 2009.

5. *Popeia phuketensis* Sumonthal, Kunya, Pauwels, Nitikul & Punnadee, 2011. Russ. J. Herp. 18(3): 186–189, figs. 1–4.

Type: Holotype, THNHM 15905, a 495 (svl) mm male (K. Kunya, 5 Oct. 2009).

Type locality: "Ban Bangrong, Thalang District, Phuket Island, Phuket Province, southwestern Thailand."

Distribution: Thailand (Phuket: Phuket Is.). Known only from type locality.

6. *Popeia popeorum* (M.A. Smith, 1937). J. Bombay Nat. Hist. Soc. 39(4): 730. (*Trimeresurus popeorum*) (*nomen corrigendum*)

Synonyms: *Coluber viridicaeruleus* Lacépède, 1789 (*nomen rejiciendum*), *Trimeresurus popeiorum* M.A. Smith, 1937 *(nomen incorrigendum)*, and *Trimeresurus popeorum* – M.A. Smith, 1943 *(nomen corrigendum)*.

Type: Lectotype, BMNH 1872.4.17.137, a 925 mm male (T.C. Jerdon, 1836–1852), designated by E.H. Taylor & Elbel (1958: 1174).

Type locality: " Khasi Hills, Assam" [= Khasi Hills, E Meghalaya State, NE India, 25°35'N, 91°38'E, elevation 1370–1675 m] via lectotype selection.

Distribution: Southeastern Asia. Northeastern India (Arunachal Pradesh, Assam, Meghalaya, Sikkim, West Bengal), Myanmar (Bago, Chin, Kayin, Mon, Taninthayi), Laos (Louangphrabang, Luang Prabang, Phongsali, Vientiane) and Thailand (Chiang Mai, Chiang Rai, Lampang, Loei, Muang Pan, Phang Nga, Tak), NSL–2100 m.

Sources: C.H. Pope & Pope, 1933, E.H. Taylor & Elbel, 1958, E.H. Taylor, 1965, Regenass & Kramer, 1981, Tweedie, 1983, G. Vogel, 1990, M.J. Cox, 1991b, David and Vogel, 1996, Malhotra & Thorpe, 1997, Orlov, 1997, M.J. Cox et al., 1998, Chan-ard et al., 1999, Orlov et al., 2000, Grismer et al., 2002b, Malkmus et al., 2002, Schleich & Kästle, 2002, Leviton et al., 2003 and Vogel et al., 2004.

Remarks: May occur in E Nepal but no specific records available (Schleich & Kästle, 2002). Records from Cambodia and Vietnam are different species *fide* Vogel et al. (2004: 24). Original orthography of *T. popeiorum* was a clerical error [corrected by the author on reprints] *fide* M.A. Smith (1943: 518) and following Art. 32.5.1.1 (ICZN, 1999: 40), the snake named after Mr. and Mrs. Pope is correctly *Pope + orum*.

7. *Popeia sabahi* (Regenass & Kramer, 1981). Rev. Suisse Zool. 89(1): 190–192, figs. 10–12. (*Trimeresurus popeorum sabahi*)

Type: Holotype, MCZ 43608, a 682 mm male (J.A. Griswold, 8 June 1937).

Type locality: "Mount Kinabulu, Kulapis River, British North Borneo."

Distribution: East Malaysia (Sabah, Sarawak) and Brunei, 200–1500 m

Sources: Vogel et al., 2004, I. Das, 2007b and David et al., 2009.

8. *Popeia toba* (David, Petri, Vogel & Doria, 2009). Ann. Mus. Civ. Stor. Nat. "Giacomo Doria," Genova 100: 325–331, figs. 1–7. (*Trimeresurus toba*)

Type: Holotype, MSNG 30988, a 798 mm female (E. Modigliani, 1891).

Type locality: " 'Si Rambie', now Sirambi, about 13 km SE of Balige, off the southern shore of Danau Toba (Indonesia, Sumatra Island, Province of Sumatera Utara)."

Distribution: Western Indonesia (N Sumatra), 300 m.

PORTHIDIUM Cope, 1871b (Viperidae)

Synonyms: *Thanatos* Posada Arango, 1889a, *Thanatophis* Posada-Arango, 1889b (*nomen substitutum*), and *Tejeraia* Sandner-Montilla, 1989

Type species: *Bothrops nasutus* Bocourt, 1868.

Distribution: Latin America.

Sources: H.M. Smith, 1941i, H.M. Smith & Taylor, 1945, J.A. Peters & Orejas-Miranda, 1970, Hoge, 1966a, Burger, 1971, Hoge & Romano-Hoge, 1971, 1981a, J.A. Campbell & Lamar, 1989, 2004, Maes, 1989, Kraus et al., 1996, Werman, 1997, Gutberlet, 1998, David & Ineich, 1999, McDiarmid et al., 1999, Parkinson, 1999, Gutberlet & Harvey, 2002, Castoe et al., 2005, Castoe & Parkinson, 2006 and Hoser, 2012d.

1. *Porthidium arcosae* Schätti & Kramer, 1993. Rev. Suisse Zool. 100(2): 264–266, fig. 4. (*Porthidium lansbergii arcosae*)

Synonym: ? *Bothrops boussingaultii* Jan, 1863b (*nomen nudum*).

Type: Holotype, QCAZ 1325, a 635 mm female (E. Ventura, Aug. 1989).

Type locality: "Agua Blanca, ca. 15km NW Salango (1°37'S x 80°48'W), Manabi, ca. 100 m ü. M., Ecuador."

Distribution: Western Ecuador (Manabi), 100 m.

Source: Coloma et al., 2000.

2. *Porthidium dunni* (Hartweg & Oliver, 1938). Occ. Pap. Mus. Zool., Univ. Michigan (390): 6, pl. 1, figs. b, d. (*Trimeresurus dunni*)

Type: Holotype, UMMZ 82732, a 404 mm female (N. Hartweg & J.A. Oliver, 7 July 1936).

Type locality: "immediate vicinity of the village of Tehuantepec, the Pacific slope of the Isthmus of Tehuantepec, Oaxaca" [= vicinity of Tehuantepec (16°21'N, 95°16'W, elevation 60 m), SE Oaxaca State, S Mexico].

Distribution: Southern Mexico (ext. W Chiapas, S Oaxaca), NSL–700 m.

Sources: Amaral, 1944b, J.D. Johnson, 1974, J.A. Campbell, 1976, Casas-Andreu et al., 1996 and Bryson et al., 2008.

3. *Porthidium hespere* (J.A. Campbell, 1976). J. Herp. 10(3): 152–156, figs. 1, 3. (*Bothrops hesperis*)

Type: Holotype, UTA 4443, a 579 mm female (S. Aviña, 15 Oct. 1973).

Type locality: "west-facing slope in the foothills ca. 12 airline km NE of Tecomán, Municipio de Ixlahuacán, Colima, western Mexico. This site lies between the Río Armería drainage and the Río Salado (a tributary of the Río Coahuayana) at an elevation of approximately 300 m." Emended to 19.2 air-line km northeast of Tecomán *fide* J.A. Campbell & Lamar (1989: 316).

Distribution: Western Mexico (SW Colima, W Michoacán), NSL–610 m.

Sources: J.A. Campbell, 1976, and Alvarado-Díaz et al., 1997 and Bryson et al., 2008.

4. *Porthidium lansbergii* (Schlegel, 1841). Mag. Zool. Anat. Comp. Palaeont. 1841: 1–3, pl. 1, 4 figs. (*Trigonocephalus lansbergii*)

Synonoms: ? *Trigonocephalus savianus* Filippi, 1848, *Trigonocephalus landsbergii* – Jan, 1859b *(nomen incorrectum), Teleuraspis castelnaui brachystoma* Cope, 1860b, *Bothrops lansbergii venezuelensis* Roze, 1959c *(nomen praeoccupatum), Bothrops lansbergii rozei* J.A. Peters, 1968 *(nomen substitutum), Bothrops lansbergii janisrozei* – Hoge & Romano-Hoge, 1971 *(nomen incorrectum),* and *Bothrops lansbergi hutmanni* Sandner-Montilla, 1989.

Type: Holotype, RMNH 1556, a 361 mm specimen (R.F. van Lansberge).

Type locality: "environs de Turbaco, en Colombie" [= Turbaco, N Bolívar Dept., N Colombia, 10°20'N, 75°25'W, elevation 175 m].

Distribution: Panama and N South America. Eastern Panama (Coclé, Colón, Darién, Panamá), NW Colombia (Antioquia, Atlántico, Bolívar, Chocó, Córdoba, La Guajira, Magdalena, Norte de Santander, Santander, Tolima) and Venezuela (Anzoátegui, Bolívar, Carabobo, Delta Amacuro, Falcón, Guárico, Lara, Monagas, Nueva Esparta, Sucre, Trujillo, Zulia, Coche and Margarita Is.), NSL–1270 m.

Sources: Nicéforo-María, 1938, Amaral, 1944b, Roze, 1966a, L.D. Wilson & Meyer, 1985, Lancini-V., 1986, Pérez-Santos, 1986, Pérez-Santos & Moreno, 1988, Schätti & Kramer, 1993, Solórzano, 1994, Mijares-Urrutia & Arends, 2000, Manzanilla & Natera, 2003, Infante-Rivero, 2009, Navarrete et al., 2009 and Ugueto & Rivas, 2010.

Remarks: Type locality correction to Tumaco [= Tumaco, W Narino Prov., W Colombia, 1°48'N, 78°46'W, elevation NSL] *fide* Amaral (1929: 21) invalid.

P

5. *Porthidium nasutum* (Bocourt, 1868). Ann. Sci. Nat. (Zool.) (5) 10: 202. (*Bothrops nasutus*)

Synonyms: *Bothriopsis proboscideus* Cope, 1875a and *Thanatos sutus* Posada-Arango, 1889a.

Type: Holotype, MNHN 1592 (M.-F. Bocourt, 1865–1866).

Type locality: "Pansos, sur les bords du Polochic (Guatémala)" [= Panzos, on the Río Polochic, ext. E Alta Verapaz Dept., E Guatemala].

Distribution: Latin America. Southeastern Mexico (NW Chiapas, ? Tabasco, SE Veracruz), S Belize (Cayo), N Guatemala (Alta Verapaz, Izabal, Petén), NW Honduras (Atlántida, Colón, Cortés, Gracias a Dios, Olancho, Santa Bárbara, Yoro), E Nicaragua (Atlántico Norte, Atlántico Sur, Jinotega, Matagalpa, Río San Juan), Costa Rica (Alajuela, Cartago, Guanacaste, Heredia, Limón), Panama (Bocas del Toro, Canal Zone, Coclé, Colón, Darién, Limón, Panamá), W Colombia (Antioquia, Cauca, Chocó, Valle) and NW Ecuador (Esmeraldas, Manabi, Pichincha), NSL–900 [1500] m.

Sources: Nicéforo-María, 1938, Pifano & Römer, 1949b, E.H. Taylor, 1951, 1954, R.W. Henderson & Hoevers, 1975, J.C. Lee, 1980, 1996, 2000, Miyata, 1980, Porras et al., 1981, Villa, 1984, L.D. Wilson & McCranie, 1984, L.D. Wilson & Meyer, 1985, Pérez-Santos, 1986, Pérez-Santos & Moreno, 1988, 1991, Solórzano, 1994, J.A. Campbell, 1998b, Savage, 2002, Lamar & Sasa, 2003, Cisneros-Heredia, 2004b, McCranie et al., 2006, McCranie, 2011a and Travers et al., 2011.

Remarks: Colombian records from 1500–1900 m rejected *fide* Porras et al. (1981: 91).

6. *Porthidium ophryomegas* (Bocourt, 1868). Ann. Sci. Nat. (Zool.) (5) 10: 201–202. (*Bothrops ophryomegas*)

Synonym: *Trimeresurus lansbergii annectens* K.P. Schmidt, 1936d.

Type: Holotype, MNHN 1593 (M.-F. Bocourt, 1865–1867).

Type locality: "versant occidental de la Cordillère. Escuintla (Guatémala)" [= S slope of mountains at Escuintla, E Escuintla Dept., S Guatemala].

Distribution: Central America. South-central Guatemala (El Progreso, Retalhuleu, Santa Rosa, Zacapa), Honduras (Choluteca, Comayagua, El Paraíso, Francisco Morazán, Gracias a Dios, Olancho, Valle, Yoro), El Salvador (Chalatenango, La Unión, Santa Ana, Sonsonate), W Nicaragua (Carazo, Chinandega, Chontales, Granada, Managua, Matagalpa, Río San Juan, Rivas) and NW Costa Rica (Alajuela, Guanacaste, Puntarenas), NSL–1400 m.

Sources: Amaral, 1944b, E.H. Taylor, 1951, 1954, Mertens, 1952c, Villa, 1984, L.D. Wilson & Meyer, 1985, Solórzano, 1994, 2004, D. Lawson, 1997, G. Köhler, 1999b, Savage, 2002, Lamar & Sasa, 2003, McCranie et al., 2006, N. Herrera et al., 2007, Bryson, et al., 2008 and McCranie, 2011a.

Remarks: Holotype erroneously listed as MNHN 1539 *fide* McDiarmid et al. (1999: 470).

7. *Porthidium porrasi* Lamar & Sasa, 2003. Rev. Biol. Trop. 51(3–4): 799–800, figs 2a–b.

Type: Holotype, UTA 30829, a 306 mm male (10 July 1990).

Type locality: "Drake Bay, Península de Osa, Puntarenas Province, Costa Rica."

Distribution: Western Costa Rica (Puntarenas), NSL–400 m.

8. *Porthidium volcanicum* Solórzano, 1995. Rev. Biol. Trop. 42(3): 696–700, figs. 2–3.

Synonym: *Bothrops volcanica* Solórzano, 1987 *(nomen nudum).*

Type: Holotype, UCR 11642, a 259 mm male (natives, 7 Jan. 1988).

Type locality: "Ujarrás de Buenos Aires (Valle del General, suroeste de la provincia de Puntarenas, Costa Rica)."

Distribution: Southwestern Costa Rica (SW Puntarenas) and Panama (Chiriquí), 400–1000 m.
Sources: Solórzano, 1995, 2004, Savage, 2002 and Dwyer & Van Den Burgh, 2012.

9. *Porthidium yucatanicum* (H.M. Smith, 1941i). Zoologica 26(1): 62–63. (*Trimeresurus yucatanicus*)

Synonym: *Bothrops yucatannicus* Hoge & Romano-Hoge, 1981a *(nomen emendatum).*
Type: Holotype, USNM 46571, a 254 mm female (E.W. Nelson and E.A. Goldman, Feb. 1901).
Type locality: "Chichen Itza, Yucatán, Mexico" [= Chichen-itza, Yucatán State, SE Mexico, 20°41'N, 88°34'W, elevation 35 m].
Distribution: Yucatán Peninsula, SE Mexico (ext. N Campeche, N Quintana Roo, Yucatán), NSL–250 m.
Sources: J.C. Lee, 1980, 1996, 2000, C.J. McCoy & Censky, 1992 and McCranie & G. Köhler, 2001.
Remarks: Belize record is misidentified *Bothrops asper.*

†*POUITELLA* Rage, 1988b (†Lapparentophiidae)

Type species: †*Pouitella pervetus* Rage, 1988b.
Distribution: Upper Cretaceous of France.
Source: Rage, 1988b.

1. †*Pouitella pervetus* Rage, 1988b. C. R. Acad. Sci. Paris (2) 307B: 1029–1032, figs. a–f.

Type: Holotype, UPVI BRZ 1, one trunk vertebra (A.L. Pouit and D. Pouit).
Type locality: "Brézé (Maine-et-Loire, France); fin du Cénomanien inférieur ou Cénomanien moyen" [= Brézé, Maine-et-Loire Dept., NW France, early or middle Cenomanian, upper Cretaceous].
Distribution: Upper Cretaceous (Cenomanian: 93.9–100.5 mya) of France. Known only from holotype.

PROATHERIS Broadley, 1996b (Viperidae)

Type species: *Vipera superciliaris* W.C.H. Peters, 1854.
Distribution: Eastern Africa.
Sources: W.C.H. Peters, 1882, Kramer, 1961a, Stevens, 1973, V.F.M. FitzSimons, 1974, Broadley, 1983, 1996b, Morgan, 1988, Golay et al., 1993, Herrmann & Joger, 1995, 1997, Spawls & Branch, 1995, David & Ineich, 1999, Herrmann et al., 1999, McDiarmid et al., 1999, D.P. Lawson et al., 2001, Lenk et al., 2001b, Spawls et al., 2002, Mallow et al., 2003, Dobiey & Vogel, 2007, Phelps, 2010 and Hoser, 2012d.

1. *Proatheris superciliaris* (W.C.H. Peters, 1854). Mber. Königl. Akad. Wiss. Berlin 1854(11): 625. (*Vipera superciliaris*)

Type: Holotype, ZMB 4121, a 570 mm male (W.C.H. Peters, April 1847).
Type locality: "Terra Querimba" [Mozambique]. Emended to Festlande, der Cap Delgado-Insel Querimba [= Quissanga, mainland opposite Querimba Island, E Cabo Delgado Prov., NE Mozambique, 12°26'S, 40°30'E, elevation NSL] *fide* W.C.H. Peters (1882: 145).
Distribution: Eastern Africa. Southern Tanzania (SW Iringa, S Mbeya), S Malawi (Southern, Nchisi Is.) and N Mozambique (Cabo Delgado, Sofala, Tete, Zambézia), NSL–800 m.

†*PROCEROPHIS* Rage, Folie, Rana, Singh, Rose & Smith, 2008 (Colubroidea *incertae sedis*)

Type species: †*Procerophis sahnii* Rage, Folie, Rana, Singh, Rose & Smith, 2008.
Distribution: Lower Eocene of India.
Source: Rage et al., 2008.

1. †*Procerophis sahnii* Rage, Folie, Rana, Singh, Rose & Smith, 2008. Acta Palaeontol. Polon. 53(3): 397–399, fig. 4a.

Type: Holotype, GU/RSR/VAS 1014, one posterior trunk vertebra.
Type locality: Vastan Lignite Mine, northeast of Surat, Gujarat, western India; Early Eocene (middle to late Ypresian) continental beds of Cambay Formation."
Distribution: Lower Eocene (Ypresian: 48.6–55.8 mya) of India. Known only from type locality.

P

†*PROPTYCHOPHIS* Whistler & Wright, 1989 (Colubridae)

Type species: †*Proptychophis achoris* Whistler & Wright, 1989.
Distribution: Upper Miocene of USA.
Sources: Whistler & Wright, 1989 and Holman, 2000a.

1. †*Proptychophis achoris* Whistler & Wright, 1989. Herpetologica 45(3): 352–355, figs. 1a–c.

Type: Holotype, MVP 120208, one trunk vertebra (D.P. Whistler and previous field crews, late 1960s–1989).
Type locality: "badlands 1 km northwest of Red Rock Canyon State Park Headquarters, SW1/4, Sec. 27, T29S, R37E (Mt. Diablo Base and Meridian), northwestern Mojave Desert, eastern Kern County, California; lower part of Member 5, Dove Spring Formation;

Late Miocene, early late Clarendonian North America Mammal Age, 10.0-10.1 MYBP."
Distribution: Upper Miocene (Clarendonian: 10.3–13.6 mya) of USA (California). Known only from type locality.

PROSYMNA Gray, 1849a (Prosymnidae)

Synonyms: *Temnorhynchus* A. Smith, 1849, *Prosymma* – J.T. Reinhardt, 1861c *(nomen incorrectum), Ligonirostra* Cope, 1863c *(nomen substitutum), Asthenophis* Boulenger, 1896g, *Pseudoprosymna* Lindholm *in* Lampe, 1902, *Stenorhabdium* F. Werner, 1909d, *Astenophis* – Calabresi, 1927 *(nomen incorrectum),* and *Proxymna* – Briceño-Rossi, 1934 *(nomen incorrectum).*
Type species: *Calamaria meleagris* J.T. Reinhardt, 1843.
Distribution: Central, E and S Africa.
Sources: Chabanaud, 1917b, Loveridge, 1957a, 1958, Broadley, 1980, 1983, Branch, 1988, Vidal et al., 2008, Kelly et al., 2009 and Chirio et al., 2011.

1. *Prosymna ambigua* Bocage, 1873a. J. Sci. Math. Phys. Nat., Lisboa 4(15): 218. (*Prosymna ambiguus*)

Synonyms: *Prosymna bocagii* Boulenger, 1897b, *Prosymna meleagris concolor* Lönnberg, 1910, *Prosymna ambigua brevis* Laurent, 1954a, *Prosymna ambigua loveridgei* Laurent, 1954a, and *Prosymna ambigua urundiensis* Laurent, 1954a.
Type: Holotype, MBL 610, a 125 mm female (Bayão, 1863), destroyed by fire 18 March 1978.
Type locality: "Duque de Bragança, Angola, Afrique occidentale" [= Duque de Bragança, W Malanje Prov., N Angola, 09°06'S, 15°57'E, elevation 1110 m].
Distribution: Central Africa. Central Cameroon (Adamaoua, Centre, Est), Central African Republic (Kemo-Gribingui, Mbomou, Ombella-Mpoko, Ouaka, Ouham, Sangha), S Congo (Bouenza, Brazzaville, Kouilou, Niari, Pool), Democratic Republic of the Congo (Bandundu, Bas-Congo, Equateur, Kasai Occidental, Kasai Oriental, Katanga, Kinshasa, Nord-Kivu, Orientale, Sud-Kivu), NE Angola (Lunda Norte, Lunda Sul, Malanje), N Zambia (N Northern, North-Western), Malawi (Central, Northern, Southern), Zimbabwe (Bulawayo, Manicaland, Mashonaland Central, Mashonaland East, Mashonaland West, Masvingo, Matabeleland North, Matabeleland South, Midlands), SE South Sudan (Eastern Equatoria), N Uganda (N Eastern, Northern), SW Kenya (Nyanza), NW Tanzania (Mwanza, Kagera), S Burundi (Bururi), 15–1260 m.
Sources: Hewitt, 1910, K.P. Schmidt, 1923, Loveridge, 1933, 1942, Witte, 1933a, 1953, 1962, 1966, Monard, 1937, Uthmöller, 1937, 1941a–b, Scortecci, 1939b, Villiers, 1952b, 1966, Laurent, 1954a, 1960, Vesey-FitzGerald, 1958, Perret, 1961, V.F.M. FitzSimons, 1962a, 1974, Laurent, 1964a, Thys van den Audenaerde, 1965, Broadley, 1971d, 1980, Pitman, 1974, Haacke and Bruton, 1978, Lanza, 1983a, Pienaar et al., 1983, Joger, 1990, Boycott & Culverwell, 1992, J.-F. Trape & Roux-Estève, 1995, Chippaux, 1999, Spawls et al., 2002, Broadley et al., 2003, Chirio & Ineich, 2006 and Chirio & LeBreton, 2007.
Remarks: Loveridge (1958: 152) erroneously listed the holotype as 225 mm.

2. *Prosymna angolensis* Boulenger, 1915a. Proc. Zool. Soc. London 85(2): 208. (*nomen substitutum*)

Synonym: *Prosymna frontalis* Bocage, 1895 (*nomen praeoccupatum*).
Type: Neotype, NMW 19275b, a 182 mm male, designated by Broadley (1980: 497).
Type locality: "Caconda, Angola" [= Caconda, N Huíla Prov., SW Angola, 13°44'S, 15°04'E, elevation 1700 m] via neotype selection.
Distribution: Southwestern Africa. Western Angola (Benguela, Cuanza Sul, Huambo, Huíla, Luanda, Mocamedes, Namibe), W Zambia (W Western), NE Namibia (Caprivi, Okavango, Oshikoto), N Botswana (North West) and W Zimbabwe (Matabeleland North), 150–1020 m.
Sources: Mertens, 1937, 1955, Monard, 1937, Bogert, 1940, V.F.M. FitzSimons, 1962a, 1974, Broadley, 1971d, Auerbach, 1987, Broadley et al., 2003, M. Griffin, 2003 and Broadley & Blaylock, 2013.
Remarks: Original description in Bocage (1895: 98–99, pl. 11, fig. 2). Holotype destroyed by fire 18 March 1978 *fide* Broadley (1980: 497).

3. *Prosymna bivittata* F. Werner, 1903a. Abh. Bayer. Akad. Wiss. 22(2): 381. (*Prosymna sundevalli bivittata*)

Synonym: *Prosymna sundevallii macrospila* F. Werner, 1910.
Type: Holotype, ZSM 1890/0, a 296 mm female (Kuhn, 1902), lost *fide* EMBL database.
Type locality: "Deutsch-Südwestafrika" [= German South West Africa or Namibia]. Emended to Windhoek, south of the ODistribution River, Namibia [= Little Namaqualand *fide* Loveridge, 1958: 136] *fide* F. Werner (1902: 339).
Distribution: Southern Africa. Northern Namibia (Khomas, Kunene, Omaheke, Otjozondjupa), Botswana (Gaborone, Ghanzi, Kgalagadi, North West, Southern), S Zimbabwe (Masvingo, S Matebeleland North, Matebeleland South), and N South Africa (NW Cape, NW Free State, KwaZulu-Natal, Transvaal), 300–1380 m.
Sources: Mertens, 1955, 1971, V.F.M. FitzSimons, 1962a, 1974, Broadley & Cock, 1975, De Waal, 1978, Pienaar et al., 1983, Auerbach, 1987, Warner & Combrint, 2010a and Broadley & Blaylock, 2013.

Remarks: Probably occurs in S Mozambique *fide* Broadley (1980: 511).

4. *Prosymna frontalis* (W.C.H. Peters, 1867c). Mber. Königl. Akad. Wiss. Berlin 1867(4): 236, pl., figs. 1, 1a–e, 2, 2–d. (*Temnorhynchus frontalis*)

Synonym: *Prosymna bergeri* Lindholm *in* Lampe & Lindholm, 1902.

Type: Lectotype, ZMB 5763a, a 300–302 mm male (Hahn), designated by Mertens (1955: 94).

Type locality: "Otjimbingue, Damaraland, Südwestafrika" [= Otjimbingwe, SE Erongo Distr., cen. Namibia, 22°21'S, 16°08'E, elevation 900 m] via lectotype selection.

Distribution: Southwestern Africa. Southwestern Angola (Benguela), Namibia (Erongo, Hardap, Karas, Khomas, Kunene, Omaheke, Oshikoto, Otjozondjupa) and W South Africa (ext. NW Northern Cape), 300–1765 m.

Sources: Sternfeld, 1910a, Metheun & Hewitt, 1914, V.F.M. FitzSimons, 1938, 1962a, 1974 and Mertens, 1955, 1971.

Remarks: Loveridge (1958: 149) clarified the erroneous length of the type reported by W.C.H. Peters (1867c: 236).

5. *Prosymna greigerti* Mocquard, 1906, Bull. Mus. Hist. Nat., Paris (4) 12(7): 466.

Synonyms: *Prosymna meleagris collaris* Sternfeld, 1908c and *Prosymna meleagris laurenti* Loveridge, 1958.

Type: Holotype, MNHN 1906.154, a 173 mm specimen (Greigert, 1897–1902).

Type locality: "region Lobi, Haute-Volta" [= region of Lobi tribe, SW Burkina Faso, NE Ivory Coast and NW Ghana].

Distribution: West and cen. Africa. Western Senegal (Saint-Louis, Kédougou, Kaolack, Kolda, Matam, Tambacounda, Thiès), Gambia (McCarthy Island), Guinea (Labé, Siguiri, Télimélé), Burkina Faso (Centre, Centre-Est, Centre-Norde, Centre-Ouest, Est, Hauts-Bassins, Nord, Sahel, Sud-Ouest, Volta-Noire), N Ghana (NE Northern), N Benin (Alibori), N Togo (Kara, Savanes), SW Niger (Dosso, Niamey, S Tahoua, N Nigeria (Bauchi, Borno, Kaduna, Plateau, Sokoto), N Cameroon (Extreme-Nord, Nord), SW Mali (Kayes, Koulikoro, Mopti, Ségou, Sikasso), SW Chad (Chari-Baguirmi, Mayo-Kebbi), S Central African Republic (Bamingui-Bangoran, Ombella-Mpoko, Ouham, Vakaga), NW Democratic Republic of the Congo (W Bandundu), E Sudan (An Nil Al Azraq, Nahr An Nil, Sennar), SE South Sudan (Central Equatoria, Eastern Equatoria), and W Ethiopia (Wollega), 10–1200 m.

Sources: Aylmer, 1922, Angel, 1933b, Villiers, 1950a, 1956b, 1975, Loveridge, 1958, Condamin & Villiers, 1962, Witte, 1962, Doucet, 1963, Papenfuss, 1969, Roux-Estève, 1969a, Hulselmans & Verheyen, 1970, Miles et al., 1978, Roman 1980, 1984, Hakansson, 1981, Joger, 1990, Chippaux, 1999, Spawls et al., 2002, Chirio & Ineich, 2006, J.-F. Trape & Mané, 2006b, Chirio & LeBreton, 2007, Chirio, 2009, Chirio et al., 2011, Segniagbeto et al., 2011, Chirio, 2013 and Hughes, 2013.

Remarks: Type locality probably in the vicinity of Bouna (13°47'N, 0°06'W, elevation 320 m), Sahel Dept., NE Burkina Faso or Diébougou (10°58'N, 3°15'W, elevation 300 m), Sud-Ouest Dept., SW Burkina Faso as those localities were Lt. Greigert's habitation. A valid species *fide* Chirio et al., 2011. Previously a synonym of *P. meleagris*. Possibly occurs in N Uganda *fide* Spawls et al., 2002.

6. *Prosymna janii* Bianconi, 1862. Mem. Accad. Sci. Ist. Bologna (2) 1: 470–472, pl. 1, 4 figs.

Type: Holotype, MZUB, a 180 mm male.

Type locality: "Mosambicana" [= Mozambique]. Restricted to Inhambane, Mozambique *fide* W.C.H. Peters (1882a: 106).

Distribution: Southern Mozambique (Gaza, Inhambane, Maputo, Bazaruto, Benguera and Inhaca Is.) and ext. NE South Africa (E KwaZulu-Natal), NSL–80 m.

Sources: Jan & Sordelli, 1876 *in* 1870–1881, W.C.H. Peters, 1882d, V.F.M. FitzSimons, 1962a, 1974, Bruton & Haacke, 1975, Bourquin, 1977, Haacke & Bruton, 1978 and Broadley, 1990a, 1992a.

Remarks: Original description reprinted in Bianconi (1850–1870: 286–288, pl. 15, 4 figs. [pagination misprinted as 386–388]). Holotype head illustrated in Broadley (1980: fig. 1).

7. *Prosymna lineata* (W.C.H. Peters, 1871b). Mber. Königl. Akad. Wiss. Berlin 1871(11): 568–569. (*Temnorhynchus lineatus*)

Type: Holotype, ZMB 7257, a 126 mm specimen (Grützner).

Type locality: "Matlale (S Ostafrika)" [= Polokwane, cen. Limpopo Prov., N South Africa, 23°54'S, 29°27'E, elevation 1250 m].

Distribution: Southern Africa Central Mozambique (Nampula, Tete), Zimbabwe (Mashonaland East, Mashonaland Central, Mashonaland West, Masvingo, Matabeleland North, Matabeleland South, Midlands, Umtali), E Botswana (Central, Gaborone, Kweneng, North East, North West, Southern) and N South Africa (N KwaZulu-Natal, Limpopo, Mpumalanga), NSL–1250 m.

Sources: Broadley, 1980, 1983, 1999c, Auerbach, 1987, M. Griffin, 2003, Marais, 2004 and Broadley & Blaylock, 2013.

8. *Prosymna meleagris* (J.T. Reinhardt, 1843). Kgl. Danske Vidensk. Selsk. Natur. Math. Afh. 10: 238–241, pl. 1, figs. 4–6. (*Calamaria meleagris*)

Type: Lectotype, ZMUC 60618, a 215 mm male (J.R. Chenon, 1826–1836), designated by F.W. Braestrup *in* Hughes & Barry (1969: 1018).

Type locality: "Guinea" [= coast of Ghana *fide* Hughes & Barry, 1969: 1018] via lectotype selection. Restricted to vicinity of Accra, Ghana *fide* J.B. Rasmussen & Hughes (1997: 14, 16).

Distribution: West Africa. Eastern Senegal (Dakar, Fatick, Louga, Saint-Louis, Tambacounda, Ziguinchor), Guinea-Bissau (Oio), Guinea (Boké, Dubréka, Gaoual, Kindia, Mali, Télimélé), Sierra Leone, Liberia, Ivory Coast (Bouaké, Dabakala, Toumodi), Ghana (Accra, Central, Eastern, Northern, Upper East, Upper West, Volta, Western), S Togo (Centrale, Maritime, Plateaux), S Benin (Collines) and S Nigeria (Benue, Lagos, Plateau), NSL–1200 m.

Sources: Aylmer, 1922, Angel, 1933b, Villiers, 1950a, 1956b, 1975, Condamin & Villiers, 1962, Witte, 1962, Doucet, 1963, Papenfuss, 1969, Roux-Estève, 1969a, Hulselmans & Verheyen, 1970, Miles et al., 1978, Roman 1980, 1984, Joger, 1990, Largen & Rasmussen, 1993, J.B. Rasmussen & Hughes, 1997, Chippaux, 1999, Spawls et al., 2002, J.-F. Trape & Mané, 2006b, Largen & Spawls, 2009, Chirio et al., 2011, Segniagbeto et al., 2011, Auliya et al., 2012, Chirio, 2013 and Hughes, 2013.

9. *Prosymna ornatissima* T. Barbour & Loveridge, 1928. Mem. Mus. Comp. Zool. 50(2): 120–121, pl. 2, fig. 2.

Type: Holotype, MCZ 23271, a 286 mm female (A. Loveridge, 7 Oct. 1926).

Type locality: "Nyange, Uluguru Mountains, Tanganyika Territory, 2,500 ft." [= Nyange, N Morogoro Region, E Tanzania, elevation 760 m].

Distribution: East-central Tanzania (Uluguru Mtns. of N Morogoro), 760 m.

Source: Spawls et al., 2002.

10. *Prosymna pitmani* Battersby, 1951. Ann. Mag. Nat. Hist. (12) 4(44): 828–829.

Type: Holotype, BMNH 1950.1.3.96, a 309 mm male (C.J.P. Ionides, 1926–1951).

Type locality: "Kilwa, Southern Province, Tanganyika Territory." Restricted to Nanguale, Kilwa District [= Lindi Region, SE Tanzania, ca. 9°00'S, 39°00'E, elevation 150–200 m] *fide* Ionides *in* Loveridge (1958: 168).

Distribution: Southeastern Tanzania (Lindi, Mtwara) and S Malawi (Southern), 150–600 m.

Sources: Loveridge, 1955b, Stevens, 1974 and Spawls et al., 2002.

P

11. *Prosymna ruspolii* (Boulenger, 1896g). Ann. Mus. Civ. Stor. Nat. Genova (2) 17(37): 12. (*Asthenophis ruspolii*)

Synonyms: *Prosymna agrestis* Scortecci, 1929 and *Prosymna ruspolii keniensis* Broadley, 1980.

Type: Holotype, MSNG CE 29057, a 198 mm male (E. Ruspoli, 1893).

Type locality: "Magala Umberto I, Ganale Doria" [= Magala Umberto I, Ganale Doria River, S Bale Prov., SE Ethiopia, 4°18'N, 42°03'E, elevation 200 m].

Distribution: Northeastern Africa. Southern Ethiopia (Bale, Gemu Gofa, Sidamo), S Somalia (Bay, Gedo, Hiraan, Jubbada Dhexe, Jubbada Hoose, Mogadishu, Shabeellaha Dhexe, Shabeellaha Hoose), NE Tanzania (Arusha) and NW Kenya (Rift Valley), NSL–1500 m.

Sources: Scortecci, 1929, 1930, 1939b, H.W. Parker, 1949, Lanza, 1983a, Largen & Rasmussen, 1993, Gravlund, 2000c, Spawls et al., 2002 and Largen & Spawls, 2010.

Remarks: H.W. Parker (1949: 65) and Loveridge (1958: 158) listed type locality as Magala, Umberto Island but Broadley (1980: 524) and Largen & Rasmeussen (1993: 402) listed it as Umberto I village on the Ganale Doria River.

12. *Prosymna semifasciata* Broadley, 1995b. Arnoldia, Zimbabwe 10(4): 30–31, fig. 1.

Type: Holotype, NMZB 13732, a 258 mm male (K.M. Howell, C.A. Msuya & W.T. Stanley, 22 July 1994).

Type locality: "Kwamgumi Forest Reserve, Muheza District, Tanga Region, Tanzania (4°56'S, 38°44'E) at approximately 230 m.a.s.l."

Distribution: Northeastern Tanzania (Tanga), 230 m. Known only from type locality.

Source: Spawls et al., 2002.

13. *Prosymna somalica* H.W. Parker, 1930d. Ann. Mag. Nat. Hist. (10) 6(35): 605–606.

Type: Holotype, BMNH 1946.1.5.42 (formerly BMNH 1930.5.9.12), a 193 mm male (R.H.R. Taylor, 1929–1930).

Type locality: "8° N. x 47° 22' E. 2000 feet, British Somaliland." Emended to The Haud, on the boundary between Italian and British Somaliland, [= The Haud, on the border between Nugaal Region, NW Somalia and Hararge Prov., E Ethiopia, 8°00'N, 47°22'E, elevation 610 m] *fide* H.W. Parker (1932b: 335, 364).

Distribution: Eastern Ethiopia (E Hararge) and N Somalia (Bari, Nugaal, Togdheer, Woqooyi Galbeed), 610–1525 m.

Sources: H.W. Parker, 1932b, 1949, Lanza, 1983a, Largen & Rasmussen, 1993 and Largen & Spawls, 2010.

14. *Prosymna stuhlmannii* (Pfeffer, 1893). Jahrb. Hamburg. Wiss. Anst. 10: 78–79, pl. 1, figs. 8–10. (*Ligonirostra stuhlmannii*)

Synonyms: *Prosymna vassei* Mocquard, 1906, *Prosymna variabilis* F. Werner, 1909d, *Stenorhabdium temporale* F. Werner, 1909d, and *Prosymna transvaalensis* Hewitt, 1910.

Types: Syntypes (3), ZMH 7910 (formerly ZMH 1646) and ZMH 521–22, longest syntype 190–200 mm (F. Stuhlmann, Sept. 1888).

Type locality: "Usambáa, Ostafrikanische" [= Usambara Mtns., N Tanga Region, NE Tanzania] *fide* Hallermenn (2006: 141).

Distribution: Eastern Africa. Extreme S Somalia (Jubbada Hoose), S Kenya (Coast, S Eastern, Rift Valley), SE Tanzania (Arusha, Dar es Salaam, Dodoma, Kilimanjaro, Lindi, Mbeya, Mtwara Piwani, Rukwa, Ruvuma, Tabora, Tanga, Kagera, Zanzibar Is.), SE Zambia (E Central, Eastern, Northern), S Malawi (Southern), Mozambique (Gaza, Inhambane, Maputo, Nampula, Sofala, Tete, Zambézia, Mozambique Is.), Zimbabwe (Manicaland, Mashonaland East, Mashonaland West, Masvingo, Matabeleland North, Matebeleland South), Swaziland and NE South Africa (KwaZulu-Natal,), NSL–1675 m.

Sources: Lanza, 1983a, 1990b, Broadley, 1992c, Broadley et al., 2003, Hallermann, 2006 and Broadley & Blaylock, 2013.

15. *Prosymna sundevallii* (A. Smith, 1849 *in* 1838–1849). Illust. Zool. So. Afr., Rept. (App.): 17–18. (*Temnorhynchus sundevallii*) (*nomen corrigendum*)

Synonyms: *Temnorhynchus sundervallii* A. Smith, 1849 *in* 1838–1849 *(nomen incorrigendum)* and *Rhinostoma cupreum* A.C.L.G. Günther, 1858.

Type: Holotype, NMSZ 1859.13.246 (formerly RSM), a 199 mm specimen (A. Smith, 1831–1834).

Type locality: "Kaffirland, to the eastward of the Cape Colony" [= KwaZulu-Natal, South Africa], (in error *fide* Broadley, 1980: 494). Corrected to Eastern or Western Cape Prov., South Africa *fide* Broadley (1980: 494).

Distribution: Southeastern Africa. Central Mozambique (Nampula, Tete), Zimbabwe (Bulawayo, Manicaland, Mashonaland Central, Mashonaland East, Mashonaland West, Masvingo, Matabeleland North, Matabeleland South, Midlands), S Botswana (Central, SW Kgalagadi), Lesotho and South Africa (Eastern Cape, Free State, Gauteng, Limpopo, Western Cape), NSL–1800 m.

Sources: V.F.M. FitzSimons, 1962a, 1974, Broadley, 1965, Broadley & Cock, 1975, De Waal, 1978, Pienaar et al., 1983, Auerbach, 1987, Spawls et al., 2002 and M. Griffin, 2003.

Remarks: Specific epithet of *sundervallii* corrected to *sundevalli* in index and errata sheet. Natal [= KwaZulu-Natal], South Africa records need confirmation *fide* Broadley (1980: 507). Probably occurs in Namibia *fide* M. Griffin 2003.

16. *Prosymna visseri* V.F.M. FitzSimons, 1959. Ann. Transvaal Mus. 23(4): 408–409.

Type: Holotype, TMP 24531, a 307 mm male (C. Koch, Sept. 1956).

Type locality: "near Caracul, S. Angola" [= vicinity of Caraculo (15°01'S, 12°40'E, elevation 480 m), Namibe Prov., ext. SW Angola].

Distribution: Southwestern Angola (Benguela, Huíla, Namibe) and NW Namibia (E Kunene), 480–1250 m.

Sources: V.F.M. FitzSimons, 1962a, Branch, 1988, Broadley, 1990b and M. Griffin, 2003.

PROTOBOTHROPS Hoge & Romano-Hoge, 1983 (Viperidae)

Synonym: *Ceratrimeresurus* Liang & Liu *in* Liang, 2003.

Type species: *Bothrops flavoviridis* Hallowell, 1861.

Fossil records: Lower Miocene of Japan.

Distribution: Southern and SE Asia.

Sources: Hoge & Romano-Hoge, 1981a, 1983, Maes, 1989, Kraus et al., 1996, David & Ineich, 1999, McDiarmid et al., 1999, Toda et al., 1999, Malhotra & Thorpe, 2000, Ziegler et al., 2001, Gumprecht et al., 2004, Herrmann et al., 2004. Castoe & Parkinson, 2006, Gao et al., 2006b, 2007, Orlov et al., 2009, Yang et al., 2011, Liu et al., 2012 and Hoser, 2012d.

Remarks: A synonym of *Trimeresurus fide* Golay et al., 1993 and McDiarmid et al., 1999.

1. *Protobothrops cornutus* (M.A. Smith, 1930b). Ann. Mag. Nat. Hist. (10) 6(36): 682–683, 1 fig. (*Trimeresurus cornutus*)

Synonym: *Ceratrimeresurus shenlii* Liang and Liu *in* Liang, 2003.

Type: Holotype, BMNH 1946.1.19.25 (formerly BMNH 1930.11.16.2), a 587 mm female (J.T. Delacour & W. Lowe [Fifth Delacour-Lowe-Greenway-Poilane Mission], late Nov.–mid-Dec. 1929).

Type locality: "Fan-si-pan mountains, Chapa, a hill-station at 5000 feet altitude, Tonkin, Indo-China." [= Mt. Fan Si Pan, Sa Pa Distr., Lai Châu Prov., N Vietnam, ca. 22°18'N, 103°46'E *fide* Herrmann et al., 2004: 214].

Distribution: Northern Vietnam (Ha Giang, Ha Tinh, Lao Cai, Nghe An, Thua Thien-Hue) and S China (Hunan), 300–2000 m.

Sources: Ziegler et al., 2001, Orlov et al., 2003, Herrmann et al., 2004, Shiryaey et al., 2007, V.S. Nguyen et al., 2009, Quyet & Ziegler, 2009 and I. Das, 2010.

Remarks: Holotype listed as male in original description. *Ceratrimeresurus shenlii* is possibly a synonym *fide* Vogel, 2006, confirmed by David et al. (2008c: 17).

2. *Protobothrops dabieshanensis* X. Huang, Pan, Han, Zhang, Hou, Yu, Zheng & Zhang, 2012. Asian Herpetol. Res. 3(3): 213–216, figs. 1a–f.

Type: Holotype, AHU 2011-EE-Hp01, an 836 mm female (X. Huang, B.-W. Zhang & D.-M. Han, July 2011).

Type locality: "Yaoluoping Nature Reserve in the Dabie Mountains (116°04'12" N, 30°58'17" E, 1210 m a.s.l.)" [= Yaoluoping Nature Reserve, Dabie Mtns., SW Anhui Prov., SE China, 30°58'N, 116°04'E, elevation 1210 m].

Distribution: Southeastern China (SW Anhui), 1210 m.

3. *Protobothrops elegans* (Gray, 1849a). Cat. Snakes Brit. Mus.: 7. (*Craspedocephalus elegans*)

Synonyms: *Trimeresurus luteus* Boettger, 1895c and *Lachesis lutea* – Boettger, 1898.

Types: Syntypes (2), BMNH 1946.1.20.2 (formerly BMNH 1847.3.4.62) and BMNH 1946.1.19.31 (formerly BMNH 1847.3.4.64), an adult and juvenile (E. Belcher, 1843–1846).

Type locality: "West coast of (North?) America?" (in error). Corrected to Ishigakishima Is., Ryukyu Arch., S Japan *fide* Stejneger (1907: 477).

Distribution: Extreme S Japan (Ryukyus: Aragusukujima, Irimoteshima, Ishigakishima, Kayamajima, Kohamajima, Kuroshima, Miyakoshima, Sakishima and Taketomijima Is.).

Sources: Maki, 1931, Keegan & Yoshino, 1959, Takara, 1962, Koba & Kikukawa, 1976, M. Mori, 1984 and Toriba, 1989a–b.

Remarks: Closely related to, and sometimes a synonym of, *Protobothrops mucrosquamatus*.

P

4. *Protobothrops flavoviridis* (Hallowell, 1861). Proc. Acad. Nat. Sci. Philadelphia (1860) 12(11): 492–493. (*Bothrops flavoviridis*)

Synonyms: *Trimeresurus riukiuanus* Hilgendorf, 1880, *Trimeresurus riukianus* – Boulenger, 1896a, *Trimeresurus fravoviridis* – Takahashi, 1922 *(nomen incorrectum),* and *Trimeresurus flavoviridis tinkhami* Gloyd, 1955b.

Type: Holotype, ANSP, a mutilated specimen (Macomb [No. Pacific Explor. Exped.], April 1855), lost *fide* Stejneger (1907a: 476).

Type locality: "Amakarima Island, (one of the Loo-Choo group)" [= Keramashima, Keramashima group, cen. Ryukyu Arch., S Japan, 26°05'N, 127°15'E].

Distribution: Extreme S Japan (Ryukyus: Amamioshima, Gishifujima, Gusukujima, Hamahigajima, Hyanzajima, Iejima, Iheyajima, Ikeijima, Kakeromajima, Korijima, Kumejima, Kuroshima, Minnajima, Miyagijima, Ohajima, Okinawajima, Sesokojima, Tokashikijima, Tokunojima, Tonakijima, Ukejima, Ukibarujima, Yabuchijima, Yagachijima and Yorojima Is.), NSL–200 m.

Sources: Stejneger, 1907a, Maki, 1931, Takara, 1962, Leviton, 1968b, Koba, 1971, M. Mori, 1982, Nishimura, 1986, Shiroma & Kamura, 1987, Toriba, 1989a–b, Nishimura et al., 1991, Nishimura & Kamura, 1992, 1993, Nobusaka et al., 1994, Shiroma, 1997 and Rodda et al., 1999.

5. *Protobothrops jerdonii* (A.C.L.G. Günther, 1875). Proc. Zool. Soc. London 43(1): 233–234, pl. 34. (*Trimeresurus jerdonii*)

Synonyms: *Trimeresurus xanthomelas* A.C.L.G. Günther, 1889, *Lachesis melli* T. Vogt, 1922, *Trimeresurus jerdonii meridionalis* Bourret, 1935b *(nomen praeoccupatum), Trimeresurus jerdonii bourreti* Klemmer, 1963 *(nomen substitutum),* and *Protobothrops jordonii* – Yao, 2012 *(nomen incorrectum).*

Types: Syntypes (3), BMNH 1946.1.18.66–68 (formerly BMNH 1872.4.17.138, BMNH 1872.4.17.242 and BMNH 1872.4.17.267), two males and a female, longest syntype 711 mm (T.C. Jerdon, 1836–1852).

Type locality: "Khassya, northern India" [= Khasi Hills, E Meghalaya State, NE India, 25°35'N, 91°38'E, elevation 1370–1675 m].

Distribution: Southeastern Asia. Eastern Nepal (Dolakha), NE India (Arunachal Pradesh, Assam, Manipur, Meghalaya, Mizoram, Nagaland), Bhutan, NW Myanmar (Chin, Kachin), S China (Gansu, Guangxi, Guizhou, Henan, Hubei, Hunan, Shaanxi, Shanxi, Sichuan, Yunnan, Xizang), Laos and N Vietnam (Lai Chau), 1200–2700 (3250) m.

Sources: A.C.L.G. Günther, 1892b, Wall, 1910i, 1925a, 1928, Bourret, 1935b, 1936b, C.H. Pope, 1935, M.A. Smith, 1935, C.-C. Liu, 1940, Kramer, 1977, Hu et al., 1980, Tian et al., 1986, Y.-X. Zhang & Tang, 1990, Zhang & Zhao, 1990, Orlov et al., 2001, 2003, Leviton et al., 2003, Ao et al., 2004, Tiwari & Shah, 2004, Whitaker & Captain, 2004, Guo et al., 2006a–b, M.F. Ahmed et al., 2009, V.S. Nguyen et al., 2009, Wangyal, 2011 and Yao, 2012.

6. *Protobothrops kaulbacki* (M.A. Smith, 1940c). Rec. Ind. Mus. 42(3): 485–486, pl. 8, fig. 5. (*Trimeresurus kaulbacki*)

Synonym: *Protobothrops kaulbachi* Kraus, Mink & Brown, 1996 *(nomen emendatum).*

Types: Syntypes (2), BMNH 1946.1.19.23–24 (formerly BMNH 1940.6.1.51–52), a 1390 mm female and 1340 mm male (R. Kaulbach, 1937–1939).

Type locality: "Trans-Himalayan region on the borders of Tibet and China, to the north of the tribal area generally known as the Triangle." Restricted to Pangnamdim, north of the Triangle, Upper Burma [= Pangnamdim, Nam Tamai Valley, Kachin Div., N. Myanmar, ca. 27°42'N, 97°54'E, elevation 1750 m] *fide* M.A. Smith (1943: 512).

Distribution: Northeastern India (Arunachal Pradesh), SW China (Xizang) and N Myanmar (Kachin), 1015–1750 m.

Sources: Kraus et al., 1996, Leviton et al., 2003 and I. Das, 2010, 2012.

7. *Protobothrops maolanensis* Yang, Orlov & Wang, 2011. Zootaxa (2936): 60–65, fig. 1.

Type: Holotype, SYS 211, an 805 mm male (S. Zheng & J.-H. Yang, 18 May 2010).
Type locality: "Maolan National Nature Reserve, Maolan town, Libo County, Guizhou, China (25°12'50.0"N, 107°59'56.04"E, 560 m a.s.l.)."
Distribution: Southern China (Guizhou), 560 m. Known only from type locality.
Source: Liu et al., 2012.

8. *Protobothrops mucrosquamatus* (Cantor, 1839a). Proc. Zool. Soc. London 7(1): 32. (*Trigonocephalus mucrosquamatus*)

Type: Lectotype, specimen illustrated in colored sketch of T.E. Cantor (1834–1837: no. 18) in RSL, designated by M.A. Smith (1943: 507).
Type locality: "Naga Hills, Assam" [= Nagaland State, NE India].
Distribution: Southeastern Asia. Northeastern India (Arunachal Pradesh, Nagaland), Bangladesh, N Myanmar (Kachin), SE China (Anhui, Fujian, Gansu, Guangdong, Guangxi, Guizhou, Hainan, Hunan, Jiangxi, Shaanxi, Sichuan, Yunnan, Zhejiang), Taiwan (Chiai, Pingtung, Taichung, Taipei), Laos (Khammouan) and N Vietnam (Bac Kan, Cao Bang, Gia Lai, Ha Giang, Hai Duong, Hai Phong, Ha Tay, Ha Tinh, Kon Tum, Lang Son, Lao Cai, Nghe An, Ninh Binh, Quang Ninh, Quang Tri, Thai Nguyen, Thua Thien-Hue, Vinh Phuc, Hon Nor Way Is.), 100–1830 m.
Sources: A.C.L.G. Günther, 1864a, Maki, 1933a–b, C.H. Pope, 1929, 1935, Bourret, 1936b, 1939d, C.S. Wang & Wang, 1956, Kuntz, 1963, Hu et al., 1980, Lue et al., 1989, Lue, 1990, M.-H. Huang, 1990, Zhang & Zhao, 1990, Orlov, 1997, Darevsky, 1999, Orlov et al., 2000, 2001, Leviton et al., 2003, Ao et al., 2004, P. Guo et al., 2006a–b, Ziegler et al., 2006, Bain et al., 2007, B.L. Stuart & Heatwole, 2008, V.S. Nguyen et al., 2009, Xiang & Li, 2009, I. Das, 2010, 2012 and Yao, 2012.
Remarks: Holotype lost *fide* A.C.L.G. Günther (1864a: 390).

9. *Protobothrops tokarensis* (Nagai, 1928). Surv. Nat. Hist. Kagoshima 3: 6, 3 pls. (*Trimeresurus tokarensis*)

Type: Holotype, KMJ (K. Nagai, 1922–1928), lost [probably during World War II] *fide* McDiarmid et al. (1999: 347).
Type locality: "Takarashima, Prov. Kagoshima" [= Takarajima, Tokara Group, N Ryukyu Arch., ext. S Japan, 29°09'N, 129°13'E].
Distribution: Extreme S Japan (Ryukyus: Kotakarajima and Takarajima Is.), NSL–100 m.
Sources: Maki, 1931, Takara, 1962, Koba & Kikukawa, 1969, 1971, Koba et al., 1970, M. Mori, 1984 and Toriba, 1989a–b.

10. *Protobothrops trungkhanhensis* Orlov, Ryabov & Nguyen, 2009. Russ. J. Herp. 16(1): 71–74, figs. 1–4.

Type: Holotype, ZISP 25351, a 733 mm male (T.T. Nguyen, N.L. Orlov & S.A. Ryabov, Sept. 2008).
Type locality: "Trung Khanh Nature Reserve, Trung Khanh district, Cao Bang Province, N-E Vietnam, elevation 600 m."
Distribution: Northern Vietnam (Cao Bang), 600 m. Known only from type locality.
Source: Ziegler & Nguyen, 2010.

11. *Protobothrops xiangchengensis* (Zhao, Jiang & Huang, 1978). Mater. Herpetol. Res. Chengdu 4: 21. (*Trimeresurus xiangchengensis*)

Type: Holotype, CIB 725049 or CIB 725050, a 741 mm male (17 Oct. 1972).
Type locality: "Sichuan Province, Xiangcheng, altitude 3100 m,"
Distribution: China (W Sichuan, Yunnan), 3000–3200 m.
Sources: Zhao, 1979a, Zhang & Zhao, 1990, Zhao & Adler, 1993, David & Tong, 1997, P. Guo et al., 1999b, 2006b and G. Peng & Fuji, 2001
Remarks: Supplemental original escription in Zhao (1979a: 422–424, fig. 1). Photographs of the holotype in Orlov et al. (2009: figs. 12–13). Holotype listed as CIB 725049 (*fide* Zhao, 1979a) and CIB 725050 (*fide* Zhao et al., 1978). Collection date of type listed as 10 Oct. 1972 *fide* David & Tong (1997: 20). Probably a synonym of *Trimeresurus* (= *Protobothrops*) *mucrosquamatus fide* McDiarmid et al. (1999: 348). P. Guo et al., 2006b indicated a close relationship to *P. jerdoni*.

PSAMMODYNASTES A.C.L.G. Günther, 1858 (Elapoidea *incertae sedis*)

Synonyms: *Psammodytes* – Gervais *in* d'Orbigny, 1847b *(nomen incorrectum), Psamodynastes* – Palacky, 1898 *(nomen incorrectum), Anisodon* Rosén, 1905a *(nomen praeoccupatum), Anisodontes* Rosén, 1905b *(nomen susbstitutum), Psalmodynastes* – E.H. Taylor, 1917 *(nomen incorrectum), Psammodinastes* – Scortecci, 1929 *(nomen incorrectum), Psammdynastes* – Maki, 1931 *(nomen incorrectum), Psamnodynastes* – Reeves, 1933 *(nomen incorrectum), Psammodynates* – Briceño-Rossi, 1934 *(nomen incorrectum), Psammodymastes* – Majupuria, 1981 *(nomen incorrectum),* and *Psamaodynastes* – W.W. Tanner & Avery, 1982 *(nomen incorrectum).*
Type species: *Psammophis pulverulenta* H. Boie *in* F. Boie, 1827.
Distribution: Southeastern Asia and East Indies.

Sources: J.G. Fischer, 1885a, Mocquard, 1887d, 1888, Rooij, 1917, C. Haas, 1950, Tweedie, 1983, Stuebing, 1991, David & Vogel, 1996, Vidal et al., 2008, Kelly et al., 2009 and I. Das, 2007b, 2010, 2012.

1. *Psammodynastes pictus* A.C.L.G. Günther, 1858. Cat. Colub. Snakes Brit. Mus.: 251–252.

Synonym: *Psammophis conjunctus* W.C.H. Peters, 1868a.

Types: Syntypes (4), BMNH 1946.1.8.54–55, two males (E. Belcher, 1843–1847) and BMNH 1946.1.8.56 and BMNH 1946.1.9.88, two females (T.S.B. Raffles, 1805–1824), longest syntype 432 mm.

Type locality: "Borneo" and "India," (latter in error *fide* A.C.L.G. Günther, 1864a: 292). Raffles syntype locality corrected to Sumatra *fide* Boulenger (1896a: 174) [= Kalimantan and Sumatra, Indonesia]. Restricted to Sumatra *fide* C. Haas (1950: 584).

Distribution: Malay Peninsula and Greater Sundas. West Malaysia (Johore, Selangor, Terengganu), Singapore, East Malaysia (Sabah, Sarawak), Brunei and W Indonesia (Bangka, Belitung, Bintan, Galang, Kalimantan, Mentawai Arch., Riau Arch., Simeulue, Sulawesi, Sumatra), NSL–300 m.

Sources: Chasen & Smedley, 1927, B.L. Lim, 1963 and K.K.P. Lim & Chou, 1990.

Remarks: India and Bangladesh records doubtful. Sulawesi specimens probably introduced *fide* I. Das (2010: 339).

2. *Psammodynastes pulverulentus* (H. Boie *in* F. Boie, 1827). Isis von Oken 20(6): 547–548. (*Psammophis pulverulenta*)

P

Synonyms: *Psammophis pulverulenta* H. Boie in Schlegel, 1826a *(nomen nudum), Psammophis pulverulenta* Schlegel, 1826b *(nomen nudum), Dipsas ferruginea* Cantor, 1839b, *Psammophis floresianus* Bleeker, 1859b (*nomen nudum*), *Lycodon bairdii* Steindachner, 1867a, *Anisodon lilljeborgi* Rosén, 1905a, and *Psammodynastes pulverulentus papenfussi* Zhao, 1995.

Type: Lectotype, RMNH 765 (H. Boie & H.C. Macklot, Dec. 1825–Sept. 1827), designated by R.F. Inger herein.

Type locality: "Java" [SW Indonesia].

Distribution: Southeastern Asia and East Indies. Northeastern India (Andhra Pradesh, Arunachal Pradesh, Assam, Meghalaya, Mizoram, Nagaland, N Orissa, Sikkim, Uttar Pradesh, N West Bengal), Nepal (Ilam, Khotang, Udayapur, Rupandehi), Bhutan, SE Bangladesh (Chittagong), Myanmar (Bago, Kachin, Mon, Sagaing, Shan, Tanintharyi, Yangon), S China (Fujian, Guangdong, Guangxi, Guizhou, Hainan, Hong Kong, Hunan, Jiangxi, Yunnan, Xizang, Lanyu Is.), Taiwan (Changhwa, Hsinchu, Pingtung, Taichung, Taipei, Botel-Tobago and Batan Is.), Vietnam (Bac Kan, Dac Lac, Dong Nai, Gia Lai, Hai Duong, Hai Phong, Ha Tinh, Hoa Binh, Ho Chi Minh City, Khanh Hoa, Lam Dong, Lang Son, Nghe An, Phu Yen, Quang Binh, Quang Ninh, Son La, Tay Ninh, Thai Nguyen, Thanh Hoa, Thua Thien-Hue, Vinh Phuc), Laos (Attapu, Champasak, Savannakhet, Vientiane, Xiangkhouang), Cambodia (Kampot, Koh Kong, Kampong Chhnang, Oddar Meanchey), Thailand (Chanthaburi, Chiang Mai, Chon Buri, Kanchanaburi, Khon Kaen, Ko Chang, Krabi, Loei, Lop Buri, Nakhon Ratchasima, Nakhon Si Thammarat, Narathiwat, Nong Khai, Pattani, Phetchabun, Phrae, Phuket, Prachin Buri, Ranong, Rayong, Sakon Nakhon, Surat Thani, Ubon Ratchathani, Chang Is.), West Malaysia (Pahang, Perak, Pinang, Perlis, Selangor, Tioman Is.), East Malaysia (Sabah, Sarawak), Brunei, Indonesia (Alor, Bali, Bawean, Buton, Bangka, Bintan, Enggano, Flores, Great Natuna, Java, Kalimantan, Komodo, Lombien, Lombok, Natuna Arch., Pador, Riau Arch., Rinca, Sangihe Arch., Siau, Selantan, Siberut, Simeulue, Sula, Sulawesi, Sumatra, Sumba, Sumbawa, Tengah, Togian Arch., Utara) and Philippines (Balabac, Basilan, Bohol, Bongao, Busuanga, Camiguin, Dinagat, Jolo, Leyte, Luzon, Mindanao, Negros, Palawan, Panay, Polillo, Sabtán, Samar, Siargao), NSL–2000 m.

Sources: Wall, 1910g, 1912c, M.A. Smith, 1922b, 1930a, 1940b, 1943, E.H. Taylor, 1922a, 1965, Kopstein, 1927, 1930, 1938, 1941, K.P. Schmidt, 1927b, Brongersma, 1929, 1933e, Cochran, 1930, Mertens, 1930, Maki, 1931–b, Smedley, 1932a, Bourret, 1935d, 1936b, 1937a–b, C.H. Pope, 1935, G.E. Shaw et al., 1941, C.-S. Wang & Wang, 1956, C.-S. Wang, 1962, Kuntz, 1963, Campden-Main, 1970a, Deuve, 1970, Saint-Girons, 1972a, J.B. Rasmussen, 1975, 1990, Auffenberg, 1980, Hu et al., 1980, Daniel, 1983, Leviton, 1983, Bosch, 1985, Tian et al., 1986, Dring et al., 1990, H.W. Greene, 1990, M.J. Cox, 1991b, Ross & Lazell, 1991, Zhao & Adler, 1993, Zhao, 1995, K. Jackson & Fritts, 1996, M.J. Cox et al., 1998, Chan-ard et al., 1999, Lazell, 1999, Das & Palden, 2000, Orlov et al., 2000, Hallermann et al., 2001, Grismer et al., 2002b, Malkmus et al., 2002, Tiwari & Shah, 2004, Whitaker & Captain, 2004, McKay, 2006, B.L. Stuart & Emmett, 2006, M.F. Ahmed et al., 2009, V.S. Nguyen et al., 2009, Xiang & Li, 2009, Mohapatra et al., 2010a, Lang et al., 2011 and T. Hartmann et al., 2013.

Remarks: Original description based on H. Boie's MS (1823–1825). Records from India (Madhya Pradesh, Tamil Nadu) and Indonsia (Timor, Lomblen, Halmahera) need confirmation.

PSAMMOPHIS H. Boie *in* Fitzinger, 1826a (*nomen substitutum*) (Psammophiidae)

Synonyms: *Macrosoma* Leach *in* Bowdich, 1819 *(nomen praeoccupatum), Psammophis* H. Boie *in* F. Boie, 1826 *(nomen nudum), Psammophis* H. Boie *in* Schlegel, 1826a *(nomen nudum), Psamnophis* – Gray *in* Griffith & Pidgeon, 1831 *(nomen incorrectum), Prammophis* – Beddome, 1862 *(nomen incorrectum), Psamonophis*

– J. Müller, 1832 *(nomen incorrectum)*, *Amphiophis* Bocage, 1872, *Anphiophis* – Hoffmann, 1890 *(nomen incorrectum)*, *Pammophis* – Boettger, 1898 *(nomen incorrectum)*, *Psammmophis* – Nikolsky, 1913 *(nomen incorrectum)*, *Psammphis* – Briceño-Rossi, 1934 *(nomen incorrectum)*, *Eipperus* Hoser, 2012u *(nomen illegitimum)*, *Elliottus* Hoser, 2012u *(nomen illegitimum)*, and *Slatteryus* Hoser, 2012u *(nomen illegitimum)*.

Type species: *Coluber sibilans* Linnaeus, 1758.

Distribution: Africa and S Asia.

Fossil records: Fossils unassigned to species include upper Miocene (Turolian, MN 13: 4.9–7.3 mya) of Spain.

Sources: H. Boie, 1826, Boulenger, 1895h, F. Werner, 1924c, Loveridge, 1940, Broadley, 1977b, 1983, 2002, Brandstätter, 1995, 1996, Hughes, 1999, Gravlund, 2001, Nagy et al., 2003, Lawson et al., 2005, Chippaux, 2006, Kelly et al., 2008, 2009, Vidal et al., 2008, Zaher et al., 2009, Hoser, 2012u and Szyndlar, 2012.

Remarks: *Rhamphiophis acutus* transferred to *Psammophis fide* Kelly et al., 2008.

1. *Psammophis acutus* A.C.L.G. Günther, 1888c. Ann. Mag. Nat. Hist. (6) 1(3): 327–328, pl. 19, fig. d.

Synonyms: *Rhamphiophis acutus wittei* Laurent, 1956a and *Rhamphiophis acutus jappi* Broadley, 1971a.

Type: Holotype, BMNH 1946.1.2.81 (formerly BMNH 1864.7.13.36), a 914 mm male (F.M.J. Welwitsch, 18 Oct. 1856–11 June 1857).

Type locality: "Pungo Andongo" [= Pedras Negras, NW Malanje, N Angola, 9°40'S, 15°35'E, elevation 1100–1265 m].

Distribution: Southern Africa. Gabon (Haut-Ogooué), S Congo, SE Democratic Republic of the Congo (Katanga), Angola (Benguela, Bié, Huambo, Huíla, Lunda Norte, Malanje), Rwanda, W Burundi (Gitega, Ruyigi), W Tanzania (Iringa, Kigoma, Mbeya, Rukwa), N Malawi (Northern) and Zambia (Luapula, Northern, North-Western, Western), 450–1815 m.

Sources: Loveridge, 1933, Witte, 1953, 1962, 1966, Laurent, 1956, Vesey-FitzGerald, 1958, Broadley & Pitman, 1960, Ashe, 1965, Broadley, 1971a, Derleyn et al., 1983, Spawls et al., 2002, Broadley et al., 2003 and Segniagbeto et al., 2011.

Remarks: Previously in the genus *Rhamphiophis*. Probably occurs in N Malawi *fide* Broadley (1971a: 3) and N Namibia *fide* M. Griffin, 2003.

2. *Psammophis aegyptius* Marx, 1958b. Fieldiana: Zool. 39(18): 194–198, figs. 30c, 31c.

Type: Holotype, FMNH 75092 (formerly CNHM 75092), an 1183 mm male (R.E. Kuntz & G. M. Malakatis).

Type locality: "near the government house at Siwa, Siwa Oasis, Western Desert Governate, Egypt" [= Siwa, Siwa Oasis, W Matrouh Govern., NW Egypt, 29°12'N, 25°31'E, elevation BSL].

Distribution: North Africa and Israel. Southeastern Algeria (Illizi, Tamanrasset), S Libya (Al Kurfah, Murzuq), N Niger (Agadez), N Chad (Kanem), Sudan (Al Khartum), Egypt (Assiut, Aswan, El-Bahr El-Ahmar, Matrouh, New Valley, Qena, Sohag, South Sinai, Suez, Giftun El Saghir, Shadwan and Tiran Is.) and ext. S Israel (S Southern), BSL–1500 m.

Sources: Kramer & Schnurrenberger, 1963, Schleich et al., 1996, Bouskila & Amitai, 2001, Baha El Din, 2006, J.-F. Tarpe & Mané, 2006b and Bar & Haimovitch, 2011.

Remarks: Previously a subspecies of *P. schokari*.

3. *Psammophis angolensis* (Bocage, 1872). J. Sci. Math. Phys. Nat., Lisboa 4(13): 82. (*Amphiophis angolensis*)

Synonym: *Ablabes homeyeri* W.C.H. Peters, 1877d.

Type: Holotype, MBL (F.A.P. Bayão, 1868), destroyed by fire 18 March 1978.

Type locality: "Dondo (intérieur d'Angola)" [= Dondo, north bank of Kwanza River, SW Cuanza Norte Distr., NW Angola, 09°41'S, 14°26'E, elevation 50 m].

Distribution: Southern Africa. Northwestern Ethiopia (Gojjam), Tanzania (Dar es Salaam, Dodoma, Iringa, Lindi, Morogoro, Mbeya, Mtwara, Mwanza, Piwani, Ruvuma, Shinyanga, Singida, Tabora, Zanzibar Is.), SE Democratic Republic of the Congo (Katanga), Angola (Benguela, Cuanza Norte, Cuanza Sul, Cunene, Luanda, Lunda Norte, Malanje, Moxico, Huíla, Zaire), Zambia (Central, Copperbelt, Eastern, Luapula, Lusaka, Northern, North-Western, Southern, Western), Malawi (Central, Northern, Southern), S Mozambique (Gaza, Inhambane, Sofala, Zambézia), N Namibia (Caprivi, Kunene, Okavango, Oshikoto), N Botswana (North West), Zimbabwe (Bulawayo, Manicaland, Mashonaland East, Mashonaland Central, Mashonaland West, Masvingo, Matebeleland North, Matebeleland South, Midlands) and NE South Africa (Limpopo, Mpumalanga), NSL–2000 m.

Sources: Bocage, 1895, Monard, 1937, Laurent, 1950c, 1954a, 1964a, Witte, 1953, 1962, Mertens, 1955, 1971, Sweeney, 1961, V.F.M. FitzSimons, 1962a, 1974, Laurent, 1964a, Noble, 1966, Bourgeois, 1967, Broadley, 1971d, Pienaar et al., 1983, Auerbach, 1987, Broadley & Howell, 1991, Largen & Rasmussen, 1993, Broadley, 2002, Spawls et al., 2002, Broadley et al., 2003, Marais, 2004, Largen & Spawls, 2010 and Broadley & Blaylock, 2013.

Remarks: Type locality has been variously cited as Danda, Loanda Distr. (Loveridge, 1940: 68 and Broadley, 1962c: 834), Donda, Loanda Distr. (V.F.M. FitzSimons, 1962a: 235 and Auerbach, 1987: 171), Dondo, Luanda Distr. (Crawford-Cabral & Mesquitela, 1989), all of which are incorrect. Loveridge (1957a: 280) and Hellmich (1957b: 42) correctly located Dondo on the north bank of the Kwanza River, Cuanza Norte

Distr., Angola. Free State, South Africa record rejected by Hewitt (1912b: 25) and Loveridge (1940: 69).

4. *Psammophis ansorgii* Boulenger, 1905d. Ann. Mag. Nat. Hist. (7) 16(92): 113, pl. 4, fig. 4.

Type: Holotype, BMNH 1946.1.8.49, a 700 mm male (W.J. Ansorge, end of 1904).

Type locality: "between Benguella and Bihé, Angola."

Distribution: Central Angola (E Benguella, Huambo), 1800 m.

Sources: Loveridge, 1940, Helmich, 1957b and Broadley, 2002, 2005.

5. *Psammophis biseriatus* W.C.H. Peters, 1881d. Sitz. Ges. Naturf. Freunde Berlin 1881(6): 88.

Type: Holotype, ZMB 9394, a male (J.M. Hilderbrandt, 1878–1881).

Type locality: "Taita (Africa orientalis)" [= Taita, S Rift Valley, S Kenya, 1°40'N, 36°10'E, elevation 665 m].

Distribution: Eastern Africa. Eritrea (Southern Red Sea: Dahlak Is.), S Ethiopia (Bale, Gemu Gofa, Hararge, Sidamo), Djibouti, Somalia (Bakool, Bay, Galguluud, Gedo, Hiraan, Jubbada Dhexe, Jubbada Hoose, Mogadishu, Mudug, Nugaal, Sanaag, Sool, Shabeellaha Dhexe, Shabeellaha Hoose, Coiama Is.), E Kenya (Coast, Eastern, North-Eastern, N Rift Valley, Patta Is.) and NE Tanzania (Arusha, Dodoma, Iringa, Kilimanjaro, Tabora), NSL–1300 m.

Sources: Stejneger, 1894b, Bezy & Drewes, 1985, Broadley & Howell, 1991, Largen & Rasmussen, 1993, Vernet, 2000, Spawls et al., 2002 and Largen & Spawls, 2010.

Remarks: Type locality (Taita Distr.) and Ukamba Distr. (1°29'S, 37°17'E) visited by Hildebrandt before traveling to Mombasa *fide* W.C.H. Peters (1878: 194), suggesting type locality is Taita rather than Taita Hills, W of Voi, Kenya.

P

6. *Psammophis brevirostris* W.C.H. Peters, 1881d. Sitz. Ges. Naturf. Freunde Berlin 1881(6): 89.

Type: Lectotype, ZMB 7256A, designated by Broadley (1977b: 19).

Type locality: "Matlale Mission, near Pietersburg, Transvaal" [= Matlale Mission, near Polokwane (23°53'S, 29°27'E, elevation 1315 m), Limpopo Prov., NE South Africa] via lectotype selection.

Distribution: Southern Africa. Southeastern Zimbabwe (Manicaland, Mashonaland East, Matebeleland South), S Mozambique (Maputo), E Botswana (Central, Ghanzi, Kweneng, North East, South East), Swaziland (Maputo) and E South Africa (Eastern Cape, Gauteng, E KwaZulu-Natal, Limpopo, Mpumalanga, North-West), NSL–2000 m.

Sources: Auerbach, 1987, Broadley, 2002, M. Griffin, 2003 and Broadley & Blaylock, 2013.

Remarks: Photograph of lectotype in Broadley (1977b: pl. 3a).

7. *Psammophis elegans* (G. Shaw, 1802). Gen. Zool., Amph. 3(2): 536. (*Coluber elegans*)

Synonym: *Psammophis elegans univittatus* Perret, 1961.

Type: Lectotype, BMNH 1946.1.8.8, a 610 mm specimen, designated by K.L. Williams & Smith (1962: 300).

Type locality: "South America," (in error) via lectotype selection. Corrected to West Africa *fide* Boulenger (1896a: 168).

Distribution: West Africa. Southern Mauritania (Hodh El Gharbi, Trarza), Senegal (Dakar, Fatick, Kédougou, Kolda, Matam, Saint-Louis, Tambacounda, Thiès, Ziguinchor), Gambia (Western), Guinea–Bissau (Biombo, Bissau, Oio, Tombali, Bolama Is.), Guinea (Boffa, Kouroussa, Nzérékoré, Télimélé), Sierra Leone (Eastern, W Northern, Southern), SW Mali (Kayes, Koulikoro, Mopti), Ivory Coast (Abidjan, Bouaké, Bouna, Fekéssédougou, Toumodi), Burkina Faso (Centre, Centre-Est, Centre-Nord, Centre-Ouest, Est, Haute-Bassins, Sahel, Sud-Ouest, Volta-Noire), Ghana (Central, Northern, Upper East, Upper West, Volta, Tumbo Is.), Togo (Kara, Plateaux), N Benin (Alibori, Atakora, Borgou, Cotonou, Zou), S Niger (Dosso, S Maradi, Niamey, S Zinder), Nigeria (Delta, Kaduna, Lagos, Plateau, Sokoto), N Cameroon (Adamaoua, Extreme-Nord, Nord) and Central African Republic (Bamingui-Bangoran, Sangha), NSL–1475 m.

Sources: Angel, 1933b, Villiers, 1950a, 1951b, 1952b, 1953, 1954a, 1956b, 1963, 1975, Perret, 1961, Doucet, 1963, Menzies, 1966, Leston & Hughes, 1968, Hughes & Barry, 1969, J Papenfuss, 1969, Roux-Estève, 1969a, W. Böhme, 1978, Roman, 1980, Joger, 1981, 1990, Hughes, 1983, 2013, Roman, 1984, Chippaux, 2006, Chirio & Ineich, 2006, J.-F. Trape & Mané, 2006b, Chirio & LeBreton, 2007, Chirio, 2009, 2013, Ullenburch et al., 2010, Segniagbeto et al., 2011 and Auliya et al., 2012.

Remarks: Original description based on Seba (1735: 59, pl. 60, fig. 1) from Nova Hispaniâ (= Mexico & Central America *fide* Wallach, 2011: 6). Official Specific Name no. 2046 *fide* Opinion 717 (ICZN, 1965a).

8. *Psammophis jallae* Peracca, 1896c. Boll. Mus. Zool. Anat. Comp. Univ. Torino 11(255): 2–4, 2 figs.

Synonyms: *Psammophis rohani* Angel, 1921b and *Psammophis longirostris* V.F.M. FitzSimons, 1932.

Types: Syntypes (2), MZUT 1633 and MZUT 1871 (formerly MZUT 1823a–b) (L. Jalla).

Type locality: "Strada da Kazungula a Buluwaio, Zambese" [= road between Kazungula (17°13'S, 25°51'E, elevation 1135 m), Southern Prov., Zambia to Bulawayo (20°10'S, 28°35'E, elevation 1380 m), Bulawayo Prov., W Zimbabwe].

Distribution: Southern Africa. Southeastern Angola (Bié, Cuando Cubango), SE Democratic Republic of the Congo (S Katanga), W Zambia (Western), Zimbabwe (Bulawayo, Mashonaland East, Matebeleland North, Matebeleland South, Midlands), NE Namibia (Okavango, Otjozondjupa), Botswana (Central, Ghanzi, Kgatleng, Kweneng, North West, Southern) and NE South Africa (Limpopo), 600–1500 m.

Sources: Mertens, 1955, 1971, V.F.M. FitzSimons, 1962a, Witte, 1962, Broadley, 1971d, Broadley & Cock, 1975, Auerbach, 1985, 1987, Broadley, 2002, Broadley et al., 2003, M. Griffin, 2003, Marais, 2004 and Broadley & Blaylock, 2013.

9. *Psammophis leightoni* Boulenger, 1902f. Proc. Zool. Soc. London 72(1): 126, pl. 12.

Type: Holotype, BMNH 1946.1.8.27, a 910 mm specimen (G. Leighton).

Type locality: "Eerste River Station, 21 miles east of Cape Town, Cape Colony" [= Eerste River Station, SW Western Cape Prov., SW South Africa, 34°05'S, 18°46'E, elevation 10 m].

Distribution: Extreme SW South Africa (SW Western Cape), NSL–185 m.

Sources: V.F.M. FitzSimons, 1962a, 1974 and Broadley, 1975, 2002.

Remarks: Possibly not a valid species *fide* Kelly et al. (2008: 1053).

10. *Psammophis leopardinus* Bocage, 1887a. J. Sci. Math. Phys. Nat., Lisboa 11(44): 206–207. (*Psammophis sibilans leopardinus*)

Type: Lectotype, MBL 1798, designated by Broadley (1977b: 18), destroyed by fire 18 March 1978.

Type locality: "Catumbela, Angola" [= Catumbela, NW Benguela Distr., W Angola, 12°26'S, 13°33'E, elevation 25 m] via lectotype selection.

Distribution: Southwestern Angola (Benguella, Huíla, Namibe) and Namibia (Erongo, Hardap, Khomas, Kunene, Okavango, Oshikoto, Otjozondjupa), 25–1500 m.

Sources: Bocage, 1887d, Broadley, 1977, 2002, Elzen, 1980, Hughes, 1999 and M. Griffin, 2003.

11. *Psammophis mossambicus* W.C.H. Peters, 1882d. Reise Mossambique, Zool. 3: 122. (*Psammophis sibilans mossambica*)

Synonyms: *Psammophis sibilans tettensis* W.C.H. Peters, 1882d and *Psammophis thomasi* Gough, 1908.

Type: Lectotype, ZMB 2468A (W.C.H. Peters, June 1843–Aug. 1847), designated by Broadley (2002: 96).

Type locality: "Insel Mossambique" [= Mozambique Is., Mozambique] via lectotype selection.

Distribution: Eastern Africa. Sudan Sudan, Uganda (Central, Eastern, Northern, Western), SW Kenya (S Coastal, S Eastern, Nyanza, Rift Valley, Western), Tanzania (Arusha, Dar es Salaam, Dodoma, Iringa, Kigoma, Kilimanjaro, Lindi, Mara, Mbeya, Morogoro, Mtwara, Mwanza, Piwani, Rukwa, Ruvuma, Shinyanga, Singida, Tabora, Tanga, Kagera), Democratic Republic of the Congo (Bandundu, Bas-Congo, Equateur, Katanga, Kinshasa, Orientale), Angola (Huíla, Lunda Norte, Moxico), Mozambique (Cabo Delgado, Gaza, Inhambane, Manica, Maputo, Nampula, Niassa, Sofala, Zambézia), Malawi (Central, Northern, Southern), Zambia (Central, Copperbelt, Eastern, Luapula, Lusaka, Northern, North-Western, Southern, Western), Zimbabwe (Bulawayo, Manicaland, Mashonaland Central, Mashonaland East, Mashonaland West, Masvingo, Matabeleland North, Matabeleland South, Midlands), N Botswana (Central, Ghanzi, North West), N Namibia (Khomas, Kunene, Okavango, Otjozondjupa), Swaziland and NE South Africa (KwaZulu-Natal, Limpopo, Mpumalanga), NSL–1500 m.

Sources: Mertens, 1955, 1971, Sweeney, 1961, Witte, 1962, Laurent, 1964a, Thys van den Audenaerde, 1965, Pienaar et al., 1983, Broadley & Howell, 1991, Boycott & Culverwell, 1992, Branch & McCartney, 1993, Hughes, 1999, Broadley, 2002, Clauss & Clauss, 2002, Spawls et al., 2002, Broadley et al., 2003, M. Griffin, 2003 and Broadley & Blaylock, 2013.

Remarks: Taxonomy of *P. sibilans* complex *(mossambicus-occidentalis-phillipsii-rukwae-sibilans)* unresolved and therefore ranges are only approximate.

12. *Psammophis namibensis* Broadley, 1975. Arnoldia, Rhodesia 7(13): 9–11, pl. 1. (*Psammophis leightoni namibensis*)

Type: Holotype, TMP 37093, a 885+ mm female (W.D. Haacke, 12 May 1969).

Type locality: "Harus in the Uri-Hauchab Mountains, Diamond Area No. 2, South West Africa" [= NW Karas Prov., SW Namibia, 25°21'S, 15°10'E, elevation 465 m].

Distribution: Southwest Africa. Southwestern Angola (Namibe), W Namibia (Erongo, Hardap, Karas, Kunene) and NW South Africa (NW Northern Cape), NSL–1500 m.

Sources: Broadley, 2002 and M. Griffin, 2003.

Remarks: Possibly not a valid species *fide* Kelly et al. (2008: 1053).

13. *Psammophis notostictus* W.C.H. Peters, 1867c. Mber. Königl. Akad. Wiss. Berlin 1867(4): 237. (*Psammophis moniliger notostictus*)

Synonym: *Psammophis sibilans stenocephalus* Bocage, 1887a.

Types: Syntypes (2), ZMB 5756a–b (C.H. Hahn, 1844–1867).

Type locality: "Otjimbingue in Südwestafrica" [= Otjimbingwe, SE Karibib Distr., cen. Namibia, 22°21'S, 16°08'N, elevation 900 m].

Distribution: Southwestern Africa. Southwestern Angola (Namibe), Namibia (Erongo, Hardap, Karas, Khomas, Kunene, Omaheke, Oshikoto, Otjozondjupa), ext. SW Botswana (SW Kgalagadi), W Lesotho and W South Africa (Eastern Cape, S Free State, Northern Cape, Western Cape), NSL–1500 m.

Sources: Bocage, 1887d, Mertens, 1955, 1971, V.F.M. FitzSimons, 1962a, 1974, Broadley, 1975, 1977b, 2002, De Waal, 1978, Elzen, 1980, Auerbach, 1987, M. Griffin, 2003, Marais, 2004 and Broadley & Blaylock, 2013.

14. *Psammophis occidentalis* F. Werner, 1919. Denkschr. Akad. Wiss. Wien, Math.-Natur. Kl. 96: 504. (*Psammophis subtaeniatus occidentalis*)

Synonym: *Psammophis irregularis* Fischer, 1856b *(nomen oblitum).*

Type: Lectotype, NMW 19245.2, a 511+ mm female, designated by Hughes & Wade (2004: 129).

Type locality: "Congo" via lectotype selection.

Distribution: West Africa. Southern Mauritania (Assaba, Guidimaka, Hodh El Gharbi), Senegal (Dakar, Diourbel, Fatick, Kaolack, Kédougou, Kolda, Louga, Matam, Saint-Louis, Tambacounda, Thiès, Ziguinchor), Gambia (MacCarthy Island, Western), Guinea-Bissau (Bissau, Cacheu, Gabú, Tombali, Bolama Is.), Guinea (Beyla, Dubréka, Kérouané, Macenta, Nzérékoré, Yomou, Los Is.), Sierra Leone (Eastern, Northern, Southern, Western, Bunce Is.), Liberia (Monrovia, Montserrado, Nimba), S Mali (S Gao, Kayes, Koulikoro, Mopti, Ségou, Sikasso, Tahoua, S Tombouctou), Ivory Coast (Abidjan, Daloa, Toumodi), Burkina Faso (Centre, Centre-Est, Centre-Nord, Centre-Ouest, Est, Haute-Bassins, Nord, Sahel, Sud-Ouest, Volta-Noire), Ghana (Accra, Northern, Volta), Togo (Centrale, Kara, Maritime, Plateaux, Savanes), Benin (Atakora, Borgou), S Niger (Agadez, Diffa, Dosso, Maradi, Niamey, Zinder), Nigeria (Akwa Ibom, Bauchi, Borno, Cross River, Delta, Gongola, Kano, Kwara, Niger, Plateau, Rivers), S Cameroon (Centre, Est, Littoral, Nord-Ouest, Ouest, Sud, Sud-Ouest), Central African Republic (Bamingui-Bangoran, Nana-Mambere, Ouham, Ouham-Pende, Vakaga), Equatorial Guinea, Gabon (Ogooué-Ivindo, Ogooué-Maritime) and S Congo (Bouenza, Brazzavillle, Kouilou), 20–2000 m.

Sources: Menzies, 1966, Hakansson, 1981, J.-F. Trape & Roux-Estève, 1995, W. Böhme et al., 1996, Hughes & Wade, 2004, Luiselli et al., 2004b, Chirio & Ineich, 2006, J.F. Trape & Mané, 2006b, Chirio & LeBreton, 2007, Kelly et al., 2008, 2011, Pauwels & Vande Weghe, 2008, Segniagbeto et al., 2011 and Auliya et al., 2012.

Remarks: A valid species *fide* Kelly et al., 2008. In accordance with Art. 23.9.2 of the Code (ICZN, 1999), *Psammophis subtaeniatus occidentalis* Werner is designated a *nomen protectum* and *Psammophis irregularis* Fischer a *nomen oblitum.* Taxonomy of *P. sibilans* complex *(mossambicus-occidentalis-phillipsii-rukwae-sibilans)* unresolved and therefore ranges are only approximate.

15. *Psammophis orientalis* Broadley, 1977b. Arnoldia, Rhodesia 8(12): 17–18. (*Psammophis subtaeniatus orientalis*)

Type: Holotype, MCZ 146965, a 972 mm female (A. Loveridge, 23 Sept. 1917).

Type locality: "Morogoro, Tanzania" [= Morogoro, Morogoro Prov., E Tanzania, 6°49'S, 37°40'E, elevation 510 m].

Distribution: Eastern Africa. Southwestern Sudan (Janub Darfur), S South Sudan (Eastern Equatoria, Western Equatoria), E Kenya (Coast, Eastern, Lamu Is.), E Tanzania (Arusha, Dar es Salaam, Dodoma, Kigoma, Kilimanjaro, Mbeya, Morogoro, Mtwara, Mwanza, Shinyanga, Singida, Tabora, Tanga, Zanzibar Is.), Mozambique (Gaza, Inhambane, Manica, Maputo, Sofala, Tete, Zambézia, Bazaruto and Benguela Is.), Malawi (Central, Northern, Southern) and E Zimbabwe (Manicaland, Masvingo), NSL–1300 m.

Sources: Steehouder, 1992b, Broadley, 2002, Spawls et al., 2002 and Broadley & Blaylock, 2013.

Remarks: A valid species *fide* Spawls et al. (2002: 405).

16. *Psammophis phillipsii* (Hallowell, 1844c). Proc. Acad. Nat. Sci. Philadelphia (1844–1845) 2(6): 169. (*Coluber phillipsii*)

Synonym: *Psammophis regularis* Sternfeld, 1908b.

Type: Lectotype, ANSP 5112, a female (S.M.E. Goheen, 1836–1838), designated by Broadley (1977b: 24); longest syntype 1006 mm.

Type locality: "Liberia, West coast of Africa," via lectotype selection.

Distribution: West Africa. Senegal (Tambacounda), Gambia (Western), Guinea-Bissau (Bubaque, Soga and Uno Is.), Guinea (Dubréka, Kouroussa, Macenta, Nzérékoré, Télimélé), Liberia (Montserrado, Margibi), S Ivory Coast (Abidjan, Daloa, Lagunes, Toumodi), S Ghana (Accra), S Togo (Centrale, Kara, Maritime, Plateaux, Savanes), S Benin (Atlantique, Borgou, Collines, Zou), S Nigeria (Akwa Ibom, Cross River, Delta, Rivers), Cameroon (Adamaoua), Gabon, Congo (Brazzaville, Cuvette, Kouilou, Plateau, Pool), Central African Republic (Bamingui-Bangoran, Gribingui, Haute-Kotto, Haute-Sangha, Kemo-Gribingui, Lobaye, Mbomou, Ombella-Mpoko, Ouaka, Ouham, Ouham-Pende, Sangha), NSL–1200 m.

Sources: Sternfeld, 1908b–c, 1909a–b, Angel, 1933b, Villiers, 1950a, 1954a, 1959, 1963, 1975, E.H. Taylor & Weyer, 1958, Broadley, 1959, 1971d, 1977b, 1983, Doucet, 1963, Knoepffler, 1966, Leston & Hughes, 1968, Hughes & Barry, 1969, Roux-Estève, 1969a,

P

Roman, 1980, 1984, Hakansson, 1981, Hughes, 1983, 1999, 2013, J.A. Butler & Reid, 1986, Joger, 1990, Gruschwitz et al., 1991, D.P. Lawson, 1993, J.-F. Trape & Roux-Estève, 1995, Akani et al., 2002b, Broadley et al., 2003, Luiselli et al., 2004b, Chirio & Ineich, 2006, J.-F. Trape & Mané, 2006b, Ullenburch et al., 2010, W. Böhme et al., 2011, Segniagbeto et al., 2011 and Chirio, 2013.

Remarks: Taxonomy of *P. sibilans* complex *(mossambicus-occidentalis-phillipsii-rukwae-sibilans)* unresolved and therefore ranges are only approximate.

17. *Psammophis pulcher* Boulenger, 1895h. Proc. Zool. Soc. London 63(3): 537–538, pl. 30, figs. 3–3a.

Type: Holotype, BMNH 1946.1.8.50 (formerly BMNH 1895.6.11.7), a 435 mm female (A. Donaldson-Smith, 24 Dec. 1894).

Type locality: "Webi Shebeli, Galla or western Somaliland" [= Webi Shebeli, Dacata River, SE Hamarro Hadad, Hararge Prov., E Ethiopia, 7°20'N, 42°17'E, elevation 500–600 m].

Distribution: Southeastern Ethiopia (Hararge), S Somalia and S Kenya (SW Coast, S Eastern), 400–820 m.

Sources: Drewes & Spawls, 1973, Spawls, 1978, Largen & Rasmussen, 1993, Spawls et al., 2002 and Largen & Spawls, 2010.

18. *Psammophis punctulatus* A.-M.-C. Duméril, Bibron & Duméril, 1854b. Erpét. Gén. 7(2): 897–898.

Synonyms: *Dendrophis furcata* Bianconi, 1859, and *Psammophis punctulatus trivirgatus* W.C.H. Peters, 1878.

Type: Holotype, MNHN 6552, a 1230 mm specimen (B. d'Arnaud [Second-White Nile Exped.], Nov. 1840–April 1841).

Type locality: "Arabie" [= Arabia] (in error). Corrected herein to Nile River, bet. Al Khartum (15°35'N, elevatuon 390 m), An Nil Al Abyad Prov., SE Sudan, and Juba (4°42'N, elevation 500 m), Upper Nile, Jonglei or Central Equatoria Prov., South Sudan].

Distribution: Eastern Africa. Extreme SE Egypt (SE El-Bahr El-Ahmar), E Sudan (An Nil Al Abyad, An Nil Al Azraq, Ash Shaqiyah, Sennar), South Sudan, Eritrea (Anseba, Gash-Barka, Northern Red Sea, Southern Red Sea), Ethiopia (Bale, Geua Gofa, Hararge, Kefa, Shoa, Sidamo), Djibouti, Somalia (Awdal, Bakool, Bay, Galguduud, Gedo, Hiraan, Jubbada Dhexe, Jubbada Hoose, Mudug, Nugaal, Shabeellaha Dhexe, Shabeellaha Hoose, Togdheer, Woqooyi Galbeed), NE Uganda (NE Northern), Kenya (Coast, Eastern, Rift Valley, North-Eastern), NE Tanzania (Arusha, Kilimanjaro), NSL–1400 m.

Sources: H,W. Parker, 1932b, 1936c, 1949, Pitman, 1974, Spawls, 1978, Lanza, 1983a, 1990b, Schätti, 1989, Broadley & Howell, 1991, Largen & Rasmussen, 1993, Spawls et al., 2002, Baha El Din, 2006 and Largen & Spawls, 2010.

Remarks: Loveridge (1940: 21) correctly suspected erroneous type locality data for *P. punctulatus* (Arabia) and *Dendrophis furcata* (Mozambique). Concerning the former, B. d'Arnaud collected on two White Nile expeditions with G. Thibaut and Sabatier (1840–1842).

19. *Psammophis rukwae* Broadley, 1966c. Arnoldia, Rhodesia 2(36): 3–4. (*Psammophis sibilans rukwae*)

Synonym: *Psammophis leucogaster* Spawls, 1983.

Type: Holotype, NMZB 4212 (formerly NMSR 4212), a 1255 mm male (L.D.E.F. Vesey-FitzGerald, 23 Nov. 1959).

Type locality: "Kafukola, Rukwa Valley, Tanganyika at 2,600 feet" [= Kafukola, S Rukwa Region, SW Tanzania, 8°05'S, 31°57'E, elevation 790 m].

Distribution: West, cen. and E Africa. Western Senegal (Dakar, Saint-Louis, Thiès), N Ghana (Northern, Upper West), Burkina Faso (Est, Nord), Togo, N Benin (Borgou), Niger (Dosso, Niamey), Nigeria (Borno, Gongola, Kaduna), S Algeria (Tamanrasset), N Cameroon (Nord), S Chad, Central African Republic, SW Sudan (Janub Darfur), S South Sudan (Eastern Equatoria), E Uganda (N Eastern), SW Kenya (S Rift Valley) and W Tanzania (Arusha, Dodoma, Mwanza, Rukwa, Shinyanga), NSL–1675 m.

Sources: W. Böhme, 1975, 1978, 1986, Joger, 1981, 1982, Spawls, 1983, Largen & Rasmussen, 1993, Chippaux, 1999 and Spawls et al., 2002.

20. *Psammophis schokari* (Forskål, 1775). Descr. Animal.: 14–15. (*Coluber schokari*)

P

Synonyms: *Coluber gemmatus* G. Shaw, 1803, *Coluber lacrymans* A. Reuss, 1834, *Psammophis punctatus* A.-M.-C. Duméril, 1853 (*nomen nudum*), *Psammophis punctatus* A.-M.-C. Duméril, Bibron & Duméril, 1854b, *Psammophis sibilans hierosolimitana* Jan, 1863b (*nomen nudum*), *Psammophis sibilans hierosolimitana* Jan, 1870 *in* Jan & Sordelli, 1866–1870, *Psammophis sindanus* Stoliczka, 1872a, and *Psammophis sibilans teknae* Valverde, 1989a.

Type: Neotype, FMNH 66153, a 763 mm male (H. Hoogstraal & R.E. Kuntz, 27 Jan. 1951), designated by Marx (1988: 13–14).

Type locality: "Hodeida (= Al Hudaydah), Yemen" [= Al Hudaydah, Al Hudaydah Prov., W Yemen, 14°48'N, 42°57'E, elevation NSL] via neotype selection.

Distribution: Northern Africa, Middel East and SW Asia. Morocco (Agadir, Azilal, Boulemane, Chechaouene, Essaouira, Figuig, Nador, Khenitra, Khemisset, Kaar es Souk, Marrakech, Nador, Oujda, Ouarzazate, Rabat, Tan Tan, Tata, Taza, Tiznit), Algeria (Adrar, Annaba, Béchar, Biskra, El Bayadh, El Oued Laghouat, M'Sila, Naama, Ouargla), S Tunisia (Gabès, Gafsa, Nabeul, Sfax, Sidi Bou Zid, Sousse, Tataouine), Libya (Ajdabiya,

Al Jabal, Al Jufrah, Al Khums, Al Kurfah, Az Zawiyah, Banghazi, Butnan, Gharyan, Ghat, Misratah, Murzuq, Nalut, Sabha, Surt, Tarabulus), Egypt (Alexandria, Aswan, Beheira, Cairo, El-Bahr El-Ahmar, Faiyum, Giza, Ismailia, Matrouh, New Valley, North Sinai, Port Said, Qena, Sharkeya, South Sinai, Suez, Shadwan Is.), Western Sahara (Rio de Oro, Saguia El Hamra), W Mauritania (W Adrar, Brakna, Inchiri, Tagant, Trarza), Senegal (Dakar, Saint-Louis), N Mali (Tombouctou), N Niger (Agadez), N Chad, N Sudan (El-Bahr El-Ahmar, Al Khartum, An Nil Al Abyad, An Nil Al Azraq, Ash Shamaliyah, Janub Kurdufan, Shamal Darfur, Shamal Kurdufan), Eritrea (Central, Gash-Barka, Northern Red Sea), Ethiopia (E Hararge), Djibouti, N Somalia (Awdal, Bari, Sanaag, Sool, Togdheer, Woqooyi Galbeed), Jordan (Ajloun, Amman, Aqaba, Irbid, Karak, Maan, Mafraq, Tafilah, Zarqa), Israel (Central, Gaza Strip, Golan Heights, Haifa, Jerusalem, Northern, Southern, Tel Aviv, West Bank), Lebanon (Beirut, El Jnoub), S Syria, Saudi Arabia (Asir, Bahah, Eastern, Ha'il, Jawf, Jazan, Madinah, Makkah, Northern, Qasim, Riyadh, Tabuk, Farasan Is.), Bahrain and Sitra Is., Kuwait, United Arab Emirates (Abu Dhabi, Dubai, Fujairah, Ras al Khaimah, Sharjah), Oman (Batinah, Janubiah, Muscat, Sharqiyah, Masirah Is.), Yemen (Abyan, Ad Dali', Aden, Al Hudaydah, Dhamar, Hadhramaut, Lahij, Sana'a, Shabwah, Ta'izz, Kamaran Is.), S Iraq (Al Anbar, Al Basrah, Maysan), Iran (Central, Esfahan, Fars, Kerman, Khuzestan, Mazandaran, Razavi Khorasan, Semnan, Sistan va Baluchestan, Yazd), S Turkmenistan, S Afghanistan (Helmand, Kandahar Nangarhar, Nimroze), Pakistan (Balochistan, F.A.T.A., Jammu & Kashmir, Khyber Pakhtunkhwa, Punjab, Sindh) and NW India (Gujarat, Maharashtra, Rajasthan), NSL–2400 m.

Sources: J. Anderson, 1898, Doumergue, 1901, Ingoldby & Procter, 1923, Wall, 1924a, Angel, 1933b, Flower, 1933, Corkill, 1935, H.W. Parker, 1941, 1949, M.A. Smith, 1943, Villiers, 1950a–b, 1954a, 1963, 1975, Tataouine Marx, 1958, 1968, 1988, Khalaf, 1959, Kramer & Schnurrenberger, 1963, Corkill & Cochrane, 1965, Bons, 1967, Leviton & Anderson, 1967, 1970a, Papenfuss, 1969, Gallagher, 1971, 1990, Arnold, 1980b, M.S. Khan, 1980, 2002, Hraoui-Bloquet, 1981, Lanza, 1983a, Disi et al., 1988, Gallagher & Arnold, 1988, Gasperetti, 1988, Marx, 1988, Osman & Sir, 1988, Clark, 1990, Latifi et al., 1990, Largen & Rasmussen, 1993, Szczerbak 1994, Bons & Geniez, 1996, Schleich et al., 1996, Bogaerts, 1998, Chippaux, 1999, Bouskila & Amitai, 2001, Disi et al., 2001, Baker et al., 2004, Geniez et al., 2004, Whitaker & Captain, 2004, J.-F. Trape & Mané, 2006b, Largen & Spawls, 2010, Amr & Disi, 2011, Bar & Haimovitch, 2011 and Masroor, 2012.

Remarks: Syntypes in ZMUC lost *fide* F.W. Braestrup *in* Kramer & Schnurrenberger (1963: 516). Photographs and illustrations of neotype in Marx (1988: figs. 2–5).

21. *Psammophis sibilans* (Linnaeus, 1758). Syst. Nat., ed. 10, 1: 222. (*Coluber sibilans*)

Synonyms: *Coluber moniliger* Daudin, 1803d, *Coluber auritus* I. Geoffroy-Saint-Hilaire, 1827, *Psammophis irregularis* J.G. Fischer, 1856b, *Psammophidion smithii* Fitzinger, 1861, *Psammophis sibilans intermedius* J.G. Fischer, 1884a, and *Psammophis silvomarginata* Brandstätter, 1995 (*nomen ineditum*).

Type: Holotype, ZMUU 42 (formerly MP 30), a 950 mm specimen.

Type locality: "Asia," (in error). Corrected to Africa *fide* J. Anderson (1898: 305) and Flower (1933: 824). Restricted to Egypt *fide* Loveridge (1953: 274).

Distribution: Northeastern Africa. Egypt (Assiut, Aswan, Beheira, Beni Suef, Cairo, Faiyum, Gharbia, Giza, Ismailia, Matrouh, Minya, North Sinai, Qena, Sohag), SE Sudan (Al Khartum, An Nil Al Abyad, An Nile Al Azraq, Sennar, South Sudan (Bahr al Ghazal, Eastern Equatoria), Eritrea (Anseba, Central, Northern Red Sea, Southern), Ethiopia (Arussi, Bale, Gemu Gofa, Gojjam, Gondar, Hararge, Ilubabor, Kefa, Shoa, Sidamo, Tigre, Wollega, Wollo), Somalia (Hiraan, Jubbada Dhexe, Jubbada Hoose, Togdheer, Sanaag, Shabeellaha Dhexe, Shabeellaha Hoose, Woqooyi Galbeed), Kenya (Central, Coast, Eastern, Nairobi, North-Eastern, Nyanza, Rift Valley, Western), Uganda (Acholi, Ankole, Bukedi, Bunyoro, Busoga, East Mengo, Karamoja, Lango, Madi, Masaka, Sebei, Teso, Toro, West Mengo, West Nile, Sese Is.), Tanzania (Arusha, Dar es Salaam, Dodoma, Kilimanjaro, Morogoro, Mtwara, Piwani, Singida, Tanga, Zanzibar, Ukwere Is.), NE Democratic Republic of the Congo (E Kivu E Orientale), Rwanda (Butare, Byumba, Kibungo) and Burundi (Muyinga, Ngozi, Ruyigi), NSL–2500 m.

Sources: Loveridge, 1936c, 1953, Marx, 1958, 1988, Broadley, 1959, 1971d, 1983, Sweeney, 1961, Witte, 1962, 1966, Blackwell, 1967, V.F.M. FitzSimons, 1974, Pitman, 1974, Spawls, 1978, Roman, 1980, 1984, Lanza, 1983a, 1990b, Boycott & Culverwell, 1992, Steehouder, 1992b, Largen & Rasmussen, 1993, Hughes, 1999, Ibrahim, 2002 and Largen & Spawls, 2010.

Remarks: Holotype illustrated in J. Anderson (1898: fig. 12) and Brandstätter (1996: fig. 25). Original description based on A. Seba (1735: 52–53, pl. 52, fig. 4; 1735: 57, pl. 56, fig. 4; 1735: 115, pl. 107, fig. 4). Taxonomy of *P. sibilans* complex (*mossambicus-occidentalis-phillipsii-rukwae-sibilans*) unresolved and therefore ranges are only approximate. Multiple records from Guinea and Benin *fide* Chirio (2013: 83) and Hughes (2013: 123).

22. *Psammophis subtaeniatus* W.C.H. Peters, 1882d. Reise Mossambique, Zool. 3: 121–122. (*Psammophis sibilans subtaeniatus*)

Synonyms: *Psammophis moniliger bilineatus* W.C.H. Peters, 1867c (*nomen rejiciendum*), *Psammophis*

P

bocagii Boulenger, 1895h, and *Psammophis transvaalensis* Gough, 1908.

Type: Lectotype, ZMB 9992A (W.C.H. Peters, June 1843–Aug. 1847), designated by Broadley (1977b: 13).

Type locality: "Tete" [= Tete, Tete Prov., W Mozambique, 16°10'S, 33°36'E, elevation 125 m] via lectotype selection.

Distribution: Southern Africa. Southwestern Angola (Benguela, Cunene, Namibe), S Zambia (Central, Eastern, Southern, Western), S Malawi (Southern), Zimbabwe (Bulawayo, Manicaland, Mashonaland Central, Mashonaland East, Mashonaland West, Masvingo, Matabeleland North, Matabeleland South, Midlands), Botswana (Central, Gaborone, Ghanzi, Kgatleng, Kweneng, North West, North East, South East, Southern), N Namibia (Caprivi, Erongo, Khomas, Kunene, Okavango, Omaheke, Oshikoto, Otjozondjupa), Mozambique (Gaza, Tete), Swaziland and NE South Africa (Limpopo, Mpumalanga, North West), 35–1500 m.

Sources: Mertens, 1955, Laurent, 1956a, Broadley, 1959, 1966c, 1971d, 1990, 1992a, Sweeney, 1961, Witte, 1962, 1975, V.F.M. FitzSimons, 1974, Pitman, 1974, Broadley & Cock, 1975, Spawls, 1978, Pienaar et al., 1983, Auerbach, 1987, W. Böhme, 1987, Broadley & Howell, 1991, Boycott & Culverwell, 1992, Chippaux, 1999, Broadley, 2002, Clauss & Clauss, 2002, Broadley et al., 2003 and Broadley & Blaylock, 2013.

Remarks: Official Specific Name no. 2801 *fide* Opinion 1210 (ICZN, 1982c).

23. *Psammophis sudanensis* F. Werner, 1919. Denkschr. Akad. Wiss. Wien, Math.-Natur. Kl. 96: 504. (*Psammophis sibilans sudanensis*)

Synonym: *Psammophis leucogaster* Spawls, 1983.

Type: Lectotype, NMW 19086, a male (F. Werner [Kordofan Zool. Exped.], 28 March 1914), designated by Tiedemann et al. (1994: 73).

Type locality: "Kasugli, S-Kordofan, Anglo-ägyptischen Sudan" [= Kadugli, Janub Kurdufan Prov., S Sudan, 11°01'N, 29°43'E, elevation 510 m] via lectotype selection.

Distribution: West, cen. and E Africa. Western Guinea (Boké, Télimélé), S Benin (Atlantique, Zou), N Cameroon (Extreme-Nord, Nord), Central African Republic (Ouham, Ouham-Pende, Vakaga), S Sudan (Janub Kurdufan), E Uganda (Eastern, E Northern), Kenya (SW Coast, Eastern, Nairobi, Nyanza, N Rift Valley, Western), and NE Tanzania (Arusha, Dar es Salaam, Dodoma, Kilimanjaro, Pwani, Singida, Tabora), NSL–2700 m.

Sources: Spawls, 1983, W. Böhme, 1986, 1987, Hughes, 1999, Spawls et al., 2002, Chirio & Ineich, 2006, Chirio & LeBreton, 2007, Ullenburch et al., 2010, Chirio, 2013 and Hughes, 2013.

24. *Psammophis tanganicus* Loveridge, 1940. Bull. Mus. Comp. Zool. 87(1): 57–60. (*Psammophis biseriatus tanganicus*)

Type: Holotype, MCZ 30380, a 633 mm female (A. Loveridge, 12 Dec. 1929).

Type locality: "Mangasini, Usandawi, central Tanganyika Territory" [= Mangasini, NW Singida Region, cen. Tanzania, ca. 4°53'S, 34°09'E, elevation 1110 m].

Distribution: Northeastern Africa. Southern Libya (Awbari), Sudan (El-Bahr El-Ahmar, Shamal Kurdufan), Eritrea (Southern Red Sea), E Ethiopia (Bale, Hararge, Shoa, Sidamo), W Somalia (Awdal, Bakool, Bari, Gedo, Nugaal, Sanaag, Togdheer, Woqooyi Galbeed), Kenya (SW Coast, Eastern, North-Eastern, N Rift Valley), E Uganda (Eastern) and NE Tanzania (Dodoma, Kilimanjaro, Singida), 1000–1700 m.

Sources: Corkill, 1935, Scortecci, 1936, Kramer & Schnurrenberger, 1963, Pitman, 1974, Spawls, 1978, Lanza, 1983a, Bezy & Drewes, 1985, Largen & Rasmussen, 1993, Spawls et al., 2002 and Largen & Spawls, 2010.

Remarks: Broadley & Howell (1991: 27) listed type locality as Kwa Mtoro, Dodoma Region, 5°14'S, 35°26'E, elevation 1185 m.

25. *Psammophis trigrammus* A.C.L.G. Günther, 1865. Ann. Mag. Nat. Hist. (3) 15(86): 95, pl. 2, fig. e.

Type: Holotype, BMNH 1946.1.8.12 (formerly BMNH 1864.7.11.4), a 1180–1219 mm male (F.F. Monteiro, 1864).

Type locality: "on the banks of the river S. Nicolao (Little Fish Bay, West Africa)," (partly in error). Corrected to Mossamedes [= Rio São Nicolau or Bentiaba River, N Namibe Distr., SW Angola, 14°16'S, 12°23'E, elevation 10–100 m] *fide* Bocage (1887c: 206).

Distribution: Southwestern Africa. Southwestern Angola (Namibe), W Namibia (Erongo, Hardap, Karas, Khomas, Kunene, Okavango, Otjozondjupa) and ext. W South Africa (W Northern Cape), NSL–1500 m.

Sources: Mertens, 1955, 1971, V.F.M. FitzSimons, 1974 and Broadley, 1983, 2002.

26. *Psammophis trinasalis* F. Werner, 1902. Verh. Zool. Bol. Ges. Wien 52(1): 340. (*Psammophis sibilans trinasalis*)

Synonym: *Psammophis moniliger furcatus* W.C.H. Peters, 1867c (*nomen praeoccupatum*).

Type: Lectotype, ZFMK 30503, a 663 mm male (F. Werner, 1900), designated by Elzen (1980: 343).

Type locality: "Windhoek, Deutsch-Südwest-Afrika" [= Windhoek, Khomas Distr., cen. Namibia, 22°34'S, 17°05'E, elevation 1665 m] via lectotype selection.

Distribution: Southern Africa. Southwestern Angola (Namibe), Namibia (Erongo, Hardap, Karas, Khomas, Kunene, Okavango, Omaheke, Oshikoto, Otjozondjupa),

Botswana (Central, Ghanzi, Kgalagadi, Kweneng, North West, Southern), and N South Africa (Eastern Cape, Free State, Gauteng, Limpopo, Mpumalanga, N Northern Cape, North West), NSL–1665 m.

Sources: Mertens, 1955, 1971, Buys & Buys, 1983, Broadley, 2002, M. Griffin, 2003 and Broadley & Blaylock, 2013.

Remarks: Photograph of lectotype in Elzen (1980: fig. 3a), who stated it to be female.

27. *Psammophis zambiensis* Hughes & Wade, 2002. Bull. Brit. Mus. Nat. Hist. (Zool.) 68(2): 75–77, figs. 1–3.

Type: Holotype, BMNH 1959.1.1.81 (formerly IRSNB), a female (H.J. Bredo, 1943–1944).

Type locality: "Abercorn (now Mbala)" (in error). Probably from Mweru Wantipa, Mporokoso Distr., N Northern Prov., Zambia *fide* Vesey-FitzGerald *in* Hughes & Wade (2002: 79).

Distribution: Southeastern Democratic Republic of Congo (Katanga) and N Zambia (Central, Luapula, Northern, North-Western), 1000–1750 m.

Sources: Broadley et al., 2003 and Broadley & Cotterill, 2004.

PSAMMOPHYLAX Fitzinger, 1843 (Psammophiidae)

Synonyms: *Cerastes* Wagler, 1830 *(nomen praeoccupatum), Trimerorhinus* A. Smith, 1847, *Psammophilax* – Doumergue, 1901 *(nomen incorrectum), Trimerhinus* – Bogert, 1942b *(nomen incorrectum),* and *Trimerirhinus*– Anthony, 1955 *(nomen incorrectum).*

Type species: *Coluber rhombeatus* Linnaeus, 1758.

Distribution: Eastern and S Africa.

Sources: V.F.M. FitzSimons, 1962a, 1966, 1974, Broadley, 1977a, 1983, Branch, 1988, Marais, 1992, Gravlund, 2002, Vidal & Hedges, 2002, Kelly et al., 2008, 2009, Vidal et al., 2008 and Zaher et al., 2009.

1. *Psammophylax multisquamis* (Loveridge, 1932a). Proc. Biol. Soc. Washington 45: 84–85. (*Trimerorhinus tritaeniatus multisquamis*)

Type: Holotype, MCZ 18213, an 804 mm female (A. Loveridge, 21 Aug. 1915).

Type locality: "Nairobi, Kenya Colony" [= Nairobi, cen. Nairobi Prov., S Kenya, 1°18'S, 36°49'E, elevation 1675 m].

Distribution: Northeast Africa. Central Ethiopia (Arussi, Bale, Gojjam, Hararge, Kefa, Shoa, Sidamo), S Kenya (Central, S Eastern, Nairobi, S Rift Valley, Western), N Rwanda (Kibungo) and N Tanzania (Arusha, Dodoma, Iringa, Mbeya, Rukwa, Shinyanga), 600–3350 m.

Sources: Spawls et al., 2002 and Largen & Spawls, 2010.

Remarks: A valid species *fide* Spawls et al., 2002 and Kelly et al., 2008. A synonym of *P. variabilis fide* Largen & Spawls, 2010.

2. *Psammophylax rhombeatus* (Linnaeus, 1758). Syst. Nat., ed. 10, 1: 220. (*Coluber rhombeatus*)

Synonyms: *Coronella tygrina* Laurenti, 1768, *Coluber aristotelis* Merrem, 1820, *Psammophylax ocellatus* Bocage, 1873a, *Psammophylax rhombeatus trilineata* Boettger, 1883b, *Psammophylax rhombeatus biseriata* F. Müller, 1893, and *Psammophis longementalis* Roux, 1907c.

Types: Syntypes (2), NHR 1232 (formerly MAFR & NHR 28), NHR 1233 (formerly MAFR & NHR 26), NHR 1234 (formerly MAFR & NHR 26), NHR 1236 (formerly MDG) and/or NHR 1240, a 600 mm and 480+ mm specimen (Mus. Drottn.).

Type locality: "Indiis" [= India] (in error). Corrected to South Africa *fide* V.F.M. FitzSimons (1962a: 210). Restricted to SW Western Cape Prov., South Africa *fide* Broadley (1977a: 17).

Distribution: Southern Africa. Southwestern Angola (Cunene, Huíla), W Namibia (Erongo, Karas, Kunene), W Swaziland, NW Lesotho and South Africa (Eastern Cape, E Free State, Gauteng, KwaZulu-Natal, Limpopo, Mpumalanga, Northern Cape, Western Cape), NSL–1500 m.

Sources: De Waal, 1978, R.M. Douglas, 1992 and Cottone & Bauer, 2010.

Remarks: Possibly occurs in Gaborone, Botswana *fide* Auerbach (1985: 35).

3. *Psammophylax togoensis* (Matschie, 1893b). Mitt. Forsch. Gelehrt. Deutsch. Schutzgeb. 6(3): 212–213. (*Psammophis togoensis*)

Synonym: *Rhamphiophis acutus garambensis* Witte, 1959.

Type: Holotype, ZMB 13249 (H.E. Kling, 26 Oct. 1891–4 Mar. 1892, and R. Büttner, 1890–1891).

Type locality: "Hinterlande von Togo," [= interior of Togo].

Distribution: West and cen. Africa. Guinea (Kouroussa), Ivory Coast (Toumodi), Ghana (Brong-Ahafo, Volta), Togo (Centrale, Plateaux), Cameroon (Adamaoua, Nord, Sud-Ouest), Congo (Cuvette, Plateau), W Central African Republic (S Ouham-Pendé or N Nana Nambere), NE Democratic Republic of the Congo (Oriental), W South Sudan (Western Equatoria) and W Uganda (S Northern, N Western), 200–1000 m.

Sources: Hughes & Barry, 1969, Broadley, 1971a, Hughes, 1983, Chirio & Ineich, 1991, J.-F. Trape & Roux-Estève, 1995, Chippaux, 2006, Chirio & LeBreton, 2007 and Segniagbeto et al., 2011.

Remarks: A valid species *fide* Segniagbeto et al. (2012: 347). Probably occurs in Burkina Faso and Nigeria *fide*

P

Hughes (1983: 353) and possibly occurs in Benin *fide* Hughes (2013: 152).

4. *Psammophylax tritaeniatus* (A.C.L.G. Günther, 1868). Ann. Mag. Nat. Hist. (4) 1(6): 423–424, pl. 19, fig. h. (*Rhagerrhis tritaeniata*)

Synonyms: *Coronella tritaenia* A.C.L.G. Günther, 1881c and *Psammophylax tritaeniatus fitzgeraldi* Broadley, 1960.

Type: Holotype, BMNH 1946.1.2.78 (formerly BMNH 1853.8.17.9), a 508 mm male (Warwick).

Type locality: "South-eastern Africa." Restricted to Rhodesia or the Transvaal [= Zimbabwe or Limpopo/ Mpumalanga, South Africa] *fide* Broadley (1977a: 17).

Distribution: Southern Africa. Southeastern Democratic Republic of the Congo (Katanga), S Tanzania (Arusha, Lindi, Mbeya, Mtwara, Rukwa, Ruvuma, ext. N Singida), S Angola (Benguella, Cuando Cubango, Cunene, Huambo, Huíla, Moxico, Namibe), Zambia (Central, Copperbelt, Eastern, Luapula, Lusaka, Northern, North-Western, Southern, Western), Malawi (Central, Northern, Southern), Mozambique (Cabo Delgado, Manica), Zimbabwe (Bulawayo, Manicaland, Mashonaland Central, Mashonaland East, Mashonaland West, Masvingo, Matebeleland North, Matebeleland South, Midlands), Botswana (Central, Gaborone, Kgatleng, Kweneng, North West, South East, Southern), Namibia (Erongo, Khomas, Kunene, Ohangwena, Okavango, Omaheke, Omasati, Oshana, Oshikoto, Otjozondjupa) and N South Africa (Free State, Gauteng, KwaZulu-Natal, Limpopo, Mpumalanga, E Northern Cape), 35–2200 m.

Sources: Mertens, 1955, 1971, Witte, 1962, Broadley, 1971d, De Waal, 1978, Elzen, 1980, Pienaar et al., 1983, Auerbach, 1987, Broadley & Howell, 1991, Clauss & Clauss, 2002, Spawls et al., 2002, Broadley et al., 2003 and Broadley & Blaylock, 2013.

5. *Psammophylax variabilis* A.C.L.G. Günther, 1892a. Proc. Zool. Soc. London 60(4): 557–558, pl. 35.

Synonyms: *Psammophylax tritaeniatus festivus* Laurent, 1956a, *Psammophylax tritaeniatus subniger* Laurent, 1956a, and *Psammophylax tritaeniatus vanoyei* Laurent, 1956a.

Types: Syntypes (4), BMNH 1946.1.2.66–69 (formerly BMNH 1892.12.31–34), three males and one female, longest syntype 622 mm (A. Whyte, 1891–1892, via H.H. Johnston).

Type locality: "Shiré Highlands south of Lake Nyassa, principally upon Mount Zomba and Mount Milanji, Nyassaland" [= Shire Highlands, mainly Mt. Zomba (ca. 15°21'S, 35°18'E) and Mt. Mulanji (ca. 16°00'S, 35°33'E), Southern Region, S Malawi, elevation 1000–2000 m].

Distribution: Southeastern Africa. Eastern Democratic Republic of the Congo (Katanga, Kivu, Orientale), S Rwanda, N Burundi (Bururi, Muramvya, Ngozi), S Tanzania (Iringa, Mbeya), Zambia (Eastern, Northern, Western), S Malawi (Southern), N Mozambique (Niassa), NE Namibia (E Caprivi) and ext. N Botswana (N North West), 1000–3500 m.

Sources: A.C.L.G. Günther, 1893b, Bogert, 1940, Laurent, 1956a, Sweeney, 1961, Witte, 1962, M.M. Stewart & Wilson, 1966, Broadley, 1971d, Stevens, 1974, Spawls, 1978, Auerbach, 1987, Broadley & Howell, 1991, Largen & Rasmussen, 1993, Spawls et al., 2002, Largen & Spawls, 2010 and Broadley & Blaylock, 2013.

PSEUDAGKISTRODON Van Denburgh, 1909 (Natricidae)

Synonyms: *Pseudoagkisrrodon* – Maki, 1933 *(nomen incorrectum)* and *Pseudoagkistrodon* – Maki, 1931 *(nomen incorrectum)*.

Type species: *Macropisthodon rudis* Boulenger, 1906a.

Distribution: Eastern Asia.

Sources: Takahashi, 1930, Maki, 1931, C.H. Pope, 1935, C.-S. Wang & Wang, 1956, Kuntz, 1963, Hu et al., 1980, Tian et al.,1986, B.-H. Li, 1987, K.-Y. Lue, 1987, 1990, McDowell, 1987, K.Y. Lue et al., 1989, M.-H. Huang, 1990, Zhao & Adler, 1993 and Xiang & Li, 2009.

1. *Pseudagkistrodon rudis* (Boulenger, 1906a). Ann. Mag. Nat. Hist. (7) 17(102): 568. (*Macropisthodon rudis*)

Synonyms: *Pseudagkistrodon carinatus* Van Denburgh, 1909, *Natrix namiei* Oshima, 1910, *Macropisthodon rudis melanogaster* Mell, 1922, and *Macropisthodon rudis multiprefrontalis* Zhao & Jiang, 1981.

Type: Holotype, BMNH 1946.1.13.82, a 590 mm male (F.J. Dymond).

Type locality: "Tongchuan fu, Yunnan" [= Dongchuan, Kunming, N Yunnan Prov., S China, 26°05'N, 103°11'E, elevation 1310 m].

Distribution: Southeastern China (Anhui, Fujian, Guangdong, Guangxi, Guizhou, Hunan, Jiangxi, Sichuan, Yunnan, Zhejiang) and Taiwan (Hsinchu, Pingtung, Taichung, Taipei), 300–2300 m.

PSEUDALSOPHIS Zaher, Grazziotin, Cadle, Murphy, Moura-Leste & Bonatto, 2009 (Xenodontidae)

Type species: *Lygophis elegans* Tschudi, 1845.

Distribution: Galapagos Arch. and W South America.

Fossil records: Holocene of Galapagos Islands (Floreana).

Sources: Steindachner, 1876, S.W. Garman, 1892a, Van Denburgh, 1912b, Steadman, 1986, Zaher et al., 2009 and Vidal et al., 2010.

P

1. *Pseudalsophis biserialis* (A.C.L.G. Günther, 1860a). Proc. Zool. Soc. London 28(1): 97–98. (*Herpetodryas biserialis*)

Synonyms: *Dromicus chamissonis habelii* Steindachner, 1876, and *Dromicus biserialis eibli* Mertens, 1960

Type: Holotype, BMNH 1946.1.4.56, a 362 mm female (C. Darwin, Sept. 1835).

Type locality: "Charles Island (Galapagos Islands)" [= Floreana Is., Galapagos Arch., Ecuador].

Distribution: Ecuador (Galapagos Arch.: Champion, Floreana, Gardner, San Cristóbal).

Sources: Van Denburgh, 1912b, Steadman et al., 1984, Steadman, 1986, Mertens, 1960 and R.A. Thomas, 1997.

Remarks: Original description reprinted in A.C.L.G. Günther (1860k: 78–79).

2. *Pseudalsophis dorsalis* (Steindachner, 1876). Schlangen Galapagos: 306–309, pl. 1, figs. 1, 1a–c. (*Dromicus chamissonis dorsalis*)

Synonyms: *Herpetodryas dorsalis* Günther *in* Steindachner, 1876 *(nomen nudum)* and *Dromicus chamissonis habelli* Steindachner, 1876.

Types Syntypes (2), NMW 25168.1–2, longest syntype 990 mm (Habel, 1868).

Type locality: "Galapagos-Inseln" [= Galapagos Is.]. Restricted to Santa Cruz Is. *fide* R.A. Thomas (1997: 36).

Distribution: Ecuador (Galapagos Is.: Baltra, Bartólome, Rábida, Santiago, Santa Cruz, Santa Fé).

Remarks: NMW catalogue lists type locality as Indefatigable (= Santa Cruz) and Jervis Is. (= Rábida) and collection date as 1875 *fide* Tiedemann et al. (1994: 65).

P

3. *Pseudalsophis elegans* (Tschudi, 1845). Arch. Naturg. 11(1): 164–165. (*Lygophis elegans*)

Synonyms: *Dryophylax fremvillei* A.-M.-C. Duméril, 1853 (*nomen nudum*), *Dryophylax freminvillei* A.-M.-C. Duméril, Bibron and Duméril, 1854a, *Dromicus rufodorsatus* A.C.L.G. Günther, 1858, *Tachymenis canilatus* Cope, 1868b, *Lygophis poecilostomus* Cope, 1875b, *Dryophylax vitellinus* Cope, 1878a, and *Tachymenis Surinamensis* Dunn, 1922.

Type: Holotype, MHNN 11 (formerly MZN 80), a 356 mm female (J.J. von Tschudi, Aug. 1838–Feb. 1841).

Type locality: "Republica Peruana" [= Peru]. Emended to Montañas von Urubamba gefunden, ein jüngeres in der Nahe von Lima, Peru [= Urubamba Mtns. in the vicinity of Lima, Peru] *fide* Tschudi (1846: 54) (partially in error and corrected to Lima, Peru *fide* Schätti, 1986: 101). Restricted to Lima *fide* K.P. Schmidt & Walker (1943b: 317).

Distribution: Coastal Ecuador (Guayas), Peru (Libertad, Lima, Piura, Tacna, San Lorenzo Is.) and ext. N Chile (Arica, Tarapacá), NSL–2400 m.

Sources: K.P. Schmidt & Walker, 1943b, Peters, 1960, Donoso-Barros, 1966, Myers & Hoogmoed, 1974 and R.A. Thomas & Ineich, 1999.

Remarks: Page 164 is mislabeled 166 in Tschudi, 1845. Supplemental original description in Tschudi (1846: 53–54, pl. 6, 2 figs.). Previously in the genus *Alsophis*.

4. *Pseudalsophis hoodensis* (Van Denburgh, 1912b). Proc. California Acad. Sci. (4) 1(4): 338–341, pl. 24. (*Dromicus hoodensis*)

Type: Holotype, CAS 11799, a 735 mm male (J.R. Slevin [CAS Galapagos Exped.], 23 June 1906).

Type locality: "Hood Island, Galapagos Archipelago" [= Española Is., Galapagos Arch., Ecuador].

Distribution: Ecuador (Galapagos Arch.: Española, Gardner).

Sources: R.A. Thomas, 1997 and Zaher, 1999.

Remarks: Previously a subspecies of *Alsophis biserialis* (A.C.L.G. Günther).

5. *Pseudalsophis occidentalis* (Van Denburgh, 1912b). Proc. California Acad. Sci. (4): 1(4): 347–349, pl. 27. (*Dromicus occidentalis*)

Synonym: *Dromicus occidentalis helleri* Van Denburgh, 1912b.

Type: Holotype, CAS 11488, a 1142 mm female (J.R. Slevin [CAS Galapagos Exped.], 18 April 1906).

Type locality: "Narborough Island, Galapagos Archipelago" [= Fernandina Is., Galapagos Arch., Ecuador].

Distribution: Ecuador (Galapagos Arch.: Fernandina, Isabela, Tortuga).

Source: Zaher, 1999.

Remarks: A valid species based on hemipeneal morphology *fide* Zaher (1999: 52).

6. *Pseudalsophis slevini* (Van Denburgh, 1912b). Proc. California Acad. Sci. (4) 1(4): 351–353, pl. 29 (upper). (*Dromicus slevini*)

Type: Holotype, CAS 12216, a 323 mm male (J.R. Slevin [CAS Galapagos Exped.], 14 Aug. 1906).

Type locality: "Duncan Island, Galapagos Archipelago" [= Pinzón Is., Galapagos Arch., Ecuador].

Distribution: Ecuador (Galapagos: Fernandina, Isabela, Pinzón).

Source: R.A. Thomas, 1997.

7. *Pseudalsophis steindachneri* (Van Denburgh, 1912b). Proc. California Acad. Sci. (4) 1(4): 353–355, pl. 30 (upper). (*Dromicus steindachneri*)

Type: Holotype, CAS 10795, a 420 mm male (J.R. Slevin [CAS Galapagos Exped.], 16 Jan. 1906).

Type locality: "Indefatigable Island, Galapagos Archipelago" [= Santa Cruz Is., Galapagos Arch., Ecuador].
Distribution: Ecuador (Galapagos: Baltra, Rábida, Santa Cruz, Santiago).
Source: R.A. Thomas, 1997.

PSEUDASPIS Fitzinger, 1843 (Pseudaspididae)

Synonyms: *Cadmus* Theobald, 1868b, *Ophirhina* Bocage, 1882a, and *Pseudaspic* – Skinner, 1973 *(nomen incorrectum).*
Type species: *Coluber canus* Linnaeus, 1758.
Distribution: Southern and E Africa.
Sources: Bocage, 1895, Loveridge, 1929, 1957a, Monard, 1937, Bogert, 1940, Witte, 1941, 1953, 1962, Mertens, 1955, 1971, Laurent, 1956a, Vesey-FitzGerald, 1958, 1975, Sweeney, 1961c, V.F.M. FitzSimons, 1962a, 1974, Broadley, 1971d, 1983, Broadley & Cock, 1975, De Waal, 1978, Spawls, 1978, Jacobsen & Haacke, 1980, Buys & Buys, 1983, Derleyn et al., 1983, Pienaar et al., 1983, Auerbach, 1987, Branch, 1988, Broadley & Howell, 1991, Boycott & Culverwell, 1992, Marais, 1992, Clauss & Clauss, 2002, Spawls et al., 2002, Broadley et al., 2003, Marais, 2004, Lawson et al., 2005, Vidal et al., 2008, Kelly et al., 2009 and Broadley & Blaylock, 2013.

1. *Pseudaspis cana* (Linnaeus, 1758). Syst. Nat., ed. 10, 1: 221–222. (*Coluber canus*)

Synonyms: *Coluber padera* Linnaeus, 1758, *Coluber elegantissimus* Laurenti, 1768, *Coluber ocellatus* Gmelin, 1789, *Coluber politus* Seetzen, 1796, *Coluber crucifer* G. Shaw, 1802, *Coluber parias* Hermann, 1804, *Cadmus cuneiformis* Theobald, 1868b, *Coronella phocarum* A.C.L.G. Günther, 1872c, and *Ophirina anchietae* Bocage, 1882a.
Type: Holotype, NHR Lin-22 (formerly MAFR), a 330 mm male (Mus. Drottn.).
Type locality: "Indiis" [= India] (in error). Corrected to Cape of Good Hope, South Africa [= Cape of Good Hope, Western Cape Prov., SW South Africa, 34°21'S, 18°28'E, elevation 20 m] *fide* Flower (1933: 818).
Distribution: Southern and E Africa. Southern Kenya (Central, S Eastern, S Rift Valley), Rwanda (Cyangugu), Burundi (Bururi, Ruyigi), Tanzania (Iringa, Kilimanjaro, Mbeya), SE Democratic Republic of the Congo (Katanga), SW Angola (Cuanza Sul, Huíla), Zambia (Central, Copperbelt, Eastern, Luapula, Lusaka, Northern, North-Western, Southern, Western), S Malawi (Southern), S Mozambique (Ganza, Inhambane, Maputo), Zimbabwe (Bulawayo, Manicaland, Mashonaland Central, Mashonaland East, Mashonaland West, Matabeleland North, Matabeleland South, Midlands), Botswana (Central, Ghanzi, Gaborone, Kgalagadi, Kgatleng, Kweneng, North East, North West, South East, Southern), Namibia (Caprivi, Erongo, Hardap, Karas, Khomas, Kunene, Okavango, Omaheke, Otjozondjupa), Swaziland, and South Africa (Eastern Cape, Free State, Gauteng, KwaZulu-Natal, Limpopo, Mpumalanga, Northern Cape, North West, Western Cape, Robben Is.), NSL–1950 m.
Remarks: Possibly extirpated in Malawi *fide* Sweeney (1961: 127). Possibly *P. anchietae* is a valid species based on hemipenial morphology *fide* Visser (2010: 3).

PSEUDECHIS Wagler, 1830 (Elapidae)

Synonyms: *Alecto* Wagler, 1830 *(nomen praeoccupatum), Pseudechys* – Spencer, 1892 *(nomen incorrectum), Peseudechis* – Brazil, 1914 *(nomen incorrectum), Pseudoechis* – Anon., 1978 *(nomen incorrectum), Cannia* Wells & Wellington, 1984, *Panacedechis* Wells & Wellington, 1985, and *Pailsus* Hoser, 1998a.
Type species: *Coluber porphyriacus* G. Shaw, 1794.
Distribution: Austro-Papua.
Fossil record: Pleistocene of Australia.
Sources: F. Werner, 1923b, R. Mackay, 1955, Cogger et al., 1983a, Storr, 1985, Longmore, 1986, Mengden et al., 1986, Shine, 1987e–f, S.K. Wilson & Knowles, 1988, Gow, 1989, Hoser, 1989, 1998a, 2001, 2012e, Covacevich & Couper, 1991, Greer, 1997, Cogger, 2000, Sanders et al., 2008 and Zaher et al., 2009.

1. *Pseudechis australis* (Gray, 1842c). Zool. Misc. 2(Apr.): 55. (*Naja australis*)

Synonyms: *Pseudechis darwiniensis* Macleay, 1878b, *Pseudechis cupreus* Boulenger, 1896a, *Pseudechis denisonioides* F. Werner, 1909e, *Pseudechis platycephalus* Thomson, 1933, *Denisonia brunnea* F.J. Mitchell, 1951, *Cannia centralis* Wells & Wellington, 1985, *Cannia weigeli* Wells & Wellington, 1987, *Cannia australis aplini* Hoser, 2001, *Cannia australis burgessi* Hoser, 2001, and *Cannia australis newmani* Hoser, 2001.
Type: Holotype, BMNH 1946.1.20.39, a female (Gilbert).
Type locality: "North coast of New Holland; Port Essington" [= ruins at tip of Cobourg Peninsula, N Northern Territory, N Australia, ca. 11°09'N, 132°00'E, elevation NSL].
Distribution: Austro-Papua. Eastern Indonesia (E Papua), SW Papua New Guinea (Western) and Australia (New South Wales, Northern Territory, Queensland, South Australia, Tasmania, ext. N Victoria, Western Australia, Angarbulumardja, Augustus, Barrow, Bernier, Bickerton, Brigadier, Carlia, Centre, Cockatoo, Comeile, Connexion, Coronation, Cotton, Dirk Hartog, Dorre, Fenelon, Finch, Fletcher, Groote Eylandt, Hidden, Inglis, Jirgan, Jungulu, Katers, Koolan, Long, Maria, Middle Osborn, Mornington, Muir, North Point, Pobassoo, Probable, Quoin, Rosemary, Sir Graham

P

Moore, South West, Sunday, Wessel Marchinbar, West, West Lewis and Wigram Is.), NSL–995 m.

Sources: F. McCoy, 1887, Roux, 1919b, L.A. Smith, 1982, F. Parker, 1982, Whitaker et al., 1982, Bush et al., 1995, O'Shea, 1996, Hoser, 1998a, Shea & Sadlier, 1999 and Storr et al., 2002.

Remarks: Composed of several unnamed species *fide* Kuch et al. (2005a: 125).

2. *Pseudechis butleri* L.A. Smith, 1982. Rec. W. Aust. Mus. 10(1): 43, fig. 1a.

Type: Holotype, WAM 22345, a 930 mm (svl) gravid female (I.C. Carnaby, 15 Oct. 1963).

Type locality: "19 km SE of Yalgoo, Western Australia, in 28°29'S, 116°49'E."

Distribution: Western Australia (W Western Australia), 260–555 m.

Sources: Maryan, 1994 and Storr et al., 2002.

3. *Pseudechis colletti* Boulenger, 1902e. Ann. Mag. Nat. Hist. (7) 10(60): 494–495.

Type: Holotype, BMNH 1946.1.20.69, a 450 mm specimen (R. Collett).

Type locality: "Queensland" [Australia].

Distribution: Northeastern Australia (Queensland), 20–320 m.

4. *Pseudechis guttatus* De Vis, 1905. Ann. Queensland Mus. (6): 49–50.

Synonyms: *Pseudechis mortonensis* De Vis, 1911, *Panacedechis worrelli* Wells & Wellington, 1985, and *Pseudechis qutatus* – Khole, 1991 *(nomen incorrectum)*.

Type: Holotype, QM J189, a 1070 mm eviscerated specimen.

Type locality: "Cecil Plains, Queensland" [= Cecil Plains, SE Queensland, E Australia, 27°32'S, 151°12'E, elevation 355 m].

Distribution: Southeastern Australia (E New South Wales, SE Queensland), 35–1240 m.

Sources: F. McCoy, 1879a, Charles et al., 1979, Gibson, 1983 and Shea & Sadlier, 1999.

5. *Pseudechis pailsi* (Hoser, 1998a). Monitor 10(1): 5–9, 31, 1 fig. (*Pailsus pailsi*) (*nomen corrigendum*)

Synonyms: *Pailsus pailsei* Hoser, 1998a *(nomen incorrigendum)* and *Pailsus pailsi* – Wüster et al., 2001b *(nomen corrigendum)*.

Type: Holotype, NMV D69704, a 1071 mm female (19 Nov. 1984).

Type locality: "East Leichardt Dam, near Mount Isa, Queensland, Lat. 20°47' [S], Long. 139°47' [E], Australia."

Distribution: Northeasterm Australia (NW Queensland), 360 m.

Sources: Hoser, 2000c and Wüster et al., 2005.

6. *Pseudechis papuanus* W.C.H. Peters & Doria, 1878. Ann. Mus. Civ. Stor. Nat. Genova 13: 409–410.

Synonym: *Panacedechis papuanus trevorhawkeswoodi* Hoser, 2009c & 2012i *(nomen illegitimum)*.

Types: Syntypes (2), MSNG 8018a–b, longest syntype 1900 mm (C.L.M. D'Albertis, 1875).

Type locality: "Monte Epa" and "coste opposte dell'IsolaYule, Nuova Guinea meridionale orientale" [= Mt. Epa (8°40'S, 148°05'E) and coast opposite Yule Island (8°49'S, 146°32'E), Central Prov., SE Papua New Guinea].

Distribution: New Guinea. Eastern Indonesia (SE Papua, Fredrik Hendrik Is.) and S Papua New Guinea (Central, Western, Yule Is.), NSL–250 m.

Sources: Roux, 1919b, K.R. Slater, 1968, Whitaker et al., 1982, F. Parker, 1982, O'Shea, 1996, Wüster et al., 2005 and Hoser, 2009c.

7. *Pseudechis porphyriacus* (G. Shaw, 1794). Zool. New Holland: 27–28, pl. 10. (*Coluber porphyriacus*)

Synonyms: *Trimeresurus leptocephalus* Lacépède, 1804, *Trimeresurus porphyreus* Merrem, 1820 *(nomen emendatum)*, *Duberria porphyriaca* – Fitzinger, 1826a, *Acanthophis tortor* Lesson, 1828 *(nomen nudum)*, *Acanthophis tortor* Lesson, 1829, *in* Duperrey, 1826–1832, *Pseudechis porphyricus* – Wagler, 1830 *(nomen incorrectum)*, *Naja porphyrica* –Schlegel, 1837 *(nomen incorrectum)*, *Trimeresurus porphyrica* – A.-M.C. Duméril, Bibron & Duméril, 1854b *(nomen incorrectum)*, *Pseudechis porphyraicus* – F. McCoy, 1867 *(nomen incorrectum)*, *Pseudechis porphyriacea* – F. Müller, 1878b *(nomen incorrectum)*, *Pseudechis porphyriachis* – Limpus, 1978 *(nomen incorrectum)*, *Pseudechis porphyrica* – Mengden, 1985 *(nomen incorrectum)*, *Pseudechis porphyriacus eipperi* Hoser, 2003d, and *Pseudechis porphyriacus rentoni* Hoser, 2003d.

Type: Holotype, not designated, lost *fide* Cogger et al. (1983: 231).

Type locality: "New Holland" [= Australia].

Distribution: Eastern Australia (Australian Capital Territory, New South Wales, E Queensland, SE South Australia, Victoria), NSL–1280 m.

Sources: F. McCoy, 1878a, M.J. Smith, 1975, Heatwole & Johnson, 1979, Shine et al., 1981, Shine, 1987d, Hoser, 2003d and J.C. Murphy & Schlager, 2003.

8. *Pseudechis rossignolii* (Hoser, 2000a). Litt. Serp. 20(6): 178–180. (*Pailsus rossignolii*)

Type: Holotype, MZB 364, a 1050 mm specimen.

Type locality: "south New Guinea, in the general region of Meraoke Lat 8° 30' Long 140° 20' and areas a short distance west of here along the coast."

P

Distribution: Eastern Indonesia (SE Papua), NSL.
Source: Wüster et al., 2005b.
Remarks: Original description reprinted in Hoser (2003b: 23–26).

PSEUDELAPHE Mertens & Rosenberg, 1943 (Colubridae)

Synonym: *Pseudoelaphe* – Dowling, 1952a *(nomen incorrectum).*
Type species: *Coluber flavirufus* Cope, 1867b.
Distribution: Southeastern Mexico and upper Central America.
Sources: H.M. Smith, 1941L, Dowling, 1952a–b, H.M. Smith & Williams, 1966b, L.D. Wilson & Hahn, 1973, L.D. Wilson & Meyer, 1985, Schulz, 1993, 1996, J.C. Lee, 1996, 2000, Schulz et al., 1997, J.A. Campbell, 1998b, G. Köhler, 1998a, 1999b, 2003, Canseco-Marquez et al., 2000, Utiger et al., 2002, Dowling et al., 2003, McCranie et al., 2005, Burbrink & Lawson, 2007, Pyron & Burbrink, 2009a and McCranie, 2011a.

1. *Pseudelaphe flavirufa* (Cope, 1867b). Proc. Acad. Nat. Sci. Philadelphia (1866) 18(4): 319. (*Coluber flavirufus*)

Synonyms: *Elaphis pardalinus* W.C.H. Peters, 1868, *Elaphis rodriguezii* Bocourt, 1887, *Elaphe flavirufa matudai* H.M. Smith, 1941L, *Elaphe flavirufa phaescens* Dowling, 1952a, *Elaphe flavirufa polysticha* H.M. Smith & Williams, 1966b, and *Elaphe flavirufa yucatanensis* Dowling *in* Schulz, 1996.
Type: Holotype, USNM 6566, a 559 mm specimen (A.C.V. Schott, 1865).
Type locality: "Yucatan" [Mexico]. Emended to Isamal, Yucatán, Mexico *fide* Cochran (1961: 165). Restricted to Chichén Itzá, Yucatán, Mexico *fide* H.M. Smith & Taylor (1950: 352), and to the vicinity of Campeche, Campeche, Mexico *fide* Dowling (1952a: 3).
Distribution: Southeastern Mexico and upper Central America. Southeastern Mexico (Campeche, Chiapas, Hidalgo, Oaxaca, Puebla, Querétaro, Quintana Roo, SE San Luis Potosí, Tabasco, S Tamaulipas, Veracruz, Yucatán), Belize (Cayo, Orange Walk), Guatemala (Alta Verapaz, Petén), N Honduras (Atlántida, Colón, Cortés, Islas de Bahías, Olancho, Yoro, Guanaja, Roatán and Utila Is.), and Nicaragua (Managua, Great Corn Is.), NSL–1500 m.
Remarks: Previously in the genus *Elaphe.*

PSEUDOBOA J.G. Schneider, 1801 (Xenodontidae)

Synonyms: *Scytalus* Fischer von Waldheim, 1813 *(nomen emendatum), Scytale* Merrem, 1820 *(nomen praeoccupatum), Olisthenes* Cope, 1860a, *Brachyrhyton* Jan, 1870 *in* Jan & Sordelli, 1870–1881, *Pseudo-boa* – Westphal-Castelnau, 1870 *(nomen illegitimum), Pseudoba* – Briceño-Rossi, 1934 *(nomen incorrectum), Pseutoboa* – Hoge, 1953b *(nomen incorrectum),* and *Psedoboa* – Villa, 1962 *(nomen incorrectum).*
Type species: *Pseudoboa coronata* J. G. Schneider, 1801.
Distribution: Panama, Amazonia and Grenada.
Sources: Bailey, 1939b, Bailey *in* J.A. Peters & Orejas-Miranda, 1970, Cunha & Nascimento, 1983a, Zaher, 1996, Vidal et al., 2000 and Zaher et al., 2008, 2009.

1. *Pseudoboa coronata* J.G. Schneider, 1801. Hist. Amph. 2: 286.

Synonyms: *Brachyrruton nuchaluteum* A.-M.-C. Duméril, 1853 *(nomen nudum), Brachyruton occipitoluteum* A.-M.-C. Duméril, Bibron & Duméril, 1854b *(nomen incorrigendum),* and *Brachyrhyton occipitoluteum* – Jan, 1870 *in* Jan & Sordelli, 1870–1881 *(nomen corrigendum).*
Type: Holotype, not designated, RMNH (formerly Mus. Heyderi, Brunovicensis), lost *fide* Hoge (1964d: 61).
Type locality: Unknown. Designated as America *fide* J.A. Peters (1960a: 535).
Distribution: Amazonia. Colombia (Amazonas, Chocó, Cundinamarca, Magdalena, Meta, Vaupés), E Venezuela (Amazonas, Anzoátegui, Bolívar, Delta Amacuro, Monagas, Sucre), Guyana (Essequibo Islands-West Demerara, Upper Demerara-Berbice), Suriname (Marowijne, Paramaribo, Saramacca), French Guyana (Cayenne, Saint-Laurent-du-Maroni), N Brazil (Acre, Amazonas, Maranhão, Pará, Rondônia), E Ecuador (Morona-Santiago, Napo, Pastaza, Sucumbios), E Peru (Loreto, Madre de Dios) and E Bolivia (Beni, La Paz, Santa Cruz), NSL–1200 m.
Sources: Beebe, 1946, Roze, 1966a, Hoge et al., 1973, Cunha and Nascimento, 1978, 1983a, 1993, Duellman, 1978, 2005, Fugler & Walls, 1978, Gasc & Rodrigues, 1980, Fugler, 1983, 1984, Abuys, 1985b, Dixon and Soini, 1986, Lancini-V, 1986, Pérez-Santos, 1986, Pérez-Santos & Moreno, 1988, 1991, Duellman & Salas, 1991, Henle & Ehrl, 1991, Gorzula & Señaris, 1998, Starace, 1998, Rivas-Fuenmayor & La Marca, 2001c, Navarrete et al., 2009 and C.J. Cole et al., 2013.

2. *Pseudoboa haasi* (Boettger, 1905). Zool. Anz. 29(11): 374–375. (*Oxyrrhopus haasi*)

Type: Lectotype, SMF 19867 (formerly SMF-B 9111, 2a), a 690 mm specimen (A. Haas, 1905), designated by Mertens (1922: 181).
Type locality: "Campos von Palmas, Parana" [= Palm country of Paraná State, S Brazil].
Distribution: Southern Brazil (Paraná, S Santa Catarina, NE Rio Grande do Sul, E São Paulo) and Argentina (Misiones), 100–600 m.
Sources: Amaral, 1926e, 1937d, Lema & Ely, 1979, Giraudo, 1999 and Marques et al., 2001.

3. *Pseudoboa martinsi* Zaher, Oliveira & Franco, 2008. Zootaxa (1674): 28–32, figs. 1a–b, 2a–d, 3a–b.

Type: Holotype, MZUSP 8549, a 1020 mm male (B. Zimmerman, Nov. 1984).
Type locality: "Fazenda Porto Alegre, a forest fragment belonging to the PDBFF (Projeto Dinâmica Biológica de Fragmenios Florestais), Manaus, State of Amazonas, Brazil (2° 25' 00" S, 59° 43' 00' W)."
Distribution: Brazil (Amazonas, Pará, Rondônia, Roraima).
Source: Martins & Oliveira, 1998.
Remarks: Possibly occurs in other Brazilian states as well as Bolivia, Colombia, Peru and Venezuela *fide* Zaher et al. (2008: 33).

4. *Pseudoboa neuwiedii* (A.-M.-C. Duméril, Bibron & Duméril, 1854b). Erpét. Gén. 7(2): 1001–1002. (*Scytale neuwiedii*)

Synonyms: *Scytale neuwiedii* A.-M.-C. Duméril, 1853 (*nomen nudum*), *Olisthenes euphaeus* Cope, 1860a, *Rhinocheilus thominotii* Bocourt and Dugès, 1887, and *Pseudoboa robinsoni* Stejneger, 1901c.
Type: Lectotype, MNHN 3779, a 900 mm specimen (L.D. de Bauperthuy, 1841), designated by Hoge & Lancini (1960a: 59).
Type locality: "Côte-Ferme" [= South America] via lectotype selection. Restricted to "Cumaná y sus alrededores" [= vicinity of Cumaná, Sucre State, NE Venezuela] *fide* Hoge & Lancini (1960a: 61).
Distribution: Panama, Amazonia and Grenada. Panama (Chiriquí, Coclé, Colón, Darién, Herrera, Panamá), Colombia (Antioquia, Atlántico, Bolívar, Boyaca, Caldas, Chocó, Córdoba, Cundinamarca, La Guajira, Magdalena, Meta, Norte de Santander, Santander, Sucre, Tolima, Vaupés, Vichada), Venezuela (Amazonas, Anzoátegui, Apure, Aragua, Barinas, Bolívar, Carabobo, Cojedes, Delta Amacuro, Distrito Federal, Falcón, Guárico, Lara, Mérida, Miranda, Monagas, Nueva Esparta, Portuguesa, Sucre, Táchira, Trujillo, Yaracuy, Vargas, Zulia, Margarita Is.), Trinidad and Tobago (Tobago, Trinidad), Grenada, Guyana (Potaro-Siparuni, Upper Demerara-Berbice, Upper Takutu-Upper Essequibo), Suriname (Saramacca), French Guiana (Cayenne, Saint-Laurent-du-Maroni), and N Brazil (Amapá, Amazonas, Distrito Federal, Goiás, Maranhão, Mato Grosso, Minas Gerais, Paraiba, Roraima), NSL–2750 m.
Sources: Amaral, 1935d–e, Shreve, 1947, Roze, 1952a, 1966a, Hoge, 1953b, 1967, Emsley, 1977, Abuys, 1985b, Lancini, 1986, Pérez-Santos, 1986, Pérez-Santos & Moreno, 1988, Gorzula & Señaris, 1998, Starace, 1998, Esqueda & La Marca, 1999, Kornacker, 2001, Markezich, 2002, Rivas-Fuenemayer & Barrio-Amorgós, 2005, Navarrete et al., 2009, Ugueto & Rivas, 2010 and C.J. Cole et al., 2013.
Remarks: Probably extirpated in Grenada *fide* Emsley (1977: 250).

P

5. *Pseudoboa nigra* (A.-M.-C. Duméril, Bibron & Duméril, 1854b). Erpét. Gén. 7(2): 1002. (*Scytale neuwiedii nigrum*)

Synonym: *Pseudoboa albimaculata* Mello, 1926.
Type: Holotype, MNHN (Lemelle-Deville, 1841–1843).
Type locality: "Bahia" [Brazil].
Distribution: Eastern Brazil (Alagoas, Bahia, Ceará, Distrito Federal, Goias, Maranhão, Mato Grosso, Mato Grosso do Sul, Minas Gerais, Pará, Paraiba, Pernambuco, Piauí, Rio de Janeiro, Rio Grande do Norte, Rondônia, São Paulo, Tocantins), E Bolivia (Beni, Santa Cruz) and N Paraguay (Alto Paraguay, Amambay, Boquerón, Concepción, Presidente Hayes), NSL–300 m.
Sources: Bailey, 1939b, Cordeiro & Hoge, 1974, Abalos & Mischis, 1975, Hoge et al., 1975, Vanzolini et al., 1980, J.D. Williams & Francini, 1991, Freitas, 1999, Cacciali, 2008 and Orofino et al., 2010.
Remarks: Official Specific Name no. 1969 *fide* Opinion 698 (ICZN, 1964a). Argentine records doubtful *fide* J.D. Williams & Francini (1991: 77).

6. *Pseudoboa serrana* Morato, Moura-Leite, Costa-Prudente & Silveira-bérnila, 1995. Biociências 3(2): 254–257, figs. 1–4.

Type: Holotype, IB 10393, a 1154 mm male (30 Jan. 1942), destroyed by fire 15 May 2010.
Type locality: "Brazil: Rio de Janeiro: Teresópolis: Sierra dos Órgãos (Parque Nacional) (22°27'S; 42°58'W)."
Distribution: Southeastern Brazil (S Minas Gerais, Rio de Janeiro, E São Paulo), NSL–650 m..
Source: Marques et al., 2001.

PSEUDOBOODON Peracca, 1897 (Lamprophiidae)

Type species: *Pseudoboodon gascae* Peracca, 1897a.
Distribution: Northeast Africa.
Sources: J.B. Rasmussen & Largen, 1992, Largen & Rasmussen, 1993, Spawls, 2004, Vidal et al., 2008, Kelly et al., 2009, Zaher et al., 2009 and Largen & Spawls, 2010.

1. *Pseudoboodon boehmei* J.B. Rasmussen & Largen, 1992. Steenstrupia 18(3): 76–79, figs. 7–8.

Type: Holotype, BMNH 1976.1668, a 620 mm male (Polytech. Cen. London & Imper. College W Ethiopia Exped., 20 Sept. 1974).
Type locality: "Godare Forest, Illubabor Province, Ethiopia (07°26'N 35°00'E, altitude 820 m)."
Distribution: Southwestern Ethiopia (Illubabor, Kefa, Wollega), 800–2150 m.

2. *Pseudoboodon gascae* Peracca, 1897a. Boll. Mus. Zool. Comp. Univ. Torino 12(273): 2–3, 1 fig.

Type: Lectotype, MZUT 1658, a 660 mm female (A. Gasca), designated by Andreone & Gavetti (2007: 91).
Type locality: "Maldi (Eritrea)" [= Maldi, S Anseba Prov., N Eritrea, 15°38'N, 38°48'E, elevation 2450 m] via lectotype selection.
Distribution: Eritrea (Anseba, S Northern Red Sea, Southern) and E Ethiopia (Hararge, Sidamo), 1500–2450 m.
Remarks: Lectotype illustrated in J.B. Rasmussen (1992: figs. 5–6) and Andreone & Gavetti (2007: fig. 10).

3. *Pseudoboodon lemniscatus* (A.-M.-C. Duméril, Bibron & Duméril, 1854a). Erpét. Gén. 7(1): 365–366. (*Boaedon lemniscatum*)

Synonyms: *Boedon lemniscatum* A.-M.-C. Duméril, 1953 (*nomen nudum*), *Lamprophis rogeri* Mocquard, 1904a, and *Pseudoboodon erlangeri* F. Werner, 1923a.
Type: Holotype, MNHN 1903.285, a 400 mm skin (E.-J. Petit).
Type locality: "Abyssinie" [= Ethiopia].
Distribution: Eritrea (S Northern Red Sea, Southern) and Ethiopia (Arussi, Bale, Gemu Gofa, Gojjam, Gondar, Hararge, Illubabor, Kefa, Shoa, Sidamo, Tigre, Wollega, Wollo), 1750–3300 m.
Sources: Bogert, 1940, H.W. Parker, 1949 and Largen & Largen, 1982.

4. *Pseudoboodon sandfordorum* Spawls, 2004. Afr. J. Herp. 53(1): 12–16, figs. 1–3.

Type: Holotype, WML 2003.47.4, a 655 mm male (natives, 23 Nov. 1997).
Type locality: "Mugher River Gorge, near Keriyo Hamlet (09° 24' N 38° 39' E, altitude 1 800 m), central highlands of Ethiopia."
Distribution: Central Ethiopia (Shoa), 1800–2400 m. Known from vicinity of type locality.

†*PSEUDOCEMOPHORA* Auffenberg, 1963 (Colubridae)

Type species: †*Pseudocemophora antiqua* Auffenberg, 1963.
Distribution: Lower to upper Miocene of USA.
Sources: Holman, 1979a and Rage, 1984b.

1. †*Pseudocemophora antiqua* Auffenberg, 1963. Tulane Stud. Zool. 10(3): 187–188, fig. 36 (paratype).

Type: Holotype, UF 6458, one trunk vertebra (W. Auffenberg, March 1954).
Type locality: "Boulder Bar, Thomas Farm, Gilchrist County (R15E, T75, Sec. 20), Florida; lower Miocene (Arikareean), Hawthorne fm."
Distribution: Lower Miocene (Arikareean) of USA (Wyoming), lower Miocene (Hemingfordian) of USA (Florida), and lower/upper Miocene (Hemingfordian/Barstovian) of USA (Wyoming).
Remarks: Possibly known from the middle Miocene (Barstovian) *fide* Rage (1984b: 51).

PSEUDOCERASTES Boulenger, 1896a (Viperidae)

Synonyms: *Pseudocerastas* – F. Werner, 1922b (*nomen incorrectum*), *Pseudoceratus* – Briceño-Rossi, 1934 (*nomen incorrectum*), *Pseudocrastes* – Leviton, 1959a (*nomen incorrectum*), and *Pseudocerates* – Marx & Rabb, 1965 (*nomen incorrectum*).
Type species: *Cerastes persicus* A.-M.-C. Duméril, Bibron & Duméril, 1854b.
Distribution: Arabia and SW Asia.
Sources: Strauch, 1873, Blanford, 1876, Nikolsky, 1913, 1916, Wall, 1925a, 1928, Flower, 1930, 1933, K.P. Schmidt, 1930, 1939b, F. Werner, 1938a, M.A. Smith, 1943, Laurent, 1948, Barash & Hoofien, 1956, G. Haas, 1957, Khalaf, 1959, Marx & Rabb, 1965, Mendelssohn, 1965, Minton, 1966, Marx, 1968, Minton et al., 1968, Leviton & Anderson, 1970a, Kochva, 1974, Baran, 1976a, Arnold & Gallagher, 1977, Groombridge, 1980, M.S. Khan, 1980, 1982b, 1990, 2002, Joger, 1984, Brodmann, 1987, R. Price, 1987a, Gasperetti, 1988, Marx et al., 1988, Gallagher, 1990, Latifi, 1991, Leviton et al., 1992a, Martens, 1993, Spawls & Branch, 1995, Nilson & Andrén, 1997, David & Ineich, 1999, Herrmann et al., 1999, McDiarmid et al., 1999, Disi et al., 2001, Lenk et al., 2001b, Mallow et al., 2003, Bostanchi et al., 2006, Wüster et al., 2008, Stümpel & Joger, 2009, Fathinia & Rastegar-Pouyani, 2010, Phelps, 2010 and Hoser, 2012d.

1. *Pseudocerastes fieldi* K.P. Schmidt, 1930. Field Mus. Nat. Hist. (Zool.) 17(6): 227–228, fig. 2.

Type: Holotype, FMNH 11061, a 726 mm male (H. Field [Marshall Field No. Arab. Desert Exped.], 9 May 1928).
Type locality: "Bair Wells, Transjordan" [= Jibal Matahat Ba'ir, Ma'an Govern, cen. Jordan, ca. 30°41'N, 36°45'E, elevation 890 m].
Distribution: Middle East and Arabia. Northeastern Egypt (North Sinai, South Sinai), S Israel (Southern), Jordan (Amman, Aqaba, Maan, Mafraq, Zarqa), S Syria (Homs), N Saudi Arabia (Jawf, Northern), W Iraq (Al Anbar) and SW Iran (Bushehr), NSL–915 m.
Sources: Gasperetti, 1988, Bouskila & Amitai, 2001, M.S. Khan, 2002, Baha El Din, 2006, Amr & Disi, 2011 and Bar & Haimovitch, 2011.

Remarks: A valid species *fide* Bostanchi et al. (2006: 445) and Fathinia & Rastegar-Pouyani (2010: 277).

2. *Pseudocerastes persicus* (A.-M.-C. Duméril, Bibron & Duméril, 1854b). Erpét. Gén. 7(2): 1443–1444, pl. 78bis, fig. 5. (*Cerastes persicus*)

Synonyms: *Cerastes persicus* A.-M.-C. Duméril, 1853 (*nomen nudum*), *Vipera persica* – Jan, 1859a, and *Pseudocerastes bicornis* Wall, 1913.

Type: Holotype, MNHN 4027 (P.M.R. Aucher-Eloy, 1830–1838).

Type locality: "Perse" [= Iran].

Distribution: Southwestern Asia. Extreme SE Turkey (Kars), NW Azerbaijan, United Arab Emirates (Ras al Khaimah), N Oman (Ad Dakhiliyah, Al Batinah, Ash Sharqiyah, Musandam, Muscat), W Iraq (Al Anbar, Dahuk, Ninawa), Iran (Central, Esfahan, Fars, Hormozghan, Kerman, Khuzestan, Lorestan, Merkazi Qum, Razavi Khorasan, Semnan, Sistan va Baluchestan, Tehran, West Azarbaijan, Yazd, Zanjan), S Afghanistan (Kandahar, Nimroze, Zabul) and Pakistan (Balochistan, Khyber Pakhtunkhwa, Sindh), 335–2200 m.

Sources: Strauch, 1873, Blanford, 1876, Nikolsky, 1913, 1916, 1928, K.P. Schmidt, 1930, 1939b, F. Werner, 1938a, Laurent, 1948, Khalaf, 1959, Marx & Rabb, 1965, Mendelssohn, 1965, Minton, 1966, Marx, 1968, Minton et al., 1968, Leviton & Anderson, 1970a, Kochva, 1974, Baran, 1976a, Arnold & Gallagher, 1977, M.S. Khan, 1980, 1982b, 1990, 2002, Joger, 1984, Brodmann, 1987, R. Price, 1987a, Gasperetti, 1988, Marx et al., 1988, Latifi, 1991, Leviton et al., 1992a, Martens, 1993, Nilson & Andrén, 1997, David & Ineich, 1999, Herrmann et al., 1999, McDiarmid et al., 1999, Disi et al., 2001, Lenk et al., 2001b, Mallow et al., 2003, Wüster et al., 2008, Stümpel & Joger, 2009, Fathinia & Rastegar-Pouyani, 2010 and Grossmann et al., 2012.

Remarks: Records from S Armenia and Afghanistan doubtful *fide* Spawls et al., 2002. *Pseudocerastes bicornis*, known only from head and neck of holotype, considered a valid species *fide* M.S. Khan (2002: 190).

3. *Pseudocerastes urarachnoides* Bostanchi, Anderson, Kami & Papenfuss, 2006. Proc. California Acad. Sci. (4) 57(14): 446–448, figs. 1–12.

Type: Holotype, FMNH 170929, a 531 mm female (D.R. Womochel and A.F. DeBlase [Second Street Exped. Iran], 27 Aug. 1968).

Type locality: "Iran: Ilam Province: 70 km SW Ilam [probably on road to Amirabad and Mehran]."

Distribution: Western Iran (Zagros Mtns. of Ilam, Kermanshah, Khuzestan), 200–300 m.

Source: Fathinia et al., 2009.

P

PSEUDOERYX Fitzinger, 1826a (Xenodontidae)

Synonyms: *Pseudoerix* Thon *in* Ersch & Gruber, 1838 (*nomen emendatum*), *Dimades* Gray, 1842d, *Pseuderyx* Fitzinger, 1843 (*nomen emendatum*), *Callopisma* A.-M.-C. Duméril, 1853 (*nomen praeoccupatum*), *Calopisma* A.-M.-C. Duméril, Bibron & Duméril, 1854a (*nomen emendatum*), and *Pseudaeryx* – Senna, 1886 (*nomen incorrectum*).

Type species: *Coluber plicatilis* Linnaeus, 1758.

Distribution: South America.

Sources: Bertoni, 1915, Serié, 1921, Hoge, 1964d, 1967, Neill, 1964, Nicéforo-María, 1964, J.A. Peters & Orejas-Miranda, 1970, Amaral, 1978, Cunha & Nascimento, 1993, Lancini, 1979, Gasc & Rodrigues, 1980, Fugler, 1983, Abuys, 1986a, Chippaux, 1987, Pérez-Santos & Moreno, 1988, 1991, Duellman & Salas, 1991, Henle & Ehrl, 1991, J.D. Williams & Francini, 1991, Giraudo, 1999, Vidal et al., 2000, Schargel et al., 2007, Navarrete et al., 2009, and Zaher et al., 2009.

1. *Pseudoeryx plicatilis* (Linnaeus, 1758). Syst. Nat., ed. 10, 1: 217. (*Coluber plicatilis*)

Synonyms: *Coluber bali* Lacépède, 1789 (*nomen rejiciendum*), *Pseudoeryx daudinii* Fitzinger, 1826a, *Pseudoeryx mimeticus* Cope, 1886a, *Pseudoeryx plicatilis anomalolepis* Bocourt, 1895 *in* Duméril, Bocourt & Mocquard, 1870–1909, *Hydrops lehmanni* Dunn, 1944e, and *Pseudoeryx plicatilis ecuadorensis* Mertens, 1965b.

Type: Lectotype, NRH Lin-57 (formerly MAFR) a 770 mm male (Mus. Drottn.), designated by Hoge (1964d: 61).

Type locality: "Ternataeis" [= Ternate, Indonesia] (in error *fide* Hoge, 1964d: 61).

Distribution: South America. Southeastern Colombia (Amazonas, Cauca, Leticia), SE Venezuela (Amazonas, Bolívar), Guyana (East Berbice-Corentyne, Pomeroon-Supenaam), Suriname (Nickerie, Para, Paramaribo, Saramacca), French Guyana (Cayenne, Saint-Laurent-du-Maroni), Brazil (Acre, Amapá, Amazonas, Maranhão, Mato Grosso, Mato Grosso do Sul, Pará, Rondônia, Roraima, São Paulo, Marajo Is.), E Ecuador (Napo), NE Peru (Loreto, Madre de Dios, Napo), E Bolivia (Beni, Santa Cruz), Paraguay (Alto Paraguay, Central, Neembucú) and NE Argentina (Chaco, Corrientes, Formosa, Misiones), NSL–410 m.

Sources: Roze, 1957b, Hoge & Maranhão Nina, 1964, Mertens, 1965b, K.L. Williams & Fugler, 1968, Cunha & Nascimento, 1978, Dixon & Soini, 1986, Pérez-Santos, 1986, Starace, 1998, Giraudo, 2001, Duellman, 2005, Scartozzoni et al., 2010, Cabral & Caballero, 2012 and C.J. Cole et al., 2013.

2. *Pseudoeryx relictualis* Schargel, Rivas-Fuenmayor, Barros-Blanco, Pèfaur & Navarrete, 2007. Herpetologica 63(2): 236–241, fig. 1.

Type: Holotype, CVULA IV 6955, a 627 mm female (F. Diaz, 11 Sept. 2003).
Type locality: "Hacienda La Trinidad, sector Montalbán, Municipio Catatumbo, Zulia state, Venezuela (09° 09' 01" N; 72° 02' 29" W)."
Distribution: Northwestern Venezuela (Zulia), NSL.

PSEUDOFERANIA Ogilby, 1890a (Homalopsidae)

Type species: *Pseudoferania macleayi* Ogilby, 1890a.
Distribution: Tidal rivers and coastal marine waters of Austro-Papua.
Sources: Kinghorn, 1964, F. Parker, 1982, S.K. Wilson & Knowles, 1988, Gow, 1989, Ehmann, 1992, O'Shea, 1996, Shea & Sadlier, 1999, Cogger, 2000 and J.C. Murphy, 2011.
Remarks: A valid genus *fide* J.C. Murphy (2011a: 234).

1. *Pseudoferania macleayi* Ogilby, 1890a. Proc. Linn. Soc. N.S.W. (2) 5(1): 51–52.

Type: Lectotype, AMS 460 (J.A. Boyd), designated by Gyi (1970: 109).
Type locality: "Ripple Creek, Herbert River District, Queensland" [Australia].
Distribution: Extreme NE Australia (NE Queensland), 90 m.
Sources: Kinghorn, 1929, Gow, 1983, 1989 and Ehmann, 1992.
Remarks: Previously a synonym of *Enhydris polylepis.*

2. *Pseudoferania polylepis* (J.G. Fischer, 1886). Abh. Ges. Naturwiss. Ver. Hamburg 9(1): 62–63. (*Hypsirhina polylepis*)

Synonym: *Pseudoferania harritosi* Wells & Wellington, 1985.
Type: Holotype, MTKD D 437, a 280 mm female (Gerrard, 1885), destroyed 13–14 Feb. 1945 during World War II.
Type locality: "Fly River, Süd-Neuguinea" [= Fly River, Western Prov., SW Papua New Guinea].
Distribution: Austro-Papua. Eastern Indonesia (S Papua), S Papua New Guinea (Western) and N Australia (N Northern Territory, N Queensland, ext. N Western Australia, Groote Eylandt and Melville Is.), NSL–280 m.
Remarks: Previously in the genus *Enhydris.*

PSEUDOFICIMIA Bocourt, 1883 *in* Duméril, Bocourt & Mocquard, 1870–1909 (Colubridae)

Type species: *Toluca frontalis* Cope, 1864.
Distribution: Western Mexico.
Sources: E.H. Taylor & Smith, 1942a, 1945, Campbell & Simmons, 1962, L.M. Hardy, 1972, 1973, 1975c and H.M. Smith & Smith, 1976, 1993.

1. *Pseudoficimia frontalis* (Cope, 1864). Proc. Acad. Nat. Sci. Philadelphia 16(3): 167. (*Toluca frontalis*)

Synonyms: *Pseudoficimia pulchra* Bocourt, 1883 *in* A.H.A. Duméril, Bocourt & Mocquard, 1870–1909, *Pseudoficimia pulcherrima* E.H. Taylor & Smith, 1942a, and *Pseudoficimia hiltoni* Bogert & Oliver, 1945.
Type: Lectotype, USNM 31425 (formerly USNM 1363), a 502 mm female (J. Xantus, 1859–1861), designated by L.M. Hardy (1972: 58).
Type locality: "Colima" [= Colima State, W Mexico] via lectotype selection.
Distribution: Western Mexico. (Colima, ext. SW Durango, Guanajuato, Guerrero, Jalisco, S México, W Michoacán, Morelos, Nayarit, Puebla, Sinaloa, ext. S Sonora), NSL–1100 m.

PSEUDOHAJE A.C.L.G. Günther, 1858 (Elapidae)

Synonyms: *Pseudohage* – Villiers, 1951a *(nomen incorrectum)* and *Pseodohaja* – Khalil, 1963 *(nomen incorrectum).*
Type species: *Pseudohaje nigra* A.C.L.G. Günther, 1858.
Distribution: West and cen. Africa.
Sources: Bogert, 1942a, 1943, Klemmer, 1963, Hughes & Barry, 1969, Villiers, 1975, Hughes, 1976b, 1983, Roman, 1976, 1984, Courtois & Chippaux, 1977, Hughes, 1983, Golay, 1985, Joger, 1990, Golay et al., 1993, Spawls & Branch, 1995, Keogh, 1998, David & Ineich, 1999, Chippaux, 2006, Dobiey & Vogel, 2007 and Zaher et al., 2009.

1. *Pseudohaje goldii* (Boulenger, 1895c). Ann. Mag. Nat. Hist. (6) 16(91): 34. (*Naia goldii*)

Synonyms: *Naia yakomae* Mocquard, 1896a, *Pseudohaje gold* – Chapman, 1968 *(nomen incorrectum),* and *Pseudohaje gouldii* – Skoczylas, 1978 *(nomen incorrectum).*
Type: Holotype, BMNH 1946.1.21.43 (formerly BMNH 1895.5.3.53), a 1750 mm male (W.H. Crosse, 1890–1895).

Type locality: "near Asaba, about 150 miles up the Niger" [= vicinity of Asaba (6°12'N, 6°41'E, elevation 50 m), Niger River, NE Delta State, S Nigeria].
Distribution: West and cen. Africa. Southeastern Ivory Coast (Abidjan, Agboville), S Ghana (Eastern, Western), S Nigeria (Akwa Ibom, Bayelsa, Bendel, Cross River, Delta, Edo, Plateau, Rivers), S Cameroon (Adamaoua, Centre, Est, Littoral, Nord-Ouest, Ouest, Sud, Sud-Ouest), S Central African Republic (Haute-Sangha, Lobaye, Ombella-Mpoko, Ouaka, Ouham, Sangha), Equatorial Guinea (Bioko Is.), Gabon (Estuaire, Ngouié, Ogooué-Maritime, Woleu-Ntem), S Congo (Kouilou), Democratic Republic of the Congo (Bandundu, Bas-Congo, Equateur, Kinshasa, Nord-Kivu, Orientale), S Uganda (Central, Western, Sanga Is.), SW Kenya (Western), N Angola (Cuanza Sul, Lunda Norte) and NE Namibia (Okavango), NSL–1525 m.
Sources: Sternfeld, 1908b, 1909a, Boulenger, 1915a, 1920c, Angel, 1933b, H.W. Parker, 1936d, Mertens, 1941a, 1955, 1964a, 1971, Laurent, 1954a, 1958, 1960, Loveridge, 1957a, E.H. Taylor & Weyer, 1958, Perret, 1961, Isemonger, 1962, Roux-Estève, 1965, Thys van den Audenaerde, 1965, Broadley, 1968a, A. Brunner, 1968, Pitman, 1974, Courtois, 1978, Spawls, 1978, Stucki-Stirn, 1979, J.-F. Trape, 1981, 1985, J.-F. Trape & Carme, 1982, J.A. Butler & Reid, 1986, J.-F. Trape & Roux-Estève, 1995, Akani et al., 2005, Chirio & LeBreton, 2007 and Pauwels & Vande weghe, 2008.
Remarks: Probably occurs in Togo *fide* Segniagbeto et al. (2011: 353) and possibly occurs in Rwanda *fide* Spawls et al.(2002: 460).

2. *Pseudohaje nigra* A.C.L.G. Günther, 1858. Cat. Colub. Snakes Brit. Mus.: 222.

Synonym: *Naia guentheri* Boulenger, 1896a.
Type: Holotype, BMNH 1946.1.21.41, a 2130–2311 mm male (Edward, 13th Earl of Derby).
Type locality: Unknown. Designated as West Africa *fide* Boulenger (1896a: 388).
Distribution: West Africa. Guinea (Nzérékoré, Télimélé), Sierra Leone (Southern, Western), Liberia (Nimba), SE Ivory Coast, S Ghana (Ashanti, Brong-Ahafo, Central, Eastern, Western), S Togo (Plateaux), and S Nigeria (Anambra), NSL–600 m.
Sources: Sternfeld, 1908c, 1909b, Barbour & Loveridge, 1930, Cansdale, 1954, Doucet, 1963, Menzies, 1966, Roman, 1976, Rödel et al., 1995, W. Böhme et al., 2011, Segniagbeto et al., 2011 and Chirio, 2013.

PSEUDOLATICAUDA Kharin, 1984 (Elapidae)

Type species: *Platurus semifasciatus* Schlegel, 1837.
Distribution: Marine waters and outlying reefs of East Asia, East Indies and SW Pacific, including Banda Sea, Flores Sea, Java Sea, Manila Bay, Sea of Japan, South China Sea, Taiwan Strait and Yellow Sea.
Sources: Kharin, 1984, Golay et al., 1993, Kharin & Czeblukov, 2006, Kharin & Hallerman, 2009a, Zaher et al., 2009 and Hoser, 2012e.

1. *Pseudolaticauda schistorhynchus* (A.C.L.G. Günther, 1874). Proc. Zool. Soc. London 42(1): 295–299, pl. 45, fig. b. (*Platurus schistorhynchus*)

Synonym: *Laticauda semifasciata schistorhyncha* – Sternfeld, 1920c.
Type: Lectotype, BMNH 1946.1.6.98 (formerly BMNH 1874.3.31.4), a juvenile male (Rolleston via A.C.L.G. Günther), designated by Kharin & Czeblukov (2006: fig. 14) and Kharin et al. (2010: 288).
Type locality: "Niue or Savage Island,"[= Niue Island, bet. 18°57'–19°09'S and 169°46–57'W, SW Pacific Ocean].
Distribution: Southwest Pacific. Northern New Guinea (East Sepik, Bertrand Is.), Tonga, Samoa and Niue.
Remarks: Photographs of lectotype in Kharin & Czeblukov (2006: fig. 14). Boulenger (1896a: 309) erroneously listed Günther's illustration of the type as fig. A.

2. *Pseudolaticauda semifasciata* (Schlegel, 1837). Essai Phys. Serp. 2: 516. (*Platurus semifasciatus*)

Synonyms: *Platutus semifasciatus* H. Boie *in* Schlegel, 1826a *(nomen nudum), Platutus semifasciatus* F. Boie, 1827 *(nomen nudum), Platurus fasciatus* Hilgendorf, 1876 *(lapsus calami), Laticauda septemfasciata* – Rizzoli, 1973 *(nomen incorrectum), Fasciata semifasciata* – Cahill, 1981 *(nomen incorrectum),* and *Laticauda semifesciata* – Culotta & Pickwell, 1993 *(nomen incorrectum).*
Type: Holotype, RMNH 1468, an 1118 mm female (C.G.C. Reinwardt, Feb. 1821–June 1822), designated by Stejneger (1907a: 409).
Type locality: "mer des Moluques" [= Maluku, E Indonesia].
Distribution: East Asia and East Indies. Southeastern Russia, S Japan (Kyushu, Amanioshima, Erabushima, Iriomotejima, Kagoshima, Miyakojima, Okinawashima, Yakujima and Yokoshima Is.), China (Fujian, Liaoning), Taiwan (Taitung, Lan Yu Is.), Indonesia (Ambon, Flores, Java, Malukku, Seram, Sulawesi) and Philippines (Bohol, Capones, Cuyo, Gato, Luzon, Monja, Negros, Palawan, Sulu Arch., Visayas).
Sources: Temminck & Schlegel, 1838, Rooij, 1917, E.H. Taylor, 1922a, Herre & Rabor, 1949, Kuntz, 1963, Guinea et al., 1983, Tamiya et al., 1983b, M. Mori, 1986, Tian et al., 1986, Toriba & Yosida, 1986, Toriba & Nakamoto, 1987, Kharin, 2009, Xiang & Li, 2009 and A.R. Rasmussen et al., 2011.

Remarks: C.G.C. Reinwardt listed as collector of type *fide* Schlegel (1837: 516) and Stejneger (1907a: 409) but Reinwardt only collected in Java and Sumatra.

PSEUDOLEPTODEIRA E.H. Taylor, 1938 (Dipsadidae)

Type species: *Hypsiglena latifasciata* A.C.L.G. Günther, 1894.
Distribution: Western Mexico.
Sources: Boulenger, 1905g, Mocquard, 1908, Dunn, 1936, E.H. Taylor, 1938, Duellman, 1954, 1958a, Shannon & Humphrey, 1964, Dowling & Jenner, 1987, García & Ceballos, 1994, H.M. Smith et al., 2000, Canseco-Márquez & Nolasco Vólez, 2008, Reyes-Velasco & Mulcahy, 2010 and Mulcahy et al., 2011.

1. *Pseudoleptodeira latifasciata* (A.C.L.G. Günther, 1894 *in* 1885–1902). Biol. Cen.-Amer., Rept. Batr.: 138, pl. 49, fig. b. (*Hypsiglena latifasciata*)

Synonym: *Leptodira guilleni* Boulenger, 1905g.
Type: Holotype, BMNH 1946.1.23.44, a 349 mm female (F.D. Godman & O. Salvin, 1861–1863).
Type locality: "Southern Mexico." Restricted to Huajintlán, Guerrero, Mexico *fide* H.M. Smith & Taylor (1950: 331).
Distribution: Southwestern Mexico (Chiapas, Colima, Guerrero, Jalisco, Michoacán, Morelos, Oaxaca, Puebla), 100–1300 m.
Remarks: Duellman (1958a: 91) erroneously cited type as BMNH 1946.1.2.44. BMNH catalogue lists BMNH 1946.1.23.43 as a syntype.

PSEUDONAJA A.C.L.G. Günther, 1858 (Elapidae)

Synonyms: *Euprepiosoma* Fitzinger, 1861, *Pseudonaia* Krefft, 1862 *(nomen emendatum)*, *Pseudonaje* – Hoffmann, 1890 *(nomen incorrectum)*, *Pseudojata* – K.N.G. Simpson, 1973 *(nomen incorrectum)*, *Dugitophis* Wells, 2002, *Notopseudonaja* Wells, 2002, *Placidaserpens* Wells, 2002, and *Hulimkai* Hoser, 2012i *(nomen illegitimum)*.
Type species: *Pseudonaja nuchalis* A.C.L.G. Gunther, 1858.
Distribution: Austro-Papua.
Fossil records: Pleistocene of Australia.
Sources: M.J. Smith, 1975, Gillam, 1979, Cogger et al., 1983a, Golay, 1985, Mengden, 1985b, Storr, 1985, S.K. Wilson & Knowles, 1988, Gow, 1989, Shine, 1989, Golay et al., 1993, Greer, 1997, David & Ineich, 1999, Cogger, 2000, Wells, 2002, Scanlon & Lee, 2004, A. Skinner, 2005, 2009, Sanders et al., 2008, Zaher et al., 2009 and Hoser, 2012e.

1. *Pseudonaja affinis* A.C.L.G. Günther, 1872b. Ann. Mag. Nat. Hist. (4) 9(49): 35–36, pl. 4, fig. c.

Synonyms: *Demansia nuchalis tanneri* Worrell, 1961b, *Pseudonaja affinis exilis* Storr, 1989, and *Pseudonaja affinis charlespiersoni* Hoser, 2009c & 2012i *(nomen illegitimum)*.
Type: Holotype, BMNH 1946.1.19.77, a 1372 mm specimen (J.L.G. Krefft, 1861–1872).
Type locality: "Australia."
Distribution: Southwestern Australia (Southern Australia, S Western Australia, Boxer, Figure of Eight and Rottnest Is.), NSL–465 m.
Sources: Storr et al., 1986 and Bush et al., 1995.

2. *Pseudonaja aspidorhyncha* (F. McCoy, 1879c). Prod. Zool. Victoria 3: 13–14, pl. 23, figs. 4–4a. (*Diemenia aspidorhyncha*)

Synonyms: *Diemenia carinata* Longman, 1915, *Demansia acutirostris* F.J. Mitchell, 1951, and *Pseudonaja gowi* Wells, 2002.
Type: Lectotype, NMV D12352 (formerly NMV 3261 & NMV 12871), a 1345 mm specimen with prepared skull, designated by Coventry (1970: 121).
Type locality: "Junction of Murray River and Darling River, Victoria, Australia" via lectotype selection.
Distribution: Southern Australia (cen. New South Wales, S Northern Territory, SW Queensland, South Australia, NW Victoria), NSL–585 m.
Source: A. Skinner, 2009.

3. *Pseudonaja guttata* (H.W. Parker, 1926a). Ann. Mag. Nat. Hist. (9) 17(102): 668–669. (*Demansia guttata*)

Synonym: *Pseudonaja guttata whybrowi* Hoser, 2009c & 2012i *(nomen illegitimum)*.
Type: Holotype, BMNH 1946.1.20.67 (formerly BMNH 1926.2.25.107), a 740 mm female (G.H. Wilkins).
Type locality: "Winton, Queensland (Lat. 22° 19' S., Long. 143° 4' E.), Australia."
Distribution: Northeastern Australia (E Northern Territory, Queensland, ext. NE South Australia), 15–455 m.
Sources: B. Miller, 1982 and Covacevich & Couper, 1991.

4. *Pseudonaja inframacula* (Waite, 1925). Rec. So. Aust. Mus. 3(1): 26–28, figs. 10–11. (*Demansia textilis inframacula*)

Types: Syntypes (2), SAMA 960a–b, 1055 mm and 960 mm specimens (E.R. Waite & F. Wood Jones, Jan. 1925), lost *fide* Cogger et al. (1983: 234) and A. Skinner (2009: 189).
Type locality: "northern end of Coffin's Bay Peninsula, on the west coast of Eyre Peninsula" [= Point Sir Isaac, S South Australia, Australia, 34°26'S, 135°12'E, elevation NSL].

Distribution: Extreme S Australia (S South Australia, North Neptunes and Wardang Is.), NSL–100 m.

5. *Pseudonaja ingrami* (Boulenger, 1908d). Ann. Mag. Nat. Hist. (8) 1(4): 334. (*Diemenia ingrami*)

Type: Holotype, BMNH 1946.1.20.32, a 1510 mm male (W. Stalker).
Type locality: "near Alexandria, in the Northern Territory of the colony of South Australia" [= vicinity of Alexandria Station, Nicholson, NE Northern Territory, N Australia, 19°05'S, 136°43'E, elevation 240 m].
Distribution: North-central Australia (cen. Northern Territory, ext. W Queensland, ext. NE Western Australia), 130–240 m.
Source: S. Phillips, 1993.

6. *Pseudonaja mengdeni* Wells & Wellington, 1985. Aust. J. Herp. (Suppl. 1): 48.

Synonym: *Pseudonaja kellyi* Wells & Wellington, 1985.
Type: Holotype, NTM 1989, a gravid female (B. Jukes).
Type locality: "2 km east of Maryvale, Northern Territory, Australia" [= 2 km E Maryvale, Titjikala, S Northern Territory, cen. Australia, 24°40'S, 134°06'E, elevation 390 m].
Distribution: Australia (W New South Wales, Northern Territory, South Australia, Western Australia), NSL–740 m.
Source: A. Skinner, 2009.

7. *Pseudonaja modesta* (A.C.L.G. Günther, 1872b). Ann. Mag. Nat. Hist. (4) 9(49): 35, pl. 3, fig. c. (*Cacophis modesta*)

P

Synonyms: *Brachysoma sutherlandi* De Vis, 1884c and *Furina ramsayi* Macleay, 1885a.
Type: Lectotype, BMNH 1946.1.18.42, a juvenile male (Duboulay), designated by Gillam (1979: 13).
Type locality: "Perth, Western Australia" via lectotype selection.
Distribution: Australia (NW New South Wales, Northern Territory, W Queensland, South Australia, Western Australia), NSL–955 m.
Source: Mengden, 1985a.
Remarks: Ingram (1979: 13) erroneously listed type figure as pl. 2, fig. C.

8. *Pseudonaja nuchalis* A. Gunther, 1858. Cat. Colub. Snakes Brit. Mus.: 227–228.

Synonyms: *Pseudelaps bancrofti* De Vis, 1911, *Pseudonaja imperitor* Wells & Wellington, 1985, *Pseudonaja jukesi* Wells & Wellington, 1985, and *Pseudonaja vanderstraateni* Wells & Wellington, 1985.
Type: Lectotype, BMNH 1946.1.20.41, skin of male, longest syntype 1626 mm (Edward, 13th Earl of Derby), designated by Wells & Wellington (1985: 48).
Type locality: "Port Essington, N.T." [= ruins at tip of Cobourg Peninsula, N Northern Territory, N Australia, ca. 11°09'N, 132°00'E, elevation NSL] via lectotype selection
Distribution: Australia (W New South Wales, Northern Territory, Queensland, South Australia, ext. N Victoria, Western Australia), NSL–820 m.
Sources: M.J. Smith, 1975, Bush, 1989 and Bush et al., 1995.

9. *Pseudonaja textilis* (A.-M.-C. Duméril, Bibron & Duméril, 1854b). Erpét. Gén. 7(2): 1242–1243. (*Furina textilis*)

Synonyms: *Furina textilis* A.-M.-C. Duméril, 1853 *(nomen nudum), Pseudoelaps superciliosus* J.G. Fischer, 1856b, *Demansia annulata* A.C.L.G. Günther, 1858, *Pseudoelaps kubingii* Jan, 1859a, e, *Pseudoelaps sordellii* Jan, 1859a, e, *Pseudoelaps kubinyi* – Jan, 1863b *(nomen incorrectum), Pseudoelaps superciliosus beckeri* Jan, 1863a *(nomen nudum), Cacophis güntherii* Steindachner, 1867a *(nomen incorrigendum), Pseudelaps beckeri* Jan, 1873 *in* Jan & Sordelli, 1870–1881, *Furina bicucullata* F. McCoy, 1879d, *Pseudechis cupreus* (*partim*) Boulenger, 1896a, *Pseudonaja ohnoi* Wells and Wellington, 1985, *Psuedonaja textilis* – Kaiser & Middlebrook, 1988 *(nomen incorrectum), Pseudonaja elliotti* Hoser, 2003b, *Pseudonaja textilis pughi* Hoser, 2003e, *Pseudonaja textilis cliveevattii* Hoser, 2009c & 2012i *(nomen illegitimum), Pseudonaja textilis jackyhoserae* Hoser, 2009c & 2012i *(nomen illegitimum), Pseudonaja textilis leswilliamsi* Hoser, 2009c & 2012i *(nomen illegitimum), Pseudonaja textilis rollinsoni* Hoser, 2009c & 2012i *(nomen illegitimum),*
Type: Holotype, MNHN 3944, a 300 mm specimen (J. Verreaux, Oct. 1846), lost *fide* Cogger et al. (1983: 233).
Type locality: "Australie" [= Australia].
Distribution: Eastern Papua New Guinea (Central, Milne Bay, Oro) and Australia (Australian Capital Territory, New South Wales, S Northern Territory, Queensland, South Australia, Victoria, ext. NE Western Australia), NSL–1235 m
Sources: F. Scott et al., 1977, Whitaker et al., 1982, K. Griffiths, 1987, Hoser, 1989, 2003b, e, 2009c, O'Shea, 1996, Watharow, 2000, J.C. Murphy & Schlager, 2003, D.J. Williams et al., 2008 and A. Skinner, 2009.
Remarks: New Guinea population possibly a separate species *fide* D.J. Williams et al. (2008: 58–59).

PSEUDOPLECTRURUS Boulenger, 1890a (Uropeltidae)

Synonym: *Pseudoplectrusus* – Briceño-Rossi, 1934 *(nomen incorrectum).*
Type species: *Silybura canarica* Beddome, 1870.
Distribution: Southern India.

Sources: A.C.L.G. Günther, 1875, Beddome, 1886, Boulenger, 1896a, Wall, 1923c, M.A. Smith,1943, Gans, 1966, Murthy, 1973, 1981c, 1982b, 1985, Mahendra, 1984 and McDiarmid et al., 1999.

1. *Pseudoplectrurus canaricus* (Beddome, 1870). Madras Month. J. Med. Sci. 2: 170. (*Silybura canarica*)

Types: Syntypes (6), BMNH 1946.1.15.95–98 (formerly BMNH 1874.4.29.940–43), BMNH 1946.1.16.71 (formerly BMNH 1872.1.2.3), and MNHN 1895.102, three males, a female and a juvenile, longest syntype 406 mm (R.H. Beddome, 1857–1870).

Type locality: "South Canara, very common on the top of the Kudra Mukh, a mountain 6,000 feet high, Madras Presidency" [= Mt. Kudremukh, SW Karnataka State, SW India, 13°08'N, 75°16'E]. Emended to near Mangalore [= 12°52'N, 74°53'E] *fide* Boulenger (1893a: 160).

Distribution: Southwestern India (Western Ghats of SW Karnataka), 1830–2000 m.

PSEUDORABDION Jan, 1862b (Calamariidae)

Synonyms: *Oxycalamus* A.C.L.G. Günther, 1864a, *Typhlogeophis* A.C.L.G. Günther, 1879a, *Thyphlogeophis* Carus, 1879 *(nomen emendatum)*, *Idiopholis* Mocquard, 1892a–b, *Pseudorhabdion* Boulenger, 1885c *(nomen emendatum)*, *Pseudorhabdium* Boulenger, 1893a *(nomen emendatum)*, *Typhlogeophus* – Casto de Elera, 1895 *(nomen incorrectum)*, and *Agrophis* F. Müller, 1895.

Type species: *Calamaria longiceps* Cantor, 1847.

Distribution: Malay Peninsula and East Indies.

Sources: Rooij, 1917, Leviton & Brown, 1959, Inger & Leviton, 1966, R.M. Brown et al., 1999, Zaher et al., 2009 and I. Das, 2010, 2012.

1. *Pseudorabdion albonuchalis* (A.C.L.G. Günther, 1896b). Ann. Mag. Nat. Hist. (6) 17(99): 229. (*Geophis albonuchalis*)

Type: Neotype, FMNH 151709 (formerly CNHM 151709), a 272 mm male (W. Hosmer, 22 March 1964), designated by Inger & Leviton (1966: 312).

Type locality: "Sungei Pesu, Fourth Division, Sarawak" [East Malaysia] via neotype selection.

Distribution: East Malaysia (Sabah, Sarawak), NSL–500 m.

Sources: Shelford, 1901b, C. Haas, 1950, Haile, 1958, Stuebing, 1991, Manthey & Grossmann, 1997 and I. Das, 2012.

Remarks: Holotype lost *fide* J.C. Battersby *in* Leviton & Brown (1959: 496). Photographs of neotype in Inger & Leviton (1966: fig. 3a–b).

2. *Pseudorabdion ater* (E.H. Taylor, 1922b). Philippine J. Sci. 21D(2): 202–203, figs. 6–7. (*Typhlogeophis ater*)

Type: Holotype, CAS 62043 (formerly EHT 1103), a 173 mm male (E.H. Taylor, 28 Sept. 1920).

Type locality: "small brook that empties into Tumugao River just above the waterwork's intake near Pasananka, Zamboanga Peninsula, Mindanao, Philippine Islands" [= Pasonanca, ext. SW Zamboanga del Sur Prov., ext. W Mindanao Is., ext. S Philippines, 6°57'N, 122°04'E, elevation 45 m].

Distribution: Extreme S Philippines (ext. SW Mindanao), 45 m. Known only from holotype.

Sources: Leviton, 1959b and Alcala, 1986.

3. *Pseudorabdion collaris* (Mocquard, 1892a). Naturaliste (2) 14(117): 35. (*Idiopholis collaris*)

Synonym: *Idiopholis everetti* Shelford, 1901.

Type: Holotype, MNHN 1891.65, a 190 mm specimen (M.A. Chaper, Nov. 1890–Jan. 1891).

Type locality: "vallée du Sebroeang (ouest de Bornéo)" [= Sebruang Valley, joining the Kapuas River NE of Sintang, cen. Kalimantan Barat Prov., W Kalimantan, Indonesia, 0°26'N, 111°53'E, elevation 300 m]. Emended to Sintang, Bornéo *fide* Mocquard (1892b: 190).

Distribution: Borneo. East Malaysia (Sabah, Sarawak) and Indonesia (W Kalimantan), 30–650 m.

Sources: Brongersma, 1946, Haile, 1959, Manthey & Grossmann, 1997, Malkmus et al., 2002 and I. Das, 2010.

Remarks: Supplemental original description in Mocquard (1892b: 191–192, pl. 7, figs. 1, 1a–c).

P

4. *Pseudorabdion eiselti* Inger & Leviton, 1961. Fieldiana: Zool. 44(5): 45–47, fig. 13.

Type: Holotype, NMW 16806, a 200 mm gravid female (J. Schild, 1899).

Type locality: "Padang, Sumatra" [= Padang, Sumatera Barat Prov., W Sumatra, W Indonesia, 0°57'S, 100°21'E, elevation NSL].

Distribution: Extreme W Indonesia (Nias, Siberut, W Sumatra), NSL.

Sources: Dring et al., 1990, David & Vogel, 1996 and Iskandar & Colijn, 2001.

5. *Pseudorabdion longiceps* (Cantor, 1847). J. Asiat. Soc. Bengal 16(182): 910, pl. 40, fig. 1. (*Calamaria longiceps*)

Synonym: *Rabdion torquatum* A.-M.-C. Duméril, Bibron & Duméril, 1854a.

Type: Holotype, BMNH 1946.1.2.13, a 156 mm male (W.T. Lewis, 1825–1847).

Type locality: "Great Hill of Pinang, Malayan Peninsula" [= Pinang Hill, Penang Is., Pinang State, NW West Malaysia, 5°25'N, 100°17'E].

Distribution: Malay Peninsula and Indonesia. Peninsular Thailand (Narathiwat, Pattani, Yala), West Malaysia (Johore, Pahang, Perak, Pinang, Selangor, Penang and Sibu Is.), Singapore, East Malaysia (Sarawak), Brunei and Indonesia (Bengkalis, Bintan, Galang, W Kalimantan, Nias, Riau Arch., Sumatra), NSL–500 m.

Sources: A.C.L.G. Günther, 1864a, Blanford, 1881, Lidth de Jeude, 1890, Flower, 1896, M.A. Smith, 1916, Chasen & Smedley, 1927, Brongersma, 1931, 1933e, Bourret, 1936b, Inger & Leviton, 1961, E.H. Taylor, 1965, Tweedie, 1983, Bosch, 1985, K.F.L. Lim & Lee, 1989, M.J. Cox, 1991b, Stuebing, 1991, K.K.P. Lim & Lim, 1992, David & Vogel, 1996, Manthey & Grossmann, 1997, Grismer, 2011 and I. Das, 2012.

Remarks: Philippine (Luzon) record doubtful *fide* Leviton & Brown (1959: 481) and Sulawesi records doubtful *fide* Lang & Vogel (2005: 247).

6. *Pseudorabdion mcnamarae* E.H. Taylor, 1917. Philippine J. Sci. 12D(6): 363, figs. 2a–c. (*Pseudorhabdium mcnamarae*)

Synonym: *Pseudorhabdium minutum* E.H. Taylor, 1922b.

Type: Holotype, CM 2606 (formerly EHT 196), a 242 mm female (H. McNamara & E.H. Taylor, 24 Dec. 1915).

Type locality: "Mount Canlaon, or Malaspina, a volcano in the north-central part of Negros Island, Occidental Negros, P.I., elevation about 900 m" [= Mount Canlaon, N Negros Occidental Prov., N Negros Is., S Philippines, 10°25'N, 123°08'E, elevation 900 m].

Distribution: Northern Philippines (Luzon, Negros, Panay), 300–1600 m.

Sources: E.H. Taylor, 1922a–b, Leviton, 1959b, Alcala, 1986, Sison & Gonzales, 1995, Ferner et al., 2001 and Gaulke, 2011.

P

7. *Pseudorabdion modiglianii* G. Doria & Petri, 2010. Ann. Mus. Civ. Stor. Nat. "Giacomo Doria," Genova 102: 193–195, fig. 2.

Type: Holotype, MSNG 55461, a 203 mm female (E. Modigliani, 1–15 April 1894).

Type locality: Indonesia, Sumatra Island, Province of Sumatera Barat, Padang" [= Padang, Sumatera Barat Prov., W Sumatra, W Indonesia, 0°57'S, 100°21'E, elevation NSL].

Distribution: Western Indonesia (W Sumatra), NSL.

8. *Pseudorabdion montanum* Leviton & Brown, 1959. Proc. California Acad. Sci. (4) 29(14): 491–494, figs. 5–6.

Type: Holotype, CAS–SU 21080 (formerly SU 21080), a 531 mm female (W.C. Brown, M. Pinero & T. Serate, 22 April 1958).

Type locality: "on the north side of the north peak of Cuernos de Negros, Negros Oriental Province, Negros Island" [= Cuernos de Negros, S Negros Oriental Prov., ext. S Negros Is., S Philippines, 9°15'N, 123°11'E, elevation ca. 1200 m].

Distribution: Central Philippines (S Negros), 1200 m.

Sources: Leviton, 1959b and Alcala, 1986.

9. *Pseudorabdion oxycephalum* (A.C.L.G. Günther, 1858). Cat. Colub. Snakes Brit. Mus.: 242–243. (*Rhabdosoma oxycephalum*)

Synonym: *Typhlogeophis brevis* A.C.L.G. Günther, 1879a.

Type: Holotype, BMNH 1946.1.1.99, a 305 mm female (H. Cuming, 1836–1840).

Type locality: "Philippines."

Distribution: Central Philippines (Negros, Panay), NSL–1200 m.

Sources: A.C.L.G. Günther, 1873b, Casto de Elera, 1895, L.E. Griffin, 1911, E.H. Taylor, 1922a–b, Leviton, 1959b, Alcala, 1986, Ferner et al., 2001 and Gaulke, 2011.

Remarks: Records from Luzon, Mindanao, Dinagat and Calamian doubtful *fide* Leviton & Brown (1959: 491) and Gaulke (2011: 309).

10. *Pseudorabdion sarasinorum* (F. Müller, 1895). Verh. Naturf. Ges. Basel (1892–1895) 10(3): 827–828, 2 figs. (*Agrophis sarasinorum*)

Type: Holotype, NMBA 1678, a 235 mm male (P. Sarasin & F. Sarasin, 1894).

Type locality: "Gunung (Vulcan) Soudara ca. 1200 m., Celebes" [= Gunung Duasudara, ext. E Sulawesi Utara Prov., ext. NE Sulawesi, 1°29'N, 125°09'E, elevation 1200 m].

Distribution: Eastern Indonesia (N Sulawesi), 1000–1720 m.

Sources: Boulenger, 1897g, Bosch, 1985 and Lang & Vogel, 2005.

Remarks: Illustration of holotype in Leviton & Brown (1959: fig. 7).

11. *Pseudorabdion saravacensis* (Shelford, 1901a). Ann. Mag. Nat. Hist. (7) 8(48): 516. (*Agrophis saravacensis*)

Type: Holotype, BMNH 1946.1.2.3 (formerly SMK), a 142 mm male.

Type locality: "Kuching, Sarawak, Borneo" [= Kuching, ext. W Sarawak, ext. W East Malaysia, 1°33'N, 110°21'E, elevation 30 m].

Distribution: Western East Malaysia (ext. W Sarawak), 30 m. Known only from holotype.

Source: Rooij, 1917.

12. *Pseudorabdion sirambense* G. Doria & Petri, 2010. Ann. Mus. Civ. Stor. Nat. 'Giacomo Doria' 102: 196–197, fig. 3.

Type: Holotype, MSNG 55463 (formerly MSNG 30372), a 98 mm male (E. Modigliani, 1890).
Type locality: Indonesia, Sumatra Island, Province of Sumatera Utara, 'Si Rambé' (now Sirambi, about 13 km SE of Balige, off the southern shore of Danau Toba)" [= Sirambe, 2 km N Naga Saribu, N Sumatra, 2°16'N, 99°07'E, elevation 1430 m].
Distribution: Western Indonesia (Sumatra), 1430 m.

13. *Pseudorabdion talonuran* R.M. Brown, Leviton & Sison, 1999. Asiatic Herpetol. Res. 8: 7–10, figs. 2–3.

Type: Holotype, PNM 2712, a 265 mm male (R.M. Brown & R. Sison, 28 May 1992).
Type locality: "western foothills of Mt. Madja-as (11°23'N, 122°09'E; elev. 1500 m), Barangay Allojipan, Municipality of Culasi, Antique Province, Panay Island, Philippines."
Distribution: Central Philippines (Panay), 1410–1500 m. Known only from type locality.
Sources: Ferner et al., 2001 and Gaulke, 2011.

14. *Pseudorabdion taylori* Leviton & Brown, 1959. Proc. California Acad. Sci. (4) 29(14): 502–504, figs. 9–10.

Type: Holotype, MCZ 25749, a 211 mm male (E.H. Taylor, 24–25 April 1923).
Type locality: "Saub, Cotabato Province, Mindanao Island" [= Saub, Maitum SW Sarangani Prov., ext. S Mindanao Island, ext. S Philippines, 6°02'N, 124°30'E, elevation NSL].
Distribution: Extreme S Philippines (ext. S Mindanao), NSL. Known only from the type series.
Sources: Leviton, 1959b and Alcala, 1986.

PSEUDOTOMODON Koslowsky, 1896b (Xenodontidae)

Synonym: *Pseudotomodon* Peracca, 1897c.
Type species: *Pelias trigonatus* Leybold, 1873.
Distribution: Western Argentina.
Sources: Koslowsky, 1896b, Serié, 1936, Abalos et al., 1964, J.A. Peters & Orejas-Miranda, 1970, Cei, 1986, J.D. Williams & Francini, 1991, Avila, 1997, Cruz et al., 1999, Giraudo & Scrocchi, 2002, Harvey & Muñoz, 2004, Scolaro, 2005, 2006, Gallardo, 2006, Avila, 2009a and Zaher et al., 2009.

1. *Pseudotomodon trigonatus* (Leybold, 1873). Pampas Arjent.: 82–83. (*Pelias trigonatus*)

Synonyms: *Pseudotomodon mendozinus* Koslowsky, 1896b and *Pseudotomodon crivellii* Peracca, 1897c.
Type: Holotype, ? MNHNC, a 400 mm specimen (F. Leybold, 18 Feb. 1871).
Type locality: "estancia de Lo-Aguirre, Pampas Arjentinas, Provincia Mendoza, Sud-América" [= Los Aguirre ranch [bet. San Carlos (33°46'S, 69°03W, elevation 960 m) and Tierra Blanca (ca. 34°02'S, 69°05'W, elervation 1300 m) *fide* Leybold map of expedition], San Carlos Dept., W Mendoza Prov., cen. Argentina].
Distribution: Western Argentina (Chubut, Córdoba, La Pampa, La Rioja, Mendoza, Neuquén, Río Negro, San Juan, San Luis, Santiago del Estero), 150–1650 m.
Remarks: Catamarca and Tucumán records need confirmation *fide* J.D. Williams & Francini (1991: 78).

PSEUDOTYPHLOPS Schlegel, 1839 (*nomen corrigendum*) (Uropeltidae)

Synonym: *Pseudo-typhlops* Schlegel, 1839 *in* 1837–1844 *(nomen incorrigendum)*, and *Pseudotyphlops* – M.A. Smith, 1943 *(nomen corrigendum)*.
Type species: *Uropeltis philippinus* J.P. Müller, 1832.
Distribution: Sri Lanka.
Sources: A.C.L.G. Günther, 1864a, Wall, 1921g, 1923h, M.A. Smith, 1943, E.H. Taylor, 1953b, Deraniyagala, 1955, Gans, 1966, 1976, Murthy, 1973, P. Silva, 1980a, A. Silva, 1990, McDiarmid et al., 1999 and Somaweera, 2006.

1. *Pseudotyphlops philippinus* (J.P. Müller, 1832). Zeits. Physiol. 4(2): 252–253, pl. 22, figs. 2a–c, 3. (*Uropeltis philippinus*)

Synonyms: *Uropeltis philippinus* Cuvier, 1829 (*nomen nudum*), *Uropeltis philippina* Wagler, 1830 (*nomen nudum*), *Uropeltis grandis* Kelaart, 1853, *Uropeltis pardalis* Kelaart, 1853, and *Uropeltis saffragamus* Kelaart, 1853.
Type: Holotype, MNHN 5621, a 205 mm specimen.
Type locality: "Philippinischen Inseln," (in error *fide* M.A. Smith, 1943: 93). Corrected to Ceylon [= Sri Lanka] *fide* A.C.L.G. Günther (1864a: 188).
Distribution: Southeastern Sri Lanka (Central, N Eastern, Sabaragamuwa, Southern, Uva), NSL–1035 m.

PSEUDOXENODON Boulenger, 1890a (Pseudoxenodontidae)

Synonyms: *Pseudocenodon* – Palacky, 1898 *(nomen incorrectum)*, *Psendoxenodon* – Phisalix, 1917a *(nomen incorrectum)*, *Pseudovenodon* – Maki, 1931

P

(nomen incorrectum), Pseudoxenedon – Chin & Chou, 1968 *(nomen incorrectum), Pseudoendon* – Shrestha, 1981 *(nomen incorrectum),* and *Pseudaxoxenodon* – Yao, 2012 *(nomen incorrectum).*

Type species: *Tropidonotus macrops* Blyth, 1854a.

Distribution: Southeastern Asia and Greater Sundas.

Sources: C.H. Pope, 1929, 1935, Maki, 1931, Zaher et al., 2009 and I. Das, 2010.

1. *Pseudoxenodon bambusicola* T. Vogt, 1922. Arch. Naturg. 88A(10): 138–139.

Synonym: *Pseudoxenodon melli* T. Vogt, 1922.

Type: Holotype, ZMB 38070, a 510 mm male (R. Mell, 1914).

Type locality: "Provinz Kuangtung, Sudchinas" [= Guangdong Prov., S China]. Emended to Grenzberge gegen Hunan und Kiangsi in Bambus- und lichtem Laubwald zwischen 600–900 m Seehöhe nicht gerade selten by Mell (1922: 118) [= N Guangdong Prov. at the Hunan-Jiangsu border, S China, 600–900 m, ca. 24°30'N, 114°00'E].

Distribution: Southeastern China (Fujian, Guangdong, Guangxi, Guizhou, Hainan, Hunan, Jiangxi, Zhejiang), Laos (Khammouan) and N Vietnam (Bac Giang, Bac Kan, Cao Bang, Ha Giang, Ha Tay, Lao Cai, Phu Tho, Thai Nguyen, Tuyen Quang, Vinh Phuc), 300–1500 m.

Sources: Mell, 1922, 1931a, M.A. Smith, 1923a, F. Werner, 1926, K.P. Schmidt, 1927a, Bourret, 1936b, 1939c–d, 1943, Gressitt, 1941b, Hu et al., 1980, Tian et al., 1986, Huang et al., 1990, Zhao & Adler, 1993, Zhao, 2006, Ziegler et al., 2006, B.L. Stuart & Heatwole, 2008, V.S. Nguyen et al., 2009 and I. Das, 2010, 2012.

2. *Pseudoxenodon baramensis* (M.A. Smith, 1921b). J. Fed. Malay. States Mus. 10(3): 199. (*Tropidonotus baramensis*)

Type: Holotype, BMNH 1946.1.13.11 (formerly FMSM & MAS 4579), a 700 mm male (natives, 1919).

Type locality: "Mount Dulit, Sarawak, North Borneo, at 1,000 metres" [= Banjaran Dulit, 3rd Division, Baram Distr., N Sarawak, cen. East Malaysia, ca. 3°15'N, 114°15'E, elevation 1000 m].

Distribution: East Malaysia (Sarawak), 1000 m.

Sources: C. Haas, 1950, Malnate, 1960b, Stuebing, 1991 and Stuebing & Inger, 1999.

Remarks: C. Haas (1950: 538) erroneously listed Robinson and Kloss as collectors.

3. *Pseudoxenodon inornatus* (H. Boie *in* F. Boie, 1827). Isis von Oken 20(6): 541. (*Xenodon inornatus*)

Synonyms: *Xenodon inornatus* H. Boie *in* Schlegel, 1826a *(noemen nudum), Xenodon inornatus* Schlegel, 1826b *(noemen nudum), Pseudoxenodon jacobsonii* Lidth de Jeude, 1922, and *Pseudoxenodon inornatus buettikoferi* Brongersma & Helle, 1951.

Type: Holotype, RMNH 233, a 464–470 mm male (H. Kuhl & J.C. van Hasselt, 1821).

Type locality: "Java" [SW Indonesia]. Restricted to region of Mt. Pangerango and Mt. Salak, West Java, Indonesia *fide* Brongersma (1948b: 14). Emended to Tjihandjawar, at the foot of Mt. Pangerango, W. Java [= Cihanyawar, Nagrak, Jawa Barat, W Java, SW Indonesia, 07°03'S, 108°40'E] *fide* Brongersma (1950: 1499).

Distribution: Western Indonesia (Java, Kalimantan, Sumatra), 600–915 m.

Sources: Schlegel, 1837, Jan & Sordelli, 1866–1870, Rooij, 1917, Dammerman, 1926, Brongersma, 1950, C. Haas, 1950 and David & Vogel, 1996.

Remarks: Original description based on H. Boie's MS (1823–1825). Authorship attributed to Kuhl by F. Boie but not so stated in H. Boie MS.

4. *Pseudoxenodon karlschmidti* C.H. Pope, 1928a. Amer. Mus. Novit. (320): 3–4.

Synonyms: *Pseudoxenodon angusticeps sikiangensis* Mell, 1931a, *Pseudoxenodon sinii* Fan, 1931, *Pseudoxenodon angusticeps banaensis* Bourret, 1936b, and *Pseudoxenodon popei* Gressitt, 1936.

Type: Holotype, AMNH 34658, a 796 mm female (C.H. Pope [Third AMNH Asiatic Exped.], April–Sept. 1926).

Type locality: "Kuatun, a village in the mountains of Ch'ungan Hsien, northwest Fukien Province, China" [= Guadun, in Wuyishan Mtns., ext. N Fujian Prov., SE China, 27°45'N, 117°50'E, elevation 900–1200 m].

Distribution: Southern China (Fujian, Guangdong, Guangxi, Guizhou, Hainan, Hunan) and ext. N Vietnam (Ha Giang, Vinh Phuc), 500–2500m.

Sources: Fan, 1931, Bourret, 1936b, 1939d, Gressitt, 1941a, Hu et al., 1980, Tian et al., 1986, Zhao & Adler, 1993, Zhao, 2006 and V.S. Nguyen et al., 2009.

Remarks: Supplemental original description in C.H. Pope (1929: 410–412, figs. 6a–c). Possibly occurs in Hong Kong, China *fide* Karsen et al. (1998: 162).

5. *Pseudoxenodon macrops* (Blyth, 1854a). J. Asiat. Soc. Bengal 23(3): 296–297. (*Tropidonotus macrops*)

Synonyms: *Tropidonotus angusticeps* Blyth, 1854a (*partim*), *Xenodon macrophthalmus* A.C.L.G. Günther, 1858 (*partim*), *Tropidonotus sikkimensis* J. Anderson, 1871b, *Pseudoxenodon sinensis* Boulenger, 1904a, *Tropidonotus handeli* F. Werner, 1922a, *Tropidonotus tigrinus niger* T. Vogt, 1922, *Pseudoxenodon fukienensis* C.H. Pope, 1928a, *Pseudoxenodon angusticeps uniformis* Bourret, 1935c, and *Pseudoxsenodon macrops sinensis* – Yao, 2012 *(nomen incorrectum).*

Types: Syntypes (4), ZSI 7506–07 (formerly IMC 7506–07), and ZS1 7556–57, longest syntype 744 mm (W.S. Sherwill, 1843–1854).

Type locality: "near Darjiling" [= vicinity of Darjeeling (27°02'N, 88°16'E, elvation 2100 m), N West Bengal State, NE India].

Distribution: Southeastern Asia. China (Fujian, Gansu, Guangdong, Guangxi, Guizhou, Hubei, Hunan, Shaanxi, Sichuan, Xizang, Yunnan), NE India (Aranachal Pradesh, Assam, Meghalaya, Mizoram, Nagaland, Sikkim, N West Bengal), Nepal (Baglung, Darchula, Dolakha, Ilam, Kanchanpur, Kaski, Kathmandu), Bhutan, Myanmar (Kachin, Rakhine, Shan, Tanintharyi, Ramri Is.), N Thailand (Chiang Mai, Loei, Mae Hong Son, Phetchabun), N Laos (Champasak, Khammouan), Vietnam (Da Nang, Ha Tinh, Kon Tum, Lam Dong, Lao Cai, Lai Chau, Nighe An, Quang Binh,Vinh Phuc), West Malaysia (N Pahang, Perak) and East Malaysia (Sarawak), 150–3300 m.

Sources: Wall, 1909d, 1909h, 1910d, 1923d, T. Barbour, 1912a, G.E. Shaw & Shebbeare, 1929, M.L.Y. Chang, 1932, Bourret, 1935c, 1936b, 1939c–d, 1943, G.E. Shaw et al., 1939, M.A. Smith, 1943, E.H. Taylor & Elbel, 1958, L.W. Swan & Leviton, 1962, E.H. Taylor, 1965, B.L. Lim, 1967b, Campden-Main, 1970a, Deuve, 1970, Fleming & Fleming, 1973, Kramer, 1977, Hu et al., 1980, Majupuria, 1981, Tweedie, 1983, Murthy, 1986, Tian et al., 1986, K.-Y. Lue, 1987, Nanhoe & Ouboter, 1987, Zhao & Adler, 1993, M.J. Cox et al., 1998, Chan-ard et al., 1999, Farkas and Fritz, 1999c, Schleich and Kästle, 2002, Ao et al., 2004, Tiwari & Shah, 2004, Whitaker & Captain, 2004, Zhao, 2006, Bain et al., 2007, Zhao, 2006, Ziegler et al., 2007, B.L. Stuart & Heatwole, 2008, V.S. Nguyen et al., 2009 and Yao, 2012.

Remarks: K.-Y. Lue (1987: 115) listed *Pseudoxenodon macrops* from Taiwan.

6. *Pseudoxenodon stejnegeri* T. Barbour, 1908. Bull. Mus. Comp. Zool. 51(12): 317–318.

Synonyms: *Xenodon macrophthalmus* A.C.L.G. Günther, 1858 (*in parte*), ? *Tropidonotus dorsalis* A.C.L.G. Günther, 1864a, *Pseudoxenodon macrops sauteri* Steindachner, 1913b, *Pseudoxenodon striaticaudatus* C.H. Pope, 1928a, and *Pseudoxenodon nothus* H.M. Smith, 1942a (*nomen substitutum*).

Type: Holotype, MCZ 7103, a 602+ mm male (A. Owston, 29 Nov. 1906).

Type locality: "Mt. Arizan, Central Formosa" [= Mt. Arishan, ext. NE Chiayi Prov., cen. Taiwan, 23°35'N, 120°49'E].

Distribution: Southeastern China (Anhui, Fujian, Guangxi, Guizhou, Hebei, Hubei, Hunan, Jiangxi, Sichuan, Yunnan, Zhejiang) and Taiwan (Chiayi), 400–2100 m.

Sources: T. Barbour, 1909, Maki, 1931–b, C.-S. Wang & Wang, 1956, Hu et al., 1980, Toriba *in* Zhao, 1983, Tian et al., 1986, B.-H. Li, 1987, K.-Y. Lue et al., 1989, Huang et al., 1990, K.-Y. Lue, 1990, Zhao & Adler, 1993, Kou, 1995, Zhao, 2006, Yang, 2008 and Xiang & Li, 2009.

Remarks: Original description reprinted in T. Barbour (1909: 67–68, pl. 7, fig. 8). T. Barbour (1908: 67) erroneously reported length of type as 470 mm.

PSEUDOXYRHOPUS A.C.L.G. Günther, 1881b (*nomen protectum*) (Pseudoxyrhophiidae)

Synonym: *Homalocephalus* Jan, 1863a (*nomen rejiciendum*).

Type species: *Homalocephalus heterurus* Jan, 1863a.

Distribution: Madagascar.

Sources: Boulenger, 1890e, Guibé, 1958, Brygoo, 1983, Raxworthy & Nussbaum, 1994, Cadle, 1999, Nagy et al., 2003, Glaw et al., 2005b, Glaw & Vences, 2007, Kelly et al., 2009 and Zaher et al., 2009.

Remarks: Official Generic Name *fide* Opinion 1789 (ICZN, 1994).

1. *Pseudoxyrhopus ambreensis* Mocquard, 1894a. C. R. Soc. Philom. Paris 1894(9): 4.

Type: Holotype, MNHN 1893.213, a 296–300 mm male (C. Alluaud and Belly, May–July 1893).

Type locality: "mont d'Ambre, Madagascar" [= Mt. Ambre, Madagascar]. Emended to Mararaomby [= Maroaomby, Montagne d'Ambre, N Antsiranana Prov., ext. N Madagascar, ca. 12°30'S, 49°11'E, elevation 900 m] *fide* Raxworthy and Nussbaum (1994: 3, 5).

Distribution: Extreme N Madagascar (N Antsiranana), 900–1200 m.

2. *Pseudoxyrhopus analabe* Nussbaum, Andreone & Raxworthy, 1998. Copeia 1998(1): 129–130, figs. 1–2.

P

Type: Holotype, MRSN 1591, a 311 mm female (F. Andreone, H. Randriamahazo & J.E. Randrianirina, 2 Feb. 1996).

Type locality: "western slope of Anjanaharibe-Sud Mountain, Valley of Analabe River, 49°26.60'E, 14°46.62'S, 1050-m elevation, Befandriana Fivondronana, Mahajunga Province, Madagascar."

Distribution: Northern Madagascar (ext. N Mahajanga), 1050 m. Known only from type locality.

3. *Pseudoxyrhopus ankafinaensis* Raxworthy & Nussbaum, 1994. Misc. Publ. Mus. Zool., Univ. Michigan (182): 6–8, figs. 5–6.

Type: Holotype, BMNH 1882.2.25.65, a 1064+ mm male (W.D. Cowan, mid-Feb.–mid-Mar., 1880).

Type locality: "Ankafana, Betsileo, Madagascar" [= Ankafina forest, 3 km W Tsarafidy, cen. Fianarantsoa Prov., cen. Madagascar, 21°12'S, 47°13'E, elevation 1550–1650 m].

Distribution: Central Madagascar (cen. Fianarantsoa), 1550–1650 m. Known only from the holotype.

4. *Pseudoxyrhopus heterurus* (Jan, 1863a). Arch. Zool. Anat. Fis. 2(2): 286. (*Homalocephalus heterurus*)

Type: Holotype, formerly MSNM, a 525 mm specimen, destroyed in 1943 during World War II.
Type locality: "Madagascar."
Distribution: Eastern Madagascar (Antsiranana, SE Fianarantsoa, N Toamasina, ext. SE Toliara, Nosy Mangabe and Sainte-Marie Is.), NSL–950 m.
Sources: Jan & Sordelli, 1860–1866, Raxworthy, 1988 and Henkel & Schmidt, 2000.
Remarks: Official Specific Name *fide* Opinion 1789 (ICZN, 1994).

5. *Pseudoxyrhopus imerinae* (A.C.L.G. Günther, 1890). Ann. Mag. Nat. Hist. (6) 5(25): 71–72. (*Liophis imerinae*)

Synonym: *Coronella microps* Boulenger, 1888a (*nomen occupatum*).
Type: Holotype, BMNH 1946.1.7.80 (formerly BMNH 1889.8.28.22), a 457 mm female (J. Wills, 1870–1889).
Type locality: "forest-district east of Imerina, Madagascar" [= E of Imerina, (ca. 18°54'S, 47°50'E, elevation 1500 m), E Antananarivo Prov., cen. Madagascar].
Distribution: Eastern Madagascar (Antananarivo, Fianarantsoa, SE Toliara), 1275–2200 m.

6. *Pseudoxyrhopus kely* Raxworthy & Nussbaum, 1994. Misc. Publ. Mus. Zool., Univ. Michigan (182): 13–15, figs. 11–12.

Type: Holotype, UMMZ 192021, a 217 mm male (C.J. Raxworthy, 4 Sept. 1989).
Type locality: "Mandena, 24°58'S, 47°2'E, 20 m altitude, Tôlanaro (Fort Dauphin) Fivondronana, Toliara (Tuléar) Province, Madagascar."
Distribution: Extreme SE Madagascar (ext. SE Toliara), NSL–25 m. Known only from type locality.

7. *Pseudoxyrhopus microps* A.C.L.G. Günther, 1881b. Ann. Mag. Nat. Hist. (5) 7(41): 359–360, 2 figs.

Type: Holotype, BMNH 1946.1.7.76 (formerly BMNH 1881.3.29.16), a 1355–1372 mm male (T. Waters).
Type locality: "district of Betsileo, Madagascar" [= N Fianarantsoa Prov., cen. Madagascar].
Distribution: Eastern Madagascar (N Antsiranana, Fianarantsoa, Toamasina, SE Toliara, Nosy Be Is.), 70–1285 m.
Source: Angel, 1935.

8. *Pseudoxyrhopus oblectator* Cadle, 1999. Bull. Mus. Comp. Zool. 155(8): 385–390, figs. 1–2, 4, 5, 17.

Type: Holotype, MCZ 182292, a 392 mm male (J.E. Cadle, T. Pierre & R. Emile, 10 Jan. 1998).
Type locality: "Mahajo Stream (a southern tributary of the Ranomena River), Ranonvafana National Park (northern sector), Fivondronana Ifanadiana, Fianarantsoa Province, Madagascar; 1,110 m [approximately 21°13'S, 47°28'E]."
Distribution: Central Madagascar (Fianarantsoa), 900–1110 m. Known only from type locality.

9. *Pseudoxyrhopus quinquelineatus* (A.C.L.G. Günther, 1881b). Ann. Mag. Nat. Hist. (5) 7(41): 359, 2 figs. (*Liophis quinquelineatus*)

Types: Syntypes (2), BMNH 1946.1.12.33–34 (formerly BMNH 1881.3.29.13–14), a 490 mm male and 427–432 mm female (T. Waters, 1873–1881).
Type locality: "District of Betsileo, Madagascar" [= N Fianarantsoa Prov., cen. Madagascar].
Distribution: Madagascar (Antananarivo, N Antsiranana, Fianarantsoa, cen. Mahajanga, S Toamasina, Toliara), NSL–1300 m.
Source: Henkel & Schmidt, 2000.

10. *Pseudoxyrhopus sokosoko* Raxworthy & Nussbaum, 1994. Misc. Publ. Mus. Zool., Univ. Michigan (182): 22–23, figs. 17–18.

Type: Holotype, UMMZ 195847, a 353 mm male (J.B. Ramanamanjato, 27 Dec. 1990).
Type locality: "Ampamakiesiny Pass 14 km northwest of Ranomafana-Sud, 24°32'S, 47°51'E, 800 m altitude, Tôlanaro (Fort Dauphin) Fivondronana, Toliara (Tuléar) Province, Madagascar."
Distribution: Extreme SE Madagascar (SE Toliara), 75–800 m.

11. *Pseudoxyrhopus tritaeniatus* Mocquard, 1894a. C.R. Soc. Philom. Paris 1894(9): 4–5.

Synonym: *Pseudoxyrhopus dubius* Mocquard, 1904a.
Type: Holotype, MNHN 1891.24, a 620–642 mm male (Catat).
Type locality: "Betsileo, Madagascar" [= N Fianarantsoa Prov., cen. Madagascar].
Distribution: Eastern Madagascar (Antsiranana, Fianarantsoa, Toamasina, Nosy Mangabe Is.), 20–1000 m.
Source: Angel, 1935.

PSEUSTES Fitzinger, 1843 (Colubridae)

Synonyms: *Thamnobius* Fitzinger, 1843, *Pseuste* – A.-M.-C. Duméril, Bibron & Duméril, 1854b *(nomen incorrectum), Phrynonax* Cope, 1862d, *Synchalinus* Cope, 1894d, *Paraphrynonax* Lutz & Mello, 1922a, *Paraphinonax* – Briceño-Rossi, 1934 *(nomen incorrectum), Paratynonax* – Briceño-Rossi, 1934 *(nomen*

P

incorrectum), Phrinonax – Briceño-Rossi, 1934 *(nomen incorrectum), Thannobsus* – Briceño-Rossi, 1934 *(nomen incorrectum),* and *Pseutes* – Fugler & Walls, 1978 *(nomen incorrectum).*

Type species: *Natrix sulphurea* Wagler *in* Spix, 1824.

Distribution: Latin America and Trinidad & Tobago.

Sources: Amaral, 1930h, J.A. Peters, 1960a, J.A. Peters & Orejas-Miranda, 1970 and Pérez-Santos & Moreno, 1988, 1991.

1. *Pseustes poecilonotus* (A.C.L.G. Günther, 1858). Cat. Colub. Snakes Brit. Mus.: 100–101. (*Spilotes poecilonotus*)

Synonyms: *Tropidodipsas lunulata* Cope, 1861b, *Ahaetulla polylepis* W.C.H. Peters, 1867d, *Spilotes fasciatus* W.C.H. Peters, 1869, *Spilotes chrysobronchus* Cope, 1875a, *Spilotes argus* Bocourt, 1888 *in* A.H.A. Duméril, Mocquard & Bocourt, 1870–1909, *Synchalinus corallioides, Phrynonax eutropis* Boulenger, 1894a, *Phrynonax guentheri* Boulenger, 1894a, *Phrynonax lyoni* Stejneger, 1901c, *Phrynonax atriceps* F. Werner, 1913, *Phrynonax poecilonatus* – Briceño-Rossi, 1934 *(nomen incorrectum),* and *Phrynonax sancheri* Briceño-Rossi, 1934 *(nomen nudum).*

Type: Lectotype, BMNH 1946.1.7.39, a 1600 mm male (D. Dyson, 1844–1858), designated by Boulenger (1894a: 20).

Type locality: "Honduras," (probably in error) via lectotype selection. Corrected to Bristish Honduras [= Beize] *fide* K.P. Schmidt (1941: 499).

Distribution: Latin America and Trinidad & Tobago. Southern Mexico (Campeche, Chiapas, Oaxaca, Quintana Roo, San Luis Potosí, Tabasco, Veracruz, Yucatán), W Belize (Belize, Cayo, Orange Walk, Stann Creek), N Guatemala (Alta Verapaz, Petén), NW Honduras (Atlántida, Colón, Cortés, Gracias a Dios, Olancho, Yoro), E Nicaragua (Atlántico Norte, Atlántico Sur, Matagalpa), Costa Rica (Alajuela, Cartago, Guanacaste, Heredia, Limón, Puntarenas, San José), Panama (Bocas del Toro, Chiriquí, Colón, Canal Zone, Darién, Panamá, Barro Colorado Is.), Colombia (Amazonas, Cauca, Chocó, Magdalena, Santander, Valle, Gorgona Is.), Venezuela (Amazonas, Aragua, Bolívar, Distrito Federal, Monagas, Sucre, Zulia), Trinidad and Tobago (Trinidad), Guyana (Cuyuni-Mazaruni, Essequibo Islands-West Demerara, Pomeroon-Supenaam, Potaro-Siparuni, Upper Demerara-Berbice, Upper Takutu-Upper Essequibo), Suriname (Para, Saramacca), French Guiana (Cayenne, Saint-Laurent-du-Maroni), E Ecuador (Napo, Pastaza), E Peru (Junín, Loreto, Madre de Dios, San Martín), N Brazil (Amapá, Pará) and Bolivia (Beni, Santa Cruz), NSL–1420 m.

Sources: Bocourt, 1888 *in* Duméril, Bocourt & Mocquard, 1870–1909, A.C.L.G. Günther, 1894 (*in* 1895–1902), Peracca, 1896c, Gomes, 1918, T. Barbour & Amaral, 1924, L.C. Stuart, 1935b, 1963, Brongersma, 1937b, K.P. Schmidt, 1941, H.M. Smith & Taylor, 1945, Beebe, 1946, E.H. Taylor, 1951, 1954, Duellman, 1963, Medem, 1965, 1979, Roze, 1966a, Hoge, 1967, R.W. Henderson & Hoevers, 1975, H.M. Smith & Smith, 1976, 1993, Amaral, 1978, Emsley, 1977, Cunha & Nascimento, 1978, 1993, Gasc & Rodrigues, 1980, J.C. Lee, 1980, 1996, Fugler, 1983, N.J. Scott et al., 1983, Villa, 1983, Cunha et al., 1985, L.D. Wilson & Meyer, 1985, Abuys, 1986a, Dixon & Soini, 1986, Lancini-V., 1986, Pérez-Santos, 1986, Savage & Villa, 1986, Chippaux, 1987, Villa et al., 1988, Pérez-Higareda & Smith, 1991, Henle & Ehrl, 1991, Pérez-Santos & Moreno, 1991, Casas-Andreu et al., 1996, J.A. Campbell, 1998b, J.C. Murphy, 1997, Rivas-Fuenmayor & Molina, 1998, Gorzula & Señaris, 1998, Lemoine & Rivas, 1998, Franklin & Franklin, 1999, Kohler, 1999b, Stafford & Meyer, 2000, Doan & Arizábal, 2002, Savage, 2002, Solórzano, 2004, Townsend et al., 2006, Navarrete et al., 2009, McCranie, 2011a and C.J. Cole et al., 2013.

Remarks: Boulenger (1894a: 20) listed holotype as female. L.C. Stuart (1963: 111) listed the lectotype as BMNH 1946.1.7.41. Manabi, Ecuador record doubtful *fide* Pérez-Santos & Moreno (1991: 291).

2. *Pseustes sexcarinatus* (Wagler *in* Spix, 1824). Serp. Brasil. Sp. Nov.: 35–36, pl. 12. (*Natrix sexcarinata*)

Type: Holotype, ZSM 1744/0, a 1509 mm male (J.B. Spix & K.F.P. von Martins, Aug. 1819–April 1820), destroyed 11 April 1945 during World War II.

Type locality: "ad flumen Amazonum, dans l'intérieur du Brésil" [= Río Amazonas, bet. Tajapuru (1°02'S, 51°02'W) and Manaus (3°06'S, 60°01'W), N Pará State, N Brazil].

Distribution: Northern Brazil (Pará).

Sources: Serié, 1915a, 1936, Hoge, 1964a, J.A. Peters & Orejas-Miranda, 1970, Hoogmoed & Gruber, 1983, J.D. Williams & Francini, 1991 and Cei, 1994.

Remarks: Probably a senior synonym of *P. poecilonotus* (A.C.L.G. Günther) *fide* J.A. Peters & Orejas-Miranda (1970: 258) and Dixon and Soini (1977: 68).

3. *Pseustes shropshirei* (T. Barbour & Amaral, 1924). Occ. Pap. Boston Soc. Nat. Hist. (1921–1931) (5): 131. (*Phrynonax shropshirei*)

Type: Holotype, MCZ 18819, a 1460 mm male (J.B. Shropshire [and/or W.S. Brooks & E. Wigglesworth], Jan.–Feb. 1924).

Type locality: "vicinity of Gatun, Canal Zone of Panama" [= vicinity of Gatun (9°17'N, 79°46'W, elevation 40 m), Colon Prov., cen. Panama]. Emended to Fort Sherman, Canal Zone, Panamá *fide* T. Barbour and Loveridge (1929a: 313).

Distribution: Lower Central America and N South America. Panama (Bocas del Toro, Canal Zone: Barro Colorado Is.; Colón, Darién, Panamá, Veraguas),

P

W Colombia (Antioquia, Atlántico, Boyacá, Cauca, Chocó, Cundinamarca, Magdalena, Meta, Norte de Santander, Santander, Tolima, Valle, Gorgona Is.), W Ecuador (Esmeraldas, Guayas), Venezuela (Aragua, Barinas, Carabobo, Cojedes, Distrito Federal, Lara, Mérida, Miranda, Táchira, Trujillo, Zulia), Trinidad and Tobago (Trinidad), N Guyana, Bolivia, Peru and Brazil, NSL–2000 m.

Sources: Brongersma, 1937b, Niceforo-Maria, 1942, C.W. Myers & Rand, 1969, Rand & Ortleb, 1969, Miyata, 1982, Pérez-Santos, 1986, Rivas-Fuenmayor & Fuentes, 1998, Gorzula & Señaris, 1998, Starace, 1998, Renjifo & Lundberg, 1999, Calcaño & Barrio-Amorós, 2003b and Navarrete et al., 2009.

Remarks: Original description listed no specific collector of holotype but Shopshire collected paratype. Introductory remarks stated Barbour had little time for field work but that Brooks & Wigglesworth collected extensively and that all specimens (including Shropshire's) were put untagged in containers from a given locality. MCZ catalogue lists Barbour, Brooks & Wigglesworth as collectors of entire series. T. Barbour & Loveridge (1929: 318) listed Shropshire as sole collector of both type and paratype. Probably a synonym of *P. poecilonotus fide* Amaral (1930h: 317) and Savage (2002: 680). E.H. Taylor's (1951, 1954) Costa Rican records were misidentified *P. poecilonotus*. Meta, Colombia record doubtful *fide* Pérez-Santos & Moreno (1988: 279).

4. *Pseustes sulphureus* (Wagler *in* Spix, 1824). Serp. Brasil. Sp. Nov.: 26–27, pl. 9. (*Natrix sulphurea*)

P

Synonyms: *Coluber caracaras* Gmelin, 1789, *Coluber poecilostoma* Wied-Neuwied, 1824b, *Dipsas dieperinkii* Schlegel, 1837, *Spilotes poecilostoma serrarius* Jan, 1863b (*nomen nudum*), *Phrynonax faucherei* Mocquard, 1903, and *Paraphrynonax versicolor* Lutz & Mello, 1922a.

Type: Lectotype, ZSM 1681/0, a 2791+ mm male (K.F.P. von Martius, Dec. 1819–Feb. 1820), designated by Hoogmoed & Gruber (1983: 329).

Type locality: "ad littora fluminis Japura" [= shore of Río Japurá, NW Amazonas State, ext. NW Brazil or E Amazonas Prov., ext. SE Colombia (3°08'S, 64°66'W to 0°24'S, 72°17'W) *fide* Vanzolini, 1981: xxvi].

Distribution: Northern South America and Trinidad. Southeastern Colombia (Amazonas, Caquetá, Meta, Vaupés), E Venezuela (Amazonas, Bolívar, Monagas), Trinidad and Tobago (Trinidad), Guyana (Cuyuni-Mazaruni, Demerara-Mahaica, Essequibo Islands-West Demerara, Upper Takutu-Upper Essequibo), Suriname (Marowijne, Nickerie, Para, Paramaribo), French Guiana (Cayenne, Saint-Laurent-du-Maroni), E Ecuador (Morona-Santiago, Napo), E Peru (Loreto, Madre de Dios, Pasco, San Martín), Brazil (Amazonas, Bahia, Ceará, Distrito Federal, Espírito Santo, Mato Grosso, Minas Gerais, Pará, Rio de Janeiro, São Paulo) and E Bolivia (Cochabamba, Santa Cruz), NSL–840 m.

Sources: Gomes, 1918, H.W. Parker, 1933a, Beebe, 1946, Hoge, 1964d, 1969c, Rossman & Williams, 1966, P. Müller, 1968b, Hoge & Romano-Hoge, 1970a, Amaral, 1978, Emsley, 1977, Cunha & Nascimento, 1978, 1993, Gasc & Rodrigues, 1980, Cunha et al., 1985, Abuys, 1986a, Dixon & Sonini, 1986, Pérez-Santos, 1986, Chippaux, 1987, Nascimento et al., 1988, Duellman & Salas, 1991, Henle & Ehrl, 1991, Riva et al., 1991, Fugler et al., 1995, Marques & Calleffo, 1997, J.C. Murphy, 1997, Starace, 1998, Freitas, 1999, Lehr, 2001, Duellman, 2005, Borges-Nojosa et al., 2006, Navarrete et al., 2009 and C.J. Cole et al., 2013.

PSOMOPHIS C.W. Myers & Cadle, 1994 (Xenodontidae)

Type species: *Rhadinaea obtusa* Cope, 1863b.

Distribution: Central and South America.

Sources: C.W. Myers and Cadle, 1994, Zaher et al., 2009 and Vidal et al., 2010.

1. *Psomophis genimaculatus* (Boettger, 1885a). Zeits. Naturwiss., Halle 58: 229–231. (*Liophis [Lygophis] genimaculata*)

Type: Lectotype, BMNH 1946.1.5.83 (formerly BMNH 1885.91.11), a 362 mm male, designated by Hoge (1958c: 222).

Type locality: "Paraguay" via lectotype selection.

Distribution: Southern South America. Eastern and S Bolivia (Beni, Santa Cruz), SW Brazil (Mato Grosso do Sul), N Paraguay (Boquerón, Central) and N Argentina (Corrientes, Formosa, Salta), 90 m.

Sources: Lions & Alvarez, 1997, Leynaud & Bucher, 1999 and Ziegler et al., 2003.

Remarks: Photographs of lectotype in Myers & Cadle (1994: figs. 5a, 6 upper). Lectotype erroneously listed as BMNH 1946.1.5.38 *fide* Hoge (1958b: 222). Records from Chaco, Misiones and Santa Fe, Argentina doubtful *fide* Giraudo & Scrocchi (2002: 35).

2. *Psomophis joberti* (Sauvage, 1884a). Bull. Soc. Philom. Paris (7) 8: 146–147. (*Enicognathus joberti*)

Type: Holotype, MNHN 5353 (formerly MNHN 1879.64), a 370 mm female (Jobert).

Type locality: "Marajo, Amazone" [= Marajó Is., mouth of Río Amazonas, E Pará State, N Brazil, ca. 0°53'S, 49°38'W, elevation 15 m].

Distribution: Eastern Brazil (Amazonas, Bahia, Ceará, Goias, Mato Grosso, Minas Gerais, Pará, Paraiba, Pernambuco, Rio Grande do Norte, São Paulo), NSL–125 m.

Sources: Freitas, 1999 and Passos & Fernandes, 2002b.

3. *Psomophis obtusus* (Cope, 1863b). Proc. Acad. Nat. Sci. Philadelphia 15(3): 101. (*Rhadinaea obtusa*)

Type: Holotype, USNM uncatalogued, a 410+ female (H.W. Kennedy), lost *fide* Myers & Cadle (1994: 23).

Type locality: "Paysondu, Uraguay" [= Paysandú, W Paysandú Prov., W Uruguay, 32°19'S, 58°04'W, elevation 25 m].

Distribution: Southern South America. Southern Brazil (Rio Grande do Sul), Paraguay (Central), Uruguay (Artigas, Canelones, Cerro Largo, Flores, Florida, Lavalleja, Maldonado, Montevideo, Paysandú, Rivera, Rocha, Salto, Soriano, Tacuarembó, Treinta y Tres) and N Argentina (NE Buenos Aires, Chaco, Corrientes, Entre Ríos, Formosa, San Juan, Santa Fe), 25–435 m.

Sources: Yanosky et al., 1996, Leynaud & Bucher, 1999, Giraudo & Scrocchi, 2002, Carreira-Vidal et al., 2005 and Di Pietro et al., 2010.

†*PTEROSPHENUS* Lucas, 1898 (†Palaeophiidae)

Synonyms: †*Moeriophis* C.W. Andrews, 1901, †*Mocriophis* – Gilmore, 1938 *(nomen incorrectum)*, †*Pterophenus* – Huene, 1959 *(nomen incorrectum)*, and †*Pterosphernus* – Terent'ev, 1961 *(nomen incorrectum)*.

Type species: †*Pterosphenus schucherti* Lucas, 1899.

Distribution: Lower Eocene of India, middle Eocene of Egypt and Uzbekistan, upper Eocene of Ecuador, Egypt, Libya and USA, and Eocene of USA.

Sources: Janensch, 1906a, Holman, 1979a, Rage, 1984b and Rage et al., 2003, 2008.

1. †*Pterosphenus biswasi* Rage, Bajpai, Thewissen & Tiwari, 2003. Geodiversitas 25(4): 705–707, fig. 4.

Type: Holotype, RUSB 2784-4, one trunk vertebra.

Type locality: "HD Pit (also called North Pit) in Panandhro Lignite Mine, District Kutch, Gujarat State, on the northwestern margin of India. Naredi Formation, Ypresian, Lower Eocene."

Distribution: Lower Eocene (Ypresian: 48.6–55.8 mya) of India. Known only from type locality.

2. †*Pterosphenus kutchensis* Rage, Bajpai, Thewissen & Tiwari, 2003. Geodiversitas 25(4): 701–704, fig. 2.

Type: Holotype, RUSB 2721-1, one trunk vertebra.

Type locality: "HD Pit (also called North Pit) in Panandhro Lignite Mine, District Kutch, Gujarat State, on the northwestern margin of India. Naredi Formation, Ypresian, Lower Eocene."

Distribution: Lower Eocene (Ypresian: 48.6–55.8 mya) of India. Known only from type locality.

3. †*Pterosphenus schucherti* Lucas, 1899. Proc. U.S. Natl. Mus. 21(1164): 637–638, pl. 45, figs. 1–5, 46.

Type: Holotype, USNM 4047, one trunk vertebra (C. Schuchert, 1896).

Type locality: "Eocene of Cocoa, Alabama." Emended to about two miles southeast of Melvin (Cocoa), Choctaw Co., Alabama, Jackson formation, Eocene *fide* Gilmore (1938: 57) [= two miles southeast of Melvin, ext. W Choctaw County, SW Alabama, USA; upper Eocene, Jackson formation].

Distribution: Upper Eocene (Chadronian: 33.9–38.0 mya) of USA (Alabama, Arkansas, Georgia, Louisiana, Mississippi, Texas) and Eocene (33.9–55.4) of USA (Florida, New Jersey).

Sources: Gilmore, 1938, H.W. Miller, 1955, Holman, 1977a, 2000a, Westgate & Ward, 1981, Westgate, 1984, Hutchison, 1985, Parmley & Case, 1988 and Holman et al., 1991.

Remarks: Rage (184b: 36) listed ca. 40 vertebrae as syntypes.

5. †*Pterosphenus schweinfurthi* (C.W. Andrews, 1901). Geol. Mag. (4) 8(10): 438–440, fig. 2. (†*Moeriophis schweinfurthi*)

Type: Holotype, CGM C10194, one trunk vertebra.

Type locality: "Lower Tertiary deposits of Egypt" [= north of Birket Qurun 8 (The Fayum), Qasr el Sagha formation, upper Bartonian (= Priabonian), upper Eocene].

Distribution: Middle Eocene (Lutetian: 37.2–48.6 mya) of Egypt and upper Eocene (Priabonian: 33.9–37.2 mya) of Egypt and Libya.

Sources: C.W. Andrews, 1906 and Hoffstetter, 1960b.

6. †*Pterosphenus sheppardi* Hoffstetter, 1958a. Bull. Soc. Géol. France (6) 8(1): 45–49, pl. 7.

Type: Holotype, BMNH 5504, five articulated trunk vertebrae with six proximal ribs (G. Sheppard, 1929).

Type locality: "Ancon se situe au S de la péninsule de Santa Elena, dans la région Côtière de la République de l'Équateur (Amérique du Sud), et correspond approximativement aux coordonnées: lat. 2°20' S, long. 80°50' W. La formation Seca, dont le nom est emprunté à la Quebrada Seca du Même district... indique clairement l'Éocène supérieur et plus précisément l'âge Jackson de l'échelle nord-américaine" [= vicinity of Ancon, ext. W Guayas Prov., SW Ecuador, 2°20'S, 80°50'W; upper Eocene, Seca Formation].

Distribution: Upper Eocene (33.9–37.2 mya) of Ecuador. Known only from type series.

Sources: Hoffstetter, 1958b, 1970.

†*PTERYGOBOA* Holman, 1976a (Boidae)

Type species: †*Pterygoboa miocenica* Holman, 1976a.
Distribution: Lower and middle Miocene of USA.
Sources: Holman, 1976a, 1977e, 1979a, 1998b.

1. †*Pterygoboa delawarensis* Holman, 1998b. Delaware Geol. Surv., Spec. Publ. (21): 143–144, figs. 1a–e.

Type: Holotype, USNM 483393, one trunk vertebra.
Type locality: "From the Pollack Farm Site (Delaware Geological Survey site Idll-a, near Cheswold, Kent County, Delaware. Cheswold sands of the Calvert Formation; lower Miocene, Hemingfordian Age."
Distribution: Lower Miocene (Hemingfordian: 16.3–20.6 mya) of USA (Delaware). Known only from holotype.

2. †*Pterygoboa miocenica* Holman, 1976a. Herpetologica 32(1): 43–44, figs. 2a–c.

Type: Holotype, SDSM 8376, one trunk vertebra.
Type locality: "Black Bear Quarry II (early Heminfordian) Bennett County, South Dakota, USA; Rosebud Formation, early Hemingfordian (lower middle Miocene)."
Distribution: Lower Miocene (Hemingfordian: 16.3–20.6 mya) of USA (South Dakota), and middle Miocene (Barstovian: 13.6–16.3 mya) of USA (Nebraska, South Dakota).

PTYAS Fitzinger, 1843 (Colubridae)

Synonyms: *Zaocys* Cope, 1861c, *Zapyrus* A.C.L.G. Günther, 1864a, *Megablabes* A.C.L.G. Günther, 1865, *Zaoccys* Theobald, 1872 *(nomen emendatum), Hemidryas* W. Peters & Doria, 1878, *Zoacys* – Steindachner, 1913b *(nomen incorrectum), Zaocis* – Kopstein, 1930 *(nomen incorrectum), Ptays* – Maki, 1931 *(nomen incorrectum)* and *Pytas* – Shrestha, 1981 *(nomen incorrectum).*
Type species: *Coluber mucosus* Linnaeus, 1758.
Distribution: Southern and E Asia and East Indies.
Fossil records: Pleistocene of India.
Sources: Rooij, 1917, Wall, 1926b, M.A. Smith, 1943, C. Haas, 1950, E.H. Taylor, 1965, Saint-Girons, 1972a, Hu et al., 1980, Tweedie, 1983, Hodges, 1993, Nagy et al., 2004 and I. Das, 2011.
Remarks: *Zaocys* a synonym *fide* Wall (1914e: 168), E.H.Taylor (1965: 738) and Leviton (1983: 200).

1. *Ptyas carinata* (A.C.L.G. Günther, 1858). Cat. Colub. Snakes Brit. Mus.: 112–113. (*Coryphodon carinatus*)

Synonym: *Zaocys tenasserimensis* Sclater, 1891a.
Type: Lectotype, BMNH 1946.1.11.35, a 3048 mm male (E. Belcher, 1843–1846), designated by A.C.L.G. Günther (1864a: 256).
Type locality: "Borneo" via type locality restriction.
Distribution: Southeast Asia and East Indies. Southwestern China (Yunnan), S Myanmar (Karen, Tanintharyi), Thailand (Chiang Mai, Lampang, Nakhon Si Thammarat, Pattani, Phang Nga), Laos, cen. Vietnam (Binh Phuoc, Da Nang, Thua Thien-Hue), Cambodia (Mandolkivi), West Malaysia (Johor, Perak, Perlis, Selangor, Tioman Is.), Singapore, East Malaysia (Sarawak), Brunei, W Indonesia (Java, W Kalimantan, Sumatra), and ext. SW Philippines (cen. Palawan), NSL–2165 m.
Sources: Lidth de Jeude, 1922, E.H. Taylor, 1922a, Kopstein, 1930, Bourret, 1934c, B.L. Lim, 1964, Campden-Main, 1970a, Leviton, 1983, Alcala, 1986, K.K.P. Lim & Chou, 1990, Stuebing, 1991, K.K.P. Lim & Lim, 1992, Zhao & Adler, 1993, David & Vogel, 1996, M.J. Cox et al., 1998, Grismer et al., 2002b, B.L. Stuart et al., 2006, I. Das, 2007b, 2010, 2012 and V.S. Nguyen et al., 2009.
Remarks: Possibly occurs in Cambodia *fide* Iskandar & Colijn (2001: 79).

2. *Ptyas dhumnades* (Cantor, 1842b). Ann. Mag. Nat. Hist. 9(59): 483. (*Coluber dhumnades*)

Synonyms: *Ablabes vittatus* A.-M.-C. Duméril, Bibron & Duméril, 1854a, *Zaocys oshimai* Stejneger, 1925, and *Zaocys dhumnades montanus* C.H. Pope, 1928b.
Type: Lectotype, BMNH 1946.1.14.89, a male (T.E. Cantor, July 1840–March 1841, via British East India Comp.), designated by Boulenger (1893a: 376), lost *fide* V.S. Nguyen et al. (2009: 335).
Type locality: "Chusan" [= Zhoushan Is., Zhoushan Arch., Zhejiang Prov., E China, bet. 29°57'–30°11'N & 121°56'–122°20'E] via lectotype selection.
Distribution: Eastern Asia. Eastern China (Anhui, Fujian, Gansu, Guangdong, Guangxi, Guizhou, Hebei, Henan, Hubei, Hunan, Jiangsu, Jiangxi, Shaanxi, Sichuan, Yunnan, Zhejiang), Taiwan (Chiai, Nantou, Pingtung, Taichung, Taipei) and ext. N Vietnam (Hoang Lien Son, Lao Cai, Vinh Phuc), 130–2000 m.
Sources: Wall, 1903, C.H. Pope, 1929, 1935, Maki, 1931, Angel & Bourret, 1933, Bourret, 1939d, C.-S.Wang & Wang, 1956, Kuntz, 1963, Mao, 1965b, Deuve, 1970, Hu et al., 1980, Tian et al., 1986, B.H. Li, 1987, K.-Y. Lue, 1987, K.-Y. Lue et al., 1989, Huang et al., 1990, Ota, 1991, Zhao & Adler, 1993, Kou, 1995, Zhao, 2006, V.S. Nguyen et al., 2009, Xiang & Li, 2009, I. Das, 2010, 2012 and Yao, 2012.

Remarks: Boulenger (1893a: 376) listed both syntypes as male. Probably occurs in N Laos *fide* Deuve (1970: 131).

3. *Ptyas dipsas* (Schlegel, 1837). Essai Phys. Serp. 1: 153, 2: 197–198. (*Herpetodryas dipsas*)

Synonyms: *Leptophis olivaceus* Duméril, Bibron & Duméril, 1854a and *Megablabes olivaceus* A.C.L.G. Günther, 1865.

Type: Holotype, MNHN 7313, an 1830–1872 mm male lacking a skull (J.R.C. Quoy & J.P. Gaimard, 1828).

Type locality: "Célèbe" [= Sulawesi, E Indonesia].

Distribution: Eastern Indonesia (Sulawesi), NSL–1100 m.

Sources: Bosch, 1985, Lazell et al., 1991, Lang & Vogel, 2005, and Koch, 2012.

Remarks: Halmahera record of Roux (1904) rejected *fide* Lazell et al. (1991: 356).

4. *Ptyas fusca* (A.C.L.G. Günther, 1858). Cat. Colub. Snakes Brit. Mus.: 112. (*Coryphodon fuscus*)

Synonym: *Herpetodryas horneri* Jan, 1863b.

Type: Lectotype, BMNH 1946.1.17.14, a juvenile male (E. Belcher, 1843–1846), designated by C. Haas (1950: 538).

Type locality: "Borneo" [= Kalimantan] via lectotype selection..

Distribution: Malay Peninsula and W Indonesia. Peninsular Thailand (Phang Nga), West Malaysia (Johor, Pinang), Singapore, East Malaysia (Sabah, Sarawak), Brunei and W Indonesia (Bangka, Belitung, Kalimantan, Natuna Besar, Nias, Sumatra), NSL–1330 m.

Sources: Lidth de Jeude, 1922, C. Haas, 1950, K.K.P. Lim & Chou, 1990, Stuebing, 1991, David & Vogel, 1996, M.J. Cox et al., 1998, Malkmus et al., 2002, Pauwels et al., 2002 and I. Das, 2010, 2012.

Remarks: Singapore record doubtful *fide* K.K.P. Lim & Lim (1992: 145).

5. *Ptyas korros* (Schlegel, 1837). Essai Phys. Serp. 1: 145, 2: 139. (*Coluber korros*)

Synonyms: *Coluber korros* H. Boie *in* Schlegel, 1826a *(nomen nudum), Coluber korros* Schlegel, 1826b *(nomen nudum), Coluber boncorage* Lesson, 1831, *Coluber korros* Lesson, 1831 *(nomen praeoccupatum), Coluber cancellatus* Oppel *in* Schlegel, 1837 *(nomen ineditum), Liopeltis libertatis* T. Barbour, 1910, *Ptyas korros chinensis* Mell, 1931a, and *Ptyas korros indicus* Mell, 1931a.

Types: Syntypes (7), RMNH 410a–d, RMNH 411 (H. Boie & H.C. Maclklot, Dec. 1825–Sept. 1827) [Java] and RMNH 414a–b (L. Horner, Feb.–Dec. 1835) [Padang], longest syntype 1630 mm.

Type locality: "îles de Java et de Sumatra" [= Java and Sumatra, Indonesia]. Restricted to Java, W Indonesia *fide* M.A. Smith (1943: 162) and RMNH catalogue.

Distribution: Southeastern Asia and W Indonesia. Northeastern India (Arunachal Pradesh, Assam, Manipur, Meghalaya, Mizoram, Nagaland, Sikkim, N West Bengal), Bangladesh (Chitagong), Bhutan, Myanmar (Kachin, Tanintharyi, Yangon), Thailand (Chainat, Chaiyaphum, Chanthaburi, Chiang Mai, Chon Buri, Chumphon, Kalasin, Khon Kaen, Krung Thep Mahanakhon, Loei, Lop Buri, Nakhon Pathom, Nakhon Ratchasima, Nakhon Si Thammarat, Narathiwat, Pattani, Phang Nga, Phatthalung, Phetchabun, Phuket, Prachin Buri, Prachuap Khiri Khan, Rayong, Saraburi, Sisaket, Songkhla, Suphan Buri, Trang, Yala), SE China (Anhui, Fujian, Guangdong, Guangxi, Guizhou, Hainan, Hong Kong, Hunan, Jiangxi, Yunnan, Zhejiang), Taiwan (Hsinchu, Nantou, Tainan, Taipai), Macau, Laos (Attapu, Bolikhansai, Champasak, Houaphan, Khammouan, Louangphrabang, Phongsali, Savannakhet, Salavan, Vientiane, Xaignabouli, Xekong, Xiangkhouang), Cambodia (Chhnang, Koh Kong, Kampong, Kampong Cham, Kampong Speu, Kandal, Siem Reap, Takeo), Vietnam (Bac Giang, Bac Kan, Bac Ninh, Binh Dinh, Binh Dong, Binh Phuoc, Ca Mau, Cao Bang, Dak Lak, Da Nang, Dong Nai, Ha Giang, Hai Duong, Ha Tay, Ha Tinh, Hoa Binh, Ho Chi Minh City, Khanh Hoa, Kien Giang, Kon Tum, Lam Dong, Lang Son, Lao Cai, Nghe An, Ninh Binh, Phu Tho, Quang Binh, Quang Nam, Quang Ninh, Quang Tri, Soc Trang, Son La, Tay Ninh, Thai Nguyen, Thanh Hoa, Thua Thien-Hue, Tuyen Quang, Vinh Phuc), West Malaysia (Perak, Perlis, Pinang, Selangor), Singapore and W Indonesia (Bali, Java, Kalimantan, Sumatra), NSL–3000 m.

Sources: Schlegel, 1844, Wall & Evans, 1900, Stejneger, 1907a, Wall, 1925d, Dunn, 1927, K.P. Schmidt, 1927a, C.H. Pope, 1929, 1935, Mell, 1930, Fan, 1931, Maki, 1931–b, Brongersma & Wehlburg, 1933, Bourret, 1934d, 1935b–d, 1937b, 1939a, 1941b, Herklots, 1934, Kopstein, 1938, G.F. Shaw et al., 1939?, Hoesel, 1948, 1959, Acharji and Kripalani, 1951, Bergman, 1953a, C.-S. Wang & Wang, 1956, Kuntz, 1963, Campden-Main, 1970a, Deuve, 1970, Saint-Girons, 1972, B.L. Lim, 1974, J.D. Romer, 1979a, Hu et al., 1980, Lue, 1987, 1989, Huang et al., 1990, K.K.P. Lim and Chou, 1990, K.-Y. Lue et al., 1990, M.J. Cox, 1991b, K.K.P. Lim & Lim, 1992, Easton & Leung-Va, 1993, Zhao & Adler, 1993, David & Vogel, 1996, M.J. Cox et al., 1998, Chan-ard et al., 1999, Orlov et al., 2000, Ahsan & Parvin, 2001, Auliya, 2002, Whitaker & Captain, 2004, Zhao, 2006, M.F. Ahmed et al., 2009, V.S. Nguyen et al., 2009, Xiang & Li, 2009, I. Das, 2010, 2012 and Wangyal, 2011.

Remarks: Schlegel listed C. Reinwardt as collector but types collected by H. Boie, H.C. Macklot & L. Horner *fide* C. Haas (1950: 546) and RMNH catalogue. Probably belongs in *Xenelaphis* based on hemipenis

(Brongersma & Wehlburg, 1933) and multicameral tracheal lung (Wallach, 1998a).

6. *Ptyas luzonensis* (A.C.L.G. Günther, 1873b). Proc. Zool. Soc. London 41(1): 169–170. (*Zaocys luzonensis*)

Type: Holotype, BMNH 1946.1.7.89, a 2500–2591 mm male (A.B. Meyer, 1870–1873).

Type locality: "Luzon, Philippine Islands." Restricted to Manila, Luzon, N Philippines *fide* A.B. Meyer (1887: 9).

Distribution: Northern and cen. Philippines (Catanduanes, Leyte, Luzon, Negros, Panay, Polillo), NSL–1100 m.

Sources: Boulenger, 1893a, Boettger, 1898, E.H. Taylor, 1922a, Leviton, 1983, Alcala, 1986, Ross et al., 1987, Ross & Gonzales, 1992 and Devan-Song & Brown, 2012.

7. *Ptyas mucosa* (Linnaeus, 1758). Syst. Nat., ed. 10, 1: 226. (*Coluber mucosus*)

Synonyms: *Coluber blumenbachii* Merrem, 1820, *Natrix shawii* Wagler, 1825 (*nomen ineditum*), *Coluber dhumna* Cantor, 1839b, *Leptophis trifrenatus* Hallowell, 1861, *Coryphodon blumenbachi carinata* Jan, 1867 *in* Jan & Sordelli, 1866–1870, *Zamenis mucosus pallidus* Abreu, 1918, and *Ptyas mucosus maximus* Deraniyagala, 1955.

Type: Holotype, NHR Lin- 61 (formerly MAFR), a 455 mm specimen (Mus. Drottn.).

Type locality: "Indiis" [= India].

Distribution: Southern Asia and W Indonesia. Southern Turkmenistan, E Iran, Afghanistan (Faryab, Herat, Kabul, Kandahar, Nangarhar, Wardak), Pakistan (SW Balochistan, F.A.T.A., Gilgit-Baltistan, Khyber Pakhtunkhwa, Punjab, Sindh), India (Andaman & Nicobars: Andaman Is., Andhra Pradesh, Assam, Bihar, Gujarat, Himachal Pradesh, Jammu and Kashmir, Karnataka, Kerala, Maharashtra, Madhya Pradesh, Manipur, Mizoram, Nagaland, Orissa, Punjab, Rajasthan, Sikkim, Tamil Nadu, Uttar Pradesh, West Bengal), Sri Lanka (Central, Eastern, North Central, Northern, North Western, Sabaragamuwa, Southern, Uva, Western), Nepal (Achham, Bajhang, Baitadi, Bajura, Banke, Bara, Bardiya, Bhojpur, Chitwan, Dailekh, Dang, Dadeldhura, Dhankuta, Dhanusa, Doti, Jhapa, Kailali, Kanchanapur, Kapilbastu, Kaski, Kathmandu, Mahottari, Makwanpur, Morang, Mugu, Nawalparasi, Nuwakot, Parsa, Rautahat, Rupandehi, Sankhuwasabha, Saptari, Sarlahi, Siraha, Sunsari, Surkhet, Udayapur), Bangladesh, Myanmar (Chin, Mandalay, Rangoon, Shan, Yangon), Thailand (Bangkok, Chaiyaphum, Kanchanaburi, Krung Thep Mahanakhon, Lampang, Loei, Lop Buri, Nakhon Pathom, Rayong, Si Saket), S China (Anhui, Fujian, Guangdong, Guangxi, Guizhou, Hainan, Hong Kong, Hubei, Hunan, Jiangxi, Sichuan, Yunnan, SE Xizang, Zhejiang), Taiwan (Chiai, Hsinchu, Nantou, Pingtung, Taiman, Taipei, Penghu Is.), Macau, Cambodia (Kandal, Kampong Chan, Kampong Chhnang, Preah Vihear, Siem Reap, Takeo), Laos (Champasak, Khammouan, Louangphrabang, Vientiane, Xiangkhouang), Vietnam (Ca Mau, Cao Bang, Dak Lak, Da Nang, Dien Bien, Dong Nai, Hanoi, Ha Tinh, Hoa Binh, Ho Chi Minh City, Kien Giang, Lang Son, Lao Cai, Nghe An, Ninh Binh, Phu Tho, Quang Binh, Quang Nam, Son La, Tay Ninh, Thai Nguyen, Thanh Hoa, Vinh Phuc, Yen Bai), West Malaysia (Pahang, Pinang) and W Indonesia (Bali, Java, Sumatra), NSL–4000 m.

Sources: Evans & Wall, 1900, Wall, 1906e, 1919b, 1921g, 1925d, Venning, 1910, Abreu, 1916, 1918, Nikolsky, 1916, Stanley, 1918, C.H. Pope, 1929, 1935, Maki, 1931, Bourret, 1934f, 1935d, 1941a, 1943, Herklots, 1934, Kopstein, 1938, 1941, G.E. Shaw et al., 1939b, Acharji, 1946, 1958, Constable, 1949, Terent'ev & Chernov, 1949, 1965, Bergman, 1953a, Deraniyagala, 1955, C.-S. Wang & Wang, 1956, Hoesel, 1959, L.S. Swan & Leviton, 1962, S.C. Anderson, 1963, Kuntz, 1963, Leviton & Anderson, 1963, 1970a, Acharji & Mukherjee, 1966, Minton, 1966, S.C. Anderson & Leviton, 1969, Majupuria, 1969, P. Silva, 1969, 1980a, Campden-Main, 1970a, Deuve, 1970, R.C. Sharma, 1971, 1974, 1982, Singh, 1972, Bannikov et al., 1977, Biswas & Acharjyo, 1977, Kramer, 1977, Sharma & Vazirani, 1977, Whitaker, 1978a, J.D. Romer, 1979a, Biswas and Sanyal, 1980, Hu et al., 1980, M.S. Khan, 1980, 1982b, 2002, Daniel, 1983, Sahi & Duda, 1986, Tiwari, 1985, Murthy, 1986, Nanhoe & Ouboter, 1987, K.-Y. Lue, 1987, 1989, M.R. Khan, 1988, Clark, 1990, Huang et al., 1990, K.-Y. Lue et al., 1990, A. Silva, 1990b, 2009, M.J. Cox, 1991b, Easton & Leung-Va, 1993, Hodges, 1993, Zhao & Adler, 1993, Szczerbak, 1994, David & Vogel, 1996, M.J. Cox et al., 1998, Zug et al., 1998, Chan-ard et al., 1999, B.L. Lim et al., 2002, Tiwari & Shah, 2004, Whitaker & Captain, 2004, S. Dutta, 2005, N. Khaire, 2006, McKay, 2006, M.F. Ahmed et al., 2009, V.S. Nguyen et al., 2009, Xiang & Li, 2009, I. Das, 2010, 2012, Vyas, 2011 and T. Hartmann et al., 2013.

Remarks: Singapore record doubtful *fide* K.K.P. Lim & Chou (1990: 54) and K.K.P. Lim & Lim (1992: 145).

8. *Ptyas nigromarginata* (Blyth, 1854a). J. Asiat. Soc. Bengal 23(3): 290–291. (*Coluber nigromarginatus*)

Type: Lectotype, ZSI 7343 (formerly IMC 7343), a 2362 mm specimen (W.S. Sherwill, 1843–1854), designated by I. Das et al. (1998: 147).

Type locality: "Vicinity of Darjiling" [= vicinity of Darjeeling (27° 02'N, 88°16'E, elevation 2100 m), N West Bengal, NE India].

Distribution: Southern Asia. Northeastern India (Arunachal Pradesh, Meghalaya, Nagaland, Sikkim,

N West Bengal), Bangladesh, Nepal, Bhutan, W China (Gansu, Guizhou, Hubei, Sichuan, Xizang, Yunnan), N Myanmar (Kachin), Laos and Vietnam (Ha Tay, Kon Tum, Lang Son, Lao Cai), 500–2350 m.

Sources: Sclater, 1891a, Annandale, 1911, Wall, 1925d, C.H. Pope, 1929, 1935, L.S. Swan & Leviton, 1962, Hu et al., 1980, Tian et al., 1986, M.A.R. Khan, 1988, Zhao & Adler, 1993, Schleich & Kästle, 2002, Ao et al., 2004, Whitaker & Captain, 2004, V.S. Nguyen et al., 2009, I. Das, 2010, 2012 and Yao, 2012.

PTYCHOPHIS Gomés, 1915 (Dipsadidae)

Synonym: *Paraptychophis* Lema, 1967: 62.
Type species: *Ptychophis flavovirgatus* Gomés, 1915.
Distribution: Southeastern Brazil.
Sources: Amaral, 1930e–f, 1977, Lema, 1967, Hoge & Romano-Hoge, 1969a, J.A. Peters & Orejas-Miranda, 1970, Porto & Caramaschi, 1988, Lema & Deiques, 1992 and Zaher et al., 2009.

1. *Ptychophis flavovirgatus* Gomés, 1915. Ann. Paul. Med. Cirurg. 4(6): 128–129, pl. 4, figs. 4–6.

Synonym: *Paraptychophis meyeri* Lema, 1967.
Type: Holotype, IB 526, a 510 mm male (A. Swarowsky, April 1914), destroyed by fire 15 May 2010.
Type locality: "São Bento, Estado de Santa Catharina, Brazil" [= São Bento do Sul, NE Santa Catarina State, S Brazil, 26°15'S, 49°23'W, elevation 835 m].
Distribution: Southeastern Brazil (ext. S Minas Gerais, E Paraná, Rio Grande do Sul, N Santa Catarina), 10–1300 m.

PYTHON Daudin, 1803b (Pythonidae)

Synonyms: *Pyphon* – Fischer von Waldheim, 1813 *(nomen incorrectum), Pylhon* – Oken, 1817 *(nomen incorrectum), Constrictor* Wagler, 1830 *(nomen praeoccupatum), Enygrus* Wagler, 1830, *Engyrus* – Gray *in* Griffith & Pidgeon, 1831 *(nomen incorrectum), Enygris* Gray, 1842a *(nomen emendatum), Heleionomus* Gray, 1842a, *Hortulia* Gray, 1842a, *Asterophis* Fitzinger, 1843, *Pijthon* Bleeker, 1856 *(nomen emendatum), Hortalia* – Chenu & Desmarest, 1857 *(nomen incorrectum), Astrophis* Fitzinger, 1861 *(nomen emendatum), Atrophis* – Fitzinger, 1867 *(nomen incorrectum), Exygrus* – A.C.L.G. Günther, 1877 *(nomen incorrectum), Emigras* – F. Müller, 1878b *(nomen incorrectum), Hortalia* Scudder, 1882 *(nomen emendatum), Aspidoboa* Sauvage, 1884, *Pyton* – Peracca, 1909 *(nomen incorrectum), Pithon* Marelli, 1931 *(nomen emendatum), Phyton* – J.F. Bonaparte, 1978 *(nomen incorrectum), Phython* – Mahendra, 1984 *(nomen incorrectum), Broghammerus* Hoser, 2004 *(nomen illegitimum),* and *Shireenhoserus* Hoser, 2004 *(nomen illegitimum).*
Type species: *Coluber molurus* Linnaeus, 1758.
Distribution: Africa, S Asia and East Indies.
Fossil records: Lower Miocene of Kenya, Namibia and Greece, middle Miocene of Italy, Greece and Morocco, upper Miocene of Pakistan, Uganda and Chad, Miocene of Myanmar, Pliocene of Tanzania, Miocene/Pliocene of Chad, Libya and Uganda, middle Pliocene of Uganda, upper Pliocene of Tanzania, Pliocene/Pleistocene of Ethiopia and Tanzania, lower Pleistocene of Tanzania, lower/middle Pleistocene of Eritrea, middle Pleistocene of Tanzania, and Pleistocene of India.
Sources: Stull, 1935, M.A. Smith, 1943, Stimson, 1969, Matz & Matz, 1971, McDowell, 1975a, Kluge, 1993b, Pearson, 1993, Walls, 1998b, McDiarmid et al., 1999, Hoser, 2003f, 2004, 2012b, J.R. Berry, 2006, Rawlings et al., 2008 and Schleip & O'Shea, 2010.

1. *Python anchietae* Bocage, 1887e. J. Sci. Math. Phys. Nat., Lisboa 12(46): 87–88.

Types: Syntypes (2), MBL T 89.1206–07, a 1050–1140 mm male and a 600 mm juvenile (J.A. d'Anchieta, 1861–1866), destroyed by fire 18 March 1978.
Type locality: "Catumbella, (Benguella)" [= Catumbela, NW Benguela Prov., W Angola, 12°26'S, 13°33'E, elevation 25 m].
Distribution: Southwestern Angola (Benguela, Huila) and N Namibia (Erongo, Khomas, Kunene, Otjozondjupa), 10–600 m.
Sources: Bocage, 1895, Monard, 1937, Bogert, 1940, Mertens, 1955, 1971, V.F.M. FitzSimons, 1962a, Finkeldey, 1963, Pitman, 1974, Patterson & Erasmus, 1978, Broadley, 1983, Branch, 1988 and Marais, 1992.

2. *Python bivittatus* Kuhl, 1820. Beitr. Zool. Anat. 1: 94.

Synonyms: *Python javanicus* Kuhl & Hasselt, 1822a, *Python molurus sondaica* F. Werner, 1899d, and *Python bivittatus progchai* Jacobs, Auliya & W. Böhme, 2009.
Type: Holotype, not designated (H. Kuhl, 1820), location unknown.
Type locality: Unknown. Restricted to Java, Indonesia *fide* Mertens (1930:287).
Distribution: Southeast Asia and Indonesia. Northeastern India (Arunachal Pradesh, Assam,Manipur, Meghalaya, Mizoram, Nagaland, Shillong, West Bengal), Bhutan, Myanmar (Bago, Kachin, Karen, Magwe, Mandalay, Mon, Rakhine, Shan, Tanintharyi, Yangon), Nepal (Banke, Bara, Bardiya, Chitwan, Dang, Dadeldhura, Dhading, Dhankuta, Dhanusa, Dolakha, Doti, Gorkha, Gulmi, Ilam, Jhapa, Kailali, Kanchanapur, Kapilbastu, Kaski, Kathmandu, Kavre, Khotang, Lalitpur, Lamjung, Mahottari, Makwanpur, Morang, Nuwakot,

P

Parsa, Rautahat, Rupandehi, Sankhuwasabha, Saptari, Sarlahi, Siraha, Sunsari, Tanahu), Thailand (Bangkok, Buriram, Chaiyaphum, Chiang Kanchaburi, Chon Buri, Chumphon, Khon Kaen, Krung Thep Mahanakhon, Loei, Lop Buri, Maha Sarakham, Mai, Nan, Nakhon, Nonthaburi, Phet Buri, Phichit, Phitsanulok, Ranong, Ratchasima, Rayong, Surin, Uthai Thani) S China (Fujian, Guangdong, Guangxi, Guizhou, Hainan, Hong Kong, Sichuan, Yunnan), Vietnam (Bac Giang, Ba Ria-Vung Tau, Binh Phuoc, Ca Mau, Can Pho, Dak Lak, Da Nang, Dong Nai, Gia Lai, Hai Duong, Ha Tinh, Hoa Binh, Ho Chi Minh City, Khanh Hoa, Kon Tum, Lai Chau, Lam Dong, Lang Son, Lao Cai, Nghe An, Ninh Binh, Phu Tho, Quang Binh, Quang Nam, Quang Ninh, Quang Tri, Son La, Tay Ninh, Thai Nguyen, Thanh Hoa, Thua Thien-Hue, Vinh Phuc, Yen Bai), Laos (Champasak, Savannakhet), Cambodia (Kampong Chhnang, Siem Reap), and Indonesia (Bali, Java, Nusa Barung, Sulawesi, Sumbawa), NSL–1200 m.

Sources: F. Werner, 1930b, Bourret, 1935d, 1936b, 1937b, C.H. Pope, 1935, 1961, E.H. Taylor, 1965, Campden-Main, 1970a, Saint-Girons, 1972a, J.D. Romer, 1979a, Hu et al., 1980, Bosch, 1985, Tian et al., 1986, M.J. Cox, 1991b, Hodges, 1993, Zhao & Adler, 1993, Zhong, 1993, M.J. Cox et al., 1998, Chan-ard et al., 1999, Orlov et al., 2000, Tiwari & Shah, 2004, M.F. Ahmed et al., 2009, Jacobs et al., 2009, V.S. Nguyen et al., 2009, Xiang & Li, 2009, Barker & Barker, 2010a and Dorcas & Willson, 2011.

Remarks: Original description based on Seba (1735: 20, pl. 19, figs. 1 and 28, pl. 27, fig. 1). Introduced and established in USA (S Florida) *fide* Dorcas & Willson, 2011. Possibly occurs in West Malaysia *fide* I. Das (2012: 142).

P

3. *Python breitensteini* Steindachner, 1880. Sitz. Akad. Wiss. Wien, Math.-Naturwiss. Kl. 82(2): 267–268.

Type: Holotype, NMW 13287, an 800 mm specimen (H. Breitenstein, 1880).

Type locality: "Teweh, Borneo" [= Muarateweh, Central Prov., cen. Kalimantan, Indonesia, 0°56'S, 143°53'E, elevation 35 m].

Distribution: Borneo. East Malaysia (Sabah, Sarawak), Brunei and Indonesia (Kalimantan), NSL–600 m.

Sources: Keogh et al., 2001, Malkmus et al., 2002, I. Das, 2010, 2012 and Zug et al., 2011.

Remarks: Type locality at junction of Barito and Teweh rivers *fide* Keogh et al. (2001: 122).

4. *Python brongersmai* Stull, 1938. Occ. Pap. Boston Soc. Nat. Hist. (8): 297–298. (*Python curtus brongersmai*)

Type: Holotype, MCZ 29779, an 1180 mm male.

Type locality: "Singapore, Malay Peninsula."

Distribution: Southeast Asia and W Indonesia. Southern Thailand (Krabi, Nakhon Si Thammarat, Pattani, Phang Nga, Ratchaburi, Songkhla, Surat Thani), Vietnam (Binh Thuan, Ca Mau, Ho Chi Minh City), West Malaysia (Johor, Kedah, Kuala Lumpur, Melaka, Pinang, Perak, Selanger), Singapore and W Indonesia (Bangka, Bintan, Sumatra), NSL–85 m.

Sources: Keogh et al., 2001, J.C. Murphy & Schlager, 2003, V.S. Nguyen et al., 2009, I. Das, 2010, 2012 and Zug et al., 2011.

Remarks: Status undetermined in Singapore *fide* K.K.P. Lim & Lim (1992: 144). Questionable records from Vietnam may be introductions.

5. *Python curtus* Schlegel, 1872. De Dierentuin Nat. Art. Mag., Rept. (4): 54, 1 fig.

Type: Holotype, RMNH 3782, a 460 mm male (S. Müller, 1833–1835).

Type locality: "Sumatra" [W Indonesia]. Restricted to between Padang and Indrapura, Sumatra *fide* Hubrecht (1879c: 245) and vicinity of Padang and in the Padang Highlands *fide* Brongersma (1947b: 667).

Distribution: Indonesia (Bangka, Beitung, Bengkulu, Lampung, Sumatra), NSL–920 m.

Sources: Hubrecht, 1879d, Blanford, 1881, Sauvage, 1884, J.G. Fischer, 1885a, Boulenger, 1889c, 1912c, 1920e, Flower, 1896, 1899, Zenneck, 1898, Ridley, 1899, Rooij, 1917, Robinson & Kloss, 1920, Stull, 1935, 1938, Westermann, 1942, Brongersma, 1947b, C. Haas, 1950, B.L. Lim, 1955, 1956, 1958, 1960, Haile, 1959, E.H. Taylor, 1965, Stimson, 1969, Tweedie, 1983, K.F.L. Lim & Lee, 1989, K.K.P. Lim & Chou, 1990, Stuebing, 1991, D.G. Barker & Barker, 1994, David & Vogel, 1996, M.J. Cox et al., 1998, Chan-ard et al., 1999, Keogh et al., 2001, I. Das, 2010, 2012 and Zug et al., 2011.

6. †*Python europaeus* Szyndlar & Rage, 2003. Booidea Oligocene Miocene Europe: 68–70, figs. 31e–g.

Type: Holotype, MNHN VCO 29, one trunk vertebra.

Type locality: "Vieux Collonges (= Mont Ceindre) (France), early/middle Miocene (MN 4/5)."

Distribution: Lower Miocene (Orleanian, MN 4–5: 13.7–16.9 mya) of France.

7. *Python kyaiktiyo* Zug, Gotte & Jacobs, 2011. Proc. Biol. Soc. Washington 124(2): 129–130, figs. 2a–b, 3.

Type: Holotype, USNM 572046, 1270 mm gravid female (W.K. Sai, Z.L. Win & N. Hla, 11 Mar. 2002).

Type locality: "Myanmar, Mon State, in Kyaiktiyo Wildlife Sanctuary at Yetagon Myaung (17°26'38"N, 97°05'58"E;~390 m elevation), approximately 5 km NE Kinmun."

Distribution: Southern Myanmar (Mon), 390 m. Known only from type locality.

8. †*Python maurus* Rage, 1976. Géol. Méditer. 3(2): 59–60, fig. 3, pl., fig. 13.

Type: Holotype, MNHN BML900, one trunk vertebra.

Type locality: "Beni Mellal, Maroc; Miocène" [= Beni Mellal, Beni Mellal Prov., cen. Morocco; middle Miocene].

Distribution: Middle Miocene (Astaracian, MN 7–8: 11.1–12.8 mya) of Morocco. Known only from type series.

9. *Python molurus* (Linnaeus, 1758). Syst. Nat., ed. 10, 1: 225. (*Coluber molurus*)

Synonyms: *Boa albicans* J.G. Schneider, 1801, *Boa castanea* J.G. Schneider, 1801, *Boa cinerea* J.G. Schneider, 1801, *Boa orbiculata* J.G. Schneider, 1801, *Boa ordinata* J.G. Schneider, 1801, *Coluber boaeformis* G. Shaw, 1802, *Boa moluroidea* Daudin, 1803b (*nomen nudum*), *Python bora* Daudin, 1803b, *Python tigris* Daudin, 1803b, *Python jamesonii* Gray, 1842a, *Astrophis tigris* Fitzinger, 1861, *Python albo-maculata* Higgins, 1873 *Python bimaculata* Higgins, 1873, *Python bora* Higgins, 1873, *Python lazuli* Higgins, 1873, *Python molurus intermedia* F. Werner, 1899d, *Python molurus ocellata* F. Werner, 1899d, and *Python molurus pimbura* Deraniyagala, 1945.

Type: Holotype, NHR Lin-5 (formerly MDG), an 1140 mm female.

Type locality: "Indiis" [= India].

Distribution: Southern Asia. Pakistan (E Balochistan, N Punjab, Sindh), India (Andhra Pradesh, Arunachal Pradesh, Goa, Gujarat, Jammu and Kashmir, Karnataka, Kerala, Madhya Pradesh, Maharashtra, Orissa, Punjab, Rajasthan, Tamil Nadu, Uttaranchal, Uttar Pradesh, West Bengal, Sikkim), Nepal (Bardiya), Bangladesh and Sri Lanka (Central, Eastern, Sabaragamuwa, Western), NSL–2500 m.

Sources: Boulenger, 1893a, Wall, 1907e, 1912a, 1921g, 1923c, 1925a , 1926a, Rooij, 1917, Mertens, 1930, F. Werner, 1930b, G.E. Shaw et al., 1938a, M.A. Smith, 1943, Deraniyagala, 1955, Daniel & Shull, 1963, Deoras, 1965, Minton, 1966, Deuve, 1970, Fleming & Fleming, 1973, Acharjyo & Mishra, 1976, 1980, R.C. Sharma, 1976a, Biswas & Acharjyo, 1977, Kramer, 1977, Whitaker, 1978, M.S. Khan, 1980, 1982, 2002, P. Silva, 1980a, Majupuria, 1981, Daniel, 1983, Matz, 1983c, Murthy, 1986, Sahi and Duda, 1986, 1986, M.A.R. Khan, 1988, Dutta & Acharjyo, 1990, A. Silva, 1990, J.C. Murphy & R.W. Henderson, 1997, Tiwari & Shah, 2004, Whitaker & Captain, 2004, N. Khaire, 2006, Bellosa, 2007, Barker & Barker, 2010b and I. Das, 2010, 2012.

10. *Python natalensis* A. Smith, 1840 *in* 1838–1849. Illustr. Zool. So. Afr., Rept.: pl. 9 and 3 pp.

Synonyms: *Python natalensis* A. Smith, 1833 (*nomen nudum*), *Heleionomus variegatus* Gray, 1842a, and *Python saxuloides* T.J. Miller & Smith, 1979.

Type: Holotype, not designated, Cape Town Zool. Soc. Mus., lost *fide* Stimson (1969: 31).

Type locality: "from the interior, eastward of Latakoo; also Port Natal" *fide* A. Smith (1833: 80) [= E of Kuruman, Northern Cape Prov., and Port Natal, KwaZula Natal Prov., E South Africa].

Distribution: Eastern Africa. Eastern Democratic Republic of the Congo (Katanga, Sud-Kivu), Burundi, S Kenya (S Central, Coast, S Eastern, S Rift Valley), Mozambique (Cabo Delgado, Gaza, Inhambane, Manica, Maputo, Nampula, Zambézia, Ukerewe and Vamizi Is.), Malawi (Central, Northern, Southern), Zambia (Central, Copperbelt, Eastern, Luapula, Lusaka, Northern, North-Western, Southern, Western), Angola (Benguela, Cunene, Huíla, Luanda, Lunda Moxico, Norte), N Namibia (Grootfontein, Karibib, Omaruru, Outjo), Botswana (Central, Kgalagadi, Kweneng, North West, South East, Southern), Zimbabwe (Bulawayo, Manicaland, Mashonaland East, Mashonaland West, Masvingo, Matebeleland North, Matebeleland South, Midlands), Swaziland and NE South Africa (Eastern Cape, Gauteng, KwaZulu-Natal, Limpopo, Northern, Mpumalanga), NSL–1800 [2200] m.

Sources: FitzSimons, 1962a, Broadley, 1983, 1999b, Auerbach, 1987, Clauss & Clauss, 2002, Spawls et al., 2002, Alexander & Marais, 2007 and Broadley & Blaylock, 2013.

11. *Python regius* (G. Shaw, 1802). Gen. Zool., Amph. 3(2): 347–348, pl. 96. (*Boa regia*)

Synonym: *Python bellii* Gray, 1842a.

Types: Syntypes, based on Seba (1734: 97–98, pl. 62, fig. 1, and 1735: 107–108, pl. 102), lost *fide* Stimson (1969: 30).

Type locality: "Africae Mosambicensi" and "Brasiliensis" [= Mozambique, Africa and Brazil] (in error). Restricted to Africa *fide* Papenfuss (1969: 304).

Distribution: West and cen. Africa. Senegal (Dakar, Fatick, Kédougou, Kolda, Tambacounda, Thiès, Ziguinchor), Gambia (MacCarthy Island, Western), Guinea–Bissau (Bissau), Guinea (Boffa, Koundara, Kouroussa, Siguiri, Télimélé), Sierra Leone (Northern, Western), Ivory Coast (Bondoukou, Bouaké, Korhogo, Lagunes, Toumodi), Ghana (Accra, Northern, Upper East, Upper West), SW Mali (Kayes, Koulikoro, Ségou, Sikasso), Burkina Faso (Centre, Centre-Est, Centre-Norte, Centre-Ouest, Est, Hauts-Bassins, Sud-Ouest, Volta-Noire), Togo (Centrale, Kara, Maritime, Plateaux, Savanes), Benin (Alibori, Atakora, Atlantique, Collines, Cotonou, Zou), SW Niger (Niamey), Nigeria (Cross River, Oyo, Rivers), SW Chad (Mayo-Kebbi), N Cameroon (Adamaoua,

Extreme-Nord, Nord, Sud-Ouest), Central African Republic (Haut-Mbomou, Ombella-Mpoko, Ouham, Ouham-Pende), Democratic Republic of the Congo (Katanga, Orientale), S Sudan (Janub Kurdufan), South Sudan (Bahr al Ghazal), and ext. W Uganda (Madi, Toro, West Nile), NSL–1000 m.

Sources: Sternfeld, 1909a, Aylmer, 1922, Angel, 1933b, Corkill, 1935, Andersson, 1937, Monard, 1940, 1951, Cansdale, 1948, 1954, 1961, Villiers, 1950a, 1953, 1954a, 1975, J.D. Romer, 1953, Witte, 1962, 1966, Doucet, 1963, Roussel & Villiers, 1965, Menzies, 1966, Blackwell, 1967, Papenfuss, 1969, Roux-Estève, 1969a, Hulselmans & Verheyen, 1970, Hulselmans et al., 1970, Pitman, 1974, W. Böhme, 1978, Miles et al., 1978, Roman, 1980, 1984, Hakansson, 1981, Hughes, 1983, 2013, J.A. Butler & Reed, 1986, Branch, 1988, Joger, 1990, Luiselli & Angelici, 1998, Chippaux, 1999, Kölpin, 2002, Spawls et al., 2002, Kirschner & Seufer, 2003, J.C. Murphy & Schlager, 2003, Leaché, 2005, McCurley, 2005, J.-F. Trape & Mané, 2006b, Chirio & LeBreton, 2007, Berry, 2010, Segniagbeto et al., 2011, Auliya et al., 2012 and Chirio, 2013.

12. *Python reticulatus* (J.G. Schneider, 1801). Hist. Amph. 2: 264–266. (*Boa reticulata*)

Synonyms: *Coluber oularsawa* Bonnaterre, 1790 (*nomen rejiciendum*), *Coluber oryzivorus* Suchow, 1797 (*nomen rejiciendum*), *Boa rhombeata* J.G. Schneider, 1801, *Boa phrygia* G. Shaw, 1802, *Coluber javanicus* G. Shaw, 1802, *Python schneideri* Merrem, 1820, *Python reticulates jampeanus* Auliya, Mausfeld, Schmitz & Böhme, 2002, *Python reticulatus saputrai* Auliya, Mausfeld, Schmitz & Böhme, 2002, *Broghammerus reticulatus euanedwardsi* Hoser, 2004, *Broghammerus reticulatus dalegibbonsi* Hoser, 2004, *Broghammerus reticulatus haydnmacphiei* Hoser, 2004, *Broghammerus reticulatus neilsonnemani* Hoser, 2004, *Broghammerus reticulatus patrickcouperi* Hoser, 2004, and *Broghammerus reticulatus stuartbigmorei* Hoser, 2004.

Type: Neotype, ZFMK 32378 (natives, 1980), a 1308 mm male, designated by Auliya et al. (2002: 206–207, fig. 3a).

Type locality: "Rengit, West Malaysia" [= Rengit, SW Johore State, S West Malaysia, 1°40'N, 103°11'E, elevation 10 m] via neotype selection.

Distribution: Southeast Asia and the East Indies. Northeastern India (West Bengal, Andaman & Nicobars: Great Nicobar Is.), Bangladesh, S Myanmar (Bago, Mon, Tanintharyi), Thailand (Kanchanaburi, Krung Thep Mahanakhon, Loei, Nakhon Nayok, Nakhon Ratchasima, Nakhon Si Thammaret, Narathiwat, Pattani, Phang Nga, Phatthalung, Phuket, Phatthalung, Prachin Buri, Ranong, Sisaket, Tak, Trang: Koh Chang Is.; Trat, Uthai Thani), Laos (Savannakhet), Cambodia (Kampong Chhnang, Oddar Meanchey), Vietnam (Ba Ria-Vung Tau, Binh Dinh, Binh Phuoc, Ca Mau, Dak Lak, Da Nang, Dong Nai, Gia Lai, Ha Tinh, Ho Chi Minh City, Khanh Hoa, Kien Giang, Kon Tum, Long An, Quang Binh, Quang Nam, Tay Ninh, Thua Thien-Hue, Cu Lao Panjang Is.), West Malaysia (Johore, Kedah, Perak, Pinang, Selangor, Penang and Tioman Is.), Singapore, East Malaysia (Sabah, Sarawak), Brunei, Indonesia (Alor, Ambon, Anambas Arch., Babi, Bacan, Bali, Banda, Banjak, Bangka, Belitung, Bintan, Boano, Bulan, Buru, Buton, Djemadja, Durian, Enggano, Flores, Galang, Halmahera, Haruku, Java, Kalimantan, Kayuadi, Krakatau, Lang, Leti, Lomblen, Lombok, Mentawai Arch., Natuna Arch., Nias, Nusa Barung, Obi, Oriba, Pantar, Riau Arch, Romang, Saparua, Seleyar, Seram, Simeulue, Sipura, Sula, Sulawesi, Sumatra, Sumbawa, Sumba, Tanimbar Arch., Ternate, Timor, Verlaten, Weh, Wetar, Yamdena), Timor-Leste and Philippines (Basilan, Bohol, Catanduanes, Cebu, Itbayat, Jolo, Leyte, Luzon, Masbate, Mindanao, Mindoro, Negros, Palawan, Panay, Polillo, Samar, Sibutu, Tawitawi), NSL–1500 m.

Sources: Flower, 1896, 1899, Rooij, 1917, E.H. Taylor, 1922a, 1965, Wall, 1923?, 1926a, Kopstein, 1926, 1927, Chasen & Smedley, 1927, Smedley, 1928, Mertens, 1928, 1930, Brongersma, 1933a, M.A. Smith, 1943, Leviton, 1959b, C.H. Pope, 1961, Deoras, 1965, Campden-Main, 1970a, Deuve, 1970, Saint-Girons, 1972a, Biswas & Sanyal, 1977, Daniel, 1983, Matz, 1983c, Tweedie, 1983, Bosch, 1985, M.A.R. Khan, 1988, M.J. Cox, 1991b, Ross & Gonzales, 1992, David & Vogel, 1996, J.C. Murphy & R.W. Henderson, 1997, M.J. Cox et al., 1998, Chan-ard et al., 1999, Darevsky, 1999, Iskandar & Colijn, 2001, Malkmus et al., 2002, J.C. Murphy & Schlager, 2003, Whitaker & Captain, 2004, I. Das, 2007b, 2010, 2012, V.S. Nguyen et al., 2009, Wöllner, 2011, O'Shea et al., 2012, C. Sanchez et al., 2012 and T. Hartmann et al., 2013.

Remarks: Description based upon specimen described and illustrated by Seba (1734: pl. 62, fig. 2) and Seba (1735: pl. 79, fig. 1, pl. 80, fig. 1). Holotype, formerly ZMG, lost *fide* Kluge (1993b: 8). Official Specific Name *fide* Opinion 1463 (ICZN, 1987). Type locality restriction of Java, Indonesia *fide* Brongersma (1972: 58) invalid.

13. †*Python sardus* (Portis, 1901). Boll. Soc. Geol. Ital. 20(2): 247–253, 1 fig. (†*Palaeopython sardus*)

Type: Holotype, ? IZUCS, one articulated palatine and pterygoid.

Type locality: "Calcare araìlloso della base di Monte Albu (Bosa), Sardegna; Miocene medio" [= Monte Bianco, W Sardegna, ext. W Italy, 45°48'N, 6°50'E; middle Miocene].

Distribution: Middle Miocene (Astaracian, MN 6–8: 11.1–13.7 mya) of Italy. Known only from holotype.

Source: Szyndlar & Rage, 2003.

P

14. *Python sebae* (Gmelin, 1789). Syst. Nat., ed. 13, 1(3): 1118. (*Coluber sebae*)

Synonyms: *Coluber speciosus* Bonnaterre, 1790, *Boa hieroglyphica* J.B. Schneider, 1801, ? *Python houttuyni* Daudin, 1803b, *Python liberiensis* Hallowell, 1845b, and *Python jubalis* Pitman, 1936 (*nomen nudum*).

Type: Lectotype, specimen described and illustrated by Seba (1735: 105, pl. 99, fig. 2) *fide* Loveridge (1957a: 248).

Type locality: "Brasiliensis, è regione Guairâ" [= Guaíra, Paraná, Brazil] (in error). Corrected to Africa *fide* V.F.M. FitzSimons (1962a: 94).

Distribution: Subsaharan Africa. Southern Mauritania (Gorgol, Guidimaka, Hodh Ech Chargui, Trarza), Senegal (Dakar, Fatick, Kédougou, Louga, Saint-Louis, Tambacounda, Thiès, Ziguinchor), Gambia (Western, MacCarthy Island), Guinea–Bissau (Bissau, Oio, Quinara, Tombali, Bubaque, Imbone, Orango and Soga Is.), Guinea (Boffa, Dubréka, Kouroussa, Macenta, Nzérékoré, Télimélé), Sierra Leone (Eastern, Northern, Southern, Western), Liberia (Bong, Grand Cape Mount, Grand Gedeh, Loffa, Montserrado), SW Mali (Bamako, Koulikoro, Ségou, Sikasso), Ivory Coast, (Abidjan, Bouaflé, Bouna, Guiglo, Ferkéssédougou, Lagunes, San Pédro, Tabou, Toumodi), Burkino Faso (Centre, Centre-Nord, Centre-Ouest, Est, Hauts-Bassins, Sahel, Sud-Ouest, Volta-Noire), Ghana (Accra, Ashanti, Central, Eastern, Northern, Upper East, Upper West, Volta, Western), N Togo (Kara, Maritime, Savanes), Benin (Alibori, Atlantique, Collines, Zou), ext. SW Niger (Dosso, Niamey), Nigeria (Cross River, Delta, Lagos, North-Eastern), Cameroon (Adamaoua, Centre, Est, Extreme-Nord, Littoral, Nord, Nord-Ouest, Ouest, Sud, Sud-Ouest), S Chad (Mayo-Kebbi Ouest, Moyen-Chari), Equatorial Guinea (Littoral, Bioko Is.), Gabon (Estuaire, Haut-Ogooué, Moyen-Ogooué, Ngounié, Ogooué-Ivindo, Ogooué-Lolo, Ogooué-Maritime), Congo (Kouilou, Likouala, Pool), Central African Republic (Bamíngui-Bangoran, Haut-Mbomou, Haute-Sangha, Lobaye, Nana Mambere, Ombella-Mpoko, Sangha, Vakaga), N Angola (Cabinda, Lunda Norte), N Namibia (Erongo, Ohangwena, Okavango, Otjozondjupa), Democratic Republic of the Congo (Bandundu, Bas-Congo, Equateur, Kasai Occidental, Kasai Oriental, Katanga, Kinshasa, Maniema, Nord-Kivu, Orientale, Sud-Kivu), Sudan (An Nil Al Abyad, An Nil Al Azraq, Al Khartum, Ash Sharqiyah, Janub Kurdufan, Sennar), SE South Sudan (Eastern Equatoria), Eritrea, Ethiopia (Gemu Gofa, Gojjam, Hararge, Illubabor, Kefa, Shoa, Sidamo, Tigre, Wollega), S Somalia (Bay, Gedo, Jubbada Dhexe, Jubbada Hoose, Mogadishu, Shabeellaha Dhexe, Shabeellaha Hoose, Alessandra Is.), Kenya (Central, Coast, Eastern, Nyanza, Nairobi, North-Eastern, Rift Valley), Uganda (Central, Eastern, Northern, Western, Sese Is.), Rwanda (Byumba, Kibungo) and Burundi, NSL–2250 m.

Sources: Sternfeld, 1909b, Aylmer, 1922, Angel, 1933b, Corkill, 1935, Andersson, 1937, Monard, 1937, 1951, Cansdale, 1948, 1961, Villiers, 1948, 1950a, 1954a, 1956a, 1965, 1966, 1975, Witte, 1953, 1962, 1966, 1975, Laurent, 1954a, 1956a, 1960, 1964a, Mertens, 1955, 1964a, 1971, Hellmich, 1957a, Condamin, 1958, Vesey-FitzGerald, 1958, Broadley, 1959, 1971d, 1983, 1984, Sweeney, 1961, Condamin & Villiers, 1962, V.F.M. FitzSimons, 1962a, 1970, Doucet, 1963, Knoepffler, 1966, Menzies, 1966, Hughes & Barry, 1969, Papenfuss, 1969, Hulselmans et al., 1970, Broadley & Cock, 1975, W. Böhme, 1978, Miles et al., 1978, Spawls, 1978, Roman, 1980, 1984, Hakansson, 1981, Trape, 1981, 1985, Hoevers & Johnson, 1982, Buys & Buys, 1983, Hughes, 1983, 2013, Lanza, 1983a, 1990b, Pienaar et al., 1983, J.A. Butler, 1986, Auerbach, 1987, Branch, 1988, Joger, 1990, Broadley & Howell, 1991, J.B. Rasmussen, 1991, Boycott & Culverwell, 1992, Marais, 1992, Haagner, 1993, Largen & Rasmussen, 1993, Bailon & Rage, 1994, J.-F. Trape & Roux-Estève, 1995, J.C. Murphy & R.W. Henderson, 1997, Chippaux, 2001, Spawls et al., 2002, Lanza & Nistri, 2005, J.-F. Trape & Mané, 2006b, Chirio & LeBreton, 2007, Largen & Spawls, 2010, Segniagbeto et al., 2011 and Chirio, 2013.

Remarks: Holotype in ZMUU *fide* Stimson (1969: 31) but not listed by Lönnberg, 1896, Vanzolini, 1969, or Wallin, 2001. *Python sebae* or *P. natalensis* are recorded in the Pliocene of Laetoli, Tanzania (Rage & Bailon, 2011).

15. *Python timoriensis* (W.C.H. Peters, 1876b). Mber. Königl. Akad. Wiss. Berlin 1876(8): 533, pl., figs. 3–3c. (*Liasis amethystinus timoriensis*)

Synonyms: *Python timorensis* Boulenger, 1893a *(nomen emendatum)*, and *Liasis petersii* Hubrecht, 1879b.

Type: Holotype, ZMB 8927, a decapitated head (De Jong) *fide* C. Haas (1950: 523).

Type locality: "Kupang (Timor)" [Indonesia] (in error *fide* McDiarmid et al., 1999: 182, confirmed *fide* H. Kaiser *in* Lang, 2011: 184).

Distribution: Lesser Sundas of S Indonesia (Adonara, Flores, Lomblen, Lombok, Pantar, Solor, Sumbawa).

Sources: Rooij, 1917, F. Werner, 1921b, Mertens, 1930, C. Haas, 1950, Brongersma, 1956b, 1968 and Lang, 2011.

Remarks: Official Specific Name no. 2425 *fide* Opinion 939 (ICZN, 1970). Records from Komodo and Timor questionable *fide* Iskandar & Colijn (2001: 31).

PYTHONODIPSAS A.C.L.G. Günther, 1868 (Pseudaspididae)

Synonyms: *Pythondipsas* – Ditmars, 1933 *(nomen incorrectum)*, *Phitodipsas* – Briceño-Rossi, 1934 *(nomen incorrectum)*, and *Pythanodipsas* – Visser, 1967 *(nomen incorrectum)*.

Type species: *Pythonodipsas carinata* A.C.L.G. Günther, 1868.

Distribution: Southwestern Africa.

Sources: Mertens, 1955, 1971, V.F.M. FitzSimons, 1962a, 1974, C.T. Stuart, 1976, Marx et al., 1982, Broadley, 1983, Schätti & McCarthy, 1987, Branch, 1988, Marais, 1992, Marais, 2004, Lawson et al., 2005, Vidal et al., 2008 and Kelly et al., 2009.

1. *Pythonodipsas carinata* A.C.L.G. Günther, 1868. Ann. Mag. Nat. Hist. (4) 1(6): 426, pl. 19, fig. k.

Type: Holotype, BMNH 1946.1.4.70 (formerly SAM), a 622 mm male (? J. Chapman, 1861–1863).

Type locality: "Zambezi" [= Zambezi River, Mozambique] (in error). Corrected to Damaraland, NW Namibia *fide* Broadley (1971c: 679).

Distribution: Southwestern Angola (Namibe) and W Namibia (Erongo, Hardap, Karas, Khomas, Kunene), NSL–870 m.

Remarks: Original description attributes collector to J. Chapman on the Zambezi River (and Broadley has shown that Chapman did not go down the Zambezi and that type locality is in Namibia); however, Günther (1868: 413–414) also listed Cape Town Mus. as source so it is more likely that the type came from SAM rather than Chapman.

P

R

RABDION A.-M.-C. Duméril, 1853 (Calamariidae)

Synonyms: *Rhabdion* Lichtenstein & Martens, 1856 *(nomen emendatum), Rhabdophidium* Boulenger, 1894a *(nomen substitutum), Rabdophidium* – Palacky, 1898 *(nomen incorrectum),* and *Rhacophidium* – Boettger, 1900 *(nomen incorrectum).*

Type species: *Rabdion forsteni* A.-M.-C. Duméril, Bibron & Duméril, 1854a.

Distribution: Eastern Indonesia.

Sources: Zaher et al., 2009 and Koch, 2012.

1. *Rabdion forsteni* A.-M.-C. Duméril, Bibron & Duméril, 1854a. Erpet. Gén. 7(1): 116–119.

Types: Syntypes (2), MNHN 7210, a 399 mm specimen, and RMNH 46 (E.A. Forsten, 1840–1842).

Type locality: "iles Célèbes" [= Sulawesi, E Indonesia].

Distribution: Eastern Indonesia (Sulawesi), 1830 m.

Sources: Bleeker, 1857b, Boettger, 1900, Rooij, 1917c, Haas, 1950, Bosch, 1985, Malkmus, 1993a, 2000, Lang & Vogel, 2005 and Koch, 2012.

RAMPHOTYPHLOPS Fitzinger, 1843 (Typhlopidae)

Synonyms: *Typhlina* Wagler, 1830 *(nomen praeoccupatum), Pseudotyphlops* Fitzinger, 1843 *(nomen praeoccupatum), Pilidion* A.-M.-C. Duméril & ad Bibron, 1844 *(nomen substitutum), Typhlinalis* Gray, 1845 *(nomen substitutum), Rhamphotyphlops* Agassiz, 1847 *(nomen emendatum), Rhinophidion* Fitzinger, 1861 *(nomen nudum), Typhlira* – Jan, 1861a *(nomen incorrectum), Pilidium* A.C.L.G. Günther, 1872a *(nomen emendatum), Tychlina* – Grzimek, 1975 *(nomen incorrectum), Ramphotypholps* – H.M. Smith & Kohler, 1978 *(nomen incorrectum), Ramphothyphlops* – Ziegler, 2002 *(nomen incorrectum), Billmacordus* Hoser, 2012am *(nomen illegitimum), Funkityphlops* Hoser, 2012am *(nomen illegitimum), Johnwilsontyphlops* Hoser, 2012am *(nomen illegitimum), Maxhoserus* Hoser, 2012am *(nomen illegitimum), Oxytyphlops* Hoser, 2012am *(nomen illegitimum),* and *Piersontyphlops* Hoser, 2012am *(nomen illegitimum).*

Type species: *Typhlops multilineatus* Schlegel, 1839 *in* 1837–1844.

Distribution: Southeastern Asia and Indo-Australia The parthenogenetic *R. braminus* has been introduced into many areas worldwide.

Sources: Robb, 1960, 1966, McDowell, 1974a, Hahn, 1980a, Cogger et al., 1983a, Storr, 1981d, 1983, Cogger, 1992, Storr et al., 1986, S.K. Wilson & Knowles, 1988, Welch, 1988, Ehmann, 1992, Wallach, 1995a, 1998b, Wallach & Ineich, 1996, Greer, 1997, M.S. Khan, 1999b, McDiarmid et al., 1999 and Vidal et al., 2010.

Remarks: Official Generic Name no. 2156 *fide* Opinion 1207 (ICZN, 1982b).

1. *Ramphotyphlops acuticaudus* (W.C.H. Peters, 1877a). Mber. Königl. Akad. Wiss. Berlin 1877(6): 416–417, pl., figs. 2, 2a–2c. (*Typhlops acuticaudus*)

Synonym: *Typhlops acuticauda* Waite, 1981.

Type: Holotype, ZMB 9127, a 123 mm specimen (Kubary via Mus. Goddefroy).

Type locality: "Palaos" [= Palau Islands, Caroline Group, Federated States of Micronesia]

Distribution: Micronesia (Palau: Angaur, Babelthaup, Koror and Palau Is.), NSL–10 m.

Source: Crombie & Pregill, 1999.

2. *Ramphotyphlops adocetus* Wynn, Reynolds, Buden, Falanruw & Lynch, 2012. Zootaxa (3172): 41–42, figs. 2, 7a.

Type: Holotype, USNM 529971, a 373 mm male (O. Sirom, 15 July 1999).

Type locality: "Pasa Island, Ant Atoll, Caroline Islands (Federated States of Micronesia)."

Distribution: Micronesia (Ant Atoll: Pasa Is.), NSL.

3. *Ramphotyphlops albiceps* (Boulenger, 1898b). Ann. Mag. Nat. Hist. (7) 1(2): 124. (*Typhlops albiceps*)

Synonyms: *Typhlops albicep* – F. Werner, 1921a *(nomen incorrectum),* and *Typhlops malaisei* Rendahl, 1937.

Type: Holotype, BMNH 1946.1.10.50 (formerly BMNH 1897.10.8.17), a 128 mm specimen (Royal Siamese Museum, Bangkok).

Type locality: "Chantaboon, Siam" [= Chanthaburi, cen. Chanthaburi Prov., SE Thailand, 12°48'N, 102°10'E, elevation NSL–105 m].

Distribution: Southeastern Asia. Southern China (Hong Kong), Thailand (Chanthaburi, Pattani, Phang Nga), S Myanmar (Kayin), West Malaysia (Kedah,

R

Pahang, Perak, Selangor, Similan and Tioman Is.) and Singapore, NSL–1370 m.

Sources: E.H. Taylor, 1965, Tweedie, 1983, Lazell & Lu, 1990, Zhao & Adler, 1993, Wallach, 1996a, Manthey & Grossmann, 1997, Karsen et al., 1998, Grismer, 2005, 2011 and I. Das, 2010, 2011, 2012.

Remarks: Occurs in Vietnam *fide* Das (2011: 348).

4. *Ramphotyphlops angusticeps* (W.C.H. Peters, 1877a). Mber. Königl. Akad. Wiss. Berlin 1877(6): 417, pl., figs. 3, 3a–3c. (*Typhlops angusticeps*)

Synonym: *Typhlops olivaceus reduncus* T. Barbour, 1921a.

Type: Holotype, ZMB 9055, a 455 mm specimen (Cox via Mus. Godeffroy).

Type locality: "Neu-Caledonien" [= New Caledonia] (in error *fide* Roux, 1913: 135 and Bauer et al., 2002: 163).

Distribution: Southern Solomons (Guadalcanal, Malaita, Rennell and San Cristobal Is.), NSL–15 m.

Sources: Waite, 1918b and V. Tanner, 1948.

5. *Ramphotyphlops becki* (V. Tanner, 1948). Great Basin Nat. 9(1–2): 12, 15, figs. 4–5. (*Typhlops becki*)

Type: Holotype, BYU 7448, a 117 mm specimen (D.E. Beck & Ramsey, 30 Nov. 1944).

Type locality: "Tenaru River, northeast portion of Guadalcanal, Solomon Islands."

Distribution: Southern Solomons (Guadacanal Is.).

Source: Shea & Wallach, 2000.

Remarks: Previously a synonym of *R. willeyi*.

6. *Ramphotyphlops braminus* (Daudin, 1803d). Hist. Nat. Rept. 7: 279–280. (*Eryx braminus*)

Synonyms: *Tortrix russelii* Merrem, 1820, *Argyrophis bramicus* Gray, 1845 *(nomen emendatum), Argyrophis truncatus* Gray, 1845, *Onychocephalus capensis* A. Smith, 1846 *in* 1838–1849, *Ophthalmidium tenue* Hallowell, 1861, *Rhinophidion nicobaricum* Fitzinger, 1861 *(nomen nudum), Typhlops accedens* Jan, 1863b, *Typhlops euproctus* Boettger, 1882, *Typhlops bramineus* – A.B. Meyer, 1887 *(nomen incorrectum), Tortrix russellii* Boulenger, 1893a *(nomen emendatum), Typhlops limbrickii* Annandale, 1906, *Typhlops braminus arenicola* Annandale, 1906, *Typhlops braminus pallidus* Wall, 1909f, *Glauconia braueri* Sternfeld, 1910c, *Typhlops fletcheri* Wall, 1919b, *Typhlops khoratensis* E.H. Taylor, 1962, and *Typhlops pseudosaurus* Dryden & Taylor, 1969.

Type: Lectotype, specimen described and illustrated in P. Russell (1796: 48, pl. 43), designated herein.

Type locality: "Vizagapatam, coast of Coromandel, India" [= Vishakhapatnam, Andhra Pradesh State, SE India, 17°41'N, 83°13'E, elevation 25 m] via lectotype selection.

R

Distribution: Worldwide distribution. Saudi Arabia (Asir, Eastern, Jazan, Makkah, Riyadh), Bahrain (Muharraq Is.), UAR (Dubai), N Oman (Al Batinah, Muscat), W Yemen, Kuwait, Iran (Fars, Kerman, Khuzestan, Sistan va Baluchestan), Iraq, Afghanistan, Pakistan (Baluchistan, Khyber Pakhtunkhwa, Punjab, Sind), India (Andhra Pradesh, Arunachal Pradesh, Assam, Bihar, Gujarat, Haryana, Kaynataka, Kerala, Madhya Pradesh, Maharashtra, Orissa, Rajasthan, Uttarakhand, Uttar Pradesh, West Bengal, Andaman & Nicobars: Car Nicobar, Chowra, Great Nicobar, Katchall, Long, North Andaman, South Andaman and Tarasa, Barkuda, Cochin Willingdon, Laccadive and Neil Is.), Bangladesh (Hatiya Is.), Sri Lanka (Central, Eastern, North-Central, Northern, North-Western, Sabaragamuwa, Southern, Uva, Western), Madagascar (Antsiranana, Mahajanga, Toamasina, Toliara, Nosy Be Is.), Comoros (Anjouan, Mayotte and Mohéli Is.), Maldives (Girawa and Miladumadulu Is.), Mascarenes (Mauritius, Aigrettes, Fourneaux, Marginay, La Passe, Plate, Round, Reunion and Rodrigues Is.), Seychelles (Aride, Assumption, Bird, Cerf, Cousin, Cousine, Curieuse, Fregate, Ilot, La Digue, Mahé, Praslin, Silhouette), Nepal (Banke, Bara, Bardiya, Chitwan, Dhankuta, Jhapa, Kailali, Kanchanpur, Kapilbastu, Kaski, Kathmandu, Lalitpur, Nawalparasi, Nuwakot, Rupandehi, Sankhuwasabha, Saptari, Siraha, Sunsari, Surkhet, Udayapur), Bhutan (Chhukha, Punakha, Samtse), Thailand (Chon Buri, Chiang Mai, Nakhon Ratchasima, Nakhon Si Thammarat, Phang Nga, Phatthalung, Phra Nakhon Si Ayutthaya, Phuket, Prachin Buri, Prachuap Kiri Khan, Saraburi, Songkhla, Thon Buri, Udon Thani, Ho Kay, Phuket, Salanga and Samet Is.), Myanmar (Ayeyarwady, Mandalay, Sagaing, Shan, Yangon), Cambodia (Kampong Chhnang, Koh Kong, Phnom Penh, Preah Vihear, Siem Reap), Laos (Champasak, Vientiane), Vietnam (Bac Giang, Dong Nai, Gia Lai, Hai Duong, Hai Phong, Hanoi, Ha Tinh, Ho Chi Minh City, Kien Giang, Lang Son, Nghe An, Quang Binh, Quang Nam, Quang Ninh, Quang Tri, Son La, Tay Ninh, Thai Binh, Thai Nguyen, Thanh Hoa, Thua Trien-Hue, Vinh Phuc, Cat Ba, Condor, Cu Lao Phon Vong and Hon Nor Way Is.), West Malaysia (Johor, Kedah, Pahang, Pinang, Selangor, Aceh, Aur, Babi Besar, Dayang, Langkawi, Penang, Perhentian Besar, Sibu, Sibu Tengah, Tioman, Tulai and Tunas Selatan Is.), Singapore (Singapore and Ubin Is.), East Malaysia (Sabah, Sarawak), Brunei, Indonesia (Papua, Ambon, Aru Arch., Bali, Bangka, Batjan, Belitung, Buru, Buton, Butung, Flores, Halmahera, Hoga, Java, Kabaena, Kalimantan, Kai Arch., Komodo, Krakatau, Lomblen, Lombok, Madura, Maluku, Nias, Niasa Tenggara, Nila, Postiljon, Riau Arch., Sabalana, Saparua, Selayer, Seram, Sertung, Sulawesi, Sumatra, Sumba, Sumbawa, Timor, Ternate, Timor, Weh), Timor-Leste, China (Fujian, Guangdong, Guangxi, Guizhou, Hainan, Hong Kong, Hubei, Jiangxi, Sichuan, Yunnan, Zhejiang, Chek Lap Kok, Cheung Chau, Guan Yu, Hei Ling Chau, Kat O Chau, Kau

Yi Chau, Lamma, Lantau, Matsu, Ma Wan, Nan Ao, Ping Yu, Po Toi, Shek Kwu Chau, Soko, Stonecutters, Tai Lei, Tung Ping Chau, Yim Tin Tsai Is.), Japan (Kyushu, Ryukyus: Agunijima, Akajima, Akusekijima, Amamioshima, Amurojima, Aragusukujima, Chichijima, Kitadaitojima, Okinodaitojima, Fukajijima, Gishifujima, Gushikawajima, Gusukujima, Hachijojima, Hamahigajima, Hatejima, Haterumajima, Hatizyozima, Hyanzajima, Iejima, Iheyajima, Ikeijima, Ikemajima, Irabujima, Iriomotejima, Ishigakijima, Izenajima, Kakeromajima, Kamiyamajima, Kayamajima, Kerumajima, Kikaijima, Kikaigashima, Kobijima, Kodakarajima, Kohamajima, Kitakojima, Minamikojima, Korijima, Kubajima, Kudakajima, Kumejima, Kurimajima, Kuroshima, Maejima, Minnajima, Miyagijima, Miyakojima, Nakajima, Nakanokamishima, Nanseisyotou, Nohojima, Ogamijima, Ogasawara, Ohajima, Ohmura, Ojima, Okierabujima, Okinawajima, Sakishima, Senkaku, Sesokojima, Shimojishima, Takarajima, Taketomijima, Tanegashima, Taramajima, Tokashikijima, Tokunoshima, Tonakijima, Tsukenjima, Ukejima, Ukibarujima, Uotsurijima, Yabuchijima, Yaguchijima, Yakabijima, Yanahajima, Yonagunijima, Yorojima, Yoronjima and Zamamijima Is.), Macau (Coloane and Taipa Is.), Taiwan (Chiayi, Changhua, Hsinchu, Hualien, Ilan, Kaohsiung, Miaoli, Pingtung, Taichung, Tainan, Taipei, Taitung, Taoyüan, Yünlin, Chihmei and Lanyu Is.), Philippines (Aguyayan, Apo, Bantayan, Barit, Basilan, Batan, Bohol, Bongao, Borocay, Busuanga, Calauit, Calamian, Camiguin, Catanduanes, Cebu, Corregidor, Dalupiri, Gigante South, Grande, Guimaras, Ibuhos, Jolo, Lapinin Chico, Leyte, Luzon, Mactan, Marinduque, Masbate, Maybag, Mindanao, Mindoro, Negros, Pacijan, Palawan, Pamilacan, Panay, Panubulon, Polillo, Ponson, Samar, Semirara, Sibay, Sibuyan and Tintiman Is.), Papua New Guinea (East Sepik, Madang, Milne Bay, Morobe, Bougainville, Gizo, Karkar, Kiriwina, New Britain, and Nusa Tupe Is.), Solomons (Boku, Buin, Guadalcanal and Tulaghi Is.), Australia (Northern Territory, Western Australia, Bathurst, Christmas, Cook, Cocos-Keeling (Panjang, Selma, West, Melville and Thursday Is.), New Zealand), American Samoa (Tau and Tutuila Is.), Fiji (Taveuni and Viti Levu Is.), Hawaii (Hawaii, Kahoolawe, Kauai, Lanai, Maui, Molokai and Oahu Is.), Kiribati (Tarawa Is.), Loyalties (Maré Is., Marianas: Agrihan, Aguiguan, Alamagan, Anatahan, Guam, Pagan, Rota, Saipan, Sarigan and Tinian Is.), Marshalls (Enewetak, Jaluit, Kwajalein, Medren and Parry Is.), Federated States of Micronesia (Bulbul, Dorooleng, Giilab, Iaar, Kosrae, Lenger, Pasa, Ponhpei, Mwahnd Peidi, Piig, Sokehs, Soong, Ulithi, and Yap Is.), Midway (Sand Is., Nauru, Caledonia: Grand Terre, Noumea, Pia and Pins Is.), Palau (Babeldaob, Carp, Koror/Oreor, Malakal, Ngcheangel, Ngeaur, Ngedbus, Ngerekebesang and Ngetmeduch Is.), Vanuatu (Ambrym, Efaté and Espiritu Santo Is.), American Samoa (Ta'u and Tutuila Is.), Canaries (Grand Canary Is.), Cape Verde Santiago Is.), S Mauritania (Traza), Senegal (Dakar), Togo (Maritime), S Benin (Cotonou), Burkina Faso (Centre), S Ivory Coast (Abidjan, Lagunes), S Nigeria (Lagos), S Cameroon (Littoral), NW Gabon (Estuaire, Ogooué-Maritime), Central African Republic (Ombell-Mpoko), Libya (Al Jabal al Gharbi), NE Egypt (Cairo, Beni Suef, Ismaili, Marsa Matrouh), Equatorial Guinea (Annobon Is.), Congo, Gabon (Estuaire, Ogooué-Maritine), Somalia (Banaadir), Kenya (S Coast), Tanzania (Dar es Salaam, Lindi, Pwani, Pemba and Zanzibar Is.), Mozambique (Inhambane, Mogadishu, Sofala), South Africa (KwaZulu-Natal, Western Cape), USA (Alabama, Arizona, California, Florida (Big Pine, Marco, Sanibel and Stock Is.), Georgia, Hawaii, Louisiana, Massachusetts, Minnesota, North Carolina, Ohio, Texas, Virginia), Mexico (Aguascalientes, Baja California Sur, Campeche, Chiapas, Distrito Federal, Durango, Guanajuato, Guerrero, Hidalgo, Jalisco, México, Michoacán, Morelos, Nayarit, Nuevo León, Oaxaca, Puebla, Querétaro, Quintana Roo, Sinaloa, Sonora, Tabasco, Veracruz), Belize (Turneffe Is.), Guatemala (Guatemala), El Salvador (San Salvador), Honduras (Santa Bárbara, Islas de la Bahía, Utila Is.), Lesser Antilles (Anguilla, Guadeloupe, St. Barthélemy, St. Martin, St. Kitts), Turks & Caicos (Grand Turk Is.), Netherland Antilles (Aruba, Curaçao), and British West Indies (Barbados, Grand Cayman Is.), NSL–2000 (2600) m.

Sources: Wall, 1918a, Kuntz, 1963, E.H. Taylor, 1965, Dixon & Hendricks, 1979, Nussbaum, 1980, Storr, 1981d, Broadley, 1983, Chen & Zheng, 1983, Bauer, 1987, Gasperetti, 1988, Ernst & Barbour, 1989, Trape, 1990, Ota et al., 1991, Zhao & Adler, 1993, Wallach, 1993b, I. Das, 1994, Gaulke, 1994b, David & Vogel, 1996, Benitez-Gálvez, 1997, Censky & Hodge, 1997, Chirio & Blanc, 1997a, Chirio & Ineich, 1997, Tennant, 1997, L.A. Thomas, 1997, M.J. Cox et al., 1998, Mancilla & Ramirez-Bautista, 1998, Chan-ard et al., 1999, Crombie & Pregill, 1999, Nevárez, 1999b, Wallach, 1999b, 2009, Bauer & Sadlier, 2000, Ferner et al., 2001, Pauwels et al., 2001, Vásquez-Díaz & Quintero-Díaz, 2001, M.S. Khan, 2002, B.A. Savitzky et al., 2002, Cedeño-Vázquez et al., 2003, Echternacht & Burton, 2003, Kraus & Allison, 2004, Whitaker & Captain, 2004, Köhler et al., 2005, N. Khaire, 2006, A. Mori et al., 2006, Solano Zavaleta et al., 2006, J.-F. Trape & Ba, 2006, J.-F. Trape & Mané, 2006b, Ziegler et al., 2006, Jensen, 2007, Quijada-Mascareños & Canseco-Marquez, 2007, Quijada-Mascareños & Enderson, 2007, Borg, 2008, Broadley & Wallach, 2009, R.W. Henderson & Powell, 2009, V.S. Nguyen et al., 2009, Vesely & G. Köhler, 2009, Xiang & Li, 2009, Cravev & Shea, 2010, Salinas & Ramírez-Bautista, 2010, I. Das, 2010, 2012, Wood & Law, 2010, Hecht-Kardesz & Nickerson, 2011, Kaiser et al., 2011, McCranie, 2011a, Pauwels et al., 2011, Reynolds et al., 2011, Hernández-Ríos & Trejo-Perez, 2012, Wangyal, 2012, T. Hartmann et al., 2013, and Hughes, 2013.

7. *Ramphotyphlops cumingii* (Gray, 1845). Cat. Lizards Brit. Mus.: 133–134. (*Onychophis cumingii*)

Synonyms: *Typhlops longicauda* E.H. Taylor, 1919, *Typhlops rugosa* E.H. Taylor, 1919, *Typhlops dendrophis* E.H. Taylor, 1922a, and *Typhlops mindanensis* E.H. Taylor, 1922a.

Types: Syntypes (3), BMNH 1946.1.11.19–20 (H. Cuming, 1836–1840), and BMNH 1946.1.10.83 (E. Belcher, 1843–1846).

Type locality: "Philippines" and "Indian Ocean."

Distribution: Philippines (Bohol, Marinduque, Mindanao, Negros, Panay, Polillo).

Source: Ferner et al., 2001.

8. *Ramphotyphlops depressus* (W.C.H. Peters, 1880a). Mber. Königl. Akad. Wiss. Berlin 1880(2): 220. (*Typhlops depressus*)

Synonyms: *Typhlops aluensis* Boulenger, 1887g, *Typhlops philococos* F. Werner, 1898b, *Typhlops philococcus* – Waite, 1918b *(nomen incorrectum), Typhlops alluensis* – Kinghorn, 1928a *(nomen incorrectum), Typhlops buehleri* Hediger, 1933, and *Typhlops alvensis* – Brattstrom & Schwenkmeyer, 1951 *(nomen incorrectum).*

Type: Holotype, ZMB 9667, a 235 mm specimen.

Type locality: "papuanischen Insel Duke of York" [= Duke of York Is., Bismarck Archipelago, Papua New Guinea, 4°10'S, 152°28'E].

Distribution: Eastern Papua New Guinea (Ambitle, Bougainville, Duke of York, Manus, Matupi, New Britian, New Ireland, Pinipel, St. Matthias, Tambiu, Woodlark) and Solomons (Guadalcanal, Kolombangara, Malaita, Mbanika, New Georgia, Nggela, Pirumeri, Ranongga, Rossel, San Cristobal, Santa Ysabel, Shortland, Simbo, Three Sisters, Tulaghi), NSL–915 m.

Sources: McDowell, 1974a, Wallach, 1996b, Bauer et al., 2002 and Kraus & Allison, 2004.

9. *Ramphotyphlops exocoeti* (Boulenger, 1887j). Proc. Zool. Soc. London 55(3): 517. (*Typhlops exocoeti*)

Synonym: *Typhlops capensis* Rendahl, 1918.

Type: Lectotype, BMNH 1946.1.11.78 (formerly BMNH 1887.4.25.3), a 350 or 230 mm specimen, designated by Wells & Wellington (1985: 40).

Type locality: "Christmas Island, Indian Ocean," via lectotype selection.

Distribution: Australia (Christmas Is.), NSL–360 m.

Sources: Cogger et al., 1983b, Wallach, 2003 and Wallach & Glaw, 2009.

10. *Ramphotyphlops flaviventer* (W.C.H. Peters, 1864a). Mber. Königl. Akad. Wiss. Berlin 1864(4): 271–272. (*Typhlops flaviventer*)

Type: Holotype, ZMB 5029, a 320 mm specimen (K.E. von Martens, March 1862–Dec. 1864).

Type locality: "Ternate" [= Ternate Is., Maluku, E Indonesia, bet. 0°45'–52'N and 127°18'–23'E].

Distribution: Eastern Indonesia (Maluku: Batjan, Halmahera, Morotai, Ternate, Tidore).

Sources: Wallach, 1996b and Wynn et al., 2012.

11. *Ramphotyphlops hatmaliyeb* Wynn, Reynolds, Buden, Falanruw & Lynch, 2012. Zootaxa (3172): 46–47, figs. 5, 7c.

Type: Holotype, USNM 558300, a 331 mm female (M. Falanruw, 4 Aug. 2007).

Type locality: "Giilab (= Gielap, Gielop) Island, Ulithi, Caroline Islands (Federated States of Micronesia)."

Distribution: Micronesia (Ulithi: Bulbul, Dorooleng, Giilab, Piig and Soong Is.).

12. *Ramphotyphlops lineatus* (Schlegel, 1839 *in* 1837–1844). Abbild. Amph.: 39, pl. 32, figs. 32–34. (*Typhlops lineatus*)

Synonyms: *Typhlops lineatus* Fitzinger, 1826a *(nomen nudum), Typhlops lineatus* F. Boie, 1827 *(nomen nudum), Acontias lineatus* Wagler, 1830 *(nomen nudum), Typhlops lineatus acontias* Gray *in* Griffith & Pidgeon, 1831 *(nomen nudum), Typhlina lineata* Fitzinger, 1843, *Typhlops lineatus sumatrana* F. Werner, 1896b, and *Typhlops lineatus numatrana* – Boulenger, 1897j *(nomen incorrectum).*

Types: Syntypes (2), RMNH 3696a–b, longest syntype 273 mm (C.G.C. Reinwardt, 1816–1821).

Type locality: "Java" [SW Indonesia]

Distribution: Southeast Asia and Greater Sundas. Cambodia, Laos, Vietnam, S Thailand (Narathiwat, Pattani), West Malaysia (Peark, Pinang, Selangor), Singapore, East Malaysia (Sarawak) and W Indonesia (Bangka, Java, Kalimantan, Nias, Sumatra), NSL–1420 m.

Sources: T. Barbour, 1912, E.H. Taylor, 1965, McDowell, 1974, David & Vogel, 1996 and I. Das, 2010, 2012.

Remarks: RMNH 3692 also a syntype *fide* Cox et al. (2013: 11).

13. *Ramphotyphlops lorenzi* (F. Werner, 1909b). Mitt. Naturhist. Mus. Hamburg 26: 209. (*Typhlops lorenzi*)

Type: Holotype, ZMH, a 337 mm specimen, (T. Lorenz, 1901), destroyed in July 1943 during World War II.

Type locality: "Insel Poeloe (Miang besar), etwas nördlich von Äquator, Ostküste von Borneo" [= Palau Miang-Besar somewhat N of the equator on the E coast of Borneo, 00°43'N, 118°00'E].

Distribution: Indonesia (Kalimantan: Miang Besar). Known only from type locality.

R

Sources: Rooij, 1917, McDowell, 1974a, Stuebing, 1991 and I. Das, 2010.

14. *Ramphotyphlops mansuetus* (T. Barbour, 1921). Proc. New England Zool. Club 7: 108–109, pl. 6. (*Typhlops cumingii mansuetus*)

Type: Holotype, MCZ 14270, a 154 mm specimen (W.M. Mann, 19 May–24 Nov. 1916).
Type locality: "Keri Keri, San Cristóbal, Solomon Islands" [= Kirakira, Makira Is., Makira-Ulawa Prov., Solomon Islands, 10°27'S, 161°54'E, elevation 15 m].
Distribution: Papua New Guinea (Bougainville Is.) and Solomons (Makira Is.), NSL–1065 m.
Sources: McDowell, 1974a, Iskandar & Colijn, 2001, Wallach, 2003 and McCoy, 2006.
Remarks: Previously a synonym of *Anilios affinis.*

15. *Ramphotyphlops marxi* (Wallach, 1993b). Raffles Bull. Zool. 41(2): 265–270, figs. 1–2. (*Typhlops marxi*)

Type: Holotype, FMNH 96520, a 180 mm female (D.S. Rabor, 1957).
Type locality: "Tarabucan (12°13'N, 124°35'E), four miles southeast of spur of Sigarag Mountains (which are 13 miles south of Catarman), Matuguinao Municipality, extreme northern Western Sámar Province, northwestern Sámar Island, northeastern Visayas Group, east-central Philippines."
Distribution: Central Philippines (Samar). Known only from type locality.
Source: Wallach, 2003.

16. *Ramphotyphlops multilineatus* (Schlegel, 1839 *in* 1837–1844). Abbild. Amph.: 40, pl. 32, figs. 39–42. (*Typhlops multilineatus*) (*nomen corrigendum*)

Synonyms: *Typlops multilineatus* Schlegel, 1839 *in* 1837–1844 *(nomen incorrigendum), Typhlops multilineatus* – A.-M.-C. Duméril & Bibron, 1844 *(nomen corrigendum), Onchophis multicarinatus* Gray, 1845, *Typhlops bibroni* Jan, 1863b, and *Typhlina multilineata* McDowell, 1974a.
Type: Holotype, MNHN 1067, a 370 mm specimen (S. Müller, 1825–1827).
Type locality: "Westküste von Neu-Guinea" [= W coast of New Guinea, Papua Prov., E Indonesia]. Emended to Port-Dory [= Manokwari, West Papua, E Indonesia] *fide* MNHN catalogue.
Distribution: New Guinea. Eastern Indonesia (N Papua, Angadi, Kai Besar, Kai Cecil, Salawati Is.) and cen. Papu New Guinea (Chimbu), NSL–870 m.
Sources: Brongersma, 1934, Haas, 1950, McDowell, 1974a and O'Shea, 1996.
Remarks: Official Specific Name no. 2798 *fide* Opinion 1207 (ICZN, 1982b).

17. *Ramphotyphlops olivaceus* (Gray, 1845). Cat. Lizards Brit. Mus.: 133. (*Onychophis olivaceus*)

Synonyms: *Typhlops ligorostris* M.A. Smith & Procter, 1921, and *Typhlina olivacea* McDowell, 1974a.
Type: Holotype, BMNH 1946.1.10.57 (H. Cuming, 1836–1839).
Type locality: "Philippines."
Distribution: East Indies. East Malaysia (Sarawak), Indonesia (Ambon, Ewaafi, Kalimantan, Misool, Sangihe, Seram, Sulawesi) and Philippines (Babuan, Samar, Dalutan), NSL–35 m.
Sources: E.H. Taylor, 1922, McDowell, 1974a, Lang & Vogel, 2005 and Koch, 2012.
Remarks: Australian record doubtful *fide* McDowell (1974a: 43).

18. *Ramphotyphlops ozakiae* Wallach *in* Niyomwan, 1999. Sp. Divers. Blind Snakes Thailand: 14, 32, 41, 42, 47, 57, 79, 96, 107, 118, 130, 140, figs. 4.1, 4.3D, 4.5D, 4.8H.

Type: Holotype, FMNH 180007 (formerly WRH 3679), a 158 mm male (W.R. Heyer, 17 Nov. 1969).
Type locality: "NE Thailand." Corrected herein to Sakaerat Experimental Station, Amphoe Pak Thong Chai, Nakhon Ratchasima Prov., SE Thailand, 14°43'N, 102°01'E, elevation 200 m.
Distribution: Thailand (Nakhon Ratchasima) and W Indonesia (Sumatra [ZMUC 52174]), 200–1100 m.
Sources: Niyomwan et al., 2001 and Wallach (pers. data).
Remarks: Uncertain if the inadvertent description by Niyomwan (mistakenly assuming the species had already been published by Wallach) is valid under the Rules.

19. *Ramphotyphlops similis* (Brongersma, 1934). Zool. Meded. 17(3–4): 192–193, figs. 22–24. (*Typhlops similis*)

Type: Holotype, RMNH 6295, a 235 mm specimen (C.B.H. von Rosenberg, 1870).
Type locality: "Doré, New Guinea" [Manokwari, N West Papua Prov., E Indonesia, 0°52' S, 134°04'E, elevation NSL].
Distribution: Eastern Indonesia (N West Papua), NSL. Known only from type locality.
Sources: O'Shea, 1996 and Wallach, 1996b.

20. *Ramphotyphlops suluensis* (E.H. Taylor, 1918a). Philippine J. Sci. 13D(5): 257–259, 265, fig. 11a–c. (*Typhlops suluensis*)

Type: Holotype, BSCM 2001, a 340 mm specimen (E.H. Taylor, 2 Oct. 1917).
Type locality: "Bunuan, Tapian Group, Sulu Archipelago, Philippine Islands" [= Bubuan Island, Sulu Archipelago, SW Philippines, 5°25'N, 120°35'E, elevation NSL].

Distribution: Philippines (Basilan, Bubuan and Sibutu Is.), NSL–20 m.

Remarks: Previously a synonym of *R. olivaceus.*

21. *Ramphotyphlops supranasalis* (Brongersma, 1934). Zool. Meded. 17(3–4): 191, fig. 25. (*Typhlops supranasalis*)

Type: Holotype, RMNH 6291, a 301 mm specimen (Bernstein, 1866).

Type locality: "Salawatti" [= Salawati Island, West Papua Prov., E Indonesia, bet. 0°53'–1°22'S and 130°38'–131°05'E].

Distribution: Eastern Indonesia (West Papua: Salawati). Known only from type locality.

Sources: O'Shea, 1996 and Wallach, 1996b.

22. *Ramphotyphlops willeyi* (Boulenger *in* Willey, 1900h). Zool. Results (5): 603–604, 4 figs. (*Typhlops willeyi*)

Type: Holotype, BMNH 1946.1.10.80 (formerly BMNH 1900.5.29.1), a 195 mm specimen (A. Willey, 1895–1897).

Type locality: "Lifu, Loyalty Islands" [= Lifou Is., Loyalty Is., New Caledonia, bet. 20°41'–21°11'S and 167°01–28'E].

Distribution: New Caledonia (Loyalties: Lifou, Maré Is.), NSL–60 m.

Sources: Roux, 1913, Wallach, 1995a, Bauer & Sadlier, 2000, Shea & Wallach, 2000 and Wynn et al., 2012.

REGINA Baird & Girard, 1853 (Natricidae)

Synonyms: *Liodytes* Cope, 1885a, *Reginaus* A.L. Herrera, 1899 *(nomen emendatum),* and *Mariolisus* Hoser, 2012h *(nomen illegitimum).*

Type species: *Coluber septemvittatus* Say, 1825.

Distribution: Extreme SE Canada and E USA.

Fossil records: Upper Pliocene to upper Pleistocene of E USA. Fossils unassigned to species include upper Pleistocene (Rancholabrean II) of USA (Georgia).

Sources: Jan, 1865d, H.M. Smith & Huheey, 1960, Rossman, 1963b, 1985, Meylan, 1982, R. Price, 1983, Ernst & Barbour, 1989, Conant & Collins, 1991, Holman, 1995a, 2000a, Dwyer & Kaiser, 1997, Alfaro & Arnold, 2001, Ernst, 2002a, Gibbons & Dorcas, 2004 and Zaher et al., 2009.

Remarks: *Regina rigida* and *R. alleni* form a monophyletic clade with *Seminatrix fide* Alfaro & Arnold, 2001, Pyron et al., 2011 and Zaher et al., 2012 but morphological characters support the four species of *Regina* as a single taxon *fide* Rossman, 1963a and 1985.

1. *Regina alleni* (S.W. Garman, 1874). Proc. Boston Soc. Nat. Hist. (1874–1875) 17(2): 92, 3 figs. (*Helicops alleni*)

Synonym: *Liodytes alleni lineapiatus* Auffenberg, 1950.

Type: Holotype, MCZ 2255, a 650 mm female (J.A. Allen, March 1869).

Type locality: "near Jacksonville, Florida" [USA] (in error). Corrected to Hawkinsville, 2 miles S of Crow's Bluff, Lake County, *fide* R.E. Hellman *in* K.P. Schmidt (1953a: 178).

Distribution: Extreme SE USA (Florida, ext. S Georgia), NSL–75 m.

Fossil records: Lower Pleistocene (Irvingtonian I) of USA (Florida), and upper Pleistocene (Rancholabrean II) of USA (Florida)

Sources: Auffenberg, 1950, 1963, Godley, 1980, 1982, Ashton & Ashton, 1988, Tennant, 1997 and Dorcas et al., 2003.

Remarks: Data tag on type reads St. Johns River, Hawkinsville and MCZ catalogue reads Fla., St. Johns River, Hawkinsville. A member of *Seminatrix fide* Zaher et al., 2012.

2. *Regina grahamii* Baird & Girard, 1853. Cat. No. Amer. Rept., Serp. 1: 47–48.

Synonym: *Natrix grohamii* – F.C. Baker & Woodruff *(nomen incorrectum).*

Type: Holotype, USNM 1376, a 273 mm specimen (J.H. Clark [U.S.-Mex. Bound. Surv.], 1850–1852, via J.D. Graham).

Type locality: "Rio Salado, Tex." [Texas, USA]. Emended to Salado Creek [= 4 miles from San Antonio, Bexar Co. *fide* Cochran, 1961: 208], Bell Co., Texas, USA *fide* K.P. Schmidt (1953a: 158).

Distribution: Central USA (Arkansas, Illinois, Iowa, E Kansas, Louisiana, W Mississippi, Missouri, ext. SE Nebraska, Oklahoma, E Texas), NSL–600 m.

Fossil records: Upper Pliocene (Blancan V) of USA (Nebraska), middle Pleistocene (Irvingtonian II) of USA (Nebraska), and upper Pleistocene (Rancholabrean I, II) of USA (Kansas).

Sources: Hall, 1969, Holman, 1972, Dundee & Rossman, 1989, Hardy, 1997 and Werler & Dixon, 2000.

Remarks: This species forms a monophyletic clade with *Nerodia cyclopion* and *N. floridana fide* Zaher et al., 2012.

3. †*Regina intermedia* Meylan, 1982. Bull. Florida St. Mus. (Biol. Sci.) 27(3): 40–43, figs. 10a–c, j.

Type: Holotype, UF 26383, fragmentary left dentary (J. Klein & R. Martin, 1967–1974).

Type locality: "Inglis IA, a sinkhole in the Ocala limestones on the north bank of the now defunct Cross Florida Barge Canal in Section 8, R.16 E., T.17 S., Citrus County, Florida; Inglis IA Local Fauna, early Irvingtonian, early Pleistocene."

Distribution: Lower Pleistocene (Irvingtonian: 0.40–1.9 mya) of USA (Florida). Known only from type locality.

4. *Regina rigida* (Say, 1825). J. Acad. Nat. Sci. Philadelphia 4(2): 239–240. (*Coluber rigidus*)

Synonyms: *Tropidonotus leberis unicolor* Lichtenstein & Martens, 1856, *Natrix rigida deltae* Huheey, 1959, and *Natrix rigida sinicola* Huheey, 1959.
Type: Holotype, ANSP 16795, a 523 mm specimen.
Type locality: "southern states" [USA]. Restricted to Charleston, South Carolina *fide* K.P. Schmidt (1953a: 158).
Distribution: Southeastern USA (S Alabama, Arkansas, N Florida, S Georgia, Louisiana, S Mississippi, SE North Carolina, ext. SE Oklahoma, E South Carolina, E Texas, SE Virginia), NSL–150 m.
Sources: Huheey, 1959, Dundee & Rossman, 1989, Palmer & Braswell, 1995, Tennant, 1997, and Werler & Dixon, 2000.
Remarks: A member of *Seminatrix fide* Zaher et al., 2012.

5. *Regina septemvittata* (Say, 1825). J. Acad. Nat. Sci. Philadelphia 4(2): 240–241. (*Coluber septemvittatus*)

Synonym: *Natrix septemvittata mabila* Neill, 1963.
Types: Syntypes (3), ANSP, one syntype 251 mm (R. Haines & T. M'Euen), and WHC, a 582 mm specimen (W. Hyde).
Type locality: "Germantown" and "near Philadelphia, Pennsylvania" [USA].
Distribution: Extreme SE Canada (SW Ontario) and E USA (Alabama, cen. Arkansas, Delaware, NW Florida, N Georgia, NE Illinois, Indiana, Kentucky, Maryland, Michigan, Mississippi, ext. S Missouri, W New Jersey, W New York, W North Carolina, Ohio, ext. SE Oklahoma, W and S Pennsylvania, NW South Carolina, Tennessee, Virginia, West Virginia, SE Wisconsin), NSL–760 m.
Fossil records: Upper Pleistocene (Rancholabrean II) of USA (Ohio, Pennsylvania, Virginia).
Sources: J. Wood, 1949, Conant, 1960, Logier & Toner, 1961, Froom, 1967, 1972, Spangler & Mount, 1969, Branson & Baker, 1973, 1974, Butler et al., 1980, Trauth, 1991a, Palmer & Braswell, 1995, J. Harding, 1997, Tennant, 1997 and Ernst, 2002b.
Remarks: Official Specific Name no. 1814 *fide* Opinion 632 (ICZN, 1962). Types not in ANSP *fide* Malnate, 1971. This species is sister taxon to *Tropidoclonion lineatum fide* Zaher et al., 2012.

RENA Baird & Girard, 1853 (Leptotyphlopidae)

Synonym: *Renaus* A.L. Herrera, 1899 *(nomen emendatum)*.
Type species: *Rena dulcis* Baird & Girard, 1853.
Distribution: Southwestern USA and Latin America.
Fossil records: Pleistocene of SW USA.
Sources: Klauber, 1940a, J.A. Peters & Orejas-Mirando, 1970, Hahn, 1980a, Wallach, 1998b, McDiarmid et al., 1999 and Pinto, 2010.

1. *Rena bressoni* (E.H. Taylor, 1939c). Copeia 1939(1): 5–6, pl. 1, figs. 7–8. (*Leptotyphlops bressoni*)

Type: Holotype, UIMNH 25076 (formerly EHT-HMS 5247 & EHT-CC 12), a 261–265 mm specimen (J.R. Bresson, 1935).
Type locality: "Hda. El Sabino, Uruapan, Michoacán, Mexico" [= Hacienda El Sabino, near Uruapan (19°25'N, 102°03'W, elevation 1625 m), Michoacán, SW México].
Distribution: Southwestern México (W Michoacán), 490–1625 m.
Sources: Klauber, 1940a, H.M. Smith & Taylor, 1945, Duellman, 1961, 1965b, Dixon & Vaughn, 2003 and Pinto, 2010.

2. *Rena dissecta* (Cope, 1896a). Amer. Nat. 30(357): 753. (*Glauconia dissecta*)

Synonym: *Leptotyphlops dulius supraocularis* W.W. Tanner, 1985.
Type: Holotype, ANSP 10752, a 235 mm specimen (E.D. Cope).
Type locality: "at the silver mines at Lake Valley, southern New Mexico" [= Lake Valley, Sierra County, S New Mexico, USA, 36°05'N, 108°09'W, elevation 1800 m].
Distribution: Southwestern USA (SE Arizona, SE Colorado, ext. S Kansas, New Mexico, Oklahoma, W Texas) and N México (Chihuahua, Coahuila, ext. NE Sonora), 215–2170 m.
Sources: Klauber, 1940a, H.M. Smith & Taylor, 1945, Hibbard, 1964, Fowlie, 1965, Hahn, 1979c, H.M. Smith & Chiszar, 1993, Degenhardt et al., 1996, Werler & Dixon, 2000, Dixon & Vaughn, 2003, Ernst & Ernst, 2003 and Lemos-Espinal & Smith, 2007a–b.
Remarks: A synonym of *Leptotyphlops myopicus fide* Pinto (2010: 87)

3. *Rena dugesii* (Bocourt, 1881). Bull. Soc. Philom. Paris (7) 5(2): 81–82. (*Catodon dugesii*)

Synonym: *Stenostoma tenuiculum* S.W. Garman, 1884.
Type: Syntypes (2), MNHN 1651a–b, longest syntype 229 mm (A.A.D. Dugès, 1853–1881).
Type locality: "Colima (Mexique)" [= Colima, Colima State, SW México, 19°15'N, 103°43'W, elevation 490 m].
Distribution: Western México (SW Chihuahua, Colima, Jalisco, Sinaloa, SE Sonora), NSL–2000 m.
Sources: Klauber, 1940a, E.H. Taylor, 1940d, Bogert & Oliver, 1945, H.M. Smith & Taylor, 1945, Hardy &

R

McDiarmid, 1969, Hahn, 1979d, Webb, 1984, García & Ceballos, 1994 and Ramírez-Bautista, 1994.

Remarks: Supplemental original description in A.H.A. Duméril et al. (1870–1909: 507–508, pl. 29, figs. 9, 9a–c, pl. 30, fig. 4).

4. *Rena dulcis* Baird & Girard, 1853. Cat. No. Amer. Rept., Serp. 1: 142–143.

Synonym: *Stenostoma rubellum* S.W. Garman, 1884.

Type: Holotype, USNM 7296, a 186–191 mm male (J.H. Clark [U.S.-Mex. Bound. Surv.], 1850–1852, via J.D. Graham).

Type locality: "Between San Pedro and Camanche Springs, Tex." [= between San Pedro camp (31°54'N, 110°12'W) or San Pedro springs (31°51'N, 110°12'W) and Comanche Springs (30°53'N, 102°53'W, elevation 900 m), Pecos Co., W Texas, USA].

Distribution: Central USA (SE Colorado, S Oklahoma, cen. Texas) and NE México (Coahuila, N Nuevo León, N Tamaulipas), 10–1525 m.

Sources: Yarrow, 1882, Klauber, 1940a, H.M. Smith & Taylor, 1945, Hahn, 1979c, H.M. Smith & Chiszar, 1993, Degenhardt et al., 1996, Bartlett & Tennant, 2000, Moeller et al., 2000, Werler & Dixon, 2000, Dixon & Vaughn, 2003, J.C. Murphy & Schlager, 2003, Lemos-Espinal & Smith, 2007 and Pinto, 2010.

Remarks: Holotype was sent to Smithsonian in 1852 by J.D. Graham, who was listed as collector by Baird & Girard (1853: 143) and Cochran (1961: 209) but corrected by Baird (1859a: 24) and Degenhardt et al., (1996: 257). Type locality restriction to Comanche, Comanche Co., Texas *fide* H.M. Smith & Taylor (1950a: 361) invalid as it is outside the route of the US-Mex. Bound. Surv. (Bartlett, 1854; Emory, 1857).

5. *Rena humilis* Baird & Girard, 1853. Cat. No. Amer. Rept., Serp. 1: 143.

R

Synonyms: *Glauconia boettgeri* F. Werner, 1899a, *Leptotyphlops humilis cahuilae* Klauber, 1931b, *Leptotyphlops humilis slevini* Klauber, 1931b, *Leptotyphlops humilis utahensis* V. Tanner, 1938, *Leptotyphlops humilis levitoni* R.W. Murphy, 1975, *Leptotyphlops humilis lindsayi* R.W. Murphy, 1975, and *Leptotyphlops humilis chihuahuaensis* W.W. Tanner, 1985.

Type: Holotype, USNM 2101, a 152 mm specimen (J.L. Le Conte, 1850).

Type locality: "Valliecitas, Cal." [= Vallecito, San Diego Co., S California, USA, 38°05'N, 120°28'W, elevation 535 m]. Restricted to vicinity of Vallecito, E San Diego Co., California, USA *fide* Klauber (1931b: 343) and to the upper Sonoran Life Zone north or west of the Vallecito area *fide* Brattstrom (1953b: 180).

Distribution: Southwestern USA (S Arizona, S California, S Nevada, SW Texas, SW Utah) and N México (Baja California Norte, Baja California Sur, SW Chihuahua, Sonora, Carmen, Cedros, Cerralvo, Danzante, San Marcos, Santa Catalina and Santa Cruz Is.), 75 m BSL–1575 m.

Fossil records: Upper Pleistocene (Rancholabrean II) of USA (Arizona).

Sources: Klauber, 1931b, 1940a, R.W. Murphy, 1975, Hahn, 1979d, Stebbins, 1985, Degenhardt et al., 1996, P.R. Brown, 1997, Recchio, 1997, H.M. Smith et al., 1998, 2005, Grismer, 1999, 2002, Bartlett & Tennant, 2000, Werler & Dixon, 2000 and Pinto, 2010.

6. *Rena iversoni* (H.M. Smith, Breukelen, Auth & Chiszar, 1998). SW Nat. 43(4): 437–439, fig. 2a. (*Leptotyphlops dulcis iversoni*)

Type: Holotype, UF 105870 (formerly UF 81JI557), a 196 mm specimen (J.B. Iverson, 11 May 1981).

Type locality: "vicinity of Hwy 101, 23.7 km SW Río San Marcos (in turn 8.1 km SW Cd. Victoria), Tamaulipas, México, 36.5 km NE of Jaumave" [= ca. 23°35'N, 99°15'W, elevation ca. 1000 m].

Distribution: Northeastern México (cen. Tamaulipas), 375–1000 m.

Sources: Dixon & Vaughn, 2003 and Pinto, 2010.

Remarks: Elevated to species status *fide* Pinto (2010: 113).

7. *Rena maxima* (Loveridge, 1932b). Proc. Biol. Soc. Washington 45: 151–152. (*Leptotyphlops maximus*)

Type: Holotype, MCZ 33604, a 289–300 mm male (W.W. Brown, Jan.–June 1932).

Type locality: "between 4,000 and 6,000 feet at Chilpancingo, State of Guerrero, Mexico" [= Chilpancingo, cen. Guerrero State, S Mexico, 17°33'N, 99°30'W, elevation 1220–1830 m].

Distribution: Southern México (Guerrero, S México, Morelos, Oaxaca, S Puebla), 90–1850 m.

Sources: E.H. Taylor, 1939c, Klauber, 1940a, H.M. Smith & Taylor, 1945, Hahn, 1980b, Dixon & Vaughn, 2003, Woolrich-Piña et al., 2005, Canseco-Márquez & Gutiérrez-Mayén, 2006, 2010, Martínez-Coronel & López-Hernández, 2008 and Pinto, 2010.

8. *Rena myopica* (S.W. Garman, 1884). Mem. Mus. Comp. Zool. (1883) 8: 130–131. (*Stenostoma myopicum*)

Type: Lectotype, FMNH 73386, a 137 mm specimen (E. Palmer, 1879), designated by Pinto (2010: 87).

Type locality: "Savineto near Tampico, Tamaulipas, México" [= ca. 22°15'N, 97°52'W, elevation 25 m] via lectotype selection.

Distribution: Northeastern México (Chihuahua, Coahuila, Hidalgo, Nuevo León, Puebla, San Luis Potosí, NE Sonora, Tamaulipas), 25–2175 m.

Sources: H.M. Smith & Taylor, 1945, E.H. Taylor, 1949, Hahn, 1979c, H.M. Smith et al., 1998, Auth et al., 2000,

Dixon & Vaughn, 2003, Pinto, 2010 and Ramírez-Bautista et al., 2010.

9. *Rena segrega* (Klauber, 1939). Trans. San Diego Soc. Nat. Hist. 9(14a): 67–68. (*Leptotyphlops humilis segregus*)

Type: Holotype, USNM 103670, a 313 mm specimen (T.F. Smith, 11 Aug. 1936).
Type locality: "Chalk Draw, Brewster County, Texas" [= 15 mi. SW Marathon (30°12'N, 103°15'W, elevation 1245 m) and 65 mi. S Alpine (30°22'N, 103°40'W, elevation 1375 m), Brewster Co., W Texas, elevation ca. 1200 m].
Distribution: Southwestern USA (SE Arizona, S New Mexico, SW Texas) and N Mexico (N Chihuahua, W Coahuila), 700–1525 m.
Sources: Klauber, 1940a, H.M. Smith & Taylor, 1945, Lemos-Espinal & Smith, 2007a–b and Pinto, 2010.

RHABDOPHIS Fitzinger, 1843 (Natricidae)

Synonyms: *Tropinotus* Kuhl & Hasselt, 1822a *(partim, nomen nudum), Tropidonotus* H. Boie, 1826 *(partim), Tropedinotus* – Holbrook, 1842a *(nomen incorrectum), Steirophis* Fitzinger, 1843, *Stirophis* Agassiz, 1847 *(nomen emendatum), Hydrophilophis* P. Schmidt, 1852, *Tropidonatus* – Higgins, 1873 *(nomen incorrectum), Trepidonotus* – Lessona, 1877 *(nomen incorrectum), Tropidenotus* – Bedriaga, 1883 *(nomen incorrectum), Bothrodytes* Cope, 1886c, *Tropidonotis* – Velasco, 1890d *(nomen incorrectum), Tropinodotus* – Phisalix, 1922 *(nomen incorrectum), Tropitonotus* – Mertens, 1921b *(nomen incorrectum), Topedonotus* – Roddy, 1928 *(nomen incorrectum), Rhabdophes* – Taub, 1965 *(nomen incorrectum), Rhadophis* – Marx & Rabb, 1972 *(nomen incorrectum), Tropidontus* – Tiedemann & Häupl, 1980 *(nomen incorrectum), Nuchisulcophis* Mahendra, 1984, and *Rabdophis* – Yao, 2012 *(nomen incorrectum).*
Type species: *Tropidonotus subminiatus* Schlegel, 1837.
Distribution: Southeastern and E Asia and East Indies.
Fossil record: Subrecent.
Sources: Stejneger, 1907a, Rooij, 1917, C.H. Pope, 1935, M.A.Smith, 1943c, Haas, 1950, Malnate, 1960a, McDowell, 1961, Zhao & Adler, 1993, I. Das, 2010, 2012 and Cadle, 2011.

1. *Rhabdophis adleri* Zhao, 1997a. Asiatic Herpetol. Res. 7: 168–169, fig. 1.

Type: Holotype, CIB 64-III-5917, a 790 mm male (native, 10 June 1964).
Type locality: "Dali Village in Mt Diaoluo, Lingshui Co., Hainan Prov., China, 225 meters."
Distribution: Southern China (Hainan), 60–800 m.
Source: Zhao et al., 2004.

2. *Rhabdophis angelii* (Bourret, 1934b). Bull. Gén. Instr. Publ. Hanoi 14(8): 151–152, fig. 1. (*Natrix angelii*)

Type: Holotype, MNHN 1938.123 (formerly RLB M.295), a 429 mm specimen (Deloustal).
Type locality: "Station d'altitude du Tam-Dao, située à 900 m d'altitude dans la province de Vinh-Yên, au bord même du delta tonkinois, sur des montagnes du Tonkin, l'Indochine française" [= Tam Dao, Vinh Phuc Prov., N Vietnam, 21°27'N, 105°39'E, elevation 900 m].
Distribution: Northern Vietnam (Thai Nguyen, Vinh Phuc), 450–900 m.
Sources: Bourret, 1935c, 1936b, 1939c–d, V.S. Nguyen et al., 2009 and I. Das, 2010.

3. *Rhabdophis auriculatus* (A.C.L.G. Günther, 1858). Cat. Colub. Snakes Brit. Mus.: 80. (*Tropidonotus auriculatus*)

Synonym: *Rhabdophis auriculata myersi* Leviton, 1970c.
Type: Holotype, BMNH 1946.1.15.16, a 470 mm female (H. Cuming, 1836–1840).
Type locality: "Philippine Islands." Restricted to Mt. Apo, Davao Prov., Mindanao Is., S Philippines *fide* Leviton (1970c: 356).
Distribution: Philippines (Basilan, Bohol, Dinagat, Leyte, Mindanao, Samar), 130–1980 m.
Sources: E.H. Taylor, 1922a, 1923, Leviton, 1970c, Ross & Lazell, 1991 and B.E. Smith, 1993.
Remarks: Luzon record needs confirmation *fide* Leviton (1970c: 357); it may represent the Mindanao population, in which case this species is restricted to S Philippines.

4. *Rhabdophis barbouri* (E.H. Taylor, 1922c). Philippine J. Sci. 21(3): 291–293. (*Natrix barbouri*)

Synonym: *Leptophis vertebralis* A.-M.-C. Duméril, Bibron & Duméril, 1854a *(nomen rejiciendum).*
Type: Holotype, CAS 61552 (formerly EHT 939), an 803 mm male (E.H. Taylor, 26 April 1920).
Type locality: "Balbalan, Kalinga, Luzon, Philippine Islands" [= Balbalan, Kalinga Subprov., Mountain Prov., Luzon Is., Philippines *fide* Leviton, 1959b: 415].
Distribution: Northern Philippines (Luzon), NSL–1500 m.
Sources: E.H. Taylor, 1922d, Leviton, 1959b and Malnate & Underwood, 1988.
Remarks: Official Specific Name no. 2106 *fide* Opinion 756 (ICZN, 1966a).

5. *Rhabdophis callichromus* (Bourret, 1934b). Bull. Gén. Instr. Publ. Hanoi 14(8): 155–156. (*Natrix chrysarga callichroma*)

Synonym: *Natrix auchenia* M.A. Smith, 1938.

Type: Holotype, MNHN 1938.122 (formerly RLB M.310), a 714 mm specimen (Jabouille, 1933).

Type locality: "station d'altitude du Mont Bavi, située à 400 m d'altitude dans la province de Son-Tây, ègalement en bordure du delta tonkinois, sur des montagnes du Tonkin, l'Indochine française" [= Bavì, Hà Son Binh Prov., N Vietnam, 21°05'N, 105°23'E, elevation 400 m].

Distribution: Northern Vietnam (Ha Tay, Son Tay, Vinh Phuc), 400 m.

Sources: Bourret, 1936b, 1939d, Orlov et al., 2000 and V.S. Nguyen et al., 2009.

Remarks: Occurs in Hainan, China *fide* V.S. Nguyen et al. (2009: 366) and Das (2011: 339).

6. *Rhabdophis callistus* (A.C.L.G. Günther, 1873). Proc. Zool. Soc. London 41(1): 170, pl. 17, fig. c. (*Tropidonotus callistus*)

Type: Holotype, BMNH 1946.1.15.42 (formerly BMNH 1872.46.124), a 295–305 mm specimen (A.B. Meyer, Dec. 1870–Nov. 1871).

Type locality: "Northern Celebes" [= N Sulawesi, E Indonesia].

Distribution: Eastern Indonesia (N Sulawesi).

Sources: Land & Vogel, 2005 and Koch, 2012.

Remarks: Photographs of the holotype in Lang & Vogel (2005: figs. 118–119). A valid species *fide* Iskandar & Colijn (2001: 104) and Lang & Vogel (2005: 172–173).

7. *Rhabdophis chrysargoides* (A.C.L.G. Günther, 1858). Cat. Colub. Snakes Brit. Mus.: 71. (*Tropidonotus chrysargoides*)

Type: Holotype, BMNH 1946.1.13.42 (formerly RMNH), an adult female (H. Schlegel exch.).

Type locality: Unknown. Designated as Java, Indonesia *fide* C. Haas (1950: 539).

Distribution: Indonesia (Buton, Java, Muna, Sangihe Arch., Siau, Sulawesi), 350–1000 m.

Sources: Kopstein, 1930, Bosch, 1985, Malkmus, 1993a, 2000 and Koch, 2012.

Remarks: Bosch (1985: 16) erroneously attributed authorship to A.-M.-C. Duméril, Bibron & Duméril, 1854.

R

8. *Rhabdophis chrysargos* (Schlegel, 1837). Essai Phys. Serp. 1: 167, 2: 312–313, pl. 12, figs. 6–7. (*Tropidonotus chrysargos*)

Synonyms: *Natrix nobilis* Wagler, 1825 (*nomen ineditum*), *Tropidonotus chrysargus* H. Boie *in* Schlegel, 1826a *(nomen nudum), Tropidonotus chrysargus* Schlegel, 1826b *(nomen nudum), Tropidonotus chrysargus* H. Boie *in* F. Boie, 1827 *(nomen nudum), Tropidonotus junceus* Cantor, 1847, *Amphiesma chrysargum* – A.-M.-C. Duméril, Bibron & Duméril, 1854a, *Leptophis albomaculatus* A.-M.-C. Duméril, Bibron & Duméril, 1854a, *Tropidonotus chrysargus* – A.-M.-C. Duméril, Bibron & Duméril, 1854a, *Tropidonotus dipsas* Blyth, 1854a, *Tropidonotus zebrinus* Blyth, 1854a, *Amphiesma chrijsargum* Bleeker, 1856 *(nomen emendatum),* and *Tropidonotus maculatus torquatus* Mocquard, 1890b.

Type: Lectotype, RMNH 10426 (formerly RMNH 1051), a 690 mm specimen (H. Kuhl & J.C. van Hasselt, Dec. 1820–Sept. 1823), designated by Iskandar & Colijn (2001: 104).

Type locality: "Magmedon, Java" [= Mt. Megamendung, West Java, SW Indonesia, 6°38'S, 106°54'E, elevation 650 m] via lectotype selection.

Distribution: Southeastern Asia and East Indies. Southern China (Guangdong, Hainan), Thailand (Chachoengsao, Chiang Mai, Krabi, Nakhon Ratchasima, Nakhon Si Thammarat, Pattani, Phang Nga, Prachuap Khiri Khan, Rayong, Rat Buri, Roi Et, Sisaket, Trang, Trat, Yala), S Myanmar (Tanintharyi, Mergui Arch.: Sullivan and King's Is.), Vietnam (Bac Giang, Dong Nai, Gai Lai, Ha Tinh, Hoa Binh, Lam Dong, Nghe An, Quang Binh, Quang Tri, Son La, Thanh Hoa, Thua Thien-Hue), N Laos (Champasak, Xiangkhouang), SW Cambodia (Kampot, Kandal, Koh Kong, Mondolkiri), West Malaysia (Johore, Kelantan, Pahlang, Perak, Pinang, Selangor, Tioman Is.), East Malaysia (Sabah, Sarawak), Brunei, Indonesia (Anambas Arch., Bali, Flores, Java, S Kalimantan, Mentawai Arch., Nias, Siberut, Simeulue, Sipora, Sumatra, Ternate) and SW Philippines (Balabac, Busuanga, Culion, Palawan), 100–1675 m.

Sources: Sclater, 1891a–b, Flower, 1899, M.A. Smith, 1916, 1923a, 1931b, Wall, 1923b, Kopstein, 1930, 1938, Bourret, 1934b, 1936b, 1939d, C.H. Pope, 1935, C. Haas, 1941, 1949, Bergman, 1959, Leviton, 1959b, E.H. Taylor, 1965, Campden-Main, 1970a, Deuve, 1970, Saint-Girons, 1972a, B. Hu et al., 1980, Tweedie, 1983, Bosch, 1985, Tian et al., 1986, Malkmus, 1987, 1989, 1992, M.J. Cox, 1991b, Stuebing, 1991, David & Vogel, 1996, Zhao, 1997, M.J. Cox et al., 1998, Chan-ard et al., 1999, Grismer et al., 2002, Malkmus et al., 2002, B.L. Stuart & Emmett, 2006, B.L. Stuart et al., 2006, I. Das, 2007b and V.S. Nguyen et al., 2009.

Remarks: Original description based on H. Boie's MS (1823–1825: pls. 27–28). C.H. Pope (1935: 102) reported two specimens from Hong Kong. Sulawesi records doubtful, probably *Rhabdophis chrysargoides, fide* Lang & Vogel (2005: 254).

9. *Rhabdophis conspicillatus* (A.C.L.G. Günther, 1872a). Proc. Zool. Soc. London 40(1): 596–597, fig. 4. (*Tropidonotus conspicillatus*)

Synonym: *Natrix conspicillata* – T. Barbour, 1912.

Types: Syntypes (3), BMNH 1946.1.23.76–77, a 406 mm male and a female [Matang] (A.H. Everett, 1869–1872), and BMNH 1946.1.23.78, a juvenile male [Borneo] (L.L. Dillwyn, 1855–1865).

Type locality: "Matang in the district of Sarawak, western Borneo" [= Kampong Matang, First Division, Kidi Distr., W Sarawak, East Malaysia, 1°35'N, 110°14'E, elevation 15 m]. Emended to "Matang" and "Borneo" *fide* Boulenger (1893a: 222).

Distribution: Malaysia and Borneo. West Malaysia (Perak), East Malaysia (Sabah, Sarawak) and W Indonesia (Great Natuna, Kalimantan, Singkep, Sumatra), 15–1000 m.

Sources: Boulenger, 1893a, T. Barbour, 1912b, Smedley, 1932b, C. Haas, 1950, Tweedie, 1983, Malnate & Underwood, 1988, Stuebing, 1991, David & Vogel, 1996, Malkmus et al., 2002 and I. Das, 2010, 2012.

Remarks: L.L. Dillwyn and J. Motley collected around Labuan (5°19'N, 115°13'E) and adjacent Sabah *fide* Motley & Dillwyn, 1855, the probable type locality of BMNH 1946.1.23.78, which Boulenger (1893a: 222) lists as a type.

10. *Rhabdophis himalayanus* (A.C.L.G. Günther, 1864a). Rept. Brit. India: 265–266, pl. 22, fig. h. (*Tropidonotus himalayanus*)

Synonyms: *Xenodon macrophthalmus* A.C.L.G. Günther, 1858 *(partim)*, *Tropidonotus himalayanus ornatus* Wall, 1908c, and *Rhabdophis speciosus* Wall, 1925c.

Types: Syntypes (2), BMNH 1946.1.13.15, a 787 mm female (B.H. Hodgson, 1842) [Nepal], and 1946.1.23.75, a female (H.R.A. von Schlagintweit, A. von Schlagintweit, & R. von Schlagintweit, 1854–1858) [Sikkim].

Type locality: "Nepal" and "Sikkim" [NE India].

Distribution: Southern Asia. Southwestern China (Xizang, Yunnan), Nepal (Ilam, Kaski, Kathmandu, Myagdi, Nuwakot, Sankhuwasabha), Bhutan, NE India (Arunachal Pradesh, Assam, Manipur, Meghalaya, Mizoram, Nagaland, North Bengal, Sikkim), Bangladesh (Sylhet) and Myanmar (Chin, Kachin, Mon, Shan), 30–2350 m.

Sources: Theobald, 1876a–b, Sclater, 1891a, Annandale, 1905b, Wall, 1908c, 1909d, 1923d, 1925c, M.A. Smith, 1938, 1940, G.E. Shaw et al., 1939, Anonymous, 1962, L. Swan & Leviton, 1962, Deuve, 1970, Fleming & Fleming, 1973, Kramer, 1977, Majupuria, 1981, Nanhoe & Ouboter, 1987, Tiwari & Shah, 2004, Whitaker & Captain, 2004, M.F. Ahmed et al., 2009 and Rahman & Ahmed, 2012.

Remarks: Probably cen. Laos *fide* Deuve (1970: 101).

11. *Rhabdophis leonardi* (Wall, 1923b). J. Bombay Nat. Hist. Soc. (1923–1924) 29(2): 466. (*Natrix leonardi*)

Synonym: *Natrix swinhonis reducta* Mell, 1931a.

Type: Holotype, formerly BNHS 466, a 660 mm specimen (P.M.R. Leonard).

Type locality: "Sinlum Kaba, Upper Burma Hills, circ. 6,000 feet (Lat. 25° long. 97°)" [= Sinlumkaba, S Kachin Div., N Myanmar, ca. 24°16'N, 97°31'E, elevation 1830 m].

Distribution: Southwestern China (Guangxi, Sichuan, Xizang, Yunnan) and NE Myanmar (Kachin), 1250–2850 m.

Sources: Wall, 1925d, Jiang & Zhao, 1983, Tian et al., 1986 and Zhao, 2006.

Remarks: Sinlum Kaba was variously reported by Wall as 6000', 24°N, 97°E (Wall, 1921b: 43), Upper Burma Hills, 6000', 24°N, 97 1/2°E (Wall, 1923d: 610), and Kachin Hills, 6500', 25°N, 97°E (Wall, 1925d: 810).

12. *Rhabdophis lineatus* (W.C.H. Peters, 1861c). Mber. Königl. Akad. Wiss. Berlin 1861(4): 686–687. (*Tropidonotus lineatus*)

Types: Syntypes (2), ZMB 3976a–b, 690 mm and 205 mm specimens (F. Jagor, July 1860).

Type locality: "Loquilocun, Insel Samar, Philippinen" [= Loquilócon, a little forest village on Loquilócon River, cen. Western Samar, cen. Samar Is., Philippines, 11°48'N, 125°06'E].

Distribution: Southern Philippines (Basilan, Bohol, Dinagat, Mindanao, Samar), 300–1065 m.

Sources: Leviton, 1959b and Ross & Lazell, 1991.

Remarks: Tiedemann & Häupl (1980: 63), Tiedemann et al. (1994: 76) and Bauer et al. (1995: 75) listed a supposed syntype (NMW 23469) with locality as E Mindanao and collector as Semper in 1874, all of which exclude it as a type specimen.

13. *Rhabdophis murudensis* (M.A. Smith, 1925b). Sarawak Mus. J. 3(8): 5–6. (*Natrix murudensis*)

Type: Holotype, BMNH 1946.1.13.26 (formerly MAS 7210), a female (E.G. Mjöberg, 1925).

Type locality: "Mt. Murud, Borneo at 5500-6000 feet altitude" [= Mt. Murud, ext. NE Sarawak, cen. East Malaysia, 3°52'N, 115°30'E, elevation 1700 m].

Distribution: East Malaysia (Sabah, Sarawak), 915–2500 m.

Sources: Stuebing & Inger, 1999, Malkmus et al., 2002 and Stuebing & Lian, 2002.

R

14. *Rhabdophis nigrocinctus* (Blyth, 1855c). Proc. Asiatic Soc. Bengal 24(7): 717. (*Tropidonotus nigrocinctus*)

Synonyms: *Tropidonotus eisenhoferi* Gyldenstolpe, 1916, and *Pseudoxenodon fruhstorferi* F. Werner, 1925.

Types: Syntypes (3), ZSI 7479–81 (formerly IMC 7479–81), longest syntype 673 mm (T.M. Berdmore, 1855).

Type locality: "Mergui and the valley of the Tenasserim river" [Myanmar] (in error). Corrected to Pegu [= Schwe Gyen, on the lower Sitang River, bet. Bago and Mon Div., S Myanmar, bet 16°58'–17°36'N, elevation < 30 m] *fide* Sclater (1891b: 37).

Distribution: Southeast Asia. Southern China (Sichuan, Yunnan), N Vietnam (Dien Bien, Nghe An, Son La), Laos (Khammouan, Phongsali, Vientiane), Cambodia (Kampot, Koh Kong), Thailand (Bangkok, Chanthaburi, Chiang Mai, Chumphon, Kanchanaburi, Loei, Mae Hong Son, Nakhon Ratchasima, Phang Nga, Ranong, Surat Thani) and S Myanmar (Bago, Tanintharyi, Mergui Arch.: King's Is.), 400–2410 m.

Sources: Sclater, 1891a, C.H. Pope, 1935, Bourret, 1936b, 1939d, E.H. Taylor, 1965, Deuve, 1970, Hu et al., 1980, Tian et al., 1986, M.J. Cox, 1991b, M.J. Cox et al., 1998, Chan-ard et al., 1999, B.L. Stuart & Emmett, 2006, Zhao, 2006, B.L. Stuart & Heatwole, 2008 and V.S. Nguyen et al., 2009.

15. *Rhabdophis nuchalis* (Boulenger, 1891a). Ann. Mag. Nat. Hist. (6) 7(39): 281. (*Tropidonotus nuchalis*)

Synonyms: *Natrix nivalis* K.P. Schmidt, 1925a, and *Natrix nuchalis collaris* T. Vogt, 1927.

Types: Syntypes (4), BMNH 1946.1.12.69–70, four females, longest syntype 640 mm (A.E. Pratt, 1887–1890).

Type locality: "China. Ichang." Emended to Ichang, Upper Yangtse-Kiang by A.C.L.G. Günther (1889: 218) and Boulenger (1893a: 218) [= Yichang, W Hubei, cen. China, 30°42'N, 111°17'E].

Distribution: Southern Asia. Northeastern India (Nagaland), S China (Gansu, Guangxi, Guizhou, Shaanxi, Xizang, Hubei, Yunnan), N Myanmar (Kachin) and N Vietnam (Cao Bang, Lao Chai, Vinh Phuc), 620–2750 m.

Sources: A.C.L.G. Günther, 1889, Boulenger, 1893a, Bourret, 1936b, 1939, B.-Q. Hu et al., 1980, Tien et al., 1986, Inger et al., 1990, Zhao, 2006, V.S. Nguyen et al., 2009, I. Das, 2010 and Yao, 2012.

R

16. *Rhabdophis pentasupralabialis* Jiang & Zhao, 1983. Acta Herpetol. Sinica (2) 2(1): 60–61, 1 fig. (*Rhabdophis nuchalis pentasupralabialis*)

Type: Holotype, CIB 80-II-0040, a male (E.-M. Zhao, 26 July 1980).

Type locality: "Jiulong County, Sichuan, alt. 2750m, China."

Distribution: Southern China (Sichuan, Yunnan), 1200–3200 m.

Sources: Zhao, 1995 and 2006.

17. *Rhabdophis sarawacensis* (A.C.L.G. Günther, 1872a). Proc. Zool. Soc. London 40(1): 596 (*Tropidonotus sarawacensis*)

Synonym: *Tropidonotus maculatus* W.C.H. Peters, 1871b.

Types: Syntypes (3), BMNH 1946.1.13.12–14, two males and one female (A.H. Everett, 1869–1872).

Type locality: "Matang in the district of Sarawak, western Borneo" [= Kampong Matang, First Division, Kidi Distr., W Sarawak, East Malaysia, 1°35'N, 110°14'E, elevation 15 m].

Distribution: Malay Peninsula and Borneo. West Malaysia (Kelantan, Pahang, Perak), East Malaysia (Sabah, Sarawak) and Indonesia (Kalimantan), 15–1760 m.

Sources: Tweedie, 1983, Stuebing, 1991, Stuebing & Inger, 1999, Malkmus et al., 2002, David & I. Das, 2003 and Inger & Tan, 2010.

18. *Rhabdophis spilogaster* (F. Boie, 1827). Isis von Oken 20(6): 535. (*Tropidonotus spilogaster*)

Synonyms: *Tropidonotus spilogaster* H. Boie *in* Schlegel, 1826a *(nomen nudum), Tropidonotus spilogaster* Schlegel, 1826b *(nomen nudum),* and *Tropidonotus crebripunctatus* Wiegmann *in* Meyen, 1834.

Types: Syntypes (5), RMNH 1048a–b (G.A. Frank, 1843) and RMNH 1049a–c (Frank, 1847).

Type locality: "Java" [W Indonesia] (in error). Corrected to Philippines *fide* RMNH catalogue and Boulenger (1893a: 257). Restricted to Luzon Is., Philippines *fide* Leviton (1959b: 439) and herein.

Distribution: Northern Philippines (Batan, Camiguin Norte, Catanduanes, Limay, Lubang, Luzon, Polillo), NSL–30 m.

Sources: L.E. Griffin, 1911, E.H. Taylor, 1922a, Leviton, 1959b, Rabor, 1981, Alcala, 1986, Ross & Gonzales, 1992, R. Brown et al., 1996, 2000 and Oliveros et al., 2011.

Remarks: Original description supposedly based on H. Boie's MS (1823–1825) but description of *T. spilogaster* is not in it. Schlegel (1837: 309) erroneously suggests that type locality is Mariana Is. based on Boie's types. Data for Negros Is. record in error *fide* Leviton (1959b: 435).

19. *Rhabdophis subminiatus* (Schlegel, 1837). Essai Phys. Serp. 1: 167–168, 2: 313–314. (*Tropidonotus subminiatus*)

Synonyms: *Tropidonotus subminiatus* H. Boie *in* Schlegel, 1826a (*nomen nudum*), *Tropidonotus subminiatus* Schlegel, 1826b (*nomen nudum*), *Tropidonotus subminiatus* Reinwardt *in* F. Boie, 1827 (*nomen nudum*), *Tropidonotus manadensis* A.C.L.G. Günther, 1873b, *Pseudoxenodon intermedius* Lönnberg, 1899, *Natrix helleri* K.P. Schmidt, 1925a, *Natrix subminiata hongkongensis* Mell, 1931a, *Natrix subminiata siamensis* Mell, 1931a, *Natrix laobaoensis* Bourret, 1934c, and *Rhabdophis subminatus* – Kochva, 1987 *(nomen incorrectum).*

Type: Lectotype, RMNH 1067, a 710 mm specimen (C.G.C. Reinwardt, 1816–1821), designated by R.F. Inger herein.

Type locality: "Java dans les champs de Sawa inondés" [= wetlands of Java, Indonesia]. Emended to West Java *fide* C. Haas (1950: 543).

Distribution: Southeastern Asia and Greater Sundas. Northeastern India (Arunachal Pradesh, Assam, Manipur, Meghalaya, Mizoram, Nagaland, North Bengal, Sikkim, West Bengal), Bangladesh, Nepal (Chitwan), Bhutan (Sarpang), S China (Fujian, Guangdong, Guangxi, Guizhou, Hainan, Hong Kong, Yunnan), Myanmar (Bago, Chin, Kachin, Rakhine, Shan, Tanintharyi, Mergui Arch.), Thailand (Bangkok, Chanthaburi, Chiang Mai, Chiang Rai, Kanchanaburi, Khon Kaen, Krabi, Krung Thep Mahanakhon, Loei, Nakhon Ratchasima, Nakhon Si Thammarat, Narathiwat, Phang Nga, Phuket, Rayong, Sisaket, Sukhothai, Surin, Trang, Ubon Ratchathani, Uttaradit), Cambodia (Kampot, Kampong Chhnang, Oddar Meanchey, Preah Vihear), Laos (Attapu, Borikhan, Champasak, Khammouan, Phongsali, Savannakhet, Vientiane, Xekong, Xiangkhouang), Vietnam (Bac Giang, Bac Kan, Bac Ninh, Binh Phuoc, Cao Bang, Dak Lak, Da Nang, Dong Nai, Gia Lai, Hai Duong, Ha Tay, Ha Tinh, Hoa Binh, Ho Chí Minh, Kien Giang, Kon Tum, Lam Dong, Lang Son, Lao Cai, Nghe An, Phu Tho, Quang Binh, Quang Nam, Quang Ninh, Quang Tri, Son La, Tay Nình, Than Hoa, Thua Thien-Hue, Vinh Phuc, Yen Bai), N West Malaysia (Pinang), Singapore and W Indonesia (Java, SE Kalimantan, Sumatra), NSL–2250 m.

Sources: A.C.L.G. Günther, 1864a, J. Anderson, 1879, Boulenger, 1890a, 1912c, Flower, 1899, Wall, 1908f, 1910a, 1923d, 1925d, 1926, T. Barbour, 1912b, Mell, 1922, 1931a, Brongersma, 1930, M.A. Smith, 1930a, Fan, 1931, Bourret, 1934b, 1936b, 1939d, Rendahl, 1937, Kopstein, 1938, Shaw et al., 1939, Gressitt, 1941, C. Haas, 1941, Bergman, 1956b, E.H. Taylor, 1965, Campden-Main, 1970a, Deuve, 1970, Saint-Girons, 1972a, Romer, 1979b, B.-Q. Hu et al., 1980, Tweedie, 1983, Tian et al., 1986, M.A.R. Khan, 1988, K.K.P. Lim & Chou, 1990, M.J. Cox, 1991b, Stuebing, 1991, I. Das, 1994, David & Vogel, 1996, M.J. Cox et al., 1998, Chan-ard et al., 1999, Schleich & Kästle, 2002, Tiwari & Shah, 2004, Whitaker & Captain, 2004, Zhao, 2006, M.F. Ahmed et al., 2009, V.S. Nguyen et al., 2009, Wangyal, 2012, T. Hartmann et al., 2013 and S.K. Sharma et al., 2013.

Remarks: Records for Ternate and Seram need confirmation *fide* C. Haas (1950: 544). Sulawesi record doubtful *fide* Lang & Vogel (2005: 255).

20. *Rhabdophis swinhonis* (A.C.L.G. Günther, 1868). Ann. Mag. Nat. Hist. (4) 1(6): 420, pl. 19, fig. f. (*Tropidonotus swinhonis*)

Synonym: *Tropidonotus swinhoei* Boettger, 1888c *(nomen emendatum).*

Type: Holotype, BMNH 1946.1.12.76, a 635 mm female (R. Swinhoe, 7 June–1 July 1858).

Type locality: "Formosa" [= Taiwan].

Distribution: Taiwan (Chiai, Taichung, Taipei), 40–850 m.

Sources: Boulenger, 1909a, Oshima, 1910, Steindachner, 1913b, Takahashi, 1930, Maki, 1931, Wang & Wang, 1956, Kuntz, 1963, Lue, 1987, 1990, Lue et al., 1989, Ota, 1991, Zhao, 2006 and Xiang & Li, 2009.

21. *Rhabdophis tigrinus* (H. Boie, 1826). Isis von Oken 18(2): 205–207. (*Tropidonotus tigrinus*)

Synonyms: *Amphiesma tigrinum* – A.-M.-C. Duméril, Bibron & Duméril, 1854a, *Tropidonotus lateralis* Berthold, 1859, *Tropidonotus orientalis* A.C.L.G. Günther, 1861i, *Natrix tigrina* – Stejneger, 1907, *Natrix tigrina formosana* Maki, 1931, *Rhabdophis tigrinus multiventralis* G. Stewart, 1970 (*nomen nudum*), and *Rhabdophis tigrina pateralis* – Yao, 2012 *(nomen incorrectum).*

Types: Syntypes (7), RMNHa–g, longest syntype 693 mm (J.C. Blomhoff, 1817–1823).

Type locality: "Decima, Japan, [= Dejima Is., previously in Nagasaki harbor, now part of Nagasaki, Japan, 32°45'N, 129°52'E, elevation 10 m].

Distribution: Eastern Asia. China (Anhui, Fujian, Gansu, Guangxi, Guizhou, Hebei, Heilongjiang, Henan, Hubei, Hunan, Jiangsu, Jiangxi, Jilin, Liaoning, Nei Monggol, Ningxia, Shanxi, Shandong, Sichuan, Xizang, Yunnan, Zhejiang), SE Russia, North Korea (Kangwon, North Hwanghae, North Pyongan, South Pyongan), South Korea (Kangwon, Kyonggi, North Chungchong, South Cholla, South Chungchong, South Kyongsang, Cheju, Jin, Namhae and Yokji Is.), Japan (Honshu, Kyushu, Ryukyus: Mageshima, Takeshima, Tanegashima and Yakujima Is.) and Taiwan (Taichung), NSL–2200 m.

Sources: Temminck & Schlegel, 1838, A.C.L.G. Günther, 1864a, Strauch, 1873, 1876, Hilgendorf, 1880, Nikolsky, 1905, Wall, 1905f, Stejneger, 1907a, Slevin, 1925, Chang & Fang, 1931, Maki, 1931, Boring, 1932, C.H. Pope, 1935, V. Tanner, 1953, Dixon, 1956, Shannon, 1956, Takara, 1962, Nakamura & Ueno, 1969, Bannikov et al., 1977, M. Mori, 1982, Ota & Mori, 1985a, Paik & Yang, 1986, Lue, 1987, 1990, Szyndlar & Hung, 1987, Lue et al., 1989, Toriba, 1992a, Xu, 2001, J.C. Murphy & Schlager, 2003, Zhao, 2006, San & Lee, 2007, Xiang & Li, 2009 and Yao, 2012.

Remarks: Original description reprinted in H. Boie (1827b: 248–252) with type illustrated in Temminck & Schlegel (1838: pl. 4, figs. 1-9). Blomhoff was stationed at Dejima Is., Nagasaki, Japan so that is likely type locality. Schlegel (1837: 315) reported many specimens sent from Japan by Siebold and Bürger. Occurs in N Vietnam *fide* Bourret (1939d: 53) and I. Das (2010: 342). Type locality possibly in error *fide* Siebold *in* Temminck & Schlegel (1838: iii), who mistakenly corrected it to East Indies.

RHABDOPS Boulenger, 1893a (Natricidae)

Synonyms: *Grotea* Theobald, 1868b *(nomen praeoccupatum), Rabdops* – Briceño-Rossi, 1934 *(nomen incorrectum),* and *Rhabodops* – Deoras, 1965 *(nomen incorrectum).*
Type species: *Ablabes olivaceus* Beddome, 1863a.
Distribution: Southern Asia.
Sources: A.C.L.G. Günther, 1864a, Boulenger, 1890a, 1893a, C.H. Pope, 1935, M.A. Smith, 1943, Murthy, 1986 and Ahmed et al., 2009.

1. *Rhabdops bicolor* (Blyth, 1854a). J. Asiat. Soc. Bengal 23(3): 289. (*Calamaria bicolor*)

Type: Holotype, ZSI 7030 (formerly IMC 7030), a 241 mm specimen (W. Robinson).
Type locality: "Asám, from the Khásya hills, or other upland country" [= probably Khasi Hills, E Meghalaya State, NE India, 25°35'N, 91°38'E, elevation 1370–1675 m].
Distribution: Southern Asia. Northeastern India (Meghalaya), N Myanmar (Kachin) and S China (W Yunnan), 900–1700 m.
Sources: Theobald, 1868a, J. Anderson, 1879, Wall, 1925a, 1926b, B.Q. Hu et al., 1980, Tian et al., 1991 and Zhao & Adler, 1993.

2. *Rhabdops olivaceus* (Beddome, 1863a). Madras Quart. J. Med. Sci. 6: 42. (*Ablabes olivaceus*)

Type: Holotype, BMNH 1946.1.12.44, a 521 mm male (R.H. Beddome, 1857–1863).
Type locality: "Manantoddy, Madras Presidency" [= Mananthavady, the Wynaad, Malabar, N Kerala, ext. SW India, 11°48'N, 76°01'E, elevation 760 m]. Emended to Manantoddy, Wynad *fide* Boulenger (1893a: 301) and Manatoddy, Malabar District *fide* M.A. Smith (1943: 328).
Distribution: Southwestern India (Western Ghats of Karnataka, Kerala, SW Maharashtra), 610–1300 m.
Sources: Wall, 1919b, Soman, 1962, Radhakrishnan, 1997, Whitaker & Captain, 2004, N. Khaire, 2006 and Ganesh et al., 2012a.

RHACHIDELUS Boulenger, 1908 (Xenodontidae)

Synonyms: *Rachidelus* – Phisalix, 1922 *(nomen incorrectum), Rhachedelis* – J.A. Peters & Orejas-Miranda, 1970 *(nomen incorrectum),* and *Rhacidelus* – Marx & Rabb, 1972 *(nomen incorrectum).*
Type species: *Rhachidelus brazili* Boulenger, 1908f.
Distribution: Southern South America.
Sources: Serie, 1915b, Amaral, 1930f, 1977, J.A. Peters & Orejas-Miranda, 1970, Lema et al., 1980, Lema, 1987, J.D. Williams & Francini, 1991, Scrocchi, 1992b, Nogueira, 2001, Fernandes & Passos, 2002, Silveira, 2004, Zaher et al., 2009 and Arruda et al., 2011.

1. *Rhachidelus brazili* Boulenger, 1908f. Ann. Mag. Nat. Hist. (8) 2(7): 31–32.

Type: Holotype, BMNH 1946.1.9.35, a 1320 mm male (V. Brazil, 1897–1908).
Type locality: "near the city of São Paulo, Brazil" [= vicinity of São Paulo, E São Paulo State, SE Brazil, 23°33'S, 46°38'W, elevation 800 m].
Distribution: Southern Brazil (Distrito Federal, Goiás, Mato Grosso, Mato Grosso do Sul, Minas Gerais, Paraná, Rio Grande do Sul, São Paulo), SE Paraguay and ext. NE Argentina (Córdoba, Misiones), 720–800 m.

RHADINAEA Cope, 1863b (Dipsadidae)

Synonyms: *Radinaea* – Sumichrast, 1882 *(nomen incorrectum), Rhadinea* S.W. Garman, 1884 *(nomen emendatum), Rhadinea* Palacky, 1898 *(nomen emendatum), Radianea* – Hilzheimer, 1912 *(nomen incorrectum), Radinoea* – Brazil, 1914 *(nomen incorrectum), Rhadinae* – Serié, 1919 *(nomen incorrectum), Rhadinia* – Briceño-Rossi, 1934 *(nomen incorrectum), Rhadinacea* – Hoge, 1964d *(nomen incorrectum), Rhabdinaea* – Duke, 1967 *(nomen incorrectum), Rhadineae* – W.W. Tanner & Avery, 1982 *(nomen incorrectum), Alexteesus* Hoser, 2012ah *(nomen illegitimum), Barrygoldsmithus* Hoser, 2012ah *(nomen illegitimum), Desmondburkeus* Hoser, 2012ah *(nomen illegitimum), Dudleyserpens* Hoser, 2012ah *(nomen illegitimum), Jockpaullus* Hoser, 2012ah *(nomen illegitimum), Robvalenticus* Hoser, 2012ah *(nomen illegitimum),* and *Wallisserpens* Hoser, 2012ah *(nomen illegitimum).*
Type species: *Taeniophis vermiculaticeps* Cope, 1860e.
Distribution: Southeastern USA and Latin America.
Fossil records: Pleistocene of SE USA.
Sources: J.A. Peters & Orejas-Miranda, 1970, C.W. Myers, 1974, 2011, Cadle, 1984b, C.W. Myers & Cadle, 1994, Zaher et al., 2009 and Mulcahy et al., 2011.
Remarks: The following genera and species have been separated from *Rhadinaea: Rhadinella* (*anachoreta, godmani, hannsteini, hempsteadae, kanalchutchan, kinkelini, lachrymans, montecristi, pegosalyta, pilonaorum, posadasi, rogerromani, schistose, serperaster, tolpanorum*), *Taeniophallus* (*affinis, bilineatus, brevirostris, nebularis, nicagus, occipitalis, persimilis, precilopogon, quadriocellatus*), and *Urotheca* (*decipiens, dumerilii, fulviceps, guentheri, lateristriga, multilineata, myersi, pachyura*).

R

1. *Rhadinaea bogertorum* C.W. Myers, 1974. Bull. Amer. Mus. Nat. Hist. 153(1): 59–65, figs. 10j, 11g.

Type: Holotype, AMNH 100907, a 448 mm male (M.R. Bogert, 27 Aug. 1967).
Type locality: "10.5 miles (16.8 kilometers) by road north of Cerro Pelón, near km. 123 on Tuxtepec Road, 6650 feet (2027 meters), Sierra de Juárez (part of the Sierra Madre del Sur), Oaxaca, Mexico."
Distribution: Mexico (Sierra de Juárez of N-cen. Oaxaca), 2025–2250 m.
Sources: Casas-Andreu et al., 1996 and Ramírez-Bautista et al., 1998.

2. *Rhadinaea calligaster* (Cope, 1875a). J. Acad. Nat. Sci. Philadelphia (1876) (2) 8(2): 146, pl. 28, fig. 12. (*Contia calligaster*)

Type: Lectotype, USNM 30679 (W.M. Gabb, 1873–1875), designated by C.W. Myers (1974: 155).
Type locality: "Pico Blanco, 5000 to 7000 feet, Limón Prov., Costa Rica" via lectotype selection. Emended to Cerro Utyum *fide* Savage (1970: 282).
Distribution: Costa Rica (Alajuela, Cartago, Heredia, Limón, Puntarenas, San José) and ext. W Panama (Bocas del Toro, Chiriquí), 1220–2440 m.
Sources: A.C.L.G. Günther 1893 *in* 1885–1902, E.H. Taylor, 1951, Savage, 1980a, 2002, N.J. Scott et al., 1983 and Solórzano, 2004.

3. *Rhadinaea cuneata* C.W. Myers, 1974. Bull. Amer. Mus. Nat. Hist. 153(1): 65–68, figs. 10f, 11i.

Type: Holotype, UAZ 26580, a 514 mm female (native via R.L. Bezy & C.J. Cole, 28 Sept. 1966).
Type locality: "Ojo del Aqua, Nacimiento del Río Atoyac, about 10 kilometers (air line) north-northeast of Córdoba,Veracruz, Mexico" [= Ojo de Agua, W-cen. Veracruz State, SE Mexico, 18°59'N, 96°54'W, elevation 1200 m].
Distribution: Southeastern Mexico (Oaxaca, cen. Veracruz), 50–1400 m.
Sources: McCoid & Dixon, 1981, Pérez-Higareda & Smith, 1991 and Casas-Andreu et al., 1996.

4. *Rhadinaea decorata* (A.C.L.G. Günther, 1858). Cat. Colub. Snakes Brit. Mus.: 35–36. (*Coronella decorata*)

Synonyms: *Enicognathus vittatus* Jan, 1863a, *Dromicus ignitus* Cope, 1871b, *Erythrolamprus longicaudus* F. Werner, 1903a, and *Erithrolamprus longicaudus* – Briceño-Rossi, 1934 *(nomen incorrectum)*.
Type: Lectotype, BMNH 1946.1.9.4, an adult female (A. Sallé, 1846–1856), designated by C.W. Myers (1974: 68).
Type locality: "Mexico" via lectotype selection. Probably Veracruz or Oaxaca State *fide* A.C.L.G. Günther (1885–1902: vi). Restricted to Potrero Viejo, Veracruz State, Mexico *fide* H.M. Smith & Taylor (1950a: 350).
Distribution: Latin America. Southeastern Mexico (Chiapas, Oaxaca, Puebla, SE San Luis Potosí, Veracruz), Belize (Toledo), N Guatemala (Alta Verapaz, El Petén), E Honduras (Gracias a Dios, Olancho), E Nicaragua (Atlántico Norte, Atlántico Sur, Río San Juan), Costa Rica (Alajuela, Cartago, Guanacaste, Heredia, Limón, Puntarenas, San José), Panama (Bocas del Toro, Canal Zone, Chiriquí, Coclé, Colón, Darién, Panama, Veraguas, and Barro Colorado and Colón Is.), W Colombia (Cauca) and NW Ecuador (Esmeraldas, Imbabura, Pichincha), NSL–1750 [2000] m.
Sources: Jan & Sordelli, 1866 *in* 1860–1881, A.C.L.G. Günther, 1893 *in* 1885–1902, Bocourt, 1886 *in* Duméril, Bocourt & Mocquard, 1870–1909, H.M. Smith, 1943a, E.H. Taylor, 1951, 1954, L.C. Stuart, 1963, J.C. Lee, 1980, 1996, Savage, 1980a, 2002, Villa, 1983, J.D. Johnson, 1984, 1989, Savage & Villa, 1986, Pérez-Santos & Moreno, 1988, 1991, J.A. Campbell, 1998b, Stafford & Meyer, 2000, McCranie, 2004a, 2011 and Ortega-Andrade et al., 2010.

5. *Rhadinaea flavilata* (Cope, 1871b). Proc. Acad. Nat. Sci. Philadelphia 23(2): 222–223. (*Dromicus flavilatus*)

Type: Holotype, ANSP 5583 (H.C. Yarrow, Nov. 1871), lost *fide* E. Böhlke *in* Malnate (1939: 359).
Type locality: "near Fort Macon, on the coast of North Carolina" [USA]. Emended to Bogue Banks some eight miles S of Fort Macon *fide* Coues & Yarrow (1878: 27) (direction in error *fide* Robertson & Tyson (1950: 137). Corrected to approximately 8 miles westward from Fort Macon, on Bogue Banks, Carteret County, North Carolina, USA *fide* C.W. Myers (1967: 51).
Distribution: Southeastern USA (ext. SW Alabama, Florida, ext. SE Georgia, ext. SE Louisiana, ext. S Mississippi, SE North Carolina, SE South Carolina), NSL–190 m.
Fossil records: Lower Pleistocene (Irvingtonian I) of USA (Florida), and upper Pleistocene (Rancholabrean II) of USA (Florida, Tennessee).
Sources: Malnate, 1939, K.P. Schmidt & Davis, 1941, Carr & Goin, 1955, A.H. Wright & Wright, 1957, C.W. Myers, 1967, Mount, 1975, Ashton & Ashton, 1988, Dundee & Rossman, 1989, Conant & Collins, 1991, Holman, 1995a, 2000a, Palmer & Braswell, 1995, Whiteman et al., 1995, Tennant, 1997 and Walley, 1999.

6. *Rhadinaea forbesi* H.M. Smith, 1942i. Proc. Biol. Soc. Washington 55: 188–189, pl. 3, fig. 3.

Type: Holotype, USNM 110365, an adult male (H.M. Smith, 23 March 1940).
Type locality: "Tequeyutepec, seven miles west of Jalapa, Veracruz, Mexico" [= Tequeyutepec, 7 miles W of

Jalapa Enríquez, Veracruz State, Mexico, 19°32'N, 97°01'W, elevation 1800 m].

Distribution: Eastern Mexico (Sierra Madre Oriental of cen. Veracruz), 520–2590 m.

Sources: H.M. Smith, 1943a, H.M. Smith & Taylor, 1945, E.H. Taylor, 1950a, Pérez-Higareda & Smith, 1991 and Montes de Oca & Mendelson III, 1997.

Remarks: Type locality erroneously listed as 10 mi. NE of Jalapa *fide* H.M.Smith & Taylor (1950b: 11).

7. *Rhadinaea fulvivittis* Cope, 1875a. J. Acad. Nat. Sci. Philadelphia (1876) (2) 8(2): 139–140.

Type: Holotype, USNM 7075, adult female (A.L.J.F. Sumichrast, 1855–1875).

Type locality: "Orizaba, Vera Cruz" [Mexico]. Emended to Alpine region of Cerro Orizaba, Veracruz, Mexico [= lower slopes of Pico de Orizaba or Volcán Citlaltépetl (19°01'N, 97°16'W), W-cen. Veracruz State, E Mexico] *fide* Cochran (1961: 210) and C.W. Myers (1974: 96–97).

Distribution: Southern Mexico (Sierra Madre de Oaxaca of SE Puebla and cen. Veracruz, Sierra Madre del Sur of W Oaxaca), [1740] 2195–3300 m.

Sources: H.M. Smith, 1943a, Pérez-Higareda & Smith, 1991 and Casas-Andreu et al., 1996.

8. *Rhadinaea gaigeae* Bailey, 1937b. Copeia 1937(2): 118–119.

Synonym: *Rhadinaea crassa* H.M. Smith, 1942j.

Type: Holotype, MCZ 24983, a 482 mm male (W.W. Brown, 9 June 1926).

Type locality: "Alvarez, San Luis Potosi, Mexico" [= Alvarez, W San Luis Potosí State, cen. Mexico, 22°03'N, 100°37'W, elevation 2680 m].

Distribution: East-central Mexico (Sierra Madre Oriental of N Hidalgo, Querétaro, E San Luis Potosí, SW Tamaulipas, Sierra Camaron of NE Guanajuato, S-cen. San Luis Potosí), 200–2835 m.

Sources: H.M. Smith, 1943a, McCoid & Dixon, 1981, Auth et al., 2000, Farr et al., 2009, Ramírez-Bautista et al., 2010 and Dixon et al., 2011.

R

9. *Rhadinaea hesperia* Bailey, 1940. Occ. Pap. Mus. Zool., Univ. Michigan (412): 8–10, pl. 2, fig. 3.

Synonyms: *Rhadinaea hesperia baileyi* H.M. Smith, 1942i, and *Rhadinaea hesperia hesperioides* H.M. Smith, 1942i.

Type: Holotype, MCZ 42661, a 264 mm (svl) male (W.W. Brown, 1936).

Type locality: "Omilteme and Sierra de Burro, Guerrero, Mexico." Corrected to Omilteme, Guerrero, Mexico [= Omiltemi, cen. Guerrero State, S Mexico, 17°33'N, 99°41'W, elevation 2150 m] *fide* T. Barbour & Loveridge (1946: 186) and MCZ catalogue. Emended to Omilteme, on the Río Balsas drainage of the Sierra Madre del Sur *fide* C.W. Myers (1974: 81).

Distribution: Western Mexico (SW Aguascalientes, Chihuahua, Colima, N Guanajuato, Guerrero, Jalisco, Michoacán, Morelos, Nayarit, Oaxaca, Puebla, S Sinaloa, S Zacatecas), 100–2560 m.

Sources: H.M. Smith, 1942i, H.M. Smith & Taylor, 1945, W.W. Tanner, 1985, Vázquez-Díaz et al., 1999, Canseco-Marquez et al., 2000, H.M. Smith et al., 2001, Lemos-Espinal et al., 2004c and Díaz & Díaz, 2005.

10. *Rhadinaea laureata* (A.C.L.G. Günther, 1868). Ann. Mag. Nat. Hist. (4) 1(6): 419, pl. 19, fig. e. (*Dromicus laureatus*)

Synonym: *Erythrolamprus grammophrys* Dugès, 1888b.

Type: Holotype, BMNH 1946.1.9.2 (formerly BMNH 1868.4.7.9), a 510–533 mm male (H. Doorman).

Type locality: "neighbourhood of the city of Mexico" [= vicinity of Ciudad de Mexico (19°26'N, 99°08'W, elevation 2245 m), Distrito Federal, S-cen. Mexico].

Distribution: Western and S-cen. Mexico (Sierra Madre Occidental of SW Chihuahua, W cen. Durango and Jalisco, Cordillera Volcánica of Distrito Federal, México, cen. Morelos, Michoacán), 1525–3080 m.

Sources: Auth et al., 1999 and R.A. Villa et al., 2012a, c.

Remarks: Probably occurs in Aguascalientes, Mexico *fide* McCranie & Wilson (2001b: 35).

11. *Rhadinaea macdougalli* H.M. Smith & Langebartel, 1950. J. Washington Acad. Sci. (1949) 39(12): 413, fig. 1 (left).

Type: Holotype, UIMNH 3775, a 294 mm male (T.B. MacDougall, 7 March 1949).

Type locality: "near Buena Vista, at crest of Sierra Madre, 4,000–4,500 feet, directly north of Río Grande, 15 miles north of Niltepec, and 1–2 miles downstream from Scarces; 50 miles northeast of Tehuantepec, Oaxaca, Isthmus of Tehuantepec, México" [= Buenavista, E Oaxaca State, ext. S Mexico, ca. 16°45'N, 94°31'W, elevation 1220–1370 m].

Distribution: Extreme S Mexico (Sierra Madre del Sur and Sierra Madre de Chiapas of E Oaxaca and S Veracruz), 900–1500 m.

Sources: Casas-Andreu et al., 1996, Ramírez-Bautista et al., 1998 and Pérez-Higareda et al., 2002.

12. *Rhadinaea marcellae* E.H. Taylor, 1949. Univ. Kansas Sci. Bull. 33(2): 197–198.

Type: Holotype, LSUMZ 270, a 290 mm male (M. Newman [LSU Exped.], 12 June 1947).

Type locality: "Xilitla region, southeastern part of the State of San Luis Potosí a few miles from the border of the State of Hidalgo and equally close to the border of Querétaro, México, between 2,000 and 3,000 feet" [= ca. 21°23'N, 98°58'W, elevation 610–915 m].

Distribution: East-central Mexico (Hidalgo, Puebla, ext. SE San Luis Potosí), 670–2130 m.

Sources: Nieto-Montes de Oca & Mendelson, 1997, Ramírez-Bautista et al., 1998, Canseco-Marquez et al., 2000 and Ramírez-Bautista et al., 2010.

Remarks: E.H. Taylor (1949: 197) erroneously listed the collection date as 12 May 1947. C.W. Myers (1974: 88) pointed out that nearby mountains exceed 2100 m elevation, so the type may have originated well above the altitude of Xilitla (670 m).

13. *Rhadinaea montana* H.M. Smith, 1944a. Field Mus. Nat. Hist. (Zool.) 29(8): 146–148.

Type: Holotype, FMNH 30826, a 580 mm female (H. Hoogstraal et al., 11 Aug. 1938).

Type locality: "Ojo de Agua, Galeana, Nuevo León, México" [= Ojo de Agua, near Galeana (24°49'N, 100°04'W, elevation 1630 m), cen. Nuevo León State, NE Mexico]. Emended to spring located 34 km NNE Galeana at the 14th de Marzo (2134 m), a village at the base of Cerro Potosí *fide* Liner (1996b: 640.2).

Distribution: Northeastern Mexico (cen. Nuevo León and Tamaulipas), 540–2135 m.

Sources: Bailey, 1940, H.M. Smith & Taylor, 1945, Liner, 1996b, Dixon et al., 2011 and García-Vázquez, 2012.

14. *Rhadinaea myersi* Rossman, 1965b. Occ. Pap. Mus. Zool., Louisiana St. Univ. (32): 1–4, fig. 1.

Type: Holotype, LSUMZ 7566 (formerly LCB 1137), a 388 mm male (L.C. Binford, 2 May 1964).

Type locality: "ecotone between pine and cloud forests (5,000 feet) 3 mi. N Pluma Hidalgo, Oaxaca, southern México" [= 3 miles N of Pluma Hidalgo, S-cen. Oaxaca State, S Mexico, ca. 15°57'N, 96°25'W, elevation 1525 m].

Distribution: Extreme S Mexico (Sierra Madre del Sur of S Oaxaca), 1300–2225 m.

Source: Ramirez-Bautista et al., 1998.

Remarks: Casas-Andreu et al., 1996 listed type locality at 1600 m.

15. *Rhadinaea omiltemana* (A.C.L.G. Günther, 1894 *in* 1885–1902). Biol. Cen.-Amer., Rept. Batr.: 113, pl. 40, fig. b. (*Dromicus omiltemanus*)

Type: Holotype, BMNH 1946.1.1.7, a 394 mm male (H.H. Smith via F.D. Godman).

Type locality: "Mexico, Omilteme in Guerrero 8000 feet" [= Omiltemi, cen. Guerrero State, S Mexico, 17°33'N, 99°41'W, elevation 2150 m].

Distribution: Southern Mexico (Sierra Madra del Sur of cen. Guerrero), 1900–2440 m.

Remarks: Boulenger (1894a: 178) listed type as female.

16. *Rhadinaea pulveriventris* Boulenger, 1896a. Cat. Snakes Brit. Mus. 3: 635–636.

Type: Holotype, BMNH 1946.1.9.6 (formerly BMNH 1895.7.13.15), a 430 mm female (C.F. Underwood).

Type locality: "Azahar de Cartago, Costa Rica."

Distribution: Central Costa Rica (Cordillera Central of Alajuela, Cartago, Heredia, San José) and W Panama (Cordillera de Talamanca of Bocas del Toro and Chiriquí), 1000–1600 m.

Sources: E.H. Taylor, 1951, Savage, 2002 and Solórzano, 2004.

17. *Rhadinaea quinquelineata* Cope, 1886b. Proc. Amer. Philos. Soc. (1885) 23(122): 277.

Type: Lectotype, USNM 31350, a 435–438 mm female (S. Bernad), designated by Bailey (1940: 11).

Type locality: "State of Hidalgo, Meixico." (sic) Emended to Teziutlán, Puebla, Mexico *fide* USNM catalogue and Bailey (1940: 11) [= Teziutlán, N Puebla State, SE Mexico, 19°49'N, 97°21'W, elevation 1940 m].

Distribution: Mexico (Sierra Madre Oriental of N Oaxaca and Puebla), 1940–2560 m.

Sources: Bailey, 1940, H.M. Smith, 1944a, McCoid & Dixon, 1981, Benitez-Gálvez, 1997 and Dixon et al., 2011.

18. *Rhadinaea sargenti* Dunn & Bailey, 1939. Bull. Mus. Comp. Zool. 86(1): 10–11.

Type: Holotype, MCZ 42788, a 307 mm male (W.M. Sergeant, 1936).

Type locality: "Pequeni-Esperanza ridge near head of Pequeni, 1800 feet, eastern Panama" [= Pequení-Esperanza ridge near head of Río Pequení, W Panamá Prov., cen. Panama, 550 m].

Distribution: Central Panama (Colón, W Panamá), 1000–2460 m.

Source: C.W. Myers & Ibañez, 2012.

19. *Rhadinaea stadelmani* L.C. Stuart & Bailey, 1941. Occ. Pap. Mus. Zool., Univ. Michigan (442): 4–6.

Type: Holotype, UMMZ 89078, a 173 mm female (R. Stadelman, Nov. 1937).

Type locality: "Todos Santos, Huehuetenango, Guatemala, at 8000 feet altitude."

Distribution: Western Guatemala (Sierra de los Cuchumatanes and Montañas del Cuilco of Huehuetenango), 2000–2660 m.

Source: Mendelson & Kizirian, 1995a.

20. *Rhadinaea taeniata* (W.C.H. Peters, 1863c). Mber. Königl. Akad. Wiss. Berlin 1863(6): 275–277. (*Dromicus taeniatus*)

Synonym: *Rhadinaea aemula* Bailey, 1940.

Type: Lectotype, ZMB 2117A, a 505–510 mm male (F. Deppe, Dec. 1824–May 1829), designated by C.W. Myers (1974: 106).

Type locality: "Mexico" via lectotype selection. Restricted to S half of México State *fide* C.W. Myers (1974: 104).

Distribution: Southern Mexico (Cordillera Volcánica of Distrito Federal, S Jalisco, México, Michoacán, Morelos and Sierra Madre del Sur of cen. Guerrero, Oaxaca), 1000–2835 m.

Source: Zaldivar-Riverón & Pérez-Ramos, 2001.

21. *Rhadinaea vermiculaticeps* (Cope, 1860e). Proc. Acad. Nat. Sci. Philadelphia 12(6): 249–250. (*Taeniophis vermiculaticeps*)

Type: Lectotype, ANSP 3535, a 340 mm male (R.W. Mitchell, 1857–1860), designated by C.W. Myers (1974: 163).

Type locality: "Cocuyos de Veraguas, Veraguas Province, Panama" via lectotyype selection. Emended to Cocuyos de Veraguas, a gold mine near the Río Concepción, N Veraguas Province, cen. Panama, ca. 8°45'N, 81°00'W, elevation 100 m *fide* C.W. Myers (1974: 166).

Distribution: Western Panama (Coclé, N Veraguas), 100–840 m.

Remarks: Type locality emendation to Cocuyas de Veraguas, New Granada (Colombia) *fide* Malnate (1971: 373) in error.

RHADINELLA H.M. Smith, 1941a (Dipsadidae)

Synonym: *Rhadinelia* – W.W. Tanner & Avery, 1982 *(nomen incorrectum).*

Type species: *Rhadinella schistosa* H.M. Smith, 1941a.

Distribution: Southern Mexico and upper Central America.

Sources: H.M. Smith, 1941a and C.W. Myers, 1974, 2011.

R

1. *Rhadinella anachoreta* (E.N. Smith & Campbell, 1994). Occ. Pap. Mus. Nat. Hist. Univ. Kansas (167): 2–6, figs. 1–2. (*Rhadinaea anachoreta*)

Type: Holotype, UTA 33051 (formerly ENS 3360), a 251 mm male (E.N. Smith, 26 June 1991).

Type locality: "north slope of Cerro del Aguacate, Aldea Negro Norte, Sierra de Caral, Municipio de Morales, Departamento de Izabal, Guatemala, 1180 m, about 15°20'49" N, 88°40'35"W."

Distribution: Guatemala (Izabal) and Honduras (Cortés, Santa Bárbara), 200–1180 m.

Sources: J.A. Campbell, 1998b, Townsend et al., 2005b, Townsend, 2006 and McCranie, 2007b, 2011a.

2. *Rhadinella godmani* (A.C.L.G. Günther, 1865). Ann. Mag. Nat. Hist. (3) 15(86): 94. (*Dromicus godmani*) (*nomen corrigendum*)

Synonyms: *Dromicus godmanni* A.C.L.G. Günther, 1865 (*nomen incorrigendum*), *Dromicus godmani* A.C.L.G. Günther, 1983 *in* 1885–1902 (*nomen corrigendum*), *Rhadinaea zilchi* Mertens, 1952a, *Rhadinaea altamontana* E.H. Taylor, 1954, and *Rhadinaea binfordi* Rossman, 1965b.

Type: Lectotype, BMNH 1946.1.9.17, a 453–457 mm male (O. Salvin & F.D. Godman, 1861–1863), designated by C.W. Myers (1974: 122).

Type locality: "Dueñas (Guatemala)" [= San Miguel Dueñas, SW Sacatepéquez Dept., S-cen. Guatemala, 14°31'N, 90°48'W, elevation 1430 m]. Emended to between the volcanos of Agua and Fuego, on an offshoot of the table lands, elevated about 4700 feet above the level of the sea *fide* Salvin (1860: 453).

Distribution: Mesoamerica. Southeastern Mexico (ext. SE Oaxaca, Chiapas), S Guatemala (El Progreso, Chimaltenango, Sacatepéquez, San Marcos), W El Salvador (Santa Ana), SW Honduras (Comayagua, Francisco Morazán, Intibucá, La Paz, Ocotepeque, Santa Bárbara, Yoro), Costa Rica (Cartago, Heredia, San José) and ext. W Panama (Chiriquí), 1000–2650 m.

Sources: Bocourt, 1886 *in* Duméril, Mocquard & Bibron, 1870–1909, A.C.L.G. Günther, 1893 *in* 1885–1902, L.C. Stuart & Bailey, 1941, Uzzell & Starrett, 1958, L.D. Wilson et al., 1979, L.D. Wilson & Meyer, 1985, McCranie & Wilson, 1992, Holm & Cruz-Díaz, 1994, Mendelson & Kizirian, 1995a, G. Köhler, 1999b, G. Köhler & McCranie, 1999b, Solórzano, 2004 and McCranie, 2006, 2011.

Remarks: Probably occurs in N Nicaragua *fide* G. Köhler (1999b: 81).

3. *Rhadinella hannsteini* (L.C. Stuart, 1949). Proc. Biol. Soc. Washington 62: 165–167. (*Trimetopon hannsteini*)

Type: Holotype, UMMZ 98756, a 324 mm male (L.C. Stuart, 14 May 1947).

Type locality: "Finca La Paz, (18 kilometers [straight line] due north of Coatepeque), Pacific slopes of southwestern highlands, Departament of San Marcos, Guatemala. Elevation, 1,450 meters" [= Finca La Paz, S San Marcos Dept., ext. SW Guatemala, 14°52'N, 91°52'W, elevation 1450 m].

Distribution: Extreme SE Mexico (SE Chiapas) and SW Guatemala (S San Marcos, Suchitepéquez), 500–2000 m.

Sources: L.C. Stuart, 1963, Landy et al., 1966, Villa et al., 1988, J.A. Campbell & Vannini, 1989 and E.N. Smith & Campbell, 1994.

4. *Rhadinella hempsteadae* (L.C. Stuart & Bailey, 1941). Occ. Pap. Mus. Zool., Univ. Michigan (442): 2–4. (*Rhadinaea hempsteadae*)

Type: Holotype, UMMZ 89080, adult male (L.C. Stuart, 26 May 1940).

Type locality: "cloud forest zone (ca. 5700 feet) above Finca Chichén, Alta Verapaz, Guatemala" [= Finca Chichén, SW Alta Verapaz Dept., cen. Guatemala, 15°23'N, 90°18'W, 1740 m].

Distribution: Central Guatemala (Cordillera de los Cuchumatanes of Huehuetenango and Montañas de Alta Verapaz of Alta Verapaz, Baja Verapaz, El Progreso, Zacapa), 1740–1900 m.

Sources: L.C. Stuart, 1954a, 1963, Holm & Cruz-Díaz, 1994 and Mendelson & Kizirian, 1995a.

5. *Rhadinella kanalchutchan* (Mendelson & Kizirian, 1995a). Herpetologica 51(3): 302–308, figs. 1–3. (*Rhadinaea kanalchutchan*)

Type: Holotype, UTEP 12233, a 490 mm male (J.D. Johnson & C. Harrison, 10 May 1989).

Type locality: "8 km (5 mi.) S San Cristóbal de Las Casas, Chiapas, Mexico" [= 8 km S San Cristóbal de Las Casas (16°44'N, 92°38'W, elevation 2135 m), cen. Chiapas State, SE Mexico].

Distribution: Southern Mexico (Chiapas), 2300–2710 m.

6. *Rhadinella kinkelini* (Boettger, 1898). Kat. Rept. Senck. Mus. 2: 68. (*Rhadinaea kinkelini*)

Synonyms: *Rhadinaea veraepacis* L.C. Stuart & Bailey, 1941, and *Rhadinaea pinicola* Mertens, 1952b.

Type: Holotype, SMF 19062 (formerly SMF-B 8178,1a), a 358 mm female (A. Kinkelin, 1897).

Type locality: "Matagalpa, Nicaragua" [= Matagalpa, cen. Matagalpa Dept., N-cen. Nicaragua, 12°53'N, 85°57'W, elevation 720 m]. Probably the humid forest above the town in the Cordillera Dariense *fide* C.W. Myers (1974: 141).

Distribution: Mesoamerica. Southern Mexico (Chiapas), Guatemala (Alta Verapaz), ext. NW El Salvador (ext. N Santa Ana), W Honduras (Copán, Cortés, Intabucá, Ocotepeque, Olancho, Yoro) and N-cen. Nicaragua (Jinotega, Matagalpa), [700] 900–2200 m.

Sources: L.D. Wilson & Meyer, 1971, 1985, Campbell, 1982, L.D. Wilson, 1983a, Villa et al., 1988, McCranie & Wilson, 1991a, E.N. Smith & Campbell, 1994, Mendelson III & Kizirian, 1995b, G. Köhler & McCranie, 1999a, G. Köhler et al., 2005 and McCranie, 2011a.

7. *Rhadinella lachrymans* (Cope, 1870). Proc. Amer. Philos. Soc. (1869–1870) 11(81): 154. (*Lygophis lachrymans*)

Type: Holotype, ANSP 5539, a 385 mm female (F. Sumichrast, 1855–1869).

Type locality: Unknown. Designated as Orizaba, Mexico *fide* F. Sumichrast *in* Cope (1900: 759) (in error). Probably Chiapas, Mexico *fide* Bailey (1940: 7) and C.W. Myers (1974: 141). Restricted to Cerro Ovando, Chiapas, Mexico *fide* H.M. Smith & Taylor (1950a: 326).

Distribution: Extreme SE Mexico (Sierra Madre of SE Chiapas), SW Guatemala (Sierra Madre of Chimaltenango, Quezaltenango, San Marcos) and Honduras (Olancho), 1050–2635 m.

Sources: A.C.L.G. Günther, 1894 *in* 1885–1902, F. Werner, 1929a, Bailey, 1940, H.M. Smith & Taylor, 1945, L.C. Stuart, 1963, McCranie & Wilson, 2001c and McCranie, 2011a.

8. *Rhadinella montecristi* (Mertens, 1952b). Zool. Anz. 149(5–6): 136–137. (*Rhadinaea montecristi*)

Type: Holotype, SMF 43188, a 408 mm male (A. Zilch, 26–27 Aug. 1951).

Type locality: "Hacienda Monte Cristo. 2200 m. H., Gebirge von Metapan, Dept. Santa Ana, El Salvador" [= Hacienda, Monte Cristo, Cordillera Metapán-Alotepeque, ext. N Santa Ana Dept., ext. NW El Salvador, 14°24'N, 89°35'W, elevation 2200 m].

Distribution: Upper Central America. Guatemala (Zacapa), ext. W Honduras (Comayagua, Copán, Cortés, Lempíra, Ocotepeque) and ext. NW El Salvador (Chalatenango, Santa Ana), 1300–2620 m.

Sources: Mertens, 1952c, Uzzell & Starret, 1958, L.D. Wilson & Meyer, 1985, McCranie & Wilson, 1991b, Holm, 1993, E.N. Smith & Campbell, 1994, G. Köhler, 1996b, G. Köhler et al., 2005, Townsend et al., 2006, L.D. Wilson et al., 2006 and McCranie, 2011a.

9. *Rhadinella pegosalyta* (McCranie, 2006). Proc. Biol. Soc. Washington 119(4): 528–532, figs. 1–2, 4–5. (*Rhadinaea pegosalyta*)

Type: Holotype, USNM 56920, a 355 mm male (F.E. Astañeda, T. Manzanares & J.R. McCranie, 18 June 2005).

Type locality: "near visitor's center of Parque Nacional El Cusuco, 15°31'N, 88°12'W, 1550 m elev., Departamento de Cortés, Honduras."

Distribution: Northwestern Honduras (Cortés). Known only from the vicinity of type locality.

Source: McCranie, 2011a.

10. *Rhadinella pilonaorum* (L.C. Stuart, 1954a). Proc. Biol. Soc. Washington 67: 176–177. (*Trimetopon piloñaorum*)

Type: Holotype, UMMZ 102635, a 300–310 mm female (A. Piloña, 25 July 1947).

Type locality: "Finca La Gloria (about 12 kilometers [straight line] northeast of Chiquimulilla), Department of Santa Rosa, Guatemala at about 950 meters" [=

Finca La Gloria, N Santa Rosa Dept., SE Guatemala, 14°05'N, 90°23'W, elevation 950 m].

Distribution: Southeastern Guatemala (N Santa Rosa) and W El Salvador (La Libertad, San Salvador), 670–1080 m.

Source: G. Köhler et al., 2005.

11. *Rhadinella posadasi* (Slevin, 1936). Proc. California Acad. Sci. (4) 23(4): 79–80. (*Trimetopon posadasi*)

Type: Holotype, CAS 66964, a 257–276 mm male (J.R. Slevin, 8 Aug. 1924).

Type locality: "Southern slope Volcan Zunil, Suchitepequez, Guatemala" [= Volcán Zunil, N Suchitepéquez Dept., SW Guatemala, 14°44'N, 91°27'W].

Distribution: Southwestern Guatemala (Pacific versant of Sierra Madre of Sololá, N Suchitepéquez), 875–1000 [1830] m.

12. *Rhadinella rogerromani* (G. Köhler & McCranie, 1999b). Senck. Biol. 79(2): 244–248, figs. 1–6. (*Rhadinaea rogerromani*)

Type: Holotype, SMF 79156, a 286 mm male (G. Köhler & E. G. Köhler, 13 April 1999).

Type locality: "from S slope of Cerro Saslaya (13°46.0' N, 85°02.3' W), 1450 m elevation, Región Autónoma Atlántico Norte, Nicaragua."

Distribution: Northeastern Nicaragua (Atlántico Norte), 1450 m. Known only from type locality.

Source: McCranie, 2006.

13. *Rhadinella schistosa* H.M. Smith, 1941a. Copeia 1941(1): 7–9, fig. 1.

Type: Holotype, FMNH 100065 (formerly EHT-HMS 23580 & EHT-CC 2116), a 186–187 mm male (E.H. Taylor, Aug. 1940).

Type locality: "Cuautlapan, Veracruz, México" [= Cuautlapán, W-cen. Veracruz State, SE Mexico, 18°52'N, 97°01'W, elevation 1000 m].

Distribution: Mexico (Sierra Madre de Oaxaca of N Oaxaca and W-cen. Veracruz), 1000 m.

Source: Pérez-Higareda & Smith, 1991.

14. *Rhadinella serperaster* (Cope, 1871b). Proc. Acad. Nat. Sci. Philadelphia 23(2): 212–213. (*Rhadinaea serperaster*)

Type: Holotype, ANSP 3739, a 176 mm male (C.H. Van Patten, 1865–1871).

Type locality: "near San José, Costa Rica" [= vicinity of San José (9°56'N, 84°05'W, elevation 1160 m), N San José Prov., cen. Costa Rica].

Distribution: Central Costa Rica (Alajuela, Cartago, Heredia, San José), 1160–2200 m.

Sources: E.H. Taylor, 1951, 1954.

15. *Rhadinella tolpanorum* (Holm & Cruz-Díaz, 1994). Herpetologica 50(1): 17–20, figs. 1–2. (*Rhadinaea tolpanorum*)

Type: Holotype, UNAH 2732, a 381 mm female (G. Dodge, T. Powell & P.A. Holm, 23 May 1991).

Type locality: "0.8 km N and 1.5 km E Cerro San Francisco in the Cordillera Nombre de Dios at 1900 m, Department of Atlántida, Honduras (15°27' N, 87°18' W)."

Distribution: Northwestern Honduras (Atlántida, N Yoro), 1500–1800 m.

Sources: G. Köhler & McCranie, 1999b, McCranie & Castaneda, 2004a and McCranie, 2011a.

RHADINOPHANES C.W. Myers & Campbell, 1981 (Dipsadidae)

Type species: *Rhadinophanes monticola* C.W. Myers & Campbell, 1981.

Distribution: Southwestern Mexico.

Sources: C.W. Myers & Campbell, 1981, Zaher et al., 2009 and Mulcahy et al., 2011.

1. *Rhadinophanes monticola* C.W. Myers & Campbell, 1981. Amer. Mus. Novit. (2708): 5–7, figs. 2, 5, 8a.

Type: Holotype, AMNH 116332, a 357 mm male (J.A. Campbell, 21 May 1974).

Type locality: "1 mile (1.6 km.) north of Puerto del Gallo, at an elevation of approximately 9000 feet (~2750 m.) on Cerro Teótepec, State of Guerrero, Mexico." Also following the above statement is "Puerto del Gallo, at nearly 2500 m elevation, is the site of an abandoned lumber camp on the southwestern slope of Cerro Teótepec, roughly 40 km. by air northwest of Atoyac and 70 km. by air north-northwest of Acapulco; 17°27'N, 100°09'W)."

Distribution: Southern Mexico (Sierra Madre del Sur of cen. Guerrero), 2600–2950 m. Known only from vicinity of type locality.

RHAGERHIS W.C.H. Peters, 1862a (Psammophiidae)

Synonyms: *Rhagerrhis* A.C.L.G. Günther, 1866 *(nomen emendation), Ragerrhis* – W.C.H. Peters, 1880b *(nomen incorrectum), Rhagechis* – Hoffmann, 1890 *(nomen incorrectum), Rhagerris* – Elter, 1981 1890 *(nomen incorrectum),* and *Scutophis* Brandstatter, 1995 (*nomen ineditum*)

Type species: *Coluber moilensis* A. Reuss, 1834.

Distribution: Middle East and N Africa.

R

Sources: Brandstatter, 1995, Vidal et al., 2008 and W. Bohme & Pury, 2011.

Remarks: Previously in the genus *Malpolon.*

1. *Rhagerhis moilensis* (A. Reuss, 1834). Mus. Senck. 1: 142–143, pl. 7, fig. 1a–b. (*Coluber moilensis*)

Synonyms: *Coelopeltis producta* Gervais, 1857, *Malpolon cordofanensis* F. Werner, 1919, ? *Rhamphiophis septentrionalis* Angel, 1923a, and *Malpolon monspessulanus saharatlanticus* Geniez, Cluchier & Haan, 2006.

Type: Holotype, SMF 20017 (formerly SMF-R III.R.1.a & SMF-B 9143c), a 930 mm female (E. Rüppell, 1826).

Type locality: "Arabien, Gegend von Moilah" [= vicinity of Al Muwaylah, Red Sea coast, Al Muqata'ah Ash Shamaliyah Prov., NW Saudi Arabia, 27°41'N, 35°27'E, elevation NSL].

Distribution: Northern Africa and SW Asia. Northern Mauritania (Adrar, Assaba, Dakhlet Nouadhibou, Hodh el Gharbi, Inchiri, Trarza, Zemmour), Western Sahara (Rio de Oro, Saguia El Hamra), S Morocco (Agadir, Ouarzazate, Safi, Tan Tan, Tata, Tiznit), Algeria (Biskra, El Bayadh, M'Sila, Tamanrasset), N Libya (Ajdabiya, Al Jufrah, Az Zawiyah, Banghazi, Surt, Tarabulus, Tripoli), S Tunisia (Ariana, Gabès, Gafsa, Jendouba, Kebili, Medenine, Nabeul, Sidi Bou Zid, Sousse, Tataouine, Tozeur), E Mali (Gao, Kayes, Ségou, Tombouctou), Niger (Agadez, Diffa, Maradi, Zinder), N and E Egypt (Alexandria, Beheira, El-Bahr El-Ahmar, Faiyum, Giza, Ismailia, Matrouh, North Sinai), N Sudan (El-Bahr El-Ahmar, Al Khartum, Ash Shamaliyah, Shamal Darfur), Eritrea (Northern Red Sea), Syria, SE Israel (E Southern), Jordan (Amman, Aqaba, Irbid, Karak, Maan, Mafraq, Tafilah, Zarqa), Saudia Arabia (Eastern, Jazan, Madinah, Makkah, Tabuk), Kuwait, United Arab Emirates (Abu Dhabi, Dubai, Abu Dhabi Is.), Oman (Al Batinah, Dhofar, Muscat, Ash Sharqiyah, Masirah Is.), Yemen (Abyan, Aden, Al Jawf, Hadhramaut, Shabwah), Iraq (Al Basrah, Baghdad, Wasit) and SW Iran (Hormozghan, Khuzestan), NSL–1885 m.

Sources: Anderson, 1898, Doumergue, 1901, Mayet, 1903, Flower, 1933, Angel & Lhote, 1938, Villiers, 1950a, 1975, Chpakowsky & Chnéour, 1954, Domergue, 1959b, Pasteur & Bons, 1960, Kramer & Schnurrenberger, 1963, Corkill & Cochrane, 1966, Mandaville, 1967, Papenfuss, 1969, Y. Werner, 1971, 1973, 1988, Gasperetti, 1988, Gruber, 1989, Gallagher, 1990, Latifi, 1991, Leviton et al., 1992, Schätti & Gasperetti, 1994, Bons & Geneiz, 1996, Brandstätter, 1996, Schleich et al., 1996, Bogaerts, 1998, Chippaux, 1999, 2006, Jongbloed, 2000, Bouskila & Amitai, 2001, Disi et al., 2001, Geniez et al., 2004, J.-F. Trape & Mané, 2006b, Egan, 2007, Largen & Spawls, 2010, Amr & Disi, 2011 and Bar & Haimovitch, 2011.

Remarks: Mertens (1967: 92) listed the type locality as Mohila, Arabia. Two specimens from S Ghana are problematic (Gallagher, 1988: 255).

RHAMNOPHIS A.C.L.G. Günther, 1862b (Colubridae)

Synonyms: *Crypsidomus* A.C.L.G. Günther, 1864c *(nomen substitutum), Chrysidomus* – Matschie, 1893a *(nomen incorrectum), Rhamphophis* – Palacky, 1898 *(nomen incorrectum), Rhamnaethiops* – F. Werner, 1923 *(lapsus calami), Rhamnnophis* – Loveridge, 1924 *(nomen incorrectum), Rhamophis* – Ditmars, 1933 *(nomen incorrectum), Rhamnopis* – Witte, 1941 *(nomen incorrectum),* and *Rhamnophi* – Stucki-Stirn, 1979 *(nomen incorrectum).*

Type species: *Rhamnophis aethiopissa* A.C.L.G. Günther, 1862b.

Distribution: West, cen. and E Africa.

Sources: Sternfeld, 1909a, Loveridge, 1944b, Monard, 1951, Laurent, 1956a, Witte, 1962, Thys van den Audenaerde, 1965, Knoepffler, 1966, Villiers, 1966, J.-F. Trape, 1981, 1985, J.-F. Trape & Roux-Estève, 1995, Broadley & Wallach, 2002, Chippaux, 2006, Chirio & LeBreton, 2007 and Pauwels & Vande weghe, 2008.

1. *Rhamnophis aethiopissa* A.C.L.G. Günther, 1862b. Ann. Mag. Nat. Hist. (3) 9(49): 129, pl. 10.

Synonyms: *Crypsidomus aethiops* Boettger, 1888e, *Thrasops splendens* Andersson, 1901, *Rhamnophis ituriensis* K.P. Schmidt, 1923, and *Rhamnophis aethiopissa elgonensis* Loveridge, 1929.

Type: Holotype, BMNH 1946.1.4.99, a 933 mm specimen.

Type locality: "West Africa."

Distribution: West, cen. and East Africa. Guinea (Macenta, Nzérékoré), Sierra Leone (Western), Liberia (Montserrado), S Ivory Coast (Abidjan, Lagunes, San Pédro), S Ghana (Accra, Eastern, Volta), S Togo (Centrale, Plateaux), S Nigeria (Akwa Ibom, Bayelsa, Lagos, Rivers), S Cameroon (Est, Centre, Littoral, Nord-Ouest, Sud, Sud-Ouest), Equatorial Guinea (Bioko Is.), Gabon (Moyen-Ogooué, Ngounié, Ogooué-Ivindo, Ogooué-Lolo, Ogooué-Maritime, Woleu-Ntem), S Congo (Brazzaville, Cuvette, Kouilou, Sangha), Central African Republic (Lobaye, Ombella-Mpoko, Sangha), Democratic Republic of the Congo (Bandundu, Bas-Congo, Equateur, Kasai Occidental, Kasai Oriental, Kinshasa, Nord-Kivu, Orientale, Sud-Kivu), N Angola (Luanda, Lunda Norte), S Uganda (Central, Western), W Kenya (S Rift Valley, Western), cen. Rwanda, W Tanzania (Kigoma, Kagera) and NW Zambia (North-Western), NSL–2000 m.

Sources: Sternfeld, 1909b, K.P. Schmidt, 1923, Loveridge, 1929, Angel, 1933b, Laurent, 1960, Doucet, 1963, Villiers, 1963, Mertens, 1965c, Roux-Estève, 1965, Menzies, 1966, Leston & Hughes, 1968, Hughes & Barry, 1969, Hulselmans et al., 1970, Pitman, 1974, Spawls, 1978, Hughes, 1983, Joger, 1990, Spawls et al., 2002, Broadley et al., 2003, Chirio & Ineich, 2006 and Segniagbeto et al., 2011.

Remarks: Probably occurs in Benin *fide* Villiers (1951a: 27) but possibly occurs in Benin *fide* Hughes (2013: 152).

2. *Rhamnophis batesii* (Boulenger, 1908g). Ann. Mag. Nat. Hist. (8) 2(7): 93. (*Thrasops batesii*)

Types: Syntypes (3), BMNH 1946.1.4.63–65, an 1800 mm specimen and two juveniles (G.L. Bates, 1895–1908).

Type locality: "South Cameroon (Efulen and Akok, Kribi River district, and Ja River district)" [= Akok (2°47'N, 10°17'E, elevation 110 m), W Sud Prov., and Efoulan (2°39'N, 12°46'E, elevation 665 m), E Sud Prov., S Cameroon].

Distribution: Central Africa. Southern Cameroon (Est, Centre, Nord-Ouest, Sud, Sud-Ouest), SW Central African Republic (Lobaye), Gabon (Ogooué-Ivindo, Ogooué-Maritime), S Congo (Kouilou) and Democratic Republic of the Congo (Bandundu, Bas-Congo, Equateur, Nord-Kivu, Orientale), 20–1300 m.

Sources: K.P. Schmidt, 1923, Knoepffler, 1966, Stucki-Stirn, 1979 and D. Lawson, 1993.

RHAMPHIOPHIS W.C.H. Peters, 1854 (Psammophiidae)

Synonyms: *Ramphiophis* – J.T. Reinhardt, 1861c *(nomen incorrectum), Rhamphioppis* – Rochebrune, 1884 *(nomen incorrectum), Rhamphophis* – Palacky, 1898 *(nomen incorrectum), Rhamphiopholis* – M.A. Smith, 1930c *(nomen incorrectum), Rhampiophis* – Anthony, 1955 *(nomen incorrectum), Rhamphiopsis* – Thistleton, 1961 *(nomen incorrectum), Rhamphiophs* – Skinner, 1973 *(nomen incorrectum),* and *Raphiophis* – Stucki-Stirn, 1979 *(nomen incorrectum).*

Type species: *Rhamphiophis rostratus* W.C.H. Peters, 1854.

Distribution: Subsaharan Africa.

Fossil record: Pliocene of East Africa.

Sources: Loveridge, 1957a, Pitman, 1974, Chirio & Ineich, 1991, Gravlund, 2001, Vidal & Hedges, 2002, Kelly et al., 2008, 2009, Vidal et al., 2008 and Zaher et al., 2009.

Remarks: Kelly et al., 2008 transferred the species *acutus* to the genus *Psammophylax.*

R

1. *Rhamphiophis maradiensis* Chirio & Ineich, 1991. Bull. Mus. Natl. Hist. Nat. (4) 13A(1–2): 226–229, figs. 1b, d, 2b, d, pl. 1, fig. b.

Type: Holotype, MNHN 1989.2739, a 573 mm female (native, Oct. 1987).

Type locality: "environs de Gari'n Bakwai, 50 km au sud-ouest de Maradi, au Niger" [= vicinity of Gari'n Bakwai, 50 km SW Maradi (13°29'N, 7°06'E, elevation 360 m) near Nigerian border, S Maradi Dept., SW Niger].

Distribution: Southern Niger (SW Maradi), 390 m.

Source: Chippaux, 2006.

2. *Rhamphiophis oxyrhynchus* (J.T. Reinhardt, 1843). Kgl. Danske Vidensk. Selsk. Natur. Math. Afh. 10: 244–246, pl. 1, figs. 10–12. (*Psammophis oxyrhynchus*)

Synonym: *Rhamphiophis connali* H.W. Parker, 1929.

Type: Lectotype, ZMUC 63125, a 462 mm specimen (J.R. Chenon, 1826–1836), designated by F.W. Braestrup *in* Hughes & Barry (1969: 1021).

Type locality: "Guinea" [= coast of Ghana *fide* Hughes & Barry, 1969: 1021] via lectotype selection. Restricted to vicinity of Accra, Ghana *fide* J.B. Rasmussen & Hughes (1997: 14, 16).

Distribution: West Africa. Senegal (Dakar, Fatick, Kaolack, Kolda, Tambacounda, Thiès, Ziguinchor), Guinea, SW Mali (Kayes, Koulikoro, Mopti, Sikasso), Ivory Coast (Daloa), Burkina Faso (Centre, Centre-Est, Centre-Nord, Centre-Ouest, Est, Nord, Hauts-Bassins, Sud-Ouest, Volta-Noire), S Ghana (Accra, Volta), Togo (Kara, Maritime, Savanes), Benin (Alibori, Atakora, Atlantique, Collines, Zou), SW Niger (Dosso, Maradi, Niamey, Tahoua), N Nigeria (Gongola, Kano), N Cameroon (Adamaoua, Extreme-Nord, Nord), S Chad (Kanem, Moyen-Chari) and Central African Republic (Bamingui-Bangoran, Ouham, Ouham-Pende), NSL–1700 m.

Sources: Sternfeld, 1909b, 1910d, Doucet, 1963, Leston & Hughes, 1968, Hughes & Barry, 1969, J Hulselmans & Verheyen, 1970, Hulselmans et al., 1970, 1971, Roman, 1980, Hughes, 1983, 2013, Chirio & Ineich, 1991, Largen & Rasmussen, 1993, J.B. Rasmussen & Hughes, 1997, Spawls et al., 2002, Chippaux, 2006, Chirio & Ineich, 2006, J.-F. Trape & Mané, 2006b, Chirio & LeBreton, 2007, Ullenburch et al., 2010 and Segniagbeto et al., 2011.

Remarks: Possibly occurs in Sudan *fide* Hughes (1983: 353).

3. *Rhamphiophis rostratus* W.C.H. Peters, 1854. Mber. Königl. Akad. Wiss. Berlin 1854(11): 624.

Synonyms: *Rhagerrhis unguiculata* A.C.L.G. Günther, 1868, and *Coelopeltis porrectus* Jan, 1870 *in* Jan & Sordelli, 1866–1870.

Types: Syntypes (4), ZMB 2484 and ZMB 9994, a 1210 mm, 1210 mm, 430 mm and 380 mm specimen (W.C.H. Peters, June 1843–Aug. 1847).

Type locality: "Tette" [= Tete, Tete Prov., W Mozambique, 16°10'S, 33°36'E, elevation 125 m]. Emended to Mesuril, Quitangonha und bei Tette *fide* W.C.H. Peters (1882d: 124) [= Mossuril District, Quitangonha Is., Nampula Prov., and Tete, Tete Prov., Moazmbique].

Distribution: Eastern Africa. Southern Sudan (Janub Kurdufan), SE South Sudan (Eastern Equatoria), Ethiopia (Awdal, Bale, Gemu Gofa, Hararge, Shoa,

Sidamo), Somalia (Bakool, Bari, Bay, Gedo, Jubbada Dhexe, Jubbada Hoose, Nugaal, Mogadishu, Mudug, Shabeellaha Hoose, Togdheer, Woqooyi Galbreed), Kenya (Central, Coast, Eastern, North-Eastern, Rift Valley, Lamu Is.), Uganda (N Eastern, N Western), E Tanzania (Arusha, Dar es Salaam, Dodoma, Kilimanjaro, Lindi, Morogoro, Mtwara, Mwanza, Piwani, Rukwa, Shinyanga, Tabora, Tanga, Kagera, Zanzibar Is.), SE Democratic Republic of the Congo (Katanga, Orientale, Sud-Kivu), SE Zambia (S Eastern, SE Southern), Malawi (Central, Northern, Southern), Mozambique (Gaza, Inhambane, Sofala, Tete, Quitangonha Is.), Zimbabwe (Manicaland, Mashonaland Central, Mashonaland East, Mashonaland West, Masvingo, Matebeleland North, Matebeleland South), ext. NE Namibia (Caprivi), N Botswana (E Central, North West) and NE South Africa (Limpopo, Mpumalanga), NSL–2135 m.

Sources: W.C.H. Peters, 1882a, Witte, 1933a, 1962, Corkill, 1935, Loveridge, 1936b, 1953, 1955b, Laurent, 1956a, Sweeney, 1961, Broadley, 1971a, Broadley & Cock, 1975, Spawls, 1978, Lanza, 1983a, Hughes, 1983, Pienaar et al., 1983, Auerbach, 1987, Largen & Rasmussen, 1993, Spawls et al., 2002, Broadley et al., 2003, Marais, 2004, Largen & Spawls, 2010 and Broadley & Blaylock, 2013.

Remarks: Original description reprinted in W.C.H. Peters (1855: 53). A subspecies of *R. oxyrhynchus fide* M Griffin (2003: 104). Possibily occurs in Ankole, Uganda *fide* Pitman (1974: 151).

4. *Rhamphiophis rubropunctatus* (J.G. Fischer, 1884a). Jahrb. Hamburg. Wiss. Anst. 1: 7–8, pl. 1, figs. 3a–c. (*Dipsina rubropunctata*)

Type: Holotype, ZMH 4179 (formerly ZMH 1192), a 485 mm specimen (G.A. Fischer [Hamburg Geog. Soc. Exped.], 1884).

Type locality: "bei Aruscha am Fuss des Kilima-Ndjaro, Massai-Gebiete (Ost-Afrika)" [= vicinity of Arusha, base of Mt. Kilimaddjaro, E Arusha Region, NE Tanzania, 3°22'S, 36°40'E, elevation 1400 m].

Distribution: Northeastern Africa. Southeastern South Sudan (Eastern Equatoria), S Ethiopia (Gemu Gofa, Hararge, Illubabor, Kefa, Shoa, Sidamo), W Somalia (Bay, Jubbada Dhexe, Jubbada Hoose, Mogadishu, Shabeellaha Dhexe, Togdheer), E Kenya (Coast, Eastern, North-Eastern, Rift Valley) and NE Tanzania (Arusha, Kilimanjaro, Tanga, Zanzibar Is.), NSL–1455 m.

Sources: Sternfeld, 1908d, Scortecci, 1929, 1939, 1943, H.W. Parker, 1949, Loveridge, 1955, Spawls, 1978, Lanza, 1983a, 1990b, Largen & Rasmussen, 1993, Spawls et al., 2002 and Largen & Spawls, 2010.

Remarks: Figure number in original description misprinted as Fig. 2.

RHINOBOTHRYUM Wagler, 1830 (Colubridae)

Synonyms: *Rhinobothrium* Fitzinger, 1843 *(nomen emendatum), Rhinobotryum* Agassiz, 1844 *(nomen emendatum), Rhinobotrion* – A.-M.-C. Duméril & Bibron, 1844 *(nomen incorrectum), Rhinobotryum* – Palacky, 1898 *(nomen incorrectum),* and *Rhinobotrium* – Maass-Berlin, 1933 *(nomen incorrectum).*

Type species: *Coluber lentiginosus* Scopoli, 1788.

Distribution: Lower Central America and N South America.

Sources: J.A. Peters & Orejas-Miranda, 1970, Kornacker, 1999 and Navarrete et al., 2009.

1. *Rhinobothryum bovallii* Andersson, 1916. Medd. Göteborgs Mus. Zool. (9) 17(5): 32–33, fig. 4. (*Rhinobothrium bovallii*)

Type: Holotype, GNM 1221, an 1190 mm specimen (C. Bovallius, 25 Sept. 1882).

Type locality: "Costa Rica, Siquirres" [= Siquirres, cen. Limón Prov., E Costa Rica, 10°06'N, 83°31'W, elevation 100 m].

Distribution: Lower Central America and NW South America. Southern Honduras (El Paraíso), E Costa Rica (NE Alajuela, Cartago, Heredia, Limón), Panama (Colón, Darién, Los Santos, Panamá), Colombia (Antioquia, Boyacá, Caldas, Cesar, Chocó, Córdoba, Risaralda, Santander, Valle del Cauca), NW Venezuela (Zulia) and NW Ecuador (Carchi, Esmeraldas, Imbabura), 70–900 (2250) m.

Sources: Dunn & Bailey, 1939, E.H. Taylor, 1951, H.M. Smith, 1958, Campbell & Howell, 1965, Pons, 1965, Medem, 1969, Miyata, 1982, L.D. Wilson & Meyer, 1985, Lancini, 1986, Pérez-Sa ntos, 1986, Savage & Villa, 1986, Pérez-Santos & Moreno, 1988, 1991, Villa et al., 1988, Savage, 2002, Solórzano, 2004, McCranie, 2011a and Rojas-Morales, 2012a.

Remarks: No precise Nicaraguan records known *fide* G. Köhler (1999b: 81). Guatemala record doubtful *fide* McCranie (2011a: 191).

2. *Rhinobothryum lentiginosum* (Scopoli, 1788). Delic. Flor. Faun. Insub. Aust. 3: 41, pl. 20, fig. 2. (*Coluber lentiginosus*)

Synonym: *Coluber macrorhinus* H. Boie *in* Schlegel, 1826a *(nomen nudum), Coluber macrorhinus* Schlegel, 1826b *(nomen nudum),* and *Rhinobothryum macrorhinum* H. Boie *in* Wagler, 1830.

Type: Holotype, not designated, a 1030 mm specimen, location unknown.

Type locality: Unknown. Designated as America *fide* Fitzinger (1843: 27). Restricted to tropical South America *fide* Boulenger (1896a: 82).

Distribution: South America. Colombia (Bolívar, Caquéta, Chocó), Venezuela (Amazonas, Bolívar), NW Guyana (Barima-Waini), Suriname (Brokopondo, Para), French Guiana (Cayenne), Brazil (Acre, Amazonas, Mato Grosso, Pará, Rondônia), E Bolivia, E Peru (Cusco, Loreto, Madre de Dios, Ucayali), and Paraguay, 10–490 m.

Sources: Duméril, Bibron & Duméril, 1854b, Amaral, 1930, Niceforo-Maria, 1942, Prado & Hoge, 1948, Cunha & Nascimento, 1976a, 1978, 1993, Gasc & Rodrigues, 1980, Abuys, 1986b, 2003, Dixon Soini, 1986, Chippaux, 1987, Roge & Sauvenet, 1987, Duellman, 1990, 2005, M. Rodriguez & Cadle, 1990, Molina R. et al., 1996, Martins & Oliveira, 1998, Starace, 1998, Lehr & Aguilar, 2000b, Lehr, 2002, Freitas, 2003, Miranda et al., 2009, and C.J. Cole et al., 2013.

Remarks: Date of Scopoli universally listed incorrectly as 1785.

RHINOCEROPHIS S.W. Garman, 1881 (Viperidae)

Type species: *Bothrops ammodytoides* Leybold, 1873.

Distribution: Southern South America.

Fossil records: Upper Pleistocene of Argentina.

Sources: J.A. Campbell & Lamar, 2004, Fenwick et al., 2009, Carrasco et al., 2010 and Hoser, 2012d.

Remarks: Taxonomic status questionable *fide* Carrasco et al., 2010. A valid genus *fide* Fenwick et al., 2009.

1. *Rhinocerophis alternatus* (A.-M.-C. Duméril, Bibron & Duméril, 1854b). Erpét. Gén. 7(2): 1512, pl. 82bis, fig. 1–1a. (*Bothrops alternatus*) (*nomen protectum*)

Synonyms: *Craspedocephalus brasiliensis* Gray, 1849a (*nomen oblitum*), *Lachesis alternata* – Boettger, 1898, and *Lachesis inaequalis* Magalhães, 1925.

Types: Syntypes (2), MNHN 7744 (F.L.N.C.L. de Castelnau & E. Deville, June 1843–Aug. 1847), and MNHN 7765 (A.C.V.M.D. d'Orbigny, 27 Dec. 1827–4 Feb. 1828), longest syntype 1150 mm, lost *fide* McDiarmid et al. (1999: 253).

Type locality: "Amérique méridionale" and "Paraguay" [= South America and Paraguay].

Distribution: Southern South America. Southern Brazil (Goiás, S Mato Grosso, Minas Gerais, Paraná, Rio de Janeiro, Rio Grande do Sul, Santa Catarina, São Paulo), Uruguay (Artigas, Canelones, Cerro Largo, Durazno, Flores, Floridae, Maldonado, Montevideo, Paysandú, Rio Negro, Rivera, Rocha, Salto, San José, Soriano, Tacuarembó, Trienta y Tres), Paraguay (Alto Paraguay, Alto Paraná, Amambay, Asunción, Boquerón, Caaguazú, Caazapá, Canindeyú, Central, Concepción, Cordillera, Guairá, Itapúa, Misiones, Neembucú, Paraguarí, Presidente Hayes, San Pedro) and N Argentina (Buenos Aires, Catamarca. Chaco, Córdoba, Corrientes, Entre Ríos, Formosa, La Pampa, Misiones, Rio Negro, San Luis, Santa Fe, Santiago del Estero, Treinta y Tres, Tucumán), NSL–700 m.

Sources: Hoge & Belluomini, 1964, Hoge, 1966a, Cei, 1994, D.R. Norman, 1994, Moro, 1996, Leynaud & Bucher, 1999, Nogueira & Valdujo, 2001, Scanferla & Nenda, 2005 Carr et al., 2007, Cacciali, 2008, Carrasco et al., 2009 and Prigioni et al., 2011.

Remarks: *Craspedocephalus brasiliensis* Gray, 1849a would be preoccupied by *Craspedocephalus brasiliensis* Lacépède, 1789 but the latter work is invalid *fide* Opinion 1463 (ICZN, 1987). Therefore, in accordance with Art. 23.9.2 of the Code (ICZN, 1999), *Bothrops alternatus* A.-M.-C. Duméril, Bibron & Duméril is designated a *nomen protectum* and *Craspedocephalus brasiliensis* Gray a *nomen oblitum*.

2. *Rhinocerophis ammodytoides* (Leybold, 1873). Pampas Arjent.: 80–81. (*Bothrops ammodytoides*)

Synonyms: *Rhinocerophis nasus* S.W. Garman, 1881, *Bothrops patagonicus* F. Müller, 1885, and *Bothrops burmeisteri* Koslowsky, 1895b.

Type: Holotype, ? MNHNC, a 750 mm specimen (F. Leybold, 18 Feb. 1871).

Type locality: "estancia de Lo-Aguirre, Pampas Arjentinas, Provincia Mendoza, Sud-América" [= Los Aguirre ranch, bet. San Carlos (33°46'S, 69°03W, elevation 960 m) and Tierra Blanca (ca. 34°02'S, 69°05'W, elervation 1300 m) *fide* Leybold map of expedition], San Carlos Dept., W Mendoza Prov., cen. Argentina].

Distribution: Southern Argentina (S Buenos Aires, Catamarca, Chubut, Córdoba, Jujuy, La Pampa, La Rioja, Mendoza, Neuguén, Rio Negro, Salta, San Juan, San Luis, Santa Cruz, Tucumán), NSL–2000 [3700] m.

Sources: Cei, 1994, Moro, 1996, J.C. Murphy & Schlager, 2003, Scolaro, 2005, 2006 and Carrasco et al., 2009, 2010.

3. *Rhinocerophis cotiara* (J. Gomés, 1913). Ann. Paul. Med. Cirurg. 1(3): 65–67, pl. 8, figs. 1–6. (*Lachesis cotiara*)

Type: Lectotype, IB 452, an 850–852 mm male (R. da Cunhae-Silva, March 1913), designated by Franco et al. (2005: 76), destroyed by fire 15 May 2010.

Type locality: "Nucleo Colonial Cruz Machado" [= Cruz Machado, S Paraná State, S Brasil, 26°01'S, 51°21'W, elevation 890 m] via lectotype selection

Distribution: Southeastern South America. Southeastern Brazil (Paraná, N Rio Grande do Sul, Santa Catarina, SE São Paulo) and NE Argentina (Misiones), NSL–1800 m.

Sources: Bousek, 1928, Lema, 1960a–b, Hoge & Belluomini, 1964, J.D. Williams & Francini, 1991, Cei, 1994 and Franco et al., 2005.

R

4. *Rhinocerophis fonsecai* (Hoge & Belluomini, 1959). Mem. Inst. Butantan (1957–1958) 28: 195–197, figs. 1–2, 3b, 4b, 5a, 9. (*Bothrops fonsecai*)

Type: Holotype, IB 7896, an 872 mm male (J. Sterce, 2 April 1933), destroyed by fire 15 May 2010.
Type locality: "Santo Antonio do Capivary, Estado do Rio de Janeiro, Brasil."
Distribution: Southeastern Brazil (ext. S Minas Gerais, S Rio de Janeiro, NE São Paulo), 1000–1600 m.
Sources: Hoge & Belluomini, 1964 and Manzani et al., 1997.

5. *Rhinocerophis itapetiningae* (Boulenger, 1907b). Ann. Mag. Nat. Hist. (7) 20(118): 338. (*Lachesis itapetiningae*)

Types: Syntypes (2), BMNH 1946.1.19.49–50 (formerly BMNH 1907.9.12.12–13), two females, longest syntype 400 mm (V. Brasil, 1897–1907).
Type locality: "Itapetininga, a city in the State of São Paulo, Brazil" [= Itapetininga, S São Paulo State, SE Brazil, 23°36'S, 48°03'W, elevation 685 m].
Distribution: South-central Brazil (S Goias, SE Mato Grosso, Mato Grosso do Sul, W Minas Gerais, NE Paraná, São Paulo), 685–1500 m.
Source: Lema, 1962a.
Remarks: Rio Grande do Sul records need verification *fide* J.A. Campbell & Lamar (2004: 389).

6. *Rhinocerophis jonathani* (Harvey, 1994). Proc. Biol. Soc. Washington 107(1): 61–64. (*Bothrops jonathani*)

Type: Holotype, MNK R1000, a 620 mm male (M.B. Harvey, 17 Jan. 1992).
Type locality: "Highway 4, approximately 35 km N (by road) of El Empalme, Provincia Carrasco, Departamento de Cochabamba, Bolivia, approximately 2800 m (17°45'S, 65°00 'W)."
Distribution: Central Bolivia (?Chuquisaca, Cochabamba, Mendez, Santa Cruz, Tarija) and Argentina (Jujuy, Salto), 1960–3220 m.
Sources: Dirksen et al., 1995, Harvey et al., 2003, 2005 and Carrasco et al., 2009.

RHINOCHEILUS Baird & Girard, 1853 (Colubridae)

Synonyms: *Rhinochilus* Yarrow, 1883 *(nomen emendatum), Rhinochilus* – Troschel, 1877 *(nomen incorrectum), Rinochilus* – Amaral, 1930f *(nomen incorrectum), Rhnocheilus* – Briceño-Rossi, 1934 *(nomen incorrectum), Rhnochelus* – Briceño-Rossi, 1934 *(nomen incorrectum), Thnocheilus* – Briceño-Rossi, 1934 *(nomen incorrectum),* and *Rhinoceilus* – Meylan, 1982 *(nomen incorrectum).*
Type species: *Rhinocheilus lecontei* Baird & Girard, 1853.
Distribution: Southwestern USA and NW Mexico.
Fossil records: Upper Miocene to upper Pleistocene of SW USA.
Sources: Bocourt, 1887, Klauber, 1941a, Medica, 1975, Stebbins, 1985, Lucas et al., 1995, Degenhardt et al., 1996, Keogh, 1996, P.R. Brown, 1997, Rodríguez-Robles & Greene, 1999, Rodríguez-Robles & Jesús-Escobar, 1999, Bartlett & Tennant, 2000, Holman, 2000a, Werler & Dixon, 2000, Grismer, 2002, Parmley, 2002, Manier, 2004, J.-F. Trape & Mané, 2006b and Cabajal-Márquez et al., 2011b.

1. *Rhinocheilus lecontei* Baird & Girard, 1853. Cat. No. Amer. Rept., Serp. 1: 120.

Synonyms: *Rhinocheilus lecontei tessellatus* S.W. Garman, 1884, *Rhinocheilus antonii* Dugès, 1886, *Rhnocheilus thenminotii* – Briceño-Rossi, 1934 *(nomen incorrectum), Rhnochelus thominotii* – Briceño-Rossi, 1934 *(nomen incorrectum), Rhinocheilus lecontei clarus* Klauber, 1941a and *Rhinocheilus lecontei etheridgei* Grismer, 1990c.
Type: Holotype, MCZ 137, a 533 mm male (J.L. Le Conte, 1849–1852).
Type locality: "San Diego, Cal." [= San Diego, San Diego Co., S California, USA].
Distribution: Southwestern USA (Arizona, California, SE Colorado, SW Idaho, SW Kansas, Nevada, New Mexico, W Oklahoma, Texas, Utah) and N Mexico (Aguascalientes, Baja California Norte, Baja California Sur, Chihuahua, Coahuila, NE Durango, Jalisco, Nayarit, Nuevo Leon, San Luis Potosí, Sinaloa, Sonora, Tamaulipas, Zacatecas, Cerralvo Is.), NSL–1645 m.
Fossil records: Upper Miocene (Hemingfordian) of USA (Texas), middle Pliocene (Blancan III) of USA (Texas), upper Pliocene (Blancan V) of USA (Kansas, Nebraska), lower/middle Pleistocene (Irvingtonian I) of USA (Texas), and upper Pleistocene (Rancholabrean I, II) of USA (Arizona, California, Nevada, New Mexico, Texas).
Remarks: *Rhinocheilus etheridgei* a valid species *fide* Grismer (1999: 460) and *R. antonii* valid *fide* Lemos-Espinal et al. (2004d: 210).

RHINOLEPTUS Orejas-Miranda, Roux-Estève & Guibé, 1970 (Leptotyphlopidae)

Type species: *Typhlops koniagui* Villiers, 1956b.
Distribution: West Africa.
Sources: Guibé et al., 1967, Orejas-Miranda et al., 1970, Villiers, 1975, Hahn, 1980a, Wallach & Ineich, 1996, Wallach, 1998b, Broadley, 1999, McDiarmid et al., 1999, J.-F. Trape & Mané, 2004, 2006 and Chirio, 2013.

R

1. *Rhinoleptus koniagui* (Villiers, 1956b). Mém. Inst. Franç. Afr. Noire (48): 151–152, figs. 1–3. (*Typhlops koniagui*)

Type: Holotype, MNHN 8730 (formerly IFAN 1955.9.58), a 460 mm specimen (Denis, Dec. 1955).

Type locality: "Guinée française: Youkounkoun" [= Youkounkoun, Koundara Prefect., N Boké Region, ext. NW Guinea, 12°32'N, 13°07'W, elevation 75 m].

Distribution: West Africa. Southern Senegal (Kédougou, Kolda, Tambacounda, Ziguinchor), N Guinea (Koundara, Kouroussa, Télimélé) and SW Mali (Kayes, Sikasso), 25–380 m.

RHINOPHIS Hemprich, 1820 (Uropeltidae)

Synonyms: *Dapatnaya* Kelaart, 1853, *Morina* Gray, 1858c, *Mitylia* Gray, 1858a, *Mytilia* Gray, 1858c *(nomen emendatum), Dapatnya* Marschall, 1873 *(nomen emendatum), Rhinopsis* – F. Müller, 1878b *(nomen incorrectum), Rinophis* – Bedriaga, 1884 *(nomen incorrectum), Rhinophus* – Wall, 1921g *(nomen incorrectum), Rhiriophis* – Briceño-Rossi, 1934 *(nomen incorrectum),* and *Rinophis* – Donoso-Barros, 1965 *(nomen incorrectum).*

Type species: *Anguis oxyrynchus* J.G. Schneider, 1801.

Distribution: Southern India and Sri Lanka.

Sources: W.C.H. Peters, 1861e, A.C.L.G. Günther, 1864a, Beddome, 1886, Wall, 1921g, 1923c, M.A. Smith, 1943, E.H. Taylor, 1950b, Deraniyagala, 1955, Gans, 1966, Murthy, 1973, 1981c, 1982b, 1985, 1986, P. Silva, 1980, Cadle et al., 1990, A. Silva, 1990, McDiarmid et al., 1999, Gower et al., 2005, Somaweera, 2006 and Gower & Maduwage, 2011.

Remarks: Original description of *Mitylia* reprinted in Gray (1858c–d and 1858f).

R

1. *Rhinophis blythii* Kelaart, 1853. J. Ceylon Br. Roy. Asiat. Soc. 22(6): 106–107.

Synonym: *Mytilia templetonii* Gray, 1858c.

Types: Syntypes (3), not designated, longest syntype 406 mm, lost *fide* Gans (1966a: 13).

Type locality: "Mountains of Ceylon...3 or 4 feet below the surface soil of Coffee plantations" [= Sri Lanka mountains].

Distribution: Southern Sri Lanka (Central, Sabaragamuwa, Southern, Uva), 300–1360 m.

Sources: Willey, 1905, Wickramasinghe et al., 2009 and Olori & bell, 2012.

Remarks: Original description reprinted in Kelaart (1854a: 14 and 1854b: 26–27). E.H. Taylor (1947a: 26) discussed dates of Kelaart's "Prodromus."

2. *Rhinophis dorsimaculatus* Deraniyagala, 1941. J. Bombay Nat. Hist. Soc. (1941–1942) 42(4): 800–801, figs. 1a–c, pl. 1, figs. a–c.

Type: Holotype, CMS 86, a 350 mm specimen (Feb. 1938 or March 1941), lost *fide* Gans (1966: 14).

Type locality: "coastal village of Marichchukate, North Western Province, Ceylon" [= Marichchukkaddi, ext. N North-Western Prov., W Sri Lanka, ca. 8°35'N, 79°56'E, elevation NSL].

Distribution: Sri Lanka (North-Western), NSL. Known only from type locality.

3. *Rhinophis drummondhayi* Wall, 1921g. Ophidia Taprobanica: 43–44

Synonym: *Rhinophis drummond-hayi* T. Barbour & Loveridge, 1929a *(nomen illegitimum).*

Types: Syntypes (13), BMNH 1946.1.16.79–81 (formerly BMNH 1920.5.5.1–4) (H.M. Drummond-Hay & Ormiston), and MCZ 14348–49 (formerly BMNH), a 275 mm and 263 mm specimen (H.M. Drummond-Hay, 1896–1921), BNMM and CMS, longest syntype 330 mm female.

Type locality: "Lennock estate in the centre of the Uva Patnas at about 4,000 to 4,200 feet" and "Kalupahani estate, Haldummulle District, about 3,000 feet, Ceylon" [= Lennock Estate (ca. 7°N, 81°E, elevation 1220–1280 m) and Kalupahana Estate (6°46'N, 80°54'E, elevation 915 m), W Uva Prov., SE Sri Lanka].

Distribution: Southern Sri Lanka (Central, Uva), 915–1525 m.

Sources: McDiarmid et al., 1999 and Wickramasinghe et al., 2009.

Remarks: Data tag for MCZ 14348 reads Ceylon, Uva Patnas, 5000'. Gans (1966a: 14) erroneously cited original description as 1921 J. Bombay Nat. Hist. Soc. 28: 43.

4. *Rhinophis erangaviraji* Wickramasinghe, Vidanapathirana, Wickramasinghe & Ranwella, 2009. Zootaxa (2044): 6–17, figs. 3, 4a–12a.

Type: Holotype, NMSL 20080601, a 214 mm (svl) male (D. Vidanapathirana, N. Ranwella & L.J.M. Wickramasinghe, 5 Dec. 2007).

Type locality: "Enselwatte Estate, Sinharaja Division (Army Camp Forest), Rakwana hills, Matara District, Southern Province (N 06° 23', E 080° 36'), 1042 m."

Distribution: Southern Sri Lanka (Southern), 1040 m.

5. *Rhinophis fergusonianus* Boulenger, 1896i. J. Bombay Nat. Hist. Soc. (1895–1897) 10(2): 236, pl.

Type: Holotype, BMNH 1946.1.16.77 (formerly BMNH 1895.7.29.1), a 320 mm specimen (J.S. Sealy).

Type locality: "Cardamon hills, Travancore, Southern India" [= Cardamon Hills, S Western Ghats, SE Kerala Prov., S India]. Emended to the High Distribution [=

High Distribution, Cardamon Hills, South Sahyadri (Western Ghats), S Kerala, ext. SW India, 9°45'N, 77°10'E] *fide* Ferguson (1895: 70).

Distribution: Southwestern India (Western Ghats of SE Kerala). Known only from holotype.

Sources: Ferguson, 1895, Murthy, 1981c, 1982b, 1990d, I. Das, 1996, Aengals, 2009 and Wickramasinghe et al., 2009.

Remarks: A synonym of *R. sanguineus fide* Mahendra (1984: 64).

6. *Rhinophis homolepis* (Hemprich, 1820). Grundr. Naturgesch.: 119. (*Anguis homolepis*)

Synonyms: *Dapatnaya trevelyanii* Kelaart, 1853, *Mitylia gerrardi* Gray, 1858a, *Mytilia gerrardi* Gray, 1858c *(nomen emendatum), Dapatnaya trevalyana* A.C.L.G. Günther, 1864a *(nomen emendatum), Rhinophis trevelyanus* A.C.L.G. Günther, 1864a *(nomen emendatum),* and *Rhinophis trevelyana* I. Das, 1996 *(nomen emendatum).*

Type: Holotype, ZMB 3827.

Type locality: Unknown. Designated as "Zeylania" [= Sri Lanka] *fide* W.C.H. Peters (1861: 14).

Distribution: Southern Sri Lanka (Central, Sabaragamuwa, Uva), 750–950 m.

Sources: Willey, 1905 and Olori & Bell, 2012.

Remarks: A valid species *fide* McDiarmid et al. (1999: 137).

7. *Rhinophis lineatus* Gower & Maduwage, 2011. Zootaxa (2881): 53–56, figs. 2–4.

Type: Holotype CAS 226024, a 286 mm female (L. Jayawickrama, 30 Oct. 1976).

Type locality: "Harasbedda (07°03'N, 80°52'E, alt. 1,460 m), near Ragala, Central Province."

Distribution: Central Sri Lanka (Central), 1460 m. Known only from vicinity of type locality.

8. *Rhinophis oxyrhynchus* (J.G. Schneider, 1801). Hist. Amph. 2: 341–342. (*Anguis oxyrhynchus*) (*nomen corrigendum*)

Synonyms: *Anguis oxyrynchus* J.G. Schneider, 1801 *(nomen incorrigendum), Anguis oxyrrhynchus* Hemprich, 1820 *(nomen emendatum), Rhinophis oxyrhynchos* Wagler, 1830 *(nomen emendatum), Pseudotyphlops oxirhynchus* Schlegel, 1839 in 1837–1844 *(nomen illegitimum), Dapatnaya lankadivana* Kelaart, 1853, *Rhinophis oxyrhincus* A.-M.-C. Duméril, Bibron & Duméril, 1854a *(nomen emendatum), Rhinophis oxyrhinchus* A.-M.-C. Duméril, Bibron & Duméril, 1854a *(nomen emendatum), Mytilia unimaculata* Gray, 1858c, and *Rhinophis oxyrhynchus* – A.C.L.G. Günther, 1864a *(nomen corrigendum).*

Types: Syntypes (2), ZMB 3825–26, a 455 mm female and 175 mm male (C.R. John via M.E. Bloch).

Type locality: "India orientali" [= E India]. Emended to Sri Lanka *fide* M.A. Smith (1943: 92).

Distribution: Northeastern Sri Lanka (Central, Eastern, Northern, North-Central), NSL–270 m.

Sources: E.H. Taylor, 1950b and Bauer, 1998.

Remarks: This epithet is derived from two Greek terms, *oxys* (sharp) + *rhynchos* (snout), and the original orthography was incorrect.

9. *Rhinophis philippinus* (Cuvier, 1829). Règne Anim., ed. 2, 2: 74. (*Typhlops philippinus*)

Synonym: *Rhinophis planiceps* W.C.H. Peters, 1861h.

Type: Holotype, MNHN 1864.94, a 217–264 mm female.

Type locality: "Philippines" (in error *fide* A.C.L.G. Günther, 1864a: 185). Corrected to Zeylania [= Sri Lanka] *fide* W.C.H. Peters (1861h: 15).

Distribution: Southern Sri Lanka (Central, Sabaragamuwa), 300–900 m.

Sources: J.P. Müller, 1832, Boulenger, 1890a and A. Silva et al., 2005.

Remarks: Holotype possibly illustrated by Jan & Sordelli (1865 *in* 1860–1866) *fide* McDiarmid et al. (1999: 139).

10. *Rhinophis porrectus* Wall, 1921g. Ophidia Taprobanica: 35.

Type: Holotype, BMNH 1946.1.16.70 (formerly BMNH 1920.8.25.1), a 355 mm specimen (A. Willey, Oct. 1903).

Type locality: "on the road between Puttalam and Chilaw, Ceylon" [= between Puttalam (8°02'N, 79°49'E) and Chilaw (7°34'N, 79°47'E), W North-Western Prov., W Sri Lanka, NSL]. Emended to Madurankuli, [7°54'N, 79°49'E] *fide* Willey (1903: 88).

Distribution: Western Sri Lanka (W North-Western), NSL. Known only from holotype.

Sources: Willey, 1903 and A. Silva, 1982.

Remarks: A valid species *fide* E.H. Taylor (1950b: 530), A. Silva (1990: 70) and McDiarmid et al. (1999: 140). A synonym of *R. punctatus* Müller *fide* M.A. Smith (1943: 92), and Gans (1966a: 15).

11. *Rhinophis punctatus* J.P. Müller, 1832. Zeits. Physiol. 4(2): 248–252, pl. 21, figs. 1–3, pl. 22, figs. 1a–f. (*Rhinophis punctata*)

Synonym: *Pseudotyphlops oxyrhynchus* Schlegel, 1837.

Type: Holotype, a 265 mm male, lost *fide* W.C.H. Peters (1861h: 13) and Gans (1966a: 15).

Type locality: "Guiana" [South America, or ? Guinea, W Africa *fide* W.C.H. Peters, 1861h: 13] (in error). Corrected to Zeylania` [= Sri Lanka] by Schlegel *in* W.C.H. Peters (1861h: 13).

Distribution: Central Sri Lanka (Central, North-Western), 300–900 m.

Source: Willey, 1903b.

12. *Rhinophis sanguineus* Beddome, 1863a. Madras Quart. J. Med. Sci. 6: 47, pl. 1, figs. 6–6a.

Synonym: *Rhinophis microlepis* Beddome, 1863a.

Types: Syntypes (2+), BMNH 1946.1.16.54 (formerly BMNH 1864.3.9.2) and BMNH 1946.1.16.76, longest syntype a 457 mm male, location of other syntypes unknown.

Type locality: "Cherambady in the Wynaud, on the Western slopes of the Neilgherries, Madras Presidency" [= Cherambadi, N Kerala State, SW India, 11°32'N, 76°17'E, elevation 1065 m]. Emended to Cherambòdy in the Wynaud (Malabar), elevation 3500 feet *fide* Beddome (1863b: 227).

Distribution: Southwestern India (Western Ghats of W Karnataka, E Kerala, SW Tamil Nadu), 750–1065 m.

Sources: Murthy, 1981c, 1982b, R. Sharma, 2003, 2007 and Aengals, 2009.

Remarks: Original description reprinted in Beddome (1863b: 227). Beddome reported several syntypes but only one was catalogued by the BMNH *fide* Gans (1966a: 15) and M. Lang (*in litt.*).

13. *Rhinophis travancoricus* Boulenger, 1893b. J. Bombay Nat. Hist. Soc. (1892–1893) 7(3): 318, pl.

Type: Holotype, BMNH 1892.10.5.2, a 170 mm specimen (H.S. Ferguson, June 1892).

Type locality: "near Trevandrum, at the 6th mile-stone towards Vambayam, Travancore, S. India" [= vicinity of Thiruvanthapuram (8°29'N, 76°57'E, elevation 20 m), ext. S Kerala, SW India].

Distribution: Southwestern India (Western Ghats of SE Kerala), NSL–1335 m.

Sources: Mahendra, 1938a, Murthy, 1981c, 1982b, Rajendran, 1985 and Aengals, 2009.

Remarks: Supplemental original description in Boulenger (1893a: 143, pl. 9, figs. 3, 3a–3b) [Boulenger 1893b published before 1893a].

R

14. *Rhinophis tricoloratus* Deraniyagala, 1975. Spolia Zeylan. 33(1–2): 535–536, figs. 1a–e. (*Rhinophis tricolorata*)

Type: Holotype, CMS 1978.5.1 (formerly PEPD), a 258 mm specimen.

Type locality: "vicinity of the rain forest of Sinha Raja to the south west of Ratnapura at an elevation of 1,500 feet" [= Sinharaja Forest, S Sabaragamuwa Prov., SW Sri Lanka, 6°23'N, 80°25'E, elevation 455 m].

Distribution: Southern Sri Lanka (Sabaragamuwa), 455 m. Known only from type locality.

Sources: Karunaratne et al., 1981 and A. Silva, 1982.

15. *Rhinophis zigzag* Gower & Maduwage, 2011. Zootaxa (2881): 59–63, figs. 6–8.

Type: Holotype, CAS 226306, a 338 mm male (C. Gans, May 1974).

Type locality: "'Biblegemma Rd.' (ca. 06°54'N, 81°08'E and approximately 1,000 in altitude), near Passara, Province of Uva, Sri Lanka."

Distribution: Central Sri Lanka (Uva), 1000 m. Known only from vicinity of type locality.

RHINOPLOCEPHALUS F. Müller, 1885 (Elapidae)

Synonyms: *Rhinhoplocephalus* Boulenger, 1896a *(nomen emendatum), Riinhoplocephalus* – Briceño-Rossi, 1934 *(nomen incorrectum), Cryptophis* Worrell, 1961c, *Unechis* Worrell, 1961c, *Rhioplocephalus* – Kramer, 1978 *(nomen incorrectum),* and *Rhinocephalus* – Obst, Richter & Jacob, 1984 *(nomen incorrectum).*

Type species: *Rhinoplocephalus bicolor* F. Müller, 1885.

Distribution: New Guinea and Australia.

Sources: Kinghorn, 1929, 1964, Cogger et al., 1983a, Shine, 1985a, Storr, 1985, Longmore, 1986, S.K. Wilson & Knowles, 1988, Gow, 1989, Hoser, 1989, 2012e, Hutchinson, 1990, Cogger, 1992, Keogh et al., 1998, Sanders et al., 2008 and Zaher et al., 2009.

1. *Rhinoplocephalus bicolor* F. Müller, 1885. Verh. Naturf. Ges. Basel (1882–1885) 7(3): 690–692, pl. 9, figs. f–i.

Type: Holotype, NMBA 2190, a 395 mm male.

Type locality: "Australia." Emended to S Australia *fide* Cogger et al. (1983a: 234).

Distribution: Southwestern Australia (SW Western Australia), NSL–320 m.

Sources: Glauert, 1960, Christensen, 1972, Shine, 1986b and Storr et al., 1986.

2. *Rhinoplocephalus boschmai* (Brongersma & Knaap-van Meeuwen, 1964). Zool. Meded. 39: 553–554, figs. a–c. (*Denisonia boschmai*)

Synonym: *Unechis boschmai crutchfieldi* Hoser, 2012ac *(nomen illegitimum).*

Type: Holotype, RMNH 10874, a 439 mm male (A.J.M. Monsanto, 10 April 1959).

Type locality: "Merauke in southern West New Guinea" [= Merauke, ext. SE Papua Prov., ext. E Indonesia, 8°30'S, 140°24'E, elevation NSL].

Distribution: Extreme E Indonesia (E Papua) and NE Australia (ext. NE New South Wales, ext. N Northern Territory, E Queensland), 30–615 m.

Source: O'Shea, 1996.

Remarks: Possibly occurs in SW Papua New Guinea *fide* F. Parker (1972: 14) and Whitaker et al. (1982: 52).

3. *Rhinoplocephalus incredibilis* (Wells & Wellington, 1985). Aust. J. Herp. (Suppl. 1): 49. (*Unechis incredibilis*)

Type: Holotype, AMS, an adult (H.G. Cogger et al.).
Type locality: "Prince of Wales Island, Torres Strait, Queensland," Australia.
Distribution: Australia (Queensland, Prince of Wales Is.), NSL.
Sources: Cameron et al., 1978, Cogger, 1996, 2000 and Shea & Sadlier, 1999.
Remarks: A valid species *fide* Shea & Sadlier (1999: 66) and Cogger (2000: 67) although the type cannot be located in AMS. Possibly occurs on N Cape York Peninsula, Queensland *fide* Wells & Wellington (1985: 49).

4. *Rhinoplocephalus nigrescens* (A.C.L.G. Günther, 1862b). Ann. Mag. Nat. Hist. (3) 9(30): 131, pl. 9, fig. 12. (*Hoplocephalus nigrescens*)

Synonyms: *Alecto melanotus* Jan, 1863b *(nomen nudum)*, *Alecto permixta* Jan, 1863b *(nomen nudum)*, *Alecto permixta* Jan, 1873 *in* Jan & Sordelli, 1870–1881, *Hoplocephalus assimilis* Macleay, 1885b, *Cryptoptis nigrescence* – Khole, 1991 *(nomen incorrectum)*, and *Cryptophis edwardsi* Hoser, 2012ac *(nomen illegitimum)*.
Type: Holotype, BMNH 1946.1.20.58, a 394 mm male (G. Krefft, 1861–1874).
Type locality: "environs of Sydney" [Australia]. Emended to rocky neighbourhood of Middle Harbour (Port Jackson) [= Port Jackson, E New South Wales, SE Australia, 33°51'S, 151°15'E, elevation NSL] *fide* Krefft (1869b: 69).
Distribution: Eastern Australia (Australian Capital Territory, E New South Wales, E Queensland, E Victoria), NSL–1395 m.
Sources: Krefft, 1869b, Shine, 1984b and Covacevich, 1991.

5. *Rhinoplocephalus nigrostriatus* (Krefft, 1864c). Proc. Zool. Soc. London 32(1): 181–182. (*Hoplocephalus nigrostriatus*) (*nomen corrigendum*)

Synonyms: *Alecto dorsalis* Jan, 1863b *(nomen nudum)*, *Hoplocephalus nigro-striatus* Krefft, 1864c *(nomen incorrigendum)*, *Alecto dorsalis* Jan, 1873 *in* Jan & Sordelli, 1870–1881, *Hoplocephalus nigrostriatus* – F. Müller, 1885 *(nomen corrigendum)*, *Denisonia nigrostriata* – Boulenger, 1896a, and *Unechis durhami* Hoser, 2012ac *(nomen illegitimum)*.
Type: Holotype, AMS 131705 (formerly AMS 6660), a 279 mm specimen (G. Salting).
Type locality: "North-east Australia, neighbourhood of Rockhampton" [= vicinity of Rockhampton, SE Queensland, E-cen. Australia, 23°23'S, 150°30'E, elevation NSL].
Distribution: Papua New Guinea (Western) and N Australia (E Queensland), NSL–855 m.
Sources: Krefft, 1869b, Covacevich, 1991 and O'Shea, 1996.
Remarks: Original description reprinted in Krefft (1865a: 67). Type locality Queensland, Australia and collector Capt. Hurley *fide* H. Ehmann *in* Shea & Sadlier (1999: 50).

6. *Rhinoplocephalus pallidiceps* (A.C.L.G. Günther, 1858). Cat. Colub. Snakes Brit. Mus.: 214. (*Hoplocephalus pallidiceps*)

Type: Lectotype, BMNH 1946.1.20.65, a 610 mm female (Earl of Derby), designated by Wells & Wellington (1985: 45).
Type locality: "Port Essington" [= Port Essington, ext. N Northern Territory, N Australia, 11°10'S, 132°08'E, elevation NSL] via lectotype selection.
Distribution: North-central Australia (N Northern Territory, ext. N Western Australia), NSL–100 m.
Sources: Krefft, 1869, Cogger & Lindner, 1974, L. Smith, 1978, Shine, 1984b and Storr et al., 1986.

RHINOTYPHLOPS Fitzinger, 1843 (Typhlopidae)

Synonyms: *Onychocephalus* A.-M.-C. Duméril & Bibron, 1844, *Onychophis* Gray, 1845, *Onycocephalus* – A.H.A. Duméril, 1856 *(nomen incorrectum)*, *Onychophalus* – Bocage, 1866a *(nomen incorrectum)*, *Onicocephalus* – Amaral, 1930f *(nomen incorrectum)*, *Onyehocephalus* – Tchang, 1932 *(nomen incorrectum)*, and *Onchocephalus* – Nussbaum, 1984 *(nomen incorrectum)*.
Type species: *Typhlops lalandei* Schlegel, 1839 *in* 1837–1844.
Distribution: Eastern and S Africa.
Sources: Witte, 1962, Pitman, 1974, Roux-Estève, 1974a, Hahn, 1977, 1980a, Broadley, 1983, Wallach, 1993a, c, 1994b, Wallach & Ineich, 1996, McDiarmid et al., 1999 and Broadley & Wallach, 2009.

R

1. *Rhinotyphlops boylei* (V.F.M. FitzSimons, 1932). Ann. Transvaal Mus. 15(1): 38. (*Typhlops boylei*)

Type: Holotype, TMP 14695 (formerly VLKE 805), a 182 male (A.M. Boyle, 28 May 1930).
Type locality: "Gemsbok Pan, Ghanzi District, Kalahari" [= West Kalahari, Botswana, 21°34'S, 21°47'E, elevation 1150 m].
Distribution: Southwestern Africa. Central Namibia (Khomas, Otjozondjupa), SW Botswana (Ghanzi, Kgalagadi) and N South Africa (Northern Cape), 915–1155 m.
Sources: Mertens, 1955, 1971, Wallach, 1994b, Branch, 1998, M. Griffin, 2003 and Broadley & Blaylock, 2013.

2. *Rhinotyphlops lalandei* (Schlegel, 1839 *in* 1837–1844). Abbild. Amph.: 38, pl. 32, figs. 17–20. (*Typhlops lalandei*)

Synonyms: *Onychocephalus delalandii* A.-M.-C. Duméril & Bibron, 1844 *(nomen emendatum), Onychophis fordii* Gray, 1845, *Onychophis franklinii* Gray, 1845, *Onychophis lalandii* Gray, 1845 *(nomen emendatum), Typhlops smithi* Jan 1860 *in* Jan & Sordelli, 1860–1865, *Typhlops lalandei paucisquamosa* Boettger, 1883b, and *Typhlops delalandei* Boettger, 1887b *(nomen emendatum).*

Types: Syntypes (9), MNHN 1492, a 235 mm male, MNHN 3214, a 215 mm male, MNHN 3214A, a 186 mm female, MNHN 6992–92A, and RMNH 3715a–d, (P.A. Delalande & H.B. van Horstolk, 1818–1820).

Type locality: "Vorgebirge der guten Hoffnung" [= Cape of Good Hope, SW Western Cape Prov., SW South Africa *fide* V.F.M. FitzSimons, 1962: 71].

Distribution: Southern Africa. Southern Namibia (Erongo, Hardap, Karas, Khomas, Omaheke), E Botswana (Central, Gaborone, Kweneng, South East), Zimbabwe (Bulawayo, Manicaland, Mashonaland West, Masvingo, Matebeleland North, Matebeleleand South, Midlands), S Mozambique (Gaza), Lesotho, Swaziland and South Africa (Eastern Cape, Free State, Gauteng, KwaZulu-Natal, Limpopo, Mpumalanga, Northern Cape, North-West, Western Cape), NSL–1670 m.

Sources: Mertens, 1955, 1971, FitzSimons, 1974, Broadley & Cock, 1975, Broadley, 1983, Pienaar et al., 1983, Meirte, 1992, M., Branch, 1998, 2001, Griffin, 2003, Warner & Combrink, 2010b and Broadley & Blaylock, 2013.

Remarks: MNHN 6992–92A not listed as syntypes *fide* Guibé & Roux-Estève (1972: 134).

3. *Rhinotyphlops schinzi* (Boettger, 1887b). Ber. Senck. Naturf. Ges. Freunde Berlin 1886–1887: 154–156, pl. 5, figs. 1a–e, 2. (*Typhlops* [*Onychocephalus*] *schinzi*)

Type: Lectotype, SMF 16665 (formerly SMF-B 7047, 1a), ? a 226 mm specimen (H. Schinz, 1884–1885), designated by Mertens (1922: 178).

Type locality: "Noi Xas bei Ghanze, Nord-Kalahari" [= Noixas, between Aus and Keetmanshoop, Great Namaqualand, Ghansi Distr., W Kalahari, Botswana, 21°34'S, 21°47'E, elevation 1155 m] via lectotype selection.

Distribution: Southwestern Africa. Namibia (Karas, Khomas, Kunene, Otjozondjupa), W Botswana (Ghanzi) and South Africa (W Northern Cape), NSL–1670 m.

Sources: Mertens, 1955, 1971, FitzSimons, 1974, Broadley, 1983, Auerbach, 1987, Branch, 1998, M. Griffin, 2003 and Broadley & Blaylock, 2013.

Remarks: Mertens (1922: 178) listed lectotype collection date as 1887, probably referring to date received in Germany.

R

RHYNCHOCALAMUS A.C.L.G. Günther, 1864d (Colubridae)

Synonyms: *Rhyncocalamus* – Tristram, 1884 *(nomen incorrectum),* and *Rhynocalamus* – Marx & Rabb, 1972 *(nomen incorrectum).*

Type species: *Homalosoma melanocephalum* Jan, 1862b.

Distribution: Southwestern Asia.

Sources: K.P. Schmidt, 1939b, C.A. Reed & Marx, 1959, Gasperetti, 1988 and Olgun et al., 2007.

1. *Rhynchocalamus arabicus* K.P. Schmidt, 1933b. Field Mus. Nat. Hist. (Zool.) 20: 9–10.

Type: Holotype, FMNH 18219, a 278 mm female (A.R.M. Rickards, 1932).

Type locality: "Aden, Arabia" [= W Aden, Yemen, ca. 13°30'N, 45°00'E].

Distribution: Southwestern Yemen (Aden). Known only from holotype.

Sources: Corkill & Cochrane, 1965, Leviton & Aldrich, 1984, Gasperetti, 1988 and Egan, 2007.

2. *Rhynchocalamus barani* Olgun, Avci, Ilgaz, N. Uzum & C.Yilmaz, 2007. Zootaxa (1399): 58–62, figs. 1–5.

Type: Holotype, ZDEU 122/2006, a 403 mm female (A.A. Vic & C. Yilmaz, 1 May 2006).

Type locality: "Amanos Mountain (36°50'N, 36°25'E; altitude 310 m a.s.l.), 34 km E of Dörtyol, Hatay Province, Turkey."

Distribution: Southern Turkey (Hatay), 310 m. Known only from type locality.

3. *Rhynchocalamus melanocephalus* (Jan, 1862b). Arch. Zool. Anat. Fis. 2(1): 34, pl. e, no. 3. (*Homalosoma melanocephalum*)

Synonyms: *Rhynchocalamus melanocephalus* Günther, 1864d *(nomen praeoccupatum), Contia satunini* Nikolsky, 1899, and *Oligodon melanocephalus septentrionalis* F. Werner, 1905.

Type: Lectotype, MHNG 1246.77, a 400 mm female (M. Sordet), designated herein.

Type locality: "Bairut" [= Beirut, Beirut Prov., W Lebanon, 33°53'N, 35°30'E, elevation NSL] via lectotype selection.

Distribution: Southwestern Asia. Southeastern Turkey (Adana, Hatay, Mardin), W Syria (Damascus, Homs, Latakia), Lebanon (Beirut, Mont-Liban, Nabatiye), W Jordan (Amman, Irbid, Jarash, Karak, Maan, Madaba, Mafraq), Israel (Central, Golan Heights, Haifa, Jerusalem, Northern, N Southern, Tel Aviv, West Bank), NE Egypt (South Sinai), SW Armenia, SW Azerbaijan, W Iran (Central, East Azarbaijan, Khuzestan) and N Iraq (Ta'min), 50–1800 m.

Sources: Boettger, 1879a, Tristram, 1884, J. Anderson, 1898, Nikolsky, 1913, 1916, T. Barbour, 1914b, Flower, 1933, Angel, 1936c, Bird, 1936, Bodenheimer, 1944, G. Haas, 1951, G. Haas & Werner, 1969, Darevsky, 1970, Baran, 1976, Bannikov et al., 1977, Basoglu & Baran, 1980, Hraoui-Bloquet, 1981, Latifi, 1991, Gasperetti, 1988, Gruber, 1989, Leviton et al., 1992, Franzen & Bischoff, 1995, Bouskila & Amitai, 2001, Disi et al., 2001, Hraoui-Bloquet et al., 2002, Baha El Din, 2006, Avci et al., 2008, Tuniyev et al., 2009, Amr & Disi, 2011 and Bar & Haimovitch, 2011.

Remarks: Illustration of the MHNG lectotype in Jan & Sordelli (1860–1866, 1[13]: pl. 3, fig. 4). Probably occurs in N Saudi Arabia *fide* Gasperetti (1988: 272).

RHYNCHOPHIS Mocquard, 1897c (Colubridae)

Synonyms: *Rhadinophis* T. Vogt, 1922, *Proboscidophis* Fan, 1931, and *Katrinahoserea* Hoser, 2012r *(nomen illegitimum).*

Type species: *Rhynchophis boulengeri* Mocquard, 1897c.

Distribution: Southeastern Asia.

Remarks: Includes *Rhadinophis frenata* and *R. prasina fide* Zaher et al., 2012.

1. *Rhynchophis boulengeri* Mocquard, 1897c. Bull. Mus. Hist. Nat. 3(6): 215.

Synonym: *Proboscidophis versicolor* Fan, 1931.

Types: Syntypes (6), MNHN 1897.94–99, males and females, longest syntype 1134 mm female (Leichtenfelder).

Type locality: "des îles Norway, de la baie d'Along, dans le golfe de Tonkin" [= Xuy Nong Chao, Ha Long Bay, Gulf of Tonkin, Quang Ninh Prov., N Vietnam, 20°37'N, 107°10'E].

Distribution: Southern China (Guangxi, Hainan) and N Vietnam (Hai Phong, Ha Tay, Ha Tinh, Quang Ninh, Son La, Thai Nguyen, Vinh Phuc, Hon Nor Way Is.), 200–1500 m.

Sources: F. Werner, 1929a, Fan, 1931, Bourret, 1934f, 1935c, 1936b, 139d, C.H. Pope, 1935, M.A. Smith, 1943, Zhao & Adler, 1993, Orlov, 1995, Darevsky, 1999, Orlov et al., 2000, T. Schmidt, 2002, Burbrink & Lawson, 2007, V.S. Nguyen et al., 2009, I. Das, 2010, 2012, Schulz et al., 2011 and Hoser, 2012r.

2. *Rhynchophis frenatus* (Gray, 1853). Ann. Mag. Nat. Hist. (2) 12(72): 390–391. (*Herpetrodryas frenatus*)

Synonyms: *Rhadinophis melli* T. Vogt, 1922, *Gonyosoma caldwelli* K.P. Schmidt, 1925a, and *Chrysopelea ornata lungchuanensis* B. Hu, Huang, He & Wei, 1958.

Type: Holotype, BMNH 1946.1.6.5 (formerly BMNH 1853.8.12.20), adult female (J. Hooker, May–Dec. 1850).

Type locality: "Khassia Mountains, East Bengal" [= Khasi Hills, Meghalaya and Assam States, NE India].

Distribution: Southern Asia. Northeastern India (Assam, Meghalaya), SE China (Anhui, Fujian, Guangdong, Guangxi, Guizhou, Hunan, Sichuan, Zhejiang) and N Vietnam (Lang Son, Lao Cai, Quang Binh, Vinh Phuc), 550–2000 m.

Sources: Boulenger, 1893a, K.P. Schmidt, 1927b, C.H. Pope, 1935, Bourret, 1936b, E.H. Taylor, 1965, B. Hu et al., 1980, Tian et al., 1986, Y. Zhang & Tang, 1990, Schulz, 1991, Zhao & Adler, 1993, Staszko & Walls, 1994, V.S. Nguyen et al., 2009, Xiang & Li, 2009 and I. Das, 2012.

Remarks: Original orthography of *Herpetrodryas* was in error for *Herpetodryas*.

3. *Rhynchophis prasinus* (Blyth, 1854a). J. Asiat. Soc. Bengal 23(3): 291. (*Coluber prasinus*)

Synonym: *Gonyosoma gramineum* A.C.L.G. Günther, 1864a.

Type: Lectotype, BMNH 1946.1.10.22 (formerly IMC 7670 & BMNH 1860.3.19.1155), longest syntype 940 mm (W. Robinson), designated by Cox et al. (2013: 221).

Type locality: "Asám, from the Khásya hills, or other upland territory" [= Khasi Hills, E Meghalaya State, NE India, 25°35'N, 91°38'E, elevation 1370–1675 m].

Distribution: Southeastern Asia. Northeastern India (Andaman & Nicobars, Assam, Arunachal Pradesh, Himachal Pradesh, Manipur, Meghalaya, Nagaland, Sikkin, N West Bengal), N Thailand (Loei), N Myanmar (Chin, Kachin, Karen, Mandalay, Sagaing, Shan), S China (Guanxi, Guizhou, Hainan, Hubei, Sichuan, Yunnan), N Laos (Champasak), Vietnam (Bac Kan, Gia Lai, Lao Cai, Nghe An, Quang Binh, Thai Nguyen, Vinh Phú), Thailand (Loei, Tak) and West Malaysia (Pahang, Perlis, Selangor), 75–2650 m.

Sources: M.A. Smith, 1922b, Bourret, 1936b, Tweedie, 1940a, 1950, Schulz, 1991, Zhao & Adler, 1993, Orlov, 1995, M.J. Cox et al., 1998, Chan-ard et al., 1999, Orlov et al., 2000, 2003, Schleich & Kästle, 2002, Vogel & Pauwels, 2004, Zhao, 2006, Yang, 2008, V.S. Nguyen et al., 2009, Thanh & Ziegler, 2009 and I. Das, 2012.

Remarks: Probably occurs in E Nepal *fide* Schleich & Kästle (2002: 873).

†*RIONEGROPHIS* Albino, 1986 (Boidae)

Type species: †*Rionegrophis madtsoioides* Albino, 1986.

Distribution: Upper Cretaceous of Argentina.

Sources: Albino, 1987b, 2011b.

1. †*Rionegrophis madtsoioides* Albino, 1986. Actas IV Congr. Argent. Paleont. Boiestr. 2: 18–19, fig. 3.

Type: Holotype, MACN-RN 32, one middle trunk vertebra.

Type locality: "Ladera sudoeste del Cerro Cuadrado, noreste de la Ea. Los Alamitos, Arroyo Verde, Dpto. Valcheta, Río Negro, Argentina. Sector medio de la Formación los Alamitos, Campaniano-Maastrichtiano temprano, cretácio tardío" [= SW slope of Cerro Cuadrado, NE of Los Alamitas (41°57'S, 66°21"W), proximal zone of Arroyo Verde (41°55'S, 66°10'W), Valcheta Dept., SE Río Negro Prov., S Argentina; middle section of Los Alamitos formation, early Campanian-Maestrichtian, upper Cretaceous].

Distribution: Upper Cretaceous (Campanian-Maastrichtian: 65.5–83.6 mya) of Argentina. Known only from type locality.

RODRIGUESOPHIS Grazziotin, Zaher, Murphy, Scrocchi, Benavides, Zhang & Bonatto, 2012 (Xenodontidae)

Type species: *Rhinostoma iglesiasi* Gomés, 1915.

Distribution: Northeastern Brazil.

1. *Rodriguesophis chui* (Rodrigues, 1993b). Pap. Avul. Zool. 38(11): 190–191, fig. 3. (*Phimophis chui*)

Type: Holotype, MZUSP 10291 (formerly MTR 90.7212), a 370 mm male (M.T. Rodrigues, 6 Oct. 1990).

Type locality: "Brazil: Bahia: Santo Ignácio" [= Santo Inácio, N Bahia State, NE Brazil, 11°07'S, 42°43'W, elevation 545 m].

Distribution: Northeastern Brazil (N Bahia), 545 m. Known only from type series.

2. *Rodriguesophis iglesiasi* (Gomés, 1915). Ann. Paul. Med. Cirurg. 4(6): 126–127, pl. 4, figs. 1–3. (*Rhinostoma iglesiasi*)

Synonym: *Rhinostoma bimaculatum* Lutz & Mello, 1922b.

Type: Holotype, IB 513, a 490 mm female (F. Iglesias), destroyed by fire 15 May 2010.

Type locality: "Estado do Piauhy, Brazil" [= Piauí State, NE Brazil].

Distribution: Eastern Brazil (Bahia, Minas Gerais, Piauí, Tocantins), 300–600 m.

Sources: Hoge et al., 1981, Freitas, 1999 and Recorder et al., 2011.

3. *Rodriguesophis scriptorcibatus* (Rodrigues, 1993b). Pap. Avul. Zool. 38(11): 188–189, figs. 1–2. (*Phimophis scriptorcibatus*)

Type: Holotype, MZUSP 10062 (formerly MTR 88.6299), a 316 mm male (M.T. Rodrigues, 26 Aug. 1988).

Type locality: "Brazil: Bahia: Ibiraba" [= Ibiraba, N Bahia State, NE Brazil, ca. 10°46'S, 42°45'W, elevation 400 m].

Distribution: Northeastern Brazil (N Bahia), 400 m. Known only from type locality.

†*ROTTOPHIS* Szyndlar & Böhme, 1996 (Tropidophiidae)

Type species: †*Tropidonotus atavus* H. Meyer, 1855.

Distribution: Upper Oligocene of Germany.

Sources: Troschel, 1859, H. Meyer, 1860, Rochebrune, 1880, Mlynarski, 1961, Rage, 1984b, 2011, W. Böhme & Lang, 1991, Szyndlar & Böhme, 1996, Szyndlar, 1997, Szyndlar & Rage, 2003 and Szyndlar et al., 2008.

1. †*Rottophis atavus* (H. Meyer, 1855). Neues Jb. Mineral. Geog. Geol. Petrefakt.-Kunde 23: 336–337. (†*Tropidonotus atavus*)

Synonyms: †*Coluber papyraceus* Troschel, 1854 (*nomen nudum*), and †*Coluber papyraceus* J.G. Fischer, 1857.

Type: Holotype, GPIBO Ro 4023 (formerly HvM 9), a nearly complete 520–535 mm skeleton (H. von Dechen, April 1855).

Type locality: "Rheinischen Braunkohle" [= brown coal of Rhine, Germany]. Emended to Braunkohle des Siebengebirges *fide* H. Meyer (1860: 232) and Romeriken-Berge Quarry, Rott near Bonn, North Rhine-Westfalia Prov., W Germany, Upper Oligocene, Arvernian *fide* Rage (1984b: 44).

Distribution: Upper Oligocene (Chattian, MP 25–30: 23.0–28.4) of Germany. Known only from type locality.

Remarks: Supplemental original description in H. Meyer (1860: 232–240, pl. 25, figs. a–c, 1–4). Photograph of holotype in W. Böhme & Koenigswald (1989: fig. 7.5) and Szyndlar & Böhme (1996: pl. 1, fig. b).

†*RUSSELLOPHIS* Rage, 1975a (†Russellophiidae)

Synonym: †*Russelophis* – Rage & Rocek, 1983 *(nomen incorrectum).*

Type species: †*Russellophis tenuis* Rage, 1975a.

Distribution: Lower Eocene of France and India.

Sources: Rage, 1983b, 1984b, 1988c, Auge et al., 1997 and Rage et al., 2008.

R

1. †*Russellophis crassus* Rage, Folie, Rana, Singh, Rose & Smith, 2008. Acta Palaeont. Polon. 53(3): 394–396, figs. 3a–b.

Type: Holotype, VAS 1039, one middle trunk vertebra.
Type locality: "Vastan Lignite Mine, northeast of Surat, Gujarat, western India; Early Eocene (middle to late Ypresian) continental beds of Cambay Formation."
Distribution: Lower Eocene (Bumbanian: 48.6–55.8 mya) of India. Known only from type locality.

2. †*Russellophis tenuis* Rage, 1975a. C.R. Soc. Géol. France 2: 46–47, figs. 1a–e.

Type: Holotype, MNHN CB 1603, one middle trunk vertebra (D.E. Russell).
Type locality: "Condé-en-Brie (Aisne, France); Yprésien, l'Eocène inférieur" [= Condé-en-Brie, S Aisne Prov., N France, lower Eocene, Ypresian].
Distribution: Lower Eocene (Ypresian, MP 7–10: 48.6–55.8 mya) of France.

S

SALOMONELAPS McDowell, 1970 (Elapidae)

Synonyms: *Salmonelaps* – Dowling *in* Dowling & Duellman, 1978 *(nomen incorrectum),* and *Solomonelaps* – O'Shea, 2007 *(nomen incorrectum).*
Type species: *Hoplocephalus par* Boulenger, 1884b.
Distribution: Papua New Guinea and Solomons.
Sources: Boulenger, 1888b, T. Barbour, 1921a, Kinghorn, 1928a, Burt & Burt, 1932, K.P. Schmidt, 1932c, Loveridge, 1945b, E.E. Williams & Parker, 1964, McDowell, 1970, Keogh, 1992, O'Shea, 1996, Zaher et al., 2009 and Hoser, 2012e.

1. *Salomonelaps par* (Boulenger, 1884b). Proc. Zool. Soc. London 52(2): 210. (*Hoplocephalus par*)

Synonyms: *Hoplocephalus melanurus* Boulenger, 1888b, *Hoplocephalus woodfordi* Boulenger, 1888b, *Denisonia melanura boulengeri*, Waite, 1899, and *Denisonia furva* Brazenor, 1947..
Type: Holotype, BMNH 1946.1.20.66 (formerly BMNH 1884.3.24.94), a 750 mm female (H.B. Guppy).
Type locality: "Faro Island, Solomon Islands."
Distribution: Eastern Papua New Guinea (North Solomons: Buka Is.) and the Solomons (Fauro, Ganongga, Giza, Guadalcanal, Kolombangara, Lavella, Malaita, New Georgia, Rendova, Santa Isabel, Shortland, Simbo, Tetipari, Vangunu and Vella Is.).

SALVADORA Baird & Girard, 1853 (Colubridae)

Synonyms: *Phimothyra* Cope, 1860e *(nomen nudum), Phimothyra* Cope, 1861c *(nomen substitutum), Phimathyra* – W.C.H. Peters, 1870a *(nomen incorrectum), Phymothyra* S.W. Garman, 1884 *(nomen emendatum), Salvador* – Briceño-Rossi, 1934 *(nomen incorrectum),* and *Aiselfakharius* Hoser, 2012aL *(nomen illegitimum).*
Type species: *Salvadora grahamiae* Baird & Girard, 1853.
Distribution: Southwestern USA and Mexico.
Fossil records: Miocene and upper Pleistocene of W USA. Fossils unassigned to species include upper Pleistocene (Rancholabrean II) of USA (Arizona, New Mexico, Texas).
Sources: Bogert, 1939a, H.M. Smith, 1939a, 1941d, H.M. Smith & Taylor, 1945, Van Devender & Bradley, 1994, Holman, 1995b, 2000a and Nagy et al., 2004.

1. *Salvadora bairdi* Jan, 1860 *in* Jan & Sordelli, 1860–1866. Icon. Gén. Ophid. 1(1): 1, pl. 3, figs. 2, 2a–c, f–g, h, n, r, v.

Synonym: *Salvadora bogerti* H.M. Smith, 1941d.
Types: Syntypes (2), USTL (A. Westphal-Castelnau), and formerly MSNM, destroyed in 1943 during World War II.
Type locality: "Mexique" [= Mexico]. Restricted to Acambaro, Guanajuato, Mexico *fide* H.M. Smith & Taylor (1950a: 330).
Distribution: Mexico (Aguascalientes, S Chihuahua, S Coahuila, Distrito Federal, Guanajuato, Guerrero, Hidalgo, Jalisco, México, Michoacán, Morelos, Nayarit, Oaxaca, Puebla, Querétaro, Sinaloa, Veracruz, Zacatecas), 1000–3200 m.
Sources: Jan, 1863b, Bocourt, 1888 *in* Bocourt, Duméril & Mocquard, 1870–1909, H.M. Smith, 1943a, Duellman, 1961, McCranie & Wilson, 1984, 2001b, W.W. Tanner, 1985, Pérez-Higareda & Smith, 1991, Camarillo & Smith, 1992 and Benitez-Gálvez, 1997.
Remarks: A synonym of *S. grahamiae fide* Bogert (1939a: 186) and W.W. Tanner (1985: 641–642).

2. *Salvadora deserticola* K.P. Schmidt, 1940. Field Mus. Nat. Hist. (Zool.) 24(12): 146–148, fig. 14. (*Salvadora hexalepis deserticola*)

Synonym: *Salvadora hexalepis celeris* H.M. Smith, 1941d.
Type: Holotype, FMNH 26615, an 854 mm specimen (T. Carney, 1935).
Type locality: "Government Spring, near Chisos Mountains, Brewster County, Texas" [USA].
Distribution: Extreme SW USA (SE Arizona, S New Mexico, W Texas) and N Mexico (Chihuahua, ext. N Coahuila, Sinaloa, E Sonora), 300–1830 m.
Sources: H.M. Smith, 1940, 1943a, Bogert & Oliver, 1945, A.H. Wright & Wright, 1957, Fowlie, 1965, Tennant, 1984, 1985, Dixon, 1987b, Degenhardt et al., 1996, Bartlett & Tennant, 2000 and Werler & Dixon, 2000.

3. *Salvadora grahamiae* Baird & Girard, 1853. Cat. No. Amer. Rept., Serp. 1: 104–105.

Synonym: *Salvadora lineata* K.P. Schmidt, 1940.
Type: Holotype, USNM 2081, a 698–724 mm male (J.H. Clark [U.S.-Mex. Bound. Surv.], 1851, via J.D. Graham).
Type locality: "Sonora, Mex." (in error). Corrected to south of the Gila River in Santa Cruz or Cochise Co., SE Arizona, USA *fide* Bogert (1939a: 198–199), Stejneger

(1940: 204) and Stickel (1943: 120). Restricted to Huachuca Mountains, Cochise Co., SE Arizona, USA *fide* Bogert (1939a: 186, 199).

Distribution: Extreme SW USA (SE Arizona, New Mexico, SW Texas) and N Mexico (Chihuahua, Coahuila, Durango, Distrito Federal, Nuevo León, Querétaro, San Luis Potosí, ext. NE Sonora, Tamaulipas, Veracruz, Zacatecas), NSL–1980 m.

Sources: A.H. Wright & Wright, 1957, Fowlie, 1965, Van Devender & Lowe, 1977, Tennant, 1984, 1985, Stebbins, 1985, Degenhardt et al., 1996, Bartlett & Tennant, 2000 and Werler & Dixon, 2000.

4. *Salvadora hexalepis* (Cope, 1867a). Proc. Acad. Nat. Sci. Philadelphia (1866) 18(4): 304. (*Phimothyra hexalepis*)

Synonyms: *Salvadora grahamiae virgultea* Bogert, 1935, *Salvadora hexalepis klauberi* Bogert, 1945, and *Salvadora hexalepis mojavensis* Bogert, 1945.

Type: Holotype, USNM 7894, a 240+ mm juvenile (E. Coues [Wheeler Surv.], 1864–1865).

Type locality: "Fort Whipple, Territory of Arizona, near the parallel of 35°" [= Fort Whipple, W Yavapai Co., W Arizona, USA].

Distribution: Southwestern USA (Arizona, S and E California, W Nevada, ext. SW Utah) and NW Mexico (Baja California Norte, Baja California Sur, W Sonora, Espíritu Santo, San Gerónimo, San José, Tiburón and Todos Santos Is.), BSL–2200 m.

Sources: Fowlie, 1965, Stebbins, 1985, Bartlett & Tennant, 2000, Germano, 2001 and Grismer, 2002.

Remarks: Original description reprinted in Coues (1875: 620).

5. *Salvadora intermedia* Hartweg, 1940. Copeia 1940(4): 256–258.

Synonym: *Salvadora intermedia richardi* H.M. Smith, 1941.

Type: Holotype, UMMZ 85733, a male (W.W. Brown, 1932–1940).

Type locality: "vicinity of Chilpancingo, Guerrero, Mexico" [= vicinity of Chilpancingo (17°33'N, 99°30'W, elevation 1270 m), Sierra Madre del Sur, cen. Guerrero State, S Mexico].

Distribution: Southern Mexico (Sierra Madre del Sur of Guerrero, Oaxaca and Puebla), 500–2700 m.

Sources: Davis & Dixon, 1957, Casas-Andreu et al., 1996 and Benitez-Gálvez, 1997 and Canseco-Márquez & Gutiérrez-Mayén, 2010.

6. *Salvadora lemniscata* (Cope, 1895c). Trans. Amer. Philos. Soc. (1895) 18(2): 203. (*Drymobius lemniscatus*)

Synonym: *Salvadora pulchra* K.P. Schmidt, 1943 (*nomen nudum*).

Types: Syntypes (?), MNHN, longest syntype 1610 mm (A.L.J.F. Sumichrast, 1855–1882).

Type locality: "Mexique occidental" [= W part of Isthmus of Tehuantepec, Oaxaca State, S Mexico *fide* Sumichrast, 1880: 162].

Distribution: Southern Mexico (Chiapas, Guerrero, Oaxaca), 50–1200 m.

Source: Bogert, 1939b.

Remarks: Sumichrast (1882: 285) listed *Coluber* sp. nov. ? and *Masticophis* sp. from near Tapana, either of which could be this species. Type locality restriction to Acapulco, Guerrero, Mexico *fide* H.M. Smith & Taylor (1950a: 351) invalid.

7. *Salvadora mexicana* (A.-M.-C. Duméril, Bibron & Duméril, 1854a). Erpét. Gén. 7(1): 695–696. (*Zamenis mexicanus*)

Type: Holotype, MNHN 3585, a 1220 mm specimen (Dubois).

Type locality: "Cap Corrientes (Mexique)" [= Cabo Corrientes, NW Jalisco State, W Mexico, ca. 20°25'N, 105°42'E, elevation NSL].

Distribution: Western Mexico (Colima, Guanajuato, Guerrero, Jalisco, México, Michoacán, Morelos, Nayarit, Oaxaca, Puebla), NSL–1200 m.

Sources: A.C.L.G. Günther, 1863c, Bocourt, 1888 *in* A.H.A. Duméril, Bocourt & Mocquard, 1870–1909, L.C. Stuart, 1933, Bogert, 1939b, Duellman, 1961, Camarillo & Smith, 1992, Casas-Andreu et al., 1996 and Benitez-Gálvez, 1997.

8. †*Salvadora paleolineata* Holman, 1973a. Contr. Mus. Paleont., Univ. Michigan 24(12): 132, figs. 5a–b.

Type: Holotype, UMMP V56292, one trunk vertebra (1967–1971).

Type locality: "one mile north of the Niobrara River on the Ranch of Franklin H. Egelhoff in the SE corner of the NE1/4 SW1/4 sec. 29, T 33 N, R 23 W, Keya Paha County, Nebraska, Upper Miocene, lower Valentine formation, Egelhoff local fauna" [= middle Miocene *fide* Holman, 2000: 188].

Distribution: Lower Miocene (Hemingfordian) of USA (Nebraska, South Dakota), lower/middle Miocene (Hemingfordian/Barstovian) of USA (Wyoming), middle Miocene (Barstovian) of USA (Nebraska, South Dakota, Texas), and upper Miocene (Hemphillian) of USA (Nebraska, Texas).

Sources: Holman, 1976a, 1977c, 1977e, 1978, 1979b, 1981a, 1982a, 1987a, 1996a, Green & Holman, 1977, Van Devender & Worthington, 1977, Holman & Sullivan, 1981, Rage, 1984b, Parmley & Holman, 1995 and Holman, 2000a.

†*SANAJEH* J.A. Wilson, Mohabey, Peters & Head, 2010 (†Madtsoiidae)

Type species: †*Sanajeh indicus* J.A. Wilson, Mohabey, Peters & Head, 2010.
Distribution: Upper Cretaceous of India.
Source: J.A. Wilson et al., 2010

1. †*Sanajeh indicus* J.A. Wilson, Mohabey, Peters & Head, 2010. PLoS Biology 8(3), 2–3, figs. 1–4.

Type: Holotype, GSI/GC/2901–2906 (cast in UMMP 14265), nearly complete skull and lower jaws and 72 trunk vertebrae plus ribs from a 3460 mm specimen (D.M. Mohabey, 1984).
Type locality: "infratrapean calcareous sandstones of the Lameta Formation exposed near Dholi village in Gujarat, western India (23°08' N, 73°23' E); Late Cretaceous (Maastrichtian)."
Distribution: Upper Cretaceous (Maastrichtian: 65.5–72.1 mya) of India. Known only from holotype.

†*SANJUANOPHIS* Sullivan & Lucus, 1988 (Boidae)

Type species: †*Sanjuanophis froehlichorum* Sullivan & Lucus, 1988.
Distribution: Lower Eocene of USA.
Sources: Sullivan & Lucas, 1988 and Longrich et al., 2012b.

1. †*Sanjuanophis froehlichorum* Sullivan & Lucus, 1988. J. Paleont. 62(4): 637, figs. 5.1–5.6, 11. (*nomen corrigendum*)

Synonyms: *Sanjuanophis froehlichi* Sullivan & Lucas, 1988 *(nomen incorrigendum)*, and *Sanjuanophis froehlichorum* – Morgan & Lucas *in* Lucas, 1999 *(nomen corrigendum)*.
Type: Holotype, NMMNH P-9867 (formerly UNM J-105), seven trunk vertebrae, two fragments of trunk vertebrae, and three vertebrae in concretions (P.H. Froehlich, 1984).
Type locality: "UNM 33A, T23N, R2W, San Juan Basin, New Mexico; San José Formation (Regina Member), Almagre local fauna. Early Eocene (Wasatchian)" [= 4 mi. W and 1.75 mi. S Regina, NW Sandoval Co., NW New Mexico, USA]. Emended to Jicarilla Indian Reservation, Sandoval County, New Mexico (L-204), Regina Member, San José Formation, early Eocene (Wasatchian) *fide* Morgan & Lucas *in* Lucas (1999: 255).
Distribution: Lower Eocene (Wasatchian: 50.3–55.4 mya) of USA (New Mexico). Known only from type locality.

S

SANZINIA Gray, 1849a (Boidae)

Synonyms: *Sganzinia* – W.C.H. Peters, 1880c *(nomen incorrectum)*, and *Sanzina* – Auffenberg, 1958 *(nomen incorrectum)*.
Type species: *Xiphosoma madagascariensis* A.-M.-C. Duméril & Bibron, 1844.
Distribution: Madagascar.
Sources: Jourdran, 1904, Guibé, 1949a, 1958, Genest-Villard, 1966, Stimson, 1969, Branch, 1980, 1986, Kluge, 1991, Wall, 1998a, Henkel & Schmidt, 2000, Vences et al., 2001, Vences & Glaw, 2003, A. Mori et al., 2006 and Glaw & Vences, 2007.

1. *Sanzinia madagascariensis* (A.-M.-C. Duméril & Bibron, 1844). Erpét. Gén. 6: 549–552. (*Xiphosoma madagascariense*)

Synonyms: *Boa manditra* Kluge, 1991 (*nomen substitutum*), and *Sanzinia madagascariensis volontany* Vences & Glaw, 2003.
Types: Syntypes (3), MNHN 43, MNHN 7329, and ? ZMB 6400 (formerly MNHN), longest syntype 1450 mm (V. Sganzin, 1831–1832, C.A.C.J. Bernier, 1831–1836, and L. Rousseau, 1841).
Type locality: "île de Madagascar" [= Madagsascar]. Restricted to Nosy Be Is., W coast of Madagascar *fide* A.-M.-C. Duméril & Duméril (1851: 219).
Distribution: Madagascar (Antsiranana, Antananarivo, Fianarantsoa, Mahajanga, Toamasina, S Toliara, Nosy Be, Nosy Boraha and Nosy Mangabe Is.), NSL–1350 m.
Remarks: Original description listed three syntypes. ZMB 6400 possibly a syntype *fide* Bauer et al. (2002: 171).

SAPHENOPHIS C.W. Myers, 1973 (Xenodontidae)

Type species: *Dromicus boursieri* Jan, 1867 *in* Jan & Sordelli, 1866–1870.
Distribution: Northwestern South America.
Sources: C.W. Myers, 1969b, 1973, J.A. Peters & Orejas-Miranda, 1970, Pérez-Santos & Moreno, 1988 and Zaher et al., 2009.

1. *Saphenophis antioquiensis* (Dunn, 1943). Caladasia 2(8): 307–308. (*Rhadinaea antioquiensis*)

Type: Holotype, MLS 386, a 665–678 mm female (Dec. 1939), destroyed by fire 9–10 April 1948.
Type locality: "San Pedro, Antioquia, Colombia, alt. 2560 m" [= San Pedro, cen. Antioquia Dept., NW Colombia, 6°28'N, 75°33'W, elevation 2560 m].
Distribution: Northwestern Colombia (Antioquia), 2560 m. Known only from type locality.

Remarks: Photographs of holotype and illustration of hemipenis in C.W. Myers (1969b: figs. 1a–b, 2a–b, 3 and 1973: fig. 2a).

2. *Saphenophis atahuallpae* (Steindachner, 1901). Anz. Akad. Wiss. Wien, Math.-Naturwiss. Kl. 38(18): 195–196. (*Liophis atahuallpae*)

Type: Holotype, ZSM 106/1926 (formerly PTB 68), a 600 mm specimen (T. von Bayern, 1898).

Type locality: "unterhalb Las Palmas am Westabhange der Anden auf dem Wege von Babahoyo nach Guaranda, West-Ecuador" [= below Las Palmas on W slope of Andes, between Babahoyo (1°49'S, 79°31'W), S Los Rías Prov., and Guaranda (1°36'S, 79°00'W), Bolívar Prov., cen. Ecuador, elevation 2500 m]. Emended to ca. 2500 m *fide* Steindachner (1902: 105).

Distribution: Central Ecuador (Bolívar), 2500 m. Known only from type locality.

Source: Pérez-Santos & Moreno, 1991.

Remarks: Supplemental original description in Steindachner (1902: 105, pl. 1, figs. 4–4a).

3. *Saphenophis boursieri* (Jan, 1867 *in* Jan & Sordelli, 1866–1870). Icon. Gén. Ophid. 2(25): 4, pl. 2, figs. 2, 2a–c, c–e, f–g, g, p, r, v, y. (*Dromicus boursieri*)

Synonym: *Coronella whymperi* Boulenger, 1882.

Type: Holotype, MNHN 858, a 590 mm female.

Type locality: "Quito" [Ecuador] (probably in error, possibly the shipping site).

Distribution: Extreme SW Colombia (Nariño) and NE Ecuador (Napo, Pastaza, Pichincha), [300] 1000–1890 m.

Sources: H.W. Parker, 1935, J.A. Peters, 1960a, 1963, C.W. Myers 1966b and Pérez-Santos & Moreno, 1988, 1991.

Remarks: Photographs of holotype in C.W. Myers (1973: figs. 5a, 6a–b). Possibly occurs in Loreto, Peru *fide* Myers (1973: 33). Lowland record from E Ecuador needs confirmation *fide* C.W. Myers (1973: 12).

4. *Saphenophis sneiderni* C.W. Myers, 1973. Amer. Mus. Novit. (2522): 22–26, figs. 5c, 9a–b.

Type: Holotype, NHR 3129a, a 422 mm female (K. von Sneidern, 1935–1936).

Type locality: "El Tambo, Department of Cauca, Colombia, about 21 kilometers westward from Popayán, between the headwaters of the Río Cauca and the Río Patía, on the eastern slopes of the Cordillera de Occidental, at an elevation of 1745 meters" [= El Tambo, cen. Cauca Dept., SW Colombia, ca. 2°45'N, 77°00'W, elevation 1745 m].

Distribution: Southwestern Colombia (cen. Cauca), 1745 m. Known only from type locality.

5. *Saphenophis tristriatus* (Rendahl & Vestergren, 1940). Arkiv Zool. 33A(1): 5–6. (*Rhadinaea tristiata*)

Type: Holotype, NHR 3119, a 655–660 mm female (G. Gerring, 1938).

Type locality: "Colombia, Cauca" [= Cauca Dept., SW Colombia].

Distribution: Southwestern Colombia (Cauca), 3200 m.

Remarks: Photograph of holotype and illustration of hemipenis in C.W. Myers (1969b: figs. 4, 7 and 1973: figs. 4b, 5b).

SCAPHIODONTOPHIS E.H. Taylor & Smith, 1943 (Sibynophiidae)

Synonyms: *Scaphiodontopsis* – Beltrán, 1953 *(nomen incorrectum), Scaphiondontophis* – K.P. Schmidt & Inger, 1957 *(nomen incorrectum),* and *Scaphodontophis* – Alvarez del Toro, 1960 *(nomen incorrectum).*

Type species: *Enicognathus annulatus* A.-M.-C. Duméril, Bibron & Duméril, 1854a.

Distribution: Latin America.

Sources: E. Morgan, 1973, R.W. Henderson, 1984, Savage & Slowinski, 1996, McCranie et al., 2006 and McCranie, 2006c, 2011a.

Remarks: Pyron et al., 2011, placed this genus in the Scaphiodontophiinae.

1. *Scaphiodontophis annulatus* (A.-M.-C. Duméril, Bibron & Duméril, 1854a). Erpét. Gén. 7(1): 335–336, pl. 80. (*Enicognathus annulatus*)

Synonyms: *Ablabes annulata* A.-M.-C. Duméril, 1853 (*nomen nudum*), *Henicognathus sumichrasti* Bocourt, 1886 *in* A.H.A. Duméril, Mocquard & Bocourt, 1870–1909 *(nomen rejiciendum), Henicognathus annulata cyclura* Cope *in* Ferrari-Pérez, 1886, *Sibynophis annulatus hondurensis* K.P. Schmidt, 1936, *Scaphiodontophis albonuchalis* E.H. Taylor & Smith, 1943, *Scaphiodontophis carpicinctus* E.H. Taylor & Smith, 1943, *Scaphiodontophis cyclurus* E.H. Taylor & Smith, 1943, and *Scaphiodontophis nothus* E.H. Taylor & Smith, 1943.

Type: Holotype, MNHN 7283, a 590+ mm male.

Type locality: "Coban (Huate-Vera-Paz)" [= Cobán, Alta Verapaz Dept., Guatemala, 15°28'N, 90°22'W, elevation 1325 m].

Distribution: Mexico and upper Central America. Southern Mexico (Chiapas, Colima, Distrito Federal, Oaxaca, Puebla, Quintana Roo, Tabasco, ext. S Tamaulipas, Veracruz, Yucatán), Belize (Belize, Cayo, Orange Walk, Stann Creek, Toledo), Guatemala (Alta Verapaz, Escuintla, Petén), El Salvador (La Libertad, San Salvador, Sonsonate) and NE Honduras (Atlántida, Comayagua, Cortés, Francisco Morazán, Gracias a Dios, Lempira, Olancho, Santa Bárbara, Yoro), NSL–1550 m.

Sources: K.P. Schmidt, 1941, E.H. Taylor & Smith, 1943, L.C. Stuart, 1948, 1963, E.H. Taylor, 1951, 1954, Mertens, 1952c, R.W. Henderson & Hoevers, 1975, J.C. Lee, 1980, 1996, N.J. Scott et al., 1983, J.D. Johnson, 1984, L.D. Wilson & Meyer, 1985, H.M. Smith et al., 1986, Casas-Andreu et al., 1996, Lamar, 1997b, J.A. Campbell, 1998b, Canseco-Marquez et al., 2000 and G. Köhler et al., 2006.

Remarks: Savage & Slowinski (1996: 151) erroneously listed holotype as MNHN 7823.

2. *Scaphiodontophis venustissimus* (A.C.L.G. Günther, 1894 *in* 1885–1902). Biol.-Cent. Amer., Rept.: 144, pl. 51, fig. c. (*Henicognathus venustissimus*)

Synonyms: *Sibynophis zeteki* Dunn, 1930, and *Scaphiodontophis dugandi* Roze, 1969.

Type: Lectotype, BMNH 1946.1.136, a 430 mm male (E. Rothschuh), designated by Savage & Slowinski (1996: 151).

Type locality: "Nicaragua: Matagalpa: Jérico" [= coffee plantation Hacienda Las Lajas, NE of San Ramón (12°55'N, 85°50'W), W Mataglapa Dept., cen. Nicaragua, ca. 13°N, 85°41'W, elevation 990 m] via lectotype selection.

Distribution: Lower Central America and Colombia. Extreme SE Honduras (Gracias a Dios), Nicaragua (Atlántico Sur, Managua, Matagalpa), Costa Rica (Alajuela, Cartago, Guanacaste, Heredia, Limón, Puntarenas, San José), Panama (Bocas del Toro, Chiriquí, Colón, Darién, Panamá) and NW Colombia (Antioquia, Boyaca, Santander), 30–990 m.

Sources: Roze, 1969, Nicéforo-María, 1970, Villa, 1983, Pérez-Santos, 1986, Savage & Villa, 1986, Pérez-Santos & Moreno, 1988, Villa et al., 1988, Savage, 2002 and McCranie, 2006c.

Remarks: Type locality of Hacienda de Jerico described by A.C.L.G. Günther, 1902 via pers. comm. to A. Voglin (10 July 1966) *fide* Mayorga (1967: 75).

S

SCAPHIOPHIS W.C.H. Peters, 1870b (Colubridae)

Synonym: *Scahphiophis* – Briceño-Rossi, 1934 *(nomen incorrectum).*

Type species: *Scaphiophis albopunctatus* W.C.H. Peters, 1870b.

Distribution: West, cen. and E Africa.

Sources: Loveridge, 1936c, 1955a, 1957a, Broadley, 1994 and Spawls et al., 2002.

1. *Scaphiophis albopunctatus* W.C.H. Peters, 1870b. Mber. Königl. Akad. Wiss. Berlin 70(8): 645–646, pl. 1, figs. 4, 4a–c.

Type: Holotype, ZMB 6945, a 352 mm specimen (Jahn).

Type locality: "Keta (Guinea)" [= Keta, ext. S Volta Region, ext. SE Ghana, 5°55'N, 0°59'E, elevation NSL].

Distribution: Central Africa. Guinea, Sierra Leone, S Ghana (Volta), S Togo (Plateau), Benin (Alibori, Atlantique, Borgou, Collines, Zou), Nigeria (Bauchi, Enugu, Kaysina, Kwara, Plateau), S Chad (Mayo-Kebbi Ouest), South Sudan (SW Eastern Equatoria), Cameroon (Adamaoua, Est, Centre, Extreme-Nord, Nord, Ouest, Sud, Sud-Ouest), Central African Republic (Bangui, Kemo-Grébizi, Ombella-Mpoko, Vakaga), Gabon, Congo (Brazzaville, Plateau), Democratic Republic of the Congo (Bandundu, Bas-Congo, Equateur, Kasai Occidental, Kasai Oriental, Katanga, Kinshasa), SE Uganda (Eastern), S Burundi, N Tanzania (Dodoma, Mbeya, Mtwara, Mwanza, Rukwa, Shinayanga, Tabora), SE Kenya (Coast, S Eastern), ext. N Mozambique (N Niassa), N Angola (Lunda Norte) and N Zambia (Luapula, Northern), NSL–1475 m.

Sources: J.G. Fischer, 1885a, Rochebrune, 1885, Bocage, 1887, Mocquard, 1887c, Sternfeld, 1910d, Boulenger, 1915, Chabanaud, 1917b, K.P. Schmidt, 1923, Scortecci, 1930, Angel, 1933b, Laurent, 1950c, Villiers, 1951b, 1952b, 1963, 1965, 1966, 1975, Witte, 1953, 1962, 1966, Vesey-FitzGerald, 1958, 1975, Perret, 1961, Doucet, 1963, Thys van den Audenaerde, 1965, Bourgeois, 1968, Leston & Hughes, 1968, Hughes & Barry, 1969, Broadley, 1971d, 1994, Spawls, 1978, Stucki-Stirn, 1979, J.-F. Trape, 1981, Joger, 1982, 1990, N. Hedges, 1983, Hughes, 1983, 2013, Roman, 1984, Largen & Rasmussen, 1993, J.-F. Trape & Roux-Estève, 1995, Broadley et al., 2003, Chippaux, 2006, Chirio & Ineich, 2006, Chirio & LeBreton, 2007, Pauwels & Vande weghe, 2008, Ullenburch et al., 2010 and Segniagbeto et al., 2011.

Remarks: Loveridge (1936c: 255) erroneously listed the type locality as Kita, Guinea, as clarified by Hughes & Barry (1969: 1018). Occurs in Mali *fide* Loveridge (1957a: 264) and Senegal and Gambia *fide* Rochebrune (1885: 163), and Guinea and Sierra Leone *fide* Chirio & LeBreton (2007: 546).

2. *Scaphiophis raffreyi* Bocourt, 1875. Ann. Sci. Nat. (Zool.) (6) 2(3): 1.

Synonym: *Scaphiophis calciatii* Scortecci, 1928.

Type: Holotype, MNHN 4331 (M.J.A. Raffray, 1874–1875).

Type locality: "Devatabar (Abyssimie), à une altitude d'environ 2500 mètres" [= Debere Tabor, Ethiopia, 11°51'N, 38°01'E, elevation 2500 m]

Distribution: Northeastern Africa. Southeastern South Sudan (Eastern Equatoria), W Eritrea (Anseba, Gash-Barka), W Ethiopia (Gemu Gofa, Gondar, Illubabor, Shoa, Wollega), NW Kenya (Rift Valley) and NE Uganda (Northern), 515–1760 (2500) m.

Sources: Scortecci, 1928, H.W. Parker, 1949, Pitman, 1974, Largen & Rasmussen, 1993, Spawls et al., 2002 and Largen & Spawls, 2010.

Remarks: Type locality (2500 m) might represent where it was dispatched rather than where it was captured *fide* Spawls et al. (2002: 369).

SCHWARTZOPHIS Zaher, Grazziotin, Cadle, Murphy, Moura-Leite & Bonatto, 2009 (Xenodontidae)

Type species: *Natrix callilaema* Gosse, 1851.
Distribution: Jamaica.
Sources: A. Schwartz & R.W. Henderson, 1991, R.W. Henderson & Powell, 2009 and Zaher et al., 2009.
Remarks: A valid genus *fide* Grazziotin et al., (2012:) but a synonym of *Hypsirhynchus fide* Hedges et al. (2009: 9) and *Darlingtonia fide* Crother, 1989.

1. *Schwartzophis callilaemum* (Gosse, 1851). Nat. Sojourn Jamaica: 384–385. (*Natrix callilaema*)

Type: Lectotype, BMNH 1946.1.5.90, a 157 mm specimen, designated by Buden (1966: 2).
Type locality: "Bluefields, Westmoreland Parish, Jamaica" [= vicinity of Bluefields (18°10'N, 78°01'W, elevation 125 m), Westmoreland Par., W Jamaica] via lectotype selection.
Distribution: Jamaica (Clarendon, Kingston & St. Andrews, Manchester, Portland, St. Catherine, St. Elizabeth, St. James, St. Mary, St. Thomas, Trelawny, Westmoreland), NSL–1000 m.
Source: Buden, 1966.

2. *Schwartzophis funereum* (Cope, 1862b). Proc. Acad. Nat. Sci. Philadelphia 14(1): 77. (*Alsophis funereus*)

Type: Neotype, USNM 12372 (formerly USNM 5778, paratype), a 378 mm male (C.B. Adams), designated by Buden (1966: 5).
Type locality: "Jamaica" via neotype selection.
Distribution: Jamaica (Clarendon, Hanover, Kingston & St. Andrews, Manchester, St. Ann, St. Catherine, St. Elizabeth, St. James, St. Mary, Trelawny, Westmoreland), NSL–610 m.
Source: Buden, 1966.
Remarks: Holotype (USNM 5779, Jamaica, C.B. Adams) lost *fide* Grant (1940: 125).

3. *Schwartzophis polylepis* (Buden, 1966). Breviora (238): 7–8. (*Dromicus polylepis*)

Type: Holotype, MCZ 81020 (formerly ASFS 13122), a 387 mm male (native collector, 19 June 1961).
Type locality: "Port Antonio, Portland Parish, Jamaica."
Distribution: Eastern Jamaica (Portland, Kingston & St. Andrews, St. Thomas), NSL–210 m.

SCOLECOPHIS Fitzinger, 1843 (Colubridae)

Synonym: *Scoleophis* – A.-M.-C. Duméril & Bibron, 1844 *(nomen incorrectum).*
Synonyms: *Pseudoelaps* Fitzinger, 1826a *(nomen oblitum), Pseudelaps* Agassiz, 1847 *(nomen emendatum), Platycranion* Jan, 1863b, and *Pseudaelaps* – Senna, 1886 *(nomen incorrectum).*
Type species: *Calamaria atrocincta* Schlegel, 1837.
Distribution: Central America.
Sources: Jan & Sordelli, 1865 *in* 1860–1866, Salvin, 1860, 1861, Bocourt, 1883 *in* A.H.A. Duméril, Bocourt & Mocquard, 1870–1909, A.C.L.G. Günther, 1895 *in* 1885–1902, E.H. Taylor, 1951, Mertens, 1952c, L.C. Stuart, 1963, J.A. Peters & Orejas-Miranda, 1970, N.J. Scott et al., 1983, L.D. Wilson, 1983c, L.D. Wilson & Meyer, 1985, Savage & Villa, 1986, L.D. Wilson et al., 1986, Villa et al., 1988, Campbell & Vannini, 1989, G. Köhler, 1999b, 2001a–b, 2008, Savage, 2002, L.D. Wilson & Williams, 2002, Solórzano, 2004, G. Köhler et al., 2005, Kolby & Fraser, 2009, Lovich et al., 2010 and McCranie, 2011a.
Remarks: In accordance with Art. 23.9.2 of the Code (ICZN, 1999), *Scolecophis* Fitzinger is designated a *nomen protectum* and *Pseudoelaps* Fitzinger a *nomen oblitum.*

1. *Scolecophis atrocinctus* (Schlegel, 1837). Essai Phys. Serp. 1: 134, 2: 47–48. (*Calamaria atrocincta*)

Synonym: *Elaps zonatus* Hallowell, 1854e.
Type: Holotype, MNHN 519, a 418–441 mm male (C. Gay, 1828–1837).
Type locality: "Chilé," (in error *fide* Bocourt, 1883 *in* A.H.A. Duméril, Bocourt & Mocquard, 1870–1909: 578). Corrected to Central America *fide* Bocourt, 1883 *in* A.H.A. Duméril, Bocourt & Mocquard (1870–1909: 578).
Distribution: Central America. Guatemala (Baja Verapaz, Sacatepéquez), El Salvador (Ahuachapán, La Libertad, San Salvador, Santa Ana, Sonsonate), SW Honduras (Comayagua, Cortés, El Paraíso, Francisco Morazán, Lempira), W Nicaragua (Boaco, Estelí, Granada, Managua, Rivas, Ometepe Is.) and NW Costa Rica (Alajuela, Cartago, Guanacaste, Heredia, Puntarenas, San José), NSL–1140 m.
Remarks: Two syntypes sent from Chile by Gay *fide* Duméril, Bibron & Duméril (1854b: 864). L.C. Stuart (1963: 115) questioned the application of this name to the Central American taxon.

SEMINATRIX Cope, 1895b (Natricidae)

Type species: *Contia pygaea* Cope, 1871b.
Distribution: Southeastern USA.

Sources: Dowling, 1950, Martof, 1956, Rossman, 1956, Mount, 1975, Martof et al., 1980, Ashton & Ashton, 1988, Ernst & Barbour, 1989, Conant & Collins, 1991, Price, 1991, Dodd, 1993, Palmer & Braswell, 1995, Tennant, 1997, Dorcas et al., 1998, Alfaro & Arnold, 2001, Gibbons & Dorcas, 2004, Zaher et al., 2009 and Rheubert et al., 2010.

1. *Seminatrix pygaea* (Cope, 1871b). Proc. Acad. Nat. Sci. Philadelphia 23(2): 223–224. (*Contia pygaea*)

Synonyms: *Seminatrix pygaea cyclas* Dowling, 1950, and *Seminatrix pygaea paludis* Dowling, 1950.

Type: Holotype, ANSP 3533, a 226–229 mm male (E. Tatnall).

Type locality: "Volusia, Florida" [= National Gardens, Volusia County, NE Florida, elevation 10 m *fide* Carr & Goin, 1942: 48].

Distribution: Southeastern USA (ext. S cen. Alabama, Florida, SE Georgia, ext. E North Carolina, E South Carolina), NSL–150 m.

SENTICOLIS Dowling & Fries, 1987 (Colubridae)

Synonym: *Senclicolis* – Schulz, 1992e *(nomen incorrectum).*

Type species: *Coluber triaspis* Cope, 1866.

Distribution: Southwestern USA and Mesoamerica.

Sources: Bocourt, 1888 *in* A.H.A. Duméril, Bibron & Mocquard, 1870–1909, K.P. Schmidt & Andrews, 1936, L.C. Stuart, 1948, 1963, E.H. Taylor, 1951, Mertens, 1952c, Dowling, 1960, Webb, 1960, H. Campbell & Howell, 1965, Duellman, 1965, J.C. Lee, 1980, 1996, Stebbins, 1985, L.D. Wilson & Meyer, 1985, Savage & Villa, 1986, Dowling & Fries, 1987, Villa et al., 1988, Pérez-Higareda & Smith, 1991, Price, 1991, Garrett & Painter, 1992, Schulz, 1992d, Staszko & Walls, 1994, Degenhardt et al., 1996, Schultz, 1996, J.A. Campbell, 1998b, G. Köhler, 1999b, Rodrígues-Robles & Jesús-Escobar, 1999, Bartlett & Tennant, 2000, McCranie & Wilson, 2001b, Lemos-Espinal et al., 2004c, G. Köhler et al., 2005, Burbrink & Lawson, 2007, Dixon & Lemos-Espinal, 2010, Lovich et al., 2010, Ramírez-Bautista et al., 2010, McCranie, 2011a and Morales-Capellan et al., 2011.

1. *Senticolis triaspis* (Cope, 1866). Proc. Acad. Nat. Sci. Philadelphia 18(2): 128. (*Coluber triaspis*)

Synonyms: *Pityophis intermedius* Boettger, 1883b, *Coluber mutabilis* Cope, 1885a, *Coluber chlorosoma* A.C.L.G. Günther, 1894 *in* 1885–1902, and *Coluber polylepis* F. Werner, 1896c.

Type: Holotype, USNM 24903, a 451–455 mm male (D.B. Parsons).

Type locality: "Belize." Restricted to the vicinity of Belize [= Belize city, E Belize District, NE Belize, 17°30'N, 88°12'W, elevation NSL] *fide* H.M. Smith & Taylor (1950a: 316)

Distribution: Southwestern USA and Mesoamerica. Extreme SW USA (SE Arizona, ext. SW New Mexico), Mexico (Aguascalientes, Campeche, Chiapas, SW Chihuahua, Colima, Durango, Guerrero, Hidalgo, Jalisco, México, Michoacán, Morelos, Nayarit, Oaxaca, Puebla, Querétaro, Quintana Roo, San Luis Potosí, Sinaloa, E Sonora, S Tamaulipas, Veracruz, Yucatán, Zacatecas), Belize (Belize, Toledo), Guatemala (Alta Verapaz, Petén, Santa Rosa), Honduras (Choluteca, Colón, Cortés, El Paraíso, Francisco Morazán, Intibucá, Lempira, Ocotepeque, Yoro), W El Salvador (Ahuachapán, Cuscatlán, La Libertad, Santa Ana, Sonsonate), Nicaragua (Estelí, Granada, Managua, Matagalpa) and Costa Rica (Alajuela, NW Cartago, Guanacaste, Heredia, Puntarenas, San José), NSL–2425 m.

Remarks: Official Specific Name no. 2106 *fide* Opinion 745 (ICZN, 1965).

SIAGONODON W.C.H. Peters, 1881b (*nomen substitutum*) (Leptotyphlopidae)

Synonyms: *Catodon* A.-M.-C. Duméril & Bibron, 1844 *(nomen praeoccupatum), Catadontus* – Hoffmann, 1890 *(nomen incorrectum), Siagondon* – Bogert, 1930 *(nomen incorrectum),* and *Evanwhittonus* Hoser, 2012am *(nomen illegitimum).*

Type species: *Anguis septemstriatus* J.G. Schneider, 1801.

Distribution: South America.

Sources: J.A. Peters & Orejas-Miranda, 1970, Hahn, 1980a, Wallach, 1998b and McDiarmid et al., 1999.

1. *Siagonodon acutirostris* Pinto & Curcio, 2011. Copeia 2011(1): 58–61, figs. 5–6.

Type: Holotype, CHUNB 35648, a 216 mm female (22 Feb. 2004).

Type locality: "Brazil, state of Tocantins, municipality of Almas, 11°31'00"S, 47°09'00"W, ca. 397 m elevation."

Distribution: Central Brazil (E Tocantins), 400–590 m.

2. *Siagonodon borrichianus* (Degerbøl, 1923). Vidensk. Medd. Dansk Naturhist. Foren. København 76: 113–114, 3 figs. (*Glauconia borrichiana*)

Type: Holotype, ZMUC R539, a 185 mm specimen (Jensen-Haarup, 1905).

Type locality: "Santa Rosa, Mendoza, Argentina" [= Santa Rosa, N Mendoza Prov., cen. Argentian, 32°54'S, 68°46'W, elevation 700 m].

Distribution: West-central Argentina (La Pampa, La Rioja, Mendoza, Río Negro, San Juan), 240–770 m.

Sources: Freiberg, 1951, Miranda & Tio-Vallejo, 1985, Cei, 1986, J.D. Williams & Francini, 1991, Scolaro, 2006 and Pinto, 2010.

3. *Siagonodon cupinensis* (Bailey & Carvalho, 1946). Bol. Mus. Nac., Zool. (2) (52): 1–3, figs. 1–3. (*Leptotyphlops cupinensis*)

Type: Holotype, MNRJ 387, a 128 mm specimen (A.L. de Carvalho, Jan. 1940).
Type locality: "near the mouth of the Tapirapé River (tributary to the Araguaya), northeastern Mato Grosso" [= vicinity of Río Tapirapé and Río Araguaia junction, E Mato Grosso State, cen. Brazil, 14°07'S, 50°51'W, elevation 250 m].
Distribution: Northeastern South America. Northeastern Brazil (Amapa, NE Mato Grosso, N Rondônia) and Suriname (Marowijne), NSL–600 m.
Sources: Amaral, 1955b, Orejas-Miranda, 1966b, 1967, Hoogmoed, 1977 and Abuys, 2003.
Remarks: Holotype listed as MNRJ 377 *fide* Orejas-Miranda (1966c: 1).

4. *Siagonodon septemstriatus* (J.G. Schneider, 1801). Hist. Amph. 2: 341. (*Anguis septemstriatus*)

Synonyms: *Stenostoma septemstriatum* Jan & Sordelli, 1860 *in* 1860–1866, *Leptotyphlops septemstriata* Amaral, 1930e, *Leptotyphlos septemstriata* – Briceño-Rossi, 1934 *(nomen incorrectum),* and *Leptotyphlops tatacua* Briceño-Rossi, 1934.
Type: Holotype, ZMB 3876 (formerly MLH) (J.B. Lampe).
Type locality: Unknown. Designated as Suriname *fide* Hoogmoed (1977: 114).
Distribution: Northern South America. Southeastern Venezuela (Amazonas, Bolívar, Zulia), Guyana (Cuyuni-Mazaruni, Essequibo Islands-West Demerara, Potaro-Siparuni), Suriname (Brokspondo, Marowijne, Nickerie, Saramacca), French Guiana (Cayenne, Saint-Laurent-du-Maroni), N Brazil (Amazonas, Maranhao, Pará, Rondônia, Roraima) and Bolivia (Tarija), 20–850 m.
Sources: Mertens, 1925, Amaral, 1955b, 1977, Roze, 1966a, Orejas-Miranda, 1967, Hoogmoed, 1977, Cunha & Nascimento, 1978, 1993, Fugler & Riva, 1990, Gorzula & Señaris, 1998, Starace, 1998, Kornacker, 1999, Bauer et al., 2002, Abuys, 2003, Börschig, 2007 and C.J. Cole et al., 2013.
Remarks: Type locality erroneously listed as Río Negro, Amazonia, Brasil *fide* Orejas-Miranda (1967: 426) and holotype erroneously listed in RMNH *fide* McDiarmid et al. (1999: 43). The holotype came from Lampe's collection, was donated by Gravenhorst to the Breslau Museum, which sent it to ZMB *fide* Hoogmoed (1977: 113).

5. *Siagonodon unguirostris* (Boulenger, 1902c). Ann. Mag. Nat. Hist. (7) 9(53): 338. (*Glauconia unguirostris*)

Type: Holotype, BMNH 1946.1.11.52 (formerly BMNH 1902.5.22.21), a 180 mm female (P.O. Simons, 1899–1901).
Type locality: "Cruz del Eje, Prov. Cordova, Argentina" [= Cruz del Eje, Córdoba Prov., Argentina, 30°43'S, 64°49'W, elevation 465 m].
Distribution: Northern Argentina (Buenos Aires, Catamarca, Chaco, Córdoba, Corrientes, Formosa, La Pampa, La Rioja, Río Negro, Salta, San Juan, San Luis, Santiago del Estero) and S Paraguay (Boquerón, Chaco), 75–1215 m.
Sources: Serie, 1915, Abalos et al., 1964, Hulse, 1977, Laurent, 1984, Miranda & Tio-Vallejo, 1985, Cei, 1986, 1994, J.D. Williams & Francini, 1991, Kretzschmar, 1996, Leynaud & Bucher, 1999, Giraudo, 2001, Cabrera, 2004, Passos et al., 2006, Scolaro, 2006, Börschig, 2007 and Pinto & Curcio, 2011.
Remarks: Possibly occurs in Bolivian Chaco *fide* Leynaud & Bucher (1999: 11).

SIBON Fitzinger, 1826a (Dipsadidae)

Synonyms: *Sibynon* Fitzinger, 1843, *Sybinon* – A.-M.-C. Duméril & Bibron, 1844 *(nomen incorrectum), Sibynum* Agassiz, 1847 *(nomen emendatum), Petalognathus* A.-M.-C. Duméril Bibron & Duméril, 1854a, *Mesopeltis* Cope, 1867b, *Pethalognathus* Marschall, 1873 *(nomen emendatum),* and *Asthenognathus* Bocourt, 1884 *(nomen rejiciendum).*
Type species: *Coluber nebulatus* Linnaeus, 1758.
Distribution: Latin America.
Sources: J.A. Peters, 1965, J.A. Peters & Orejas-Miranda, 1970, Kofron, 1980, 1985a, Cadle, 1984b, Villa et al., 1988, Savage & McDiarmid, 1992, Wallach, 1995b, McCranie, 2006, Cadle, 2007, Harvey et al., 2009, Zaher et al., 2009, Mulcahy et al., 2011 and Rovito et al., 2012.

1. *Sibon annulatus* (A.C.L.G. Günther, 1872b). Ann. Mag. Nat. Hist. (4) 9(49): 30. (*Leptognathus annulatus*)

Synonym: *Leptognathus pictiventris* Cope, 1875a.
Type: Holotype, BMNH 1946.1.20.81 (formerly BMNH 1871.11.22.14), a 440–445 mm specimen (native, 1871).
Type locality: "elevated country of Costa Rica, near Cartago" [= vicinity of Cartago (9°51'N, 83°55'W, elevation 1375 m), W Cartago Prov., cen. Costa Rica].
Distribution: Lower Central America and NW South America. Eastern Honduras (Colón, Gracias a Dios), N Nicaragua (Jinotega), E Costa Rica (Alajuela, E Cartago, Heredia, Limón, NW Puntarenas), Panama

(Bocas del Toro, Coclé, Darién,Veraguas) and Colombia (César), 15–1250 m.

Sources: A.C.L.G. Günther, 1894 *in* 1885–1902, Savage & Villa, 1986, Kofron, 1990, Savage & McDiarmid, 1992, G. Köhler, 1998b, 1999b, G. Köhler & Seipp, 1998, Savage, 2002, McCranie et al., 2003, 2007, G. Köhler et al., 2010, Moreno-Arias, 2010 and McCranie, 2011a.

Remarks: A synonym of *S. dimidiata* (Günther) *fide* Kofron (1990: 209).

2. *Sibon anthracops* (Cope, 1868b). Proc. Acad. Nat. Sci. Philadelphia 20(2): 108. (*Leptognathus anthracops*)

Synonym: *Sibynomorphus ruthveni* T. Barbour & Dunn, 1921.

Type: Holotype, ANSP 10135, a 499 mm female (R. Bridges via E.D. Cope).

Type locality: "Central America."

Distribution: Central America. Guatemala (Baja Verapaz), El Salvador (Cuscatlán, San Salvador), N Honduras (Francisco Morazán, Yoro), W Nicaragua (Carazo, Estelí, Managua) and NW Costa Rica (Alajuela, Guanacaste, N Puntarenas, San José), NSL–1300 m.

Sources: A.C.L.G. Günther, 1894 *in* 1885–1902, E.H. Taylor, 1954, Heyer, 1967, J.R. Meyer & Wilson, 1971, 1983, Hidalgo, 1979, N.J. Scott et al., 1983, L.D. Wilson & Meyer, 1985, Kofron, 1987, Savage, 2002, McCranie, 2007, 2011a and G. Köhler et al., 2010.

Remarks: Supplemental original description in Cope (1868c: 136). Holotype listed as ANSP 10134 *fide* Kofron (1987: 220). Colombia (Valle) record doubtful *fide* Harvey & Embert (2008: 105).

3. *Sibon argus* (Cope, 1875a). J. Acad. Nat. Sci. Philadelphia (1876) (2) 8(2): 130, pl. 27, fig. 1, pl. 28, fig. 7. (*Leptognathus argus*)

Type: Holotype, USNM 30656, a 345 mm male (W.M. Gabb, 1873–1875).

Type locality: "Sipurio, Costa Rica" [= Sipurio, SE Limón Prov., SE Costa Rica, 9°32'N, 82°55'W, elevation 200 m].

Distribution: Southeastern Costa Rica (Cartago, S Limón) and Panama (Coclé, Darién), NSL–800 m.

Sources: T. Barbour & Dunn, 1921, Kofron, 1987, Savage & McDiarmid, 1992, Savage, 2002, Solórzano, 2004 and G. Köhler et al., 2010.

Remarks: A synonym of *S. dimidiata* (A.C.L.G. Günther) *fide* Kofron (1990: 209).

4. *Sibon carri* (Shreve, 1951). Copeia 1951(1): 52. (*Tropidodipsas carri*)

Type: Holotype, MCZ 49797, a 411+ mm male (A.F. Carr, Jr., 16 Aug. 1945).

Type locality: "Escuela Agricola Panamericana, near Tegucigalpa, Honduras." Emended to Escuela Agricola Panamericana at El Zamorano, 66.5 km S of Tegucigalpa, S Francisco Morazán, S Honduras [14°01'N, 87°02'W, elevation 865 m] *fide* Hahn (1971: 111).

Distribution: Upper Central America. Guatemala (Baja Verapaz, Guastatoya, El Progreso, Zacapa), El Salvador (Cuscatlán, San Salvador) and S Honduras (Choluteca, Francisco Morazán, Intibucá, Valle), NSL–1200 m.

Sources: Mertens, 1952c, Hahn, 1971, J.R. Meyer & Wilson, 1971, L.D. Wilson, 1983c, Kofron, 1985b, L.D. Wilson & Meyer, 1985, G. Köhler, 1996b, G. Köhler et al., 2005 and McCranie, 2011a.

5. *Sibon dimidiatus* (A.C.L.G. Günther, 1872b). Ann. Mag. Nat. Hist. (4) 9(1): 31. (*Leptognathus dimidiatus*)

Synonyms: *Petalognathus multifasciatus* Jan *in* Bocourt, 1884, and *Leptognathus grandoculis* F. Müller, 1887.

Type: Holotype, BMNH 1946.1.20.97, a 432 mm female (1868–1871).

Type locality: "Mexico."

Distribution: Southern Mexico and Central America. Southern Mexico (Chiapas, Oaxaca, S Veracruz), Belize (Cayo, Toledo), Guatemala (Izabal, Petén, San Marcos, Suchitepéquez), El Salvador (Morazán), W Honduras (Copán, Cortés, Francisco Morazán, Gracias a Dios, Olancho, Yoro), cen. Nicaragua (Matagalpa) and W Costa Rica (Alajuela, Cartago, Guanacaste, Limón, Puntarenas, San José), NSL–1600 m.

Sources: A.C.L.G. Günther, 1894 *in* 1888–1902, L.C. Stuart, 1935b, H.M. Smith, 1943a, L.D. Wilson, 1983c, L.D. Wilson & Meyer, 1985, J.A. Campbell & Vannini, 1989, Kofron, 1990, Pérez-Higareda & Smith, 1991, Savage & McDiarmid, 1992, J.C. Lee, 1996, J.A. Campbell, 1998b, Savage, 2002, McCranie, 2007, 2011a, G. Köhler et al., 2010 and Marczak et al., 2010

6. *Sibon dunni* J.A. Peters, 1957b. Copeia 1957(2): 110–111.

Type: Holotype, UMMZ 92068, a 402 mm male (P. Hershkovitz, 10 June 1934).

Type locality: "Pimanpiro, San Nicholas, Province of Imbabura, Ecuador" [= Pimampiro, E Imbabura, N Ecuador, 0°23'N, 77°56'W, elevation 2150 m].

Distribution: Northern Ecuador (Cordillera Oriental of E Imbabura), 2150 m. Known only from type locality.

Sources: Duellman, 1979 and Kofron, 1990.

Remarks: Photograph of holotype in J.A. Peters (1960: pl. 7, fig. d).

7. *Sibon lamari* Solórzano, 2001. Rev. Biol. Trop. 49(3–4): 1111–1118, figs. 1–2.

Type: Holotype, UCR 13983, a 589 mm male (M. Solano & N. Solano, 7 Aug. 1998).

S

Type locality: "Guayacán de Siquirres, al noreste de la provincia de Limón (83°32'30"N y 10°02'10"NE), Costa Rica."
Distribution: Northeastern Costa Rica (Limón) and Panama (Boca del Toro, Colón), NSL–650 m.
Sources: Solórzano, 2004, McCranie, 2007, Jaramillo et al., 2008 and G. Köhler et al., 2010.

8. *Sibon linearis* Pérez-Higareda, López-Luna & Smith, 2002. Bull. Maryland Herp. Soc. 38(2): 62–64, figs. 1–4.

Type: Holotype, UNAM-LT 1796, a 311 mm female (G. Pérez-Higareda, Aug. 1983).
Type locality: "Balzapote, municipality of San Andrés Tuxtla, Veracruz, Mexico, 100 m" [= Balzapote, city of San Andres Tuxtla, S Veracruz, E Mexico, 18°27'N, 95°13'W, elevation 100 m].
Distribution: Eastern Mexico (S Veracruz), 100 m. Known only from type locality.
Source: McCranie, 2007.

9. *Sibon longifrenis* (Stejneger, 1909). Proc. U.S. Natl. Mus. 36(1681): 457–458. (*Mesopeltis longifrenis*)

Synonym: *Dipsas costaricensis* E.H. Taylor, 1951.
Type: Holotype, USNM 38750, a 488 mm female (F.E. Read).
Type locality: "Bocas del Toro, on a small island at one of the entrances to the Bay of Chiriqui, on the Atlantic side and near the western extremity of the Republic of Panama" [= Bocas del Toro Is., NW Bocas del Toro Prov., W Panama, ca. 9°23'N, 82°15'W, elevation NSL].
Distribution: Lower Central America. Honduras (Colón, Gracias a Dios, Olancho), Nicaragua (Jinotega), E Costa Rica (Alajuela, Cartago, Heredia, Limón) and Panama (Bocas del Toro, Coclé, Darién), NSL–800 m.
Sources: E.H. Taylor, 1951, Kofron, 1990, Savage & McDiarmid, 1996, McCranie et al., 2001, Savage, 2002, Solórzano, 2004, McCranie, 2006, 2011a, G. Köhler et al., 2010 and Travers et al., 2011.
Remarks: A synonym of *S. dimidiata* (A.C.L.G. Günther) *fide* Kofron (1990: 209).

10. *Sibon manzanaresi* McCranie, 2007a. Herpetologica 63(2): 213–216, figs. 1–2.

Type: Holotype, USNM 563364, a 407 mm male (T. Manzanares & J.R. McCranie, 27 May 2005).
Type locality: "San San Hil, 14°56.809'N, 84°30.715'W, 300 m elevation, Departamento de Gracias a Dios, Honduras."
Distribution: Northeastern Honduras (Gracias a Dios), 40–300 m.
Source: McCranie, 2011a.

11. *Sibon merendonensis* Rovito, Papenfuss & Vásquez-Almazán, 2012. Zootaxa (3206): 65–67, figs. 2–3.

Type: Holotype, MVZ 263880, a 687 mm female (S.M. Rovito & T.J. Papenfuss, 16 Jan. 2008).
Type locality: "Finca la Bendición, 4.0 km SE (by road) of soccer field at junction with road to Pinalito, then 7.6 km E (by road) of La Trementina, 14.93042°N, 89.4167°W (WGS84 Datum), 1364 m, Departamento Zacapa, Guatemala."
Distribution: Eastern Guatemala (Zacapa), 1365 m. Known only from holotype.

12. *Sibon miskitus* McCranie, 2006. J. Herp. 40(1): 16–18.

Type: Holotype, UF 142076, a 477 mm female (T. Manzanares & J.R. McCranie, 14 May 2004).
Type locality: "Warunta Tingni Kiamp, 14°55.559'N, 84°41.471'W, 150 m elevation, Departamento de Gracias a Dios, Honduras."
Distribution: Northeastern Honduras (Gracias a Dios), 35–300 m.
Source: McCranie, 2007, 2011a.

13. *Sibon nebulatus* (Linnaeus, 1758). Syst. Nat., ed. 10, 1: 222. (*Coluber nebulatus*)

Synonyms: *Coluber sibon* Linnaeus, 1758, *Coluber variegatus* Hallowell, 1845a, *Leptognathus brevis* A.-M.-C. Duméril, Bibron & Duméril, 1854a, *Leptognathus affinis* J.G. Fischer, 1879, *Leptognathus leucomelas* Boulenger, 1896b, *Sibon nebulata hartwegi* J.A. Peters, 1960b, and *Sibon nebulata popayanensis* J.A. Peters, 1960b.
Type: Lectotype, NHR Lin-60 (formerly MAFR), a 640 mm specimen (Mus. Drottn.), designated by Andersson (1899: 19).
Type locality: "America" via lectotype selection. Restricted to Jicaltepec, Veracruz, Mexico *fide* H.M. Smith & Taylor (1950a: 349).
Distribution: Latin America. Southern Mexico (Campeche, Chiapas, Colima, Guerrero, Jalisco, México, Michoacán, Nayarit, Oaxaca, Quintana Roo, Tabasco, Veracruz, Yucatán), Guatemala (Alta Verapaz, Petén), Belize (Belize, Cayo, Orange Walk, Stann Creek, Toledo), El Salvador (Ahuachapán, Santa Ana), N Honduras (Atlántida, Copán, Cortés, Gracias a Dios, Olancho, Yoro), E Nicaragua (Atlántico Norte, Atlántico Sur, Boaco, Jinotega, Matagalpa, Río San Juan, Great Corn Is.), Costa Rica (Alajuela, Cartago, Guanacaste, Heredia, Limón, Puntarenas, San José), Panama (Bocas del Toro, Chiriquí, Coclé, Colón, Darién, Panamá, Veraguas), Colombia (Antioquia, Bolívar, Boyaca, Cauca, Chocó, Cundinamarca, Magdalena, Meta, Nariño, Norte de Santander, Risaralda, Santander, Tolima, Valle, Vaupés), Venezuela (Amazonas, Anzoátegui, Apure, Aragua,

Barinas, Bolívar, Carabobo, Cojedes, Delta Amacuro, Distrito Federal, Falcón, Guárico, Lara, Mérida, Miranda, Monagas, Nueva Esparta, Portuguesa, Sucre, Táchira, Trujillo, Yaracuy, Vargas, Zulia, Margarita Is.), Trinidad and Tobago (Huevos, Tobago, Trinidad), Guyana (Barimi-Waini, Demerara-Mahaica, Essequibo Islands-West Demerara, Potaro-Siparuni, Upper Demerara-Berbice), Suriname (Para, Paramaribo, Saramacca, Suriname), French Guiana (Cayenne), NW Ecuador (Carchi, Esmeraldas, Guayas, Imbabura, Los Rios, Napo, Pichincha) and N Brazil (Alagoas, Amapá, Amazonas, Bahia, Ceará, Maranhão, Pará, Roraima), NSL–2630 m.

Sources: Jan & Sordelli, 1870–1881, A.H.A. Duméril, Bocourt & Mocquard, 1870–1909, A.C.L.G. Günther, 1894 *in* 1885–1902, E.H. Taylor, 1940c, 1951, L.C. Stuart, 1948, 1963, Aleman, 1953, Roze, 1958d, 1966a, Zweifel, 1959a, Duellman, 1963, Neill, 1965, Heyer, 1967, Hoge, 1967, C.J. McCoy, 1970, Villa, 1972, R.W. Henderson & Hoevers, 1975, Emsley, 1977, Cunha & Nascimento, 1978, 1993, Hoogmoed, 1979, J.C. Lee, 1980, 1996, Miyata, 1980, 1982, Hidalgo, 1981, N.J. Scott et al., 1983, L.D. Wilson, 1983, L.D. Wilson & Meyer, 1985, Abuys, 1986b, Lancini, 1986, Pérez-Santos, 1986, Savage & Villa, 1986, Chippaux, 1987, Pérez-Santos & Moreno, 1988, 1991, Nascimento & Lima-Verde, 1989, Weyer, 1990, Pérez-Higareda & Smith, 1991, García & Ceballos, 1994, J.A. Campbell, 1998b, Freire, 1998, Starace, 1998, G. Köhler, 1999b, Stafford & Meyer, 2000, Markezich, 2002, Savage, 2002, G. Köhler et al., 2005, 2010, Rivas-Fuenemayer & Barrio-Amorgós, 2005, Navarrete et al., 2009, Ortega-Andrade et al., 2010, Ugueto & Rivas, 2010, McCranie, 2011a and Travers et al., 2011.

14. *Sibon noalamina* Lotzkat, Hert & G. Köhler (2012). Zootaxa (3485): 27–32, figs. 1a–c, 2a–c, 3a–b.

Type: Holotype, SMF 91539 (formerly SL 775), a 546 mm male (A. Hertz & S. Lotzkat, 10 Aug. 2010).

Type locality: "headwaters of Río Chiriquí Malí, approximately 6.4 km NW Fortuna dam (8.7891°N, 82.2155°W, 1050 m), Bosque Palo Seco, Comarca Ngöbe-Buglé (formerly province of Bocas del Toro), Panama."

Distribution: Western Panama (Bocas del Toro, Veraguas), 1050–1260 m.

S

15. *Sibon perissostichon* G. Köhler, Lotzkat & Hertz, 2010. Herpetologica 66(1): 81–83, figs. 1–4.

Type: Holotype, SMF 88716, a 668 mm female (S. Lotzkat, 14 May 2008).

Type locality: "near Lost and Found Ecohostel, Reserva Forestal La Fortuna, 8°40.47' N, 82°12.97' W, 1434 m elevation, western Panama."

Distribution: Western Panama (Chiriquí), 1435 m.

16. *Sibon sanniolus* (Cope, 1867b). Proc. Acad. Nat. Sci. Philadelphia (1866) 18(4): 318. (*Mesopeltis sanniolus*)

Synonym: *Sibon neilli* R.W. Henderson, Hoevers & Wilson, 1977.

Type: Holotype, USNM 6564, a 279 mm (svl) male (A.C.V. Schott, 1865–1866).

Type locality: "Yucatan, Mexico."

Distribution: Yucanán Peninsula. Southeastern Mexico (N Campeche, E Quintana Roo, Yucatán), Belize (Belize, Corozal, Cayo, Toledo) and NE Guatemala (E Petén), NSL–750 m.

Sources: H.M. Smith, 1943c, Duellman, 1965a, R.W. Henderson & Hoevers, 1975, J.C. Lee, 1980, Kofron, 1983, C.J. McCoy, 1986, J.A. Campbell & Vannini, 1989, Weyer, 1990, J.A. Campbell, 1998b, Stafford & Meyer, 2000 and McCranie, 2007.

SIBYNOMORPHUS Fitzinger, 1843 (Dipsadidae)

Synonyms: *Anholodon* A.-M.-C. Duméril, Bibron & Duméril, 1854a, *Cochliophagus* A.-M.-C. Duméril, Bibron & Duméril, 1854a, *Cochleophagus* – Marschall, 1873 *(nomen incorrectum), Pseudopareas* Boulenger, 1896a, *Sybinomorphus* – Amaral, 1926 *(nomen incorrectum), Pseudoparcas* – Gee, 1930 *(nomen incorrectum), Pseudopares* – Briceño-Rossi, 1934 *(nomen incorrectum), Sibinomorphus* – Briceño-Rossi, 1934 *(nomen incorrectum), Sibonnymorphus* – Dunn, 1944 *(nomen incorrectum),* and *Sibonomorphus* – Mattison, 1986 *(nomen incorrectum).*

Type species: *Dipsas mikanii* Schlegel, 1837.

Distribution: South America.

Sources: Ihering, 1911, J.A. Peters, 1960b, J.A. Peters & Orejas-Miranda, 1970, Carillo de Espinoza, 1974, Orcés & Almendáriz, 1989a, Scrocchi et al., 1993, Cadle, 2007, Harvey et al., 2009 and Zaher et al., 2009.

1. *Sibynomorphus inaequifasciatus* (A.-M.-C. Duméril, Bibron & Duméril, 1854a). Erpét. Gén. 7(1): 480–482. (*Cochliophagus inaequifasciatus*)

Type: Holotype, MNHN 3436, a 445 mm specimen (J.P. Verreaux, 1823–1854).

Type locality: "Amérique méridionale," possibly "provenant du Brésil" [= South America, possibly Brazil]. MNHN catalogue entry reads Amérique du Sud [= South America].

Distribution: South America (? Brazil). Known only from holotype.

Sources: Jan, 1863 and Jan & Sordelli, 1870–1881.

Remarks: Ihering (1911: 334) and J.A. Peters (960b: 147) suggested the type was an aberrant *S. mikani* (Schlegel) or *S. vagrans* Dunn.

2. *Sibynomorphus lavillai* Scrocchi, Porto & Rey, 1993. Rev. Brasil. Biol. 53(2): 200–204, figs. 1, 2b–c, 3a, 4a.

Type: Holotype, FML 2053 (formerly PT 441), a 554 mm female (F. Cruz, G. Perotti, O. Donadío & L. Fitzgerald, 16 Jan. 1988).
Type locality: "10 kilómetros al norte de Joaquín V. Gonzálezy 14.7 kilómetros al este del cruce de la ruta nacional 16 y la ruta provincial 30, sobre ruta 30. Departamento Anta. SALTA. Argentina" [= NE of Joaquín V. González, cen. Salta prov., N Argentina, ca. 25°02'S, 64°04'W, elevation 375 m].
Distribution: Southern South America. Brazil (Mato Grosso do Sul), Bolivia (Santa Crux, Tarija), N Paraguay (Alto Paraguay, Boquerón) and N Argentina (Chaco, Formosa, Jujuy, Salta, Santiago del Estero), 100–1190 m.
Sources: Lions & Alvarez, 1996a, Leynaud & Bucher, 1999, Cacciali, 2006 and Ferreira & Ávila, 2009.
Remarks: Probably occurs in Chaco of Paraguay *fide* Leynaud & Bucher (1999: 34).

3. *Sibynomorphus mikanii* (Schlegel, 1837). Essai Phys. Serp. 1: 162, 2: 277–278. (*Dipsas mikanii*)

Synonyms: *Leptognathus garmani* Cope, 1887c, and *Sibynomorphus mikani septentrionalis* Cunha, Nascimento & Hoge, 1980.
Type: Lectotype, NMW, specimen with 167 V + 46 SC (J. Natterer, 1817–1835), designated by J.A. Peters (1960b: 153).
Type locality: "Brésil" [= Brazil]. Restricted to SE Brazil *fide* J.A. Peters (1960b: 153).
Distribution: Southeastern Brazil (Bahia, Goiás, Maranhão, Mato Grosso, Minas Gerais, Pará, Paraná, Rio Grande do Norte, Rio Grande Grande do Sul, São Paulo), E Paraguay (Alto Paraná, Amambay) and NE Argentina (Misiones), 200–600 m.
Sources: Amaral, 1930a, 1977, Rosenberg, 1955, Barth, 1957, Hoge et al., 1975, 1981b, Cunha et al., 1980, Laporta-Ferreira et al., 1986, Lema, 1987, Freitas, 1999, Franca, 2003 and Cacciali, 2006, 2008.

4. *Sibynomorphus neuwiedi* (Ihering, 1911). Rev. Mus. Paulista 8: 333. (*Cochliophagus mikani neuwiedi*)

Synonym: *Sibynomorphus mikanii fasciatus* Amaral, 1930a.
Type: Holotype, not designated, MPSP.
Type locality: "Est. S. Paulo, Piquete, São Sebastião; Est. Espírito Santo, Rio Doce, Brazil" [= Piquete (22°37'S, 45°11'W, elevation 640 m) and São Sebastião (23°46'S, 45°25'W, elevation 40 m) E São Paulo State, and Rio Doce (21°06'S, 41°32'W, elevation 270 m) S Espírito Santo State, SE Brazil].
Distribution: Southeastern Brazil (Bahia, Espírito Santo, Minas Gerais, Paraná, Rio de Janeiro, Rio Grande do Sul, Santa Catarina, São Paulo, Cananéia, Cardoso, Grande, Santo Amaro, São Sebastião and São Vicente Is.), NSL–640 m.
Sources: Cunha et al., 1980, Hoge et al., 1981b, Laporta-Ferreira et al., 1986, Lema, 1987 and Cicchi et al., 2007.

5. *Sibynomorphus oligozonatus* Orcés & Almendáriz, 1989a. Politecnica, Biol. 14(3): 63–66, fig. 3.

Type: Holotype, EPN 3612, a 358 mm male (A. Almendáriz and field party).
Type locality: "Comuna Zhila (2.250 m.s.n.m.) (79°17' 26" W; 03°39'45" S.), parroquia Oña, Cantón Girón, provincia Azuay, Ecuador."
Distribution: Southwestern Ecuador (Azuay, Loja) and N Peru (Piura), 1440–2250 m.
Sources: Cadle, 2005, 2007.
Remarks: Possibly a synonym of *S. vagrans* (Dunn) *fide* Scrocchi et al. (1993: 205).

6. *Sibynomorphus oneilli* Rossman & Thomas, 1979. Occ. Pap. Mus. Zool., Louisiana St. Univ. (54): 1–5, figs. 1 (upper), 2 (upper).

Type: Holotype, LSUMZ 33736 (formerly JPO 5224), a 233 mm male (J.P. O'Neill, 8 Aug. 1975).
Type locality: "semiarid brushland (*ca.* 1645 m) NNE Balsas on the road to Abra Chanchillo, Departamento de Amazonas, Perú" [= between Balsas and Abra Chanchille, S Amazonas, NW Peru, ca. 6°50'S, 78°01'W, elevation 1645 m].
Distribution: Northwestern Peru (S Amazonas, Ancash, Cajamarca, La Libertad), 900–3500 m.
Sources: Rossman & Kizirian, 1993 and Cadle, 2005, 2007.

7. *Sibynomorphus petersi* Orcés-V. & Almendáriz, 1989a. Politecnica, Biol. 14(3): 58–62, fig. 1.

Type: Holotype, EPN 2659, a 763 mm male (A. Almendáriz et al.).
Type locality: "la comuna Zhila (2.250 m.s.n.m.,) 79°17'26" W, 03°39'45" S), perteneciente a la parroquia Oña, cantón Girón provincia Azuay, Ecuador."
Distribution: Southwestern Ecuador (Azuay, Loja) and N Peru (Ancash, Cajamarca), 1250–3120 m.
Sources: Cadle, 2005, 2007.

8. *Sibynomorphus turgidus* (Cope, 1868c). Proc. Acad. Nat. Sci. Philadelphia 20(2): 108, 136–137. (*Leptognathus turgida*)

Synonyms: ? *Leptognathus atypicus* Cope, 1874b, and *Tropidodipsas spilogaster* L.E. Griffin, 1916.
Types: Syntypes (2), USNM 5815a–b, 406 mm and 408 mm males (T.J. Page, Nov.–Dec. 1853).
Type locality: "Northern part of the Paraguay river" [= upper Paraguay River, Paraguay]. Emended to boundary of Brazil and Paraguay *fide* Cochran (1961: 192) and Paraguay *fide*

USNM catalogue and data tags [= upper Paraguay River, Presidente Hayes and Alto Paraguay Dept., N Paraguay, 20–24°S, 58°W, elevation 60–90 m].

Distribution: Southern South America. Bolivia (Cochabamba, La Paz, Santa Cruz, Tarija), S Brazil (S Mato Grosso, S Minas Gerais, Rio de Janeiro, Rio Grande do Sul, São Paulo), Paraguay (Alto Paraguay, Amambay, Boquerón, Caazapá, Central, Concepción, Itapúa, Paraguarí, Presidente Hayes, San Pedro, Modesto and Yacyretá Is.), N Argentina (Chaco, Córdoba, Corrientes, Entre Ríos, Formosa, Jujuy, Misiones, Salta, Santa Fe, Santiago del Estero, Tucumán) and NW Uruguay (Artigas, Paysandú, Rio Negro, Salto), 50–300 m.

Sources: Peracca, 1897, 1904a, Berg, 1898, 1901, Boulenger, 1898h, Koslowsky, 1898a, Lönnberg, 1902, Amaral, 1926, 1977, Serie, 1936, Freiberg, 1939, Vanzolini, 1948, Barth, 1957, Orejas-Miranda, 1958, Achaval, 1976, Fabián-Beurmann, 1977, Melgarejo, 1980, Meneghel & Melgarejo, 1984, Lema & Lema, 1987, Cabrera & Merlini, 1989, J.D. Williams & Francini, 1991, D.R. Norman, 1994, Yanosky et al., 1996, Leynaud & Bucher, 1999, Leynaud & Silmara-Carvantes, 2004, Carreira-Vidal et al., 2005 and Cacciali, 2006, 2008.

Remarks: Page Exped. traveled up the Parana as far as Corumba, Brazil (19°S, 57°45'W). *Leptognathus atypicus* Cope, 1874b is listed as *incertae sedis fide* Cadle (see 2007: 255–257 for discussion).

9. *Sibynomorphus vagrans* (Dunn, 1923b). Proc. Biol. Soc. Washington 36: 187–188. (*Pseudopareas vagrans*)

Type: Holotype, MCZ 17284, a 557–560 mm female (G.K. Noble [Harvard Peru. Exped.], 19–29 Sept. 1916).

Type locality: "Bellavista, in the low, broad and arid valley of the Marañon River, northwestern Peru" [= Bellavista, N Cajamarca Dept., NW Peru, 5°40' S, 78°41' W, elevation 500 m].

Distribution: Northwestern Peru (N Cajamarca), 500 m. Known only from type locality.

Sources: Duellman, 1979, Rossman & Thomas, 1979 and Cadle, 2005, 2007.

Remarks: Type locality listed as 1900 m *fide* Duellman (1979: 456).

10. *Sibynomorphus vagus* (Jan, 1863b). Elenco Sist. Ofidi: 101. (*Leptognathus vagus*)

Type: Holotype, formerly MSNM, a female, destroyed in 1943 during World War II.

Type locality: "Hong Kong" [China] (in error). Restricted to South America *fide* Cope (1868c: 136). Restricted to NW South America *fide* Dunn (1923b: 187).

Distribution: Northwestern Peru (Cajamarca, Lima, E Piura), 1810–1980 [3270] m.

Sources: Jan & Sordelli, 1870 *in* 1870–1881, Dunn, 1923b, Rossman & Thomas, 1979, Rossman & Kizirian, 1993, Aguilar et al., 2007 and Cadle, 2007.

Remarks: Holotype illustrated in Jan & Sordelli (1870 in 1870–1881, 3[37]: pl. 6, fig. 2).

11. *Sibynomorphus ventrimaculatus* (Boulenger, 1885b). Ann. Mag. Nat. Hist. (5) 16(92): 87–88. (*Leptognathus ventrimaculatus*)

Synonyms: *Leptognathus intermedius* Steindachner, 1903b, and *Heterorhachis poecilolepis* Amaral, 1923 & 1926a.

Types: Syntypes (2), BMNH 1946.9.7.29–30 (formerly BMNH 1885.6.26.32–33), a male and female, longest syntype 450 mm (H.F.I. von Ihering, 1880–1885).

Type locality: "S. Lorenzo, on the southern border of the Lagoa dos Patos, Province Rio Grande do Sul, Brazil." Emended to colony of São Lorenzo, Serra dos Tapes, Río Grande do Sul Province, Brazil [= São Lourengo do Sul, SE Rio Grande do Sul, ext. S Brazil, 31°22'S, 51°58'W, elevation NSL] *fide* J.A. Peters (1960b: 165). BMNH catalogue entry reads "San Lorenzo" and "Rio Grande do Sul."

Distribution: Southern South America. Southern Brazil (S Mato Grosso, Paraná, Rio Grande do Sul, São Paulo), S Paraguay (Amambay, Alto Paraná, Central, Cordillera, Itapúa, Misiones, Yacyretá Is.) and N Argentina (Corrientes, Formosa, Misiones, Santa Fe), NSL–630 m

Sources: Berg, 1901, Steindachner, 1903b, Bertoni, 1913, Amaral, 1930, 1977, Lema & Fabián-Beurmann, 1977, Lema, 1987, J.D. Williams & Francini, 1991, Lions & Alvarez, 1996b, Fernandes et al., 1998, Giraudo, 1999 and Cacciali, 2006, 2008.

12. *Sibynomorphus williamsi* Carrillo de Espinoza, 1974. Publ. Mus. Hist. Nat. Javier Prado, Zool. (24): 3–5, figs. 1–7.

Synonym: *Sibynomorphus williamsi* Carrillo-Tarazona, 1973 *(nomen ineditum)*.

Type: Holotype, MHNJP 2170 (formerly OF 492), a 530–545 mm male (A. Hoempler, 3 June 1969).

Type locality: "Jicamarca, Radio-Observatorio, Distrito Lurigancho, la región árida de la costa central, en la lomas cercanas, Provincia Lima, Departamento Lima, Peru, a una altitud de 300-500 m. sobre el nivel del mar" [= Jicamarca, cen. Lima Prov., W Peru, 11°57'S, 76°52'W, elevation 510 m].

Distribution: West-central Peru (Ancash, Lima), 300–2990 m.

Sources: Cadle, 2005, 2007.

Remarks: Photograph of holotype in Cadle (2007: fig. 40).

SIBYNOPHIS Fitzinger, 1843 (Sibynophiidae)

Synonyms: *Sybinophis* – A.-M.-C. Duméril & Bibron, 1844 *(nomen incorrectum), Enicognathus* A.-M.-C. Duméril, Bibron & Duméril, 1854a *(nomen praeoccupatum), Henicognathus* Cope, 1868 *(nomen praeoccupatum), Polyodontophis* Boulenger, 1890a *(nomen substitutum), Polyofontophis* – Sollas, 1906 *(nomen incorrectum), Sibonophis* – Steindachner, 1913b *(nomen incorrectum), Sibinophis* – Ditmars, 1931 *(nomen incorrectum), Polydontophis* – Maass-Berlin, 1933 *(nomen incorrectum), Sibynophys* – Ochomogo, 1957 *(nomen incorrectum), Enicognatus* – Guibé, 1958 *(nomen incorrectum), Sybynophis* – Picado, 1976 *(nomen incorrectum),* and *Sybynophis* – Nutaphand, 1991b *(nomen incorrectum).*

Type species: *Coluber geminatus* H. Boie, 1826.

Distribution: Southern, SE and E Asia and the East Indies.

Sources: Bourret, 1936b, M.A. Smith, 1943, Leviton & Munsterman, 1956, Morgan, 1973 and I. Das, 2010, 2012.

1. *Sibynophis bistrigatus* (A.C.L.G. Günther, 1868). Ann. Mag. Nat. Hist. (4) 1(6): 417–418. (*Ablabes bistrigatus*)

Type: Holotype, BMNH 1946.1.15.61, a 267 mm male (W. Theobald, Jr., 1862–1873).

Type locality: "Pegu" [= Bago, S Bago Div., S Myanmar, 17°20'N, 96°30'E, elevation 15 m].

Distribution: Eastern India (Andaman & Nicobars: Nicobar Is.) and S Myanmar (Bago), 15–300 m.

Sources: Theobald, 1868b, Boulenger, 1890a, Sclater, 1891b, Annandale, 1905a, Wall, 1923d, Biswas & Sanyal, 1980 and I. Das, 1994, 1999.

Remarks: Nicobar Is. record requires confirmation *fide* I. Das (2010: 300).

2. *Sibynophis bivittatus* (Boulenger, 1894d). Ann. Mag. Nat. Hist. (6) 14(80): 82–83. (*Polyodontophis bivittatus*)

Types: Syntypes (2), BMNH 1946.1.1.48–49, two males, longest syntype 490 mm (A.H. Everett, Nov. 1877–Jan. 1878).

Type locality: "Palawan" [= Palawan Is., SW Philippines].

Distribution: Southwestern Philippines (Busuanga, Culion, Dumaran, Palawan).

Sources: Boettger, 1898, L.E. Griffin, 1909b, 1911, J.C. Thompson, 1913a, E.H. Taylor. 1922a, Leviton, 1964a, Alcala, 1986 and Schoppe & Cervancia, 2009.

3. *Sibynophis chinensis* (A.C.L.G. Günther, 1889). Ann. Mag. Nat. Hist. (6) 4(21): 220–221. (*Ablabes chinensis*)

Synonyms: *Henicognathus sumichrasti* Bocourt, 1886 *(nomen suppressum), Ablabes sinensis* Boulenger, 1893 *(nomen emendatum), Polyodontophis grahami* Boulenger, 1904a, *Sibynophis hainanensis* K.P. Schmidt, 1925a, *Sibynophis collaris formosensis* Maki, 1931, *Sibynophis collaris parkeri* Mell, 1931a, and *Sibynophis chinensis miyiensis* Zhao & Kou *in* Zhao, 1987a.

Type: Holotype, not designated, a 500 mm specimen (A.E. Pratt, 1887–1889), location unknown.

Type locality: "Ichang, Upper Yangtsze-Kiang" [= Yichang, W Hubei Prov., cen. China, 30°42'N, 111°17'E, elevation 95 m].

Distribution: Southeastern Asia. Southern China (Anhui, Fujian, S Gansu, Guangdong, Guangxi, Guizhou, Hainan, Henan, Hong Kong, Hubei, Hunan, Jiangsu, Jiangxi, Shaanxi, Sichuan, Yunnan, Zhejiang, Lantau Is.), ext. S South Korea (Cheju Is.), Taiwan (Chiai, Pingtung, Taichung, Taipei) and N Vietnam (Bac Kan, Cao Bang, Gia Lai, Hoa Binh, Lao Cai, Ninh Binh, Quang Tri, Vinh Phuc), 95–2000 m.

Sources: Maki, 1931–b, Wu et al., 1985, Chen, 1991, Karsen et al., 1998, Lue et al., 1999, Orlov et al., 2000, Zhao, 2002, Captain et al., 2004, Zhao, 2006, San & Lee, 2007, Yang, 2008, V.S. Nguyen et al., 2009, Xiang & Li, 2009, Zhang, 2009, Li, 2011 and Yao, 2012.

Remarks: Official Specific Name no. 2092 fide Opinion 744 (ICZN, 1965b). A synonym of *S. collaris fide* Morgan (1973: 36).

4. *Sibynophis collaris* (Gray, 1853). Ann. Mag. Nat. Hist. (2) 12(72): 390. (*Psammophis collaris*)

Synonym: *Coluber colubrinus* Blyth *in* Sclater, 1891 (*nomen nudum*),

Type: Lectotype, BMNH 1946.1.1.51A (formerly BMNH 1853.8.12.34), an adult male (J.D. Hooker, May–Dec. 1850), designated by Kramer (1977: 747).

Type locality: "Khasi Hills" [= Khasi Hills, Meghalaya State, NE India, ca. 25°30'N, 92°00'E, elevation 1370–1675 m] via lectotype selection.

Distribution: Southeastern Asia. Southwestern China (Guangdong, SE Xizang, Yunnan), N India (Arunachal Pradesh, Assam, Himachal Pradesh, Manipur, Meghalaya, Nagaland, Sikkim, West Bengal), Nepal (Chitwan, Ilam, Kaski, Kathmandu, Makwanpur, Mustang, Myagdi, Sankhuwasabha), Bhutan, Myanmar (Bago, Rakhine, Sagaing, Shan, Yangon), Cambodia, N Laos (Houaphan, Phongsali, Xiangkhouang), S Vietnam (Bac Kan, Dien Bien, Dong Nai, Gia Lai, Ha Tinh, Lam Dong, Nghe An, Ninh Binh, Quang Binh, Quang Tri, Son Lai, Thua Thien-Hue), Thailand (Chiang Mai, Kanchanaburi, Loei, Nakhon Ratchasima, Phetchaburi, Trat, Udon Thani, Koh Chang Is.) and West Malaysia (N Pahang), NSL–3280 m.

Sources: Mocquard, 1905, Wall, 1908c, 1909d, 1925d, 1926b, Stejneger, 1910, 1925, M.A. Smith, 1922, 1930, K.P. Schmidt, 1925, 1927, Angel, 1929, Chang & Fang, 1931, Maki, 1931, Bourret, 1935c, 1937a, b, 1939a, c, d, C.H. Pope, 1935, G.E. Shaw et al., 1939, Wang & Wang, 1956, Swan & Leviton, 1962, E.H. Taylor, 1965, Sonderberg, 1966, 1967, Campden-Main, 1970, Deuve, 1970, Fleming & Fleming, 1974, Romer, 1979, Hu et al., 1980, Majupuria, 1981, Paik, 1982, Tweedie, 1983, Karsen et al., 1986, Tian et al., 1986, Lu, 1987, Nanhoe & Ouboter, 1987, Szyndlar & Hung, 1987, Zhao, 1987, Lu et al., 1989, Huang et al., 1990, M.J. Cox, 1991b, Ota, 1991, Zhao & Adler, 1993, M.J. Cox et al., 1998, Zug et al., 1998, Orlov et al., 2000, Schleich & Kästle, 2002, Ao et al., 2004, Captain et al., 2004, 2007, Whitaker & Captain, 2004, Yang, 2008, V.S. Nguyen et al., 2009, Li et al., 2010, Li, 2011 and Wangyal, 2011.

Remarks: Record from Hokkaido, Japan (Morgan, 1973) needs confirmation. Some of the above geographic records may belong to *S. chinensis* and/or *S. triangularis*.

5. *Sibynophis geminatus* (H. Boie, 1826). Isis von Oken 19(2): 211. (*Coluber geminatus*)

Synonyms: *Sibynophis geminatus maculatus* Angel, 1941.

Types: Syntypes (3), MNHN 3392–93, two males, and RMNH 687, a female (H. Kuhl & J.C. van Hasselt, Dec. 1820–Sept. 1823).

Type locality: "Java" [SW Indonesia]. Restricted to region of Mt. Pangerango and Mt. Salak, West Java, Indonesia *fide* Brongersma (1948b: 14). Emended to Tjihandjawar, at the foot of Mt. Pangerango, W. Java [= Cihanyawar, Nagrak, Jawa Barat, W Java, SW Indonesia, 07°03'S, 108°40'E] *fide* Brongersma (1950: 1499).

Distribution: Southwestern Indonesia (Bali, Java, Lombok, SE Sumatra), NSL–1300 m.

Sources: Gaulke, 1993a, 1994, David & Vogel, 1996 and McKay, 2006.

Remarks: Original description reprinted in H. Boie (1827b: 21–22).

S

6. *Sibynophis melanocephalus* (Gray, 1835 *in* Gray & Hardwicke, 1830–1835). Illust. Indian Zool. 2(19–20): pl. 83, fig. 2. (*Lycodon melanocephalus*)

Synonyms: *Lycodon catenatus* Gray, 1835 *in* Gray & Hardwicke, 1830–1835, *Herpetodryas prionotus* Cantor, 1839b, *Enicognathus javanicus* Bleeker, 1859b, *Enicognathus ornatus* Jan, 1863a, *Ablabes flaviceps* A.C.L.G. Günther, 1866, and *Sibynophis geminatus insularis* Mertens, 1927b.

Type: Holotype, BMNH 1946.1.17.76, a male (T. Hardwicke, 1756–1823).

Type locality: Unknown. Designated as Singapore ? *fide* Boulenger (1893a: 186) and Singapore *fide* Morgan (1973: 22).

Distribution: Southeast Asia and East Indies. Laos, Vietnam (Gia Lai), peninsular Thailand (Narathiwat, Pattani, Phang Nga), West Malaysia (Melaka, Pahang, Perak, Selangor, Terengganu, Tioman Is.), Singapore, East Malaysia (Sabah, Sarawak), Brunei, W Indonesia (Bulan, Bangka, Belitung, W Java, Natuna Arch., Nias, Siberut, S Sumatra, W Kalamantan, Weh) and ext. SW Philippines (Tawitawi), 100–1300 m

Sources: Flower, 1899, Ridley, 1899, T. Barbour, 1912, Rooij, 1917, M.A. Smith, 1926b, 1930a, Chasen & Smedley, 1927, Mertens, 1930, Bourret, 1935b, 1936b, 1939c, d, C. Haas, 1950, B.L. Lim, 1955, 1956, E.H. Taylor, 1965, Tweedie, 1983, K.F.L. Lim & Lee, 1989, K.K.P. Lim & Chou, 1990, M.J. Cox, 1991b, Stuebing, 1991, K.K.P. Lim & Lim, 1992, Gaulke, 1993a, David & Vogel, 1996, M.J. Cox et al., 1998, Malkmus et al., 2002, J.L. Grismer et al., 2003a, Orlov et al., 2003, Dehling, 2006 and V.S. Nguyen et al., 2009.

Remarks: Type locality designation by A.C.L.G. Günther (1858: 28) in error.

7. *Sibynophis sagittarius* (Cantor, 1839b). Proc. Zool. Soc. London 7(1): 49. (*Calamaria sagittaria*)

Synonym: *Enicognathus braconnieri* Jan, 1863b.

Type: Lectotype, BMNH 1860.3.19.1268, a male, designated by Kramer (1977: 747).

Type locality: "Tirhoot, Bengal" [= Tirhut, Muzaffarpur, N Bihar State, NE India, 26°07'N, 85°23'E, elevation 60 m] via lectotype selection.

Distribution: Southern Asia. Pakistan (Jammu & Kashmir), N India (Andhra Pradesh, Bihar, Himachal Pradesh, Kerala, Madhya Pradesh, Maharashtra, Orissa, Kerala, Tamil Nadu, Uttar Pradesh, West Bengal), Bhutan, Nepal (Bardiya, Chitwan, Dang, Mahottari, Nawalparasi, Parsa, Rupandehi) and Bangladesh, NSL–500 m.

Sources: A.C.L.G. Günther, 1864a, Wall, 1907d, 1921g, 1924, Swan & Leviton, 1962, Murthy, 1972c, 1985, 1986, 1990, Kramer, 1977, Murthy & Sharma, 1977b, M.A.R. Khan, 1988, Malhotra & Davis, 1991, Das & Palden, 2000, M.S. Khan, 2002, Captain et al., 2004, Whitaker & Captain, 2004 and Ahsan et al., 2009.

Remarks: Morgan (1973: 63) listed holotype as BMNH 1947.3.6.53 from Pinang, India.

8. *Sibynophis subpunctatus* (A.-M.-C. Duméril, Bibron & Duméril, 1854a). Erp. Gén. 7(1): 58–59. (*Oligodon subpunctatum*)

Synonyms: *Coluber humberti* H. Boie in Schlegel, 1826a *(nomen nudum)*, *Coluber humberti* Schlegel, 1826b *(nomen nudum)*, *Oligodon spinaepunctatus* Jan, 1862b, *Enicognathus humberti* Jan, 1863b, *Enicognathus grayi* Jan, 1866 *in* Jan & Sordelli, 1860–1877, and *Sibynophis subpunctatus ceylonicus* P. Silva, 1969.

Type: Holotype, MNHN 3240 (J.-J. Dussumier, 1816–1840).

Type locality: "Malabar" [= India].

Distribution: Southern India (Andhra Pradesh, Madhya Pradesh, Maharashtra) and Sri Lanka (Central, Eastern, North-Western, Sabaragamuwa, Southern, Uva, Western), NSL–500 m.
Sources: K.P. Schmidt, 1926b, Deraniyagala, 1955, P. Silva, 1969, 1980a, A. Silva, 1990b, 2001, 2009, Captain et al., 2004, Whiatker & Captain, 2004, N. Khaire, 2006, Somaweera, 2006, Nande & Deshmukh, 2007 and Ahsan et al., 2009.
Remarks: A synonym of *S. sagittarius fide* Morgan (1973: 63). A valid species *fide* Captain et al. (2004: 92). Occurs in Bangladesh fide Ahsan et al. (2009: 148).

9. *Sibynophis triangularis* E.H. Taylor & Ebel, 1958. Univ. Kansas Sci. Bull. 31(13): 1130–1134, fig. 23. (*Sibynophis collaris triangularis*)

Type: Holotype, KU 33520, a 327 mm specimen (B. Lekagul, 18 Aug. 1953).
Type locality: "Nongko (village), Siracha (district), Chon Buri (province), Thailand."
Distribution: Eastern Thailand (Chanthaburi, Chaiyaphum, Chon Buri, Nakhon Ratchasima, Phetchaburi, Saraburi) and Cambodia (Mandolkiri), 500 m.
Sources: Captain et al., 2004, B.L. Stuart et al., 2006 and I. Das, 2010.

†*SIMOLIOPHIS* Sauvage, 1880b (†Pachyophiidae)

Synonyms: †*Symoliophis* Rochebrune, 1880 *(nomen emendatum)*, †*Siniliophis* – Sauvage, 1880b *(nomen incorrectum)*, †*Symoliopis* – Hoffmann, 1890 *(nomen incorrectum)*, †*Symoliopsis* – Cope, 1898 *(nomen incorrectum)*, †*Symoleophis* – Williston, 1925 *(nomen incorrectum)*, †*Cimoliophis* Hoffstetter, 1960a *(nomen emendatum)*, and †*Kimoliophis* Hoffstetter, 1960a *(nomen emendatum)*.
Type species: †*Simoliophis rochebrunei* Sauvage, 1880a.
Distribution: Upper Cretaceous of Egypt, France, Libya, Morocco, Portugal and Spain.
Sources: Rage, 1984b, 1987, Nessov et al., 1998, Rage & Escuillié, 2003a–b, Bardet et al., 2008 and Vullo et al., 2011.
Remarks: Probably marine.

1. †*Simoliophis libycus* Nessov, Zhegallo & Averianov, 1998. Ann. Paléont. 84(3–4): 5, fig. 2, pl. 1, fig 2a–e.

Type: Holotype, ZIN PC 2/31, middle trunk vertebra.
Type locality: "Draa Ubari at the southern border of the Hamada Al-Khamra, southeast of the Gadames Oasis in western Libya. 'Unnamed' beds between the Thala and Mazuzah Members of the Mizdah Formation, Upper Cretaceous, Santonian-Campanian."
Distribution: Upper Cretaceous (Cenomanian: 93.9–100.5 mya) of Libya.
Source: Bardet et al., 2008.
Remarks: Probably Cenomanian age *fide* Rage & Cappetta, 2002.

2. †*Simoliophis rochebrunei* Sauvage, 1880a. C.R. Acad. Sci. 91: 672.

Synonym: †*Simoliophis delgadoi* Sauvage, 1898.
Type: Holotype, several middle trunk vertebrae from a 3000 mm specimen (T. de Rochebrune), lost *fide* Rage (1984b: 11).
Type locality: "île d'Aix, dans la Charente; l'étage carentonien, l'époque cénomanienne" [= Aix Island, W Charente-Maritime Prov., W France; upper Cretaceous, Cenomanian].
Distribution: Upper Cretaceous (Cenomanian: 93.9–100.5 mya) of France and Portugal.
Source: Vullo et al., 2011.
Remarks: Original description reprinted in Sauvage (1880b: 313 and 1881: 206). Vertebra illustrated in Vullo et al. (2011: figs. 6a–e).

SIMOPHIS W.C.H. Peters, 1860d (Colubridae)

Synonyms: *Rhinaspis* Fitzinger, 1843 *(nomen rejiciendum)*, and *Simphis* – Briceño-Rossi, 1934 *(nomen incorrectum)*.
Type species: *Heterodon rhinostoma* Schlegel, 1837.
Distribution: Brazil and Paraguay.
Sources: Bertoni, 1915, J.A. Peters & Orejas-Miranda, 1970 and Cacciali et al., 2009.

1. *Simophis rhinostomus* (Schlegel, 1837). Essai Phys. Serp. 1: 141, 2: 100–103, pl. 3, figs. 17–19. (*Heterodon rhinostoma*)

Synonyms: *Rhinostoma proboscidea* Fitzinger, 1826a (*nomen rejiciendum*), *Rhinostoma proboscidea* Fitzinger, 1843 (*nomen rejiciendum*), *Rhinostoma schlegelii* A.C.L.G. Günther, 1858, *Rhinaspis probocideus* Jan, 1863a, and *Rhinaspis rohdei* Boettger, 1885a.
Type: Holotype, ZMB 10602 (formerly MPB 249), a 935 mm specimen (J. Natterer, 1817–1835).
Type locality: "intérieur du Brésil" [= cen. Brazil].
Distribution: Southeastern Brazil (Bahia, Distrito Federal, Goiás, Mato Grosso, Mato Grosso do Sul, Minas Geráis, São Paulo, São Sebastião Is.) and Paraguay (Alto Paraná, Caaguazú), 180–1065 m.
Sources: Koslowsky, 1898, Vanzolini, 1948, P. Müller, 1968a, Amaral, 1978, Santos-Jordäo & Fernandes-Bizerra, 1996, Argôlo, 1998c and Freitas, 1999.
Remarks: Official Specific Name no. 1970 *fide* Opinion 698 (ICZN, 1964). Holotype erroneously listed as ZISP 6658 *fide* Cacciali et al. (2009: 699). Cacciali et al., 2009 synonymized *S. rohdei* with *S. rhinostomus*.

SIMOSELAPS Jan, 1859a (Elapidae)

Synonyms: *Rhynchoelaps* Jan, 1858 (*nomen nudum*), *Rhynchelaps* Boulenger, 1896a *(nomen emendatum), Rhynchaelaps* – Longman, 1930 *(nomen incorrectum), Rimchelaps* – Briceño-Rossi, 1934 *(nomen incorrectum), Melwardia* Worrell, 1960, *Narophis* Worrell, 1961c, and *Simonelaps* – Mattison, 1986 *(nomen incorrectum).*

Type species: *Elaps bertholdi* Jan, 1859a.

Distribution: Australia.

Sources: Thomson, 1934, Kinghorn, 1964, Storr, 1968, 1985, 1986, Cogger et al., 1983a, Shine, 1984a, 1985a, Golay, 1985, Longmore, 1986, Scanlon & Shine, 1988, S.K. Wilson & Knowles, 1988, Gow, 1989, Hoser, 1989, 2012e, Hutchinson, 1990, Cogger, 1992, Clarke & How, 1995, Horner, 1998, Keogh et al., 1998, How & Shine, 1999, Scanlon & Lee, 2004, Sanders et al., 2008 and Zaher et al., 2009.

1. *Simoselaps anomalus* (Sternfeld, 1919). Senckenbergiana (1918) 1(3): 77–78. (*Rhynchelaps anomalus*)

Synonym: *Simoselaps anomala* – Hutchinson, 1990.

Type: Lectotype, SMF 20557 (formerly SMF-B 9362, 1a) (M. von Leonhardi, 1908), designated by Mertens (1922: 182).

Type locality: "Hermannsburg Central-Australien" [= Hermannsburg Mission, upper Finke River S of Maconnell Distributions, S Northern Territory, cen. Australia, 23°57'S, 132°46'E, elevation 575 m].

Distribution: Western Australia (SW Northern Territory, NW South Australia, N Western Australia), NSL–790 m.

Sources: Sternfeld, 1925, Waite, 1928, Thomson, 1934 and Storr, 1968, 1979.

2. *Simoselaps bertholdi* (Jan, 1859a). Rev. Mag. Zool. (2) 11: 123. (*Elaps [Simoselaps] bertholdi*)

Synonyms: *Elaps bertholdi* Jan, 1858b *(nomen nudum), Elaps mattozoi* Bethencourt-Ferreira, 1891, and *Furina robusta* De Vis, 1905.

Type: Holotype, ZMG 191A, a 200 mm specimen.

Type locality: "Australie" [= Australia]. Emended to Ile Adélaïde (Australie méridionale) [= Adelaide, SE South Australia, SE Australia, 34°56'S, 138°36'E, elevation NSL] *fide* Jan (1858: 518) and Perth, Western Australia (31°57'S, 115°51'E, elevation 15 m) *fide* Storr (1968: 82–83).

Distribution: Southwestern Australia (ext. S Northern Territory, South Australia, S Western Australia, Wallabi Is.), NSL–810 m.

Sources: Storr, 1965, 1979, Bush et al., 1995, 2007, Storr et al., 2002 and Bush & Maryan, 2011.

Remarks: Supplemental original description in Jan (1859e: pl. b, fig. no. 7).

S

3. *Simoselaps bimaculatus* (A.-M.-C. Duméril, Bibron & Duméril, 1854b). Erpét. Gén. 7(2): 1240–1241. (*Furina bimaculata*) (*nomen corrigendum*)

Synonyms: *Furina bimaculata* A.-M.-C. Duméril, 1853 (*nomen nudum*), *Furina bi-maculata* A.-M.-C. Duméril, Bibron & Duméril, 1854b *(nomen incorrigendum), Furina bimaculata* – A.-M.-C. Duméril, Bibron & Duméril, 1854c *(nomen corrigendum),* and *Brachysoma bimaculatum* – A.C.L.G. Günther, 1858.

Type: Holotype, MNHN 3942, a 340 mm specimen (J.P. Verreaux, 1844).

Type locality: "Tasmanie" [= Tasmania, Australia] (in error *fide* Storr, 1968: 85). Corrected to Perth, Australia *fide* Storr (1968: 85). Restricted to W Australia *fide* Boulenger (1896a: 407).

Distribution: Southwestern Australia (W South Australia, S Western Australia), 120–205 m.

Sources: Bush et al., 1995 and Storr et al., 2002.

Remarks: Basal sister taxon to *Brachyurophis* and *Simoselaps fide* Metzger et al. (2010: 409).

4. *Simoselaps littoralis* (Storr, 1968). J. Roy. Soc. West. Aust. (1967) 50(3): 84. (*Vermicella bertholdi littoralis*)

Type: Holotype, WAM 12770, a 110–199 mm male (R. Vollprecht, 16 Jan. 1958).

Type locality: "coastal dunes 7 miles south of Geraldton, Western Australia, in 28° 52' S, 114° 38' E, Australia."

Distribution: Extreme W Australia (ext. W Western Australia, Bernier, Dorre, Dirk Hartog, East Wallabi, Faure and West Wallabi Is.), NSL–220 m.

Sources: Storr, 1968 and Storr et al., 2002.

5. *Simoselaps minimus* (Worrell, 1960). West. Aust. Nat. 7(5): 132–133, figs. 1, 2a–c. (*Melwardia minima*)

Type: Holotype, AMS 16494 (formerly MW), a 175 mm specimen (M. Ward, 1945).

Type locality: "Broome in Western Australia" [= Broome, NW Western Australia, NW Australia, 17°57'S, 122°14'E, elevation NSL].

Distribution: Extreme NW Australia (NW Western Australia), 10–20 m.

Source: Storr, 1979.

SINOMICRURUS Slowinski, Boundy & Lawson, 2001 (Elapidae)

Synonym: *Funkelapidus* Hoser, 2012n *(nomen illegitimum).*

Type species: *Elaps macclellandi* J.T. Reinhardt, 1844.

Distribution: Southeastern and E Asia.

Sources: Shrestha & Majupuria, 1977, Golay et al., 1993, David & Ineich, 1999, Slowinski et al., 2001, Castoe et al., 2007a, Zaher et al., 2009 and Hoser, 2012e, n.

1. *Sinomicrurus hatori* (Takahashi, 1930). Terr. snakes Japan: 139–141 (unnumbered), 1 fig., col. pl. for species 65. (*Calliophis hatori*)

Type: Holotype, not designated, (formerly "Taiwan Library"), a 550 mm specimen (J. Hatori, 1905), lost *fide* Oshima (1920: 2).
Type locality: "Giran, Formosa" [= Yilan, Yilan Co., NE Taiwan, 24°45'N, 121°45'E, elevation 10 m].
Distribution: Northern and E Taiwan (Yilan), 10–1500 m.
Sources: J. Hatori, 1905, Oshima, 1920, Maki, 1931–b, Ota et al., 1999 and Xiang & Li, 2009.

2. *Sinomicrurus japonicus* (A.C.L.G. Günther, 1868). Ann. Mag. Nat. Hist. (4) 1(6): 428, pl. 17, fig. c. (*Callophis japonicus*)

Synonyms: *Callophis boettgeri* Fritze, 1894, and *Hemibungarus japonicus takarai* Ota, Ito & Lin, 1999.
Type: Holotype, BMNH 1946.1.17.97, a 241 mm specimen (Whitely).
Type locality: "Nagasaki" [Japan] (in error *fide* Boulenger, 1896a: 395). Corrected to Ryukyu Is. *fide* Stejneger (1907: 389).
Distribution: Southern Japan (Ryukyus: Akajima, Amamioshima, Amurojima, Fukajijima, Gerumajima, Gishifujima, Gushikawajima, Gusukujima, Hatejima, Iejima, Iheyajima, Izenajima, Kakeromajima, Kerumajima, Korijima, Kubajima, Kumejima, Kuroshima, Maejima, Nakajima, Nohojima, Okinawajima, Tokashikijima, Tokunoshima, Tonakijima, Ukejima, Yaguchijima, Yakabijima, Yanahajima, Yorojima and Zamamijima Is.).
Sources: Stejneger, 1907a, Maki, 1931, Takara, 1962, Nakamura & Ueno, 1963, Ota, 1991, Ota et al., 1999, Uchiyama et al., 2002 and Goris & Maeda, 2004.

3. *Sinomicrurus kelloggi* (C.H. Pope, 1928a). Amer. Mus. Novit. (320): 6. (*Hemibungarus kelloggi*)

Synonyms: *Calliophis wongii* Fan, 1931, and *Calliophis wongi tonkinensis* Bourret, 1935c.
Type: Holotype, AMNH 33744, a 774 mm female (C.H. Pope, 12 June–20 July 1925).
Type locality: "Kuatun, a village in the mountains of Ch'ungan Hsien, northwest Fukien Province, China" [= Guadun, in Wuyishan Mtns., ext. N Fujian Prov., SE China, 27°45'N, 117°50'E, elevation 900–1200 m].
Distribution: Southeastern China (Fujian, Guangdong, Guangxi, Guizhou, Hainan, Jiangxi, Sichuan), N Laos and N Vietnam (Cao Bang, Hai Duoag, Lai Chau, Lang Son, Son La, Vinh Phuc), 300–1500 m.
Sources: Zhao, 1998, 2006, Orlov et al., 2000, 2003, V.S. Nguyen et al., 2009 and I. Das, 2010, 2012.
Remarks: Supplemental original description in C.H. Pope (1929: 466–467, fig. 19).

4. *Sinomicrurus macclellandi* (J.T. Reinhardt, 1844). Calcutta J. Nat. Hist. 4: 532–534. (*Elaps macclellandi*)

Synonyms: *Elaps personatus* Blyth, 1854a, *Elaps univirgatus* A.C.L.G. Günther, 1858, *Calliophis annularis* A.C.L.G. Günther, 1864, *Callophis macclellandi nigriventer* Wall, 1909k, *Callophis macclellandi gorei* Wall, 1910c, *Callophis formosensis* J.C. Thompson, 1912, *Calliophis swinhoei* Van Denburgh, 1912a, *Calliophis macclellandi concolor* Wall, 1925d, *Calliophis macclellandi formosensis* Maki, 1931, *Calliophis macclellandi iwasakii* Maki, 1935, *Calliophis maclellandii univirgatus* – Kramer, 1977 *(nomen incorrectum), Calliophis macclellandi nepalensis* Shrestha & Majupuria, 1977, and *Calliophis maccellandi* – Shrestha & Majupuria, 1977 *(nomen incorrectum).*
Type: Holotype, ZMUC 65339, a 637 mm female (J. McClelland, 1835–1836).
Type locality: "Assam" [= Arunachal Pradesh, Assam, Manipur, Meghalaya, Mizoram or Nagaland State, NE India].
Distribution: Northeastern India (Arunachal Pradesh, Assam, Himachal Pradesh, Manipur, Meghalaya, Mizoram, Nagaland, North Bengal, Sikhim, West Bengal), Nepal (Chitwan, Ilam, Kaski, Kathmandu), Bhutan, Bangladesh (Sylhet), Myanmar (Bago, Kachin, Mandalay, Yangon), Thailand (Chachaengsao, Chiang Mai, Loei, Nakhon Ratchasima, Tak), Laos, Vietnam (Cao Bang, Dong Nai, Hai Duong, Ha Tay, Ha Tinh, Kon Tum, Lam Dong, Lao Cai, Phu Tho, Quang Binh, Quang Tri, Son Lai, Thanh Hoa, Thua Thien-Hue, Vinh Phuc), S China (Fukien, Gansu, Guangdong, Guangxi, Hainan, Hong Kong, Jiangsu, Jiangxi, Shaanxi, Lantau Is.), Taiwan, ext. S Japan (Ryukyus: Iriomotejima, Ishigakijima and Kohamajima Is.), 45–2485 m.
Sources: Stejneger, 1907a, Wall, 1918c, Takara, 1962, Kuntz, 1963, Nakamura & Ueno, 1963, E.H. Taylor, 1965, Deuve, 1970, Kramer, 1977, J. Lin, 1982, Karsen et al., 1986, 1998, M.J. Cox et al., 1998, Lue et al., 1998, Zhao, 1998, 2006, Zou & Chen, 1998, Orlov et al., 2000, Schleich & Kästle, 2002, Leviton et al., 2003, Ao et al., 2004, Tiwari & Shah, 2004, Whitaker & Captain, 2004, Zhao, 2006, M.F. Ahmed et al., 2009, V.S. Nguyen et al., 2009, Xiang & Li, 2009, I. Das, 2010, 2012, Wangyal, 2011, Rahman, 2012d, Yao, 2012 and S.K. Sharma et al., 2013.

5. *Sinomicrurus sauteri* (Steindachner, 1913a). Anz. Akad. Wiss. 50(12): 219. (*Oligodon sauteri*)

Synonym: *Hemibungarus matsudai* Oshima, 1920.
Type: Holotype, NMW 23456, a 510 mm specimen (H. Sauter, 1912).

Type locality: "Formosa" [= Taiwan]. "Suisharío" [= Shui-she-liao] *fide* Steindachner (1913b: 338) and NMW catalogue.
Distribution: Southern and cen. Taiwan, 180–300 m.
Sources: Maki, 1931–b, Wang & Wang, 1956, Kuntz, 1963, Tiedemann et al., 1994, Lue et al., 1998, Zhao, 1998, 2006, Ota et al., 1999 and Xiang & Li, 2009.

SINONATRIX Rossman & Eberle, 1977 (Natricidae)

Type species: *Tropidonotus annularis* Hallowell, 1856a.
Distribution: Southeastern Asia.
Sources: Stejneger, 1925, C.H. Pope, 1935, Rossman & Eberle, 1977, Hu et al., 1980, Tian et al., 1986, Zhao & Adler, 1993, Rao & Yang, 1998, Zhao, 1998, 2006 and I. Das, 2010, 2012.

1. *Sinonatrix aequifasciata* (T. Barbour, 1908). Bull. Mus. Comp. Zool. 51(12): 317. (*Natrix aequifasciata*)

Types: Syntypes (2), MCZ 7101a–b (A. Owston, 1906 via T. Barbour).
Type locality: "Mt. Wuchi, central Hainan" [= Mt. Wuzhishan, S Hainan Is., Hainan Prov., S China, 18°54'N, 109°42'E, elevation 1000–1800 m].
Distribution: Southern China (Fujian, Guangdong, Guangxi, Guizhou, Hainan, Hong Kong, Hunan, Jiangxi, Yunnan, Zhejiang), Myanmar, Laos (Bolikhamxay, Khammouan) and Vietnam (Bac Giang, Cao Bang, Ha Giang, Ha Tinh, Lang Son, Lao Cai, Naghe An, Vinh Phuc), 40–2000 m.
Sources: Stejneger, 1925, K.P. Schmidt, 1927, C.H. Pope, 1929, Fan, 1931, Bourret, 1936b, Romer, 1978, 1979a, Hu et al., 1980, Karsen et al., 1986, Tian et al., 1986, Huang et al., 1991, Ziegler & Quang, 2005, Bain et al., 2007, B.L. Stuart & Heatwole, 2008, V.S. Nguyen et al., 2009 and I. Das, 2010.
Remarks: Supplemental original description in T. Barbour (1909: 66–67, pl. 7, figs. 1–4).

S

2. *Sinonatrix annularis* (Hallowell, 1856a). Proc. Acad. Nat. Sci. Philadelphia (1856–1857) 8(4): 151–152. (*Tropidonotus annularis*)

Synonyms: *Tropidonotus chinensis* Jan *in* Berthold, 1859, *Tropidonotus semifasciatus* Berthold, 1859, and *Tropidonotus habereri* F. Werner, 1903a.
Type: Holotype, ANSP 6640, a 775 mm specimen (D.B. McCartee, 1844–1856).
Type locality: "Ningpo, China" [= Ningbo, E Zhejiang Prov., E China, 29°52'N, 121°33'E, elevation 10 m].
Distribution: Southeastern Asia. Myanmar, SE China (Anhui, Fujian, Guangdong, Guangxi, Hainan, Hebei, Hubei, Hunan, Jiangsu, Jiangxi, Hainan, Sichuan, Zhejiang) and Taiwan (Taichung, Taoyuan, Tainan, Taipei, Taitung), 10–1000 m.
Sources: A.C.L.G. Günther, 1864a, Wall, 1903, Stejneger, 1907a, Maki, 1931–b, Wang & Wang, 1956, Kuntz, 1963, B. Li, 1987, Lue, 1987, 1990, Lue et al., 1989, Zhao, 2006, Xiang & Li, 2009 and I. Das, 2010.
Remarks: Publication date Dec. 1856 *fide* Proc. ANSP, 1857, vol. 9(1) *contra* Malnate (1971: 362) and Zhao & Adler (1993: 374) who cited 1857.

3. *Sinonatrix percarinata* (Boulenger, 1899b). Proc. Zool. Soc. London 67(1): 163–164, pl. 17, fig. 2. (*Tropidonotus percarinatus*)

Synonym: *Natrix suriki* Maki, 1931.
Type: Holotype, BMNH 1946.1.15.15, a 500 mm male (J.D. La Touche, spring 1896 or 1898).
Type locality: "Kuatun, a village about 270 miles from Foochow, in the mountains at the North-west of the Province of Fokien, at an altitude of 3000 to 4000 feet or more, China" [= Guadun, in Wuyishan Mtns., ext. N Fujian Prov., SE China, 27°45'N, 117°50'E, elevation 900–1200 m].
Distribution: Eastern Asia. Northeastern India (Arunachal Pradesh), N Myanmar (Kachin), N Thailand (Chiang Mai, Loei), Laos (Bolikhamsai, Champasak, Huaphan, Phongsali, Xekong) Vietnam (Bac Giang, Bac Kan, Bac Ninh, Cao Bang, Dong Nai, Gia Lai, Ha Tinh, Hoa Binh, Lao Cai, Ha Giang, Lang Son, Phu Tho, Quang Binh, Quang Ninh, Son La, Thai Nguyen, Thua Thien-Hue, Vinh Phuc), SE China (Anhui, Fujian, S Gansu, Guangdong, Guangxi, Guizhou, Hainan, Henan, Hong Kong, Hubei, Hunan, Jiangsu, Jiangxi, Shaanxi, Sichuan, Yunnan, Zhejiang) and Taiwan (Pingtung), 90–2000 m.
Sources: Wall, 1903, Mell, 1922, M.A. Smith, 1923a, 1940, 1943, H.W. Parker, 1925a, Fan, 1931, Maki, 1931, Chang, 1932, 1934, Bourret, 1934a, 1934f, 1935b–c, 1936b, 1937a–b, 1939a–d, 1941b, 1944, Mao, 1965a, E.H. Taylor, 1965, Romer, 1974a, 1979a, Lue, 1987, 1990, Lue et al., 1989, M.J. Cox, 1991b, Captain & Patel, 1998, B.L. Stuart & Heatwole, 2008, V.S. Nguyen et al., 2009, Xiang & Li, 2009, I. Das, 2012 and Yao, 2012.

4. *Sinonatrix yunnanensis* Rao & Yang, 1998. Russ. J. Herp. 5(1): 70–71, fig. 1.

Type: Holotype, KIZ 75II0191, a 440–498 mm specimen (native, 11 May 1975).
Type locality: "Dingpa of Jingdong County, Yunnan, altitude 1700 m, China."
Distribution: Southeastern Asia. Northern Myanmar (Kachin), N Thailand (Chiang Rai) and S China (Yunnan), 450–2000 m.
Sources: Pauwels et al., 2009 and I. Das, 2010, 2012.

SINOVIPERA P. Guo & Wang, 2011 (Viperidae)

Type species: *Sinovipera sichuanensis* P. Guo & Wang, 2011.

Distribution: China.

1. *Sinovipera sichuanensis* P. Guo & Wang, 2011. Zootaxa (2918): 9–12, figs. 3, 4a–c.

Type: Holotype, YBU 71077, an 1080 mm female (native, Aug. 2007).
Type locality: Hejiang County, Sichuan Province, China. It was collected near a small river in the forest, elevation 1000 m."
Distribution: Southwestern China (Sichuan), 1000 m. Known only from type locality.
Remarks: Probably occurs in Chongquing and Guizhou, China *fide* P. Guo & Wang (2011: 12).

SIPHLOPHIS Fitzinger, 1843 (Xenodontidae)

Synonyms: *Lycodon* H. Boie *in* F. Boie, 1827 (*nomen praeoccupatum*), *Tripanurgos* Fitzinger, 1843, *Siphlopis* – A.-M.-C. Duméril & Bibron, 1844 *(nomen incorrectum)*, *Trypanurgos* Tschudi, 1845 *(nomen emendatum)*, *Tripanurgus* Agassiz, 1847 *(nomen emendatum)*, *Lycognathus* A.-M.-C. Duméril, 1853, *Lycognatus* – Rochebrune, 1885 *(nomen incorrectum)*, *Tympanurgas* – Palacky, 1898 *(nomen incorrectum)*, *Tympanurgos* – Palacky, 1898 *(nomen incorrectum)*, *Licognathus* – Briceño-Rossi, 1934 *(nomen incorrectum)*, *Tripanurgo* – Briceño-Rossi, 1934 *(nomen incorrectum)*, *Callopistria* Amaral, 1935a, *Alleidophis* Prado, 1940a, *Calliopistria* – Hoge, 1964c *(nomen incorrectum)*, *Lycognthus* – Hoge, 1964c *(nomen incorrectum)*, *Lyucognathus* – Hoge, 1964c *(nomen incorrectum)*, *Sophlophis* – Hoge, 1964c *(nomen incorrectum)*, and *Syphlophis* – Leviton, 1965 *(nomen incorrectum)*.
Type species: *Coronella cervina* Laurenti, 1768.
Distribution: Lower Central America and South America.
Sources: Hoge, 1964c, Bailey *in* J.A. Peters & Orejas-Miranda, 1970, Zaher & Prudente, 1999, Costa-Prudente et al., 1998 and Zaher et al., 2009.
Remarks: *Tripanurgos* a valid genus *fide* Vidal et al., 2000.

1. *Siphlophis cervinus* (Laurenti, 1768). Synop. Rept.: 88. (*Coronella cervina*)

Synonyms: *Coluber zeylonicus* Gmelin, 1789, *Coluber audax* Daudin, 1803c, *Coluber maximiliani* Merrem, 1820, *Natrix pardalis* Wagler, 1825 (*nomen ineditum*), *Lycognathus scolopax* A.-M.-C. Duméril, Bibron & Duméril, 1854b, *Clelia euprepa* L.E. Griffin, 1916, and *Drepanodon attenuatus* T. Barbour & Noble, 1921.
Type: Lectotype, specimen described and illustrated by Seba (1735: 84, pl. 79, fig. 3), designated herein.
Type locality: "America" via lectotype selection.
Distribution: Panama and N South America. Eastern Panama (Canal Zone, Colón, Darién, Panamá, Barro Colorado Is.), Colombia (Amazonas, Chocó, Vaupés), Trinidad and Tobago (Trinidad), SE Venezuela (Amazonas, Bolívar), Guyana (Demerara-Mahaica), Suriname (Brokopondo, Saramacca), French Guiana (Cayenne), Brazil (Amapá, Amazonas, W Maranhão, Mato Grosso, Pará, Rondônia, Roraima, São Paulo), E Ecuador (Napo, Pastaza, Sucumbios), E Peru (Cusco, Loreto, Madre de Dios) and E Bolivia (Santa Cruz), NSL–915 m.
Sources: Prado, 1940a, Beebe, 1946, Amaral, 1948, 1977, Hoge & Maranhão-Nina, 1964, Hoge, 1967, Emsley, 1977, Cunha & Nascimento, 1978, 1993, Duellman, 1978, 2005, Hoogmoed, 1979, Gasc & Rodriguez, 1980, Abuys, 1986b, Dixon & Soini, 1986, Pérez-Santos, 1986, Chippaux, 1987, Nascimento et al., 1987, Pérez-Santos & Moreno, 1988, 1991, Rodriguez & Cadle, 1990, Zimmerman & Rodriguez, 1990, Duellman & Salas, 1991, Lamar, 1997b, J.C. Murphy, 1997, Barrio-Amorgós et al., 1998, Starace, 1998, Boos, 2001, Navarrete et al., 2009, Sajdak, 2010 and C.J. Cole et al., 2013.

2. *Siphlophis compressus* (Daudin, 1803c). Hist. Nat. Rept. 6: 247–248. (*Coluber compressus*)

Synonym: *Coluber leucocephalus* Mikan, 1820 *in* 1820–1825.
Type: Holotype, MNHN 3730.
Type locality: "la colonie hollandaise de Suriname" [= Suriname].
Distribution: Lower Central America and Amazonia. Costa Rica (Puntarenas, San José), Panama (Canal Zone, Chiriquí), Trinidad and Tobago (Trinidad), Colombia (Amazonas, Boyacá, Casanare, Cauca, Chocó, Meta, Putumayo, Santander, Valle de Cauca, Vaupés), Venezuela (Amazonas, Anzoátegui, Bolívar, Delta Amacuro, Monagas, Sucre), Guyana (Barima-Waini, Cuyuni-Mazaruni, Essequibo Islands-West Demerara, Potaro-Siparuni, Upper Demerara-Berbice), Suriname (Para), SE French Guiana (S Cayenne), Ecuador (Esmeraldas, Napo), Brazil (Amapá, Amazonas, Bahia, Espinto Santo, Mato Grosso, Pará, Rio de Janeiro, Rondônia, Roraima, São Paulo, Sergipe), Peru (Amazonas, Loreto, Madre de Dios, San Martin) and Bolivia (Santa Cruz), NSL–190 m.
Sources: Roze, 1966a, Bailey *in* J.A. Peters & Orejas-Miranda, 1970, Cunha & Nascimento, 1978, 1993, Lancini, 1979, Cadle, 1984a, Chippaux, 1987, Rogé & Sauvenet, 1987, Pérez-Santos & Moreno, 1988, Villa et al., 1988, Reynolds et al., 1997, Starace, 1998, Freitas, 1999, Lehr & Fernandez, 2000b, Marques et al., 2001, 2004, Savage, 2002, Abuys, 2003, Rivas-Fuenmayor & Molina, 2003, Kaiser et al., 2003, Toral-C., et al., 2004, Navarrete et al., 2009 and C.J. Cole et al., 2013.
Remarks: Transferred from *Tripanurgos fide* Zaher & Prudente, 1999.

3. *Siphlophis leucocephalus* (A.C.L.G. Günther, 1863a). Ann. Mag. Nat. Hist. (3) 11(61): 23. (*Leptodira leucocephala*)

Type: Holotype, BMNH 1946.9.7.25, a 796 mm male (O.E.H. Wucherer, 1859–1860).
Type locality: "Bahia" [= Salvador, E Bahia, E Brazil, 12°59'S, 38°31'W, elevation NSL].
Distribution: Eastern Brazil (Bahia, Goiás, Minas Gerais), NSL–10 m.

4. *Siphlophis longicaudatus* (Andersson, 1901). Bihang Kon. Svenska Vet.-Akad. Handl. (4) 27(5): 17–18, pl. 2, figs. 9–11. (*Tropidodipsas longicaudata*)

Synonym: *Siphlophis cinereus* Lema, 1964.
Type: Holotype, NHR, a 920 mm female (Touzet, 1862 via Hyltén-Cavallius).
Type locality: "Brasilia" [= Brazil].
Distribution: Southeastern Brazil (Espírito Santo, Minas Gerais, E Rio de Janeiro, Rio Grande do Sul, E São Paulo), 65–900 m.
Sources: Marques et al., 2001, 2004 and Prudente & Feio, 2001.

5. *Siphlophis pulcher* (Raddi, 1820). Mem. Math. Fis. Soc. Ital. Sci. Modena 18(2): 337–338. (*Coluber pulcher*)

Synonyms: *Lycognathus geminatus* A.-M.-C. Duméril, Bibron & Duméril, 1854b, *Oxyrhopus rhombeatus* W.C.H. Peters, 1863c, and *Callopistria rubrovertebralis* Amaral, 1935a.
Type: Holotype, IRMC (G. Raddi, 1817–1818).
Type locality: "vicinanze di Rio-Janeiro, Brasile" [= Rio de Janeiro, S Rio de Janeiro State, SE Brazil, 22°54'S, 43°13'W, elevation 15 m].
Distribution: Southeastern Brazil (Bahia, S Minas Gerais, Paraná, Rio de Janeiro, Rio Grande do Sul, Santa Catarina, E São Paulo, Búzios, Cannanéia, Grande, Santo Amaro, São Sebastião and São Vicente Is.), NSL–50 m.
Sources: Amaral, 1935a, 1977, Freitas, 1999, Marques et al., 2001, 2004, Duarte Roche & Van Sluys, 2006 and Cicchi et al., 2007.

6. *Siphlophis worontzowi* (Prado, 1940a). Mem. Inst. Butantan (1939) 13: 5–6, pl., 4 figs. (*Alleidophis worontzowi*)

Type: Holotype, IB 10062, an 885 mm male (C. Worontzow, Feb. 1937), destroyed by fire 15 May 2010.
Type locality: "rio Amana, na confluencia deste com o rio Parauarí, a sudeste do Estado do Amazonas, Brasil" [= Amana River, near its confluence with the Parauarí River (4°23'S, 57°35'W, elevation 25 m), E Amazonas, N Brazil].
Distribution: Northwestern Brazil (Amazonas, Mato Grosso, Pará, Rondônia), 25–240 m.
Sources: A.P. Santos, 2003 and Costa et al., 2010a.

SISTRURUS S.W. Garman, 1884 (Viperidae)

Synonyms: *Crotalophorus* Gray, 1825, *Caudisona* Fitzinger, 1826a, †*Neurodromicus* Cope, 1873, *Crotolophorus* – W.H. Smith, 1882 *(nomen incorrectum)*, *Systrurus* – Cope, 1887c *(nomen incorrectum)*, †*Neurodromus* – Hoffmann, 1890 *(nomen incorrectum)*, *Sistruras* – Crouse, 1902 *(nomen incorrectum)*, and *Sisturus* – Gharpurey, 1962 *(nomen incorrectum)*.
Type species: *Crotalus miliarius* Linnaeus, 1766.
Distribution: Central and E North America and N Mexico.
Fossil records: Upper Miocene to upper Pleistocene of E USA. Fossils unassigned to species include upper Miocene (Clarendonian) of USA (Nebraska).
Sources: Gloyd, 1940, Brattstrom, 1964, Klauber, 1972, Harris & Simmons, 1978b, Stille, 1987, J.A. Campbell & Lamar, 1989, Maes, 1989, Minton, 1992, A. Knight et al., 1993, Rubio, 1998, David & Ineich, 1999, McDiarmid et al., 1999, Parkinson, 1999, Holman, 2000a, Castoe & Parkinson, 2006, Parmley & Holman, 2007, Parmley & Hunter, 2010, Ernst & Ernst, 2011, Kubatko et al., 2011 and Hoser, 2012f.
Remarks: *Sistrurus ravus* is now in the genus *Crotalus*.

1. *Sistrurus catenatus* (Rafinesque-Schmaltz, 1818b). Amer. Month. Mag. Crit. Rev. 4(1): 41. (*Crotalinus catenatus*)

Synonyms: *Crotalus tergeminus* Say *in* E. James, 1823, *Crotalophorus edwardsii* Baird & Girard, 1853, *Crotalophorus consors* Baird & Girard, 1853, †*Neurodromicus dorsalis* Cope, 1873, and *Caudisona tergemina* – Cope, 1875c.
Type: Neotype, USNM 526, an 830 mm specimen, designated by Crother et al. (2011a: 273) in petition to ICZN.
Type locality: "Poland, Mahoning County, Ohio, USA" via neotype selection.
Distribution: Western and S USA (ext. SE Arizona, SE Colorado, SW Iowa, Kansas, N and E Missouri, SE Nebraska, S New Mexico, cen. Oklahoma, Texas) and ext. NE Mexico (N Coahuila, S Nuevo León), NSL–1650 m.
Fossil records: Middle Pliocene (Blancan II, III) of USA (Kansas, Texas), upper Pliocene (Blancan V) of USA (Kansas, Nebraska), middle Pleistocene (Irvingtonian II) of USA (Nebraska), middle/upper Pleistocene (Irvingtonian II) of USA (West Virginia), and upper Pleistocene (Rancholabrean I) of USA (Kansas).
Sources: Gloyd, 1955a, P. Anderson, 1965, Fowlie, 1965, Webb, 1970, Stebbins, 1985, B. Johnson & Menzies, 1993, Hammerson, 1999, Werler & Dixon, 2000, Holycross et al., 2008, C.D. Anderson et al., 2009, Gibbs et al., 2011 and Kubatko et al., 2011.

Remarks: Holotype lost *fide* Holycross et al. (2008: 421). Type locality restrictions to vicinity of Kansas City, Missouri [USA] by K.P. Schmidt (1953a: 226) and to "flood plains of the Missouri River, between the mouth of the Platte River and Nebraska City, Nebraska," *fide* Holycross et al. (2008: 422–423) invalid. Possibly occurs in NE Sonora, Mexico *fide* Bogert and Oliver (1945: 417) and NW Chihuahua and N Tamaulipas, Mexico *fide* Minton (1983: 332.1). Durango, Mexico record needs confirmation *fide* Campbell & Lamar (1989: 374).

2. *Sistrurus messasaugus* (Kirtland *in* Mather, 1838). Second Ann. Rep. Geol. Surv. State Ohio: 190. (*Crotalus messasaugus*)

Synonyms: *Crotalophorus kirtlandi* Holbrook, 1842a, and *Crotalus messaugus* – Golay, Smith, Broadley, Dixon, McCarthy, Rage, Schätti & Toriba, 1993 *(nomen incorrectum).*

Type: Holotype, not designated, a 665 mm specimen, location unknown.

Type locality: "Ohio" [USA].

Distribution: Extreme SE Canada (SE Ontario), cen. and NE USA (Illinois, N Indiana, E Iowa, Michigan, ext. SE Minnesota, W and cen. New York, N and cen. Ohio, ext. W Pennsylvania, S Wisconsin), NSL–1650 m.

Sources: B. Wright, 1941, Breckenridge, 1944, Logier & Toner, 1961, P.W. Smith, 1961, P. Anderson, 1965, H.M. Smith et al., 1965, Froom, 1972, Minton, 1972, 1983, R.C. Vogt, 1981, Cook, 1984, Ernst & Barbour, 1989, Harding, 1997 and Gibbs et al., 2011.

3. *Sistrurus miliarius* (Linnaeus, 1766). Syst. Nat., ed. 12, 1: 372. (*Crotalus miliarius*)

Synonyms: *Crotalus miliaris* Palisot de Beauvois, 1799 *(nomen emendatum), Caudisona miliaria* – Fitzinger, 1826a, *Sistrurus miliarius barbouri* Gloyd, 1935, and *Sistrurus miliarius streckeri* Gloyd, 1935.

Type: Holotype, a male (A. Garden, 1760–1761), lost *fide* Palmer (1978: 220.1).

Type locality: "Carolina" [USA]. Restricted to vicinity of Charleston, South Carolina *fide* Klauber (1948: 6).

Distribution: Southeastern and S cen. USA (Alabama, Arkansas, Florida, Georgia, ext. SW Kentucky, Louisiana, Mississippi, S Missouri, S and SE North Carolina, E Oklahoma, South Carolina, W Tennessee, E Texas), NSL–500 m.

Fossil records: Lower Pleistocene (Irvingtonian I) of USA (Florida), and upper Pleistocene (Rancholabrean II) of USA (Florida).

Sources: Martof, 1956, Mount, 1975, Palmer, 1978, Christman, 1980, Martof et al., 1980, T.R. Johnson, 1987, 2000, Ashton & Ashton, 1988, Dundee & Rossman, 1989, Ernst & Barbour, 1989, Conant & Collins, 1991, Palmer & Braswell, 1995, Werler & Dixon, 2000, Kubatko et al., 2011 and Snellings, 2012.

Remarks: *Sistrurus barbouri* possibly a valid species *fide* Kubtko et al., 2011.

†*SIVAOPHIS* Head, 2005 (Colubridae)

Type species: †*Sivaophis downsi* Head, 2005.

Distribution: Miocene of Pakistan.

Source: Head, 2005.

1. †*Sivaophis downsi* Head, 2005. Palaeont. Electron. 8(1): 20–21, figs. 10a.1–5.

Type: Holotype, H-GSP 49661, one trunk vertebra (E.H. Linday & W.R. Downs III).

Type locality: "Y-227 (9.29 Ma), Potwar Plateau of north-central Pakistan; Dhok Pathan Formation, Winnewala, Miocene."

Distribution: Middle Miocene (6.3–13.6 mya) of Pakistan.

SONORA Baird & Girard, 1853 (Colubridae)

Synonym: *Procinura* Cope, 1879.

Type species: *Sonora semiannulata* Baird & Girard, 1853.

Distribution: Southwestern USA and N Mexico.

Fossil records: Upper Pleistocene and Holocene of SW USA.

Sources: Stickel, 1943, H.M. Smith & Taylor, 1945, Heringhi, 1966, Frost & Van Devender, 1979, Holman, 1995b, 2000a and C.L. Cox et al., 2012.

Remarks: *Procinura* a synonym of *Sonora fide* C.L. Cox et al., 2012.

1. *Sonora aemula* (Cope, 1879). Proc. Amer. Philos. Soc. (1878–80) 18: 262. (*Procinura aemula*)

Type: Holotype, ANSP 11641, a 364 mm male (E. Wilkinson, Jr., 1873–1875).

Type locality: "Batopilas, in Southern Chihuahua, on the western side of the water shed of the Cordilleras on the upper waters of a tributary of the Rio Fuerte" [= Batopilas, SW Chihuahua, NW Mexico, 27°01'N, 107°44'W, elevation 575 m].

Distribution: Northwestern Mexico (ext. SW Chihuahua, Sinaloa, SE Sonora), NSL–1680 m.

Sources: Bogert & Oliver, 1945, Zweifel & Norris, 1955, McDiarmid et al., 1976, Webb, 1984 and H.M. Smith et al., 2006.

Remarks: This species should be in the genus *Procinura fide* Lemos-Espinal et al., 2004b.

2. *Sonora michoacanensis* (Dugès *in* Cope, 1885b). Proc. Amer. Philos. Soc. (1884) 22: 178–179. (*Contia michoacanensis*)

Synonym: *Sonora erythrura* E.H. Taylor, 1937b.

Type: Neotype, BMNH 1946.1.14.65 (formerly BMNH 1903.3.21), a 195 mm male (A.A.D. Dugès, 1853–1885), designated by Stickel (1943: 113).

Type locality: "Michoacan, Mexico," via neotype selection. Restricted to Apatzingán, Michoacán, Mexico *fide* H.M. Smith & Taylor (1950a: 335).

Distribution: Southwestern Mexico (Colima, Guerrero, Michoacán, Morelos, Puebla), NSL–2500 m.

Sources: E.H. Taylor, 1937b, Zweifel, 1959b, Duellman, 1961, Echternacht, 1973, Benitez-Gálvez, 1997 and Lamar, 1997b.

Remarks: Holotype lost *fide* H.M. Smith & Necker (1943: 185).

3. *Sonora mutabilis* Stickel, 1943. Proc. Biol. Soc. Washington 56: 116–17. (*Sonora michoacanensis mutabilis*)

Synonym: *Sonora aequalis* H.M. Smith & Taylor, 1945.

Type: Holotype, FMNH 105257 (formerly EHT-HMS 4661), a male (H.M. Smith, 29 July 1935)

Type locality: "near Magdalena, Jalisco, Mexico" [= Magdalena, N Jalisco State, W Mexico. 20°55'N, 103°59'W, elevation 1380 m].

Distribution: Mexico (Aguacalientes, Jalisco, Nayarit, S Zacatecas, ext. S Sinaloa), 400–1800 m.

Sources: E.H. Taylor, 1937b and C.L. Cox et al., 2012.

4. *Sonora semiannulata* Baird & Girard, 1853. Cat. No. Amer. Rept., Serp. 1: 117.

Synonyms: *Lamprosoma episcopum* Kennicott *in* Baird, 1859a, *Contia isozona* Cope, 1867a, *Contia episcopa torquata* Cope, 1880, *Contia taylori* Boulenger, 1894a, *Contia nuchalis* Schenkel, 1901, *Sonora semiannulata blanchardi* Stickel, 1938, *Sonora semiannulata gloydi* Stickel, 1938, *Sonora miniata miniata* Stickel, 1938, *Sonora miniata linearis* Stickel, 1938, *Sonora mosaueri* Stickel, 1938, and *Sonora bancroftae* Klauber, 1943a.

S

Type: Holotype, USNM 2109, a 241–243 mm male (J.H. Clark [U.S.-Mex. Bound. Surv.], 1851, via J.D. Graham).

Type locality: "Sonora, Mex." [= Sonora State, Mexico] (in error). Corrected to south of the Gila River in Santa Cruz or Cochise Co., SE Arizona, USA *fide* Bogert (1939a: 198–199) and Stejneger (1940: 204). Restricted to the vicinity of the Santa Rita Mountains, Santa Cruz Co., SE Arizona, USA by Stickel (1943: 119).

Distribution: Western USA (ext. NW Arkansas, Arizona, SE California, SE Colorado, SW Idaho, S Kansas, SW Missouri, New Mexico, Oklahoma, ext. SE Oregon, W Texas, ext. SW Utah) and N Mexico (Baja California Norte, Baja California Sur, Chihuahua, Coahuila, NE Durango, Nuevo León, ext. N Sonora, N Tamaulipas, San José and San Marcos Is.), NSL–2080 m.

Fossil records: Upper Pleistocene (Rancholabrean I, II) of USA (Arizona, Kansas, New Mexico, Texas).

Sources: Klauber, 1943a, Kassing, 1961, Frost & Van Devender, 1979, Frost, 1983, Stebbins, 1985, Van Devender & Bradley, 1994, H. Brown et al., 1995, Degenhardt et al., 1996, P.R. Brown, 1997, Bartlett & Tennant, 2000, Werler & Dixon, 2000, Grismer, 2002, Trauth et al., 2004 and Milanovich et al., 2005.

SORDELLINA Procter, 1923a (Xenodontidae)

Type species: *Xenodon punctatus* W.C.H. Peters, 1880a.

Distribution: Southern Brazil.

Sources: F. Werner, 1909b, Procter, 1923a, Amaral, 1926a, 1930, 1936, 1977, Hoge, 1958a, J.A. Peters & Orejas-Miranda, 1970, Hoge & Romano-Hoge, 1978b, Marques et al., 2001, Cicchi et al., 2007, Zaher et al., 2009 and Miranda & Fernandes, 2012.

1. *Sordellina punctata* (W.C.H. Peters, 1880a). Mber. Königl. Akad. Wiss. Berlin 1880(2): 221–222, pl., fig. 3. (*Xenodon punctatus*)

Synonyms: *Liophis rehi* F. Werner, 1909b, *Sordellina brandonjonesii* Procter, 1923a, and *Sordellina pauloensis* Amaral, 1923 & 1926a.

Type: Holotype, ZMB 9647, a 170 mm specimen (Baden).

Type locality: "Brasilien" [= Brazil]. Restricted to Poá, São Paulo State, Brazil *fide* Hoge & Romano (1978b: 64).

Distribution: Southern Brazil (E Paraná, E Santa Catarina, E São Paulo, Comprida Is.), NSL–960 m.

Remarks: Type locality erroneously restricted to São Bento, São Paulo, Brazil *fide* Hoge (1958a: 52). Mato Grosso record in error *fide* Miranda & Fernandes (2012: 7).

SPALEROSOPHIS Jan *in* Filippi, 1865a (Colubridae)

Synonyms: *Chilolepis* Fitzinger, 1843 *(nomen rejiciendum), Sphalerosophis* – Jan *in* Filippi, 1865 *(nomen incorrectum), Sphalerosophis* Jan, 1867 *in* Jan & Sordelli, 1866–1870 *(nomen rejiciendum), Loxodon* Jan, 1867 *in* Jan & Sordelli, 1866–1870 *(nomen substitutum), Toxodon* Jan, 1867 *in* Jan & Sordelli, 1866–1870 *(nomen incorrectum), Spalerophis* – Cope, 1887c *(nomen incorrectum), Spelaerosophis* – Loveridge, 1957 *(nomen incorrectum), Sphaerosophis* – Leviton, 1959a *(nomen incorrectum), Spalesorophis* – Majupuria, 1981 *(nomen incorrectum), Sphaelerosophis* – Hughes, 1983 *(nomen incorrectum),* and *Spalersophis* – Yaron, 1985 *(nomen incorrectum).*

Type species: *Spalerosophis microlepis* Jan *in* Filippi, 1865a.

Distribution: North Africa, Arabia and SW Asia.

Sources: M.A. Smith, 1943, Marx, 1959, Lanza, 1964, Minton, 1966, Marx et al., 1982, Sahi & Duda, 1986, Schätti & McCarthy, 1987, Schätti, 1993a, Nagy et al., 2004, Baig & Masroor, 2008, Baig et al., 2008 and Schätti et al., 2009.

Remarks: Official Generic Name no. 1741 *fide* Opinion 794 (ICZN, 1966c). *Spalerosophis* consists of six species (*i.e.*, *arenarius, atriceps, cliffordii, diadema, josephscortecci, microlepis*) with *dolichosphilus* a subspecies of *S. cliffordii fide* Ratagat Masroor (pers. comm., unpub. data).

1. *Spalerosophis arenarius* (Boulenger, 1890a). Fauna Brit. India, Rept. Batr.: 329. (*Zamenis arenarius*)

Type: Lectotype, BMNH 1946.1.14.62, head and anterior body of a 1200 mm specimen (J.A. Murray, 1871–1886), designated by Marx (1959: 356).

Type locality: "Karachi" [= Karachi, SW Sindh, S Pakistan, 24°54'N, 67°02'E, elevation 25 m] via type locality restriction.

Distribution: Southeastern Pakistan (ext. S Balochistan, S Punjab, S Sindh) and NW India (N Gujaret, S Jammu and Kashmir, Rajasthan), NSL–1650 m.

Sources: Boulenger, 1893a, K.P. Schmidt, 1930, Mertens, 1969a, M.S. Khan, 1982a–b, 2002, Whitaker & Captain, 2004 and Baig & Masroor, 2008.

2. *Spalerosophis atriceps* (J.G. Fischer, 1885b). Jahrb. Hamburg. Wiss. Anst. 2: 102–103. (*Zamenis diadema atriceps*)

Synonym: *Zamenis diadema melanoides* Wall, 1914f.

Type: Holotype, SMF 41134 (formerly NMB 7796), a 1475 mm female (F. Krüger).

Type locality: "Himalaya" [= Himalaya Mtns. of NE Pakistan, N India, Nepal or Bhutan].

Distribution: Pakistan (Balochistan, Gilgit-Baltistan, Jammu & Kashmir, Khyber Pakhtunkhwa, Punjab, Sindh) and N India (Delhi, Gujarat, Haryana, Hamadral Pradesh, Jammu and Kashmir, Rajasthan, Uttar Pradesh) and Nepal (Chitwan, Kathmandu, Saptari), NSL–2000 m.

Sources: Wall, 1914f, Sharma, 1974, Vyas, 1989, M.S. Khan, 2002, Schleich & Kästle, 2002, Tiwari & Shah, 2004, Whitaker & Captain, 2004 and Masroor, 2012.

Remarks: A valid species *fide* Baig & Masroor, 2008 and Masroor (2012: 140). *Spalerosophis melanoides* (Wall) possibly a valid species *fide* Sharma (1974: 22) A synonym of *S. diadema* (Schlegel) *fide* M.A. Smith (1943: 173), Mertens (1969a: 69) and M.S. Khan (1982b: 237). *Spalerosophis atriceps* an ontogenetic phase of *S. diadema fide* Mahendra (1984: 293).

3. *Spalerosophis cliffordii* (Schlegel, 1837). Essai Phys. Serp. 1: 148–149, 2: 163–164, pl. 6, figs. 13–14. (*Coluber cliffordii*) (*nomen protectum*)

Synonyms: *Coluber guttatus* Forskal *in* Niebuhr 1775 (*nomen oblitum*), *Coluber cahirinus* Gmelin, 1789 (*nomen oblitum*), *Coluber versicolor* Wagler, 1830 (*nomen praeoccupatum*), *Periops parallelus* A.-M.-C. Duméril, Bibron & Duméril, 1854a

Type: Lectotype, RMNH 467, an 888 mm female (J.H.F. Clifford), designated by L.D. Brongersma *in* Kramer & Schnurrenberger (1963: 524).

Type locality: "États barbaresques" [= North African countries of Morocco, Algeria, Tunisia and Libya] via lectotype selection. Restricted to Tripoli, Libya *fide* Marx (1959: 350).

Distribution: North Africa and Middle East. Southern Morocco (Tan Tan), N Western Sahara (Saguia El Hamra), W Mauritania (Adrar, Tiris Zemmour, Trarza), NW Senegal, Mali (Mopti, S Tombouctou), Algeria (Adrar, Béchar, Biskra, Ouargla, S Tamanrasset), Niger (W Agadez, Maradi, Niamey, S Tahoua, Tamanrasset, Zinder), Tunisia (Gabès, Gafsa, Kasserine, Kebili, Medenine, Sidi Bou Zid, Tatahouine, Tozeur, Tunis), Libya (Ajdabiya, Al Fatih, Al Khums, Al Kurfah, Al Nuqat Al Khams, Al Wahah, Awbari, Banghazi, Damah, Murzuq, Sabha, Sawfajjin, Surt, Tarabulus, Tarhunah), N Chad, N Sudan (El-Bahr El-Ahmar, Al Khartum, An Nil Al Azraq, Ash Shamaliyah, Ash Shaqiyah, Sennar, Shamal Kurdufan), E Egypt (Alexandria, Assiut, Aswan, Beheira, Beni Suef, Cairo, El-Bahr El-Ahmar, Faiyum, Giza, Ismailia, Matrouh, Menufia, Minya, New Valley, North Sinai, Port Said, Qena, Sharkeya, Sohag, South Sinai, Suez), S Israel (Central, Gaza Strip, Jerusalem, Southern, West Bank), cen. Syria (Homs), SW Jordan (Amman, Aqaba, Irbid, Karak, Maan, Mafraq, Tafilah), Saudi Arabia (Asir, Bahah, Eastern, Ha'il, Jazan, Makkah, Northern, Qasim, Riyadh, Tabuk), Kuwait, E United Arab Emirates (Ras al Khaimah), Oman (Ad Dakhiliyah, Ad Dhahirah, Al Batinah, Dhofar, Muscat), Yemen (Abyan, Aden, Hadhramaut, Hajjah, Sana'a, Shabwah), Iraq (Babil, Baghdad, Dhi Qar, Maysan, Ninawa), ext. SE Turkey (Adryaman, Sanliurfa, Urfa) and ext. W Iran (Bushehr, Central, Fars, Kerman, Khuzestan, Razavi Khorasan, Semnan, Sistan va Baluchestan, Zanjhan), NSL–2400 m.

Sources: Olivier, 1894, 1896, J. Anderson, 1896, 1898, Doumergue, 1901, H.W. Parker, 1931, 1938, Angel, 1933, Flower, 1933, Villiers, 1950a–b, Piguet, 1955, Dekeyser & Villiers, 1956, Mertens, 1956, Domergue, 1959b, 1962, Pasteur & Bons, 1960, Kramer & Schnurrenberger, 1963, Corkill & Cochrane, 1965, Bons, 1967, 1972, Gauthier, 1967, Pasteur, 1967, Papenfuss, 1969, Gasperetti, 1988, Le Berre, 1989, Bons & Geniez, 1996, Bouskila & Amitai, 2001, Geniez et al., 2004, Chippaux, 2006, Amr & Disi, 2011 and Bar & Haimovitch, 2011.

Remarks: A valid species *fide* Latifi, 1991. *Coluber guttatus* Forskal and *Coluber cahirinus* Gmelin are *nomina oblita* according to Article 23.9.2 of the Code (ICZN

1999) *fide* Schätti et al. (2009: 43). Possibly occurs in Sahrawi Republic *fide* Monteil (1951: 82) and Lebanon *fide* Gruber (1989: 149).

4. *Spalerosophis diadema* (Schlegel, 1837). Essai Phys. Serp. 1: 146, 2: 148. (*Coluber diadema*)

Synonyms: *Natrix oppelii* Wagler, 1825 (*nomen ineditum*), *Coluber geoffroyii* Gray *in* Griffith & Pidgeon, 1831, *Coluber oppellii* Blyth, 1854a *(nomen nudum), Periops parallelus schirasiana* Jan 1863b, and *Spalerosophis maximus* Clark & Inger, 1942 (*nomen nudum*).

Type: Lectotype, a 1016 mm specimen described and illustrated by P. Russell (1801: 34–35, pl. 30) (Bruce via Scott), designated herein.

Type locality: "environs de Bombay" [India] (in error) via lectotype selection. Specified as Buchier, [= Bushehr, W Bushehr Prov., SW Iran, 28°59'N, 50°50'E, elevation NSL] *fide* P. Russell (1801: 35, 56).

Distribution: Southwestern Asia. Iran (Bushehr, Central, Fars, Kerman, Khuzestan, Razavi Khorasan, Semnan, Sistan va Baluchestan, Zanjhan), SW Kazakhstan, Uzbekistan, S Turkmenistan, Krygyzstan, Tadzhikistan, Afghanistan (Helmand, Nangarhar, Nimruz), Pakistan (Balochistan, Khyber Pakhtunkhwa, Punjab, Sindh) and NW India, NSL–2400 m.

Sources: Murray, 1886, Wall, 1914f, 1923, K.P. Schmidt, 1930, 1939b, 1941, Zavattari, 1930, 1934, 1937, Mosauer, 1934, Andersson, 1935, Corkill, 1935, Terent'ev & Chernov, 1949, 1965, G. Haas, 1951, 1975, Monteil, 1951, Chpakowsky & Chnéour, 1953, Piquet, 1955, Minton, 1956, 1969, Sochurek, 1956, 1979b, Khalaf, 1959, Kramer & Schnurrenberger, 1963, Minton, 1966, Dmi'el, 1967, Mandaville, 1967, 1974, Král, 1969, Mertens, 1969a, Eiselt, 1970, Leviton & Anderson, 1970a, Y. Werner, 1971, 1973, 1988, Dmi'el and Borut, 1972, Hoofien, 1972, Kochva, 1974, Sharma, 1974, Baran, 1976a, Matz, 1976a, Arnold & Gallagher, 1977, Bannikov et al., 1977, Arnold, 1980b, Basoglu & Baran, 1980, M.S. Khan, 1982a–b, 2002, Daniel, 1983, Murthy, 1985, 1986, Gasperetti, 1988, Osman & El Sir, 1988, Gruber, 1989, Gallagher, 1990, Latifi et al., 1992, Leviton et al., 1992, Szczerbak, 1994, Lamar, 1997b, Franzen, 1999, Disi et al., 2001, Baker et al., 2004 and Baran et al. 2004.

Remarks: *Spalersophis diadema schirasianus* a synonym *fide* Baig & Masroor (2008: 113) but a valid species *fide* M.S. Khan (2002: 125) and possibly valid *fide* Schätti et al. (2009: 131).

5. *Spalerosophis dolichospilus* (F. Werner, 1923a). Ann. Naturhist. Mus. Wien 36: 166 (*Zamenis diadema dolichospila*)

Synonym: *Coluber choumowitchi* Domergue, 1954.

Type: Holotype, NMW 23418, a 1020 mm male (Schönbrunn).

Type locality: "Indien" [= India] (in error). Corrected to Ain Sefra, Heimat die westalgerische (und wohl auch die marokkanische) Sahara [= Ain Sefra, Naama Wilaya, NW Algeria, 32°45'N, 0°35'W, elevation 1080 m] *fide* Werner (1929b: 7).

Distribution: Northwestern Africa. Northern Western Sahara (Saguia El Hamra), Morocco (Agadir, Al Hoceima, Boulemane, Fes, Kaar Es Souk, Khenitra, Ouarzazate, Tan Tan, Tata), Algeria (Naama, Souk Ahras, Tamanrasset) and W Tunisia (Chott el Gharsa), 80–1800 m.

Source: Geniez et al., 2004.

Remarks: A valid species *fide* Pasteur (1967: 444) and Geneiz et al. (2004: 150). A subspecies of *S. cliffordii fide* Ratagat Masroor (pers. comm., unpublished data).

6. *Spalerosophis josephscorteccii* Lanza, 1964. Monit. Zool. Ital. 72(1–2): 47–54, figs. 1, 2a–d, 3a–d, 4a–g. (*Sphalerosophis josephscorteccii*)

Type: Holotype, MZUF 2587, a 998 mm male (G. Scortecci, autumn 1953).

Type locality: "L'oasi di Gàlgalo (49°03'E X 11°00'N) circa a 35 km a SSW di Bosaso (Bender Cassim), sulle montagne calcaree dell'Ahl Medò, ad un'altitudine approssimativa di 875 m (Somalia settentrionale)" [= Gàlgalo Oasis, NW Bari Region, NE Somalia, 10°58'N, 49°05'E, elevation 875 m].

Distribution: Northeastern Somalia (NW Bari), 875 m.

Source: Lanza, 1983a.

7. *Spalerosophis microlepis* Jan *in* Filippi, 1865a. Viaggio Persia 1: 356–357.

Synonym: *Spalerosophis macrolepis* – Elter, 1982 (*nomen incorrectum*).

Type: Lectotype, MZUT 1843 (formerly MZUT 4024), a 1230 mm female (G. Doria, 1862–1864), designated by Kramer & Schnurrenberger (1963: 526), lost *fide* Schätti et al. (2009: 122).

Type locality: "Laristan, Persia meridionale" [= Larestan region, W Hormozghan Prov., SW Iran, ca. 27°25N, 5606'°E] via lectotype selection.

Distribution: Southwestern Iran (Central, Esfahan, Fars, Hamadan, Hormozghan, Khuzestan, Lorestan), 1530–2400 m.

Sources: Jan & Sordelli, 1867 *in* 1866–1870, K.P. Schmidt, 1939b, Latifi, 1991, Leviton et al., 1992 and Schätti et al., 2009.

Remarks: Official Specific Name no. 2176 *fide* Opinion 794 (ICZN, 1966c). Original generic orthography *Spalerosophis* and *Sphalerosophis,* the former selected by Marx (1959: 348) as first reviewer. Possibly occurs in Zagros Mountains, Iraq *fide* Leviton et al. (1992: 104).

SPILOTES Wagler, 1830 Colubridae)

Synonyms: *Spiletes* Swainson, 1839 *(nomen emendatum), Agriotes* Jan, 1863b, *Agriotis* – F. Werner, 1909b *(nomen incorrectum), Spilote* – Briceño-Rossi, 1934 *(nomen incorrectum), Spillotes* – Tomas-Rodas, 1938 *(nomen incorrectum),* and *Spilotus* – Moonen, Eriks & Deursen, 1979 *(nomen incorrectum).*

Type species: *Coluber pullatus* Linnaeus, 1758.

Distribution: Latin America.

Source: Amaral, 1930g.

1. *Spilotes megalolepis* A.C.L.G. Günther, 1865. Ann. Mag. Nat. Hist. (3) 15(86): 93–94.

Type: Holotype, BMNH 1946.1.13.2, a 2362 mm female.

Type locality: "South America."

Distribution: Western Ecuador (W Carchi, E Esmeraldas, El Oro, Pichinicha), 100–500 m.

Sources: Boulenger, 1894a, Miyata, 1980, Orcés & Almendáriz, 1989b and Almendáriz, 1991.

2. *Spilotes pullatus* (Linnaeus, 1758). Syst. Nat., ed. 10, 1: 225–226. (*Coluber pullatus*)

Synonyms: *Cerastes coronatus* Laurenti, 1768, *Cerastes mexicanus* Laurenti, 1768, *Natrix humana* Laurenti, 1768, *Coluber novae-hispaniae* Gmelin, 1789 *(nomen incorrigendum), Coluber variabilis* Merrem, 1790, *Coluber plutonius* Daudin, 1803c, *Natrix canininana* Merrem, 1820, *Coluber variabilis kuhii* Wied-Neuwied, 1825, *Natrix mycthemera* Wagler, 1825 *(nomen ineditum), Spilotes pullatus auribundus* Cope, 1861e, *Spilotes salvini* A.C.L.G. Günther, 1862b, *Herpetodryas incertus* Jan, 1863b, *Spilotes pullatus anomalepis* Bocourt, 1888 *in* A.H.A. Duméril, Bocourt & Mocquard, 1870–1909, *Coluber novae-hispaniae* – Boulneger, 1894a *(nomen illegitimum), Spilotes microlepis* F. Werner, 1903a *(partim), Spilotes pullatus ater* Sternfeld, 1920b, *Spilotes pullatus argusiformis* Amaral, 1930g, *Spilotes pullatus maculatus* Amaral, 1930g, and *Tryria pullata* – Briceño-Rossi, 1934 *(nomen incorrectum).*

Types: Syntypes (2), ZMUU 40 (formerly MP) (A. Fredrik), and NHR Lin-30 (formerly MAFR), a 580 mm specimen (Mus. Drottn.).

Type locality: "Asia" (in error). Corrected to Regno Mexicano [= Mexico] *fide* Seba (1735: 21).

Distribution: Latin America. Southeastern Mexico (Campeche, Chiapas, ? Colima, ? Guerrero, Hidalgo, E Oaxaca, San Luis Potosí, E Oaxaca, Querétaro. Quintana Roo, Tabasco, ext. S Tamaulipas, Veracruz, Yucatán), Belize (Belize, Cayo, Corozal, Orange Walk, Stann Creek, Toledo), Guatemala (Alta Verapaz, Izabal, Petén, Sololá), Honduras (Atlántida, Choluteca, Colón, Comayagua, Cortés, El Paraíso, Francisco Morazán, Gracias a Dios, Ocotepeque, Olancho, Yoro), El Salvador (Ahuachapán, Cuscatlán, La Libertad, Morazán, San Miguel, Santa Ana, Sonsonate), Nicaragua (Atlántico Norte, Atlántico Sur, Granada, Jinotega, Managua, Matagalpa, Rio San Juan, Ometepe Is.), Costa Rica (Alajuela, Cartago, Guanacaste, Heredia, Limón, Puntarenas, San José), Panama (Bocas del Toro, Canal Zone, Chiriquí, Coclé, Colón, Darién, Herrera, Panamá, Veraguas, Barro Colorado Is.), Colombia (Amazonas, Antioquia, Atlántico, Boyaca, Caquetá, Cauca, Chocó, Córdoba, Cundinamarca, Magdalena, Meta, Nariño, Norte de Santander, Risaralda, Santander, Sucre, Tolima, Valle, Vaupés), Venezuela (Amazonas, Anzoátegui, Apure, Aragua, Barinas, Bolívar, Carabobo, Cojedes, Delta Amacuro, Distrito Federal, Falcón, Guárico, Lara, Mérida, Miranda, Monagas, Nueva Esparta, Portuguesa, Sucre, Táchira, Trujillo, Yaracuy, Vargas, Zulia, Margarita Is.), Trinidad and Tobago (Tobago, Trinidad), Guyana (Barima-Waini, East Berbice-Corentyne, Mahaica-Berbice, Potaro-Siparuni, Upper Demerara-Berbice, Upper Takutu-Upper Essequibo), Suriname (Brokopondo, Nickerie, Para, Paramaribo, Saramacca), French Guiana (Cayenne, Saint-Laurent-du-Maroni), E Ecuador (Napo, Pastaza), E Peru (Loreto, Madre de Dios), Brazil (Acre, Amapá, Amazonas, Bahia, Ceará, Goiás, Mato Grosso, Minas Gerais, Pará, E Paraná, Piauí, Rio Grande do Norte, Pernambuco, Rio de Janeiro, Rio Grande do Sul, Rondônia, Santa Catarina, São Paulo, Sergipe, Cananéia, Cardoso, Comprida, Grande, Marajó, Porchat, Santo Amaro, São Sebastião and São Vicente Is.), Bolivia (Beni, La Paz), S Paraguay (Alto Paraná, Amambay, Asunción, Caaguazú, Caazapá, Canindeyú, Central, Concepción, Cordillera, Guairá, Itapúa, Misiones, Neembucú, Paraguarí, San Pedro) and ext. N Argentina (Formosa), NSL–2000 (2630) m.

Sources: A.H.A. Duméril, Bocourt & Mocquard, 1888 *in* 1870–1909, A.C.L.G. Günther, 1894 *in* 1885–1902, Bertoni, 1915, Sternfeld, 1920b, K.P. Schmidt & Andrews, 1936, K.P. Schmidt, 1941, Niceforo-Maria, 1942, H.M. Smith & Taylor, 1945, Beebe, 1946, Shreve, 1947, L.C. Stuart, 1948, 1963, Dunn, 1949, E.H. Taylor, 1951, 1954, Mertens, 1952c, Aléman, 1953, Roze, 1953a, 1966a, Hoge & Maranhao-Nina, 1964, Duellman, 1965, 1979, Sexton & Heatwole, 1965, Hoge, 1967, P. Muller, 1968b, Medem, 1969, Carillo de Espinoza, 1970, J.A. Peters & Orejas-Miranda, 1970, R.W. Henderson & Hoevers, 1975, Abe & Fernandes, 1977, Emsley, 1977, Amaral, 1978, Cunha & Nascimento, 1978, 1993, Gasc & Rodriguez, 1980, J.C. Lee, 1980, 1996, Vanzolini et al., 1980, Abuys, 1983a, J.D. Johnson, 1984, Cunha et al., 1985, L.D. Wilson & Meyer, 1985, Dixon & Soini, 1986, Fugler, 1986, Lancini-V., 1986, Pérez-Santos, 1986, Savage & Villa, 1986, Vanzolini, 1986, Chippaux, 1987, Lema, 1987, Villa et al., 1988, Pérez-Santos & Moreno, 1988, 1991, J.A. Campbell & Vannini, 1989, Weyer, 1990, Nascimento et al., 1991, J.D. Williams & Francini, 1991, Flores-Villela, 1993, J.A. Campbell, 1998b, Gorzula & Señaris, 1998, Starace, 1998, Freitas, 1999, G. Köhler, 1999b, Kornacker, 1999, La Marca & Soriano, 2000, Mendoza-Quijano & Hernández-Hernández, 2001, Doan & Arizábal, 2002, Savage, 2002, Cicchi et al., 2007, Cacciali, 2008, Navarrete et al., 2009, Ugueto & Rivas, 2010, McCranie, 2011a and C.J. Cole et al., 2013.

Remarks: Original description based on Seba (1735: 21, pl. 20, fig. 1), Linnaeus (1746: 581 & 1749: 35, pl. 20, fig.

3), and Gronovius (1756: 56), the latter all citing Seba. Amaral (1930g: 277) incorrectly listed type locality as Tojuquá Is., Río de Prata based on Seba (1735: pl. 105, fig. 4), which Linnaeus did not cite.

STEGONOTUS A.-M.-C. Duméril, Bibron & Duméril, 1854a (Colubridae)

Synonyms: *Lielaphis* A.C.L.G. Günther, 1863f, *Zamenophis* A.C.L.G. Günther, 1872b, *Liclaphis* – Marschall, 1873 *(nomen incorrectum), Pseudolycodon* W.C.H. Peters, 1876b, *Herbertophis* Macleay, 1884a, and *Stegonothus* – Obst, Richter & Jacob, 1984 *(nomen incorrectum).*

Type species: *Stegonotus muelleri* A.-M.-C. Duméril, Bibron & Duméril, 1854a.

Distribution: Sino-Australia.

Sources: Rooij, 1917, C. Haas, 1950, McDowell, 1972b, F. Scott et al., 1977, Whitaker et al., 1982 and Hoser, 2012s.

1. *Stegonotus batjanensis* (A.C.L.G. Günther, 1865). Ann. Mag. Nat. Hist. (3) 15(86): 93, pl. 2, fig. a. (*Coluber batjanensis*)

Synonym: *Leptophis batjanensis* Bleeker, 1857a *(nomen nudum).*

Type: Holotype, BMNH 1946.1.11.36, a 1549 mm female (J.G.T. Bernelot-Moens via P. de Bleeker).

Type locality: "Batjan" [= Bacan Is., Maluku, E Indonesia, bet. 0°18–52'S and 127°17–54'E].

Distribution: Eastern Indonesia (Malukus: Ambon, Bacan, Batanta, Halmahera, Morotai, Salawati, Ternate).

Source: V.M. Tanner, 1950.

Remarks: Source of type listed as Bleeker *fide* A.C.L.G. Günther (1865: 93) and Boulenger (1893a: 368). Possibly a synonym of *S. cucullatus* (A.-M.-C. Duméril, Bibron & Duméril) *fide* McDowell (1972b: 22).

S

2. *Stegonotus borneensis* Inger, 1967. Fieldiana: Zool. 51(5): 82–83, fig. 1.

Type: Holotype, FMNH 164746, a 955 mm female (F.W. King, 24 May 1963).

Type locality: "Nanga Tekalit, Kapit District, Third Division, Sarawak, Borneo (1°38'N, 113°35'E)."

Distribution: East Malaysia (Sabah, Sarawak), 1370–1800 m.

Sources: Malkmus, 1987, 1989, 1992, Malkmus et al., 2002 and I. Das, 2006, 2010–2012.

Remarks: Possibly occurs in S Thailand *fide* Das (2011: 302 & 2012: 147).

3. *Stegonotus cucullatus* (A.-M.-C. Duméril, Bibron & Duméril, 1854a). Erpét. Gén. 7(1): 376. (*Lycodon cucullatum*)

Synonyms: *Zamenophis australis* A.C.L.G. Günther, 1872b, *Lycodon magnus* A.B. Meyer, 1874, *Lycodon keyensis* G. Doria, 1875, *Lycodon darnleyensis* Macleay, 1877, *Herbertophis plumbeus* Macleay, 1884a, and *Stegonotus reticulatus* Boulenger, 1895b.

Type: Holotype, MNHN 3412, a 1051 mm specimen (P. Garnot & R.-P. Lesson [Coquille or Astrolabe Voy.], 1822–1829).

Type locality: "Nouvelle Guinée" [= New Guinea]. Emended to Dore, Japen Is., Indonesia by Cogger et al. (1983: 213). [= Doré, Yapen Is., N West Papua Prov., E Indonesia, 1°40'S, 136°12'E, elevation 30 m].

Distribution: Austro-Papua. Eastern Indonesia (West Papua, Biak and Yapen Is.), Papua New Guinea (Central, Chimbu, Eastern Highlands, East Sepik, Gulf, Madang, Milne Bay, Morobe, Northern, Western Highlands, West Sepik, Bougainville, Daru, Fergusson, Karkar, Rossel, Samarai and Woodlark Is.) and N Australia (N Northern Territory, N Queensland, Darnley, Groote Eylandt, Melville, Morington and Murray Is.), NSL–1220 m.

Sources: Worrell, 1963c, Kinghorn, 1964, F. Parker, 1982, Cogger et al., 1983a, McDowell, 1984, S.K. Wilson & Knowles, 1988, Gow, 1989, Hoser, 1989, Cogger, 1992, O'Shea, 1996, Kraus & Allison, 2004 and Trembath & Lloyd, 2005.

Remarks: Bougainville record needs confirmation *fide* McDowell (1972: 23).

4. *Stegonotus diehli* Lindholm, 1905b. Jahrb. Nassau. Ver. Naturk. 58: 236–237.

Synonyms: *Stegonotus dorsalis* F. Werner, 1924a, and *Stegonotus adelynhoserae* Hoser, 2012s *(nomen illegitimum).*

Type: Holotype, MWNH 1244 (formerly SMW) (W. Diehl, July 1905).

Type locality: "Bogadjim an der Astrolabe Bai (Deutsch-Neuguinea)" [= Bogadjim, Astrolabe Bay, Madang Distr., NE Papua New Guinea, 5°26'S, 145°44'E, elevation NSL].

Distribution: New Guinea. Eastern Indonesia (West Papua, Mansinam Is.) and Papua New Guinea (East Sepik, Central, Chimbu, Gulf, Madang, Milne Bay, Morobe, Northern, Western, West Sepik, Daru and Fak-Fak Is.), NSL–1100 m.

Sources: Jong, 1927, Loveridge, 1948, F. Parker, 1982 and McDowell, 1984.

5. *Stegonotus florensis* (Rooij, 1917). Rept. Indo-Aust. Arch. 2: 110–111, figs. 46–47. (*Lycodon florensis*)

Synonym: *Stegonotus sutteri* Forcart, 1953.

Type: Holotype, ZMA 4080, a 568 mm specimen (G.A.J. van der Sande, 14 July 1908).

Type locality: "Flores (near Kotting!)." Emended to near Kotting along upper Maumeri River, Flores, Indonesia, altitude 220 m [= vicinity of Koting along upper Manumere River, E Flores, Lesser Sundas, S Indonesia, ca. 8° 40'S, 122°12'E, elevation 200 m] *fide* Daan & Hillenius (1966: 138).

Distribution: Lesser Sundas of S Indonesia (Flores, Sumba), 200–800 m.

Sources: Mertens, 1930, Forcart, 1954 and Lang, 2011.

Remarks: Photographs of holotype in Lang (2011: figs. 75–77).

6. *Stegonotus guentheri* Boulenger, 1895b. Ann. Mag. Nat. Hist. (6) 16(91): 31.

Types: Syntypes (3), BMNH 1946.1.11.37–39, two females and one male, longest syntype 1150 mm (A.S. Meek, 1894–1895).

Type locality: "Ferguson Island, D'Entrecasteaux Group, British New Guinea" [= Fergusson Island, Milne Bay Distr., SE Papua New Guinea, bet. 9°20–43'S and 150°26–57'E].

Distribution: Southeastern Papua New Guinea (Milne Bay, Fergusson, Goodenough, Kiriwina, Kitava and Normanby Is.).

Source: O'Shea, 1996.

7. *Stegonotus heterurus* Boulenger, 1893a. Cat. Snakes Brit. Mus. 1: 367, pl. 25, fig. 1.

Types: Syntypes (3), BMNH 1946.1.14.95 and 1946.1.15.10, two females (G. Brown) and 1946.1.14.91 (formerly Mus. Godeffroy), a female.

Type locality: "Duke of York Island" and "New Britain" [= Duke of York Is., East New Britain Distr., 4°10'S, 152°28'E, and New Britain Is., West or East New Britian Distr., E Papua New Guinea].

Distribution: Northeastern Papua New Guinea (East New Britain, New Ireland, E New Britain, New Ireland and Duke of York Is.), NSL–60 m.

Sources: F. Werner, 1899b, Wandolleck, 1911 and O'Shea, 1996.

Remarks: Probably a subspecies of *S. modestus* (Schlegel) *fide* McDowell (1972b: 16).

8. *Stegonotus modestus* (Schlegel, 1837). Essai Phys. Serp. 1: 143, 2: 119–120, pl. 4, figs. 16–17. (*Lycodon modestus*)

Synonyms: *Alopecion annulifer* A.-M.-C. Duméril, 1853 (*nomen nudum*), *Lycodon lividum* A.-M.-C. Duméril, Bibron & Duméril, 1854a, *Lycodon modestum* – A.-M.-C. Duméril, Bibron & Duméril, 1854a, *Ablabes greineri* Bleeker, 1860, *Coronella rosenbergii* Bleeker, 1860e, *Coluber holochrous* A.C.L.G. Günther, 1863f, *Lycodon aruensis* G. Doria, 1875, and *Stegonotus lenihoseri* Hoser, 2012s (*nomen illegitimum*).

Type: Lectotype, RMNH 324, a 910 mm female (S. Müller, Feb. 1828–Nov. 1829), designated by A.-M.-C. Duméril, Bibron & Duméril (1854a: 381) via type locality restriction.

Type locality: "Amboine" [= Ambon Is., cen. Maluku, E Indonesia, 3°40'S, 128°10'E] via type locality restriction.

Distribution: New Guinea. Eastern Indonesia (N Papua, Ambon, Aru Arch., Buru, Gag, Kai Arch., Misool, Nusa Laut, Salawati, Samao, Seram, Supiori Biak) and N Papua New Guinea (East Sepik, Madang, Manus, West Sepik, Manus Is.), NSL–460 m.

Source: O'Shea, 1996.

9. *Stegonotus muelleri* A.-M.-C. Duméril, Bibron & Duméril, 1854a. Erpét. Gén. 7(1): 682–683. (*nomen corrigendum*)

Synonyms: *Stegonotus mülleri* A.-M.-C. Duméril, Bibron & Duméril, 1854a (*nomen incorrigendum*), *Spilotes samarensis* W.C.H. Peters, 1861c, and *Stegonotus muelleri* – Boulenger, 1896a (*nomen corrigendum*).

Type: Holotype, MNHN 848 (formerly RMNH), a 1032 mm specimen (S. Müller via RMNH, 1845).

Type locality: "Java" [W Indonesia] (in error). Corrected and restricted to Samar Is., Philippines [bet. 10°54'–12°36'N and 124°15'–125°51'E] *fide* Leviton (1959b: 586).

Distribution: Southern Philippines (Leyte, Mindanao, Samar), 10–590 m.

Sources: L.E. Griffin, 1911, E.H. Taylor, 1922a, Leviton, 1959b, Inger, 1967, Alcala, 1986 and B.E. Smith, 1993.

Remarks: Possibly a synonym of *S. cucullatus* (Duméril, Bibron & Duméril) *fide* McDowell (1972b: 23). Lectotype designation via type locality restriction by Leviton (1959b: 586) invalid (unpublished).

10. *Stegonotus parvus* (A.B. Meyer, 1874). Mber. Königl. Akad. Wiss. Berlin 1874(2): 137. (*Lycodon parvus*)

Synonym: *Stegonotus sammacdowelli* Hoser, 2012s (*nomen illegitimum*).

Types: Syntypes (2), MTKD 876a–b (A.B. Meyer, 1873), destroyed 13–14 Feb. 1945 during World War II.

Type locality: "Neu-Guinea, Jobi" [= Yapen Is., West Papua Prov., E Indonesia, bet. 1°37–58'S and 135°26'–136°54'E].

Distribution: Austro-Papua. Eastern Indonesia (West Papua, Yapen Is.), Papua New Guinea (East New Britain, East Sepik, Eastern Highlands, Madang, Morobe, Western, Western Highlands, West New Britain, West Sepik, Daru, Duke of York, Manam and New Britrain Is.) and ext. N Australia (ext. N Queensland, Murray Is.), 460–915 m.

Sources: F. Parker, 1982, McDowell, 1984 and O'Shea, 1996.

Remarks: Publication date 1875 *fide* McDowell (1972b: 16). Duke of York, East New Britain and West New Britain, Papua New Guinea population possibly an undescribed species *fide* McDowell (1972: 18–19).

STENOPHIS Boulenger 1896a (Pseudoxyrhophiidae)

Synonyms: *Parastenophis* Domergue, 1995, and *Phisalixella* Domergue, 1995.
Type species: *Dipsas gaimardii* Schlegel, 1837.
Distribution: Madagascar.
Sources: Domergue, 1995, Hallermann, 2000, Nagy et al., 2005, Glaw & Vences, 2007, Vidal et al., 2008, Kelly et al., 2009 and Zaher et al., 2009.
Remarks: A synonym of *Lycodryas fide* Zaher et al., 2012.

1. *Stenophis arctifasciatus* (A.-M.-C. Duméril, Bibron & Duméril, 1854b). Erpét. Gén. 7(2): 1176–1677. (*Heterurus arctifasciatus*) (*nomen corrigendum*)

Synonyms: *Heterurus arctifasciatus* A.-M.-C. Duméril, 1853 (*nomen nudum*), *Heterurus arcti-fasciatus* A.-M.-C. Duméril, Bibron & Duméril, 1854b *(nomen incorrigendum), Heterurus arctifasciatus* – Jan, 1871 *in* Jan & Sordelli, 1870–1881 *(nomen corrigendum)*, and *Stenophis jaosoloa* Domergue, 1995.
Type: Holotype, MNHN 3870, a 431 mm specimen (J.R.C. Quoy & J.P. Gaimard, 1817–1829).
Type locality: "Madécasse" [= Madagascar].
Distribution: Eastern Madagascar (Antananarivo, Antsiranana, Toamasina, SE Toliara, Nosy Be Is.), NSL–1285 m.
Sources: Raxworthy, 1988 and Henkel & Schmidt, 2000.

2. *Stenophis betsileanus* (A.C.L.G. Günther, 1880b). Ann. Mag. Nat. Hist. (5) 6(33): 238. (*Dipsas betsileana*)

Type: Holotype, BMNH 1946.1.2.51, a 394–400 mm specimen (? J. Kirk & C.E. Bewsher).
Type locality: "S.E. Betsileo, Madagascar" [= N Fianarantsoa Prov., cen. Madagascar].
Distribution: Eastern Madagascar (Antananarivo, Antsiranana, Fianarantsoa, E SE Toliara, Nosy Mangabe Is.), 80–1325 m.
Source: Boulenger, 1896a.

S

3. *Stenophis carleti* Domergue, 1995. Arch Inst. Pasteur Madagascar (1994) 61(2): 121.

Type: Holotype, MNHN 1988.321, a 770+ mm specimen.
Type locality: "Côte sud-est de Manakara à Tolagnaro, Madagascar."
Distribution: Southeastern Madagascar (SE Toliara), NSL–570 m.
Remarks: Taxonomic status uncertain *fide* Glaw & Vences (2007: 414).

4. *Stenophis citrinus* Domergue, 1995. Arch. Inst. Pasteur Madagascar (1994) 61(2): 122.

Type: Holotype, MNHN 1978.2790, a 415 mm specimen.
Type locality: "forêt caducifolée d'Analabe-Beroboka, dans le Menabe, à mi-distance entre Morondava et Belo-sur-Tsirbihina, Madagascar."
Distribution: Western Madagascar (S Mahajanga, N Toliara), 125–380 m.

5. *Stenophis gaimardii* (Schlegel, 1837). Essai Phys. Serp. 1: 165, 2: 293–294. (*Dipsas gaimardii*)

Type: Holotype, MNHN 3868, a 550–565 mm male (J.R.C. Quoy & J.P. Gaimard [Coquille or Astrolabe Voy.], 1817–1829).
Type locality: "Madagascar."
Distribution: Madagascar (Antananarivo, Antsiranana, Nosy Be, S Toliara).

6. *Stenophis granuliceps* (Boettger, 1877). Abh. Senckenb. Naturf. Ges. 11: 14–16, pl. 1, fig. 3. (*Dipsas* [*Heterurus*] *gaimardi granuliceps*)

Synonym: *Stenophis capuroni* Domergue, 1995.
Type: Lectotype, SMF 19612 (formerly SMF-B 9041, a), a 1011 mm female (A. Stumpff, 1877), designated by Mertens (1922: 181),
Type locality: "Nossibé (= Nosy Bé), Madagascar" [= Nosy Be Is., N Antsiranana Prov., ext. N Madagascar] via lectotype selection.
Distribution: Northern Madagascar (Antsiranana, Nosy Be Is.), 400–800 m.

7. *Stenophis guentheri* Boulenger, 1896a. Cat. Snakes Brit. Mus. 3: 40–41, pl. 4, fig. 1.

Type: Holotype, BMNH 1946.1.2.53, a 435 mm female (Last).
Type locality: "S.W. Madagascar."
Distribution: Southern Madagascar (S Fianarantsoa, Toliara), 130–1345 m.

8. *Stenophis iarakaensis* Domergue, 1995. Arch. Inst. Pasteur Madagascar (1994) 61(2): 122.

Type: Holotype, MNHN 1988.320, a 295 mm juvenile.
Type locality: Madagascar. Emended to Iaraka, Madagascar [= Maroantsetra, N Toamasina Prov., NE Madagascar, 15°26'S, 49°44'E, elevation NSL] *fide* Glaw & Vences (1994: 346).
Distribution: Madagascar (ext. N Toamasina), NSL. Known only from holotype.

9. *Stenophis inopinae* Domergue, 1995. Arch. Inst. Pasteur Madagascar (1994) 61(2): 122.

Type: Holotype, MNHN 1950.160, a 620 mm specimen.
Type locality: Madagascar. Emended to Antongonbato, Madagascar *fide* Glaw & Vences (1994: 346).
Distribution: Extreme N Madagascar (N Antsiranana), 180–500 m.

10. *Stenophis inornatus* Boulenger, 1896a. Cat. Snakes Brit. Mus. 3: 42.

Type: Holotype, BMNH 1946.1.7.9, a 550 mm male (Last).
Type locality: "Madagascar."
Distribution: Extreme S Madagascar (S Toliara), 140 m.

11. *Stenophis pseudogranuliceps* Domergue, 1995. Arch. Inst. Pasteur Madagascar (1994) 61(2): 121–122.

Type: Holotype, MNHN 1978.42, a 733 mm specimen.
Type locality: "forêts caducifolées de la région ouest, entre l'Ankarafantsika et Toliara, Madagascar."
Distribution: Western and S Madagascar (Mahajanga, Toliara), 100–200 m.
Source: A. Mori et al., 2006.

12. *Stenophis variabilis* Boulenger, 1896a. Cat. Snakes Brit. Mus. 3: 43–44, pl. 4, figs. 3–3a.

Synonym: *Stenophis tulearensis* Domergue, 1995.
Types: Syntypes (2), BMNH 1946.1.7.10–11, a 1098–1100 mm male and a juvenile.
Type locality: "Madagascar."
Distribution: Western and S Madagascar (Antsiranana, Mahajanga, Toliara), NSL–800 m.
Sources: A. Mori et al., 2006 and Sajdak, 2010.

STENORRHINA A.-M.-C. Duméril, 1853 (Colubridae)

Synonyms: *Stenorhina* A.-M.-C. Duméril, Bibron & Duméril, 1854b *(nomen emendatum), Microphis* Hallowell, 1854b, *Bergenia* Steindachner, 1867a, *Sthenorhina* – Alvarez del Toro, 1972 *(nomen incorrectum).*
Type species: *Calamaria degenhardtii* Berthold, 1845.
Distribution: Latin America.
Sources: E.H. Taylor & Smith, 1939, H.M. Smith, 1943a, H.M. Smith & Taylor, 1945, E.H. Taylor, 1951, L.C. Stuart, 1963, J.A. Peters & Orejas-Miranda, 1970, J.C. Lee, 1980, 1996, L.D. Wilson & Meyer, 1985, Villa et al., 1988, Pérez-Higareda & Smith, 1991, J.A. Campbell, 1998b, G. Köhler, 1999b, Stafford & Meyer, 2000, Savage, 2002 and McCranie, 2011a.

1. *Stenorrhina degenhardtii* (Berthold, 1845). Nachr. Georg-Augusts-Univ. Königl. Ges. Wiss. Göttingen (3): 39–40, pl. 1, figs. 3–4. (*Calamaria degenhardtii*)

Synonyms: *Stenorrhina ventralis* A.-M.-C. Duméril, 1853 *(nomen nudum), Stenorhina ventralis* A.-M.-C. Duméril, Bibron & Duméril, 1854b, *Stenorhina kennicottiana* Cope, 1860c, *Bergenia mexicana* Steindachner, 1867a, and *Stenorhina degenhardti ocellata* Jan, 1876 *in* Jan & Sordelli, 1870–1881.
Type: Holotype, ZFMK 36828 (formerly ZIMG 44/256), a 200 mm specimen (C. Degenhardt, 1839–1845).
Type locality: "Provinz Popayan, Etwa 2°N.B. und 301° L., Colombien, Neu-Granada" [= W Colombia, probably on the Pacific versant *fide* C.W. Myers & Böhme, 1996: 17].
Distribution: Southeastern Mexico (Chiapas, Oaxaca, Veracruz), N Guatemala (Alta Verapaz, Péten), Belize (Cayo), El Salvador (Ahuachapán, Cuscatlán, La Libertad, Morazán, San Miguel, San Salvador, Santa Ana, Sonsonate, Usulután), W Honduras (Atlántida, Comayagua, Copán, Cortés, Francisco Morazán, Lempira, Ocotepeque, Olancho, Santa Bárbara, Yoro), Nicaragua (Chontales, Jinotega, Matagalpa), Costa Rica (Alajuela, Cartago, Guanacaste, Heredia, Limón, Puntarenas, San José), Panama (Canal Zone: Barro Colorado Is.; Chiriquí, Colón, Darién, Panamá, Veraguas), W Colombia (Antioquia, Atlántico, Bolívar, Boyacá, Cauca, Chocó, Cundinamarca, La Guajira, Nariño, Norte de Santander, Tolima, Valle), W Ecuador (Esmeraldas, Guayas, Manabí) and NW Venezuela (Aragua, Barinas, Carabobo, Distrito Federal, Falcón, Miranda,Táchira, Yaracuy), NSL–2410 (2750) m.
Sources: A.C.L.G. Günther, 1895 *in* 1885–1902, E.H. Taylor, 1954, Roze, 1966a, J.D. Johnson, 1976, Lancini., 1986, Pérez-Santos, 1986, Pérez-Santos & Moreno, 1988, 1991, Casas-Andreu et al., 1996, Stafford, 1996, J.C. Lee, 2000, Barrio-Amorgós & Calcaño, 2001, G. Köhler et al., 2005, Navarrete et al., 2009 and McCranie & Orellana, 2011.
Remarks: Supplemental original description in Berthold (1846: 8–9, pl. 1, figs. 3–4). See C.W. Myers & Böhme, 1996 for summary of Berthold papers. Type locality designation of near Popayán, Cauca Dept., Colombia, 2°27'N, 76°37'W, elevation 1800 m *fide* Dunn & Stuart (1951: 56) in error *fide* C.W. Myers & Böhme (1996: 17).

2. *Stenorrhina freminvillei* A.-M.-C. Duméril, Bibron & Duméril, 1854b. Erpét. Gén. 7(2): 868–869 & 1854d [Atlas]: pl. 70, figs. 1–2. (*Stenorhina freminvillei*) (*nomen corrigendum*)

Synonyms: *Stenorhina freminvillei* A.-M.-C. Duméril, 1853 *(nomen nudum), Stenorhina freminvillii* A.-M.-C. Duméril, Bibron & Duméril, 1854b *(nomen incorrigendum), Microphis quinqueliniatus* Hallowell, 1854b, *Stenorhina lactea* Cope, 1861e, *Stenorhina degenhardtii apiata* Cope, 1875a, and

Geophis multitorques yucatanicus T. Barbour & Cole, 1906.

Type: Holotype, MNHN 816, a 463 mm specimen (C.P. de la Poix Chevalier de Fréminville, 1824–1831).

Type locality: "Mexique" [= Mexico]. Restricted to Totalapán, Oaxaca Mexico *fide* H.M. Smith & Taylor (1950: 341).

Distribution: Latin America. Southern Mexico (Chiapas, Campeche, Guerrero, Michoacán, Oaxaca, Quintana Roo, Yucatán), Belize (Belize, Cayo, Corozal, Orange Walk), Guatemala (Petén), Honduras (Choluteca, Comayagua, Copán, Cortés, El Paraíso, Francisco Morazán, Gracias a Dios, Intibucá, Lempira, Santa Bárbara, Valle, Yoro), Nicaragua (Atlántico Norte, Chinandega, Granada, Madriz, Managua, Matagalpa, Rivas: Ometepe Is.), Costa Rica (Alajuela, E Cartago, Guanacaste, Limón, Puntarenas, NW San José), W Panama (Chiriquí) and Colombia (Boyacá), NSL–2200 m.

Sources: Duméril, Bocourt & Mocquard, 1870–1909, R.W. Henderson & Hoevers, 1975, Weyer, 1991, J.C. Lee, 2000, J.D. Johnson & Webb, 2001 and Hernández-Roz, 2005.

Remarks: This taxon was named in honor of Capt. Fréminville; since A.-M.-C. Duméril (1853: 490) and A.-M.-C. Duméril, Bibron & Duméril (1854c: 359) used the orthography of *freminvillei* the A.-M.-C. Duméril, Bibron & Duméril (1854b: 868) usage of *freminvillii* is considered a *nomen incorrigendum*. Probably occurs in Veracruz *fide* Pérez-Higareda & Smith (1991: 44).

STOLICZKIA Jerdon, 1870 (Xenodermatidae)

Synonyms: *Stoliczkaia* Boulenger, 1890a *(nomen praeoccupatum), Stolickaia* – Palacky, 1898 *(nomen incorrectum), Stolickaja* – Palacky, 1898 *(nomen incorrectum), Estoliczkaia* – Briceño-Rossi, 1934 *(nomen incorrectum), Stolzickia* – Taub, 1967 *(nomen incorrectum),* and *Stoliczkai* – Murthy & Pillai *in* Majupuria, 1986 *(nomen incorrectum).*

Type species: *Stoliczkia khasiensis* Jerdon, 1870.

Distribution: Southeastern Asia.

Source: Zaher et al., 2009.

S

1. *Stoliczkia borneensis* Boulenger, 1899c. Ann. Mag. Nat. Hist. (7) 4(24): 452. (*Stoliczkaia borneensis*)

Type: Holotype, BMNH 1946.1.15.58 (formerly RM), a 750 mm female (R. Hanitsch, March 1899).

Type locality: "Mount Kina Balu, 4200 feet, North Borneo" [= Mt. Kinabalu, N Sabah, NE East Malaysia, 6°05'N, 116°33'E, elevation 1280 m].

Distribution: East Malaysia (Sabah, Sarawak), 800–1800 m.

Sources: C. Haas, 1950, Stuebing, 1991, Malkmus et al., 2002 and I. Das, 2006, 2010–2012.

Remarks: Supplemental original descripation in Boulenger *in* Hanitsch (1900: 71–72, pl. 1, fig. 2).

2. *Stoliczkia khasiensis* Jerdon, 1870. Proc. Asiatic Soc. Bengal 1870(3): 81–82.

Type: Holotype, BMNH 1946.1.15.67, a 670–673 mm female (T.C. Jerdon, 1836–1852).

Type locality: "Khasi hills" [= Khasi Hills, E Meghalaya State, NE India, 25°35'N, 91°38'E, elevation 1370–1675 m].

Distribution: Northeastern India (Meghalaya), 1370–1675 m.

Sources: Boulenger, 1890a, Annandale, 1904, Wall, 1923d and M.A. Smith, 1938, 1943.

STORERIA Baird & Girard, 1853 (Natricidae)

Synonyms: *Ischnognathus* A.-M.-C. Duméril, Bibron & Duméril, 1854a, *Stoneria* – Wied-Neuwied, 1865 *(nomen incorrectum),* and *Hemigenius* Dugès, 1888a.

Type species: *Coluber dekayi* Holbrook, 1839.

Distribution: Eastern North America and Mesoamerica.

Fossil records: Middle Pliocene to upper Pleistocene of E USA. Fossils unassigned to species include middle Pliocene (Blancan II) of USA (Nebraska), middle Pleistocene (Irvingtonian II) of USA (Nebraska), middle/upper Pleistocene (Irvingtonian II) of USA (Kansas, Maryland, Texas, West Virginia), and upper Pleistocene (Rancholabrean II) of USA (Alabama, Missouri, West Virginia).

Sources: Trapido, 1944, H.M. Smith & Taylor, 1945, Conant, 1951, A.H. Wright & Wright, 1957, Ernst & Barbour, 1989, Rossman & Myers, 1990, Conant & Collins, 1991, Holman, 1995b, 2000a, Palmer & Braswell, 1995, Alfaro & Arnold, 2001, Zaher et al., 2009 and Ernst, 2012.

1. *Storeria dekayi* (Holbrook, 1839). No. Amer. Herp. 1 [Second version]: 121–122, pl. 25. (*Coluber dekayi*)

Synonyms: *Coluber dorsalis* De Kay *in* Holbrook, 1839 (*nomen nudum*), *Storeria tropica* Cope, 1885a, *Storeria dekayi anomala* Dugès, 1888a, *Storeria victa* Hay, 1892b, *Storeria dekayi temporalineata* Trapido, 1944, *Storeria dekayi texana* Trapido, 1944, *Storeria dekayi wrightorum* Trapido, 1944, and *Storeria dekayi limnetes* P.K. Anderson, 1961.

Type: Lectotype, ANSP 5832, a 382 mm female (C.F. Pickering, 1825–1826), designated by Trapido (1944: 47).

Type locality: "Massachusetts" via lectotype selection. Restricted to Cambridge, Massachusetts, USA *fide* K.P. Schmidt (1953a: 165).

Distribution: Eastern North America and Mesoamerica. Extreme SE Canada (SE Ontario, SW Quebec), E USA (Alabama, Arkansas, Connecticut, Delaware, District of Columbia, Florida: Big Pine, Little Torch, Middle Torch, No Name and Sugarloaf Is.; Georgia, Illinois, Indiana, Iowa, E Kansas, Kentucky, Louisiana, S Maine, Maryland, Massachusetts, Michigan, S Minnesota, Mississippi, Missouri, SE Nebraska, New Hampshire, New Jersey, New York: Long Island; North Carolina, Ohio, Oklahoma, Pennsylvania, Rhode Island, South Carolina, Tennessee, E Texas, Vermont, Virginia, West Virginia, Wisconsin), E Mexico (Chiapas, Hidalgo, Nuevo León, Oaxaca, Puebla, Tamaulipas, Veracruz), Guatemala (Alta Verapaz, Péten) and W Honduras (Cortés, Francisco Morazán, Yoro), NSL–2035 m.

Fossil records: Upper Pleistocene (Rancholabrean I, II) of USA (Florida, Georgia, Kansas, Pennsylvania, Tennessee, Texas, Virginia).

Sources: Ditmars, 1936, 1939, McCauley, 1945, L.C. Stuart, 1954a, Carr & Goin, 1955, P.K. Anderson, 1961, P.W. Smith, 1961, Fouquette & Rossman, 1963, Sabath & Sabath, 1969, Minton, 1972, Lazell, 1976, Martof et al., 1980, Christman, 1982, Tennant, 1984, L.D. Wilson & Meyer, 1985, Green & Pauley, 1987, Ashton & Ashton, 1988, Dundee & Rossman, 1989, Allen, 1992, Collins, 1993, J.C. Lee, 1996, Harding, 1997, Tennant, 1997, J.A. Campbell, 1998b, Fitch, 1999, Werler & Dixon, 2000 and McCranie, 2011a.

Remarks: Official Specific Name no. 1816 *fide* Opinion 632 (ICZN, 1962). Original description reprinted in Holbrook (1842b: 53–55, pl. 14). Original plate mislabeled as pl. 25 (correct in text as pl. 24). The specimen in Holbrook (1838: pl. 14), was collected by De Kay on Long Island, New York *fide* Adler (1976: xxxiii). Malnate (1971: 365) erroneously referred to ANSP 5832 as a holotype. Type locality restriction of north coast of Long Island *fide* Christman (1982: 2) invalid.

2. *Storeria hidalgoensis* E.H. Taylor, 1942. Herpetologica 2(2): 78.

Type: Holotype, UIMNH 25061 (formerly EHT-HMS 16145 & EHT-CC 3082), a 282 mm specimen (E.H. Taylor, 13 Aug. 1938).

Type locality: "near Zacualtipan, Hidalgo, Mexico" [= vicinity of Zacualtipan, Hidalgo State, Mexico, 20°39'N, 98°39'W, elevation 1990 m].

Distribution: Northeastern Mexico (Coahuila, Hidalgo, México, Nuevo León, Querétaro, San Luis Potosí, Tamaulipas), 200–3500 m.

Sources: Dixon & Lemos-Espinal, 2010 and Ramírez-Bautista et al., 2010.

3. *Storeria occipitomaculata* (Storer, 1839). Rept. Massachusetts: 230. (*Coluber occipitomaculatus*) (*nomen corrigendum*)

Synonyms: *Coluber leberis* Linnaeus, 1758 *(nomen rejiciendum), Coluber occipito-maculatus* Storer, 1839 *(nomen incorrigendum), Coluber venustus* Hallowell, 1847, *Storeria occipitomaculata* – S.W. Garman, 1884 *(nomen corrigendum), Storeria occipitomaculata obscura* Trapido, 1944, and *Storeria occipitomaculata pahasapae* H.M. Smith, 1963a.

Type: Holotype, not designated, a 267 mm specimen (Adams), lost *fide* Trapido (1944: 21).

Type locality: "Amherst, Massachusetts,"[USA].

Distribution: Eastern North America. Southeastern Canada (ext. SE Alberta, S Manitoba, New Brunswick, Nova Scotia, S Ontario, S Quebec, ext. SE Saskatchewan, Prince Edward Is.) and E USA (Alabama, Arkansas, Connecticut, Delaware, District of Columbia, N Florida, Georgia, ext. E Kansas (a relic population in W), Kentucky, Illinois, Indiana, Iowa, Louisiana, Maine, Maryland, Massachusetts, Michigan, Minnesota, Mississippi, Missouri, cen. Nebraska, New Hampshire, New Jersey, New York, North Carolina, E North Dakota, Ohio, E Oklahoma, Pennsylvania, Rhode Island, South Carolina, ext. W and ext. E South Dakota, Tennessee, ext. E Texas, Vermont, Virginia, West Virginia, Wisconsin, ext. NE Wyoming). NSL–1500 m.

Fossil records: Upper Pleistocene (Rancholabrean II) of USA (Pennsylvania, Virginia).

Sources: Hay, 1902, Fowler, 1907, Blanchard, 1937, Breckenridge, 1944, Bleakney, 1958, Logier & Toner, 1961, H.M. Smith, 1963a, Webb, 1970, Behler & King, 1979, Rossman & Erwin, 1980, Preston, 1982, De Graaf & Rudis, 1983, Ashton & Ashton, 1988, Ernst & Barbour, 1989, Klemens, 1991, Harding, 1997, Tennant, 1997, Fitch, 1999, Werler & Dixon, 2000 and Ernst, 2000c.

Remarks: Official Specific Name no. 1816 *fide* Opinion 632 (ICZN, 1962). Original description reprinted in Storer (1840: 33–34).

4. *Storeria storerioides* (Cope, 1865). Proc. Acad. Nat. Sci. Philadelphia 17(4): 190–191. (*Tropidoclonium storerioides*)

Synonym: *Hemigenius variabilis* Dugès, 1888b.

Types: Syntypes (4), USNM 24987, a 318–341 mm female, USNM 24988, a 308 mm male, USNM 24989, a 268 mm specimen, and USNM 24990, a 258 mm male (C. Sartorius).

Type locality: "Mexican plateau between the eastern Distribution and the valley of Mexico." Restricted to

Tres Cumbres, Morelos, Mexico *fide* H.M. Smith & Taylor (1950a: 336).

Distribution: Central Mexico (SW Chihuahua, SW Coahuila, Distrito Federal, Durango, Guanajuato, Guerrero, Hidalgo, Jalisco, México, Michoacán, Morelos, Oaxaca, Querétaro, San Luis Potosí, Zacatecas), 1200–4000 m. Probably occurs in Aguascalientes, Mexico *fide* McCranie & Wilson, 2001b.

Sources: A.H.A. Duméril, Bibron & Mocquard, 1870-1909, S.W. Garman, 1884, A.C.L.G. Günther, 1894 *in* 1885–1902, E.H. Taylor & Smith, 1939, H.M. Smith, 1944a, W.W. Tanner & Robison, 1960, P.K. Anderson, 1961, Duellman, 1961, Hardy & McDiarmid, 1969, W.W. Tanner, 1985, Flores-Villela, 1993, Casas-Andreu et al., 1996, Reyes-Velasa et al., 2008, Dixon & Lemos-Espinal, 2010 and Ramírez-Bautista et al., 2010.

SUTA Worrell, 1961c (Elapidae)

Type species: *Hoplocephalus sutus* W.C.H. Peters, 1863a.

Distribution: Australia.

Sources: Cogger et al., 1983a, Mengden, 1985a, Schwaner et al., 1985, Shine, 1985a, 1988, Wallach, 1985, Longmore, 1986, Storr et al., 1986, S.K. Wilson & Knowles, 1988, Gow, 1989, Hutchinson, 1990, Cogger, 1992, Keogh et al., 1998, Sanders et al., 2008, Zaher et al., 2009 and Hoser, 2012e.

1. *Suta fasciata* (Rosén, 1905). Ann. Mag. Nat. Hist. (7) 15(86): 179. (*Denisonia fasciata*)

Type: Lectotype, ZMUL 129, longest syntype 410 mm (N. Holst, 1897), designated by Wells & Wellington (1985: 45).

Type locality: "West Australia" [= Western Australia, Australia].

Distribution: Western Australia (W Western Australia), NSL–790 m.

Source: L.A. Smith, 1980.

S

2. *Suta ordensis* (Storr, 1984a). Rec. West. Aust. Mus. 11(3): 254–256. (*Denisonia ordensis*)

Type: Holotype, WAM 58877, a 385–662 mm (svl) specimen (W.H. Butler, 9 Jan. 1972).

Type locality: "Argyle Downs, W.A., in 16°17'S, 128°47'E."

Distribution: Northern Australia (ext. NW Northern Territory, ext. NE Western Australia), 75–395 m.

3. *Suta punctata* (Boulenger, 1896a). Cat. Snakes Brit. Mus. 3: 341–342, pl. 18, fig. 4. (*Denisonia punctata*)

Synonym: *Rhinoplocephalus punctatus* – Storr, 1984a.

Type: Holotype, BMNH 1946.1.20.7, a 350 mm female (Beckett).

Type locality: "Port Walcott, north-western Australia" [= Port Walcott, W Western Australia, W Australia, 20°35'S, 117°11'E, elevation NSL].

Distribution: Northwestern Australia (Northern Territory, W Queensland, N Western Australia), NSL–760 m.

Sources: L.A. Smith, 1980, Shine, 1983b and Covacevich, 1991.

4. *Suta suta* (W.C.H. Peters, 1863a). Mber. Königl. Akad. Wiss. Berlin 1863(4): 234. (*Hoplocephalus sutus*)

Synonyms: *Hoplocephalus frenatus* W.C.H. Peters, 1870b, *Hoplocephalus carpentariae* Macleay, 1887a, *Hoplocephalus frontalis* Ogilby, 1890b, *Hoplocephalus stirlingi* Lucas & Frost, 1896, *Denisonia frontalis propinqua* De Vis, 1905, and *Denisonia forresti* Boulenger, 1906c.

Type: Holotype, ZMB 4715, a 325 mm specimen (R.H. Schomburgk, 1849–1863).

Type locality: "aus Buchsfelde bei Adelaide in Südaustralien" [= vicinity of Adelaide (34°56'S, 138°36'E, elevation 45 m), SE South Australia, S Australia]. Emended to Loos, near Gawler [34°36'S 138°45'E, elevation 55 m], South Australia *fide* Bauer et al. (1995: 76).

Distribution: Australia (New South Wales, Northern Territory, Queensland, South Australia, N Victoria, E Western Australia), NSL–895 m.

Sources: Coventry, 1971, L.A. Smith, 1978, Shine, 1983b, Storr, 1984a, Hoser, 1989 and Covacevich, 1991.

SYMPHIMUS Cope, 1870 (Colubridae)

Synonyms: *Symphymus* – Hoffmann, 1890 *(nomen incorrectum),* and *Symphinus* F. Werner, 1929a *(nomen emendatum).*

Type species: *Symphimus leucostomus* Cope, 1870.

Distribution: Southern Mexico.

Sources: Rossman & Schaefer, 1974, Cundall, 1981 and G. Köhler, 2001, 2008.

1. *Symphimus leucostomus* Cope, 1870. Proc. Amer. Philos. Soc. (1869–1870) 11(81): 150.

Type: Lectotype, USNM 30310 (formerly SMI 240), a 799–803 mm male (A.L.J.F. Sumichrast, 1855–1869), designated by H.M. Smith & Taylor (1945: 133).

Type locality: "Chihuitan, Tehuantepec, Western Mexico" [= Santo Domingo Chihuitán, SE Oaxaca, SE Mexico, 16°35'N, 95°10'W] via lectotype selection.

Distribution: Southern Mexico (W Chiapas, Colima, Jalisco, Michoacán, E Oaxaca), 200–1000 m.
Sources: Hartweg & Oliver, 1940, Casas-Andreu et al., 1996 and Reyes-Velasco et al., 2009.

2. *Symphimus mayae* (Gaige, 1936). Publ. Carnegie Inst. (457): 300–301. (*Eurypholis mayae*)

Type: Holotype, UMMZ 73082, a 314 mm specimen (A.S. Pearse & E. Creaser, 3 July 1932).
Type locality: "Dzitas-Chichen Itza road, Dzitas, Yucatan, Mexico" [= between Dzitás (20°51'N, 88°31'W) and Chichén Itzá (20°40'N, 88°34'W), E Yucatán State, ext. SE Mexico]. Emended to NE Dzitás in UMMZ catalogue *fide* Kluge (984: 57).
Distribution: Yucatán Peninsula. Southeastern Mexico (Quintana Roo, Yucatán) and Belize (Belize, Orange Walk), NSL–300 m.
Sources: J.C. Lee, 1980, 1996, 2000, J.A. Campbell, 1998b and Stafford & Meyer, 2000.
Remarks: Collection date of type 13 July 1932 *fide* Kluge (1984: 57).

SYMPHOLIS Cope, 1862a (Colubridae)

Synonyms: *Cheilorhina* Filippi *in* Jan, 1862b, and *Chilorhina* Cope, 1887c *(nomen emendatum)*.
Type species: *Sympholis lippiens* Cope, 1862a.
Distribution: Western Mexico.
Sources: H.M. Smith & Taylor, 1945, Humphrey & Shannon, 1958, Zweifel, 1959b, Hensley, 1966, Hardy & McDiarmid, 1969, W.W. Tanner, 1985 and Lemos-Espinal et al., 2004c.

1. *Sympholis lippiens* Cope, 1862a. Proc. Acad. Nat. Sci. Philadelphia (1861) 13: 524.

Synonyms: *Cheilorhina villarsii* Jan, 1862b, and *Sympholis lippiens rectilimbus* Hensley, 1966.
Types: Syntypes (3), USNM 6092, and USNM 31345–46, two females, longest syntype 527 mm (J.J. Major), 6092 lost *fide* Zweifel (1959b: 3).
Type locality: "Guadalaxara, Mexico" [= Guadalajara, N-cen. Jalisco, mid-western Mexico, 20°40'N, 103°20'W].
Distribution: Western Mexico (ext. SW Chihuahua, cen. Jalisco, Nayarit, Sinaloa, S Sonora), 435–915 (1585) m.

SYNOPHIS Peracca, 1896d (Dipsadidae)

Synonym: *Sinophis* – Briceño-Rossi, 1934 *(nomen incorrectum)*.
Type species: *Synophis bicolor* Peracca, 1896d.
Distribution: Colombia and Ecuador.
Sources: Nicéforo-Maria, 1950, Bogert, 1964, J.A. Peters & Orejos-Miranda, 1970, Hillis, 1990, Pérez-Santos & Moreno, 1991, Sheil, 1998, Sheil & Grant, 2001 and Zaher et al., 2009.

1. *Synophis bicolor* Peracca, 1896d. Boll. Mus. Zool. Anat. Comp. Univ. Torino 11(266): 1–2, 1 fig.

Type: Holotype, MZUT 257, a juvenile (1887).
Type locality: "America meridionale" [= South America].
Distribution: Colombia (Antioquia) and N Ecuador (Chimborazo, Cotopaxi, Pastaza, Pichincha), 1200–2000 m.
Sources: Nicéforo-Miria, 1970 and Sheil & Grant, 2001.
Remarks: Possibly two invariable allopatric species *fide* Sheil & Grant (2001: 208).

2. *Synophis calamitus* Hillis, 1990. Occ. Pap. Mus. Nat. Hist., Univ. Kansas (135): 2–4, fig. 1.

Type: Holotype, KU 197107, a 223 mm female (D.M. Hillis, 2 April 1984).
Type locality: "4 km SE Tandayapa, Pichincha Province, Ecuador, elevation 1890 m" [= 4 km SE Tandayapa (0°01'S, 78°46'W), NE Pichincha Prov., NW Ecuador, elevation 1890 m].
Distribution: Northwestern Ecuador (Pichincha), 1890–2150 m.

3. *Synophis lasallei* (Nicéforo-Maria, 1950). Rev. Acad. Colomb. Cienc. Exec. Fis. Nat. 7(28): 517, figs. 1–2, pl. (*Diaphorolepis lasallei*)

Type: Holotype, MLS (formerly CSJP), a 470 mm male.
Type locality: "Paraje situado al norte de albán, Cundinamarca, Cordillera Oriental (vertiente occidental), a unos 2.200 metros de altitud y aproximadamente a 60 kilómetros al noroeste de Bogotá" [= N of Albán (4°53N, 74°27'W), cen. Cundinamarca Dept., cen. Colombia, elevation 2200 m].
Distribution: Central Colombia (Cundinamarca, Meta) and E Ecuador (Morona-Santiago, Napo, Pastaza), 460–2200 m.
Sources: J.A. Peters, 1960a and Pérez-Santos & Moreno, 1988.
Remarks: Bogert (1964: 518) and Pérez-Santos & Moreno (1988: 320) listed type locality as northwest of Albán.

4. *Synophis plectovertebralis* Sheil & Grant, 2001. J. Herp. 35(2): 204–207, figs. 1–2a–b.

Type: Holotype, UV 11858, a 272.3 mm female (L.A. Ortega, 31 March 1995).
Type locality: "Hacienda San Pedro, about 6 km south El Queremal, Municipio Dagua, Departamento del Valle del Cauca, Colombia, 03°29'N, 76°42'W,

at 1800 m on the Pacific versant of the Cordillera Occidental."

Distribution: Western Colombia (Valle del Cauca), 1800 m. Known only from type locality.

†*SZYNDLARIA* Rage & Augé, 2010 (Tropidophiidae)

Type species: †*Szyndlaria aureomontensis* Rage & Augé, 2010.

Distribution: Middle Eocene of France.

Source: Rage & Augé. 2010.

1. †*Szyndlaria aureomontensis* Rage & Augé, 2010. Géobios 43: 260–262, fig. 3, no. 1.

Type: Holotype, MHNL Li 331, one middle trunk vertebra.

Type locality: "near the village of Lissieu ("department" or administrative region, of Rhône), central eastern France; MP 14, middle Eocene."

Distribution: Middle Eocene (Lutetian, MP 14: 37.2–42.7 or 48.6 mya) of France. Known only from type locality.

Remarks: Type locality no longer exists as the fissure was mined out by end of the 19th Century *fide* Rage & Augé (2010: 254).

T

TACHYMENIS Wiegmann, 1834 (Xenodontidae)

Synonyms: *Tachymenes* – Girard, 1855 (*nomen incorrectum*), *Zacholomorphus* Fitzinger, 1861, *Tachydermis* – A.C.L.G. Günther, 1881 (*lapsus calami*), *Colubellus* Philippi, 1899 (*nomen nudum*), *Aspidula* Philippi *in* Quijada, 1916, *Colubellus* Philippi *in* Quijada, 1916, Trachymenis– Escomel, 1929 (*nomen incorrectum*), *Tachymensis* – Maass-Berlin, 1933 (*nomen incorrectum*), *Tachymeris* – Neave, 1940b (*nomen incorrectum*), *Tachimenus* – Vuletin, 1960 (*nomen incorrectum*), and *Tachymenys* – Shine *in* Cogger & Zweifel, 1992 (*nomen incorrectum*).

Type species: *Tachymenis peruviana* Wiegmann, 1834a.

Distribution: Southern South America.

Sources: W. Walker, 1945, Donoso-Barros, 1966, J.A. Peters & Orejas-Miranda, 1970, Zaher et al., 2009 and Vidal et al., 2010.

Remarks: Congeneric with *Thamnodynastes fide* Bailey & Thomas, 1991.

1. *Tachymenis affinis* Boulenger, 1896a. Cat. Snakes Brit. Mus. 3: 119, pl. 7, fig. 1.

Type: Holotype, BMNH 1946.1.2.80, a 520 mm male (W. Davis).

Type locality: "Muña, Peru." Restricted to the upper Río Huallaga, Huanuco, Peru *fide* W. Walker (1945: 22).

Distribution: Southern Peru (Cuzco, Huanuco), 2000–2500 m.

2. *Tachymenis attenuata* W. Walker, 1945. Bull. Mus. Comp. Zool. 96(1): 24–25, pl. 2, fig. 3, pl. 4, figs. 13–15.

Synonym: *Tachymenis attenuata boliviana* W. Walker, 1945.

Type: Holotype, FMNH 40071 (formerly UADZ), a 582 mm male (J. Perea).

Type locality: "Peru." Restricted to Madre de Dios Dept. *fide* W. Walker (1945: 24).

Distribution: Southern Peru (Madre de Dios) and Bolivia (Cochabamba, La Paz), 2500–4115 m.

Source: Harvey & Muñoz, 2004.

3. *Tachymenis chilensis* (Schlegel, 1837). Essai Phys. Serp. 1: 136, 2: 70. (*Coronella chilensis*)

Synonyms: *Zacholomorphus brevicaudatus* Fitzinger, 1861, *Zaccholomorphus chilensis* Fitzinger, 1861, *Tachymenis peruviana catenata* F. Werner, 1898c, *Tachymenis peruviana coronellina* F. Werner, 1898c, *Tachymenis peruviana vittata* F. Werner, 1898c, *Coronella leucognatha* Philippi, 1899 (*nomen nudum*), *Aspidula celeris* Philippi *in* Quijada, 1916 (*nomen nudum*), *Aspidula vittata* Philippi *in* Quijada, 1916 (*nomen nudum*), *Coronella atrovittata* Philippi *in* Quijada, 1916 (*nomen nudum*), *Coronella campestris* Philippi *in* Quijada, 1916 (*nomen nudum*), *Coronella concolor* Philippi *in* Quijada, 1916 (*nomen nudum*), *Coronella lepida* Philippi *in* Quijada, 1916 (*nomen nudum*), *Coronella melanogastra* Philippi *in* Quijada, 1916 (*nomen nudum*), *Coronella plumbea* Philippi *in* Quijada, 1916 (*nomen nudum*), and *Coronella montana* Philippi *in* Quijada, 1916 (*nomen nudum*).

Types: Syntypes (2), RMNH 150a–b (formerly MNHPB), a 450 mm and 388 mm specimen (R.-P. Lesson & P. Garnot [Coquille Voy.], Aug. 1822–March 1825).

Type locality: "Chilé."

Distribution: Central Chile (Aconcagua, Arauco, Atacama, Bio-Bio, Cautin, Chiloe, Colchagua, Concepción, Coquimbo, Llanquihue, Malleco, Nuble, Osorno, Santiago, Talca, Valdivia) and cen. Argentina (Chubut, Neuquén, Río Negro), NSL–2000 m.

Sources: Donoso-Barros, 1966, Ortiz, 1973, Troncoso & Ortiz, 1987, J.D. Williams & Francini, 1991, Valenzuela-Dellarossa et al., 2010 and Avila et al., 2012.

Remarks: The subspecies *chilensis* and *coronellina* possibly sympatric and possibly represent separate species *fide* Valenzuela-Dellarossa et al., 2010.

4. *Tachymenis elongata* Despax, 1910. Bull. Mus. Hist. Nat., Paris 16(7): 373–374.

Type: Holotype, MNHN 1906.254, a 450 mm specimen (P. Rivet, 1902–1906).

Type locality: "Tablazo de Paita (Pérou)" [Dept. Piura, Peru]. Corrected to Tablazo de Payta (30 m) *fide* Despax (1911: 34).

Distribution: Coastal Peru (Ancash, Lima, Piura), 30–155 m.

Source: Carrillo de Espinoza & Icochea, 1995.

Remarks: Original description reprinted in Despax (1911: 33–35, pl. 1, figs. 2–2a).

5. *Tachymenis peruviana* Wiegmann, 1834a. Nova Acta Acad. Caesar. Leop.-Carol. 17(1): 252–253, pl. 20, fig. 1.

Synonyms: *Ophis peruana* Tschudi, 1945 (*nomen nudum*), *Psammophylax assimilis* Jan, 1863a, *Tachymenis peruviana dorsalis* F. Werner, 1901c, *Leimadophis andicolus* T. Barbour, 1915b, *Tachymenis chilensis melanura* W. Walker, 1945, and *Tachymenis peruviana yutoensis* Miranda & Couturier, 1981.
Type: Holotype, ZMB, probably a female (F.J.F. Meyen, 1830–1832).
Type locality: Not stated. "Peru" by implication. Restricted to Puno, Puno Dept., Peru *fide* W. Walker (1945: 15).
Distribution: Peru (Ancash, Arequipa, Ayachucho, Cuzco, Huancavelica, Huanuco, Junin, La Libertad, Lima, Madre de Dios, Moquegua, Puno, Tacana), Bolivia (Cochabamba, La Paz, Oruro, Potosi, Santa Cruz), NW Argentina (Jujuy) and Chile (Aconcaqua, Antofagasta, Atacama, Coquimbo, Santiago, Tarapaca), 900–4250 m.
Sources: Carrillo de Espinoza, 1970, Ortiz, 1973, Pefaur et al., 1978, Henle & Ehrl, 1991, J.D. Williams & Francini, 1991, Cei, 1994, Lehr & G. Köhler, 2000, Harvey & Muñoz, 2004 and Valenzuela-Dellarossa et al., 2010.

6. *Tachymenis tarmensis* W. Walker, 1945. Bull. Mus. Comp. Zool. 96(1): 21–22, pl. 4, figs. 10–11.

Type: Holotype, FMNH 5698, a 393 mm female (M.P. Anderson, 19 May 1914).
Type locality: "Tarma in the departament of Junin, Peru."
Distribution: Central Peru (Junin). Known only from type locality.

TAENIOPHALLUS Cope, 1895c (Xenodontidae)

Type species: *Lygophis nicagus* Cope, 1868c.
Distribution: South America.
Sources: C.W. Myers, 1974, C.W. Myers & Cadle, 1994, Schargel et al., 2005, Zaher et al., 2009 and Vidal et al., 2010.

1. *Taeniophallus affinis* (A.C.L.G. Günther, 1858). Cat. Colub. Snakes Brit. Mus.: 128–129. (*Dromicus affinis*)

Synonyms: *Enicognathus melanocephalus* A.-M.-C. Duméril, Bibron & Duméril, 1854a (*partim*), and *Coronella iheringii* Boulenger, 1885a.
Type: Lectotype, BMNH 1946.1.5.80, a juvenile male (G. Busk), designated by C.W. Myers (1974: 197).
Type locality: "Rio de Janeiro, Brazil" [= Rio de Janeiro, Rio de Janeiro State, SE Brazil, 22°54'S, 43°14'W, elevation NSL] via lectotype selection.
Distribution: Southeastern Brazil (Alagoas, Bahia, Ceará, Espírito Santo, SE Minas Gerais, Paraná, Rio de Janeiro, N and NE Rio Grande do Sul, Santa Catarina, São Paulo), NSL–940 m.
Sources: Prado, 1943, Di-Bernardo & Lema, 1988, Argólo, 1998b, Freire & Silva, 2000, Loebmann, 2008a and Paula et al., 2011.
Remarks: Head and body pattern of holotype illustrated in C.W. Myers (1974: figs. 45f (upper), 46a).

2. *Taeniophallus bilineatus* (J.G. Fischer, 1885b). Jahrb. Hamburg. Wiss. Anst. 2: 98–99, pl. 3, figs. 5a–c. (*Enicognathus bilineatus*)

Type: Holotype, BMNH 1946.1.5.77 (formerly JGF 858 & BMNH 1886.5.15.19), a 325 mm male (commercial dealer).
Type locality: "Santos" [Brazil] [= Santos, E São Paulo State, SE Brazil, 23°57'S, 46°20'W, elevation 15 m].
Distribution: Southeastern Brazil (Espírito Santo, S Minas Gerais, Paraná, S Rio de Janeiro, NE Rio Grande Do Sul, Santa Catarina, São Paulo, Cardoso Is.), 15–1075 m.
Sources: Di-Bernardo & Lima, 1991, Hiert, 2006 and Cicchi et al., 2007.

3. *Taeniophallus brevirostris* (W.C.H. Peters, 1863c). Mber. Königl. Akad. Wiss. Berlin 1863(6): 280–281. (*Dromicus brevirostris*)

Synonyms: *Enicognathus taeniolatus* Jan, 1863a, and *Dromicus viperinus* A.C.L.G. Günther, 1868.
Type: Holotype, ZMB 3869, a 315 mm specimen (purchased), lost *fide* G. Peters in C.W. Myers (1974: 202) but extant *fide* Bauer et al. (1995: 71).
Type locality: "Quito" [Ecuador] (in error). Probably lowland locality in Ecuador *fide* J.A. Peters (1955: 347), but considered unknown *fide* C.W. Myers (1974: 202).
Distribution: Northern South America. Southern Colombia (Amazonas, Putumayo, Vaupés), ext. W Venezuela (Zulia), Guyana, French Guiana (Cayenne), Suriname (Brokopondo, Para), E Ecuador (Chimborazo, Morona-Santiago, Napo, Pastaza, Sucumbios, Tungurahua, Zamora-Chinchipa), E Peru (Amazonas, Cuzco, Huánuco, Loreto, Madre de Dios, San Martín), N Brazil (Acre, Amapá, Amazonas, Pará, Rondônia) and N Bolivia (Beni, La Paz), 30–1500 (2000) m.
Sources: Boettger, 1888b, Shreve, 1934, H.W. Parker, 1935, Hoge & Belluomini, 1960, Cunha & Nascimento, 1978, 1993, Duellman, 1978, Hoogmoed, 1979, Gasc & Rodrigues, 1980, Abuys, 1986a, Lancini, 1986, Pérez-Santos, 1986, Chippaux, 1987, Pérez-Santos & Moreno, 1988, 1991, Duellman & Mendelson, 1995, Starace, 1998 and Yuki, 1999.

4. *Taeniophallus nebularis* Schargel, Rivas-Fuenmayor & Myers, 2005. Amer. Mus. Novit. (3484): 4–11, figs. 1–6.

Type: Holotype, MHNLS 15743, a 492 mm male (G. Rivas, C.L. Barrio-Amorgós, & W.S. Schargel, 2 June 2002).
Type locality: "trail between Las Melenas and Cerro Humo, 800 m elevation (10°39'N, 62°34'W), Peninsula de Paria, Estado Sucre, Venezuela."
Distribution: Northeastern Venezuela (Sucre), 800 m. Known only from type locality.
Source: Navarrete et al., 2009.

5. *Taeniophallus nicagus* (Cope, 1868c). Proc. Acad. Nat. Sci. Philadelphia 20(2): 132–133. (*Lygophis nicagus*)

Type: Holotype, MNHN 55, a 367 mm male.
Type locality: "Neotropical" [region]. Brazil *fide* Jan and Sordelli (1866 *in* 1860–1866: 11).
Distribution: Northern South America. Suriname, Guyana and N Brazil (Amapá).
Remarks: Holotype illustrated in Jan & Sordelli, 1866 *in* 1860–1866(16): pl. 1, fig. 4.

6. *Taeniophallus occipitalis* (Jan, 1863a). Arch. Zool. Anat. Fis. 2(2): 266–267. (*Enicognathus occipitalis*)

Synonyms: *Dromicus wuchereri* A.C.L.G. Günther, 1863f, and *Dromicus miolepis* Boettger, 1891.
Type: Lectotype, ZMH 4423, a 340 mm specimen (Schramm, 1861), designated herein.
Type locality: "Bahia" [= Bahia State, NE Brazil] via lectotype selection.
Distribution: South America. Southeastern Colombia (Amazonas), NE Peru (Loreto, Madre de Dios), E Brazil (Bahia, Ceará, Distrito Federal, Goiás, Pará, Paraiba, Paraná, Piauí, Rio Grande do Sul, Rondônia, São Paulo, Sergipe), Bolivia (Beni, La Paz, Santa Cruz, Tarija), Paraguay (Alto Paraguay, Boquerón, Central, Concepción, Neembucú, Presidente Hayes), Uruguay (Artigas, Cerro Largo, Paysandú, Rivera, Salto, Treinta y Tres) and N Argentina (Córdoba, Corrientes, Entre Ríos, Formosa, Jujuy, Misiones, Salta, Santiago del Estero, Tucumán), NSL–600 m.
Sources: Bertoni, 1913, Devincenzi, 1925, Amaral, 1934b, Prado, 1946a, Vaz-Ferreira & Sierra de Soriano, 1960, Achaval, 1976, Lema & Fabián-Beurmann, 1977, Dixon & Soini, 1986, Fugler, 1986, Pérez-Santos, 1986, Pérez-Santos & Moreno, 1988, Duellman & Salas, 1991, J.D. Williams & Francini, 1991, Cunha & Nascimento, 1993, Reati, 1996, Yanosky et al., 1996, Fugler et al., MS, Leynaud & Bucher, 1999, Nogueira, 2001, Giraudo & Scrocchi, 2002, A.P. Santos & Lema, 2004, Carreira-Vidal et al., 2005, Duellman, 2005, Cacciali, 2008, Santos, et al., 2008, Morato et al., 2011 and Prigioni et al., 2011.
Remarks: ZMH 4423, which was illustrated in Jan & Sordelli, 1866 *in* 1860–1866(16): pl. 1, fig. 1, would "make the best lectotype" *fide* C.W. Myers (1974: 209).

7. *Taeniophallus persimilis* (Cope, 1869b). Proc. Acad. Nat. Sci. Philadelphia (1868) 20: 308. (*Liophis persimilis*)

Synonyms: *Liophis insignissimus* Amaral, 1926a, and *Rhadinaea beui* Prado, 1944b.
Type: Holotype, MCZ 436, a 286 mm specimen (J.L.R. Agassiz, April 1865–June 1866), lost *fide* T. Barbour & Loveridge (1929a: 297) and E.E. Williams *in* C.W. Myers (1974: 212).
Type locality: "Rio de Janeiro, Brazil" [Rio de Janeiro, Rio de Janeiro State, SE Brazil, 22°54'S, 43°14'W, elevation NSL].
Distribution: Southeastern Brazil (Espírito Santo, Paraná, Rio de Janeiro, Santa Catarina, São Paulo).
Source: Di-Bernardo & Lema, 1986.
Remarks: Possibly occurs in NE Rio Grande do Sul *fide* Di-Bernardo & Lema (1986: 114).

8. *Taeniophallus poecilopogon* (Cope, 1863b). Proc. Acad. Nat. Sci. Philadelphia 15(3): 100–101. (*Rhadinaea poecilopogon*)

Synonyms: *Enicognathus elegans* Jan, 1863a (*nomen oblitum*), and *Dromicus melanocephalus* W.C.H. Peters, 1863.
Type: Holotype, USNM 31278, a 1008 mm male (H.W. Kennedy).
Type locality: "Paysondu, Uraguay" [= Paysandú, ext. SW Paysandú Dept., W Uruguay, 32°19'S, 58°05'W, elevation 25 m].
Distribution: Southeastern South America. Southern Brazil (S Mato Grosso, S Minas Gerais, Paraná, Rio Grande do Sul, Santa Catarina, São Paulo), Uruguay (Cerro Largo, Montevideo, Paysandú, Rocha, Treinta y Tres), S Paraguay (Asunción) and ext. NE Argentina (Buenos Aires, Corrientes, Entre Ríos, La Paz, Misiones), 25–65 m.
Sources: Di-Bernardo & Lema, 1987, J.D. Williams & Francini, 1991, Carreira-Vidal et al., 2005, Etchepare & Zaracho, 2009, Nenda & Di Pietro, 2009 and Falcione et al., 2010.
Remarks: In accordance with Art. 23.9.2 of the Code (ICZN, 1999), *Rhadinaea poecilopogon* Cope is designated a *nomen protectum* and *Enicognathus elegans* Jan a *nomen oblitum*.

9. *Taeniophallus quadriocellatus* Santos, Di-Bernardo & Lema, 2008. J. Herp. 42(3): 420–423, figs. 1–2.

Type: Holotype, MPEG 20144, a 434 mm male (E.F. Santos, 11 Feb. 2002).

Type locality: "Floresta Nacional de Caxiulanã, Estação Cientifica Ferreira Penna (01°42'30"S and 51°31'45"W), Melgaço municipality, Pará, Brazil."
Distribution: Northeastern Brazil (Pará, Marajó Is.).

†*TALLAHATTAOPHIS* Holman & Case, 1988 (Boidae)

Type species: †*Tallahattaophis dunni* Holman & Case, 1988.
Distribution: Lower Eocene of USA.
Sources: Holman & Case, 1988 and Holman, 2000a.

1. †*Tallahattaophis dunni* Holman & Case, 1988. J. Vert. Paleont. 8(3): 329–331, figs. 3a–e.

Type: Holotype, MSUVP 1189, one trunk vertebra (W.E. Dunn & G. Ziegenhine).
Type locality: "Point 'A' Dam Site, about 123–182 m south of the McWilliams Power Plant at the Dam, Conecuh River, 1.6 km west of Andalusia, Covington County, Alabama, NW1/4, SW1/4, Sec. 35, T5N, R15E. (Early Middle Eocene (Bridgerian); Claiborne Group, Tallahatta formation."
Distribution: Lower Eocene (Bridgerian: 46.2–50.3 mya) of USA (Alabama). Known only from type locality.

TANTALOPHIS Duellman, 1958b (Dipsadidae)

Type species: *Leptodeira discolor* A.C.L.G. Günther, 1860g.
Distribution: Southern Mexico.
Sources: A.C.L.G. Günther, 1885–1902, A.H.A. Duméril, Bocourt & Mocquard, 1870–1909, E.H. Taylor, 1938, H.M. Smith & Taylor, 1945, Duellman, 1958a–b, C.W. Myers & Campbell, 1981, Casas-Andreu et al., 1996, Zaher et al., 2009 and Mulcahy et al., 2011.

1. *Tantalophis discolor* (A.C.L.G. Günther, 1860g). Proc. Zool. Soc. London 28(1): 317–318. (*Leptodeira discolor*)

Types: Syntypes (2), BMNH 1946.1.23.67–68, a 535 mm and 509–536 mm and male (A. Sallé, 1846–1856).
Type locality: "Oaxaca (Mexico)."
Distribution: Southern Mexico (Sierra Madre del Sur of W Oaxaca), 2400–2520 m.

TANTILLA Baird & Girard, 1853 (Colubridae)

Synonyms: *Homalocranion* A.-M.-C. Duméril, 1853, *Lioninia* Hallowell, 1861, *Homalocranium* A.C.L.G. Günther, 1863c (*nomen emendatum*), *Homolocranium* – Dugès, 1869 (*nomen incorrectum*), *Tortilla* – Westphal-Castelnau, 1870 (*nomen incorrectum*), *Microdromus* A.C.L.G. Günther, 1872b, *Homulocranion* – Locard, 1875 (*nomen incorrectum*), *Homolacranion* – Velasco, 1890b (*nomen incorrectum*), *Pogonaspis* Cope, 1894d, *Hamalocranion* – E.B. Branson, 1904 (*nomen incorrectum*), *Pagonaspis* – Amaral, 1930f (*nomen incorrectum*), and *Tortix* – Marcus, 1937 (*nomen incorrectum*).
Type species: *Tantilla coronata* Baird & Girard, 1853.
Distribution: Southern USA and Latin America.
Fossil records: Lower Pleistocene (Irvingtonian I) of USA (Florida), middle/upper Pleistocene (Irvingtonian II) of USA (Texas), and upper Pleistocene (Rancholabrean I, II) of USA (California, Kansas, New Mexico, Texas).
Sources: F. Werner, 1924c, H.M. Smith & Taylor, 1945, J.A. Peters & Orejas-Miranda, 1970, L.D. Wilson, 1982a–b, 1999, Villa et al., 1988, Flores-Villela, 1993, L.D. Wilson & McCranie, 1999 and Holman, 2000a.

1. *Tantilla albiceps* T. Barbour, 1925. Occ. Pap. Boston Soc. Nat. Hist. (1921–1931) (5): 156.

Type: Holotype, MCZ 20600, a 212 mm male (Indian boy via T. Barbour, Feb. 1925).
Type locality: "near the laboratory building, Barro Colorado Island, Gatun Lake, Canal Zone of Panama."
Distribution: Central Panama (Canal Zone), 100–160 m.
Source: L.D. Wilson, 1985g.

2. *Tantilla alticola* (Boulenger, 1903b). Ann. Mag. Nat. Hist. (7) 12(69): 353. (*Homalocranium alticola*)

Synonyms: *Homalocranium coralliventre* Boulenger, 1913d, and *Tantilla costaricensis* E.H. Taylor, 1954.
Types: Syntypes (3), BMNH 1946.1.8.63–65, (formerly BMNH 1898.10.27.7–9), first two males, third a juvenile female, longest syntype 330 mm (A.E. Pratt, 1891–1901).
Type locality: "Santa Rita, north of Medellin, Colombia, altitude 9000 feet" [= Antioquia Dept.].
Distribution: Lower Central America and ext. NW South America. Nicaragua (Jinotega, Matagalpa), Costa Rica (Alajuela, Cartago, Guanacaste, Heredia, Puntarenas), Panama (Chiriquí) and NW Colombia (Antioquia, Chocó, Nariño), 50–2745 m.
Sources: L.D. Wilson, 1986b, 1987c, Castillo et al., 1990, Solis, 1991, G. Köhler, 1999b, Solórzano, 2004 and Mueses-Cisneros & Cepeda Quilindo, 2006.

3. *Tantilla andinista* L.D. Wilson & Mena, 1980. Mem. San Diego Soc. Nat. Hist. (11): 21–22, figs. 7, 12a.

Type: Holotype, KU 135209, a 285 mm female (T.H. Fritts, 9 June 1970).
Type locality: "5 km E Alausí, Prov. Chimborazo, Ecuador, elevation 2600–2750 m."

Distribution: Central Ecuador (Chimborazo), 2600–2750 m. Known only from holotype.
Sources: L.D. Wilson, 1985i, 1987c.

4. *Tantilla armillata* Cope, 1875a. J. Acad. Nat. Sci. Philadelphia (1876) (2) 8(2): 143.

Type: Holotype, ANSP 5363 (C.H. Van Patten, 1865–1875).
Type locality: "Middle Costa Rica."
Distribution: Central America. Guatemala (Sacatepéquez, Santa Rosa), El Salvador (San Salvador), Honduras (Choluteca, Comayagua, Francisco Morazán, Gracias a Dios, Olancho, Valle, Yoro), Nicaragua (Chontales, Managua, Matagalpa, Rivas), Costa Rica (Alajuela, Cartago, Guanacaste, Heredia, Limón, Puntarenas, San José) and possibly W Panama, NSL–1400 m.
Sources: Savage, 2002, Solórzano, 2004, G. Köhler et al., 2005 and McCranie, 2011a.
Remarks: A valid species *fide* Savage (2002: 692–693) but see Remarks under *T. melanocephala* and *T. ruficeps.*

5. *Tantilla atriceps* (A.C.L.G. Günther, 1895 *in* 1885–1902). Biol. Cen.-Amer., Rept. Batr.: 146–147, pl. 52, fig. b. (*Homalocranium atriceps*)

Types: Syntypes (2), BMNH 1946.1.8.81–82 (formerly BMNH 1889.7.3.36–37), a 173 mm and 168 mm male, respectively (W. Taylor, 1889).
Type locality: "Mexico, Nuevo Leon."
Distribution: Southern USA (S Texas) and N Mexico (SE Chihuahua, S Coahuila, Durango, NW Nuevo León, San Luis Potosí, Tamaulipas, NE Zacatecas), 150–2400 m.
Sources: Cole & Hardy, 1981, 1983a, Conant & Collins, 1998, Werler & Dixon, 2000, Gotte & Wilson, 2005 and Lazcano-Villarreal et al., 2010.

6. *Tantilla bairdi* L.C. Stuart, 1941b. Occ. Pap. Mus. Zool., Univ. Michigan (452): 1–3.

Type: Holotype, UMMZ 89223, a 342 mm female (L.C. Stuart, 17 May 1940).
Type locality: "Two km. northeast of Finca Chichén (10 km. south of Cobán [straight line]) on Chamelco trail, Alta Verapaz, Guatemala. Altitude, about 1550 m."
Distribution: Northern Guatemala (Alta Verapaz, Baja Verapaz), 1350–1750 m.
Sources: L.D. Wilson, 1985a, k.

7. *Tantilla bocourti* (A.C.L.G. Günther, 1895 *in* 1885–1902). Biol. Cen.-Amer., Rept. Batr.: 149. (*Homalocranium bocourti*)

Type: Lectotype, BMNH 1946.1.8.70 (formerly MNHN 3694 & BMNH 1894.10.2.1), a 241 mm male (A.A.D. Dugès), designated by Boulenger (1896a: 224).
Type locality: "Guanajuato" [= Guanajuato, Guanajuato State, cen. Mexico, 20°55'N, 101°10'W, elevation 2000 m] via lectotype selection.
Distribution: Mexico (Aguacalientes, Colima, Distrito Federal, S Durango, Guanajuato, Guerrero, Hidalgo, Jalisco, México, Michoacán, Morelos, Nayarit, Oaxaca, Puebla, S Querétaro, NE Sinaloa, cen. Veracruz, W Zacatecas, María Cleofas Is.), (400) 1500–2510 m.
Sources: McDiarmid et al., 1976, McDiarmid & Folke, 1991, McDiarmid, 1992, Benitez-Gálvez, 1997, L.D. Wilson, 1999 and Ponce-Campos & Huerta-Ortega, 2000.

8. *Tantilla boipiranga* Sawaya & Sazima, 2003. Herpetologica 59(1): 119–124, figs. 1a, 2–3 (left).

Type: Holotype, ZUEC 1840, a 375 mm male (J.R. Meyer & R.J. Sawaya, 4 May 1995).
Type locality: "'the serrote' in montane fields ("campos rupestres") of the Serra do Cipó (19°17' S; 43°36' W; approximate elevation 1200 m), Santana do Riacho, Minas Gerais, southeastern Brazil."
Distribution: Southeastern Brazil (Minas Gerais), 450–1300 m.
Sources: Cassimiro, 2003 and Lema, 2004.

9. *Tantilla brevicauda* Mertens, 1952b. Zool. Anz. 149(5–6): 137–138.

Type: Holotype, SMF 43243, a 182 mm female (A. Zilch, 12 Nov. 1951).
Type locality: "El Grito, Finca Los Angeles, 1510 m. H., Cumbre de Jayaque, Dept. La Libertad, El Salvador."
Distribution: El Salvador (Ahuachapán, La Libertad, Morazán, San Salvador, Morazán, La Libertad, San Vicente) and S Guatemala (Chimaltenango), 750–1750 m.
Sources: L.D. Wilson, 1988a, L.D. Wilson et al., 1999 and G. Köhler et al., 2005.

10. *Tantilla briggsi* Savitzky & Smith, 1971. J. Herp. 5(3–4): 167–169, fig. 1.

Type: Holotype, UCM 40000, a 301 mm male (T. MacDougall, July –Sept. 1968).
Type locality: "12 de Julio, an ejido 3 leagues west of Donají (formerly Tolosa) on the trans-isthmus railway, Caribbean drainage, Oaxaca, Mexico."
Distribution: Southern Mexico (S Oaxaca), 90–500 m. Known only from holotype.
Sources: L.D. Wilson, 1985b, Casas-Andreu et al., 1996 and L.D. Wilson et al., 1999.

11. *Tantilla calamarina* Cope, 1867b. Proc. Acad. Nat. Sci. Philadelphia (1866) 18(4): 320.

Synonyms: *Tantilla bimaculata* Cope, 1875a, *Tantilla martindelcampoi* E.H. Taylor, 1937a, and *Geophis gertschi* Bogert & Porter, 1966a.
Type: Holotype, USNM 6600, a 196 mm female (J.J. Major).

Type locality: "Guadalaxara, Mexico" [= Guadalajara, Jalisco, Mexico, 20°40'N, 103°21'W, elevation 1550 m] (probably in error *fide* J.A. Peters (1954: 31). Corrected to Colima State, Mexico *fide* Zweifel (1959a: 6).
Distribution: Mexico (Colima, Distrito Federal, cen. Guerrero, Jalisco, Michoacán, Morelos, Nayarit, ? Puebla, Sinaloa, María Madre Is.), NSL–1655 m.
Sources: E.H. Taylor, 1937a, H.M. Smith, 1942b, J.A. Peters, 1954, Zweifel, 1960, Duellman, 1965b, L. Hardy & McDiarmid, 1969, McDiarmid et al., 1976, L.D. Wilson & Meyer, 1981, L.D. Wilson, 1988b, H.M. Smith et al., 1993, 1997, García & Ceballos, 1994, Ramirez-Bautista, 1994, Benitez-Gálvez, 1997, L.D. Wilson & Campbell, 2000 and Canseco-Márquez et al., 2007.
Remarks: Cope (1867b: 320) incorrectly reported the collector of the type as I.I. Major.

12. *Tantilla capistrata* Cope, 1875b. J. Acad. Nat. Sci. Philadelphia (1876) (2) 8(2): 181.

Type: Holotype, ANSP 11581, an 136 mm female (J. Orton, 1867–1877).
Type locality: "Valley of Jequetepeque, Peru" [= La Libertad Dept.].
Distribution: Southern Ecuador (El Oro, Loja, Zamora-Chinchipe) and W Peru (Cajamarca, La Libertad, Lima, Piura), NSL–1830 m.
Sources: Lema, 1978c, L.D. Wilson & Mena, 1980, L.D. Wilson, 1990a, Pérez-Santos & Moreno, 1991 and Aguilar et al., 2007.

13. *Tantilla cascadae* L.D. Wilson & Meyer, 1981. Milwaukee Pub. Mus. Contr. Biol. Geol. (42): 13–15, figs. 4, 5a.

Type: Holotype, AMNH 107389, a female (D.F. Brand, June 1939).
Type locality: "Tzaráracua Falls (= Cascada la Tzaráracua), S of Uruapán [10.5 km south, according to Duellman, 1961], Michoacán, Mexico."
Distribution: Southwestern Mexico (Michoacán), 1430 m. Known only from holotype.
Sources: L.D. Wilson, 1988e and L.D. Wilson & Campbell, 2000.
Remarks: Length of type not determined due to dessication *fide* L.D. Wilson & Meyer (1981: 13).

14. *Tantilla ceboruca* Canseco-Marquez, Smith, Ponce-Campos, Flores-Villela & Campbell, 2007. J. Herp. 41(2): 221–223, figs. 1–2.

Type: Holotype, MZFC 17048, a 195 mm male (E.N. Smith, P. Ponce-Campos & J. Malone, 17 July 2003).
Type locality:"Volcán Ceboruco, Nayarit, Mexico. Carretera Jala-Cerro microhonl. Das, 2094 m. (21.13199°N, 104.50462°W)."
Distribution: Western Mexico (Colima, Nayarit), 1700–2095 m.
Source: Reyes-Velasco et al., 2012.

15. *Tantilla coronadoi* Hartweg, 1944. Occ. Pap. Mus. Zool., Univ. Michigan (486): 4–5.

Type: Holotype, UMMZ 85697, adult ? female (W.W. Brown, 1932–1944).
Type locality: "vicinity of Chilpancingo, Guerrero" [= vicinity of Chilpancingo [17°33'N, 99°30'W, elevation 1270 m], Sierra Madre del Sur, Guerrero, Mexico].
Distribution: Southwestern Mexico (cen. Guerrero), 1400–1525 m.
Sources: W. Davis & Dixon, 1959, L.D. Wilson & Meyer, 1981, L.D. Wilson, 1990f, L.D. Wilson & Campbell, 2000 and Reyes-Velasco et al., 2012.

16. *Tantilla coronata* Baird & Girard, 1853. Cat. No. Amer. Rept., Serp. 1: 131.

Synonyms: *Homalocranium wagneri* Jan, 1862b, and *Tantilla coronata mitrifer* A. Schwartz, 1953.
Type: Holotype, USNM 1875, a 222 mm female (D.C. Lloyd).
Type locality: "Kemper Co., Miss." [Mississippi, USA].
Distribution: Southeastern USA (Alabama, NW Florida, Georgia, ext. S Indiana, W Kentucky, SE Louisiana, Mississippi, North Carolina, South Carolina, Tennessee, S Virginia), NSL–460 m.
Sources: A. Schwartz, 1953, Telford, 1966, 1982, Mount, 1975, Christman, 1980, Ernst & Barbour, 1989, Ashton & Ashton, 1988, Aldridge, 1992, Aldridge & Semlitsch, 1992a–b, J.C. Mitchell, 1994, Palmer & Braswell, 1995, Tennant, 1997, Conant & Collins, 1998 and Werler & Dixon, 2000.
Remarks: Holotype erroneously listed as USNM 1876 *fide* Telford (1966: 262, 285).

17. *Tantilla cucullata* Minton, 1956. Fieldiana: Zool. 34(39): 449–452, fig. 98.

Synonym: *Tantilla diabola* Fouquette & Potter, 1961.
Type: Holotype, FMNH 74384, a 336 mm male (S.A. Minton, Jr., 1 July 1955).
Type locality: "six miles south-southeast of Alpine, Brewster County, Texas" [USA].
Distribution: Southern USA (SW Texas).
Fossil records: Upper Pleistocene (Rancholabrean II) of USA (Texas).
Sources: Dixon et al., 2000 and Werler & Dixon, 2000.
Remarks: Elevated to species status *fide* Werler & Dixon (2000: 20).

18. *Tantilla cuniculator* H.M. Smith, 1939b. Field Mus. Nat. Hist. (Zool.) 24(3): 32–34. (*Tantilla moesta cuniculator*)

Type: Holotype, FMNH 19408, a 101 mm female (E. Blackburn & E.W. Andrews, 1934).

Type locality: "Mérida, Yucatán, México."
Distribution: Southeastern Mexico (Quintana Roo, Yucatán), N Belize (Belize, Orange Walk) and N Guatemala (Petén), NSL–200 m.
Sources: L.D. Wilson, 1985d, J.C. Lee, 1980, 1996, 2000, J.A. Campbell, 1998b, L.D. Wilson et al., 1999, Stafford & Meyer, 2000 and Stafford, 2004.

19. *Tantilla deppii* (Bocourt, 1883) *in* A.H.A. Duméril, Bocourt & Mocquard, 1870–1909. Miss. Sci. Mex. Amér. Cen., Rept. 3(9): 584–585, pl. 36, figs. 11, 11a–d. (*Homalocranion deppii*)

Type: Lectotype, MNHN 54, a 198 mm male (A.B. Ghiesbreght), designated by L.D. Wilson & Meyer (1981: 18).
Type locality: "México" via lectotype selection. Restricted to vicinity of Huitzilac, Morelos, Mexico *fide* W. Davis & Smith (1953: 138).
Distribution: Southern Mexico (Guerrero, N Morelos, NW Oaxaca), 1525–2440 m.
Sources: L.D. Wilson & Meyer, 1981, L.D. Wilson, 1988f, Flores-Villela & Hernandez-García, 1989, Flores-Villela et al., 1991, L.D. Wilson & Campbell, 2000 and Canseco-Márquez et al., 2007.

20. *Tantilla flavilineata* H.M. Smith & Burger, 1950. Herpetologica 6(5): 117–119.

Type: Holotype, UIMNH 6321, a 300 mm female (J. Burger & W.L. Burger, 23 Aug. 1949).
Type locality: "8 miles southeast of Nochixtlan, Oaxaca, Mexico" [= 13 km SE Nochixtlan, Oaxaca State, Mexico, 17°25'N, 97°11'W, elevation 2090 m].
Distribution: Southern Mexico (cen. Oaxaca), 2190–2340 m.
Sources: L.D. Wilson & Meyer, 1971, L.D. Wilson, 1985e, Casas-Andreu et al., 1996 and L.D. Wilson et al., 1999.

21. *Tantilla gracilis* Baird & Girard, 1853. Cat. No. Amer. Rept., Serp. 1: 132.

Synonym: *Tantilla hallowelli* Cope, 1860d.
Type: Holotype, USNM 2040, a 184 mm specimen (J.H. Clark [U.S.-Mex. Bound. Surv.], 1850–1852, via J.D. Graham), lost *fide* Gotte & Wilson (2005: 597–599).
Type locality: "Indianola, Texas" [USA].
Distribution: Central USA (Arkansas, ext. SW Illinois, E Kansas, N Louisiana, Missouri, Oklahoma, Texas) and N Mexico (NE Coahuila), NSL–610 m.
Sources: Baird, 1859, Kirn et al., 1941, Webb, 1970, Cole & Hardy, 1981, Tennant, 1984, T. R. Johnson, 1987, 2000, Dundee & Rossman, 1989, Ernst & Barbour, 1989, Trauth, 1991b, Collins, 1993, Conant & Collins, 1998, Fitch, 1999, Werler & Dixon, 2000 and Gotte & Wilson, 2005.
Remarks: UMMZ 3781 (formerly USNM 4500) erroneously listed as holotype *fide* Kirn et al. (1949: 240), J.A. Peters (1952: 51), and Kluge (1984: 74).

22. *Tantilla hendersoni* Stafford, 2004. J. Herp. 38(1): 44–47, figs. 1, 2 (upper), 3.

Type: Holotype, BMNH 2002.3, a 272 mm female (P. Stafford, S. McMurry & T. Rainwater, 13 Sept. 2002).
Type locality: "0.5 km east of Las Cuevas on trail to Monkey Tail River, Cayo District (GR 1643'95"N, 8859'17"W), Belize."
Distribution: Southern Belize (Cayo), 580 m. Known only from type locality.
Source: Stafford et al., 2010.
Remarks: Holotype possibly a juvenile *T. impensa* and additional material should demonstrate *T. hendersoni* is a synonym of *T. impensa fide* Stafford et al. (2010: 385).

23. *Tantilla hobartsmithi* E.H. Taylor, 1937a. Trans. Kansas Acad. Sci. (1936) 39: 340–342, fig. 2.

Synonym: *Tantilla utahensis* Blanchard, 1938.
Type: Holotype, UIMNH 25066 (formerly EHT-HMS 4558 & EHT-CC 326), adult male (E.H. Taylor, 3 July 1934).
Type locality: "near La Posa, 10 mi. northwest of Guaymas, Sonora" [Mexico]. Further specified as two miles north of La Posa, back of the first low Distribution on the bank of a dry stream bed at night *fide* E.H. Taylor (1938: 497).
Distribution: Southwestern USA (Arizona, S California, ext. W-cen. Colorado, SW Nevada, S New Mexico, SW Texas, S Utah) and N Mexico (E Chihuahua, N Coahuila, W Sonora), 150–1980 m.
Sources: W.W. Tanner & Banta, 1962b, Fowlie, 1965, Cole & Hardy, 1981, 1983b, Tennant, 1984, Van Devender & Bradley, 1994, M. Williamson et al., 1994, Degenhardt et al., 1996, P.R. Brown, 1997, Hammerson, 1999, Bartlett & Tennant, 2000, Werler & Dixon, 2000, Gotte & Wilson, 2005 and Lazcano-Villarreal et al., 2010.

24. *Tantilla impensa* J.A. Campbell, 1998a . Sci. Pap. Nat. Hist. Mus., Univ. Kansas (7): 6–9, fig. 3.

Type: Holotype, UTA 38196 (formerly ENS 5199), a 642 mm female (E.N. Smith, 2 July 1994).
Type locality: "Aldea San Miguelito, Sierra de Caral, Municipio de Morales, Izabal, Guatemala, 460 m (15°22' N, 88°43'W)."
Distribution: Southern Mexico and upper Central America. Southeastern Mexico (Chiapas), E Guatemala (Alta Verapaz, Huehuetenango, Izabal) and W Honduras (Copán, Cortés), 10–1600 m.
Sources: L.D. Wilson & McCranie, 1999, Stafford, 2004, Stafford et al., 2010 and McCranie, 2011a.
Remarks: Possibly includes *T. hendersoni fide* Stafford et al., (2010: 385).

25. *Tantilla insulamontana* L.D. Wilson & Mena, 1980. Mem. San Diego Soc. Nat. Hist. (11): 24–25, figs. 10, 12d.

Type: Holotype, KU 152207, a 249 mm male (A.C. Echternacht, 12 June 1971).
Type locality: "Río Minas, 15.1 km W Santa Isabel, Prov. Azuay, Ecuador, elevation ~ 1250 m."
Distribution: Southern Ecuador (Azuay), 1250–2100 m.
Sources: L.D. Wilson, 1987c, 1990g and Pérez-Santos & Moreno, 1991.

26. *Tantilla jani* (A.C.L.G. Günther, 1895 *in* 1885–1902). Biol. Cen.-Amer., Rept. Batr.: 148, pl. 52, fig. d. (*Homalocranium jani*)

Synonym: *Tantilla cuesta* L.D. Wilson, 1982a.
Type: Lectotype, BMNH 1946.1.8.68, a 241 mm female (possibly G.C. Champion, 1879–1883, via F.D. Godman), designated by H.M. Smith (1942b: 37).
Type locality: "Guatemala" via lectotype selection.
Distribution: Southeastern Mexico (S Chiapas, E Oaxaca) and S Guatemala (San Marcos, Suchitepéquez), 100–1050 m.
Sources: L.C. Stuart, 1963, L.D. Wilson & Meyer, 1971, L.D. Wilson, 1982a, 1985b–c, f, Casas-Andreu et al., 1996, J.A. Campbell, 1998b and L.D. Wilson et al., 1999.
Remarks: L.D. Wilson & Meyer (1971: 25) examined both *T. jani* syntypes and confirmed H.M. Smith's (1942d: 32) selection of the lectotype, as the status of other syntype (BMNH 1946.1.8.71) is uncertain. If it eventually proves to be *T. jani,* the Distribution would include Nicaragua (Matagalpa).

27. *Tantilla johnsoni* L.D. Wilson, Vaughan & Dixon, 1999. J. Herp. 33(1): 1–3, figs. 1–2.

Type: Holotype, CM 51741, a 337 mm male (E.C. Welling, 11 Aug. 1968).
Type locality: "Musté, Municipio Motozintla, Chiapas, México, approximate elevation 450 m."
Distribution: Extreme SE Mexico (SE Chiapas), 450 m.
Remarks: Locality and elevation of the holotype questioned *fide* J.D. Johnson et al., 2010.

T

28. *Tantilla lempira* L.D. Wilson & Mena, 1980. Mem. San Diego Soc. Nat. Hist. (11): 25–26, figs. 11, 12e.

Type: Holotype, LSUMZ 26093, a 258 mm female (E.A. Liner, 3 June 1968).
Type locality: "41 km NW Tegucigalpa, Depto. Francisco Morazán, Honduras."
Distribution: Southern Honduras (Comayagua, El Paraíso, Francisco Morazán, La Paz), 1450–1730 m.
Sources: L.D. Wilson, 1982a, 1990b, L.D. Wilson & Meyer, 1985 and McCranie, 2011a.

29. *Tantilla marcovani* Lema, 2004. Acta Biol. Leopold. 26(2): 269–271, figs. 1–2.

Type: Holotype, MNRJ 6525, a 283+ mm male (M. Porto, 14 April 1990).
Type locality: "Pico do Jabre, highest point of Paraiba State, Brazil at 50 km SW from Teixeira city (7.13 S; 37.15 W), Maturéia locality. The region is located in highlands within Caatinga, with 1090 m of altitude."
Distribution: Extreme E Brazil (Paraiba), 1090 m. Known only from type locality.

30. *Tantilla melanocephala* (Linnaeus, 1758). Syst. Nat., ed. 10, 1: 218. (*Coluber melanocephalus*)

Synonyms: *Coluber capiteniger* Lacépède, 1789 (*nomen rejiciendum*), *Elaps melanocephalus* Wagler *in* Spix, 1824, *Homalocranion semicinctum* A.-M.-C. Duméril, 1853 (*nomen nudum*), *Elapomorphus mexicanus* A.C.L.G. Günther, 1862a, *Homolocranion melanocephalum fuscum* Bocourt, 1883 *in* A.H.A. Duméril, Bocourt & Mocquard, 1870–1909, *Tantilla pallida* Cope, 1887b, *Homalocranium melanocephalum fraseri* A.C.L.G. Günther, 1895 *in* 1885–1902, *Homalocranium melanocephalum pernambucense* A.C.L.G. Günther, 1895 *in* 1885–1902, *Homalocranium longifrontale* Boulenger, 1896b, *Homalocranium hoffmanni* F. Werner, 1909b, *Elapomorphus nuchalis* T. Barbour, 1914d, and *Tantilla equatoriana* L.D. Wilson & Mena, 1980.
Type: Lectotype, NHR Lin-86 (formerly MAFR), a 210 mm specimen (Mus. Drottn.), designated by Andersson (1899: 10–11).
Type locality: "America" via lectotype selection.
Distribution: Panama and South America. Panama (Chiriquí, Coclé, Colón, Darién, Panamá, Veragua, Barro Colorado Is.), Trinidad and Tobago (SE Tobago, Trinidad), Colombia (Amazonas, Antioquia, Atlántico, Bolívar, Caldas, Cauca, Chimborazo, Chocó, Cundinamarca, Loreto, Magdalena, Meta, Morona, Napo, Nariño, Pastaza, Putamayo, Santander, Santiago, Valle, Vaupés, Gorgona Is.), Venezuela (Amazonas, Anzoátegui, Aragua, Barinas, Bolívar, Carabobo, Cojedes, Distrito Federal, Falcón, Lara, Merída, Miranda, Monagas, Portuguesa, Sucre, Zulia) Guyana (Barima-Waini, Essequibo Islands-West Demerara, Upper Demerara-Berbice, Upper Takutu-Upper Essequibo), Suriname (Suriname), French Guiana (Cayenne, Saint-Laurent-du-Maroni), Ecuador (Cotopaxi, Esmeraldas, Guayas, Inbabura, Loja, Morena-Santiago, Napo, Pastaza, Pichincha), E Peru (Amazonas, Ayacucho, Cuzco, Huanuco, Junín, Loreto, Madre de Dios, Pasco, San Martin), Brazil (Amazonas, Bahia, Ceara, Goiás, Mato Grosso, Pará, Rio Grande do Norte, São Paulo, Tocantins), Bolivia (Cochabamba, La Paz, Santa Cruz), Uruguay (Artigas, Maldonado, Rivera, Salto, Tacuarembó), Paraguay (Cordillera) and N Argentina (Corrientes, Entre Ríos, Misiones), NSL–3080 m.

Sources: Roze, 1966a, Achaval & Vaz-Ferreira, 1974, Emsley, 1977, Cunha & Nascimento, 1978, 1993, L.D. Wilson & Mena, 1980, Vanzolini et al., 1980, L.D. Wilson & Meyer, 1985, Lancini, 1986, Chippaux, 1987, L.D. Wilson, 1987c, f, 1988g, 1992a, Peréz-Santos & Moreno, 1988, 1991, Lancini & Kornacker, 1989, Duellman & Salas, 1991, Gorzula & Señaris, 1998, Montanelli & Alvarez, 1998, Marques & Puorto, 1998, Starace, 1998, Freitas, 1999, Schaeffer & Alvarez, 2000, Giraudo & Scrocchi, 2002, Sawaya & Sazima, 2003, Greenbaum et al., 2004, Lema, 2004, Carreira-Vidal et al., 2005, Duellman, 2005, G. Köhler et al., 2005, Rivas-Fuenmayor & Barrio-Amorás, 2005, R.W. Henderson & Powell, 2006, 2009, Rojas-Runjaic et al., 2007, Navarrete et al., 2009 and C.J. Cole et al., 2013.

Remarks: Population in Panama questionable *fide* by L.D. Wilson (pers. comm.) and a review of *T. armillata, T. melanocephala,* and *T. ruficeps* of lower Central America is needed. Records for Mustique and Union Is., St Vincent & Grenadines, are probable introductions from Trinidad *fide* R.W. Henderson & Powell (2006: 501).

31. *Tantilla miyatai* L.D. Wilson & Knight *in* L.D. Wilson, 1987c. Milwaukee Pub. Mus. Contr. Biol. Geol. (68): 12–15, fig. 7.

Type: Holotype, MCZ 166541, a 375 mm male (G. Onore, Sept. 1983).

Type locality: "Puerto Quito (0°10'N, 79°16'W), Prov. Pichincha, Ecuador."

Distribution: Northern Ecuador (Pichincha), 750 m. Known only from type locality.

Sources: L.D. Wilson, 1987c, 1990c.

32. *Tantilla moesta* (A.C.L.G. Günther, 1863c). Ann. Mag. Nat. Hist. (3) 12(71): 352. (*Homalocranium moestum*)

Type: Holotype, BMNH 1946.1.9.74 (formerly BMNH 1864.1.26.119), a 322 mm female (O. Salvin & F.D. Godman, 1861–1863).

Type locality: "Province of Peten" [Guatemala]. Restricted to Flores, El Petén, Guatemala *fide* H.M. Smith & Taylor (1950a: 318).

Distribution: Southeastern Mexico (Quintana Roo, Yucatán) and N Guatemala (N Petén), NSL–200 m.

Sources: J.C. Lee, 1980, 1996, 2000, L.D. Wilson, 1982a, 1988h, J.A. Campbell, 1998 and Canseco-Márquez et al., 2002.

33. *Tantilla nigra* (Boulenger, 1914b). Proc. Zool. Soc. London 84(3): 816, pl. 2, figs. 2–2a. (*Homalocranium nigrum*)

Type: Holotype, BMNH 1946.1.8.69, a 171–175 mm female (H.G.F. Spurrell, 1912–1913).

Type locality: "near Peña Lisa, Condoto, Choco, Colombia" [= vicinity of Peña Lisa, Río Condoto, Chocó Dept., W Colombia, 5°06'N, 76°37'W, elevation 90 m].

Distribution: Northwestern Colombia (Antioquia, Chocó), 90–1970 m.

Sources: L.D. Wilson, 1987c, 1992b, Pérez-Santos & Moreno, 1988 and H. Sanchez, et al., 1995.

34. *Tantilla nigriceps* Kennicott, 1860. Proc. Acad. Nat. Sci. Philadelphia 12(8): 328.

Synonyms: *Scolecophis fumiceps* Cope, 1861a, *Homalocranion praeoculum* Bocourt, 1883 *in* A.H.A. Duméril, Bocourt & Mocquard, 1870–1909, and *Tantilla kirnia* Blanchard, 1938.

Type: Neotype, UTEP 1349, a 257 mm male (A.J. Ward, E. Rogers & B. Kinniburgh, 15 June 1971), designated by Gotte & Wilson (2005: 600).

Type locality: "on Hueco Tanks [State Park] Road (Farm Road 2775), 0.25 miles [= 0.4 kilometers] N junction with United States highway 62–180" via neotype selection.

Distribution: South-central USA (SE Arizona, E Colorado, Kansas, SW Nebraska, W Oklahoma, New Mexico, Texas, E Wyoming) and N Mexico (E Chihuahua, ext. NW Coahuila, N Durango, NE Nuevo León, N Tamaulipas), NSL–2285 m.

Sources: Fowlie, 1965, Webb, 1970, Cole & Hardy, 1981, Tennant, 1984, W. Tanner, 1985, Hammerson, 1986, 1999, Collins, 1993, M. Williamson et al., 1994, Degenhardt et al., 1996, Conant & Collins, 1998, Nevárez, 1999, L.D. Wilson, 1999, Bartlett & Tennant, 2000, Werler & Dixon, 2000, Gotte & Wilson, 2005, Lazcano-Villarreal et al., 2010 and H.M. Smith et al., 2005.

Remarks: Photographs of neotype in Gotte & Wilson (2005: figs. 1–2). Syntypes (USNM 2046 & 4491) lost *fide* Gotte *in* L.D. Wilson (1999: 12). One syntype erroneously listed as USNM 2040 *fide* H.M. Smith & Taylor (1945: 140). Type locality restriction of Fort Bliss, El Paso Co., Texas *fide* H.M. Smith & Taylor (1950a: 362) invalid.

35. *Tantilla oaxacae* L.D. Wilson & Meyer, 1971. Herpetologica 27(1): 26–27, fig. 3, 7b.

Type: Holotype, UIMNH 40910, a 216 mm male (T. MacDougall, Feb. 1965).

Type locality: "Santo Tomas Teipan, Oaxaca, México" [= Santo Tomás Teipan, Oaxaca State, S Mexico, 17°06'N, 96°44'W, 1720 m]."

Distribution: Southern Mexico (S Oaxaca), 1720–1745 m.

Sources: L.D. Wilson, 1990h, Casas-Andreu et al., 1996 and L.D. Wilson et al., 1999.

36. *Tantilla oolitica* Telford, 1966. Bull. Florida St. Mus. (Biol. Sci.) 10(7): 281–282, fig. 7b.

Type: Holotype, UF 17326, a 215 mm male (D.R. Paulson, April 1955).

Type locality: "Miami, Dade County, Florida in a vacant lot on southwest 27th Avenue near 24th Street" [USA].

Distribution: Extreme SE USA (ext. S Florida, Grassy Key and Key Largo Is.), NSL–10 m.

Sources: Porras & Wilson, 1979, Christman, 1980, Telford, 1980a, Ashton & Ashton, 1988, Ernst & Barbour, 1989, Lazell, 1989, H. Campbell & Moler *in* Moler, 1992, Tennant, 1997, Conant & Collins, 1998 and Hines, 2011.

37. *Tantilla petersi* L.D. Wilson, 1979. Herpetologica 35(3): 274–275, figs. 1–2.

Type: Holotype, UMMZ 92074, a 443 mm female (P. Hershkovitz, 10 June 1934).

Type locality: "San Nicolás (this is probably one of several haciendas San Nicolás in the vicinity of Pimampiro) Pimanpiro (= Pimampiro, 0°26'N, 77°58'W), Provincia Imbabura, Ecuador."

Distribution: Northern Ecuador (E Imbabura), ca. 2100–2150 m.

Sources: L.D. Wilson, 1987c, 1991a and Pérez-Santos & Moreno, 1991.

Remarks: L.D. Wilson (1997: 274) erroneously reported the collecting date as 15 April 1934.

38. *Tantilla planiceps* (Blainville, 1835). Nouv. Ann. Mus. Hist. Nat. Paris (3) 4(3): 294–295, pl. 27, figs. 3, 3a–b. (*Coluber planiceps*)

Synonyms: *Tantilla eiseni* Stejneger, 1896, and *Tantilla eiseni transmontana* Klauber, 1943b.

Type: Holotype, MNHN 818, a 238–240 mm male (M.P.E. Botta, 1827–1829).

Type locality: "La Californie" [= Baja California, Mexico]. Restricted to Cabo San Lucas, Baja California Sur, Mexico *fide* H.M. Smith & Taylor (1950a: 322).

Distribution: Extreme SW USA (SW California) and NW Mexico (Baja California Norte, Baja California Sur, Carmen Is.), NSL–840 m.

Sources: Cole & Hardy, 1981, 1983c, R.W. Murphy & Ottley, 1984, Stebbins, 1985, P.R. Brown, 1997, Bartlett & Tennant, 2000 and Grismer, 2002.

39. *Tantilla psittaca* McCranie, 2011b. Zootaxa (3037): 38–40, figs. 1–3.

Type: Holotype, USNM 578389, a 354 mm male (J.R. McCranie, 15 July 2008).

Type locality: "Rawa Kiamp (15°06'N, 84°26'W, 60 m) along Rio Siksatingni, department of Gracias a Dios, Honduras."

Distribution: Northeastern Honduras (Colón, Gracias a Dios), NSL–420 m.

40. *Tantilla relicta* Telford, 1966. Bull. Florida St. Mus. (Biol. Sci.) 10(7): 270–271, fig. 7c.

Synonyms: *Tantilla relicta neilli* Telford, 1966, and *Tantilla relicta pamlica* Telford, 1966.

Type: Holotype, UF 12421, a 171 mm female (S.R. Telford, Jr., 26 Dec. 1960).

Type locality: "south side of Babson Park, Polk County, Florida" [USA].

Distribution: Southeastern USA (cen. Florida, S Georgia), NSL–45 m.

Fossil records: Upper Pleistocene (Rancholabrean II) of USA (Florida).

Sources: Telford, 1966, 1980b, Christman, 1980, Ashton & Ashton, 1988, Ernst & Barbour, 1989, Holman, 1996c, 2000a, Tennant, 1997, Conant & Collins, 1998, Elliott & McGhee, 2000 and Holman, 2000a.

41. *Tantilla reticulata* Cope, 1860d. Proc. Acad. Nat. Sci. Philadelphia 12(2): 77.

Synonyms: *Microdromus virgatus* A.C.L.G. Günther, 1872b, and *Homalocranion sexfasciatum* J.G. Fischer, 1882.

Type: Holotype, ANSP 3361, a 254 mm specimen (R.W. Mitchell, 1857–1860), lost *fide* E.V. Malnate *in* L.D. Wilson (1982a: 47).

Type locality: "Cocuyas de Veraguas, New Grenada" [= Cocuyas de Veraguas, a gold mine near the Río Concepción, N Veraguas Prov., cen. Panama, ca. 8°45'N, 81°00'W, elevation 100 m].

Distribution: Lower Central America and Colombia. Extreme SE Nicaragua (SE Río San Juan), NE Costa Rica (Alajuela, Cartago, Heredia, Limón, Puntarenas), cen. Panama (Panamá, Veraguas) and W Colombia (Chocó, Nariño), NSL–1430 m.

Sources: E.H. Taylor, 1951, L.D. Wilson & Meyer, 1971, Savage, 1980a, 2002, L.D. Wilson, 1982a, 1985g, 1987c, Pérez-Santos & Moreno, 1988 and L.D. Wilson et al., 1999.

42. *Tantilla robusta* Canseco-Márquez, Mendelson & Gutiérrez-Mayén, 2002. Herpetologica 58(4): 493–495, figs. 1–2.

Type: Holotype, EBUAP 1031, a 426 female (natives, 4 March 1998).

Type locality: "Octimaxal Norte, 930 m, Municipio de Cuetzalan del Progreso, Sierra Norte de Puebla, Puebla, Mexico (20° 02.743' N, 97° 30.103' W)."

Distribution: Southern Mexico (Puebla), 930 m. Known only from type locality.

43. *Tantilla rubra* Cope, 1875a. J. Acad. Nat. Sci. Philadelphia (1876) (2) 8(2): 144.

Synonyms: *Tantilla miniator* Cope, 1863b, *Homalocranium boulengeri* A.C.L.G. Günther, 1895 *in* 1885–1902, and *Tantilla morgani* Hartweg, 1944.

Type: Holotype, USNM 26500, a 310 mm (svl) specimen (A.L.J.F. Sumichrast, 1855–1869).
Type locality: "Japana, Tehuantepec" [= San Pedro Tapanatepec, Oaxaca State, Mexico, 16°22'N, 94°12'W, elevation 55 m].
Distribution: Eastern and S Mexico (Chiapas, Hidalgo, México, S Nuevo León, cen. Oaxaca, Puebla, Querétaro, San Luis Potosí, Tamaulipas, cen. Veracruz) and SW Guatemala (Huehuetenango), 55–2200 m.
Sources: H.M. Smith, 1942b, L.C. Stuart, 1963, H.M. Smith & Werler, 1970, Van Devender & Bradley, 1994, Casas-Andreu et al., 1996, Auth et al., 1997b, Conant & Collins, 1998, Auth et al., 2000, Dixon et al., 2000, Werler & Dixon, 2000, Canseco-Márquez et al., 2002, Lazcano-Villarreal et al., 2010 and Ramírez-Bautista et al., 2010.
Remarks: Petition to conserve the name *rubra* Cope, 1875 has been sent to ICZN *fide* Dixon et al. (2000: 141).

44. *Tantilla ruficeps* (Cope, 1894d). Proc. Acad. Nat. Sci. Philadelphia 46: 204–205. (*Pogonaspus ruficeps*)

Type: Holotype, (MNCR 141), a 223 mm (svl) specimen (A. Faro).
Type locality: "Jimenez, Costa Rica" [= Jiménez, Limón Prov., E Costa Rica, 10°13'N, 83°45'W, elevation 230 m].
Distribution: Costa Rica (Alajuela, Cartago, Heredia, Limón, Puntarenas, San José) and W Panama, NSL–1600 m.
Sources: Savage, 2002, Solórzano, 2004 and Sunver & G. Köhler, 2005.
Remarks: Holotype not in ANSP or USNM. A valid species *fide* Savage (2002: 694–695) but L.D. Wilson (pers. comm.) believes that status in Panama and relationship to Costa Rican material needs review.

45. *Tantilla schistosa* (Bocourt 1883) *in* A.H.A. Duméril, Bocourt & Mocquard, 1870–1909. Miss. Sci. Mex. Amér. Cen., Rept. 3(9): 584, pl. 36, figs. 10, 10a–e. (*Homalocranion schistosum*)

Synonyms: *Tantilla phrenitica* H.M. Smith, 1942b, and *Tantilla schistosa taylori* H.M. Smith, 1962.
Type: Lectotype, MNHN 1883.506, a 191 mm specimen (M. F. Bocourt, 1864–1866), designated by H.M. Smith (1942b: 39).
Type locality: "Alta Vera Paz, Guatemala" via lectotype selection.
Distribution: Mesoamerica. Southeastern Mexico (Oaxaca, Quintana Roo, Veracruz), Guatemala (Alta Verapaz, Petén, Quiché), Belize (Belize, Cayo, Orange Walk), NW Honduras (Copán, Cortés, Olancho, Santa Bárbara, Yoro), cen. Nicaragua (Matagalpa), Costa Rica (Alajuela, Cartago, Guanacaste, Heredia, Limón, Puntarenas, San José) and Panama (Chiriqui, Cocle, Darién, Panamá), 40–1700 m.
Sources: E.H. Taylor, 1951, 1954, H.M. Smith, 1962, L.C. Stuart, 1963, R.W. Henderson & Hoevers, 1975, L.D. Wilson & Meyer, 1985, L.D. Wilson, 1987a, J.C. Lee, 1996, 2000, J.A. Campbell, 1998b, Stafford & Meyer, 2000, Canseco-Márquez et al., 2002, Savage, 2002, Solórzano, 2004, Luja, 2006, Pérez-Higareda et al., 2007, McCranie, 2011a and McCranie & Orellana, 2011.

46. *Tantilla semicincta* (A.-M.-C. Duméril, Bibron & Duméril, 1854b). Erpét. Gén. 7(2): 862–363. (*Homalocranion semicinctum*) (*nomen corrigendum*)

Synonyms: *Elapsoïdes pleii* A.-M.-C. Duméril, Bibron & Duméril, 1854b (*nomen nudum*), *Homalocranion semicinctum* A.-M.-C. Duméril, Bibron & Duméril, 1854b (*nomen incorrigendum*), *Homalocranium laticeps* A.C.L.G. Günther, 1860f, *Homalocranion semicinctum* – Jan, 1862b (*nomen corrigendum*), *Homalocranion lineatum* J.G. Fischer, 1883, and *Homalocranium senicintum* – Briceño-Rossi, 1934 (*nomen incorrectum*).
Types: Syntypes (2), MNHN 3695, a 404 mm female, and RMNH 77 (A. Plée, 1823–1825).
Type locality: "Colombie" [= Colombia] and "Martinique," (in error *fide* L.D. Wilson, 1976: 44), respectively.
Distribution: Northern Colombia (Atlántico, Bolívar, César, Magdalena, Norte de Santander) and NW Venezuela (Aragua, Carabobo, Distrito Federal, Falcón, Mérida, Trujillo, Zulia), NSL–455 m.
Sources: Roze, 1966a, L.D. Wilson, 1976, 1987c, 1990d, Pérez-Santos & Moreno, 1988, Lancini & Kornacker, 1989 and Navarrete et al., 2009.
Remarks: MNHN 3695 erroneously listed as holotype *fide* L.D. Wilson (1976: 44). Designation of type locality as Colombia by J.A. Peters & Orejas-Miranda (1970) erroneously considered unjustified *fide* L.D. Wilson (1987c: 19). Unconfirmed records from Panama and Ecuador *fide* Roze (1966: 226).

47. *Tantilla sertula* L.D. Wilson & Campbell, 2000. Proc. Biol. Soc. Washington 113(3): 821–823, fig. 1.

Type: Holotype, UTA 38145, a 99 mm female (J.A. Campbell, July 1978).
Type locality: "0.8 km NNE of the junction of Mexico highway 200 on the road to La Unión (17°59'N, 101°49'W), Guerrero, México, approximate elevation slightly above 150 m."
Distribution: Southwestern Mexico (Guerrero, Nayarit), 150 m.
Source: Canseco-Márquez et al., 2007.

48. *Tantilla shawi* E.H. Taylor, 1949. Univ. Kansas Sci. Bull. 33(2): 207–209.

Type: Holotype, LSUMZ 306, a 369 mm specimen (C. Shaw, 28 Aug. 1947).

Type locality: "Xilitla (Miramar), San Luis Potosí, Mexico."
Distribution: Eastern Mexico (ext. SE San Luis Potosí, N Veracruz), 670–1400 m.
Sources: L.D. Wilson, 1991b and J.A. Campbell et al., 1995.

49. *Tantilla slavensi* Pérez-Higareda, Smith & Smith, 1985. J. Herp. 19(2): 290–291, figs. 1–2.

Type: Holotype, UNAM-LT 1668, a 346 mm female (G. Pérez-Higareda, 17 April 1983).
Type locality: "Cerro Chochobi, elevation 800 m, El Acuyal area, 8 km NW Catemaco, Veracruz, Mexico."
Distribution: Southeastern Mexico (S Veracruz), 50–800 m.
Sources: Pérez-Higareda & Smith, 1991, L.D. Wilson et al., 1999 and Pérez-Higareda et al., 2007.

50. *Tantilla striata* Dunn, 1928a. Amer. Mus. Novit. (314): 3.

Type: Holotype, AMNH 19745, adult male (P.D.R. Rüthling).
Type locality: "Mixtequillo, Oaxaca, Mexico" [= Santa María Mixtequilla, Oaxaca State, S Mexico, 16°22'N, 95°16'W, elevation 55 m].
Distribution: Southern Mexico (S Oaxaca), NSL–1000 m.
Sources: H.M. Smith, 1942b, L.D. Wilson & Meyer, 1971, L.D. Wilson, 1990i, Casas-Andreu et al., 1996 and L.D. Wilson et al., 1999.

51. *Tantilla supracincta* (W.C.H. Peters, 1863c). Mber. Königl. Akad. Wiss. Berlin 1863(6): 272–273. (*Homalocranion supracinctum*)

Synonym: *Tantilla annulata* Boettger, 1892b.
Type: Holotype, ZMB 4791, a 530 mm female (C. Reiss, 1860).
Type locality: "Umgebung von Guayaquil" [Ecuador] [= vicinity of Guayaquil, Guayas Prov., SW Ecuador, 2°12'S, 78°54'W, elevation 10 m].
Distribution: Lower Central America and N South America. Southeastern Nicaragua (Río San Juan), Costa Rica (Alajuela, Cartago, probably Guanacaste, Heredia, Limón, Puntarenas, San José), Panama (Chiriquí, Coclé, Colón, Darién, Panamá), Ecuador (Esmeraldas, Guayas) and Venezuela (Mérida), NSL–900 m.
Sources: E.H. Taylor, 1951, J.A. Peters, 1960a, L.D. Wilson, 1982a, 1985j, 1987c, Pérez-Santos & Moreno, 1991, Esqueda & La Marca, 1999, Savage, 2002 and Solórzano, 2004.

T

52. *Tantilla taeniata* (Bocourt, 1883) *in* A.H.A. Duméril, Bocourt & Mocquard, 1870–1909. Miss. Sci. Mex. Amér. Cen., Rept. 3(9): 587, pl. 37, figs. 3, 3a–e. (*Homalocranion taeniatum*)

Synonym: *Homalocranium trivittatum* F. Müller, 1885.
Type: Holotype, MNHN 1666, a 341–356 mm male (A. Bouvier).
Type locality: "Guatemala."
Distribution: Mexico and upper Central America. Southern Mexico (S Oaxaca), Guatemala (Guatemala, Santa Rosa), El Salvador (La Union, Sonsonate, Usulután), Honduras (Atlântida, ? Colón, Cortés, El Paraíso. Francisco Morazán, Olancho, Yoro) and Nicaragua (Matagalpa), 650–1600 m.
Sources: L.C. Stuart, 1963, Savitzky & Smith, 1971, L.D. Wilson & Meyer, 1971, 1985, L.D. Wilson, 1982a, 1983b, 1985a, J.A. Campbell, 1998b, L.D. Wilson & McCranie, 1999, L.D. Wilson et al., 1999, Stafford, 2004, G. Köhler et al., 2005, McCranie et al., 2006, McCranie, 2011a–b and McCranie et al., 2012a.
Remarks: *Tantilla taeniata* may represent one or more undescribed species *fide* McCranie (2011b: 42).

53. *Tantilla tayrae* L.D. Wilson, 1983a. J. Herp. 17(1): 54–57, fig. 1.

Type: Holotype, MVZ 159203, a 317 mm male (R.L. Seib, 30 July 1978).
Type locality: "Finca San Jerónimo, 7.5 km N (by rd.) Cacaohatán [= Cacahoatán or Cacahuatán], elevation 760 m, Volcán Tacaná, Municipio de Unión Juárez, Chiapas, México."
Distribution: Southeastern Mexico (Volcán Tacaná of Chiapas), 760–960 m. Known only from vicinity of type locality.
Sources: J.D. Johnson, 1984, L.D. Wilson, 1990e and L.D. Wilson et al., 1999.
Remarks: Closely related to *T. jani:* additional specimens needed to clarify its status *fide* J.A. Campbell (1998: 10–11).

54. *Tantilla tecta* J.A. Campbell & Smith, 1997. Proc. Biol. Soc. Washington 110(3): 333–337, figs. 1–2.

Type: Holotype, UTA 41160 (formerly UVG 1742), a 222 mm female (C. Granizo, 29 June 1992).
Type locality: "slope flanking the NE side of Laguna Yaxhá, Petén, Guatemala (17°03'43"N, 89°23'12"W), at about 220 m."
Distribution: Northeastern Guatemala (Petén), 220 m. Known only from type locality.
Sources: L.D. Wilson et al., 1999, J.C. Lee, 2000 and Stafford, 2004.

55. *Tantilla trilineata* (W.C.H. Peters, 1880a). Mber. Königl. Akad. Wiss. Berlin 1880(2): 221, pl., figs. 2, 2a–c. (*Leptocalamus trilineatus*)

Type: Holotype, ZMB 9648, a 230 mm female (H. Boeckmann).
Type locality: "Brasilien" [= Brazil] (in error *fide* Savitzky & Smith, 1971: 170).
Distribution: Unknown. Known only from the holotype.
Sources: F. Werner 1925, Amaral, 1929, Savitzky & Smith, 1971, L.D. Wilson & Meyer, 1971, L.D. Wilson, 1974, 1999 and L.D.Wilson & McCranie, 1999.
Remarks: Status of this taxon is unclear *fide* L.D. Wilson (pers. comm.).

56. *Tantilla triseriata* H.M. Smith & Smith, 1951. Proc. Biol. Soc. Washington 64: 97–99.

Type: Holotype, UIMNH 20198, a 203 mm female (T. MacDougall, 9 Oct. 1949).
Type locality: "Coatlán, Oaxaca, slopes of the Isthmus of Tehuantepec, Mexico" [= Coatlán, Oaxaca State, S Mexico, 16°18'N, 96°46'W, elevation 1725 m].
Distribution: Southern Mexico (Oaxaca), 1635–1725 m.
Source: H.M. Smith et al., 1998.

57. *Tantilla tritaeniata* H.M. Smith & Williams, 1966c. Southwest. Nat. 11(4): 483–485, fig. 1.

Type: Holotype, BMNH 1894.12.28.23, a 273 mm female (? Gaumer via F.D. Godman).
Type locality: "Bonacca Island, Bay Islands, Honduras."
Distribution: Northern Honduras (Islas de la Bahía: Guanaja Is.), NSL.
Sources: L.D. Wilson & McCranie, 1999, McCranie et al., 2005 and McCranie, 2011a.

58. *Tantilla vermiformis* (Hallowell, 1861). Proc. Acad. Nat. Sci. Philadelphia (1860) 12(11): 484. (*Lioninia vermiformis*)

Type: Lectotype, USNM 32338 (formerly USNM 5792), a 129 mm female (C. Wright [No. Pacific Explor. Exped.], 1853–1856), designated by R.W. Van Devender & Cole (1977: 2).
Type locality: "Nicaragua" via lectotype selection.
Distribution: Central America. El Salvador (San Salvador), W Nicaragua (León, Managua) and NW Costa Rica (Guanacaste), 50–700 m.
Sources: R.W. Van Devender & Cole, 1977, L.D. Wilson, 1982a, 1987b, L.D. Wilson et al., 1999, L.D. Wilson & Campbell, 2000, Dueñas et al., 2000, G. Köhler et al., 2005 and Canseco-Márquez et al., 2007.

59. *Tantilla vulcani* J.A. Campbell, 1998a. Sci. Pap. Mus. Nat. Hist. Univ. Kansas (7): 11–13, figs. 4–5.

Type: Holotype, UTA 21772 (formerly JAC 12410), a 211 mm female (C. Mirón, April–May 1986).
Type locality: "Finca El Carmen, Km 197.5 on Ca-2, 518 m, Quezaltenango, Guatemala."
Distribution: Southern Mexico (Oaxaca) and S Guatemala (Escuintla, Quezaltenango, San Marcos, Suchitepéquez), NSL–610 m.

60. *Tantilla wilcoxi* Stejneger, 1902b. Proc. U.S. Natl. Mus. 25: 156–157.

Synonyms: *Tantilla deviatrix* T. Barbour, 1916b, and *Tantilla wilcoxi rubricata* H.M. Smith, 1942b.
Type: Holotype, USNM 19674, a 184+ mm male (T.E. Wilcox, 1892).
Type locality: "Fort Huachuca, Arizona" [= Huachuca Mtns., Cochise Co., USA].
Distribution: Extreme SW USA (ext. SE Arizona) and N Mexico (W Chihuahua, ext. SE Coahuila, Durango, S Nuevo León, W San Luis Potosí, NE Sinaloa, E Sonora, Tamaulipas, Zacatecas), 915–2440 m.
Sources: H.M. Smith, 1942b, Fowlie, 1965, C. Shaw & Campbell, 1974, Cole & Hardy, 1981, Liner, 1983, Stebbins, 1985, McDiarmid, 1992, H.M. Smith & Smith, 1993, Bartlett & Tennant, 2000, Farr et al., 2009 and Lazcano-Villarreal et al., 2010.

61. *Tantilla yaquia* H.M. Smith, 1942b. Zoologica 27(7): 41.

Synonym: *Tantilla bogerti* Hartweg, 1944.
Type: Holotype, MCZ 43274, a 240 mm female (H.S. Gentry, Aug. 1936).
Type locality: "Guasaremos, Río Mayo, Chihuahua, Mexico" [= Guasaremos (27°40'N, 108°42'W, elevation 880 m), Río Mayo, ext. W Chihuahua State, NW Mexico].
Distribution: Southwestern USA (SE Arizona, ext. SW New Mexico) and NW Mexico (ext. W Chihuahua, Nayarit, Sinaloa, E Sonora), 60–1585 m.
Sources: Fowlie, 1965, McDiarmid, 1968, 1977, Cole & Hardy, 1981, W.W. Tanner, 1985, M. Williamson et al., 1994, Degenhardt et al., 1996, Bartlett & Tennant, 2000, Lemos-Espinal et al., 2004b and Bogan et al., 2009.

TANTILLITA H.M. Smith, 1941f (Colubridae)

Type species: *Tantilla lintoni* H.M. Smith, 1940.
Distribution: Mesoamerica.
Sources: H.M. Smith, 1941f, H.M. Smith & Taylor, 1945, J.A. Peters & Orejas-Miranda, 1970, Villa et al., 1988, L.D. Wilson, 1988i and J.A. Campbell & Vannini, 1989.

1. *Tantillita brevissima* (E.H. Taylor, 1937a). Trans. Kansas Acad. Sci. (1936) 39: 344–345, fig. 4. (*Tantilla brevissima*)

Synonym: *Tantilla excubitor* L.D. Wilson, 1982a.
Type: Holotype, UIMNH 25075 (formerly EHT-HMS 4557 & EHT-CC 4623), a 151 mm specimen (E.H. Taylor, 27–28 Aug. 1935).
Type locality: "Tonalá, Chiapas (on top of low mountain near the city)" [= mountaintop near Tonalá (16°05'N, 93°45'W, elevation 50 m), Chiapas State, SE Mexico].
Distribution: Southeastern Mexico (Chiapas, SE Oaxaca) and S Guatemala (Escuintla). 200–1700 m.
Sources: L.D. Wilson, 1982a, 1988j and J.D. Johnson, 1984, 1989.

2. *Tantillita canula* (Cope, 1875a). J. Acad. Nat. Sci. Philadelphia (1876) (2) 8(2): 144. (*Tantilla canula*)

Synonym: *Homalocranium breve* A.C.L.G. Günther, 1895 *in* 1885–1902.
Types: Syntypes (2), USNM 24881–82, a male and female, longest syntype 172 mm (A.C.V. Schott, 1865–1866).
Type locality: "Yucatan" [Mexico]. Restricted to Chichén Itzá, Yucatán *fide* H.M. Smith & Taylor (1950a: 352).
Distribution: Mesoamerica. Southeastern Mexico (Campeche, Quintana Roo, Yucatán), Guatemala (Petén) and Belize (Cayo, Orange Walk), NSL–300 m.
Sources: R.W. Van Devender & Cole, 1977, L.D. Wilson, 1982a, 1986a, 1988c, H.M. Smith et al., 1993, J.C. Lee, 1996, 2000 and Stafford & Meyer, 2000.

3. *Tantillita lintoni* (H.M. Smith, 1940). Proc. Biol. Soc. Washington 53: 61–62, fig. 1. (*Tantilla lintoni*)

Synonym: *Tantillita lintoni rozellae* Pérez-Higareda, 1985.
Type: Holotype, USNM 108603 (formerly HMS 7460), a 207 mm female (H.M. Smith, 24 May 1939).
Type locality: "Piedras Negras, Guatemala, at a locality some 45 miles by trail from Tenosique, Tabasco, Mexico, on the bank of Usumacinta River" [= ext. NW Petén, Guatemala, 17°19'N, 91°17'W, elevation 70 m].
Distribution: Southeastern Mexico (Tabasco, Quintana Roo, S Veracruz), N Guatemala (Petén), Belize (Toledo), NW Honduras (Atlántida, Cortés, Gracias a Dios, Islas de la Bahía, Santa Bárbara and Utila Is.) and N Nicaragua (Jinotega), NSL–900 m.
Sources: L.C. Stuart, 1963, L.D. Wilson, 1982a, Pérez-Higareda, 1985, L.D. Wilson & Meyer, 1985, J.C. Lee, 1996, 2000, J.A. Campbell, 1998b, G. Köhler, 1999a, López-Luna & Torres-Laranca, 2000, Stafford & Meyer, 2000, Cedeño-Vázquez et al., 2003, Martinez, 2011 and McCranie, 2011a.

TAPHROMETOPON Brandt, 1838 (Psammophiidae)

Synonyms: *Chorisodon* A.-M.-C. Duméril, 1853, *Monodiastema* Bibron *in* A.-M.-C. Duméril, Bibron & Duméril, 1854, *Phayrea* Theobald, 1868b, *Taphrometopum* Blanford, 1876 (*nomen emendatum*), *Tafrometopon* – Palacky, 1898 (*nomen incorrectum*), *Mike* F. Werner, 1924a, *Taphrometopom* – Briceño-Rossi, 1934 (*nomen incorrectum*), *Tephrometopon* – Terent'ev & Chernov, 1965 (*nomen incorrectum*), and *Rayhammondus* Hoser, 2012u (*nomen illegitimum*).
Type species: *Coluber lineolatus* Brandt, 1838.
Distribution: Africa and Asia.
Sources: M.A. Smith, 1943, Leviton & Anderson, 1970a, M.S. Khan, 2002, 2006, Whitaker & Captain, 2004 and Hoser, 2012u.
Remarks: A valid genus *fide* Zaher et al., 2012; previously a synonym of *Psammophis*. Arrangement of Hoser, 2012u, who used the name *Phayrea*.

1. *Taphrometopon condanarum* (Merrem, 1820). Syst. Amph.: 107. (*Coluber condanarus*)

Synonyms: *Leptophis bellii* Jerdon, 1854, *Psammophis taeniata* A.C.L.G. Günther, 1862b, *Psammophis indicus* Beddome, 1863a, *Phayrea isabellina* Theobald, 1868b, *Psammophis sibilans quadrilineata* Jan, 1863b (*nomen nudum*), *Psammophis sibilans quadrilineata* Jan, 1870 *in* Jan & Sordelli, 1866–1870, and *Mike elegantissima* F. Werner, 1924a.
Type: Lectotype, a 724 mm specimen described and illustrated in P. Russell (1796: 32–33, pl. 27) (Snodgrass, Oct. 1788), designated herein.
Type locality: "Ganjam" [= Ganjam, Ganjam Distr., E Orissa State, E India, 19°23'N, 85°03'E, elevation NSL] via lectotype selection.
Distribution: Southern Asia. Eastern Pakistan (Khyber Pakhtunkhwa, Punjab, Sindh), N India (Andhra Pradesh, Bihar, Gujarat, Himachal Pradesh, Karnataka, Madhya Pradesh, Maharashtra, Orissa, Punjab, Uttaranchal, Uttar Pradesh, S West Bengal) and Nepal (Chitwan, Sunsari), NSL–2000 m.
Sources: J. Anderson, 1871a, Murray, 1884, Boulenger, 1890a, Wall, 1911a, 1924a, Minton, 1966, Kramer,1977, Murthy, 1986, Prasad, 1992, Sanyal et al., 1993, M.S. Khan, 2002, 2006, Tiwari & Shah, 2004, Whitaker & Captain, 2004 and N. Khaire, 2006.
Remarks: Original description based on P. Russell (1796: 32–33, pl. 27).

2. *Taphrometopon crucifer* (Daudin, 1803d). Hist. Nat. Rept. 7: 189–190. (*Coluber crucifer*)

Type: Holotype, not designated, a 236 mm specimen, location unknown.

Type locality: "Indes orientales" [= Southern Asia and East Indies] (in error). Corrected to South Africa *fide* V.F.M. FitzSimons (1962a: 239).

Distribution: Southern Africa. Eastern Zimbabwe (Manicaland), S Mozambique (W Manica), NW Swaziland, W Lesotho and South Africa (Eastern Cape, Free State, Gauteng, KwaZulu-Natal, Limpopo, Mpumalanga, W Northern Cape, North West, Western Cape), NSL–2300 m.

Sources: V.F.M. FitzSimons, 1962a, 1974, Broadley & Cock, 1975, De Waal, 1978, Branch, 1988, Boycott & Culverwell, 1992, Broadley, 2002, Marais, 2004, Cottone & Bauer, 2010 and Broadley & Blaylock, 2013.

Remarks: Description based on Merrem (1790: 13, pl. 3) from "Ostindien." Forms a basal sister clade to *Taphrometopon fide* Kelly et al. (2008: 1049).

3. *Taphrometopon indochinensis* (M.A. Smith, 1943). Fauna Brit. India, Rept. Amph. 3: 365. (*Psammophis candanarus indochinensis*)

Types: Syntypes (9), BMNH 1868.4.3.26–29, BMNH 1914.5.11.9–10, and BMNH 1969.1920–22 (M.A. Smith).

Type locality: "Indo-China south of lat. 21° (Taungyi, Pegu, Lopburi, Bangkok, Phan-rang in Annam)" [= Taungyi (SW Shan Div., 20°47'N, 97°02'E, elevation 1415 m) and Bago (Bago Div., 17°25'S, 96°40'E, elevation 10 m), S Myanmar; Lopburi (Lop Buri Prov., 15°03'N, 100°53'E, elevation 95 m) and Bangkok (Bangkok Prov., 13°45'N, 100°30'E, elevaton NSL), S Thailand; Phan Rang (Ninh Thuan Prov., 11°35'N, 108°59'E, elevation NSL), S Vietnam].

Distribution: Southeast Asia. S Myanmar (Bago, Irawaddy, Magwe, Mandalay, Sagaing, S Shan), Thailand (Bangkok, Chaiyaphum, Chiang Mai, Krung Thep Mahanakhon, Lop Buri, Nakhon Sawan, Phra Nakhon Si Ayutthaya, Ratchaburi, Saraburi), Laos (Champasak), Cambodia, S Vietnam (Binh Thuan, Ninh Thuan) and Indonesia (Bali, Java), NSL–2000 m.

Sources: M.A. Smith, 1943, Bourret, 1936b, E.H. Taylor, 1965, Campden-Main, 1970a, M.J. Cox et al., 1998, Hughes, 1999, Chan-ard et al., 1999, Orlov et al., 2003, Teynié & David, 2007, Stuart & Heatwole, 2008, V.S. Nguyen et al., 2009 and I. Das, 2010, 2012.

Remarks: A valid species *fide* Hughes (1999: 65). BMNH 77 listed as type *fide* Cox et al. (2013: 522).

4. *Taphrometopon leithii* (A.C.L.G. Günther, 1869). Proc. Zool. Soc. London 37(1): 505–506, pl. 39. (*Psammophis leithii*)

Type: Holotype, BMNH 1946.1.2.82, a 650 mm male (A.H. Leith).

Type locality: "Sindh" [= Sindh Prov., SE Pakistan].

Distribution: Southern Asia. Eastern Afghanistan (Nangarhar), Pakistan (Balochistan, Khyber Pakhtunkhwa, Punjab, Sindh) and NW India (Gujarat, Jammu and Kashmir, Maharashtra, Punjab, Rajasthan, Uttar Pradesh, Dwarka Is.), NSL–800 m.

Sources: Boulenger, 1890a, Wall, 1907e, 1911d, 1924b, M.A. Smith, 1943, Leviton, 1959a, Minton, 1966, Král, 1969, Leviton & Anderson, 1970a, M.S. Khan, 1982, 2006, R.C. Sharma, 1982, Murthy, 1986, Sahi & Duda, 1986, Whitaker & Captain, 2004 and N. Khaire, 2006.

5. *Taphrometopon lineolatum* (Brandt, 1838). Bull. Acad. Imp. Sci. St. Pétersbourg (1837) 3(16): 243–244. (*Coluber [Taphrometopon] lineolatus*)

Synonyms: *Coluber caspius* Lichtenstein, 1823 (*nomen praeoccupatum*), *Chorisodon sibericum* A.-M.-C. Duméril, 1853 (*nomen nudum*), *Chorisodon sibericum* A.-M.-C. Duméril, Bibron & Duméril, 1854, *Psammophis doriae* Jan *in* Filippi, 1865, and *Psammophis triticeus* Wall, 1912b.

Type: Holotype, ZISP 2042, a 584 mm specimen (G.S. Karelin, 1837).

Type locality: "côte orientale de la mer Caspienne, dans la Turcomanie et dans le nord de la Perse, notamment dans les environs d'Astrabad" [= eastern shore of Caspian Sea, within Turkey and northern Iran, notably the vicinity of Gorgan, Golestan Prov., N Iran]. Restricted to region of Krasnovodsk, W Turkmenistan *fide* Strauch (1873: 193).

Distribution: Central Asia. Extreme S Azerbaijan, ext. NW Iran (East Azarbaijan, West Azarbaijan), Kazakhstan (Barsakel'mes Is.), Kyrgyzstan, W Tadzhikistan, Turkmenistan, Uzbekistan (Vozrozhdenya Is.), NW China (Gansu, Nei Monggol, Ningxia, Xingjiang), S Mongolia (Bayan Hongar, Dornogovi, Dundgovi, Govi-Altay, Hovd, Omnigovi, Suhbaatar), E Iran (Central, Esfahan, Fars, Kerman, Khuzestan, Lorestan, Manzandaran, Razavi Khorasan, Sennan, Sistan va Baluchestan), Afghanistan (Balkh, Farah, Faryab, Helmand, Herat, Jowzjan, Kabul, Kunduz, Nangarhar) and Pakistan (Balochistan, SW Punjab), 150–2440 m.

Sources: W.C.H. Peters, 1861g, Strauch, 1873, Blanford, 1876, Alcock & Finn, 1896, Nikolsky, 1916, Wall, 1924a, C.H. Pope, 1935, M.A. Smith, 1943, Leviton, 1959a, Minton, 1966, S.C. Anderson & Leviton, 1969, Král, 1969, Leviton & Anderson, 1970a, Bannikov et al., 1977, M.S. Khan, 1980, 1982b, 1986, 2002, Tien et al., 1986, Borkin et al., 1990a–b, Clark, 1990, Latifi, 1991, Toktosunov et al., 1992, Szczerbak, 1994, Xu, 2001, M.S. Khan, 2002, Baig et al., 2008 and Tuniyev et al., 2009.

6. *Taphrometopon longifrons* (Boulenger, 1890a). Fauna Brit. India, Rept. Batr.: 366–367. (*Psammophis longifrons*)

Type: Holotype, BMNH 1946.1.8.44, head and neck of a 1200 mm specimen (R.H. Beddome, 1857–1882).

Type locality: "Cuddapah Hills, British India" [= Kadapa, S Andhra Pradesh State, S India, 14°28'N, 78°49'E, elevation 135 m] (in error *fide* Wall, 1924a: 875 & 601 and M.A. Smith, 1943: 366).

Distribution: Central India (Gujarat, Madhya Pradesh, Maharashtra), NSL–310 m.

Sources: Dreckmann, 1893, Gleadow, 1894, Abreu, 1913, 1916, Wall, 1924a, M.A. Smith, 1943, Gharpurey, 1962, Whitaker, 1978a, Murthy, 1986, R.C. Sharma, 2003, 2007, Whitaker & Captain, 2004 and N. Khaire, 2006.

Remarks: Authorship erroneously attributed to Boulenger, 1896a.

†*TAUNTONOPHIS* Parmley & Walker, 2003 (Colubridae)

Type species: †*Tauntonophis morganorum* Parmley & Walker, 2003.

Distribution: Pliocene of USA.

Source: Parmley & Walker, 2003.

1. †*Tauntonophis morganorum* Parmley & Walker, 2003. J. Herp. 37(2): 240–242, fig. 2a–e.

Type: Holotype, BMUW 87325, one trunk vertebra.

Type locality: "Taunton Fossil Site, 0.5 km east of Taunton, Adams Co., Washington. Ringold Formation: Pliocene (Blancan Land Mammal Age), at 2.85–2.95 MYBP" [USA].

Distribution: Upper Pliocene (Blancan: 2.85–2.95 mya) of USA (Washington). Known only from type locality.

TELESCOPUS Wagler, 1830 (Colubridae)

Synonyms: *Tarbophis* Fleischmann, 1831, *Trigonophis* Eichwald, 1831, *Ailurophis* Fitzinger *in* Bonaparte, 1837, *Ailurops* Michahelles *in* Bonaparte, 1837, *Aelurophis* Agassiz, 1847 (*nomen emendatum*), *Aelurops* Agassiz, 1847 (*nomen emendatum*), *Tarbophus* – Briceño-Rossi, 1934 (*nomen incorrectum*), *Tarbophys* – Buresch & Zonkow, 1934 (*nomen incorrectum*), *Tharbophis*– Scortecci, 1934 (*nomen incorrectum*), *Migiurtinophis* Scortecci, 1935, *Tarborhis* – Chernov, 1939 (*nomen incorrectum*), *Torborhis* – Chernov, 1939 (*nomen incorrectum*), *Torbophis* – Radovanovic & Martino, 1950 (*nomen incorrectum*), *Pseudotarbophis* Domergue, 1955, *Tarbophia* – Stanek, 1962 (*nomen incorrectum*), *Tlescopus* – Baran, 1976 (*nomen incorrectum*), and *Telescopis* – Mahendra, 1984 (*nomen incorrectum*).

Type species: *Coluber obtusus* A. Reuss, 1834.

Distribution: Africa, SE Europe, Middle East, Arabia and SW Asia.

Fossil records: Lower Miocene to middle Pleistocene of Europe. Fossils unassigned to species include lower Miocene (Orleanian, MN 4–5: 13.7–16.9 mya) of Czech Republic and Germany, upper Miocene (Turolian, MN 13: 4.9–7.3 mya) of Hungary and Italy, and middle Pleistocene (Ionian: 0.13–0.78 mya) of Bulgaria.

Sources: W. Böhme, 1977b, Zinner, 1977, J.B. Rasmussen, 1979, Szyndlar, 1991a, 2005, 2009, Broadley, 1994a, Venczel & Sen, 1994, Venczel, 1998 and Delfino, 2002.

1. *Telescopus beetzii* (T. Barbour, 1922). Proc. Biol. Soc. Washington 35: 230. (*Tarbophis beetzii*)

Type: Holotype, MCZ 16728, a 251 mm specimen (W. Beetz)

Type locality: "Kolmanskop, about seventy miles south of Lüderitzbucht, Southwest Protectorate (formerly German Southwest Africa), Africa" [= Kolmanskop, W Karas Distr., SW Namibia, 26°42'S, 15°14'W, elevation 165 m].

Distribution: Southern Namibia (Hardap, Karas) and W South Africa (W Free State, Northern Cape, N Western Cape), 165–1040 m.

Sources: Mertens, 1955, 1971, FitzSimons, 1962, De Waal, 1978 and Broadley, 1983.

2. †*Telescopus bolkayi* Szyndlar, 2005. Palaeontogr. Ital. (2004) 90: 36–37, figs. 4d–h.

Synonym: †*Telescopus bolkayi* Szyndlar, 2000 (*nomen nudum*).

Type: Holotype, GIH V.22182, one trunk vertebra.

Type locality: Rudabánya (locality R3), northeast Hungary; Late Miocene, early Vallesian (biozone MN 9)."

Distribution: Upper Miocene (Vallesian, MN 9: 9.7–11.1 mya) of Hungary and Italy. Known only from type locality.

Source: Szyndlar, 2012.

Remarks: Closely related to *T. fallax fide* Szyndlar (2005: 38).

3. *Telescopus dhara* (Forskal *in* Niebuhr, 1775). Descr. Anim.: 14. (*Coluber dhara*)

Synonyms: *Dipsas aegypticus* Schlegel, 1837, *Tarbophis guentheri* J. Anderson, 1895, and *Tarbophis obtusus nubamontana* F. Werner, 1919.

Type: Holotype, none designated, a 450 mm specimen (P.S. Forskal, Dec. 1762–July 1763), lost *fide* M. Andersen *in* Crochet et al. (2008: 37).

Type locality: "Yemen."

Distribution: Egypt, Middle East and Arabia. Northeastern Egypt (South Sinai), W Jordan (Aqaba, Balqa, Karak, Maan, Madaba), E Israel (S Northern, E Southern, West Bank), Saudi Arabia (Asir, Bahah, Jawf, Jazan, Madinah, Makkah, Riyadh, Tabuk), United Arab Emirates, Oman (Al Batinah, Al Dakhiliyah, Al Hudaydah, Al Wusta, Dhofar, Musandam, Muscat, Ash Sharqiyah, Masirah Is.) and Yemen (Abyan, Ad Dali', Aden, Hadhramaut, Ibb, Lahij, Sana'a), 400 m BSL–2100 m.

Sources: Gasperetti, 1988, Leviton et al., 1992, Bouskila & Amitai, 2001, Disi et al., 2001, Baker et al., 2004, Baha El Din, 2006, Crochet et al., 2008, Soorae et al., 2009, Amr & Disi, 2011, Bar & Haimovitch, 2011 and Grossmann et al., 2012.

Remarks: A valid, clearly diagnosable, allopatric taxon *fide* Crochet et al. (2008: 38).

4. *Telescopus fallax* (Fleischmann, 1831). Dalmatiae Nova Serp. Gen.: 18–24, pl. 1, figs. a–e. (*Tarbophis fallax*)

Synonyms: *Coluber vivax* Fitzinger, 1826a (*nomen nudum*), *Trigonophis iberus* Eichwald, 1831, *Coluber carneus* Dwigubski, 1832, *Ailurophis vivax* Bonaparte, 1837 *in* 1832–1841, *Tarbophis vivax syriaca* Boettger, 1880b, *Tarbophis savignyi* Boulenger, 1896a, *Tarbophis fallax cyprianus* T. Barbour & Amaral, 1927, *Tarbophis fallax mcewani* K.P. Schmidt, 1939b, *Tarbophis barnumbrowni* Bogert, 1940, *Telescopus fallax pallidus* Stepanek, 1944, *Telescopus fallax rhodicus* Wettstein, 1953a, *Telescopus fallax multisquamatus* Wettstein, 1953a, *Telescopus fallax christianus* Buchholz, 1955, and *Telescopus intermedius* Gruber, 1974.

Type: Holotype, none designated, an 864 mm specimen, location unknown.

Type locality: "Dalmatiae, Istriae agrique Tergestinensis muris et petris" [= Istria Peninsula, Croatia, Slovenia and Italy]. Restricted to Trieste Prov., Italy *fide* Mertens & Müller (1928: 50).

Distribution: Southeastern Europe, Middle East and SW Asia. Extreme NE Italy (Friuli-Venezia Giulia, Trieste), Slovenia, Croatia (Ada, Arbe, Brac, Cres, Hvar, Korcula, Kornat, Krk, Pag, Rab and Silba, Svetac, Solta and Vis Is.), extr. S Bosnia-Herzegovina, S Montenegro, Bosnia-Herzegovina, W Albania, Macedonia, SW Bulgaria (Blagoevgrad, Haskovo), Greece (Aitolia and Akarnania, Andros, Amorgós, Antikithira, Antiparos, Arkadhia, Akhaia, Argolis, Arta, Attica, Despotikó, Drama, Elasa, Evritania, Evros, Euboia, Florina, Fokis, Fthiotis, Gávdos, Grevena, Ilia, Imathia, Ioannina, Kéalymnos, Kardhitsa, Kássos, Kastoria, Kéa, Kefaloniá, Kerkira, Khalkidhiki, Kilkis, Kíthira, Kithnos, Korinthia, Koufoníssi, Kozani, Kristiani, Lakonia, Larisa, Lesvos, Levkas, Magnisia, Messinia, Míkonos, Milos, Náxos, Páros, Pella, Pieria, Preveza, Rodhopi, Santorini, Serrai, Sérifos, Skiros, Spétses, Strofádes, Thesprotia, Thessaloniki, Thíra, Thirassia, Tinos, Trikala, Voiotia, Xanthi, Zákinthos), Crete, Cyprus, Malta, Rhodes, Turkey (Adana, Alyon, Aydin, Balikesir, Bursa, Denizli, Eskisehir, Famagusta, Gaziantep, Hatay, Icel, Isparta, Istanbul, Izmir, Konya, Manisa, Maras, Mugla, Nicosia, Sanliurfa, Siirt, Urfa, Usak), N Syria (Homs), Lebanon (S Beqaa, Mont-Liban, Nabatiye), N Israel (Central, Gaza, Haifa, Jerusalem, Northern, Tel Aviv, West Bank), NE Egypt (Sinai), SW Russia (Dagestan), S Georgia, S Armenia, Azerbaijan, Turkmenistan, Iraq and Iran (Central, East Azarbaijan, Esfahan, Hamadan, Semnan, Sistan va Baluchestan, West Azarbaijan, Zanjhan), NSL–2000 m.

Sources: A.C.L.G. Günther, 1880a, Wettstein, 1953a, Bannikov et al., 1977, Bruno & Magueri, 1977, Kock, 1979, Basoglu & Baran, 1980, Hraoui-Bloquet, 1981, Bruno, 1984, Ilani, 1983a, Disi et al., 1988, Bruno & Maugeri, 1990, J.B. Rasmussen, 1990, Latifi, 1991, Leviton et al., 1992, Valakos & Mylonas, 1992, Ferri, 1994, Gasc et al., 1997, Bouskila & Amitai, 2001, Beshkov & Nanev, 2002, Demetropoulos & Ioannides, 2002, Hraoui-Bloquet et al., 2002, Ananjeva et al., 2004, Dimaki, 2004, Sindaco et al., 2006, Göçman et al., 2007, Kreiner, 2007, Trapp, 2007, Franzen et al., 2008, Valakos et al., 2008, Baier et al., 2009, Tuniyev et al., 2009, Arakelyan et al., 2011, Bar & Haimovitch, 2011 and Stojanov et al., 2011.

Remarks: Type locality restriction to southern Syria and adjacent Lower Egypt *fide* T. Barbour & Amaral (1927: 27) invalid but restriction to Trieste *fide* Mertens & Müller (1928: 50) valid because Trieste was a part of Dalmatia in 19th century. Introduced into Malta *fide* Arakelyan et al. (2011: 97). Population from mainland Greece forms separate clade from those of the E Aegean islands, Cyprus, Turkey and Jordan *fide* Poulakakis et al. (2013: 625).

5. *Telescopus gezirae* Broadley, 1994a. J. Afr. Zool. 108(2): 205–206, fig. 3.

Type: Holotype, NMZB 11807, a 500 mm male (G.R.C. Van Someren, late 1966).

Type locality: "Blue Nile, 30 km North from Sennar, Sudan" [= Blue Nile River, 30 km N Sennar, N Sennar Prov., SE Sudan, 13°47'N, 33°35'E, elevation 415 m].

Distribution: Southeastern Sudan (Al Jazirah, N Sennar), 400–415 m.

Source: Crochet et al., 2008.

Remarks: Photographs of holotype in Crochet et al. (2008: fig. 7a).

6. *Telescopus hoogstraali* K.P. Schmidt & Marx, 1956. Fieldiana: Zool. 39(4): 33–34, figs. 5–6.

Type: Holotype, FMNH 72027, a 993 mm male (H. Hoogstraal, 15 May 1953).

Type locality: "Wadi el Sheikh, St. Catherine's Monastery area, Sinai Peninsula. Collected at about 5,000 feet altitude, among rocks on a hillside" [= Wadi el Sheikh, South Sinai Govern., NE Egypt, 28°33'N, 33°59'E, elevation 1585 m].

Distribution: Africa and Middle East. Northeastern Egypt (North Sinai, South Sinai), S Israel (Southern, West Bank) and W Jordan (Amman, Maan), 50–1585 m.

Sources: Zinner, 1977, Y. Werner, 1983, Bouskila & Amitai, 2001, Disi et al., 2001, Baha El Din, 2006, Amr & Disi, 2011 and Bar & Haimovitch, 2011.

Remarks: A valid species *fide* Y. Werner, 1983; a subspecies of *T. fallax fide* Zinner, 1977.

7. *Telescopus nigriceps* (Ahl, 1924). Arch. Naturg. 90A(5): 246. (*Tarbophis nigriceps*)

Type: Holotype, ZMB 23787, a 482 mm specimen (von Oppenheim), lost *fide* R. Günther *(in litt.)*.
Type locality: "Central-Mesopotamien" [= central Iraq].
Distribution: Southwestern Asia. Southeastern Turkey (Homs, Kilis, Sanliurfa), S Syria (Homs, Suweida), Lebanon, W Jordan (Amman, Irbid, Jarash, Karak, Maan) and Iraq (Al Anbar, Baghdad), 300–1000 m.
Sources: Disi et al., 1988, 2001, Leviton et al., 1992, Amr & Disi, 2011 and Ilgaz et al., 2007.
Remarks: A valid species *fide* K.P. Schmidt, 1939b, J.B. Rasmussen, 1979, Martens, 1993, Göçmen et al., 2007b and Amr & Disi, 2011. Occurs in Israel *fide* Amr & Disi (2011: 233). Two taxa masquerading in Jordan under *T. nigriceps fide* Disi et al. (2001: 315).

8. *Telescopus obtusus* (A. Reuss, 1834). Mus. Senck. 1: 137–139. (*Coluber obtusus*)

Synonyms: *Coluber geoffr.* Wagler, 1830 (*nomen illegitimum*), *Telescopus geoffroanus* Fitzinger, 1843, and *Telescopus geoffreanus* A.-M.-C. Duméril, Bibron & Duméril, 1854b (*nomen emendatum*).
Type: Lectotype SMF 19664 (formerly SMF-R III.V.10.a & SMF-B 9053, a), a female (E. Rüppell, 1832), designated by Mertens (1922: 181).
Type locality: "Ägypten" [= Egypt] via lectotype selection.
Distribution: Northeastern Africa. Egypt (Aswan, Assiut, Beheira, Beni Suef, Cairo, S El-Bahr El-Ahmar, Giza, Ismailia, Matrouh, Minya, Qena, Sohag, Suez), Eritrea, Djibouti (Djibouti, Obock), Ethiopia (Bale, Gemu Gofa, Gondar, Harerge, Kefa, Sidamo), Somalia (Bakool, Bay, Galguduud, Gedo, Hiraan, Jubbada Dhexe, Jubbada Hoose, Mogadishu, Mudug, Nugaal, Shabeellaha Dhexe, Shabeellaha Hoose, Togdheer, Woqooyi Galbeed), Sudan (El-Bahr El-Ahmar, Janub Darfur), E South Sudan (Eastern Equatoria, Jonglei), NE Uganda (E Northern), Kenya (Coast, Eastern, North-Eastern, Rift Valley) and NE Tanzania (Arusha, Kilimanjaro), NSL–1400 m.
Sources: Lanza, 1983a, 1990, W. Böhme et al., 1989, Broadley, 1994, Leviton et al., 1996, Schleich et al., 1996, Spawls et al., 2002, Crochet et al., 2008, and Largen & Spawls, 2010.
Remarks: A valid, clearly diagnosable, allopatric taxon *fide* Crochet et al. (2008: 39). A valid species fide Broadley (1994: 207) and Schleich et al. (1996: 524) but a subspecies of *T. dhara fide* Gasperetti (2008: 281), Leviton et al. (1992: 107), and Baha El Din (2006: 274).

9. *Telescopus pulcher* (Scortecci, 1935). Ann. Mus. Civ. Stor. Nat. 'Giacomo Doria,' Genova (1935–1937) 59: 2–5, 3 figs. (*Migiurtinophis pulcher*)

Type: Holotype, MSNG, a 410 mm male (N.M. Bronzi, 28 June 1930).
Type locality: "Migiurtinia (località imprecisata)" [= Bari and E Nugaal Region, NE Somalia].
Distribution: Northern Somalia (Bari, Nugaal, Togdheer) and E Ethiopia (E Harerge), 1065–1100 m.
Sources: H.W. Parker, 1949, Lanza, 1983a, 1990, Largen & Rasmussen, 1993 and Largen & Spawls, 2010.
Remarks: Holotype was supposed to be returned to MSNG from Prof. O. De Beaux *fide* Scortecci but apparently was not as it is not listed by Capoccacia, 1961.

10. *Telescopus rhinopoma* (Blanford, 1874). Ann. Mag. Nat. Hist. (4) 14(79): 34. (*Dipsas rhinopoma*)

Synonym: *Dipsadomorphus jollyi* Wall, 1914d.
Types: Syntypes (2), BMNH 1946.1.5.10, a 1207 mm male (W.T. Blanford), and ZSI 3500 (formerly IMC).
Type locality: "Carmania, Persia" [= Kerman Prov., SE Iran]. Restricted to Kerman, Kerman Prov., S Iran *fide* W. Böhme (1977a: 203, fig. 1.4).
Distribution: Southwestern Asia. Southern Turkmenistan, S Iran (Central, Fars, Razavi Khorasan, Semnan, Sistan va Baluchestan), SW Afghanistan (Qandahar) and S Pakistan (Balochistan, Sindh), 985–1985 m.
Sources: M.A. Smith, 1943, Mertens, 1969a, Bannikov et al., 1977, W. Böhme, 1977b, Rai, 1978, Mahendra, 1984, Kudrjavtsev & Mamet, 1989a, Latifi, 1991, Leviton et al., 1992, Szczerbak, 1994, M.S. Khan, 2002, 2006 and Ananjeva et al., 2004.
Remarks: W. Böhme (1977b: 203) indicated that *T. rhinopoma* and *T. fallax* may represent a cline and thus be conspecific.

11. *Telescopus semiannulatus* A. Smith, 1849 *in* 1838–1849. Illust. Zool. So. Africa, Rept.: pl. 72, figs. a–c, 2 pp.

Synonyms: *Tarbophis semiannulatus meridionalis* F. Werner, 1898d, and *Telescopus semiannulatus polystictus* Mertens, 1954b.
Type: Holotype, BMNH 1946.1.2.55 (formerly BMNH 1871.4.21.1), a 743–749 mm specimen (A. Smith, 1825–1837).
Type locality: Not stated. Designated as Loanda, South Africa [= La Locanda, E Western Cape Prov., SW South Africa, 33°51'S, 22°31'E, elevation 350 m] by inference *fide* Loveridge (1957a: 270).
Distribution: Southeastern Africa. Southern Democratic Republic of the Congo (Bandundu, Bas-Congo, Katanga, Kinshasa), NW Angola (Luanda), SE Kenya (Coast, Eastern), Tanzania (Dodoma, Iringa, Lindi, Mbeya, Morogoro, Mtwara, E Piwani, Rukwa, Ruvuma, Singida, Tabora, E Tanga), Zambia (Central, Copperbelt, Eastern, Luapula, Lusaka, Northern, North-Western, Southern, Western), Malawi (Central, Northern, Southern), Mozambique (Gaza, Inhambane, Manica, Maputo, Sofala, Tete), Zimbabwe (Bulawayo, Manicaland, Mashonaland Central, Mashonaland East, Mashonaland West, Masvingo, Matabeleland North,

Matabeleland South, Midlands), Botswana (Central, Gaborone, Ghanzi, Kgalagadi, Kweneng, North West, North East, South East, Southern, Tuli), Namibia (Caprivi, Erongo, Hardap, Karas, Khomas, Kunene, Okavango, Oshikoto, Otjozondjupa), Swaziland and South Africa (Eastern Cape, Gauteng, KwaZulu-Natal, Limpopo, Mpumalanga, Northern Cape, North West), NSL–1700 m.

Sources: Mertens, 1954b, 1955, 1971, Sweeney, 1961, FitzSimons, 1962a, 1974, Witte, 1962, Thys van den Audenaerde, 1965, Broadley, 1983, Pienaar et al., 1983, Auerbach, 1987, Branch, 1988, Clauss & Clauss, 2002, Spawls et al., 2002, Broadley et al., 2003, M. Griffin, 2003, Marais, 2004 and Broadley & Blaylock, 2013.

Remarks: Holotype not in BMNH or RSM *fide* V.F.M. FitzSimons (1937: 262). Record from East London, South Africa needs confirmation *fide* Broadley, 1983.

12. *Telescopus somalicus* (H.W. Parker, 1949). Zool. Verh. (6): 88. (*Tarbophis dhara somalicus*)

Type: Lectotype, BMNH 1946.1.23.52 (formerly BMNH 1893.11.21.64), a male (J.W. Gregory, 1892–1893), designated by Crochet et al. (2008: 40).

Type locality: "Ngatana, Kenya" via lectotype selection. Further specified as the old village Ngatana on the lower Tana River valley in Coast Prov., Kenya, ca. 1.7 km NW of the new village Wema (2°11'S, 40°12'E, elevation 15 m) and 36 km NW Witu (2°23'S, 40°26'E, elevation 20 m) *fide* Broadley & Wallach (2007a: 28) and Crochet et al. (2008: 40).

Distribution: Southern Somalia (Jubbada Dhexe) and Kenya (Coast, S Rift Valley), NSL–600 m.

Sources: Spawls et al., 2002 and Crochet et al., 2008.

Remarks: A valid, clearly diagnosable, allopatric taxon *fide* Crochet et al. (2008: 40).

13. *Telescopus tessellatus* (Wall, 1908i). J. Bombay Nat. Hist. Soc. (1907–1908) 18(4): 802–803. (*Tarbophis tessellatus*)

Synonym: *Tarbophis martini* K.P. Schmidt, 1939b.

Type: Holotype, BNHS 2665, ca. 457 mm specimen in five pieces (A.T. Wilson, 10 July 1908).

Type locality: "Maidan Mihaftan 30 m. E. of Shushtar in S.W. Persia" [= Maidan-e-Naftun, Khuzistan Prov., SW Iran, 31°59'N, 49°19'E, elevation 300 m *fide* Schätti et al., 2009b: 121].

Distribution: Southwestern Iran (Central, Esfahan, Hamadan, Khuzestan, Kordestan) and Iraq (Bagdad), 20–2245 m.

Sources: W. Böhme, 1977b, Latifi, 1991 and Leviton et al., 1992.

14. *Telescopus tripolitanus* (F. Werner, 1909a). Zool. Jahrb. (Abt. Syst. Ökol. Geog. Tiere) 27(6): 619–620. (*Leptodira tripolitana*)

Synonyms: *Tarbophis guidimakaensis* Chabanaud, 1916b, *Tarbophis gabesi* Domergue, 1955, and *Pseudotarbophis gabesiensis,* Domergue, 1959 (*nomen emendatum*).

Type: Holotype, NMW, a 740 mm male (R. Storch via W. Schlüter), lost *fide* Kramer & Schnurrenberger (1963: 528).

Type locality: "Tripolis" [= Tripoli, Tripoli Govern., NW Libya, 32°53'N, 13°11'E, elevation 20 m]. Specified as NW Libya or S Tunisia *fide* Crochet et al. (2008: 35).

Distribution: North Africa. Morocco (Agadir, Meknes, Ouarzazate, Tan Tan, Tata), Algeria (Tamanrasset, Tindouf), S Tunisia (Gabès, Gafsa), Libya (Nalut), Western Sahara (NE Rio de Oro, Saguia El Hamra), Mauritania (Guidimaka, Tagant, Trarza), N Senegal (Fatick, Kaolack, Louga, Matam, Saint-Louis, Tambacounda, Thiès), Mali (Gao, Koulikoro, Mopti, Tombouctou), Niger (S Agadez, Dosso, Maradi, Niamey, Tahoua, Zinder), N Burkina Faso (Nord, Sahel), N Nigeria (Bauchi, Sokoto), Central African Republic (Vakaga) and Sudan (Shamal Kurdufan), NSL–1135 m.

Sources: Kramer & Schnurrenberger, 1963, Domergue, 1966, Roman, 1980, W. Böhme & Nikolaus, 1989, W. Böhme et al., 1989, Geniez et al., 2004, Chippaux, 2006, J.-F. Trape & Mané, 2006b, Chirio & LeBreton, 2007 and Crochet et al., 2008.

Remarks: *Telescopus guidimakaensis* a valid species *fide* Geniez et al. (2004: 155). Probably occurs in ext. N Cameroon *fide* Chirio & LeBreton (2007: 548).

15. *Telescopus variegatus* (J.T. Reinhardt, 1843). Kgl. Danske Vidensk. Selsk. Natur. Math. Afh. 10: 249–251, pl. 1, figs. 15–17. (*Dipsas variegata*)

Synonym: *Leptodira pobequini* Mocquard, 1902b.

Type: Holotype, ZMUC 63126, a 588 mm specimen (J.R. Chenon, 1826–1836).

Type locality: "Guinea" [= coast of Ghana *fide* Hughes & Barry, 1969: 1027]. Restricted to vicinity of Accra, Ghana *fide* J.B. Rasmussen & Hughes (1997: 14, 16).

Distribution: West Africa. Southern Senegal (Dakar, Kédougou, Tambacounda), Gambia (Western), Guinea (Kouroussa), S Mali (Kayes, Koulikoro), Sierra Leone (Northern), Ivory Coast (Bouna, Daloa, Toumodi), Burkina Faso (Centre-Est, Centre-Nord, Centre-Ouest, Est, Hauts-Bassins, Nord, Sud-Est, Volta-Noire), N Ghana (Upper East, Upper West), S Togo (Centrale, Plateaux), Benin (Alibori, Atakora, Collines), Nigeria (Gongola, Kwara), S Niger (Niamey, Tahoua), SW Chad (Chari-Baguirmi, Mayo-Kebbi), N Cameroon (Extreme-Nord, Nord) and N Central African Republic (Ouham, Vakaga), 270–1000 m.

Sources: Doucet, 1963, Leston & Hughes, 1968, Hughes & Barry, 1969, Papenfuss, 1969, Roman, 1977, Roman, 1980, Hughes, 1983, 2013, Rödel et al., 1995, Greenbaum

& Carr, 2005, Chippaux, 2006, Chirio & Ineich, 2006, J.-F. Trape & Mané, 2006b, Chirio & LeBreton, 2007, Ullenbruch et al., 2010 and Segniagbeto et al., 2011.

TERETRURUS Beddome, 1886 (Uropeltidae)

Type species: *Plectrurus sanguineus* Beddome, 1867.
Distribution: Southwestern India.
Sources: M.A. Smith, 1943, Constable, 1949, Gans, 1966, Murthy, 1981c, 1982b, Cadle et al., 1990, McDiarmid et al., 1999 and Pradhan, 2002.

1. *Teretrurus sanguineus* (Beddome, 1867). Madras Quart. J. Med. Sci. 11: 14, pl. 1, fig. 2. (*Plectrurus sanguineus*)

Synonyms: *Platyplectrurus hewstoni* Beddome, 1876, *Plectrurus scabricauda* Theobald, 1876, and *Teretrurus travancoricus* Beddome, 1886.
Types: Syntypes (12), BMNH 1946.1.16.57–62 (R.H. Beddome, 1857–1867), MNHN 1895.116a–c (R.H. Beddome, 1857–1867), and NMW 2161.1–3 (R.H. Beddome, 1857–1867, via F. Steindachner, 1883).
Type locality: "Anamallay forests; 4,000 feet elevation, Madras Presidency" [= Anamalai Hills, S Western Ghats, E Kerala /W Tamil Nadu States, SW India, ca. 10°22'N, 77°08'E, 1220 m].
Distribution: Southwestern India (Western Ghats of E Kerala, SW Tamil Nadu), 220–2800 m.
Remarks: Type locality above Ponachi *fide* BMNH catalogue. Only a single specimen described originally so one syntype is lectotype (with 144 V + 7 SC).

TETRACHEILOSTOMA Jan, 1861b (Leptotyphlopidae)

Synonym: *Eucephalus* Fitzinger, 1843.
Type species: *Typhlops bilineatus* Schlegel, 1839 *in* 1837–1844.
Distribution: Lesser Antilles.
Sources: Hedges, 2008 and R.W. Henderson & Powell, 2009.

1. *Tetracheilostoma bilineata* (Schlegel, 1839 *in* 1837–1844). Abbild. Amph.: 36, pl. 32, figs. 5–8. (*Typhlops bilineatus*)

Synonyms: *Stenostoma bilineatum* A.-M.-C. Duméril & Bibron, 1844, *Glauconia bilineata* Boulenger, 1893a, and *Leptotyphlops bilineatus* Hahn, 1980a.
Types: Syntypes (3), RMNH 3704a–b and MNHN 3234, longest syntype 101 mm (A. Plée, 1823–1825).
Type locality: "Martinique" [= Lesser Antilles, West Indies].
Distribution: Martinique and Chancel Is., NSL–90 m.
Sources: Barbour, 1915a, R. Thomas, 1965a, Hahn, 1980a, R. Thomas et al., 1985, A. Schwartz & R.W. Henderson, 1991, Wallach, 1998b, McDiarmid et al., 1999, Breuil, 2003 and Hedges, 2008.
Remarks: Guadeloupe record doubtful, but see Lazell *in* Thomas (1965a: 5).

2. *Tetracheilostoma breuili* (Hedges, 2008). Zootaxa (1841): 9–11, figs. 1b, 2a (middle), 2b (middle). (*Leptotyphlops breuili*)

Type: Holotype, USNM 564810 (formerly SBH 267731), a 102–113 mm male (S.B. Hedges & C.A. Hass, 13 June 2006).
Type locality: "north slope of Maria Major Island, St. Lucia, 60 m (13° 43.430' N, 60° 55.897' W)."
Distribution: St Lucia (Maria Major Is.), NSL–60 m.

3. *Tetracheilostoma carlae* (Hedges, 2008). Zootaxa (1841): 5–7, figs. 1a, 2a (lower), 2b (right). (*Leptotyphlops carlae*)

Type: Holotype, USNM 564819 (formerly SBH 267708), a 93–104 mm female (S.B. Hedges & C.A. Hass, 9 June 2006).
Type locality: "near Bonwell, St. Joseph Parish, Barbados, 280 m (13° 11.196' N, 59° 32.445' W)."
Distribution: Barbados (St. John, St. Joseph), 100–280 m.
Source: Emsley, 1966.

TETRALEPIS Boettger, 1892a (Colubroidea *incertae sedis*)

Type species: *Tetralepis fruhstorferi* Boettger, 1892a.
Distribution: Western Indonesia.
Sources: Hodges, 1993 and I. Das, 2010, 2012.
Remarks: This genus unassignable to family *fide* Zaher et al. (2009: 132).

1. *Tetralepis fruhstorferi* Boettger, 1892a. Ber. Offenb. Ver. Naturk. (1887–1891) 29–32: 125–126.

Type: Lectotype, SMF 17891 (formerly SMF-B 7320, a), a 502 mm female (H. Fruhstorfer, May 1891), designated by Mertens (1922: 179).
Type locality: "Tengger-Gebirge, 1200 m H., Ost-Java" [= Tengger Mtns., E Java, W Indonesia, elevation 1200 m] via lectotype selection.
Distribution: Western Indonesia (E Java), 1200 m. Known only from type series.

†*TEXASOPHIS* Holman, 1977b (Colubridae)

Type species: †*Texasophis fossilis* Holman, 1977b.

Distribution: Middle Oligocene to middle/upper Miocene of W Europe and lower Oligocene to upper Miocene of cen. USA. Fossils unassigned to species include lower Miocene (Arikareean) of USA (Nebraska), and upper Miocene (Clarendonian) of USA (Nebraska).

Sources: Holman, 1979b, 1984a, 1984b, 2000a, Rage, 1984b and Parmley & Hunter, 2010.

Remarks: *Texasophis* might represent an extant Asian genus *fide* Szyndlar (1994: 33).

1. †*Texasophis bohemiacus* Szyndlar, 1987. J. Vert. Paleont. 7(1): 62–64, fig. 7a–e.

Type: Holotype, CUP 1238, one trunk vertebra.

Type locality: "Lower Miocene (early Orleanian or Ottnangian; European Land Mammalian Biozone MN 4 = upper Hemingfordian in North America) locality of Dolnice near to the town of Cheb in westernmost Bohemia (Czechoslovakia)." [= Czech Republic].

Distribution: Lower Oligocene (Rupelian, MP 22: 28.4–33.9 mya) of Germany, and lower Miocene (Orleanian, MN4: 16.0–16.9 mya) of Czech Republic.

Sources: Szyndlar, 1994 and Szyndlar, 2012.

2. †*Texasophis fossilis* Holman, 1977b. Herpetologica 33(4): 397–398, figs. 9a–b.

Type: Holotype, SMPSMU 63691, one trunk vertebra.

Type locality: "Town Bluff site, eastern-central Tyler County Texas" [= Fleming Formation, Early Barstovian, middle Miocene].

Distribution: Lower Miocene (Hemingfordian: 16.3–20.6 mya) of USA (Colorado), and middle Miocene (Barstovia: 13.6–16.3 mya) of USA (Texas).

Sources: Holman, 2000a, 2003.

3. †*Texasophis galbreathi* Holman, 1984a. J. Vert. Paleont. 3(4): 223–225, fig. 1.

Type: Holotype, MSUVP 1038, one trunk vertebra (E.C. Galbreath).

Type locality: "Flats Area; SW 1/4 sec. 12, T. 11 N., R. 54 W., Logan County, Colorado. Scenic member of Brule Formation, White River Group; Orella Land Mammal Age, middle Oligocene." Emended to Flats, E. 1/2 Sec. 12, T11N, R54W, Logan Co., Colorado, Scenic Member of the Brule Formation. White River Group: early Orellan to Whitneyan age; middle to late Oligocene *fide* E.C. Galbreath *in* Sullivcan & Holman (1996: 366).

Distribution: Lower Oligocene (Orellan: 33.3–33.9 mya) of USA (Colorado). Known only from type locality.

Sources: Sullivan & Holman, 1996 and Holman, 2000a.

4. †*Texasophis hecki* M. Böhme, 2008. Cour. Forsch.-Inst. Senck. (260): 170, figs. 3a–c, pl. 4, figs. 10a–e.

Type: Holotype, SMF 5014, one trunk vertebra.

Type locality: "Oberleichtersbach, Rhön Mountains, Northern Bavaria, Germany; uppermost Oligocene, MP 30."

Distribution: Upper Oligocene (Chattian, MP 30: 23.0–28.4 mya) of Germany. Known only from type locality.

5. †*Texasophis meini* Rage & Holman, 1984. Géobios 17(1): 91–93, figs. 2a–e.

Type: Holotype, UCBL 285002, one trunk vertebra.

Type locality: "La-Grive-Saint-Alban (Isère) fissure M, Sansan, France, zone MN 7, Astaracien" [= French Miocene].

Distribution: Lower to middle Miocene (Orleaniain-Astaracian, MN 4–8: 11.1–16.9 mya) of France, Germany, Hungary and Ukraine.

Sources: Holman, 1984b, Augé & Rage, 2000 and Szyndlar, 1987, 1991, 2012.

Remarks: Congeneric with *Boiga fide* Zerova, 1987.

6. †*Texasophis wilsoni* Holman, 1984b. Copeia 1984(3): 660–661, fig. 1.

Type: Holotype, FHSC 3640, one trunk vertebra (R.J. Zakrzewski).

Type locality: "Ogallala Formation, middle or late Clarendonian Age (late medial Miocene) near Wakeeney, Trego County, Kansas" [= middle or upper Miocene].

Distribution: Middle/upper Miocene (Clarendonian: 10.3–13.6 mya) of USA (Kansas).

Source: Holman, 2000a.

THALASSOPHINA M.A. Smith, 1926a (Elapidae)

Synonyms: *Thalassiophis* – A.-M.-C. Duméril, Bibron & Duméril, 1854b (*nomen incorrectum*), *Praescutata* Wall, 1921g (*nomen ineditum*), *Praescuta* – Deraniyagala, 1949 (*nomen incorrectum*), *Praecutata* – Nutapahnd & Tumvipart, 1982 (*nomen incorrectum*), and *Thalasophina* – W.W. Tanner & Avery, 1982 (*nomen incorrectum*).

Type species: *Thalassophis viperina* P. Schmidt, 1852.

Distribution: Marine waters of South, SE and E Asia, including Andaman Sea, Arabian Sea, Bay of Bengal, Gulf of Oman, Gulf of Thailand, Hainan Strait, Indian Ocean, Java Sea, Persian Gulf, Strait of Malacca and South China Sea.

Sources: Wall, 1909a, 1921g, Rooij, 1917, M.A. Smith, 1926a, 1943, Cochran, 1930, Tchang, 1932, Bourret, 1934g, C.H. Pope, 1935, Volsoe, 1939, C. Haas, 1950, Deraniyagala, 1955, E.H. Taylor, 1965, Minton, 1966, 1975, Barme, 1968, Halstead, 1970, McDowell, 1972a, Murthy, 1977b, 1986, Voris, 1977, J.D. Romer, 1979a, M.S. Khan, 1982b, 2002, Tweedie, 1983, Joger, 1984, Kharin, 1984d, Zhou, 1985, Tian et al., 1986, Gasperetti,

1988, A.R. Rasmussen, 1989a–b, K.K.P. Lim & Chou, 1990, B.L. Lim, 1991, Stuebing, 1991, Leviton et al., 1992, Golay et al., 1993, Zhao & Adler, 1993, Xiang & Li, 2009, Zaher et al., 2009, I. Das, 2010, 2012, A.R. Rasmussen et al., 2011, A. Silva et al., 2011 and Hoser, 2012e.

Remarks: *Praescutata* was not proposed *fide* errata sheet in Wall, 1921g (Golay & McCarthy *in* Golay et al., 1993: 247). A synonym of *Lapemis fide* McDowell *in* Golay et al. (1993: 247) or *Hydrophis fide* Sanders et al., 2013.

1. *Thalassophina viperina* (P. Schmidt, 1852). Abh. Geb. Naturw. Ver. 2(2): 79–81, pl. 3. (*Thalassophis viperina*)

Synonyms: *Disteira praescutata* A.-M.-C. Duméril, 1853 (*nomen nudum*), *Disteira praescutata* A.-M.-C. Duméril, Bibron & Duméril, 1854, *Hydrophis doliata* J.G. Fischer, 1856a (*nomen praeoccupatum*), *Hydrophis nigra* J. Anderson, 1872, *Hydrophis obscurus* Jan, 1859b (*nomen praeoccupatum*), *Hydrophis jayakari* Boulenger, 1887d, *Hydrophis plumbea* Murray, 1887, *Hydrophis viperinus* – Boettger, 1888c, *Thallassophina viperina* – Corkill, 1932 (*nomen incorrectum*), *Thalassophinae viperina* – Maegraith, 1958 (*nomen incorrectum*), *Praescutata viperine* – Sayed, 1972 (*nomen incorrectum*), *Paraescutata viperina* – M.S. Khan, 1982 (*nomen incorrectum*), *Enhydrus plumbea* – Culotta & Pickwell, 1993 (*lapsus calami*), and *Thalassophina veperina* – Culotta & Pickwell, 1993 (*nomen incorrectum*).

Type: Holotype, ZMH 404, a 650 mm male (Meyer & Werner, 1847), destroyed in July 1943 during World War II.

Type locality: "Küste von Java" [= coast of Java, Greater Sundas, SW Indonesia].

Distribution: Asia. United Arab Emirates (Sir Abu Nu'ayr Is.), Oman (Muscat), S Iran (Hormozghan, Sistan va Baluchestan), Pakistan (Balochistan, Sindh), India (Karnataka, Kerala, Maharashtra, Orissa, Tamil Nadu, West Bengal), Sri Lanka (Northern, North-Western, Vaduga Bank, Western), Myanmar (Tanintharyi, Yangon), S Thailand (Pattani, Phatthalung, Phuket, Prachuap Khiri Khan, Singgora, Songkhla, Trat, Phuket Is.), Cambodia (Koh Kong Is.), Vietnam (Dong Nai, Thuan Hai), West Malaysia, Singapore, East Malaysia (Sarawak), W Indonesia (Java, Sumatra), SE China (Fujian, Guangdong, Guangxi, Hainan, Hong Kong) and Taiwan.

T

THALASSOPHIS P. Schmidt, 1852 (Elapidae)

Synonyms: *Thalassiophis* – A.-M.-C. Duméril, Bibron & Duméril, 1854 (*nomen incorrectum*), *Thallassophis* – Sleen, 1932 (*nomen incorrectum*), *Thallassophis* – Maass-Berlin, 1933 (*nomen incorrectum*), *Talassophis* – Boquet, 1964 (*nomen incorrectum*), and *Thalosophis* – B.L. Lim & Sawai, 1975 (*nomen incorrectum*).

Type species: *Thalassophis anomala* P. Schmidt, 1852.

Distribution: Marine waters of SE Asia and Indonesia.

Sources: M.A. Smith, 1926a, C.P.J. Haas, 1950, Bergman, 1954, E.H. Taylor, 1965, B.L. Lim & Sawai, 1975, Tweedie, 1983, Golay et al., 1993, A.R. Rasmussen, 1997, David & Ineich, 1999, Zaher et al., 2009, I. Das, 2010, 2012, A.R. Rasmussen et al., 2011 and Hoser, 2012e.

1. *Thalassophis anomalus* P. Schmidt, 1852. Abh. Geb. Naturw. Ver. Hamburg 2(2): 81–82, pl. 4, figs. 1–5. (*Thalassophis anomala*)

Synonym: *Thalassophis anomalous* – Culotta & Pickwell, 1993 (*nomen incorrectum*).

Type: Holotype, ZMH 3342 (formerly ZMH 402), a 720 mm male (Meyer & Werner, 1847).

Type locality: "auf der Rhede von Samarang, an der Küste von Java" [= off the port of Semarang (6°58'S, 110°25'E), north coast of Java, Indonesia].

Distribution: Southeast Asia and Indonesia. Thailand, Cambodia, Vietnam, West Malaysia (Johore, Pahang), East Malaysia (Sabah, Sarawak) and Indonesia (Java, Moluccas, Sumatra).

THAMNODYNASTES Wagler, 1830 (Xenodontidae)

Synonyms: *Dryophylax* Wagler, 1830, *Dryophilax* Bonaparte, 1831 (*nomen emendatum*), *Dryophilas* A.-M.-C. Duméril & Bibron, 1844 (*nomen emendatum*), *Thamnodynaster* – Rüppell, 1845 (*nomen incorrectum*), *Dryiophylax* – A.-M.-C. Duméril, Bibron & Duméril, 1854b (*nomen incorrectum*), *Mesotes* Jan, 1863a, *Dryophyllax* – Hoffmann, 1890 (*nomen incorrectum*), *Tamnodynastes* – Brazil, 1914 (*nomen incorrectum*), *Dryophylar* – Amaral, 1935a (*nomen incorrectum*), and *Thamnodinastes* – Abalos & Bucher, 1970 (*nomen incorrectum*).

Type species: *Coluber pallidus* Linnaeus, 1758.

Distribution: South America.

Sources: Amaral, 1944i, Bailey *in* J.A. Peters & Orejas-Miranda, 1970, Pérez-Santos & Moreno, 1988, Franco et al., 2003, Bailey et al., 2005, Zaher et al., 2009 and Vidal et al., 2010.

Remarks: A *Thamnodynastes* sp. is listed for Trinidad by Boos, 2001. A specimen listed as *Thamnodynastes* by Salles et al., 2010, is likely a new species.

1. *Thamnodynastes almae* Franco & Ferreira, 2003. Phyllomedusa 1(2): 60–62, figs. 1–2.

Type: Holotype, IB 52135, a 657 mm male (CHESF, 20 April 1988), destroyed by fire 15 May 2010.

Type locality: "Usina Hidroelétrica Luiz Gonzaga (entre 09°06'S - 38°20'W, 280 m e 08°51'S - 38°45'W, 300 m), no rio São Francisco, municipio de Rodelas, estado da Bahia, Brasil."

Distribution: Eastern Brazil (Alagoas, E Bahia, Ceará, Paraíba, Pernambuco, Rio Grande do Norte), 135–300 m.

Sources: Roberto et al., 2009a, Guedes, 2010 and Jorge & Freire, 2011.

2. *Thamnodynastes ceibae* Bailey & Thomas, 2007. Mem. Fund. La Salle Cien. Nat. (2006) 166: 16–17, figs. 8a–b.

Type: Holotype, UMMZ 218579 (formerly MBUCV 8165), a 583 mm male (J. Racenis & J.A. Roze, 19 May 1963).

Type locality: "La Ceiba, Trujillo State, Venezuela, near sea level."

Distribution: Western Venezuela (Trujillo), NSL. Known only from type locality.

Source: Navarrete et al., 2009.

3. *Thamnodynastes chaquensis* Bergna & Alvarez, 1993. Facena 10: 7–11, fig. 2a, pl. 1, figs. 1a–b.

Type: Holotype, UNNE-CHC 339, a 480 mm male (S. Bergna, 15 Nov. 1992).

Type locality: "Colonia Las Mercedes, Departamento San Fernando, Chaco, nordeste Argentino."

Distribution: Southern South America. Central Brazil (Mato Grosso), Bolivia (Santa Cruz, Tarija), SW Paraguay (Central, Misiones, Neembucú, S Presidente Hayes), Uruguay (Colonia) and NE Argentina (Chaco, NW Corrientes, Formosa, E Salta, N Santa Fe), 50–1300 m.

Sources: Giraudo, 1996, Leynaud & Bucher, 1999, Carreira-Vidal, 2003, Cacciali, 2008 and Jensen, 2008.

4. *Thamnodynastes chimanta* Roze, 1958c. Acta Biol. Venez. 2(25): 305–307.

Type: Holotype, FMNH 74042, a 340 mm male (J.A. Steyermark & J. Wurdack, 10 Feb. 1955).

Type locality: "Chimantá Tepui, Estado Bolivar, Venezuela. Colectado sobre elsuelo en la sabana de la cumbre Apacara Tepui; 2.200 metros de altura."

Distribution: Eastern Venezuela (Bolívar), 1900–2600.

Sources: Lancini, 1986, C.W. Myers & Donnelly, 1996, Gorzula & Señaris, 1998 and Navarrete et al., 2009.

5. *Thamnodynastes corocoroensis* Gorzula & Ayarzagüena, 1996. Publ. Asoc. Amigos Doñana (1995) (6): 6–8, fig. 1.

Type: Holotype, MHNLS 11376, a 258 mm female (S. Gorzula, 10 Nov. 1987).

Type locality: "Tepuy Corocoro, Estado Bolívar, Venezuela (05°46'N-66°11'W), 2.150 m.n.s.m.," (partly in error). Corrected to in Estado Amazonas *fide* Gorzula & Señaris (1998: 180, 255) (locality H-017).

Distribution: Southern Venezuela (N Amazonas/W Bolívar), 2150 m. Known only from type locality.

Sources: Gorzula & Señaris, 1998, C.W. Myers & Donnelly, 2001 and Navarrete et al., 2009.

6. *Thamnodynastes dixoni* Bailey & Thomas, 2007. Mem. Fund. La Salle Cien. Nat. (2006) 166: 13–16, figs. 5a–b, 6–7.

Type: Holotype, TCWC 44896, a 432 mm female (J.R. Dixon & M.A. Staton, 17 Oct. 1973).

Type locality: "Hato La Guanota, 4 km west of San Fernando, Apure State, Venezuela."

Distribution: Southern Venezuela (Amazonas, Apure, Bolívar, Guárico) and SE Colombia (Arauca), NSL–500 m.

7. *Thamnodynastes duida* C.W. Myers & Donnelly, 1996. Amer. Mus. Novit. (3172): 48–51, figs. 29a, 30a, 31a.

Type: Holotype, AMNH 36608, a 455 mm male (Sidney F. Tayler Jr. [Duida Exped.], 27 Jan. 1929).

Type locality: "the south end of the summit of Cerro Duida, valley between peaks 7 and 16, about 6610 ft. (2015m) elevation, Amazonas, Venezuela."

Distribution: Southern Venezuela (Amazonas), 2015 m. Known only from type locality.

Sources: C.W. Myers & Donnelly, 2001 and Navarrete et al., 2009.

8. *Thamnodynastes gambotensis* Pérez-Santos & Moreno, 1989b. Boll. Mus. Reg. Sci. Nat. Torino 7(1): 1–4, figs. 1–4.

Type: Holotype, MNCN 11187, a 505 mm male (C. Pérez-Santos, 11 July 1987).

Type locality: "2 km de Gambote sobre la carretera de Cartagena de Indias, Dep. de Bolivar, Colombia." Elevation 5 m *fide* García-Díez & González-Fernández (2013: 57).

Distribution: Northern Colombia (Atlántico, Bolívar, Cesar), NSL–150 m.

Source: Bailey & Thomas, 2007.

Remarks: Photographs of holotype in García-Díez & González-Fernández (2013: figs. 3 left–right).

9. *Thamnodynastes hypoconia* (Cope, 1860e). Proc. Acad. Nat. Sci. Philadelphia 12(6): 247–248. (*Tachymenis hypoconia*)

Type: Holotype, ANSP 3747 (A. Kennedy).

Type locality: "Buenos Ayres" [= Buenos Aires, Buenos Aires Prov., NE Argentina, 34°36'S, 58°23'W, elevation 15 m].

Distribution: Southern South America. Southern Brazil (Goiás, Minas Gerais, Paraná, Rio de Janeiro, Rio Grande do Sul, Santa Catarina, São Paulo, Tocantins), Uruguay (Canelones, Colonia, Durazno, Flores, Florida, Lavalleja, Maldonado, Montevideo, Rivera, Rocha, San José, Treinta y Tres), Paraguay (Alto Paraguay, Alto Paraná, Amambay, Asunción, Caaguazú, Caazapá, Canindeyú, Central, Concepción, Cordillera, Guairá, Itapúa, Misiones, Neembucú, Paraguarí, Presidente Hayes, San Pedro) and Argentina (Buenos Aires, Córdoba, Santa Fe), NSL–650 m.

Sources: J.D. Williams & Francini, 1991, Bergna & Alvarez, 1993, Bellagamba & Vega, 1996, Cacivio, 1997, Leynaud & Bucher, 1999, Marques et al., 2001, Carreira-Vidal et al., 2005, Cacciali, 2008, Farlani et al., 2010 and Bellini et al., 2013.

10. *Thamnodynastes lanei* Bailey, Thomas & Silva, 2005. Phyllomedusa 4(2): 93–95, figs. 14a–b.

Type: Holotype, UMMZ 109081, a 648 mm male (J.R. Bailey, 24 Jan. 1941).

Type locality: "Salobra, Mato Grosso do Sul, Brazil" [= Salobra, W Mato Grosso do Sul State, SW Brazil, 21°14'S, 57°08;'W, elevation 200 m].

Distribution: Central South America. Brazil (Alagoas, Mato Grosso, Mato Grosso do Sul, Pará), N and E Bolivia (Beni, Santa Cruz), Peru (Loreto), Paraguay (San Pedro) and N Argentina (Formosa), NSL–200 m.

11. *Thamnodynastes longicaudus* Franco, Ferreira, Marques & Sazima, 2003. Zootaxa (334): 2–5, figs. 1–3.

Type: Holotype, IB 59545, a 574 mm male (M.D. Galanta, 22 Dec. 1998), destroyed by fire 15 May 2010.

Type locality: "Brazil, São Paulo, São Lourenço da Sena, (23°51'S—46°56'W, 690 m a.s.l.)."

Distribution: Southeastern Brazil (Espírito Santo, Rio de Janeiro, São Paulo), 520–905 m.

Source: Lima-Silveira, 2006.

T

12. *Thamnodynastes marahuaquensis* Gorzula & Ayarzagüena, 1996. Publ. Asoc. Amigos Doñana (1995) (6): 8–11, fig. 2.

Type: Holotype, MHNLS 12507, a 396 mm female (J. Serrano, 11 Dec. 1992).

Type locality: "Tepuy Marahuaca Norte, Estado Amazonas, Venezuela (3°45'N-65°30'W), 2.500 m n.s.m."

Distribution: Southern Venezuela (Amazonas), 2500 m. Known only from type locality.

Sources: C.W. Myers & Donnelly, 2001 and Navarrete et al., 2009.

13. *Thamnodynastes pallidus* (Linnaeus, 1758). Syst. Nat., ed. 10 1: 221. (*Coluber pallidus*)

Synonyms: *Coluber strigilis* Thunberg, 1787, and *Natrix punctatissima* Wagler *in* Spix, 1824, *Coluber lineolatus* Wied-Neuwied, 1825c, *Coluber nattereri* Mikan, 1828, *Thamnodynastes puntatissimus* – A.-M.-C. Duméril, Bibron & Duméril, 1854b, *Mesotes obstrusus* Jan, 1863a, *Mesotes obstrusus plataensis* Jan, 1863a (*nomen nudum*), *Mesotes obstrusus plataensis* Jan, 1866 *in* Jan & Sordelli, 1866–1870, and *Thamndodynastes nattereri laevis* Boulenger, 1885a.

Type: Neotype, ZMUU 292, a 509 mm male, designated by Bailey et al. (2005: 87).

Type locality: Unknown via neotype selection.

Distribution: Northwesrtern South America. Colombia (Atlántico, Guajira, Magdalena), S Venezuela (Amazonas, Apure, Bolívar, Cojedes, Delta Amacuro, Zulia), N Guyana (Essequibo Islands-West Demerara), N Suriname (Commewijne, Paramaribo, Wanica), N French Guinea (N Cayenne), Ecuador, Brazil (Bahia, Pará, Rio de Janeiro, São Paulo, Comprida and Grande Is.) and Peru, NSL–500 m.

Sources: Roze, 1966a, Lancini, 1986, Vanzolini et al., 1980, Chippaux, 1987, Bailey & Thomas, 1991, Cunha & Nascimento, 1993, Reati, 1996, Yanosky et al., 1996, C.W. Myers & Donnelly, 1997, Gorzula & Señaris, 1998, Starace, 1998, Freitas, 1999, Abuys, 2003, Franco & Ferreira, 2003, Harvey & Muñoz, 2004, Bailey et al., 2005, Duarte Roche & Van Sluys, 2006, Navarrete et al., 2009 and C.J. Cole et al., 2013.

Remarks: Photographs of neotype in Bailey et al. (2005: figs. 2a–c). Type lost *fide* Andersson (1896: 25).

14. *Thamnodynastes paraguanae* Bailey & Thomas, 2007. Mem. Fund. La Salle Cien. Nat. (2006) 166: 9–12, figs. 1a–b.

Type: Holotype, MBUCV 1468, a 477 mm male (Toledo, 21 Nov. 1964).

Type locality: "Norte de Paraguaná, Falcón State, Venezuela."

Distribution: Extreme N Colombia (La Guajira) and W Venezuela (Falcón, Lara, Zulia), NSL–50 m.

Source: Navarrete et al., 2009.

15. *Thamnodynastes ramonriveroi* Manzanilla & Sánchez, 2005. Mem. Fund. La Salle Cien. Nat. (2004) 161–162: 63–71, figs. 1a–c, 2.

Type: Holotype, EBRG 3957, a 469 mm male (J. Manzanilla & D. Sánchez, 25 Sept. 1999).

Type locality: "Cerro La Laguna (10°01'27"N-64°07'0) macizo del Turimiquire, municipio Freites, Estado Anzoátegui, Venezuela, 2130 m s.n.m."

Distribution: Northern South America. Venezuela (Anzoátegui, Bolívar, Delta Amacuro, Monagas, Sucre), Trinidad and Tobago (Trinidad), Guyana (Essequibo Islands-West Demerara, Upper Demerara-Berbice),

Suriname (Nickeria, Paramaribo) and W Brazil (Acre), NSL–2130 m.

Sources: Bailey & Thomas, 2007, Navarrete et al., 2009 and C.J. Cole et al., 2013.

16. *Thamnodynastes rutilus* (Prado, 1942b). Ciencia 3(7): 204, figs. 1–2. (*Dryophylax rutilus*)

Type: Holotype, IB 10337, adult female (T. Beu, 8 April 1942), destroyed by fire 15 May 2010.

Type locality: "Galia, S. Paulo" [= Gália, São Paulo State, SE Brazil, 22°17'S, 49°33'W, elevation 550 m].

Distribution: Southeastern Brazil (São Paulo), 550 m.

17. *Thamnodynastes sertanejo* Bailey, Thomas & Silva, 2005. Phyllomedusa 4(2): 90–92, figs. 8a–b, 9–11.

Type: Holotype IB 26401, a 544 mm female, destroyed by fire 15 May 2010.

Type locality: "Caraiba (Mina Caraiba in the original record), at the head of the Rio Curaçá, Bahia, Brazil" [= Caraiba, N Bahia State, E Brazil, 9°30'S, 38°55'W, elevation 425 m].

Distribution: Eastern Brazil (Bahia, Ceará, Minas Gerais, Pernambuco), 425 m.

Sources: Silveira & Cottu, 2006 and Roberto et al., 2009b.

18. *Thamnodynastes strigatus* (A.C.L.G. Günther, 1858). Cat. Colub. Snakes Brit. Mus.: 52–53. (*Tomodon strigatus*)

Type: Holotype, BMNH 1946.1.2.70, a 584 mm female (Warwick).

Type locality: "India," (in error). Corrected to South America *fide* Hoge (1953a: 159).

Distribution: Southern South America. Southeastern Brazil (Paraná, Rio de Janeiro, Santa Catarina, São Paulo), Paraguay (Alto Paraná, Itapúa), Uruguay (Artigas, Cerro Largo, Colonia, Rio Negro, Salto, Treinta y Tres) and Argentina (Buenos Aires, NE Corrientes, Entre Ríos, Misiones, Treinta y Tres), NSL–1000 m.

Sources: Hoge, 1953a, Carreira-Vidal et al., 2005, Harvey & Muñoz, 2004, Cacciali, 2008 and Prigioni et al., 2011.

19. *Thamnodynastes yavi* C.W. Myers & Donnelly, 1996. Amer. Mus. Novit. (3172): 41–45, fig. 26, fig. 27 (upper right).

Type: Holotype, EBRG 3124 (formerly CWM 19665), a 340 mm female (AMNH-Terramar Exped., 19 Feb. 1995).

Type locality: "summit of Cerro Yavi, 2150 m elevation, Amazonas, Venezuela."

Distribution: Southern Venezuela (Amazonas), 2150 m. Known only from type locality.

Sources: C.W. Myers & Donnelly, 2001 and Navarrete et al., 2009.

THAMNOPHIS Fitzinger, 1843 (Natricidae)

Synonyms: *Eutainia* Baird *in* Baird & Girard, 1853, *Eutaenia* Baird, 1854 (*nomen emendatum*), *Eutania* – Kennicott, 1855 (*nomen incorrectum*), *Prymnomiodon* Cope, 1861c, *Tamnophis* – Fitzinger, 1967 (*nomen incorrectum*), *Eulainia* – Higgins, 1873 (*nomen incorrectum*), *Prymniodon* Marschall, 1873 (*nomen emendatum*), *Chilopoma* Cope, 1875c (*nomen nudum*), *Chilopoma* Cope *in* Yarrow, 1875, *Eutaemia* – A.M. Ross, 1878 (*nomen incorrectum*), *Atomarchus* Cope, 1883c, *Stypocemus* Cope, 1885b (*nomen substitutum*), *Phamnophis* – Ferrari-Pérez, 1886 (*nomen incorrectum*), *Tamnosophis* – Dugès, 1887 (*nomen incorrectum*), *Thamnosophis* – Herrera, 1890 (*nomen incorrectum*), *Prymnomidion* – Hoffmann, 1890 (*nomen incorrectum*), *Eutoenia* – Velasco, 1891 (*nomen incorrectum*), *Eustoenia* – Velasco, 1895 (*nomen incorrectum*), *Lutainia* – C.H.T. Townsend, 1897 (*nomen incorrectum*), *Eutainiaus* A.L. Herrera, 1899 (*nomen emendatum*), *Thamnopis* – McLain, 1899 (*nomen incorrectum*), *Eutenia* – A.J. Carlson & Luckhardt, 1920 (*nomen incorrectum*), *Thomnophis* – Haltom, 1931 (*nomen incorrectum*), *Prinmomiodon* – Briceño-Rossi, 1934 (*nomen incorrectum*), *Thanmophis* – Viosca, 1949 (*nomen incorrectum*), *Thamniphis* – Z. Vogel, 1966 (*nomen incorrectum*), *Thamnoplis* – W.W. Tanner, 1970 (*nomen incorrectum*), *Thamnophsis* – Saint-Aubain, 1984 (*nomen incorrectum*), *Brucerogersus* Hoser, 2012y (*nomen illegitimum*), *Gregswedoshus* Hoser, 2012y (*nomen illegitimum*), *Neilsonnemanus* Hoser, 2012y (*nomen illegitimum*), *Pughus* Hoser, 2012y (*nomen illegitimum*), and *Whybrowus* Hoser, 2012y (*nomen illegitimum*).

Type species: *Coluber saurita* Linnaeus, 1766.

Distribution: North America and Central America.

Fossil records: Middle Miocene to upper Pleistocene of USA. Fossils unassigned to species include Miocene (Barstovian) of USA (Nebraska), middle/upper Miocene (Clarendonian) of USA (Kansas; Nebraska), upper Miocene (Hemphillian) of USA (Arizona), upper Miocene/lower Pliocene (Hemphillian) of USA (Indiana), lower Pliocene (Hemphillian) of USA (Kansas, Oklahoma), middle Pliocene (Blancan II, III) of USA (Idaho, Kansas), middle/upper Pliocene (Blancan III–V) of USA (Arizona), upper Pliocene (Blancan) of USA (Washington), lower Pleistocene (Irvingtonian I [?]) of USA (Pennsylvania), lower/middle Pleistocene (Irvingtonian I) of USA (Arizona, Texas), middle/upper Pleistocene (Irvingtonian II) of USA (Kansas, Maryland, Texas, West Virginia), and upper Pleistocene (Rancholabrean II) of USA (Indiana, Nevada, Utah, Wisconsin).

Sources: Jan, 1865d, Ruthven, 1908, Van Denburgh & Slevin, 1918, Fitch, 1940, H.M. Smith, 1942g, H.M. Smith & Taylor, 1945, K.P. Schmidt, 1953a, Rossman *in* J.A. Peters & Orejas-Miranda, 1970, N.J. Rossman et al., 1982, Webb, 1982, R. Lawson, 1985, 1987, Chiasson & Lowe, 1989, Queiroz & Lawson, 1994, Holman, 1995a, 2000a, Mutschmann, 1995, Rossman et al., 1996, Alfaro & Arnold, 2001, Hallmen & Chlebowy, 2001, Bourguignon, 2002, Queiroz et al., 2002, Parmley & Walker, 2003, Zaher et al., 2009, Parmley & Hunter, 2010 and Hoser, 2012y.

1. *Thamnophis atratus* (Kennicott *in* J.G. Cooper, 1860). Rep. Explor. Surv. Railroad, 47th & 49th Par. 12(2): 296, pl. 15, fig. 1. (*Eutainia atrata*)

Synonyms: *Eutaenia infernalis vidua* Cope, 1892a, *Thamnophis ordinoides hydrophila* Fitch, 1936, *Thamnophis elegans aquaticus* W. Fox, 1951a, and *Thamnophis atratus zaxanthus* Boundy, 1999.
Type: Lectotype, USNM 970A, a 629 mm male (R.D. Cutts [Pacific Railroad Exped.], 1854), designated by Fitch (1940: 89).
Type locality: "San Francisco, California" [USA] via lectotype selection. Corrected to Santa Cruz Mtns. or S San Francisco Peninsula *fide* Boundy (1999: 330).
Distribution: Northwestern USA (coastal cen. California, SW Oregon), NSL–1920 m.
Sources: Bellemin & Stewart, 1977, Fitch, 1984, Stebbins, 1985, Rossman & Stewart, 1987, H. Brown et al., 1995, P.R. Brown, 1997, Boundy, 1999 and Bartlett & Tennant, 2000.
Remarks: Elevated to specific status *fide* Rossman & Stewart (1987: 19), who also suggested that *T. aquaticus* and *T. hydrophilus* are valid subspecies.

2. *Thamnophis bogerti* Rossman & Burbrink, 2005. Occ. Pap. Mus. Nat. Sci., Louisiana St. Univ. (79): 22–24, fig. 8.

Type: Holotype, AMNH 93239, a 396 mm male (C.M. Bogert, 11 Oct. 1964).
Type locality: "Mexico, Oaxaca, El Tejocote, 2377 m elevation" [= El Tejocote, Oaxaca State, S Mexico, 17°21'N, 98°09'W,elevation 2375 m].
Distribution: Southern Mexico (Mesa del Sur of Oaxaca), 2195–2745 m.
Source: Casas-Andreu et al., 1996.

T

3. *Thamnophis brachystoma* (Cope, 1892b). Amer. Nat. 26: 964–965, 1 fig. (*Eutaenia brachystoma*)

Type: Holotype, ANSP 10751, a 286 mm male (A.M. Brown, 1894).
Type locality: "near Franklin, Venango County, Pennsylvania, on the Alleghany River" [USA].
Distribution: Northeastern USA (S New York, NW Pennsylvania), 270–730 m.
Fossil records: Upper Pleistocene (Rancholabrean II) of USA (Pennsylvania, West Virginia).
Sources: Bothner, 1976, Ernst & Barbour, 1989, Conant & Collins, 1991, J.H. Harding, 1997 and Holman, 2000a.

4. *Thamnophis butleri* (Cope, 1889b). Proc. U.S. Natl. Mus. (1888–1889) 11(729): 399–400. (*Eutaenia butleri*)

Type: Holotype, PUL 264, lost *fide* Minton (1980: 258.1).
Type locality: "Richmond, Ind." [Indiana, USA].
Distribution: South-central Canada (S Ontario) and N-cen. USA (NE Indiana, E Michigan, NW Ohio, SE Wisconsin), 150–460 m.
Sources: Conant, 1950, Minton, 1980, Ernst & Barbour, 1989, Conant & Collins, 1991 and J.H. Harding, 1997.

5. *Thamnophis chrysocephalus* (Cope, 1885a). Proc. Amer. Philos. Soc. (1884) 22: 173–174. (*Eutaenia chrysocephala*)

Synonym: *Thamnophis eburatus* E.H. Taylor, 1940a.
Type: Holotype, USNM 30494, a 430 mm specimen (A.L.J.F. Sumichrast, 1855–1882).
Type locality: "Orizaba, Vera Cruz" [= Orizaba, cen. Veracruz State, E Mexico, 18°51'N, 97°06'W, elevation 1235 m].
Distribution: Mexico (E Guerrero, Oaxaca, SE Puebla, cen. Veracruz), 1220–3080 m.

6. *Thamnophis conanti* Rossman & Burbrink, 2005. Occ. Pap. Mus. Nat. Sci., Louisiana St. Univ. (79): 29–31, fig. 12.

Type: Holotype, LSUMZ 75985 (formerly USL 23933), a 415 mm female (T. Hardaway, 15 Aug. 1975).
Type locality: "Mexico, Puebla-Veracruz state line at Mexico Highway 125 [number in error; it is Highway 150 that extends north from Tehuacan, Puebla, to the state line at Puerto del Aire]."
Distribution: Mexico (interface of Mesa Central and Sierra Madre Oriental in Puebla), 2135–2900 m.
Source: Canseco-Márquez & Gutiérrez-Mayén, 2010.

7. *Thamnophis couchii* (Kennicott *in* Baird, 1859b). Explor. Railroad Rte. Sacramento Val.-Columbia Riv. (1855) 10: 10–11. (*Eutaenia couchii*)

Type: Holotype, USNM 866, an adult female (J.S. Newberry [Pacific Railroad Exped.], 1855).
Type locality: "Bank of Pit river, California." Specified as near mouth of Hat Creek *fide* Fitch (1940: 59).
Distribution: Western USA (E California, W Nevada), 90–2440 m.
Sources: Lawson & Dessauer, 1979, Fitch, 1984, Stebbins, 1985, Rossman & Stewart, 1987, Blaustein et al., 1995, P.R. Brown, 1997 and Bartlett & Tennant, 2000.

8. *Thamnophis cyrtopsis* (Kennicott, 1860). Proc. Acad. Nat. Sci. Philadelphia 12(8): 333–334. (*Eutaenia cyrtopsis*)

Synonyms: *Thamnophis cyrtopsis cyclides* Cope, 1861e, *Tropidonotus collaris* Jan, 1863b (*nomen nudum*), *Tropidonotus collaris* Jan, 1865c, *Eutaenia cyrtopsis ocellata* Cope, 1880, *Eutaenia aurata* Cope, 1892a, *Thamnophis vicinus* H.M. Smith, 1942h, and *Thamnophis sumichrasti salvini* H.M. Smith, Nixon & Smith, 1950.

Type: Holotype, USNM 8067 (formerly SIM 930), a 622–626 mm male (D.N. Couch and/or Webb, 1853).

Type locality: "Rinconada Coahuila, Mexico," (in error). Corrected to just across Coahuila state line in Nuevo León, ca. 33 km NE Ramos Arizpe [= Rinconada, ext. W-cen. Nuevo León, NE Mexico, 25°39'N, 00°46'W, elevation 1250 m] *fide* Conant *in* Webb (1966: 58).

Distribution: Southwestern USA and Mesoamerica. Southwestern USA (Arizona, ext. S Colorado, New Mexico, SW and S Texas, SE Utah), Mexico (Aguascalientes, Chiapas, Chihuahua, Coahuila, Colima, Distrito Federal, W Durango, Guanajuato, Guerrero, Hidalgo, Jalisco, México, Michoacán, Morelos, Nayarit, Nuevo León, Oaxaca, Puebla, Querétaro, San Luis Potosí, Sinaloa, Sonora, SW Tamaulipas, Tlaxcala, W Veracruz, S Zacatecas) and SW Guatemala, NSL–2700 m.

Fossil records: Upper Pleistocene (Rancholabrean II) of USA (New Mexico, Texas).

Sources: Milstead, 1953, Duellman, 1961, Webb, 1966, 1978, 1980, J.D. Johnson, 1984, Stebbins, 1985, Villa et al., 1988, Rossman, 1992, 1996, Van Devender & Bradley, 1994, Degenhardt et al., 1996, Auth et al., 2000, Bartlett & Tennant, 2000, Holman, 2000a, Werler & Dixon, 2000, McCranie & Wilson, 2001b and Persons & Rosen, 2001.

Remarks: Collector listed as Webb on specimen tag *fide* Conant (1969: 85).

9. *Thamnophis elegans* (Baird & Girard, 1853). Cat. Rept. No. Amer., Serp. 1: 34–35. (*Eutainia elegans*)

Synonyms: *Eutainia vagrans* Baird & Girard, 1853, *Tropidonotus trivittatus* Hallowell, 1853a, *Eutaenia biscutata* Cope, 1883d, *Eutaenia henshawi* Yarrow, 1883a, *Eutaenia vagrans plutonia* Yarrow, 1883a, *Eutaenia elegans brunnea* Cope, 1892a, *Eutaenia elegans lineolata* Cope, 1892a, *Thamnophis ordinoides hueyi* Van Denburgh & Slevin, 1923, *Thamnophis elegans nigrescens* M. L. Johnson, 1947, *Thamnophis elegans terrestris* W. Fox, 1951a, *Coluber nutkensis* Mociño & Maldonado *in* Arias-Divito, 1968 (*nomen nudum*), *Thamnophis elegans arizonae* W.W. Tanner & Lowe, 1989, and *Thamnophis elegans vascotanneri* W.W. Tanner & Lowe, 1989.

Type: Holotype, USNM 882, a 597 mm male (C.C. Boyle).

Type locality: "El Dorado Co., Cal." [= California, USA].

Distribution: Western North America and NW Mexico. Western Canada (Alberta, British Colombia, SW Saskatchewan, Vancouver Is.), W USA (N Arizona, N California, Colorado, Idaho, Montana, W Nebraska, Nevada, New Mexico, ext. W Oklahoma, Oregon, W South Dakota, Utah, Washington, Wyoming) and NW Mexico (Baja California Norte, NW Chihuahua), NSL–3660 m.

Fossil records: Middle/upper Pleistocene (Irvingtonian II) of USA (Colorado).

Sources: Webb, 1976, Bellemin & Stewart, 1977, Lawson & Dessauer, 1979, Rossman, 1979, Fitch, 1980a, 1983, McCoy & Flores-Villela, 1988, W.W. Tanner & Lowe, 1989, H. Brown et al., 1995, Degenhardt et al., 1996, P.R. Brown, 1997, Bartlett & Tennant, 2000, Holman, 2000a, Bronikowski & Arnold, 2001, Grismer, 2002 and Lemos-Espinal et al. , 2004.

Remarks: Elevation of 3990 m possibly erroneous *fide* Wise et al. (2008: 360).

10. *Thamnophis eques* (A. Reuss, 1834). Mus. Senck. 1: 152–155, pl. 8, figs. 2a–b. (*Coluber eques*)

Synonyms: *Coluber subcarinata* Gray *in* Richardson et al., 1839, *Eutaenia macrostemma* Kennicott, 1860, *Eutaenia megalops* Kennicott, 1860, *Eutaenia flavilabris* Cope, 1867a, *Eutaenia insigniarum* Cope, 1885a, *Thamnophis stejnegeri* McLain, 1899, *Thamnophis eques virgatenuis* Conant, 1963b, *Thamnophis eques carmenensis* Conant, 2003, *Thamnophis eques cuitzeoensis* Conant, 2003, *Thamnophis eques diluvialis* Conant, 2003, *Thamnophis eques insperatus* Conant, 2003, *Thamnophis eques obscurus* Conant, 2003 (*nomen praeoccupatum*), *Thamnophis eques patzcuaroensis* Conant, 2003, *Thamnophis eques scotti* Conant, 2003, and *Thamnophis eques kiwi* Boundy, 2007 (*nomen substitutum*).

Type: Lectotype, SMF 17179 (formerly SMF-R III.O.9.a & SMF-B 7209, a), 615 mm specimen (F.A. Dillenburger, 1832), designated by Mertens (1922: 179).

Type locality: "Mexico." Restricted to El Limón, Totalco, Veracruz, Mexico by H.M. Smith & Taylor (1950a: 351).

Distribution: Southwestern USA (S Arizona, Nevada, SW New Mexico) and Mexico (Aguascalientes, Chihuahua, ext. SW Coahuila, Distrito Federal, Durango, Guanajuato, Hidalgo, Jalisco, N Michoacán, México, Morelos, Nayarit, S Nuevo León, cen. Oaxaca, Puebla, Querétaro, San Luis Potosí, NE Sonora, Tlaxcala, W-cen. Veracruz, Zacatecas), 55–2590 m.

Sources: H.M. Smith, 1951, Conant, 1961, 1963b, 2000, 2003, W.W. Tanner, 1985, Degenhardt et al., 1996, Queiroz & Smith, 1996, Benitez-Gálvez, 1997, Auth et al., 2000, Bartlett & Tennant, 2000, McCranie & Wilson, 2001b, Lemos-Espinal et al., 2004 and Blo, 2012.

Remarks: A 1911 Nevada record probably represents occurrence in Nevada and possibly Arizona where now extinct *fide* Queiroz & Smith (1996: 155).

11. *Thamnophis errans* H.M. Smith, 1942h. Zoologica 27: 112–113. (*Thamnophis ordinoides errans*)

Type: Holotype, USNM 46336, a 545 mm female (E.W. Nelson & E.A. Goldman, July 1899).
Type locality: "Colonia Garcia, Chihuahua" [= Colonia García, NW Chihuahua State, NW Mexico, 29°58'N, 108°21'W, elevation 2240 m].
Distribution: Northwestern Mexico (W Chihuahua, W Durango, ext. N Nayarit, ext. SW Zacatecas), 1860–2545 m.
Sources: Webb, 1976, Fitch, 1980a and Queiroz & Lawson, 1994.

12. *Thamnophis exsul* Rossman, 1969. Occ. Pap. Mus. Zool., Louisiana St. Univ. (39): 1–2, fig. 1.

Type: Holotype, USNM 166423, a 411 mm female (M.D. Sabath & R.W. Axtell, 6 July 1961).
Type locality: "dry pine forest (ca. 9,100 feet), 11 mi. E, 3.5 mi. S San Antonio de las Alazanas, Coahuila, northern México" [= 18 km E and 6 km S San Antonio de las Alazanas (25°16'N, 100°33'W, elevation 2275 m), ext. SE Coahuila State, NE Mexico].
Distribution: Northeastern Mexico (SE Coahuila, SE Nuevo León, SW Tamaulipas), 2275–3240 m.
Sources: Rossman et al., 1989, Liner, 1992 and Farr et al., 2007.

13. *Thamnophis fulvus* (Bocourt, 1893) *in* A.H.A. Duméril, Bocourt & Mocquard, 1870–1909. Miss. Sci. Mex. Rept. (13): 777–778, pl. 57, figs. 2, 2a–d. (*Eutaenia cyrtopsis fulvus*)

Type: Holotype, MNHN 94, a 475 mm male (Comm. Sci. Haute Vera Paz, Mex.).
Type locality: "département de la Haute Vera Paz (Guatemala)" [= Alta Verapaz Dept., N Guatemala]. Restricted to vicinity of Cobán, Alta Verapaz, Guatemala *fide* L.C. Stuart (1948: 91–92). Restricted to Cobán *fide* H.M. Smith & Taylor (1950a: 317).
Distribution: Southeastern Mexico (Chiapas), Guatemala (Alta Verapaz), N El Salvador (Chalatenango) and SW Honduras (Francisco Morazán, Intibucá, Ocotepeque), 1410–3355 m.
Sources: L.C. Stuart, 1948, Rossman, 1965b, Webb, 1982, L.D. Wilson & Meyer, 1985, Villa et al., 1988, G. Köhler et al., 2005 and McCranie, 2011a.
Remarks: Holotype listed as MNHN 94.94 *fide* L.C. Stuart (1963: 120).

T

14. *Thamnophis gigas* Fitch, 1940. Univ. California Publ. Zool. 44(1): 69–73. (*Thamnophis ordinoides gigas*)

Type: Holotype, MVZ 5428, an adult female (H.C. Bryant, 16 May 1914).
Type locality: "Gadwall, Merced County, California" [USA].
Distribution: Western USA (cen. California), NSL–120 m.
Sources: Rossman & Stewart, 1987, Brode, 1988, P.R. Brown, 1997, Bartlett & Tennant, 2000 and Halstead et al., 2010.
Remarks: Elevated to specific status *fide* Rossman & Stewart (1987: 19). Extiprated in southern part of the Distribution *fide* Rossman & Stewart, 1987.

15. *Thamnophis godmani* (A.C.L.G. Günther, 1894 *in* 1885–1902). Biol. Cen.-Amér., Rept. Batr.: 133. (*Tropidonotus godmani*)

Type: Lectotype, BMNH 1946.1.21.81, a 555 mm female (H.H. Smith), designated by Rossman & Burbrink (2005: 20–21).
Type locality: "Mexico, Omilteme" [= Omiltemi, cen. Guerrero State, SW Mexico, 17°33'N, 99°41'W, elevation 2185 m] via lectotype selection.
Distribution: Southwestern Mexico (cen. Guerrero), 1770–2440 m.
Source: Rossman & Burbrink, 2005.

16. *Thamnophis hammondii* (Kennicott, 1860). Proc. Acad. Nat. Sci. Philadelphia 12(8): 332. (*Eutaenia hammondii*)

Synonym: *Tropidonotus digueti* Mocquard, 1899.
Type: Lectotype, USNM 894, a male (J.F. Hammond), designated herein.
Type locality: "San Diego, San Diego Co., California, USA" via lectotype selection.
Distribution: Southwestern USA (SW California) and NW Mexico (NW Baja California Norte, Baja California Sur), NSL–2130 m.
Fossil records: Upper Pleistocene (Rancholabrean II) of USA (California).
Sources: Rossman & Stewart, 1987, McGuire & Grismer, 1993, P.R. Brown, 1997, Bartlett & Tennant, 2000, Holman, 2000a and Grismer, 2002.
Remarks: Elevated to specific status *fide* Rossman & Stewart (1987: 19). Three syntypes collected by J. Xantus from Fort Tejon (USNM 5496a–b and ANSP 6522) are identified as *T. couchii fide* Rossman & Stewart (1987: 16).

17. *Thamnophis lineri* Rossman & Burbrink, 2005. Occ. Pap. Mus. Nat. Sci., Louisiana St. Univ. (79): 25–29, fig. 10 (left).

Type: Holotype, UTA 12482, a 404 mm male (J.A. Campbell, 8 June 1983).

Type locality: "Mexico, Oaxaca, Llano de las Flores, 2786 m elevation."

Distribution: Southern Mexico (Mesa del Sur of Oaxaca), 2670–3050 m.

18. *Thamnophis marcianus* (Baird & Girard, 1853). Cat. No. Amer. Rept., Serp. 1: 36–37. (*Eutainia marciana*)

Synonyms: *Eutaenia nigrolateris* A. Brown, 1889, *Eutaenia praeocularis* Bocourt, 1892b, *Thamnophis arabdotus* Andrews, 1937, *Thamnophis ruthveni* Hartweg & Oliver, 1938, *Thamnophis bovallii* Dunn, 1940a, and *Thamnophis rozellae* H.M. Smith, 1940.

Type: Holotype, USNM 844, an 864 mm specimen (R.B. Marcy [Red River Louisiana Exped.], 20 May 1852).

Type locality: "Red River, Ark.," [USA] (partially in error). Specified as between Camp 5 and Red River, on the open prairie *fide* Marcy (1854: 221). Restricted to vicinity of Slough Creek, east of Hollister, Tillman Co., Oklahoma *fide* Mittleman (1949: 243). Further specified as Indian Territory between Camp 5 and Red River, Tillman Co., near Hollister, Oklahoma *fide* H.A. Dundee *in* Cochran (1961: 182) and specimen label of holotype.

Distribution: Southwestern USA and Mesoamerica. Southwestern USA (S Arizona, ext. SE California, ext. SW Kansas, New Mexico, W Oklahoma, Texas), Mexico (ext. NE Baja California Norte, Campeche, Chihuahua, Coahuila, N Durango, Hidalgo, Nuevo León, SE Oaxaca, Quintana Roo, ext. N San Luis Potosí, Sonora, Tabasco, Tamaulipas, Veracruz, Yucatán), Belize (Belize, Cayo), Guatemala (Izabal), Honduras (Atlântida, Cortés, Santa Bárbara), Nicaragua (Chontales, Granada, Managua, Masaya, Rivas) and Costa Rica (N Alajuela, Limón), NSL–1640 m.

Fossil records: Middle Pliocene (Blancan III) of USA (Texas), and upper Pleistocene (Rancholabrean II) of USA (Texas).

Sources: Mittleman, 1949, Shreve & Gans, 1958, Rossman, 1971, Stebbins, 1985, L.D. Wilson & Meyer, 1985, Villa et al., 1988, Degenhardt et al., 1996, J.C. Lee, 1996, 2000, P.R. Brown, 1997, Hollingsworth & Prosser, 1997, J.A. Campbell, 1998b, Bartlett & Tennant, 2000, Holman, 2000a, Stafford & Meyer, 2000, Werler & Dixon, 2000, Grismer, 2002, Ramírez-Bautista et al., 2010, McCranie, 2011a and McCranie et al., 2012b.

Remarks: G.B. McClellan also listed as collector in original description. Type locality suggested as Cache Creek, Oklahoma *fide* Ruthven (1908: 58).

19. *Thamnophis melanogaster* (W.C.H. Peters, 1864b). Mber. Königl. Akad. Wiss. Berlin 1864(6): 389–390. (*Tropidonotus melanogaster*)

Synonyms: *Tropidonotus melanogaster* Deppe, 1830 (*nomen nudum*), *Tropidonotus mesomelanus* Jan, 1863b (*nomen oblitum*), *Tropidonotus mesomelanus* Jan, 1865c, *Tropidonotus baronis mülleri* – Troschel *in* J.W. Müller, 1865 (*nomen incorrigendum*), *Thamnophis melanogaster canescens* H.M. Smith, 1942b, *Thamnophis melanogaster linearis* H.M. Smith, Nixon & Smith, 1950, and *Thamnophis melanogaster chihuahuaensis* W.W. Tanner, 1959.

Types: Syntypes (2), ZMB 1946 and ZMB 1996, two females.

Type locality: "Mexico." Restricted to Xochimilco, Distrito Federal, Mexico *fide* H.M. Smith & Taylor (1950a: 330).

Distribution: Mexico (Aguascalientes, SW Chihuahua, ext. SW Coahuila, Distrito Federal, Durango, Guanajuato, Hidalgo, Jalisco, W México, N Michoacán, Morelos, Nayarit, Queretaro, SW San Luis Potosí, Zacatecas), (435) 850–2830 m.

Sources: H.M. Smith et al., 1950, W.W. Tanner, 1959, Conant, 1963b, Auth et al., 2000, McCranie & Wilson, 2001b, H.M. Smith & Chiszar, 2001a, Webb, 2002, Lemos-Espinal et al., 2004 and González Hernández et al., 2011.

Remarks: Type locality Amerika in ZMB catalogue *fide* R. Günther *(in litt)*.

20. *Thamnophis mendax* C. F. Walker, 1955. Copeia 1955(2): 110–111, fig. 1 (middle).

Type: Holotype, UMMZ 104044, a 368 mm male (W.Z. Lidicker, 28 July 1951).

Type locality: "near La Joya de Salas, Tamaulipas, ± 6,000 ft." [= Joya de Salas, S Tamaulipas State, NE Mexico, 23°12'N, 99°17'W, elevation 1700 m].

Distribution: Northeastern Mexico (S Tamaulipas), 1050–2120 m.

Sources: Dowling et al., 1983 and Rossman, 1992.

21. *Thamnophis nigronuchalis* F.G. Thompson, 1957. Occ. Pap. Mus. Zool., Univ. Michigan (584): 1–3, figs. 1 (right), 2 (right).

Type: Holotype, UMMZ 113611, a 543 mm female (R. Parker, 22 July 1955).

Type locality: "San Luis, Durango, 9000 feet elevation" [= San Luis, S Durango State, NW Mexico, 24°03'N, 104°35'W, elevation 2745 m].

Distribution: Northwestern Mexico (Aguascalientes, SW Durango), 2195–2745 m.

Sources: W.W. Tanner, 1985, Rossman, 1995 and Wood et al., 2011.

Remarks: A valid species *fide* Wood et al., 2011.

22. *Thamnophis ordinoides* (Baird & Girard, 1852c). Proc. Acad. Nat. Sci. Philadelphia (1852–1853) 6(5): 176. (*Tropidonotus ordinoides*)

Synonyms: *Eutainia leptocephala* Baird & Girard, 1853, *Eutainia cooperi* Kennicott, 1859 (*nomen nudum & nomen ineditum*), *Eutainia cooperi* Kennicott *in* J.G. Cooper, 1860, *Thamnophis leptocephala olympia* Meek *in* Meek & Elliot, 1899, and *Thamnophis rubristriata* Meek *in* Meek & Elliot, 1899.

Type: Holotype, none designated (C. Wilkes [U.S. Explor. Exped.], 1841), lost *fide* Kirk, (1979: 233.1).

Type locality: "Puget Sound, Western coast of America" [Washington, USA].

Distribution: Extreme SW Canada (SW British Columbia, Vancover Is.) and NW USA (ext. NW California, W Oregon, W Washington), NSL–1500 m.

Sources: Fitch, 1940, W. Fox, 1948, Kirk, 1979, K. Lawson & Dessauer, 1979, Stebbins, 1985, H. Brown et al., 1995, P.R. Brown, 1997 and Bartlett & Tennant, 2000.

Remarks: A neotype should be designated in next systematic revision *fide* Kirk (1979: 233.2).

23. *Thamnophis postremus* H.M. Smith, 1942g. Zoologica 27(3–4): 109–110. (*Thamnophis eques postremus*)

Type: Holotype, FMNH 120235 (formerly EHT-HMS 5275), a 575 mm female (H.M. Smith).

Type locality: "El Sabino, Michoacán, Mexico" [= El Sabino, N Michoacán, SW Mexico, 20°04'N, 101°44'W, elevation 1750 m].

Distribution: Southwestern Mexico (N Michoacán), 235–1065 m.

Sources: Milstead, 1953, Duellman, 1961, Webb, 1978 and Rossman, 1992.

24. *Thamnophis proximus* (Say *in* E. James, 1823). Exped. Pittsburgh Rocky Mts. 1: 187. (*Coluber proximus*)

Synonyms: *Tropidonotus trivittatus* Deppe, 1830 (*nomen nudum*), *Eutainia faireyi* Baird & Girard, 1853, *Eutaenia rutiloris* Cope, 1885b, *Thamnophis proximus alpinus* Rossman, 1963a, *Thamnophis proximus diabolicus* Rossman, 1963a, *Thamnophis proximus orarius* Rossman, 1963a, and *Thamnophis proximus rubrilineatus* Rossman, 1963a.

Type: Holotype, none designated, ? ANSP, a 794 mm specimen (T. Say [Long Rocky Mtn. Exped.], Sept. 1819–June 1820), lost *fide* H.M. Smith & Taylor (1945: 166).

Type locality: "stone quarry below Camp Missouri" [= Nebraska stone quarry on W side Missouri River, 5 km above mouth of Boyers River, Iowa *fide* H.M. Smith & Taylor (1945: 166), ca. 5 km ENE Fort Calhoun, Washington Co., Nebraska *fide* Rossman (1963a: 109), or 8.5 km S of Fort Calhoun, Nebraska *fide* Dundee, 1996: 79].

T

Distribution: Central USA and Mesoamerica. Central USA (Arkansas, ext. SE Colorado, Kansas, Illinois, Indiana, E Iowa, ext. W Kentucky, Louisiana, W Mississippi, Missouri, ext. S and E Nebraska, E New Mexico, Oklahoma, Texas, S Wisconsin), Mexico (Campeche, Chiapas, E Coahuila, S Guerrero, Hidalgo, Nuevo León, E Oaxaca, Quintana Roo, San Luis Potosí, Tabasco, Tamaulipas, Veracruz, Yucatán), Central America: Guatemala (Petén and SE coast), Belize (Belize, Orange Walk, Stann Creek, Toledo), El Salvador (Cabañas, La Paz, La Unión, Usulután), Honduras (Choluteca, Comayagua, Francisco Morazán, Gracias a Dios, Valle), Nicaragua (Managua, Masaya, Matagalpa) and Costa Rica (Alajuela, Cartago, Guanacaste, Heredia, Puntarenas, San José), NSL–2440 m.

Fossil records: Middle/upper Pliocene (Blancan V) of USA (Kansas), upper Pliocene (Blancan V) of USA (Nebraska), lower/middle Pleistocene (Irvingtonian I) of USA (Kansas), middle Pleistocene (Irvingtonian II) of USA (Nebraska), and upper Pleistocene (Rancholabrean I, II) of USA (Kansas, Missouri, New Mexico, Texas).

Sources: Rossman, 1962, 1963a, 1970a, Stebbins, 1985, L.D. Wilson & Meyer, 1985, Holman, 1987b, Villa et al., 1988, Ernst & Barbour, 1989, Degenhardt et al., 1996, J.C. Lee, 1996, 2000, Harding, 1997, J.A. Campbell, 1998b, G. Köhler, 1999b, Bartlett & Tennant, 2000, Holman, 2000a, Stafford & Meyer, 2000, Werler & Dixon, 2000, Savage, 2002, Solórzano, 2004, G. Köhler et al., 2005, Ramírez-Bautista et al., 2010 and McCranie, 2011a.

Remarks: Rossman (1963: 109) erroneously cites original description as p. 339.

25. *Thamnophis pulchrilatus* (Cope, 1885a). Proc. Amer. Philos. Soc. 22(118): 174. (*Eutaenia pulchrilatus*)

Type: Holotype, USNM 9899, a 465 mm specimen (A.A. Dugès, 1853–1885).

Type locality: "probably Guanajuato" [Mexico].

Distribution: Southern Mexico (Aguascalientes, Distrito Federal, S Durango, Guanajuato, E Hidalgo, S Jalisco, México, E Michoacán, Morelos, Nayarit, SE Nuevo León, N Oaxaca, NE Pueblo, N Querétaro, SW Tamaulipas, Tlaxcala, W-cen. Veracruz, S Zacatecas), 1370–2805 m.

Sources: Webb, 1966, Rossman, 1992 and García-Vázquez, 2008.

26. *Thamnophis radix* (Baird & Girard, 1853). Cat. No. Amer. Rept., Serp. 1: 34. (*Eutainia radix*)

Synonyms: *Eutainia haydenii* Kennicott, 1860, *Tropidonotus glaphyros* Jan, 1863b, *Tropidonotus kennicotti* Jan, 1863b, *Eutaenia radix twiningi* Coues &

Yarrow, 1878, and *Eutaenia radix melanotaenia* Cope, 1889b.

Type: Holotype, USNM 719, a 578 mm specimen (P.R. Hoy).

Type locality: "Racine, Wisconsin" [USA].

Distribution: Southwestern Canada (S Alberta, S Manitoba, S Saskatchewan) and USA (Arkansas [MCZ 58891], E Colorado, Illinois, NW Indiana, Iowa, Kansas, Minnesota, N Missouri, Montana, Nebraska, NE New Mexico, North Dakota, N Ohio, ext. W Oklahoma, South Dakota, ext. N Texas, S Wisconsin, ext. E Wyoming), 120–2290 m.

Fossil records: Middle Pliocene (Blancan II) of USA (Nebraska), upper Pliocene (Blancan V) of USA (Nebraska), lower/middle Pleistocene (Irvingtonian I) of USA (Kansas), middle Pleistocene (Irvingtonian II) of USA (Nebraska), and upper Pleistocene (Rancholabrean I, II) of USA (Kansas, Oklahoma).

Sources: A.G. Smith, 1949, Conant, 1950, P.W. Smith, 1961, Bullock & Tanner, 1966, Stebbins, 1985, Ernst & Barbour, 1989, Conant & Collins, 1991, Degenhardt et al., 1996, J. Harding, 1997, Holman, 2000a, Werler & Dixon, 2000, Walley et al., 2003 and Hallmen, 2004a.

Remarks: Type locality not in error as suggested by K.P. Schmidt (1953: 172) *fide* Tekiela (2004a: 57) and Rossman (pers. comm.) so restriction to vicinity of Chicago, Illinois *fide* K.P. Schmidt (1953a: 172) invalid.

27. *Thamnophis rossmani* Conant, 2000. Occ. Pap. Mus. Nat. Sci, Louisiana St. Univ. (76): 2–3.

Type: Holotype, LSUMZ 30390, a 643 mm male (D.A. Rossman, W.G. Eberle, D. Morizot & A. Varkey, 26 May 1969).

Type locality: "open roadside ditch 1.4 km NW of the small settlement of San Cayetano, Nayarit, Mexico" [= 1.4 km NW San Cayetano, S Nayarit, W Mexico, 21°27'N, 104°49'W, elevation 950 m].

Distribution: Western Mexico (S Nayarit), 950 m.

28. *Thamnophis rufipunctatus* (Cope *in* Yarrow, 1875). Rep. Geog. Geol. Expl. Surv. W 100th Mer. 5: 544, pl. 20, figs. 1–1a. (*Chilopoma rufipunctatum*)

Synonyms: *Chilopoma rufopunctatum* Cope, 1875c (*nomen nudum*), and *Atomarchus multimaculatus* Cope, 1883c.

Type: Holotype, USNM 8600 (formerly SIM 1097), a 257 mm specimen (H.W. Henshaw, Oct. 1874).

Type locality: "Southern Arizona," USA.

Distribution: Southwestern USA (cen. Arizona, SW New Mexico), 600–2430 m.

Sources: Ruthven, 1908, Lowe, 1955, F.G. Thompson, 1957, Conant, 1963b, Van Devender & Lowe, 1977, W.W. Tanner, 1985, 1990, P.C. Rosen & Schwalbe, 1988, Rossman, 1995, Degenhardt et al., 1996, H.M. Smith & Chiszar, 2003 and Wood et al., 2011.

Remarks: Collection date of type Sept., 1874 *fide* Cochran (1961: 164).

29. *Thamnophis sauritus* (Linnaeus, 1766). Syst. Nat., ed. 12, 1: 385. (*Coluber saurita*)

Synonyms: *Eutaenia sackenii* Kennicott, 1859a, *Prymnomiodon chalceus* Cope, 1861c, *Eutaemia saurita* – A.M. Ross, 1878 (*nomen incorrectum*), *Thamnophis sauritus nitae* Rossman, 1963a, and *Thamnophis sauritus septentrionalis* Rossman, 1963a.

Type: Neotype, FMNH 73119 (formerly CMR 53.136.5), a 590 mm female (J. Quinby, 20 July 1953), designated by K.P. Schmidt & Conant (1956: 191).

Type locality: "15 miles northeast of Charleston, Berkeley County, South Carolina" [USA] via neotype selection.

Distribution: Southeastern Canada (Nova Scotia, SE Ontario) and E USA (Alabama, Conneticut, Delaware, Florida, Georgia, Kentucky, SE Illinois, Indiana, ext. SE Louisiana, S Maine, Maryland, Massachusetts, Michigan, Mississippi, S New Hamshire, New Jersey, New York, North Carolina, Ohio, Pennsylvania, Rhode Island, South Carolina, W Tennessee, S Vermont, Virginia, Washington D.C., West Virginia, E Wisconsin), NSL–610 m.

Fossil records: Lower Pleistocene (Irvingtonian I) of USA (Florida), and upper Pleistocene (Rancholabrean II) of USA (Georgia, Pennsylvania, Virginia).

Sources: K.P. Schmidt & Conant, 1956–1957, A.H. Wright & Wright, 1957, Rossman, 1963a, 1970b, Christman, 1980, Ernst & Barbour, 1989, Conant & Collins, 1991, Palmer & Braswell, 1995, J. Harding, 1997, Tennant, 1997, Buckner & Franz, 1998a and Holman, 2000a.

Remarks: Official Specific Name No. 677 *fide* Opinion 385 (ICZN, 1956a). Photograph of neotype in K.P. Schmidt & Conant (1956: pl. 2). Introduced in the Bahama Is. (New Providence) *fide* Buckner & Franz (1998a: 55). Holotype lost *fide* K.P. Schmidt & Conant (1956: 191) and Rossman (1963: 145).

30. *Thamnophis scalaris* Cope, 1861a. Proc. Acad. Nat. Sci. Philadelphia (1860) 12: 369.

Type: Neotype, ANSP 11694, a 335 mm male, designated by Rossman & Lara-Gongora (1997: 11).

Type locality: "Jalapa, Veracruz, México" [= Jalapa, Veracruz State, E Mexico, 19°33'N, 96°55'W, elevation 1450 m] via neotype selection.

Distribution: Central Mexico (Distrito Federal, SE Hidalgo, SW Jalisco, México, N-cen. Michoacán, N Morelos, N Puebla, Querétaro, Tlaxcala, W-cen. Veracruz), 1450–4275 m.

Sources: H.M. Smith, 1942h, Rossman & Lara-Gongora, 1997, Rossman, 1992, Casas-Andreu et al., 1996 and Benítez-Gálvez, 1997.

Remarks: Holotype lost *fide* H.M. Smith & Taylor (1945: 167).

31. *Thamnophis scaliger* (Jan, 1865c). Arch. Zool. Anat. Fis. 3(2): 214. (*Tropidonotus* [*Eutainia*] *scaliger*)

Synonym: *Tropidonotus scaliger* Jan, 1863b (*nomen nudum*).

Type: Neotype, MZFC 2001, a 490 mm female, designated by Rossman & Lara-Gongora (1997: 11).

Type locality: "Pedregal de San Angel, Ciudad Universitaria, Delegación Coyoacán, Distrito Federal, Mexico" via neotype selection.

Distribution: Central Mexico (Aguascalientes, Distrito Federal, cen. Guanajuato, Hidalgo, ext. NE Jalisco, México, N Michoacán), 1800–3400 (3800) m.

Sources: Lara-Gongora, 1991, Rossman & Lara-Gongora, 1997, Quintero-Díaz et al., 1999, Mociño-Deloya et al., 2009 and Ramírez-Bautista et al., 2010.

Remarks: Type locality designated as Mexico City *fide* H.M. Smith & Taylor (1950a: 325).

32. *Thamnophis sirtalis* (Linnaeus, 1758). Syst. Nat., ed. 10, 1: 222. (*Coluber sirtalis*)

Synonyms: *Coluber ordinatus* Linnaeus, 1766, *Coluber taenia* Schöpf, 1788, *Coluber terordinatus* Lacépède, 1789 (*nomen rejiciendum*), *Coluber borealis* Suckow, 1798, *Coluber quincuncialis* Suckow, 1797, *Coluber terordinatus* Latreille, 1801, *Coluber ibibe* Daudin, 1803d, *Coluber gravitate* Rafinesque-Schmaltz, 1819b, *Coluber trivittata* Rafinesque-Schmaltz, 1819b, *Coluber kentuckensis* Rafinesque-Schmaltz, 1820, *Coluber pennanti* Merrem, 1820, *Coluber similis* Rafinesque-Schmaltz, 1820, *Natrix pennanti* Merrem, 1820, *Coluber parietalis* Say *in* E. James, 1823, *Coluber infernalis* Blainville, 1835, *Tropidonotus bipunctatus* Schlegel, 1837, *Tropidonotus concinnus* Hallowell, 1852b, *Eutaenia dorsalis* Baird & Girard, 1853, *Eutainia pickeringii* Baird & Girard, 1853, *Tropidonotus jauresi* A.-M.-C. Duméril, Bibron & Duméril, 1854, *Eutaenia ornata* Baird, 1859a, *Eutaenia sirtalis obscura* Cope *in* Yarrow, 1875, *Eutaenia sirtalis tetrataenia* Cope *in* Yarrow, 1875, *Coluber lineatus* Mellish, 1876, *Eutaenia imperialis* Coues & Yarrow, 1878 (*nomen nudum*), *Eutaenia sirtalis graminea* Cope, 1889b, *Eutaenia sirtalis melanota* Smith *in* Higley, 1889, *Eutaenia sirtalis trilineata* Cope, 1892a, *Eutaenia sirtalis semifasciata* Cope, 1892a, *Thamnophis sirtalis pallidula* Allen, 1899, *Tropidonotus obalskii* Mocquard, 1903, *Thamnophis sirtalis annectens* B. Brown, 1950, *Thamnophis sirtalis fitchi* W. Fox, 1951, *Thamnophis sirtalis similis* Rossman, 1965a (*nomen praeoccupatum*), and *Thamnophis sirtalis lowei* W.W. Tanner, 1988.

Type: Holotype, not designated, ANSP, a 686 mm specimen, now lost.

Type locality: "Pennsylvania" [USA].

Distribution: North America. Canada (Alberta, British Columbia, S Manitoba, New Brunswick, ext. S Northern Territories, Nova Scotia, S Ontario, Prince Edward Island, S Quebec, S Saskatchewan, Vancouver Is.), USA (Alabama, ext. SE Alaska, Arkansas, California, NE Colorado, Conneticut, Delaware, District of Colombia, Florida, Georgia, Idaho, Illinois, Indiana, Iowa, Kansas, Kentucky, Louisiana, Maine, Maryland, Massachusets, Michigan, Minnesota, Mississippi, Missouri, Montana, Nebraska, NW Nevada, New Hampshire, New Jersey, New Mexico, New York, North Carolina, North Dakota, Ohio, Oklahoma, Oregon, Pennsylvania, Rhode Island, South Carolina, South Dakota, Tennessee, E Texas, NE Utah, Vermont, Virginia, Washington, West Virginia, Wisconsin, E Wyoming) and N Mexico (N Chihuahua), NSL–2540 m.

Fossil records: Middle Pliocene (Blancan II, III) of USA (Nebraska, Texas), upper Pliocene (Blancan V) of USA (Nebraska), lower Pleistocene (Irvingtonian I) of USA (South Dakota), lower/middle Pleistocene (Irvingtonian I) of USA (Kansas), middle Pleistocene (Irvingtonian II) of USA (Nebraska), middle/upper Pleistocene (Irvingtonian II) of USA (Arkansas, Colorado), and upper Pleistocene (Rancholabrean II) of USA (Alabama, California, Florida, Georgia, Missouri, Ohio, Pennsylvania, Tennessee, Texas, Virginia, West Virginia).

Sources: Inger, 1946, K.P. Schmidt & Conant, 1956–1957, Fitch & Maslin, 1961, Fitch, 1965, 1980b, Conant, 1975, Chiras, 1978, Benton, 1980ab, Christman, 1980, Stebbins, 1985, W.W. Tanner, 1988, Ernst & Barbour, 1989, Conant & Collins, 1991, H. Brown et al., 1995, Palmer & Braswell, 1995, Boundy & Rossman, 1995, Gregory & Larsen, 1996, P.R. Brown, 1997, J. Harding, 1997, Lamar, 1997b, Tennant, 1997, Fitch, 1999, Holman, 2000a, Werler & Dixon, 2000, J.C. Murphy & Schlager, 2003, Hallmen, 2004b, 2007, Placyk et al., 2007 and Mooi et al., 2011.

Remarks: Official Specific Name no. 676 *fide* Opinion 385 (ICZN, 1956a). Neotype selection of FMNH 73660 (formerly ROM 7167, a 498 mm male, G.M. Bureau, 1942) by K.P. Schmidt & Conant (1956: 228) rejected *fide* Opinion 771 (ICZN, 1966b). Type locality restriction of vicinity of Quebec *fide* Inger (19846: 254) and Quebec, Canada *fide* K.P. Schmidt (1953: 174) invalid. Photograph of neotype in K.P. Schmidt & Conant (1956: pl. 1). Record from the Bahama Is. (Abaco) possibly introduced *fide* Buckner & Franz (1998b: 55).

33. *Thamnophis sumichrasti* (Cope, 1867a). Proc. Acad. Nat. Sci. Philadelphia (1866) 18: 306. (*Eutaenia sumichrasti*)

Synonyms: *Eutaenia phenax* Cope, 1868c, and *Thamnophis halophilus* E.H. Taylor, 1940a.

Types: Syntypes (2), USNM 26501–02 (formerly SIM 45) (A.L.J.F. Sumichrast, 1855–1867).

Type locality: "Orizava, Mexico" [= Orizaba, cen. Veracruz, E Mexico, 18°51'N, 97°06'W, elevation 1235 m].

Distribution: Central Mexico (N Hidalgo, Jalisco, E Puebla, N Querétaro, SE San Luis Potosí, cen. Veracruz), 1235–2400 m.

Sources: Conant, 1965, Rossman, 1965c, 1966, 1992 and Casas-Andreu et al., 1996.

34. *Thamnophis unilabialis* W.W. Tanner, 1985. Great Basin Nat. 45(4): 648–649. (*Thamnophis rufipunctatus unilabialis*)

Type: Holotype, BYU 14217, a 508 mm female (W.W. Tanner & W.G. Robison, 11 July 1958).

Type locality: "from .5 mi SW of Bocoyna, Chihuahua, Mexico."

Distribution: Central Mexico (Chihuahua, ext. SW Coahuila, NE Durango, ext. E-cen. Sonora), (700) 1200–2530 m.

Source: Woods et al., 2011.

Remarks: Elevated to specific status *fide* Woods et al., 2011.

35. *Thamnophis validus* (Kennicott, 1860). Proc. Acad. Nat. Sci. Philadelphia 12(8): 334–335. (*Regina valida*)

Synonyms: *Tropidonotus celaeno* Cope, 1860f, *Tropidonotus tephropleura* Cope, 1860f, *Tropidonotus quadriserialis* J.G. Fischer, 1879, *Natrix valida isabelleae* Conant, 1953, and *Natrix valida thamnophisoides* Conant, 1961.

Type: Holotype, USNM 1309, an adult female (D.N. Couch, 1846–1854).

Type locality: "Durango, Mexico," (in error). Corrected to Río Presidio or coastal plain of southern Sonora, Mexico *fide* Conant (1969: 83).

Distribution: Western Mexico (ext. S Baja California Sur, ext. SW Chihuahua, Colima, W Guerrero, W Jalisco, ext. SW Michoacán, Nayarit, ext. SW Sonora, Sinaloa), NSL–1200 m.

Sources: Conant, 1946a, 1953, 1961, 1969, R. Lawson, 1987, McCranie & McAllister, 1988, Queiroz & Lawson, 1994, Grismer, 2002 and Lemos-Espinal et al., 2002a.

THAMNOSOPHIS Jan, 1863b (Pseudoxyrhophiidae)

Synonym: *Bibilava* Glaw, Nagy, Franzen & Vences, 2007.

Type species: *Leptophis lateralis* A.-M.-C. Duméril, Bibron & Duméril, 1854a.

Distribution: Madagascar.

Sources: Glaw et al., 2007, 2009, Cadle & Ineich, 2008 and Zaher et al., 2009.

1. *Thamnosophis epistibes* (Cadle, 1996a). Bull. Mus. Comp. Zool. 154(5): 384–391, fig. 7. (*Liopholidophis epistibes*)

Synonym: *Leptophis lateralis* A.-M.-C. Duméril, Bibron & Duméril, 1854a *(partim)*.

Type: Holotype, MCZ 180322 (formerly JEC 11460), a 753 mm female (J.E. Cadle, 20–26 Dec. 1991).

Type locality: "Talatakely, Ranomafana National Park, 950–1,000 m, Fivondronana Ifanadiana, Fianarantsoa Province, Madagascar [21°16'S, 47°25'E]."

Distribution: Southeastern Madagascar (Fianarantsoa), 950–1000 m.

2. *Thamnosophis infrasignatus* (A.C.L.G. Günther, 1882). Ann. Mag. Nat. Hist. (5) 9(52): 263–264, 1 fig. (*Ptyas infrasignatus*)

Synonyms: *Dromicus baroni* Boulenger, 1888a, and *Liopholidophis thieli* Domergue, 1973.

Type: Lectotype, BMNH 1946.1.7.57 (formerly BMNH 1882.2.25), a 920 mm female (W.D. Cowan, 1874–1881), designated by Cadle (1996a: 424).

Type locality: "Arkafana, eastern Betsileo" via lectotype selection.

Distribution: Eastern Madagascar (Antsiranana, Fianarantsoa, Toamasina, Toliara), 100–1600 m.

Source: Henkel & Schmidt, 2000.

Remarks: Photographs of lectotype in Cadle (1996a: figs. 27–28). Type locality corrected to Ankafana, Betsileo *fide* Boulenger (1893a: 247).

3. *Thamnosophis lateralis* (A.-M.-C. Duméril, Bibron & Duméril, 1854a). Erpét. Gén. 7(1): 544–545. (*Leptophis lateralis*)

Synonym: *Dromicus madagascariensis* A.C.L.G. Günther, 1872b.

Type: Lectotype, MNHN 1999.8232, a 791 mm female, designated by Cadle & Ineich (2008: 287).

Type locality: "Madagascar" via lectotype selection.

Distribution: Madagascar (Antsiranana, Atananarivo, Fianarantsoa, Mahajanga, Toamasina, Toliara), NSL–1565 m.

Remarks: Photographs of lectotype in Cadle & Ineich (2008: fig. 1 upper & lower).

4. *Thamnosophis martae* (Glaw, Franzen & Vences, 2005). Salamandra 41(1–2): 84–89, figs. 1–3. (*Liopholidophis martae*)

Type: Holotype, ZSM 253/2004 (formerly FGZC 492), an 847 mm male (F. Glaw, M. Pruente & R.D. Randrianiaina, 23 Feb. 2004).

Type locality: "Montagne des Français (12°19'34"S, 49°20'09"E, ca. 300 m above sea level), Antsiranana Province, northern Madagascar."

Distribution: Northern Madagascar (Antsiranana, Mahajanga), 170–500 m.

5. *Thamnosophis mavotenda* Glaw, Nagy, G. Köhler, Franzen & Vences, 2009. Organ. Divers. Evol. 9: 15–18, figs. 2–3, 4a–b, 5a, 6.

Type: Holotype, ZSM 127/2006, an 846 mm male (F. Glaw, J.G. Köhler, F. Bora & H. Enting, 26 March 2006).
Type locality: "western Madagascar, Mahajanga Province, Tsingy de Bemaraha National Park, Bendrao forest, 18°47'04"S, 44°51'37"E, 427 m a.s.l."
Distribution: Western Madagascar (Mahajanga), 425 m. Known only from type locality.

6. *Thamnosophis stumpffi* (Boettger, 1881a). Zool. Anz. 4(87): 358. (*Dromicus stumpffi*)

Synonym: *Liophidium gracile* Mocquard, 1908.
Type: Lectotype, SMF 17576 (formerly SMF-B 7242, a), a 502–750 mm female (A. Stumpff, 1881), designated by Mertens (1922: 179).
Type locality: "Nossibé" [=Nosy Be Is., W Antsiranana Prov., ext. NW Madagascar, bet. 13°12'–25'S and 48°10'–22'E, elevation < 225 m] via lectotype selection.
Distribution: Northern Madagascar (Antsiranana, Nosy Be Is.), NSL–360 m.
Source: Boettger, 1881d.

†*THAUMASTOPHIS* Rage, Folie, Rana, Singh, Rose & Smith, 2008 (Caenophidia *incertae sedis*)

Type species: †*Thaumastophis missiaeni* Rage, Folie, Rana, Singh, Rose & Smith, 2008.
Distribution: Eocene of India.
Source: Rage et al., 2008.

1. †*Thaumastophis missiaeni* Rage, Folie, Rana, Singh, Rose & Smith, 2008. Acta Palaeontol. Pol. 53(3): 399–400, figs. 5a (1–8).

T

Type: Holotype, GU/RSR/VAS 1017, one middle trunk vertebra.
Type locality: "Vastan Lignite Mine, northeast of Surat, Gujarat, western India; Early Eocene (middle to late Ypresian) continental beds of Cambay Formation."
Distribution: Lower Eocene (Ypresian, MP 8–10: 48.6–55.8 mya) of India. Known only from type locality.

THELOTORNIS A. Smith, 1849 *in* 1838–1849 (Colubridae)

Synonyms: *Cladophis* A.H.A. Duméril, 1859, *Theletornis* – Phisalix, 1922 (*nomen incorrectum*), *Thelotormis* – Briceño-Rossi, 1934 (*nomen incorrectum*), *Thelotomis* – Hedges, 1983 (*nomen incorrectum*), and *Theltornis* – Carpenter, 1986 (*nomen incorrectum*).
Type species: *Thelotornis capensis* A. Smith, 1849 *in* 1838–1849.
Distribution: Central S and E Africa.
Sources: Loveridge, 1957a, V.F.M. FitzSimons, 1962a, Witte, 1962, Pitman, 1974, Broadley, 1979, 2001 and Broadley & Wallach, 2002.

1. *Thelotornis capensis* A. Smith, 1849 *in* 1838–1849. Illust. Zool. So. Africa, Rept. (App.): 19–20.

Synonym: *Dryiophis oatesii* A.C.L.G. Günther, 1881c.
Type: Holotype, not designated, a 940 mm female, lost *fide* Broadley (2001: 60).
Type locality: "Kaffirland and the country towards Port Natal" [= KwaZulu-Natal, South Africa *fide* Loveridge, 1944b: 161]. Restricted to vicinity of Durban, KwaZulu-Natal Prov., South Africa *fide* David & Ineich (1999: 100).
Distribution: Southern Africa. Southern Angola (Lunda Norte, Lunda Sul), SE Democratic Republic of the Congo (Katanga), Malawi (Central, Northern, Southern), Mozambique (Gaza, Inhambane, Manica, Maputo, Tete, Zambézia), Zambia (Central, Copperbelt, Eastern, Luapula, Lusaka, Northern, North-Western, Southern, Western), Zimbabwe (Bulawayo, Manicaland, Mashonaland Central, Mashonaland East, Mashonaland West, Masvingo, Matabeleland North, Matabeleland South, Midlands), Namibia (Caprivi, Kunene, Okavango, Omaheke, Oshikoto, Otjozondjupa), E Botswana (Central, Gaborone, Kweneng, North East, North West, South East), Swaziland and NE South Africa (Gauteng, KwaZulu-Natal, Limpopo, Mpumalanga), NSL–1830 m.
Sources: Sweeney, 1961, Laurent, 1954a, 1964a, Mertens, 1955, 1971, Branch, 1982, 1988, Broadley, 1983, Buys & Buys, 1983, Pienaar et al., 1983, Clauss & Clauss, 2002, Spawls et al., 2002, Broadley et al., 2003, M. Griffin, 2003, Marais, 2004, Alexander & Marais, 2007 and Broadley & Blaylock, 2013.
Remarks: Holotype not in RSM or BMNH *fide* V. FitzSmons (1937: 274).

2. *Thelotornis kirtlandii* (Hallowell, 1844a). Proc. Acad. Nat. Sci. Philadelphia (1844–1845) 2: 62. (*Leptophis kirtlandii*)

Synonyms: *Tragops rufulus* A.-M.-C. Duméril, 1853 (*nomen nudum*), *Dryophis kirtlandi* Hallowell, 1854c, *Oxybelis lecomtei* A.-M.-C. Duméril, Bibron & Duméril, 1854b, *Tragops rufulus* A.-M.-C. Duméril, Bibron & Duméril, 1854b, *Dryiophis pelii* Lichtenstein, 1856, *Oxybelis violacea* J.G. Fischer, 1856b, and *Thelotomis kirtlandi* – Khole, 1991 (*nomen incorrectum*).

Type: Holotype, ANSP 5271, a 657 mm specimen (E. Blanding via W. Blanding).
Type locality: "Liberia, W. Africa."
Distribution: West and cen. Africa. Guinea-Bissau (Bubaque Is.), Guinea (Macenta, Nzérékoré, Télimélé), Sierra Leone (Eastern, Southern, Western), Liberia (Bong, Grand Bassa, Montserrado, Margibi), S Ivory Coast (Abidjan, Dalao, Guiglo, Lagunes, San Pédro, Tabou, Toumodi), S Ghana (Ashanti, Eastern), Togo (Plateaux), S Benin (Atakora, Atlantique, Plateau), S Nigeria (Akwa Ibom, Bayelsa, Cross River, Lagos, Ondo, Rivers), S Cameroon (Est, Centre, Littoral, Nord-Ouest, Sud, Sud-Est), Equatorial Guinea (Bioko Is.), Gabon (Estuaire, Haut-Ogooué, Ngounié, Ogooué-Lolo, Ogooué-Maritime, Woleu-Ntem, Elobey Is.), Congo (Brazzaville, Cuvette, Kouilou, Lekoumou, Likouala, Pool, Sangha), SW Central African Republic (Haute-Sangha, Kemo-Gribingui, Lobaye, Mbomou, Ombella-Mpoko, Sangha), Democratic Republic of the Congo (Bandundu, Bas-Congo, Equateur, Katanga, Kinshasa, Maniema, Nord-Kivu, Orientale, Sud-Kivu), N Angola (Cuanza Norte, Cuanza Sul, Lunda Norte, Bena-Mai Is.) SE South Sudan (Eastern Equatoria), SW Uganda (Central, Western, Sese Is.), W Rwanda, NW Burundi, S Tanzania (Iringa, Kigoma) and W Zambia (North-Western), NSL–2200 m.
Sources: Villiers, 1951, Mertens, 1955, 1964a, E.H. Taylor & Weyer, 1958, Doucet, 1963, Laurent, 1964a, Menzies, 1966, Leston & Hughes, 1968, Hughes & Barry, 1969, J Roux-Estève, 1969, Leston, 1970, V.F.M. FitzSimons, 1974, Pitman, 1974, J.A. Butler & Reid, 1986, D. Lawson, 1993, J.-F. Trape & Roux-Estève, 1995, Akani et al., 2002a, Spawls et al., 2002, Broadley et al., 2003, Chippaux, 2006, Chirio & Ineich, 2006, Chirio & LeBreton, 2007, Pauwels & Vande weghe, 2008, Segniagbeto et al., 2011, Auliya et al., 2012, Chirio, 2013 and Hughes, 2013.
Remarks: Edmund Blanding, William Blanding's brother, was a missionary in Liberia.

3. *Thelotornis mossambicanus* (Bocage, 1895). Herp. Angola Congo:119–120. (*Dryiophis kirtlandii mossambicana*)

Synonym: *Thelotornis capensis schilsi* Derleyn, 1978.
Type: Lectotype, MBL 1843 (A. Ennes), designated by Broadley (1979: 129), destroyed by fire 18 March 1978.
Type locality: "Manica, Mozambique" [= Mozambique], via lectotype selection.
Distribution: Eastern Africa. Southern Somalia (Jubbada Dhexe, Jubbada Hoose, Mogadishu, Alessandra Is.), SE Kenya (Coast), SE Tanzania (Arusha, Dodoma, Iringa, Kigoma, Lindi, Manyara, Mbeya, Morogoro, Mtwara, Piwani, Rukwa, Ruvuma, Tanga), SE Burundi, Morogoro, Mtwara, Mwanza, Pwani, Rukwa, Tanga, Mafia Is.), cen. Mozambique (Inhambane, Manica, Niassa, Sofala, Zambézia, Bazaruto and Benguela Is.), Malawi (Central, Northern, Southern), NE Zambia (Northern) and E Zimbabwe (Manicaland), NSL–1500 m.
Sources: Barros e Cunha, 1935, Derleyn, 1978, Broadley, 1990b, Broadley et al., 2003 and Broadley & Blaylock, 2013.

4. *Thelotornis usambaricus* Broadley, 2001. Afr. J. Herp. 50(2): 58–59, fig. 1.

Type: Holotype, NMZB 16182 (formerly KMH 16325), a 1060 mm male (Frontier-Tanzania team, 18 Feb. 1999).
Type locality: "Amani Nature Reserve (Kwan Koro/ Kwemsambia Forest Reserve), East Usambara Mountains (05° 07' S; 38° 39' E)" [= Tanga Prov., Tanzania].
Distribution: Northeastern Tanzania (Usambara Mtns. of Tanga) and SE Kenya (S Coast), 135–1715 m.

THERMOPHIS Malnate, 1953 (Xenodontidae)

Synonym: *Fhermophis* – E.-M. Zhao, 1980 (*nomen incorrectum*).
Type species: *Tropidonotus baileyi* Wall, 1907b.
Distribution: Central Asia.
Sources: Zaher et al., 2009 and X. Sun et al., 2011.

1. *Thermophis baileyi* (Wall, 1907b). J. Bombay Nat. Hist. Soc. (1906–1907) 17(3): 617–618, pl., 6 figs. (*Tropidonotus baileyi*)

Type: Lectotype, BNHS 1602 A (formerly BNHS 90 or BNHS 91), an 840 mm male (F.M. Bailey, 15 May 1906), designated by Malnate (1953: 94).
Type locality: "Tibet, at about 14,000 feet elevation" [= Xizang Prov., SW China, elevation 4265 m] via lectotype selection.
Distribution: Southwestern China (Xizang), 3660–4400 m.
Sources: Malnate, 1953, HU et al., 1980, Zhao & Adler, 1993, Zhao et al., 1998, Sung et al., 2002, Zhao, 2006, P. Guo et al., 2009, Huang et al., 2009, and Li et al., 2010.

2. *Thermophis zhaoermii* P. Guo, Liu, Feng & He, 2008. Sichuan J. Zool. 27(3): 321, 1 pl.

Type: Holotype, SCUM 44804, an adult female (native, Aug. 2004).
Type locality: "Litang County, Sichuan, China. Elevation 3700 m."
Distribution: China (W Sichuan), 3700 m. Known only from type locality.
Source: P. Guo et al., 2009a.

THRASOPS Hallowell, 1857b (Colubridae)

Synonyms: *Tharsops* – J.A. Butler & Reid *in* Rocek, 1986 (*nomen incorrectum*), and *Trasops* – J.D. Williams & Francini, 1991 (*nomen incorrectum*).
Type species: *Dendrophis flavigularis* Hallowell, 1852b.
Distribution: West and cen. Africa.
Sources: Loveridge, 1944b, 1957a, Pitman, 1974, Hughes, 1983 and Broadley & Wallach, 2002.

1. *Thrasops flavigularis* (Hallowell, 1852b). Proc. Acad. Nat. Sci. Philadelphia (1852–1853) 6(5): 205. (*Dendrophis flavigularis*)

Synonyms: *Hapsidophrys niger* A.C.L.G. Günther, 1872b, *Thrasops pustulatus* Buchholz & Peters *in* W.C.H. Peters, 1875, and *Thrasops flavigularis stirnensis* Stucki-Stirn, 1979.
Type: Holotype, ANSP 5243, a 1925 mm specimen (H.A. Ford), lost *fide* Malnate (1971: 366).
Type locality: "Liberia, Western Africa," (in error). Corrected to Gabon *fide* Loveridge (1944b: 132).
Distribution: Central Africa. Extreme SE Nigeria (Cross River), S Cameroon (Adamaoua, Est, Centre, Littoral, Nord-Ouest, Ouest, Sud, Sud-Ouest), Equatorial Guinea (Bioko Is.), SW Central African Republic (Sangha), Gabon (Estuaire, Haut-Ogoouée, Ogouué-Ivindo, Ogooué-Maritime), S Congo (Bouenza, Brazzaville, Kouilou, Lekoumou, Niari, Pool), Democratic Republic of the Congo (Bandundu, Bas-Congo, Kasai Occidental, Kasai Oriental, Kinshasa, Orientale) and N Angola, NSL–2000 m.
Sources: Witte, 1962, Mertens, 1965c, J.A. Butler & Reid, 1986, Lawson, 1993, J.-F. Trape & Roux-Estève, 1995, Villiers & Condamin, 2005, Chippaux, 2006, Chirio & Ineich, 2006, Chirio & LeBreton, 2007 and Pauwels & Vande weghe, 2008.

2. *Thrasops jacksonii* A.C.L.G. Günther, 1895a. Ann. Mag. Nat. Hist. (6) 15(90): 528–529.

Synonyms: *Thrasops rothschildi* Mocquard, 1905b, and *Thrasops mossambicus* Uthmoller, 1941a (*nomen nudum*).
Type: Holotype, BMNH 1946.1.4.8, a 1651 mm skin (F.J. Jackson).
Type locality: "Kavirondo" [= S Western and Nyanza Prov., SW Kenya].
Distribution: Central Africa. Southern Cameroon (Centre, Est, Ouest), Central African Republic (Haut-Mbomou, Lobaye, Mbomou, Nana-Mambere, Ombella-Mpoko, Ouaka, Ouham, Sangha), S Congo (Brazzaville, Lekoumou, Pool), Democratic Republic of the Congo (Bandundu, Bas-Congo, Equateur, Kasai Oriental, Katanga, Kinshasa, Nord-Kivu, Orientale, Sud-Kivu), S Uganda (Central, Eastern, Western), W Rwanda, S Kenya (Eastern, Nairobi, S Rift Valley, Western), NW Tanzania (Mwanza, Kagera, Rubondo Is.) and N Zambia (North-Western), 200–2400 m.
Sources: Witte, 1953, 1962, J.-F. Trape & Roux-Estève, 1995, Spawls et al., 2002, Broadley et al., 2003, Chippaux, 2006, Chirio & Ineich, 2006 and Chirio & LeBreton, 2007.

T

3. *Thrasops occidentalis* H.W. Parker, 1940. Ann. Mag. Nat. Hist. (11) 5(27): 273, figs. 1, 2a.

Type: Holotype, BMNH 1946.1.5.66 (formerly BMNH 1911.6.30.2), a 1488 mm female (W.P. Lowe).
Type locality: "Axim, Gold Coast" [= Axim, S Western Prov., SW Ghana, 4°52'N, 2°13'W, elevation 25 m].
Distribution: West Africa. Southwestern Senegal (Ziguinchor), Guinea-Bissau (Bubaque Is.), Guinea (Dalaba, Labé, Macenta, Mali, Nzérékoré, Télimélé), Sierra Leone (Northern, Southern), Liberia (Montserrado), Ivory Coast (Abidjan, Adiopodoumé, Daloa, Ferkéssédougou, Guiglo, Lagunes, Tabou), S Ghana (Ashanti, Central, Eastern, Western), S Togo (Plateau), SW Nigeria (Oyo) and W Cameroon (Northwest), 25–1135 m.
Sources: Villiers, 1950, 1951, 1963, Doucet, 1963, Menzies, 1966, Leston & Hughes, 1968, Hughes & Barry, 1969, Leston, 1970, Roux-Estève, 1969, Hulselmans et al., 1970, Stucki-Stirn, 1979, Hughes, 1983, Chippaux, 2006, J.-F. Trape & Mané, 2006b, W. Böhme et al., 2011, Segniagbeto et al., 2011, Auliya et al., 2012 and Chirio, 2013.
Remarks: Total length erroneously listed as 1085 mm *fide* Loveridge (1944b: 132). Possibly occurs in Benin *fide* Hughes (2013: 152).

4. *Thrasops schmidti* Loveridge, 1936a. Proc. Biol. Soc. Washington 49: 63–64. (*Thrasops jacksonii schmidti*)

Type: Holotype, MCZ 9276, a 1065 mm male (G.M. Allen, 21–26 Aug. 1909).
Type locality: "Meru Forest, northeast of Mount Kenya, Kenya Colony."
Distribution: Central Kenya (Eastern, Nairobi), 1600–2000 m.
Sources: Loveridge, 1957, Spawls, 1978 and Broadley & Wallach, 2002.

†*TITANOBOA* Head, Bloch, Hastings, Bourque, Cadena, Herrera, Polly & Jaramillo, 2009 (Boidae)

Type species: †*Titanoboa cerrejonensis* Head, Bloch, Hastings, Bourque, Cadena, Herrera, Polly & Jaramillo, 2009.
Distribution: Upper Paleocene of Colombia.
Source: Head et al., 2009.

1. †*Titanoboa cerrejonensis* Head, Bloch, Hastings, Bourque, Cadena, Herrera, Polly & Jaramillo, 2009. Nature 457(7230): 715–716, figs. 1a–d.

Type: Holotype, INIGM (formerly UF/IGM 1), one trunk vertebra (A. Rincon, M. Carvalho, J.I. Bloch, A.K. Hastings, E.A. Cadena, F.A. Herrera & Jaramillo, 2003–2008).

Type locality: "La Puente Pit, Cerrejón Coal Mine, Guajira Peninsula, northeastern Colombia (palaeolatitude 5.5°N); middle segment of the Cerrejón Formation, middle-late Paleocene epoch (58-60 Myr ago), palynological zone Cu-02."

Distribution: Upper Paleocene (58.0–60.0 mya) of Colombia. Known only from type locality.

TOMODON A.-M.-C. Duméril, 1853 (Xenodontidae)

Synonyms: *Machirodon* A.-M.-C. Duméril, Bibron & Duméril, 1854a, *Opisthoplus* W.C.H. Peters, 1883, *Aproterodon* Vanzolini, 1947, and *Tomadon* – Auffenberg, 1958 (*nomen incorrectum*).

Type species: *Tomodon dorsatum* A.-M.-C. Duméril, Bibron & Duméril, 1854b.

Distribution: Southern South America.

Sources: Hoge, 1959a, Bailey, 1967, J.A. Peters & Orejas-Miranda, 1970, Giraudo & Scrocchi, 2002, Harvey & Muñoz, 2004 and Zaher et al., 2009.

1. *Tomodon dorsatum* A.-M.-C. Duméril, Bibron & Duméril, 1854b. Erpét. Gén. 7(2): 934–936.

Synonyms: *Tomodon dorsatum* A.-M.-C. Duméril, 1853 (*nomen nudum*), *Opisthoplus degener* W.C.H. Peters, 1883, and *Aproterodon clementei* Vanzolini, 1947.

Type: Holotype, MNHN 3737 (formerly MHNMF), a 620 mm specimen.

Type locality: "Américaine," probably "Brésil" [= America, probably Brazil]. Amérique Meridionale [= South America] *fide* MNHN catalogue.

Distribution: Southern South America. Central and SE Brazil (Espírito Santo, Paraná, Rio Grande do Sul, Santa Catarina, São Paulo, Cananéia, Cardoso, Porchat and São Vicente Is.), Uruguay (Cerro Largo, Rocha), Paraguay (Itapúa) and N Argentina (Misiones), NSL–250 m.

Sources: Zanini-Cechin, 1989, Giraudo, 1997, Starace, 1998, Harvey & Muñoz, 2004, Carreira-Vidal & Lumbardo, 2006, Cicchi et al., 2007, and Carreira-Vidal et al., 2012b.

2. *Tomodon ocellatus* A.-M.-C. Duméril, Bibron & Duméril, 1854b. Erpét. Gén. 7(2): 938–940. (*Tomodon ocellatum*)

Synonym: *Tomodon dorsatum* A.-M.-C. Duméril, 1853 (*nomen nudum*).

Type: Holotype, ? RMNH, a 425 mm specimen.

Type locality: "Brésil" [= Brazil].

Distribution: Southern South America. Southern Brazil, Paraguay (Itapúa), Uruguay (Artigas, Canelones, Cerro Largo, Colonia, Durazino, Flores, Lavalleja, Montevideo, Paysandú, Rio Negro, Salto, San José, Soriano, Treinta y Tres) and Argentina (Buenos Aires, Córdoba, Corrientes, Entre Ríos, La Pampa, Santa Fe), NSL–190 m.

Sources: Gallardo, 1972, Aquino et al., 1996, Tiranti & Avila, 1997 and Carreira-Vidal et al., 2005.

Remarks: Holotype not in MNHN. Misiones, Argentina record in error *fide* Giraudo & Scrocchi (2002: 40).

3. *Tomodon orestes* Harvey & Muñoz, 2004. Herpetologica 60(3): 365–369, figs 1–2.

Type: Holotype, CBF 2316 (formerly AM-S 400), a 515 mm male (J. Muñoz S., Jan. 2000).

Type locality: "close to the Río Erquis, Reserva de Sama, Méndez province, Tarija, Bolivia, 2754 m (21° 28' 56" S, 64° 50' 32" W)."

Distribution: Southern Bolivia (Chuquisaca, Cochabamba, Tarija) and Argentina (Salto), 1705–3300 m.

Source: Akmentins & Vaira, 2010.

†*TOTLANDOPHIS* Holman & Harrison, 1998b (Boidae)

Type species: †*Totlandophis thomasae* Holman & Harrison, 1998b.

Distribution: Upper Eocene of United Kingdom, and lower Oligocene of USA.

Sources: Holman & Harrison, 1998b, 2001.

T

1. †*Totlandophis americanus* Holman & Harrison, 2001. Acta Zool. Cracov. 44(1): 26–29, figs. 1–3.

Type: Holotype, UF 190844, one trunk vertebra (UF field crews, 1966–1967).

Type locality: "I-75 Local Fauna (Florida Natural History Museum Locality AL018), 1.5 km WSW Gainesville, Alachua County, Florida; Early Oligocene (Whitneyan)" [USA].

Distribution: Lower Oligocene (Whitneyan: 30.8–33.3 mya) of USA (Florida).

2. †*Totlandophis thomasae* Holman & Harrison, 1998b. Acta Zool. Cracov. 41(1): 30–33, figs. 1–2.

Type: Holotype, MSUVP 1458, one trunk vertebra (D.L. Harrison, 1994).
Type locality: "Hordle Cliff, Hampshire, England. Rodent Bed in upper part of Totland Bay member, MP17, of the Headon Hill Formation (Upper Eocene)."
Distribution: Upper Eocene (Priabonian, MP 17: 33.9–37.2 mya) of United Kingdom (England).

TOXICOCALAMUS Boulenger, 1896e (Elapidae)

Synonyms: *Apistocalamus* Boulenger, 1898, *Pseudapistocalamus* Lönnberg, 1900, *Vanapina* De Vis, 1905, *Apisthocalamus* Boulenger, 1896a (*nomen emendatum*), *Pseudapisthocalamus* Boulenger, 1896a (*nomen emendatum*), *Ultrocalamus* Sternfeld, 1913b, *Opistocalamus* – Boquet, 1948 (*nomen incorrectum*), *Pseudopistocalamus* – Boquet, 1948 (*nomen incorrectum*), *Tocicocalamus* – McDowell, 1967 (*nomen incorrectum*), *Utocalamus* – J.H. Brown, 1973 (*nomen incorrectum*), and *Ultocalamus* – W.W. Tanner & Avery, 1982 (*nomen incorrectum*).
Type species: *Toxicocalamus longissimus* Boulenger, 1896e.
Distribution: New Guinea and adjacent islands.
Sources: Loveridge, 1946, McDowell, 1967, 1969a, O'Shea, 1996, Keogh, 1998, David & Ineich, 1999, Scanlon & Lee, 2004, Sanders et al., 2008, Zaher et al., 2009 and Hoser, 2012e.

1. *Toxicocalamus buergersi* (Sternfeld, 1913b). Sitz. Ges. Naturf. Freunde Berlin 1913(8): 388–389. (*Ultrocalamus bürgersi*)

Synonym: *Ultrocalamus latisquamatus* Schüz, 1929.
Type: Holotype, ZMB 25232, a 365 mm specimen (Bürgers).
Type locality: "Deutsch-Neuguinea" [= German New Guinea, now Papua New Guinea].
Distribution: Papua New Guinea (East Sepik, West Sepik), NSL–300 m.

T

2. *Toxicocalamus grandis* (Boulenger, 1914d). Trans. Zool. Soc. London 20(5): 265, pl. 30, figs. 3–3a. (*Apistocalamus grandis*)

Type: Holotype, BMNH 1946.1.18.34, a 940–960 mm female (A.F.R. Wollaston [2nd Wollaston Exped.], 1912).
Type locality: "Launch Camp, Setekwa R., Dutch New Guinea" [= Launch Camp, Setekwa River, Papua Prov., E Indonesia].
Distribution: Eastern Indonesia (S Papua). Known only from holotype.

3. *Toxicocalamus holopelturus* McDowell, 1969a. J. Zool. London 159: 467–469, figs. 1b, 3b, pl. 2.

Type: Holotype, AMNH 76660, a 620 mm male (R.F. Peterson & L.J. Brass).
Type locality: "Mount Rossel at 700 m, Rossel Island, Milne Bay Division, Territory of Papua" [= Rossel Is., E Milne Bay Prov., SE Papua New Guinea, bet. 11°18'–26'S and 153°58'E–154°17'E].
Distribution: Southeastern Papua New Guinea (Milne Bay: Rossel Is.), 700 m. Known only from holotype.

4. *Toxicocalamus longissimus* Boulenger, 1896e. Ann. Mag. Nat. Hist. (6) 18(104): 152.

Synonym: *Vanapina lineata* De Vis, 1905.
Type: Lectotype, BMNH 1946.1.18.93 (formerly BMNH 1896.7.8.18), a 650 mm female (A.S. Meek), designated by McDowell (1969a: 473).
Type locality: "Woodlark Island, Territory of Papua New Guinea" via lectotype selection.
Distribution: Southeastern Papua New Guinea (Milne Bay: Fergusson and Woodlark Is.).
Sources: Boulenger, 1903c and Ingram, 1989.

5. *Toxicocalamus loriae* (Boulenger, 1898i). Ann. Mus. Civ. Stor. Nat. Genova (2) 18(38): 705–706, pl. 8, fig. 1. (*Apistocalamus loriae*)

Synonyms: *Pseudapistocalamus nymani* Lönnberg, 1900, *Apistocalamus pratti* Boulenger, 1904b, *Apisthocalamus loennbergii* Boulenger, 1908c, and *Apisthocalamus lamingtoni* Kinghorn, 1928b.
Type: Holotype, MSNG 29141, a 580 mm male (L. Loria, 1889–1892).
Type locality: "Haveri, British New Guinea" [= Papua New Guinea].
Distribution: New Guinea. Eastern Indonesia (Papua) and Papua New Guinea (Central, Eastern Highlands, Madang, Milne Bay, Morobe, Oro, Simbu, Southern Highlands, Western, Western Highlands, West Sepik, Fergusson, Karkar and Normanby Is.), NSL–1490 m.
Sources: McDowell, 1984 and Kraus & Allison, 2004.

6. *Toxicocalamus misimae* McDowell, 1969a. J. Zool. London 159: 470–473, figs. 3c, 7.

Type: Holotype, AMNH 76684, a 468 mm male (R.F. Peterson & L.J. Brass [5th Archbold New Guinea Exped.], 14 July–13 Aug. 1956).
Type locality: "Misima Island, Milne Bay Division, Territory of Papua, New Guinea" [= Papua New Guinea].
Distribution: Papua New Guinea (Central, Milne Bay: Misima Is.).

7. *Toxicocalamus preussi* (Sternfeld, 1913b). Sitz. Ges. Naturf. Freunde Berlin 1913(8): 388, figs. 1–2. (*Ultrocalamus preussi*)

Synonym: *Ultrocalamus preussi angusticinctus* Bogert & Matalas, 1945.

Type: Holotype, ZMB 23948, a 600 mm specimen (P. Preuss, 1903).

Type locality: "Insel Seleo (bei Berlinhafen)" [= Seleo Is., West Sepik Prov., NW Papua New Guinea, 3°09'S, 142°30'E, elevation NSL–15 m].

Distribution: New Guinea. Eastern Indonesia (N Papua) and Papua New Guinea (East Sepik, Gulf, Southern Highlands, Western, West Sepik, Seleo Is.), NSL–850 m.

Source: Bogert & Matalas, 1945.

8. *Toxicocalamus spilolepidotus* McDowell, 1969a. J. Zool. London 159: 464–465, pl. 1.

Type: Holotype, AMNH 85745, a 770 mm female (L.J. Brass & H.M. van Deusen [6th Archbold New Guinea Exped.], 25 Sept. 1959).

Type locality: "Purosa (near Okapa), Eastern Highlands Division, Territory of New Guinea" [= Purosa camp, 27 km by road SSW Okapa Patrol Post, Eastern Highlands Prov., E Papua New Guinea, 6°33'S, 145°37'E, elevation 1950 m].

Distribution: Central Papua New Guinea (Eastern Highlands), 1500–1950 m.

9. *Toxicocalamus stanleyanus* Boulenger, 1903c. Proc. Zool. Soc. London 1903: 128–129, pl. 13, fig. 3.

Type: Holotype, BMNH 1946.1.7.55 (formerly BMNH 1903.3.10.18), a 610 mm female (A.E. Pratt, 1902–1903).

Type locality: "Dinawa, Owen Stanley Distribution, 4000 feet, British New Guinea" [= Papua New Guinea].

Distribution: New Guinea. Eastern Indonesia (ext. West Papua) and Papua New Guinea (Central, Gulf, Madang, Southern Highlands, Western, West Sepik), 100–1300 m.

Sources: De Rooij, 1917, Bogert & Matalas, 1945, Loveridge, 1948, McDowell, 1967 and Kraus, 2010.

TOXICODRYAS Hallowell, 1857b (Colubridae)

Synonyms: *Toxicordryas* – Welch, 1982 (*nomen incorrectum*) and *Toxidryas* – Broadley, 1983 (*nomen incorrectum*).

Type species: *Dipsas blandingii* Hallowell, 1844c.

Distribution: West and cen. Africa.

Sources: Witte, 1962, Laurent, 1964a, Leston & Hughes, 1968, Hughes & Barry, 1969, Hughes, 1983, 2013, J.B. Rasmussen, 1997, Chippaux, 2006, Chirio & LeBreton, 2007, Pauwels & Vande weghe, 2008, W. Böhme et al., 2011 and Segniagbeto et al., 2011.

Remarks: A valid genus *fide* Wallach (unpubl. data).

1. *Toxicodryas blandingii* (Hallowell, 1844c). Proc. Acad. Nat. Sci. Philadelphia 2: 170–171. (*Dipsas blandingii*)

Synonyms: *Triglyphodon fuscum* A.-M.-C. Duméril, Bibron & Duméril, 1854b, *Dipsas fasciata* J.G. Fischer, 1856, *Dipsas globiceps* J.G. Fischer, 1856 *Dipsas valida* J.G. Fischer, 1856, *Dipsas fischeri* Jan *in* A.H.A. Duméril, 1859, *Triglyphodon fuscum obscurum* A.H.A. Duméril, 1861, *Dipsas regalis* Jan, 1871 *in* Jan & Sordelli, 1871–1881, *Dipsas globiceps tumboensis* F. Müller, 1885, *Boiga blandingii occidentalis* Stucki-Stirn, 1979, and *Boiga blandingi subfulva* Stucki-Stirn, 1979.

Type: Holotype, not designated (formerly ? ANSP), a 1670 mm specimen (E. Blanding via W. Blanding), lost *fide* Hughes & Barry (1969: 1021).

Type locality: "Liberia, West Africa."

Distribution: West and cen. Africa. Southwestern Senegal (Ziguinchor), Guinea (Nzérékoré), Guinea Bissau (Biombo, Bissau, Bubaque and Soga Is.), Guinea (Kouroussa, Macenta, Nzérékoré, Télimélé), Sierra Leone (Eastern, Northern, Southern, Western, Sherbo Is.), Liberia (Bong, Gbarpolu, Loffa, Montserrado, Margibi, Nimba), Ivory Coast (Abidjan, Agnebi, Bouaflé, Bouaké, Bouna, Daloa, Guiglo, Lagunes, Tabou, Toumodi), Ghana (Brong-Ahafo, Eastern), Togo (Kara, Maritime, Plateaux), S Benin (Atlantique, Plateau, Zou), Nigeria (Akwa Ibom, Bayelsa, Cross River, Delta, Kaduna, Kwara, Niger, Rivers), S Cameroon (Est, Centre, Littoral, Nord-Ouest, Ouest, Sud, Sud-Ouest), Equatorial Guinea (Bioko Is.), Central African Republic (Bamingui-Bangoran, Haut-Mbomou, Haute-Sangha, Lobaye, Ombella-Mpoko, Ouaka, Sangha), Gabon (Estuaire, Haut-Ogooué, Moyen-Ogooué, Ngounié, Ogooué-Ivindo, Ogooué-Lolo, Ogooué-Maritime, Woleu-Ntem), Congo (Brazzaville, Kouilou, Pool, Sangha), Democratic Republic of the Congo (Bandundu, Bas-Congo, Equateur, Kasai Occidental, Kasai Oriental, Katanga, Kinshasa, Maniema, Nord-Kivu, Orientale, Sud-Kivu), NE Angola (Lunda Norte), S South Sudan (SW Eastern Equatoria), S Uganda (Central, Western), W Kenya (Western), W Tanzania (Kigoma, Kagera, Rubondo Is.) and N Zambia (Northern), NSL–2200 m.

Sources: Villiers, 1951, Doucet, 1963, Mertens, 1965c, Menzies, 1966, J.A. Butler & Reid, 1986, Luiselli et al., 1998b, Spawls et al., 2002, Broadley et al., 2003, Greenbaum & Carr, 2005, Chippaux, 2006, J.-F. Trape & Mané, 2006b, W. Böhme et al., 2011, Auliya et al., 2012 and Chirio, 2013.

Remarks: Edmund Blanding, William Blanding's brother, was a missionary in Liberia.

2. *Toxicodryas pulverulentus* (J.G. Fischer, 1856b). Abh. Ges. Naturw. Ver. Hamburg 3: 81–83, pl. 3, figs. 1a–c. (*Dipsas pulverulenta*)

Synonym: *Dipsadomorphus boueti* Chabanaud, 1917a.

Type: Lectotype, ZMH 4376 (formerly ZMH 339), a 790 mm specimen (Davis), designated herein.
Type locality: "Edina, Grand Bassa County in Liberia (West-Afrika)" [= Edina, Grand Bassa Co., Liberia, 5°55'N, 10°05'W, elevation 10 m] (in error *fide* Ladiges *in* Hughes & Barry, 1969: 1020). Corrected to São Tomé Is., São Tomé and Príncipe, Gulf of Guinea *fide* Ladiges *in* Hughes & Barry (1969: 1020).
Distribution: West and cen. Africa. Southeastern Guinea (Guéckédou, Macenta, Nzérékoré), Sierra Leone (Northern, Southern, Liberia, S Ivory Coast (Abidjan, Daloa, Guiglo, Lagunes, Tabou), Ghana (Eastern, Volta), Togo (Centrale, Plateaux), S Benin (Cotonou, Plateau), S Nigeria (Cross River, Delta), S Cameroon (Est, Centre, Littoral, Nord-Ouest, Sud, Sud-Ouest), Gabon (Moyen-Ogooué, Ogooué-Ivindo, Ogooué-Maritime), Equatorial Guinea, São Tomé and Príncipe, Central African Republic (Bamingui-Bangoran, Haute-Sangha, Lobaye, Ombella-Mpoko), Congo (Brazzaville, Cuvette, Kouilou, Lekoumou, Sangha), Democratic Republic of the Congo (Bandundu, Bas-Congo, Equateur, Kasai Occidental, Kasai Oriental, Katanga, Kinshasa, Nord-Kivu, Orientale, Sud-Kivu), S Uganda (Central, Western), SW Kenya (Western) and N Angola, NSL–1050 m.
Sources: Doucet, 1963, Menzies, 1966, J.A. Butler & Reid, 1986, Spawls et al., 2002 and Chippaux, 2006.
Remarks: Second syntype lost *fide* Hallermann (1998: 202).

TRACHISCHIUM A.C.L.G. Günther, 1858 (Colubroidea *incertae sedis*)

Synonyms: *Trachyschium* Marschall, 1873 (*nomen emendatum*), *Trachyischium* Berg, 1901 (*nomen emendatum*), *Eminophis* F. Werner, 1924a, and *Erninophis* – A.S. Romer, 1956 (*nomen incorrectum*).
Type species: *Calamaria fusca* Blyth, 1854a.
Distribution: Southern Asia.
Sources: Boulenger, 1890a, 1893a, M.A. Smith, 1943, Kramer, 1977, R.C. Sharma, 2003, Shah & Tiwari, 2004 and Zaher et al., 2009.

T

1. *Trachischium fuscum* (Blyth, 1854a). J. Asiat. Soc. Bengal 23(3): 288. (*Calamaria fusca*)

Synonyms: *Calamaria fusca* Blyth, 1854a, *Calamaria obscuro-striata* Blyth, 1854a (*nomen incorrigendum*), *Trachischium rugosum* A.C.L.G. Günther, 1858, *Calamaria obscurostriata* – A.C.L.G. Günther, 1864a (*nomen corrigendum*), *Ablabes gilgiticus* Annandale, 1905b, and *Eminophis lineolata* F. Werner, 1924a.
Type: Lectotype, ZSI 7044, a 305 mm specimen (W.S. Sherwill, 1843–1854), designated by Talukdar et al. (1980: 94).
Type locality: "Darjeeling (Darjeeling District, West Bengal, India)" [= 27°02'N, 88°16'E, elevation 1600–2250 m] via lectotype selection.
Distribution: Southern Asia. Northeastern Pakistan (Gilgit-Balistan), India (Arunachal Pradesh, Kashmir, Sikkim, Uttar Pradesh, West Bengal), Nepal (Bhojpur, Dolakha, Ilam, Kaski, Kathmandu, Panchthar, Sankhuwasabha, Taplejung, Tehrathum) and Bhutan border (Teesta; MCZ 7514), 920–2590 m.
Sources: J. Anderson, 1871a, M.A. Smith, 1928b, Talukdar et al., 1980, Nanhoe & Ouboter, 1987, Schleich & Kästle, 2002, Tiwari & Shah, 2004, Ahmed et al., 2009 and Agarwal et al., 2010.
Remarks: Photographs of lectotype in Talukdar et al. (1980: pl. 2, figs. a–c). Myanmar record probably in error *fide* Dowling & Jenner (1988: 11).

2. *Trachischium guentheri* Boulenger, 1890a. Fauna Brit. India, Rept. Batr.: 285–286.

Types: Syntypes (4), BMNH 1946.1.12.29, male (V. Schlagintweit), BMNH 1946.1.12.36, juvenile (R.H. Beddome), and BMNH 1946.1.12.49 and 1946.1.12.50, male and female, one is 280 mm (W.T. Blanford).
Type locality: "Sikhim" [= Sikkim State, NE India].
Distribution: Southern Asia. Northern India (West Bengal), cen. Nepal (Kathmandu) and Bhutan, 1400–2500 m.
Sources: Bauer & Günther, 1992, Schleich & Kästle, 2002, Tiwari & Shah, 2004 and Chettri et al., 2009.
Remarks: Occurs in Bangladesh *fide* Ahsan et al. (2009: 149).

3. *Trachischium laeve* Peracca, 1904b. Rev. Suisse Zool. 12(3–4): 665–666.

Synonym: *Trachischium quinquelabialis* Wall, 1911h.
Types: Syntypes (2), MHNG 667.46, a 502 mm female (H. de Saussure), and NHMB 911, a 337 mm male.
Type locality: "Indes orientales" [= Southern Asia and East Indies].
Distribution: Southern Asia. Northern India (Uttar Pradesh), cen. Nepal (Kaski) and Bhutan, 1540–2700 m.
Sources: Tillack & Shah, 2002, Tiwari & Shah, 2004, Whitaker & Captain, 2004 and Wangyal, 2011.
Remarks: Photograph of syntype, mislabeled as MHNG 677.40, in Tillack & Shah (2002: fig. 5).

4. *Trachischium monticola* (Cantor, 1839b). Proc. Zool. Soc. London 1839(7): 50. (*Calamaria monticola*)

Synonyms: *Cyclophis rubriventris* Jerdon, 1870, and *Ablabes albiventer* A.C.L.G. Günther, 1875.
Type: Holotype, BMNH 1946.1.12.32 (T.E. Cantor, 1834–1839).
Type locality: "Naga Hills in Asám" [= Naga Hills, Nagaland State, NE India].
Distribution: Northern India (Arunachal Pradesh, Assam, West Bengal) and SW China (Xizang), 600–2000 m.
Sources: Zhao & Adler, 1993, Zhao, 2006, Agarwal et al., 2010 and Li et al., 2010.

Remarks: Occurs in Bangladesh *fide* Ahsan et al. (2009: 149) and Nepal *fide* Tiwari & Shah, 2004.

5. *Trachischium tenuiceps* (Blyth, 1854a). J. Asiat. Soc. Bengal 23(3): 288. (*Calamaria tenuiceps*)

Type: Holotype, ZSI 7181 (formerly IMC), a 355 mm specimen (W.S. Sherwill, 1843–1854).
Type locality: "vicinity of Darjiling" [= vicinity of Darjeeling (27°02'N, 88°16'E), West Bengal State, NE India, elevation 1600–2250 m].
Distribution: Southern Asia. Northern India (Arunachal Pradesh, Sikkim, West Bengal) Bangladesh, Nepal (Bhaktapur, Dolakha, Ilam, Kathmandu, Kavre, Makwanpur, Nuwakot, Panchthar) and SW China (Xizang), 800–2440 m.
Sources: Schleich & Kästle, 2002, Tiwari & Shah, 2004, Zhao, 2006, Ahsan et al., 2009, Agarwal et al., 2010 and Li et al., 2010.

TRACHYBOA W.C.H. Peters, 1860b (Tropidophiidae)

Synonyms: *Trachiboa* – Senna, 1886 (*nomen incorrectum*), and *Xrachyboa* – Briceño-Rossi, 1934 (*nomen incorrectum*).
Type species: *Trachyboa gularis* W.C.H. Peters, 1860b.
Distribution: Panama and NW South America.
Sources: Brongersma, 1951, Stimson, 1969, Zaher, 1994, Walls, 1998a and McDiarmid et al., 1999.

1. *Trachyboa boulengeri* Peracca, 1910. Ann. Mus. Zool. Univ. Napoli (2) 3(12): 1–2, 1 fig.

Type: Holotype, not designated, a 371 mm specimen, location unknown.
Type locality: "regioni tropicali del Sud America" [= tropical South America].
Distribution: Eastern Panama (Darien), W Colombia (Magdalena) and NW Ecuador (Esmeraldas), NSL–750 m.
Sources: Dunn & Bailey, 1939, H.M. Smith, 1958, Lehmann, 1970, 1974, 1993a–b, Pérez-Santos, 1986, Villa et al., 1988 and Cisneros-Heredia, 2004d.
Remarks: Holotype not in SZN *fide* Andreone & Gavetti (2009: 94).

2. *Trachyboa gularis* W.C.H. Peters, 1860b. Mber. Königl. Akad. Wiss. Berlin 1860(4): 200–202, pl., figs. 1, 1a–d.

Synonym: *Trachyboa gularis multimaculata* Rosén, 1905.
Types: Syntypes (3), ZMB 3770–71 and ZMB 4494, longest syntypes 395 mm and 285 mm (C. Reiss, 1860).
Type locality: "Guayaquil" [Ecuador].
Distribution: Southwestern Ecuador (Guayas), 10–50 m.
Remarks: A possible syntype = RMNH 1297 (ZMB exchange), a 289 mm male. Possibly occurs in Brazil *fide* J.A. Peters & Orejas-Miranda (1970: 305) but doubtful *fide* McDiarmid et al. (1999: 215).

†*TREGOPHIS* Holman, 1975 (Boidae)

Type species: †*Tregophis brevirachis* Holman, 1975.
Distribution: Middle-upper Miocene of USA.
Sources: Holman, 1975, 1979b, 2000a, Rage, 1984b and Parmley & Hunter, 2010.

1. †*Tregophis brevirachis* Holman, 1975. Univ. Michigan Paps. Paleont. (12): 61, fig. 2f.

Type: Holotype, MSUVP 783, one trunk vertebra (J.A. Holman, 1969–1973).
Type locality: "locality UM-K6-59 on the Lowell Hillman Ranch 2350–2550 ft S and 75 ft E of the NW corner, Sec. 22, R. 22W, T.11 S, elevation 2255 ft., Wakeeney, Kansas" [Gallala formation, Clarendonian, lower or middle Miocene].
Distribution: Midddle Miocene (Clarendonian: 10.3–13.6 mya) of USA (Kansas, Nebraska).
Remarks: Revised diagnosis in Holman (2000a: 92–93).

TRETANORHINUS A.-M.-C. Duméril, Bibron & Duméril, 1854a (Dipsadidae)

Synonyms: *Tretanorhinos* – Jan, 1857 (*nomen incorrectum*), *Tretanorhine* – Marschall, 1873 (*nomen incorrectum*), *Tetranorhinus*– Senna, 1886 (*nomen incorrectum*), *Tretatorhinus* – Hoffmann, 1890 (*nomen incorrectum*), and *Tretanorphinus* – Grant, 1946a (*nomen incorrectum*).
Type species: *Tretanorhinus variabilis* A.-M.-C. Duméril, Bibron & Duméril, 1854a.
Distribution: Latin America and West Indies.
Sources: Dunn, 1939a, J.A. Peters & Orejas-Miranda, 1970, Villa et al., 1988, Pinou & Dowling, 1994 and Zaher et al., 2009.

T

1. *Tretanorhinus mocquardi* Bocourt, 1891. Naturaliste (2) 13(101): 122.

Types: Syntypes (5), ANSP 11656 and MNHN 3675a–b, longest syntype 730 mm (Criado), location of other syntypes unknown.
Type locality: "Amérique intertropicale" [= America between Tropics of Cancer and Capricorn, i.e. 23.5°N and 23.5°S]. Restricted to Panama *fide* Bocourt (1895: 798–799, pl. 52, fig. 5 *in* A.H.A. Duméril et al., 1870–1909) and specimen label. Further restricted to Panama City, Panama *fide* Dunn (1939a: 214).

Distribution: Panama (Panama, Colon) and Ecuador (Esmeraldas), NSL–50 m.
Sources: A.H.A. Duméril et al., 1870–1909, Dunn, 1939a and Cisneros-Heredia, 2005b.
Remarks: Five syntypes originally *fide* Bocourt (1895: 799 *in* A.H.A. Duméril et al., 1870–1909).

2. *Tretanorhinus nigroluteus* Cope, 1861e. Proc. Acad. Nat. Sci. Philadelphia 13(7): 298–299.

Synonyms: *Tretanorhinos mexicanus* Jan, 1857 (*nomen nudum*), *Helicops agassizi* Jan, 1863b, *Helicops bifrenatus* Bocourt, 1884, *Tretanorhinus lateralis* Bocourt, 1891, *Tretanorhinus intermedius* Rosén, 1905, *Tretanorhinus nigroluteus mertensi* H.M. Smith & Gillespie *in* H.M. Smith, 1965, *Tretanorhinus nigroluteus dichromaticus* Villa, 1969, and *Tretanorhinus nigroluteus obscurus* Villa, 1969.
Type: Holotype, USNM 5568 (formerly SIM), a 635 mm (svl) specimen (probably J. Rowell, 1857).
Type locality: "Greytown, Nicaragua," (in error *fide* Dunn, 1939a: 216). Corrected to Aspinwall, Panama *fide* USNM catalogue (Dunn, 1939a: 216 & Cochran, 1961: 220).
Distribution: Mesoamerica. Southeastern Mexico (Campeche, Oaxaca, Quintana Roo, Tabasco, Veracruz), N Guatemala (Alta Verapaz, Petén), Belize (Belize, Cayo, Orange Walk, Stann Creek), Honduras (Atlántica, Colón, Cortés, Francisco Morazán, Gracias a Dios, Islas de la Bahía, Olancho, Yoro, Guanaja, Roatán and Utila Is.), Nicaragua (Atlántico Norte, Atlántico Sur, Jinotega, Matagalpa, Rio San Juan, Zelaya, Corn Is.), Costa Rica (Heridia, Limón), Panama and Colombia (Atlántico), NSL–1260 m.
Sources: Villa, 1968, 1970a, Alarcón-Pardo, 1978, R.W. Henderson & Hoevers, 1979, L.D, Wilson & Meyer, 1985, Pérez-Santos & Moreno, 1988, J.C. Lee, 1996, 2000, G. Köhler, 1996a, 1999, J.A. Campbell, 1998b, Auth et al., 2000, Stafford & Meyer, 2000, Savage, 2002, McCranie et al., 2005, McCranie, 2011a and Travers et al., 2011.
Remarks: Collector (Dr. Caldwell) of type and its locality in error *fide* Dunn (1939a: 216).

T

3. *Tretanorhinus taeniatus* Boulenger, 1903b. Ann. Mag. Nat. Hist. (7) 12(69): 350.

Type: Holotype, BMNH 1946.1.15.40, a 570 mm female (W.F.H. Rosenberg, 1896–1903).
Type locality: "Rio Sapayo, N.W. Ecuador, altitude 450 feet."
Distribution: Western Colombia (Valle del Cauca) and NW Ecuador (Esmeraldas), NSL–150 m.
Sources: Pérez-Santos & Moreno, 1988, 1991.

4. *Tretanorhinus variabilis* A.-M.-C. Duméril, Bibron & Duméril, 1854a. Erpét. Gén. 7(1): 349–351, pl. 80, fig. 4.

Synonyms: *Helicops wagleri* Jan, 1863b, *Tretanorhinus variabilis adnexus* Jan, 1863b (*nomen nudum*), *Tretanorhinus variabilis adnexus* Bocourt, 1895 *in* A.H.A. Duméril, Bocourt & Mocquard, 1870–1909, *Tretanorhinus variabilis cubanus* Bocourt, 1895 *in* A.H.A. Duméril, Bocourt & Mocquard, 1870–1909, *Tretanorhinus insulaepinorum* T. Barbour, 1916a, *Tretanorhinus variabilis lewisi* Grant, 1940c, *Tretanorhinus gaigeae* Grant, 1946a, and *Tretanorhinus variabilis binghami* A. Schwartz & Ogren, 1956.
Types: Syntypes (2), MNHN 7161 and MNHN 7346, longest syntype 511 mm.
Type locality: Unknown.
Distribution: Cuba (Camagüey, Cienfuegos, Ciego de Aviula, Granma, Guantánamo, Habana, Las Tunas, Matanzas, Pinar del Río, Sancti Spíritus, Villa Clara, Juventad Is.) and Cayman Is. (Grand Cayman).
Sources: Petzold, 1967, A. Schwartz & R.W. Henderson, 1988, 1991, Seidel & Franz, 1994, Rodríguez-Robles & Greene, 1996, Rodríguez-Schettino, 2003 and R.W. Henderson & Powell, 2009.

TRICERATOLEPIDOPHIS Ziegler, Herrmann, David, Orlov & Pauwels, 2001 (Viperidae)

Type species: *Triceratolepidophis sieversorum* Ziegler, Herrmann, David, Orlov & Pauwels, 2001.
Distribution: Indo-China.
Sources: Ziegler et al., 2001, Herrmann et al., 2002, 2004, Orlov et al., 2003, 2009, Gumprecht et al., 2004, Shiryaev et al., 2007 and V.S. Nguyen et al., 2009.

1. *Triceratolepidophis sieversorum* Ziegler, Herrmann, David, Orlov & Pauwels, 2001. Russ. J. Herp. (2000) 7(3): 201–206, figs. 1–8.

Type: Holotype, ZFMK 71262, a 1255 mm male (native, Jan.–June 1999).
Type locality: "Phong Nha (village), Nature Reserve, Quang Binh Province, Vietnam, ~100 m a.s.l."
Distribution: Central Laos (Khammouan) and cen. Vietnam (Quang Binh), 100–210 m.

TRICHEILOSTOMA Jan & Sordelli, 1860 *in* 1860–1866 (Leptotyphlopidae)

Synonym: *Guinea* Hedges, Adalsteinsson & Branch *in* Adalsteinsson, Branch, Trape, Vitt & Hedges, 2009.
Type species: *Stenostoma bicolor* Jan & Sordelli, 1860 *in* 1860–1866.
Distribution: West Africa.
Sources: Hahn, 1980a, Wallach, 1998b and McDiarmid et al., 1999.

1. *Tricheilostoma bicolor* (Jan & Sordelli, 1860 *in* 1860–1866). Icon. Gén. Ophid. 1(1): 1, pl. 5, figs. 15a–b, f, x. (*Stenostoma bicolor*)

Synonyms: *Stenosoma gracile* Jan & Sordelli, 1860 *in* 1860–1866, *Stenostoma brevicauda* Bocage, 1887a, *Glauconia bicolor gruveli* Chabanaud, 1917a, and *Leptotyphlops brevicaudus* Hahn, 1980a.

Type: Holotype, RMNH 3682, a 100 mm specimen.

Type locality: "Boutry, Côte-d'Or" [= Butre, Western Region, SW Ghana, 04°49'N, 01°55'W, elevation 5 m].

Distribution: West Africa. Southern Mali (Koulikoro, Mopti, Sikasso), Burkina Faso (Centre, Centre-Nord, Centre-Ouest, Est, Hauts-Bassins, Sud-Ouest, Volta-Noire), SW Niger (Dosso, Niamey), SW Chad (Mayo Kebbi), Guinea (Kouroussa), Ivory Coast (Ferkéssédougou, Katiola, Tiassalé, Toumodi), Ghana (Ashanti, Eastern, Northern, Upper East, Western), S Togo (Centrale, Kara, Maritime, Plateaux), Benin (Alibori, Atlantique, Plateau) and Nigeria (Benue, Kano, Kwara, Oyo, Plateau), NSL–690 m.

Sources: Jan, 1864, Bocage, 1887b, Boulenger, 1893a, 1920c, Angel, 1933, Leeson, 1950, Villiers, 1951a–b, 1952b, 1965, Doucet, 1963, Roussel & Villiers, 1965, Leston & Hughes, 1968, Hughes & Barry, 1969, Papenfuss, 1969, Roux-Estève, 1969, Leston, 1970, Roman, 1980, 1984, Hughes, 1983, 2013, 1988, Hallermann & Rödel, 1995, Wallach & Hahn, 1997, Hahn & Wallach, 1998, LeBreton, 1999, Wallach & Boundy, 2005, J.-F. Trape & Mané, 2006b, Chirio, 2009 and Segniagbeto et al., 2011.

Remarks: Hughes & Barry (1969: 1008) listed holotype as RMNH 3682; Hahn & Wallach (1998: 58) listed a pair of syntypes (RMNH 3683). Possibly occurs in Cameroun *fide* Hughes (1983: 350).

2. *Tricheilostoma broadleyi* (Wallach & Hahn, 1997). Afr. J. Herp. 46(2): 104–106, fig. 1. (*Leptotyphlops broadleyi*)

Type: Holotype, MNHN 1977.103, a 110 mm specimen (M. Lamotte, 1961–1969).

Type locality: "vicinity of the Tropical Ecology station, Lamto, 300 m North of the Bandama River, Toumodi Department, Ivory Coast, 6°13' N, 5°02' W, ca. 300 m elevation."

Distribution: Central Ivory Coast (Toumodi), 300 m. Known only from vicinity of type locality.

Source: Wallach & Boundy, 2005.

3. *Tricheilostoma dissimilis* (Bocage, 1886). J. Sci. Math. Phys. Nat., Lisboa 11(43): 174. (*Stenostoma dissimile*)

Type: Holotype, MBL, a 104 mm specimen (Peteani de Steinberg), destroyed by fire in 18 March 1978.

Type locality: "Afrique centrale, Nil-blanc" [= White Nile, An Nil Al Abyad Prov., SE Sudan].

Distribution: Southeastern Sudan (An Nil Al Abyad). Known only from type locality.

Sources: Corkill, 1935, Hahn, 1978 and Broadley & Wallach, 2007a.

Remarks: *Incertae sedis fide* Broadley & Wallach (2007a: 49); possibly related to *L. bicolor* and/or *L. broadleyi*.

4. *Tricheilostoma greenwelli* (Wallach & Boundy, 2005). Ann. Carnegie Mus. 74(1): 39–42, figs. 1a–c. (*Leptotyphlops greenwelli*)

Type: Holotype, CM 92653, a 91 mm specimen (R.R. Golding, 19 Feb. 1965).

Type locality: "Vicinity of University of Ibadan Zoological Garden, city of Ibadan, Ibadan Province, Oyo State, Western Region, southwestern Nigeria, 7°23'N, 3°54'E, elevation ca. 150 m."

Distribution: Southwestern Nigeria (Western), 150 m. Known only from type locality.

5. *Tricheilostoma sundewalli* (Jan, 1861b). Arch. Zool. Anat. Fis. 1(2): 191. (*Stenostoma sundewalli*)

Synonyms: *Glauconia sundevallii* Boulenger, 1890d *(nomen emendatum)*, *Glauconia gestri* Boulenger, 1906f, and *Leptotyphlops sundevalli* – Villiers, 1950.

Type: Holotype, formerly MSNM, destroyed in 1943 during World War II.

Type locality: Unknown. Designated as West Africa *fide* Jan & Sordelli (1861 *in* 1860–1866: 2).

Distribution: West Africa. Ghana, Togo (Plateaux), SE Cameroon (Est, Sud), Equatorial Guinea (Bioko Is.) and SW Central African Republic (Kemo-Gribingui, Lobaye, Ombella-Mpoko, Ouham, Sangha), 300–1200 m.

Sources: Jan & Sordelli, 1861 *in* 1860–1866, Boulenger, 1893a, 1920c, Sternfeld, 1909b, Angel, 1933, Leeson, 1950, Villiers, 1951a, Perret, 1961, Mertens, 1965c, Roux-Estève, 1965, Hughes & Barry, 1969, Hughes, 1983, 1988, Roman, 1984, Joger, 1990, Hallermann & Rödel, 1995, Hahn & Wallach, 1998, LeBreton, 1999, Herrmann et al., 2004a, Wallach & Boundy, 2005, Chirio & Ineich, 2006, Chirio & LeBreton, 2007 and Segniagbeto et al., 2011.

Remarks: Records for Burkina Faso and Chad need verification. Possibly occurs in Benin *fide* Hughes (2013: 152).

TRILEPIDA Hedges, 2011 (Leptotyphlopidae)

Type species: *Stenostoma macrolepis* W.C.H. Peters, 1857b.

Distribution: South America.

Sources: Hahn, 1980a, Wallach, 1998b, McDiarmid et al., 1999, Hedges, 2011 and Pinto & Fernandes, 2012.

1. *Trilepida anthracina* (Bailey, 1946). Occ. Pap. Mus. Zool., Univ. Michigan (492): 1–5. (*Leptotyphlops anthracinus*)

Type: Holotype, UMMZ 90816, a 214–246 mm female (W. Clarke-MacIntyre, Aug. 1939).

Type locality: "at an elevation of 1800 meters near Baños, in eastern Ecuador" [= Baños de Agua Santa, Provincia Tungurahua, E Ecuador, 01°24'S, 78°25'W, elevation 1800 m].

Distribution: Ecuador (Azuay, Bolívar, Guayas, Pastaza, Tungurahua, Zamora-Chinchipe), 1000–1800 m.

Sources: J.A. Peters, 1955, 1960, Orejas-Miranda, 1967, Miyata, 1982, Cisneros-Heredia, 2008, Pinto, 2010 and Salazar-Valenzuela et al., 2010.

Remarks: Pinto (2010: 164) corrected the sex of the holotype.

2. *Trilepida brasiliensis* (Laurent, 1949). Bull. Inst. Roy. Sci. Nat. Belgique 25(9): 4–5, figs. 7–9. (*Leptotyphlops brasiliensis*)

Type: Holotype, IRSNB 2049 (formerly IG 12594 & IRSNB 2081), a 253–255 mm specimen (Abbaye de Saint André don. [= Saint Andrews Abbey, Bruges], 16 July 1939).

Type locality: "Brésil" [= Brazil]. Restricted to Barreiras, W Bahía State, Brazil *fide* Wallach (1996a: 10).

Distribution: Southeastern Brazil (W Bahía, Ceará, Goiás, Maranhão, Mato Grosso, Mato Grosso do Sul, W Minas Gerais, SE Piauí, Tocantins), 85–850 m.

Sources: Vanzolini et al., 1980, Lang, 1990, Rodrigues & Puorto, 1994, Curcio et al., 2002, Borges-Nojosa et al., 2009, Pinto, 2010, Silveira, 2010 and Pinto & Curcio, 2011.

Remarks: IRSNB 12594 erroneously cited as holotype number *fide* Rodrigues & Puorto (1994: 93) and Curcio et al. (2002:101).

3. *Trilepida brevissima* (Shreve, 1964). Breviora (211): 1–4. (*Leptotyphlops brevissima*)

Synonym: *Leptotyphlops brevissimus* J.A. Peters & Orejas-Miranda, 1970.

Type: Holotype, MLS 1311, a 66–68 mm male (H. Nicéforo-María, 10 Feb. 1951).

Type locality: "Florencia, Departamento de Caquetá, Colombia" [= 01°37'N, 75°37'W, elevation 560 m].

Distribution: Colombia (Antioquia, W Caquetá, Distrito Capital), 560–2800 m.

Sources: Orejas-Miranda, 1967, Pérez-Santos & Moreno, 1988, Passos et al., 2005, Pinto, 2010 and Pinto et al., 2010.

4. *Trilepida dimidiata* (Jan, 1861b). Arch. Zool. Anat. Fis. 1(2): 188. (*Stenostoma dimidiatum*)

Synonyms: *Glauconia dimidiata* Boulenger, 1893a, and *Leptotyphlops dimidiatus* Orejas-Miranda, 1967.

Type: Neotype, IB 24011, a 222 mm female (A.R. Hoge, 27 May 1964), designated by Pinto & Fernandes (2012: 38).

Type locality: "Brazil, State of Roraima, Boa Vista Municipality, São Marcos, 03°05' N, 60°25' W, elevation 75 m" via neotype selection.

Distribution: Northern South America. Southeastern Venezuela, Guyana (Demerara-Mahaica, Upper Takutu-Upper Essequibo), Suriname (N Brokopondo, Para, Suriname), French Guiana (Cayenne) and N Brazil (Minas Gerais, Roraima, Maracá Is.), 15–1100 m.

Sources: Jan & Sordelli, 1861 *in* 1860–1866, Orejas-Miranda, 1967, Amaral, 1978, Hoogmoed, 1977, Lancini & Kornacker, 1989, Kornacker, 1999, Abuys, 2003, Silveira et al., 2004b, Passos et al., 2005, 2006, Lima-Silveira, 2007b, Pinto, 2010, Pinto & Fernandes, 2012 and C.J. Cole, et al., 2013.

Remarks: The holotype, MSNM (Voyage de M. Peteani de Steinberg), was destroyed in 1943 during World War II. Neotype selection based upon type locality restriction of Peters & Orejas-Miranda (1970: 169) of "São Marcos, near confluence of Rios Uriracuera and Tacutu, both tributaries of Rio Branco, Territorio de Roraima, Brazil." Occurrence in Venezuela needs confirmation (Pinto & Fernandes, 2012 and Rivas-Fuenmayor et al., 2012).

5. *Trilepida dugandi* (Dunn, 1944c). Caldasia 3(11): 52–53. (*Leptotyphlops dugandi*)

Type: Holotype, CBB (unnumbered), a 143–165 mm male (H. Hildeberto), lost *fide* Pinto et al. (2010: 15).

Type locality: "Juanmina, 20 m. alt., about 11 km. S. W. of Barranquilla (Departamento del Atlántico), Colombia" [= Juan Mina, Barranquilla, Atlántico Dept., Colombia, 10°57'N, 74°53'W, elevation 30 m].

Distribution: Northern Colombia (Atlántico, Magdalena), NSL–30 m.

Sources: Pérez-Santos & Moreno, 1988, Passos et al., 2006 and Pinto, 2010.

6. *Trilepida fuliginosa* (Passos, Caramaschi & Pinto, 2006). Amphibia-Reptilia 27(3): 351–355, figs. 3a–b, 4a–c. (*Leptotyphlops fuliginosus*)

Type: Holotype, MNRJ 10034, a 256 mm female (R. Carvalho, Jr., 18 July 2003).

Type locality: "in the Rio Preto during the fauna rescue of the Queimados Hydroelectric Plant, between Municipalities of Luziânia (16°15'S, 46°57'W), State of Goiás, and Unaí (16°21'S, 46°54'W), State of Minas Gerais, Brazil" [elevation ca. 850 m].

Distribution: Southeastern Brazil (Bahia, Goiás, W Minas Gerais, Tocantis), 190–975 m.

Source: Pinto, 2010.

Remarks: Type locality coordinates are 16°15'S, 47°56'W, elevation 855 m (Luziânia) and 16°21'N, 46°55'W, elevation 595 m (Unaí) *fide* Google Earth.

7. *Trilepida guayaquilensis* (Orejas-Miranda & Peters, 1970). Mitt. Zool. Mus. Berlin (1969) 46(2): 439–441, fig. 1. (*Leptotyphlops guayaquilensis*)

Type: Holotype, ZMB 4508, a 170 mm specimen (C. Reiss, 1860).
Type locality: "Guayaquil an der Pazifikküste von Ecuador" [= Guayaquil, Guayas, Ecuador, 02°10'S, 79°54'W, elevation 10 m].
Distribution: Western Ecuador (Guayaquil), 10 m.
Sources: Miyata, 1980, 1982, Pérez-Santos & Moreno, 1991, Cisneros-Heredia, 2008 and Pinto, 2010.
Remarks: Ecuadorian specimens from El Oro, Esmeraldas, Guayas and Manabí provinces possibly referrable to this species (Cisneros-Heredia, 2008: 179).

8. *Trilepida jani* (Pinto & Fernandes, 2012). Copeia, 2012(1), 40–45, figs. 3–4. (*Tricheilostoma jani*)

Type: Holotype, MNRJ 4263, a 187 mm male (G. Kisteumacher, 1985).
Type locality: "Brazil, State of Minas Gerais, Belo Horizonte Municipality, Parque das Mangabeiras, 19° 55' S, 43° 56' W, elevation ca. 1000 m."
Distribution: Southeastern Brazil (Minas Gerais), 1000–1750 m.
Source: Pinto, 2010.

9. *Trilepida joshuai* (Dunn, 1944c). Caldasia 3(11): 53, figs. 9–10. (*Leptotyphlops joshuai*)

Type: Holotype, MLS 13, a 259–270 mm male, destroyed by fire 9–10 April 1948.
Type locality: "Jericó, Departamento de Antioquia (1967 m.), Colombia" [= Jericó, Antioquia Dept., Colombia, 05°48'N, 75°47'W, elevation 1965 m].
Distribution: Colombia (Antioquia, Caldas, Quindío, Risaralda, Valle del Cauca), 1600–2835 m.
Sources: Pérez-Santos & Moreno, 1988, Passos et al., 2006, Pinto, 2010, Pinto et al., 2010, Rojas-Morales & González-Durán, 2011 and Rojas-Morales, 2012b.

10. *Trilepida koppesi* (Amaral, 1955c). Mem. Inst. Butantan (1954) 26: 203–204, figs. 4–6. (*Leptotyphlops koppesi*)

Type: Holotype, IB 8883, a 144–145 mm male (S.J. Koppes, 10 Oct. 1934), destroyed by fire 15 May 2010.
Type locality: "Terenos, Mato Grosso, Brasil" [= Municipality of Terenos, Mato Grosso do Sul, SE Brazil, 20°26'S, 54°51'W, elevation 400 m].
Distribution: Brazil (Goiás, Mato Grosso, Mato Grosso do Sul, Minas Gerais, São Paulo, Tocantins), 270–880 m.
Sources: Amaral, 1978, Nogueira, 2001, Valdujo & Nogueiro, 2001, Sousa & Silva, 2002b, Passos et al., 2005, 2006 and Pinto, 2010.

11. *Trilepida macrolepis* (W.C.H. Peters, 1857b). Mber. Kønigl. Akad. Wiss. Berlin 1857(8): 402. (*Stenostoma macrolepis*)

Synonyms: *Stenostoma affine* Boulenger, 1884a, *Glauconia affinis* – Boulenger, 1893a, *Leptotyphlops affinis* – Amaral, 1930f, *Leptotyphlops ihlei* Brongersma, 1933a, and *Leptotyphlops afinis* – Briceño-Rossi, 1934 (*nomen incorrectum*).
Type: Lectotype, ZMB 1434, a 280 mm specimen (C.F. Appun), designated via restriction by Orejas-Miranda (1967: 430).
Type locality: "Puerto Cabello" [= Município do Puerto Cabello, Carabobo State, Venezuela, 10°28'N, 68°01'W, elevation 10 m], by type locality restriction and designation.
Distribution: Northern South America. Panamá (Panamá), Colombia (Antioquia, Boyacá, Caldas, Chocó, Córdoba, Huila, La Guajira, Magdalena, Norte de Santander, Santander, Tolima, Valle del Cauca, Vaupés), Venezuela (Acosta, Aragua, Carabobo, Cojedes, Distrito Federal, Falcón, Mérida, Miranda, Portuguesa, Táchira, Vargas, Yaracuy, Zulia), Guyana (Potaro-Siparuni, Upper Takutu-Upper Essequibo), Suriname (Brokopondo, Saramacca), French Guiana (Cayenne, Saint-Laurent-du-Maroni), N Brazil (Amazonas, Bahia, Maranhão, Pará, Rio de Janeiro, Rondônia) and E Perú (Loreto), NSL–1930 (2000) m.
Sources: Boulenger, 1893a, Nicéforo María, 1942, Shreve, 1964, Roze, 1966a, Orejas-Miranda, 1967, Amaral, 1978, Hoogmoed, 1977, Cunha & Nascimento, 1978, 1993, Gasc & Rodrigues, 1980, Chippaux, 1987, Pérez-Santos & Moreno, 1988, Lancini & Kornacker, 1989, Cunha & Nascimento, 1993, Bauer et al., 1995, Starace, 1998, Freitas, 1999, 2003, Kornacker, 1999, Markezich, 2002, Abuys, 2003, Argôlo, 2004, Rivas-Fuenemayer & Barrio-Amorgós, 2005, Passos et al., 2005, 2006, Navarrete et al., 2009, Pinto, 2010 and Pinto et al., 2010 and C.J. Cole et al., 2013.
Remarks: *Stenostoma affine* Boulenger is a synonym *fide* Pinto (2010: 121).

12. *Trilepida nicefori* (Dunn, 1946b). Caladasia 4(17): 121–122. (*Leptotyphlops nicefori*)

Type: Holotype, MLS 17, a 90 mm male (H. Nicéforo-María), destroyed by fire 9–10 April 1948.
Type locality: "Mogotes, Santander, Colombia, 1746 m. elevation" [= Mogótes, Santander Dept., NE Colombia, 06°29'N, 72°58'W, elevation 1685 m].
Distribution: Northeastern Colombia (Santander), 1685–1825 m.
Sources: Pinto, 2010 and Pinto et al., 2010.

13. *Trilepida salgueiroi* (Amaral, 1955c). Mem. Inst. Butantan (1954) 26: 203, figs. 1–3. (*Leptotyphlops salgueiroi*)

Type: Holotype, IB 8876, a 297–301 mm female (W.S. Salgueiro, 30 Nov. 1934), destroyed by fire 15 May 2010.

Type locality: "Itá, Espírito Santo, Brasil" [= Baixo Guandu, State of Espírito Santo, SE Brazil, 20°26'S, 54°51'W, elevation 75 m].

Distribution: Southeastern Brazil (S Bahia, Espírito Santo, E Minas Gerais, Rio de Janeiro), 20–515 m.

Sources: Amaral, 1978, Bilate & Ribeiro-Costa, 2005, Passos et al., 2005, Costa et al., 2009, Pinto, 2010 and Figueiredo-de-Andrade et al., 2011.

Remarks: Pinto (2010: 198) corrected the sex of the holotype.

TRIMERESURUS Lacépède, 1804 (Viperidae)

Synonyms: *Trimeresura* Fleming, 1822 (*nomen emendatum*), *Trimesurus* Jarocki, 1822 (*nomen emendatum*), *Trimereresurus* – Meckel, 1829 (*nomen incorrectum*), *Trimensurus* – J. Müller, 1832 (*nomen incorrectum*), *Bothrophis* Fitzinger, 1843, *Cryptelytrops* Cope, 1860b, *Tremeresurus* – Kulagin, 1888 (*nomen incorrectum*), *Botrophis* – Hoffmann, 1890 (*nomen incorrectum*), *Cryptelythrops* – Hoffmann, 1890 (*nomen incorrectum*), *Trimaesurus* – Stanley, 1917 (*nomen incorrectum*), *Trimererusus* – Takahashi, 1930 (*nomen incorrectum*), *Trineresurus* – Horikawa, 1941 (*nomen incorrectum*), *Trimerasurus* – Leviton, 1961b (*nomen incorrectum*), *Trimersurus* – H.-F. Huang, Chen, Min & Ho, 1965 (*nomen incorrectum*), *Ttimeresurus* – Deuve, 1970 (*nomen incorrectum*), *Trimerusurus* – Frith, 1977 (*nomen incorrectum*), *Trimetresurus* – Murthy, 1977c (*nomen incorrectum*), *Trimeresus* – Majupuria, 1981 (*nomen incorrectum*), *Trimcresurus* – Murthy & Pillai *in* Majupuria, 1986 (*nomen incorrectum*), *Trimerisurus* – Cogger & Zweifel, 1992 (*nomen incorrectum*), and *Adelynhoserea* Hoser, 2012o (*nomen illegitimum*).

Type species: *Trimeresurus insularis* Kramer, 1977.

Distribution: Asia and Indonesia.

Sources: Maslin, 1942, M.A Smith, 1943, Maslin, 1945, Klemmer, 1963, Leviton, 1964d, W. Burger, 1971, Kramer, 1977, Hoge & Romano-Hoge, 1981a, Malhotra & Davis, 1991, Golay et al., 1993, Zhao & Adler, 1993, Kraus et al., 1996, David & Vogel, 1998, David & Ineich, 1999, McDiarmid et al., 1999, Parkinson, 1999, Malhotra & Thorpe, 1997, 2000, 2004c, Ziegler et al., 2001, Giannasi et al., 2001a, G. Peng & Fuji, 2001, David et al., 2002, 2009, 2011, Gumprecht et al., 2004c, Castoe & Parkinson, 2006, Grismer et al., 2006a, 2008b, Sanders et al., 2006, Dawson et al., 2008, Ganesh et al., 2008, Orlov et al., 2008, 2009, I. Das, 2010, 2012, David et al., 2011c, P. Guo & Wang, 2011, Sumontha et al., 2011 and Hoser, 2012d, o.

Remarks: The following genera and species have been separated from *Trimeresurus*, based on Malhotra & Thorpe 2004: *Craspedocephalus (gramineus, andalasensis, borneensis, brongersmai, malabaricus, puniceus, strigatus, trigonocephalus, wiroti), Garthius (chaseni), Himalayophis (tibetanus), Ovophis (convictus, gracilis, monticola, okinavensis, tonkinensis, zayuensis), Parias (flavomaculatus, hageni, malcolmi, mcgregori, schultzei, sumatranus), Peltopelor (macrolepis), Popeia (barati, buniana, fucata, inornata, nebularis, popeorum, sabahi), Protobothrops (cornutus, elegans, flavoviridis, jerdonii, kaulbacki, mucrosquamatus, tokarensis trungkhanensis, xiangchengensis), Triceratolepidophis (sieversorum), Tropidolaemus (huttoni, laticinctus, wagleri), Viridovipera (gumprechti, medoensis, stejnegeri, truongsonensis, vogeli, yunnanensis), and Zhaoermia (mangshanensis)*, David et al., 2011c, relegated these genera to subgenera of *Trimeresurus*.

1. *Trimeresurus albolabris* Gray, 1842b. Zool. Misc. 2(May): 48. (*nomen corrigendum*)

Synonyms: *Trimesurus albolabris* Gray, 1842b (*nomen incorrigendum*), *Trimeresurus purpureomaculatus bicolor* Boulenger, 1890 (*nomen praeoccupatum*), and *Trimeresurus albolabris* – C.H. Pope & Pope, 1933 (*nomen corrigendum*).

Type: Lectotype, BMNH 1946.1.19.85 (formerly BMNH I.2.2b), a juvenile male (J. Reeves), designated by Regenass & Kramer (1981: 168).

Type locality: "China." Restricted to Hong Kong, China *fide* Iskandar & Colijn (2001: 157).

Distribution: Southeastern Asia. Southern China (Fukien, Guangdong, Guangxi, Hainan, Hong Kong, Hunan, Jiangxi, Nan Ao Is.), Macau (Coloane Is.), NE India (Assam, Arunachal Pradesh, Meghalaya, Nagaland, North Bengal, Sikkim), Bhutan (Sarpang), Thailand (Bangkok, Buriram, Chachoengsao, Chaiyaphum, Chiang Mai, Chiang Rai, Chon Buri, Chumphon, Kalasin, Kamphang Phet, Kanchanaburi, Khiri Khan, Khon Kaen, Krabi, Loei, Lop Buri, Maha Sarakham, Nakhon Nayok, Nakhon Pathom, Nakhon Phanom, Nakhon Ratchasima, Nakhon Sawan, Nakhon Si Thammarat, Nan, Nong Khai, Nonthaburi, Phet Buri, Phetchabun, Phetchaburi, Phitsanulok, Prachin Buri, Prachuap Khiri Khan, Rayong, Roi Et, Sakhon Nakhon, Samut Sakhon, Saraburi, Satun, Sisaket, Songkhla, Surat Thani, Surin, Trang, Ubon Ratchathani, Udon Thani), Myanmar (Bago, Kachin, Mandalay, Mon), Cambodia (Kampot, Koh Kong, Mondulkiri), Laos (Xiangkhouang) and S Vietnam (Bac Giang, Bac Kan, Binh Phuoc, Ca Mau, Cao Bang, Dak Lak, Da Nang, Dong Nai, Gia Lai, Hai Duong, Hao Binh, Ha Tay, Ha Tinh, Khanh Hoa, Kien Giang, Lam Dong, Lang Son, Lao Cai, Nghe An, Quang Binh, Quang Nam, Quang Tri, Son La, Tay Ninh, Thai Nguyen, Thanh Hoa, Thua Thien-Hue, Vinh-Phu, Cat Ba Is.) and W Indonesia (Sumatra), 50–3050 m.

Sources: C.H. Pope & Pope, 1933, C.H. Pope, 1935, E.H. Taylor, 1965, Saint-Girons, 1972a, Regenass & Kramer, 1981, Zhang & Zhao, 1990, M.J. Cox, 1991b, Easton & Leung-Va, 1993, Zhao & Adler, 1993, Malhotra & Thorpe, 1997, Orlov, 1997, M.J. Cox et al., 1998, David & Vogel, 1998, Karsen et al., 1998, Chan-ard et al., 1999, Darebsky, 1999, Lazell, 1999, David & Gernot, 2000, Malhotra & Thorpe, 2000, Giannasi et al., 2001b, G. Peng & Fuji, 2001, Gumprecht, 2004f, B.L. Stuart & Emmett, 2006, B.L. Stuart et al., 2006, Vijayakumar & David, 2006, Zhao, 2006, Ziegler et al., 2007, Datong, 2008, M.F. Ahmed et al., 2009, V.S. Nguyen et al., 2009, I. Das, 2012 and Wangyal, 2012.

Remarks: C.H. Pope & Pope (1933: 10) erroneously credited with lectotype designation but they did not meet the requirements of Art. 74.6. *Trimeresurus albolabris* not included in Hainan fauna *fide* Shi, 2011. Malaysia, Borneo and Andaman & Nicobar Is. records rejected *fide* David & Ineich (1999: 281) and Das (1994: 48). Presence on Sumatra (Indonesia) confirmed *fide* David & Vogel (2000: 225).

2. *Trimeresurus andersoni* Theobald, 1868b. J. Asiat. Soc. Bengal (1866) 37: 75–76.

Type: Holotype, ZSI 3057.

Type locality: Unknown. Designated as Andaman Is., Andaman & Nicobars, E India *fide* Regenass & Kramer (1981: 195).

Distribution: Eastern India (Andaman & Nicobars: Little Andaman, Car Nicobar, Cinque, Long, Nicobar, Preparis and South Andaman Is.).

Sources: Biswas & Sanyal, 1980, Malhotra & Thorpe, 1997, I. Das, 1994, 1996, Whitaker & Captain, 2004, Vijayakumar & David, 2006 and Harikrishnan et al., 2012.

Remarks: A synonym of *T. purpureomaculatus fide* McDiarmid et al. (1999: 342).

3. *Trimeresurus cantori* Blyth, 1846. J. Asiat. Soc. Bengal 15(173): 377.

Synonym; *Trimeresurus cantoris* Stoliczka, 1870c–d (*nomen emendatum*).

Type: Lectotype, ZSI 2959, a 762 mm specimen (Lewis & Barbe), designated by Sclater (1891: 71).

Type locality: "Nicobar Islands,"[Andaman & Nicobars, E India, Bay of Bengal].

Distribution: Eastern India (Andaman & Nicobars: Nicobar Is.).

Sources: Biswas & Sanyal, 1980, Malhotra & Thorpe, 1997 and Whitaker & Captain, 2004.

Remarks: Original description included a 762 mm adult and a 394 mm juvenile. Andaman Is. record needs confirmation *fide* Whitaker & Captain (2007: 448).

4. *Trimeresurus cardamomensis* (Malhotra, Thorpe, Mrinalini & Stuart, 2011). Zootaxa (2757), 3–8, fig. 3. (*Cryptelytrops cardamomensis*)

Type: Holotype, FMNH 259191, a 347 mm female (B.L. Stuart and S. Platt, 25 Aug. 2000).

Type locality: "Phnom Chan Mountain, Cardamom Mountains, Sre Ambei District, Koh Kong Province, Cambodia (11.44 N, 103.79 E), 100–200 m elevation."

Distribution: Southeastern Thailand (Chantaburi) and SW Cambodia (Koh Kong), NSL–800 m.

Remarks: Probably occurs on offshore islands like Koh Chang *fide* Malhotra et al. (2011: 9).

5. *Trimeresurus erythrurus* (Cantor, 1839a). Proc. Zool. Soc. London 7(1): 31–32. (*Trigonocephalus erythrurus*)

Synonym: *Trimesurus bicolor* Gray, 1853.

Type: Holotype, BMNH 1946.1.19.99 (formerly BMNH 1894.6.25.15), a juvenile female illustrated in colored sketch of (T.E. Cantor (1834–1837: no 17) *fide* M.A. Smith (1943: 522).

Type locality: "Delta Gangeticum" [= Ganges Delta, Bangladesh and West Bengal State, NE India].

Distribution: Southern Asia. Northeastern India (Arunachal Pradesh, Assam, Manipur, ? Meghalaya, ? Mizoram, Nagaland, Sikkim, West Bengal), cen. Nepal (Kaski), Bangladesh and Myanmar (Kachin, Rakhine, Yangon), 25–2500 m.

Sources: C.H. Pope & Pope, 1933, E.H. Taylor, 1965, Regenass & Kramer, 1981, Toriba et al., 1990, G. Vogel, 1990, Toriba, 1994, Malhotra & Thorpe, 1997, David & Vogel, 1998, Zug et al., 1998, Leviton et al., 2003, Ao et al., 2004, Tiwari & Shah, 2004 and Whitaker & Captain, 2004.

Remarks: Old holotype number listed as BMNH 94.6.25.5 *fide* Regenass & Kramer (1981: 179).

6. *Trimeresurus fasciatus* (Boulenger, 1896d). Ann. Mag. Nat. Hist. (6) 18(103): 63–64. (*Lachesis fasciatus*)

Type: Holotype, BMNH 1896.4.29.46, a 455 mm female (A.H. Everett, 1896).

Type locality: "Jampea Island" [= Tanahjampea Island, SE Indonesia, bet. 7°00'–09'S and 120°36'–47'E]. Specified as Jampea Island, between Celebes [= Sulawesi] and Flores *fide* BMNH catalogue.

Distribution: Southern Indonesia (Tanahjampea Is.).

Sources: Auffenberg, 1980, David et al., 2003 and Koch, 2012.

Remarks: Photograph of holotype in David et al. (2003: fig. 1).

7. *Trimeresurus honsonensis* (Grismer, Ngo & Grismer, 2008b). Zootaxa (1715): 61–64, fig. 2 (upper). (*Cryptelytrops honsonensis*)

Type: Holotype, UNS 353, a 648 mm female (V.T. Ngo, 29 July 2007).
Type locality: "Hon Son island in the Kien Hai District of the Kien Giang Province, Vietnam, 09° 47' 95.2" N, 104° 37' 85.6" E at 100 m a.s.l."
Distribution: Vietnam (Kien Giang: Hon Son Is.), 100 m. Known only from type locality.
Source: V.S. Nguyen et al., 2009.

8. *Trimeresurus insularis* Kramer, 1977. Rev. Suisse Zool. 84(3): 755. (*Trimeresurus albolabris insularis*)

Synonym: *Trimeresurus viridis* Lacépède, 1804 (*nomen oblatum*).
Type: Holotype, MHNG 12773, a 532 mm male (A. Bühler, 1935).
Type locality: "Soe, Timor" [= Soe, cen. Timor, SE Indonesia, 9°52'S, 124°17'E, elevation 800 m].
Distribution: Lesser Sundas of S Indonesia (Adonara, Alor, Bali, Flores, Java, Kisar, Komodo, Lembata, Lomblem, Lombok, Padar, Pantar, Rinca, Romang, Roti, Semau, Sumba, Sumbawa, Timor, Wetar) and Timor-Leste, NSL–1200 m.
Sources: David & Vogel, 1996, Giannasi et al., 2001b, David et al., 2003, 2011c, Vijayakumar & David, 2006, I Das, 2010, Kaiser et al., 2011, Lang, 2011 and O'Shea et al., 2012.
Remarks: Previously considered a subspecies of *T. albolabris*. *Trimeresurus viridis* Lacépède is type species of *Trimeresurus fide* David et al. (2011c: 35) and a *nomen oblitum fide* Art. 23.9.2 of the Code (ICZN, 1999). Sulawesi record doubtful *fide* Lang & Vogel (2005: 264).

9. *Trimeresurus kanburiensis* M.A. Smith, 1943. Fauna Brit. India, Rept. Amph. 3: 519.

Type: Holotype, BMNH 1946.1.8.91 (formerly BMNH 1927.8.10.1), a 405 mm female (M.A. Smith).
Type locality: "limestone hills near Kanburi, south-western Siam" [= vicinity of Kanchanaburi (14°40'N, 99°01'E, elevation 525 m), Kanchanaburi Prov., W-cen. Thailand].
Distribution: Central Thailand (Kanchanaburi), 200–1000 m.
Sources: E.H. Taylor, 1965, M.J. Cox 1991b, 1991, Vogel, 1991, Warrell et al., 1992, Malhotra & Thorpe, 1997, 2004a, M.J. Cox et al., 1998, Chan-ard et al., 1999, Orlov et al., 2003, 2004 and David et al., 2004.
Remarks: Type locality listed as "S.W. China" *fide* BMNH catalogue. Probably occurs in adjacent Myanmar.

T

10. *Trimeresurus labialis* Fitzinger *in* Steindachner, 1867a. Reise Öster. Novara 1: 86–87, pl. 3, figs. 1–2.

Synonyms: *Vipera acontia* Laurenti, 1768 (*nomen oblitum*), *Bothrophis labialis* Fitzinger, 1861 (*nomen nudum*), and *Trimeresurus mutabilis* Stoliczka, 1870a–d.
Types: Syntypes (3), NMW 18813.1–3 (G. von Frauenfeld [Novara Exped.], 23 Feb.–26 Mar. 1858).
Type locality: "Nikobaren" [= Nicobar Is., Andaman & Nicobars, E India, Bay of Bengal].
Distribution: Eastern India (Andaman & Nicobars: Nicobar Is.).
Sources: Biswas & Sanyal, 1980, David & Vogel, 1998, Whitaker & Captain, 2004 and Vijayakumar & David, 2006.
Remarks: McDiarmid et al. (1999: 337–338) gives Steindachner as author of this name, but Steindachner in the citation on the plates clearly indicates Fitzinger (*in litt.*). A single record from the Andaman Is. needs confirmation *fide* Whitaker & Captain (2007: 466). In accordance with Art. 23.9.2 of the Code (ICZN, 1999), *Trimeresurus labialis* Fitzinger is designated a *nomen protectum* and *Vipera acontia* Laurenti a *nomen oblitum*.

11. *Trimeresurus macrops* Kramer, 1977. Rev. Suisse Zool. 84(3): 757, pl. 5, figs. 27–29.

Type: Holotype, MHNG 1400.85, a 639–761 mm male.
Type locality: "Bangkok (Thailand)" [= Bangkok, Bangkok Prov., Thailand, 13°43'N, 100°29'E, elevation NSL].
Distribution: Indo-China. Thailand (Ang Thong, Ayutthaya, Bangkok, Chaiyaphum, Chiang Mai, Kanchanaburi, Khan Kaen, Lampang, Nakhon Phanom, Nonthaburi, Ratchasima, Pathon Thani, Samut Prakan), Laos (Bolikhamxai, Champasak, Khammouan) and ext. N Cambodia (Kampot, Koh Kong, Mondulkiri, Stung Treng), NSL–800 m.
Sources: Regenass & Kramer, 1981, M.J. Cox, 1991b, M.J. Cox et al., 1998, Chan-ard et al., 1999, Orlov et al., 2003, B.L. Stuart & Emmett, 2006, B.L. Stuart et al., 2006, Teynié & David, 2007, V.S. Nguyen et al., 2009 and Malhotra et al., 2011b.

12. *Trimeresurus purpureomaculatus* (Gray, 1832 *in* Gray & Hardwicke, 1830–1835). Illust. Indian Zool. 1(10): pl. 81. (*Trigonocephalus purpureomaculatus*) (*nomen corrigendum*)

Synonyms: *Trigonocephalus purpureo-maculatus* Gray, 1832 *in* Gray & Hardwicke, 1830–1835 (*nomen incorrigendum*), *Trimesurus carinatus* Gray, 1842b, *Trimesurus purpureus* Gray, 1842b (*nomen substitutum*), *Trimeresurus porphyraceus* Blyth, 1860, *Trimeresurus obscurus* Theobald, 1868, *Trimeresurus purpureomaculatus* – Boulenger, 1896a (*nomen* *corrigendum*), and *Trimeresurus acutimentalis* F. Werner, 1927.

Type: Holotype, BMNH 1946.1.19.54 (formerly BMNH I.2.4a & BMNH 1894.6.25.15), an 820 mm female illustrated by Hardwicke (1756–1823, 1: fig. 158) (T. Hardwicke, 1756–1823).

Type locality: "Singapore."

Distribution: Southeastern Asia and W Indonesia. India (West Bengal), Bangladesh, S Myanmar (Mon, Karen, Tanintharyi), Singapore, Thailand (Kanchanaburi, Ko Surin, Nakhon Si Thammarat, Phang Nga, Phuket, Ranong, Trang, Koh Surin and Phuket Is.), West Malaysia (Johore, Perak, Pinang, Pangkor Is.) and Indonesia (Sumatra, Bengkalis and Riau Arch.), NSL–1500 m.

Sources: C.H. Pope & Pope, 1933, E.H. Taylor, 1965, Whitaker, 1978a, 1978c, Regenass & Kramer, 1981, Tweedie, 1983, M.J. Cox, 1991b, Lim & Lim, 1992, David & Vogel, 1996, Malhotra & Thorpe, 1996, 1997, 2000, M.J. Cox et al., 1998, Chan-ard et al., 1999, Leviton et al., 2003 and Onn et al., 2010.

Remarks: Original color sketch of holotype of Hardwicke (1756–1823) (1: fig. 158) in BMNH. Date of publication 1829 *fide* Regenass & Kramer (1981: 195).

13. *Trimeresurus septentrionalis* Kramer, 1977. Rev. Suisse Zool. 84(3): 755–756, pl. 5, figs. 24–26. (*Trimeresurus albolabris septentrionalis*)

Type: Holotype, MHNG 1404.31, a 773 mm male (H. Schnurrenberger, 1962–1964).

Type locality: "Nepal 83° 55' 28° 15' 1500 m (Nähe Pokhara)" [= vicinity of Pokhara, Kaski District, Western Region, Nepal, 28°15'E, 83°55'N, elevation 1500 m].

Distribution: Southern Asia. Northern India (Arunachal Pradesh, Meghalaya, Nagaland, Sikkim, West Bengal), Bangladesh and Nepal (Baglung, Chitwan, Dadeldhura, Dhading, Kaski, Kathmandu, Lalitpur, Lamjung, Manang, Mustang, Myagdi, Rasuwa, Sankhuwasabha, Sindhupalchok), 200–3050 m.

Sources: Nanhoe & Ouboter, 1987, Giannasi et al., 2001b, Shah & Tiwari, 2004, Whitaker & Captain, 2004, Vijayakumar & David, 2006, Ahmed et al., 2009, Ahsan et al., 2009 and S.K. Sharma et al., 2013.

14. *Trimeresurus venustus* Vogel, 1991. Sauria 13(1): 23–26, 2 figs. (paratype), 1 fig. (hemipenis).

Type: Holotype, ZMB 48045, a 580 mm male.

Type locality: "Thung Song, Provinz Nakhon Si Thammarat, Süd-Thailand" [= Thung Song, Nakhon Si Thammarat Prov., S Thailand, 8°10'N, 99°41'E, elevation 70 m].

Distribution: Peninsular Thailand (Krabi, Nakhon Si Thammarat, Surat Thani), 70–700 m.

Sources: David et al., 2004, Malhotra & Thorpe, 2004a, Orlov et al., 2004 and Ziegler et al., 2006.

TRIMETOPON Cope, 1885a (Dipsadidae)

Synonym: *Trimepodon* – Amaral, 1930f (*nomen incorrectum*).

Type species: *Ablabes gracilis* A.C.L.G. Günther, 1872b.

Distribution: Lower Central America.

Sources: Dunn, 1937b, E.H. Taylor, 1951, 1954, J.A. Peters & Orejas-Miranda, 1970, C.W. Myers, 1974, Villa et al., 1988, Savage, 2002, Solórzano, 2004 and Zaher et al., 2009.

1. *Trimetopon barbouri* Dunn, 1930. Occ. Pap. Boston Soc. Nat. Hist. (1921–1931) (5): 331–332.

Type: Holotype, MCZ 23877, a 260 mm male (Canal Zone Health Dept).

Type locality: "Pedro Miguel, Canal Zone, Panama" [= Pedro Miguel, Canal Zone, cen. Panama, 9°01'N, 79°37'W, elevation 25 m].

Distribution: Central Panama (Canal Zone, Barro Colorado Is.), NSL–60 m.

Source: H.M. Smith & Grant, 1958b.

2. *Trimetopon gracile* (A.C.L.G. Günther, 1872b). Ann. Mag. Nat. Hist. (4) 9(49): 18, pl. 3, fig. d. (*Ablabes gracilis*)

Type: Holotype, BMNH 1946.1.1.1 (formerly BMNH 1871.11.22.19 or 20), a 290–305 mm male.

Type locality: "elevated country of Costa Rica, near Cartago" [= vicinity of Cartago, Cartago Prov., cen. Costa Rica, 9°51'N, 83°55'W, elevation 1435 m].

Distribution: Central Costa Rica (Cartago, Heredia, Puntarenas), 1250–2285 m.

Remarks: Although Günther stated "one specimen," Boulenger (1894a: 185) and Dunn (1937: 214) reported two male syntypes.

3. *Trimetopon pliolepis* Cope, 1894d. Proc. Acad. Nat. Sci. Philadelphia 46: 201–202.

Type: Holotype, AMNH 17302 (formerly MNCR 451), a 287 mm specimen (P. Biolley, 1890–1894).

Type locality: "San José, Costa Rica" [= San José, San José Prov., cen. Costa Rica, 9°56'N, 84°05'W, elevation 1160 m].

Distribution: Costa Rica (Cartago, Heredia, Limón, Puntarenas, San José, Bonita Is.), 60–1600 m.

4. *Trimetopon simile* Dunn, 1930. Occ. Pap. Boston Soc. Nat. Hist. (1921–1931) (5): 331.

Type: Holotype, MCZ 15263 (formerly MNCR), a 155 mm specimen.

Type locality: "Reventazon, Costa Rica (= La Junta)" [= Junta, Río Reventazon, cen. Limón Prov., E Costa Rica,

10°07'N, 83°32'W, elevation 100 m] (possibly in error). Probably Siquirres [= cen. Limón Prov., E Costa Rica, 10°06'N, 83°31'W, elevation 100 m] *fide* Dunn (1937: 21).

Distribution: Costa Rica (Cartago, Limón, N Puntarenas), 60–1500 m.

5. *Trimetopon slevini* Dunn, 1940b. Proc. Acad. Nat. Sci. Philadelphia 92: 117–118.

Type: Holotype, CAS 78938, a 220 mm male (J.R. Slevin, 20 July 1939).

Type locality: "Near Boquete, Chiriquí Prov., Panama, altitude 4000 feet" [= vicinity of Boquete, N Chiriquí Prov., W Panama, 8°46'N, 82°26'W, elevation 1090 m].

Distribution: Costa Rica (Cartago, Heredia, Puntarenas) and W Panama (Chiriquí), 120–1825 m.

6. *Trimetopon viquezi* Dunn, 1937b. Copeia 1937(4): 215.

Type: Holotype, ERD (formerly MNCR), a 209 mm male (C. Viquez-Segrada).

Type locality: "Siquirres, Costa Rica" [= Siquirres, Limón Prov., E-cen. Costa Rica, 10°06'N, 83°31'W, elevation 60 m].

Distribution: Eastern Costa Rica (Limón), 60 m. Known only from type locality.

Remarks: Possibly occurs in Heredia, Costa Rica *fide* Savage (2002: 640).

TRIMORPHODON Cope, 1861e (Colubridae)

Synonyms: *Trimorphon* – Boulenger, 1896a (*nomen incorrectum*), *Trimopbodon* – Dugès, 1897b (*nomen incorrectum*), *Trimorphondon* – Woodbury, 1931 (*nomen incorrectum*), and *Mimorphodon* – Anthony, 1955 (*nomen incorrectum*).

Type species: *Lycodon lyrophanes* Cope, 1860f.

Distribution: Southwestern USA and Mesoamerica.

Fossil records: Pleistocene of SW USA.

Sources: E.H. Taylor, 1939a, Klauber, 1940b, H.M. Smith, 1940o, Gehlbach, 1971, N.J. Scott & McDiarmid, 1984a–b, Casas-Andreu et al., 1996, Devitt, 2006 and Devitt et al., 2008.

T

1. *Trimorphodon biscutatus* (A.-M.-C. Duméril, Bibron & Duméril, 1854b). Erpét. Gén. 7(2): 1153–1154. (*Dipsas biscutata*)

Synonyms: *Trimorphodon major* Cope, 1870, and *Trimorphodon biscutatus semirutus* H.M. Smith, 1943a

Type: Holotype, MNHN 5900, a 580 mm specimen.

Type locality: "Mexique" [= Mexico]. Restricted to Tehuantepec, Oaxaca State, Mexico *fide* H.M. Smith & Taylor (950a: 340)

Distribution: Southwestern Mexico (Chiapas, Colima, Guerrero, Jalisco, Michoacán, Oaxaca, Puebla), NSL–1500 (2400) m.

Sources: N.J. Scott & McDiarmid, 1984b and Holman, 2000a.

2. *Trimorphodon lambda* Cope, 1886b. Proc. Amer. Philos. Soc. (1885) 23(122): 286–287.

Type: Holotype, USNM 13487, a 304 mm specimen (H.F. Emerich, 1883).

Type locality: "Guaymas, Sonora" [= Guaymas, Sonora State, Mexico, 27°55'N, 110°54'W, elevation 20 m].

Distribution: Southwestern USA (Arizona, SE California, S Nevada, SW New Mexico, ext. SW Utah) and NW Mexico (W Chihuahua, N Sinaloa, Sonora), NSL–800 m.

Fossil records: Upper Pleistocene (Rancholabrean II) of USA (Arizona, New Mexico).

Sources: Klauber, 1928, Davis & Dixon, 1957, N.J. Scott & McDiarmid, 1984b, Stebbins, 1985, Degenhardt et al., 1996, P.R. Brown, 1997, Bartlett & Tennant, 2000, Holman, 2000a and Ernst & Ernst, 2003.

3. *Trimorphodon lyrophanes* (Cope, 1860f). Proc. Acad. Nat. Sci. Philadelphia 12(8): 343–344. (*Lycodon lyrophanes*)

Synonym: *Trimorphodon vandenburghi* Klauber, 1924.

Types: Syntypes (4), USNM 4680a–b, a male and female, longest syntype 711 mm, broken in half and partially macerated, ANSP 10146 (formerly USNM 4680c), and ANSP (formerly USNM 4680d) (J. Xantus, May 1859), both ANSP types lost *fide* Malnate *in* N.J. Scott & McDiarmid (1984b: 353.2).

Type locality: "Cape St. Lucas, Lower California" [= Cabo San Lucas, ext. S Baja California Sur State, NW Mexico, 22°53'N, 109°55'W, elevation 25 m].

Distribution: Southwestern USA (SW California) and NW Mexico (Baja Californnia Norte, Baja California Sur, Cerralvo, Danzante, San José, San Marcos and Tiburon Is.), NSL–1800 m.

Sources: R.W. Murphy & Ottley, 1984, N.J. Scott & McDiarmid, 1884b, Mahrdt & Grismer, 1995, Hollingsworth et al., 2000 and Grismer, 2002.

4. *Trimorphodon paucimaculatus* E.H. Taylor, 1938. Univ. Kansas Sci. Bull. (1936) 24(26): 527–529, pl. 46, fig. 1.

Type: Holotype, UIMNH 25072 (formerly EHT 709), an 880 mm specimen (E.H. Taylor, 24 July 1934).

Type locality: "Mazatlan, Sin., Mexico" [= Mazatlan, S Sinaloa State, NW Mexico, 23°14'N, 106°25'W, elevation 10 m].

Distribution: Northwestern Mexico (Jalisco, Nayarit, Sinaloa), NSL–800 m.

Sources: Davis & Dixon, 1957 and N.J. Scott & McDiarmid, 1884b.

5. *Trimorphodon quadruplex* H.M. Smith, 1941o. Proc. U.S. Natl. Mus. 91(3130): 157–159. (*Trimorphodon biscutatus quadruplex*)

Type: Holotype, USNM 89476, a female (J.H. Ivy, 1932).
Type locality: "Esteli, Nicaragua" [= Estelí, Estelí Dept., cen. Nicaragua, 13°07'N, 86°25'W, elevation 830 m].
Distribution: Central America. Southeastern Guatemala (Huehuetenango), El Salvador (Ahuachapán, Cuscatlán, La Libertad, La Unión, San Miguel, San Salvador, San Vicente, Sonsonate), Honduras (Choluteca, Comayagua, Copán, Francisco Morazán, Valle, Yoro). Nicaragua (Carazo, Chinandega, Chontales, Estelí, Granada, Managua, Matagalpa, Nueva Segovia) and Costa Rica (Alajuela, Guanacaste, Puntarenas), NSL–1500 m.
Sources: E.H. Taylor, 1951, L.D. Wilson & Meyer, 1985, Villa et al., 1988, G. Köhler, 1999b, Savage, 2002, G. Köhler et al., 2005, N. Herrera et al., 2007, Jadin & Smith, 2010 and McCranie, 2011a.

6. *Trimorphodon tau* Cope, 1870. Proc. Amer. Philos. Soc. (1869–1870) 11(81): 151–152.

Synonyms: *Dipsas biscutata latifascia* W.C.H. Peters, 1870a, *Trimorphodon upsilon* Cope, 1870, *Trimorphodon collaris* Cope, 1875a, *Trimorphodon forbesi* H.M. Smith, 1941o, and *Trimorphodon fasciolata* H.M. Smith, 1941o.
Type: Holotype, USNM 30338 (formerly SIM 236), a 236 mm specimen (A.L.J.F. Sumichrast, 1855–1870).
Type locality: "western part of the Isthmus of Tehuantepec, Mexico," (in error *fide* McDiarmid & Scott, 1970: 29). Designated as near Quiotepec [= Coyotepec, cen. Oaxaca State, S Mexico, 16°57'N, 96°42'W, elevation 1530 m], between Tehuacán and Oaxaca, Oaxaca State, Mexico *fide* Sumichrast (1892: 44).
Distribution: Mexico (Aguascalientes, ext. SW Chihuahua, Colima, Distrito Federal, Durango, Guanajuato, Guerrero, Hidalgo, Jalisco, México, Michoacán, Morelos, Nayarit, Nuevo León, Oaxaca, Puebla, Querétaro, San Luis Potosí, Sinaloa, Sonora, Tamaulipas, cen. Veracruz, Zacatecas), 50–2600 m.
Sources: E.H. Taylor, 1939a, Davis & Dixon, 1957, McDiarmid & Scott, 1970, N.J. Scott & McDiarmid, 1984c, Benítez-Gálvez, 1997, Dundee & Liner, 1997, Auth et al., 2000, McCranie & Wilson, 2001b, Lemos-Espinal et al., 2004a–b, H.M. Smith et al., 2005 and Lemos-Espinal & Smith, 2007a.
Remarks: Type locality listed as Tuiatepec, Mexico *fide* USNM catalogue.

7. *Trimorphodon vilkinsonii* Cope, 1886b. Proc. Amer. Philos. Soc. (1885) 23(122): 285–286.

Synonym: *Trimorphodon wilkinsonii* Cope, 1887c (*nomen emendatum*).
Type: Holotype, USNM 14268, a 272 mm male (E. Wilkinson).
Type locality: "City of Chihuahua" [= vicinity of Chihuahua, Chihuahua State, NW Mexico, 28°38'N, 106°04'W, elevation 1440 m].
Distribution: Southwestern USA (S New Mexico, SW Texas) and N Mexico (N Chihuahua), 900–1850 m.
Sources: Gehlbach, 1971, Degenhardt et al., 1996, Werler & Dixon, 2000, LaDuc & Johnson, 2003 and Lemos-Espinal & Smth, 2007a.
Remarks: Probably occurs in NW Coahuila, Mexico *fide* Lemos-Espinal & Smith (2007a: 377).

TROPIDECHIS A.C.L.G. Günther, 1863c (Elapidae)

Synonyms: *Trophidechis* – De Vis, 1911 (*nomen incorrectum*), *Tropidechia* – Maass-Berlin, 1933 (*nomen incorrectum*), and *Tropedechis* – Boquet, 1948 (*nomen incorrectum*).
Type species: *Hoplocephalus carinatus* Krefft, 1863.
Distribution: Eastern Australia.
Sources: Beard, 1979, Cogger et al., 1983a, Hutchinson, 1990, Cogger, 1992, David & Ineich, 1999, Hoser, 2003c, 2012e, Sanders et al., 2008 and Zaher et al., 2009.

1. *Tropidechis carinatus* (Krefft, 1863). Proc. Zool. Soc. London 31(1): 86, 3 figs. (*Hoplocephalus carinatus*)

Synonyms: *Tropidechis carinata* – A.C.L.G. Günther, 1863c, and *Tropidechis sadlieri* Hoser, 2003c (*nomen illegitimum*).
Type: Holotype, a 965 mm specimen, (J.J. Wilcox), lost *fide* Cogger et al. (1983: 237).
Type locality: "near Grafton, in the Clarence River District" [= vicinity of Grafton, NE New South Wales, Australia, 29°42'S, 152°56'E, elevation NSL].
Distribution: Eastern Australia (ext. NE New South Wales, ext. SE Queensland), NSL–1395 m.
Sources: Shine & Charles, 1982 and Covacevic & Couoper, 1991.

TROPIDOCLONION Cope, 1860d (Natricidae)

Synonyms: *Microps* Hallowell, 1856c (*nomen praeoccupatum*), *Tropidoclonium* Cope, 1864d (*nomen emendatum*), *Tropidoclonio* – A.H. Wright & Wright, 1952 (*nomen incorrectum*), and *Tropidocloniom* – Marx & Rabb, 1972 (*nomen incorrectum*).
Type species: *Microps lineatus* Hallowell, 1856c.
Distribution: North and central USA
Fossil records: Upper Miocene to upper Pleistocene of USA. Fossils unassigned to species include upper Miocene (Clarendonian) of USA (Nebraska).
Sources: Bocourt, 1892a, Ramsey, 1953, P.W. Smith & Smith, 1963, Ernst & Barbour, 1989, Conant & Collins,

1991, H.M. Smith & Chiszar, 1994, Degenhardt et al., 1996, Fitch, 1999, Holman, 1984c, 2000a, Werler & Dixon, 2000, Alfaro & Arnold, 2001, Ernst & Ernst, 2003, Zaher et al., 2009 and Parmley & Hunter, 2010.

1. *Tropidoclonion lineatum* (Hallowell, 1856c). Proc. Acad. Nat. Sci. Philadelphia 8(5): 241. (*Microps lineatus*)

Synonyms: *Tropidoclonion lineatum iowae* Call, 1891, *Tropidoclonion lineatum annectens* Ramsey, 1953, *Tropidoclonion lineatum texanum* Ramsey, 1953, and *Tropidoclonion lineatum mertensi* H.M. Smith, 1965.

Type: Holotype, ANSP 5922, a 230 mm specimen (W.A. Hammond, 1847–1856).

Type locality: "Kansas" [USA]. Restricted to vicinity of Kansas City, Kansas, USA *fide* K.P. Schmidt (1953a: 176) and to the vicinity of Fort Leavenworth, Kansas, USA *fide* L.W. Ramsay (1953: 9).

Distribution: Central USA (E Colorado, cen. Illinois, Iowa, Kansas, ext. SW Minnesota, Missouri, E Nebraska, E New Mexico, Oklahoma, SE South Dakota, Texas), NSL–2015 m.

Fossil records: Lower/middle Pleistocene (Irvingtonian I) of USA (Texas), middle/upper Pleistocene (Irvingtonian II) of USA (Texas), and upper Pleistocene (Rancholabrean I, II) of USA (Kansas, Texas).

TROPIDODIPSAS A.C.L.G. Günther, 1858 (Dipsadidae)

Synonyms: *Galedon* Jan, 1863b, *Tropidogeophis* F. Müller, 1878a, *Dipeltophis* Cope, 1887c, *Geatractus* Dugès, 1898a, *Tropidodipsa* – Briceño-Rossi, 1934 (*nomen incorrectum*), *Exelencophis* H.M. Smith, 1942b, and *Excelencophis* – Obst, Richter & Jacob, 1984 (*nomen incorrectum*).

Type species: *Tropidodipsas fasciata* A.C.L.G. Günther, 1858.

Distribution: Mesoamerica.

Sources: Cadle, 1984b, 2007, Kofron, 1980, 1985a, 1987, 1988b, Wallach, 1995b, Harvey et al., 2008 and Zaher et al., 2009.

1. *Tropidodipsas annulifera* Boulenger, 1894a. Cat. Snakes Brit. Mus. 2: 297, pl. 14, fig. 1.

Synonyms: *Geophis tecpanecus* Dugès, 1898a–b, *Tantilla nelsoni* Slevin, 1926, and *Tropidodipsas malacodryas* Shannon & Humphrey, 1959b.

Type: Holotype, BMNH 1946.1.6.44, a 430 mm male (Zool. Soc. London).

Type locality: Unknown. Designated as W Mexico *fide* N.J. Scott (1967: 283).

Distribution: Western Mexico (Colima, Guerrero, Jalisco, Michoacán, Nayarit, Sinaloa, María Madre Is.), NSL–2000 m.

Sources: H.M. Smith& Taylor, 1945, H. Campbell & Simmons, 1962, Greer, 1965, N.J. Scott, 1967, Hardy & McDiarmid, 1969 and Reyes-Velasco et al., 2008.

2. *Tropidodipsas fasciata* A.C.L.G. Günther, 1858. Cat. Colub. Snakes Brit. Mus.: 181.

Synonyms: *Leptognathus subannulatus* F. Müller, 1887, and *Tropidodipsas guerreroensis* E.H. Taylor, 1939.

Type: Lectotype, BMNH 1946.1.23.66, a 610 mm female (Zool. Soc. London), designated by A.C.L.G. Günther (1894: 139).

Type locality: "Central America," (in error) via lectotype selection. Unknown *fide* Boulenger (1894a: 295). Restricted to Chichén Itzá, Yucatán, Mexico *fide* H.M. Smith & Taylor (1950a: 352).

Distribution: Eastern Mexico (Campeche, Chiapas, Guerrero, Nuevo León, Oaxaca, Quintana Roo, San Luis Potosí, Tamaulipas, Veracruz, Yucatán) and Guatemala, NSL–2000 m.

Sources: Kofron, 1987, J.C. Lee, 1996, Auth et al., 2000 and Farr et al., 2009.

3. *Tropidodipsas fischeri* Boulenger, 1894a. Cat. Snakes Brit. Mus. 2: 296. (*nomen substitutum*)

Synonyms: *Virginia fasciata* J.G. Fischer, 1885b (*nomen praeoccupatum*), *Tropidoclonion annulatum* Bocourt, 1892a (*nomen praeoccupatum*), and *Tropidodipsas kidderi* L.C. Stuart,1942.

Types: Syntypes (2), BMNH 1946.1.1.81 (formerly SMNS 2454b), a 540 mm male (F.C. Sarg, 1867–1894), and SMNS 2454a, the latter destroyed 25 Aug. 1940–19 April 1945 during World War II.

Type locality: "Guatemala." Restricted (by implication) to Tecpam, Guatemala *fide* Dunn & Saxe (1951: 61).

Distribution: Mesoamerica. Southern Mexico (Chiapas, Oaxaca), S Guatemala, W El Salvador (Santa Ana) and SW Honduras (Intibucá, Ocotepeque), 1000–3800 m.

Sources: Kofron, 1985b, L.D. Wilson & Meyer, 1985, Casas-Andreu et al., 1996, G. Köhler et al., 2006 and McCranie, 2011a.

Remarks: Type locality restriction of Volcán Atitlán, Sololá Dept., Guatemala *fide* H.M. Smith & Taylor (1950a: 319) rejected *fide* Dunn & Saxe (1951: 60).

4. *Tropidodipsas philippii* (Jan, 1863b). Elenco Sist. Ofidi: 101. (*Leptognathus philippii*)

Synonyms: *Leptognathus albocinctus* J.G. Fischer, 1885b, *Tropidodipsas occidentala* Oliver, 1937, and *Tropidodipsas freiae* Shannon & Humphrey, 1959b.

Types: Syntypes (2), MSNM and ZMH, both destroyed in 1943 during World War II.

Type locality: "Chili" [= Chile] (in error) and "Mazatlan" [= Mazatlan, S Sinaloa State, NW Mexico, 23°14'N, 106°25'W, elevation 10 m].

Distribution: Western Mexico (Jalisco, W Michoacán, Nayarit, W Oaxaca, Sinaloa), NSL–1100 m.
Sources: N.J. Scott, 1962, Hardy & McDiarmid, 1969, Kofron, 1987 and García & Ceballos, 1994.

5. *Tropidodipsas repleta* H.M. Smith, Lemos-Espinal, Hartman & Chiszar, 2005. Bull. Maryland Herp. Soc. 41(1): 39–40.

Type: Holotype, UBIPRO 11635, a 383 mm male (J.A. Lemos-Espinal & D. Hartman, 29 Sept. 2003).
Type locality: "km 236.2 hwy 16 Chihuahua-Hermosillo in Sonora (28°26'12.5"N, 109°10'5.7"W), 1643 m."
Distribution: Northwestern Mexico (SW Chihuahua, SE Sonora), 1500–1645 m.
Sources: H.M. Smith & Lemos-Espinal, 2006c and Lemos-Espinal & Smith, 2007a.

6. *Tropidodipsas sartorii* Cope, 1863b. Proc. Acad. Nat. Sci. Philadelphia 15(3): 100.

Synonyms: *Leptognathus dumerili* Jan, 1863b, *Galedon annularis* Jan, 1863b, *Geophis annulatus* W.C.H. Peters, 1870b, *Leptognathus leucostomus* Bocourt, 1884, *Leptognathus semicinctus* Bocourt, 1884, *Leptognathus sexscutatus* Bocourt, 1884, *Leptognathus bernoullii* F. Müller, 1887, *Leptognathus cuculliceps* F. Müller, 1887, and *Tropidodipsas macdougalli* H.M. Smith, 1943c.
Type: Holotype, USNM (formerly SIM 20), a 1422 mm specimen (C. Sartorius), lost *fide* H.M. Smith & Taylor (1945: 152).
Type locality: "Mirador, Vera Cruz" [= Mirador, Veracruz, E Mexico, 20°57'N, 97°27'W, elevation 35 m].
Distribution: Mesoamerica. Eastern Mexico (Campeche, Chiapas, Hidalgo, Nuevo León, Oaxaca, Puebla, Quintana Roo, Querétaro, San Luis Potosí, Tabasco, Tamaulipas, Veracruz, Yucatán), Belize (Cayo, Corozal, Orange Walk, Stann Creek, Toledo), Guatemala (Petén), El Salvador (La Libertad, Morazán, San Salvador, Santa Ana), Honduras (Atlántida, Comayagua, Copán, Cortés, Gracias a Dios, Santa Bárbara, Yoro), Nicaragua (Atlántico Norte, Managua) and NW Costa Rica (N Guanacaste), NSL–2440 m.
Sources: Lara & Flores Villeto, 1978, Hidalgo, 1981a, L.D. Wilson & Meyer, 1985, Casas-Andreu et al., 1996, J.C. Lee, 1996, 2000, J.A. Campbell, 1998b, G. Köhler, 1999b, Auth et al., 2000, Marineros, 2000, Stafford & Meyer, 2000, Solórzano, 2004, G. Köhler et al., 2006, McCranie, 2006, 2011a and Travers et al., 2011.

7. *Tropidodipsas zweifeli* Liner & Wilson, 1970. Copeia 1970(4): 787. (*nomen susbstitutum*)

Synonym: *Chersodromus annulatus* Zweifel, 1954a (*nomen praeoccupatum*).
Type: Holotype, MVZ 45030, a 477 mm male (W.W. Brown, Aug. 1942–Aug. 1943).
Type locality: "near Chilpancingo, Guerrero, México" [= vicinity of Chilpancingo, Guerrero State, S Mexico, 17°33'N, 99°30'W, elevation 1260 m].
Distribution: Southern Mexico (Guerrero, Morelos), 1020–1800 m.
Source: Zweifel, 1954a.

TROPIDODRYAS Fitzinger, 1843 (Xenodontidae)

Synonyms: *Galeophis* Berthold, 1859, *Geleophis* Berthold, 1859 (*nomen incorrectum*), *Teleolepis* Cope, 1870, and *Teleolapsis* – Osborn, 1930 (*nomen incorrectum*).
Type species: *Herpetodryas serra* Schlegel, 1837.
Distribution: Southeastern Brazil.
Sources: R.A. Thomas & Dixon, 1977, Cicchi et al., 2007, Zaher et al., 2009 and Vidal et al., 2010.
Remarks: Previously a synonym of *Philodryas*.

1. *Tropidodryas serra* (Schlegel, 1837). Essai Phys. Serp. 1: 150–151, 2: 180–181, pl. 7, figs. 1–2. (*Herpetodryas serra*)

Synonyms: *Geleophis jani* – Berthold, 1859 (*nomen incorrectum*), *Galeophis jani* Berthold, 1859, and *Philodryas serra laevis* Jan, 1863b (*nomen nudum*).
Type: Lectotype, RMNH 624, an adult female (G.H. von Langsdorff, 1822–1828), designated by R.A. Thomas & Dixon (1977: 7).
Type locality: "Brésil" [= Brazil] via lectotype selection.
Distribution: Southeastern Brazil (Bahia, Espírito Santo, Minas Gerais, Paraná, Rio de Janeiro, Santa Catarina, São Paulo, Cardoso and Santo Amaro Is.), NSL–100 m.
Sources: Argôlo, 1999b and Marques et al., 2001.
Remarks: RMNH catalogue cards indicate the collector of the lectotype to be H. Boie.

2. *Tropidodryas striaticeps* (Cope, 1870). Proc. Amer. Philos. Soc. (1869–1870) 11(81): 153–154. (*Teleolepis striaticeps*)

Synonym: *Philodryas pseudo-serra* Amaral, 1939a (*nomen incorrigendum*).
Type: Holotype, MCZ 2909, a 295–300 mm female (G. Sceva [Thayer Exped.], 1864–1865).
Type locality: "Brazil."
Distribution: Southeastern Brazil (Bahia, Espírito Santo, Minas Gerais, Paraná, Rio de Janeiro, Rio Grande do Sul, Santa Catarina, São Paulo, Cardoso and Santo Amaro Is.), 235–1050 m.
Sources: Barbour & Loveridge, 1929, Argôlo, 1999c, Freitas, 1999, Hofstadler-Deiques et al., 2004, Hamdan & Lira-da-Silva, 2007 and Guedes & Marques, 2011.
Remarks: Holotype erroneously listed as MCZ 909 *fide* Cope (1870: 154).

TROPIDOLAEMUS Wagler, 1830 (Viperidae)

Synonyms: *Tropidolesmus* – A.-M.-C. Duméril & Bibron, 1844 *(nomen incorrectum), Tropidolepius* – Chenu & Desmarest, 1857 *(nomen incorrectum), Tropidolamus* – Cope, 1862c *(nomen incorrectum), Tropidolemus* – Senna, 1886 *(nomen incorrectum),* and *Tropidolenus* – Senna, 1886 *(nomen incorrectum).*

Type species: *Cophias wagleri* H. Boie *in* F. Boie, 1827.

Distribution: India, Malay Peninsula and East Indies.

Sources: Maslin, 1942, Brattstrom, 1964, Leviton, 1964d, E.H. Taylor,1965, W. Burger, 1971, Hoge & Romano-Hoge, 1981a, Tweedie, 1983, K.F.L. Lim & Lee, 1989, Maes, 1989, M.J. Cox, 1991b, Gaulke, 1994b, David & Ineich, 1999, McDiarmid et al., 1999, Parkinson, 1999, Malhotra & Thorpe, 2000, Ziegler et al., 2001, Gumprecht et al., 2004. Koch et al., 2007, Vogel et al., 2007 and Hoser, 2012d.

Remarks: At least two more revisions of the genus in progress *fide* Vogel et al. (*in litt.*).

1. *Tropidolaemus huttoni* (M.A. Smith, 1949). J. Bombay Nat. Hist. Soc. 48(3): 596. (*Trimeresurus huttoni*)

Synonym: *Trimeresurus huttoni* Hutton, 1949 (*nomen nudum*).

Type: Holotype, BMNH 1948.1.8.75, a 136–138 mm specimen (A.F. Hutton, 1946–1948).

Type locality: "High Wavy Mountains, Madura District, South India, altitude 5,200 feet."

Distribution: Southwestern India (Maharashtra, Tamil Nadu), 1585 m.

Sources: David & Vogel, 1998, Boundy, 2008 and Hutton & David, 2009.

Remarks: Photographs of holotype in David & Vogel (1998: figs. 1–3).

2. *Tropidolaemus laticinctus* Koch, Gumprecht & Melaun, 2007. Zootaxa (1446): 3–17, figs. 1–5, 6a, 7a.

T

Type: Holotype, BMNH 1896.12.9.80, a 416 mm male (P. Sarasin & F. Sarasin, 1893–1896).

Type locality: "between L. Posso and Tomini Bay, Celebes" [= between Lake Poso and Tomini Bay, North Sulawesi Prov., E Indonesia].

Distribution: Eastern Indonesia (Sulawesi), NSL–100 m.

Sources: Koch, 2008a, 2012.

3. *Tropidolaemus philippensis* (Gray, 1842b). Zool. Misc. 2(Mar.): 48. (*Trimesurus philippensis*)

Synonyms: *Tropidolaemus hombroni* Guichenot *in* Jacquinot & Guichenot, 1853, *Tropidolaemus hombronii* A.-M.-C. Duméril, 1853 *(nomen nudum), Tropidolaemus philippinensis* W.C.H. Peters, 1861 *(nomen emendatum), Tropidolaemus schlegeli* Theobald, 1868b *(nomen nudum),* and *Bothrops philippinensis* Maslin, 1942 *(nomen emendatum).*

Type: Holotype, BMNH 1946.1.17.67 (H. Cuming, 1836–1840).

Type locality: "Philippine Islands."

Distribution: Southern Philippines (Mindanao), 195 m.

Remarks: Photographs of holotype of *T. philippensis* in Vogel et al. (2007: figs. 25–26) and photographs of holotype of *T. hombroni* in Vogel et al. (2007: figs. 27–30).

4. *Tropidolaemus subannulatus* (Gray, 1842b). Zool. Misc. 2(Mar.): 48. (*Trimesurus subannulatus*)

Synonyms: *Trigonocephalus wagleri celebensis* Gray, 1849a, *Tropidolaemus schlegelii* Bleeker, 1859b (*nomen nudum*), *Trimeresurus subannulatus immaculatus* W.C.H. Peters, 1872d, and *Trimeresurus wagleri alboviridis* E.H. Taylor, 1917.

Types: Syntypes (2), BMNH 1946.1.19.32–33, a female and juvenile (H. Cuming, 1836–1840).

Type locality: "The Philippines."

Distribution: East Indies. East Malaysia (Sabah, Sarawak), Brunei, cen. Indonesia (Belitung, Buton, Kalimantan, Sangihe Arch., Sulawesi) and Philippines (Balabac, Basilan, Bohol, Dinagat, Jolo, Leyte, Luzon, Mindanao, Negros, Palawan, Panay, Samar, Sibutu, Tumindao), NSL–100 m.

Sources: Card, 1994, Vogel, 2006, Kuch et al., 2007, I. Das, 2010, 2012 and Koch, 2012.

Remarks: Photographs of syntypes in Vogel et al. (2007: figs. 12–16). This species is a complex under study *fide* Kuch et al., (2007: 2).

5. *Tropidolaemus wagleri* (H. Boie *in* F. Boie, 1827). Isis von Oken 20(6): 561. (*Cophias wagleri*)

Synonyms: *Cophias wagleri* H. Boie, 1826 (*nomen nudum*), *Trimesurus maculatus* Gray, 1842b, *Trimeresurus sumatranus* Gray, 1842b, *Trigonocephalus wagleri sumatrenis* Gray, 1849a, and *Trimesurus formosus* Gray, 1849a,

Type: Neotype, MNHN 5784 (formerly MNHN 1879.708), an adult female (Rochet), designated by Vogel et al. (2007: 8).

Type locality: "Deli: rivière de Bedagei River (about 3°30'N, 99°13'E), Sumatra Barat Province, Sumatra, Indonesia" via neotype selection.

Distribution: Southeast Asia and W Indonesia. Thailand (Nakhon Si Tammarat, Narathiwat, Pattani, Phang Nga, Surat Thani, Yala), Vietnam (Minh Hai, Song Be), West Malaysia (Pahang, Perak, Perlis, Pinang, Selangor, Aceh Is.), Singapore and W Indonesia (Bangka, Durian, Mentawai Arch., Natuna, Riau Arch., Sumatra), NSL–1300 m.

Sources: Card, 1994, Inger & Tan, 1996, M.J. Cox et al., 1998, Chan-ard et al., 1999, Ferner et al., 2001,

Malkmus et al., 2002, Orlov et al., 2003, Gumprecht & Toenjes, 2007, V.S. Nguyen et al., 2009, I. Das, 2010, 2012 and Grismer, 2011.

Remarks: Original syntypes based on specimen figured in Seba (1735, pl. 68, fig. 4) and RMNH 1518 and RMNH 1522 (2 specimens) and perhaps RMNH 5667, subsequently described and illustrated by Schlegel, 1837 *fide* McDiarmid et al. (1999: 351). Photographs of neotype in Vogel et al. (2007: figs. 4–7).

TROPIDONOPHIS Jan, 1863b (Natricidae)

Synonyms: *Styporhynchus* W.C.H. Peters, 1863d, *Katophis* Macleay, 1877, *Macropophis* Boulenger, 1893a, *Styporhinchus* – Capocaccia, 1961 *(nomen incorrectum), Alanbrygelus* Hoser, 2012ag *(nomen illegitimum), Desburkeus* Hoser, 2012ag *(nomen illegitimum), Kirnerea* Hoser, 2012ag *(nomen illegitimum),* and *Oxynatrix* Hoser, 2012ag *(nomen illegitimum).*

Type species: *Tropidonotus picturatus* Schlegel, 1837.

Distribution: East Indies and Australasia.

Sources: McDowell, 1961, Malnate & Underwood, 1988, Kraus & Allison, 2004a and Zaher et al., 2009.

1. *Tropidonophis aenigmaticus* Malnate & Underwood, 1988. Proc. Acad. Nat. Sci. Philadelphia 140(1): 165–166, fig. 30.

Type: Holotype, AMNH 76633, a 506 mm male.

Type locality: "between Agamoia and Ailuluai, Fergusson Island, D'Entrecasteaux Islands, Milne Bay District, Papua New Guinea."

Distribution: Papua New Guinea (E Eastern Highlands, Milne Bay, Morobe, Fergusson, Goodenough, Misima, Normanby and Trobriand Is.), 700–1705 m.

Source: O'Shea, 1996.

Remarks: Holotype listed as AMNH 76638 in Malnate & Underwood (1988: Fig. 30).

2. *Tropidonophis dahlii* (F. Werner, 1899b). Zool. Anz. 22(596): 373. (*Tropidonotus* [*Macropophis*] *dahlii*)

Synonym: *Tropidonotus melanocephalus* F. Werner, 1925.

Type: Lectotype, NMW 23488, an 820–882 female (K.T.F. Dahl, 1896–1897), designated by Malnate & Underwood (1988: 169).

Type locality: "Herbertshöhe, Neu-Pommern [New Britain]" [= Kokopo, N East New Britain, Bismarck Arch., NE Papua New Guinea, 4°21'S, 152°16'E, elevation 50 m] via lectotype selection.

Distribution: Northeastern Papua New Guinea (East New Britain, West New Britain, New Britain Is.), 45–1070 m.

Source: O'Shea, 1996.

Remarks: F. Werner's (1900c: 85) report of holotype in ZMB erroneous *fide* G. Peters *in* Malnate & Underwood (1988: 169).

3. *Tropidonophis dendrophiops* (A.C.L.G. Günther, 1883). Ann. Mag. Nat. Hist. (5) 11(62): 136–137, 1 fig. (*Tropidonotus dendrophiops*)

Type: Holotype, BMNH 1946.1.15.41 (formerly BMNH 1882.11.25.13), a 914 mm female (H.M.S. Challenger Voy. [Stn. 201 or 212], 23–25 Oct. 1874 or 30 Jan.–5 Feb. 1875).

Type locality: "Zamboanga (Philippine Islands)" [= Zamboanga, SW Mindanao Is., Philippines, 7°01'N, 122°11'E, elevation 10 m].

Distribution: Philippines (Basilan, Bohol, Camiguin, Dinagat, Leyte, Mindanao, Samar), NSL–825 m.

Sources: Alcala, 1986, Ross & Lazell, 1991 and B.E. Smith, 1993.

Remarks: Possibly occurs on Luzon, Philippines *fide* Iskandar & Colijn (2001: 106).

4. *Tropidonophis dolasii* Kraus & Allison, 2004a. Proc. Biol. Soc. Washington 117(3): 303–306, figs. 1a–b, 2.

Type: Holotype, BPBM 16539 (formerly FK 6118), an 1145 mm female (D. Salepuna, 23 Aug. 2002).

Type locality: "E slope of Oya Tabu (Mt. Kilkerran), 9.4555°S, 150.7857°E, 1090 m, Fergusson Island, Milne Bay Province, Papua New Guinea."

Distribution: Southeastern Papua New Guinea (Milne Bay: Fergusson and Goodenough Is.), 900–1090 m.

5. *Tropidonophis doriae* (Boulenger, 1898i). Ann. Mus. Civ. Stor. Nat. Genova (2) 18(38): 704. (*Tropidonotus doriae*)

Synonym: *Tropidonophis doriae desburkei* Hoser, 2012ag *(nomen illegitimum).*

Types: Syntypes (18), BMNH 1946.1.12.84 (formerly BMNH 1897.12.10.116), a female [Moroka], 1946.1.12.61–62 (formerly BMNH 1897.12.10.117–118), two males [Haveri], MSNG CE 36101a–c [Haveri], and MSNTO 946a–b (formerly MZUT 2561a–b), longest syntype 990 mm (L. Loria, 1889–1892), location of other syntypes unknown.

Type locality: "Haveri, Moroka, Bartholomew Distribution, 2300 feet, South-Eastern British New Guinea" [= Haveri, 700 m, and Moroka, 1280 m, Bartholomew Mtns., Central Prov., SE Papua New Guinea]. Restricted to Haveri, SE Papua New Guinea *fide* Haas (1950: 539).

Distribution: New Guinea. Eastern Indonesia (Papua, Aru Is.: Kobroor) and Papua New Guinea (Central, Gulf, Morobe, Western, Western Highlands, West Sepik), NSL–1300 m.

Sources: F. Parker, 1982 and O'Shea, 1996.

Remarks: Malnate & Underwood (1988: 102) listed just three syntypes from BMNH and erroneously cited collection date as 1893. Eight syntypes are known; perhaps Boulenger's (1898i: 704) listing of 18 was a typo for 8?

6. *Tropidonophis elongatus* (Jan, 1865c). Arch. Zool. Anat. Fis. 3(2): 228. (*Tropidonotus picturatus elongatus*)

Synonyms: *Tropidonotus picturatus elongatus* Jan, 1863b (*nomen nudum*), *Coryphodon carinatus* Bleeker *in* Boulenger, 1893a *(nomen ineditum)*, and *Tropidonophis alanbrygeli* Hoser, 2012ag *(nomen illegitimum), Tropidonophis alanbrygeli sammywatsonae* Hoser, 2012ag *(nomen illegitimum)*, and *Tropidonophis smythi* Hoser, 2012ag *(nomen illegitimum)*.

Type: Holotype, RMNH 1085, a 1651 mm female (E.A. Forsten, 1838–1843).

Type locality: "Molucche" [= Maluku, E Indonesia]. Restricted to Ambon, *fide* Malnate & Underwood (1988: 96).

Distribution: Eastern Indonesia (Papua, Ambon, Biak, Halmabera, Noemfor, Salawati, Seram), 90–500 m.

Source: O'Shea, 1996.

7. *Tropidonophis halmahericus* (Boettger, 1895a). Zool. Anz. 18(479): 130–131. (*Tropidonotus halmahericus*)

Type: Lectotype, SMF 17750 (formerly SMF-B 7264, 1a), an 1102–1294 mm male (W. Kükenthal, 1894), designated by Mertens (1922: 179).

Type locality: "Soa Konorra, Nord-Halmaheira, Molukken" [= Soa Konora, N Halmahera, E Indonesia].

Distribution: Eastern Indonesia (Maluku: Batjan, Halmahera, Ternate).

Remarks: Lectotype listed as SMF-B 7264, 1a *fide* SMF catalogue and Mertens (1922: 179) but SMF-B 7264, 1b *fide* Boettger (1898: 28). SMF-B 7264, 1b is paralectotype SMF 17751.

T

8. *Tropidonophis hypomelas* (A.C.L.G. Günther, 1877). Proc. Zool. Soc. London 45(1): 130–131, fig. 1. (*Tropidonotus hypomelas*)

Type: Holotype, BMNH 1946.1.7.59 (formerly BMNH 1872.2.24.24), a 813 mm male (G. Brown).

Type locality: "Duke of York Island" [= Bismarck Arch., NE Papua New Guinea].

Distribution: Papua New Guinea (East New Britain, New Ireland, Umboi, West New Britain, Didao of Yoda Is.), 950 m.

Source: O'Shea, 1996.

9. *Tropidonophis mairii* (Gray *in* Grey, 1841b). J. Two Exped. Australia 2: 442. (*Tropidonotus mairii*)

Synonyms: *Tropidonotus australis* Gray, 1842c, *Tropidonotus semicinctus* A.-M.-C. Duméril, Bibron & Duméril, 1854a, *Kalophis plumbea* Macleay, 1877, *Tropidonotus angusticeps* Macleay, 1884a, *Tropidonotus ater* Macleay, 1885b, and *Tropidonophis mairii brongersmai* Malnate & Underwood, 1988.

Type: Neotype, BMNH 1946.1.13.52, a 605 mm male (J. Gilbert, 1838–1841), designated by Malnate & Underwood (1988: 110).

Type locality: "Port Essington, Northern Territory, Australia" [= ruins at tip of Cobourg Peninsula, N Northern Territory, N Australia, ca. 11°09'N, 132°00'E, elevation NSL] via neotype selection.

Distribution: Austro-Papua. Eastern Indonesia (SE Papua), Papua New Guinea (Central, Western) and N Australia (ext. NE New South Wales, N Northern Territory, N and E Queensland, ext. N Western Australia, Bathurst, Croker, Groote Eylandt, Melville, Milingimbi, Mornington, Prince of Wales and Wessel Marchinbar Is.), NSL–1500 m.

Sources: F. Parker, 1982, Cogger et al., 1983a, Hoser, 1989, Shea, 1990, Cogger, 1992, O'Shea, 1996, Greer, 1997, Storr et al., 2002 and Swan et al., 2004.

Remarks: Holotype (FPM 39) destroyed during World War II *fide* Malnate & Underwood (1988: 110).

10. *Tropidonophis mcdowelli* Malnate & Underwood, 1988. Proc. Acad. Nat. Sci. Philadelphia 140(1): 142–145, fig. 23.

Type: Holotype, BMNH 1913.11.1.100, a 650 mm male (Wollaston Exped., 1910–1913).

Type locality: "Camp III, Otakwa River, Irian Jaya, 2500 feet (767 meters)" [E Indonesia].

Distribution: New Guinea. Eastern Indonesia (Papua) and Papua New Guinea (East Sepik, West Sepik), 580–1885 m.

Sources: O'Shea, 1996 and Kraus & Allison, 2006.

11. *Tropidonophis montanus* (Lidth de Jeude, 1911b). Nova Guinea (Zool.) 9(2): 271. (*Tropidonotus montanus*)

Synonym: *Natrix polytaenia* Laurent, 1948.

Type: Holotype, RMNH 4704 (formerly RMNH 323), an 800 mm male (H.A. Lorentz [First Dutch New Guinea Exped.], 14 Oct. 1909).

Type locality: "Biwak V. (1360 m.), Nova Guinea" [= Bivak V., Hellwig Mtns., on Lorentz River, S Papua Prov., E Indonesia, ca. 4°32'S, 138°41'E, elevation 1360 m].

Distribution: Eastern Indonesia (Papua, Biak, Numfor, Salawati), 1260–2200 m.

Sources: F. Parker, 1982, McDowell, 1984 and O'Shea, 1996.

12. *Tropidonophis multiscutellatus* (Brongersma, 1948a). Proc. Kon. Ned. Akad. Wetensch. 51(3): 372–378. (*Natrix mairii multiscutellata*)

Synonym: *Tropidonophis multiscutellatus cottoni* Hoser, 2012ag *(nomen illegitimum).*

Type: Holotype, RMNH 8669, a male (H.A. Lorentz [Second Dutch New Guinea Exped.], 1909).

Type locality: "Alkmaar, Lorentz River, Dutch New Guinea" [= Camp Alkmaar, on Lorentz River, S Irian Jaya, E Indonesia, 4°40'S, 138°42'E, elevation 215 m].

Distribution: New Guinea. Eastern Indonesia (Papua, Batanta, Biak, Salawati) and Papua New Guinea (Central, Eastern Sepik, Gulf, Madang, Milne Bay, Morobe, Northern, Western Sepik, Western, Western Highlands, Karkar, Liki and Manam Is.), 15–1440 m.

Sources: Brongersma, 1948a, McDowell, 1984 and O'Shea, 1996.

Remarks: Malnate & Underwood (1988: 120) listed the reprint pagination for original description (p. 3).

13. *Tropidonophis negrosensis* (E.H. Taylor, 1917). Philippine J. Sci. 12D(6): 356–358, fig. 1. (*Natrix dendrophiops negrosensis*)

Type: Holotype, CM 2261 (formerly EHT 128), a 730 mm male (E.H. Taylor, Dec. 1915).

Type locality: "Canlaon Volcano, Occidental Negros P.I." [= Philippine Islands].

Distribution: Central Philippines (Cebu, Masbate, Mindoro, Negros, Panay, Pan de Azucar, Sicogen), NSL–1200 m.

Sources: Leviton, 1963b and Ferner et al., 2001.

Remarks: Holotype illustrated in Malnate & Underwood (1988: fig. 1).

14. *Tropidonophis novaeguineae* (Lidth de Jeude, 1911b). Nova Guinea (Zool.) 9(2): 271. (*Tropidonotus novaeguineae*) (*nomen corrigendum*)

Synonyms: *Tropidonotus novae-guineae* Lidth de Jeude, 1911b *(nomen incorrigendum), Natrix novaeguineae* – C.P.J. Haas *(nomen corrigendum),* and *Tropidonophis novaeguineae trioani* Hoser, 2012ag *(nomen illegitimum).*

Type: Holotype, RMNH 4702, a 600 mm female (H.A. Lorentz [Second New Guinea Exped.], 17 Dec. 1909).

Type locality: "Alkmaar, Nova Guinea" [= Camp Alkmaar, on Lorentz River, S Irian Jaya, E Indonesia, 4°40'S, 138°42'E, elevation 215 m].

Distribution: New Guinea. Eastern Indonesia (Papua, Misol) and Papua New Guinea (Central, Chimbu, Gulf, Western), 75–1065 m.

Sources: F. Parker, 1982 and O'Shea, 1996.

15. *Tropidonophis parkeri* Malnate & Underwood, 1988. Proc. Acad. Nat. Sci. Philadelphia 140(1): 161–162, fig. 28.

Type: Holotype, CAS 139588, a 504 mm male (F.S. Parker).

Type locality: "Wahgi River (specific collection site unknown), Eastern Highlands Province, Papua New Guinea," (partially in error). Corrected to Wahgi River, Western Highlands or Chimbu provinces *fide* F.S. Parker, *in litt.*

Distribution: Central Papua New Guinea (Chimbu, Western Highlands), 1065–2135 m.

Source: O'Shea, 1996.

16. *Tropidonophis picturatus* (Schlegel, 1837). Essai Phys. Serp. 2: 314–315, pl. 12, figs. 8–9. (*Tropidonotus picturatus*)

Synonym: *Tropidonophis picturatus pillotti* Hoser, 2012ag *(nomen illegitimum).*

Type: Lectotype, RMNH 1076, a female (S. Müller & H. Macklot, Aug. 1828), designated by Malnate & Underwood (1988: 135).

Type locality: "Lobo Bay, Netherlands New Guinea [Irain Jaya]" [E Indonesia] via lectotype selection.

Distribution: New Guinea. Eastern Indonesia (Papua, Misol, Salawati, Weigeau) and Papua New Guinea (Central, Chimbu, Milne Bay, Morobe, Western, West Sepik), 200–1545 m.

Sources: F. Parker, 1982 and O'Shea, 1996.

17. *Tropidonophis punctiventris* (Boettger, 1895a). Zool. Anz. 18(471–472): 129–130. (*Tropidonotus punctiventris*)

Type: Holotype, SMF 17195 (formerly SMF-B 7216a), a 342 mm male (W. Kükenthal, 1894).

Type locality: "Soa Konorra, Nord-Halmaheira, Molukken" [E Indonesia.].

Distribution: Eastern Indonesia (Maluku: N Halmahera). Known only from type locality.

Remarks: Holotype illustrated in Malnate & Underwood (1988: fig. 7).

18. *Tropidonophis statisticus* Malnate & Underwood, 1988. Proc. Acad. Nat. Sci. Philadelphia 140(1): 155–156, fig. 27.

Type: Holotype, MCZ 142422, a 514 mm male (F.S. Parker, 8 July 1964).

Type locality: "Orumba, Eastern Highlands Province, Papua New Guinea."

Distribution: New Guinea. Eastern Indonesia (SE Papua) and Papua New Guinea (Central, Eastern Highlands, East Sepik, Enga, Milne Bay, Morobe, Chimbu, Southern Highlands, Western, Western Highlands), 500–2135 m.

Sources: O'Shea, 1996 and Kraus & Allison, 2004.

19. *Tropidonophis truncatus* (W.C.H. Peters, 1863d). Mber. Königl. Akad. Wiss. Berlin 1863(10): 399–403. (*Styporhynchus truncatus*)

Types: Syntypes (2), ZMB 4883a–b, two females, longest syntype 425 mm (K.E. von Martens, March 1862–Dec. 1864).

Type locality: "der mitte der Insel Djololo" [= Dodinga, Jailolo, Halmahera Is., North Maluku State, Indonesia, 1°04'N, 127°28'E, elevation 15 m].

Distribution: Eastern Indonesia (Papua, Batjan, Halmahera, Ternate, Salawati), NSL–15 m.

Remarks: Syntype illustrated in Malnate & Underwood (1988: fig. 6).

TROPIDOPHIS Bibron, 1840 *in* Ramón de la Sagra, 1838–1843 (Tropidophiidae)

Synonyms: *Leionotus* Bibron *in* Ramón de la Sagra, 1840 *in* 1838–1843, *Ungalia* Gray, 1842a, *Erycopsis* Fitzinger, 1843, ? *Tropidophis* A.-M.-C. Duméril & Bibron, 1844, *Notophis* Hallowell, 1856b, *Notopholis* – J.T. Reinhardt & Lütken, 1862 *(nomen incorrectum), Lionotus* W.C.H. Peters, 1864b *(nomen emendatum), Ungualia* Cope, 1868c *(nomen emendatum), Leionothus* – Hoffmann, 1890 *(nomen incorrectum), Tripidophis* – Briceño-Rossi, 1934 *(nomen incorrectum), Trophidophis* – H.W. Parker, 1963 *(nomen incorrectum),* and *Tropilophis*– W.W. Tanner & Avery, 1982 *(nomen incorrectum).*

Type species: *Boa melanura* Schlegel, 1837.

Distribution: West Indies and N South America.

Fossil record: Pleistocene of Cayman Is. and USA (Florida).

Sources: Stull, 1928, Bailey, 1937c, Brongersma, 1951, A. Schwartz & Marsh, 1960, Stimson, 1969, A. Schwartz & Thomas, 1975, Hedges & Garrido, 1992b, 2002, Tolson & R.W. Henderson, 1993, Meylan, 1995, R.M. Burger, 1996, Rodríguez-Robles & Greene, 1996, Walls, 1998a, Hedges et al., 1999, McDiarmid et al., 1999, Holman, 2000b, Hedges, 2002, Rodríguez-Schettino, 2003, M.S.Y. Lee et al., 2007 and R.W. Henderson & Powell, 2009.

T

1. *Tropidophis battersbyi* Laurent, 1949. Bull. Inst. Roy. Sci. Nat. Belg. 25(9): 6–7, figs. 10–12.

Type: Holotype, IRSNB 2050 (formerly IG 3701 & IRSNB Reg. no. 739), a 347 mm male (de Ville, 20 Nov. 1875).

Type locality: "Equateur" [= Ecuador, or equatorial South America].

Distribution: Unknown, possibly Ecuador. Known only from holotype.

Source: Curcio et al., 2012.

Remarks: Photographs of holotype in Curcio et al. (2012: figs. 2a–b, 3a–d). Type locality probably Ecuador, through which the Equator extends.

2. *Tropidophis bucculentus* (Cope, 1868c). Proc. Acad. Nat. Sci. Philadelphia 20(2): 129. (*Ungalia pardalis bucculenta*)

Types: Syntypes (4), USNM 12377a–c (formerly SIM), a 635–655 mm female, 399 mm male and 342 mm male, and USNM 12370 (W.J. Rasin), the latter rotted and destroyed 9 April 1958 *fide* Powell (1999: 10).

Type locality: "Navassa Id" [= Navassa Is., West Indies, bet. 18°24'–25'N & 75°00'–02'W, elevation NSL–80 m].

Distribution: Navassa Island, < 80 m.

Sources: Bailey, 1937, R. Thomas, 1966c, Powell, 1999, 2002 and Hedges, 2002.

Remarks: ANSP 10281 was questionably considered a syntype *fide* Bailey (1937: 45) but it has no data and E.R. Dunn's label reads "probably the type of *sancticrucis* Cope" plus the fourth of Cope's syntypes is more likely to have been USNM 12370. A valid species *fide* Powell (1999: 10); previously a subspecies of *T. melanurus.* Probably extinct *fide* Powell (1999: 11) and Hedges (2002: 88).

3. *Tropidophis canus* (Cope, 1868c). Proc. Acad. Nat. Sci. Philadelphia 20(2): 129. (*Ungalia cana*)

Types: Syntypes (6), USNM 7111a–e (formerly SIM) (T. Blain), and USNM 26763 (formerly SIM), longest syntype 349 mm.

Type locality: "Bahama Island of Inagua" [= Great Inagua Is., Bahamas, West Indies]. Designated as S side of Inagua *fide* Stimson (1969: 33).

Distribution: Bahamas (Great Inagua Is.).

Source: Tolson & R.W. Henderson, 1993.

4. *Tropidophis caymanensis* Battersby, 1938. Ann. Mag. Nat. Hist. (11) 1(5): 558–559. (*Tropidophis melanurus caymanensis*)

Type: Holotype, BMNH 1946.1.1.53 (formerly BMNH 1912.7.18.1), a 448 mm female (T.M. Savage).

Type locality: "Grand Cayman Island" [Cayman Islands, West Indies].

Distribution: Caymans (Grande Cayman and Little Cayman Is.).

Sources: R. Thomas, 1963 and Seidel & Franz, 1994.

5. *Tropidophis celiae* Hedges, Estrada & Diaz, 1999. Copeia 1999(3): 376–379, figs. 1–2.

Type: Holotype, MNHNCU 4474, a 385 mm female (A.R. Estrada & L.M. Diaz, 7 June 1996).

Type locality: "northern (coastal) side of Loma Canasi, at the mouth of Río Canasi, Santa Cruz del Norte

Municipality, La Habana Province, Cuba, 3 m elevation, 23°08'37"N, 81°46'40"W."
Distribution: Northwestern Cuba (Habana), NSL.
Source: Hedges et al., 2001.

6. *Tropidophis curtus* (S.W. Garman, 1887b). Proc. Amer. Philos. Soc. 24(126): 279. (*Ungualia curta*)

Synonyms: *Tropidophis pardalis androsi* Stull, 1928, and *Tropidophis pardalis barbouri* Bailey, 1937c.
Type: Holotype, MCZ 6114.
Type locality: "Cuba" [Greater Antilles] (in error *fide* Stull, 1928: 31).
Distribution: Bahamas (Andros, Bimini, Eleuthera, Eleuthera Cays, Exuma, Gun Cay, New Providence, South Bimini, Cat, Cay Sal, Long, New Providence, Ragged).
Source: Hedges, 2002.
Remarks: A. Schwartz & Thomas (1975: 192) reported a specimen (AMNH 2946) from Camagüey, Cuba, either a waif or support for S.W. Garman's type locality.

7. *Tropidophis feicki* A. Schwartz, 1957. Amer. Mus. Novit. (1839): 3–6, fig. 1 (upper).

Type: Holotype, AMNH 76224 (formerly JRF 1267), a 474 mm female (J.R. Feick, 20 Dec. 1956).
Type locality: "Cueva de los Indios, San Vicente, Pinar del Río Province, Cuba."
Distribution: Western Cuba (Habana, Mantanzas, Pinar del Río), NSL–125 m.
Sources: Hedges & Garrido, 1999 and J.C. Murphy & Schlager, 2003.
Remarks: Record from Granma Prov., Cuba needs confirmation *fide* A. Schwartz & Henderson (1991: 633).

8. *Tropidophis fuscus* Hedges & Garrido, 1992b. Copeia 1992(3): 820–823, figs. 1, 3 (middle).

Type: Holotype, MNHNCU 2705 (formerly SBH 190300), a 338 mm female (S.B. Hedges, 27 July 1989).
Type locality: "Minas Amores (21.7 km NW, 7.7 km SE Baracoa, by road), Guantánamo Province, Cuba, 76 m."
Distribution: Extreme E Cuba (NE Guantánamo), 75–700 m.
Sources: Fong, 2002, 2005.

9. *Tropidophis galacelidus* A. Schwartz & Garrido, 1975. Proc. Biol. Soc. Washington 88(9): 81–82. (*Tropidophis pilsbryi galacelidus*)

Type: Holotype, CZACC 4.3435 (formerly IZAC 4052), a 213 mm male (P. Alayo, 9 July 1973).
Type locality: "near Cafetal de Gaviña, Sierra de Trinidad, Las Villas Province, Cuba" [= SW Sancti Spíritus Prov., cen. Cuba].
Distribution: Central Cuba (S Sancti Spíritus), 500 m.
Sources: Hedges & Garrido, 2002 and Chamizo-Lara et al., 2005.
Remarks: A valid species *fide* Hedges & Garrido (2002: 160).

10. *Tropidophis grapiuna* Curcio, Sales-Nuñes, Suzart-Argôlo, Skuk & Rodrigues, 2012. Herpetol. Monogr. (26): 86–91, figs. 4a–b, 5a–h.

Type: Holotype, MNRJ 19593 (formerly CZGB 3820), a 324 mm female (J.R. Martins, 15 June 1995).
Type locality: "'Serra de Pedra Lascada', region of Itajuípe, state of Bahia, Brazil."
Distribution: Northeastern Brazil (Bahia), 725–750 m.

11. *Tropidophis greenwayi* T. Barbour & Shreve, 1936. Proc. New England Zool. Club 16: 2. (*Tropidophis pardalis greenwayi*)

Synonym: *Tropidophis greenwayi lanthanus* A. Schwartz, 1963.
Type: Holotype, MCZ 42051, a 359 mm male (G. Greenway, March 1936).
Type locality: "Ambergris Cay, Caicos Islands, Bahamas."
Distribution: Turks and Caicos (Ambergris Cay, Big Ambergris Cay, Long Cay, Middle Caicos, Middleton Cay, North Caicos, South Caicos), NSL.
Sources: A. Schwartz, 1963, Iverson, 1986, R. Burger, 1997 and Reynolds et al., 2010.
Remarks: Probably occurs on Providencia *fide* Iverson (1986: 192).

12. *Tropidophis haetianus* (Cope, 1879). Proc. Amer. Philos. Soc. 18(104): 273. (*Ungualia haetiana*)

Synonyms: *Tropidophis conjunctus* J.G. Fischer, 1888c, *Tropidophis haetianus hemerus* A. Schwartz, 1975, and *Tropidophis haetianus tiburonensis* A. Schwartz, 1975.
Types: Syntypes (3), USNM 10164, a 680–690 mm specimen (J.J. Brown) [Port-au-Prince], USNM 10169, ca. 400 mm specimen broken into several pieces (J.J. Brown) [Gonave], and USNM 10275, a 410 mm specimen (C.A. Fraser, 1878) [Puerto Plata].
Type locality: "Port-au-Prince and Gonave Island" [= Port-au-Prince and Gonave Is., Ouest Dept., SW Haiti] and "Puerto Plata, Santo Domingo" [= Dominican Republic].
Distribution: Haiti (Artibonite, Centre, Grande'Anse, Nord, Nord Ouest, Ouest, Sud, Sud-Est, Gonâve and Tortue Is.), Dominican Republic (Azua, Barahona, Dajabón, Distrito Nacional, Duarte, El Seibo, Espaillat, Hato Mayor, Independencia, La Altagracia, La Vega, María Trinidad Sánchez, Monseñor Nouel, Monte Cristi, Monte Plata, Pedernales, Puerto Plata, Salcedo, Samaná, San Cristóbal, San Juan, Santiago, Santiago Rodríguez) and E Cuba (Oriente), NSL–820 m.
Sources: A. Schwartz, 1975 and Hedges & Garrido, 2002.

Remarks: Stimson (1969: 35) and A. Schwartz (1975:308) listed only two syntypes (USNM 10164 & 10169).

13. *Tropidophis hardyi* A. Schwartz & Garrido, 1975. Proc. Biol. Soc. Washington 88(9): 86–87. (*Tropidophis nigriventris hardyi*)

Type: Holotype: USNM 138510, a 351 mm male (J.D. Hardy, 5–10 Sept. 1956).
Type locality: "10 mi (16 km) W Trinidad, Las Villas Province, Cuba" [= SW Sancti Spíritus Prov., cen. Cuba].
Distribution: Central Cuba (Cienfuegos, S Sancti Spíritus), NSL–90 m.
Source: Domínguez et al., 2006.
Remarks: A valid species *fide* Hedges & Garrido (2002: 160).

14. *Tropidophis hendersoni* Hedges & Garrido, 2002. J. Herp. 36(2): 158–159, fig. 1.

Type: Holotype, MCZ 47896, an 287 mm (svl) specimen (W.J. Clench & Alayo, 14 Aug. 1945).
Type locality: "Guarda la Vaca, Holguín Province, Cuba, 21°07'35"N, 77°49'55"W, probably at sea level."
Distribution: Eastern Cuba (N Holguín), NSL. Known only from type locality.

15. *Tropidophis jamaicensis* Stull, 1928. Occ. Pap. Mus. Zool. Univ. Michigan (195): 12–13. (*Tropidophis maculatus jamaicensis*)

Type: Holotype, MCZ 12090, a female (M. Graham & T. Barbour, 1917).
Type locality: "Kingston, Jamaica" [= Kingston, St. Andrew Parish, Jamaica, 18°00'N, 76°48'W, elevation 55 m].
Distribution: Southern Jamaica (Manchester, St. Andrews, St. Catherine, St. Elizabeth, St. Thomas), 55 m.
Source: Hedges & Garrido, 2002.
Remarks: T. Barbour & Loveridge (1929: 353) misspelled collector of type as M. Garbham.

T

16. *Tropidophis maculatus* (Bibron, 1840 *in* Ramón de la Sagra, 1838–1843). Hist. Fís. Polít. Nat. Cuba 2(2): 127–128, pl. 24. (*Leionotus maculatus*)

Synonyms: *Tropidophis distinctus* Jan, 1863b, and *Ungalia dipsadina* Cope, 1868c.
Type: Holotype, MNHN 7184, a 472 mm specimen (D. Ramón de la Sagra, 1824–1835).
Type locality: "Cuba."
Distribution: Western Cuba (Habana, Matanzas, Pinar del Río, Juventud Is.).
Source: Hedges & Garrido, 1999.
Remarks: Original description reprinted in Bibron, 1840 *in* Ramón de la Sagra (1841–1843: 212–214, pl. 25).

17. *Tropidophis melanurus* (Schlegel, 1837). Essai Phys. Serp. 1: 177, 2: 399–401. (*Boa melanura*)

Synonyms: *Notophis bicarinatus* Hallowell, 1856b, *Tropidophis melanurus dysodes* A. Schwartz & Thomas, 1960, and *Tropidophis melanurus ericksoni* A. Schwartz & Thomas, 1960.
Types: Syntypes (2), MNHN 1325 and MNHN 3284, longest syntype 480 mm (Ricord via D. Ramón de la Sagra), lost *fide* A. Schwartz & Thomas (1975: 194),
Type locality: "île de Cuba" [Greater Antilles].
Distribution: Cuba (Camagüey, Cienfuegos, Ciego de Aviula, Granma, Guantánamo, Habana, Holguín, Las Tunas, Matanzas, Pinar del Río, Sancti Spíritus, Santiago de Cuba, Villa Clara, Juventud Is.), 10–800 m.
Sources: Hallowell, 1857c, A. Schwartz & Thomas, 1960, R. Thomas, 1966c, Estrada, 1994 and Estrada, 1998.
Remarks: Holotype erroneously listed as ANSP 10308 (= *Notophis bicarinatus* Hallowell type) *fide* Rodriguez-Schettino (2000: 25).

18. *Tropidophis morenoi* Hedges, Garrido & Diaz, 2001. J. Herp. 35(4): 615–617.

Type: Holotype, CZACC 2943 (formerly IBAC 2943), a 337 mm female (J. Salas, 1969).
Type locality: "Dolinas de Cueva de Humboldt, Caguanes, Villa Clara Province, Cuba, 22°50'04"N, 80°12'02"W, 25 m.
Distribution: Central Cuba (Ciego de Ávila, Cocokey, Sabina-Camagüey Arch., Sancti Spíritus, Villa Clara), 25 m.
Sources: Chamizo-Lara et al., 2005 and Domínguez & Parada, 2009.

19. *Tropidophis nigriventris* Bailey, 1937c. Proc. New England Zool. Club 16: 45–46.

Type: Holotype, UMMZ 70888, a 232 mm female (A. Greenhall, 30 July 1931).
Type locality: "six miles east of Marti, Camaguey, Cuba."
Distribution: Eastern Cuba (E Camagüey).
Source: A. Schwartz & Garrido, 1975.
Remarks: A. Schwartz & Garrido (1975: 89) erroneously listed the holotype as UMMZ 70889.

20. *Tropidophis pardalis* (Gundlach, 1840). Arch. Naturg. 6(1): 359–361. (*Boa pardalis*)

Types: Syntypes (2), none designated, a 523 mm and 270 mm specimen (Ricord, 1839), lost *fide* Stimson (1969: 37) and Rodríguez-Schettino (2000: 24).
Type locality: "Cuba."
Distribution: Cuba (Camaguey, Ciego de Avila, Cienfuegos, Holguín, Habana, Matanas, Pinar del Río, Santiago de Cuba, Villa Clara, Cayo Coco, Cayo Parédon Grande and Juventud Is.), NSL–130 m.
Sources: Estrada, 1998 and Domínguez et al., 2006.

Remarks: Holotype not ZMB 8043 *fide* A. Schwartz & Thomas (1975: 195).

21. *Tropidophis parkeri* Grant, 1940c. Bull. Inst. Jamaica (Sci.) (2): 44–46.

Type: Holotype, formerly CG 2596, a 385+ mm female (C. Grant, 10 April 1938), location unknown.
Type locality: "Little Cayman, Cayman Islands."
Distribution: Caymans (Little Cayman Is.).
Source: Hedges, 2002.

22. *Tropidophis paucisquamis* (F. Müller *in* Schenkel, 1901). Verh. Naturf. Ges. Basel (1901–1902) 13(1): 154. (*Ungalia paucisquamis*)

Synonym: *Ungalia brasiliensis* Andersson, 1901.
Type: Holotype, NMBA 426, a 330 mm male.
Type locality: "Continent von trop. Amerika" [= tropical Mesoamerica or South America]. Restricted to EBB, municipality of Salesópolis, state of São Paulo, Brazil *fide* Curcio et al. (2012: 93).
Distribution: Southeastern Brazil (Espírito Santo, Rio de Janeiro, São Paulo), 500–1250 m.
Sources: Amaral, 1930k and Curcio et al., 2012.
Remarks: Photographs of holotype in Curcio et al. (2012: figs. 8a–b, 9a–d). Record from Peru doubtful *fide* Walls (1998a: 242).

23. *Tropidophis pilsbryi* Bailey, 1937c. Proc. New England Zool Club 16: 42–43. (*Tropidophis maculatus pilsbryi*)

Type: Holotype, ANSP 20822, a 243 mm male (H.A. Pilsbry, July 1928).
Type locality: "Cayo del Rey, near Miranda, Oriente Province, Cuba" [= Lomas de Cayo del Rey, Santiago de Cuba Prov., E Cuba, 20°25'N, 75°55'W, elevation 115 m.]
Distribution: Eastern Cuba (Guantánamo, Holguín, Santiago de Cuba), 115–800 m.
Sources: A. Schwartz & Garrido, 1975 and Estrada, 1994.

24. *Tropidophis preciosus* Curcio, Sales-Nuñes, Suzart-Argôlo, Skuk & Rodrigues, 2012. Herpetol. Monogr. (26): 101–108, figs. 15a–b, 16a–h, 19.

Type: Holotype, MZUSP 17957, a 429 mm female (G. Skuk, Sept. 1989).
Type locality: "Conselheiro Mata, small village at approximately 40 km east from the municipality of Diamantina, state of Minas Gerais, Brazil."
Distribution: Southeastern Brazil (E Minas Gerais), 980–1100 m.

25. *Tropidophis schwartzi* R. Thomas, 1963. Breviora (195): 3–6, figs. 1, 3. (*Tropidophis caymanensis schwartzi*)

Type: Holotype, MCZ 69618, a 435 mm male (native via A. Schwartz, 25 Aug. 1961).
Type locality: "The Creek, 8 mi. NE of West End, Cayman Brac" [Cayman Islands].
Distribution: Caymans (Cayman Brac), NSL–35 m.
Source: Hedges, 2002.

26. *Tropidophis semicinctus* (Gundlach & Peters *in* W.C.H. Peters, 1864b). Mber. Königl. Akad. Wiss. Berlin 1864(6): 388. (*Ungalia maculata semicincta*)

Synonym: *Tropidophis moreletii* Bocourt, 1885.
Types: Syntypes (2), ZMB 5076a–b (J.C. Gundlach, 1839–1864).
Type locality: "Cuba."
Distribution: Western Cuba (Cienfuegos, Habana, Matanzas, Pinar del Río, Sacti Spíritus, Villa Clara).
Sources: F. Werner, 1901b, Petzold, 1969, A. Schwartz & Garrido 1975, Hedges & Garrido, 1999 and Hedges et al., 2001.

27. *Tropidophis spiritus* Hedges & Garrido, 1999. J. Herp. 33(3): 436–440, figs. 1a–b, 3a.

Type: Holotype, MNHNCU 4085, a 351 mm male (O. Jíminez, Feb. 1993).
Type locality: "Canal Zaza, Cacerio Chorrera Brava, Sancti Spíritus Province, Cuba, 21°47'07"N, 79°21'38"W."
Distribution: Central Cuba (S Sancti Spíritus).
Source: Hedges et al., 2001.

28. *Tropidophis stejnegeri* Grant, 1940a. Jamaica To-Day: 157. (*Tropidophis pardalis stejnegeri*)

Type: Holotype, MCZ 44769 (formerly CG 770), a 273 mm female (C. Grant, 29 April 1937).
Type locality: "north slope, Jamaica." Specified as Boston Bay, Jamaica *fide* Grant (1940b: 119).
Distribution: Northern Jamaica (Portland, St. Ann, N St. Elizabeth, St. James, Trelawny, Westmoreland).
Source: Hedges, 2002.
Remarks: Original description reprinted in Grant (1940b: 119–120).

29. *Tropidophis stullae* Grant, 1940a. Jamaica To-Day: 157. (*Tropidophis maculatus stullae*) (*nomen corrigendum*)

Synonyms: *Tropidophis maculatus stulli* Grant, 1940a *(nomen incorrigendum),* and *Tropidophis haetianus stullae* – Schwartz & Thomas, 1975 *(nomen corrigendum).*

Type: Holotype, MCZ 44870 (formerly CG 384), a 318 mm male (C. Grant, 20 April 1937).
Type locality: "Portland Point, Jamaica" [= S Clarendon Parish, S Jamaica].
Distribution: Southern Jamaica (ext. S Clarendon).
Source: Hedges, 2002.
Remarks: Original description reprinted in Grant (1940b: 121–122).

30. *Tropidophis taczanowskyi* (Steindachner, 1880a). Sitz. Akad. Wiss. Wien, Math.-Natur. Kl. (1879) 80(1): 522–525, pl., figs. 1, 1a–c, 2a–b. (*Ungalia taczanowskyi*)

Type: Lectotype, NMW 14858, a 321 mm female (Stolzmann, 1880), designated by Curcio et al. (2012: 116).
Type locality: "Tambillo, department of Cajamarca, Peru" [= Tambillo, Cajamarca Dept., cen. Peru, 6°10'S, 78°45'W, elevation 1770 m] via lectotype selection.
Distribution: Southern Ecuador (Loja, Zamora-Chinchipe) and N Peru (Cajamarca, Ucayali), 840–2900 m.
Sources: J.A. Peters, 1960a and Curcio et al., 2012.
Remarks: Photographs of lectotype in Curcio et al. (2012: figs. 20a–b, 21a–d), who listed authorship as Steindachner, 1870. Records from Guayaquil, Ecuador and Amazonas, Brazil need confirmation *fide* Curcio et al. (2012: 117).

31. *Tropidophis wrighti* Stull, 1928. Occ. Pap. Mus. Zool., Univ. Michigan (195): 38–39, pl. 2, fig. 7.

Type: Holotype, USNM 12420, a 348 mm male.
Type locality: "East Cuba."
Distribution: Eastern Cuba (Camagüey, Granma, Sancti Spíritus, Santiago de Cuba), 300–730 m.
Sources: Estrada, 1994, Hedges & Garrido, 1999 and Hedges et al., 2001.

32. *Tropidophis xanthogaster* Domínguez, Moreno & Hedges, 2006. Amphibia-Reptilia 27(3): 427–429, fig. 1a.

T

Type: Holotype, CZACC 4.9465, a 279 mm male (A. Rodríguez & R. Alonso, 4 May 2000).
Type locality: "Las Perlas Cave, La Bajada, Guanahacabibes Peninsula, Pinar del Río Province (21°55'53.38"N 84°28'48.90"W), Cuba."
Distribution: Western Cuba (Pinar del Río), NSL–15 m.

†*TUSCAHOMAOPHIS* Holman & Case, 1992 (Alethinophidia *incertae sedis*)

Type species: †*Tuscahomaophis leggetti* Holman & Case, 1992.
Distribution: Upper Paleocene of USA.
Sources: Holman & Case, 1992 and Holman, 2000a.

1. †*Tuscahomaophis leggetti* Holman & Case, 1992. Ann. Carnegie Mus. 61(3): 198–201, figs. 1a–e.

Type: Holotype, CM 47872, one trunk vertebra (K.C. Beard, A.R. Tabrum, C.R. Case, & J.J. Leggett, April 1990).
Type locality: "site CM loc. 517 (coordinates on file at CM), Meridian, Lauderdale County, Mississippi; Latest Paleocene, Tuscahoma Formation" [USA].
Distribution: Upper Paleocene (Clarkforkian: 55.4–56.8 mya) of USA (Mississippi). Known only from type series.

TYPHLOPHIS Fitzinger, 1843 (Anomalepididae)

Synonyms: *Cephalolepis* A.-M.-C. Duméril & Bibron, 1844, *Typhlopis* – Palacky, 1898 *(nomen incorrectum)*, and *Thlophis* – Briceño-Rossi, 1934 *(nomen incorrectum)*.
Type species: *Typhlops squamosus* Schlegel, 1839 *in* 1837–1844.
Distribution: Northern South America.
Sources: Hahn, 1980a and McDiarmid et al., 1999.

1. *Typhlophis squamosus* (Schlegel, 1839 *in* 1837–1844). Abbild. Amph.: 36, pl. 32, figs. 9–12. (*Typhlops squamosus*)

Synonyms: *Cephalolepis leucocephalus* A.-M.-C. Duméril & Bibron, 1844, and *Typhlophis ayarzaguenai* Señaris, 1998.
Type: Holotype, RMNH 3685, a 145 mm specimen (1835).
Type locality: "Cayenne" [= Cayenne, N Cayenne Dept., N French Guiana, 04°55'N, 52°20'W, elevation 10 m].
Distribution: Eastern Venezuela (Bolívar), Guyana (Cuyuni-Mazaruni), Surinam (Nickerie), French Guiana (Cayenne, Saint-Laurent-du-Maroni) and N Brazil (Amazonas, Pará), NSL–650 m.
Sources: J.A. Peters & Orejas-Miranda, 1970, Cunha & Nascimento, 1978, 1993, Chippaux, 1987, Señaris, 1998, Starace, 1998, Abuys, 2003, Kok & Rivas-Fuenmayor, 2008, Navarrete et al., 2009 and C.J. Cole et al., 2013.
Remarks: Hahn (1980a: 5) erroneously reported the type to be lost from MNHN.

TYPHLOPS Oppel, 1811a (Typhlopidae)

Synonyms: *Tiphlops* Rafinesque-Schmaltz, 1815 *(nomen emendatum)*, *Typhlus* Koch *in* Sturm, 1828 *(nomen emendatum)*, *Typhlos* Bonaparte, 1831 *(nomen emendatum)*, *Ophthalmidion* A.-M.-C. Duméril & Bibron, 1844, *Argyrophis* Gray, 1845, *Meditoria* Gray, 1845, *Typhylops* – Bianconi, 1848 *(nomen incorrectum)*,

Thyphlops Gistel, 1849a *(nomen emendatum), Ophtalmidion* – Bianconi, 1849c *(nomen incorrectum), Diaphorotyphlops* Jan, 1860 *in* Jan & Sordelli, 1860–1866, *Diophorotyphlops* – Jan, 1861a *(nomen incorrectum), Tychlpos* – Steindachner, 1867a *(nomen incorrectum), Tyophlops* – Cope, 1887c *(nomen incorrectum), Cyphlops* – Cardew, 1897 *(nomen incorrectum), Typlops* – Bertoni, 1915 *(nomen incorrectum), Tychlops* – Sternfeld, 1917 *(nomen incorrectum), Typhops* – M.A. Smith, 1922b *(nomen incorrectum), Typhlopus* – Nakamura, 1938 *(nomen incorrectum), Ophthalmidium* Loveridge, 1957a *(nomen emendatum), Typhtoelps* – B.D. Sharma & Sharma, 1976 *(nomen incorrectum), Typhlophs* – R.C. Sharma, 1976 *(nomen incorrectum), Typlhops* – J.T.F. Chen & Yu, 1984 *(nomen incorrectum), Tyhhlops* – Mahendra, 1984 *(nomen incorrectum), Tylops* – L.D. Wilson & Meyer, 1985 *(nomen incorrectum), Typholops* – Bhamrah & Juneja, 1989 *(nomen incorrectum), Typhlop* – Nutaphand, 1991a *(nomen incorrectum), Acetyphlops* Hoser, 2012am, *(nomen illegitimum), Altmantyphlops* Hoser, 2012am *(nomen illegitimum), Arnoldtyphlops* Hoser, 2012am *(nomen illegitimum), Carrytyphlopea* Hoser, 2012am *(nomen illegitimum), Copelandtyphlops* Hoser, 2012am *(nomen illegitimum), Crottytyphlops* Hoser, 2012am *(nomen illegitimum), Cyrilhoserus* Hoser, 2012am *(nomen illegitimum), Dannytyphlops* Hoser, 2012am *(nomen illegitimum), Edwardstyphlops* Hoser, 2012am *(nomen illegitimum), Eilliotttyphlopea* Hoser, 2012am *(nomen illegitimum), Freudtyphlops* Hoser, 2012am *(nomen illegitimum), Goldsteintyphlops* Hoser, 2012am *(nomen illegitimum), Katrinahosertyphlops* Hoser, 2012am *(nomen illegitimum), Lenhosertyphlops* Hoser, 2012am *(nomen illegitimum), Mosestyphlops* Hoser, 2012am *(nomen illegitimum), Nintyphlops* Hoser, 2012am *(nomen illegitimum), Rentontyphlops* Hoser, 2012am *(nomen illegitimum), Rolyburrellus* Hoser, 2012am *(nomen illegitimum), Ronhoserus* Hoser, 2012am *(nomen illegitimum), Smythtyphlops* Hoser, 2012am *(nomen illegitimum), Swiletyphlops* Hoser, 2012am *(nomen illegitimum),* and *Woolftyphlops* Hoser, 2012am *(nomen illegitimum).*

Type species: *Anguis lumbricalis* Linnaeus, 1758.

Distribution: Mexico, Latin America, West Indies, SE Europe, Africa, Madagascar, S Asia and East Indies.

Fossil records: Oligocene of USA, Miocene of USA, upper Miocene of Hungary, lower Miocene of Spain; upper Pliocene of Greece, Pleistocene of Bahamas, Cayman Is. and Greece, and Holocene of Antigua and Jamaica. Records for Scolecophidia indet. are listed in Wallach (2002: 114).

Sources: Jan, 1861a, 1864c, de Rooij, 1917, E.H. Taylor, 1922a, F. Werner, 1921a, M.A. Smith, 1943, Roux-Estève, 1974a, 1974b, R. Thomas, 1976, Dixon & Hendricks, 1979, Mahendra, 1984, Hahn, 1980a, Wallach & Ineich, 1996, M.S. Khan, 1999b, McDiarmid et al., 1999, Wallach, 2004, R.W. Henderson & Powell, 2009, Wallach & Glaw, 2009 and Vidal et al., 2010.

Remarks: See Zhao & Adler (1993: 219) and McDiarmid et al. (1999: 89-90) for authorship of the generic name *Typhlops.* The following genera and species have been separated from *Typhlops: Acutotyphlops (banaorum, infralabialis kunuaensis, solomonis, subocularis), Afrotyphlops (angeli, angolensis, bibronii, blanfordii, congestus, elegans, fornasini, gierrai, kaimosae, liberiensis, lineolatus, manni, nanus, nigrocandidus, punctatus, rondoensis, schmidti, steinhausi, tanganicanus, usambaricus), Anilios (affinis, ammodytes, aspina, australis, batillus, bicolor, bituberculatus, broomi, centralis, chamodracaena, diversus, endoterus, erycinus, ganei, grypus, guentheri, hamatus, howi, kimberleyensis, leptosomus, leucoproctus, ligatus, longissimus, margaretae, micrommus, minimus, nema, nigrescens, nigricaudus, nigroterminatus, pilbarensis, pinguis, polygrammus, proximus, robertsi, silvia, splendidus, tovelli, troglodytes, unguirostris, waitii, wiedii, yampiensis, yirrikalae), Grypotyphlops (acutus), Letheobia (acutirostrata, ataeniata, caeca, crossii, debilis, decorosa, episcopus, erythraea, feae, gracilis, graueri, jubana, kibarae, largeni, leucosticta, lumbriciformis, newtoni, obtusus, pallida, pauwelsi, pembana, praeocularis, rufescens, scorteccii, simoni, somalica, stejnegeri, sudanensis, swahilica, toritensis, uluguruensis, unitaeniata, wittei), Megatyphlops (anomalus, brevis, mucruso, schlegelii), Ramphotyphlops (acuticaudus, albiceps, angusticeps, becki, braminus, cumingii, depressus, exocoeti, flaviventer, lineatus, lorenzi, macrurus, mansuetus, marxi, multilineatus, olivaceus, ozakiae, similis, suluensis, supranasalis, willeyi),* and *Xenotyphlops (grandidieri).*

1. *Typhlops agoralionis* R. Thomas & Hedges, 2007. Zootaxa (1400): 9–10, fig. 3c.

Type: Holotype, USNM 564777 (formerly USNM-FN 191925), a 139–193 mm specimen (S.B. Hedges, M. Leal, N. Plummer & R. Thomas, 28 May 1991).

Type locality: "8.0 km S Marché Léon, 435 m, Dépt. de la Grande Anse, Haiti."

Distribution: Extreme SW Haiti (Grand'Anse), 435 m. Known only from type locality.

Source: R.W. Henderson & Powell, 2009.

2. *Typhlops albanalis* Rendahl, 1918. Arkiv. Zool. 11(17): 3–5, figs. 4–6.

Type: Holotype, NHR 2574 (formerly NRM ex. A), a 276 mm specimen (W. Kauden, Nov. 1906).

Type locality: "Kaplande" [= South Africa] (in error). Corrected to Madagascar *fide* V.F.W. FitzSimons (1962: 63).

Distribution: Unknown (not Africa or Madagascar), but possibly Indonesia.

Remarks: A synonym of *Typhlops ocularis fide* Roux-Estève (1974: 250). Based upon Kaudern's label, which came from his collection in Madagascar in 1906, it was assumed the origin was Madagascar. However, neither Andersson, 1911, who reported on Kaudern's Malagasy snake collection from 1906–1907, nor Kaudern, 1922, who reported on his 1906–07 and 1911–12 Malagasy collections, mentioned this species.

3. *Typhlops amoipira* Rodrigues & Junca, 2002. Pap. Avul. Zool. 42(13): 327–330, figs. 1–3.

Type: Holotype, MZUSP 12298 (formerly MZUSP-FN 200170), a 208 mm specimen (F.A. Juncá, 5 March 2000).
Type locality: "sand-dune area of the left bank of Río São Francisco, Ibiraba (10°48'S, 42°50'W): state of Bahía: Brazil," (partly in error). Corrected to 11°06'06"S, 43°09'07"W, elevation 435 m *fide* Fernandes et al. (2010: 268).
Distribution: Eastern Brazil (W Bahía, NW Minas Gerais), 425–550 m.
Source: Fernandes et al., 2010.

4. *Typhlops anchaurus* R. Thomas & Hedges, 2007. Zootaxa (1400): 22–23, fig. 8j.

Type: Holotype, MNHNCU 4553 (formerly USNM-FN 190563), a 240 mm specimen (R. Thomas, 31 July 1989).
Type locality: "within 1 km of Cueva de Agua (2.5 km airline NW Maisí), Guantánamo Province, Cuba" [= 20°15'N, 74°10'W, elevation 10 m].
Distribution: Extreme E Cuba (Guantánamo), NSL–10 m. Known only from type locality.
Source: Domínguez & Moreno, 2009.

5. *Typhlops andasibensis* Wallach & Glaw, 2009. Zootaxa (2294): 24–28, fig. 3, figs. 1–2, 4 (paratype).

Type: Holotype, ZSM 54/2005, a 325 mm female, (NGO "Mitsinjo" & R. Dolch, 23–26 Feb. 2005).
Type locality: "near Andasibe (= Perinet), 18°56.169'S, 48°24.734'E, 939 m elevation, central eastern Madagascar."
Distribution: Eastern Madagascar (Toamasina), 940 m. Known only from type locality.

T

6. *Typhlops annae* Breuil, 1999. Bull. Mens. Soc. Linn. Lyon 68(2): 32–34, fig. 1.

Type: Holotype, MNHN 1997.6058, a 110 mm specimen (M. Breuil, 12 Aug. 1996).
Type locality: "sud du Morne Vitet, à 100 m à l'ouest du point de vue, à 5 m au-dessus du niveau de la mer, Saint-Barthélemy, département de la Guadeloupe, Petites Antilles" [= S of Morne de Vitet, Saint Barthélemy Is., Lesser Antilles, 17°54'N, 62°48'W, elevation 5 m].
Distribution: Lesser Antilles (Guadeloupe: St-Barthélemy Is.), NSL. Known only from type locality.
Sources: Breuil, 2002, 2003.

7. *Typhlops anousius* R. Thomas & Hedges, 2007. Zootaxa (1400): 20, fig. 8g, 9c.

Type: Holotype, MNHNCU 4550 (formerly USNM-FN 191295), a 109–197 mm female (S.B. Hedges, R. Thomas & D. McCallister, 28 June 1990).
Type locality: "3.5 km E Tortuguilla, Guantánamo Province, Cuba, 10 meters elevation" [= 19°59'N, 74°56'W].
Distribution: Extreme E Cuba (Guantánamo), 10 m. Known only from type locality.
Source: Domínguez & Moreno, 2009.

8. *Typhlops arator* R. Thomas & Hedges, 2007. Zootaxa (1400): 19–20, figs. 8f, 9b.

Type: Holotype, MNHNCU 191, a 460 mm female (R. Regalado & C. Wotzkow, 20 April 1988).
Type locality: "'at El Narigon and La India', near Puerto Escondido, La Habana, Province, Cuba" [= 23°09'N, 81°43'W, elevation NSL].
Distribution: Northwestern Cuba (Habana), NSL. Known only from type locality.
Source: Domínguez & Moreno, 2009.

9. *Typhlops arenarius* (Grandidier, 1872). Ann. Sci. Nat. (Zool.) (5) 15(20): 9. (*Onychocephalus arenarius*)

Type: Holotype, MNHN 1895.216, a 180 mm specimen (A. Grandidier, 1870).
Type locality: "Mouroundava, Madagascar" [= Morondava, Toliara Prov., SW Madagascar, 20°17'S, 44°19'E, elevation 5 m].
Distribution: Western Madagascar (Mahajanga, Toliara), NSL–985 m.
Sources: Guibe, 1958, Glaw & Vences, 1994, A. Mori et al., 2006 and Wallach & Glaw, 2009.

10. *Typhlops biminiensis* Richmond, 1955. Amer. Mus. Novit. (1734): 2–4, fig. 1.

Type: Holotype, CM 32604, a 257 mm specimen (M.G. Netting & N.D. Richmond, 27 March 1954).
Type locality: "near Nixon's Harbor, along trail to 'Buck Lands', South Bimini Islands, Bahama Islands" [= between Nixon's Harbor and "Black Lands," South Bimini Is., Bahamas, ca. 25°43'N, 79°18'W, elevation NSL].
Distribution: Bahamas (Andros, Berry, Elbow Cay, Little Ragged, New Providence, North Bimini, South Bimini), NSL–70 m.
Sources: R. Thomas, 1968, A. Schwartz & R.W. Henderson, 1991 and R. Thomas & Hedges , 2007.

11. *Typhlops bipartitus* Sauvage, 1879. Bull. Soc. Philom. Paris (7) 3: 59–60.

Types: Syntypes (2), MNHN 1878.214–15, a 250 and 175 mm specimen (M.J.A. Raffray & M. Maindron, Dec. 1876).
Type locality: "'ile de Tidore, Nouvelle-Guinée" [= Tidore Is., Maluku, Indonesia, 00°41'N, 127°24'E].
Distribution: Eastern Indonesia (Maluku: Tidore). Known only from type locality.
Sources: Boulenger, 1993a and Rooij, 1917.
Remarks: Possibly a synonym of *Ramphotyphlops polygrammicus fide* Wallach *in* McDiarmid et al. (1999: 93). Possibly occus in Papua, Indonesia *fide* Iskandar & Colijn (2001: 17).

12. *Typhlops boettgeri* Boulenger, 1893a. Cat. Snakes Brit. Mus. 1: 39, pl. 2, figs. 6a–c.

Types: Syntypes (6), BMNH 1946.1.11.98–99 and BMNH 1946.1.12.1–4 (formerly BMNH 1892.11.26.1–6), longest syntype 220 mm (Last).
Type locality: "S. W. Madagascar."
Distribution: Southwestern Madagascar (Toliara), NSL–1040 m.
Sources: Guibe, 1958 and Wallach & Glaw, 2009.
Remarks: Previously considered a synonym of *T. arenarius.*

13. *Typhlops bothriorhynchus* A.C.L.G. Günther, 1864a. Rept. Brit. India: 174, pl. 16, fig. g.

Type: Holotype, BMNH 1946.1.11.59 (formerly BMNH 1860), a 180 mm specimen (T. Cantor, 1834–1855).
Type locality: "Pinang" [= Penang Is., Pinang State, West Malaysia, 05°22'N, 100°14'E].
Distribution: Southeastern Asia. Northeastern India (Assam), N Myanmar (Shan), West Malaysia (Pinang) and W Indonesia (Java), NSL–1510 m.
Sources: Boulenger, 1893a, Wall, 1923c, I. Das, 1996 and Schleich & Kästle, 2002.
Remarks: Type locality listed as Pinang, Malaya *fide* BMNH catalogue but Wall (1923c: 350) and I. Das (2012: 153) questioned type locality. Possibly occurs in Nepal *fide* Schleich & Kästle, 2002.

14. *Typhlops brongersmianus* Vanzolini, 1976. Pap. Avul. Zool. 29(24): 247. (*nomen substitutum*)

Synonyms: *Typhlops brongersmai* Vanzolini, 1972 *(nomen praeoccupatum), Altmantyphlops kirnerae* Hoser, 2012am *(nomen illegitimum),* and *Altmantyphlops kirnerae wellingtoni* Hoser, 2012am *(nomen illegitimum).*
Type: Holotype, MZUSP 5218, a 214 mm specimen (E.E. Williams & P.E. Vanzolini, 14 Dec. 1968).
Type locality: "Barra de Itaipe, Ilheus, Bahia, Brasil" [= Barra de Itaipe, Bahia Prov., E Brazil, 14°49'S, 39°02'W, elevation NSL].
Distribution: South America. Southeastern Colombia (Amazonas, Meta, Santa Sophia Is.), Venezuela (Delta Amacuro, Monagas, Sucre), Trinidad and Tobago (N Trinidad), Guyana (Upper Demerara-Berbice), Suriname (Saramacca), Brazil (Bahia, Ceará, Espírito Santo, Goias, Mato Grosso, Mato Grosso do Sul, Minas Gerais, Pará, Paraíba, Pernambuco, Rio de Janeiro, Rio Grande do Sul, São Paulo, Sergipe), Ecuador, NE Peru (Loreto), Bolivia (Beni, La Paz, Potosí, Santa Cruz), Paraguay (Alto Paraná, Amambay, Asunción, Caaguazú, Caazapá, Canindeyú, Central, Concepción, Cordillera, Guairá, Itapúa, Misiones, Neembucú, Paraguarí, Presidente Hayes, San Pedro) and Argentina (Buenos Aires, Chaco, Córdoba, Corrientes, Entre Ríos, Formosa, Misiones, Salta, Santa Fe, Tucumán), NSL–835 m.
Sources: Vanzolini, 1972, Dixon & Hendricks, 1979, Lema, 1982, Dixon et al., 1993, Cunha & Nascimento, 1993, Fugler et al., 1995, J.C. Murphy, 1997, Freitas, 1999, Leynaud & Bucher, 1999, Boos, 2001, Giraudo, 2001, Giraudo & Scrocchi, 2002, Cacciali, 2008, Loebmann, 2008b, Navarrete et al., 2009, Martins et al., 2010, Arruda et al., 2011, Caicedo-Portilla, 2011 and Morato et al., 2011 and C.J. Cole et al., 2013.
Remarks: Original description in Vanzolini (1972: 27–29). La Rioja, Argentina record rejected *fide* Giraudo & Scrocchi (2002: 4).

15. *Typhlops calabresii* Gans & Laurent, 1965. Ann. Mus. Roy. Afr. Cen., Tervuren (134): 54–55, pl. 6. (*Typhlops cuneirostris calabresii*)

Type: Holotype, MCZ 74465, a 147 mm specimen (C.F. Hemming, Dec. 1961).
Type locality: "56 miles from Hargeisa, on Berbera road, Northern Territory, Somalia Republic" [= mile 56 on Hargeisa-Berbera road, Woqooyi Galbeed Region, NW Somalia, 44°35'E, 09°52'N, elevation 945 m].
Distribution: Northeastern Africa. Southeastern Ethiopia (Bale, Hararage, Sidamo), Somalia (Bay, Mudug, Woqooyi Galbeed), and Kenya (Coast, North-Eastern), NSL–1065 m.
Sources: Spawls et al., 2002, Broadley & Wallach, 2009 and Largen & Spawls, 2010.

16. *Typhlops canlaonensis* E.H. Taylor, 1917. Philippine J. Sci. 12D(6): 354–355.

Type: Holotype, CM 2666 (formerly EHT 241), a 122 mm specimen (E.H. Taylor, 25 Dec. 1915).
Type locality: "Canlaon Volcano, Negros, P. I; elevation about 750 meters" [= Mt. Kanlaon, Negros Occidental Prov., Negros Is., Philippines, 10°25'N, 123°08'E, elevation 750 m].
Distribution: Central Philippines (Negros), 750 m.

17. *Typhlops capitulatus* Richmond, 1964. Breviora (202): 2–3, fig. 1.

Synonym: *Typhlops capitulata* R. Thomas, 1966a.
Type: Holotype, MCZ 62636, a 205 mm specimen (A.S. Rand & J. D. Lazell, 10 Aug. 1960).
Type locality: "Manneville, Haiti, at the northwest end of Lake Saumâtre" [= Ouest Dept., SW Haiti, 18°39'N, 72°03'W, elevation 15 m].
Distribution: Southwestern Haiti (Ouest, Sud), NSL–40 m.
Sources: Richmond, 1964, R. Thomas, 1965b, 1974b, 1989, A. Schwartz & R.W. Henderson, 1988, 1991 and R. Thomas & Powell, 1995a.

18. †*Typhlops cariei* Hoffstetter, 1946. Bull. Mus. Natl. Hist. Nat. (2) 18(3): 313–314, fig. 2.

Types: Syntypes (7), MNHN MAU 1000, seven trunk vertebrae (P. Carié).
Type locality: "la Mare aux Songes (Ile Maurice), subfossile" [= La Mare aux Songes, Mon Trésor Sugar Estate, SE Mauritius Is., Indian Ocean].
Distribution: Holocene (0.001–0.01 mya) of Mauritius. Known only from type locality.
Source: Rage, 1984a.

19. *Typhlops castanotus* Wynn & Leviton, 1993. Proc. Biol. Soc. Washington 106(1): 35–41, figs. 1–2.

Type: Holotype, CAS-SU 27940 (formerly SU 27940), a 224 mm male (A.C. Acala & crew, 23 May 1967).
Type locality: "8 km west of Pulupandan, Inampulugan Island, Negros Occidental Province, Philippines."
Distribution: Central Philippines (Borocay, Inampulugan, Negros, Panay), NSL–175 m.
Source: Ferner et al., 2001.

20. *Typhlops catapontus* R. Thomas, 1966b. Rev. Biol. Trop. (1965) 13(2): 190–191, fig. 4d (paratype). (*Typhlops richardi catapontus*)

Type: Holotype, MCZ 77220, a 265 mm specimen (R. Thomas, 18 Aug. 1964).
Type locality: "island of Anegada, British Virgin Islands, in the vicinity of the Settlement" [= Anegada Is., British Virgin Is., 18°43'N, 64°19'W, elevation NSL].
Distribution: British Virgin Islands (Anegada, Necker), NSL–245 m.
Source: Lazell, 2006.

T

21. *Typhlops caymanensis* Sackett, 1940b. Not. Nat. (48): 1–2.

Synonym: *Typhlops richardi capapontus* – Hahn, 1980a *(nomen incorrectum).*
Type: Holotype, ANSP 22123, a 180 mm specimen (J.T. Sackett, 26 Feb. 1940).
Type locality: "between Pedro Point and North Sound, Grand Cayman Island, West Indies" [= ca. 19°18'N, 81°14'W, elevation NSL].
Distribution: Caymans (Grand Cayman Is.), NSL.
Sources: R. Thomas, 1968, A. Schwartz & R.W. Henderson, 1991 and R. Thomas & Hedges, 2007.

22. *Typhlops coecatus* Jan, 1863b. Elenco Sist. Ofidi: 10.

Synonyms: *Typhlops caecatus* Jan, 1861b *(nomen nudum),* and *Typhlops caecatus* Jan, 1864 *(nomen emendatum).*
Type: Holotype, NMBA 310.
Type locality: "Costa d'oro" [= Gold Coast, presumably Ghana].
Distribution: Southern Ghana (Eastern, Volta) and S Ivory Coast (Aboisso, Dabakala, Toumodi), NSL–590 m.
Sources: Angel, 1933, Leston & Hughes, 1968, Hughes & Barry, 1969, Roux-Estève, 1974, Kramer, 1978, Hughes, 1983 and Wallach, 1993b.
Remarks: For disscussion of the name *coecatus* see McDiarmid et al. (1999: 96).

23. *Typhlops collaris* Wynn & Leviton, 1993. Proc. Biol. Soc. Washington 106(1): 41–44, fig. 3.

Type: Holotype, UF 55123, a 226 mm male (W. Auffenberg, 29 July 1982).
Type locality: "Anuling Mt. (150 m elev.), Caramoan Municipality, Camarines Sur, Luzon Island, Philippines."
Distribution: Northern Philippines (SE Luzon), 150–250 m.
Source: Auffenberg, 1988.

24. *Typhlops comorensis* Boulenger, 1889b. Ann. Mag. Nat. Hist. (6) 4(23): 361–362.

Type: Holotype, BMNH 1946.1.11.92 (formerly BMNH 1879.11.12.7), a 245 mm male (J. Kirk, Feb.–Nov. 1861).
Type locality: "Comoro Islands" [= ca. 11°42'S, 43°20'E, Indian Ocean].
Distribution: Comoros (Grande Comore Is.).
Sources: Boulenger, 1983a, Wallach, 1993b, 2009 and Wallach & Glaw, 2009.

25. *Typhlops conradi* W.C.H. Peters, 1874b. Mber. Kønigl. Akad. Wiss. Berlin 1874(2): 162–163, pl., fig. 1a–c.

Synonym: *Typhlops conradti* – Boulenger, 1897g *(nomen incorrectum).*
Type: Holotype, ZMB 7934, a 175 mm specimen (Conrad via Mus. Godeffroy).
Type locality: "Nord-Celebes" [= N Sulawesi, cen. Indonesia].
Distribution: Central Indonesia (ext. NE Sulawesi).
Sources: McDowell, 1974a, Bosch & Ineich, 1994, Wallach, 1993b, Bauer et al., 2002, Lang & Vogel, 2005 and Koch, 2012.

Remarks: Holotype illustrated in Bosch & Ineich (1994: fig. 3), who incorrectly listed holotype as ZMB 7834.

26. *Typhlops contorhinus* R. Thomas & Hedges, 2007. Zootaxa (1400): 21–22, figs. 8i, 9f.

Type: Holotype, MNHNCU 4552 (formerly USNM-FN 191296), a 316 mm male (R. Thomas, S. B. Hedges, E. Alfaro & D. McCallister, 29 June 1990).
Type locality: "4.9 km S La Tinta, Guantánamo Province, Cuba, 5 meters elevation" [= 20°05'N, 74°20'W].
Distribution: Extreme SE Cuba (Oriente), NSL. Known only from type locality.
Source: Domínguez & Moreno, 2009.

27. *Typhlops costaricensis* Jiménez & Savage, 1963. Rev. Biol. Trop. (1962) 10(2): 199–201.

Type: Holotype, LACM 26767 (formerly MNCR 1960), a 369 mm specimen (J. Walter, 6–13 March 1960).
Type locality: "Quaker colony at Monteverde, Sierra de Tilarán, Provincia de Puntarenas, Costa Rica, elevation 1500 m" [= 10°18'N, 84°48'W].
Distribution: Central America. Southern Honduras (Francisco Morazán, Gracias a Dios), Nicaragua (Chontales, Matagalpa) and N Costa Rica (Alajuela, Puntarenas), 150–1600 m.
Sources: Dixon & Hendricks, 1979, Vanzolini, 1986, Villa, 1988b, Villa et al., 1988, Savage, 2002, Solórzano, 2004, McCranie et al., 2006, Townsend et al., 2008 and McCranie, 2011a.
Remarks: Holotype is on permanent loan to LACM. Hahn (1980a: 55) erroneously reported the type on permanent loan to USC.

28. *Typhlops cuneirostris* W.C.H. Peters, 1879. Mber. Königl. Akad. Wiss. Berlin 1879(8): 775–776, pl., figs. 4, 4a–4c.

Type: Holotype, ZMB 9531, a 146 mm specimen (G.A. Fischer).
Type locality: "Barawa im Somalilande" [= Brava, Shabeellaha Hoose Region, S Somalia, 01°07'N, 44°02'E, elevation 15 m].
Distribution: Southern Somalia (Mogadishu, Shabeellaha Hoose), NSL–200 m.
Sources: Lanza, 1983a, 1990b and Broadley & Wallach, 2009.

29. *Typhlops decorsei* Mocquard, 1901. Bull. Mus. Hist. Nat., Paris 7(6): 255.

Type: Holotype, MNHN 1901.242, a 455 mm specimen (G.-J. Decorse, 1898–1901).
Type locality: "Ambovombé, Madagascar" [= Ambovombe, SE Toliara Prov., S Madagascar, 25°11'S, 46°05'E, elevation 130 m].
Distribution: Western and S Madagascar (Mahajanga, Toliara), 10–200 m.
Sources: Guibé, 1958, Glaw & Vences, 1994, A. Mori et al., 2006 and Wallach & Glaw, 2009.

30. *Typhlops diardii* Schlegel, 1839 *in* 1837–1844. Abbild. Amph.: 39.

Synonyms: *Argyrophis horsfieldii* Gray, 1845, *Typhlops striolatus* W.C.H. Peters, 1861e, *Typhlops barmanus* Stoliczka, 1872c, *Typhlops kapaladua* Annandale, 1906, *Typhlops tephrosoma* Wall, 1908c, *Typhlops diardi cinereus* Wall, 1909f *(nomen praeoccupatum)*, *Typhlops labialis* Waite, 1918a, and *Typhlops diardi platyventris* M.S. Khan, 1998.
Type: Holotype, MNHN 1065, a 372–420 mm specimen (P.-M. Diard, 1824–1825).
Type locality: "Cochinchina" [= S Vietnam]. Designated as Indes orientales [= Southern Asia and East Indies] *fide* A.-M.-C. Duméril & Bibron (1844: 301).
Distribution: Southeastern Asia. Northeastern Pakistan (Jammu & Kashmir), India (Arunachal Pradesh, Assam, Manipur, Meghalaya, Mizoram, Nagaland, Orissa, Tripura, Uttar Pradesh, West Bengal), S China (Hainan, Yunnan), S Nepal (Bardiya, Chitwan), Bangladesh, N Myanmar, Thailand (Chaiyaphum, Chanthaburi, Chiang Mai, Chon Buri, Krung Thep Mahanakhon, Loei, Nakhon Panom, Nakhon Ratchasima, Narathiwat, Prachin Buri, Tak, Trang, Uthai Thani), Laos (Houaphan, Xiangkhouang), Cambodia (Pursat) and N Vietnam (Bac Giang, Gia Lai, Ha Tinh, Lam Dong, Nghe An, Quang Binh, Quang Nak, Quang Ninh, Tay Ninh, Thanh Hoa, Cu Lao Ba Mun and Cu Lao Cham Is.), 140–1525 m.
Sources: Wall, 1918a, E.H. Taylor, 1965, Deuve. 1970, McDowell, 1974a, Tweedie, 1983, Zhao & Adler, 1993, M.J. Cox et al., 1998, M.S. Khan, 1998, 2002, 2006, Chan-ard et al., 1999, Darevsky, 1999, Nabhitabhata, 2000, Wallach, 2001, 2004, Schleich & Kästle, 2002, Shah & Tiwari, 2004, Whitaker & Captain, 2004, Ao et al., 2005, Nabhitabhata & Chan-ard, 2005, Grismer et al., 2008a, M.F. Ahmed et al., 2009, V.S. Nguyen et al., 2009, I. Das, 2010, 2012 and Haitao, 2011.
Remarks: BMNH 1946.1.11.50 from India listed as type *fide* Cox et al. (2013: 19) and BMNH catalogue.

31. *Typhlops domerguei* Roux-Estève, 1980. Bull. Mus. Natl. Hist. Nat. (4) 2A(1): 321–323, fig. 1.

Type: Holotype, MNHN 1977.1650 (Brunhes, Feb. 1967).
Type locality: "Ranomafana (47°26'E-21°18'S), village situé à 50 km à l'est de Fianarantsoa sur la route de Fianarantsoa à Mananjary, Madagascar" [Fianarantsoa Prov., elevation 625 m].
Distribution: Eastern Madagascar (Fianarantsoa), 455–625 m.
Sources: Glaw & Vences, 1994, Wallach, 1993b and Wallach & Glaw, 2009.

32. *Typhlops dominicanus* Stejneger, 1904. Rep. U.S. Natl. Mus. 1902: 687. (*Typhlops dominicana*)

Synonym: *Typhlops platycephalus* Boulenger, 1893a *(nomen praeoccupatum).*
Types: Syntypes (10), BMNH 1865.5.4.177, BMNH 1898.8.14.1–1898.8.14.1–8 (G.A. Ramage, June 1889), and BMNH 1891.5.11.2 (H A.A. Nichols).
Type locality: "island of Dominica" [Lesser Antilles].
Distribution: Dominica (Saint David, Saint Joseph, Saint Mark, Saint Patrick, Saint Peter).
Sources: Boulenger, 1893a, Richmond, 1966, A. Schwartz & R.W. Henderson, 1991, Malhotra & Thorpe, 1999, Breuil, 2002 and Rudman & Powell, 2010.
Remarks: Record from Barbados is doubtful *fide* Richmond (1966: 1310).

33. *Typhlops epactius* R. Thomas, 1968. Copeia 1968(4): 715–717, fig. 2c. (*Typhlops biminiensis epactia*)

Type: Holotype, MCZ 92048, a 209 mm male (R. Thomas, 24 March 1967).
Type locality: "5.4 miles E of West End, southern coastal platform, Cayman Brac, British West Indies" [= ca. 19°43'N, 79°48'W, elevation NSL],
Distribution: Caymans (Cayman Brac), NSL.
Sources: Richmond, 1955, R. Thomas, 1989, A. Schwartz & R.W. Henderson, 1991 and R. Thomas & Hedges, 2007.
Remarks: A valid species *fide* R. Thomas (1989: 409).

34. *Typhlops eperopeus* R. Thomas & Hedges, 2007. Zootaxa (1400): 12–13, figs. 3e, 4b.

Type: Holotype, USNM 564785 (formerly USNM-FN 266250), a 140–281 mm female (R. Thomas, 30 July 1999).
Type locality: "3.9 km airline SSW Barahona (4.5 km S Barahona along coast road and 2.8 km inland), 18°9.854' N, 71°5.497' W, 305 m."
Distribution: Southwestern Dominican Republic (Barahona, Independencia), NSL–700 m.

T

35. *Typhlops etheridgei* Wallach, 2002. Hamadryad 27(1): 109–111, fig. 1.

Type: Holotype, MHNG 1326.62 (formerly EK 11443), a 220 mm female (J. Garzoni, 1967).
Type locality: "Between Atar (20°31'N, 13°03'W) and Choûm (21°20'N, 12°59'W) along Route Nationale 1, a distance of ca. 85 air km through the Amsâga region, western Adrar Region, west-central Mauritania, elevation ca. 200 m."
Distribution: Western Mauritania (Adrar), 200 m. Known only from type locality.

36. *Typhlops exiguus* Jan, 1864 *in* Jan & Sordelli. 1860–1866. Icon. Gén. Ophid. 1(3): 3, pl. 4, fig. 11, pl. 5, figs. 11a–c, g, n, p, r, u, x.

Synonym: *Typhlops exignus* – F. Werner, 1921a *(nomen incorrectum).*
Type: Holotype, SMNS, a 135 mm specimen, destroyed in Sept. 1944 during World War II.
Type locality: "Indies orientales" [Southern Asia and East Indies].
Distribution: Southwestern India (Karnataka), 765 m.
Sources: Jan, 1854c and Wallach, 2004.

37. *Typhlops filiformis* A.-M.-C. Duméril & Bibron, 1844. Erpét. Gén. 6: 307–309.

Type: Holotype, MNHN 929, a 121–135 mm specimen.
Type locality: Unknown.
Distribution: Probably Asia.
Source: Jan, 1864.
Remarks: Listed as *incertae sedis* by Hahn (1980a: 76) and McDiarmid et al. (1999: 125). Either a valid member of the *Typhlops porrectus* group or a senior synonym of *T. porrectus* (Wallach, unpubl. data). Currently being studied by A.H. Wynn.

38. *Typhlops fuscus* (A.-M.-C. Duméril & A.H.A. Duméril, 1851). Cat. Méth.: 203. (*Ophthalmidion fuscum*)

Type: Holotype, MNHN 1062, a 210–214 mm specimen (S. Müller, 1825–1827 or 1831–1832).
Type locality: "Java" [SW Indonesia].
Distribution: Indonesia (Java, Maluku).
Source: Bleeker, 1858f.
Remarks: A valid species *fide* Wallach (unpubl. data). A synonym of *Typhlops platycephalus* (= *T. dominicana*) *fide* Boulenger (1893a: 30). Hahn (1980a: 76) and McDiarmid et al. (1999: 126) listed as *incertae sedis.*

39. *Typhlops geotomus* R. Thomas, 1966a. Proc. Biol. Soc. Washington 79: 260–261. (*Typhlops monastus geotomus*)

Type: Holotype, MCZ 81115, a 196 mm specimen (R. Thomas, 8 Aug. 1965).
Type locality: "Approximately 1 mi. N Carlisle, St. Mary's Parish, Antigua, British West Indies" [ca. 17°02'N, 61°50'W, elevation 50 m].
Distribution: Lesser Antilles. Antigua (Saint George, Saint Mary, Saint Paul, Great Bird Is.), Barbuda, St. Kitts & Nevis (Nevis: Christ Church Nichola Town, Saint Thomas Middle Island, St. Kitts: Saint George Gingerland, Saint James Windward), NSL–50 m.
Sources: Malhotra & Thorpe, 1999 and R.W. Henderson & Powell, 2009.

40. *Typhlops giadinhensis* Bourret, 1937b. Bull. Gén. Instr. Pub. Hanoi 5(4): 68–69, figs. 7a–c.

Type: Holotype, MNHN 1938.118 (formerly RLB M.559) (Inst. Pasteur de Saigon).
Type locality: "Giadinh, (Colchinchine)" [= Ho Chi Minh City, Ho Chi Minh City Prov., S Vietnam, 10°48'N, 106°38'E, elevation NSL].
Distribution: Southern Vietnam (Ho Chi Minh City), NSL. Known only from type locality.
Source: Wallach, 2001.
Remarks: A synonym of *Typhlops diardii fide* V.S. Nguyen (2007: 28) and V.S. Nguyen et al. (2009: 281); however, it differs from *T. diardii* in having 22 scale rows (vs. 24–48), a narrow rostral 0.3 head width (vs. 0.4–0.5), an L/W ratio of 26 (vs. 29–48), and a pair of longitudinal dorsal stripes (vs. unstriped).

41. *Typhlops golyathi* Domínguez & Moreno, 2009. Zootaxa (2028): 61–63, fig.. 1.

Type: Holotype CZACC 4.5385, a 371 mm male (R. Alonso, April 2003).
Type locality: "Valle de San Vicente, Viñales Municipality, Pinar del Río Prov., Cuba, 83°43'00" N, 22°41'00" W, 112 m elevation."
Distribution: Western Cuba (Pinar del Río), 110 m.

42. *Typhlops gonavensis* Richmond, 1964. Breviora (202): 3–4, fig. 2.

Type: Holotype, YPM 3003, a 189 mm specimen (P.S. Humphrey & S. Van Vleek, 9 April 1959).
Type locality: "Point à Raquette, on the south shore of Gonave Island, Haiti" [= 18°47'N, 73°04'W, elevation 18 m].
Distribution: Western Haiti (Gonâve Is.), NSL. Known only from type locality.
Sources: R. Thomas, 1989 and R. Thomas & Powell, 1995b.
Remarks: Previously considered a subspecies of *T. capitulatus*.

43. *Typhlops granti* Ruthven & Gaige, 1935. Occ. Pap. Mus. Zool., Univ. Michigan (307): 2–3, fig. 1 (paratype).

Type: Holotype, UMMZ 76669, a 154 mm specimen (C. Grant, 20 April 1931).
Type locality: "Caja de Muertos (eight miles off Ponce, Puerto Rico)" [= 17°54'N, 66°31'W, elevation NSL].
Distribution: Southwestern Puerto Rico (Lajas, Ponce, Caja de Muertos Is.).
Sources: Hedges & R. Thomas, 1991 and Rivero, 1998.

44. †*Typhlops grivensis* Hoffstetter, 1946. Bull. Mus. Natl. Hist. Nat., Paris (2) 18(3): 311–313, fig. 1.

Types: Syntypes (3), MNHN LGA 53–55, three trunk vertebrae (R. Hoffstetter).
Type locality: "le Vindobonien moyen de la Grive-Saint-Alban (Isère), France," Astaraciàn, middle Miocene."
Distribution: Middle Miocene (Astaracian, MP 6–8: 11.1–13.7 mya) of France. Known only from type series.
Sources: Rage, 1984 and Wallach, 2002.

45. *Typhlops guadeloupensis* Richmond, 1966. Herpetologica 22(2): 129–130, fig. 1.

Synonym: *Typhlops cinereus* J.G. Schneider, 1801.
Type: Holotype, CM 41216 (formerly ASFS X4988), a 162 mm specimen (A. Schwartz, 19 Jan. 1963).
Type locality: "Guadelupe, from Grandterre part of the island, 2 km. SW Port-Blanc" [= 16°14'N, 61°27'W, elevation 50–75 m].
Distribution: Lesser Antilles. Guadeloupe, 50–75 m.
Source: Breuil, 2002.

46. *Typhlops hectus* R. Thomas, 1974a. Proc. Biol. Soc. Washington 87(2): 12–14.

Type: Holotype, MCZ 81149 (formerly ASFS V9145), a 192 mm specimen (R. Thomas, 12 March 1966).
Type locality: "Martineau, ca. 9 km (airline) W Jérémie, Dépt. du Sud, Haiti" [= Grand'Anse Dept., ca. 18°39'N, 74°07'W, elevation 10 m].
Distribution: Southern Haiti (Grand'Anse, Sud, Gônave Is.), NSL–870 m.
Sources: R. Thomas, 1976, 1989, R. Thomas & Powell, 1992 and R. Thomas & Hedges, 2007.

47. *Typhlops hypogius* Savage, 1950. Proc. California Zool. Club 1(10): 52–53, figs. 1 a–c. (*Typhlops hypogia*)

Type: Holotype, CAS-SU 12347 (formerly SU 12347), a 176 mm specimen (A.W. Herre, 23 Aug. 1940).
Type locality: "Cebu, Cebu, Philippine Islands" [= Cebu City, Cebu Is., cen. Philippines, 10°22'N, 123°52'E, elevation 370 m].
Distribution: Central Philippines (Cebu), 370 m.
Sources: McDowell, 1974b, Alcala, 1986 and Ferner et al., 2001.
Remarks: Illustrations of holotype mislabeled as figs. 2a–c in Savage (1950: 49) (but corrected in reprint).

48. *Typhlops hypomethes* Hedges & Thomas, 1991. Herpetologica 47(4): 452–454, figs. 4b, 6 (upper left, lower right).

Type: Holotype, USNM 300574, a 201 mm female (R. Thomas, 19 Nov. 1986).

Type locality: "the University of Puerto Rico Campus (faculty housing), Río Piedras, San Juan, Puerto Rico" [= 18°24'N, 66°03'W, elevation 25 m].

Distribution: Eastern Puerto Rico (Arecibo, Barceloneta, Cataño, Cidra, Fajardo, Guayama, Loiza, Manati, Maulnabo, Río Grande, San Juan, Santa Isabel, Vega Baja, Palominos, Piñeros and Vieques Is.), NSL–435 m.

Sources: Hedges & Thomas, 1991 and Rivero, 1998.

49. *Typhlops hypsobothrius* F. Werner, 1917a. Mitt. Zool. Mus. Hamburg (1916) 34: 34–35.

Types: Syntypes (2), ZMH, longest syntype 285 mm, destroyed in July 1943 during World War II.

Type locality: "Sumatra" [W Indonesia].

Distribution: Western Indonesia (Sumatra). Known only from syntypes.

Sources: Haas, 1950, David & Vogel, 1996 and Manthey & Grossmann, 1997.

50. *Typhlops jamaicensis* (G. Shaw, 1802). Gen. Zool., Amph. 3(2): 588, pl. 133. (*Anguis jamaicensis*)

Synonyms: *Meditoria nasuta* Gray, 1845, and *Anilios leachii* Gray, 1845.

Type: Holotype, not designated, location unknown.

Type locality: "Jamaica" [Greater Antilles, West Indies].

Distribution: Jamaica (Clarendon, Kingston and St. Andrews, Manchester, St. Ann, St. Elizabeth, St James, St. Mary, St. Thomas, Portland, Trelawny, Westmoreland), NSL–700 m.

Sources: Barbour, 1910, Ruthven & Gaige, 1935, R. Thomas, 1966b, A. Schwartz & R.W. Henderson, 1991, H.M. Smith & David, 1999 and Dominguez & Díaz, 2011b.

Remarks: Original description based partly upon *Amphisbena subargentea* P. Browne (1756: 460, pl. 44, fig. 1) and partly upon *Serpens Caecilia ex. Mauritaniâ* Seba (1934: 137, pl. 86, fig. 2).

51. *Typhlops jerdoni* Boulenger, 1890a. Fauna Brit. India, Rept. Batr.: 238, pl. 34.

Synonym: *Typhlops diversiceps* Annandale, 1912.

Types: Syntypes (3), BMNH 1946.1.10.66–68, longest syntype 229 mm (T.C. Jerdon, 1836–1852).

Type locality: "Khási Hills" [= Meghalaya State, NE India, ca. 25°35'N, 91°38E].

Distribution: Southern Asia. Northeastern India (Arunachal Pradesh, Assam, Meghalaya, Nagaland, Sikkim, N West Bengal), E Nepal (Ilam), S Bhutan (Sarpang) and N Myanmar (Bago, Shan), 150–2125 m.

Sources: Boulenger, 1893a, Wall, 1923a, Dowling & Jenner, 1988, Bauer & Günther, 1992, Wallach, 1993b, Das & Palden, 2000, Schleich & Kästle, 2002, R. Sharma, 2003, 2007, Ao et al., 2004 and I. Das, 2010, 2012.

52. *Typhlops klemmeri* E.H. Taylor, 1962. Univ. Kansas Sci. Bull. 43(7): 253–254.

Type: Holotype, FMNH 178238 (formerly EHT-HMS M176 & EHT-CC 2307), a 151 mm specimen (E.H. Taylor).

Type locality: "Koh Phai, near Kuala Lumpur, Malaya" [= Koh Phai, Selangor State, West Malaysia, ca. 03°08'N, 101°40'E, elevation 75 m].

Distribution: West Malaysia (Kuala Lumpur), 75 m. Known only from type locality.

Source: Wallach, 2001.

53. *Typhlops koekkoeki* Brongersma, 1934. Zool. Meded. 17(3–4): 186–187, figs. 13–15.

Type: Holotype, RMNH 11077, a 445 mm specimen (U. Driebergen, 1924).

Type locality: "Boenjoe Id., N Borneo," [= Bunju Is., E Kalimantan, Indonesia, 03°31'N, 117°49'E, elevation NSL].

Distribution: Borneo. East Malaysia (Sarawak [FMNH 71579]) and Indonesia (E Kalimantan: Bunju Is.), NSL–230 m.

Sources: Stuebing, 1991 and Stuebing & Inger, 1999.

54. *Typhlops koshunensis* Oshima, 1916a. Zool. Mag., Tokyo 28(328): 85. (*nomen substitutum*)

Synonyms: *Typhlops leucoproctus* Oshima, 1910 *(nomen praeoccupatum)*, and *Typhlops koshunnensis* – Hahn, 1980a *(nomen incorrectum)*.

Type: Lectotype, ISM 4 (M. Oshima, 29 May 1909), designated by Hahn (1980a: 61), lost *fide* Ota (1991: 69).

Type locality: "Kosun, Formosa" [= Kuraru (= Kueitzuchiao), Koshun (= Hengchun County), Pingtung Township, ext. S Taiwan *fide* Zhao & Adler, 1993: 219].

Distribution: Southern Taiwan (Kaohsiung, Pingtung), 10–1000 m.

Sources: Oshima, 1910, Takahashi, 1930, Maki, 1931, 1933, Hu et al., 1980, 1989, Zhao & Adler, 1993, Wallach, 2001, 2004, Zhao, 2006 and Xiang & Li, 2009.

Remarks: Original description in Oshima (1910: 187).

55. *Typhlops kraalii* G. Doria, 1875. Ann. Mus. Civ. Stor. Nat. Genova (1874) 6: 347, pl. 12, fig. f.

Type: Holotype, MSNG CE 29962, a 250 mm specimen (O. Beccari, 1873), lost *fide* McDowell (1974b: 12).

Type locality: "Isole Kei" [= Great Kei Is., Maluku Prov., E Indonesia].

Distribution: Eastern Indonesia (Maluku: Kei, Seram), 915 m.

Sources: Rooij, 1917, E.H. Taylor, 1922, Brongersma, 1934, Haas, 1950 and McDowell, 1974b.

56. *Typhlops lankaensis* E.H. Taylor, 1947b. Univ. Kansas Sci. Bull. 31(13): 287–289, fig. 1.

Type: Holotype, FMNH 100066 (formerly EHT-HMS 30062 & EHT-CC C90) (E.H. Taylor, 13 Sept. 1944).
Type locality: "12 miles north of Trincomalee, Ceylon, 3–5 m" [= near Nilaveli, 18 km N Trincomalee, N Eastern Prov., NE Sri Lanka, ca. 8°43'N, 81°11'E, elevation NSL *fide* Bauer & Silva, 2007: 5].
Distribution: Northeastern Sri Lanka (N Eastern), NSL. Known only from type locality.
Sources: P. Silva, 1980a, A. Silva, 1990, Wallach, 1993b, 2009 and Somaweera, 2006.

57. *Typhlops lazelli* Wallach & Pauwels, 2002. Breviora (512): 4–8, fig. 1.

Type: Holotype, MCZ 173290 (formerly JDL 362.66), a 158 mm female (S. Brown, 27 May 1988).
Type locality: "High West, Pokfulan, Hong Kong Island, Hong Kong Territory, China, ca. 22°15'30"N, 114°08'30"E" [elevation 160 m].
Distribution: China (Hong Kong), 160 m. Known only from type locality.
Source: Zhao, 2006.

58. *Typhlops lehneri* Roux, 1926. Rev. Suisse Zool. 33(4): 298–299.

Type: Holotype, NMBA 9020, a 115–185 mm specimen (E. Lehner).
Type locality: "El Pozon, district de Falcón, Venézuéla" [= El Pozon, Falcón State, N Venezuela, 12°06'N, 70°00'W, elevation 10 m].
Distribution: Northern Venezuela (N Falcón), NSL–10 m.
Sources: Roux, 1927, Roze, 1966a, R. Thomas, 1974b, Dixon & Hendricks, 1979, Lancini & Kornacker, 1989, Mijares-Urrutia & Arends, 2000 and Navarrete et al., 2009.

59. *Typhlops leucocephalus* H.W. Parker, 1930d. Ann. Mag. Nat. Hist. (10) 6(35): 605.

Type: Holotype, BMNH 1946.1.11.96 (formerly BMNH 1930.5.9.11), a 212 mm specimen (R.H.R. Taylor, 1925–1930).
Type locality: "Las Anod (8° 28'N. x 47° 21'E.), 2400 ft., British Somaliland" [= Laascaanood, Nogal Valley, Nugaal Region, NE Somalia, 8°29'N, 47°22'E, elevation 730 m].
Distribution: Northern Somalia (SW Nugaal), 730 m. Known only from type locality.
Sources: Parker, 1932, 1949, Roux-Estève, 1974 and Lanza, 1983a, 1990b.
Remarks: Transferred to *Rhinotyphlops fide* Roux-Estève (1974: 183). Probably occurs in E Ethiopia *fide* Largen & Rasmussen (1993: 320).

60. *Typhlops leucomelas* Boulenger, 1890a. Fauna Brit. India, Rept. Batr.: 237–238.

Type: Holotype, BMNH 1946.1.10.46 (formerly BMNH 1874.4.29.223), a 127 mm specimen (R.H. Beddome, 1857–1882).
Type locality: "Haycock Mountain (2000 ft.) in Ceylon, 40 miles from Point de Galle" [= Haycock, 40 mi. NNE of Galle, Southern Prov., Sri Lanka, 06°20'N, 80°18'E, elevation 610 m].
Distribution: Southern Sri Lanka (Central, Sabaragamuwa, Southern), 610–1360 m.
Sources: Boulenger, 1893a, Wall, 1921, E.H. Taylor, 1947b and P. Silva, 1980a, R. Sharma, 2003, 2007, Das & Silva, 2005 and Somaweera, 2006.

61. *Typhlops longissimus* (A.M.C. Dumeril & Bibron, 1844). Erp. Gén. 6: 263–265. (*Ophthalmidion longissimum*)

Type: Holotype, MNHN 1061, a 346–353 mm specimen (F.L.N.C.L. de Castelnau, 1837–1843)
Type locality: "Amérique septentrionale" [= North America] (in error).
Distribution: Unknown, probably Asia.
Sources: Jan & Sordelli, 1864 *in* 1860–1866, Boulenger, 1893a and Loveridge, 1957.
Remarks: Listed as *incertae sedis* by Hahn (1980a: 77) and McDiarmid et al. (1999: 126). Possibly congeneric with and a senior synonym of *Typhlops tenuicollis* Peters, 1864a (Wallach, unpublished data).

62. *Typhlops loveridgei* Constable, 1949. Bull. Mus. Comp. Zool. 103(2): 110–111.

Type: Holotype, MCZ 2283, a 203 mm specimen (M.M. Carleton, 1873).
Type locality: "probably from North India" [probably Ambala, Haryana State, 30°21'N, 76°50'E, elevation 275 m, or Kulu, Himachal Pradesh State, 31°58N, 77°66E, elevation 1215 m, N India].
Distribution: Probably N India (Haryana or Himachal Pradesh). Known only from holotype.
Source: M.S. Khan, 1999b.
Remarks: Carleton collected material in 1873 from Ambala, Ambaia District and Kulu (Valley), Kangra District, in the Punjab (now Haryana and Himachal Pradesh states) but the type was donated with a specimen of *Typhlops diardii* with the same questionable locality data *fide* Constable (1949: 110, 113).

T

63. *Typhlops lumbricalis* (Linnaeus, 1758). Syst. Nat., ed. 10, 1: 228. (*Anguis lumbricalis*)

Synonym: *Anguis Surinamensis* Gravenhorst, 1807.
Type: Neotype, KU 273756 (R. Thomas, 7 Oct. 1965), designated by Dominguez & Diaz (2011b: 3).
Type locality: "4 miles N and 0.5 miles E of Rock Sound (76°11'02"N, 24°58'02"W, 20 m elevation, datum WGS

84), Elenthera Island, Bahama Islands" via neotype selection.

Distribution: Bahamas (Little Bahama & Great Bahama Banks: Andros, Cat, Eleuthera, Great Abaco, Great Exuma, Great Harbour Cay, Great Ragged, Little Abaco, Little Exuma, Little Ragged, Long Island, New Providence, Pipe Cay, South Bimini, Staniel Cay, Water Cay), NSL–20 m.

Sources: R. Thomas, 1974b, 1989 and Domínguez & Diaz, 2011a–b.

64. *Typhlops luzonensis* E.H. Taylor, 1919. Philippine J. Sci. (1918) 14D(1): 105–106.

Type: Holotype, CM 2653 (formerly EHT 109), a 260 mm specimen (E.H. Taylor, 12 May 1915).

Type locality: "Mount Maquiling, Laguna, Luzon, Philippine Islands" [= Mt. Makiling, Laguna Prov., Luzon Is., N Philippines, 14°08'N, 121°12'E].

Distribution: Philippines (Luzon, Negros, Marinduque), 250–300 m.

Sources: Savage, 1950, McDowell, 1974b, Alcala, 1986, Wynn & Leviton, 1993 and Brown et al., 2000.

65. *Typhlops madagascariensis* Boettger, 1877. Abh. Senck. Naturf. Ges. 11(1): 3–5, pl. 1, figs. 1a–f.

Type: Holotype, SMF 16604 (formerly SMF-B 7025, 1a), a 410 mm specimen (K. Ebenau, 1875).

Type locality: "Nossi-bé, bei Madagascar" [= Nosy Be Is., W Antsiranana Prov., N Madagascar, bet. 13°12'–25'S and 48°10'–22'E].

Distribution: Northern Madagascar (Antsiranana). Known only from type locality.

Source: Glaw & Vences, 1994.

66. *Typhlops madgemintonae* M.S. Khan, 1999b. Russ. J. Herp. 6(3): 233–236, figs. 1–2. (*nomen corrigendum*)

Synonyms: *Typhlops madgemintoni* M.S. Khan, 1999a *(nomen nudum), Typhlops ahsanai* M.S. Khan, 1999b *(nomen incorrigendum), Typhlops ahsanuli* – M.S. Khan, 1999b *(nomen corrigendum), Typhlops madgemintonai* M.S. Khan, 1999b *(nomen incorrigendum), Typhlops madgemintoni shermanai* M.S. Khan, 1999b *(nomen incorrigendum), and Typhlops madgemintonae shermani* – M.S. Khan, 1999b *(nomen corrigendum).*

Type: Holotype, AMNH 144213 (formerly MSK 904.93), a 200 mm male (M.S. Khan, 17 July 1993).

Type locality: "near house of Master Muhammad Sadiq, Goi Madan, District Kotli, Azad Kashmir, 33°30'N and 74°00'E, elevation 1315 m."

Distribution: Pakistan (Jammu & Kashmir), 1315–1495 m.

Sources: Wallach, 2000b, 2004 and M.S. Khan, 2002, 2006.

Remarks: *Typhlops madgemintonae* and *T. ahsanuli* both *incertae sedis fide* Wallach (2000b: 140).

67. *Typhlops malcolmi* E.H. Taylor, 1947b. Univ. Kansas Sci. Bull. 31(13): 291–292.

Type: Holotype, EHT-HMS 30072 (formerly EHT-CC 912), a 107 mm specimen (E.H. Taylor, Nov. 1944), lost *fide* Hahn (1980a: 64).

Type locality: "12 miles north of Trincomalee, Eastern Province, Ceylon" [= near Nilaveli, 18 km N Trincomalee, N Eastern Prov., NE Sri Lanka, ca. 8°43'N, 81°11'E, elevation NSL *fide* Bauer & Silva, 2007: 5].

Distribution: Northeastern Sri Lanka (N Eastern), NSL. Known only from type locality.

Sources: P. Silva, 1980a, A. Silva, 1990, Wallach, 1993b, 2009 and Somaweera, 2006..

68. *Typhlops manilae* E.H. Taylor, 1919. Philippine J. Sci. (1918) 14D(1): 106–108.

Type: Holotype, STUM (unnumbered), a 280 mm specimen.

Type locality: "Filipinas" [= Philippines].

Distribution: Philippines (? Luzon). Known only from holotype.

Remarks: Type locality probably Luzon, Philippines *fide* E.H. Taylor (1922a: 56).

69. *Typhlops meszoelyi* Wallach, 1999a. Herpetologica 55(2): 185–187, fig. 1a–b.

Type: Holotype, FMNH 191889 (formerly BBS 18 & ZM 18), a 162 mm male (B.B. Sinha, 5 Oct. 1971).

Type locality: "Darjeeling, 27°02'N, 88°16'E, 2285 m, West Bengal Province, India."

Distribution: Northeastern India (West Bengal), 2285 m. Known only from type locality.

Source: M.S. Khan, 1999b.

70. *Typhlops microcephalus* F. Werner, 1909d. Jahr. Vereins Vaterl. Naturkd. Württemberg 65: 60.

Type: Holotype, SMNS 6720 (formerly SMNS 4233), a 235 mm female (H. Rolle, 1905).

Type locality: "Madagaskar" [= Madagascar]. Restricted herein to Amber Mountain, NW Antsiranana Prov., N Madagascar, 12°37'S, 49°09'E, elevation 1100 m.

Distribution: Northern Madagascar (Antsiranana), 650–1200 m.

Sources: Glaw & Vences, 2007, Wallach, 2009 and Wallach & Glaw, 2009.

Remarks: A valid species *fide* Wallach *in* McDiarmid et al. (1999: 60). Previously considered a synonym of *Ramphotyphlops braminus* by Loveridge (1957: 244), Hahn (1980a: 39) and Brygoo (1987: 23).

71. *Typhlops microstomus* Cope, 1866. Proc. Acad. Nat. Sci. Philadelphia 18(2): 125–126.

Type: Holotype, USNM 61064 (formerly SIM 716), a 275 mm specimen (A.C.V. Schott, 1865–1866).
Type locality: "Yucatan" [Mexico].
Distribution: Yucatán Peninsula. Southeastern Mexico (S Campeche, Quintana Roo, Veracruz, Yucatán), Belize (Belize, Cayo, Corozal, Orange Walk) and N Guatemala (Petén), NSL–200 m.
Sources: K.P. Schmidt, 1936, Dixon & Hendricks, 1979, Villa et al., 1988, Meerman, 1992, Garel & Matola, 1995, J.C. Lee, 1996, 2000, J.A. Campbell, 1998b, Pérez-Higareda & Smith, 1999 and Townsend et al., 2008.

72. *Typhlops minuisquamus* Dixon & Hendricks, 1979. Zool. Verh. (173): 21–23, figs. 1b, 2b.

Type: Holotype, TCWC 39130, a 230 mm female (P. Soini, 17 March 1972).
Type locality: "Mishana, Loreto, Peru" [03°57'S, 73°37'W, elevation 125 m].
Distribution: Amazonia. Southeastern Colombia (Meta, Vaupés), Venezuela (Amazonas), Guyana (Cuyuni-Mazruni, Essequibo Islands-West Demerara, Mazaruni-Potaro, Upper Demerara-Berbice), Brazil (Amazonas) and NE Peru (N Loreto), NSL–200 m.
Sources: Lancini & Kornacker, 1989, Caicedo-Portilla, 2011 and C.J. Cole et al., 2013.

73. *Typhlops monastus* R. Thomas, 1966a. Proc. Biol. Soc. Washington 79: 257–260, figs. 1–2.

Type: Holotype, MCZ 81112, a 184 mm male (R. Thomas, 5 Aug. 1965).
Type locality: "between Lawyers River and Cassava Ghaut, St. Peter's Parish, Montserrat, British West Indies" [ca. 16°45'N, 62°13'W, elevation 75–100 m].
Distribution: Lesser Antilles. Montserrat (St. Anthony, St. Peter), NSL–400 m.
Source: A. Schwartz & R.W. Henderson, 1991.
Remarks: Fossil record from Holocene of Antigua *fide* Pregill et al. (1988: 14).

74. *Typhlops monensis* K.P. Schmidt, 1926a. Field Mus. Nat. Hist. (Zool.) 12(12): 157–158, fig. 1.

Type: Holotype, ZMH 1582, a 182 mm specimen, destroyed in July 1943 during World War II.
Type locality: "Mona Island, West Indies" [= Puerto Rico, 18°05'N, 67°54'W, elevation NSL–60 m].
Distribution: Puerto Rico (Mona Is.), NSL–60 m.
Sources: A. Schwartz & R.W. Henderson, 1991 and Rivero, 1998.

75. *Typhlops mucronatus* Boettger, 1880a. Zool. Anz. 3(57): 279.

Type: Lectotype, SMF 16626 (formerly SMF-B 7037, 2a), a 378 mm specimen (C. Ebenau & A. Stumpff, 1880), designated by Mertens (1922: 178).
Type locality: "Loucoubé auf Nossibé" [= Lokobe, Nosy Be Is., N Antsiranana Prov., N Madagascar, ca. 13°22'S, 48°19'E] via lectotype selection.
Distribution: Northern Madagascar (N Antsiranana, Nosy Be Is.), NSL–650 m.
Sources: Boettger, 1881d, Glaw & Vences, 2007 and Wallach & Glaw, 2009.

76. *Typhlops muelleri* Schlegel, 1839 *in* 1837–1844. Abbild. Amph.: 39, pl. 32, figs. 25–28. *(nomen corrigendum)*

Synonyms: *Typhlops mülleri* Schlegel, 1839 *in* 1837–1844 *(nomen incorrigendum)*, *Typhlops nigro-albus* A.-M.-C. Duméril & Bibron, 1844 *(nomen incorrigendum)*, *Typhlops nigroalbus* – Jan, 1863b *(nomen corrigendum)*, *Argyrophis bicolor* Gray, 1845, *Pilidion dimidiatum* Bleeker, 1860b, *Typhlops schneideri* Jan, 1863b *(nomen nudum)*, *Typhlops schneideri* Jan, 1864, *Typhlops muelleri* – Boulenger, 1893a *(nomen corrigendum)*, and *Typhlops fusconotus* Brongersma, 1934.
Type: Holotype, RMNH 3718, a 321 mm specimen (S. Müller, June 1833–Dec. 1835).
Type locality: "Padang auf Sumatra" [= Padang, Sumatra, W Indonesia, 00°57'S, 100°21'E, elevation 20 m].
Distribution: Southeast Asia. Southern Myanmar, Thailand (Bangkok, Chanthaburi, Chumphon, Nakhon Ratchasima, Nakhon Si Thammarat, Pathum Thani, Pattani, Phang Nga, Phattalung, Surat Thani, Yala), Cambodia (Kampot, Kompong Speu, Pursat), Laos, S Vietnam (Dong Nai, Gia Lai, Lam Dong, Quang Nam, Tay Ninh), West Malaysia (Kuala Lumpur, Pinang, Perlis, Selangor), Singapore and Indonesia (Bangka, Kalimantan, Nias, Padang, Sumatra, Weh, Webb), 20–1675 m.
Sources: E.H. Taylor, 1965, David & Vogel, 1996, Saint Girons, 1972, M.J. Cox et al., 1998, Nabhitabhata, 2000, Wallach, 2001, Pauwels et al., 2002, Pauwels & Sumontha, 2003, Nabhitabhata & Chan-ard, 2005, B.L. Stuart & Emmett, 2006, and V.S. Nguyen et al., 2009.
Remarks: Original color sketches of holotype of *Argyrophis bicolor* by Harwicke (1756–1823) (1: figs. 119–120) in BMNH. Record for E Indonesia (Papua) needs confirmation (McDowell, 1974a).

77. *Typhlops naugus* R. Thomas, 1966b. Rev. Biol. Trop. (1965) 13(2): 192, fig. 4e. (*Typhlops richardi naugus*)

Type: Holotype, MCZ 77221, a 243 mm male (D.C. Leber, 11 Aug. 1964).

Type locality: "on the island of Virgin Gorda, British Virgin Islands, on the hillside above Pond Bay" [= 18°28'N, 64°25'W].
Distribution: British Virgin Islands (Virgin Gorda), NSL–245 m.

78. *Typhlops notorachius* R. Thomas & Hedges, 2007. Zootaxa (1400): 21, figs. 8h, 9d–e.

Type: Holotype, MNHNCU 4551 (formerly USNM-FN 191322), a 282–301 mm male (S. B. Hedges, R. Thomas & D. McCallister, 1 July 1990).
Type locality: "9.4 km W Imias, Guantánamo Province, Cuba, 5 meters" [= 20°04'N, 74°37'W].
Distribution: Extreme E Cuba (Guantánamo), NSL.
Source: Domínguez & Moreno, 2009.

79. *Typhlops oatesii* Boulenger, 1890a. Fauna Brit. India, Rept. Batr.: 238.

Types: Syntypes (2), BMNH 1946.1.10.51–52 (formerly BMNH 1889.4.24.13–14), longest syntype 200 mm (E.W. Oates).
Type locality: "Table Island, Cocos Group" [= Myanmar, Bay of Bengal, 14°12'N, 93°22'E, elevation NSL].
Distribution: Eastern India (Andaman & Nicobars: Middle Andaman Is.) and S Myanmar (Cocos: Table Is.), NSL–15 m.
Sources: Boulenger, 1893a, Murthy & Chakrapany, 1983 and Wallach, 2001.
Remarks: Three syntypes listed *fide* Boulenger (1893a: 238).

80. *Typhlops ocularis* H.W. Parker, 1927b. Ann. Mag. Nat. Hist. (9) 19(111): 379–380, 2 figs.

Type: Holotype, BMNH 1946.1.11.25 (formerly BMNH 1926.8.18.3), a 342 mm female.
Type locality: "Antongil Forest, Maroantsetra, N.E. Madagascar" [= Maroantsetra Forest, Antongil Bay, N Toamasina Prov., Madagascar, 15°26'S, 49°44'E, elevation ca. 10 m].
Distribution: Eastern Madagascar (Antsiranana, ext. N Toamasina, ext. SE Toliara, Nosy Be Is.), NSL–400 m.
Sources: Guibé, 1958, Glaw & Vences, 2007 and Wallach and Glaw, 2009.

T

81. *Typhlops oxyrhinus* Domínguez & Diaz, 2011. Herpetologica 67(2): 204–205, figs. 1b, 2b.

Type: Holotype, CTR 2230 (formerly IES), a 244 mm female (L. Lobaina, mid–1900s).
Type locality: "Bayate (75°21'47"N, 20°20'07"W, 338 m elevation), El Salvador Municipality, Guantánamo Province, Cuba."
Distribution: Extreme E Cuba (Guantánamo), NSL–400 m.

82. *Typhlops pachyrhinus* Domínguez & Diaz, 2011. Herpetologica 67(2): 206–207, figs. 1c, 2c.

Type: Holotype, CZACC 4.11933, a 211 mm male (E. Elso, June 1966).
Type locality: "'El Liberal' coffee plantation (coordinates not known). Cabañas Municipality, Sierra del Rosario, Pinar del Río Province, Cuba" [= ca. 22°48'N, 83°15'W, elevation ca. 400 m].
Distribution: Western Cuba (Pinar del Río), 400 m. Known only from type locality.

83. *Typhlops pammeces* A.C.L.G. Günther, 1864a. Rept. Brit. India:176, pl. 16, fig. c. (*nomen susbstitutum*)

Synonyms: *Typhlops tenuis* A.C.L.G. Günther, 1864a (*nomen praeoccupatum*), *Typhlops psammophilus* Annandale, 1906, and *Typhlops psammeces* M.A. Smith, 1943 *(nomen emendatum)*.
Type: Holotype, BMNH 1946.1.11.34, a 127 mm specimen (T.C. Jerdon, 1836–1852).
Type locality: "Madras, British India" [= Chennai, Tamil Nadu State, E India, ca. 13°00'N, 80°12'E, elevation 15 m].
Distribution: Eastern India (Tamil Nadu, Uttar Pradesh), NSL–650 m.
Sources: Wallach, 1993b, 2009 and R. Sharma, 2003, 2007.

84. *Typhlops paradoxus* R. Thomas, 1968. Copeia 1968(4): 715, fig. 2b (paratype). (*Typhlops biminiensis paradoxus*)

Type: Holotype, MCZ 92993, a 245 mm specimen (R. Forbes via T. Laska, 10 Feb. 1967).
Type locality: "7.5 miles N of Mathew Town, Great Inagua, southeastern Bahamas" [= 21°02'N, 73°39'W, elevation NSL].
Distribution: Bahamas (Great Inagua Is.), NSL.
Remarks: A valid species *fide* Thomas & Hedges, 2007.

85. *Typhlops paucisquamus* Dixon & Hendricks, 1979. Zool. Verh. (173): 23–25, figs. 1a, 2a.

Type: Holotype, MCZ 147336 (formerly MCZ 1444b), a 133 mm specimen (B. Burnet, don. 21 Dec. 1863).
Type locality: "state of Pernambuco, Brasil" [= Brazil].
Distribution: Northeastern Brazil (Pernambuco), 10 m.

86. *Typhlops perimychus* R. Thomas & Hedges, 2007. Zootaxa (1400): 15–18, figs. 8e, 9a.

Type: Holotype, USNM 564800 (formerly USNM-FN 161982), a 130–280 mm specimen (S.B. Hedges & C.A. Hass, 22 March 1987).
Type locality: "Blue Beach, U. S. Naval Base Guantánamo Bay, Cuba, five meters elevation" [= 19°54'N, 75°08'W].

Distribution: Extreme E Cuba (Guantánamo), NSL. Known only from type locality.
Source: Domínguez & Moreno, 2009.

87. *Typhlops platycephalus* A.-M.-C. Duméril & Bibron, 1844. Erpét. Gén. 6: 293–295.

Synonyms: *Typhlops platycephala* R. Thomas, 1966b, and *Typhlops brachycephalus* Slowinski & Lawson, 2002 (*nomen nudum*).
Type: Holotype, MNHN 1066, a 295 mm specimen (A. Plée, 1823–1825).
Type locality: "Martinique," (in error). Corrected to Puerto Rico *fide* Stejneger (1904: 557). Restricted to eastern, southern or interior Puerto Rico *fide* Hedges and Thomas (1991: 457).
Distribution: Puerto Rico (Arecibo, Carolina, Dorado, Guánica, Río Grande, San Germán, Palominos and Vieques Is.), NSL–650 m
Sources: K. Schmidt, 1928, R. Thomas, 1989, Hedges & Thomas, 1991, A. Schwartz & R.W. Henderson, 1991, Rivero, 1998 and Torres et al., 2000.

88. *Typhlops platyrhynchus* Sternfeld, 1910c. Mitt. Zool. Mus. Berlin 5(1): 69.

Types: Syntypes (4), ZMB 20729, a 245 mm specimen, ZMB 63442–43 (formerly ZMB 20729b–c), a 218 mm and 195 mm specimen, and MCZ 39798 (formerly ZMB 20729d), a 270–273 mm specimen (Martienssen).
Type locality: "Tanga, Deutsch-Ostafrika" [= Tanga, E Tanga Prov., NE Tanzania].
Distribution: Northeastern Tanzania (Tanga). Known only from type locality.
Sources: Roux-Estève, 1974 and Bauer et al., 2002..

89. *Typhlops porrectus* Stoliczka, 1871. J. Asiat. Soc. Bengal 40(2): 426–428, pl. 25, figs. 1–4.

Synonyms: *Typhlops mackinnoni* Wall, 1910b, *Typhlops venningi* Wall, 1913e, *Typhlops porrecta* Constable, 1949, and *Typhlops ductuliformes* M.S. Khan, 1999a.
Types: Syntypes (8), NMW 15357.1–2 and NMW 15358.1–2, 150–285 mm specimens (F. Stoliczka, 1862–1871), other syntypes lost *fide* M.A. Smith (1943: 46).
Type locality: "Hurdwár" [= Haridwar, Uttarakhand State, 29°57'N, 78°10'E, elevation 285 m], "in the neighbourhood of Calcutta" [= Kolkata, West Bengal State, 22°34N, 88°22E, elevation 10 m], "at the foot of the Parisnáth hill (in western Bengal)" [= 23°58'N, 86°08'E, elevation 275 m], "the base of the Rangnu valley below Darjíling" [= 27°02'N, 88°15'E, elevation 2125], and "south of Agra" [= Uttar Pradesh State, 27°10'N, 78°00'E, elevation 170 m].
Distribution: Southern Asia. Pakistan (Khyber Pakhtunkhwa, Punjab. Sindh), India (Bihar, Gujarat, Haryana, Himachal Pradesh, Karnataka, Kerala, Maharashtra, Orissa, Rajasthan, Tamil Nadu, Uttarakhand, Uttar Pradesh, West Bengal), E Nepal (Morang), Sri Lanka (Central, North Central, Uva), Bangladesh, Nepal (Kosi), N Myanmar, Thailand (Chiang Mai) and Singapore, NSL–2300 m.
Sources: Wall, 1911i, 1923c, M.A. Smith, 1943, Mertens 1969a, P. Silva, 1980a, Murphy, 1986, A Silva, 1990, I. Das, 1994, M.J. Cox & Nabhitabhata, 1997, M.S. Khan, 1999a–b, 2002, 2006, Wallach, 2000b, 2001, 2004, Shrestha, 2001, Schleich & Kästle, 2002, R. Sharma, 2003, 2007, Tiwari & Shah, 2004, Somaweera, 2006 and Baig et al., 2008.
Remarks: Hahn (1980: 67) and McDiarmid et al. (1999: 114) listed "Roorkee" [Uttarakhand State, 29°52'N, 77°54'E, elevation 270] as the locality for NMW 15357. *Typhlops venningi* Wall and *T. mackinnoni* Wall are possibly valid species *fide* A.H. Wynn *in* Wallach (1999: 185).

90. *Typhlops proancylops* R. Thomas & Hedges, 2007. Zootaxa (1400): 7–9, fig. 3b.

Type: Holotype, KU 272267, a 127–243 mm male (native, 13 July 1979).
Type locality: "Soliette, 5 km airline NW Fond Verettes, 363 m, Dépt. du Sud-Est, Haiti" [= 18°24'N, 71°51'W].
Distribution: Southeastern Haiti (Ouest, Sud-Est) and SW Dominican Republic (Independencia), 385–800 m.

91. *Typhlops pusillus* T. Barbour, 1914a. Mem. Mus. Comp. Zool. 44(2): 323, 2 figs.

Synonym: *Typhlops pusilla* R. Thomas, 1966a.
Type: Holotype, MCZ 8719, a 127 mm specimen (W.M. Mann, Mar.–Nov. 1913).
Type locality: "Cape Haitien, Haiti" [= Cap Haitien, Nord Dept., N Haiti, 19°45'N, 72°12'W, elevation NSL].
Distribution: Hispaniola. Haiti (Artibonite, Centre, Grand'Anse, Nord, Nord-Est, Nord-Ouest, Ouest, Sud, Sud-Est: Gonave, Grande Cayemite and Grande Tortue Is.) and Dominican Republic (Azua, Barahona, Dajabón, Distrito Nacional, Duarte, Elías Piña, Espaillat, Hato Mayor, Independencia, La Altagracial, La Ramona, La Vega, Monseñor Nouel, Monte Cristi, Peravia, Puerto Plata, Salcedo, Samaná, Sánchez Ramirez, San Cristóbal, San Juan, Santiago, Santiago Rodríguez, Valverde, Catalina and Saona Is.), NSL–730 m.
Sources: Richmond, 1964, R. Thomas, 1974b, A. Schwartz & R.W. Henderson, 1991, R. Thomas & Powell, 1994a and Domínguez & Diaz, 2011a.

92. *Typhlops rajeryi* Renoult & Raselimanana, 2009. Zootaxa (2290): 65–67, fig. 1.

Type: Holotype, ZSM 1589/2008 (formerly FGZC 3165), a 269 mm specimen (J.P. Renoult, 27 Dec. 2007).

Type locality: "Madagascar: Vatovavy Fitovinany Region, the main entrance of Ranomafana National Park, 21°15.388'S, 47°25.308'E, alt.: 918 m."
Distribution: Southeastern Madagascar (N Fianarantsoa), 920 m. Known only from type locality.

93. *Typhlops reticulatus* (Linnaeus, 1758). Syst. Nat., ed. 10, 1: 228. (*Anguis reticulata*)

Synonyms: *Anguis rostralis* Weigel, 1782, *Anguis nasutus* Gmelin, 1789, *Anguis rostratus* Latreille in Sonnini & Latreille, 1801 *(nomen emendatum)*, *Anguis crocotatus* J.G. Schneider, 1801 *(nomen substitutum)*, *Typhlops leucogaster* Wied-Neuwied, 1825, *Ophthalmidion crassum* Duméril *in* A.-M.-C. Duméril & Duméril, 1851, *Onychocephalus troscheli* Jan, 1861b *(nomen nudum)*, *Typhlops reticulatus troscheli* Jan, 1863b *(nomen nudum)*, *Typhlops reticulatus nigrolactea* Jan, 1864 *in* Jan & Sordelli, 1860–1866, *Typhlops reticulatus troscheli* Jan, 1864 *in* Jan & Sordelli, 1860–1866, *Typhlopus retisculatus* – Nakamura, 1938 *(nomen incorrectum)*, and *Altmantyphlops reticulates wellsi* Hoser, 2012am *(nomen illegitimum)*.
Type: Lectotype, specimen illustrated in Scheuchzer (1735: pl. 747, fig. 4), designated by Bauer (2012: 68).
Type locality: "America" via lectotype selection.
Distribution: Amazonia. Colombia (Amazonas, Antioquia, Ayauca, Casanare, Cauca, Coquetá, Córdoba, Guaviare, Meta, Putumayo, Santander, Vaupés), E Ecuador (Esmeraldas, Napo, Sucumbios), Venezuela (Amazonas, Barinas, Bolívar, Carabobo, District Federal, Falcón, Monagas, Portuguesa, Yaracuy, Zulia), Guyana (Barima-Waini, Cuyuni-Mazaruni, Essequibo Islands-West Demerara, Potaro-Siparuni, Upper Demerara-Berbice, Santa Rosa Is.), Suriname (Marowijne, Paramaribo, Saramacca), French Guiana (Cayenne, Saint-Laurent-du-Maroni), N Brazil (Acre, Amapa, Amazonas, Mato Grosso, Pará, Paraiba, Rondônia, Marajo Is.), Peru (Loreto, Madre de Dios) and Bolivia (Beni, La Paz, Santa Cruz), NSL–750 m.
Sources: Roze, 1966, Cunha & Nascimento, 1978, 1993, Dixon & Hendricks, 1979, Lema, 1980, Starace, 1998, Mijares-Urrutia & Arends, 2000, Rivas-Fuenmayor, 2001, Rivas-Fuenmayor et al., 2001, Doan & Arizábal, 2002, Carvajal-Cogollo, 2006, Börschig, 2007, Rojas-Runjaic et al., 2007, Navarrete et al., 2009, Caicedo-Portilla, 2011, Bauer, 2012 and C.J. Cole et al., 2013.

94. *Typhlops reuteri* Boettger, 1881c. Zool. Anz. 4(99): 650.

Synonym: *Typhlops lenzi* Boettger, 1882.
Type: Holotype, SMF 16589 (formerly SMF-B 7021 a), a 95 mm specimen (A. Stumpff, 1881).
Type locality: "insula Nossi-Bé, Madagascariensis," Emended to Lukube auf Nossibe [= Lokobe, Nosy Be Is., N Antsiranana, N Madagascar, ca. 13°22'S, 48°19'E] *fide* Boettger (1898: 2).
Distribution: Northern Madagascar (N Antsiranana, Nosy Be Is.), NSL–360 m.
Sources: Guibé, 1958, Glaw & Vences, 1994 and Wallach, 2009.
Remarks: Original description cited C. Reuter as collector; corrected to A. Stumpff *fide* Boettger (1898: 2) and SMF specimen label.

95. *Typhlops richardii* A.-M.-C. Duméril & Bibron, 1844. Erpét. Gén. 6: 290–293.

Types: Syntypes (2), MNHN 3220 und RMNH 3713, longest syntype 258 mm (L.C.M. Richard, 1786).
Type locality: "île Saint-Thomas, l'une des petites Antilles" [= St Thomas, U.S. Virgin Islands, ca. 18°20'N, 64°56'W].
Distribution: Puerto Rico bank (Cayo Diablo, Cayo Icacos, Cayo Norte, Culebra, Surprise Cay), U.S. Virgin Islands (St. Croix, St. John, St. Thomas) and British Virgin Islands (Guana, Tortola).
Sources: R. Thomas, 1966a, Hedges & Thomas, 1991 and Rivero, 1998.

96. *Typhlops rostellatus* Stejneger, 1904. Rep. U.S. Natl. Mus. 1902: 686, figs. 145–147.

Synonym: *Typhlops rostellata* R. Thomas, 1966a.
Type: Holotype, USNM 25463, a 200 mm specimen (A.B. Baker, 26 Jan. 1899).
Type locality: "Lares, Porto Rico, West Indies" [= Lares, Lares Prov., N Puerto Rico, 18°18'N, 66°53'W, elevation 325 m].
Distribution: Northern Puerto Rico (Arecibo, Bayamón, Caguas, Isabela, Rio Grande, San Juan, Utuado, Vega Baja), NSL–325 m.
Sources: K.P. Schmidt, 1928, A. Schwartz & R.W. Henderson, 1991 and Rivero, 1998.

97. *Typhlops roxaneae* Wallach, 2001. Raffles Bull. Zool. 49(1): 40–42, fig. 1.

Type: Holotype, MCZ 177984 (formerly TNRC 523-549), a 231 mm female (native, 2 Oct. 1963).
Type locality: "Thailand: Phra Nakhon Si Ayutthaya Province: Bangkok, Bangkok International School, 14°20'N, 100°20'E, elevation 1–2 m," (in error). Corrected herein to Bangkok International School, Nonthaburi Prov., S Thailand, 13°54'N, 100°32'E, elevation 10 m.
Distribution: Southern Thailand (Nonthaburi), NSL–10 m. Known only from type locality.

98. *Typhlops ruber* Boettger, 1897. Zool. Anz. 20(531): 164.

Synonym: *Typhlops rubra* Savage, 1950.

Type: Holotype, SMF 16616 (SMF-B 7031, 1a), a 225 mm specimen (O.F. von Moellendorff & O. Koch, 1897).
Type locality: "Insel Samar, Philippinen" [= Samar Island, cen. Philippines].
Distribution: Philippines (Camotes: Pacijan; Cebu, Luzon, Marinduque, Mindanao, Mindoro, Negros, Samar).
Sources: McDowell, 1974a, Alcala, 1986 and Wynn & Leviton, 1993.

99. *Typhlops ruficaudus* (Gray, 1845). Cat. Lizards Brit. Mus.: 136. (*Anilios ruficauda*)

Synonyms: *Typhlops jagorii* W.C.H. Peters, 1861c, *Typhlops dichromatus* Jan, 1863b (*nomen nudum*), *Typhlops dichromatus* Jan, 1864 *in* Jan & Sordelli, 1860–1866, and *Typhlops petersii* Steindachner, 1867d.
Types: Syntypes (3), BMNH 1946.1.11.4–6, longest syntype 250 mm (H. Cuming, Jan. 1836–June 1840).
Type locality: "Philippine Islands."
Distribution: Philippines (Luzon, Marinduque, Negros, Romblon, Sibuyan, Tablas), NSL–1065 m.
Sources: McDowell, 1974b, Alcala, 1986, A.H. Wynn & Leviton, 1993 and R. Brown et al., 1996. Sulawesi record doubtful *fide* Lang & Vogel (2005: 263).

100. *Typhlops satelles* R. Thomas & Hedges, 2007. Zootaxa (1400): 23–24, fig. 8k.

Type: Holotype, SMF 61303, a 303–350 mm specimen (Z. Vogel, 22 April 1966).
Type locality: "a public beach (0 m elevation) east of inlet to the Bahia de Cienfuegos, about 12 km airline S Cienfuegos, Cienfuegos Province, in south-central Cuba" [= 22°02'N, 80°25'W].
Distribution: Central Cuba (Cienfuegos), NSL. Known only from type locality.
Source: Domínguez & Moreno, 2009.

101. *Typhlops schmutzi* Auffenberg, 1980. Bull. Florida St. Mus. (Biol. Sci.) 25(2): 114–115, fig. 36 (left).

Type: Holotype, UF 29507, a 127 mm female (W. Auffenberg & field crew, Jan. 1970).
Type locality: "Loho Lavi, Komodo, 12 m" [= Loho Lavi, E Komodo, Lesser Sundas, SE Indonesia, 08°36'S, 119°27'E].
Distribution: Southern Indonesia (N Flores, Komodo, Sumbawa), NSL–50 m.
Sources: Wallach, 2004 and McKay & Lilley, 2012.

102. *Typhlops schwartzi* R. Thomas *in* Woods, 1989. Biogeog. West Indies: 412–414, fig. 1a.

Type: Holotype, KU 208752, a 265 mm female (natives, 11 Aug. 1973).
Type locality: "1.5 km W Jayaco, La Vega Province, República Dominicana, 244 m."
Distribution: Eastern Dominican Republic (Azua, Duarte, La Altagracia, La Vega, Puerto Plata, Samaná, San Pedro de Macoris), 30–275 m.
Sources: A. Schwartz & R.W. Henderson, 1991 and R. Thomas & Powell, 1994b.

103. *Typhlops siamensis* A.C.L.G. Günther, 1864a. Rept. Brit. India: 175, pl. 16, fig. d.

Type: Holotype, BMNH 1946.1.11.69 (formerly BMNH 1859.7.1.17), a 216 mm specimen (H. Mouhot, 1858–1860).
Type locality: "Siam" [= Thailand]. Restricted to E Thailand *fide* M.A. Smith (1923: 53).
Distribution: Indo-China. Eastern Thailand (Ko Llao Is.), Cambodia (Trapeang Chan), Laos and S Vietnam (Binh Duong, Dong Nai, Nam Bo), NSL–105 m.
Sources: E.H. Taylor, 1965, Saint Girons, 1972a, Wallach, 2001, V.S. Nguyen, 2007, V.S. Nguyen et al., 2009 and Geissler et al., 2011.

104. *Typhlops silus* Legler, 1959a. Herpetologica 15(2): 105–107, fig. 1.

Type: Holotype, KU 47469, a 126 mm specimen (C. Grant, April 1956).
Type locality: "Banes, Oriente Province, Cuba" [= Banes, Holguín Prov., E Cuba, 75°43'N, 20°58'W, elevation 53 m].
Distribution: Eastern Cuba (Granma, Guantánamo, Holguín, Santiago de Cuba), 20–570 m.
Remarks: A valid species (previously a synonym of *T. lumbricalis*) *fide* Dominguez & Diaz (2010a: 11).

105. *Typhlops socotranus* Boulenger, 1889b. Ann. Mag. Nat. Hist. (6) 4(23): 362.

Types: Syntypes (2), BMNH 1946.1.11.72–73 (formerly BMNH 1881.3.30.16–17), longest syntype 200 mm (I.B. Balfour, 1879–1880).
Type locality: "Socotra" [Yemen].
Distribution: Yemen (Socotra Is.), 170–550 m.
Sources: Boulenger, 1893a, Steindachner, 1903, Roux-Estève, 1974, Schätti & Desvoignes, 1999, Rösler & Wranik, 2004 and Razzetti et al., 2011.

106. *Typhlops stadelmani* K.P. Schmidt, 1936d. Proc. Biol. Soc. Washington 49: 48.

Type: Holotype, MCZ 38701, a 112 mm specimen (R.E. Stadelman, 26 Jan. 1833).
Type locality: "Subirana Valley, 2800 feet altitude, Yoro, Honduras" [ca. 15°11'N, 87°28'W, elevation 855 m].
Distribution: Northwestern Honduras (Atlántida, NW Copán, Cortés, SW Yoro), 320–1370 m.

Sources: McCranie & Wilson, 2001a, McCranie & Castañeda, 2005, Townsend et al., 2008 and McCranie, 2011a.

Remarks: A valid species (previously a synonym of *T. tenuis* Salvin) *fide* McCranie & Wilson, 2001a.

107. *Typhlops sulcatus* Cope, 1868c. Proc. Acad. Nat. Sci. Philadelphia 20(2): 128.

Synonyms: *Typhlops haitiensis* Richmond, 1964, and *Typhlops sulcata* R. Thomas, 1966a.

Type: Holotype, USNM 12371, a 168 mm specimen (W.J. Rasin).

Type locality: "Navassa Island, southwest from St. Domingo" [West Indies, 18°24'N, 75°01'W].

Distribution: Hispaniola. Southern Haiti (Ouest, Sud, Gonáve and Grande Cayemite Is.), SW Dominican Republic (Baharona, Independencia, Pedernales, Peravea and Alto Velo Is.) and Navassa Is., NSL–25 m.

Sources: Richmond, 1964, R. Thomas, 1966c, A. Schwartz & R.W. Henderson, 1991, R. Thomas & Powell, 1994c and Powell, 1999.

108. *Typhlops sylleptor* R. Thomas & Hedges, 2007. Zootaxa (1400): 10–12, fig. 3d.

Type: Holotype, USNM 564804 (formerly USNM-FN 192317), a 118–214 mm specimen (M. Leal & R. Thomas, 7 June 1991).

Type locality: "8.0 km WSW Baradères, Dépt. de la Grande Anse, Haiti, 420 m" [= 18°30'N, 73°39'W].

Distribution: Extreme SW Haiti (Grand'Anse), 375–420 m.

109. *Typhlops syntherus* R. Thomas, 1965b. Copeia 1965(4): 436–437, figs. 1, 2b.

Type: Holotype, MCZ 77215, a 200 mm specimen (D.C. Leber & R. Thomas, 25 June 1964).

Type locality: "5 km north of Pedernales, Pedernales Province, Dominican Republic" [= 18°05'N, 71°44'W, elevation 90 m].

Distribution: Southwestern Dominican Republic (Pedernales), 15–90 m.

Sources: R. Thomas, 1974b, A. Schwartz & R.W. Henderson, 1991 and L. White et al., 1992.

110. *Typhlops tasymicris* R. Thomas, 1974b. Occ. Pap. Mus. Zool., Louisiana St. Univ. (46): 1–2, fig. 1a.

Type: Holotype, UF 21547, a 181 mm female (F.G. Thompson, 9 May 1968).

Type locality: "1 mi. E Vincennes, St. David Parish, Grenada, Lesser Antilles" [= 12°03'N, 61°41'W, elevation 170 m].

Distribution: Lesser Antilles (Grenada: St David) and St. Vincent and Grenadines (Union), 15–180 m.

Sources: Wallach, 2000a, Yorks et al., 2003 and Rivera-Roriguez et al., 2011.

111. *Typhlops tenebrarum* E.H. Taylor, 1947b. Univ. Kansas Sci. Bull. 31(13): 292–294, fig. 2.

Type: Holotype, EHT-HMS 30063 (formerly EHT-CC C186), a 111 mm specimen (E.H. Taylor, 5 Oct. 1944), lost *fide* Hahn (1980a: 73).

Type locality: "12 miles north of Trincomalee, Ceylon" [= near Nilaveli, 18 km N Trincomalee, N Eastern Prov., NE Sri Lanka, ca. 8°43'N, 81°11'E, elevation NSL *fide* Bauer & Silva, 2007: 5].

Distribution: Sri Lanka (N Eastern, Western), NSL.

Sources: P. Silva, 1980a, A. Silva, 1990, Wallach, 1993b, 2009 and Somaweera, 2006.

112. *Typhlops tenuicollis* (W.C.H. Peters, 1864a). Mber. Kønigl. Akad. Wiss. Berlin 1864(4): 272–273, pl., figs. 2, 2a–c. (*Onychocephalus tenuicollis*)

Synonym: *Typhlops theobaldanus* Stoliczka, 1871.

Type: Holotype, ZMB 5042, a 365 mm specimen.

Type locality: "Himalaya" [= mountains in Nepal, India or Bhutan].

Distribution: Northeastern India.

Sources: M.A. Smith, 1943 and Bauer et al., 2002.

113. *Typhlops tenuis* Salvin, 1860. Proc. Zool. Soc. London 28(1): 454.

Synonyms: *Typhlops basimaculatus* Cope, 1867b, *Typhlops perditus* W.C.H. Peters, 1869, *Typhlops praelongus* F. Müller, 1885, and *Typhlops praelongis* E.H. Taylor, 1940b *(nomen emendatum).*

Type: Holotype, BMNH 1946.1.11.71 (formerly BMNH 1864.1.26.47), a 305 mm specimen (native).

Type locality: "Coban in Guatemala" [= Cobán, Alta Verapaz Dept., N Guatemala, 15°28'N, 90°22'W, elevation 1330 m].

Distribution: Southern Mexico (Puebla, Tabasco, cen. Veracruz) and Guatamala (Alta Verapaz, Baja Verapaz), 95–1520 m.

Sources: E.H. Taylor, 1940b, Stuart, 1963, Dixon & Hendricks, 1979, L.D. Wilson & Meyer, 1985, Villa et al., 1988, McCranie & Wilson, 2001a, Townsend et al., 2008 and García-Vázquez et al., 2009.

114. *Typhlops tetrathyreus* R. Thomas *in* Woods, 1989. Biogeog. West Indies: 414–415, fig. 1b.

Type: Holotype, KU 208793, a 240 mm male (native, 28 Nov. 1970).

Type locality: "3 mi. (4.8 km) N Pétionville, Dépt. de l'Ouest, Haiti" [= 18°33'N, 72°17'W, elevation ca. 100 m].

Distribution: Southern Haiti (Centre, Ouest), NSL–545 m.

Sources: A. Schwartz & R.W. Henderson, 1991, R. Thomas & Powell, 1994d and Domínguez & Diaz, 2011a.

115. *Typhlops thurstoni* Boettger, 1890b. Ber. Senck. Naturf. Ges. 1890: 297–298.

Synonyms: *Typhlops walli* Procter, 1924, and *Typhlops hurstonii* – Bourret, 1936 *(nomen incorrectum).*
Type: Holotype, BMNH (formerly MGM), a 240 mm specimen (F. Thurston).
Type locality: "Nilgiri Hills, Brit. Ostindien" [= Nilgiri Hills, cen. Westsrn Ghats, W Tamil Nadu/SE Kerala States, SW India].
Distribution: Southwestern India (Western Ghats of Kerala, Tamil Nadu), NSL–1300 m.
Sources: Procter, 1924, Murthy, 1990 and R. Sharma, 2003, 2007
Remarks: Holotype erroneously listed in SMF *fide* R. Sharma (2007: 39).

116. *Typhlops titanops* R. Thomas *in* Woods, 1989. Biogeog. West Indies: 415–417, fig. 1c.

Type: Holotype, KU 208812, a 166 mm female (R. Thomas, 5 Aug. 1975).
Type locality: "El Mulito on Río El Mulito, 18 km N Pedernales, Pedernales Province, República Dominicana, 500 feet (152 m)."
Distribution: Hispaniola. Southeastern Haiti (Ouest) and SW Dominican Republic (Pedernales), 150–730 m.
Sources: A. Schwartz & R.W. Henderson, 1991, R. Thomas & Powell, 1995c and Domínguez & Diaz, 2011a.

117. *Typhlops trangensis* E.H. Taylor, 1962. Univ. Kansas Sci. Bull. 43(7): 251–253, fig. 14.

Type: Holotype, FMNH 178238 (formerly EHT-HMS 35754), a 155 mm specimen (E.H. Taylor, 20 May 1958).
Type locality: "Khao Chong Forest Experimental Station, Trang province, Thailand" [= ca. 07°33'N, 99°47'E].
Distribution: Peninsular Thailand (Trang). Known only from type locality.
Sources: E.H. Taylor, 1965 and Wallach, 2001.
Remarks: Taxonomic status uncertain *fide* Wallach (2001: 46).

118. *Typhlops trinitatus* Richmond, 1965. Proc. Biol. Soc. Washington 78: 121–123, fig. 1.

Type: Holotype, AMNH 89820, a 240 mm specimen (J.A. Oliver & J. Tee-Van, 1 April 1956).
Type locality: "Trinidad, in log beside Arima Road, 3 miles above Simla" [= William Beebe Tropical Research Station, Trinidad & Tobago, 10°41'N, 61°17'W, elevation 240 m].
Distribution: Trinidad & Tobago (Tobago, Trinidad), 30–240 m.
Sources: R. Thomas, 1974b, Emsley, 1977, Dixon & Hendricks, 1979, J.C. Murphy, 1997 and Boos, 2001.

119. *Typhlops tycherus* Townsend, Wilson, Ketzler & Luque-Montes, 2008. Zootaxa (1932): 20–23, figs. 2–3.

Type: Holotype, UF 151797 (formerly JHT 2376), a 371 mm female (L.E. Chakerian & L. P. Ketzler, 3 Feb. 2008).
Type locality: "on road between San José de Los Andes and El Cedral (14°54.241'N. 88°05.489'W), 1550 m elevation, western slope of Montaña de Santa Bárbara, Parque Nacional Montaña, Depto. Santa Bárbara, Honduras," (partly in error). Corrected to eastern slope of Montaña de Santa Bárbara *fide* McCranie (2011: 58).
Distribution: Western Honduras (Santa Bárbara), 1550 m.
Source: McCranie, 2011a.

120. *Typhlops unilineatus* (A.-M.-C. Duméril & Bibron, 1844). Erp. Gén. 6: 278–278. (*Onychocephalus unilineatus*)

Type: Holotype, MNHN 1064, a 310 mm specimen (Ms. Richard via Rivoire).
Type locality: "Cayenne" [= French Guiana, 04°55'N, 52°20'W, elevation 10 m].
Distribution: Northern French Guiana (N Cayenne), NSL.
Sources: Jan, 1864 *in* Jan & Sordelli, 1860–1866, Boulenger, 1893a, Peters & Orejas-Miranda, 1970 and Dixon & Hendricks, 1979.
Remarks: Suriname record needs confirmation *fide* A.H. Wynn *in* McDiarmid et al. (1999: 123).

121. *Typhlops veddae* E.H. Taylor, 1947b. Univ. Kansas Sci. Bull. 31(13): 294–295, fig. 3.

Type: Holotype, FMNH 100033 (formerly EHT-HMS 30089 & EHT-CC 50), a 91 mm specimen (E.H. Taylor, 29 Sept. 1944).
Type locality: "12 mi. north of Trincomalee, Ceylon" [= near Nilaveli, 18 km N Trincomalee, N Eastern Prov., NE Sri Lanka, ca. 8°43'N, 81°11'E, elevation NSL *fide* Bauer & Silva, 2007: 5].
Distribution: Northeastern Sri Lanka (N Eastern), NSL. Known only from type locality.
Sources: P. Silva, 1980a, A. Silva, 1990, Wallach, 1993b, 2009 and Somaweera, 2006.

122. *Typhlops vermicularis* Merrem, 1820. Tent. Syst. Amph.: 158.

Synonyms: *Anguis lumbricalis* Daudin, 1803d, *Typhlops flavescens* Bibron & Bory de Saint-Vincent *in* Bory de Saint-Vincent, 1833, *Typhlops syriacus* Jan, 1863b, *Typhlops reticulatus flavescens* Jan 1864 *in* Jan &

Sordelli, 1860–1866, *Typhlops persicus* Blanford, 1874, and *Typhlops vermicalis* – Brehm, 1878 *(nomen incorrectum)*.

Type: Holotype, not designated, lost *fide* Hahn (1980: 76).

Type locality: "Archipelago, Asia." Restricted to Greek Islands *fide* Mertens & Müller (1928: 49).

Distribution: Southeastern Europe, SW Asia and Egypt. Croatia (Dugi Otok Is.), S Serbia, SW Montenegro, SE Kosovo, W Albania, SE Macedonia, S Bulgaria (Blagoevgrad, Burgas, Haskovo, Kardzhali, Pazardzhij, Sveti Ivan Is.), Greece (Aitolia Kai Akarnania, Akhaia, Argolis, Arkadhia, Attiki, Dráma, Évros, Grevená, Ilía, Imathia, Ioánnina, Korinthía, Lakonia, Lárisa, Magnisia, Messinia, Préveza, Thessaloniki, Voiotia, Andros, Chios, Epidavros, Euboea, Kalymnos, Kérkyra, Kithira, Korfu, Kos, Kreta, Lefkas, Leros, Lésvos, Megisti, Náxos, Nidri, Patmos, Rhodos, Salamis, Samos, Skiros and Thásos Is.), Cyprus, SW Turkey (Adana, Adiyaman, Antalya, Aydin, Canakkale, Denizli, Diyarbakir, Gaziantep, Hakkari, Isparta, Izmir, Kilis, Konya, Manisa, Mardin, Mersin, Mugla, Sanliurfa, Siirt, Sirnak, Urfa, Bozcaada, Ciplak, Dereköy, Fener, Gazipasa, Gökçeada, Hekim, Kalem, Karantina, Kekova, Koyun, Marmara, Orak, Pasalimani, Silvan, Sirnak, and Yilan Is.), SW Russia (E Dagestan), SE Georgia, S Armenia, Azerbaijan, W Uzbekistan, SW Turkmenistan, SW Tajikastan, Syria (Damascus, Halab), Jordan (Ajloun, Amman, Balqa, Irbid, Jarash, Mafraq, Tafilah), W Lebanon (Beirut, S Beqaa, Mont- Liban, Nabatiye), Israel (Central, Gaza Strip, Golan Heights, Haifa, Jerusalem, Northern, Tel Aviv, West Bank), NE Egypt (Beheira, Cairo, Faiyum, North Sinai), W Saudi Arabia (Jazan), Iraq (Baghdad, Dhi Qar), Iran (Central, East Azarbaijan, Fars, Golestan, Ilam, Kerman, Kermanshah, Khuzestan, Lorestan, Markazi, Mazandaran, Razavi Khorasan, Tehran, West Azarbaijan, Kaboudan Is.) and NW Afghanistan (Herat), NSL–1900 m.

Sources: Nikolsky, 1916, Corkill, 1932, Werner, 1939, Leviton & Anderson, 1970, Latifi, 1991, Bannikov et al., 1977, Hraoui-Bloquet, 1981, Ilani, 1983a, Baran & Atatür, 1987, Bruno & Maugeri, 1990, Leviton et al., 1992, Grillitsch & Grillitsch, 1993, Szczerbak, 1994, Gasc et al., 1997, Bouskila & Amitai, 2001, Disi et al., 2001, Arnold, 2002, Beshov & Nanev, 2002, Hraoui-Bloquet et al., 2002, Wallach, 2002, Ananjeva et al., 2004, Baran, 2005, Baha El Din, 2006, Kreiner, 2007, Trapp, 2007, Franzen, 2008, Valakos et al., 2008, Baier et al., 2009, Tuniyev et al., 2009, Amr & Disi, 2011, Bar & Haimovitch, 2011, Kornilios et al., 2011, 2012, Stojanov et al., 2011 and Afroosheh et al., 2012.

Remarks: Records from Crete, Kazakhstan and Pakistan need confirmation *fide* Bruno & Maugeri (2001: 77). *Typhlops syriacus* Jan a valid species *fide* Kornilios et al. (2012: 867).

123. *Typhlops violaceus* E.H. Taylor, 1947b. Univ. Kansas Sci. Bull. 31(13): 289–290.

Type: Holotype, FMNH 100068 (formerly EHT-HMS 30091 & EHT-CC C170), a 111 mm specimen (E.H. Taylor, 5 Oct. 1944).

Type locality: "12 miles north of Trincomalee, Ceylon" [= near Nilaveli, 18 km N Trincomalee, N Eastern Prov., NE Sri Lanka, ca. 8°43'N, 81°11'E, elevation NSL *fide* Bauer & Silva, 2007: 5].

Distribution: Northeastern Sri Lanka (N Eastern), NSL. Known only from type locality.

Sources: P. Silva, 1980a, A. Silva, 1990, Wallach, 1993b, 2009 and Somaweera, 2006.

124. *Typhlops wilsoni* Wall, 1908i. J. Bombay Nat. Hist. Soc. (1907–1908) 18(4): 796.

Type: Holotype, formerly BMNH, a 343 mm specimen (A.T. Wilson), lost *fide* Hahn (1980: 76).

Type locality: "Maidan Mihaftan, 30 m. E. of Shushtar in S.W. Persia" [= Khuzestan, Iran, 32°02'N, 49°23'E, elevation ca. 300 m].

Distribution: Southwestern Iran (Khuzestan), ca. 300 m. Known only from type locality.

125. *Typhlops yonenagae* Rodrigues, 1991. Pap. Avul. Zool. 37(22): 343–345, figs. 1–2.

Type: Holotype, MZUSP 10086, a 101 mm specimen (M.T. Rodrigues, 7 Feb. 1989).

Type locality: "Brasil: Bahia: Santo Ignácio (11°06'S, 42° 44'W)" [= Santo Inácio, elevation 530 m].

Distribution: Eastern Brazil (Bahia), 530 m. Known only from type locality.

126. *Typhlops zenkeri* Sternfeld, 1908a. Sitz. Ges. Naturf. Freunde Berlin 1908(4): 92.

Synonym: *Typhlops vermis* Boulenger, 1914c.

Type: Holotype, ZMB 11091, a 135 mm specimen (Morgan).

Type locality: "Kribi (Kamerun)" [= Kribi, ext. W Sud Prov., SW Cameroon, 02°56'N, 09°55'E, elevation 20 m].

Distribution: Southern Cameroon (Sud), 20–620 m.

Sources: Wallach, 1993b and Chirio & LeBreton, 2007.

U

UMBRIVAGA Roze, 1964 (Xenodontidae)

Synonym: *Umbravaga* – Dixon & Soini, 1986 *(nomen incorrectum).*
Type species: *Umbrivaga mertensi* Roze, 1964.
Distribution: Northern South America.
Sources: J.A. Peters & Orejas-Miranda, 1970, Markezich & Dixon, 1979, Zaher et al., 2009 and Vidal et al., 2010.
Remarks: A synonym of *Erythrolamprus fide* Grazziotin et al., 2012.

1. *Umbrivaga mertensi* Roze, 1964. Senck. Biol. 45(3–5): 536–539, figs. 1–5.

Type: Holotype, MBUCV 3046, a 345 mm male (L. Aristeguieta, April 1953).
Type locality: "Parque Nacional Henri Pittier (Rancho Grande), estado Aragua, Venezuela, in a cloud forest, approximately 1000 meters above sea level."
Distribution: Northern Venezuela (Aragua, Distrito Federal), 1000 m.
Sources: Roze, 1966a and Navarrete et al., 2009.

2. *Umbrivaga pyburni* Markezich & Dixon, 1979. Copeia 1979(4): 698–700, figs. 1, 2 (upper).

Synonym: ? *Liophis leucogaster* Jan, 1863a.
Type: Holotype, UTA 3861, a 304 mm male (J.K. Salser, 20 Oct. 1973).
Type locality: "Loma Linda, Departmento de Meta, Colombia" [= Loma Linda, cen. Meta Dept., Colombia, 3°18'N, 73°22'W, elevation 285 m].
Distribution: Central Colombia (Meta), 285 m. Known only from type locality.
Source: Pérez-Santos & Moreno, 1988.

3. *Umbrivaga pygmaea* (Cope, 1868b). Proc. Acad. Nat. Sci. Philadelphia 20(2): 103. (*Liophis pygmaeus*)

Synonym: ? *Liophis leucogaster* Jan, 1863a [fide Dixon, 1980: 10, 20].
Type: Holotype, ANSP 6668, a 176 mm specimen (J. Orton, winter 1867–1868).
Type locality: "Napo, or neighboring pt. of Maranon" [Peru].
Distribution: Amazonia. Colombia, French Guiana, N Brazil (Amazonas, Pará), Ecuador (Napo, Pastaza) and E Peru (Loreto), 50–100 m.
Sources: Dixon, 1980, Pérez-Santos & Moreno, 1988, 1991, Duellman & Mendelson III, 1995, Martins & Oliveira, 1998, Fernandes et al., 1999, Vigle, 2008 and Kawashita-Ribeiro et al., 2011.
Remarks: Transferred to *Umbrivaga fide* Markezich & Dixon, 1979.

UNGALIOPHIS F. Müller, 1880 (Ungaliophiidae)

Synonyms: *Peropodum* F. Müller, 1878b *(nomen illegitimum), Peropodum* Bocourt, 1882 *in* A.H.A. Duméril, Bocourt & Mocquard, 1870–1909, *Ungualiophis* Cope, 1887c *(nomen emendatum),* and *Muzaliophis* – Briceño-Rossi, 1934 *(nomen incorrectum).*
Type species: *Ungaliophis continentalis* F. Müller, 1880.
Distribution: Mesoamerica and Colombia.
Sources: L.C. Stuart, 1954b, Bogert, 1968a, Stimson, 1969, Villa & Wilson, 1990, Zaher, 1994, Walls, 1998a, McDiarmid et al., 1999, M.S.Y. Lee et al., 2007 and G. Köhler, 2008.

1. *Ungaliophis continentalis* F. Müller, 1880. Verh. Naturf. Ges. Basel (1882) 7(1): 142.

Synonym: *Peropodum guatemalensis* Bocourt, 1882 *in* A.H.A. Duméril, Bocourt & Mocquard, 1870–1909.
Type: Holotype, NMBA 427, a 760 mm female (G. Bernoulli, 1877).
Type locality: "Guatemala." Specified as Retaluléu, costa grande v. Guatémala *fide* F. Müller (1878: 59 & 652) [= Retalhuleu, Retalhuleu Dept., SW Guatemala, 14°22'N, 91°45'W, elevation 65 m].
Distribution: Mesoamerica. Southeastern Mexico (E Chiapas), SW Guatemala (Retalhuleu), Honduras (Atlántida, Francisco Morazán) and Nicaragua (Matagalpa), 65–2300 m.
Sources: Conant, 1966, Kramer, 1978, L.D. Wilson & Meyer, 1985, R.M. Burger, 1995, McCranie, 1996, 2011 and G. Köhler, 1997.
Remarks: Original description in F. Müller (1878: 591, 652–654, pl. 1, figs. a, a1–a3). Holotype erroneously listed as NMBA 426 *fide* L. Stuart (1963: 85).

2. *Ungaliophis panamensis* K.P. Schmidt, 1933a. Smithson. Misc. Coll. 89(1): 12–13.

Synonym: *Ungaliophis danieli* Prado, 1941e.
Type: Holotype, USNM 54059, a 500 mm specimen (E.A. Goldman [Smithsonian Biol. Surv.], 6 June 1911).
Type locality: "Cerro Brujo, Panama, altitude 2,000 feet."

Distribution: Lower Central America and Colombia. Southern Nicaragua (Atlántico Sur), Costa Rica (Alajuela, Cartago, Guanacaste, Heredia, Limón, Puntarenas, San José), Panama (Colón) and W Colombia (Antioquia), NSL–2100 m.

Sources: M.J. Corn, 1974, G. Köhler, 1999b, Savage, 2002 and J.C. Murphy & Schlager, 2003.

URAEUS Wagler, 1830 (Elapidae)

Synonyms: *Uracus* – J. Müller, 1832 *(nomen incorrectum), Otracma* – A.-M.-C. Duméril & Bibron, 1844 *(lapsus calami),* and *Wellsus* Hoser, 2009c *(nomen illegitimum).*

Type species: *Coluber haje* Linnaeus, 1758.

Distribution: Africa and Arabia.

Sources: F. Werner, 1923b, Bogert, 1943, Klemmer, 1968, Broadley, 1968c, 1974, Minton et al., 1970, Golay, 1985, Szyndlar & Rage, 1990, Ulber, 1995, Golay et al., 1993, Keogh, 1998, David & Ineich, 1999, Slowinski & Keogh, 2000, Dobiey & Vogel, 2007, Wüster et al., 2007, Hoser, 2009c, 2012a, e, Wallach et al., 2009 and Zaher et al., 2009, 2012.

1. *Uraeus anchietae* (Bocage, 1879). J. Sci. Math. Phys. Nat., Lisboa (1879–1882) 7(26): 89–90. (*Naja anchietae*)

Synonyms: *Naia anchietae barotseensis* Angel, 1921a, and *Naja anchieta* – Bethencourt-Ferreira, 1930 *(nomen incorrectum).*

Type: Lectotype, MBL 1987, designated by Broadley & Wüster (2004: 111), destroyed by fire 18 March 1978.

Type locality: "Caconda, Angola" [= Caconda, Huíla Distr., NW Angola, 13°44'S, 15°04'E, elevation 1700 m].

Distribution: Southern Africa. Angola (Bie, Cuando Cubango, Cuanza Norte, Huambo, Huíla, Lunda, Namibe), SE Democratic Republic of the Congo (Katanga), W Zambia (Central, Luapula, Southern, Western, Chilubi Is.), N Namibia (Caprivi, Kunene, Ohangwena, Okavango, Omaheke, Omusati, Oshana, Oshikoto, Otjozondjupa), N Botswana (Central, Ghanzi, North West) and NW Zimbabwe (Matabeleland North), NSL–2045 m.

Sources: Mertens, 1955, 1971, Laurent, 1964a, Buys & Buys, 1983, Auerbach, 1985, 1987, Broadley, 1995a, Broadley et al., 2003, M. Griffins, 2003, Broadley & Wüster, 2004, J.-F. Trape et al., 2009 and Broadley & Blaylock, 2013.

2. *Uraeus annuliferus* (W.C.H. Peters, 1854). Mber. Königl. Akad. Wiss. Berlin 1854(11): 624–625. (*Naja haje annulifera*)

Type: Holotype, ZMB 2813, a 1650 mm specimen (W.C.H. Peters, June 1843–Aug. 1847).

Type locality: "Tette" [= Tete, Tete Prov., Mozambique, 16°10'S, 33°36'E, elevation 125 m].

Distribution: Southeastern Africa. Zambia (Northern, Southern, Western), S Malawi (Southern), Mozambique (Gaza, Inhambane, Manica, Safala, Tete, Zambézia), Zimbabwe (Bulawayo, Manicaland, Mashonaland Central, Mashonaland East, Mashonaland West, Masvingo, Matabeleland North, Matabeleland South, Midlands), SE Botswana (Central, Gaborone, Kgalagadi, Kgatleng, North East, Southern), Swaziland and NE South Africa (Gauteng, KwaZulu-Natal, Limpopo, Mpumalanga, North West), NSL–1375 m.

Sources: Auerbach, 1985, 1987, Broadley & Cock, 1975, Broadley, 1995a, Clauss & Clauss, 2002, Broadley et al., 2003, Broadley & Wüster, 2004, J.-F. Trape et al., 2009 and Broadley & Blaylock, 2013.

Remarks: Original description reprinted in W.C.H. Peters (1855: 54).

3. *Uraeus arabicus* (Scortecci, 1932a). Atti Soc. Ital. Sci. Nat. 71: 47–49. (*Naja haie arabica*)

Types: Syntypes (5), ISMF, a 1620 mm male, 1535 mm, 1095 mm, and 900 mm females, and a 685 mm juvenile (E. Dubbiosi, Aug. 1928–July 1929).

Type locality: "dintorni di Sana, Yemen, a circa 2400 metri sul livello del mare" [= vicinity of Sana'a, Amanat Al Asimah Govern., W Yemen, 15°21'N, 44°12'E, elevation ca. 2400 m].

Distribution: Southwestern Arabia. Southwestern Saudi Arabia (Asir, Bahah, Jazan, Madinah, Makkah, Qizan), W Oman (Dhofar) and Yemen (Abyan, Ad Dali', Al Bayda, Dhamar, Lahij, Sana'a, Shabwah), NSL–2400 m.

Sources: Gasperetti, 1988 and J.-F. Trape et al., 2009.

Remarks: Gasperetti (1974: 57 & 1988: 427) listed a holotype (without number) among the type series. A highland species that is washed down to lowlands during floods *fide* Gasperetti (1988: 293). A valid species *fide* J.-F. Trape et al., 2009.

4. *Uraeus haje* (Linnaeus, 1758). Syst. Nat., ed. 10, 1: 225. (*Coluber haje*)

Synonyms: *Cerastes candidus* Laurenti, 1768, *Coluber candidissimus* Lacépède, 1789 (*nomen rejiciendum*), *Naja haje viridis* W.C.H. Peters, 1873b, *Naja haje nigra* F. Müller, 1885, *Naja haie* Boulenger, 1896a *(nomen emendatum), Naja haje negra* Bethencourt-Ferreira, 1930 *(nomen emendatum),* and *Naja haje legionis* Valverde, 1989b.

Type: Holotype, NHR (formerly MAFR), a 2000 mm specimen, lost *fide* Kramer & Schnurrenburger (1963: 536).

Type locality: "Aegypto inferiore" [= Lower Egypt].

Distribution: Africa. Western Sahara (N Rio de Oro, Saguia El Hamra), S Morocco (Agadir, Essaouira, Figuig, Kaar es Souk, Marrakech,

U

Oujda, Ouarzazate, Safi, Tan Tan, Tiznit), N Algeria (Béchar, El Oued, Naama), S Tunisia (Bizerte, Gabès, Gafsa, Kairouan, Kasserine, Medenine, Sfax, Sidi Bou Zou, Sousse, Tebessa), N Libya (Al Jabal Al Akhdar, Al Khums, Al Murqub, Banghazi, Darnah, Jabal al-Akhdar, Tripoli), Egypt (Alexandria, Assiut, Aswan, Beheira, Beni Suef, Cairo, Faiyum, Giza, Kafr El-Shaikh, Menufia, N Matrouh, Minya, Qena, Sohag), Senegal (Dakar, Fatick, Kédougou, Louga, Tambacounda, Thiès), Guinea-Bissau (Bolama Is.), Guinea, SW Mali (Kayes, Koulikoro, Mopti, Ségou, Sikasso, S Tomboucto), Burkina Faso (Centre, Centre-Nord, Centre-Ouest, Est, Haute-Bassins, Sahel, Sud-Ouest, Volta-Noire), N Ghana (Upper West), Niger (W Agadez, Dosso, Maradi, Tahoua, Zinder), Nigeria (Borno, Kaduna, Kwara), N Cameroon (Extreme-Nord, Nord), SW Chad (Mayo-Kebbi), N Central African Republic (Vakaga), NE Democratic Republic of the Congo (Orientale), Sudan (Al Jazirah, Al Khartum, An Nil Al Azraq, Ash Shaqiyah, Janub Kurdufan, Nahr An Nil, Sennar), SE South Sudan (Eastern Equatoria), Eritrea (Anseba, Cenral, Gash-Barka, Northern Red Sea, Southern), Ethiopia (Arussi, Gemu Gofa, Hararge, Shoa, Sidamo), N Somalia (Awdal, Bari, Sanaag, Sool, Togdheer, Woqooyi Galbeed), Kenya (Central, S Coast, S Eastern, Nairobi, S Rift Valley, Western), N Uganda (N Eastern, Northern, N Western) and N Tanzania (Arusha, Dodoma, Morogoro, Shinyanga, Singida, Tanga, Zanzibar Is.), NSL–2000 m.

Sources: J. Anderson, 1898, Doumergue, 1901, Mayet, 1903, Loveridge, 1923, 1928, 1953, 1957a, Scortecci, 1928, 1939a–b, Zavattari, 1930a, Flower, 1933, Corkill, 1935, Foley & Parrot, 1935, Bogert, 1940, 1943, Villiers, 1950a–b, 1953, 1975, Chpakowsky & Chnéour, 1954, Mertens, 1955a, 1971, Laurent, 1956a, Saint-Girons, 1956, Condamin, 1958, Domergue, 1959a–b, Sweeney, 1961, Witte, 1962, Kramer & Schnurrenberger, 1963, Roussel & Villiers, 1965, Corkill & Cochrane, 1965, Bons, 1967, 1972, Broadley, 1968c, Marx, 1968, Hughes & Barry, 1969, Papenfuss, 1969, Roman, 1969, 1973b, 1976, 1980, Visser, 1972, V.F.M. FitzSimons, 1974, Pitman, 1974, Stevens, 1974, Broadley & Cock, 1975, Spawls, 1978, Visser & Chapman, 1978, Arnold 1980a, Manaças, 1982, Hughes, 1983, Lanza, 1983a, 1990b, Pienaar et al., 1983, Joger, 1984, MacKay & MacKay, 1985, Patterson, 1987, Branch, 1988, Gasperetti, 1988, Le Berre, 1989, Schätti & Gasperetti, 1994, Broadley, 1995, Spawls & Branch, 1995, Bons & Geniez, 1996, Vernet, 2000, Spawls et al., 2002, Geniez et al., 2004, Chippaux, 2006, Chirio & Ineich, 2006, J.-F. Trape & Mané, 2006b, Chirio, 2009, J.-F. Trape et al., 2009, Largen & Spawls, 2010 and Auliya et al., 2012.

5. *Uraeus niveus* (Linnaeus, 1758). Syst. Nat., ed. 10, 1: 223. (*Coluber niveus*)

Synonyms: *Vipera flava* Merrem, 1820, *Naja gutturalis* A. Smith, 1838, *Naja intermixta* A.-M.-C. Duméril, Bibron & Duméril, 1854b, and *Naja haje capensis* Jan, 1863b (*nomen nudum*).

Type: Holotype, NHR Lin-91 (formerly MDG), a 975 mm male.

Type locality: "Africa." Restricted to Cape of Good Hope, South Africa *fide* V.F.M. FitzSimons (1962a: 297).

Distribution: Southwestern Africa. Southern Namibia (Erongo, Hardap, Karas, Khomas, Omaheke), SW Botswana (Central, Ghanzi, Kgalagadi), W Lesotho and W South Africa (Eastern Cape, W Free State, Northern Cape, North West, Western Cape), NSL–600 m.

Sources: Mertens, 1955, 1971, V.F.M. FitzSimons, 1962, 1974, Visser, 1972, De Waal, 1978, Visser & Chapman, 1978, Broadley, 1983, 1990, Buys & Buys, 1983, Auerbach, 1987, Patterson, 1987, Branch, 1988, Spawls & Branch, 1995, Bauer & Branch, 2001, and Clauss & Clauss, 2002,

Broadley & Wüster, 200 and Broadley & Blaylock, 20134.

6. *Uraeus senegalensis* (Trape, Chirio & Wüster *in* J.F. Trape, Chirio, Broadley & Wüster, 2009). Zootaxa (2236): 12–18, figs. 4–6. (*Naja senegalensis*)

Type: Holotype, MNHN 2008.74 (formerly IRD S-8549), a 1430 mm male (B. N'Dao, Sept. 2008).

Type locality: "near Dielmo (13°43'N, 16°25'W), West Africa" [= vicinity of Dielmo, Fatick Region, SW Senegal, 13°43'N, 16°25'W, elevation 30 m].

Distribution: West Africa. Senegal (Dakar, Fatick, Kaolack, Kédougou, Tambacounda, Thiès), NE Guinea (Kankan), SW Mali (Kayes, Koulikoro, Mopti, Ségou, Sikasso), Burkina Faso (Centre, Centre-Nord, Centre-Ouest, Est, Hauts-Bassins, Sahel, Sud-Ouest, Volta-Noire), N Ghana (Upper West), N Benin (Alibori), Niger (S Niamey, Maradi) and Nigeria (Kwara, North-Eastern), NSL–370 m.

Sources: Chirio, 2009, J.-F. Trape et al., 2009, Wallach et al., 2009, and Hughes, 2013.

UROMACER A.-M.-C. Duméril, Bibron & Duméril, 1854a (Xenodontidae)

U

Synonyms: *Megalocercus* A.H.A. Duméril, 1859, *Uromacre* – Marschall, 1873 (*nomen incorrectum*), *Uromaces* – Palacky, 1898 (*nomen incorrectum*), and *Neomacer* – Cope, 1900 (*nomen incorrectum*).

Type species: *Dendrophis catesbyi* Schlegel, 1837.

Distribution: Hispaniola.

Sources: H.S. Horn, 1969, A. Schwartz & R.W. Henderson, 1984a, 1988, 1991, Rodríguez-Robles & Greene, 1996, R.W. Henderson & Powell, 2009 and Zaher et al., 2009.

1. *Uromacer catesbyi* (Schlegel, 1837). Essai Phys. Serp. 1: 156–157, 2: 226–227. (*Dendrophis catesbyi*)

Synonyms: *Uromacre castesbyi* – A.-M.-C. Duméril, Bibron & Duméril, 1854b *(nomen incorrectum), Uromacer scandax* Dunn, 1920a, *Uromacer catesbyi cereolineatus* A. Schwartz, 1970a, *Uromacer catesbyi frondicolor* A. Schwartz, 1970a, *Uromacer catesbyi hariolatus* A. Schwartz, 1970a, *Uromacer catesbyi inchausteguii* A. Schwartz, 1970a, *Uromacer catesbyi insulaevaccarum* A. Schwartz, 1970a, and *Uromacer catesbyi pampineus* A. Schwartz, 1970a.

Types: Syntypes (2), MNHN 8670–71, an 1165 mm specimen and a 1095 mm specimen lacking a skull (A. Ricord, 1826–1834).

Type locality: "île de St. Domingue" [= Dominican Republic, Hispaniola].

Distribution: Haiti (Artibonite, Nord, Nord-Ouest, Ouest, Sud, Gonâve, Petite Cayemite, Grande Cayemite, Tortuga and Vache Is.) and Dominican Republic (Baoruco, Barahona, Dajabón, Distrito Nacional, Elías Piña, El Seibo, Espillat, Hato Mayor, Independencia, La Altagracia, La Estrelleta, La Romana, La Vega, María Trinidad Sánchez, Monte Christi, Pedernales, Peravia, Puerto Plata, Salcedo, Samaná, San Cristóbal, San Pedro de Macorís, Santiago, Catalina and Saona Is.), NSL–1525 m.

Sources: A. Schwartz, 1970a and A. Schwartz & R.W. Henderson, 1984b.

2. *Uromacer frenatus* (A.C.L.G. Günther, 1865). Ann. Mag. Nat. Hist. (3) 15(86): 94–95, pl. 2, fig. b. (*Ahaetulla frenata*)

Synonyms: *Uromacer inornatus* S.W. Garman, 1887b, *Uromacer dorsalis* Dunn, 1920a, *Uromacer wetmorei* Cochran, 1931, and *Uromacer frenatus chlorauges* Schwartz, 1976.

Type: Holotype, BMNH 1946.1.6.70, a 1320 mm male.

Type locality: Unknown. Designated as Haiti *fide* Cochran (1941: 335).

Distribution: Southwestern Haiti (Grand'Anse, Ouest, Sud, Sud-Est, Gônave, Grande Cayemite, Grosse Caye and Vache Is.) and SW Dominican Republic (Barahona, Independencia, Pedernales, Beata Is.), NSL–915 m.

Sources: Horn, 1969, A. Schwartz, 1976, 1979a and R.W. Henderson & Schwartz, 1984b.

U

3. *Uromacer oxyrhynchus* A.-M.-C. Duméril, Bibron & Duméril, 1854a. Erpét. Gén. 7(1): 722–724, pl. 83, fig. 1.

Type: Holotype, MNHN 8672, a 1228 mm specimen.

Type locality: "Senegal," (in error).

Distribution: Haiti (Nord, Nord-Est, Nord-Ouest, Ouest, Sud-Est, Tortue Is.) and Dominican Republic (Azua, Dajabón, Distrito Nacional, Duarte, El Seibo, Espillat, Hato Mayor, La Altagracia, La Estrelleta, La Romana, La Vega, María Trinidad Sánchez, Monseñor Nouel, Monte Christi, Pedernales, Peravia, Puerto Plata, Salcedo, Samaná, San Cristóbal, San Juan, San Pedro de Macorís, Santiago, Santiago Rodriguez, Valverde, Catalina and Saona Is.), NSL–1220 m.

Source: A. Schwartz & R.W. Henderson, 1984c.

UROMACERINA Amaral, 1930d (Xenodontidae)

Type species: *Uromacer ricardinii* Peracca, 1897d.

Distribution: Brazil.

Sources: Hoge, 1959c, Lema, 1973, P. Müller & Ritter, 1978, Cunha & Nascimento, 1982b, 1993, Zamprogno, 1997, Argôlo, 2001, Barcelos-Ribeiro et al., 2004 and Zaher et al., 2009.

1. *Uromacerina ricardinii* (Peracca, 1897d). Boll. Mus. Zool. Anat. Comp. Univ. Torino 12(282): 1–2, 2 figs. (*Uromacer ricardinii*)

Type: Holotype, MZUT 1769 (formerly MZUT 2057), a 720 mm specimen (Ricardini).

Type locality: "S. Paulo (Brasile)" [= São Paulo, E São Paulo State, SE Brazil, 23°33'S, 46°38'W, elevation 800 m].

Distribution: Brazil (Bahia, Espírito Santo, Minas Gerais, Pará, Paraná, Rio de Janeiro, Rio Grande do Sul, Santa Catarina, São Paulo), 600–850 m.

UROPELTIS Cuvier, 1829 (Uropeltidae)

Synonyms: *Uropeltes* – J. Müller, 1832 *(nomen incorrectum), Urapeltis* – Swainson, 1839 *(nomen incorrectum), Siluboura* Gray, 1845 *(nomen emendatum), Coloburus* A.H.A. Duméril *in* A.-M.-C. Duméril & Bibron, 1851c, *Uropelta* Hoeven, 1852 *(nomen emendatum), Crealia* Gray, 1858c, *Silybura* W.C.H. Peters, 1861d *(nomen emendatum), Euroleptis* – Hoffmann, 1890 *(nomen incorrectum), Sylybura* – Briceño-Rossi, 1934 *(nomen incorrectum), Silubura* – Deraniyagala, 1955 *(nomen incorrectum), Uroaeltis* – Murthy, 1990d *(nomen incorrectum), Uroeltisp* – Murthy, 1990d *(nomen incorrectum),* and *Uropeltls* – Murthy, 1990d *(nomen incorrectum).*

Type species: *Uropeltis ceylanicus* Cuvier, 1829.

Distribution: India and Sri Lanka.

Sources: Jan, 1865b, M.A. Smith, 1943, Gans, 1966, Murthy, 1981c, 1982b, Mahendra, 1984, Rajendran, 1985, Cadle et al., 1990, McDiarmid et al., 1999, Gower et al., 2005, 2008 and Aengals, 2009.

Remarks: Gender of *Uropeltis* is feminine *fide* Lanza & Boscherini (2000: 328).

1. *Uropeltis arcticeps* (A.C.L.G. Günther, 1875). Proc. Zool. Soc. London 43(1): 229, fig. 1. (*Silybura arcticeps*)

Synonyms: *Silybura madurensis* Beddome, 1878c, and *Silybura nilgherriensis picta* Beddome, 1886.

Types: Syntypes (2), BMNH 1946.1.16.11–12, a 194 mm and 148 mm specimen.

Type locality: "Tinnevelly" [= Tinnevelly Hills, N Tirunelveli Distr., S Tamil Nadu State, SE India, 1200–1800 m].

Distribution: Southern India (Western Ghats of SE Kerala, SW Tamil Nadu), NSL–1830 m.

Source: Chandramouli & Ganesh, 2010.

2. *Uropeltis beddomii* (A.C.L.G. Günther, 1862a). Ann. Mag. Nat. Hist. (3) 9(50): 56–57. (*Silybura beddomii*)

Synonym: *Silybura beddomei* Beddome, 1886 *(nomen emendatum).*

Types: Syntypes (4), BMNH 1946.1.16.13–15, three 250–279 mm specimens, and MNHN 1895.90 (formerly BMNH) (R.H. Beddome, 1857–1861).

Type locality: "Anamallay Hills" [= Anamalai Hills, S Western Ghats, E Kerala /W Tamil Nadu States, SW India, ca. 10°22'N, 77°08'E].

Distribution: Southern India (Western Ghats of SE Kerala, SW Tamil Nadu).

Source: Whitaker & Captain, 2004.

3. *Uropeltis bicatenata* (A.C.L.G. Günther, 1864a). Rept. Brit. India: 191, pl. 17, figs. h, h'. (*Silybura bicatenata*)

Synonym: *Silybura bicatenata* A.C.L.G. Günther, 1863c (*nomen nudum*).

Type: Holotype, BMNH 1946.1.16.8 (formerly BMNH 1860.3.19.1277), a 248 mm male (W.H. Sykes, 1824–1831).

Type locality: "Deccan, British India" [= Deccan Plateau, India].

Distribution: Western India (Western Ghats of W Maharashtra), 740–1140 m.

Sources: Beddome, 1886 and Gower et al., 2008.

Remarks: Illustrations of holotype in Gower et al. (2008: figs. 2–3). A valid species (previously a synonym of *Uropeltis ceylanica) fide* Gower et al. (2008: 67). McDiarmid et al. (1999: 144) questioned type locality but Sykes worked in the Deccan as a statistical reporter for many years. Type locality restriction of Wynad, Malabar, S India, 1065 m *fide* Gans (1966: 18) invalid.

4. *Uropeltis broughami* (Beddome, 1878c). Proc. Zool. Soc. London 46(1): 800–801. (*Silybura broughami*)

Synonym: *Silybura levingii* Beddome, 1878c.

Types: Syntypes (2), BMNH 1946.1.16.29 (formerly BMNH 1883.1.12.23), a female, and MNHN 1895.89 (formerly BMNH) (R.H. Beddome).

Type locality: "Sirumullay hills (Madura district), 5500 feet elevation" [= Tirumalai Hills, Madura Distr., Eastern Ghats, N Tamil Nadu State, SE India, elevation 1675 m].

Distribution: Southern India (Western and Eastern Ghats of SE Kerala, Tamil Nadu), 1230–1680 m.

5. *Uropeltis ceylanica* Cuvier, 1829. Règne Anim., ed. 2, 2: 76, pl. 7, fig. 2. (*Uropeltis ceylanicus*)

Synonyms: *Uropeltis ceylonica* Wagler, 1830 *(nomen emendatum), Pseudo-typhlops ceylanicus* Schlegel, 1839 *in* 1837–1844 *(nomen illegitimum),* ? *Uropeltis affinis* Jerdon, 1854, *Silybura brevis* A.C.L.G. Günther, 1862a, *Silybura shorttii* Beddome, 1863a, *Silybura nilgherriensis* Beddome, 1863a, and *Silybura nilgherriensis annulata* Beddome, 1886.

Type: Lectotype, MNHN 39, a 161 mm specimen (L. de Lateur, July 1822), designated by Gans (1966a: 18).

Type locality: "Ceylan" [= Sri Lanka] (in error *fide* Constable, 1949: 119) via lectotype selection.

Distribution: Southwestern India (Western Ghats of W Karnataka, E Kerala, SW Tamil Nadu), 500–1500 m.

Sources: Cocteau, 1833, Murthy, 1990, Malhotra & Davis, 1991 and Pradhan, 2008.

6. *Uropeltis dindigalensis* (Beddome, 1877a). Proc. Zool. Soc. London 45(1): 167. (*Silybura dindigalensis*)

Types: Syntypes (4), BMNH 1946.1.16.2 (formerly BMNH 1883.1.12.5), BMNH 1946.1.16.4 (formerly BMNH 1883.1.12.7), BMNH 1946.1.16.37 (formerly BMNH 1877.8.10.3), and MNHN 1895.88 (formerly BMNH), 305–381 mm specimens (R.H. Beddome, 1875–1877).

Type locality: "Sirumullay hills, near Dindigul, at 4000–5000 feet elevation, Southern India" [= Tirumalai Hills, near Dindigul (10°22'N, 77°59'E), Eastern Ghats, N Tamil Nadu State, SE India, elevation 1200–1525 m].

Distribution: Southeastern India (Eastern Ghats of Tamil Nadu), 1200–1700 m.

Source: Chandramouli & Ganesh, 2010.

7. *Uropeltis ellioti* (Gray, 1858c). Proc. Zool. Soc. London 26(1): 262, fig. 1. (*Siloboura ellioti*)

Synonyms: *Silybura elliotti* A.C.L.G. Günther, 1864a *(nomen emendatum),* and *Silybura punctata* A.C.L.G. Günther, 1875.

Type: Holotype, BMNH 1946.1.16.6 (W. Elliot, 1845–1858).

Type locality: "Madras" [= Chennai, NE Tamil Nadu State, SE India].

Distribution: Southern India (Western and Eastern Ghats of Andhra Pradesh, Goa, S Gujarat, W Karnataka, E Kerala, Orissa, Tamil Nadu), 100–1800 m.

Source: Rajendran, 1985, Sanyal, 1993, Sanyal et al., 1993, R.C. Sharma, 2000, Whitaker & Captain, 2004, N. Khaire, 2006 and Pradhan, 2008.

Remarks: Original description reprinted in Gray (1858f: 377–378).

8. *Uropeltis liura* (A.C.L.G. Günther, 1875). Proc. Zool. Soc. London 43(1): 228, pl. 31, fig. b. (*Silybura liura*)

Type: Holotype, BMNH 1946.1.16.7 (formerly BMNH 1874.4.29.1206), a male (R.H. Beddome, 1857–1875).

Type locality: "Malabar" [= N Kerala State, SW India]. Specified as Madura and Tinnevelly Hills *fide* BMNH catalogue and M.A. Smith (1943: 84).

Distribution: Southern India (Eastern Ghats of Tamil Nadu), 920–1520 m.

Sources: Rajendran, 1985 and Whitaker & Captain, 2004.

9. *Uropeltis macrolepis* (W.C.H. Peters, 1861d). Mber. Königl. Akad. Wiss. Berlin 1861(10): 904–905. (*Silybura macrolepis*)

Synonym: *Uropeltis macrolepis mahableshwarensis* Chari, 1955.

Type: Holotype, BMNH 1946.1.15.99 (formerly BMNH 1888.4.20.4), a 265 mm specimen.

Type locality: "Ceylon?" (in error). Unknown *fide* BMNH catalogue *fide* M.A. Smith (1943: 79). Designated as Bombay Hills between lats. 18° 7' and 19° 7' N, British India *fide* M.A. Smith (1943: 80).

Distribution: Western India (Western Ghats of Goa, S Gujarat, W Maharashtra, Salsette Is.), NSL–1370 m.

Sources: Beddome, 1886, Chari, 1955, R.C. Sharma, 2000, Whitaker & Captain, 2004, N. Khaire, 2006 and Pradhan, 2008.

10. *Uropeltis macrorhyncha* (Beddome, 1877a). Proc. Zool. Soc. London 45(1): 167–168. (*Silybura macrorhyncha*)

Synonym: *Silybura macrorhynchus* Boulenger, 1893a *(nomen emendatum)*.

Type: Holotype, BMNH 1946.9.7.45 (formerly BMNH 1883.1.12.24), a female (R.H. Beddome, 1857–1877).

Type locality: "Anamullay Mountains, 4000 feet elevation, southern India" [= Anamalai Hills, S Western Ghats, E Kerala/W Tamil Nadu States, SW India, ca. 10°22'N, 77°08'E, elevation 1220 m].

Distribution: Southern India (Western Ghats of SE Kerala, SW Tamil Nadu), 100–1335 m.

11. *Uropeltis maculata* (Beddome, 1878a). Proc. Zool. Soc. London 46(1): 154–155. (*Silybura maculata*)

Synonym: *Uropeltis maculatus* M.A. Smith, 1943.

Types: Syntypes (8), BMNH 1946.1.16.63–65, BMNH 1946.1.16.83–84, MNHN 1895.81 (formerly BMNH), and NMW 21598.1–2 (formerly BMNH) (R.H. Beddome, 1857–1878, via G. Edwards & F. Steindachner, 1883).

Type locality: "Southern India." Specified as Anamallays, higher ranges 6000 to 7000 feet elevation *fide* Beddome (1886: 22).

Distribution: Southern India (Western Ghats of SE Kerala, SW Tamil Nadu), 1065–2400 m.

Sources: Rajendran, 1985, Pradhan, 2002 and Hutton & David, 2009.

12. *Uropeltis melanogaster* (Gray, 1858c). Proc. Zool. Soc. London 26(1): 264–265, fig. 5. (*Mytilia melanogaster*)

Synonym: *Plectrurus ceylonicus* W.C.H. Peters, 1859c.

Types: Syntypes (4), BMNH 1946.1.16.94–97 (G.H.K. Thwaites, 1854).

Type locality: "Ceylon" [= Sri Lanka]. Specified as Meniakanda Group, Mousakanda Estate and Gammaduwa, in the East Matale Hills, Sri Lanka *fide* A. Silva (2009: 38).

Distribution: Central Sri Lanka (Central, Uva), 700–1300 m.

Sources: Wall, 1921, P. Silva, 1980, A. Silva, 1990, 2009, A. Silva et al., 2005, Somaweera, 2006 and Olori & Bell, 2012.

Remarks: Original description reprinted in Gray (1858f: 380). Only two syntypes mentioned by Gray (1858c: 265).

13. *Uropeltis myhendrae* (Beddome, 1886). Ann. Mag. Nat. Hist. (5) 17(97): 15–16. (*Silybura nilgherriensis myhendrae*)

Types: Syntypes (2), BMNH 1946.1.16.9 (formerly BMNH 1883.1.12.49), a 334 mm specimen (R.H. Beddome, 1857–1882), and MNHN 1895.95 (formerly BMNH) (R.H. Beddome, 1857–1882).

Type locality: "South Travancore, on the Myhendra Mountain" [= Mahendragiri, ext. S Tamil Nadu State, S India, 8°23'N, 77°31'E, elevation 1200 m].

Distribution: Southwestern India (Western Ghats Goa, E Kerala, SW Tamil Nadu), 600–1200 m.

Sources: Rajendran, 1985 and Pradhan, 2008.

Remarks: MNHN 1897.255 considered a possible type *fide* Das *in* McDiarmid et al. (19999: 147) but only two syntypes are mentioned *fide* Beddome (1886: 16) and MNHN 1897.255 is from Punga Hills, India.

14. *Uropeltis nitida* (Beddome, 1878a). Proc. Zool. Soc. London 46(1): 154. (*Silybura nitida*)

Synonym: *Uropeltis nitidus* M.A. Smith, 1943.

Types: Syntypes (7), BMNH 1946.1.16.30–31, BMNH 1946.1.13.95–97, MNHN 1895.87a–b (formerly BMNH) (R.H. Beddome, 1857–1878), and NMW

U

21597 (R.H. Beddome, 1857–1878, via G. Edwards & F. Steindachner, 1883).

Type locality: "Anamallays, 4000 to 5000 ft., Southern India" [= Anamalai Hills, S Western Ghats, SE Kerala State, SW India]. Specified as Anamallays, about the Nelliampady estates on the Cochin side [= vicinity of W Nelliyampathy Hills, Palghat Distr., Kerala State, S India, ca. 10°32'N, 76°42'E, 1200–1525 m] *fide* Beddome (1886: 19).

Distribution: Southwestern India (Western Ghats of SE Kerala), 1335–1700 m.

Source: Murthy, 1990d.

15. *Uropeltis ocellata* (Beddome, 1863a). Madras Quart. J. Med. Sci. 6: 46, pl. 1, fig. 4. (*Silybura ocellata*)

Synonyms: *Silybura dupeni* Beddome, 1878a, *Silybura ochracea* Beddome, 1878a, and *Uropeltis ocellatus* M.A. Smith, 1943.

Type: Lectotype, BMNH 1946.1.15.59 (R.H. Beddome, 1857–1863), designated by Gans (1966a: 21).

Type locality: "Walaghat on the Western slopes of the Neilgherries in the dense forests at an elevation of 3,500 feet, ['in the dense moist forests'], India" via lectotype selection.

Distribution: Southwestern India (Western Ghats of W Karnataka, E Kerala, Madhya Pradesh, SW Tamil Nadu), 610–1980 m.

Sources: Wall, 1918c, Constable, 1949, Rajendran, 1985, Murthy, 1990d, Radhakrishna & Palot, 2007, Sur et al., 2007, Chandra, 2009 and Hutton & David, 2009.

Remarks: See McDiarmid et al. (1999: 148) for discussion of types. Original description reprinted in Beddome (1853b: 226). Longest syntype 368 mm.

16. *Uropeltis petersi* (Beddome, 1878a). Proc. Zool. Soc. London 46(1): 154. (*Silybura petersi*)

Types: Syntypes (6), BMNH 1946.1.1.7–9 and BMNH 1946.1.16.75 (formerly BMNH 1878.1.11.3a–d), and MNHN 1895.80a–b (formerly BMNH) (R.H. Beddome, 1857–1878).

Type locality: "Anamallays, 4000 ft." [= Anamalai Hills, S Western Ghats, E Kerala/W Tamil Nadu States, SW India, ca. 10°22'N, 77°08'E, elevation 1220 m]. Restricted to forests above Ponache, India *fide* Beddome (1886: 22).

Distribution: Southern India (Western Ghats of SE Kerala, SW Tamil Nadu), 1220–1700 m.

Remarks: McDiarmid et al. (1999: 149) listed only four syntypes.

17. *Uropeltis phillipsi* (G. Nicholls, 1929b). Ceylon J. Sci. 15B: 153–155, pl. 32, figs. 1–4, 7. (*Silybura phillipsi*)

Type: Holotype, BMNH 1946.1.17.2 (formerly BMNH 1929.2.5.1), a 222 mm specimen (W.W.A. Phillips).

Type locality: "Meniakanda Group, Gammaduwa, in the East Matale Hills, Ceylon" [= Sri Lanka].

Distribution: Central Sri Lanka (Central), 360–1115 m.

Sources: A. Silva, 1990, A Silbva et al., 2005 and Somaweera, 2006.

18. *Uropeltis phipsonii* (G.E. Mason, 1888). Ann. Mag. Nat. Hist. (6) 1(3): 184–185. (*Silybura phipsonii*)

Types: Syntypes (2), BMNH 1946.1.16.33–34 (formerly BMNH 1897.7.19.2–3), a 277 mm and 218 mm specimen (H.M. Phipson, 1883–1888).

Type locality: "Bombay ghats, Bombay Presidency" [= N Western Ghats, Maharashtra State, W India].

Distribution: Southwestern India (Western Ghats of Goa, W Karnataka, E Kerala, W Maharashtra, SW Tamil Nadu), NSL–500 m.

Sources: Tiwari & Sharma, 1970, Rajendran, 1985, J.C. Murphy & Schlager, 2003, Whitaker & Captain, 2004, N. Khaire, 2006, Radhakrishna & Palot, 2007 and Pradhan, 2008.

Remarks: Previously a synonym of *U. rubrolineatus.*

19. *Uropeltis pulneyensis* (Beddome, 1863b). Proc. Zool. Soc. London 31(1): 228, pl. 25, fig. 2. (*Plectrurus pulneyensis*)

Synonym: *Silybura guentheri* Beddome, 1878a.

Types: Syntypes (9), BMNH 1946.1.17.6, MNHN 1895.78, and ZSI 4381 [unknown], ZSI 6948 [Palni], ZSI 6950 [unknown], and ZSI 6972–75 [Madras] (R.H. Beddome, 1857–1863).

Type locality: "Pulney Hills, 7000 to 8000 feet, Southern India" [= Palani Hills, W Tamil Nadu State, S India, 2135–2440 m].

Distribution: Southern India (Western Ghats of SE Kerala, SW Tamil Nadu), 915–2440 m.

Sources: Rajendran, 1985 and Pradhan, 2002.

Remarks: I. Das et al. (1998: 145) listed seven syntypes from ZSI.

20. *Uropeltis rubrolineata* (A.C.L.G. Günther, 1875). Proc. Zool. Soc. London 43(1): 228. (*Silybura rubrolineata*)

Synonym: *Uropeltis rubrolineatus* M.A. Smith, 1943.

Type: Lectotype, BMNH 1946.1.15.53 (formerly BMNH 1874.4.29.804), a 177 m specimen (R.H. Beddome, 1857–1875), designated by Beddome (1886: 14).

Type locality: "Tranvancore hills" [= Tirunelveli Hills, SW Tamil Nadu State, S India] via lectotype selection.
Distribution: Southern India (Western Ghats of SE Kerala, SW Tamil Nadu), NSL–835 m.
Sources: Rajendran, 1985 and Murthy, 1990.
Remarks: See McDiarmid et al. (1999: 150) for details on types.

21. *Uropeltis rubromaculata* (Beddome, 1867). Madras Quart. J. Med. Sci. 11: 15, pl. 2, fig. 3. (*Silybura rubromaculata*) (*nomen corrigendum*)

Synonyms: *Silybura rubro-maculata* Beddome, 1867 *(nomen incorrigendum), Silybura rubromaculata* – A.C.L.G. Günther, 1875 *(nomen corrigendum),* and *Uropeltis rubromaculatus* M.A. Smith, 1943.
Types: Syntypes (6), BMNH 1946.1.15.51–52, a 365 mm and 368 mm specimen, BMNH 1946.1.15.82–84, a 343 mm, 349 mm and 256 mm specimen, MNHN 1895.97 (formerly BMNH) (R.H. Beddome, 1857–1867), and NMW 18932 (R.H. Beddome, 1857–1867, via G. Edwards & F. Steindachner, 1883).
Type locality: "Anamallay forests; 4,000 feet elevation, Madras Presidency" [= Anamalai Hills, S Western Ghats, E Kerala/W Tamil Nadu States, SW India, ca. 10°22'N, 77°08'E, elevation 1220 m].
Distribution: Southern India (Western Ghats of SE Kerala, SW Tamil Nadu), 1065–1820 m.
Sources: Rajendran, 1985, Murthy, 1990 and Hutton & David, 2009.
Remarks: Plate in original description erroneously listed as pl. 1 *fide* Beddome (1867: 15).

22. *Uropeltis ruhunae* Deraniyagala, 1954. Proc. Tenth Ann. Ceylon Assoc. Advan. Sci. (1): 24.

Synonym: *Uropeltis ruhanae* Gans, 1966 *(nomen emendatum).*
Type: Holotype, CMS RS 552, a 250–268 mm specimen (P. Abrahams).
Type locality: "Galle, Ceylon" [= Galle, Southern Prov., S Sri Lanka].
Distribution: Southern Sri Lanka (Southern). Known only from holotype.
Sources: Deraniyagala, 1955, P. Silva, 1980, A. Silva, 1990 and Somaweera, 2006.
Remarks: Taxonomic status doubtful *fide* A. Silva (2001: 64).

23. *Uropeltis smithi* Gans, 1966. Das Tierreich (84): 22–24. (*nomen substitutum*)

Synonym: *Rhinophis grandis* Beddome, 1867 *(nomen praeoccupatum).*
Types: Syntypes (2), BMNH 1946.1.16.82 and MNHN 1895.79.
Type locality: "Anamallay forests in moist woods at 4,000 feet elevation, Madras Presidency" [= Anamalai Hills, S Western Ghats, E Kerala/W Tamil Nadu States, SW India, ca. 10°22'N, 77°08'E].
Distribution: Southern India (Western Ghats of SE Kerala, SW Tamil Nadu), 1220–2155 m.
Remarks: Original description in Beddome (1867: 15, pl. 2, fig. 4).

24. *Uropeltis woodmasoni* (Theobald, 1876). Cat. Rept. Brit. India: 135. (*Silybura woodmasoni*) (*nomen corrigendum*)

Synonyms: *Silybura wood-masoni* Theobald, 1876 *(nomen incorrigendum), Silybura melanogaster* A.C.L.G. Günther, 1875 *(nomen praeoccupatum), Silybura nigra* Beddome, 1878a, and *Silybura woodmasoni* – Sclater, 1891 *(nomen corrigendum).*
Type: Lectotype, ZSI 8760, a 204 mm specimen.
Type locality: "Palney Hills, British India" [= Palani Hills, W Tamil Nadu State, S India] via lectotype selection.
Distribution: Southern India (Western and Eastern Ghats of SE Kerala, Tamil Nadu), 1860–2100 m.
Sources: Constable, 1949 and Rajendran, 1985.

UROTHECA Bibron, 1843 *in* Ramón de la Sagra, 1838–1843 (Dipsadidae)

Synonym: *Uroteca* Gundlach, 1867 *(nomen emendatum).*
Type species: *Calamaria dumerilii* Bibron, 1840 *in* Ramón de la Sagra, 1838–1843.
Distribution: Central America and N South America.
Sources: J.A. Peters & Orejas-Miranda, 1970, C.W. Myers, 1974, Villa et al., 1988, Savage & Crother, 1989, Savage & Lahanas, 1989 and Zaher et al., 2009.
Remarks: *Pliocercus* recognized as distinct from *Urotheca fide* C.W. Myers & Cadle (1999) but Savage & Crother, 1989, allocated *Pliocercus* and seven species formerly in *Rhadinaea* to *Urotheca.*

1. *Urotheca decipiens* (A.C.L.G. Günther, 1893 *in* 1885–1902). Biol. Cen.-Amer., Rept. Batr.: 105, pl. 37, fig a. (*Ablabes decipiens*)

Synonym: *Rhadinaea albiceps* Amaral, 1924b.
Types: Syntypes (3), BMNH 1946.1.3.95–97, longest syntype 457 mm (H. Rogers via F.D. Godman, 1861–1889).
Type locality: "Costa Rica, Irazu" [= Volcán Irazú, N Cartago or E San José Prov., Costa Rica, ca. 9°59'N, 83°51'W]
Distribution: Lower Central America and NW South America. Honduras (Gracias a Dios), S Nicaragua (Río San Juan), Costa Rica (Alajuela, Cartago, Guanacaste, Heredia, Limón, Puntarenas, San José), W Panama, Colombia (Valle de Cauca), Ecuador (Esmeraldas), NSL–1800 m.

Sources: Pérez-Santos & Moreno, 1988, Savage, 2002, McCranie et al., 2003, 2006, Solórzano, 2004, Castro-Herrera & Vargas-Salinas, 2007, McCranie, 2011a and Salazar & Barquero, 2012.

Remarks: Record for Medellin, Colombia needs confirmation *fide* C.W. Myers (1974: 176). Ecuador record is tentatively identified as *U. decipiens fide* Ortega-Andrade et al., 2010.

2. *Urotheca dumerilii* (Bibron, 1840 *in* Ramón de la Sagra, 1838–1843). Hist. Fís. Polít. Nat. Cuba 2(2): pl. 26, figs. 1–5. (*Calamaria dumerilii*)

Type: Holotype, MNHN 733, a 344 mm male (D. Ramón de la Sagra, 1824–1835).

Type locality: "Cuba," (in error). Corrected to Colombia *fide* C.W. Myers (1974: 176).

Distribution: Northwestern Colombia (Antioquia, Chocó), 150–1540 m.

Sources: Maglio, 1970 and Pérez-Santos & Moreno, 1988.

Remarks: Plate 26 of Bibron (1840) was labelled *Calamaria dumerilii,* but the text (1843) employed *Urotheca dumerilii.* Supplemental original description in Bibron, 1843 *in* Ramón de la Sagra (1838–1843: 131–132). Original description reprinted in Bibron, 1843 *in* Ramón de la Sagra (1841–1843: 218, pl. 26, figs. 1–5).

3. *Urotheca fulviceps* (Cope, 1886b). Proc. Amer. Philos. Soc. (1885) 23(122): 279–280. (*Rhadinaea fulviceps*)

Type: Holotype, USNM 61187 (formerly SIM 14118), a 341–347 mm female (G.W. Nelson, 1880–1885).

Type locality: "Panama, tropical America" [= Panama city, Panama Dept., cen. Panama, 9°00'N, 79°31'W, elevation 15 m].

Distribution: Lower Central America and NW South America. Southwestern Costa Rica (Puntarenas, San José), Panama, NW Ecuador (Esmeraldas, Pichincha), cen. Colombia (Caldas, Cundinamarca, Santander) and W Venezuela (Zulia), NSL–700 m.

Sources: Pérez-Santos & Moreno, 1988, 1991, Manzanilla et al., 1998, Rojas-Runjaic et al., 2008 and Navarrete et al., 2009.

Remarks: High elevation records of 800–2000 m doubtful *fide* C.W. Myers (1974: 180).

4. *Urotheca guentheri* (Dunn, 1938b). Copeia 1938(4): 198. (*Rhadinaea guentheri*) (*nomen substitutum*)

Synonyms: *Tachymenis decipiens* A.C.L.G. Günther, 1895 *in* 1895–1902 *(nomen praeoccupatum), Rhadinaea persimilis* Dunn, 1938b *(nomen praeoccupatum), Rhadinaea güntheri* Dunn, 1938b *(nomen incorrigendum)* and *Rhadinaea guentheri* – J.A. Peters & Orejas-Miranda, 1970 *(nomen corrigendum).*

Type: Lectotype, BMNH 1946.1.8.15, adult female (H. Rogers), designated by C.W. Myers (1974: 182). Longest syntype 483 mm (svl).

Type locality: "Irazu, Costa Rica" [= Volcán Irazú, N Cartago or E San José Prov., Costa Rica, ca. 9°59'N, 83°51'W] via lectotype selection.

Distribution: Central America. Eastern Honduras (Colón, Gracias a Dios), E Nicaragua (Atlántico Norte, Chontales, Jinotega), N Costa Rica (Alajuela, Cartago, Heredia, Guanacaste, Limón, Puntarenas, San José) and W Panama (Bocas del Toro, Coclé, Panamá, Veraguas), NSL–2100 m.

Sources: G. Köhler, 1999, 2001, 2008, McCranie et al., 2001, 2006, Lotzkat et al., 2010, McCranie, 2011a, Bartuano & Ibañez, 2012 and Salazar & Barquero, 2012.

Remarks: Original description in A.C.L.G. Günther (1895 *in* 1895–1902: 163, pl. 53, fig. a).

5. *Urotheca lateristriga* (Berthold, 1859). Nachr. Georg.-Augusts-Univ. Königl. Ges. Wiss. Göttingen (17): 180–181. (*Liophis lateristriga*)

Synonyms: *Dromicus temporalis* Cope, 1861a, *Dromicus frenatus* W.C.H. Peters, 1863a, *Dromicus nuntius* Jan, 1863b, *Urotheca coronata* Steindachner, 1901, *Erythrolamprus labialis* F. Werner, 1909b, and *Liothis lateristriga* – Briceño-Rossi, 1934 *(nomen incorrectum).*

Type: Holotype, not designated, (formerly ZMUG) (C. Degenhardt, 1839–1845), lost *fide* Medem (1965: 304).

Type locality: "Nova-Granada" [= Colombia]. Specified as Popayan, New Granada [= W Colombia, probably on the Pacific versant *fide* C.W. Myers & Böhme, 1996: 17] *fide* Jan, 1866 *in* 1866–1870, 2(18): pl. 5, fig 2.

Distribution: Northwestern South America. Colombia (Antioquia, Magdalena, Santander), W Ecuador (Guayas, Los Rios, Pichincha) and Peru, NSL–2000 m.

Sources: Pérez-Santos & Moreno, 1988, 1991 and C.W. Myers & Böhme, 1996.

Remarks: Type locality designation of near Popayán, Cauca Dept., Colombia, 2°27'N, 76°37'W, elevation 1800 m *fide* Dunn & Stuart (1951: 56) in error *fide* C.W. Myers & Böhme (1996: 17).

U

6. *Urotheca multilineata* (W.C.H. Peters, 1863c). Mber. Königl. Akad. Wiss. Berlin 1863(6): 279–280. (*Dromicus multilineatus*)

Type: Lectotype, ZMB 2108, a 315 mm specimen (Appun), designated by C.W. Myers (1974: 190).

Type locality: "Puerto Cabello und Caracas, Venezuela" via lectotype selection. Restricted to Caracas, Venezuela *fide* Dunn (1944f: 479).

Distribution: Venezuela (Aragua, Carabobo, Distrito Federal, Falcón, Lara, Miranda), 800–2000 m.

Sources: Mijares-Urrutia, 2000 and Navarrete et al., 2009.

Remarks: Lectotype is specimen referred to as variety "A" by W.C.H. Peters (1863c: 280). C.W. Myers (1974: 190) reported the lectotype (ZMB 2108) to be lost but Bauer et al. (1995: 71) discovered three syntypes, including ZMB 2108.

7. *Urotheca myersi* Savage & Lahanas, 1989. Copeia 1989(4): 892–894, fig. 1 (paratype).

Type: Holotype, LACM 137604 (formerly CRE 1036), a 305 mm male (J. Hernandez-Camacho, 11 June 1961).

Type locality: "Cordillera de Talamanca, near Finca de Jardin (2255 m), approx. 9.6 km south by road of Villa Mills, just south of Division, on the Carretera Interamericana (Ruta 2), Canton Perez-Zeledon, Provincia de San Jose, Costa Rica."

Distribution: Central Costa Rica (Puntarenas, San José), 1500–2255 m.

Sources: Lips, 1993b, Savage, 2002 and Solórzano, 2004.

8. *Urotheca pachyura* (Cope, 1875a). J. Acad. Nat. Sci. Philadelphia (1876) (2) 8(2): 145–146. (*Contia pachyura*)

Synonym: *Rhadinaea decipiens rubricollis* E.H. Taylor, 1954.

Type: Holotype, USNM 30618, a 335 mm male (W.M. Gabb, 1873–1875).

Type locality: "Sipurio, Costa Rica" [= Sipurio, near Suretka (9°34'N, 82°56'W, elevation 50 m), S Limón Prov., E Costa Rica].

Distribution: Lower Central America. Extreme S Nicaragua (W Río San Juan), E Costa Rica (Alajuela, Cartago, Heredia, Limón) and W Panamá, NSL–1360 m.

Sources: Savage, 2002, Solórzano, 2004 and Salazar & Barquero, 2012.

V

†*VECTOPHIS* Rage & Ford, 1980 (Alethinophidia *incertae sedis*)

Type species: †*Vectophis wardi* Rage & Ford, 1980.
Distribution: Upper Eocene of United Kingdom.
Source: Rage & Ford, 1980.
Remarks: Alethinophidia, *incertae sedis fide* Rage (1984b: 58).

1. †*Vectophis wardi* Rage & Ford, 1980. Tertiary Res. 3(2): 54–56, fig. 4, pl. 1, figs. 14–17.

Type: Holotype, MNHN CGB 27, one middle trunk vertebra.
Type locality: "Totland Bay, Isle of Wight, (Great Britain)" [United Kingdom, lower Headon Beds, Middle Headonian (= upper Bartonian), upper Eocene].
Distribution: Upper Eocene (Priabonian, MP 17–20: 33.9–37.2 mya) of United Kingdom (England).

VERMICELLA Gray *in* A.C.L.G. Günther, 1858 (Elapidae)

Synonyms: *Homaloselaps* Jan, 1858b, *Vermicalla* – A.C.L.G. Günther, 1858, and *Neelaps* A.C.L.G. Günther, 1863a.
Type species: *Calamaria annulata* Gray *in* Grey, 1841b.
Distribution: Australia.
Sources: F. McCoy, 1881, Storr, 1968, 1979, Cogger et al., 1983a, Storr, 1985, S.K. Wilson & Knowles, 1988, Hoser, 1989, 2012e, Hutchinson, 1990, Cogger, 1992, Keogh & Smith, 1996, Keogh, 1998, Keogh et al., 1998, Storr et al., 2002, Scanlon & Lee, 2004, Sanders et al., 2008 and Zaher et al., 2009.
Remarks: *Neelaps* a synonym *fide* Wallach, 1985 and Zaher et al., 2012.

1. *Vermicella annulata* (Gray *in* Grey, 1841b). J. Two Exped. Australia 2: 443–444. (*Calamaria annulata*)

Synonyms: *Elaps occipitalis* A.-M.-C. Duméril, Bibron & Duméril, 1854b, *Vermicella lunulata* Krefft, 1869b, and *Rhynchelaps latizonatus* De Vis, 1905.
Type: Holotype, BMNH 1946.1.19.21 (Lewis).
Type locality: "New Holland" [= Australia]. Specified as Swan River, Australia *fide* BMNH catalogue.
Distribution: Eastern Australia (Australian Capital Territory, New South Wales, Northern Territory, Queensland, SE South Australia, N Victoria), NSL–1325 m.
Sources: J. White, 1790, Shine, 1980a and Longmore, 1986.

2. *Vermicella calonotus* (A.-M.-C. Duméril, Bibron & Duméril, 1854b). Erpét. Gén. 7(2): 1241–1242, pl. 75 bis. (*Furina calonotus*) (*nomen corrigendum*)

Synonyms: *Furina calonotos* A.-M.-C. Duméril, 1853 (*nomen nudum*), *Furina calonotos* A.-M.-C. Duméril, Bibron & Duméril, 1854c (*nomen incorrigendum*), *Furina calonotus* – A.-M.-C. Duméril, Bibron & Duméril, 1854c (*nomen corrigendum*), *Neelaps caledonicus* Hoffmann, 1890 (*lapsus calami*), *Neelaps neocaledonicus* Palacky, 1898 (*lapsus calami*), and *Furina calonota* – Boulenger, 1896a.
Type: Lectotype, MNHN 3943a, the larger of the two syntypes (J.P. Verreaux, 1844), designated by Wells & Wellington (1985: 46),
Type locality: "Tasmanie" [= Tasmania, Australia] (in error). Corrected to Nouvelle-Hollande [= Australia] *fide* Duméril (1853: 517).
Distribution: Extreme SW Australia (ext. SW Western Australia), NSL–35 m.
Source: Bush et al., 1995.
Remarks: A.C.L.G. Günther (1863a: 24) emended the specific epithet to *calonotus,* which was considered valid *fide* Cogger et al. (1983: 228).

3. *Vermicella intermedia* Keogh & Smith, 1996. J. Zool., London 240(4): 688–691, pl. 1, fig. a.

Type: Holotype, NTM 17196, a 360 mm female (G. Gow, 1985).
Type locality: "Humpty Doo District of the Northern Territory (12°38'S, 131°15'E)," Australia.
Distribution: Northern Australia (N Northern Territory, N Western Australia), NSL–320 m.

4. *Vermicella multifasciata* (Longman, 1915). Mem. Queensland Mus. 3: 30. (*Furina multifasciata*)

Type: Holotype, QM J2019 (formerly QM J14/2019), a 362–365 mm specimen (G.F. Hill).
Type locality: "Port Darwin, Northern Territory" [= Port Darwin, Darwin, ext. N Northern Territory, N Australia, 12°28'S, 130°51'E, elevation NSL].
Distribution: Northern Australia (ext. N Western Australia, NW Northern Territory, Melville Is.), 15–300 m.

Source: Clarke & How, 1995.

5. *Vermicella snelli* Storr, 1968. J. Roy. Soc. West. Australia (1967) 50(2): 82. (*Vermicella annulata snelli*)

Type: Holotype, WAM 19203, a 271–504 mm female, (C. Snell, early 1963).

Type locality: "Mundiwindi, Western Australia, in 23° 51' S, 120° 09' E," Australia.

Distribution: Western Australia (NW Western Australia), 45–755 m.

6. *Vermicella vermiformis* Keogh & Smith, 1996. J. Zool., London 240(4): 695–696, pl. 1, fig. e.

Type: Holotype, NTM 34484, a 494 mm male (6 June 1972).

Type locality: "Alice Springs, Northern Territory (23°42'S, 133°53'E)," Australia.

Distribution: Central Australia (Northern Territory), 550–725 m.

Remarks: Holotype listed as NTM 34485 *fide* Keogh & Smith (1996: 691).

VIPERA Laurenti, 1768 (Viperidae)

Synonyms: *Echidna* Link, 1806 *(nomen praeoccupatum), Chersea* Fleming, 1822, *Ammodytes* Bonaparte, 1831b (*nomen rejiciendum*), *Rhinaspis* Bonaparte, 1834 *(nomen nudum), Rhinechis* Fitzinger, 1843 *(nomen praeoccupatum), Rinechis* – Vandoni, 1914 *(nomen incorrectum), Rhinochis* – Caziot, 1922 *(nomen incorrectum), Mesovipera* A.F.T. Reuss, 1927a, *Teleovipera* A.F.T. Reuss, 1927a *(nomen illegitimum), Latastea* A.F.T. Reuss, 1929a, *Latasteopara* A.F.T. Reuss, 1935a, *Rhinophis* – Guillon, 1936 *(lapsus calami)* and *Latasteopara* A.F.T. Reuss, 1937 *(nomen praeoccupatum).*

Type species: *Coluber aspis* Linnaeus, 1758.

Distribution: Europe and N Africa.

Fossil records: Lower Miocene to upper Pleistocene of Europe. Fossils unassigned to species include lower Miocene (Agenian, MN 1–2: 20.0–23.0 mya) of France, lower Miocene (Orleanian, MN 3–5: 13.7–20.0 mya) of Austria, France and Spain, middle Miocene (Astaracian, MN 6–8: 11.1–13.7 mya) of Germany, Poland and Slovakia, upper Miocene (Vallesian, MN 9: 9.7–11.1 mya) of Austria, Hungary and Ukraine, upper Miocene (Turolian, MN 11–13: 4.9–8.7 mya) of Spain, lower Pliocene (Ruscinian, MN 14: 4.2–4.9 mya) of Poland and Ukraine, middle Pliocene (Ruscinian, MN 15: 3.2–4.2 mya) of Russia and Ukraine, middle/upper Pliocene (Ruscinian/Villanyian, MN 15–16: 2.6–4.2 mya) of Spain, upper Pliocene (Villanyian, MN 16–17: 1.8–3.2 mya) of Austria, France, Greece and Ukraine, lower Pleistocene (Calabrian: 0.8–1.8 mya) of France and Greece, middle Pleistocene (Ionian: 0.13–0.78 mya) of Croatia, and upper Pleistocene (Tarantian: 0.01–0.13 mya) of Bulgaria, France and Ukraine.

Sources: Schwarz, 1936, F.A. Reuss, 1939b, Wittmann, 1954, Groombridge, 1980, Jirousek, 1983, Obst, 1983, W. Böhme & Joger, 1984, Bozhansky & Kudryavcev, 1986, Nilson & Andrén, 1986, 1988a–b, Brodmann, 1987, Marx & Nörstrom, 1988, Ashe & Marx, 1988, Golay et al., 1993, Bergman, 1994, Leenders, 1994, Venczel & Sen, 1994, David & Ineich, 1999, Herrmann, et al., 1999, McDiarmid et al., 1999, Lenk et al., 2001, Szyndlar & Rage, 2002, Mallow et al., 2003, Garrigues et al., 2005, Dobiey & Vogel, 2007, Stümpel & Joger, 2009, Phelps, 2010 and Hoser, 2012d.

Remarks: Bailon et al., 2010, reported *Vipera* sp. from the Pliocene of Majorca, Spain. The following genera and species have been separated from *Vipera: Daboia (deserti, †gedulyi, †maghrebiana, mauritanica, †maxima, palaestinae, russelii, †sarmatica, siamensis), Macrovipera (†burgenlandica, †kuchurganica, lebetina, schweizeri), Montivipera (albicornuta, albizona, bornmulleri, bulgardaghica, latifi, raddei, wagneri, xanthina),* and *Pelias (altaica, anatolica, barani, berus, darevskii, dinniki, ebneri, eriwanensis, kaznakovi, lotievi, magnifica, nikolskii, olguni, orlovi, pontica, renardi, sachalinensis, seoanei, ursinii).*

1. †*Vipera aegertica* Augé & Rage *in* Ginsburg, 2000. Mém. Mus. Natl. Hist. Nat. 183: 303–305, fig. 29.

Type: Holotype, MNHN Sa 23714, one middle trunk vertebra.

Type locality: "Sansan, Gers, France; Astaracien (MN 6), Miocène moyen."

Distribution: Middle Miocene (Astaracian, MN 6: 12.8–13.7 mya) of France. Known only from type series.

Source: Szyndlar & Rage, 2002.

2. *Vipera ammodytes* (Linnaeus, 1758). Syst. Nat., ed. 10, 1: 216–217. (*Coluber ammodytes*)

Synonyms: *Vipera illyrica* Laurenti, 1768, *Vipera illyrica fusca* Laurenti, 1768, *Vipera illyrica pallido-caerulescens* Laurenti, 1768 *(nomen incorrigendum), Vipera monoceros* Bonaparte, 1840 *(nomen nudum), Vipera amodytes* Erber, 1863 *(nomen emendatum), Vipera ammodyta* – Canton, 1895 *(nomen incorrectum), Vipera ammodytes steindachneri* F. Werner, 1897d, *Vipera ammodytes meridionalis* Boulenger, 1903d, *Vipera ammodytes montandoni* Boulenger, 1904e, *Vipera ammodytes transcaucasiana* Boulenger, 1913a, *Vipera ammodytes transversovirgata* Tzarevsky, 1915, *Vipera meridionalis connectens* Bolkay, 1919, *Teleovipera transversovirgata* A.F.T. Reuss, 1927d *(nomen illegitimum), Vipera aspis balcanica* Buresch & Zonkow, 1934, *Rhinaspis ammodytes velebitensis* A.F.T. Reuss, 1932, *Rhinaspis illyrica litoralis* A.F.T.

V

Reuss, 1935a, *Rhinaspis illyrica melanura* A.F.T. Reuss, 1937, *Vipera ammodytes ruffoi* Bruno, 1968, *Vipera ammodytes gregorwallneri* Sochurek, 1974, *Vipera ammodytes montondoni* – Weinstein, Minton & Wilde, 1985 *(nomen incorrectum)*, and *Vipera ammodytes montadori* – Sota, Pérez & Minton, 1988 *(nomen incorrectum)*.

Type: Holotype, ZMUU 95 (E. Carleson via C. Grill., 1747).

Type locality: "Oriente" [= Asia]. Restricted to Zara [= Zara, Croatia] *fide* Schwarz (1936: 225). Restricted to Illyria [= Croatia] *fide* Mertens & Müller (1928: 51).

Distribution: Southeastern Europe. Northeastern Italy (Friuli-Venezia Giulia, Trentino-Alto Adige, Veneto), S Austria (Carinthia, E Tyrol), W Hungary, Romania, Bulgaria (Blagoevgrad, Burgas, Dobrich, Gabrovo, Grand Sofiya, Haskovo, Kardzhali, Lovech, Kyustendil, Montana, Pazardzhik, Pernik, Pleven, Plovdiv, Razgrad, Rousse, Shumen, Silistra, Sliven, Smolyan, Sofiya, Stara Zagora, Targovishte, Varna, Veliko Tarnovo, Vidin, Vratsa, Yambol), Slovenia, Croatia (Ada, Brag, Hvar, Korcula, Krk, Mljet, Pag, Solta, and Vis Is.), Bosnia-Herzegovina, Serbia, Montenegro, Kosovo, Macedonia, Albania, Greece (Aitolia & Akarnania, Akhaia, Argolis, Arkhadia, Arta, Attica, Evritania, Evvoia, Fthiotis, Ilia, Ioannina, Kastoria, Kavala, Kilkis, Laconia, Magnisia, Messalina, Pieria, Rhodhopi, Thesprotia, Thessaloniki, Voiotia, Xanthi, Andros, Ano-Kufonissi, Antíparos, Délos, Euboea, Herákleia, Ios, Itháki, Kefalloniá, Kérkira, Lefkáda, Mykonos, Náxos, Páros, Sikinos, Syros, Stróngylo, and Tinos Is.), W Turkey and Georgia, NSL–2500 m.

Fossil records: Upper Miocene (Turolian, MN 13: 4.9–7.3 mya) of Hungary, lower Pliocene (Ruscinian, MN 14: 4.2–4.9 mya) of Hungary, middle Pliocene (Ruscinian, MN 15: 3.2–4.2 mya) of Poland, upper Pliocene (Villanyian, MN 16: 2.6–3.2 mya) of Poland, lower Pleistocene (Calabrian: 0.8–1.8 mya) of Austria, Czech Republic, Poland and Spain, middle Pleistocene (Ionian: 0.13–0.78 mya) of Greece, and upper Pleistocene (Tarantian: 0.01–0.13 mya) of Serbia.

Sources: Tzarevsky, 1915, Bolkay, 1920, Bruno, 1968, 1984, Eiselt & Baran, 1970, Lotze, 1973, Sochurek, 1976, Bannikov et al., 1977, Szyndlar, 1981, 1984a, 1991b, Billing et al., 1990, Bruno & Maugeri, 1990, Biella & Blattler, 1992, Biella,1995, Herrmann & Joger, 1997, Nilson & Andrén, 1997, Holman, 1998a, Kutrup, 1999, Szyndlar & Rage, 1999, Cabela et al., 2001, Demetropoulos & Ioannides, 2002, Szyndlar & Rage, 2002, Tomovic et al., 2002, Tomovíc & Dzukic, 2003, Sindaco et al., 2006, Smedt, 2006, Tomovic, 2006, Ivanov, 2007, Trapp, 2007, Ursenbacher et al., 2008, Valakos et al., 2008 and Stojanov et al., 2011.

Remarks: Original description based on Linnaeus (1749b: 517–519, pl., fig. 2) with type locality of Portam Ottomanicam [= ? Ottoman seaport, now Balkans or Turkey]. *Vipera transcaucasiana* a valid species *fide* Tuniyev et al. (2009: 140).

3. †*Vipera antiqua* Szyndlar, 1987. J. Vert. Paleont. 7(1): 68–69, figs. 11a–f.

Type: Holotype, CUP 4538, one trunk vertebra.

Type locality: "Lower Miocene (MN 4) locality of Dolnice (west Czechoslovakia) near to the town of Cheb in westernmost Bohemia."

Distribution: Lower Miocene (Agenian, MN 1: 22.4–23.0 mya) of Germany, and lower Miocene (Orleanian, MN 4: 16.0–16.9 mya) of Czech Republic and Germany.

Sources: Szyndlar & Schleich, 1993, Nilson & Andrén, 1997, Szyndlar & Rage, 1999, 2002 and Augé & Rage, 2000.

4. *Vipera aspis* (Linnaeus, 1758). Syst. Nat., ed. 10, 1: 218. (*Coluber aspis*)

Synonyms: *Vipera francisci redi* Laurenti, 1768 *(nomen incorrigendum)*, *Vipera mosis charas* Laurenti, 1768, *Coluber redi* Gmelin, 1789 (*nomen substitutum*), *Coluber charasii* G. Shaw, 1802, *Vipera vulgaris* Latreille *in* Sonnini & Latreille, 1801a, *Vipera ocellata* Latreille *in* Sonnini & Latreille, 1801a, *Vipera aspis pallasii* Merrem, 1820, *Vipera atra* Meisner, 1820, *Vipera communis* Millet, 1828, *Vipera hugyi* Schinz, 1833, *Vipera hugii* – Schinz, 1833 *(nomen incorrectum)*, *Vipera aspis fusca* Bonaparte, 1834 *in* 1832–1841, *Vipera aspis nigra* Bonaparte, 1834 *in* 1832–1841, *Vipera aspis ocellata* Bonaparte, 1834 *in* 1832–1841, *Vipera aspis rubriventris* Bonaparte, 1834 *in* 1832–1841, *Vipera aspis rufa* Bonaparte, 1834 *in* 1832–1841, *Vipera cholobia* Bonaparte, 1840 *(nomen nudum)*, *Vipera heegeri* Schinz *in* Bonaparte, 1840 *(nomen nudum)*, *Vipera aspis cinerea* Betta, 1853 (*nomen nudum*), *Vipera aspis rufescens* Betta, 1853 (*nomen nudum*), *Vipera aspis plumbea* Massalongo, 1854, *Vipera aspis vulgaris* Massalongo, 1854, *Vipera aspis brunnea* Betta, 1857, *Vipera aspis cinerascens* Betta, 1857 *(nomen substitutum)*, *Vipera aspis cinerea* Betta, 1857, *Vipera aspis fulva* Betta, 1857, *Vipera aspis fusca-plumbeiventris* Betta, 1857 *(nomen incorrigendum)*, *Vipera aspis immaculata* Betta, 1857, *Vipera aspis isabellina* Betta, 1857, *Vipera aspis rufiventris* Betta, 1857, *Vipera heegeri* Schreiber, 1875 *(nomen nudum)*, *Vipera aspis immaculata* Calderini, 1878, *Vipera aspis calderinii* Betta, 1879 *(nomen substitutum)*, *Vipera aspis immaculata* Betta, 1879, *Vipera aspis infernalis* F. Müller, 1885, *Vipera berus aspis hugyi* Camerano, 1889 *(nomen illegitimum)*, *Vipera aspis lineata* Dürigen, 1897, *Vipera aspis delalande* Phisalix, 1902, *Vipera aspis typus* Calabresi, 1924, *Vipera aspis rudolphi-italica* A.F.T. Reuss, 1924a *(nomen incorrigendum)*, *Mesovipera alpina* Reuss, 1927a *(nomen nudum)*, *Mesovipera maculata* A.F.T. Reuss, 1930c (*nomen nudum*), *Mesovipera maculata* A.F.T. Reuss, 1935a *(nomen praeoccupatum)*, *Mesovipera morathi* A.F.T. Reuss, 1938a *(nomen nudum)*, *Mesovipera morathi leopontica* A.F.T.

V

Reuss, 1938a *(nomen nudum), Mesovipera morathi* A.F.T. Reuss, 1938b, *Mesovipera morathi leopontica* A.F.T. Reuss, 1938b, *Mesovipera stemmler-morathi alpina* A.F.T. Reuss, 1939 *(nomen incorrigendum), Mesovipera stemmler-morathi despaxi* A.F.T. Reuss, 1939 *(nomen incorrigendum), Vipera aspis montecristi* Mertens, 1956, *Vipera aspis zinnikeri* Kramer, 1958, *Vipera aspis delalandei* Phisalix, 1968 *(nomen emendatum), Vipera aspis francisciredi* – Kramer, 1971 *(nomen corrigendum), Vipera aspis montechristi* – Saint-Girons *in* H.W. Parker, Bellairs & Saint-Girons, 1971 *(nomen incorrectum), Vipera aspis zinnikeri bilineata* Saint-Girons, 1978 *(nomen illegitimum), Vipera aspis heinzdischeki* Sochurek, 1979a, *Vipera aspis francisredi* – Obst, 1983 *(nomen incorrectum), Vipera aspis franciscia-edi* – González, 1991 *(nomen incorrectum),* and *Vipera aspis typus* Golay, Smith, Broadley, Dixon, McCarthy, Rage, Schätti & Toriba, 1993 *(nomen nudum).*

Type: Holotype, not designated, a male, lost *fide* Golay et al. (1993: 280).

Type locality: "Gallia" [= France]. Restricted to Poitou, France *fide* Schwarz (1936: 213).

Distribution: Southern Europe. Northeastern Spain (Álava, Andorra, Barcelona, Burgos, Gerona, Huesca, Lérida, Navarra, Santander, N Zaragossa), France (Ain, Allier, Alpes-Maritimes, Ardèche, Ariège, Aube, Aude, Aveyron, Basses-Alpes, Basses-Pyrénées, Belfort, Cantal, Cher, Charente, Charente-Maritime, Corrèze, Corsica, Côte-d'Or, Creuse, Deux-Sèvres, Dordogne, Doubs, Drôme, Essonne, Eure-et-Loir, Gard, Gers, Gironde, Haut-Rhin, Haute-Garonne, Haute-Loire, Haute-Marne, Haute-Saône, Haute-Savoie, Haute-Vienne, Hautes-Alpes, Hautes-Pyrénées, Hérault, Indre, Indre-et-Loire, Isère, Jura, Landes, Loire, Loire-Atlantique, Loiret, Loir-et-Cher, Lot, Lot-et-Garonne, Lozère, Maine-et-Loire, Manche, Mayenne, Meuse, Meurthe-Moselle, Moselle, Nièvre, Orne, Puy-de-Dôme, Pyrénées-Orientales, Rhône, Saône-et-Loire, Sarthe, Savoie, Seine-et-Marne, Tarn, Tarn-et-Garonne, Val-d'Oise, Var, Vaucluse, Vendée, Vienne, Vosges, Yonne, Yvelines), ext. SW Germany (S Baden-Württemberg), S Switzerland (Aargau, Basel-Landschaft, Basel-Stadt, Bern, Fribourg, Genève, Graubungen, Jura, Solothurn, Neuchâtel, Ticino, Valais, Vaud), Monaco, Elba and Monte Cristo Is., Italy (Abruzzi, Basilicata, Calabria, Campania, Emilia-Romagna, Friuli-Venezia Giulia, Lazio, Liguria, Lombardia, Marche, Molise, Piemonte, Puglia, Toscana, Trentino-Alto Adige, Umbria, Valle d'Aosta, Veneto, Elba and Sicily Is.), San Marino, SE Bulgaria and NW Slovenia, NSL–2930 m.

Fossil records: Middle Pleistocene (Ionian: 0.13–0.78 mya) of France.

Sources: Kramer, 1958, P. Müller, 1969c, Eiselt & Baran, 1970, Bruno, 1975, 1977, 1984, Saint-Girons, 1980b, Naulleau, 1987b, 1997, Billing et al., 1990, Duguy, 1995, R. Günther, 1996, Herrmann & Joger, 1997, Nilson & Andrén, 1997, Holman, 1998a, Szyndlar & Rage, 1999, Zuffi & Bonnet, 1999, Bologna et al., 2000, Hofer et al., 2001, Zuffi et al., 2000, Zuffi, 2002, 2003, Filippi, 2003, Aprea et al., 2006, Sindaco et al., 2006, Smedt, 2006, Ursenbacher et al., 2006, Krecsák, 2007, Renner & Vitzthum, 2007, Golay et al., 2008, Vacher & Geniez, 2010 and Stojanov et al., 2011.

Remarks: Former subspecies *V. a. atra, V. a. hugyi,* and *V. a. zinnikeri* are valid species *fide* Zuffi et al. (2000: 191).

5. *Vipera latastei* Boscá, 1878. Bull. Soc. Zool. France 3: 116–121, pl. 4, figs. 1–1a, 4. (*nomen corrigendum*)

Synonyms: *Vipera latasti* Boscá, 1878 *(nomen incorrigendum), Vipera latastei* – Boscá, 1879 *(nomen corrigendum), Vipera berus aspis latastei* Camerano, 1889 *(nomen illegitimum), Vipera latastii* Boulenger, 1896a *(nomen emendatum), Rhinaspis latastei nigricaudata* A.F.T. Reuss, 1933a, and *Vipera latastei gaditana* Saint-Girons, 1977.

Type: Lectotype, BMNH 1946.1.21.5 (formerly BMNH 1920.1.20.252), a 518 mm female (V.-F. Lataste, Nov. 1878), designated by Saint-Girons (1977: 605–606).

Type locality: "Ciudad Real" [= Ciudad Real, Castille La Mancha Prov., cen. Spain, 38°58'N, 3°52'E, elevation 635 m] via lectotype selection.

Distribution: Iberian Peninsula and ext. NW Africa. Portugal (Beja, Braga, Bragansa, Coimbra, Évora, Faro, Lisboa, Portalegre, Santarém, Viseu), Spain (Alicante, Almería, Aragon, Badajoz, Cáceres, Cádiz, Castellón, Catalonia, Ciudad Real, Córdoba, Granada, Guadalajara, Huelva, Huesca, Jaén, Léon, Lérida, Lugo, Madrid, Málaga, Melilla, Navarra, Orense, Salamanca, Segovia, Soria, Tarragona, Teruel, Toledo, Valencia, Zamora), ext. N Morocco (Nador, N Oujda, Tangier, Tétouan), N Algeria (Algiers, Annaba, Tipasa) and N Tunisia (Jendouba), NSL–2780 m.

Sources: Chpakowsky & Chnéour, 1954, A. Davidson, 1964, Saint-Girons, 1977, 1980b, Spawls & Branch, 1995, Bons & Geniez, 1996, Schleich et al., 1996, Herrmann & Joger, 1997, Pleguezuelos, 1997, Holman, 1998a, A. Salvador, 1998, Szyndlar & Rage, 1999, Amat, 2004, Malkmus, 2005, Brito et al., 2006 and Smedt, 2006.

Fossil records: Middle to upper Pleistocene (Ionian-Tarantian: 0.01–0.78 mya) of Spain.

Remarks: Original description reprinted in Boscá (1879: 85–86, pl. 4). Type locality restriction of Valencia *fide* Mertens (1925: 105) and Valencia del Cid *fide* Mertens & Müller (1928: 52) invalid.

6. †*Vipera meotica* Zerova *in* Szyndlar & Zerova, 1992. Neues Jb. Geol. Paläont. Abh. 184(1): 94–96, figs. 19–21.

Type: Holotype, ZIKP 45-5023, one basisphenoid.

Type locality: "lower layer of Cherevichnoie (Ukraine, USSR), located on the left bank of the estuarine lake Khadzibeyskiy, between villages Morozovka and Cherevichnoie (Belyaevka district in Odessa oblast'), approximately 20 km north from Odessa; upper Miocene, early Meotian, i.e., middle Turolian or MN 12."

Distribution: Upper Miocene (Turolian, MN 12: 7.3–7.8 mya) of Ukraine. Known only from type series.

Source: Nilson & Andrén, 1997.

7. *Vipera monticola* Saint-Girons, 1954. Bull. Soc. Zool. France 78(5–6): 475. (*Vipera latastei monticola*) (*nomen susbstitutum*)

Synonym: *Vipera latastei montana* Saint-Girons, 1953 *(nomen praeoccupaum).*

Types: Syntypes (10), MHNG 1353.7 (F. Proust, 1951), MNHN 8410, MNHN 8958–59, MNHN 1973.79–80, SMF 42001, ZMA 11154, MHNR 492 R, and ISCM (H. Saint-Girons, 1952), latter lost *fide* Golay et al. (1993: 287).

Type locality: "Haut-Atlas (Massif du Toubkal), Maroc, 2.500–3.100 m" [= Toubkal Massif, High Atlas Mtns., E Marrakech and W Ouarzazate Prov., cen. Morocco, elevation 2500–3100 m].

Distribution: Central Morocco (Azilal, Beni Mellal, Fès, Marrakech, Meknes, Ouarzazate), 1200–3100 (3900) m.

Sources: Beerli et al., 1986, Pillet, 1994, Spawls & Branch, 1995, Bons & Geniez, 1996, Schleich et al., 1996 and Brito et al., 2006.

Remarks: Original description in H. Saint-Girons (1953: 24–27). Proust collected three syntypes, Saint-Girons the other seven. A valid species *fide* Beerli et al., 1986.

8. †*Vipera natiensis* Bailon, García-Porta & Quintana-Cardona, 2002. C. R. Palevol. 1: 228, 230–231, figs. 1a–h.

Type: Holotype, IMCS Nati-12 no. 1362, one trunk vertebra.

Type locality: "Punta Nati 12, northwestern Minorca (Balearic Islands, Spain), probably Pliocene."

Distribution: Pliocene (1.8–4.9 mya) of Spain. Known only from type locality.

VIRGINIA Baird & Girard, 1853 (Natricidae)

Synonyms: *Potamophis* Fitzinger, 1843 *(nomen praeoccupatum), Haldea* Baird & Girard, 1853, *Conocephalus* A.-M.-C. Duméril, Bibron & Duméril, 1854a, *Falcóneria* Theobald, 1868b, *Amphiardis* Cope, 1889a, *Virginea* Hoffmann, 1890 *(nomen incorrectum), Amphardis* Cope, 1895c *(nomen emendatum),* and *Haldae* – Gehlbach, 1965a *(nomen incorrectum).*

Type species: *Virginia valeriae* Baird & Girard, 1853.

Distribution: Southeastern USA.

Fossil records: Upper Pleistocene of E USA.

Sources: Zillig, 1958, D. Clark, 1964, Pisani, 1972, Pisani & Collins, 1972, Ernst & Barbour, 1989, Conant & Collins, 1991, Rossman & Wallach, 1991, Holman, 1995a, 2000a, Palmer & Braswell, 1995, Tennant, 1997, Werler & Dixon, 2000, Alfaro & Arnold, 2001, Ernst & Ernst, 2003 and Zaher et al., 2009.

1. *Virginia striatula* (Linnaeus, 1766). Syst. Nat., ed. 12, 1: 375–376. (*Coluber striatulus*)

Synonyms: *Falcóneria bengalensis* Theobald, 1868b, and *Virginia inornata* S.W. Garman, 1884.

Type: Holotype, not designated (A. Garden, 1760–1761), location unknown.

Type locality: "Carolina" [USA]. Restricted to the vicinity of Charleston, South Carolina *fide* Klauber (1948: 6).

Distribution: Southeastern USA (S Alabama, Arkansas, N Florida, Georgia, SE Kansas, Louisiana, Mississippi, S Missouri, North Carolina, E Oklahoma, South Carolina, SW Tennessee, Texas, SE Virginia), NSL–350 m.

Fossil records: Upper Pleistocene (Rancholabrean II) of USA (Georgia, Texas, Virginia).

Source: Powell et al., 1994.

2. *Virginia valeriae* Baird & Girard, 1853. Cat. No. Amer. Rept., Serp. 1: 127–128.

Synonyms: *Carphophis harperti* A.-M.-C. Duméril, Bibron & Duméril, 1854a, *Virginia elegans* Kennicott, 1859a, and *Haldea valeriae pulchra* Richmond, 1954.

Type: Holotype, USNM 1962, a 238 mm specimen (V. Blaney, 1847).

Type locality: "Kent Co., Md" [= Kent Co., Maryland, USA].

Distribution: Eastern USA (Alabama, Arkansas, Delaware, District of Colombia, N Florida, Georgia, E Kansas, Kentucky, S Illinois, S Indiana, S Iowa, Louisiana, Maryland, Missouri, Mississippi, New Jersey, North Carolina, S Ohio, E Oklahoma, Pennsylvania, South Carolina, Tennessee, E Texas, Virginia, West Virginia), NSL–885 m.

Fossil records: Upper Pleistocene (Rancholabrean II) of USA (Georgia, Pennsylvania, Tennessee, Virginia).

Sources: Powell et al., 1992, B. Miller, 1995 and Fitch, 1999.

VIRIDOVIPERA Malhotra & Thorpe, 2004c (Viperidae)

Type species: *Trimeresurus stejnegeri* K.P. Schmidt, 1925a.

Distribution: Southeastern Asia.

Sources: Maki, 1931, 1931b, Maslin, 1942, M.A Smith, 1943, Maslin, 1945, Klemmer, 1963, Leviton, 1964d, W. Burger, 1971, Kramer, 1977, Hoge & Romano-Hoge, 1981a, Maes, 1989, Malhotra & Davis, 1991, Golay et al., 1993, Zhao & Adler, 1993, Kraus et al., 1996, David & Vogel, 1998, David & Ineich, 1999, McDiarmid et al., 1999, Parkinson, 1999, Malhotra & Thorpe, 1997, 2000, 2004c, Ziegler et al., 2001, Giannasi et al., 2001a, G. Peng & Fuji, 2001, David et al., 2002, 2009, 2011, Gumprecht et al., 2004c, Castoe & Parkinson, 2006, Grismer et al., 2006a, 2008b, Sanders et al., 2006, Dawson et al., 2008, Ganesh et al., 2008, Orlov et al., 2008, 2009, I. Das, 2010, 2012, David et al., 2011c, P. Guo & Wang, 2011, Sumontha et al., 2011 and Hoser, 2012d, o.

1. *Viridovipera gumprechti* (David, Vogel, Pawels & Vidal, 2002). Nat. Hist. J. Chulalongkorn Univ. 2(1): 7–10, figs. 4–8. (*Trimeresurus gumprechti*)

Type: Holotype, MNHN 1999.9072, a 726 mm male (T. Chan-ard, May 1996).

Type locality: "Phu Luang Wildlife Research Station, about 15 km northwest from the city of Phu Luang and 45 km southwest from the city of Loei, Loei Province, Thailand."

Distribution: Southeast Asia. Northeastern India (Assam, Meghalaya West Bengal), S China, Myanmar, N Thailand (Loei, Phitsanulok), Laos and Vietnam (Lai Chau, Lao Cai), 800–1200 m.

Sources: Malhotra & Thorpe, 2004b, David & Mathew, 2005, V.S. Nguyen et al., 2009 and I. Das, 2012.

Remarks: Holotype born in captivity May 1996 and died May 1998. Probably occurs in W Laos.

2. *Viridovipera medoensis* (Zhao *in* Zhao & Jiang, 1977). Acta Zool. Sinica 23(1): 66–67, 71, pl. 2, figs. 9.1–9.5. (*Trimeresurus medoensis*)

Type: Holotype, SBRI 73-II-5208, an adult male (E.-M. Zhao & Y-A. Gao, 3 Aug. 1973).

Type locality: "near A-nie Bridge, Medo Xian, Xizang, alt. 1,200 m., China."

Distribution: Northern India (Arunachal Pradesh), SW China (Xizang) and N Myanmar (Kachin), (150) 1000–1400 m.

Sources: Zhao & Adler, 1993, David & Tong, 1997, G. Peng & Fuji, 2001, Leviton et al., 2003, David et al., 2002, Whitaker & Captain, 2004, David & Mathew, 2005, Zhao, 2006, I. Das, 2012 and Li et al., 2010.

3. *Viridovipera stejnegeri* (K.P. Schmidt, 1925a). Amer. Mus. Novit. (157): 4. (*Trimeresurus stejnegeri*)

Synonyms: *Trimeresurus gramineus formosensis* Maki, 1931–b, *Trimeresurus gramineus kodairai* Maki, 1931–b, *Trimeresurus stejnegeri makii* Klemmer, 1963, and *Trimeresurus stejnegeri chenbihuii* Zhao, 1995.

Type: Holotype, AMNH 21054, a 688 mm male (R.C. Andrews & E. Heller [Third AMNH Cen. Asiatic Exped.], summer 1916).

Type locality: "Shaowu, Fukien Province, China" [= Shaowu, N Fujian Prov., SE China, 27°20'N, 117°30'E, elevation 245 m].

Distribution: Indo-China. Southeastern China (Chekiang, Fujian, Gansu, Guangdong, Guangxi, Guizhou, Hainan, Jiangxi, Jilin, Sichuan, Yunnan, Zhejiang), Taiwan, Laos and Vietnam (Bac Thai, Cao Bang, Da Nang, Gia Lai, Ha Giang, Ha Tinh, Hoa Binh, Lao Cai, Nghe An, Quang Nang, Vinh Phu), 90–2845 m.

Sources: C.H. Pope & Pope, 1933, C.H. Pope, 1935, Mao, 1962, Kuntz, 1963, E.H. Taylor, 1965, Fleming & Fleming, 1973, Zhao & Yan, 1979, Regenass & Kramer, 1981, Jiang, 1983, Zhang & Zhao, 1990, Zhao & Adler, 1993, Castellano et al., 1994, Zhao, 1995, David & Tong, 1997, Malhotra & Thorpe, 1997, 2004b, Orlov, 1997, M.J. Cox et al., 1998, David & Vogel, 1998, Xie et al., 1998, Orlov et al., 2000, Creer et al., 2001, David et al., 2001, G. Peng & Fuji, 2001, Tsai & Tu, 2001, Ziegler, 2002, P. Guo et al., 2006a, V.S. Nguyen et al., 2009, Xiang & Li, 2009, I. Das, 2010, 2012 and Yao, 2012.

Remarks: Supplemental original description in K.P. Schmidt (1927: 546–547, fig. 22). Nepalese records are *T. tibetanus fide* Schleich & Kästle, 2002. Occurs in Cambodia and Laos *fide* Ziegler (2002: 257).

4. *Viridovipera truongsonensis* (Orlov, Ryabov, Bui & Ho, 2004). Russ. J. Herp. 11(2): 140–144, col. figs. 1–6. (*Trimeresurus truongsonensis*)

Type: Holotype, ZISP 22931 (formerly NLO 3995), a 595 mm male (T.C. Ho, N.L. Orlov, S.A. Ryabov & N.T. Bui, Sept. 2003).

Type locality "Phong Nha-Ke Bang National Park, Quang Binh province, Vietnam, elevation 600 m."

Distribution: Central Vietnam (Quang Binh, Quang Tri), 500–600 m. Known only from vicinity of type locality.

Sources: Ziegler et al., 2006, Dawson et al., 2008, Grismer et al., 2008b, V.S. Nguyen et al., 2009 and I. Das, 2010, 2012.

Remarks: This population was originally reported as *Trimeresurus* (= *Cryptelytrops*) *kanburiensis*.

5. *Viridovipera vogeli* (David, Vidal & Pauwels, 2001). Russ. J. Herp. 8(3): 206–211, figs. 1–5, 7. (*Trimeresurus vogeli*)

Synonym: *Trimeresurus koratensis* H.J. Peters, 1997 (*nomen nudum*).

Type: Holotype, MNHN 1999.9036, an 819 mm male (A. Gumprecht, K.-D. Schulz & A. Wlodarczyk, Aug. 1996).

Type locality: "Headquarters area, Khao Yai National Park (ca 14°21'N, 101°31'E), Nakhon Ratchasima Province, Thailand."

Distribution: Indo-China. Thailand (Chanthaburi, Krabi, Nakhon Ratchasima, Prachin Buri, Trat), SW Cambodia (Kampot, Koh Kong, Monulkiri, Pursat), Laos (Champasak) and Vietnam (Dak Lak, Da Nang, Gia Lai, Ha Tinh, Kon Tum, Quang Binh, Quang Nam, Thua Thien-Hue), 55–1200 m.

Sources: Orlov et al., 2003, Malholtra & Thorpe, 2004b, B.L. Stuart & Emmett, 2006, B.L. Stuart et al., 2006, Bain et al., 2007 and V.S. Nguyen et al., 2009.

Remarks: Probably occurs in Chon Buri and Suraburi, Thailand *fide* Orlov et al. (2003: 237).

6. *Viridovipera yunnanensis* (K.P. Schmidt, 1925a). Amer. Mus. Novit. (157): 4–5. (*Trimeresurus yunnanensis*)

Type: Holotype, AMNH 21058, a 701–707 mm male (R.C. Andrews & E.H. Heller [Third AMNH Cen. Asiatic Exped.], 18 May 1917).

Type locality: "Tengyueh, Yunnan Province, China" [= Tengchong, W Yunnan Prov., SW China, 25°01'N, 98°29'E, elevation 1810 m].

Distribution: Northeastern India (Arunachal Pradesh, Meghalaya, Nagaland, Sikkim, West Bengal), Nepal (Kathmandu, Sindhupalchok), China (Anhui, Fujian, Gansu, Guangdong, Guangxi, Guizhou, Hubei, Hunan, Jiangsu, Jiangxi, Jilin, Sichuan, Yunnan, Zhejiang), Myanmar (Chin, Kachin, Mandalay) and ? Laos, 1205–2845 m.

Sources: Regenass & Kramer, 1981, Zhao & Adler, 1993, Zhao, 1995, 2006, Leviton et al., 2003, Ao et al., 2004, Malhotra & Thorpe, 2004b, Tiwari & Shah, 2004, P. Guo et al., 2006a and Yang, 2008.

Remarks: Supplemental original description in K.P. Schmidt (1927: 547–548). David & Ineich, 1999, agreed with Zhao, 1999, in the elevation to specific status. Not listed by V.S. Nguyen et al., 2009 for Vietnam.

W

†*WAINCOPHIS* Albino, 1987a (Boidae)

Type species: †*Waincophis australis* Albino, 1987a.
Distribution: Lower Eocene of Argentina, and middle Paleocene of Brazil.
Sources: Albino, 1987a, 1996 and Rage, 2001.

1. †*Waincophis australis* Albino, 1987a. Ameghiniana 24(1–2): 62–63, fig. 1.

Type: Holotype, MACN CH 1008, one trunk vertebra.
Type locality: "Rocas Gemelas, Ea. El Sol, Cañadón Hondo, Provincia del Chubut, Argentina." Formacion Cañadón Hondo. Edad mamifero Casa mayorense (Eoceno temprano).
Distribution: Lower Eocene (48.0–54.0 mya) of Argentina.
Source: Albino, 1996a.

2. †*Waincophis cameratus* Rage, 2001. Palaeovertebrata 30(3–4): 130–132, fig. 11.

Type: Holotype, DGM 1343-R, one middle trunk vertebra.
Type locality: "São José de Itaboraí, State of Rio de Janeiro, Brazil; middle Palaeocene."
Distribution: Middle Paleocene (Itaboraian: 57.0–59.0 mya) of Brazil.
Source: Rage, 2011.

3. †*Waincophis pressulus* Rage, 2001. Palaeovertebrata 30(3–4): 126–129, fig. 8.

Type: Holotype, DGM 1337-R, one middle trunk vertebra.
Type locality: "São José de Itaboraí, State of Rio de Janeiro, Brazil; middle Paleocene."
Distribution: Middle Paleocene (Itaboraian: 57.0–59.0 mya) of Brazil .
Source: Rage, 2011.

WALTERINNESIA Lataste, 1887 (Elapidae)

Synonyms: *Warterinnesia* – Palacky, 1898 *(nomen incorrectum), Valterinnesia* – Phisalix, 1922 *(nomen incorrectum), Walterinesia* – Nopsca, 1928 *(nomen incorrectum), Walterinneria* – F. Werner, 1929a *(nomen incorrectum),* and *Walternnesia* – Tuck, 1974b *(nomen incorrectum).*
Type species: *Walterinnesia aegyptia* Lataste, 1887.
Distribution: Egypt and Middle East.
Sources: J. Anderson, 1898, F. Werner, 1923b, Bogert, 1943, Klemmer, 1968, Broadley, 1968c, 1974, Minton et al., 1970, Ilani, 1983b, Golay, 1985, Gasperetti, 1988, Szyndlar & Rage, 1990, Ulber, 1995, Golay et al., 1993, Keogh, 1998, David & Ineich, 1999, Slowinski & Keogh, 2000, Bouskila & Amitai, 2001, Dobiey & Vogel, 2007, Wüster et al., 2007, Amr, 2009b, Hoser, 2009c, 2012e, Wallach et al., 2009, Zaher et al., 2009, Amr & Disi, 2011 and Bar & Haimovitch, 2011.

1. *Walterinnesia aegyptia* Lataste, 1887. Naturaliste 9(52): 412–413.

Synonym: *Walterinnesia aegyptica* – Thompson, 1913 *(nomen incorrectum), Walterinnesia aegypti* – G. Haas, 1957 *(nomen incorrectum), Walterinnesia aegyptea* – Mohamed & Kamel, 1963 *(nomen incorrectum), Walterinessia aegypta* – Mohamed, El-Serougi & Kamel, 1964 *(nomen incorrectum),* and *Walterinneasia aegyptia* – Latifi, 1984 *(nomen incorrectum),* and *Walterinesia aegyptia* – Salhab, 1989 *(nomen incorrectum).*
Type: Holotype, BMNH 1946.1.21.42 (formerly BMNH 1892.6.28.1 & BMNH 1920.1.20.3375), a 1075–1187 mm female (native via W.F. Innes).
Type locality: "Égypte" [= Egypt]. Restricted to vicinity of Cairo (30°03'N, 31°14'E, elevation 25 m), Cairo Govern., NE Egypt] *fide* W.F. Innes *in* J. Anderson (1898: 325).
Distribution: Egypt and Middel East. Northeastern Egypt (Cairo, Ismailia, North Sinai, South Sinai, Suez), S Israel (Southern, West Bank), Jordan (Aqaba, Irbid, Karak, Maan, Madaba, Mafraq, Tafilah, Zarqa) and W Saudi Arabia (Jazan, Madinah, Tabuk), 500–1200 m.
Sources: Marx, 1953a, G. Haas & Werner, 1969, Zinner, 1971, Gasperetti, 1976a, Joger, 1984, Dmi'el et al., 1990, Abdeen, 1991, Leviton et al., 1992, Spawls & Branch, 1995, Y.L. Werner, 1995, Disi et al., 2001, Ugurtas et al., 2001, Baha El Din, 2006, Nilson & Rastegar-Pouyani, 2007, Amr & Disi, 2011 and Bar & Haimovitch, 2011.
Remarks: Illustration of holotype in J. Anderson (1898: pl. 46).

2. *Walterinnesia morgani* (Mocquard, 1905a). Bull. Mus. Hist. Nat., Paris 11(2): 78–79. (*Naja morgani*)

Synonym: *Atractaspis wilsoni* Wall, 1908L.
Types: Syntypes (5), MNHN 1904.562–66, three males and two females, longest male 902 mm (Morgan).

Type locality: "Perse (Arabistan)" [= Kuzestan Prov., W Iran].

Distribution: Arabia and SW Asia. Southeastern Turkey (Urfa), Iraq (Al Anbar, Al Basrah, Baghdad, Diyalá, Kirkuk, Ninawa), W Iran (Fars, Kermanshah, Khuzestan, Kurdistan, Mazandaran), E Saudi Arabia (Eastern, Najran, Riyadh), Kuwait and United Arab Emirates, NSL–2000 m.

Sources: J.C. Thompson, 1913c, Marx, 1953a, Raï, 1970, Gasperetti, 1976a, 1988, Joger, 1984, Latifi, 1991, Leviton et al., 1992, Spawls & Branch, 1995, Baran et al., 2006 and Nilson & Rastegar-Pouyani, 2007.

Remarks: A valid species *fide* Nilson & Rastegar-Pouyani (2007: 11).

†*WONAMBI* M.J. Smith, 1976 (†Madtsoiidae)

Type species: †*Wonambi naracoortensis* M.J. Smith, 1976.

Distribution: Oligocene to Pleistocene of Australia.

Sources: Rage, 1984b, Barrie, 1990, Scanlon, 1992, 1997, 2003, 2006, Scanlon & Lee, 2002, La Duke et al., 2010 and J.A. Wilson et al., 2010.

1. †*Wonambi barriei* Scanlon *in* Scanlon & Lee, 2000. Nature 403(6768): 417, fig. 2e.

Type: Holotype, QMF F23038, one middle trunk vertebra.

Type locality: "WW Site, Riversleigh (northwest Queensland, Australia); Late Oligocene to Early Miocene of Riversleigh, northwest Queensland."

Distribution: Upper Oligocene to lower Miocene (16.3–25.0 mya) of Australia.

2. †*Wonambi naracoortensis* M.J. Smith, 1976. Trans. Roy. Soc. So. Australia 100(1): 41–43, figs. 2a–e, 3a–c.

Type: Holotype, SAMA P16168, one trunk vertebra (J. Barre).

Type locality: "Fossil Chamber, Victoria Cave, Naracoorte, S. Aust." [= South Australia, Australia].

Distribution: Pliocene and Pleistocene (0.01–4.5 mya) of Australia.

Sources: M.J. Smith, 1985 and Scanlon, 1995.

†*WOUTERSOPHIS* Rage, 1980 (†Nigerophiidae)

Type species: †*Woutersophis novus* Rage, 1980.

Distribution: Middle Eocene of Belgium.

Sources: Rage, 1980, 1983b, 1984b and LaDuke et al., 2010.

1. †*Woutersophis novus* Rage, 1980. C.R. Acad. Sci., Paris 291D(5): 469–471, figs. a–b.

Type: Holotype, MNHN CBL 1, one middle trunk vertebra (G. Wouters).

Type locality: "sablière Van Pachtenbeke, à la limite de Wolu Weh-Saint-Étienne et Wodu Weh-Saint-Lambert (Est de Bruxelles), Belgique; Bruxellien (= Lutétien inférieur), Éocène moyen."

Distribution: Middle Eocene (Lutetian, MP 11–15: 40.4–48.6 mya) of Belgium. Known only from type locality.

X

XENELAPHIS A.C.L.G. Günther, 1864a (Colubridae)

Synonyms: *Henelaphis* – Higgins, 1873 *(nomen incorrectum), Xenalaphis* – Nutaphand, 1991b *(nomen incorrectum),* and *Senelapsis* – Nutaphand, 2001 *(nomen incorrectum).*

Type species: *Coluber hexagonotus* Cantor, 1847.

Distribution: Southeast Asia and Indonesia.

Sources: Rooij, 1917, M.A. Smith, 1943, C. Haas, 1950, Tweedie, 1983, Welch, 1988, David & Vogel, 1996, Iskandar & Colijn, 2001 and I. Das, 2010, 2012.

1. *Xenelaphis ellipsifer* Boulenger, 1900f. Proc. Zool. Soc. London 69(2): 184, pl. 16.

Type: Holotype, BMNH 1946.1.7.38 (formerly SMK), a 2000 mm male.

Type locality: "Head-waters of Sarawak River, Borneo" [= Sarawak, East Malaysia].

Distribution: West Malaysia (Perak, Selangor), East Malaysia (Sabah, Sarawak) and W Indonesia (Sumatra), (150) 800–1100 m.

Sources: Rooij, 1917, Malkmus et al., 2002, Auliya & Shepard, 2009 and Grismer et al., 2010.

Remarks: Record from Mt. Kinabalu needs confirmation *fide* Stuebing & Inger (1999: 79).

2. *Xenelaphis hexagonotus* (Cantor, 1847). J. Asiat. Soc. Bengal 16(182): 921–922. (*Coluber hexagonotus*) (*nomen corrigendum*)

Synonyms: *Coluber sublutescens* H. Boie *in* Schlegel, 1826a *(nomen nudum), Coluber sublutescens* Schlegel, 1826b *(nomen nudum), Coluber hexahonotus* Cantor, 1847 *(nomen incorrigendum), Coluber hexagonotus* Cantor, 1847 *(nomen corrigendum), Coluber sublutescens* A.-M.-C. Duméril, 1853 *(nomen nudum), Coryphodon sub-lutescens* A.-M.-C. Duméril, Bibron & Duméril, 1854a *(nomen incorrigendum), Dendrophis dumerilii* Bleeker, 1858a *(nomen nudum), Coryphodon hexanotus* – A.C.L.G. Günther, 1858 *(nomen incorrectum),* and *Coryphodon sublutescens* – Boulenger, 1894a *(nomen corrigendum).*

Type: Holotype, BMNH 1946.1.14.86, a 387 mm specimen (W. Norris via T. Cantor).

Type locality: "Great Hill of Pinang, Prince of Wales Island (Pulo Pinang), 5° 25' N. L. 100° 19'E." [= Great Hill, Penang Is., Pinang State, NW West Malaysia, 5°25'N, 100°17'E].

Distribution: Southeast Asia and Indonesia. S Myanmar, S Vietnam (Ho Chi Minh City), S Thailand (Narathiwat, Phattalung, Surat Thani), West Malaysia (Perlis, Pinang, Penang Is.), Singapore, East Malaysia (Sabah, Sarawak), Brunei and Indonesia (Bangka, Belitung, Java, Kalimantan, Riau Arch., Sumatra), NSL–1200 m.

Sources: K.F.L. Lim & Lee, 1989, K.K.P. Lim & Lim, 1992, Manthey & Grossmann, 1997, Chan-ard et al., 1999, B.L. Lim et al., 2002, Auliya, 2006, I. Das, 2007b, V.S. Nguyen et al., 2009 and Onn et al., 2009.

Remarks: Possibly occurs in Cambodia *fide* Manthey & Grossmann (1997: 396). Myanmar record doubtful *fide* Dowling & Jenner (1988: 11).

XENOCALAMUS A.C.L.G. Günther, 1868 (Atractaspididae)

Synonym: *Micaela* F. Werner *in* Michaelsen, 1915.

Type species: *Xenocalamus bicolor* A.C.L.G. Günther, 1868.

Distribution: Southern Africa.

Sources: Witte & Laurent, 1947, Broadley, 1971c, 1983, V.F.M. FitzSimons, 1974, Branch, 1988, Wallach, 1991, Underwood & Kochva, 1993, Nagy et al., 2005, Kelly et al., 2009 and Zaher et al., 2009.

1. *Xenocalamus bicolor* A.C.L.G. Günther, 1868. Ann. Mag. Nat. Hist. (4) 1(6): 415, pl. 19, fig. a.

Synonyms: *Micaela pernasuta* F. Werner *in* Michaelsen, 1915, *Xenocalamus bicolor maculatus* V.F.M. FitzSimons, 1932, *Xenocalamus bicolor concavorostralis* A. Hoffman, 1940, *Xenocalamus bicolor australis* V.F.M. FitzSimons, 1946b, and *Xenocalamus bicolor machadoi* Laurent, 1954a.

Type: Holotype, BMNH 1946.1.2.89 (formerly SAM), a 430–432 mm female (? J. Chapman, 1861–1863).

Type locality: "on the Zambeze" [= Zambezi River, Mozambique] (in error). Corrected to Damaraland, NW Namibia *fide* Broadley (1971c: 679).

Distribution: Southern Africa. Southeastern Democratic Republic of the Congo (Katanga), Angola (Benguela, Bié, Lunda Norte), Zimbabwe (Mashonaland Central, Mashonaland West, Masvingo, Matebeleland North), S Mozambique (Gaza, Inhambane, Maputo, Sofala), Namibia (Erongo, Khomas, Okavango, Omaheke, Otjozondjupa), Botswana (Gaborone, Ghanzi, Kgalagadi, Kweneng, North West) and NE South Africa (Free State, KwaZulu-Natal, Limpopo, Mpumalanga, North-West), 30–1700 m.

Sources: Laurent, 1954, Mertens, 1955, 1971, Broadley, 1959, Witte, 1962, Broadley & Cock, 1975, Auerbach, 1987, Bates, 1991, Clauss & Clauss, 2002, M. Griffin, 2003, J.C. Murphy & Schlager, 2003, Marais, 2004, Alexander & Marais, 2007 and Broadley & Blaylock, 2013.

Remarks: Original description attributes collector to J. Chapman on the Zambezi River (and Broadley has shown that Chapman did not go down the Zambezi and that type locality is in Namibia); however, A.C.L.G. Günther (1868: 413–414) also listed Cape Town Mus. as source so it is more likely that the type came from SAM rather than Chapman.

2. *Xenocalamus lineatus* Roux, 1907a. Rev. Suisse Zool. 15(1): 79–80. (*Xenocalamus bicolor lineatus*)

Type: Holotype, MHNN 16 (formerly MZN 66), a 220 mm specimen (H.A. Junod).

Type locality: "Rikatla, colline sablonneuse, située dans le pays de Mozambique, au nord de Lourenço-Marquès à 50 mètres d'altitude et à 5 à 6 kilom. de la mer" [= Ricatla, Maputo Prov., S Mozambique, 25°26'S, 32°31'E, elevation 50 m].

Distribution: Southern Mozambique (Gaza, Maputo) and ext. S Zimbabwe (S Masvingo, S Matebeleland South), 50–460 m.

Remarks: A valid species *fide* Broadley & Blaylock (2013: 252).

3. *Xenocalamus mechowii* W.C.H. Peters, 1881e. Sitz. Ges. Naturf. Freunde Berlin 1881(9): 147–148.

Synonym: *Xenocalamus mechowii inornatus* Witte & Laurent, 1947.

Type: Holotype, ZMB 10044, female (M.-F.W.A. von Mechow, 1873–1881).

Type locality: "Malange, West-Afrika" [= Malanje, Malanje Prov., cen. Angola, 9°33'S, 16°21'E, elevation 1130 m].

Distribution: Southern Africa. Southern Congo (Brazzaville, Pool), S Democratic Republic of the Congo (Bandundu, Bas-Congo, Kasai Occidental, Kasai Oriental, Katanga, Kinshasa), E Angola (Cuando Cubango, Lunda Norte, Malanje), W Zambia (North-Western, Southern, Western), W Zimbabwe (Matebeleland North), N Namibia (Caprivi, Okavango, Oshikoto) and Botswana (Kweneng, North West), 285–1245 m.

Sources: Laurent, 195a4, Mertens, 1955, 1971, Thys van den Audenaerde, 1965, Broadley & Cock, 1975, Auerbach, 1987, Branch & McCartney, 1993, J.-F. Trape & Roux-Estève, 1995, Broadley et al., 2003, Chippaux, 2006 and Broadley & Blaylock, 2013.

4. *Xenocalamus michelli* L. Müller, 1911. Zool. Anz. 38(14–15): 359–360.

Type: Holotype, ZSM 524/1911, a 540 mm specimen (Michell), lost *fide* Franzen & Glaw (2007: 247).

Type locality: "Kituri, Katangadistrikt, Kongostaat" [= Lukaga, Katanga Prov., SE Democratic Republic of the Congo].

Distribution: Southeastern Democratic Republic of the Congo (E Katanga), 685–730 m.

5. *Xenocalamus sabiensis* Broadley, 1971c. Occ. Pap. Natl. Mus. Rhodesia 4B(33): 673–675, fig. 17.

Type: Holotype, NMZB 1448 (formerly UM 1448), a 424+ mm male (D. Bredenkamp, Nov. 1961).

Type locality: "Birchenough Bridge, Sabi River, Rhodesia, at an altitude of 450 metres" [= Birchenough Bridge, Zimbabwe, 19°58'S, 32°21'E, elevation 490 m].

Distribution: Southeastern Zimbabwe (S Manicaland, Masvingo) and S Mozambique (Manica), 240–1100 m.

Sources: Broadley & Cock, 1975 and Broadley & Blaylock, 2013.

6. *Xenocalamus transvaalensis* Methuen, 1919. Proc. Zool. Soc. London 89(3–4): 350–351, fig. 1b.

Type: Holotype, TMP 10121 (formerly TM 1689), a 414 mm specimen (Noomé & Roberts, Sept. 1913).

Type locality: "north of the Zoutpansbergen, Northern Transvaal, near the Ingelel River, within 25 miles of the Limipopo" [= N of Soutpansberg (23°02'S, 29°48'E, elevation 965 m), near Nzhelele River, N Limpopo Prov., NE South Africa *fide* Broadley, 1983: 205].

Distribution: Southern Africa. Eastern Botswana (E Central), NE South Africa (NE KwaZulu-Natal, N Limpopo) and ext. S Mozambique (Maputo, Bazaruto Is.), NSL–965 m.

Sources: Auerbach, 1987, Broadley, 1990b, Marais, 2004 and Alexander & Marais, 2007.

XENOCHROPHIS A.C.L.G. Günther, 1864a (Natricidae)

Synonyms: *Fowlea* Theobald, 1868b, *Diplophallus* Cope, 1893a, *Diplophatulus* – Cope, 1900 *(nomen incorrectum), Fowleya* – Král, 1969 *(nomen incorrectum), Xenochropis* – Negi, 1992 *(nomen incorrectum),* and *Rentonus,* Hoser, 2012ab *(nomen illegitimum).*

Type species: *Psammophis cerasogaster* Cantor, 1839b.

Distribution: Southern Asia and Indonesia.

Sources: M.A. Smith, 1943, Malnate, 1960a, McDowell, 1961, Kramer, 1977, Welch, 1988, Wallach, 1998, Zaher et al., 2009, I. Das, 2010, 2012, Cadle, 2011 and Hoser, 2012ab.

Remarks: An undescribed species occurs in Sri Lanka *fide* Vogel & David (2006: 245 & 2012: 50–52).

1. *Xenochrophis asperrimus* (Boulenger, 1891a). Ann. Mag. Nat. Hist. (6) 7(39): 281–282. (*Tropidonotus asperrimus*)

Synonym: *Tropidonotus quincunciatus unicular* F. Müller, 1887.

Types: Syntypes (12), BMNH 1946.1.7.60–61, BMNH 1946.1.12.43, BMNH 1946.1.12.74–75, BMNH 1946.1.12.85, BMNH 1946.1.13.33–34, BMNH 1946.1.15.17, BMNH 1946.1.15.22–23, and BMNH 1946.1.15.30, longest syntype 820 mm.

Type locality: "Ceylon" [= Sri Lanka].

Distribution: Sri Lanka (Central, Eastern, North-Central, Northern, North-Western, Sabaragamuwa, Southern, Uva, Western), NSL–1220 m.

Sources: Wall, 1910k, E.H. Taylor, 1950b, Deraniyagala, 1954, P. Silva, 1969, A. Silva, 1990, I. Das, 2005 and Vogel & David, 2006, 2007, 2012.

Remarks: A valid species *fide* Silva (1990: 98). A sympatric undescribed species also on Sri Lanka *fide* Vogel & David (2006: 243).

2. *Xenochrophis bellulus* (Stoliczka, 1871). J. Asiat. Soc. Bengal 40(2): 432–434, pl. 26, figs. 2–2b. (*Tropidonotus bellulus*)

Type: Holotype, a 419 mm specimen (S. Kurz), lost *fide* M.A. Smith (1943: 298).

Type locality: "between Prome and Tonghoo, in N. Western Pegu, Burma" [= between Pyay (18°49'N, 95°13'E, elevation 40 m), and Toungoo (18°56'N, 96°26'E, elevation 55 m), N Bago Div., S Myanmar].

Distribution: Southern Myanmar (Bago, Yangon), 30–300 m.

Sources: Theobald, 1876, Boulenger, 1890a, Wall, 1926b, M.A. Smith, 1943, Dowling & Jenner, 1988, Malnate & Underwood, 1988, D.-Q. Rao & Yang, 1998 and Pauwels et al., 2001.

3. *Xenochrophis cerasogaster* (Cantor, 1839b). Proc. Zool. Soc. London 7(1): 52–53. (*Psammophis cerasogaster*)

Synonyms: *Amphiesma schistaceum* Jan, 1864a, and *Xenochrophis cirasogaster* – R.C. Sharma, 2003 *(nomen incorrectum)*.

Type: Lectotype, specimen illustrated in colored sketch of T.E. Cantor (1834–1837: no. 30) in RSL, designated by M.A. Smith (1943: 317).

Type locality: "near Calcutta (22°34'N, 88°22'E, elevation 10 m), S West Bengal State, NE India" via lectotype selection.

Distribution: South Asia. Southeastern Pakistan (Sindh), NE India (Assam, Uttar Pradesh, West Bengal), Bangladesh (Barisol), and Nepal (Rupandehi), NSL–200 m.

Sources: A.C.L.G. Günther, 1864a, Malnate & Minton, 1965, Kramer, 1977, M.S. Khan, 2002, 2006, Schleich & Kästle, 2002, Tiwari & Shah, 2004, Ahsan et al., 2009, Grismer et al., 2010 and Sourav & Purkayastha, 2011.

Remarks: Holotype lost *fide* Malnate & Minton (1965: 24). Penang, Malaysia record rejected *fide* Tweedie (1935: 126).

X

4. *Xenochrophis flavipunctatus* (Hallowell, 1861). Proc. Acad. Nat. Sci. Philadelphia (1860) 12(11): 503–504. (*Amphiesma flavipunctatum*)

Synonym: *Tropidonotus piscator ornata* Wall, 1907c.

Type: Neotype, USNM 7387, a 398 mm male (J.M. Brooke [No. Pacific Explor. Exped.], May 1854), designated by Zug et al. (2006: 165) and Vogel & David (2012: 17).

Type locality: "Island of Hong Kong, China" via lectotype/neotype selection.

Distribution: Southeastern Asia. Eastern India, Nepal (Lumbini), SE China (Anhui, Fujian, Guizhou, Guangxi, Guangdong, Hainan, Hong Kong, Hubei, Hunan, Jiangsu, Jiangxi, Shaanxi, Yunnan, Xizang, Zhejiang), Taiwan, Myanmar (Ayeyarwady, Bago, S Kachin, Sagaing, Yangon), Thailand (Bangkok, Chaiyaphum, Chanthaburi, Chiang Mai, Chiang Rai, Chomphun, Chon Buri, Kanchanaburi, Khan Kaen, Lampang, Nakhon Ratchasima, Nakhon Si Thammarat, Phatthalung, Prachuap Khiri Khan, Lop Buri, Phang Nga, Phetchabun, Phuket, Prachuap Khirikhan, Satun, Si Sa Ket, Surat Thani, Tak, Ubon Ratchathani, Udon Thani, Uthai Thani), Cambodia (Banteay Meanchey, Battambang, Kampong Cham, Kampong Chhang, Kampong Speu, Kampong Thom, Kampot, Kandal, Kep, Koh Kong, Kratié, Mondulkiri, Oddar Meancheay, Pailin, Phnom Penh, Preah Vihear, Prey Veng, Pursat, Ratanakiri, Siem Reap, Sihanoukville, Stung Treng, Svay Rieng, Takéo), Laos (Attapu, Borikhan, Champasak, Houaphan, Khammouan, Louangphrabang, Phongsali, Salavan, Savannakhet, Vientiane, Xaignabouli, Xekong, Xiangkhouang), Vietnam (An Giang, Bac Giang, Bac Kan, Bac Lieu, Bac Ninh, Ba Ria-Vung Tau, Ben Tre, Binh Dinh, Binh Duong, Binh Phuoc, Binh Thuan, Ca Mau, Can Tho, Cao Bang, Dak Lak, Dak Nong, Da Nang, Dien Bien, Dong Nai, Dong Thap, Gia Lai, Ha Giang, Hai Duong, Hai Pohng, Ha Nam, Hanoi, Ha Tay, Ha Tinh, Hau Giang, Hoa Binh, Ho Chi Minh City, Hung Yen, Khanh Hoa, Kien Giang, Kon Tum, Lai Chau, Lam Dong, Lang Son, Lao Cai, Long An, Nam Dinh, Nghe An, Ninh Binh, Ninh Thuan, Phu Tho, Phu Yen, Quang Binh, Quang Nam, Quang Ngai, Quang Ninh, Quang Tri, Soc Trang, Son La, Tai Ninh, Thai Binh, Thai Nguyen, Thanh Hoa, Thua Thien-Hue, Tien Giang, Tra Vinh, Tuyen Quang, Vinh Phuc, Vinh Long, Yen Bai, Con Dao Is.), West Malaysia (Johore, Kedah, Kelantan, Melaka, Negeri Sembilan, Pahang, Perak, Perlis, Pinang, Selangor, Terengganu) and Singapore, NSL–1200 m.

Sources: K.P. Schmidt, 1927, H. Saint-Girons, 1972, Zhao & Adler, 1993, Darevsky, 1999, Schleich & Kästle, 2002, Vogel & David, 2006, 2012, Zug et al., 2006,

V.S. Nguyen et al., 2009, Xiang & Li, 2009, I. Das, 2010, Cadle, 2011 and T. Hartmann et al., 2013.

Remarks: A valid species *fide* Zug et al. (2006: 164) and Vogel & David (2006: 244). Photographs of neotype in Vogel & David (2012: figs. 9–11). Original description based on ANSP 6616 (Canton River, China, 553 mm female, Ruschenberger coll.) *fide* Vogel & David (2012: 17), lost *fide* Zug et al. (2006: 164). Hallowell (1861: 504) clearly had two syntypes and, although basing the description on ANSP specimen, he listed type locality, date and collector of other syntype. Cochran (1961: 158) erroneously listed USNM 7387 as holotype. Nias Is., Indonesia record doubtful *fide* Vogel & David (2012: 30).

5. *Xenochrophis maculatus* (Edeling, 1864a). Nat. Tijds. Ned. Indïe (6) 26(1): 488–489. (*Tropidonotus maculatus*)

Synonyms: *Tropidonotus sundanensis* Bleeker, 1857a *(nomen nudum), Amphiesma substolatum* Jan, 1863b *(nomen nudum), Amphiesma substolatum* Jan, 1865c, *Tropidonotus sundanensis* A.C.L.G. Günther, 1865 *(nomen praeoccupatum),* and *Tropidonotus sarawacensis* J.G. Fischer, 1885a.

Type: Holotype, not designated, a 685 mm specimen (Benjamins), lost *fide* Iskandar & Colijn (2001: 109).

Type locality: "Martapoera, près de Bandjermasin" [= Martapura, near Bandjermasin, S Kalimantan, cen. Indonesia, 3°25'S, 114°51'E, elevation 10 m].

Distribution: Malay Peninsula and Indonesia. West Malaysia (Johore, Kedah, Kelantan, Pahang, Perak, Selangor, Terengganu), Singapore, East Malaysia (Sabah, Sarawak), Brunei and Indonesia (Bangka, Belitung, Kalimantan, Natuna Arch., Riau Arch., Sumatra), 10 m.

Sources: Tweedie, 1983 and David & Vogel, 1996, Stuebing & Inger, 1999, Malkmus et al., 2002, Auliya, 2006 and I. Das, 2006, 2007b.

Remarks: Original description reprinted in Edeling (1865: 203–204).

6. *Xenochrophis melanzostus* (Gravenhorst, 1807). Vergleich. Linne. Zool. Syst.: 402. (*Coluber melanzostus*)

Synonyms: *Tropidonotus melanozostus* F. Boie, 1827 *(nomen emendatum), Coluber lippus* A. Reuss, 1834, and *Tropidonotus striolatus* Blyth *in* Blanford, 1863b. [or 1868]

Type: Neotype, MNHN 60, an 805 mm female (Méder), designated by Vogel & David (2012: 10).

Type locality: "'Batavia', now Jakarta, Java Island, Indonesia" [= 6°13'S, 106°51'E, elevation 15 m] via neotype selection.

Distribution: Southern Indonesia (Bali, Java, Sumatra), NSL–15 m.

Sources: David & Vogel, 1996 and Vogel & David, 2006, 2012.

Remarks: A valid species *fide* Das (1996: 59). Syntypes (3+) in BM destroyed [9 Sept. 1944 during World War II] *fide* M. Rybecky *in* Vogel & David (2012: 10). Photographs of neotype in Vogel & David (2012: 10, figs. 2–4). Holotype erroneously listed in RMNH *fide* Iskandar & Colijn (2001: 109). Possibly introduced in Bali *fide* Kusuma et al., 2010. Sulawesi records doubtful *fide* Lang & Vogel (2005: 258).

7. *Xenochrophis piscator* (J.G. Schneider, 1799). Hist. Amph. 1: 247–248. (*Hydrus piscator*)

Synonyms: *Hydrus palustris* J. Schreider, 1799, *Coluber anastomosatus* Daudin, 1803d, *Coluber braminus* Daudin, 1803d, *Coluber dora* Daudin, 1803d, *Coluber mortuarius* Daudin, 1803d, *Coluber umbratus* Daudin, 1803d, *Natrix sagittata* Wagler, 1825 *(nomen ineditum), Coluber funebris* F. Boie, 1827 *(nomen nudum), Coluber bengalensis* Gray, 1834, *Coluber rectangulus* Gray, 1834, *Tropidonotus quincunciatus* Schlegel, 1837, *Coluber sinuatus* Schlegel, 1837 *(nomen nudum), Tropidonotus piscator lateralis* Wall, 1907c, *Tropidonotus piscator obscurus* Wall, 1907c, *Tropidonotus piscator punctatus* Wall, 1907c, and *Tropidonotus piscator unicolor* Wall, 1907c.

Type: Holotype, BMNH 1904.7.24.32, skin no. 32 of a female (P. Russell,

Type locality: "Indiae orientalis" [= E India]. Restricted to Coast of Coromandel, N Andhra Pradesh State, E India *fide* Vogel & David (2012: 6).

Distribution: Southern and SE Asia. Pakistan (E Balochistan, Gilgit-Baltistan, Jammu and Kashmir, Khyber Pakhtunkhwa, Punjab, Sindh), India (Andhra Pradesh, Arunachal Pradesh, Assam, Bihar, Gujarat, Haryana, Himachal Pradesh, Jammu & Kashmir, Kerala, Karnataka, Madhya Pradesh, Maharashtra, Manipur, Meghalaya, Mizoram, Nagaland, Orissa, Pondicherry, Punjab, Rajasthan, Sikkim, Tamil Nadu, Tripura, Uttar Pradesh, West Bengal), Bangladesh, Nepal (Bara, Bardiya, Chitwan, Dhanusa, Doti, Jhapa, Kanchanpur, Kaski, Kathmandu, Mahottari, Morang, Parsa, Rupandehi, Saptari, Sunsari, Udayapur), Bhutan, S China (Guangxi, Yunnan), Myanmar (Ayeyarwaddy, Bago, Chin, Kachin, Kayah, Kayin, Magwe, Mandalay, Mon, Rakhine, Sagaing, Shan, Yangon), Thailand (Chiang Mai, Loei, Udon Thani) and Laos (Louangphrabang, Vientiane, Xaignabouri), 150–2100 m.

Sources: Wall, 1907c, Bergman, 1958b, Deuve, 1963a, c, Kuntz, 1963, Kramer, 1977, Tweedie, 1983, Inger et al., 1984, M.S, Khan, 1984a, 2002, Nanhoe & Ouboter, 1987, K.F.L. Lim & Lee, 1989, A. Silva, 1990, Zhao & Adler, 1993, I. Das, 1994, David & Vogel, 1996, M.J. Cox et al., 1998, Chan-ard et al., 1999, Farkas & Fritz, 1999d, Xie et al., 1998, Hallermann et al., 2001, Ao et al., 2004, Tiwari & Shah, 2004, Whitaker & Captain,

2004, N. Khaire, 2006, M.F. Ahmed et al., 2009, Vyas, 2011 and Vogel & David, 2012.

Remarks: Original description based on P. Russell (1796: 38, pl. 33). Photograph of holotype in Vogel & David (2012: fig. 1). See Kramer (1977: 750) for discussion of type. Probably occurs in Nepal *fide* Schleich & Kästle, 2002.

8. *Xenochrophis punctulatus* (A.C.L.G. Günther, 1858). Cat. Colub. Snakes Brit. Mus.: 247–248. (*Tropidonotus punctulatus*)

Synonym: *Fowlea peguensis* Theobald, 1868.

Type: Holotype, BMNH 1946.1.13.59, a 559 mm female (Zool. Soc. London).

Type locality: Unknown.

Distribution: Southeast Asia. Southern Myanmar (Bago, Mon, Taninthayi, Yangon) and Thailand (Mae Hong Son, Ranong), NSL–30 m.

Sources: Pauwels et al., 2001, 2004.

Remarks: A valid species *fide* Pauwels et al. (2001: 247). Manipur, India record doubtful *fide* Pauwels et al. (2001: 250).

9. *Xenochrophis sanctijohannis* (Boulenger, 1890a). Fauna Brit. India, Rept. Batr.: 350. (*Tropidonotus sanctijohannis*) (*nomen corrigendum*)

Synonyms: *Tropidonotus sancti-johannis* Boulenger, 1890a *(nomen incorrigendum),* and *Xenochrophis piscator sanctijohannis* – Kramer, 1977 *(nomen corrigendum).*

Type: Holotype, BMNH 1946.1.13.61 (formerly BMNH 1889.11.8.1), a 406 mm specimen (O.B. St. John).

Type locality: "Kashmir" [= Azad Kashmir Prov., NE Pakistan or Jammu & Kashmir Prov., NW India].

Distribution: South Asia. Extreme E Afghanistan, NE Pakistan (Jammu & Kashmir, S Punjab), N India (Arunachal Pradesh, Himachal Pradesh, Jammu & Kashmir, West Bengal), and Nepal, 405–1400 (2200) m.

Sources: M.S. Khan, 2002, Tiwari & Shah, 2004, Whitaker & Captain, 2004, Masroor, 2012 and Vogel & David, 2012.

Remarks: A valid species *fide* Tiwari & Shah (2004: 200) or separate *X. piscator* morph *fide* Vogel & David (2012: 45).

10. *Xenochrophis schnurrenbergeri* Kramer, 1977. Rev. Suisse Zool. 84(3): 750, figs. 16–17. (*Xenochrophis flavipunctatus schnurrenbergeri*)

X

Type: Holotype, MHNG 1377.44, a 480 mm male (H. Schnurrenberger, Oct. 1962).

Type locality: "Nepal. 83° 28'/27° 30'/150 m" [= vicinity of Bhairahawa, Lumbini Zone, Rupandehi Prov., cen. Nepal, 27°30'N, 83°28'E, elevation 150 m].

Distribution: South Asia. Southeastern Pakistan (Sindh), S Nepal (Mahottari, Rupandehi) and N India (Assam, Orissa, Uttar Pradesh, West Bengal), NSL–150 m.

Sources: Tiwari & Shah, 2004, Vogel & David, 2006, 2012 and Mohapatra et al., 2010b.

Remarks: A valid species *fide* Vogel & David (2006: 244). Photographs of holotype in Vogel & David (2012: figs. 17–18). Sympatric with *X. piscator* in northern Distribution and parapatric with *X. flavipunctatus fide* Vogel & David (2006: 244).

11. *Xenochrophis trianguligerus* (F. Boie, 1827). Isis von Oken 20(6): 535. (*Tropidonotus trianguligerus*)

Synonyms: *Tropidonotus trianguligerus* H. Boie *in* Schlegel, 1826a *(nomen nudum), Tropidonotus trianguligerus* Schlegel, 1826b *(nomen nudum), Tropidonotus umbratus* Cantor, 1847 *(nomen praeoccupatum),* and *Tropidonotus trianguligerus annularis* J.G. Fischer, 1885a.

Type: Lectotype, RMNH 1102 (H. Boie & H.C. Macklot, 1825–1832), designated by R.F. Inger *in* Koch (2012: 284).

Type locality: "Java" [W Indonesia] via lectotype selection. Restricted to region of Mt. Pangerango and Mt. Salak, West Java, Indonesia *fide* Brongersma (1948b: 14). Emended to Tjihandjawar, at the foot of Mt. Pangerango, W. Java [= Cihanyawar, Nagrak, Jawa Barat, W Java, SW Indonesia, 07°03'S, 108°40'E] *fide* Brongersma (1950: 1499).

Distribution: Southeast Asia and Borneo. Eastern India (Andamans & Nicobars: Great Nicobar and Little Nicobar Is.), Myanmar, Thailand (Nakhon Si Thammarat, Pattani, Phang Nga, Phatthalung, Phuket, Ranong, Satan), SW Cambodia (Kampot, Koh Kong), Laos, Vietnam (Quang Nam, Vinh Phuc), West Malaysia (Melaka, Perlis), Singapore, East Malaysia (Sabah, Sarawak), Brunei and Indonesia (Bali, Bangka, Belitung, Buton, Java, Kabaena, Kalimantan, Mentawai Arch., Nias, Riau Arch., Sangihe Arch., Sulawesi, Sumatra, Ternate), NSL–1370 m.

Sources: Bergman, 1959, E.H. Taylor, 1965, Tweedie, 1983, M.J. Cox, 1991b, I. Das, 1994, 2006, 2007, David & Vogel, 1996, Inger & Tan Fui Lian, 1996, Manthey & Grossmann, 1997, M.J. Cox et al., 1998, Rao & Yang, 1998, Malkmus et al., 2002, Whitaker & Captain, 2004, Lang & Vogel, 2005, B.L. Stuart & Emmett, 2006, Vijayakuman & David, 2006, Auliya, 2006, V.S. Nguyen et al., 2009 and Koch, 2012.

Remarks: Original description based on Boie's MS (1823–1825) so although F. Boie credited authorship to Reinwardt, he cannot be the author. Type locality listed as Bogor, West Java, Indonesia *fide* Iskandar & Colijn (2001: 110).

12. *Xenochrophis tytleri* (Blyth, 1863). Proc. Asiat. Soc. Bengal 32(1): 88–89. (*Tropidonotus tytleri*) (*nomen corrigendum*)

Synonyms: *Tripidonotus tytleri* Blyth, 1863 *(nomen incorrigendum), Tropidonotus tytleri* – Theobald, 1876 *(nomen corrigendum),* and *Tropidonotus striolatus* Theobald, 1868a *(nomen substitutum).*

Type: Holotype, ZSI 7402 (formerly ASB 46a), an 876 mm male (R.C. Tytler, April 1862–1863).

Type locality: "Port Blair" [= South Andaman Is., Andaman & Nicobars, E India, Bay of Bengal].

Distribution: Eastern India (Andaman & Nicobars: Baratang, Long, North Andaman, Ross and South Andaman Is.).

Sources: Biswas & Sanyal, 1980, Talukdar et al., 1989, I. Das et al., 1998, Whitaker & Captain, 2004, Vogel & David, 2006, 2012 and Harikrishnan et al., 2012.

Remarks: Holotype illustrated in Vogel & David (2012: 19–20, figs. 12–13). Possibly occurs in the Nicobars, India *fide* Whitaker & Captain (2004: 226) although doubtful *fide* Vogel & David (2012: 31).

XENODERMUS J.C.H. Reinhardt, 1836 (Xenodermatidae)

Synonyms: *Gonionotus* Gray, 1846a, *Xenoderma* Lichtenstein & Martens, 1856 *(nomen emendatum),* and *Xenordermus* – Lim & Sharef bin Kanarudin, 1975 *(nomen incorrectum).*

Type species: *Xenodermus javanicus* J.C.H. Reinhardt, 1836.

Distribution: Southeastern Asia and Greater Sundas.

Sources: Duméril, Bibron & Duméril, 1854c, Rooij, 1917, Kopstein, 1926, 1930, 1938a–b, 1941, E.H. Taylor, 1965, Tweedie, 1983, M.J. Cox, 1991, J.C. Murphy et al., 1994, David & Vogel, 1996, J.B. Rasmussen & Hughes, 1997, Keller, 2008, I. Das, 2006, 2010 and Zaher et al., 2009.

1. *Xenodermus javanicus* J.C.H. Reinhardt, 1836. Over. Kon. Dansk Vidensk. Selsk. Forhand. (1836–1837) 3: 6–7.

Synonyms: *Gonionotus plumbeus* Gray, 1846, and *Xenoderma gonyonotus* Lichtenstein & Martens, 1856.

Type: Holotype, ZMUC 5481, a 637 mm specimen (E. Lange, 1834).

Type locality: "Java" [SW Indonesia]

Distribution: Malay Peninsula and Greater Sundas. Myanmar, ext. S Thailand (Yala), West Malaysia (Johor), and Indonesia (Java, Kalimantan, Sumatra), NSL–1300 m.

XENODON H. Boie *in* Fitzinger, 1826a (Xenodontidae)

Synonyms: *Ophis* Wagler *in* Spix, 1824, *Xenodon* H. Boie *in* Schlegel, 1826, *Acanthophallus* Cope, 1893a, *Procteria* F. Werner, 1924 *(nomen praeoccupatum), Zenodon* Swainson, 1839 *(nomen praeoccupatum), Xenedon* – Brazil, 1914 *(nomen incorrectum), Lenodon* – Phisalix, 1917a *(nomen incorrectum), Senodon* – Briceño-Rossi, 1934 *(nomen incorrectum), Waglerophis* Romano & Hoge, 1973, *Xenoon* – Cunha & Nascimento, 1978 *(nomen incorrectum),* and *Thalesius* Ferreira-Yuki, 1993.

Type species: *Coluber severus* Linnaeus, 1758.

Distribution: Latin America.

Sources: Wucherer, 1861b, J.A. Peters & Orejas-Miranda, 1970, Amaral, 1978, Cadle 1984a, 1984c, Villa et al., 1988, Ferreira-Yuki, 1993, Lúcia-Ferreira, 1997, Wallach, 1998, Vidal et al., 2000, Gravlund, 2001, M.S.Y. Lee et al., 2007, Zaher et al., 2009 and Vidal et al., 2010.

Remarks: *Waglerophis* a synonym of *Xenodon fide* Zaher et al. (2009: 146) and Grazziotin et al., 2012.

1. *Xenodon guentheri* Boulenger, 1894a. Cat. Snakes Brit. Mus. 2: 147–148, pl. 7, fig. 1.

Type: Holotype, BMNH 1946.1.5.93, a 430 mm male (Michaëlis).

Type locality: "Lagos, Sta. Catharina, southern Brazil."

Distribution: Southern Brazil (Paraná, Santa Catarina).

Source: Amaral, 1978.

2. *Xenodon merremii* (Wagler *in* Spix, 1824). Serp. Brasil. Sp. Nov.: 47–48, pl. 17. (*Ophis merremii*)

Synonyms: *Xenodon aeneus* H. Boie *in* Schlegel, 1826a *(nomen nudum), Xenodon ocellatus* H. Boie *in* Schlegel, 1826a *(nomen nudum), Xenodon aeneus* Schlegel, 1826b *(nomen nudum), Xenodon ocellatus* Schlegel, 1826b *(nomen nudum), Xenodon aeneus* F. Boie *in* H. Boie, 1827, *Xenodon ocellatus* F. Boie *in* H. Boie, 1827, *Xenodon irregularis* A.C.L.G. Günther, 1863c, *Trigonocephalus alternatus binocularis* Bacque, 1906, and *Trigonocephalus flavescens* Bacque, 1906.

Type: Holotype, a 737 mm specimen (J.B. von Spix & K.F.P. von Martius, 10 Nov.–10 Dec. 1818), lost before 1859 *fide* Hoogmoed & Gruber (1983: 334).

Type locality: "urbem Bahiae, Brésil" [= Salvador, Bahia State, Brazil, 12°58'S, 38°31'W, elevation 40 m].

Distribution: South America. Southeastern Venezuela (Bolívar), Guyana (Cuyuni-Mazaruni), Suriname (Sipaliwini), French Guiana (Cayenne, Saint-Laurent-du-Maroni), Brazil (Bahia, Brasilia, Ceará, Goias, Mato Grosso, Pará, Paraiba, Parana, Pernambuco, Rondônia, São Paulo, Tocantins, Santo Amaro Is.), E Bolivia (Cochabamba, Santa Cruz, Tarija), Paraguay (Alto Paraguay, Alto Paraná, Amambay, Asunción, Boquerón, Caaguazú, Caazapá, Canindeyú, Central, Concepción, Cordillera, Guairá, Itapúa, Misiones,

Neembucú, Paraguarí, Presidente Hayes, San Pedro), N Uruguay (Artigas) and N Argentina (Buenos Aires, Catamarca, Chaco, Córdoba, Corrientes, Entre Ríos, Formosa, Jujuy, Li Rioja, Misiones, Salta, San Juan, San Luis, Santa Fe, Santiago del Estero, Tucumán), NSL–600 m.

Sources: G. Schreiber, 1950, Anthony & Serra, 1951, Amaral, 1978, Vanzolini et al., 1980, Hoogmoed, 1985, Abuys, 1986d, Chippaux, 1987, D.R. Norman, 1994, Reati, 1996, Vuoto, 1996, Lúcia-Ferreira, 1997, Starace, 1998, Fugler et al., 1995, Freitas, 1999, Leynaud & Bucher, 1999, Moura-Leite & Bernarde, 1999, Giraudo, 2001, Giraudo & Scrocchi, 2002, Abuys, 2003, Carreira-Vidal et al., 2005, Cicchi et al., 2007, Cacciali, 2008, Navarrete et al., 2009, Zaher et al., 2009 and C.J. Cole et al., 2013.

3. *Xenodon neuwiedii* A.C.L.G. Günther, 1863c. Ann. Mag. Nat. Hist. (3) 12(71): 354–355, pl. 5, fig. c.

Synonyms: *Xenodon neovidii* Cope, 1868c *(nomen emendatum),* and *Xenodon hemileucurus* Lutz & Mello, 1922a.

Types: Syntypes (16 ?), BMNH 1946.1.4.29, BMNH 1946 .1.4.37–40, BMNH 1946.1.4.45, and BMNH 1946.1.5.94–96 (A. Fry), location of other sytypes unknown.

Type locality: "Rio Janeiro" [= Rio de Janeiro, S Rio de Janeiro State, SE Brazil, 22°54'S, 43°13'W, elevation 15 m].

Distribution: Southeastern Brazil (Bahia, Paraná, Rio de Janeiro, Santa Catarina, São Paulo, Cananéia, Cardoso, Grande, Santo Amaro and São Sebastião Is.), Paraguay and ext. NE Argentina (Misiones), 15–800 m.

Sources: Argólo, 1999, Freitas, 1999, Giraudo, 2001, Giraudo & Scrocchi, 2002, Duarte-Rocha & Van Sluys, 2006 and Cicchi et al., 2007.

Remarks: Record from Corrientes, Argentina doubtful *fide* Giraudo & Scrocchi (2002: 40).

4. *Xenodon rabdocephalus* (Wied-Neuwied, 1824). Isis von Oken 14(6): 668. (*Coluber rabdocephalus*)

Synonyms: *Vipera xaxalhua* Gravenhorst, 1807 *(nomen oblitum), Coluber rhapdocephalus* – Wied-Neuwied, 1828 *(nomen incorrectum), Xenodon rhabdocephalus* A.-M.-C. Duméril, Bibron & Duméril, 1854a *(nomen emendatum), Xenodon colubrinus* A.C.L.G. Günther, 1858, *Ophis maximiliani* Fitzinger, 1861, *Xenodon bertholdi* Jan, 1863a, *Xenodon angustirostris* W.C.H. Peters, 1864b, *Xenodon suspectus* Cope, 1868c, *Xenodon bipraeocularis* Cope, 1886a, and *Xenodon mexicanus* H.M. Smith, 1940.

X

Type: Holotype, AMNH 3609 (A.P.M. zu Wied-Neuwied, Sept. 1815–May 1817), lost *fide* J.A. Peters (1960a: 541).

Type locality: "Brasiliens" [= Brazil]. Specified as Sertâo von Bahia, Brazil *fide* Wied-Neuwied (1825c: 351).

Distribution: Latin America. Mexico (Chiapas, Guerrero, Oaxaca, Quintana Roo, Veracruz, Yucatán), Belize (Belize, Cayo, Orange Walk, Stann Creek), Guatemala (Alta Verapaz, Petén), Honduras (Atlántida, Colón, Cortés, El Paraiso, Gracias a Dios, Olancho), Nicaragua (Atlántico Norte, Atlántico Sur, Jinotega, Matagalpa, Río San Juan), Costa Rica (Alajuela, Cartago, Guanacaste, Heredia, Limón, Puntarenas, San José), Panama, Colombia (Cundinamarca), Venezuela (Amazonas, Barinas, Bolívar, Zulia), Guyana (Essequibo Islands-West Demerara), N Suriname (Wanica), French Guiana (Cayenne, Saint-Laurent-du-Maroni), N Brazil (Amazonia, Bahia, Pará, Rondônia), Ecuador (Carchi, Cotopaxi, Esmeraldas, Los Rios, Manabi, Morona-Santiago, Napo, Pastaza, Pichincha), E Peru (Madre de Dios, Pasco) and E Bolivia (Beni, Santa Cruz), NSL–1900 m.

Sources: Dixon, 1983c, L.D. Wilson & Meyer, 1985, Abuys, 1986d, 2003, Chippaux, 1987, Pérez-Santos & Moreno, 1988, Cunha & Nascimento, 1993, Casas-Andreu et al., 1996, J.C. Lee, 1996, Chippaux, 1987, J.A. Campbell, 1998b, Gorzula & Señaris, 1998, Starace, 1998, Freitas, 1999, G. Köhler, 1999b, Lehr, 2001, Doan & Arizábal, 2002, Savage, 2002, Abuys, 2003, McCranie et al., 2006, Rojas-Runjaic et al., 2007, Jansen, 2008, Navarrete et al., 2009, McCranie, 2011a and C.J. Cole et al., 2013.

Remarks: In accordance with Art. 23.9.2 of the Code (ICZN, 1999), *Coluber rabdocephalus* Wied-Neuwied is designated a *nomen protectum* and *Vipera xaxalhua* Gravenhorst a *nomen oblitum.*

5. *Xenodon severus* (Linnaeus, 1758). Syst. Nat., ed. 10, 1: 219. (*Coluber severus*)

Synonyms: *Coluber guineensis* Bonnaterre, 1790, *Coluber exalbidus* Suckow, 1797, *Coluber breviceps* G. Shaw, 1802, *Coluber versicolor* Merrem, 1820, *Coluber saurocephalus* Wied-Neuwied, 1824, and *Senodon severus* – Briceño-Rossi, 1934 *(nomen incorrectum).*

Type: Holotype, NHR Lin-48 (formerly MAFR), a 280 mm specimen (Mus. Drottn.).

Type locality: "Asia," (in error). Corrected to South America *fide* A.C.L.G. Günther (1863c: 353).

Distribution: Amazonia. Colombia (Amazonas), Venezuela (Amazonas, Aragua, Bolívar, Carabobo, Cojedes, Delta Amacuro, Falcón, Guárico, Lara, Miranda, Monagas, Zulia), Guyana (Barima-Waini, Cuyuni-Mazaruni, Demerara-Mahaica, Essequibo Islands-West Demerara, Pomeroon-Supenaam, Upper Demerara-Berbice, Santa Rosa Is.), Suriname, French Guiana (Cayenne, Saint-Laurent-du-Maroni), Ecuador (Carchi, Imbabura, Loja, Morona-Santiago, Napo, Pastaza), Peru (Madre de Dios, Pasco), N Brazil (Pará) and E Bolivia (Santa Cruz), NSL–2000 m.

Sources: Roze, 1966a, Lancini, 1986, Pérez-Santos & Moreno, 1988, 1991, Duellman & Salas, 1991, Cunha & Nascimento, 1993, Gorzula & Señaris, 1998, Starace, 1998, Mijares-Urrutia & Arends, 1999, 2000, Lehr, 2001, Duellman, 2005, Rivas-Fuenemayer &

Barrio-Amorgós, 2005, Navarrete et al., 2009 and C.J. Cole et al., 2013.

6. *Xenodon werneri* Eiselt, 1963. Ann. Naturhist. Mus. Wien 66: 280–281, fig. 1. (*nomen substitutum*)

Synonym: *Procteria viridis* F. Werner, 1924 (*nomen praeoccupatum*).

Type: Holotype, RMNH 17119, a 605 mm female (F.H. Rolle, 1910).

Type locality: "Tsumeb, Deutsch-Südwestafrika" [= Tsumeb, Namibia] (in error *fide* Bogert, 1940: 13, and Mertens, 1955: 12).

Distribution: Suriname (Marowijne, Nickerie) and French Guiana (Cayenne, Saint-Laurent-du-Maroni), 35–200 m.

Sources: F. Werner, 1924a, Gasc & Rodrigues, 1980, Hoogm 2oed, 1985, Ferreira-Yuki, 1993, Starace, 1998, Zaher, 1999, Abuys, 2003 and Sajdak, 2010.

Remarks: Original and supplemental description in F. Werner (1924a: 1 & 1924b: 48, figs. 1a–b). Redescription and illustrations of holotype in Eiselt (1963: 280–281, figs. 1a–h). F.H. Rolle was a natural history dealer from Institut Kosmos, Berlin, so it is not surprising that locality data for *Procteria viridis* were incorrect. Possibly occurs in Brazil *fide* Starace (1998: 261).

XENOPELTIS F. Boie, 1827 (Xenopeltidae)

Synonyms: *Xenopeltis* F. Boie, 1826 *(nomen nudum)*, *Xenopeltis* Schlegel, 1826a *(nomen nudum)*, *Zenopeltis* Swainson, 1839 *(nomen emendatum)*, *Zenopeltes* – Theobald, 1868a *(nomen incorrectum)*, and *Xenopeltus* – Porter, 1972 *(nomen incorrectum)*.

Type species: *Xenopeltis unicolor* F. Boie, 1827.

Distribution: Southeastern Asia and East Indies.

Sources: Mahendra, 1938, M.A. Smith, 1943, G. Haas, 1955, McDowell, 1975a, Underwood, 1976b, B. Hu et al., 1980, Leviton, 1983, Zhao & Adler, 1993, Walls, 1998b, McDiarmid et al., 1999, Orlov, 2000, V.S. Nguyen et al., 2009, I. Das, 2010, 2012 and Rage, 2011.

1. *Xenopeltis hainanensis* S.-Q. Hu & Zhao *in* Zhao, 1972. Mater. Herpetol. Res. 1(1): 36.

Synonym: *Xenopeltis hainanensis jidamingae* Zhao, 1995.

Type: Holotype, CIB 64-III-6016, a 628 mm male (native, 15 June 1964).

Type locality: "Dali of Diaulo Shan, Hainan, altitude 200 m," China.

Distribution: Southeastern China (Fujian, Guangdong, Guangxi, Hainan, Hunan, Jiangxi, Zhejiang) and N Vietnam (Cao Bang, Hai Dung, Ha Tinh, Lang Son, Quang Binh, Vinh Phu, Yen Bai), 200–2000 m.

Sources: Hu et al., 1975, Zhong, 1981, Jiang, 1987, M.-H. Huang, 1990, Zhao, 1995, 1997b, 2006, Zhao & Adler, 1993, Orlov et al., 2000, Kizirian et al., 2003, D.-T. Yang, 2008, Y.-X. Zhang, 2009, Li, 2011 and Shi, 2011.

Remarks: English translation of original description by Koshikawa (1982: 4, 9); holotype erroneously listed as SBRI 66-III-6016 on p. 9. Possibly occurs in Hong Kong, China *fide* Karsen et al. (1998: 162).

2. *Xenopeltis unicolor* F. Boie, 1827. Isis von Oken 20(6): 564.

Synonyms: *Coluber alvearius* Oppel *in* F. Boie, 1826 (*nomen nudum*), *Xenopeltis concolor* F. Boie, 1827, *Xenopeltis leucocephala* F. Boie, 1827, and *Tortrix xenopeltis* Schlegel, 1837 *(nomen substitutum)*.

Type: Holotype, not designated, formerly MNHN (L.T. Leschenault, 1805), lost *fide* Koch (2012: 304).

Type locality: "Java" [SW Indonesia].

Distribution: Southeastern Asia and East Indies. Eastern India (Andaman & Nicobars: Nicobar Is.), S China (Guangdong, Guangxi, Yunnan), Myanmar, Thailand (Bangkok, Chaiyaphum, Chiang Mai, Chiang Rai, Chon Buri, Krabi, Lampang, Loei, Nakhon Ratchasima, Phang Nga, Phitsanulok, Phrae, Phuket, Prachuap Khiri Khan, Ranong, Saraburi, Surin), Cambodia (Kampong Chhnang, Koh Kong, Preah Vihear, Stung Treng), Laos (Champasak), Vietnam (Binh Phuoc, Ca Mau, Can Tho, Da Nang, Dong Nai, Hai Duong, Hanoi, Ha Tinh, Hoa Binh, Ho Chi Minh City, Kien Giang, Kon Tum, Lam Dong, Lang Son, Nghe An, Ninh Binh, Ninh Thuan, Quang Binh, Quang Nam, Quang Ninh, Quang Tri, Son La, Tay Ninh, Thai Nguyen, Thanh Hoa, Tien Giang, Vinh Phuc, Yen Bai), Singapore, West Malaysia (Pahang, Ubin Is.), East Malaysia (Sabah, Sarawak), Brunei, Indonesia (Banka, Belitung, Java, Kalimantan, Mentawai Arch., Natuna Besar, Nias, Riau Arch., Siberut, Simeulue, Sipura, Sulawesi, Sumatra, We) and SW Philippines (Balabac, Bongao, Jolo, Palawan, Sanga Sanga), NSL–2000 m.

Sources: Wall, 1909c, E.H. Taylor, 1922, 1965, Bourret, 1936a–b, Mertens, 1943, Bergman, 1955e, Hoesel, 1959, Deuve, 1962a, Hoge, 1964b, Leviton, 1963b, 1983, Dowling & Gibson, 1970b, Biswas & Sanyal, 1980, Tweedie, 1983, Mahendra, 1984, K.F.L. Lim & Lee, 1989, M. Tiwari, 1992, I. Das, 1994, Gaulke, 1994b, David & Vogel,1996, Wallach & Jones, 1997, M.J. Cox et al., 1998, Zug et al., 1998, Chan-ard et al., 1999, I. Das, 1999, 2010, Stuebing & Inger, 1999, Orlov et al., 2000, Malkmus et al., 2002, J.C. Murphy & Schlager, 2003, Whitaker & Captain, 2004, B.L. Stuart & Emmett, 2006, D.-T. Yang, 2008, Koch, 2012 and Hartmann et al., 2013.

Remarks: Original description based on Boie's MS (1823–1825) so although F. Boie credited authorship to Reinwardt, he cannot be the author. Holotype in MNHN *fide* McDiarmid et al. (1999: 159) or RMNH *fide* C.P.J. Haas (1950: 527) and V.S. Nguyen et al. (2009: 286).

XENOPHIDION R. Günther & Manthey, 1995 (Xenophidiidae)

Type species: *Xenophidion acanthognathus* R. Günther & Manthey, 1995.
Distribution: Malaysia.
Sources: Wallach & Günther, 1998, R. Lawson et al., 2004 and I. Das, 2010, 2012.

1. *Xenophidion acanthognathus* R. Günther & Manthey, 1995. Amphibia-Reptilia 16(3): 231–233, fig. 1.

Type: Holotype, FMNH 235170, a 337 mm female (R.F. Inger, 20 Nov. 1987).
Type locality: "Sipitang Dist., Mendolong, Sabah (Borneo)" [East Malaysia].
Distribution: East Malaysia (Sabah). Known only from holotype.

2. *Xenophidion schaeferi* R. Günther & Manthey, 1995. Amphibia-Reptilia 16(3): 234–238, figs. 2–7.

Type: Holotype, ZMB 50534, a 263 mm male (C. Schäfer, 5 Nov. 1989).
Type locality: "Primary rain forest ca. 20 km north of Kuala Lumpur on the border of Templer Park, Province Selangor, Malay Peninsula" [West Malaysia].
Distribution: West Malaysia (Selangor).
Source: Chan-ard et al., 1999.

XENOPHOLIS W.C.H. Peters, 1869 (Dipsadidae)

Synonyms: *Gerrhosteus* Cope, 1874b, *Genhosteus* – Cope, 1874b *(nomen incorrectum), Paroxyrhopus* Schenkel, 1901, *Sympeltophis* F. Werner, 1925, *Paraoxyrhopus* – Dunn, 1928b *(nomen incorrectum), Gerrosteus* – T. Barbour & Loveridge, 1929 *(nomen incorrectum), Paraxirhopus* – Briceño-Rossi, 1934 *(nomen incorrectum),* and *Xenophis* – Elter, 1981 *(nomen incorrectum).*
Type species: *Elapomorphus scalaris* Wucherer, 1861c.
Distribution: Amazonia.
Sources: Bogert, 1964, J.A. Peters & Orejas-Miranda, 1970, Hoge & Federsoni, 1975, Sheil & Grant, 2001, Jansen et al., 2009, Zaher et al., 2009 and Vidal et al., 2010.

1. *Xenopholis scalaris* (Wucherer, 1861c). Proc. Zool. Soc. London 29(1): 325–326. (*Elapomorphus scalaris*)

Synonyms: *Xenopholis braconnieri* W.C.H. Peters, 1869, *Gerrhosteus prosopis* Cope, 1874b, and *Sympeltophis ungalioides* F. Werner, 1925.
Types: Syntypes (2), BMNH 1946.1.8.60, a 250 mm specimen, and OEHW, a 320 mm damaged specimen (M.C. Gayleard).
Type locality: "Cañavieras, Matta de S. João, a few leagues south from the city of Bahia, Brazil" [= Canavieiras, SE Bahia State, E Brazil, 15°37'S, 38°55'W, elevation NSL (or Mata de São João, E Bahia State, 12°40'S, 38°17'E, elevation 40 m)].
Distribution: Amazonia. Southern Colombia, S Guyana (Upper Takutu-Upper Essequibo), NW French Guiana (N Saint-Laurent-du-Maroni), E Ecuador (Napo, Pastaza), E Peru (Amazonas, Loreto, Madre de Dios, Pasco), Bolivia (Beni, La Paz, Pastaza, Santa Cruz) and Brazil (Bahia, Maranhão, Pará, Rondônia), NSL–1500 m.
Sources: M.A. Smith, 1928b, Cunha & Nascimento, 1978, 1993, Pérez-Santos & Moreno, 1988, 1991, Duellman & Salas, 1991, Duellman & Mendelson III, 1995, Starace, 1995, 1998, Freitas, 1999, Lehr, 2001, Duellman, 2005 and C.J. Cole et al., 2013.
Remarks: Original description reprinted in Wucherer (1862: 318–319). Type locality appears as two different locations with Canavieiras to the south of Bahia (= Salvador) and Mata de São João to the north.

2. *Xenopholis undulatus* (A.S. Jensen, 1900). Vidensk. Medd. Naturhist. Foren. Kjöbenhaven 1900: 106–107, fig. 2. (*Oxyrhopus undulatus*)

Synonyms: *Paroxyrhopus reticulatus* Schenkel, 1901, *Oxyrhopus latifrontalis* F. Werner, 1913, and *Paroxyrhopus atropurpureus* Amaral, 1923 & 1926a.
Types: Syntypes (2), ZMUC, longest specimen 465 mm (J.T. Reinhardt, 1854–1856 and E. Warming, 1863–1866).
Type locality: "Lagoa Santa er en lille Landsby, der ligger i den brasilianske Stat Minas geraës under 19°40' S. B., altsaa godt 3 Breddegrader nordligere end Rio de Janeiro og N. t.V. for denne By. Højden over Havet er ca. 835 Metre" [= Logoa Santa, Minas Gerais Sate, Brazil, ca. 19°38'S, 43°53'W, elevation 835 m].
Distribution: Brazil (Ceará, Distrito Federal, Goiás, Mato Grosso, Maranhão, Mato Grosso do Sul, Minas Gerais, Pará, Paraná, Pernambuco, Rondônia, São Paulo, Tocantins) and Paraguay, 115–895 m.
Sources: Hoge & Federsoni, 1975, Nogueira, 2001, Cacciali & Wüest, 2009, Loebmann, 2009a and Ribeiro et al., 2011.
Remarks: Photographs of holotype of *Paroxyrhopus reticulatus* Schenkel in M. Jansen et al. (2009: figs. 9a–b). The *P. reticulatus* type locality of Bermalcue (= *lapsus calami* for Bernal Cué) is now Cerrito, Paraguay *fide* Cacciali & Wüest (2009: 383).

3. *Xenopholis werdingorum* M. Jansen, Álvarez & G. Köhler, 2009. Zootaxa (2222): 33–38, figs. 2–4.

Type: Holotype, MNKR 4700, a 390 mm female (M. Jansen & A. Schulze, 25 Oct. 2006).
Type locality: "Hacienda San Sebastián (S 16°21.676', W 62°00.017', 550 m), Province of Ñuflo de Chávez, Department of Santa Cruz, Bolivia."
Distribution: Eastern Bolivia (Santa Cruz), 550 m.

X

Remarks: Probably occurs in Mato Grosso, Brazil *fide* M. Jansen et al. (2009: 41).

XENOTYPHLOPS Wallach & Ineich, 1996 (Xenotyphlopidae)

Type species: *Typhlops grandidieri* Mocquard, 1905b.
Distribution: Madagascar.
Sources: Wallach, 1993c, Glaw & Vences, 1994, Wallach & Ineich, 1996, McDiarmid et al., 1999, Wallach et al., 2007, Wallach & Glaw, 2009 and Wegener et al., in press.

1. *Xenotyphlops grandidieri* (Mocquard, 1905b). Bull. Mus. Hist. Nat., Paris 11(5): 287. (*Typhlops grandidieri*)

Synonym: *Xenotyphlops mocquardi* Wallach, Mercurio & Andreone, 2007.
Type: Lectotype, MNHN 1905.271, a 257 mm female (M. de Rothschild, 1904–1905), designated by Wallach & Ineich (1996: 368).
Type locality: "Madagascar," via lectotype selection. Restricted to Baie de Sakalava, N Antsiranana Prov., ext. N Madagascar *fide* Wegener et al. (in press, p. 11).
Distribution: Northern Madagascar (N Antsiranana), NSL–40 m.

XYELODONTOPHIS Broadley & Wallach, 2002 (Colubridae)

Type species: *Xyelodontophis uluguruensis* Broadley & Wallach, 2002.
Distribution: Southeast Africa.
Source: Broadley & Wallach, 2002.

1. *Xyelodontophis uluguruensis* Broadley & Wallach, 2002. Bull. Nat. Hist. Mus. (Zool.) 68(2): 66–69, figs. 1f, 2a, 4a.

Type: Holotype, NMZB 7443 (formerly KMH 2636), an 1147 mm female (J. Lovett, Nov. 1983)
Type locality: "Lupanga Peak, Uluguru Mountains, Tanzania (06°52'S: 37°43'E)" [= Lupanga Peak, Uluguru Mtns., Morogoro Prov., NE Tanzania, 6°52'S, 37°43'E, max. elevation 2150 m].
Distribution: Northeastern Tanzania (Morogoro), 1650–2000 m.

XYLOPHIS Beddome, 1878b (Elapoidea *incertae sedis*)

Synonyms: *Platypteryx* A.-M.-C. Duméril, 1853, *Platypterix* A.-M.-C. Duméril, Bibron & Duméril, 1854a *(nomen emendatum), Sylophis* – A.S. Romer, 1956 *(nomen incorrectum), Xilophis* – Briceño-Rossi, 1934 *(nomen incorrectum),* and *Xylphis* – B.D. Sharma *in* B.D. Sharma & Kumari, 1998a *(nomen incorrectum).*
Type species: *Geophis stenorhynchus* A.C.L.G. Günther, 1875.
Distribution: Southern India.
Sources: M.A. Smith, 1943, Mahendra, 1984, Whitaker & Captain, 2004 and Zaher et al., 2009.

1. *Xylophis captaini* Gower & Winkler, 2007. Hamadryad 31(2): 316–322, figs. 1–3, 4 (paratype), 5.

Type: Holotype, BNHS 3375, a 137 mm male (O.V. Oommen and field crew, 8 Aug. 2000).
Type locality: "Kannam, Kottayam District, in the state of Kerala at approximately 9° 32' N, 76° 41' E and 110 m a.s.l., India."
Distribution: Southwestern India (S Kerala, Tamil Nadu), 50–300 m.
Source: Ganesh et al., 2012a.

2. *Xylophis perrotetii* (A.-M.-C. Duméril, 1853). Mém. Acad. Sci. Inst. France, Paris 23: 468. (*Platypteryx perrotetii*)

Synonym: *Rabdosoma microcephalum* A.C.L.G. Günther, 1858.
Type: Holotype MNHN 7300, a 488 mm specimen (G.S. Perrotet, 1834–1853).
Type locality: "monts Nilgherry" [= Nilgiri Hills, cen. Western Ghats, SE Kerala/W Tamil Nadu States, SW India].
Distribution: Southwestern India (Western Ghats of E Kerala, W Tamil Nadu), NSL–2000 m.
Sources: Wall, 1919b, 1923a and Whitaker & Whitaker, 1977.
Remarks: *Platypteryx perrotetii* A.-M.-C. Duméril, 1853 is not a *nomen nudum* (*sensu* Williams & Wallach, 1989: 116) but a valid name as the monotypic genus *Platypteryx* was described in 1853 with *P. perrotetii* as its only species.

3. *Xylophis stenorhynchus* (A.C.L.G. Günther, 1875). Proc. Zool. Soc. London 43(1): 230–231. (*Geophis stenorhynchus*)

Synonyms: *Xylophis indicus* Beddome, 1878b, and *Xylophis stenorynchus* Wall, 1923a *(nomen emendatum).*
Types: Syntypes (3), BMNH 1946.1.14.13, a 159 mm male, BMNH 1946.1.14.14, a 165 mm male, and BMNH 1946.1.14.15, a 185–191 mm female (R.H. Beddome, 1857–1875).
Type locality: "Travancore, Southern India" [= S Kerala and SW Tamil Nadu States, SW India].
Distribution: Southern India (Western and Eastern Ghats of S Kerala, S Tamil Nadu), 1075–1525 m.
Source: Inger et al., 1984.

Y

†*YURLUNGGUR* Scanlon, 1992 (†Madtsoiidae)

Type species: †*Yurlunggur camfieldensis* Scanlon, 1992.
Distribution: Upper Oligocene to Pleistocene of Australia.
Sources: Scanlon, 1997, 2003b, 2006, Scanlon & Lee, 2002, La Duke et al., 2010 and J.A. Wilson et al., 2010.

1. †*Yurlunggur camfieldensis* Scanlon, 1992. The Beagle 9(1): 50–55, figs. 1a–f.

Type: Holotype, NTM P894-5, -3, -5, -21, -22, -23, -24, -25, -27, -28, -245; P8692-28; P8695-23, -25, -127, -128, -169, -243, -244, -245, -246, -247, -260, -261, -262, -263, -264; P901-1, -2, -4; P87103-22, -23, -24; P8792-23, -24; P87108-17, -18, -19, -20, -21; P87115-1, a partial skeleton of 29 trunk vertebrae and 11 rib fragments (P. Murray, D. Megirian & K. Roth).
Type locality: "'Blast Site', Bullock Creek Local Fauna, Camfield Beds, north central Northern Territory, Australia; middle Miocene, approx. 12 MYBP."
Distribution: Upper Oligocene to Pleistocene (0.01–25.0 mya) of Australia.

Z

ZAMENIS Wagler, 1830 (Colubridae)

Synonyms: *Rhinechis* Michahelles *in* Wagler, 1833, *Calopeltis* Fitzinger, 1834 *(nomen nudum), Callopeltis* Fitzinger *in* Bonaparte 1834, *Calopeltis* Fitzinger, 1843 *(nomen emendatum), Callipelta* Agassiz, 1847 *(nomen emendatum), Lamenis* – Tristram, 1860 *(nomen incorrectum), Tysia* – Carus & Engelmann, 1861 *(nomen incorrectum), Tyria* Cope, 1862c *(nomen praeoccupatum), Zameuis* – Rochebrune, 1885 *(nomen incorrectum), Famenis* – F. Werner, 1924c *(nomen incorrectum), Zamensis* – Roddy, 1928 *(nomen incorrectum), Zeamnis* – Roddy, 1928 *(nomen incorrectum), Zamonis* – Maass-Berlin, 1933 *(nomen incorrectum), Caolepellis* – Briceño-Rossi, 1934 *(nomen incorrectum), Tryria* – Briceño-Rossi, 1934 *(nomen incorrectum), Rhinophis* Guillon, 1936 *(nomen praeoccupatum), Aamenis* – Sundersingh, 1960 *(nomen incorrectum), Lamenius* – Yakovleva, 1964 *(nomen incorrectum), Zameni* – Wu, Li & & Liu, 1985 *(nomen incorrectum),* and *Richardwellsus* Hoser, 2012m *(nomen illegitimum).*

Type species: *Natrix longissima* Laurenti, 1768.

Distribution: Eurasia.

Fossil records: Lower Miocene to upper Pleistocene of Europe.

Sources: Schulz, 1996, Lenk et al., 2001a, Utiger et al., 2002, 2005, and Burbrink & Lawson, 2007.

1. †*Zamenis algorensis* (Szyndlar, 1985a). Estud. Geol. 41(5–6): 448–451, figs. 2.1–2.3. (*Elaphe algorensis*)

Type: Holotype, MNCN Alg. I-3, one basiparasphenoid (B. Sanchíz).

Type locality: "Fossil Karstic breccia from Algora (province of Guadalajara, Spain); Uppermost Miocene (European Continental biozone NM 13)."

Distribution: Upper Miocene (Turolian, MN 13: 4.9–7.3 mya) of Spain. Known only from type locality.

Source: Szyndlar, 2012.

Remarks: Similar to †*Z. kohfidischi* and *Z. longissima fide* Szyndlar, 1985a.

2. *Zamenis hohenackeri* (Strauch, 1873). Mém. Acad. Imp. Sci. St.-Pétersbourg (7) 21(4): 69–73, pl. 2, figs. a–b. (*Coluber hohenackeri*)

Synonyms: *Coluber rubriventer* Dwigubsky, 1832 *(nomen oblitum),* and *Coluber tauricus* F. Werner, 1898a.

Type: Lectotype, ZISP 1642 (formerly AIS) (R.F. Hohenacker, 1838), designated by Bischoff *in* W. Böhme (1993: 322).

Type locality: "Kolonie Helenendorf = Chanlar/ Aserbaidzan" [= Goygol, Azerbaijan, 40°34'N, 46°21'E, elevation 700 m] via lectotype selection.

Distribution: Middle East and Asia Minor. Southern Turkey (Adana, Agri, Amasya, Antalya, Icel, Konya, Siirt, Trabzon), SW Syria (Damascus), Lebanon (Beqaa, Liban-Nord, Mont-Liban, Nabatiye), ext. N Israel (N Golan Heights), SW Russia (Dagestan), Georgia, Armenia, Azerbaijan, NE Iraq (Erbil, Sulaymaniyah) and NW Iran (East Azarbaijan, Isfahan, West Azarbaijan), 180–3000 m.

Sources: Nikolsky, 1916, Baran, 1976a, Bannikov et al., 1977, Basoglu & Baran, 1980, Nilson & Andrén, 1984b, Schätti & Baran, 1988, Bischoff, 1993, Hraoui-Bloquet et al., 2002, Franzen et al., 2008, Tuniyev et al., 2009, Afrasiab & Mohamad, 2011 and Bar & Haimovitch, 2011.

Remarks: In accordance with Art. 23.9.2 of the Code (ICZN, 1999), *Coluber hohenackeri* Strauch is designated a *nomen protectum* and *Coluber rubriventer* Dwigubsky a *nomen oblitum.*

3. †*Zamenis kohfidischi* (Bachmayer & Szyndlar, 1985). Ann. Naturhist. Mus. Wien 87A: 80–81, figs. 1.1–1.5, pl. 1, figs. 1–3. (†*Elaphe kohfidischi*)

Type: Holotype, NMWGP 1984/96, one middle trunk vertebra.

Type locality: "Kohfidisch fissures, Burgenland, Austria; Late Pannonian (= Pontian S. Str.) (MN 11); Early Pliocene or Upper Miocene."

Distribution: Lower Miocene (Orleanian, MN 5: 13.7–16.0 mya) of Austria, and upper Miocene (Turolian, MN 11–12: 7.3–8.7 mya) of Austria and Ukraine.

Sources: Bachmayer & Szyndlar, 1987, Szyndlar, 1991a, 2012 and Szyndlar & Zerova, 1992.

4. †*Zamenis kormosi* (Bolkay, 1913). Mitt. Jb. Kgl. Ungar. Geol. Reichsanst. 21(7): 224, pl. 12, figs. 4–8. (†*Coluber kormosi*)

Types: Syntypes (14), HGIM Ob-4465/Vt. 75: Ob-4465/a, a basioccipital, Ob-4465/b, a right quadrate, Ob-4465/c, a left ectopterygoid, Ob-4465/d, a parabasisphenoid, Ob-4465/e, plus a premaxilla, a palatine, a second ectopterygoid, a second and third quadrate, and five maxillary fragments (T. Kormos).

Type locality: "Polgárdi, Hungary; Pannonian, intermixed with the remains of Hipparion fauna, Praeglacial (Interglacial ?) periode."

Distribution: Upper Miocene to lower Pliocene (Turolian, MN 12–14: 4.2–7.8 mya) of Hungary.

Sources: Rage, 1984b, Szyndlar, 1991a, 2012 and Venczel, 1994.

Remarks: A valid species *fide* Mlynarski (1961: 26). Publication date listed as 1913/1915 *fide* Mlynarski (1961: 41).

5. *Zamenis lineatus* (Camerano, 1891a). Mem. Reale Accad. Sci. (Fis. Mat. Nat.), Torino (2) 41: 458. (*Callopeltis longissimus lineata*)

Type: Holotype, MSNM 942 (formerly MZUT 434), an 1140 mm male.

Type locality: "contorni di Napoli, Italia" [= vicinity of Naples, Napoli Prov., Campania Region, S Italy, 40°51'N, 14°16'E, elevation 20 m].

Distribution: Southern Italy (Basilicata, Calabria, Campania, Sardegna, Sicilia), NSL–900 m.

Source: Lenk & Wüster, 1999.

Remarks: Original description is reprinted in Camerano (1891b: 58–59).

6. †*Zamenis lobsingensis* (Heller, 1960). Erlanger Geol. Abh. (34): 27–28, pl. 3, fig. 14. (†*Coluber lobsingensis*)

Type: Holotype, SPLUE H Lo 58–59, one parietal and articulated parabasisphenoid (F.X. Schnittmann, Jan. 1956).

Type locality: "Dolomitsteinbruch Lobsing bei Neustadt/Donau; Gerinne-Verfüllungen jungpleistozänen Alters (letztes Interglazial)" [= dolomites of Lobsing near Neustadt on the Danube River, Bayern Prov., S Germany; Pleistocene, Riss-Würm interglacial *fide* Rage, 1984b: 46].

Distribution: Pleistocene (0.01–1.9 mya) of Germany. Known only from type series.

Sources: Mlynarski, 1961 and Rage, 1984b.

7. *Zamenis longissimus* (Laurenti, 1768). Synop. Rept.: 74. (*Natrix longissima*)

Synonyms: *Coluber aesculapii* Lacépède, 1789 (*nomen rejiciendum*), *Coluber flavescens* Gmelin, 1789, *Coluber asclepiadeus* Donndorff, 1798, *Coluber leprosus* Donndorff, 1798, *Coluber pannonicus* Donndorff, 1798, *Coluber sellmanni* Donndorff, 1798, *Coluber romanus* Suckow, 1798, *Coluber leprosus* Bechstein, 1802, *Natrix scopolii* Merrem, 1820, *Coluber aesculapii niger* Fitzinger, 1826a, *Coluber fugax* Eichwald, 1831, *Coluber elegantulus* Bonaparte, 1840 (*nomen nudum*), *Coluber isabellinus* Freyer, 1842, *Coluber flavescens nigrescens* Massalongo, 1853, *Zamenis aesculapii nigra* Fitzinger, 1853 (*nomen nudum*), *Coluber flavescens nigrescens* Massalongo, 1854, *Coluber atratus* Costa, 1871 (*nomen nudum*), *Coluber flavescens carbonarius* Pirona, 1871 (*partim*), *Callopeltis longissma lineata* Camerano, 1891, *Coluber longissimus deubeli* Dürigen, 1897 (*nomen nudum*), *Coluber aesculapii niger* Dürigen, 1897, *Coluber aesculapii ventrimaculatus* Dürigen, 1897 (*nomen substitutum*), *Coluber aesculapii virgatus* Dürigen, 1897d (*nomen substitutum*), *Coluber longissimus subgrisea* F. Werner, 1897, *Elaphe rechingeri* F. Werner, 1932, and *Coluber longissimus lineata* Balli, 1944 (*nomen praeoccupatum*).

Type: Holotype, not designated, location unknown.

Type locality: "Austria." Restricted to Vienna, Austria *fide* Mertens & Müller (1928: 47).

Distribution: Southern Europe and SW Asia. Northern Spain (Barcelona, Gerona, Guipúzcoa, Huesca, Lérida, Navarra, W Santander, Vizcaya), France (Ain, Allier, Alpes-Maritimes, Ardèche, Ariège, Aude, Aveyron, Basses-Alpes, Basses-Pyrénées, Bouchés-du-Rhône, Cantal, Cher, Charente, Charente-Maritime, Corrèze, Corsica, Côte-d'Or, Creuse, Deux-Sèvres, Dordogne, Drôme, Essonne, Eure-et-Loir, Finistère, Gard, Gers, Gironde, Haut-Rhin, Haute-Garonne, Haute-Loire, Haute-Marne, Haute-Saône, Haute-Savoie, Haute-Vienne, Hautes-Alpes, Hautes-Pyrénées, Hérault, Ille-et-Vilaine, Indre-et-Loire, Isère, Jura, Landes, Loire, Loire-Atlantique, Loiret, Loir-et-Cher, Lot, Lot-et-Garonne, Lozère, Maine-et-Loire, Manche, Mayenne, Morbihan, Nièvre, Orne, Puy-de-Dôme, Pyrénées-Orientales, Rhône, Saöne-et-Loire, Sarthe, Savoie, Seine-et-Marne, Tarn, Tarn-et-Garonne, Val-d'Oise, Var, Vaucluse, Vendée, Vienne, Yonne, Yvelines), S Germany (E N Baden-Württemberg, Bavaria, S Hesse), SW Poland (Lubelskie, Malopolskie, Podkarpackie, Slaskie), Czech Republic (Jihomoravsky, Zapadocesky), Slovakia, Switzerland (Bern, Genève, Graubunden, St. Gallen, Ticino, Valais, Vaud), Austria (Burgenland, Carinthia, Lower Austria, Salzburg, Styria, Tyrol, Upper Austria, Vienna), Hungary, Italy (Abruzzi, Basilicata, Calabria, Campania, Emilia-Romagna, Friuli-Venezia Giulia, Lazio, Liguria, Lombardia, Marche, Molise, Piemonte, Puglia, Toscana, Trentino-Alto Adige, Umbria, Valle d'Aosta, Veneto, Sardinia and Sicily Is.), S Romania, SW Ukraine, Moldavia, Slovenia, Croatia, Bosnia-Herzegovina, Montenegro, Serbia, Macedonia, Albania, Bulgaria (Blagoevgrad, Burgas, Dobrich, Gabrovo, Grand Sofiya, Haskovo, Kardzhali, Lovech, Kyustendil, Montana, Pazardzhik, Pernik, Pleven, Plovdiv, Razgrad, Rousse, Shumen, Silistra, Sliven, Smolyan, Sofiya, Stara Zagora, Targovishte, Varna, Veliko Tarnovo, Vidin, Vratsa, Yambol), Greece (Aitolia and Akarnania, Arkadhia, Akhaia, Argolis, Arta, Attica, Drama, Evritania, Evros, Evvoia, Florina, Fokis, Fthiotis, Grevena, Ilia, Imathia, Ioannina, Kardhitsa, Kastoria, Kefallinia, Kerkira, Khalkidhiki, Kilkis, Korinthia, Kozani, Lakonia, Larisa, Levkas, Magnisia, Messinia, Pella, Pieria, Preveza, Rodhopi,

Serrai, Thesprotia, Thessaloniki, Trikala, Voiotia, Xanthi, Zakinthos), N Turkey (Giresun, Istanbul, Trabzon, Corfu, Paxos), W Georgia, SW Azerbaijan, SW Russia (Dagestan), and N Iran (East Azarbaijan, Hamadan, West Azarbaijan), NSL–1700 (2000) m.

Fossil records: Lower Pliocene (Ruscinian, MN 14: 4.2–4.9 mya) of Bosnia & Herzegovina and Hungary, middle Pliocene (Ruscinian, MN 15: 3.2–4.2 mya) of Moldavia and Slovakia, upper Pliocene (Villanyian, MN 16: 2.6–3.2 mya) of Austria, France, Hungary and Moldavia, lower Pleistocene (lower, Calabrian: 0.8–1.8 mya) of Austria, Czech Republic, Poland and Romania, middle Pleistocene (Ionian: 0.13–0.78 mya) of Austria, France, Germany, Greece, Romania and United Kingdom, and upper Pleistocene (Tarantian: 0.01–0.13 mya) of Croatia, France, Germany, Greece, Poland and Serbia.

Sources: Schreiber, 1875, Steinheil, 1914, Roillant, 1934, Mertens & Wermuth, 1960, Fuhn & Vancea, 1961, Capocaccia, 1964, 1965, Steward, 1971, Lotze, 1975, Bannikov et al., 1977, Bruno, 1977, 1984, Basoglu & Baran, 1980, Nilson & Adrén, 1984b, Szyndlar, 1984c, Wütschert, 1984, König, 1985, Bailon, 1989, Gillette, 1990, Latifi, 1991, Szyndlar, 1991a, 2012, W. Böhme, 1993d, Clark, 1994, Lenk & Joger, 1994, Schweiger, 1994, R. Günther, 1996, Borkin et al., 1997, Holman, 1998a, Cabela et al., 2001, Hofer et al., 2001, Mikátová et al., 2001, F.P. Pérez, 2001, Gomille, 2002, Glowacinski & Rafinski, 2003, Ratnikov, 2004, Sindaco et al., 2006, Ivanov, 2007, Kreiner, 2007, Renner & Vitzthum, 2007, Trapp, 2007, Valakos et al., 2008, Tuniyev et al., 2009, Bezman-Moseyko, 2010, Vacher & Geniez, 2010, Stojanov et al., 2011 and Jablonski et al., 2012.

Remarks: *Zamenis romani* possibly a valid species related to *Z. persica fide* Lenk & Joger (1994: 371) but later a synonym of *Z. longissima fide* Lenk & Wüster, 1999.

8. †*Zamenis paralongissimus* (Szyndlar, 1984a). Acta Zool. Cracov. 28(1): 67–71, figs. 26.1–26.5. (†*Elaphe paralongissima*)

Type: Holotype, IZPAN R III/11, one middle trunk vertebra.

Type locality: "Weze II (W II) at Zelce Hill near Dzialoszyn (Northern Kraków—Wielun Upland in central Southern Poland); Upper Pliocene, MN biozone presumably 16."

Distribution: Upper Pliocene (Villanyian MN 16: 2.6–3.2 mya) of Poland, and lower Pleistocene (Calabrian: 0.8–1.8 mya) of Austria and Czech Republic.

Sources: Mlynarski et al., 1984, Szyndlar, 1991a and Ivanov, 2007.

9. *Zamenis persicus* (F. Werner, 1913). Mitt. Naturhist. Mus. Hamburg (1912) 30: 23–24. (*Coluber longissimus persica*) (*nomen protectum*)

Synonyms: *Zamenis persicus* Jan, 1863b (*nomen nudum*), *Zamenis persicus* Jan, 1867 *in* Jan & Sordelli, 1866–1870 (*nomen oblitum*), and *Coluber longissimus nigra* Nikolsky, 1907 *(nomen praeoccupatum)*.

Type: Holotype, ZMH, a 640 mm male (F. Bruns, 20–26 July 1909), destroyed in July 1943 during World War II.

Type locality: "Barferusch, Provinz Mazanderan, Persien" [= Babol, Mazandaran Prov., N Iran, 36°33'N, 52°41'E, elevation 10 m].

Distribution: Southeastern Azerbaijan and N Iran (Azarbayjan Khavari, Gilan, Mazandaran), 25 m BSL–1565 (4000) m.

Sources: Nilson & Andren, 1984b, Tuniyev et al., 2009 and Schätti et al., 2010a.

10. †*Zamenis praelongissima* (Venczel, 1994). Acta Zool. Cracov. 37(1): 12–16, figs. 6a–c. (†*Elaphe praelongissima*)

Type: Holotype, HGIM V.19004/Vt.149, one basiparasphenoid (1984–1994).

Type locality: "Polgárdi 4 'Lower' limestone quarry of the Somló-Hill and Koszár-Hill, located near the village of Polgárdi in western Hungary; Uppermost Miocene, Pontian or Upper Turolian (MN 13)."

Distribution: Upper Miocene to lower Pliocene (Turolian-Ruscinian, MN 13–14: 4.2–7.3 mya) of Hungary.

Source: Venczel, 1998 and Szyndlar, 2012.

11. *Zamenis scalaris* (Schinz *in* Cuvier, 1822). Das Thierreich 2: 123. (*Coluber scalaris*)

Synonyms: *Coluber marmoratus* Gravenhorst, 1807 (*nomen oblitum*), *Rhinechis agassizii* Michahelles *in* Wagler, 1833, *Coluber bilineatus* Schinz, 1833, *Xenodon michahelles* Schlegel, 1837, *Coluber bitaeniatus* Duméril *in* Bonaparte, 1840 *(nomen nudum)*, *Coluber dorsalis* Bonaparte, 1840 *(nomen nudum)*, *Coluber meiffreni* Bonaparte, 1840 *(nomen nudum)*, *Coluber meifreni* Schinz, 1840, *Simus danmanni* Schinz, 1840, *Coluber hermannii* Lesson, 1841, *Simus baumanni* Fitzinger, 1867, and *Coluber agassirii* – Westphal-Castelnau, 1870 *(nomen incorrectum)*.

Type: Holotype, not designated, location unknown.

Type locality: "Südlichen Frankreich" [= S France].

Distribution: Southwestern Europe. Portugal (Aveiro, Beja, Braga, Bragansa, Castelo Branco, Coimbra, Évora, Faro, Guarda, Leiria, Lisboa, Portalegre, Porto, Santarém, Sertubal, Viana de Castelo, Vila Real, Viseu), Spain (Álava, Alicante, Almería, Andorra, Aragon, Ávila, Badajoz, Burgos, Cáceres, Cádiz, Castellón, Catalonia, Ciudad Real, Córdoba, Cuenca, Gerona, Granada, Guadalajara, Guipúzcoa, Huelva, Huesca, Jaén, La Coruña, Léon, Lérida, Logroño, Lugo, Madrid, Málaga, Murcia, Navarra, Orense,

Oviedo, Palencia, Pontevedra, Salamanca, Segovia, Seville, Soria, Tarragona, Teruel, Toledo, Valencia, Valladolid, Zamora, Zaragossa, Minorca Is.), S France (Alpes-Maritimes, Ardèche, Aude, Basses-Alpes, Bouchés-du-Rhône, Drôme, Gard, Hérault, Pyrénées-Orientales, Vaucluse, Var, Hyères Is.), NSL–2100 m.

Fossil records: Lower Pleistocene (Calabrian: 0.8–1.8 mya) of Spain, and middle Pleistocene (Ionian: 0.13–0.78 mya) of France and Spain.

Sources: Mertens & Wermuth, 1960, Rage & Truc, 1970, Sanz & Sanchíz, 1980, Bruno, 1984, Salvador, 1985, Calmonte & Ferri, 1987, Cheylan & Guillaume, 1993 and Blázquez, 1994, Schulz, 1996, Gasc et al., 1997, Holman, 1998a, Szyndlar, 1999, Hilfenberger, 2001, Lenk et al., 2001a, Utiger et al., 2002, 2005, Burbrink & Lawson, 2007, Feriche et al., 2008 and Vacher & Geniez, 2010.

Remarks: A synonym of *Zamenis fide* Zaher et al., 2012. In accordance with Art. 23.9.2 of the Code (ICZN, 1999), *Coluber scalaris* Schinz is designated a *nomen protectum* and *Coluber marmoratus* Gravenhorst a *nomen oblitum*.

12. *Zamenis situla* (Linnaeus, 1758). Syst. Nat., ed. 10, 1: 223. (*Coluber situla*)

Synonyms: *Coluber quadrilineatus* Pallas, 1814 *(nomen praeoccupatum), Coluber trilineatus* Metaxa, 1823, *Coluber leopardinus* Bonaparte, 1840 *in* 1832–1841, *Coluber metaxa* Bonaparte, 1837 *in* 1832–1841 (*nomen substitutum*), *Coluber cruentatus* Steveni, 1835, *Callopeltis leopardinus metaxa* Bonaparte, 1834 *in* 1832–1841, *Coluber leopardinus striata* Dehne, 1856, *Coronella quadrilineata punctulata* Jan, 1863, *Coelopeltis leopardinus otthianus* Gistel, 1868, *Coluber quadrilineatus pallasii* Bedriaga, 1881 *in* 1881–1882, and *Coluber leopardinus sparsus* Schreiber, 1912.

Type: Holotype, NHR Lin-32, an 890 mm specimen (F. Hasselquist, 1749–1752, via Mus. Drottn.).

Type locality: "Aegypto" [= Egypt] (in error, possibly for Asia Minor *fide* Flower, 1933: 815). Corrected to ruins of Ephesus, 3 km SW Selçuk, Izmir Prov., W Turkey *fide* Bruno (1992: 116) and to Izmir, Izmir Prov., Turkey *fide* Obst et al. (1993: 443).

Distribution: Southeastern Europe and Asia Minor. Extreme S Italy (Basilicata, Puglia, E Sicily Is.), Malta, Bosnia-Herzegovina, Montenegero, Croatia, Ada, Cres, Korcula, Krk, Lissa, Losinj, Mljet, Sipan and Vis Is., Macedonia, Albania, S Bulgaria (Blagoevgrad, Burgas, Pazardzhik), ext. S Ukraine, Greece (Aitolia and Akarnania, Arkadhia, Akhaia, Argolis, Arta, Attica, Drama, Evritania, Evros, Evvoia, Florina, Fokis, Fthiotis, Grevena, Ilia, Imathia, Ioannina, Kardhitsa, Kastoria, Kefallinia, Kerkira, Khalkidhiki, Kilkis, Korinthia, Kozani, Lakonia, Larisa, Lesvos, Levkas, Magnisia, Messinia, Pella, Pieria, Poros, Preveza, Rodhopi, Serrai, Thesprotia, Thessaloniki, Trikala, Voiotia, Xanthi, Zakinthos, Aegina, Andros, Argos, Chios, Ermione, Epidavros, Euboea, Kea, Kefaloniá, Kérkira, Kíthnos, Kríti, Konitsa, Lésvos, Mílos, Páros, Ródos, Salamis, Sámos, Sérifos, Skíros, Skópelos, Sífnos, Síros, Spetasi, Thíra and Tínos Is.) and Turkey (Aydin, Balikesir, Denizli, Istanbul, Izmir, Kocaeli, Trabzon), NSL–1600 m.

Fossil records: Lower Pleistocene (Calabrian: 0.8–1.8 mya) of Austria and Czech Republic, and middle Pleistocene to upper Pleistocene (Ionian-Tarantian: 0.01–0.78 mya) of Bulgaria and Greece.

Sources: F. Werner, 1938a, Buchholz, 1955, Pozzi, 1966, Clark, 1968, Bruno, 1969, 1977, 1984, Baran, 1976, Bannikov et al., 1977, Goran, 1986, Bruno & Maugeri, 1990, Szyndlar, 1991a, Obst et al., 1993, Holman, 1998a, Farkas & Toth, 1999, Petkovski et al., 2000, Ratnikov, 2004, Valakos et al., 2004, Sindaco et al., 2006, Ivanov, 2007, Naumov & Tomovic, 2007, Trapp, 2007, Franzen et al., 2008, Valakos et al., 2008, Tuniyev et al., 2009 and Stojanov et al., 2011.

Remarks: Type locality restriction to Misir, Egypt *fide* Baran (1976: 82) invalid.

13. †*Zamenis szyndlari* (Venczel, 1998). Acta Zool. Cracov. 41(1): 12–14, figs. 8a–c. (†*Elaphe szyndlari*)

Type: Holotype, HGIM V.20628, one basiparasphenoid.

Type locality: "Polgárdi 4 'Lower' limestone quarries of Somló-Hill and Koszár-Hill, near the village of Polgárdi (W-Hungary); Pontian or Upper Turolian (mammalian biozone MN 13), Late Miocene."

Distribution: Upper Miocene (Turolian, MN 13: 4.9–7.3 mya) of Hungary. Known only from type locality.

Source: Szyndlar, 2012.

†*ZELCEOPHIS* Szyndlar, 1984a (Colubridae)

Type species: †*Zelceophis xenos* Szyndlar, 1984a.

Distribution: Upper Miocene of Poland.

Source: Szyndlar, 1984a.

1. †*Zelceophis xenos* Szyndlar, 1984a. Acta Zool. Cracov. 28(1): 38–39, figs. 12a, d., l, p, v.

Type: Holotype, ZPW IZ-6/R/1, one trunk vertebra.

Type locality: "Mala Cava (MA) at Zelce Hill near Dzialoszyn (Northern Kraków—Wielún Upland in central Southern Poland); Upper Miocene (Turolian/Pontian), layer 7+8 (= Early/Middle Pliocene)."

Distribution: Upper Miocene (Turolian, MN 11–13: 4.9–9.7 mya) of Poland.

ZHAOERMIA Gumprecht & Tillack, 2004 (*nomen substitutum*) (Viperidae)

Synonym: *Ermia* F.-J. Zhang *in* Zhao et al., 1993. *(nomen praeoccupatum)*

Type species: *Trimeresurus mangshanensis* Zhao *in* Zhao & Chen, 1990.

Distribution: China.

Sources: F.-J. Zhang, 1993, David & Tong, 1997, P. Guo et al., 1999, Gumprecht & Tillack, 2004, Gumprecht et al., 2004 and Zaher et al., 2012.

1. *Zhaoermia mangshanensis* (Zhao *in* Zhao & Chen, 1990). Sichuan J. Zool. 9(1): 11–12. (*Trimeresurus mangshanensis*)

Type: Holotype, CIB 8901 (formerly ZS 8901), a 447 mm female (G.-H. Chen & Y.L. Tao, 15–30 Sept. 1989).

Type locality: "Pingkeng District, Mt. Mang (Mangshan), Yizhang County, Hunan Province, altitude 700–900 m" [China].

Distribution: Southern China (Hunan), 700–900 m. Known only from type locality.

Source: Guo et al., 2012.

Remarks: Holotype listed as male in original description. Holotype "considerably damaged" with skull removed *fide* P. Guo et al. (2012: 340).

Type: Neotype designation of CIB 98485 (a 370 mm male from Pingkeng in Mt. Mang, Yizhang, Hunan, China) by P. Guo et al. (2012: 340–341, figs. 1–2) invalid as an extracted skull and poor preservation are insufficient reasons for a neotype designation.

Genera and Species Inquirenda (*incertae sedis, nomina dubia, nomina nuda, nomina inedita and nomina rejicienda*)

GENERA

1. *ACRODERUS* Haworth, 1825. Philos. Mag. J. 65: 372 (*nomen nudum*).

Type species: Not designated.

2. *ATRYPTODON* Agassiz, 1844. Nomencl. Zool.: 5 (*nomen nudum*).

Synonym: *Athryptodon* Agassiz, 1847 *(nomen emendatum).*
Type species: Not designated.

3. *BORANUS* Rafinesque-Schmaltz, 1815. Anal. Nat. Tabl.: 77 (*nomen nudum*).

Type species: Not designated.

4. †*BOTROPHIS* Rochebrune, 1880. Nouv. Arch. Mus. Hist. Nat. (2) 3: 280.

Synonyms: †*Bothrophis* Zittel, 1889 *(nomen praeoccupatum),* and †*Borthrophis* – Huene, 1956 *(nomen incorrectum).*
Type species: †*Botrophis gaudryi* Rochebrune, 1880.
Distribution: Lower Miocene of France.
Remarks: Boidae or Viperidae *fide* Rage (1984b: 33).

5. *CERCOPSIS* A.-M.-C. Duméril & Bibron, 1844. Erpét. Gén. 6: 63 (*nomen nudum*).

Type species: Not designated.

6. *CHALINOPHORUS* Jan, 1857. Cenni Mus. Civ. Milano: 50 (*nomen nudum*).

Type species: *Chalinophorus semidoliatus* Jan, 1857.
Type locality: "Giava" [= Java, W Indonesia].

7. *CORALLIOPHIS* Fitzinger, 1843. Syst. Rept.: 25 (*nomen nudum*).

Synonym: *Corralliophis* Fitzinger *in* A.-M.-C. Duméruil, Bibron & Duméril, 1854b *(nomen emendatum).*
Type species: *Coronella spixii* Fitzinger, 1826a.

8. *DOCHIS* Jan, 1857. Cenni Mus. Civ. Milano: 44 (*nomen nudum*).

Type species: *Dochis strigata* Jan, 1857.

9. *DRACONISCUS* Philippi, 1899. Anal. Univ. Chile 104: 720 (*nomen nudum*).

Type species: Not designated.

10. *EUPELTIS* Fitzinger, 1843. Syst. Rept.: 25 (*nomen nudum*).

Type species: *Eupeltis pulchra* Fitzinger, 1843.

11. *EUPREPES* Philippi, 1899. Anal. Univ. Chile 104: 720 (*nomen nudum*).

Type species: Not designated.
Remarks: Possibly *Dromicus (= Liophis) fide* Donoso-Barros & Cardenas (1964: 232).

12. *GONYOGASTER* A.-M.-C. Duméril & Bibron, 1844. Erpét. Gén. 6: 64 (*nomen nudum*).

Type species: Not designated.

13. †*HETEROPYTHON* Rochebrune, 1880. Nouv. Arch. Mus. Hist. Nat. (2) 3: 290. (Boidae)

Synonyms: †*Heterophyton* – Hoffmann, 1890 *(nomen incorrectum),* †*Heteropyton* – Huene, 1959 *(nomen incorrectum),* and †*Heterophyton* – P.H.D.H. Silva, 1980 *(nomen incorrectum).*
Type species: †*Python euboeicus* Roemer, 1870.
Distribution: Southern France.
Remarks: A *nomen dubium fide* Rage (1984b: 33).

14. †*HOMEOTYPHLUS* Palacky, 1898. Mém. Soc. Zool. France 11: 104 (*nomen nudum*).

Type species: Not designated.

15. *HYDRECHIS* Ritgen, 1828. Nova Acta Acad. Caesar Leop.-Carol. 14: 262 (*nomen nudum*).

Type species: Not designated.

16. *HYPNALE* Rafinesque-Schmaltz, 1815. Anal. Nat. Tabl.: 77 (*nomen nudum*).

Type species: Not designated.

17. *IPHISA* Rafinesque-Schmaltz, 1815: Anal. Nat. Tabl.: 77 (*nomen nudum*).

Type species: Not designated.

18. *LAMANDA* Oken, 1816. Lehrbuch Naturg.: 274 (*nomen nudum*).

Type species: Not designated.

19. †*LAOPHIS* Owen, 1857. Quart. J. Geol. Soc. London 13: 196. (? Viperidae)

Type species: †*Laophis crotaloides* Owen, 1857.
Distribution: Upper Miocene or lower Pliocene of Greece.
Sources: Kuhn, 1939a, 1963, Hoffstetter, 1955 and Golay et al., 1993.
Remarks: A *nomen dubium fide* Rage (1984b: 58).

20. *LEPIDON* Swainson, 1839. Cabinet Cyclo. Nat. Hist. 2: 345 (*nomen nudum*).

Type species: Not designated.
Remarks: Possibly *Elapsoidea* or *Pseudohaje fide* K.L. Williams & Wallach (1989: 83).

21. *LEPTASPIS* Fitzinger, 1861. Sitz. Akad. Wiss. Wiss., Wien Math-Natur. Klass (1860) 42(25): 410 (*nomen nudum*).

Type species: *Leptaspis inornata* Fitzinger, 1861.
Remarks: Probably an Australian elapid.

22. †*LESTOPHIS* Marsh, 1885. Amer. J. Sci. (3) 29(170): 169. (*nomen substitutum*).

Synonyms: †*Limnophis* Marsh, 1871 *(nomen praeoccupatum)*, and *Paleoboa* K.P. Schmidt, 1927c *(nomen substitutum)*.
Type species: †*Limnophis crassus* Marsh, 1871.
Remarks: Not a snake but an amphisbaenid *fide* Estes (1983: 202).

23. *LIOSTEIRA* Fitzinger, 1843. Syst. Rept.: 26 (*nomen nudum*).

Synonym: *Liostira* Agassiz, 1847 *(nomen emendatum)*.
Type species: *Tropidonotus meleagris* Wiegmann.

24. *MACROBOTHRIUM* Fitzinger, 1843. Syst. Rept.: 26 (*nomen nudum*).

Type species: *Coronella narica* Fitzinger, 1826a.

25. *MEGACEPHALUS* Fitzinger, 1843. Syst. Rept.: 27 (*nomen nudum*).

Type species: *Coluber similis* Fitzinger, 1843.

26. †*OPHIDION* Pomel, 1853. Cat. Méth. Vert. Fossil.: 128 (*nomen nudum*). (Boidae)

Synonyms: †*Ophiodon* – Mlynarski, 1961 *(nomen incorrectum)*, and †*Ophidioniscus* Kuhn *in* Westphal, 1963 *(nomen substitutum)*.
Type species: *Ophidion antiquus* Pomel, 1853.
Remarks: Official Rejected Name *fide* Direction 56 (ICZN, 1956d).

27. †*OPHIDIONISCUS* Kuhn *in* Westphal, 1963. Fossil. Cat., Animalia 1: 35 (*nomen substitutum*). (Boidae)

Type species: †*Ophidion antiquus* Pomel, 1853.
Remarks: A *nomen dubium fide* Rage (1984b 32).

28. †*OPHIDIUM* Burmeister, 1885: Anal. Mus. Nac. Buenos Aires 3(2): 152.

Type species: †*Ophidium incertum* Burmeister, 1885.
Distribution: Upper Miocene or Pliocene of Argentina.
Source: Rage, 1984b.
Remarks: A *nomen dubium fide* Rage, 1984b.

29. *OPHIS* Goldfuss, 1831. Nova Acta Phys.-Med. Caesar. Leop.-Carol. Nat. Cur. 15: 127 (*nomen nudum*).

Type species: *Ophis dubius* Goldfuss, 1831.
Remarks: A *nomen nudum fide* Rage (1984b: 61).

30. *OPHRIAS* Rafinesque-Schmaltz, 1815. Anal. Nat. Tabl.: 77 (*nomen nudum*).

Type species: Not designated.

31. *OPHTHALMOPHIS* Fitzinger, 1843. Syst. Rept.: 27 (*nomen nudum*).

Type species: *Coluber partschii* Fitzinger, 1826a.

32. †*PALAELAPHIS* Rochebrune, 1884. Mém. Soc. Sci. Nat. Saone-et-Loire 5: 156. (Boidae)

Synonym: †*Paleoelaphis* – Palacky, 1898 *(nomen incorrectum)*.
Type species: †*Palaelaphis antiquus* Rochebrune, 1884.
Distribution: Upper Eocene or Oligocene of France.
Remarks: A *nomen dubium fide* Rage (1984b: 32).

33. *PETROLUS* Rafinesque-Schmaltz, 1815. Anal. Nat. Tabl.: 77 (*nomen nudum*).

Type species: Not designated.

34. *PHENACOPHIS* Fitzinger, 1861. Sitz. Akad. Wiss. Wien, Math-Natur. Kl. (1860) 42(25): 407 (*nomen nudum*).

Type species: *Phenacophis zelebori* Fitzinger, 1861.

35. *PRODIPLUS* Rafinesque-Schmaltz, 1819a. J. Phys. Chim. Hist. Nat. Arts., Paris 88: 418 (*nomen nudum*).

Type species: Not designated.

36. †*PROTROPIDONOTUS* Schlosser, 1916. Abh. Bayer. Akad. Wiss. Math.-Phys. Kl. 28(6): 34. (Colubridae)

Type species: †*Protropidonotus neglectus* Schlosser, 1916.
Distribution: Middle Miocene of Germany.
Source: Rage, 1984b.
Remarks: A *nomen dubium fide* Szyndlar & Böhme (1993: 388–389).

37. †*PROVIPERA* Kinkelin, 1892. Zool. Anz. 15(385): 94. (Viperidae)

Type species: †*Provipera boettgeri* Kinkelin, 1892.
Type: Holotype, SMF R 4186, one fang.
Type locality: "Hessler, Hessen Prov., Germany; Agenian (= Aquitanian), Lower Miocene" *fide* Rage (1984b: 58).
Distribution: Lower Miocene of Germany.
Remarks: A *nomen dubium fide* Rage (1984b: 58).

38. *PSAMMOPHIDION* Fitzinger, 1861. Sitz. Akad. Wiss. Wien, Math.-Natur. Kl. (1860) 42(25): 407 (*nomen nudum*).

Type species: *Psammophidion smithii* Fitzinger, 1861.

39. *PYTHONISCUS* Philippi, 1899. Anal. Univ. Chile 104: 720 (*nomen nudum*).

Type species: Not designated.
Remarks: Probably an Argentinian Xenodontidae *fide* Donoso-Barros & Cardenas (1964: 232).

40. †*RAGEOPHIS* Wallach, 1986. J. Herp. 20(3): 448. (*nomen substitutum*). (Boidae)

Synonyms: †*Scytalophis* Rochebrune, 1880 *(nomen praeoccupatum),* and †*Scytalopus* – Hoffmann, 1890 *(nomen incorrectum).*
Type species: †*Coluber lafonti* Filhol, 1877.
Distribution: Upper Eocene or Oligocene of France.
Sources: Rage, 1974, 1984b, 1988a and Wallach, 1986.
Remarks: A *nomen dubium fide* Rage (1984b: 32).

41. *RHINOCLEIS* Jan, 1857. Cenni Mus. Civ. Milano: 46 (*nomen nudum*).

Type species: *Rhinocleis gradatus* Jan, 1857.

42. *RHINODIPSAS* Fitzinger, 1843. Syst. Rept.: 27 (*nomen nudum*).

Type species: *Dipsas maculata* Wiegmann.
Remarks: A New World colubrid.

43. *RHINOPELTIS* Wagler *in* Agassiz, 1844. Nomen. Zool.: 39 (*nomen nudum*).

Type species: Not designated.
Remarks: Colubridae *fide* Agassiz (1844: 39). A valid genus *fide* Schulze et al. (1954 *in* 1936–1954: 3034).

44. †*SAUROPHIS* Brunner, 1957. Neues Jb. Geol. Paläont. 1957(7–8): 376–377, fig. 15. (Natricidae)

Type species: †*Saurophis crassus* Brunner, 1957.
Distribution: Upper Pleistocene of Germany (Bavaria).
Remarks: A *nomen dubium fide* Estes (1983: 211).

45. †*SCAPTOPHIS* Rochebrune, 1880. Nouv. Arch. Mus. Hist. Nat. (2) 3: 279. (? Colubridae)

Synonyms: †*Scaptophus* – Hoffmann, 1890 *(nomen incorrectum),* and †*Scatophis* – Williston, 1925 *(nomen incorrectum).*
Type species: †*Scaptophis miocenicus* Rochebrune, 1880.
Distribution: Southern France.
Remarks: A *nomen dubium fide* Rage (1981: 541) and Augé & Rage *in* Ginsburg (2000: 301).

46. *STENODEIROS* Philippi, 1899. Anal. Univ. Chile 104: 722 (*nomen nudum*).

Type Species: Not designated.

47. †*SUFFOSIO* Standhardt, 1986. Vert. Paleont. Cretaceous/Tertiary Texas: 124 (*nomen ineditum*). (Anomalepididae)

Type species: †*Suffosio praedatrix* Standhardt, 1986. *(nomen ineditum)*
Remarks: A *nomen dubium fide* Rage (pers. comm.).

48. †*TACHYOPHIS* Rochebrune, 1884. Mém. Soc. Sci. Nat. Saone-et-Loire 5: 159. (Boidae)

Synonym: †*Tachiophis* – Palacky, 1898 *(nomen incorrectum).*
Type species: †*Tachyophis nitidus* Rochebrune, 1884.

Distribution: Upper Eocene or Oligocene of France.
Sources: Hoffstetter, 1955 and Rage, 1974, 1984b.
Remarks: A *nomen dubium fide* Rage (1984b: 33).

49. *TAMACUILLA* Brüggemann & Wigand, 1838. Naturg. Abbild. Besch. Amph. 1: 98 (*nomen nudum*).

Type species: *Tamacuilla huilia* Brüggemann & Wigand, 1838.

50. †*THORACOPHIS* Trochel, 1854. Verh. Naturhist. Ver. Pruess. Rheinlande Westfalens 11(1): xix.

Type species: †*Thoracophis rugosus* Trochel, 1854 *(nomen dubium).*
Distribution: Upper Oligocene of Germany.
Remarks: A *nomen dubium fide Rage (1984b*: 61). Not a snake but an anguid lizard *fide* Szyndlar & Böhme (1993: 390).

51. *TRACHELOGERRHON* Fitzinger, 1843. Syst. Rept.: 27 (*nomen nudum*).

Synonyms: *Trachelogerron* A.-M.-C. Duméril & Bibron, 1844. Erpét. Gén. 6: 64 *(nomen nudum),* and *Trachelogerrhum* Agassiz, 1847 *(nomen emendatum).*
Type species: *Dipsas trachelogerrhos* Hemprich *in* Fitzinger, 1843.

52. *TRIMACULATUS* Higgins, 1873. Ophidians: 80 (*nomen nudum*).

Type species: *Trimaculatus ceylonensis* Higgins, 1873.

53. *TROPIDOPELTIS* Fitzinger, 1843. Syst. Rept.: 25 (*nomen nudum*).

Synonym: *Tropidopellis* Gervais, 1848 *(nomen incorrectum).*
Type species: *Pseudoeryx strigiventris* Fitzinger, 1826.
Type locality: "? Asia." Emended to Brazil *fide* Jan (1857: 47).

SPECIES

1. *Ablabes annulata* A.-M.-C. Duméril, 1853. Mém. Acad. Sci. Inst. France 23: 454 (*nomen nudum*).

Type: Holotype, MNHN, not designated.
Type locality: "Vera-Paz, Guatemala."

2. *Ablabes punctatus impunctata* Jan, 1857. Cenni Mus. Civico Milano: 46 (*nomen nudum*).

Type: Holotype, MSNM, not designated, destroyed in 1943 during World War II.
Type locality: "America sett." [= North America].

3. *Ablabes raimondii* Jan, 1863a. Arch. Zool. Anat. Fis. 2(2): 72–73 (*nomen nudum*).

Type: Holotype, MSNM (Raimondi), destroyed in 1943 during World War II.
Type locality: "Lima" [Peru.].

4. *Ablabes sexlineata* A.-M.-C. Duméril, 1853. Mém. Acad. Sci. Inst. France 23: 454 (*nomen nudum*).

Type: Holotype, not designated, location unknown.
Type locality: Unknown.

5. *Alecto wiebeli* Jan, 1863b. Elenco Sist. Ofidi: 117 (*nomen nudum*).

Type: Holotype, ZMB, not designated.
Type locality: "Australia."

6. *Amphiesma schistaceum* Jan, 1863b. Elenco Sist. Ofidi: 73 (*nomen nudum*).

Type: Holotype, MZN, not designated.
Type locality: "Indie orientali" [= E India].

7. *Anguis brachyura* Hermann, 1804. Observ. Zool.: 288.

Type: Holotype, not designated, location unknown.
Type locality: "Ramsayanis."
Remarks: A *nomen dubium.*

8. *Anguis brunneus* Seetzen, 1795. Intelligenz. Allgel. Lit.-Zeitung: 320.

Type: Holotype, not designated, location unknown.
Type locality: Unknown.
Remarks: A *nomen dubium.*

9. *Anguis cinereus* J.G. Schneider, 1801. Hist. Amph. 2: 341.

Type: Holotype, ZMB, not designated.
Type locality: Unknown. Emended to "Brésil et les Antilles" [Brazil and the Antilles] *fide* Guérin-Méneville (1838: 13).
Remarks: *Incertae sedis fide* McDiarmid et al. (1999: 125) but possibly a synonym of *Typhlops richardii fide* A.-M.-C. Duméril & Bibron (1844: 293) or *T. dominicanus fide* Hahn (1980a: 57). MNHN 3219 listed as holotype *fide* Hahn (1980a: 57) but this is probably the specimen reported by Cuvier, 1829, and illustrated by Guérin-Méneville, 1838 *in* 1829–1844.

10. *Anguis gronovianus* Schneider, 1801. Hist. Amphib.: 338.

Type: Holotype, not designated, location unknown.
Type locality: "Guineam" [= West Africa].
Remarks: A *nomen dubium.*

11. *Anguis miguel* Lacépède, 1789. Hist. Nat. Serp. 2: 134, 445 (*nomen rejiciendum*).

Type: Holotype, MNHN, not designated (King of France's cabinet).
Type locality: "Paraguay."
Remarks: Official Rejected Name *fide* Opinion 1463 (ICZN, 1987).

12. *Anguis sagitta* Lacépède, 1789. Hist. Nat. Serp. 2: 134 (*nomen rejiciendum*).

Type: Holotype, not designated, location unknown.
Type locality: Unknown.
Remarks: Official Rejected Name *fide* Opinion 1463 (ICZN, 1987).

13. *Boa flavicans* Bonnaterre, 1790: Tabl. Ency. Méth., Ophiol.: 8–9.

Type: Holotype, not designated, location unknown.
Type locality: "Guinée" [= West Africa].
Remarks: A *nomen dubium.*

14. *Boa isobequensis* Bonnaterre, 1790: Tabl. Ency. Méth., Ophiol.: 9.

Type: Holotype, not designated, location unknown.
Type locality: "Amérique septentrionale" [= North America].
Remarks: A *nomen dubium.*

15. *Boa merremii* Seetzen, 1796. Meyer's Zool. Arch. 2: 53.

Type: Holotype, not designated, location unknown.
Type locality: Unknown.
Remarks: A *nomen dubium fide* McDiarmid et al. (1999: 193).

16. *Boa moluroidem* Schneider, 1801. Hist. Amph. 1: 279–280.

Type: Holotype, MJU, not designated.
Type locality: Unknown.
Remarks: A *nomen dubium fide* J.C. Murphy et al. (2012a: 15).

17. †*Botrophis gaudryi* Rochebrune, 1880. Nouv. Arch. Mus. Hist. Nat. (2) 3: 280–281, pl. 12, figs. 8a–c.

Type: Holotype, MNHN FP 3112, one poorly preserved trunk vertebra.
Type locality: "Faluns de Pont-Levoy (Loir-et-Cher)" [= Pontlevoy, Loir-et-Cher Prov., France, lower Miocene (Orleanian) *fide* Rage, 1984b: 33].
Distribution: Lower Miocene (Orleanian MN 4: 16.0–16.9 mya) of France (Loir-et-Cher). Known only from type locality.
Remarks: A *nomen dubium fide* Rage (1984b: 33).

18. *Brachyorrhos torquatus* F. Boie, 1827. Isis von Oken 20(6): 540 (*nomen nudum*).

Type: Holotype, not designated, location unknown.
Type locality: "indischen Archipelagus" [= East Indies].

19. *Brachyruton plumbeum maculatum* Jan, 1863b. Elenco Sist. Ofidi: 92 (*nomen nudum*).

Type: Holotype, SMF, not designated.
Type locality: "Buenos Ayres" [Argentina].

20. *Bungarus ferrum-equinum* Fitzinger, 1826a. Neue Class. Rept.: 61 (*nomen nudum*).

Type: Holotype, not designated (H. Kuhl), location unknown.
Type locality: "Asia, Insula Java" [= Java, W Indonesia].

21. *Bungarus filum* Oppel *in* Schlegel, 1837. Essai Phys. Serp. 2: 231 (*nomen nudum*).

Type: Holotype, not designated, location unknown.
Type locality: Unknown.
Remarks: Probably *Dendrelaphis pictus* or *D. tristis.*

22. *Bungarus venosus* Oppel, 1811a. Ann. Mus. Hist. Nat. (1810) 16: 392 (*nomen nudum*).

Type: Holotype, not designated, location unknown.
Type locality: Unknown.

23. *Calamaria occipitalis* Jan, 1862b. Arch. Zool. Anat. Fis. 2(1): 9–10 (*nomen nudum*).

Types: Syntypes (2), formerly MSNM, a 500 mm and 490 mm specimen, destroyed in 1943 during World War II.
Type locality: "Giava" [= Java, Indonesia].
Remarks: A *nomen dubium fide* Inger & Marx (1965: 240): *cf. Calamaria lumbricoidea* but with a divided anal shield.
Sources: Jan & Sordelli, 1865 *in* 1860–1866, Rooij, 1917, Haas, 1950 and Inger & Marx, 1965.

24. *Callophis pectolineatus* Beddome *in* F. Müller, 1887: 276 (*nomen nudum*).

Type: Holotype, NMBA, not designated (F. Müller).
Type locality: "Malabar" [SW India].

25. *Chalinophorus semidoliatus* Jan, 1857. Cenni Mus. Civico Milano: 50 (*nomen nudum*).

Type: Holotype, MSNM, not designated, destroyed in 1943 during World War II.
Type locality: "Giava" [= Java, W Indonesia].

26. *Chilolepis domestica* Fitzinger, 1853. Sitz. Akad. Wiss. Wien, Math.-Natur. Klass. 10(4): 658 (*nomen dubium*).

Types: Syntypes (7), NMW (formerly HNC) (Lautner, 1851 [6 spec., Egypt] and Kotschy, 1839 [1 spec., Tripoli]), probably destroyed 6 Nov. 1944 during World War II.
Type locality: "Ägypten, Tripolis" [= Egypt and Tripoli].

27. *Chionactis palarostris seri* Blake, 1970. Distr. and var. *C. palarostris*. (*nomen ineditum*).

Type: Holotype, not designated, location unknown.
Type locality: Unknown.

28. *Chironius leuckarti* Fitzinger, 1826a. Neue Class. Rept.: 60 (*nomen nudum*).

Type: Holotype, not designated (Leuckart), location unknown.
Type locality: Unknown.

29. *Coelopeltis productus* Jan, 1863b. Elenco Sist. Ofidi: 89 (*nomen nudum*).

Type: Holotype, formerly MSNM, destroyed in 1943 during World War II.
Type locality: "Algeri" [= Algeria].

30. *Coluber 5-lineatus* Raddi, 1820. Mem. Mat. Fis. Soc. Ital. Sci., Modena 18: 339–340) (*nomen dubium*).

Type: Holotype, not designated (G. Raddi, 1817–1818), location unknown.
Type locality: "contorni di Rio-Janeiro, Brasile" [= vicinity of Rio de Janeiro, Rio de Janeiro State, SE Brazil].
Remarks: Possibly an *Apostolepis, Elapomorphus* or *Phalotris.*

31. *Coluber acantia* Rystedt, 1823. Fauna Amer. Merid.: 3 (*nomen nudum*).

Type: Holotype, not designated, location unknown.
Type locality: "Amer. Merid." [= South America].

32. *Coluber africanus* Bonnaterre, 1790. Tabl. Ency. Meth., Ophiol.: 49.

Type: Holotype, not designated, location unknown.
Type locality: "Afrique" [= Africa].
Remarks: A *nomen dubium.*

33. *Coluber angustulus* Hermann, 1804. Observ. Zool.: 286.

Type: Holotype, not designated, location unknown.
Type locality: Unknown.
Remarks: A *nomen dubium.*

34. *Coluber argentea* Rafinesque-Schmaltz, 1818a. Amer. Month. Mag. Critical Rev. 3(1): 416.

Type: Holotype, not designated, location unknown.
Type locality: "Western region of the United States."
Remarks: A *nomen dubium.* Possibly a *Rena.*

35. *Coluber asiaticus* Lacépède, 1789. Hist. Nat. Serp. 2: 94, 249 (*nomen rejiciendum*).

Type: Holotype, MNHN, not designated (King of France's cabinet).
Type locality: "isle de Ceylan" [= Sri Lanka].
Remarks: Official Rejected Name *fide* Opinion 1463 (ICZN, 1987).

36. *Coluber asiaticus* Latreille *in* Sonnini & Latreille, 1801b. Hist. Nat. Rept. 4: 118–119.

Type: Holotype, ? MNHN, not designated.
Type locality: Unknown.
Remarks: A *nomen dubium.*

37. *Coluber atro-reticulata* Bonnaterre, 1790. Tabl. Ency. Meth., Ophiol.: 29.

Type: Holotype, not designated, location unknown.
Type locality: "Guinée" [= West Africa].
Remarks: A *nomen dubium.*

38. *Coluber azureus* Merrem, 1820. Tent. Syst. Amph.: 129.

Type: Holotype, MNHN, not designated.
Type locality: "Promontorio viridi" [= Cape Verde].
Remarks: A *nomen dubium.*

39. *Coluber baetaen* Forskal *in* Niebuhr, 1775. Descr. Anim.: 15.

Type: Holotype, not designated (P.S. Forskal, Dec. 1762–July 1763), location unknown.
Type locality: "Arab."
Remarks: A *nomen dubium.*

40. *Coluber bechsteinii* Merrem, 1820. Tent. Syst. Amph.: 110–111.

Type Holotype, not designated, location unknown.
Type locality: Unknown.
Remarks: A *nomen dubium.*

41. †*Coluber beggiatoi* Zigno, 1890. Atti Mem. Accad. Sci. Lett. Arti Padova (1889–1890) (2) 6: 112–113, pl. figs. 1–8.

Type: Holotype, cranial bones and axial skeleton (F. Beggiato), destroyed *fide* Rage (1984b 52).
Type locality: "colle di Lonigo, Veneto; terreni Terziari" [= Colle di Lonigo, Veneto Prov., NE Italy; upper Eocene (in error *fide* Rage, 1984a: 52].
Distribution: Italy (Veneto).
Remarks: A *nomen dubium fide* Rage (1984b 52) but probably a Recent colubrid.

42. *Coluber bilis* Ekstrand, 1823: Fauna Brasliensis: 6 (*nomen nudum*).

Type: Holotype, not designated, location unknown.
Type locality: "Brasiliensis" [= Brazil].

43. *Coluber bimaculatum* Lacépède, 1789. Hist. Nat. Serp. 2: 78, 222 (*nomen rejiciendum*).

Type: Holotype, MNHN, not designated (King of France's cabinet).
Type locality: Unknown.
Remarks: Official Rejected Name *fide* Opinion 1463 (ICZN, 1987).

44. *Coluber blochii* Merrem, 1820. Tent. Syst. Amph.: 107.

Type: Holotype, not designated, location unknown.
Type locality: Unknown.
Remarks: A *nomen dubium.*

45. *Coluber boa* Duvernoy, 1832. Ann. Sci. Nat. (Zool.): 128 (*nomen nudum*).

Type: Holotype, not designated, location unknown.
Type locality: Unknown.

46. *Coluber braminus* Raddi, 1823. Mem. Mat. Fis. Soc. Ital. Sci., Modena 19: 64–65 (*nomen dubium*).

Type: Holotype, not designated (G. Raddi. 1817–1818), location unknown.
Type locality: "Rio-janeiro, Brasile" [= Rio de Janeiro, Rio de Janeiro State, SE Brazil].
Remarks: Possibly a *Helicops.*

47. *Coluber capitatus* Lacépède, 1789. Hist. Nat. Serp. 2: 90, 280, pl. 14, fig. 1 (*nomen rejiciendum*).

Type: Holotype, MNHN, not designated (King of France's cabinet).
Type locality: "Amérique" [= America].
Remarks: Official Rejected Name *fide* Opinion 1463 (ICZN, 1987).

48. *Coluber cenchrus* Bonnaterre, 1790. Tabl. Ency. Méth., Ophiol.: 22, pl. 41, fig. 2.

Type: Holotype, MNHN, not designated (Cabinot du Roi).
Type locality: "Asie" [= Asia].
Remarks: A *nomen dubium.*

49. *Coluber cineraceus* Lacépède, 1789. Hist. Nat. Serp. 2: 94 (*nomen rejiciendum*).

Type: Holotype, MNHN, not designated (d'Aubenton).
Type locality: "les Indes" [= East Indies].
Remarks: Official Rejected Name *fide* Opinion 1463 (ICZN, 1987).

50. *Coluber conspersus* Gravenhorst, 1807. Linn. Zool. Syst.: 406.

Type: Holotype, not designated, location unknown.
Type locality: Unknown.
Remarks: A *nomen dubium.*

51. *Coluber cyanoleucus* Gravenhorst, 1807. Linn. Zool. Syst.: 401–402.

Type: Holotype, not designated, location unknown.
Type locality: Unknown.

Remarks: A *nomen dubium.*

52. *Coluber edentulus* Gravenhorst, 1807. Linn. Zool. Syst.: 409.

Type: Holotype, not designated, location unknown.
Type locality: Unknown.
Remarks: A *nomen dubium.*

53. *Coluber enzelii* Fitzinger, 1826a. Neue Class. Rept.: 59 (*nomen nudum*).

Type: Holotype, not designated (Enzel), location unknown.
Type locality: "America septentionali" [= North America].

54. *Coluber ephippium* Seetzen, 1796a. Meyer's Zool. Archiv 2: 54.

Type: Holotype, not designated, location unknown.
Type locality: Unknown.
Remarks: A *nomen dubium.*

55. *Coluber flagellarius* Hermann, 1894. Observ. Zool.: 286.

Type: Holotype, MZUS (formerly MH), destroyed in 1943–1944 during World War II.
Type locality: Unknown.
Remarks: A *nomen dubium.*

56. *Coluber flexosus* Fischer von Waldheim, 1832. Bull. Soc. Impér. Nat. Moscou 4(3): 575

Type: Holotype, not designated, location unknown.
Type locality: "Caucasus" [SW Asia].
Remarks: A *nomen dubium.* Data given are zigzag yellow lateral stripe, keeled scales, 177 ventrals and 75 subcaudals, length 4 feet.

57. *Coluber frenatus* Gravenhorst, 1807. Linn. Zool. Syst.: 403.

Type: Holotype, not designated, location unknown.
Type locality: Unknown.
Remarks: A *nomen dubium.*

58. *Coluber funestus* Fitzinger, 1826a. Neue Class. Rept.: 58 (*nomen nudum*).

Type: Holotype, MV, not designated.
Type locality: "America."

59. *Coluber fuscescens* Fitzinger, 1826a. Neue Class. Rept.: 57 (*nomen nudum*).

Type: Holotype, not designated (Bräunlicher), location unknown.
Type locality: Unknown.

60. *Coluber gesneri* Merrem, 1820. Tent. Syst. Amph.: 120.

Type: Holotype, not designated, location unknown.
Type locality: Unknown.
Remarks: A *nomen dubium.*

61. *Coluber gliricapa* Bonnaterre, 1790. Tabl. Ency. Méth., Ophiol.: 29.

Type: Holotype, not designated, location unknown.
Type locality: "Surinam, Ceylan" [= Suriname and Sri Lanka].
Remarks: A *nomen dubium.* Possibly a *Psammophis.*

62. *Coluber gracilis* Gravenhorst, 1807. Linn. Zool. Syst.: 410.

Type: Holotype, not designated, location unknown.
Type locality: Unknown.
Remarks: A *nomen dubium.*

63. *Coluber hermanni* Merrem, 1820. Tent. Syst. Amph.: 94.

Type: Holotype, not designated, location unknown.
Type locality: "Tranquebaria" [= Tharangambadi, Tamil Nadu State, SE India].
Remarks: A *nomen dubium.*

64. *Coluber hoelleik* Forskal *in* Niebuhr, 1775. Descr. Anim.: 15.

Type: Holotype, not designated (P.S. Forskal, Dec. 1762–July 1763), location unknown.
Type locality: "Arab." [= Arabia].
Remarks: A *nomen dubium.*

65. *Coluber jaspideus* Hermann, 1804. Observ. Zool.: 275.

Type: Holotype, not designated, location unknown.
Type locality: Unknown.
Remarks: A *nomen dubium.*

66. *Coluber jenisonii* Gistel, 1850. Handbuch Naturg. Reiche: 328 (*nomen nudum*).

Type: Holotype, not designated, location unknown.
Type locality: Unknown.

67. †*Coluber kargii* H. Meyer, 1845. Fauna Vorwelt Fossil.: 41, pl. 6, fig. 2. (? Viperidae)

Type: Holotype, MNHN BML 905, one right maxilla; or syntypes (2), PIMUZ A/III 163 and A/III 164, an embedded complete skeleton on opposing slabs.

Type locality: "Molasse-Mergel Steinbrüche von Öningen" [= Oeningen, Baden-Württemberg State, SW Germany; Astaracian, middle Miocene].
Distribution: Middle Miocene (Astaracian, MN 7–8: 11.1–12.8 mya) of Germany. Known only from holotype.
Sources: Szyndlar, 1992, 1997.
Remarks: A *nomen dubium fide* Szyndlar, 1992, and Szyndlar & Böhme, 1993.

68. *Coluber lateralis* Gravenhorst, 1807. Linn. Zool. Syst.: 403.

Type: Holotype, not designated, location unknown.
Type locality: Unknown.
Remarks: A *nomen dubium.*

69. *Coluber laticapitantus* Lacépède, 1789. Hist. Nat. Serp. 2: 84, 336–337 (*nomen rejiciendum*).

Type: Holotype, MNHN, not designated (King of France's cabinet).
Type locality: "Amérique méridionale" [= South America].
Remarks: Official Rejected Name *fide* Opinion 1463 (ICZN, 1987).

70. *Coluber laticapitatus* Merrem, 1820. Tent. Syst. Amph.: 99.

Type: Holotype, not designated, location unknown.
Type locality: "America meridionali" [= South America].
Remarks: A *nomen dubium.*

71. *Coluber latreillei* Merrem, 1820. Tent. Syst. Amph.: 112.

Type: Holotype, not designated, location unknown.
Type locality: Unknown.
Remarks: A *nomen dubium.*

72. *Coluber leucomelas* Gmelin, 1789. Syst. Nat., ed. 13, 1(3): 1090–1091.

Type: Holotype, not designated, location unknown.
Type locality: Unknown.
Remarks: A *nomen dubium.*

73. *Coluber longicauda* Fitzinger *in* Ritter, 1823b. Archiv. Ges. Stat. Lit. Kunst (84–85): 440 (*nomen nudum*).

Type: Holotype, NMW, not designated (C. Ritter, 1819–1821).
Type locality: "St. Domingo" [= Santo Domingo, Dominican Republic].

74. *Coluber lumbricosa* Lesson, 1828. Ann. Sci. Nat. (Zool.) 13: 372 (*nomen nudum*).

Type: Holotype, not designated, location unknown.
Type locality: Unknown.
Remarks: Type locality Brazil *fide* Lesson (1830: 4).

75. *Coluber lunulatus* Hermann, 1804. Observ. Zool.: 274–275.

Types: Syntypes (2), MZUS (formerly MH), destroyed in 1943–1944 during World War II.
Type locality: Unknown.
Remarks: A *nomen dubium.*

76. *Coluber margaritiferus* Bonnaterre, 1790. Tabl. Ency. Méth., Ophiol.: 56, pl. 41, fig. 1.

Type: Holotype, MNHN, not designated (Cabint du Roi).
Type locality: Unknown.
Remarks: A *nomen dubium.* Possibly a *Psammophis.*

77. *Coluber melanogramus* Fitzinger *in* Ritter, 1823b. Archiv. Ges. Stat. Lit. Kunst (84–85): 440 (*nomen nudum*).

Type: Holotype, NMW, not designated (C. Ritter, 1819–1821).
Type locality: "St. Domingo" [= Santo Domingo, Dominican Republic].

78. *Coluber melanoleucos* Donndorff, 1798. Zool. Beyträge 3: 209.

Type: Holotype, not designated, a male, location unknown.
Type locality: Unknown.
Remarks: A *nomen dubium.*

79. *Coluber monochromis* F. Boie, 1827. Isis von Oken 20(6): 538.

Type: Holotype, not designated, location unknown.
Type locality: "Africa."
Remarks: A *nomen dubium.*

80. *Coluber naevius* Gmelin, 1789. Syst. Nat., ed. 13, 1(3): 1095.

Type: Holotype, not designated, location unknown.
Type locality: Unknown.
Remarks: A *nomen dubium.*

81. *Coluber nasica* Bonnaterre, 1790. Tabl. Ency. Méth., Ophiol.: 36.

Type: Holotype, not designated, location unknown.
Type locality: Unknown.
Remarks: A *nomen dubium.*

82. *Coluber neumayeri* Fitzinger, 1826a. Neue Class. Rept.: 57 (*nomen nudum*).

Type: Holotype, not designated (Neumayer), location unknown.
Type locality: "Europa, Dalmatia" [= Croatia].

83. *Coluber nicandri* Merrem, 1820. Tent. Syst. Amph.: 97.

Type: Holotype, not designated, location unknown.
Type locality: Unknown.
Remarks: A *nomen dubium*.

84. *Coluber nigerrimus* Hermann, 1804. Observ. Zool.: 287.

Type: Holotype, not designated, location unknown.
Type locality: Unknown.
Remarks: A *nomen dubium*.

85. *Coluber nocturnes* Donndorff, 1798. Zool. Beyträge 3: 209.

Type: Holotype not designated, a male, location unknown.
Type locality: Unknown.
Remarks: A *nomen dubium*.

86. *Coluber nuchimacula* Fitzinger, 1826a. Neue Class. Rept.: 58 (*nomen nudum*).

Type: Holotype, not designated, location unknown.
Type locality: Unknown.

87. †*Coluber ombonii* Zigno, 1890. Atti Mem. Accad. Sci. Lett. Arti Padova (1889–1890) (2) 6: 113, fig. 9.

Type: Holotype, IGP 8360, a 55 mm imprint of posterior body and tail section.
Type locality: "marne superiori di M. Bolca, Veneto; terreni Terziari" [= Pesciara, Monte Bolca, Veneto Prov., NE Italy; Ypresian, lower Eocene *fide* Rage, 1984a: 58].
Distribution: Lower Eocene (Ypresian) of Italy. Known only from the holotype.
Remarks: *Incertae sedis fide* Rage (1984b: 58).

88. *Coluber oppelii* Fitzinger, 1826a. Neue Class. Rept.: 57 (*nomen nudum*).

Type: Holotype, not designated, location unknown.
Type locality: "America septentrionali" [= North America].

89. *Coluber ovivorus* Linnaeus, 1758. Syst. Nat., ed. 10, 1: 223.

Type: Holotype, not designated, location unknown.
Type locality: "America."
Remarks: A *nomen dubium*. Possibly *Coluber constrictor* or *Pantherophis vulpinus*.

90. †*Coluber owenii* H. Meyer, 1844a. Neues Jb. Min. Geog. Geol. Petrefak.-Kunde 12: 330–331. (Colubridae)

Type: Holotype, LNK TE/58, an incomplete axial skeleton (ca. 900 mm with 202 vertebrae) embedded in the matrix (H. Seyfried).
Type locality: "Molasse-Mergel Steinbrüche von Öningen" [= Oeningen, Baden-Württemberg State, SW Germany; Astaracian, middle Miocene].
Distribution: Middle Miocene (Astaracian, MN 7–8: 11.1–12.8 mya) of Germany. Known only from holotype.
Sources: Rage, 1984b, Szyndlar & Böhme, 1993 and Szyndlar, 2012.
Remarks: Supplemental original description in H. Meyer (1845: 40–41, pl. 7, fig. 1). Identification below family level impossible *fide* Szyndlar & Böhme (1993: 389).

91. *Coluber pallasii* Merrem, 1820. Tent. Syst. Amph.: 109.

Type: Holotype, not designated, location unknown.
Type locality: Unknown.
Remarks: A *nomen dubium*.

92. *Coluber panamensis* Gmelin, 1789. Syst. Nat., ed. 13, 1(3): 1095.

Type: Holotype, not designated, location unknown.
Type locality: "Panama."
Remarks: A *nomen dubium*.

93. *Coluber papyraceus* Troschel, 1854. Verh. Naturhist. Ver. Preuß. Rheinl. Westfaalens 11(1): xix (*nomen nudum*).

Type: Holotype, lost *fide Rage (1984b*: 60).
Type locality: "Braunkohle von Rott."
Distribution: Upper Oligocene of W Germany (Rott).
Remarks: A *nomen nudum fide* Rage (1984b: 60).

94. *Coluber partschii* Fitzinger, 1826a. Neue Class. Rept.: 57 (*nomen nudum*).

Type: Holotype, not designated (Partsch), location unknown.
Type locality: Unknown.

95. *Coluber phorbeia* Gravenhorst, 1807. Linn. Zool. Syst.: 410.

Type: Holotype, not designated, location unknown.
Type locality: Unknown.
Remarks: A *nomen dubium*.

96. *Coluber planiventer* Seetzen, 1796. Meyer's Zool. Archiv 2: 56.

Type: Holotype, not designated, location unknown.
Type locality: Unknown.
Remarks: A *nomen dubium.*

97. †*Coluber plioagellus* R.L. Wilson, 1968. Contri. Mus. Paleont. Univ. Michigan 22(7): 102–103, figs. 10f–h. (Colubridae)

Type: Holotype, UMMP V 55711, one trunk vertebra.
Type locality: "UM-K6-59, Ogallala Formation, Lower Pliocene, Wakeeney local fauna, Lowell Hillman Ranch, northwest of Ogallah, Trego County, Kansas, elevation 2255 feet," Clarendonian, upper Miocene.
Distribution: USA (Kansas). Known only from type locality.
Sources: Holman, 1975 and Rage, 1984b.
Remarks: A *nomen dubium fide* Rage (1984b: 52) but *Coluber* or *Masticophis fide* Holman (1975: 63).

98. †*Coluber podolicus* H. Meyer, 1844b. Nueues Jb. Min. Geog. Geol. Petrefak.-Kunde 12: 565. (? Natricidae)

Type: Holotype, one trunk vertebra, lost *fide* Rage (1984b: 52).
Type locality: "Holuzubiniec" *fide* Pusch (1837: 168) or "Hotozubince" *fide* Pusch (1842: 179) and Szyndlar (1991b: 265).
Distribution: Ukraine (Khmel'nyts'ka). Known only from holotype.
Remarks: Rage (1984b: 52) listed Bolurubince, near Kamenets Podo'sky, Ukrainian S.S.R., USSR, age unknown. A *nomen dubium fide* Rage (1984b: 52) but a *Natrix fide* Szyndlar (1997: 213–214).

99. *Coluber polyzimias* Donndorff, 1798. Zool. Beyträge 3: 202.

Type: Holotype, not designated, a 325 mm specimen (King of France's cabinet).
Type locality: Unknown.
Remarks: A *nomen dubium.*

100. *Coluber porosus* Fitzinger *in* Ritter, 1823b. Archiv. Ges. Stat. Lit. Kunst (84–85): 440 (*nomen nudum*).

Type: Holotype, NMW, not designated (C. Ritter, 1819–1821).
Type locality: "St. Domingo" [= Santo Domingo, Dominican Republic].

101. *Coluber praetexata* Gravenhorst, 1807. Linn. Zool. Syst.: 414.

Type: Holotype, not designated, location unknown.
Type locality: Unknown.
Remarks: A *nomen dubium.*

102. *Coluber precatorius* Donndorff, 1798. Zool. Beyträge 3: 205.

Type: Holotype, not designated, a 473 mm specimen (King of France's cabinet), based on Lacépède (1789: pl. 12, fig 1).
Distribution: Unknown.
Remarks: A *nomen dubium.* Possibly a *Psammophis.*

103. *Coluber pseudechidna* Hermann, 1804. Observ. Zool.: 276–278.

Types: Syntypes (7), not designated (D. Hecht, 1791), location unknown.
Type locality: "Pontarlier, Aliae minores ex Salins" [= Pontarlier and Salins-les-Bains, France].
Remarks: A *nomen dubium.*

104. *Coluber pulchellus* Hermann, 1804. Observ. Zool.: 286.

Type: Holotype, not designated, location unknown.
Type locality: Unknown.
Remarks: A *nomen dubium.*

105. *Coluber pulcher* Raddi, 1820. Mem. Mat. Fis. Soc. Ital. Sci., Modena 18: 337–338.

Type: Holotype, not designated (G. Raddi, 1817–1818), location unknown.
Type locality: "vicinanze di Rio-Janeiro, Brasile" [= vicinity of Rio de Janeiro, Rio de Janeiro State, SE Brazil].
Remarks: A *nomen dubium.* Possibly an *Apostolepis, Elapomorphus* or *Phalotris.*

106. *Coluber pusillus* Bonnaterre, 1790. Tabl. Ency. Méth., Ophiol.: 61–62.

Type: Holotype, not designated, location unknown.
Type locality: "Inde" [= India].
Remarks: A *nomen dubium.*

107. *Coluber quinquefasciatus* Fitzinger, 1826a. Neue Class. Rept.: 58.

Type: Holotype, not designated (C.G.C. Reinwardt), location unknown.
Type locality: "Asia, Insula Java" [= Java, W Indonesia].
Remarks: A *nomen dubium.*

108. *Coluber rufofuscus* Fitzinger *in* Ritter, 1823b. Archiv. Ges. Stat. Lit. Kunst (84–85): 440 (*nomen nudum*).

Type: Holotype, NMW, not designated (C. Ritter, 1819–1821).
Type locality: "St. Domingo" [= Santo Domingo, Dominican Republic].

109. *Coluber schneideri* Merrem, 1820. Tent. Syst. Amph.: 96.

Type: Holotype, not designated, location unknown.
Type locality: Unknown.
Remarks: A *nomen dubium.*

110. *Coluber schytha* Bonnaterre, 1790. Tabl. Ency. Méth., Ophiol.: 15.

Type: Holotype, not designated (P.S. Pallas, 1768–1774).
Type locality: "Sibérie" [= N Russia].
Remarks: A *nomen dubium.*

111. *Coluber seetzenii* Merrem, 1820. Tent. Syst. Amph.: 97 (*nomen nudum*).

Type: Holotype, RMNH, not designated.
Type locality: "Africa."

112. *Coluber seriatus* Hermann, 1804. Observ. Zool.: 273.

Type: Holotype, MZUS (formerly MH), destroyed in 1943–1944 during World War II.
Type locality: Unknown.
Remarks: A *nomen dubium.*

113. *Coluber similis* Fitzinger, 1843. Syst. Rept.: 27 (*nomen nudum*).

Type: Holotype, not designated, location unknown.
Type locality: "As. Am." [= Asia, America].

114. *Coluber sitala* Latreille *in* Sonnini & Latreille, 1801b. Hist. Nat. Rept. 4: 162.

Type: Holotype, not designated, location unknown.
Type locality: "Egypte" [= Egypt].
Remarks: A *nomen dubium.*

115. *Coluber splendidus* Bonnaterre, 1790. Tabl. Ency. Méth., Ophiol.: 14–15.

Type: Holotype, not designated, location unknown.
Type locality: Unknown.
Remarks: A *nomen dubium.*

116. *Coluber subcaerulous* Lacépède, 1789. Hist. Nat. Serp. 2: 104, 288 (*nomen rejiciendum*).

Type: Holotype, not designated, location unknown.
Type locality: "Amérique" [= America].
Remarks: Official Rejected Name *fide* Opinion 1463 (ICZN, 1987).

117. *Coluber subulatus* Gravenhorst, 1807. Linn. Zool. Syst.: 404.

Type: Holotype, not designated, location unknown.
Type locality: Unknown.
Remarks: A *nomen dubium.*

118. *Coluber thoracicus* Hermann, 1804. Observ. Zool.: 276.

Type: Holotype, MZUS (formerly MH), destroyed in 1943–1944 during World War II.
Type locality: Unknown.
Remarks: A *nomen dubium.*

119. *Coluber tiedemanni* Merrem, 1820. Tent. Syst. Amph.: 126.

Type: Holotype, not designated, location unknown.
Type locality: Unknown.
Remarks: A *nomen dubium.*

120. *Coluber toxicon* Bonnaterre, 1790. Tabl. Ency. Méth. Ophiol.: 62.

Type: Holotype, not designated, location unknown.
Type locality: "Inde" [= India].
Remarks: A *nomen dubium.*

121. *Coluber trifasciatus* Shaw, 1802. Gen. Zool. 3(2): 542.

Type: Holotype, WHM, not designated.
Type locality: Unknown.
Remarks: A *nomen dubium.*

122. *Coluber urens* Bonnaterre, 1790. Tabl. Ency. Méth., Ophiol.: 62.

Type: Holotype, not designated, location unknown.
Type locality: "Inde" [= India].

123. *Coluber variegatus* Bonnaterre, 1790. Tabl. Ency. Méth., Ophiol.: 39.

Type: Holotype, not designated, location unknown.
Type locality: Unknown.
Remarks: A *nomen dubium.*

124. *Coluber virgata* Bonnaterre, 1790. Tabl. Ency. Méth. Ophiol.: 49–50.

Type: Holotype, not designated, location unknown.
Type locality: "Surinam."
Remarks: A *nomen dubium.*

125. *Coluber weigelii* Seetzen, 1796. Meyer's Zool. Archiv 2: 57.

Type: Holotype, not designated, location unknown.
Type locality: Unknown.
Remarks: A *nomen dubium.*

126. *Coluber xequipeles* Bonnaterre, 1790. Tabl. Ency. Méth., Ophiol.: 52.

Type: Holotype, not designated, location unknown.
Type locality: "Surinam, l'Amerique" [= Surinam, South America].
Remarks: A *nomen dubium.*

127. *Coronella agnellii* Fitzinger, 1826a. Neue Class. Rept.: 59 (*nomen nudum*).

Type: Holotype, not designated (Agnello), location unknown.
Type locality: "Africa."

128. *Coronella anguiformis* Laurenti, 1768. Synop. Rept.: 85.

Type: Holotype, Museo Turriano (? MZUT).
Type locality: Unknown.
Remarks: A *nomen dubium.*

1292. *Coronella cerastoides* Laurenti, 1768. Synop. Rept.: 86.

Type: Holotype, Museo Turriano (? MZUT).
Type locality: Unknown.
Remarks: A *nomen dubium.*

130. *Coronella geoffroana* Fitzinger, 1826a. Neue Class. Rept.: 59. (*nomen nudum*).

Type: Holotype, not designated (Geoffroy), location unknown.
Type locality: Unknown.

131. *Coronella latirostra* Laurenti, 1768. Synop. Rept.: 86.

Type: Holotype, Museo Turriano (? MZUT).
Type locality: Unknown.
Remarks: A *nomen dubium.*

132. *Coronella lichtensteinii* Fitzinger, 1826a. Neue Class. Rept.: 59 (*nomen nudum*).

Type: Holotype, not designated (Lichtenstein), location unknown.
Type locality: Unknown.

133. *Coronella lineolata* Fitzinger, 1826a. Neue Class. Rept.: 59. (*nomen nudum*).

Type: Holotype, not designated, location unknown.
Type locality: Unknown.

134. *Coronella macroura* Fitzinger, 1826a. Neue Class. Rept.: 59. (*nomen nudum*).

Type: Holotype, not designated, location unknown.
Type locality: Unknown.

135. *Coronella narica* Fitzinger, 1826a. Neue Class. Rept.: 59 (*nomen nudum*).

Type: Holotype, MV, not designated.
Type locality: "America, Brasilia" [= Brazil, South America].

136. *Coronella notata* Blyth, 1859. J. Asiat. Soc. Bengal 28(3): 297. (*nomen nudum*).

Type: Holotype, not designated, location unknown.
Type locality: Unknown.

137. *Coronella pohlii* Fitzinger, 1826a. Neue Class. Rept.: 59 (*nomen nudum*).

Type: Holotype, not designated (Pohl), location unknown.
Type locality: "America, Brasilia" [= Brazil, South America].

138. *Coronella russellii* Blyth, 1854a. J. Asiat. Soc. Bengal 23(3): 293 (*nomen nudum*).

Type: Holotype, not designated, location unknown.
Type locality: Unknown.

139. *Coronella scherei* Fitzinger, 1826a. Neue Class. Rept.: 59 (*nomen nudum*).

Type: Holotype, not designated (Scherer), location unknown.
Type locality: Unknown.

140. *Coronella spixii* Fitzinger, 1826a. Neue Class. Rept.: 59 (*nomen nudum*).

Synonym: *Coralliophis spixii* Fitzinger, 1843.
Type: Holotype, MV, not designated.
Type locality: "America, Brasilia" [= Brazil, South America].

141. *Coronella treuensteinii* Fitzinger, 1826a. Neue Class. Rept.: 59. (*nomen nudum*).

Type: Holotype, MV, not designated.
Type locality: "America, Brasilia" [= Brazil, South America].

142. †*Crotalus potterensis* Brattstrom, 1953. Copeia 1953(3): 175–176, fig. 1. (Viperidae)

Type: Holotype, UCMP 3129, two middle trunk vertebrae, part of a pterygoid, rib fragments and three bone fragments.
Type locality: "Potter Creek Cave, Shasta County, California, Late Pleistocene."
Distribution: Upper Pleistocene of USA (California). Known only from type locality.
Remarks: A *nomen dubium fide* Rage (1984b: 56).

143. *Crotalus tesselatus* Hermann, 1804. Observ. Zool.: 271. (Viperidae)

Type: Holotype, not designated, location unknown.
Type locality: Unknown.
Sources: Klauber, 1936 and McDiarmid et al., 1999.
Remarks: Resembles *Crotalus durrisus* and *C. adamanteus* in size and pattern and *C. triseriatus* in scutellation *fide* Klauber (1936: 197). A *nomen dubium*.

144. *Cylindrophis macroscelis* Jerdon, 1853. J. Asiat. Soc. Bengal 22(6): 527. (Cylidrophiidae)

Type: Holotype, not designated, location unknown.
Type locality: "India."
Remarks: A *nomen dubium*. Probably referable to *Cylindrophis*.

145. *Dendrophis fuscus* Jan, 1863b. Elenco Sist. Ofidi: 86 (*nomen nudum*).

Types: Syntypes (3), MCZ, MNHN, and ZMH, not designated.
Type locality: "Australia" and "Sydney" [Australia].

146. *Dipsas bertholdi* Jan, 1863b. Elenco Sist. Ofidi: 103 (*nomen nudum*).

Type: Holotype, ZMG, not designated.
Type locality: Unknown.

47. *Dipsas infrenalis* Rosén, 1905. Ann. Mag. Nat. Hist. (7) 15(86): 150. (Dipsadidae)

Type: Holotype, ZMUL, a 300 mm specimen.
Type locality: Unknown.
Remarks: *Incertae sedis fide* J.A. Peters (1960: 214).

148. *Dipsas interruptus* Duvernoy, 1831. J. Roy. Inst. Great Britain (5): 158 (*nomen nudum*).

Type: Holotype, not designated, location unknown.
Type locality: Unknown.

149. *Dipsas lineatus* A.-M.-C. Duméril, 1853. Mém. Acad. Sci. Inst. France 23: 508 (*nomen nudum*).

Type: Holotype, not designated, location unknown.
Type locality: Unknown.

150. *Dipsas valdiviensis* Jan, 1857. Cenni Mus. Civico Milano: 49 (*nomen nudum*).

Type: Holotype, MSNM, not designated, destroyed in 1943 during World War II.
Type locality: "Valdivia" [Chile].

151. *Dipsas trachelogerrhos* Hemprich *in* Fitzinger, 1843. Syst. Rept.: 27 (*nomen nudum*).

Type: Holotype, not designated, location unknown.
Type locality: "As.?" [= Asia].

152. *Dochis strigata* Jan, 1857. Cenni Mus. Civico Milano: 44 (*nomen nudum*).

Type: Holotype, MSNM, not designated, destroyed in 1943 during World War II.
Type locality: "Venezuela."

153. *Dromicus capensis* Jan, 1863b. Elenco Sist. Ofidi: 66 (*nomen nudum*).

Type: Holotype, formerly MSNM, destroyed in 1943 during World War II.
Type locality: "Capo di B. Sp.'" [= Cape of Good Hope, South Africa].

154. *Dromicus infrenatus* Jan, 1857. Cenni Mus. Civico Milano: 46 (*nomen nudum*).

Type: Holotype, MSNM, not designated, destroyed in 1943 during World War II.
Type locality: Unknown.

155. *Dromicus modestus* Jan, 1863b. Elenco Sist. Ofidi: 66 (*nomen nudum*).

Type: Holotype, MNHN, not designated.
Type locality: "Brasile" [= Brazil].

156. *Dryophis lophorhynchus* Fitzinger, 1826a. Neue Class. Rept.: 61 (*nomen nudum*).

Type: Holotype, not designated, location unknown.
Type locality: Unknown.

157. *Duberria amoena* Fitzinger, 1826a. Neue Class. Rept.: 55 (*nomen nudum*).

Type: Holotype, not designated, location unknown.
Type locality: Unknown.

158. *Duberria badia* Fitzinger, 1826a. Neue Class. Rept.: 55 (*nomen nudum*).

Type: Holotype, not designated, location unknown.
Type locality: Unknown.

159. *Duberria bivittata* Fitzinger, 1826a. Neue Class. Rept.: 56 (*nomen nudum*).

Type: Holotype, MV, not designated.
Type locality: "America, Brasilia" [= Brazil, South America].

160. *Duberria boiei* Fitzinger, 1826a. Neue Class. Rept.: 55 (*nomen nudum*).

Type: Holotype, not designated (Boie), location unknown.
Type locality: Unknown.

161. *Duberria cinerascens* Fitzinger, 1826a. Neue Class. Rept.: 56 (*nomen nudum*).

Type: Holotype, MV, not designated.
Type locality: "America, Brasilia" [= Brazil, South America].

162. *Duberria dumerilii* Fitzinger, 1826a. Neue Class. Rept.: 55 (*nomen nudum*).

Type: Holotype, not designated (Duméril), location unknown.
Type locality: Unknown.

163. *Duberria hemprichii* Fitzinger, 1826a. Neue Class. Rept.: 55 (*nomen nudum*).

Type: Holotype, not designated (Hemprich), location unknown.
Type locality: "America septentrionali" [= North America].

164. *Duberria maximiliani* Fitzinger, 1826a. Neue Class. Rept.:56 (*nomen nudum*).

Type: Holotype, not designated (Maximilian), location unknown.
Type locality: "America septentrionali" [= North America].

165. †*Elaphis boulei* De Stefano, 1905. Bol. Soc. Geol. Ital. 24: 41–42, pl. 4, figs. 3–6. (Boidae)

Types: Syntypes (?), MNHN, several vertebrae and one dentary (1893), lost *fide* Rage (1984b: 33).
Type locality: "Quercy" [= Phosphorites of Quercy, on the Quercy Plateau, from the departments of Lot, Tarne-et-Garonne, and Aveyron in SW France; upper Eocene to Oligocene].
Distribution: Upper Eocene or Oligocene of France. Known only from syntypes.
Remarks: A *nomen dubium fide* Rage (1984b 33).

166. *Elaphis poecilurus* Jan, 1863b. Elenco Sist. Ofidi: 61 (*nomen nudum*).

Type: Holotype, formerly MSNM, destroyed in 1943 during World War II.
Type locality: Unknown.

167. *Elaps anulatus* Schneider, 1801. Hist. Amph.: 305.

Type: Holotype, MB, not designated.
Type locality: Unknown.
Remarks: A *nomen dubium.*

168. *Elaps corallinus gastrostictus* Jan, 1863b. Elenco Sist. Ofidi: 113 (*nomen nudum*).

Type: Holotype, NMW, not designated.
Type locality: "Colombia."

169. *Elaps obtusicaudus* Wagler, 1825. Amphibia: xii. (*nomen ineditum*).

Type: Holotype, not designated, location unknown.
Type locality: Unknown.

170. *Elaps obtusirostris* Wagler, 1825. Amphibia: xii. (*nomen ineditum*).

Type: Holotype, not designated, location unknown.
Type locality: "Brasilia" [= Brazil].

171. *Elaps schneideri* Jan, 1863b. Elenco Sist. Ofidi: 113 (*nomen nudum*).

Type: Holotype, formerly MSNM, destroyed in 1943 during World War II.
Type locality: "Amer. merid." [= South America].

172. *Enicognathus sagittifer* Berthold, 1859. Nach. Ges. Wiss. Georg-Augustus-Univ. Göttingen (17): 180.

Type: Holotype, not designated, ZMG.
Type locality: "Surinamo" [= Surinam].
Remarks: A *nomen dubium.*

173. *Erythrolamprus mentalis* F. Werner, 1909b. Mitt. Naturhist. Mus. Hamburg 26: 238. (Xenodontidae)

Type: Holotype, ZMH, destroyed in July 1943 during World War II.
Type locality: "Guatemala."
Remarks: A *nomen dubium.* Possibly referable to *Rhadinaea.*

174. *Eupeltis pulchra* Fitzinger, 1843. Syst. Rept.: 25 (*nomen nudum*).

Type: Holotype, not designated, location unknown.
Type locality: "Af.?" [= Africa].

175. *Euprepes multilineatus* Philippi *in* Quijada, 1916. Anal. Univ. Chile 104: 46 (*nomen nudum*).

Type: Holotype, not designated, location unknown.
Type locality: Unknown.

176. *Eutaenia angustirostris* Kennicott, 1860. Proc. Acad. Nat. Sci. Philadelphia 12(8): 332–333. (Natricidae)

Type: Lectotype, USNM 959 (D.N. Couch, 1846–1854), designated by F.G. Thompson (1957: 6).
Type locality: "Parras, Coahuila, Mexico."
Distribution: Northern Mexico (Coahuila).
Source: F.G. Thompson, 1957.
Remarks: A *nomen dubium fide* Rossman et al. (1996: 245).

177. *Furina bilineata* Jan, 1859a. Rev. Mag. Zool. (2) 11: 123, 125–126. (Elapidae)

Type: Holotype, MSNM, not designated, destroyed in 1943 during World War II.
Type locality: "Nouvelle-Holland?" [= Australia].
Remarks: Supplemental original description in Jan (1859c: pl. 9, fig. 5). A *nomen dubium.*

178. *Herpetodryas irregularis* Jan, 1863b. Elenco Sist. Ofidi: 81. (*nomen nudum*).

Type: Holotype, MSNM, not designated, destroyed in 1943 during World War II.
Type locality: "Surinam."

179. *Herpetodryas lineolatus* Jan, 1863b. Elenco Sist. Ofidi: 81. (*nomen nudum*).

Type: Holotype, MSNM, not designated, destroyed in 1943 during World War II.
Type locality: Unknown.

180. *Herpetodryas poiteaui* Jan, 1857. Cenni Mus. Civico Milano: 45 (*nomen nudum*).

Type: Holotype, MSNM, not designated, destroyed in 1943 during World War II.
Type locality: "Cayenne" [French Guiana].

181. *Herpetodryas xanthogaster* Jan, 1863a. Elenco Sist. Ofidi: 80.

Type: Holotype, MSNM, not designated, destroyed in 1943 during World War II.
Type locality: "Brasile" [= Brazil].
Remarks: A *nomen dubium.*

182. *Heterodon vipera* Fitzinger, 1826a. Neue Class. Rept.: 56. (*nomen nudum*).

Type: Holotype, not designated, location unknown.
Type locality: "America septentrionali" [= North America].

183. *Homalopsis 5-vittata* A.-M.-C. Duméril, 1853. Mém. Acad. Sci. Inst. France 23: 499 (*nomen nudum*).

Type: Holotype, ? MNHN, not designated (Morelet).
Type locality: "Brésil ? de Peten."
Remarks: Possibly a *Helicops.*

184. *Homalopsis python* Fitzinger, 1826a. Neue Class. Rept.: 55.(*nomen nudum*).

Type: Holotype, not designated, location unknown.
Type locality: Unknown.

185. *Homalopsis robustus* Jan, 1863b. Elenco Sist. Ofidi: 77 (*nomen nudum*).

Type: Holotype, MSNM, not designated, destroyed in 1943 during World War II.
Type locality: "Brasile" [= Brazil].

186. *Homalopsis rufotaeniatus* Fitzinger, 1826a. Neue Class. Rept.: 55 (*nomen nudum*).

Type: Holotype, not designated, location unknown.
Type locality: Unknown.

187. *Hydrus atricapillus* F. Boie, 1827. Isis von Oken 20(6): 554 (*nomen nudum*).

Type: Holotype, not designated (C.G.C. Reinwardt), location unknown.
Type locality: "moluckischen Inseln" [= Maluku, Indonesia].

188. *Imantodes ancoatl* Sumichrast, 1873. Arch. Sci. Phys. Nat. 46: 246 (*nomen nudum*).

Type: Holotype, not designated, location unknown.
Type locality: "Etat de Vera-Cruz, Mexique" [= Veracruz State, Mexico].
Distribution: Southern Mexico (Veracruz), 1000–1200 m.

189. *Lachesis wettsteini* F. Werner, 1923a. Ann. Naturhist. Mus. Wien 36: 166. (Viperidae)

Type: Holotype, NMW 23398 *fide* Tiedemann & Häupl (1980: 56).
Type locality: Unknown.
Remarks: A *nomen dubium*. Probably referable to *Trimeresurus*.

190. †*Laophis crotaloides* Owen, 1857. Quart. J. Geol. Soc. London 13: 196–198, pl. 4, figs. 2–3. (Viperidae)

Types: Syntypes (13), 13 vertebrae (Spratt), lost *fide* Rage (1984b: 58).
Type locality: "near the Promontory of Karabournou, on the eastern coast of the Gulf of Salonica" [= Macedonia, Greece]. Miocene *fide* Rochebrune (1880: 292).
Distribution: Upper Miocene or lower Pliocene of Greece. Known only from type series.
Remarks: A *nomen dubium*.

191. *Leptaspes atacamensis* Philippi *in* Quijada, 1916. Cat. Sist. Rept. Chil.: 23 (*nomen nudum*).

Type: Holotype, not designated, location unknown.
Type locality: Unknown.
Remarks: Probably a *Chironius fide* Donoso-Barros & Cardenas (1964: 230).

192. *Leptaspis inornata* Fitzinger, 1861. Sitz. Akad. Wiss. Wien Math.-Natur. Kl. (1860) 42(25): 410 (*nomen nudum*).

Type: Holotype, not designated, location unknown.
Type locality: "Neu-Holland" [= Australia].

193. *Leptophis chenonii heuglini* Jan, 1857. Cenni Museo Civico di Milano: 45 (*nomen nudum*).

Type: Holotype, MSNM, not designated, destroyed in 1943 during World War II.
Type locality: "Sennar" [Sudan].

194. *Leptophis nilagiricus* Jerdon, 1853. J. Asiat. Soc. Bengal 22(6): 529. (Colubridae)

Type: Holotype, not designated, location unknown.
Type locality: "grassy hills of the Neelgherries" [SW India].
Source: Ganesh & Chalndramouli, 2011.
Remarks: A *nomen dubium*. Possibly a synonym of *Ahaetulla perroteti*.

195. *Leptophis tatacua* Briceõ-Rossi, 1934. Bol. Min. Salub. Agric. Cria, Caracas 2(14): 1130, 1 fig (*nomen nudum*).

Type: Holotype, not designated, location unknown.
Type locality: Unknown.
Remarks: Possibly an Amphisbaenia.

196. *Liopeltis brevicauda* Jan, 1863b. Elenco Sist. Ofidi: 82.

Type: Holotype, MSNM, not designated, destroyed in 1943 during World War II.
Type locality: Unknown.
Remarks: A *nomen dubium*.

197. *Liophis poecilogyrus californica* Jan, 1863a. Arch. Zool. Anat. Fis. 2(2): 292.

Types: Syntypes (2), MNHN, (Freyburg [California]).
Type locality: "California, Brasile" [California, USA and Brazil].
Remarks: A *nomen dubium*.

198. *Liophis poecilogyrus doliata* Jan, 1863a. Arch. Zool. Anat. Fis. 2(2): 292.

Type: Holotype, MZN, not designated.
Type locality: "Brasile" [= Brazil].
Remarks: A *nomen dubium*.

199. *Lycodon assimilis* Jerdon, 1853. J. Asiat. Soc. Bengal 22(6): 528–529. (Colubridae)

Type: Holotype, not designated, location unknown.
Type locality: Unknown.
Remarks: A *nomen dubium*. Probably referable to *Lycodon*.

200. *Lycodon aulicum subfasciatum* Jan, 1863b. Elenco Sist. Ofidi: 96 (*nomen nudum*).

Type: Holotype, ZMG, not designated.
Type locality: Unknown.

201. *Lycodon aulicum transversalis* Jan, 1863b. Elenco Sist. Ofidi: 96 (*nomen nudum*).

Type: Holotype, MHNG, not designated.
Type locality: Unknown.

202. *Lycodon bicolor* Jan, 1863b. Elenco Sist. Ofidi: 97 (*nomen nudum*).

Type: Holotype, MM, not designated.
Type locality: Unknown.

203. *Lycodon linckii* F. Boie, 1827. Isis von Oken 20(6): 528 (*nomen nudum*).

Type: Holotype, not designated, location unknown.
Type locality: Unknown.

204. *Lycodon mordax* F. Boie, 1827. Isis von Oken 20(6): 534 (*nomen nudum*).

Type: Holotype, not designated, location unknown.
Type locality: Unknown.

205. †*Naja sauvagei* Rochebrune, 1880. Nouv. Arch. Mus. Hist. Nat. (2) 3: 293, pl. 12, figs. 14a–c.

Type: Holotype, MNHN MCF 5, one trunk vertebra.
Type locality: "Brèches de Coudes, près Issoire, terrain Récent" [= Coudes, Pay-de-Dôme Dept., cen. France; Pleistocene].
Distribution: Middle Pleistocene of France. Known only from type specimen.
Remarks: A Recent juvenile colubrid (perhaps *Natrix*) *fide* Rage (1984b: 54).

206. †*Naja suevica* (Fraas, 1870). Jahresh. Ver. Vaterl. Naturk. Württemberg 26: 291. (Colubridae)

Type: Lectotype, SMNS 51026a, one trunk vertebra, designated by Rage (1984b: 45).
Type locality: Steinheim, Baden-Württemberg, Germany; Astaracian, middle Miocene.
Distribution: Middle Miocene (Astaracian: MN6–8: 11.1–12.8 mya) of Germany and Czech Republic.
Sources: Rage, 1984b and Ivanov, 2002.
Remarks: *Coluber incertae sedis fide* Rage (1984b: 45).

207. *Naja vera* Fitzinger, 1826a. Neue Class. Rept.: 61 (*nomen nudum*).

Type: Holotype, not designated (Hemprich), location unknown.
Type locality: "Asia, India."

208. *Natrix argus* Laurenti, 1768. Synop. Rept.: 78.

Type: Holotype, Museo Turriano (? MZUT).
Type locality: Unknown.
Remarks: A *nomen dubium*.

209. *Natrix callocephala* Wagler, 1825. Amphibia: lvi. (*nomen ineditum*).

Type: Holotype, not designated, location unknown.
Type locality: "Java" [= Java, W Indonesia].

210. *Natrix cinnamomeus* Wagler *in* Spix, 1824. Serp. Brasil. Sp. Nov.:20–21, pl. 6, fig. 1. (Colubridae)

Synonyms: *Chironius cinnamomeus* – J.A. Peters & Orejas-Miranda, 1970, and *Pseustes cinnamomoeus* – Claessen, 2003a.
Type: Holotype, formerly ZSM, a 1480 mm specimen (J.B. von Spix, 21 Aug. 1819–3 Feb. 1820), lost *fide* Hoogmoed & Gruber (1983: 328).
Type locality: "sylvis flumini Amazonum confinibus, l'intérieur du Brésil" [= rainforest of Amazion River, Pará or Amazonas State, N Brazil].
Distribution: Northern South America. Colombia, N Venezuela (Monagas), Brazil (Amazonas) and Suriname.
Sources: F. Werner, 1898, Hoge, 1964d, J.A. Peters & Orejas-Miranda, 1970, Lancini & Kornacker, 1989 and Claessen, 2003a.
Remarks: Holotype is ZSM 1679/0 *fide* F. Werner (1897a: 207). A *nomen dubium fide* Hoogmoed & Gruber (1983: 328).

211. *Natrix humana* Laurenti, 1768. Synop. Rept.: 80.

Types: Syntypes (2), not designated, ? RMNH (formerly MG).
Type locality: "Amboina & Mexico" [= Ambon, Indonesia and Mexico].
Remarks: A *nomen dubium*.

212. *Natrix maculiventris* Wagler, 1825. Amphibia: xlvii. (*nomen ineditum*).

Type: Holotype, not designated, location unknown.
Type locality: "Brasilia" [= Brazil].

213. *Natrix moniliformis* Wagler, 1825. Amphibia: xlvii. (*nomen ineditum*).

Type: Holotype, not designated, location unknown.
Type locality: "Brasilia" [= Brazil].

214. *Natrix ourdia* Wagler, 1825. Amphibia: liv–lv (*nomen ineditum*).

Type: Holotype, not designated, location unknown.
Type locality: "India orientalis" [= E India].

215. *Natrix punctata* Laurenti, 1768. Synop. Rept.: 80.

Synonyms: *Coluber puntulatus* Gmelin, 1789, and *Coluber farinosus* Bonneterre, 1790.
Type: Holotype, ? RMNH (formerly MG), not designated.
Type locality: Unknown.
Remarks: A *nomen dubium*.

216. *Natrix rhomboigera* Wagler, 1825. Amphibia: xlvii. (*nomen ineditum*).

Type: Holotype, not designated, location unknown.
Type locality: Unknown.
Remarks: Type locality "Brasilia" [= Brazil]. Possibly either *Xenodon rabdocephalus* or *Waglerophis merremii.*

217. †*Natrix rusticus* G. Brunner, 1957. Neues Jb. Geol. Paläont., Stuttgart 1957(7–8): 377, fig. 16.5.

Type: Holotype, BSM 1982 X, one dentary (H. Brand).
Type locality: "middle Pleistocene of Breitenberghöhle near Gössweinstein" [= Breitenberghöhle ca. 1 km E Gößweinstein, Bavaria, W Germany; middle Pleistocene].
Distribution: Middle Pleistocene of Germany. Known only from the holotype.
Remarks: A *nomen dubium fide* Rage (1984b 52) and Szyndlar & Böhme (1993: 390). Possibly a Recent species *fide* Rage (1984b 52).

218. †*Ogmophis kansasensis* R.L. Wilson, 1968. Contr. Mus. Paleont., Univ. Michigan 22(7): 99–100, 9a–c. (Colubridae)

Type: Holotype, UMMZP V55687, one precaudal vertebra.
Type locality: "UM-K6-59. Ogallala Formation, Lower Pliocene, Trego County, Kansas" [USA].
Distribution: Lower Pliocene of USA (Kansas). Known only from type locality.
Remarks: A *nomen dubium* fide Rage (1984b 28) and Holman (2000a: 79), who referred it to the Colubridae.

219. †*Ogmophis parvus* K.L. Rogers, 1976. Publ. Mus. Michigan St. Univ. (Paleont.) 1(5): 186–188, figs. 16a–c. (Colubridae)

Type: Holotype, MSUVP 9284, one middle trunk vertebra (W.W. Dalquest, 1967–1972).
Type locality: "on the property of Mr. Jimmie Beck, 17 kilometers E. Snyder, SW corner of NE quarter of NE quarter, Sec. 62, HTC Survey, Scurry County, Texas, at an elevation of 518 meters; Beck Ranch local fauna, Blancan age, Upper Pliocene."
Distribution: Upper Pliocene (Blancan: 1.9–4.9 mya) of Texas. Known only from type locality.
Remarks: A *nomen dubium fide* Holman (2000a: 254).

220. †*Ogmophis pauperrimus* Vanzolini, 1952. J. Paleont. 26(3): 453–454, pl. 55, figs. 1–5.

Type: Holotype, MCZVP 1976, one trunk vertebra.
Type locality: "Thomas Farm, Gilchrist County, Florida; Arikareean, lower Miocene."
Distribution: Lower Miocene (Hemingfordian) of Florida. Known only from type locality.
Source: Auffenberg, 1963.
Remarks: A *nomen dubium fide* Holman (2000a: 79).

221. *Oligodon dorsale* Berthold, 1859. Nach. Ges. Wiss. Georg-Augustus-Univ. Göttingen (17): 179.

Synonym: *Oligodon bertholdi* F. Werner, 1929a *(nomen substitutum).*
Type: Holotype, not designated, ZMG.
Type locality: "Bengala" [= Bengal, India].
Remarks: A *nomen dubium.*

222. †*Ophidion antiquus* Pomel, 1853. Cat. Méth. Vert. Foss.: 128 (*nomen rejiciendum*). (Boidae)

Type: Holotype, not designated, location unknown.
Type locality: "Terrain tertiare à Langy, le basin hydrographique supérieur de la Loire" [France].
Distribution: Miocene of France.
Remarks: A *nomen dubium fide* Rage (1984b: 32). Official Rejected Name no. 806 *fide* Direction 56 (ICZN, 1956e).

223. †*Ophidium incertum* Burmeister, 1885. Anal. Mus. Nac. Buenos Aires 3(2): 152.

Type: Holotype, MACN, one fragmentary quadrate (D.J. José de Urquiza, 18 Oct. 1855), lost *fide* Rage (1984b: 60).
Type locality: "los terrenas marinos Terciarios, de las cercanías del Paraná... la base de la meseta sobre la cual se levanta la ciudad del Paraná, capital de la Confederacion Argentina, á una altura media de 35 metros sobre el nivel ordinario de las aguas del rio; la Formacion Terciaria, el Terreno Mioceno" [= Proximity of Paraná, W Entre Rios Prov., NE Argentina; Tertiary, Miocene].
Distribution: Upper Miocene or Pliocene of Argentina. Known only from holotype.
Remarks: A *nomen dubium fide* Rage (1984b: 60).

224. †*Ophis dubius* Goldfuss, 1831. Nova Acta Acad. Phys.-Med. Caesar. Leop.-Carol. Nat. Cur. 15: 127–128, pl. 13, fig. 8 (*nomen dubium*).

Type: Holotype, GPIBO Goldfuss 1345, an imprint of skin, lost *fide* Szyndlar & Böhme (1993: 388).
Type locality: "Braunkhle des Siebengebirges" [= Rott, Nordhein-Westfalen Prov., W Germany; Arvenian, upper Oligocene *fide* Rage (1984b: 61)].
Distribution: Upper Oligocene (Arvernian, MP 30) of Germany. Known only from holotype.
Remarks: A *nomen dubium fide* Rage (1984b: 61) and Szyndlar & Böhme (1993: 388).

225. *Oxybelis punctulatus* A.-M.-C. Duméril, 1853. Mém. Acad. Sci. Inst. France 23: 487 (*nomen nudum*).

Type: Holotype, ? MNHN, not designated.
Type locality: Unknown.

226. *Oxybelis venulatus* Jan, 1863b. Elenco Sist. Ofidi: 88.

Type: Holotype, MSNM, not designated, destroyed in 1943 during World War II.
Type locality: Unknown.
Remarks: A *nomen dubium*.

227. *Oxyrhopus doliatus problematicus* Jan, 1863b. Elenco Sist. Ofidi: 93 (*nomen nudum*).

Type: Holotype, ZMH, not designated.
Type locality: "Brasile" [= Brazil].

228. *Oxyrhopus petolarius versatilis* Jan, 1863b. Elenco Sist. Ofidi: 94 (*nomen nudum*).

Type: Holotype, MNHN, not designated.
Type locality: "Corrientes" [Argentina].

229. *Oxyrhopus petolarius amoenus* Jan, 1863b. Elenco Sist. Ofidi: 94 (*nomen nudum*).

Type: Holotype, MSNM, not designated, destroyed in 1943 during World War II.
Type locality: "Venezuela."

230. †*Palaelaphis antiquus* Rochebrune, 1884. Mém. Soc. Sci. Nat. Saone-et-Loire 5: 156–157, pl. 1, figs. 3a, 3b, pl. 2, figs. 7–7c. (Boidae)

Type: Lectotype, MNHN QU 16339, one anterior trunk vertebra, designated by Rage (1984b: 32).
Type locality: "Phosphorites du Quercy, Eocène" [= Phosphorites of Quercy, on the Quercy Plateau, from the departments of Lot, Tarne-et-Garonne, and Aveyron in SW France; upper Eocene to Oligocene] via lectotype selection.
Distribution: Upper Eocene or Oligocene of France. Known only from type series.
Remarks: A *nomen dubium fide* Rage (1984b: 32).

231. †*Palaelaphis robustus* Rochebrune, 1884. Mém. Soc. Sci. Nat. Saone-et-Loire 5: 157, pl. 2, fig. 8.

Type: Holotype, MNHN QU 16341, one dentary.
Type locality: "Phosphorites du Quercy, Eocène" [= Phosphorites of Quercy, on the Quercy Plateau, from the departments of Lot, Tarne-et-Garonne, and Aveyron in SW France; upper Eocene to Oligocene].
Distribution: Upper Eocene or Oligocene of France. Known only from holotype.
Remarks: A *nomen dubium fide* Rage (1984b: 32).

232. †*Palaeophis giganteus* Pomel *in* Graves, 1847. Essai Topogr. Géog. Oise: 586 (*nomen nudum*).

Type: Holotype, several vertebrae (Lévesque), lost *fide Rage (1984b*: 60).
Type locality: "Sables glauconieux moyens: Cuise-Lamotte" [France].
Distribution: Lower Eocene (Ypresian) of France (Cuise-la-Motte).
Remarks: A *nomen nudum fide* Rage (1984b: 60).

233. †*Palaeophis halidanus* Cope, 1869a. Proc. Acad. Nat. Sci. Philadelphia (1868) 20(4): 234–235.

Type: Holotype, YPM 2763, one fragmentary vertebra (O.B. Kinney).
Type locality: "Squankum, Monmouth Co., N.J., a few miles south of Shark River. The horizon is Eocene." Emended to Manasquan Marl, 3 miles S of Farmingdale, New Jersey *fide* Gilmore (1938: 50) and Ypresian, lower Eocene *fide* Rage (1984b: 34).
Distribution: Lower Eocene (Ypresian) of USA (New Jersey). Known only from holotype.
Source: Cope, 1869d.
Remarks: A *nomen dubium fide* Rage (1984b: 36).

234. †*Palaeophis longus* Owen, 1850. Palaeontogr. Soc. Monogr. (1849) 3(2): 66, pl. 14, figs. 35–37, 45–46.

Types: Syntypes (?), BMNH, several vertebrae (Colchester), lost *fide* Rage (1984b: 36).
Type locality: "Eocene formation underlying the Red Crag at Kyson or Kingston in Suffolk" [= Kingston, Suffolk, England, UK; Suffolk Pebble Beds, Sparnacian, lower Eocene *fide* Rage, 1984b: 36].
Distribution: Lower Eocene (Ypresian) of UK (England). Known only from type specimens.
Remarks: A *nomen dubium fide* Rage (1983b: 226 & 1984b: 36).

235. †*Paleryx cayluxi* De Stefano, 1905. Boll. Soc. Geol. Ital. 24: 48–49, pl. 4, figs. 9–10, pl. 5, figs. 1–2.

Types: Syntypes, MNHN QU 16327, one dentary, MNHN QU 16328, one pterygoid, and two articulated trunk vertebrae, the latter lost *fide* Rage (1984b: 22).
Type locality: "Phosphorites du Quercy" [= Phosphorites of Quercy, on the Quercy Plateau, from the departments of Lot, Tarne-et-Garonne, and Aveyron in SW France; upper Eocene to Oligocene].
Distribution: Upper Eocene or Oligocene of France. Known only from type locality.

Remarks: Similar to *Paraepicrates fide* Rage (1984b: 22) but without vertebrae impossible to assign to genus.

236. *Phenacophis zelebori* Fitzinger, 1861. Sitz. Akad. Wiss. Wien, Math.-Natur. Kl. (1860) 42(25): 407 (*nomen nudum*).

Type: Holotype, not designated, location unknown.
Type locality: "Chile."

237. *Philodendros agnellii* Fitzinger, 1853. Sitz. Akad. Wiss. Wien, Math-Natur. Klasse 10(4): 658.

Types: Syntypes (4), NMW (formerly HNC) (Lautner, 1851), probably destroyed 6 Nov. 1944 during World War II.
Type locality: "Ägypten" [= Egypt].
Remarks: A *nomen dubium.*

238. *Pilidion montanum* Jerdon, 1853. J. Asiat. Soc. Bengal 22(6): 527. (? Uropeltidae)

Type: Holotype, not designated, location unknown.
Type locality: "Neelgherries" [= Nilgiri Hills, cen. Western Ghats, W Tamil Nadu/SE Kerala States, SW India].
Remarks: A *nomen dubium.*

239. †*Protropidonotus steinheimensis* (Fraas, 1870). Jahresh.Ver. Vaterländ Naturk. Württemberg 26(2–3): 291. (†*Coluber steinheimensis*)

Synonym: †*Protropidonotus neglectus* Schlosser, 1916.
Type: Lectotype, SMNS 51027a, one posterior trunk vertebra, designated by Rage (1984b: 51).
Type locality: "Steinheim, bei Heidenheim, eine tertiäre Oase des Steinheimer Beckens; miocänen, tertiären Localitäten namentlich nicht mit Bohnerzlocalitäten" [= Steinheim am Albuch, NE Baden-Württemberg State, SW Germany; Miocene, Bohnerz locality].
Distribution: Middle Miocene (Astaracian, MN 7–8: 11.1–12.8 mya) of Germany.
Sources: Rage, 1984b and Szyndlar, 2012.
Remarks: A *nomen dubium fide* Szyndlar & Böhme (1993: 388–389).

240. *Psammophidion smithii* Fitzinger, 1861. Sitz. Akad. Wiss. Wien, Math.-Natur. Kl (1860) 42(25): 407 (*nomen nudum*).

Type: Holotype, not designated, location unknown.
Type locality: "Cap" [= Cape of Good Hope, South Africa].

241. *Psammophis oblitus* Jan, 1863b. Elenco Sist. Ofidi: 90 (*nomen nudum*).

Type: Holotype, MSNM, not designated, destroyed in 1943 during World War II.
Type locality: Unknown.

242. *Psammophis sibilans amoena* Jan, 1863b. Elenco Sist. Ofidi: 90. (Psammophiidae)

Type: Holotype, MSNM, not designated, destroyed in 1943 during World War II.
Type locality: "Capo" [= Cape of Good Hope, South Africa].
Remarks: A *nomen dubium.* Probably a *Psammophis.*

243. *Pseudoboa sapa* Briceño-Rossi, 1934. Bol. Min. Salub. Agric. Cria, Caracas 2(15)25–29, 1 fig. (Xenodontidae)

Type: Holotype, not designated, a 1200 mm specimen, location unknown.
Type locality: "Rio de Oro y Catatumbo, Venezuela" [= Rio de Oro, Sucre State, and Catatumbo, Zulia State, N Venezuela].
Distribution: Northern Venezuela (Sucre, Zulia).
Remarks: A black opisthoglyph with 23 smooth scale rows, divided anal, 42 subcaudals, 8 (4-5) supralabials, 2 preoculars, 2 postoculars and asubocular.

244. *Pseudoelaps cingulatus* Fitzinger, 1826a. Neue Class. Rept.: 56 (*nomen nudum*).

Type: Holotype, not designated, location unknown.
Type locality: Unknown.

245. *Pseudoeryx pyrrhogrammus* Fitzinger, 1826a. Neue Class. Rept.: 55 (*nomen nudum*).

Type: Holotype, not designated, location unknown.
Type locality: Unknown.

246. *Pseudoeryx strigiventris* Fitzinger, 1826a. Neue Class. Rept.: 55 (*nomen nudum*).

Type: Holotype, not designated, location unknown.
Type locality: Unknown.

247. †*Pterosphenus muruntau* Averianov, 1997. Russ. J. Herp. 4(2): 138, figs. 7a–h.

Type: Holotype, ZISP PC 1/34, one anterior trunk vertebra (A.O. Averianov, L.A. Nessov, or A.V. Panteleev, 1987 or 1989).
Type locality: "Dzheroi 2, white quartz sands in a large hollow near the eastern Muruntau-Tamdy road, Central Kyzylkum Desert, Bukhara Viloyat, Uzbekistan."
Distribution: Middle Eocene (Bartonian) of Uzbekistan. Known only from type locality.
Remarks: A *nomen dubium fide* Rage et al. (2003: 701).

248. †*Pylmophis gracilis* Rochebrune, 1884. Mém. Soc. Sci. Nat. Saone-et-Loire 5: 158, pl. 1, figs. 4–4a, pl. 2, figs. 9, 9a–c.

Type: Lectotype, MNHN QU 16335, two articulated posterior trunk vertebrae, designated by Rage (1984a: 32).
Type locality: "Phosphorites du Quercy" [= Phosphorites of Quercy, on the Quercy Plateau, from the departments of Lot, Tarne-et-Garonne, and Aveyron in SW France; upper Eocene-Oligocene] via lectotype selection.
Distribution: Upper Eocene or Oligocene of SW France. Known only from the type series.
Remarks: A *nomen dubium fide* Rage (1984a: 32).

249. †*Python euboeicus* Roemer, 1870. Zeits. Deutsch. Geol. Ges. 22(13): 582, pl. 13, 3 figs.

Type: Holotype, ZPWM, ca. 25 trunk vertebrae, ribs and one dentary (Beinert, 1852), lost *fide* Rage (1984b: 33).
Type locality: "eines Braunkohlenlagers von Kumi auf der Insel Euboea... in tertïaren Schichten Griechenlands" [= Kími, E Évvoia Is., E Greece; lower Miocene].
Distribution: Lower Miocene of Greece. Known only from type locality.
Sources: Rochebrune, 1880, Kuhn, 1939b, 1963, Rage, 1984b, Szyndlar, 1991a and Szyndlar & Rage, 2003.
Remarks: A *nomen dubium fide* Rage (1984b: 33).

250. †*Rageophis lafonti* (Filhol, 1877). Ann. Sci. Géol. 8(1): 558, pl. 25, figs. 413–418. (†*Coluber lafonti*)

Type: Lectotype, QU 16342, mummified skin with attached vertebrae and ribs (H. Filhol, 1865–1876), designated by Rage (1988a: 460).
Type locality: "Les Phosphorites du Quercy, sur le plateau du Quercy, sur divers points des départements du Lot, du Tarn-et-Garonne, et de l'Aveyron" [= Phosphorites of Quercy, on the Quercy Plateau, from the departments of Lot, Tarne-et-Garonne, and Aveyron in SW France; upper Eocene to Oligocene] via lectotype selection.
Distribution: Upper Eocene or Oligocene of France. Known only from type series.
Remarks: A *nomen dubium fide* Rage (1984a: 32).

251. *Rhadinaea steinbachi* Boulenger, 1905b. Ann. Mag. Nat. Hist. (7) 15(89): 454–455.

Synonym: *Aporophis melanocephalus* L.E. Griffin, 1916.
Types: Syntypes (2), BMNH, a 540 mm female, and NMW 23106, a juvenile male (J. Steinbach, 1904–1905).
Type locality: "Province Sara, Department Santa Cruz de la Sierra" [= Ñuflo de Chávez Prov., Santa Cruz Dept., SE Bolivia, 16°50'S, 63°55'W].
Distribution: Eastern Bolivia (Santa Cruz), 250 m. Known only from syntypes.
Remarks: *Incertae sedis fide* C.W. Myers (1974: 22) and Dixon (1980: 15).

252. *Rhinocleis gradatus* Jan, 1857. Cenni Mus. Civico Milano: 46. (*nomen nudum*).

Type: Holotype, MSNM, not designated, destroyed in 1943 during World War II.
Type locality: Unknown.

253. *Rhinostoma rufo-fusca* Fitzinger, 1826a. Neue Class. Rept.: 56 (*nomen nudum*).

Type: Holotype, MV, not designated.
Type locality: "America, Insula St. Dominici" [= Dominican Republic].

254. †*Saurophis crassus* Brunner, 1957. Neues Jb. Geol. Paläont. 1957(7–8): 376–377, fig. 15. (Natricidae)

Type: Holotype, PCBN, one left dentary, lost *fide* Estes (1983: 211).
Type locality: "Die Breitenberghöhle bei Gössweinstein (Ofr.)." [= Breitenberghöhle ca. 1 km E Gößweinstein, Bavaria, W Germany; middle Pleistocene].
Distribution: Upper Pleistocene of Germany (Bavaria).
Remarks: A *nomen dubium fide* Estes (1983: 211).

255. †*Scaptophis miocenicus* Rochebrune, 1880. Nouv. Arch. Mus. Hist. Nat. (2) 3: 279–280, pl. 12, figs. 7a–c. (Colubridae)

Type: Holotype, MNHN SA 9880, one trunk vertebra.
Type locality: "Colline de Sansan. Miocène," (partly in error). Corrected to Sansan, Gers Dept., S France, Recent *fide* Hoffstetter (1955: 657), Mlynarski (1961: 29), and Rage (1981: 541).
Distribution: Recent of France (Gers). Known only from holotype.
Sources: Hoffstetter, 1955, Mlynarski, 1961 and Rage, 1981, 1984b.
Remarks: A *nomen dubium fide* Rage (1981: 541) and Augé & Rage *in* Ginsburg (2000: 301).

256. *Scolecophis cingulata* Fitzinger *in* Jan, 1857. Cenni Mus. Civico Milano: 44 (*nomen nudum*).

Type: Holotype, not designated, location unknown.
Type locality: Unknown.

257. *Sepedon brongniardii* Fitzinger, 1826a. Neue Class. Rept.: 62 (*nomen nudum*).

Type: Holotype, not designated, location unknown.
Type locality: Unknown.

258. *Simotes aspersus* A.-M.-C. Duméril, 1853. Mém. Acad. Sci. Inst. France 23: 472 (*nomen nudum*).

Type: Holotype, ? MNHN, not designated.
Type locality: "Indes orientales" [= East Indies].
Remarks: Probably an *Oligodon*.

259. †*Suffosio praedatrix* Standhart, 1986. Vert. Paleont. Cretaceous/Tertiary Texas: 124–127, figs. 36a–f. (Anomalepididae) (*nomen ineditum*)

Type: Holotype, LSUMG V-797, one vertebra.
Type locality: "Dogie (VL-108), Upper Javelina Member, Tornillo Formation, Big Bend National Park, Texas. Puercan (early Paleocene)."
Distribution: Lower Paleocene of USA (Texas).
Remarks: A *nomen dubium fide* Rage (pers. comm.).

260. †*Tachyophis nitidus* Rochebrune, 1884. Mém. Soc. Sci. Nat. Saone-et-Loire 5: 159–160, pl. 1, figs. 5–5a, pl. 2, figs. 10–10b.

Type: Lectotype, MNHN QU 16331, four articulated trunk vertebrae, designated by Rage (1984b: 32).
Type locality: "Phosphorites du Quercy, France; l'epoque de l'Eocène supérieur" [= Phosphorites of Quercy, on the Quercy Plateau, from the departments of Lot, Tarne-et-Garonne, and Aveyron in SW France; upper Eocene] via lectotype selection.
Distribution: Upper Eocene of France. Known only from type specimens.
Remarks: A *nomen dubium fide* Rage (1984b: 33).

261. *Tamacuilla huilia* Brüggemann & Wigand, 1838. Naturg. Abbild. Besch. Amph. 1: 98 (*nomen nudum*).

Type: Holotype, not designated, location unknown.
Type locality: Unknown.

262. *Thamnophis sumichrasti cerebrosus* H.M. Smith, 1942g. Zoologica 27(3–4): 99.

Type: Holotype, USNM 12734, a female specimen (H.J. Stuart).
Type locality: "Escuintla, Guatemala."
Distribution: Guatemala (Escuintla).
Source: Rossman, 1991.
Remarks: *Incertae sedis fide* Rossman et al. (1996: 160).

263. *Thamnosophis delesserti* Jan, 1863b. Elenco Sist. Ofidi: 82.

Type: Holotype, MNHN, not designated.
Type locality: Unknown.
Remarks: A *nomen dubium*.

264. *Thanatophis patoquilla* Posada Arango, 1889. Anal. Acad. Med. Medellin 2: 47.

Type: Holotype, not designated, lost *fide* Niceforo María (1938: 417).
Type locality: "las regiones bajas y cálidas del Departamento especialmente hacia el nordeste" [= Medellín, Colombia].
Distribution: Colombia. Known only from holotype.
Source: McDiarmid et al., 1999.
Remarks: A synonym of *Porthidium lansbergi fide* Daniels (1949: 329) but *incertae sedis fide* J.A. Peters and Orejas-Miranda (1970: 329) and McDiarmid et al. (1999: 324).

265. †*Thoracophis rugosus* Trochel, 1854. Verh. Naturhist. Ver. Pruess. Rheinlande Westfalens 11(1): xix (*nomen dubium*).

Type: Holotype, lost *fide* Rage (1984a: 61).
Type locality: "Braunkohle von Rott" [Germany].
Distribution: Upper Oligocene of Germany.
Remarks: A *nomen dubium fide* Rage (1984a: 61). Not a snake but an anguid lizard *fide* Szyndlar & Böhme (1993: 390).

266. *Tortrix australis* Gray *in* Grey, 1841b. J. Two Exped. Australia 2: 444.

Type: Holotype, not designated, location unknown.
Type locality: "Australia."
Remarks: A *nomen dubium fide* Cogger et al. (1983a: 264).

267. *Trimaculatus ceylonensis* Higgins, 1873. Ophidians: 80 (*nomen nudum*).

Type: Holotype, not designated, location unknown.
Type locality: "Ceylon" [= Sri Lanka].

268. *Trimeresurus colubrinus* Fitzinger, 1826a. Neue Class. Rept.: 61 (*nomen nudum*).

Type: Holotype, not designated, location unknown.
Type locality: Unknown.

269. *Tropidolaemus bivittatus* Kröyer *in* J.T. Reinhardt, 1861a. Vidensk. Medd. Naturhist. Foren. Kjöbenhavn (1860) 22: 228 (*nomen nudum*).

Type: Holotype, not designated, location unknown.
Type locality: Unknown.

270. *Tropidolaemus mystax* Kröyer *in* J.T. Reinhardt, 1861a. Vidensk. Medd. Naturhist. Foren. Kjöbenhavn (1860) 22: 228 (*nomen nudum*).

Type: not designated, location unknown.
Type locality: Unknown.
Remarks: Collector was Hombron and type locality Philippines.

271. †*Tropidonotus atavus* H. Meyer, 1855. Neues Jahrb. Min. Geog. Geol. Petrefak. 23: 337. (Colubridae)

Type: Holotype, IPUB 9, skull and incomplete axial skeleton embedded in matrix.
Type locality: "Grube Romerikenberg" [= Romeriken-Berge Quarry, Rott, Bavaria Distr., W Germany; Arvernian, upper Oligocene *fide* Rage, 1984b: 44].
Distribution: Upper Oligocene (Arvenian) and lower Miocene of Germany (Bavaria).
Remarks: A *nomen dubium fide* Ragc (1984b:44).

272. *Tropidonotus catenulatus* Jan, 1857. Cenni Mus. Civico Milano: 45 (*nomen nudum*).

Type: Holotype, MSNM, not designated, destroyed in 1943 during World War II.
Type locality: "Messico" [= Mexico].

273. †*Tropidonotus elongatus* Troschel, 1854. Verh. Naturhist. Ver. Pruess. Rheinlande Westfalens 11(1): xix (*nomen nudum*).

Type: Holotype, lost *fide* Rage (1984b: 60).
Type locality: "Braunkohle von Rott." [Germany].
Distribution: Upper Oligocene of W Germany (Rott).
Remarks: A *nomen nudum fide* Rage (1984b: 60). Not a snake but an anguid lizard *fide* Szyndlar & Böhme (1993: 390)

274. *Tropidonotus glaphyros* Jan, 1865c. Arch. Zool. Anat. Fis. 3(2): 214.

Synonym: *Tropidophis glaphyros* Jan, 1863b *(nomen nudum).*
Type: Holotype, MSNM, not designated, destroyed in 1943 during World War II.
Type locality: "Amer. settentr." [= North America].
Remarks: A *nomen dubium.*

275. *Tropidonotus incertus* Jan, 1863b. Elenco Sist. Ofidi: 72 (*nomen nudum*).

Type: Holotype, MSNM, not designated, destroyed in 1943 during World War II.
Type locality: Unknown.

276. *Tropidonotus intermedius* Jan, 1863b. Elenco Sist. Ofidi: 69 (*nomen nudum*).

Type: Holotype, not designated, location unknown.
Type locality: "Indie orientali ?" [= E India].

277. *Tropidonotus meleagris* Wiegmann *in* Fitzinger, 1843. Syst. Rept.: 26 (*nomen nudum*).

Type: Holotype, not designated, location unknown.
Type locality: "Am." [= America].

278. *Tropidophis distinctus* Jan, 1863b. Elenco Sist. Ofidi: 22 (*nomen nudum*).

Type: Holotype, MSNM, not designated, destroyed in 1943 during World War II.
Type locality: "Charlestown" [= Charleston, South Carolina, USA].

279. *Typhlops cubae* Bibron, 1838/39 *in* Ramon de la Sagra, 1838–1843. Hist. Fis. Pol. Nat. Isla Cuba 2(2): 122–123, pl. 22, figs. 1–5.

Types: Syntypes (2), MNHN 3218, a 215 mm specimen, and MNHN 1999.8218 (formerly MNHN 3218a), a 197 mm specimen (D. Ramón de la Sagra, 1824–1835).
Type locality: "Cuba."
Source: Domínguez & Diaz, 2011a.

Remarks: Previously a synonym of *Typhlops lumbricalis*. A *nomen nudum fide* R. Thomas (1968: 714).

280. *Typhlops disparilis* Jan, 1860 *in* Jan & Sordelli, 1860–1866. Icon. Gen. Ophid. 1(1): 1, pls. 5, fig. 6, pl. 6, fig. 6.

Type: Holotype, MSNM, not designated, a 300 mm specimen, destroyed in 1943 during World War II.
Type locality: Unknown.
Sources: W.C.H. Peters, 1881 and Boulenger, 1893a.
Remarks: Listed as *incertae sedis* by Hahn (1980a: 76).

281. *Typhlops inconspicuus* Jan, 1863b. Elenco Sist. Ofidi: 11 (*nomen nudum*).

Type: Holotype, MNHN, not designated.
Type locality: "Madagascar."

282. *Typhlops tenuis* Jan, 1863b. Elenco Sist. Ofidi: 10.

Type: Holotype, MSNM, not designated, destroyed in 1943 during World War II.
Type locality: Unknown.
Remarks: A *nomen dubium*.

283. *Typhlops textilis* Jan, 1863b. Elenco Sist. Ofidi: 11 (*nomen nudum*).

Type: Holotype, MNHN, not designated.
Type locality: Unknown.

284. *Typhlops troscheli* Jan, 1863b. Elenco Sist. Ofidi: 11 (*nomen nudum*).

Type: Syntypes, MNHN and MSNM, not designated, the latter destroyed in 1943 during World War II.
Type locality: Unknown.

285. *Tyria cloquetii* Fitzinger, 1826a. Neue Class. Rept.: 60 (*nomen nudum*).

Type: Holotype, not designated (Cloquet), location unknown.
Type locality: Unknown.

286. *Tyria kollari* Fitzinger, 1826a. Neue Class. Rept.: 60 (*nomen nudum*).

Type: Holotype, not designated (Kollar), location unknown.
Type locality: Unknown.

287. *Tyria schreibersii* Fitzinger, 1826a. Neue Class. Rept.: 60 (*nomen nudum*).

Type: Holotype, not designated (Schreibers), location unknown.
Type locality: Unknown.

288. *Vipera cobra* Laurenti, 1768. Synop. Rept.: 103.

Type: Holotype, Museo Turriano (? MZUT).
Type locality: Unknown.
Remarks: A *nomen dubium.*

289. *Vipera lanceolata* Raddi, 1823. Mem. Mat. Fis. Soc. Ital. Sci., Modena 19: 63–64. (Viperidae)

Type: Holotype, not designated (G. Raddi, 1817–1818), location unknown.
Type locality: "provincia di Rio-janeiro" [= Rio de Janeiro State, SE Brazil].
Remarks: A *nomen dubium*. Probably a crotaline like *Bothrops*.

290. †*Vipera sansaniensis* Lartet, 1851. Notice Colline Sansan: ?? (*nomen nudum*).

Type: Holotype, tubular fangs, lost *fide* Rage (1984b: 60).
Type locality: "Sansan, du terrain tertiare Miocène dans le Bassin Sous-Pyrènéen" [= Gers Prov., France].
Distribution: Upper Miocene of France.
Remarks: A *nomen nudum fide* Rage (1984b: 60).

291. *Vipera urens* F. Boie, 1827. Isis von Oken 20(6): 559 (*nomen nudum*).

Type: Holotype, not designated, location unknown.
Type locality: Unknown.

292. *Xenodon reinhardtii* Kröyer *in* J.T. Reinhardt, 1861a. Vidensk. Medd. Naturhist. Foren. Kjöbenhavn (1860) 22: 222 (*nomen nudum & nomen corrigendum*).

Synonym: *Xenodon einhardii* Kröyer *in* J.T. Reinhardt, 1861a (*nomen incorrigendum*).
Type: Holotype, not designated, location unknown.
Type locality: Unknown.

293. *Xenodon treuensteini* Jan, 1857. Cenni Mus. Civico Milano: 46 (*nomen nudum*).

Type: Holotype, NMW, not designated.
Type locality: "Brasile" [= Brazil].

294. *Xenodon uniporus* Fitzinger, 1826a. Neue Class. Rept.: 57 (*nomen nudum*).

Type: Holotype, not designated, location unknown.
Type locality: "America, Brasilia" [= Brazil].

295. *Xenopeltis trivirgatus* Jerdon, 1853. J. Asiat. Soc. Bengal 22 (6): 527.

Type: Holotype, not designated, location unknown.

Type locality: "Neelgherries" [= Nilgiri Hills, cen. Western Ghats, W Tamil Nadu/SE Kerala States, SW India].

Remarks: A *nomen dubium*. Possibly referable to *Calamaria*.

296. *Zamenis argentinus* Brèthes, 1917. Physis 3: 92–94, 2 figs.

Type: Holotype, not designated, location unknown.

Type locality: "Argentina."

Remarks: *Incertae sedis fide* J.A. Peters & Orejas-Miranda (1970: 129). Referable to *Philodryas fide* F. Werner (1924a: 34).

Geographical References

World

GENERAL: Adler & Halliday, 1986; Badger, 1999; Bauchot, 1994; Boulenger, 1893a, 1894a; Braßeler, 1989, 1991, 1994; Brazaitis & Watanabe, 1993; J.A. Burton, 1991; Coborn, 1991, 1994; Cogger & Zweifel, 1992; Delhey, 2009; Ditmars, 1931; Du, 2004; Gower et al., 2012; Greene, 1997; Hahn, 1980a (Scolecophidia); D. Hall, 2004; Heatwole, 1999 (sea snakes); Ineich, 2004 (sea snakes); Kundert, 1974, 1984; Lamar, 1997b; Marais, 1997; Mattison, 1995, 1999; Mehrtens, 1987; McDiarmid et al., 1999 (Scolecophidia, Henophidia, Acrochoridae); Moncuit & Daoues, 2002; J.B. Murphy & Schlager, 2003; Obst et al., 1984, 1988; Perrier, 1928; O'Shea, 2005 (Henophidia); Rage, 1984b (fossils); D. Schmidt, 1996a (Booidea), 1996b (Natricidae), 1998 (Boiginae), 2001; K.P. Schmidt & Inger, 1957; Sengoku, 1993a–b; B. Song & Shi, 2002; Stafford, 2000b; Terent'ev, 1961; Trutnau, 2002; Weidensaul, 1991; Welch, 1994b (Scolecophidia & Henophidia); K.L. Williams & Wallach, 1988 (genera)

VENOMOUS SNAKES: Boulenger, 1896a; J.H. Brown, 1973; Bücherel et al., 1968; Bucherel & Buckley, 1971; Calmette, 1908; David & Ineich, 1999; Golay, 1985 (elapids); Golay et al., 1993; K.A. Harding & Welch, 1980; Jaroniewski, 1992; Klemmer, 1963; R. Kraus & Werner, 1931; McDiarmid et al., 1999 (vipers); Minton et al., 1968; Noguchi, 1909; O'Shea, 2007; Phelps, 1981, 2010 (Old World vipers); Phisalix, 1922; Trutnau, 2004; Ulber, 1993–1995; Welch, 1994a

Africa & Madagascar

GENERAL: Bogert, 1940; Boulenger, 1910a, 1915a–d, 1920c–d; Broadley & Minshull, 1986; Brogard, 2005; Dobiey & Vogel, 2007 (venomous); Isemonger, 1983; Lacarriere & Ineich, 1992; Meirte, 1992; Spawls & Branch, 1995; Sternfeld, 1912, 1917; Welch, 1982; Wilkey, 2002a–b (venomous)

NORTH AFRICA: Andersson, 1935; Angel & Lhote, 1938; Doumergue, 1901; Grenot & Vernet, 1972; Gruber, 1992; Lambert, 1984; Le Berre, 1989; Olivier, 1899; Papenfuss, 1969; Saint-Girons, 1951; Sauer, 1989; Schleich et al., 1996; Sochurek, 1979b; Venchi & Sindaco, 2006; Vernet & Grenot, 1972

WEST AFRICA: Angel, 1933b; Cansdale, 1961; Chippaux, 1999, 2001, 2006; Chirio, 2009; Condamin, 1958; Dekeyser & Derivot, 1960 (venomous); Hughes, 1983; Joger, 1981; Papenfuss, 1969; Spawls, 1992; J.-F. Trape & Mané, 2006b; Villiers, 1950–1975

EAST AFRICA: Branch, 2005; N.G. Hedges, 1983; Howell, 1993; Lanza, 1978; Loveridge, 1929, 1957a; A. MacKay & MacKay, 1985; Spawls et al., 2002, 2006; Tornier, 1896

ALGERIA: Andersson, 1935; Angel, 1944; Boulenger, 1891h; Chananaud, 1920b; Chirio & Blanc, 1997b–d; Doumergue, 1901; Gauthier, 1967; Geniez et al., 2004 (Tindouf area); Guichenot, 1850; Hartert, 1913; Olivier, 1894; Sergent, 1919 (venomous); Strauch, 1862; Witte, 1930c

ANGOLA: Angel, 1923c; Bocage, 1895a; Bogert, 1940; Ferreira, 1898, 1903a–b, 1906; Hellmich, 1957, 1975a–b; Laurent, 1950c 1954a, 1964a; Manaças, 1982; Monard, 1935, 1937; H.W. Parker, 1936d; K.P. Schmidt, 1933d; Thys van der Audenaerde, 1966

BENIN (Dahomey): Bocage, 1887b; Chirio, 2009; Hughes, 2013; Roman, 1984; Ullenbruch et al., 2010; Villiers, 1951a

BOTSWANA: Auerbach, 1985, 1987; Broadley & Blaylock, 2013; Clauss & Clauss, 2002

BURKINA FASO (Upper Volta): Chirio, 2009; Roman, 1969, 1976, 1980, 1984; J.-F. Trape, 2005

BURUNDI (German East Africa): Derleyn, 1977, 1978b; Derleyn et al., 1983; Laurent, 1956a; Rosselot, 1978; Spawls et al., 2002; Witte, 1962

CAMEROON: Angel, 1940; Böhme, 1975, 1982; Chirio & Le Breton, 2007; Eisentraut, 1973; Gauduin, 1970; Joger, 1982; D.P. Lawson, 1993; Le Breton, 1999; Mertens, 1938, 1940b; Monard, 1951; L. Muller, 1910; Ota et al., 1987; Perret, 1961; Perret & Mertens, 1957a–b; Sjöstedt, 1897; Sternfeld, 1908a, 1909a, 1917; Stucki-Stirn, 1979

CENTRAL AFRICAN REPUBLIC: Chirio & Ineich, 2006; Joger, 1990; Roux-Estève, 1965; Sternfeld, 1917

CHAD: Roussel & Villiers, 1965; Wake & Kluge, 1961

COMORO ISLANDS: Boulenger, 1915b; Henkel & Schmidt, 2000; Henkel et al., 1995

CONGO (Brazzaville, People's Republic of the Congo): Bocage, 1887a; Bourgeois, 1964; Mocquard, 1887c, 1896b; J.B. Rasmussen, 1991; J.F. Trape, 1981, 1985; J.-F. Trape & Carme, 1982; J.-F. Trape & Roux-Estève, 1995; Villiers, 1966

DEMOCRATIC REPUBLIC OF CONGO (Belgian Congo, Zaire): Bouillon, 1965; Boulenger, 1909c; Bourgeois, 1963a (Katanga); Chifundera, 1990; Curry-Lindahl, 1956, 1961; Laurent, 1953, 1956a, 1960; Loveridge, 1942; K.P. Schmidt, 1923; Schwetz, 1934; Thys van den Audenaerde, 1965; Witte, 1933b, 1941, 1953, 1962, 1966, 1975

DJIBOUTI: Ineich, 1999; Schätti, 1989; Sternfeld, 1908d

EGYPT (Sinai): J. Anderson, 1896, 1898, 1903; Baha El Din, 2006; T. Barbour, 1914b; Eiselt, 1962; Flower,

1933; Marx, 1956, 1968; Saleh, 1997; K.P. Schmidt & Marx, 1956; Stein & Helmy, 1994; Y. Werner, 1973, 1983; Y. Werner *in* Scott, 1982

EQUATORIAL GUINEA (Spanish Guinea, Río Muni, Fernando Po, Bioko, Annobón): Capoccacia, 1961b; Eisentraut, 1973; Manaças, 1982; Mertens, 1941a, 1964a

ERITREA: Boulenger, 1896f; Largen, 1997; Largen & Spawls, 2010

ETHIOPIA: Boulenger, 1891i, 1896h, 1897i, 1898j; Lanza, 1972 (Awash Natl. Pk.); Largen & Largen, 1982; Largen & Rasmussen, 1993; Largen & Spawls, 2010; Rehák, 1994; Roux, 1936 (Omo); Scortecci, 1930a–b; Sternfeld, 1908d; Zavattari, 1930a, 1933

GABON: Boulenger, 1900g; A.H.A. Duméril, 1857; Girardin, 1978; A. Günther, 1896a; Knoepffler, 1966, 1968; Pauwels & Vande weghe, 2008

GAMBIA: Andersson, 1937; L.K. Barnett et al., 2001; Condamin & Villiers, 1962; Gruschwitz et al., 1991; Hakansson, 1981; Miles et al., 1978; Pauwels & Meikte, 1996

GHANA (Gold Coast): Cansdale, 1948, 1949a–b, 1954; Hughes, 1988; Hughes & Barry, 1969; Leeson, 1950; Leston, 1970; Leston & Hughes, 1968; Orton & Morrison, 1946; Raxworthy & Attuguayefio (Muni Lagoon), 2000

GUINEA (French Guinea): Bedriaga, 1882, 1892; Boulenger, 1905h; Chirio, 2013; Condamin, 1959; Ineich, 2003; Villiers, 1956a

GUINEA–BISSAU (Portuguese Guinea): Auliya et al., 2012; Bocage, 1896; Manaças, 1957, 1982; Monard, 1940; Villiers, 1948

IVORY COAST (Côte d'Ivoire): Angel et al., 1954; Barbault, 1971; Chippaux, 1980; Chippaux & Bressy, 1981; Courtois & Chippaux, 1977; Doucet, 1963; Ineich, 2003; Rodel et al., 1995, 1999; Roman, 1984; Roux-Estève, 1969a

KENYA: Battersby, 1954, 1955; Bogert, 1942b; Lötters et al., 2007 (Kakamega Forest); Loveridge, 1935, 1936c; 1942; H.W. Parker, 1932c; H.A. Skinner, 1973; Spawls, 1978; Spawls et al., 2002; Vesey-FitzGerald, 1975

LESOTHO: Bates, 1991; Bourquin, 1989; Lambiris, 1993; Mouton & Wyk, 1993

LIBERIA: T. Barbour & Loveridge, 1930; Briscoe, 1949; Ineich, 2003; Johnsen, 1962; Loveridge, 1938b, 1941a; E.H. Taylor & Weyer, 1958

LIBYA: Ibrahim, 2008 (southern); Kramer & Schnurrenberger, 1959, 1963; Le Berre, 1989; Schleich, 1984; Scortecci, 1937, 1939a; Vinciguerra, 1931; F. Werner, 1909a; Zavattari, 1922, 1929, 1930b, 1936

MADAGASCAR (Malagasy Republic): Andersson, 1911; Angel, 1931; T. Barbour, 1918; Blanc, 1972, 1986; Boettger, 1877, 1878, 1879b; Brygoo, 1983, 1987; Domergue, 1969; Glaw & Vences, 1992, 1996, 2007; Guibé, 1958; Henkel & Schmidt, 2000; Henkel et al., 1995; Jourdran, 1904; Mertens, 1933, 1972c; Mocquard, 1895a–b; 1909; Nagy et al., 2003; Raxworthy, 1990; Raxworthy & Nussbaum, 1996

MALAWI (Nyasaland): Boulenger, 1891e; De Fonseca & Jocque, 1979; Loveridge, 1953; Royle, 1989; Stevens, 1974; M. Stewart, 1968, 1969; Stewart & Wilson, 1966; Sweeney, 1961; Tonge & Morgan, 1984

MALI (French Soudan): Angel, 1922, 1933b; Papenfuss, 1969; Schätti, 1986c; Villiers, 1965c

MAURITANIA: J. Anderson, 1892 (Barbary); Angel, 1938; Angel & Lhote, 1938; Dekeyser & Villiers, 1956; Geniez et al., 2004 (Sahara); Villiers, 1950b

MASCARENE ISLANDS (Mauritius, Round): Boulenger, 1915b; Bullock, 1977, 1980 (Round Is.); Cole, 2009; Henkel & Schmidt, 2000; Owadally & Lambert, 1988; Tirvengadum & Bour, 1985; Vinson, 1949

MOROCCO (Maroc): Aellen, 1951; Boettger, 1874, 1883a; Bons, 1958, 1967, 1972, 1973; Bons & Geniez, 1996; Bons & Girot, 1962; Dollfus & Beaurieaux, 1928; Geniez et al., 1991, 2004 (Sahara); Harris et al., 2010; Hediger, 1935; Mellado & Mateo, 1992; Pasteur & Bons, 1960; Pellegrin, 1925, 1927; Saint Girons, 1956; J.R. Schouten & Thevenot, 1988; Schweiger, 1992; Stemmler, 1972a–b, d–e; Stemmler & Hotz, 1973; F. Werner, 1931; Zulueta, 1908, 1909

MOZAMBIQUE (Portuguese East Africa): Angel, 1920d; J.G. Barros & Cunha, 1935; Bocage, 1896; Broadley, 1990, 1992a; Cott, 1928, 1935; Loveridge, 1953; Manaças, 1959, 1961, 1982; W.C.H. Peters, 1882d; Thermido, 1941

NAMIBIA (South West Africa): Branch, 1994; Buys & Buys, 1983; Fischer, 1888b; M. Griffin, 2003; Lambiris, 1993; Mertens, 1937, 1955, 1971; H.W. Parker, 1936d; Sternfeld, 1910c–d; C.T. Stuart, 1980; Elzen, 1980, 1983

NIGER: Chirio, 2009; Roman, 1974a, 1976, 1984; Villiers, 1950c

NIGERIA: Akinpeln & Areo, 2007 (Osun Grove, Osogbo); Blackwell, 1967; J.A. Butler & Reid, 1986, 1990; Cansdale, 1971 (Gold Coast); Dunger, 1971–1973; Luiselli et al., 1998; Romer, 1953b (Port Harcourt)

RWANDA (German East Africa): Fischer & Hinkel, 1992; Laurent, 1956a; Spawls et al., 2002; Witte, 1962

SÃO TOMÉ AND PRÍNCIPE ISLANDS: Manaças, 1958; Schätti & Loumont, 1992

SENEGAL: Böhme, 1978; Joger & Lambert, 2002; Miles et al., 1978; Rochebrune, 1885; J.-F. Trape & Mané, 2000, 2004 (Sénégal Oriental)

SEYCHELLES ISLANDS: Gaymer, 1968; Gerlach, 2006; Henkel & Schmidt, 2000; Henkel et al., 1995; Honeggar, 1966; Nussbaum, 1984; H.W. Parker, 1936e; Rendahl, 1939; Vesey-FitzGerald, 1948

SIERRA LEONE: Aylmer, 1922; Menzies, 1966; Zug, 1987

SOMALIA: Calabresi, 1927; Cherchi, 1958 (venomous); Gans & Laurent, 1965; Hoevers & Johnson, 1982 (Middle Juba); Lanza, 1983a, 1990b; Mocquard, 1888; H.W. Parker, 1932b, 1949; Scortecci, 1934, 1939b; Sternfeld, 1908d

SOUTH AFRICA: Alexander, 1990 (Durban); Alexander & Marais, 2007; Bates, 1991b (Golden Highlands Natl. Pk.), 1996 (Free State); Bauer & Branch, 2001 (N Cape);

Boulenger, 1887j; Bourgeois, 1964 (Kasapa); Bourquin & Channing, 1980 (Natal Drakenberg); Bourquin & Sowler, 1980 (Vernon Crookes Nat. Res.); Bourquin et al., 1971 (Hluhluwe Game Res.); Branch, 1978–1979, 1982, 1988, 1991, 1993, 1998, 2001, 2005; Broadley, 1983; Bruton & Haacke, 1980 (Maputaland); De Waal, 1978 (Orange Free State); F.W. FitzSimons, 1912; V.F.M. FitzSimons, 1930, 1935a–b, 1938, 1956, 1962a, 1966, 1974; Gough, 1908; Haacke, 1984 (Mapelane); Haagner & Els, 1986; Hawthorne, 1998; Hewitt, 1912, 1937 (E Cape); Hewitt & Power, 1913; Jacobsen, 2005; Jacobsen & Haacke, 1980 (Transvaal); Lambiris, 1993; Marais, 1985, 1992, 2004, 2007; Patterson, 1987; Pienaar et al., 1983 (Kruger Natl. Pk.); Pringle, 1954; Reitz, 1978; W. Rose, 1929, 1950, 1955, 1962; Sternfeld, 1910c; Ulber, 1999; Visser, 1967, 1972, 1979b; Visser & Chapman, 1978

SUDAN: Angel & Lhote, 1938; Colley, 1946 (Geneina); Corkill, 1935; Le Berre, 1989; Loveridge, 1955a; Osman & El Sir, 1988

SWAZILAND: Boycott & Culverwell, 1992

TANZANIA (German East Africa): T. Barbour & Loveridge, 1928; Broadley & Howell, 1991; Cott, 1928; Dollo, 1886; Ionides, 1950; Loveridge, 1924b, 1933, 1942, 1951a, 1955b, 1957b, 1959; Mocquard, 1888; Pakenham, 1947 (Zanzibar Is.), 1983 (Zanzibar & Pemba Is.); J.B. Rasmussen, 1981, 1982b (both Usambaras); Razzetti & Msuya, 2002 (Arusha Natl. Pk.); Rodgers & Homewood, 1982; H.A. Skinner, 1973; Spawls et al., 2002; Sternfeld, 1910c; Uthmöller, 1937–1938 (both Kilimanjaro), 1941a–b; Vesey-FitzGerald, 1958, 1975

TOGO: Hulselmans & Verheyen, 1970; Hulselmans et al., 1970, 1971; Matschie, 1890, 1893a–b; Roman, 1984; Segniagbeto et al., 2011; Sternfeld, 1908c, 1909b; Villiers, 1951a

TUNISIA (Tunis): Blanc, 1988; Blanc & Nouira, 1988; Bogaerts, 1998; Boulenger, 1920d; Chpakowsky & Chneour, 1954; A. Davidson, 1964; C. Domergue, 1959a–b; F. Doumergue, 1901; Mayet, 1903; Mertens, 1929b; Mosauer, 1934; Olivier, 1896a–b

UGANDA: Loveridge, 1936c, 1942; H.W. Parker, 1932c; Pitman, 1938, 1974; H.A. Skinner, 1973; Spawls et al., 2002

WESTERN SAHARA (Rio de Oro, Sahrawi Republic): Geniez et al., 2004; A.C.L.G. Günther, 1903; Monteil, 1951; Salvador & Perls, 1975

ZAMBIA (Northern Rhodesia): Angel, 1923c; Boulenger, 1907e; Broadley, 1971d; Broadley & Pitman, 1960; Broadley et al., 2003; C. Doria & Nyirenda, 1995; Johnsen, 1962; Simbotwe & Mubamba, 1993; Vesey-FitzGerald, 1939, 1958; V.J. Wilson, 1965

ZIMBABWE (Southern Rhodesia): Broadley, 1959, 1988; Broadley & Blake, 1979; Broadley & Blaylock, 2013; Broadley & Cock, 1975; Lambiris, 1993

Europe & Western Asia

GENERAL: Arnold, 2002; Arnold & Burton, 1978; Bannikov et al., 1977; Blum, 1976; Brodmann, 1987 (vipers); Boulenger, 1913e; Böhme, 1993; Maugeri, 1990; Cihar, 1979; Corbett, 1989; David & Vogel, 2010b (vipers); De Smedt, 2001, 2006 (vipers); Diesener et al., 1986, 1992; Dottrens & Aellen, 1963; Engelmann et al., 1985; Ferri, 1994, 2002; Frommhold, 1959; Gasc et al., 1997; Gruber, 2009; G. Hecht, 1929a; Hediger, 1937; Hellmich, 1956, 1962; Honeggar, 1981; Kreiner, 2007; Kwet, 2005; Lanka & Vit, 1985; Lanza, 1973; Matz, 1983a–b; Mertens, 1964b; Mertens & Muller, 1928, 1940; Mertens & Wermuth, 1960; Mezzena & Dolce, 1977; Parent, 1983; Radovanovic, 1941; Radovanovic & Martino, 1950; Saint-Girons, 1952a (S Europe), 1978 (vipers); Sauer, 1989; Schreiber, 1875, 1912; Spellerberg, 2002 (NE Europe); Speybroeck et al., 2010; Steinheil, 1913–1914; Steinheil & Müller, 1927, 1928, 1931; Sternfeld, 1913c; Sternfeld & Steiner, 1952; Steward, 1971; Street, 1979; Trutnau, 1975; Venchi & Sindaco, 2006; Welch, 1983b

ALBANIA: Bruno, 1988a, 1989; Fejérváry, 1922; Haxhiu, 1998; Kopstein & Wettstein, 1920

ARMENIA: Bannikov et al., 1977; Chernov, 1937, 1939; Arakelyan et al., 2011; Tuniyev, 1995; Tuniyev et al., 2009 (Caucasus)

AUSTRIA: Cabela, 1982; Cabela & Tiedemann, 1985; Cabela et al., 1992 (Carinthia), 2001; Eiselt, 1961; Fabian et al., 2007; Galán-Regalado & Fernández-Arias, 1993; Grillitsch & Cabela, 1992; Grillitsch & Grillitsch, 1984; Grillitsch et al., 1983; Knauer, 1875; Luttenberger, 1978; Tiedemann et al., 1990; Weissmair, & Moser, 2008; F. Werner, 1897a

AZERBAIJAN: Bannikov et al., 1977; Eichwald, 1842; Sobolevsky, 1929; Tuniyev, 1995

BELGIUM: Parent, 1979, 1984; Sparreboom, 1981; Vacher & Geniez, 2010; Witte, 1948

BELARUS: Bannikov et al., 1977; Pikulik, 1996

BOSNIA & HERZEGOVINA (Yugoslavia): Veith, 1991a–b

BULGARIA: Beshkov, 1984; Beshkov & Beron, 1964; Beshkov & Nanev, 2002, 2006; Buresch & Zonkow, 1934; L. Müller, 1932a, 1939; Stojanov et al., 2011; Sura, 1981b

CRETE: Sowig, 1985; Stepánek, 1934

CROATIA (Yugoslavia): Tóth et al., 2006 (Cres); Vogrin, 1997 (Lake Vrana)

CYPRUS: Atatür & Göçmen, 2001 (northern); Baier et al., 2009; Boettger, 1880b; Boulenger, 1888g; Demetropoulos & Ioannides, 2002; Schätti & Sigg, 1989

CZECH REPUBLIC: Barus & Oliva, 1992; Mikátová et al., 2001; Necas et al., 1997; Stepánek, 1949; Vozenilek, 1966

DENMARK: Hvass, 1945; Pfaff, 1950; Sarauw, 1903

ESTONIA: Bannikov et al., 1977

FINLAND: Langerwerf, 1975; Malmberg, 1872; Terhivuo, 1993

FRANCE: Angel, 1946; Bedriaga, 1883 (Corsica); Bertrand & Cochet, 1992 (Ariege); Cantuel, 1949; Castanet & Guyetant, 1989; Cheylan, 1992 (Corsica); Delaugerre & Cheylan, 1992 (Corsica); Deliry, 2002 (Rhône-Alpes); Fretey, 1975; Lataste, 1876a–b; Lescure & Massary, 2012; Lescure et al., 2010 (Seine-Saint-Denis); Matz, 1976b; Mertens, 1957b; Naulleau, 1987a; Parent, 1982; Pinston et al., 2000 (Franche-Comté); Portevin, 1942; Pottier, 2004, 2008 (Pyrenees); Renner & Vitzthum, 2007 (Lorraine); Rolliant, 1934; Saint-Girons, 1952b (vipers); Thiriet & Vacher, 2010 (Alsace); Thirion et al., 2002; Vacher & Geniez, 2010

GEORGIA: Bannikov et al., 1977; Eichwald, 1842; Muskhelishvili, 1970a–b; Tuniyev, 1995

GERMANY: Bitz & Veith, 1992 (Rhineland-Pfalz); Blauscheck, 1985, 1989; Bühler, 1981; Durigen, 1897; Feldmann, 1981; Franke, 1881; Fröhlich et al., 1987; Frommhold, 1952; R. Günther, 1996; Hölzinger & Schmid, 1987 (Baden-Württemberg); Mertens, 1947a, 1968c; Mey & Schmidt, 2002 (Thüringen); Quetz, 2003 (Stuttgart); Schiemenz & Günther, 1994

GIBRALTAR: Cortés, 1982; Mateo et al., 2003

GREECE: Bedriaga, 1881–82; Bird, 1935; Boettger, 1888a; Buchholz, 1955; Buchholz & Schultze-Westrum, 1964; Buttle, 1986, 1989, 1995, 1997; Cattaneo, 1989 (Serfos, Sifnos & Mitos), 1990; Chondropoulos, 1989; R.J. Clark, 1967b, 1989; Cyrén, 1941; Demetropoulos & Ioannides, 2002; Grillitsch & Tiedemann, 1984; Keymar, 1986; Mertens, 1961a; Ondrias, 1968; Paphiles & Valakos, 2012; Rödel, 1994; B. Schneider, 1975, 1983, 1986; Stepánaek, 1934, 1944; Trapp, 2007 (mainland); Valakos & Papapanagiotou, 1985; E. Valakos et al., 2008; S. Valakos et al., 2004; F. Werner, 1930a & 1935 (both Aegean Is.), 1935; Wettstein, 1953a–b, 1957

HUNGARY: Benke, 2000; Dely, 1978, 1981, 1987; Fejérváry-Langh, 1943; Marián, 1988; Méhely, 1991

ITALY: Abram & Menegon, 1994; Andreone & Sindaco, 1998 (Posta Valley); Barbieri et al., 1996, 2000 (Vicenza), 2004; Bernini et al., 2004 (Lombardia); Betta, 1874; Bologna et al., 2000 (Lazio); Bruno, 1970 (Sicily), 1977 (Monte della Tolfa), 1980 (Veneto), 1984, 1985 (vipers), 1988a (Cres, Krk & Ada Is.), 1988b (Sicily), 1998; Bruno & Cesare, 1991 (SE Peligna); Bruno & Magueri, 1977, 1990; Camerano, 1889 (vipers), 1891 (colubrids); Capoccacia, 1958, 1959, 1968; Corsetti & Capula, 1992 (Lepini Mts.); Corti et al., 1997 (Sicilian Is.); Gherardini, 1988; Ghidini, 1958; Kasapidis et al., 1996 (Lesbos); Lanza, 1968, 1983b-c, 1987, 1996; Lapini, 1983; Lo Valvo & Longo, 2001 (Sicily); Luiselli & Rugiero, 1990 (Tolfa Mts.); Marchi, 1901; Mariani, 1988; Mazzotti & Stagni, 1993 (Emilia-Romagna); Menegon, 1996; Morisi, 1983 (Cuneo); Sindaco et al., 2006; Stemmler, 1959 (Sardinia); Tedaldi, 1998 (Emilio-Romagna); Tortonese, 1942, 1953; Tortonese & Lanza, 1968; Vandoni, 1914

LATVIA: Ananjeva et al., 2004; Bannikov et al., 1977

LITHUANIA: Ananjeva et al., 2004; Bannikov et al., 1977

LUXEMBURG: Fontaine, 1870; G. Hecht, 1930b; Sparreboom, 1981; Vacher & Geniez, 2010

MACEDONIA: Petkovski et al., 2000

MALTA: Mertens, 1921

THE NETHERLANDS: Bergmans & Zuiderwijk, 1980, 1986; Bund, 1964; Creemers & Delft, 2009; Floericke, 1920; Sparreboom, 1981; Van Kampen & Heimans, 1927

NORWAY: Ahlén et al., 1992; Dolmen, 1986; Gislén & Kauri, 1959

POLAND: Berger, 1975, 2000; Berger et al., 1969; Gillett, 1990; Glowacinski & Rafinski, 2003; Herczek & Gorczyca, 1999; Lanka & Vit, 1993; Mlynarski, 1961, 1966, 1971

PORTUGAL: Andrada, 1985; Aragón-Rebollo et al., 2006 (Iberica); Barbadillo-Escriva & Martínez de Castilla, 1987; Barbadillo-Escriva et al., 1999; Bethencourt-Ferreira, 1892; Crespo, 1972, 1973, 1975; Crespo & Oliviera, 1989; Crespo & Sampaio, 1994; Ferreira & Seabra, 1911; Malkmus, 1979a–b, 1982, 1983, 1987, 1991, 1993b–c, 1995b, 2004, 2005; Martínez-Rica, 1989; Olivier, 1896; Pleguezuelos, 1997; Salvador, 1985, 1998 (Iberica); Thermido, 1942

ROMANIA: Fuhn, 1969; Fuhn & Vancea, 1961; Vasiliu & Sova, 1968

SLOVENIA (Yugoslavia): Krofel et al., 2009; Mrsic, 1997; Tome, 2002; Vogrin, 1997

SPAIN: Alberdi et al., 1981 (Murcia); Andrada, 1980, 1985; Aragón-Rebollo et al., 2006 (Iberica); Barbadillo-Escriva & Martínez de Castilla, 1987; Barbadillo-Escriva et al., 1999; Busack, 1977; Buttle, 1986; Ferri, 1994; Gabaldón, 1997 (Valencia); Galán-Regalado & Fernández-Arias, 1993 (Galicia); García-París et al., 1989 (Madrid); González de la Vega, 1989 (Huelva); Lizana et al., 1995; Llorente et al., 1995 (Andorra, Catalunya); Maluquer, 1917; Martínez-Rica, 1989; Massip-Gibert, 1993 (Catalunya); Mayol-Serra, 2003 (Balears); Mertens, 1925; Monzó-Giménez, 2001, 2002 (Pinoso), 2002; F.P. Pérez, 2001 (Bizkaia); Pinta & Carretero, 2011 (Balearic Is.); Pleguezuelos, 1997; Pleguezuelos & Fericha, 2003; Salvador, 1974, 1985, 1998 (Iberica); Salvador & Pleguezuelos, 2002; Silva-Rubio, 1995; Vega, 1989 (Huelva); Vives-Balmaña, 1984 (Catalunya)

SWEDEN: Curry-Lindahl, 1975; Gislén & Kauri, 1959

SWITZERLAND: Dossenbach & Ganz, 1978; Fatio, 1872; Grossenbachner & Brand, 1973; Hofer, 1991 (Kantons Bern); Hofer et al., 2001; Hotz & Broggi, 1982; Meier & Schnieper, 1986; A. Meyer, 2001; Stemmler, 1967; Stemmler et al., 1971; Vacher & Geniez, 2010

UKRAINE (Crimea): Bannikov et al., 1977; Kurilenko & Verves, 1999; Szczerbak, 1966; Tarashchuk, 1959

UNITED KINGDOM: E.N. Arnold, 1995; E.N. Arnold & Burton, 1978; H.R. Arnold, 1973; Beebee & Griffiths, 2000; Bell, 1838; Bernhard, 1999 (Hampshire & Isle of Wight); E.A. Ellis, 1979 (Great Britain); Griffiths, 2000; Clemons, 1990; Cortes, 1982; D. Frazer, 1983, 1989; J.F.D. Frazer, 1949; Halfpenny, 1978 (Staffordshire); Hand et al., 2006 (Herefordshire); Inns,

2009; Leighton, 1901; M.A. Smith, 1954; Smout, 1995; Smout & Pritchard, 1995 (Fife); R. Taylor, 1948, 1963; Tonge, 1986; Wareham, 2008 (Dorset); Wycherley & Anstis, 2001 (Surrey)

YUGOSLAVIA (Albania, Balkans, Bosnia & Herzegovina, Croatia, Kosovo, Macedonia, Montenegro, Serbia, Slovenia, Vojvodina): Brelih & Dzukic, 1974; Karaman, 1939; Pozzi, 1966; Radovanovic, 1951, 1964; Rödel & Bussmann, 1992; Sklenár & Rocek, 1979; Wettstein, 1928

Southwest Asia

GENERAL: David & Vogel, 2010b (elapids & vipers); Dept. of Defense, 1991; Egan, 2007 (Arabia); Gumprecht et al., 2004 (pitvipers); Leviton et al., 1992; K.P. Schmidt, 1939b; Vogel, 2006 (venomous); Welch, 1983b

AFGHANISTAN: S.C. Anderson & Leviton, 1969; Boulenger, 1889d; Brück, 1968; R.J. Clark, 1990; R.J. Clark et al., 1969; Král, 1969; Leviton, 1959a; Leviton & Anderson, 1963, 1970a; Murray, 1892; M.A. Smith, 1940a

BAHRAIN: Gallagher, 1971; Loveridge, 1955a

IRAN: S.C. Anderson, 1963; Andreas & Werner, 1917; Blanford, 1876; Firouz, 2000, 2005; Forcart, 1950; Latifi, 1985, 2000; Latifi et al., 1991; Mertens, 1940a; Nikolsky, 1907; Nilson & Andren, 1981b; Rai, 1965 (colubrids); Schleich, 1977; K.P. Schmidt, 1955b, Tuck, 1971, 1974a–b; Volsoe, 1939; Wall, 1908i, F. Werner, 1916b (Fars)

IRAQ: Boulenger, 1920g; Corkill, 1932a–b; Khalaf, 1959; Leviton & Anderson, 1967; Mahdi & Georg, 1969; C.A. Reed & Marx, 1959

ISRAEL: Bar & Haimovitch, 2011; Barash & Hoofien, 1956, 1961; Berger-Dell'mour, 1986; Boettger, 1880b; Bouskila & Amitai, 2001; Dawydow, 1898; G. Haas, 1951, 1952b; Hoofien, 1972; Kochva, 1974, 1990; Mendelssohn, 1963, 1965; Sivan & Werner, 1992; Theodor, 1955; Tristram, 1884; Warburg, 1964; Y.L. Werner, 1965, 1995

JORDAN (Transjordan): Al-Quran, 2009; Amr & Disi, 2011; Amr & Al-Oran, 1995; Amr & Amr, 1983; Amr et al., 1994; Disi, 1983, 1985, 1987, 1990; Disi et al., 1988, 2001; El Oran et al., 1994; El Shafië, 1995; Sindaco et al., 1989; Tarawneh et al., 1999; Y. Werner, 1971

KUWAIT: Eissa & El-Assy, 1975

LEBANON: Hraoui-Bloquet, 1981; Hraoui-Bloquet et al., 2002; Ilani, 1983a; L. Müller & Wettstein, 1933; Zinner, 1967

OMAN: Arnold, 1980b; Arnold & Gallagher, 1977; Gallagher, 1990; Gallagher & Arnold, 1988; A.S. Gardner, 1999, 2009, 2013; Wilms & Hulbert, 2000

SAUDIA ARABIA: Al-Sadoon et al., 1991; J. Anderson, 1896; Corkill & Cochrane, 1965; Cunningham, 2010; Egan, 2007; Gasperetti, 1974, 1977a, 1988; G. Haas, 1957, 1961; Schätti & Gasperetti, 1994

SYRIA: T. Barbour, 1914b; Boettger, 1880b; Disi & Böhme, 1996; Lortet, 1883; Martens, 1997; F. Werner, 1939

TURKEY (Asia Minor): Baran, 1976, 2005; Baran & Atatür, 1997; Basoglu & Baran, 1980; Bird, 1936; Bodenheimer, 1944; Budak et al., 2000 (Datca); R.J. Clark & Clark, 1973; Daszak & Cawthraw, 1991; Eiselt, 1970, 1976; Eiselt & Baran, 1970; Mertens, 1952d; Nelson & Flärdh, 1988 (vipers); Sindaco et al., 2000 (Anatolia); Van der Winden et al., 1997 (Gokan Delta)

UNITED ARAB EMIRATES (UAE): Gasperetti et al., 1964, 1967; A.S. Gardner, 2009, 2013; Jongbloed & Brown, 2000; Soorae et al., 2009

YEMEN (Aden, South Yemen, Yemen Arab Republic, Socotra): Abadi, 1998; Cherchi & Spano, 1966; Corkill & Cochrane, 1965 (Socotra); Gasperetti, 1988; Obadi, 1997; Nagy et al., 2003 (Socotra); H.W. Parker, 1941; Razzetti et al., 2011; Schätti & Desvoignes, 1999; Schätti & Gasperetti, 1994; Scortecci, 1932a; K.P. Schmidt, 1953d; Steindachner, 1903a

Central & Eastern Asia

GENERAL: Bannikov et al., 1977; David & Vogel, 2010b (elapids & vipers); Gumprecht et al., 2004 (pitvipers); L.W. Swan, 1947; Szczerbak, 2003; Vogel, 2006 (venomous)

CHINA: Anonymous, 1977, 1979; Blanford, 1878b (SW Xinjiang); Boettger, 1888c; Boring et al., 1932 (Anhui, Gansu, Hebei, Henan, Jiangxi, Shaanxi, Shandong, Shanxi), 1932b (Fujian); Chang, 1932, 1933 (both Sichulan), 1936 (Jiangsu), 1947 (Guangxi); Chang & Fang, 1931; B. Chen, 1991 (Anhui); Deng et al., 1983 (Sichuan); Deng & Zheng, 1974 (Fujian); Fan, 1931 (Guangxi, Jiangxi); Fan, 1998 (Shanxi); Feng, 1981 (Gansu); Z.F. Gao, 1984 (Sichuan); Gee, 1930; Glass, 1946; Gressitt, 1941a (Guangdong, Hainan, Zhejiang), 1941b (Fujian, Jiangxi); Gu et al., 1982 (Zhejiang); B.-C. Hu et al., 1958, 1959 (both Zhejiang); S.-Q. Hu, 1966 (Sichuan), 1980; M.-H. Huang et al., 1990 (Zhejiang); Z.-Y. Huang, 1998; Inger et al., 1990 (Sichuan); Ji et al., 1987 (Liaoning); Jiang & Hu, 1979 (Sichuan); Kudrjavtsev & Mamet, 1989; Lazell, 1988, 2002 (Shek Kwu Chou); B.-H. Li, 1987 (Jiuhua Mt.); D.-J. Li, 1989 (Guizhou); D.J. Li et al., 1985 (Guizhan), H.-F. Li, 1986 (Tianjin); P.P. Li et al., 2010 (Tibet); Z.-C. Li, 2009 (Dinghushan, Guangdong), 2011 (Guangdong); Lui & Hu, 1962 (Kwangsi); Maslin, 1950 (Jiangsu); Matsui & Ota, 1995; Mell, 1922, 1929, 1931a–b; Okada, 1935 (Hopeh, Inner Mongolia, Liaoning); Pavloff, 1926 (Heilongjiang, Liaoning); Pope, 1929 (Fujian), 1935; Reeves, 1933 (NE China); Schammakov et al., 1993; K.P. Schmidt, 1927a–b (Hainan); Shi, 2011 (Hainan); Shuntov, 1966 (E Indian Ocean & South Sea); M. Smith, 1923a (Hainan); M.-T. Song, 1987a–b (Shaanxi); Song & Fang, 1982 (Shaanxi); Sowerby, 1930 (Liaoning), 1943 (Zhejiang); Stanley, 1915; Stejneger, 1925; Sung et al., 2002; Tchang, 1932;

Tian et al., 1986; Wall, 1903; Wen, 1983 (Guangxi); Wettstein, 1938 (Tibet); Wu et al., 1985, 2009 (Hebei); Xie et al., 1998; Xu, 2001 (Inner Mongolia); Yang & Inger, 1986; D.-T. Yang et al., 1983 (Yunnan); Yuan, 1983c (Yunnan), 1985 (both Shaanxi), 2008 (Yunnan); Y.-X. Zhang, 2009 (Guangxi); Zhao, 1990b, 1998, 2003 (Sichuan), 2006, 2008 (Heilongjiang); Zhao & Adler, 1993; Zhao & Huang, 1982 (Liaoning); Zhao & Jiang, 1979 (N Xinjiang), 1986; Zhao & Li, 1985, 1987 (both are on Mt. Namjagbarwa, Tibet); Zhao & Yang, 1997 (Hengduan Mts.); Zhao & Zhao, 1994; Zhao et al., 1986 (Sichuan, Xizang Autonomous Region), 1998, 2000; Zhong, 1983, 1985, 1990 (all Jiangxi); Zou & Chen, 1998 (Jiangsu), Zhong & Wu, 1981 (Jiangxi)

HONG KONG (China): Bogadek & Lau, 1997; Chan et al., 2006; Herklotts, 1933–1938; Karsen et al., 1986, 1998; J.D. Romer, 1959, 1961, 1965, 1970, 1972, 1974b, 1979a, 1983

JAPAN (Loo Choo Is.; Ryukyus): Fukada, 1992; Gans, 1949; Goris, 1966; Goris & Maeda, 2004; Hikida et al., 1992 (Ryukyus); Hilgendorf, 1876; Koba, 1955–1973, 1974 (venomous, Ryukyus); Maki, 1931a–b, 1933; T. Matsui, 1985; M. Mori, 1982, 1984, 1986; Nakamura & Uéno, 1963; Okada, 1938a–b; Oshima, 1920; Ota, 1986 (Ryukyus); Sengoku, 1979; Shibata, 1960, 1981; Stejneger, 1907a; Takahashi, 1922, 1930 (venomous); Takara, 1962 (Ryukyus); Toriba & Sawai, 1990; Toriba et al., 1983; Uchiyama et al., 2002; Utsonomiya et al., 1996 (Hiroshima); Wall, 1903, 1905f

KAZAKHSTAN: Ananjeva et al., 2004; Paraskev, 1956

KYRGYSTAN: Ananjeva et al., 2004; Yakovleva, 1961, 1964

MACAO: L. Barros, 1978; Easton & Leung, 1993

MONGOLIA: Ananjeva et al., 1997; Borkin et al., 1990a–b; Emelianov, 1929 (N Mongolia); T. Mori, 1928; Obst, 1963; Strauch, 1876; Tzarewsky, 1926

NORTH KOREA (Democratic People's Republic of Korea): San & Lee, 2007; Shannon, 1956; G.D. Stewart, 1954; Szyndlar, 1984b; Szyndlar & Hung, 1987; Toriba, 1990b

RUSSIA (USSR): Alenizin, 1876; Ananjeva et al., 1998, 2004; Anufriev & Bobretsov, 1996 (NW Russia); Bannikov et al., 1977; Bassarukin, 1983; Bedriaga, 1912; Bobrov & Aleshchenko, 2001; Boettger, 1888f; 1965; Bol'shakov & Vershinin, 2005 (Urals); Borkin & Darevsky, 1987; Borkin et al., 1981, 1997 (Moldavia); Chernov, 1939; Darevsky, 1987; Dunaev & Orlova, 2003; Eichwald, 1842 (Caspio-Caucasia); Emelianov, 1929 (Far Eastern); Gardin, 1983 (Volga-Kana); Ivanter & Korosov, 2002 (Karelia); Kashchenko, 1902 (Tomsk); Kharin, 2008b (sea snakes); Klimov et al., 1999 (Lipetsk); Kudrjavstev & Mamet, 1989a, 1998 (vipers); Kuznetsov et al., 2000 (Moldavia); Mikhailovsky, 1904 (Transcapia); Nikolsky, 1905, 1913, 1916; Orlova & Semenov, 1999; Pavlov & Zamaletdinov, 2002 (Tatarstan); Pestov, 1999 (Nizhny Novgorod); Pestov et al., 2001 (Nizhny Novgorod); Rustanov & Szczerbak, 1986; Shkatulova, 1978 (Transbaikalia); Strauch, 1873; Szczerbak, 2003; Terent'ev & Chernov, 1949; Yushkov & Voronev, 1994 (Perm)

SOUTH KOREA (Republic of Korea): Dixon, 1956; Hahn, 1960; San & Lee, 2007; Shannon, 1956; Toriba, 1990b; Webb et al., 1962

TADZHIKISTAN: Chernov, 1959; Said-Aliyev, 1979

TAIWAN (Formosa): J. Chen & Yu, 1984; Du, 2004; Fet & Atamuradov, 1994; Hatori, 1905; Horikawa, 1941; Kuntz, 1963; T.-H. Lee, 2005; Lue, 1987, 1990; Lue et al., 1984 (Kenting), 1989, 1999; Maki, 1931a–b; Mao, 1993, Mao & Chen, 1980; Oshima, 1914, 1916b (both sea snakes); Ota, 1991; Steindachner, 1913a–b; C.-S. Wang, 1962; C.-S. Wang & Wang, 1956; Xiang, 2007 (Shei-Pa Natl. Pk.); Xiang & Li, 2009; Yu, 1996; Zhang, 1987

TURKMENISTAN: Ataev et al., 1994; Blanford, 1875; Bogdanov, 1962; Boulenger, 1891g; Fet & Atamuradov, 1994; Nikolsky, 1899b, 1908; Shammakov, 1981 (lowlands); Shammakov & Ataev, 1971; Shammakov et al., 1993; Szczerbak, 1994; Tuniyev et al., 1998, 1999

UZBEKISTAN: Ananjeva et al., 2004

South & Southeast Asia

GENERAL: Bain & Hurley, 2011 (Indochina); Bourret, 1935e, 1936a–b; H.W. Campbell & Christman, 1972; I. Das, 1996, 2002b, 2010, 2012; Koba, 1973; Gumprecht et al., 2004 (pitvipers); Leviton, 1963b; B.L. Lim, 1971 (venomous); K.F.L. Lim & Lee, 1989; Mahendra, 1984; Murray, 1884, 1886; Oshima, 1944; Reitinger & Lee, 1978; M.A. Smith, 1943; L.W. Swan, 1947; Vogel, 2006 (venomous); Wall, 1923c–d, 1924a–b, 1925a, 1928; Werler & Keegen, 1963

ANDAMAN & NICOBAR ISLANDS (India): Biswas & Sanyal, 1980; I. Das, 1994; M.A. Smith, 1941; Tikader & Das, 1985

BANGLADESH (East Bengal, East Pakistan): Ashan et al., 2009; M.A.R. Khan, 1988, 1992; Mahony et al., 2009; Montaquim et al., 1980

BHUTAN: Bauer & Günther, 1992; Biswas, 1975; I. Das & Palden, 2000; Wangyal, 2011

CAMBODIA (Indochina, Kampuchea): Bain & Hurley, 2011, Bezuijen et al., 2009 (Mekong Valley); I. Das, 2010, 2012; Grismer et al., 2008a; Saint-Girons, 1971b, 1972a; Saint-Girons & Pfeffer, 1972; B.L. Stuart et al., 2006 (E Cambodia); Tirant, 1885a, 1904

INDIA: Acharji, 1951 (Manipur), 1958 (Bihar); Acharji & Kripalani, 1951 (Assam); Acharji & Mukherjee, 1966; Aengals, 2009 (Tamil Nadu); Agarwal et al., 2010 (Arunachal Pradesh); Agrawal, 1970 (Madhya Pradesh); S. Ahmed & Dasgupta, 1991 (Bihar), 1992 (West Bengal); M.F. Ahmed et al., 2009 (Assam); Annandale, 1915, 1921 (both Chilka Lake); Behura, 1966 (Orissa); Bahuguna, 2010 (Uttarakhand); Bethencourt-Ferreria, 1897; Biswas & Sanyal, 1977 (Rajasthan), 1980 (Andaman & Nicobars Is.); Blanford, 1870; Cazaly, 1914; Constable, 1949; J.C. Daniel, 1983; J.C. Daniel & Shull, 1963 (Gujarat); I. Das,

1994 (Andaman Is.), 1996, 2002, 2003; Dasgupta & Raha, 2004 (Bihar), 2006 (Nagaland); Deoras, 1965, 1978, 1981, 1990; Dutta & Acharjyo, 1990 (Orissa); Dutta et al., 2009 (Orissa); Ewart, 1878 (venomous); Fayrer, 1872 (venomous); Ferguson, 1895 (Kerala), 1902; A.G.L. Fraser, 1936, 1937a–b; Gharpurey, 1962; A. Günther, 1864a; Hora & Jayaram, 1949; Husain & Tilak, 1995 (Uttarakhand); Hutton, 1949 (Tamil Nadu); Inglis et al., 1920 (Bengal); Jha & Thapa, 2002 (Sikkim); Jayaram, 1949, 1974; N. Khaire, 1996, 2006 (Goa, Karnataka, Maharashtra), 2010; Kumar et al., 2010 (Rajasthan); Lamba, 1987 (Uttar Pradesh); Lindberg, 1932, 1939 (both Deccan); Mahajan & Agrawal, 1976 (Himachal Pradesh); Mahendra, 1984; Malhotra & Davis, 1991; Mathew, 1983; Mathew, 1992 (Meghalaya), 1995 (Meghalaya), 2007 (Mizoram); McCann, 1938; Murthy, 1972–1986, 1987 & 1988 (Orissa), 1990a–d, 1992a (Karnutalca), 1995 (Tamil Nadu), 2001 (Kerala), 2007 (marine reptiles), 2009; Murthy & Rao, 1988 (Orissa); Murthy et al., 1979 (Jammu, Kashmir), 1993; Nande & Deshmukh, 2007 (Maharashtra); Negi, 1992; Nicholson, 1874; Pradhan, 2004, 2005, 2006 (Maharashtra), 2008 (Goa); Prater, 1926; Rajendran, 1986; Roux, 1928; Sahi & Duda, 1986 (Jammu & Kashmir); Sanyal, 1993 (Andhra Pradesh, Orissa); Sanyal & Gayen, 1985 (Aranchal Pradesh); Sanyal et al., 1993 (Andhra Pradesh), 2002 (Tripura), 2006 (Sikkim); Sclater, 1891a–b; V.M. Satishkumar, 2007 (Tamil Nadu); B. Sharma, 1976a (Goa, Jammu, Kashmir), 1998; R.C. Sharma, 1974 (Rajasthan), 1976a, 1977 (Rajasthan), 1982 (Gujarat), 2000 (Gujarat), 2003, 2007; G.E. Shaw & Shebbeare, 1928–1931 (Bengal, Sikkim); G.E. Shaw et al., 1928–1931, 1938–1947, 1999; Sights, 1949 (Bengal); Singh & Pradhan, 1992 (Maharashtra); M.A. Smith, 1940d, 1941 (Andamans), 1943; Srinivasulu & Das, 2008 (Eastern Ghats); Sundersingh, 1960; Sur et al., 2007 (Madhya Pradesh); Thermido, 1941 (Goa); Theobald, 1868a, 1876; Tikader & Das, 1985 (Andaman Is.); K.K. Tiwari & Sharma, 1970 (Maharashtra), 1971; R. Vyas, 1993, 1996, 1998, 2008 (all Gujarat); Wall, 1905–1919, 1923–1925; Whitaker, 1978a, 1978b (Andamar Is.), 1982 (Tamil Nadu), 1999 (Kodagu); Whitaker & Captain, 2004; Whitaker & Whitaker, 1986

LAOS (Indochina): Bain & Hurley, 2011; I. Das, 2010, 2012; Deuve, 1961b, 1962a–c, 1970; B.L. Stuart, 1999

MALDIVE & LACCADIVE ISLANDS: Laidlaw, 1902

MYANMAR (Burma, Cocos Is.): Aye, 1990; Boulenger, 1893g; Cazaly, 1914; I. Das, 2010, 2012; Dowling & Jenner, 1988; Leviton et al., 2003 (venomous); F. Mason, 1852; Rendahl, 1937; M.A. Smith, 1940c; Theobald, 1880; Wall, 1925d, 1926b; Wall & Evans, 1900, 1901; Wogan et al., 2008

NEPAL: Fleming & Fleming, 1973; A. Günther, 1860d, 1861h (Himalayan Mts.); Kramer, 1977; Majupuria, 1981; Mrsíc, 1980; Nanhoe & Ouboter, 1987; Negi, 1992; O'Shea, 1998; Schleich & Kästle, 2002; Schleich & Maskey, 1992; S.K. Sharma et al., 2013 (venomous); Shaw & Tiwari, 2004; T.K. Shrestha, 2001; R. Shrestha & Shah, 1984; L.W. Swan & Leviton, 1962; Wall, 1907f; Zug & Mitchell, 1995

PAKISTAN (West Pakistan, Punjab, Sindh): Akram & Qureshi, 1995; Annandale, 1906; Baig et al., 2008 (Cholistan Desert); Ingoldby & Procter, 1923; M.S. Khan, 1977, 1982a, 1983, 1985b, 1986, 1987, 1993, 1994, 2002, 2006; M.S. Khan & Baig, 1988; Mahendra, 1984; Masroor, 2012 (N Pakistan); Mertens, 1969a: Minton, 1962, 1966; Murray, 1886; Siddiqui, 1969

SRI LANKA (Ceylon): Abercromby, 1910, 1913; I. Das & A. Silva, 2005; Deraniyagala, 1955; W. Ferguson, 1877; Fletcher, 1908; Haly, 1891; Holtzinger-Tenever, 1917; Jayawickrama, 1994; Karunarathna & Amarasinghe, 2011 (Nilgala Forest); Kelaart, 1852; Mahendra, 1984; Méhely, 1897; Nicholls, 1929a; Pyron et al., 2013; Sarasin, 1910; A. Silva, 1990a–b, 1994, 1996, 1998, 2001, 2004, 2009; N.S. Silva, 2000; P.H.D.H. Silva, 1980a–b; Somaweera, 2006; Somaweera & Somaweera, 2009 (sea snakes); E.H. Taylor, 1950b, 1953b; Wall, 1905g, 1921e, 1923h, 1924d; Willey, 1903b, 1906

THAILAND (Siam): Baiget et al., 2008 (Cholistan desert); Bussarawit et al., 1989 (sea snakes of Phuket Is.); Chanard et al., 1999; Cochran, 1930; M.J. Cox, 1991b; M.J. Cox et al., 1998, 2013; I. Das, 2010, 2012; Flower, 1899; C.B. Frith, 1977a, 1978 (Phuket Is.); Grismer et al.. 2008 (Cardamom region); A. Günther, 1860b, 1860d, 1861h; Gyldenstolpe, 1916; Jayasuriya, 2001; Jintakune & Chanhome, 1995; Nabhitabhata, 2004; Nabhitabhata & Chan-ard, 2005; Nabhitabhata et al., 2004; Nutaphand, 1971, 2001; Nutaphand & Tumvipart, 1982; Pauwels et al., 2002 (Phlang-nga); Purananda, 1957; M.A. Smith, 1914b–c, 1915b–c, e, 1923b; Soderberg, 1967a–b, 1973; Suvatti, 1950; E.H. Taylor, 1965; E.H. Taylor & Elbel, 1958; Tu, 1974

VIETNAM (Indochina, North Vietnam, South Vietnam): Bain & Hurley, 2011; Barme, 1968; Bourret, 1927, 1935c, 1936a–b, 1938, 1939d; Campden-Main, 1970a; Darevsky, 1990; I. Das, 2010, 2012; Dowling, 1966; Kharin, 2006 (sea snakes); T. Kien & Nguyen, 1995 (venomous); Koukoulis, 1967; Mocquard, 1904b; V.S. Nguyen, 2007; V.S. Nguyen & Ho, 1996, 2005; V.S. Nguyen et al., 2009; Orlov et al., 2000, 2010 (Fansipan Mts.); Ota & Weidenhöfer, 1990; H.W. Parker, 1925a; Saint-Girons, 1972b; Szyndlar & Nguyen, 1996; Tien, 1957 (Quang-Tri), 1960; Tran & Nguyen, 1995; Viravan et al., 1992; Vo & Nguyen, 1993 (venomous); Ziegler, 2002; Ziegler et al., 2006a–b (Truong Son, Quang Binh); Ziegler & Nguyen, 2010

East Indies

GENERAL: T. Barbour, 1912b; Brongersma, 1934; Burt & Burt, 1932c; I. Das, 2002b; C.J.P. Haas, 1950; Leviton, 1968b; Loveridge, 1945b; Mertens, 1930; Oshima, 1944; W.C.H. Peters & Doria, 1878; Rooij, 1917; Roux, 1913a; Steindachner, 1867a; Sternfeld, 1914, 1920c; Werler & Keegen, 1963

ARU & KEI ISLANDS: Roux, 1910b

BRUNEI (Negara Brunei Darussalam): I. Das, 2007b; I. Das et al., 2008b; Stuebing & Inger, 1999

EAST MALAYSIA (Sabah, Sarawak, North Borneo): E. Bartlett, 1895a–b, 1896a–b; I. Das, 2006c, 2010, 2012; I. Das & Hee, 2008 (Pulau Banggi Is.); Grismer, 2011 (Seribuat Arch.); Inger & Tan Fui Lian, 1996; Malkmus, 1987, 1989, 1992; Malkmus et al., 2002 (Mt. Kinabalu, Sabah); J.C. Murphy et al., 1994; M.A. Smith, 1931b; Stuebing, 1991, 1994a; Stuebing & Inger, 1999

INDONESIA (Netherland/Dutch Indies): Boulenger, 1885 (Nias), 1894c (Mentawei), 1897d; Brongersma, 1931, 1933b (Aklu-Weh), 1933d (Weh), 1933f (Wetar); I. Das, 1998b, 2006a (Borneo), 2010, 2012; Dring et al., 1990 (Mentawei); Ferreira, 1898 (Timor); Focart, 1953 (Sundas); Inger & Voris, 2001 (Sundaland); Iskandar & Coliljn, 2001; Jong, 1926 (Burkana), 1927 (Dutch New Guinea), 1928, 1930a; Kaiser et al., 2011 (Timor-Leste); Lang, 2011a–b (Lesser Sundas); Manthey & Grossmann, 1997; Maradjo, 1977; McKay, 2006 (Bali); Mertens, 1927b, 1930, 1957a (all Sundas); Roux, 1913b, 1931; Supriatna, 1981 (venomous); Westermann, 1942

JAVA: Brongersma, 1929, 1930; Capoccacia, 1976; I. Das, 2010, 2012; C.J.P. Haas, 1941; Hodges, 1993; Hoesel, 1959; Kopstein, 1938; Mertens, 1957a; Rooij, 1922; Whitten & McCarthy, 1993

KALIMANTAN (Borneo, Indonesian Borneo): Auliya, 2006; E. Bartlett, 1895a–b, 1896a–b; A.E. Brown, 1902a; I. Das, 2002b, 2006a, 2010, 2012; Haile, 1959; Malkmus, 1987, 1989; Mocquard, 1890a–b; Shelford, 1901b; M. Smith, 1925a–b; Stuebing, 1991, 1994a; Stuebing & Inger, 1999

KOMODO ISLAND: Auffenberg, 1980; Darevsky, 1964

NATUNA ISLAND: A. Günther, 1895b

PAPUA NEW GUINEA (Dutch New Guinea): Brongersma, 1956d; Hediger, 1934; Jong, 1927; Lidth de Jeude, 1911a–b; Loveridge, 1948; McDowell, 1974a, 1975a–b, 1979, 1984; Méhely, 1895, 1898; O'Shea, 1990, 1996; F. Parker, 1983, H.W. Parker, 1936a; Peters & Doria, 1878; Room, 1974; Roux, 1919c; Scott et al., 1977; K.R. Slater, 1968; F. Werner, 1900c; Whitaker et al., 1982

PALAU ISLANDS (Belau): Crombie & Pregill, 1999

PHILIPPINE ISLANDS: Alcala, 1986; Boettger, 1886c; W.C. Brown & Alcala, 1964 (S Negros), 1970, 1986; R.M. Brown et al., 1996, 2000 (Luzon); Casto de Elera, 1895; Devan-Song & Brown, 2012 (Luzon); Ferner et al., 2001 (Panay); Gaulke, 1996 (Sibutu), 2011 (Panay); L.E. Griffin, 1909b (Palawan), 1910 (Palilb), 1911; Grismer, 1996; Leviton, 1959b, 1961b, 1963b; McLeod et al., 2011 (Luzon); Mocquard, 1890a–b; Rabor, 1981; Ross & Gonzales, 1992 (Catanduanes); Ross & Lazell, 1991 (Dinagat & Siargao); Siler et al., 2011 (Luzon); B.E. Smith, 1993 (E Mindanao); E.H. Taylor, 1918a, 1922–1925; Toriba, 1990c

SINGAPORE: Chan-ard et al., 1999; M.J. Cox et al., 1998; I. Das, 2010, 2012; K.K.P. Lim & Chou, 1990; Ridley, 1899; Sworder, 1922, 1924

SULAWESI (Celebes): Bosch, 1985; Boulenger, 1897g; Iskandar & Nio, 1996; Jong, 1926; Koch, 2012; Lang & Vogel, 2005; Malkmus, 1993a

SUMATRA: Baumann, 1913; Boulenger, 1920e; Brongersma & Wehlburg, 1933; Cohn, 1905; I. Das, 2010, 2012; David & Vogel, 1996; Edeling, 1870; Holtzinger-Tenever, 1917; Lange & Rooij, 1912; Lidth de Jeude, 1922; Lönnberg & Rendahl, 1925; Matsui et al., 1984; Robinson & Kloss, 1920; Volz, 1904; F. Werner, 1900b; Westermann, 1942

WEST MALAYSIA (Malay Peninsula, Penang): Batchelor, 1958; Boulenger, 1903f, 1912c; Cantor, 1847; Chan-ard et al., 1999; M.J. Cox et al., 1998; I. Das, 2010, 2012; Dring, 1979; Flower, 1896, 1899; Grandison, 1978; Grismer, 2005 (Tioman Arch.), 2011 (Seribuat Arch.), Grismer et al., 2002b, 2006 (Seribuat Arch.), 2010 (Banjaran Bintang); Grismer & Khang Aun, 2008; Heang, 1987; Hendrickson, 1966; Inger & Voris, 2001; Koba, 1971, 1973; Kopstein, 1938, 1941; B.-L. Lim, 1963, 1982, 1990; Manthey & Grossmann, 1997; Onn et al., 2010 (Palau Pangkor); Ridley, 1899; Robinson & Kloss, 1916, 1920; Smedley, 1928–1932; M.A. Smith, 1927, 1930a; Tweedie, 1940–1983

Australia & Pacific Islands

GENERAL: Kinghorn & Kellaway, 1943; Werler & Keegen, 1963

AUSTRALIA: Barker & Barker, 1994; Barrett, 1950; R. Bennett, 1997; Bush, 1981; Bush & Maryan, 2011 (the Pilbara); Bush et al., 1995, 2007; Cadle, 1985 (biogeography); Cameron & Cogger, 1992; Cermak, 2008; Cogger, 1975, 1975, 1981 (Arnhem Land), 1983, 1988, 1992, 2000; Cogger et al., 1983a–b, 1993; Covacevich, 1981, 1987; Covacevich et al., 1987; Covacevich & Couper, 1991; Coventry & Robertson, 1991; Edwards, 2009; Ehmann, 1992; Emmott & Wilson, 2009; C. Frith & Frith, 1987; Glauert, 1957; Gow, 1976, 1977b (Darwin area), 1983, 1989; Greer, 1997; Griffiths, 1987, 1996, 2012 (Sydney); Hines & Hoser, 1987; Hunt, 1947; Hutchinson, 1990; Ingram & Covacevich, 1981; Jenkins & Bartell, 1980 (high country); Kend, 1997 ("pythons"); Kinghorn, 1964; Krefft, 1866c (Sidney area); 1869b; Leggett, 1996; Longmore, 1986, 2002; Loveridge, 1934, 1938c; Low, 1978; Lunney & Ayers, 1993; F. McCoy, 1867 (Victoria); McDowell, 1985; Mengden, 1983; Mincham, 1970; Mirtschin & Davis, 1983; Mirtschin et al., 1990; F.J. Mitchell, 1951, 1955, 1964 (Arnhem Land) 1966; Sadlier, 1990 (Northern Territory); Schmida, 1985; Schwaner, 1985 (So. Australia); Shine, 1991; Shuntov, 1971 (sea snakes of N Australian shelf); P. Slater, 1997; L.A. Smith, 1976 (Barrow Is.); Storr & Hanlon, 1980; Storr & Harold, 1978, 1980, 1985; Stanger et al., 1998; Storr, 1985 (West. Australia); Storr et al., 1981, 1983, 1986, 2002 (West. Australia); Sutherland, 1981; G. Swan, 1990, 1995, 2000, 2005 (Queensland); G. Swan & Watharow, 2005 (Victorian Mallee, 1966); G. Swan et al., 2004 (New South Wales); M. Swan, 2007 (pythons); Swanson, 2008, 2012; Torr, 2000 (pythons); Waite, 1898, 1918a, 1928; Weigel, 1990; Wells, 2007 (sea snakes); S.K. Wilson, 2005 (Queensland); S.K. Wilson & Swan, 2003; S.K. Wilson & Knowles, 1988; Worrell, 1953, 1961d, 1963c, 1967, 1970

CHRISTMAS ISLAND: Boulenger, 1887i, 1900i; Maclear, 1887; Cogger et al., 1983

COCOS ISLANDS (Keeling): Cogger et al., 1983b

EASTER ISLAND: Garman, 1908

FIJI: Morrison, 2003

NEW CALEDONIA: Bauer & Sadlier, 2000; Bauer & Vindum, 1990; Roux, 1913a

NEW ZEALAND: Gill, 1986; McCann, 1966; Robb, 1980; West, 1979

SAMOA: Gill, 1993

SOLOMON ISLANDS (Bougainville): Boulenger, 1886b, 1887g, 1888f; Kinghorn, 1928a; M. McCoy, 1980, 2006; McDowell, 1970; V. Tanner, 1951

South America

GENERAL: Bartlett & Bartlett, 2003 (Amazonia); J.A. Campbell & Lamar, 1989, 2004 (both venomous); Dobiey & Vogel, 2012 (venomous); Duellman, 1979; Freiberg, 1954, 1982; Jenner, 1981; J.C. Murphy, 1996; J.A. Peters & Orejas-Miranda, 1970; Tipton, 2005; Vanzolini, 1977, 1978, 1980, 1986a

ARGENTINA: Abalos & Mischis, 1975; Abalos et al., 1964; Acosta et al., 1994; Albino & Quintana, 1992 (Buenos Aires); Álvarez et al., 2009 (Chaco, Formosa); Avila et al., 1998; Berg, 1898; Cabrera, 2004 (central Argentina); Cei, 1986, 1993; Cei et al., 1978; Cruz et al., 1992; Freiberg, 1939, 1968, 1982; Gallardo, 1977a–b; Giraudo, 2001; Giraudo & Arzamendia, 1997a; Giraudo & Scrocchi, 2002; Kacoliris et al., 2006 (Chaco); Koslowsky, 1898b; Laurent & Teran, 1981; Lavilla & Scrocchi, 1991; Lavilla et al., 2000; Leynaud & Bucher, 1999; Marelli, 1924, 1931; Milstead, 1956; Miranda et al., 1983; Orrego-Aravena, 1971; Reati, 1996; Scolaro, 2005 (S Patagonia), 2006 (N Patagonia); Serié, 1915b, 1916, 1921, 1936; Tiranti & Avila, 1997; J.D. Williams & Francini, 1991; J.D. Williams & Scrocchi, 1994; Yanosky et al., 1993, 1997

BOLIVIA: Embert, 2007; Fugler, 1983, 1984, 1986; Fugler & Riva, 1990; Fugler et al., 1995; Gans, 1960; Kempff-Mercado, 1975; Lavilla et al., 1995 (Salta); Mijaria-Uriona, 2001 (Amazonia); Montero et al., 1995; Riva et al., 1991; Sandoval, 1991

BRAZIL: Amaral, 1925b (Mato Grosso), 1930e–f, 1934d, 1935d (Goias, Paraiba, Pernambuco, Rio Grande do Norte), 1937c–d, 1948 (Mato Grosso), 1949 (Pará), 1977; Araujo et al., 2010 (São Paulo); Argôlo, 2004 (Bahia); Avila-Pires, 2005 (Guiana shield); Barth, 1957 (Minas Gerais, Rio de Janeiro); Boulenger, 1885a, 1886a (both Rio Grande do Sul); Brazil, 1914; Carvallho et al., 2007 (Manaus area); Cicchi et al., 2007 (São Paulo); Cordeiro & Hoge, 1974 (Pernambuco); H.C. Costa et al., 2010b (Minas Gerais); Cunha & Nascimento, 1978 (Pará), 1980 (Roraima), 1993; Cunha et al., 1985 (Pará); Domingos Cintra et al., 2009 (Goiás); Duellman, 1990 (Amazonia); Forlani et al., 2010 (Parque Estadual Carlos Botelho, São Paulo); Freitas, 1999 (Bahia), 2003; Freitas & Pavie, 2002 (Bahia); Gomes, 1916, 1918 (Ceará, Pará); Guichenot, 1855; Hamdan & Lira-da-Silva, 2012; Hoge, 1953b–d (Amazonia, Mato Grosso), 1967 (Amapá); Hoge & Belluomini, 1960 (Pará); Hoge & Maranhao-Nina, 1964 (Amazonia); Hoge & Romano-Hoge, 1981b; Hoge et al., 1973 (Amazonia), 1975 (São Paulo), 1978b (Maranhão), 1981a (Piauí); Ihering, 1911; Jorge de Silva et al., 2009 (Mato Grosso, Mato Grosso do Sul); Koslowsky, 1898a (Mato Grosso); Lema, 1977a, 1994; Lema & Fabián-Beurmann, 1977 (Rio Grande do Sul), 1987; Lema, 1973b, 1980 (Rio Grande do Sul), 1983 (Pampeana, Limitropies), 1987 (Rio Grande do Sul); Lema et al., 1980 (Rio Grande do Sul); Lima-Verde, 1972 (Ceará, Rio Grande do Norte); Loebmann & Hadded, 2010 (Ceará); Machado, 1944, 1945a–c; Marques et al., 2001 (Mata Atlántica), 2004 (Sierra do Mar range), 2005 (Pantanal); Martins & Oliveira, 1998 (Manaus region); Mercadante, 2005; L. Müller, 1927 (Pernambuco); P. Müller, 1968a, 1968c (both São Paulo); Nascimento & Lima-Verde, 1989 (Ceará); Nascimento et al., 1987 (Pará), 1988 (Mato Grosso, Rondônia), 1991 (Pará); O'Shea, 1989 (Roraima); Prudente et al., 2010 (Coari, Amazonas); Recoder et al., 2011 (Tocantins); Ribeiro et al., 2012 (Ceará); Rocha & Pontes, 2008 (Rio de Janeiro); Santos, 1941, 1955, 1981; Salles et al., 2010 (Rio de Janeiro); F.M. Silva, 2011 (Barcarena area, Pará); M. Silva, 1956; Spix & Martius, 1823; Suzart-Argôlo, 2004 (Bahia); Thermido, 1945; Valdujo et al., 2009 (central Brazil); Vanzolini, 1948 (São Paulo), 1986b (Rondônia); Vanzolini et al., 1980; Vas-Silva et al., 2007 (Goiás); Vrcibradic et al. (Rio de Janeiro); Yuki & M. de Santos, 1996 (Pará); Zimmerman & Rodrigues, 1990

CHILE: Donoso-Barros, 1965, 1966, 1970; Gay, 1848; Gigoux, 1940; Nuñez, 1992; Quijada, 1916; Pflaumer, 1943; Troncoso & Ortiz, 1987; Veloso & Navarro, 1988

COLOMBIA: Aléman, 1953; Amaral, 1932b, 1935b; T. Barbour, 1905; Barriga–bonilla et al., 1969; H. Daniel, 1949, 1955; Dugand, 1975; Dunn, 1944f–h, 1945, 1957; García, 1896; Medem, 1965, 1969, 1979; Moreno-Arias et al., 2010 (Cundinamarca, Yacopí); Nicéforo Maria, 1930–1975; H.W. Parker, 1926b (Gorgona Is.); Peracca, 1912; Pérez-Santos, 1983, 1986; Pérez-Santos & Moreno, 1986, 1988, 1989a; Rendahl & Vestergren, 1940; Ruthven, 1922; Sanchez et al., 1995; Tinoco, 1976; Touzet, 1986

ECUADOR: Albuja et al., 1980; Almendáriz, 1987 (Pastaza), 1991; Duellman, 1978 (Napo), 1990; Freire, 1982; Fugler & Walls, 1978; Miyata, 1980, 1982 (both Esmeraldas, Los Ruis, Pichicha); Orcés, 1942, 1943, 1948 (venomous); Ortega-Andrade et al., 2010 (Esmeraldas Bilsa Biol. Station); H.W. Parker, 1938b; Pérez-Santos & Moreno, 1991; J.A. Peters, 1960a; Rendahl & Vestergren, 1941; Vigle, 2008 (Estación Biol. Jatun Jacha, Napo)

FRENCH GUIANA: Abuys, 2003; Ávila-Pires, 2005 (Guiana Shield); Born, 1996; Chippaux, 1987; Claessen, 2001–2006b (Guiana Shield); Gasc & Rodriques, 1980; Hoogmoed, 1979; Hoogmoed & Avila-Pires, 1991; Starace, 1998

GALAPAGOS (Ecuador): S. Garman, 1892a; Mertens, 1960a; Slevin, 1935; Steindachner, 1876; Swash & Still, 2000; Van Denbrugh, 1912c

GUYANA (British Guiana): Ávila-Pires, 2005 (Guiana Shield); Beebe, 1919, 1946; Claessen, 2001–2006b (Guiana Shield); Cole et al., 2013; Hoogmoed, 1979, 1983; MacCulloch & Reynolds, 2012 (Paramakatoi & Kato); H.W. Parker, 1935; Quelch, 1898a; Reynolds & MacCulloch, 2012 (Baramita)

PARAGUAY: Aquino et al., 1996; Bertoni, 1918; Boettger, 1885a; Cacciali, 2008; Canese, 1966 (venomous); Gatti, 1955; Norman, 1994 (Chaco); Schouten, 1937; Scott et al., 1975; Serié, 1915c

PERU: Carrillo de Espinoza, 1966, 1970, 1977, 1983, 1989; Carrillo de Espinoza & Icochea, 1995; Dixon & Soini, 1977, 1986; Duellman, 1990, 2005 (Cusco Amazonico); Duellman & Mendelson, 1995; Duellman & Salas, 1991; Henle & Ehrl, 1991; Lamar, 1997a; Lehr, 2002; Meneses, 1974; Morales & McDiarmid, 1996 (Madre de Dios); Pefaur et al., 1978; Prado & Hoge, 1948; Reynolds et al., 1997 (Madre de Dios); Rodriguez & Cadle, 1990; K.P. Schmidt & Walker, 1943a–c, 1945; Yarlequé-Chocas, 2000; Young & León, 1999

SURINAME (Dutch Guiana): Abuys, 1982–1989, 2003; Brongersma, 1967; Claessen, 2001–2006b (Guiana Shield); Hoge, 1964d; Hoogmoed, 1979; Moonen et al., 1979

URUGUAY: Achaval, 1976, 2001; Achaval & Olmos, 2003; Carreira-Vidal et al., 2005, 2012a; Carreira-Vidal, 2002; Devincenzi, 1925, 1939; A. Ferreira, 1949; Gudynas, 1989; Lema & Fabián-Beurmann, 1977; Vaz-Ferreira & Sierra de Soriano, 1960

VENEZUELA: Aleman, 1952; Ávila-Pires, 2005 (Guiana Shield); Beebe, 1946; Fuentes & Rodriguez-Acosta, 1997; Gorzula & Señaris, 1998; Kornacker, 1999; La Marca & Soriano, 2004; Lancini, 1979, 1983, 1986; Lancini-V. & Kornacker, 1989; Mijares-Urrutia & Arends R., 2000; Navarrete et al., 2009; Péfaur, 1992; Péfaur & Rivero, 2000; Rivas-Fuenmayer & Oliveros, 1998; Rivas-Fuenmayer et al., 2012; Roze, 1954 (upper Orinoco), 1956 (Roques and Orchila Is.), 1957a, 1958b–c, 1962a–b, 1965 (Margarita Is.), 1966a; Sandner-Montilla, 1965, 1975; Stejneger, 1901c; Test et al., 1966a–b; Ugueto & Rivas, 2010 (Nuevá Esparta); Vellard, 1941; Vetencourt, 1960 (venomous)

West Indies

WEST INDIES: T. Barbour, 1914a, 1930, 1935, 1937; Crother, 1999a–b; A.C.L.G. Günther, 1888d (Dominica); S.B. Hedges, 1996; Henderson & Powell, 1999, 2009; Maclean et al., 1977; Maglio, 1970; J.C. Murphy, 1996; Pregill & Crother, 1999; K.P. Schmidt, 1921a (Santo Domingo); A. Schwartz & Henderson, 1985, 1988, 1991; A. Schwartz & Thomas, 1975; A. Schwartz et al., 1978; Underwood, 1962

CAYMAN ISLANDS: Grant, 1940b; Seidel & Franz, 1994

CUBA: Alayo Dalmau, 1955; T. Barbour & Ramsden, 1919; Buide, 1966 (Hicacos), 1967; Estrada & Ruibal, 1999; Estrada, 1994; Fong & Navarro, 2001 (Sagua–Baracoa Mts.); Garrido & Jaume, 1984; Gundlach, 1867, 1875, 1880; Lando & Williams, 1969; Rodríguez-Schettino, 2000, 2003

DOMINICAN REPUBLIC (Hispaniola): Cochran, 1941; Henderson & Schwartz, 1984a; Henderson et al., 1984; Mertens, 1939; Powell et. al.,1999

HAITI (Hispaniola): Cochran, 1924, 1941; Franz & Cordier, 1986; Henderson & Schwartz, 1984a; Henderson et al., 1984; Mertens, 1939; Powell et al., 1999

JAMAICA: Grant, 1940a, 1940c; Lambert, 1986; Lynn & Grant, 1940; Crombie, 1999

LESSER ANTILLES (Leeward Is., Windward Is.): Breuil, 2002, 2003 (Guadeloupe, Saint Martin, Saint Barthélemy), 2009 (Martinique); Brongersma, 1940, 1959; Censky & Kaiser, 1999; Hodge et al., 2003 (Anguilla); Powell, 1999 (Navassa); Powell et al., 1996, 2005 (Dutch Westward Is.); K.P. Schmidt, 1921b (Navassa)

NETHERLAND ANTILLES (Dutch Antilles): Brongersma, 1940, 1959; Borg, 2008 (Curaçao); Buurt, 2001, 2005; Malhotra & Thorpe, 1999; Ruthven, 1923

PUERTO RICO: Grant, 1931, 1932a–b; Rivero, 1978, 1998; K.P. Schmidt, 1920, 1928a; Stejneger, 1904; R. Thomas & Powell, 1999; R. Thomas & Joglar, 1996

SAN ANDRES & PROVIDENCIA ISLANDS: Dunn & Saxe, 1950; Valdivieso & Tamsitt, 1963b

TRINIDAD & TOBAGO: T. Barbour, 1916c; Boos, 1967 (Huevos Is.), 1983b, 2001; Boos & Quesnel, 1968; Brongersma, 1956a; Emsley, 1963a, 1977; J.D. Hardy, 1982; Mertens, 1972b; Mole, 1924; Mole & Ulrich, 1894; J.C. Murphy, 1997; K.P. Schmidt (venomous); Wehekind, 1960

VIRGIN ISLANDS: Boos & Quesnel, 1968; Grant, 1932c, 1937; Maclean, 1982; K.P. Schmidt, 1928a

Mesoamerica

CENTRAL AMERICA: Bolaños, 1983 (venomous), 1984; J.A. Campbell & Lamar, 1989, 2004 (venomous); Duellman, 1966a; J.D. Johnson, 1984, 1989; J.D. Johnson et al., 2001; G. Köhler, 2001a, 2003, 2008; J.A. Peters & Orejas-Miranda, 1970; Savage, 1966, 1982; Tipton, 2005; Vanzolini, 1986a; Villa et al., 1988; L.D. Wilson & Johnson, 2010; L.D. Wilson & Townsend, 2010; L.D. Wilson et al., 2010

BELIZE (British Honduras): J.A. Campbell, 1998b; J.A. Campbell & Vannini, 1988; Garel & Matola, 1995; Henderson & Hoevers, 1975; J.C. Lee, 1996, 2000; Neill, 1965; Neill & Allen, 1959; S.G. Platt et al., 1999;

K.P. Schmidt, 1941; Stafford & Meyer, 2000; Stafford et al., 2010; Weyer, 1990

COSTA RICA: A. Alfaro, 1985; Bolaños, 1984 (venomous); R.M. Burger, 2001 (Caño Palma Biol. Stn.); Cope, 1875a; Duellman, 1990; Guyer, 1994 (La Selva); Guyer & Donnelly, 2005 (La Selva & Caribbean); M.P. Hayes et al., 1989 (Monteverde); Leenders, 2001; McConnell, 2013; Picado, 1931, 1976; Sasa & Solórzano, 1995 (Santa Rosa Natl. Pk.); Sasa et al., 2010; Savage, 1980a–b, 2002; Savage & Bolaños, 2009; Savage & Villa, 1986; N.J. Scott, 1969; N.J. Scott et al., 1983; Solórzano, 2004; E.H. Taylor, 1951, 1954, 1955; R. Taylor et al., 1974 (venomous); R. Van Devender, 1980 (Monteverde); Verhoeven, 1995 (Rara Avis); Wettstein, 1934

EL SALVADOR: Dueñas et al., 2001; Greenbaum & Komar, 2010; G. Köhler, 1996b; G. Köhler et al., 2006; Mertens, 1952a–c; Rand, 1957; K.P. Schmidt, 1928b; Uzzell & Starett, 1958

GUATEMALA: Acevedo, 2006; Acevedo et al., 2010; J.A. Campbell, 1982, 1998b, 2001; J.A. Campbell & Vannini, 1988; Duellman, 1963 (Peten); J.C. Lee, 1996, 2000; K.P. Schmidt & Stuart, 1941 (Baja Verapaz); Slevin, 1939; L.C. Stuart, 1934–1963; Tomas Rodas, 1938

HONDURAS: Carr, 1950; Cruz-Diaz, 1987; Dunn & Emlen, 1932; Espinal et al., 2001; Marineros, 2000; McCranie, 2011a, McCranie & Castañeda, 2005 (Parque Nacional Pico); McCranie et al., 2005 (Bay Is. & Cayos Cochinos), 2006 (Mosquita); Meyer, 1969; Townsend & Wilson, 2008 (Cusco Natl. Pk.), 2010a–b; L.D. Wilson, 1985; L.D. Wilson & Cruz-Diaz, 1993 (Cayo Cochinos); L.D. Wilson & Hahn, 1973 (Bay Is.); L.D. Wilson & McCranie, 1994, 2003, 2004a–b; L.D. Wilson & Meyer, 1985; L.D. Wilson et al., 2001, 2010

MEXICO: Alvarez del Toro, 1960, 1973, 1982 (all Chiapas); Benítez-Gálvez, 1997 (Puebla); Bogert & Oliver, 1945 (Sonora); P.R. Brown & Wright, 1994 (deserts); Calderon et al., 2003 (Campeche); Camarillo & Smith, 1992 (state of México); J.A. Campbell, 1998b (Yucatán); Canseco-Márquez & Gutiérrez-Mayén, 2006 (Puebla), 2010 (Tehuacán-Cuicatlán Valley); Canseco-Márquez et al., 2000 (Puebla); Cárdenas et al., 1988 (Baja California); Casas-Andreu & McCoy, 1979; Casas-Andreu et al., 1996 (Oaxaca); Castro-Franco & Bustos-Zagal, 1994 (Morelos); Castro-Franco & Gaviño de la Torre, 1990 (Nayarit); Conant, 1965; Cruz-Sáenz et al., 2009 (Jalisco), 2011 (Jalisco: Huaxtla); Dixon & Lemos-Espinal, 2010 (Querétaro); Dixon et al., 1972 (Querétaro); Duellman, 1954 (Michoacán), 1958c (Colima), 1961 (Michoacán), 1965a (Yucatán), 1965b (Michoacán); Dugès, 1897b; Dundee et al., 1986 (Yucatán); Enderson et al., 2009 (Sonora); Flores-Villela, 1991, 1993; Flores-Villela et al., 1987, 2010 (Central Highlands); Fugler, 1953 (San Luis Potosí); Fugler & Dixon, 1961 (Sinaloa); García & Ceballos, 1994 (Jalisco); Grismer, 1993, 1994, 1999, 2002 (all Baja California); L.M. Hardy & McDiarmid, 1969 (Sinaloa); Hensley & Lannom, 1966 (Colima); Himmelstein, 1981 (Quintana Roo); J.D. Johnson, 1976 (Chiapas), 1984, 1989; J.D. Johnson et al., 2010 (SE México); Lascano-Villarreal & Contreras, 1995 (Nuevo León); Lascano-Villarreal et al., 2010 (Nuevo León); Lavin-Murcio & Lazcano-Villarreal, 2010 (N Mexico); Lazcano-Villarreal et al., 2004, 2010 (both Nuevo León); J.C. Lee, 1980, 1996, 2000 (all Yucatán); Lemos-Espinal & Smith, 2007a–c (Chihuahua, Coahuila, Sonora); Lemos-Espinal et al., 2002a–b (Chihuahua), 2004c-e (Chihuahua, Sonora, Zacatecas); Leviton & Banta, 1964 (Baja California); Liner, 1994, 2007 (checklist); Linsdale, 1932 (Baja California); Martin, 1958 (Tamaulipas); McCranie & Wilson, 1987 (Sierra Madre Occidental), 2001b (Aguascalientes); Mendez-de La Cruz et al., 1992 (Distrito Federal); Mendoza-Q., 1990 (Hidalgo); Mocquard, 1899 (Baja California); Morafka, 1977 (Chihuahuan desert); Muñoz-Alonso, 1988 (Chiapas, Guerrero); R.W. Murphy & Méndez de la Cruz, 2010 (Baja California); R.W. Murphy & Ottley, 1984 (Baja California); Pérez-Higareda & Smith, 1991 (Veracruz); Pérez-Higareda et al., 1987, 2007 (both Veracruz); Peréz-Ramos et al., 2000 (Guerrero); J.A. Peters, 1954 (Michoacán); Ramírez-Bautista, 1994 (Jalisco); Ramírez-Bautista et al., 1991 (state of México), 2010 (Hidalgo); Rossi & Rossi, 2003; K.P. Schmidt, 1922 (Baja California), 1936a (Yucatán); K.P. Schmidt & Andrews, 1936 (Yucatán); K.P. Schmidt & Owens, 1944 (N Coahuila); N.J. Scott, 1951 (Sinoloa), 1962 (Sinaloa); H.M. Smith, 1943a; H.M. Smith & Shannon, 1947 (Michoacán); H.M. Smith & Smith, 1973, 1976, 1993; H.M. Smith & Taylor, 1945; Soule & Sloan, 1966 (Baja California); W.W. Tanner, 1985 (Chihuahua); E.H. Taylor, 1936b (Sinaloa), 1940c, 1949–1953 (San Luis Potosí); Toro & Smith, 1956 (Chiapas); Van Denburgh, 1895a (Baja California); Vásquez-Díaz & Quintero-Díaz, 1997a–b, 2005 (Aguascalientes); Vega-López & Alvarez-S., 1992 (Puebla); Webb, 1984 (Durango, Sinaloa); Welsh, 1976 (Baja California); Woolrich-Piña et al., 2005 (Puebla); Zweifel, 1958 (Baja California), 1959a, 1959b (Nayarit), 1960 (Tres Marias Is.)

NICARAGUA: Brattstrom & Howell, 1954; Cope, 1871c; G. Köhler, 1999b, 2001b; Ochomogo, 1957 (venomous); Ruiz-Pérez, 1996; Sunyer & Köhler, 2010; Vences et al., 1998; Villa, 1962, 1972 (Corn Is.), 1983, 1984 (venomous)

PANAMA: Auth, 1994; T. Barbour, 1906; Breder, 1946 (Darien); Duellman, 1990; Dunn, 1931, 1942, 1943a, 1944f–g, 1949; Dunn & Bailey, 1939; Grocott & Sadler, 1958; Ibañez & Solis, 1991; Ibanez-D. et al., 1994, 2001; Jaramillo et al., 2010; C.W. Myers & Rand, 1969; Peracca, 1896b; Perez-Santos et al., 1993; Rand & Myers, 1990; Slevin, 1942; H.M. Smith, 1958; Solis et al., 1996

North America

GENERAL: Baird & Girard, 1853; Ballinger & Lynch, 1983; Banks et al., 1987; Behler & King, 1979; Blanchard, 1925; Cagle, 1968; Cochran & Goin, 1970; Cope, 1975c, 1900; Ditmars, 1907, 1936, 1939; Ernst, 1992 (venomous); Ernst & Ernst, 2003, 2011 (venomous); Garman, 1884; Gehlbach, 1965a (incl. Pliocene and Pleistocene); Leviton, 1972; Perkins, 1949b; Powell et al., 2012; Rossi & Rossi, 2003; K.P. Schmidt, 1953a; K.P. Schmidt & Davis, 1941; Shupe, 2004 (venomous); H.M. Smith & Brodie, 1982; Stejneger, 1895 (venomous); Stejneger & Barbour, 1943; Tipton, 2005; Wright & Wright, 1957; Yarrow, 1883b

WESTERN USA: Adler, 1978; R.D. Bartlett & Tennant, 2000; P.R. Brown & Wright, 1994; Collins & Taggart, 2009; Conant et al., 1956; Crother et al., 2012; Hanson & Hanson, 1997; Heymann, 1975; Nussbaum et al., 1983; Rossi & Rossi, 1995; C.E. Shaw & Campbell, 1974; Slevin, 1934; Stebbins, 1954, 1985; St. John, 2002; Stoops & Wright, 1993 (Southwest); W.W. Tanner & Banta, 1966 (Great Basin); Van Denburgh, 1897, 1922; Woodhouse, 1853

EASTERN USA: R.D. Bartlett & Bartlett, 2005; Collins & Taggart, 2009; Conant, 1957 (NE USA), 1975; Conant & Collins, 1991, 1998; Conant et al., 1956; Crother et al., 2012; N.S. Davis & Rice, 1883a; DeGraaf & Rudis, 1981, 1983; Ernst & Barbour, 1989; Gibbons & Dorcas, 2005 (SE USA); Gray, 2005; Hulse et al., 2001 (NE USA); J.J. Jackson, 1983 (SE USA); Knobel, 1896 (New England); Minter, 2005; Reichling, 2008; Rossi, 1992; Tennant, 2003; Tennant & Bartlett, 2000

ALABAMA: Boschung, 1977; Chermock, 1952; Gibbons & Dorcas, 2005; Haltom, 1931; Linzey, 1979; Löding, 1922; Mount, 1975

ALASKA: Hodge, 1976; Logier & Toner, 1961; St. John, 2002

ARIZONA: Brennan & Hollycross, 2005 (Maricopa Co.), 2006; Fowlie, 1965; Gloyd, 1937; Grater, 1981 (Lake Mead); Lazaroff et al., 2006 (Sabino Canyon); Lowe, 1964; Lowe et al., 1986; D.M. Miller et al., 1982; Ruthven, 1907; Stejneger, 1902b; Van Denburgh & Slevin, 1913

ARKANSAS: Dellinger & Black, 1938; Dowling, 1957b; Hurter & Strecker, 1909; K.J. Irwin, 2004; Schwardt, 1938; Trauth et al., 2004; Vance, 1982, 1985

CALIFORNIA: P.R. Brown, 1997; J.G. Cooper, 1870; Delisle et al., 1987 (Santa Monica Mts.); Dixon, 1967; Glaser, 1970 (Riverside Co.); Hill, 1948 (Los Angeles Co.); Jennings, 1983b, 1987; Jennings & Hayes, 1994; D.H. Johnson et al., 1948; Klauber, 1939d; Lemm, 2006 (San Diego Co.); McKeown, 1974; J.C. Murphy, 1990 (southern); Perkins, 1949a; Pickwell, 1947; Schoenherr, 1976 (San Gabriel Mtns.); Stebbins, 1959, 1972; Stebbins & McGinnis, 2012; St. John, 2002; Zeiner et al., 1998

COLORADO: M.M. Ellis & Henderson, 1913; Hammerson, 1986, 1999; Jones-Burdick, 1949; Maslin, 1959; H.M. Smith et al., 1965

CONNECTICUT: Klemens, 1991; Krulikowski, 2004; Lamson, 1935; Peterson & Fritsch, 1986

DELAWARE: Arndt, 1985; Conant, 1945; McCauley, 1945; C.F. Reed, 1956b, 1958b; J.E. White & White, 2002

DISTRICT OF COLUMBIA: Harris, 1975; Hay, 1902; McCauley, 1945

FLORIDA: R. Anderson, 1983, 1985, 1989; Ashton, 1977; Ashton & Ashton, 1988; Bartlett & Bartlett, 2003; Bartolotti & Bartolotti, 2010; Blaney, 1971; P. Carmichael & Williams, 1991; Carr, 1940; Carr & Goin, 1955; Cope, 1889a; Duellman & Schwartz, 1959; Enge & Dodd, 1992; Gibbons & Dorcas, 2005; Haast & Anderson, 1981; Laerm et al., 1980; Lazell, 1989; Meshaka et al., 2004; Moler, 1992; Tennant, 1997, 2003; Truitt, 1962; L.D. Wilson & Porras, 1983

GEORGIA: Gibbons & Dorcas, 2005; Gibbons & Semlitsch, 1991; Gibbons & West, 1998; Jensen & Moulis, 1999; Jensen et al., 2008; Martof, 1956; Neill, 1949b; Vitt & Laerm, 1982; Wharton, 1963; G.K. Willamson & Moulis, 1979, 1994

HAWAII: McKeown, 1996; Oliver & Shaw, 1953

IDAHO: Linder & Fichter, 1977; Nussbaum et al., 1983; W.W. Tanner, 1941

ILLINOIS: Cagle, 1941; N.S. Davis & Rice, 1883b; H. Garman, 1892; J.H. Harding, 1997 (Great Lakes); Lueth, 1941, 1949; Necker, 1939; Parmalee, 1955; Phillips et al., 1999; Pope, 1944, 1966; Redmer & Ballard, 1995; P.W. Smith, 1961

INDIANA: J.H. Harding, 1997 (Great Lakes); Hay, 1892b; MacGowan & Kingsbury, 2001; Minton, 1972, 2001

IOWA: Christiansen & Bailey, 1990; Guthrie, 1923, 1926

KANSAS: Branson, 1904; Busby et al., 1994; R.F. Clarke, 1959, 1980; Collins, 1982b, 1993; Collins & Collins, 2008; Collins et al., 2010, 2011 (Cimarron Natl. Grassland); Fitch, 1999; Karns et al., 1974; KSU Coop. Ext. Serv., 1974; H.M. Smith, 1950, 1956; E.H. Taylor, 1929

KENTUCKY: R. Barbour, 1971; Dury & Williams, 1933; Meade, 1991, 2005

LOUISIANA: Beyer, 1900; Boundy, 1997, 2006; Cagle, 1952; R.F. Clark, 1949; Dundee & Rossman, 1989; Gibbons & Dorcas, 2005; Keiser, 1971; Keiser & Wilson, 1979; Viosca, 1944

MAINE: Fogg, 1862; Hulse, et al., 2001; Hunter et al., 1999

MARYLAND: Brady, 1937; Conant, 1945; Harris, 1975; Kelly, 1936; Mansueti, 1941; McCauley, 1945; J.C. Mitchell & Anderson, 1994; C.F. Reed, 1956a–b, 1957a, 1957c–d, 1958a–b; J.E. White & White, 2002

MASSACHUSETTS: Cardoza & Mirick, 1979; Herscu, 2004; Jackson & Mirick, 1993; Krulikowski, 2004; Lazell, 1972, 1974, 1976

MICHIGAN: J.H. Harding, 1997 (Great Lakes); Holman, 2012; Holman & Harding, 2006; Holman et al., 1989; Notestein, 1905; Pentecost & Vogt, 1976; Ruthven,

1912; Ruthven et al., 1928; Sheldon, 2006; W. Smith, 1879; Takiela, 2004b

MINNESOTA: Blasus, 1997; Breckenridge, 1944; J.H. Harding, 1997 (Great Lakes); Moriarty & Jones, 1997; Oldfield & Moriarty, 1994; Perry & Dexter, 1989; Sheldon, 2006; Takiela, 2003

MISSISSIPPI: Cliburn, 1965; F.A. Cook, 1962; Gibbons & Dorcas, 2005; Keiser, 1982a; Lohoefener & Altig, 1983

MISSOURI: P. Anderson, 1965; Boyer & Heinze, 1934; Hurter, 1911; T.R. Johnson, 1980, 1987, 2000

MONTANA: Coues & Yarrow, 1878; Maxwell et al., 2003; J.K. Werner et al., 2004

NEBRASKA: Ballinger et al., 2010; Fogell, 2011; Hudson, 1942; Lynch, 1985; W.E. Taylor, 1892

NEVADA: Banta, 1965; Linsdale, 1940

NEW HAMPSHIRE: Klemens, 1995; J. Taylor, 1993

NEW JERSEY: Fowler, 1907; Hulse et al., 2001; V. Schwartz & Golden, 2002; Trapido, 1937

NEW MEXICO: Degenhardt et al., 1996; Gehlbach, 1965b; Ruthven, 1907; Van Denburgh, 1924; M.A. Williamson et al., 1994

NEW YORK: Baird, 1854; DeKay, 1842; Eckel, 1902; Gibbs et al., 2007; J.H. Harding, 1997 (Great Lakes); Hulse et al., 2001

NORTH CAROLINA: Beane et al., 2010; Brimley, 1939–1943; Brothers, 1992 (Dismal Swamp); Depoe et al., 1961; Dorcas, 2004; Gibbons & Dorcas, 2005; Huheey & Stupka, 1967 (Smoky Mts.); Martof et al., 1980; Palmer, 1974; Palmer & Braswell, 1995; Tilley & Huheey, 2004 (Smoky Mts.)

NORTH DAKOTA: Wheeler, 1947, 1954; Wheeler & Wheeler, 1966

OHIO: Conant, 1938b, 1951; J.H. Harding, 1997 (Great Lakes); Morse, 1904; W. Smith, 1882; D.E. Wynn & Moody, 2006

OKLAHOMA: Carpenter & Krupa, 1989; Engbretson, 1974; Lardie, 1982; Ortenburger, 1927, 1930; Sievert & Sievert, 1988; Van Vleet, 1902; Webb, 1970

OREGON: Blaustein et al., 1995; H.A. Brown et al., 1995; Csuti et al., 1997; K. Gordon, 1939; Nussbaum et al., 1983; Pickwell, 1947; Storm et al., 1995

PENNSYLVANIA: W.B. Allen, 1992; J.H. Harding, 1997 (Great Lakes); Hulse et al., 2001; C.J. McCoy, 1980, 1982; Netting, 1932, 1936a; Roddy, 1928; Shaffer, 1992; Surface, 1906

RHODE ISLAND: Drowne, 1905; Hopkins, 1918

SOUTH CAROLINA: Beane et al., 2010; Gibbons & Dorcas, 2005; Gibbons & Semlitsch, 1991; Gibbons & West, 1998; Martof et al., 1980

SOUTH DAKOTA: Over, 1923; S.W. Thompson, 1982

TENNESSEE: Gibbons & Dorcas, 2005; Huheey & Stupka, 1967 (Smoky Mts.); R.M. Johnson, 1964; Jones, 1928; J.A. Klein, 1973, 1989; Redmond et al., 1990; Rhoads, 1895; Shoup, 1974; Sinclair et al., 1970; Snyder, 1992; Tilley & Huheey, 2001 (Smoky Mts.)

TEXAS: B.C. Brown, 1950; Dixon, 1987b, 1993, 1996a–b; Dixon & Werler, 2005; J.D. Mitchell, 1903; Price, 1998, 2009; Raun, 1965; Raun & Gehlbach, 1972; Strecker, 1915; Tennant, 1984, 1985, 1998, 2003; R.A. Thomas, 1976b; Vermersch & Kuntz, 1986; Werler, 1970, 1978; Werler & Dixon, 2000

UTAH: D.C. Cox & Tanner, 1995; Pack, 1930; Shofner, 2007; Stansbury, 1852, 1854; V.M. Tanner, 1929; Van Denburgh & Slevin, 1915; Wauer, 1964; Woodbury, 1931

VERMONT: J.S. Andrews, 2002

VIRGINIA: Beane et al., 2010; Brothers, 1992 (Dismal Swamp); Ernst et al., 1997 (northern); Gibbons & Dorcas, 2005; Linzey & Clifford, 1981; Martof et al., 1980; J.C. Mitchell, 1974, 1994; J.C. Mitchell & Anderson, 1994 (Assateague & Chincoteague Is.); J.C. Mitchell & Reay, 1999; Pinder & Mitchell, 2002; C.F. Reed, 1957a (Northern Neck), 1957b (Anna Arundel Co.), 1957c (Accomac & Northhampton Co.); Tobey, 1985; J.E. White & White, 2002

WASHINGTON: Blaustein et al., 1995; H.A. Brown et al., 1995; Carlson, 1967; M.L. Johnson, 1942, 1954; Nussbaum et al., 1983; Paulson, 1971, 1972; Pickwell, 1947; Storm et al., 1995

WEST VIRGINIA: N.B. Green, 1953; N.B. Green & Pauley, 1987

WISCONSIN: Casper, 1996, 2008; Christoffel et al., 2008; Dickinson, 1949; J.H. Harding, 1997 (Great Lakes); Higley, 1889; Hoy, 1883; T.E.B. Pope & Dickinson, 1928; Sheldon, 2006; Takiela, 2004a, R.C. Vogt, 1981; Watermolen, 1992

WYOMING: Baxter & Stone, 1980; Boyer & Heinze, 1934; P.S. Corn et al., 1984; Koch & Peterson, 1995; Lewis, 2011; Turner, 1955

CANADA: Bider & Matte, 1996 (Québec); Bleakney, 1958 (eastern); Carl, 1944 (British Columbia); Christie, 1997 (Ontario); F.R. Cook, 1967 (Prince Edwards Island), 1977a (Saskatchewan), 1977b; 1980, 1984; Cowan, 1936 (British Columbia); P. Cox, 1903 (Maritimes); Desroches & Rodrigue, 2004 (Québec); Ernst & Ernst, 2003; C. Fisher et al., 2007; Froom, 1972, 1981 (Ontario); Gilhen, 1984 (Nova Scotia); Gorham, 1970b; Gregory & Campbell, 1984 (British Columbia); J.H. Harding, 1997 (Great Lakes); Hodge, 1976 (Yukon & Northwest Terr.); B. Johnson, 1989 (Ontario); Logier, 1939, 1958 (both Ontario); Logier & Toner, 1961; MacCulloch, 2002 (Ontario); Matsuda et al., 2006 (British Columbia); R.C. Mills, 1948; Powell et al., 2012; Preston, 1982 (Manitoba); A.P. Russell & Bauer, 2000 (Alberta); St. John, 2002 (W Canada)

Literature Cited

Abadi, N. (1998) *Fauna of Yemen. Part 2. Amphibians and reptiles.* Abadi Center for Studies and Publishing, Sana'a, 151 pp., 59 figs. (in Arabic)

Abalos, J.W. & Baez, E.C. (1963) Variaciones del diseño en *Bothrops neuwiedii meridionalis* de Santiago del Estero. *Acta Zoologica Lilloana* 19, 479–486.

Abalos, J.W. & Bucher, E. (1970) Zoo-epidemiologia del ofidismo en Santiago del Estero. *Boletin de la Academia Nacional di Ciencias,* Córdoba 47(2–4), 259–272, 4 figs.

Abalos, J.W. & Mischis, C.C. (1975) Elenco sistemático de los ofidios Argentinos. *Boletin de la Academia Nacional de Ciencias,* Córdoba 51(1–2), 55–76.

Abalos, J.W. & Nader, R. (1962) La coral *Micrurus lemniscatus frontalis* de Santiago del Estero. *Anales del Instituto Nacional de Microbiologia,* Buenos Aires 1, 83–94.

Abalos, J.W., Baez, E.C. & Nader, R. (1964) Serpientes de Santiago del Estero. *Acta Zoologica Lilloana* 20, 211–283, 23 figs.

Abbott, C.C. (1869) Appendix E. Catalogue of vertebrate animals of New Jersey. *In:* Cook, G.H. (Ed.), *Geology of New Jersey.* State Geological Report (1868), Newark, pp. 751–830.

Abdála, V. (1990) Morfometria en dos especies del genero *Phimophis* Cope (Ophidia, Colubridae). *Acta Zoologica Lilloana* 39(2), 85–89, 8 figs.

Abdeen, A.M. (1991a) Further studies on the ophidian cranial osteology, the skull of the Egyptian black cobra, *Walterinnesia aegyptia* (family Elapidae). *Journal of the Egyptian Gerontological Society, Zoology* 5, 367–389, 4 figs.

Abdeen, A.M. (1991b) Further studies on the ophidian cranial osteology, the skull of the Egyptian blind snake *Leptotyphlops cairi* (family Leptotyphlopidae). II. The lower jaw and the hyoid apparatus. *Journal of the Egyptian Gerontological Society, Zoology* 5, 457–467, 2 figs.

Abdeen, A.M., Abo-Taira, A.M. & Zaher, M.M. (1991) Further studies on the ophidian cranial osteology, the skull of the Egyptian blind snake *Leptotyphlops cairi* (family Leptotyphlopidae). I. The cranium. A, the median dorsal bones of the upper jaw, circumorbital Series and occipital ring. *Journal of the Egyptian Gerontological Society, Zoology* 5, 417–437, 3 figs.

Abdulali, H. (1935) An addition to the list of snakes of Bombay and Salsette, *Coronella brachyura. The Journal of the Bombay Natural History Society* (1935–1936) 38(1), 197.

Abe, A.S. & Fernandes, W. (1977) Polymorphism in *Spilotes pullatus anomalepis* Bocourt (Reptilia, Serpentes, Colubridae). *Journal of Herpetology* 11(1), 98–100, illust.

Abes, S.S. & Ferrarezzi, H. (2003) Geographic distribution: Serpentes: *Apostolepis goiasensis. Herpetological Review* 34(2), 167–168.

Abercromby, A.F. (1910) *The snakes of Ceylon.* Murray, London, 89 pp.

Abercromby, A.F. (1913) Distribution of snakes in Ceylon. *Spolia Zeylanica,* Colombo (1912) 8(32), 304–305.

Abram, S. & Menegon, M. (1994) *Vipere e altri serpenti Italiani.* Siste Edizioni, Trento, 176 pp., figs., col. photos.

Abreu, E.A. de. (1913) Occurrence of the snake (*Psammophis longifrons*) at Nagpur. *The Journal of the Bombay Natural History Society* (1913–1914) 22(3), 634.

Abreu, E.A. de. (1916) The snakes of Nagpur. *Records of the Nagpur Museum* (1), 1–47, 3 figs.

Abreu, E.A. de. (1918) On an undescribed colour variety of the snake (*Zaocys mucosus*) from the Central Province. *The Journal of the Bombay Natural History Society* (1917–1918) 25(4), 753–754.

Abuys, A. (1982a) De slangen van Suriname, Deel I, De wormslangen. *Litteratura Serpentium,* Utrecht 2(2), 64–81, 33 figs., 2 photos.

Abuys, A. (1982b) De slangen van Suriname, Deel II, De Families Aniliidae en Boidae. *Litteratura Serpentium,* Utrecht 2(3), 112–133, 2 figs., 4 photos.

Abuys, A. (1982c) De slangen van Suriname, Deel III, De Familie Colubridae (algemene gegevens en het genus *Chironius*). *Litteratura Serpentium,* Utrecht 2(5), 226–245, 1 fig., 2 photos.

Abuys, A. (1982d) De slangen van Suriname, Deel IV, Onderfamilie Colubrinae (genera *Dendrophidion, Drymarchon, Drymobius,* en *Drymoluber*). *Litteratura Serpentium,* Utrecht 2(26), 274–287, 2 photos.

Abuys, A. (1983a) The snakes of Surinam, Part V, Subfamily Colubrinae (genera *Geophis, Leptophis, Mastigodryas,* and *Spilotes*). *Litteratura Serpentium,* Utrecht (English ed.) 3(1), 22–31, 3 photos.

Abuys, A. (1983b) The snakes of Surinam, Part VI, Subfamily Xenodontinae (general data and genus *Atractus*). *Litteratura Serpentium,* Utrecht (English ed.) 3(2–3), 66–81, 33 figs., 1 photo.

Abuys, A. (1983c) The snakes of Surinam, Part VII, Subfamily Xenodontinae (genera *Clelia* and *Dipsas*). *Litteratura Serpentium,* Utrecht (English ed.) 3(4), 111–120, 2 photos.

Abuys, A. (1983d) The snakes of Surinam, Part VIII, Subfamily Xenodontinae (genera *Erythrolamprus, Helicops,* and *Hydrodynastes*). *Litteratura Serpentium,* Utrecht (English ed.) 3(6), 203–212, 4 photos.

Abuys, A. (1984a) The snakes of Surinam, Part IX, Subfamily Xenodontinae (genera *Hydrops, Imantodes,* and *Leimadophis*). *Litteratura Serpentium,* Utrecht (English ed.) 4(2), 63–73.

Abuys, A. (1984b) The snakes of Surinam, Part X, Subfamily Xenodontinae (genera *Leptodeira, Liophis,* and *Lygophis*). *Litteratura Serpentium,* Utrecht (English ed.) 4(5–6), 160–172.

Abuys, A. (1985a) The snakes of Surinam, Part XI, Subfamily Xenodontinae (genera *Ninia, Oxybelis,* and *Oxyrhopus*). *Litteratura Serpentium,* Utrecht (English ed.) 5(3), 96–109, 5 photos.

Abuys, A. (1985b) The snakes of Surinam, Part XII, Subfamily Xenodontinae (genera *Philodryas, Phimophis,* and *Pseudoboa*). *Litteratura Serpentium,* Utrecht (English ed.) 5(5), 188–198, 4 photos.

Abuys, A. (1986a) The snakes of Surinam, Part XIII, Subfamily Xenodontinae (genera *Pseudoeryx, Pseustes,* and *Rhadinaea*). *Litteratura Serpentium,* Utrecht (English ed.) 6(1), 19–30, 4 photos.

Abuys, A. (1986b) The snakes of Surinam, Part XIV, Subfamily Xenodontinae (genera *Rhinobothryum, Sibon,* and *Siphlophis*). *Litteratura Serpentium,* Utrecht (English ed.) 6(2), 66–73, 2 photos.

Abuys, A. (1986c) The snakes of Surinam, Part XV, Subfamily Xenodontinae (genera *Tantilla*, *Thamnodynastes*, and *Tripanurgos*). *Litteratura Serpentium,* Utrecht (English ed.) 6(3), 107–116, 4 photos.

Abuys, A. (1986d) The snakes of Surinam, Part XVI, Subfamily Xenodontinae (genera *Waglerophis*, *Xenodon*, and *Xenopholis*). *Litteratura Serpentium,* Utrecht (English ed.) 6(5), 181–192, 2 figs., 3 photos.

Abuys, A. (1987a) The snakes of Surinam, Part XVII, Family Elapidae, subfamily Micrurinae (genera *Leptomicrurus* and *Micrurus*). *Litteratura Serpentium,* Utrecht (English ed.) 7(4), 185–194, 1 photo.

Abuys, A. (1987b) A new coral snake from Surinam. *Litteratura Serpentium,* Utrecht (English ed.) 7(5), 215–220, illust.

Abuys, A. (1987c) The snakes of Surinam, Part XVIII, Family Elapidae, subfamily Micrurinae (genera *Leptomicrurus* and *Micrurus*). *Litteratura Serpentium,* Utrecht (English ed.) 7(5), 221–242, 1 fig., 5 photos.

Abuys, A. (1987d) The snakes of Surinam, Part XIX, Family Viperidae, subfamily Crotalinae (genus *Crotalus*). *Litteratura Serpentium,* Utrecht (English ed.) 7(6), 282–296, 4 figs., 2 photos.

Abuys, A. (1988a) The snakes of Surinam, Part XX, Family Viperidae, subfamily Crotalinae (the genus *Lachesis*). *Litteratura Serpentium,* Utrecht (English ed.) 8(1), 41–49, 2 photos.

Abuys, A. (1988b) The snakes of Surinam, Part XXI, Family Viperidae, subfamily Crotalinae (the genus *Bothrops*). *Litteratura Serpentium,* Utrecht (English ed.) 8(2), 73–88, 6 photos.

Abuys, A. (1989) Additional notes to the article The snakes of Suriname, part 2. The genus Aniliidae and Boidae. *Litteratura Serpentium,* Utrecht (English ed.) 9(3), 126–128.

Abuys, A. (2003) *De slangen van Suriname en de andere Guyana's.* Gopher Publishers, Groningen, 592 pp., col. illust.

Acevedo, M. (2006) Anfibios y reptiles de Guatemala: una breve sintesis con bibliografia. *In:* Cano, E. (Ed.), *Biodiversidad de Guatemala.* Universidad del Valle de Guatemala, Guatemala, pp. 487–524.

Acevedo, M., Wilson, L.D., Cano, E.B. & Vásquez-Almazán, C. (2010) Diversity and conservation status of the Guatemalan herpetofauna. *In:* Wilson, L.D., Townsend, J.H. & Johnson, J.A. (Eds.), *Conservation of Mesoamerican amphibians and reptiles.* Eagle Mountain Publishing, Eagle Mountain, Utah, pp. 406–435, 2 figs.

Acharji, M.N. (1946) A note on some snakes of Benares (U.P.). *The Journal of the Bombay Natural History Society* (1946–1947) 46(2), 344–347.

Acharji, M.N. (1958) On a collection of chelonians and snakes from Chota Nagpur, Bihar. *Records of the Indian Museum,* Calcutta (1955) 53(3–4), 383–392.

Acharji, M.N. (1964) Zoological results of the Indian Cho-Oyu Expedition (1958) in Nepal. Part 4. Reptilia. *Records of the Indian Museum,* Calcutta 59, 238–240, pl.

Acharji, M.N. & Kripalani, M.B. (1951) Contributions to the fauna of the Manipur State, Assam. Part IV. Reptilia. *Records of the Indian Museum,* Calcutta (1950) 48(2), 93–105.

Acharji, M.N. & Mukherjee, A.K. (1966) Report on a collection of snakes from Lower Bengal (Reptilia, Ophidia). *Journal of the Zoological Society of India,* Calcutta 16(1–2), 76–81.

Acharji, M.N. & Ray, H.C. (1936) A new species of *Oligodon* from the United Provinces (India). *Records of the Indian Museum,* Calcutta 38(4), 519–520, figs. a–b.

Acharjyo, L.N. & Mishra, R. (1976) Aspects of reproduction and growth of the Indian python, *Python molurus molurus* in captivity. *British Journal of Herpetology* 5, 562–565.

Acharjyo, L.N. & Mishra, R. (1980) Growth rate of the Indian python *Python molurus molurus* (Serpentes: Boidae) in captivity with special reference to age at first egg-laying. *The Journal of the Bombay Natural History Society* 77(2), 344–353.

Acharjyo, L.N. & Murthy, T.S.N. (1983) Studies on the king cobras of Orissa, India. *The Snake* 15(1), 22–31.

Acharjyo, L.N., Mohapatra, S. & Mishra, B. (1978) On skin sloughing of king cobras, *Ophiophagus hannah* (Cantor) in captivity. *The Journal of the Bombay Natural History Society* 75(1), 234–236, illust.

Achaval, F. (1973) El genero *Clelia* en el Uruguay. *Trabalhos del V Congreso Latinoamericano de Zoología,* Montevideo 1, 17–29, 3 pls.

Achaval, F. (1976) Reptiles. *In:* Langguth, A. (Ed.), *Lista de las especies de vertebrados del Uruguay.* Museo Nacional de Historia Naturale, Monteovideo, pp. 26–29.

Achaval, F. (2001) Actualización sistemática y mapas de distribución de los reptiles del Uruguay. *Smithsonian Herpetological Information Service* (129), 21 pp.

Achaval, F. & Olmos, A. (2003) *Anfibios y reptiles del Uruguay.* Second edition. Graphis Impresora, Montevideo. 136 pp., col. illust. [first edition, 1997, Barreiro y Ramos, Montevideo, 128 pp., illust.]

Achaval, F. & Vaz-Ferreira, R. (1974) La presencia de *Tantilla melanocephala melanocephala* (L.) en la fauna Uruguaya. *Neotropica,* La Plata 20(62), 107–111, 6 figs.

Acosta, S., Giraudo, A. & Montanelli, S. (1994) Composicion de la ofidiofauna (Reptilia, Serpentes) del Parque Nacional Iguazu, Provincia de Misiones, Argentina. *Boletín de la Asociación Herpetólogica Argentina* 10(1), 6–9.

Adalsteinsson, S.A., Branch, W.R., Trape, S., Vitt, L.J. & Hedges, B. (2009) Molecular phylogeny, classification, and biogeography of snakes of the family Leptotyphlopidae (Squamata, Scolecophidia). *Zootaxa* (2244), 1–50, 13 figs.

Adcock, F. (1922) Ecological survey of the fauna of Lake Knox. *Transactions of the Illinois State Academy of Science* 15, 186–199.

Adiyodi, K.G. (1963) First record of Beddome's wolf-snake *Lycodon travancoricus* (Beddome) from the Laccadive Archipelago. *The Journal of the Bombay Natural History Society* 60(1), 261–262.

Adler, K. (Ed.). (1978) *Herpetological explorations of the great American West.* 2 volumes. Arno Press, New York, xx + 719 pp., 153 pls.

Adler, K. (1989) *Contributions to the history of herpetology.* Society for the Study of Amphibians and Reptiles, Ithaca, 202 pp., illust.

Adler, K. (2007) *Contributions to the history of herpetology. Volume 2.* Society for the Study of Amphibians and Reptiles, Saint Louis, 389 pp., illust.

Adler, K. (2012) *Contributions to the history of herpetology. Volume 3.* Society for the Study of Amphibians and Reptiles, Vancouver, 564 pp., illust.

Adler, K. & Halliday, T.R. (1986) *Reptiles & amphibians.* Torstar Books, New York, 160 pp., col. illust.

Adler, K. & Zhao, E. (1995) The proper generic name for the Asian wolf snakes, *Lycodon* (Serpentes, Colubridae). *Sichuan Journal of Zoology* 14(2), 74–75.

Adler, K., Zhao, E. & Darevsky, I.S. (1992) First records of the pipe snake (*Cylindrophis*) in China. *Asiatic Herpetological Research,* Berkeley 4, 37–41, 2 figs.

Adnagulov, E.V., Tarasov, T. & Gorobeiko, V.V. (2000) New data on amphibians and reptiles distribution in the Russian Far East. *Russian Journal of Herpetology* 7(2), 139–154, 7 figs.

Aellen, V. (1951) Contribution a l'herpétologie du Maroc. *Bulletin de la Société des Sciences Naturelles et Phyiques du Maroc*, Rabat 31, 153–199.

Aengals, R. (2009) Reptilia. *In:* Alfred, J.R.B. (Ed.), *Fauna of Tamil Nadu. State Fauna Series* 17. Zoological Survey of India, Kolkata, pp. 187–198, 3 pls.

Afrasiab, S.R. & Mohamad, S.I. (2011) First record of the rat snake, *Zamenis hohenackeri* (Strauch, 1873), from north-eastern Iraq with notes on other colubrid snakes. *Zoology in the Middle East* 54, 19–22, 2 figs.

Afroosheh, M., Rastegar-Pouyani, N. & Kami, H.G. (2012) Comments on the distribution and external morphology of *Typhlops vermicularis* (Ophidia: Typhlopidae) in Iran. *Hamadryad* 36(1), 12–16, 3 figs.

Afsar, M. & Afsar, B. (2009) A new locality for *Vipera* (*Pelias*) *kaznakovi* Nikolsky, 1909 (Reptilia, Viperidae) in the north-eastern Anatolia. *Russian Journal of Herpetology* 16(2), 155–158, 2 figs.

Agarwal, I., Mistry, V.K. & Athreya, R. (2010) A preliminary check-list of the reptiles of Eaglenest Wildlife Sanctuary, West Kameng District, Arunachal Pradesh, India. *Russian Journal of Herpetology* 17(2), 81–93, 9 figs.

Agassiz, L. (1844–1845) *Nomenclator zoologicus, continens nomina systematica generum animalium tam viventium quam fossilium, secundum ordinem alphabeticum disposita, adjectis auctoribus, libris, in quibus reperiuntur, anno editionis, etymologia et familiis, ad quas pertinent, in singulis classibus. Fasc. VI–VII.* Jent et Gassmann, Soloduri, vii + 48 + 8 pp.

Agassiz, L. (1847) *Nomenclatoris zoologici index universalis, continens nomina systematica classium, ordinum, familiarum et generum animalium omnium, tam viventium quam fossilium, secundum ordinem alphabeticum unicum disposita, adjectis homonymiis plantarum.* Jent and Gassmann, Soloduri, x + 1135 pp.

Agasyan, A.L. (1987) The snake *Pseudocyclophis persicus* (Anderson, 1872)—a first record for the Caucasian fauna. *Proceedings of the Zoological Institute, Leningrad, USSR Academy of Science* (1986) 158, 177–179. (in Russian)

Agostini, F.M. de. (1998) Variação da folidose de *Philodryas aestivus* (Duméril, Bibron et Duméril, 1854) (Serpentes: Colubridae: Xenodontinae: Philodryadini). *Biociências* 6(1), 169–182.

Agostini, F.M. de, Lema, T. de & Ribeiro, S.L.B. (2003) The status of *Drymobius reissii* Peters, 1868 (Serpentes, Colubridae). *Phyllolmedusa* 2(2), 115–116, 1 fig.

Agrawal, H.P. (1970) On a collection of snakes from Jabalpur (Reptilia: Ophidia). *Journal of the Zoological Society of India* 22(1–2), 41–46.

Agrimi, U. & Luiselli, L. (1992) Feeding strategies of the viper *Vipera ursinii ursinii* (Reptilia, Viperidae) in the Apennines. *The Herpetological Journal*, London 2, 37–42, 4 figs.

Agrimi, U. (1994) Ecology of the snake *Coronella girondica* (Reptilia, Colubridae) in central Italy. *Vie et Milieu*, Paris 44(3–4), 203–210, 3 figs.

Aguilar, C., Lundberg, M., Siu-Ting, K. & Jiménez, M.E. (2007) Nuevos registros para la herpetofauna del departmento de Lima, descripción del renacuajo de *Telmatobius rimac* Schmidt, 1954 (Anura, Ceratophryidae) y una clave de los anfibios. New records for the herpetofauna of Lima, description of *Telmatobius rimac* Schmidt, 1954 tadpole (Anura, Ceratophryidae) and a key to the amphibians. *Revista Peruana de Biologia* 14(2), 209–216, 4 figs.

Ahl, E. (1924) Neue Reptilien und Batrachier aus dem Zoologischen Museum Berlin. *Archiv für Naturgeschichte*, Berlin 90A(5), 246–254.

Ahl, E. (1925) Uber zwei neue Schlangen des Zoolog. Museums Berlin. *Zoologischer Anzeiger*, Leipzig 63(9–10), 271–272.

Ahl, E. (1927) Zwei neue Korallenottern der Gattung *Elaps*. *Zoologischer Anzeiger*, Leipzig 70(9–10), 251–252.

Ahl, E. (1933a) Zur Kenntnis der afrikanischen Wühlschlangen der Gattung *Eryx*. *Sitzungsberichte der Gesellschaft Naturforschender Freunde zu Berlin* 1933, 324–326, 6 figs.

Ahl, E. (1933b) Ergebnisse der Celebes-und Halmaheira-Expedition Heinrich 1930–32. 1. Reptilien und Amphibien. *Mitteilungen aus dem Zoologischen Museum in Berlin* 19, 577–581.

Ahlén, I., Andrén, C. & Nilson, G. (1992) *Sveriges grodor, ödlor och ormar: fältguide och faktasamling.* Naturskyddsföreningen, Stockholm, 48 pp., 2 figs., 38 col. photos.

Ahmed, M.F., Das, A. & Dutta, S.K. (2009) *Amphibians and reptiles of Northeast India: a photographic guide.* Aaranyak, Guwahati, 170 pp., col. illust.

Ahmed, S. (1978) Sea-snakes of the Indian Ocean in the collections of the Zoological Survey of India together with remarks on the geographical distribution of all Indian Ocean species. *Journal of the Marine Biology Association of India*, Madras (1975) 17(1), 73–81.

Ahmed, S. & Dasgupta, G. (1991) Studies on the lizards and snakes of North Bihar. *Records of the Zoological Society of India* 88(1), 75–80.

Ahmed, S. & Dasgupta, G. (1992) Reptilia. *In:* Alfred, J.R.B. (Ed.), *Fauna of West Bengal. Part 2 (Reptilia, Amphibia, fishes, Hemichordata and Archaeozoology). State Fauna Series* 3. Zoological Survey of India, Calcutta, pp. 1–65.

Ahsan, M.F. & Parvin, S. (2001) The first record of *Ptyas korros* (Colubridae) from Bangladesh. *Asiatic Herpetological Research* 9, 23–24, 1 fig.

Ahsan, M.F., Asmat, G.S.M., Islam, M.A. Khan, M.M.H., Muzaffar, S.B., Kamruzzaman, M. & Chakma, S. (2009) Amphibians and reptiles. *In:* Kabir, S.M.H., Ahmed, M., Ahmed, A.T.A., Rahman, A.K.A., Ahmed, Z.U., Begum, Z.N.T., Hassan, M.A. & Khondker, M. (Eds.), *Encyclopedia of flora and fauna of Bangladesh. Volume 25.* Asiatic Society of Bangladesh, Dhaka, 204 pp., col. illust.

Ahumada-Carrillo, I.T. & Vazquez-Huizar, O. (2012a) Geographic distribution: Serpentes: *Micrurus distans*. *Herpetological Review* 43(1), 106.

Ahumada-Carrillo, I.T. & Vázquez-Huizar, O. (2012b) Geographic distribution: Serpentes: *Drymarchon melanurus*. *Herpetological Review* 43(3), 448.

Ahumada-Carrillo, I.T., Vázquez-Huizar, O., Vázquez-Diaz, J. & Garcia-Vázquez, U.O. (2011) Noteworthy records of amphibians and reptiles from Zacatecas, México. *Herpetological Review* 42(3), 397–398, 1 fig.

Aiyar, T.V.R. (1906) Notes on some sea-snakes caught at Madras. *Journal and Proceedings of the Asiatic Society of Bengal* (Series 2) 2, 69–72.

Akani, G.C. & Luiselli, L. (2001a) Aspects of the natural history of *Natriciteres* (Serpentes, Colubridae) in Nigeria, with special reference to *N. variegata* and *N. fuliginoides*. *Herpetological Natural History* (1999–2000) 7(2), 163–168.

Akani, G.C. & Luiselli, L. (2001b) Ecological stuidies on a population of the water snake *Grayia smythii* in a rainforest swamp of the Niger Delta, Nigeria. *Contributions to Zoology* 70(3), 139–146, 5 figs.

Akani, G.C., Luiselli, L., Tooze, Z., Angelici, F.M., Corti, C. & Zuffi, M.A.L. (2001) The ecological distribution of *Causus* Wagler 1830 (Viperidae) in Nigeria, with special reference to *C. resimus* (Peters 1862) and *C. lichtensteini* (Jan 1859), two species rarely recorded from this country. *Tropical Zoology*, Firenze 14(1): 185–195, 1 fig.

Akani, G.C., Luiselli, L. & Angelici, F.M. (2002a) Diet of *Thelotornis kirtlandii* (Serpentes: Colubridae: Dispholidini) from southern Nigeria. *Herpetological Journal* 12, 179–182.

Akani, G.C., Eniang, E.A., Ekpo, I.J., Angelici, F.M. & Luiselli, L. (2002b) Thermal and reproductive ecology of the snake *Psammophis phillipsi* from the rainforest region of southern Nigeria. *Herpetological Journal* 12, 63–67, 4 figs.

Akani, G.C., Angelici, F.M. & Luiselli, L. (2005) Ecological data on the Goldie's tree cobra, *Pseudohaje goldii* (Elapidae) in southern Nigeria. *Amphibia-Reptilia*, Leiden 26, 382–387, 1 fig.

Akinpelu, A. & Arco, A. (2007) The snakes of Osun Grove: a World Heritage site in Osogbo, Nigeria. *Revista de Biologia Tropical* 55(2), 717–721.

Akmentins, M.S. & Vaira, M. (2010) Reptilia, Squamata, Dipsadidae, *Tomodon orestes* Harvey and Muñoz, 2004: distribution extension, new country record. *Check List* 6(2), 248–249, 2 figs.

Akram, S.M. & Qureshi, J.I. (1995) Snakes of Faisalabad, Punjab, Pakistan. *The Snake* 27(1), 25–29, 1 fig.

Alarcón-Pardo, H. (1978) Primer registro de *Tretanorhinus nigroluteus nigroluteus* Cope (Reptilia, Serpentes, Colubridae) para Colombia. *Lozania* (27), 1–4.

Alayo-Dalmau, P. (1955) *Lista de los reptiles de Cuba.* Museo Charles T. Ramsden, Santiago de Cuba, 31 pp.

Alberdi, M.T., Morales, J., Maya, S. & Sanchiz, B. (1981) Macrovertebrados (Reptilia y Mammalia) del Yacimiento finimioceno de Librilla (Murcia). *Estudios Geologicos*, Madrid 37, 307–312, figs.

Albino, A.M. (1986) Nuevos Boidae Madtsoiinae en el Cretácico tardío de Patagonia (Formación Los Alamitos, Río Negro, Argentina). *Actas IV Congreso Argentino de Paleontología y Bioestratigrafía*, Mendoza 2, 15–21, 3 figs.

Albino, A.M. (1987a) Un nuevo Boidae (Reptilia, Serpentes) del Eoceno temprano de la provincia del Chubut, Argentina. *Ameghiniana*, Buenos Aires 24(1–2), 61–66, 1 fig.

Albino, A.M. (1987b) The late Cretaceous fauna of Los Alamitos, Patagonia, Argentina. Part V—the ophidians. *Revista del Museo Argentino de Ciencias Naturales "Bernardino Rivadavia" (Paleontología)*, Buenos Aires 3(3), 141–145, 1 fig.

Albino, A.M. (1993) Snakes from the Paleocene and Eocene of Patagonia (Argentina): paleoecology and coevolution with mammals. *Historical Biology* 7, 51–69, 5 figs.

Albino, A.M. (1994) Una nueva serpiente (Reptilia) del Cretácico tardío de Patagonia, Argentina. *Pesquisas*, Porto Alegre 21(1), 58–63, 4 figs.

Albino, A.M. (1996a) Snakes from the Miocene of Patagonia (Argentina). Part I. The Booidea. *Neues Jahrbuch für Geologie und Palaeontologie. Abhandlungen*, Stuttgart 199(3), 417–434, 7 figs.

Albino, A.M. (1996b) Snakes from the Miocene of Patagonia (Argentina). Part II. The Colubroidea. *Neues Jahrbuch für Geologie und Palaeontologie. Abhandlungen*, Stuttgart 200(3), 353–360

Albino, A.M. (2000) New records of snakes from the Cretaceous of Patagonia (Argentina). *Geodiversitas* 22(2), 247–253, 3 figs.

Albino, A. (2011a) Morfología vertebral de *Boa constrictor* (Serpentes: Boidae) y la validez del género Mioceno *Pseudoepicrates* Auffenberg, 1923. *Ameginiana* 48(1), 53–62, 4 figs.

Albino, A. (2011b) Evolution of Squamata reptiles in Patagonia based on the fossil record. *Biological Journal of the Linnean Society*, London 103(2), 441–457, 7 figs.

Albino, A.M. & Caldwell, M.W. (2003) Hábitos de vida de la serpiente Cretácica *Dinilysia patagonica* Woodward. *Ameghiniana*, Buenos Aires 40(3), 407–414, 2 figs.

Albino, A.M. & Carlini, A.A. (2008) First record of *Boa constrictor* (Serpentes, Boidae) in the Quaternary of South America. *Journal of Herpetology* 42(1), 82–88, 3 figs.

Albino, A.M. & Quintana, C.A. (1992) Los Colubridae (Reptilia, Serpentes) del Chapadmalalense y Uquiense (Plioceno Tardio–Pleistoceno Temprano?) de la provincia de Buenos Aires, Argentina. *Ameghiniana*, Buenos Aires 29(2), 125–133, 3 figs.

Albuja, L., Ibarra, M., Urgilés, J. & Barriga, R. (1980) *Estudio preliminar de los vertebra ecuatorianos.* Escuela Politécnica Nacional, Departamento de Ciencias Biológicas, Quito, 143 pp.

Albuquerque, N.R. (2000) The status of *Hydrops martii* (Wagler, 1824) (Serpentes, Colubridae). *Boletim do Museu Paraense Emilio Goeldi (Zoological Series)* 16, 153–161.

Albuquerque, N.R. (2001) Geographic distribution: Serpentes: *Hydrops triangularis. Herpetological Review* 32(1), 60.

Albuquerque, N.R. (2002) Osteologia craniana, morfologia do hemipênis e o posicionamento sistemático do gênero *Hydrops* Wagler, 1830 (Serpentes, Colubridae). *Comunicações do Museu de Ciências e Tecnologia da PUCRS (Zoologia)*, Pôrto Alegre 1, 21–32.

Albuquerque, N.R. (2008) Taxonomic revision of the Neotropical water snake *Hydrops triangularis* (Serpentes, Colubridae). *Zootaxa* (1685), 55–66, 6 figs.

Albuquerque, N.R. (2009) New records of *Leptophis ahaetulla* (Serpentes, Colubridae) for Venezuela, Colombia and the placement of *L. a. copei* into the synonymy of *L. a. ahaetulla. Biota Neotropica* 9(4), 293–297, 2 figs.

Albuquerque, N.R. de & Lema, T. de. (2012) Description of the second known specimen of *Apostolepis intermedia* (Serpentes, Colubridae, Xenodontinae). *Zootaxa* (3325), 53–58, 3 figs.

Albuquerque, N.R. & McDiarmid, R.W. (2010) Redescription of *Leptophis cupreus* (Cope) (Serpentes, Colubridae), a rare South American colubrine snake. *Papéis Avulsos de Zoologia* 50(23), 375–384, 7 figs.

Albuquerque, N.R. de, Passos, P. & Gotte, S.W. (2012) *Leptophis santamartensis* (Serpentes, Colubridae), a junior synonym of *Leptophis ahaetulla occidentalis. Journal of Herpetology* 46(2), 248–252, 1 fig.

Alcala, A.C. (1986) *Guide to Philippine flora and fauna. Vol. X. Amphibians and reptiles.* Natural Resources Management Center, Ministry of Natural Resources, University of the Philippines, Manila, 195 pp., 172 figs.

Alcock, A. & Finn, F. (1897) An account of the Reptilia collected by Dr. F.P. Maynard, Captain A.H. McMahon, C.I.E., and the members of the Afghan-Baluch Boundary Commission of 1896. *Journal of the Asiatic Society of Bengal (Natural History)*, Calcutta (1896) 65(4), 550–566, pls. 11–15.

Aldridge, R.D. (1992) Oviductal anatomy and seasonal sperm storage in the southeastern crowned snake (*Tantilla coronata*). *Copeia* 1992(4), 1102–1106, 2 figs.

Aldridge, R.D. & Semlitsch, R.D. (1992a) Female reproductive biology of the southeastern crowned snake (*Tantilla coronata*). *Amphibia-Reptilia*, Leiden 13(3), 209–218, 5 figs.

Aldridge, R.D. (1992b) Male reproductive biology of the southeastern crowned snake (*Tantilla coronata*). *Amphibia-Reptilia*, Leiden 13(3), 219–225, 4 figs.

Aleman, G.C. (1952) Apuntes sobre reptiles y anfibios de la region Baruta-El Hatillo. *Memorias de la Sociedad de Ciencias Naturales "La Salle,"* Caracas 12, 11–30.

Aleman, G.C. (1953) Contribucion al estudio de los reptiles y batrocios de la Sierra de Perija. *Memorias de la Sociedad de Ciencias Naturales "La Salle,"* Caracas 13(35), 205–225, 5 figs.

Alenizin, V. (1876) Reptiles of the islands and shores of the Aral Sea. *Travaux de la Société des Naturalistes de St.-Pétersbourg* 1876(3), 1–64. (in Russian)

Alexander, G.J. (1990) Reptiles and amphibians of Durban. *Durban Museum Novitates* 15, 1–41, illust.

Alexander, G.J. (2007) Thermal biology of the Southern African python (*Python natalensis*), does temperature limit its distribution? *In:* Henderson, R.W. & Powell, R. (Eds.), *Biology of the Boas and Pythons*. Eagle Mountain Publishing, Eagle Mountain, pp. 51–75, 14 figs., col. illust.

Alexander, G. & Marais, J. (2007) *A guide to the reptiles of southern Africa.* Struik Publishers, Cape Town, 408 pp., col. illust.

Alfaro, A. (1985) Catálogo de los reptiles de Costa Rica. *Gaceta Médica Costa Rica* 10(4), 78–81.

Alfaro, M.E. & Arnold, S. J. (2001) Molecular systematics and evolution of *Regina* and the thamnophiine snakes. *Molecular Phylogenetics and Evolution* 21(3), 408–423, 7 figs.

Alfaro, M.E., Karns, D.R., Voris, H.K., Brock, C.D. & Stuart, B.L. (2008) Phylogeny, evolutionary history, and biogeography of Oriental-Australian rear-fanged water snakes (Colubroidea: Homalopsidae) inferred from mitochondrial and nuclear DNA sequences. *Molecular Phylogenetics and Evolution* 46(2), 576–593, 6 figs.

Allen, G.M. (1899) Notes on amphibians and reptiles of Intervale, New Hampshire. *Proceedings of the Boston Society of Natural History* 29, 63–75.

Allen, W.B., Jr. (1992) The snakes of Pennsylvania. *Reptile & Amphibian Magazine* 33 pp., col. illust.

Allsteadt, J., Savitzey, A.H., Petersen, C.E. & Naik, D.N. (2006) Geographic variation in the morphology of *Crotalus horridus* (Serpentes, Viperidae). *Herpetological Monographs* (20), 1–63, 14 figs.

Almaça, C. & Neves, A.M. (1987) The Museu Bocage and the new Series of its Arquivos. *Arquivos do Museu Bocage* 1(1), 1–8.

Almendáriz, A. (1987) Contribucion al conocimiento de la herpetofauna centroriental Ecuatoriana. *Politecnica (Biologia 1)*, Quito 12(4), 77–133, 29 figs.

Almendáriz, A. (1991) Anfibios y reptiles *in* Lista de vertebrados del Ecuador (Peces de agua dulce, anfibios, reptiles y mamíferos). *Politecnica (Biologia 3)*, Quito 16(3), 89–162.

Al-Mohanna, S., Jaman, S. & Meakins, R. (1997) First report of a melanistic Whip Snake, *Coluber jugularis* Linnaeus, 1758 from the Arabian Peninsula. *Zoology in the Middle East* 14, 73.

Al-Oran, R.M. (1994) Snakes of southern Jordan. *Bollettino di Zoologia* 61, 359–367, 1 fig.

Al-Oran, R.M. & Amr, Z.S. (1995) First record of the mole viper, *Atractaspis microlepidota engaddensis* from Jordan. *Zoology in the Middle East* 11, 47–49.

Al-Quran, S. (2009) The herpetofauna of the southern Jordan. *American-Eurasian Journal of Agriculture & Environmental Science* 6(4), 385–391.

Al-Sadoon, M.K., Al-Farraj, S.A. & Abdo, N.M. (1991) Survey of the reptilian fauna of the Kingdom of Saudi Arabia. III. An ecological survey of the lizard, amphisbaenian and snake fauna of Al-Zulfi Area. *Bulletin of the Maryland Herpetological Society* 27(1), 1–22, 10 figs.

Alström, P. & Olsson, U. (1999) The golden-spectacled warbler: a complex of sibling species, including a previously undescribed species. *Ibis* 141, 545–568.

Alvarado-Diaz, J. & Campbell, J.A. (2004) A new montane rattlesnake (Viperidae) from Michoacán, Mexico. *Herpetologica* 60(2), 281–286, 3 figs.

Alvarado-Diaz, J., Huacuz-Elias, D.C. & Suazo-Ortuno, I. (1997) Geographic distribution: Serpentes: *Porthidium hespere*. *Herpetological Review* 28(2), 98.

Alvarado-Diaz, J., Estrada-Virgen, A. & Suazo-Ortuno, I. (2007) Natural history notes on *Crotalus tancitarensis* (Serpentes, Viperidae). *Herpetological Review* 38(2), 155–157, 1 col. fig.

Alvarez, B.B., Rey, L. & Cei, J.M. (1992) A new subspecies of the *reticulatus* group, genus *Atractus*, from southeastern South America (Serpentes, Colubridae). *Bollettino del Museo Regionale di Scienze Naturali Torino* 10(2), 249–256, 3 figs.

Alvarez, B.B., Garcia, J.A.R., Céspedez, J.A., Hernando, A.B., Zaracho, V.H., Calamante, C.C. & Aguirre, R.H. (2009) Herpetofauna, provinces of Chaco and Formosa, Chaco Oriental region, north-eastern Argentina. *Check List* 5(1), 74–82, 3 figs.

Alvarez del Toro, M. (1982) *Los reptiles de Chiapas*. Third edition. Instituto Zoologico del Estado, Tuxtla Gutierrez, 248 pp., 168 figs. [first edition, 1960, 204 pp., figs.; second edition, 1972, 178 pp., 154 figs.]

Alvarez del Toro, M. & Smith, H.M. (1956) Notulae herpetologicae Chiapasiae I. *Herpetologica* 12(1), 3–17.

Alves, F.Q. & Argôlo, A.J.S. (1998) *Dipsas indica petersi*. Geographic distribution: Serpentes. *Herpetological Review* 29(3), 176.

Amaral, A. do. (1921a) Contribuição para o conhecimento dos ofídios do Brasil – A. Parte I. Quatro novas espécies de serpents brasileiras. *Anexos de Memorias do Instituto de Butantan (Ofiologia)*, São Paulo 1(1), 5–37, 49–81, 4 pls.

Amaral, A. do. (1921b) Contribution towards the knowledge of snakes in Brazil – A. Parte II. Biologia da nova espécie, *Lachesis insularis*. *Anexos de Memorias do Instituto de Butantan (Ofiologia)*, São Paulo 1(1), 39–44, 83–88, pls. 5–17.

Amaral, A. do. (1921c) Ultimos trabalhos inéditos de J. Florencio Gomés, duos novas espécies de Colubrideos opisthoglyphos brasileiros (*Philodryas oligolepis* e *Apostolepis longicaudata*). *Anais Paulistas de Medicina e Cirurgia*, São Paulo 9(7–8), 110–113, pl. A.

Amaral, A. do. (1923) New genera and species of snakes. *Proceedings of the New England Zoological Club* 8, 85–105.

Amaral, A. do. (1924a) *Helminthophis*. *Proceedings of the New England Zoological Club* 9, 25–30.

Amaral, A. do. (1924b) New genus and species of South American snakes contained in the United States National Museum. *Journal of the Washington Academy of Science* 14(9), 200–202.

Amaral, A. do. (1925a) South American snakes in the collection of the United States National Museum. *Proceedings of the United States National Museum* (1925–1926) 67(2596), 1–30.

Amaral, A. do. (1925b) Ophidios de Matto Grosso (Contribuição II para o conhecimento dos ophidios do Brasil). *Commissão de Linhas Telegraphicas Estrategicas de Matto Grosso ao Amazonas (Annexo 5, Historia Natural, Zoologia)*, São Paulo (84), 1–29, 1 pl.

Amaral, A. do. (1925c) A general consideration of snake poisoning and observations on Neotropical pit-vipers. *Contributions of the Harvard Institute for Tropical Biology and Medicine* 2, 1–64, 16 col. pls.

Amaral, A. do. (1926a) Novos generos e espécies de ophídios brasileiros. (Contribuição III para o conhecimento dos ophidios do Brasil). *Archivos do Museu Nacional*, Rio de Janeiro 26, 95–121, 4 pls.

Amaral, A. do. (1926b) Ophidia from South America in the Carnegie Museum, a critique of Dr. L.E. Griffin's "Catalog of the Ophidia from South America at present (June, 1916) contained in the Carnegie Museum." *Annals of the Carnegie Museum* (1925–1926) 16(2), 319–323.

Amaral, A. do. (1926c) Notas de ophiologia. 2a. Nota de nomenclatura ophiologica. Sobre o emprego do nome generico *Sibynomorphus* em vez de *"Leptognathus," "Cochliophagus," "Stremmatoganthus," "Anholodon,"* etc. *Revista do Museu Paulista*, São Paulo 14, 1–3.

Amaral, A. do. (1926d) Notas de Ophiologia. Sobre a invalidez de um genero e algumas especies de ophidios sul-americanos. *Revista do Museu Paulista*, São Paulo 14, 17–40.

Amaral, A. do. (1926e) Estudo comparativo da evolução ontogenetica de *Pseudoboa cloelia* (Daudin, 1803) e *Ps. haasi* (Boettger, 1906). *Revista do Museu Paulista*, São Paulo 15, 105–110, pl. A.

Amaral, A. do. (1927a) Studies of neotropical Ophidia. 4. A new form of Crotalidae from Bolivia. *Bulletin of the Antivenin Institute of America* 1(1), 5–6.

Amaral, A. do. (1927b) Studies of neotropical Ophidia. VI. A new genus of snakes from Honduras. *Bulletin of the Antivenin Institute of America* 1(1), 28–29, 1 fig.

Amaral, A. do. (1927c) Variações das marcas dorsaes de *Crotalus terrificus* Laurenti, 1768. Collectanea Ophiologica 12. *Revista do Museu Paulista* (1926) 15, 89–91.

Amaral, A. do. (1929a) Studies of nearctic Ophidia. V. On *Crotalus confluentus* Say, 1823, and its allied forms. *Bulletin of the Antivenin Institute of America* 2(4), 86–97.

Amaral, A. do. (1929b) Studies of neotropical Ophidia. XIII. A new colubrine snake in the collection of the Vienna Museum. *Bulletin of the Antivenin Institute of America* 3(2), 40.

Amaral, A. do. (1930a) Studies of neotropical Ophidia. XVI. Two new snakes from central Colombia. *Bulletin of the Antivenin Institute of America* 4(2), 27–28.

Amaral, A. do. (1930b) Studies of neotropical Ophidia. XXIV. A new Brazilian snake. *Bulletin of the Antivenin Institute of America* 4(3), 65.

Amaral, A. do. (1930c) A new race of *Bothrops neuwiedii*. *Bulletin of the Antivenin Institute of America* 4(3), 65–67, 1 fig.

Amaral, A. do. (1930d) Estudos sobre ophídios neotrópicas. XVII. Valor systematico de varias formas de ophídios neotrópicos. *Memórias do Instituto Butantan*, São Paulo (1929) 4, 3–68.

Amaral, A. do. (1930e) Contribuição dos ophídios do Brasil. IV. Lista remissiva dos ophidios do Brasil. *Memórias do Instituto Butantan*, São Paulo (1929) 4, 71–125 + i–viii.

Amaral, A. do. (1930f) Estudos sobre ophídios neotrópicos. XVIII. Lista remissiva dos ophidios da região neotropica. *Memórias do Instituto Butantan*, São Paulo (1929) 4, 129–271.

Amaral, A. do. (1930g) Estudos sobre ophídios neotrópicos. XIX. Revisão do genero *Spilotes* Wagler, 1830. *Memórias do Instituto Butantan*, São Paulo (1929) 4, 275–298, 8 figs.

Amaral, A. do. (1930h) Estudos sobre ophídios neotrópicos. XX. Revisão do genero *Phrynonax* Cope, 1862. *Memórias do Instituto Butantan*, São Paulo (1929) 4, 301–320, 6 figs.

Amaral, A. do. (1930i) Estudos sobre ophídios neotrópicos. XXI. Revisão do genero *Drymarchon* Fitzinger, 1843. *Memórias do Instituto Butantan*, São Paulo (1929) 4, 323–330, 3 figs.

Amaral, A. do. (1930j) Estudos sobre ophídios neotrópicos. XXII. Sobre a especie *Coluber dichrous* (Peters) Boulenger, 1894. *Memórias do Instituto Butantan*, São Paulo (1929) 4, 333–337, 1 fig.

Amaral, A. do. (1930k) Uma raridade ophidica do Brasil. *Boletim do Museu Nacional,* Rio de Janeiro 6(1), 3–8.

Amaral, A. do. (1931a) Studies of neotropical Ophidia. XXIII. Additional notes on Colombian snakes. *Bulletin of the Antivenin Institute of America* 4(4), 85–89.

Amaral, A. do. (1931b) Studies of neotropical Ophidia. XXVI. Ophidia of Colombia. *Bulletin of the Antivenin Institute of America* 4(4), 89–94.

Amaral, A. do. (1932a) Studies of neotropical Ophidia. XXVII. On two small collections of snakes from Central Colombia. *Bulletin of the Antivenin Institute of America* 5(3), 66–68.

Amaral, A. do. (1932b) Novas notas sobre especies da Colombia. *Memórias do Instituto Butantan*, São Paulo 7, 105–123.

Amaral, A. do. (1933a) Contribuição ao conhecimento dos ophidios do Brasil. V. Uma nova raça de *Bothrops neuwiedii. Memórias do Instituto Butantan*, Sao Paulo (1932) 7, 97–98, 1 pl.

Amaral, A. do. (1933b) Contribuição ao conhecimento dos ophídios do Brasil. VI. Uma nova espécie de Colubrideo opisthoglypho, do genero *Chlorosoma* Wagler, 1930. *Memórias do Instituto Butantan*, São Paulo (1932) 7, 99–101, 4 figs.

Amaral, A. do. (1933c) Estudos sobre ophidios neotropicos. XXIX. Novas notas sobre especies da Colombia. *Memórias do Instituto Butantan*, São Paulo (1932) 7, 105–123.

Amaral, A. do. (1934a) Estudos sobre ophídios neotropicos. XXX. Novo genero e espécie de colubrideo na fauna da Colombia. *Memórias do Instituto Butantan*, São Paulo (1933–1934) 8, 157–159.

Amaral, A. do. (1934b) Collecta herpetologica no nordeste do Brasil. *Memórias do Instituto Butantan*, São Paulo (1933–1934) 8, 183–192.

Amaral, A. do. (1935a) Contribuição ao conhecimento dos ophídios do Brasil. VII. Novos generos e espécies de colubrideos opisthoglyphos. *Memórias do Instituto Butantan*, São Paulo 9, 203–206, 2 figs.

Amaral, A. do. (1935b) Estudos sobre ophidios neotropicos. XXXII. Apontamentos sobre a fauna da Colombia. *Memórias do Instituto Butantan*, São Paulo 9, 209–216.

Amaral, A. do. (1935c) Estudos sôbre ophídios neotrópicos. XXXIII. Novas especies de ophidios da Colombia. *Memórias do Instituto Butantan.*, São Paulo 9, 219–223, 1 pl.

Amaral, A. do. (1935d) Collecta herpetologica no Nordeste do Brasil (Contribuição II). *Memórias do Instituto Butantan*, São Paulo 9, 227–232.

Amaral, A. do. (1935e) Collecta herpetologica no centro do Brasil. *Memórias do Instituto Butantan*, São Paulo 9, 235–246.

Amaral, A. do. (1937a) New genera and species of opisthoglyphous colubrids from Brazil. *Comptes Rendus du XII Congrès de l'International Zoologie*, Lisbonne (1935) 3, 1708–1711, pl. 82.

Amaral, A. do. (1937b) New species of ophidians from Colombia. *Comptes Rendus du XII Congrès de l'International Zoologie*, Lisbonne (1935) 3, 1762–1767, figs.

Amaral, A. do. (1937c) Check-list of the Ophidia of Brazil. *Comptes Rendus du XII Congrès de l'International Zoologie*, Lisbonne (1935) 3, 1744–1761.

Amaral, A. do. (1937d) Contribuição ao conhecimento dos ophidios do Brasil. VIII. Lista remissiva dos ophidios do Brasil. 2.a edição. *Memórias do Instituto Butantan*, São Paulo (1935–1936) 10, 87–162 + i–xix.

Amaral, A. do. (1938a) Contribuição ao conhecimento dos ophídios do Brasil. 9. Nova espécie de colubrideo opisthoglypho, confundivel com *Philodryas serra* (Schlegel, 1837). *Memórias do Instituto Butantan*, São Paulo (1937) 11, 205–211.

Amaral, A. do. (1938b) Estudos sôbre ophídios neotrópicos. 34. Novas notas sobre a fauna da Colombia e descripção de uma espécie nova de colubrideo aglypho. *Memórias do Instituto Butantan*, São Paulo (1937) 11, 231–240.

Amaral, A. do. (1944a) Notas sôbre a ofiologia neotrópica e brasilica. I. *Papeis Avulsos do Departamento de Zoologia*, São Paulo 5, 1–6.

Amaral, A. do. (1944b) Notas sôbre a ofiologia neotrópica e brasilica. II. Sobre *Bothrops lansbergii* (Schlegel, 1841); *Trimeresurus dunni* Hartweg & Oliver, 1938; *T. lansbergii annectens* Schmidt, 1936. *Papeis Avulsos do Departamento de Zoologia*, São Paulo 5, 7–12.

Amaral, A. do. (1944c) Notas sôbre a ofiologia neotrópica e brasilica. III. Sobre a aplicação do nome genérico *Trimeresurus* em vez de *Bothrops*, a serpentes neotrópicas. *Papeis Avulsos do Departamento de Zoologia*, São Paulo 5, 13–18.

Amaral, A. do. (1944d) Notas sôbre a ofiologia neotrópica e brasilica. IV. Da invalidez da especie *Bothrops pessoai* A. Prado, 1939. *Papeis Avulsos do Departamento de Zoologia*, São Paulo 5, 19–28.

Amaral, A. do. (1944e) Notas sôbre a ofiologia neotrópica e brasilica. V. Sobre a invalidez especifica de *Crotalus unicolor*. *Papeis Avulsos do Departamento de Zoologia*, São Paulo 5, 29–40.

Amaral, A. do. (1944f) Notas sôbre a ofiologia neotrópica e brasilica. VI. Formas de boíneos de recente registro. *Papeis Avulsos do Departamento de Zoologia*, São Paulo 5, 41–48.

Amaral, A. do. (1944g) Notas sôbre a ofiologia neotrópica e brasilica. VII. Formas de colubríneos de recente registro. *Papeis Avulsos do Departamento de Zoologia*, São Paulo 5, 49–60.

Amaral, A. do. (1944h) Notas sôbre a ofiologia neotrópica e brasilica. VIII. Formas de dipsadíneos de recente registro. *Papeis Avulsos do Departamento de Zoologia*, São Paulo 5, 61–64.

Amaral, A. do. (1944i) Notas sôbre a ofiologia neotrópica e brasilica. IX. Formas de boigíneos de recente registro. *Papeis Avulsos do Departamento de Zoologia*, São Paulo 5, 65–74.

Amaral, A. do. (1944j) Notas sôbre a ofiologia neotrópica e brasilica. X. Distribuição geográfica e racial de *Leimadophis poecilogyrus* (Wied). *Papeis Avulsos do Departamento de Zoologia*, São Paulo 5, 75–82.

Amaral, A. do. (1944k) Notas sôbre a ofiologia neotrópica e brasilica. XI. Subespécies de *Micrurus lemniscatus* (L.) e suas afindades com *M. frontalis* (Dm. & Bibr.). *Papeis Avulsos do Departamento de Zoologia*, São Paulo 5, 83–94.

Amaral, A. do. (1948) Ofidio de Mato Grosso (Contribuição II para o conhecimento dos ofidios do Brasil). *Commissão de Linhas Telegraphicas Estrategicas de Matto Grosso ao Amazonas (Annexo 5, Historia Natural, Zoologia)*, São Paulo, 43 pp.

Amaral, A. do. (1949) Ofidios do Para. *Boletim do Museu Paraense Emilio Goeldi (Zoologia)*, Belém (1948) 10, 149–159.

Amaral, A. do. (1955a) Contribuição ao conhecimento dos ofidios do Brasil. 12. Notas a respeito de *Helminthophis ternetzii* Boulenger 1896. *Memórias do Instituto Butantan*, São Paulo (1954) 26, 191–195, 2 figs.

Amaral, A. do. (1955b) Contribuição ao conhecimento dos ofidios do Brasil. 13. Observações a próposito de "cobras–cegas" (fam. Typhlopidae e fam. Leptotyphlopidae). *Memórias do Instituto Butantan*, São Paulo (1954) 26, 197–202, 4 figs.

Amaral, A. do. (1955c) Contribuição ao conhecimento dos ofidios do Brasil. 14. Descrição de duas espécies novas de "cobra–cega" (fam. Leptotyphlopidae). *Memórias do Instituto Butantan*, São Paulo (1954) 26, 203–205, 6 figs.

Amaral, A. do. (1955d) Contribuição ao conhecimento dos ofidios do Brasil. 15. Situação taxonómica de algumas formas de Crotalidae Lachesinae, recentemente descritas. *Memórias do Instituto Butantan*, São Paulo (1954) 26, 215–220.

Amaral, A. do. (1955e) Contribuição ao conhecimento dos ofidios neotrópicos. XXXVI. Redescrição da espécie *Bothrops hyoprora* Amaral, 1935. *Memórias do Instituto Butantan*, São Paulo (1954) 26, 221–225.

Amaral, A. do. (1955f) Contribuição ao conhecimento dos ofídios neotrópicos. XXXVII. Sub-espécies de *Epicrates cenchria* (Lineu, 1758). *Memórias do Instituto Butantan*, São Paulo (1954) 26, 227–247.

Amaral, A. do. (1963) Herpetological note. *Copeia* 1963(3), 580.

Amaral, A. do. (1978) *Serpentes do Brasil, Iconografia colorida (Brazilian snakes, a color iconography)*. Second edition. Universidade de São Paulo, São Paulo, 247 pp., 164 col. pls. [first edition, 1977]

Amat, F. (2004) Analysis of the morphological variation of *Vipera latastei*, taxonomic and biogeographic implications. *Russian Journal of Herpetology* 11(3), 198–202, 3 figs.

Ambu, S. & Ng, C.S. (1991) The first record of the green cat snake (*Boiga cyanea*) from peninsular Malaysia. *The Snake* 23(1), 40–41, 3 figs.

Amerson, A.B., Jr., Whistler, W.A. & Schwaner, T.D. (1982) *Wildlife and wildlife habitat of American Samoa. II. Accounts of flora and fauna*. United States Department of the Interior, Fish and Wildlife Service, Washington, 151 pp., 26 figs.

Amr, Z.S. (2009a) Reptiles and amphibians of Iraq. *Nature Iraq*, Sulaimani, 10 pp.

Amr, Z.S. (2009b) Venomous snakes of Iraq. *Nature Iraq*, Sulaimani, 22 pp., 8 col. photos.

Amr, Z.S. & Amr, S.S. (1983) Snakebites in Jordan. *The Snake* 15(2), 81–85, 1 fig.

Amr, Z.S. & Al-Oran, R. (1995) *Venomous animals in Jordan. The snakes of Jordan*. Mutah University, Mu'tah, 88 pp., 33 figs., 41 col. photos. (in Arabic)

Amr, Z.S., Al-Oran, R. & Disi, A.M. (1994) Reptiles of southern Jordan. *The Snake* 26(1), 41–49.

An, J., Park, D., Lee, J. Kim, K.-S., Lee, H. & Min, M.-S. (2010) No genetic differentiation of *Elaphe schrenckii* subspecies in Korea based on 9 microsatellite loci. *Korean Journal of Systematic Zoology* 26(1), 15–19.

Ananjeva, N.B., Munkhbayar, K., Orlov, N.L., Orlova, V.F., Semenov, D.V. & Terbish, K. (1997) *Amphibians and reptiles of Mongolia*. KMK Publishing, Moscow, 416 pp., 105 figs. (in Russian)

Ananjeva, N.B., Borkin, L.J., Darevsky, I.S. & Orlov, N. (1998) *Encyclopedia of Russian nature: amphibians and reptiles*. ABF, Moscow, 574 pp., 35 figs., 64 pls., 58 col. photos. (in Russian)

Ananjeva, N.B., Orlov, N.L., Khalikov, R.G., Darevsky, I.S., Ryabov, S.A. & Barabanov, A.V. (2006). *The reptiles of northern Eurasia: taxonomic diversity, distribution, conservation status*. Pensoft Publishers, Sofia, 245 pp., col. illust. [English trans. of Ananjeva et al., 2004, "Colored atlas of the reptiles of the North Eurasia (taxonomic diversity, distribution, conservation status)," Zoological Institute, Russian Academy of Sciences, St.-Petersburg, 230 pp., col. illust.]

Anderson, C.D., Gibbs, H.L., Douglas, M.E. & Holycross, A.T. (2009) Conservation genetics of the desert massasauga rattlesnake (*Sistrurus catenatus edwardsii*). *Copeia* 2009(4), 740–747, 3 figs.

Anderson, C.G. & Greenbaum, E. (2012) Phylogeography of northern populations of the black-tailed rattlesnake *(Crotalus molossus* Baird and Girard, 1853), with the revalidation of *C. ornatus* Hallowell, 1854. *Herpetological Monographs* (26), 19–57, 11 figs.

Anderson, J. (1871a) On some Indian reptiles. *Proceedings of the Zoological Society of London* 39(1), 149–211.

Anderson, J. (1871b) A list of the reptilian accessions to the Indian Museum, Calcutta, from 1865 to 1870, with a description of some new species. *Journal of the Asiatic Society of Bengal (Natural History),* Calcutta 40(1), 12–39.

Anderson, J. (1872) On some Persian, Himalayan, and other reptiles. *Proceedings of the Zoological Society of London* 40(1), 371–404, 9 figs.

Anderson, J. (1879) *Anatomical and zoological researches, comprising an account of the zoological results of the two expeditions to western Yunnan in 1868 and 1875; and a monograph of the two cetacean genera,* Platanista *and* Orcella. First volume–text. Second volume–plates. Bernard Quaritch, London (1878), 985 pp., 84 pls.

Anderson, J. (1892) On a small collection of mammals, reptiles, and batrachians from Barbary. *Proceedings of the Zoological Society of London* 60(1), 3–24, 1 pl.

Anderson, J. (1893) On a new species of *Zamenis* and a new species of *Bufo* from Egypt. *Annals and Magazine of Natural History*, London (Series 6) 12(72), 439–440.

Anderson, J. (1895) On a collection of reptiles and batrachians made by Colonel Yerbury at Aden and its neighbourhood. *Proceedings of the Zoological Society of London* 63(3), 635–663, pls. 36–37.

Anderson, J. (1896) *A contribution to the herpetology of Arabia. With a preliminary list of the reptiles and batrachians of Egypt.* R.H. Porter, London, 122 pp.

Anderson, J. (1898) *Zoology of Egypt. Volume First. Reptilia and Batrachia.* Bernard Quaritch, London, xv + 371 pp., 50 pls. [reprinted in 1965 by Wheldon & Wesley, Codicote, and J. Cramer, Weinheim]

Anderson, J. (1900) On the presence of *Echis coloratus* Günther, in Africa. *Annals and Magazine of Natural History*, London (Series 7) 6(34), 419–425, 4 figs.

Anderson, J.D. (1962) A new subspecies of the ridge-nosed rattlesnake, *Crotalus willardi*, from Chihuahua, Mexico. *Copeia* 1962(1), 160–163.

Anderson, P. (1965) *The reptiles of Missouri*. University of Missouri Press, Columbia, xxiii + 330 pp., 169 figs.

Anderson, P.K. (1961) Variation in populations of brown snakes, genus *Storeria*, bordering the Gulf of Mexico. *American Midland Naturalist* 66(1), 235–249.

Anderson, R. (1983) *Guide to Florida poisonous snakes*. Second edition. Anderson's Nature Guides, Port Richey, 60 pp., 40 figs., 22 col. photos.

Anderson, R. (1985) *Guide to Florida nonpoisonous snakes.* Winner Enterprises, Altamonte Springs, 56 pp., 7 figs., 16 col. photos.

Anderson, R. (1989) *The great outdoors book of Florida snakes.* Great Outdoors Publishing Company, St. Petersburg, 129 pp., figs., 33 col. photos.

Anderson, S.C. (1963) Amphibians and reptiles from Iran. *Proceedings of the California Academy of Sciences* (Series 4) 31(16), 417–498, 15 figs.

Anderson, S.C. & Leviton, A.E. (1969) Amphibians and reptiles collected by the Street Expedition to Afghanistan, 1965. *Proceedings of the California Academy of Sciences* (Series 4) 37(2), 25–56, 8 figs.

Andersson, L.G. (1898) List of the reptiles and batrachians collected by the Swedish Expedition to Tierra del Fuego (1895–1896) under direction of Dr. Otto Nordenskiöld. *Öfversigt af Kongliga Vetensk-Akademiens Förhandlingar*, Stockholm 55(7), 457–462.

Andersson, L.G. (1899) Catalogue of Linnean type-specimens of snakes in the Royal Museum in Stockholm. *Bihang till Kongliga Svenska Vetenskaps-Akademiens. Handlingar*, Stockholm (Series 4) 24(6), 1–35.

Andersson, L.G. (1901) Some new species of snakes from Cameroon and South America, belonging to the collections of the Royal Museum in Stockholm. *Bihang till Kongliga Svenska Vetenskaps-Akademiens. Handlingar*, Stockholm (Series 4) 27(5), 1–26, 2 pls.

Andersson, L.G. (1903) List of reptiles and batrachians collected by the Swedish Zoological Expedition to Egypt, the Soudan and the Sinaitic Peninsula. *In:* Jägerskiöld, L.A. (Ed.), *Results of the Swedish Zoological Expedition to Egypt and the White Nile 1901 under the direction of L. A. Jägerskiöld.* Royal University Library, Uppsala (4), 1–12.

Andersson, L.G. (1911) Reptiles and batrachians from the north-west of Madagascar collected by V. Kaudern, 1906–1907. *Arkiv för Zoologi*, Stockholm (1910) 7(7), 1–15.

Andersson, L.G. (1916) Notes on the reptiles and batrachians in the Zoological Museum at Gothenburg. *Meddelanden från Götesborgs Musei Zoologiska* (Series 9) 17(5), 1–41, 4 figs.

Andersson, L.G. (1923) Some reptiles and batrachians from central-Borneo. *Nytt Magazin for Naturvidenskaberne*, Kristiania 61(7), 119–125.

Andersson, L.G. (1935) Reptiles and batrachians from the central Sahara. *Göteborgs Kungliga Vetenskaps-och Vitterhets-Samhälles Handlingar* 4B(10), 1–19.

Andersson, L.G. (1937) Reptiles and batrachians collected in the Gambia by Gustav Svensson and Birger Rudebeck (Swedish Expedition 1931). *Arkiv för Zoologi*, Stockholm 29A(16), 1–28, 8 figs.

Andrada, J. (1980) *Guia de Campo de los anfibios y reptiles de la peninsula Iberica.* Ediciones Omega, Barcelona, 159 pp., 8 figs., 94 col. pls.

Andrén, C., Nilson, G., Höggren, M. & Tegelström, H. (1997) Reproductive strategies and sperm competition in the adder, *Vipera berus*. *In:* Thorpe, R.S., Wüster, W. & Malhotra, A. (Eds.), *Ecology, evolution and snakebite of venomous snakes.* Clarendon Press, Oxford, pp. 129–141, 1 fig.

Andreone, F. & Gavetti, E. (2007) *The life and herpetological contributions of Mario Giacinto Peracca (1861–1923).* Society for the Study of Amphibians and Reptiles, and, Societa Herpetologia Italica, 570 pp., illust.

Andreone, F. & Raxworthy, C.T. (1998) The colubrid snake *Brygophis coulangesi* (Domergue 1988) rediscovered in north-eastern Madagascar. *Tropical Biology* 11(2), 249–257, illust.

Andreone, F. & Sindaco, R. (1998) Erpetologia del Piemonte e della Valle d'Aosta. Atlante degli anfibi e dei rettili. *Museo Regionale di Scienze Naturali Torino, Monografias* (26), 1–283, illust.

Andrews, C.W. (1901) Preliminary note on some recently discovered extinct vertebrates from Egypt. (Part II.). *Geological Magazine*, London (Series 2) 8(10), 436–444, 3 figs.

Andrews, C.W. (1906) *A descriptive catalogue of the Tertiary Vertebrata of the Fayum, Egypt. Based on the collection of the Egyptian government in the Geological Museum, Cairo, and on the collection in the British Museum (Natural History), London*. British Museum (Natural History), London, 324 pp., 26 pls.

Andrews, C.W. (1924) Notes on some ophidian vertebrae from Nigeria, Appendix. *In:* Russ, W. (Ed.), The phosphate deposits of Abeokuta Province. *Bulletin of the Geological Survey of Nigeria* 7, 39–43, 2 figs.

Andrews, E.W. (1937) Notes on snakes from the Yucatan Peninsula. *Field Museum of Natural History Publications, Zoological Series* 20, 355–359.

Andrews, J.S. (2002) *The atlas of the reptiles and amphibians of Vermont*. James S. Andrews, Middlebury, 90 pp., 49 figs.

Andrzejowsky, A. (1832) Amphibia nostratia, seu enumeration sauriorum, ophidiorum nec non sireniorum in excursionibus per Volhyniam, Podoliam Guberniumque Chersonens e usque ad euxinum observatorum. *Nouveaux Mémoires de la Société Impériale des Naturalistes de Moscou* 2(2), 321–346.

Angarita-Sierra, T.G. (2009) Variación geográfica de *Ninia atrata* en Colombia (Colubridae: Dipsadinae). *Papéis Avulsos de Zoologia* 49(22), 277–288, 4 figs.

Angel, F. (1920a) Liste de reptiles récemment déterminés et entrés dans les collections et description d'une nouvelle espèce du genre *Amblycephalus*. *Bulletin du Museum d'Histoire Naturelle*, Paris 26(2), 112–114.

Angel, F. (1920b) Sur un collection de Reptilia et de batraciens de l'île de San Thomé et de l'île du Prince et description d'une espèce nouvelle du genre *Typhlops*. *Bulletin du Museum d'Histoire Naturelle*, Paris 26, 197–199, 3 figs.

Angel, F. (1920c) Sur deux ophidiens nouveaux de la collection du muséum. *Bulletin du Museum d'Histoire Naturelle*, Paris 26, 291–294, 4 figs.

Angel, F. (1920d) Liste de reptiles du Haut-Zambèze et de l'Afrique australe. Description d'une espèce nouvelle du genre *Monopeltis*. *Bulletin du Museum d'Histoire Naturelle*, Paris 26, 614–617.

Angel, F. (1921a) Reptiles du Haut-Zambèze et de l'Afrique australe. Description d'une espèce et d'une variété nouvelles. *Bulletin du Museum d'Histoire Naturelle*, Paris 27, 42–44, 3 figs.

Angel, F. (1921b) Description d'un ophidien nouveau de l'Angola appartenant au genre *Psammophis*. *Bulletin de la Société Zoologique de France*, Paris 46(8–10), 116–118, 1 fig.

Angel, F. (1921c) Sur des reptiles de la region du Gribingui. *Bulletin du Museum d'Histoire Naturelle*, Paris 27, 141–142.

Angel, F. (1922) Sur une collection de reptiles et de batrachiens, recueillis au Soudan Français par le Mission, du Dr. Millet Horsin. *Bulletin du Museum d'Histoire Naturelle*, Paris 28, 39–41.

Angel, F. (1923a) Reptiles du Sahara, rapportés par la Mission du Colonel Hovart. Description d'un ophidien nouveau du genre *Rhamphiophis*. *Bulletin du Museum d'Histoire Naturelle*, Paris 29, 205–208, 3 figs.

Angel, F. (1923b) Sur un genre nouveau de serpent aglyphe du Congo Français. *Bulletin du Museum d'Histoire Naturelle*, Paris 29, 348–350, 4 figs.

Angel, F. (1923c) Reptiles. *In: Mission Rohan-Chabot Angola et Rhodesia (1912–1914)*. Imprimérie Nationale, Paris, pp. 157–169, 12 figs., 1 pl.

Angel, F. (1927) Liste des reptiles et des batraciens rapports de l'Indo-Chine par M.P. Chevey. Description d'une variété nouvelle de *Simotes violaceous* Cantor. *Bulletin du Museum d'Histoire Naturelle*, Paris 33, 496–498.

Angel, F. (1929) Liste des reptiles et batraciens du Haut-Laos recueillis par M. Delacour. Description d'un genre, de deux espèces et d'une variété d'ophidiens. *Bulletin du Museum d'Histoire Naturelle*, Paris (Série 2) 1(1), 75–81, 4 figs.

Angel, F. (1931) *Contribution à l'étude de la fauna de Madagascar. Reptilia et Batrachia*. Faune des Colonies Francaises, Paris 4, 495–556, figs., pls.

Angel, F. (1932) Sue quelques reptiles et batraciens du Sahara. *Bulletin du Museum d'Histoire Naturelle*, Paris (Série 2) 4, 385–387.

Angel, F. (1933a) Une vipère nouvelle de l'Annam. *Bulletin du Museum d'Histoire Naturelle*, Paris (Série 2) 5(4), 277–278, 1 fig.

Angel, F. (1933b) *Les serpents del'Afrique Occidentale Française*. Larose Éditeur, Paris, 246 pp., 83 figs.

Angel, F. (1934a) Description d'un vipéridé nouveau, du Congo belge, et de deux batraciens de Madagascar. *Bulletin de la Société Zoologique de France*, Paris 59(3–4), 169–172.

Angel, F. (1934b) Remarques sur le genre *Oophilositum* Parker (Colubridé aglyphe) et description d'une espèce nouvelle. *Bulletin de la Société Zoologique de France*, Paris 59(5), 417–419.

Angel, F. (1935) Sur la validité spécifique de deux couleuvres de Madagascar. *Archives du Museum d'Histoire Naturelle*, Paris (Série 6) 12, 465–467.

Angel, F. (1936a) Materiaux erpetologiques recueillis a Madagascar par M. Roger Heim, Chargé de Mission. Description de deux formes nouvelles. *Bulletin du Museum d'Histoire Naturelle*, Paris (Série 2) 8(2), 125–139.

Angel, F. (1936b) Sur quelques formes nouvelles de reptiles et de batrachiens du Sahara central. *Bulletin de la Société Zoologique de France*, Paris 61(4), 273–277.

Angel, F. (1936c) Reptiles et batraciens de Syrie et de Mésopotamie récoltés par M.P. Pallary. *Bulletin de l'Institut d'Égypte* 18(1), 107–116.

Angel, F. (1938) Liste de reptiles de Mauritanie recueillis par la Mission d'études de la biologie des Acridiens en 1936 et 1937. Description d'une sous–espèce nouvelle d'*Eryx muelleri*. *Bulletin du Museum d'Histoire Naturelle*, Paris (Série 2) 10(5), 485–487.

Angel, F. (1939) Reptiles et batraciens de Madagascar et de la Réunion. Description d'un serpent nouveau du genre *Alluaudina*. *Bulletin du Museum d'Histoire Naturelle*, Paris (Série 2) 11(6), 536–538.

Angel, F. (1940) Reptiles et amphibiens recueillis au Cameroun par la Mission P. Lepesme R. Paulian et A. Villiers. *Bulletin du Museum d'Histoire Naturelle*, Paris (Série 2) 12(7), 390–399.

Angel, F. (1941) Sur quelques reptiles de Java. *Bulletin du Museum d'Histoire Naturelle*, Paris (Série 2)13(5), 411–413.

Angel, F. (1944) Contribution a l'étude de la faune herpétologique du Sahara central. *Bulletin du Museum d'Histoire Naturelle*, Paris (Série 2) 16(6), 418–419.

Angel, F. (1946) *Faune de France. Reptiles et amphibiens*. Paul Lechevalier, Paris, 204 pp., 83 figs.

Angel, F. (1947) Liste des reptiles et amphibiens de l'Inde envoyés au Muséum par M. le Dr. K. Lindberg. *Bulletin du Museum d'Histoire Naturelle*, Paris (2) 19(1), 57–59.

Angel, F. (1950) *Vie et moeurs des serpents*. Payot, Paris, 319 pp., 122 figs.

Angel, F. & Bourret, R.L. (1933) Sur une petite collection de serpents du Tonkin. Descriptions d'espèces nouvelles. *Bulletin de la Société Zoologique de France*, Paris 58(3–4), 129–140.

Angel, F. & Lhote, M.H. (1938) Reptiles et amphibiens du Sahara Central et du Soudan. *Bulletin du Comité d'Études Historiques et Scientifiques de l'Afrique Occidental Française* 21(3), 345–381.

Angel, F., Guibé, J., Lamotte, J. & Roy, R. (1954) La reserve naturelle intégrale du Mont Nimba. Serpents. *Memoirs Institut Française d'Afrique Noire*, Dakar 40(2), 381–402, 1 fig.

Annandale, N. (1904) Additions to the collection of Oriental snakes in the Indian Museum. *Journal of the Asiatic Society of Bengal (Natural History),* Calcutta 73(5), 207–211, pl. 9.

Annandale, N. (1905a) Additions to the collection of Oriental snakes in the Indian Museum, Part 2. Specimens from the Andamans and Nicobars. *Journal of the Asiatic Society of Bengal (Natural History),* Calcutta (Series 2) 1(7), 173–176.

Annandale, N. (1905b) Additions to the collection of Oriental snakes in the Indian Museum, Part 3. *Journal of the Asiatic Society of Bengal (Natural History),* Calcutta (Series 2) 1(8), 208–214, 3 figs.

Annandale, N. (1906) Notes on the fauna of a desert tract in southern India. Part I. Batrachians and reptiles, with remarks on the reptiles of the desert region of the North-West Frontier. *Memoirs of the Asiatic Society of Bengal*, Calcutta (1905) 1(10), 183–222, pls. 9–10.

Annandale, N. (1907) Miscellanea. Reptiles and Batrachia. *Records of the Indian Museum*, Calcutta 1, 397.

Annandale, N. (1911) Contributions to the fauna of Yunnan based on collections made by J. Coggin Brown, B.Sc., 1909–1910. Part IV. Batrachia and reptiles. *Records of the Indian Museum*, Calcutta 6(4), 215–218.

Annandale, N. (1912) Zoological results of the Abor Expedition, 1911–1912. II. Reptilia. *Records of the Indian Museum*, Calcutta 8, Part 1(2), 37–55, pl. 5.

Annandale, N. (1913) Description of a new sand-boa from the Persian Gulf. *Records of the Indian Museum*, Calcutta 9, 217–218, 4 figs.

Annandale, N. (1915) Fauna of the Chilka Lake. Mammals, reptiles and batrachians. *Memoirs of the Indian Museum*, Calcutta 5, 163–174.

Annandale, N. (1921) The fauna of an island in the Chilka Lake. The reptiles and batrachians. *Records of the Indian Museum*, Calcutta 22, Pt. 4(23), 331–333.

Anonymous. (1902) Les serpents de mer et leur venin. *La Revue Scientifique de la France et de l'Étranger Revue des Cours Scientifiques,* Paris (Série 4) 18(7), 219.

Anonymous. (1944) *Poisoning by snakes, plants and fish.* Arctic, Desert & Tropic Information Center (Medical Series), New York 1(4), 1–17.

Anonymous. (1958) *A list of zoological and botanical types preserved in collections in southern and East Africa. Volume 1—Zoology. Part 1.* Transvaal Museum, Pretoria, 147 pp.

Anonymous. (1961–1969) Type specimens in the Western Australian Museum. *Museum and Art Gallery of Western Australia, Annual Report* Part (2), 35–39 (1960–1961); (Part 3), 38–40 (1962–1963); (Part 5), 34–35 (1963–1964); (Part 6), 38–42 (1965–1966); (Part 8), 59–60 (1966–1967); (Part 9), 62–64 (1967–1968); (Part 10), 58–60 (1968–1969).

Anonymous. (1972a) Les ophidiens. *Les Animaux Découvrir,* Paris (131), 1–20, 26 col. illust.

Anonymous. (1972b) Les vipéridés. *Les Animaux Découvrir,* Paris (132), 1–20, 26 col. illust.

Anonymous. (1972a) Les hydrophiidés, les élapidés. *Les Animaux Découvrir,* Paris (133), 1–20, 24 col. illust.

Anonymous. (1972a) Les colubridés. *Les Animaux Découvrir,* Paris (134), 1–20, 28 col. illust.

Anonymous. (1976) Reptilia. *In: The Zoological Record* (1972) 109(17), i–xviii + 1–202.

Anonymous. (1977) *A synopsis of Chinese reptiles*. Science Press Beijing, 111 pp., 82 figs. (In Chinese)

Anonymous. (1978) The Japan Snake Institute (JSI) three-year report (January 1, 1975–December 31, 1977). *The Snake* 10(2), 168–193, 16 figs.

Anonymous. (1979) *Chinese poisonous snakes and treatment of their bites.* Second edition. Chengdu Institute of Biology, Shanghai, 390 pp., figs., 26 col. illust.

Anonymous. (1993) The Japan Snake Institute (JSI) three-year report (January 1, 1990–December 31, 1992). *The Snake* 25(1), 77–89, 3 figs.

Anthony, J. (1955) Essai sur l'évolution anatomique de l'appareil venimeux des ophidiens. *Annales des Sciences Naturelles (Zoologie),* Paris (Série 11) 17, 7–53, 16 figs.

Anthony, J. & Guibé, J. (1952) Les affinités anatomiques de *Bolyeria* et de *Casarea* Boidés). *Memoires de l'Institut Scientifique de Madagascar* (*Biologique Animales*), Tananarive 7A(2), 189–201, 5 figs.

Anthony, J. & Serra, R.G. (1951) Anatomie de l'appareil de la morsure chez *Xenodon merremii* B., serpent aglyphe de l'Amérique, tropicale. *Arquivos du Museu Nacional,* Rio de Janeiro 42, 21–47, 15 figs.

Anufriev, V.W. & Bobretsov, A.V. (1996) *Fauna of the European North-Eastern part of Russia. Vol. 4. Amphibians and reptiles.* Nauka, St. Petersburg, 129 pp., 42 figs. (in Russian)

Ao, J.M., David, P., Bondoloi, S. & Ohler, A. (2004) Notes on a collection of snakes from Nagaland, northeast India, with 19 new records for the state. *Russian Journal of Herpetology* 11(2), 155–165, 4 col. pl.

Apesteguía, S. & Zaher, H. (2006) A Cretaceous terrestrial snake with robust hindlimbs and a sacrum. *Nature* 440(7087), 1037–1040, 2 figs.

Aplin, K.P. (1998) Three new blindsnakes (Squamata, Typhlopidae) from northwestern Australia. *Records of the Western Australian Museum* 19, 1–12, 8 figs.

Aplin, K.P. & Donnellan, S.C. (1993) A new species of blindsnake, genus *Ramphotyphlops* (Typhlopidae, Squamata), from northwestern Western Australia, with a redescription of *R. hamatus*, Storr 1981. *Records of the Western Australian Museum*, Perth 16(2), 243–256, 4 figs.

Aplin, K.P. & Donnellan, S.C. (1999) An extended description of the Pilbara death adder, *Acanthophis wellsi* Hoser (Serpentes: Elapidae), with notes on the desert death adder, *A. pyrrhus* Boulenger, and identification of a possible hybrid zone. *Records of the Western Australian Museum* 19, 277–298.

Applegate, R. (2007) *Kingsnakes & milksnakes in captivity.* ECO Herpetological Publishing, Lansing, 92 pp., col. illust.

Aprea, G., Odierna, G., Andreone, F., Glaw, F. & Vences, M. (2003) Unusual karyotype in the Malagasy colubrid snake *Mimophis mahfalensis*. *Amphibia-Reptilia,* Leiden 24(2), 215–219, 2 figs.

Aprea, G., Gentilli, A., Zuffi, M.A.L. & Odierna, G. (2006) The karyology of *Vipera aspis*, *V. atra*, *V. hugyi*, and *Cerastes vipera*. *Amphibia-Reptilia,* Leiden 27(1), 113–119, 3 figs.

Aquino, A.L., Scott, N.J. & Motte, M. (1996) Lista de los anfibios y reptiles del Museo Nacional de Historia Natural del Paraguay (Marzo de 1980–Septiembre de 1995). *In:* Martinez, O.R. (Ed.), *Colecciones de flora y fauna del Museo Nacional de Historia Natural del Paraguay*. Ministerio de Agricultura y Ganadería, Asunción, pp. 331–400.

Aragón-Rebollo, T., Pierna-Chamorro, J., Aragón-Rebollo, D. & Hernández-Estévez, J.A. (2006) *Anfibios y reptiles de la Península Ibérica.* Ediciones Jaguar, Madrid, 495 pp., col. illust.

Aramata, H. (1990). *Atlas anima. Volume 3. Amphibia and Reptilia.* Heibonsha Publishers, Tokyo, 375 pp., col. illust. (in Japanese)

Arambourg, C. (1952) Les vertébrés fossiles des gisements de phosphates (Maroc-Algérie-Tunisie). *Notes et Mémoires du Service Géologique du Maroc*, Paris (92), 1–372, 44 pls.

Araujo, C.O., Condez, T.H., Bovo, R.P., Cruz-Centeno, F. da & Lutz, A.M. (2010) Amphibians and reptiles of the Parque Estadual Turistico do Alto Ribeira (PETAR), SP: an Atlantic forest remnant of southeastern Brazil. *Biota Neotropica* 10(4), 257–274, 4 figs.

Archer, M. & Wade, M. (1976) Results of the Ray E. Lemley Expedition, Part I. The Allingham Formation and a new Pliocene vertebrate fauna from northern Australia. *Memoirs of the Queensland Museum* 17, 379–397.

Arenas-Monroy, J.C., García-Balderas, C.M. & Carbajal-Márquez, R.A. (2012) Geographic distribution: Serpentes: *Geophis dugesii. Herpetological Review* 43(4), 620.

Argôlo, A.J.S. (1998a) Geographic distribution: Serpentes: *Chironius quadricarinatus quadricarinatus. Herpetological Review* 29(3), 175.

Argôlo, A.J.S. (1998b) Geographic distribution: Serpentes: *Echinanthera affinis. Herpetological Review* 29(3), 176.

Argôlo, A.J.S. (1998c) Geographic distribution: Serpentes: *Simophis rhinostoma. Herpetological Review* 29(3), 179.

Argôlo, A.J.S. (1999a) Geographic distribution: Serpentes: *Liophis maryellenae. Herpetological Review* 30(1), 54.

Argôlo, A.J.S. (1999b) Geographic distribution: Serpentes: *Tropidodryas serra. Herpetological Review* 30(1), 55.

Argôlo, A.J.S. (1999c) Geographic distribution: Serpentes: *Tropidodryas striaticeps. Herpetological Review* 30(1), 56.

Argôlo, A.J.S. (1999d) Geographic distribution: Serpentes: *Xenodon neuwiedii. Herpetological Review* 30(1), 56.

Argôlo, A.J.S. (2001a) Geographic distribution: Serpentes: *Oxyrhopus clavatus. Herpetological Review* 32(1), 61.

Argôlo, A.J.S. (2001b) Geographic distribution: Serpentes: *Uromacerina ricardinii. Herpetological Review* 33(2), 196–197.

Argôlo, A.J.S. (2002a) Geographic distribution: Serpentes: *Lystrophis nattereri. Herpetological Review* 33(2), 150.

Argôlo, A.J.S. (2002b) Geographic distribution: Serpentes: *Philodryas aestivus. Herpetological Review* 33(3), 229.

Argôlo, A.J.S. (2004) *As Serpentes dos Cacauais do sudeste da Bahia.* Editoria da UESC, Ilhéus, 259 pp., 13 figs., 61 photos.

Argôlo, A.J.S. (2008) Geographic distribution: Serpentes: *Echinanthera cephalostriata. Herpetological Review* 39(1), 111.

Argôlo, A.J.S. & Alves, F.Q. (2002) Geographic distribution: Serpentes: *Dipsas indica indica. Herpetological Review* 33(4), 323–324.

Argôlo, A.J.S. & Freitas, M.A. de. (2000) Geographic distribution: Serpentes: *Oxyrhopus rhombifer rhombifer. Herpetological Review* 31(1), 57.

Argôlo, A.J.S. & Freitas, M.A. de. (2002) Geographic distribution: Serpentes: *Philodryas aestivus. Herpetological Review* 33(3), 229.

Argôlo, A.J.S., Jesus, J.A. de, Vila Nova, M.F. & Sousa-Costa, C.A. (2011) Geographic distribution: Serpentes: *Atractus potschi. Herpetological Review* 42(3), 392.

Arias-Divito, J.C. (1968) *Las expediciones cientificas Españolas el Siglo XVIII, Expedición botánica de Nueva España.* Editiones Cultura Hispánica, Madrid, 427 pp.

Arikan, H., Göçmen, B., Mermer, A. & Bahar, H. (2005) An electrophoretic comparison of the venom of a colubrid and various viperid snakes from Turkey and Cyprus, with some taxonomic and phylogenetic implications. *Zootaxa* (1038), 1–10.

Armstrong, B.L. & Murphy, J.B. (1979) The natural history of Mexican rattlesnakes. *Special Publication, University of Kansas Museum of Natural History* (5), 1–88, illust.

Armstrong, J.G. (1976) *Amphibia and Reptilia of the late Cretaceous Fruitland Formation, northwestern New Mexico.* Master's Thesis, Wayne State University, Detroit.

Armstrong, M.P., Frymire, D. & Zimmerer, E.J. (2001) Analysis of sympatric populations of *Lampropeltis triangulum syspila* and *Lampropeltis triangulum elapsoides*, in western Kentucky and adjacent Tennessee with relation to the taxonomic status of the scarlet king snake. *Journal of Herpetology* 35(4), 688–693, 2 figs.

Armstrong-Ziegler, J.G. (1978) An aniliid snake and associated vertebrates from the Campanian of New Mexico. *Journal of Paleontology* 52(2), 480–483, 1 fig.

Arndt, R.G. (1985) Distributional notes on some Delaware snakes, with two first records for the state. *Bulletin of the Maryland Herpetological Society* 21(2), 67–73.

Arnold, E.N. (1980a) Recently extinct reptile populations from Mauritius and Réunion, Indian Ocean. *Journal of Zoology*, London 191(1), 33–47.

Arnold, E.N. (1980b) The reptiles and amphibians of Dhofar, southern Arabia. *In:* The scientific results of the Oman flora and fauna survey 1977 (Dhofar). *Journal of Oman Studies Special Report* (2), 273–332, pls. 19–24.

Arnold, E.N. (1982) Reptiles of Saudi Arabia. A new semaphore gecko (*Pristurus*, Gekkonidae) and a new dwarf snake (*Eirenis*, Colubridae) from southwestern Arabia. *Fauna of Saudi Arabia*, Riyadh 4, 468–477, 3 figs.

Arnold, E.N. (1995) *Atlas of amphibians and reptiles in Britain.* Institute of Terrestrial Ecology, Research Publication (10), 1–40, illust.

Arnold, E.N. (2002) *Reptiles and amphibians of Europe*, second edition, Princeton University Press, Princeton 288 pp., 256 figs., 49 col. pls.

Arnold, E.N. & Burton, J.A. (1978) *A field guide to the reptiles and amphibians of Britain and Europe.* Collins, London, 272 pp., 256 figs., 40 col. pls.

Arnold, E.N. & Gallagher, M.D. (1977) Reptiles and amphibians from the mountains of northern Oman with special reference to the Jebel Akhdar region. *In:* The scientific results of the Oman flora and fauna survey 1975. *The Journal of Oman Studies, Special Report* (1), 59–80, illust.

Arnold, E.N., Robinson, M.D. & Carranza, S. (2009) A preliminary analysis of phylogenetic relationships and biogeography of the dangerously venomous carpet vipers, *Echis* (Squamata, Serpentes, Viperidae) based on mitochondrial DNA sequence. *Amphibia-Reptilia,* Leiden 30(2), 273–282, 2 figs.

Arnold, H.R. (1973) *Provisional atlas of the amphibians and reptiles of the British Isles.* Biological Records Centre, Monks Wood Experimental Station, 4 pp.

Arnoult, J. & Fourmanoir, P. 1967. Sur une colleccion de poissons marins et de reptiles, provenant de l'Ile de Nossi–Bé (Madagascar). *Bulletin du Museum National d'Histoire Naturelle*, Paris 39, 134–142.

Arruda, L.A.G., Avila, R.M., Morais, D.H. & Kawashita-Ribeiro, R.A. (2011) *Rhachidelus brazili* Boulenger, 1908 (Squamata: Serpentes: Dipsadidae): new records, range extension, and distribution map in Mato Grosso and Mato Grosso do Sul states, Brazil. *Check List* 7(6), 854–855, 2 figs.

Arruda, M.P. de. (2011) First record in Midwestern region of the state of Säo Paulo, Brazil of *Typhlops brongersmianus* Vanzolini, 1976 (Squamata: Typhlopidae). *Check List* 7(4), 571–573, 2 figs.

Arseniev, A.S., Pashkov, V.S., Pluzhnikov, K.A., Rochat, H. & Bystrov, V.F. (1981) The 1H nuclear-magnetic-resonance spectra and spatial structure of *Naja mossambica mossambica* neurotoxin III. *European Journal of Biochemistry* 118, 453–462.

Arth, S. & Baus, S. (2006) *Le Python Vert,* Morelia viridis. Natur und Tier-Verlag, Münster, 63 pp., col. illust.

Arzamendia, V. (1999) Geographic distribution: Serpentes: *Philodryas baroni*. *Herpetological Review* 30(1), 55.

Arzamendia, V. & Giraudo, A.R. (1999) Geographical distribution: *Liophis sagittifer modestus*. *Herpetological Review* 30(1), 54.

Arzamendia, V. & Giraudo, A.R. (2002) Geographic distribution: Serpentes: *Liophis frenatus*. *Herpetological Review* 33(3), 228.

Ashe, J.O.P. (1965) First record of snake from Uganda. *Journal of the East Africa Natural History Society and National Museum*, Nairobi 25(2), 162.

Ashe, J.O.P. (1968) A new bush viper. *Journal of the East Africa Natural History Society and National Museum*, Nairobi 27(1), 53–59, 2 figs., 2 pls.

Ashe, J.S. & Marx, H. (1988) Phylogeny of the viperine snakes (Viperinae): Part II. Cladistic analysis and major lineages. *Fieldiana: Zoology* (Series 2) (52), i–iii + 1–23, 3 figs.

Ashton, K.G. & Queiroz, A. de. (2001) Molecular systematics of the western rattlesnake, *Crotalus viridis* (Viperidae), with comments on the utility of the D-loop in phylogenetic studies of snakes. *Molecular Phylogenetics and Evolution* 21(2), 176–189, 5 figs.

Ashton, R.E., Jr. (1977) *Identification manual to the reptiles and amphibians in Florida*. Florida State Museum, Gainesville, 65 pp.

Ashton, R.E., Jr. & Ashton, P.S. (1988) *Handbook of reptiles and amphibians of Florida. Part one. The Snakes*. Windward Publishing, Miami, 176 pp., col. illust. [first edition, 1981, 176 pp., col. illust.]

Astiba, H., Buffetaut, E., Buscalioni, A.D., Cappetta, H., Corral, C., Estes, R., Garcia-Garmilla, F., Jaeger, J.J., Jimenez-Fuentes, E., Le Loeuff, J., Mazin, J.M., Orue-Etxebarria, X., Pereda-Suberbiola, J., Powell, J.E., Rage, J.-C., Rodriguez-Lazaro, J., Sanz, J.L. & Tong, H. (1990) The fossil vertebrates from Laño (Basque country, Spain); new evidence on the composition and affinities of the late Cretaceous continental faunas of Europe. *Terra Nova* 2, 460–466, 6 figs.

Ataev, C., Rustamov, A.K. & Shammakov, S. (1994) Reptiles of Kopetdagh. *In:* Fet, V. & Atamuradov, K.I. (Eds.), *Biogeography and ecology of Turkmanistan*. Kluwer Academic Publishers, Dordrecht, pp. 329–349.

Atatür, M.K. & Göçmen, B. (2001) *Amphibians and reptiles of northern Cyprus*. Ege Üniversitesi Basimevi (170), 64 pp.

Atsatt, S.R. (1923) Behavior of the leaf-nosed snake, *Phyllorhynchus decurtatus*. *University of California Publications in Zoology* 21(9), 303–312.

Auerbach, R.D. (1985) *The reptiles of Gaborone. A guide to the reptiles of the south-eastern Hardveld of Botswana*. Botswana Book Centre, Gaborone, 47 pp., 16 col. pls.

Auerbach, R.D. (1987) *The amphibians and reptiles of Botswana*. Okwepa Consulate, Botswana, v + 295 pp., text figs., 19 pls.

Auffenberg, W. (1950) A new subspecies of the mud snake, *Liodytes alleni*. *Herpetologica* 6(1), 13–16, 3 figs.

Auffenberg, W. (1955) A reconsideration of the racer, *Coluber constrictor*, in eastern United States. *Tulane Studies in Zoology* 2(6), 89–155, 17 figs.

Auffenberg, W. (1958) A new genus of colubrid snake from the upper Miocene of North America. *American Museum Novitates* (1874), 1–16, 3 figs.

Auffenberg, W. (1959) *Anomalophis bolcensis* (Massalongo), a new genus of fossil snake from the Italian Eocene. *Breviora* (114), 1–16, 4 figs.

Auffenberg, W. (1963) The fossil snakes of Florida. *Tulane Studies in Zoology* 10(3), 131–216, 51 figs.

Auffenberg, W. (1980) The herpetofauna of Komodo, with notes on adjacent areas. *Bulletin of the Florida State Museum, Biological Science* 25(2), 39–156, 52 figs.

Auffenberg, W. (1988) *Gray's monitor lizard*. University Press of Florida, Gainesville, xi + 419 pp., illust.

Auffenberg, W. & Babbitt, L.H. (1953) A new subspecies of *Coluber constrictor* from Florida. *Copeia* 1953(1), 44.

Auffenberg, W. & Rehman, H. (1991) Studies on Pakistan reptiles. Part I. The genus *Echis*. *Bulletin of the Florida Museum of Natural History, Biological Science* 35(5), 263–314, illust.

Augé, M. & Rage, J-C. (2000) Les squamates (Reptilia) du Miocéne moyen de Sansan. *In:* Ginsburg, L. (Ed.), *La faune Miocène de Sansan et son environnement*. Mémoires du Museum National d'Histoire Naturelle, Paris 183, 263–313, 30 figs.

Augé, M., Duffaud, S., La Broin, F., Rage, J.-C. & Vasse, D. (1997) Les amphibiens et les reptiles de Prémontré (Cuisien, Bassin parisien), une herpétofaune de référence pour l'Eocène inférieur. *Géologie de la France* (1), 23–33, 9 figs.

Auliya, M. (2002) Rediscovery of the Indochinese rat snake *Ptyas korros* (Schlegel, 1837) (Serpentes, Colubridae) in Borneo. *Raffles Museum Bulletin* 50(1), 197–198.

Auliya, M. & Shepherd, C.R. (2009) New data on the poorly known snake, *Xenelaphis ellipsifer* Boulenger, 1900 (Squamata: Colubridae). *Hamadryad* 34(1), 168–171, 5 figs.

Auliya, M., Mausfeld, P., Schmitz, A. & Böhme, W. (2002). Review of the reticulated python (*Python reticulatus* Schneider, 1801) with the description of new subspecies from Indonesia. *Naturwissenschaften* 89, 201–213, 3 figs.

Auliya, M., Wagner, P. & Böhme, W. (2012) The herpetofauna of the Bijagós Archipelago, Guinea-Bissau (West Africa) and a first country-wide checklist. *Bonn Zoological Bulletin* 61(2), 255–281, 31 figs.

Austin, C. (2000) Molecular phylogeny and historical biogeography of Pacific Island boas (*Candoia*). *Copeia* 2000(2), 341–352, 2 figs.

Auth, D.L. (1994) Checklist and bibliography of the amphibians and reptiles of Panama. *Smithsonian Herpetological Information Service* (98), 1–59.

Auth, D.L, Auffenberg, K. & Dorman, D.K. (1990) Geographic distribution: Serpentes: *Gerarda prevostiana* (Gerard's water snake). *Herpetological Review* 21(2), 41.

Auth, D.L., Chiszar, D., Smith, H.M. & Breukelen, F. van (1997a) Geographic distribution: Serpentes: *Masticophis mentovarius mentovarius*. *Herpetological Review* 28(4), 210.

Auth, D.L., Smith, H.M., Breukelen, F. van & Chiszar, D. (1997b) Geographic distribution: Serpentes: *Tantilla rubra*. *Herpetological Review* 28(4), 211.

Auth, D.L., Chiszar, D., Smith, H.M. & Breukelen, F. van (1998a) Geographic distribution: Serpentes: *Masticophis mentovarius mentovarius*. *Herpetological Review* 29(1), 54.

Auth, D.L., Mariaux, J., Clary, J., Chiszar, D., Breukelen, F. van & Smith, H.M. (1998b) The report of the snake genus *Conophis* in South America is erroneous. *Bulletin of the Maryland Herpetological Society* 34(4), 107–112, 2 figs.

Auth, D.L., Brown, B.C., Smith, H.M. & Chiszar, D. (1999) Geographic distribution: Serpentes: *Rhadinaea laureata*. *Herpetological Review* 30(4), 236.

Auth, D.L., Smith, H.M., Brown, B.C., Lintz, D. & Chiszar, D. (2000a) Further observations on Iverson's blind snake in Tamaulipas, Mexico. *Bulletin of the Maryland Herpetological Society* 36(1), 1–4, 1 fig.

Auth, D.L., Smith, H.M., Brown, B.C. & Lintz, D. (2000b) A description of the Mexican amphibian and reptile collection of the Strecker Museum. *Bulletin of the Chicago Herpetological Society* 35(4), 65–85.

Avci, A. & Olgun, K. (2011) On additional specimens of *Eirenis (Pediophis) coronelloides* (Serpentes: Colubridae) collected from southeastern Anatolia, Turkey, with a discussion of its status. *Russian Journal of Herpetology* 18(1), 73–79, 4 figs.

Avci, A., Dinçaslan, Y.E., Ulgaz, Ç. & Üzüm, N. (2008) Contribution to the distribution and morphology of *Rhynchocalamus melanocephalus melanocephalus* (Jan 1862) (Reptilia, Colubridae) in Turkey. *North-Western Journal* 4(1), 161–166, 3 figs.

Avci, A., Ilgaz, Ç., Baskaya, S., Baran, I. & Kumlutas, Y. (2010) Contribution to the distribution and morphology of *Pelias darevskii* (Vedmederja, Orlov et Tuniyev 1986) (Reptilia: Squamata: Viperidae) in northeastern Anatolia. *Russian Journal of Herpetology* 17(1): 1–7, 4 col. figs.

Averianov, A.O. (1997) Paleogene sea snakes from the eastern part of Tethys. *Russian Journal of Herpetology* 4(2), 128–142, 9 figs.

Avila, L.J. (1996) Geographic distribution: Serpentes: *Philodryas aestivus subcarinatus*. *Herpetological Review* 27(3), 154.

Avila, L.J. (1997) Geographic distribution: Serpentes: *Pseudotomodon trigonatus*. *Herpetological Review* 28(2), 98.

Avila, L.J. (2009a) Reptilia, Squamata, Dipsadidae, *Pseudotomodon trigonatus* (Leybold, 1873): Distribution extension. *Check List* 5(3), 391–393, 2 figs.

Avila, L.J. (2009b) Reptilia, Serpentes, Colubridae, *Liophis sagittifer sagittifer*: Distribution extension. *Check List* 5(3), 712–713, 2 figs.

Avila, L.J. & Morando, M. (1999) Geographic distribution: Serpentes: *Oxyrhopus rhombifer bachmanni*. *Herpetological Review* 30(2), 114.

Avila, L.J., Acosta, J.C. & Murúa, A.F. (1998) Herpetofauna de la provincia de San Juan, Argentina: lista comentada y distribución geográphica. *Cuadernos de Herpetología* 12(1), 11–29, 1 fig.

Avila, L.J., Morando, M. & Perez, D.R. (2001) New records and natural history notes for lizards and snakes from Patagonia, Argentina. *Herpetological Review* 32(1), 64–65.

Avila, L.J., Morando, M. & Belver, L.C. (2012) *Tachymenis chilensis* Schlegel, 1837 (Reptilia: Squamata: Dipsadidae). New record and geographic distribution map. *Cuadernos de Herpetología* 26(2), 103–104, 2 figs.

Avila, R.W. & Kawashita-Ribeiro, R.A. (2011) Herpetofauna of Sao Joao da Barra Hydroelectric Plant, state of Mato Grosso, Brazil. *Check List* 7(6), 750–755, 4 figs.

Avila, R.W., Ferreira, V.L. & Souza, V.B. (2006) Biology of the blindsnake *Typhlops brongersmianus* (Typhlopidae) in a semidecidous forest from central Brazil. *Herpetological Journal* 16, 403–405.

Avila-Pires, F.D. de. (1965) The type specimens of Brazilian mammals collected by Prince Maximilian zu Wied. *American Museum Novitates* (2209), 1–21.

Ávila-Pires, T.C.S. de. (2005) Reptiles. *In:* Hollowell, T. & Reynolds, R.P. (Eds.), Checklist of the terrestrial vertebrates of the Guiana Shield. *Bulletin of the Biological Society of Washington* (13), 25–42.

Aye, M.M. (1990) Venomous snakes of medical importance in Burma. *In:* Gopalakrishnakone, P. & Chou, L.M. (Eds.), *Snakes of medical importance (Asia-Pacific region)*. National University of Singapore, Singapore, pp. 211–241, 8 figs.

Ayerbe-González, S. & López-López, F.J. (2005) Descripción de una nueva especie de serpiente coral (Elapidae: *Micrurus*). *Novedades Colombianas, Museo de Historia Natural Universidad del Cauca*, Popayán 8(1), 41–43, 1 fig.

Ayerbe, S., Tidwell, M.A. & Tidwell, M. (1990) Observaciones sobre la biología y comportamiemnto de la serpiente coral "rabo de aji" (*Micrurus mipartitus*). Descripción de une subespecie nueva. *Novedades Colombianas, Museo de Historia Natural Universidad del Cauca*, Popayán 2, 30–41.

Aylmer, C. (1922) The snakes of Sierra Leone. *Sierra Leone Studies* 5, 7–37, 8 figs.

Azevedo, A.C.P. (1963) Variaçãos cromáticas em *Micrurus corallinus* (Wied, 1820) Serpentes, Elapidae. *Iheringia (Zoologia)*, Pôrto Alegre 34, 1–16.

Babocsay, G. (2003) Geographic variation in *Echis coloratus* (Viperidae, Ophidia) in the Levant with the description of a new subspecies. *Zoology in the Middle East* 29,13–32.

Babocsay, G. (2004) A new species of saw-scales viper of the *Echis coloratus* complex (Ophidia, Viperidae) from Oman, eastern Arabia. *Systematics and Biodiversity* 1(4),503–514.

Bachman, E.S. (1985) Distribution and variability of the Sri Lankan pipe snake (*Cylindrophis maculatus*). *The Journal of the Bombay Natural History Society* 82(2), 322–327, 1 fig.

Bachmayer, F. (1987) A second contribution to the ophidian fauna (Reptilia, Serpentes) of Kohfidisch, Austria. *Annalen des Naturhistorischen Museums in Wien* 88A, 25–39, 6 figs., 1 pl.

Bachmayer, F. & Szyndlar, Z. (1985) Ophidians (Reptilia, Serpentes) from the Kohfidisch fissures of Burgenland, Austria. *Annalen des Naturhistorischen Museums in Wien* 87A, 79–100, 6 figs., 2 pls.

Bacqué, A. (1905) Trois trigonocéphales du Paraguay. *Revista del Museo de La Plata* 12(1), 111–119, 6 figs.

Badger, D. (1999) *Snakes*. Voyager Press, Stillwater, 144 pp., col. illust.

Baha El Din, S. (2006) *A guide to the reptiles and amphibians of Egypt*. American University of Cairo Press, Cairo, xi + 359 pp., 127 col. pls.

Bahuguna, A. (2010) Reptilia. *In:* Alfred, J.R.B. (Ed.), *Fauna of Uttarakhand, (Part-1) Vertebrates. State Fauna Series* 18. Zoological Survey of India, Kolkata, pp. 445–503.

Baier, F., Sparrow, D.J. & Wiedl, H.-J. (2009) *The amphibians and reptiles of Cyprus*. Edition Chimaira, Frankfurt-am-Main, 383 pp., 239 col. figs.

Baig, K.J. & Masroor, R. (2008) The snakes of the genus *Spalerosophis* Jan, 1865 in Indo-Pakistan and Iran (Squamata: Serpentes: Colubridae). *Herpetozoa* (2007) 20(3–4), 109–115, 7 figs.

Baig, K.J., Masroor, R. & Arshad, M. (2008) Biodiversity and ecology of the herpetofauna of Cholistan Desert, Pakistan. *Russian Journal of Herpetology* 15(3), 193–205, 11 figs.

Bailey, J.R. (1937a) New forms of *Coniophanes* Hallowell, and the status of *Dromicus clavatus* Peters. *Occasional Papers of the Museum of Zoology, University of Michigan* (362), 1–6.

Bailey, J.R. (1937b) A new species of *Rhadinaea* from San Luis Potosí. *Copeia* 1937(2), 118–119.

Bailey, J.R. (1937c) A review of some recent *Tropidophis* material. *Proceedings of the New England Zoological Club* 16, 41–52.

Bailey, J.R. (1939a) A systematic revision of the snakes of the genus *Coniophanes*. *Papers of the Michigan Academy of Science, Arts and Letters* 24(2), 1–48, 5 figs., 3 pls.

Bailey, J.R. (1939b) *Relationships and distributions of the snakes allied to the genus* Pseudoboa. Doctoral Dissertation, University of Michigan, Ann Arbor, 239 pp., 13 pls.

Bailey, J.R. (1940) The Mexican snakes of the genus *Rhadinaea*. *Occasional Papers of the Museum of Zoology, University of Michigan* (412), 1–19, 2 pls.

Bailey, J.R. (1946) *Leptotyphlops anthracinus*, a new blind snake from eastern Ecuador. *Occasional Papers of the Museum of Zoology, University of Michigan* (492), 1–5.

Bailey, J.R. (1955) The snakes of the genus *Chironius* in southeastern South America. *Occasional Papers of the Museum of Zoology, University of Michigan* (571), 1–21, 1 fig.

Bailey, J.R. (1966) A redescription of the snake *Calamodontophis paucidens*. *Copeia* 1966(4), 885–886.

Bailey, J.R. (1967) The synthetic approach to colubrid classification. *Herpetologica* 23, 155–161.

Bailey, J.R. (1986) The *Oxyrhopus petola-petolarius* question continued. *Bulletin of the Maryland Herpetological Society* 22(3), 144–145.

Bailey, J.R & de Carvalho, A.L. (1946) A new *Leptotyphlops* from Mato Grosso, with notes on *Leptotyphlops tenella* Klauber. *Boletim do Museu Nacional (Zoologia),* Rio de Janeiro (Series 2) (52), 1–7, 4 figs.

Bailey, J.R. & Thomas, R.A. (1991) Snakes related to *Thamnodynastes* updated. *Abstract, Society for the Study of Amphibians and Reptiles and Herpetologists League Meetings*, 1 p.

Bailey, J.R. & Thomas, R.A. (2007) A revision of the South American snake genus *Thamnodynastes* Wagler, 1830 (Serpentes, Colubridae, Tachymenini). II. Three new species from northern South America, with further descriptions of *Thamnodylnastes gambotensis* Pérez-Santos and Moreno and *Thamnodynastes ramonriveroi* Manzanilla and Sánchez. *Memoria de la Fundación La Salle de Ciencias Naturales* (2006) 166, 7–27, 10 figs.

Bailey, J.R., Thomas, R.A. & Silva, N.J. da, Jr. (2005) A revision of the South American snake genus *Thamnodynastes* Wagler, 1830 (Serpentes, Colubridae, Tachymenini). I. Two new species of *Thamnodynastes* from central Brazil and adjacent areas, with a redefinition of and neotype designation for *Thamnodynastes pallidus* (Linnaeus, 1758). *Phyllomedusa* 4(2), 83–101, 14 figs.

Bailon, S. (1988) Un Aniliidé (Reptilia, Serpentes) dans le Pliocène supérieur européen. *Comptes Rendus de l'Académie des Sciences*, Paris (Series 2) 306, 1255–1258, figs. a–e.

Bailon, S. (1989) Les amphibiens et les reptiles du Pliocene superior de Balaruc II (Herault, France). *Palaeovertebrata*, Montpellier 19(1), 7–28, 7 figs.

Bailon, S. (1991) Le genre *Malpolon* (Serpentes, Colubridae) dans les gisements française. *Bulletin de la Société Herpétologique de France* (58), 1–10, 4 figs.

Bailon, S. (2000) Amphibiens et reptiles du Pliocène terminal d'Ahl al Oughlam (Casablanca, Maroc). *Geodiversitas* 22, 539–558.

Bailon, S. & Aouraghe, H. (2002) Amphibiens, chéloniens et squamates du Pléistocène supérieur d'El Harhoura 1 (Témara, Maroc). *Geodiversitas* 24, 821–830.

Bailon, S. & Rage, J.-C. (1994) Squamates Néogènes et Pléistocénes du Rift Occidental, Ouganda. *In: Geology and Palaeobiology of the Albertine Rift Valley, Uganda-Zaire. Vol. II. Palaeobiology*. CIFEG Occasional Papers, Orleans, pp. 129–135, 4 figs.

Bailon, S., Garcia-Porta, J. & Quintana-Cardona, J. (2002) Première découverte de Viperidae (Reptilia, Serpentes) dans les îles Baléares (Espagne): des vipers du Néogène de Minorque. Description d'une nouvelle espèce du Pliocène. *Comptes Rendus Palevol* 1, 227–234.

Bailon, S., Bover, P., Quintana-Cardona, J. & Alvover, J.A. (2010) First fossil record of *Vipera* "oriental vipers complex" (Serpentes: Viperidae) from the early Pliocene of the western Mediterranean islands. *Comptes Rendus Palevol* 9, 147–154, 2 figs.

Bain, R.H., Truong, N.Q. & Kien, D. van. (2007a) Geographic distribution: Serpentes: *Calloselasma rhodostoma*. *Herpetological Review* 38(1), 102.

Bain, R.H. (2007b) Geographic distribution: Serpentes: *Lycodon laoensis*. *Herpetological Review* 38(1), 104.

Bain, R.H. (2007c) New herpetofauna records from Vietnam. *Herpetological Review* 38(1), 107–117.

Bain, R.H. & Hurley, M.M. (2011) A biogeographic synthesis of the amphibians and reptiles of Indochina. *Bulletin of the American Museum of Natural History* (360), 1–138, 9 figs.

Baird, S.F. (1853) Reptiles. Order Ophidia. *In:* Sitgreaves, L. (Ed.), *Report of an expedition down the Zuni and Colorado Rivers*. Robert Armstrong, Washington, pp. 130–141, pls. 10–18.

Baird, S.F. (1854) *On the serpents of New York; with a notice of a species not hitherto included in the fauna of the state*. Appendix. C. Van Benthuysen, Albany (50), 97–124.

Baird, S.F. (1859a) Reptiles of the boundary, with notes by the naturalists of the survey. *In: United States and Mexican Boundary Survey, under the orders of Lieut. Col. W.H. Emory, Major First Cavalry, and United States Commissioner.* Government Printing Office, Washington (1858) 2(2), 1–36, 41 pls.

Baird, S.F. (1859b) Zoological report. No. 4. Report on reptiles collected on the survey. *In:* Williamson, R.S. & Abbot, H.L. (Eds.), *Reports of Explorations and Surveys, to ascertain the most practicable and economical route for a railroad from the Mississippi River to the Pacific Ocean, 1853-6. Explorations for a railroad route, from the Sacramento Valley to the Columbia River.* Government Printing Office, Washington, D.C. (1855) 10(4), 9–13, pls. 11, 28, 30, 44.

Baird, S.F. & Girard, C. (1852a) Characteristics of some new reptiles in the museum of the Smithsonian Institution. *Proceedings of the Academy of Natural Sciences of Philadelphia* (1852–1853) 6(1), 68–70.

Baird, S.F. & Girard, C. (1852b) Appendix C. Reptiles. *In:* Stansbury, H. (Ed.), *Exploration and survey of the valley of the Great Salt Lake of Utah, including a reconnaissance of a new route through the Rocky Mountains*. Lippincott, Grambo & Co., Philadelphia, pp. 336–353, 6 pls. [reprinted in 1853 by House of Representatives, in 1855 with different title and frontispiece, and in 1978 by K. Adler, Arno Press]

Baird, S.F. & Girard, C. (1852c) Descriptions of new species of reptiles, collected by the U.S. Exploring Expedition under the command of Capt. Charles Wilkes, U.S.N. First Part. Including the species from the western coast of America. *Proceedings of the Academy of Natural Sciences of Philadelphia* (1852–1853) 6(5), 174–177.

Baird, S.F. & Girard, C. (1853) *Catalogue of North American reptiles in the museum of the Smithsonian Institution. Part I. Serpents.* Smithsonian Institution, Washington, xvi + 172 pp.

Baker, F.C. & Woodruff, F.M. (1897) Note on *Natrix grohamii* B. & G. *Science* (Series 2) 5(115), 447.

Baker, M.A., Rifai, L., Jogel, U., Nagy, Z., Wink, M. & Amr, Z. (2002) Occurrence of *Coluber* (*Hierophis*) *schmidti* Nikolsky, 1909 in Jordan. *Herpetozoa* 15(1–2), 29–36, 3 figs.

Baker, M.A., Qarqaz, M., Rifai, L., Hamidan, N., Al Omari, K., Modry, D. & Amr, Z. (2004) New locality records extending the distribution of some ophidians in southeastern Anatolia. *Russian Journal of Herpetology* 11(1), 1–5, 1 fig.

Balk, L. (1749, "1746") Museum Adolpho-Fridericianum—"Museum Principis." *In:* Linnaeus, C. (Ed.), *Amoenitates academicae, seu dissertationes variae physicae, medicae, botanicae antehac seorsim editae nunc collectae et auctae cum tabulis aeneis*. Godofredum Kiesewetter, Holmiae et Lipsiae, pp. 557-610.

Ballinger, R.E. & Lynch, J.D. (1983) *How to know the amphibians and reptiles*. Wm. C. Brown, Dubuque, 229 pp., 488 figs.

Ballinger, R.E., Lynch, J.D. & Smith, G.R. (2010) *Amphibians and reptiles of Nebraska*. Rusty Lizard Press, Oro Valley, 400 pp., 308 col. figs.

Baloutch, M. (1976) Première récolte de *Lytorhynchus maynardi* (Alcock et Finn, 1896) en Iran (Serpentes, Colubridae). *Bulletin du Museum National d'Histoire Naturelle*, Paris (Série 3) (417), 1377–1378, 1 fig.

Balozet, L. (1957) La vipère lébétine et son venin. *Archives de l'Institut Pasteur l'Algèrie* 35(3), 220–295.

Banerjee, R.N. & Siddiqui, Z.A. (1978) Epidemiological study of snake-bite in India. *In:* Rosenberg, P. (Ed.), *Toxins: animal, plant and microbial*. Pergamon Press, Oxford, pp. 439–446, 4 figs.

Banerji, P. (1929) *Handbook of snake-bite*. Paresh Bannerji, Mihijam, xvii + 430 pp., illust., 23 pls.

Banks, R.C., McDiarmid, R.W. & Gardner, A.L. (1987) *Checklist of the Vertebrates of the United States, the U.S. Territories, and Canada*. United States Department of the Interior, Washington (166), 1–79.

Bannerman, W.B. & Pocha, J.B. (1905) On the distribution of the varieties of cobra (*Naja tripudians*) in India. *The Journal of the Bombay Natural History Society* (1904–1905) 16(4), 638–643, 4 figs.

Bannerman, W.B. (1907) A further note on the distribution of the varieties of cobra in India. *The Journal of the Bombay Natural History Society* (1906–1907) 17(4), 1031–1032.

Bannikov, A.G., Darevskii, I.S., Ishchenko, V.G., Rustamov, A.K. & Szczerbak, N.N. (1977) *Guide to amphibians and reptiles of the USSR fauna*. Second edition. Enlightenment Publications, Moskva, 414 pp., 108 figs., 32 col. pls. (in Russian) [first edition, Bannikov, Darevsky & Rustamov, 1971, 303 pp., 62 figs, 32 col. pls.]

Banta, B.H. (1965) An annotated chronological bibliography of the herpetology of the state of Nevada. *Wasmann Journal of Biology* 23(1–2), 1–224.

Banta, B.H. & Leviton, A.E. (1963) Remarks on the colubird genus *Chilomeniscus* (Serpentes, Colubridae). *Proceedings of the California Academy of Sciences* (4)31(11), 309–327, 10 figs.

Bar, A. & Haimovitch, G. (2011) *A field guide to reptiles and amphibians of Israel*. Aviad Bar and Guy Haimovitch, Herzliya, 245 pp., col. illust.

Barabanov, A. (2002) Taxonomic status of *Coluber ventromaculatus bengalensis* Khan et Khan, 2000 (Reptilia, Squamata, Colubridae). *Russian Journal of Herpetology* 9(3), 255.

Baran, I. (1976) Revision and Verbreitung der Schlangen der Türkei. *Türkiye Bilimsel ve Teknik Arastirma Kurumu Yayinlari*, Ankara (Series 9) (309), 177 pp., 49 figs., 11 pls. (in Turkish)

Baran, I. (1982) Zur Taxonomie der Schlangen in Südost- und Ost-Anatolien. *Spixiana* 5, 51–59.

Baran, I. (1986) On an island population of *Eirenis modestus* in the eastern Mediterranean Sea. *Zoology in the Middle East* 1, 80–83, illust.

Baran, I. (2005) *Türkiye amphibi ve sürüngenleri*. Popüler Bilim Kitaplari, Tübitak, 165 pp., col. illust.

Baran, I. & Atatür, M.K. (1997) *Turkish herpetofauna (amphibians and reptiles)*. Ministry of Environmental Publications, Ankara, 214 pp., 122 figs.

Baran, I., Joger, U., Kutrup, B. & Türkozan, O. (2001) On new specimens of *Vipera barani* Böhme and Joger, 1983, from northeastern Anatolia, and implications for the validity of *Vipera pontica* Billing, Nilson & Sattler, 1990 (Reptilia, Viperidae). *Zoology in the Middle East* 23, 47–53.

Baran, I., Kumlutas, Y., Ilgaz, C., Türkozan, O. & Aver, A. (2004) New locality records extending the distribution of some ophidians in southeastern Anatolia. *Russian Journal of Herpetology* 11(1), 6–9.

Baran, I., Kumlutas, Y., Avci, A. & Ilzag, C. (2006) Second record of *Walterinnesia aegyptia* Lataste, 1887 from southeastern Anatolia. *Herpetozoa* 19(1–2), 1–4, 7 figs.

Barash, A. & Hoofien, J.H. (1956) *Reptiles of Israel: a guide and a key*. Hakibutz Hamenchad Publishing House, Tel-Aviv, 180 pp., 80 figs.

Barash, A. (1961) *Reptiles of Israel. A guide and a key*. (First revision). Hakibutz Hameuchad Publishing House, Tel-Aviv, pp. 177–179.

Barbadillo-Escriva, L.J. & Martinez de Castilla, A. (1987) *La guia de incafo de los anfibios y reptiles de la peninsula Iberica, Islas Baleares y Canarias*. Incafo, Madrid, 694 pp., 99 figs., 81 col. pls.

Barbadillo-Escriva, L.J., Lacomba, J.I., Pérez-Mellado, V., Sancho, V. & López-Jurado, L.F. (1999) *Anfibios y reptiles de la peninsula Ibérica, Baleares y Canarias*. Editorial GeoPlaneta, Barcelona, 419 pp., col. illust.

Barbault, R. (1971) Les peuplemente d'ophidiens des savanes de Lamto (Cote-d'Ivoire). *Annals de l'Université d'Abidjan (Écologie)* 4E(1), 133–193.

Barbieri, F., Doria, G. & Sindaco, R. (Eds.). (1996) Atlante provisorio degli anfibi e dei rettili Italiani. *Annali della Museo Civico di Storia Naturale "Giacamo Doria,"* Genova 91, 95–178, 62 figs.

Barbieri, F., Doria, G. & Sindaco, R. (Eds.). (2000) *Atlante degli anfibi e dei rettili della provincia di Vecenza*. GilbertoPadovan Editore, Vicenza, 203 pp., col. illust.

Barbour, R.W. (1971) *Amphibians and reptiles of Kentucky*. University Press of Kentucky, Lexington, x + 334 pp., 64 pls.

Barbour, T. (1905) The Vertebrata of Gorgona Island, Colombia. V. Reptilia and Amphibia. *Bulletin of the Museum of Comparative Zoology* 46, 98–102.

Barbour, T. (1906) Reptilia and Amphibia. *In:* Bangs, O. (Ed.), Vertebrata from the savanna of Panama. *Bulletin of the Museum of Comparative Zoology* 46(12), 224–229.

Barbour, T. (1908) Some new reptiles and amphibians. *Bulletin of the Museum of Comparative Zoology* 51(12), 315–325.

Barbour, T. (1909) Notes on Amphibia and Reptilia from eastern Asia. *Proceedings of the New England Zoological Club* 4, 53–78, pl. 7.

Barbour, T. (1910) A new colubrine snake from Java. *Proceedings of the Biological Society of Washington* 23, 169–170.

Barbour, T. (1912a) Amphibia and Reptilia. *In:* Henshaw, S. (Ed.), Some Chinese Vertebrates. *Memoirs of the Museum of Comparative Zoology* 40(4), 125–136, 2 pls.

Barbour, T. (1912b) A contribution to the zoogeography of the East Indian Islands. *Memoirs of the Museum of Comparative Zoology* 44(1), 203 pp., 8 pls.

Barbour, T. (1913a) Reptiles collected by the Yale Peruvian expedition of 1912. *Proceedings of the Academy of Natural Sciences of Philadelphia* 65(3), 505–507, pl. 17.

Barbour, T. (1913b) Reptiles and amphibians from eastern Sudan. *Proceedings of the Biological Society of Washington* 26, 145–149.

Barbour, T. (1914a) A contribution to the zoögeography of the West Indies, with especial reference to amphibians and reptiles. *Memoirs of the Museum of Comparative Zoology* 44(2), 209–359, 1 pl.

Barbour, T. (1914b) Notes on some reptiles from Sinai and Syria. *Proceedings of the New England Zoological Club* 5, 73–92, 2 figs.

Barbour, T. (1914c) Some new reptiles. *Proceedings of the New England Zoological Club* 4, 95–98.

Barbour, T. (1914d) A new snake from northern Brazil. *Proceedings of the Biological Society of Washington* 27, 199–200.

Barbour, T. (1914e) On some Australasian reptiles. *Proceedings of the Biological Society of Washington* 27, 201–206.

Barbour, T. (1915a) Recent notes regarding West Indian reptiles and amphibians. *Proceedings of the Biological Society of Washington* 28, 71–78.

Barbour, T. (1915b) A new snake from southern Peru. *Proceedings of the Biological Society of Washington* 28, 149–150.

Barbour, T. (1916a) The reptiles and amphibians of the Isle of Pines. *Annals of the Carnegie Museum* 10(1–2), 297–308, 4 figs.

Barbour, T. (1916b) A new *Tantilla* from Mexico. *Proceedings of the Biological Society of Washington* 29, 93–94.

Barbour, T. (1916c) Amphibians and reptiles from Tobago. *Proceedings of the Biological Society of Washington* 29, 221–224.

Barbour, T. (1917a) A most regretable tangle of names. *Occasional Papers of the Museum of Zoology, University of Michigan* (44), 1–9.

Barbour, T. (1917b) Notes on the herpetology of the Virgin Islands. *Proceedings of the Biological Society of Washington* 30, 97–104.

Barbour, T. (1918) Amphibia; Reptilia. *In:* Agassiz, G.R. (Ed.), Vertebrata from Madagascar. *Bulletin of the Museum of Comparative Zoology* 61(14), 475–516, 2 pls.

Barbour, T. (1919) Another new race of the kingsnake. *Proceedings of the New England Zoological Club* 7, 1–3, 1 pl.

Barbour, T. (1921a) Reptiles and amphibians from the British Solomon Islands. *Proceedings of the New England Zoological Club* 7, 91–112, pls. 2–6.

Barbour, T. (1921b) The Florida pine snake. *Proceedings of the New England Zoological Club* 7, 117–118.

Barbour, T. (1922) A new snake from southwest Africa. *Proceedings of the Biological Society of Washington* 35, 229–230.

Barbour, T. (1923) Notes on reptiles and amphibians from Panama. *Occasional Papers of the Museum of Zoology, University of Michigan* (129), 1–16.

Barbour, T. (1925) A new frog and a new snake from Panamá. *Occasional Papers of the Boston Society of Natural History* (5), 155–156.

Barbour, T. (1927) Two new Bornean snakes. *Proceedings of the Biological Society of Washington* 40, 127–128.

Barbour, T. (1930) A list of Antillean reptiles and amphibians. *Zoologica* 11(4), 61–116.

Barbour, T. (1935) A second list of Antillean reptiles and amphibians. *Zoologica* 19(3), 77–141.

Barbour, T. (1937) Third list of Antillean reptiles and amphibians. *Bulletin of the Museum of Comparative Zoology* 82(2), 77–166.

Barbour, T. (1941) A new boa from the Bahamas. *Proceedings of the New England Zoological Club* 18, 61–65.

Barbour, T. (1943) A new water snake from North Carolina. *Proceedings of the New England Zoological Club* 22, 1–2, 1 pl.

Barbour, T. & Amaral, A. do. (1924) Notes on some Central American snakes. *Occasional Papers of the Boston Society of Natural History* (1921–1931) (5), 129–132.

Barbour, T. & Amaral, A. do. (1926) A new North American snake. *Proceedings of the New England Zoological Club* 9, 79–81.

Barbour, T. & Amaral, A. do. (1927) Studies on African Ophidia. *Bulletin of the Antivenin Institute of America* 1(1), 25–27.

Barbour, T. & Amaral, A. do. (1928) A new elapid from western Panamá. *Bulletin of the Antivenin Institute of America 1*(4), 100.

Barbour, T. & Carr, A.F., Jr. (1940) Notes on *Elaphe* and a new species. *Occasional Papers of the Boston Society of Natural History* (8), 337–342.

Barbour, T. & Cole, L. (1906) Vertebrata from Yucatán. Reptilia, Amphibia, and Pisces. *Bulletin of the Museum of Comparative Zoology* 50(5), 146–155, 2 pls.

Barbour, T. & Dunn, E.R. (1921) Herpetological novelties. *Proceedings of the Biological Society of Washington* 34, 157–162.

Barbour, T. & Engels, W.T. (1942) Two interesting new snakes. *Proceedings of the New England Zoological Club* 20, 101–104, pl. 16.

Barbour, T. & Loveridge, A. (1928) A comparative study of the herpetological faunae of the Uluguru and Usambara Mountains, Tanganyika Territory, with descriptions of new species. *Memoirs of the Museum of Comparative Zoology* 50(2), 87–265, 4 pls.

Barbour, T. & Loveridge, A. (1929a) Typical reptiles and amphibians. *Bulletin of the Museum of Comparative Zoology* 69(10), 205–360.

Barbour, T. & Loveridge, A. (1929b) On some Honduranian and Guatemalan snakes with the description of a new arboreal pit viper of the genus *Bothrops*. *Bulletin of the Antivenin Institute of America* 3(1), 1–3, 1 fig.

Barbour, T. & Loveridge, A. (1930) Reptiles and amphibians from Liberia. *In:* Strong, R.P. (Ed.), *The African report of Liberia and the Belgian Congo based on the observations made and material collected during the Harvard African Expedition 1926–1927.* Contribution from the Department of tropical Medicine and the Institute for Tropical Biology and Medicine No. V. Harvard University Press, Cambridge, pp. 769–786, illust.

Barbour, T. & Loveridge, A. (1946) First supplement to typical reptiles and amphibians. *Bulletin of the Museum of Comparative Zoology* 96(2), 59–214.

Barbour, T. & Noble, G.K. (1921) Amphibians and reptiles from southern Peru collected by the Peruvian expedition of 1914–1915 under the auspices of Yale University and the National Geographic Society. *Proceedings of the United States National Museum* (1920–1921), 58(2352), 609–620. [6 Jan. 1921]

Barbour, T. & Ramsden, C.T. (1919) The herpetology of Cuba. *Memoirs of the Museum of Comparative Zoology* 47(2), 71–213, 15 pls.

Barbour, T. & Shreve, B. (1935) Concerning some Bahamian reptiles, with notes on the fauna. *Proceedings of the Boston Society of Natural History* 40(5), 347–366.

Barbour, T. & Shreve, B. (1936) New races of *Tropidophis* and of *Ameiva* from the Bahamas. *Proceedings of the New England Zoological Club* 16, 1–3.

Barcelos Ribeiro, S.L. de, d'Agostini, F.M. & Lema, T. de. (2004) Geographic distribution: Serpentes: *Uromacerina ricardinii. Herpetological Review* 35(1), 85.

Bardet, N., Houssaye, A., Rage, J.-C. & Suberbiola, X.P. (2008) The Cenomanian-Turonian (late Cretaceous) radiation of marine squamates (Reptilia): the role of the Mediterranean Tethys. *Bulletin de la Société Géologique de France* 179(6), 605–622, 4 figs.

Barker, B.S., Henderson, R.W. & Powell, R. (2009) Geographic distribution: Serpentes: *Epicrates monensis granti. Herpetological Review* 40(4), 455–456.

Barker, D.G. (1992) Variation, infraspecific relationships, and biogeography of the ridgenose rattlesnake, *Crotalus willardi. In:* Campbell, J.A. & Brodie, E.D., Jr. (Eds.), *Biology of the pitvipers.* Selva, Tyler, pp. 89–105, 6 figs.

Barker, D.G. & Barker, T.M. (1994) *Pythons of the world. Vol. I, Australia.* Advanced Vivarium Systems, Lakeside, 171 pp., 40 figs., 80 col. pls.

Barker, D.G. & Barker, T.M. (2006) *Pythons of the world. Vol. II. Ball pythons: the history, natural history, care, and breeding.* Vida Preciosa International Library, Boerne, 321 pp., col. illust.

Barker, D.G. & Barker, T.M. (2010a) The distribution of the Burmese python, *Python bivittatus,* in China. *Bulletin of the Chicago Herpetological Society* 45(5), 86–88.

Barker, D.G. & Barker, T.M. (2010b) A critique of the analysis used to predict the climate space of the Burmese python in the United States by Rodda et al. (2008, 20009) and Reed and Rodda (2009). *Bulletin of the Chicago Herpetological Society* 45(6), 97–106.

Barme, M. (1968) Venomous sea snakes (Hydrophiidae). *In:* Bücherl, W., Buckley, E.E. & Deulofen, V. (Eds.), *Venomous animals and their venoms. Volume I. Venomous vertebrates.* Academic Press, New York, pp. 285–308.

Barnard, S.M. (1996) *Reptile keeper's handbook.* Krieger Publishing Company, Malabar, 252 pp., 119 figs.

Barnes, B. (1927) Eine eozäne Wirbeltier-Fauna aus der Braunkohle des Geiseltales. *Jahrbuch der Halleschen Verbandes für die Erforschung der Mitteldeutschen Bodenschaetz und ihrer Verwertung* (Series 2) 6, 5–24, 7 figs., 3 pls.

Barnett, B. (1999a) The eastern children's or spotted python *(Antaresia maculosus)* in captivity. *Monitor,* Ardeer 10(2–3), 14–23, 3 figs.

Barnett, B. (1999b) Keeping and breeding the coastal taipan *(Oxyuranus scutellatus). Monitor,* Ardeer 10(2–3), 38–45, 1 fig.

Barnett, L.K., Emms, C. & Santoni, C. (2001) The herpetofauna of Abuko Nature Reserve, The Gambia. *Herpetological Bulletin* (77), 5–14, illust.

Barragan-Vázquez, R., Castillo-Sánchez, I.M. & Frias-Aguilar, J.R. (2004) Geographic distribution: Serpentes: *Conophis lineatus. Herpetological Review* 35(1), 82–83.

Barrett, C. (1950) *Reptiles of Australia, crocodiles, snakes and lizards.* Cassell & Company, London, 168 pp., 41 illust.

Barrie, J. (1990) Skull elements and additional remains of the Pleistocene boid snake *Wonambi naracoortensis. Memoirs of the Queensland Museum,* Brisbane 28(1), 139–151, 13 figs.

Barriga-Bonilla, E., Hernandez-Camacho, J., Jaramillo, I., Jaramillo-Mejia, R., Mora-Osejo, L.E., Pinto-Escobar, P. & Ruiz-Carranza, P.M. (1969) *La isla de San Andres, contribuciones al conocimiento de su ecologia, flora, fauna y fesca.* Universidade Nacional de Colombia, Bogota, 152 pp.

Barrio, A. & Miranda-Ribeiro, M.E. do. (1967) Estudio comparativo morfológico e inmunológico entre las diferentes entidades del género *Micrurus* Wagler (Ophidia, Elapidae) de la Argentina. *Memórias do Instituto Butantan,* São Paulo (1966) 33(3), 869–880, 8 figs.

Barrio, A., Laurent, R.F. & Thomas, R.A. (1977) The status of *Philodryas subcarinatus* Boulenger (Reptilia, Serpentes, Colubridae). *Journal of Herpetology* 11(2), 230–231.

Barrio-Amorgós, C.L. (2001) Geographic distribution: Serpentes: *Micrurus dumerilii carinicauda. Herpetological Review* 32(4), 279.

Barrio-Amorgós, C.L. & Brewer-Carias, C. (2008) Herpetological results of the 2002 expedition to Sarisariñama, a tepui in Venezuelan Guayana, with the descriptions of five new species. *Zootaxa* (1942), 1–68, 19 figs.

Barrio-Amorgós, C.L. & Calcaño, D. (2001) Geographic distribution: Serpentes: *Stenorrhina degenhardti. Herpetological Review* 33(4), 280.

Barrio-Amorgós, C.L. & Pascual, A.D. (2008) Reptilia, Boidae, *Epicrates cenchria cenchria*: Distribution extension. *Check List* 4(3), 244–247, 1 fig.

Barrio-Amorgós, C.L. & Navarrete, L.F. (1999) Geographic distribution: Serpentes: *Lampropeltis triangulum andesiana. Herpetological Review* 30(3), 174–175.

Barrio-Amorgós, C.L., Navarrete, L.F., Fuentes, O. & Mattei, R. (1998) *Siphlophis cervinus* (Serpentes: Colubridae) en Venezuela. *Acta Biologica Venezuelica* 18, 49–53.

Barrio-Amorgós, C.L., Brewer-Carías, C. & Fuentes, O. (1999) Geographic distribution: Serpentes: *Atractus torquatus. Herpetological Review* 30(1), 53.

Barrio-Amorgós, C.L., Chacón, A. & de Pascual, A.D. (2002) Geographic distribution: Serpentes: *Anilius scytale. Herpetological Review* 33(1), 66–67.

Barro, F.E., Grazziotin, F.G., Sazima, J., Martins, M. & Sawaya, R.J. (2012) A new and threatened insular species of lancehead from southeastern Brazil. *Herpetologica* 68(3), 418–429, 6 figs.

Barros, L. (1978) *Guia ilustrado de cobras venenosas de macau e das ilhas da Taipa e Coloane.* Centro de Informacao e Turismo de Macau, Missao, 48 pp.

Barros-Blanco, T.R. (2000) Una nueva especie de *Atractus* (Serpentes, Colubridae) de la Sierra de Perija, estado Zulia, Venezuela. *Anartia* (11), 1–9, 3 figs.

Barros-Blanco, T.R. & Pirela, R. (2000) Geographic distribution: Serpentes: *Oxybelis fulgidus. Herpetological Review* 31(1), 57.

Barros-Blanco, T.R., Jadin, R.C., Caicedo-Portilla, J.R. & Rivas-Fuenemayer, G.A. (2012) Discovery of a rare snail-eater snake in Venezuela (Dipsadinae, *Dipsas pratti*), with additions to its natural history and morphology. *Zoosystematics and Evolution* 88(1), 125–134.

Barros e Cunha, J.G. de. (1935) Ofidios de Massangulo, Distrito do Niassa, coligidos por A. F. Gomes e Sousa. *Memórias e Estudos de Museu Zoologico da Universidade Coimbra* (83), 1–16.

Barth, R. (1957) A faune do Parque Nacional do Itatiaia. *Boletim do Parque Nacional do Itatiaia* (6), 1–150, 39 figs.

Bartlett, E. (1895a) Notes on the snakes of Borneo and the adjacent islands. Part I. *Sarawak Note Book* (1), 68–76. [reprinted from 2 Sept. 1895 Sarawak Gazette, 25, 160–162]

Bartlett, E. (1895b) Notes on the snakes of Borneo and the adjacent islands. Part II. *Sarawak Note Book* (1), 78–85. [reprinted from 1 Oct. 1895 Sarawak Gazette, 25, 182–184]

Bartlett, E. (1896a) Notes on the snakes of Borneo and the adjacent islands. Part III. *Sarawak Note Book* (2), 100–114. [reprinted from Aug.-Oct. 1896 Sarawak Gazette, 26, 153–157]

Bartlett, E. (1896b) Notes on the snakes of Borneo and the adjacent islands. Part IV. Addenda. *Sarawak Note Book* (2), 115–116. [reprinted from 1896 Sarawak Gazette, 26, 241]

Bartlett, J.B. (1854) *Personal narrative or explorations and incidents in Texas, New Mexico, California, Sonora, and Chihuahua, connected with the United States and Mexican Boundary Commission, during the years 1850, '51, '52. and '53. Vol. I.* D. Appleton, New York, xii + 596 pp., 40 pls.

Bartlett, R.D. & Tennant, A. (2000) *Snakes of North America, western region.* Gulf Publishing, Houston, xvi + 311 pp., illust., 186 col. pls.

Bartolotti, K. & Bartolotti, G. (2010) *A field guide to snakes of Hernando, Hillsborough, Manatee, Pasco, and Pinellas counties in Florida.* Robertson Publishing, Los Gatos, 39 pp., col. illust.

Bartram, W. (1791) *Travels through North & South Carolina, Georgia, East & West Florida, the Cherokee country, the extensive territories of the Muscogulges, or Creek Confederacy, and the country of the Chactaws; containing an account of the soil and natural productions of those regions, together with observations on the manners of the Indians.* James & Johnson, Philadelphia. 522 pp.

Barts, M., Schneider, C., Boone, J., Marais, J., Branch, W.R. & Haacke, W.D. (2012) *Lamprophis fiskii* Boulenger, 1887 (Ophidia, Colubroidea, Lamprophiidae: Lamprophiinae), eine selten gefundene Hausschlange aus Südafrika. *Sauria,* Berlin 34(2), 41–51, 10 col. figs.

Bartuano, A.S. & Ibañez D. (2012) Geographic distribution: Serpentes: *Urotheca guentheri. Herpetological Review* 43(2), 210.

Barus, V. & Oliva, A. (1992) *Plazi—Reptilia. Fauna CSFR.* Academia, Praha, 26, 222 pp., 32 figs.,15 pls.

Basoglu, M. (1947) On some varieties of *Vipera berus* from the extreme northeastern Anatolia. *Revue de la Faculté des Sciences de l'Université d'Istanbul* 12B(3), 182–190, 1 photo.

Basoglu, M. (1970) On two specimens of *Eirenis lineomaculata* Schmidt (Colubridae, Ophidia) from Turkey. *Scientific Reports of the Faculty of Science, Ege University* (110), 1–7, 1 pl.

Basoglu, M. & Baran, I. (1980) *Türkiye Sürüngenleri Kisim II. Yilanlar. The reptiles of Turkey. Part II. The snakes (Taxonomy and distribution, key for identification).* Ege Üniversitesi Matbaasi, Bornova–Izmir, ix + 205 pp., 91 figs., 10 pls.

Bassarukin, A.M. (1983) *Cadastrum of the distribution of amphibians and reptiles of the Sakhalin region.* Yuzhno-Sakhalinsk, 29 pp., 11 figs. (in Russian)

Basu, D. (1989) Range extension of *Chrysopelea ornata* Shaw (Reptilia, Colubridae) with comments on the distribution of some snakes in north India. *The Journal of the Bombay Natural History Society* 86(1), 110–111.

Baszio, S. (2004) *Messelophis variatus* n. gen. n. sp., from the Eocene of Messel: a tropidopheine snake with affinities to Erycinae (Boidae). *Courier Forschungsinstitut Senckenberg,* Frankfurt-am-Main (252), 47–66.

Batchelor, D.M. (1958) Some notes on the snakes of Asaban, Malacca. *Malay Nature Journal,* Kuala Lumpur 12, 103–112.

Bates, M.F. (1991a) A re-evaluation of the taxonomic status of *Xenocalamus bicolor concavorostralis* Hoffman, 1940 (Serpentes, Atractaspidinae). *South African Journal of Zoology,* Pretoria 26(2), 78–81, illust.

Bates, M.F. (1991b) A provisional check list of the reptiles and amphibians of Golden Highlands National Park. *Koedoe* 34(2), 153–155.

Bates, M.F. (1996) New reptile distribution records for the Free State Province of South Africa. *Navorsinge van die Nasionale Museum Bloemfontein* 12(1), 1–47.

Battersby, J.C. (1938) Some snakes of the genus *Tropidophis. Annals and Magazine of Natural History,* London (Series 11) 1(5), 557–560.

Battersby, J.C. (1950) A new amphisbaenid lizard from Tanganyika Territory and notes on the rare snake *Chilorhinophis. Annals and Magazine of Natural History,* London (Series 12) 3(29), 413–417, 2 figs.

Battersby, J.C. (1951) Description of a new species of snake *Prosymna pitmani,* sp. n., from Tanganyika Territory and notes on the genus. *Annals and Magazine of Natural History,* London (Series 12) 4(44), 828–829.

Battersby, J.C. (1953) On a further record of *Micrelaps mulleri* Boettger from Syria. *Annals and Magazine of Natural History,* London (Series 12) 6(68), 640.

Battersby, J.C. (1954) Reptiles and amphibians collected in the northern Frontier Division of Kenya in 1951. *Annals and Magazine of Natural History,* London (Series 12) 7(76), 241–248.

Battersby, J.C. (1955) On further amphibians and reptiles from the Northern Frontier Division of Kenya. *Annals and Magazine of Natural History,* London (Series 12) 8(86), 149–150.

Battersby, J.C. & Swinton, W.E. (1960) Reptilia. *In: The Zoological Record* (1957) 94(16B), 1–94.

Bauchot, R. (Ed.). (1994) *Snakes: a natural history.* Sterling Publishing Company, New York, 220 pp., col. illust.

Bauer, A.M. (1987) Geographic distribution: Serpentes: *Ramphotyphlops braminus. Herpetological Review* 18(2), 41.

Bauer, A.M. (1998) Some Asian herpetological specimens of historical note in the Zoological Museum, Berlin. *Hamadryad* 23(2), 133–149.

Bauer, A.M. (2000) Comments on the types and type localities of South African reptiles collected by Heinrich Bergius and Ludwig Krebs. *African Journal of Herpetology* 49, 53–60.

Bauer, A.M. (Ed.). (2004) Studies in the history of herpetological exploration. *Bonner Zoologische Beiträge* 52(3–4), 181–335.

Bauer, A.M. & Branch, W.R. (2001) The herpetofauna of the Richtersveld National Park and the adjacent northern Cape Province, Republic of South Africa. *Herpetological Natural History* 8(2), 111–160, 39 col. figs.

Bauer, A.M. & Das, I. (1999) North Carolina reptiles and amphibians of historical note in the collection of the Zoological Survey of India. *The Journal of the Elisha Mitchell Scientific Society* 115(1), 1–7.

Bauer, A.M. & Günther, R. (1992) A preliminary report on the reptile fauna of the Kingdom of Bhutan with the description of a new species of scincid lizard (Reptilia, Scincidae). *Asiatic Herpetological Research,* Berkeley 4, 23–36, 2 figs.

Bauer, A.M. & Sadlier, R.A. (2000) *The herpetofauna of New Caledonia.* Society for the Study of Amphibians and Reptiles, xii + 310 pp., 63 figs., 24 col. pls.

Bauer, A.M. & Vindum, J. (1990) A checklist and key to the herpetofauna of New Caledonia, with remarks on biogeography. *Proceedings of the California Academy of Sciences* 47(2), 17–45, 5 figs.

Bauer, A.M., Günther, R. & Klipfel, M. (1995) The herpetological contributions of Wilhelm Peters (1815–1883). *In:* Peters, W., *The herpetological contributions of Wilhelm Peters (1815–1883).* Society for the Study of Amphibians and Reptiles, Ithaca, pp. 5–87

Bauer, A.M., Wallach, V. & Günther, R. (2002) An annotated type catalogue of the scolecophidian, alethinophidian, and macrostomatan snakes in the collection of the Museum für Naturkunde der Humboldt-Universität zu Berlin. *Mitteilungen aus dem Museum für Naturkunde in Berlin* 78(1), 157–176.

Baumann, F. (1913) Reptilien und Batrachier des Berner Naturhistorischen Museums aus dem Battak-Gebirge von West-Sumatra. *Zoologische Jahrbücher. Abteilung für Systematik, Ökologie und Geographie der Tiere*, Jena (1912–1913) 34(3), 257–278, figs. a–e.

Bavay, A. (1869) Catalogue des reptiles de la Nouvelle-Calédonie et description d'espèces nouvelles. *Mémoires de la Société Linnéenne de Normandie* 15(5), 1–37.

Baxter, G.T. & Stone, M.D. (1980) *Amphibians and reptiles of Wyoming*. Wyoming Game & Fish Department, Cheyenne, vi + 137 pp., illust.

Baynham, T. (2010) Geographic distribution: Reptilia: Squamata: Elapidae: *Dendroaspis polylepis polylepis. African Herp News* (51), 25–27.

Beaman, K.R. & Dugan, E.A. (2006) *Crotalus ruber. Catalogue of American amphibians and reptiles,* pp. 840.1–840.17, 2 figs.

Beaman, K.R. & Grismer, L.L. (1994) *Crotalus enyo* (Cope) Baja California rattlesnake. *Catalogue of American amphibians and reptiles*, pp. 589.1–589.6, 3 figs.

Beaman, K.R. & Spencer, C.L. (2004) *Crotalus tortugensis. Catalogue of American amphibians and reptiles*, pp. 798.1–798.5, 1 fig,

Beaman, K.R. & Wong, N. (2001) *Crotalus catalinensis. Catalogue of American amphibians and reptiles*, pp. 733.1–733.4, 2 col. figs.

Beane, J.C., Braswell, A.L., Mitchell, J.C., Palmer, W.M. & Harrison, J.R. (2010) *Amphibians and reptiles of the Carolinas and Virginia*. Second edition. University of North Carolina Press, Chapel Hill, x + 274 pp., col. photos. [first edition, Martof et al., 1980, 264 pp., col. photos]

Beard, D.J. (1979) Rough-scaled snake, *Tropidechis carinatus. Herpetofauna*, Sydney 10(2), 26–28.

Bechstein, J.M. (1801) *Herrn de La Cépède's Naturgeschischte der Amphibien oder der enerlegenden vierfüssingen Thiere und der Schlangen. Eine Forsekung von Buffon's Naturgeschichte aus dem Französischen übersetzt und mit Anmerkungen und Zusatzen versehen.* Dritter Band. Industrie Comptoir's, Weimar, xxvi + 454 pp., 30 pls.

Bechstein, J.M. (1802a) *Herrn de La Cépède's Naturgeschischte der Amphibien oder der enerlegenden vierfüssigen Thiere und der Schlangen. Eine Fortsekung von Buffon's Naturgeschichte aus dem Französischen übersetzt und mit Anmerkungen und Zusatzen versehen.* Vierter Band. Industrie Comptoir's, Weimar, xx + 298, 48 pls.

Bechstein, J.M. (1802b) *Herrn de La Cépède's Naturgeschischte der Amphibien oder der enerlegenden vierfüssigen Thiere und der Schlangen. Eine Fortsekung von Buffon's Naturgeschichte aus dem Französischen übersetzt und mit Anmerkungen und Zusatzen versehen.* Fünfter Band. Industrie Comptoir's, Weimar, xii + 200, 21 pls

Beddard, F.E. (1903) On the trachea, lungs and other points in the anatomy of the hamadryad snake (*Ophiophagus bungarus*). *Proceedings of the Zoological Society of London* 73(1), 319–328, figs.

Beddard, F.E. (1906) Contributions to the anatomy of the Ophidia. *Proceedings of the Zoological Society of London* 76(1–2), 12–44, figs. 2–11.

Beddard, F.E. (1908) A comparison of the Neotropical species of *Corallus, C. cookii*, with *C. madagascariensis*; and on some points in the anatomy of *Corallus caninus. Proceedings of the Zoological Society of London* 78(2), 135–158, figs. 21–27.

Beddome, R.H. (1862) Notes upon the land and fresh-water snakes of the Madras Presidency. *Madras Quarterly Journal of Medical Science* 5, 1–31, pl. 2.

Beddome, R.H. (1863a) Further notes upon the snakes of the Madras Presidency; with descriptions of new species. *Madras Quarterly Journal of Medical Science* 6, 41–48, 2 pls.

Beddome, R.H. (1863b) Descriptions of new species of the family Uropeltidae from southern India, with notes on other little known species. *Proceedings of the Zoological Society of London* 31(1), 225–229, 3 col. pls.

Beddome, R.H. (1864) Description of a new species of *Elaps* from Malabar. *Proceedings of the Zoological Society of London* 32(1), 179–180.

Beddome, R.H. (1867) Descriptions and figures of five new snakes from the Madras Presidency. *Madras Quarterly Journal of Medical Science* 11, 14–16, 2 pls.

Beddome, R.H. (1870) Descriptions of new reptiles from the Madras Presidency. *Madras Monthly Journal of Medical Science* (Series 2) 2, 169–176.

Beddome, R.H. (1871) Descriptions of new reptiles from the Madras Presidency. *Madras Monthly Journal of Medical Science* (Series 2) 4, 401–404.

Beddome, R.H. (1876) Description of a new species of Indian snake of the genus *Platyplectrurus* from the Wynad. *Proceedings of the Zoological Society of London* 39(1), 701.

Beddome, R.H. (1877a) Description of three new snakes of the family Uropeltidae from southern India. *Proceedings of the Zoological Society of London* 45(1), 167–168.

Beddome, R.H. (1877b) Description of new reptiles from the Madras Presidency. *Proceedings of the Zoological Society of London* 45(1), 685–686.

Beddome, R.H. (1878a) Descriptions of new Uropeltidae from southern India, with remarks on some previously described species. *Proceedings of the Zoological Society of London* 46(1), 154–155.

Beddome, R.H. (1878b) Description of a new genus of snakes of the family Calamaridae, from southern India. *Proceedings of the Zoological Society of London* 46(1), 576.

Beddome, R.H. (1878c) Description of six new species of snakes of the genus *Silybura*, family Uropeltidae, from the Peninsula of India. *Proceedings of the Zoological Society of London* 46(1), 800–802.

Beddome, R.H. (1880) Description of a new snake of the genus *Plectrurus* from Malabar. *Proceedings of the Zoological Society of London* 48(2), 182.

Beddome, R.H. (1886) An account of the earth-snakes of the peninsula of India and Ceylon. *Annals and Magazine of Natural History*, London (Series 5) 17(97), 3–33.

Bedriaga, J. von. (1881–1882) Die Amphibien und Reptilien Griechenlands. *Bulletin de la Société Impériale des Naturalistes de Moscou* 56(2), 242–310 (1881), 56(3), 43–103 (1881) & 56(4), 278–344 (1882).

Bedriaga, J. von. (1883) Beiträge zur Kenntniss der Amphibien und Reptilien der fauna von Corsika. *Archiv für Naturgeschichte*, Berlin 49(1), 124–273, pls. 3–5.

Bedriaga, J. von. (1884) *Amphisbaena cinerea* Vand. und *A. strauchi* v. Bedr. Erster Beitrag zur Kenntniss der Doppelschleichen. *Archiv für Naturgeschichte,* Berlin 50(1), 23–75, pl. 4.

Bedriaga, J. von. (1892) Notes sur les amphibiens et reptiles recuéillis par M. Adolphe F. Moller aux Iles de la Guinée. *O Instituto*, Coïmbre 40, 316–320, 432–447.

Bedriaga, J. von. (1907) Verzeichnis der von der Central-Asiatischen Expedition unter Stabs-Kapitan W. Roborowski in den jahren 1893–1895 gesammelten reptilian. *Annuaire de Musée Zoologique de l'Académie Impériale des Sciences de St. Pétersbourg* 10, 159–200.

Bedriaga, J. von. (1912) *Wissenschaftliche Resultate der von N.M. Przewalski nach Central-Asien unternommenen Reisen. Auf kosten einer von seiner kaiserlichen hoheit dem Grossfursten Thronfolger Nikolai Alexandrowitsch, seiner Majestat dem jetzt regierenden summe heransgegeben von der Kaiserlichen Akademie der Wissenschaften. Zoologischer Theil. Band III, Abtheilung 1. Amphibien und Reptilien.* Kaiserlichen Akademie der Wissenschaften, St. Pétersbourg, 769 pp., 10 pls.

Beebe, W. (1919) The higher vertebrates of British Guiana, with special reference to the fauna of Bartica District. List of Amphibia, Reptilia and Mammalia. *Zoologica* 2(7), 205–227.

Beebe, W. (1946) Field notes on the snakes of Kartabo, British Guiana, and Caripito, Venezuela. *Zoologica* 31(4), 11–52, 4 figs., 13 pls.

Beebee, T.J.C. & Griffiths, R.A. (2000) *Amphibians and reptiles: a natural history of the British herpetofauna*. HarperCollins Publishers, London, 270 pp., 8 col. pls., 80 b&w photos.

Beerli, P., Billing, H. & Schätti, B. (1986) Taxonomischer status von *Vipera latasti monticola* Saint Girons, 1953 (Serpentes, Viperidae). *Salamandra*, Frankfurt-am-Main 22(2–3), 101–104, illust.

Begg, R.J. & Martin, K. (1980) Capture of a further specimen of *Python oenpelliensis*. *Herpetofauna*, Sydney 11(2), 11.

Behler, J.L. & King, F.W. (1979) *National Audubon Society field guide to North American reptiles and amphibians*. Alfred A. Knopf, New York, 742 pp., 657 col. photos.

Behura, B.K. (1966) *Odisara sapa (Snakes of Orissa)*. Zoological Society of Orissa, Orissa, 244 pp., figs. (in Oriya)

Beirn, J.T. (1979) Sea snakes of Gato Island: an exotic resource of the Philippines. *Explorer's Journal* 57(2), 90–91.

Belding, L. (1887) Reptiles of the Cape region of Lower California. *The West-American Scientist*, San Diego 3(24), 97–99.

Bell, C.J. & Mead, J.I. (1996) *Charina* Gray, 1849 (Boidae, Erycinae) from the Pleistocene of California. *Herpetological Natural History* 4(2), 161–168, 1 fig.

Bell, L.N. (1952) A new subspecies of the racer *Coluber constrictor*. *Herpetologica* 8(2), 21.

Bell, T. (1825) On Leptophina, a group of serpents comprising the genus *Dryinus* of Merrem, and a newly formed genus proposed to be named *Leptophis*. *Zoological Journal*, London 2(7), 322–329, 1 pl.

Bell, T. (1826) Sur les Leptophina, groupe de serpents qui comprend les genres *Dryinus* de Merrem, et le nouveau genre récemment nommé *Leptophis*. *Bulletin des Sciences Naturelles*, Paris 9, 104–105.

Bell, T. (1830) Uber Leptophina, eine Schlangengruppe, Taf. 12., welche *Dryinus* Merr. und eine neue Sippe *Leptophis* begreift. *Isis von Oken*, Leipzig 23(10), 1035–1038.

Bell, T. (1849) *A history of British reptiles*. Second edition. John Van Voorst, London, 159 pp., 40 figs. [first edition, 1839, 142 pp., 40 figs.]

Bellagamba, P.J. & Vega, L.E. (1996) Geographic distribution: Serpentes: *Thamnodynastes hypoconia*. *Herpetological Review* 27(1), 36.

Bellairs, A. (1969) *Die reptilien*. Editions Rencontre, Lausanne, 768 pp, col. illust.

Bellemin, J.M. & Stewart, G.R. (1977) Diagnostic characters and color convergence of the garter snakes *Thamnophis elegans terrestris* and *Thamnophis couchii atratus* along the central California coast. *Bulletin of the Southern California Academy of Sciences* 76(2), 73–84, 3 figs.

Bellini, G.P., Arzamendia, V. & Giraudo, A.R. (2013) Ecology of *Thamnodynastes hypocopnia* in subtropical-tempertae South America. *Herpetologica* 69(1), 67–79, 2 figs.

Bellosa, H. (2007) Python molurus, *der Tigerpython*. Natur und Tier-Verlag, Münster, 111 pp., col. illust.

Bellosa, H. & Bisplinghof, H. (2012) *Rainbow boas: natural history & captive husbandry*. Edition Chimaira, Frankfurt-am-Main, 223 pp., col. photos.

Belluomini, H.E., Maranhão-Nina, A.F. & Hoge, A.R. (1960) Contribuição à biologia do genero *Eunectes* Wagler, 1830. (Serp. Boidae). Estudo de seis ninhadas de "sucuris." *Memórias do Instituto Butantan*, São Paulo (1959) 29, 165–174, 2 figs.

Belluomini, H.E., Veinert, T., Dissmann, F., Hoge, A.R. & Penha, A.M. (1978) Notas biológicas a respeito do genero *Eunectes* Wagler, 1830 "sucuris." [Serpentes, Boinae]. *Memórias do Instituto Butantan*, São Paulo (1976–1977) 40–41, 79–115.

Bels, V. (1987) Observations of the courtship and mating behavior in the snake *Hydrodynastes gigas*. *Journal of Herpetology* 21(4), 350–352, 2 figs.

Belt, P.J., Malhotra, A., Thorpe, R.S., Warrell, D.A. & Wüster, W. (1997) Russell's viper in Indonesia, snakebite and systematics. *In:* Thorpe, R.S., Wüster, W. & Malhotra, A. (Eds.), *Venomous Snakes, Ecology, evolution and snakebite*. Zoological Society of London, Clarendon Press, Oxford, pp. 219–234, 2 figs.

Belt, T. (1874) *The naturalist in Nicaragua*. J.M. Dent & Sons, London, xxxiv + 306 pp., illust.

Beltrán, E. (1953) *Vida Silvestre y recursos naturales alo largo de la carretera Panamericana*. Instituto Mexicano de Recursos Naturales Renovables, México City, vii + 228 pp.

Bendiscioli, G. (1826) Monografia dei serpenti della Provincia di Mantova. *Giornale di Fisica, Chimica e Storia Naturale Medicina ed Arti*, Pavia 9(6), 413–434.

Beneden, P.-J. van. (1871) Les reptiles fossiles en Belgique; notice. *Bulletins de la Académie Royale de Belgique*, Bruxelles (Série 2) 31(1), 21–28.

Benítez-Gálvez, J.E. (1997) *Los ofidios de Puebla*. Comisión Estattal Forestal y de la Fauna Silvestre, Puebla, 206 pp., col. illust.

Benke, Z. (2000) *Magyarország kígyói: as üldözött oslakosok*. Alexandra, Alkotás, 88 pp., col. illust. (in Hungarian)

Bennett, D.K. (1979) The fossil fauna from Lost and Found Quarries (Hemphillian: latest Miocene), Wallace County, Kansas. *Occasional Papers of the Museum of Natural History, University of Kansas* (79), 1-24, illust.

Bennett, G.F. (1862) The snakes of New South Wales. *The Medical Record of Australia* 2(8), 92–93.

Bennett, R. (1997) *Reptiles & frogs of the Australian Capital Territory*. National Parks Association of the ACT, Woden, 86 pp., col. illust.

Benton, M.J. (1980a) Geographic variation in the garter snakes (*Thamnophis sirtalis*) of the north-central United States, a multivariate study. *Zoological Journal of the Linnean Society*, London 68(4), 307–323, figs.

Benton, M.J. (1980b) Geographic variation and the validity of subspecies names for the eastern garter snake, *Thamnophis sirtalis*. *Bulletin of the Chicago Herpetological Society* 15(3), 57–69, figs.

Beolens, B., Watkins, M. & Grayson, M. (2011) *The eponym dictionary of reptiles*. Johns Hopkins University Press, Baltimore, 296 pp.

Berg, C. (1895) Dos reptiles nuevos. *Anales del Museo Nacional de Buenos Aires* 4, 189–194.

Berg, C. (1898) Contribuciones al conocimiento de la fauna erpetológica argentina y de los países limitrofes. *Anales del Museo Nacional de Buenos Aires* (1898–1899) 6, 1–32.

Berg, C. (1901) Herpetological notes. *Comunicaciones del Museo Nacional de Buenos Aires* 1(8), 289–291.

Berger, C. (2007) *Venomous snakes*. Stackpole Books, Mechanicsburg, 104 pp., col. illust.

Berger, L. (1975) Gady i plazy (Reptilia et Amphibia). *Fauna Slodkowodna Polski* (4), 1–108, 46 figs.

Berger, L. (2000) *Plazy I gady Polski: klucz do oznaczania*. Wydawnictwo Naukowe PWN, Warszawa, 146 pp., 51 figs., 28 col. photos.

Berger, L, Jaskowska, J. & Mlynarski, M. (1969) Plazy i Gady. *Katalog Fauny Polski* 39(14), 1–73.

Berger-Dell'mour, H. (1986) Zur herpetofauna des Golan. *Annalen des Naturhistorischen Museums in Wien* 87B, 59–67.

Bergman, J. & Norström, M. (1994) Some notes on the genus *Vipera* (Serpentes, Viperidae, Viperinae) in Anatolia, Turkey. *Litteratura Serpentium* 14(6), 165–175, 8 pls.

Bergman, R.A.M. (1949) The anatomy of *Lapemis hardwickii* Gray. *Proceedings of the Koninklijke Nederlandse Akademie van Wetenschappen* 52C, 882–898.

Bergman, R.A.M. (1950a) The life of *Natrix vittata*. *Zoologische Mededelingen*, Leiden 31(1), 1–11, 2 figs.

Bergman, R.A.M. (1950b) The anatomy of *Natrix vittata* (L.). *Zoologische Mededelingen*, Leiden 31(2), 13–24, 1 fig.

Bergman, R.A.M. (1953a) L'anatomie du genre *Ptyas* a Java. *Rivista di Biologia Coloniale*, Roma (1952) 12, 5–42, 12 figs.

Bergman, R.A.M. (1953b) The anatomy of *Cylindrophis rufus* (Laur.). I. *Proceedings of the Koninklijke Nederlandse Akademie van Wetenschappen*, Amsterdam 56C(5), 650–666, 8 figs.

Bergman, R.A.M. (1954) *Thalassophis anomalus* Schmidt. *Koninklijke Instituut voor de Tropen*, Amsterdam, 16 pp., 5 figs.

Bergman, R.A.M. (1955a) L'anatomie de *Enhydris enhydris*. *Rivista di Biologia Coloniale*, Roma 15, 5–28, 7 figs., 1 pl.

Bergman, R.A.M. (1955b) L'anatomie de *Cerberus rhynchops*. *Archives Néerlandaises de Zoologie*, Leiden 11(1), 113–126, 7 figs.

Bergman, R.A.M. (1955c) L'anatomie de *Enhydrina schistosa* D. *Archives Néerlandaises de Zoologie*, Leiden 11(1), 127–142, 7 figs.

Bergman, R.A.M. (1955d) *Dendrophis pictus*. *Proceedings of the Koninklijke Nederlandse Akademie van Wetenschappen*, Amsterdam 58C(2), 206–218, 9 figs.

Bergman, R.A.M. (1955e) The anatomy of *Xenopeltis unicolor*. *Zoologische Mededelingen*, Leiden 33(22), 209–225, 7 figs.

Bergman, R.A.M. (1956a) The anatomy of *Dryophis prasinus*. I. *Proceedings of the Koninklijke Nederlandse Akademie van Wetenschappen*, Amsterdam 59C(2), 263–279, 10 figs.

Bergman, R.A.M. (1956b) The anatomy of *Natrix subminiata*. *Biologisch Jaarboek*, Gent 23, 306–326, 7 figs.

Bergman, R.A.M. (1958a) L'anatomie de *Elapoides fuscus* B. *Rivista di Biologia Coloniale*, Roma (1956–1958) 16, 9–32, 11 figs.

Bergman, R.A.M. (1958b) The anatomy of *Natrix piscator*. *Biologisch Jaarboek*, Gent 26, 77–79, 8 figs.

Bergman, R.A.M. (1958c) The anatomy of the Acrochordinae. I–IV. *Proceedings of the Koninklijke Nederlandse Akademie van Wetenschappen*, Amsterdam 61C(2), 145–184.

Bergman, R.A.M. (1959) *Natrix chrysarga*, *Natrix trianguligera*. *Biologisch Jaarboek*, Gent 27, 73–79, 4 figs.

Bergman, R.A.M. (1960) The anatomy of some Homalopsinae. *Biologisch Jaarboek*, Gent 28, 119–138, 10 figs.

Bergman, R.A.M. (1961a) The anatomy of *Coluber radiatus* and *Coluber melanurus*. *Pacific Science*, Honolulu 15(1), 144–154, 7 figs.

Bergman, R.A.M. (1961b) The anatomy of some Viperidae (I and II). *Acta Morphologica Neerlando-Scandinavica*, Amsterdam 4(3), 195–230, 8 figs.

Bergman, R.A.M. (1962a) Die anatomie der Elapinae. *Zeitschrift für Wissenschaftliche Zoology*, Leipzig 167(3–4), 291–337, 18 figs.

Bergman, R.A.M. (1962b) The anatomy of *Hydrophis fasciatus atriceps*. *Biologisch Jaarboek*, Gent 30, 389–416, 9 figs.

Bergman, R.A.M. (1963) The anatomy of *Ablabes baliodeira*, a colubrid snake from Java. *Journal of the Ohio Herpetological Society* 4(1–2), 1–14, 6 figs.

Bergman, R.A.M. (1965) The anatomy of *Calamaria multipunctata* (Boie). *Bulletin of the National Museum*, Singapore 33(7), 35–56, 10 figs.

Bergmans, W. & Zuiderwijk, A. (1980) Amphibieën en reptielen in Nederland. *Wetenschappelijke Mededelingen van de Koninklijke Nederlandse Natuurhistorische Vereniging* (139), 1–74, 101 figs.

Bergmans, W. & Zuiderwijk, A. (1986) *Atlas van de Nederlandse Amfibieën en Reptielen en hun Bedreiging. Vijfde Herpetogeografisch verslag*. Koninklijke Nederlandse Natuurhistorische Vereniging, Hoogwoud, 177 pp.

Bergna, S. & Alvarez, B. (1993) Descripción de una nueva especie de *Thamnodynastes* (Reptilia, Serpentes, Colubridae) del nordeste argentino. *Facena* 10, 5–18, 2 figs., 1 pl.

Bernal-Carlo, A. (1994) A new subspecies of the colubrid snake *Liophis epinephelus* from Sierra Nevada de Santa Marta, Colombia. *Bulletin of the Maryland Herpetological Society* 30(4), 186–189.

Bernal-Carlo, A. (1997) Snake genus *Atractus* (Colubridae) from Sierra Nevada de Santa Marta, Colombia, with a description of a new species. *Bulletin of the Maryland Herpetological Society* 33(4), 165–170.

Bernal-Carlo, A. & Roze, J.A. (1994) A new *Leptophis* (Serpentes, Colubridae) from Sierra Nevada de Santa Marta, Colombia. *Bulletin of the Maryland Herpetological Society* 30(1), 46–49.

Bernarde, P.S. & Moura-Leite, J.C. de (1999) Geographic distribution: Serpentes: *Hydrodynastes gigas*. *Herpetological Review* 30(1), 54.

Bernarde, P.S., Souza do Amaral, E. de & Vale, A.D. do. (2011) Squamata, Serpentes, Viperidae, *Bothrocophias hyoprora* (Amaral, 1935): distribution extension in the state of Acre, northern Brazil. *Check List* 7, 813–814, 3 figs.

Bernardo, P.H., Machado, F.A., Murphy, R.W. & Zaher, H. (2012) Redescription and morphological variation of *Oxyrhopus clathratus* Duméril, Bibron and Duméril, 1854 (Serpentes, Dipsadidae, Xenodontinae). *South American Journal of Herpetology* 7(2), 134–148, 6 figs.

Bernhard, T. (1999) *The reptiles and amphibians of Hampshire and the Isle of Wight*. Hampshire & Isle of Wight Naturalists' Trust, Romsey, 12 pp., illust.

Bernheimer, A.W., Weinstein, S.A. & Linder, R. (1986) Isoelectric analysis of some Australian elapid snake venoms with special reference to phospholipase B and hemolysis. *Toxicon* 24(8), 841–849, 3 figs.

Bérnils, R.S. & Gilardi, J. (2000) Geographic distribution: Serpentes: *Imantodes cenchoa*. *Herpetological Review* 31(1), 55–56.

Bernini, F., Bonini, L., Ferri, V., Gentilli, A., Razzetti, E. & Scali, S. (2004) *Atlante degli anfibi e dei rettili della Lombardia.* Monographie di Pianura (5), 1–254, figs., col. illust.

Berry, J.R. (2010) *Designer morphs. Volume one: ball python & boa constrictor. A guide to mutations and variations.* Second edition. Edition Chimaira, Frankfurt-am-Main, 348 pp., col. photos. [first edition, 2006 by John R. Berry, Palm Beach, 323 pp., 275 col. photos]

Berry, J.R. (2012) *Designer-morphs, western hognose snakes: a complete guide to care, breeding morphs and mutations.* Big Reptile World, Bristol, UK, 198 pp., col. illust.

Berry, P.S.M. (1963) The Gaboon viper of the plateau forests of northern Rhodesia. *Puku*, Chilanga (1), 71–74.

Berry, P.S.M. (1970) A note on Shreve's tree snake, *Dipsadaboa shrevei* (Loveridge) in Zambia. *Journal of the Herpetological Association of Africa* (6), 20–21.

Berthold, A.A. (1842) Ueber verschiedene neue oder seltene Amphibienarten. *Sitzungsberichte königlich Gesellschaft der Wissenschaften zu Göttingen* (18), 1–28, 1 pl.

Berthold, A.A. (1845) Ueber verschiedene neue oder seltene Reptilien aus Neu-Granada und Crustaceen aus China. *Nachrichten von der Gesellschaft der Wissenschaften und der Georg-Augustus-Universität zu Göttingen* (3), 37–48, 3 pls.

Berthold, A.A. (1846) Über verschiedene neue oder seltene Reptilien aus Neu-Granada und Crustaceen aus China. *Sitzungsberichte königlich Gesellschaft der Wissenschaften zu Göttingen* 3, 1–32, 3 pls. [reprint of Berthold, 1845]

Berthold, A.A. (1859) Einige neue Reptilien des Akad. Zoolog. Museums in Göttingen. *Nachrichten von der Gesellschaft der Wissenschaften und der Georg-Augustus-Universität zu Göttingen* (17), 179–181.

Bertke, E.M., Watt, D.D. & Tu, T. (1966) Electrophoretic patterns of venoms from species of Crotalidae and Elapidae snakes. *Toxicon* 4(1), 73–76, 2 figs.

Bertoni, A.W. (1913) Fauna Paraguaya. Catálogos sistemáticos de los vertebrados del Paraguay. *In: Descripción Fisica y Económica del Paraguay.* Moisés S. Bertoni, Asunción, 86 pp.

Bertoni, A.W. (1915) Fauna Paraguaya. Catálogos sistemáticos de los vertebrados del Paraguay. Peces, batracios, reptiles, aves y mamiferos conocides hasta 1913. *Revista de la Sociedad Cientifico de Paraguay* 4(4), 1–59.

Bertoni, A.W. (1918) Adiciones a los vertebrados del Paraguay. *Anales Científicas Paraguayos* (2) 3, 233–241.

Bertrand, A. & Cochet, P.A. (1992) *Amphibiens et reptiles d'Ariege.* Société Herpétologique de France, Paris, 139 pp., illust.

Beshkov, V.A. (1984) The effect of the Balkan Range on the distribution of the herpetofauna in Bulgaria. *Acta Zoologica Bulgarica*, Sofia 25, 9–15, 3 figs. (in Russian)

Beshkov, V.A. (1986) Striped and non-striped type of colouring of the ringed snake *Natrix natrix* (L.) (Colubridae, Serpentes) in Bulgaria. *Acta Zoologica Bulgarica*, Sofia 31, 32–36, 1 fig. (in Russian)

Beshkov, V. & Beron, P. (1964) Catalogue et bibliographie des amphibiens et des reptiles en Bulgarie. *Editions de l'Academie Bulgare des Sciences*, Sofia, 39 pp., 4 figs.

Beshkov, V. & Nanev, K. (2002) *Amphibians and reptiles in Bulgaria.* Pensoft, Sofia, 120 pp., col. illust. (in Russian)

Bethencourt-Ferreira, J. (1891) Sur quelques espèces du genre *Elaps* déposés au Muséum de Lisbonne. *Jornal de Sciencias, Mathematicas, Physicas e Naturaes*, Lisboa (Série 2) 2(6), 89–95.

Bethencourt-Ferreira, J. (1892) Revisao dos reptis e batrachios de Portugal. *Jornal de Sciencias, Mathematicas, Physicas e Naturaes*, Lisboa (Série 2) 2(8), 268–290.

Bethencourt-Ferreira, J. (1897a) Reptis da India do Museu de Lisboa. *Jornal de Sciencias, Mathematicas, Physicas e Naturaes*, Lisboa (Série 2) 4(16), 212–234.

Bethencourt-Ferreira, J. (1897b) Sobre alguns reptis ultimamente enviados á secção zoological do Museu de Lisbon. *Jornal de Sciencias, Mathematicas, Physicas e Naturaes*, Lisboa (Série 2) 5(18), 111–114.

Bethencourt-Ferreira, J. (1898a) Sobre alguns reptis ultimamente enviados á secção zoological do Museu de Lisbon. *Jornal de Sciencias, Mathematicas, Physicas e Naturaes*, Lisboa (Série 2) 5(18), 111–116.

Bethencourt-Ferreira, J. (1898b) Reptis de Timôr no Museu de Lisboa. *Jornal de Sciencias, Mathematicas, Physicas e Naturaes*, Lisboa (Série 2) 5(19), 151–156.

Bethencourt-Ferreira, J. (1903) Reptis de Angola da regiao norte do Quanza da collecao Pereira do Nascimento. *Jornal de Sciencias, Mathematicas, Physicas e Naturaes*, Lisboa (Série 2) 7(25), 9–16.

Bethencourt-Ferreira, J. (1905) Reptis e amphibios de Angola da regiao ao norte do Quanza (colleccao Newton—1903). *Jornal de Sciencias, Mathematicas, Physicas e Naturaes*, Lisboa (1904) (Série 2) 7(26), 111–117.

Bethencourt-Ferreira, J. (1906) Alcumas especies novas ou pouco conhecidas de amphibios e reptis de Angola (Collecção Newton—1903–1904). *Jornal de Sciencias, Mathematicas, Physicas e Naturaes*, Lisboa (Série 2) 7(27), 159–169, 1 pl.

Bethencourt-Ferreira, J. (1930) Some notes on the habits and peculiarities of the African snakes of the genus *Naja*, their taxonomic status and geographical distribution. *Bulletin of the Antivenin Institute of America* 4(3), 53–57.

Bethencourt-Ferreira, J. & Seabra, A.F. de. (1911) Catalogue systematique des Vertébrés du Portugal. *Bulletin de la Société Portugaise des Sciences Naturelles*, Lisbon 5(3), 97–128.

Betta, F.E. de. (1853) *Catalogus systematicus rerum naturalium in Museo Extantium. Sectio I. Reptilia Europae.* J. Antonelli, Verona, 29 pp.

Betta, F.E. de. (1857) *Erpetologia delle provincie venete e del Tirolo meridionale. Opera premiata dall'accademia di agricoltura, arti e commercio di Verona e formante il vol. XXXV de suoi atti.* Vicentini e Franchini, Verona, 365 pp.

Betta, F.E. de. (1865) Sui serpenti Italiani del genere *Tropidonotus* Kuhl, osservazioni critiche. *Atti del Reale Istituto Veneto di Scienze, Lettere ed Arti* (Series 3) 10, 513–542.

Betta, F.E. de. (1874) *Fauna d'Italia divisa in due branche: vertebrati e invertebrati. Vertebrati. Parte quarta. Rettili ed anfibi.* Dottor Francesco Vallardi, Milano 107 pp.

Betta, F.E. de. (1879) Sulla vipera ammodyte (*Vipera ammodytes* Latr.) nell'Italia e sulla sua distribuzione geografica. *Atti del Reale Istituto Veneto di Scienze, Lettere ed Arti* (Series 5) 5, 589–615.

Betz, T.W. (1963) Neonatal *Natrix cyclopion floridana. Copeia* 1963(3), 575–576.

Beyer, G.E. (1900) Louisiana Herpetology, with a check-list of the batrachians and reptiles of the state. *Proceedings of the Louisiana Society of Naturalists* 1897–1899, 25–46.

Bezman-Moseyko, O.S. (2010) New records of three rare snake species from the left bank of Dniester River (Pridnestrovie). *Russian Journal of Herpetology* 17(4), 280–284, 11 figs.

Bezuijen, M.R., Vinn, B. & Seng, L. (2009) A collection of amphibians and reptiles from the Mekong river, north-eastern Cambodia. *Hamadryad* 34(1), 135–164, 28 figs.

Bezy, R.L. & Drewes, R.C. (1985) Specific status of the east African colubrid snake, *Psammophis tanganicus. Journal of Herpetology* 19(2), 246–253, 4 figs.

Bhamrah, H.S. & Juneja, K. (1989) *Recent trends in chordates.* Anmol Publications, New Delhi, 186 pp., illust.

Bhati, D.P.S. & Wadhawan, B.S. (1974) Sexual dimorphism in the Indian sand snake, *Eryx johni* (Russell). *Annals of Zoology,* Agra 10(3), 71–76, figs. 27–28.

Bhatnagar, A.N. (1959) Studies on the structure and behavior of chromosomes of *Oligodon arnensis* Shaw (Colubridae, Ophidia). *Cytologia* 24, 459–465, 20 figs.

Bhatnagar, A.N. (1960) Chromosomes of *Bungarus caeruleus* Schneider (Elapidae, Ophidia). *Cytologia* 25, 173–177, 19 figs.

Bhatnagar, R.K. (1969) Extension of range of copperhead snake, *Elaphe radiata* Schlegel (Ophidia, Colubridae) to Doon Valley and Doon Siwaliks. *The Journal of the Bombay Natural History Society* 66(2), 383.

Bhatnagar, R.K. (1975) Hemipeneal structure in *Elaphe radiata* Schlegel, with a diagnostic key based on hemipenis. *The Journal of the Bombay Natural History Society* 72(2), 560–562.

Bianconi, G.G. (1849a) Lettera al Dottore Filippo de–Filippi, Professore di Zoologia a Torino sopra alcune nuove specie di rettili del Mozambico. *Nuovi Annali dell Scienze Naturali,* Bologna (1848) (Series 2) 10, 106–109.

Bianconi, G.G. (1849b) La sua memoria di turno che intitola Spicilegia Zoologica Mozambicana. Fasciculus III. *Nuovi Annali delle Scienze Naturali* (Bologna), (Series 2) 10, 202–208.

Bianconi, G.G. (1849c) Giuseppe Bianconi…legge la sua Memoria… Spicilegia Zoologia Mozambicana, fasciculus III. *Rendiconto delle Sessioni dell'Accademia delle Scienze dell'Istituto di Bologna* (1848–1849), 33–39.

Bianconi, G.G. (1850) Specimina zoologica Mosambicana. Fasciculus II. De Reptilibus. *Memorie della Accademia delle Scienze dell'Istituto di Bologna* 1, 171–186, pls. 6–8.

Bianconi, G.G. (1852) Specimina zoologica Mosambicana. Fasciculus XII. De Reptilibus. *Memorie della Accademia delle Scienze dell'Istituto di Bologna* 6, 89–115, pl. 12.

Bianconi, G.G. (1859) Specimina zoologica Mosambicana. Fasciculus XII. De Reptilibus. *Memorie della Accademia delle Scienze dell'Istituto di Bologna* 10, 501, pl. 26.

Bianconi, G.G. (1862) Specimina zoologica Mosambicana. Fasciculus XIII. De Reptilibus. *Memorie della Accademia delle Scienze dell'Istituto di Bologna* (Series 2) 1, 469–472, pl. 1.

Bianconi, G.G. (1850–1870) *Specimina zoologica Mosambicana quibis vel novae vel minus notae animalium species illustrantur. Fasc. I–XX.* Typ. Academiae Scientiarum, Bononiae, 363 pp., 48 pls.

Bider, J.R. & Matte, S. (1996) *The atlas of amphibians and reptiles of Québec.* St. Lawrence Valley Natural History Society, Sainte-Anne-de-Bellevue, 106 pp., illust.

Biella, H.-J. (1995) *Die sandotter: Vipera ammodytes.* Second edition. Westarp Wissenschaften [Neue Brehm-Bücherei Nr. 558], Magdeburg, 84 pp., 54 figs., 2 col. pls. [first edition, 1983]

Biella, H.-J. & Blattler, E. (1992) Zur geographischen Variation eidonomischer Merkmale bei der Sandviper (*Vipera ammodytes*). *Mitteilungen aus dem Zoologischen Museum in Berlin* 68(2), 249–258, illust.

Bilate, M. & Ribeiro-Costa, E. (2005) Geographic distribution: Serpentes: *Leptotyphlops salgueiroi. Herpetological Review* 36(3), 339.

Billing, H. & Schätti, B. (1984) Vorläufige Mitteilung zum Subspezies-Problem bei *Vipera lebetina* (Linnaeus, 1758) (Serpentes, Viperidae). *Salamandra,* Frankfurt-am-Main 20(2–3), 65–69, 4 figs.

Billing, H., Nilson, G. & Sattler, U. (1990) *Vipera pontica* sp. n., a new viper species in the *kaznakovi* group (Reptilia, Viperidae) from northeastern Turkey and adjacent Transcaucasia. *Zoologica Scripta,* Oxford 19(2), 227–231, 4 figs.

Binder, S. (2002) *Boa constrictor.* Natur und Tier-Verlag, Münster. 96 pp., 89 col. illust.

Bird, C.G. (1935) The reptiles and amphibians of the Cyclades. *Annals and Magazine of Natural History,* London (Series 10) 16(92), 274–284.

Bird, C.G. (1936) The distribution of reptiles and amphibians in Asiatic Turkey, with notes on a collection from the Vilayets of Adana, Gaziantep and Malatya. *Annals and Magazine of Natural History,* London (Series 10) 18(104), 257–281, pls.

Bischoff, W. (1993) *Elaphe hohenackeri* (Strauch, 1873) – Transkaukasische kletternatter. *In:* Böhme, W. (Ed.), *Handbuch der Reptilien und Amphibien Europas. Band 3/I. Schlangen (Serpentes) I. (Typhlopidae, Boidae, Colubridae 1, Colubrinae).* AULA-Verlag, Wiesbaden, pp. 317–329, figs. 60–62.

Bischoff, W. & Schmidtler, J. (1997) Nach 115 Jahren in Syrien wiedergefunden, *Micrelaps muelleri* Böttger, 1880 (Serpentes, Atractaspididae). *Salamandra,* Frankfurt-am-Main 33(1), 25–32, 4 figs.

Bisplinghof, H. & Bellosa, H. (2007) *Anakondas.* Edition Chimaira, Frankfurt-am-Main, 127 pp., 119 col. figs.

Biswas, S. (1975) Reptilia from Bhutan with description of a new species of *Calotes* Rafinesque. *The Journal of the Bombay Natural History Society* 72(3), 774–777.

Biswas, S. (1978) A new species of krait of the genus *Bungarus* Daudin, 1803 (Serpentes, Elapidae) from the Andaman Island. *The Journal of the Bombay Natural History Society* 75(1), 179–183, 4 figs.

Biswas, S. (1980) A report on the Reptilia fauna of Andaman and Nicobars Islands in the collection of Zoological Survey of India. *Records of the Zoological Survey of India,* Calcutta 77(1–4), 255–292, 3 figs.

Biswas, S. & Acharjyo, L.N. (1977) Notes on the ecology and biology of some reptiles occurring in and around Nandankanan Biological Park, Orissa. *Records of the Zoological Survey of India,* Calcutta 73(1–4), 95–109, illust.

Biswas, S. & Sanyal, D.P. (1965) A new species of Wolf-snake of the genus *Lycodon* Boie (Reptilia, Serpentes, Colubridae) from the Andaman and Nicobar Islands. *Proceedings of the Zoological Society of Calcutta* 18(2), 137–141, pl. 5.

Biswas, L.N. & Sanyal, D.P. (1977a) Notes on the Reptilia collection from the Great Nicobar Island during the Great Nicobar Expedition in 1966. *Records of the Zoological Survey of India* 72(1–4), 107–124.

Biswas, L.N. & Sanyal, D.P. (1977b) Fauna of Rajasthan, India. Part: Reptilia. *Records of the Zoological Survey of India* 73(1–4), 247–269.

Biswas, S. & Sanyal, D.P. (1980) A report on the Reptilia fauna of Andaman and Nicobars Islands in the collection of Zoological Survey of India. *Records of the Zoological Survey of India* 77, 255–292.

Biswas, S., Acharjyo, L.N. & Misra, R. (1976) On the feeding habits of the king cobra *Ophiophagus hannah* (Cantor) at Nandankanan Biological Park, Orissa. *The Journal of the Bombay Natural History Society* 73(1), 225–226.

Bitz, A. & Veith, M. (Eds.). (1992) Herpetologie in Rheinland-Pfalz – Faunistik, Schutz und Forschung. *Zeitschrift für Naturschutz,* Nassau (6), 1–191, illust.

Blackwell, K. (1967) A preliminary survey of the reptiles of the Ibadan area. *British Journal of Herpetology*, London 3(12), 307–310.

Blainville, H.M.D. de. (1818) Sur un nouveau genre de serpent, *Scoliophis*, et le serpent de mer vu en Amerique en 1817. *Journal de Physique, de Chimie, d'Histoire Naturelle et des Arts*, Paris 86, 297–304, 1 pl.

Blainville, H.M.D. de. (1835) Description de quelques espèces de reptiles de la Californie, précedée de l'analyse d'un système generale d'herpétologie et d'amphibiologie. *Nouvelles Annales du Muséum National d'Histoire Naturelle de Paris* (Série 3) 4(3), 232–296, pls. 24–27.

Blair, K.B., Killebrew, F.C., Smith, H.M. & Chiszar, D. (1997) Mexican amphibians and reptiles in the West Texas A & M University Natural History Museum. *Bulletin of the Chicago Herpetological Society* 32(8), 174–177.

Blake, R.A. (1970) *The distribution and variation of the shovel-nosed snake Chionactis palarostris with the description of a new subspecies from coastal Sonora, Mexico*. Master's Thesis, University of Arizona, Tucson, 40 pp.

Blanc, C.P. (1972) Les reptiles de Madagascar et des îles voisines. *In:* Battistini, R. & Richard-Vinard, G. (Eds.). *Biogeography and ecology in Madagascar*. Dr. W. Junk, The Hague, pp. 501–614, 22 figs., 6 pls.

Blanc, C.P. (1986) The Reptiles. *In:* Jolly, A., Oberle, P. & Albignac, R. (Eds.), *Madagascar*. Pergammon Press, Oxford, pp. 105–114, illust.

Blanc, C.P. (1988) Faune herpétologique des îles Kerkennah: inventaire, distribution et zoogéographie. *Bulletin d'Ecologie* 19(2–3), 255–258, illust.

Blanc, C.P. & Nouira, S. (1988) Biogeographie des reptiles des iles Zembra et Zembretta. *Bulletin d'Ecologie* 19(2–3), 255–258, illust.

Blanchard, F.N. (1919) Two new snakes of the genus *Lampropeltis*. *Occasional Papers of the Museum of Zoology, University of Michigan* (70), 1–11, 1 pl.

Blanchard, F.N. (1920a) Three new snakes of the genus *Lampropeltis*. *Occasional Papers of the Museum of Zoology, University of Michigan* (81), 1–10, 1 pl.

Blanchard, F.N. (1920b) A synopsis of the king snakes, genus *Lampropeltis* Fitzinger. *Occasional Papers of the Museum of Zoology, University of Michigan* (87), 1–7.

Blanchard, F.N. (1921) A revision of the king snakes, genus *Lampropeltis*. *Bulletin of the United States National Museum* (114), vi + 260 pp., 78 figs.

Blanchard, F.N. (1923a) A new North American snake of the genus *Natrix*. *Occasional Papers of the Museum of Zoology, University of Michigan* (140), 1–6.

Blanchard, F.N. (1923b) Comments on ring-necked snakes (genus *Diadophis*), with diagnoses of new forms. *Occasional Papers of the Museum of Zoology, University of Michigan* (142), 1–9.

Blanchard, F.N. (1924a) A new snake of the genus *Arizona*. *Occasional Papers of the Museum of Zoology, University of Michigan* (150), 1–5.

Blanchard, F.N. (1924b) A new name for the black *Pituophis* from Alabama. *Papers of the Michigan Academy of Science, Arts and Letters* 4, 531–532.

Blanchard, F.N. (1925) A key to the snakes of the United States, Canada, and Lower California. *Papers of the Michigan Academy of Science, Arts and Letters* (1924) 4(2), xiii + 65 pp., 78 figs.

Blanchard, F.N. (1933) Eggs and young of the smooth green snake, *Liopeltis vernalis* (Harlan). *Papers of the Michigan Academy of Science, Arts and Letters* 17, 493–508, pl. 51–56.

Blanchard, F.N. (1937) Data on the natural history of the red-bellied snake, *Storeria occipito-maculata* (Storer), in northern Michigan. *Copeia* 1937, 151–162.

Blanchard, F.N. (1938) Snakes of the genus *Tantilla* in the United States. *Field Museum of Natural History Publications, Zoological Series* 20(28), 369–376.

Blanchard, F.N. (1942) The ring-neck snakes, genus *Diadophis*. *Bulletin of the Chicago Academy of Sciences* 7(1), 1–144, 26 figs.

Blaney, R.M. (1971) An annotated check list and biogeographical analysis of the insular herpetofauna of the Apalachicola region, Florida. *Herpetologica* 27(4), 406–430.

Blaney, R.M. (1973) *Lampropeltis*. *Catalogue of American amphibians and reptiles*, pp. 150.1–150.2.

Blaney, R.M. (1977) Systematics of the common kingsnake, *Lampropeltis getulus* (Linnaeus). *Tulane Studies in Zoology and Botany* 19(3–4), 47–103, 37 figs.

Blaney, R.M. (1979a) The status of the Outer Banks kingsnake, *Lampropeltis getulus sticticeps* (Reptilia, Serpentes, Colubridae). *Brimleyana* (1), 125–128, 1 fig.

Blaney, R.M. (1979b) *Lampropeltis calligaster*. *Catalogue of American amphibians and reptiles*, pp. 229.1–229.2.

Blaney, R.M & Blaney, P.K. (1978) Notes on three species of *Micrurus* (Serpentes, Elapidae) from Mexico. *Herpetological Review* 9(3), 92.

Blaney, R.M & Blaney, P.K. (1979a) Variation in the coral snake, *Micrurus diastema*, in Quintana Roo, México. *Herpetologica* 35(3), 276–278, 1 fig.

Blaney, R.M & Blaney, P.K. (1979b) The *Nerodia sipedon* complex of water snakes in Mississippi and southeastern Louisiana. *Herpetologica* 35(4), 350–359, 9 figs.

Blanford, W.T. (1870) Notes on some Reptilia and Amphibia from central India. *Journal of the Asiatic Society of Bengal (Natural History)*, Calcutta 39(4), 335–376, pls. 14–16.

Blanford, W.T. (1874) Descriptions of new Reptilia and Amphibia from Persia and Baluchistán. *Annals and Magazine of Natural History*, London (Series 4) 14(79), 31–35.

Blanford, W.T. (1875) List of Reptilia and Amphibia collected by the late Dr. Stoliczka in Kashmir, Ladak, eastern Turkestan, and Wakhán, with descriptions of new species. *Journal of the Asiatic Society of Bengal (Natural History)*, Calcutta 44(3), 191–196.

Blanford, W.T. (1876) *Eastern Persia. An account of the journeys of the Persian Boundary Commission 1870–71–72. Vol. II. The Zoology and Geology*. Macmillan, London , 516 pp., 28 pls.

Blanford, W.T. (1878a) Notes on Reptilia from the Himalayas and Assam (Abstract.). *Proceedings of the Asiatic Society of Bengal (Natural History)*, Calcutta 1878(6), 141–142. [June, 1878]

Blanford, W.T. (1878b) Notes on some Reptilia from the Himalayas and Burma. Part II. *Journal of the Asiatic Society of Bengal (Natural History)*, Calcutta 47(3), 125–131. [Oct. 1878]

Blanford, W.T. (1878c) *Scientific results of the second Yarkand Mission; based upon the collections and notes of the late Ferdinand Stoliczka, Ph.D. Reptilia and Amphibia*. Government of India, Calcutta, 26 pp., 2 pls.

Blanford, W.T. (1879a) Notes on reptiles. (Abstract.). *Proceedings of the Asiatic Society of Bengal (Natural History)*, Calcutta 1879(8): 216–217. [Aug. 1879]

Blanford, W.T. (1879b) Notes on Reptilia. *Journal of the Asiatic Society of Bengal (Natural History)*, Calcutta 48(3), 127–132. [Nov. 1879]

Blanford, W.T. (1881) On a collection of reptiles and frogs chiefly from Singapore. *Proceedings of the Zoological Society of London* 49(1), 215–226, pls. 20–21.

Blanford, W.T. (1897) No title (on *Eristicophis macmahoni*). *Proceedings of the Zoological Society of London* 65(1), 295.

Blasus, R.E. (1997) Amphibian & Reptile time table for Minnesota. A guide to the seasonal activity of Minnesota's amphibians & reptiles. *Minnesota Herpetological Society Occasional Paper* (4), 1–30.

Blauscheck, R. (1985) *Amphibien und Reptilien Deutschlands.* Landbuch-Verlag, Gmbh, Hannover, 158 pp., 59 figs.

Blauscheck, R. (1989) *Amphibien und Reptilien Deutschlands.* Second edition. Landbuch-Verlag, Gmbh, Hannover, 158 pp., 59 figs.

Blaustein, A.R., Beatty, J.J., Olson, D.H. & Storm, R.M. (1995) *The biology of amphibians and reptiles in old-growth forests in the Pacific Northwest.* United States Department of Agriculture, General Technical Report, Washington, 98 pp., 25 figs.

Blázquez, M.C. (1994) Notes on the reproduction of *Elaphe scalaris* (Schinz, 1822) in southern Spain. *The Herpetological Journal* 4(3), 109–111.

Bleakney, J.S. (1958) A zoogeographical study of the amphibians and reptiles of eastern Canada. *Bulletin of the National Museum of Canada*, Ottawa (155), 1–119 pp., 1 fig.

Bleeker, P. de. (1856) *Reis door de Minahassa en den Moluskschen Archipel gedaan in de maanden September en Oktober 1855 in het gevolg van den gouverneur generaal Mr. A.J. Duymaer van Twist. Eerste deel.* Lange & Co., Batavia, viii + 288 pp.

Bleeker, P. de. (1857a) Berigt omtrent eenige Reptiliën van Sumatra, Borneo, Batjan en Borneo. *Natuurkundig Tijdschrift voor Nederlandsch Indië*, Batavia (Série 3) 13(3), 470–475.

Bleeker, P. de. (1857b) Opsomming der soorten van Reptiliën, tot dus verre van het eiland Java bekend geworden. *Natuurkundig Tijdschrift voor Nederlandsch Indië*, Batavia (Série 3) 14(4), 235–239.

Bleeker, P. de. (1857c) Een groot gedeelte dezer soorten bevindt zich reeds in mijn kabinet, hoezeer ik niet alle van Java zelf ontving. *Natuurkundig Tijdschrift voor Nederlandsch Indië*, Batavia (Série 3) 14(4), 239–243.

Bleeker, P. de. (1858a) Eenige slangen en visschen, gevangen in de omstreken van Montrado de gewezen hoofdplaats der Chinezen ter westkust van Borneo. *Natuurkundig Tijdschrift voor Nederlandsch Indië*, Batavia (Série 4) 15(1), 196–198.

Bleeker, P. de. (1858b) Opsomming der tot dus verre het eiland Sumatra bekend gewordene Reptilën. *Natuurkundig Tijdschrift voor Nederlandsch Indië*, Batavia (Série 4) 15(1), 260–263.

Bleeker, P. de. (1858c) Eine Zeeslang (*Aipysurus margaritophorus*). *Natuurkundig Tijdschrift voor Nederlandsch Indië*, Batavia (1858–1859) (Série 4) 16(1), 49–50.

Bleeker, P. de. (1858d) Reptiliën en visschen van Sinkawang (westkust van Borneo). *Natuurkundig Tijdschrift voor Nederlandsch Indië*, Batavia (1858–1859) (Série 4) 16(2), 188–189.

Bleeker, P. de. (1858e) *Gonyosoma jansenii* Blkr., eene nieuwe slang van Manado. *Natuurkundig Tijdschrift voor Nederlandsch Indië*, Batavia (1858–1859) (Série 4) 16(2), 242.

Bleeker, P. de. (1858f) Reptiliën van Oost-Java. *Natuurkundig Tijdschrift voor Nederlandsch Indië*, Batavia (1858–1859) (Série 4) 16(2), 310–314.

Bleeker, P. de. (1859a) Reptiliën van Nieuw-Guinea. *Natuurkundig Tijdschrift voor Nederlandsch Indië*, Batavia (1858–1859) (Série 4) 16(2), 420–423.

Bleeker, P. de. (1859b) Verslag van eene verzameling reptiliën en visschen van westelijk Borneo. *Natuurkundig Tijdschrift voor Nederlandsch Indië*, Batavia (1858–1859) (Série 4) 16(2), 433–440.

Bleeker, P. de. (1859c) Reptiliën van Sintang. *Natuurkundig Tijdschrift voor Nederlandsch Indië*, Batavia (1859–1860) (Série 4) 20(6), 200–201.

Bleeker, P. de. (1859d) Reptiliën uit de omstreken van Tandjong. *Natuurkundig Tijdschrift voor Nederlandsch Indië*, Batavia (1859–1860) (Série 4) 20(6), 220–221.

Bleeker, P. de. (1859e) Reptiliën van Agam. *Natuurkundig Tijdschrift voor Nederlandsch Indië*, Batavia (1859–1860) (Série 4) 20(6), 325–329.

Bleeker, P. de. (1860a) Heer P.L. Bloemen Waanders, van Banjoewangi, aanbiedende een slang van Bali. *Natuurkundig Tijdschrift voor Nederlandsch Indië*, Batavia (1859–1860) (Série 4) 20(6), 470–471.

Bleeker, P. de. (1860b) Over de Reptiliën-Fauna van Sumatra. *Natuurkundig Tijdschrift voor Nederlandsch Indië*, Batavia (Série 5) 21(1), 284–298.

Bleeker, P. de. (1860c) Soorten van reptilian van het eiland Banka. *Natuurkundig Tijdschrift voor Nederlandsch Indië*, Batavia (Série 5) 21(1), 331–334.

Bleeker, P. de. (1860d) Over de Reptiliën-Fauna van Ceram. *Natuurkundig Tijdschrift voor Nederlandsch Indië*, Batavia (Série 5) 22(2), 35–38.

Bleeker, P. de. (1860e) Over de Reptiliën-Fauna van Amboina. *Natuurkundig Tijdschrift voor Nederlandsch Indië*, Batavia (Série 5) 22(2), 39–43.

Bleeker, P. de. (1860f) Vereeniiging de heer J.T. Van Bloemen Waanders, van de Bonische expeditie teruggekeerd. *Natuurkundig Tijdschrift voor Nederlandsch Indië*, Batavia (Série 5) 22(2), 81–85.

Bleeker, P. de. (1860g) Reptiliën van het eiland Timor. *Natuurkundig Tijdschrift voor Nederlandsch Indië*, Batavia (Série 5) 22(2), 86–88.

Blem, C.R. (1981) *Heterodon platyrhinos. Catalogue of American amphibians and reptiles,* pp. 282.1–282.2.

Blo, S. (2012) De herontdekking van *Thamnophis eques insperatus* (de Zacapu Mexicaanse Kousebandslang) in Zacapu, Michoacám, Mexico. *Lacerta* 70(2), 48–63.

Blum, J. (1976) *European reptiles and amphibians. A pocket identification guide.* Foulis Spectrum Books, Somerset, 64 pp., illust.

Blumenbach, J.F. (1788) Beytrag zur Naturgeschichte der Schlangen. *J.H. Voigt's Magazin für das Neueste aus der Physik und Naturgeschichte,* Gotha 5(1), 1–13, pl. 1.

Blyth, E. (1846) Notes on the fauna of the Nicobar Islands. Reptilia. *Journal of the Asiatic Society of Bengal (Natural History),* Calcutta 15(173), 376–377.

Blyth, E. (1854a) Notices and descriptions of various reptiles, new or little known. *Journal of the Asiatic Society of Bengal (Natural History),* Calcutta 23(3), 287–302.

Blyth, E. (1854b) Report of Curator, Zoological Department, for September Meeting. *Proceedings of the Asiatic Society of Bengal (Natural History),* Calcutta 23(7), 729–740. [? 1855 *fide* Zhao & Adler, 1993: 353]

Blyth, E. (1855a) Report of Curator, Zoological Department, for March Meeting. *Proceedings of the Asiatic Society of Bengal (Natural History),* Calcutta 24(2), 187–188.

Blyth, E. (1855b) Report of Curator, Zoological Department, for May Meeting, 1855. *Proceedings of the Asiatic Society of Bengal (Natural History),* Calcutta 24(4), 359–363.

Blyth, E. (1855c) Report of Curator, Zoology Department, for October Meeting, 1855. *Proceedings of the Asiatic Society of Bengal (Natural History)*, Calcutta 24(7), 711–723.

Blyth, E. (1859) Report of Curator, Zoological Department, for February to May Meetings, 1859. *Proceedings of the Asiatic Society of Bengal (Natural History)*, Calcutta 28(3), 271–298.

Blyth, E. (1860) Report of Curator, Zoological Department. *Proceedings of the Asiatic Society of Bengal (Natural History)*, Calcutta 29(1), 87–115.

Blyth, E. (1863) Report of the Curator, Zoology Department. Reptilia. *Proceedings of the Asiatic Society of Bengal (Natural History)*, Calcutta 32(1), 80–90.

Boback, S.M. (2006) A morphometric comparison of island and mainland boas (*Boa constrictor*) in Belize. *Copeia* 2006(2), 261–267, 2 figs.

Boback, S.M. & Carpenter, M. (2007) Body size and head shape of island *Boa constrictor* in Belize, environmental versus genetic contributions. *In:* Henderson, R.W. & Powell, R. (Eds.), *Biology of the boas and pythons*. Eagle Mountain Publishing, Eagle Mountain, pp. 103–117, 2 figs., col. illust.

Bobrov, V.V. & Aleshchenko, G.M. (2001) Herpetogeographical regionalizations of Russia and adjacent countries. *Russian Journal of Herpetology* 8(3), 223–238, 3 figs.

Bocage, J.V. Barboza du. (1866a) Lista dos reptis das possessões portuguezas d'Africa occidental que existem no Museu de Lisboa. *Jornal de Sciencias, Mathematicas, Physicas e Naturaes*, Lisboa 1(1), 37–56.

Bocage, J.V. Barboza du. (1866b) Reptiles nouveaux ou peu connus recueillis dans les possessions portugaiese de l'Afrique occidentale, que se trouvent au Muséum de Lisbonne. *Jornal de Sciencias, Mathematicas, Physicas e Naturaes*, Lisboa 1(1), 57–78, pl. 1.

Bocage, J.V. Barboza du. (1867a) Segunda lista dos reptis das possessões portuguezas d'Africa occidental que existem no Museu de Lisboa. *Jornal de Sciencias, Mathematicas, Physicas e Naturaes*, Lisboa 1(3), 217–225.

Bocage, J.V. Barboza du. (1867b) Diagnoses de quelques reptiles nouveaux de l'Afrique occidentale. *Jornal de Sciencias, Mathematicas, Physicas e Naturaes*, Lisboa 1(3), 229–232, pl. 3.

Bocage, J.V. Barboza du. (1872) Diagnoses de quelques espèces nouvelles de reptiles d'Afrique occidentale. *Jornal de Sciencias, Mathematicas, Physicas e Naturaes*, Lisboa 4(13), 72–82.

Bocage, J.V. Barboza du. (1873a) Mélanges erpétologiques. II. Sur quelques reptiles et batraciens nouveaux, rares ou peu connus d'Afrique occidentale. *Jornal de Sciencias, Mathematicas, Physicas e Naturaes*, Lisboa 4(15), 209–227.

Bocage, J.V. Barboza du. (1873b) Reptiles nouveaux de l'intérieur de Mossamedes. *Jornal de Sciencias, Mathematicas, Physicas e Naturaes*, Lisboa 4(15), 247–253, 4 figs., 1 pl.

Bocage, J.V. Barboza du. (1879) Subsidios para a fauna das possessoes portuguezas d'Africa occidental. *Jornal de Sciencias, Mathematicas, Physicas e Naturaes*, Lisboa (1879–1882) 7(26), 85–96.

Bocage, J.V. Barboza du. (1882a) Reptiles rares ou nouveaux d'Angola. *Jornal de Sciencias, Mathematicas, Physicas e Naturaes*, Lisboa 8(32), 299–304.

Bocage, J.V. Barboza du. (1882b) Notice sur les espèces du genre "*Philothamnus*" qui se trouvent au Muséum de Lisbonne. *Jornal de Sciencias, Mathematicas, Physicas e Naturaes*, Lisboa 9(33), 1–19, 6 figs.

Bocage, J.V. Barboza du. (1886) Typhlopiens nouveaux de la faune africaine. *Jornal de Sciencias, Mathematicas, Physicas e Naturaes*, Lisboa 11(43), 171–174.

Bocage, J.V. Barboza du. (1887a) Mélanges erpétologiques. I. Reptiles et batraciens du Congo. *Jornal de Sciencias, Mathematicas, Physicas e Naturaes*, Lisboa 11(44), 177–208.

Bocage, J.V. Barboza du. (1887b) Mélanges erpétologiques. II. Reptiles de Dahomey. *Jornal de Sciencias, Mathematicas, Physicas e Naturaes*, Lisboa 11(44), 192–197.

Bocage, J.V. Barboza du. (1887c) Mélanges erpétologiques. III. Reptiles de l'Ile du Prince. *Jornal de Sciencias, Mathematicas, Physicas e Naturaes*, Lisboa 11(44), 198–201.

Bocage, J.V. Barboza du. (1887d) Mélanges erpétologiques. IV. Reptiles du dernier voyage de MM. Capello et Ivens à travers l'Afrique. *Jornal de Sciencias, Mathematicas, Physicas e Naturaes*, Lisboa 11(44), 201–208.

Bocage, J.V. Barboza du. (1887e) Sur un *Python* nouveau d'Afrique. *Jornal de Sciencias, Mathematicas, Physicas e Naturaes*, Lisboa 12(46), 87–88.

Bocage, J.V. Barboza du. (1888) Mélanges erpétologiques. VI. Espèces du genre *Dendraspis*. *Jornal de Sciencias, Mathematicas, Physicas e Naturaes*, Lisboa 12(47), 138–147, 11 figs.

Bocage, J.V. Barboza du. (1889) Sur une vipère apparemment nouvelle d'Angola. *Jornal de Sciencias, Mathematicas, Physicas e Naturaes*, Lisboa (Série 2) 1(2), 127–128, 1 fig.

Bocage, J.V. Barboza du. (1890) Sur une espèce nouvelle à ajoutter à la faune erpétologique de St. Thomé et Rolas. *Jornal de Sciencias, Mathematicas, Physicas e Naturaes*, Lisboa (Série 2) 2(5), 61–62.

Bocage, J.V. Barboza du. (1893a) Mammiferos, Aves e Reptis da Ilha de Anno-Bom. *Jornal de Sciencias, Mathematicas, Physicas e Naturaes*, Lisboa (Série 2) 3(9), 43–46.

Bocage, J.V. Barboza du. (1893b) Diagnoses de deux nouveaux reptiles de l'ile de Anno-Bom. *Jornal de Sciencias, Mathematicas, Physicas e Naturaes*, Lisboa (Série 2) 3(9), 47–48.

Bocage, J.V. Barboza du. (1893c) Diagnoses de quelques nouvelles espéces de reptiles et batraciens d'Angola. *Jornal de Sciencias, Mathematicas, Physicas e Naturaes*, Lisboa (Série 2) 3(10), 115–121.

Bocage, J.V. Barboza du. (1895) *Herpetologie d'Angola et du Congo*. Lisbonne, xx + 203 pp., 14 pls.

Bocage, J.V. Barboza du. (1896) Reptis de algumas possessões Portuguezas d'Africa que existem no Museum de Lisboa. *Jornal de Sciencias, Mathematicas, Physicas e Naturaes*, Lisboa (Série 2) 4(14), 65–103, 2 pls.

Bocage, J.V. Barboza du. (1897) Mammiferos, reptis e batrachios d'Africa de que existen exemplars typicos no Museu de Lisboa. *Jornal de Sciencias, Mathematicas, Physicas e Naturaes*, Lisboa (Série 2) 4(16), 187–206.

Bocourt, M.-F. (1866) Notes sur les reptiles, les batraciens et les poissons recueillis pendant un voyage dans le Royaume de Siam. *Nouvelles Archives du Muséum National d'Histoire Naturelle*, Paris 2(1), 4–9. 1 pl.

Bocourt, M.-F. (1868) Descriptions de quelques crotaliens nouveaux appartenant au genre *Bothrops*, recuellis dans le Guatémala. *Annales des Sciences Naturelles (Zoologie)*, Paris (Série 5) 10, 201–202.

Bocourt, M.-F. (1875) Note sur une nouvelle espèce d'ophidien. *Annales des Sciences Naturelles (Zoologie)*, Paris (Série 6) 2(3), 1.

Bocourt, M.-F. (1876) Note sur une nouvelle espèce d'ophidien du Mexique. *Annales des Sciences Naturelles (Zoologie)*, Paris (Série 6) 4(7), 1.

Bocourt, M.-F. (1881) Description d'un ophidien opoterodonte appurtenant au genre *Catodon*. *Bulletin de la Société Philomathique de Paris* (Série 7) 5, 81–82.

Bocourt, M.-F. (1884) Note sur quelques ophidiens nouveaux, provenant de l'Amérique inter-tropicale. *Bulletin de la Société Philomathique de Paris* (Série 7) 8, 133–142.

Bocourt, M.-F. (1885) Note sur un boidien nouveaux, provenant du Guatemala. *Bulletin de la Société Philomathique de Paris* (Série 7) 9, 112–115.

Bocourt, M.-F. (1887) Note sur un ophidien nouveau appartenant au groupe des colubriens provenant du Guatemala. *Le Naturaliste* (Série 2) 9(14), 168.

Bocourt, M.-F. (1891) Note sur quelques ophidiens de l'Amérique intertropicale appartenant au genre *Tretanorhinus*. *Le Naturaliste*, Paris (Série 2) 13(101), 121–122 & 13(108), 208.

Bocourt, M.-F. (1892a) Note sur un petit ophidien. Appartenant au genre *Tropidoclonion*. *Le Naturaliste*, Paris (Série 2) 14(126), 132.

Bocourt, M.-F. (1892b) Note sur un ophidien appartenant au genre *Eutaenia*. *Le Naturaliste*, Paris (Série 2) 14(138), 278–279.

Bocourt, M.-F. (1894) Note sur un ophidien de la Guyane. *Le Naturaliste*, Paris (Série 2) 16(176), 155.

Bocourt, M.-F. & Dugès, A.A.D. (1887) Sur deux ophidiens de la familie des Coronellidae, provenant de l'Amérique intertropicale. *Le Naturaliste*, Paris (Série 2) 9(4), 45–46, 4 figs.

Boddaert, P. (1783) Specimen novae methodi distinguendi Serpentia. *Nova Acta Physico-Medica Academiae Caesareae Leopoldino-Carolinae Naturae Curiosorum* 7(2), 12–27.

Bodenheimer, F.S. (1944) Introduction into the knowledge of the Amphibia and Reptilia of Turkey. *Revue de la Faculté des Sciences de l'Université d'Istanbul* 9(1), 1–93, 10 pls.

Boettger, O. (1869) Beitrag zur Kenntniss der Reptilien Spaniens und Portugals. *Berichte des Offenbacher Vereins für Naturkunde* (1868–1869) 9, 50–59, 1 pl.

Boettger, O. (1873) *Calamaria iris* n. sp. neue Schlange von Sumatra. *Berichte des Offenbacher Vereins für Naturkunde* (1871–1872) 13, 1–7, 1 pl.

Boettger, O. (1874) Reptilien von Marocco und von den canarischen Inseln. *Abhandlungen der Senckenbergischen Naturforschenden Gesellschaft*, Frankfurt-am-Main 9, 121–191, 1 pl.

Boettger, O. (1876) Bemerkungen über einige Reptilien von Griechenland und von der Insel Chios. *Berichte des Offenbacher Vereins für Naturkunde* (1873–1875) 15–16, 55–64.

Boettger, O. (1877) Die Reptilien und Amphibien von Madagascar. *Abhandlungen der Senckenbergischen Naturforschenden Gesellschaft*, Frankfurt-am-Main 11(1), 1–55, 1 pl.

Boettger, O. (1878) Die Reptilien und Amphibien von Madagascar. Erster Nachtrag. *Abhandlungen der Senckenbergischen Naturforschenden Gesellschaft im Frankfurt-am-Main* 11(2–3), 269–282, 1 pl.

Boettger, O. (1879a) Reptilien und Amphibien aus Syrien. *Berichte der Senckenbergischen Naturforschenden Gesellschaft im Frankfurt-am-Main* 1878–1879, 57–69.

Boettger, O. (1879b) Die Reptilien und Amphibien von Madagascar, Zweiter Nachtrag. *Abhandlungen der Senckenbergischen Naturforschenden Gesellschaft im Frankfurt-am-Main* 11(4), 457–497, 1 pl.

Boettger, O. (1880a) Diagnoses Reptilium et Batrachiorum Novorum a Carolo Ebenau in insula Nossi-Bé Madagascariensi lectorum. *Zoologischer Anzeiger*, Leipzig 3(57), 279–283.

Boettger, O. (1880b) Die Reptilien und Amphibien von Syrien, Palaestina und Cypern. *Berichte der Senckenbergischen Naturforschenden Gesellschaft im Frankfurt-am-Main* 1879–1880, 132–219, pl. 3.

Boettger, O. (1881a) Diagnoses Reptilium et Batrachiorum Novorum ab illust. Antonio Stumpff in insula Nossi-Bé Madagascariensi lectorum. *Zoologischer Anzeiger*, Leipzig 4(87), 358–359.

Boettger, O. (1881b) Diagnoses Reptilium Novorum Maroccanorum. *Zoologischer Anzeiger*, Leipzig 4, 570–572.

Boettger, O. (1881c) Diagnoses Reptilium et Batrachiorum Novorum insulae Nossi-Bé Madagascariensis. *Zoologischer Anzieger*, Leipzig 4, 650–651.

Boettger, O. (1881d) Die Reptilien und Amphibien von Madagascar. Dritter Nachtrag. *Abhandlungen der Senckenbergischen Naturforschenden Gesellschaft*, Frankfurt-am-Main 12(3–4), 435–558, 5 pls.

Boettger, O. (1882) Diagnoses Reptilium et Batrachiorum Novorum insulae Nossi-Bé Madagascariensis. *Zoologischer Anzieger*, Leipzig 5, 478–480.

Boettger, O. (1883a) Die Reptilien und Amphibien von Marocco. II. *Abhandlungen der Senckenbergischen Naturforschenden Gesellschaft*, Frankfurt-am-Main 13, 93–146, 1 pl.

Boettger, O. (1883b) Herpetologische Mittheilungen. *Berichte des Offenbacher Vereins für Naturkunde* (1880–1882) 22–23, 147–156.

Boettger, O. (1885a) Liste von Reptilien und Batrachiern aus Paraguay. *Zeitschrift für Naturwissenschaften*, Halle 58, 213–248.

Boettger, O. (1885b) Materialien zur herpetologischen Fauna von China I. 1. Liste der ersten Moellendorff schen Sendung Sudchinesischer Kriechthiere. *Berichte des Offenbacher Vereins für Naturkunde* (1882–1884) 24–25, 115–170.

Boettger, O. (1886a) Beiträge zur Herpetologie und Malakozoologie Sudwest-Afrikas. *Berichte der Senckenbergischen Naturforschenden Gesellschaft in Frankfurt-am-Main* 1886, 1–29, 1 pl.

Boettger, O. (1886b) Diagnoses Reptilium Novorum ab illust. viris O. Herz et Consule Dr. O. Fr. de Moellendorff in Sina meridionali repertorum. *Zoologischer Anzieger*, Leipzig 9(231), 519–520.

Boettger, O. (1886c) Aufzählung der von den Philippinen bekannten Reptilien und Batrachier. *Berichte der Senckenbergischen Naturforschenden Gesellschaft in Frankfurt-am-Main* 1886, 91–134, 2 pls.

Boettger, O. (1887a) Diagnosis Reptilium Novorum ab ill. Paul Hesse in finibus fluminis Congo repertorum. *Zoologischer Anzieger*, Leipzig 10(267), 649–651.

Boettger, O. (1887b) Zweiter Beitrag zur Herpetologie Südwest- und Süd-Afrikas. *Berichte der Senckenbergischen Naturforschenden Gesellschaft in Frankfurt-am-Main* 1886–1887, 135–173, pl. 5.

Boettger, O. (1887c) Diagnoses reptilium novorum ab illust. viro Paul Hesse infinibus fluminis Congo repertorum. *Zoologischer Anzieger*, Leipzig 10(267), 649–651.

Boettger, O. (1888a) Verzeichniss der von Hrn. E. von Oertzen aus Griechenland und aus Kleinasien mitgebrachten Batrachier und Reptilien. *Berichte der Senckenbergischen Naturforschenden Gesellschaft in Frankfurt-am-Main* 1888, 139–186.

Boettger, O. (1888b) Beitrag zur Reptilfauna des oberen Beni in Bolivia. *Berichte der Senckenbergischen Naturforschenden Gesellschaft in Frankfurt-am-Main* 1888, 191–199, 3 figs.

Boettger, O. (1888c) Materialien zur herpetologischen Fauna von China II. 1. Liste der zweiten von Moellendorff'schen und der Herz'schen und Schmacker'schen Sendungen chinesischer Kriechtriere. *26, 27, und 28. Berichte des Offenbacher Vereins für Naturkunde* (1884–1887) 26–28, 53–102, 2 pls..

Boettger, O. (1888d) Über die Reptilien und Batrachier Transcaspiens. *Zoologischer Anzieger*, Leipzig 11(279), 259–263.

Boettger, O. (1888e) Materialien zur Fauna des unteren Congo II. Reptilien und Batrachier. *Berichte der Senckenbergischen Naturforschenden Gesellschaft in Frankfurt-am-Main* 108 pp., 2 pls.

Boettger, O. (1888f) Die Reptilien und Batrachier Transkaspiens. *Zoologische Jahrbücher, Abteilung für Systematik, Okologie und Geographie der Tiere*, Jena 3, 871–972, pl. 34.

Boettger, O. (1890a) Eine neue viper aus Armenien. *Zoologischer Anzieger*, Leipzig 13(327), 62–64.

Boettger, O. (1890b) Neue Schlange aus Ostindien. *Berichte der Senckenbergischen Naturforschenden Gesellschaft in Frankfurt-am-Main* 1890, 297–298.

Boettger, O. (1891) Reptilien und Batrachier aus Bolivia. *Zoologischer Anzieger*, Leipzig 14(374), 343–347.

Boettger, O. (1892a) Herpetologische Mitteilungen. Listen von Kriechtieren und Lurchen aus dem tropischen Asien u. au Paupuasien. *Berichte des Offenbacher Vereins für Naturkunde* (1887–189) 29–32, 61–164.

Boettger, O. (1892b) Drei neue colubriforme Schlangen. *Zoologischer Anzieger*, Leipzig 15(405), 417–420.

Boettger, O. (1893a) Kriechtiere der Kaukasusländer, gesammelt durch die Radde-Valentin'sche Expedition nach dem Karabagh und durch die Herren Dr. J. Valentin und P. Reibisch. *Berichte der Senckenbergischen Naturforschenden Gesellschaft in Frankfurt-am-Main* 1893, 131–150.

Boettger, O. (1893b) Übersicht der von Prof. C. Keller anlässlich der Ruspoli'schen Expedition nach den Somaliländern gesammelten Reptilien und Batrachier. *Zoologischer Anzieger*, Leipzig 16(416), 113–119.

Boettger, O. (1893c) Übersicht der von Prof. C. Keller anlässlich der Ruspoli'schen Expedition nach den Somaliländern gesammelten Reptilien und Batrachier. *Zoologischer Anzieger*, Leipzig 16(417), 129–132.

Boettger, O. (1893d) Neue Reptilien und Batrachien aus West-Java. *Zoologischer Anzieger*, Leipzig 16(427), 334–340.

Boettger, O. (1894) Materialien zur herpetologischen Fauna von China III. *Berichte der Senckenbergischen Naturforschenden Gesellschaft in Frankfurt-am-Main* 1894, 129–152, 1 pl.

Boettger, O. (1895a) Liste der Reptilien und Batrachier der insel Halmaheira nach den Sammlungen Prof. Dr. W. Kükenthal's. *Zoologischer Anzieger*, Leipzig 18(471–472), 129–138.

Boettger, O. (1895b) Zwei neue Reptilien vom Sambesi. *Zoologischer Anzieger*, Leipzig 18(468), 62–63.

Boettger, O. (1895c) Neue Frösche und Schlangen von den Liukiu-Inseln. *Zoologischer Anzieger*, Leipzig 18(479), 266–270.

Boettger, O. (1895d) Beitrag zur herpetologischen Kenntniss der Calamianen, Philippinische Inseln. *Abhandlungen und Berichte der königliches Zoologisches und Anthropologisch-Ethnographisches Museums*, Dresden (1894–1895) (7), 1–5.

Boettger, O. (1895e) Neue Frösche und Schlangen von den Liukiu-Inseln. *Bericht des Offenbacher Vereins für Naturkunde* (1891–1895) 101–117.

Boettger, O. (1897) Neue Reptilien und Batrachier von den Philippinen. *Zoologischer Anzieger*, Leipzig 20(531), 161–166.

Boettger, O. (1898) *Katalog der Reptilien-Sammlung im Museum der Senckenbergischen Naturforschenden Gesellschaft in Frankfurt-am-Main. Il. Teil (Schlangen)*. Gebrüder Knauer, Frankfurt-am-Main, ix + 160 pp.

Boettger, O. (1900) Die Reptilien und Batrachier. *Abhandlungen der Senckenbergischen Naturforschenden Gesellschaft*, Frankfurt-am-Main 25, 321–403, pls. 14–15.

Boettger, O. (1905) Reptilien aus dem Staate Parana. *Zoologischer Anzieger*, Leipzig 29(11), 373–375.

Boettger, O. (1913) Reptilien und Amphibien von Madagascar, den Inseln und dem Festland Ostafrikas (Sammlung Voeltzkow 1889–1895 und 1903–1905). *In:* Voeltzkow, A. (Ed.), *Reise in Ostafrika in den Jahren 1903–1905 mit Mitteln der Hermann und Elise geb. Heckmann Wentzel-Stiftung ausgefuhrt. Wissenschaftliche Ergebnisse. Band III. Systematische Arbeiten.* Heft IV. E. Schweizerbart'sche Verlagsbuchhandlung, Nägele & Dr. Sproesser, Stuttgart, pp. 269–375, 1 fig., pls. 23–30.

Boettger, O. & Walter, A. (1888) *Wissenschaftliche Ergebnisse der im Jahre 1886 in Transkaspien von Dr. G. Radde, Dr. A. Walter und A. Konschin ausgefuhrten Expedition. Band I. Zoologische Abtheilung. 3. Lieferung. Die Reptilien und Amphibien*. Gustav Fischer, Jena, 116 pp., pl. 34. [reprint of Boettger, 1888d]

Bogadek, A. & Lau, M.W.N. (1997) A revised checklist of Hong Kong amphibians and reptiles. *Memoirs of the Hong Kong Natural History Society* 21, 173–188.

Bogaerts, S. (1998) Herpetofauna in Tunisia. *Lacerta* 56(5), 154–161, illust.

Bogan, M.T., Findley, L.T. & Enderson, E.F. (2009) Geographic distribution: Serpentes: *Tantilla yaquia*. *Herpetological Review* 40(4), 458.

Bogdanov, O.P. (1962) *Reptiles of the Turkmenistan*, Academy of Sciences of the Turkmenistan Soviet Socialist Republic, Ashabad, 235 pp., 44 figs. (in Russian)

Bogdanov, O.P. (1965) *The ecology of the reptiles of Central Asia.* Akademii Nauk Uzbekistan SSR, Instituta Zoologii i Parazitologii, 258 pp., 10 figs., 36 pls.

Bogert, C.M. (1930) An annotated list of the amphibians and reptiles of Los Angeles County, California. *Bulletin of the Southern California Academy of Science,* Los Angeles 29(1), 3–14.

Bogert, C.M. (1935) *Salvadora grahamiae virgultea*, a new subspecies of the patch-nosed snake. *Bulletin of the Southern California Academy of Sciences* 34(1), 88–94, pls. 13–14.

Bogert, C.M. (1939a) A study of the genus *Salvadora* the patch–nosed snakes. *Publications of the University of California at Los Angeles in Biological Sciences* 1(10), 177–236, 2 figs., pls. 3–7.

Bogert, C.M. (1939b) Notes on snakes of the genus *Salvadora* with a redescription of a neglected Mexican species. *Copeia* 1939(3), 140–149, 1 fig.

Bogert, C.M. (1940) Herpetological results of the Vernay Angola Expedition, with notes on African reptiles in other collections. Part I. Snakes, including an arrangement of African Colubridae. *Bulletin of the American Museum of Natural History* 77(1), 1–107, 18 figs., 1 pl.

Bogert, C.M. (1942a) *Pseudohaje* Günther, a valid genus for two West African arboreal cobras. *American Museum Novitates* (1174), 1–9, 8 figs.

Bogert, C.M. (1942b) Snakes secured by the Snyder East African Expedition in Kenya Colony and Tanganyika Territory. *American Museum Novitates* (1178), 1–5.

Bogert, C.M. (1943) Dentitional phenomena in cobras and other elapids with notes on adaptive modifications of fangs. *Bulletin of the American Museum of Natural History* 81(3), 285–360, 73 figs., pls. 48–51.

Bogert, C.M. (1945) Two additional races of the patch-nosed snake, *Salvadora hexalepis. American Museum Novitates* (1285), 1–14, 10 figs.

Bogert, C.M. (1947) The status of the genus *Leptodrymus* Amaral, with comments on modifications of colubrid premaxillae. *American Museum Novitates* (1352), 1–14, 6 figs.

Bogert, C.M. (1961) Book review: Die Amphibien und Reptilien Europas, by Robert Mertens and Heinz Wermuth. *Copeia* 1961(2), 252–256

Bogert, C.M. (1964) Snakes of the genera *Diaphorolepis* and *Synophis* and the colubrid subfamily Xenoderminae. *Senckenbergiana Biologica*, Frankfurt-am-Main 45(3–5), 509–531, 48 figs.

Bogert, C.M. (1968a) The variations and affinities of the dwarf boas of the genus *Ungaliophis*. *American Museum Novitates* (2340), 1–26, 8 figs.

Bogert, C.M. (1968b) A new arboreal pit viper of the genus *Bothrops* from the Isthmus of Tehuantepec, Mexico. *American Museum Novitates* (2341), 1–14, 3 figs.

Bogert, C.M. (1968c) A new genus and species of dwarf boa from southern México. *American Museum Novitates* (2354), 1–38, 12 figs.

Bogert, C.M. & Degenhardt, W.G. (1961) An addition to the fauna of the United States, the Chihuahua ridge-nosed rattlesnake in New Mexico. *American Museum Novitates* (2064), 1–15, 3 figs.

Bogert, C.M. & Duellman, W.E. (1963) A new genus and species of colubrid snake from the Mexican state of Oaxaca. *American Museum Novitates* (2162), 1–15, 7 figs.

Bogert, C.M. & Matalas, B.L. (1945) Results of the Archbold Expeditions. No. 53. A review of the elapid genus *Ultrocalamus* of New Guinea. *American Museum Novitates* (1284), 1–8, 10 figs.

Bogert, C.M. & Oliver, J.A. (1945) A preliminary analysis of the herpetofauna of Sonora. *Bulletin of the American Museum of Natural History* 83(6), 297–426, 13 figs., pls. 30–37.

Bogert, C.M. & Porter, A.P. (1966a) A new species of *Geophis* (Serpentes, Colubridae) from the state of Colima, México. *American Museum Novitates* (2260), 1–10, 1 fig.

Bogert, C.M. & Porter, A.P. (1966b) The differential characteristics of the Mexican snakes related to *Geophis dubius* (Peters). *American Museum Novitates* (2277), 1–19, 1 fig.

Böhme, M. (2008) Ectothermic vertebrates (Teleostei, Allocaudata, Urodela, Anura, Testudines, Choristodera, Crocodylis, Squamata) from the Upper Oligocene of Oberleichtersbach (Northern Bavaria, Germany). *Courier Forschungsinstitut Senckenberg*, Frankfurt-am-Main (260), 161–183, 4 figs., 4 pls.

Böhme, W. (1974) Die Typusexemplare der Herpetologischen Sammlung des Museums Alexander Koenig, Bonn. *Bonner Zoologische Beiträge*, Bonn 25, 165–176.

Böhme, W. (1975) Zur herpetofaunistik Kameruns, mit Beschreibung eines neuen Scinciden. *Bonner Zoologische Beiträge*, Bonn 26(1–3), 2–48, 10 figs., 1 pl.

Böhme, W. (1977a) Eine neue Art der Gattung *Bitis* (Serpentes Viperidae) aus Äthiopien. *Monitore Zoologico Italiano*, Firenze Supplement 9(3), 59–68, 3 figs.

Böhme, W. (1977b) Further specimens of the rare cat snake, *Telescopus rhinopoma* (Blanford, 1874) (Reptilia, Serpentes, Colubridae). *Journal of Herpetology* 11(2), 201–205, 1 fig.

Böhme, W. (1978) Zur Herpetofaunistik des Senegal. *Bonner Zoologische Beiträge*, Bonn 29(4), 360–417, illust.

Böhme, W. (1986) Preliminary note on the taxonomic status of *Psammophis leucogaster* Spawls, 1983 (Colubridae, Psammophini). *Litteratura Serpentium* (English ed.) 6(5), 171–180, 2 figs.

Böhme, W. (1987) Seltene Reptilien aus dem Uzungwe-Gebirge, Sud-Tansania. *Herpetofauna* 9(48), 27–34, illust.

Böhme, W. (1993a) *Handbuch der Reptilien und Amphibien Europas. Schlangen (Serpentes) I. (Typhlopidae, Boidae, Colubridae 1, Colubrinae)*. AULA-Verlag, Wiesbaden 3(1), 1–479, 85 figs.

Böhme, W. (1993b) *Coluber gyarosensis* Mertens, 1968—Gyaros-Pfeilnatter. *In:* Böhme, W. (Ed.), *Handbuch der Reptilien und Amphibien Europas, Schlangen (Serpentes) I. (Typhlopidae, Boidae, Colubridae 1, Colubrinae)*. AULA-Verlag, Wiesbaden, pp. 111–114, fig. 22.

Böhme, W. (1993c) *Coluber ravergieri* Ménétriés, 1832—Ravergiers Zornnatter. *In:* Böhme, W. (Ed.), *Handbuch der Reptilien und Amphibien Europas, Schlangen (Serpentes) I. (Typhlopidae, Boidae, Colubridae 1, Colubrinae)*. AULA-Verlag, Wiesbaden, pp. 145–153, figs. 34–35.

Böhme, W. (1993d) *Elaphe longissima* (Laurenti, 1768)—Askulapnatter. *In:* (Böhme, W. (Ed.), *Handbuch der Reptilien und Amphibien Europas, Schlangen (Serpentes) I. (Typhlopidae, Boidae, Colubridae 1, Colubrinae)*. AULA-Verlag, Wiesbaden, pp. 331–372, figs. 63–69.

Böhme, W. (2010) A list of the herpetological type specimens in the Zoologisches Forschungsmuseum Alexander Koenig, Bonn. *Bonn Zoological Bulletin* 59, 79–108.

Bohme, W. (2011) Type list of amphibians and reptiles in the Zoologisches Forschungsmuseum A. Koenig, Bonn: corrections and additions. *Bonn Zoological Bulletin* 60(1), 61–71.

Böhme, W. & Bischoff, W. (1984) Amphibien und Reptilien. *In:* Rheinweld, G. (Ed), Die Wirbeltier sammlungen des Museums Alexander Koenig. *Bonner Zoolologische Monographien*, Bonn (19), 151–213.

Böhme, W. & Joger, U. (1984) Eine neue Art des *Vipera berus*-Komplexes aus der Türkei. *Amphibia-Reptilia*, Leiden (1983) 4(2–4), 265–271, 3 figs.

Böhme, W. & Lang, M. (1991) The reptilian fauna of the Late Oligocene locality Rott near Bonn (Germany) with special reference to the taxonomic assignment of "*Lacerta*" *rottensis* von Meyer, 1856. *Neues Jahrbuch für Geologie und Paläontologie, Monatshefte*, Stuttgart 1991, 515–525, 6 figs.

Böhme, W. and Nikolaus, G. (1989) Herpetological specimens from the Gotel Mountains and Mambilla Plateau, Nigeria. *Tauraco Research Report* (1), 28–30, illust.

Böhme, W. & Pury, S. de. (2011) A note on the generic allocation of *Coluber moilensis* Reuss, 1834 (Serpentes: Psammophiidae). *Salamandra* 47(2), 120–123.

Böhme, W. & Szczerbak, N. (1993) *Elaphe quatuorlineata* (Lacépède, 1789)—Vierstreifennatter. *In:* Böhme, W. (Ed.), *Handbuch der Reptilien und Amphibien Europas, Schlangen (Serpentes) I. (Typhlopidae, Boidae, Colubridae 1, Colubrinae)*. AULA-Verlag, Wiesbaden, pp. 373–396, figs. 70–72.

Böhme, W., Schmitz, G. & Messer, J. (1989) Erster Nachweis der Gattung *Telescopus* für die Schlangenfauna Marokkos. *Salamandra*, Frankfurt-am-Main 25(2), 73–76, 1 fig.

Böhme, W., Smith, H.M., Rybak, J., van Breukelen, F. & Chiszar, D. (1998) The lectotype and type locality of *Candoia carinata* (Reptilia, Serpentes). *Contemporary Herpetology* 1998(2), 6-9.

Böhme, W., Rödel, M.-O., Brede, C. & Wagner, P. (2011) The reptiles (Testudines, Squamata, Crocodylia) of the forested southeast of the Republic of Guinea (Guinée forestière), with a country-wide checklist. *Bonn Zoological Bulletin* 60(1), 35–61, 30 figs.

Boie, F. (1826) Generalübersicht der Familien und Gattungen der Ophidier. *Isis von Oken*, Jena 19(10), 981–982.

Boie, F. (1827) Bemerkungen über Merrem's Versuch eines Systems der Amphibien. Marburg, 1820. 1te Lieferung, Ophidier. *Isis von Oken*, Jena 20(6), 508–566.

Boie, F (1828) Bemerkungen über die Abtheilungen im Natürlichen systeme und deren characteristik. *Isis von Oken*, Leipzig 21(4), 351–363.

Boie, H. (MS) (1823–1825) *Erpétologie de Java*. Unpublished manuscript deposited in the Leiden Museum; snake section with 30 pp.

Boie, H. (1826a) Notice sur l'erpétologie de l'ile de Java (ouvrage manuscrit). *Bulletin des Sciences Naturelles et de Géologie*, Paris 9, 233–240.

Boie, H. (1826b) Merkmale einiger japanischen Lurche. *Isis von Oken*, Jena 18(2), 203–216.

Boie, H. (1827a) Caractères de quelques espèces de reptiles du Japan; par le Dr. Boié, à Leyde. *Bulletin des Sciences Naturelles et de Géologie*, Paris (Férussac) 10, 160–162. [partial reprint of H. Boie, 1826]

Boie, H. (1827b) Kenteekenen van eenige Japansche Amphibien. *Bijdragen tot de Natuurkundige Wetenschappen, Amsterdam* 2(1), 243–272. [reprint of H. Boie, 1826b]

Boie, H. (1831) Briefs von Heinrich Boie geschrieben aus Ostindien und auf der Reise dahin. *Neues Staatsbürgerliches Magazin*, Holstein 1(1), 126–218, 1(2), 440–500.

Bokermann, W.C.A. (1957) Atualização do itinerario da viagem do Principe de Wied ao Brasil (1815–1817). *Arquivos de Zoologia*, São Paulo (1956–1957) 10(3), 210–251.

Bolaños, R. (1983) Serpientes venenosas de Centro America. Distribución características y patrones cariológicos. *Memórias do Instituto Butantan*, São Paulo (1982) 46, 275–291, 11 figs.

Bolaños, R. (1984) *Serpientes venenos y ofidismo en Centroamérica*. Editorial Universidad de Costa Rica, San José 7, 136 pp., illust.

Bolkay, S.J. (1913) Additions to the fossil herpetology of Hungary from the Pannonian and Praeglacial Periode. *Mitteilungen aus dem Jahrbuch der königlich Ungarischen Geologische Reichsanstalt* 21(7), 217–230, 5 figs., pls. 11–12.

Bolkay, S.J. (1919) Additions to the herpetology of the western Balkan Peninsula. *Glasnik Zemaljskog Muzeja u Bosni I Hercegovini*, Sarajevo 31, 1–38, 5 pls. (in Czech)

Bolkay, S.J. (1920) On the phytogenetical Series *Vipera gedulyi* By., *Vipera ammodytes* L., *Vipera meridionalis* Blgr. *Glasnika Zemaljskog Muzeja u Bosni I Hercegovini*, Sarajevo 32, 1–12. (in Czech)

Bolkay, S.J. (1925) *Mesophis Nopcsai* n. g. n. sp., ein neues, schlangenähnliches Reptil aus der unteren Kreide (Neocom) von Bilek–Selista (Ost–Hercegovina). *Glasnik Zemaljskog Muzeja u Bosni I Hercegovini*, Sarajevo 37, 125–134, 5 figs. 3 pls.

Bologna, M.A., Capula, M. & Carpaneto, G.M. (2000) *Anfibi e rettili del Lazio*. Fratelli Palombi Editori, Roma, 159 pp., col. illust.

Bol'shakov, V.N. & Vershinin, V.L. (2005) *Amphibians and reptiles of the Middle Urals*. Ural Division of the Russian Academy of Sciences, Ekaterinburg, 179 pp., 34 figs. (in Russian)

Bonaparte, C.L. (1831a) Saggio d'una distribuzione metodica degli animali vertebrati a sangue freddo. *Giornale Arcadico di Scienze, Lettere, ed Arti*, Roma 52, 129–209.

Bonaparte, C.L. (1831b) *Saggio di una distribuzione metodica degli animali*. Presso Antonio Boulzaler, Rome 144 pp. [reprint of Bonaparte, 1831a]

Bonaparte, C.L. (1832–1841) *Iconografia della fauna Italica per le quattro classi degli animali vertebrati. Tomo II. Amfibi*. Tipografie Salviucci, Roma, 65 pp., 153 pls. [issued in 30 fasc.: 1 (1832), 2–5 (1833), 6–11 (1834), 12–13 (1835), 14–18 (1835/36), 19–21 (1837), 22–23 (1838), 24–26 (1839), 27–29 (1840) & 30 (1841)]

Bonaparte, C.L. (1833) Versuch einer methodischen vertheilung der Wirbelthiere mit Kaltem Blut von Carl Lucian Bonaparte. *Isis von Oken*, Leipzig 26, 1183–1229.

Bonaparte, C.L. (1840) Amphibia Europaea ad systema nostrum vertebratorum ordinata. *Memorie della Reale Accademia delle Scienze di Torino* (1839) (Série 2) 2, 385–456.

Bonaparte, C.L. (1849) On the lorine genus of parrots, with the description of a new species, *Electus Cornelia*. *Proceedings of the Zoological Society of London* 17(1), 142–146, pl. 11.

Bonaparte, J.F. (1978) *El Mesozóica de América del Sur y sus tetrapodos*. Opera Lilloana, Tucumán, 596 pp.

Bonaparte, J.F. (1991) Los vertebrados fosiles de la formación Río Colorado, de la ciudad de Neuquen y Cercanias, Cretacico superior, Argentina. *Revista del Museo Argentino de Cièncias Naturales "Bernardino Rivadavia,"* Buenos Aires 4(3), 1–123, 30 figs.

Bonino, N.A. (1987) Aspectos morfologicos de *Philodryas psammophideus* (Serpentes, Colubridae) de la Prov. de Cordoba, Rep. Argentina. *Anales, Museo Nacional de Historia Naturale*, Valparaíso 18, 123–130, 9 figs.

Bonnaterre, J.P. (1789) Erpétologie. *In: Tableau encyclopédique et méthodique des trois règnes de la nature*. Panckoucke, Paris, xxviii + 70 pp., 35 pls.

Bonnaterre, J.P. (1790) Ophiologie. *In: Tableau encyclopédique et méthodique des trois règnes de la nature*. Panconoke, Paris, xliv + 76 pp., 43 pls. [partial reprint of Bonnaterre, 1789]

Bonny, K. (2007) *Die Gattung Boa: Taxonomie und Fortpflanzung*. Kirschner & Seufer Verlag, Rheinstetten, 262 pp., 417 col. photos.

Bons, J. (1958) Contribution a l'étude de l'herpétofaune Marocaine (Reptiles de la région d'Ifrane). *Bulletin de la Société des Sciences Naturelles et Physiques du Maroc*, Rabat 38, 167–182, illust.

Bons, J. (1962) Notes sur trois couleuvres Africaines, *Coluber algirus*, *Coluber florulentus* et *Coluber hippocrepis*. Description de *Coluber algirus villiersi* subsp. nov. *Bulletin de la Société des Sciences Naturelles et Physiques du Maroc*, Rabat 42, 61–86, 4 figs., 2 pls.

Bons, J. (1967) *Recherches sur la biogeographie et la biologie des amphibiens et des reptiles du Maroc*. Doctoral Diss., Universite de Montpellier, Montpellier, 321 pp., 32 figs., 20 pls.

Bons, J. (1972) Herpetologie Marocaine. I. Liste commentée des amphibiens et reptiles du Maroc. *Bulletin de la Société des Sciences Naturelles et Physiques du Maroc*, Rabat 52(3–4), 107–126.

Bons, J. (1973) Herpetologie Marocaine. 2. Origines, évolution et particularités du peuplement herpétologique du Maroc. *Bulletin de la Société des Sciences Naturelles et Physiques du Maroc*, Rabat 53(1–2), 63–110, 6 figs.

Bons, J. & Geniez, P. (1996) *Amphibiens et reptiles du Maroc (Sahara Occidental compris)*. Atlas biogéographique. Asociación Herpetologia Española, Barcelona, 310 pp., col. photos.

Bons, J. & Girot, B. (1962) Cle illustrée des reptiles du Maroc. *Travaux de l'Institut Scientifique Chérifien (Zoologique)*, Rabat (26), 1–62, 15 figs.

Bonta, I.L., Vargaftig, B.B., Bhargava, N. & Vos, C.J. de. (1970) Method for study of snake venom induced hemorrhages. *Toxicon* 8(1), 3–10, 5 figs.

Boos, H. (1967) Reptiles on Huevos. *Living World, Journal of the Trinidad Field Naturalist's Club*, Port-of-Spain 1967, 15–18.

Boos, H. (1983a) The terrestrial reptiles of Monos Island. *Living World, Journal of the Trinidad Tobago Field Naturalist's Club*, Port-of-Spain 1983, 14–18.

Boos, H. (1983b) A consideration of the terrestrial reptile fauna on some offshore islands north west of Trinidad. *Living World, Journal of the Trinidad Tobago Field Naturalist's Club*, Port-of-Spain 1983, 19–26.

Boos, H. (1992) A note on the 18.5 ft. *Boa constrictor* from Trinidad. *British Herpetological Society Bulletin*, London 40, 15–17.

Boos, H. (2001) *The snakes of Trinidad & Tobago*. Texas A & M University, College Station, xvx + 270 pp., 50 figs., 48 col. pls.
Boos, H. & Quesnel, V. (1968) *Reptiles of Trinidad and Tobago*. Ministry of Education and Culture, Port-of-Spain, 39 pp., col. illust.
Booth, E.S. (1959) Amphibians and reptiles collected in Mexico and Central America from 1952 to 1958 by the Walla Walla College of Natural History. *Walla Walla College Publications of the Department of Biological Sciences and Biological Station*, College Place (24), 1–9.
Boquet, P. (1948) *Venins de serpents et antivenins*. Éditions Médicales Flammarion, Paris, 157 pp.
Boquet, P. (1964) Venins de serpents (1ère partie). Physio-pathologie de l'envenimation et propriétés biologiques des venins. *Toxicon* 2(1), 5–41.
Borg, J. ter. (2008) Reptielen en amfibieën in Curaçao. *Lacerta*, Utrecht 66(6), 216–235, c25 ol. photos.
Borges-Martins, M., Rossetti, J.G. & Anés, A.C. (2005) Geographic distribution: Serpentes: *Clelia hussami*. *Herpetological Review* 36(3), 338.
Borges-Nojosa, D.M. & Lima-Verde, J.S. (1999) Geographic distribution: Serpentes: *Lachis muta rhombeata*. *Herpetological Review* 30(4), 235.
Borges-Nojosa, D.M., Loebmann, D., Lima, D.C., Lima-Melo, J.C. & Giacometti-Mai, A.C. (2006) Reptilia, Colubridae, *Pseustes sulphureus*, distribution extension, new state record. *Check List* 2(3), 79–81, 4 figs.
Borges-Nojosa, D.M., Dourado de Mesquita, P.C.M. & Fernandes-Ferreira, H. (2008) Geographic distribution: Serpentes: *Helicops angulatus*. *Herpetological Review* 39(4), 485.
Borges-Nojosa, D.M., Alencar Oliveira, C. de & Bezerra Maia, A.P. (2009) Geographic distribution: Serpentes: *Leptotyphlops brasiliensis*. *Herpetological Review* 40(2), 238.
Boring, A.M. (1932) A list of Fukien Amphibia and reptiles. *First Annual Report of the Marine Biology Association of China*, 99–124.
Boring, A.M, Liu, C.-C. & Chou, S.-C. (1932) Handbook of North China amphibia and reptiles (Herpetology of North China). *Peking Natural History Bulletin, Handbook* (3), 64 pp., figs.
Borkin, L.J. & V. Cherlin (1995) On the validity of the name *Echis coloratus* and the type locality of *Echis megalocephalus*. *Zoologichesky Zhurnal* 74, 134–136.
Borkin, L.J. & Darevsky, I.S. (1987) List of amphibians and reptiles of the fauna of USSR. *In:* Kriukova, E.P. (Ed.), *Amphibians and reptiles of conservation areas*. Chief Administration of Hunting and Conservation Areas, Moscow. pp. 128–143. (in Russian)
Borkin, L.J., Velizhanin, A.G., Korotkova, E.B. & Korotkov, Y.M. (1981) On the herpetofauna of islands in Gulf of Peter the Great. *In:* Borkin, L.J. (Ed.), *Herpetological investigations in Siberia and the Far East*. Proceedings of the Zoological Institute, Leningrad, USSR Academy of Science, pp. 24–27. (in Russian)
Borkin, L.J., Munkhbayar, K., Orlov, N.L., Semenov, D.V. & Terbish, K. (1990a) Distribution of reptiles in Mongolia. *Proceedings of the Zoological Institute, Akademia Nauk SSR*, Leningrad 207, 22–138, 20 figs. (in Russian)
Borkin, L.J., Vorobeva, E.I., Darevkii, I.S., Kuzmin, S.L., Munkhbayar, K. & Semenov, D.V. (1990b) *Amphibians and reptiles of Mongolian People's Republic, general problems, amphibians*. Nauka, Moscow, 1–248, illust. (in Russian)
Borkin, L.J., Litvinchuk, S.N. & Rosanov, Y.M. (1997) Amphibians and reptiles of Moldavia, additions and corrections, with a list of species. *Russian Journal of Herpetology* 4(1), 50–63, 5 figs.
Born, M. (1996) *Reptiles and amphibians of Nouragues, French Guiana*. Marga Born, Amsterdam, 55 pp., 92 col. photos.
Börschig, C. (2007) *Taxonomie und Zoogeographie der subterranean Squamatenfamilien Boliviens (Amphisbaenidae, Typhlopidae, Leptotyphlopidae)*. Master's Thesis, Rheinischen Friedrich-Wilhelms-Universität, Bonn, 267 pp.
Bory de Saint-Vincent, J.B.G.M. (1822–1831) *Dictionnaire classique d'histoire naturelle, par Messieurs Audouin, Isid. Bourdon, Ad. Brongniart, De Candolle, Daudebard de Férussac, A. Desmoulins, Drapiez, Edwards, Flourens, Geoffroy de Saint-Hilaire, A. De Jussieu, Kunth, G. De Larosse, Lamouroux, Latreille, Lucas fils, Presle–Duplessis, C. Prévost, A. Richard, Thiébaut de Berneaud, et Bory de Saint–Vincent. Ouvrage dirigé par ce dernier collaborateur.* 17 vols. Rey et Gravier, Baudouin Frères, Paris.
Bory de Saint-Vincent, J.B.G.M. (1832) *In:* Geoffroy-Saint-Hilaire, E., Goeffroy-Saint-Hilarie, I., Deshayes, G.P., Bibron, G. & Bory de Saint-Vincent, J.B.G.M. (Eds.), *Expedition scientifique de Morée. Tome III.—1.re partie. Zoologie. Premiere section. Animaux vertébrés, mollusques et polypiers*. F. G. Levrault, Paris, 400 pp., 58 pls.
Boscá, E. (1878) Note sur une forme nouvelle ou peu connue de vipère. *Bulletin de la Société Zoologique de France*, Paris 3, 116–121, pl. 4.
Boscá, E. (1879) Las viboras de Espana. *Anales de la Sociedad Española de Historia Naturale*, 8, 65–86.
Bosch, H. (1994) *Boa constrictor*. Terrarien Bibliothek, Münster, 88 pp., 31 figs., 19 col. illust.
Bosch, H.A.J. in den. (1985) Snakes of Sulawesi, checklist, key and additional biogeographical remarks. *Zoologische Verhandelingen*, Leiden (217), 1–50, 2 figs.
Bosch, H.A.J. in den & Ineich, I. (1994) The Typhlopidae of Sulawesi (Indonesia), a review with description of a new genus and a new species (Serpentes, Typhlopidae). *Journal of Herpetology* 28(2), 206–217, 10 figs.
Boschung, H. (1977) How to know the poisonous snakes of Alabama. *Nature Notebook*, Alabama Museum of Natural History (3), 1–7, illust.
Bostanchi, H., Anderson, S.C., Kami, H.G. & Papenfuss, T.J. (2006) A new species of *Pseudocerastes* with elaborate tail ornamentation from western Iran (Squamata, Viperidae). *Proceedings of the California Academy of Sciences* 57(14), 443–450, 13 figs.
Bostic, D.L. (1971) Herpetofauna of the Pacific Coast of north central Baja California, Mexico, with a description of a new subspecies of *Phyllodactylus xanti*. *Transactions of the San Diego Society of Natural History* 16(10), 237–264, 8 figs.
Botha, S.A. (1986) Geographic distribution, *Bitis atropos*. *Journal of the Herpetological Association of Africa*, Stellenbosch 32, 31.
Bothner, R. (1976) *Thamnophis brachystoma*. *Catalogue of American amphibians and reptiles*, pp. 190.1–190.2.
Boucard, A. (1884) Notice sur François Sumichrast. *Bulletin de la Société Zoologique de France,* Paris 9, 305–312.
Bouillon, A. (1965) Serpents, danger! *Zooleo*, Leopoldville (2), 1–76, 52 figs.
Boulenger, G.A. (1880) Reptiles et batraciens recueillis par M. Émile de Ville dans les Andes de l'Équateur. *Bulletin de la Société Zoologique de France* 5, 41–48.
Boulenger, G.A. (1882) Account of the reptiles and batrachians collected by Mr. Edward Whymper in Ecuador in 1879–80. *Annals and Magazine of Natural History*, London (Series 5) 9(54), 457–467.

Boulenger, G.A. (1883) Report on a collection of reptiles and batrachians from the Timor Laut Islands, formed by Mr. H.O. Forbes. *Proceedings of the Zoological Society of London* 51(1), 386–388, pls. 41–42.

Boulenger, G.A. (1884a) Descriptions of new species of reptiles and batrachians in the British Museum. Part II. *Annals and Magazine of Natural History*, London (Series 5) 13(77), 396–398.

Boulenger, G.A. (1884b) Diagnoses of new reptiles and batrachians from the Solomon Islands, collected and presented to the British Museum by H.B. Guppy Esq., M.B., H.M.S. "Lark." *Proceedings of the Zoological Society of London* 52(2), 210–213.

Boulenger, G.A. (1885a) A list of reptiles and batrachians from the Province Rio Grande do Sul, Brazil, sent to the Natural History Museum by Dr. H. von Ihering. *Annals and Magazine of Natural History*, London (Series 5) 15(87), 191–196.

Boulenger, G.A. (1885b) Second list of reptiles and batrachians from the Province Rio Grande do Sul, Brazil, sent to the Natural-History Museum by Dr. H. von Ihering. *Annals and Magazine of Natural History*, London (Series 5) 16(92), 85–88.

Boulenger, G.A. (1885c) A list of reptiles and batrachians from the island of Nias. *Annals and Magazine of Natural History*, London (Series 5) 16(95), 388–389.

Boulenger, G.A. (1886a) A synopsis of the reptiles and batrachians of the Province of Rio Grande do Sul, Brazil. *Annals and Magazine of Natural History*, London (Series 5) 18(108), 423–445.

Boulenger, G.A. (1886b) On the reptiles and batrachians of the Solomon Islands. *Transactions of the Zoological Society of London* 12(1), 35–61, pls. 7–13.

Boulenger, G.A. (1887a) Description of a new snake, of the genus *Calamaria*, from Borneo. *Annals and Magazine of Natural History*, London (Series 5) 19(111), 169–170, 3 figs.

Boulenger, G.A. (1887b) On new reptiles and batrachians from North Borneo. *Annals and Magazine of Natural History*, London (Series 5) 20(116), 95–97.

Boulenger, G.A. (1887c) List of reptiles and batrachians from Cyprus. *Annals and Magazine of Natural History*, London (Series 5) 20(119), 344–345.

Boulenger, G.A. (1887d) A list of the reptiles and batrachians obtained near Muscat, Arabia, and presented to the British Museum by Surgeon-Major A.S.G. Jayakar. *Annals and Magazine of Natural History*, London (Series 5) 20(120), 407–408.

Boulenger, G.A. (1887e) Description of a new snake from Afghanistan. *Annals and Magazine of Natural History*, London (Series 5) 20(120), 413–414.

Boulenger, G.A. (1887f) A synopsis of the snakes of South Africa. *The Zoologist*, London (Series 3) 11(125), 171–182.

Boulenger, G.A. (1887g) On a collection of reptiles and batrachians made by Mr. H. Pryer in the Loo Choo Islands. *Proceedings of the Zoological Society of London* 55(1), 146–150, pl. 17–18.

Boulenger, G.A. (1887h) Second contribution to the herpetology of the Solomon Islands. *Proceedings of the Zoological Society of London* 55(2), 333–340, pl. 28.

Boulenger, G.A. (1887i) On a new snake of the genus *Lamprophis* living in the Society's Gardens. *Proceedings of the Zoological Society of London* 55(3), 397–398, pl. 34.

Boulenger, G.A. (1887j) Report on a zoological collection made by the officers of H.M.S. "Flying Fish" at Christmas Island, Indian Ocean. III. Reptiles. *Proceedings of the Zoological Society of London* 55(3), 516–517.

Boulenger, G.A. (1888a) Descriptions of new reptiles and batrachians from Madagascar. *Annals and Magazine of Natural History*, London (Series 6) 1(2), 101–107, pls. 5–6.

Boulenger, G.A. (1888b) Descriptions of new reptiles and batrachians obtained by Mr. H.O. Forbes in New Guinea. *Annals and Magazine of Natural History*, London (Series 6) 1(5), 343–346.

Boulenger, G.A. (1888c) Description of two new snakes from Hongkong, and note on the dentition of *Hydrophis viperina*. *Annals and Magazine of Natural History*, London (Series 6) 2(7), 43–45.

Boulenger, G.A. (1888d) On new or little-known South-African reptiles. *Annals and Magazine of Natural History*, London (Series 6) 2(8), 136–141.

Boulenger, G.A. (1888e) Description of a new snake from Muscat, Arabia. *Annals and Magazine of Natural History*, London (Series 6) 2(12), 508–509.

Boulenger, G.A. (1888f) Third contribution to the herpetology of the Solomon Islands. *Proceedings of the Zoological Society of London* 57(2), 88–90.

Boulenger, G.A. (1888g) Second list of reptiles and batrachians from Cyprus. *Proceedings of the Zoological Society of London* 57(4), 505–506.

Boulenger, G.A. (1888h) An account of the Reptilia obtained in Burma, north of Tenasserim, by M.L. Fea, of the Genoa Civic Museum. *Annali del Museo Civico di Storia Naturale di Genova* (Series 2) 6, 593–604, pls. 5–7.

Boulenger, G.A. (1889a) Description of a new snake and two new fishes obtained by Dr. H. von Ihering in Brazil. *Annals and Magazine of Natural History*, London (Series 6) 4(22), 265–267.

Boulenger, G.A. (1889b) Descriptions of new Typhlopidae in the British Museum. *Annals and Magazine of Natural History*, London (Series 6) 4(23), 360–363.

Boulenger, G.A. (1889c) Note on *Python curtus*. *Proceedings of the Zoological Society of London* 57(1), 432–433, pl. 45.

Boulenger, G.A. (1889d) The zoology of the Afghan Delimitation Commission. Reptiles and batrachians. *Transactions of the Linnaean Society of London (Zoology)*, (Series 2) 5(3), 94–106, pls. 8–11.

Boulenger, G.A. (1890a) *The fauna of British India, including Ceylon and Burma. Reptilia and Batrachia*. Taylor and Francis, London, xviii + 541 pp. 142 figs.

Boulenger, G.A. (1890b) Fourth contribution to the herpetology of the Solomon Islands. *Proceedings of the Zoological Society of London* 58(1), 30–31, pl. 2.

Boulenger, G.A. (1890c) List of the reptiles, batrachians, and freshwater fishes collected by Professor Moesch and Mr. Iversen in the district of Deli, Sumatra. *Proceedings of the Zoological Society of London* 58(1), 31–40.

Boulenger, G.A. (1890d) Description of a new snake of the genus *Glauconia*, Gray, obtained by Dr. Emin Pasha on the Victoria Nyanza. *Annals and Magazine of Natural History*, London (Series 6) 6(31), 91–93.

Boulenger, G.A. (1890e) On the ophidian genus *Pseudoxyrhopus*, Gthr. *Annals and Magazine of Natural History*, London (Series 6) 6(34), 311–314.

Boulenger, G.A. (1891a) Descriptions of new oriental reptiles and batrachians. *Annals and Magazine of Natural History*, London (Series 6) 7(39), 279–283.

Boulenger, G.A. (1891b) Remarks on the herpetological fauna of Mount Kina Baloo, North Borneo. *Annals and Magazine of Natural History*, London (Series 6) 7(40), 341–345.

Boulenger, G.A. (1891c) On new or little-known Indian and Malayan reptiles and batrachians. *Annals and Magazine of Natural History*, London (Series 6) 8(46), 288–292.

Boulenger, G.A. (1891d) Remarks on the genus *Heterolepis*, Smith. *Annals and Magazine of Natural History*, London (Series 6) 8(47), 344–346.

Boulenger, G.A. (1891e) On the state of our knowledge of the reptiles and batrachians of British Central Africa. *Proceedings of the Zoological Society of London* 59(2), 305–309.

Boulenger, G.A. (1891f) On reptiles, batrachians, and fishes from the Lesser West Indies. *Proceedings of the Zoological Society of London* 59(3), 351–357.

Boulenger, G.A. (1891g) Notes on Transcaspian reptiles. *Proceedings of the Zoological Society of London* 59(4), 628–633.

Boulenger, G.A. (1891h) Catalogue of the reptiles and batrachians of Barbary (Morocco, Algeria, Tunisia), based chiefly upon the notes and collections made in 1880–1884 by M. Fernand Lataste. *Transactions of the Zoological Society of London* 13(1), 93–164, pls. 13–18.

Boulenger, G.A. (1891i) On some reptiles collected by Sig. L. Bricchetti Robecchi in Somaliland. *Annali del Museo Civico di Storia Naturale di Genova* (Series 2) 12(32), 5–13, pl. 1.

Boulenger, G.A. (1892a) Description of a new snake from Nubia. *Annals and Magazine of Natural History*, London (Series 6) 9(49), 74–76.

Boulenger, G.A. (1892b) Descriptions of new reptiles and batrachians from the Loo Choo Islands. *Annals and Magazine of Natural History*, London (Series 6) 10(58), 302–304.

Boulenger, G.A. (1892c) Reptilia and Batrachia. *In:* Distant, W.L. (Ed.), *A Naturalist in the Transvaal.* H.H. Porter, London, pp. 174–176, 3 figs.

Boulenger, G.A. (1892) Reptilia and Batrachia. *In: Zoological Record* (1891) 28(4), 1–24.

Boulenger, G.A. (1893a) *Catalogue of the snakes in the British Museum (Natural History). Volume I, containing the families Typhlopidae, Glauconiidae, Boidae, Ilysiidae, Uropeltidae, Xenopeltidae and Colubridae aglyphae, part.* British Museum (Natural History), London, xiv + 448 pp., 26 figs., 28 pls.

Boulenger, G.A. (1893b) Description of a new earth-snake from Travancore. *The Journal of the Bombay Natural History Society* (1892–1893) 7(3), 318, 1 pl.

Boulenger, G.A. (1893c) Description of a new snake from the Gold Coast. *Annals and Magazine of Natural History*, London (Series 6) 12(70), 273.

Boulenger, G.A. (1893d) Descriptions of new reptiles and batrachians obtained in Borneo by Mr. A. Everett and Mr. C. Hose. *Proceedings of the Zoological Society of London* 61(1), 522–528, pls. 42–44.

Boulenger, G.A. (1893e) On a little-known European viper, *Vipera ursinni*, Bonap. *Proceedings of the Zoological Society of London* 61(1), 596–599, pl. 51.

Boulenger, G.A. (1893f) On *Vipera renardi*, Christoph. *Proceedings of the Zoological Society of London* 61(1), 757–760, pl. 64.

Boulenger, G.A. (1893g) Viaggio di Leonardo Fea in Birmania e regioni vicine. LII. Concluding report on the reptiles and batrachians obtained in Burma by Signor L. Fea, dealing with the collection made in Pegu and the Karin Hills in 1887–88. *Annali del Museo Civico di Storia Naturale di Genova* (Series 2) 13, 304–347, pls. 7–12.

Boulenger, G.A. (1894a) *Catalogue of the snakes in the British Museum (Natural History). Volume II, containing the conclusion of the Colubridae aglyphae.* British Museum (Natural History), London, xii + 382 pp., 25 figs., 20 pls.

Boulenger, G.A. (1894b) Description of a new snake found in Travancore by Mr. S. Dighton. Pirmaad. *The Journal of the Bombay Natural History Society* (1893–1894) 8(4), 528, 1 pl.

Boulenger, G.A. (1894c) List of reptiles and batrachians collected by Dr. J. Bohls near Asuncion, Paraguay. *Annals and Magazine of Natural History*, London (Series 6) 13(76), 342–348.

Boulenger, G.A. (1894d) On the herpetological fauna of Palawan and Balabac. *Annals and Magazine of Natural History*, London (Series 6) 14(80), 81–90.

Boulenger, G.A. (1894e) A list of the reptiles and batrachians collected by Dr. E. Modigliani on Sereinu (Sipora), Mentawei Islands. *Annali del Museo Civico di Storia Naturale di Genova* (Series 2) 14(34), 613–618.

Boulenger, G.A. (1895a) Esplorazione del Giuba e dei suoi affluenti compiuta dal Cap. V. Bottego durante gli anni 1892–93 sotto gli auspicii della Società Geografica Italiana. Risultati zoologici. II. Rettili e Batraci. *Annali del Museo Civico di Storia Naturale di Genova* (Series 2) 15(35), 7–18, 4 pls.

Boulenger, G.A. (1895b) On a collection of reptiles and batrachians from Ferguson Island, D'Entrecasteaux Group, British New Guinea. *Annals and Magazine of Natural History*, London (Series 6) 16(91), 28–32.

Boulenger, G.A. (1895c) On some new or little-known reptiles obtained by W.H. Crosse, Esq., on the Niger. *Annals and Magazine of Natural History*, London (Series 6) 16(91), 32–34.

Boulenger, G.A. (1895d) Descriptions of two new snakes from Usambara, German East Africa. *Annals and Magazine of Natural History*, London (Series 6)16(92), 171–173.

Boulenger, G.A. (1895e) Descriptions of two new reptiles obtained by Mr. A.S. Meek in the Trobriand Islands, British New Guinea. *Annals and Magazine of Natural History*, London (Series 6) 16(95), 408–409.

Boulenger, G.A. (1895f) Descriptions of two new snakes of the genus *Calamaria*. *Annals and Magazine of Natural History*, London (Series 6) 16(96), 481.

Boulenger, G.A. (1895g) On a new *Typhlops* previously confounded with *Typhlops unguirostris* Peters. *Proceedings of the Linnaean Society of New South Wales*, Sydney (1894), (Series 2) 9(4), 718–719.

Boulenger, G.A. (1895h) An account of the reptiles and batrachians collected by Dr. A. Donaldson Smith in western Somali-land and the Galla country. *Proceedings of the Zoological Society of London* 63(3), 530–540, pls. 29–30.

Boulenger, G.A. (1895i) On the type specimen of *Boulengerina stormsi*, an elapoid snake from central Africa. *Proceedings of the Zoological Society of London* 63(4), 865–866, pl. 48.

Boulenger, G.A. (1895j) Descriptions of a new snake and a new frog from North Australia. *Proceedings of the Zoological Society of London* 63(4), 867, pl. 49.

Boulenger, G.A. (1895k) Description of a new snake from Borneo. *Annals and Magazine of Natural History*, London (Series 6) 15(88), 329.

Boulenger, G.A. (1896a) *Catalogue of the snakes in the British Museum (Natural History). Volume III, containing the Colubridae (Opisthoglyphae and Proteroglyphae), Amblycephalidae, and Viperidae.* British Museum (Natural History), London, xiv + 727 pp., 37 figs., 25 pls.

Boulenger, G.A. (1896b) Descriptions of new reptiles and batrachians from Colombia. *Annals and Magazine of Natural History*, London (Series 6) 17(97), 16–21.

Boulenger, G.A. (1896c) Descriptions of new reptiles and batrachians collected in Celebes by Drs. P. and F. Sarasin. *Annals and Magazine of Natural History*, London (Series 6) 17(101), 393–395.

Boulenger, G.A. (1896d) Descriptions of new reptiles and batrachians obtained by Mr. Alfred Everett in Celebes and Jampea. *Annals and Magazine of Natural History*, London (Series 6) 18(103), 62–64.

Boulenger, G.A. (1896e) Description of a new genus of elapine snakes from Woodlark Island, British New Guinea. *Annals and Magazine of Natural History*, London (Series 6) 18(104), 152.

Boulenger, G.A. (1896f) A list of the reptiles and batrachians collected by Dr. Ragazzi in Shoa and Eritrea. *Annali del Museo Civico di Storia Naturale di Genova* (Series 2) 16(36), 545–554.

Boulenger, G.A. (1896g) A list of the reptiles and batrachians collected by the late Prince Eugenio Ruspoli in Somaliland and Gallaland in 1893. *Annali del Museo Civico di Storia Naturale di Genova* (Series 2) 17(37), 5–14.

Boulenger, G.A. (1896h) Report on Capt. Bottego's second collection of reptiles and batrachians from Somaliland. *Annali del Museo Civico di Storia Naturale di Genova* (Series 2) 17(37), 15–23, pl. 1.

Boulenger, G.A. (1896i) Description of a new earth-snake from Travancore (*Rhinophis fergusonianus*). *The Journal of the Bombay Natural History Society* (1895–1897) 10(2), 236, 1 pl.

Boulenger, G.A. (1897a) Description of a new snake from Sierra Leone. *Annals and Magazine of Natural History*, London (Series 6) 19(110), 154.

Boulenger, G.A. (1897b) A list of reptiles and batrachians from the Congo Free State, with descriptions of two new snakes. *Annals and Magazine of Natural History*, London (Series 6) 19(111), 276–281, 2 figs.

Boulenger, G.A. (1897c) Description of a new snake from Usambara, German East Africa. *Annals and Magazine of Natural History*, London (Series 6) 19(111), 281.

Boulenger, G.A. (1897d) A list of the reptiles and batrachians collected by Mr. Alfred Everett in Lombok, Flores, Sumba, and Savu, with descriptions of new species. *Annals and Magazine of Natural History*, London (Series 6) 19(113), 503–509.

Boulenger, G.A. (1897e) Description of a new snake from the Andes of Colombia. *Annals and Magazine of Natural History*, London (Series 6) 20(120), 523.

Boulenger, G.A. (1897f) A new krait from Sind (*Bungarus sindanus*). *The Journal of the Bombay Natural History Society* (1897–1898) 11(1), 73–74, 1 pl.

Boulenger, G.A. (1897g) A catalogue of the reptiles and batrachians of Celebes, with special reference to the collections made by Drs. P. & F. Sarasin in 1893–1896. *Proceedings of the Zoological Society of London* 65(1), 193–237, pls. 7–16.

Boulenger, G.A. (1897h) A list of the reptiles and batrachians collected in northern Nyasaland by Mr. Alex. Whyte, F.Z.S., and presented to the British Museum by Sir Harry H. Johnston, K.C.B.; with descriptions of new species. *Proceedings of the Zoological Society of London* 65(3), 800–803, pl. 46.

Boulenger, G.A. (1897i) A list of the reptiles and batrachians of Somaliland and Gallaland. *Annali del Museo Civico di Storia Naturale di Genova* (Series 2) 17(37), 275–280.

Boulenger, G.A. (1897j) Reptilia and Batrachia. *In: Zoological Record* (1896) 3(4)3, 1–38.

Boulenger, G.A. (1898a) A list of reptiles and batrachians from Ombaai, East Indian Archipelago. *Annals and Magazine of Natural History*, London (Series 7) 1(2), 122–124.

Boulenger, G.A. (1898b) Description of two new blind snakes. *Annals and Magazine of Natural History*, London (Series 7) 1(2), 124.

Boulenger, G.A. (1898c) Description of a new genus of aglyphous colubrine snakes from Sumatra. *Annals and Magazine of Natural History*, London (Series 7) 2(7), 73–74, 4 figs.

Boulenger, G.A. (1898d) Description of a new death-adder (*Acanthophis*) from central Australia. *Annals and Magazine of Natural History*, London (Series 7) 2(7), 75.

Boulenger, G.A. (1898e) On a second collection of reptiles made by Mr. E. Lort-Phillips in Somaliland. *Annals and Magazine of Natural History*, London (Series 7) 2(8), 130–133.

Boulenger, G.A. (1898f) Descriptions of two new snakes from Queensland. *Annals and Magazine of Natural History*, London (Series 7) 2(11), 414.

Boulenger, G.A. (1898g) Description of a new sea-snake from Borneo. *Proceedings of the Zoological Society of London* 66(1), 106–107, pl. 9.

Boulenger, G.A. (1898h) An account of the reptiles and batrachians collected by Mr. W.F.H. Rosenberg in western Ecuador. *Proceedings of the Zoological Society of London* 66(1), 107–126, pls. 10–18.

Boulenger, G.A. (1898i) An account of the reptiles and batrachians collected by Dr. L. Loria in British New Guinea. *Annali del Museo Civico di Storia Naturale di Genova* (Series 2) 18(38), 694–710, pls. 6–8.

Boulenger, G.A. (1898j) Concluding report on the late Capt. Bottego's collection of reptiles and batrachians from Somaliland and British East Africa. *Annali del Museo Civico di Storia Naturale di Genova* (Series 2) 18(38), 715–723, pls. 9–10.

Boulenger, G.A. (1898k) A list of the reptiles and batrachians collected by the late Prof. L. Balzan in Bolivia. *Annali del Museo Civico di Storia Naturale di Genova* (Series 2) 19(39), 128–133.

Boulenger, G.A. (1898l) On a little-known sea-snake from the South Pacific. *In:* Willey, A. (Ed.), *Zoological results, based on material from New Britain, New Guinea, Loyalty Islands and elsewhere collected during the years 1895, 1896 and 1897.* Cambridge University, Cambridge, 1, 57–58, pl. 5.

Boulenger, G.A. (1899a) Descriptions of the new species of reptiles collected by Dr. H.O. Forbes and Mr. W.R. Ogilvie Grant in the islands of Abd-el-Kuri and Socotra. *Bulletin of the Liverpool Museum* 2, 4–7.

Boulenger, G.A. (1899b) On a collection of reptiles and batrachians made by Mr. J.D. La Touche in N.W. Fokien, China. *Proceedings of the Zoological Society of London* 67(1), 159–172, 2 figs., pls. 16–19.

Boulenger, G.A. (1899c) Descriptions of three new reptiles and a new batrachian from Mount Kina Balu, North Borneo. *Annals and Magazine of Natural History*, London (Series 7) 4(24), 451–454.

Boulenger, G.A. (1899d) A new sea snake of the genus *Distira*, from Kurrachee. *The Journal of the Bombay Natural History Society* (1898–1899) 12(4), 642, 1 pl.

Boulenger, G.A. (1899e) Reptilia and Batrachia. *In: Zoological Record* (1898) 35(4), 1–29.

Boulenger, G.A. (1900a) Descriptions of new reptiles from Perak, Malay Peninsula. *Annals and Magazine of Natural History*, London (Series 7) 5(27), 306–308.

Boulenger, G.A. (1900b) Descriptions of new batrachians and reptiles collected by Mr. P.O. Simons in Perú. *Annals and Magazine of Natural History*, London (Series 7) 6(32), 181–186.

Boulenger, G.A. (1900c) Description of a new snake from the Perak Hills. *The Journal of the Bombay Natural History Society* (1900–1902) 13(2), 336.

Boulenger, G.A. (1900d) Descriptions of new batrachians and reptiles from the Larut Hills, Perak. *Annals and Magazine of Natural History*, London (Series 7) 6(32), 186–193.

Boulenger, G.A. (1900e) Description of a new snake of the genus *Ablabes* from Burma. *Annals and Magazine of Natural History*, London (Series 7) 6(34), 409.

Boulenger, G.A. (1900f) Descriptions of new reptiles and batracians from Borneo. *Proceedings of the Zoological Society of London* 69(2), 182–187, pls. 14–17.

Boulenger, G.A. (1900g) A list of the batrachians and reptiles of the Gaboon (French Congo), with descriptions of new genera and species. *Proceedings of the Zoological Society of London* 69(3), 433–466, pls. 27–32.

Boulenger, G.A. (1900h) On a new blind snake from Lifu, Loyalty Islands. *In:* Willey, A. (Ed.), *Zoological results, based on material from New Britain, New Guinea, Loyalty Islands and elsewhere collected during the years 1895, 1896 and 1897.* Cambridge University, Cambridge (5), 603–604, 4 figs.

Boulenger, G.A. (1900i) Reptilia. *In:* Andrews, C.W. (Ed.), *A monograph of Christmas Island (Indian Ocean), physical features and geology, with descriptions of the fauna and flora by numerous contributors.* British Museum (Natural History), London, pp. 51–54, pl. 7.

Boulenger, G.A. (1901a) Matériaux pour la faune du Congo. Reptiles nouveaux. *Annales du Musée du Congo (Zoologie)*, Bruxelles 2(1), 7–14, pls. 3–5.

Boulenger, G.A. (1901b) Description of a new snake of the genus *Ablabes* from Burma. *The Journal of the Bombay Natural History Society* (1900–1901) 13(3), 553.

Boulenger, G.A. (1902a) Descriptions of new batrachians and reptiles from north-western Ecuador. *Annals and Magazine of Natural History*, London (Series 7) 9(49), 51–57.

Boulenger, G.A. (1902b) Description of new fishes and reptiles discovered by Dr. F. Silvestri in South America. *Annals and Magazine of Natural History*, London (Series 7) 9(52), 284–288.

Boulenger, G.A. (1902c) List of the fishes, batrachians, and reptiles collected by the late Mr. P.O. Simons in the provinces of Mendoza and Cordova, Argentina. *Annals and Magazine of Natural History*, London (Series 7) 9(53), 336–339.

Boulenger, G.A. (1902d) Descriptions of new batrachians and reptiles from the Andes of Peru and Bolivia. *Annals and Magazine of Natural History*, London (Series 7) 10(59), 394–402.

Boulenger, G.A. (1902e) Description of a new snake of the genus *Pseudechis* from Queensland. *Annals and Magazine of Natural History*, London (Series 7) 10(60), 494–495.

Boulenger, G.A. (1902f) Description of a new snake of the genus *Psammophis*, from Cape Colony. *Proceedings of the Zoological Society of London* 72(1), 126, pl. 12.

Boulenger, G.A. (1903a) On some batrachians and reptiles from Venezuela. *Annals and Magazine of Natural History*, London (Series 7) 11(65), 481–484.

Boulenger, G.A. (1903b) Descriptions of new snakes in the collection of the British Museum. *Annals and Magazine of Natural History*, London (Series 7) 12(69), 350–354.

Boulenger, G.A. (1903c) Descriptions of new reptiles from British New Guinea. *Proceedings of the Zoological Society of London* 73(1), 125–129, pls. 12–13.

Boulenger, G.A. (1903d) On the geographical variations of the sand-viper, *Vipera ammodytes*. *Proceedings of the Zoological Society of London* 73(1), 185–186, figs. 27–28.

Boulenger, G.A. (1903e) Description of a new sea-snake from Rangoon. *The Journal of the Bombay Natural History Society* (1902–1903) 14(4), 719, 1 pl.

Boulenger, G.A. (1903f) Report on the batrachians and reptiles. *Fasciculi Malayenses*, London 1, 131–176, 3 figs., pls. 5–10.

Boulenger, G.A. (1903g) The reptiles of Sokotra. *In:* Forbes, H.O. (Ed.), *Natural History of Sokotra and Abd-el-Kuri.* Liverpool Museum, Liverpool, pp. 75–104, pls. 8–11.

Boulenger, G.A. (1904a) Descriptions of new frogs and snakes from Yunnan. *Annals and Magazine of Natural History*, London (Series 7) 13(74), 130–134.

Boulenger, G.A. (1904b) Descriptions of three new snakes. *Annals and Magazine of Natural History*, London (Series 7) 13(78), 450–452.

Boulenger, G.A. (1904c) Descriptions of two new Elapine snakes from the Congo. *Annals and Magazine of Natural History*, London (Series 7) 14(79), 14–15.

Boulenger, G.A. (1904d) Description of a new snake. *Spolia Zeylanica*, Colombo 2(7), 95–96, 1 pl.

Boulenger, G.A. (1904e) On the sand-viper of Roumania (*Vipera ammodytes* var. *Montandoni*). *Annals and Magazine of Natural History*, London (Series 7) 14(80), 134–135.

Boulenger, G.A. (1905a) Description of a new snake of the genus *Atractaspis* from Mount Kenya, British East Africa. *Annals and Magazine of Natural History*, London (Series 7) 15(86), 190.

Boulenger, G.A. (1905b) Descriptions of new snakes in the collection of the British Museum. *Annals and Magazine of Natural History*, London (Series 7) 15(89), 453–456.

Boulenger, G.A. (1905c) Description of a new snake from Venezuela. *Annals and Magazine of Natural History*, London (Series 7) 15(90), 561.

Boulenger, G.A. (1905d) A list of the batrachians and reptiles collected by Dr. W.J. Ansorge in Angola, with descriptions of new species. *Annals and Magazine of Natural History*, London (Series 7) 16(92), 105–115, pl. 4.

Boulenger, G.A. (1905e) Descriptions of three new snakes discovered in south Arabia by Mr. G.W. Bury. *Annals and Magazine of Natural History*, London (Series 7) 16(92), 178–180.

Boulenger, G.A. (1905f) Descriptions of two new snakes from Upper Burma. *The Journal of the Bombay Natural History Society* (1904–1905) 16(2), 235–236, 1 pl.

Boulenger, G.A. (1905g) Descriptions of new reptiles discovered in Mexico by Dr. H. Gadow, F.R.S. *Proceedings of the Zoological Society of London* 75(2), 245–247, pls. 6–7.

Boulenger, G.A. (1905h) Reptiles de la Guinée Espagnole. *Memorias de la Real Sociedad Española de. Historia Natural*, Madrid 1(8), 183–186.

Boulenger, G.A. (1906a) Descriptions of new reptiles from Yunnan. *Annals and Magazine of Natural History*, London (Series 7) 17(102), 567–568.

Boulenger, G.A. (1906b) Description of a new tree-viper from Mount Ruwenzoni. *Annals and Magazine of Natural History*, London (Series 7) 18(103), 37–38.

Boulenger, G.A. (1906c) Description of a new lizard and a new snake from Australia. *Annals and Magazine of Natural History*, London (Series 7) 18(108), 440–441.

Boulenger, G.A. (1906d) Description of a new snake of the genus *Glauconia*, from Somaliland. *Annals and Magazine of Natural History*, London (Series 7) 18(108), 441.

Boulenger, G.A. (1906e) Additions to the herpetology of British East Africa. *Proceedings of the Zoological Society of London* 76(3–4), 570–573, figs. 96–98.

Boulenger, G.A. (1906f) Report on the reptiles collected by the late L. Fea in West Africa. *Annali dell Museo Civico di Storia Naturale di Genova* (Series 3) 2(42), 196–216, 9 figs.

Boulenger, G.A. (1907a) Descriptions of three new snakes discovered by Mr. G.L. Bates in south Cameroon. *Annals and Magazine of Natural History*, London (Series 7) 19(112), 324–326.

Boulenger, G.A. (1907b) Description of a new pit-viper from Brazil. *Annals and Magazine of Natural History*, London (Series 7) 20(118), 338.

Boulenger, G.A. (1907c) Description of a new African snake of the genus *Simocephalus*. *Annals and Magazine of Natural History*, London (Series 7) 20(120), 489.

Boulenger, G.A. (1907d) Description of a new snake from Nepal. *Records of the Indian Museum*, Calcutta 1(3), 217.

Boulenger, G.A. (1907e) On a collection of fishes, batrachians, and reptiles, made by Mr. S.A. Neave in Rhodesia, north of the Zambesi, with field notes by the collector. *Memoirs and Proceedings of the Manchester Literary and Philosophical Society* (1906–1907) 51(12), 1–12.

Boulenger, G.A. (1908a) Descriptions of new South-American reptiles. *Annals and Magazine of Natural History*, London (Series 8) 1(1), 111–115.

Boulenger, G.A. (1908b) Note on the ophidian genus *Emydocephalus*. *Annals and Magazine of Natural History*, London (Series 8) 1(3), 231.

Boulenger, G.A. (1908c) Description of a new elapine snake of the genus *Apisthocalamus*, Blg., from New Guinea. *Annals and Magazine of Natural History*, London (Series 8) 1(3), 248–249.

Boulenger, G.A. (1908d) Description of a new elapine snake from Australia. *Annals and Magazine of Natural History*, London (Series 8) 1(4), 333–334.

Boulenger, G.A. (1908e) Note on the type specimen of a blind snake, *Helminthophis wilderi* (Garman), from Brazil. *Annals and Magazine of Natural History*, London (Series 8) 1(4), 334–335, 3 figs.

Boulenger, G.A. (1908f) On a new genus of snake from Brazil. *Annals and Magazine of Natural History*, London (Series 8) 2(7), 31–32.

Boulenger, G.A. (1908g) Descriptions of three new snakes from Africa. *Annals and Magazine of Natural History*, London (Series 8) 2(7), 93–94.

Boulenger, G.A. (1908h) Descriptions of a new frog and a new snake from Formosa. *Annals and Magazine of Natural History*, London (Series 8) 2(8), 221–222.

Boulenger, G.A. (1908i) Description of a new snake from Yunnan. *Annals and Magazine of Natural History*, London (Series 8) 2(8), 244.

Boulenger, G.A. (1908j) Descriptions of new batrachians and reptiles discovered by Mr. M.G. Palmer in south-western Colombia. *Annals and Magazine of Natural History*, London (Series 8) 2(12), 515–522.

Boulenger, G.A. (1908k) On a collection of fresh-water fishes, batrachians, and reptiles from Natal and Zululand, with descriptions of new species. *Annals of the Natal Museum*, Pietermaritzburg 1(3), 219–235, pls. 35–36.

Boulenger, G.A. (1909a) Descriptions of four new frogs and a new snake discovered by Mr. H. Sauter in Formosa. *Annals and Magazine of Natural History*, London (Series 8) 4(24), 492–495.

Boulenger, G.A. (1909b) On the ophidian genus *Grayia*. *Proceedings of the Zoological Society of London* 79(4), 944–952, figs. 295–299.

Boulenger, G.A. (1909c) Ruwenzori Expedition reports. 15. Pisces, Batrachia, and Reptilia. *Transactions of the Zoological Society of London* 19(3), 237–252, pls. 8–9.

Boulenger, G.A. (1910a) A revised list of the South African reptiles and batrachians, with synoptic tables, special reference to the specimens in the South African Museum, and descriptions of new species. *Annals of the South African Museum*, Cape Town 5, 455–538.

Boulenger, G.A. (1910b) Descriptions of four new African snakes in the British Museum. *Annals and Magazine of Natural History*, London (Series 8) 5(30), 512–513.

Boulenger, G.A. (1911a) Descriptions of new reptiles from the Andes of South America, preserved in the British Museum. *Annals and Magazine of Natural History*, London (Series 8) 7(37), 19–25.

Boulenger, G.A. (1911b) Descriptions of three new snakes discovered by Mr. G. L. Bates in South Cameroon. *Annals and Magazine of Natural History*, London (Series 8) 8(45), 370–371.

Boulenger, G.A. (1912a) Missione per la frontiera Italo-Etiopica sotto il comando del Capitano Carlo Citerni. Risultati zoologici. List of the reptiles and batrachians. *Annali del Museo Civico di Storia Naturale di Genova* (Series 3) 5(45), 329–332.

Boulenger, G.A. (1912b) Descriptions of new reptiles from the Andes of South America, preserved in the British Museum. *Annals and Magazine of Natural History*, London (Series 8) 10(58), 420–424.

Boulenger, G.A. (1912c) *A vertebrate fauna of the Malay Peninsula from the Isthmus of Kra to Singapore including the adjacent islands. Reptilia and Batrachia*. Taylor and Francis, London, xiii + 294 pp., 79 figs.

Boulenger, G.A. (1913a) On the geographical races of *Vipera ammodytes*. *Annals and Magazine of Natural History*, London (Series 8) 11(63), 283–287, pl. 5.

Boulenger, G.A. (1913b) Description of a new snake discovered by Mr. A. E. Pratt in eastern Peru. *Annals and Magazine of Natural History*, London (Series 8) 12(67), 72.

Boulenger, G.A. (1913c) Description de deux reptiles nouveaux provenant du Katanga. *Revue Zoologique Africaine*, Tervuren 3(1), 103–105, 2 figs.

Boulenger, G.A. (1913d) On a collection of batrachians and reptiles made by Dr. H.G.F. Spurrell, F.Z.S., in the Choco, Colombia. *Proceedings of the Zoological Society of London* 83(4), 1019–1038, figs. 174–178, pls. 102–108.

Boulenger, G.A. (1913e) *The snakes of Europe*. Methuen, London, xi + 269 pp., 42 figs., 14 pls.

Boulenger, G.A. (1913f) Description of a new lizard and a new snake from South America. *Annali del Museo Civico di Storia Naturale di Genova* (Series 3) 6(46), 49–50.

Boulenger, G.A. (1913g) A list of the reptiles obtained by Mr. H. Stevens in Upper Assam and the Eastern Himalayas. *Records of the Indian Museum*, Calcutta 9(5), 337–338.

Boulenger, G.A. (1914a) Description of a new snake of the genus *Coluber* from northern China. *Annals and Magazine of Natural History*, London (Series 8) 13(78), 576.

Boulenger, G.A. (1914b) On a second collection of batrachians and reptiles made by Dr. H.G.F. Spurrell, F.Z.S., in the Choco, Colombia. *Proceedings of the Zoological Society of London* 84(3), 813–817, 2 pls.

Boulenger, G.A. (1914c) Descriptions of new species of snakes in the collection of the British Museum. *Annals and Magazine of Natural History*, London (Series 8) 14(84), 482–485.

Boulenger, G.A. (1914d) An annotated list of the batrachians and reptiles collected by the British Ornithologists' Union Expedition and the Wollaston Expedition in Dutch New Guinea. *Transactions of the Zoological Society of London* 20(5), 247–274, pls. 27–30.

Boulenger, G.A. (1914e) Descriptions of new reptiles from Siam. With notes by Malcolm Smith, M.R.C.S., L.R.C.P. *Journal of the Natural History Society of Siam*, Bangkok 1(2), 67–70, 6 figs.

Boulenger, G.A. (1915a) A list of the snakes of the Belgian and Portuguese Congo, northern Rhodesia, and Angola. *Proceedings of the Zoological Society of London* 85(2), 193–223, 3 figs.

Boulenger, G.A. (1915b) A list of the snakes of Madagascar, Comoro, Mascarenes, and Seychelles. *Proceedings of the Zoological Society of London* 85(3), 369–382.

Boulenger, G.A. (1915c) A list of the snakes of East Africa, north of the Zambesi and south of the Soudan and Somaliland, and of Nyasaland. *Proceedings of the Zoological Society of London* 85(4), 611–640.

Boulenger, G.A. (1915d) A list of the snakes of North-east Africa, from the Tropic to the Soudan and Somaliland, including Socotra. *Proceedings of the Zoological Society of London* 85(4), 641–658.

Boulenger, G.A. (1915e) Description of a new Amphisbaena and a new snake discovered by Dr. H.G.F. Spurrell in southern Colombia. *Proceedings of the Zoological Society of London* 85(4), 659–661, 2 figs.

Boulenger, G.A. (1916) Description of a new snake of the genus *Coluber* from northern China. *Annals and Magazine of Natural History*, London (Series 8) 17(99), 243–244.

Boulenger, G.A. (1918) Description of a new snake of the genus *Oligodon* from Upper Burma. *Proceedings of the Zoological Society of London* 89(1), 9–10, 1 fig.

Boulenger, G.A. (1919a) Batraciens et reptiles recueillis par le Dr C. Christy au Congo Belge dans les districts de Stanleyville, Haut-Uelé et Ituri en 1912–1914. *Revue Zoologique Africaine*, Tervuren 7(1), 1–29.

Boulenger, G.A. (1919b) Descriptions d'un ophidien et d'un batracien nouveaux du Congo. *Revue Zoologique Africaine*, Tervuren 7(2), 186–187.

Boulenger, G.A. (1920a) Descriptions of a new gecko and a new snake from Sumatra. *Annals and Magazine of Natural History*, London (Series 9) 5(27), 281–283.

Boulenger, G.A. (1920b) Descriptions of four new snakes in the collection of the British Museum. *Annals and Magazine of Natural History*, London (Series 9) 6(31), 108–111.

Boulenger, G.A. (1920c) A list of the snakes of West Africa, from Mauritania to the French Congo. *Proceedings of the Zoological Society of London* 90(3), 267–298, 2 figs.

Boulenger, G.A. (1920d) A list of the snakes of North Africa. *Proceedings of the Zoological Society of London* 90(3), 299–307.

Boulenger, G.A. (1920e) Reptiles and batrachians collected in Korinchi, West Sumatra, by Messrs. H.C. Robinson and C. Boden Kloss. *Journal of the Federated Malay States Museums*, Kuala Lumpur 8(2), 285–296, pl. 8.

Boulenger, G.A. (1920f) Description of a new snake of the genus *Contia*, B. & G., from Persia. *The Journal of the Bombay Natural History Society* (1918–1920) 26(4), 1024.

Boulenger, G.A. (1920g) A list of snakes from Mesopotamia. Collected by members of the Mesopotamian Expeditionary Force, 1915–1919. *The Journal of the Bombay Natural History Society* (1920–1921) 27(2), 347–350.

Boulenger, G.A. (1920h) Description of a new snake of the genus *Zamenis* from Persia. *The Journal of the Bombay Natural History Society* (1920–1921) 27(2), 251.

Boundy, J. (1999) Systematics of the garter snake *Thamnophis atratus* at the southern end of its range. *Proceedings of the California Academy of Sciences* 51(6), 311–336, 8 figs.

Boundy, J. (2006) *Snakes of Louisiana*. Second edition. Louisiana Department of Wildlife & Fisheries, Baton Rouge, 40 pp., 80 col. figs. [first edition, 1997, 33 pp., 66 col. figs.]

Boundy, J. (2007) A replacement name for *Thamnophis eques obscurus* Conant. *Bulletin of the Maryland Herpetological Society* 43(3), 134.

Boundy, J. (2013) Description of a second specimen of *Leptotyphlops parkeri* (Squamata: Leptotyphlopidae), with comments on its generic placement. *Zootaxa* (3637), 493–497, 3 figs.

Boundy, J. & Wallach, V. (2008) The identity of the leptotyphlopid snake *Glauconia unicolor*, Werner, 1913 (Squamata: Serpentes: Leptotyphlopidae). *Mitteilungen aus dem Hamburgischen Zoologischen Museum und Institute* 105, 53–56, 1 fig.

Bour, R. (1993) Les voyages de Peter Simon Pallas et l'origine de *Coluber halys* (Serpentes Viperidae). *Bulletin Mensuel de la Société Linnéenne de Lyon* 62(9), 328–340, illust.

Bourgeois, M. (1963a) Contribution à l'herpétologie du Haut-Katanga. *Revue de Zoologie et de Botanique Africaines*, Tervuren 68(1–2), 117–134,

Bourgeois, M. (1963b) Note sur *Atractaspis irregularis conradsi* Sternfeld (Viperidae). *Annales de la Société Royal Zoologique de Belgique* 93(1), 159–169, 3 figs.

Bourgeois, M. (1963c) Note sur le crane de *Miodon* (Colubridae). *Annales de la Société Royal Zoologique de Belgique* 93(1), 171–178, 3 figs.

Bourgeois, M. (1964) Les reptiles de la Kasapa (Elisábethville). *Publication de l'Université de l'État à Elisabethville* 7, 65–89, 4 figs.

Bourgeois, M. (1967) Note sur les cranes de *Psammophis angolensis* (Bocage) et de *Psammophis sibilans* (Linne). *Publications de l'Universitè Officielle du Congo à Lubumbashi* 17, 37–48, 4 figs.

Bourgeois, M. (1968) Contribution à la morphologie comparée du crane des ophidiens de l'Afrique Centrale. *Publications de l'Universitè Officielle du Congo à Lubumbashi* 18, 1–293, 116 figs.

Bourguignon, T. (2002) *Strumofbandnattern: Herkunft, Pflege, Arten*. Ulmer, 125 pp., 82 col. photos.

Bourquin, O. (1977) Some reptile records from north-eastern Zululand. *Lammergeyer* 23, 46–48.

Bourquin, O. (1989) Herpetofauna in the Natal Parks Board reserves and resorts. *Journal of the Herpetological Society of Africa* 36, 19–25, illust.

Bourquin, O. (1991) A new genus and species of snake from the Natal Drakensberg, South Africa. *Annals of the Transvaal Museum*, Pretoria 35(12), 199–203, 5 figs.

Bourquin, O. & Channing, A. (1980) Herpetofauna of the Natal Drakenberg, an annotated checklist. *Lammergeyer*, Pietermaritzburg (30), 1–20.

Bourquin, O. & Sowler, S.G. (1980) The vertebrates of Vernon Crookes Nature Reserve, 1. *Lammergeyer*, Pietermaritzburg (28), 20–32.

Bourquin, O., Vincent, J. & Hitchens, P.M. (1971) The vertebrates of the Hluhluwe Game Reserve-Corridor (State-land)-Umfolozi Game Reserve complex. *Lammergeyer*, Pietermaritzburg (14), 5–58.

Bourret, R.L. (1927) *Inventaire général de l'Indochine. La fauna de l'Indochine. Vertébrés*. Société de Géographie de Hanoi, Hanoi, 453 pp.

Bourret, R.L. (1934a) Notes herpétologiques sur l'Indochine française. I. Ophidiens de Chapa. *Bulletin Générale de l'Instruction Publique*, Hanoi 14(7), 129–138, 7 figs.

Bourret, R.L. (1934b) Notes herpétologiques sur l'Indochine française. II. Sur que quelques serpents des montagnes du Tonkin. *Bulletin Générale de l'Instruction Publique*, Hanoi 14(8), 149–157, 3 figs.

Bourret, R.L. (1934c) Notes herpétologiques sur l'Indochine française. III. Ophidiens d'Annam et du moyen Laos. *Bulletin Générale de l'Instruction Publique*, Hanoi 14(9), 167–176, 4 figs.

Bourret, R.L. (1934d) Notes herpétologiques sur l'Indochine française. IV. Sur une collection d'ophidiens de Cochinchine et du Cambodge. *Bulletin Générale de l'Instruction Publique*, Hanoi 14(1), 5–17, 5 figs.

Bourret, R.L. (1934e) Notes herpétologiques sur l'Indochine française. V. Sur *Liopeltis major* et ses alliés. *Bulletin Générale de l'Instruction Publique*, Hanoi 14(1), 18–22.

Bourret, R.L. (1934f) Notes herpétologiques sur l'Indochine française. VI. Sur diverses collections de serpents appartenant a l'Université de Hanoi. *Bulletin Générale de l'Instruction Publique*, Hanoi 14(4), 73–83, 2 figs. (*Amphiesma atemporalis*)

Bourret, R.L. (1934g) *Les serpents marins de l'Indochine française*. Institut Océanographique de l'Indochine, Hanoi, 69 pp., 10 pls.

Bourret, R.L. (1935a) Notes herpétologiques sur l'Indochine française. VIII. Sur les *Achalinus* d'Indochine. *Bulletin Générale de l'Instruction Publique*, Hanoi 15(5), 101–104, 4 figs.

Bourret, R.L. (1935b) Notes herpétologiques sur l'Indochine française. IX. Les serpents de Chapa. *Bulletin Générale de l'Instruction Publique*, Hanoi 15(7), 239–254, 2 figs.

Bourret, R.L. (1935c) Notes herpétologiques sur l'Indochine française. X. Les Serpents de la station d'altitude du Tam-dao. *Bulletin Générale de l'Instruction Publique*, Hanoi 15(8), 259–271, 4 figs.

Bourret, R.L. (1935d) Notes herpétologiques sur l'Indochine française. XI. Sur quelques serpents recoltés en 1934. *Bulletin Générale de l'Instruction Publique*, Hanoi 15(9), 289–296.

Bourret, R.L. (1935e) Comment déterminer un serpent d'Indochine. *Bulletin Générale de l'Instruction Publique*, Hanoi 15(3), 1–28, 14 figs.

Bourret, R.L. (1936a) *Les serpents de Indochine. Tome I. Études sur la fauna*. Henri Basuyau, Toulouse, 141 pp., 14 figs.

Bourret, R.L. (1936b) *Les serpents de Indochine. Tome II. Catalogue systématique descriptif.* Henri Basuyau, Toulouse, 505 pp., 189 figs.

Bourret, R.L. (1937a) Notes herpétologiques sur l'Indochine française. XIII. Serpents récemment récoltés au Tonkin et en Annam. *Bulletin Générale de l'Instruction Publique*, Hanoi 17(9), 27–36.

Bourret, R.L. (1937b) Notes herpétologiques sur l'Indochine française. XV. Lézards et serpents reçus au Laboratoire des Sciences Naturelles de l'Université au cours de l'année 1937. Descriptions de deux espèces et de deux variétés nouvelles. *Bulletin Générale de l'Instruction Publique*, Hanoi 17(4), 57–80, 7 figs.

Bourret, R.L. (1938) Les serpents venimeux en Indochine. *Bulletin Générale de l'Instruction Publique*, Hanoi 18(9), 1–24, 12 figs.

Bourret, R.L. (1939a) Notes herpétologiques sur l'Indochine française. XVII. Reptiles et batraciens reçus au Laboratoire des Sciences Naturelles de l'Université au cours de l'année 1938. Descriptions de trois espèces nouvelles. *Bulletin Générale de l'Instruction Publique*, Hanoi 19(6), 13–34.

Bourret, R.L. (1939b) Notes herpétologiques sur l'Indochine française. XVIII. Reptiles et batraciens reçus au Laboratoire des Sciences Naturelles de l'Université au cours de l'année 1939. Descriptions de quatre espèces et d'une variété nouvelles. *Bulletin Générale de l'Instruction Publique*, Hanoi 19(4), 5–39, 8 figs.

Bourret, R.L. (1939c) Notes herpétologiques sur l'Indochine française. XIX. La faune herpétologique des stations d'altitude du Tonkin. *Bulletin Générale de l'Instruction Publique*, Hanoi 19(4), 41–47.

Bourret, R.L. (1939d) Notes herpétologiques sur l'Indochine française. XX. Liste de reptiles et batraciens actuellement connus en Indochine française. *Bulletin Générale de l'Instruction Publique*, Hanoi 19(4), 49–60.

Bourret, R.L. (1941a) Notes herpétologiques sur l'Indochine française. XXI. Reptiles et batraciens reçus au Laboratoire des Sciences Naturelles de l'Université au cours de l'année 1940. Description d'une espèce fossile nouvelle. *Bulletin Générale de l'Instruction Publique*, Hanoi 20, 1–29.

Bourret, R.L. (1941b) Notes herpétologiques sur l'Indochine française. XXII. Reptiles et batraciens reçus au Laboratoire des Sciences Naturelles de l'Université au cours de l'année 1941. Description d'une espèce et d'une variété nouvelles. *Bulletin Générale de l'Instruction Publique*, Hanoi 21, 5–29.

Bourret, R.L. (1943) Notes herpétologiques sur l'Indochine française. XXIV. Reptiles et batraciens reçus au Laboratoire de Zoologie de l'École Supérieur des Sciences au cours de l'année 1942. *Notes et Travaux de l'École Supérieur des Sciences de la Université Indochinoise*, Hanoi 22(2), 1–17.

Bourret, R.L. (1944) Notes herpétologiques sur l'Indochine française. XXV. Reptiles et batraciens reçus au Laboratoire de Zoologie de l'École Supérieur des Sciences au cours de l'année 1943. *Notes et Travaux de l'École Supérieur des Sciences de la Université Indochinoise*, Hanoi 23(3), 37–43.

Bousek, R.M. (1928) *Trimeresurus* (*Lachesis*) *cotiara* Gom. *Herpetologica*, Tábor (1), 1–8, 4 figs., 1 pl. (in Polish)

Bouskila, A. & Amitai, P. (2001) *Handbook of amphibians & reptiles of Israel.* Keter Publishing House, Jerusalem, 345 pp., col. illust. (in Hebrew)

Bovo, R.P., Marques, O.A.V. & Andrade, D.V. (2012) When basking is not an option: thermoregulation of a viperid snake endemic to a small island in the South Atlantic of Brazil. *Copeia* 2012(3), 408–418, 4 figs.

Bowdich, T.E. (1819) Descriptions par M. le Dr. W. E. Leach, de quelques nouveaux genres et espèces d'animaux decouverts en Afrique. *Journal de Physique, de Chimie, d'Histoire Naturelle et des Arts* 88, 258–260.

Boycott, R.C. (1995) Notes on the distribution and ecology of *Amblyodipsas concolor* (A. Smith, 1849) in Swaziland (Serpentes, Colubridae). *African Journal of Ecology* 22, 417–419.

Boycott, R.C. & Culverwell, J.B. (1992) Swaziland herpetofauna, a preliminary synthesis. *Journal of the Herpetological Association of Africa*, Stellenbosch 40, 38–41.

Boyer, D.A. & Heinze, A.A. (1934) An annotated list of the amphibians and reptiles of Jefferson County, Missouri. *Transactions of the Academy of Sciences of Saint Louis* 28(4), 185–201.

Bozhanskij, A.T. & Kudryavtsev, S.V. (1986) The sand viper *Vipera ammodytes*. *Priroda*, Moskow 1986(2), 46–47, illust. (in Russian)

Brady, M.K. (1932) A new snake from Florida. *Proceedings of the Biological Society of Washington* 45, 5–7.

Brady, M.K. (1937) Natural history of Plummer's Island, Maryland. VI. Reptiles and amphibians. *Proceedings of the Biological Society of Washington* 50, 137–139.

Brana, F. & Bas, S. (1983) *Vipera seoanei cantabrica* ssp. n. *Munibe* 38(1–2), 87–88, illust.

Branch, W.R. (1976) The wolf snakes *Lycophidion capense* and *Lycophidion variegatum* (Reptilia, Serpentes, Colubridae) in South Africa. *Journal of Herpetology* 10(1), 1–11, 6 figs.

Branch, W.R. (1978) The venomous snakes of southern Africa. Part I. Introduction and Viperidae. *The Snake* 9(2), 67–86, 15 figs.

Branch, W.R. (1979) The venomous snakes of southern Africa. Part 2. Elapidae and Hydrophidae. *The Snake* 11(2), 199–225, 20 figs. [reprinted in 1981 by *Bulletin of the Maryland Herpetological Society* 17(1), 1–47, 20 figs.]

Branch, W.R. (1980) Chromosome morphology of the Madagascar tree boa *Sanzinia madagascariensis*. *South African Journal of Zoology*, Pretoria 15(2), 79–82, 1 fig.

Branch, W.R. (1982) Venomous snakes of southern Africa. 3. Concluding part, Colubridae. *The Snake* 14(1), 1–17, 12 figs.

Branch, W.R. (1986) Hemipenial morphology of African snakes, A taxonomic review. Part 1. Scolecophidia and Boidae. *Journal of Herpetology* 20(3), 285–299, 5 figs.

Branch, W.R. (1988) *Bill Branch's field guide to the snakes and other reptiles of southern Africa*. Ralph Curtis, Sanibel Island, 326 pp., 96 col. pls.

Branch, W.R. (1989) A new adder (*Bitis*) from the Cederberg, and the status of *Bitis inornata* (A. Smith, 1849), preliminary observations. *Journal of the Herpetological Association of Africa*, Stellenbosch 36, 64–67.

Branch, W.R. (1991) *Everyone's guide to snakes of southern Africa*. Central News Agency, Johannesburg, 112 pp., col. illust.

Branch, W.R. (1993) *Southern African snakes and other reptiles. A photographic guide*. Ralph Curtis, London, 144 pp., illust.

Branch, W.R. (1994) Herpetofauna of the Sperrgebeit region of southern Namibia. *Herpetological Natural History* 2(1), 1–11, 1 fig.

Branch, W.R. (1997) A new adder (*Bitis*; Viperidae) from the Western Cape Province, South Africa. *South African Journal of Zoology* 32(2), 37–42, 2 figs.

Branch, W.R. (1998) *Field guide to snakes and other reptiles of southern Africa*. Third edition Ralph Curtis Books, Sanibel Island, 399 pp., 112 pls.

Branch, W.R. (1999a) Dwarf adders of the *Bitis cornuta-inornata* complex (Serpentes, Viperidae) in southern Africa. *In:* Joger, U. (Ed.), Phylogeny and systematics of the Viperidae. *Kaupia, Darmstädter Beiträge zur Naturgeschichte* 8, 39–63, 6 figs.

Branch, W.R. (1999b) *Bitis arietans peghullae* Stewart 1973 (Serpentes, Viperidae), a valid race of puff adder? *African Journal of Herpetology* 48(1–2), 15–19, illust.

Branch, W.R. (2001) *A photographic guide to snakes and other reptiles of southern Africa*. Second edition. Struik, Cape Town, 144 pp., col. illust.

Branch, W.R. (2005) *A photographic guide to snakes and other reptiles of East Africa*. Struik, Cape Town, 144 pp., col. illust.

Branch, W.R. & Bauer, A.M. (2005) *The herpetological contributions of Sir Andrew Smith, with an introduction, concordance of names, and annotated bibliography*. Society for the Study of Amphibians and Reptiles, Villanova, 80 pp.

Branch, W.R. & Bayliss, J. (2009) A new species of *Atheris* (Serpentes: Viperidae) from northern Mozambique. *Zootaxa* (2113), 41–54, 9 figs.

Branch, W.R & Haagner, G.V. (1991) New reptile records from three southern national parks. *Koedoe*, Pretoria 35(2), 59–60.

Branch, W.R. & McCartney, C.J. (1993) A report on a small collection of reptiles from southern Angola. *Journal of the Herpetological Association of Africa*, Stellenbosch (1992) (41), 1–3.

Branch, W.R., Haagner, G.V. & Bourquin, O. (1993) Further specimens of the cream-spotted mountain snake *Montaspis gilvomaculata* from Natal. *Lammergeyer*, Pietermaritzburg 42, 50–52.

Branch, W.R., Rödel, M.-O. & Marais, J. (2005) Herpetological survey of the Niassa Game Reserve, northern Mozambique – part I: reptiles. *Salamandra* 41(4), 195–214.

Brandley, M.C., Guiher, T.J., Pyron, R.A. Winne, C.T. & Burbrink, F.T. (2010) Does dispersal across an aquatic geographic barrier obscure phylogeographic structure in the diamondback water snake (*Nerodia rhombifer*)? *Molecular Phylogenetics and Evolution* 57(2), 552–566, 3 figs.

Brandstätter, F. (1995) *Eine revision der gattung Psammophis mit berücksicht der Schwesterngattungen innerhalf der Tribus Psammophini (Colubridae, Lycodontinae)*. Doctoral Dissertation, Universität Saarlandes, Saarbrücken, 480 pp.

Brandstätter, F. (1996) *Die Sandrennattern: Gattung Psammophis*. Westarp Wissenschaften [Die Neue Brehm-Büchereri Nr. 636], Magdeburg, 142 pp., 30 figs., 2 col. pls.

Brandt, J.F. (1838) Note sur quatre nouvelles espèces de serpents de la côte occidentale de la mer Caspienne et de la Perse septentrionale, découvertes par M. Kareline. *Bulletin de l'Académie Impériale des Sciences de St. Pétersbourg* (1837) 3(16), 241–244.

Branson, B.A. (1974) An ecological study of the queen snake *Regina septemvittata* (Say) in Kentucky. *Tulane Studies in Zoology and Botany* 18(4), 153–171, 6 figs.

Branson, B.A. & Baker, E.C. (1973) Sexual dimorphism in Kentucky queen snakes (*Regina septemvittata*) based on scute counts. *Transactions of the Kentucky Academy of Sciences* 34, 57–58.

Branson, E.B. (1904) Snakes of Kansas. *Kansas University Science Bulletin* 2(13), 353–430.

Brass, L.D. (1959) Results of the Archbold expeditions. No. 79. Summary of the Fifth Archbold Expedition to New Guinea (1956–1957). *Bulletin of the American Museum of Natural History* 118(1), 1–69, illust.

Braßeler, H. (1989) Bibliography on herpetological bibliographies I: Amphibia, Reptilia. *Courier Forschungsinstitut Senckenberg*, Frankfurt-am-Main (116), 1–214.

Braßeler, H. (1991) Bibliography on herpetological bibliographies II: Amphibia, Reptilia. *Courier Forschungsinstitut Senckenberg*, Frankfurt-am-Main (140), 1–129.

Braßeler, H. (1994) Bibliography on herpetological bibliographies III: Amphibia, Reptilia. *Courier Forschungsinstitut Senckenberg*, Frankfurt-am-Main (167), 1–139.

Brattstrom, B.H. (1953a) Records of Pleistocene reptiles from California. *Copeia* 1953(3), 174–179, 2 figs.

Brattstrom, B.H. (1953b) An ecological restriction of the type locality of the western worm snake, *Leptotyphlops h. humilis*. *Herpetologica* 8(4), 180–181.

Brattstrom, B.H. (1954) The fossil pit-vipers of North America. *Transactions of the San Diego Society of Natural History* 12(3), 31–46.

Brattstrom, B.H. (1955a) New snakes and lizards from the Eocene of California. *Journal of Paleontology* 29(1), 145–149, 1 fig.

Brattstrom, B.H. (1955b) Pliocene and Pleistocene amphibians and reptiles from southeastern Arizona. *Journal of Paleontology* 29(1), 150–154, 2 figs.

Brattstrom, B.H. (1955c) Records of some Pliocene and Pleistocene reptiles and amphibians from México. *Bulletin of the Southern California Academy of Sciences* 54(1), 1–4, 1 pl.

Brattstrom, B.H. (1958) New records of Cenozoic amphibians and reptiles from California. *Bulletin of the Southern California Academy of Sciences* 57(1), 5–12, pls. 3–4.

Brattstrom, B.H. (1964) Evolution of the pit vipers. *Transactions of the San Diego Society of Natural History* 13(11), 185–267, 41 figs.

Brattstrom, B.H. & Howell, T.R. (1954) Notes on some collections of reptiles and amphibians from Nicaragua. *Herpetologica* 10(2), 114–123, 1 fig.

Brattstrom, B.H. & Warren, J.W. (1953) A new subspecies of racer, *Masticophis flagellum*, from the San Joaquin Valley of California. *Herpetologica* 9(4), 177–179.

Brazaitis, P. & Watanabe, M.E. (1993) *The world of snakes.* Tetra Press, Blacksburg, 176 pp., col. illust.

Brazenor, C.W. (1947) A new snake from Torokina, Bougainville Island. *Memoirs of the National Museum of Victoria*, Melbourne (15), 128–130, figs.

Brazil, V. (1914) *La defense contre l'ophidisme.* Second edition. Pocai Weiss & Co., São Paulo, 319 pp., 28 pls. [first edition, 1911]

Breckenridge, W.J. (1944) *Reptiles and amphibians of Minnesota.* University of Minnesota Press, Minneapolis, xiii + 202 pp., 52 figs.

Breder, C.M., Jr. (1946) Amphibians and reptiles of the Rio Chucunaque drainage, Darien, Panama, with notes on their life histories and habits. *Bulletin of the American Museum of Natural History* 86(8), 377–435.

Brehm, A.E. (1878) *Brehms Thierleben, Allgemeine Kunde des Thierreichs. Dritte Abtheilung. Kriechthiere, Lurche und Fische. Tome VII.* Second edition. Bibliographischen Institut, Leipzig, xiv + 673 pp., 158 figs., 16 pls.

Breithaupt, B.H. & Duvall, D. (1986) The oldest record of serpent aggregation. *Lethaia*, Oslo 19, 181–185, 1 fig.

Brelih, S. & Dzukic, G. (1974) Reptilia. Catalogus Faunae Jugoslaviae. *Consiliorum Academiarum Scientiarum rei Publicae Socialisticae Foederativae Jugoslaviae*, Ljubljana 4(2), 1–32.

Brennan, K.E.C., Morley, T., Hutchinson, M. & Donnellan, S. (2012) Redescription of the western desert taipan, *Oxyuranus temporalis* (Serpentes: Elapidae), with notes on its distribution, diet and genetic variation. *Australian Journal of Zoology* 59(4), 227–235.

Brennan, T.C. & Hollycross, A. (2004) Geographic distribution: Serpentes: *Crotalus oreganus concolor. Herpetological Review* 35(2), 190-191.

Brennan, T.C. & Hollycross, A. (2005) *A field guide to amphibians and reptiles of Maricopa County.* Arizona Game and Fish Department, Phoenix, iv + 68 pp., illust.

Brennan, T.C. & Hollycross, A. (2006) *A field guide to amphibians and reptiles in Arizona.* Arizona Game and Fish Department, Phoenix, iv + 150 pp., illust.

Brèthes, J. (1917) Description d'un nouveau Colubridae aglypha de la République Argentine, "*Zamenis argentinus*" Brèthes, n. sp. *Physis*, Buenos Aires 3, 92–94, 2 figs.

Breuil, M. (1999) Nouvelle espèce du genre *Typhlops* (Serpentes, Typhlopidae) de l'île de St.-Barthélemy, comparaison avec les autres espèces des Petites Antilles. *Bulletin Mensuel de la Société Linnéenne de Lyon* 68, 30–40.

Breuil, M. (2002) *Histoire Naturelle des amphibiens et reptiliens de l'Archipel Guadeloupéen: Guadeloupe, Saint-Martin, Saint-Barthélemy.* Patrimoines Naturels (54), 1–339.

Brueil, M. (2003) *A la découverte des amphibiens & reptiles des Antilles.* PLB Éditions, Gosier, Guadeloupe, 64 pp., col. illust.

Breuil, M. (2009) The terrestrial herpetofauna of Martinique: past, present, future. *Applied Herpetology* 6, 123–149, 4 figs.

Briceño-Rossi, A.L. (1934) El problema del ofidismo en Venezuela. *Boletin del Ministerio de Salubridad y de Agricultura y Cria*, Caracas 2(14), 1079–1177 & 2(15), 15–103, illust.

Brickell, J. (1805) On two species of *Crotalus*. *Philadelphia Medical and Physical Journal* 2(1, sec. 3): 164.

Briggs, P. & McKeown, S. (1994) Natural and captive-selected color phases of California kingsnakes *(Lampropeltis getula californiae). The Vivarium* 6(4), 26–29, 44–46, 28 col. photos.

Brimley, C.S. (1944) *Amphibians and reptiles of North Carolina. The snakes (suborder Serpentes or Ophidia of order Squamata).* Carolina Biological Supply Company, Elon College. [reprinted from *Carolina Tips,* 1941, 4(5), 18–19, 4(6), 22–23, 4(7), 26–27; 1942, 5(1), 2–3, 5(2), 6–7, 5(3), 10–11, 5(4), 14–15, 5(5), 18–19]

Briscoe, M.S. (1949) Notes on snakes collected in Liberia. *Copeia* 1949(1), 16–18.

Brito, J.C., Santos, X., Pleguezuelos, J.M., Fahd, S., Llorente, G.A. & Parellada, X. (2006) Morphological variability of the Lataste's viper (*Vipera latastei*) and the Atlas dwarf viper (*Vipera monticola*), patterns of biogeographical distribution and taxonomy. *Amphibia-Reptilia,* Leiden 27(2), 219–240, 4 figs.

Broad, A.J., Sutherland, S.K., Tanner, C. & Covacevich, J. (1979) Electrophoretic, enzyme, and preliminary toxicity studies of the venom of the small-scaled snake, *Parademansia microlepidota* (Serpentes, Elapidae), with additional data on its distribution. *Memoirs of the Queensland Museum*, Brisbane 19(3), 319–329, 2 pls.

Broadley, D.G. (1958) Serpentes, Colubridae, *Duberria lutrix rhodesiana* n. subsp. *Occasional Papers of the National Museum of Southern Rhodesia*, Salisbury 3B(22), 215–216.

Broadley, D.G. (1959) The herpetology of southern Rhodesia. Part I. Snakes. *Bulletin of the Museum of Comparative Zoology* 120(1), 1–100, 6 pls.

Broadley, D.G. (1960) Serpentes, Colubridae, *Psammophylax tritaeniatus fitzgeraldi* n. subsp. *Occasional Papers of the National Museum of Southern Rhodesia*, Salisbury 3B(24), 431–432.

Broadley, D.G. (1962a) Snakes of southern Rhodesia. 12. The common boomslang (*Dispholidus typus typus*). *African Wild Life* 16(2), 109–111, 1 fig.

Broadley, D.G. (1962b) Serpentes, Colubridae, *Natriciteres olivacea bipostocularis* n. subsp. *Occasional Papers of the National Museum of Southern Rhodesia*, Salisbury 3B(26), 785–786.

Broadley, D.G. (1962c) On some reptile collections from the north-western and north-eastern districts of Southern Rhodesia 1958–61, with descriptions of four new lizards. *Occasional Papers of the National Museum of Southern Rhodesia*, Salisbury 3B(26), 787–843, 2 figs.

Broadley, D.G. (1965) A revision of the *Prosymna sundevalli* group (Serpentes, Colubridae). *Arnoldia (Rhodesia)*, Salisbury 2(5), 1–6.

Broadley, D.G. (1966a) A review of the Natal green snake, *Philothamnus natalensis* (A. Smith), with a description of a new subspecies. *Annals of the Natal Museum*, Pietermaritzburg 18(2), 417–423, 1 fig.

Broadley, D.G. (1966b) A review of the genus *Natriciteres* Loveridge (Serpentes, Colubridae). *Arnoldia (Rhodesia)*, Salisbury 2(35), 1–11, 1 fig.

Broadley, D.G. (1966c) A review of the African stripe-bellied sand-snakes of the genus *Psammophis*. *Arnoldia (Rhodesia)*, Salisbury 2(36), 1–9, 3 figs.

Broadley, D.G. (1967a) The scientific name of the African puff adder. *Copeia* 1967(4), 864–865.

Broadley, D.G. (1967b) A review of the genus *Lycodonomorphus* Fitzinger (Serpentes, Colubridae) in southeastern Africa, with a key to the genus. *Arnoldia (Rhodesia)*, Salisbury 3(16), 1–9, 1 fig., 2 pls.

Broadley, D.G. (1968a) The venomous snakes of central and south Africa. *In:* Bucherl, W., Buckley, E.E. & Deulofeu, V. (Eds.), *Venomous animals and their venoms. Volume I. Venomous vertebrates.* Academic Press, New York, pp. 403–435, 20 figs.

Broadley, D.G. (1968b) A new species of *Crotaphopeltis* (Serpentes, Colubridae) from Barotseland, Zambia. *Fieldiana: Zoology* 51(10), 135–139, 2 figs.

Broadley, D.G. (1968c) A review of the African cobras of the genus *Naja* (Serpentes, Elapinae). *Arnoldia (Rhodesia)*, Salisbury 3(29), 1–14, 5 figs.

Broadley, D.G. (1968d) A revision of *Aspidelaps scutatus* (A. Smith) (Serpentes, Elapinae). *Arnoldia (Rhodesia)*, Salisbury 4(2), 1–9, 3 figs.

Broadley, D.G. (1968e) The African wolf snakes (Serpentes, Colubridae), a new genus and species from Mozambique. *Arnoldia (Rhodesia)*, Salisbury 4(10), 1–5, 3 figs.

Broadley, D.G. (1969) A new species of *Lycophidion* from Rhodesia (Serpentes, Colubridae). *Arnoldia (Rhodesia)*, Salisbury 4(27), 1–8, 2 figs., 2 pls.

Broadley, D.G. (1971a) A review of *Rhamphiophis acutus* (Günther) with the description of a new subspecies from Zambia. *Arnoldia (Rhodesia)*, Salisbury 5(8), 1–8, fig. 1.

Broadley, D.G. (1971b) A revision of the African snake genus *Elapsoidea* Bocage (Elapidae). *Occasional Papers of the National Museum of Rhodesia*, Salisbury 4B(32), 577–626, 12 figs., 2 pls.

Broadley, D.G. (1971c) A revision of the African snake genera *Amblyodipsas* and *Xenocalamus*. *Occasional Papers of the National Museum of Rhodesia*, Salisbury 4B(33), 629–697, 22 figs., 4 pls.

Broadley, D.G. (1971d) The reptiles and amphibians of Zambia. *Puku*, Chilanga 6, 1–143, illust.

Broadley, D.G. (1974) A review of the cobras of the *Naja nigricollis* complex in southwestern Africa (Serpentes, Elapidae). *Cimbebasia*, Windhoek 2A(14), 155–162, 1 fig.

Broadley, D.G. (1975) A review of *Psammophis leightoni* and *Psammophis notostictus* in southern Africa (Serpentes, Colubridae). *Arnoldia (Rhodesia)*, Salisbury 7(13), 1–17, 6 figs., 1 pl.

Broadley, D.G. (1977a) A revision of the African snakes of the genus *Psammophylax* Fitzinger (Colubridae). *Occasional Papers of the National Museum of Rhodesia*, Salisbury 6B(1), 1–44, 12 figs.

Broadley, D.G. (1977b) A review of the genus *Psammophis* in southern Africa (Serpentes, Colubridae). *Arnoldia (Rhodesia)*, Salisbury 8(12), 1–29, 5 figs., 4 pls.

Broadley, D.G. (1979) Problems presented by geographical variation in the African vine snakes, genus *Thelotornis*. *Suid-Afrikaanse Tydskrif vir Dierkunde*, Pretoria 14, 125–131, 4 figs.

Broadley, D.G. (1980) A revision of the African snake genus *Prosymna* Gray (Colubridae). *Occasional Papers of the National Museum of Rhodesia*, Bulawayo 6B(7), 481–556, 18 figs.

Broadley, D.G. (1983) *FitzSimons' Snakes of southern Africa*. Second edition. Delta Books, Johannesburg, 376 pp., 209 figs., 84 col. pls.

Broadley, D.G. (1984) A review of geographical variation in the African python, *Python sebae* (Gmelin). *British Journal of Herpetology*, London 6, 359–367, 2 figs.

Broadley, D.G. (1988) A check list of the reptiles of Zimbabwe, with synoptic keys. *Arnoldia, Zimbabwe*, Bulawayo 9(30), 369–430, 22 figs.

Broadley, D.G. (1989) *Atheris* Cope, 1862 (Reptilia, Serpentes), proposed conservation, and proposed confirmation of *Vipera chlorechis* Pel, (1851) as the valid name of the type species. *Bulletin of Zoological Nomenclature*, London 46(4), 264–266.

Broadley, D.G. (1990a) The herpetofaunas of the islands off the coast of South Moçambique. *Arnoldia, Zimbabwe*, Bulawayo 9(35), 469–493, 4 figs.

Broadley, D.G. (1990b) *FitzSimons' Snakes of southern Africa*. Third edition. Jonathan Ball and Ad. Donker Publishers, Parklands, 387 pp., 211 figs., 84 col. pls. [first edition by V. FitzSimons, 1962a; second edition by Delta Books, Johannesburg, 376 pp., 209 figs., 84 col. pls.]

Broadley, D.G. (1991a) A review of the Namibian snakes of the genus *Lycophidion* (Serpentes, Colubridae), with the description of a new species. *Annals of the Transvaal Museum*, Pretoria 35(14), 209–215, 6 figs.

Broadley, D.G. (1991b) A review of the southern African stiletto snakes of the genus *Atractaspis* A. Smith. *Arnoldia, Zimbabwe*, Bulawayo 9(36), 495–517, 5 figs.

Broadley, D.G. (1992a) Reptiles and amphibians from the Bazaruto Archipelago, Mozambique. *Arnoldia, Zimbabwe*, Bulawayo 9(38), 539–548.

Broadley, D.G. (1992b) The taxonomy and zoogeography of the genus *Lycophidion* (Serpentes, Colubridae). *Journal of the Herpetological Association of Africa*, Stellenbosch (40), 30–36, 5 figs.

Broadley, D.G. (1994a) A collection of snakes from eastern Sudan, with the description of a new species of *Telescopus* Wagler, 1830 (Reptilia, Ophidia). *Journal of African Zoology*, Tervuren 108, 201–208, 3 figs.

Broadley, D.G. (1994b) A revision of the African genus *Scaphiophis* Peters (Serpentes, Colubridae). *The Herpetological Journal*, London 4, 1–10, 2 figs.

Broadley, D.G. (1995a) The snouted cobra, *Naja annulifera*, a valid species in southern Africa. *Journal of the Herpetological Association of Africa*, Stellenbosch 44(2), 26–32, 5 figs.

Broadley, D.G. (1995b) A new species of *Prosymna* Gray (Serpentes, Colubridae) from coastal forest in northeastern Tanzania. *Arnoldia, Zimbabwe*, Bulawayo 10(4), 29–32, 1 fig.

Broadley, D.G. (1995c) A small collection of reptiles and amphibians from central and southern Malawi. *African Herp News* 24, 16–18.

Broadley, D.G. (1996a) A revision of the genus *Lycophidion* Fitzinger (Serpentes, Colubridae) in Africa south of the Equator. *Syntarsus*, Bulawayo (3), 1–33, 23 figs.

Broadley, D.G. (1996b) A review of the tribe Atherini (Serpentes, Viperidae), with the descriptions of two new genera. *African Journal of Herpetology* 45(2), 40–48, 6 figs.

Broadley, D.G. (1998a) A review of the African *Elapsoidea semiannulata* complex (Serpentes, Elapidae). *African Journal of Herpetology* 47(1), 13–23, 2 figs.

Broadley, D.G. (1998b) A review of the genus *Atheris* Cope, (Serpentes, Viperidae), with the description of a new species from Uganda. *The Herpetological Journal* 8, 117–135, 13 figs.

Broadley, D.G. (1999a) A new species of worm snake from Ethiopia (Serpentes, Leptotyphlopidae). *Arnoldia, Zimbabwe*, Bulawayo 12(14), 141–144, 1 fig.

Broadley, D.G. (1999b) The southern African python, *Python natalensis* A. Smith 1840, is a valid species. *African Herp News* (29), 31–32.

Broadley, D.G. (1999c) Geographical distribution. Reptila, Serpentes: *Prosymna lineata* (Peters, 1871) – Lined Shovel-snout. *African Herp News* 30, 36–37.

Broadley, D.G. (2001) A review of the genus *Thelotornis* A. Smith in eastern Africa, with the description of a new species from the Usambara Mountains (Serpentes, Colubridae, Dispholidini). *African Journal of Herpetology* 50(2), 53–70, 2 figs.

Broadley, D.G. (2002) A review of the species of *Psammophis* Boie found south of latitude 12° S (Serpentes, Psammophiinae). *African Journal of Herpetology* 51(2), 83–119, 6 figs.

Broadley, D.G. & Baldwin, A.S. (2006) Taxonomy, natural history, and zoogeography of the southern African shield cobras, genus *Aspidelaps* (Serpentes, Elapidae). *Herpetological Natural History* 9(2), 163–176, 3 figs.

Broadley, D.G. & Bates, M.F. (2009) New range extensions for *Dasypeltis atra* Sternfeld in Tanzania (Serpentes: Colubridae), with a review of the distribution of colour phases. *African Journal of Herpetology* 58(1), 50–55, 3 figs.

Broadley, D.G. & Blake, D.K. (1979) A check list of the reptiles of the national parks and other conservation areas of Zimbabwe Rhodesia. *Arnoldia (Rhodesia)*, Salisbury 8(35), 1–15, illust.

Broadley, D.G. & Blaylock, R. (2013) *The snakes of Zimbabwe and Botswana*. Edition Chimaira, Frankfurt-am-Main, 387 pp., 101 figs., 346 col. photos.

Broadley, D.G. & Broadley, S. (1999) A review of the African worm snakes from south of latitude 12°S (Serpentes: Leptotyphlopidae). *Syntarsus* (5), 1–36.

Broadley, D.G. & Cock, E.V. (1975) *Snakes of Rhodesia*. Longman Rhodesia, Salisbury, iv + 152 pp., 12 figs., 64 pls. [reprinted in 1989]

Broadley, D.G. & Cotterill, F.P.D. (2004) The reptiles of southeast Katanga, an overlooked "hot spot." *African Journal of Herpetology* 53(1), 35–61.

Broadley, D.G. & Howell, K.H. (1991) A check list of the reptiles of Tanzania, with synoptic keys. *Syntarsus*, Bulawayo (1), 1–70, 64 figs.

Broadley, D.G. & Hughes, B. (1993) A review of the genus *Lycophidion* (Serpentes, Colubridae) in northeastern Africa. *The Herpetological Journal*, London 3, 8–18, 4 figs.

Broadley, D.G. & Hughes, B. (2000) A revision of the African genus *Hemirhagerrhis* Boettger 1893 (Serpentes, Colubridae). *Syntarsus*, Bulawayo 6, 3–17, 6 figs.

Broadley, D.G. & Minshull, J.I. (1986) A gazetteer of African countries, their constituent parts, and their synonyms. *Arnoldia, Zimbabwe*, Bulawayo 9(26), 333–342.

Broadley, D.G. & Parker, R.H. (1976) Natural hydridization between the puffadder and Gaboon viper in Zululand (Serpentes, Viperidae). *Durban Museum Novitates* 11(3), 77–83, 2 figs., 1 pl.

Broadley, D.G. & Pitman, C.R.S. (1960) On a collection of snakes taken in northern Rhodesia by Monsieur H. J. Bredo. *Occasional Papers of the National Museum of Southern Rhodesia*, Salisbury 3B(24), 437–451.

Broadley, D.G. & Schätti, B. (2000) A new species of *Coluber* from northern Namibia (Reptilia, Serpentes). *Madoqua*, Windhoek (1997) 19(2), 172–174, 5 figs.

Broadley, D.G. & Stevens, R.A. (1971) A review of *Chamaetortus aulicus* Günther, with the description of a new subspecies from Malawi (Serpentes, Colubridae). *Arnoldia (Rhodesia)*, Salisbury 5(11), 1–11, 2 figs., 2 pls.

Broadley, D.G. & Wallach, V. (1996) Remarkable new worm snake (Serpentes, Leptotyphlopidae) from the East African coast. *Copeia* 1996(1), 162–166, 2 figs.

Broadley, D.G. & Wallach, V. (1997a) A review of the genus *Leptotyphlops* (Serpentes, Leptotyphlopidae) in KwaZulu-Natal, South Africa, with the description of a new forest-dwelling species. *Durban Museum Novitates* 22, 37–42, 3 figs.

Broadley, D.G. & Wallach, V. (1997b) A review of the worm snakes of Mozambique (Serpentes, Leptotyphlopidae) with the description of a new species. *Arnoldia, Zimbabwe*, Bulawayo 10(11), 111–119, 1 fig.

Broadley, D.G. & Wallach, V. (2000) A new blind snake (Serpentes, Typhlopidae) from montane forests of the Eastern Arc Mountains in Tanzania. *African Journal of Herpetology* 49(2), 165–168, 1 fig.

Broadley, D.G. & Wallach, V. (2002) Review of the Dispholidini, with the description of a new genus and species from Tanzania (Serpentes, Colubridae). *Bulletin of the Natural History Museum of London (Zoology)* 66(2), 57–74, 6 figs.

Broadley, D.G. & Wallach, V. (2007a) A revision of the genus *Leptotyphlops* in northeastern Africa and southwestern Arabia (Serpentes, Leptotyphlopidae). *Zootaxa* (1408), 1–78, 12 pls.

Broadley, D.G. & Wallach, V. (2007b) A review of east and central African species of *Letheobia* Cope, revived from the synonym of *Rhinotyphlops* Fitzinger, with descriptions of five new species (Serpentes, Typhlopidae). *Zootaxa* (1515), 31–68, 13 figs.

Broadley, D.G. & Wallach, V. (2009) A review of the eastern and southern African blind-snakes (Serpentes: Typhlopidae), excluding *Letheobia* Cope, with the description of two new genera and a new species. *Zootaxa* (2255), 1–100, 19 figs.

Broadley, D.G. & Watson, G. (1976) A revision of the worm snakes of southeastern Africa (Serpentes, Leptotyphlopidae). *Occasional Papers of the National Museum of Rhodesia*, Salisbury 5B(8), 465–510, 17 figs.

Broadley, D.G. & Wüster, W. (2004) A review of the southern African "non-spitting" cobras (Serpentes, Elapidae, *Naja*). *African Journal of Herpetology* 53(2), 101–122, 5 figs.

Broadley, D.G., Doria, C.T. & Wigge, J. (2003) *Snakes of Zambia. An atlas and field Guide*. Edition Chimaira, Frankfurt-am-Main, 280 pp., 56 figs., 172 col. photos.

Brocchi, P. (1876) Sur un calamarien provenant des iles Fidji. *Bulletin de la Société Philomathique de Paris* (Series 6) 13, 92–94.

Brock, V.E. (1942) A new snake of the genus *Drymarchon* from the Tres Marías Islands. *Copeia* 1942(4), 249–250.

Brode, J.M. (1988) Natural history of the giant garter snake (*Thamnophis couchii gigas*). *In:* De Lisle, H.F., Brown, P.R., Kaufman, B. & McGurty, B.M. (Eds.), Proceedings of the conference on California herpetology. *Southwest Herpetological Society Special Publication* (4), 25–28, 3 figs.

Brode, J.M. (1991) Reptiles of California. *In:* Williams, D.F., Toth, E., Evans, W. Brode, J.M. & Reed, S. (Eds.), *Checklist of birds, mammals, fish, reptiles and amphibians present on forest and rangelands in California*. California Department of Fish & Game, Sacramento, pp. 29–31.

Brodmann, P. (1987) *Die Giftschlangen Europas und die Gattung Vipera in Afrika and Asien*. Kümmerly & Frey, Bern, 148 pp., 25 figs., col. pls.

Brogard, J. (2005) *Zoogeographical checklist of reptiles: Volume 1. Afrotropical and Palearctic realms*. Jacques Brogard, Condé-sur-Noireau, 301 pp.

Broin, F. de, Buffetaut, E., Koeniguer, J.-C., Rage, J.-C., Russell, D., Taquet, P., Vergnaud-Grazzini, C. & Wenz, S. (1974) La faune de vertébrés continentaux du gisement d'In Beceten (Sénonien du Niger). *Comptes Rendus de l'Académie des Sciences*, Paris 279D(6), 469–472.

Brongersma, L.D. (1928) Neue Reptilien aus dem Zoologischen Museum Amsterdam. *Zoologischer Anzieger*, Leipzig 75, 251–257, 3 figs.

Brongersma, L.D. (1929) A list of the reptiles from Java. *In:* Dammerman, K.W. (Ed.), On the zoogeography of Java. *Treubia*, Buitenzorg 11(1), 64–68.

Brongersma, L.D. (1930) Notes on the list of reptiles of Java. *Treubia*, Buitenzorg 12(3–4), 299–303.

Brongersma, L.D. (1931) Résultats scientifiques du Voyage aux Indes Orientales Néerlandaises de LL. AA. RR. le Prince et la Princesse Léopold de Belgique. Reptiles. *Memoires du Musée Royal d'Histoire Naturelle de Belgique* (Série 2) 5(2), 1–39, 4 pls.

Brongersma, L.D. (1933a) A new species of *Leptotyphlops* from Surinam. *Zoologische Mededelingen*, Leiden 15(3–4), 175–176.

Brongersma, L.D. (1933b) Herpetological notes. II. On some snakes from Pulu Babi. *Zoologische Mededelingen*, Leiden 16(1–2), 2–3.

Brongersma, L.D. (1933c) Herpetological notes. III. Snakes from the Mentawei-Islands. *Zoologische Mededelingen*, Leiden 16(1–2), 3–12.

Brongersma, L.D. (1933d) Herpetological notes. V. The herpetological fauna of Pulu Weh. *Zoologische Mededelingen*, Leiden 16(1–2), 12–17.

Brongersma, L.D. (1933e) Herpetological notes. VII. New locality-records. *Zoologische Mededelingen*, Leiden 16(1–2), 21–25.

Brongersma, L.D. (1933f) Herpetological notes. VIII. On some reptiles from Wetar. *Zoologische Mededelingen*, Leiden 16(1–2), 25–27.

Brongersma, L.D. (1933g) Herpetological notes. IX. Contribution to the herpetology of the Babber-Islands. *Zoologische Mededelingen*, Leiden 16(1–2), 27–29.

Brongersma, L.D. (1934) Contributions to Indo-Australian herpetology. *Zoologische Mededelingen*, Leiden 17(3–4), 161–251, 47 figs., 2 pls.

Brongersma, L.D. (1937a) Herpetological notes. XIV. The types of *Psammophis antillensis* Schlegel, 1837. *Zoologische Mededelingen*, Leiden 20(1–2), 1–5.

Brongersma, L.D. (1937b) Herpetological notes. XV. The type of *Dipsas dieperinkii* Schlegel 1837, and the genus *Pseustes* Fitzinger, 1843. *Zoologische Mededelingen*, Leiden 20(1–2), 5–6, 4 figs.

Brongersma, L.D. (1940) Snakes from the Leeward group, Venezuela and eastern Colombia. *In:* Hummelinck, P.W. (Ed.), *Studies on the fauna of Curaçao, Aruba, Bonaire and the Venezuelan Islands. Vol. II.* Martinus Nijhoff, The Hague, pp. 115–137, fig. 15, pl. 12.

Brongersma, L.D. (1946) Note on *Idiopholis collaris* Mocquard. *Zoologische Mededelingen*, Leiden 26(1–4), 249–250, 1 fig.

Brongersma, L.D. (1947a) On the identity of *Maticora intermedia* Westermann and *Dipsadoides decipiens* Annandale. *Proceedings of the Koninklijke Nederlandse Akademie van Wetenschappen*, Amsterdam 50C(4), 419–425.

Brongersma, L.D. (1947b) On the subspecies of *Python curtus* Schlegel occurring in Sumatra. *Proceedings of the Koninklijke Nederlandse Akademie van Wetenschappen*, Amsterdam 50C(6), 666–671.

Brongersma, L.D. (1948a) A new subspecies of *Natrix mairii* (Gray) from Dutch New Guinea. *Proceedings of the Koninklijke Nederlandse Akademie van Wetenschappen*, Amsterdam 51C(3), 372–381.

Brongersma, L.D. (1948b) Notes on *Maticora birvigata* (Boie) and on *Bungarus flaviceps* Reinh. *Zoologische Mededelingen*, Leiden 30(1), 1–29.

Brongersma, L.D. (1950) Notes on *Pseudoxenodon inornatus* (Boie) and *Pseudoxenodon jacobsonii* Lidth. *Proceedings of the Koninklijke Nederlandse Akademie van Wetenschappen*, Amsterdam 53C(9), 1498–1505.

Brongersma, L.D. (1951) Some notes upon the anatomy of *Tropidophis* and *Trachyboa* (Serpentes). *Zoologische Mededelingen*, Leiden 31(11), 107–124, 8 figs.

Brongersma, L.D. (1952) Zoologisch onderzoek met de M. L.D. Brass in Nieuw-Guinea. *Alle Hens, Ministerie van Defensie* 6(5), 4–5.

Brongersma, L.D. (1953) Notes on New Guinean reptiles and amphibians. II. *Proceedings of the Koninklijke Nederlandse Akademie van Wetenschappen*, Amsterdam 56C(3), 317–325, 1 fig., 1 pl.

Brongersma, L.D. (1956a) On some reptiles and amphibians from Trinidad and Tobago, B.W.I. I–II. *Proceedings of the Koninklijke Nederlandse Akademie van Wetenschappen*, Amsterdam 59C(2), 165–188, 4 figs.

Brongersma, L.D. (1956b) On two species of boid snakes from the Lesser Sunda Islands. *Proceedings of the Koninklijke Nederlandse Akademie van Wetenschappen*, Amsterdam 59C(3), 290–300, 3 figs.

Brongersma, L.D. (1956c) Notes on New Guinea reptiles and amphibians, IV. *Proceedings of the Koninklijke Nederlandse Akademie van Wetenschappen*, Amsterdam 59C(4), 447–453, 3 figs.

Brongersma, L.D. (1956d) Notes on New Guinean reptiles and amphibians. V. *Proceedings of the Koninklijke Nederlandse Akademie van Wetenschappen*, Amsterdam 59C(5), 599–610, 6 figs.

Brongersma, L.D. (1958) Note on *Vipera russelii* (Shaw). *Zoologische Mededelingen*, Leiden 36(4), 55–76, 3 pls.

Brongersma, L.D. (1959) Some snakes from the Lesser Antilles. *Studies on the fauna of Curaçao and Caribbean Islands* 9(37), 50–60, fig. 83, pls. 4–5.

Brongersma, L.D. (1966) Note on *Leptomicrurus collaris* (Schlegel) (Reptilia, Serpentes). *Zoologische Mededelingen*, Leiden 41(17), 243–254.

Brongersma, L.D. (1967) Poisonous snakes of Surinam. *Memórias do Instituto Butantan*, São Paulo (1966) 33(1), 73–79.

Brongersma, L.D. (1968) Proposal to suppress *Python timoriensis* Müller, 1844, and *Python timoriensis* Müller, 1857, and to add *Python timoriensis* (Peters, 1876) to the Official List (Reptilia, Serpentes, Boidae). *Bulletin of Zoological Nomenclature*, London 25(1), 55–59.

Brongersma, L.D. (1969) Note on *Liasis boeleni* Brongersma. *Proceedings of the Koninklijke Nederlandse Akademie van Wetenschappen*, Amsterdam 72C(2), 124–128, 2 pls.

Brongersma, L.D. (1972) On the "Histoire naturelle des serpens" by de la Cépède, 1789 and 1790, with a request to reject this work as a whole, and with proposals to place seven names of snakes, being nomina oblita, on the Official Index of rejected and invalid names in zoology, and to place three names of snakes on the Official List of specific names in zoology (class Reptilia). *Bulletin of Zoological Nomenclature* 29(1), 44–61.

Brongersma, L.D. & Helle, W. (1951) Notes on Australian snakes, I. *Proceedings of the Koninklijke Nederlandse Akademie van Wetenschappen*, Amsterdam 54C(1), 1–8, 1 fig.

Brongersma, L.D. & Knaap-van Meeuwen, M.S. (1964) On a new species of *Denisonia* (Reptilia, Serpentes) from New Guinea. *Zoologische Mededelingen*, Leiden 39, 550–554, 3 figs.

Brongersma, L.D. & Wehlburg, C. (1933) Notes on some snakes from Sumatra. *Miscellanea Zoologica Sumatrana*, Medan (74), 1–6, 1 pl.

Brongniart, A. (1800) Essai d'une classification naturelle des reptiles. Iere. partie. Etablissement des orders. *Bulletin des Sciences, par la Société Philomatique,* Paris 2(35), 81–82.

Bronikowski, A.M. & Arnold, S.J. (2001) Cytochrome b phylogeny does not match subspecific classification in the western garter snake, *Thamnophis elegans*. *Copeia* 2001(2), 508–513, 2 figs.

Brooke, R.K. (1992) The date of publication of certain names of reptiles and amphibians proposed by Andrew Smith. *Journal of the Herpetological Association of Africa* (41), 9.

Brothers, D.R. (1992) *An introduction to snakes of the Dismal Swamp region of North Carolina and Virginia*. Edgewood PROBES, Boise, 139 pp., 32 col. pls.

Brown, A.E. (1889) Description of a new species of *Eutaenia*. *Proceedings of the Academy of Natural Sciences of Philadelphia* 41, 421–422.

Brown, A.E. (1890) On a new genus of Colubridae from Florida. *Proceedings of the Academy of Natural Sciences of Philadelphia* 42, 199–200.

Brown, A.E. (1893) Notes on some snakes from tropical America lately living in the collection of the zoological society of Philadelphia. *Proceedings of the Academy of Natural Sciences of Philadelphia* 45(3), 429–435.

Brown, A.E. (1902a) A collection of reptiles and batrachians from Borneo and the Loo Choo Islands. *Proceedings of the Academy of Natural Sciences of Philadelphia* 53(1), 175–186.

Brown, A.E. (1902b) A new species of *Coluber* from western Texas. *Proceedings of the Academy of Natural Sciences of Philadelphia* (1901) 53(3), 492–495, pl. 29.

Brown, A.E. (1902c) A new species of *Ophibolus* from western Texas. *Proceedings of the Academy of Natural Sciences of Philadelphia* (1901) 53(3), 612–613, pl. 34.

Brown, B.C. (1950) *An annotated check list of the reptiles and amphibians of Texas*. Baylor University Press, Waco, xii + 257 pp.

Brown, B.C. & Smith, H.M. (1942) A new subspecies of Mexican coral snake. *Proceedings of the Biological Society of Washington* 55, 63–66.

Brown, C.W. & Ernst, C.H. (1986) A study of variation in eastern timber rattlesnakes, *Crotalus horridus* Linnae (Serpentes, Viperidae). *Brimleyana* 12, 57–74, illust.

Brown, F.M. (1941) A gazetteer of entomological collecting stations in Ecuador. *Annals of the Entomological Society of America* 34, 809–851.

Brown, H.A., Bury, R.B., Darda, D.M., Diller, L.V., Peterson, C.R. & Storm, R.M. (1995) *Reptiles of Washington and Oregon*. Seattle Audubon Society, viii + 176 pp., col. pls.

Brown, J.H. (1973) *Toxicology and pharmacology of venoms from poisonous snakes.* Charles C. Thomas, Springfield, xiv + 184 pp., 12 col. pls.

Brown, J.H. & Bowles, M.E. (1966) Studies on the phospholipase A2 activity of *Crotalus atrox* venom. *Toxicon* 3(3), 205–212, 4 figs.

Brown, P.R. (1997) *A field guide to snakes of California*. Gulf Publishing, Houston, viii +209 pp., 6 figs., 37 col. pls.

Brown, P.R. & Wright, J.W. (Eds.). (1994) *Herpetology of the North American deserts, proceedings of a symposium.* Southwestern Herpetological Society, Serpent's Tale Books, Excelsior, iv + 311 pp., illust.

Brown, R.M., Ferner, J.W., Sison, R.V., Gonzales, P.C. & Kennedy, R.S. (1996) Amphibians and reptiles of the Zambales Mountains of Luzon Island, Republic of the Philippines. *Herpetological Natural History* 4(1), 1–22, 20 figs.

Brown, R.M., Leviton, A.E. & Sison, R.V. (1999) Description of a new species of *Pseudorabdion* (Serpentes, Colubridae) from Panay Island, Philippines with a revised key to the genus. *Asiatic Herpetological Research* 8, 7–12, 4 figs.

Brown, R.M., McGuire, J.A., Ferner, J.W., Icarangal, N., Jr. & Kennedy, R.S. (2000) Amphibians and reptiles of Luzon Island, II. Preliminary report on the herpetofauna of Aurora Memorial National Park, Philippines. *Hamadryad* 25(2), 175–195.

Brown, R.M., Leviton, A.E., Ferner, J.W. & Sison, R.V. (2001) A new snake of the genus *Hologerrhum* Günther (Reptilia, Squamata, Colubridae) from Panay Island, Philippines. *Asiatic Herpetological Research* 9, 9–22, 7 figs.

Brown, T.W. & Lillywhite, H.B. (1992) Autecology of the Mojave desert sidewinder, *Crotalus cerastes cerastes*, at Kelso dunes, Mojave desert, California, USA. *In:* Campbell, J.A. & Brodie, E.D., Jr. (Eds.), *Biology of the pitvipers*. Selva, Tyler, pp. 279–308, 8 figs.

Brown, W.C. & Alcala, A.C. (1964) Relationship of the herpetofauna of the non-dipterocarp communities to that of the dipterocarp forest on southern Negros Island, Philippines. *Senckenbergiana Biologica*, Frankfurt-am-Main 45(3–5), 591–611, 8 figs.

Brown, W.C. & Alcala, A.C. (1970) The zoogeography of the herpetofauna of the Philippine Islands, a fringing archipelago. *Proceedings of the California Academy of Sciences* 38, 105–130, 2 figs.

Brown, W.C. & Alcala, A.C. (1986) Comparison of the herpetofaunal species richness on Negros and Cebu islands, Philippines. *Silliman Journal*, Dumaguete City 33(1–4), 74–86, 2 figs.

Brown, W.C. & Fehlmann, A. (1958) A new genus and species of arboreal scincid lizards from the Palau Islands. *Occasional Papers of the Natural History Museum of Stanford University,* Palo Alto (6), 1–7, 7 figs..

Brown, W.C. & Leviton, A.E. (1961) Discovery of the snake genus *Opisthotropis* in the Philippine Islands, with description of a new species. *Occasional Papers of the Natural History Museum of Stanford University* (8), 1–5, 1 fig.

Brown, W.S. (1993) Biology, status, and management of the timber rattlesnake (*Crotalus horridus*), a guide for conservation. Society for the Study of Amphibians and Reptiles *Herpetological Circular* (22), vi + 78 pp., 16 col. figs.

Browne, P. (1756) *Civil and natural history of Jamaica. In three parts. Containing, I. An accurate description of that island, its situation and soil; with a brief account of its former and present state, government, revenues, produce, and trade. II. A history of the natural productions, including the various sorts of native fossils; perfect and imperfect vegetables; quadrupedes, birds, fishes, reptiles and insects; with their properties and uses in mechanics, diet, and physic. III. An account of the nature of climates in general, and their different effects upon the human body; with a detail of the diseases arising from this source, particularly within the tropics*. Patrick Browne, London, 503 pp., 50 pls.

Bruck, G. (1968) Zur herpetofauna Afghanistans. *Vestnik Ceskoslovenske Spolecnosti Zoologicke, Acta Zoologicae Bohemoslovacae* 32, 201–208.

Brüggemann, C. & Wigand, O. (1838) *Die Naturgeschichte in getreuen Abbildungen und mit ausfuhrlicher Beschreibung derselben. Band I. Amphibien.* Verlag von Carl Brüggemann und Otto Wigand, Halberstadt, 100 pp., 74 pls.

Brugnière, M. (1784) Description d'une espèce particulière de serpent à Madagascar. *Observations sur la Physique, sur l'Histoire Naturelle et sur les Arts* (*Journal Physique*), Paris 24(1), 132–133, pl. 2.

Bruner, J.C. (1991) A catalogue of type specimens of fossil vertebrates in the Field Museum of Natural History. Classes Amphibia, Reptilia, Aves, and Ichnites. *Fieldiana: Geology* (22), 1–51.

Bruner, K. (1908) Die cettische Ringelnatter, *Tropidonotus natrix* subspec. *cettii* (Gené). *Blätter für Aquarien- und Terrarienkunde*, Stuttgart 19, 3–7, 1 pl.

Brunner, A. (1968) Über die black forest cobra, *Pseudohaje goldii*. *Salamandra*, Frankfurt-am-Main 4(2–3), 56–59, 2 figs.

Brunner, G. (1954) Das Fuchsloch bei Siegmannsbrunn (Oberfr.). *Neues Jahrbuch für Geologie und Paläontologie Abhandlung*, Stuttgart 100(1), 83–118, 20 figs.

Brunner, G. (1957) Die Breitenberghöhle bei gössweinstein Ofr. Eine Mindel-Riss-und eine postglaziale Mediterran-Fauna. *Neues Jahrbüch für Geologie und Paläontologie, Monatshefte*, Stuttgart 1957(7–8), 352–378, 17 figs.

Bruno, S. (1966) Sulle specie del genere *Coronella* Laurenti viventi in Italia (Terzo contriubto alla conoscenza degli Ofidi italiani). *Atti della Accademia Gioenia di Scienze Naturali in Catania* (Series 6) 18, 1–117, 5 figs.

Bruno, S. (1967) La *Vipera ursinii* (Bonaparte 1853) in Italia (Quinto contributa alla conoscenza degli Ofidi itlaiani). *Memorie del Museo Civico di Storia Naturale*, Verona 15, 105–125, fig. 1, 1 pl.

Bruno, S. (1968a) I serpenti Europei della collezione Edoardo De Betta conservata nel Civico Museo di Storia Naturale di Verona. Considerazioni sui serpenti d'Italia. *Memorie del Museo Civico di Storia Naturale*, Verona (1967) 15, 173–211, 1 fig.

Bruno, S. (1968b) Sulla *Vipera ammodytes* (Linnaeus 1758) in Italia (VI contributo alla conoscenza degli Ofidi italiani). *Memorie del Museo Civico di Storia Naturale*, Verona (1967) 15, 269–336, 6 figs., 4 pls.

Bruno, S. (1969) Morfologia distribuzione e biologie di *Elaphe situla* (Linnaeus 1758) (Reptilia, Serpentes, Colubridae) (Studi sulla Fauna Erpetologica Italiena. X). *Atti della Accademia Gioenia di Scienze Naturali in Catania* (Series 7) 1, 1–44, 2 pls.

Bruno, S. (1970) Anfibi e rettili di Sicilia (Studi sulla Fauna Erpetologica Italiana. XI). *Atti della Accademia Gioenia di Scienze Naturali in Catania* (Series 7) 2, 1–144, 38 figs., 1 pl.

Bruno, S. (1975) Note riassuntive sull'erpetofauna dell'isola di Montecristo (Archipelago Toscano, Mare Tirreno). *Lavori della Societa Italiana di Biogeografia* (Série 2) (1974) 5, 1–98, 45 figs.

Bruno, S. (1977) Gli anfibi e rettili dei Monti della Tolfa (Antiappennino Laziale). *Quaderni dell'Accademia Nazionale dei Lincei*, Roma 227, 89–124.

Bruno, S. (1980) I serpenti del Veneto (Italia NE). I—Morfologia, tassonomia, geonemia (Studi sulla fauna erpetologica Italiana. XXXV). *Lavori-Società Veneziana di Scienze Naturali*, Venezia 5 (Supplement), 1–70, 22 figs.

Bruno, S. (1984) *Guida ai serpenti d'Italia*. Giuinti Martello, Firenze, 191 pp., 78 figs.

Bruno, S. (1985) *Le vipere d'Italia e d'Europa*. Edagricole, Bologna, 269 pp., 123 figs., 12 pls.

Bruno, S. (1988a) L'erpetofauna delle isole di Cres, Krk e Ada (Jugoslavia-Albania). *Bulletin d'Ecologie*, Paris 19(2–3), 265–281, illust.

Bruno, S. (1988b) Considerazioni sull'erpetofauna della Sicilia. *Bulletin d'Ecologie*, Paris 19(2–3), 283–303, illust.

Bruno, S. (1989) Introduction to a study of the herpetofauna of Albania. *British Herpetological Society Bulletin*, London (29), 16–41, illust.

Bruno, S. (1998) *Serpenti*. Giunti Gruppo Editoriale, Firenze, 255 pp., 89 figs.

Bruno, S. & Cesare, E. (1991) The herpetofauna of the south-east Peligna region (Abruzzo, Italy). *British Herpetological Society Bulletin*, 35, 20–34, illust.

Bruno, S. & Hotz, H. (1976) *Coluber hippocrepis* auf der insel Sardinien. *Salamandra* 12(2), 69–86.

Bruno, S. & Maugeri, S. (1977) *Rettili d'Italia. Serpenti*. Aldo Martello, Giunti Editore, Firenze, 207 pp., 125 figs.

Bruno, S. & Maugeri, S. (1990) *Serpenti d'Italia e d'Europa*. Editoriale Giorgio Mondadori, Milano, 223 pp., unnumbered figs. & col. illust.

Bruton, M.N. & Haacke, W.D. (1975) New reptile records from the tropical transition zone of south-east Africa. *Lammergeyer* 22, 23–32.

Bruton, M.N. & Haacke, W.D. (1980) The reptiles of Maputaland. *In:* Bruton, M.N. & Cooper, K.H. (Eds.), *Studies on the ecology of Maputaland*. Rhodes University & Natal Branch of the Wildlife Society of South Africa, pp. 251–287, illust..

Bryant, L.J. (1989) Non-dinosaurian lower vertebrates across the Cretaceous-Tertiary boundary in northeastern Montana. *University of California Publications in Geological Science* 134, i–xi + 1–107, illust.

Brygoo, E.R. (1977) Bibliographie de la zoologie parasitaire des vertébrés de Madagascar. *Archives de l'Institut Pasteur de Madagascar* 44(1), 245–288.

Brygoo, E.R. (1983) Les ophidiens de Madagscar. *Memórias do Instituto Butantan*, São Paulo (1982) 46, 19–58

Brygoo, E.R. (1987) L'endemisme des reptiles de Madagascar. *Bulletin de la Société Zoologique de France*, Paris 112(1–2), 5–38.

Bryson, R.W., Jr. (2006) Geographic distribution: Serpentes: *Crotalus triseriatus triseriatus*. *Herpetological Review* 37(4), 497.

Bryson, R.W., Jr. & Mendoza-Quijano, F. (2007) Cantils of Hidalgo and Veracruz, Mexico, with comments on the validity of *Agkistrodon bilineatus lemosespinali*. *Journal of Herpetology* 41(3), 536–539, fig. 1.

Bryson, R.W., Jr. & Mueller, J.M. (2001) Geographic distribution: Serpentes: *Crotalus lepidus lepidus*. *Herpetological Review* 32(2), 122.

Bryson, R.W., Jr., Salmon, G.T. & Lazcano-Villarreal, D. (2001) Geographic distribution: Serpentes: *Lampropeltis mexicana*. *Herpetological Review* 32(2), 123.

Bryson, R.W., Jr., Dixon, J.R. & Lazcano-Villarreal, D. (2005) New species of *Lampropeltis* (Serpentes, Colubridae). *Journal of Herpetology* 39(2), 207–214, 4 figs.

Bryson, R.W., Jr., Pastorini, J., Burbrink, F.T. & Forstner, M.R.J. (2007) A phylogeny of the *Lampropeltis mexicana* complex (Serpentes, Colubridae) based on mitochondrial DNA sequences suggests evidence for species-level polyphyly within *Lampropeltis*. *Molecular Phylogenetics and Evolution* 43(2), 674–684, 4 figs.

Bryson, R.W., Jr., Nieto-Montes de Oca, A. & Velasco, J.R. (2008) Phylogenetic position of *Porthidium hespere* (Viperidae, Crotalinae) and phylogeography of arid-adapted hognosed pitvipers based on mitochondrial DNA. *Copeia* 2008(1), 172–178, 2 figs.

Bryson R.W., Jr., Murphy, R.W., Lathrop, A. & Lazcano-Villarreal, D. (2011) Evolutionary drivers of phylogeographical diversity in the highlands of Mexico: a case study of the *Crotalus triseriatus* species group of montane rattlesnakes. *Journal of Biogeography* 38, 697–710, 6 figs.

Bücherel, W. & Buckley, E.E. (Eds.). (1971) *Venomous animals and their venoms. Volume II. Venomous vertebrates.* Academic Press, New York, 687 pp., illust.

Bücherel, W., Buckley, E.E. & Deulofeu, V. (Eds.). (1968) *Venomous animals and their venoms. Volume I. Venomous vertebrates.* Academic Press, New York, 707 pp., illust.

Buchholz, K.F. (1955) Herpetologische Ergebnisse zweier Cycladenreisen. *Bonner Zoologische Beiträge*, Bonn 6(1–2), 95–110, 1 fig.

Buchholz, K.F. (1961) Über *Elaphe quatuorlineata* von den Zykladen und das *Elaphe*—Taxon von der Insel Amorgos (Reptilia, Colubridae). *Bonner Zoologische Beiträge*, Bonn 12(1–2), 68–86, 1 fig.

Buchholz, K.F. & Schultze-Westrum, T. (1964) Zur Kenntnis der Schlangenfauna der Nördlichen Sporaden. *Zoologischer Anzieger*, Leipzig 173(2), 127–136, 3 figs.

Buckley, E.E. & N. Porges (1956) *Venoms*. Government Printing Office, Washington, xii + 467 pp.

Buckner, S.D. (1998b) Geographic distribution: Serpentes: *Thamnophis sirtalis sirtilis. Herpetological Review* 29(1), 55.

Buckner, S.D. & Franz, R. (1998a) Geographic distribution: Serpentes: *Thamnophis sauritus sackenii. Herpetological Review* 29(1), 55.

Budak, A., Atatür, M.K. & Tok, C.V. (2000) *Amphibians and reptiles of Resadiye (Datca) Peninsula.* Ege Üniversitesi Fen Fakültesi Biyoloji Bölümü, Bornova (165), 1–45, 29 col. pls.

Buden, D.W. (1966) An evaluation of Jamaican *Dromicus* (Serpentes, Colubridae) with the description of a new species. *Breviora* (238), 1–10, 3 figs.

Buden, D.W. (1975) Notes on *Epicrates chrysogaster* (Serpentes, Boidae) of the southern Bahamas, with description of a new subspecies. *Herpetologica* 31(2), 167–177, 4 figs.

Bühler, W. (1981) *Kennst du mich? Band 4. Amphibien und Reptilien.* Verlag Sauerländer, Aarau, 127 pp., illust.

Buide, M.S. (1966) Reptiles de la Península Hicacos. *Poeyana* A(21), 1–21.

Buide, M.S. (1967) Lista de los anfibios y reptiles de Cuba. *Torreia*, Habana (Série 2) (1), 1–60.

Bullock, D.J. (1977) Round Island—a tale of destruction. *Oryx*, London 14(1), 51–58, 3 figs.

Bullock, D.J. (1986) The ecology and conservation of reptiles on Round Island and Gunner's Quoin, Mauritius. *Biological Conservation* 37(2), 135–156, 1 fig.

Bullock, R.E. & Tanner, W.W. (1966) A comparative osteological study of two species of Colubridae (*Pituophis* and *Thamnophis*). *Brigham Young University Science Bulletin (Biological Series)* 8(3), 1–29, 11 figs.

Bullock, T.H. & Barrett, R. (1968) Radiant heat reception in snakes. *Communications in Behavioral Biology* 1A(1), 19–29, 4 figs.

Bulmer, R.N.H., Menzies, J.I. & Parker, F.S. (1975) Kalan classification of reptiles and fishes. *Journal of Polynesian Society* 84(3), 267–308.

Bund, C.F. van den (1964) De verspreiding van de reptilen en amphibieen in Nederland. *Lacerta*, Gravenhage 22, 1–72.

Buongerminip, E. & Waller, T. (1998) Geographic distribution: Serpentes: *Hydrops triangularis. Herpetological Review* 29(2), 113.

Buongerminip, E. & Waller, T. (1999) Geographic distribution: Serpentes: *Bothrops moojeni. Herpetological Review* 30(1), 53.

Burbrink, F.T. (2001) Systematics of the eastern ratsnake complex (*Elaphe obsoleta*). *Herpetological Monographs* (15), 1–53, 16 figs.

Burbrink, F.T. (2002) Phylogeographic analysis of the cornsnake (*Elaphe guttata*) complex as inferred from maximum likelihood and Bayesian analysis. *Molecular Phylogenetics and Evolution* 25(3), 465–476, 4 figs.

Burbrink, F.T. (2005) Inferring the phylogenetic position of *Boa constrictor* among the Boinae. *Molecular Phylogenetics and Evolution* 34(1), 167–180, 5 figs.

Burbrink, F.T. & Lawson, R. (2007) How and when did Old World ratsnakes disperse into the New World? *Molecular Phylogenetics and Evolution* 43(1), 173–189, 4 figs.

Burbrink, F.T., Lawson, R. & Slowinski, J.B. (2000) Mitochondrial DNA phylogeography of the polytypic North American rat snake (*Elaphe obsoleta*), a critique of the subspecies concept. *Evolution* 54(6), 2107–2118, 5 figs.

Burbrink, F.T., Fontanella, F., Pyron, R.A., Guiher, T.J. & Jimenez, C. (2008) Phylogeography across a continent, the evolutionary and demographic history of the North American racer (Serpentes, Colubridae, *Coluber constrictor*). *Molecular Phylogenetics and Evolution* 57(1), 274–288, 7 figs.

Burbrink, F.T., Yao, H., Ingrasci, M., Bryson Jr., R.W., Guiher, T.J. & Ruane, S. (2011) Speciation at the Mogollon Rim in the Arizona mountain kingsnake (*Lampropeltis pyromelana*). *Molecular Phylogenetics and Evolution* 60, 445–454, 4 figs.

Burbrink, F.T., Ruane, S. & Pyron, A. (2012) When are adaptive radiations replicated in areas? Ecological opportunity and unexceptional diversification in West Indian dipsadine snakes (Colubridae: Alsophiini). *Journal of Biogeography* 39, 465–475.

Burchell, W.J. (1822–1824) *Travels in the interior of southern Africa. 2 vols.* Longman, Hurst, Rees, Orme, Brown and Green, London, viii + 582 + 648 pp., 10 pls.

Buresch, I. & Zonkow, J. (1934) Untersuchungen über die Verbreitung der Reptilien und Amphibien in Bulgarien und auf der Balkanhalbinsel. II. Teil. Schlangen (Serpentes). *Mitteilungen aus den königlich Naturwissenschaftlichen Instituten in Sofia* 7, 106–188, 37 figs.

Burger, R.M. (1995) An arboreal burrower: the dwarf boa *Ungaliophis. The Vivarium* 7(2), 46–49, 6 col. photos.

Burger, R.M. (1996) Natural history and husbandry of the Caribbean dwarf boas of the genus *Tropidophis. The Vivarium* 8(3), 6–11, 8 col. photos.

Burger, R.M. (1997) Miscellaneous notes on the Caicos Islands dwarf boa, *Tropidophis greenwayi:* reproduction, locomotion and maximum size. *Bulletin of the Chicago Herpetological Society* 32(2), 26–27, illust.

Burger, R.M. (2001) The herpetofauna of Caño Palma Biological Station, Tortuguero, Costa Rica. *Bulletin of the Chicago Herpetological Society* 36(12), 243–253.

Burger, W.L. (1947) A taxonomic and statistical study of the keeled green snake, *Opheodrys aestivus. Bulletin of the Ecological Society of America* 28(5), 54.

Burger, W.L. (1950) A preliminary study of the subspecies of the jumping viper, *Bothrops nummifer. Bulletin of the Chicago Academy of Sciences* 9(3), 59–67, pl. 1.

Burger, W.L. (1955) A new subspecies of the coral snake, *Micrurus lemniscatus*, from Venezuela, British Guiana and Trinidad, and a key for the identification of associated species of coral snakes. *Boletín del Museo de Ciencias Naturales*, Caracas 1(2), 1–18.

Burger, W.L. (1971) *Genera of pitvipers (Serpentes, Crotalidae).* Doctoral Dissertation, University of Kansas, Lawrence, 186 pp., 9 figs.

Burger, W.L. & Natsuno, T. (1974) A new genus for the Arafura smooth seasnake and redefinitions of other seasnake genera. *The Snake* 6(2), 61–75.

Burger, W.L. & Robertson, W.B. (1951) A new subspecies of the Mexican mocassin, *Agkistrodon bilineatus. University of Kansas Science Bulletin* 34(5), 213–218.

Burger, W.L & Werler, J.E. (1954) The subspecies of the ring-necked coffee snake, *Ninia diademata*, and a short biological and taxonomic account of the genus. *University of Kansas Science Bulletin* 36(10), 643–672, 2 figs.

Burkett, R.D. (1966) Natural history of cottonmouth moccasin, *Agkistrodon pisciviorus* (Reptilia). *University of Kansas Publications, Museum of Natural History*, Lawrence 17(9), 435–491, 7 figs.

Burmeister, C.H.C. (1861) Reise durch die La Plata-Staaten, mit besonderer Rücksicht auf die physische Beschaffenheit und den Culturzustand der Argentinischen Republik. Zweiter Band. Ausgeführt in den Jahren 1857, 1858, 1859 und 1860. H.W. Schmidt, Halle, v + 538 pp.

Burmeister, G.H.C. (1885) Examen critico de los mamiferos y reptiles fosiles denoinados por Dr. Augusto Bravard y mencionados en su obra precendente. *Anales del Museo Nacional de Buenos Aires* 3(2), 95–173, 4 figs., pl. 1.

Burns, G. & Heatwole, H. (1998) Home range and habitat use of the olive sea snake, *Aipysurus laevis*, on the Great Barrier Reef, Australia. *Journal of Herpetology* 32(3), 350–358, 3 figs.

Burns, G. & Heatwole, H. (2000) Growth, sexual dimorphism, and population biology of the olive sea snake, *Aipysurus laevis*, on the Great Barrier Reef of Australia. *Amphibia-Reptilia,* Leiden 21(3), 289–300, 2 figs.

Burt, C.E. (1936) The nomenclature of western coral king snakes *Lampropeltis zonata* L. versus *L. multicincta. Copeia* 1936, 94–98.

Burt, C.E. (1946) Description of *Elaphe quivira*, a new rat-snake from Kansas. *Transactions of the Kansas Academy of Sciences* 49, 116.

Burt, C.E. & Burt, M.D. (1932) Herpetological results of the Whitney South Sea Expedition. VI. Pacific Island amphibians and reptiles in the collection of the American Museum of Natural History. *Bulletin of the American Museum of Natural History* 63(5), 461–597, 38 figs.

Burton, E. St. J. (1933) Faunal horizons of the Barton Beds in Hampshire. *Proceedings of the Geologists' Association* 44, 131–167.

Burton, J.A. (1991) *The book of snakes.* Australian Book Centre, Summer Park, 144 pp., col. illust.

Busack, S.D. (1966) Notes on a herpetological collection from the Azuero Peninsula, Panama. *Copeia* 1966(2), 371.

Busack, S.D. (1977) Zoogeography of amphibians and reptiles in Cadiz Province, Spain. *Annals of the Carnegie Museum* (1976–1977) 46(17), 285–316, 30 figs.

Busack, S.D. & McCoy, C.J. (1990) Distribution, variation and biology of *Macroprotodon cucullatus* (Reptilia, Colubridae, Boiginae). *Annals of the Carnegie Museum* 59(4), 261–285, 4 figs.

Busby, W.H., Parmelee, J.R., Dwyer, C.M., Hooper, E.D. & Irwin, K.J. (1994) *A survey of the herpetofauna on the Fort Riley Military Reservation, Kansas.* Report of the State Biological Survey of Kansas, University of Kansas, Lawrence, 78 pp., 8 figs.

Bush, B. (1981) *Reptiles of the Kalgoorlie-Esperance region.* Vanguard Press, Perth, 46 pp., 52 col. figs.

Bush, B. (1989) Ontogenetic colour change in the gwardar, *Pseudonaja nuchalis. Western Australian Naturalist* 18(2), 25–29, illust.

Bush, B., Browne-Cooper, R. & Robinson, D. (1995) *A guide to the reptiles and frogs of the Perth region.* University of Western Australian Press, Nedlands, xiv + 226 pp., col. pls.

Bushar, L.M., Maliga, M. & Reinert, H.K. (2001) Cross-species and amplification of *Crotalus horridus* microsatellites and their application in phylogenetic analysis. *Journal of Herpetology* 35(3), 532–537, 1 fig.

Bussarawit, S., Rasmussen, A.R. & Andersen, M. (1989) A preliminary study on sea snakes (Hydrophiidae) from Phuket Harbour, Phuket Island, Thailand. *Natural History Bulletin of the Siam Society*, Bangkok 37(2), 209–225, 13 figs.

Butler, A.L. (1899a) Note on the white snake (*Coluber taeniurus*), Cope. *The Journal of the Bombay Natural History Society* (1898–1899) 12(2), 424–425.

Butler, A.L. (1899b) Further notes on the varieties of *Coluber taeniurus. The Journal of the Bombay Natural History Society* (1898–1899) 12(2), 425–426.

Butler, A.W. (1887) Department of zoological miscellany: herpetology. *Journal of the Cincinnati Society of Natural History* 9, 263–265.

Butler, J.A. & Reid, J.C. (1986) Habitat preferences of snakes in the southern Cross River State, Nigeria. *In:* Rocek, Z. (Ed.), *Studies in herpetology.* Charles University, Prague, pp. 483–488, 1 fig.

Butler, J.A. & Reid, J.C. (1990) Records of snakes from Nigeria. *Nigerian Field*, Woodchester 55(1–2), 19–40, illust.

Butler, J.A., Dalrymple, G.H. & Hinko, A. (1980) Early spring feeding habits and comments on the physiological ecology of the queen snake, *Regina septemvittata*, in central Ohio. *Bulletin of the New York Herpetological Society* 16(1), 3–9, 1 fig.

Buttle, D. (1986) Amphibians and reptiles on the Spanish island of Mallorca. *British Herpetological Society Bulletin*, London (18), 12–15, 1 fig.

Buttle, D. (1989) Notes on reptiles and amphibians of northeastern Greece and the island of Samothraki. *British Herpetological Society Bulletin*, London (29), 49–53, 1 fig.

Buttle, D. (1995) An introduction to reptiles and amphibians of the Greek islands. *International Reptilian Magazine* 3(7), 15–25.

Buttle, D. (1997) Observations on reptiles and amphibians of Andros (Cyclades, Greece). *British Herpetological Society Bulletin* 60, 5–12, illust.

Buurt, G. van. (2001) *De Amfibieen en Reptielen van Aruba, Curacao en Bonaire*. Uitgegeven door de schrijver, Curacao. [not seen]

Buurt, G. van. (2005) *Field guide to the amphibians and reptiles of Aruba, Curaçao and Bonaire*. Edition Chimaira, Frankfurt-am-Main, 137 pp., 80 col. figs.

Buys, P.J. & Buys, P.J.C. (1983) *Snakes of South West Africa.* Gamsberg Publishers, Windhoek, 64 pp., col. illust.

Cabela, A. (1982) *Catalogus faunae Austriae. Teil XXI ab. Amphibia, Reptilia*. Österreichischen Akademie der Wissenschaften, Wien, 17 pp.

Cabela, A. & Tiedemann, F. (1985) Atlas der Amphibien und Reptilien Osterreichs. *Neue Denkschrift der Naturhistorischen Museums in Wien* 4, 1–80, illust.

Cabela, A., Grillitsch, H., Happ, H., Happ, F. & Kollar, R. (1992) Die Kriechtiere Kärntens. *Carinthia II*, Klagenfurt 182, 195–316, 56 figs.

Cabela, A., Grillitsch, H. & Tiedemann, F. (Eds.). (2001) *Atlas zur Verbreitung und Ökologie der Amphibien und Reptilien in Österreich.* Naturhistorischen Museums, Wien, 880 pp., col. illust.

Cabral, H. & Caballero, A. (2012) Geographic distribution: Serpentes: *Pseudoeryx plicatilis*. *Herpetological Review* 43(4), 622.

Cabrera, M.R. (2004) *Las serpientes de Argentina Central.* Publicaciones Universidad Nacional de Córdoba, 108 pp., 65 figs.

Cabrera, M.R. & Merlini, H.O. (1989) Nota científica, a new record of *Sibynomorphus turgidus* (Cope, 1868) from southern South America (Serpentes, Colubridae). *Iheringia (Zoologia)*, Pôrto Alegre 69, 151–153.

Cacciali, P. (2006) Las serpientes caracoleras (Colubridae: Dipsadini) de Paraguay. *Revista Española de Herpetología* 20, 71–85.

Cacciali, P. (2008) *Guía para la identificación de 60 serpientes del Paraguay.* Asociación Guyra Paraguay, Asunción, 218 pp., col. illust.

Cacciali, P. & Brusqueti, F. (2005) Geographic distribution: Serpentes: *Leptotyphlops unguirostris*. *Herpetological Review* 36(2), 203.

Cacciali, P. & Wüest, U. (2009) Reptilia, Squamata, Colubridae, *Lygophis meridionalis*: Type locality. *Check List* 5(3), 383–385, 1 fig.

Cacciali, P., Fernández, S. & Ramírez, F. (2005) Geographic distribution: Serpentes: *Drymoluber brazili*. *Herpetological Review* 36(3), 338.

Cacciali, P., Villalba, R. & Yanosky, A.A. (2007a) A new species of *Atractus* (Serpentes, Colubridae, Dipsadinae) from the Atlantic Forest of Alto Paraná, Paraguay. *South American Journal of Herpetology* 2(2), 83–88, 3 figs.

Cacciali, P., Carreira, S. & Scott, N. (2007b) Redescription of *Phalotris nigrilatus* Ferrarezzi, 1993 (Serpentes, Colubridae, Xenodontinae). *Herpetologica* 63(4), 552–559, 4 figs.

Cacciali, P., Scott, N.J., Guenther, R., Sawaya, R.J., Brusquetti, F. & Bauer, F. (2009) Taxonomic status of the false coral snake genus *Simophis* (Peters, 1860) (Serpentes: Colubridae: Colubrinae) from Paraguay and Brazil. *Journal of Herpetology* 43(4), 698–703, 3 figs.

Cacciali, P., Espinola, D, Vinales, S.C., Espinola, I.G. & Cabral, H. (2011) Squamata, Serpentes, *Micrurus silviae* Di-Bernardo, Borges-Martins and Silva, 2007: presence confirmation in Paraguay. *Check List* 7(4), 809–810, 3 figs.

Cacivio, P.M. (1997) Geographic distribution: Serpentes: *Thamnodynastes hypoconia*. *Herpetological Review* 28(3), 160.

Cacivio, P.M. (1999) Geographic distribution: Serpentes: *Clelia bicolor*. *Herpetological Review* 30(3), 174.

Cadle, J.E. (1982) *Evolutionary relationships among advanced snakes.* Doctoral Dissertation, University of California, Berkeley, 274 pp., 13 figs.

Cadle, J.E. (1983) Problems and approaches in the interpretation of the evolutionary history of venomous snakes. *Memórias do Instituto Butantan*, São Paulo (1982) 46, 255–274, 1 fig.

Cadle, J.E. (1984a) Molecular systematics of neotropical xenodontine snakes, I. South American xenodontines. *Herpetologica* 40(1), 8–20, 3 figs.

Cadle, J.E. (1984b) Molecular systematics of neotropical xenodontine snakes, II. Central American xenodontines. *Herpetologica* 40(1), 21–30, 1 fig.

Cadle, J.E. (1984c) Molecular systematics of neotropical xenodontine snakes. III. Overview of xenodontine phylogeny and the history of New World snakes. *Copeia* 1984(3), 641–652.

Cadle, J.E. (1985) The neotropical colubrid snake fauna (Serpentes, Colubridae), lineage components and biogeography. *Systematic Zoology* 34(1), 1–20, 5 figs.

Cadle, J.E. (1988) Phylogenetic relationships among advanced snakes: a molecular perspective. *University of California Publications in Zoology*, Berkeley (119), x + 77 pp., 7 figs.

Cadle, J.E. (1989) A new species of *Coniophanes* (Serpentes, Colubridae) from northwestern Perú. *Herpetologica* 45(4), 411–424, 7 figs.

Cadle, J.E. (1992) Phylogenetic relationships among vipers, immunological evidence. *In:* Campbell, J.A. & Brodie, E.D., Jr. (Eds.), *Biology of the pitvipers.* Selva, Tyler, pp. 41–48, 1 fig.

Cadle, J.E. (1994) The colubrid radiation in Africa (Serpentes, Colubridae), phylogenetic relationships and evolutionary patterns based on immunological data. *Zoological Journal of the Linnean Society*, London 110(2), 103–140, 3 figs.

Cadle, J.E. (1996a) Snakes of the genus *Liopholidophis* (Colubridae) from eastern Madagascar, new species, revisionary notes, and an estimate of phylogeny. *Bulletin of the Museum of Comparative Zoology* 154(5), 369–464, 42 figs.

Cadle, J.E. (1996b) Systematics of snakes of the genus *Geodipsas* (Colubridae) from Madagascar, with descriptions of new species and observations on natural history. *Bulletin of the Museum of Comparative Zoology* 155(2), 33–87, 17 figs.

Cadle, J.E. (1998) The identity of *Leptophis varius* Fischer, 1884, and placement of *Liopholidophis pinguis* Parker, 1925, in its synonymy. *Journal of Herpetology* 32(3), 434–437, 2 figs.

Cadle, J.E. (1999) The dentition, systematics, and phylogeny of *Pseudoxyrhopus* and related genera from Madagascar (Serpentes, Colubridae), with descriptions of a new species and a new genus. *Bulletin of the Museum of Comparative Zoology* 155(8), 381–443, 19 figs.

Cadle, J.E. (2005) Systematics of snakes of the *Dipsas oreas* complex (Colubridae: Dipsadinae) in western Ecuador and Peru with revalidation of *D. elegans* (Boulenger) and *D. ellipsifera* (Boulenger). *Bulletin of the Museum of Comparative Zoology* 158(3), 67–136, 23 figs.

Cadle, J.E. (2007) The snake genus *Sibynomorphus* (Colubridae: Dipsadinae: Dipsadini) in Peru and Ecuador with comments on the systematics of Dipsadini. *Bulletin of the Museum of Comparative Zoology* 158(5), 183–284, 46 figs.

Cadle, J.E. (2010) Systematics, natural history, and hemipenial morphology of *Dendrophidion brunneum* (Günther) (Serpentes, Colubridae), a poorly known snake from the Andes of Ecuador and Peru. *Zootaxa* (2433), 1–24, 6 figs.

Cadle, J.E. (2011) Hemipenial morphology in the North American snake genus *Phyllorhynchus* (Serpentes: Colubridae), with a review of and comparisons with natricid hemipenes. *Zootaxa* 3092, 1–25, 12 figs.

Cadle, J.E. (2012a) Cryptic species within the *Dendrophidion vinitor* complex in Middle America (Serpentes: Colubridae). *Bulletin of the Museum of Comparative Zoology* 160(4), 183–240, 23 figs.

Cadle, J.E. (2012b) Systematics of the Neotropical snake *Dendrophidion percarinatum* (Serpentes: Colubridae), with descriptions of two new species from western Colombia and Ecuador and supplementary data on *D. brunneum*. *Bulletin of the Museum of Comparative Zoology* 160(6), 259–344, 46 figs.

Cadle, J.E. & Myers, C.W. (2003) Systematics of snakes referred to *Dipsas variegata* in Panama and western South America, with revalidation of two species and notes on defensive behaviors in the Dipsadini (Colubridae). *American Museum Novitates* (3409), 1–47, 21 figs.

Cadle, J.E. & Sarich, V.M. (1981) An immunological assessment of the phylogenetic position of New World coral snakes. *Journal of Zoology*, London 195(2), 157–167, 1 fig.

Cadle, J.E. & Savage, J.M. (2012) Systematics of the *Dendrophidion nuchale* complex (Serpentes: Colubridae) with the description of a new species from Central America. *Zootaxa* (3513), 1–50, 29 figs.

Cadle, J.E., Dessauer, H.C., Gans, C. & Gartside, D.F. (1990) Phylogenetic relationships and molecular evolution in uropeltid snakes (Serpentes, Uropeltidae), allozymes and albumin immunology. *Biological Journal of the Linnean Society*, London 40(3), 293–320, 5 figs.

Cagle, F.R. (1941) *A key to the reptiles and amphibians of Illinois*. Museum of Natural and Social Sciences, Southern Illinois University, Carbondale, iv + 32 pp.

Cagle, F.R. (1952) *A key to the amphibians and reptiles of Louisiana*. Tulane University, New Orleans, 42 pp., 65 figs., 3 pls.

Cagle, F.R. (1968) Reptiles. *In:* Blair, W.F., Blair, A.P., Brodkorb, P., Cagle, F.R. & Moore, G.A. (Eds.), *Vertebrates of the United States*. Second editon. McGraw Hill, New York, pp. 213–268, 108 figs. [first edition, 1957]

Cahill, M., Heatwole, H.F. & Goldman, B. (1973) Ecology of the reefs. *Occasional Papers of the Department of Geography, University of Papua New Guinea*, Port Moresby 1, 209–229.

Calabresi, E. (1924) Ricerche sulla variazioni della *Vipera aspis* auct. in Italia. *Bollettino dell'Istituto di Zoologia*, Reale Università di Roma 2, 78–126.

Calabresi, E. (1927) Anfibi e Rettili raccolti nella Somalia dai Proff. G. Stefanini e N. Puccioni. *Atti della Società Italiana di Scienze Naturali e del Museo Civico di Storia Naturali di Milano* 66, 14–60, pl. 1.

Calcaño, D. & Barrio-Amorós, C.J. (2002) Geographic distribution: Serpentes: *Drymobius rhombifer*. *Herpetological Review* 33(1), 67.

Calcaño, D. & Barrio-Amorós, C.J. (2003a) Geographic distribution: Serpentes: *Dipsas variegata*. *Herpetological Review* 34(4), 388.

Calcaño, D. & Barrio-Amorós, C.J. (2003b) Geographic distribution: Serpentes: *Pseustes shropshirei*. *Herpetological Review* 34(4), 389–390.

Calcaño, D. & Barrio-Amorós, C.L. (2010) Geographic distribution: Serpentes: *Oxyrhopus leucomelas*. *Herpetological Review* 41(3), 381.

Calderon, R., Cedeño-Vázquez, J.R. & Pozo, C. (2003) New distributional records for amphibians and reptiles from Campeche, México. *Herpetological Review* 34(3), 269–272.

Caldwell, M.W. (2000) On the phylogenetic relationships of *Pachyrhachis* within snakes: a response to Zaher (1998). *Journal of Vertebrate Paleontology* 20, 187–190.

Caldwell, M.W. & Albino, A.M. (2001) Palaeoenvironment and palaeoecology of three Cretaceous snakes: *Pachyophis, Pachyrhachis,* and *Dinilysia*. *Acta Palaeontologica Polonica* 46(2), 203–218, 4 figs.

Caldwell, M.W. & Albino, A.M. (2002) Exceptionally preserved skeletons of the Cretaceous snake *Dinilysia patagonica* Woodward, 1901. *Journal of Vertebrate Paleontology* 22(4), 861–866, 3 figs.

Caldwell, M.W. & Lee, M.S.Y. (1997) A snake with legs from the marine Cretaceous of the Middle East. *Nature* 386, 705–709, 4 figs.

Call, R.E. (1891) On a new serpent from Iowa. *American Journal of Science*, New Haven (Series 3) 41, 297–299.

Calmette, A. (1908) *Venoms, venomous animals, and antivenomous serum-therapeutics*. John Bale, Sons & Danielsson, London, 403 pp., 125 figs.

Calmonte, T. & Ferri, V. (1987) Un serpent nuovo per la fauna italiana, il colubro scalare, *Elaphe scalaris* (Schinz, 1822) (Reptilia, Colubridae). *Atti della Società Italiana di Scienze Naturali e del Museo Civico di Storia Naturali di Milano* 128(3–4), 314–316, illust.

Camarillo, J.L. (1995) Distribution records for some amphibians and reptiles from Mexico. *Bulletin of the Maryland Herpetological Society* 31(4), 195–197.

Camarillo, J.L. & Campbell, J.A. (1993) A second confirmed population of the rare Mexican rattlesnake, *Crotalus transversus* (Serpentes: Viperidae). *Texas Journal of Science* 45, 178–179.

Camarillo, J.L. & Campbell, J.A. (2002) Observaciones sobre la historia natural de *Crotalus transversus* (Squamata: Viperidae). *Boletin de la Sociedad Herpetológica Mexicana* 10(1), 7–9.

Camarillo, J.L. & Smith, H.M. (1992) A handlist of the amphibians and reptiles of the state of México, Mexico. *Greater Cincinnati Herpetological Society, Contributions to Herpetology* 1992, 39–41.

Camerano, L. (1889) Monografia degli ofidi Italiani. Parte Prima. Viperidi. *Memorie della Reale Accademia delle Scienze (Fisiche, Matematiche e Naturali)*, Torino (1888) (Série 2) 39, 195–243, 2 pls., 77 figs. [reprinted in 1888 by Ermanno Loescher, Torino, 51 pp., 2 pls., 77 figs.]

Camerano, L. (1891) Monografia degli ofidi Italiani. Parte Seconda. Colubridi e monografia dei cheloni Italiani. *Memorie della Reale Accademia delle Scienze, (Fisiche, Matematiche e Naturali)*, Torino (Série 2) 41, 403–481, 2 pls., 57 figs. [reprinted in 1891 by Carlo Clausen, Torino, 81 pp., 2 pls., 57 figs.]

Cameron, E.E. & Cogger, H.G. (1992) The herpetofauna of the Weipa region, Cape York Peninsula. *Technical Report of the Australian Museum* (7), 1–200, 4 figs., col. pls.

Campbell, H.W. & Christman, S.P. (1972) Dangerous land snakes of southeast Asia. *In:* Pickwell, G.V & Evans, W.E. (Eds.), *Handbook of dangerous animals for field personnel*. Naval Undersea Center, San Diego, pp. 27–84, 43 figs.

Campbell, H.W. & Howell, T.R. (1965) Herpetological records from Nicaragua. *Herpetologica* 21(2), 130–140, 1 fig.

Campbell, H.W. & Simmons, R.S. (1962) Notes on some reptiles and amphibians from western Mexico. *Bulletin of the Southern California Academy of Sciences* 61(4), 193–203.

Campbell, J.A. (1976) A new terrestrial pit viper of the genus *Bothrops* (Reptilia, Serpentes, Crotalidae) from western Mexico. *Journal of Herpetology* 10(3), 151–160, 4 figs.

Campbell, J.A. (1977) *The distribution, variation, and natural history of the Middle American highland pitvipers, Bothrops barbouri and Bothrops godmani*. Master's Thesis, University of Texas at Arlington, Arlington, 125 pp.

Campbell, J.A. (1978) A new rattlesnake (Reptilia, Serpentes, Viperidae) from Jalisco, Mexico. *Transactions of the Kansas Academy of Sciences* 81(4), 365–370, 3 figs.

Campbell, J.A. (1982) *The biogeography of the cloud forest herpetofauna of Middle America, with special reference to the Sierra de las Minas of Guatemala*. Doctoral Dissertation, University of Kansas, Lawrence, 322 pp.

Campbell, J.A. (1985) A new species of highland pitviper of the genus *Bothrops* from southern México. *Journal of Herpetology* 19(1), 48–54, 3 figs.

Campbell, J.A. (1988a) *Crotalus transversus*. *Catalogue of American amphibians and reptiles*, pp. 450.1–450.3.

Campbell, J.A. (1988b) The distribution, variation, natural history, and relationships of *Porthidium barbouri* (Viperidae). *Acta Zoologica Mexicana*, México (Series 2) (26), 1–32, 8 figs.

Campbell, J.A. (1989) A new species of colubrid snake of the genus *Coniophanes* from the highlands of Chiapas, Mexico. *Proceedings of the Biological Society of Washington* 102(4), 1036–1044, 3 figs.

Campbell, J.A. (1998a) Comments on the identities of certain *Tantilla* (Squamata, Colubridae) from Guatemala, with the descriptions of two new species. *Scientific Papers, Natural History Museum, The University of Kansas* (7), 1–14, 5 figs.

Campbell, J.A. (1998b) *Amphibians and reptiles of northern Guatemala, the Yucatán, and Belize*. University of Oklahoma Press, Norman, xix + 380 pp., 189 figs.

Campbell, J.A. (2000) A new species of venomous coral snake (Serpentes, Elapidae) from high desert in Puebla, Mexico. *Proceedings of the Biological Society of Washington* 113(1), 291–297, 3 figs.

Campbell, J.A. (2001) The herpetofauna of the mesic upland forest of the Sierra de Las Minas and Montañas del Mico of Guatemala. *In:* Johnson, J.D., Webb, R.G. & Flores-Villela, OA. (Eds.), *Mesoamerican herpetology, systematics, zoogeography, and conservation*. Univ. Texas El Paso, pp. 80–92, 3 figs.

Campbell, J.A. & Armstrong, B.L. (1979) Geographic variation in the Mexican pygmy rattlesnake, *Sistrurus ravus*, with the description of a new subspecies. *Herpetologica* 35(4), 304–317, 8 figs.

Campbell, J.A. & Brodie, E.D., Jr. (1988) A new colubrid snake of the genus *Adelphicos* from Guatemala. *Herpetologica* 44(4), 416–422, 3 figs.

Campbell, J.A. & Brodie, E.D., Jr. (Eds.), (1992) *Biology of the pitvipers*. Selva, Tyler, xi + 467 pp., 17 pls.

Campbell, J.A. & Flores-Villela, O. (2008) A new long-tailed rattlesnake (Viperidae) from Guerrero, Mexico. *Herpetologica* 64(2), 246–257, 5 figs.

Campbell, J.A. & Ford, L.S. (1982) Phylogenetic relationships of the colubrid snakes of the genus *Adelphicos* in the highlands of Middle America. *Occasional Papers of the Museum of Natural History*, University of Kansas (100), 1–22, 7 figs.

Campbell, J.A. & Lamar, W.W. (1989) *The venomous reptiles of Latin America*. Comstock Publishing Associates, Cornell University Press, Ithaca, 425 pp., 443 figs.

Campbell, J.A. & Lamar, W.W. (1992) Taxonomic status of miscellaneous neotropical viperids, with the description of a new genus. *Occasional Papers of the Museum of Texas Tech University* (153), 1–31, 10 figs.

Campbell, J.A. & Lamar, W.W. (2004) *The venomous reptiles of the Western Hemisphere*. Cornell University Press, Ithaca, 2 vols., 870 pp., 281 figs., 1500 col. pls.

Campbell, J.A. & Murphy, J.B. (1977a) A new species of *Geophis* (Reptilia, Serpentes, Colubridae) from the Sierra de Coalcomán, Michoacán, Mexico. *Journal of Herpetology* 11(4), 397–403, 4 figs.

Campbell, J.A. & Murphy, J.B. (1977b) Miscellaneous notes on the reproductive biology of reptiles. 1. Two colubrid snake species from the Malagasy Republic, *Leioheterodon madagascariensis* and *Madagascariophis colubrina* (Reptilia, Serpentes, Colubridae). *Journal of Herpetology* 11(2), 228–230, 2 figs.

Campbell, J.A. & Murphy, J.B. (1984) Reproduction in five species of Paraguayan colubrids. *Transactions of the Kansas Academy of Sciences* 87(1–2), 63–65.

Campbell, J.A. & Smith, E.N. (1997) A new species of *Tantilla* (Serpentes, Colubridae) from northeastern Guatemala. *Proceedings of the Biological Society of Washington* 110(3), 332–337, 2 figs.

Campbell, J.A. & Smith, E.N. (1998) A new genus and species of colubrid snake from the Sierra de Las Minas of Guatemala. *Herpetologica* 54(2), 207–220, 5 figs.

Campbell, J.A. & Smith, E.N. (2000) A new species of arboreal pitviper from the Atlantic versant of northern Central America. *Revista de Biologia Tropical* 48(4), 1001–1013, 4 figs.

Campbell, J.A. & Solorzano, A. (1992) The distribution, variation, and natural history of the Middle American montane pitviper, *Porthidium godmani*. *In:* Campbell, J.A. & Brodie, E.D., Jr. (Eds.), *Biology of the pitvipers*. Selva, Tyler, pp. 223–250, 18 figs.

Campbell, J.A. & Vannini, J.P. (1988) Preliminary checklist of the herpetofauna of Finca El Faro, El Palmar, Quezaltenango, Guatemala. *Fundación Interamericana de Investigación Tropical, Publicación Ocasional*, Guatemala City (1), 1–10.

Campbell, J.A. & Vannini, J.P. (1989) Distribution of amphibians and reptiles in Guatemala and Belize. *Proceedings of the Western Foundation of Vertebrate Zoology*, Los Angeles 4(1), 1–21, 3 figs.

Campbell, J.A. & Whitmore, D.H., Jr. (1989) A comparison of the skin keratin biochemistry in vipers with comments on its systematic value. *Herpetologica* 45(2), 242–249, 4 figs.

Campbell, J.A., Ford, L.S. & Karges, J.P. (1983) Resurrection of *Geophis anocularis* Dunn with comments on its relationships and natural history. *Transactions of the Kansas Academy of Sciences* 86(1), 38–47, 2 figs.

Campbell, J.A., Camarillo, J.L. & Ustach, P.C. (1995) Redescription and rediagnosis of *Tantilla shawi* (Serpentes, Colubridae) from the Sierra Madre Oriental of Mexico. *Southwestern Naturalist* 40(1), 120–123, 1 fig.

Campbell, T. & Campbell, K.R. (1999) Geographic distribution: Serpentes: *Arrhyton exiguum exiguum*. *Herpetological Review* 30(2), 112.

Campden-Main, S.M. (1969a) The subspecies of *Calliophis maculiceps* (Günther). *British Journal of Herpetology*, London 4(3), 49–50.

Campden-Main, S.M. (1969b) The status of *Oligodon taeniatus* (Günther), 1861, and *Oligodon mouhoti* (Boulenger), 1914 (Serpentes, Colubridae). *Herpetologica* 25(4), 295–299.

Campden-Main, S.M. (1970a) *A field guide to the snakes of South Vietnam*. Smithsonian Institution, Washington (1969), vii + 114 pp., illust.

Campden-Main, S.M. (1970b) The identity of *Oligodon cyclurus* (Cantor, 1839) and revalidation of *Oligodon brevicauda* (Steindachner, 1867) (Serpents, Colubridae). *Proceedings of the Biological Society of Washington* 82(58), 763–766. (*Oligodon analepticos*)

Camper, J.D. (1996a) *Masticophis bilineatus*. *Catalogue of American amphibians and reptiles*, pp. 637.1–637.4, 2 figs.

Camper, J.D. (1996b) *Masticophis schotti*. *Catalogue of American amphibians and reptiles*, pp. 638.1–638.2, 2 figs.

Camper, J.D. (1996c) *Masticophis taeniatus*. *Catalogue of American amphibians and reptiles*, pp. 639.1–639.6, 2 figs.

Camper, J.D. & Dixon, J.R. (1990) High incidence of melanism in *Masticophis taeniatus girardi* (Reptilia, Colubridae), from the Cuatro Cienegas Basin of Coahuila, Mexico. *Texas Journal of Science* 42(2), 202–204.

Camper, J.D. & Dixon, J.R. (1994) Geographic variation and systematics of the striped whipsnakes (*Masticophis taeniatus* complex; Reptilia; Serpentes, Colubridae). *Annals of the Carnegie Museum* 63(1), 1–48, 20 figs.

Campos-Nogueira, C. de. (2001) New records of squamate reptiles in central Brazilian Cerrado. *Herpetological Review* 32(4), 285–287.

Campos-Rodríguez, J.I. & Mendoza-Quijano, F. (2004) Geographic distribution: Serpentes: *Crotalus ravus*. *Herpetological Review* 35(3), 291.

Campos-Rodríguez, J.I., Chambert, L., del Carmen-Díaz, M., Elizalde, C., Lopez-Vidal, J.C., Hernández-Arciga, R., Santos-Babbera, G. & Mendoza-Quijano, F. (2004) Geographic distribution: Serpentes: *Leptodeira septentrionalis septentrionalis*. *Herpetological Review* 35(3), 292–293.

Canese, A. (1966) Animales venenosos del Paraguay. *Revista Paraguaya Microbiologica*, Asuncion 1(1), 56–72.

Cansdale, G.S. (1948) Field notes on some Gold Coast snakes. *Nigerian Field*, Woodchester 13(2), 43–50, 14 photos.

Cansdale, G.S. (1949a) Further notes on Gold Coast snakes. *Nigerian Field*, Woodchester 14(2), 52–54, 3 photos.

Cansdale, G.S. (1949b) Further notes on Gold Coast snakes. *Nigerian Field*, Woodchester 14(3), 106–113, 8 photos.

Cansdale, G.S. (1954) Gold Coast snakes—A complete list. *Nigerian Field*, Woodchester 19(3),118–132, illust.

Cansdale, G.S. (1961) *West African snakes*. Longmans, London, vi + 74 pp., 15 col. pls.

Cansdale, G.S. (1971) Field notes on some Gold Coast snakes. *Nigerian Field Special Issue*, Woodchester, 57–58, 2 photos.

Canseco-Márquez, L. & Austin, C.C. (2005) Geographic distribution: Serpentes: *Geophis blanchardi*. *Herpetological Review* 36(1), 82.

Canseco-Márquez, L. & Campbell, J.A. (2003) Variation in Zapotitlán coralsnake, *Micrurus pachecogili* (Serpentes, Elapidae). *Southwestern Naturalist* 48(4), 705–707.

Canseco-Márquez, L. & Flores-Villela, O. (1995) Geographic distribution: Serpentes: *Chersodromus liebmanni* (Liebmann's earth runner). *Herpetological Review* 26(2), 109.

Canseco-Márquez, L. & Gutiérrez-Mayén, M.G. (2006) *Guía de campo de los anfibios y reptiles del Valle de Zapotitlán, Puebla*. Sociedad Herpetológica Mexicana, Puebla, 78 pp., col. illust.

Canseco-Márquez, L. & Gutiérrez-Mayén, M.G. (2010) *Anfibios y reptiles del Valle de Tehuacán-Cuicatlán*. Comisión Nacional para el Conocimiento y uso de la Biodiversidad, Mexico City, 301 pp., 188 col. pls.

Canseco-Márquez, L. & Nolasco Vélez, A.L. (2008) Geographic distribution: Serpentes: *Pseudoleptodeira latifasciata*. *Herpetological Review* 39(3), 373.

Canseco-Márquez, L., Gutierrez-Mayen, G. & Salazar-Arenas, J. (2000) New records and range extensions for amphibians and reptiles from Puebla, México. *Herpetological Review* 31(4), 259–263, 1 fig.

Canseco-Márquez, L., Mendelson, J.R., III & Gutiérrez-Mayén, G. (2002) A new species of large *Tantilla* (Squamata, Colubridae) from the Sierra Madre Oriental of Puebla Mexico. *Herpetologica* 58(4), 492–497, 3 figs.

Canseco-Márquez, L., Ramos-Torres, A. & Flores-Villela, O. (2004) Geographic distribution: Serpentes: *Geophis blanchardi*. *Herpetological Review* 35(2), 191–192.

Canseco-Márquez, L., Mendoza-Hernández, A.A. & Nieto-Montes de Oca, A. (2005) Geographic distribution: Serpentes: *Enulius flavitorques*. *Herpetological Review* 36(3), 339.

Canseco-Márquez, L., Smith, E.N., Ponce-Campos, P., Flores-Villela, O. & Campbell, J.A. (2007) A new species of *Tantilla* (Squamata, Colubridae) of the *calamarina* group from Volcán Ceboruco, Nayarit, Mexico. *Journal of Herpetology* 41(2), 220–224, 4 figs.

Canton, E. (1895) *Conferencías sobre zoologia médica*. Second edition. Imprenta Helvetia, Buenos Aires, 306 pp.

Cantor, T.E. (1834–1837) *Drawings of animals*. Vols. 1–3 (vol. 1–harmless serpents, 35 drawings, vol. 2–venomous serpents, 23 drawings, vol 3–Indian serpents, 240 pp.) [deposited in Radcliffe Science Library, University of Oxfoed, Oxford].

Cantor, T.E. (1836a) Sketch of an undescribed hooded serpent, with fangs and maxillary teeth. *Asiatic Researches*, Calcutta 19(1), 87–94, pls. 10–12.

Cantor, T.E. (1836b) Sketches of two undescribed venomous serpents, with fangs behind the maxillar teeth ("A crochets postèrieurs."). *Transactions of the Medical and Physical Society of Calcutta* 8, 137–142.

Cantor, T.E. (1838) A notice of the *Hamadryas*, a genus of hooded serpents with poisonous fangs and maxillary teeth. *Proceedings of the Zoological Society of London* 6(1), 72–75.

Cantor, T.E. (1839a) Spicilegium Serpentium Indicorum. A. Venomous serpents. *Proceedings of the Zoological Society of London* 7(1), 31–34. [reprinted in 1839 by Richard and John E. Taylor, London, 5 pp.]

Cantor, T.E. (1839b) Spicilegium Serpentium Indicorum. B. Innocuous serpents. *Proceedings of the Zoological Society of London* 7(1), 49–55. [reprinted in 1839 by Richard and John E. Taylor, London, 6 pp.]

Cantor, T.E. (1842a) *Zoology of Chusan*. J. Bennett, Calcutta, 32 pp., 13 col. pls.

Cantor, T.E. (1842b) General features of Chusan, with remarks on the flora and fauna of that island. Animals observed at Chusan. *Annals and Magazine of Natural History*, London 9(58), 265–278, 9(59), 361–370, 9(60), 481–493.

Cantor, T.E. (1847) Catalogue of reptiles inhabiting the Malayan Peninsula and islands, collected or observed by Theodore Cantor, Esq., M.D. Bengal Medical Service. Ophidia. *Journal of the Asiatic Society of Bengal (Natural History)*, Calcutta 16(180), 607–656, 16(182), 897–952, & 16(183), 1026–1078, pl. 40. [reprinted in 1886 by Trübner & Co., London and in 1981 by Cosmo Publications, New Delhi, with author listed as T.E. Canton]

Cantor, T.E. (1886) Catalogue of reptiles inhabiting the Malayan Peninsula and islands, collected or observed by Theodore Cantor, Esq., M.D. Bengal Medical Service. Ophidia. *Miscellaneous Papers Relating to Indo-China*. 2, 158–238. [reprint of Cantor, 1847]

Cantuel, P. (1949) *Faune des vertébrés du Massif Central de la France. Contribution à l'étude de la biologie des régions altitudinaires*. Paul Lechevalier, Paris, 404 pp., illust.

Capocaccia, L. (1955) Contributo allo studio della distribuzione geographica di *Bitis lachesis somalica* Parker (Ophidia, Viperidae). *Doriàna*, Genova 2(60), 1–7, 2 figs.

Capocaccia, L. (1958) I serpenti della Liguria. Parte I, Viperidi. *Annali del Museo Civico di Storia Naturale "Giacomo Doria,"* Genova 70, 173–192, 2 pls.

Capocaccia, L. (1959) I serpenti della Liguria. Parte II, Colubridi. *Annali del Museo Civico di Storia Naturale "Giacomo Doria,"* Genova 71, 248–275.

Capocaccia, L. (1961a) Catalogo dei tipi di rettili del Museo Civico di Storia Naturale di Genova. *Annali del Museo Civico di Storia Naturale "Giacomo Doria,"* Genova 72, 86–111.

Capocaccia, L. (1961b) Contributo allo studio dei serpenti delle isole del Golfo di Guinea. *Annali del Museo Civico di Storia Naturale "Giacomo Doria,"* Genova 72, 285–309, 1 pl.

Capocaccia, L. (1964) Variabilita' e sottospecie di *Elaphe longissima* (Laur.) in Italia (Serpentes). *Annali del Museo Civico di Storia Naturale "Giacomo Doria,"* Genova 74, 353–387, 15 figs.

Capocaccia, L. (1965) Intorno a *Elaphe longissima* (Laur.) della Sardegna (Serpentes). *Doriàna*, Genova 4(161), 1–4.

Capocaccia, L. (1968) *Anfibi e rettili*. Arnoldo Mondadori, Milano, 159 pp., illust.

Capocaccia, L. (1976) Contributo allo studio dei serpenti dell'isola di Giava. *Annali del Museo Civico di Storia Naturale "Giacomo Doria,"* Genova 81, 51–95, 4 figs.

Capolongo, D. (1984) Note sull'erpetofauna pugliese. *Atti della Società Italiana di Scienze Naturali e del Museo Civico di Storia Naturali di Milano* 125(3–4), 189–200, illust.

Captain, A. (1999) On the identification of *Lycodon flavomaculatus* Wall, 1907. *The Journal of the Bombay Natural History Society* 96(2), 323–327, 2 pls.

Captain, A. & Bhatt, B.B. (2002) First record of *Amphiesma venningi* (Wall, 1910) (Serpentes, Colubridae, Natricinae) from India, with remarks on its subspecies. (2001). *Hamadryad* 26, 354–358.

Captain, A. & Patel, A. (1998) *Sinonatrix*, a new genus for India. *Hamadryad* 22(2), 114–115, 2 figs.

Captain, A. & Thakur, S. (1999) Aberrant banded racers *Argyrogena fasciolatus. The Journal of the Bombay Natural History Society* 96(1), 147–151.

Captain, A., Gower, D.J., David, P. & Bauer, A.M. (2004) Taxonomic status of the colubrid snake *Sibynophis subpunctatus* (Duméril, Bibron and Duméril, 1854). *Hamadryad* 28(1–2), 90–94.

Captain, A., Tillack, F., Gumprecht, A. & Dandge, P. (2005) First record of *Elachistodon westermanni* Reinhardt, 1863 (Serpentes, Colubridae, Colubrinae) from Maharashtra State, India. *Russian Journal of Herpetology* 12(2), 155–158, 3 figs.

Capula, M., Luiselli, L. & Rugiero, L. (1995) Ecological correlates of reproductive mode in reproductively bimodal snakes of the genus *Coronella. Vie et Milieu* 45, 165–175.

Carbajal-Márquez, R.A., Quintero-Díaz, G.E. & Domínguez-de La Riva, M.A. (2011a) Geographic distribution: Serpentes: *Coniophanes melanocephalus. Herpetological Review* 42(2), 242.

Carbajal-Márquez, R.A., Quintero-Díaz, G.E. & González-Saucedo, Z.Y. (2011b) Geographic distribution: Serpentes: *Rhinocheilus lecontei. Herpetological Review* 42(2), 244.

Carbajal-Marquez, R.A., Arenas-Monroy, J.C., Gonzalez-Saucedo, Z.Y., Jones, J.M. & Roman, L.G. (2012) Geographic distribution: Serpentes: *Conopsis lineata. Herpetological Review* 43(1), 105.

Card, W. (1994) Notes on the natural history and husbandry of the temple viper *(Tropidolaemus wagleri). The Vivarium* 5(5), 22–25, 10 col. photos.

Cardenas, S.A., Tessaro, P.G., Romero, A.G. & Rubio, A.O. (1988) Herpetofauna. *In:* Arriaga, L. & Ortega, A. (Eds.), *La Sierra de la Laguna de Baja California Sur*. Centro de Investigaciones Biologicas de Baja California Sur A.C. Publications (1), 167–184.

Cardew, A.G. (1897) A rough key to the identification of Indian Ophidia. *Journal of the Bombay Natural History Society* (1895–1897) 10(4), 585–596.

Cardoza, J.E. & Mirick, P.G. (1979) List of the reptiles and amphibians of Massachusetts. *Massachusetts Division of Fish and Wildlife, Fauna of Massachusetts Series* (3), 1–5.

Cardwell, M.D., Gotte, S.W., McDiarmid, R.W., Gilmore, N. & Poindexter, J.A. II. (2013) Type specimens of *Crotalus scutulatus* (Chordata: Reptilia: Squamata: Viperidae) re-examined, with new evidence after more than a century of confusion. *Proceedings of the Biological Society of Washington* 126(1), 11–16, 3 figs.

Carl, G.C. (1944) The reptiles of British Columbia (popular). *Handbook of the British Columbia Provincial Museum* 3, 60 pp., illust.

Carlleyle, A.C.L. (1869) Description of two new species belonging to the genera *Varanus* and *Feranioides* respectively from near Agra. *Journal of the Asiatic Society of Bengal (Natural History),* Calcutta 38(2), 192–200.

Carlson, A.J. & Luckhardt, B. (1920) Studies on the visceral sensory nervous system. III. Lung automatism and lung reflexes in Reptilia (Turtles: *Chrysemys elegans* and *Malacoclemmys lesueurii.* Snake: *Eutenia elegans). American Journal of Physiology* 54, 261–306, figs. 22–24.

Carlson, D.S. (1967) A checklist of amphibians and reptiles of Washington. *Bulletin of the Pacific Northwest Herpetological Society* 2(1), 7–8.

Carlsson, M. (2003) *Phylogeography of the adder,* Vipera berus. Doctoral Dissertation, Uppsala University, Uppsala, 32 pp., 5 figs.

Carmichael, C.K. (2007) *Phylogeography of the Indonesian water python, Liasis mackloti ssp. (Squamata: Boidea: Pythoninae): a comparative approach toward resolving phylogeny*. Doctoral Dissertation, University of Southern Mississippi, xii + 138 pp.

Carmichael, P. & Williams, W. (1991) *Florida's fabulous reptiles & amphibians: snakes, lizards, alligators, frogs, and turtles.* World Publications, Tampa, 120 pp., col. illust.

Carpenter, C.C. (1986) An inventory of combat rituals in snakes. *Smithsonian Herpetological Information Service* (69), 1–18.

Carpenter, C.C. & Krupa, J.J. (1989) *Oklahoma herpetology, an annotated bibliography*. University of Oklahoma Press, Norman, vii + 258 pp.

Carr, A. (1940) A contribution to the herpetology of Florida. *University of Florida Publications, Biological Science Series* 3(1), 1–118.

Carr, A. (1950) Outline for a classification of animal habitats in Honduras. *Bulletin of the American Museum of Natural History* 94(10), 567–594, 6 figs., 33 pls.

Carr, A. & Goin, C.J. (1942) Rehabilitation of *Natrix sipedon taeniata* Cope. *Proceedings of the New England Zoological Club* 21, 47–54.

Carr, A. & Goin, C.J. (1955) *Guide to the reptiles, amphibians and fresh-water fishes of Florida*. University of Florida Press, Gainesville, 341 pp., 30 figs., 67 pls.

Carranza, S., Arnold, E.N., Wade, E. & Fahd, S. (2004) Phylogeography of the false smooth snakes, *Macroprotodon* (Serpentes, Colubridae): mitochondrial DNA sequences show European populations arrived recently from northwest Africa. *Molecular Phylogenetics and Evolution* 33(3), 523–532, 2 figs.

Carranza, S., Arnold, E.N. & Pleguezuelos, J.M. (2006) Phylogeny, biogeography, and evolution of two Mediterranean snakes, *Malpolon monspessulanus* and *Hemorrhois hippocrepis* (Squamata, Colubridae), using mtDNA sequences. *Molecular Phylogenetics and Evolution* 40(2), 532–546, 3 figs.

Carrasco, P.A., Harvey, M.B. & Muñoz Saravia, A. (2009) The rare Andean pitviper *Rhinocerophis jonathani* (Serpentes: Viperidae: Crotalinae): redescription with comments on its systematics and biogeography. *Zootaxa* (2283), 1–15, 8 figs.

Carrasco, P.A., Leynaud, G.C. & Scrocchi, G.J. (2010) Redescription of the southernmost snake species, *Bothrops ammodytoides* (Serpentes: Viperidae: Crotalinae). *Amphibia-Reptilia,* Leiden 31(3), 323–338, 9 figs.

Carreira-Vidal, S. (2002) Alimentación de los ofidios de Uruguay. *Monografia de Herpetologia,* Barcelona 6, 1–127, col. figs.

Carreira-Vidal, S. (2003) Geographic distribution: Serpentes: *Thamnodynastes chaquensis. Herpetological Review* 34(4), 390.

Carreira-Vidal, S. (2004) Geographic distribution: Serpentes: *Liotyphlops ternetzii. Herpetological Review* 35(4), 411–412.

Carreira-Vidal, S. (2006) *Calamodontophis. C. paucidens. Catalogue of American amphibians and reptiles,* pp. 819.1–819.3, 1 fig.

Carreira-Vidal, S. & Lombardo, I. (2006) Geographic distribution: Serpentes: *Tomodon dorsatus. Herpetological Review* 37(1), 110.

Carreira-Vidal, S. & Lombardo, I. (2007a) Geographic distribution: Serpentes: *Chironius bicarinatus. Herpetological Review* 38(1), 102.

Carreira-Vidal, S. & Lombardo, I. (2007b) Geographic distribution: Serpentes: *Liophis flavifrenatus. Herpetological Review* 38(4), 487–488.

Carreira-Vidal, S. & Lombardo, I. (2008) The hemipenis and external morphology of a new specimen of *Calamodontophis paucidens* (Amaral, 1935) (Serpentes: Colubridae, Dipsadinae, Tachymenini) from Uruguay. *Herpetological Review* 39(3), 282–283, 1 fig.

Carreira-Vidal, S., Meneghel, M. & Achaval, F. (2004) Geographic distribution: Serpentes: *Atractus reticulatus. Herpetological Review* 35(2), 189.

Carreira-Vidal, S., Meneghel, M., & Achaval, F. (2005) *Reptiles de Uruguay. University de la Republica,* Facultad de Ciencias, Montevideo, 637 pp., 132 figs., 125 col. pls.

Carreira-Vidal, S., Brazeiro, A., Camargo, A., Rosa, I. de, Canavero, A. & Maneyro, R. (2012a) Diversity of reptiles of Uruguay: knowledge and information gaps. *Boletín de la Sociedad Zoológica del Uruguay* 21(1–2), 9–29, 4 figs.

Carreira-Vidal, S., Lombardo, I. & Achaval-Coppes, F. (2012b) Nuevos registros de ofidios poco conocidos en Uruguay (Reptilia: Squamata: Serpentes). *Boletín de la Sociedad Zoológica del Uruguay* 21(1–2), 70–74, 1 fig.

Carrera, M. & Javier-Avila, L. (2008) Geographic distribution: Serpentes: *Oxyrhopus rhombifera bachmanni. Herpetological Review* 39(2), 242–243.

Carrillo de Espinoza, N. (1966) Contribucion al conocimiento de los boideos Peruanos (Boidae, Ophidia, Reptilia). *Publicaciones del Museo de Historia Natural "Javier Prado" (Zoología),* Lima (21A), 1–51, 20 figs.

Carrillo de Espinoza, N. (1970) Contribucion al conocimiento de los reptiles del Peru (Squamata, Crocodylia, Testudinata, Reptilia). *Publicaciones del Museo de Historia Natural "Javier Prado" (Zoología),* Lima (22A), 1–64.

Carrillo de Espinoza, N. (1974) *Sibynomorphus williamsi* nov. sp. (Serpentes, Colubridae). *Publicaciones del Museo de Historia Natural "Javier Prado" (Zoología),* Lima (24A), 1–16, 8 figs.

Carrillo de Espinoza, N. (1977) Arañas y serpientes muy venenosas en el Departamento do Lima. *Universidad Nacional Mayor de San Marcos (Série de Divulgación)* (8), 1–8, figs. 3–4.

Carrillo de Espinoza, N. (1983) Contribucion al conocimiento de las serpientes venenosas del Peru de las familias Viperidae, Elapidae e Hydrophiidae. *Publicaciones del Museo de Historia Natural "Javier Prado" (Zoología),* Lima (30A), 1–55, 31 figs.

Carrillo de Espinoza, N. (1989) *List of Peruvian Colubridae.* Departamento de Herpetología, Museo de Historia Natural, Universidad Nacional Mayor de San Marcos, Lima, 4 pp.

Carrillo de Espinoza, N. & Icochea, J. (1995) Lista taxonomica preliminar de los reptiles vivientes del Perú. *Publicaciones de Museo de Historia Natural, Universidad Nacional Mayor de San Marcos,* Lima (A) 49, 1–27.

Carus, J.V. & Engelmann, W. (1861) *Bibliotheca zoological. Verzeichniss der Schriften über zoologie, welche in den Periodischen Werken enthalten und vom Jahre 1846–1860 selbständig erschienen sind. Zweiter Band.* Verlag von Wilhelm Engelmann, Leipzig, x + 969 pp.

Carvajal-Cogollo, J.E. (2006) Geographic distribution: Serpentes: *Typhlops reticulatus. Herpetological Review* 37(4), 501–502.

Carvajal-Cogollo, J.E., Bernal-Gonzalez, V.A. & Nonzoque, L. (2011a) Geographic distribution: Serpentes: *Dipsas temporalis. Herpetological Review* 42(3), 393.

Carvajal-Cogollo, J.E., Bernal-Gonzalez, V.A. & Nonzoque, L. (2011b) Geographic distribution: Serpentes: *Drymobius margaritiferus. Herpetological Review* 42(3), 393.

Carvajal-Cogollo, J.E., Bernal-Gonzalez, V.A. & Nonzoque, L. (2011c) Geographic distribution: Serpentes: *Oxyrhopus formosus. Herpetological Review* 42(3), 395.

Carvalho, C.M. de. (2002) Descrição de uma nova espécie de *Micrurus* do estado de Roraima, Brasil (Serpentes, Elapidae). *Papéis Avulsos de Zoologia, Museu de Zoologia da Universidade de São Paulo* 42, 183–192.

Carvalho, C.M. de, Souza-Alencar, I.C. de & Carvalho-Vilar, J. (2007) Serpentes da região Manaus, Amazônia central, Brasil. *Biologia Geral e Experimental,* São Cristóvão 7(2), 41–59, 7 figs.

Casado-B. R. & Zoltan, S. (1996) Geographic distribution: Serpentes: *Boa constrictor constrictor. Herpetological Review* 27(2), 88.

Casais e Silva, L.L. (1996) Geographic distribution: Serpentes: *Micrurus lemniscatus. Herpetological Review* 27(2), 88.

Casas-Andreu, G. & López-Forment, W. (1978) Notas sobre *Micrurus browni taylori* Schmidt y Smith, en Guerrero, México. *Anales del Instituto de Biología, Universidad Nacional Autónomica de México (Zoología)* 49(1), 291–294.

Casas-Andreu, G. & McCoy, C.J. (1979) *Anfibios y reptiles de México. Claves ilustradas para su identificación.* Ed. Limusa, México, 87 pp., 111 figs.

Casas-Andreu, G., Mendez-de la Cruz, F.R. & Camarillo, J.L. (1996) Anfibios y reptiles de Oaxaca. Lista, distribucion y conservacion. *Acta Zoologica Mexicana* 69, 1–35, 3 figs.

Casper, G.S. (1996) *Geographic distributions of the amphibians and Reptiles of Wisconsin.* Milwaukee Public Museum, Milwaukee, 87 pp., illust.

Casper, G.S. (2008) An annotated checklist of the amphibians and reptiles of Wisconsin. *Bulletin of the Chicago Herpetological Society* 43(5), 73–79.

Casper, G.S. & Burnham, J. (2001) Geographic distribution: Serpentes: *Dryocalamus subannulatus. Herpetological Review* 32(3), 195.

Cassimiro, T. (2003) Geographic distribution: Serpentes: *Tantilla boipiranga. Herpetological Review* 34(4), 390.

Castanet, J. & Guyetant, R. (Eds.). (1989) *Atlas de repartition des amphibians et reptiles de France.* Société Herpetologique de France, Paris, 191 pp., illust.

Castellani, A. & Chalmers, A.J. (1913) *Manual of tropical medicine.* Second edition. Baillère, Tindall & Cox, London, xxxii + 1747 pp.

Castellano, S., Malhotra, A. & Thorpe, R.S. (1994) Within-island geographic variation of the dangerous Taiwanese snake, *Trimeresurus stejnegeri*, in relation to ecology. *Biological Journal of the Linnean Society*, London 52(4), 365–375, 2 figs.

Castillo, D., Sousa, F. & Pedrol, J. (1990) *Tantilla alticola* (Boulenger, 1903) Colubridae, primer reporte para Panamá. *Scientia (Panamá)* 5(1), 73–84.

Casto de Elera, R.P.F. (1895) *Catálogo sistemático de toda la fauna de Filipinas conocida hasta el presente, y á la vez el de la colección zoológica del museo de Pp. Dominicos del Colegio-Universidad de Sto. Tomás de Manila, escrito con motiro de la Exposición Regional Filipina. Volumen I. Vertebrados*. Colegio de Santo Tomás, Manila, viii + 701 pp.

Castoe, T.A. & Parkinson, C.L. (2006) Bayesian mixed models and the phylogeny of pitvipers (Viperidae: Serpentes). *Molecular Phylogenetics and Evolution* 39(1), 91–110, 3 figs.

Castoe, T.A., Chippindale, P.T., Campbell, J.A., Ammerman, L.K. & Parkinson, C.L. (2003) Molecular systematics of the Middle American jumping pitvipers (genus *Atropoides*) and phylogeography of the *Atropoides nummifer* complex. *Herpetologica* 59(5), 420–431, illust.

Castoe, T.A., Sasa, M.M. & Parkinson, C.L. (2005) Modeling nucleotide evolution at the mesoscale: the phylogeny of the Neotropical pitvipers of the *Porthidium* group (Viperidae, Crotalinae). *Molecular Phylogenetics and Evolution* 37(3), 881–898, 3 figs.

Castoe, T.A., Smith, E.N., Brown, R.M. & Parkinson, C.L. (2007a) Higher-level phylogeny of Asian and American coralsnakes, their placement within the Elapidae (Squamata), and the systematic affinities of the enigmatic Asian coralsnake *Hemibungarus calligaster* (Wiegmann, 1834). *Zoological Journal of the Linnean Society*, London 151(4), 809–831, 5 figs.

Castoe, T.A., Spencer, C.L. & Parkinson, C.L. (2007b) Phylogeographic structure and historical demography of the western diamondback rattlesnake (*Crotalus atrox*): a perspective on North American desert biogeography. *Molecular Phylogenetics and Evolution* (2006) 42(1), 193–212, 6 figs.

Castoe, T.A., Daza, J.M., Smith, E.N., Sasa, M.M., Kuch, U., Campbell, J.A., Chippindale, P.T. & Parkinson, C.I. (2009) Comparative phylogeography of pitvipers suggests a consensus of ancient Middle American highland biogeography. *Journal of Biogeography* 36(1), 88–103, 6 figs.

Castro-Franco, R. & Bustos-Zagal, M.G. (1994) List of reptiles of Morelos, Mexico, and their distribution in relation to vegetation types. *Southwestern Naturalist* 39(2), 171–175.

Castro-Franco, R. & Bustos-Zagal, M.G. (2004) Additional records and range extentions of reptiles from Morelos, México. *Herpetological Review* 35(2), 196–197.

Castro-Franco, R. & Gaviño de la Torre, G. (1990) Reptiles de la Isla La Peña, Nayarit, México. *Anales del Instituto Biología, Universidad Nacional Autónomica de México (Zoología)* 61(1), 175–187.

Castro-Herrera, F. & Vargas-Salinas, F. (2007) Reptilia, Squamata, Serpentes, Colubridae, *Urotheca decipiens*: distribution extension. *Check List* 3(3), 185–189, 3 figs.

Catesby, M. (1743) *The natural history of Carolina, Florida, and the Bahama Islands, containing the figures of birds, beasts, fishes, serpents, insects and plants, not hitherto described, or very incorrectly figured by authors. Together with their descriptions in English and French. To which are added, observations on the air, soil, and waters, with remarks upon agriculture, grain, pulse, roots, &c. To the whole is prefixed a new and correct map of the countries treated of. Vol. II.* Mark Catesby, London, 100 pp., 100 col. pls., App., 20 pp., 20 col. pls.

Cattaneo, A. (1985) Il colubro ferro di cavallo dell'isola di Pantelleria: *Coluber hippocrepis nigrescens* subsp. nova. *Atti della Società Italiana di Scienze Naturali e del Museo Civico di Storia Naturale di Milano* 126, 165–184.

Cattaneo, A. (1989) Note erpetologiche sulle isole greche di Serifos, Sifnos e Mitos (Cicladi occidentali). *Atti della Società Italiana di Scienze Naturali e del Museo Civico di Storia Naturale di Milano* 130(2), 57–76, 1 fig.

Cattaneo, A. (1990) I serpenti della isole greche di Kythnos e Kea (Cicladi occidentali). *Atti della Società Italiana di Scienze Naturali e del Museo Civico di Storia Naturale di Milano* 131(11), 209–219, 3 figs.

Cattaneo, A. (1999) Variabilità e sottospecie di *Elaphe quatuorlineata* (Lacépède) nelle picole isole Egee (Serpentes: Colubridae). *Atti della Società Italiana di Scienze Naturali e del Museo Civico di Storia Naturale di Milano* 140, 119–139.

Cattaneo, A. (2012) Il colubrogola rossa dell'Arcipelagodi Rodi: *Dolichophis jugularis zinneri* subsp. nov. (Reptilia Serpentes). *Naturalista Siciliano* 36(1), 77–103.

Cazaly, W.H. (1914) *The common snakes of India and Burma and how to recognize them*. International Book Distributors, Dehra Dun, India, iii + 60 + iv pp., 11 figs.

Caziot, C. (1922) Les reptiles du Département des Alpes-Maritimes. *In: Congrès de Montpellier.* Association Française pour l'avancement des sciences, pp. 428–432.

Cedeño-Vázquez, J.R., Calderón, R. & Pozo, C. (2003) New distributional records for amphibians and reptiles from Quintana Roo on the Yucatán Peninsula, México. *Herpetological Review* 34(4), 393–395.

Cei, J.M. (1984) Una nueva especie de *Elapomorphus* de la region de Cuyo (Colubridae, Serpentes). *Boletín del Museo de Ciencias Naturales y Anthropológicas Juan Cornelio Moyano*, Mendoza (4), 47–50.

Cei, J.M. (1986) Reptiles del centro, centro-oeste y sur de la Argentina. Herpetofauna de las zonas áridas y semiáridas. *Monografia, Museo Regionale di Scienze Naturali*, Torino (4), 1–527, 77 figs., 48 pls.

Cei, J.M. (1994) Reptiles del noroeste, nordeste y este de la Argentina. *Monografia, Museo Regionale di Scienze Naturali*, Torino (1993) (14), 1–949, 112 figs., 126 col. pls.

Cei, J.M., Castro, L.P., Ormeno, E.A. & Arratia, G.F. (1978) Atlas de los vertebrados inferiores de la region de Cuyo. *Publicaciones Ocasionales del Instituto Biología Animal, Universidad Nacional de Cuyo, Series Cientificas* (2), 1–10, 27 pls.

Cekalovic-K.T. & Artigas, J.N. (1981) Catalogo de los tipos depositados en la coleccion del Departamento de Zoologia de la Universidad de Concepcion, Chile. (Parte 3). *Boletin de la Sociedad de Biología de Concepción* 51(2), 75–107, illust.

Celsi, C.E., Monserrat, A.L. & Kocoliris, F.P. (2008) Reptilia, Colubridae, *Philodryas aestivus*: distribution extension. *Check List* 4(1), 12–14, 3 figs.

Censky, E.J. & Hodge, K. (1997) Geographic distribution: Serpentes: *Ramphotyphlops braminus*. *Herpetological Review* 28(4), 210.

Censky, E.J. & Kaiser, H. (1999) The Lesser Antillean fauna. *In:* Crother, B.I. (Ed.), *Caribbean amphibians and reptiles*. Academic Press, San Diego, pp. 181–222, 2 figs.

Centeno, F.C., Sawaya, R.J. & Germano, V.J. (2010) A new species of *Liotyphlops* (Serpentes: Anomalepididae) from the Atlantic coastal forest in southeastern Brazil. *Herpetologica* 66(1), 86–91, 3 figs.

Cermak, M. (2008) *Spectacular snakes of Australia.* CSIRO Publishing, Collingwood, 120 pp., col. illust.

Chabanaud, P. (1916a) Serpents d'Afrique Occidentale recueillis par M. Gruvel. *Bulletin du Museum d'Histoire Naturelle*, Paris 22(2), 75–76.

Chabanaud, P. (1916b) Description d'un serpent nouveau de Mauritanie saharienne. *Bulletin du Museum d'Histoire Naturelle*, Paris 22(2), 77–78.

Chabanaud, P. (1917a) Enumération des ophidiens non encore étudies de l'Afrique occidentale, appartenant aux collections du Muséum, avec la description des espèces et des variétés nouvelles. *Bulletin du Museum d'Histoire Naturelle*, Paris (1916) 22(7), 362–382, 23 figs.

Chabanaud, P. (1917b) Revision du genre *Prosymna* Gray. *Bulletin du Museum d'Histoire Naturelle*, Paris (1916) 22(8), 433–440, 2 figs.

Chabanaud, P. (1917c) Note complémentaire sur les ophidiens de l'Afrique occidentale, avec la description d'une espèce nouvelle. *Bulletin du Museum d'Histoire Naturelle*, Paris 23(1), 7–14, 3 figs.

Chabanaud, P. (1917d) Descriptions de trois espèces nouvelles de reptiles de l'Afrique. *Bulletin du Museum d'Histoire Naturelle*, Paris 23(4), 219–225, 9 figs.

Chabanaud, P. (1918a) Revision de quelques reptiles d'Afrique et description de trois espèces nouvelles. *Bulletin du Museum d'Histoire Naturelle*, Paris (1917) 23(7), 442–454, 7 figs.

Chabanaud, P. (1918b) Étude complémentaire de deux *Agama* de l'Afrique Occidentale et description de quatre espèces nouvelles de reptiles de la même région. *Bulletin du Museum d'Histoire Naturelle*, Paris 24(2), 104–112.

Chabanaud, P. (1918c) Étude d'une collection de reptiles de l'Afrique Occidentale Française, récemment donnée au Muséum d'Histoire Naturelle de Paris par le Dr. G. Bouet, avec la description de deux espèces nouvelles. *Bulletin du Museum d'Histoire Naturelle*, Paris 24(3), 160–166.

Chabanaud, P. (1920a) Description d'un *Typhlops* nouveau découvert au Togo par le D'Millet-Horsin. *Bulletin du Museum d'Histoire Naturelle*, Paris 26(6), 463–464.

Chabanaud, P. (1920b) Contribution à l'étude de la faune herpétologique de l'Afrique occidentale. *Bulletin du Comité d'Études Historiques et Scientifiques de l'Afrique Occidental Française* 4, 489–497.

Chamizo-Lara, A.R., Moreno-Garcia, L.V. & Rodríguez Schettino, L. (2005) Holotypes of reptiles in the Herpetological collections of the Institute of Ecology and Systematics, Cuba. *Smithsonian Herpetological Information Service* (135), 10 pp.

Chan, K.O., Grismer, L.L., Sharma, D.S., Belabut, D. & Ahmad, N. (2009) New herpetofaunal records for Perlis State Park and adjacent areas. *Malayan Nature Journal* 61(4), 277–284.

Chan, S.K.-F., Cheung, K.-S., Ho, C.-Y., Lan, F.-M., Tang, W.-S. & Man, J.L.T. (2006) *A field guide to the venomous land snakes of Hong Kong.* Agriculture, Fisheries and Conservation Department, Hong Kong, 196 pp., col. illust.

Chan-ard, T. & Makchai, S. (2011) Catalogue of type specimens of amphibians and reptiles in the Natural History Museum, National Science Museum, Thailand. *The Thailand Natural History Museum Journal* 5(2), 113–123.

Chan-ard, T., Grossmann, W., Gumprecht, A. & Schulz, K.D. (1999) *Amphibians and reptiles of Peninsular Malaysia and Thailand.* Bushmaster Publishing, Wuerselen, 240 pp., col. illust.

Chan-ard, T., Stuart, B.L. & Wüster, W. (2000) First record of Indochinese spitting cobra *Naja siamensis* Laurenti (Serpentes, Elapidae) from Laos, with comments on the genus in the country. *Natural History Bulletin of the Siam Society* 48, 149–152, 2 figs.

Chandramouli, S.R. & Ganesh, S.R. (2010) Herpetofauna of southern Western Ghats, India – reinvestigation after decades. *Taprobanica,* Homagama 2(2), 72–85, 4 col. pls.

Chandramouli, S.R., Baiju, Sebastien, J.J. & Ganesh, S.R. (2012) Expanded description of *Enhydris dussumierii* (Duméril, Bibron & Duméril, 1854) (Reptilia: Colubridae: Homalopsinae). *Taprobanica,* Homagama 4(1), 42–47, 3 figs.

Chang, M.L.Y. (1932) Notes on a collection of reptiles from Szechuan. *Contributions from the Biological Laboratory of the Science Society of China (Zoological Series)*, Shanghai 8(2), 9–95, 28 figs.

Chang, M.L.Y. (1933) Herpetological notes on Szechuan. *Transactions of the Science Society of China*, Nanking 8(1), 57–59.

Chang, M.L.Y. (1934) Preliminary report on some reptiles from Chekiang. *China Journal of Shanghai* 21(5), 251–253.

Chang, M.L.Y. (1936) Snakes of Luschan (Kiangsi) collected by R.C. Ching. *Contributions from the Biological Laboratory of the Science Society of China (Zoological Series)*, Shanghai 11(10), 317–344, 6 figs.

Chang, M.L.Y. (1947) Herpetological notes on Kwangsi. *Transactions of the Chinese Association for the Advancement of Science*, Taipei 9, 85–120.

Chang, T.H. & Fang, P.W. (1931) A study of the ophidians and chelonians of Nanking. *Contributions from the Biological Laboratory of the Science Society of China (Zoological Series)*, Shanghai 7(8), 249–288, 19 figs.

Chanhome, L., Pauwels, O.S.G., Jintakune, P. & David, P. (2001) Catalogue of the herpetological collection of the Queen Saovabha Memorial Institute, Thai Red Cross Society, Bangkok. Part 1. Snakes (except Elapidae and Viperidae). *Bulletin of the Maryland Herpetological Society* 37(2), 49–72.

Chapman, D.S. (1968) The symptomatology, pathology, and treatment of the bites of venomous snakes of Central and southern Africa. *In:* Bücherl, W., Buckley, E.E. & Deulofeu, V. (Eds.), *Venomous animals and their venoms. Volume I. Venomous vertebrates.* Academic Press, New York, pp. 463–529, 7 figs.

Chari, V.K. (1955) A new form of the burrowing snake *Uropeltis macrolepis* (Peters) from Mahableshwar (*U. m. mahableshwarensis*, ssp. nov.). *The Journal of the Bombay Natural History Society* (1954–1955) 52(4), 901.

Charles, N., Whitaker, P. & Shine, R. (1979) Oviparity and captive breeding in the spotted blacksnake, *Pseudechis guttatus* (Serpentes, Elapidae). *Australian Zoologist*, Mosman 20(2), 361–364, illust.

Chasen, F.N. & Smedley, N. (1927) A list of reptiles from Pulau Galang and other islands of the Rhio Archipelago. *Journal of the Malay Straits Branch of the Royal Asiatic Society* 5, 351–355.

Chatman, V.B. & DiMari, S.J. (1974) Use of sulphopropyl-Sephadex chromatography and disc electrophoresis for the fractionation and identification of elapid venoms. *Toxicon* 12, 405–414.

Cheh, Z.-C., Lin, Y.-L. & Lingbing. (1987) A discussion on the specific categories of *Hydrophis cyanocinctus* and *H. melanocephalus* (Hydrophiidae). *Wuyi Science Journal* 7, 141–144.

Cheke, A.S. (1973) Snakes at Chiang Mai University. *Natural History Bulletin of the Siam Society* 24(3–4), 469–471.

Cheke, A.S. (2010) Is the enigmatic blind-snake *Cathetorhinus melanocephalus* (Serpentes: Typhlopidae) an extinct species from Mauritius? *Hamadryad* 35(1), 101–104.

Chen, B. (Ed.), (1991) *The amphibian and reptilian fauna of Anhui*. Anhui Publishing House, Science and Technology, Hefei, 408 pp., 151 figs., 4 col. pls. (in Chinese)

Chen, B. & Zheng, G. (1983) A snake record new to Sichuan Province. *Acta Herpetologica Sinica*, Chengdu (Series 2) 2(1), 78–79. (in Chinese)

Chen, J.T.F. & Yu, M.-J. (1984) A synopsis of the vertebrates of Taiwan. Vol. III. *Commercial Press*, Taipei, 3 + 633 pp., 8 pls. (in Chinese)

Chen, Y.-C., Wu, X.-F. & Zhao, E.N. (1984) Classification of *Agkistrodon* species in China. *Toxicon*, London 22(1), 53–61, 8 figs.

Chenu, J.C. & Desmarest, E. (1857) *Encyclopédie d'histoire naturelle ou traité complet de cette science d'apres les travaux des naturalists, les plus éminents de tous les pays et de toutes les époques: Buffon, Daubenton, Lacépède, G. Cuvier, F. Cuvier, Geoffroy Saint-Hilaire, Latreille, de Jussieu, Brongniart, etc., etc. Ouvrage resumant les observations des auteurs anciens et comprenant toutes les descouvertes modernes jusqu'a nos jours. Tome troisème. Reptiles et poissons.* Marescq et Compagne, Paris, 360 pp., 48 pls.

Cherchi, M.A. (1958) Note sugli ofidi velenosi Somali. *Bolletino dei Musei e degli Instuti Biologici dell'Università di Genova* 18(169), 79–86.

Cherchi, M.A. and S. Spano. (1966) Viperidi ed elapidi dello Yemen. (Missione Scortecci 1965). *Bolletino dei Musei e degli Instiuti Biologici dell'Università di Genova* 34(210), 139–148, 4 figs.

Cherlin, V.A. (1981) The new saw-scaled viper *Echis multisquamatus* sp. nov. from south-western and Middle Asia. *Trudy Zoologicheskogo Instituta Akademii Nauk S.S.S.R.*, Leningrad 101, 92–95, pl. 14. (in Russian)

Cherlin, V.A. (1983) New facts on taxonomy of snakes from the *Echis* genus. *Vestnik Zoologii*, Kiev 1983 (2), 42–46, 3 figs. (in Russian) [English translation by F. Owusu published in 1984 by *Smithsonian Herpetological Information Service* (61), 1–7, 3 figs.]

Cherlin, V.A. (1990) Taxonomic revision of the snake genus *Echis* (Viperidae). II. An analysis of taxonomy and description of new forms. *Proceedings of the Zoological Institute, USSR Academy of Science*, Leningrad 207, 193–223, 7 figs., pls. 4–6. (in Russian)

Cherlin, V.A. & Borkin, L.J. (1990) Taxonomic revision of the snake genus *Echis* (Viperidae). I. An analysis of the history of study and synonymy (in Russian). *Proceedings of the Zoological Institute, Leningrad, USSR Academy of Science* 207, 175–192.

Chermock, R.L. (1952) A key to the amphibians and reptiles of Alabama. *Geological Survey of Alabama Museum Papers* (33), 1–88.

Chernov, S.A. (1934) On the systematics and distribution of *Agkistrodon* (Ophidia) in the Soviet Union. *Doklady Akademii Nauk S.S.R.*, Leningrad (new Series) 1934, 350–356 (in Russian).

Chernov, S.A. (1935) New material on the distribution of the Indian snake *Lycodon striatus* (Shaw) (Ophidia, Colubridae) in Soviet Middle Asia. *Comptes Rendus (Doklady) de l'Académie des Sciences de l'URSS* 3(8), 189–192.

Chernov, S.A. (1937) *Check-list of snakes, lizards, and turtles of Armenia*. Biological Institute, Armenian Branch, Academy of Science U.S.S.R., Erevan, pp.1–54, 27 figs. (In Russian)

Chernov, S.A. (1939) On the herpetological fauna of Armenia and Nachichevan ARSS. *Zoological Paper, Biological Institute, Armenian Branch, Academy of Scence, USSR*, Erevan 1, 77–194, 37 figs. (in Russian).

Chernov, S.A. (1940) *In:* Terentyev, P.V. & Chernov, S.A. *Opredelitel Presmykayushchikhsya i Zemmovodnykh*. Soviet Nauka Press, Moscow, 340 pp. (in Russian)

Chernov, S.A. (1948) Key to Paleoarctic *Contia*. *Proceedings of the Zoological Institute*, Leningrad 7, 118–123, 2 figs. (in Russian)

Chernov, S.A. (1957) The systematic position of the poisonous snake *Ancistrodon rhodostoma* (Boie) (Serpentes, Crotalidae) based on its cranial osteology. *Zoologischeskii Zhurnal*, Moscow 36(5), 790–792. (in Russian) [English translation by L. Ewbank in 1968 by *Smithsonian Herpetological Information Service* (9), 1–2]

Chernov, S.A. (1959) *The fauna of Tadzhikistan. Vol. 18. Reptiles. In:* Proceedings of the Institute of Zoology and Parasitology of SSR Turkmenia Academy of Sciences, Stalinabad 98, 1–202, 13 figs. (in Russian)

Chernov, S.A. (1962) On one still unknown make of the genus *Calamaria*. (Scientific results of the Chinese-Soviet expeditions in 1955–1957 to south western China). *Trudy Zoologicheskogo Instituta Akademii Nauk S.S.S.R.*, Leningrad 30, 382–384, 1 fig. (in Russian)

Chettri, B., Bhupathy, S. & Acharya, B.K. (2009) Morphometry and aspects of breeding biology of *Trachischium guentheri* Boulenger, 1890 (Serpentes, Colubridae) in north Sikkim, eastern Himalaya, India. *Russian Journal of Herpetology* 16(3), 177–182, 6 figs.

Cheylan, M. & Guillaume, C.-P. (1993) *Elaphe scalaris* (Schinz, 1822) – Treppennatter. *In:* Böhme, W. (Ed.), *Handbuch der Reptilien und Amphibien Europas. Band 3/I. Schlangen (Serpentes) I. (Typhlopidae, Boidae, Colubridae 1, Colubrinae).* AULA-Verlag, Wiesbaden, pp. 397–429, figs. 73–81.

Chhapgar, B.F. & Kewalramani, H.G. (1967) Mating and other observations on sea snakes in captivity. *The Journal of the Bombay Natural History Society* 64(3), 563–564, 1 pl.

Chiari, Y., Ravelonkasina, D. & Faliarisoa, E. (2006) Geographic distribution: Serpentes: *Liopholidophis sexlineatus*. *Herpetological Review* 37(3), 363.

Chiasson, R.B. & Lowe, C.H. (1989) Ultrastructural scale pattern in *Nerodia* and *Thamnophis*. *Journal of Herpetology* 23(2), 109–118, 8 figs.

Chiasson, R.B., Bentley, D.L. & Lowe, C.H. (1989) Scale morphology in *Agkistrodon* and closely related crotaline genera. *Herpetologica* 45(4), 430–438, 5 figs.

Chifundera, K. (1990) Snakes of Zaire and their bites. *African Study Monographs* 10(3), 137–157.

Chikin, Y.A. (1991) The geographic variations of non-metric characteristics of colour pattern in *Vipera lebetina turanica* (Reptilia, Viperidae). *Herpetological Research*, Leningrad (1), 5–11, 3 figs.

Chikin, Y.A. & Szczerbak, N.N. (1992) New subspecies of *Vipera lebetina cernovi* ssp. nov. (Reptilia, Viperidae) from Middle Asia. *Vestnik Zoologii*, Kiev 1992(6), 45–49. (in Russian)

Chin, Y.-L. & Chou, Y.-T. (1968) Research data of order Squamata of class Reptilia in Wu-i-shan, Fukien Province. *In: Communist Chinese Scientific Abstracts (30th Annual China Zoological Society Conference, December 1965). No. 2,* Joint Publications Research Service, Washington, p. 274.

Chippaux, J.-P. (1980) *Epidemiologie des morsures de Serpents en Cote-d' Ivoire*. Doctoral Dissertation, Faculte de Medicine de Marseille, Marseille, 154 pp.

Chippaux, J.-P. (1987) *Les serpents de la Guyane française*. Office de la Recherche Scientifique et Technique Outre-Mer, Paris (1986) (27), 1–165, 62 figs.

Chippaux, J.-P. (2006) *Les serpents d'Afrique occidentale et centrale*. Third edition. IRD Éditions, Paris, 311 pp., figs., 58 col. photos. [first edition, 1999, 278 pp., figs., 55 col. photos; second edition, 2001, 292 pp., figs., 55 col. photos]

Chippaux, J.-P. & Bressy, C. (1981) L'endémie ophidienne de plantations de Côte d'Ivoire. *Bulletin, Société de Pathologie Exotique* 74, 458–467.

Chippaux, J.-P. & Goyffon, M. (1983) Producers of antivenomous sera. *Toxicon* 21(6), 739–752.

Chippindale, P. (1989) Geographic distribution: Serpentes: *Gonionotophis klingi*. *Herpetological Review* 20(3), 75.

Chiras, S. (1978) Description of a multicolor *Thamnophis sirtalis sirtalis* and her offspring. Herp, *Bulletin of the New York Herpetological Society* 14(2), 25, illust.

Chirio, L. (2009) Inventaire des reptiles de la région de la Réserve de Biosphere Transfrontalière du W (Niger/Bénin/Burkina Faso: Afrique de l'Ouest). *Bulletin de la Société Herpétologique de France* (132), 13–41, 9 figs.

Chirio, L. (2012) Geographic distribution: Serpentes: *Myriopholis narirostris*. *Herpetological Review* 43(3), 449.

Chirio, L. (2013) Inventaire des reptiles de la region de Sangarédi (Guinée maritime). *Bulletin de la Société Herpétologique de France* (2012) (144), 67–100, 6 figs., 1 photo.

Chirio, L. & Blanc, C.P. (1997a) Geographic distribution: Serpentes: *Ramphotyphlops braminus*. *Herpetological Review* 28(1), 52

Chirio, L. & Blanc, C.P. (1997b) Analyse biogeographique du peuplement reptilian de l'Aures (Algerie). *Biogeographica*, Paris 73(1), 13–22, illust.

Chirio, L. & Blanc, C.P. (1997c) Analyse de la distribucion des reptiles dan le Massif de'l'Aures (Algeria). *Ecologie* 28(4), 281–292, illust.

Chirio, L. & Blanc, C.P. (1997d) Statut et distribution des reptiles dans le massif de l'Aures (Algerie). *Journal of African Zoology* 111(3), 205–233, illust.

Chirio, L. & Ineich, I. (1991) Les genres *Rhamphiophis* Peters, 1854 et *Dipsina* Jan, 1863 (Serpentes, Colubridae), revue des taxons reconnus et description d'une espèce nouvelle. *Bulletin du Museum National d'Histoire Naturelle*, Paris (Series 4) 13A (1–2), 217–235, 3 figs., 1 pl.

Chirio, L. & Ineich, I. (1993) Geographic distribution: Serpentes: *Hemirhagerrhis nototaenia*. *Herpetological Review* 24(4), 156.

Chirio, L. & Ineich, I. (1997) Geographic distribution: Serpentes: *Ramphotyphlops braminus*. *Herpetological Review* 28(1), 52.

Chirio, L. & Ineich, I. (2006) Biogeography of the reptiles of the Central African Republic. *African Journal of Herpetology* 55(1), 23–59, 3 figs.

Chirio, L., Ineich, I., Schmitz, A. & Trape, J.-F. (2011) Note sur la systématique de quelques du genre *Prosymna* Gray, 1849 en Afrique au Nord de l'équateur (Serpentes, Prosymnidae). *Revue Suisse de Zoologie* 118(1), 157–173, 4 figs.

Chiszar, D. & Smith, H.M. (2004) Geographic distribution: Serpentes: *Crotalus basiliscus*. *Herpetological Review* 35(1), 83.

Chiu, K.W. & Wong, C.C. (1974) Testicular cycles of the striped racer, *Elaphe taeniura*. *General and Comparative Endocrinology* 23(2), 242–244, 1 fig., 1 pl.

Chondropoulos, B.P. (1989) A checklist of Greek reptiles. II. Snakes. *Herpetozoa*, Wien 2(1–2), 3–36.

Chpakowsky, N. & Chénour, A. (1954) Les serpents de Tunisie. *Bulletin de la Société des Sciences Naturelles de Tunis* (1953) 6(1–4), 125–146, 4 pls.

Christensen, P. (1972) New record of Mueller's snake, *Rhinoplocephalus bicolor*. *Western Australian Naturalist*, Perth 12(4), 88–89.

Christiansen, J.L. & Bailey, R.M. (1990) *The snakes of Iowa*. Iowa Conservation Commission, Des Moines, 15 pp., 28 col. photos.

Christiansen, J.L. & Leclere, J.B. (2002) *Nerodia erythrogaster neglecta* in Iowa with comments on ontogenesis of color and pattern. *Herpetological Review* 33(2), 152–154, 2 figs.

Christie, P. (1997) *Reptiles and amphibians of Prince Edward County, Ontario*. Natural Heritage Books, Toronto, 143 pp.

Christman, S.P. (1974) Recent amphibians and reptiles. *In:* Gilbert, C.R. (Ed.), Catalogue of type specimens in the Department of Natural Sciences, Florida State Museum. *Bulletin of the Florida State Museum, Biological Science* 18(2), 101–120.

Christman, S.P. (1980) Patterns of geographic variation in Florida snakes. *Bulletin of the Florida State Museum, Biological Science* 25(3), 157–256, 112 figs.

Christman, S.P. (1982) *Storeria dekayi*. *Catalogue of American amphibians and reptiles*, pp. 306.1–306.4.

Christoffel, R., Hay, R., Paloski, R. & Ramirez, L. (2008) *Snakes of Wisconsin*. Second edition. Wisconsin Department of Natural Resources, Madison, 32 pp., 6 figs., 29 col. photos. [first edition, 2000, 32 pp., 6 figs., 29 col. photos]

Christoph, H. (1861) *Pelias renardi* mihi. *Bulletin de la Société Impériale des Naturalistes de Moscou* 34(4), 599–606.

Chuaynkern, Y. & Makchal, S. (2006) Geographic distribution: Serpentes: *Boiga drapiezii*. *Herpetological Review* 37(4), 496.

Cicchi, P.J.P., Sena, M.A. de, Peccinini-Seale, D.M. & Duarte, M.R. (2007) Snakes from coastal islands of the state of São Paulo, southeastern Brazil. *Biota Neotropica* 7(2), 1–26, 44 figs.

Cihar, J. (1979) *A colour guide to familiar amphibians and reptiles*. Octopus Books, London, 192 pp., 11 figs., 64 col. pls.

Cisneros-Heredia, D.E. (2004a) Geographic distribution: Serpentes: *Bothrops brazili*. *Herpetological Review* 35(2), 189.

Cisneros-Heredia, D.E. (2004b) Geographic distribution: Serpentes: *Porthidium nasutum*. *Herpetological Review* 35(3), 293.

Cisneros-Heredia, D.E. (2004c) Geographic distribution: Serpentes: *Erythrolamprus mimus micrurus*. *Herpetological Review* 35(4), 410–411.

Cisneros-Heredia, D.E. (2004d) Geographic distribution: Serpentes: *Trachyboa boulengeri*. *Herpetological Review* 35(4), 413.

Cisneros-Heredia, D.E. (2005a) Rediscovery of the Ecuadorian snake *Atractus dunni* Savage, 1955 (Serpentes, Colubridae). *Casopis Nórodniho Muzeo, Rada Prirodovedna* 174(1–4), 87–114, 3 figs.

Cisneros-Heredia, D.E. (2005b) Geographic distribution: Serpentes: *Tretanorhinus mocquardi*. *Herpetological Review* 36(3), 340.

Cisneros-Heredia, D.E. (2007) Distribution and natural history of the Ecuadorian snake, *Dipsas andiana* (Boulenger, 1896) (Colubridae, Dipsadinae) with considerations on its conservation status. *Russian Journal of Herpetology* 14, 199–202.

Cisneros-Heredia, D.E. (2008) Reptilia, Squamata, Leptotyphlopidae, *Leptotyphlops*, re-evaluation of the species cited for the country. *Check List* 4(2), 178–181.

Cisneros-Heredia, D.E. & Touzet, J.-M. (2007) Ecuadorian distribution of snakes of the genera *Oxybelis* Wagler, 1830 and *Xenoxybelis* Machado, 1993. *Herpetozoa* 19(3–4), 188–189.

Cisneros-Heredia, D.E., Boria, M.O., Proaño, D. & Touzet, J.-M. (2006) Distribution and natural history of the Ecuadorian toad-headed pitvipers of the genus *Bothrocophias* (Squamata, Serpentes, Viperidae, Crotalinae). *Herpetozoa* 19(1–2), 17–26, 1 fig.

Cisneros-Heredia, D.E., Kuch, U., Freire-Lascano, A. & Wüster, W. (2007) Reptilia, Squamata, Colubridae, *Clelia clelia*, range extension and new provincial records from Ecuador. *Check List* 3(3), 280–281.

Claessen, H. (2001) De Slangen van de Guyana's. Deel I. *Lacerta* 59(6), 221–230, 4 col. photos.

Claessen, H. (2002a) De Slangen van de Guyana's. Deel II. *Lacerta* 60(2), 55–63, 8 col. photos.

Claessen, H. (2002b) De Slangen van de Guyana's. Deel III. *Lacerta* 60(6), 213–220, 8 col. photos.

Claessen, H. (2003a) De Slangen van de Guyana's. Deel IV. *Lacerta* 61(1), 19–27, 5 col. photos.

Claessen, H. (2003b) De Slangen van de Guyana's. Deel V. *Lacerta* 61(4), 124–133, 8 col. photos.

Claessen, H. (2003c) De Slangen van de Guyana's. Deel VI. *Lacerta* 61(5), 177–187, 7 col. photos.

Claessen, H. (2003d) De Slangen van de Guyana's. Deel VII. *Lacerta* 61(6), 221–234, 13 col. photos.

Claessen, H. (2005a) De Slangen van de Guyana's. Deel VIII. *Lacerta* 63(4), 136–151, 12 col. photos.

Claessen, H. (2005b) De Slangen van de Guyana's. Deel IX. *Lacerta* 63(6), 228–241, 6 col. photos.

Claessen, H. (2006a) De Slangen van de Guyana's. Deel X. *Lacerta* 64(3), 92–100, 3 col. photos.

Claessen, H. (2006b) De Slangen van de Guyana's. Deel XI. *Lacerta* 64(4), 136–146, 9 col. photos.

Clark, A.M., Moler, P.E., Possardt, E.E., Savitzky, A.H., Brown, W.S. & Brown, B.W. (2003) Phylogeography of the timber rattlesnake (*Crotalus horridus*) based on mtDNA sequences. *Journal of Herpetology* 37(1), 145–154, 3 figs.

Clark, D.R., Jr. (1964) The structures of the hemipenis as systematic characters in the genus *Virginia* Baird and Girard. *Herpetologica* 20(1), 33–37.

Clark, D.R., Jr. (1968) A proposal of specific status for the western worm snake, *Carphophis amoenus vermis* (Kennicott). *Herpetologica* 24(2), 104–112, 3 figs.

Clark, D.R., Jr. (1970) Ecological study of the worm snake *Carphophis vermis* (Kennicott). *University of Kansas Publications, Museum of Natural History*, Lawrence 19(2), 85–194, 48 figs., 4 pls.

Clark, P.J. & Inger, R.F. (1942) Scale reductions in snakes. *Copeia* 1942(3), 163–170.

Clark, R.F. (1949) Snakes of the Hill Parishes of Louisiana. *Journal of the Tennessee Academy of Sciences* 24, 244–261.

Clark, R.J. (1967a) Comments on the subspecies of the snake *Elaphe quatuorlineata* in Greece. *Copeia* 1967(4), 868–871, 2 figs.

Clark, R.J. (1967b) Herpetofauna of the islands of the Argo-Saronic Gulf, Greece. *Proceedings of the California Academy of Sciences* (Series 4) 35(2), 23–36, 2 figs.

Clark, R.J. (1968) A collection of snakes from Greece. *British Journal of Herpetology*, London 4(3), 45–48.

Clark, R.J. (1989) A check list of the herpetofauna of the Argo-Saronic Gulf district, Greece. *British Herpetological Society Bulletin*, London 28, 8–24, illust.

Clark, R.J. (1990) A report on herpetological observations in Afghanistan. *British Herpetological Society Bulletin*. (33), 20–42, illust.

Clark, R.J. (1994) New observations on the *Elaphe* snakes from Amorgos (Cyclades, Greece) and the validity of *Elaphe rechingeri* Werner as an endemic species. *The Herpetological Journal*, London 4(3), 91–97, 2 figs.

Clark, R.J. & Clark, E.D. (1973) Report on a collection of amphbians and reptiles from Turkey. *Occasional Papers of the California Academy of Sciences* (104), 1–62, 9 figs.

Clark, R.J., Clark, E.D., Anderson, S.C. & Leviton, A.E. (1969) Report on a collection of amphibians and reptiles from Afghanistan. *Proceedings of the California Academy of Sciences* (Series 4)36, 279–315, 5 figs.

Clarke, R.F. (1959) Poisonous snakes of Kansas. *Kansas School Naturalist* 5(3), 3–15, 15 figs., 2 col. pls.

Clarke, R.F. (1980) Snakes in Kansas. *Kansas School Naturalist* 26(3), 3–15, 21 figs.

Clarke, V.L.R. & How, R.A. (1995) Body dimensions in *Simoselaps* and *Vermicella* (Elapidae), a method for determining sex in natural populations. *Records of the Western Australian Museum* 17, 69–72.

Clary, J. & Martelli, J.-L. (1995) Catalogue des collections de reptiles du Museum d'Histoire Naturelle de Lyon. *Nouvelles Archives du Muséum d'Histoire Naturelle de Lyon* 33, 69–83.

Clauss, B. & Clauss, R. (2002) *Common amphibians and reptiles of Botswana: more than just creepy-crawlies.* Gamsberg Macmillan, Windhoek, 114 pp., illust.

Clay, W.M. (1938a) A new water snake of the genus *Natrix* from Mexico. *Annals of the Carnegie Museum* (1938–1939) 27(15), 251–253, pl. 25, 2 figs.

Clay, W.M. (1938b) A synopsis of the North American water snakes of the genus *Natrix*. *Copeia* 1938(4), 173–182.

Clemons, J. (1990) *Provisional atlas of the amphibians and reptiles of Warwickshire.* Warwickshire Museum Service, Warwick, 20 pp.

Cliburn, J.W. (1960) *The phylogeny and zoogeography of North American* Natrix. Doctoral Dissertation, University of Alabama, Tuscaloosa, 319 pp., 12 figs.

Cliburn, J.W. (1965) *A key to the amphibians and reptiles of Mississippi with guides to their study.* 3rd rev. ed., State Wildlife Museum, Jackson, 63 pp.

Cliff, F.S. (1954) Snakes of the islands in the Gulf of California, Mexico. *Transactions of the San Diego Society of Natural History* 12(5), 67–98, 1 fig., pls. 6–7.

Cloquet, H. (1818) Couleuvre. *In:* Levrault, F.G. (Ed.), 1816–1830, *Dictionnaire des sciences naturelles. Tome II. Zoologie.* F.G. Levrault, Paris, pp. 170–216, pls. 21–32.

Coborn, J. (1991) *The atlas of snakes of the world.* T.F.H. Publications, Neptune City, 591 pp., 1462 col. photos.

Coborn, J. (1994) *The mini-atlas of snakes of the world.* T.F.H. Publications, Neptune City, 736 pp., 950 col. photos.

Cochran, D.M. (1923) Two new species of *Calamaria* from Borneo. *Proceedings of the Biological Society of Washington* 36, 91–92.

Cochran, D.M. (1924) Notes on the herpetological collections made by Dr. W. L. Abbott on the island of Haiti. *Proceedings of the United States National Museum* (1924–1926) 66(2543), 1–15.

Cochran, D.M. (1927) New reptiles and batrachians collected by Dr. Hugh M. Smith in Siam. *Proceedings of the Biological Society of Washington* 40, 179–191.

Cochran, D.M. (1930) The herpetological collections made by Dr. Hugh M. Smith in Siam from 1923 to 1929. *Proceedings of the United States National Museum* 77(2834), 1–39, 6 figs.

Cochran, D.M. (1931) New reptiles from Beata Island, Dominican Republic. *Proceedings of the Biological Society of Washington* 44, 89–92.

Cochran, D.M. (1932) A new snake, *Ialtris parishi*, from the Republic of Haiti. *Proceedings of the Biological Society of Washington* 45, 189–190.

Cochran, D.M. (1934) Herpetological collections made in Hispaniola by the Utowana Expedition, 1934. *Occasional Papers of the Boston Society of Natural History* (8), 163–188.

Cochran, D.M. (1935) New reptiles and amphibians collected in Haiti by P.J. Darlington. *Proceedings of the Boston Society of Natural History* 40(6), 367–375.

Cochran, D.M. (1938) Reptiles and amphibians from the Lesser Antilles collected by Dr. S.T. Danforth. *Proceedings of the Biological Society of Washington* 51, 147–155.

Cochran, D.M. (1941) The herpetology of Hispaniola. *Bulletin of the United States National Museum* (177), vii + 397 pp., 117 figs., 12 pls.

Cochran, D.M. (1946) Notes on the herpetology of the Pearl Islands, Panamá. *Smithsonian Miscellaneous Collections* 106(4), 1–8.

Cochran, D.M. (1961) Type specimens of reptiles and amphibians in the U.S. National Museum. *Bulletin of the United States National Museum* (220), 1–291.

Cochran, D.M. & Goin, C.J. (1970) *The new field book of reptiles and amphibians*. G. P. Putnam's Sons, New York, xxii + 359 pp., 100 figs., 16 col. pls.

Cockeran, A. & Haagner, G.V. (1993) Life history notes. Serpentes, Colubridae, *Dasypeltis inornata*, brown egg eater, size, reproduction and sperm retention. *Journal of the Herpetological Association of Africa*, Stellenbosch (1992) (41), 44.

Cocteau, T. (1833) Sur le genre de reptiles ophidiens nommé *Uropeltis* par Cuvier, et description d'une espèce de ce genre. *Magasin de Zoologie*, Paris 3(2), 1–7, pl. 2.

Codoceo, M. (1957) Serpientes marinas. *Noticiario Mensual del Museo Nacional de Historia Natural,* Santiago 1(6), 7.

Cogger, H.G. (1975) Sea snakes of Australia and New Guinea. *In:* Dunson, W.A. (Ed.), *The biology of sea snakes*. University Park Press, Baltimore, pp. 59–139.

Cogger, H.G. (1979) Type specimens of reptiles and amphibians in the Australian Museum. *Records of the Australian Museum*, Sydney 32(4), 163–210.

Cogger, H.G. (1981) A biogeographic study of the Arnhem Land herpetofauna. *In: Proceedings of the Melbourne Herpetological Symposium*. Dominion Press, Blackburn, pp. 148–155, 3 figs.

Cogger, H.G. (2000) *Reptiles & amphibians of Australia*. Sixth edition. Comstock, Cornell, Ithaca, 808 pp., col. illust. [first edition, 1975, 630 pp., 786 figs.; second edition, 1979, 608 pp., 798 figs.; third edition, 1983, 630 pp., 882 figs.; fourth edition, 1986, 630 pp., 882 figs.; fifth edition, 1992, 775 pp., col. illust.]

Cogger, H.G. & Heatwole, H. (2006) *Laticauda frontalis* (de Vis, 1905) and *Laticauda saintgironsi* n. sp. From Vanuatu and New Caledonia (Serpentes: Elapidae: Laticaudinae)—a new lineage of sea kraits? *Records of the Australian Museum* 58(2), 245–256.

Cogger, H.G. & Lindner, D.A. (1974) Frogs and reptiles. Fauna survey of the Port Essington District, Cobourg Peninsula, Northern Territory of Australia. *Technical Paper of the Wildlife Survey Section CSIRO* (28), 61–107, 22 figs.

Cogger, H.G. & Zweifel, R.G. (1992) *Reptiles & amphibians*. Smithmark Publishers, New York, 240 pp., col. illust.

Cogger, H.G., Cameron, E.E. & Cogger, H.M. (1983a) *Zoological Catalogue of Australia. Volume 1. Amphibia and Reptilia*. Australian Government Publishing Service, Canberra, vi + 313 pp.

Cogger, H.G., Sadlier, R. & Cameron, E. (1983b) The terrestrial reptiles of Australia's island territories. *Australian National Parks and Wildlife Service, Special Publication* (11), 1–80, 37 figs.

Cogger, H.G., Heatwole, H., Ishikawa, Y., McCoy, M., Tamiya, N. & Teruuchi, T. (1987) The status and natural history of the Rennell Island sea krait, *Laticauda crockeri* (Serpentes, Laticaudidae). *Journal of Herpetology* 21(4), 255–266, illust.

Cogger, H.G., Cameron, E.E., Sadlier, R.A. & Eggler, P. (1993) The action plan for Australian reptiles. *Australian Nature Conservation Agency* (124), 4 + 254 pp.

Cohn, L. (1905) Schlangen aus Sumatra. *Zoologischer Anzieger*, Leipzig 29(17), 540–548.

Colares-Canto, A.L. (2000) Geographic distribution: Serpentes: *Eunectes deschauenseei*. *Herpetological Review* 31(4), 255.

Cole, C.J. & Hardy, L.M. (1981) Systematics of North American colubrid snakes related to *Tantilla planiceps* (Blainville). *Bulletin of the American Museum of Natural History* 171(3), 199–284, 39 figs.

Cole, C.J. & Hardy, L.M. (1983a) *Tantilla atriceps*. *Catalogue of American amphibians and reptiles*, pp. 317.1–317.2, 2 figs.

Cole, C.J. & Hardy, L.M. (1983b) *Tantilla hobartsmithi*. *Catalogue of American amphibians and reptiles*, pp. 318.1–318.2, 2 figs.

Cole, C.J. & Hardy, L.M. (1983c) *Tantilla planiceps*. *Catalogue of American amphibians and reptiles*, pp. 319.1–319.2, 2 figs.

Cole, C.J., Townsend, C.R., Reynolds, R.P., MacCulloch, R.D. & Lathrop, A. (2013) Amphibians and reptiles of Guyana, South America: illustrated keys, annotated species accounts, and a biogeographic synopsis. *Proceedings of the Biological Society of Washington* 125(4), 317–620, 40 col. pls..

Cole, L.R. (1967) The snake *Miodon acanthias* found with *Geotrypetes seraphini* (Amphibia, Caeciliidae) as prey. *Copeia* 1967(4), 862.

Cole, N. (2009) *A field guide to the reptiles and amphibians of Mauritius*. Mauritius Wildlife Foundation, Vacoas. iii + 80 pp., col. illust.

Colley, W.M. (1946) Snakes in Geneina. *Sudan Notes*, Khartoum 27, 213–220.

Colli, G.R. & Péres, A.K., Jr. (1997) Geographic distribution: Serpentes: *Chironius flavolineatus*. *Herpetological Review* 28(3), 158.

Colli, G.R. & Péres, A.K., Jr. (1998) Geographic distribution: Serpentes: *Chironius flavolineatus*. *Herpetological Review* 29(1), 53.

Collins, J.T. (1982) *Crotalus stejnegeri*. *Catalogue of American amphibians and reptiles*, pp. 303.1–303.2, 1 fig.

Collins, J.T. (1991) Viewpoint, a new taxonomic arrangement for some North American amphibians and reptiles. *Herpetological Review* 22(2), 42–43.

Collins, J.T. (1993) *Amphibians and reptiles in Kansas*. Third edition. University of Kansas Natural History Museum, Lawrence, 397 pp., 50 figs., 96 col. figs. [first edition, 1974, 283 pp., illust.; second edition, 1982, 356 pp., 50 figs., illust.]

Collins, J.T. & Collins, S.L. (2006) *A pocket guide to Kansas snakes*. Great Plains Nature Center, Wichita, 67 pp., col. illust.

Collins, J. & Knight, J.L. (1980) *Crotalus horridus*. *Catalogue of American amphibians and reptiles*, pp. 253.1–253.2.

Collins, J.T. & Taggart, T.W. (2008a) An alternative classification of the New World rat snakes (genus *Pantherophis* [Reptilia: Squamata: Colubridae]). *Journal of Kansas Herpetology* (26), 16–18, 1 fig.

Collins, J.T. & Taggart, T.W. (2008b) A proposal to retain *Masticophis* as the generic name for the coachwhip and whipsnakes. *Journal of Kansas Herpetology* (27), 12.

Collins, J.T. & Taggart, T.W. (2009) *Standard common and current scientific names for North American amphibians, turtles, reptiles & crocodilians*. Sixth edition. The Center for North American Herpetology, Lawrence, 44 pp. [first edition (SSAR Herp. Circ. 7), 36 pp.; second edition, 1982 (SSAR Herp. Circ. 12), 28 pp.; third edition, 1990 (SSAR Herp. Circ. 19), 41 pp.; fourth edition, 1997 (SSAR Herp. Circ. 25), 40 pp.; fifth edition, 2002, 44 pp.]

Collins, J.T., Collins, S.L. & Taggart, T.W. (2010) *Amphibians, reptiles, and turtles in Kansas*. Eagle Mountain Publishing, Eagle Mountain, xvi + 312 pp., 65 figs., 497 col. pls.

Collins, J.T., Collins, S.L. & Taggart, T.W. (2011) *Amphibians, reptiles and turtles of the Cimarron National Grassland Kansas*. United States Forest Service, Elkhart, 66 pp., col. illust.

Coloma, L.A., Quiguana, A. & Ron, S. (2000) *Reptiles de Ecuador, lista de especies y distribución. Crocodylia, serpenties, y testudines*. Version 1.1.25 Mayo 2000. Museo de Zoologia, Pontificia Universidad Católica del Ecuador. access, http://www.puce.edu.ec/zoologia/repecua.htm.

Colston, T.J., Grazziotin, F.G., Shepard, D.B., Vitt, L.J., Colli, G.R., Henderson, R.W., Hedges, S.B., Bonatto, S., Zaher, H., Noonan, B.P. & Burbrink, F.T. (2013) Molecular systematics and historical biogeography of tree boas *(Corallus* spp.). *Molecular Phylogenetics and Evolution* 66, 953–959, 2 figs.

Conant, R. (1937a) *Alsophis* from New Islands with the description of a new subspecies. *Proceedings of the New England Zoological Club* 16, 81–83.

Conant, R. (1937b) The snakes of the Philadelphia region. *Frontiers*, Los Angeles 1, 118–123.

Conant, R. (1938a) Note on the eggs and young of *Lioheterodon madagascariensis* (D. and B.). *Zoologica*, New York 23, 389–391, 1 pl.

Conant, R. (1938b) The reptiles of Ohio. *American Midland Naturalist* 20, 1–200.

Conant, R. (1940) A new subspecies of the fox snake *Elaphe vulpina* Baird and Girard. *Herpetologica* 2(1), 1–14, 1 fig.

Conant, R. (1943) Studies on North American water snakes—I. *Natrix kirtlandii* (Kennicott). *American Midland Naturalist* 29(2), 313–341, 2 figs., 1 pl.

Conant, R. (1945) *An annotated checklist of the amphibians and reptiles of the Del-Mar-Va Peninsula*. Society of Natural History of Delaware, Wilmington, 8 pp.

Conant, R. (1946) Studies in North American water snakes—II. The subspecies of *Natrix valida*. *American Midland Naturalist* 35(1), 250–275, 2 figs.

Conant, R. (1949) Two new races of *Natrix erythrogaster*. *Copeia* 1949(1), 1–15, 1 col. pl.

Conant, R. (1950) On the taxonomic status of *Thamnophis butleri* (Cope). *Bulletin of the Chicago Academy of Sciences* 9(4), 71–77.

Conant, R. (1951) *The reptiles of Ohio*. Second edition University of Notre Dame Press, Notre Dame, 284 pp.

Conant, R. (1953) Three new water snakes of the genus *Natrix* from Mexico. *Natural History Miscellanea*, Chicago (126), 1–9.

Conant, R. (1956) A review of two rare pine snakes from the Gulf Coastal Plain. *American Museum Novitates* (1781), 1–31, 12 figs.

Conant, R. (1957) *Reptiles and amphibians of the northeastern states*. Third edition. Zoological Society of Philadelphia, Philadelphia, 41 pp., illust. [first edition, 1951; second edition, 1952, 41 pp., illust.]

Conant, R. (1960) The queen snake, *Natrix septemvittata*, in the interior highlands of Arkansas and Missouri, with comments upon similar disjunct distributions. *Proceedings of the Academy of Natural Sciences of Philadelphia* 112(2), 25–40, 12 figs.

Conant, R. (1961) A new water snake from Mexico, with notes on anal plates and apical pits in *Natrix* and *Thamnophis*. *American Museum Novitates* (2060), 1–22, 2 figs.

Conant, R. (1963a) Evidence for the specific status of the water snake *Natrix fasciata*. *American Museum Novitates* (2122), 1–38, 13 figs.

Conant, R. (1963b) Semiaquatic snakes of the genus *Thamnophis* from the isolated drainage system of the Río Nazas and adjacent areas in México. *Copeia* 1963(3), 473–499, 10 figs.

Conant, R. (1963c) Another new water snake of the genus *Natrix* from the Mexican Plateau. *Proceedings of the Biological Society of Washington* 76, 169–172.

Conant, R. (1965) Miscellaneous notes and comments on toads, lizards, and snakes from Mexico. *American Museum Novitates* (2205), 1–38, 13 figs.

Conant, R. (1966) A second record for *Ungaliophis continentalis* from México. *Herpetologica* 22(2), 157–160.

Conant, R. (1969) A review of the water snakes of the genus *Natrix* in México. *Bulletin of the American Museum of Natural History* 142(1), 1–140, 3 figs., 22 pls.

Conant, R. (1975) *A field guide to reptiles and amphibians of eastern and central North America*. Second edition. Houghton Mifflin Co., Boston, xviii + 429 pp., 105 figs., 48 col. pls. [first edition, 1958, 366 pp., 61 figs., 40 col. pls.]

Conant, R. (1982) *Agkistrodon* in Europe (Reptilia, Serpentes, Viperidae). *Salamandra*, Frankfurt-am-Main 18(3–4), 191–195, 1 fig.

Conant, R. (1984) A new subspecies of the pit viper, *Agkistrodon bilineatus* (Reptilia, Viperidae) from Central America. *Proceedings of the Biological Society of Washington* 97(1), 135–141, 1 fig.

Conant, R. (1986) Phylogeny and zoogeography of the genus *Agkistrodon* in North America. *In:* Rocek, Z. (Ed.), *Studies in herpetology*. Charles University, Prague, pp. 89–92.

Conant, R. (1990) The fossil history of the genus *Agkistrodon* in North America. *In:* Gloyd, H.K. & Conant, R. *Snakes of the Agkistrodon complex. A monographic review*. Society for the Study of Amphibians and Reptiles, pp. 539–543.

Conant, R. (2000) A new species of garter snake from western Mexico. *Occasional Papers of the Museum of Natural Science, Louisiana State University* (76), 1–7, fig. 3.

Conant, R. (2003) Observations on garter snakes of the *Thamnophis eques* complex in the lakes of Mexico's Transvolcanic belt, with description of new taxa. *American Museum Novitates* (3406), 1–64, 20 figs.

Conant, R. & Clay, W.M. (1937) A new subspecies of water snake from islands in Lake Erie. *Occasional Papers of the Museum of Zoology, University of Michigan* (346), 1–9, 3 pls.

Conant, R. & Clay, W.M. (1963) A reassessment of the taxonomic status of the Lake Erie water snake. *Herpetologica* 19(3), 179–184.

Conant, R. & Collins, J.T. (1991) *A field guide to reptiles and amphibians*. Eastern and central North America. Houghton Mifflin Co., Boston, xviii + 450 pp., 120 figs., 48 pls.

Conant, R. & Collins, J.T. (1998) *A field guide to reptiles and amphibians of eastern and central North America*. Third edition. Houghton Mifflin Co., Boston, 616 pp., col. figs., 48 col. pls.

Conant, R., Cagle, F.R., Goin, C.J., Lowe, C.H., Jr., Neill, W.T., Netting, M.G., Stebbins, R. & Bogert, C.M. (1956) Common names for North American amphibians and reptiles. *Copeia* 1956(3), 172–185.

Conant, R. & Lazell, J.D., Jr. (1973) The Carolina salt marsh snake, a distinct form of *Natrix sipedon*. *Breviora* (400), 1–13, 2 figs.

Condamin, M. (1958) La collection de serpents de l'I.F.A.N. (acquisitions 1956). *Bulletin de l'Institut Française d'Afrique Noire*, Dakar 20A(1), 243–262, 2 figs.

Condamin, M. (1959) Serpents récoltés à Sérédon (Guinée) par R. Pujol. *Bulletin de l'Institut Française d'Afrique Noire*, Dakar 21A(4), 1351–1366, 1 fig.

Condamin, M. & Villiers, A. (1962) Contribution à l'étude de la faune de la basse Casamance. II. Reptiles. *Bulletin de l'Institut Française d'Afrique Noire*, Dakar 24A(3), 897–908, 5 figs.

Condrea, E. & Vries, A. de. (1965) Venom phospholipase A: a review. *Toxicon* 2(4), 261–273.

Constable, J.D. (1949) Reptiles from the Indian Peninsula in the Museum of Comparative Zoology. *Bulletin of the Museum of Comparative Zoology* 103(2), 59–160.

Cook, F.A. (1962) *Snakes of Mississippi*. Second edition. Wildlife Museum, Jackson, ii + 45 pp. [first edition, 1954, ii + 40 pp.]

Cook, F.R. (1964) Communal egg laying in the smooth green green snake. *Herpetologica* 20(3), 206.

Cook, F.R. (1967) An analysis of the herpetofauna of Prince Edward Island. *Bulletin of the Natlional Museum of Canada* (212), i–vii + 1–60, 26 figs.

Cook, F.R. (1977a) Amphibians and reptiles of Saskatchewan. *Saskatchewan Museum of Natural History, Popular Series*, Regina (13), 1–40, illust.

Cook, F.R. (1977b) Review of the Canadian herpetological scene. *In: Canada's threatened species and habitats. Proceedings of the Symposium on Canada's Threatened Species and Habitats, co-sponsored by the Canadian Nature Federation and the World Wildlife Fund (Canada).* Canadian Nature Federation, Ottawa (1976), pp. 117–121, illust.

Cook, F.R. (1980) Checklist of amphibians and reptiles of Canada. *Canadian Amphibian and Reptile Conservation Society*, Toronto 18(2), 1–6.

Cook, F.R. (1984) *Introduction to Canadian amphibians and reptiles*. National Museum of Natural Science, National Museum of Canada, Ottowa, 200 pp.

Cooper, J.G. (1860) Zoological report. No. 4. Report upon the reptiles collected on the survey. *In:* Stevens, I.I. (Ed.), *Reports of explorations and surveys, to ascertain the most practicable and economical route for a railroad from the Mississippi River to the Pacific Ocean, made under the direction of the Secretary of War, in 1853-5. Route near the forty-seventh and forty-ninth parallels.* 12(2), 292–306, pls. 12–16, 19–22, 29, 31.

Cooper, J.G. (1868) Zoology. *In:* Cronise, T.F. (Ed.), *The natural wealth of California*. H.H. Bancroft & Company, San Francisco, pp. 434–501.

Cooper, J.G. (1870) The fauna of California and its geographical distribution. *Proceedings of the California Academy of Sciences,* San Francisco (1868–1872) 4(2), 61–81.

Cope, E.D. (1860a) Notes and descriptions of foreign reptiles. *Proceedings of the Academy of Natural Sciences of Philadelphia* (1859) 11(10), 294–297.

Cope, E.D. (1860b) Catalogue of the venomous serpents in the Museum of the Academy of Natural Sciences of Philadelphia, with notes on the families, genera and species. *Proceedings of the Academy of Natural Sciences of Philadelphia* (1859) 11(12), 332–347.

Cope, E.D. (1860c) Supplement to "A catalogue of the venomous serpents in the museum of the Academy," etc. *Proceedings of the Academy of Natural Sciences of Philadelphia* 12(2), 72–74.

Cope, E.D. (1860d) Catalogue of Colubridae in the Museum of the Academy of Natural Sciences of Philadelphia. I. Calamarinae. *Proceedings of the Academy of Natural Sciences of Philadelphia* 12(2), 74–79.

Cope, E.D. (1860e) Catalogue of the Colubridae in the Museum of the Academy of Natural Sciences of Philadelphia, with notes and descriptions of new species. Part 2. *Proceedings of the Academy of Natural Sciences of Philadelphia* 12(6), 241–266.

Cope, E.D. (1860f) Notes and descriptions of new and little known species of American reptiles. *Proceedings of the Academy of Natural Sciences of Philadelphia* 12(8), 339–345.

Cope, E.D. (1861a) Descriptions of reptiles from tropical America and Asia. *Proceedings of the Academy of Natural Sciences of Philadelphia* (1860) 12(9), 368–374.

Cope, E.D. (1861b) Description of new species of the reptilian genera *Hyperolius*, *Liuperus* and *Tropidodipsas*. *Proceedings of the Academy of Natural Sciences of Philadelphia* (1860) 12(12), 517–518.

Cope, E.D. (1861c) Catalogue of the Colubridae in the Museum of the Academy of Natural Sciences of Philadelphia. Part 3. *Proceedings of the Academy of Natural Sciences of Philadelphia* (1860) 12(12), 553–566.

Cope, E.D. (1861d) "Some remarks defining...species of Reptilia Squamata..." *Proceedings of the Academy of Natural Sciences of Philadelphia* 13(3), 73–77.

Cope, E.D. (1861e) Contributions to the ophiology of Lower California, Mexico and Central America. *Proceedings of the Academy of Natural Sciences of Philadelphia* 13(7), 292–306.

Cope, E.D. (1862a) "Permission was granted to Mr. E.D. Cope to state to the Academy a few observations..." *Proceedings of the Academy of Natural Sciences of Philadelphia* (1861) 13(9), 522–524.

Cope, E.D. (1862b) Synopsis of the species of *Holcosus* and *Ameiva*, with diagnoses of new West Indian and South American Colubridae. *Proceedings of the Academy of Natural Sciences of Philadelphia* 14(1), 60–82.

Cope, E.D. (1862c) Notes upon some reptiles of the Old World. *Proceedings of the Academy of Natural Sciences of Philadelphia* 14(5), 337–344.

Cope, E.D. (1862d) Catalogues of the reptiles obtained during the explorations of the Parana, Paraguay, Vermejo and Uraguay Rivers, by Capt. Thos. J. Page, U.S.N.; and of those procured by Lieut. N. Michler, U.S. Top. Eng., Commander of the expedition conducting the survey of the Atrato River. I. The Paraguay collection. *Proceedings of the Academy of Natural Sciences of Philadelphia* 14(5), 346–359.

Cope, E.D. (1862e) Errata and addenda. *Proceedings of the Academy of Natural Sciences of Philadelphia* 14(7), 594.

Cope, E.D. (1862f) On *Carphoptis harpesti*. *Proceedings of the Academy of Natural Sciences of Philadelphia* 14(4), 249.

Cope, E.D. (1863a) On a species of *Vipera* hitherto unknown. *Proceedings of the Zoological Society of London* 31(1), 229–230, 1 fig.

Cope, E.D. (1863b) Descriptions of new American Squamata, in the Museum of the Smithsonian Institution, Washington. *Proceedings of the Academy of Natural Sciences of Philadelphia* 15(3), 100–106.

Cope, E.D. (1863c) On part II of Prof. G. Jan's Prodromo della iconografia generale degli ofidi. *The American Journal of Science and Arts,* New Haven (Series 2) 35(105), 455–458.

Cope, E.D. (1864) Contributions to the herpetology of tropical America. *Proceedings of the Academy of Natural Sciences of Philadelphia* 16(3), 166–181.

Cope, E.D. (1865) Third contribution to the herpetology of tropical America. *Proceedings of the Academy of Natural Sciences of Philadelphia* 17(4), 185–198.

Cope, E.D. (1866) Fourth contribution to the herpetology of tropical America. *Proceedings of the Academy of Natural Sciences of Philadelphia* 18(2), 123–132.

Cope, E.D. (1867a) On the Reptilia and Batrachia of the Sonoran Province of the Neartic Region. *Proceedings of the Academy of Natural Sciences of Philadelphia* (1866) 18(4), 300–314.

Cope, E.D. (1867b) Fifth contribution to the herpetology of tropical America. *Proceedings of the Academy of Natural Sciences of Philadelphia* (1866) 18(4), 317–323.

Cope, E.D. (1868a) Observations on some vertebrates presented by Wm. M. Gabb, of San Franciso, which were procured by him in western Nevada and the northern part of Lower California. *Proceedings of the Academy of Natural Sciences of Philadelphia* 20(1), 2.

Cope, E.D. (1868b) An examination of the Reptilia and Batrachia obtained by the Orton Expedition to Equador and the Upper Amazon, with notes on other species. *Proceedings of the Academy of Natural Sciences of Philadelphia* 20(2), 96–119.

Cope, E.D. (1868c) Additional descriptions of Neotropical Reptilia and Batrachia not previously known. *Proceedings of the Academy of Natural Sciences of Philadelphia* 20(2), 119–140.

Cope, E.D. (1868d) On the vertebrae of a serpent from the Greensand of New Jersey. *Proceedings of the Academy of Natural Sciences of Philadelphia* 20(2), 147.

Cope, E.D. (1869a) On some Cretaceous reptiles. *Proceedings of the Academy of Natural Sciences of Philadelphia* (1868) 20(4), 233–242.

Cope, E.D. (1869b) Sixth contribution to the herpetology of tropical America. *Proceedings of the Academy of Natural Sciences of Philadelphia* (1868) 20(5), 305–313.

Cope, E.D. (1869c) Observations on reptiles of the Old World. Art. II. *Proceedings of the Academy of Natural Sciences of Philadelphia* (1868) 20(5), 316–323.

Cope, E.D. (1869d) Synopsis of the extinct Batrachia, Reptilia, and Aves of North America. Transactions of the American Philosophical Society (Series 2) 14, 1-252, pls.

Cope, E.D. (1870) Seventh contribution to the herpetology of tropical America. *Proceedings of the American Philosophical Society* (1869–1870) 11(81), 147–169, pls. 9–11.

Cope, E.D. (1871a) Eighth contribution to the herpetology of tropical America. *Proceedings of the American Philosophical Society* (1869–1870) 11(81), 553–559.

Cope, E.D. (1871b) Ninth contribution to the herpetology of tropical America. *Proceedings of the Academy of Natural Sciences of Philadelphia* 23(2), 200–224, pl. 17.

Cope, E.D. (1871c) Catalogue of Batrachia and Reptilia obtained by J.A. McNiel in Nicaragua. *Second and Third Annual Reports of the Trustees of the Peabody Academy of Science*, Salem, pp. 80–82.

Cope, E.D. (1872) Third account of new Vertebrata from the Bridger Eocene of Wyoming Territory. *Proceedings of the American Philosophical Society* (1871–1872) 12(86), 469–472.

Cope, E.D. (1873) *Synopsis of new Vertebrata from the Tertiary of Colorado, obtained during the summer of 1873*. Government Printing Office, Washington 19 pp. [partial preprint of Cope, 1874a]

Cope, E.D. (1874a) Report on the vertebrate paleontology of Colorado. *In:* Hayden, F.V. (Ed.), *7th Annual report of the United States Geological and Geographical Survey of the Territories, embracing Colorado, being a report of progress of the exploration for the year 1873*. Government Printing Office, Washington, pp. 427–533, 8 pls.

Cope, E.D. (1874b) Description of some species of reptiles obtained by Dr. John F. Bransford, assistant surgeon United States Navy, while attached to the Nicaraguan Surveying Expedition in 1873. *Proceedings of the Academy of Natural Sciences of Philadelphia* 26(1), 64–72.

Cope, E.D. (1875a) On the Batrachia and Reptilia of Costa Rica. With notes on the herpetology and ichthyology of Nicaragua and Peru. *Journal of the Academy of Natural Sciences of Philadelphia* (1876) (Series 2) 8(2), 93–154, pls. 23–28. [preprint in 1875]

Cope, E.D. (1875b) Report on the reptiles brought by Professor James Orton from the middle and upper Amazon, and western Peru. *Journal of the Academy of Natural Sciences of Philadelphia* (1876) (Series 2) 8(2), 159–183. [preprint in 1875]

Cope, E.D. (1875c) *Check-list of North American Batrachia and Reptilia, with a systematic list of the higher groups, and an essay on Geographic distribution: Serpentes: Based on the specimens in the U.S. National Museum*. Government Printing Office, Washington, 104 pp.

Cope, E.D. (1878a) Synopsis of the cold-blooded Vertebrata, procured by Prof. James Orton during his expedition of Peru in 1876–1877. *Proceedings of the American Philosophical Society* (1877–1878) 17(101), 33–49.

Cope, E.D. (1878b) Tenth contribution to the herpetology of tropical America. *Proceedings of the American Philosophical Society* (1877–1878) 17(101), 85–97.

Cope, E.D. (1879) Eleventh contribution to the herpetology of tropical America. *Proceedings of the American Philosophical Society* (1878–1880) 18(104), 261–277.

Cope, E.D. (1880) On the zoological position of Texas. *Bulletin of the United States National Museum* (17), 1–51.

Cope, E.D. (1883a) First addition to the fauna of the Puerco Eocene. *Proceedings of the American Philosophical Society* (1882–1883) 20(113), 545–563. [reprinted in 1883 by *Paleontological Bulletin* 36]

Cope, E.D. (1883b) New Mammalia from the Puerco Eocene. *The American Naturalist* 17(2), 191.

Cope, E.D. (1883c) A new snake from New Mexico. *The American Naturalist* 17(12), 1300–1301.

Cope, E.D. (1884a) Notes on the geographical distribution of Batrachia and Reptilia in western North America. *Proceedings of the Academy of Natural Sciences of Philadelphia* (1883) 35, 10–35.

Cope, E.D. (1884b) The Vertebrata of the Tertiary formations of the West. Book I. Part First. The Puerco, Wasatch, and Bridger faunae. Part Second. The White River and John Day faunae. *In:* Hayden, F.V. (Ed.), *Reports of the United States Geological Survey of the Territories. Vol. III.* Government Printing Office, Washington, xxxv + 1009 pp., 75 pls.

Cope, E.D. (1885a) Twelfth contribution to the herpetology of tropical America. *Proceedings of the American Philosophical Society* 22(118), 167–194, 1 pl.

Cope, E.D. (1885b) A contribution to the herpetology of Mexico. *Proceedings of the American Philosophical Society* 22(120), 379–404.

Cope, E.D. (1886a) Catalogue of the species of batrachians and reptiles contained in a collection made at Pebas, Upper Amazon, by John Hauxwell. *Proceedings of the American Philosophical Society* 23(121), 94–103.

Cope, E.D. (1886b) Thirteenth contribution to the herpetology of tropical America. *Proceedings of the American Philosophical Society* 23(122), 271–287.

Cope, E.D. (1886c) An analytical table of the genera of snakes. *Proceedings of the American Philosophical Society* 23(124), 479–499.

Cope, E.D. (1887a) Brazilian Reptilia. *American Naturalist* 21, 388–389.

Cope, E.D. (1887b) Synopsis of the Batrachia and Reptilia obtained by H.H. Smith, in the province of Mato Grosso, Brazil. *Proceedings of the American Philosophical Society* 24(125), 44–60.

Cope, E.D. (1887c) Catalogue of batrachians and reptiles of Central America and Mexico. *Bulletin of the United States National Museum* (32), 1–98.

Cope, E.D. (1887d) On a new species of *Tropidonotus* found in Washington. *Proceedings of the United States National Museum* (1887–1888) 10(615), 146.

Cope, E.D. (1888) On a new species of *Charina* from California. *Proceedings of the United States National Museum* (1888–1889) 11(689), 88, pl. 36, fig. 2.

Cope, E.D. (1889a) On the snakes of Florida. *Proceedings of the United States National Museum* (1888–1889) 11(727), 381–394, pl. 36, figs. 3–4. [7 July 1889]

Cope, E.D. (1889b) On the Eutaeniae of southeastern Indiana. *Proceedings of the United States National Museum* (1888–1889) 11(729), 399–401. [7 July 1889]

Cope, E.D. (1891a) The epiglottis in colubrine snakes. *The American Naturalist* 25(291), 156–157, 1 fig.

Cope, E.D. (1891b) Snakes in banana bunches. *American Naturalist* 25, 742.

Cope, E.D. (1892a) A critical review of the characters and variations of the snakes of North America. *Proceedings of the United States National Museum* (1891–1892) 14(882), 589–694.

Cope, E.D. (1892b) A new species of *Eutaenia* from western Pennsylvania. *The American Naturalist* 26, 964–965.

Cope, E.D. (1892c) Remarks on the communication "Ein fossiler Giftzahn," by Dr. F. Kinkelin. *Zoologischer Anzieger*, Leipzig 15(393), 224.

Cope, E.D. (1893a) Prodromus of a new system of the non-venomous snakes. *The American Naturalist* 27(317), 477–483.

Cope, E.D. (1893b) The color variation of the milk snake. *The American Naturalist* 27(324), 1066–1071, pl. 24–28.

Cope, E.D. (1894a) Second addition to the knowledge of the Batrachia and Reptilia of Costa Rica. *Proceedings of the American Philosophical Society* (1893) 31(142), 333–347.

Cope, E.D. (1894b) On the species of *Himantodes* D.& B. *The American Naturalist* 28(331), 612–614.

Cope, E.D. (1894c) The classification of snakes. *The American Naturalist* 28(334), 831–844, pls. 27–28.

Cope, E.D. (1894d) Third addition to a knowledge of the Batrachia and Reptilia of Costa Rica. *Proceedings of the Academy of Natural Sciences of Philadelphia* 46, 194–206, pl. 10.

Cope, E.D. (1895a) On a collection of Batrachia and Reptilia from the island of Hainan. *Proceedings of the Academy of Natural Sciences of Philadelphia* (1894) 46, 423–428.

Cope, E.D. (1895b) On some new North American snakes. *The American Naturalist* 29(343), 676–680.

Cope, E.D. (1895c) The classification of the Ophidia. *Transactions of the American Philosophical Society* (Series 2) (1894) 18(3), 186–219, pls. 14–23.

Cope, E.D. (1896a) On a new *Glauconia* from New Mexico. *The American Naturalist* 30(357), 753.

Cope, E.D. (1896b) Boulenger's Catalogue of Snakes in the British Museum. *The American Naturalist* 30(360), 1029–1031.

Cope, E.D. (1898) *Syllabus of lectures on the Vertebrata.* University of Pennsylvania, Philadelphia, 135 pp.

Cope, E.D. (1899a) Vertebrate remains from Port Kennedy bone deposit. *Journal of the Academy of Natural Sciences of Philadelphia* (Series 2) 11, 193–286.

Cope, E.D. (1899b) Contributions to the herpetology of New Granada and Argentina, with descriptions of new forms. On a collection of Batrachia and Reptilia from New Granada. *Philadelphia Commercial Museum Scientific Bulletin* (1), 1–22, 4 pls. [published posthumously and edited by J.P. Moore]

Cope, E.D. (1900) The crocodilians, lizards, and snakes of North America. *Report of the United States National Museum* 1898, 153–1270, 347 figs., 36 pls. [published posthumously]

Corbett, K. (Ed.). (1989) *The conservation of European reptiles and amphibians.* Christopher Helm, London, 274 pp., illust.

Cordeiro, C.L. & Hoge, A.R. (1974) Contribuição ao conhecimento das serpentes do Estado de Pernambuco. *Memórias do Instituto Butantan*, São Paulo (1973) 37, 261–290.

Corkill, N.L. (1932a) *Snakes and snake bite in Iraq. A handbook for medical officers.* Bailliere, Tindall and Cox, London, 50 pp., 20 pls.

Corkill, N.L. (1932b) Snakes of Iraq. *The Journal of the Bombay Natural History Society* (1931–1932) 35(3), 552–572.

Corkill, N.L. (1935) Notes on Sudan snakes. A guide to the species represented in the collection in the Natural History Museum Khartoum. *Sudan Government Museum (Natural History)*, Khartoum (3), 1–41.

Corkill, N.L. (1956) Snake poisoning in the Sudan. *In:* Buckley, E.E. & Porges, N. (Eds.), *Venoms.* American Association for the Advancement of Science, Washington, D.C., pp. 331–339.

Corkill, N.L. & Cochrane, J.A. (1965) The snakes of the Arabian Peninsula and Socotra. *The Journal of the Bombay Natural History Society* 62(3), 475–506.

Corn, M.J. (1974) Report on the first certain collection of *Ungaliophis panamensis* from Costa Rica. *Caribbean Journal of Science*, Mayagüez 14(3–4), 167–175, 2 figs.

Corn, P.S. & Bury, R.B. (1986) Morphological variation and zoogeography of racers (*Coluber constrictor*) in the central Rocky Mountains. *Herpetologica* 42(2), 258–264, 3 figs.

Corn, P.S., Bury, R.B. & Welch, H.H. (1984) Selected bibliography of Wyoming amphibians and reptiles. *Smithsonian Herpetological Information Service* (59), 1–34.

Cornelissen, T. (1970) *Erpeton tentaculatum*, een vissende slang (Serpentes, Colubridae, Homalopsinae). *Lacerta*, Gravenhage 28, 88–92 & 29, 3–6, 3 figs., 17 illust.

Corsetti, L. & Capula, M. (1992) The amphibians and reptiles of the Lepini Mountains (Latium, central Italy), checklist and provisional atlas. *British Herpetological Society Bulletin*, London (39), 8–15, 6 figs.

Cortés, J.E. (1982) The herpetofauna of Gibraltar—status, recent history and current research. *British Journal of Herpetology*, London 6, 273–275.

Corti, C. & Luiselli, L. (2001) *Macroprotodon cucullatus* on Lampedusa Island (Mediterranean Sea); notes on its natural history, morphometrics, and conservation. *Amphibia-Reptilia,* Leiden 22(1), 129–134.

Corti, C., Lo-Cascio, P., Vanni, S., Turrisi, G. & Vaccaro, A. (1997) Amphibians and reptiles of the circumsicilian islands: new data and some considerations. *Bollettino del Museo Regionale di Scienze Naturali*, Torino 15(1), 179–211, illust.

Corti, C., Luiselli, L., Filippi, E. & Capula, M. (2000) Distribution, natural history and morphometrics of the critically endangered *Coluber hippocrepis* populations of Sardina, a review, with additional data and conservation implications. *Amphibia-Reptilia,* Leiden 21(3), 279–287, 1 fig.

Costa, H.C., Pinto, R.R. & Santana, D.J. (2009) Reptilia, Leptotyphlopidae, *Leptotyphlops salgueiroi* Amaral, 1954: distribution extension and geographic variation. *Check List* 5(4), 783–786, 2 figs.

Costa, H.C., São-Pedro, V. de A. & Feio, R.N. (2010a) A new record of the poorly known Amazonian snake *Siphlophis worontzowi*. *Herpetology Notes* 3, 197–100, 3 figs.

Costa, H.C., Pantoia, D.L., Pontes, J.L. & Feio, R.N. (2010b) Snakes do Municipio de Viçosa, Mata Atlántica do Sudeste do Brasil. *Biota Neotropica* 10(3), 352–377, 8 col. figs.

Costa, H.C., Barros, A.B., Sueiro, L.R. & Feio, R.N. (2010c) The blunt-headed vine snake, *Imantodes cenchoa* (Linnaeus, 1758), in Minas Gerais, southeastern Brazil. *Biotemas* 23(4), 173–176, 1 fig.

Costa-Prudente, A.L. da & Brandão, R.A. (1998) Geographic distribution: Serpentes: *Gomesophis brasiliensis*. *Herpetological Review* 29(2), 112.

Costa-Prudente, A.L. & dos Santos-Costa, M.C. (2004) Geographic distribution: Serpentes: *Atractus major*. *Herpetological Review* 35(2), 189.

Costa-Prudente, A.L. & dos Santos-Costa, M.C. (2006) A new species of *Atractus* Wagler, 1828 (Serpentes, Colubridae) from eastern Amazonia, Brazil. *Zootaxa* (1285), 21–29, figs.

Costa-Prudente, A.L., de Moura-Leite, J.C. & Morato, S.A.A. (1998) Alimentação das espécies de *Siphlophis* Fitzinger (Serpentes, Colubridae, Xenodontinae, Pseudoboini). *Revista Brasiliera de Zoologia* 15(2), 375–383.

Cota, M., Mendyk, R.W., Chan-ard, T., Makchai, S. & Meewattana, P. (2011) Range extension, new records and natural history notes on *Lepturophis albofuscus* (Duméril, Bibron & Duméril, 1854) (Squamata: Colubridae) in Thailand. *The Thailand Natural History Journal* 5(2), 151–155.

Cott, H.B. (1928) Report on the Zoological Society's expedition to the Zambesi, 1927. *Proceedings of the Zoological Society of London* 98(4), 923–961, 4 pls.

Cott, H.B. (1935) The Zoological Society's expedition to the Zambesi, 1927, No. 6. On a collection of snakes from Portuguese East Africa. *Proceedings of the Zoological Society of London* 98(4), 963–975, 1 pl.

Cottone, A.M. & Bauer, A.M. (2010) Sexual dimorphism, diet, reproduction, and their geographic variation in sympatric psammophiids, *Psammophis crucifer* and *Psammophylax rhombeatus rhombeatus*, from southern Africa. *Copeia* 2010(4), 578–590, 2 figs.

Coues, E. (1875) Synopsis of the reptiles and batrachians of Arizona; with critical and field notes, and an extensive synonymy. *In:* Wheeler, G.M. (Ed.), *Report upon geographical and geological explorations and surveys west of the One Hundredth Meridian. Vol. 5.–Zoology.* Government Printing Office, Washington, pp. 585–633, pls. 16–25.

Coues, E. & Yarrow, H.C. (1878) *Notes on the herpetology of Dakota and Montana.* Department of the Interior, United States Geological and Geographical Survey, Washington, 259–294.

Couper, J., Chiszar, D. & Smith, H.M. (1996) Geographic distribution: Serpentes: *Drymarchon couperi*. *Herpetological Review* 27(3), 154.

Couper, P.J., Covacevich, J.A. & Wilson, S.K. (1998) Two new species of *Ramphotyphlops* (Squamata: Typhlopidae) from Queensland. *Memoirs of the Queensland Museum* 42(2), 459–464.

Courtois, B. (1978) Note sur la description de deux specimens de *Pseudohaje goldii* (Boulenger) trouvés en Côte d'Ivoire (Serpentes, Elapidae). *Bulletin de l'Institut Fondamental d'Afrique Noire*, Dakar 40A(1), 212–214.

Courtois, B. (1979) A propos de deux noureaux spécimens d'*Atractaspis corpulenta* (Hallowell) trouvés en Côte d'Ivoire. *Bulletin de l'Institut Fondamental d'Afrique Noire*, Dakar 41A(1), 206–209, 2 figs.

Courtois, B. & Chippaux, J.-P. (1977) *Serpentes venimeux en Côte d'Ivoire.* Réalisation, Maison Carree, 91 pp., 4 figs., 7 illust.

Couterier, G. & Faivovich, J. (1996) *Clelia bicolor* (Peracca) en la Provincia de Santa Fe. *Cuadernos de Herpetologia* 10, 71.

Covacevich, J.A. (1971) Amphibians and reptile type-specimens in the Queensland Museum. *Memoirs of the Queensland Museum*, Brisbane 16(1), 49–68.

Covacevich, J.A. (1981) The snakes of Brisbane (rev. ed.). *Queensland Museum Booklet,* Brisbane (4), 1–20, 10 figs., plus unnumbered illust.

Covacevich, J.A. (1987) Reptiles of far northern Cape York Peninsula. *Queensland Naturalist* 28(1–4), 22–30, illust.

Covacevich, J.A. & Couper, P.J. (1991) The reptile records. *In:* Ingram, J.G. & Raven, R.J. (Eds.), *An atlas of Queensland's frogs, reptiles, birds and mammals.* Queensland Museum, Brisbane, pp. 45–140.

Covacevich, J.A. & McDonald, K.R. (1993) Distribution and conservation of frogs and reptiles of Queensland rainforests. *Memoirs of the Queensland Museum* 34(1), 189–199, illust.

Covacevich, J. & Wombey, J. (1976) Recognition of *Parademansia microlepidotus* (McCoy) (Elapidae), a dangerous Australian snake. *Proceedings of the Royal Society of Queensland*, Brisbane 87, 29–32, 1 fig., 2 pls.

Covacevich, J.A., McDowell, S.B., Jr., Tanner, C. & Mengden, G.A. (1981) The relationship of the taipan, *Oxyuranus scutellatus*, and the small-scaled snake, *Oxyurnaus microlepidotus* (Serpentes, Elapidae). *In:* Banks, C.B. & Martin, A.A. (Eds.), *Proceedings of the Melbourne Herpetological Symposium.* Dominion Press, Blackburn, pp. 160–168, 3 pls.

Covacevich, J.A., Davie, P. & Pearn, J. (Eds.). (1987) *Toxic plants & animals. A guide for Australia.* Old Museum, Brisbane, viii + 501 pp., illust.

Coventry, A.J. (1970) Reptile and amphibian type specimens housed in the National Museum of Victoria. *Memoirs of the National Museum of Victoria*, Melbourne 31, 115–124.

Coventry, A.J. (1971) Identification of the black-headed snakes (*Denisonia*) within Victoria. *The Victorian Naturalist* 88, 304–306, 1 fig.

Coventry, A.J. & Rawlinson, P.A. (1980) Taxonomic revision of the elapid genus *Drysdalia* Worrell 1961. *Memoirs of the National Museum of Victoria*, Melbourne 41, 65–78, pl. 12.

Coventry, A.J. & Robertson, P. (1991) *The snakes of Victoria.* Victorian Department of Conservation and Environment, East Melbourne, 70 pp.

Cowan, I.M. (1936) *A review of the reptiles and amphibians of British Columbia.* Report for the Year, Provincial Museum of Natural History and Anthropology, Victoria 1936, 16–25.

Cowan, I.M. & Hick, W.B.M. (1951) A comparative study of the mycology of the head region in three species of *Thamnophis* (Reptilia, Ophidia). *Transactions of the Royal Society of Canada* (Series 3) 45(5), 19–60, 10 figs.

Cox, C.L., Davis-Rabosky, A.R., Reyes-Velasco, J., Ponce-Campos, P., Smith, E.N., Flores-Villela, O. & Campbell, J. (2012). Molecular systematics of the genus *Sonora* (Squamata: Colubridae) in central and western Mexico. *Systematics and Biodiversity* 10(1), 93–108, 14 figs.

Cox, D.C. & Tanner, W.W. (1995) *Snakes of Utah.* Brigham Young University, Provo, 92 pp.

Cox, M.J. (1991a) Geographic distribution: Serpentes: *Lycodon butleri* (Butler's wolf snake). *Herpetological Review* 22(1), 26.

Cox, M.J. (1991b) *The snakes of Thailand and their husbandry.* Robert E. Krieger, Malabar, xxxviii + 526 pp., 36 figs. 164 pls.

Cox, M.J. (1995) Geographic distribution: Serpentes: *Opisthotrophis boonsongi* (Boonsong's stream snake). *Herpetological Review* 26(3), 157.

Cox, M.J., Hoover, M.F., Chanhome, L. & Thirakhupt, K. (2013) *The snakes of Thailand.* Chulalongkorn University, Bangkok (2012), 845 pp.

Cox, M.J. & Nabhitabhata, J. (1997) Geographic distribution: Serpentes: *Typhlops porrectus*. *Herpetological Review* 28(1), 53.

Cox, M.J., van Dijk, P.P., Nabhitabhata, J. & Thirakhupt, K. (1998) *A photographic guide to snakes and other reptiles of Peninsular Malaysia, Singapore and Thailand.* Ralph Curtis Books, Sanibel Island, 144 pp., col. illust.

Cox, M.J., Hoover, M.F., Chanhome, L. & Thirakhupt, K. (2013) *The snakes of Thailand.* Chulalongkorn University Museum of Natural History, Bangkok (2012), 845 pp., 179 col. photos.

Cox, P. (1903) The snakes of the Maritime provinces of Canada. *Proceedings of the Miramichi Natural History Association* (3), 11–20.

Cranston, T. (1991) Notes on the natural history, husbandry, and breeding of the gray-banded kingsnake (*Lampropeltis alterna*). *The Vivarium* 3(2), 7–10, 5 illust.

Cranston, T. (1993a) The natural history and captive husbandry of the Trans-Pecos rat snake (*Bogertophis subocularis*). *The Vivarium* 4(5), 18–20, 3 col. photos.

Cranston, T. (1993b) Notes on the natural history and captive husbandry of the Sonoran mountain kingsnake (*Lampropeltis pyromelana*). *The Vivarium* 5(6), 15–17, 35, 4 col. photos.

Cranston, T. (1994a) Common kingsnakes (*Lampropeltis getula*). *The Vivarium* 5(4), 30–33, 46–47, 54, 8 col. photos.

Cranston, T. (1994b) Natural history of the Sierra Mountain kingsnake (*Lampropeltis zonata multicincta*). *The Vivarium* 6(3), 38–43, 47, 11 col. photos.

Cranston, T. & Cranston, E. (1996) Natural history and captive care of Baird's rat snake, *Elaphe bairdi*. *The Vivarium* 8(1), 60–62, 5 col. photos.

Cranston, T. & Loza, E. (1995) Captive care and propagation of the California kingsnake (*Lampropeltis getula californiae*). *The Vivarium* 6(5), 50–54, 10 col. photos.

Cranwell, J.A. (1942) Consideraciones sobre "*Rhadinaea elegantissima*" de Koslowsky. *Revista Argentina de Zoogeografía*, Buenos Aires 2(3), 143–146.

Craven, R. & Shea, G.M. (2010) Geographic distribution: Serpentes: *Ramphotyphlops braminus*. *Herpetological Review* 41(4), 518.

Crawford-Cabral, J. & Mesquitela, L.M. (1989) *Índice toponímico de colheitas zoológicas em Angola.* Instituto de Investigação Cientifica Tropical, Centro de Zoologia, Lisbon, 206 pp.

Creemers, R.C.M. & Delft, J.J.C.W. van. (2009) *De amfibieën en reptielen van Nederland.* Nationaal Natuurhistorisch Museum Naturalis, Leiden, 476 pp., col. illust.

Creer, S., Malhotra, A., Thorpe, R.S. & Chou, W.-H. (2001) Multiple causation of phylogeographical pattern as revealed by nested clade analysis of the bamboo viper (*Trimeresurus stejnegeri*) within Taiwan. *Molecular Ecology* 10(8), 1967–1981, 4 figs.

Creer, S., Malhotra, A. & Thorpe, R.S. (2003) Assessing the phylogenetic utility of four mitrochondrial genes and a nuclear intron in the Asian pit viper genus, *Trimeresurus*, simultaneous, and conditional data combination analyses. *Molecular Biology and Evolution* 20(8), 1240–1251, 2 figs.

Crespo, E.G. (1972) Répteis de Portugal continental das colecções do Museo Bocage. *Arquivos do Museu Bocage*, Lisboa (Série 2) 3(17), 447–612, 13 figs.

Crespo, E.G. (1975) Aditamento dos catálogos dos répteis e anfibios de Portugal continental das colecções do Museu Bocage. *Arquivos do Museu Bocage*, Lisboa (Series 2) 5(3), 479–497.

Crespo, E.G. & Oliveira, M.E. (1989) *Atlas da distribuição dos anfibios and répteis de Portugal continental.* SNPRCN, Lisboa, 98 pp.

Crespo, E.G. & Sampaio, L. (1994) *Serpentes de Portugal.* Instituto de Conservação da Natureza, Lisboa, 40 pp.

Crippen, R.C. (1962) Holotype specimens of the amphibians and reptiles in the Museum of Vertebrate Zoology, University of California, Berkeley. *Herpetologica* 18(3), 187–194.

Crochet, P.-A., Rasmussen, J.B., Wilms, T., Geniez, P., Trape, J.-F. & Böhme, W. (2008) Systematic status and correct nomen of the western North African cat snake, *Telescopus tripolitanus* (Werner, 1909) (Serpentes, Colubridae), with comments on the other taxa in the *dhara-obtusus* group. *Zootaxa* (1703), 25–46, 7 figs.

Crombie, R.I. (1999) Jamaica. *In:* Crother, B.I. (Ed.), *Caribbean amphibians and reptiles.* Academic Press, San Diego, pp. 63–92, 2 figs.

Crombie, R.I. & Pregill, G.K. (1999) A checklist of the herpetofauna of the Palou Islands (Republic of Belau). *Herpetological Monographs* (13), 29–80.

Cross, J.K. (1979) Multivariate and univariate character geography in *Chionactis* (Reptilia, Serpentes). *Dissertation Abstracts International* 40B(2), 538.

Crother, B.I. (1989) A redescription of the hemipenis of *Hydromorphus concolor* (Colubridae) with comments on its tribal allocation. *Copeia* 1989(1), 227–229, 2 figs.

Crother, B.I. (Ed.). (1999a) *Caribbean amphibians and reptiles.* Academic Press, San Diego, xxx + 495 pp., illust., 8 col. pls.

Crother, B.I. (1999b) Evolutionary relationships. *In:* Crother, B.I. (Ed.), *Caribbean amphibians and reptiles.* Academic Press, San Diego, pp. 269–334, 18 figs.

Crother, B.I., Campbell, J.A. & Hillis, D.M. (1992) Phylogeny and historical biogeography of the palm-pitvipers, genus *Bothriechis*, biochemical and morphological evidence. *In:* Campbell, J.A. & Brodie, E.D., Jr. (Eds.), *Biology of the pitvipers.* Selva, Tyler, pp. 1–19, 21 figs.

Crother, B.I., Savage, J.M. & Holycross, A.T. (2011a) *Crotalinus catenatus* Rafinesque, 1818 (currently *Sistrurus catenatus)* and *Crotalus tergeminus* Say in James, 1822 (currently *sistrurus tergeminus*; Reptilia, Serpentes): proposed conservation of usage by designation of neotypes for both species. *Bulletin of Zoological Nomenclature* 68(4), 271–274.

Crother, B.I., White, M.E., Savage, J.M., Eckstut, M.E., Graham, M.R. & Gardner, D.W. (2011b) A reevaluation of the status of the foxsnakes *Pantherophis gloydii* Conant and *P. vulpinus* Baird and Girard (Lepidosauria). *International Scholarly Research Network* 2011, 1–15, 5 figs.

Crother, B.I., Boundy, J., Burbrink, F., Campbell, J.A., Queiroz, K. de, Frost, D.R., Green, D.M., Highton, R., Iverson, J.B., Kraus, F., McDiarmid, R.W., Mendelson, J.R., III, Meylan, P., Pyron, R.A., Reeder, T.W., Seidel, M.E., S.G. & Wake, D.B. (2012) *Scientific and standard English names of amphibians and reptiles of North America north of Mexico, with comments regarding confidence in our understanding.* Seventh edition. Society for the Study of Amphibians and Reptiles, Herpetological Circular (39), 92 pp. [fifth edition, 2000, (Herp. Circ. 29), 82 pp.; sixth edition, 2008, (Herp. Circ. 37), 84 pp.]

Crouse, H.W. (1902) *The venomous snakes and spiders of Texas.* Transactions of State Medical Association of Texas, 39 pp., 9 figs.

Crumly, C. (1990) Type catalogues of herpetological collections, an annotated list of lists. *Society for the Study of Amphibians and Reptiles, Herpetological Circular* (18), 1–50.

Cruz, F.B., Gabriela-Perotti, M. & Fitzgerald, L.A. (1992) Lista de anfibios y reptiles colectados en una localidad del Chaco Salteño. *Acta Zoologica Lilloana*, Tucumán 42(1), 101–107, 1 fig.

Cruz, F.B., Schulte, J.A., II & Bellagamba, P. (1999) New distributional records and natural history notes for reptiles from southern Argentina. *Herpetological Review* 30(3), 182–183.

Cruz-Díaz, G. (1987) *Serpientes venenosas de Honduras*. Editorial Universitaria, Tegucigalpa, 160 pp.

Cruz-Sáenz, D., Guerrero, S., Lazcano-Villarreal, D. & Téllez-López, J. (2009) Notes on the herpetofauna of western Mexico 1: an update on the herpetofauna of the state of Jalisco, Mexico. *Bulletin of the Chicago Herpetological Society* 44(7), 105–113.

Cruz-Sáenz, D., Lazcano-Villarreal, D. & Ontiveros-Esqueda, L. (2011) Notes on the herpetofauna of western Mexico 5: an update of herpetofauna in the oak forest of Huaxtla, Zapopan, Jalisco, Mexico. *Bulletin of the Chicago Herpetological Society* 46(7), 81–87, 7 figs.

Csiki, E. (1903) Holdhaus, K., Beiträge zur Kenntniss der Koleopterengattung Atomaria Steph. (review). *Rovartani Lapok*, Budapest 10, 198 (footnote).

Csiki, E. (1904) Über einige Gattungsnamen. *Zoologischer Anzieger*, Leipzig 28(7), 266–267.

Csuti, B., Kimerling, A., Jon, A., O'Neil, T.A., Shaughnessy, M.M., Gaines, E.P. & Huso, M.M.P. (1997) *Atlas of Oregon wildlife, distribution, habitat, and natural history*. Oregon State University, Corvallis, xiv + 492 pp., illust.

Cullings, K.W., Morafka, D.J., Hernandez, J. & Roberts, J.W. (1997) Reassessment of phylogenetic relationships among pitviper genera based on mitochondrial cytochrome b gene sequences. *Copeia* 1997(2), 429–432, 2 figs.

Culotta, W.A. & Pickwell, G.V. (1993) *The venomous sea snakes: a comprehensive bibliography*. Krieger Publishing Company, Malabar, xx + 504 pp.

Cundall, D. (1981) Cranial osteology of the colubrid snake genus *Opheodrys*. *Copeia* 1981(2), 351–371, 16 figs.

Cundall, D. (1986) Variations of the cephalic muscles in the colubrid snake genera *Entechinus*, *Opheodrys*, and *Symphimus*. *Journal of Morphology* 187, 1–21, 21 figs.

Cundall, D. & Irish, F.J. (1986) Aspects of locomotor and feeding behavior in the Round Island boa *Casarea dussumieri*. *Dodo, Journal of the Jersey Wildlife Preserve Trust* 23, 108–111, 1 fig., 1 pl.

Cundall, D. & Irish, F.J. (1989) The function of the intramaxillary joint in the Round Island boa, *Casarea dussumieri*. *Journal of Zoology*, London 217(4), 569–598, 13 figs.

Cundall, D. & Rossman, D.A. (1984) Quantitative comparisons of skull form in the colubrid snake genera *Farancia* and *Pseudoeryx*. *Herpetologica* 40(4), 388–405, 13 figs.

Cundall, D. & Rossman, D.A. (1994) Cephalic anatomy of the rare Indonesian snake *Anomochilus weberi*. *Zoological Journal of the Linnean Society*, London (1993) 109(3), 235–273, 25 figs.

Cundall, D., Wallach, V. & Rossman, D.A. (1994) The systematic relationships of the snake genus *Anomochilus*. *Zoological Journal of the Linnean Society*, London (1993) 109(3), 275–299, 4 figs.

Cunha, O.R. da. (1967) Ofídios da Amazônia. I. A ocorrência de *Bothrops bilineatus* (Wied, 1825) nas matas dos arredores da didade de Belém, Pará. *Boletim do Museu Paraense Emílio Goeldi (Zoologia)*, Belém (66), 1–12.

Cunha, O.R. da & do Nascimento, F.P. (1970) Ofidios da Amazônia. II. *Liophis miliaris* (Linnaeus, 1758) na Amazônia norte oriental (Território Federal do Amapá), (Ophidia, Colubridae). *Boletim do Museu Paraense Emílio Goeldi (Zoologia)*, Belém (70), 1–6.

Cunha, O.R. da & do Nascimento, F.P. (1972) Ofidios de Amazônia. III. A ocorrência de *Bothrops lichenosus* Roze, 1958, no Brasil (Ophidia, Crotalidae). *Revista Brasileira do Biologia*, Rio de Janeiro 32(1), 27.

Cunha, O.R. da & do Nascimento, F.P. (1973) Ofidios de Amazônia. IV. As cobras corais (gênero *Micrurus*) da região leste do Pará (Ophidia, Elapidae) Nota preliminar. *Publicaciones Avulsas do Museu Paraense Emílio Goeldi (Zoologia)*, Belém (20), 273–286, 3 figs.

Cunha, O.R. da & do Nascimento, F.P. (1975a) Ofidios da Amazônia. V. *Bothrops lichenosus* Roze, 1958, sinônimo de *Bothrops castelnaudi* Duméril, Bibron & Duméril, com nova descrição e comentários. *Boletim do Museu Paraense Emílio Goeldi (Zoologia)*, Belém (80), 1–14, illust.

Cunha, O.R. da & do Nascimento, F.P. (1975b) Ofidios da Amazônia. VI. *Liotyphlops ternetzii* (Boulenger, 1896) ofidio raro e da hábitos subterrâneos, na região leste do Pará (Ophidia, Anomalepididae). *Boletim do Museu Paraense Emílio Goeldi (Zoologia)*, Belém (82), 1–8.

Cunha, O.R. da & do Nascimento, F.P. (1975c) Ofídios da Amazônia. VII—As serpents peçonhentas do gênero *Bothrops* (jararacas) e *Lachesis* (surucucu) da região leste do Pará (Ophidia, Viperidae). *Boletim do Museu Paraense Emílio Goeldi (Zoologia)*, Belém (83), 1–42, 4 figs., 2 pls.

Cunha, O.R. da & do Nascimento, F.P. (1976a) Ofídios da Amazônia. VIII. A ocorrência de *Rhinobothryum lentiginosum* (Scopoli, 1785) nas proximidades de Belém, Pará (Opidia, Viperidae). *Boletim do Museu Paraense Emílio Goeldi (Zoologia)*, Belém (84), 1–6, illust.

Cunha, O.R. da & do Nascimento, F.P. (1976b) Ofídios da Amazônia. IX. O gênero *Liophis* Wagler, 1830, na região leste do Pará (Ophidia, Colubridae). *Boletim do Museu Paraense Emílio Goeldi (Zoologia)*, Belém (85), 1–32.

Cunha, O.R. da & do Nascimento, F.P. (1978) Ofídios da Amazônia. X—As Cobras da região leste do Pará. *Publicaciones Avulsas do Museu Paraense Emílio Goeldi (Zoologia)*, Belém (32), 1–218, 41 pls.

Cunha, O.R. da & do Nascimento, F.P. (1980) Ofídios da Amazônia. XI. Ofidios de Roraima e notas sobre *Erythrolamprus bauperthrusii* Duméril, Bibron and Duméril, 1854, sinônimo de *Erythrolamprus aesculapii* (Linnaeus, 1758). *Boletim do Museu Paraense Emílio Goeldi (Zoologia)*, Belém (102), 1–21.

Cunha, O.R. da & do Nascimento, F.P. (1981) Ofídios da Amazônia. XIII. Observações sobre a viviparidade em ofidios do Pará e Maranhão (Ophidia, Aniliidae, Boidae, Colubridae e Viperidae). *Boletim do Museu Paraense Emílio Goeldi (Zoologia)*, Belém (109), 1–20.

Cunha, O.R. da & do Nascimento, F.P. (1982a) Ofidios da Amazônia. XIV. As espécies de *Micrurus*, *Bothrops*, *Lachesis* e *Crotalus* do sul do Pará e oeste do Maranhão, incluindo áreas de cerrado desse Estado (Ophidia, Elapidae e Viperidae). *Boletim do Museu Paraense Emílio Goeldi (Zoologia)*, Belém (112), 1–58.

Cunha, O.R. da & do Nascimento, F.P. (1982b) Ofídios da Amazonia. XVI. A espécie *Uromacerina ricardini* (Peracca, 1897) na Amazônia oriental, leste do Pará (Ophidia, Colubridae). *Boletim do Museu Paraense Emílio Goeldi (Zoologia)*, Belém (113), 1–9, 4 figs.

Cunha, O.R. da & do Nascimento, F.P. (1982c) Ofídios da Amazônia. XVII. Revalidação de *Micrurus ornatissimus* (Jan, 1858) diferenciada de *M. Langsdorffi* (Wagler, 1824) e distribuição geográfica das duas espécies (Ophidia, Elapidae). *Boletim do Museu Paraense Emílio Goeldi* (*Zoologia*), Belém (116), 1–19.

Cunha, O.R. da & do Nascimento, F.P. (1982d) Ofídios da Amazônia. XVIII. O gênero *Chironius* Fitzinger, na Amazônia oriental (Ophidia, Colubridae). *Boletim do Museu Paraense Emílio Goeldi* (*Zoologia*), Belém (119), 1–17.

Cunha, O.R. da & do Nascimento, F.P. (1983a) Ofídios da Amazônia. XIX. As espécies de *Oxyrhopus* Wagler, com uma subespécie nova, e *Pseudoboa* Schneider na Amazônia oriental e Maranhão (Ophidia, Colubridae). *Boletim do Museu Paraense Emílio Goeldi* (*Zoologia*), Belém (122), 1–47.

Cunha, O.R. da & do Nascimento, F.P. (1983b) Ofídios da Amazonia. XX. As espécies de *Atractus* Wagler, 1828, na Amazônia Oriental e Maranhão (Ophidia, Colubridae). *Boletim do Museu Paraense Emílio Goeldi* (*Zoologia*), Belém (123), 1–38, 3 figs., 2 pls.

Cunha, O.R. da & do Nascimento, F.P. (1983c) Ofídios da Amazônia. XV. As espécies de *Chironius* da Amazônia oriental (Pará, Amapá e Maranhão). *Memórias do Instituto Butantan*, São Paulo (1982) 46, 139–172.

Cunha, O.R. da & do Nascimento, F.P. (1984) Ofídios da Amazônia. XXI. *Atractus zidoki* no região leste do Pará e notas sobre *A. alphonsehogei* e *A. schach* (Ophidia, Colubridae). *Boletim do Museu Paraense Emílio Goeldi* (*Zoologia*), Belém 1(2), 219–228, 2 pls.

Cunha, O.R. da & do Nascimento, F.P. (1991) Ofídios da Amazônia. XXII. Revalidação e redescrição de *Micrurus albicinctus* Amaral, de Rondônia, e sobre a validade de *Micrurus waehnerorum* Meise, do Amazonas (Ophidia, Elapidae). *Boletim do Museu Paraense Emílio Goeldi* (*Zoologia*), Belém 7(1), 43–52, 2 figs.

Cunha, O.R. da & do Nascimento, F.P. (1993) Ofídios da Amazônia. As cobras da região leste do Pará. *Boletim do Museu Paraense Emílio Goeldi* (*Zoologia*). Belém 9(1), 1–191, 87 figs.

Cunha, O.R. da, do Nascimento, F.P. & Hoge, A.R. (1980) Ofídios da Amazônia. XII. Uma subspécie de *Sibynomorphus mikani* do noroeste do Maranhão (Ophidia, Colubridae, Dipsadinae). *Boletim do Museu Paraense Emílio Goeldi* (*Zoologia*), Belém (103), 1–15.

Cunha, O.R. da, do Nascimento, F.P. & Avila-Pires, T.C.S. (1985) Os Répteis da área de Carajás, Pará, Brasil (Testudines e Squamata). I. *Publicaciones Avulsas do Museu Paraense Emílio Goeldi* (*Zoologia*), Belém (40), 1–92, 6 pls.

Cunningham, P.L. (2010) Checklist of terrestrial reptiles in three protected areas in the Kingdom of Saudi Arabia. *Herpetological Review* 41(1), 25–28, 1 fig.

Cuny, G., Jaeger, J.-J., Mahboubi, M. & Rage, J.-C. (1990) Les plus anciens serpents (Reptilia, Squamata) connus. Mise au point sur l'âge géologique des serpents de la partie moyenne du Crétacé. *Comptes Rendus de l'Académie des Sciences*, Paris (Série 2) 311, 1267–1272, 2 figs.

Curcio, F.F., Zaher, H. & Rodrigues, M.T. (2002) Rediscovery of the blind-snake *Leptotyphlops brasiliensis* Laurent, 1949 (Serpentes, Leptotyphlopidae) in the wild. *Phyllomedusa* 1(2), 101–104, 2 figs.

Curcio, F.F., Piacentini, V. de & Fernandes, D.S. (2009a) On the status of the snake genera *Erythrolamprus* Boie, *Liophis* Wagler and *Lygophis* Fitzinger (Serpentes, Xenodontinae). *Zootaxa* (2173), 66–68.

Curcio, F.F., Sánchez-Pacheco, S.J., Mueses-Cisneros, J.J. & Rodrigues, M.T. (2009b) Notes on distribution, variation and characterization of *Erythrolamprus pseudocorallus* Roze, 1959 (Serpentes: Colubridae) with the first records from Colombia. *Zootaxa* (2045), 33–42, 3 figs.

Curcio, F.F., Sales-Nuñes, P.M., Harvey, M.B. & Rodrigues, M.T. (2011) Redescription of *Apostolepis longicaudata* (Serpentes: Xenodontinae) wth comments on its hemipenial morphology and natural history. *Herpetologica* 67(3), 318–331, 7 figs.

Curcio, F.F., Sales-Nuñes, P.M., Suzart-Argôlo, A.J., Skuk, G. & Rodrigues, M.T. (2012) Taxonomy of the South American dwarf boas of the genus *Tropidophis* Bibron, 1840, with the description of two new species from the Atlantic forest (Serpentes: Tropidophiidae). *Herpetological Monographs* (26), 80–121, 26 figs.

Curry-Lindahl, K. (1956) Ecological studies on mammals, birds, reptiles and amphibians in the eastern Belgian Congo. *Annales du Musée Royal du Congo Belge* (8, *Science Zoologique*), Tervuren 42, 1–78, 23 pls.

Curry-Lindahl, K. (1961) *Exploration du Parc National Albert et du Parc National de la Kagera. II. Mission K. Curry-Lindahl (1951–1952, 1958–1959). Fascicule 1. Contribution à l'étude des vertébrés terrestres en Afrique tropicale.* Institut des Parcs Nationaux du Congo et Ruanda-Urundi, 331 pp., 22 pls.

Curry-Lindahl, K. (1975) *Groddjur och kräldjur i färg.* Awe/Gebers, Stockholm, 198 pp.

Cuvier, G. (1817) *Le Règne animal distribue d'apres son organisation, pour servir de base a l'histoire naturelle des animaux et d'introduction a l'anatomie comparee. Tome II, contenant les reptiles, les poissons, les mollusques et les annelides.* Deterville, Paris, xviii + 532 pp.

Cuvier, G. (1822) *Das Thierreich eingetheilt nach den bau der Thiere als Grundlage ihrer Naturgeschichte und der vergleichenden Anatomie von dem herrn Ritter von Cuvier Staatsrath von Franfreich und bestandiger Secretar der Academie der Wissenschaften u.s.w. Aus dem Franzosischen... H.R. Schinz, Med. Dr. Zweiter Band. Reptilien, Fische, Weichthiere, Ringelwurmer.* J.G. Gotta'schen, Stuttgart und Tubingen, 835 pp.

Cuvier, G. (1829) *Le Règne animal distribue d'apres son organisation, pour servir de base a l'histoire naturelle des animaux et d'introduction a l'anatomie comparee. Tome II. Nouvelle edition, revue et augmentee.* Deterville, Paris, xv + 406 pp.

Cuvier, G. & Voigt, F.S. (1832) *Das Thierreich, geordnet nach seiner Organisation: als Grundlage der Naturgeschichte der Thiere und Einleitung in die vergleichende Anatomie. Nach der zweiten, vermehrten Ausgabe übersetzt und durch Zusätze erweitert von F.S. Voigt. Zweiter Band, die Reptilien und Fische enthaltend.* F.A. Brockhaus, Leipzig, xvi + 539 pp.

Cyrén, C.A.O. (1934) *Ormar: I fantasi och verklighet.* Bockforlaget Natur och Kultur, Stockholm, 249 pp.

Cyrén, C.A.O. (1941) Beiträge zur Herpetologie der Balkanhalbinsel. *Bulletin des Institutions Royales d'Histoire Naturelle à Sofia-Bulgarien* 14, 36–139, 27 figs., 6 pls.

Daan, S. & Hillenius, D. (1966) Catalogue of type specimens of amphibians and reptiles in the Zoological Museum. *Beaufortia*, Amsterdam 13(158), 117–144.

Dalman, J.W. (1823) *Ahalecta entomologica.* Bokförlaget Natur och Kultur, Stockholm, 249 pp.

Daltry, J.C. & W. Wüster. (2002) A new species of wolf snake (Serpentes, Colubridae, *Lycodon*) from the Cardamom Mountains, southwestern Cambodia. *Herpetologica* 58(4), 498–504, 4 figs.

Daltry, J.C., Ponnudural, G., Shin, C.K., Tan, N.-H., Thorpe, R.S. & Wüster, W. (1996) Electrophoretic profiles and biological activities: infraspecific variation in the venom of the Malayan pit viper (*Calloselasma rhodostoma*) *Toxicon* 34(1), 67–79, 2 figs.

Daltry, J.C., Wüster, W. & Thorpe, R.S. (1997) The role of ecology in determining venom variation in the Malayan pitviper (*Calloselasma rhodostoma*). *In:* Thorpe, R.S., Wüster, W. & Malhotra, A. (Eds.), *Ecology, evolution and snakebite venomous snakes*. Zoological Society of London, Clarendon Press, Oxford, pp. 155–171, 4 figs.

Daltry, J.C., Bloxam, Q., Cooper, G., Day, M.L., Hartley, J., Henry, M., Lindsay, K. & Smith, B.L. (2001) Five years of conserving the "world's rarest snake," the Antiguan racer *Alsophis antiguae*. *Oryx* 35(2), 119–127, 1 fig., 2 pls.

Dammermann, K.W. (1926) List of reptiles, and of Amphibia, from the Rhio-Lingga Archipelago. *Treubia*, Batavia 8(3–4), 323.

Dandge, P.H. (2008) Food and feeding habits of *Elachistodon westermanni* Reinhardt, 1863. *Hamadryad* 32(1), 75–77, 1 fig.

Daniel, H. (1949) Las serpientes en Colombia. *Revista, Facultad Nacional de Agronomía*, Medellín 10(36), 301–333.

Daniel, H. (1955) Algunos aspectos de la lucha biologica. IV. *Revista, Facultad Nacional de Agronomía*, Medellín 17(48), 38–88.

Daniel, J.C. (1962) Extension of the known range of the catsnake *Boiga forsteni* (Dum. & Bibr.). *The Journal of the Bombay Natural History Society* 59(3), 966–967.

Daniel, J.C. (1983) *The book of Indian reptiles*. Bombay Natural History Society, x + 141 pp., figs., pls.

Daniel, J.C. & Shull, E.M. (1963) A list of the reptiles and amphibians of the Surat Dangs, South Gujarat. *The Journal of the Bombay Natural History Society* 60(3), 737–743.

Danilov, I.G. & Averianov, A.O. (1999) A new species of *Calamagras* Cope, 1873 (Serpentes, Boidae, Erycinae) from the early Eocene of Kirghizia. *Geodiversitas* 21(1), 85–91, 3 figs.

Darevsky, I.S. (1956) A new species of the venomous snake, *Vipera kaznakovi* Nikolsky, for the fauna of Armenia. *Bulletin of the Academy of Sciences of Armenia*, Yerevan 9(12), 127–130.

Darevsky, I.S. (1964) Die Reptilien der Inseln Komodo, Padar und Rintja im Kleinen Sunda-Archipel, Indonesien. *Senckenbergiana Biologica*, Frankfurt-am-Main 45(3–5), 563–576.

Darevsky, I.S. (1970) Systematic status of *Rhynchocalamus melanocephalus satunini* Nik. (Serpentes, Colubridae) previously included in the genus *Oligodon*. *Zoologicheskii Zhurnal*, Moscow 49(7), 1685–1690, 4 figs.

Darevsky, I.S. (1987) Aleksandr Filippovich Laister (1884–1941) and his role in the investigation of the Caucasian herpetological fauna. *Trudy Zoologicheskovo Instituta, Akademiya Nauk SSSR* 158, 4–19, illust. (in Russian)

Darevsky, I.S. (1990) Notes on the reptiles (Squamata) of some offshore islands along the coast of Vietnam. *In:* Peters, G. & Hutterer, R. (Eds.), *Vertebrates in the tropics*. Museum Alexander Koenig, Bonn, pp. 125–129, 1 fig.

Darevsky, I.S. (1993) *Eryx miliaris* (Pallas, 1773) - Östliche Sandboa. *In:* Böhme, W. (Ed.), *Handbuch der Reptilien und Amphibien Europas. Band 3/I. Schlangen (Serpentes) I (Typhlopidae, Boidae, Colubridae 1, Colubrinae)*. AULA-Verlag, Wiesbaden, pp. 55–66, figs. 10–12.

Darevsky, I.S. (1999) The herpetofauna of some offshore islands of Vietnam, as related to that of the adjacent mainland. *In:* Ota, H. (Ed.), *Tropical island herpetofauna: origin, current diversity, and conservation*. Elsevier, Amsterdam, pp. 27–42, 10 figs.

Darevsky, I.S. & Bakradze, M.A. (1982) The taxonomic status of *Contia collaris* var. *macrospilota* Werner, 1909 (Reptilia, Serpentes, Colubridae). *Amphibia-Reptilia*, Wiesbaden 3(2–3), 283–286, 2 figs.

Darevsky, I.S. & Orlov, N.L. (1992) A new subspecies of the dwarf snake *Calamaria lowi ingermarxi* ssp. nov. (Serpentes, Colubridae) from southern Vietnam. *Asiatic Herpetological Research*, Berkeley 4, 13–17, 5 figs.

Darevsky, I.S. & Orlov, N.L. (1994) The systematic position of the slender racer *Coluber najadum* (Eichwald) from south-east Azerbaijan, and some remarks on the herpetological fauna of the region. *Russian Journal of Herpetology*, Moscow 1(2), 93–97, 4 figs.

Darevsky, I.S. & Szczerbak, N.N. (1993) *Coluber najadum* (Eichwald, 1831) - Schlanknatter. *In:* Böhme, W. (Ed.), *Handbuch der Reptilien und Amphibien Europas. Band 3/I. Schlangen (Serpentes) I (Typhlopidae, Boidae, Colubridae 1, Colubrinae)*. AULA-Verlag, Wiesbaden, pp. 131–144, figs. 27–33.

Dart, R.C., O'Brien, P.C. & McNally, J. (1990) Neutralization of *Micrurus distans* venom by antivenin *(Micrurus fulvius)*. *Toxicon* 28(6), 606.

Das, A., Mohapatra, P.P., Purkayasthla, J., Sengupta, S., Dutta, S.K., Ahmed, M.F. & Tillack, F. (2010) A contribution to *Boiga gokool* (Gray, 1835) (Reptilia: Squamata: Colubridae). *Russian Journal of Herpetology* 17(3), 161–178, 7 figs.

Das, I. (1991) A new species of *Eryx* (Boidae, Serpentes, Squamata) from south-western India. *The Journal of the Bombay Natural History Society* 88(1), 92–97, 3 figs.

Das, I. (1993) Annandale's seasnake, *Kolpophis annandalei* (Laidlaw, 1901): a new record for Borneo (Reptilia, Serpentes, Hydrophiidae). *Raffles Bulletin of Zoology*, Singapore 41(2), 359–361, 1 fig.

Das, I. (1994) A check-list of the amphibians and reptiles of Andaman and Nicobar islands. *Journal of the Andaman Science Association* 10(1–2), 44–49.

Das, I. (1996) *Biogeography of the reptiles of south Asia*. Krieger Publishing Company, Malabar, 87 pp., 36 col. pls.

Das, I. (1997) Herpetological explorations in the Andaman and Nicobar Islands, India. *In:* Roçek, Z. & Hart, S. (Eds.), *Abstracts of the Third World Congress of Herpetology*. Third World Congress of Herpetology, Prague, p. 50.

Das, I. (1998a) A new species of *Boiga* (Serpentes, Colubridae) from the Nicobar Archipelago. *Journal of South Asian Natural History* 3(1), 59–67, 3 figs.

Das, I. (1998b) *Herpetological bibliography of Indonesia*. Krieger Publishing Company, Malabar, 92 pp.

Das, I. (1999a) The dates of publication of amphibian and reptile names by Blanford and Stoliczka in the Journal and Proceedings of the Asiatic Society of Bengal. *Asiatic Herpetological Research* 8, 18–24.

Das, I. (1999b) Biogeography of the amphibians and reptiles of the Andaman and Nicobar Islands, India. *In:* Ota, H. (Ed.), *Tropical island herpetofauna: origin, current diversity, and conservation*. Elsevier, Amsterdam, pp. 43–77.

Das, I. (2002a) *Snakes and other reptiles of India*. Ralph Curtis Books, Sanibel Island, 144 pp., col. photos.

Das, I. (2002b) *An introduction to the amphibians and reptiles of tropical Asia*. Natural History Publications (Borneo), Kota Kinabalu, 207 pp., 68 col. pls.

Das, I. (2003) Growth of knowledge on the reptiles of India, with an introduction to systematics, taxonomy and nomenclature. *The Journal of the Bombay Natural History Society* 100(2–3), 446–501.

Das, I. (2004a) On neotype designations for *Coronella cyclura* Cantor, 1839 (Serpentes: Colubridae). *Hamadryad* 29(1), 133–134.

Das, I. (2004b) On the identity of *Amphiesma venningi* (Wall, 1910) reported from Meghalaya, India. *Hamadryad* 29(1), 134–135.

Das, I. (2006a) *Snakes & other reptiles of Borneo*. Ralph Curtis Books, Sanibel Island, 144 pp., col. illust.

Das, I. (2006b) Geographic distribution: Serpentes: *Stoliczkia borneensis*. *Herpetological Review* 37(4), 500–501.

Das, I. (2006c) The reptile fauna. *In:* Abang, F. & Das, I. (Eds.), *The biodiversity of a peat swamp forest in Sarawak*. Universiti Malaysia Sarawak, Kota Samarahan, pp. 115–128.

Das, I. (2007a) Geographic distribution: Serpentes: *Dendrelaphis kopsteini*. *Herpetological Review* 38(2), 220.

Das, I. (2007b) *Amphibians and reptiles of Brunei*. Natural History Publication (Borneo), Kota Kinabalu, 200 pp., col. illust.

Das, I. (2010) *A field guide to the reptiles of South-East Asia*. New Holland Publishers, London, 376 pp., 74 col. pls.

Das, I. (2012) *A naturalist's guide to the snakes of South-East Asia: Malaysia, Singapore, Thailand, Myanmar, Borneo, Sumatra, Java and Bali*. John Beaufoy Publishing, Oxford, 176 pp., 300 col. photos.

Das, I. & Chandra, K. (1994) Two snakes new to Andaman and Nicobar islands. *Journal of the Andaman Science Association* 10(1–2), 114–115.

Das, I. & Gayen, N.C. (2004) Addenda and corrigenda to the catalogue of reptile types in the collection of the Zoological Survey of India. *Hamadryad* 28(1–2), 95–97.

Das, I. & Hee, K.B. (2008) Herpetofauna of the Pulau Banggi group of islands off northeastern Borneo. *Herpetological Review* 39(3), 296–298, 1 fig.

Das, I. & Lim, K.K.P. (2001) Catalogue of herpetological types in the collection of the Raffles Museum of Biodiversity Research, National University of Singapore. *The Raffles Bulletin of Zoology* 49(1), 7–11.

Das, I. & Palden, J. (2000) A herpetological collection from Bhutan, with new country records. *Herpetological Review* 31(4), 256–258.

Das, I. & de Silva, A. (2005) *A photographic guide to snakes and other reptiles of Sri Lanka*. Ralph Curtis Books, Sanibel Island, 144 pp., col. illust.

Das, I. & Whitaker, R. (1996) Bibliography of the king cobra (*Ophiophagus hannah*). *Smithsonian Herpetological Information Service* (108), 1–24.

Das, I., Dattagupta, B. & Gayen, N.C. (1998). History and catalogue of reptile types in the collection of the Zoological Survey of India. *Journal of South Asian Natural History* 3(2), 121–172, illust.

Das, I., Lakin, M., Lim, K.K.P. & Hui T.H. (2008a) New species of *Anomochilus* from Borneo (Squamata: Anomochilidae). *Journal of Herpetology* 42(3), 584–591, 4 figs.

Das, I., Nyawa, S.B. & Charles, J.K. (Eds.). (2008b) *A guide to the amphibians and reptiles of Tasek Merimbun Heritage Park, Brunei Darussalam*. Brunei Museums Department, Bandar Seri Begawan, 92 pp., col. illust.

Das, I., Breuer, H. & Shonleben, S. (2013) *Gerarda prevostiana* (Eydoux and Gervais, 1837) (Squamata: Serpentes: Homalopsidae), a new snake for Borneo. *Asian Herpetological Research* 4(1), 76–78, 4 figs.

Dasgupta, G. & Raha, S. (2004) Reptilia. *In:* Alfred, J.R.B. (Ed.), *Fauna of Bihar (including Jharkhand), Part-1. State Fauna Series* 11. Zoological Survey of India, Kolkata, pp. 143–179.

Dasgupta, G. & Raha, S. (2006) Reptilia. *In:* Alfred, J.R.B. (Ed.), *Fauna of Nagaland. State Fauna Series* 12. Zoological Survey of India, Kolkata, pp. 433–460, 21 col. figs.

Daszak, P. & Cawthraw, S. (1991) A review of the reptiles and amphibians of Turkey, including a literature survey and species checklist. *British Herpetological Society Bulletin*, London (36), 14–26, 3 figs., 6 pls.

Daudin, F.M. (1803a) Erpétologie. Caractères des vingt-trois genres qui composent l'ordre des ophidiens. *Magasin Encyclopédique, ou Journal des Sciences, des Lettres et des Arts,* Paris, Année 8 5(20), 433-438.

Daudin, F.M. (1803b) *Histoire naturelle, génerale et particulière des reptiles; Ouvrage faisant suite aux Oeuvres de Leclerc de Buffon, et partie du cours complet d'histoire naturelle rédigé par C.S. Sonnini, membre de plusieurs sociétés savantes. Tome cinquième.* F. Dufart, Paris, 365 pp., pls. 59–70.

Daudin, F.M. (1803c) *Histoire naturelle, génerale et particulière des reptiles; Ouvrage faisant suite aux Oeuvres de Leclerc de Buffon, et partie du cours complet d'histoire naturelle rédigé par C.S. Sonnini, membre de plusieurs sociétés savantes. Tome sixième.* F. Dufart, Paris, 447 pp., pls. 71–80.

Daudin, F.M. (1803d) *Histoire naturelle, génerale et particulière des reptiles; Ouvrage faisant suite aux Oeuvres de Leclerc de Buffon, et partie du cours complet d'histoire naturelle rédigé par C.S. Sonnini, membre de plusieurs sociétés savantes. Tome septième.* F. Dufart, Paris, 436 pp., pls. 81–92.

Daudin, F.M. (1803e) *Histoire naturelle génerale et particulière des reptiles; Ouvrage faisant suite aux Cuvres de Leclerc de Buffon, et partie du cours complet d'histoire naturelle rédigé par C.S. Sonnini, membre de plusieurs sociétés savantes. Tome huitième.* F. Dufart, Paris, 439 pp., pls. 93–100.

David, P. (1995) Geographic distribution: Serpentes: *Ovophis tonkinensis* (Tonkin pitviper). *Herpetological Review* 26(3), 157.

David, P. (2009) Les collections herpétologiques de Jean Deuve. *Bulletin de la Société Herpétologique de France* (129), 21–24, 1 fig.

David, P. & Das, I. (2003) A new species of the snake genus *Amphiesma* (Serpentes, Colubridae, Natricinae) from western Sumatra, Indonesia. *Raffles Bulletin of Zoology* 51(2), 413–419, 4 figs.

David, P. & DuBois, A. (2005) Découverte et redescription de l'holotype d' *Ahaetulla pulverulenta* (Duméril, Bibron & Duméril, 1854) (Reptilia, Serpentes, Colubridae), avec une remarque sur le status de *Dryinus fuscus* Duméril, Bibron & Duméril, 1854. *Zoosystema* 27(1), 163–178, 1 fig.

David, P. & Ineich, I. (1999) Les serpents venimeux du monde, systématique et répartition. *Dumerilia*, Paris 3, 1–499.

David, P. & Pauwels, O.S.G. (2004) A re-evaluation of the taxonomy of *Macrocalamus lateralis* Günther, 1864 (Serpentes, Colubridae), with the descriptions of two new species. *Raffles Bulletin of Zoology* 52(2), 635–645, 9 figs.

David, P. & Tong, H. (1997) Translations of recent descriptions of Chinese pitvipers of the *Trimeresurus*-complex (Serpentes, Viperidae), with a key to the complex in China and adjacent areas. *Smithsonian Herpetological Information Service* (112), 1–31.

David, P. & Vogel, G. (1996) *The snakes of Sumatra, an annotated checklist and key with natural history notes*. A.S. Brahm, Ed. Chimaira, Frankfurt-am-Main, 260 pp., col. illust.

David, P. & Vogel, G. (1998) Redescription of *Trimeresurus huttoni* Smith, 1949 (Serpentes, Crotalinae), with a descussion of its relationships. *Hamadryad* 22(2), 73–87, 3 figs.

David, P. & Vogel, G. (2010a) A new species of the natricine snake genus *Amphiesma* from Borneo (Squamata: Natricidae). *Russian Journal of Herpetology* 17(2), 121–127, 4 figs.

David, P. & Vogel, G. (2010b) *Venomous snakes of Europe, northern, central, and eastern Asia.* Edition Chimaira, Frankfurt-am-Main, 160 pp., 580 col. photos.

David, P. & Vogel, G. (2012a) A new species of the genus *Oligodon* Fitzinger, 1826 (Squamata: Colubridae) from Pulau Nias, Indonesia. *Zootaxa* (3201), 58–68, 3 figs.

David, P. & Vogel, G. (2012b) On the status of *Trimeresurus monicola meridionalis* Bourret, 1935 (Squamata: Viperidae). *Zootaxa* (3304), 43–53, 2 figs.

David, P., Vogel, G. & Pauwels, O.S.G. (1998) *Amphiesma optata* Hu & Djao, 1966, (Serpentes, Colubridae): an addition to the snake fauna of Vietnam, with a list of the species of the genus *Amphiesma* and a note on its type species. *Journal of the Taiwan Museum* 15(2), 83–92, 3 figs..

David, P., Ananjeva, N.B., Das, I. & Golubev, M. (1999) Translation of the original description of *Vipera obtusa* Dwigubsky, 1832 (Serpentes, Viperidae, Viperinae), with designation of a neotype. *Russian Journal of Herpetology* 6(3), 193–198, 2 col. figs.

David, P., Vidal, N. & Pauwels, O.S.G. (2001) A morphological study of Stejneger's pitviper *Trimeresurus stejnegeri* (Serpentes, Viperidae, Crotalinae), with the description of a new species from Thailand. *Russian Journal of Herpetology* 8(3), 205–222, 7 col. figs.

David, P., Vogel, G., Pauwels, O.S.G. & Vidal, N. (2002) Description of a new species of the genus *Trimeresurus* from Thailand, related to *Trimeresurus stejnegeri* Schmidt, 1925 (Serpentes, Crotalidae). *The Natural History Journal of Chulalongkorn University*, Bangkok 2(1), 5–19, 8 figs.

David, P., Vogel, G. & Vidal, N. (2003) On *Trimeresurus fasciatus* (Boulenger, 1896) (Serpentes, Crotalidae), with a discussion on its relationships based on morphological, and molecular data. *Raffles Bulletin of Zoology* 51(2), 149–157, 5 figs.

David, P., Vogel, G., Sumontha, M., Pauwels, O.S.G. & Chanhome, L. (2004) Expanded descriptions of the poorly known pitviper *Trimeresurus kanburiensis* Smith, 1943, with confirmation of the validity of *Trimeresurus venustus* G. Vogel, 1991 (Reptilia, Serpentes, Crotalidae). *Russian Journal of Herpetology* 11(2), 81–90, 4 col. figs.

David, P., Vogel, G., Vijayakumar, S.P. & Vidal, N. (2006) A revision of the *Trimeresurus puniceus*-complex (Serpentes, Viperidae, Crotalinae) based on morphological and molecular data. *Zootaxa* (1293), 1–78, 31 figs.

David, P., Bain, R.H., Nguyen, T.Q., Orlov, N.L., Vogel, G., Thanh, V.N. & Ziegler, T. (2007) A new species of the natricine snake genus *Amphiesma* from the Indochinese region (Squamata, Colubridae, Natricinae). *Zootaxa* (1462), 41–60, 12 figs.

David, P., Vogel, G. & Pauwels, O.S.G. (2008a) A new species of the genus *Oligodon* Fitzinger, 1826 (Squamata: Colubridae) from southern Vietnam and Cambodia. *Zootaxa* (1939), 19–37, 16 figs.

David, P., Vogel, G. & de Rooijen, J. van. (2008b) A revision of the *Oligodon taeniatus* (Günther, 1861) group (Squamata: Colubridae), with the description of three new species from the Indochinese region. *Zootaxa* (1965), 1–49, 34 figs.

David, P., Tong, H., Vogel, G. & Tiang, M. (2008c) On the status of the Chinese pitviper *Ceratrimeresurus shenii* Liang and Liu *in* Liang, 2003 (Serpentes, Viperidae), with the addition of *Protobothrops cornutus* (Smith, 1930) to the Chinese snake fauna. *Asiatic Herpetological Research* 11, 17–23, 5 figs.

David, P., Petri, M., Vogel, G. & Doria, G. (2009) A new species of pitviper of the genus *Trimeresurus* (*Popeia*) from northern Sumatra (Reptilia, Squamata, Viperidae). *Annali del Museo Civico di Storia Naturale "Giacomo Doria,"* Genova 100, 323–346, 11 figs.

David, P., Pham, C.T., Nguyen, T.Q. & Ziegler, T. (2011a) A new species of the genus *Opisthotropis* Günther, 1872 (Squamata: Natricidae) from the highlands of Kon Tum Province, Vietnam. *Zootaxa* (2758), 43–56, 5 figs.

David, P., Das, I. & Vogel, G. (2011b) On some taxonomic and nomenclatural problems in Indian species of the genus *Oligodon* Fitzinger, 1826 (Squamata: Colubridae). *Zootaxa* (2799), 1–14, 5 figs.

David, P., Vogel, G. & Dubois, A. (2011c) On the need to follow rigorously the Rules of the Code for the subsequent designation of a nucleospecies (type species) for a nominal genus which lacked one: the case of the nominal genus *Trimereserus* Lacépède, 1864 (Reptilia: Squamata: Viperidae). *Zootaxa* (2992), 1–51.

David, P., Nguyen, T.Q., Nguyen, T.T., Jiang, K., Chen, T.-B., Teynié, A. & Ziegler, T. (2012) A new species of the genus *Oligodon* Fitzinger, 1826 (Squamata: Colubridae) from northern Vietnam, southern China and central Laos. *Zootaxa* (3498), 45–62, 6 figs.

Davidson, A. (1964) *Snakes and scorpions found in the land of Tunisia. A handbook of the twenty two snakes and four scorpions, with notes on recognition points and advice on what to do if bitten.* Alan Davidson, Tunis, 26 pp., 20 figs.

Davis, D.D. (1946) Observations on the burrowing behavior of the hog-nosed snake. *Herpetologica* 1946(2), 75–78, 1 fig.

Davis, J., Bigelow, J. & Gray, F.C. (1817) *Report of a committee of the Linnaean Society of New England relative to a large marine animal, supposed to be a sea-serpent seen near Cape Ann, Massachusetts, in August, 1817.* Cummings and Hilliard, Boston, 1-52, 2 pls.

Davis, N.S. & Rice, F.L. (1883a) Descriptive catalogue of North American Batrachia and Reptilia, found east of Mississippi River. *Illinois State Laboratory Natural History Bulletin* (5), 1–66.

Davis, N.S. & Rice, F.L. (1883b) List of Batrachia and Reptilia of Illinois. *Bulletin of the Chicago Academy of Sciences* 1(3), 25–32.

Davis, W.B. & Dixon, J.R. (1957) Notes on Mexican snakes (Ophidia). *Southwestern Naturalist* 2(1), 19–27.

Davis, W.B. & Dixon, J.R. (1959) Snakes of the Chilpancingo region, Mexico. *Proceedings of the Biological Society of Washington* 72, 79–92.

Davis, W.B. & Smith, H.M. (1953) Snakes of the Mexican state of Morelos. *Herpetologica* 8(4), 133–143.

Dawson, K., Malhotra, A., Thorpe, R.S., Guo, P., Mrinalini & Ziegler, T. (2008) Mitochondrial DNA analysis reveals a new member of the Asian pitviper genus *Viridovipera* (Serpentes: Viperidae: Crotalinae). *Molecular Phylogenetics and Evolution* 49(1), 356–361, 1 fig.

Dawydow, K.N. (1898) Materiaux pour une faune herpetologique de la Palastine sud-est. *Annuaire du Musée Zoologique de l'Académie Impériale des Sciences de St. Pétersbourg* 3, 136–154.

Day, D. (1989) *The doomsday book of animals. A natural history of vanished species.* Viking Press, New York, 288 pp., illust.

Daza, J.M. (2005) Geographic distribution: Serpentes: *Corallus annulatus*. *Herpetological Review* 36(2), 202–203.

Daza, J.M. & Henderson, R.W. (2005) Geographic distribution: Serpentes: *Corallus caninus*. *Herpetological Review* 36(3), 338–339.

Daza, J.M., Quintana, J.R. & Otero, R. (2005) Geographic distribution: Serpentes: *Bothrops punctatus*. *Herpetological Review* 36(3), 338.

Daza, J.M., Smith, E.N., Páez, V.P. & Parkinson, C.L. (2009) Complex evolution in the Neotropics: the origin and diversification of the widespread genus *Leptodeira* (Serpentes: Colubridae). *Molecular Phylogenetics and Evolution* 53(3), 653–667, 8 figs.

Dean, B. (1938) Note on the sea-snake, *Pelamas platurus* (Linnaeus). *Science*, New York 88(2276), 145.

Deepak, V., Harikrishnan, S., Vasudevan, K. & Smith, E.N. (2010) Redescription of Bibron's coral snake, *Calliophis bibroni* Jan 1858 with notes and new records from south of the Palghat and Shencottah Gaps of the Western Ghats, India. *Hamadryad* 35(1), 1–10, 4 figs.

De Fonseca, P. & Jocqué, R. (1979) A herpetological survey of the Chintheche area in Malawi. *Revue de Zoologie Africaine 93*(4), 797–809, illust.

Degenhardt, W.G., Painter, C.W. & Price, A.H. (1996) *Amphibians and reptiles of New Mexico*. University of New Mexico Press, Albuquerque, 431 pp., 40 figs., 123 col. pls.

Degerbol, M. (1923) Description of a new snake of the genus *Glauconia*, from Mendoza. *Videnskabelige Meddelelser fra Dansk Naturhistorisk Forening i Kjøbenhavn* 76, 113–114, 3 figs.

DeGraaf, R.M. & Rudis, D.D. (1983) *Amphibians and reptiles of New England*. University of Massachusetts Press, Amherst, 83 pp., illust.

Dehling, J.M. (2006) Geographic distribution: Serpentes: *Sibynophis melanocephalus*. *Herpetological Review* 37(4), 500.

Dehling, J.M. & Das, I. (2006) Geographic distribution: Serpentes: *Dryocalamus tristrigatus*. *Herpetological Review* 37(4), 497.

Dehne, A. (1856) Verzeichniss derjenigen Reptilien, welche Herr Dr. Rabenhorst im Jahre 1847 in Italien gefunden und mir gütigst überlassen hat. *Allgemeine Deutsche Naturhistorische Zeitung*, Dresden 2, 212–221.

Dejesus, J.E. (2007). Geographic distribution: Serpentes: *Bothriechis supraciliaris*. *Herpetological Review* 38(4), 485.

DeKay, J.A. (1842) *Zoology of New-York, or the New-York fauna; comprising detailed descriptions of all the animals hitherto observed within the state of New-York, with brief notices of those occasionally found near its borders, and accompanied by appropriate illustrations. Part III. Reptiles and Amphibia*. 2 volumes. W. & A. White & J. Visscher, Albany, vi + 98 pp., 23 pls.

Dekeyser, P.L. & Derivot, J. (1960) Serpents et venins; morsures et traitements. *Notes Africaines*, Dakar (85), 1–35, 48 + 19 figs.

Dekeyser, P.L. & Villiers, A. (1954) Essai sur le peuplement zoologique terrestre de l'Ouest Africain. *Bulletin de l'Institut Français d'Afrique Noire,* Dakar 16, 957–970.

Dekeyser, P.L. & Villiers, A. (1956) Contribution a l'étude du peuplement de la Mauritanie. Notations ecologiques et biogeographiques sur la faune de l'Adrar. *Mémoires de l'Institut Français d'Afrique Noire (Sciences Naturelles),* Dakar (44), 1–222, 25 pl.

Delfino, M. (2002) *Erpetofaune Italiane del Neogene e del Quaternario*. Doctoral Dissertation, Università degli Studi di Modena e Reggio Emilia, Modena, 382 pp.

Delfino, M. & Szyndlar, Z. (2003) *Coluber etruriae* Portis, 1890 (Serpentes: Colubridae) a redescription. *Neues Jahrbuch fur Geologie und Palaeontologie Monatshefte* 2003(2), 65–72.

Dellinger, S.C. & Black, J.D. (1938) Herpetology of Arkansas. Part One. The reptiles. *Occasional Papers of the University of Arkansas* (1), 1–47.

Deliry, C. (Ed.). (2002) *Reptiles et amphibiens de Rhône-Alpes: atlas préliminaire*. Centre Ornithologique Rhône-Alps, Lyon, 146 pp., illust.

Dely, O.G. (1978) Reptilia. *Fauna Hungariae* 130, 1–120, illust. (in Hungarian)

Dely, O.G. (1981) Amphibians and reptiles of the Hortobagy. *Natural History of the National Parks of Hungary* 1, 387–390.

Dely, O.G. (1983) Wann ist die Beschreibung von Mehely uber *Vipera ursinii rakosiensis* erschienen? *Allattani Közle-menyek* 7(1–4), 117.

Dely, O.G. (1987) Amphibians and reptiles of the Kiskunsag. *Natural History of the National Parks of Hungary* 5, 404–426, illust.

Dely, O.G. & Stohl, G. (1984) Weitere Beitrage zur Kenntnis des *Vipera ursinii*-Formenkreises (Viperidae). *Vertebrata Hungarica*, Budapest 22, 15–46, 4 pls.

Demetropoulos, A. & Ioannides, Y. (2002) *Reptiles of Greece and Cyprus*. Goulandres Natural History Museum, Athens, 275 pp., 110 col. pls. (in Greek)

Demidoff, A. de (1840) *Voyage dans la Russie méridionale et la Crimée, par la Hongrie, la Válachie et la Moldavie, exécuté en 1837, sous la direction de M. Anatole de Demidoff, par MM. de Sainson, le Play, Huot, Leveille, Raffet, Rousseau, de Nordmann et du Ponceau. Tome troisième.* Ernest Bourdin, Paris, 756 pp. (Atlas, 1842, 87 pls.)

Deng, H.-B. & Zheng, J. (1974) Snakes of Fujian Province. *Publications of the Department of biology, Fujian Normal University*, Fuzhou (6), 93 pp., 37 pl. (in Chinese)

Deng, H.-B., Zhong, J., Liu, L.-B., Shi, P., Liu, G.-F., Zhao, J.-C. & Lin., J.-C. (1978) *Poisonous snakes of Fujian Province, snake-bites and treatment*. Published by authors, Quanzhou, 82 pp. (in Chinese)

Deng, Q.-X., Yu, Z.-W., Hu, J.-Z., Chen, H.-X. & Li, J.-G. (1983) A herpetological survey of some cities and counties in Sichuan. *Acta Herpetologica Sinica*, Chengdu (Series 2) 2(3), 68–69.

Densmore, L.D., Rose, F.L. & Kain, S.J. (1992) Mitochondrial DNA evolution and speciation in water snakes (genus *Nerodia*) with special reference to *Nerodia harteri*. *Herpetologica* 48(1), 60–68, 4 figs.

Deoras, P.J. (1990) *Snakes of India*. Fourth edition. National Book Trust, New Delhi, xv + 160 pp., 42 B & W pls., 13 col. pls. [first edition, 1965, xv + 148 pp., 50 figs., 12 col. pls.; second edition, 1978, xv + 156 pp., 50 figs., 12 col. pls.; third edition, 1981, xv + 152 pp., 13 col. pls.]

Department of Defense. (1991) *Venomous snakes of the Middle East*. Department of Defense Intelligence Agency, 160 pp.

Depéret, C. (1897) Les animaux pliocènes du Roussillon. *Mémoires de la Société Géologique de France (Paléontologie)*, Paris 7(3): 165–194, pl. 18.

Depoe, C.E., Funderburg, J.B. & Quoy, T.L. (1961) The reptiles and amphibians of North Carolina: (a preliminary check list and bibliography). *Journal of the Elisha Mitchell Scientific Society* 77(2), 125–136.

Deraniyagala, P.E.P. (1936) The snake *Oligodon albiventer* (Günther). *Ceylon Journal of Science*, Colombo 20B(1), 89–91.

Deraniyagala, P.E.P. (1939) A new colour variety of cobra from Ceylon & south India. *Ceylon Journal of Science*, Colombo 21B(3), 233–235, pl.21.

Deraniyagala, P.E.P. (1942) A new fossorial snake (*Rhinophis dorsimaculatus*) from Ceylon. *The Journal of the Bombay Natural History Society* (1940–1942) 42(4), 800–802, 1 fig., 1 pl.

Deraniyagala, P.E.P. (1945) Some new races of the *Python*, *Chrysopelea*, binocellate cobra, and Tith-Polonga inhabiting Ceylon and India. *Spolia Zeylanica*, Colombo 24(2), 103–112, 2 figs., pl. 13.

Deraniyagala, P.E.P. (1949) Some vertebrate animals of Ceylon. Vol. 1. *The National Museums of Ceylon Pictorial Series,* Colombo, 119 pp., 96 pls.

Deraniyagala, P.E.P. (1951) Some new races of the snakes *Eryx*, *Callophis* and *Echis*. *Spolia Zeylanica*, Colombo 26(2), 147–150.

Deraniyagala, P.E.P. (1954) Two new snakes from Ceylon. *Proceedings of the Tenth Annual Session of the Ceylon Association for the Advancement of Science (25th-27th Nov., 1954),* Colombo (1), 24.

Deraniyagala, P.E.P. (1955) *A colored atlas of some vertebrates from Ceylon. Volume three. Serpentoid Reptilia.* Colombo National Museum, Colombo, xii + 121 pp., 36 figs., 6 pls., col. pls. 36–49.

Deraniyagala, P.E.P. (1960a) A new red color variety *leucohaematus* of the cat snake *Boiga forsteni*. *Spolia Zeylanica*, Colombo 29(1), 37, 1 pl.

Deraniyagala, P.E.P. (1960b) The taxonomy of the cobras of southeastern Asia. *Spolia Zeylanica*, Colombo 29(1), 41–63.

Deraniyagala, P.E.P. (1961a) Some of Linne's reptiles and a mammal from Ceylon and elsewhere that are in Sweden. *Spolia Zeylanica*, Colombo 29(1), 8.

Deraniyagala, P.E.P. (1961b) The taxonomy of the cobras of southeastern Asia. Part 2. *Spolia Zeylanica*, Colombo 29(2), 205–232, 2 figs., 2 pls.

Deraniyagala, P.E.P. (1963) A new record of the sand boa from Ceylon. *Spolia Zeylanica*, Colombo 30(1), 25–26, 1 pl.

Deraniyagala, P.E.P. (1975) A new fossorial snake of the genus *Rhinophis*. *Spolia Zeylanica*, Colombo 33(1–2), 535–536, 1 fig.

Deraniyagala, P.E.P. (1977) The history of marine herpetology of the Indian and Pacific Oceans. *Journal of the Marine Biological Association of India,* Ernakulam (1974) 16(3), 793–806.

Derleyn, P.P. (1977) Notes sur les serpents du Burundi. Les genres *Thrasops*, *Dendroaspis* et *Bitis*. *Revue de Zoologie Africaine*, Tervuren 91(3), 525–537.

Derleyn, P.P. (1978a) Notes sur les serpents du Burundii. (2me partie) Les genres *Geodipsas* et *Thelotornis*. *Revue de Zoologie Africaine*, Tervuren 92(1), 208–218, illust.

Derleyn, P.P. (1978b) Notes sur les serpents du Burundi (3me partie). *Revue de Zoologie Africaine*, Tervuren 92(3), 728–737, illust.

Derleyn, P.P., Taverne, L., Ndabakubije, V., Madodo, G. & Karonkano, J. (1983) Etude sur les serpents des vallées de la Kayongozi et de la Ruvubu (Burundi). *Revue de Zoologie Africaine*, Tervuren 97(4), 770–800.

Despax, R. (1910) Mission géodésique de l'Équateur. Collections recueillies par M. le Dr. Rivet. Liste des ophidiens et description des espèces nouvelles. (Note préliminaire). *Bulletin du Muséum National d'Histoire Naturelle*, Paris 16(7), 368–376.

Despax, R. (1911) Reptiles et batraciens de l'Equateur recueillis par M. le Dr. Rivet. *In: Arc de Méridien Équatorial en Amérique du Sud sous le contrôle scientifique de l'Académie des Sciences, 1899–1906. Tome 9. Zoologie. Fascicule 2.-Reptiles-poissons-batraciens.* Gauthier-Villars, Paris, pp. 1–38, 2 pls.

Despax, R. (1912) Sur trois collections de reptiles et de batraciens provenent de l'Archipel Malais. *Bulletin du Muséum National d'Histoire Naturelle*, Paris 18, 198–205.

Despax, R. (1913) Sur une collections de reptiles et de batraciens rassembléee. *Bulletin du Muséum National d'Histoire Naturelle*, Paris 19(4), 179–184.

Desquez, M. (1884) Note sur l'alimentation de l'*Heterodon madagascariensis*. *Bulletin de la Société Philomathique Paris* (Série 7) 8, 180–182.

Desroches, J.-F. & Rodrigue, D. (2004) *Amphibiens et reptiles du Québec et des Maritimes.* Éditions Michel Quintin, Waterloo, 288 pp., col. illust.

Dessauer, H.C. (1967) Molecular approach to the taxonomy of colubrid snakes. *Herpetologica* 23(2), 148–155, 4 figs.

De Stefano, G. (1905) Appunti sui Batraci e sui Rettili del Quercy appartenenti alla collezione Rossignol. Parte Terza. Coccodrilli-Serpenti-Tartarughe. *Bollettino della Società Geologica Italiana*, Roma 24, 17–63, pls. 3–5.

Deufel, A. & Cundall, D. (2003) Feeding in *Atractaspis* (Serpentes: Atractaspididae): a study in conflicting functional constraints. *Zoology* 106, 43–61, 15 figs.

Deuve, J. (1961a) Observations sur les serpents du Laos, avec description d'une variété nouvelle. *Bulletin du Muséum National d'Histoire Naturelle*, Paris (1960) (Série 2) 32(5), 376–379.

Deuve, J. (1961b) Liste annotée des serpents du Laos. *Bulletin de la Société Royale des Sciences Naturelles du Laos*, Vientiane (1), 5–32.

Deuve, J. (1962a) Notes herpétologiques. *Bulletin de la Société Royale des Sciences Naturelles du Laos*, Vientiane (2), 83–88, 5 figs.

Deuve, J. (1962b) Notes herpétologiques [II]. *Bulletin de la Société Royale des Sciences Naturelles du Laos*, Vientiane (3), 13–20, 9 figs.

Deuve, J. (1962c) Clef d'identification des serpents du Laos. *Bulletin de la Société Royale des Sciences Naturelles du Laos*, Vientiane (4), 28–63, 20 figs.

Deuve, J. (1962d) Note sur un *Amblycephalus moellendorffii* Boettger. *Bulletin de la Société Royale des Sciences Naturelles du Laos*, Vientiane (5), 13–16, 4 figs.

Deuve, J. (1962e) Note sur *Dryophis prasinus* Boie. *Bulletin de la Société Royale des Sciences Naturelles du Laos*, Vientiane (5), 65–68, 6 figs.

Deuve, J. (1963a) Comparison entre les plaques anterieurs cephaliques d'*Holarchus purpurascens* Schlegel et d'*Holarchus taeniatus* Gunther. *Bulletin de la Société Royale des Sciences Naturelles du Laos*, Vientiane (6), 57–59, 1 fig.

Deuve, J. (1963b) *Holarchus purpurascens* Schlegel. Observations sur l'ornementation (exemplaires adultes vivants au fraichement captures). *Bulletin de la Société Royale des Sciences Naturelles du Laos*, Vientiane (6), 73–79, 18 figs.

Deuve, J. (1963c) *Natrix piscator* Schneider. *Bulletin de la Société Royale des Sciences Naturelles du Laos*, Vientiane (7), 25–35, 19 figs.

Deuve, J. (1963d) Observations sur la tendance a la division de certaines plaques chez *Natrix piscator piscator* Schneider. *Bulletin de la Société Royale des Sciences Naturelles du Laos*, Vientiane (7), 39–44, 16 figs.

Deuve, J. (1963e) Notes herpétologiques III. *Bulletin de la Société Royale des Sciences Naturelles du Laos*, Vientiane (8), 29–42, 18 figs.

Deuve, J. (1964) Note sur un jeune *Boiga cynodon* Boié. *Bulletin de la Société Royale des Sciences Naturelles du Laos*, Vientiane (10), 39.

Deuve, J. (1970) Serpents du Laos. *Mémoirs O.R.S.T.O.M.,* Paris (39), 1–252, 18 figs., 32 pls.

Devan-Song, A. & Brown, R.M. (2012) Amphibians and reptiles of Luzon Island, Philippines, VI: the herpetofauna of the Subic Bay area. *Asian Herpetological Research* (Series 2) 3(1), 1–20, 38 col. figs.

Devincenzi, G.J. (1925) Fauna erpetológica del Uruguay. *Anales del Museo de Historia Natural de Montevideo* (Series 2) 2(1), 1–65, 4 pls.

Devincenzi, G.J. (1939) *Ofidios del Uruguay*. Sociedad Linneana, Montevideo, 53 pp., illust.

De Vis, C.W. (1884a) On a new species of *Hoplocephalus*. *Proceedings of the Royal Society of Queensland*, Brisbane 1(2), 100, pl. 15.

De Vis, C.W. (1884b) Descriptions of new snakes. *Brisbane Courier* (Sept. 13) 39(8324), 5.

De Vis, C.W. (1884c) Descriptions of new snakes with a synopsis of the genus *Hoplocephalus*. *Proceedings of the Royal Society of Queensland*, Brisbane 1(3), 138–140.

De Vis, C.W. (1889) List of birds, lizards, and snakes collected at Cambridge Gulf, and descriptions of two new vertebrates in Mr. Saville-Kent's collection. *Proceedings of the Royal Society of Queensland*, Brisbane 6, 236–239, pl. 14.

De Vis, C.W. (1905) A new genus of lizard. *Annals of the Queensland Museum*, Brisbane (6), 46–52, pl. 15.

De Vis, C.W. (1911) Descriptions of snakes apparently new. *Annals of the Queensland Museum*, Brisbane (10), 21–25.

Devitt, T.J. (2006) Phylogeography of the western lyre snake (*Trimorphodon biscutatus*), testing aridland biogeographical hypotheses across the Nearctic-Neotropical transition. *Molecular Ecology* 15(14), 4387–4407, 5 figs.

Devitt, T.J., LaDuc, T.J. & McGuire, J.A. (2008) The *Trimorphodon biscutatus* (Squamata: Colubridae) species complex revisited: a multivariate statistical analysis of geographic variation. *Copeia* 2008(2), 370–387, 4 figs.

De Waal, S.W.P. (1978) The Squamata (Reptilia) of the Orange Free State, South Africa. *Memoirs of the National Museum*, Bloemfontein 11, 1–160, 2 figs.

Di-Bernardo, M. (1992) Revalidation of the genus *Echinanthera* Cope, 1894, and its conceptual amplification (Serpentes, Colubridae). *Comunicaciones do Museo Ciências e Tecnologia PUCRS (Zoologia)*, Pôrto Alegre 5(13), 225–256, 17 figs.

Di-Bernardo, M. (1994) Uma nova espécie de *Echinanthera* Cope (Serpentes, Colubridae) do nordeste do Brasil. *Biociències*, Porto Alegre 2(2), 75–81, 1 fig.

Di-Bernardo, M. (1996) A new species of the neotropical snake genus *Echinanthera* Cope, 1894 from southeastern Brazil (Serpentes, Colubridae). *The Snake* 27(2), 120–126, 3 figs.

Di-Bernardo, M. & Lema, T. de. (1986) O gênero *Rhadinaea* Cope, 1863, no Brasil Meridional. II. *Rhadinaea persimilis* (Cope, 1869) (Serpentes, Colubridae). *Acta Biologica Leopoldensia* 8(1), 101–122, 7 figs.

Di-Bernardo, M. & Lema, T. de. (1987) O gênero *Rhadinaea* Cope, 1863, no Brasil meridional. I. *Rhadinaea poecilopogon* Cope, 1863 (Serpentes, Colubridae). *Acta Biologica Leopoldensia* 9(2), 203–224, 7 figs.

Di-Bernardo, M. & Lema, T. de. (1988) O gênero *Rhadinaea* Cope, 1863, no Brasil meridional. III. *Rhadinaea affinis* (Günther, 1858) (Serpentes, Colubridae). *Acta Biologica Leopoldensia* 10(2), 223–252, 10 figs.

Di-Bernardo, M. & Lema, T. de. (1991) O gênero *Rhadinaea* Cope, 1863 no Brasil meridional. IV. *Rhadinaea bilineata* (Fischer, 1885) (Serpentes, Colubridae). *Acta Biologica Leopoldensia* (1990) 12(2), 359–392, 11 figs.

Di-Bernardo, M., Borges-Martins, M. & Jorge de Silva, N., Jr. (2007) A new species of coralsnake (*Micrurus*, Elapidae) from southern Brazil. *Zootaxa* (1447), 1–26, 10 figs.

Dickinson, W.E. (1949) *Field guide to the lizards and snakes of Wisconsin*. Milwaukee Public Museum, Milwaukee, 70 pp., illust.

Dimaki, M. (2004) Geographic distribution: Serpentes: *Telescopus fallax*. *Herpetological Review* 35(3), 294.

Dimitropoulos, A. (1987) Some notes on observations of the Ottoman Viper, (*Vipera xanthina xanthina*, Gray, 1849) in the Greek islands of Leros, Dodecanes (S.E. Aegean) and Chios (E. Aegean). *Herptile* 12(2), 101–104, illust.

Dimitropoulos, A. (1990) A new locality record of Ottoman viper, *Vipera xanthina* (Serpentes, Viperidae) from the Greek island of Oenousses, N.E. Aegean. *Annales Musei Goulandris* 8, 245–249, illust.

Di Pietro, D.O., Nenda, S.J. & Williams, J.D. (2010) New records of *Psomophis obtusus* (Cope, 1863) (Serpentes: Dipsadidae) in Argentina and Uruguay. *Cuadernos de Herpetología* 24(2), 125–127.

Dirksen, L. (2002) *Anakondas, monographische Revision der Gattung Eunectes Wagler, 1830 (Serpentes, Boidae)*. Natur und Tier Publ. House, Münster, 189 pp., 124 figs.

Dirksen, L. & Böhme, W. (1998a) Studien an Anakondas. 1. Indizilien für natürliche Bastardierung zwischen der Grossen Anakonda *Eunectes murinus* und der Paraguayanakonda *Eunectes notaeus* in Bolivien, mit Bemerkungen zur Taxonomie der Gattung *Eunectes*. *Zoologische Abhandlungen, Staatliches Museum für Tierkunde*, Dresden 4, 45–58.

Dirksen, L. & Böhme, W. (1998b) Studien an Anakondas. II. Zum taxonomischen Status von *Eunectes murinus gigas* (Latreille, 1801) (Serpentes, Boidae). *Salamandra* 34(4), 359–374.

Dirksen, L. & Böhme, W. (2005) Studies on Anacondas III. A reappraisal of *Eunectes beniensis* Dirksen, 2000, from Bolivia, and a key to the species of the genus *Eunectes* Wagler, 1830 (Serpentes, Boidae). *Russian Journal of Herpetology* 12(3), 223–229, 6 col. figs.

Dirksen, L. & Henderson, R.W. (2002) *Eunectes deschauenseei*. *Catalogue of American amphibians and reptiles*, pp. 755.1–755.3, 3 figs.

Dirksen, L., Ibisch, P.L., Köhler, J. & Böhme, W. (1995) Zur herpetofauna der semihumiden Samaipata-Region, Bolivien. 2. Reptilien. *Herpetofauna*, Weinstadt 17(99), 15–28.

Disi, A.M. (1983) A contribution to the herpetofauna of Jordan. 1. Venomous snakes. *DIRASAT (Natural Science)* 10(2), 167–180, 5 pls.

Disi, A.M. (1985) A contribution to the herpetofauna of Jordan. 2. New records and a systematic list of snakes from Jordan. *The Snake* 17(1), 31–42, 1 fig.

Disi, A.M. (1987) Environmental factors affecting snake distribution in Jordan. *In:* Krupp, F., Schneider, W. & Kinzelbach, R. (Eds.), *Proceedings of the symposium on the fauna and zoogeography of the Middle East*. Dr. Ludwig Reichert, Wiesbaden, pp. 296–310.

Disi, A.M. (1990) Venomous snakes in Jordan. *In:* Gopalakrishnakone, P. & Chou, L.M. (Eds.), *Snakes of medical importance (Asia-Pacific region)*. National University of Singapore, Singapore, pp. 349–376.

Disi, A.M. (1993). A contribution to the herpetofauna of Jordan, V. New records of three colubrid snakes from Jordan. *The Snake* 25(2), 109–113.

Disi, A.M. & Böhme, W. (1996) Zoogeography of the amphibians and reptiles of Syria, with additional new records. *Herpetozoa* 9(1–2), 63–70.

Disi, A.M., Amr, Z.S. & Defosse, D. (1988) Contribution to the herpetofauna of Jordan. III. Snakes of Jordan. *The Snake* 20(1), 40– 51, 2 figs.

Disi, A.M., Modry, D., Necas, P. & Rifai, L. (2001) *Amphibians and reptiles of the Hashemite Kingdom of Jordan. An atlas and field guide*. Edition Chimaira, Frankfurt-am-Main, 408 pp., 260 col. (mostly) figs.

Ditmars, R.L. (1905) A new species of rattlesnake. *Ninth Annual Report, New York Zoological Society* (1904), 197–200.

Ditmars, R.L. (1907) *The reptile book: a comprehensive, popularized work on the structure and habits of the turtles, tortoises, crocodilians, lizards and snakes which inhabit the United States and northern Mexico*. Doubleday, Page & Company, New York, 472 pp., 136 pls.

Ditmars, R.L. (1931) *Snakes of the world*. The Macmillan Company, New York, 207 pp., 84 pls.

Ditmars, R.L. (1936) *The reptiles of North America. A review of the crocodilians, lizards, snakes, turtles and tortoises inhabiting the United States and northern México*. Second edition. Doubleday & Company, New York, xvi + 476 pp., 135 pls.

Ditmars, R.L. (1939) *A field book of North American snakes*. Doubleday, Doran & Company, New York, xii + 305 pp., 48 pls., 1 col. pl.

Dixon, J.R. (1956) A collection of amphibians and reptiles from west central Korea. *Herpetologica* 12(1), 50–56, 1 fig.

Dixon, J.R. (1960) A new name for the snake, *Arizona elegans arizonae*. *Southwestern Naturalist* 5(4), 226.

Dixon, J.R. (1965) A taxonomic re-evaluation of the night snake *Hypsiglena ochrorhyncha* and relatives. *Southwestern Naturalist* 10(2), 125–131, 2 figs.

Dixon, J.R. (1967) Amphibians and reptiles of Los Angeles County California. *Los Angeles County Museum of Natural History, Scientific Series* 23, Zoology (10), 1–63, illust.

Dixon, J.R. (1980) The neotropical colubrid snakes genus *Liophis*. The generic concept. *Milwaukee Public Museum, Contributions in Biology and Geology* (31), 1–40, 2 figs.

Dixon, J.R. (1981) The neotropical colubrid snake genus *Liophis*, the eastern Caribbean complex. *Copeia* 1981(2), 296–304, 2 figs.

Dixon, J.R. (1983a) The *Liophis cobella* group of the neotropical colubrid snake genus *Liophis*. *Journal of Herpetology* 17(2), 149–165, 4 figs.

Dixon, J.R. (1983b) Systematics of *Liophis reginae* and *L. williamsi* (Serpentes, Colubridae), with a descriptions of a new species. *Annals of the Carnegie Museum* 52(6), 113–138, 6 figs.

Dixon, J.R. (1983c) Taxonomic status of the Brazilian colubrid snake, *Xenodon suspectus* Cope. *Texas Journal of Science* 35(3), 257–260, 1 fig.

Dixon, J.R. (1983d) Taxonomic status of the South American snakes *Liophis miliaris*, *L. amazonicus*, *L. chrysostomus*, *L. mossoroensis* and *L. purpurans* (Colubridae, Serpentes). *Copeia* 1983(3), 791–802, 3 figs.

Dixon, J.R. (1983e) Systematics of the Latin American snake, *Liophis epinephelus* (Serpentes, Colubridae). *In:* Rhodin, A.G.J. & Miyata, K. (Eds.), *Advances in herpetology and evolutionary biology*. Museum of Comparative Zoology, Cambridge, pp. 132–149, 8 figs.

Dixon, J.R. (1984) Taxonomic status of the Venezuelan macagua, *Bothrops colombiensis*. *Journal of Herpetology* 18(3), 329–332, 1 fig.

Dixon, J.R. (1985a) A review of *Liophis anomalus* and *Liophis elegantissimus*, and the description of a new species (Serpentes, Colubridae). *Copeia* 1985(3), 565–573, 4 figs.

Dixon, J.R. (1985b) A new species of the colubrid snake genus *Liophis* from Brazil. *Proceedings of the Biological Society of Washington* 98(2), 295–302, 3 figs.

Dixon, J.R. (1987) Taxonomy and geographic variation of *Liophis typhlus* and related "green" species of South America (Serpentes, Colubridae). *Annals of the Carnegie Museum* 56(8), 173–191, 4 figs.

Dixon, J.R. (1989) A key and checklist to the neotropical snake genus *Liophis* with country lists and maps. *Smithsonian Herpetological Information Service* (79), 1–28.

Dixon, J.R. (1991) Geographical variation and taxonomy of *Liophis almadensis* (Wagler) (Serpentes, Colubridae), and description of a new species of *Liophis* from Argentina and Bolivia. *Texas Journal of Science* 43(3), 225–236, 4 figs.

Dixon, J.R. (1993) Supplement to the literature for the "amphibians and reptiles of Texas" 1987. *Smithsonian Herpetological Information Service* (94), 1–43.

Dixon, J.R. (1996a) Texas herpetological literature: 1828–1996. *Texas Herpetological Society Special Publication* (1), 1–201.

Dixon, J.R. (1996b) Ten year supplement to Texas herpetological county records published in Amphibians and Reptiles of Texas, 1987. *Texas Herpetological Society Special Publication* (2), 1–64.

Dixon, J.R. (2000a) Ecuadorian, Peruvian, and Bolivian snakes of the *Liophis taeniurus* complex with descriptions of two new species. *Copeia* 2000(2), 482–490, 3 figs.

Dixon, J.R. (2000b) *Amphibians and reptiles of Texas*. Second edition. Texas A&M University Press, College Station, vii + 421 pp., 25 figs. [first edition, 1987, xii + 434 pp., 18 figs., 25 figs.]

Dixon, J.R. & Dean, R.H. (1986) Status of the southern populations of the night snake *(Hypsiglena*, Colubridae) exclusive of California and Baja California. *Southwestern Naturalist* 31(3), 307–318, 6 figs.

Dixon, J.R. & Fleet, R.R. (1976) *Arizona A. elegans*. *Catalogue of American amphibians and reptiles*, pp. 179.1–179.4.

Dixon, J.R. & Hendricks, F.S. (1979) The wormsnakes (family Typhlopidae) of the Neotropics, exclusive of the Antilles. *Zoologische Verhandelingen*, Leiden (173), 1–39, 10 figs.

Dixon, J.R. & Ketchersid, C.A. (1969) The status of the Mexican snake genus *Schmidtophis* Taylor (Colubridae). *Journal of Herpetology* 3(3–4), 163–165, 2 figs.

Dixon, J.R. & Kofron, C.P. (1984) The Central and South American anomalepid snakes of the genus *Liotyphlops*. *Amphibia-Reptilia*, Leiden (1983) 4(2–4), 241–264, 6 figs.

Dixon, J.R. & Lemos Espinal, J.A. (2010) *Anfibios y reptiles del estado de Querétaro, México. Amphibians and reptiles of the state of Querétaro, Mexico*. Texas A & M University, College Station and the Universidad Nacional Autónoma de México, 428 pp., 240 col. pls.

Dixon, J.R. & Lieb, C.S. (1972) A new night snake from México (Serpentes, Colubridae). *Contributions in Science, Natural History Museum of Los Angeles County* (222), 1–7, 2 figs.

Dixon, J.R. & Markezich, A.L. (1979) Rediscovery of *Liophis taeniurus* Tschudi (Reptilia, Serpentes, Colubridae) and its relationship to other Andean colubrid snakes. *Journal of Herpetology* 13(3), 317–320, 1 fig.

Dixon, J.R. & Markezich, A.L. (1992) Taxonomy and geographic variation of *Liophis poecilogyrus* (Wied) from South America (Serpentes, Colubridae). *Texas Journal of Science* 44(2), 130–166, figs.

Dixon, J.R. & Michaud, E.J. (1992) Shaw's black-backed snake (*Liophis melanotus*) (Serpentes, Colubridae) of northern South America. *Journal of Herpetology* 26(3), 250–259, 3 figs.

Dixon, J.R. & Soini, P. (1986) *The reptiles of the Upper Amazon Basin, Iquitos region, Perú. Part. 1. Lizards and amphisbaenians. Part 2. Crocodilians, turtles and snakes*. Second edition.

Milwaukee Public Museum, Milwaukee, 154 pp., 41 figs. [first edition, 1977, *Milwaukee Public Museum, Contributions in Biology and Geology* (12), 1–91, 5 figs.]

Dixon, J.R. & Thomas, R.A. (1982) The status of the Argentine colubrid snakes *Liophis sagittifer* and *L. trifasciatus*. *Herpetologica* 38(3), 389–395, 2 figs.

Dixon, J.R. & Thomas, R.A. (1984) A dichromatic population of the snake *Geophis latifrontalis*, with comments on the status of *Geophis semiannulatus*. *Journal of Herpetology* 8(3), 269–271, 1 fig.

Dixon, J.R. & Thomas, R.A. (1985) A new species of South American water snake (genus *Liophis*) from southeastern Brazil. *Herpetologica* 41(3), 259–262, 1 fig.

Dixon, J.R. & Tipton, B.L. (2003) *Liophis miliaris intermedius* (Henle and Ehrl, 1991) is actually *Liophis reginae* (Serpentes, Colubridae). *Journal of Herpetology* 37(1), 191.

Dixon, J.R. & Tipton, B.L. (2004) *Dryadophis* versus *Mastigodryas* (Ophidia, Colubridae): a proposed solution. *Herpetological Review* 35(4), 347–348.

Dixon, J.R. & Vaughan, R.K. (2003) The status of Mexican and southwestern United States blind snakes allied with *Leptotyphlops dulcis* (Serpentes: Leptotyphlopidae). *Texas Journal of Science* 55(1), 3–24.

Dixon, J.R., Ketchersid, C.A. & Lieb, C.S. (1972) The herpetofauna of Queretaro, Mexcio, with remarks on taxonomic problems. *Southwestern Naturalist* 16(3–4), 225–237, 1 fig.

Dixon, J.R., Thomas, R.A. & Greene, H.W. (1976) Status of the Neotropical snake *Rhabdosoma poeppigi* Jan, with notes on variation in *Atractus elaps* (Günther). *Herpetologica* 32(2), 221–227, 2 figs.

Dixon, J.R., Yanosky, A.A. & Mercolli, C. (1993) *Typhlops brongersmianus* Vanzolini and *Liophis almadensis* (Wagler): two new records for the snake fauna of the province of Formosa, Argentina. *The Herpetological Journal*, London 3(2), 72.

Dixon, J.R., Wiest, J.A., Jr. & Cei, J.M. (1995) Revision of the Neotropical snake genus *Chironius* Fitzinger (Serpentes, Colubridae). *Museo Regionale di Scienze Naturali, Torino, Monografia* XIII (1993), 268 pp., 64 figs., 21 col. pls. [15 Jan. 1995]

Dixon, J.R., Vaughan, R.K. & Wilson, L.D. (2000) The taxonomy of *Tantilla rubra* and allied taxa (Serpentes, Colubridae). *Southwestern Naturalist* 45(2), 141–153, 2 figs.

Dixon, J.R., Lazcano-Villarreal, D., Banda-Leal, J. & Chávez-Cisneros, J.A. (2011) Notes on Mexican herpetofauna 17: The status of *Rhadinaea montana*, *R. gaigeae* and *R. quinquelineata* of the Sierra Madre Oriental of eastern Mexico. *Bulletin of the Chicago Herpetological Society* 46(6), 69–73, 1 fig., 3 photos.

Dmi'el, R. (1967) Studies on reproduction, growth, and feeding in the snake *Spalerosophis cliffordi* (Colubridae). *Copeia* 1967(2), 332–346.

Dmi'el, R. & Borut, A. (1972) Thermal behavior, heat exchange, and metabolism in the desert snake *Spalerosophis cliffordi*. *Physiological Zoology* 45, 78–94, 8 figs.

Dmi'el, R., Perry, G. & Mendelssohn, H. (1990) Sexual dimorphism in *Walterinnesia aegyptia* (Reptilia, Ophidia, Elapidae). *The Snake* 22(1), 33–35, 3 figs.

Doan, T.M. & Arizábal, W. (2002) Microgeographic variation in the species composition of the herpetofaunal communities of Tambopata region, Peru. *Biotropica* 34, 101–117.

Doan, T.M. & Arriaga, W.A. (2000) New herpetological records for the Tambopata Province, Department of Madre de Dios, Perú. *Herpetological Review* 31(3), 188–189.

Dobiey, M. & Vogel, G. (2007) *Venomous snakes of Africa*. Edition Chimaira, Frankfurt-am-Main, 148 pp., 480 col. photos.

Dobiey, M. & Vogel, G. (2012) *Venomous snakes of South America*. Edition Chimaira, Frankfurt-am-Main, 144 pp., 450 col. photos. [not seen]

Dodd, C.K., Jr. (1993) Population structure, body mass, activity, and orientation of an aquatic snake (*Seminatrix pygaea*) during a drought. *Canadian Journal of Zoology*, Ottawa 71(7), 1281–1288, 7 figs.

Dollfus, R.P. & Beaurieux, C. (1928) Tableau pour la determination facile des Serpents du Maroc. *Variétés Scientifiques de la Société des Sciences Naturelles du Maroc*, Rabat 1, 1–27, 2 pls.

Dollo, L. (1886) Notice sur les reptiles et Batraciens recueillis par M. le capitaine Em. Storms dans la region du Tanganyka. *Bulletin du Musée Royal d'Histoire Naturelle de Belgique*, Bruxelles 4, 151–160, 5 figs.

Dolmen, D. (1986) Norwegian amphibians and reptiles; current situation 1985. *In:* Rocek, Z. (Ed.), *Studies in Herpetology. Proceedings of the European Herpetological Meeting Prague 1985*. Charles University, Prague, pp. 743–746.

Domergue, C.A. (1954) Note sur une nouvelle espèce ophidienne de Tunisie et l'Afrique du Nord *Coluber choumowitchi* Domergue. *Bulletin de la Société des Sciences Naturelles du Tunisie*, Tunis 7, 37–49, 15 figs.

Domergue, C.A. (1955) Note sur un serpent nouveau, *Pseudotarbophis gabesi* nov. sp. *Bulletin de la Société des Sciences Naturelles du Tunisie*, Tunis (1954–1955) 8(1–2), 119–123, pls. 29–31.

Domergue, C.A. (1959a) Liste des Ophidiens de Tunisie. *Archives de l'Institut Pasteur de Tunis* 36(2), 157–161.

Domergue, C.A. (1959b) Clé de determination des serpents de Tunisie. *Archives de l'Institut Pasteur de Tunis* 36(2), 163–172, 9 figs., 22 pls.

Domergue, C.A. (1962) Observations sur les pénis des ophidies (deuxième note). *Bulletin de la Société des Sciences Naturelles et Physiques du Maroc*, Rabat 42(1–2), 87–105, 14 figs., 1 pl.

Domergue, C.A. (1966) *Pseudotarbophis gabesi*—Domergue 1955. (Le serpent de Teboulbou). *Bulletin de la Société des Sciences Naturelles du Tunisie*, Tunis (1956–1966) 9–10, 85–95.

Domergue, C.A. (1969) Clé simplifiee pour la determination sur le terrain des serpents communs de Madagascar. *Bulletin de l'Academie Malagche*, Tananarive (1967) 45(2), 13–26, 10 figs.

Domergue, C.A. (1970) Notes sur les serpents de la région Malagache *Lycodryas maculatus* (Günther, 1858), espèce des Comores—description de deux femelles. *Bulletin du Muséum National d'Histoire Naturelle*, Paris (Série 3) 42, 449–451.

Domergue, C.A. (1973) Étude de trois serpents malgaches, *Liopholidophis lateralis* (D. & B.), *L. stumpffi* (Boettger) et *L. thielli* n. sp. *Bulletin du Muséum National d'Histoire Naturelle*, Paris (Série 3) 45, 1397–1412, 8 figs.

Domergue, C.A. (1984a) Notes sur les serpents de la région Malgache III. Description de trois espèces nouvelles rapportees au genre *Liophidium* Boulenger, 1896. *Bulletin du Muséum National d'Histoire Naturelle*, Paris (1983) (Série 4) 5(4), 1109–1122, 6 figs.

Domergue, C.A. (1984b) Notes sur les serpents de la région Malgache IV. Le genre *Pararhadinaea* Boettger, 1898. Descriptions d'une espèce et d'une sous-espèce nouvelles. *Bulletin du Muséum National d'Histoire Naturelle*, Paris (Série 4) 6A(1), 149–157, 3 figs.

Domergue, C.A. (1984c) Notes sur les Serpents de la région Malgache V. Le genre *Alluaudina* Mocquard, 1894. *Bulletin du Muséum National d'Histoire Naturelle*, Paris (Série 4) 6A(2), 537–549, 10 figs.

Domergue, C.A. (1986) Notes sur les serpentes de la région Malgache VI. Le genre *Ithycyphus* Günther, 1873; description de deux espèces nouvelles. *Bulletin du Muséum National d'Histoire Naturelle*, Paris (Série 4) 8A(2), 409–434, 8 figs.

Domergue, C.A. (1987) Notes sur les Serpents de la région Malgache VII. Révision du genre *Madagascarophis* Mertens, 1952. *Bulletin du Muséum National d'Histoire Naturelle*, Paris (Série 4) 9A(2), 455–489, 6 figs.

Domergue, C.A. (1988) Notes sur les serpents de la région Malgache VIII. Colubridae nouveaux. *Bulletin du Muséum National d'Histoire Naturelle*, Paris (Série 4) 10A(1), 135–146, 4 figs.

Domergue, C.A. (1994) Serpents de Madagascar, note preliminaire sur des especes nouvelles du genre *Stenophis* Boulenger, 1896 (Colubridae-Boiginae). *Archives de l'Institut Pasteur de Madagascar* 61(2), 121–122.

Domergue, C.A. & Bour, R. (1988) *Brygophis* nom nouveau por *Perinetia* Domergue, 1988, preemploye (Reptilia, Colubridae). *Bulletin du Muséum National d'Histoire Naturelle*, Paris (Série 4) 10A(4), 805–806.

Domingos-Cintra, C.E., Rodrigues Silva, H.L. & Silva Jr., N.J. da. (2009) Herpetofauna, Santa Edwiges I and II hydroelectric power plants, state of Goiás, Brazil. *Check List* 5(3), 570–576, 2 figs.

Domínguez, M. & Diaz, Jr., R.E. (2011a) Taxonomy of the blind snakes associated with *Typhlops lumbricalis* (Linnaeus, 1758) (Scolecophidia, Typhlopidae) from the Bahama Islands and Cuba. *Herpetologica* 67(2), 194–211, 4 figs.

Domínguez, M. & Díaz, Jr., R.E. (2011b) *Anguis jamaicensis* Shaw, 1802 (currently *Typhlops jamaicensis*: Reptilia, Serpentes): proposed conservation of the specific name by a ruling that it is not to be treated as a replacement name for *A. lumbricalis* Linnaeus, 1758 (currently *T. lumbricalis*) and the designation of neotypes for both taxa. *Bulletin of Zoological Nomenclature* 68(3), 1–7, 1 fig.

Domínguez, M. & Moreno, L.V. (2009) Taxonomy of the Cuban blind snakes (Scolecophidia, Typhlopidae), with the description of a new large species. *Zootaxa* (2028), 59–66, 2 figs.

Domínguez, M. & Parada, A. (2009) Geographic distribution: Serpentes: *Tropidophis morenoi*. *Herpetological Review* 40(4), 458.

Domínguez, M., Moreno, L.V. & Hedges, S.B. (2006) A new snake of the genus *Tropidophis* (Tropidophiidae) from the Guanahacabibes Peninsula of western Cuba. *Amphibia-Reptilia*, Leiden 27(3), 427–432, 2 figs.

Donndorff, J.A. (1798) *Zoologische [Amphibiologische und Ichthyologische] Beyträge zur XIII. Ausgabe des Linneischen Natursystems. Dritter Band:Amphibien und Fische. Amphibien und Fische*. Weidmannschen Buchhandlung, Leipzig, vi + 980 pp.

Donnelly, M.A. & Myers, C.W. (1991) Herpetological results of the 1990 Venezuelan expedition to the summit of Cerro Guaiquinima, with new tepui reptiles. *American Museum Novitates* (3017), 1–54, 33 figs.

Donoso-Barros, R. (1954) Consideraciones sobre la ecologia de los reptiles del sur de Coquimbo (Santiago). *Zooiatría* 3(11), 3–5.

Donoso-Barros, R. (1962) *Los ofidios chilenos*. Noticiario Mensuel de Museo Nacional de Historia Natural, Santiago (66), unpaginated.

Donoso-Barros, R. (1965) *Las serpientes*. Museo Nacional de Historia Natural, Santiago de Chile, 24 pp.

Donoso-Barros, R. (1966) *Reptiles de Chile*. La Universidad de Chile, Santiago de Chile, cxlvi + 458 pp., 32 col. pls.

Donoso-Barros, R. (1969a) Contribución al conocimiento del genero *Chironius* (Serpentes, Colubridae). *Boletin de la Sociedad de Biología de Concepción* 41, 189–197.

Donoso-Barros, R. (1969b) Consieraciones sobre un reflejo postural de *Tachymenis peruviana chilensis*. *Boletin de la Sociedad de Biología de Concepción* 41, 157–160, 17 figs.

Donoso-Barros, R. (1970) Catalogo herpetologico chileno. *Boletin del Museo Nacional de Historia Natural de Santiago* 31, 49–124, 8 pls.

Donoso-Barros, R. (1974) Descripcion de una culebra del Perú. *Neotropica*, La Plata 20(61), 14–16, 7 figs.

Donoso-Barros, R. & Candiani, H. (1950) Reptiles de la Provincia de Santiago. *Revista de la Academia Colombiana de Ciencias*, Bogotá 7(28), 482–489, 2 pls.

Donoso-Barros, R. & Cardenas, S. (1964) Los tipos de serpientes de R.A. Philippi. *Anais do Segundo Congresso Latino-Americano de Zoología*, São Paulo 3, 225–234.

d'Orbigny, C. (1843) *Dictionnaire universel d'histoire naturelle résumant et complétant tous les faits présentés par les encyclopédies les anciens dictionnaires scientifiques, les Oeuvres completes de Buffon, et les meilleurs traits spéciaux sur les diverses branches des sciences naturelles; –Donnant la description des ètres et des divers phénomènes de la nature, l'étymologie et la definition des noms scientifiques, et les principales applications des corps organiques et inorganiques à l'agriculture, à la medicine, aux artys industriels, etc. Tome troisiéme (CAA–CLA)*. Bureau Principal des Éditeurs, Paris, 744 pp.

d'Orbigny, C. (1847a) *Dictionnaire universel d'histoire naturelle résumant et complétant tous les faits présentés par les encyclopédies les anciens dictionnaires scientifiques, les Oeuvres completes de Buffon, et les meilleurs traits spéciaux sur les diverses branches des sciences naturelles; –Donnant la description des ètres et des divers phénomènes de la nature, l'étymologie et la definition des noms scientifiques, et les principales applications des corps organiques et inorganiques à l'agriculture, à la medicine, aux artys industriels, etc. Tome neuvième (OIE–PHO)*. Éditeurs Mm. Renard, Martinet et Cie., Paris, 776 pp.

d'Orbigny, C. (1847b) *Dictionnaire universel d'histoire naturelle résumant et complétant tous les faits présentés par les encyclopédies les anciens dictionnaires scientifiques, les Oeuvres completes de Buffon, et les meilleurs traits spéciaux sur les diverses branches des sciences naturelles; –Donnant la description des ètres et des divers phénomènes de la nature, l'étymologie et la definition des noms scientifiques, et les principales applications des corps organiques et inorganiques à l'agriculture, à la medicine, aux artys industriels, etc. Tome dixième (PHO–REP)*. Bureau Principal des Éditeurs, Paris, 760 pp.

d'Orbigny, C. (1848) *Dictionnaire universel d'histoire naturelle résumant et complétant tous les faits présentés par les encyclopédies les anciens dictionnaires scientifiques, les Oeuvres completes de Buffon, et les meilleurs traits spéciaux sur les diverses branches des sciences naturelles; –Donnant la description des ètres et des divers phénomènes de la nature, l'étymologie et la definition des noms scientifiques, et les principales applications des corps organiques et inorganiques à l'agriculture, à la medicine, aux artys industriels, etc. Tome douzième (STE–VAN)*. Éditeurs Mm. Renard, Martinet et Cie., Paris, 816 pp.

Dorcas, M.E. (1992) Relationships among montane populations of *Crotalus lepidus* and *Crotalus triseriatus*. *In:* Campbell, J.A. & Brodie, E.D., Jr. (Eds.), *Biology of the pitvipers*. Selva, Tyler, pp. 71–87, 13 figs.

Dorcas, M.E. (2004) *A guide to the snakes of North Carolina*. Michael E. Dorcas, Charlotte, 40 pp., col. illust.

Dorcas, M.E. & Willson, J.D. (2011) *Invasive pythons in the United States: ecology of an introduced predator.* University of Georgia Press, Athens, 156 pp., col. illust.

Dorcas, M.E., Gibbons, J.W. & Dowling, H.G. (1998) *Seminatrix, S. pygaea. Catalogue of American amphibians and reptiles*, pp. 679.1–679.5, 2 figs.

Dorcas, M.E., Poppy, S.M., Ernst, C.H. & Gibbons, J.W. (2003) *Regina alleni. Catalogue of American amphibians and reptiles*, pp. 778.1–778.4, 1 col. fig.

Doria, C. & Nyirenda, P. (1995) *A guide to the snakes of Luangwa Valley.* Wildlife Conservation Society of Zambia, Chipata Branch, 107 pp., 60 figs. [not seen]

Doria, G. (1875) Enumerazione dei Rettili raccolti dal Dott. O. Beccari in Amboina, alle Isole Aru ed alle Isole Kei durante gli anni 1872–73. *Annali del Museo Civico di Storia Naturale di Genova* (1874) 6, 325–357, pls. 11–12.

Doria, G. & Petri, M. (2010) *Pseudorabdion* in the museum of Genova, with description of two new species from Sumatra and a revised key to the genus (Reptilia, Serpentes, Colubridae, Calamariinae). *Annali del Museo Civico di Storia Naturale "Giacomo Doria,"* Genova 102, 187–201, 3 figs.

Dossenbach, H.D. & Ganz, A. (1978) Die Reptilien der Schweiz. *Neujahrsblatt der Naturforschenden Gesellschaft in Schaffhausen* (30), 1–32, illust.

Dotsenko, I.B. (1985) A revision of the genus *Eirenis* (Reptilia, Colubridae). Communication 1. Resurrection of the genus *Pseudocyclophis* Boettger, 1888. *Vestnik Zoologii*, Kiev 1985(4), 41–44, 1 fig. (in Russian)

Dotsenko, I.B. (1989) A review of the genus *Eirenis* (Reptilia, Colubridae). *Vestnik Zoologii*, Kiev 1989(5), 23–29. (in Russian)

Dotsenko, I.B. (1999) Snakes from Vietnam at the National Museum of Natural History of Ukrainian National Academy of Sciences stock collection. Communication 1. The sea snakes-Hydrophiidae. *Vestnik Zoologii*, Kiev 33(6), 39–51, illust. (In Russian)

Dottrens, È. & Aellen, V. (1963) *Batraciens et reptiles d'Europe*. Éditions Lacháux & Niestlé, Neuchatel, 261 pp., 59 figs., 52 pls.

Doucet, J. (1963) Les serpents de la Republique de Cote d'Ivoire. Ire partie. Generalites et serpents non venimeux. *Acta Tropica*, Abidjan 20(3), 201–259, 20(4), 297–340, 57 figs., 10 pls.

Doughty, P., Maryan, B., Donnellan, S.C. & Hutchinson, M.N. (2007) A new species of taipan (Elapidae, *Oxyuranus*) from central Australia. *Zootaxa* (1422), 45–58, 5 figs.

Douglas, M.E., Douglas, M.R., Schuett, G.W., Porras, L.W. & Holycross, A.T. (2002) Phylogeography of the western rattlesnake (*Crotalus viridis*) complex, with emphasis on the Colorado plateau. *In:* Schuett, G.W., Höggren, M., Douglas, M.E. & Greene, H.W. (Eds.), *Biology of the vipers*. Eagle Mountain Press, Salt Lake City, pp. 11–50, 8 figs.,

Douglas, M.E., Douglas, M.R., Schuett, G.W. & Porras, L.W. (2006) Evolution of rattlesnakes (Viperidae; *Crotalus*) in the warm deserts of western North America shaped by Neogene vicariance and Quaternary climate change. *Molecular Ecology* 15(11), 3353–3374, 7 figs.

Douglas, M.E., Douglas, M.R., Schuett, G.W., Porras, L.W. & Thomason, B.L. (2007) Genealogical concordance between mitochondrial and nuclear DNAs supports species recognition of the Panamint rattlesnake (*Crotalus mitchellii stephensi*). *Copeia* 2007(4), 920–932, 4 figs.

Douglas, M.E., Douglas, M.R., Schuett, G.W. & Porras, L.W. (2009) Climate change and evolution of the New World pitviper species *Agkistrodon* (Viperidae). *Journal of Biogeography* 36, 1164–1180, 6 figs.

Douglas, R.M. (1992) Microscopic investigation of the digestive tract contents of spotted grass snakes, *Psammophylax rhombeatus rhombeatus* (Reptilia, Colubridae). *Revue de Zoologie Africaine*, Tervuren 106(5), 401–411.

Douglass, E. (1903) New vertebrates from the Montana Tertiary. *Annals of the Carnegie Museum* (1903–1904) 2(2), 145–199, 13 figs.

Doumergue, F. (1901) *Essai sur la faune erpetologique de l'Oranie. Avec des tableaux analytiques et des notions pour determination de tous les reptiles & batraciens du Maroc, de l'Algerie et de la Tunisie.* L. Fouque, Oran, 404 pp., 217 figs., 27 pls. [reprint in 1972 by Linnaeus Press, Amsterdam]

Dowling, H.G. (1950) Studies of the black swamp snake, *Seminatrix pygaea* (Cope), with descriptions of two new subspecies. *Miscellaneous Publications, Museum of Zoology, University of Michigan* (76), 1–38, 6 figs.

Dowling, H.G. (1951) A taxonomic study of the ratsnakes, genus *Elaphe* Fitzinger. I. The status of the name *Scotophis laetus* Baird and Girard (1853). *Copeia* 1951(1), 39–44.

Dowling, H.G. (1952a) A taxonomic study of the ratsnakes, genus *Elaphe* Fitzinger. II. The subspecies of *Elaphe flavirufa* (Cope). *Occasional Papers of the Museum of Zoology, University of Michigan* (540), 1–14, 1 pl.

Dowling, H.G. (1952b) A taxonomic study of the ratsnakes, genus *Elaphe* Fitzinger, IV. A check list of the American forms. *Occasional Papers of the Museum of Zoology, University of Michigan* (541), 1–12.

Dowling, H.G. (1957a) A taxonomic study of the ratsnakes, genus *Elaphe* Fitzinger. V. The *rosalie* section. *Occasional Papers of the Museum of Zoology, University of Michigan* (583), 1–22, 2 pls.

Dowling, H.G. (1957b) A review of the amphibians and reptiles of Arkansas. *Occasional Papers of the University of Arkansas Museum* (3), 4–51.

Dowling, H.G. (1958) A taxonomic study of the ratsnakes. VI. Validation of the genera *Gonyosoma* Wagler and *Elaphe* Fitzinger. *Copeia* 1958(1), 29–40, 3 figs.

Dowling, H.G. (1960) A taxonomic study of the ratsnakes, genus *Elaphe* Fitzinger. VII. The *triaspis* section. *Zoologica*, New York 45(6), 53–80, 1 pl.

Dowling, H.G. (1966) Poisonous snakes of Vietnam. *Animal Kingdom*, New York 69(2), 34–43, illust.

Dowling, H.G. (1990) Taxonomic status and relationships of the genus *Lycognathophis*. *Herpetologica* 46(1), 60–66, 2 figs.

Dowling, H.G. & Duellman, W.E. (1978) *Systematic herpetology: a synopsis of families and higher categories.* Herpetological Information Search Service, New York, vii + 240 + viii pp., 118 figs.

Dowling, H.G. & Fries, I. (1987) A taxonomic study of the ratsnakes. VIII. A proposed new genus for *Elaphe triaspis* (Cope). *Herpetologica* 43(2), 200–207, 5 figs.

Dowling, H.G. & Gibson, F.W. (1970a) Relationship of the neotropical snakes *Hydrodynastes bicinctus* and *Cyclagras gigas*. *Herpetological Review* 2(2), 37–38, 3 figs.

Dowling, H.G. & Gibson, F.W. (1970b) Relationships of the Asian sunbeam snake, *Xenopeltis unicolor*. *Herpetological Review* 2(3), 51–52, 2 figs.

Dowling, H.G. & Gilboa, I. (1975) A revised list of herpetological type lists. *In: Yearbook of herpetology (1974)*. Herpetological Information Search Systems, New York, 171–173.

Dowling, H.G. & Jenner, J.V. (1987) Taxonomy of American xenodontine snakes. II. The status and relationships of *Pseudoleptodeira*. *Herpetologica* 43(2), 190–200, 2 figs.

Dowling, H.G. & Jenner, J.V. (1988) Snakes of Burma. Checklist of reported species & Bibliography. *Smithsonian Herpetological Information Service* (76), 1–19.

Dowling, H.G. & Maxson, L.R. (1990) Genetic and taxonomic relations of the short-tailed snakes, genus *Stilosoma*. *Journal of Zoology*, London 221(1), 77–85, 2 figs.

Dowling, H.G. & Price, R.M. (1988) A proposed new genus for *Elaphe subocularis* and *Elaphe rosaliae*. *The Snake* 20(1), 52–63, 14 figs.

Dowling, H.G., Gilboa, I. & Zweifel, R.G. (1970) A list of herpetological type lists. *Herpetological Review* 2(3), 53–54.

Dowling, H.G., Highton, R., Maha, G.C. & Maxson, L.R. (1983) Biochemical evaluation of colubrid snake phylogeny. *Journal of Zoology*, London 201(3), 309–329, 2 figs.

Downs, F.L. (1961) Generic reallocation of *Tropidodipsas leucomelas* Werner. *Copeia* 1961(4), 383–387, 1 fig.

Downs, F.L. (1967) Intrageneric relationships among colubrid snakes of the genus *Geophis* Wagler. *Miscellaneous Publications, Museum of Zoology, University of Michigan* (131), 1–193, 23 figs.

Dreckmann, F. (1893) A rare snake (*Psammophis longifrons*). *The Journal of the Bombay Natural History Society* (1892–1893) 7(3), 406–407.

Drewes, R.C. (1972) Report on a collection of reptiles and amphibians from the Ilemi Triangle, southwestern Sudan. *Occasional Papers of the California Academy of Sciences* (100), 1–14, 2 figs.

Drewes, R.C. & Sacherer, J.M. (1974) A new population of carpet vipers *Echis carinatus* from northern Kenya. *Journal of the East Africa Natural History Society and National Museum*, Nairobi (145), 1–7, 2 figs.

Drewes, R.C. & Spawls, S. (1973) The occurrence of the colubrid snake, *Psammophis pulcher*, in Kenya. *Herpetologica* 29(4), 306–307.

Dring, J.C.M. (1979) Amphibians and reptiles from northern Trengganu, Malaysia, with descriptions of two new geckos, *Cnemaspis* and *Cyrtodactylus*. *Bulletin of the British Museum of Natural History* (*Zoology*) 34(5), 181–241, illust.

Dring, J.C.M., McCarthy, C.J. & Whitten, A.J. (1990) The terrestrial herpetofauna of the Mentawai Islands, Indonesia. *Indo-Malayan Zoology* (1989) 6(1), 119–132, illust.

Drowne, F.P. (1905) The reptiles and amphibians of Rhode Island. *Monograph Roger Williams Park Museum*, Providence (15), 1–15.

Dryden, G.L. & Taylor, E.H. (1969) Reptiles from the Mariana and Caroline Islands. *University of Kansas Science Bulletin* 58(8), 269–279.

Du, M.-Z. (2004) *Amazing snakes: big surprises*. Yuan-Lion Publishing Company, Taipei, 279 pp. col. illust. (in Chinese)

Duarte, M.R., Puorto, G. & Franco, F.L. (1995) A biological survey of the pitviper *Bothrops insularis* Amaral (Serpentes, Viperidae): an endemic and threatened offshore island snake of southeastern Brazil. *Studies on Neotropical Fauna and Environment* 30(1), 1–13, 3 figs.

Duarte-Rocha, C.F. & Van Sluys, M. (2006) New records of reptiles from Ilha Granda Island in Rio de Janeiro State, Brazil. *Herpetological Review* 37(1), 112–114.

Dubey, B., Meganstthlan, P.R., Vidal, N. & Haque, I. (2012) Molecular evidence for the nonmonophyly of the Asian natricid genus *Xenochrophis* (Serpentes, Colubroidea) as inferred from mitochondrial and nuclear genes. *Journal of Herpetology* 46(2), 263–268, 3 figs.

DuBourdieu, D.J., Kawaguchi, H. & Shier, W.T. (1987) Molecular weight variations in the diversity of phospholipase A2 forms in reptile venoms. *Toxicon* 25(3), 333–343, 4 figs.

Duckett, J.E. (1987) The common cobra (*Naja-naja*) and its hood markings in Malaysia. *Malayan Nature Journal*, Kuala Lumpar 40(3–4), 46–47, 4 figs.

Duellman, W.E. (1954) The amphibians and reptiles of Jorullo Volcano, Michoacán, Mexico. *Occasional Papers of the Museum of Zoology, University of Michigan* (560), 1–24, 1 fig., 2 pls.

Duellman, W.E. (1956) A new snake of the genus *Leptotyphlops* from Michoacán, México. *Copeia* 1956(1), 93–94, 1 fig.

Duellman, W.E. (1958a) A monographic study of the colubrid snake genus *Leptodeira*. *Bulletin of the American Museum of Natural History* 114(1), 1–152, 25 figs., 31 pls.

Duellman, W.E. (1958b) Systematic status of the colubrid snake, *Leptodeira discolor* Günther. *University of Kansas Publications, Museum of Natural History* 11(1), 1–9, 4 figs.

Duellman, W.E. (1958c) A preliminary analysis of the herpetofauna of Colima, Mexico. *Occasional Papers of the Museum of Zoology, University of Michigan* (589), 1–22.

Duellman, W.E. (1959) Two new snakes, genus *Geophis*, from Michoacán, México. *Occasional Papers of the Museum of Zoology, University of Michigan* (605), 1–9, 2 figs., 1 pl.

Duellman, W.E. (1960) A taxonomic study of the Middle American snake, *Pituophis deppei*. *University of Kansas Publications, Museum of Natural History* 10(10), 599–610, 1 fig., pl. 45.

Duellman, W.E. (1961) The amphibians and reptiles of Michoacán, México. *University of Kansas Publications, Museum of Natural History* 15(1), 1–148, 11 figs., 6 pls.

Duellman, W.E. (1963) Amphibians and reptiles of the rainforests of southern El Petén, Guatemala. *University of Kansas Publications, Museum of Natural History* 15(5), 205–249, 6 figs., pls. 7–10.

Duellman, W.E. (1965a) Amphibians and reptiles from the Yucatán Peninsula, México. *University of Kansas Publications, Museum of Natural History* 15(12), 577–614, 1 fig.

Duellman, W.E. (1965b) A biogeographic account of the herpetofauna of Michoacán, México. *University of Kansas Publications, Museum of Natural History* 15 (14), 627–709, 5 figs.

Duellman, W.E. (1966a) The Central American herpetofauna: an ecological perspective. *Copeia* 1966(4), 700–719, 8 figs.

Duellman, W.E. (1966b) Remarks on the systematic status of certain Mexican snakes of the genus *Leptodeira*. *Herpetologica* 22(2), 97–106.

Duellman, W.E. (1978) The biology of an equatorial herpetofauna in Amazonian Ecuador. *Miscellaneous Publications, University of Kansas Museum of Natural History* (65), 1–352, 198 figs.

Duellman, W.E. (1979) The South American herpetofauna: its origin, evolution, and dispersal. *Monographs, University of Kansas Museum of Natural History* (7), 1–485, 172 figs.

Duellman, W.E. (1990) Herpetofaunas in neotropical rainforests, comparative composition, history and resource use. *In:* Gentry, A.H. (Ed.), *Four neotropical rainforests*. Yale University Press, New Haven, pp. 455–505, illust.

Duellman, W.E. (2005) *Cusco Amazónica. The lives of amphibians and reptiles in an Amazonian rainforest.* Cornell University Press, Ithaca, xv + 433 pp., 236 col. pls.

Duellman, W.E. & Berg, B. (1962) Type specimens of amphibians and reptiles in the Museum of Natural History, the University of Kansas. *University of Kansas Publications, Museum of Natural History* 15(4), 183–204.

Duellman, W.E. & Mendelson, J.R., III. (1995) Amphibians and reptiles from northern Departamento Loreto, Peru, taxonomy and biogeography. *University of Kansas Science Bulletin* 55(10), 329–376, 25 figs.

Duellman, W.E. & Salas, A.W. (1991) Annotated checklist of the amphibians and reptiles of Cuzco Amazonico, Peru. *Occasional Papers of the Museum of Natural History, University of Kansas* (142), 1–13, illust.

Duellman, W.E. & Schwartz, A. (1958) Amphibians and reptiles of southern Florida. *Bulletin of the Florida State Museum (Biological Sciences)* 3(5), 181–324, 28 figs.

Duellman, W.E. & Werler, J.E. (1955) Variation and relationships of the colubrid snake *Leptodeira frenata. Occasional Papers of the Museum of Zoology, University of Michigan* (570), 1–9, 2 pls.

Dueñas, C., Wilson, L.D. & McCranie, J.R. (2000) A list of the amphibians and reptiles of El Salvador, with notes on additions and delections. *In:* Johnson, J.D., Webb, R.G. & Flores-Villela, O.A. (Eds.), *Mesoamerican herpetology: systematics, zoogeography, and conservation.* University of Texas at El Paso, El Paso, pp. 93–99.

Duffaud, S. & Rage, J.-C. (1997) Les remplissages karstiques polyphasés (Éocène, Oligocène, Pliocène) de Saint-Maximin (Phosphorites du Gard) et leur apport à la connaissance des faunes Européennes, notamment pour l'Éocène moyen (MP 13). 2. Systématique, amphibiens et reptiles. *Mémoires et Travaux de l'Institut de Montpellier, E.P.H.E.*, 21, 729–735, 1 fig.

Dugand, A. (1975) Serpentifauna de la llanura costera del Caribe. *Caldasia*, Bogotá 11(53), 61–82.

Dugès, A.A.D. (1827) Recherches anatomiques et physiologiques sur la deglutition dans les reptiles. *Annales des Sciences Naturelles (Zoologie),* Paris (Série 1) 12, 337–395, pl.

Dugès, A.A.D. (1865) Du *Liophis janii. Mémoires de l'Académie des Sciences et Lettres du Montpellier* 6, 32–33.

Dugès, A.A.D. (1869) Catálogo de animals vertebrados observados en la República Mexicana. *La Naturaleza*, México City 1, 137–145.

Dugès, A.A.D. (1877) Apuntes para la monografia de los crótalos de México. *La Naturaleza*, México City 4, 1–29.

Dugès, A.A.D. (1884) Dos reptiles de Mexico. *La Naturaleza,* México City 6(23), 359–362, pl. 9.

Dugès, A.A.D. (1886) Sur le *Rhinocheilus antonii. Proceedings of the American Philosophical Society* 23(122), 290–291.

Dugès, A.A.D. (1887) *Adelophis copei*, *Argas sanchezi* y *Ornitomyia villadae. La Naturaleza*, México City (Serie 2) 1, 18–21, pl. 3.

Dugès, A.A.D. (1888a) Description of *Storeria dekayi* var. *anomala. Proceedings of the United States National Museum* 11(1), 9–10.

Dugès, A.A.D. (1888b) Sur deux espèces nouvelles des ophidiens de Mexique. *Proceedings of the American Philosophical Society* 25(128), 181–183, 2 figs.

Dugès, A.A.D. (1890) Dos nuevas especies de ophidios mexicanos. *La Naturaleza,* México City (Serie 2) 1, 402–403, figs. 13–14, pl. 27.

Dugès, A.A.D. (1891) *Elaps diastema*, var. *michoacanensis*, A. Dug. *La Naturaleza,* México City (Serie 2) 1, 487, pl. 32 (col.).

Dugès, A.A.D. (1897a) Description d'un ophidien nouveau du Mexique (*Oreophis boulengeri*, g. et sp. n.). *Proceedings of the Zoological Society of London* 65(1), 284–285.

Dugès, A.A.D. (1897b) Reptiles y batracios de los E.U. Mexicanos. *La Naturaleza,* México City (Serie 2) 2, 479–485.

Dugès, A.A.D. (1898a) Un nuevo género de ofidio. *La Naturaleza,* México City (Serie 2) 3, 52.

Dugès, A.A.D. (1898b) Description d'un genre nouveau d'ophidiens: *Geatractus. Proceedings of the Zoological Society of London* 66(3), 539–540.

Dugès, A.A.D. (1905) Description d'un ophidien nouveau do Mexique (*Morenoa orizabensis* g. et sp. n.). *Proceedings of the Zoological Society of London* 75(4), 517–518.

Duguy, R. (1970) Numbers of blood cells and their variation. *In:* Gans, C. & Parsons, T.S. (Eds.), *Biology of the Reptilia. Volume 3 (Morphology C).* Academic Press, London, pp. 93–110.

Duguy, R. & Saint-Girons, H. (1993) Repartition et frequency du phenotype "raye" chez *Natrix maura* (Linne, 1758) dans les departments de Charente-Marit. *Annales de la Société Sciences Naturelles de la Charente-Maritime* 8(2), 133–138, illust.

Duke, J.A. (1967) *Herpetological dietary.* Battelle Memorial Institute, Columbus, 35 pp.

Duméril, A.H.A. (1856) Note sur les Reptiles du Gabon. Ordre des ophidiens. *Revue et Magasin de Zoologie*, Paris (Série 2) (8), 460–476.

Duméril, A.H.A. (1857) *Notes pour servir a l'histoire de l'erpétologie de l'Afrique occidentale et en particulier de la côte du Gabon, précédées de considérations générales sur les reptiles de ces contrées.* Bouchard-Huzard, Paris, 36 pp., 2 pls. [reprint of A.H.A. Duméril, 1856 with different title]

Duméril, A.H.A. (1859) Reptiles et Poissons de l'Afrique occidentale. Etude precedee de considerations generales sur leur distribution geographique. *Archives du Muséum National d'Histoire Naturelle*, Paris (1858–1861) 10, 137–240, 7 pls. [Royal Soc. London catalog cites 1860 as date *fide* Loveridge, 1958, 184]

Duméril, A.H.A., Bocourt, M.-F. & Mocquard, F. (1870–1909) *Mission Scientifique au Mexique et dans l'Amérique Centrale. Recherches zoologiques pour servir à l'histoire de la faune de l'Amérique Centrale et du Mexique. Troisième partie.—1re section. Etudes sur les reptiles.* Imprimerie Nationale, Paris, 1012 pp., 98 pls. [issued in 17 livr.: 1, 1–32, pls. 1–7, 9, 11, 12 (1870); 2, 33–112, pls. 8, 10, 13–15 (1873); 3, 113–192, pls. 16–18 (1874); 4, 193–280, pls. 19, 20A–C, 23 (1874); 5, 281–360, pls. 20, 20D–G, 21A–C (1878); 6, 361–440, pls. 21–22, 22A–D (1879); 7, 441–488, pls. 22E–J (1881); 8, 489–528, pls. 27–30 (1882); 9, 529–592, pls. 31–35 (1883); 10, 593–666, pls. 36–41 (1886); 11, 667–696, pls. 42–47 (1888); 12, 697–732, pls. 25–26, 48–51 (1890); 13, 733–780, pls. 52–57 (1893); 14, 781–828, pls. 24, 58–62 (1895); 15, 829–860, pls. 63–68 (1897); 16, 861–932, pls. 69–74 (1908); 17, 933–1012, pls. 75–77 (1909)]

Duméril, A.M.C. (1806) *Zoologie analytique, ou méthode naturelle de classification des animaux, rendue plus facile à l'aide de tableaux synoptiques.* Allais Libraire, Paris, 344 pp.

Duméril, A.M.C. (1853) Prodrome de la classification des reptiles ophidiens. *Mémoires de l'Académie de Sciences de l'Institut de France,* Paris 23, 399–536, 2 pls., 15 figs. [reprinted in 1853 by Firmin Didot Frères, Paris, 139 pp., 2 pls.]

Duméril, A.M.C. & Bibron, G. (1844) *Erpétologie générale ou histoire naturelle complète des reptiles. Tome sixième, comprenant l'histoire générale des ophidiens, la description des genres et des espèces de serpents non venimeux, savoir, la totalité des vermiformes ou des scolecophides, et partie des circuriformes ou azémiophides; en tout vingt-cinq genres et soixante-cinq espèces*. Librairie Encyclopédique de Roret, Paris, xii + 609 pp. [reprinted in 2012 by Society for the Study of Amphibians and Reptiles]

Duméril, A.M.C. & Duméril, A.H.A. (1851) *Catalogue methodique de la collection des reptiles du Museum d'Histoire naturelle de Paris*. Gide et Baudry, Paris, iv + 224 pp.

Duméril, A.M.C., Bibron, G. & Duméril, A.H.A. (1854a) *Erpétologie générale ou histoire naturelle complète des reptiles. Tome septième—Premiere partie. Comprenant l'histoire des serpents non venimeux*. Librairie Encyclopédique de Roret, Paris, vii + 780 pp. [reprinted in 2012 by Society for the Study of Amphibians and Reptiles]

Duméril, A.M.C., Bibron, G. & Duméril, A.H.A. (1854b) *Erpétologie générale ou histoire naturelle complète des reptiles. Tome septième—Deuxième partie. Comprenant l'histoire des serpents venimeux*. Librairie Encyclopédique de Roret, Paris, xii + pp. 781–1536. [reprinted in 2012 by Society for the Study of Amphibians and Reptiles]

Duméril, A.M.C., Bibron, G. & Duméril, A.H.A. (1854c) *Erpétologie générale ou histoire naturelle complète des reptiles. Tome neuvième. Comprenant l'histoire des batraciens urodèles. Plus, un repertoire ou catalogue méthodique de tous les reptiles décrits dans les neuf volumes; avec des tables generales pour le texte et pour les figures*. Librairie Encyclopédique de Roret, Paris, xx + 440 pp. [reprinted in 2012 by Society for the Study of Amphibians and Reptiles]

Duméril, A.M.C., Bibron, G. & Duméril, A.H.A. (1854d) *Erpétologie générale ou histoire naturelle complète des reptiles. Tome neuvième. Comprenant l'histoire des batraciens urodèles. Plus, un répertoire ou catalogue méthodique de tous les reptiles décrits dans les neuf volumes; avec des tables générales pour le texte et pour les figures. Ouvrage accompagné de planches*. Librarie Encyclopédique de Roret, Paris, xx + 440 pp., 120 pls. [reprinted in 2012 by Society for the Study of Amphibians and Reptiles]

Dunaev, E.A. & Orlova, V.F. (2003) *Snake diversity (from materials of the exposition of the Zoological Museum of Moscow State University)*. Moscow University, Moscow, 376 pp., 209 figs.

Dundee, H.A. (1992) New data on hatchling mud snakes, *Farancia abacura reinwardtii* Schlegel. *Bulletin of the Maryland Herpetological Society* 28(3), 78–82.

Dundee, H.A. (1996a) Geographical distribution. *Typhlops ater*. *Herpetological Review* 27(3), 155.

Dundee, H.A. (1996b) Some reallocations of type localities of reptiles and amphibians described from the Major Stephen H. Long expedition to the Rocky Mountains, with comments on some of the statements made in the account written by Edwin James. *Tulane Studies in Zoology and Botany* 30, 75–89.

Dundee, H.A. & Liner, E.A. (1997) Geographical distribution. *Trimorphodon tau tau*. *Herpetological Review* 28(4), 211.

Dundee, H.A. & Rossman, D.A. (1989) *The amphibians and reptiles of Louisiana*. Louisiana State University Press, Baton Rouge, xi + 300 pp., 10 figs., 24 pls.

Dundee, H.A., White, D.A. & Rico-Gray, V. (1986) Observations on the distribution and biology of some Yucatan Peninsula amphibians and reptiles. *Bulletin of the Maryland Herpetological Society* 22(2), 37–50.

Dunger, G.T. (1966) A new species of the colubrid genus *Mehelya* from Nigeria. *American Museum Novitates* (2268), 1–8, 4 figs.

Dunger, G.T. (1971a) The snakes of Nigeria. Part 1, The file snakes of Nigeria. *Nigerian Field*, Woodchester 36(2), 54–71, 4 figs., 6 pls.

Dunger, G.T. (1971b) The snakes of Nigeria. Part 2, The house snakes of Nigeria. *Nigerian Field*, Woodchester 36(4), 151–163, 7 pls.

Dunger, G.T. (1972) The snakes of Nigeria. Part 3, The harmless water and marsh snakes of Nigeria. *Nigerian Field*, Woodchester 37(1), 21–38, 10 pls.

Dunger, G.T. (1973) The snakes of Nigeria. Part 4, The harmless green snakes of Nigeria. *Nigerian Field*, Woodchester 38(4), 158–179, 1 fig., 13 pls.

Dunn, E.R. (1919) Two new crotaline snakes from western México. *Proceedings of the Biological Society of Washington* 32, 213–216.

Dunn, E.R. (1920a) On the Haitian snakes of the genera *Leimadophis* and *Uromacer*. *Proceedings of the New England Zoological Club* 7, 37–44.

Dunn, E.R. (1920b) A new *Geophis* from México. *Proceedings of the Biological Society of Washington* 33, 127–128.

Dunn, E.R. (1922) Two new South American snakes. *Proceedings of the Biological Society of Washington* 35, 219–220.

Dunn, E.R. (1923a) On a collection of reptiles from Sarawak. *Journal of the Malayan Branch of the Royal Asiatic Society* 1(1), 1–4.

Dunn, E.R. (1923b) Some snakes from northwestern Peru. *Proceedings of the Biological Society of Washington* 36, 185–188.

Dunn, E.R. (1925) *Amastridium*: a neglected genus of snakes. *Proceedings of the United States National Museum* (1924) 65(2524), 1–3.

Dunn, E.R. (1927) Results of the Douglas Buden Expedition to the island of Komodo. II. Snakes from the East Indies. *American Museum Novitates* (287), 1–7.

Dunn, E.R. (1928a) New Central American snakes in the American Museum of Natural History. *American Museum Novitates* (314), 1–4.

Dunn, E.R. (1928b) A tentative key and arrangement of the American genera of Colubridae. *Bulletin of the Antivenin Institute of America* 2(1), 18–24.

Dunn, E.R. (1930) New snakes from Costa Rica and Panamá. *Occasional Papers of the Boston Society of Natural History* (1921–1931) (5), 329–332

Dunn, E.R. (1931a) Preliminary list of the reptiles and amphibians of the Canal Zone and the provinces of Panama and Colon, R.P. *Seventh Annual Report of Barro Colorado Island Biology Laboratory*, 7, 15–18.

Dunn, E.R. (1931b) Some Central American snake genera. *Copeia* 1931(4), 163.

Dunn, E.R. (1933a) Amphibians and reptiles from El Valle de Anton, Panama. *Occasional Papers of the Boston Society of Natural History* (8), 65–79.

Dunn, E.R. (1933b) A new snake from Panama. *Copeia* 1933(4), 193–194.

Dunn, E.R. (1935) The snakes of the genus *Ninia*. *Proceedings of the National Academy of Sciences*, Washington 21(1), 9–12.

Dunn, E.R. (1936) Notes on North American *Leptodeira*. *Proceedings of the National Academy of Sciences*, Washington 22(12), 689–698.

Dunn, E.R. (1937a) Notes on tropical *Lampropeltis*. *Occasional Papers of the Museum of Zoology, University of Michigan* (353), 1–11.

Dunn, E.R. (1937b) New or unnamed snakes from Costa Rica. *Copeia* 1937(4), 213–215.

Dunn, E.R. (1938a) The snake genus *Enulius* Cope. *Proceedings of the Academy of Natural Sciences of Philadelphia* 89, 415–418.

Dunn, E.R. (1938b) A new *Rhadinaea* from Central America. *Copeia* 1938(4), 197–198.

Dunn, E.R. (1939a) Mainland forms of the snake genus *Tretanorhinus*. *Copeia* 1939(4), 212–217.

Dunn, E.R. (1939b) A new pit viper from Costa Rica. *Proceedings of the Biological Society of Washington* 52, 165.

Dunn, E.R. (1940a) Notes on some American lizards and snakes in the Museum of Goteborg. *Herpetologica* 1(8), 189–194.

Dunn, E.R. (1940b) New and noteworthy herpetological material from Panama. *Proceedings of the Academy of Natural Sciences of Philadelphia* 92, 105–122.

Dunn, E.R. (1941) Notes on the snake genus *Anomalepis. Bulletin of the Museum of Comparative Zoology* 87(7), 511–526, 5 figs.

Dunn, E.R. (1942) New or noteworthy snakes from Panama. *Notulae Naturae*, Philadelphia (108), 1–8.

Dunn, E.R. (1943) Notes on Colombian herpetology: I, A new snake of the genus *Rhadinaea. Caldasia*, Bogotá 2(8), 307–308.

Dunn, E.R. (1944a) The snake genus *Dendrophidion* in Colombia. *Caldasia*, Bogotá 2(10), 474–477.

Dunn, E.R. (1944b) A review of the Colombian snakes of the families Typhlopidae and Leptotyphlopidae. *Caldasia*, Bogotá 3(11), 47–55, 10 figs.

Dunn, E.R. (1944c) *Dugandia*, a new snake genus for *Coluber bicinctus* Herrmann. *Caldasia*, Bogotá 3(11), 69–70.

Dunn, E.R. (1944d) A new snake of the genus *Hydrops* from Colombia. *Caldasia*, Bogotá 3(11), 71–72.

Dunn, E.R. (1944e) Los genero de anfibios y reptiles de Colombia. III. Tercera parte. Reptiles: orden de las serpientes. *Caldasia*, Bogotá 3(12), 155–224.

Dunn, E.R. (1944f) A revision of the Colombian snakes of the genera *Leimadophis, Lygophis, Liophis, Rhadinaea* and *Pliocercus*, with a note on Colombian *Coniophanes. Caldasia,* Bogotá 2(10), 479–495.

Dunn, E.R. (1944g) Herpetology of the Bogotá area. *Revista de la Academia Colombiana de Ciencias Exactas, Fisicas y Naturales*, Bogotá 6(21), 68–81.

Dunn, E.R. (1945) The amphibians and reptiles of the Colombian Caribbean islands San Andres and Providencia. *Caldasia*, Bogotá 3(14), 363–365.

Dunn, E.R. (1946a) *Atractus sanctaemartae*, a new species of snake from the Sierra Nevada de Santa Marta, Colombia. *Occasional Papers of the Museum of Zoology, University of Michigan* (493), 1–6

Dunn, E.R. (1946b) A new snake from the eastern Andes of Colombia. *Caldasia*, Bogotá 4(17), 121–122.

Dunn, E.R. (1949) Relative abundance of some Panamanian snakes. *Ecology* 30(1), 39–57.

Dunn, E.R. (1957) *Contributions to the herpetology of Colombia, 1943–1946. Appendix: Herpetological publications of Emmett Reid Dunn, 1915–1957.* M.T. Dunn, iv + 296 pp. [reprinted from *Revista de la Academia Colombiana de Ciencias Exactas, Fisicas y Naturales* (1944) and *Caldasia* (1943–1946)]

Dunn, E.R. & Bailey, J.R. (1939) Snakes from the uplands of the Canal Zone and of Darién. *Bulletin of the Museum of Comparative Zoology* 86(1), 1–22.

Dunn, E.R. & Conant, R. (1936) Notes on anacondas, with descriptions of two new species. *Proceedings of the Academy of Natural Sciences of Philadelphia* 88, 503–506, pl. 14.

Dunn, E.R. & Dowling, H.G. (1957) The Neotropical snake genus *Nothopsis* Cope. *Copeia* 1957(4), 255–261, 1 pl.

Dunn, E.R. & Emlen, J.T., Jr. (1932) Reptiles and amphibians from Honduras. *Proceedings of the Academy of Natural Sciences of Philadelphia* 84, 21– 32.

Dunn, E.R. & Saxe, L.H., Jr. (1950) Results of the Catherwood-Chaplin West Indies Expedition, 1948. Part V. Amphibians and Reptiles of San Andres and Providencia. *Proceedings of the Academy of Natural Sciences of Philadelphia* 102, 141–165.

Dunn, E.R. & Stuart, L.C. (1935) A new race of *Leptodeira* from northern Central America. *Occasional Papers of the Museum of Zoology, University of Michigan* (313), 1–3.

Dunn, E.R. & Stuart, L.C. (1951) Comments on some recent restrictions of type localities of certain South and Central American amphibians and reptiles. *Copeia* 1951(1), 55–61.

Dunn, E.R. & Wood, G.C. (1939) Notes on eastern snakes of the genus *Coluber. Notulae Naturae*, Philadelphia (5), 1–4.

Dunson, W.A. (1979) Occurrence of partially striped forms of the mangroove snake *Nerodia fasciata compressicauda* Kennicott and comments on the status of *N. f. taeniata* Cope. *Florida Scientist* 42(2), 102–112, figs.

Dunson, W.A. & Minton, S.A. (1978) Diversity, distribution, and ecology of Philippine marine snakes (Reptilia, Serpentes). *Journal of Herpetology* 12(3), 281–286.

Duong, H.-M. (1962) Contribution a l'étude du venin de *Trimeresurus gramineus* du Sud-Vietnam. *Bulletin de la Société de Pathologie Exotique* 55(1), 147–153.

Durigën, B. (1897) *Deutschlands Amphibien und Reptilien.* Schluss, Magdeburg, 369–576, pl. 11.

Dury, R. & Williams, R.S. (1933) Notes on some Kentucky amphibians and reptiles. *Bulletin of the Baker-Hunt Museum*, Covington (1), 1–22.

Dusej, G. (1993) *Coronella girondica* (Daudin, 1803)—Girondische Glatt- oder Schlingnatter. *In:* Böhme, W. (Ed.), *Handbuch der Reptilien und Amphibien Europas. Band 3/I. Schlangen (Serpentes) I. (Typhlopidae, Boidae, Colubridae 1, Colubrinae).* AULA-Verlag, Wiesbaden, pp. 247–264, figs. 48–50b.

Dutta, S. (2005) *Common rat snake of India,* Ptyas mucosus: *their functional morphology and bionomics, including deadly venomous snakes of India.* Asiatic Publishing House, Delhi. 334 pp., illust.

Dutta, S.K. & Acharjyo, L.N. (1990) Checklist of the herpetofauna of Orissa, India. *The Snake* 22(1), 36–43.

Dutta, S.K. & Acharjyo, L.N. (1993) Additions to the herpetofauna of Orissa, India. *The Snake* 25(2), 149–150.

Dutta, S.K., Nair, M.V., Mohapatra, P.P. & Mahapatra, A.K. (2009) *Amphibians and reptiles of Similipal Biosphere Reserve.* Regional Plant Resource Centre, Bhubaneswar, 173 pp., col. illust.

Duvernoy, G. L. (1831) On the distinguishing marks of venomous serpents. *The Journal of the Royal Institution of Great Britain* (5), 157-159.

Duvernoy, G.L. (1832) Mémoire sur les caractères tirés de l'anatomie pour distinguer les serpents venimeux des serpents non venimeux. *Annales des Sciences Naturelles (Zoologie),* Paris 26, 113–160, pls. 5–10.

Duvernoy, G.L. (1833a) Fragmens d'anatomie sur l'organisation des serpens. *Annales des Sciences Naturelles (Zoologie),* Paris 30, 5-32, 4 pls.

Duvernoy, G.L. (1833b) Ueber die anatomischen kennzeichen zur unterschreibung der giftschlangen gen von den ungiftigen. *Isis von Oken*, Leipzig 26, 626–642, pls. 18–19. (Partial translation of 1832)

Dwigubsky, J.A. (1832) *Opyt estestvennoi Istorii Gady.* [*Concerning the natural history of the Russian Empire*]. *Vol. 3.* Rossiskoj Imperi., Moscova, 46 pp. (in Russian) [not seen]

Dwyer, C.M. & Kaiser, H. (1997) Relationships between skull form and prey selection in the Thamnophiine snake genera *Nerodia* and *Regina. Journal of Herpetology* 31(4), 463–475, 4 figs.

Dwyer, Q. & Perez, M. (2009) Geographic distribution: Serpentes: *Lachesis melanocephala. Herpetological Review* 40(1), 114.

Dwyer, Q. & Van Den Burgh, J. (2012) Geographic distribution: Serpentes: *Porthidium volcanicum*. *Herpetological Review* 43(2), 308.

Easton, E.R. & Leung-Va. (1993) An annotated list of the reptiles and amphibians of Macau, Southeast Asia. *Herpetological Review* 24(4), 158–160.

Echternacht, A.C. (1973) The color pattern of *Sonora michoacanensis* (Dugès) (Serpentes, Colubridae) and its bearing on the origin of the species. *Breviora* (410), 1–18, 5 figs.

Echternacht, A.C. & Burton, F.J. (2003) Geographic distribution: Serpentes: *Ramphotyphlops braminus*. *Herpetological Review* 34(3), 265–266.

Eckel, E.C. (1902) Serpents of northeastern United States. *In:* Eckel, E.C. & Paulmier, F.C. (Eds.), Catalogue of New York reptiles and batrachians. *Bulletin of the New York State Museum* 51, 355–386, 24 figs.

Eckstein, H.-P. (1993) Untersuchungen zur Ökologie der Ringelnatter (*Natrix natrix* Linnaeus 1758). *Jahrbuch für Feldherpetologie* 4, 1–145, 69 figs.

Edeling, A.C.J. (1864a) Recherches sur la faune erpétologique de Borneo. *Natuurkundig Tijdschrift voor Nederlandsch Indïe*, Batavia (Série 6) 1(1), 482–490.

Edeling, A.C.J. (1864b) Dezelfde vertoont eene hagedis met gespleten start, door hem ontvangen van den heer A.H.G. Blokzeijl op Balie. *Natuurkundig Tijdschrift voor Nederlandsch Indïe*, Batavia (Série 6) 2(2), 387–389.

Edeling, A.C.J. (1865) Koninklijke natuurkundige vereeniging van Nederlandsch Indië. Zoologie. Recherches sur la fauna erpétologique de Bornéo. *Nederlandsch Tijdschrift voor Dierkunde*, Amsterdam 2(4–6), 200–204. [reprint of Edeling, 1864a]

Edeling, A.C.J. (1870) Recherches sur la faune erpétologique de Sumatra. *Natuurkundig Tijdschrift voor Nederlandsch Indïe*, Batavia (Série 7) 31(1), 376–388.

Edgren, R.A. (1952a) Biogeographical and behavioral considerations of the snakes of the genus *Heterodon*. Doctoral Dissertation, Northwestern University, Evanston, ix + 232 pp.

Edgren, R.A. (1952b) A synopsis of the snakes of the genus *Heterodon*, with the diagnosis of a new race of *Heterodon nasicus* Baird and Girard. *Natural History Miscellanea*, Chicago (112), 1–4.

Edgren, R.A. (1953) On the identity of *Coluber simus* Linnaeus. *Copeia* 1953(1), 64–65.

Edwards, B. (2009) *Australian snakes: a pictorial guide.* The Five Mile Press, Scoresby, 80 pp., col. illust.

Efrati, P. (1969) Clinical manifestations and treatment of viper bite in Israel. *Toxicon* 7(1), 29–31.

Egan, D. (2007) *Snakes of Arabia: a field guide to the snakes of the Arabian Peninsula and its shores.* Motivate Publishing, Dubai, 208 pp., col. illust.

Ehmann, H. (1992) *Encyclopedia of Australian Animals. Reptiles.* Angus & Robertson, Pymble, xv + 405 pp., col. illust.

Eichwald, K.E.I. (1831) *Zoologia specialis quam expositis animalibus tum vivis, tum fossilibus potissimum Rossiae in universum, et Poloniae in specie, in usum lectionum publicarum in Universitate Caesarea Vilnensi habendarum. Pars posterior [vol.3], specialem expositionem spondylozoorum continens.* Josephi Zawadzki, Vilnae, 369 pp., 1 pl.

Eichwald, K.E.I. (1837) *Reise auf dem Caspischen Meere und in den Kaukasas (Unternommen in den Jahren 1825–1826).* Erster Band. J.G. Cotta'schen Buchhandlung, Stuttgart, 894 pp., 2 pls.

Eichwald, K.E.I. (1839) De duabus novis amphibiorum speciebus. *Bulletin de la Société Impériale des Naturalistes du Moscou* 12, 303–307.

Eichwald, K.E.I. (1841) *Fauna Caspio-Caucasia nonnullis observationibus novis.* Litteris Typographiae Diarii Gall. Politic. Petropolensis, Petropoli, 233 pp.

Eichwald, K.E.I. (1842) *Fauna Caspio-Caucasia nonnullis observationibus novis, illustravit. [Atlas]* Literis Typographiae Diarii Gall. Politic. Petropolensis, Petropoli, 128 pp, 40 pls.

Einfalt, P. (2012) *Die Brillenschlange,* Naja naja. Natur und Tier-Verlag, Münster, 62 pp., col. illust.

Eiselt, J. (1961) Amphibia, Reptilia. *In: Ein systematisches Verzeichnis aller auf osterreichischem-Gebiet festgestellten Tierarten.* Part XXI. Springer-Verlag, Wien, 21 pp.

Eiselt, J. (1962) Ergelbnisse der zoologischen Nubien-Expedition 1962 Teil II. Amphibien und Reptilien. *Annalen des Naturhistorischen Museums in Wien* 65, 281–296.

Eiselt, J. (1963) Zur Kenntnis der colubriden Schlangengattungen *Procteria* und *Xenodon*. *Annalen des Naturhistorischen Museums in Wien* 66, 279–282, 1 fig.

Eiselt, J. (1970) Ergebnisse zoologischer Sammelreisen in der Türkei: Bemerkenswerte Funde von Reptilien, I. *Annalen des Naturhistorischen Museums in Wien* 74, 343–355.

Eiselt, J. (1971) *Eirenis rechingeri* n. sp. (Colubridae, Serpentes) aus dem Iran. *Annalen des Naturhistorischen Museums in Wien* 75, 375–381, 2 pls.

Eiselt, J. (1976) Ergebnisse zoologischer Sammelreisen in der Türkei: Bemerkenswerte Funde von Reptilien, II. *Annalen des Naturhistorischen Museums in Wien* 80, 803–814, 5 figs., 2 pls.

Eiselt, J. & Baran, I. (1970) Ergebnisse zoologischer Sammelreisen in der Türkei: Viperidae. *Annalen des Naturhistorischen Museums in Wien* 74, 357–369, 3 pls.

Eisentraut, M. (1973) Die Wirbeltierfauna von Fernando Poo und Westkamerun. Unter besonderer Berücksichtigung der Bedentung der pleistozänen Klimaschwankungen für die heutige Faunenverteilung. *Bonnischer Zoologische Mongraphien*, Bonn (3), 3–428, illust.

Eissa, S.M. & El-Assy, Y.S. (1975) Record of certain reptilian species found in Kuwait. *Journal of the University of Kuwait (Science)* 2, 123–145.

Ekstrand, C.H. (1823) *D. D. Fauna Brasliensis quam Venia. Exp. Facult. Med. Upsal. praeside C.P. Thunberg. VII Junii MDCCCXXIII.* Excudebant Palmblad et C., Upsaliae. iii + 10 pp.

Elliot, W. (1840) Description of a new species of naga or cobra de capello [*Naia vittata*]. *Madras Journal of Literature and Science* 11, 39–41, pl. 1.

Elliott, M.J. & McGhee, M.E. (2000) Geographic distribution: Serpentes: *Tantilla relicta relicta*. *Herpetological Review* 31(4), 256.

Ellis, E.A. (1979) *British amphibians and reptiles: a fully illustrated introduction to species found in the British Isles.* Jarrold Colour Publications, Norwich, 32 pp., 35 col. photos.

Ellis, M.M. & Henderson, J. (1913) *The Amphibia and Reptilia of Colorado. Part I.* University of Colorado Bulletin 10(2), 39–129, 8 pls.

Elpatjewsky, W.S. (1902) Sur la corrélation des espèces *Contia modesta* Mart. et *Contia collaris* Mén. *Annuaire du Musée Zoologique de l'Académie Impériale des Sciences de St. Pétersbourg* 7, 223–234.

Elpatjewsky, W.S. & Sabanejew, L.L. (1906) Ergänzungen zur herpetologischen Fauna des Russischen Reichs. *Zoologischer Jahrbücher (Abteilung fü Systematic, Ökologie und Geographie der Tiere)*, Jena 24(4), 247–261, pls. 18–19.

El Shafië, D.M. (1995) *Snakes in Jordan.* Yarmook University, Museum of Natural History, Irbid, 82 pp., 56 col. photos. (in Arabic)

Elter, O. (1982) *Cataloghi V—La collezione Erpetologica del Museo di zoologia dell'Universita di Torino.* Museo Regionale di Scienze Naturale, Torino, 116 pp., 2 figs., 1 pl.

Elvin, D.W. (1963) Variation and distribution of the shovel-nosed snakes (*Chionactis occipitalis*) in the northern Majave Desert, California and Nevada. *Herpetologica* 19(2), 73–76, 2 figs.

Elzen, P. van den. (1980) Die Schlangen Südwest-Afrikas in der Herpetologischen Abteilung des Zoologischen Forschungsinstitutes und Museums Alexander Koenig in Bonn. *Bonner Zoologische Beiträge,* Bonn 31(3–4), 333–356, 5 figs.

Embert, D. (2002a) Geographic distribution: Serpentes: *Chironius multiventris multiventris. Herpetological Review* 33(3), 226.

Embert, D. (2002b) Geographic distribution: Serpentes: *Chironius quadricarinatus maculoventris. Herpetological Review* 33(3), 227.

Embert, D. (2007) *Distribution, diversity and conservation status of Bolivian reptiles.* Doctoral Dissertation, Rheinischen Friedrichs-Wilhelms-Universität Bonn, Bonn, 429 pp.

Embert, D., Gonzales, L. & Montaño, R. (2006) Geographical distribution. *Atractus latifrons. Herpetological Review* 27(1), 108.

Emelianov, A.A. (1929) Snakes of the Far Eastern District. *Memoires Vladivostok Section, Russian State Geographical Society* 3(1), 1–208, 62 figs., 7 pls. (in Russian)

Emelianov, A.A. (1937) On a new species of *Ancistrodon* (Ophidia) in the Far East. (*Ancistrodon saxatilis* sp. nov.). *Bulletin of the Far Eastern Branch of the Academy of Sciences of the USSR* 1937(24), 19–40, 4 figs. (in Russian)

Emmott, A. & Wilson, S.G. (2009) *Snakes of western Queensland.* Desert Channels Queensland, Longreach, 136 pp., col. illust.

Emmrich, D. (1997) New data for the rare horned bush-viper, *Atheris ceratophora*, Werner, 1895. *Mitteilungen aus dem Naturhistorischen Museum in Berlin* 73(1), 131–151.

Emory, W.H. (1857) *Report on the United States and Mexican boundary survey, made under the direction of the Secretary of the Interior, by William H. Emory, Major First Cavalry, and the United States commissioner.* C. Windell, Washington, xvi + 615 pp., 21 pls.

Emsley, M.G. (1963a) A consideration of the list of snakes recorded from Trinidad. *Copeia* 1963(3), 576–577.

Emsley, M.G. (1963b) The rediscovery of Cope's *Liophis perfuscus* in Barbados. *Copeia* 1963(3), 577–579, 1 fig.

Emsley, M.G. (1966a) The status of the snake *Erythrolamprus aesculapii ocellatus* Peters. *Copeia* 1966(1), 128–129.

Emsley, M.G. (1966b) The mimetic significance of *Erythrolamprus aesculapii ocellatus* Peters from Tobago. *Evolution* 20, 663–664.

Emsley, M.G. (1977) Snakes of Trinidad and Tobago. *Bulletin of the Maryland Herpetological Society* 13(4), 201–304, illust.

Enderson, E.F. & Bezy, R.L. (2007a) Geographic distribution: Serpentes: *Geophis dugesii aquilonaris. Herpetological Review* 38(1), 103.

Enderson, E.F. & Bezy, R.L. (2007b) Geographic distribution: Serpentes: *Leptodeira splendida ephippiata. Herpetological Review* 38(2), 220–221.

Enderson, E.F., Quijada-Mascareñas, A., Turner, D.S., Rosen, P.C. & Bezy, R.L. (2009) The herpetofauna of Sonora, Mexico, with comparisons to adjoining states. *Check List* 5(3), 632–672, 8 figs.

Enge, K.M. & Dodd, C.K., Jr. (1992) An indexed bibliography of the herpetofauna of Florida. *Florida Game and Freshwater Fish Commission, Nongame Wildlife Program* (11), iv + 231 pp.

Engelmann, W.-E. (1993) *Coronella austriaca* (Laurenti, 1768) - Schlingnatter, Glatt- oder Haselnatter. *In:* Böhme, W. (Ed.), *Handbuch der Reptilien und Amphibien Europas. Band 3/I. Schlangen (Serpentes) I. (Typhlopidae, Boidae, Colubridae 1, Colubrinae).* AULA-Verlag, Wiesbaden, pp. 200–245, figs. 42–47.

Engelmann, W.-E. & Obst, F.J. (1981) *Snakes: biology, behavior and relationship to man.* Exeter Books, New York, 222 pp., 45 figs., 156 photos.

Engelmann, W.-E., Fritzsche, J., Günther, J. & Obst, F.J. (1985) *Lurche und Kriechtiere Europas.* Neumann, Leipzig, 420 pp.

Engbretson, G. (1974) Amphibians and reptiles of Oklahoma. *In: Field Guide to Oklahoma.* Oklahoma Biological Survey, Norman, pp. 103–118.

Engkamat, L., Stuebing, R.B. & Voris, H.K. (1991) A population size estimate of the yellow-lipped sea krait, *Laticauda colubrina*, on Kalampunian Damit Island, Sabah, Malaysia. *Copeia* 1991(4), 1139–1142.

Erber, J. (1863) Beobachtungen an Amphibien in der Gefangenschaft. *Verhandlungen der Kaiserlich-Königlichen Zoologisch-Botanischen Gesellschaft in Wien* 13, 129–132.

Erhard, D. (1858) *Die Fauna der Cykladen, 1. Teil, Die Wirbelthiere der Cykladen.* Voigt & Günther, Leipzig, 116 pp.

Ernst, C.H. (1992) *Venomous reptiles of North America.* Smithsonian Institution Press, Washington, ix + 236 pp., 55 col. pls.

Ernst, C.H. (2002a) *Regina. Catalogue of American amphibians and reptiles*, pp. 756.1–756.4.

Ernst, C.H. (2002b) *Regina septemvittata. Catalogue of American amphibians and reptiles*, pp. 757.1–757.5, 2 col. figs.

Ernst, C.H. (2002c). *Storeria occipitomaculata. Catalogue of American amphibians and reptiles*, pp. 759.1–769.8, 2 col. figs.

Ernst, C.H. (2012) *Storeria. Catalogue of American amphibians and reptiles,* pp. 900.1–900.14, 4 figs.

Ernst, C.H. & Barbour, R.W. (1989) *Snakes of eastern North America.* George Mason University Press, Fairfax, 282 pp., 78 figs., 64 col. pls. [first edition, 1989, 282 pp., 78 figs., 64 col. pls.]

Ernst, C.H. & Ernst, E.M. (2003) *Snakes of the United States and Canada.* Smithsonian Books, Washington, xi + 668 pp., col. photos.

Ernst, C.H. & Ernst, E.M. (2011) *Venomous reptiles of the United States, Canada, and northern Mexico. Vol. 1. Heloderma, Micruroides, Micrurus, Pelamis, Agkistrodon, Sistrurus.* John Hopkins University Press, Baltimore, xviii + 352 pp., 16 col. pls.

Ernst, C.H. & Ernst, E.M. (2012) *Venomous reptiles of the United States, Canada, and northern Mexico. Vol. 2. Crotalus.* John Hopkins University Press, Baltimore, xiv + 391 pp., 16 col. pls.

Ernst, C.H., Belfit, S.C., Sekscienski, S.W. & Laemmerzahl, A.F. (1997) The amphibians and reptiles of Ft. Belvoir and northern Virginia. *Bulletin of the Maryland Herpetological Society* 33(1), 1–62, 1 fig.

Ernst, C.H., Orr, J.M. & Creque, T.R. (2003a) *Carphophis. Catalogue of American amphibians and reptiles*, pp. 773.1–773.2.

Ernst, C.H., Orr, J.M. & Creque, T.R. (2003b) *Carphophis amoenus. Catalogue of American amphibians and reptiles*, pp. 774.1–774.7, 2 col. figs.

Ernst, C.H., Orr, J.M. & Creque, T.R. (2003c) *Carphophis vermis. Catalogue of American amphibians and reptiles*, pp. 775.1–775.4, 1 col. fig.

Ernst, R. & Rodel, M.-O. (2002) A new *Atheris* species (Serpentes, Viperidae), from Tai National Park, Ivory Coast. *Herpetological Journal* 12(2), 55–61.

Escomel, E. (1929) Fauna de Arequipa. *Obras Cientificas,* Lima 1, 367–418, 8 pls.
Espinal, M.R., McCranie, J.R. & Wilson, L.D. (2001) The herpetofauna of Parque Nacional La Muralla, Honduras. *In:* Johnson, J.D., Webb, R.G. & Flores-Villela, O.A. (Eds.), *Mesoamerican herpetology: systematics, zoogeography, and conservation.* University of Texas at El Paso, El Paso, pp. 100–108, 1 fig.
Esqueda, L.F. & La Marca, E. (1999) New reptilian species records from the Cordillera de Mérida, Andes of Venezuela. *Herpetological Review* 30(4), 238–240.
Esqueda, L.F. & La Marca, E. (2004) Geographic distribution: Serpentes: *Corallus ruschenbergerii. Herpetological Review* 35(2), 190.
Esqueda, L.F. & La Marca, E. (2005) Revisión taxonómica y biogeográfica (con descripción de cinco nuevas especies) del género *Atractus* (Colubridae: Dipsadinae) en los Andes de Venezuela. *Herpetotropicos* 2(1), 1–32, 27 figs.
Esqueda, L.F., La Marca, E., Natera, M. & Battiston, P. (2001) Noteworthy reptilian state records and a lizard species to the herpetofauna of Venezuela. *Herpetological Review* 32(3), 198–200.
Esqueda, L.F., La Marca, E. & Bazo, S. (2005a) Un nuevo colúbrido semifossorial del género *Atractus* (Dipsadinae) de la vertiente lacustre de Los Andes de Venezuela. *Herpetotropicos* 2(2), 87–93, 5 figs.
Esqueda, L.F., Natera, M., La Marca, E. & Ilija-Fistar, M. (2005b) Nueva especie de serpiente (Reptilia: Colubridae: *Liophis*) de un bosque tropical relictual en el Estado Barinas, Venezuela. *Herpetotropicos* 2(2), 95–103, 5 figs.
Esterbauer, H. (1992) Some notes on *Daboia lebetina obtusa. Lacerta,* Gravenhage 50(6), 198–204, illust.
Estes, R. (1983) Sauria terrestria, Amphisbaenia. *Handbuch der Paläoherpetologie. Teil 10A.* Gustav Fischer Verlag, Stuttgart, 249 pp., 69 figs.
Estes, R. & Báez, A. (1985) Herpetofauna of North and South America during the late Cretaceous and Cenozoic: evidence for interchange? *In:* Stehli, F.G. & Webb, S.D. (Eds.), *The great American biotic interchange.* Plenum Press, New York, pp. 139–197, 5 figs.
Estes, R., Frazzetta, T.H. & Williams, E.E. (1970) Studies on the fossil snake *Dinilysia patagonica* Woodward, Part I. Cranial morphology. *Bulletin of the Museum of Comparative Zoology* 140(2), 25–74, 12 figs., 5 pls.
Estrada, A.R. (1994) Herpetofauna de la Cuenca Banao-Higuanojo, Sancti Spíritus, Cuba. *Revista de la Academia Colombiana de Ciencias Exactas, Físicas y Naturales,* Bogotá 19(73), 353–360, 2 figs.
Estrada, A.R. (1998) Anfibios y reptiles encontrados durante 1988 y 1989 en Cayo Paredón Grande, Archipiélago de Sabana-Camagüey, Cuba. *Caribbean Journal of Science* 34(1–2), 106–112.
Estrada, A.R. & Ruibal, R. (1999) A review of Cuban herpetology. *In:* Crother, B.I. (Ed.), *Caribbean amphibians and reptiles.* Academic Press, San Diego, pp. 31–62, 4 figs.
Etchepare, E.G. & Zaracho, V.H. (2009) Serpentes, Colubridae, *Taeniophallus poecilopogon* (Cope, 1863): Rediscovery in Corrientes, Argentina, and natural history. *Check List* 5(4), 770–772, 4 figs.
Evans, G.H. (1905) *Simotes splendidus. The Journal of the Bombay Natural History Society* (1904–1905) 16(2), 362.
Evans, G.H. & Wall, F. (1900) Incubation of a brood of *Zamenis mucosus. The Journal of the Bombay Natural History Society* (1900–1901) 13(1), 189–190.
Evans, H.E. (1947) Notes on Panamanian reptiles and amphibians. *Copeia* 1947(3), 166–170.
Eversmann, E.F. (1823) *Reise von Orenburg nach Buchara...nebst einem Wortverzeichniss aus der Afghanischen Sprache begleitet von einem naturhistorischen Anhange und einer Vorrede von H. Lichenstein.* E.H.G. Christiani, Berlin, viii + 150 pp., 2 pls.
Ewart, J. (1878) The poisonous snakes of India. For the use of the Officials and others residing in the Indian Empire. J. & A. Churchill, London, 64 pp., 2 pls. + 19 col. pls. [a partial reprint of Fayrer, 1872]
Eydoux, F.-J.-T. & Gervais, P. (1837) Reptiles. *Magasin de Zoologie,* Paris 7(3), 1–10, pls. 12–16.
Eydoux, F.-J.-T. & Souleyet, L.-F.-A. (1842) Reptiles. *In:* Vaillant, A.N. (Ed.), *Voyage autour du monde exécuté pendant les années 1836 et 1837 sur la corvette La Bonite commandée par M. Vaillant, Capitaine de vaisseau, publié par ordre du Roi sous les auspices du Département de la Marine. Zoologie. Tome premier. Seconde partie.* Arthur Bertrand, Paris (1841), pp. 135–154 [Atlas], 9 pls.
Falcione, C., Cajade, R., Barrasso, D.A. & Nenda, S.J. (2010) *Taeniophallus poecilopogon* (cope, 1863) (Serpentes: Dipsadidae): filling a gap on its known distribution in Argentina. *Cuadernos de Herpetología* 24(2), 137–140, 2 figs.
Falk, K. (1923) *Die Giftschlangen in Südwest, ihr Biss und dessen Heilung.* Swakopmunder Buchhandlung, Windhoek, 23 pp.
Fan, L. (1998) *The Amphibia and Reptilia Fauna of Shanxi.* China Forestry Publishing House, Beijing, 206 pp., 96 figs. (in Chinese)
Fan, T.H. (1931) Preliminary report of reptiles from Yaoshan, Kwangsi, China. *Bulletin of the Deptartment of Biology, Sun Yatsen University* (11), 1–154, 11 figs., 9 pls.
Fang, R.-S. & Wang, T.-Z. (1983) Three new records of snakes from Shaanxi. *Acta Herpetologica Sinica,* Chengdu (Series 2) 2(2), 75–76.
Farang, A.A. & Banaja, A.A. (1980) Amphibians and reptiles from the western region of Saudi Arabia. *Bulletin, Faculty of Science, King Abdul Aziz University,* Jeddah 4, 5–29.
Farkas, B. & Fritz, U. (1999a) Geographic distribution: Serpentes: *Boiga multomaculata. Herpetological Review* 30(3), 174.
Farkas, B. & Fritz, U. (1999b) Geographic distribution: Serpentes: *Enhydris plumbea. Herpetological Review* 30(3), 174.
Farkas, B. & Fritz, U. (1999c) Geographic distribution: Serpentes: *Pseudoxenodon macrops. Herpetological Review* 30(3), 175.
Farkas, B. & Fritz, U. (1999d) Geographic distribution: Serpentes: *Xenochrophis piscator. Herpetological Review* 30(3), 175.
Farkas, B. & Tóth, T. (1999) Geographic distribution: Serpentes: *Elaphe situla. Herpetological Review* 30(2), 113.
Farr, W.L., Lavin-Murcio, P.A. & Luzcano, D. (2007) New distributional records for amphibians and reptiles from the state of Tamaulipas, México. *Herpetological Review* 38(2), 226.
Farr, W.L., Lazcano-Villarreal, D., & Murcino, P.A.L. (2009) New distributional records for amphibians and reptiles from the state of Tamaulipas, México II. *Herpetological Review* 40(4), 459–467.
Fathinia, B. & Rastegar-Pouyani, N. (2010) On the species of *Pseudocerastes* (Ophidia: Viperidae) in Iran. *Russian Journal of Herpetology* 17(4), 275–279, 7 figs.
Fathinia, B., Anderson, S.C., Rastegar-Pouyani, N., Jahani, H. & Mohamadi, H. (2009) Notes on the natural history of *Pseudocerastes urarachnoides* (Squamata: Viperidae). *Russian Journal of Herpetology* 16(2), 134–138, 12 figs.

Fatio, V.P. (1872) *Fauna des vertébrés de la Suisse. Vol. III. Histoire naturelle des Reptiles et des batraciens.* H. George, Genève, 603 pp., 6 pls.

Fayrer, J. (1872) *The Thanatophidia of India being a description of the venomous snakes of the Indian Peninsula, with an account of the influence of their poison on life; and a Series of experiments.* J. and A. Churchill, London, x + 131 pp., 31 col. pls. [second edition in 1874]

Fayrer, J. (1877) Venomous animals. *Edinburgh Medical Journal* 23(27), 97–117.

Fearn, S. (1993) The tiger snake *Notechis scutatus* (Serpentes: Elapidae) in Tasmania. *Herpetofauna,* Sydney 23(2), 17–29, 1 fig., 6 col. pls.

Fearn, S. & Munday, B. (1995) The copperhead or superb snake in captivity. *Monitor,* Ardeer 7(1), 9–11, col. photos 3–6.

Federsoni, P.A., Jr. (1981) Novo artefato para a extração de venenos de serpentes do gênero *Micrupus* Wagler. *Memórias do Instituto Butantan,* São Paulo (1978–1979) 42–43, 171–174.

Feitosa, D.T., Passos, P.P. & Prudente, A.L.C. (2007a) Taxonomic status and geographic variation of the slender coralsnake, *Micrurus filiformis* (Günther, 1859) (Serpentes, Elapidae). *South American Journal of Herpetology* 2(3), 149–156, 5 figs.

Feitosa, D.T., Prudente, A.L.C. & Lima, A.C. de. (2007b) Redescription and variation of *Micrurus paraensis* Cunha & Nascimento 1973 (Serpentes: Elapidae). *Zootaxa* (1470), 35–45.

Fejérváry, G.J. (1922) The batrachians and reptiles collected by Mr. E. Csiki in the northern parts of central Albania and in Servia. *Magyar Tudományos Akademie Balkán-Kutatáainak Tudományos Eredményei,* Budapest 1, 64 pp., pls. 2–3.

Fejérváry-Lángh, A.H. (1943) Beitrage und Berichtigungen zum Reptilien—Teil des ungarischen Faunenkataloges. *Fragmenta Faunistica Hungarica* 6, 81–98.

Feldman, C.R. & Hoyer, R.F. (2010) A new species of snake in the genus *Contia* (Squamata: Colubridae) from California and Oregon. *Copeia* 2010(2), 254–267, 4 figs.

Feldman, C.R. & Spicer, G.S. (2002) Mitochondrial variation in sharp-tailed snakes (*Contia tenuis*): evidence of a cryptic species. *Journal of Herpetology* 36(4), 648–655, 2 figs.

Feldmann, R. (1981) Die Amphibien und Reptilien Westfalens. *Abhandlungen aus dem Landesmuseum für Naturkunde zu Munster* 43(4), 4–161, 23 figs.

Felger, R.S. & Moswer, M.B. (1985) *People of the desert and sea: ethnobotany of the Seri Indians.* University of Arizona Press, Tucson, xv + 480 pp., illust.

Felipe-Esqueda, L. & La Marca, E. (2004) Geographic distribution: Serpentes: *Corallus ruschenbergii. Herpetological Review* 35(2), 190.

Feng, X.-Y. (1981) The snakes of Gansu Province. *Acta Herpetologica Sinica,* Chengdu 5(5), 29–44.

Feng, X.-Y. (1986) Two new records of venomous snakes from Gansu. *Acta Herpetologica Sinica,* Chengdu (Series 2) 5(2), 158.

Fenwick, A.M., Gutberlet R.L., Jr., Evans, J.A. & Parkinson, C.L. (2009) Morphological and molecular evidence for phylogeny and classification of South American pitvipers, genera *Bothrops, Bothriopsis,* and *Bothrocophias* (Serpentes: Viperidae). *Zoological Journal of the Linnean Society,* London 156(2), 617–640, 2 figs.

Ferguson, D., Mathieson, M. & Eyre, T. (2012) Southerly range extension of the poorly known, Queensland endemic yellow-naped snake *Furina barnardi* (Squamata: Elapidae) into the Mulga Lands. *Memoirs of the Queensland Museum* 56(1), 9–12, 1 fig.

Ferguson, G.W. (1977) Variation and evolution of stereotyped behavior in reptiles. Part II. Social displays of reptiles: communications value, ultimate causes of variation, taxonomic significance and heritability of population differences. *In:* Gans, C. & Tinkle, D.W. (Eds.), *Biology of the Reptilia. Volume 7 (Ecology and behavior A).* Academic Press, London, pp. 405–554, 10 figs.

Ferguson, H.S. (1895) List of snakes taken in Travancore from 1888–1895. *The Journal of the Bombay Natural History Society* (1895–1897) 10(1), 68–77.

Ferguson, H.S. (1902) Travancore snakes. *The Journal of the Bombay Natural History Society* (1902–1903) 14(2), 386–387.

Ferguson, W. (1876) Description of a new snake of the genus *Aspidura* from Ceylon. *Proceedings of the Zoological Society of London* 44(1), 819–820.

Ferguson, W. (1877) *Reptile fauna of Ceylon. Letter on a collection sent to the Colombo Museum.* William Henry Herbert, Colombo, 42 pp. [reprinted in 2001 by Society for the Study of Amphibians and Reptiles]

Feriche, M., Pleguezuelos, J.M. & Santos, X. (2008) Reproductive ecology of the Montpellier snake, *Malpolon monspessulanus* (Colubridae), and comparison with other sympatric colubrids in the Iberian Peninsula. *Copeia* 2008(2), 279–285, 4 figs.

Fernandes, D.S. & Passos, P. (2002) Geographic distribution: Serpentes: *Rachidelus brazili. Herpetological Review* 33(2), 150.

Fernandes, D.S., Franco, F.L. & Germano, V.J. (1999) Geographic distribution: Serpentes: *Umbrivaga pygmaea. Herpetological Review* 30(3), 175.

Fernandes, D.S., Germano, V.J., Fernandes, R. & Franco, F.L. (2002) Taxonomic status and geographic distribution of the lowland species of the *Liophis cobella* group with comments on the species from the Venezuelan Tepuis (Serpentes, Colubridae). *Boletim do Museu Nacional,* Rio de Janeiro (481), 1–14, 12 figs.

Fernandes, D.S., Franco, F.L. & Fernandes, R. (2004) Systematic revision of the genus *Lachesis* Daudin, 1803 (Serpentes, Viperidae). *Herpetologica* 60(2), 245–260, 4 figs.

Fernandes, D.S., Marques, O.A.V. & Argólo, A.J.S. (2010) A new species of *Dipsas* from the Atlantic forest of Brazil (Serpentes: Dipsadidae). *Zootaxa* (2691), 57–66. 6 figs.

Fernandes, R. (1995a) A new species of snake in the genus *Atractus* (Colubridae, Xenodontinae) from northeastern Brazil. *Journal of Herpetology* 29(3), 416–419, 2 figs.

Fernandes, R. (1995b) Variation and taxonomy of the *Atractus reticulatus* complex (Serpentes, Colubridae). *Comunicações do Museu de Ciências e Tecnologia da PUCRS (Zoologia),* Porto Alegre 8, 37–53, 4 figs.

Fernandes, R. & Argôlo, A.J.S. (1999) Rediscovery of *Atractus guentheri* (Wucherer, 1861) (Serpentes, Colubridae) in southeastern Bahia, Brazil. *Boletim do Museu Nacional, Zoologia* (397), 1–5, 1 fig.

Fernandes, R. & Puorto, G. (1994) A new species of *Atractus* from Brazil and the status of *A. guentheri* (Serpentes, Colubridae). *Memórias do Instituto Butantan,* São Paulo (1993) 55(1), 7–14, 5 figs.

Fernandes, R., Porto, M. & Caramaschi, U. (1998) The taxonomic status of *Heterorhachis poecilolepis* Amaral, 1923. *Journal of Herpetology* 32(1), 139–141, 1 fig.

Fernandes, R., Freire, E.M.X. & Puorto, G. (2000) Geographic variation of the Brazilian Atlantic Rainforest snake *Atractus maculatus* (Gunther, 1858) with the revalidation of *Rhabdosoma zebrinum* Jan, 1862 (Serpentes: Colubridae). *Boletim do Museu Nacional,* Rio de Janeiro (419), 1–8.

Fernandes, V.D., Moura, M.R. de, Dayrell, J.S., Santana, D.J. & Lima, L.H.R. (2010) Reptilia, Squamata, Serpentes, Typhlopidae, *Typhlops amoipira* Rodrigues and Juncá, 2002: range extensions and new state record. *Check List* 6(2), 268–269, 2 figs.

Fernandéz, J.A., Barrios-Quiroz, G., Sanchez-Aguilar, G. & Minor, M. (2003) Geographic distribution: Serpentes: *Crotalus molossus nigrescens*. *Herpetological Review* 34(4), 387.

Fernandez, W. & Abe, A. (1991) An electrophoretic approach to the relationships among the subspecies of the lancehead *Bothrops neuwiedi* (Serpentes, Viperidae). *Zoologischer Anzeiger*, Leipzig 226(3–4), 195–201, illust.

Ferner, J.W., Brown, R.M., Sison, R.V. & Kennedy, R.S. (2001) The amphibians and reptiles of Panay Island, Philippines. *Asiatic Herpetological Research* 9, 34–70, 56 col. figs.

Ferrarezzi, H. (1993) *Sistemática filogenética de Elapomorphus, Phalotris e Apostolepis (Serpentes, Colubridae, Xenodontinae)*. M.S. Thesis, Instituto de Biociências, Universidade de São Paulo, São Paulo.

Ferrarezzi, H. (1994a) Nota sobre o gênero *Phalotris* com revisão do grupo *nasutus* e descrição de três novas espécies (Serpentes, Colubridae, Xenodontinae). *Memórias do Instituto Butantan* (1993) 55(1), 21–38, 6 figs.

Ferrarezzi, H. (1994b) Uma sinopse dos gêneros e classifiçaõ das Serpentes (Squamata), I. Scolecophidia e Alethinophidia não colubrídeos. *Herpetologia no Brasil* 1, 69–80.

Ferrarezzi, H. (1994c) Uma sinopse dos gêneros e classifiçaõ das Serpentes (Squamata), II. Família Colubridae. *Herpetologia no Brasil* 1, 81–91.

Ferrarezzi, H. & Freire, E.M.X. (2001) New species of *Bothrops* Wagler, 1824, from the Atlantic forest of northeastern Brazil (Serpentes, Viperidae, Crotalinae). *Boletim do Museu Nacional, Zoologia* (440), 1–10, figs.

Ferrarezzi, H., Erritto-Barbo, F., España-Albuquerque, C. (2005a) Phylogenetic relationships of a new species of *Apostolepis* from Brazilian cerrado with notes on the *assimilis* group (Serpentes, Colubridae, Xenodontinae, Elapomorphini). *Papeis Avulsos do Departamento de Zoologia, Universidade de São Paulo* 45(16), 215–229, 12 figs.

Ferrarezzi, H., Velloso-Calleffo, M.E., Castellar, T.M. & Lauro, M. de. (2005b) Geographic distribution: Serpentes: *Phalotris reticulatus*. *Herpetological Review* 36(4), 420.

Ferrari-Pérez, F. (1886) Catalogue of animals collected by the Geographical and Exploring Commission of the Republic of México. *Proceedings of the United States National Museum* (1886–1887) 9(11), 125–199.

Ferreira, A. (1949) *Ofidismo en el Uruguay*. Talleres Gráficos de los Establecimièntos Galien, Montevideo 191 pp., 23 figs.

Ferreira, V.L. & Ávila, R.W. (2009) Reptilia, Squamata, Serpentes, Dipsadidae, *Sibynomorphus lavillai* Scrocchi, Porto and Rey 1993: New country record and geographic distribution map. *Check List* 5(4), 773–775, 2 figs.

Ferreira, V.L., Wang, E. & Himmelstein, J.A. (2005) Geographic distribution: Serpentes: *Phalotris nasutus*. *Herpetological Review* 36(2), 204.

Ferreira-Yuki, V.L. (1993) Realocação genérica de *Xenodon werneri* Eiselt, 1963 (Serpentes, Colubridae). *Comunicações do Museu de Ciências da PUCRS (Zoologia)*, Pôrto Alegre (53), 39–47, 4 figs.

Ferri, V. (1994) *Las serpientes de España y Europa*. Editorial de Vecchi, Barcelona, 158 pp., col. illust.

Fet, V. & Atamuradov, K.I. (1994) Biogeography and ecology of Turkmenistan. *Monographiae Biologicae* (72), Kluwer Academy Publishers, Dordrecht, 650 pp., illust.

Ffolliott, P. & Liversidge, R. (1971) *Ludwig Krebs: Cape naturalist to the King of Prussia, 1792–1844*. A.A. Balkema, Cape Town, iv + 304 pp.

Figueiredo-de-Andrade, C.A., Oliveira-Nogueira, C.H. & Pereira, C.A., Jr. (2011) Filling gaps on the distribution of *Tricheilostoma salgueiroi* (Amaral, 1955) in the state of Rio de Janeiro, Brazil (Serpentes: Leptotyphlopidae). *Check List* 7(4), 409–410.

Filho, P.L. & Aoki, C. (2011) Geographic distribution: Serpentes: *Helicops infrataeniatus*. *Herpetological Review* 42(3), 394.

Filhol, H. (1877) Recherches sur les phosphorites du Quercy. Étude des fossiles qu'on y recontre et spécialement des mammifères. *Annales des Sciences Géologiques*, Paris 8(1), 1–562, 55 pls.

Fillipi, E. (2003) On the proposed elevation of *Vipera aspis* subspecies to full species rank: a cautionary note. *Amphibia-Reptilia*, Leiden 24(2), 235–238.

Filippi, F. de. (1840) Catalogo ragionato e descrittivo della raccolta de' serpenti del Museo dell'I.R. Università di Pavia. *Biblioteca Italiana*, Milano 99, 163–187, 306–343, 21 pls.

Filippi, F. de. (1848) Nota sopra tre specie di serpenti. *Nuovi Annali dell Scienze Naturali*, Bologna (Série 2) 9, 87–89, pl. 4.

Filippi, F. de. (1865) *Note di un viaggio in Persia nel 1862. Volume Unico*. G. Daelli & C. Editori, Milano, xii + 398 pp., 7 figs.

Filippi, L.S. & Garrido, A.C. (2012) Nuevo registro del género *Dinilysia* (Squamata, Serpentes) para la formación Anacleto (Campaniano inferior–medio), Rincón de los Sauces, Nequén, Argentina. *Ameghiniana*, Buenos Aires 49(1), 132–136, 2 figs.

Finkeldey, H. (1963) *Python anchietae* Bocage. *Cimbebasia*, Windhoek (6), 23–28, 5 figs.

Firouz, E. (2000) *A guide to the fauna of Iran*. Iran University Press, Tehran. vi + 491 pp. (in Persian)

Firouz, E. (2005) *The complete fauna of Iran*. I.B. Taurus, London. 322 pp., col. illust.

Fischer, E. & Hinkel, H. (1992) *La nature du Rwanda*. Johannes Scientifique et Technologique, Butare, 452 pp., 349 col. figs.

Fischer, J.G. (1855) *Die Familie der Seeschlangen systematisch beschrieben von J.G. Fischer Dr*. Johann August Meissner, Hamburg, 68 pp., 3 pls.

Fischer, J.G. (1856a) Die Familie der Seeschlangen. *Abhandlungen aus dem Gebiete der Naturwissenschaften herausgegeben von dem Naturwissenschaftlichen Verein in Hamburg* 3, 1–78, 3 pls.

Fischer, J.G. (1856b) Neue Schlangen des Hamburgischen Naturhistorischen Museums. *Abhandlungen aus dem Gebiete der Naturwissenschaften herausgegeben von Naturwissenschaftlichen Verein in Hamburg* 3, 79–116, 3 pls.

Fischer, J.G. (1857) *De serpentibus quibusdem fossilibus*. Dissertatio Zoologico, Bonn, 35 pp.

Fischer, J.G. (1879) Neue oder wenig bekannte Reptilien. *Abhandlungen vom Naturwissenschaftlichen Verein zu Hamburg* 1879, 76–103, 5 pls.

Fischer, J.G. (1880) Neue Amphibien und Reptilien. *Archiv für Naturgeschichte*, Berlin 46(1), 215–227, pls. 8–9.

Fischer, J.G. (1881) Beschreibung neuer Reptilien. *Archiv für Naturgeschichte*, Berlin 47, 225–238, pls. 11–12.

Fischer, J.G. (1882) Herpetologische Bemerkungen vorzugsweise über Stücke der Sammlung des Naturhistorischen Museums in Bremen. *Abhandlungen vom Naturwissenschaftlichen Verein zu Bremen* 7, 225–238, pls. 14–17.

Fischer, J.G. (1883) Beschreibungen neuer Reptilien. *Oster-Programm des Akademischen Gymnasiums*, Hamburg 1883, 1–16, 1 pl.

Fischer, J.G. (1884a) Über einige afrikanische Reptilien, Amphibien und Fische des Naturhistorischen Museums. I. Über die von Herrn Dr. G.A. Fischer im Massai-Gebiete (Ost-Afrika) auf seiner in Veranlassung der Geographischen Gesellschaft in Hamburg unternommenen Expedition gesammelten Reptilien, Amphibien und Fische. *Jahrbuch der Hamburgischen Wissenschaftlichen Anstalten*, Hamburg 1, 3–32, 2 pls.

Fischer, J.G. (1884b) Über einige afrikanische Reptilien, Amphibien und Fische des naturhistorischen Museums. II. Über einige Reptilien von Nossi-Bé und Madagaskar. *Jahrbuch der Hamburgischen Wissenschaftlichen Anstalten*, Hamburg 1, 33–38, pl. 3.

Fischer, J.G. (1884c) Herpetologische Bemerkungen. *Abhandlungen vom Naturwissenschaftlichen Verein zu Hamburg* 8(1), 1–11, pl. 7.

Fischer, J.G. (1885a) Ueber eine Kollektion von Amphibien und Reptilien aus Südost-Borneo. *Archiv für Naturgeschichte*, Leipzig 51(1), 41–72, pls. 4–5.

Fischer, J.G. (1885b) Ichthyologische und herpetologische Bemerkungen. V. Herpetologische Bemerkungen. *Jahrbuch der Hamburgischen Wissenschaftlichen Anstalten*, Hamburg 2, 82–121, 4 pls.

Fischer, J.G. (1886a) Ueber eine Kollektion Reptilien und Amphibien der Insel Nias. *Abhandlungen aus dem Gebiete der Naturwissenschaften herausgegeben vom Naturwissenschaftlichen Verein in Hamburg* 9(1), 1–9.

Fischer, J.G. (1886b) Herpetologische Notizen. *Abhandlungen aus dem Gebiete der Naturwissenschaften herausgegeben vom Naturwissenschaftlichen Verein in Hamburg* 9(1), 51–67, 2 pls.

Fischer, J.G. (1888a) Herpetologische Mitteilungen. I. Über zwei neue Schlangen und einen neuen Laubfrosch von Kamerun. *Jahrbuch der Hamburgischen Wissenschaftlichen Anstalten*, Hamburg (1887) 5, 3–10, 2 pls.

Fischer, J.G. (1888b) Herpetologische Mitteilungen. II. Über eine Kollektion Reptilien von Angra Pequenna. *Jahrbuch der Hamburgischen Wissenschaftlichen Anstalten*, Hamburg (1887) 5, 11–17.

Fischer, J.G. (1888c) Herpetologische Mitteilungen. IV. Über eine Kollektion Reptilien und Amphibien von Hayti. *Jahrbuch der Hamburgischen Wissenschaftlichen Anstalten*, Hamburg (1887) 5, 23–43, pl. 3.

Fischer von Waldheim, G. (1808) *Tableaux synoptiques de Zoognosie. Publiés à l'usage de ses éléves à l'Université Impériale de Moscou...Tome III. Reptiles. Editio primus.* Imprimerie de l'Université Impériale, Moscou, iv + 186 pp., 6 pls.

Fischer von Waldheim, G. (1813) *Zoognosia tabulis synopticis illustrata, in usum praelectionum Academiae Imperialis Medico-Chirurgicae Mosquensis edita. Editio tertia, classium, ordinum, generum illustratione perpetua aucta. Volumen primum, tabulis synopticas generales et comparativas, nec non characteram quorundam explicationem iconographicam continens. Editio tertia, classium, ordinum, generum illustratione perpetua aucta. Volumen primum, tabulis synopticas generales et comparativas, nec non characterum quorundam explicationem iconographicam continens. Parte tertia (Reptiles, poisons).* Typis Nicolai Sergeidis Vsevolozsky, Mosquae, pp. 57–117.

Fischer von Waldheim, G. (1832) Notices sur les nouvelles acquisitions de la Sociéte de l'année 1832. *Bulletin de la Société Impériale des Naturalistes de Moscou* 4(3), 572–576.

Fisher, C., Joynt, A. & Brooks, R.J. (2007) *Reptiles and amphibians of Canada.* Lone Pine Publishing, Edmonton, 208 pp., col. figs.

Fisher, C.B. 1968. *A systematic revision of the species* Coniophanes fissidens *(Günther) (Serpentes, Colubridae).* Master's Thesis, Northwestern State University, Natchitoches, 65 pp., 9 pls.

Fitch, H.S. (1936) Amphibians and reptiles of the Rogue River Basin, Oregon. *American Midland Naturalist* 17(3), 634–652.

Fitch, H.S. (1940) A biogeographical study of the *ordinoides* artenkreis of garter snakes (genus *Thamnophis*). *University of California Publications in Zoology* 44(1), 1–150, 22 figs., 7 pls.

Fitch, H.S. (1960) Autecology of the copperhead. *University of Kansas Publications, Museum of Natural History* 13(4), 85–288, 26 figs., pls. 13–20.

Fitch, H.S. (1963) Natural history of the racer *Coluber constrictor*. *University of Kansas Publications, Museum of Natural History* 15(8), 351–468, 20 figs., pls. 19–22.

Fitch, H.S. (1965) An ecological study of the garter snake, *Thamnophis sirtalis*. *University of Kansas Publications, Museum of Natural History* 15(10), 493–564, 14 figs., pls. 23–25.

Fitch, H.S. (1975) A demographic study of the ring neck snake (*Diadophis punctatus*) in Kansas. *Miscellaneous Publications, University of Kansas Museum of Natural History* (62), 4–53, 19 figs.

Fitch, H.S. (1980a) Remarks concerning certain western garter snakes of the *Thamnophis elegans* complex. *Transactions of the Kansas Academy of Sciences* 83(3), 106–113.

Fitch, H.S. (1980b) *Thamnophis sirtalis*. *Catalogue of American amphibians and reptiles*, pp. 270.1–270.4.

Fitch, H.S. (1983) *Thamnophis elegans*. *Catalogue of American amphibians and reptiles*, pp. 320.1–320.4.

Fitch, H.S. (1984) *Thamnophis couchii*. *Catalogue of American amphibians and reptiles*, pp. 351.1–351.3.

Fitch, H.S. (1999) *A Kansas snake community: composition and changes over 50 years*. Krieger Publ. Co., Malabar, 165 pp., 37 figs., 16 col. pls.

Fitch, H.S. & Maslin, T.P. (1961) Occurrence of the garter snake, *Thamnophis sirtalis*, in the Great Plains and Rocky Mountains. *University of Kansas Publications, Museum of Natural History* 13(5), 493–564.

Fitzinger, L.J.F.J. (1823) Einiges über gehörnte Schlangen. *Archiv für Geschichte, Statistik, Literatur und Kunst,* Wien (59), 301–303.

Fitzinger, L.J.F.J. (1824) Sur les reptiles d'Autriche. *Bulletin des Sciences Naturelles et de Géologie,* Paris 2, 294–297.

Fitzinger, L.J.F.J. (1826a) *Neue Classification der Reptilien nach ihren Natürlichen Verwandtschaften. Nebst einer Verwandtschafts-Tafel und einem Verzeichnisse der Reptilien-Sammlung des k. k. zoologischen Museums zu Wien.* J. G. Huebner, Wien, viii, 66 pp.

Fitzinger, L.J.F.J. (1826b) Gritische Bemerkungen über J. Wagler's Schangenmert. *Isis von Oken*, Jena 19(9), 881–909.

Fitzinger, L.J.F.J. (1827) Über die Hydren oder Wasserschlangen. *Isis von Oken*, Jena 20, 731–741.

Fitzinger, L.J.F.J. (1828) Erwiderung an Herrn Schlegel und Herrn Wagler in Betreff ihrer Angriffe gegen meine Classification der Reptilien. *Isis von Oken,* Jena 21(1), 3–24.

Fitzinger, L.J.F.J. (1833) *Elaphe parreyssii* — Parreyss's *Elaphe*. *In:* Wagler, J. (Ed.), *Description et icones Amphibiorum* (3). J.G. Cotta'sche Buchhandlung, München, Stuttgart, Tübingen, 3–4 (Taf. 27).

Fitzinger, L.J.F.J. (1843) *Systema reptilium. Fasciculus primus.* Amblyglossae. Braumüller et Seidel Bibliopolas, Vindobonae, vi + 106 pp.

Fitzinger, L.J.F.J. (1853) Versuch einer Geschichte der Menagerie des Österreichisch- Kaiserlichen Hofes. *Sitzungsberichte der Akademie der Wissenschaften in Wien, Mathematische-Naturwissenschaftliche Klasse*, Wien 10(4), 626–710.

Fitzinger, L.J.F.J. (1861) Die Ausbeute der österreichischen Naturforscher an Säugethieren und Reptilien während der Weltumsegelung Sr. Majestät Fregatte Novara. Novara. *Sitzungsberichte der Akademie der Wissenschaften in Wien, Mathematische-Naturwissenschaftliche Klasse*, Wien (1860) 42(25), 383–416.

Fitzinger, L.J.F.J. (1867) *Bilder-Atlas zur wissenschaftlich-populären Naturgeschichte der Wirbelthiere*. Verlag der Kaiserlich-königlichen Hof- und Staatsdruckerei, Wien, 102 pp., 200 figs.

FitzSimons, F.W. (1912) *The snakes of South Africa. Their venom and the treatment of snake bite*. T. Maskew Miller, Cape Town, xvi + 547 pp., 162 figs., figs. a–z.

FitzSimons, V.F.M. (1930) Descriptions of new South African Reptilia and Batrachia, with distributional records of allied species in the Transvaal Museum collections. *Annals of the Transvaal Museum*, Pretoria 14(1), 20–48, 20 figs.

FitzSimons, V.F.M. (1932) Preliminary descriptions of new forms of S. African Reptilia and Amphibia from the Vernay-Lang Kalahari Expedition, 1930. *Annals of the Transvaal Museum*, Pretoria 15(1), 35–40.

FitzSimons, V.F.M. (1935a) Notes on a collection of reptiles and amphibians made in the southern Kalahari, Bushmanland and Little and Great Namaqualand. *Annals of the Transvaal Museum*, Pretoria 15(4), 519–550, 2 figs.

FitzSimons, V.F.M. (1935b) Scientific results of the Vernay-Lang Kalahari Expedition, March to September, 1930. Reptilia and Amphibia. *Annals of the Transvaal Museum*, Pretoria 16(2), 295–397, 30 figs., pls. 10–11.

FitzSimons, V.F.M. (1937) Notes on the reptiles and amphibians collected and described from South Africa by Andrew Smith. *Annals of the Transvaal Museum*, Pretoria 17(4), 259–274, pl. 10.

FitzSimons, V.F.M. (1938) Transvaal Museum expedition to South-West Africa and Little Namaqualand, May to August 1937. Reptiles and amphibians. *Annals of the Transvaal Museum*, Pretoria 19(2), 153–209, 9 figs., pls. 2–3.

FitzSimons, V.F.M. (1939) An account of the reptiles and amphibians collected on an expedition to South-eastern Rhodesia during December 1937 and January 1938. *Annals of the Transvaal Museum*, Pretoria 20(1), 17–46.

FitzSimons, V.F.M. (1946a) An account of the reptiles and amphibians collected on an expedition to the Cape Province, October to December, 1940. *Annals of the Transvaal Museum*, Pretoria 20(4), 351–377, 13 figs., pls. 14–17.

FitzSimons, V.F.M. (1946b) Notes on some South African snakes, including a description of a new subspecies of *Xenocalamus*. *Annals of the Transvaal Museum*, Pretoria 20(4), 379–393, 20 figs.

FitzSimons, V.F.M. (1947) Descriptions of new species and subspecies of reptiles and amphibians from Natal, together with notes on some other little known species. *Annals of the Natal Museum*, Pietermaritzburg 11(1), 111–137, 6 figs., 3 pls.

FitzSimons, V.F.M. (1956) Reptiles: Serpentes & Sauria. *In:* Hanström, B., Brinck, P. & Rudebeck, G. (Eds.), *South African animal life. Results of the Lund University Expedition in 1950-1951. Vol. IV.* Almqvist & Wiksell, Stockholm, pp. 385–405.

FitzSimons, V.F.M. (1958) On a small collection of reptiles and amphibians from the Inyanga District, S. Rhodesia. *Occasional Papers of the National Museum of Southern Rhodesia*, Salisbury 3B(22), 204–214, 1 fig.

FitzSimons, V.F.M. (1959) Some new reptiles from southern Africa and southern Angola. *Annals of the Transvaal Museum*, Pretoria 23(4), 405–409.

FitzSimons, V.F.M. (1962a) *Snakes of southern Africa*. Purnell and Sons, Cape Town, 423 pp., 106 figs., 43 pls., 75 col. pls.

FitzSimons, V.F.M. (1962b) A new worm-snake (*Leptotyphlops*) from South-West Africa. *Annals of the Transvaal Museum*, Pretoria 24(2–3), 239–240, 1 fig.

FitzSimons, V.F.M. (1964) A new subspecies of water-snake from Kruger National Park. *Koedoe*, Pretoria (7), 26–29, 1 fig.

FitzSimons, V.F.M. (1966) A check-list, with synoptic keys, to the snakes of southern Africa. *Annals of the Transvaal Museum*, Pretoria 25(3), 35–79.

FitzSimons, V.F.M. (1974) *A field guide to the snakes of southern Africa*. Second edition. Cape and Transvaal Printers, Cape Town, 221 pp., 7 figs., 32 pls. [first edition, 1970, 221 pp., 7 figs., 32 pls.]

FitzSimons, V.F.M. & Brain, C.K. (1958) Serpentes, Colubridae, *Calamelaps ventrimaculatus websteri* n. subsp. *Occasional Papers of the National Museum of Southern Rhodesia*, Salisbury 3B(22), 202–203, 1 fig.

Flagle, A.R. & Stoops, E.D. (2009) *Black python* Morelia boeleni. Edition Chimaira, Frankfurt-am-Main, 157 pp., 175 col. figs.

Fleet, R.R. & Dixon, J.R. (1971) Geographic variation within the long-tailed group of the glossy snake, *Arizona elegans* Kennicott. *Herpetologica* 27(3), 295–302, 4 figs.

Fleischmann, F.L. (1831) *Dissertatio inauguralis sistens Dalmatiae Nova Serpentum Genera quam gratiosi medicorum ordinis auctoritate in Regia Literarum Universitate Friderico–Alexandrina pro impetrandis Doctoris Medicinae, Chirurgiae atque Artis obstetriciae Summis honoribus et privileglis eruditorum examini submittit*. Commissis C. Heyderi, Erlangae, 35 pp., 2 col. pls.

Fleming, J. (1822) *The philosophy of zoology; or a general view of the structure, functions, and classification of animals. Vol. II.* Archibald Constable & Co., Edinburgh, 618 pp.

Fleming, J. (1828) *History British Animals, exhibiting the descriptive characters and systematical arrangement of the genera and species of quadrupeds, birds, reptiles, fishes, Mollusca, and Radiata of the United Kingdom; including the indigenous, extirpated, and extinct kinds, together with periodical and occasional visitants*. Bell & Bradfute, Edinburgh, xxiii + 565 pp. [reprinted in 1842 by Duncan & Malcolm, London]

Fleming, R.L., Jr. & Fleming, R.L., Sr. (1973) Some snakes from Nepal. *The Journal of the Bombay Natural History Society* 70(3), 426–437.

Fletcher, T.B. (1908) Notes on snakes from Diyatalawa, Ceylon. *Spolia Zeylanica*, Colombo 5(18), 98–101.

Floericke, K. (1912) *Kriechtiere und Lurche fremder Länder.* Kosmos, Stuttgart 3, 1–100, 20 figs.

Floericke, K. (1920) *Kriechtiere und Lurche Deutschlands*. Kosmos, Stuttgart, 112 pp, 31 figs.

Flores-Benabib, J. & Flores-Villela, O. (2008) Nuevo registro estatal de *Leptotyphlops goudotii* en Tamaulipas, México. *Boletin de la Sociedad Herpetologica Mexicana* 16(1), 13–14.

Flores-Villela, O. (1991) *Analisis de la distribucion de la herpetofauna de Mexico*. Tesis Doctoral, Facultad de Sciencias, Universidad Nacional Autonoma de Mexico, Mexico City, 269 pp., 55 figs.

Flores-Villela, O. (1993) Herpetofauna Mexicana. *Carnegie Museum of Natural History, Special Publication* (17), iv + 72 pp.

Flores-Villela, O. & Hernández-García, E. (1989) New state records from northern Guerrero, Mexico. *Herpetological Review* 20(1), 15–16.

Flores-Villela, O. & Smith, E.N. (2009) A new species of *Coniophanes* (Squamata: Colubridae), from the coast of Michoacán, Mexico. *Herpetologica* 65(4), 404–412, 4 figs.

Flores-Villela, O., Perez-Higareda, G., Vogt, R.C. & Muñoz, M.P. (1987) *Clavas para los generos y las especies de anfibios y reptiles de la Region de Las Tuxtlas*. Universidad Nacional Autonoma de Mexico, i–xxvii.

Flores-Villela, O., Hernández-García, E. & Montes de Oca, A.N. (1991) Catalogo de anfibios y reptiles del Museo de Zoologia, Falcultad de Ciencias, Universidad Nacional Autonoma de Mexico. *Série Catalogos del Museo de Zoología "Alfonso L. Herrera," México* (3), 1–222.

Flores-Villela, O., Loeza-Corichi, A. & Pérez-Nuñez, A. (1995) Geographic distribution: Serpentes: *Geophis nigrocinctus* (Sierra Coalcoman earth snake). *Herpetological Review* 26(2), 109.

Flores-Villela, O., Reynoso, V.H. & Magaña-Cota, G. (2010) More notes on Alfredo Dugès' names for Mexican taxa. *Zootaxa* (2600), 66–68.

Flower, S.S. (1896) Notes on a collection of reptiles and batrachians made in the Malay Peninsula in 1895–96; with a list of the species recorded from that region. *Proceedings of the Zoological Society of London* 64(4), 856–914, pls. 44–46.

Flower, S.S. (1899) Notes on a second collection of reptiles made in the Malay Peninsula and Siam, from November 1896 to September 1899, with a list of the species recorded from these countries. *Proceedings of the Zoological Society of London* 68(3), 600–696.

Flower, S.S. (1926) Reptilia and Batrachia. *In: Zoological Record* (1925) 62(14), 1–61.

Flower, S.S. (1929) *List of vertebrated animals exhibited in the gardens of he Zoological Society of London, 1828–1927. Vol. 3. Reptiles.* Zoological Society of London, London, 272 pp.

Flower, S.S. (1930) The occurrence of *Pseudocerastes* in Sinai. *Annals and Magazine of Natural History*, London (Series 10) 6(32), 224.

Flower, S.S. (1931) The occurrence of the "black and white cobra," *Naja melanoleuca*, in the Sudan. *Annals and Magazine of Natural History*, London (Series 10) 7(41), 499–500.

Flower, S.S. (1933) Notes on the recent reptiles and amphibians of Egypt, with a list of the species recorded from that kingdom. *Proceedings of the Zoological Society of London* 103(3), 735–851.

Flury, A. (1950) A new kingsnake from Trans-Pecos Texas. *Copeia* 1950(3), 215–217.

Fogel, D. (1997) *Captive husbandry and propagation of the boa constrictors and related boas.* Krieger Publishing Company, Malabar, 98 pp., 13 figs., 60 col. pls.

Fogell, D.D. (2010) *A field guide to the amphibians and reptiles of Nebraska.* Institude of Agriculture & Natural Resources, University of Nebraska, Lincoln, vi + 158 pp., 155 col. figs.

Fogg, B.F. (1862) A list of the reptiles and amphibians of Maine. *Proceedings of the Portland Society of Natural History* 1, 86.

Foley, H. & Parrot, L. (1935) Sur la répartition du *Naja haje* L. (Ophidiens protéroglyphes) en Algérie. *Bulletin de la Société d'Histoire Naturelle d'Afrique Nord*, Algiers 26, 166–168.

Fong, A. (2002) Geographic distribution: Serpentes: *Tropidophis fuscus*. *Herpetological Review* 33(4), 326.

Fong, A. (2005) Variation in and natural history notes on *Tropidophis fuscus* (Serpentes, Tropidophiidae) from Cuba. *Herpetological Review* 36(2), 118–119.

Fong, A. & Navarro, N. (2001) Checklist of amphibians and reptiles of the Sagua-Baracoa Mountains, eastern Cuba. *Smithsonian Herpetological Information Service* (130), 13 pp.

Fons, R., Saint-Girons, H., Salotti, M., Cheylan, M. & Clara, J.P. (1991) Contribution à la faune herpétologique des Iles Méditerranéennes: présence de la couleuvre vipérine *Natrix maura* (Reptilia, Colubridae) en Corse. *Bonnischer Zoologische Beiträge*, Bonn 42(2), 181–186, 1 fig.

Fontaine, A. de la. (1870) *Faune du pays de Luxembourg ou manuel de zoologie contenant la description des animaux vertébrés observés dans le pays de Luxembourg. III. Reptiles.* V. Buck, Luxembourg, ii + 49 pp.

Fontanella, F.M., Feldman, C.R., Siddal, M.B. & Burbrink, F.T. (2008) Phylogeography of *Diadophis punctatus*: extensive diversity and repeated patterns of historical demography in a trans-continental snake. *Molecular Phylogenetics and Evolution* 46(3), 1049–1070, 5 figs.

Forcart, L. (1950) Amphibien und Reptilien von Iran. *Verhandlungen der Naturforchenden Gesellschaft in Basel* 61, 151–156.

Forcart, L. (1953) Die Amphibien und Reptilien von Sumba, ihre zoogeographischen Bezichungen und Revision der Unterarten *Typhlops polygrammicus*. *Verhandlungen der Naturforchenden Gesellschaft in Basel* 64(2), 356–388, 3 figs.

Forcart, L. (1954) Die taxonomie von *Lycodon florensis* Rooij und *Stegonotus sutteri* Forcart (Ophidia, Colubridae). *Verhandlungen der Naturforchenden Gesellschaft in Basel* 65(1), 7–8.

Forlani, M.C., Bernardo, P.H., Haddad, C.B.F. & Zaher, H. (2010) Herpetofauna du Parque Estadual Carlos Botelho, São Paulo, Brasil. *Biota Neotropica* 10(3), 265–309, 14 figs.

Forskal, P. (1775) *Descriptiones animalium avium, amphibiorum, piscium, insectorum, vermium; quae in itinere orientali.* Officina Mölleri Typographi apud Heineck et Faber, Hauniae, xxxiv + 164 pp.

Forsman, A. (1997) Growth and survival of *Vipera berus* in a variable environment. *In:* Thorpe, R.S., Wüster, W. & Malhotra, A. (Eds.), *Ecology, evolution and snakebite venomous snakes.* Zoological Society of London, Clarendon Press, Oxford, pp. 143–154, 4 figs.

Forster, J.R. (1771) *A catalogue of the animals of North America. In:* Bossu, J.B., *Travels through that part of North America formerly called Louisiana. Vol. I.* T. Davies, London, vii + 407 pp.

Forstner, M.R.J., Hilsenbeck, R.A. & Scudday, J.F. (1997) Geographic variation in whole venom profiles from the mottled rock rattlesnake (*Crotalus lepidus lepidus*) in Texas. *Journal of Herpetology* 31(2), 277–287, 4 figs.

Foufopoulos, J. (2001) Geographic distribution: Serpentes: *Typhlops depressiceps*. *Herpetological Review* 32(1), 61–62.

Fouin, C. (1969) Le crane d'*Elapsoidea guntheri* Bocage. Contribution à la phylogénèse des Colubroidea. *Publications de l'Universitè Officielle du Congo à Lubumbashi* 19, 61–91, 6 figs.

Fouquette, M.J. & Potter, F. (1961) A new black-headed snake (*Tantilla*) from southwestern Texas. *Copeia* 1961(2), 144–148, 4 figs.

Fouquette, M.J. & Rossman, D.A. (1963) Noteworthy records of Mexican amphibians and reptiles in the Florida State Museum and the Texas Natural History Collection. *Herpetologica* 19(3), 185–201.

Fowler, H.W. (1907) Amphibians and reptiles of New Jersey. *Annual Report of the New Jersey State Museum* (1906) (2), 25–250, 351–384, 402–408, 69 pls.

Fowlie, J.A. (1965) *The snakes of Arizona.* Azul Quinta Press, Fallbrook, iv + 164 pp., illust.

Fox, H. (1977) The urogenital system of reptiles. *In:* Gans, C. & Parsons, T.S. (Eds.), *Biology of the Reptilia. Volume 6 (Morphology E).* Academic Press, London, pp. 1–157, 51 figs.

Fox, R.C. (1975) Fossil snakes from the upper Milk River formation (upper Cretaceous), Alberta. *Canadian Journal of Earth Science*, Ottawa 12, 1557–1563, 2 figs.
Fox, W. (1948) The relationships of the garter snake, *Thamnophis ordinoides*. *Copeia* 1948(2), 113–120.
Fox, W. (1951a) Relationships among the garter snakes of the *Thamnophis elegans* rassenkreis. *University of California Publications in Zoology* 50, 485–530.
Fox, W. (1951b) The status of the gartersnake *Thamnophis sirtalis tetrataenia*. *Copeia* 1951(4), 257–267.
Fraas, O. (1870) Die Fauna von Steinheim. Mit Rücksicht auf die miocänen Säugethier- und vogelreste des Steinlheimer Beckens. *Jahresheft des Vereins für Vaterländische Naturkunde in Württemberg* 26(2–3), 145–306, pls. 4–13.
França, F.G.R. (2003a) Geographic distribution: Serpentes: *Philodryas psammophideus*. *Herpetological Review* 34(2), 170.
França, F.G.R. (2003b) Geographic distribution: Serpentes: *Sibynomorphus mikanii*. *Herpetological Review* 34(3), 266.
França, F.G.R. (2004) Geographic distribution: Serpentes: *Liophis maryellenae*. *Herpetological Review* 35(1), 84.
França, F.G.R., Mesquita, D.O. & Garda, A.A. (2005) Geographic distribution: Serpentes: *Phalotris labiomaculatus*. *Herpetological Review* 36(1), 83.
Francisco, B.C.S., Pinto, R.R. & Frenandes, D.S. (2012) Taxonomy of *Epictia munoai* (Orejas-Miranda, 1961) (Squamata: Serpentes: Leptotyphlopidae). *Zootaxa* (3512), 42–52, 3 figs.
Franco, F.L. (2012) A coleção herpetological do Instituto Butantan: da sua origem ao incêndio ocorrido em 15 de Maio de 2010. *Herpetologica Brasileira* 1(1), 22–31, 4 figs.
Franco, F.L. & Gancev-Ferreira, T. (2003) Descrição de uma nova Espécie de *Thamnodynastes* Wagler, 1830 (Serpentes, Colubridae) do nordeste brasileiro, com comentários sobre o gênero. (2002). *Phyllomedusa* 1(2), 57–74, 4 figs.
Franco, F.L. & Pinto, R.R. (2009) *Stenostoma albifrons* Wagler *in* Spix, 1824 as *nomen dubium* and recognition of the name *Leptotyphlops tenellus* Klauber, 1939 (Serpentes: Leptotyphlopidae). *Salamandra* 45(4), 239–244.
Franco, F.L., Marques, O.A.V. & Puorto, G. (1997) Two new species of colubrid snakes of the genus *Clelia* from Brazil. *Journal of Herpetology* 31(4), 483–490, 5 figs.
Franco, F.L., Salomao, E.L., Martins, M.B., Di-Bernardo, M., Meneghel, M.D. & Carreira, S. (2001) New records of *Calamodontophis paucidens* (Serpentes, Colubridae, Xenodontinae) from Brazil and Uruguay. *Cuadernos de Herpetología* 14(2), 155–159, 3 figs.
Franco, F.L., Ferreira, T.G., Marques, O.A.V. & Sazima, I. (2003) A new species of hood-displaying *Thamnodynastes* (Serpentes, Colubridae) from the Atlantic forest in southeast Brazil. *Zootaxa* (334), 1–7, 3 figs.
Franco, F.L., Germano, V.J. & Martão, L.R. (2005) Notas sobre o Lectótipo de *Bothrops cotiara* (Gomes, 1913) (Serpentes, Viperidae). *Phyllomedusa* 4(1), 75–77.
Franco, F.L., de Carvalho, L.A. & Lema, T. de.. (2006) A new species of *Calamodontophis* Amaral, 1963 (Serpentes, Colubridae, Xenodontinae) from southern Brazil. *South American Journal of Herpetology* 1(3), 218–226, 4 figs.
Franco, F.L., Fernandes, D.S. & Bentim, B.M. (2007) A new species of *Hydrodynastes* Fitzinger, 1843 from central Brazil (Serpentes: Colubridae: Xenodontinae). *Zootaxa* (1613), 57–65, 7 figs.
Franke, A. (1881) *Die Reptilien und Amphibien Deutschlands*. Veit, Leipzig, x + 174 pp.
Franklin, C.J. (2001) Checklist of the amphibians and reptiles of the Madre Selva II Biological Reserve of Loreto, Peru. *Bulletin of the Chicago Herpetological Society* 36(7), 137–143.
Franklin, C.J. & Franklin, J. (1999) Geographic distribution: Serpentes: *Pseustes poecilonotus*. *Herpetological Review* 30(2), 114.
Franz, R. & Cordier, D. (1986) *The herpetofauna of the proposed national parks in southern Haiti*. USAID/Haiti Report, University of Florida, Gainesville, 73 pp.
Franzen, M. (1999) A record of *Spalerosophis diadema* (Reptilia: Colubridae) from Adiyaman Province, Turkey. *Zoology in the Middle East* 19, 33.
Franzen, M. & Bischoff, W. (1995) Erstnachweis von *Rhynchocalamus melanocephalus melanocephalus* für die Turkei. *Salamandra* 31(2), 107–122.
Franzen, M. & Glaw, F. (2007) Type catalogue of reptiles in the Zoologische Staatssammlung München. *Spixiana* 30(2), 201–274, 23 figs.
Franzen, M. & Heckes, V. (2000) *Vipera barani* Böhme & Joger, 1983 aus dem östlichen Pontusgebirge, Turkei: Differentialmerkmale, Verbreitung, Habitate. *Spixiana* 23(1), 61–71.
Franzen, M. & Sigg, H. (1989) Bemerkungen zu einigen Schlangen Ostanatoliens. *Salamandra* 25(3–4), 203–212.
Franzen, M. & Wallach, V. (2002) A new *Rhinotyphlops* from southeastern Turkey (Serpentes, Typhlopidae). *Journal of Herpetology* 36(2), 176–184, 3 figs.
Franzen, M., Bussmann, M., Kordges, T. & Thiesmeier, B. (2008) *Die Amphibien und Reptilien des Südwest-Türkei*. Laurenti Verlag, Bielefeld, 328 pp., col. illust.
Franzen, M., Jones, J., Raselimanana, A.P., Nagy, Z.T., Craze, N.D., Glaw, F. & Vences, M. (2009) A new black-bellied snake (Pseudoxyrhophiinae: *Liophidium*) from western Madagascar, with notes on the genus *Pararhadinaea*. *Amphibia-Reptilia*, Leiden 30(2), 173–183, 6 figs.
Fraser, A.G.L. (1936) The snakes of Deolali. Part I. *The Journal of the Bombay Natural History Society* (1936–1937) 39(1), 58–82, illust.
Fraser, A.G.L. (1937a) The snakes of Deolali. Part II. *The Journal of the Bombay Natural History Society* (1936–1937) 39(2), 264–291, illust.
Fraser, A.G.L. (1937b) The snakes of Deolali. Part III. *The Journal of the Bombay Natural History Society* (1936–1937) 39(3), 464–501, illust.
Fraser, D.F. (1964) *Micrurus limbatus*, a new coral snake from Veracruz, México. *Copeia* 1964(3), 570–573, 3 figs.
Fraser, D.F. (1973) Variation in the coral snake, *Micrurus diastema*. *Copeia* 1973(1), 1–17.
Frazer, D. (1989) *Reptiles and amphibians in Britian*. Second edition. Collins New Nature Series, Bloomsbury Books, London, 254 pp., 49 figs., 16 pls. [first edition, 1983, 256 pp., 16 pls.]
Frazer, J.F.D. (1949) The reptiles and amphibia of the Channel Isles and their distribution. *British Journal of Herpetology*, London 2, 51–53.
Frazier, J.A., Montgomery, C.E., Boback, S.M. & Reed, R.A. (2007) Geographic distribution: Serpentes: *Leptophis mexicanus*. *Herpetological Review* 38(2), 221.
Frazzetta, T.H. (1970) Studies on the fossil snake *Dinilysia patagonica* Woodward. Part II. Jaw machinery in the earliest snakes. *Forma et Functio*, Brunswick 3, 205–221, 5 figs.
Freiberg, M.A. (1939) Enumeracion sistematica de los reptiles en entre Rios y lista de los ejemplares que los representan en el Museo de Entre Rios. *Memorias del Museo de Entre Rios, Zoología* (11), 1–28.

Freiberg, M.A. (1951) Nuevo hallazgo de *Leptotyphlops borrichiana* Degerbol en Argentina. *Physis*, Buenos Aires 20(58), 260–262, 2 figs.

Freiberg, M.A. (1954) *Vida de batracios y reptiles Sudamericanas*. Ed. Cesarini Hnos, Buenos Aires, 192 pp., 45 pls.

Freiberg, M.A. (1968) Ofidios ponzonosos de la Argentina. *Ciencia y Investigación*, Buenos Aires 24(8), 338–353.

Freiberg, M.A. (1982) Snakes of South America. T.F.H. Publ., Neptune, 189 pp., illust.

Freiberg, M.A. & Orejas-Miranda, B.R. (1968) Un nuevo Leptotyphlopidae de la Republica Argentina (Reptilia, Ophidia). *Physis*, Buenos Aires 28(76), 145–147, 2 figs.

Freire, E.M.X. (1998) Geographical distribution. *Sibon nebulata*. *Herpetological Review* 29(3), 178.

Freire, E.M.X. (1999a) Geographic distribution: Serpentes: *Dendrophidion dendrophis*. *Herpetological Review* 30(1), 53.

Freire, E.M.X. (1999b) Geographical distribution. *Oxyrhopus guibei*. *Herpetological Review* 30(1), 55.

Freire, E.M.X. (2000) Geographic distribution: Serpentes: *Drymoluber dichrous*. *Herpetological Review* 31(1), 55.

Freire, E.M.X. (2001) Geographic distribution: Serpentes: *Micrurus corallinus*. *Herpetological Review* 32(1), 60.

Freire, E.M.X. & da Silva, S.D. (2000) Geographic distribution: Serpentes: *Taeniophallus affinis*. *Herpetological Review* 31(3), 187.

Freire, E.M.X., Caramaschi, U. & Argôlo, A.J.S. (2007) A new species of *Liotyphlops* (Serpentes, Anomalepididae) from the Atlantic rain forest of northeastern Brazil. *Zootaxa* (1393), 19–26, 2 figs.

Freire, E.M.X., Caramaschi, U. & Gonçalves, U. (2010) A new species of *Dendrophidion* (Serpentes: Colubridae) from the Atlantic rain forest of northeastern Brazil. *Zootaxa* (2719), 62–68, 4 figs.

Freire-Lascano., A. (1982) *Serpientes en los Guaduales*. Segundo Simposio Latinoamericano del BAMBU, Guayaquil, 16 pp.

Freire-Lascano, A. (1991) Dos nuevas especies de *Bothrops* en el Ecuador (Serpientes venenosas). *Publicaciones Trabajos Cientificos del Ecuador, Universidad Tecnica de Máchala* 2, 1–10, 8 figs.

Freire-Lascano, A., Garcia, F., Wüster, W. & Kuch, U. (2003) Geographic distribution: Serpentes: *Bothriopsis taeniata* or *Bothrops taeniatus*. *Herpetological Review* 34(3), 263.

Freitas, M.A. de. (1999) *Serpentes da Bahia e do Brasil. Suas características e hábitos*. Ed. Dall, Feira de Santana, 78 pp., 107 col. photos.

Freitas, M.A. de. (2003) *Serpentes Brasileiras*. Proquigel Quimica, Camaçari, 120 pp., 300 col. photos.

Freitas, M.A. de. & Pavie, I. (2002) *Guia de Répteis da região metropolitana de Salvador e litoral norte da Bahia*. Malha-de-Sapo-Publicações, Lauro de Freitas, 71 pp., 80 col. photos.

Freitas, M.A. de & Santos, T.F. (2011) Geographic distribution: Serpentes: *Helicops modestus*. *Herpetological Review* 42(3), 394.

Freitas, M.A. de, França, D.P.F. de & Verissimo, D. (2012a) Distribution of *Drymoluber brazili* (Gomes, 1918) (Serpentes: Colubridae) for the state of Piauí, Brazil. *Check List* 8(1), 168–160, 1 fig.

Freitas, M.A. de, França, D.P.F. de. & Bernarde, P.S. (2012b) Squamata, Serpentes, Dipsadidae, *Philodryas viridissima* (Linnaeus, 1758): first record in the state of Acre, northern Brazil. *Check List* 8(2), 258–259, 1 fig.

Freitas, M.A. de, Franca, D.P.F. de, Graboski, R., Uhlig, V. & Veréissimo, D. (2012c) Notes on the conservation status, geographic distribution and ecology of *Bothrops muriciensis* Ferrarezzi & Freire, 2001 (Serpentes, Viperidae). *North-Western Journal of Zoology* 8(2), 338–343, 3 figs.

Fretey, J. (1975) *Guide des reptiles et batraciens de France*. Hatier, Paris, 239 pp., illust., 32 col. pls.

Freyer, H. (1842) *Fauna der in Krain bekannten Säugethiere, Vögel, Reptilien und Fische. Nach Cuvier's System geordnet*. Egers'chen Guberniol-Buchdruckerei, Laibach, viii + 90 pp.

Frith, C. & Frith, D. (1987) *Australian tropical reptiles & frogs*. Tropical Australia Graphics, Townsville, 70 pp., col. photos.

Frith, C.B. (1977a) A survey of the snakes of Phuket Island and the adjacent mainland areas of peninsular Thailand. *Natural History Bulletin of the Siam Society*, Bangkok 26(3–4), 263–316, 3 figs.

Frith, C.B. (1977b) The sea snake *Hydrophis spiralis* (Shaw); a new species of the fauna of Thailand. *Natural History Bulletin of the Siam Society*, Bangkok 26, 339–341, pl. 11.

Frith, C.B. (1978) Additions of the snake fauna of Phuket Island, peninsular Thailand. *Natural History Bulletin of the Siam Society*, Bangkok 27, 181–186, 1 fig.

Frith, C.B. & Boswall, J. (1978) Cantor's water snake, *Cantoria violacea* Girard; a vertebrate new to the fauna of Thailand. *Natural History Bulletin of the Siam Society*, Bangkok 27, 187–188, pl. 2.

Frith, C.B. & MacIver, D. (1978) The crab-eating water snake, *Fordonia leucobalia* (Schleg.); another new snake to Thailand. *Natural History Bulletin of the Siam Society*, Bangkok 27, 189–191, pl. 3.

Fritts, T.H. (1988) The brown tree snake, *Boiga irregularis*: a threat to Pacific islands. *Fish & Wildlife Service, United States Department of the Interior, Biological Report* 88(31), vi + 36 pp., 10 figs.

Fritts, T.H. (1993) The common wolf snake, *Lycodon aulicus capucinus*: a recent colonist of Christmas Island in the Indian Ocean. *Wildlife Research* 20, 261–266.

Fritts, T.H. & Smith, H.M. (1969) A new genus and species of snake from western Ecuador. *Transactions of the Kansas Academy of Sciences* 72(1), 60–66, 2 figs.

Fritts, T.H., Scott, N.J., Jr. & Savidge, J.A. (1987) Activity of the arboreal brown tree snake (*Boiga irregularis*) on Guam as determined by electrical outages. *The Snake* 19(1), 51–58.

Fritz, U. (2002) Herpetology and herpetological type specimens at the Museum für Tierkunde Dresden with a bibliography of herpetological contributions by Fritz Jürgen Obst (Amphibia, Reptilia). *Faunistiche Abhandlungen Staatliches Museum für Tierkunde Dresden* 23(1), 3–34.

Fritze, A. (1894) Die Fauna der Liu-Kiu Insel, Okinawa. *Zoologische Jahrbücher. Abteilung für Systematik, Ökologie und Geographie der Tiere*, Jena 7, 852–926.

Fröhlich, G., Oertner, J. & Vogel, S. (1987) *Schützt Lurche und Kriechtiere*. Deutscher Landwirtschaftsverlag, Berlin, 324 pp., col. illust.

Frommhold, E. (1952) *Heimische Lurche und Kriechtiere*. Akademische Verlagsgesellschaft Geest & Portig K.-G., Leipzig, 122 pp., 56 figs.

Frommhold, E. (1959) *Wir bistimmen Lurche und Kriechtierre Mitteleuropas*. Neumann, Radebeul, 219 pp., 40 pls.

Frommhold, E. (1964) *Die Kreuzotter*. A. Ziemsen Verlag, Wittenberg Lutherstadt, 76 pp., 9 figs., 32 pls.

Froom, B. (1981) *Ontario snakes*. Second edition. Ministry of Natural Resources, Ontario, 36 pp., 18 figs. [first edition, 1967, Department of Lands and Forests, 36 pp., 18 figs.]

Froom, B. (1972) *Snakes of Canada*. McClelland and Steward, Toronto, 128 pp., 39 illust., 8 col. pls.
Frost, D.R. (1983) *Sonora semiannulata. Catalogue of American amphibians and reptiles*, pp. 333.1–333.4, 1 fig.
Frost, D.R. & Collins, J.T. (1988) Nomenclatural notes on reptiles of the United States. *Herpetological Review* 19(4), 73–74.
Frost, D.R. & Van Devender, T.R. (1979) The relationship of the ground snakes *Sonora semiannulata* and *S. episcopa* (Serpentes, Colubridae). *Occasional Papers of the Museum of Zoology, Louisiana State University* (52), 1–9, 1 fig.
Frota, J.G. da. (2000a) Geographic distribution: Serpentes: *Atractus snethlageae*. *Herpetological Review* 31(4), 254.
Frota, J.G. da. (2000b) Geographic distribution: Serpentes: *Oxyrhopus melanogenys melanogenys*. *Herpetological Review* 31(4), 255–256.
Frota, J.G. da. (2001a) Geographic distribution: Serpentes: *Liophis breviceps breviceps*. *Herpetological Review* 32(1), 60.
Frota, J.G. da. (2001b) Geographic distribution: Serpentes: *Liophis cobella taeniogaster*. *Herpetological Review* 32(2), 123.
Frota, J.G. da. (2005) Nova espécie de *Helicops* Wagler, 1830 (Serpentes, Colubridae) do rio Tapajós, Amazônia, Brasil. *Phyllomedusa* 4(1), 61–68, 4 figs.
Frota, J.G. da & dos Santos, A.P., Jr. (2002) Geographic distribution: Serpentes: *Liophis miliaris chrysostomus*. *Herpetological Review* 33(1), 68.
Frota, J.G. da & dos Santos, A.P., Jr. (2003) Geographic distribution: Serpentes: *Liophis taeniogaster*. *Herpetological Review* 34(4), 388–389.
Frota, J.G. da & Yuki, R.N. (2001) Geographic distribution: Serpentes: *Liophis festae*. *Herpetological Review* 33(4), 278.
Frota, J.G. da & Yuki, R.N. (2002) Geographic distribution: Serpentes: *Chironius exoletus*. *Herpetological Review* 33(1), 67.
Fry, B.G., Wickramaratna, J.C., Jones, A., Alewood, P.F. & Hodgson, W.C. (2001) Species and regional variations in the effectiveness of antivenom against the *in vitro* neurotoxicity of death adder (*Acanthophis*) venoms. *Toxicology and Applied Pharmacology* 175, 140–148, 5 figs.
Fry, B.G., Wickramaratna, J.C., Hodgson, W.C., Alewood, P.F., Kini, R.M., Ho, H. & Wüster, W. (2002) Electrospray liquid chromatography/mass spectrometry fingerprinting of *Acanthophis* (death adder) venoms: taxonomic and toxinological implications. *Rapid Communications in Mass Spectrometry* 16, 600–608, 4 figs.
Fry, D.B. (1915) Herpetological notes. *Proceedings of the Royal Society of Queensland*, Brisbane 27, 60–95, figs.
Frymire, D. & Scott, A.F. (2012) Geographic distribution: Serpentes: *Clonophis kirtlandii*. *Herpetological Review* 43(3), 447.
Fuentes-R., O. (2003) Geographic distribution: Serpentes: *Liophis miliaris miliaris*. *Herpetological Review* 34(3), 265.
Fuentes-R., O. & Barrio-Amorgós, C.L. (1999a) Geographic distribution: Serpentes: *Erythrolamprus pseudocorallus*. *Herpetological Review* 30(1), 53.
Fuentes-R., O. & Barrio-Amorgós, C.L. (1999b) Geographic distribution: Serpentes: *Liophis poecilogyrus*. *Herpetological Review* 30(1), 54.
Fuentes-R., O. & Rodriguez-Acosta, A. (1997) Sobre los generos *Bothriechis*, *Bothriopsis*, *Bothrops* y *Porthidium* (Serpentes, Crotalidae) existentes en Venezuela. Claves para su identificacion de interes biomedico. *Acta Biologica Venezuelica* 17(3), 31–38, illust.
Fugler, C.M. (1953) *The distribution of the amphibians and reptiles of the state of San Luis Potosi*. Master's Thesis, Louisiana State University, Baton Rouge, 139 pp.
Fugler, C.M. (1983) Lista preliminar de los anfibios y reptiles de Tumi Chucua, Provincia Vaca Diez, Departamento de Beni. *Comunicación, Museo Nacional de Historia Natural*, La Paz 2, 4–11.
Fugler, C.M. (1984) Tercera contribucion a la fauna herpetologica del Oriente Boliviano. *Ecología en Bolivia*, La Paz (5), 63–72.
Fugler, C.M. (1986) Una lista preliminar de las serpientes de Bolivia. *Ecologia en Bolivia*, La Paz (8), 45–72.
Fugler, C.M. & Dixon, J.R. (1961) Notes on the herpetofauna of the El Dorado area of Sinaloa, Mexico. *Publications of Michigan State University* 2(1), 1–23.
Fugler, C.M. & Riva, I. de la. (1990) Herpetologia boliviana: lista provisional de las serpientes conocidas en el país. *Comunicación, Museo Nacional de Historia Natural,* La Paz 9, 22–53.
Fugler, C.M. & Walls, A.B. (1978) Snakes of the Upano Valley of Amazonian Ecuador. *Journal of the Tennessee Academy of Sciences* 53(3), 81–87, 1 fig.
Fugler, C.M., Riva, I. de la & Cabot, J. (1995) Herpetologica Boliviana: una lista comentada de las serpientes de Bolivia con datos sobre su distribución. *Ecología en Bolivia*, La Paz (24), 40–87.
Fuhn, I.E. (1969) *Natura si Omul. Broaste, Serpi, Sopirle*. Bucuresti, 246 pp., 6 pls.
Fuhn, I.E. & Vancea, S. (1961) Fauna Republicii Populare Romine. Reptilia (Testoase, Sopirle, Serpi). *Academia Republicii Populare Romine* 14(2), 338 pp., 244 figs.
Fukada, H. (1975) A list of Tanzanian snakes collected by Dr. Masakazu Kawabata. *Kyoto University African Studies* 9, 33–43, 11 figs.
Fukada, H. (1992) *Snake life history in Kyoto.* Impact Shuppankai Co., Tokyo, 171 pp., 54 figs.
Funk, R.S. (1967) A new colubrid snake of the genus *Chionactis* from Arizona. *Southwestern Naturalist* 12(2), 180–188, 3 figs.
Furman, J. (2007) *Timber rattlesnakes in Vermont & New York: biology, history, and the fate of an endangered species.* University Press of New England, Hanover, 207 pp., 9 figs., 11 col. pls.
Fyfe, G. (1994) The central carpet python *Morelia spilota bredli* in the field and in captivity. *Monitor,* Ardeer 6(2), 70–72, 1 fig., col. photos 3–6.
Fyfe, G. & Green, D. (2003) *Keeping children's pythons.* Australian Reptile Keeper Publications, Bendigo, 37 pp., illust., 8 col. pls.
Gabaldón, F.S. (Ed.). (1997) *Guía de os anfibios y reptiles de la Comunidad Valenciana.* Sociedad Herpetologica Valenciana, Valencia, 184 pp., col. photos.
Gachet 1829. Description d'une espèce inedited? De Couleuvre observée aux environs de Bordeaux. *Bulletin d'Histoire Naturelle de la Société Linnéenne de Bordeaux* 3, 255–259.
Gaige, H.T. (1936) Some reptiles and amphibians from Yucatán and Campeche, México. *Publications, Carnegie Institution of Washington* (457), 289–304.
Gaige, H.T., Hartweg, N. & Stuart, L.C. (1937) Notes on a collection of amphibians and reptiles from eastern Nicaragua. *Occasional Papers of the Museum of Zoology, University of Michigan* (357), 1–18.
Gail, R. & Rageau, J. (1958) Introduction a l'étude des serpents marins (ophidiens Hydrophiidae) en Nouvelle-Calédonie. *Bulletin de la Société Pathologie Exotique* 51(3), 448–459, 13 figs.
Galán-Regalado, P. & Fernández-Arias, G. (1993) *Anfibios e réptiles de Galicia.* Edicíons Xerais de Galicia, Vigo, 501 pp., 26 figs., 48 col. pls.
Gallagher, M.D. (1971) *The amphibians and reptiles of Bahrain.* Michael D. Gallagher, Bahrain, 37 pp., 15 figs.

Gallagher, M.D. (1993) *Snakes of the Arabian Gulf and Oman*. Second edition. Michael D. Gallagher, Muscat, 16 pp., 5 figs., 37 col. photos [first edition, 1990, 16 pp., 5 figs., 37 col. photos]

Gallagher, M.D. & Arnold, E.N. (1988) Reptiles and amphibians from the Wahiba Sands, Oman. *Journal of Oman Studies, Special Report* (3), 405–413.

Gallardo, G.A. (2006) Geographic distribution: Serpentes: *Pseudotomodon trigonatus*. *Herpetological Review* 37(2), 243.

Gallardo, J.M. (1972) Observations biologicas sobre una falsa yarara, *Tomodon ocellatus* Duméril, Bibron et Duméril. *Neotropica*, La Plata 18(56), 58–63.

Gallardo, J.M. (1977a) Composition, distribution and origin of the Chacoan herpetofauna. *Herpetological Review* 8(3), 8–9.

Gallardo, J.M. (1977b) Estudio ecologico sobre los anifibios y reptiles de la Depresion del Salado, Provincia de Buenos Aires, Argentina. *Revista del Museo Argentino de Ciencias Naturales Bernardino Rivadavia* (*Ecologia*) (1976) 2(1), 1–26.

Ganesh, S.R. (2011) On a rare, South Indian burrowing snake *Platyplectrurus trilineatus* (Beddome, 1867). *Taprobanica*, Homagama 3(1), 11–14, 1 pl., 5 figs.

Ganesh, S.R. & Chandramouli, S.R. (2011a) On the nomenclature and taxonomy of the south Indian colubrid snake *Ahaetulla perroteti* (Duméril, Bibron & Duméril, 1854). *Herpetological Bulletin* (117), 19–24, 1 fig.

Ganesh, S.R. & Chandramouli, S.R. (2011b) Report on some noteworthy specimens and species of herpetofauna from South-East India. *Taprobanica*, Homagama 3(1), 5–10, 5 figs.

Ganesh, S.R., Asokan, S. & Kannan, P. (2008) Record length of large scaled pit viper (*Trimeresurus macrolepis*) Beddome, 1862, with notes on its ecological and conservation significance. *Cobra* 2(3), 17–22, 1 fig.

Ganesh, S.R., Asokan, S. & Kannan, P. (2009) Record of *Oligodon travancoricus* Beddome, 1877 (Serpentes, Colubridae) from Grizzled Squirrel Sanctuary, Western Ghats, Tamil Nadu, India. *Hereptological Bulletin* (109), 25–28.

Ganesh, S.R., Chandramouli, S.R. & Shankar, P.G. (2012a) Record lengths of two endemic caenophidian snakes from the Western Ghats Mountains, India. *Hamadryad* 36(1), 44–46, 2 figs.

Ganesh, S.R., Shankar, P.G. & Sreekar, R. (2012b) First record of Wayanad shieldtail, *Melanophidium wynaudense* (Beddome, 1863), from the central Western Ghats, India. *Hamadryad* 36(1), 53–56, 2 figs.

Gans, C. (1949) A bibliography of the herpetology of Japan. *Bulletin of the American Museum of Natural History* 93(6), 389–496.

Gans, C. (1957) *Dasypeltis medici lamuensis*, a new race of egg-eating snake (Ophidia, Reptilia) from coastal east Africa. *Breviora* (79), 1–13.

Gans, C. (1959) A taxonomic revision of the African snake genus *Dasypeltis* (Reptilia, Serpentes). *Annales du Musée Royal du Congo Belge* (*Série in Octavo, Science Zoologique*), Tervuren 74, ix + 237 pp., 36 figs., 13 pls.

Gans, C. (1960) Notes on a herpetological collecting trip through the south-eastern lowlands of Bolivia. *Annals of the Carnegie Museum* (1957–1960) 35(13), 283–313, 11 photos.

Gans, C. (1964a) Further comments on the forms of the African snake genus *Dasypeltis* (Reptilia, Serpentes). *Revue de Zoologie et Botanique Africaine*, Bruxelles 69(3–4), 279–295, 5 figs.

Gans, C. (1964b) A redescription of and geographic variation in *Liophis miliaris* Linné, the common water snake of south-eastern South America. *American Museum Novitates* (2178), 1–58.

Gans, C. (1966) Liste der rezenten Amphibien und Reptilien: Uropeltidae. *Das Tierreich*, Berlin (84), i–viii + 1–29.

Gans, C. (1976) Aspects of the biology of uropeltid snakes. *In:* Bellairs, A.D. & Cox, C.B. (Eds.), *Morphology and biology of reptiles*. Linnaean Society Symposium Series 3, Academic Press, London, pp. 191–204.

Gans, C. (1978) Reptilian venoms: some evolutionary considerations. *In:* Gans, C. & Gans, K.A. (Eds.), *Biology of the Reptilia. Volume 8 (Physiology B)*. Academic Press, London, pp. 1–42.

Gans, C. & Fetcho, J.R. (1982) The Sri Lanka genus *Aspidura* (Serpentes, Reptilia, Colubridae). *Annals of the Carnegie Museum* 51(14), 271–316, 13 figs.

Gans, C. & Laurent, R.F. (1965) Snakes. *In:* Gans, C., Laurent, R.F. & Pandit, H. (Eds.), Notes on a herpetological collection from the Somali Republic. *Annales du Musée Royal de l'Afrique Centrale* (*Science Zoologique*), Tervuren (134), 47–70, figs. 10–12.

Gans, C. & Maderson, P.F. (1970) Sound-producing mechanism in Recent reptiles: review and comment. *American Zoologist* 13, 1195–1203, 1 fig.

Gans, C. & Oshima, M. (1952) Adaptations for egg eating in the snake *Elaphe climacophora* (Boie). *American Museum Novitates* (1571), 1–15, 8 figs.

Gans, C. & Williams, E.E. (1954) Present knowledge of the snake *Elachistodon westermanni* Reinhardt. *Breviora* (36), 1–17, 4 figs.

Gao, W. (1987) A new record of the snakes of Beijing—*Elaphe carinata*. *Acta Herpetologica Sinica*, Chengdu (Series 2) 6(2), 75.

Gao, Z.-F. (1984) Snakes of Anxian County, Sichuan. *Acta Herpetologica Sinica*, Chengdu (Series 2) 3(3), 69–71.

Garanin, V.I., Pavlov, A. & Bakiev, A.G., 2004. Stepnaia gadyuka, ili Gadyuka Renarda *Vipera renardi* (Christoph, 1861). *In:* Bakiev, A.G., Garanin, V.I., Litvinov, N.A., Pavlov, A.V. & Ratnikov, V.Ju. (Eds.), *Snakes of Volzhsko-Kamskij Region*. Publishing House of Samara Scientific Center, Russian Academy of Sciences, Samara, pp. 61–90.

García, A. & Ceballos, G. (1994) *Guia de campo de los reptiles y anfibios de la costa de Jalisco, Mexico. Field guide to the Reptiles and amphibians of the Jalisco coast, Mexico*. Fundación Ecológica de Cuixmala, Instituto de Biol.ogía, Uiversidad Nacional Autónoma de México, 184 pp., 14 col. figs.

García, A. & Valtierra-Azotla, M. (1996) Geographic distribution: Serpentes: *Ficimia publia*. *Herpetological Review* 27(2), 88.

García, E. (1896) *Los ofidios venenosas del Cauca. Métodos empíricos y racionales empleados contra los accidentes producidos por la mordedura de esos reptiles*. Libreria Colombiana, Cali, xv + 102 pp., 15 pls.

Garcia, R.L., Huang, S.-Y. & Perez, J.C. (1989) Cytotoxic activity of HPLC fractionated venom on hyvridoma cells. *Toxicon* 27(1), 46–47.

García-Balderas, C.M. & Quintero-Díaz, G.E. (2012) Geographic distribution: Serpentes: *Geophis dugesii*. *Herpetological Review* 43(4), 621.

García-Cubas, A. (1893) *México: its trade, industries and resources*. Oficina del Departamento de Fomento, Colónia y Industría, México City, xviii + 463 + iv pp.

García-Díez, T. & Gonzáles-Fernández, J.E. (2013) The reptile type specimens preserved in the Museo Nacional de Ciencias Naturales (CSIC) of Madrid, Spain. *Zootaxa* (3619), 46–58.

García-París, M., Martín, C., Dorda, J. & Esteban, M. (1989) *Los anfibios y reptiles de Madrid*. Ministerio de Agricultura Pesca y Alimentación, Madrid, 243 pp., 137 col. figs.

García-Pérez, J.E. (1995) Una nueva especie de cascabel (Serpentes: Crotalidae) para el bolsón árido de Lagunillas, Cordillera de Mérida, Venezuela. *Revista de Zoologia Latinoamericana* 3(2), 7–12.

García-Vázquez, U.O. (2008) Geographic distribution: Serpentes: *Thamnophis pulchrilatus*. *Herpetological Review* 39(4), 487.

García-Vázquez, U.O. (2012) Geographic distribution: Serpentes: *Rhadinaea montana*. *Herpetological Review* 43(2), 209.

García-Vázquez, U.O., Pérez-Ramos, E. & Durán-Fuentes, I. (2008a) Geographic distribution: Serpentes: *Conophis acuta*. *Herpetological Review* 39(4), 484.

García-Vázquez, U.O., Solano-Zavaleta, I. & Nieto-Montes de Oca, A. (2008b) Geographic distribution: Serpentes: *Crotalus transversus*. *Herpetological Review* 39(4), 484.

García-Vázquez, U.O., Canseco-Márquez, L., Aguilar-López, J.L., Solano-Zavaleta, I. & Maceda-Cruz, R. J. (2009) Noteworthy records of amphibians and reptiles from Puebla, México. *Herpetological Review* 40(4), 467–470.

Gärdenfors, U. (1980) Ein Nachweis von *Vipera xanthina* in Griechenland (Reptilia, Serpentes, Viperidae). *Salamandra*, Frankfurt-am-Main 16(4), 270.

Gardner, A.S. (1999) The reptiles and amphibians of Oman: herpetological history and current status. *In:* Fisher, M., Ghazanfar, S.A. & Spalton, A. (Eds.), *The natural history of Oman: a festschrift for Michael Gallagher*. Backhuys Publishers, Leiden, pp. 65–88.

Gardner, A.S. (2013) *The amphibians and reptiles of Oman and the UAE*. Edition Chimaira, Frankfurt-am-Main, 460 pp., 300 col. photos.

Gardner, J.D. & Cifelli, R.L. (1999) A primitive snake from the Cretaceous of Utah. *Special Papers in Palaeontology* 60, 87–100.

Gardner, S.A. & Mendelson, J.R., III. (2004) Taxonomy and geographic variation in the leaf-nosed snake *Phyllorhynchus decurtatus* (Squamata, Colubridae). *Journal of Herpetology* 38(2), 187–196, 3 figs.

Garel, T. & Matola, S. (1995) *A field guide to the snakes of Belize*. Belize Zoo and tropical Education Center, Belize, 147 pp., col. illust.

Garman, H. (1892) A synopsis of the reptiles and amphibians of Illinois. *Bulletin of the Illinois State Laboratory of Natural History* 3(13), 215–388 + 8.

Garman, S.W. (1874) Description of a new species of North American serpent. *Proceedings of the Boston Society of Natural History* (1874–1875) 17(2), 92–94, 3 figs.

Garman, S.W. (1881) New or little-known reptiles and fishes in the Museum collections. *Bulletin of the Museum of Comparative Zoology* 8(3), 85–93.

Garman, S.W. (1883) On certain reptiles from Brazil and Florida. *Science Observer*, Boston 4(5–6), 47–48.

Garman, S.W. (1884) The reptiles and batrachians of North America. *Memoirs of the Museum of Comparative Zoology* (1883) 8(3), i–xxvii + 1–185, 9 pls.

Garman, S.W. (1887a) On the reptiles and batrachians of Grand Cayman. *Proceedings of the American Philosophical Society* 24(126), 273–277.

Garman, S.W. (1887b) On West Indian reptiles in the Museum of Comparative Zoology at Cambridge, Mass. *Proceedings of the American Philosophical Society* 24(126), 278–286.

Garman, S.W. (1887c) Reptiles and batrachians from Texas and Mexico. *Bulletin of the Essex Institute* 19, 119–138.

Garman, S.W. (1888) Reptiles and batrachians from the Caymans and from the Bahamas, collected by Prof. C.J. Maynard for the Museum of Comparative Zoology at Cambridge. *Bulletin of the Essex Institute* 20, 101–116.

Garman, S.W. (1892a) The reptiles of the Galapagos Islands. From the collection of Dr. Geo. Baur. *Bulletin of the Essex Institute* 24, 73–87.

Garman, S.W. (1892b) On reptiles collected by Dr. Geo. Baur near Guayaquil. *Bulletin of the Essex Institute* 24, 88–95.

Garman, S.W. (1901) Some reptiles and batrachians from Australasia. *Bulletin of the Museum of Comparative Zoology* 39(1), 1–14, 2 pls.

Garman, S.W. (1908) The reptiles of Easter Island. *Bulletin of the Museum of Comparative Zoology* 52(1), 3–14, 1 pl.

Garrett, C.M. & Painter, C.W. (1992) Geographic distribution: Serpentes: *Senticolis triaspis intermedia* (green rat snake). *Herpetological Review* 23(4), 124.

Garrido, O.H. (1973) Nuevas subespecies de reptiles para Cuba. *Torreia* 30, 1–31, 3 figs.

Garrido, O.H. & Jaume, M.L. (1984) Catalogo descriptivo de los anifibios y reptiles de Cuba. *Doñana, Acta Vertebratica*, Sevilla 11(2), 5–128.

Garrigues, T., Dauga, C., Ferquel, E. Choumet, V. & Failloux, A.-B. (2005) Molecular phylogeny of *Vipera* Laurenti, 1768 and the related genera *Macrovipera* (Reuss, 1927) and *Daboia* (Gray, 1842), with comments about neurotoxic *Vipera aspis aspis* populations. *Molecular Phylogenetics and Evolution* 35(1), 35–47, 4 figs.

Garstka, W.R. (1982) Systematics of the *mexicana* species group of the colubrid genus *Lampropeltis* with an hypothesis of mimicry. *Breviora* (466), 1–35, 8 figs.

Garton, J.S., Harris, E.W. & Brandon, R.A. (1970) Descriptive and ecological notes on *Natrix cyclopion* in Illinois. *Herpetologica* 26(4), 454–461.

Garza, M. de la. (1984) *El universo sagrado de la serpiente entre los Mayos*. Universidad Nacional Autonóma de México, México, 462 pp., 78 figs.

Gasc, J.-P. & Rodrigues, M.T. (1979a) Une nouvelle espèce du genre *Atractus* (Colubridae, Serpentes) de la Guyane française. *Bulletin du Muséum National d'Histoire Naturelle*, Paris (Série 4) 1A(2), 547–557, 6 figs.

Gasc, J.-P. & Rodrigues, M.T. (1979b) Sur la présence du genre *Geophis* (Colubridae, Serpentes) dans la région guyanaise. Description d'une espèce de Guyane française. *Bulletin du Muséum National d'Histoire Naturelle*, Paris (Série 4) 1A(4), 1121–1130, 4 figs., pl. 1.

Gasc, J.-P. & Rodrigues, M.T. (1980) Liste préliminaire des serpents de la Guyane française. *Bulletin du Muséum National d'Histoire Naturelle*, Paris (Série 4) 2A(2), 559–598, 12 figs.

Gasc, J.-P., Cabela, A., Crnobrnja-Isailovic, J., Dolmen, D., Grossenbacher, K., Haffner, P., Lescure, J., Martens, H., Martínez-Rica, J.P., Maurin, H., Oliveira, M.E., Sofianidou, T.S., Veith, M. & Zuiderwijk, A. (Eds). (1997) *Atlas of amphibians and reptiles in Europe*. Societas Europaea Herpetologica and Musée National d'Histoire Naturelle, Paris, 496 pp., illust.

Gasperetti, J. (1974) A preliminary sketch of the snakes of the Arabian Peninsula. *Journal of the Saudi Arabian Natural History Society*, Jiddah (1973) (12), 1–72, 30 figs.

Gasperetti, J. (1976a) *Walterinnesia aegyptia* Lataste, a rare black snake. *Journal of the Saudi Arabian Natural History Society*, Jiddah (16), 2–8, 3 figs.

Gasperetti, J. (1976b) A very rare snake, *Coluber elegantissimus*. *Journal of the Saudi Arabian Natural History Society*, Jiddah (18), 13–15, 3 figs.

Gasperetti, J. (1977a) Snakes in Arabia. *Journal of the Saudi Arabian Natural History Society*, Jiddah (19), 3–16.

Gasperetti, J. (1977b) An interesting snake from the USGS camp near Madha. *Journal of the Saudi Arabian Natural History Society*, Jiddah (20), 23–24, 3 figs.

Gasperetti, J. (1988) Snakes of Arabia. *In:* Buttiker, W. & Krupp, F. (Eds.), *Fauna of Saudi Arabia*, Riyadh 9, 169–450, 135 figs., 29 pls.

Gasperetti, J., Leviton, A.E. & Anderson, S.C. (1967) Survey of reptiles of the Sheikdom of Abu Dhabi, Arabian peninsula. Part 1. A geographical sketch of the Sheikdom of Abu Dhabi. Part 2. Systematic account of the collection of reptiles made in the Sheikdom of Abu Dhabi by John Gasperetti. *Proceedings of the California Academy of Sciences* 35(8–9), 141–192, 14 figs.

Gatti, C. (1955) Las culebras venenosas del Paraguay. *Revista Médica del Paraguay* 1(2), 81–100, 8 figs.

Gauduin, P. (1970) Contributions à la faune de la région de Yaoundé. III. Notes sur les colubridés (reptiles ophidiens). *Annales de la Faculté des Sciences du Cameroun*, Yaoundé (4), 73–99.

Gaulke, M. (1993a) First record of the polyodont snake *Sibynophis geminatus geminatus* (Boie, 1822) from the Philippines, with a discussion of *Sibynophis bivittatus* (Boulenger, 1894). *Herpetological Journal,* London 3, 151–152.

Gaulke, M. (1993b) Zur Taxonomie und Biologie von *Oligodon meyerinkii* (Steindachner, 1891). *Sauria* 15(3), 3–6, 3 col. figs.

Gaulke, M. (1994) Eine neue Unterart des Malaysischen Baumschnüfflers, *Ahaetulla prasina suluensis* n. subsp. *Senckenbergiana Biologica*, Frankfurt-am-Main 73(1–2), 45–47.

Gaulke, M. (1994) Contribution to the snake fauna of the Sulu Archipelago, with the description of a new subspecies of *Dendrelaphis caudolineatus* (Gray, 1834). *Herpetological Journal*, London 4, 136–144, 3 figs.

Gaulke, M. (1995) Observations on arboreality in a Philippine blind snake. *Asiatic Herpetological Research* 6, 45–48.

Gaulke, M. (1996) Die Herpetofauna von Sibutu Island (Philippinen), unter Berücksichtigung zoogeographischer und ökologischer Aspekte. *Senckenbergiana Biologica*, Frankfurt-am-Main 75(1–2), 45–56.

Gaulke, M. (1999) Die herpetofauna von Calauit Island (Calamianes-Inseln, Provinz Palawan, Philippinen) (Amphibia et Reptilia). *Faunistiche Abhandlungen, Staatliches Museum für Tierkdunde in Dresden* 21(19), 273-282.

Gaulke, M. (2002) A new species of *Lycodon* from Panay Island, Philippines (Reptilian, Serpentes, Colubridae). *Spixiana* 25(1), 85–92.

Gaulke, M. (2011) *The herpetofauna of Panay Island, Philippines.* Edition Chimaira, Frankfurt-am-Main, 390 pp., 258 col. figs.

Gaulke, M. & Altenbach, A.V. (1994) Contribution to the knowledge of the snake fauna of Masbate (Philippines) (Squamata, Serpentes). *Herpetozoa*, Wien 7(1–2), 63–66, 1 fig.

Gaulke, M. & Vogel, G. (2005) Verbreitungsnachweis und Längenrekord für *Calamaria bitorques* Peters, 1872. *Sauria* 27(1), 19–23, 6 col. figs.

Gaulke, M., Demegillo, A. & Vogel, G. (2004) Eine neue Unterart der Mangroven-Nachtbaumnatter von dem Philippinen. *Herpetofauna* 25(143), 5–16, 10 figs.

Gauthier, J.A., Kearney, M., Maisano, J.A., Rieppel, O. & Behlke, A.D.B. (2012) Assembling the squamate tree of life: perspectives from the phenotype and the fossil record. *Bulletin of the Peabody Museum of Natural History* 53(1), 3–308, 610 figs.

Gauthier, R. (1967) Ecologie et ethologie des reptiles du Sahara nord-occidental. *Annales du Musée Royal de l'Afrique Centrale (Série in Octavo, Science Zoologique)* 155, 1–83, 6 figs., 4 pls.

Gay, C. (1848) *Historia fisica y politica de Chile segun documentos adquiridos en esta Republica durante doce años de residencia en ella y publicada baja los auspicios del supremo gobierno. Zoologia 2.* Museo de Historia Natural de Santiago, 372 pp.

Gayen, N.C. (2002) A record of *Amphiesma venningi* (Wall, 1910) (Serpentes: Colubridae) from Meghalaya state, India. *Hamadryad* (2001) 26(2), 375.

Gaymer, R. (1968) Amphibians and reptiles of the Seychelles. *British Journal of Herpetology*, London 4, 24–28.

Gee, N.G. (1930) A contribution toward a preliminary list of reptiles recorded from China. *Bulletin of the Department of Biology, Yenching University* (1929–1930) 1(1), 53–84.

Gehlbach, F.R. (1965a) Amphibians and reptiles from the Pliocene and Pleistocene of North America: a chronological summary and selected bibliography. *Texas Journal of Science* 17(1), 56–70.

Gehlbach, F.R. (1965b) Herpetology of the Zuni Mountains region, northwestern New Mexico. *Proceedings of the United States National Museum* (1964–1965) 116(3505), 243–332, 10 figs.

Gehlbach, F.R. (1967) *Lampropeltis mexicanus. Catalogue of American amphibians and reptiles*, pp. 55.1–55.2.

Gehlbach, F.R. (1971) Lyre snakes of the *Trimorphodon biscutatus* complex: a taxonomic resume. *Herpetologica* 27(2), 200–211.

Gehlbach, F.R. (1974) Evolutionary relations of southwestern ringneck snakes (*Diadophis punctatus*). *Herpetologica* 30(2), 140–148, fig. 1.

Gehlbach, F.R. & Baker, J.K. (1962) Kingsnakes allied with *Lampropeltis mexicana*: Taxonomy and natural history. *Copeia* 1962(2), 291–300.

Gehlbach, F.R. & McCoy, C.J. (1965) Additional observations on variation and distribution of the gray-banded kingsnake, *Lampropeltis mexicana* (Garman). *Herpetologica* 21(1), 35–38.

Geissler, P., Krohn, A.R. & Rennert, D. (2011a) Herpetofaunal records in coastal dune areas, Binh Thuan Province, southern Vietnam, with the rediscovery of *Oligodon macrurus* Angel, 1927. *Russian Journal of Herpetology* 18(4), 317–3247 figs.

Geissler, P., Nguyen, T.Q., Poyarkov, N.A. & Böhme, W. (2011b) New records of snakes from Cat Tien National Park, Dong Nai and Lam Dong provinces, southern Vietnam. *Bonn Zoological Bulletin* 60(1), 9–16, 3 figs.

Gené, J. (1834) Description d'un reptile mal connu et d'un poisson nouveau. *Memorie dell'Accademie di Scienze di Torino, Classe di Scienze, Fiziche, Matematiche e Naturali* (1833), 299–305.

Gené, J. (1839) Synopsis Reptilium Sardiniae indigenorum. *Memorie della Reale Accademia di Scienze di Torino, Classe di Scienze, Fiziche, Matematiche e Naturali* (Series 2) 1, 257–286, fig. 1.

Genest-Villard, H. (1966) Développement du crâne d'un Boïdé, *Sanzinia madagascariensis. Mémoires du Muséum National d'Histoire Naturelle*, Paris (*Zoologie*) 40A(5), 207–262, 15 figs., 6 pls.

Geniez, P. & Gauthier, Y. (2008) On the distribution of *Platyceps saharicus* (Reptilia: Colubridae) in the Sahara. *Salamandra*, Frankfurt-am-Main 44(3), 179–180, 1 fig.

Geniez, P., Geniez, M., Boissinot, S., Beaubrun, P.-C. & Bons, J. (1991) Nouvelles observations sur l'herpetofaune Marocaine. 2. *Bulletin de la Société Herpétologique de France* (59), 19–27, illust.

Geniez, P., Mateo, J.A., Geniez, M. & Pether, J. (2004) *The amphibians and reptiles of the western Sahara.* Edition Chimaira, Frankfurt-am-Main, 229 pp., 212 col. photos.

Geoffroy-Saint-Hilaire, E., Geoffrey-Saint-Hilaire, I., Deshayes, G.P., Bibron, G. & Bory de Saint-Vincent, J.B.G.M. (1832) *Expedition scientifique de Morée. Tome III.—1.re partie. Zoologie. Premiere section. Animaux vertébrés, mollusques et polypiers.* F. G. Levrault, Paris, 400 pp., 58 pls.

Geoffroy-Saint-Hilaire, I. (1827) Description des reptiles qui se trouvent en Égypte. *In:* Savigny, M.J.-C.L. de (Ed.), *Description de l'Égypte, ou recueil des observations et des recherches qui ont été faites en Égypte pendant l'expédition de l'armée Française. Histoire naturelle. Tome premier. Partie premier.* Imprimerie Impériale, Paris (1809–1829), pp. 121–160.

Georgi, J.G. (1801) *Geographisch-Physikalische und naturhistorische Beschreibung des Russischen Reichs, zur uebersicht bisheriger Kenntnisse von demselben. Des dritten Theils sechster Band.* Friedrich Nicolovius, Königsberg 7(3), 1463–2122.

Gerlach, J. (2006) *The vertebrates of Seychelles: a field guide.* Justin Gerlach, 79 pp., 31 col. pls.

Germano, D.J. (2001) Geographic distribution: Serpentes: *Salvadora hexalepis. Herpetological Review* 32(1), 61.

Gervais, P. (1836) Énumération de quelques espèces de reptiles provenant de Barbarie. *Annales des Sciences Naturelles (Zoologie)*, Paris (Série 3) 6, 308–313.

Gervais, P. (1848) *Considérations générales sur la distribution géographique et la classification des reptiles vivants et fossiles.* Paris, 14 pp.

Gervais, P. (1857) Sur quelques ophidiens sur l'Algérie. *Mémoires de la Section des Sciences, Académie de Sciences et Lettres de Montpellier* 3, 511–512, pl.

Gervais, P. & Beneden, P.-J. van. (1859) *Zoologie médicale. Exposé méthodique du Régne Animal base sur l'anatomie, l'embryogénie et la paleontology comprenant la description des espèces employees en medicine de celles qui sont venimeuses et de celles qui sont parasites de l'homme et des animaux.* J.-B. Baillière et Fils, Paris, 455 pp.

Gharpurey, K.G. (1962) *The snakes of India and Pakistan.* Fifth edition. Popular Book Depot, Bombay, x + 156 pp., 29 figs., 16 pls. [first edition, 1935, x + 165 pp., 76 figs.; second edition, 1937; third edition, 1944, ix + 178 pp., 76 figs.; fourth edition, 1954, x + 154 pp.,76 figs.]

Ghate, H.V. & Yazdani, G.M. (1991) First record of the pied-belly uropelt *Melanophidium punctatum* Beddome (Serpentes, Uropeltidae) from Maharashtra. *The Journal of the Bombay Natural History Society* 87(3), 463, 1 fig.

Gherardini, F. (1988) *Serpenti Italiani velenosi e innocui.* Editoriale Olimpia, Firenze, 119 pp., col. illust.

Ghidini, G.M. (1958) *Serpenti d'Italia edi altri paesi.* La Scuola, Brescia, 60 pp., illust.

Ghodke, S. & Andrews, H.V. (2002) Recent record of *Cantoria violacea* (Girard, 1857) from North and Middle Andaman Islands, India with a note on its bite. (2001). *Hamadryad* 26(2), 371–373.

Giannasi, N., Malhotra, A. & Thorpe, R.S. (2001a) Nuclear and mtDNA phylogenies of the *Trimeresurus* complex: implications of the gene versus species tree debate. *Molecular Phylogenetics and Evolution* 19(1), 57–66.

Giannasi, N., Thorpe, R.S. & Malhotra, A. (2001b) The use of amplified fragment length polymorphism in determining species trees at fine taxonomic levels: analysis of a medically important snake, *Trimeresurus albolabris. Molecular Ecology* 10(2), 419–426, 5 figs.

Gibbons, J.W. & Dorcas, M.E. (2004) *North American watersnakes. A natural history.* University of Oklahoma Press, Norman, xxvi + 438 pp., col. photos.

Gibbons, J.W. & Dorcas, M.E. (2005) *Snakes of the Southeast.* University of Georgia Press, Athens, 253 pp., col. illust.

Gibbons, J.W. & Semlitsch, R.D. (1991) *Guide to the reptiles and amphibians of the Savannah River site.* University of Georgia Press, Athens, xii + 131 pp., 17 figs.

Gibbons, J.W. & West, P.J. (1998) *Snakes of Georgia and South Carolina.* Second edition. University of Georgia, Athens, 28 pp., col. illust.

Gibbons, J.W., Coker, J.W. & Murphy, T.M., Jr. 1977. Selected aspects of the life history of the rainbow snake (*Farancia erytrogramma*). *Herpetologica* 33(3), 276–281, 1 fig.

Gibbs, H.L., Corey, S.J., Blouin-Demers, G., Prior, K.A. & Weatherheads, P.J. (2006) Hybridization between mtDNA-defined phylogeographic lineages of black rat snakes (*Pantherophis* sp.). *Molecular Ecology* 15, 3755–3767, 4 figs.

Gibbs, H.L., Murphy, M. & Chiucchi, J.E. (2011) Genetic identity of endangered massasauga rattlesnakes (*Sistrurus* sp.) in Missouri. *Conservation Genetics* 12, 433–439, 3 figs.

Gibbs, J.P., Breisch, A.R., Ducey, P.K. & Behler, J. (2007) *The amphibians and reptiles of New York State.* Oxford Univ. Press, 422 pp., 62 col. pls.

Gibson, E. (1983) *The gamma-taxonomy of* Pseudechis guttatus *and considerations of affinity with* P. colletti *(Squamata, Serpentes, Elapidae).* Bachelor's Thesis, University of New England, Australia, 81 pp.

Giebel, C. (1862) Literatur: Zoologie. *Zeitschrift für die Gesammten Naturwissenschaften,* Berlin 20, 275–285.

Giglioli, E.H. (1873) Ricerche intorno alla distribuzione geografica generale o corologia degli animali vertebrati. *Stabilimento Giuseppe Civelli,* Roma, 200 pp.

Gigoux, E.E. (1940) Los ofidios Chilenos. *Boletin del Museo Nacional de Historia Natural,* Santiago 18, 5–7.

Gilhen, J. (1984) *Amphibians and reptiles of Nova Scotia.* Nova Scotia State Museum, Halifax, 162 pp., 162 figs., 8 col. pls.

Gill, B.J. (1993) The land reptiles of Western Samoa. *Journal of the Royal Society of New Zealand* 23(2), 79–89, illust.

Gill, B.J. (1995) Notes on the land reptiles of Wallis and Futuna, south-west Pacific. *Records of the Auckland Museum* 32, 55–61, 1 fig.

Gillam, M.W. (1979) The genus *Pseudonaja* (Serpentes, Elapidae) in the Northern Territory. *Territory Parks and Wildlife Commission, Research Bulletin* (1), 1–28, 15 figs.

Gillett, L. (1990) The reptiles and amphibians of Poland. *British Herpetological Society Bulletin,* London (31), 36–38.

Gillespie, G.R., Howard, S., Lockie, D., Scroggie, M. & Boeadi. (2005) Herpetofaunal richness and community structure of offshore islands of Sulawesi, Indonesia. *Biotropica* 37(2), 279–290, 6 figs.

Gillispie, C.C. (1992) Palisot de Beauvois on the Americans. *Proceedings of the American Philosophical Society* 136(1), 33–50.

Gilmore, C.W. (1938) Fossil snakes of North America. *Geological Society of America, Special Paper* (9), vii + 96 pp., 38 figs., 4 pls.

Gilmore, C.W. (1943) Fossil lizards of Mongolia. *Bulletin of the American Museum of Natural History* 81(4), 361–384, pl. 52.

Ginsburg, L. (1970) Les reptiles fossils. *In:* Grassé, P.-P. (Ed.), *Traité de zoologie: anatomie, systématique, biologie. Tome XIV (Reptiles). Fascicule III (Glands endocrines, embryologie, systématique, paleontologie).* Masson et Cie., Éditeurs, Paris, pp. 1161–1332.

Girard, C. (1854) Abstract of a report to Lieut. James M. Gilliss, U.S.N., upon the reptiles collected during the U.S.N. Astronomical Expedition to Chili. *Proceedings of the Academy of Natural Sciences of Philadelphia* (1854–1855) 7(6), 226–227.

Girard, C. (1855) Reptiles. *In:* Gilliss, J.M. (Ed.), *Report of the U.S. Naval Astronomical Expedition to the Southern Hemisphere, during the years 1849–'50–'51–'52. Appendix F. Zoology. Volume II.* A.O.P. Nicholson, Washington, pp. 207–229, pls. 24–27.

Girard, C. (1858a) Descriptions of some new reptiles, collected by the U.S. Exploring Expedition, under the command of Capt. Charles Wilkes, U.S.N. Third Part. Including the species of ophidians, exotic to North America. *Proceedings of the Academy of Natural Sciences of Philadelphia* (1857) 9(6), 181–182. [April 1858]

Girard, C. (1858b) Ophidia. 1855. *In:* Wilkes, C. (Ed.), *United States Exploring Expedition. During the years 1838, 1839, 1840, 1841, 1842. Under the command of Charles Wilkes, U.S.N. [Vol. XX.] Herpetology. With a folio atlas.* J.B. Lippincott, Philadelphia, pp. 103–190, pls. 7–15. [reprinted in 1978 by Arno Press New York]

Girardin, N. (1978) *Les serpents du Gabon.* Ministere de l'Education Nationale, Section des Sciences Naturelles, Libreville, 76 pp., figs.

Giraudo, A.R. (1994) Comentarios sobre las especies del genero *Liotyphlops* Peters (Serpentes, Anomalepididae) presentes en la Provincia de Misiones (Republica Argentina). *Cuadernos de Herpetología* 8(2), 229–233.

Giraudo, A.R. (1996) Geographic distribution: Serpentes: *Thamnodynastes chaquensis. Herpetological Review* 27(4), 215.

Giraudo, A.R. (1997) Geographic distribution: Serpentes: *Lystrophis matogrossensis. Herpetological Review* 28(3), 159.

Giraudo, A.R. (1999) New records of snakes from Argentina. *Herpetological Review* 30(3), 179–181.

Giraudo, A.R. (2002) *Serpientes de la Selva Paraense y del Choco Humedo.* Monograph 17, Literature of Latin America, Buenos Aires (2001) 285 pp., 54 figs., 28 col. pls.

Giraudo, A.R. & Arzamendia, V. (1997a) Clave para identificación de serpientes (Reptilia, Serpentes) de la Provincia de Santa Fe, Argentina. *Natura Neotropicalis* 28(2), 131–145, illust.

Giraudo, A.R. & Arzamendia, V. (1997b) Geographic distribution: Serpentes: *Clelia rustica. Herpetological Review* 28(3), 158–159.

Giraudo, A.R. & Scrocchi, G.J. (1998) A new species of *Apostolepis* (Serpentes, Colubridae) and comments on the genus in Argentina. *Herpetologica* 54(4), 470–476, 2 figs.

Giraudo, A.R. & Scrocchi, G.J. (2000) The genus *Atractus* (Serpentes: Colubridae) in north-eastern Argentina. *Herpetological Journal* 10, 81–90, 6 figs.

Giraudo, A.R. & Scrocchi, G.J. (2002) Argentinian snakes: an annotated checklist. *Smithsonian Herpetological Information Service* (132), 53 pp.

Giraudo, A.R., Couturier, G.A. & Di Bernardo, M. (1996) *Echinanthera cyanopleura* (Cope, 1885), a new record for the ophidiofauna of Argentina (Serpentes, Colubridae). *Cuadernos de Herpetología* 10(1–2), 72.

Giraudo, A.R., Arzamendia, V. & Cacciali, P. (2006) Geographic variation and taxonomic status of the southernmost population of *Liophis miliaris* (Linnaeus, 1758) (Serpentes: Colubridae). *Herpetological Journal,* London 16: 213–220.

Gisbert, J. & Garcia-Perea, R. (1986) Nuevas citas para la distribución de *Macroprotodon cucullatus* (Geoffroy, 1827) en la Península Ibérica. *Revista Española de Herpetología* 1, 177–185.

Gislen, T. & Kauri, H. (1959) Zoogeography of the Swedish amphibians and reptiles with notes on their growth and ecology. *Acta Vertebratica,* Stockholm 1(3), 193–397, 69 figs.

Gistel, J.N.F.X. (1848) *Naturgeschichte des Thierieichs. Für höhere schulen bearbeitet. Atlas.* Hoffaman'sche Verlags–Buchhandlung, Stuttgart, xvi + 216 pp, 32 pls.

Gistel, J.N.F.X. (1868) Die lurche Europas. Ein Beitrag zur Lehre von der geographischen Verbreitung derselben. *In:* Gistel, J. (Ed.), *Blicke in das Leben der Natur und des Menschen.* Verlag von Ed. Wartig, Leipzig, pp. 144–167.

Gistel, J.N.F.X. & Bromme, G. (1850) *Handbuch der Naturgeschichte aller dreireiche für lehrer und lernende, für Schule und Haus.* Hoffmann'sche Verlags–Buchhandlung, Stuttgart, 1037 pp.

Glaser, H.S.R. (1970) *The distribution of amphibians and reptiles in Riverside County, California.* Riverside Museum Press, Riverside, 40 pp., 13 figs.

Glass, B.P. (1946) A collection of reptiles from Hunan and Anhwei Provinces, China. *Copeia* 1946(4), 249–252.

Glauert, L. (1948) A western tiger snake, *Notechis scutatus occidentalis* subsp. nov. *Western Australian Naturalist,* Perth 1, 139–141.

Glauert, L. (1954) Herpetological miscellanea. III A new burrowing snake from north-western Australia. *Western Australian Naturalist,* Perth 4(4), 85, 1 fig.

Glauert, L. (1967) *A handbook of the snakes of Western Australia.* Third edition. Western Australia Naturalists' Club, Perth, 62 pp., 29 figs., 1 col. pl. [first edition, 1950, 50 pp., 15 figs., 1 col. pl.; second edition, 1957, 58 pp.]

Glaw, F. & Vences, M. (2007) *A field guide to the amphibians and reptiles of Madagascar.* Third edition. Vences & Glaw Verlags, Cologne, 496 pp., 1500 col. photos. [first edition, 1992, 331 pp., 340 figs., 127 col. pls.; second edition, 1994, 480 pp., 552 figs., 448 col. figs.]

Glaw, F., Franzen, M. & Vences, M. (2005a) A new species of colubrid snake (*Liopholidophis*) from northern Madagascar. *Salamandra* 41(1–2), 83–90, 5 figs.

Glaw, F., Vences, M. & Nussbaum, R.A. (2005b) A new species of *Heteroliodon* (Reptilia, Squamata, Colubridae) from Montagne des Français, far northern Madagascar. *Herpetologica* 61(3), 275–280, 2 figs.

Glaw, F., Nagy, Z.T., Franzen, M. & Vences, M. (2007a) Molecular phylogeny and systematics of the pseudoxyrhophiine snake genus *Liopholidophis* (Reptilia, Colubridae): evolution of its exceptional sexual dimorphism and descriptions of new taxa. *Zoologica Scripta* 36(4), 291–300, 6 figs.

Glaw, F., Nagy, Z.T. & Vences, M. (2007b) Phylogenetic relationships and classification of the Malagasy pseudoxyrhophiine snake genera *Geodipsas* and *Compsophis* based on morphological and molecular data. *Zootaxa* (1517), 53–62, 4 figs.

Glaw, F., Nagy, Z.T., Köhler, J., Franzen, M. & Vences, M. (2009) Phylogenetic relationships of a new species of pseudooxyrhophiine snake (Reptilia: Lamprophiidae: *Thamnosophis*) suggest a biogeographical link between western and northern Madagascar. *Organisms, Diversity & Evolution* 9, 13–22, 6 figs.

Gleadow, F. (1894) *Psammophis longifrons. The Journal of the Bombay Natural History Society* (1893–1894) 8(4), 553.

Glodek, G.S. & Voris, H.K. (1982) Marine snake diets: prey composition, diversity and overlap. *Copeia* 1982(3), 661–666.

Glowacinski, Z. & Rafinski, J. (2003) *Atlas of the amphibians and reptiles of Poland: status – distribution – conservation.* Biblioteka Monitoringu Srodowiska, Warszawa, 152 pp., 4 col. pls.

Gloyd, H.K. (1935) The subspecies of *Sistrurus miliarius. Occasional Papers of the Museum of Zoology, University of Michigan* (322), 1–7.

Gloyd, H.K. (1936a) A Mexican subspecies of *Crotalus molossus* Baird and Girard. *Occasional Papers of the Museum of Zoology, University of Michigan* (325), 1–5.

Gloyd, H.K. (1936b) The subspecies of *Crotalus lepidus*. *Occasional Papers of the Museum of Zoology, University of Michigan* (337), 1–6, 1 pl.

Gloyd, H.K. (1937) A herpetological consideration of faunal areas in southern Arizona. *Bulletin of the Chicago Academy of Sciences* 5(5), 79–136, 22 figs.

Gloyd, H.K. (1940) The rattlesnakes, genera *Sistrurus* and *Crotalus*. A study in zoogeography and evolution. *Chicago Academy of Science, Special Publication* (4), 1–270, 31 pls., 10 figs. [reprinted in 1978 by Society for the Study of Amphibians and Reptiles]

Gloyd, H.K. (1948) Description of a neglected subspecies of rattlesnake from Mexico. *Natural History Miscellanea*, Chicago (17), 1–4.

Gloyd, H.K. (1955a) A review of the massasaugas, *Sistrurus catenatus*, of the southwestern United States (Serpentes, Crotalidae). *Bulletin of the Chicago Academy of Sciences* 10(6), 83–98, 1 fig.

Gloyd, H.K. (1955b) A new crotalid snake from Kume Shima, Riu Kiu Islands. *Bulletin of the Chicago Academy of Sciences* 10(8), 123–134, 5 figs.

Gloyd, H.K. (1957) Distribution of the Mexican moccasin: a correction. *Copeia* 1956(4), 259.

Gloyd, H.K. (1969) Two additional subspecies of North American crotalid snakes, genus *Agkistrodon*. *Proceedings of the Biological Society of Washington* 82, 219–232, 6 figs.

Gloyd, H.K. (1972a) A subspecies of *Agkistrodon bilineatus* (Serpentes, Crotalidae) on the Yucatán peninsula, Mexico. *Proceedings of the Biological Society of Washington* 84, 327–334.

Gloyd, H.K. (1972b) The Korean snakes of the genus *Agkistrodon* (Crotalidae). *Proceedings of the Biological Society of Washington* 85(49), 557– 577, 6 figs.

Gloyd, H.K. (1977) Descriptions of new taxa of crotalid snakes from China and Ceylon (Sri Lanka). *Proceedings of the Biological Society of Washington* 90(4), 1002–1015.

Gloyd, H.K. (1979) A new generic name for the hundred-pace viper. *Proceedings of the Biological Society of Washington* 91(4), 963–964.

Gloyd, H.K. (1990) The Palearctic species of *Agkistrodon*. *In:* Gloyd, H.K. & Conant, R. (Eds.), *Snakes of the Agkistrodon complex. A monographic review*. Society for the Study of Amphibians and Reptiles, pp. 545– 561.

Gloyd, H.K. & Conant, R. (1934) The broad-banded copperhead, a new subspecies of *Agkistrodon mokasen*. *Occasional Papers of the Museum of Zoology, University of Michigan* (283), 1–6.

Gloyd, H.K. & Conant, R. (1938) The subspecies of the copperhead, *Agkistrodon mokasen* Beauvois. *Bulletin of the Chicago Academy of Sciences* 5, 163–166.

Gloyd, H.K. & Conant, R. (1943) A synopsis of the American forms of *Agkistrodon* (copperheads and moccasins). *Bulletin of the Chicago Academy of Sciences* 7, 147–170.

Gloyd, H.K. & Conant, R. (1982) The classification of the *Agkistrodon halys* complex. *Japanese Journal of Herpetology*, Kyoto 9, 75–78.

Gloyd, H.K. & Conant, R. (1990) *Snakes of the Agkistrodon complex. A monographic review.* Society for the Study of Amphibians and Reptiles, vi + 521 pp., 59 figs., 52 pls.

Gloyd, H.K. & Kauffeld, C.F. (1940) A new rattlesnake from Mexico. *Bulletin of the Chicago Academy of Sciences* 6(2), 11–14, 2 figs.

Glückselig, A.M. (1832) *Synopsis reptilium et amphibiorum Bohemiae.* Typis Filiorum Theophili Haase, Pragae, 68 pp.

Gmelin, J.F. (1789) *Caroli a Linne...Systema naturae per regna tria natural, secundum classes, ordines, genera, species, cum characteribus differentilis, synonymis, locis. Tomus I, Editio decima tertia, aucta, reformata. Pars III. Amphibia et Pisces.* Georg. Emanuel Beer, Lipsiae, 1038–1516.

Gmelin, J.F. (1799) Beschreibung zweyer neuer Schlangenarten. *Der Naturforscher*, Halle 28, 154–171.

Göçmen, B., Arikan, H., Mermer, A., Langerwerf, B. & Bahar, H. (2006) Morphological, hemipenial and venom electrophoresis. Comparisons of the Levantine viper, *Macrovipera lebertina* (Linnaeus, 1758), from Cyprus and southern Anatolia. *Turkish Journal of Zoology* 30, 225–234, 5 figs.

Göçmen, B., Arikan, H., Çícek, K. & Yildiz, Z. (2007a) A serological comparison of the populations of the Levantine viper, *Macrovipera lebetina* (Linnaeus, 1758) in Cyprus and southern Turkey. *South-Western Journal of Zoology* 3(2), 75–82, figs.

Göçmen, B., Nilson, G., Yildiz, M.Z., Arikan, H., Yalçinkaya, D. & Akman, B. (2007b) On the occurrence of the black cat snake, *Telescopus nigriceps* (Ahl, 1924) (Serpentes: Colubridae) from the southeastern Anatolia, Turkey with some taxonomical comments. *North-Western Journal of Zoology,* Oradea 3(2), 81–95, 4 figs.

Goddard, P. (1980) Limited movement areas and spatial behavior in the smoothsnake *Coronella austriaca* in southern England. *Proceedings of the European Herpetological Symposium 1980*, 25–40, 9 figs.

Goddard, P. (1984) Morphology, growth, food habits and population characteristics of the smooth snake *Coronella austriaca* in Britain. *Journal of Zoology*, London 204(2), 241–257, illust.

Godley, J.S. (1980) Foraging ecology of the striped swamp snake, *Regina alleni*, in southern Florida. *Ecological Monographs* 50(4), 411–436, 11 figs.

Godley, J.S. (1982) Predation and defensive behavior of the striped swamp snake (*Regina alleni*). *Florida Field Naturalist* 10(2), 31–36.

Goff, C.C. (1936) Distribution and variation of a new subspecies of water snake, *Natrix cyclopion floridana*, with a discussion of its relationships. *Occasional Papers of the Museum of Zoology, University of Michigan* (327), 1–9, pl. 1.

Golay, P. (1985) *Checklist and keys to the terrestrial proteroglyphs of the world (Serpentes, Elapidae - Hydrophiidae).* Elapsoïdea, Geneva, 90 pp., 4 figs.

Golay, P., Smith, H.M., Broadley, D.G., Dixon, J.R., McCarthy, C., Rage, J.-C., Schätti, B. & Toriba, M. (1993) *Endoglyphs and other major venomous snakes of the world. A checklist.* Azemiops Herpetological Data Center, xv + 478 pp.

Golay, P., Monney, J.-C., Conelli, A., Durand, T., Thiery, G., Zuffi, M.A.L. & Ursenbacher, S. (2008) Systematics of the Swiss asp viper, some implications for the European *Viper aspis* (Linnaeus, 1758) complex (Serpentes, Viperidae): a tribute to Eugen Kramer. *Amphibia-Reptilia,* Leiden 29(1), 71–83, 5 figs.

Golder, F. (1984) Haltung und Zucht von *Leptodeira nigrofasciata* Günther, 1868 (Serpentes, Colubridae). *Salamandra*, Frankfurt-am-Main 20(1), 3–10, 3 figs.

Goldberg, S.R. (1996) Reproduction in the saddled leafnose snake, *Phyllorhynchus browni*, and the spotted leafnose snake, *Phyllorhynchus decurtatus*, from Arizona. *Journal of Herpetology* 30(2), 280–282.

Goldberg, S.R. (1997) Reproduction in the western coral snake, *Micruroides euryxanthus* (Elapidae), from Arizona and Sonora, Mexico. *Great Basin Naturalist* 57(4), 363–365.

Goldfuss, G.A. (1820) Zweite Ordung Ophidii, Schlangen. *In: Handbuch der Zoologie. Erste Abtheilung*. Johann Leonhaard Schrag, Nürnberg, pp. 135–151.

Goldfuss, G.A. (1831) Beiträge zur Kenntniss verschiedener Reptilien der Vorwelt. *Nova Acta Physico-Medica Caesareae Leopoldino-Carolinae Naturae Curiosorum*, Halle 15, 61–128, pls. 7–13.

Goldman, J., Hill, L. & Stanbury, P.J. (1969) Type specimens in the Macleay Museum, University of Sydney. II. Amphibians and reptiles. *Proceedings of the Linnaean Society of New South Wales*, Sydney 93(3), 427–438.

Gomés, J.F. (1913) Uma nova cobra venenosa do Brasil. *Anais Paulistas de Medicina e Cirurgia*, São Paulo 1(3), 65–67, pl. 8.

Gomés, J.F. (1915) Contribuição para o conhecimento dos ofidios do Brasil. Descrição de quatro especies novas e um novo género de opistoglifos. *Anais Paulistas de Medicina e Cirurgia.*, São Paulo 4(6), 121–129, pls. 3–4.

Gomés, J.F. (1916) Ofidios. *In*: Neiva, A. & Penna, B.V. (Ed.), Viajem cientifica pelo norte da Bahia, sudoeste de Pernambuco, sul do Piauhí e de norte a sul de Goiáz. *Memorias do Instituto Oswaldo Cruz*, Rio de Janeiro 8(3), 101–103.

Gomés, J.F. (1918) Contribuição para o conhecimento dos ofidios do Brasil. III (1). 1. Ofidios do Muse Paraense. 2. Descrição de duas especies novas. *Memórias do Instituto Butantan,* São Paulo 1(1), 57–83, pl. 14.

Gomes, N. & Puorto, G. (1993) Atlas anatômico de *Bothrops jararaca* Wied, 1824 (Serpentes, Viperidae). *Memórias do Instituto Butantan,* São Paulo 55, 69–100, 24 figs.

Gómez, R.O., Báez, A.M. & Rougier, G.W. (2008) An anilioid snake from the Upper Cretaceous of northern Patagonia. *Cretaceous Research* 29(3), 481–488, 6 figs.

Gomille, A. (2002) *Die "Askulapnatter" Elaphe longissima.* Edition Chimaira, Frankfurt-am-Main, 158 pp., 96 figs.

Gonzales, L. & Embert, D. (2008) Geographic distribution: Serpentes: *Atractus snethlageae. Herpetological Review* 39(3), 370.

Gonzáles-Sponga, M.A. (1971) *Atractus emigdioi* (Serpentes, Colubridae) nueva especie para los Andes de Venezuela. *Instituto Pedagógico de Caracas, Monografias Ciencias* (3), 3–11, 5 figs.

González, D. (1991) Snakebite problems in Europe. *In:* Tu, A.T. (Ed.), *Handbook of natural toxins. Volume 5. Reptile venoms and toxins.* Marcel Dekker, New York, pp. 687–751, 29 figs.

González de la Vega, J.P. (1989) *Anfibios y reptiles de la Provincia de Huelva.* Second edition. Juan Pablo González de la Vega, Huelva, 238 pp., 209 figs., 287 photos. [first edition, 1988]

Goode, G.B. (1883) Bibliographies of American naturalists. I. The published writings of Spencer Fullerton Baird, 1843–1882. *Bulletin of the United States National Museum* (20), i–xvi + 1–377.

Goran, D. (1986) Zum vorkommen der Leopardnatter *Elaphe situla* (Linnaeus, 1758) auf Kea (NW-Cycladen) (Serpentes, Colubridae). *Salamandra*, Frankfurt-am-Main 22(2–3), 213–214, 1 fig.

Gordon, K. (1939) The Amphibia and Reptilia of Oregon. *Oregon State Monographs, Studies in Zoology* (1), 1–82, 54 figs.

Gordon, R.E. 1957. Remarks on the range, behavior and habitat of *Abastor erytrogrammus* (Latreille). *Proceedings of the Louisiana Academy of Sciences* 20, 35–37.

Gorham, S.W. (1970a) The rare Fiji snake *Ogmodon vitianus* Peters. *Agricultural Journal, Fiji Islands Department of Agriculture*, Suva 32, 49–51, 2 pls.

Gorham, S.W. (1970b) *Amphibians and reptiles of New Brunswick.* New Brunswick Museum, Halifax, 30 pp., illust.

Goris, R.C. (1966) *The reptiles of Japan*. Shogakkan, Tokyo, 127 pp, 24 pl. (in Japanese)

Goris, R.C. (1967) The reptiles and amphibians of Hachijojima Island. *Acta Herpetologica Japonica*, Kyoto 2, 25–30, 2 figs.

Goris, R.C. & Maeda, N. (2004) *Guide to the amphibians and reptiles of Japan.* Krieger Publishing, Malabar, 285 pp., 139 col. figs.

Goris, R.C. & Naganuma, K. (1979) Albino *Elaphe climacophora* occurring in southwestern Kanto. *Japanese Journal of Herpetology*, Kyoto 8(1), 1–7, 3 figs.

Gorman, G.C. (1965) The distribution of *Lichanura trivirgata* and the status of the species. *Herpetologica* 21(4), 283–287, 1 fig.

Gorman, G.C. (1981) The chromosomes of *Laticauda* and a review of karyotypic evolution in the Elapidae. *Journal of Herpetology* 15(2), 225–233, 2 figs.

Gorman, G.C., Licht, P. & McCollum, F. (1981) Annual reproductive patterns in three species of marine snakes from the central Philippines. *Journal of Herpetology* 15(3), 335–354, 5 figs.

Gorzula, S. & Ayarzagüena, J. (1996) Dos nuevas especies del género *Thamnodynastes* (Serpentes, Colubridae) de los tepuyes de la Guayana venezolana. *Publicaciones de la Asociación de los Amigos de Doñana* (1995)(6), 1–17, 2 figs.

Gorzula, S. & Señaris, J.C. (1998) Contribution to the herpetofauna of the Venezuelan Guayana. I. A data base. *Scientia Guaianae* (8), xviii + 268 pp., 18 figs., 129 col. pls.

Gosner, K.L. (1987) Observations on Lesser Antillean pit vipers. *Journal of Herpetology* 21(1), 78–80, 1 fig.

Gosse, P.H. (1851) A *naturalist's sojourn in Jamaica.* Longman, Brown, Green, and Longmans, London, xxvi + 508 pp.

Gotte, S.W. & Wilson, L.D. (2005) Commentary on the type material of *Tantilla gracilis* Baird and Girard, 1853 and *Tantilla nigriceps* Kennicott, 1860 (Reptilia, Squamata), with a neotype designation for *T. nigriceps. Proceedings of the Biological Society of Washington* 118(3), 596–604, 2 figs.

Gough, L.H. (1902) New snakes in the collections of the Zoological Institute of the University Strassburg. *Zoologischer Anzeiger*, Leipzig 25(681), 645–647.

Gough, L.H. (1907a) Description of a new species of *Atractaspis*, collected at Serowe, north eastern Kalahari. *Records of the Albany Museum*, Grahamstown 2, 178–179, 1 fig.

Gough, L.H. (1907b) Description of a new snake from the Transvaal. *Zoologischer Anzeiger*, Leipzig 32(15–16), 454–455.

Gough, L.H. (1908) South African snakes in the collections of the Transvaal Museum, Pretoria, the Albany Museum, Grahamtown, and the State Museum, Bloemfontein. *Annals of the Transvaal Museum*, Pretoria 1, 17–40, illust.

Gould, R.T. (1930) *The case for the sea-serpent.* Philip Allan, Detroit, 291 pp.

Gow, G.F. (1977a) A new species of *Python* from Arnhem Land. *Australian Zoologist*, Mosman 19(2), 133–139, 4 figs.

Gow, G.F. (1977b) *Snakes of the Darwin area.* Museums and Art Galleries Board of the Northern Territory, Darwin, 29 pp., 6 figs., 31 pls.

Gow, G.F. (1981) A new species of *Python* from central Australia. *Australian Journal of Herpetology*, Sydney 1(1), 29–34, 5 figs. [reprinted in 1994 by the Monitor, 6(2), 64–69, 5 figs.]

Gow, G.F. (1983) *Snakes of Australia.* Second edition Angus & Robertson, London, vii + 118 pp., 11 figs., 48 col. pls. [first edition, 1976, 88 pp., 5 figs., 83 col. pls.]

Gow, G.F. (1989) *Graeme Gow's complete guide to Australian snakes.* Angus & Robertson, North Ryde, 171 pp., figs., col. pls.

Gowanloch, J.N. (1943) Poisonous snakes of Louisiana. *In:* Gowanloch, J.N. & Brown, C.A. (Eds.), *Poisonous snakes, plants and black widow spider in Louisiana.* Louisiana Department of Conservation, New Orleans, pp. 7–62, 23 figs.

Gower, D.J. & Maduwage, K. (2011) Two new species of *Rhinophis* Hemprich (Serpentes: Uropeltidae) from Sri Lanka. *Zootaxa* (2881), 51–68, 8 figs.

Gower, D.J., Garrett, K. & Stafford P.J. (2012) *Snakes.* Firefly Books, Buffalo, 144 pp., col. illust.

Gower, D.J., Vidal, N., Spinks, J.N. & McCarthy, C.J. (2005) The phylogenetic position of Anomochilidae (Reptilia: Serpentes): first evidence from DNA sequences. *Journal of Zoological Systematics and Evolutionary Research* 43(4), 315–320, 2 figs.

Goyenechea, I. (1995) *Revisión taxonómica de los géneros Conophis Günther y Toluca Kennicott (Reptilia, Colubridae).* Master's thesis, Facultad de Ciencias, Universidad Nacional Autónoma de México, Mexico, 61 pp.

Goyenechea, I. & Flores-Villela, O. (2000) Designation of a neotype for *Conophis nasus* (Serpentes, Colubridae). *Copeia* 2000(1), 285–287, 1 fig.

Goyenechea, I. & Flores-Villela, O. (2002) Taxonomic status of the snake genera *Conopsis* and *Toluca* (Colubridae). *Journal of Herpetology* 36(1), 92–95, fig. 1.

Goyenechea, I. & Flores-Villela, O. (2006) Taxonomic summary of *Conopsis*, Günther, 1858 (Serpentes, Colubridae). *Zootaxa* (1271), 1–27, 7 figs.

Gradstein, F.M., Ogg, J.G. & Smith, A.G. (Eds.). (2004) *A geologic time scale 2004.* Cambridge University Press, Cambridge, 589 pp., illust.

Graham, G.L. (1977) The karyotype of the Texas coral snake, *Micrurus fulvius tenere. Herpetologica* 33(3), 345–348.

Graham, J.B., Rubinoff, I. & Hecht, M.K. (1971) Temperature physiology of the sea snake *Pelamis platurus*: an index of its colonization potential in the Atlantic Ocean. *Proceedings of the National Academy of Sciences* 68, 1360–1363.

Grande, L. (1980) Paleontology of the Green River formation with a review of the fish fauna. *Bulletin of the Geological Survey of Wyoming* 63, 1–334, figs.

Grandidier, A. (1867) Liste des reptiles nouveaux découverts, en 1866, sur la côte sud-ouest de Madagascar. *Revue et Magasin de Zoologie*, Paris (Série 2) 19(7), 232–234.

Grandidier, A. (1872) Descriptions de quelques reptiles nouveaux découverts a Madagascar en 1870. *Annales des Sciences Naturelles (Zoologie)*, Paris (Série 5) 15(20), 6–11.

Grandison, A.G.C. (1972) The Gunong Benom Expedition 1967. 5. Reptiles and amphibians of Gunong Benom with a description of a new species of *Macrocalamus. Bulletin of the British Museum of Natural History (Zoology)*, London 23(4), 43–101, 4 figs.

Grandison, A.G.C. (1978) Snakes of West Malaysia and Singapore. *Annalen des Naturhistorischen Museums in Wien* 81, 283–303.

Grant, C. (1931) A revised list of the herpetological fauna of Culebra Island. *Journal of the Department of Agriculture of Porto Rico*, San Juan 15(3), 215.

Grant, C. (1932a) The herpetology of St. John and adjacent keys, U.S. Virgin Islands. *Journal of the Department of Agriculture of Puerto Rico*, San Juan 16(3), 331–334.

Grant, C. (1932b) Herpetology of Tortola, notes on Anegada and Virgin Gorda, British Virgin Islands. *Journal of the Department of Agriculture of Puerto Rico*, San Juan 16(3), 339–346.

Grant, C. (1932c) The growth of herpetology in the Puerto Rico and Virgin Island area. *Journal of the Department of Agriculture of Puerto Rico*, San Juan 16(4), 401–404.

Grant, C. (1937) Herpetological notes with new species from the American and British Virgin Islands, 1936. *Journal of the Department of Agriculture of Puerto Rico*, San Juan 21(4), 503–522, pls. 12–15.

Grant, C. (1940a) Notes on the reptiles and amphibians of Jamaica, with diagnosis of new species and subspecies. *Jamaica To-Day*, P.M. Sherlock, London, pp. 151–157.

Grant, C. (1940b) The herpetology of Jamaica. II. The reptiles. *Bulletin of the Institute of Jamaica (Science Series)*, Kingston (1), 61–148, 3 figs.

Grant, C. (1940c) The herpetology of the Cayman Islands. *Bulletin of the Institute of Jamaica (Science Series)*, Kingston (2), 1–65, 6 pls.

Grant, C. (1946a) Notes on *Tretanorhinus* of Cuba and the Isle of Pines. *Journal of Agriculture of the University of Puerto Rico*, San Juan 30(2), 102–117, 4 figs.

Grant, C. (1946b) A new name for *Alsophis antillensis. Journal of Agriculture of the University of Puerto Rico*, San Juan 30(2), 118–125, figs. 5–6.

Grant, C. (1959a) New subspecies of *Alsophis* from Oriente, Cuba. *Herpetologica* 15(1), 58–60, 2 figs.

Grant, C. (1959b) The status of snakes of the genus *Arrhyton* in Cuba. *Herpetologica* 15(3), 129–133, 4 figs.

Grant, C., Smith, H.M. & Dalmau, P.A. (1959) The status of snakes of the genus *Arrhyton* in Cuba. *Herpetologica* 15(2), 129–133, 4 figs.

Grantsau, R. (1991) *Die Giftschlangen Brasiliens.* Bandeirante S.A. Gráfica e Editora, São Bernardo do Campo, 101 pp., 22 figs., 75 col. figs., 36 col. pls. [Portuguese edition, 1991]

Gras, U. (1988) Neuer Beleg fur *Lamprophis fuliginosus* (Boie, 1827) in Sudmarokko (Serpentes, Colubridae). *Salamandra*, Frankfurt-am-Main 24(1), 69–71, illust.

Grater, R.K. (1981) *Snakes, lizards & turtles of the Lake Mead region.* Southwest Parks and Monuments Association, Globe, 48 pp., col. illust.

Gravenhorst, J.L.C.C. (1807) *Vergleichende Uebersicht des Linneischen und einiger neuern zoologischen Systeme nebst dem eingeschalteten Verzeichnisse der zoologischen Sammlung des Versassers und den Beschreibungen neuer Thierarten, die in derselben vorhanden sind.* Heinrich Dieterich, Göttingen, viii + 476 pp.

Gravenhorst, J.L.C.C. (1832) *Das Zoologische Museum der Universität Breslau.* Gedrucht bei Grass, Barth und Co., Breslau, xviii + 188 pp.

Graves, L. (1847) *Essai sur la topographie géognostique du département de l'Oise.* Achille Desjardins, Beauvois, 804 pp.

Gravlund, P. (2000a) Geographic distribution: Serpentes: *Gastropyxis smaragdina. Herpetological Review* 31(1), 55.

Gravlund, P. (2000b) Georgraphic distribution. *Lycophidion taylori. Herpetological Review* 31(1), 56.

Gravlund, P. (2000c) Geographic distribution: Serpentes: *Prosymna ruspolii keniensis. Herpetological Review* 31(1), 57.

Gravlund, P. (2000d) Geographic distribution: Serpentes: *Telescopus obtusus. Herpetological Review* 31(1), 58.

Gravlund, P. (2001) Radiation within the advanced snakes (Caenophidia) with special emphasis on African opistoglyph colubrids, based on mitochondrial sequence data. *Biological Journal of the Linnean Society,* London 72(1), 99–114, 3 figs.

Gravlund, P., Berggreen, A. & Andersen, M. (1998) Geographic distribution, *Dasypeltis atra. Herpetological Review* 29(1), 53.

Gray, J.E. (1825) A synopsis of the genera of reptiles and Amphibia, with a description of some new species. *Annals of Philosophy*, London (Series 2) 10, 193–217.

Gray, J.E. (1827) Appendix B. Reptilia. *In:* King, P.P., *Narrative of a survey of the intertropical and western coasts of Australia performed between the years 1818 and 1822. With an appendix, containing various subjects relating to hydrography and natural history. Vol. 2.* John Murray, London, pp. 424–434.

Gray, J.E. (1831) A synopsis of the species of the class Reptilia. *In:* Griffith, E. & Pidgeon, E. (Eds.), *The animal kingdom arranged in conformity with its organization, by the Baron Cuvier, member of the Institute of France, &c. &c. &c. With additional descriptions of all the species hitherto named, and of many not before noticed. Volume the ninth. The class Reptilia arranged by the Baron Cuvier, with specific descriptions.* Whittaker, Treacher, London, pp. 1–110 (Appendix).

Gray, J.E. (1839) Reptiles. *In:* Richardson, J., Vigors, N.A., Lay, G.T., Bennett, E.T., Owen, R., Gray, J.E., Buckland, W. & Sowerby, G.B. (Eds.), *The zoology of Captain Beechey's voyage; compiled from the collections and notes made by Captain Beechey, the officers and naturalist of the expedition, during a voyage to the Pacific and Behring's Straits performed in His Majesty's ship Blossom under the command of Captain F.W. Beechey, R.N., F.R.S., &c. &c. in the years 1825, 26, 27, and 28.* Henry G. Bohn, London, pp. 93–94, pls. 29–32.

Gray, J.E. (1840) *Synopsis of the contents of the British Museum.* 42nd edition. G. Woodfall, London, 370 pp.

Gray, J.E. (1841a) Description of some new species and four new genera of reptiles from W. Australia, discovered by Mr. J. Gould. *Annals and Magazine of Natural History*, London 7(42), 86–91.

Gray, J.E. (1841b) Appendix E. A catalogue of the species of reptiles and Amphibia hitherto described as inhabiting Australia, with a description of some new species from Western Australia, and some remarks on their geographical distribution. *In:* Grey, G., *Journals of two expeditions of discovery in north-west and western Australia, during the years 1837, 38, and 39, under the authority of Her Majesty's Government, describing many newly discovered, important, and fertile districts, with observations on the moral and physical condition of the aboriginal inhabitants, &c. &c. Vol. II.* T. and W. Boone, London, pp. 422–449, 8 pls.

Gray, J.E. (1842a) Synopsis of the species of prehensile-tailed snakes, or family Boidae. *The Zoological Miscellany*, John E. Gray, London 2(March), 41–46.

Gray, J.E. (1842b) Synopsis of the species of rattle-snakes, or family of Crotalidae. *The Zoological Miscellany*, John E. Gray, London 2(March), 47–51.

Gray, J.E. (1842c) Description of some hitherto unrecorded species of Australian reptiles and batrachians. *The Zoological Miscellany*, John E. Gray, London 2(April), 51–55.

Gray, J.E. (1842d) Monographic synopsis of the water snakes, or the family Hydridae. *The Zoological Miscellany*, John E. Gray, London 2(May), 59–68.

Gray, J.E. (1842e) Monographic synopsis of the vipers, or the family Viperidae. *The Zoological Miscellany*, John E. Gray, London 2(May), 68–71.

Gray, J.E. (1843) Description of two new species of reptiles from the collections made during the voyages of H.M.S. Sulphur. *Annals and Magazine of Natural History*, London 11(68), 46.

Gray, J.E. (1845) *Catalogue of the specimens of lizards in the collections of the British Museum.* British Museum (Natural History), London, xxvii + 289 pp., 28 pls.

Gray, J.E. (1846a) Appendix. Descriptions of some new Australian reptiles. *In:* Stokes, J.L., *Discoveries in Australia; with an account of the coasts and rivers explored and surveyed during the voyage of H.M.S. Beagle, in the years 1837–38–39–40–41–42–43. Vol. I.* T. and W. Boone, London, pp. 498–504, 4 pls.

Gray, J.E. (1846b) A new genus of sea-snake from Port Essington. *Annals and Magazine of Natural History*, London 18(119), 284.

Gray, J.E. (1847) Description of a new genus of snakes, Appendix. *In:* Jukes, J.B., *Narrative of the surveying voyage of H.M.S. Fly, commanded by Captain F.P. Blackwood, R.N. in Torres Strait, New Guinea, and other islands of the eastern Archipelago, during the years 1842–1846. Vol. II.* T. W. Broone, London, pp. 332–334, 1 pl.

Gray, J.E. (1849a) *Catalogue of the specimens of snakes in the collection of the British Museum.* British Museum (Natural History), London, xv + 125 pp.

Gray, J.E. (1849b) Description of three new genera and species of snakes. *Annals and Magazine of Natural History*, London (Series 2) 4(22), 246–248.

Gray, J.E. (1853) Descriptions of some undescribed species of reptiles collected by Dr. Joseph Hooker, in the Khasia Mountains, E Bengal and Sikkim, Himalaya. *Annals and Magazine of Natural History*, London (Series 2) 12(72), 386–392.

Gray, J.E. (1858a) Notice of a new genus of Uropeltidae from Ceylon, in the collection of the British Museum. *Proceedings of the Zoological Society of London* 26(1), 57–58.

Gray, J.E. (1858b) Description of a new genus of Boidae from Old Calabar, and a list of W. African reptiles. *Proceedings of the Zoological Society of London* 26(1), 154–167, pl. 14.

Gray, J.E. (1858c) On a new genus and several new species of Uropeltidae, in the collection of the British Museum. *Proceedings of the Zoological Society of London* 26(1), 260–265.

Gray, J.E. (1858d) Notice of a new genus of Uropeltidae from Ceylon, in the collection of the British Museum. *Annals and Magazine of Natural History*, London (Series 3) 1(6), 440–441.

Gray, J.E. (1858e) Description of a new genus of Boidae from Old Calabar. *Annals and Magazine of Natural History*, London (Series 3) 2(10), 300–301.

Gray, J.E. (1858f) On a new genus and several new species of Uropeltidae, in the collection of the British Museum. *Annals and Magazine of Natural History*, London (Series 3) 2(11), 376–381.

Gray, J.E. (1858g) Appendix (2.) to the Catalogue of the specimens of snakes in the collection of the British Museum. First Part (Crotalidae, Viperidae, Hydridae and Boidae). *In:* Günther, A.C.L.G., *Catalogue of colubrine snakes in the collection of the British Museum.* British Museum (Natural History), London, pp. 265–281.

Gray, J.E. (1860a) Description of a new genus of Boidae discovered by Mr. Bates on the Upper Amazon. *Proceedings of the Zoological Society of London* 28(1), 132–133, pl. 24.

Gray, J.E. (1860b) Description of a new genus of Boidae discovered by Mr. Bates on the upper Amazon. *Annals and Magazine of Natural History*, London (Series 3) 6(32), 131–132.

Gray, J.E. & Hardwicke, T. (1830–1835) *Illustrations of Indian Zoology; chiefly selected from the collection of Major-General Hardwicke.* 2 Vols. Treuttel, Wurtz, Treuttel, Jun and Richter, London, 202 col. pls. [Publication dates of parts are as follows, 1–4 (1830), 5–9 (1831), 10–12 (1832), 13–18 (1834), 19–20 (1835) *fide* Sherborn, 1922, Kinnear, 1925, Dawson, 1946, and Sawyer, 1953].

Gray, J.E. & Neill, J. (1845) Catalogue of reptiles and fishes, found at King George's Sound. *In:* Eyre, E.J., *Journals of expeditions of discovery into central Australia and overland from Adelaide to King George's Sound in the years 1840–41*. T. & W. Boone, London 1, 412–418.

Grazziotin, F.G., Zaher, H., Murphy, R.W., Scrocchi, G., Benavides, M.A., Zhangg, Y.-P. & Bonatto, S.L. (2012) Molecular phylogeny of the New World Dipsadidae (Serpentes: Colubroidea): a reappraisal. *Cladistics* 28(5), 437–459, 3 figs.

Green, M. & Holman, J.A. (1977) A late Tertiary stream channel fauna from South Bijou Hill, South Dakota. *Journal of Paleontology* 51(3), 543–547, 1 fig.

Green, M.D. (2010) *Molecular phylogeny of the snake genus Oligodon (Serpentes: Colubridae), with an annotated checklist and key*. Master's Thesis, University of Toronto, Toronto, 161 pp.

Green, M.D., Orlov, N.L. & Murphy, R.W. (2010) Toward a phylogeny of the Kukri snakes, genus *Oligodon*. *Asian Herpetological Research* (Series 2) 1(1), 1–21, 10 figs.

Green, N.B. (1953) Snakes. *West Virginia Conservation* 1953, 8–10.

Green, N.B. & Pauley, T.K. (1987) *Amphibians and reptiles in West Virginia*. University of Pittsburg Press, Pittsburg, xi + 284 pp., 85 col. pls.

Greenbaum, E. & Komar, O. (2010) A conservation assessment of Salvadoran protected areas: priorities and recommendations based on amphibian and reptile distribution. *In:* Wilson, L.D., Townsend, J.H. & Johnson, J.A. (Eds.), *Conservation of Mesoamerican amphibians and reptiles*. Eagle Mountain Publishing, Eagle Mountain, pp. 436–459, 1 fig.

Greenbaum, E., Komar, O. & Bolaños, R. (2002) Geographic distribution: Serpentes: *Ninia espinali*. *Herpetological Review* 33(4), 325.

Greenbaum, E., Cellou-Baldé, M. & Carr, J.L. (2003) Geographic distribution: Serpentes: *Dendroaspis polylepis*. *Herpetological Review* 34(2), 168.

Greenbaum, E., Carr, J.L. & Almendáriz, A. (2004) Taxonomic status of *Tantilla equatoriana* Wilson and Mena 1980 (Serpentes, Colubridae). *Southwestern Naturalist* 49(4), 457–464, 3 figs.

Greene, B.D., Wilkinson, R.F., Jr. & Powell, R. (1998) *Micrurus limbatus*. *Catalogue of American amphibians and reptiles*, pp. 678.1–678.3, 2 figs.

Greene, B.T., Powell, R. & Henderson, R.W. (2003) *Mastigodryas bruesi*. *Catalogue of American amphibians and reptiles*, pp. 777.1–777.3, 1 col. fig.

Greene, H.W. (1969) Unusual pattern and coloration in snakes of the genus *Pliocercus* from Veracruz. *Journal of Herpetology* 3(1–2), 27–31, 2 figs.

Greene, H.W. (1989) Ecological, evolutionary, and conservation implications of feeding biology in Old World cat snakes, genus *Boiga* (Colubridae). *Proceedings of the California Academy of Sciences* 46(8), 193–207, 2 figs.

Greene, H.W. (1990) Defensive behavior and feeding biology of the Asian mock viper, *Psammodynastes pulverulentus* (Colubridae): a specialized predator on scincid lizards. *Chinese Herpetological Research*, Berkeley (1989) 2(2), 21–32, 3 figs.

Greene, H.W. (1997) *Snakes: the evolution of mystery in nature*. University of California Press, Berkeley, 351 pp., 4 figs., 313 col. photos.

Greene, H.W. & Burghardt, G.M. (1978) Behavior and phylogeny: constriction in ancient and modern snakes. *Science* 200(4337), 74–77.

Greene, H.W. & McDiarmid, R.M. (1981) Coral snake mimicry: does it occur? *Science* 213(4513), 1207–1212, 2 figs.

Greer, A.E. (1965a) A new subspecies of *Clelia clelia* (Serpentes, Colubridae) from the island of Grenada. *Breviora* (223), 1–6.

Greer, A.E. (1965b) Distribution notes for *Leptodeira punctata*, *Tropidodipsas philippii*, and *Tropidodipsas malacodryas* in northwestern Mexico. *Herpetologica* 21(3), 237.

Greer, A.E. (1997) *The biology and evolution of Australian snakes*. Surrey Beatty & Sons, Chipping Norton, xii + 358 pp., 181 figs.

Gregory, P.T. & Campbell, R.W. (1984) *The reptiles of British Columbia*. British Columbia Provincial Museum Handbook, Victoria, viii + 93 pp., illust.

Gregory, P.T. & Larsen, K.W. (1996) Are there any meaningful correlations of geographic life-history variation in the garter snake, *Thamnophis sirtalis*? *Copeia* 1996(1), 183–189, 1 fig.

Grenot, C. & Vernet, R. (1972) Place des reptiles dans l'ecosystème du désert pierreux, au Sahara occidental. *Bulletin de l'Association des Naturalistes Orléanais*, Orléans (Series 3) (5), 25–47, 2 figs., 13 pls.

Gressitt, J.L. (1936) New reptiles from Formosa and Hainan. *Proceedings of the Biological Society of Washington* 49, 117–122.

Gressitt, J.L. (1937) A new snake from southeastern China. *Proceedings of the Biological Society of Washington* 50, 125–128.

Gressitt, J.L. (1941a) On a collection of amphibians and reptiles made on Hainan Island. *Peking Natural History Bulletin* (1940–1941) 15(3), 175–193, 1 fig.

Gressitt, J.L. (1941b) Amphibians and reptiles from southeastern China. *The Philippine Journal of Science*, Manila 75(1), 1–59, 1 pl.

Griffin, L.E. (1909a) Two new species of snakes found in the Philippine Islands. *The Philippine Journal of Science*, Manila 4A(1), 55–56.

Griffin, L.E. (1909b) A list of snakes found in Palawan. *The Philippine Journal of Science*, Manila 4A(6), 595–601.

Griffin, L.E. (1910) A list of snakes from the island of Polillo, P. I., with descriptions of a new genus and two new species. *The Philippine Journal of Science*, Manila 5D(4), 211–215, 1 pl.

Griffin, L.E. (1911) A check-list and key of Philippine snakes. *The Philippine Journal of Science*, Manila 6D(5), 253–269, 2 figs.

Griffin, L.E. (1916) A catalog of the Ophidia from South America at present (June, 1916) contained in the Carnegie Museum with descriptions of some new species. *Memoirs of the Carnegie Museum* 7(3), 163–228, pl. 28.

Griffin, L.E. (1917) *Leptodeira albofusca* (Lacépède) a synonym of *Leptodeira annulata* (Linnaeus). *Annals of the Carnegie Museum* 11(1–2), 321–326.

Griffin, M. (2003) *Annotated checklist and provisional national conservation status of Namibian reptiles*. Namibia Science Society, Windhoek, 169 pp.

Griffith, E.J., Ibáñez-D.R. & Jaramillo, C.A. (2008) Geographic distribution: Serpentes: *Bothriechis lateralis*. *Herpetological Review* 39(3), 370.

Griffiths, K. (1987) *Reptiles of the Sydney region*. Three Sisters Productions, Winmalee, 120 pp., col. illust.

Griffiths, K. (1996) *Australia's reptiles and frogs: a guide to identification and captive care*. Topmill Pty. Limited, Marrickville, 80 pp., col. illust.

Griffiths, K. (2012) *Frogs and reptiles of the Sydney region*. Second edition. New Holland Publishers, French's Forest, 128 pp., col. illust. [first edition, 1997, University of New South Wales Press, Sydney, 128 pp., col. illust.]

Grillitsch, B. & Grillitsch, H. (1984) Zur Verbreitung der Amphibien und Reptilien im Gebiet des Neusiedlersees (Burgenland, Osterreich) unter besonderer Berucksichtigung der Verhaltnisse im westlichen Schifgurtel. *Annalen des Naturhistorischen Museums in Wien* (1982) 86, 29–64, illust.

Grillitsch, B. & Grillitsch, H. (1993) *Typhlops vermicularis* Merrem 1820 — Wurmschlange oder Blödange. *In:* Böhme, W. (Ed.), *Handbuch der Reptilien und Amphibien Europas. Band 3/I. Schlangen (Serpentes) I. (Typhlopidae, Boidae, Colubridae 1, Colubrinae).* AULA-Verlag, Wiesbaden, pp. 15–32, 3 figs.

Grillitsch, H. & Cabela, A. (1992) Die Arealgrenzen der Reptilien in Karnten (Osterreich), dargestellt durch den Verlauf ausgewahlter klimatischer Iso- und Grenzlinien. *Herpetozoa*, Wien 5(1–2), 41–49, illust.

Grillitsch, H. & Tiedemann, F. (1984) Zur Herpetofauna der griechischlen Inseln Kea, Spanopoula, Kithnos, Sifnos, Kitriani (Cycladen), Alonissos und Piperi (nordliche Sporaden). *Annalen des Naturhistorischen Museums in Wien* (1982) 86, 7–28, illust.

Grismer, L.L. (1990a) Relationships, taxonomy, and biogeography of the *Masticophis lateralis* complex in Baja California, México. *Herpetologica* 46(1), 66–77, 4 figs.

Grismer, L.L. (1990b) *Masticophis aurigulus* (Cope). *Catalogue of American amphibians and reptiles*, pp. 499.1–499.2, 2 figs.

Grismer, L.L. (1990c) A new long-nosed snake (*Rhinocheilus lecontei*) from Isla Cerralvo, Baja California Sur, México. *Proceedings of the San Diego Society of Natural History* (4), 1–7, 3 figs.

Grismer, L.L. (1993) The insular herpetofauna of the Pacific Coast of Baja California, México. *Herpetological Natural History* 1(2), 1–10, 4 figs.

Grismer, L.L. (1994) Geographic origins for the reptiles on islands in the Gulf of California, México. *Herpetological Natural History* 2(2), 17–40, 13 figs.

Grismer, L.L. (1997) The distribution of *Pituophis melanoleucus* and *P. vertebralis* in northern Baja California, México. *Herpetological Review* 28(2), 68–70, fig. 1.

Grismer, L.L. (1999) An evolutionary classification of reptiles on islands in the Gulf of California, México. *Herpetologica* 55(4), 446–469, 1 fig.

Grismer, L.L. (2001a) Comments on the taxonomy of gopher snakes from Baja California, a reply to Rodríguez-Robles and de Jesús-Escobar. *Herpetological Review* 32(2), 81–83, 1 fig.

Grismer, L.L. (2001b) An evolutionary classification and checklists of amphibians and reptiles on the Pacific islands of Baja California, México. *Bulletin of the Southern California Academy of Sciences* 100(1), 12–23, 1 fig.

Grismer, L.L. (2002) *Amphibians and reptiles of Baja California including its Pacific islands and the islands in the Sea of Cortes.* University of California Press, Berkeley, xii + 399 pp., col. illust.

Grismer, L.L. (2011) *Amphibians and reptiles of the Seribuat Archipelago (Peninsular Malaysia) – A field guide.* Second edition. Edition Chimaira, Frankfurt-am-Main, 239 pp, 218 col. photos. [first edition, 2005, Forestry Department Peninsular Malaysia, Kuala Lumpur, 215 pp., 172 col. photos]

Grismer, L.L & Hollingsworth, B.D. (2000) Geographic distribution: Serpentes: *Lampropeltis getula. Herpetological Review* 31(1), 56.

Grismer, L.L. & Khang Aun, P. (2008) Diversity, endemism and conservation of the amphibians and reptiles of southern Peninsular Malaysia and its offshore islands. *Herpetological Review* 39(3), 270–281, 25 figs.

Grismer, L.L. & Mahrdt, C.R. (1996) Geographic distribution: Serpentes: *Masticophis lateralis lateralis. Herpetological Review* 27(1), 34.

Grismer, L.L., McGuire, J.A. & Hollingsworth, B.D. (1994) A report on the herpetofauna of the Vizcaino Peninsula, Baja California, Mexico, with a discussion of its biogeographic and taxonomic implications. *Bulletin of the Southern California Academy of Sciences* 93(2), 45–80.

Grismer, L.L., Wong, H. & Galina-Tessaro, P. (2002a) Geographic variation and taxonomy of the sand snakes, *Chilomeniscus* (Squamata, Colubridae). *Herpetologica* 58(1), 18–31, 6 figs.

Grismer, L.L., McGuire, J.A., Sosa, R.A. & Kaiser, H. (2002b) Revised checklist and comments on the terrestrial herpetofauna of Pulau Tioman, Peninsular Malaysia. *Herpetological Review* 33(1), 26–29.

Grismer, L.L., Rassmusen, C. & McGuire, J.A. (2003a) Geographic distribution: Serpentes: *Sibynophis melanocephalus. Herpetological Review* 34(3), 266.

Grismer, L.L., Das, I. & Leong, T.M. (2003b) A new species of *Gongylosoma* (Squamata, Colubridae) from Pulau Tioman, West Malaysia. *Herpetologica* 59(4), 565–572, 4 figs.

Grismer, L.L., Kaiser, H. & Yaskob, N.S. (2004) A new species of reed snake of the genus *Calamaria* H. Boie, 1827, from Pulau Tioman, Pahang, West Malaysia. *Hamadryad* 28(1–2), 1–6, 4 figs.

Grismer, L.L., Grismer, J.L. & McGuire, J.A. (2006a) A new species of the genus *Popeia* (Squamata, Viperidae) from Pulau Tioman, Pahang, West Malaysia. *Zootaxa* (1305), 1–19, 4 figs.

Grismer, L.L., Youmans, T.M., Wood, P.L., Jr. & Grismer, J.L. (2006b) Checklist of the herpetofauna of the Seribuat Archipelago, West Malaysia with comments on biogeography, natural history, and adaptive types. *Raffles Bulletin of Zoology* 54(1), 157–180, 7 figs.

Grismer, L.L., Neang, T., Chau, T., Wood, P.L., Jr., Oaks, J.R., Holden, J., Grismer, J.L., Szutz, T.R. & Youmans, T.M. (2008a) Additional amphibians and reptiles from the Phnom Sankos Wildlife Sanctuary in northwestern Cardamon Mountains, Cambodia, with comments on their taxonomy and the discovery of three new species. *Raffles Bulletin of Zoology* 56(1), 161–175, 20 col. figs.

Grismer, L.L., Ngo, V.T. & Grismer, J.L. (2008b) A new species of insular pitviper of the genus *Cryptelytrops* (Squamata, Viperidae) from southern Vietnam. *Zootaxa* (1715), 57–68, 2 figs.

Grismer, L.L., Onn, C.K., Grismer, J.L., Wood, P.L. & Norhayati, A. (2010) A checklist of the herpetofauna of the Banjaran Bintang, Peninsular Malaysia. *Russian Journal of Herpetology* 17(2), 147–160, 1 fig.

Gritis, P. & Voris, H.K. (1990) Variability and significance of parietal and ventral scales in the marine snakes of the genus *Lapemis* (Serpentes, Hydrophiidae), with comments on the occurrence of spiny scales in the genus. *Fieldiana: Zoology* (Series 2) (56), i–iii + 1–13, 8 figs.

Grobman, A.B. (1941) A contribution to the knowledge of variation in *Opheodrys vernalis* (Harlan), with the description of a new subspecies. *Miscellaneous Publications, Museum of Zoology, University of Michigan* (50), 1–38.

Grobman, A.B. (1984) Scutellation variation in *Opheodrys aestivus. Bulletin of the Florida State Museum, Biological Sciences* 29(3), 153–170, 2 figs.

Grobman, A.B. (1991) Does the smooth green snake occur in Missouri? *Transactions of the Missouri Academy of Sciences* 25, 1–3.

Grobman, A.B. (1992) Metamerism in the snake *Opheodrys vernalis*, with a description of a new subspecies. *Journal of Herpetology* 26(2), 175–186, 2 figs.

Grocott, R.G. & Sadler, G.G. (1958) *The poisonous snakes of Panama*. Panama City, 38 pp., 24 figs.

Grodzinski, Z. (1928) Rozwój naczyn krwionosnych u zaskronca. — Die Blut-gefässentwicklung bei der Natter, *Tropidonotus natrix* L. *Mémoires de l'Académie Polonaise des Sciences et des Lettres* (1), 1–110, 5 figs., 10 pls.

Groen, J. (2008) Het *Boiga*-genus. *Lacerta* 66(1–3), 64–79, 10 col. photos.

Groen, J. (2010a) De cobra's van het genus *Naja* in Azie. Deel 1. *Lacerta* 68(1–2), 4–27, 16 col. photos.

Groen, J. (2010b) De cobra's van het genus *Naja* in Azie. Deel 2. *Lacerta* 68(5), 182–203, 20 col. photos.

Grognet, J.M., Gatineau, E. Bougis, P.E., Harvey, A.L., Coudere, J., Fromageot, P. & Ménez, A. (1986) Two neutralizing monoclonal antibodies specific for *Naja nigricollis* cardiotoxin: preparation, characterization and localization of the epitopes. *Molecular Immunology* 23(12), 1329–1337, 4 figs.

Gronovius, L.T. (1756) *Musei Ichthyologici, Tomus secundus sistens Piscium indigenorum & nonnullorum exoticorum, quorum maxima pars in Museo Laurentii Theodori Gronovii, J.U.D., adservantur, nec non quorumdam in aliis Museis observatorum descriptiones. Accedunt nonnullorum exoticorum Piscium icones aeri incisae, et Amphibiorum Animalium, Historia Zoologica*. Theodorum Haak, Lugduni Batavorum, 8 + 88 pp., pls. 5–7.

Gronovius, L.T. (1763) *Zoophylacii Gronoviani fasciculus primus exhibens Animalia Quadrupeda, Amphibia atque Pisces, quae in Museo suo adservat, rite examenavit, systematice disposuit, descripsit, atque iconibus illustravit*. Theodorum Haak, Leiden, iv + 136 pp., 13 pls.

Groombridge, B.C. (1979) On the vomer in Acrochordidae (Reptilia, Serpentes), and its cladistic significance. *Journal of Zoology*, London 189(4), 559–567, 3 figs.

Groombridge, B.C. (1980) *A phyletic analysis of viperine snakes*. Doctoral Dissertation, London Polytechnic, London, 249 pp., 178 figs.

Groombridge, B.C. (1986) Phyletic relationships among viperine snakes. *In:* Rocek, Z. (Ed.), *Studies in Herpetology*. Charles University, Prague, pp. 219–222.

Grosselet, O., Vauche, M., Gupta, A. & Gupta, S. (2004) *Bungarus niger* Wall, 1908 (Reptilia, Serpentes, Elapidae): extension of range to Cachar District, Assam, India. *Russian Journal of Herpetology* 11(1), 10–11.

Grossenbacher, K. & Brand, M. (1973) *Schüssel zur Bestimmung der Amphibien und Reptilien der Schweiz*. Naturhistorisches Museum, Bern, 55 pp.

Grossmann, W. & Tillack, F. (2003) On the taxonomic status of *Asthenodipsas tropidonotus* (Van Lidth de Jeude, 1923) and *Pareas vertebralis* (Boulenger, 1909) (Serpentes, Colubridae, Pareatinae). *Russian Journal of Herpetology* 10(1), 175–190, 9 figs.

Grossmann, W., Kowalski, T., Zwanzig, B.M. & Zilger, H.-J. (2012) Ergänzende herpetologische Beobachtungen auf dem Said-Plateau und im Jebel al-Akhdar, Sultanat Oman. *Sauria*, Berlin 34(4), 3–18, 14 figs.

Groves, J.D. (1995) Reproduction and feeding behavior of *Oxybelis wilsoni*, a new species of vine snake (Serpentes, Colubridae). *Revista de Biologia Tropical*, San José 43(1–3), 307–309.

Gruber, U. (1974) Zur Taxonomic und Ökologie der Reptilien von der Insel Antikythira. *Salamandra*, Frankfurt-am-Main 10(1), 31–41, 1 fig., 5 pls.

Gruber, U. (2009) *Die Schlangen Europas alle Arten Europas und es Mittelmeerraums*. Second edition. Franckh-Kosmos Verlags-GmbH, Stuttgart, 264 pp., 217 col. photos. [first edition, 1989, by Franckh'sche Verlagshandlung, Stuttgart, 248 pp., 37 figs., 197 col. photo; French translation in 1992 by Delachaux et Niestlé, Lausanne]

Gruber, U. & Pace, H. (1978) *Lebende Schlangen: Riesenschlangen, Nattern, Vipern aus funf Erdteilen*. Zoologischen Staatssammlung, München, 48 pp., illust.

Gruschwitz, M. (1978) Untersuchungen zu Vorkommen und Lebensweise der Würfelnatter (*Natrix t. tessellata*) im Bereich der Flüsse Mosel und Lahn (Rheinland-Pfalz) (Reptilia, Serpentes, Colubridae). *Salamandra*, Frankfurt-am-Main 14(2), 80–89, 5 figs.

Gruschwitz, M. (1986) Notes on the ecology of the dice snake, *Natrix tessellata* Laur. in West Germany. *In:* Rocek, Z. (Ed.), *Studies in herpetology*. Charles University, Prague, pp. 499–502.

Gruschwitz, M., Lenz, S. & Böhme, W. (1991) Zur Kenntnis der Herpetofauna von Gambia (Westafrika) Teil 2, Schlangen (Reptilia, Serpentes) herpetofaunistische Bewertung. *Herpetofauna*, Weinstadt 13(75), 27–34, illust.

Gu, H.-Q., Jin, Y.-L. & Geng, J.-S. (1982) A herpetological survey on the isles belonging to Dinghai County, Zhejiang. *Acta Herpetologica Sinica*, Chengdu (Series 2) 1(1), 89–91.

Gu, H.-Q., Ji, X. & Fan, C.-L. (1987) A new record of the snakes of Zhejiang—*Calamaria pavimentata*. *Acta Herpetologica Sinica*, Chengdu (Series 2) 6(2), 76.

Guardado, J. & Gatica-Colima, A. (2009) Geographic distribution: Serpentes: *Micruroides euryxanthus*. *Herpetological Review* 40(4), 456–457.

Gudynas, E. (1979) Notes on the ecology of *Lystrophis dorbignyi* in Uruguay. *ASRA Journal* 1, 24–33.

Gudynas, E. (1989) Amphibians and reptiles of a coastal periurban ecosystem (Solymar, Uruguay): list, preliminary analysis of community structure, and conservation. *Bulletin of the Maryland Herpetological Society* 25(3), 85–123, illust.

Guedes, T.B. (2010) Geographic distribution: Serpentes: *Thamnodynastes almae*. *Herpetological Review* 41(2), 245.

Guedes, T.B. & Marques, O.A.V. (2011) Reptilia, Squamata, Serpentes, Dipsadidae, *Tropidodryas striaticeps* (Cope, 1869): latitudinal and altitudinal extension and geographic distribution map. *Check List* 7(1), 78–82, 2 figs.

Guérin-Méneville, F.-E. (1829–1844) *Iconographie du Règne animal de G. Cuvier ou représentation d'après nature de l'une des espèces les remarquables et souvent non encore figurées, de chaque genre d'animaux. Avec un texte descriptif mis au courant de la science. Tome 1. Planches des animaux vertebrés*. J.B. Baillière, Paris, 215 pls.

Guibé, J. (1948) Sur le dimorphisme sexuel des espèces du genre *Langaha* (Ophidiens). *Comptes Rendus de l'Académie des Sciences*, Paris 226, 1219–1220.

Guibé, J. (1949a) Révision des Boides de Madagascar. *Mémoires de l'Institut Scientifique de Madagascar*, Tananarive 3A(1), 95–105, 3 figs.

Guibé, J. (1949b) Révision du genre *Langaha* (Ophidiens): le dimorphisme ssexual, ses consequences taxonomiques. *Mémoires de l'Institut Scientifique de Madagascar*, Tananarive 3A(2), 147–155, 2 figs.

Guibé, J. (1952) *Typhlops angeli* (Serpent), espèce nouvelle Mont Nimba. *Bulletin du Muséum National d'Histoire Naturelle*, Paris (Série 2) 24(1), 79.

Guibé, J. (1954) Étude de *Liopholidophis lateralis* (D. et B.) et description d'une espece nouvelle [Reptiles]. *Mémoires de l'Institut Scientifique de Madagascar*, Tananarive 9A, 241–246, 2 figs.

Guibé, J. (1957) Reptiles d'Iran récoltés par M. Francis Petter. Descriptions d'un vipéridé nouveau, *Pseudocerastes latirostris* n. sp. *Bulletin du Muséum National d'Histoire Naturelle* Paris (Série 2)29, 136–142.

Guibé, J. (1958) Les serpents de Madagascar. *Mémoires de l'Institut Scientifique de Madagascar*, Tananarive 12A, 189–260, 42 figs.

Guibé, J. (1970) La systématique des reptiles actuels. *In:* Grassé, P.-P. (Ed.), *Traité de zoologie: anatomie, systématique, biologie. Tome XIV (Reptiles). Fascicule III (Glands endocrines, embryologie, systématique, paleontologie).* Masson & Cie., Éditeurs, Paris, pp. 1054–1160, figs. 707–808.

Guibé, J. & Roux-Estevé, R. (1972a) Les types de Schlegel (ophidiens), présents dans les collections du Muséum National d'Histoire Naturelle de Paris. *Zoologische Mededelingen*, Leiden 47(9), 129–134.

Guibé, J. & Roux-Estevé, R. (1972b) Les espèces ouest-africaines du genre *Lycophidion* (Serpentes, Colubridae). *Zoologische Mededelingen*, Leiden 47(29), 391–400, 5 figs., 2 pls.

Guibé, J., Roux-Estève, R. & Villiers, A. (1967) *Typhlops koniagui* = *Leptotyphlops koniagui* (Serpentes). *Bulletin du Muséum National d'Histoire Naturelle*, Paris (Série 2) 39(3), 452–453.

Guichenot, A. (1850) *Exploration scientifique de l'Algérie pendant les années 1840, 1841, 1842 publiée par ordre du gouvernement et avec le concours d'une Commission Académique. Sciences physiques, Zoologie. V. Histoire naturelle des reptiles et des poissons de l'Algérie. Atlas.* Imprimerie National, Paris, 144 pp., 12 pls.

Guichenot, A. (1855) *Animaux nouveaux ou rares recueillis pendant l'expédition dans les parties centrales de l'Amérique du Sud, de Rio de Janeiro aà Lima, et de Lima au Pará; executee par ordre du gouvernement Français pendant les années 1843 à 1847 sous la direction du Compte Francis de Castelnau. Reptiles.* P. Bertrand, Paris, 96 pp., 18 col. pls.

Guicking, D., Lawson, R., Joger, U. & Wink, M. (2006) Evolution and phylogeny of the genus *Natrix* (Serpentes: Colubridae). *Biological Journal of the Linnean Society*, London 87(1), 127–143, 5 figs.

Guicking, D., Joger, U. & Wink, M. (2008) Molecular phylogeography of the viperine snake (*Natrix maura*). *Organisms, Diversity and Evolution* 8, 130–145.

Guicking, D., Joger, U. & Wink, M. (2009) Cryptic diversity in a Eurasian water snake (*Natrix tessellata*, Serpentes: Colubridae): evidence from mitochondrial sequence data and nuclear ISSR-PCR fingerprinting. *Organisms, Diversity and Evolution* 9, 201–214.

Guiher, T.J. & Burbrink, F.T. (2008) Demographic and phylogeographic histories of two venomous North American snakes of the genus *Agkistrodon*. *Molecular Phylogenetics and Evolution* 48(2), 543–553, 5 figs.

Guillon, M. (1936) Observations sur quelques batraciens, lézards et serpents en captivité. *Bulletin de la Société Nationale d'Acclimatation de France*, Paris 83(9–10), 196–197.

Guimarães-Guedes, A. (2000) Geographic distribution: Serpentes: *Micrurus frontifasciatus*. *Herpetological Review* 31(4), 255.

Guimarães-Guedes, A. (2003) Geographic distribution: Serpentes: *Micrurus lemniscatus carvalhoi*. *Herpetological Review* 34(2), 169.

Guimarães-Guedes, A., Rodrigues-Maia, A. & Yuki, R.N. (2002) Geographic distribution: Serpentes: *Hydrops triangularis neglectus*. *Herpetological Review* 33(4), 324.

Guinea, M.L., Tamiya, N. & Cogger, H.G. (1983) The neurotoxins of the sea snake *Laticauda schistorhynchus*. *Biochemical Journal* 213, 39–41, 1 fig.

Gumprecht, A. (2002) Die Bambusottern der Gattung *Trimeresurus* Lacépède. Teil VI: Die philippinischen Bambusottern des *Trimeresurus flavomaculatus*-Komplexes. II: Die Batanes-Bambusotter *Trimeresurus mcgregori* Taylor, 1919. *Sauria*, Berlin 24(3), 31–44, 21 col. figs.

Gumprecht, A. (2004a) *Die Blumennatter,* Orthriophis moellendorffi (Elaphe moellendorffi). Natur und Tier-Verlag, Münster, 61 pp., col. illust.

Gumprecht, A. (2004b) *Die Königs-Kletternatter,* Elaphe carinata. Natur und Tier-Verlag, Münster, 61 pp., col. illust.

Gumprecht, A. (2004c) *Die Mandarinnatter,* Euprepiophis mandarinus (Elaphe mandarina). Natur und Tier-Verlag, Münster, 63 pp., col. illust.

Gumprecht, A. (2004d) *Die Schönnatter,* Orthriophis taeniurus (Elaphe taeniura). Natur und Tier-Verlag, Münster, 61 pp., col. illust.

Gumprecht, A. (2004e) *Die Spitzkopfnattern, Die Gattung* Gonyosoma. Natur und Tier-Verlag, Münster, 63 pp., col. illust.

Gumprecht, A. (2004f) *Die Weißlippen-Bambusotter,* Cryptelytrops albolabris (Trimeresurus albolabris). Natur und Tier-Verlag, Münster, 63 pp., col. illust.

Gumprecht, A. & Tillack, F. (2004) Proposal for a replacement name of the snake genus *Ermia* Zhang, 1993. *Russian Journal of Herpetology* 11(1), 73–76, 3 col. figs.

Gumprecht, A. & Toenjes, E. (2007) *Die Tempelotter,* Tropidolaemus wagleri. Natur und Tier-Verlag, Münster, 63 pp., col. illust.

Gumprecht, A., Tillack, F., Orlov, N.L., Captain, A. & Ryabov, S. (2004) *Asian pitvipers*. Geitje Books, Berlin, 368 pp., col. illust.

Gundlach, J. (1840) Ueber zwei von mir gesammelte Boen (*Boa pardalis*) Wiegmann. *Archiv für Naturgeschichte*, Berlin 6(1), 359–361.

Gundlach, J. (1867) Revista y catálogo de los reptiles Cubanos. *Anuario de la Sección de Ciencias Físicas y Naturales del Liceo de Matanzas* (1866) 2, 102–122.

Gundlach, J. (1875) Catálogo de los reptiles cubanos. *Anales de la Sociedad Española de Historia Natural*, Madrid 4, 347–368.

Gundlach, J. (1880) *Contribucion a la erpetología cubana*. G. Montiel, La Habana, 98 pp.

Günther, A.C.L.G. (1858) *Catalogue of colubrine snakes in the collection of the British Museum.* British Museum (Natural History), London, xvi + 281 pp.

Günther, A.C.L.G. (1859a) Description of a new genus of west African snakes, and revision of the South American *Elaps*. *Annals and Magazine of Natural History*, London (Series 3) 4(21), 161–174, pl. 4.

Günther, A.C.L.G. (1859b) On the genus *Elaps* of Wagler. *Proceedings of the Zoological Society of London* 27(1), 79–89, pls. 16–18.

Günther, A.C.L.G. (1859c) Second list of cold-blooded Vertebrata collected by Mr. Fraser in the Andes of western Ecuador. *Proceedings of the Zoological Society of London* 27(1), 402–422, pl. 20.

Günther, A.C.L.G. (1860a) On a new snake from the Galapagos Islands. *Proceedings of the Zoological Society of London* 28(1), 97–98.

Günther, A.C.L.G. (1860b) On the reptiles of Siam. *Proceedings of the Zoological Society of London* 28(1), 113–117, 1 pl.

Günther, A.C.L.G. (1860c) Contributions to a knowledge of the reptiles of the Himalaya Mountains. I. Descriptions of the new species. *Proceedings of the Zoological Society of London* 28(1), 148–166, pls. 25–28.

Günther, A.C.L.G. (1860d) Contributions to a knowledge of the reptiles of the Himalaya Mountains. III. List of the Himalayan reptiles according to their altitudinal distribution, and remarks on it. *Proceedings of the Zoological Society of London* 28(1), 167–175.

Günther, A.C.L.G. (1860e) Third list of cold-blooded Vertebrata collected by Mr. Fraser in Ecuador. *Proceedings of the Zoological Society of London* 28(1), 233–240, pl. 10.

Günther, A.C.L.G. (1860f) Description of *Homalocranium laticeps*, a new snake from Carthagena. *Proceedings of the Zoological Society of London* 28(1), 240–241.

Günther, A.C.L.G. (1860g) On new reptiles and fishes from Mexico. *Proceedings of the Zoological Society of London* 28(1), 316–319.

Günther, A.C.L.G. (1860h) On a West-African genus of snakes (*Meizodon*). *Proceedings of the Zoological Society of London* 28(1), 427–430, 3 figs.

Günther, A.C.L.G. (1860i) Description of *Uriechis microlepidotus*, a new snake from South Africa. *Annals and Magazine of Natural History*, London (Series 3) 5(27), 168–169, pl. 9.

Günther, A.C.L.G. (1860j) Description of *Leptodeira torquata*, a new snake from Central America. *Annals and Magazine of Natural History*, London (Series 3) 5(27), 169–171, pl. 10.

Günther, A.C.L.G. (1860k) On a new snake from the Galapagos Islands. *Annals and Magazine of Natural History* (Series 3) 6(31), 78–79.

Günther, A.C.L.G. (1860l) Note on *Psammophis perroteti*. *Annals and Magazine of Natural History*, London (Series 3) 6(36), 428–429, pl. 7, figs. 5–7.

Günther, A.C.L.G. (1861a) On a West-African genus of snakes (*Meizodon*). *Annals and Magazine of Natural History*, London (Series 3) 7(39), 223–226, 6 figs.

Günther, A.C.L.G. (1861b) Account of the reptiles sent by Dr. Wucherer from Bahia. *Annals and Magazine of Natural History*, London (Series 3) 7(42), 412–417, 3 figs.

Günther, A.C.L.G. (1861c) On the ophidian genus *Helicops*. *Annals and Magazine of Natural History*, London (Series 3) 7(42), 425–428, 1 fig.

Günther, A.C.L.G. (1861d) On a new species of the family Boidae. *Annals and Magazine of Natural History*, London (Series 3) 8(44), 189.

Günther, A.C.L.G. (1861e) Account of the reptiles sent by Dr. Wucherer from Bahia. *Proceedings of the Zoological Society of London* 29(1), 12–17, 3 figs. [reprint of A. Günther, 1861b]

Günther, A.C.L.G. (1861f) On a new species of the family Boidae. *Proceedings of the Zoological Society of London* 29(1), 142, pl. 23.

Günther, A.C.L.G. (1861g) Second list of Siamese reptiles. *Proceedings of the Zoological Society of London* 29(1), 187–189.

Günther, A.C.L.G. (1861h) List of the cold-blooded Vertebrata collected by B.H. Hodgson, Esq., in Nepal. *Proceedings of the Zoological Society of London* 29(1), 213–227.

Günther, A.C.L.G. (1861i) [non-titled]. *Proceedings of the Zoological Society of London* 29(1), 390–391.

Günther, A.C.L.G. (1862a) On new species of snakes in the collection of the British Museum. *Annals and Magazine of Natural History*, London (Series 3) 9(49), 52–59, pl. 9.

Günther, A.C.L.G. (1862b) On new species of snakes in the collection of the British Museum. *Annals and Magazine of Natural History*, London (Series 3) 9(50), 124–132, pls. 9–10.

Günther, A.C.L.G. (1862c) Eine neue Art von *Coronella* mit Furchenzahn. *Archiv für Naturgeschichte*, Berlin 28(1), 48–52.

Günther, A.C.L.G. (1863a) On new species of snakes in the collection of the British Museum. *Annals and Magazine of Natural History*, London (Series 3) 11(61), 20–25, pl. 3.

Günther, A.C.L.G. (1863b) On some species of tree-snakes (*Ahaetulla*). *Annals and Magazine of Natural History*, London (Series 3) 11(64), 283–287.

Günther, A.C.L.G. (1863c) Third account of new species of snakes in the collection of the British Museum. *Annals and Magazine of Natural History*, London (Series 3) 12(71), 348–365, pls. 5–6.

Günther, A.C.L.G. (1863d) On *Atheris burtonii*, a new snake from West Africa. *Proceedings of the Zoological Society of London* 31(1), 16–17, pl. 3.

Günther, A.C.L.G. (1863e) Note on *Diemennia superciliosa*. *Proceedings of the Zoological Society of London* 31(1), 17–18.

Günther, A.C.L.G. (1863f) Addition to Dr. Wucherer's article on the ophidians of Bahia. *Proceedings of the Zoological Society of London* 31(1), 56–58. 3 figs.

Günther, A.C.L.G. (1863g) Contribution to the herpetology of Ceram. *Proceedings of the Zoological Society of London* 31(1), 58–60, 2 figs.

Günther, A.C.L.G. (1864a) *The reptiles of British India*. Ray Society, London, xxvii + 452 pp., 28 pls.

Günther, A.C.L.G. (1864b) On the reptiles and fishes obtained by Capt. Speke during the East-African expedition. *Proceedings of the Zoological Society of London* 32(1), 115, pl. 15.

Günther, A.C.L.G. (1864c) Report on a collection of reptiles and fishes made by Dr. Kirk in the Zambesi and Nyassa regions. *Proceedings of the Zoological Society of London* 32(1), 303–314, pls. 26–27.

Günther, A.C.L.G. (1864d) Report on a collection of reptiles and fishes from Palestine. *Proceedings of the Zoological Society of London* 32(1), 488–493.

Günther, A.C.L.G. (1865) Fourth account of new species of snakes in the collection of the British Museum. *Annals and Magazine of Natural History*, London (Series 3) 15(86), 89–98, pls. 2–3.

Günther, A.C.L.G. (1866) Fifth account of new species of snakes in the collection of the British Museum. *Annals and Magazine of Natural History*, London (Series 3) 18(103), 24–29, pls. 6–7.

Günther, A.C.L.G. (1867) Additions to the knowledge of Australian reptiles and fishes. *Annals and Magazine of Natural History*, London (Series 3) 20(115), 45–68.

Günther, A.C.L.G. (1868) Sixth account of new species of snakes in the collection of the British Museum. *Annals and Magazine of Natural History*, London (Series 4) 1(6), 413–429, pls. 17–19.

Günther, A.C.L.G. (1869) Report on two collections of Indian reptiles. *Proceedings of the Zoological Society of London* 37(1), 500–507, pls. 38–40.

Günther, A.C.L.G. (1870) Letter to L. Bouton [regarding identification of Mauritian snake as *Typhlops flavoterminatus,* in error for *Ramphotyphlops braminus]. Transactions of the Royal Society of Arts and Sciences of Mauritius* (Series 2) 4, 139.

Günther, A.C.L.G. (1872a) On the reptiles and amphibians of Borneo. *Proceedings of the Zoological Society of London* 40(1), 586–600, 5 figs., pls. 35–40.

Günther, A.C.L.G. (1872b) Seventh account of new species of snakes in the collection of the British Museum. *Annals and Magazine of Natural History*, London (Series 4) 9(49), 13–37, pls. 3–6.

Günther, A.C.L.G. (1872c) Note on the black snake of Robben Island, South Africa. *Proceedings of the Zoological Society of London* 40(1), 836–837.

Günther, A.C.L.G. (1873a) Description of a new snake from Madagascar. *Annals and Magazine of Natural History*, London (Series 4) 11(65), 374–375.

Günther, A.C.L.G. (1873b) Notes on some reptiles and batrachians obtained by Dr. Adolf Bernhard Meyer in Celebes and the Philippine Islands. *Proceedings of the Zoological Society of London* 41(1), 165–172, 1 fig., pls. 17–18.

Günther, A.C.L.G. (1874a) A contribution to the fauna of Savage Island. *Proceedings of the Zoological Society of London* 42(1), 295–297, pl. 45.

Günther, A.C.L.G. (1874b) Descriptions of some new or imperfectly known species of reptiles from the Camaroon Mountains. *Proceedings of the Zoological Society of London* 42(1), 442–445, pls. 56–57.

Günther, A.C.L.G. (1875) Second report on collections of Indian reptiles obtained by the British Museum. *Proceedings of the Zoological Society of London* 43(1), 224–234, 5 figs., pls. 30–34.

Günther, A.C.L.G. (1876) Descriptions of new species of reptiles from Australia, collected by Hr. Dämel for the Godeffroy Museum. *Journal des Museum Godeffroy*, Hamburg 5(12), 45–47.

Günther, A.C.L.G. (1877) On a collection of reptiles and fishes from Duke-of-York Island, New Ireland, and New Britian. *Proceedings of the Zoological Society of London* 45(1), 127–132, pls. 20–21.

Günther, A.C.L.G. (1878) On reptiles from Midian collected by Major Burton. *Proceedings of the Zoological Society of London* 46(1), 977–978, pl. 62.

Günther, A.C.L.G. (1879a) List of the mammals, reptiles, and batrachians sent by Mr. Everett from the Philippine Islands. *Proceedings of the Zoological Society of London* 47(1), 74–79, pl. 4.

Günther, A.C.L.G. (1879b) Notice of a collection of reptiles from islands of Torres Strait. *Annals and Magazine of Natural History*, London (Series 5) 3(13), 84–87, 2 figs.

Günther, A.C.L.G. (1879c) On mammals and reptiles from Johanna, Comoro Islands. *Annals and Magazine of Natural History*, London (Series 5) 3(15), 215–219.

Günther, A.C.L.G. (1880a) On the occurrence of *Tachymenis vivax* in Cyprus. *Annals and Magazine of Natural History*, London (Series 5) 5(29), 436.

Günther, A.C.L.G. (1880b) Description of new species of reptiles from eastern Africa. *Annals and Magazine of Natural History*, London (Series 5) 6(33), 234–238, 5 figs.

Günther, A.C.L.G. (1880c) Description of *Ophites japonicus*, a new snake from Japan. *Annals and Magazine of Natural History*, London (Series 5) 6(36), 462.

Günther, A.C.L.G. (1881a) Descriptions of the amphisbaenians and ophidians collected by Prof. I. Bayley Balfour in the island of Socotra. *Proceedings of the Zoological Society of London* 49(2), 461–463, 1 fig, pls. 40–41.

Günther, A.C.L.G. (1881b) Seventh contribution to the knowledge of the fauna of Madagascar. *Annals and Magazine of Natural History*, London (Series 5) 7(41), 357–360, 3 figs, pl. 19

Günther, A.C.L.G. (1881c) Appendix III. Herpetology. *In:* Oates, F., *Matabele Land and the Victoria Falls*. C. Kegan Paul, London, pp. 330–331, pls. c, d. [second edition, 1889, pp. 336–337, pls. 3–4]

Günther, A.C.L.G. (1882) Ninth contribution to the knowledge of the fauna of Madagascar. *Annals and Magazine of Natural History*, London (Series 5) 9(52), 262–266, 3 figs.

Günther, A.C.L.G. (1883) Descriptions of two snakes from the "Challenger" collections. *Annals and Magazine of Natural History*, London (Series 5) 11(62), 136–137, 1 fig.

Günther, A.C.L.G. (1888a) Report on a collection of reptiles and batrachians sent by Emin Pasha from Monbuttu, Upper Congo. *Proceedings of the Zoological Society of London* 56(1), 50–51.

Günther, A.C.L.G. (1888b) On a collection of reptiles from China. *Annals and Magazine of Natural History*, London (Series 6) 1(3), 165–172, pl. 12.

Günther, A.C.L.G. (1888c) Contribution to the knowledge of snakes of tropical Africa. *Annals and Magazine of Natural History*, London (Series 6) 1(3), 322–335, pls. 18–19.

Günther, A.C.L.G. (1888d) Notes on reptiles and frogs from Dominica, West Indies. *Annals and Magazine of Natural History*, London (Series 6) 2(11), 362–366.

Günther, A.C.L.G. (1889) Third contribution to our knowledge of reptiles and fishes from the upper Yangtsze-Kiang. *Annals and Magazine of Natural History*, London (Series 6) 4(21), 218–229.

Günther, A.C.L.G. (1890) Tenth contribution to the knowledge of the fauna of Madagascar. *Annals and Magazine of Natural History*, London (Series 6) 5(25), 69–72, pl. 6.

Günther, A.C.L.G. (1892a) Report on a collection of reptiles and batrachians transmitted by Mr. H.H. Johnston, C.B., from Nyassaland. *Proceedings of the Zoological Society of London* 60(4), 555–558, pls. 33–35.

Günther, A.C.L.G. (1892b) Appendix II. List of the species of reptiles and fishes collected by Mr. A.E. Pratt on the Upper Yangtze-kiang and in the province Szechuen, with description of the new species. *In:* Pratt, A.E., *To the snows of Tibet through China*. Longmans, Green, London, pp. 238–250, pl. 1.

Günther, A.C.L.G. (1894a) Report on the collection of reptiles and fishes made by Dr. J.W. Gregory during his expedition to Mount Kenia. *Proceedings of the Zoological Society of London* 62(1), 84–91, pls. 8–11.

Günther, A.C.L.G. (1894b) Second report on the reptiles and batrachians transmitted by Mr. H.H. Johnston, C.C. from British Central Africa. *Proceedings of the Zoological Society of London* 62(1), 616–628.

Günther, A.C.L.G. (1894c) Descriptions of the reptiles and fishes collected by Mr. E. Coode-Hore on Lake Tanganyika. *Proceedings of the Zoological Society of London* (1893) 62(1), 628–632, 1 fig., pl. 58.

Günther, A.C.L.G. (1895a) Notice of reptiles and batrachians collected in the eastern half of tropical Africa. *Annals and Magazine of Natural History*, London (Series 6) 15(90), 523–529, pl. 21.

Günther, A.C.L.G. (1895b) The reptiles and batrachians of the Natuna Islands. *Novitates Zoologicae*, London 2(4), 499–502.

Günther, A.C.L.G. (1896a) Report on the collections of reptiles, batrachians and fishes made by Messrs. Potanin and Berezowski in the Chinese provinces Kansu and Sze-chuen. *Annuaire du Musée Zoologique de l'Académie Impériale des Sciences de St.-Pétersbourg* 1, 199–219, 2 pls.

Günther, A.C.L.G. (1896b) Descriptions of two new species of snakes from Sarawak. *Annals and Magazine of Natural History*, London (Series 6) 17(99), 228–229.

Günther, A.C.L.G. (1896c) Report on a collection of reptiles and fishes made by Miss M.H. Kingsley during her travels on the Ogowe River and in Old Calabar. *Annals and Magazine of Natural History*, London (Series 6) 17(100), 261–285, pls. 13–15.

Günther, A.C.L.G. (1885–1902) Reptilia and Batrachia. *In:* Godman, F.D. (Ed.), *Biologia Centrali-Americana. Vol. 7.* Taylor & Francis, London, xx + 326 pp., 76 pls.

Günther, A.C.L.G. (1903) Reptiles from Rio de Oro, Western Sahara. *Novitates Zoologicae*, London 10(2), 298–299, 2 figs.

Günther, R. (1996) *Die Amphibien und Reptilien Deutschlands.* Gustav Fischer, Jena, 825 pp., 328 figs.

Günther, R. & Manthey, U. (1995) *Xenophidion*: a new genus with two new species of snakes from Malaysia (Serpentes, Colubridae). *Amphibia-Reptilia*, Leiden 16(3), 229–240, 7 figs.

Guo, K. & Deng, X. (2009) A new species of *Pareas* (Serpentes: Colubridae: Pareatinae) from the Gaoligong Mountains, south-western China. *Zootaxa* (2008), 53–60, 3 figs.

Guo, P. & Wang, Y. (2011) A new genus and species of cryptic Asian green pitviper (Serpentes: Viperidae: Crotalinae) from south-west China. *Zootaxa* (2918), 1–14, 7 figs.

Guo, P., Zhang, F.-J. & Chen, Y.-Y. (1999a) The hemipenes of Chinese species of *Deinagkistrodon* and *Gloydius* (Serpentes, Crotalinae). *Asiatic Herpetological Research* 8, 38–42, 7 figs.

Guo, P., Zhang, F.-J. & Chen, Y.-Y. (1999b) Catalogue of type specimens of reptiles in the herpetological collections of Chengdu Institute of Biology, the Chinese Academy of Sciences. *Asiatic Herpetological Research* 8, 43–47.

Guo, P., Lu, S., Huang, S., Zhao, H. & Zhao, E. (2006a) Hemipenial morphology of five Asian pitvipers, with a discussion on their taxonomy. *Amphibia-Reptilia,* Leiden 27(1), 19–23, 1 fig.

Guo, P., Pang, J., Zhang, Y. & Zhao, E. (2006b) A re-analysis of the phylogeny of the genus *Protobothrops* (Reptilia, Viperidae), with particular reference to the systematic position of *P. xiangchengensis. Amphibia-Reptilia,* Leiden 27(3), 433–439, 1 fig.

Guo, P., Malhotra, A., Li, P.P., Pook, C.E. & Creer, S. (2007) New evidence on the phylogenetic position of the poorly known Asia pitviper *Protobothrops kaulbacki* (Serpentes: Viperidae: Crotalinae) with a redescription of the species and a revision of the genus *Protobothrops. Herpetological Journal,* London 17, 237–246, 4 figs.

Guo, P., Liu, S.Y., Feng, J.-C. & He, M. (2008) The description of a new species of *Thermophis* (Serpentes: Colubridae). *Sichuan Journal of Zoology* 27(3), 231, 1 pl.

Guo, P., Liu, S.Y., Huang, S., He, M., Sun, Z.Y., Feng, J.C. & Zhao, E.M. (2009a) Morphological variation in *Thermophis* Malnate (Serpentes: Colubridae), with an expanded description of *T. zhaoermii. Zootaxa* (1973), 51–60, 3 figs.

Guo, P., Malhotra, A., Creer, S. & Pook, C.E. (2009b) An evaluation of the systematic value of skull morphology in the *Trimeresurus* radiation (Serpentes: Viperidae: Crotalinae) of Asian pitvipers. *Journal of Zoological Systematics and Evolutionary Research* 47(4), 378–384, 8 figs.

Guo, P., Liu, Q., Xu, Y., Jiang, K., Hou, M., Ding, L., Pyron, R.A. & Burbrink, F.T. (2012) Out of Asia: Natricine snakes support the Cenezoic Beringian Dispersal Hypothesis. *Molecular Phylogenetics and Evolution* 63(2), 825–833, 2 figs.

Guo, P., Li, J.-T., Chen, Y.-Y. & Wang, Y.-Z. (2012) Designation of a neotype for *Protobothrops mangshanensis* (Zhao, 1990). *Asian Herpetological Research* (Series 2) 3(4), 340–341, 2 figs.

Guo, Y., Wu, Y., He, S., Shi, H. & Zhao, E. (2011) Systematics and molecular phylogenetics of Asian snail-eating snakes (Pareatidae). *Zootaxa* (3001), 57–64, 2 figs.

Gurney, J.H. (1860) *Uriëchis microlepidotus. Annals and Magazine of Natural History*, London (Series 3) 5(28), 342.

Gutberlet, R.L., Jr. (1995) A new locality for Rowley's palm pitviper, *Bothriechis rowleyi* (Serpentes, Viperidae): a Mexican relict. *Southwestern Naturalist* 40(1), 124–125.

Gutberlet, R.L., Jr. (1998) The phylogenetic position of the Mexican black-tailed pitviper (Squamata, Viperidae, Crotalinae). *Herpetologica* 54(2), 184–206, 10 figs.

Gutberlet, R.L., Jr. & Campbell, J.A. (2001) Generic recognition for a neglected lineage of South American pitvipers (Squamata, Viperidae, Crotalinae), with the description of a new species from the Colombian Chocó. *American Museum Novitates* (3316), 1–15, 6 figs.

Gutberlet, R.L., Jr. & Harvey, M.B. (2002) Phylogenetic relationships of New World pitvipers as inferred from anatomical evidence. *In:* Schuett, G.W., Höggren, M., Douglas, M.E. & Greene, H.W. (Eds.), *Biology of the vipers.* Eagle Mountain Press, Salt Lake City, pp. 51–68.

Guthrie, J.E. (1926) The snakes of Iowa. *Bulletin of the Agricultural Experiment Station,* Ames (239), 147–192, 21 figs., 5 pls.

Gutsche, A. (2003) Geographic distribution: Serpentes: *Enulius flavitorques. Herpetological Review* 34(3), 265.

Guyer, C. (1994) The reptile fauna: diversity and ecology. *In:* McDade, L.A., Bawa, K.S., Hespenheide, H.A. & Hartshorn, G.S. (Eds.), *La Selva: ecology and natural history of a Neotropical rain forest.* University of Chicago Press, Chicago, pp. 210–216, 382–383, 1 fig..

Guyer, C. & Donnelly, M.A. (2005) *Amphibians and reptiles of La Selva, Costa Rica, and the Caribbean slope. A comprehensive guide.* Chicago University Press, Chicago, vii + 299, 207 col. pls.

Gvozdík, V., Jandzik, D., Cordos, B., Rehák, I. & Kotlík, P. (2012) A mitochondrial DNA phylogeny of the endangered vipers of the *Vipera ursinii* complex. *Molecular Phylogenetics and Evolution* 62, 1019–1024, 2 figs.

Gyi, K.K. (1970) A revision of colubrid snakes of the subfamily Homalopsinae. *University of Kansas Publications, Museum of Natural History* 20(2), 47–223, 38 figs.

Gyldenstolpe, N. (1916) Zoological results of the Swedish Zoological expeditions to Siam, 1911–1912 and 1914–1915. I. Snakes. *Kungliga Svenska Vetenskapsakademiens Handlingar*, Stockholn 95(3), 1–28, 2 figs.

Haacke, W.D. (1975) Description of a new adder (Viperidae, Reptilia) from southern Africa, with a discussion of related forms. *Cimbebasia*, Windhoek 4A(5), 115–128, 6 figs., 3 pls.

Haacke, W.D. (1981) The file-snakes of the genus *Mehelya* in southern Africa with special reference to South West Africa/Namibia. *Madoqua*, Windhoek 12(4), 217–224, 2 figs., 3 pls.

Haacke, W.D. (1982) Further notes on reproduction and distribution of the Natal purple gloss snake *Amblyodipsas concolor* (A. Smith). *Journal of the Herpetological Association of Africa*, Stellenbosch 28, 24–25, illust.

Haacke, W.D. (1984) The herpetology of the southern Kalahari domain. *Koedoe* (*Supplement*), Pretoria 1984, 171–186.

Haacke, W.D. (1985) Occurance of the spotted bush-snake (*Philothamnus semivariegatus semivariegatus*) in the arid south-west of southern Africa. *Journal of the Herpetological Association of Africa*, Stellenbosch 31, 7–9, 1 fig.

Haacke, W.D. & Bruton, M.N. (1978) On two little known snakes from the tropical subtraction zone of south-eastern Africa. *Annals of the Transvaal Museum*, Pretoria 31(5), 43–50, illust.

Haacke, W.D. & Finkeldey, H. (1967) The Angola garter snake *Elapsoidea sundevallii semiannulata* Bocage: a new record from southern Africa. *South-West Africa Science Society*, Windhoek, *Special Publication* (8), 1–12, 3 pls.

Haagner, G.V. (1990) A first record of the eastern bark snake *Hemirhagerrhis nototaenia* in Natal. *Lammergeyer*, Pietermaritzburg 41, 37–39, illust.

Haagner, G.V. (1993) The husbandry and captive propagation of the southern rock python, *Python sebae natalensis*. *British Herpetological Society Bulletin*, London (1992–1993) (42), 30–41, 8 figs.

Haagner, G.V. & Branch, W.R. (1994) A taxonomic revision of the dusky-bellied water snake, *Lycodonomorphus laevissimus* (Serpentes: Colubridae). *Journal of African Zoology* 108, 237–250, 4 figs.

Haagner, G.V. & Cockeran, A. (1993) Life history notes. Serpentes, Colubridae, *Lycodonomorphus rufulus*, brown watersnake: size, reproduction and sperm retention. *Journal of the Herpetological Association of Africa*, Stellenbosch (1992) (41), 43.

Haagner, G.V. & Els, R.A. (1986) The reptiles of Mapelane. *Lammergeyer*, Pietermaritzburg 37, 14–21, illust.

Haas, C.P.J., de. (1941) Some notes on the biology of snakes and on their distribution in two districts of West Java. *Treubia*, Bogor 18(2), 327–375, 8 figs.

Haas, C.P.J., de. (1949) The genus *Natrix* in the collection of the Raffles Museum and its distribution in the Malay Peninsula. *Bulletin of the Raffles Museum*, Singapore (19), 78–97.

Haas, C.P.J., de. (1950) Checklist of the snakes of the Indo-Australian Archipelago (Reptilia, Ophidia). *Treubia*, Bogor 20(3), 511–625.

Haas, G. (1950) A new *Atractaspis* (mole viper) from Palestine. *Copeia* 1950(1), 52–53, 1 fig.

Haas, G. (1951) On the present state of our knowledge of the herpetofauna of Palestine. *Bulletin of the Research Council of Israel* (*Zoology*), Jerusalem 1(3), 67–95.

Haas, G. (1952a) Two collections of reptiles from Iraq with descriptions of two new forms. *Copeia* 1952(1), 20–22.

Haas, G. (1952b) Remarks on the origin of the herpetofauna of Palestine. *Revue de la Faculté des Sciences de l'Université d'Istanbul, Sciences Naturelles* 17B(2), 95–105.

Haas, G. (1955) The systematic position of *Loxocemus bicolor* Cope (Ophidia). *American Museum Novitates* (1748), 1–8, 2 figs.

Haas, G. (1957) Some amphibians and reptiles from Arabia. *Proceedings of the California Academy of Sciences* (Series 3) 29(3), 47–86, 12 figs.

Haas, G. (1961) On a collection of Arabian reptiles. *Annals of the Carnegie Museum* (1961–1964) 36(3), 19–28, 2 figs.

Haas, G. (1973) Muscles of the jaws and associated structures in the Rhynchocephalia and Squamata. *In:* Gans, C. & Parsons, T.S. (Eds.), *Biology of the Reptilia. Volume 4 (Morphology D)*. Academic Press, New York, pp. 285–490, 169 figs.

Haas, G. (1979) On a new snakelike reptile from the lower Cenomanian of Ein Jabrud, near Jerusalem. *Bulletin du Muséum National d'Histoire Naturelle*, Paris (Série 4) 1C(1), 51–64, 6 figs.

Haas, G. (1980a) *Pachyrhachis problematicus* Haas: snakelike reptile from the Lower Cenomanian, ventral view of the skull. *Bulletin du Muséum National d'Histoire Naturelle*, Paris (Série 4) 2C(2), 87–104, 9 figs.

Haas, G. (1980b) Remarks on a new ophiomorph reptile from the lower Cenomanian of Ein Jabrud, Israel. *In:* Jacobs, L.L. (Ed.), *Aspects of vertebrate history*. Museum of Northern Arizona Press, Flagstaff, pp. 177–192, figs. 10.1–10.6

Haas, G. & Werner, Y.L. (1969) Lizards and snakes from southwestern Asia, collected by Henry Field. *Bulletin of the Museum of Comparative Zoology* 138(6), 327–405, 13 figs., 21 pls.

Haast, W.E. & Anderson, R. (1981) *Complete guide to the snakes of Florida*. Phoenix Publishing Co., Miami, 139 pp., 37 col. pls.

Hadley, W.F. & Gans, C. (1972) Convergent ontogenetic change of color pattern in *Elaphe climacophora* (Colubridae, Reptilia). *Journal of Herpetology* 6(1), 75–78, 1 fig.

Hahn, D.E. (1960) Collecting notes on central Korean reptiles and amphibians. *Journal of the Ohio Herpetological Society* 2(4), 16–25, 4 figs.

Hahn, D.E. (1971) Noteworthy herpetological records from Honduras. *Herpetological Review* 3(4), 111.

Hahn, D.E. (1977) Book review: "Révue systématique des Typhlopidae d'Afrique, Reptilia-Serpentes." By Rolande Roux-Estève 1974. *Journal of the Herpetological Association of Africa* 15, 34–39.

Hahn, D.E. (1978) A brief review of the genus *Leptotyphlops* (Reptilia, Serpentes, Leptotyphlopidae) of Asia, with description of a new species. *Journal of Herpetology* 12(4), 477–489, 8 figs.

Hahn, D.E. (1979a) The identity of the blind snake *Stenostoma Signatum* Jan, 1861 (Serpentes, Leptotyphlopidae). *Herpetologica* 35(1), 57–60, 1 fig.

Hahn, D.E. (1979b) Leptotyphlopidae, *Leptotyphlops*. *Catalogue of American amphibians and reptiles*, pp. 230.1–230.4.

Hahn, D.E. (1979c) *Leptotyphlops dulcis*. *Catalogue of American amphibians and reptiles*, pp. 231.1–231.2.

Hahn, D.E. (1979d) *Leptotyphlops humilis*. *Catalogue of American amphibians and reptiles*, pp. 232.1–232.4.

Hahn, D.E. (1980a) Liste der rezenten Amphibien und Reptilien: Anomalepididae, Leptotyphlopidae, Typhlopidae. *Das Tierreich*, Berlin (101), 1–93.

Hahn, D.E. (1980b) *Leptotyphlops maximus*. *Catalogue of American amphibians and reptiles*, p. 244.1.

Hahn, D.E. & Wallach, V. (1998) Comments on the systematics of Old World *Leptotyphlops* (Serpentes, Leptotyphlopidae), with description of a new species. *Hamadryad* 23(1), 50–62, 7 figs.

Haile, N.S. (1959) The snakes of Borneo, with a key to the species. *Sarawak Museum Journal*, Kuching 8, 743–771, 2 figs., pls. 22–23.

Hakansson, N.T. (1981) An annotated checklist of reptiles known to occur in the Gambia. *Journal of Herpetology* 15(2), 155–161, 1 fig.

Hale, S.F. (1977) Observations on additional specimens of the Mexican snail-eating snake, *Dipsas gaigeae* (Reptilia, Serpentes, Colubridae). *Journal of Herpetology* 11(3), 374–377.

Halfpenny, G. (1978) *Atlas of the amphibians and reptiles of Staffordshire*. The City Museum & Art Gallery, Stoke-on-Trent, 48 pp., illust.

Hall, C.W. (1951) Notes on a small herpetological collection from Guerrero. *University of Kansas Science Bulletin* 34(4), 201–212, pl. 24.

Hall, D. (2004) *The ultimate guide to snakes & reptiles*. Chartwell Books, Edison, 443 pp., col. photos.

Hall, R.J. (1969) Ecological observations on Graham's water snake, (*Regina grahami* Baird and Girard). *American Midland Naturalist* 81(1), 156–163, 3 figs.

Hallermann, J. (1998) Annotated catelogue of the type specimens of the herpetological collection in the Zoological Museum of the University of Hamburg. *Mitteilungen aus dem Hamburgisches Zoologisches Museum und Institut* 95, 197–223.

Hallermann, J. (2000) Korrekturen zur Liste der Neubeschreibungen von Reptilien des Jahres 1999 in Elaphe 8 (2000) Heft 2. *Elaphe* 8(3), 46–47.

Hallermann, J. & Ródel, M.O. (1995) A new species of *Leptotyphlops* (Serpentes, Leptotyphlopidae) of the longicaudus-group from West Africa. *Stuttgarter Beiträge zur Naturkunde*, Stuttgart (Series A) (532), 1–6, 2 figs.

Hallermann, J., Ródel, M.O., Ananjeva, N.B. & Orlov, N.L. (2001) On a remarkable collection of reptiles and amphibians collected by the German Indian Expedition 1955–1958. *Russian Journal of Herpetology* 8(1), 59–68, 4 col. figs.

Halliday, T.R. & Adler, K. (1986) *All the world's animals: reptiles & amphibians.* Torstar Books, New York, 160 pp., col. illust.

Hallmen, M. (2004a) *Die Prärie-Strumpfbandnatter,* Thamnophis radix. Natur und Tier-Verlag, Münster, 61 pp., col. illust.

Hallmen, M. (2004b) *Die Strumpfbandnatter,* Thamnophis sirtalis. Natur und Tier-Verlag, Münster, 61 pp., col. illust.

Hallmen, M. (2007) *Die San-Francisco-Strumpfbandnatter,* Thamnophis sirtalis tetrataenia. Natur und Tier-Verlag, Münster, 62 pp., col. illust.

Hallmen, M. & Chlebowy, J. (2001) *Strumpfbandnattern.* Natur und Tier-Verlag, Münster, 191 pp., 125 col. photos.

Hallowell, E. (1842) Description of a new genus of serpents from western Africa. *Journal of the Academy of Natural Sciences of Philadelphia* 8(2), 336–338, pl. 19.

Hallowell, E. (1844a) Descriptions of new species of African reptiles. *Proceedings of the Academy of Natural Sciences of Philadelphia* (1844–1845) 2(3), 58–62

Hallowell, E. (1844b) Description of new species of reptiles from Africa. *Proceedings of the Academy of Natural Sciences of Philadelphia* (1844–1845) 2(5), 118–120.

Hallowell, E. (1844c) Description of new species of African reptiles. *Proceedings of the Academy of Natural Sciences of Philadelphia* (1844–1845) 2(6), 169–172.

Hallowell, E. (1845a) Description of reptiles from South America, supposed to be new. *Proceedings of the Academy of Natural Sciences of Philadelphia* (1844–1845) 2(9), 241–247.

Hallowell, E. (1845b) Description of new species of African reptiles. *Proceedings of the Academy of Natural Sciences of Philadelphia* (1844–1845) 2(9), 247–250.

Hallowell, E. (1847) Description of a new species of *Coluber* inhabiting the United States. *Proceedings of the Academy of Natural Sciences of Philadelphia* 3(11), 278–280.

Hallowell, E. (1848) Description of two new species of *Onychocephalus* from the western coast of Africa. *Proceedings of the Academy of Natural Sciences of Philadelphia* (1848–1849) 4, 59–61, 1 pl.

Hallowell, E. (1849) Description of a new species of *Eryx* from Madras. *Proceedings of the Academy of Natural Sciences of Philadelphia* (1848–1849) 4, 184–185, 1 pl.

Hallowell, E. (1852a) Descriptions of new species of reptiles inhabiting North America. *Proceedings of the Academy of Natural Sciences of Philadelphia* (1852–1853) 6(5), 177–182.

Hallowell, E. (1852b) Descriptions of new species of reptiles from Oregon. *Proceedings of the Academy of Natural Sciences of Philadelphia* (1852–1853) 6(5), 182–183.

Hallowell, E. (1852c) On a new genus and two new species of African serpents. *Proceedings of the Academy of Natural Sciences of Philadelphia* (1852–1853) 6(6), 203–205.

Hallowell, E. (1852d) On some new reptiles from Oregon and the western coast of Africa. *Journal of the Academy of Natural Sciences of Philadelphia* (Series 2) 2, 301–306, pls. 28–29.

Hallowell, E. (1853a) On some new reptiles from California. *Proceedings of the Academy of Natural Sciences of Philadelphia* (1852–1853) 6(7), 236–238.

Hallowell, E. (1853b) Reptiles. Order Ophidia. *In:* Sitgreaves, L. (Ed.), *Report of an expedition down the Zuni and Colorado Rivers.* Robert Armstrong, Washington, pp. 106–147, 20 pls.

Hallowell, E. (1854a) Descriptions of new reptiles from California. *Proceedings of the Academy of Natural Sciences of Philadelphia* (1854–1855) 7(3), 91–97.

Hallowell, E. (1854b) On a genus and species of serpent from Honduras, presumed to be new. *Proceedings of the Academy of Natural Sciences of Philadelphia* (1854–1855) 7(3), 97.

Hallowell, E. (1854c) Remarks on the geographical distribution of reptiles, with descriptions of several species supposed to be new, and corrections of former papers. *Proceedings of the Academy of Natural Sciences of Philadelphia* (1854–1855) 7(3), 98–105.

Hallowell, E. (1854d) Contributions to South American herpetology. *Journal of the Academy of Natural Sciences of Philadelphia* (1855–1858) (Series 2) 3(1), 33–36, pls. 3–4. [preprint in 1854]

Hallowell, E. (1855a) Notices of new reptiles from Texas. *Proceedings of the Academy of Natural Sciences of Philadelphia* (1854–1855) 7(5), 192–193.

Hallowell, E. (1855b) Descriptions of new reptiles from Guinea. *Proceedings of the Academy of Natural Sciences of Philadelphia* (1854–1855) 7(5), 193–194.

Hallowell, E. (1856a) Notes on the reptiles in the collection of the Museum of the Academy of Natural Sciences. *Proceedings of the Academy of Natural Sciences of Philadelphia* 8(4), 146–153.

Hallowell, E. (1856b) On several new species of reptiles in the collection of the Academy of Natural Sciences. *Proceedings of the Academy of Natural Sciences of Philadelphia* (1856–1857) 8(4), 153–156.

Hallowell, E. (1856c) Notice of a collection of reptiles from Kansas and Nebraska, presented to the Academy of Natural Sciences, by Dr. Hammond, U.S.A. *Proceedings of the Academy of Natural Sciences of Philadelphia* (1856–1857) 8(5), 238–253.

Hallowell, E. (1857a) Description of a new genus of colubriform serpents from California. *Proceedings of the Academy of Natural Sciences of Philadelphia* (1856–1857) 8(6), 310–311.

Hallowell, E. (1857b) Notice of a collection of reptiles from the Gaboon country, West Africa, recently presented to the Academy of Natural Sciences of Philadelphia, by Dr. Henry A. Ford. *Proceedings of the Academy of Natural Sciences of Philadelphia* (1857–1858) 9(3), 48–72.

Hallowell, E. (1857c) On a new genus of Boidae from Cuba. *Transactions of the American Philosophical Society* (Series 2) 11(3), 65–69, pl. 2.

Hallowell, E. (1859a) Zoological report. No. 1. Report upon reptiles collected on the survey. *In:* Williamson, R.S. (Ed.), *Reports of Explorations and Surveys, to ascertain the most practicable and economical route for a railroad from the Mississippi River to the Pacific Ocean, 1853-6. Report of explorations in California for railroad routes to connect with the routes near the 35th and 32nd parallels of north latitude.* Government Printing Office, Washington, D.C. (1853) 10(1), 1–27, 10 pls.

Hallowell, E. (1859b) Zoological report. No. 2. Report upon reptiles collected on the survey. *In:* Parke, J.G. (Ed.), *Reports of Explorations and Surveys, to ascertain the most practicable and economical route for a railroad from the Mississippi River to the Pacific Ocean, 1853-6. Report of explorations for a railroad route near the 32nd Parallel of North Latitude, lying between Doña Ana, on the Río Grande, and Pimas villages, on the Gila.* Government Printing Office, Washington, D.C. (1855) 10(2), 23–24, pl. 2.

Hallowell, E. (1861) Report upon the Reptilia of the North Pacific Exploring Expedition, under command of Capt. John Rogers, U.S.N. *Proceedings of the Academy of Natural Sciences of Philadelphia* (1860) 12(11), 480–510. [published posthumously and edited by E.D. Cope]

Halloy, M. (1985a) Note sur *Liophis sagittifer* (Jan, 1861) (Colubridae, Serpentes). *Revue Français d'Aquariologie*, Nancy 12(1), 27–30, illust.

Halloy, M. (1985b) *Liophis poecilogyrus* (Wied). *Revue Français d'Aquariologie*, Nancy 12(2), 2 pp., illust.

Halloy, M. & Laurent, R. (1984) *Clelia rustica* (Cope). *Revue Français d'Aquariologie*, Nancy 11(1), 2 pp., illust.

Halpin, Z.T. (1983) Naturally occurring encounters between black-tailed prairie dogs *(Cynomys ludovicianus)* and snakes. *American Midland Naturalist* 109(1), 50–54.

Halstead, B.J., Wylie, G.D. & Casazza, M.L. (2010) Habitat suitability and conservation of the giant gartersnake (*Thamnophis gigas*) in the Sacramento Valley of California. *Copeia* 2010(4), 591–599, 3 figs

Halstead, B.W. (1970) *Poisonous and venomous marine animals of the world. Volume three—vertebrates. Class Reptilia.* United States Government Printing Office, Washington, pp. 617–756, 36 col. pls.

Haltom, W.L. (1931) Alabama reptiles. *Alabama Museum of Natural History, Paper* (11), vi + 145 pp., 57 figs., 39 pls.

Haly, A. (1887) On a remarkable sea snake from Colombo (*Hydrophis taprobanica* Haly). *Taprobanian,* Bombay 2(4), 107–108.

Haly, A. (1888) Two new Ceylon snakes. *Taprobanian*, Bombay 3, 51.

Haly, A. (1891) *Report on the collection of reptiles and Batrachia in the Colombo Museum.* Colombo Museum, Colombo, 39 pp.

Hamdan, B.S. & Lira-da-Silva, R.M. (2007) Geographic distribution: Serpentes: *Tropidodryas striaticeps. Herpetological Review* 38(1), 107.

Hamdan, B.S. & Lira-Da-Silva, R.M. (2008) Geographic distribution: Serpentes: *Oxyrhopus rhombifer rhombifer. Herpetological Review* 39(3), 372–373.

Hamdan, B.S. & Lira-da-Silva, R.M. (2012) The snakes of Bahia State, northeastern Brazil: species richness, composition and biogeographical notes. *Salamandra* 48(1), 31–50, 4 figs.

Hamilton, B.T., Hart, R. & Sites, J.W., Jr. (2012) Feeding ecology of the milksnake *(Lampropeltis triangulum,* Colubridae) in the western United States. *Journal of Herpetology* 46(4), 515–522, 5 figs.

Hammar, A.G. (1908) Note on the type specimen of a blind snake, *Helminthophis wilderi* (Garman), from Brazil. *Annals and Magazine of Natural History*, London (Series 8) 1(4), 334–335.

Hammerson, G.A. (1999) *Amphibians and reptiles in Colorado, a Colorado field guide.* Second edition. University of Colorado Press & Colorado Division of Wildlife, Niwot, xxvi + 484 pp., 26 figs., 225 pls. [first edition, 1986, vii + 131 pp., illust.]

Han, K.H., Stuebing, R.B. & Voris, H.K. (1991) Population structure and reproduction in the marine snake, *Lapemis hardwickii* Gray, from the west coast of Sabah. *Sarawak Museum Journal*, Kuching 42(63), 463–475, 10 figs.

Hancock, J. (1830) Observations on the snake called yellow tail *(Coluber flavicolis,* Linn., belonging to the division of *Cerberus* of Cuvier), and on the supposed power of fascination in serpents. *The Edinburgh New Philosophical Journal* 9, 165–170.

Hand, N., Watson, W. & King, P. (2006) *Amphibians and reptiles of Herefordshire.* Herefordshire Biological Records Centre, Hereford, 72 pp., col. photos.

Hanitsch, R. (1900) An expedition to Mount Kina Balu, British North Borneo. *Journal of the Straits Branch of the Royal Asiatic Society*, Singapore (34), 49–88, 4 pls.

Hankinson, T.L. (1917) Amphibians and reptiles of the Charleston region. *Transactions of the Illinois Academy of Science* 10, 322–330.

Hansen, R.W. & Bryson, Jr., R.W. (2009) Geographic distribution: Serpentes: *Lampropeltis mexicana. Herpetological Review* 40(1), 114.

Hansen, R.W., Rodriguez, C.M. & Grünwald, C.I. (2011) Geographic distribution: Serpentes: *Lampropeltis mexicana. Herpetological Review* 42(2), 242.

Hanson, J. & Hanson, R.B. (1997) *50 common reptiles and amphibians of the Southwest.* Western National Parks Association, Tucson, 59 pp., col. photos.

Hanvivatvong, O., Ohanuphak, P., Lowcharoenkul, C., Benyajati, C. & Sakulramrung, R. (1987) Serum and urine levels of cobra venom in non-antivenin treated patients. *In:* Gopalakrishnakone, P. & Tan, C.K. (Eds.), *Progress in venom and toxin research,* National University of Singapore, Singapore, pp. 69–75, 1 fig.

Happ, F. (1985) *Coronella austriaca austriaca* — eine Mimikry-Schlange? Ein Erfahrungsbericht uber die Sonderstellung der *Coronella austriaca austriaca* unter den Karntner Schlangen. *Carinthia*, Klagenfurt 175(95), 285–290, illust.

Hardaway, T.E. & Williams, K.L. (1976) Costal cartilage in snakes and their phylogenetic significance. *Herpetologica* 32(4), 378–387, 6 figs.

Harding, J.H. (1997) *Amphibians and reptiles of the Great Lakes region.* University of Michigan Press, Ann Arbor, xvi + 378 pp., col. photos.

Harding, K.A. (1995) A new species of tree snake of the genus *Leptophis* Bell 1825 from Mount Aripo, Trinidad. *Tropical Zoology*, Firenze 8(2), 221–226, 2 figs.

Harding, K.A. & Welch, K.R.G. (1980) *Venomous snakes of the World. A checklist.* Pergamon Press, Oxford, x + 188 pp.

Hardwicke, T. (1756–1823) *Water-colour sketches of snakes.* 2 vols. Unpublished, deposited in British Museum (Natural History), London, 263 col. pls.

Hardy, J.D., Jr. (1982) Biogeography of Tobago, West Indies, with special reference to amphibians and reptiles: a review. *Bulletin of the Maryland Herpetological Society* 18(2), 37–142, 29 figs.

Hardy, J.D., Jr. & Boos, H.A.E. (1995) Snakes of the genus *Erythrolamprus* (Serpentes, Colubridae) from Trinidad and Tobago, West Indies. *Bulletin of the Maryland Herpetological Society* 31(3), 158–190, 10 figs.

Hardy, L.M. (1963) Description of a new species of snake (genus *Dryadophis*) from México. *Copeia* 1963(4), 669–672, 1 fig.

Hardy, L.M. (1964) A replacement name for *Dryadophis fasciatus* Hardy. *Copeia* 1964(4), 714.

Hardy, L.M. (1972) A systematic revision of the genus *Pseudoficimia* (Serpentes, Colubridae). *Journal of Herpetology* 6(1), 53–69, 8 figs., 4 pls.

Hardy, L.M. (1973) *Pseudoficimia. P. frontalis. Catalogue of American amphibians and reptiles*, pp. 146.1–146.2.

Hardy, L.M. (1975a) A systematic revision of the colubrid snake genus *Gyalopion. Journal of Herpetology* 9(1), 107–132, 11 figs.

Hardy, L.M. (1975b) A systematic revision of the colubrid snake *Ficimia. Journal of Herpetology* 9(2), 133–168, 24 figs.

Hardy, L.M. (1975c) Comparative morphology and evolutionary relationships of the colubrid snake genera *Pseudoficimia*, *Ficimia*, and *Gyalopion*. *Journal of Herpetology* 9(4), 323–336, 11 figs.

Hardy, L.M. (1976a) *Ficimia streckeri*. *Catalogue of American amphibians and reptiles*, pp. 181.1–181.2.

Hardy, L.M. (1976b) *Gyalopion. G. canum, G. quadrangularis*. *Catalogue of American amphibians and reptiles*, pp. 182.1–182.4.

Hardy, L.M. (1978) *Ficimia olivacea*. *Catalogue of American amphibians and reptiles*, pp. 219.1– 219.2.

Hardy, L.M. (1979) *Ficimia ramirezi*. *Catalogue of American amphibians and reptiles*, p. 228.1.

Hardy, L.M. (1980a) *Ficimia ruspator*. *Catalogue of American amphibians and reptiles*, p. 243.1.

Hardy, L.M. (1980b) *Ficimia publia*. *Catalogue of American amphibians and reptiles*, pp. 254.1– 254.2.

Hardy, L.M. (1980c) *Ficimia variegata*. *Catalogue of American amphibians and reptiles*, pp. 269.1–269.2.

Hardy, L.M. (1989) The karyotype of *Exiliboa placata* Bogert (Tropidopheidae), and comparisons with the family Boidae (Reptilia, Serpentes). *Proceedings of the Biological Society of Washington* 102(4), 1045–1049, 3 figs.

Hardy, L.M. (1990) *Ficimia*. *Catalogue of American amphibians and reptiles*, pp. 471.1–471.5, 9 figs.

Hardy, L.M. & McDiarmid, R.W. (1969) The amphibians and reptiles of Sinaloa, México. *University of Kansas Publications, Museum of Natural History* 18(3), 39–252, 91 figs.

Harlan, R. (1827) Genera of North America Reptilia, and a synopsis of the species. *Journal of the Academy of Natural Sciences of Philadelphia* (1827–1829) 6(1), 7–37, 317–372.

Harlan, R. (1835) Genera of North American Reptilia, and a synopsis of the species. Second edition. *In: Medical and physical researches, or original memoirs in medicine, surgery, physiology, geology, zoology, and comparative anatomy*. Lydia R. Bailey, Philadelphia, pp. 84–160.

Harlow, P., & Shine, R. (1992) Food habits and reproductive biology of the Pacific island boas (*Candoia*). *Journal of Herpetology* 26(1), 60–66, 4 figs.

Harman, A.J.E. (1961) A collection of snakes from Singapore. *Malayan Nature Journal*, Kuala Lumpur 15(3–4), 181–183.

Harper, F. (1940) Some works of Batram, Daudin, Latreille, and Sonnini, and their bearing upon North American herpetological nomenclature. *American Midland Naturalist* 23(3), 692–723.

Harris, D.J., Perera, A., Barata, M., Tarroso, P. & Salvi, D. (2010) New distributional notes for terrestrial herpetofauna from Morocco. *North-Western Journal of Zoology*, Oradea 6(2), 309–315, 2 figs.

Harris, H.S., Jr. (1975) Distributional survey (Amphibia/Reptilia): Maryland and the District of Columbia. *Bulletin of the Maryland Herpetological Society* 11(3), 73–167, 3 figs.

Harris, H.S., Jr. & Simmons, R.S. (1967) Another *Dipsas gaigeae* (Oliver) from Colima, Mexico. *Herpetologica* 23(3), 234–235, 1 fig.

Harris, H.S., Jr. & Simmons, R.S. (1972a) A checklist of the rattlesnake (*Crotalus durissus* group) of South America. *Bulletin of the Maryland Herpetological Society* 8(1), 28–31.

Harris, H.S., Jr. & Simmons, R.S. (1972b) Keys to the Neotropical species and subspecies of the *Crotalus durissus* group. *Bulletin of the Maryland Herpetological Society* 8(1), 32–34, 1 fig.

Harris, H.S., Jr. & Simmons, R.S. (1974) The New Mexican ridge-nosed rattlesnake. *National Parks and Conservation Magazine* 48(3), 22–24.

Harris, H.S., Jr. & Simmons, R.S. (1975) An endangered species: the New Mexican ridge-nosed rattlesnake. *Bulletin of the Maryland Herpetological Society* 11(1), 1–7, 8 figs.

Harris, H.S., Jr. & Simmons, R.S. (1976) The paleogeography and evolution of *Crotalus willardi*, with a formal description of a new subspecies from New Mexico, United States. *Bulletin of the Maryland Herpetological Society* 12(1), 1–22, 5 figs.

Harris, H.S., Jr. & Simmons, R.S. (1977) A preliminary account of insular rattlesnake populations, with special reference to those occurring in the Gulf of California and off the Pacific coast. *Bulletin of the Maryland Herpetological Society* 13(2), 92–110, 11 figs.

Harris, H.S., Jr. & Simmons, R.S. (1978a) A new subspecies of *Crotalus durissus* (Serpentes, Crotalidae) from the Rupununi Savanna of southwestern Guyana. *Memórias do Instituto Butantan*, São Paulo (1976–1977) 40–41, 305–311, 2 figs.

Harris, H.S., Jr. & Simmons, R.S. (1978b) A preliminary account of the rattlesnakes with the descriptions of four new subspecies. *Bulletin of the Maryland Herpetological Society* 14(3), 105–211, 88 figs.

Harrison, C.R. (1992) *A taxonomic revision of the snakes in the* Coniophanes piceivittis *species group*. Master's Thesis, University of Texas at El Paso, El Paso, 65 pp., 4 figs.

Hartert, E. (1913) Reptiles and batrachians. Expedition to Central Western Sahara by E. Hartert. *Novitates Zoologicae*, London 20, 76–84.

Hartman, F.A. (1911) Description of a little-known rattlesnake, *Crotalus willardi*, from Arizona. *Proceedings of the United States National Museum* (1910–1911) 39(1800), 569–570.

Hartmann, M.T., Hartmann, P.A., Cechin, S.Z. & Martins, M. (2005) Feeding habits and habitat use in *Bothrops pubescens* (Viperidae, Crotalinae) from southern Brazil. *Journal of Herpetology* 39(4), 664–669, 2 figs.

Hartmann, T., Ihlow, F., Edwards, S., Sovath, S., Handschuh, M. & Böhme, W. (2013) A preliminary annotated checklist of the amphibians and reptiles of the Kulen Promtep Wildlife Sanctuary in northern Cambodia. *Asian Herpetological Research* 4(1), 36–55, 7 figs.

Hatori, J. (1905) *Report on the Thanatophidia of Formosa*. Commission of Investigation of Endemic and Infectious Diseases of Formosan Government, Taipei, 20+ pp., 9 figs.

Hartweg, N. (1932) The status of *Apostolepis tenuis* Ruthven. *Copeia* 1932(2), 103.

Hartweg, N. (1940) Description of *Salvadora intermedia*, new species, with remarks on the *grahamiae* group. *Copeia* 1940(4), 256–259.

Hartweg, N. (1944) Remarks on some Mexican snakes of the genus *Tantilla*. *Occasional Papers of the Museum of Zoology, University of Michigan* (486), 1–9.

Hartweg, N. (1959) A new colubrid snakes of the genus *Geophis* from Michoacán. *Occasional Papers of the Museum of Zoology, University of Michigan* (601), 1–5, 1 fig., 1 pl.

Hartweg, N. & Oliver, J.A. (1938) A contribution to the herpetology of the Pacific slope. *Occasional Papers of the Museum of Zoology, University of Michigan* (390), 1–8, 1 pl.

Hartweg, N. & Oliver, J.A. (1940) A contribution to the herpetology of the Isthmus of Tehuantepec. IV. *Miscellaneous Publications, Museum of Zoology, University of Michigan* (47), 7–31.

Harvey, M.B. (1994) A new species of montane pitviper (Serpentes, Viperidae, *Bothrops*) from Cochabamba, Bolivia. *Proceedings of the Biological Society of Washington* 107(1), 60–66, 3 figs.

Harvey, M.B. (1999) Revision of Bolivian *Apostolepis* (Squamata, Colubridae). *Copeia* 1999(2), 388–409, 11 figs.

Harvey, M.B. (2008) New and poorly known *Dipsas* (Serpentes: Colubridae) from northern South America. *Herpetologica* 64(4), 422–451, 8 figs.

Harvey, M.B. & Embert, D. (2009) Review of Bolivia *Dipsas* (Serpentes: Colubridae) with comments on other South American species. *Herpetological Monographs* (2008) (22), 54–105, 14 figs.

Harvey, M.B. & Muñoz-S, A. (2004) A new species of *Tomodon* (Serpentes, Colubridae) from high elevations in the Bolivian Andes. *Herpetologica* 60(3), 364–372, 3 figs.

Harvey, M.B., Barker, D.G., Ammerman, L.K. & Chippindale, P.T. (2000) Systematics of pythons of the *Morelia amethistina* complex (Serpentes, Boidae) with the description of three new species. *Herpetological Monographs* (14), 139–185, 11 figs.

Harvey, M.B., Gonzales, L. & Scrocchi, G.J. (2001) New species *Apostolepis* (Squamata, Colubridae) from the Gran Choco in southern Bolivia. *Copeia* 2001(2), 501–507, 2 figs.

Harvey, M.B., Aparicio, J. & Gonzalez, L. (2003). Revision of the venomous snakes of Bolivia. Part 1. The coralsnakes (Elapidae: *Micrurus*). *Annals of the Carnegie Museum* 72(1), 1–52.

Harvey, M.B., Aparicio, J. & Gonzales, L. (2005) Revision of the venomous snakes of Bolivia. II: the pitvipers (Serpentes: Viperidae). *Annals of the Carnegie Museum* 74(1), 1–37, 14 figs.

Harvey, M.B., Fuenmayor, G.E., Partilla, J.R.C. & Rueda-Almonacid, J.V. (2008) Systematics of the enigmatic dipsadine snake *Tropidodipsas perijanensis* Alemán (Serpentes: Colubridae) and review of morphological characters of Dipsadini. *Herpetological Monographs* (22), 106–132, 11 figs.

Hasselquist, F. (1762) *Iter Palaestinum. Reise nach Palästina in der Jahren 1749 bis 1752. Auf befehl ihro Majestät der koniginn von Schweden herausgegeben von Carl Linnäus. Aus dem Swedischen.* Johann Christian Koppe, Rostock, xiii + 606 pp.

Hasselt, A.W.M. (1857) Aanteekening over en nadere beschrijving van een individu der grootste tot nu bekende gift-slangen uit het geslacht der *Naja's. Verslagen en Mededeelingen der Koninklijke Akademie van Afdeeling Natuurkunde* 7, 200–208, 1 fig.

Hasselt, A.W.M. (1862) De Slangen-beker van Ceylon. *Verslagen der Mededeelingen der Koninklijke Akademie van Wetenschappen*, Amsterdam 13, 116–128.

Hasselt, A.W.M. (1882) Eine monster-*Naja*. *Verslagen en Mededeelingen der Koninklijke Akademie van Wetenschappen*, Amsterdam (Series 2) 17, 140–143.

Hati, R.N., Mandal, M. & Hati, A.K. (1990) Active immunization of rabbit with gamma irradiated Russell's viper venom toxoid. *Toxicon* 28(8), 895–902, 2 figs.

Hatooka, K. (1996) Catalogue of the type specimens of amphibians and reptiles preserved in Osaka Museum of Natural History. *Osaka Museum of Natural History* (50), 17–43.

Hatori, J. (1905) Formosa thanatophidia. *Transactions of the Natural History Society of Formosa* 3, 53–58. (in Chinese)

Hatori, S. (1913) Newly discovered poisonous snakes and their venoms. *Transactions of the Natural History Society of Formosa* 3(10), 53–58. (in Chinese)

Hattori, Z., Inoue, T. & Okonogi, T. (1974) In front of a pile of sea snake carcasses. *The Snake* 6(2), 94–98, 5 figs. (in Japanese)

Hawlitschek, O., Nagy, Z.T. & Glaw, F. (2012) Island evolution and systematic revision of Comoran snakes: why and when subspecies still make sense. *Public Library of Science (PLoS) One* 7(8): 1–19, 7 figs.

Haworth, A.H. (1825) A binary arrangement of the Class Amphibia. *Philosophical Magazine and Journal* 65, 372, 1 pl.

Hawthorne, T. (1998) *First field guide to snakes & other reptiles of southern Africa.* Struik Publishers, Cape Town, 56 pp., col. illust.

Haxhiu, I. (1997) The Reptilia of Albania: species composition, distribution, habitats. *Bonnischer Zoologische Beiträge* 48(1), 35–57, illust.

Hay, C. (1998) Captive husbandry and reproduction of the water python *Liasis fuscus* (Dumeril & Bibron, 1844). *Monitor,* Ardeeer 9(2), 14–19, 3 figs., 5 col. photos.

Hay, O.P. (1887) A preliminary catalogue of the Amphibia and Reptilia of the state of Indiana. *Journal of the Cincinnati Society of Natural History* 10, 59–69.

Hay, O.P. (1892a) Description of a supposed new species of *Storeria* from Florida, *Storeria victa. Science* 19(479), 199.

Hay, O.P. (1892b) The batrachians and reptiles of the state of Indiana. *Annual Report, Indiana Department of Geology and Natural Resources* (1891) 17, 409–602, 3 pls.

Hay, W.P. (1902) A list of the batrachians and reptiles of the District of Columbia and vicinity. *Proceedings of the Biological Society of Washington* 15, 121–145.

Hayes, F.E. (1991) Notes on reproduction in the false water cobra (*Hydrodynastes gigas*). *Bulletin of the Maryland Herpetological Society* 27(3), 166–170.

Hayes, M.P. (1975) *The taxonomy and evolution of* Lampropeltis zonata. Master's Thesis, California State University, Chico, 128 pp.

Hayes, M.P., Pounds, J.A. & Timmerman, W.W. (1989) An annotated list and guide to the amphibians and reptiles of Monteverde, Costa Rica. *Society for the Study of Amphibians and Reptiles, Herpetological Circular* (17), 1–67, 32 figs.

Hayes, W.K., Beaman, K.R., Cardwell, M.D. & Bush, S.P. (Eds.). (2008) *The biology of rattlesnakes.* Loma Linda University Press, Loma Linda, xv + 590 pp., 20 col. pls.

He, X.-R. (1983) Preliminary studies on biology of *Amphiesma octolineata. Acta Herpetologica Sinica*, Chengdu (Series 2) 2(2), 11–17.

Head, J.J. (2005) Snakes of the Siwalik group (Miocene of Pakistan): systematics and relationships to environmental change. *Palaeontologia Electronica* 8(1), 1–31, 15 figs.

Head, J.J. (2006) Fossil snakes from the Neogene of Venezuela (Falcón state). *Journal of Systematic Palaeontology* 4(3), 233–240, 1 fig.

Head, J.J., Mohabey, D.M. & Wilson, J.A. (2007) *Acrochordus* Hornstedt (Serpentes, Caenophidia) from the Miocene of Gujarat, western India: temporal constraints on dispersal of a derived snake. *Journal of Vertebrate Paleontology* 27(3), 720–723, 2 figs.

Head, J.J., Bloch, J.I., Hastings, A.K., Bourque, J.R., Cadena, E.A., Herrera, F.A., Polly, P.D. & Jaramillo, C.A. (2009) Giant boid snake from the Paleocene neotropics reveals hotter past equatorial temperatures. *Nature* 457(7230), 715–717, 3 figs.

Head, J.J., Rincon, A.F., Suarez, C., Montes, C. & Jaramillo, C.A. (2012) Fossil evidence for earliest Neogene American faunal interchange: *Boa* (Serpentes, Boinae) from the early Miocene of Panama. *Journal of Vertebrate Paleontology* 32(6), 1328–1334, 3 figs.

Heang, K.B. (1987) An annotated checklist of the herpetofauna of Ulu Endau, Johore, Malaysia. *Malayan Nature Journal*, Kuala Lumpur 41(2–3), 413–423.

Heatwole, H. (1981) Role of the saccular lung in the diving of the sea krait, *Laticauda colubrina* (Serpentes, Laticaudidae). *Australian Journal of Herpetology*, Sydney 1(1), 11–16, figs.

Heatwole, H. (1999) *Sea snakes*. Second edition Krieger Publishing Company, Malabar, 148 pp., illust., 29 col. pls. [first edition by New South Wales University Press, 1987, 85 pp., illust.]

Heatwole, H. (2010) *Distribution and geographic variation in sea kraits in the* Laticauda colubrina *complex (Serpentes, Elapidae, Laticaudinae)*. Doctoral Dissertation, James Cook University, Townsville, xxiv + 285 pp., 53 figs.

Heatwole, H. & Johnson, C.R. (1979) Thermoregulation in the red-bellied blacksnake, *Pseudechis porphyriacus* (Elapidae). *Zoological Journal of the Linnean Society*, London 65(1), 83–101.

Heatwole, H., Busack, S. & Cogger, H. (2005) Geographic variation in sea kraits of the *Laticauda colubrina* complex (Serpentes, Elapidae, Hydrophinae, Laticaudini). *Herpetological Monographs* (19), 1–136, 40 figs.

Hecht, G. (1929a) Zur Kenntnis der Nordgrenzen mitteleuropäischen Reptilien. *Mitteilungen aus dem Naturhistorischen Museum in Berlin* 14, 501–596, pl. 3.

Hecht, G. (1929b) Zur Kenntnis der Schlangengattung *Chlorophis* Hallow. *Zoologischer Anzeiger*, Leipzig 81(11–12), 329–335.

Hecht, G. (1930a) Systematik, Ausbreitungsgeschichte und Oekologie der europäischen Arten der Gattung *Tropidonotus* (Kuhl) H. Boie. *Mitteilungen aus dem Naturhistorischen Museum in Berlin* 16(2), 244–393, pls. 2–3.

Hecht, G. (1930b) *Luxemburger Reptilien und Amphibien: ihre geographischen Beziehungen und ihre Einwanderungsgeschichte*. Luxemburg, 26 pp.

Hecht, M.K. (1959) Amphibians and reptiles. *In:* McGrew, P.O. (Ed.), The geology and paleontology of the Elk Mountain and Tabernacle Butte area, Wyoming. *Bulletin of the American Museum of Natural History* 117(3), pp. 130–146, pls. 52–57.

Hecht, M.K. (1982) The vertebral morphology of the Cretaceous snake, *Dinilysia patagonica* Woodward. *Neues Jahrbuch für Geologie und Paläontologie, Monatshheft*, Stuttgart 1982(9), 523–532, illust.

Hecht, M.K. & LaDuke, T.C. (1988) Bolyerine vertebral variation: a problem for paleoherpetology. *Acta Zoologica Cracoviensia*, Kraków 31(24), 605–614, 2 figs.

Hecht, M.K., Kropach, C. & Hecht, B.M. (1974) Distribution of the yellow-bellied sea snake, *Pelamis platurus*, and its significance to the fossil record. *Herpetologica* 30(4), 387–396.

Hecht-Kardasz, K. & Nickerson, M.A. (2011) Geographic distribution: Serpentes: *Ramphotyphlops braminus*. *Herpetological Review* 42(3), 395–396.

Hedges, N.G. (1983) *Reptiles and amphibians of East Africa*. Kenya Literature Bureau, Nairobi, ix + 139 pp., 112 figs.

Hedges, S.B. (1996) The origin of West Indian amphibians and reptiles. *In:* Powell, R. & Henderson, R. (Eds.), *Contributions to West Indian herpetology: a tribute to Albert Schwartz*. Society for the Study of Amphibians and Reptiles, pp. 95–128, 5 figs.

Hedges, S.B. (2002) Morphological variation and the definition of species in the snake genus *Tropidophis*. *Bulletin of the Natural History Museum of London (Zoology)* 66(2), 83–90.

Hedges, S.B. (2008) At the lower size limit in snakes: two new species of threadsnakes (Squamata: Leptotyphlopidae: *Leptotyphlops*) from the Lesser Antilles. *Zootaxa* (1841), 1–30, 7 figs.

Hedges, S.B. (2011) The type species of the threadsnake genus *Tricheilostoma* Jan revisited (Squamata, Leptotyphlopidae). *Zootaxa* (302), 63–64.

Hedges, S.B. & Garrido, O.H. (1992a) Cuban snakes of the genus *Arrhyton*: two new species and a reconsideration of *A. redimitum* Cope. *Herpetologica* 48(2), 168–177, 4 figs.

Hedges, S.B. & Garrido, O.H. (1992b) A new species of *Tropidophis* from Cuba (Serpentes, Tropidophiidae). *Copeia* 1992(3), 820–825, 4 figs.

Hedges, S.B. & Garrido, O.H. (1999) A new snake of the genus *Tropidophis* (Tropidophiidae) from central Cuba. *Journal of Herpetology* 33(3), 436–441, 3 figs.

Hedges, S.B. & Garrido, O.H. (2002) A new snake of the genus *Tropidophis* (Tropidophiidae) from eastern Cuba. *Journal of Herpetology* 36(2), 157–161, 1 fig.

Hedges, S.B. & Thomas, R. (1991) Cryptic species of snakes (Typhlopidae, *Typhlops*) from the Puerto Rico Bank detected by protein electrophoresis. *Herpetologica* 47(4), 448–459, 6 figs.

Hedges, S.B., Estrada, A.R. & Diaz, L.M. (1999) New snake (*Tropidophis*) from western Cuba. *Copeia* 1999(2), 376–381, 3 figs.

Hedges, S.B., Garrido, O.H. & Diaz, L.M. (2001) A new banded snake of the genus *Tropidophis* (Tropidophiidae) from north-central Cuba. *Journal of Herpetology* 35(4), 615–617, 1 fig.

Hedges, S.B., Couloux, A. & Vidal, N. (2009) Molecular phylogeny, classification, and biogeography of West Indian racer snakes of the tribe Alsophiini (Squamata, Dipsadidae, Xenodontinae). *Zootaxa* (2067), 1–28, 4 figs.

Hediger, H. (1933) Über die von Herrn Dr. A. Bühler auf der Admiralitäts-Gruppe und einigen benachbarten Inseln gesammelten Reptilien und Amphibien. *Verhandlungen der Naturforchenden Gesellschaft in Basel* 44(2), 28–52, 2 figs.

Hediger, H. (1934) Beitrag zur herpetologie und zoogeographie Neu Britanniens und einiger umliegender Gebiete. *Zoologische Jahrbücher (Abteilung für Systematic, Ökologie und Geographie der Tiere)*, Jena 65, 441–582.

Hediger, H. (1935) Herpetologische Beobachtungen in Marokko. *Verhandlungen der Naturforchenden Gesellschaft in Basel* 46, 1–49.

Hediger, H. (1937) *Die Schlangen Mitteleuropas*. Benno Schwabe, Basel, 54 pp., 40 figs.

Heilprin, A. (1887) Explorations on the west coast of Florida and in the Okeechobee wilderness. *Transactions of the Wagner Free Institute of Science*, Philadelphia 1, vii + 134 pp.

Heimes, P. (1993) *Coluber viridiflavus* Lacépède, 1789 — Gelbgrüne Zornnatter. *In:* Böhme, W. (Ed.), *Handbuch der Reptilien und Amphibien Europas. Band 3/I. Schlangen (Serpentes) I. (Typhlopidae, Boidae, Colubridae 1, Colubrinae)*. AULA-Verlag, Wiesbaden, pp. 177–198, figs. 40–41.

Heise, P.J., Maxson, L.R., Dowling, H.G. & Hedges, S.B. (1995) Higher-level snake phylogeny inferred from mitochondrial DNA sequences of 12S rRNA and 16S rRNA genes. *Molecular Biology and Evolution* 12(2), 259–265, 1 fig.

Helfenberger, N. (2001) Phylogenetic relationships of Old World ratsnakes based on visceral organ topography, osteology, and alloyme variation. *Russian Journal of Herpetology (Supplement)*, 62 pp., 19 figs.

Helfenberger, N. & Schätti, B. (1998) Morphological adaptations for egg-eating in the snake *Elaphe davidi* Sauvage 1884. *Russian Journal of Herpetology* 5(1), 36–42, 3 figs.

Helfenberger, N., Shan, K.B., Orlov, N.L. & Guen, G.-D. (2000) Eine seltene Natter aus Nepal, *Elaphe cantoris* (Boulenger, 1894) (Squamata, Serpentes, Colubridae). *Sauria*, Berlin 22(2), 3–10, 9 col. figs.

Heller, F. (1960) Würmeiszeitliche und letztinterglaziale Faunenreste von Lobsing bei Neustadt/Donau. *Erlanger Geologische Abhandlungen* (34), 19–33, pls. 2–3.

Hellmich, W. (1953) Contribuciones al conocimiento de los ofidios de Venezuela. 2. Sobre la subespecie venezalana de *Coluber (Masticophis) mentovarius*. *Acta Biologica Venezuelica*, Caracas 1, 146–154.

Hellmich, W. (1956) *Die Lurche und kriechtiere Europas*. Carl Winters, Heidelberg, 26, 1–166, 9 figs., 68 pls. [English translation in 1962 by Blandford Press, London]

Hellmich, W. (1957a) Herpetologische Ergebnisse einer Forschungsreise in Angola. *Veroffentlichungen der Zoologische Staatssammlung München* 5, 1–92, 3 figs., 8 pls.

Hellmich, W. (1957b) Die Reptilienausbeute der Hamburgischen Angola-Expedition. *Mitteilungen aus dem Hamburgisches Zoologisches Museum und Institut*, Hamburg 55, 39–80.

Hellmich, W. (1962) *Reptiles and amphibians of Europe*. Blandford Press, London, viii + 160 pp., 68 pls.

Hemprich, C. (1827) Wilhelm Friedrich Hemprich. *Neuer Nekrolog der Deutschen* 3, 774–793.

Hemprich, W. (1829) *Grundriss der Naturgeschichte für höhere Lehranstalten. Zweite Auflage*. August Rücker, Berlin, x + 489 pp. [first edition, 1820, viii + 432 pp.]

Henderson, J.T. & Bieber, A.L. (1986) Antigenic relationships between Mojave toxin subunits, Mojave toxin and some crotalid venoms. *Toxicon* 24(5), 473–479, 3 figs.

Henderson, R.W. (1976) A new insular subspecies of the colubrid snake *Leptophis mexicanus* (Reptilia, Serpentes, Colubridae) from Belize. *Journal of Herpetology* 10(4), 329–331.

Henderson, R.W. (1984) *Scaphiodontophis* (Serpentes: Colubridae): natural history and test of a mimicry-related hypothesis. *In:* Seigel, R.S., Hunt, L.E., Knight, J.L, Malaret, L. & Zuschlag, N.L. (Eds.), *Vertebrate Ecology and Systematics: a tribute to Henry S. Fitch.* University of Kansas Museum of Natural History, 185–194.

Henderson, R.W. (1990) A new subspecies of *Alsophis antiguae* (Serpentes, Colubridae) from Great Bird Island (Antigua), Lesser Antilles. *Caribbean Journal of Science*, Mayagüez 25(3–4), 119–122, figs.

Henderson, R.W. (1991) Distribution and preliminary interpretation of geographic variation in the Neotropical tree boa *Corallus enydris*: a progress report. *Bulletin of the Chicago Herpetological Society* 26(5), 105–110, 6 figs.

Henderson, R.W. (1993a) *Corallus* Daudin. *Catalogue of American amphibians and reptiles*, pp. 572.1–572.2.

Henderson, R.W. (1993b) *Corallus annulatus* (Cope). *Catalogue of American amphibians and reptiles*, pp. 573.1–573.3, 2 figs.

Henderson, R.W. (1993c) *Corallus caninus* (Linnaeus). *Catalogue of American amphibians and reptiles*, pp. 574.1–574.4, 3 figs.

Henderson, R.W. (1993d) *Corallus enydris* (Linnaeus). *Catalogue of American amphibians and reptiles*, pp. 576.1–576.6, 7 figs.

Henderson, R.W. (1997) A taxonomic review of the *Corallus hortulanus* complex of neotropical tree boas. *Caribbean Journal of Science* 33(3–4), 198–221, 10 figs.

Henderson, R.W. (2002) *Neotropical treeboas. Natural history of the* Corallus hortulanus *complex*. Krieger Publishing Company, Malabar, xiv + 197 pp., figs., 32 col. pls.

Henderson, R.W. & Arias-B., A. (2001) *Epicrates angulifer. Catalogue of American amphibians and reptiles*, pp. 734.1–734.4, 5 figs.

Henderson, R.W. & Hedges, S.B. (1995) Origin of West Indian populations of the geographically widespread boa *Corallus enydris* inferred from mitochondrial DNA sequences. *Molecular Phylogenetics and Evolution* 4(1), 88–92, 3 figs.

Henderson, R.W. & Hoevers, L.G. (1975) A checklist and key to the amphibians and reptiles of Belize, Central America. *Contributions in Biology and Geology, Milwaukee Public Museum* (5), 1–63, 1 fig.

Henderson, R.W. & Hoevers, L.G. (1979) Variation in the snake *Tretanorhinus nigroluteus lateralis* in Belize with notes on breeding tubercles. *Herpetologica* 35(3), 245–248.

Henderson, R.W. & Pauers, M.J. (2012) On the diets of Neotropical treeboas (Squamata: Boidae: *Corallus). South American Journal of Herpetology* 7(2), 172–180, 2 figs..

Henderson, R.W. & Powell, R. (1996a) *Alsophis ater. Catalogue of American amphibians and reptiles*, pp. 633.1–633.2, 1 fig.

Henderson, R.W. & Powell, R. (1996b) *Alsophis sanctaecrucis. Catalogue of American amphibians and reptiles*, pp. 634.1–634.2, 1 fig.

Henderson, R.W. & Powell, R. (1996c) *Chironius vincenti. Catalogue of American amphibians and reptiles*, pp. 635.1–635.2, fig. 1.

Henderson, R.W. & Powell, R. (1999) West Indian herpetoecology. *In:* Crother, B. (Ed.), *Caribbean amphibians and reptiles.* Academic Press, San Diego, pp. 223–268, 4 figs.

Henderson, R.W. & Powell, R. (2002) *Epicrates gracilis. Catalogue of American amphibians and reptiles*, pp. 754.1–754.4, 3 figs.

Henderson, R.W. & Powell, R. (2004) *Epicrates fordii. Catalogue of American amphibians and reptiles*, pp. 800.1–800.3, 2 figs.

Henderson, R.W. & Powell, R. (2006) Geographic distribution: Serpentes: *Tantilla melanocephala. Herpetological Review* 37(4), 501.

Henderson, R.W. & Powell, R. (Eds.). (2007) *Biology of the boas and pythons.* Eagle Mountain Publishing, Salt Lake City, x + 438 pp., illust.

Henderson, R.W. & Powell, R. (2009) *Natural history of West Indian Reptiles and Amphibians.* University Press of Florida, Gainesville, xxiv + 495 pp., 37 figs.

Henderson, R.W. & Puorto, G. (1993) *Corallus cropanii* (Hoge). *Catalogue of American amphibians and reptiles*, pp. 575.1–575.2, 1 fig.

Henderson, R.W. & Sajdak, R.A. (1996) Diets of West Indian racers (Colubridae, *Alsophis*), composition and biogeographic implications. *In:* Henderson, R.W. & Powell, R. (Eds.), *Contributions to West Indian herpetology: a tribute to Albert Schwartz.* Society for the Study of Amphibians and Reptiles, pp. 327–338, 5 figs.

Henderson, R.W. & Schwartz, A. (1984a) A guide to the identification of the amphibians and reptiles of Hispaniola. *Special Publications in Biology and Geology, Milwaukee Public Museum* (4), 1–70, illust.

Henderson, R.W. & Schwartz, A. (1984b) *Uromacer frenatus. Catalogue of American amphibians and reptiles*, pp. 357.1–357.2.

Henderson, R.W., Hoevers, L.G. & Wilson, L.D. (1977) A new species of *Sibon* (Reptilia, Serpentes, Colubridae) from Belize, Central America. *Journal of Herpetology* 11(1), 77–79, 1 fig.

Henderson, R.W., Schwartz, A. & Inchaustegui, S.J. (1984) Guia para la identificacion de anfibios y reptiles de la Hispaniola. *Museo Nacional de Historia Natural, Monografias*, Santo Domingo (1), 1–128.

Henderson, R.W., Sajdak, R.A. & Henderson, R.M. (1988) The rediscovery of the West Indian colubrid snake *Chironius vincenti. Amphibia-Reptilia*, Leiden 9(4), 415–416.

Henderson, R.W., Micucci, T.W.P., Puorto, G. & Bourgeois, R.W. (1995) Ecological correlates and patterns in the distribution of Neotropical boines (Serpentes, Boidae): a preliminary assessment. *Herpetological Natural History* 3(1), 15–27, 3 figs.

Henderson, R.W., Powell, R., Daltry, J.C. & Day, M.L. (1996a) *Alsophis antiguae. Catalogue of American amphibians and reptiles*, pp. 632.1–632.3, 1 fig.

Henderson, R.W., Winstel, R.A. & Friesch, J. (1996b) *Corallus hortulanus* (Serpentes, Boidae) in the Post-Columbian West Indies: new habitats, new prey species, and new problems. *In:* Henderson, R.W. & Powell, R. (Eds.), *Contributions to West Indian herpetology: a tribute to Albert Schwartz*. Society for the Study of Amphibians and Reptiles, pp. 417–423, 3 figs.

Henderson, R.W., Höggren, M., Lamar, W.W. & Porras, L.W. (2001) Distribution and variation in the treeboa *Corallus annulatus* (Serpentes: Boidae). *Studies on Neotropical Fauna and Environment* 36(1), 39–47.

Henderson, R.W., Passos, P. & Feitoso, D. (2009) Geographic variation in the emerald treeboa, *Corallus caninus* (Squamata: Boidae). *Copeia* 2009(3), 572–582, 5 figs.

Hendrickson, J.R. (1966) The reptiles. *In:* Observations on the fauna of Pulau Tiomàn and Pulau Tulai. *Bulletin of the National Museum*, Singapore (34), 374–377.

Henkel, F.-W. & Schmidt, W. (2000) *Amphibians and reptiles of Madagascar and the Mascarene, Seychelles, and Comoro Islands*. Krieger Publishing Company, Malabar, viii + 316 pp., col. illust.

Henle, J. (1839) *Vergleichend-Anatomische beschreibung des Kehlkopfs mit besonderer berücksichtigung des Kehlkopfs der Reptilien.* Verlag von Leopold Voss, Leipzig, 83 pp., 5 pls.

Henle, K. (1993) *Coluber gemonensis* Laurenti, 1768—Balkanzornnatter. *In:* Böhme, W. (Ed.), *Handbuch der Reptilien und Amphibien Europas. Band 3/I. Schlangen (Serpentes) I. (Typhlopidae, Boidae, Colubridae 1, Colubrinae).* AULA-Verlag, Wiesbaden, pp. 97–110, figs. 19–21.

Henle, K. & Ehrl, A. (1991) Zur reptilienfauna Perus nebst Beschreibung eines neuen *Anolis* (Iguanidae) und zweier neuer Schlangen (Colubridae). *Bonnischer Zoologische Beiträge*, Bonn 42(2), 143–180, 10 figs.

Hensel, R. (1868) Beiträge zur Kenntniss der Wirbelthiere Südbrasiliens. *Archiv für Naturgeschichte*, Berlin 34(1), 323–375.

Hensley, M.M. (1950) Results of a herpetological reconnaissance in extreme southwestern Arizona and adjacent Sonora, with description of a new subspecies of the Sonoran whipsnake, *Masticophis bilineatus*. *Transactions of the Kansas Academy of Sciences* 53(2), 270–288, 3 figs.

Hensley, M.M. (1966) A new subspecies of the Mexican snake, *Sympholis lippiens* Cope. *Herpetologica* 22(1), 48–55, 2 figs.

Hensley, M.M. & Lannom, J.R. (1966) Noteworthy snake records for the Mexican states of Colima, Jalisco, and Nayarit. *Herpetologica* 22(3), 231–235.

Herczek, A. & Gorczyca, J. (1999) *Plazy I gady Polski: atlas I klucz.* Wydawnictwo Kubajak, Lesna, 104 pp., col. illust.

Herklots, G.A.C. (1933) Land snakes of Hong Kong. *Hong Kong Naturalist* 4(2), 115–126, 2 figs., pl. 17.

Herklots, G.A.C. (1934) Land snakes of Hong Kong. Part II. *Hong Kong Naturalist* 5(1), 23–30, 3 figs.

Herklots, G.A.C. (1935) Land snakes of Hong Kong. Part III. *Hong Kong Naturalist* 6(3–4), 192–206, 3 figs., pl. 18.

Herklots, G.A.C. (1938) Land snakes of Hong Kong. Part IV. *Hong Kong Naturalist* 8(3–4), 189–200, 3 figs., pls. 17–18.

Herman, J.S., McGowan, R.Y. & Swinney, G.N. (1990) *Catalogue of the type specimens of recent vertebrates in the National Museums of Scotland. National Museums of Scotland.* Online at http://hdl.handle.net/10068/584605.

Hermann, J. (1804) *Observationes zoologicae quibus novae complures, aliaeque animalium species describuntur et illustrantur, opus posthumus edidit Fridericus Ludovicus Hammer. Pars prior, observationum quatuor centurias continens.* Amandum Koenig, Parisiis, viii + 332 pp.

Hermann, R. (1989a) *Lamprophis fuliginosus* (Boie 1827): die affrikanische Hausschlange. *Sauria,* Berlin 11(1), 3–8, illust.

Hermann, R. (1989b) *Lamprophis aurora* (Linnaeus 1758): eine selten gehaltene Hausschlange. *Sauria,* Berlin 11(3), 25–30, illust.

Hermann, R. (1991) Seltene Hausschlangen Sudafrikas. Teil 1. *Lamprophis fuscus* Boulenger 1893 und *Lamprophis guttatus* (Smith 1843). *Sauria*, Berlin 13(2), 25–30, illust.

Hermann, R. (1993) Seltene Hausschlangen Sudafrikas. Teil 2. *Lamprophis inornatus* (Duméril, Bibron & Duméril, 1854) und *Lamprophis swazicus* Schaefer, 1970. *Sauria*, Berlin 15(4), 3–8, illust.

Hernández, M. & Rada, M.D. de. (1992) Contribucion al conocimiento del genero *Mastigodryas* (Serpentes, Colubridae) en Venezuela. *Acta Biologica Venezuelica*, Caracas 13(3–4), 67–81, 7 figs.

Hernandez, R.A. (2004) Geographic distribution: Serpentes: *Oxybelis aeneus*. *Herpetological Review* 35(1), 84.

Hernández-Ibarra, X., Ramírez-Bautista, A., Torres-Cervantes, R. & Hernández-Macías, H. (1999) Geographic distribution: Serpentes: *Ficimia hardyi*. *Herpetological Review* 30(4), 235.

Hernández-Ríos, A. & Trejo-Perez, R.G. (2012) Geographic distribution: Serpentes: *Ramphotyphlops braminus*. *Herpetological Review* 43(4), 622.

Hernández-Ruiz, E.J. (2005) Geographic distribution: Serpentes: *Stenorrhina freminvillei*. *Herpetological Review* 36(2), 204.

Hernández-Salinas, U., Cruz-Elizalde, R., Berriosabal-Islas, C. & Goyenechea, I. (2011) Geographic distribution: Serpentes: *Rena (=Leptotyphlops) dulcis*. *Herpetological Review* 42(2), 244.

Herre, A.W.C.T. (1942) Notes on Philippine sea-snakes. *Copeia* 1942(1), 7–9.

Herre, A.W.C.T. & Rabor, D.S. (1949) Notes on Philippine sea snakes of the genus *Laticauda*. *Copeia* 1949(4), 282–284.

Herrera, A.L. (1899) *Sinonimia vulgar y cientifíca de los principales vertebrados Mexicanos.* Oficina Tipográfica de la Secretaría de Fomento, Méjico, 31 pp.

Herrera, A.L. (1890) Notas acerca de los vertebrados del Valle de México. *La Naturaleza,* México City (Serie 2) 1, 299–342.

Herrera, N., Hendriquez, V. & Greenbaum, E. (2007) New country and department records for amphibians and reptiles from El Salvador. *Herpetological Review* 38(2), 222–226.

Herrmann, H.-W. & Joger, U. (1995) Molecular data on the phylogeny of African vipers-preliminary results. *In:* Llorente et al. (Eds.), *Scientia Herpetologica*. Asociación Herpetológica Española, Barcelona, pp. 92–97, 1 fig.

Herrmann, H.-W. & Joger, U. (1997) Evolution of viperine snakes. *In:* Thorpe, R.S., Wüster, W. & Malhotra, A. (Eds.), *Ecology, evolution and snakebite venomous snakes*. Zoological Society of London, Clarendon Press, Oxford, pp. 43–61, 7 figs.

Herrmann, H.-W., Joger, U. & Nilson, G. (1992) Phylogeny and systematics of viperine snakes. III: resurrection of the genus *Macrovipera* (Reuss, 1927) as suggested by biochemical evidence. *Amphibia-Reptilia*, Leiden 13(4), 375–392, 5 figs.

Herrmann, H.-W., Joger, U., Lenk, P. & Wink, M. (1999) Morphological and molecular phylogenies of Viperines: conflicting evidence? *In:* Joger, U. (Ed.), *Phylogeny and systematics of the Viperidae.* Kaupia, Darmstädter Beiträge zur Naturgeschichte, 21–30, 5 figs.

Herrmann, H.-W., Ziegler, T., Stuart, B.L. & Orlov, N.L. (2002) New findings on the distribution, morphology, and natural history of *Triceratolepidophis sieversorum* (Serpentes, Viperidae). *Herpetological Natural History* 9(1), 89–94, 4 figs.

Herrmann, H.-W., Herrmann, P.A. & Wallach, V. (2004a) Geographic distribution: Serpentes: *Leptotyphlops sundewalli gestri*. *Herpetological Review* 35(4), 411.

Herrmann, H.-W., Ziegler, T., Malhotra, A., Thorpe, R.S. & Parkinson, C.L. (2004b) Redescription and systematics of *Trimeresurus cornutus* (Serpentes, Viperidae) based on morphology and molecular data. *Herpetologica* 60(2), 211–221, 4 figs.

Herscu, M. (2004) *Snakes of the Pioneer Valley.* Apple Blossom Press, Amherst, 144 pp., col. illust.

Herter, K. (1960) *Das Thierreich. VII. Chordatiere. 4. Kreichtiere.* Walter de Gruyter, Leipzig, 200 pp., 142 figs.

Hes, O., Duda, Z., Hnízdo, J., Pantchev, N., Bulantová, J. & Vrabec, V. (2007) *Boa constrictor: Biologie – Pflege – Zucht – Erkrankungen.* Herpeton, Verlag Elke Köhler, Offenbach, 270 pp., col. illust.

Hess, S. (1980) *Rattlers & rattler hunters.* Stan Hess, Pennsylvania, 112 pp., illust.

Hewitt, J. (1910) Description of a new snake, *Prosymna transvaalensis* from the Transvaal. *Annals of the Transvaal Museum*, Pretoria 2(3), 73.

Hewitt, J. (1912) Notes on the specific characters and distribution of some South Africa Ophidia and Batrachia. *Records of the Albany Museum*, Grahamstown 2(4), 264–281.

Hewitt, J. (1937) *A guide to the vertebrate fauna of the eastern Cape Province, South Africa. Part II. Reptiles, amphibians, and freshwater fishes.* Albany Museum, Grahamstown, vii + 141 pp., 34 pls.

Hewitt, J. & Power, J.H. (1913) A list of South African Lacertilia, Ophidia, and Batrachia in the McGregor Museum, Kimberley; with field-notes on various species. *Transactions of the Royal Society of South Africa*, Cape Town 3(1), 147–176.

Heyer, W.R. (1967) A herpetological study of an ecological transect through the Cordillera de Tilaran, Costa Rica. *Copeia* 1967(2), 259–287.

Heymann, M.M. (1975) *Reptiles and amphibians of the American Southwest.* Doubleshoe Publishers, Scottsdale, 77 pp., col. photos.

Hidalgo, H. (1979) Range extension of the snake *Sibon anthracops* (Cope) in El Salvador. *Herpetological Review* 10(3), 103.

Hidalgo, H. (1980a) *Enyaliosaurus quinquecarinatus* (Gray) and *Leptodeira nigrofasciata* Günther in El Salvador. *Herpetological Review* 11(2), 42–43.

Hidalgo, H. (1980b) Occurrence of *Pelamis platurus* (Linnaeus) in El Salvador. *Herpetological Review* 11(4), 117.

Hidalgo, H. (1981a) Additions to the reptile fauna of El Salvador. *Transactions of the Kansas Academy of Sciences* 84(1), 55–58.

Hidalgo, H. (1981b) Additions to the snake fauna of El Salvador. *Herpetological Review* 12(2), 67–68.

Hiert, C. (2006) Geographic distribution: Serpentes: *Echinanthera bilineata*. *Herpetological Review* 37(1), 108.

Higgins, S.B. (1873) *Ophidians: zoological arrangement of the different genera.* Boericke and Tafel, New York, 239 pp.

Highton, R. (1956) Systematics and variation of the endemic Florida snake genus *Stilosoma*. *Bulletin of the Florida State Museum, Biological Sciences* 1(2), 73–96, 3 figs.

Highton, R. (1976) *Stilosoma. S. extenuatum. Catalogue of American amphibians and reptiles*, pp. 183.1–183.2.

Higley, W.K. (1889) Reptilia and Batrachia of Wisconsin. *Transactions of the Wisconsin Academy of Sciences* 7, 155–176.

Hikida, T., Ota, H. & Toyama, M. (1992) Herpetofauna of an encounter zone of Oriental and Palearctic elements: amphibians and reptiles of the Tokara Group and adjacent islands in the northern Ryukyus. *Japanese Biological Magazine*, Okinawa, 30, 29–43, 2 figs.

Hilgendorf, F. (1876) Die japanischen Schlangen. *Mitteilungen der Deutschen Gesellschaft für Natur- und Völkerkunde Ostasiens* 1(10), 29–34.

Hilgendorf, F. (1880) Bemerkungen über die von ihm in Japan gesammelten Amphibien nebst Beschreibung zweier neuer Schlangenarten. *Sitzungsberichte der Gesellschaft Naturforschender Freunde zu Berlin* 1880(8), 111–121, 1 pl.

Hilken, G. & Schleppler, R. (1998) Der *Lampropeltis mexicana*-Komplex (Serpentes, Colubridae): Naturgeschichte und Terrarienhaltung. *Salamandra*, Reinbach 34(2), 91–124, 6 figs.

Hill, H.R. (1948) Amphibians and reptiles of Los Angeles County. *The Los Angeles County Museum Science Series* (12), Zoology (5), 1–30, 30 illust.

Hillis, D.M. (1990) A new species of Xenodontine colubrid snake of the genus *Synophis* from Ecuador and the phylogeny of the genera *Synophis* and *Emmochliophis*. *Occasional Papers of the Museum of Natural History, University of Kansas* (135), 1–9, 2 figs.

Hilzheimer, M. (1912) Reptilien. *In:* Hilzheimer, M. & Haempel, O. (Eds.), *Handbuch der Biologie der Wirbeltiere. 1. Hälfte. Fische. – Amphibien. – Reptilien.* Verlag von Ferdinand Enke, Stuttgart, pp. 231–374, figs. 108–245.

Himmelstein, J. (1973a) The development of a northern pine snake embryo-outside of the egg. *Bulletin of the New York Herpetological Society* (1972) 9(1–2), 22–25, 4 figs.

Himmelstein, J. (1973b) Hybridization between a female Florida pine snake and a male San Diego gopher snake. *Bulletin of the New York Herpetological Society* 10(1–2), 19–24, 7 figs.

Himmelstein, J. (1981) Observations and distributions of amphibians and reptiles in the state of Quintana Roo, Mexico. *Bulletin of the New York Herpetological Society* 16(2), 18–34, 3 figs.

Hines, K.N. (2011) Status and distribution of the Rim Rock crowned snake, *Tantilla oolitica*. *Herpetological Review* 42(3), 352–356, 2 figs.

Hirschkorn, D. & Skubowius, B. (2011) *Bullennattern, gophernattern, Kiefernnattern: die Gattung* Pituophis. Natur und Tier-Verlag, Münster, 175 pp., col. illust.

Hnízdo, J. (2000) Neue Erkenntnisse zur Fortpflanzungs biologie von *Ahaetulla prasina* (Boie, 1827). *Sauria*, Berlin 22(2), 41–44, 4 col. figs.

Hoard, R.S. (1939) A new subspecies of snake of the genus *Chilomeniscus*. *Pomona College Journal of Entomology and Zoology* 31(4), 45–46.

Hodge, R.P. (1976) *Amphibians and reptiles in Alaska, the Yukon and Northwest Territories.* Alaska Northwest Publishing Company, Anchorage, xi + 89 pp., illust.

Hodges, R. (1993) Snakes of Java with special reference to East Java Province. *British Herpetological Society Bulletin*, London 43, 15–32, illust.

Hoesch, W. (1960) Von der Baumschlange *Dispholidus typus* und ihren Beutetieren. *Natur und Volk*, Frankfurt-am-Main 90(6), 177–185, 12 figs.

Hoesel, J.K.P. van. (1948) Snake hunting. *Treubia*, Buitenzorg 19(3), 525–538, 7 figs.

Hoesel, J.K.P. van. (1959) *Ophidia Javanica.* Museum Zoologicum Bogoriense, Bogor, 188 pp., 133 figs.

Hoevers, L.G. & Johnson, P.M. (1982) Notes on a collection of snakes from the middle Juba region, Somalia. *Monitore Zoologico Italiano, Supplemento*, Firenze 16(8), 173–203, 6 figs.

Hoffman, A.C. (1940) A new species of *Xenocalamus* from Bloemfontein. *Navorsinge van die Nasionale Museum*, Bloemfontein 1(11), 111–113.

Hoffstetter, R. (1939) Contribution a l'étude des Elapidae actuels et fossiles et de l'ostéologie des ophidiens. *Archives du Muséum d'Histoire Naturelle*, Lyon 15(3), 1–78, 13 figs., 2 pls.

Hoffstetter, R. (1946) Les Typhlopidae fossiles. *Bulletin du Muséum National d'Histoire Naturelle*, Paris (Série 2) 18(3), 309–315, 2 figs.

Hoffstetter, R. (1955) Squamates de type moderne. *In:* Piveteau, J. (Ed.), *Traite de Paléontologie, Vol. 5.* Masson, Paris, pp. 606–662, 26 figs.

Hoffstetter, R. (1958a) Un serpent marin du genre *Pterosphenus* (*Pt. sheppardi* nov. sp.) dan l'Éocène supérieur de l'Équateur (Amérique du Sud). *Bulletin de la Société Géologique de France*, Paris (Série 6) 8(1), 45–50, pl. 7.

Hoffstetter, R. (1958b) Una serpiente marina del genero *Pterosphenus* en el Eoceno superior de Ancon (Ecuador de America). *Boletin de Informaciones Cientificas Nacionales*, Quito, 87, 239–250, 1 fig.

Hoffstetter, R. (1959) Un dentaire de *Madsoia* (Serpent géant du Paléocène de Patagonie). *Bulletin du Muséum National d'Histoire Naturelle*, Paris (Série 2) 31(4), 379–386, 2 figs.

Hoffstetter, R. (1960a) Un serpent terrestre dans le Crétacé inférieur du Sahara. *Bulletin de la Société Géologique de France*, Paris (1959) (Série 7) 1(9), 897–902, 2 figs.

Hoffstetter, R. (1960b) Présence de *Pterosphenus* (serpent paléophidé) dans l'Eocène supèrieur du bord occidental du désert libyque. *Comptes Rendus Sommaires de la Société Géologique de France*, Paris 1960(2), 41, 1 fig.

Hoffstetter, R. (1961a) Nouveaux restes d'un serpent Boïdé (*Madtsoia madagascariensis* nov. sp.) dans le Crétacé supérieur de Madagascar. *Bulletin du Muséum National d'Histoire Naturelle*, Paris (Série 2) 33(2), 152–160, 3 figs.

Hoffstetter, R. (1961b) Nouvelles récoltes de serpents fossiles dans l'Éocène supérieur de désert Libyque. *Bulletin du Muséum National d'Histoire Naturelle*, Paris (Série 2) 33(3), 326–331, 1 fig.

Hoffstetter, R. (1961c) Le gisement de vertébrés miocènes de Beni Mellal (Maroc): Squamates. *Notes et Mémoires, Service Géologique du Maroc* (155), 95–101.

Hoffstetter, R. (1964) Les serpents du Néogène du Pakistan (couches des Siwaliks). *Bulletin de la Société Géologique de France*, Paris (Série 7) 6(4), 467–474, 5 figs.

Hoffstetter, R. (1970) Vertebrados Cenozoicos del Ecuador. *Actas IV Congreso Latino Amererican de Zoología, Caracas* 2, 955–969.

Hoffstetter, R. & Gayrard, Y. (1965) Observations sur l'ostéologie et la classification des Acrochordidae (Serpentes). *Bulletin du Muséum National d'Histoire Naturelle*, Paris (Série 2) 36(5), 677–696, 6 figs.

Hoffstetter, R. & Rage, J.-C. (1972) Les Erycinae fossiles de France (Serpentes, Boidae) compréhension et histoire de la sous-famille. *Annales de Paléontologie (Vertébrés)*, Paris 58(1), 81–124, 11 figs., 2 pls.

Hoffstetter, R. & Rage, J.-C. (1977) Le gisement de vertébrés miocènes de La Venta (Colombie) et sa faune de serpents. *Annales de Paléontologie (Vertébrés)*, Paris 63(2), 161–190, 7 figs., 1 pl.

Hofstadler-Deiques, C. & Lema, T. de. (2005) On the cranial morphology of *Elapomorphus*, *Phalotris* and *Apostolepis* (Serpentes, Colubridae), and its phylogenetic significance. *Zootaxa* (1042), 1–26, 15 figs.

Hofstadler-Deiques, C., Schmitt, P. & Di-Bernardo, M. (2004) Geographic distribution: Serpentes: *Tropidodryas striaticeps*. *Herpetological Review* 35(2), 193.

Hofstadler-Deiques, L.R. & Cechin, S.Z. (1992) O status de *Helicops carinicaudus* (Wied, 1825) (Serpentes: Colubridae). *Acta Biológica Leopoldensia* 12(2), 313–325, illust.

Hoge, A.R. (1948) Nota erpetológicas. 3. Uma nova especie de *Trimeresurus*. *Memórias do Instituto Butantan,* São Paulo (1947) 20, 193–202, 6 figs., 2 pls.

Hoge, A.R. (1949) Notas erpetológicas. 6. Sôbre a ocorrência de *Trimeresurus hyoprora* (Amaral) no Brasil. *Boletim do Museu Paraense Emílio Goeldi* (*Zoologia*), Belém (1948) 10, 325–329.

Hoge, A.R. (1950) Notas erpétologicas. 7. Fauna erpetológica da Ilha da Queimada Grande. *Memórias do Instituto Butantan,* São Paulo (1949) 22, 151–171, 7 pls.

Hoge, A.R. (1953a) Notas erpetológicas. Revalidação de *Thamnodynastes strigatus* (Günther, 1858). *Memórias do Instituto Butantan,* Sào Paulo (1952) 24(2), 157–171.

Hoge, A.R. (1953b) Notas erpetológicas. 1a. Contribuição ao conhecimento dos ofidios do Brasil Central. *Memórias do Instituto Butantan,* São Paulo (1952) 24(2), 179–214, 9 pls.

Hoge, A.R. (1953c) Notas erpetológicas. 2a. Contribuição ao conhecimento dos ofidios do Brasil Central. *Memórias do Instituto Butantan,* São Paulo (1952) 24(2), 215–224.

Hoge, A.R. (1953d) Snakes from Uaupás region. *Memórias do Instituto Butantan,* São Paulo (1952) 24(2), 225–230.

Hoge, A.R. (1953e) Notas erpetológicas. Revalidasão de *Bothrops lanceolata* (Lacépède). *Memórias do Instituto Butantan,* São Paulo (1952) 24(2), 231–236, 7 figs.

Hoge, A.R. (1953f) Notas erpetológicas. Uma nova subspécie de *Leimadophis reginae*. *Memórias do Instituto Butantan,* Sao Paulo (1952) 24(2), 241–244.

Hoge, A.R. (1953g) Notes on *Lygophis* Fitzinger with revalidation of two subspecies. *Memórias do Instituto Butantan,* São Paulo (1952) 24(2), 245–267, 3 pls.

Hoge, A.R. (1954a) A new *Bothrops* from Brazil. *Bothrops brazili*, sp. nov. *Memórias do Instituto Butantan,* São Paulo (1953) 25(1), 15–22, 7 figs.

Hoge, A.R. (1954b) A new genus and species of Boinae from Brazil. *Xenoboa cropanii*, gen. nov., sp. nov. *Memórias do Instituto Butantan,* São Paulo (1953) 25(1), 27–34, 8 figs.

Hoge, A.R. (1955) Eine neue Schlange der Gattung *Elapomorphus* aus Brasilien. *Senckenbergiana Biologica*, Frankfurt-am-Main 36(5–6), 301–304, pls. 27–29.

Hoge, A.R. (1957) Uma nova espécie de *Micrurus* (Serp. Elap.) do Brasil. *Memórias do Instituto Butantan,* Sao Paulo (1955–1956) 27, 67–72, 6 figs.

Hoge, A.R. (1958a) Die Systematische stellung von *Xenodon punctatus* Peters 1880 und *Philodryas taeniatus* Hensel 1868. *Mitteilungen aus dem Naturhistorischen Museum in Berlin* 34, 49–56, 7 figs., 2 pls.

Hoge, A.R. (1958b) Tres notas sobre serpentes brasileiras. 1. Sobre a posição generica de *Coluber bicinctus* Hermann, 1804 e *Xenodon gigas* Duméril, 1853 (Colubridae). *Papeis Avulsos do Departamento de Zoologia*, São Paulo 13(17), 221–222.

Hoge, A.R. (1958c) Tres notas sobre serpentes brasileiras. 2. Sobre a posição sistematica de *Enicognathus joberti* Sauvage, 1884 (Colubridae). *Papeis Avulsos do Departamento de Zoologia*, São Paulo 13(17), 222–223.

Hoge, A.R. (1958d) Tres notas sobre serpentes brasileiras. 3. Dimorfismo sexual em *Micrurus s. surinamensis* (Cuvier, 1817) (Elapidae). *Papeis Avulsos do Departamento de Zoologia*, São Paulo 13(17), 223–224, 4 figs.

Hoge, A.R. (1959a) Notes sur la position systematique de *Opisthoplus degener* Peters, 1882 et *Leimadophis reginae macrosoma* Amaral, 1935 (Serpentes). *Memórias do Instituto Butantan,* São Paulo (1957–1958) 28, 67–72, 2 figs.

Hoge, A.R. (1959b) Étude sur *Apostolepis coronata* (Sauvage 1877) et *Apostolepis quinquelineata* Boulanger [sic] 1896 (Serpentes). *Memórias do Instituto Butantan,* São Paulo (1957–1958) 28, 73–76, 2 figs.

Hoge, A.R. (1959c) Étude sur *Uromacerina ricardinii* (Peracca) (Serpentes). *Memórias do Instituto Butantan,* São Paulo (1957–1958) 28, 77–82, 3 figs.

Hoge, A.R. (1959d) Note sur la position systematique de *Trigonocephalus* (*Bothrops*) *pubescens* Cope 1869. *Memórias do Instituto Butantan,* São Paulo (1957–1958) 28, 83–84.

Hoge, A.R. (1959e) Sur la position systematique de *Coluber quinquelineatus* Raddi 1820. *Memórias do Instituto Butantan,* São Paulo (1957–1958) 28, 267–274, 5 figs.

Hoge, A.R. (1964a) Posição sistemática de *Natrix sexcarinatus* Wagler, 1824, Serpentes. *Memórias do Instituto Butantan,* São Paulo (1960–1962) 30, 27–30, 2 figs.

Hoge, A.R. (1964b) Note sôbre *Xenopeltis unicolor* Reinwardt 1827, Serpentes. *Memórias do Instituto Butantan,* São Paulo (1960–1962) 30, 31–33, 4 figs.

Hoge, A.R. (1964c) Sur la position systemtique de quelques serpents du genre *Siphlophis* Fitzinger 1843 Serpentes. *Memórias do Instituto Butantan,* São Paulo (1960–1962) 30, 35–50, 23 figs.

Hoge, A.R. (1964d) Serpentes da fundação "Surinaam Museum." *Memórias do Instituto Butantan,* São Paulo (1960–1962) 30, 51–64.

Hoge, A.R. (1964e) Notes sur *Xenodon schottii* Schlegel, Serpentes. *Memórias do Instituto Butantan,* São Paulo (1960–1962) 30, 65–70.

Hoge, A.R. (1966a) Preliminary account on Neotropical Crotalinae (Serpentes Viperidae). *Memórias do Instituto Butantan,* São Paulo (1965) 32, 109–184, 20 pls.

Hoge, A.R. (1966b) Notes on *Hydrodynastes* [Serpentes-Colubridae]. *Ciência e Cultura*, São Paulo 18(2), 143.

Hoge, A.R. (1967) Serpentes do Territorio Federal do Amapá. *Atas do Simpósio sóbre a Biota Amazônica*, Rio de Janeiro (*Zoologia*) 5, 217–223.

Hoge, A.R. (1969a) *Chironius scurulus* (Wagler) recorded from Venezuela. *Memórias do Instituto Butantan,* São Paulo 34, 85.

Hoge, A.R. (1969b) Notes on the holotype of *Dipsas indica cisticeps* (Boettger) (Serpentes-Dipsadinae). *Memórias do Instituto Butantan,* São Paulo 34, 87–88.

Hoge, A.R. (1969c) Notes on *Pseustes dieperinckii* (Schlegel) Serpentes. *Memórias do Instituto Butantan,* São Paulo 34, 89–92, 2 figs.

Hoge, A.R. (1969d) Um nôvo *Trimeresurus* de Simalur, Sumatra (Serpentes Viperidae). *Ciência e Cultura*, São Paulo 21(2), 459–460.

Hoge, A.R. & Belluomini, H.E. (1959) Uma nova espécie de *Bothrops* do Brasil (Serpentes). *Memórias do Instituto Butantan,* São Paulo (1957–1958) 28, 195–206, 7 figs.

Hoge, A.R. & Belluomini, H.E. (1960) Serpentes coletadas em Jacaréacanga, Estado do Pará, Brasil. *Memórias do Instituto Butantan,* São Paulo (1959) 29, 15–16.

Hoge, A.R. & Belluomini, H.E. (1964) Notes sobre *Bothrops fonsecai* Hoge e Belluomini, *Bothrops alternans* Duméril, Bibron et Duméril e *Bothrops cotiara* Gomes. *Memórias do Instituto Butantan,* São Paulo (1960–1962) 30, 97–102.

Hoge, A.R. & Federsoni, P.A., Jr. (1975) Notes on *Xenopholis* Peters and *Paroxyrhopus* Schenkel (Serpentes, Colubridae). *Memórias do Instituto Butantan,* São Paulo (1974) 38, 137–145, 16 figs.

Hoge, A.R. & Gans, C. (1965) A first record for an Amazonian snake similar to the eastern Brazilian *Liophis miliaris* (Linné). *Copeia* 1965(4), 511–512, 2 figs.

Hoge, A.R. & Garcia, A. (1949) Notas erpetológicas. 5. Notas sôbre *Elapomorphus nasutus* Gomes, 1915. *Memórias do Instituto Butantan,* São Paulo 21, 67–76, 4 figs.

Hoge, A.R. & Lancini, A.R. (1960a) Notas sobre la ubicación de la terre typica de varias especies de "serpentes" colectadas por M. Bauperthuis en la "cote Ferme" y en la "Province de Venezuela". *Boletin del Muséo de Ciencias Naturales,* Caracas 6(1–4), 58–62.

Hoge, A.R. & Lancini, A.R. (1960b) Nota sobre *Micrurus surinamensis nattereri* Schmidt e *Micrurus pyrrhocryptus* Cope. *Memórias do Instituto Butantan,* São Paulo (1959) 29(1), 9–13, 2 figs.

Hoge, A.R. & Lancini, A.R. (1962) Sinopsis de las serpientes venenosas de Venezuela. *Publicaciones Ocasionales del Muséo de Ciencias Naturales*, Caracas (*Zoología*) (1), 1–24, 10 figs.

Hoge, A.R. & Lima-Verde, J.S. (1973) *Liophis mossoroensis* nov. sp. do Brasil [Serpentes, Colubridae]. *Memórias do Instituto Butantan,* São Paulo (1972) 36, 215–220, 2 figs.

Hoge, A.R. & Maranhão-Niña, A.C. (1964) Serpentes coletadas pelo Instituto Nacional de Pesquisas da Amazonia. *Memórias do Instituto Butantan,* São Paulo (1960–1962) 30, 71–96, 11 pls.

Hoge, A.R. & Mertens, R. (1955) Eine neue Schlange der Gattung *Pseudoboa* aus dem Staate Matto Grosso, Brasilien. *Senckenbergiana Biologica*, Frankfurt-am-Main 36(5–6), 305–308, pls. 30–31.

Hoge, A.R. & Mertens, R. (1959) Eine neue Gattung opisthoglypher Nattern aus Brasilien. *Senckenbergiana Biologica*, Frankfurt-am-Main 40(5–6), 241–243, 2 figs.

Hoge, A.R. & Romano-Hoge, S.A.R.W.L. (1966) *Leptomicrurus* in Brasil [Serpentes-Elapidae]. *Memórias do Instituto Butantan,* São Paulo (1965) 32(1), 1–8, 12 figs.

Hoge, A.R. & Romano-Hoge, S.A.R.W.L. (1969a) Nota sôbre *Siphlophis* e *Ptychophis*. *Ciencia e Cultura*, São Paulo 21(2), 454.

Hoge, A.R. & Romano-Hoge, S.A.R.W.L. (1969b) Espécies registradas para o Brasil (Serpentes). *Ciencia e Cultura*, São Paulo 21(2), 454.

Hoge, A.R. & Romano-Hoge, S.A.R.W.L. (1970a) Notes on *Pseustes dieperinckii* (Schlegel) Serpentes. *Memórias do Instituto Butantan,* São Paulo (1969) 34, 89–92, 2 figs.

Hoge, A.R. & Romano-Hoge, S.A.R.W.L. (1970b) A new species of *Chironius* (Serpentes-Colubridae). *Memórias do Instituto Butantan,* São Paulo (1969) 34, 93–95, 2 pls.

Hoge, A.R. & Romano-Hoge, S.A.R.W.L. (1971) Neotropical pit vipers, sea snakes, and coral snakes. *In:* Bücherl, W. & Buckley, E.E. (Eds.), *Venomous animals and their venoms. Volume II. Venomous vertebrates.* Academic Press, New York, pp. 211–293.

Hoge, A.R. & Romano-Hoge, S.A.R.W.L. (1972) *Micrurus hemprichii hemprichii* recorded from Brasil (Serpentes Elapidae). *Memórias do Instituto Butantan,* São Paulo (1971) 35, 107–109.

Hoge, A.R. & Romano-Hoge, S.A.R.W.L. (1973) Sinopse das Serpentes peçonhentas do Brasil, Serpentes, Elapidae e Viperidae. *Memórias do Instituto Butantan,* São Paulo (1972) 36, 109–208, 57 figs.

Hoge, A.R. & Romano-Hoge, S.A.R.W.L. (1975a) Notes on *Trimeresurus brongersmai* Hoge 1969 (Serpentes, Viperidae, Crotalinae). *Memórias do Instituto Butantan,* São Paulo (1974) 38, 147–158, 14 figs.

Hoge, A.R. & Romano-Hoge, S.A.R.W.L. (1975b) A new subspecies of *Dipsas indica* from Brazil [Serpentes, Colubridae, Dipsadinae]. *Memórias do Instituto Butantan,* São Paulo 39, 51–60, 15 figs.

Hoge, A.R. & Romano-Hoge, S.A.R.W.L. (1978a) Description of a new subsp. of *Oxyrhopus* Wagler (Serpentes, Colubridae). *Memórias do Instituto Butantan,* São Paulo (1976–1977) 40–41, 55–62, figs.

Hoge, A.R. & Romano-Hoge, S.A.R.W.L. (1978b) Redescription and range of *Sordellina punctata* (Peters) [Serpentes, Colubridae]. *Memórias do Instituto Butantan,* São Paulo (1976–1977) 40–41, 63–70, 8 figs.

Hoge, A.R. & Romano-Hoge, S.A.R.W.L. (1981a) Poisonous snakes of the world. Part I. Check list of the pit vipers. Viperoidea, Viperidae, Crotalinae. *Memórias do Instituto Butantan,* São Paulo (1978–1979) 42–43, 179–310, 66 figs.

Hoge, A.R. & Romano-Hoge, S.A.R.W.L. (1981b) Sinopse das Serpentes peconhentas do Brasil. (2d. ed.). *Memórias do Instituto Butantan,* São Paulo (1978–1979) 42–43, 373–496, 193 figs.

Hoge, A.R. & Romano-Hoge, S.A.R.W.L. (1983) Notes on micro and ultrastructure of "Oberhautschen" in Viperoidea. *Memórias do Instituto Butantan,* São Paulo (1980–1981) 44–45, 81–118, 84 figs.

Hoge, A.R., Belluomini, H.E., Schreiber, G. & Penha, A.M. (1960) Sexual abnormalities in *Bothrops insularis* (Amaral) 1921 (Serpentes). *Memórias do Instituto Butantan,* São Paulo (1959) 29, 17–88, 18 figs., 15 pls.

Hoge, A.R., Santos, N.P., Heitor, C., Lopes, L.A. & de Souza, I.M. (1973) Serpentes coletadas pelo Projeto Rondon VII em Iauareté, Brasil. *Memórias do Instituto Butantan,* São Paulo (1972) 36, 221–232.

Hoge, A.R., Romano-Hoge, S.A.R.W.L., Federsoni, P.A., Jr. & Cordeiro, C.L. (1975) Lista das espécies de serpentes coletadas na região da usina hidroelética de Ilha Solteira-Brasil. *Memórias do Instituto Butantan,* São Paulo (1974) 38, 167–178.

Hoge, A.R., Cordeiro, C.L. & Romano-Hoge, S.A.R.W.L. (1976a) Posição taxonômica de *Lystrophis nattereri* (Steindachner) [Serpentes, Colubridae]. *Memórias do Instituto Butantan,* São Paulo (1975) 39, 37–50, 19 figs.

Hoge, A.R., Cordeiro, C.L. & Romano-Hoge, S.A.R.W.L. (1976b) A new species of *Micrurus* from Brasil. *Ciencia e Cultura,* São Paulo 28(7), 417–418.

Hoge, A.R., Belluomini, H.E. & Fernandes, W. (1978a) Variação do número de placas ventrais de *Bothrops jararaca* em função dos climas [Viperidae, Crotalinae]. *Memórias do Instituto Butantan,* São Paulo (1976–1977) 40–41, 11–17, 7 figs.

Hoge, A.R., Romano-Hoge, S.A.R.W.L. & Cordeiro, C.L. (1978b) Contribuição ao conhecimento das serpentes do Maranhão, Brasil (Serpentes, Boidae, Colubridae e Viperidae). *Memórias do Instituto Butantan,* São Paulo (1976–1977) 40–41, 37–52, 1 fig.

Hoge, A.R., Cordeiro, C.L. & Romano-Hoge, S.A.R.W.L. (1978c) Redescription of *Micrurus donosoi* Hoge, Cordeiro et Romano (Serpentes, Elapinae). *Memórias do Instituto Butantan,* São Paulo (1976–1977) 40–41, 71–73.

Hoge, A.R., Russo, C.R., Santos, M.C. & Furtado, M.F.D. (1981a) Snakes collected by "Projeto Rondon XXII" do Piauí, Brasil. *Memórias do Instituto Butantan,* São Paulo (1978–1979) 42–43, 87–94.

Hoge, A.R., Laporta, I.L. & Romano-Hoge, S.A.R.W.L. (1981b) Notes on *Sibynomorphus mikanii* Schlegel 1837. *Memórias do Instituto Butantan,* São Paulo (1978–1979) 42–43, 175–178.

Höggren, M. (1995) *Mating strategies and sperm competition in the adder* (Vipera berus). Comprehensive Summaries of Uppsala Dissertations from the Faculty of Science and Technology, Uppsala (163), 27 pp.

Höggren, M., Nilson, G., Andrén, C., Orlov, N.L. & Tuniyev, B.S. (1993) Vipers of the Caucasus: natural history and systematic review. *Herpetological Natural History* 1(2), 11–19, 7 figs.

Holbrook, J.E. (1836) *North American herpetology; or, a description of the reptiles inhabiting the United States. Vol. I* [First version]. J. Dobson, Philadelphia, 120 pp., 23 pls.

Holbrook, J.E. (1838) *North American herpetology; or, a description of the reptiles inhabiting the United States. Vol. II* [First version]. J. Dobson, Philadelphia, 125 pp., 28 pls.

Holbrook, J.E. (1839) *North American herpetology; or, a description of the reptiles inhabiting the United States. Vol. I* [Second version]. J. Dobson, Philadelphia, 132 pp., 27 pls.

Holbrook, J.E. (1840) *North American herpetology; or, a description of the reptiles inhabiting the United States. Vol. IV* [First version]. J. Dobson, Philadelphia, 126 pp., 28 pls.

Holbrook, J.E. (1842a) *North American herpetology; or, a description of the reptiles inhabiting the United States. Vol. III.* Second edition. J. Dobson, Philadelphia, 128 pp., 30 pls.

Holbrook, J.E. (1842b) *North American herpetology; or, a description of the reptiles inhabiting the United States. Vol. IV.* Second edition. J. Dobson, Philadelphia, 138 pp., 35 pls.

Hollingsworth, B.D. (1997) Geographic distribution: Serpentes: *Thamnophis marcianus marcianus. Herpetological Review* 28(4), 211.

Hollingsworth, B.D., Galina-Tessaro, P., Castellanos-Vera, A. & Cota, A. (2000) Geographical distribution. *Trimorphodon biscutatus. Herpetological Review* 31(1), 58.

Hollis, J.L. (2006) Phylogenetics of the genus *Chironius* Fitzinger, 1826 (Serpentes, Colubridae) based on morphology. *Herpetologica* 62(4), 435–453, 6 figs.

Holm, A. (1957) Specimina Linnaeana i Uppsala bevarade zoologiska samlingar fran Linnés tid. *Uppsala Universitets Arsskrift* 6, 1–68, 14 figs., 1 pl.

Holm, P.A. (1993) Geographic distribution: Serpentes: *Rhadinaea montecristi. Herpetological Review* 24(2), 68.

Holm, P.A. (1995) *Leptophis modestus* Morphology. *Herpetological Review* 26(1), 39.

Holm, P.A. & Cruz-Díaz, G.A. (1994) A new species of *Rhadinaea* (Colubridae) from a cloud forest in northern Honduras. *Herpetologica* 50(1), 15–23, 3 figs.

Holman, J.A. (1964) Fossil caves from the Valentine formation of Nebraska. *Copeia* 1964(4), 631–637, 3 figs.

Holman, J.A. (1967) Neural spine expansions of some boid snakes. *British Journal of Herpetology* 35(12), 300–303, 1 fig.

Holman, J.A. (1968) Upper Pliocene snakes from Idaho. *Copeia* 1968(1), 152–158, 3 figs.

Holman, J.A. (1970) Herpetofauna of the Wood Mountain formation (upper Miocene) of Saskatchewan. *Canadian Journal of Earth Sciences*, Ottawa 7(5), 1317–1325, 2 figs.

Holman, J.A. (1972) Herpetofauna of the Calf Creek local fauna (lower Oligocene, Cypress Hills formation) of Saskatchewan. *Canadian Journal of Earth Sciences*, Ottawa 9(12), 1612–1631, 7 figs.

Holman, J.A. (1973a) Reptiles of the Egelhoff local fauna (upper Miocene) of Nebraska. *Contributions from the Museum of Paleontolgy, The University of Michigan* 24(12), 125–134, 5 figs.

Holman, J.A. (1973b) A new Pliocene snake, genus *Elaphe*, from Oklahoma. *Copeia* 1973(3), 574–580, 3 figs.

Holman, J.A. (1973c) Herpetofauna of the Mission local fauna (lower Pliocene) of South Dakota. *Journal of Paleontology* 47(3), 462–464.

Holman, J.A. (1975) Herpetofauna of the Wakeeney local fauna (lower Pliocene, Clarendonian) of Trego County, Kansas. *In:* Studies on Cenozoic paleontology and stratigraphy, in honor of Claude W. Hibbard. *University of Michigan Papers in Paleontology* (12), 49–66, 2 figs.

Holman, J.A. (1976a) Snakes from the Rosebud formation (middle Miocene) of South Dakota. *Herpetologica* 32(1), 41–48, 4 figs.

Holman, J.A. (1976b) Snakes of the Gering formation (lower Miocene) of Nebraska. *Herpetologica* 32(1), 88–94, 4 figs.

Holman, J.A. (1976c) Snakes of the Split Rock formation (middle Miocene), central Wyoming. *Herpetologica* 32(4), 419–426, 1 fig.

Holman, J.A. (1977a) Upper Eocene snakes (Reptilia, Serpentes) from Georgia. *Journal of Herpetology* 11(2), 141–145, 3 figs.

Holman, J.A. (1977b) Amphibians and reptiles from the Gulf Coast Miocene of Texas. *Herpetologica* 33(4), 391–403, 9 figs. (†*Texasophis fossilis*)

Holman, J.A. (1977c) Additional snakes from the Miocene of western Nebraska. *Herpetologica* 33(4), 443–446, 1 fig.

Holman, J.A. (1977d) Snakes of the Monroe Creek formation (lower Miocene, Arikareean) of Wyoming. *Copeia* 1977(1), 193–194.

Holman, J.A. (1977e) Upper Miocene snakes (Reptilia, Serpentes) from southeastern Nebraska. *Journal of Herpetology* 11(3), 323–335, 8 figs.

Holman, J.A. (1978) Herpetofauna of the Bijou Hills local fauna (late Miocene, Barstovian) of South Dakota. *Herpetologica* 34(3), 253–257, 1 fig.

Holman, J.A. (1979a) A review of North American Tertiary snakes. *Publications of the Museum, Michigan State University, Paleontological Series* 1(6), 203– 260, 33 figs.

Holman, J.A. (1979b) Paleontology and geology of the Badwater Creek area, central Wyoming. Part 17. The late Eocene snakes. *Annals of the Carnegie Museum* 48(6), 103–110, 2 figs.

Holman, J.A. (1981a) A review of North American Pleistocene snakes. *Publications of the Museum, Michigan State University, Paleontological Series* 1(7), 263–306, 18 figs.

Holman, J.A. (1981b) A herpetofauna from an eastern extension of the Harrison formation (early Miocene, Arikareean), Cherry County, Nebraska. *Journal of Vertebrate Paleontology* 1(1), 49–56, 10 figs.

Holman, J.A. (1982a) New herpetological species and records from the Norden Bridge fauna (Miocene, late Barstovian) of Nebraska. *Transactions of the Nebraska Academy of Sciences* 10, 31–36, 6 figs.

Holman, J.A. (1982b) A fossil snake (*Elaphe vulpina*) from a Pliocene ash bed in Nebraska. *Transactions of the Nebraska Academy of Sciences* 10, 37–42, 3 figs.

Holman, J.A. (1982c) *Palaeophis casei*, new species: a tiny palaeophid snake from the early Eocene of Mississippi. *Journal of Vertebrate Paleontology* 2(2), 163–166, 8 figs.

Holman, J.A. (1982d) *Geringophis* (Serpentes, Boidae) from the middle Oligocene of Nebraska. *Herpetologica* 38(4), 489–492, 4 figs.

Holman, J.A. (1982e) The Pleistocene (Kansan) herpetofauna of Trout Cave, West Virginia. *Annals of the Carnegie Museum* 51(20), 391–404, 6 figs.

Holman, J.A. (1983) A new species of *Helagras* (Serpentes) from the middle Oligocene of Nebraska. *Journal of Herpetology* 17(4), 417–419, 2 figs.

Holman, J.A. (1984a) *Texasophis galbreathi*, new species: the earliest New World colubrid snake. *Journal of Vertebrate Paleontology* 3(4), 223–225, 3 figs.

Holman, J.A. (1984b) *Texasophis* (Reptilia, Serpentes): an addition to the Miocene (Clarendonian) of North America. *Copeia* 1984(3), 660–661, 1 fig.

Holman, J.A. (1984c) Herpetofauna of the Duck Creek and Williams local faunas (Pleistocene, Illinoian) of Kansas. *In:* Genoways, H.H. & Dawson, M.R. (Eds.), Contributions in Quaternary vertebrate paleontology, a volume in memorial to John E. Guilday. *Carnegie Museum of Natural History, Special Publication* (8), pp. 20–38, 13 figs.

Holman, J.A. (1985) Herpetofauna of the late Pleistocene fissures near Ightham, Kent. *British Herpetological Journal*, London 1, 26–32, 6 figs.

Holman, J.A. (1987a) Herpetofauna of the Egelhoff site (Miocene, Barstovian) of north-central Nebraska. *Journal of Vertebrate Paleontology* 7(2), 109–120, 6 figs.

Holman, J.A. (1987b) Climatic significance of a late Illinoian herpetofauna from southwestern Kansas. *Contributions from the Museum of Paleontology, University of Michigan* 27(5), 129–141, 7 figs.

Holman, J.A. (1993) A new genus of primitive colubroid snake from the Upper Eocene, Isle of Wight, England. *Tertiary Research*, Leiden 14(4), 151–154, 1 fig.

Holman, J.A. (1995a) *Pleistocene amphibians and reptiles in North America*. Oxford University Press, New York, 243 pp., 43 figs.

Holman, J.A. (1995b) A late medial Miocene herpetofauna from northeastern Nebraska. *Herpetological Natural History* 3(2), 143–150, 2 figs.

Holman, J.A. (1996) The large Pleistocene (Sangamonian) herpetofauna of the Williston IIIA site, north-central Florida. *Herpetological Natural History* 4(1), 35–47, 4 figs.

Holman, J.A. (1996a) Herpetofauna of the Trinity River Local Fauna (Miocene, early Barstovian), San Jacinto County, Texas, USA. *Tertiary Research*, Leiden 17(1–2), 5–10, 1 fig.

Holman, J.A. (1996b) A new genus of diminutive boid snake from the upper Eocene of Hordle Cliff, Hampshire, England. *Tertiary Research*, Leiden 17(1–2), 11–13, 2 figs.

Holman, J.A. (1998a) Pleistocene amphibians and reptiles in Britain and Europe. *Oxford Monographs in Geology and Geophysics* (36), 254 pp., 46 figs.

Holman, J.A. (1998b) Reptiles of the lower Miocene (Hemingfordian) Pollack Farm fossil site, Delaware. *In:* Benson, R.N. (Ed.), Geology and Paleontology of the lower Miocene Pollack Farm fossile site, Delaware. *Delaware Geological Survey Special Publication* (21), pp. 141–147, 2 figs.

Holman, J.A. (2000a) *Fossil snakes of North America: origin, evolution, distribution, paleoecology*. Indiana University Press, Bloomington, xi + 357 pp., 152 figs.

Holman, J.A. (2000b) New herpetological record from the middle Pleistocene Boxgrove Hominid site, England. *Cranium* 17(2), 112–119, 1 fig.

Holman, J.A. (2000c) Early Oligocene (Whitneyan) snakes from Florida (USA): the second oldest colubrid snake in North America. *Acta Zoologica Cracoviensa* (1999) 42(3), 447–454, 3 figs.

Holman, J.A. (2003) Early Hemingfordian (early Miocene) squamate reptiles from the Quarry A Local Fauna, Logan County, Colorado. *Michigan Academician* 23, 477–480.

Holman, J.A. (2012) *The amphibians and reptiles of Michigan.* Wayne State University Press, Detroit, xix + 291 pp., col. illust.

Holman, J.A. & Case, G.R. (1988) Reptiles from the Eocene Tallahatta formation of Alabama. *Journal of Vertebrate Paleontology* 8(3), 328–333, 6 figs.

Holman, J.A. & Case, G.R. (1992) A puzzling new snake (Reptilia, Serpentes) from the Late Paleocene of Mississippi. *Annals of the Carnegie Museum* 61(3), 197–205, 4 figs.

Holman, J.A. & Harding, J.H. (2006) *Michigan snakes: a field guide and pocket reference.* Second edition. Michigan State University, Lansing, 74 pp., col. illust.

Holman, J.A. & Harrison, D.L. (1998a) A new genus of snake (Serpentes, Boidae) from the upper Eocene of Hordle Cliff, Hampshire, England. *Acta Zoologica Cracoviensa*, Kraków 41(1), 23–27, 2 figs.

Holman, J.A. & Harrison, D.L. (1998b) A new genus of small boid snake from the upper Eocene of Hordle Cliff, Hampshire, England. *Acta Zoologica Cracoviensa*, Kraków 41(1), 29–33, 2 figs.

Holman, J.A. & Harrison, D.L. (2000) Early Oligocene (Whitneyan) snakes from Florida (USA): a unique booid. *Acta Zoologica Cracoviensa*, Kraków 43(1–2), 127–134, 5 figs.

Holman, J.A. & Harrison, D.L. (2001) Early Oligocene (Whitneyan) snakes from Florida (USA): remaining boids, indeterminate colubroids summary and discussion of the I-75 Local Fauna snakes. *Acta Zoologica Cracoviensa*, Kraków 44(1), 25–36.

Holman, J.A. & Richards, R.L. (1981) Late Pleistocene occurrence in southern Indiana of the smooth green snake, *Opheodrys vernalis. Journal of Herpetology* 15(1), 123–125, 1 fig.

Holman, J.A. & Sullivan, R.M. (1981) A small herpetofauna from the type section of the Valentine formation (Miocene, Barstovian), Cherry County, Nebraska. *Journal of Paleontology* 55(1), 138–144, illust.

Holman, J.A., Harding, J.H., Hensley, M. & Dudderar, G.R. (1989) *Michigan snakes.* Michigan State University Extension, East Lansing, 72 pp., col. photos.

Holman, J.A., Stuart, A.J. & Clayden, J.D. (1990) A middle Pleistocene herpetofauna from Cudmore Grove, Essex, England, and its paleogeographic and paleoclimatic implications. *Journal of Vertebrate Paleontology* 10(1), 86–94, illust.

Holman, J.A., Dockery, D.T., III & Case, G.R. (1991) Paleogene snakes of Mississippi. *Mississippi Geology*, Jackson (1), 1–12, 8 figs.

Holmstrom, W.F. & Behler, J.L. (1981) Post-parturient behavior of the common anaconda, *Eunectes murinus. Zoologische Garten*, Jena 51(5–6), 353–356.

Holtzinger-Tenever, H. (1916) *Calamaria borneensis* Blkr. nov. subsp. *Zoologischer Anzeiger*, Leipzig 48(2), 33–34.

Holtzinger-Tenever, H. (1917) Verzeichniss der von Herrn H. Schoede auf Ceylon und Sumatra gesammelten Reptilien. *Mitteilungen aus dem Zoologischen Museum in Berlin* 58(3), 427–444.

Holtzinger-Tenever, H. (1920) Herpetologische Mitteilungen aus dem Zoologischen Museum in Berlin. Die von H. Mertens in Indonesien gesammelten Reptilien. *Archiv für Naturgeschichte*, Berlin 85A(11), 99–111.

Holycross, A.T., Anton, T.C., Douglass, M.E. & Frost, D.R. (2008) The type locality of *Sistrurus catenatus* and *Crotalus viridis* (Serpentes: Viperidae), with the unraveling of a most unfortunate tangle of names. *Copeia* 2008(2), 421–424.

Hölzinger, J. & Schmid, G. (Eds.). (1987) *Die Amphibien und Reptilien Baden-Württembergs.* Institut für Ökologie und Naturschutz, Karlsruhe, 500 pp., illust.

Hombron, C.B. & Jacquinot, H. (1842) *Voyage au pole sud et dans l'océanie sur les corvettes l'Astrolabe et la Zélée pendent les années 1837–1838–1839–1840 sous le commandement de M. Dumont-d'Urville, capitaine de vaisseau. Zoologie. Atlas (Ophidiens).* Gide et J. Boudry, Paris, 3 pls.

Homma, M. (1974) Pathology of venomous snake bite in Asia, with special reference to local tissue damage. *The Snake* 6(1), 1–9, 14 figs. (in Japanese)

Honegger, R.E. (1966) Beobachtungen an der Herpetofauna der Seychellen. *Salamandra*, Frankfurt-am-Main 2(1–2), 21–36, 2 figs.

Honegger, R.E. (1981) List of amphibians and reptiles either known or thought to have become extinct since 1600. *Biological Conservation* 19, 141–158.

Hoofien, J.H. (1972) *A taxonomic list of the reptiles of Israel and its administered area according to the status on May 31st, 1972.* Department of Zoology, Tel-Aviv University , 4 pp., 1 fig.

Hoogmoed, M.S. (1977) On a new species of *Leptotyphlops* from Surinam, with notes on the other Surinam species of the genus (Leptotyphlopidae, Serpentes). Notes on the herpetofauna of Surinam V. *Zoologische Mededelingen*, Leiden 51(7), 99–123, 10 figs., 2 pls.

Hoogmoed, M.S. (1979) The herpetofauna of the Guianan region. *In:* Duellman, W.E. (Ed.), The South American herpetofauna: its origin, evolution, and dispersal. *Monograph, Museum of Natural History, University of Kansas* (7), 241– 279, 13 figs.

Hoogmoed, M.S. (1980) Revision of the genus *Atractus* in Surinam, with the resurrection of two species (Colubridae, Reptilia). Notes on the herpetofauna of Surinam VII. *Zoologische Verhandelingen*, Leiden (175), 1–47, 17 figs., 6 pls.

Hoogmoed, M.S. (1982) Nomenclatural problems relating to *Atractus trilineatus* Wagler, 1828. *Zoologische Mededelingen*, Leiden 56(10), 131–138.

Hoogmoed, M.S. (1983) Snakes of the Guianan region. *Memórias do Instituto Butantan,* São Paulo (1982) 46, 219–254, 9 figs.

Hoogmoed, M.S. (1985) *Xenodon werneri* Eiselt, a poorly known snake from Guiana, with notes on *Waglerophis merremii* (Wagler) (Reptilia, Serpentes, Colubridae). Notes on the herpetofauna of Surinam IX. *Zoologische Mededelingen*, Leiden 59(8), 79–88, 4 figs.

Hoogmoed, M.S. (1997) Rediscovery of a forgotten snake in an unexpected place and remarks on a small herpetological collection form southeastern Brazil. *Zoologische Mededelingen,* Leiden 71(9), 63–81, 8 figs.

Hoogmoed, M.S. & Avila-Pires, T.C.S. (1991) Annotated checklist of the herpetofauna of Petit Saut, Sinnamary River, French Guiana. *Zoologische Mededelingen*, Leiden 65(5), 53–88, 8 figs.

Hoogmoed, M.S. & Gruber, U. (1983) Spix and Wagler type specimens of reptiles and amphibians in the Natural History Musea in Munich (Germany) and Leiden (the Netherlands). *Spixiana, Supplement*, München (9), 319–415.

Hoogmoed, M.S. & Prudente, A.L.C. (2003) A new species of *Atractus* (Reptilia, Ophidia, Colubridae, Dipsadinae) from the Amazon forest region in Brazil. *Zoologische Mededelingen*, Leiden 77(9), 425–439, 6 figs.

Hopkins, J. (1918) The reptiles of Rhode Island. Part II. Snakes. *Park Museum Bulletin*, Providence 10(2), 85–88, figs. 24–26.

Hopkins, P.W. (1974) Sobre la herpetofauna de la Sierra de Estrella (Portugal) con especial referencia a *Coronella austriaca austriaca* y *Vipera latasti*. *Doñana, Acta Vertebrata*, Sevilla 1, 11–17, 1 fig.

Hopley, C.G. (1882) *Snakes: curiosities and wonders of serpent life*. Griffith and Farran, New York, 614 pp., illust.

Hora, S.L. & Jayaram, K.C. (1949) Remarks on the distribution of snakes of peninsular India with Malayan affinities. *Proceedings of the National Institute of Science of India* 15(8), 399–402.

Horikawa, Y. (1941) *The snakes of Taiwan*. Kotaheira Seisakusho, Taipei, 77 pp. (in Japanese)

Horn, H.S. (1969) Polymorphism and evolution of the Hispaniolan snake genus *Uromacer* (Colubridae). *Breviora* (324), 1–23, 2 figs.

Horner, P. (1998) *Simoselaps morrisi* sp. nov. (Elapidae), a new species of snake from the Northern Territory. *The Beagle, Records of the Museums and Art Galleries of the Northern Territory* 14, 63–70, 3 figs.

Horner, P. (1999) Type specimens of terrestrial vertebrates in the Museum and Art Gallery of the Northern Territory — 1973–1999. *The Beagle, Records of the Museums and Art Galleries of the Northern Territory* 15, 55–74.

Hornstedt, C.F. (1787) Beskrifning po en Ny Orm fron Java. *Kongliga Svenska Vetenskaps-Akadamiens Handlingar*, Stockholm 8, 306–308.

Hornstedt, C.F. (1788) Description d'un nouveau serpent de l'ile de Java. *Journal de Physique, de Chimie, d'Histoire Naturelle et des Arts* 32(1), 284–285. [French translation of Hornstedt, 1787]

Hoser, R.T. (1989) *Australian reptiles & frogs*. Pierson, Sydney, 238 pp., 613 figs.

Hoser, R.T. (1991) Further notes on hybrid Australian pythons. *Herptile*, Dudley 16(3), 110–115, illust.

Hoser, R.T. (1995) Australia's death adders — genus *Acanthophis*. *Reptilian Magazine* 3(4), 7–21, and 3(5), 27–34.

Hoser, R.T. (1998a) A new snake from Queensland, Australia (Serpentes, Elapidae). *Monitor*, Ardeer 10(1), 5–9, 31, 6 figs., 2 col. photos.

Hoser, R.T. (1998b) Death adders (genus *Acanthophis*): an overview, including descriptions of five new species and one new subspecies. *Monitor*, Ardeer 9(2), 20–41, 24 figs., 17 col. photos.

Hoser, R.T. (1999a) Australia's dwarf pythons – genus *Antaresia*. *Monitor*, Ardeer 10(2–3), 24–32, 8 figs.

Hoser, R.T. (1999b) Ant-hill pythons *(Anataresia perthensis)* in the wild and captivity. *Monitor*, Ardeer 19(2–3), 33–37, 3 figs.

Hoser, R.T. (2000a) A new species of snake (Serpentes, Elapidae) from Irian Jaya. *Litteratura Serpentium* 20(6), 178–186. [reprinted in 2003 by Alberta Reptile and Amphibian Society Bulletin 16(8), 23–26]

Hoser, R.T. (2000b) A revision of the Australasian pythons. *Ophidia Review* 1(1), 7–21.

Hoser, R.T. (2000c) Australia's dwarf pythons (genus, *Antaresia*). *Monitor*, Ardeer 10(2–3), 24–32.

Hoser, R.T. (2000d) Comments on comments on *Pailsus pailsei*. *Monitor*, Ardeer 10(2–3), 120–122.

Hoser, R.T. (2001) A current assessment of the status of the snakes of the genus *Cannia* and *Pailsus*, including descriptions of three new subspecies from the Northern Territory and Western Australia, Australia. *Boydii, Journal of the Herpetological Society of Queensland Incorporated* (July), 26–60.

Hoser, R.T. (2002a) Death adders (genus *Acanthophis*): an updated overview, including descriptions of 3 new island species and 2 new Australian subspecies. *Crocodilian, Journal of the Victorian Association of Amateur Herpetologists* 4(1), 5–11, 16–22, 24–30, front and back covers.

Hoser, R.T. (2002b) An overview of the taipans, genus, (*Oxyuranus*) (Serpentes, Elapidae) including the description of a new subspecies. *Crocodilian, Journal of the Victorian Association of Amateur Herpetologists* 3(1), 43–50.

Hoser, R.T. (2003a) Five new Australian pythons. *Newsletter of the Macarthur Herpetological Society* (40), 4–9.

Hoser, R.T. (2003b) A new species of elapid (Serpentes, Elapidae) from western New South Wales. *Crocodilian, Journal of the Victorian Association of Amateur Herpetologists* 4(2), 19–26.

Hoser, R.T. (2003c) The rough-scaled snakes, genus *Tropidechis* (Serpentes, Elapidae), including the description of a species from far north Queensland, Australia. *Crocodilian, Journal of the Victorian Association of Amateur Herpetologists* 4(2), 11–14.

Hoser, R.T. (2003d) A re-assessment of the taxonomy of the red-bellied black snakes (genus *Pseudechis*) with the description of two new subspecies. *Boydii, Journal of the Herpetological Society of Queensland Incorporated* (May), 15–18.

Hoser, R.T. (2003e) A new subspecies of elapid (Serpentes, Elapidae) from New Guinea. *Boydii, Journal of the Herpetological Society of Queensland Incorporated* (Autumn), 2–4.

Hoser, R.T. (2003f) A reclassification of the Pythoninae, including the descriptions of two new genera, two new species and nine new subspecies. *Crocodilian, Journal of the Victorian Association of Amateur Herpetologists* 4(3), 31–37.

Hoser, R.T. (2004) A reclassification of the Pythoninae, including the descriptions of two new genera, two new species and nine new subspecies. *Crocodilian, Journal of the Victorian Association of Amateur Herpetologists* 4(4), 21–40.

Hoser, R. (2009a) One or two mutations doesn't make a new species…The taxonomy of copperheads (*Austrelaps*) (Serpentes: Elapidae). *Australasian Journal of Herpetology*, Doncaster 1, 1–28.

Hoser, R. (2009b) Creationism and contrived science: a review of recent *Python* systematics papers and the resolution of issues of taxonomy and nomenclature. *Australasian Journal of Herpetology*, Doncaster 2, 1–34.

Hoser, R. (2009c) Eight new taxa in the genera *Pseudonaja* Gunther, 1858, *Oxyuranus* Kinghorn, 1923 and *Panacedechis* Wells and Wellington, 1985 (Serpentes: Elapidae). *Australasian Journal of Herpetology*, Doncaster 4, 1–27.

Hoser, R. (2009d) A reclassification of the rattlesnakes; species formerly exclusively referred to the genera *Crotalus* and *Sistrurus*. *Australasian Journal of Herpetology*, Doncaster 6, 1–21.

Hoser, R. (2009e) A reclassification of the true cobras; species formerly referred to the genera *Naja, Boulengerina* and *Paranaja*. *Australasian Journal of Herpetology*, Doncaster 7, 1–15.

Hoser, R.T. (2012a) Exposing a Fraud! *Afronaja* Wallach, Wüster and Broadley 2009 is a junior synonym of *Spacklandus* Hoser, 2009! *Australasian Journal of Herpetology*, Doncaster 9, 1–41.

Hoser, R.T (2012b) An updated review of the pythons including resolution of issues of taxonomy and nomenclature. *Australasian Journal of Herpetology*, Doncaster 10, 2–32.

Hoser, R.T. (2012c) A new genus of jumping pitviper from Middle America (Serpentes: Viperidae). *Australasian Journal of Herpetology*, Doncaster 10, 33–34.

Hoser, R.T. (2012d) A reassessment of the higher taxonomy of the Viperidae. *Australasian Journal of Herpetology*, Doncaster 10, 35–48.

Hoser, R.T. (2012e) A reassessment of the higher taxonomy of the Elapidae. *Australasian Journal of Herpetology,* Doncaster 10, 49–63.

Hoser, R.T. (2012f) A reclassification of the rattlesnakes: species formerly exclusively referred to the genera *Crotalus* and *Sistrurus* and a division of the elapid genus *Micrurus. Australasian Journal of Herpetology,* Doncaster 11, 2–24.

Hoser, R.T. (2012g) A new genus of pitviper (Serpentes: Viperidae) from South America. *Australasian Journal of Herpetology,* Doncaster 11, 25–27

Hoser, R.T. (2012h) Two new genera of water snakes from North America. The subdivision of the genera *Regina* Baird and Girard, 1853 and *Nerodia* Baird and Girard, 1853 (Serpentes: Colubridae: Natricidae). *Australasian Journal of Herpetology,* Doncaster 11, 29–32.

Hoser, R.T. (2012i) The description of a new genus of west Australian snake and eight new taxa in the genera *Pseudonaja* Gunther, 1858, *Oxyuranus* Kinghorn, 1923 and *Panacedechis* Wells and Wellington, 1985 (Serpentes: Elapidae). *Australasian Journal of Herpetology,* Doncaster 11, 32-50.

Hoser, R.T. (2012j) A new genus of Asian pitviper (Serpentes: Viperidae). *Australasian Journal of Herpetology,* Doncaster 11, 51–52.

Hoser, R.T. (2012k) A taxonomic revision of the *Vipera palaestinae* Werner, 1938 species group, with the creation of a new genus and a new subgenus. *Australasian Journal of Herpetology,* Doncaster 11, 53–55.

Hoser, R.T. (2012L) A reassessment of the burrowing asps, *Atractaspis* Smith, 1849 with the erection of a new genus and two tribes (Serpentes: Atractaspidae). *Australasian Journal of Herpetology,* Doncaster 11, 56–58.

Hoser, R.T. (2012m) A taxonomic revision of the colubrinae genera *Zamenis* and *Orthriophis* with the creation of two new genera (Serpentes: Colubridae). *Australasian Journal of Herpetology,* Doncaster 11, 59–64.

Hoser, R.T. (2012n) A new genus of coral snake from Japan (Serpentes: Elapidae). *Australasian Journal of Herpetology,* Doncaster 12, 3–5.

Hoser, R.T. (2012o) A revision of the Asian pitvipers, referred to the genus *Cryptelytrops* Cope, 1860, with the creation of a new genus *Adelynhoserea* to accommodate six divergent species (Serpentes: Viperidae: Crotalinae). *Australasian Journal of Herpetology,* Doncaster 12, 6–8.

Hoser, R.T. (2012p) A division of the south-east Asian ratsnake genus *Coelognathus* (Serpentes: Colubridae). *Australasian Journal of Herpetology,* Doncaster 12, 9–11.

Hoser, R.T. (2012q) A new genus of Asian snail-eating snake (Serpentes: Pareatidae). *Australasian Journal of Herpetology,* Doncaster 12, 12–15.

Hoser, R.T. (2012r) The dissolution of the genus *Rhadinophis* Vogt, 1922 (Serpentes: Colubridae). *Australasian Journal of Herpetology,* Doncaster 12, 16–17.

Hoser, R.T. (2012s) Three new species of *Stegonotus* from New Guinea (Serpentes: Colubridae). *Australasian Journal of Herpetology,* Doncaster 12, 18–22.

Hoser, R.T. (2012t) A new genus and new subgenus of snakes from the South African region (Serpentes: Colubridae). *Australasian Journal of Herpetology,* Doncaster 12, 23–25.

Hoser, R.T. (2012u) A division of the African genus *Psammophis* Boie, 1825 into 4 genera and four further subgenera (Serpentes: Psammophiinae). *Australasian Journal of Herpetology,* Doncaster 12, 26–31.

Hoser, R.T. (2012v) A division of the African tree viper genus *Atheris* Cope, 1860 into four subgenera (Serpentes: Viperidae). *Australasian Journal of Herpetology,* Doncaster 12, 32–35.

Hoser, R.T. (2012w) A new subgenus of giant snakes (Anaconda) from South America (Serpentes: Boidae). *Australasian Journal of Herpetology,* Doncaster 12, 36–39.

Hoser, R.T. (2012x) A review of the South American snake genera *Leptodeira* and *Imantodes* including three new genera and two new subgenera (Serpentes: Dipsadidae: Imantodini). *Australasian Journal of Herpetology,* Doncaster 12, 40–47.

Hoser, R.T. (2012y) A review of the North American garter snakes genus *Thamnophis* Fitzinger, 1843 (Serpentes: Colubridae). *Australasian Journal of Herpetology,* Doncaster 12, 48–53.

Hoser, R.T. (2012z) A three-way division of the New World genus *Lampropeltis* Fitzinger, 1843 (Serpentes: Colubridae). *Australasian Journal of Herpetology,* Doncaster 12, 54–57.

Hoser, R.T. (2012aa) A review of the taxonomy of the European colubrid snake genera *Natrix* and *Coronella,* with the creation of three new monotypic genera (Serpentes: Colubridae). *Australasian Journal of Herpetology,* Doncaster 12, 58–62.

Hoser, R.T. (2012ab) Divisions of the Asian colubrid snake genera *Xenochrophis, Dendrelaphis* and *Boiga* (Serpentes: Colubridae). *Australasian Journal of Herpetology,* Doncaster 12, 65–76.

Hoser, R.T. (2012ac) Some new small-eyed snakes from Australia and New Guinea (Serpentes: Elapidae). *Australasian Journal of Herpetology,* Doncaster 13, 3–7.

Hoser, R.T. (2012ad) A division of the Asian forest ratsnakes genus *Euprepiophis* Boie, 1826 (Serpentes: Colubridae). *Australasian Journal of Herpetology,* Doncaster 13, 8–9.

Hoser, R.T. (2012ae) A three-way division of African centipede eating snakes, *Aparallactus* Smith, 1849 (Serpentes: Lamprophiidae: Aparallactinae) and a new subgenus of wolf snakes *Lycophidion* Fitzinger, 1843 (Serpentes: Lamprophiidae: Lamprophiinae). *Australasian Journal of Herpetology,* Doncaster 13, 10–14.

Hoser, R.T. (2012af) A review of kukri snakes, currently referred to the genus *Oligodon* Fitzinger, 1826, with a division into twelve genera, four further subgenera and the creation of a tribe to accommodate them (Serpentes: Colubridae). *Australasian Journal of Herpetology,* Doncaster 13, 15–34.

Hoser, R.T. (2012ag) A review of natricine genera *Tropidonophis* Jan, 1863 and *Amphiesma* Dumeril, Bibron and Dumeril, 1854 (Serpentes: Colubridae: Natricinae). *Australasian Journal of Herpetology,* Doncaster 13, 35–46.

Hoser, R.T. (2012ah) A division of the Neotropical genus *Rhadinaea* Cope, 1863 (Serpentes: Colubridae). *Australasian Journal of Herpetology,* Doncaster 13, 47–54.

Hoser, R.T. (2012ai) A division of Central American snake genera, *Coniophanes* Hallowell in Cope, 1860 into six subgenera and *Conophis* Peters, 1860 into two genera (Serpentes: Colubridae: Dipsadinae). *Australasian Journal of Herpetology,* Doncaster 13, 55–60.

Hoser, R.T. (2012aj) A division of the Asian reed snakes, genus *Calamaria* Boie, 1827 (Serpentes: Colubridae: Calamariinae). *Australasian Journal of Herpetology,* Doncaster 13, 61–64.

Hoser, R.T. (2012ak) *Yeomansus*: a new genus for the slender racer (Serpentes: Colubridae). *Australasian Journal of Herpetology,* Doncaster 14, 3–5.

Hoser, R.T. (2012aL) A division of the patch-nosed snakes, genus *Salvadora* Baird and Girard, 1853 (Serpentes: Colubridae: Colubrinae). *Australasian Journal of Herpetology,* Doncaster 14, 6–8.

Hoser, R.T. (2012am) A review of the extant Scolecophidians ("blind-snakes") including the formal naming and diagnosis of new tribes, genera, subgenera, species and subspecies for divergent taxa. *Australasian Journal of Herpetology,* Doncaster 15, 1–64.

Hosmer, W. (1952) The broad-headed snake, *Hoplocephalus bungaroides* (Boie). *Northern Queensland Naturalist* 20(100), 14–16.

Hotz, H. & Broggi, M.F. (1982) *Rote Liste des gefahrdeten und seltenen Amphibien und Reptilien der Schweiz.* Schweizerischer Bund fur Naturschutz, Basel, 112 pp., illust.

Houssaye, A. (2010) Rediscovery and description of the second specimen of the hind-limbed snake *Pachyophis woodwardi* Nopsca, 1923 (Squamata, Ophidia) from the Cenomanian of Bosnia Herzegovina. *Journal of Vertebrate Paleontology* 30(1), 276–279, 3 figs.

Houssaye, A., Xu, F., Helfen, L., Buffrenil, V. de, Baumbach, T. & Tafforeau, P. (2011) Three-dimensional pelvis and limb anatomy of the Cenomanian hind-limbed snake *Eupodophis descouensi* (Squamata, Ophidia) revealed by synchron-radiation computed laminography. *Journal of Vertebrate Paleontology* 31(1), 2–7, 4 figs.

Houston, D. & Shine, R. (1993) Sexual dimorphism and niche divergence: feeding habits of the Arafura filesnake. *Journal of Animal Ecology* 62(4), 737–748.

Houston, D. & Shine, R. (1994) Population demography of Arafura filesnakes (Serpentes, Acrochordidae) in tropical Australia. *Journal of Herpetology* 28(3), 273–280, 3 figs.

Houston, T.F. (1976) Vertebrate type-specimens in the South Australian Museum. III. Reptiles. *Records of the South Australian Museum,* Adelaide 17(9), 181–187.

Houttuyn, M. (1764) *Natuurlyke historie of uitvoerige Beschryving der dieren, planten en mineraalen, volgens het Samenstel van den Heer Linnaeus.* Eerste deels, Zesde stuk. F. Houttuyn, Amsterdam, viii + 558 + vi pp., pls. 50–56.

How, R.A. & Shine, R. (1999) Ecological traits and conservation biology of five fossorial "sand-swimming" snake species (*Simoselaps*: Elapidae) in southwestern Australia. *Journal of Zoology,* London 249(3), 269–282.

Howard, K.G., Eaton, J.M., Larimer, S.C. & Townsend, J.H. (2001) Geographic distribution: Serpentes: *Alsophis rijgersmaei. Herpetological Review* 32(2), 121.

Howard, S.D. & Gillespie, G.R. (2007) Two new *Calamaria* (Serpentes) species from Sulawesi, Indonesia. *Journal of Herpetology* 41(2), 237–242, 5 figs.

Howell, K. (1993) Herpetofauna of the eastern African forests. *In:* Lovett, J.C. & Wasser, S.K. (Eds.), *Biogeography and ecology of the rain forests of eastern Africa.* Cambridge University Press, Cambridge, pp. 173–201.

Hoy, P.R. (1883) Catalogue of the cold-blooded vertebrates of Wisconsin. *Survey of Geology of Wisconsin* (1873–1879) 1(2), 422–435.

Hoyer, R.F. (2001) Discovery of a probable new species in the genus *Contia. Northwestern Naturalist* 82(3), 116–122.

Hoyt, D.L. (1964) The rediscovery of the snake *Leptophis modestus. Copeia* 1964(1), 214–215.

Hraoui-Bloquet, S. (1981) Les reptiles du Liban. 1. Nomenclature et note écologique. *Ecologia Mediterranea* 7(2), 94–101, 2 figs.

Hraoui-Bloquet, S., Sadek, R.A., Sindaco, R. & Venchi, A. (2002) The herpetofauna of Lebanon: new data on distribution. *Zoology in the Middle East* 27, 35–46, 3 figs.

Hsiou, A.S. & Albino, A.M. (2009) Presence of the genus *Eunectes* (Serpentes, Boidae) in the Neogene of southwestern Amazonia, Brazil. *Journal of Herpetology* 43(4), 612–619, 3 figs.

Hsiou, A.S., Albino, A.M. & Ferigolo, J. (2010) Reappraisal of the South American Miocene snakes of the genus *Colombophis,* with description of a new species. *Acta Palaeontologica Polonica* 55(3), 365–379, 5 figs.

Hu, B.-C., M., Huang, M.-H., Ho, S.-S. & Wei, K.-C. (1958) New records of snakes from Chekiang. *Acta Zoologica Sinica,* Beijing 10(2), 113–122, 2 pls.

Hu, B.-Q., Huang, M.-H., Hé, S.-X. & Wei, K.-J. (1959) *The snakes of Zhejiang.* Science Press, Peking, 47 pp., 69 figs. (in Chinese)

Hu, B.-Q., Huang, M.-H., Xie, Z.-T., Zhao, E.-M., Chiang, Y.-M., Huang, C.-Y. & Ma, J. (Eds.). (1980) *Atlas of the snakes of China.* Shanghai Science & Technology Press, Shanghai, 166 pp., 184 figs., 112 col. pls. (in Chinese).

Hu, S.-Q. & Zhao, E.-M. (1966) Three new species of reptiles from Szechuan. *Acta Zootaxonomica Sinica,* Peking 3(2), 158–164, 2 pls. (in Chinese)

Hu, S.-Q., Zhao, E.-M. & Liu, C.-C. (1966) A herpetological survey of the Tsinling and Ta-Pa Shan region. *Acta Zoologica Sinica,* Beijing 18(1), 57–90, 2 pls.

Hu, S.-Q., Zhao, E.-M. & Liu, C.-C. (1973) A survey of amphibians and reptiles in Kweichow Province, including a herpetofaunal analysis. *Acta Zoologica Sinica,* Beijing 19(2), 149–178, 3 pls.

Hu, S.-Q., Zhao, E.-M. & Huang, C.-C. (1975) Three new species of reptiles from Hainan Island, Guangdong Province. *Acta Zoologica Sinica,* Beijing 21(4), 379–384, 1 pl.

Huang, H.-F., Chen, F.-K., Min, C.-L. & Ho, C.-T. (1965) A preliminary survey of Amphibia and Reptilia in Ch'in-ling and Ta-pa-shan. *In: Communist Chinese Scientific Abstracts (30th Annual China Zoological Society Conference, December 1965). No. 2,* Joint Publications Research Service, Washington, pp. 253–254.

Huang, M.-H. (1990) *Fauna of Zhejiang. Amphibia. Reptilia.* Zhejiang Science and Technology Publishing House, Zhejiang, 306 pp., 288 figs., 30 col. pls. (in Chinese)

Huang, M.-H. & Qu, Y.-F. (1992) *Deinagkistrodon acutus.* Second edition. Scientific Publishing House, Beijing, 278 pp., 17 col. pls. (in Chinese) [first edition 1983, 196 pp., illust.]

Huang, M.-H., Yang, Y.-J., Xie, X. & Qu, Y.-F. (1998) Karyotype of the Chinese endemic species *Boiga kraepelini* Stejneger, (Colubridae). *Russian Journal of Herpetology* 5(1), 43, 1 fig.

Huang, M.-P. (1984) An estimate on the population of *Agkistrodon shedaoensis* on Shedao Island. *Acta Herpetologica Sinica* 3(4), 17–22. (in Chinese)

Huang, Q.-Y. (2004) *Pareas chinensis* (Barbour, 1912) should be a junior synonym of *Pareas formosensis* (Van Denburgh, 1909). *Sichuan Journal of Zoology* 23(3), 209–210.

Huang, S., Liu, S.-Y., Guo, P., Zhang, Y.-P. & Zhao, E. (2009) What are the closest relatives of the hot-spring snakes (Colubridae, *Thermophis*), the relict species endemic to the Tibetan Plateau? *Molecular Phylogenetics and Evolution* 51(3), 438–446, 2 figs.

Huang, S., Ding, L., Burbrink, F.T., Yang, J., Huang, J.-T., Ling, C., Chen, X. & Zhang, Y.-P. (2012) A new species of the genus *Elaphe* (Squamata: Colubridae) from Zoige County, Sichuan, China. *Asian Herpetological Research* (Series 2) 3(1), 38–45, 5 figs.

Huang, W.-S. (1995) Notes on egg clutch size of the black burrowing snake, *Achalinus niger. Journal of the Taiwan Museum* 48(2), 105–108.

Huang, X., Pan, T., Han, D.-M., Zhang, L.-A., Hou, Y.-X., Yu, L, Zheng, H. & Zhang, B.-W. (2012) A new species of the genus *Protobothrops* (Squamata: Viperidae: Crotalinae) from the Dabie Mountains, Anhui, China. *Asian Herpetological Research* (Series 2) 3(3), 213–218, 3 figs.

Huang, Z.-J., Zheng, J. & Fang, J.J. (1978) New species of snakes. *Journal of the Fujian Normal University, Natural Science Series*, Fuzhou 2, 91–93. (in Chinese)

Huang, Z.-Y. (1982) A new species of the Crotalidae snake from Tibetan. *Acta Fudan University (Natural Sciences)*, Shanghai 21(1), 115–118, 5 figs.

Huang, Z.-Y. (Ed.). (1998) *Unique reptiles of China.* Fu Da University, Shanghai 116 pp., 53 figs., 10 photos, 83 col. figs. (in Chinese)

Hubbs, B. (2004) *Mountain kings. A collective natural history of California, Sonoran, Durango and Queretaro mountain kingsnakes.* Tricolor Books, Tempe, 319 pp., illust.

Hubbs, B. (2009) *Common kingsnakes. A natural history of* Lampropeltis getula. Tricolor Books, Tempe, 412 pp., illust.

Hubbs, B. & O'Connor, B. (2009) *A guide to the rattlesnakes of the United States.* Tricolor Books, Tempe, 89 pp., 101 col. photos.

Hubrecht, A.A.W. (1879a) Note III. On a new genus and species of Pythonidae from Salawatti. *Notes from the Leyden Museum* 1(1), 14–15.

Hubrecht, A.A.W. (1879b) Note IV. *Liasis petersii* n. sp. *Notes from the Leyden Museum* 1(1), 16–18.

Hubrecht, A.A.W. (1879c) Note V. On the geographical range of *Erebophis asper* Gthr. *Notes from the Leyden Museum* 1(1), 19–20.

Hubrecht, A.A.W. (1879d) Note XLVI. Contributions to the herpetology of Sumatra. II. On *Python curtus* Schlegel, MS. *Notes from the Leyden Museum* 1(3), 244–245.

Hubrecht, A.A.W. (1882) Note XI. List of reptiles and amphibians brought from British India by Mr. Francis Day. *Notes from the Leyden Museum* 4(2), 138–144.

Hubrecht, A.A.W. (1886) Natuurlijke historie. Tweede afdeeling. Kruipende dieren en visschen. Systematische lijst. I. Reptilia. *Midden Sumatra (Natural History)* 13(2), 1–8, 1 pl.

Hudson, G.E. (1942) The amphibians and reptiles of Nebraska. *Nebraska Conservation Bulletin* (24), 1–146, 20 pls.

Huene, F.R.F. von. (1956) *Paläontologie und Phylogenie der Niederen Tetrapoden.* Gustav Fischer Verlag, Jena, 716 pp., 689 figs.

Hughes, B. (1968) An unusual rhinocerus viper, *Bitis nasicornis*, from Ghana, West Africa. *Zoologische Mededelingen*, Leiden 43(9), 371–383, 8 figs.

Hughes, B. (1976a) Notes on African carpet vipers, *Echis carinatus, E. leucogaster* and *E. ocellatus* (Viperidae, Serpentes). *Revue Suisse de Zoologie*, Genève 83(2), 359–371, 3 figs.

Hughes, B. (1976b) Zoogeography of West Africa false cobras (*Pseudohaje* spp.). *Bulletin de l'Institut Fondamental d'Afrique Noire*, Dakar 38A(2), 457–466, 1 fig.

Hughes, B. (1978) Latitudinal clines and ecogeography of the West African night adder, *Causus maculatus* (Hallowell, 1842), Serpentes, Viperidae. *Bulletin de l'Institut Fondamental d'Afrique Noire*, Dakar (1977) 39A(2), 358–384.

Hughes, B. (1983) African snake faunas. *Bonnischer Zoologische Beiträge*, Bonn 34(1–3), 311–356, 6 figs.

Hughes, B. (1985) Progress on a taxonomic revision of the African green tree snakes (*Philothamnus* spp.). *In:* Schuchmann, K.-L. (Ed.), *Proceedings of the International Symposium on African Vertebrates: systematics, phylogeny and evolutionary ecology*. Zoologisches Forschungsinstitut und Museum Alexander Koenig, Bonn, pp. 511–530, 17 figs.

Hughes, B. (1988) Herpetology in Ghana (west Africa). *British Herpetological Society Bulletin*, London (25), 29–38, illust.

Hughes, B. (1997) *Dasypeltis scabra* and *Lamprophis fuliginosus*—two pan-African snakes in the Horn of Africa: a tribute to Don Broadley. *African Journal of Herpetology* 46, 68–77.

Hughes, B. (1999) Critical review of a revision of *Psammophis* (Linnaeus 1758) (Serpentes, Reptilia) by Frank Brandstätter. *African Journal of Herpetology* 48(1–2), 63–70.

Hughes, B. (2003) *Polemon barthi* Jan 1858 and *P. bocourti* Mocquard 1897, two West African snakes which are allopatric, not sympatric. *African Journal of Herpetology* 52(2), 113–117, 1 fig.

Hughes, B. (2004) Misidentification of *Dromophis lineatus* (Duméril & Bibron, 1854) as *Psammophis sibilans* (Linné 1758) and the perpetuation of error. *African Journal of Herpetology* 53(1), 63–76, 3 figs.

Hughes, B. (2013) Snakes of Bénin, West Africa. *Bulletin de la Société Herpétologique de France* (2012) (144), 101–159, 2 figs., 4 photos.

Hughes, B. & Barry, D.H. (1969) The snakes of Ghana, a checklist and key. *Bulletin de l'Institut Fondamental d'Afrique Noire*, Dakar 31A(3), 1004–1041.

Hughes, B. & Wade, E. (2002) On the African leopard whip snake, *Psammophis leopardinus* Bocage, 1887 (Serpentes, Colubridae), with the description of a new species from Zambia. *Bulletin of the Natural History Museum (Zoology Series)* 68(2), 75–81, 5 figs.

Hughes, B. & Wade, E. (2004) Is *Psammophis sibilans occidentalis* Werner, 1919 a junior synonym of *P. phillipsi* (Hallowell, 1844)? *Herpetozoa* 16(3–4), 127–132, 1 fig.

Huheey, J.E. (1959) Distribution and variation in the glossy water snake, *Natrix rigida* (Say). *Copeia* 1959(4), 303–311, 4 figs.

Huheey, J.E. & Stupka, A. (1967) *Amphibians and reptiles of Great Smoky Mountains National Park.* University of Tennessee Press, Knoxville, ix + 98 pp., 72 pls.

Hulse, A.C., McCoy, C.J. & Censky, E. (2001) *Amphibians and reptiles of Pennsylvania and the Northeast.* Cornell University Press, Ithaca, 419 pp., 42 figs., 133 col. photos.

Hulselmans, J.L.J. & Verheyen, W.N. (1970) Contribution à l'herpétologie de la République du Togo. 2. List préliminaire des serpents récoltés par la deuxième Mission zoologique belge au Togo. *Revue de Zoologie et Botanique Africaine*, Bruxelles 82, 200–203.

Hulselmans, J.L.J., de Roo, A. & de Vree, F. (1970) Contribution à l'herpétologie de la République du Togo. 1. Liste préliminaire des serpents récoltés par la première Mission zoologique belge au Togo. *Revue de Zoologie et Botanique Africaine,* Bruxelles 81, 193–196.

Hulselmans, J.L.J., de Vree, F. & van der Straeten, E. (1971) Contribution à l'herpétologie de la République du Togo. 3. Liste préliminaire des serpents récoltés par la troisième Mission zoologique belge au Togo. *Revue de Zoologie et Botanique Africaine*, Bruxelles 84(1–2), 47–49.

Humboldt, F.W.H. A. von. (1813) Sur deux nouvelles espèces de crotales. *In:* Humboldt, A. & Bonpland, A. *Recuell d'observations de zoologie et d'anatomie comparée faites dans l'Océan Atlantique, dans l'intérieur du Nouveau continent et dans la Mer du Sud pendant les années 1799, 1800, 1801, 1802 et 1803. Tome. II.* (1811) F. Schoell and G. Dufour, Paris, pp. 1–8. [reprinted in 1833 by J. Smith & Gide, Paris]

Humphrey, F.L. & Shannon, F.A. (1958) A state record and range extension for the snake *Sympholis lippiens* (Serpentes, Colubridae). *Herpetologica* 13(4), 257–260.

Hunt, R.A. (1947) A key to the identification of Australian snakes (other than blind snakes, rock snakes and pythons, file snakes, and sea snakes). *Victorian Naturalist*, Springvale 64, 153–167.

Hunter, M.L., Jr., Albright, J. & Arbuckle, J. (Eds.). (1992) The amphibians and reptiles of Maine. *Bulletin, Maine Agricultural Experiment Station*, Orono (838), 1–188, 70 figs.

Hunter, M.L., Jr., Calhoun, A.J.K. & McCollough, M. (Eds.). (1999) *Maine amphibians and reptiles*. Second edition. University of Maine Press, Orono, 252 pp., 36 col. pls.

Hurter, J., Jr. (1893) Catalogue of reptiles and batrachians found in the vicinity of St. Louis, Mo. *Transactions of the St. Louis Academy of Sciences* 6(11), 251–261.

Hurter, J., Jr. (1911) Herpetology of Missouri. *Transactions of the Academy of Science of St. Louis* 20(5), 59–274, 24 pls.

Hurter, J., Jr. & Strecker, J.K. (1909) The amphibians and reptiles of Arkansas. *Transactions of the Academy of Sciences of St. Louis* 18(2), 11–27.

Husain, A. & Ray, P. (1995) Reptilia. *In:* Alfred, J.R.B. (Ed.), *Fauna of western Himalaya. Part 1. Uttar Pradesh. Himalayan Ecosystem Series*. Zoological Survey of India, Calcutta, pp. 159–167.

Husain, A. & Tilak, R. (1995) Snakes (Reptilia: Serpentes). *In:* Alfred, J.R.B. (Ed.), *Fauna of Rajaji National Park. Fauna of Conservation Areas* 5. Zoological Survey of India, Calcutta, pp. 91–113, 3 pls.

Husain, A. (1997) Reptilia. *In:* Alfred, J.R.B. (Ed.), *Fauna of Delhi. State Fauna Series* 6. Zoological Survey of India, Calcutta, pp. 665–698.

Hutchinson, M.N. (1990) The generic classification of the Australian terrestrial elapid snakes. *Memoirs of the Queensland Museum*, Brisbane 29(2), 397–405.

Hutchison, J.H. (1985) *Pterosphenus* cf. *P. schucherti* Lucas (Squamata, Palaeophidae) from the late Eocene of peninsular Florida. *Journal of Vertebrate Paleontology* 5(1), 20–23.

Hutton, A.F. (1949) Notes on the snakes and mammals of the high Wavy Mountains, Madura district, S. India. Part I. Snakes. *The Journal of the Bombay Natural History Society* (1948–1949) 48(3), 454–460, fig.

Hutton, A.F. & David, P. (2009) Note on a collection of snakes from South India, with emphasis on the snake fauna of the Meghamalai Hills (High Wavy Mountains). *The Journal of the Bombay Natural History Society* (2008) 105(3), 299–316, 4 figs.

Hvass, H. (1945) De Danske krybdyrs udbredelsesforhold. *Flora og Fauna* 51, 97–126.

Ibáñez, R. & Solis, F.A. (1991) Las serpientes de Panama: lista de especies, comentarios taxonomicos y bibliografia. *Scientia* (Panamá) 6(2), 27–52.

Ibáñez, R., Arosemena, F.A., Solis, F.A. & Jaramillo, C.A. (1994) Anfibios y Reptiles de la Serrania Piedras-Pacora, Parque Nacional Chagres. *Scientia* (Panamá) 9(1), 17–31, fig. 1.

Ibáñez, R., Solis, F.A., Jaramillo, C.A. & Rand, A.S. (2001) An overview of the herpetology of Panama. *In:* Johnson, J.D., Webb, R.G. & Flores-Villela, O.A. (Eds.), *Mesoamerican herpetology: systematics, zoogeography, and conservation*. University of Texas El Paso, El Paso, pp. 159–170.

Ihering, R. von (1911) As cobras do Brazil. *Revista do Museu Paulista*, São Paulo 8, 273–379, 25 figs.

Ibrahim, A.A. (2001) Geographic distribution: Serpentes: *Malpolon monspessulanus insignatus*. *Herpetological Review* 32(2), 123.

Ibrahim, A.A. (2002) Geographic distribution: Serpentes: *Psammophis sibilans sibilans*. *Herpetological Review* 33(1), 69.

Ibrahim, A.A. (2008) Contribution to the herpetology of southern Libya. *Acta Herpetologica* 3(1), 35–49, 1 fig.

Ibrahim, S.A. & Masr, A.R.M. (1975) Action of phospholipase A from black mamba *(Dendroaspis polyelepis)* venom on the phospholipids of human blood. *Toxicon* 13(1), 99.

ICZN [International Commission of Zoological Nomenclature]. (1999) *International code of zoological nomenclature*. Fourth edition. International Trust for Zoological Nomenclature, London, 306 pp. [known as the "Code" or "Rules"]

ICZN Opinion 92. (1926) Sixteen generic names of Piscers, Amphibia, and Reptilia placed in the Official List of Generic Names. Smithsonian Miscellaneous Collections 73(4), 3–4.

ICZN Opinion 188. (1945) Suppression of the name *Cobra* Laurenti, 1768, and suspension of the rules for *Bitis* Gray, 1842 (Class Reptilia, Order Squamata). *Opinions and Declarations of the International Commission on Zoological Nomenclature*, London 3(7), 77–89.

ICZN Opinion 328. (1955a) Suppression under the Plenary Powers, of the specific name "*caesius*" Cloquet, 1818, as published in the combination "*Coluber caesius*," for the purpose of validating the specific name "*irregularis*" Leach, 1819, as published in the combination "*Coluber irregularis*" (Class Reptilia, Order Squamata). *Opinions and Declarations of the International Commission on Zoological Nomenclature*, London 9(20), 299–308.

ICZN Opinion 339. (1955b) Acceptance of the name concolor Woodbury, 1929, as published in the combination *Crotalus concolor*, as the name for the yellow rattlesnake of the Colorado River Basin. *Opinions and Declarations of the International Commission on Zoological Nomenclature*, London 10(6), 181–200.

ICZN Opinion 365. (1955c) Suppression under the Plenary Powers of the specific name "*cinereous*" LeConte, 1852, as published in the combination "*Crotalus cinereous*," for the purpose of rendering the specific name "*atrox*" Baird and Girard, 1853, as published in the combination "*Crotalus atrox*," the oldest available name for the western diamond rattlesnake (Class Reptilia, Order Squamata). *Opinions and Declarations of the International Commission on Zoological Nomenclature*, London 11(15), 215–230.

ICZN Opinion 366. (1955c) Suppression under the Plenary Powers of the specific name "*multimaculata*" Jan, 1863, as published in the combination "*Crotalus lugubris* var. *multimaculata*," for the purpose of rendering the specific name "*Polysticta*," Cope, 1865, as published in the combination "*Caudisona polysticta*," the oldest available name for the Mexican lance-headed rattlesnake (Class Reptilia, Order Squamata). *Opinions and Declarations of the International Commission on Zoological Nomenclature*, London 11(16), 231–244.

ICZN Opinion 385. (1956a) Determination under the Plenary Powers of the method to be adopted in interpreting the nominal species "*Coluber sirtalis*" Linnaeus, 1758, the common garter snake of the eastern United States, and approval of the neotype designated for the foregoing species and of that designated for the ribbon snake of North America. *Opinions and Declarations of the International Commission on Zoological Nomenclature*, London 12(6), 191–230, 2 pls.

ICZN Opinion 387. (1956b) Suppression under the Plenary Powers of the generic name *Anodon* Smith, 1829 for the purpose of validating the generic name *Dasypeltis* Wagler, 1830, for the African egg-eating snake (Class Reptilia). *Opinions and Declarations of the International Commission on Zoological Nomenclature*, London 12(8), 241–250.

ICZN Opinion 390. (1956c) Determination under the Plenary Powers of the taxon to which shall be applicable the specific name *simus* Linnaeus, 1766 as published in the combination *Coluber simus* (Class Reptilia). *Opinions and Declarations of the International Commission on Zoological Nomenclature*, London 12(13), 273–284.

ICZN Direction 37. (1956d) Completion of the entries relating to the names of certain genera of the order Decapoda (class Crustacea) placed on the Official List of Generic Names in Zoology in the period up to the end of 1936. *Opinions and Declarations of the International Commission on Zoological Nomenclature*, London 1D(2), 47–82.

ICZN Direction 56. (1956d) Completion and in certain cases correction of entries relating to the names of genera belonging to the classers Pisces, Amphibia and Reptilia made in the Official List of Generic Names in Zoology in the period up to the end of 1936. *Opinions and Declarations of the International Commission on Zoological Nomenclature*, London 1D(17), 337–364.

ICZN Opinion 490. (1957) Action under the Plenary Powers to preserve for use in its accustomed sense the generic name "*Elaphe*" Fitzinger, 1833 (Class Reptilia). *Opinions and Declarations of the International Commission on Zoological Nomenclature*, London 17(12), 179–196.

ICZN Opinion 524. (1958a) Interpretation of the nominal species "*Coluber ahaetulla*" Linnaeus, 1758, and addition to the "Official list of generic names in zoology" of the generic name "*Ahaetulla*" Link, 1807, with "*Ahaetulla mycterizans*" Link, 1807, as type species (Class Reptilia). *Opinions and Declarations of the International Commission on Zoological Nomenclature*, London 19(11), 261–276.

ICZN Opinion 525. (1958b) Determination of precedence to be accorded to the names "*Coluber pictus*" Gmelin (J.F.), 1789, and "*Coluber boiga*" Lacépède, 1789 (Class Reptilia). *Opinions and Declarations of the International Commission on Zoological Nomenclature*, London 19(12), 277–290.

ICZN Opinion 632. (1962) *Regina* Baird and Girard, 1853 (Reptilia); designation of a type-species under the Plenary Powers. *Bulletin of Zoological Nomenclature*, London 19(3), 145–147.

ICZN Opinion 644. (1963a) *Coluber atratus* Hallowell, 1845 (Reptilia); validation under the Plenary Powers. *Bulletin of Zoological Nomenclature*, London 20(1), 26–28.

ICZN Opinion 651. (1963b) *Anilius* Oken, 1816 (Reptilia); validation under the Plenary Powers. *Bulletin of Zoological Nomenclature*, London 20(2), 111–113.

ICZN Opinion 661. (1963c) *Cerastes* Laurenti, 1768 (Reptilia): designation of a type-species under the Plenary Powers. *Bulletin of Zoological Nomenclature*, London 20(3), 191–192.

ICZN Opinion 698. (1964a) *Lystrophis* Cope 1885 (Reptilia): validated under the Plenary Powers. *Bulletin of Zoological Nomenclature*, London 21(2), 101–103.

ICZN Opinion 709. (1964b) *Dendraspis* Fitzinger, 1843 (Reptilia): suppressed under the Plenary Powers. *Bulletin of Zoological Nomenclature*, London 21(3), 210–211.

ICZN Opinion 717. (1965a) *Arizona elegans* Kennicott, 1859 (Reptilia): validated under the Plenary Powers. *Bulletin of Zoological Nomenclature*, London 22(1), 19–21.

ICZN Opinion 744. (1965b) *Ablabes chinensis* Günther, 1889 (Reptilia): validated under the Plenary Powers. *Bulletin of Zoological Nomenclature*, London 22(3), 180–181.

ICZN Opinion 745. (1965c) *Coluber subocularis* Brown, 1901 (Reptilia): validated under the Plenary Powers. *Bulletin of Zoological Nomenclature*, London 22(3), 182–183.

ICZN Opinion 746. (1965d) *Spilotes melanurus* Duméril, Bibron & Duméril, 1854 (Reptilia): validated under the plenary powers. *Bulletin of Zoological Nomenclature*, London 22(3), 184–185.

ICZN Opinion 756. (1966a) *Leptophis vertebralis* Duméril, Bibron & Duméril, 1854 (Reptilia): suppressed under the Plenary Powers. *Bulletin of Zoological Nomenclature*, London 22(5–6), 284–285.

ICZN Opinion 771. (1966b) *Thamnophis sirtalis* Linnaeus, 1758 (Reptilia): rejection under the Plenary Powers of the neotype specimen designated for that species by Opinion 385. *Bulletin of Zoological Nomenclature*, London 23(1), 38–40.

ICZN Opinion 794. (1966c) *Spalerosophis* Jan, 1865 (Reptilia): validated under the Plenary Powers. *Bulletin of Zoological Nomenclature*, London 23(5), 229–231.

ICZN Opinion 804. (1967) *Coluber doliatus* Linnaeus, 1766 (Reptilia): suppressed under the Plenary Powers. *Bulletin of Zoological Nomenclature*, London 24(1), 18–19.

ICZN Opinion 939. (1970) *Python timorensis* Müller, 1844 (Reptilia): specific name suppressed under the Plenary Powers. *Bulletin of Zoological Nomenclature*, London 27(3–4), 178–179.

ICZN Opinion 963. (1971) *Amplycephalus* Kuhl & van Hasselt, 1822 (Reptilia): suppressed under the Plenary Powers. *Bulletin of Zoological Nomenclature*, London 28(1–2), 44–45.

ICZN Opinion 1035. (1975) *Mastigodryas* Amaral, 1934 (Reptilia, Serpentes): refusal to use the Plenary Powers for suppression. *Bulletin of Zoological Nomenclature*, London 32(1), 36–37.

ICZN Opinion 1045. (1976) *Boa* Linnaeus, 1758 (Reptilia, Serpentes): placed on the Official List. *Bulletin of Zoological Nomenclature*, London 32(4), 218–221.

ICZN Opinion 1165. (1981a) *Scytale niger* Daudin, 1803 (Reptilia, Serpentes): suppressed. *Bulletin of Zoological Nomenclature*, London 38(1), 61–63.

ICZN Opinion 1176. (1981b) *Echis colorata* [*sic*] Guenther, 1878 (Reptilia, Serpentes) given nomenclatural precedence over *Echis froenata* [*sic*] Duméril, Bibron & Duméril, 1854. *Bulletin of Zoological Nomenclature*, London 38(2), 110–111.

ICZN Opinion 1201. (1982a) Elapidae Boie, 1827 (Reptilia, Serpentes): ruling to stabilize nomenclature of taxa in this family (and others). *Bulletin of Zoological Nomenclature*, London 39(1), 21–26.

ICZN Opinion 1207. (1982b) *Leptotyphlops* and *Ramphotyphlops* Fitzinger, 1843 (Reptilia, Serpentes), conserved. *Bulletin of Zoological Nomenclature*, London 39(2), 106–108.

ICZN Opinion 1210. (1982c) *Psammophis sibilans subtaeniata* Peters, 1882 (Reptilia, Serpentes), conserved. *Bulletin of Zoological Nomenclature*, London 39(2), 114–116.

ICZN Opinion 1211. (1982d) *Philodryas nattereri* Steindachner, 1870 (Reptilial, Serpentes), conserved. *Bulletin of Zoological Nomenclature*, London 39(2), 117–118.

ICZN Opinion 1214. (1982e) *Chlorophis carinatus* Andersson, 1901 given conditional precedence over *Philothamnus nigrofasciatus* Buchholz & Peters, 1875 (Reptilia, Serpentes). *Bulletin of Zoological Nomenclature*, London 39(2), 124–125.

ICZN Opinion 1246. (1983) *Herpetodryas margaritiferus* Schlegel, 1837 (Reptilia, Serpentes), conserved. *Bulletin of Zoological Nomenclature*, London 40(2), 75–76.

ICZN Opinion 1374. (1986a) *Boiga* Fitzinger, 1826 (Reptilia, Serpentes), conserved. *Bulletin of Zoological Nomenclature*, London 43(1), 25–26.

ICZN Opinion 1384. (1986b) *Dromophis* Peters, 1869 (Reptilia, Serpentes), conserved. *Bulletin of Zoological Nomenclature*, London 43(2), 123–124.

ICZN Opinion 1463. (1987) De Lacépède 1788–1789, Histoire Naturelle de Serpens and later editions, rejected as a non-binominal work. *Bulletin of Zoological Nomenclature*, London 44(4), 265–267.

ICZN Opinion 1514. (1988) *Liasis* Gray, 1844 (Reptilia, Serpentes): *Liasis mackloti* Duméril & Bibron, 1844 designated as the type species. *Bulletin of Zoological Nomenclature*, London 45(3), 244.

ICZN Opinion 1634. (1991) *Atheris* Cope, 1862 (Reptilia, Serpentes), conserved. *Bulletin of Zoological Nomenclature*, London 48(1), 84.

ICZN Opinion 1686. (1992) *Natrix gemonensis* Laurenti, 1768 (currently *Coluber gemonensis*), *Coluber viridiflavus* Lacépède, 1789 and *Coluber helveticus* Lacépède, 1789 (currently *Natrix natrix helvetica*) (Reptilia, Serpentes), specific names conserved. *Bulletin of Zoological Nomenclature*, London 49(2), 174–175.

ICZN Opinion 1789. (1994) *Pseudoxyrhopus* Günther, 1881 (Reptilia, Serpentes), conserved. *Bulletin of Zoological Nomenclature*, London 51(3), 289.

ICZN Opinion 1832. (1996a) *Coluber poecilogyrus* Wied-Neuwied [1824] (currently *Liophis poecilogyrus*) (Reptilia, Serpentes): specific name conserved. *Bulletin of Zoological Nomenclature* 53(1), 72–73.

ICZN Opinion 1856. (1996b) *Lycognathophis* Boulenger, 1893 (Reptilia, Serpentes), conserved. *Bulletin of Zoological Nomenclature* 53(3), 226–227.

ICZN Opinion 1867. (1997) *Phyllophis carinata* Günther, 1864 (currently *Elaphe carinata*, Reptilia, Serpentes): specific name conserved. *Bulletin of Zoological Nomenclature* 54(1), 75–76.

ICZN Opinion 1939. (1999a) *Trigonocephalus pulcher* Peters, 1862 (currently *Bothrops pulcher, Bothriechis pulchjer* or *Bothriopsis pulchra;* Reptilia, Serpentes): defined by the holotype, and not a neotype; *Bothrops campbelli* Freire Lascano, 1991: specific name placed on the Official List. Bulletin of Zoological Nomenclature 56(3), 221–222.

ICZN Opinion 1940. (1999b) *Hoplocephalus vestigiatus* De Vis, 1884 (Reptilia, Serpentes): specific name placed on the Official List. Bulletin of Zoological Nomenclature 56(3), 218–220.

ICZN Opinion 1960. (2000a) *Crotalus ruber* Cope, 1892 (Reptilia, Serpentes): specific name given precedence over that of *Crotalus exsul* Garman, 1884. *Bulletin of Zoological Nomenclature* 57(3), 189–190.

ICZN Opinion 1960. (2000b) *Coluber infernalis* Blainville, 1835 and *Eutainia sirtalis tetrataenia* Cope *in* Yarrow, 1875 (currently *Thamnophis sirtalis infernalis* and *T. s. tetrataenia*; Reptilia, Serpentes): subspecific names conserved by the designation of a neotype for *T. s. infernalis*. *Bulletin of Zoological Nomenclature* 57(3), 191–192.

ICZN Opinion 2196. (2008a) *Calamaria* H. Boie in F. Boie, (1827) (Reptilia, Serpentes): usage conserved by designation of *C. linnaei* H. Boie in F. Boie, 1827 as the type species. *Bulletin of Zoological Nomenclature* 65(1), 79–80.

ICZN Opinion 2210. (2008b) *Atractus* Wagler, 1828 and *Atractus trilineatus* Wagler, 1828 (Reptilia, Serpentes): conserved. *Bulletin of Zoological Nomenclature* 65(3), 239–240.

ICZN Opinion 2282. (2011) *Coluber nummifer* Reuss, 1834 (currently *Hemorrhois nummifer*; Reptilia, Serpentes): specific name conserved. *Bulletin of Zoological Nomenclature* 68(3), 228–229.

Ilani, G. (1983a) Wild animals in Lebanon. *Israel – Land and Nature,* Tel Aviv 8(3), 126.

Ilani, G. (1983b) Willife news. *Israel – Land and Nature,* Tel Aviv 9(1), 37, 2 figs.

Ilani, G. (1986) Willife news. *Israel – Land and Nature,* Tel Aviv 11(2), 102.

Ilgaz, C., Baran, I. & Kumlutas, Y. (2005) First record of *Natrix megalocephala* Orlov and Tuniyev, 1986 (Reptilia, Ophidia, Colubridae) from Turkey. *Russian Journal of Herpetology* 12(2), 111–112, 2 figs.

Ilgaz, C., Avei, A., Kumlutas, Y., Baran, I. & Ozdemir, A. (2007) New locality records extended the distribution of *Telescopus nigriceps* (Ahl, 1924), black headed cat snake, in southeastern Anatolia, Turkey. *Russian Journal of Herpetology* 14(3), 207–210, 3 figs.

Imazumi, K., Kuroda, N. & Okada, Y. (1950) *Mammals, birds, reptiles, amphibians.* Japan Publishing Company, Tokyo, 410 pp. (in Japanese)

Ineich, I, (1986) Histoire naturelle du serpent marin *Pelamis platurus* (Linne, 1766). *Bulletin de la Société Études Océaniennes* (236), 1–10.

Ineich, I. (1988) Le serpent marin *Pelamis platurus* (Elapidae, Hydrophiinae): Bilan des connaissances sur sabiologie et sa distribution; situation en Polynesie Orientale. *L'Annee Biologique* (Series 4) 27(2), 93–117.

Ineich, I. (1996) Geographic distribution: Serpentes: *Hydrophis torquatus diadema. Herpetological Review* 27(3), 154.

Ineich, I. (1999) *Reptiles & amphibians de la République de Djibouti.* Rapport sur la mission herpétologique de Ivan Ineich, Ministére de l'Urbanisme de l'Environnement et de l'Aménagement du Territoire, Djibouti, 60 pp.

Ineich, I. (2003) Contributions à la connaissance de la biodiversité des régions Afro-montagnardes: les reptiles du mont Nimba. *In:* Lamotte, M. & Roy, R. (Eds.), Le peuplement animal du Mont Nimba (Guinée, Côte d'Ivoire, Liberia). *Mémoires du Muséum National d'Histoire Naturelle*, Paris 190, 597–637.

Ineich, I. (2004) *Les serpents marins.* Institut Océanographique, Paris, 320 pp., 58 figs., 16 col. pls.

Ineich, I. & Borsa, P. (2003) Geographic distribution: Serpentes: *Hydrophis coggeri. Herpetological Review* 34(4), 388.

Ineich, I. & Laboute, P. (2002) *Sea snakes of New Caledonia.* Institut de Recherche pour le Développement, Paris. 302 pp., col. illust.

Ineich, I. & Rasmussen, A.R. (1997) Sea snakes from New Caledonian and the Loyalty Islands (Elapidae, Laticaudinae and Hydrophiinae). *Zoosystema* 19, 185–192.

Ineich, I. & Tellier, J.M. (1992) Une glande supralabiale a debouche externe chez le genre *Echis* (Reptilia, Viperidae), cas unique chez les serpents. *Comptes Rendus de l'Académie des Sciences*, Paris (Série 3) 315(2), 49–53, illust.

Ineich, I., Martelli, J.-L. & Clary, J. (2002) Catalogue des collections de reptiles du Muséum d'Histoire Naturelle de Lyon. *Cahiers Scientifiques*, Lyon 2002(2), 83–96.

Ineich, I., Fretey, T., Colyn, M. & Barrière, P. (2007) Geographic distribution: Serpentes: *Gonionotophis brussauxi. Herpetological Review* 38(1), 104.

Infante-Rivero, E. (2009) Primer registro de *Porthidium lansbergii lansbergii* Schlegel, 1841, en La Guajira Venezolana, con comentarios sobre el género y la distribución en el estado Zulia. *Herpetotropicos* 5(1), 25–28, 2 figs.

Infante-Rivero, E., Delgadound-F., P.V. & Rojas-Runjaic, J.M. (2005) Geographic distribution: Serpentes: *Drymarchon caudomaculatus. Herpetological Review* 36(2), 203.

Infante-Rivero, E., Natera-Mumaw, M. & Marcano, A. (2008) Extension of the distribution of *Eunectes murinus* (Linnaeus, 1758) and *Helicops angulatus* (Linnaeus, 1758) in Venezuela, with notes on ophiophagia. *Herpetotropicos* 4(1), 39, 1 fig.

Inger, R.F. (1946) Restriction of the type locality of *Thamnophis sirtalis. Copeia* 1946(4), 254.

Inger, R.F. (1967) A new colubrid snake the genus *Stegonotus* from Borneo. *Fieldiana: Zoology* 51(5), 77–83, fig. 13.

Inger, R.F. & Clark, P.J. (1943) Partition of the genus *Coluber. Copeia* 1943(3), 141–145.

Inger, R.F. & Leviton, A.E. (1961) A new colubrid snake the genus *Pseudorabdion* from Sumatra. *Fieldiana: Zoology* 44(5), 45–47, fig. 13.

Inger, R.F. & Leviton, A.E. (1966) The taxonomic status of Bornean snakes of the genus *Pseudorabdion* Jan and of the nominal genus *Idiopholis* Mocquard. *Proceedings of the California Academy of Sciences* (Series 4) 34(4), 307–314, 3 figs.

Inger, R.F. & Lian, T.F. (2010) *The natural history of amphibians and reptiles in Sabah.* Second edition. Natural History Publications, Kota Kinabalu, vii + 121 pp., 174 col. figs. [first edition, 1996, v + 101 pp., 130 col. figs.]

Inger, R.F. & Marx, H. (1962) Variation of hemipenis and cloaca in the colubrid snake *Calamaria lumbricoidea. Systematic Zoology* 11(1), 32–38, 3 figs.

Inger, R.F. & Marx, H. (1965) The systematics and evolution of the Oriental colubrid snakes of the genus *Calamaria. Fieldiana: Zoology* 49, 1–304, 73 figs.

Inger, R.F. & Voris, H.K. (2001) The biogeographical relations of the frogs and snakes of Sundaland. *Journal of Biogeography* 28, 863–891, 4 figs.

Inger, R.F., Shaffer, H.B., Koshy, M. & Bakde, R. (1984) A report on a collection of amphibians and reptiles from the Ponmudi, Kerala, south India. *The Journal of the Bombay Natural History Society* 81(3), 551–570, pls. 4–8.

Inger, R.F., Zhao, E., Shaffer, H.B. & Wu, G. (1990) Report on a collection of amphibians and reptiles from Sichuan, China. *Fieldiana: Zoology* (Series 2) (58), i–iii + 1–24, 10 figs.

Ingle, M. & Sarsavan, A. (2011) A new locality record of *Coronella brachyura* (Günther, 1866) (Serpentes, Colubridae, Colubrinae) from Madhya Pradesh, India, with notes on its distribution and natural history. *Sauria* 33(2), 59–61.

Inglis, C.K., Travers, W.L., O'Donel, H.V. & Shebbeare, E.O. (1920) A tentative list of vertebrates of the Jalpaiguri district. *The Journal of the Bombay Natural History Society* (1920–1921) 27(1), 158–159.

Ingoldby, C.M. & Procter, J.B. (1923) Notes on a collection of Reptilia from Waziristan and the adjoining portion of the N.W. Frontier Province. *The Journal of the Bombay Natural History Society* (1923–1924) 29(1), 117–130, 4 pls.

Ingram, G.J. (1989) *Vanapina lineata* de Vis 1905 is a junior synonym of the New Guinea snake *Toxicocalamus longissimus* Boulenger, 1896. *Copeia* 1989(3), 753–755.

Ingram, G.J. (1990) The works of Charles Walter deVis alias "Devis," alias "Thickthorn." *Memoirs of the Queensland Museum*, Brisbane 28(1), 1–34, illust.

Ingram, G.J. & Covacevich, J.A. (1981) Frog and reptile type specimens in the Queensland Museum, with a checklist of frogs and reptiles in Queensland. *Memoirs of the Queensland Museum*, Brisbane 20(2), 291–306.

Ingram, G.J. & Covacevich, J.A. (1993) Two new species of striped blindsnakes. *Memoirs of the Queensland Museum*, Brisbane 34(1), 181–184, 3 figs.

Inns, H. (2009) *Britain's reptiles and amphibians: a field guide to the reptiles and amphibians of Great Britain, Ireland and the Channel Isalnds.* WildGuides, Hampshire, 164 pp., 245 col. photos.

Ionides, C.J.P. (1950) Snakes of the Southern Province. *Tanganyika Notes and Records*, Dar-es-Salaam (29), 98–108, 8 figs.

Ionides, C.J.P. & Pitman, C.R.S. (1965) Notes on three East African venomous snake populations. *Puku*, Chilanga (3), 87–95.

Irish, J. (1985) Zoological types in the State Museum. *Cimbebasia*, Windhoek (Series A) (8), 107–132.

Irwin, K.J. (2001) Geographic distribution: Serpentes: *Heterodon nasicus. Herpetological Review* 32(1), 59–60.

Irwin, K.J. (2004) *Arkansas snake guide.* Arkansas Game & Fish Commission, Little Rock, 50 pp., col. photos.

Irwin, R.L., Oliver, K.L., Mohamed, A.H. & Haast, W.E. (1970) Toxicity of Elapidae venoms and an observation in relation to geographical location. *Toxicon* 8(1), 51–54.

Isemonger, R.M. (1962) *Snakes of Africa: southern, central and east.* Thomas Nelson and Sons, Cape Town, 236 pp., 8 pls., illust.

Isemonger, R.M. (1983) *Snakes of Africa.* Books of Africa, Cape Town, vi + 263 pp., 10 pls., illust.

Iskandar, D.T. (1979) A second specimen of the Matanna water snake, *Enhydris matannensis* (Blgr.), from Raha, Muna Island, Indonesia. *British Journal of Herpetology*, London 5, 849–850.

Iskandar, D.T. (1986) On the occurrence of *Trimeresurus monticola* in Sumatra. *The Snake* 18(2), 126–128, 1 fig.

Iskandar, D.T. (1987) The occurrence of *Enhydris alternans* at Java. *The Snake* 19(1), 72–73.

Iskandar, D.T. (2004) *The amphibians and reptiles of Malinau region, Bulungan Research Forest, East Kalimantan: annotated checklist with notes on ecological preferences of the species and local utilization.* Center for International Forestry Research, Jakarta, iv + 27 pp.

Iskandar, D.T. & Tjan, K.N. (1996) The amphibians and reptiles of Sulawesi, with notes on the distribution and chromosomal number of frogs. *In:* Kitchener, D.J. & Suyanto, A. (Eds.), *Proceedings of the first international conferences on eastern Indonesia-Australian vertebrate fauna, Manado, Indonesia, Novenmber 22-26, 1994.* Western Australian Museum, Perth, pp. 39–46.

Iskandar, D.T. & Colijn, E. (2001) *A checklist of Southeast Asian and New Guinean reptiles. Part I: Serpentes.* Biodiversity Conservation Project, Djakarta, 195 pp.

Isogawa, K., Moriya, A. & Mitsui, S. (1994) A new snake of the genus *Agkistrodon* (Serpentes, Viperidae) from Tsushima Island, Nagasaki Prefecture, Japan. *Japanese Journal of Herpetology*, Kyoto 15(30), 101–111, 5 figs.

Ivanov, M. (2000) Snakes of the lower/middle Miocene transition at Vieux Collonges (Rhône, France), with comments on the colonization of western Europe by colubroids. *Geodiversitas* 22(4), 559–588.

Ivanov, M. (2002) The oldest known Miocene snake fauna from central Europe, Merkur-North locality, Czech Republic. *Acta Palaeontologica Polonica* 47(3), 513–534, 12 figs.

Ivanov, M. (2007) Herpetological assemblages from the Pliocene to middle Pleistocene in central Europe: palaeoecological significance. *Geodiversitas* 29(2), 297–316, 9 figs.

Ivanov, M. & Böhme, M. (2011) Snakes from Griesbeckerzell (Langhian, Early Badenian), North Alpine Foreland Basin (Germany), with comments on the evolution of snake faunas in central Europe during the Miocene climatic optimum. *Geodiversitas* 33(3), 411–449, 15 figs.

Ivanter, E.V. & Korosov, A.V. (2002) *Amphibians and reptiles of Karelia.* Petrozavodsk State University Press, Petrozavodsk, 160 pp., illust. (in Russian)

Iverson, J.B. (1986) Notes on the natural history of the Caicos Islands dwarf boa, *Tropidophis greenwayi. Caribbean Journal of Science* 22(2–3), 191–198.

Jablonski, D., Jandzik, D. & Gvozdík, V. (2012) New records and zoogeographic classification of amphibians and reptiles from Bosnia and Herzegovina. *North-Western Journal of Zoology*, Oradea 8(2), 324–337, 4 figs.

Jackson, D.R. & Franz, R. (1981) Ecology of the eastern coral snake (*Micrurus fulvius*) in northern peninsular Florida. *Herpetologica* 37, 213–228.

Jackson, J.J. (1983) *Snakes of the southeastern United States*. Publications Section, Georgia Extension Service, Athens, 112 pp., 104 col. photos.

Jackson, K. & Fritts, T.H. (1996) Observations of a grooved fang in *Psammodynastes pulverulentus*: does the mock viper resemble a protoelapid? *Journal of Herpetology* 30(1), 128–131, 5 figs.

Jackson, K., Underwood, G., Arnold, E.N. & Savitzky, A.H. (1999) Hinged teeth in the enigmatic colubrid, *Iguanognathus werneri Copeia* 1999(3), 815–818, 2 figs.

Jackson, S. & Mirick, P. (1993) *Massachusetts snakes: a guide*. University of Massachusetts, Amherst, 20 pp., illust.

Jacobs, H.J., Auliya, M. & Böhme, W. (2009) Zur taxonomie des Dunklen Tigerpythons, *Python molurus bivittatus* Kuhl, 1820, speziell der Population von Sulawesi. *Sauria* (Berlin) 31(3), 5–16, 10 figs.

Jacobsen, N.H.G. (1986) A new subspecies of *Amblyodipsas microphthalma* (Bianconi, 1850) (Serpentes, Colubridae) from the Transvaal. *Annals of the Transvaal Museum*, Pretoria 34(5), 123–127, 3 figs.

Jacobsen, N.H.G. (2005) *Remarkable reptiles of South Africa*. Briza Publications, Pretoria, 152 pp., col. illust.

Jacobsen, N.H.G. & Haacke, W.D. (1980) Harmless snakes of the Transvaal. *Transvaal Nature Conservation Division*, Pretoria (2), 1–64, illust.

Jacquinot, H. & Guichenot, M.A. (1853) *Voyage au Pole Sud et dans l'Océanie sur les corvettes l'Astrolabe et la Zélée; exécuté par ordre du Roi pendant les années 1837–1838–1839–1840, sous le commandement de M. J. Dumont-d'Urville, Capitaine de vaisseau; publié par ordonnance de sa majesté, sous la direction supérieure de M. Jacquinot, Capitaine de vaisseau, Commandant de la Zélée. Zoologie. Tome troisième. Reptiles et poisons*. Gide et J. Baudry, Paris, pp. 1–56, 6 pls.

Jadhav, S.S. & Kapre, S.V. (1991) Antivenin production in India. *In:* Tu, A.T. (Ed.), *Handbook of natural toxins. Volume 5. Reptile venoms and toxins*. Marcel Dekker, New York, pp. 584–610, 8 figs.

Jadin, R.C. (2010a) *Cerrophidion tzotzilorum. Catalogue of American amphibians and reptiles*, pp. 880.1–880.2.

Jadin, R.C. (2010b) Morphological phylogeny of the montane pitvipers (Crotalinae: *Cerrophidion*) with comments on the conservation of Mesoamerican highlands. *In:* Wilson, L.D., Townsend, J.H. & Johnson, J.D. (Eds.), *Conservation of Mesoamerican amphibians and reptiles*. Eagle Mountain Publications, Eagle Mountain, pp. 675–685, 4 figs.

Jadin, R.C. & Smith, E.N. (2010) The hemipenis of *Trimorphodon quadruplex*. *Herpetological Review* 41(2), 157–159, 2 figs.

Jadin, R.C., Gutberlet, R.L., Jr. & Smith, E.N. (2010a) Phylogeny, evolutionary morphology, and hemipenis of the Middle American jumping pitvipers (Serpentes: Crotalinae: *Atropoides*). *Journal of Zoological Systematics and Evolutionary Research* 48(4), 360–365, 3 figs.

Jadin, R.C. Reyes-Velasco, J. & Smith, E.N. (2010b) Hemipenes of the long-tailed rattlesnakes (Serpentes: Viperidae) from Mexico. *Phyllomedusa* 9(1), 69–73, 1 fig.

Jadin, R.C., Smith, E.N. & Campbell, J.C. (2011) Unravelling a tangle of Mexican serpents: a systematic revision of highland vipers. *Zoological Journal of the Linnean Society*, London 163(3), 943–958.

Jadin, R.C., Townsend, J.H., Castoe, T.A. & Campbell, J.C. (2012) Cryptic diversity in disjunct populations of Middle American montane pitvipers: a systematic reassessment of *Cerrophidion godmani*. *Zoologica Scripta* 41(5): 455–470.

Jakobsen, A. (1997) A review of some East African members of the genus *Elapsoidea* Bocage with the description of a new species from Somalia and a key for the genus (Reptilia, Serpentes, Elapidae). *Steenstrupia* 22, 59–82, 13 figs.

James, E. (1823) *Account of an expedition from Pittsburgh to the Rocky Mountains, performed in the years 1819 and 1820, by order of the Hon. J.C. Calhoun, Sec'y of War, under the command of Major Stephen H. Long. From the notes of Major Long, Mr. T. Say, and other gentlemen of the exploring party. Vol. I.* H.C. Carey and I. Lea, Philadelphia, 503 pp.

James, J.F. (1887) Catalogue of the mammals, birds, reptiles, batrachians and fishes in he collection of the Cincinnati Society of Natural History. Class III. Reptilia (the reptiles). *Journal of the Cincinnati Society of Natural History* 10, 34–48.

Jan, G. (1857) *Cenni sul Museo Civico di Milano ed indice sistematico dei rettili ed anfibi esposti nel medesimo*. Luigi di Giacomo Pirola, Milano, 61 pp.

Jan, G. (1858a) Plan d'une iconographie descriptive des ophidiens et description sommaire de nouvelles espèces de serpents. *Revue et Magasin de Zoologie*, Paris (Série 2) 10(Oct.), 438–449. [volume listed as 11 in signature header]

Jan, G. (1858b) Plan d'une iconographie descriptive des ophidiens, et description sommaire de nouvelles espèces de serpents. *Revue et Magasin de Zoologie*, Paris (Série 2) 10(Dec.), 514–527. [volume listed as 11 in signature header]

Jan, G. (1859a) Plan d'une iconographie descriptive des ophidiens, et description sommaire de nouvelles espèces de serpents. *Revue et Magasin de Zoologie*, Paris (Série 2) 11(Mar.), 122–130. [volume listed as 12 in signature header]

Jan, G. (1859b) Plan d'une iconographie descriptive des ophidiens, et description sommaire de nouvelles espèces de serpents. *Revue et Magasin de Zoologie*, Paris (Série 2) 11(Apr.), 148–157. [volume listed as 12 in signature header]

Jan, G. (1859c) Additions et rectifications aux Plan et prodrome de l'iconographie descriptive des ophidiens. *Revue et Magasin de Zoologie*, Paris (Série 2) 11(Dec.), 505–512, pls. 4, 5, 9. [volume listed as 12 in signature header]

Jan, G. (1859d) Spix' Serpentes brasiliensis beurtheilt nach Autopsie der Originalexemplare und auf die Nomenclatur von Dumeril und Bibron zerúckgefúhrt. *Archiv für Naturgeschichte* 25, 272–276.

Jan, G. (1859e) *Prodrome d'une iconographie descriptive des ophidiens et description sommaire de nouvelles espèces de serpents venimeux*. Imprimerie de Mme. Ve. Bouchard-Huzard, Paris, 32 pp., pls. 4–6, A–E. [reprint of Jan, 1858a–b and 1859a–c]

Jan, G. (1861a) Bemerkungen über die Typhlopiden. *Archiv für Naturgeschichte*, Berlin 27, 1–7.

Jan, G. (1861b) Note sulla famiglia dei tiflopidi sui loro generi e sulle specie del genere *Stenostoma* relative alle tav. V e Vi dei 1. ed alle tav. V e VI dell 2. fascicolo dell'Iconographie generale des Ophidiens. *Archivio per la Zoologia, l'Anatomia, e la Fisiologia*, Genova 1(2), 178–199, 2 pls.

Jan, G. (1861c) Ueber die Familie Homalopsidae. *Archiv für Naturgeschichte,* Berlin 27, 87–95.

Jan, G. (1862a) Ueber die Familien der Eryciden und Tortriciden. *Archiv für Naturgeschichte*, Berlin 28, 238–252.

Jan, G. (1862b) Enumerazione sistematica delle specie d'ofidi del gruppo Calamaridae. *Archivio per la Zoologia, l'Anatomia, e la Fisiologia*, Genova 2(1), 1–76 pp., 7 pls.

Jan, G. (1862c) *Prodromo della iconografia generale degli ofidi. II. Parte. V. Gruppo Calamaridae*. R.I. de'Sordo-Muti, Genova, xii + 76 pp., 7 pls. [reprint of Jan, 1862b]

Jan, G. (1863a) Enumerazione sistematica degli ofidi appartenenti al gruppo Coronellidae. *Archivio per la Zoologia, l'Anatomia, e la Fisiologia*, Genova 2(2), 213–330.

Jan, G. (1863b) *Elenco sistematico degli ofidi descritti e disegnati per l'Iconografia générale*. A. Lombardi, Milano, vii + 143 pp.

Jan, G. (1863c) *Prodromo della Iconographie générale degli ofidi II. Parte. VI. Gruppo. Coronellidae*. Eredi Soliani, Modena, 120 pp. [reprint of Jan, 1863a]

Jan, G. (1864) *Iconographie générale des ophidiens. Première famille, les typhlopiens*. J. B. Baillière et fils, Paris, 42 pp.

Jan, G. (1865a) Rettili. *In:* de Filippi, F. (Ed.), *Note di un Viaggio in Persia nel 1862. Vol. 1*. G. Daelli, Milano, pp. 352–357.

Jan, G. (1865b) *Iconographie générale des ophidiens. Deuxième famille, les uropeltiens; troisième famille, les tortriciens; quatrième famille, les boidiens*. J. B. Baillière et fils, Paris, ii + pp. 43–100.

Jan, G. (1865c) Enumerazione sistematica degli ofidi appartenenti al gruppo Potamophilidae. *Archivio per la Zoologia, l'Anatomia, e la Fisiologia*, Genova (1864) 3(2), 201–265.

Jan, G. (1865d) *Prodromo della iconografia generale degli ofidi. VIII. Gruppo Potamophilidae*. Erede Soliani, Modena (1864), 65 pp.

Jan, G. & Sordelli, F. (1860–1866) *Iconographie générale des ophidiens. Tome premier (livrais. 1 à 17), contenant cent deux planches*. Georges Jan and Ferdinand Sordelli, Milan, 102 pls. [livraison 1 (1860), 2 (1861), 3–8 (1864), 9–14 (1865), 15–17 (1866)]

Jan, G. & Sordelli, F. (1866–1870) *Iconographie générale des ophidiens. Tome second (livrais. 18 à 34), contenant cent deux planches*. Georges Jan and Ferdinand Sordelli, Milan, 102 pls. [livrasons 18–19 (1866), 20–25 (1867), 26–30 (1868), 31–33 (1869), 34 (1870)]

Jan, G. & Sordelli, F. (1870–1881) *Iconographie générale des ophidiens. Tome troisième (livrais. 35 à 50), contenant quatre vingt seize planches*. Georges Jan and Ferdinand Sordelli, Milan, 96 pls. [livaisons 35–37 (1870), 38 (1871), 39–42 (1872), 43–44 (1873), 45–46 (1874), 47 (1875), 48 (1876), 49 (1879), 50 (1881)]

Jan, G. & Sordelli, F. (1961) *Iconographie générale des ophidiens. Thome I-III*. J. Cramer, Weinheim, 300 pls. [reprint of Jan & Sordelli, 1860–1881]

Janecek, J. (1976) An exceptionally large puff adder brood. *International Zoo Yearbook* 16, 85–86.

Janensch, W. (1904a) Über eine fossile Schlange aus dem Eocän des Monte Bolca. *Zeitschrift der Deutschen Geologischen Gesellschaft*, Berlin 56, 54–56.

Janensch, W. (1904b) Ueber eine fossile Schlange aus dem Eocän des Monte Bolca. *Sitzunsberichte der Gesellschaft Naturforschender Freunde zu Berlin* 1904(6), 133–135.

Janensch, W. (1906a) *Pterosphenus schweinfurthi* Andrews und die Entwicklung der Palaeophiden. *Archiv für Biontologie*, Berlin 1(3), 307–350, pls. 26–27.

Janensch, W. (1906b) Über *Archaeophis proavus* Mass.: eine Schlange aus dem Eocän des Monte Bolca. *Beiträge zur Paläontologie und Geologie Österrich-Ungarns und des Orients* 19, 1–33, 2 pls.

Jansen, M. (2006) Primeros registros de *Bothrops pauloensis* Amaral, 1925 (Serpentes: Viperidae) en Bolivia. *Kempffiana* 2006, 66–71, 4 figs.

Jansen, M. (2008) New and noteworthy records of snakes from Bolivia. *Herpetological Review* 39(3), 374–375.

Jansen, M. & Köhler, G. (2008) A new species of *Phalotris* from the eastern lowlands of Bolivia (Reptiles, Squamata, Colubridae). *Senkenbergiana Biologica*, Frankfurt-am-Main 88(1), 103–110, 3 figs.

Jansen, M., Álvarez, L.G. & Köhler, G. (2009) Description of a new species of *Xenopholis* (Serpentes: Colubridae) from the Cerrado of Bolivia, with comments on *Xenopholis scalaris* in Bolivia. *Zootaxa* (2222), 31–45, 10 figs.

Janssen, M., Freyvogel, T.A. & Meier, J. (1990) Antigenic relationship between the venom of the night adder *Causus maculates* and venoms of other viperids. *Toxicon* 28(8), 975–983, 1 fig.

Jaquet, M. (1896) Sur la présence d'un *Typhlops* en Algérie. *Bibliographie Anatomique*, Paris 4(2), 79–81, 7 figs.

Jaramillo, C.A. & Ibáñez-D., R. (2003) Geographic distribution: Serpentes: *Leptophis nebulosus*. *Herpetological Review* 34(3), 265.

Jaramillo, C.A., Ibáñez-D, R., Jaramillo, F. & Solís, F.A. (2008) Geographic distribution: Serpentes: *Sibon lamari*. *Herpetological Review* 39(3), 373–374.

Jaramillo, C.A., Wilson, L.D., Ibáñez-D., R. & Jaramillo, F. (2010) The herpetofauna of Panama. *In:* Wilson, L.D., Townsend, J.H. & Johnson, J.D. (Eds.), *Conservation of Mesoamerican amphibians and reptiles*. Eagle Mountain Publications, Eagle Mountain, pp. 604-671, 2 figs.

Jaroniewski, W. (1992) *Jadowite weze swiata*. Wydawnictwa Szkolne I Pedagogiczne, Warszawa, 144 pp., 30 figs., 32 col. pls.

Jayaram, K.C. (1974) Ecology and distribution of fresh-water fish, Amphibia and reptiles. *In:* Mani, M.S. (Ed.), *Ecology and biogeography in India*. W. Junk, The Hague, pp. 517–584.

Jayasuriya, D.P. (2001) *Snakes of Sri Lanka: identification, venom and treatment*. D.P. Jayasuriya, Kandy, 49 pp., 25 col. pls. (in Sinhalese)

Jayawickrama, A. (1994) Snakes observed in Gampola, Sri Lanka. *Lyriocephalus*, Gampola 1(1–2), 63–65.

Jayne, B.C., Ward, T.J. & Voris, H.K. (1995) Morphology, reproduction, and diet of the marine homalopsine snake *Bitia hydroides* in peninsular Malaysia. *Copeia* 1995(4), 800–808, 7 figs.

Jenkins, R. & Bartell, R. (1980) *A field guide to reptiles of the Australian high country*. Inkata Press, Melbourne, 278 pp., col. illust.

Jenner, J. (1981) *A zoogeographic study and taxonomy of the Xenodontine colubrid snakes*. Ph.D. Diss., New York University, New York, 354 pp., 210 figs.

Jennings, M.R. (1983a) *Masticophis lateralis*. *Catalogue of American amphibians and reptiles*, pp. 343.1–343.2.

Jennings, M.R. (1983b) *An annotated check list of the amphibians and reptiles of California*. California Fish & Game 69(3), 151–171, illust.

Jennings, M.R. (1987) Annotated check list of the amphibians and reptiles of California. Second edition. *Southwestern Herpetological Society Special Publication* (3), 1–48.

Jennings, M.R. & Hayes, M.P. (1994) *Amphibian and reptile species of special concern in California*. California Department of Fish and Game, Rancho Cordova, 255 pp., 49 figs., 16 pls.

Jensen, A.S. (1900) Lagoa Santa Egnens Slanger. Et Bidrag til det indre Brasiliens Herpetologi. (With description of three new species.). *Videnskabelige Meddelelser fra Dansk Naturhistorisk Forening i Kjöbenhavn* 1900, 99–111.

Jensen, J.B. (2007) *Ramphotyphlops braminus*. Geographic distribution: Serpentes: *Herpetological Review* 38(4), 480.

Jensen, J.B. & Moulis, R.A. (1999) The Georgia Herp Atlas: year II county records. *Herpetological Review* 30(4), 240–247.

Jensen, J.B., Camp, C.D., Gibbons, W. & Elliott, M.J. (Eds.). (2008) *Amphibians and reptiles of Georgia.* University of Georgia, Athens, xvii + 575 pp., 1 fig., 478 col. photos.

Jerdon, T.C. (1854) Catalogue of reptiles inhabiting the peninsula of India. *Journal the Asiatic Society of Bengal (Natural History),* Calcutta (1853) 22(6), 522–534.

Jerdon, T.C. (1870) Notes on Indian herpetology. *Proceedings of the Asiatic Society of Bengal (Natural History),* Calcutta 1870(3), 66–85.

Jesus, J., Nagy, Z.T., Branch, W.R., Wink, M., Brehm, A. & Harris, D.J. (2009) Phylogenetic relationships of African green snakes (genera *Philothamnus* and *Hapsidophrys*) from São Tomé, Príncipe and Annobon islands based on mtDNA sequences, and comments on their colonization and taxonomy. *Herpetological Journal* 19, 42–48, 2 figs.

Jha, A. & Thapa, K. (2002) *Reptiles & amphibians of Sikkim.* Mrs. Shila Jha, Chattisgarh, 100 pp., illust.

Ji., D., Chang, W., Wen, S., Liu, M., Qin, Y. & Zhou, Y. (1985) Studies on *Elaphe schrenckii* of China. *Acta Herpetologica Sinica* (Series 2) 4(3), 239–246.

Ji, D.-M. & Wen, S.-S. (1997) *Elaphe anomala* is an effective species (Serpentiformes). *Acta Zootaxonomica Sinica*, Beijing 22(4), 436–440, 2 figs.

Ji, D.-M., Liu, M.-Y., Liu, Z.-Y., Zhou, Y.-F., Huang, K.-C., Wen, S.-S. & Zou, B.-Z. (1987) *Fauna Liaoningica, Amphibia Reptilia.* Liaoning Science and Technology Publishing House, Shenyang, 170 pp., 9 col. pls. (in Chinese)

Ji, D.-M., Weng, S. & Liu, M. (1989) Classification of *Agkistrodon halys* in northeast China. *In:* Matsui, M., Hikida, T. & Goris, R.C. (Eds.), *Current herpetology in East Asia.* Herpetological Society of Japan, Kyoto, pp. 185–192, 5 figs.

Jia, Z.-X. (1988) *Elaphe davidi*—a record new to Shandong. *Acta Herpetologica Sinica*, Chengdu (Series 3) 1(1), 66.

Jiang, F. & Zhao, E.-M. (2009) *Gloydius lijianlii,* a new species from the northern coastal islands along Shandong Peninsula (Reptilia, Squamata, Viperidae). *Acta Zoologica Sinica* 34(3), 642–646, 4 figs.

Jiang, K., Chen, T.-B., David, P., Vogel, G., Hou, M.-A., Yuan, Z.-Y., Meng, Y.-J. & Che, J. (2012) On the occurrence of *Oligodon joynsoni* (Smith, 1917) in China (Squamata: Colubridae). *Asian Herpetological Research* (Series 2) 3(4), 316–321, 5 figs.

Jiang, Y.-F. (2000) A summary of snake types in Chinese collections. *Sichuan Journal of Zoology* 19(3), 176–177.

Jiang, Y.-M. (1987) On *Xenopeltis hainanensis* Hu et Zhao. *Acta Herpetologica Sinica*, Chengdu (Series 2) 6(4), 74–75, 1 fig. (in Chinese).

Jiang, Y.-M. (2004) *Pareas macularius* Theobald, 1868 should be a junior synonym of *Pareas margaritophorus* (Jan, 1866). *Sichuan Journal of Zoology* 23(3), 207–208.

Jiang, Y.-M. & Hu, Q.-X. (1979) On collections of reptiles from Youyang, Xiushan and Gulin, Sichuan, China. *Acta Herpetologica Sinica*, Chengdu 2(2), 1–7.

Jiang, Y.-M. & Huang, Q.-Y. (1984) A new record of Chinese opisthoglyphic snakes, *Boiga cyanea* Dumeril et Bibron. *Acta Herpetologica Sinica*, Chengdu (Series 2) 3(2), 14.

Jiang, Y.-M. & Zhao, F. (1983) Studies on amphibians and reptiles of Mt. Gongga region, Sichuan, China. 3. A study of species-group *nuchalis*, genus *Rhabdophis*. *Acta Herpetologica Sinica*, Chengdu (Series 2) 2(1), 59–62, figs.

Jiménez, A. & Savage, J.M. (1962) A new blind snake (genus *Typhlops*) from Costa Rica. *Revista de Biologia Tropical*, San José 10(2), 199–203, 1 fig.

Jintakune, P. & Chanhome, L. (1995) *The venomous snakes of Thailand.* Thai Red Cross Society, Bangkok, 175 pp., 284 col. photos. (in Thai)

Jirousek, V. (1983) Systematic survey of adders of genus *Vipera. Fauna Bohemiae Septentrionalis*, Ústí nad Labem (8), 71–94, illust. (in Czech)

Joger, U. (1981) Zur Herpetofaunistik Westafrikas. *Bonnischer Zoologische Beiträge*, Bonn 32(3–4), 297–340, illust.

Joger, U. (1982) Zur herpetofaunistik Kameruns (II). *Bonnischer Zoologische Beiträge*, Bonn 33(2–4), 313–342, 10 figs.

Joger, U. (1984) *The venomous snakes of the Near and Middle East.* L.R. Verlag, Wiesbaden, 115 pp., 21 figs.

Joger, U. (1990) The herpetofauna of the Central African Republic, with description of a new species of *Rhinotyphlops* (Serpentes, Typhlopidae). *In:* Peters, G. & Hutterer, R. (Eds.), *Vertebrates in the tropics.* Museum Alexander Koenig, Bonn, pp. 85–102, 7 figs.

Joger, U. (Ed.), (1999) Phylogeny and systematics of the Viperidae. *Kaupia, Darmstädter Beiträge zur Naturgeschichte* 8, 1–148, illust.

Joger, U. (2003) Reptiles and amphibians of southern Tunisia. *Kaupia* 12, 71–88, 15 figs.

Joger, U. & Courage, K. (1999) Are Palaearctic "rattlesnakes" (*Echis* and *Cerastes*) monophyletic? *In:* Joger, U. (Ed.), Phylogeny and systematics of the Viperidae. *Kaupia, Darmstädter Beiträge zur Naturgeschichte* 8, 63–81, 14 figs., 4 pls.

Joger, U. & Lambert, M.R.K. (2002). Inventory of amphibians and reptiles in SE Senegal, including the Niokola-Koba National Park, with observations on factors influencing diversity. *Tropical Zoology*, Firenze 15(2), 165–185.

Joger, U., Teynié, A. & Fuchs, D. (1988) Morphological characterization of *Vipera wagneri* Nilson & Andren, 1984 (Reptilia: Viperidae), with first description of the males. *Bonner Zoologische Beiträge,* Bonn 39(2–3), 221–228.

Joger, U. & Wollesen, R. (Eds.). (2004) Distribution, ecology and conservation of the adder (*Vipera berus* [Linnaeus, 1758]). *Mertensiella (Supplement)* (15), 1–309, illust.

Johann, H. (1973) Ein Fundpunkt von *Cerastes cerastes* in Nordtunesien (Serpentes, Viperidae). *Salamandra*, Frankfurt-am-Main 9(3–4), 160.

Johnsen, P. (1962) Notes on African snakes mainly from northern Rhodesia and Liberia. *Videnskabelige Meddelelser fra Dansk Naturhistorisk Forening i Kjobenhavn* 124, 115–130.

Johnson, B. (1989) *Familiar amphibians and reptiles of Ontario.* Natural Heritage/Natural History, Toronto, 168 pp., illust.

Johnson, B. & Menzies, V. (1993) *International symposium and workshop on the conservation of the eastern massasauga rattlesnake,* Sistrurus catenatus catenatus. Metropolitan Toronto Zoo, Toronto, 136 pp., illust.

Johnson, C.R. (1975) Head-body thermal control, thermal preferenda, and voluntary maxima in the taipan, *Oxyuranus scutellatus* (Serpentes, Elapidae). *Zoological Journal of the Linnean Society*, London 56(1), 1–12, 4 figs.

Johnson, D.H., Bryant, M.D. & Miller, A.H. (1948) Vertebrate animals of the Providence Mountains area of California. *University of California Publications in Zoology* 48, 221–374, illust.

Johnson, J.D. (1974) New records of reptiles and amphibians from Chiapas, Mexico. *Transactions of the Kansas Academy of Sciences* 76(3), 223–224.

Johnson, J.D. (1977) The taxonomy and distribution of the Neotropical whipsnake *Masticophis mentovarius* (Reptilia, Serpentes, Colubridae). *Journal of Herpetology* 11(3), 287–309, 16 figs.

Johnson, J.D. (1978a) *Conopsis nasus nasus* (Reptilia, Serpentes, Colubridae) in Chihuahua, México. *Journal of Herpetology* 12(1), 108–109.

Johnson, J.D. (1978b) First record of *Manolepis putnami* (Serpentes, Colubridae) from Chiapas, Mexico. *Southwestern Naturalist* 23(3), 538.

Johnson, J.D. (1979) Taxonomic status and distribution of *Geophis laticinctus* (Colubridae) in southern Mexico. *Southwestern Naturalist* 24(4), 698–701, 1 fig.

Johnson, J.D. (1982) *Masticophis mentovarius. Catalogue of American amphibians and reptiles*, pp. 295.1–295.4.

Johnson, J.D. (1984) *A biogeographic analysis of the herpetofauna of northwestern nuclear Central America.* Doctoral Dissertation, Texas A & M University, College Station, 127 pp.

Johnson, J.D. (1989) A biogeographic analysis of the herpetofauna of northwestern nuclear Central America. *Milwaukee Public Museum, Contributions in Biology and Geology* (76), 1–66, 29 figs.

Johnson, J.D. & Dixon, J.R. (1984) Taxonomic status of the Venezuelan macagua, *Bothrops colombiensis. Journal of Herpetology* 18(3), 329–332, 1 fig.

Johnson, J.D. & Webb, R.G. (2001) Geographic distribution: Serpentes: *Stenorrhina freminvillii. Herpetological Review* 33(4), 280.

Johnson, J.D., Ely, C.A. & Webb, R.G. (1976) Biogeographical and taxonomic notes on some herpetozoa from the northern highlands of Chiapas, México. *Transactions of the Kansas Academy of Sciences* 79 (3–4), 131–139, 2 figs.

Johnson, J.D., Webb, R.G. & Flores-Villela, O.A. (2001) *Mesoamerican herpetology: systematics, zoogeography, and conservation.* University of Texas El Paso, El Paso, iv + 200 pp.

Johnson, J.D., Mata-Silva, V. & Ramirez-Bautista, A. (2010) Geographic distribution and conservation of the herpetofauna of southeastern Mexico. *In:* Wilson, L.D., Townsend, J.H. & Johnson, J.D. (Eds.), *Conservation of Mesoamerican amphibians and reptiles.* Eagle Mountain Publications, Eagle Mountain, pp 322–369, 6 figs.

Johnson, M.L. (1942) A distributional check-list of the reptiles of Washington. *Copeia* 1942(1), 15–18.

Johnson, M.L. (1947) The status of the *elegans* subspecies of *Thamnophis* with descriptions of a new subspecies from Washington state. *Herpetologica* 3(5), 159–165.

Johnson, M.L. (1995). Reptiles of the state of Washington. *Northwest Fauna* (3), 1–80. [unpublished manuscript from 1954]

Johnson, R.H. (1902) Axial bifurcation in snakes. *Transactions of the Wisconsin Academy of Science, Arts, and Letters,* Madison (1901) 13(2), 523–538, 8 pls.

Johnson, R.M. (1964) *The herpetofauna of the Oak Ridge area.* Oak Ridge National Laboratory (Doc. ORNL–3653), 29 pp.

Johnson, T.R. (1980) *Snakes of Missouri.* Missouri Department of Conservation, Jefferson City, 12 pp., illust.

Johnson, T.R. (2000) *The amphibians and reptiles of Missouri.* Second edition, Missouri Department of Conservation, Jefferson City, 400 pp., illust., 288 col. photos. [first edition, 1987, xi + 368 pp., illust., col. photos]

Johnston, G.R. (1987) Reproduction and growth in captive death adders *Acanthophis antarcticus* (Squamata, Elapidae). *Transactions of the Royal Society of South Australia,* Adelaide 111(1–2), 123–125, illust.

Johnston, G.R. & Ellins, P. (1979) The reptiles of the Sir Joseph Banks Islands, South Australia. *Herpetofauna* 10(2), 9–12.

Jones, J.P. (1928) A few facts regarding some Tennessee snakes. *Journal of the Tennessee Academy of Sciences* 3(2), 14–24.

Jones-Burdick, W.H. (1949) Guide to the snakes of Colorado. Second edition. *University of Colorado Museum,* Boulder (5), 1–13, 8 pls. [first edition, 1939, (1), 1–12]

Jong, J.K. de. (1926a) *Heurnia ventro-maculata* n. g. n. sp. und *Cantoria annulata* n. sp., Zwei Neue Schlangen von Neu-Guinea. *Zoologischer Anzeiger,* Leipzig 67(11–12), 302–304.

Jong, J.K. de. (1926b) Fauna Bukana. Reptiles. *Treubia,* Buitenzorg 7(2), 85–96.

Jong, J.K. de. (1927) Reptiles from Dutch New Guinea. *Nova Guinea (Zoology),* Leiden 15(3), 296–318, 3 figs.

Jong, J.K. de. (1928) Beiträge zur Kenntnis der Reptilien-Fauna von Niederländisch Ost-Indien. *Treubia,* Buitenzorg 10(2–3), 145–151.

Jong, J.K. de. (1930a) Notes on some reptiles from the Dutch-East-Indies. *Treubia,* Buitenzorg 12(1), 115–119.

Jong, J.K. de. (1930b) List of reptiles collected by Prof. Dr. W. Docters van Leeuwen during the North New-Guinea Expedition 1926. *Nova Guinea (Zoology),* Leiden 15(4), 405–408.

Jongbloed, M. (2000) *Wild about reptiles. Field guide to the reptiles and amphibiens of the United Arab Emirates.* Barkers Trident Communications, London, 116 pp., col. illust.

Jooris, R. & Fourmy, R. (1996) An analysis of scutellation in populations of *Cerastes vipera* (Linnaeus, 1758): scale characters co-vary with environmental temperature. *African Journal of Herpetology* 45(2), 59–67, 4 figs.

Jorge, J.S. & Freire, E.M.X. (2011) Geographic distribution: Serpentes: *Thamnodynastes almae. Herpetological Review* 42(3), 396.

Jourdan, A.J.L. (1833) Description d'un appareil dentaire particular appartenant a la colonne vertebrale (*Rachiodon scaber*). *Le Temps,* Paris 13 June 1833, and *Froriep, Notizen (1833/1834)* 41, 104–105. [not seen]

Jourdan, A.J.L. (1834) Description d'un appareil dentaire particulier appartenant a la colonne vertebrale. *L'Institut* 2(60), 214.

Jourdran, E. (1904) *Les ophidiens de Madagascar.* A. Michalon, Paris (1903), 62 pp., 37 pls., 85 figs.

Joy, W.D. (1992) World record *Boa constrictor* "de-discovered." *Notes from NOAH* 19(10), 21.

Juliá-Zertuche, J. (1982) Una nueva subespecie de "vibora de cascabel pigmea" de la Sierra de Sinaloa. *Resúmenes VI Congreso Nacional de Zoología,* Mazatlán, p. 52.

Juliá-Zertuche, J. & Treviño-Saldaña, C.H. (1978) Una nueva subespecie de *Crotalus lepidus* encontrada en Nuevo León. (Abstract) *Congreso Nacional de Zoología,* México 2, 60.

Juliá-Zertuche, J. & Treviño-Saldaña, C.H. (1980) Una nueva subespecie de *Crotalus lepidus* en Nuevo León. *Memorias del II Congreso Nacional de Zoología* (1978) 1, 286–293.

Juliá-Zertuche, J. & Varela-J., M. (1978) Una *Bothrops* de México, nueva para la ciencia. *Memoria del Primer Congreso Nacional de Zoología, Escuela Nacional de Agricultura, Universidad Autonoma de Chapingo,* Chapingo (1977), 209–210.

Kabara, J.J. & Fischer, G.A. (1972) Methodology for the isolation of low molecular weight toxins from snake venoms. *Toxicon* 10(3), 227–238, 10 figs.

Kacoliris, F.P., Berkunsky, I. & Williams, J. (2006) Herpetofauna of the Argentinean impenetrable Great Chaco. *Phyllomedusa* 5(2), 149–157, 2 figs.

Kaiser, H., Barrio-Amorós, C.L., Fuenmayor, G.R. & Grismer, L.L. (2003) Geographic distribution: Serpentes: *Siphlophis compressus. Herpetological Review* 34(2), 170.

Kaiser, H., Lopes-Carvalho, V., Ceballos, J., Freed, P., Heacox, S., Lester, B., Richards, S.J., Trainor, C.R., Sanchez, C. & O'Shea, M. (2011) The herpetofauna of Timor-Leste: a first report. *ZooKeys* 109, 19–86, 30 col. photos.

Kaiser, H., Crother, B.I., Kelly, C.M.R., Liselli, L., O'Shea, M., Ota, H., Passos, P., Schleip, W.D. & Wüster, W. (2013) Best practices: in the 21st Century, taxonomic decisions in herpetology are acceptable only when supported by a body of evidence and published via peer-review. *Herpetological Review* 44(1), 8–23.

Kallert, E. (1929) Hamburger Importe. *Blätter für Aquarien- und Terrarienkunde*, Stuttgart 38(18), 372. [not seen]

Kalyabina-Hauf, S., Schweiger, S., Joger, U., Mayer, W., Orlov, N. & Wink, M. (2004) Phylogeny and systematics of adders (*Vipera berus* complex). *Mertensiella* 15, 7–16, 3 figs.

Kamel, A. (1974) Fractionation of Egyptian cobra venom. *Toxicon* 12(5), 495–499, 7 figs.

Kammel, W. (1992a) Zur Situation der Wiesenotter, *Vipera ursinii rakosiensis* (Mehely, 1894) (Squamata, Serpentes, Viperidae), in Niederosterreich. *Herpetozoa*, Wien 5(1–2), 3–11, illust.

Kammel, W. (1992b) Zur Situation der Wiesennotter, *Vipera ursinii rakosiensis* (Mehely, 1894), under der Pannonischen Bergeidechse, *Lacerta vivipara pannonica* Lac & Kluch, 1968, im Burgenland (Osterreich) (Squamata, Serpentes, Viperidae; Sauria, Lacertidae). *Herpetozoa*, Wien 5(3–4), 109–118, illust.

Kandamby, D.S. (1997) Herpetological types reposed in the National Museum Colombo, Sri Lanka. *Lyriocephalus*, Colombo 3(1), 31–33.

Kanishika, D., Ukuwela, B., Sanders, K.L. & Fry, B. (2012) *Hydrophis donaldi* (Elapidae, Hydrophiinae), a highly distinctive new species of sea snake from northern Australia. *Zootaxa* (3201), 45–57.

Karaman, S. (1939) Über die Verbreitung der Reptilien in Jugoslawien. *Annales Musei Serbiae Meridionalis*, Skopje 1, 1–20.

Karaman, S. (1955) Eine seltene Schlange, *Eryx jaculus turcicus* Oliv. *Fragmenta Balcanica Musei Macedonia Scientiarum Naturalium*, Skopje 1(21), 181–184, 4 figs. (in Russian)

Kardon, A. (1995) Treasure of the Sierra Madre Orientals: the northern Mexican pine snake, *Pituophis deppei jani. The Vivarium* 7(1), 18–21, 6 col. photos.

Kardon, A. & Holmback, E.R. (1996) Husbandry and reproduction of the giant Central American black milksnake, *Lampropeltis triangulum gaigeae. The Vivarium* 7(4), 14–16, 4 col. photos.

Kardong, K.V. (1990) General skull bone, and muscle variation in *Agkistrodon* and related genera. *In:* Gloyd, H.K. & Conant, R., *Snakes of the* Agkistrodon *complex*. Society for the Study of Amphibians and Reptiles, pp. 573–581, 4 figs.

Karges, J.P. & McDaniel, V.R. (1982) The nomenclatural history and type specimens of *Farancia abacura. Journal of Herpetology* 16(2), 183–185.

Karns, D.R., Ashton, R.E., Jr. & Swearington, T. (1974) Illustrated guide to amphibians and reptiles in Kansas an identification manual. *University of Kansas Museum of Natural History, Public Education Series* (2), 24 pp., illust.

Karns, D.R., Voris, H.K., Chan-ard, T., Goodwin, J.C. & Murphy, J.C. (2001) The spacial ecology of the rainbow water snake, *Enhydris enhydris* (Homalopsinae) in southern Thailand. *Herpetological Natural History* (1999–2000) 7(2), 97–115, 7 figs.

Karns, D.R., Lukoschek, V., Osterhage, J., Murphy, J.C. & Voris, H.K. (2010) Phylogeny and biogeography of the *Enhydris* clade (Serpentes: Homalopsidae). *Zootaxa* (2452), 18–30, 3 figs.

Karsen, S.J., Lau, M.W. & Bogadek, A. (1998) *Hong Kong amphibians and reptiles*. Second edition. Provisional Urban Council, Hong Kong, 186 pp., illust. [first edition, 1986, 136 pp., col. illust.]

Karunarathna, D.M.S.S. & Amarasinghe, A.A. T. (2011) A preliminary survey of the reptile fauna in Nilgala Forest and its vicinity, Monaragala District, Sri Lanka. *Taprobanica,* Homagama 3(2), 69–73, 3 pls.

Karunaratne, P.B., Peiris, T. & Raheem, R. (1981) A research Project in the Sinharaja forest. *Loris* 15(6), 326–327.

Kasapidis, P., Provatidou, S., Maragou, P. & Valakos, E.D. (1996) Neue Daten über die Herpetofauna von Lesbos (Ägäische Inseln, Griechenland) und einige biogeographische Bemerkungen über die Inseln des nordöstlichen ägäischen Archipels. *Salamandra* 32(3), 171–180, 3 figs.

Kashchenko, N.F. (1902) *Review of the reptiles of the Tomsk region*. M. N. Kononov, Tomsk. 24 pp. (in Russian)

Kashyap, H.V. & Sohoni, P.R. (1973) The heart and arterial system of *Acrochordus granulatus* (Schneider). *Journal of the University of Bombay (Science)* 42(69), 34–52, 7 figs.

Kassing, E.F. (1961) A life history study of the Great Plains ground snake, *Sonora episcopa episcopa* (Kennicott). *Texas Journal of Science* 13(2), 185–203, 5 figs.

Kathariner, L. (1900) Die Nase der in Wasser lebenden Schlangen als Luftweg und Geruchsorgan. *Zoologische Jahrbücher (Systematik, Geographie und Biologie der Thiere)* 13, 415–442, pls. 27–28.

Kaudern, W. (1922) Sauropsiden aus Madagascar. *Zoologische Jahrbücher, Abteilung für Systematik, Ökologie und Geographie der Tiere* 45, 395–458, 4 figs.

Kaup, F. (1827) Zoologische monographien. *Isis von Oken* 20, 610–626.

Kawashita-Ribeiro, R.A., Carvalho, V.T. de, Lima, A.C. de, Ávila, R.W. & Fraga, R. de. (2011) Morphology and geographical distribution of the poorly known snake *Umbrivaga pygmaea* (Serpentes: Dipsadidae) in Brazil. *Phyllomedusa* 10(2), 177–182, 4 figs.

Keegan, H.L. (1956) Antivenins available for treatment of envenomation by poisonous snakes, scorpions, and spiders. *In:* Buckley, E.E. & Porges, N. (Eds.), *Venoms*. American Association for the Advancement of Science, Washington, D.C., pp. 413–438.

Keegan, H.L. (1958) *Some venomous and noxious animals of the Far East*. Department of Entomology, 406th Medical General Laboratory, U.S. Army Medical Command, Japan, Bio-Medical Report, 33 pp., 60 figs., 1 col. pl.

Keegan, H.L. & Matsui, T. (1964) *Observations on the pit viper* Trimeresurus gracilis *Oshima, 1920*. Department of Entomology, 406th Medical General Laboratory, U.S. Army Medical Command, Japan, Bio-Medical Research Report, 14 pp., 7 figs. [reprinted in 1964 by *Bulletin of the Institute of Zoology, Academia Sinica* 3: 9–18]

Keegan, H.L. & Yoshino, K. (1959) *Trimeresurus elegans* (Gray, 1849), the Sakishima habu, a venomous pit viper from the Ryukyu Islands. *American Journal of Tropical Medicine and Hygeine,* Baltimore 8(2), 124–133, 7 figs.

Keegan, H.L., Weaver, R.E., Toshioka, S. & Matsui, T. (1964) *Some venomous and noxious animals of East and Southeast Asia.* Third edition. Department of Entomology, 406th Medical General Laboratory, U.S. Army Medical Command, Japan, Bio-Medical Report (7), 43 pp., 52 figs. [first edition, 1958; second edition, 1960,

Keegan, H.L., Snover, G.E., Matsui, T. & Lloyd, H.H. (1965) *The Malayan pit viper* Agkistrodon rhodostoma *(Boie, 1827).* Department of Entomology, 406th Medical General Laboratory, U.S. Army Medical Command, Japan, Bio-Medical Report (10), 15 pp., 4 figs.

Keim, T.D. (1914) Amphibians and reptiles at Jennings, Maryland. *Copeia* 1914(2), [2].

Keiser, E.D., Jr. (1969) Evidence of a dichromatic population of the vine snake *Oxybelis fulgidus* (Daudin) on the Islas de la Bahia of Honduras. *Caribbean Journal of Science* 9(1–2), 31–32.

Keiser, E.D., Jr. (1971) *The poisonous snakes of Louisiana and the emergency treatment of their bites*. Lafayette Natural History Museum, Lafayette, 16 pp., col. illust.

Keiser, E.D., Jr. (1974) A systematic study of the neotropical vine snake *Oxybelis aeneus* (Wagler). *Bulletin of the Texas Memorial Museum*, Austin 22, 1–51, 3 figs.

Keiser, E.D., Jr. (1982a) The poisonous snakes of Mississippi with suggestions for the emergency treatment of their bite. *Mississippi Outdoors* (July/Aug.), 14 pp., col. illust.

Keiser, E.D., Jr. (1982b) *Oxybelis aeneus*. *Catalogue of American amphibians and reptiles*, pp. 305.1– 305.4.

Keiser, E.D., Jr. (1989) *Oxybelis boulengeri* Procter, a valid species of vine snake from South America. *Copeia* 1989(3), 764–768, 3 figs.

Keiser, E.D., Jr. (1991) Bibliography of the genus *Oxybelis* Wagler (Serpentes, Colubridae). *Smithsonian Herpetological Information Service*, Washington (86), 1–45.

Keiser, E.D., Jr. & Wilson, L.D. (1979) Checklist and key to the amphibians and reptiles of Louisiana. Second edition. *Lafayette Natural History Museum* (1), 1–49, 7 figs. [first edition, 1969, 51 pp., 7 figs.]

Kelaart, E.F. (1852) *Prodromus faunae Zelanicae; being contributions to the zoology of Ceylon*. Colombo, 56 pp.

Kelaart, E.F. (1853) Description of new or little-known species of reptiles collected in Ceylon. *Journal of the Ceylon Branch of the Royal Asiatic Society*, Colombo 22(6), 102–116.

Kelaart, E.F. (1854a) *Prodromus faunae Zeylanicae; being contributions to the zoology of Ceylon. Vol. II. Part I.* Government Press, Colombo (1853), 34 pp.

Kelaart, E.F. (1854b) Descriptions of new or little-known species of reptiles collected in Ceylon. *Annals and Magazine of Natural History*, London (Series 2) 13(1), 25–31.

Kellaway, C.H. & Thomson, D.F. (1932) Observations on the venom of a melanotic insular variety of the tiger snake (*Notechis scutatis*). *Australian Journal of Experimental Biology and Medical Science*, Adelaide 10(1), 35–48, 16 figs.

Keller, A. (2008a) Geographic distribution: Serpentes: *Parias sumatranus*. *Herpetological Review* 39(3), 373.

Keller, A. (2008b) Geographic distribution: Serpentes: *Xenodermis javanicus*. *Herpetological Review* 39(3), 374.

Kelly, C.M.R., Barker, N.P. & Villet, M.H. (2003) Phylogenetics of advanced snakes (Caenophidia) based on four mitochondrial genes. *Systematic Zoology* 52(4), 439–459, 6 figs.

Kelly, C.M.R., Barker, N.P., Viller, M.H., Broadley, D.C. & Branch, W.R. (2008) The snake family Psammophiidae (Reptilia: Serpentes): phylogenetics and species delimitation in the African sand snakes (*Psammophis* Boie, 1825) and allied genera. *Molecular Phylogenetics and Evolution* 47(3), 1045–1060, 4 figs.

Kelly, C.M.R., Barker, N.P., Villet, M.H. & Broadley, D.G. (2009) Phylogeny, biogeography and classification of the snake superfamily Elapoidea: a rapid radiation in the late Eocene. *Cladistics* 25(1), 38–63, 3 figs.

Kelly, C.M.R., Branch, W.R., Broadley, D.C., Barker, N.P. & Villet, M.H. (2011) Molecular systematics of the African snake family Lamprophiidae Fitzinger, 1843 (Serpentes: Elapoidea), with particular focus on the genera *Lamprophis* Fitzinger 1843 and *Mehelya* Csiki 1903. *Molecular Phylogenetics and Evolution* 58(2), 415–426, 1 fig.

Kelly, H.A. (1936) *Snakes of Maryland*. Natural History Society of Maryland, Baltimore, 103 pp., 33 figs., 10 pls.

Kempff-Mercado, N. (1975) *Ofidios de Bolivia*. Academia Nacional de Ciencias de Bolivia, La Paz, 46 pp., illust.

Kend, B.A. (1997) *Pythons of Australia*. Canyonlands Publishing Group, Provo, 206 pp., 90 figs.

Kennedy, W.P. (1937) Some addition to the fauna of Iraq. *The Journal of the Bombay Natural History Society* (1936–1937) 39(4), 745–749.

Kennicott, R. (1855) Catalogue of animals observed in Cook County, Illinois. *Transactions of the Illinois State Agricultural Society* (1853–1854) 1, 577–595.

Kennicott, R. (1856) Description of a new snake from Illinois. *Proceedings of the Academy of Natural Sciences of Philadelphia* 8(2), 95–96.

Kennicott, R. (1859) Notes on *Coluber calligaster* of Say, and a description of new species of serpents in the collection of the North Western University of Evanston, Illust. *Proceedings of the Academy of Natural Sciences of Philadelphia* 11(3), 98–100.

Kennicott, R. (1860) Descriptions of new species of North America serpents in the Museum of the Smithsonian Institution, Washington. *Proceedings of the Academy of Natural Sciences of Philadelphia* 12(8), 328–338.

Kennicott, R. (1861) On three new forms of rattlesnakes. *Proceedings of the Academy of Natural Sciences of Philadelphia* 13(6), 206–208.

Keogh, J.S. (1996) Evolution of the colubrid snake tribe Lampropeltini: a morphological perspective. *Herpetologica* 52(3), 406–416, 4 figs.

Keogh, J.S. (1998) Molecular phylogeny of elapid snakes and a consideration of their biogeographic history. *Biological Journal of the Linnean Society,* London 63(1), 177–203, 4 figs.

Keogh, J.S., Branch, W.R. & Shine, R. (2000) Feeding ecology, reproduction and sexual dimorphism in the colubrid snke *Crotaphopeltis hotamboeia* in southern Africa. *African Journal of Herpetology* 49(2), 129–137, 4 figs.

Keogh, J.S. & Smith, S.A. (1996) Taxonomy and natural history of the Australian bandy-bandy snakes (Elapidae, *Vermicella*) with a description of two new species. *Journal of Zoology*, London 240(4), 677–701, 7 figs.

Keogh, J.S., Shine, R. & Donnellan, S. (1998) Phylogenetic relationships of terrestrial Australo-Papuan elapid snakes (Subfamily Hydrophiinae) based on cytochrome b and 16 S rRNA sequences. *Molecular Phylogenetics and Evolution* 10(1), 67–81, 5 figs.

Keogh, J.S., Scott, I.A.W. & Scanlon, J.D. (2000) Molecular phylogeny of viviparous Australian elapid snakes, affinities of *Echiopsis atriceps* (Storr, 1980) and *Drysdalia coronata* (Schlegel, 1837), with description of a new genus. *Journal of Zoology,* London 252(3), 317–326.

Keogh, J.S., Barker, D.G. & Shine, R. (2001) Heavily exploited but poorly known: systematics and biogeography of commercially harvested pythons (*Python curtus* group) in southeast Asia. *Biological Journal of the Linnean Society,* London 73(1), 113–129, 4 figs.

Keogh, J.S., Scott, I.A.W., Fitzgerald, M. & Shine, R. (2003) Molecular phylogeny of the Australian venomous snake genus *Hoplocephalus* (Serpentes, Elapidae) and conservation genetics of the threatened *H. stephensii*. *Conservation Genetics* 4(1), 57–65.

Keogh, J.S., Scott, I.A.W. & Hayes, C. (2005) Rapid and repeated origin of insular gigantism and dwarfism in Australian tiger snakes. *Evolution* 59(1), 226–233.

Kerr, J.G. (1926) *Bothrodon pridii*, an extinct serpent of gigantic dimensions. *Proceedings of the Royal Society of Edinburgh* 46, 314–315, 1 fig.

Keymar, P.F. (1986) Vorlaufige Ergebnisse herpetologischer Aufsammlungen auf den Ionischen Inseln, 2. Zakynthos und Marathonisi. *Annalen des Naturhistorischen Museums in Wien* 90B, 17–25, illust.

Khaire, A. & Khaire, N. (1986) A report on the occurrence of hybrid between *Eryx conicus* (Schneider) and *Eryx johnii* (Russell). *The Snake* 18(2), 114–117, 6 figs.

Khaire, A. & Khaire, N. (1993) Occurrence of brown whip snake *Ahaetulla pulverulenta* (Dum. & Bibr.) in Pune, India. *The Snake* 25(2), 147–148, 1 fig.

Khaire, N. (1996) *Indian snakes.* Indian Herpetological Society, Pune, 106 pp., illust.

Khaire, N. (2006) *A guide to the snakes of Maharashtra, Gioa & Karnataka.* Indian Herpetological Society, Pune, 129 pp., col. photos.

Khaire, N. (2010) *Snakes.* Jyotsna Prakashan, Shaniwar Peth, 104 pp., col. illust.

Khalaf, K.T. (1959) *Reptiles of Iraq with some notes on the amphibians.* Ar-Rabitta Press, Baghdad, v + 96 pp., 40 figs.

Khalil, F. (1963) Taxonomy, ecology and zoogeography. *In: A review of the natural resources of the African continent.* UNESCO and Columbia University Press, New York, pp. 277–316.

Khan, M.A.R. (1988) An updated list of the snakes of Bangladesh. *Hamadryad*, Madras 13(1), 5–7.

Khan, M.A.R. (1992) *Snakes of Bangladesh.* Bangla Academy, Dhaka, 227 pp., figs. (In Bengali)

Khan, M.A.R. (2004) Checklist of the herpetofauna of Bangladesh. *Cobra* 57, 1–31.

Khan, M.S. (1977) Checklist and key to the snakes of Jhang District, Punjab, Pakistan. *Biologia*, Lahore 23(2), 145–157, illust.

Khan, M.S. (1980) Affinities and zoogeography of herptiles of Pakistan. *Biologia*, Lahore 26(1–2), 113–171, illust.

Khan, M.S. (1982a) Some notes on snakes in Pakistan. *Hamadryad*, Madras 7(1), 4–5.

Khan, M.S. (1982b) An annotated checklist and key to the reptiles of Pakistan, Part III, Serpentes (Ophidia). *Biologia*, Lahore 28(2), 215–254.

Khan, M.S. (1983) Venomous terrestrial snakes of Pakistan. *The Snake* 15(2), 101–105, 5 figs.

Khan, M.S. (1984a) Validity of the natricine taxon *Natrix sanctijohannis* Boulenger. *Journal of Herpetology* 18(2), 198–200.

Khan, M.S. (1984b) Rediscovery and validity of *Bungarus sindanus* Boulenger. *The Snake* 16(1), 43–48, 1 fig.

Khan, M.S. (1985a) Taxonomic notes on *Bungarus caeruleus* (Schneider) and *Bungarus sindanus* Boulenger. *The Snake* 17(1), 71–78, 1 fig.

Khan, M.S. (1985b) An interesting collection of amphibians and reptiles from Cholistan desert, Punjab, Pakistan. *The Journal of the Bombay Natural History Society* 82(1), 144–148, figs.

Khan, M.S. (1986) A noteworthy collection of amphibians and reptiles from north-western Punjab, Pakistan. *The Snake* 18(2), 118–125, 4 figs.

Khan, M.S. (1987) Checklist, distribution and zoogeographical affinities of amphibians and reptiles of Baluchistan. *Proceedings of the 7th Pakistan Congress of Zoology* 1987, 105–112, 1 fig.

Khan, M.S. (1990) Venomous terrestrial snakes of Pakistan and snake bite problems. *In:* Gopalakrishnakone, P. & Chou, L.M. (Eds), *Snakes of medical importance (Asia-Pacific region).* National University of Singapore, Singapore, pp. 419–445, 19 figs.

Khan, M.S. (1993) *Snakes of Pakistan.* Urdu Science Board, Lahore, 208 pp., 48 figs., 50 photos. (in Bengali)

Khan, M.S. (1994) Key for identification of amphibians and reptiles of Pakistan. *Pakistan Journal of Zoology* 26(3), 249–255.

Khan, M.S. (1997) Taxonomic notes on Pakistani snakes of the *Coluber karelini-rhodorachis-ventromaculatus* species complex: a new approach to the problem. *Asiatic Herpetological Research* 7, 51–60, 6 figs.

Khan, M.S. (1998) Notes on *Typhlops diardi* Schlegel, 1839, with description of a new subspecies (Squamata, Serpentes, Scolecophidia). *Pakistan Journal of Zoology* 30(3), 213–221, 5 figs.

Khan, M.S. (1999a) *Typhlops ductuliformes* a new species of blind snakes from Pakistan and a note on *T. porrectus* Stoliczka, 1871 (Squamata, Serpentes, Scolecophidia). *Pakistan Journal of Zoology* 31(4), 385–390, 2 figs.

Khan, M.S. (1999b) Two new species and a subspecies of blind snakes of genus *Typhlops* from Azad Kashmir and Punjab (Serpentes, Typhlopidae). *Russian Journal of Herpetology* 6(3), 231–240, 4 figs.

Khan, M.S. (2002) *A guide to the snakes of Pakistan.* Edition Chimaira, Frankfurt-am-Main, 265 pp., 155 col. figs.

Khan, M.S. (2006) *Amphibians and reptiles of Pakistan.* Krieger Publishing Company, Malabar, xvi + 231 pp., 176 pls.

Khan, M.S. & Baig, K.J. (1988) Checklist of the amphibians and reptiles of district Jhelum, Punjab, Pakistan. *The Snake* 20(2), 156–161, 1 fig.

Khan, M.S. & Khan, A.Q. (2000) Three new subspecies of snakes of the genus *Coluber* from Pakistan. *Pakistan Journal of Zoology* 32(1), 49–52, 3 figs.

Khan, M.S. & Tasnim, R. (1986a) Balling and caudal luring in young *Bungarus caeruleus. The Snake* 18(1), 42–45.

Khan, M.S. & Tasnim, R. (1986b) A note on tail injury in *Eryx johnii. The Snake* 18(1), 57–58.

Khan, M.S. & Tasnim, R. (1986c) Notes on the Himalayan pit viper, *Agkistrodon himalayanus* (Gunther). *Litteratura Serpentium*, Utrecht 6, 46–55.

Khan, M.S. & Tasnim, R. (1987) Observations on distress behavior of an injured *Eryx johnii. The Snake* 19(2), 144–145, 4 figs.

Kharin, V.E. (1981) A review of sea snakes of the genus *Aipysurus* (Serpentes, Hydrophiidae). *Zoologischeskii Zhurnal*, Moscow 60(2), 257–264, 2 figs.

Kharin, V.E. (1983) A new species of the genus *Hydrophis sensu lato* (Serpentes, Hydrophiidae) from the North Australian shelf. *Zoologischeskii Zhurnal*, Moscow 62(11), 1751–1753, 3 figs.

Kharin, V.E. (1984a) Sea snakes of the genus *Hydrophis sensu lato* (Serpentes, Hydrophiidae). On the taxonomic status of the New Guinea *H. obscurus. Zoologischeskii Zhurnal*, Moscow 63(4), 630–632, 4 figs.

Kharin, V.E. (1984b) A review of sea snakes of the group *Hydrophis sensu lato* (Serpentes, Hydrophiidae). 3. The genus *Leioselasma. Zoologischeskii Zhurnal*, Moscow 63(10), 1535–1546, 4 figs.

Kharin, V.E. (1984c) Revision of sea snakes of subfamily Laticaudinae Cope, 1879 *sensu lato* (Serpentes, Hydrophiidae). *In:* Borkin, L.J. (Ed.), *Ecology and faunistics of amphibians and reptiles of the USSR and adjacent countries.* Proceedings of the Zoological Institute, Leningrad, USSR Academy of Science, pp. 124–139. (in Russian)

Kharin, V.E. (1984d) The first record of these species of sea snake from Vietnam with a note on rare variety of *Praescutata viperina. Biologiya Morya*, Vladivostok 1984(2), 26–30.

Kharin, V.E. (1985) A new species of sea snakes of the genus *Enhydrina* (Serpentes, Hydrophiidae) from waters of New Guinea. *Zoologischeskii Zhurnal*, Moscow 64(5), 785–787, 3 figs.

Kharin, V.E. (1989) A new sea snake species of the genus *Disteira* (Serpentes, Hydrophiidae) from the waters of the Malay Archipelago. *Vestnik Zoologii*, Moscow 1989(1), 29–32.

Kharin, V.E. (2004a) On the taxonomic status of the sea snake *Hydrophis caerulescens* (Shaw, 1802). *Biologiya Morya* 30(3), 227–229.

Kharin, V.E. (2004b) A review of sea snakes of the genus *Hydrophis sensu stricto* (Serpentes, Hydropheidae). *Biologiya Morya* 30(6), 442–454.

Kharin, V.E. (2005a) Distribution of a little-known sea snake *Chitulia belcheri* (Gray, 1849) and new records of rare species of the genus *Leioselasma* Lacépède, 1804 (Serpentes: Hydrophiidae). *Russian Journal of Marine Biology* 31(3), 159–163, 4 figs.

Kharin, V.E. (2005b) On new findings of a rare sea snake *Leioselasma czeblukovi* Kharin, 1984, with remarks on species composition and taxonomical position of the genus *Leioselasma* Lacépède, 1804, (Serpentes: Hydrophiidae). *Russian Journal of Marine Biology* 31(5), 269–272, 1 fig.

Kharin, V.E. (2006) An annotated checklist of sea snakes of Vietnam, with notes on a new record of the yellow-lipped sea krait, *Laticauda colubrina* (Schneider, 1799) (Laticaudidae, Hydrophiidae). *Russian Journal of Marine Biology* 32(4), 223–228, 2 figs.

Kharin, V.E. (2007) On the second record of yellow-bellied sea snake *Pelamis platurus* (Linnaeus, 1766) from Russia. *Russian Journal of Herpetology* 14(1), 45–49, 1 fig.

Kharin, V.E. (2008a) On the taxonomic status of a mangrove sea snake *Hydrelaps darwiniensis* Boulenger, 1896 (Serpentes, Hydrophiidae). *Russian Journal of Herpetology* 15(2), 103–109, 1 fig.

Kharin, V.E. (2008b) Reptilians. *In: Biota of the Russian waters of the Sea of Japan.* Russian Academy of Sciences, Far Eastern Branch, A.V. Zhirmunsky Institute of Marine Biology, Vladivostok, 7, 1–169, 24 pls. (in Russian)

Kharin, V.E. (2009) Redescription of a Russian finding of the Erabu sea krait *Pseudolaticauda semifasciata* (Reinwardt *in* Schegel, 1837), with remarks about species composition of sea snakes (Serpentes: Laticaudidae, Hydrophiidae) in Russian and adjacent waters. *Russian Journal of Marine Biology* 35(1), 8–14, 4 figs.

Kharin, V.E. (2011) Rare and little-known snakes of the north-eastern Eurasia. 3. On the taxonomic status of the slender racer *Hierophis spinalis* (Serpentes: Colubridae). *Contemporary Herpetology* 11(3–4), 173–179, 6 figs. (in Russian)

Kharin, V.E. (2012) Sea snakes of the genus *Chitulia* Gray, 1849 (Serpentes: Hydrophiidae). The taxonomic status of intrageneric complexes. *Russian Journal of Marine Biology* 38(6), 456–458.

Kharin, V.E. & Czeblukov, V.P. (2005) On the new record of rare snake *Chitulia inornata* Gray, 1849 (Serpentes, Hydrophiidae) from Australia. *Russian Journal of Herpetology* 12(2), 99–101, 2 figs.

Kharin, V.E. & Czeblukov, V.P. (2006a) A new revision of sea kraits of family Laticaudidae Cope, 1879 (Serpentes, Colubroidea). *Russian Journal of Herpetology* 13(2), 227–241, 14 figs.

Kharin, V.E. & Czeblukov, V.P. (2006b) On a new record of a poorly known and rare sea snake *Aipysurus tenuis* Lönnberg et Andersson, 1913 (Serpentes: Hydrophiidae) in the waters of Australia. *Russian Journal of Marine Biology* 32(3), 194–197, 3 figs.

Kharin, V.E. & Czeblukov, V.P. (2007) On first reliable record of the sea snake *Chitulia belcheri* (Gray, 1849) from Australian waters, with notes on species composition and taxonomic status of the genus *Chitulia* (Serpentes, Hydrophiidae). *Russian Journal of Marine Biology* 33(3), 161–165, 3 figs.

Kharin, V.E. & Czeblukov, V.P. (2009) A revision of the sea snakes of subfamily Hydrophiinae. 1. Tribe Disteirini nov. (Serpentes: Hydrophiidae). *Russian Journal of Herpetology* 16(3), 183–202, 21 figs.

Kharin, V.E. & Dotsenko, I.B. (2012) The first record of the sea snake *Chitulia inornata* Gray, 1849 from the Arabain Sea with notes on the composition of the genus *Chitulia* Gray, 1849 (Serpentes: Hydrophiidae). *Biologiya Morya* 38(1), 35–42, 3 figs. (in Russian)

Kharin, V.E. & Hallermann, J. (2009a) Distribution of a little-known sea krait *Pseudolaticauda schistorhynchus* (Günther, 1874) (Serpentes: Laticaudidae). *Russian Journal of Marine Biology* 35(5), 437–440, figs. a–f.

Kharin, V.E. & Hallermann, J. (2009b) Annotated catalogue of sea kraits (Laticaudidae) and sea snakes (Hydrophiidae) of the herpetological collection of the Zoological Museum, University of Hamburg. *Mitteilungen aus dem Hamburgischen Zoologischen Museum und Institute* 106, 51–67, 9 figs.

Kharin, V.E. & Hallermann, J. (2010) A remote new distribution record and a redescription of a rare and little-known sea snake *Leioselasma czeblukovi* Kharin, 1984 (Hydrophiidae). *Herpetology Notes* 3, 315–319, 3 figs.

Kharin, V.E. & Timentzev, A.A. (2009). Rare and little-known snakes in north-eastern Eurasia. 2. Records of two-colored yellow-bellied sea snake *(Pelamis platura,* Hydrophiidae) in the Russian waters. *Contemporary Herpetology* 9(3–4), 122–129, 2 figs.

Kharin, V.E., Rödel, M.-O. & Hallermann, J. (2010) New records and distribution of a little-known sea krait *Laticauda frontalis* (Devis, 1905) (Serpentes, Laticaudidae). *Russian Journal of Herpetology* 17(4), 285–289, 3 figs.

Kharin, V.E., Orlov, N.L. & Ananjeva, N.B. (2011) New records and redescription of rare and little-known elapid snake *Bungarus slowinskii* (Serpentes: Elapidae: Bungarinae). *Russian Journal of Herpetology* 18(4), 284–294, 17 figs.

Khole, V. (1991) Snakebite problems in Europe. *In:* Tu, A.T. (Ed.), *Handbook of natural toxins. Volume 5. Reptile venoms and toxins.* Marcel Dekker, New York, pp. 405–470.

Kieffer, M.C. (1998) Geographical distribution. *Pseudablabes agassizii. Herpetological Review* 29(1), 54.

Kien, D. van & Truong, N.Q. (2009) Geographic distribution: Serpentes: *Bungarus slowinskii. Herpetological Review* 40(3), 364.

Kien, T. & Nguyen, Q.T. (1995). *Venomous snakes of Vietnam and treatment of snakebites.* Second edition. Science and Technics Publishing House, Hanoi, 204 pp., 83 figs. (in Vietnamese) [first edition, 1980, 150 pp., 82 figs.]

Kinghorn, J.R. (1921) Studies in Australian reptiles. 2. *Records of the Australian Museum*, Sydney 13(4), 143–154, figs.

Kinghorn, J.R. (1923) A new genus of elapine snake from north Australia. *Records of the Australian Museum*, Sydney 14(1), 42–45.

Kinghorn, J.R. (1926) A new genus of sea snake and other reptiles collected by H.M.A.S. Geranium during a cruise to North Australia. *Proceedings of the Zoological Society of London* 96(1), 71–74, illust.

Kinghorn, J.R. (1928a) Herpetology of the Solomon Islands. *Records of the Australian Museum*, Sydney 16(3), 58–178, 22 figs., pl. 14.

Kinghorn, J.R. (1928b) Notes on some reptiles and batrachians from the Northern Division of Papua, with description of new species of *Apisthocalamus* and *Lygosoma*. *Records of the Australian Museum*, Sydney 16(6), 289–293.

Kinghorn, J.R. (1929) Two new snakes from Australia. *Records of the Australian Museum*, Sydney 17(4), 190–193.

Kinghorn, J.R. (1931) Herpetological notes. No. 3. *Records of the Australian Museum*, Sydney 18(5), 267–269, 2 figs.

Kinghorn, J.R. (1939) Two Queensland snakes. *Records of the Australian Museum*, Sydney 20(4), 257–259, 6 figs.

Kinghorn, J.R. (1942) Herpetological notes. No. 4. *Records of the Australian Museum*, Sydney 21(2), 118–121.

Kinghorn, J.R. (1948) A new species of *Typhlops* from New Britain. *Records of the Australian Museum*, Sydney 22(1), 67–69, 3 figs.

Kinghorn, J.R. (1955) Herpetological notes. No. 5. 3. A new generic name for two nominal species of *Pseudechis*. *Records of the Australian Museum*, Sydney 23(5), 283–286, 2 figs., 1 pl.

Kinghorn, J.R. (1964) *The snakes of Australia*. Third edition. Angus and Robertson, Sydney, 197 pp., 113 figs., 9 pls. [first edition, 1929, 198 pp., illust.; second edition, 1956, 197 pp., illust.]

Kinghorn, J.R. & Kellaway, C.H. (1943) *The dangerous snakes of the South-West Pacific area*. Victorian Railways Printing, Melbourne, 43 pp., 11 figs.

Kingsley, J.S. (Ed.). (1888) *The Riverside natural history. Volume III. Lower vertebrates*. Houghton, Mifflin and Company, Boston, 488 pp., 2200 figs., 168 pls.

Kinkelin, F. (1892) Ein fossiler Giftzahn. *Zoologischer Anzeiger*, Leipzig 15(385), 93–94.

Kinkelin, F. (1896) Einige seltere Fossilien des Senckenbergischen Museums. IV. Ein fossiler Giftzahn aus den untermiocänen Hydrobienschichten vom Hessler bei Mosbach-Biebrich. *Abhandlungen der Senckenbergischen Naturforschenden Gesellschaft*, Frankfurt-am-Main 20(1), 36–40, 3 figs.

Kirchberg, J.S. & Davidson, T.M. (1991) Envenomation by the colubrid snake *Atracaspis bibronii;* a case report. *Toxicon* 29(3), 379–381.

Kirk, J.J. (1979) *Thamnophis ordinoides*. *Catalogue of American amphibians and reptiles*, pp. 233.1–233.2.

Kirn, A.J., Burger, W.L. & Smith, H.M. (1949) The subspecies of *Tantilla gracilis*. *American Midland Naturalist* 42(1), 239–251, 5 figs.

Kirschner, A. & Seufer, H. (2003) *The ball python: care, breeding and natural history*. Kirschner & Seufer Verlag, Keltern-Weiler, 94 pp., 100 col. photos. [first edition, 1995, *Der Königspython: Pflege und Zucht*, 71 pp., 17 col. photos]

Kirtland, J.P. (1838) Report on the zoology of Ohio. *In:* Mather, W.W. (Ed.), *Second annual report on the geological survey of the state of Ohio*. Samuel Medary, Columbus, pp. 157–200.

Kitchell, Jr., K. & Dundee, H.A. (1994) A trilogy on the herpetology of Linnaeus's Systema Naturae X. *Smithsonian Herpetological Information Service* (100), 1–61.

Kivit, R. & Wiseman, S. (2005) *Grüner Baumpython und Grüne Hundskopfboa: Pflege, Zucht und Lebensweise*. Second edition. Kirschner & Seufer Verlag, Keltern-Weiler, 174 pp., 259 col. photos. [first edition, 2000, *The green tree python & emerald tree boa: their caprive husbandry and reproduction,* 125 pp., 122 col. photos]

Kizirian, D., Eriksson, A. & Truong, N.Q. (2002) Geographic distribution: Serpentes: *Bungarus bungaroides*. *Herpetological Review* 33(4), 323.

Kizirian, D., Truong, N.Q., Eriksson, A. & Vora, N. (2003) Taxonomic status of *Xenopeltis hainanensis jidamingae* Zhao. *Journal of Herpetology* 37(2), 438–440, 2 figs.

Klaczko, J., Machado, F. de A., Scrocchi, G. & H. Zaher. (2010) Taxonomic status of *Chironius multiventris* and *Chironius cochranae* (Serpentes). *Herpetologica* 66(4), 476–484, 4 figs.

Klarsfeld, J.D. (1999) A new pattern morph of the Louisiana milk snake, *Lampropeltis triangulum amaura*. *The Vivarium* 10(6), 23–26, 3 col. photos.

Klauber, L.M. (1924) Notes on the distribution of snakes in San Diego County, California. *Bulletin of the Zoological Society of San Diego* (1), 1–23, 5 figs.

Klauber, L.M. (1927) Some observations on the rattlesnakes of the extreme Southwest. *Bulletin of the Antivenin Institute of America* 1(1), 7–21, 9 figs.

Klauber, L.M. (1928) The *Trimorphodon* (lyre snake) of California, with notes on the species of the adjacent areas. *Transactions of the San Diego Society of Natural History* 5(11), 183–194, pls. 22–23.

Klauber, L.M. (1930a) New and renamed subspecies of *Crotalus confluentus* Say, with remarks on related species. *Transactions of the San Diego Society of Natural History* 6(3), 95–144, pls. 9–12.

Klauber, L.M. (1930b) Differential characteristics of southwestern rattlesnakes allied to *Crotalus atrox*. *Bulletin of the Zoological Society of San Diego* (6), 1–72, 6 pls.

Klauber, L.M. (1931a) A new subspecies of the California boa, with notes on the genus *Lichanura*. *Transactions of the San Diego Society of Natural History* 6(20), 305–318, pl. 21.

Klauber, L.M. (1931b) Notes on the worm snakes of the southwest, with descriptions of two new subspecies. *Transactions of the San Diego Society of Natural History* 6(23), 333–352.

Klauber, L.M. (1931c) *Crotalus tigris* and *Crotalus enyo*, two little known rattlesnakes of the southwest. *Transactions of the San Diego Society of Natural History* 6(24), 353–370, pl. 23.

Klauber, L.M. (1933) Notes on *Lichanura*. *Copeia* 1933(4), 214–215.

Klauber, L.M. (1935a) A new subspecies of *Crotalus confluentus*, the prairie rattlesnake. *Transactions of the San Diego Society of Natural History* 8(13), 75–90, pl. 8.

Klauber, L.M. (1935b) *Phyllorhynchus*, the leaf-nosed snake. *Bulletin of the Zoological Society of San Diego* (12), 4–31, 4 figs.

Klauber, L.M. (1936a) *Crotalus mitchellii*, the speckled rattlesnake. *Transactions of the San Diego Society of Natural History* 8(19), 149–184, 3 figs., pls. 19–20.

Klauber, L.M. (1936b) A statistical study of the rattlesnakes, I. Introduction; II. Sex ratio in rattlesnake populations; III Birth rate. *Occasional Papers of the San Diego Society of Natural History* (1), 1–24, 2 figs.

Klauber, L.M. (1936c) A key to the rattlesnakes with summary of characteristics. *Transactions of the San Diego Society of Natural History* 8(20), 185–276, 112 figs.

Klauber, L.M. (1937a) A statistical study of the rattlesnakes, IV. The growth of the rattlesnake. *Occasional Papers of the San Diego Society of Natural History* (3), 1–56, figs. 3–6.

Klauber, L.M. (1937b) A new snake of the genus *Sonora* from Mexico. *Transactions of the San Diego Society of Natural History* 8(27), 363–366.

Klauber, L.M. (1938) A statistical study of the rattlesnakes. V. Head dimensions. *Occasional Papers of the San Diego Society of Natural History* (4), 1–53, figs. 7–16.

Klauber, L.M. (1939a) A statistical study of rattlesnakes. VI. Fangs. *Occasional Papers of the San Diego Society of Natural History* (5), 1–41, figs. 17–46.

Klauber, L.M. (1939b) Three new worm snakes of the genus *Leptotyphlops*. *Transactions of the San Diego Society of Natural History* 9(14), 59–66, 3 figs.

Klauber, L.M. (1939c) A new subspecies of the western worm snake. *Transactions of the San Diego Society of Natural History* 9(14), 67–68.

Klauber, L.M. (1939d) Studies of reptile life in the arid southwest. *Bulletin of the Zoological Society of San Diego* (14), 1–100.

Klauber, L.M. (1939e) A further study of pattern dimorphism in the California king snake. *Bulletin of the Zoological Society of San Diego* (15), 1–23, 7 figs.

Klauber, L.M. (1940a) The worm snakes of the genus *Leptotyphlops* in the United States and northern México. *Transactions of the San Diego Society of Natural History* 9(18), 87–162, 3 figs., pl. 6.

Klauber, L.M. (1940b) The lyre snakes (genus *Trimorphodon*) of the United States. *Transactions of the San Diego Society of Natural History* 9(19), 163–194, pl. 7.

Klauber, L.M. (1940c) Two new subspecies of *Phyllorhynchus*, the leaf- nosed snake, with notes on the genus. *Transactions of the San Diego Society of Natural History* 9(20), 195–214, pl. 8.

Klauber, L.M. (1940d) A statistical study of the rattlesnakes. VII. The rattle, Part I. *Occasional Papers of the San Diego Society of Natural History* (6), 1–62, figs. 47–73.

Klauber, L.M. (1941a) The long-nosed snakes of the genus *Rhinocheilus*. *Transactions of the San Diego Society of Natural History* 9(29), 289–332, pls. 12–13.

Klauber, L.M. (1941b) A new species of rattlesnake from Venezuela. *Transactions of the San Diego Society of Natural History* 9(30), 333–336.

Klauber, L.M. (1943a) A new snake of the genus *Sonora* from Lower California, México. *Transactions of the San Diego Society of Natural History* 10(4), 69–70.

Klauber, L.M. (1943b) A desert subspecies of the snake *Tantilla eiseni*. *Transactions of the San Diego Society of Natural History* 10(5), 71–74.

Klauber, L.M. (1943c) The coral king snakes of the Pacific coast. *Transactions of the San Diego Society of Natural History* 10(6), 75–82.

Klauber, L.M. (1943d) The subspecies of the rubber snake, *Charina*. *Transactions of the San Diego Society of Natural History* 10(7), 83–90.

Klauber, L.M. (1944) The sidewinder, *Crotalus cerastes*, with description of a new subspecies. *Transactions of the San Diego Society of Natural History* 10(8), 91–126, 1 fig., pls. 5–6.

Klauber, L.M. (1946a) The glossy snake, *Arizona*, with descriptions of new subspecies. *Transactions of the San Diego Society of Natural History* 10(17), 311–398, pls. 7–8.

Klauber, L.M. (1946b) The gopher snakes of Baja California, with descriptions of new subspecies of *Pituophis catenifer*. *Transactions of the San Diego Society of Natural History* 11(1), 1–40, 2 pls.

Klauber, L.M. (1946c) A new gopher snake (*Pituophis*) from Santa Cruz Island, California. *Transactions of the San Diego Society of Natural History* 11(2), 41–48, pl. 3.

Klauber, L.M. (1947) Classification and ranges of the gopher snakes of the genus *Pituophis* in the western United States. *Bulletin of the Zoological Society of San Diego* (22), 6–81, 6 figs.

Klauber, L.M. (1948) Some misapplications of the Linnaean names applied to American snakes. *Copeia* 1948(1), 1–14.

Klauber, L.M. (1949a) The relationship of *Crotalus ruber* and *Crotalus lucasensis*. *Transactions of the San Diego Society of Natural History* 11(5), 57–60.

Klauber, L.M. (1949b) Some new and revived subspecies of rattlesnakes. *Transactions of the San Diego Society of Natural History* 11(6), 61–116, pls. 4–6.

Klauber, L.M. (1949c) The subspecies of the ridge-nosed rattlesnake, *Crotalus willardi*. *Transactions of the San Diego Society of Natural History* 11(8), 121–140.

Klauber, L.M. (1951) The shovel-nosed snake, *Chionactis*, with descriptions of two new subspecies. *Transactions of the San Diego Society of Natural History* 11(9),141–204, pls. 9–10.

Klauber, L.M. (1952) Taxonomic studies of the rattlesnakes of mainland México. *Bulletin of the Zoological Society of San Diego* (26), 1–143, 12 figs.

Klauber, L.M. (1956a) *Rattlesnakes: their habits, life histories and influence on mankind*. 2 vols. University of California Press, Berkeley, 1476 pp., illust. [second edition, 1972, 1533 pp., illust.; third edition, 1997, 1533 pp., illust.]

Klauber, L.M. (1956b) *Agkistrodon* or *Ancistrodon*? *Copeia* 1956(4), 258–259.

Klauber, L.M. (1963) A new insular subspecies of the speckled rattlesnake. *Transactions of the San Diego Society of Natural History* 13(5), 73–80, 2 figs.

Klawe, W.L. (1964) Food of the black-and-yellow sea snake, *Pelamis platurus*, from Ecuadorian coastal waters. *Copeia* 1964(4), 712–713.

Klein, J.A. (1973) Snakes of the Oak Ridge area, Tennessee. *Bulletin of the New York Herpetological Society* 10(1–2), 29–32.

Klein, J.A. (1989) A checklist of the reptiles and amphibians on the Department of Energy Oak Ridge Reservation, Anderson and Roane counties, Tennessee. *Journal of the Tennessee Academy of Sciences* 64(4), 228–230, illust.

Klein, R.G. (1972) Preliminary report on the July through September 1970 excavations at Nelson Bay Cave, Plettenberg Bay (Cape Province, South Africa). *Palaeoecology of Africa* 6, 177–208, 9 figs.

Klemens, M.W. (1991) Checklist of the amphibians and reptiles of Connecticut with notes on uncommon species. *Connecticut Department of Environmental Protection Bulletin*, Hartford (14), 24 pp., 50 col. photos.

Klemens, M.W. (1995) New distributional records of amphibians and reptiles from New Hampshire. *Herpetological Review* 26, 49.

Klemmer, K. (1963) Liste der rezenten Giftschlangen: Elapidae, Hydropheidae, Viperidae und Crotalidae. *In: Die Giftschlangen der Erde: Wirkungen und Antigenität der Gifte Therapie von Giftschlangenbissen*. N.G. Elwert Universitäts- und Verlags-Buchhandlung, Marburg, pp. 255–464, 37 col. pls.

Klemmer, K. (1967) Observations on the sea snake, *Laticauda laticaudata* in captivity. *International Zoo Yearbook* 7, 229–231.

Klemmer, K. (1968) Classification and distribution of European, North African, and north and west Asiatic venomous snakes. *In:* Bücherl, W., Buckley, E.E. & Deulofeu, V. (Eds.), *Venomous animals and their venoms. Volume I. Venomous vertebrates*. Academic Press, New York, pp. 309–325.

Klesius, T. (2009) *Die Vipernatter,* Natrix maura. Natur und Tier-Verlag, Münster, 62 pp., col. illust.

Klesius, Y. & Klesius, T. (2010) *Die Raue Grasnatter,* Opheodrys aestivus. Natur und Tier-Verlag, Münster, 62 pp., col. illust.

Kletecki, E., Lanszki, J., Trócsányi, B., Muzinic, J. & Purger, J.J. (2009) First record of *Dolichophis caspius* (Gmelin, 1789), (Reptilia: Colubridae) on the island of Olib, Croatia. *Croatian Natural History Museum, Demetrava*, Croatia 18(2), 437–442, 5 figs.

Klimov, S.M., Klimova, N.I. & Aleksandrov, V.N. (1999) *Amphibians and reptiles of the Lipetsk region*. Lipetsk State Pedagogic Institute, Lipetsk, 81 pp., 19 figs. (in Russian)

Kluge, A.G. (1984) Type-specimens of reptiles in the University of Michigan Museum of Zoology. *Miscellaneous Publications, Museum of Zoology, University of Michigan* (167), 1–85.

Kluge, A.G. (1988a) Relationships of the Cenozoic boine snakes *Paraepicrates* and *Pseudoepicrates*. *Journal of Vertebrate Paleontology* 8(2), 299–230.

Kluge, A.G. (1988b) Parsimony in vicariance biogeography: a quantitative method and a Greater Antillean example. *Systematic Zoology* 37(4), 315–328, 10 figs.

Kluge, A.G. (1989) A concern for evidence and a phylogenetic hypothesis of relationships among *Epicrates* (*Boidae*, Serpentes). *Systematic Zoology* 38(1), 7–25, 8 figs.

Kluge, A.G. (1991) Boine snake phylogeny and research cycles. *Miscellaneous Publications, Museum of Zoology, University of Michigan* (178), i–iv, 1–58, 14 figs.

Kluge, A.G. (1993a) *Calabaria* and the phylogeny of erycine snakes. *Zoological Journal of the Linnean Society*, London 107(4), 293–351, 19 figs.

Kluge, A.G. (1993b) *Aspidites* and the phylogeny of pythonine snakes. *Records of the Australian Museum*, Melbourne (Supplement 19), 1–77, 28 figs.

Knauer, F. (1875) *Beobachtungen an Reptilien und Amphibien in der Gefangenschaft*. Wien, 54 pp.

Knight, A. & Mindell, D.P. (1993) Substitution bias, weighting of DNA sequence evolution, and the phylogenetic position of Fea's viper. *Systematic Biology* 42(1), 18–31, 3 figs.

Knight, A., Densmore, III, L.D. & Rael, E.D. (1992) Molecular systematics of the *Agkistrodon* complex. *In:* Campbell, J.A. & Brodie, Jr., E.D. (Eds.), *Biology of the pitvipers*. Selva, Tyler, pp. 49–69, 8 figs.

Knight, A., Styer, D., Pelikan, S., Campbell, J.A., Densmore, III, L.D. & Mindell, D.P. (1993) Choosing among hypotheses of rattlesnake phylogeny, a best-fit rate test for DNA sequence data. *Systematic Zoology* 42(3), 356–367, 5 figs.

Knight, J.L. (1986) Variation in snout morphology in the North American snake *Pituophis melanoleucus* (Serpentes, Colubridae). *Journal of Herpetology* 20(1), 77–79, 1 fig.

Knobel, E. (1896) *The turtles, snakes, frogs and other reptiles and amphibians of New England and the North*. Bradlee Whidden, Boston, 47 pp., illust.

Knoepffler, L.-P. (1966) Fauna du Gabon (amphibiens et reptiles). I. Ophidiens de l'Ogooué-Ivindo et du Woleu N'tem. *Biologia Gabonica*, Makoukou 2(1), 3–23.

Knoepffler, L.-P. (1968) Clef de détermination des serpents actuellement connus du Gabon et das contrées limitrophes (Moyen-Congo, Rio Muni, Cabinda et Sud Cameroun forestier). *Biologia Gabonica*, Makoukou 4(2), 183–194, 5 figs.

Knoepffler, L.-P. & Sochurek, E. (1955) Neues über die Rasun der Wiesenotter (*Vipera ursinii* Bonap.). *Burgenländische Heimatblätter*, Eisenstadt 17(4), 185–188, 1 fig.

Koba, K. (1938) Some notes on *Agkistrodon halys* (Pallas) from Syoryuzan Island, Kwangtung, South Manchuria (I). *Zoological Magazine*, Tokyo 50(5), 245–264, 8 figs., pls. 3–6. (In Japanese)

Koba, K. (1955) Herpetological fauna of Japan. *Bulletin of the Biogeographical Society of Japan*, Tokyo 16–19, 345–354. (in Japanese)

Koba, K. (1956) Herpetofauna of the Amami Group of the Loo Choo Islands, I. *Memoirs, Faculty of General Education, Kumamoto University*, Kumamoto 4, 148–164, 2 pls. (in Japanese)

Koba, K. (1958) Herpetofauna of the Anami group of the Loo Choo Islands (II). *Memoirs, Faculty of General Education, Kumamoto University*, Kumamoto 6, 173–185, 7 figs., pl. 2. (in Japanese)

Koba, K. (1959) Herpetofauna of the Anami group of the Loo Choo Islands (III). *Memoirs, Faculty of General Education, Kumamoto University*, Kumamoto 7, 187–202, 7 figs., pl. 1. (in Japanese)

Koba, K. (1960) Herpetofauna of the Anami group of the Loo Choo Islands (IV). *Memoirs, Faculty of General Education, Kumamoto University*, Kumamoto 8, 181–191, 5 figs., 1 pl. (in Japanese)

Koba, K. (1961) Food of *Trimeresurus flavoviridis flavoviridis* and *T. okinavensis* in the Amami Group of the Loo Choo Islands. *Memoirs, Faculty of General Education, Kumamoto University*, Kumamoto 9, 220–229, 1 fig., 2 pls. (in Japanese)

Koba, K. (1963) Food of *Trimeresurus flavoviridis flavoviridis* and *T. okinavensis* in the Amami Group of the Loo Choo Islands (Supplementary notes, I). *Memoirs, Faculty of General Education, Kumamoto University*, Kumamoto 11, 35–40, 2 figs., 2 pls. (in Japanese)

Koba, K. (1968) Food of *Trimeresurus flavoviridis* and *T. okinavensis* in the Amami Group of the Loo Choo Islands (Supplementary notes, II). *Memoirs, Faculty of General Education, Kumamoto University*, Kumamoto 16, 69–73. (in Japanese)

Koba, K. (1971) Natural history of the habu, *Trimeresurus flavoviridis* (Hallowell). *The Snake* 3(2), 75–96, 13 figs., 2 pls. (in Japanese)

Koba, K. (1973) The venomous snakes of southeast Asia. *The Snake* 5(2), 77–115, 30 figs. (in Japanese)

Koba, K. (1974) The venomous terrestrial snakes of the Nansei (Ryukyu) Islands. *Japanese Journal of Tropical Medicine and Hygiene*, Nagasaki (1973) 1(2), 103–104, 168–169. (in Japanese)

Koba, K. & Kikukawa, D. (1969) A taxonomic study of the Tokara-habu, *Trimeresurus tokarensis* from the Tokara Islands, Japan. *Bulletin of the Biogeographical Society of Japan*, Tokyo 25(1), 1–8, 1 figs., 2 pls. (in Japanese)

Koba, K. & Kikukawa, D. (1971) A taxonomic study of the Tokara-habu, *Trimeresurus tokarensis* Nagai. *The Snake* 3(1), 39–52, 12 figs. (in Japanese)

Koba, K. & Kikukawa, D. (1976) The morphology of *Trimeresurus elegans* (Serpentes, Viperidae). *Bulletin of Ginkyo College of Medicine and Technology*, Karamoto (1), 19–32, 9 figs. (in Japanese)

Koba, K. & Tanaka, K. (1973) Taxonomic notes on the hime-habu, *Trimeresurus okinavensis* of the Amami group, the Ryukyu islands. *Japanese Journal of Herpetology*, Kyoto 5(2), 30.

Koba, K., Tanaka, K., Yoshizaki, K. & Nakamoto, E. (1970) Eggs and hatching of the hime-habu, *Trimeresurus okinavensis* Boulenger. *The Snake* 2(2), 111–121, 3 figs., 2 pls.

Koba, K., Kikukawa, D., Fukuda, T. & Tanaka, K. (1977) A taxonomic study of *Calliophis japonicus* (Serpentes, Elapidae) of the Nansei Islands, Japan. *Bulletin of Ginkyo College of Medicine and Technology*, Karamoto (2), 7–30, 11 figs. (in Japanese)

Koch, A. (2008a) On the type specimens of the recently described Sulawesi pitviper *Tropidolaemus laticinctus* Kuch, Gumprecht and Melaun 2007 (Squamata: Viperidae). *Zootaxa* (1721), 65–68, 1 fig.

Koch, A. (2008b) Discovery of new specimens of Sarasin's keelback, *Amphiesma sarasinorum* (Boulenger, 1896), endemic to SW Sulawesi, Indonesia. *Herpetozoa* 20(3–4), 178–182, 3 figs.

Koch, A. (2012) *Discovery, diversity, and distribution of the amphibians and reptiles of Sulawesi and its offshore islands*. Edition Chimaira, Frankfurt-am-Main, 374 pp., 114 col. photos.

Koch, A., Arida, E., McGuire, J.A., Iskandar, D.T. & Böhme, W. (2009) A new species of *Calamaria* (Squamata: Colubridae) similar to *C. ceramensis* de Rooij, 1913, from the Banggai Islands, east of Sulawesi, Indonesia. *Zootaxa* (2196), 19–30, 9 figs.

Koch, E.D. & Peterson, C.R. (1995) *Amphibians and reptiles of Yellowstone and Grand Teton National Parks*. University of Utah Press, Salt Lake City, xviii + 188 pp., col. photos.

Kochva, E. (1974) *Snakes*. George S. Wise Center for the Life Sciences, Tel-Aviv 29 pp., 7 figs., 34 illust. (in Hebrew)

Kochva, E. (1987) The origin of snakes and evolution of the venom apparatus. *Toxicon* 25(1), 65–106, 20 figs.

Kochva, E. (1990) Venomous snakes of Israel. *In:* Gopalakrishnakone, P. & Chou, L.M. (Eds.), *Snakes of medical importance (Asia-Pacific region)*. National University of Singapore, Singapore, pp. 311–321, 7 figs.

Kochva, E. & Wollberg, M. (1970) The salivary glands of Apparallactinae (Colubridae) and the venom glands of *Elaps* (Elapidae) in relation to the taxonomic status of this genus. *Zoological Journal of the Linnean Society*, London 49(3), 217–224, 4 pls.

Kock, D. (1979) Zwei Schlangen neu für Kephalinia, Ionische Inseln, Griechenland (Reptilia, Serpentes, Colubridae). *Senckenbergiana Biologica*, Frankfurt-am-Main 60(1–2), 7–11, illust.

Kofron, C.P. (1980) *A revision of the Central American slug-eating snakes in the* Tropidodipsas *complex of the genus* Sibon *(Serpentes, Colubridae)*. Doctoral Dissertation, Texas A & M University, College Station, 177 pp.

Kofron, C.P. (1982a) The identities of some dipsadine snakes: *Dipsas elegans*, *D. ellipsifera* and *Leptognathus andrei*. *Copeia* 1982(1), 46–51, figs.

Kofron, C.P. (1982b) A review of Mexican snail-eating snakes, *Dipsas brevifacies* and *Dipsas gaigeae*. *Journal of Herpetology* 16(3), 270–286, 9 figs.

Kofron, C.P. (1983) Female reproductive cycle of the Neotropical snail-eating snake *Sibon sanniola* in northern Yucatan, Mexico. *Copeia* 1983(4), 963–969, 4 figs.

Kofron, C.P. (1985a) Systematics of the Neotropical gastropod-eating snake genera, *Tropidodipsas* and *Sibon*. *Journal of Herpetology* 19(1), 84–92.

Kofron, C.P. (1985b) Review of the Central American colubrid snakes, *Sibon fischeri* and *S. carri*. *Copeia* 1985(1), 164–174, figs.

Kofron, C.P. (1987) Systematics of Neotropical gastropod-eating snakes: the *fasciata* group of the genus *Sibon*. *Journal of Herpetology* 21(3), 210–225, 5 figs.

Kofron, C.P. (1988a) The Central and South American blindsnakes of the genus *Anomalepis*. *Amphibia-Reptilia*, Leiden 9(1), 7–13, 2 figs.

Kofron, C.P. (1988b) Systematics of Neotropical gastropod-eating snakes: the *sartorii* group of the genus *Sibon*. *Amphibia-Reptilia*, Leiden 9(2), 145–168, 6 figs.

Kofron, C.P. (1990) Systematics of Neotropical gastropod-eating snakes: the *dimidiata* group of the genus *Sibon*, with comments on the *nebulata* group. *Amphibia-Reptilia*, Leiden 11(3), 207–233, 5 figs.

Köhler, G. (1996a) Additions to the known herpetofauna of Isla de Utila (Islas de la Bahia, Honduras) with the description of a new species of *Norops*. *Senckenbergiana Biologica*, Frankfurt-am-Main 76(1–2), 19–28, 2 figs.

Köhler, G. (1996b) Notes on a collection of reptiles from El Salvador collected between 1951 and 1956. *Senckenbergiana Biologica*, Frankfurt-am-Main 76(1–2), 29–38, 5 figs.

Köhler, G. (1997) Geographic distribution: Serpentes: *Ungaliophis continentalis*. *Herpetological Review* 28(4), 211.

Köhler, G. (1998a) Further additions to the known herpetofauna of Isla de Utila (Islas de la Bahia, Honduras) with notes on other species and a key to the amphibians and reptiles of the island. *Senckenbergiana Biologica*, Frankfurt-am-Main 77(2), 139–145, 7 figs.

Köhler, G. (1998b) Geographic distribution: Serpentes: *Sibon annulatus*. *Herpetological Review* 29(4), 249.

Köhler, G. (1999a) Geographic distribution: Serpentes: *Tantillita lintoni*. *Herpetological Review* 30(1), 55.

Köhler, G. (1999b) The amphibians and reptiles of Nicaragua: a distributional checklist with keys. *Courier Forschungsinstitut Senckenberg*, Frankfurt-am-Main (213), 1–121, 135 figs., 10 pls.

Köhler, G. (2001a) *Reptilien und Amphibien Mittleamerikas Band 2, Schlangen*. Herpeton, Verlag Elke Köhler, Offenbach, 174 pp., 213 col. photos.

Köhler, G. (2001b) *Anfibios y reptiles de Nicaragua*. Herpeton, Verlag Elke Köhler, Offenbach, 208 pp., 277 figs.

Köhler, G. (2008) *Reptiles of Central America*. Second edition. Herpeton, Verlag Elke Köhler, Offenbach, 400 pp., 816 col. figs. [first edition, 2003, 367 pp., 748 col. figs.]

Köhler, G. & Berg, P. (2005) *Kornnattern: Lebensweise, Pflege, Zucht, Erkrankungen*. Herpeton, Verlag Elke Köhler, Offenbach, 142 pp., 194 col. photos.

Köhler, G. & Ferrari, J.A. (2009) Geographic distribution: Serpentes: *Leptophis depressirostris*. *Herpetological Review* 40(4), 456.

Köhler, G. & McCranie, J.R. (1999a) Taxonomic status of *Rhadinaea pinicola* Mertens (Serpentes, Colubridae). *Copeia* 1999(2), 529–530.

Köhler, G. & McCranie, J.R. (1999b) A new species of colubrid snake of the *Rhadinaea godmani* group from Cerro Saslaya, Nicaragua. *Senckenbergiana Biologica*, Frankfurt-am-Main 79(2), 243–249, 7 figs.

Köhler, G. & Seipp, R. (1998) Eine expedition in den Randbereich des Biosphärenreservats Bosawas, Nicaragua. *Natur und Museum* 128(6), 170–175, illust.

Köhler, G., Wilson, L.D. & McCranie, J.R. (2001) A new genus and species of colubrid snake from the Sierra de Omoa of northwestern Honduras (Reptilia, Squamata, Colubridae). *Senckenbergiana Biologica*, Frankfurt-am-Main 81(1–2), 269–276, 3 figs.

Köhler, G., Quintana, A.Z., Bustiago, F. & Diethert, H. (2004) New and noteworthy records of amphibians and reptiles from Nicaragua. *Salamandra*, Frankfurt-am-Main 40(1), 15–24.

Köhler, G., Vesely, M. & Greenbaum, E. (2006) *The amphibians and reptiles of El Salvador*. Krieger Publishing Company, Malibar (2005) ix + 238 pp., 140 col. pls.

Köhler, G., Lotzkat, S. & Hertz, A. (2010) A new species of *Sibon* (Squamata: Colubridae) from western Panama. *Herpetologica* 66(1), 80–85, 4 figs.

Köhler, J. & Güsten, R. (2007) Herpetological type specimens in the natural history collections of the museums in Darmstadt and Wiesbaden, Germany. *Spixiana* 30(2), 275–288, 9 figs.

Kohno, H. & Ota, H. (1991) Reptiles in a seabird colony: herpetofauna of Nakanokamishima Island of the Yaeyama Group, Ryukyu Archipelago. *Island Studies in Okinawa* (9), 73–89, 4 figs.

Kok, P.J.R. (2006) A new snake of the genus *Atractus* Wagler, 1828 (Reptilia, Squamata, Colubridae) from Kaieteur National Park, Guyana, northeastern South America. *Zootaxa* (1378), 19–35, 6 figs.

Kok, P.J.R. (2010) A new species of *Chironius* Fitzinger, 1826 (Squamata: Colubridae) from the Pantepui region, northeastern South America. *Zootaxa* (2611), 31–44, 8 figs.

Kok, P.J.R. & Fuenmayor, G.R. (2008) *Typhlophis ayarzaguenai* Señaris, 1998 is a junior synonym of *Typhlophis squamosus* (Schlegel, 1839). *Amphibia-Reptilia*, Leiden 29(4), 555–558, 1 fig.

Kok, P.J.R., Fuenmayor, G.A.R. & Pauwels, O.S.G. (2007) The taxonomic status of the Venezuelan snakes *Atractus matthewi* and *A. nororientalis* (Serpentes, Colubridae). *Zootaxa* (1493), 66–68.

Kolby, J.E. & Fraser, D. (2009) Geographic distribution: Serpentes: *Scolecophis atrocinctus. Herpetological Review* 40(3), 366.

Kölpin, T. (2002) Python regius, *der Königspython.* Natur und Tier-Verlag, Münster, 94 pp., 82 col. photos.

König, D. (1985) Langjährige Beobachtungen an der Äskulapnatter *Elaphe longissima* (Laurenti, 1768) Serpentes, Colubridae. *Salamandra*, Frankfurt-am-Main 21(1), 17–39, 10 figs.

Kopstein, F. (1926) Herpetologische Notizen. V. Reptilien von der Molukken und den Benachbarten Iseln. *Zoologische Mededelingen*, Leiden 9(2–3), 71–112.

Kopstein, F. (1927) Die reptilienfauna der Sula-Inseln. *Treubia*, Buitenzorg 9(4), 437–446.

Kopstein, F. (1930) Herpetologische Notizen. III. Reptilien des östlichen Preanger (West-Java). *Treubia*, Buitenzorg 12(3–4), 273–276.

Kopstein, F. (1931) Herpetologische Notizen. IV. *Fordonia leucobalia* Schlegel und *Cerberus rynchops* Schmidt. *Treubia* 13(1), 1-4.

Kopstein, F. (1932a) Herpetologische Notizen. V. *Bungarus javanicus*, eine neue Gifschlange von Java. *Treubia*, Buitenzorg 14(1), 73–77, pl.

Kopstein, F. (1932b) Die Gifttiere Javas und ihre Bedeutung für den Menschen. *Mededelingen van de Dienst den Volksgezondheld in Nederlandsch-Indië* 21, 222–256.

Kopstein, F. (1936) Herpetologische Notizen. XIII. Ueber *Vipera russellii* von Java. *Treubia*, Buitenzorg 15(3), 259–264.

Kopstein, F. (1937) Schlangen von Enggano. *Treubia*, Buitenzorg 16(2), 239–244, 1 pl.

Kopstein, F. (1938a) Ein Beitrag zur Eierkunde und zur Fortpflanzung der Malaiischen Reptilien. *Bulletin of the Raffles Museum, Straits Settlement*, Singapore (14), 81–167, pls. 6–27, 75 figs.

Kopstein, F. (1938b) Ein Beitrag zur Morphologie, Biologie und Ökologie von *Xenodermus javanicus* Reinhardt. *Bulletin of the Raffles Museum, Straits Settlement,* Singapore (14), 168–174, pl. 28.

Kopstein, F. (1941) Über sexualdimorphismus bei Malaiischen Schlangen. *Temminckia*, Leiden 6, 109–185.

Kopstein, F. & Wettstein, O. (1920) Reptilien und Amphibien aus Albanien. *Verhandlungen der Zoologisch-Botanischen Gesellschaft in Wien* 70, 387–409.

Kordikova, E.G. (1998) Herpetofaunas of Kazakhstan in the Plaeozoic, Mesozoic and Cenozoic. *Proceedings of the Third Asian Herpetological Meetings, Al-Farabi Kazakh State National University,* Almati (6), pp. 61–109, 9 figs.

Kordikova, E.G., Brushko, Z.K. & Kubykin, R.A. (1998) On distribution of some fossil amphibians and reptiles in Kazakhstan and their extant geographical range. *Proceedings of the Third Asian Herpetological Meetings, Al-Farabi Kazakh State National University,* Almati (6), pp. 35–54, 6 figs.

Kornacker, P.M. (1999) *Checklist and key to the snakes of Venezuela.* PaKo-Verlag, Rheinbach, 269 pp., 16 figs., 90 col. photos.

Kornacker, P.M. (2001) Geographic distribution: Serpentes: *Pseudoboa neuwiedii. Herpetological Review* 32(2), 124.

Korneva, L.G. (1969) Embryonic development of the water snake — *Natrix tessellata. Zoologischeskii Zhurnal*, Moscow 48(1), 110–120, 4 figs. (in Russian)

Kornilios, P., Ilgaz, Ç., Kumlutas, Y., Giokas, S., Fraguedakis-Tsolis, S. & Chondropoulos, B. (2011) The role of Anatolian refugia in herpetofaunal diversity: an mtDNA analysis of *Typhlops vermicularus* Merrem, 1820 (Squamata, Typhlopidae). *Amphibia-Reptilia* 32(3), 351–363, 3 figs.

Kornilios, P., Ilgaz, Ç., Kumlutas, Y., Lymberakis, P., Moravec, J., Sindaco, R., Rastegar-Pouyani, N., Afroosheh, M., Giokas, S., Fraguedakis-Tsolis, S. & Chondropoulos, B. (2012) Neogene climatic oscillations shape the biogeography and evolutionary history of the Eurasian blindsnake. *Molecular Phylogenetics and Evolution* 62, 856–873, 5 figs.

Kortlang, S. & Green, D. (2001) *Keeping carpet pythons.* Australian Reptile Keeper Publications, Bendigo, 41 pp., illust., 8 col. pls.

Koslowsky, J. (1895a) Batracios y reptiles de Rioja y Catamarca (República Argentina) recogidos durante los meses de Febrero á Mayo de 1895 (expedición del Director del Museo). *Revista del Museo de La Plata* (1894–1895) 6(2), 357–370, 4 pls.

Koslowsky, J. (1895b) Reptiles y batracios de la Sierra de la Ventana (Provincia de Buenos Aires). *Revista del Museo de La Plata* (1895–1896) 7(1), 149–156, 1 pl.

Koslowsky, J. (1896) Sobre algunos reptiles de Patagonia y otras regiones Argentinas. *Revista del Museo de La Plata* (1895–1896) 7(2), 447–457, 4 pls.

Koslowsky, J. (1898a) Ofidios de Mato-Grosso (Brasil). *Revista del Museo de La Plata* 8(1), 25–32, 1 pl.

Koslowsky, J. (1898b) Enumeración sistemática y distribucion geográfica de los reptiles Argentinos. *Revista del Museo de La Plata* 8, 161–200.

Kotenko, T.I., Vakarenko, V.I. & Sirenko, V.A. (1986) New data on *Coluber jugularis caspius* in the Ukraine. *Vestnik Zoologii*, Kiev 1986(6), 37–43. (in Russian)

Kou, Z.-T. (1985) *Amphiesma venningi*, a record new to China. *Acta Herpetologica Sinica*, Chengdu (Series 2) 4(2), 160–161, 1 fig.

Kou, Z.-T. (1995) An investigation of the herpetological fauna from the northeast of Yunnan, China. *Abstracts of the Second Asian Herpetological Meeting*, pp. 34–35.

Kou, Z.-T. & Wu, J.-Y. (1993) A new species of *Oligodon* from Yunnan. *Acta Zootaxonomica Sinica*, Beijing 18(3), 379–381, 4 figs.

Koukoulis, A. (1967) *Poisonous snakes of Vietnam.* Ross Allen's Reptile Institute, Silver Springs, 2 pp.

Král, B. (1969) Notes on the herpetofauna of certain provinces of Afghanistan. *Zoologiché Listy* 18(1), 55–66.

Kramer, E. (1958) Eine neue Rasse der Aspisviper aus den südwestlichen Frankreich *Vipera aspis zinnikeri* n. subsp. *Vierteljahrsschrift der Naturforschenden Gesellschaft in Zurich* 103, 321–326, 2 figs.

Kramer, E. (1961a) Über zwei afrikanische Zwergpuffottern, *Bitis hindii* (Boulenger, 1910) und *Bitis superciliaris* (Peters, 1854). *Vierteljahrsschrift der Naturforschenden Gesellschaft in Zurich* 106, 409–423.

Kramer, E. (1961b) Variation, Sexualdimorphismus, Wachstum und Taxonomie von *Vipera ursinii* (Bonaparte, 1835) und *Vipera kaznakovi* Nikolsky, 1909. *Revue Suisse de Zoologie*, Genève 68(4), 627–725, 8 figs.

Kramer, E. (1971) Revalidierte und neue Rassen der euopäischen Schlangenfauna. *Lavori della Società Italiana di Biogeographia*, Forli 1, 667–678, 20 figs.

Kramer, E. (1977) Zur Schlangenfauna Nepals. *Revue Suisse de Zoologie*, Genève 84(3), 721–761, 5 figs.

Kramer, E. (1978) Typenkatalog der Schlangen im Naturhistorischen Museum Basel (BM), Stand 1977. *Revue Suisse de Zoologie*, Genève 85(3), 657–665.

Kramer, E. & Schnurrenberger, H. (1959) Zur systematik Libyscher Schlangen. *Mitteilungen der Naturforschenden Gesellschaft in Bern* 17, 1–17, 1 pl.

Kramer, E. & Schnurrenberger, H. (1963) Systematik, Verbreitung und Okologie der Libyschen Schlangen. *Revue Suisse de Zoologie*, Genève 70(27), 453–568, 13 figs., 4 pls.

Krapp, F. & Böhme, W. (1978) *Natrix tessellata* in der Voreifel (Reptilia, Serpentes, Colubridae, Natricinae). *Salamandra*, Frankfurt-am-Main 14(3), 157–159, 1 fig.

Krassawzeff, B.A. (1932) Die Torfmoos-Viper, eine neue Varietät (*Vipera berus sphagnosa*, var. nova). *Zoologischer Anzeiger*, Leipzig 101(3–4), 80–81.

Kratzer, H. (1973) Über die Verstreifennatter (*Elaphe quatuorlineatus*) von Antimilos und die Problematik der Untgerarten dieser Species im Raum der südlichen Ägais. *Salamandra*, Frankfurt-am-Main 9(3–4), 103–117, 1 fig., 8 pls.

Kraus, F. (2005) New species of blindsnake from Rossel Island, Papua New Guinea. *Journal of Herpetology* 39(4), 591–595, 2 figs.

Kraus, F. (2009) New species of *Toxicocalamus* (Squamata: Elapidae) from Papua New Guinea. *Herpetologica* 65(4), 460–467, 4 figs.

Kraus, F. (2010) More range extensions for Papuan reptiles and amphibians. *Herpetological Review* 41(2), 246–248.

Kraus, F. & Allison, A. (2004a) A new species of *Tropidonophis* (Serpentes: Colubridae: Natricinae) from the D'Entrecasteaux Islands, Papua New Guinea. *Proceedings of the Biological Society of Washington* 117(3), 303–310, 3 figs.

Kraus, F. & Allison, A. (2004b) New records for reptiles and amphibians from Milne Bay Province, Papua New Guinea. *Herpetological Review* 35(4), 413–418.

Kraus, F. & Allison, A. (2006) Range extensions for reptiles and amphibians along the northern versant of Papua New Guinea. *Herpetological Review* 37(3), 364–368.

Kraus, F. & Shea, G.M. (2005) Additional reptile and amphibians range extensions for Milne Bay Province, Papua New Guinea. *Herpetological Review* 36(4), 471–473.

Kraus, F., Mink, D.G. & Brown, W.M. (1996) Crotaline intergeneric relationships based on mitochondrial DNA sequence data. *Copeia* 1996(4), 763–773, 5 figs.

Kraus, R. & Werner, F. (1931) Giftschlangen und die Serumbehandlung der Schlangenbisse. *Verlag von Gustav Fischer,* Jena, 220 pp., illust.

Krauss, P. (1994) Woma husbandry and captive breeding. *Monitor,* Ardeer 6(1), 9–12, 1 fig., 6 col. pls.

Krecsák, L. (2007a) An account of the generic and specific names, and type specimens of viperid taxa described by Albert Franz Theodor Reuss (Squamata, Viperidae). *Zootaxa* (1514), 1–36, 27 figs.

Krecsák, L. (2007b) Re-description of the holotype of *Vipera eriwanensis (*Reuss, 1933) (Serpentes: Viperidae). *Revue Suisse de Zoologie* 114(3), 655–662.

Krecsák, L. & Wahlgren, R. (2008) A survey of the Linnaean type material of *Coluber berus, Coluber chersea* and *Coluber prester* (Serpentes, Viperidae). *Journal of Natural History*, London 42(33–36), 2343–2377.

Krecsák, L. & Zamfirescu, S. (2001) Ecological situation and morphological characteristics of *Vipera ursinii moldavica* in the "Valea Lui David" Natural Reserve. *Russian Journal of Herpetology* 8(1), 69–73, 4 figs.

Krecsák, L., Zamfirescu, S. & Korsós, Z. (2003) An updated overview of the distribution of the Moldavian Steppe viper (*Vipera ursinii moldavica* Nilson, Andrén et Joger, 1993). *Russian Journal of Herpetology* 10(3), 199–206, 10 figs.

Krefft, G. (1863) Description of a new species of *Hoplocephalus* with keeled scales (*H. carinatus*). *Proceedings of the Zoological Society of London* 31(1), 86, 3 figs.

Krefft, G. (1864a) Description of *Aspidiotes melanocephalus*, a new snake from Port Denison, N.E. Australia. *Proceedings of the Zoological Society of London* 32(1), 20–22, 2 figs.

Krefft, G. (1864b) Description of *Aspidiotes melanocephalus*, a new snake from Port Denison, N.E. Australia. *Annals and Magazine of Natural History*, London (Series 3) 14(81), 225–226, 2 figs.

Krefft, G. (1864c) Descriptions of three new species of Australian snakes. *Proceedings of the Zoological Society of London* 32(1), 180–182, 7 figs..

Krefft, G. (1865a) Description of three new species of Australian snakes. *Annals and Magazine of Natural History*, London (Series 3) 15(85), 66–68, 7 figs.

Krefft, G. (1865b) On the reptiles found near Sydney with remarks upon their habits and geographical range. *Transactions of the Philosophical Society of New South Wales* (1866) 1862–1865, 34–60. [preprint in 1865]

Krefft, G. (1866a) Descriptions of three species of snakes of the genus *Hoplocephalus. Proceedings of the Zoological Society of London* 34(1), 370–371.

Krefft, G. (1866b) On snakes observed in the neighbourhood of Sydney. *In: Two papers on the Vertebrata of the Lower Murray and Darling; and on the snakes of Sydney*. Reading and Wellbank, Sydney (1865), pp. 34–60. [reprint of Krefft, 1865b]

Krefft, G. (1869a) Descriptions of new Australian snakes. *Proceedings of the Zoological Society of London* 37(1), 318–322, 7 figs.

Krefft, G. (1869b) *The snakes of Australia; an illustrated and descriptive catalogue of all the known species.* Thomas Richards, Sydney, xxv + 100 pp., 12 col. pls. [reprinted in 1984 by Lookout Publications, Brisbane]

Kreiner, G. (2007) *The snakes of Europe; all species from west of the Caucasus Mountains.* Edition Chimaira, Frankfurt-am-Main, 317 pp., 345 col. photos.

Kretz, J. (1971) Ueber *Vipera kaznakowi* Nikolsky 1909 aus Nordostanotolien (Reptilia, Viperidae). *Jahrbuch, Naturhistorisches Museum der Stadt Bern* 4, 125–134, 2 pls.

Kretzschmar, S. (1996) Nuevos datos sobre *Leptotyphlops unguirostris* (Boulenger, 1902) (Serpentes, Leptotyphlopidae). *Acta Zoologica Lilloana* 43(2), 275–279.

Kretzschmar, S. (1998) Geographic distribution: Serpentes: *Liotyphlops ternetzii. Herpetological Review* 29(2), 114.

Kretzschmar, S. (2006) Revision histórica y redescripción de *Leptotyphlops albipunctus* (Serpentes, Leptotyphlopidae). *Cuadernos de Herpetología* 19(2), 43–56, 3 figs.

Krishna, D. & Dave, K.C. (1956) On the distribution of reptiles in the desert of Rajasthan. *Proceedings of the 43rd Session, Indian Science Congress Association* 4, 34–35.

Krofel, M., Cafuta, V., Planinc, G., Sopotnik, M., Salamun, A., Tome, S., Vamberger, M. & Zagar, A. (2009) Distribution of reptiles in Slovenia: a review of data collected until 2009. *Natura Sloveniae* 11(2), 61–99. (In Slovac)

Kroon, C. (1973) A new colubrid snake (*Boiga*) from southeastern Asia. *Copeia* 1973(3), 580–586, 4 figs.

Kropach, C. (1972) *Pelamis platurus* as a potential colonizer of the Caribbean Sea. *Bulletin of the Biological Society of Washington* 2, 267–269.

Kropach, C. (1975) The yellow-bellied sea snake, *Pelamis*, in the eastern Pacific. *In:* Dunson, W.A. (Ed.), *The biology of sea snakes.* University Park Press, Baltimore, pp. 185–213, 3 figs.

Krulikowski, L. (2004) *Snakes of New England: photographic and natural history study.* LuvLife Publishing, Old Lyme, xi + 308 pp., col. illust.

Krynicky, J. (1837) Observationes quaedam de reptilibus indigenis. *Bulletin de la Société Impériale des Naturalistes de Moscou* 10(3), 46–69, 2 pls.

Krysko, K.L. & Judd, W.S. (2006) Morphological systematics of kingsnakes, *Lampropeltis getula* complex (Serpentes, Colubridae), in the eastern United States. *Zootaxa* (1193), 1–39, 12 figs.

Krysko, K.L., Enge, K.M. & Moler, P.E. (2011) *Atlas of amphibians and reptiles in Florida.* iv + 527 pp., 200 col. photos.

Kubatko, L.S., Gibbs, H.L. & Bloomquist, E.W. (2011) Inferring species-level phylogenies and taxonomic distinctiveness using multilocus data in *Sistrurus* rattlesnakes. *Systematic Zoology* 60(1), 1–17, 5 figs.

Kubykin, R.A. (1994) Present distribution and population density of four-lined snake *Elaphe quaturolineata* (Reptilia, Colubridae) in Kazakhstan. *Selevinia* 2(1), 61–64, 1 fig. (in Russian)

Kubykin, R.A. & Brushko, Z.K. (1998) Contemporary spreading and information on *Agkistrodon halys caraganus* Eichwald, 1831 (Reptilia, Crotalidae) numbers in Kazakhstan. *Proceedings of the Third Asian Herpetological Meetings, Al-Farabi Kazakh State National University,* Almati (6), pp. 9–13, 1 fig.

Kuch, U. (1990) *Hydrodynastes bicinctus* (Serpentes, Colubridae, Xenodontinae)—Eine Kurzportrait. *Sauria*, Berlin 12(4), 11–13, illust.

Kuch, U. (1991) Abnormal colouration in the banded krait, *Bungarus fasciatus* (Schneider, 1801). *The Snake* 23(1), 25–28, 2 figs.

Kuch, U. (1997) Comments on the proposed conservation of the specific and subspecific names of *Trigonocephalus pulcher* Peters 1863 [recte 1862] and *Bothrops albocarinatus* Shreve, 1934 (Reptilia, Serpentes) by the designation of a neotype for *T. pulcher*. *Bulletin of Zoological Nomenclature* 54(4), 245–249.

Kuch, U. & Freire, A. (1993) Life history notes. Serpentes. *Dendrophidion brunneum* (NCN). Reproduction and size. *Herpetological Review* 24(3), 106.

Kuch, U. & Freire, A. (1995a) Bemerkungen zur geographischen Verbreitung und Variabilität von Schlegels Palmen-Lanzenotter, *Bothriechis schlegelii* (Berthold, 1846), in Ecuador (Squamata, Serpentes, Viperidae). *Herpetozoa* 8(1–2), 49–58, 1 fig.

Kuch, U. & Freire, A. (1995b) Notes on morphology, reproduction and medical importance of the poorly known small-eyed lancehead, *Bothrops microphthalmus* Cope, 1876, in Ecuador. *Herpetozoa* 8(1–2), 81–83, 1 fig.

Kuch, U. & Freire, A. (2001) Geographic distribution: Serpentes: *Bothriechis schlegelii*. *Herpetological Review* 32(1), 58.

Kuch, U. & Götzke, A. (2000) Eine Freilandbeobachtung des Kinabalu-Kraits, *Bungarus flaviceps baluensis* Loveridge, 1938 (Serpentes, Elapidae). *Sauria*, Berlin 22(2), 19–22, 3 figs.

Kuch, U. & Mebs, D. (2007) The identity of the javan krait, *Bungarus javanicus* Kopstein, 1932 (Squamata, Elapidae): evidence from mitochondrial and nuclear DNA sequence analyses and morphology. *Zootaxa* (1426), 1–26, 17 figs.

Kuch, U., Keogh, J.S., Weigel, J., Smith, L.A. & Mebs, D. (2005a) Phylogeography of Australia's king brown snake (*Pseudechis australis*) reveals Pliocene divergence and Pleistocene dispersal of a top predator. *Naturwissenschaften* 92, 121–127, 2 figs.

Kuch, U., Kizirian, D., Truong, N.Q., Lawson, R., Donnelly, M.A. & Mebs, D. (2005b) A new species of krait (Squamata, Elapidae) from the Red River system of northern Vietnam. *Copeia* 2005(4), 818–833, 5 figs.

Kuch, U., Gumprecht, A. & Melaun, C. (2007) A new species of temple pitviper (*Tropidolaemus* Wagler, 1830) from Sulawesi, Indonesia (Squamata, Viperidae, Crotalinae). *Zootaxa* (1446), 1–26, 12 figs.

Kucharzewski, C. & Tillack, F. (2007) Redescription of the type specimen of *Naja* (*Hamadryas*?) *fasciata* Peters, 1861, a junior synonym of *Hemibungarus calligaster calligaster* (Wiegmann, 1834) (Reptilia, Squamata, Elapidae). *Mitteilungen aus dem Museum für Naturkunde in Berlin, Zoologische Reihe* 83, 52–55, 1 fig.

Kucharzewski, C. & Tillack, F. (2008) The identity of *Cochliophagus isolepis* Müller, 1924 (Serpentes: Colubridae). *Salamandra* 44(1), 43–49, 3 col. figs.

Kucharzewski, C., Tillack, F., & Fuhrmann, K. (2009) Rediscovery and redescription of the type specimens of *Ophites ruhstrati* Fischer, 1886 (Squamata: Serpentes: Colubridae). *Herpetozoa* 22(3–4), 181–187, 3 figs.

Kudrjavstev, S.V. & Mamet, S.V. (1989a) Terrestrial snakes of the Soviet Union. *The Snake* 21(1), 29–35.

Kudrjavstev, S.V. & Mamet, S.V. (1989b) The finding of *Dinodon rufozonatum* (Colubridae, Squamata) in the Primorski Territory. *Zoologischeskii Zhurnal*, 68(1), 153–154.

Kudrjavstev, S.V. & Mamet, S.V. (1998) *The venomous vipers.* Moscow Zoopark, Moscow, 44 pp., col. illust. (in Russian)

Kuhl, H. (1820) *Beiträge zur Zoologie und vergleichenden Anatomie, Erste Abtheilung. II. Ophidier.* Hermannschen Buchhandlung, Frankfurt-am-Main, 78–100.

Kuhl, H. (1824) Sur les reptiles de Java. *Bulletin des Sciences Naturelles et de Géologie,* Paris 2, 79–83.

Kuhl, H. & Hasselt, J.C. van. (1822a) Uittreksel uit de brieven van Heeren Kuhl en van Hasselt, aan de Heeren C.J. Temminck, Th. van Swinderen en W. de Haan. *Algemeene Konst- en Letter-Bode*, Haarlem 1(12), 180–184.

Kuhl, H. & Hasselt, J.C. van. (1822b) Aus einem schreiben von Dr. Kuhl und Dr. van Hasselt aus Java. *Isis von Oken* 10(4), 472–476.

Kuhn, O. (1939a) Squamata, Lacertilia et Ophidia. *In:* Quenstedt, W. (Ed.), *Fossilium catalogus. I, Animalia.* W. Junk, Gravenhage 86, 1–33.

Kuhn, O. (1939b) Die Schlangen (Boidae) aus dem Mitteleozän des Geiseltales. *Nova Acta Leopoldina* 7, 119–133, 3 figs.

Kuhn, O. (1963) Serpentes (Supplementum I). *In:* Westphal, F. (Ed.), *Fossilium catalogus. I, Animalia.* W. Junk, Gravenhage 103, 1–45.

Kulagin, N.M. (1888) Lists and descriptive matter of material to be found in the zoological collections of the Imperial Moscow University: reptiles and batrachians. *Transactions of the Laboratory of the Zoological Museum,* Impérial Moscow University 4(2), 1–39. (in Russian)

Kumar, A. (2010) A tragic loss: fire at Instituto Butantan, Brazil. *Toxicon* 56, 1528–1529.

Kumar, A.B. & Captain, A. (2011) Recent records of the endemic Kerain mud snake, *Enhydris dussumierii* (Duméril, Bibron & Duméril, 1854) from India. *Current Science,* Bangalore 100(6), 928–932, 7 figs.

Kumar, A.B., Sanders, K.L., George, S. & Murphy, J.C. (2012) The status of *Eurostus dussumierii* and *Hypsirhina chinensis* (Reptilia, Squamata, Serpentes): with comments on the origin of salt tolerance in homalopsid snakes. *Systematics and Biodiversity* 10(4), 479–489, 4 figs.

Kumar, S., Gaur, S. & Khandal, D. (2010). Reptilia. *In:* Alfred, J.R.B. (Ed.), *Fauna of Ranthambhore National Park, Rajasthan. Conservation Area Series* 43. Zoological Survey of India, Kolkata, pp. 163–179.

Kundert, F. (1974) *Fascination: snakes and lizards.* Verlag F. Kundert, Spreitenbach, 201 pp., col. illust.

Kundert, F. (1984) *Das neue Schlangenbuch in Farbe. Second edition.* Albert Müller Verlag, Rüschlikon-Zürich, 39 pp., 196 col. pls.

Kuntz, R.E. (1963) Snakes of Taiwan. *Quarterly Journal of the Taiwan Museum*, Taipei 16(1–2), 1–79, col. illust. [reprinted in 1963 by U.S. Naval Medical Research Unit No. 2, Taipei]

Kuntz, R.E. & Ming, D.Z. (1970) Vertebrates of Taiwan taken for parasitological and biomedical studies by U.S. Naval Medical Research Unit No. 2, Taipei, Taiwan, Republic of China. *Quarterly Journal of the Taiwan Museum*, Taipei 23(1–2), 1–37.

Kunz, K. (2004) *Die Kornnatter,* Pantherophis guttatus. Natur und Tier-Verlag, Münster, 62 pp., col. illust.

Kunz, K. (2006) *Die Hausschlange,* Lamprophis lineatus. Natur und Tier-Verlag, Münster, 61 pp., col. illust.

Kurfess, J.F. (1967) Mating, gestation and growth rate in *Lichanura r. roseofusca*. *Copeia* 1967(2), 477–479, 1 fig.

Kurilenko, V.E. & Verves, Y.G. (1998) *Amphibians and reptiles of the fauna of Ukraine: handbook and identification tables.* Bidavnitstbo Geneza, Kiev, 206 pp., 62 figs., 32 col. pls. (in Russian)

Kutrup, B. (1999) The morphology of *Vipera ammodytes transcaucasiana* (Reptilia, Viperidae); specimens collected from Murgul (Artvin, Turkey). *Turkish Journal of Zoology* 23, 433–438.

Kutrup, B. (2003) The identification of new specimens of *Vipera* from Trabzon, Turkey with affinities to *Vipera barani* and *V. pontica*. *Herpetological Review* 34(1), 28–31, 2 figs.

Kwet, A. (2005) *Reptilien und Amphibien Europas.* Franckh-Kosmos Verlags, Stuttgart, 252 pp., 317 col. photos.

Lacarrière, J. & Ineich, I. (1992) *Le serpent: images et rituel.* Editions du Perron, Alleur–Liège, 205 pp., col. illust.

Lacépède, B.G.E. (1789) *Histoire naturelle des serpens. Tome second.* Hotel de Thon, Paris 2(1), 1–19, 2(2), 1–144, 2(3), 1–527 pp., 22 pls.

Lacépède, B.G.E. (1801) Sur un nouveau genre de serpent. *Bulletin de la Société Philomathique de Paris* 2(46), 169.

Lacépède, B.G.E. (1803) Observations sur un genre de serpent qui n'a pas encore été décrit. *Annales du Musée d'Histoire Naturelle de Paris* 2, 280–284, pl. 50.

Lacépède, B.G.E. (1804) Mémoire. Sur plusieurs animaux de la Nouvelle-Hollande dont la description n'a pas encore été publiée. *Annales du Musée d'Histoire Naturelle de Paris* 4, 184–211, pls. 55–58.

Lacépède, B.G.E. (1825) *Histoire naturelle des quadrupèdes ovipares, serpens, poissons et cétcés, pour faire suite aux oeuvres de Buffon: nouvelle edition.* A. Eymerey, Paris 1(4), 620 pp.

Lacépède, B.G.E. (1832) *Oeuvres du comte de Lacépède, comprenant l'histoire naturelle des quadrupèdes ovipares, des serpents, des poissons et des cétacés; accompagnérs du portrait de l'auteur et d'environ 400 figures, exécutés sur acier pour cette édition par les meilleurs artistes. Tome IV. Serpentes.* F. D. Pillot, Paris, 428 pp.

Lacerda, J.B. de. (1884) *Leçons sur le venin des serpents du Brésil et sur la méthode de traitement des morsures venimeuses par le permanganate de potasse.* Lombaerts, Rio de Janeiro, xv + 194 pp., 3 pls.

Lada, G.A. & Nedosekin, V.Y. (1997) The first record of tesselated snake, *Natrix tesselata* Laurenti, 1768 in the central Chernozem Territory of Russia and some other results of the herpetological research in the Upper Don. *Russian Journal of Herpetology* 4(2), 192–194, 2 figs.

Ladiges, W. (1939) Herpetologische Beobachtungen auf Sumatra. *Zoologischer Anzeiger,* Leipzig 128(9–10), 235–249, 10 figs.

LaDuc, T.J. (1995) The nomenclatural status and gender of *Adelphicos*. *Journal of Herpetology* 29(1), 141–142.

LaDuc, T.J. & Johnson, J.D. (2003) A taxonomic revision of *Trimorphodon biscutatus vilkinsonii* (Serpentes, Colubridae). *Herpetologica* 59(3), 364–374, 6 figs.

LaDuke, T.C., Krause, D.W., Scanlon, J.D. & Kley, N.J. (2010) A late Cretaceous (Maastrichtian) snake assemblage from the Maevarano Formation, Mahajanga Basin, Madagascar. *Journal of Vertebrate Paleontology* 30(1), 109–138, 11 figs.

Laerm, J., Freeman, B.J., Vitt, L.J., Meyers, J.M. & Logan, L. (1980) Vertebrates of the Okefenokee Swamp. *Brimleyana* (4), 47–73.

Laidlaw, F.F. (1902) List of a collection of snakes, crocodiles, and chelonians from the Malay Peninsula, made by members of the "Skeat Expedition," 1899–1900. *Proceedings of the Zoological Society of London* (1901) 71(2), 575–583, pl. 35. [April, 1902]

Laidlaw, F.F. (1902) Amphibia and Reptilia. *In:* Gardiner, J.S. (Ed.), *The fauna and geography of the Maldive and Laccadive Archipelagoes. Being the account of the work carried on and of the collections made by an expedition during the years of 1899 and 1900. Volume I. Part II.* Cambridge University Press, Cambridge, pp. 119–122.

Laita, M. (2013) *Serpentine.* Abrams, New York, 400 pp., 182 col. photos.

Lamar, W.W. (1997a) Checklist and common names of the reptiles of the Peruvian lower Amazon. *Herpetological Natural History* 5(1), 73–76.

Lamar, W.W. (1997b) *The world's most spectacular reptiles & amphibians.* World Publications, Tampa, 208 pp., col. photos.

Lamar, W.W. (2003) A new species of slender coralsnake from Colombia, and its clinal and ontogenetic variation (Serpentes, Elapidae, *Leptomicrurus*). *Revista de Biologia Tropical* 51(3–4), 805–810, 2 col. figs.

Lamar, W.W. & Sasa, M. (2003) A new species of hognose pitviper, genus *Porthidium*, from the southwestern Pacific of Costa Rica (Serpentes, Viperidae). *Revista de Biologia Tropical* 51(3–4), 797–804, 3 col. figs.

La Marca, E. (2000) Geographical distribution. *Spilotes pullatus*. *Herpetological Review* 31(1), 57–58.

La Marca, E. & Soriano, P.J. (1995) Geographic distribution: Serpentes: *Epicrates cenchria cenchria* (Rainbow boa; tornasol). *Herpetological Review* 26(2), 109.

La Marca, E. & Soriano, P.J. (2004) *Reptiles de los Andes de Venezuela.* BIOGEOS, Mérida, 173 pp., 77 col. figs.

La Marca, E., Soriano, P.J. & Casado-B., R. (1995) Geographic distribution: Serpentes: *Drymarchon corais melanurus* (Palancacoate). *Herpetological Review* 26(2), 109.

La Marca, E., Esqueda, L.F. & Manzanilla, J. (2004) Geographic distribution: Serpentes: *Ninia atrata*. *Herpetological Review* 35(2), 192.

Lamare-Picquot, P. (1835) Rapport fait, le 19 mars 1832, par M. Duméril, au nom d'une commission composée de MM. Latreille, F. Cuvier et Duméril, sur un Mémoire de M. Lamarepicquot, relatif aux Serpens des Indes et a leur venin. *Annales des Sciences Naturelles (Zoologie)*, Paris (Série 2) 3, 35–38.

Lamba, B.S. (1987) *Fauna of Corbett National Park (mammals, birds & reptiles).* Zoological Survey of India, Madras, 86 pp., 58 pls.

Lambe, L.M. (1908) The Vertebrata of the Oligocene of the Cypress Hills, Saskatchewan. *Contributions to Canadian Palaeontology*, Ottawa 3(4), 1–65, 12 figs., 8 pls.

Lambert, M.R.K. (1984) Amphibians and reptiles. *In:* Cloudsley-Thompson, J. (Ed.), *Sahara Desert*. Pergamon Press, Oxford, pp. 205–227.

Lambert, M.R.K. (1985) A few of the Herpetofauna in the commonwealth (Oriental and Australasian zones). *Bulletin of the British Herpetological Society*, London (12), 19–27, illust.

Lambert, M.R.K. (1986) Herpetology in Jamaica with general reference to conservation. *British Herpetological Society Bulletin*, London (18), 3–8.

Lambiris, A.J.L. (1993) Annual report on the herpetological collection of A.J.L. Lambiris. *Annual Report of the Lambiris Collection* 9(1), 1–82.

Lambiris, A.J.L. (1997) Description of the hemipenis of *Montaspis gilvomaculata* Bourquin, 1991 (Serpentes, Colubridae). *Annals of the Natal Museum* 38, 1–3, 2 figs.

Lampe, E. (1911) Erster Nachtrag zumcatalog der Reptilien- und Amphibien-Sammlung des Naturhistorischen Museums der Stadt Wiesbaden. *Jahrbücher des Nassauischen Vereins Naturkunde,* Wiesbaden 64, 137–236.

Lampe, E. & Lindholm, W.A. (1902) Catalog der Reptilien- und Amphibian-Sammlung (Schlangen; Frosch-, Schwanz- und Schleichenlurche) des Naturhistorischen Museums zu Wiesbaden, mit Bemerkungen. *Jahrbuch des Nassauischen Vereins für Naturkunde*, Wiesbaden 55, 1–66.

Lamson, G.H. (1935) The reptiles of Connecticut. *Connecticut State Geological and Natural History Survey Bulletin, Hartford* (54), 1–35, 12 pls.

Lancini, A.R. (1962a) Una nueva espécie de serpiente coral (Serpentes, Elapidae) del Peru. *Publicaciones Ocasionales del Museo de Ciencias Naturales*, Caracas (2), 1–3, 1 fig.

Lancini, A.R. (1962b) Un cambio de nombre para una serpiente coral (Elapidae, *Micrurus*) del Peru. *Publicaciones Ocasionales del Museo de Ciencias Naturales*, Caracas (3), 1.

Lancini, A.R. (1964) Contribucion al conocimiento de la distribución en Venezuela de *Helicops scalaris* y descripción de *Helicops hogei* (Serpentes, Colubridae), nueva especie de serpiente semiacuatica para la ciencia. *Publicaciones Ocasionales del Museo de Ciencias Naturales*, Caracas (7), 1–3, 3 figs.

Lancini, A.R. (1966) Notas sôbre as diferenças existentes entre as sub-espécies Venezuelanas de *Coluber* (*Masticophis*) *mentovarius* (Dum. Bibr. et Dum. 1854). *Memórias do Instituto Butantan*, São Paulo (1965) 32, 9–14, 2 figs.

Lancini, A.R. (1967) *Crotalus vegrandis* Klauber: redescripción y distribución. *Memórias do Instituto Butantan*, São Paulo (1966) 33(3), 725–734, 4 figs.

Lancini, A.R. (1969) *Atractus mariselae*: una nueva especie de serpiente minadora de los Andes de Venezuela (Serpentes, Colubridae). *Publicaciones Ocasionales del Museo de Ciencias Naturales*, Caracas (15), 1–6, 3 figs.

Lancini, A.R. (1983) Serpientes de Venezuela. Distribución geografica y altitudinal de generos de serpientes en Venezuela. *Memórias do Instituto Butantan*, São Paulo (1982) 46, 95–103.

Lancini, A.R. (1986) *Serpientes de Venezuela*. Second edition. Talleres Graficas Armitano, Caracas, 262 pp., 119 figs. [first edition, 1979, 262 pp., 119 figs.]

Lancini, A.R. & Kornacker, P.M. (1989) *Die Schlangen von Venezuela*. Third edition. Armitano Editores, Caracas, 381 pp., 207 figs. & col. pls. [first edition, 1979, and second edition, 1986, by Lancini, Talleres Graficas Armitano, Caracas, 262 pp., 119 figs.]

Lando, R.V. & Williams, E.E. (1969) Notes on the herpetology of the U.S. naval base at Guantánamo Bay, Cuba. *Studies on the Fauna of Curaçao and other Caribbean Islands*, The Hague 31(116), 159–201.

Landy, M.J., Langabartel, D.A., Moll, E.O. & Smith, H.M. (1966) A collection of snakes from Volcán Tacaná, Chiapas, Mexico. *Journal of the Ohio Herpetological Society* 5(3), 93–101, 1 fig.

Lang, M. (1990) Annotated catalogue of the type-specimens from the herpetological collection of the Royal Belgian Institute of Natural Sciences, Brussels. *Documents de Travail, Institut Royal du Science Naturelle de Belgique*, Bruxelles 59, 1–36.

Lang, M. (1993) *Coluber algirus* (Jan, 1863)—Algerische Zornnatter. *In:* Böhme, W. (Ed.), *Handbuch der Reptilien und Amphibien Europas. Band 3/I. Schlangen (Serpentes) I. (Typhlopidae, Boidae, Colubridae 1, Colubrinae).* AULA-Verlag, Wiesbaden, pp. 75–82, figs. 13–14.

Lang, R. de. (2011a) *Snakes of the Lesser Sunda Islands (Nusa Tenggara), Indonesia. A field guide to the terrestrial and semi-aquatic snakes with identification key*. Edition Chimaira, Frankfurt-am-Main, 349 pp., 220 col. photos.

Lang, R. de (2011b) The snakes of the Lesser Sunda Islands (Nusa Tenggara), Indonesia. *Asian Herpetological Research* (Series 2) 2(1), 46–54, 8 figs.

Lang, R. de & Vogel, G. (2005) *The snakes of Sulawesi. A field guide to the land snakes of Sulawesi*. Edition Chimaira, Frankfurt-am-Main, 312 pp., 189 figs.

Lange, D. & de Rooij, N. (1912) Amphibien und Reptilien. *Alfred Maass Durch Zentral-Sumatra* 2, 1–8.

Langerwerf, B. (1975) Reptilien en anfipbeén in Finland. *Lacerta* 34(1), 4–7.

Langhammer, J.K. (1983) A new subspecies of boa constrictor, *Boa constrictor melanogaster*, from Ecuador (Serpentes, Boidae). *Tropical Fish Hobbyist* 32(4), 68–79, 7 figs.

Lanka, V. (1978) Variabilität und Biologie der Würfelnatter (*Natrix tessellata*). *Acta Universitatis Carolinae, Biologica*, Prague (1975–1976), 167–207, 9 figs.

Lanka, V. & Vit, Z. (1985) *Amphibians and reptiles*. Hamlyn, Feltham, 224 pp., illust.

Lanza, B. (1954) Su un esemaplare di *Pelamis platurus* (L.) catturato in un fiume della piccola Andaman (Serpentes; Hydrophiidae). *Monitore Zoologico Italiano*, Firenze 62(2), 67–70, 1 fig.

Lanza, B. (1958) Notizie su due popolazioni insulari di *Lacerta muralis* e sulla *Vipera ursinii* in Italia. *Annali dell Museo Civico di Storia Naturale di Genova* 70, 305–322, 2 figs.

Lanza, B. (1963) Il genere *Coluber* in Somalia e descrizione di una nuova specie (Reptilia, Serpentes). *Atti della Società Italiana di Scienze Naturali e del Museo Civico di Storia Naturali di Milano* 102(4), 379–396, 5 figs.

Lanza, B. (1964) Il genere *Sphalerosophis* e descrizione di una nuova specie (Reptilia, Serpentes). *Monitore Zoologico Italiano*, Firenze 72(1–2), 47–64, 6 figs.

Lanza, B. (1966) Il genere *Brachyophis* e descrizione di una nuova forma (Reptilia, Serpentes, Colubridae). *Monitore Zoologico Italiano*, Firenze 74(1–3), 30–48, 2 figs., pls. 15–20.

Lanza, B. (1968) Rettili. *In:* Tortonese, E. & Lanza, B. (Eds.), *Piccola fauna Italiana. Pesci, anfibi e rettili*. Aldo Martello, Milano, pp. 133–174.

Lanza, B. (1972) The reptiles of the Awash National Park (Shoa, Ethiopia). *Monitore Zoologico Italiano*, Firenze (New Series, Supplement) 4(7), 159–184, 18 figs.

Lanza, B. (1973) Gli anfibi e i rettili delle isole circumsiciliane. *Lavori della Società Italiana di Biogeographia*, Forlì (1972) (Series 2) 3, 755–804, 16 figs.

Lanza, B. (1978) On some new or interesting East African amphibians and reptiles. *Monitore Zoologico Italiano*, Firenze (New Series, Supplement) (14), 229–297, 41 figs.

Lanza, B. (1979) *Elapsoidea chelazzii*, a new elapid snake from Somalia. *Monitore Zoologico Italiano*, Firenze (New Series, Supplement) 12(12), 237–245, 5 figs.

Lanza, B. (1983a) A list of the Somali amphibians and reptiles. *Monitore Zoologico Italiano*, Firenze (New Series, Supplement) 18(8), 193–247.

Lanza, B. (1983b) *Guide per il riconoscimento della Specie animali delle acque interne Italiane. 27. Anfibi, rettili (Amphibia, Reptilia).* Consiglio Nazionale delle Ricerche, 196 pp., 72 figs.

Lanza, B. (1983c) Ipotesi sulle origini del popolamento erpetologico della Sardegna. *Lavori della Società Italiana di Biogeographia*, Forlì (1980) (Series 2) 8, 723–744.

Lanza, B. (1987) Erpetologia. Tutti i serpenti Italiani. *Silva*, Milano 2, 48–69, 128, 29 figs., 34 illust.

Lanza, B. (1990a) Rediscovery of the Malagasy colubrid snake *Alluaudina mocquardi* Angel 1939. *Tropical Zoology*, Firenze 3(2), 219–223, 2 figs.

Lanza, B. (1990b) Amphibians and reptiles of the Somali Democratic Republic: check list and biogeography. *Biogeographia* (1988) 14, 407–465, 1 fig.

Lanza, B. (1996) Atlante Provvisorio degli anfibi e dei rettili Italiani. *Annali dell Museo Civico di Storia Naturale di Genova* 91, 95–163, 47 figs.

Lanza, B. (1999) A new species of *Lycodon* from the Philippines, with a key to the genus (Reptilia Serpentes Colubridae). *Tropical Zoology* 12(1), 89–104, 3 figs.

Lanza, B. & Boscherini, S. (2000) The gender of the genera *Podarcis* Wagler 1830 (Lacertidae), *Pelamis* Daudin 1803 (Hydrophiidae) and *Uropeltis* Cuvier 1829 (Uropeltidae). *Tropical Zoology*, Firenze 13(2), 327–329.

Lanza, B. & Bruzzone, C.L. (1960) Reptilia. *In:* Zavattari (Ed.), Biogeografia della isole Pelagie. Fauna, Vertebrati. *Rendiconti dell'Accademia Nazionale dei XL*, Roma (Series 4) 11(83), pp. 288–328, pls. 29–42.

Lanza, B. & Nistri, A. (2005) Somali Boidae (genus *Eryx* Daudin 1803) and Pythonidae (genus *Python* Daudin 1803) (Reptilia Serpentes). *Tropical Zoology*, Firenze 18(1), 67–136, 30 figs.

Lanza, B. & Broadley, D.G. (2013) A review of the genus *Gonionotophis* in north-eastern Africa (Squamata: Lamprophiidae). *Acta Herpetologica,* Firenze 8(1), 10 pp., 4 figs. (in press).

Lapini, L. (1983) *Anfibi e rettili.* Carlo Lorenzini Editore, Tricesimo, 142 pp., col. illust.

Laporta-Ferreira, I.L., Salomão, M.G. & Sawaya, P. (1986) Biologia de *Sibynomorphus* (Colubridae-Dipsadinae)—reprodução e hábitos alimentares. *Revista Brasileira de Biologia*, Rio de Janeiro 46(4), 793–799, 8 figs.

Lara-Góngora, G. & Flores-Villela, O. (1980) Primer catálogo de una coleción herpetológia depositada en la facultad de Ciencias. *Memorias del II Congreso Nacional de Zoología* 2, 366–374.

Lardie, R.L. (1982) A preliminary checklist of the amphibians and reptiles of northwestern Oklahoma (excluding the Oklahoma Panhandle). *Bulletin of the Oklahoma Herpetological Society* 7, 36–78.

Largen, M.J. (1978) The status of the family Typhlopidae (Reptilia Serpentes) in Ethiopia. *Monitore Zoologico Italiano*, Firenze (New Series) 10(2), 43–68, 11 figs.

Largen, M.J. (1997) An annotated checklist of the amphibians and reptiles of Eritrea, with keys for their identification. *Tropical Zoology,* Firenze 10(1), 63–115, illust.

Largen, M.J. & Largen, S.E. (1982) The snakes of the Addis Ababa region. *Walia* (*Journal of the Ethiopian Natural History Society*), Addis Ababa (8), 10–15, 1 fig.

Largen, M.J. & Rasmussen, J.B. (1993) Catalogue of the snakes of Ethiopia (Reptilia Serpentes), including identification keys. *Tropical Zoology*, Firenze 6(2), 313–434.

Largen, M.J. & Spawls, S. (2010) *The amphibians and reptiles of Ethiopia and Eritrea*. Edition Chimaira, Frankfurt-am-Main, 693 pp., 380 col. photos.

Lartet, E. (1851) *Notice sur la colline de Sansan, suivie d'une recapitulation des diverses espèces d'animaux vertébrés fossiles, trouvés soit à Sansan, soit sans d'autres gisements du terrain tertiare miocene dans le Bassin Sous-Pyèenéen.* J.-A. Portes, Auch, 42 pp.

Lataste, F. (1876a) Catalogue des batraciens et reptiles des environs de Paris et distribution géographique des batraciens et reptiles de l'Ouest de la France. *Actes de la Société Linnéenne de Bordeaux* 31, 5–29.

Lataste, F. (1876b) Aperçu de la faune herpétologique du Plateau Central de la France. *Bulletin de la Société Zoologique de France*, Paris 1, 204–212.

Lataste, F. (1879) Diagnose d'une vipère nouvelle d'espagne. *Bulletin de la Société Zoologique de France*, Paris 4, 132.

Lataste, F. (1887) Description d'un nouveau genre et d'une nouvelle espèce d'ophidien protéroglyphe d'égypte. *Le Naturaliste*, Paris 9(52), 411–413.

Lataste, F. (1888) Description d'un ophidien diacrantérien nouveau (*Periops Dorri*, n. sp.) originaire du Haut-Sénégal. *Le Naturaliste*, Paris (Série 2) 10(38), 227–228.

Latifi, M. (1978) Commercial production of anti-snakebite serum (antivenin). *In:* Gans, C. & Gans, K.A. (Eds.), *Biology of the Reptilia. Volume 8 (Physiology B).* Academic Press, London, pp. 561–588, 7 figs.

Latifi, M. (1984) Variation in yield and lethality of venoms from Iranian snakes. *Toxicon* 22(3), 373–380, 2 figs.

Latifi, M. (1991) *The snakes of Iran.* Second edition. Society for the Study of Amphibians and Reptiles, Oxford, viii + 159 pp., 23 figs., 24 col. pls.

Latifi, M. (2000) *Snakes of Iran*. Third edition. Department of the Environment, Tehran, Iran, 378 pp., 14 figs., 26 col. pls. (in Farsi) [first edition, 1985, 221 pp., 12 figs., 25 col. pls.; second edition, 1991, by Society for the Study of Amphibians and Reptiles, 159 pp., 23 figs., 24 col. pls.]

Laurent, R.F. (1945) Contribution a la connaissance du genre *Atractaspis* Smith. *Revue de Zoologie et Botanique Africaine*, Tervuren 38(3–4), 312–343.

Laurent, R.F. (1947) Notes sur quelques reptiles appartenant a la collection du Musée Royal d'Histoire Naturelle de Belgique. I. Formes Africaines. *Bulletin du Musée Royal d'Histoire Naturelle de Belgique*, Bruxelles 23(16), 1–12.

Laurent, R.F. (1948) Notes sur quelques reptiles appartenant a la collection du Musee Royal d'Histoire naturelle de Belgique. II. Formes asiatiques et neo guineennes. *Bulletin du Musée Royal d'Histoire Naturelle de Belgique*, Bruxelles 24(17), 1–12, 13 figs.

Laurent, R.F. (1949) Notes sur quelques reptiles appartenant a la collection de l'Institut Royal des Sciences Naturelles de Belgique. III. Formes américaines. *Bulletin du Musée Royal d'Histoire Naturelle de Belgique*, Bruxelles 25(9), 1–20, 22 figs.

Laurent, R.F. (1950a) Revision du genre *Atractaspis* A. Smith. *Mémoires du Institut Royal des Science Naturelle de Belgique*, Bruxelles (Série 2) 38, 1–49, 18 figs.

Laurent, R.F. (1950b) Reptiles nouveaux des Kundelungu. *Revue de Zoologie et Botanique Africaine*, Tervuren 43(4), 349–352.

Laurent, R.F. (1950c) Subsídios para o estudo da biologia na Lunda, Museo do Dundo. Reptiles et batraciens de la région de Dundo (Angola du Nord-Est) (Première note). *Publicações Culturais da Companhia de Diamantes de Angola*, Dundo (6), 5–17.

Laurent, R.F. (1951) Remarques à propos des genres *Dipsadoboa* Gunther et *Crotaphopeltis* Fitzinger. *Revue de Zoologie et Botanique Africaine*, Tervuren 44(3), 210–212.

Laurent, R.F. (1952) Reptiles et batraciens nouveaux de la région des Grands Lacs africains. *Revue de Zoologie et Botanique Africaine*, Tervuren 46, 269–279.

Laurent, R.F. (1953) Reptiles et batraciens récemment parvenus au Musée royal du Congo belge. *Bulletin du Cercle Zoologique Congolais*, Tervuren 21, 21–29.

Laurent, R.F. (1954a) Subsídios para estudo da biologia na Lunda, Museo do Dundo. Reptiles et batraciens de la région de Dundo (Angola). (Deuxième note). *Publicações Culturais da Companhia de Diamantes de Angola*, Dundo (23), 35–84, 23 figs.

Laurent, R.F. (1954b) Apercu de la biogeographie des batraciens et des reptiles de la region des Grands Lacs. *Bulletin de la Société Zoologique de France*, Paris 79(4), 290–310.

Laurent, R.F. (1955) Diagnoses préliminares de quelques serpents venimeux. *Revue de Zoologie et Botanique Africaine*, Tervuren 51(1–2), 127–139.

Laurent, R.F. (1956a) Contribution a l'herpétologie de la région des Grands Lacs de l'Afrique centrale. I. Generalites—II. Cheloniens—III. Ophidiens. *Annales du Musée Royal du Congo Belge*, Tervuren (48), 1–390, 50 figs., 31 pls.

Laurent, R.F. (1956b) Notes herpétologiques africaines I. *Revue de Zoologie et Botanique Africaine*, Tervuren 53(3–4), 229–256.

Laurent, R.F. (1958) Notes herpétologiques africaines II. *Revue de Zoologie et Botanique Africaine*, Tervuren 58(1–2), 115–128.

Laurent, R.F. (1960) Notes complémentaires sur les chéloniens et les ophidiens du Congo Oriental. *Annales du Musée Royal du Congo Belge*, Tervuren (84), 1–86, 5 figs.

Laurent, R.F. (1964a) Subsídios para o estudo da biologia na Lunda, Museo do Dundo. Reptiles et amphibiens de l'Angola (Troisième contribution). *Publicações Culturais da Companhia de Diamantes de Angola*, Dundo (67), 17–165, 40 figs.

Laurent, R.F. (1964b) A revision of the *punctatus* group of African *Typhlops* (Reptilia, Serpentes). *Bulletin of the Museum of Comparative Zoology* 130(6), 387–444, 9 figs.

Laurent, R.F. (1968a) On some *Typhlops* from south-eastern Africa. *Arnoldia (Rhodesia)*, Salisbury 3(31), 1–2.

Laurent, R.F. (1968b) A re-examination of the snake genus *Lycophidium* Duméril and Bibron. *Bulletin of the Museum of Comparative Zoology* 136(12), 461–482, 12 figs.

Laurent, R.F. (1973a) Le concept de sous-espèce et la variation géographique chez *Naja melanoleuca* et *Naja nigricollis* (Reptilia, Ophidia). *Revue de Zoologie et Botanique Africaine*, Tervuren 87(1), 198–204.

Laurent, R.F. (1973b) Una nueva subespecies de culebra del noroeste Argentino y de Bolivia. *Acta Zoologica Lilloana*, Tucumán 26(20), 291–298.

Laurent, R.F. (1984) El genero *Leptotyphlops* en la coleccion de la Fundacion Miguel Lillo. *Acta Zoologica Lilloana*, Tucumán 38(1), 29–34, figs. a–c.

Laurent, R.F. & Teran, E.M. (1981) Lista de los anfibios y reptiles de la provincia de Tucumán. *Fundación Miguel Lillo, Miscelanea*, Córdoba (71), 5–15.

Laurenti, J.N. (1768) *Austriaci viennensis Specimen medicum, exhibens synopsin reptilium emendatam cum experimentis circa venena et antidota reptilium austriacorum*. Joan. Thom. Nob de Trattnern, Viennae, 214 pp., 5 pls.

Lavilla, E.O. & Scrocchi, G.J. (1991) Aportes a la herpetologia del Chaco argentino. 1. Lista comentada de los taxa colectados por la Expedicion PRHERP 1985. *Acta Zoologica Lilloana*, Tucumán 40(1), 21–32, illust.

Lavilla, E.O. & Scrocchi, G. (1999) Anfibios y reptiles de Tariquía. *In:* González, J.A., Scrocchi, G.J., Lavilla & E.O. (Eds), *Relevamiento de la biodiversidad de la Reserva Nacional de Flora y Fauna (Tarija, Bolivia)*. Fundación Miguel Lillo, Tucumán, pp. 83–92.

Lavilla, E.O., Cruz, F.B. & Scrocchi, G.J. (1995) Amphibians et reptiles de la Station Biologique "Los Colorados" dans la province de Salta, Argentina (2nd parte). *Revue Française d'Aquariologie*, Nancy 22, 117–128, col. figs. 25–54.

Lavilla, E.O., Richard, E. & Scrocchi, G.J. (Eds.). (2000) *Categorización de los anfibios y reptiles de la República Argentina*. Asociación Herpetológica Argentina, iii + 97 pp.

Lavin-Murcio, P.A. & Dixon, J.R. (2004) A new species of coral snake (Serpentes, Elapidae) from the Sierra de Tamaulipas, Mexico. *Phyllomedusa* 3(1), 3–7, 1 fig.

Lavín-Murcio, P.A. & Lazcano-Villarreal, D. (2010) Geographic distribution and conservation of the herpetofauna of northern Mexico. *In:* Wilson, L.D., Townsend, J.H. & Johnson, J.A. (Eds.), *Conservation of Mesoamerican amphibians and reptiles*. Eagle Mountain Publishing, Eagle Mountain, pp. 275-301, 2 figs.

Lawson, D.P. (1993) The reptiles and amphibians of Korup National Park Project, Cameroon. *Herpetological Natural History* 1(2), 27–90, 50 figs.

Lawson, D.P. (1997) Sexual dimorphism and reproduction of the pitviper *Porthidium ophryomegas* (Serpentes: Viperidae) in Guatemala. *Revista de Biologia Tropical* 44–45, 671–674.

Lawson, D.P. (1999) A new species of arboreal viper (Serpentes, Viperidae, *Atheris*) from Cameroon, Africa. *Proceedings of the Biological Society of Washington* 112(4), 791–803, 5 figs.

Lawson, D.P. & Ustach, P.C. (2000). A redescription of *Atheris squamigera* (Serpentes, Viperidae) with comments on the validity of *Atheris anisolepis*. *Journal of Herpetology* 34(3), 386–389, 1 fig.

Lawson, D.P., Noonan, B.P. & Ustach, P.C. (2001) *Atheris subocularis* (Serpentes, Viperidae) revisited: molecular and morphological evidence for the resurrection of an enigmatic taxon. *Copeia* 2001(3), 737–744, 4 figs.

Lawson, R. (1985) *Molecular studies of thamnophiine snakes*. Doctoral Dissertation, Louisiana State University, Baton Rouge, 183 pp., 4 figs.

Lawson, R. (1987) Molecular studies of Thamnophine snakes: 1. The phylogeny of the genus *Nerodia*. *Journal of Herpetology* 21(2), 140–157, 4 figs.

Lawson, R. & Dessauer, H.C. (1979) Biochemical genetics and systematics of garter snakes of the *Thamnophis elegans-couchii-ordinoides* complex. *Occasional Papers, Museum of Zoology, Louisiana State University* (56), 1–24, 4 figs.

Lawson, R. & Dessauer, H.C. (1981) Electrophoretic evaluation of the colubrid genus *Elaphe* Fitzinger. *Isozyme Bulletin* 14, 83.

Lawson, R. & Lieb, C.S. (1990) Variation and hybridization in *Elaphe bairdi* (Serpentes, Colubridae). *Journal of Herpetology* 24(3), 280–292, 7 figs.

Lawson, R., Meier, A.J., Frank, P.G. & Moler, P.E. (1991) Allozyme variation and systematics of the *Nerodia fasciata-Nerodia clarkii* complex of water snakes (Serpentes, Colubridae). *Copeia* 1991(3), 638–659, 5 figs.

Lawson, R., Slowinski, J.B. & Burbrink, F.T. (2004) A molecular approach to discerning the phylogenetic placement of the enigmatic snake *Xenophidion schaeferi* among the Alethinophidia. *Journal of Zoology*, London 263(3), 285–294, 5 figs.

Lazcano-Villarreal, D. & Contreras-Balderas, A. (1995) Lista revisada de los anfibios y reptiles de Nuevo León, México. *In: Listado preliminar de la fauna a silvestre del Estado de Nuevo León, México*, Consejo Consultivo Estatal para la Preservación y Fomento de la Flora y Fauna Silvestre de Nuevo León, pp. 57–70.

Lazcano-Villarreal, D., Contreras-Balderas, A., Gonzales-Rojas, J.I., Castañeda, G., García de la Peña, C. & Solis-Rojas, C. (2004) Notes on herpetofauna 6: herpetofauna of Sierra San Antonio Peña Nevada, Zaragoza, Nuevo León, México, preliminary list. *Bulletin of the Chicago Herpetological Society* 39(10), 181–187, 1 fig.

Lazcano-Villarreal, D., Leal, J.B. & Jacobo-Galvan, R.D. (2010) *Serpientes de Nuevo León*. Universidad Autónoma de Nuevo León, Monterrey, 502 pp., col. illust.

Lazcano-Barrero, M.A. & Gongora-Arones, E. (1988) A note on the reproduction of *Bothrops tzotzilorum*. *Bulletin of the Maryland Herpetological Society* 24(1), 7–8.

Lazell, J.D., Jr. (1964) The Lesser Antillean representatives of *Bothrops* and *Constrictor*. *Bulletin of the Museum of Comparative Zoology* 132(3), 245–273, 5 figs.

Lazell, J.D., Jr. (1967) Wiederentdeckung von zwei angeblich ausgestorbenen Schlangenarten der westindischen Inseln. *Salamandra*, Frankfurt-am-Main 3(3), 91–97, 2 figs.

Lazell, J.D., Jr. (1974) *Reptiles & amphibians in Massachusetts*. Second edition James D. Lazell, Jr., 34 pp., 1 fig., 5 photos. [first edition by Massachusetts Audubon Society, Lincoln, 34 pp., 2 figs., 16 photos]

Lazell, J.D., Jr. (1976) *This broken archipelago. Cape Cod and the islands: amphibians and reptiles*. Demeter Press, New York, xi + 260 pp., illust.

Lazell, J.D., Jr. (1988) Herpetology in South China. *Herpetological Review* 19(3), 49–51, 1 fig.

Lazell, J.D., Jr. (1989) *Wildlife of the Florida Keys: a natural history*. Island Press, Washington, 250 pp., illust.

Lazell, J.D., Jr. (1999) The origins and evolution of the herpetofaunas of the islands on the continental shelf of South China. *In:* Ota, H. (Ed.), *Tropical island herpetofauna: origin, current diversity, and conservation*. Elsevier, Amsterdam, pp. 79–96, 3 figs.

Lazell, J.D., Jr. (2002) The herpetofauna of Shek Kwu Chau, South China Sea, with descriptions of two new colubrid snakes. *Memoirs of the Hong Kong Natural History Society* 25, 1–81.

Lazell, J.D., Jr. & Liao, W. (1986) Contribution to the herpetofauna of Dinghushan. *Acta Herpetologica Sinica*, Beijing 5(1), 70–71.

Lazell, J.D., Jr. & Lu, W.-H. (1990) Four remarkable reptiles from South China Sea islands, Hong Kong Territory. *Asiatic Herpetological Research*, Berkeley 3, 64–66, 2 figs.

Lazell, J.D., Jr. & Musick, J.A. (1973) The kingsnake, *Lampropeltis getulus sticticeps*, and the ecology of the Outer Banks of North Carolina. *Copeia* 1973(3), 497–503.

Lazell, J.D., Jr., Keirans, J.E. & Samuelson, G.A. (1991) The Sulawesi black racer, *Coluber* (*Ptyas*) *dipsas*, and a remarkable ectoparasitic aggregation. *Pacific Science* 45(4), 355–361, 2 figs.

Leach, W.E. (1814) *The zoological miscellany; being descriptions of new or interesting animals, illustrated with coloured figures, drawn from nature, by R.P. Nodder. Vol. I.* E. Nodder & Son, London (1815), 144 pp., 60 pls.

Leach, W.E. (1817) *The zoological miscellany; being descriptions of new or interesting animals, illustrated with coloured figures, drawn from nature, by R.P. Nodder. Vol. III.* E. Nodder & Son, London, 151 pp., 30 pls.

Leach, W.E. (1818) Appendix. No. IV. A general notice of the animals taken by Mr. John Cranch, during the expedition to explore the source of the river Zaire. *In:* Tuckey, J.K. (Ed.), *Narrative of an expedition to explore the river Zaire, usually called the Congo, in South Africa, in 1816, under the direction of Captain J.K. Tuckey, R.N. To which is added, the journal of Professor Smith; some general observations on the country and its inhabitants; and an appendix, containing the natural history of that part of the kingdom of Congo through which the Zaire flows*. John Murray, London, pp. 407–419.

Leach, W.E. (1819) Appendix IV. Dr. Leach's notice of reptiles, insects, &c. *In:* Bowdich, T.E. (Ed.), *Mission from Cape Coast Castle to Ashantee, with...geographic notices of other parts of the Interior of Africa*. John Murray, London, pp. 493–496.

Leaché, A.D. (2005) Results of a herpetological survey in Ghana and a new country record. *Herpetological Review* 36(1), 16–19, 2 figs.

Le Berre, M. (1989) *Fauna du Sahara. 1. Poissons-Amphibiens, reptiles*. Editions Raymond Chabaud-Lechevalier, Paris, 336 pp., 6 pls., 130 col. pls.

LeBreton, M. (1999) *A working checklist of the herpetofauna of Cameroon, with localities for species occuring in southern Cameroon and a list of herpetofauna for the Dja Faunal Reserve*. Netherlands Committee for the IUCN, Amsterdam, 160 pp., illust.

LeBreton, M. & Chirio, L. (2004) Geographic distribution: Serpentes: *Dendroaspis polylepis*. *Herpetological Review* 35(2), 191.

Lee, B. & Koh, P.K. (2005) Geographic distribution: Serpentes: *Pareas margaritophorus*. *Herpetological Review* 36(3), 340.

Lee, J.C. (1980) An ecogeographic analysis of the herpetofauna of the Yucatán Peninsula. *Miscellaneous Publications, University of Kansas Museum of Natural History* (67), 1–75, 22 figs., 27 pls.

Lee, J.C. (1996) *Amphibians and reptiles of the Yucatán Peninsula*. Cornell University Press, Ithaca, 512 pp., 196 figs., 211 col. photos.

Lee, J.C. (2000) *A field guide to the amphibians and reptiles of the Maya world. The lowlands of Mexico, northern Guatemala, and Belize*. Cornell University Press, Ithaca, xii + 402 pp., 216 figs., 210 col. figs.

Lee, M.S.Y. (1997) The phylogeny of varanoid lizards and the affinities of snakes. *Philosophical Transactions of the Royal Society of London*, 352B, 53–91, 15 figs.

Lee, M.S.Y. & Caldwell, M.W. (1998) Anatomy and relationships of *Pachyrhachis problematicus*, a primitive snake with hindlimbs. *Philosophical Transactions of the Royal Society of London*, 353B, 1521–1552.

Lee, M.S.Y. & Scanlon, J.D. (2002) Snake phylogeny based on osteology, soft anatomy amd ecology. *Biological Review* 77, 333–401, 12 figs.

Lee, M.S.Y., Caldwell, M.W. & Scanlon, J.D. (1999) A second primitive marine snake: *Pachyophis woodwardi* from the Cretaceous of Bosnia-Herzegovina. *Journal of Zoology* 248, 509–520.

Lee, M.S.Y., Hugall, A.F., Lawson, R. & Scanlon, J.D. (2007) Phylogeny of snakes (Serpentes): combining morphological and molecular data in likelihood, Bayesian and parsimony analyses. *Systematics and Biodiversity* 5(4), 371–389, 10 figs.

Lee, T.-H. (2005) Ecological patterns of distribution on gradients of elevation and species diversity of snakes in southern Taiwan. *Amphibia-Reptilia*, Leiden 26(3), 325–332, 2 figs.

Leenders, T. (1994) The changing face of the genus *Vipera*. *Litteratura Serpentium* 14(5), 145–152, 1 fig., 1 photo.

Leenders, T. (2001) *A guide to amphibians and reptiles of Costa Rica*. Zona Tropical, Miami, 305 pp., 5 figs., 86 pls.

Leeson, F. (1950) *Identification of snakes of the Gold Coast*. Crown Agents, London, x + 142 pp., 32 pls.

Legler, J.M. (1959a) A new blind snake (genus *Typhlops*) from Cuba. *Herpetologica* 15(1), 105–112, 1 fig.

Legler, J.M. (1959b) Notes on the snake *Leptotyphlops columbi* Klauber. *Herpetologica* 15(1), 112.

Legler, J.M. (1959c) A new snake of the genus *Geophis* from Chihuahua, México. *University of Kansas Publications, Museum of Natural History* 11(4), 327–334, 2 figs.

Lehmann, H.D. (1970) Beobachtungen bei der Haltung und Aufzucht von *Trachyboa boulengeri* (Serpentes, Boidae). *Salamandra*, Frankfurt-am-Main 6, 32–42, 7 pls.

Lehmann, H.D. (1974) *Trachyboa boulengeri* (Serpentes, Boidae) frisst Fische. *Salamandra*, Frankfurt-am-Main 10(3–4), 132–133, 1 fig.

Lehmann, H.D. (1993a) Observations on the husbandry and reproduction of *Trachyboa boulengeri* (Serpentes, Boidae). *Bulletin of the Chicago Herpetological Society* 28(2), 25–30, 7 figs.

Lehmann, H.D. (1993b) *Trachyboa boulengeri* (Serpentes, Boidae) eats fish. *Bulletin of the Chicago Herpetological Society* 28(2), 30–31, 1 fig.

Lehr, E. (2000) Geographic distribution: Serpentes: *Leptotyphlops albifrons*. *Herpetological Review* 31(3), 186.

Lehr, E. (2001) New records for amphibians and reptiles from departamentos Pasco and Ucayali, Peru. *Herpetological Review* 32(2), 130–132.

Lehr, E. & Aguilar, C. (2000a) Geographical distribution. *Oxybelis aeneus*. *Herpetological Review* 31(1), 56.

Lehr, E. & Aguilar, C. (2000b) Geographical distribution. *Rhinobothryum lentiginosum*. *Herpetological Review* 31(1), 57.

Lehr, E. & Fernandez, R. (2000a) Geographical distribution. *Ninia hudsoni*. *Herpetological Review* 31(1), 56.

Lehr, E. & Fernandez, R. (2000b) Geographic distribution: Serpentes: *Tripanurgos compressus*. *Herpetological Review* 31(3), 187.

Lehr, E. & Köhler, G. (2000) Geographic distribution: Serpentes: *Tachymenis peruviana*. *Herpetological Review* 31(3), 187.

Lehr, E., Wallach, V., Köhler, G. & Aguilar, C. (2002) New species of tricolor *Leptotyphlops* (Reptilia, Squamata, Leptotyphlopidae) from central Peru. *Copeia* 2002(1), 131–136, 3 figs.

Lehr, E., Carrillo de Espinoza, N. & Hocking, P. (2004) New species of *Drymoluber* (Reptilia, Squamata, Colubridae) from southeastern Peru. *Copeia* 2004(1), 46–52, 4 figs.

Leighton, G.R. (1901) *The life-history of British serpents and their local distribution in the British Isles*. William Blackwood and Sons, Edinburg, xvi + 383 pp., 49 figs.

Le Loeuff, J. (1991) The campano-Maastrichtian vertebrate fauna from southern Europe and their relationships with other faunas in the world; palaeobiogeographical implications. *Cretaceous Research* 12, 93–114, 3 figs.

Lema, T. de. (1958) Notas sôbre os répteis do Estado Rio Grande do Sul, Brasil. II. Nota prévia sôbre a espécie de colubrideo aquatico *Helicops carinicauda* (Wied, 1825) do Estado do Rio Grande do Sul, Brasil. *Iheringia* (*Zoologia*), Pôrto Alegre (10), 19–24.

Lema, T. de. (1960a) Notas sôbre os répteis do Estado do Rio Grande do Sul, Brasil. V. Sôbre a espécie *Bothrops cotiara* (Gomés, 1913) e sua ocorréncia no Estado do Rio Grande do Sul (Serpentes, Crotalidae). *Iheringia* (*Zoologia*), Pôrto Alegre (13), 6–10, figs. 13–14.

Lema, T. de. (1960b) Notas sôbre os répteis do Estado do Rio Grande do Sul, Brasil. VI. Ampliação da descrição original de *Bothrops cotiara* (Gomés, 1913) (Serpentes, Crotalidae). *Iheringia* (*Zoologia*), Pôrto Alegre (13), 11–19, figs. 15–17.

Lema, T. de. (1961a) Notas sôbre os répteis do Estado do Rio Grande do Sul, Brasil. IX. Presenca de *Helicops carinicauda carinicauda* (Wied) no Estato do Rio Grande do Sul (Serpentes, Colubridae). *Iheringia* (*Zoologia*), Pôrto Alegre (17), 7–10, 3 figs.

Lema, T. de. (1961b) Notas sôbre os répteis do Estado do Rio Grande do Sul, Brasil. X. Estudo das marcas vertebrais de *Lystrophis dorbignyi* (Dum., Bibr. & Dum.) e descrição de um exemplar anomalo (Serpentes, Colubridae). *Iheringia* (*Zoologia*), Pôrto Alegre (17), 11–14, figs. 4–15.

Lema, T. de. (1962a) Sôbre a espécie *Bothrops itapetiningae* (Boulenger, 1907) e sua ocurrência no Estado do Rio Grande do Sul, Brasil (Serpentes, Crotalidae). *Iheringia* (*Zoologia*), Pôrto Alegre (21), 1–12, 4 pls., 7 figs.

Lema, T. de. (1962b) Ocorrência de *Philodryas arnaldoi* (Amaral, 1932) no Estado do Rio Grande do Sul, Brasil. (Serpentes, Colubridae). *Iheringia* (*Zoologia*), Pôrto Alegre (22), 1–4, 2 pls., 4 figs.

Lema, T. de. (1964) Uma nova espécie de serpente do genero *Siphlophis* Fitzinger, 1843 do Brasil meridional (Colubridae, Xenodontinae). *Revista Brasileira de Biologia*, Rio de Janeiro 24, 221–228, 13 figs.

Lema, T. de. (1967) Nôvo gênero e espécie de serpente opistoglifodonte do Brasil meridional ("Colubridae," "Colubrinae"). *Iheringia* (*Zoologia*), Pôrto Alegre (35), 61–74, 10 figs.

Lema, T. de. (1970) Sobre o status de *Elapomorphus bilineatus* Duméril, Bibron & Duméril, 1854, curiosa serpente subterrânea. *Iheringia* (*Zoologia*), Pôrto Alegre (38), 89–118, 7 figs.

Lema, T. de. (1972) Sobre *Micrurus putumayensis* Lancini, 1962 e sua ocurrência no Brasil (Serpentes, Elapidae). *Iheringia* (*Zoologia*), Pôrto Alegre (41), 35–58, 15 figs.

Lema, T. de. (1973) Ocorrência de *Uromacerina ricardinii* (Peracca 1897) no Rio Grande do Sul e contribuição ao conhecimento dessa rara serpente (Ophidia, Colubridae). *Iheringia* (*Zoologia*), Pôrto Alegre (44), 64–73, 8 figs.

Lema, T. de. (1977) Variação geografica de *Elapomorphus lemniscatus* Duméril, Bibron & Duméril, 1854, e contribuição ao conhecimento da zoogeografia da região da Bacia do Rio de la Plata. (Ophidia, Colubridae). *Resúmenes VII Congreso Latinoamericano de Zoología*, Tucumán, p. 104.

Lema, T. de. (1978a) O status de *Elapomorphus suspectus* Amaral 1924 (Ophidia, Colubridae). *Comunicacoes do Museu de Ciencias da PUCRGS*, Pôrto Alegre (16), 1–10, 3 figs.

Lema, T. de. (1978b) Invalidação de *Elapomorphus bollei* Mertens 1954 e o status de *Elapomorphus spegazzini* Boulenger 1913 (Ophidia, Colubridae). *Comunicacoes do Museu de Ciencias da PUCRGS*, Pôrto Alegre (17), 11–16.

Lema, T. de. (1978c) Ocorrência de *Tantilla melanocephala* (L., 1758) no Rio Grande do Sul, Brasil, e o "status" de *Tantilla capistrata* Cope, 1876 (Ophidia, Colubridae). *Comunicacoes do Museu de Ciencias da PUCRGS*, Pôrto Alegre (18), 1–26, 16 figs.

Lema, T. de. (1978d) Novas espécies de opistoglifontes do gênero *Apostolepis* Cope 1861 do Paraguai (Ophidia, Colubridae). *Comunicacoes do Museu de Ciencias da PUCRGS*, Pôrto Alegre (19), 27–48, 14 figs.

Lema, T. de. (1979a) Sobre a validade dos nomes *Elapomorphus bilineatus* Duméril, Bibron & Duméril, 1854 e *E. lemniscatus* Duméril, Bibron & Duméril, 1854 (Ophidia, Colubridae). *Iheringia* (*Zoologia*), Pôrto Alegre (54), 77–81.

Lema, T. de. (1979b) *Elapomorphus punctatus*: nova espécie de Colubridae para a Argentina (Ophidia). *Revista Brasileira de Biologia*, Rio de Janeiro 39(4), 835–853, 26 figs.

Lema, T. de. (1980) Ocorrência de *Typhlops reticulatus* (L., 1766) no Estado do Rio Grande do Sul (Brasil) e comentários sôbre a espécie (Ophidia, Typhlopidae). *Iheringia (Zoologia)*, Pôrto Alegre (56), 7–14, 5 figs.

Lema, T. de. (1982) Sôbre a ocorrência de *Typhlops brongersmianus* Vanzolini, 1972, no Estado do Rio Grande do Sul e regioes adjacentes. (Serpentes, Typhlopidae). *Iheringia (Zoologia)*, Pôrto Alegre (61), 3–7.

Lema, T. de. (1983) Fauna de Serpentes da Província Pampeana e inter-relações com as Província Limitrofes. *Memórias do Instituto Butantan*, São Paulo (1982) 46, 173–182, 2 figs.

Lema, T. de. (1984) Sôbre o gênero *Elapomorphus* Wiegmann, 1843 (Serpentes, Colubridae, Elapomorphinae). *Iheringia (Zoologia)*, Pôrto Alegre (64), 53–86, 18 figs.

Lema, T. de. (1985a) Melanismo em serpentes do gênero *Lygophis* Fitzinger, 1843 (Serpentes, Colubridae), no nordeste do Brasil. *Iheringia (Zoologia)*, Pôrto Alegre (65), 31–40, 17 figs.

Lema, T. de. (1985b) Aspectos biólogicos de *Elapomorphus (Phalotris) lemniscatus* Duméril, Bibron et Duméril, 1854 (Serpentes, Colubridae, Elapomorphinae). *Iheringia (Zoologia)*, Pôrto Alegre (65), 57–64, 8 figs.

Lema, T. de. (1986) *Apostolepis dimidiata* (Jan, 1862), nova combinação, e a validade de *A. erythronota* (Peters, 1880) e *A. ventrimaculata* (Serpentes, Colubridae, Elapomorphinae). *Memórias do Instituto Butantan*, São Paulo (1983–1984) 47–48, 71–80, 3 figs.

Lema, T. de. (1987) Lista preliminar das serpentes registradas para o estado do Rio Grande do Sul (Brasil meridional) (Reptilia, Lepidosauria, Squamata). *Acta Biológica Leopoldensia* 9(2), 225–240.

Lema, T. de. (1989a) Notas sobre a biologia de duas espécies de *Elapomorphus* Wiegmann, 1843 (Serpentes, Colubridae, Elapomorphinae). *Iheringia (Zoologia)*, Pôrto Alegre (69), 61–69.

Lema, T. de. (1989b) Serpentes do complexo *Liophis lineatus* (Linnaeus, 1758) no Brasil nordeste (Serpentes, Colubridae, Colubrinae). *Acta Biológica Leopoldensia* 11(2), 251–271, 6 figs.

Lema, T. de. (1992) Presença de *Elapomorphus quinquelineatus* (Raddi) no extreme-sul do Brasil e a ocorrencia de rara anomalia. (Serpentes, Colubridae, Xenodontinae, Elapomorphini). *Comunicações do Museu de Ciências da PUCRS (Zoologia)*, Pôrto Alegre 5(1), 1–7, 4 figs.

Lema, T. de. (1993) Polimorfismo em *Apostolepis dimidiata* (Jan, 1862) com a invalidação de *Apostolepis villaricae* Lema, 1978 e *Apostolepis barrioi* Lema, 1978 (Serpentes, Colubridae, Xenodontinae, Elapomorphini). *Acta Biológica Leopoldensia* 15(1), 35–52, 10 figs.

Lema, T. de. (1994) Lista comentada dos repteis ocorrentes no Rio Grande do Sul, Brasil. *Comunicações do Museu de Ciências e Tecnologia da PUCRS (Zoologia),* Pôrto Alegre 7(1), 41–150, 3 figs.

Lema, T. de. (1997) A redescription of the tropical Brazilian serpent *Apostolepis nigrolineata* (Peters, 1869) (Colubridae, Elapomorphinae), synonymous with *A. pymi* Boulenger. *Studies on Neotropical Fauna and Environment* 32(4), 193–199, illust.

Lema, T. de. (2001) Fossorial snake genus *Apostolepis* from South America (Serpentes, Colubridae, Elapomorphinae). *Cuadernos de Herpetología* 15(1), 29–43.

Lema, T. de. (2002a) Redescription of *Apostolepis sanctaeritae* Werner, 1924 (Serpentes, Colubridae), and a comparison with related species. *Copeia* 2002(4), 1147–1151.

Lema, T. de. (2002b) Nova espécie de *Apostolepis* Cope do grupo *dimidiata* do Cerrado do Brasil (Serpentes, Elapomorphinae). *Comunicações do Museu de Ciências e Tecnologia da PUCRS (Zoologia),* Pôrto Alegre 15(2), 227–238, 1 fig.

Lema, T. de. (2002c) New species of *Phalotris* from northern Brazil with notes on the *nasutus* group (Serpentes: Elapomorphinae). *Comunicações do Museu de Ciências e Tecnologia da PUCRS (Zoologia),* Pôrto Alegre 15(2), 201–214.

Lema, T. de. (2002d) Nova espécie de *Apostolepis* do grupo *lineatea* do sudoeste do Brasil (Serpentes, Elapomorphinae). *Facena* 18, 41–52, 4 pls.

Lema, T. de. (2003a) Descrição de nova espécie de *Apostolepis* Cope do Cerrado do Brasil, pertencente ao grupo *dimidiata* (Serpentes, Elapomorphinae). *Acta Biológica Leopoldensia* 25(1), 123–131, 4 figs.

Lema, T. de. (2003b) Reencontro de *Apostolepis goiasensis* Prado, 1942, com a redescrição da espécie (Serpentes, Elapomorphinae). *Comunicações do Museu de Ciências e Tecnologia da PUCRS (Zoologia),* Pôrto Alegre 16(2), 199–209, 3 figs.

Lema, T. de. (2004a) New species of *Apostolepis* Cope, 1861 from Caatinga belonging to *assimilis* group (Serpentes, Elapomorphinae). *Comunicações do Museu de Ciências e Tecnologia da PUCRS (Zoologia),* Pôrto Alegre 17(1), 3–7, 1 fig.

Lema, T. de. (2004b) Nova espécie de *Apostolepis* Cope do estado de Rondônia, Brasil (Serpentes, Elapomorphinae). *Comunicações do Museu de Ciências e Tecnologia da PUCRS (Zoologia),* Pôrte Alegre 17(2), 81–89, 1 fig.

Lema, T. de. (2004c) New species of *Tantilla* Baird & Girard from northeastern Brazil (Serpentes, Colubrinae). *Acta Biológica Leopoldensia* 26(2), 267–284, 8 figs.

Lema, T. de. (2005) New occurrence of *Atractus trihedrurus* Amaral (Serpentes, Xenodontinae). *Comunicações do Museu de Ciências e Tecnologia da PUCRS (Zoologia),* Pôrte Alegre 18(1), 57–58.

Lema, T. de & Albuquerque, N.R. (2010) The identity of *Apostolepis pymi* and placement of *A. quinquelineata* in the synonymy of *A. nigrolineata* (Serpentes, Xenodontinae). *Biota Neotropica* 10(1), 343–346, 2 figs.

Lema, T. de & Azevedo, A.C.P. (1969) Ocorrência de *Micrurus decoratus* (Jan) no Rio Grande do Sul, Brasil, (Serpentes, Elapidae). *Iheringia (Zoologia)*, Pôrto Alegre (37), 113–117.

Lema, T. de & Cappellari, L.H. (2001) Geographic distribution: Serpentes: *Apostolepis quirogai. Herpetological Review* 32(2), 121.

Lema, T. de & Cappellari, L.H. (2002) Geographic distribution: Serpentes: *Apostolepis quirogai. Herpetological Review* 33(4), 277.

Lema, T. de & Ely, L.A.M. (1979) Considerações sôbre *Pseudoboa haasi* (Boettger, 1905) no extremo sul do Brasil (Ophidia, Colubridae). *Iheringia (Zoologia)*, Pôrto Alegre (54), 53–56.

Lema, T. de & Fabián-Beurmann, M.E. (1977) Levantamento preliminar dos répteis da região da fronteira Brasil-Uruguai. *Iheringia (Zoologia)*, Pôrto Alegre (50), 61–92.

Lema, T. de. & Fernandes, R. (1997) The status of *Apostolepis sanctaeritae* Werner, 1924, and its revalidation (Serpentes, Colubridae, Elapomorphinae). O status de *Apostolepis sanctaeritae* Werner, 1924 e sua revalidacao (Serpentes, Colubridae, Elapomorphinae). *Acta Biológica Leopoldensia* 19(1), 51–59, illust.

Lema, T. de & Hofstadler-Deiques, C.H. (1992) Contribuição ao conhecimento da "cobra espada d'agua," *Ptychophis flavovirgatus* Gomes, (Serpentes, Colubridae, Xenodontinae, Tachymenini). *Comunicações do Museu de Ciências e Tecnologia da PUCRS (Zoologia),* Pôrto Alegre 5(6), 55–83, 12 figs.

Lema, T. de & Hofstadler-Deiques, C.H. (1995) Estudo revisivo de *Elapomorphus lepidus* Reinhardt com a invalidação de *E. wuchereri* Gunther, *E. accedens* Jan e *E. coronata* Sauvage mediante analise tipologica e a osteologia craniana (Serpentes, Colubridae, Xenodontinae, Elapomorphini). *Biociências*, Pôrto Alegre 3(1), 91–143.

Lema, T. de & Hofstadler-Deiques, C. (2010) Description of a new genus for allocation of *Elapomorphus lepidus* and the status of *Elapomorphus wuchereri* (Serpentes: Dipsadidae: Xenodontinae: Elapomorphini). *Neotropical Biology and Conservation* 5(2), 113–119, 10 figs.

Lema, T. de & Leitão de Araujo, M. (1980) Sôbre *Bothrops jararacussu* Lacerda, 1884 do extremo sul do Brasil e sua ocorrência no Estado do Rio Grande do Sul (Ophidia, Viperidae). *Iheringia (Zoologia)*, Pôrto Alegre (56), 63–70.

Lema, T. de & Renner, M.F. (2004) New species of *Apostolepis* from Alto Tocantins, Brazil, with comments on the striped pattern species (Serpentes, Elapomorphinae). *Biociências*, Pôrto Alegre 12(2), 139–145, fig. 1.

Lema, T. de & Renner, M.F. (2006a) Contribution to the knowledge of *Apostolepis cearensis* Gomes, 1915 (Serpentes: Colubridae: Elapomorphinae), with comments on similar species from northeastern Brazil. *Comunicações do Museu de Ciências e Tecnologia da PUCRS (Zoologia),* Pôrto Alegre 18(2), 129–140, 6 figs.

Lema, T. de & Renner, M.F. (2006b) A new species of *Apostolepis* with striped pattern from Mato Grosso, Brazil (Serpentes, Elapomorphinae). *Ciência em Movimento* 8, 13–18, 4 figs.

Lema, T. de & Renner, M.F. (2007) Contribuição ao conhecimento de *Apostolepis ammodites* (Serpentes, Colubridae, Elapomorphinae). *Biociências*, Puerto Alegre 15(1), 126–142, 9 figs.

Lema, T. de, Fabían-Beurmann, M.E., Leitão de Araujo, M., Machado-Alves, M.L. & Ibarra-Vieira, M. (1980) Lista de répteis encontrados na região da Grande Põrto Alegre, Estado do Rio Grande do Sul, Brasil. *Iheringia (Zoologia)*, Pôrto Alegre (55), 27–36, 1 fig.

Lema, T. de, Cappellari, L.H. & Maurer d'Agostini, F. (2004) Contributions to the knowledge of *Apostolepis quirogai* Giraudo & Scrocchi, 1998, with description of the skull (Serpentes, Elapomorphinae). *Comunicações do Museu de Ciências e Tecnologia da PUCRS (Zoologia),* Pôrto Alegre 17(2), 67–79, 4 figs.

Lema, T. de, D' Agostini, F.M. & Cappellari, L.H. (2005a) Nova espécie de *Phalotris*, redescrição de *P. tricolor* e osteologia craniana (Serpentes, Elapomorphinae). *Iheringia (Zoologia)* 95(1), 65–78, 22 figs.

Lema, T. de, Ferret-Renner, M. & Nascimento-Silva, C. (2005b) Contribution to the knowledge of *Apostolepis ambiniger* (Peters, 1869) (Serpentes, Elapomorphinae). *Comunicações do Museu de Ciências e Tecnologia da PUCRS (Zoologia),* Pôrto Alegre 18(1), 21–36, 9 figs.

Lemm, J.M. (2006) *Field guide to amphibians and reptiles of the San Diego region.* University of California Press, Berkeley, 326 pp., 121 pls.

Lemoinek, K. & Rivas-Fuenmayor, G.A. (1998) Geographic distribution: Serpentes: *Pseustes poecilonotus polylepis. Herpetological Review* 29(2), 115.

Lemos-Espinal, J.A. & Smith, H.M. (2007a) *Amphibians and reptiles of the state of Chihuahua, Mexico.* Universidad Nacional Autónoma de México, Mexico City, xii + 613 pp., 129 figs., 249 photos.

Lemos-Espinal, J.A. & Smith, H.M. (2007b) *Amphibians and reptiles of the state of Coahuila, Mexico.* Universidad Nacional Autónoma de México, Mexico City, xii + 550 pp., 79 figs., 285 photos.

Lemos-Espinal, J.A. & Smith, H.M. (2009) *Keys to the amphibians and reptiles of Sonora, Chihuahua and Coahuila, Mexico.* Universidad Nacional Autónoma de México, Mexico City, and University of Colorado, Boulder, 348 pp., 675 col. photos.

Lemos-Espinal, J.A., Chiszar, D. & Smith, H.M. (1994) The distribution of the prairie rattlesnake (*Crotalus v. viridis*) in Mexico. *Bulletin of the Maryland Herpetological Society* 30(4), 143–148.

Lemos-Espinal, J.A., Smith, H.M. & Chiszar, D. (2001) Geographic distribution: Serpentes: *Boa constrictor imperator. Herpetological Review* 33(4), 277.

Lemos-Espinal, J.A., Auth, D., Chiszar, D. & Smith, H.M. (2002a) Year 2000 snakes from Chihuahua, Mexico. *Bulletin of the Chicago Herpetological Society* 37(3), 51–55.

Lemos-Espinal, J.A., Auth, D., Chiszar, D. & Smith, H.M. (2002b) Year 2001 snakes from Chihuahua, Mexico. *Bulletin of the Chicago Herpetological Society* 37(10), 180–182.

Lemos-Espinal, J.A., Chiszar, D. & Smith, H.M. (2004a) Geographic distribution: Serpentes: *Drymobius margartiferus fistulosus. Herpetological Review* 35(3), 291.

Lemos-Espinal, J.A., Chiszar, D. & Smith, H.M. (2004b) Variation in *Procinura aemula*, the file-tailed groundsnake of México. *Bulletin of the Maryland Herpetological Society* 40(2), 61–69, 1 fig.

Lemos-Espinal, J.A., Chiszar, D., Ingrasci, M.J. & Smith, H.M. (2004c) Year 2002. Turtles and snakes from Chihuahua, Mexico. *Bulletin of the Chicago Herpetological Society* 39(5), 82–87.

Lemos-Espinal, J.A., Smith, H.M., Chiszar, D. & Woolrich-Piña, G. (2004d) Snakes from Chihuahua and adjacent states of Mexico. *Bulletin of the Chicago Herpetological Society* 39(11), 206–213.

Lemos-Espinal, J.A., Smith, H.M. & Chiszar, D. (2004e) *Introduction to the amphibians and reptiles of the state of Chihuahua, Mexico.* Conabio, 128 pp., 132 figs.

Lemos-Espinal, J.A., Smith, H.M. & Chiszar, D. (2006a) Recent herpetological acquisitions from the Chinipas region of southwestern Chihuahua. *Bulletin of the Chicago Herpetological Society* 41(10), 185–186.

Lemos-Espinal, J.A., Smith, H.M. & Chiszar, D. (2006b) *Agkistrodon b. bilineatus* in western Chihuahua. *Bulletin of the Maryland Herpetological Society* 42(4), 173–175, 1 fig.

Lenk, P. & Joger, U. (1994) Genetic relationships between populations and intraspecific subdivision of *Elaphe longissima* (Laurenti, 1768) as suggested by plasma protein electrophoresis and DNA fingerprinting. *Amphibia-Reptilia*, Leiden 15(4), 363–373, 4 figs.

Lenk, P. & Wüster, W. (1999) A multivariate approach to the Systematics of Italian rat snakes of the *Elaphe longissima* complex (Reptilia, Colubridae): revalidation of Camerano's *Callopeltis longissimus* var. *lineata*. *Herpetological Journal* 9, 153–162, 7 figs.

Lenk, P., Herrmann, H.-W., Joger, U. & Wink, M. (1999) Phylogeny and taxonomic subdivision of *Bitis* (Reptilia, Viperidae) based on molecular evidence. *In:* Joger, U. (Ed.), Phylogeny and systematics of the Viperidae. *Kaupia, Darmstädter Beiträge zur Naturgeschichte* 8, 31–38, 3 figs.

Lenk, P., Joger, U. & Wink, M. (2001a) Phylogenetic relationships among European ratsnakes of the genus *Elaphe* Fitzinger based on mitochondrial DNA sequence comparisons. *Amphibia-Reptilia,* Leiden 22(3), 329–339, 2 figs.

Lenk, P., Kalyabina, S., Wink, M. & Joger, U. (2001b) Evolutionary relationships among the true vipers (Reptilia: Viperidae) inferred from mitochondrial DNA sequences. *Molecular Phylogenetics and Evolution* 19(1), 94–104, 2 figs.

Lenz, H.O. (1832) *Schlangenkunde.* Bedersche, Gotha, xiv + 559 pp., 29 pls.

Léon-Soler, A. & Solórzano, A. (2000) The blind snake *Leptotyphlops goudotii* (Serpentes: Leptotyphlopidae), in Murciélagos Islands, Costa Rica. *Revista de Biologia Tropical* 48(4), 1049.

Leonard, W.P. & Ovaska, K. (1998) *Contia, C. tenuis. Catalogue of American amphibians and reptiles*, pp. 677.1–677.7, 2 figs.

Leonardi, M., Quaroni, A., Rigato, F. & Scali, S. (1995) La collezioni del Museo Civico di Storia Naturale di Milano. *Atti della Socièta Italiana di Scienze Naturali e del Museo Civico di Storia Naturale di Milano* 135(1), 1–296.

Leong, T.M. (2004) Geographic distribution: Serpentes: *Calamaria lowi gimletti. Herpetological Review* 35(3), 290.

Leong, T.M. (2005) Geographic distribution: Serpentes: *Dendrelaphis cyanochloris. Herpetological Review* 36(1), 81.

Leong, T.M. & Diesmos, A.C. (2004) Geographic distribution: Serpentes: *Amphiesma sanguinea. Herpetological Review* 35(3), 290.

Leong, T.M. & Grismer, L.L. (2004) A new species of kukri snake, *Oligodon* (Colubridae), from Pulau Tioman, West Malaysia. *Asiatic Herpetological Research* 10, 12–16.

Leong, T.M. & Lim, B.L. (2002) Geographic distribution: Serpentes: *Collorhabdium williamsoni. Herpetological Review* 33(4), 277–278.

Leong, T.M. & Tan, H.H. (2000) Geographic distribution: Serpentes: *Enhydris enhydris. Herpetological Review* 31(2), 114.

Lescure, J. & Massary, J.-C. de. (Eds.). (2012) *Atlas des amphibiens et reptiles de France.* Editions Biotope, 272 pp., col. illust.

Lescure, J., Massary, J.-C. de & Oger, F. (2010) *Atlas des amphibiens et reptiles de la Seine-Saint-Denis.* Editions Biotope, 146 pp., col. photos.

Lesson, R.-P. (1828) Observations générales sur les reptiles observes dans le voyage autour du monde de la corvette *la Coquille. Annales des Sciences Naturelles (Zoologie),* Paris 13, 369–394.

Lesson, R.-P. (1829) Reptiles. *In:* Duperrey, L.I. (1826–1832), *Voyage autour du monde execute par ordre du Roi, sur la corvette de la Majeste, la Coquille, execute pendant les années 1822, 1823, 1824 et 1825, sous le ministerre et conformément aux instructions de S.E.M. la Marquis de Clermont-Tonnerre, Ministre de la Marine; et publié sous les auspices de son excellence Mgr. le Cte. de Chabrol, Ministre de la Marine et des Colonies. Histoire naturelle, Zoologie. Atlas.* Arthus Bertrand, Paris, 7 pls.

Lesson, R.-P. (1830) Description de quelques reptiles nouveaux ou peu connus. *In:* Duperrey, L.I. (1826–1832), *Voyage autour du monde execute par ordre du Roi, sur la corvette de la Majeste, la Coquille, execute pendant les années 1822, 1823, 1824 et 1825, sous le ministerre et conformément aux instructions de S.E.M. la Marquis de Clermont-Tonnerre, Ministre de la Marine; et publié sous les auspices de son excellence Mgr. le Cte. de Chabrol, Ministre de la Marine et des Colonies. Zoologie. Tome second – 1re partie.* Arthus Bertrand, Paris, pp. 34–65.

Lesson, R.-P. (1831) Catalogue des reptiles qui font partie d'une collection zoologique recueillie dans l'Indie continentale ou en Afrique et apportée en France par M. Lamare-Piquot. Catalogue dressé (Juillet 1830) *Bulletin des Sciences Naturelles et de Géologie*, Paris 25(4), 119–123.

Lesson, R.-P. (1832) Reptiles. *In:* Bélanger, C. (Ed.) (1831–1834), *Voyage aux Indes-Orientales, par le nord de l'Europe, les provinces du Caucase, la Géorgie, l'Arménie et la Perse, suivi de details topographiques, statistiques et autres sur les Pégou, les iles de Java, de Maurice et de Bourbon, sur le Cap-de-Bonne-Espérance et Sainte-Hélène, pendant années 1825, 1826, 1827, 1828 et 1829. Dédié au Roi. Zoologie par MM. Bélanger, Isidore Geoffroy-Saint-Hilaire, Lesson, Valenciennes, Deshayes et Guérin. Zoologie et Atlas.* Arthus Bertrand, Paris, pp. 289–336, 7 pls.

Lesson, R.-P. (1841) Catalogue d'une faune du Département de la Charente–inférieure. *Actes de la Société Linnéenne de Bordeaux* 12, 1–64, pls.

Lessona, M. (1877) Nota di un axolotl del Museo Zoologico di Torino. *Atti della Reale Accademia delle Scienze di Torino* (1876–1877) 13, 137–139.

Leston, D. (1970) Some snakes from the forest zone of Ghana. *British Journal of Herpetology*, London 4, 141–144.

Leston, D. & Hughes, B. (1968) The snakes of Tafo, Dakar, a forest cocoa-farm locality in Ghana. *Bulletin de l'Institut Fondamental d'Afrique Noire*, Dakar 30A(2), 737–770, 4 figs.

Leviton, A.E. (1952) A new Philippine snake of the genus *Calamaria. Journal of the Washington Academy of Sciences* 42(7), 239–240, 1 fig.

Leviton, A.E. (1953a) A new snake of the genus *Oligodon* from Annam. *Journal of the Washington Academy of Sciences* 43(12), 422–424, 1 fig.

Leviton, A.E. (1953b) Catalogue of the amphibians and reptile types in the Natural History Museum of Stanford University. *Herpetologica* 8(4), 121–132.

Leviton, A.E. (1955) Systematic notes on the Asian snake *Lycodon subcinctus. The Philippine Journal of Science,* Manila 84(2), 195–203.

Leviton, A.E. (1957) A review of the Philippine snakes of the genus *Oxyrhabdium* (Serpentes, Colubridae). *Wasmann Journal of Biology* 15(2), 285–303, 2 figs.

Leviton, A.E. (1959a) Report on a collection of reptiles from Afghanistan. *Proceedings of the California Academy of Sciences* (Series 4) 29(12), 445–463.

Leviton, A.E. (1959b) *Systematics and zoogeography of Philippine snakes.* Doctoral Dissertation, Stanford University, Palo Alto, xx + 865 pp.

Leviton, A.E. (1960) Notes on the second specimen of the snake *Oligodon annamensis* Leviton. *Wasmann Journal of Biology* 18, 305–307.

Leviton, A.E. (1961a) Description of a new subspecies of the Philippine snake *Dendrelaphis caudolineatus. Occasional Papers of the Natural History Museum of Stanford University* (9), 1–6.

Leviton, A.E. (1961b) Keys to the dangerously venomous terrestrial snakes of the Philippine Islands. *Silliman Journal*, Dumaguete City 8(2), 98–106.

Leviton, A.E. (1963a) Contributions to a review of Philippine snakes, I. The snakes of the genus *Oligodon. The Philippine Journal of Science*, Manila (1962) 91(4), 459–484, 1 fig.

Leviton, A.E. (1963b) Remarks on the zoogeography of Philippine terrestrial snakes. *Proceedings of the California Academy of Sciences* (Series 4) 31(15), 369–416, 1 fig.

Leviton, A.E. (1964a) Contributions to a review of Philippine snakes, II. The snakes of the genera *Liopeltis* and *Sibynophis*. *The Philippine Journal of Science*, Manila (1963) 92(3), 367–381.

Leviton, A.E. (1964b) Contributions to a review of Philippine snakes, III. The genera *Maticora* and *Calliophis*. *The Philippine Journal of Science*, Manila (1963) 92(4), 523–550, fig. 1.

Leviton, A.E. (1964c) Contributions to a review of Philippine snakes, IV. The gerera *Chrysopelea* and *Dryophiops*. *The Philippine Journal of Science*, Manila 93(1), 131–145, 2 figs.

Leviton, A.E. (1964d) Contributions to a review of Philippine snakes, V. The snakes of the genus *Trimeresurus*. *The Philippine Journal of Science*, Manila 93(2), 251–276, 2 figs.

Leviton, A.E. (1965a) Contributions to a review of Philippine snakes, VI. The snakes of the genus *Oxyrhabdium*. *The Philippine Journal of Science*, Manila (1964) 93(3), 407–422, 1 fig.

Leviton, A.E. (1965b) Contributions to a review of Philippine snakes, VII. The snakes of the genera *Naja* and *Ophiophagus*. *The Philippine Journal of Science*, Manila (1964) 93(4), 531–550, 2 figs.

Leviton, A.E. (1965c) Contributions to a review of Philippine snakes, VIII. The snakes of the genus *Lycodon* H. Boie. *The Philippine Journal of Science*, Manila 94(1), 117–140, 2 figs.

Leviton, A.E. (1967) Contributions to a review of Philippine snakes, IX. The snakes of the genus *Cyclocorus*. *The Philippine Journal of Science*, Manila (1965) 94(4), 519–533, 4 figs.

Leviton, A.E. (1968a) Contributions to a review of Philippine snakes, X. The snakes of the genus *Ahaetulla*. *The Philippine Journal of Science*, Manila 96(1), 73–90, 1 fig.

Leviton, A.E. (1968b) The venomous terrestrial snakes of East Asia, India, Malaya, and Indonesia. *In:* Bücherl, W., Buckley, E.E. & Deulofeu, V. (Eds.), *Venomous animals and their venoms. Volume I. Venomous vertebrates.* Academic Press, New York, pp. 529–576.

Leviton, A.E. (1970a) Contribution to a review of Philippine snakes. XI. The snakes of the genus *Boiga*. *The Philippine Journal of Science*, Manila 97(3), 291–314, 2 figs.

Leviton, A.E. (1970b) Contributions to a review of Philippine snakes, XII. The Philippine snakes of the genus *Dendrelaphis* (Serpentes, Colubridae). *The Philippine Journal of Science*, Manila 97(4), 371–396, 2 figs.

Leviton, A.E. (1970c) Description of a new subspecies of *Rhabdophis auriculata* in the Philippines, with comments on the zoogeography of Mindanao Island. *Proceedings of the California Academy of Sciences* (Series 4) 38(18), 347–362, 8 figs.

Leviton, A.E. (1972) *Reptiles and amphibians of North America.* Doubleday, New York, 250 pp., illust.

Leviton, A.E. (1977) A new lytorhynchid snake. *Journal of the Saudi Arabia Natural History Society*, Jiddah (19), 16–25, 1 fig.

Leviton, A.E. (1979) Contributions to a review of Philippine snakes, XIII. The snakes of the genus *Elaphe*. *The Philippine Journal of Science*, Manila (1977) 106(3–4), 99–128, 5 figs.

Leviton, A.E. (1983) Contributions to a review of Philippine snakes, XIV. The snakes of the genera *Xenopeltis*, *Zaocys*, *Psammodynastes* and *Myersophis*. *The Philippine Journal of Science*, Manila 112(3–4), 195–223, 4 figs.

Leviton, A.E. (1986) Description of a new species of *Coluber* (Reptilia, Serpentes, Colubridae) from the southern Tihama of Saudi Arabia, with comments on the biogeography of southwestern Arabia. *Fauna of Saudi Arabia* 8, 436–446, 3 figs.

Leviton, A.E. & Aldrich, M.L. (1984) Introduction: John Anderson (1833–1900): A zoologist in the Victorian Period; check list of the amphibians and reptiles of the Arabian Peninsula; Bibliography of John Anderson. *In:* Anderson, J., *Herpetology of Arabia.* Society for the Study of Amphibians and Reptiles, v-xxxv.

Leviton, A.E. & Anderson, S.C. (1963) Third contribution to the herpetology of Afghanistan. *Proceedings of the California Academy of Sciences* (Series 4) 31(12), 329–339, 5 figs.

Leviton, A.E. & Anderson, S.C. (1967) Survey of the reptiles of the Sheikdom of Abu Dhabi, Arabian Peninsula. Part II. Systematic account of the collection of reptiles made in the Sheikdom of Abu Dhabi by John Gasperetti. *Proceedings of the California Academy of Sciences* (Series 4) 35(9), 157–192, 12 figs.

Leviton, A.E. & Anderson, S.C. (1970a) The amphibians and reptiles of Afghanistan: a checklist and key to the herpetofauna. *Proceedings of the California Academy of Sciences* (Series 4) 38(10), 163–206.

Leviton, A.E. & Anderson, S.C. (1970b) Review of the snakes of the genus *Lytorhynchus*. *Proceedings of the California Academy of Sciences* (Series 4) 37(7), 249–274, 25 figs.

Leviton, A.E. & Banta, B.H. (1956) Catalogue of the amphibian and reptile types in the Natural History Museum of Stanford University supplement number 1. *Herpetologica* 12(2), 213–219.

Leviton, A.E. & Banta, B.H. (1961) Description of a new species of snake of the colubrid genus *Gyalopion* from the state of Zacatecas, México. *Occasional Papers of the California Academy of Sciences* (26), 1–4, 3 fig.

Leviton, A.E. & Banta, B.H. (1964) Midwinter reconnaissance of the herpetofauna of the Cape region of Baja California, México. *Proceedings of the California Academy of Sciences* (Series 4) 30(7), 127–156, 11 figs.

Leviton, A.E. & Brown, W.C. (1959) A review of the snakes of the genus *Pseudorabdion* with remarks on the status of the genera *Agrophis* and *Typhlogeophis* (Serpentes, Colubridae). *Proceedings of the California Academy of Sciences* (Series 4) 29(14), 475–508, 10 figs.

Leviton, A.E. & Munsterman, H.E. (1956) The generic status and subfamily relationships of the colubrid snakes of the genus *Sibynophis* in Madagascar. *Occasional Papers of the Natural History Museum of Stanford University* (4), 1–11, 10 figs.

Leviton, A.E. & Tanner, W.W. (1960) The generic allocation of *Hypsiglena slevini* Tanner (Serpentes, Colubridae). *Occasional Papers of the California Academy of Sciences* (27), 1–7, 2 figs.

Leviton, A.E., Myers, G.S. & Swan, L.W. (1956) Zoological results of the California Himalayan Expedition to Makalu, eastern Nepal. I. Amphibians and reptiles. *Occasional Papers of the Natural History Museum of Stanford University* (1), 1–18, 9 figs.

Leviton, A.E., Gibbs, R.H., Jr., Heal, E. & Dawson, C.E. (1985) Standards in herpetology and ichthyology, Part I. Standard symbolic codes for institutional resource collections in herpetology and ichthyology. *Copeia* 1985(3), 802–832.

Leviton, A.E., Anderson, S.C., Adler, K. & Minton, S.A. (1992) *Handbook to Middle East amphibians and reptiles*. Society for the Study of Amphibians and Reptiles, vii + 252 pp., 23 figs., 33 pls.

Leviton, A.E., Wogan, G.O.U., Koo, M.S., Zug, G.R., Lucas, R.S. & Vindum, J.V. (2003) The dangerously venomous snakes of Myanmar: illustrated checklist with keys. *Proceedings of the California Academy of Sciences* (Series 4) 54(24), 407–462, 80 figs.

Lewis, D. (2011) *A field guide to the amphibians and reptiles of Wyoming*. The Wyoming Naturalist, Douglas, viii + 183 pp., col. illust.

Leybold, F. 1873. *Escursion a las Pampas Arjentinas. Hojas de mi diario, Febrero de 1871, seguido de tablas de observaciones barometricas, i un boceto de la ruta tomada*. Imprenta Nacional, Santiago, 107 pp.

Leynaud, G.C. & Bucher, E.H. (1999) La fauna de serpientes del Chaco Sudamericano: diversidad, distribution geografica y estado de conservation. *Academia Nacional de Ciencias* (Córdoba, Argentina) (98), 1–46, 2 figs., 1 col. pl.

Leynaud, G.C. & Chiaraviglio, M. (1996) Sobre la presencia de *Phimophis guerini* (Duméril, Bibron & Duméril) (Serpentes, Colubridae) en la Provincia de Córdoba. *Cuadernos de Herpetología* 9(2), 111–112.

Leynaud, G.C. & Silmara-Cervantes, R. (2004) Geographic distribution: Serpentes: *Sibynomorphus turgidus*. *Herpetological Review* 35(3), 294.

Leynaud, G.C., Cabrera, M.R. & Carrasco, P. (2005) A survey of The southernmost representatives of the *tricolor* species group, genus *Phalotris* (Serpentes, Colubridae). *Phyllomedusa* 4(2), 103–110, 1 fig.

Li, B.-H. (1987) The herpetofauna of Jiuhua Mountain. *Acta Herpetologica Sinica*, Chengdu (Series 2) 6(2), 78–80.

Li, C., Liu, Q. & Guo, P. (2010) Expanded description of a Chinese endemic snake *Opisthotropis cheni* (Serpentes: Colubridae: Natricinae). *Asian Herpetological Research* (Series 2) 1(1), 57–60, 4 figs.

Li, D.-J. (1989) A survey of reptiles in Leigongshan area. *In:* Matsui, M., Hikida, T. & Goris, R.C. (Eds.), *Current herpetology in East Asia*. Herpetological Society of Japan, Kyoto, pp. 269–275, 3 figs.

Li, D.-J., Wang, J.-A. & Zhao, Z.-F. (1985) Reptilian survey of the karst in Libo, Guizhou. *Acta Herpetologica Sinica*, Chengdu (Series 2) 4(2), 140–143.

Li, H.-F., Su, X.-Y. & Li, S.-Y. (1986) The preliminary investigation of the herpetofauna in Tianjin District, China. *Acta Herpetologica Sinica*, Chengdu (Series 2) 5(4), 299–300.

Li, J.-L. (1995) *China snake Island*. Liaoning Science & Technology Press, Shenyang. 136 pp., col. illust.

Li, J.-L. (1999) Infraspecific classification of two species of *Gloydius* (Serpentes: Crotalinae). *Acta Zootaxonomica Sinica* 24(4), 454-460. [not seen]

Li, M., Fry, B.G. & Kini, R.M. (2005) Putting the brakes on snake venom evolution: the unique molecular evolution patterns of *Aipysurus eydouxii* (marbled sea snake) phospholipase A2 toxins. *Molecular Biology and Evolution* 22(4), 934–941, 3 figs.

Li, P.P., Zhao, E.-M. & Dong, B.-J. (2010) *Amphibians and reptiles of Tibet*. Science Press, Beijing, xii + 251 pp., 143 pls. (in Chinese)

Li, S.-Q. (1983) A new record of Chinese colubrid—*Elaphe hodgsoni* Smith. *Acta Herpetologica Sinica*, Chengdu (Series 2) 1(2), 78.

Li, Z.-C. (2011) Amphibians and reptiles of Guangdong. China Scientific Book Services, Beijing, 266 pp., col. illust. (in Chinese)

Li, Z.-C., Xiao, Z., Qing, N., Lu, W.-H. & Lazell, J. (2009) Amphibians and reptiles of Dinghushan in Guangdong Province, China's oldest nature reserve. *IRCF Reptiles & Amphibians* 16(3), 131–151, 71 figs.

Liais, E (1872) *Climats, géologie, faune, et géographie botanique du Brésil*. Garnier Frères, Paris, viii + 640 pp.

Liang, Q. (2003) Reptiles of Guangdong Nanling Nature Reserve. *In:* Pang, X. (Ed.), *Studies in biodiversity of the Guangdong Nanling National Nature Reserve*. Guangdong Science and Technology Press, Guangzhou, pp. 408–417, 1 pl. (in Chinese)

Lichtenstein, M.H.C. (1823) *Verzeichniss der Doubletten des zoologischen Museums der Königl. Universität zu Berlin nebst Beschreibung vieler bisher unbekannten Arten von Saugethieren, Vögeln, Amphibien und Fischen*. T. Trautwein, Berlin, x + 118 pp.

Lichtenstein, M.H.C. & Martens, E. von. (1856) *Nomenclator reptilium et amphibiorum Musei Zoologici Berolinensis*. Königlichen Akademie der Wissenschaften, Berlin, iv + 48 pp.

Lidth de Jeude, T.W. van. (1886) On *Cophias wagleri*, Boie and *Coluber sumatranus*, Raffles. *Notes from the Leyden Museum* 8, 43–54, pl. 2.

Lidth de Jeude, T.W. van. (1887) On a collection of reptiles and fishes from the West-Indies. *Notes from the Leyden Museum* 9, 129–139.

Lidth de Jeude, T.W. van. (1890) Reptilia from the Malay Archipelago. II. Ophidia. *In:* Weber, M. (Ed.), *Zoologische Ergebnisse einer Reise in Niederlandisch Ost-Indien*. Erster Band. Heft 2. E.J. Brill, Leiden, pp. 178–192, pls. 15–16.

Lidth de Jeude, T.W. van. (1891) List of reptiles brought from Siam by Mr. R.C. Keun. *Notes from the Leyden Museum* 13, 255–256.

Lidth de Jeude, T.W. van. (1893) On reptiles from North Borneo. *Notes from the Leyden Museum* 15, 250–257, 1 fig. (*Calamaria prakkei*)

Lidth de Jeude, T.W. van. (1901) On a *Typhlops* new to Java. *Notes from the Leyden Museum* 23(1), 32.

Lidth de Jeude, T.W. van. (1911a) Reptilien (Schlangen). *In: Nova Guinea. Résultats de l'expédition scientifique néerlandaise à la Nouvelle-Guinée* 5(4), Zoologie. E.J. Brill, Leiden, pp. 519–530.

Lidth de Jeude, T.W. van. (1911b) Reptilien (Schlangen). *In: Nova Guinea. Résultats de l'expédition scientifique néerlandaise à la Nouvelle-Guinée* 9(2), Zoologie. E.J. Brill, Leiden, pp. 265–287, pl. 8.

Lidth de Jeude, T.W. van. (1922) Snakes from Sumatra. *Zoologische Mededelingen*, Leiden 6(4), 239–253.

Lidth de Jeude, T.W. van. (1923) A new *Amblycephalus* from Sumatra. *Zoologische Mededelingen*, Leiden 7(3–4), 243.

Lieb, C.S. (1988) Systematic status of the neotropical snakes *Dendrophidion dendrophis* and *D. nuchalis* (Colubridae). *Herpetologica* 44(2), 162–175, 5 figs.

Lieb, C.S. (1991a) *Dendrophidion nuchale. Catalogue of American amphibians and reptiles*, pp. 520.1–520.2, 1 fig.

Lieb, C.S. (1991b) *Dendrophidion paucicarinatum. Catalogue of American amphibians and reptiles*, pp. 521.1–521.2, 2 figs.

Lieb, C.S. (1991c) *Dendrophidion vinitor. Catalogue of American amphibians and reptiles*, pp. 522.1–522.2, 1 fig.

Lieb, C.S. (1996) *Dendrophidion percarinatum. Catalogue of American amphibians and reptiles*, pp. 636.1–636.2, 1 fig.

Liem, K.F., Marx, H. & Rabb, G.B. (1971) The viperid snake *Azemiops*: its comparative cephalic anatomy and phylogenetic position in relation to Viperinae and Crotalinae. *Fieldiana: Zoology* 59(2), 63–126, 16 figs., 1 pl.

Lim, B.L. (1955) Snakes collected near Kuala Lumpur. *Malayan Nature Journal*, Kuala Lumpur 9(4), 122–125.

Lim, B.L. (1956) The natural food of some Malayan snakes. *Malayan Nature Journal*, Kuala Lumpur 10(4), 139–144.

Lim, B.L. (1958) The Indo-Chinese wolf snake. *Malayan Nature Journal* 13, 70–72, 1 pl.

Lim, B.L. (1963a) *Macrocalamus tweediei*, a new species of reed snake from Malay. *Bulletin of the National Museum*, Singapore (32), 99–102, 1 fig., pl. 2.

Lim, B.L. (1963b) Comments on some rare snakes. *Federation Museum Journal*, Kuala Lumpur 8, 60–64.

Lim, B.L. (1964) Notes on the elephant's trunk snake and the puff-faced water snake in Kuala Lumpur. *Malayan Nature Journal*, Kuala Lumpur 18(4), 179–183, pls. 23–24.

Lim, B.L. (1967a) Snakes collected in Gunong Brinchang, Cameron Highlands, Pehang. *Malayan Nature Journal*, Kuala Lumpur 20, 121–127.

Lim, B.L. (1967b) Further comments on rare snakes. *Federation Museum Journal*, Kuala Lumpur 12, 123–126.

Lim, B.L. (1971) Venomous snakes of southeast Asia. *Southeast Asian Journal of Tropical Medicine and Public Health* 2(1), 56–64, 3 figs.

Lim, B.L. (1972) A genus of snakes (*Oreocalamus*) new to Malaya. *Sarawak Museum Journal* 18, 410–411.

Lim, B.L. (1974) Snakes as natural predators of rats in an oil palm estate. *Malayan Nature Journal*, Kaula Lumpur 27(3–4), 114–117.

Lim, B.L. (1990) Venomous land snakes of Malaysia. *In:* Gopalakrishnakone, P. & Chou, L.M. (Eds.), *Snakes of medical importance (Asia-Pacific region).* National University of Singapore, Singapore, pp. 387–417, 29 figs.

Lim, B.L. (1991) *Poisonous snakes of peninsular Malaysia. Third edition.* Malayan Nature Society, Kuala Lumpur, 74 pp., 52 figs. [first edition1979; second edition 1982, 74 pp., 52 figs.]

Lim, B.L. & Balasingam, E. (1969) A collection of sea snakes from Port Swetteniam, Selangor and Tumpat, Kelantan. *Federation Museum Journal*, Kuala Lumpur 14, 123–126.

Lim, B.L. & Ratnam, L. (1996) The green cat snake *Boiga cyanea*: a new record for Malaysia. *Malayan Nature Journal* 50(1), 43–46, illust.

Lim, B.L. & Sawai, Y. (1975) Ecology and distribution of some seasnakes in Malay Peninsula. *The Snake* 7(1), 28–31, 2 figs..

Lim, B.L. & Sharef bin Kanarudin, M. (1975) Notes on new locality records of some rare snakes in peninsular Malaysia. *Malayan Nature Journal*, Kuala Lumpur 29(1), 23–27.

Lim, B.L., Ratnam, L. & Hussein, N.A. (1995) Snakes examined from the Sungai Singgor area of Temengor, Hulu Perak, Malaysia. *Malayan Nature Journal* 48(3–4), 357–364.

Lim, B.L., Leong, T.M. & Oh, K.S. (2002) Some rare and interesting snake species from Genting Highlands, Pahang. *Journal of Wildlife and Parks* 20, 67–74.

Lim, F.L.K. (1991) *Tales and scales.* Graham Brash, Singapore, 94 pp., 2 figs., 32 col. pls.

Lim, K.K.P. & Chou, L.M. (1990) The herpetofauna of Singapore. *In:* Chou, L.M. & Ng, P.K.L. (Eds.), *Essays in zoology: papers commemorating the 40th annioversary of the Department of Zoology, National University of Singapore.* National University of Singapore, Singapore, pp. 49–59.

Lim, K.K.P. & Lim, F.L.K. (1992) *A guide to the amphibians & reptiles of Singapore.* Singapore Science Center, Singapore, 160 pp., 27 figs., 134 col. figs.

Lim, K.F.L. & Lee, T.M.M. (1989) *Fascinating snakes of southeast Asia—an introduction.* Tropical Press Sdn. Bhd., Kuala Lumpur, xvii + 124 pp., illust.

Lima, A.C. de & Costa-Prudente, A.L. da. (2009) Morphological variation and systematics of *Dipsas catesbyi* (Sentzen, 1796) and *Dipsas pavonina* Schlegel, 1837 (Serpentes: Dipsadinae). *Zootaxa* (2203), 31–48, 10 figs.

Lima-Duarte, M.T., Soares, M. & da Silva-Souto, A. (2003) Geographic distribution: Serpentes: *Bothrops leucurus. Herpetological Review* 34(2), 168.

Lima-Silveira, A. (2004) Geographic distribution: Serpentes: *Rhachidelus brazili. Herpetological Review* 35(4), 412.

Lima-Silveira, A. (2006) Geographic distribution: Serpentes: *Thamnodynastes longicaudus. Herpetological Review* 37(2), 243.

Lima-Silveira, A. (2008) Geographic distribution: Serpentes: *Oxyrhopus guibei. Herpetological Review* 39(2), 242.

Lima-Silveira, A. & Agostini-Cotta, G. (2006) Geographic distribution: Serpentes: *Clelia quimi. Herpetological Review* 37(2), 242.

Lima-Silveira, A., Pires, M.R.S & Agostini-Cotta, G. (2004a) Geographic distribution: Serpentes: *Echinanthera melanostigma. Herpetological Review* 35(4), 410.

Lima-Silveira, A., Pires, M.R.S & Agostini-Cotta, G. (2004b) Geographic distribution: Serpentes: *Leptotyphlops dimidiatus. Herpetological Review* 35(4), 411.

Lima-Silveira, A., Costa, E.R. & Salles, L.M. (2004c) Geographic distribution: Serpentes: *Lystrophis nattereri. Herpetological Review* 35(4), 411.

Limburg, R., Montoya, B. & Merker, G. (2011) *Rosy boas: patterns in time.* ECO Herpetological Publishing, Rodeo, 184 pp., col. illust.

Limpus, C.J. (1975) Coastal sea snakes of subtropical Queensland waters (23° to 28° south latitude). *In:* Dunson, W.A. (Ed.), *The biology of sea snakes.* University Park Press, Baltimore, pp. 173–182, 1 fig.

Limpus, C.J. (1978) The venom apparatus and venom yields of subtropical Queensland Hydrophiidae. *In:* Rosenberg, P. (Ed.), *Toxins: animal, plant and microbial.* Pergamon Press, Oxford, pp. 39–70, 19 figs.

Lin, J. (1982) Is it *Calliophis macclelandii*? *Malayan Naturalist* 35(1–2), 191, illust.

Lin, M.W. (1975) *Venomous snakes of Southeast Asia and some physiological and biochemical aspects of their venoms.* Master's Thesis, Northeastern Illinois University, Chicago, 65 pp.

Lindaker, J.T. (1791) Systematisches Verzeichnis der böhmischen Amphibien mit Beobachtungen. *Neuere Abhandlungen der Böhmischen Gesellschaft der Wissenschaften in Prag* 1, 109–128.

Lindemann, W. (1898) Ueber die Sekretionersheinungen der Giftdruse der Kreuzotter. *Archiv für Mikroskopische Anatomie und Entwicklungsgeschichte,* Bonn 53, 313–321, pl. 16.

Lindberg, K. (1932) Snakes on the Barsi light railway (Deccan). *The Journal of the Bombay Natural History Society* (1931–1932) 35(3), 690–697.

Lindberg, K. (1940) Notes sur une collection de serpents du Deccan (Inde). *Bulletin de la Société Zoologique de France*, Paris 64, 328–336.

Lindell, L.E. (1995) *Causes and consequences of variability in growth rate, energy storage and survival of adders, Vipera berus.* Doctoral Dissertation, Uppsala University, Uppsala, 43 pp.

Linder, A.D. & Fichter, E. (1977) *The amphibians and reptiles of Idaho.* Idaho State University Press, Pocatello, 78 pp., 73 figs.

Lindholm, W.A. (1905a) Beschreibung einen neuen Schlangenart. (*Dipsadophidium weileri* nov. gen. et. nov. sp.). *Jahrbuch des Nassauischen Vereins für Naturkunde*, Wiesbaden 58, 185–187.

Lindholm, W.A. (1905b) Über einige Eidechsen und Schlangen aus Deutsch-Neuguinea. *Jahrbuch des Nassauischen Vereins für Naturkunde*, Wiesbaden 58, 229–240.

Liner, E.A. (1960) A new subspecies of false coral snake (*Pliocercus elapoides*) from San Luis Potosí, Mexico. *Southwestern Naturalist* 5(4), 217–220.

Liner, E.A. (1983) *Tantilla wilcoxi*. *Catalogue of American amphibians and reptiles*, pp. 345.1–345.2.

Liner, E.A. (1992) *Thamnophis exsul* Rossman. Montane Garter Snake, Culebra de Agua Nomada de Montaña. *Catalogue of American amphibians and reptiles*, pp. 549.1–549.2, 2 figs.

Liner, E.A. (1994) *Scientific and common names for the amphibians and reptiles of Mexico in English and Spanish*. Society for the Study of Amphibians and Reptiles, *Herpetological Circular* (23), 1–113.

Liner, E.A. (1996a) Herpetological type material from Nuevo León, México. *Bulletin of the Chicago Herpetological Society* 31(9), 168–171.

Liner, E.A. (1996b) *Rhadinaea montana*. *Catalogue of American amphibians and reptiles*, pp. 640.1–640.2, 1 fig.

Liner, E.A. (2000) Type descriptions and type publications of Hobart M. Smith, 1933 through June, 1999. *Smithsonian Herpetological Information Service* (127), 1–25.

Liner, E.A. (2007) A checklist of the amphibians and reptiles of México. *Occasional Papers, Museum of Natural Science, Louisiana State University* (80), 1–60.

Liner, E.A. & Casas-Andreu, G. (2008) *Standard Spanish, English and scientific names of the amphibians and reptiles of Mexico*. Second edition. Society for the Study of Amphibians and Reptiles, *Herpetological Circular* (38), 1–162.

Liner, E.A. & Wilson, L.D. (1970) Changes in the name and generic status of the Mexican snake *Chersodromus annulatus* Zweifel (Colubridae). *Copeia* 1970(4), 786–788.

Ling, C., Liu, S.-Y., Huang, S., Burbrink, F.T., Guo, P., Sun, Z.-Y. & Zhao, J. (2010) Phylogenetic analyses reveal a unique species of *Elaphe* (Serpentes, Colubridae) new to science. *Asian Herpetological Research* (Series 2) 1(2), 1–7, 1 fig.

Link, H.F. (1806) Ueber die in Meklenburg einheimischen giftigen Schlangen. *Magazin für den Neuesten Zustand der Naturfunde* 12, 289–300, 1 pl.

Link, H.F. (1807) *Beschreibung der Naturalien-Sammlung der Universität zu Rostock*. Zweite Abtheilung. Gedruckt bei Adlers Erben, Rostock, pp. 51–100.

Linnaeus, C. (1754) *Museum Regis Adolphi Friderici regis svecorum, gothorum, vandalorumovce, haer. norv. duc. slesv. hols. storm. ditm. com. oldenb. del menhorstiae. etc. etc. Nimalia rariora imprimis, et exotica, quadrupedia, aves, Amphibia, Pisces, Insecta, Vermes describuntur et determinantur, latine et svetice cum iconibus*. Kongliga Tryckeriet, Holmie, 4 + xxx + 95 + 14 + xxxi pp.

Linnaeus, C. (1758) *Systemae naturae per regna tria naturae, secundum classes, ordines, genera, species, cum characteribus, differentiis, synonymis, locis*. Tomus I. Editio decima, Reformata. Laurentii Salvii, Holmiae, 824 pp.

Linnaeus, C. (1761) *Fauna svecica sistens animalia Sveciae Regni: mammalian, aves, amphibian, pisces, insecta, vermes. Distributa per classes & ordines, genera & species, cum differentiis speciarum, synonymis auctorum, nominibus incolarum, locis natalium, descriptionibus insectorum*. Laurentii Salvii, Stockholmiae, 48 + 578 pp., 1 pl.

Linnaeus, C. (1766) *Systema naturae per regna tria naturae, secundum classes, ordines, genera, species, cum characteribus, differentiis, synonymis, locis*. Tomus I. Editio duodecima, Reformata. Laurentii Salvii, Holmiae, 532 pp.

Linsdale, J.M. (1932) Amphibians and reptiles from lower California. *University of California Publications in Zoology* 38(6), 345–386.

Linsdale, J.M. (1940) Amphibians and reptiles in Nevada. *Proceedings of the American Academy of Arts and Sciences* 73(8), 197–257.

Linzey, D.W. (1979) *Snakes of Alabama*. Strode Publishers, Huntsville, 136 pp., illust.

Linzey, D.W. & Clifford, M.J. (1981) *Snakes of Virginia*. University Press of Virginia, Charlottesville, xiv + 173 pp., 37 figs., 53 col. pls.

Lions, M.L. & Alvarez, B.B. (1996a) Geographic distribution: Serpentes: *Sibynomorphus lavillai*. *Herpetological Review* 27(4), 214.

Lions, M.L. & Alvarez, B.B. (1996b) Geographic distribution: Serpentes: *Sibynomorphus ventrimaculatus*. *Herpetological Review* 27(4), 214.

Lions, M.L. & Alvarez, B.B. (1997) Geographic distribution: Serpentes: *Psomophis genimaculatus*. *Herpetological Review* 28(1), 52.

Lips, K.R. (1993a) Geographic distribution: Serpentes: *Enulius sclateri* (NCN). *Herpetological Review* 24(3), 109.

Lips, K.R. (1993b) Geographic distribution: Serpentes: *Urotheca myersi* (NCN). *Herpetological Review* 24(3), 111.

Lips, K.R. & Savage, J.M. (1994) A new fossorial snake of the genus *Geophis* (Reptilia, Serpentes, Colubridae) from the Cordillera de Talamanca of Costa Rica. *Proceedings of the Biological Society of Washington* 107(2), 410–416, 2 figs.

Littlejohn, M.J. (1962) Zoology of the High Plains, Part I—Ichthyology and herpetology. *Proceedings of the Royal Society of Victoria*, Melbourne 75(2), 311–313.

Liu, C.-C. (1940) Life history of *Trimeresurus jerdonii*. *Peking Natural History Bulletin*, Beijing 14(4), 245–252, 1 pl.

Liu, C.-C. & Hu, S.-C. (1962) A herpetological report of Kwangsi. *Acta Zoologica Sinica*, Beijing 14(Supplement), 75–104.

Liu, L.-B., Chen, Z.-C., Liu, Y.-L. & Shu, P. (1985) Studies on the sea snakes of the middle section of Fujian's coastal waters and their feeding habits. *Acta Herpetologica Sinica*, Chengdu (Series 2) 4(4), 341–343.

Liu, N., Song, Z. & Luo, W. (1989) A new subspecies of *Agkistrodon halys*. *Journal of Lanzhou University* (*Natural Science*) 25(1), 114–117, illust. (in Chinese)

Lizana, M., del Arco, C., Morales, J.J., Bosch, J., Cejudo, D., Lopez-Gonzales, F.J., Gutiérrez, J. & Martin-Sánchez, R. (1995) Atlas provisional de la herpetofauna en el Sistema Central Segoviano. *Revista Española de Herpetología* 9, 113–132, 10 figs.

Llorente, G.A., Montori, A., Santos, X. & Carretero, M.A. (1995) *Atlas dels amfibis i reptiles de Catalunya i Andorra*. Edicions El Brau, Alvarez de Castro, 190 pp., 27 figs., col. photos.

Lobo, F. & Scrocchi, G. (1994) Osteología cranial del género *Philodryas* (Serpentes: Colubridae). *Cuadernos de Herpetología* 8(1), 104–111, 9 figs.

Locard, A. (1875) *Guide aux collections de zoologie, géologie et minéralogie (Muséum d'Histoire Naturelle de Lyon)*. Imprimérie Pitrat Ainé, Lyon, 170 pp.

Lockington, W.N. (1877) Description of a new genus and species of colubrine snake. *Proceedings of the California Academy of Sciences* (1876) 7(1), 52–53.

Lockington, W.N. (1886) The form of the pupil in snakes. *Proceedings of the Academy of Natural Sciences of Philadelphia* 38, 300.

Löding, H.P. (1922) A preliminary catalog of Alabama reptiles and amphibians. *Alabama Museum of Natural History Museum Paper* (5), 1–59.

Loebmann, D. (2008a) Geographic distribution: Serpentes: *Echinanthera affinis*. *Herpetological Review* 39(2), 241.

Loebmann, D. (2008b) Geographic distribution: Serpentes: *Typhlops brongersmianus*. *Herpetological Review* 39(2), 244.

Loebmann, D. (2009a) Geographic distribution: Serpentes: *Xenopholis undulatus*. *Herpetological Review* 40(1), 117.

Loebmann, D. (2009b) Reptilia, Serpentes, Scolecophidia, Anomalepididae, *Liotyphlops* cf. *ternetzii* (Boulenger, 1896): first family record for the state of Ceará, Brazil. *Check List* 5(2), 249–250, 1 fig.

Loebmann, D. (2009c) Reptilia, Squamata, Serpentes, Viperidae, *Bothrops lutzi*: distribution extension, geographic distribution map. *Check List* 5(3), 373–375, 2 figs.

Loebmann, D. & Haddad, C.F.B. (2010) Amphibians and reptiles from a highly diverse area of the Caatinga domain: composition and conservation implications. *Biota Neotropica* 10(3), 227–255, 18 col. figs.

Loebmann, D. & Roberto, I.J. (2009) Geographic distribution: Serpentes: *Oxyrhopus melanogenys orientalis*. *Herpetological Review* 40(3), 366.

Loebmann, D., Ribeiro, S.C., Lima Sales, D. & Oliveira Almeida, W. de (2009) New records of *Atractus ronnie* (Serpentes, Colubridae) in relictual forests from the state of Ceará, Brazil, and comments on meristic and morphometric data. *Biotemas* 22(1), 169–173, 2 figs.

Loeffler, C.W. (2001) Geographic distribution: Serpentes: *Pituophis catenifer deserticola*. *Herpetological Review* 32(2), 124.

Logier, E.B.S. (1939) The reptiles of Ontario. *Royal Ontario Museum Zoology Handbook* (4), ii + 63 pp., 8pls.

Logier, E.B.S. (1958) *The snakes of Ontario*. University of Toronto Press, Toronto, x + 94 pp., figs., 2 pls.

Logier, E.B.S. & Toner, G.C. (1961) Check list of the amphibians and reptiles of Canada and Alaska. *Contributions, Life Sciences Division, Royal Ontario Museum*, Toronto (53), ix + 93 pp.

Lohmeyer, C. (1882) *Systematische Uebersicht der Arten der Reptilien und Amphibien des Museums der naturforschenden Gesellschaft zu Emden*. Druck von Th. Hahn, Emden, 19 pp.

Lohoefener, R. & Altig, R. (1983) Mississippi herpetology. *Mississippi State University Research Center and National Space Technology Laboratory Bulletin* (1), vi + 66 pp., 4 figs.

Lombard, R.E., Marx, H. & Rabb, G.B. (1986) Morphometrics of the ectopterygoid in advanced snakes (Colubroidea): a concordance of shape and phylogeny. *Biological Journal of the Linnean Society*, London 27(2), 133–164, 13 figs.

Lombardi, M. (1985) *Variation and systematics of the elapid snake genus* Austrelaps *(Worrell) in South Australia*. Bachelor's Thesis, University of Adelaide, Adelaide, ii + 50 pp., 6 figs.

Longman, H.A. (1913) Herpetological notes. I. Systematic. Including the description of one new species. Part II. Ethological. *Memoirs of the Queensland Museum*, Brisbane 2, 39–45.

Longman, H.A. (1915) Reptiles from Queensland and the Northern Territory. *Memoirs of the Queensland Museum*, Brisbane 3, 30–34, 2 pls.

Longman, H.A. (1930) *The vertebrates of Queensland: reptiles and amphibians*. Handbook for Queensland, Brisbane, pp. 67–69.

Longmore, R. (Ed.). (1986) Atlas of elapid snakes of Australia. *Australian Flora and Fauna Series*, Canberra (7), 1–115, 24 figs.

Longrich, N.R., Bhullar, B.-A. & Gauthier, J.A. (2012a) A transitional snake from the late Cretaceous period of North America. *Nature* 488(7410), 205–208, 5 figs.

Longrich, N.R., Bhullar, B.-A. & Gauthier, J.A. (2012b) Mass extinction of lizards and snakes at the Cretaceous-Paleogene boundary. *Proceedings of the National Academy of Sciences* 109(52), 21396–21401, 6 figs.

Lönnberg, E. (1894) Notes on reptiles and batrachians collected in Florida in 1892 and 1893. *Proceedings of the United States National Museum* (1894–1895) 17(1003), 317–339.

Lönnberg, E. (1896) Linnean type-specimens of birds, reptiles, batrachians and fishes in the Zoological Museum of the R. University in Upsala. *Kongliga Svenska Vetenskaps-Akadamiens, Handlingar*, Stockholm (Series 4) 22(1), 1–45.

Lönnberg, E. (1899) On a small collection of Javanese reptiles containing a new species of snake. *Zoologischer Anzeiger*, Leipzig 22(581), 108–111.

Lönnberg, E. (1900) Reptiles and batrachians collected in German New Guinea by the late Dr. Erik Nyman. *Annals and Magazine of Natural History*, London (Series 7) 6(36), 574–582.

Lönnberg, E. (1902) On a collection of snakes from north-western Argentina and Bolivia containing new species. *Annals and Magazine of Natural History*, London (Series 7) 10(60), 457–462.

Lönnberg, E. (1907) *Wissenschaftliche ergebnisse der Schwedischen Zoologischen Expedition nach dem Kilimandjaro, dem Meru und den umbebenden Massaisteppen 1905–1906 unter leitung von Prof. Dr. Yngve Sjöstedt. 4. Reptilia and Amphibia*. Königlich Schwedischen Akademie der Wissenschaften, Uppsala, 19 pp., 1 pl.

Lönnberg, E. (1910) Two new snakes from lower Congo. *Arkiv für Zoologi*, Stockholm 7(8), 1–8, 3 figs.

Lönnberg, E. & Andersson, L.G. (1913a) Results of Dr. E. Mjöberg's Swedish scientific expeditions to Australia 1910–13. III. Reptiles. *Kongliga Svenska Vetenskaps-Akademiens, Handlingar*, Stockholm 52(3), 1–17, 5 figs.

Lönnberg, E. & Andersson, L.G. (1913b) On a collection of reptiles from Kismayu. *Arkiv für Zoologi*, Stockholm 8(20), 1–6, 1 fig.

Lönnberg, E. & Rendahl, H. (1925) Dr. E. Mjoberg's zoological collections from Sumatra. 2. Reptiles and batrachians. *Arkiv für Zoologi*, Stockholm 17A(23), 3 pp.

Lopez, M.S. & Giraudo, A.R. (2008) Ecology of the snake *Philodryas patagoniensis* (Serpentes, Colubridae) from northeast Argentina. *Journal of Herpetology* 42(3), 474–480, 2 figs.

Lopez, T.J. & Maxson, L.R. (1995) Mitochondrial DNA sequence variation and genetic differentiation among colubrine snakes (Reptilia, Colubridae, Colubrinae). *Biochemical Systematics and Ecology* 23(5), 487–505, 2 figs.

Lopez, T.J. & Maxson, L.R. (1996) Albumin and mitochondrial DNA evolution: phylogenetic implications for colubrine snakes (Colubridae, Colubrinae). *Amphibia-Reptilia*, Leiden 17(3), 247–259, 3 figs.

Lopez, T.J., Maxson, L.R. & Dowling, H.G. (1993) Phylogenetic relationships of the African egg-eating snake *Dasypeltis scabra*. *Amphibia-Reptilia*, Leiden 14(3), 223–236, 3 figs.

López-Luna, M.A. & de la Torre-Loranca, M.A. (2000) Geographic distribution: Serpentes: *Tantillita lintoni rozellae*. *Herpetological Review* 31(3), 187.

López-Luna, M.A., Vogt, R.C. & de la Torre-Loranca, M.A. (1999) A new species of montane pitviper from Veracruz, México. *Herpetologica* 55(3), 382–389, 10 figs.

López-Luna, M.A., Pérez-Higareda, G. & Smith, H.M. (2001) Variation in *Drymobius chloroticus* from central western Veracruz, Mexico (Reptilia, Serpentes, Colubridae). *Bulletin of the Maryland Herpetological Society* 37(2), 73–76, 1 fig.

Lord, C.E. & Scott, H.H. (1924) *A synopsis of the vertebrate animals of Tasmania*. Oldham, Beddome & Meredith, Hobart, 340 pp.

Lord, J.K. (1866) *The naturalist in Vancouver Island and British Columbia. Volume II*. Richard Bentley, London, 375 pp.

Lortet, L. (1883) Poissons et reptiles du Lac de Tiberiade et de quelques autres parties de la Syrie. *Archives du Museum d'Histoire Naturelle de Lyon* 3, 99–194.

Lötters, S., Wagner, P., Bwong, B.A., Schick, S., Malonza, P.K., Muchai, V., Wasonga, D.V. & Veith, M. (2007) *A fieldguide to the amphibians and reptiles of the Kakamega Forest.* National Museums of Kenya, Nairobi, 112 pp., 102 col. photos.

Lotze, H.U. (1973) Die Schlangen der Erimonìssia in den Kykladen (Griechenland). *Salamandra*, Frankfurt-am-Main 9(2), 58–70, 9 figs.

Lotze, H.U. (1975) Zum Paarungsverhalten der Äskulapnatter, *Elaphe longissima. Salamandra*, Frankfurt-am-Main 11(2), 67–76, 4 figs.

Lotzkat, S., Natera-Mumaw, M., Hertz, A., Sunyer, J. & Mora, D. (2008) New state records of *Dipsas variegatus* (Duméril, Bibron and Duméril, 1854) (Serpentes: Colubridae) from northern Venezuela, with comments on natural history. *Herpetotropicos* 4(1), 25–29, 2 figs.

Lotzkat, S., Hertz, A., Stadler, L., Hamad, N., Arrizo, A. & Köhler, G. (2010) Noteworthy distribution records of reptiles from western Panamá. *Herpetological Review* 41(4), 520–523, 1 fig.

Lotzkat, S., Hertz, A. & Köhler, G. (2012) A new species of *Sibon* (Squamata: Colubroidea: Dipsadidae) from the Cordillera Central of western Panama, with comments on other species of the genus in the area. *Zootaxa* (3485), 26–40, 5 figs.

Lo Valvo, F. & Longo, A.M. (2001) *Anfibi e rettili in Sicilia.* Società Siciliana Scienze Naturali, Palermo, 85 pp., col. illust.

Love, W.B. (1993) Stalking the elusive twig mimic snake *(Langaha nasuta)* with preliminary notes on its behavior in captivity. *The Vivarium* 5(2), 15–17, 6 col. photos.

Loveridge, A. (1918) Notes on snakes in East Africa. *Journal of the East Africa and Uganda Natural History Society* 13, 315–338.

Loveridge, A. (1922) New reptiles from Tanganyika Territory. *Proceedings of the Zoological Society of London* 92(2), 313–315.

Loveridge, A. (1923) Notes on East African snakes collected 1918–1923. *Proceedings of the Zoological Society of London* 93(4), 871–897.

Loveridge, A. (1924a) A new snake of the genus *Lampropeltis. Occasional Papers of the Boston Society of Natural History* (1921–1931) (5), 137–139.

Loveridge, A. (1924b) The snakes of Tanganyika Territory. *Journal of the Tanganyika European Civil Servants' Association* (1923), December, 10–28.

Loveridge, A. (1924c) Check list of the Reptilia recorded from the British territories in East Africa. *Journal of the East Africa and Uganda Natural History Society,* Special Suppl. (3), 1–16.

Loveridge, A. (1928) Notes on snakes and snake bites in East Africa. 1. The Elapine snakes. II. The Hydrophine snakes. *Bulletin of the Antivenin Institute of America* 1, 106–117.

Loveridge, A. (1929) East African reptiles and amphibians in the United States National Museum. *Bulletin of the United States National Museum* (151), 1–135, 1 pl.

Loveridge, A. (1930) Preliminary description of a new tree viper of the genus *Atheris* from Tanganyika Territory. *Proceedings of the New England Zoological Club* 11, 107–108.

Loveridge, A. (1931) A new snake of the genus *Typhlops* from the Belgian Congo. *Copeia* 1931(3), 92–93, 2 figs.

Loveridge, A. (1932a) New opisthoglyphous snakes of the genera *Crotaphopeltis* and *Trimerorhinus* from Angola and Kenya Colony. *Proceedings of the Biological Society of Washington* 45, 83–85.

Loveridge, A. (1932b) A new worm snake of the genus *Leptotyphlops* from Guerrero, México. *Proceedings of the Biological Society of Washington* 45, 151–152.

Loveridge, A. (1932c) New reptiles and amphibians from Tanganyika Territory and Kenya Colony. *Bulletin of the Museum of Comparative Zoology* 72(10), 375–387.

Loveridge, A. (1933) Reports on the scientific results of an expedition to the southwestern highlands of Tanganyika Territory. VII. Herpetology. *Bulletin of the Museum of Comparative Zoology* 74(7), 197–416, 3 pls.

Loveridge, A. (1934) Australian reptiles in the Museum of Comparative Zoology Cambridge, Massachusetts. *Bulletin of the Museum of Comparative Zoology* 77(6), 243–383, 1 pl.

Loveridge, A. (1935) Scientific results of an expedition to rain forest regions in eastern Africa. I. New reptiles and amphibians from East Africa. *Bulletin of the Museum of Comparative Zoology* 79(1), 1–19.

Loveridge, A. (1936a) New tree snakes of the genera *Thrasops* and *Dendraspis* from Kenya Colony. *Proceedings of the Biological Society of Washington* 49, 63–66.

Loveridge, A. (1936b) African reptiles and amphibians in Field Museum of Natural History. *Zoological Series, Field Museum of Natural History* 22(1), 1–111.

Loveridge, A. (1936c) Scientific results of an expedition to rain forest regions in eastern Africa. V. Reptiles. *Bulletin of the Museum of Comparative Zoology* 79(5), 209–337.

Loveridge, A. (1938a) New snakes of the genera *Calamaria, Bungarus* and *Trimeresurus* from Mount Kinabalu, North Borneo. *Proceedings of the Biological Society of Washington* 51, 43–46.

Loveridge, A. (1938b) On a collection of reptiles and amphibians from Liberia. *Proceedings of the New England Zoological Club* 17, 49–74.

Loveridge, A. (1938c) On some reptiles and amphibians from the central region of Australia. *Transactions of the Royal Society of South Australia* 62, 183–191.

Loveridge, A. (1938d) A new freshwater snake (*Chersydrus granulatus luzonensis*) from the Philippines. *Proceedings of the Biological Society of Washington* 51, 209–210.

Loveridge, A. (1939) Revision of the African snakes of the genus *Mehelya* and *Gonionotophis. Bulletin of the Museum of Comparative Zoology* 86(3), 131–162.

Loveridge, A. (1940) Revision of the African snakes of the genera *Dromophis* and *Psammophis. Bulletin of the Museum of Comparative Zoology* 87(1), 1–69.

Loveridge, A. (1941a) Report on the Smithsonian-Firestone Expedition's collection of reptiles and amphibians from Liberia. *Proceedings of the United States National Museum* (1941–1942) 91(3128), 113–140.

Loveridge, A. (1941b) New geckos (*Phelsuma & Lygodactylus*), snake (*Leptotyphlops*), and frog (*Phrynobatrachus*) from Pemba Island, East Africa. *Proceedings of the Biological Society of Washington* 54, 175–178.

Loveridge, A. (1942) Scientific results of a fourth expedition to forested areas in east and central Africa. IV. Reptiles. *Bulletin of the Museum of Comparative Zoology* 91(2), 237–373, 6 pls.

Loveridge, A. (1944a) A new elapid snake of the genus *Maticora* from Sarawak, Borneo. *Proceedings of the Biological Society of Washington* 57, 105–106.

Loveridge, A. (1944b) Further revisions of African snake genera. *Bulletin of the Museum of Comparative Zoology* 95(2), 121–247.

Loveridge, A. (1945a) A new blind snake (*Typhlops tovelli*) from Darwin, Australia. *Proceedings of the Biological Society of Washington* 58, 111–112.
Loveridge, A. (1945b) *Reptiles of the Pacific world*. Macmillan, New York, xii + 259 pp., 7 pls.
Loveridge, A. (1948) New Guinean reptiles and amphibians in the Museum of Comparative Zoology and United States National Museum. *Bulletin of the Museum of Comparative Zoology* 101(2), 305–430.
Loveridge, A. (1950) The green and black mambas of East Africa. *Journal of the East Africa Natural History Society and National Museum*, Nairobi 19(5), 251–252.
Loveridge, A. (1951a) On reptiles and amphibians from Tanganyika Territory collected by C.J.P. Ionides. *Bulletin of the Museum of Comparative Zoology* 106(4), 177–204, 1 fig.
Loveridge, A. (1951b) Synopsis of the African green snakes (*Philothamnus* inc. *Chlorophis*) with the description of a new form. *Bulletin, Institut Royal des Sciences Naturelles de Belgique*, Bruxelles 27(37), 1–12.
Loveridge, A. (1953) Zoological results of a fifth expedition to East Africa. III. Reptiles from Nyasaland and Tete. *Bulletin of the Museum of Comparative Zoology* 110(3), 143–322, 5 pls.
Loveridge, A. (1955a) On snakes collected in the Anglo-Egyptian Sudan by J.S. Owen, Esq. *Sudan Notes and Records*, Khartoum 36, 2–20.
Loveridge, A. (1955b) On a second collection of reptiles and amphibians taken in Tanganyika Territory by C.J.P. Ionides, Esq. *Journal of the East Africa Natural History Society and National Museum*, Nairobi 22(5), 168–198.
Loveridge, A. (1957a) Check list of the reptiles and amphibians of East Africa (Uganda; Kenya; Tanganyika; Zanzibar). *Bulletin of the Museum of Comparative Zoology* 117(2), 153–362 + xxxvi.
Loveridge, A. (1957b) On a third collection of reptiles taken in Tanganyika by C.J.P. Ionides, Esq. *Tanganyika Notes and Records*, Dar-es-Salaam (1956) (43), 1–19.
Loveridge, A. (1958) Revision of five African snake genera. *Bulletin of the Museum of Comparative Zoology* 119(1), 1–198.
Loveridge, A. (1959) On a fourth collection of reptiles mostly taken in Tanganyika Territory by Mr. C.J.P. Ionides. *Proceedings of the Zoological Society of London* 133(1), 29–44.
Loveridge, A. (1961) Status of new vertebrates described or collected by Loveridge. *Journal of the East Africa Natural History Society and National Museum,* Nairobi 23(7), 250–280.
Lovich, R.E., Akre, T., Ryan, M., Nuñez, S., Cruz, G., Borjas, G., Scott, N.J., Flores, S., Cid, W.D., Flores, A., Rodriguez, C., Luque-Montes, I.R. & Ford, R. (2010) New herpetofaunal records from southern Honduras. *Herpetological Review* 41(1), 112–115.
Low, T. (1978) The reptiles of Magnetic Island, North Queensland. *Herpetofauna*, Sydney 9(2), 10–14.
Lowe, C.H., Jr. (1955) Generic status of the aquatic snake *Thamnophis angustirostris*. *Copeia* 1955(4), 307–309.
Lowe, C.H., Jr. (1964) *The vertebrates of Arizona*. University of Arizona Press, Tucson, vii + 259 pp., 66 figs.
Lowe, C.H., Jr. & Norris, K.S. (1954) Analysis of the herpetofauna of Baja California, Mexico. *Transactions of the San Diego Society of Natural History* 12(4), 47–64, 2 figs., pls. 4–5.
Lowe, C.H., Jr. & Norris, K.S. (1955) Analysis of the herpetofauna of Baja California, Mexico. III. New and revived reptilian subspecies of Isla de San Estaban, Gulf of California, Sonora, Mexico, with notes on other satellite islands of Isla Tiburon. *Herpetologica* 11(2), 89–96.
Lowe, C.H., Jr. & Woodin, W.H., III. (1954) A new racer (genus *Masticophis*) from Arizona and Sonora, Mexico. *Proceedings of the Biological Society of Washington* 67, 247–250.
Lowe, C.H., Jr., Schwalbe, C.R. & Johnson, T.B. (1986) *The venomous reptiles of Arizona*. Arizona Fish and Game Department, Phoenix, 115 pp., 24 col. photos.
Lowe, W.P. (1932) *The trail that is always new.* Gurney & Jacobson, London, xviii + 271 pp., 18 illust.
Luard, C.E. (1917) The bronze-backed tree snake (*Dendrolaphis tristis*) in central India. *The Journal of the Bombay Natural History Society* (1917–1918) 25(2), 306.
Lucas, A.H.S. & Frost, C. (1896) Reptilia. *In:* Spencer, B. (Ed.), *Report on the work of the Horn Scientific Expedition to Central Australia. Part II. Zoology.* Dulau and Co., London, pp. 112–151, pls. 8–12.
Lucas, F.A. (1899) A new snake from the Eocene of Alabama. *Proceedings of the United States National Museum* (1898–1899) 21(1164), 637–638, pls. 45–46.
Lucas, S.G., Heckert, A.B. & Sealey, P.L. (1995) A fossil specimen of the long-nosed snake *Rhinocheilus* from the Pliocene of southern New Mexico. *Texas Journal of Science* 47(1), 9–12, 1 fig.
Lue, K.-Y. (1987) *Snakes of Taiwan.* Second edition. Qiu Lei, Taipei, 249 pp., 23 figs., 138 col. photos. (in Chinese) [first edition, 1970]
Lue, K.-Y. (1990) *The amphibians and reptiles of Taiwan.* Council of Agriculture, Taipei, 124 pp., 3 figs., 113 col. photos. (in Chinese)
Lue, K.-Y. & Lai, J.-S. (1992) Note about the clutch size of *Pareas formosensis* (Colubridae: Squamata). *Journal of the Taiwan Museum* 45(1), 61–63.
Lue, K.-Y., Chen, S.-H. & Yeh, K.K.-C. (1984) *The amphibians and reptiles of Kenting National Park.* Taiwan, 126 pp., 96 col. photos. (in Chinese)
Lue, K.-Y., Chen, S.-H. & Chen, S.-L. (1989) *Taiwan snakes.* Taiwan Provincial Department of Education, Wufeng, 148 pp., col. photos. (in Chinese)
Lue, K.-Y., Du, M.-Z. & Xiang, G.-S. (1999) *A field guide to the reptiles and amphibians of Taiwan.* Society of Wildlife and Nature, Taipei, 343 pp., 466 col. photos. (in Chinese)
Lueth, F.X. (1941) *Manual of Illinois snakes, including a description of their most important characteristics.* Illinois Department of Conservation, Springfield, 48 pp., illust.
Lueth, F.X. (1949) *The manual of Illinois snakes*. Second edition. Illinois Department of Conservation, Springfield, 35 pp., 9 figs., 31 unnumbered illust.
Lugaro, G. (1957) Elenco sistematico dei rettili italiani conservati nella collezione di studio esistente press il Museo di Storia Naturale di Milano, con Brevi note critiche ed esplicative. *Atti della Società Italiana di Scienze Naturali e del Museo Civico di Storia Naturale de Milano* 96(1–2), 20–36.
Luiselli, L. & Akani, G.C. (1998a) Aspects of the ecology of *Calabaria reinhardtii* (Serpentes, Booidea) in southeastern Nigeria. *Herpetological Natural History* 6(1), 65–71, 1 fig.
Luiselli, L. & Akani, G.C. (1998b) Observations on the ecology of *Bitis gabonica* and *Bitis nasicornis* (Viperidae) in the rain forests of south-eastern Nigeria. *Bulletin de la Société Herpétologique de France* (87–88), 25–34, 2 figs.
Luiselli, L. & Akani, G.C. (2003) Diet of sympatric Gaboon vipers (*Bitis gabonica*) and nose-horned vipers (*Bitis nasicornis*) in southern Nigeria. *African Journal of Herpetology* 52(2), 101–106.

Luiselli, L. & Angelici, F.M. (1998) Sexual size dimorphism and natural history traits are correlated with intersexual dietary divergence in royal pythons *(Python regius)* from the rainforests of southeastern Nigeria. *Italian Journal of Zoology* 65, 183–185.

Luiselli, L. & Angelici, F.M. (2000) Ecological relationships in two Afrotropical cobra species *(Naja melanoleuca* and *Naja nigricollis). Canadian Journal of Zoology* 78, 191–198, 4 figs.

Luiselli, L. & Rugiero, L. (1990) On habitat selection and phenology in six species of snakes in Canale Monterano (Tolfa Mountains, Latium, Italy) including data on reproduction and feeding in *Vipera aspis francisciredi* (Squamata, Viperidae). *Herpetozoa*, Wien 2(3–4), 107–115, illust.

Luiselli, L., Akani, G.C. & Capizzi, D. (1998a) Food resource partitioning of a community of snakes in a swamp rainforest of south-eastern Nigeria. *Journal of Zoology*, London 246(2), 125–133.

Luiselli, L., Akani, G.C. & Barieenee, I.F. (1998b) Observations on habitat, reproduction and feeding of *Boiga blandingi* (Colubridae) in south-eastern Nigeria. *Amphibia-Reptilia*, Leiden 19(4), 430–436.

Luiselli, L., Akani, G.C., Angelici, F.M. & Barieenee, I.F. (1998c) Reproductive strategies of sympatric *Bitis gabonica* and *Bitis nasicornis* (Viperidae) in the Niger Delta (Port Harcourt, Nigeria): preliminary data. *Amphibia-Reptilia*, Leiden 19(20, 223–229.

Luiselli, L., Akani, G.C., Otonye, L.D., Ekanem, J.S. & Capizzi, D. (1999) Additions to the knowledge of the natural history of *Bothrophthalamus lineatus* (Colubridae) from the Port Harcourt region of Nigeria. *Amphibia-Reptilia*, Leiden 20(3), 318–326.

Luiselli, L., Angelici, F.M. & Akani, G.C. (2000) Large elapids and arboreality: the ecology of Jameson's green mamba *(Dendroaspis jamesoni)* in an Afrotropical forested region. *Contributions to Zoology* 69(3), 147–155, 1 fig.

Luiselli, L., Akani, G.C. & Angelici, F.M. (2001) Diet and foraging behaviour of three ecologically little-known African forest snakes: *Meizodon coronatus, Dipsadoboa duchesnei* and *Hapsidophrys lineatus. Folia Zoologica*, Praha 50(2), 151–158.

Luiselli, L., Effah, C., Angelici, F.M., Odegbune, E., Inyang, M.A., Akani, G.C. & Politano, E. (2002) Female breeding frequency, clutch size and dietary habits of a Nigerian population of Calabar ground python, *Calabaria reinhardtii. Herpetological Journal* 12, 127–129, 1 fig.

Luiselli, L., Akani, G.C., Angelici, F.M., Politano, E., Ude, L. & Wariboko, S.M. (2003) Diet of the semi-aquatic snake, *Afronatrix anoscopus* (Colubridae) in southern Nigeria. *African Journal of Herpetology* 52(2), 123–126.

Luiselli, L., Akani, G.C., Angelici, F.M., Politano, E., Ude, L. & Wariboko, S.M. (2004a) Aspects of the ecology of a population of *Causus maculatus* (Reptilia: Viperidae) from southern Nigeria. *Amphibia-Reptilia*, Leiden 25(1), 99–104.

Luiselli, L., Akani, G.C., Angelici, F.M., Eniang, E.A., Ude, L. & Politano, E. (2004b) Local distribution, habitat use, and diet of two supposed "species" of the *Psammophis "phillipsi"* complex (Serpentes: Colubridae), sympatric in southern Nigeria. *Amphibia-Reptilia*, Leiden 25(4), 415–423, 1 fig..

Luiselli, L., Akani, G.C., Angelici, F.M., Ude, L. & Wariboko, S.M. (2005) Seasonal variation in habitat use in sympatric Afrotropical semi-aquatic snakes, *Grayia smythii* and *Afronatrix anoscopus* (Colubridae). *Amphibia-Reptilia*, Leiden 26(3), 372–376, 1 fig.

Luja, V.H. (2006) Geographic distribution: Serpentes: *Tantilla schistosa. Herpetological Review* 37(4), 501.

Luna-Reyes, R. & Suárez-Velázquez, A. (2008) *Reptiles venenosos de Chiapas: reconocimiento, primeros auxilios y tratamiento medico en caso de mordedura.* Instituto de Historia Natural, Tuxtla Gutiérrez. 90 pp., col. illust.

Lungu, A.N., Zerova, G.A. and Chkhikvadze, V.M. (1989) Large lizards and venomous snakes from the middle Sarmatian of Moldavia and the significance of the fossil herpetofauna for palaeoclimatology. *In:* Lungu, A.N. (Ed.), *The fauna and flora of the Mesozoic and Caenozoic of the Russian platform region.* Shtiintsa, Kishinev, pp. 59–69, illust. (in Russian)

Lunney, D. & Ayers, D. (Eds.) (1993) *Herpetology in Australia: a diverse discipline.* Royal Zoological Society of New South Wales, Mosman, 414 pp., illust.

Luque-Montes, I.R. & Townsend, J.H. (2009) Geographic distribution: Serpentes: *Ninia espinali. Herpetological Review* 40(4), 457.

Lütken, C. (1863) Nogle nye krybodyr og padder. *Videnskabelige Meddelelser fra den Naturhistoriske Forening i Kjobenhavn* (1862–1863) 24(20–22), 292–331, 2 pls.

Luttenberger, F. (1978) *Die Schlangen Österreichs: Bestimmung, Ökologie, Verhalten, Fortpflanzung, Schlangenbiss.* Facultas-Verlag, Wien, 67 pp., 48 figs.

Lutz, A. & Mello, O. (1922) *Elaps ezequieli* e *Rhinostoma bimaculatum*, cobras novas do Estado de Minas Geraes. *Memorias do Instituto Oswaldo Cruz*, Rio de Janeiro 15(1), 235–239, 7 figs., pl. 31,

Lutz, A. & Mello, O. (1923) Duas novas especies de colubridios Brasileiros. *Folha Médica*, Rio de Janeiro 4(1), 1–4.

Luykx, P., Slowinski, J.B. & McCranie, J.R. (1992) The karyotype of the coral snake *Micrurus ruatanus. Amphibia-Reptilia*, Leiden 13(3), 289–292, 2 figs.

Lydekker, R. (1888) *Catalogue of fossil Reptilia and Amphibia in the British Museum (Natural History); Cromwell Road, S.W. Part I. Containing the orders Ornithosauria, Crocodilia, Dinosauria, Squamata, Rhynchocephalia, and Proterosauria.* British Museum (Natural History), London, 309 pp.

Lydekker, R. (1901) Reptiles. *In:* Lydekker, R. (Ed.), *The royal natural history. Volume 5.* Gebbie & Co., Philadelphia, pp. 2357–2962.

Lynch, J.D. (1985) Annotated checklist of the amphibians and reptiles of Nebraska. *Transactions of the Nebraska Academy of Sciences* 13, 33–57, illust.

Lynch, J.D. (2009) Snakes of the genus *Oxyrhopus* (Colubridae: Squamata) in Colombia: taxonomy and geographic variation. *Papéis Avulsos de Zoologia* 49(25), 319–337, 9 figs.

Lynch, J.D. & Smith, H.M. (1965) New or unusual amphibians and reptiles from Oaxaca, Mexico. I. *Herpetologica* 21(3), 168–177.

Lynch, J.D. & Smith, H.M. (1966) New or unusual amphibians and reptiles from Oaxaca, II. *Transactions of the Kansas Academy of Sciences* 69(1), 58–75, 3 figs.

Lynn, W.G. (1934) A new snake (*Paleophis virginianus*) from the Eocene of Virginia. *Johns Hopkins University Studies in Geology* 11, 245–249, 1 pl.

Ma, J. & Zong, Y. (1984) A new species of the genus *Elaphe–Elaphe maculata. Acta Herpetologica Sinica*, Chengdu (Series 2) 3(3), 59–60, figs. (in Chinese)

Maass-Berlin, T.A. (1933) Kennzeichen der sogenennten ungiftigen und verdächtigen Schlangen mit Gift-Apparat. *Tabulae Biologicae Periodicae*, W. Junk 9, 105–195.

MacCulloch, R.D. (2002) *The ROM field guide to amphibians and reptiles of Ontario.* Royal Ontario Museum, Toronto, 168 pp., col. illust.

MacCulloch, R.D. & Lathrop, A. (2004a) *Micrurus ibiboboca* (Serpentes, Elapidae) is not a Guiana Shield species. *Phyllomedusa* 3(2), 141–144.

MacCulloch, R.D. & Lathrop, A. (2004b) A new species of *Dipsas* (Squamata: Colubridae) from Guyana. *Revista de Biologia Tropical* 52(1), 239–247.

MacCulloch, R.D. & Reynolds, R.P. (2012) Amphibians and reptiles from Paramakatoi and Kato, Guyana. *Check List* 8(2), 207–210, 3 figs.

MacCulloch, R.D., Lathrop, A., Kok, P.J.R., Ernst, R. & Kalamandeen, M. (2009) The genus *Oxyrhopus* (Serpentes: Dipsadidae: Xenodontinae) in Guyana: morphology, distributions and comments on taxonomy. *Papéis Avulsos de Zoologia* 49(36), 487–495, 4 figs.

Macey, J.R., Papenfuss, T. & Zhao, E.-M. (1988) The snakes of Ningxia Hui Autonomous Region as an indication of a herpetological corridor. *Asiatic Herpetological Research* 2, 4–5.

MacGowan, B. & Kingsbury, B. (2001) *Snakes of Indiana.* Purdue University, West Lafayette, 51 pp., col. illust.

Machado, O. (1944) Observações sobre ofidios do Brasil. *Boletim do Instituto Vital Brasil*, Rio de Janeiro 4(27), 61–64, 2 pls.

Machado, O. (1945a) Estudos comparativos sôbre ofídios do Brasil. *Boletim do Instituto Vital Brasil*, Rio de Janeiro 5(1), 17–36, 2 pls.

Machado, O. (1945b) Estudo comparativo das Elapideas do Brasil. *Boletim do Instituto Vital Brasil*, Rio de Janeiro 5(2), 37–46, 1 pl.

Machado, O. (1945c) Estudo comparativo das crotalideas do Brasil. *Boletim do Instituto Vital Brasil*, Rio de Janeiro 5(1), 47–66, 4 pls.

Machado-Filho, P.R., Duarte, M.R., Carmo, L.F. & Franco, F.L. (2011) New record of *Corallus cropanii* (Boidae: Boinae): a rare snake from the Vale do Ribeira, State of São Paulo, Brazil. *Salamandra* 47(2), 112–115, 6 figs.

Mack, G. & Gunn, S.B. (1953) De Vis' types of Australian snakes. *Memoirs of the Queensland Museum*, Brisbane 13(1), 58–70.

MacKay, A. & MacKay, J. Privately printed. (1985) *Poisonous snakes of eastern Africa and the treatment of their bites.* 1–95, illust.

Mackay, R.D. (1949) The Australian coral snake. *Proceedings of the Royal Society of New South Wales*, Sydney (1947–1948), 36–37.

Mackay, R.D. (1950) Notes on the diamond snake and carpet snake. (*Morelia argus* and *Morelia argus variegata*). *Proceedings of the Royal Society of New South Wales*, Sydney (1948–1949), 35–36, 1 photo.

Mackay, R.D. (1952) The taipan. *Proceedings of the Royal Society of New South Wales*, Sydney (1950–1951), 24–25, 1 fig.

Mackay, R.D. (1955) A revision of the genus *Pseudechis*. *Proceedings of the Royal Society of New South Wales*, Sydney (1953–1954), 15–23.

Mackay, R.D. (1956) The status of *Denisonia maculata* Steindachner and related species (Ophidia, Elapidae). *Proceedings of the Royal Society of New South Wales*, Sydney (1954–55), 29–33.

Maclean, W.P. (1982) *Reptiles and amphibians of the Virgin Islands.* Macmillan Caribbean, 54 pp.

Maclean, W.P., Kellner, R. & Dennis, H. (1977) Islands lists of West Indian amphibians and reptiles. *Smithsonian Herpetological Information Service* (40), 1–47, illust.

Maclear, J.P. (1887) Report on Christmas Island. *Proceedings of the Zoological Society of London* 55(3), 508–510.

Macleay, W. (1876) Notes on a new species of *Dendrophis* from Cleveland Bay. *Proceedings of the Linnaean Society of New South Wales*, Sydney (1876–1877) 1(1), 15–16.

Macleay, W. (1877) The ophidians of the Chevert Expedition. *Proceedings of the Linnaean Society of New South Wales*, Sydney (1877–1878) 2(1), 33–41.

Macleay, W. (1878a) On a new species of *Hoplocephalus*, from Sutton Forest. *Proceedings of the Linnaean Society of New South Wales*, Sydney (1877–1878) 2(1), 52–54.

Macleay, W. (1878b) Notes on a collection of snakes from Port Darwin. *Proceedings of the Linnaean Society of New South Wales*, Sydney (1877–1878) 2(3), 219–222.

Macleay, W. (1882) Description of two new species of snakes. *Proceedings of the Linnaean Society of New South Wales*, Sydney (1881–1882) 6(4), 811–813.

Macleay, W. (1884a) Notes on some reptiles from the Herbert River, Queensland. *Proceedings of the Linnaean Society of New South Wales*, Sydney (1883–1884) 8(4), 432–436.

Macleay, W. (1884b) Census of Australian snakes with descriptions of two new species. *Proceedings of the Linnaean Society of New South Wales*, Sydney (1884–1885) 9(3), 548–568.

Macleay, W. (1885a) On a new snake from the Barrier Range. *Proceedings of the Linnaean Society of New South Wales*, Sydney (1885–1886) 10(1), 61–62.

Macleay, W. (1885b) On some Reptilia lately received from the Herbert River District, Queensland. *Proceedings of the Linnaean Society of New South Wales*, Sydney (1885–1886) 10(1), 64–68.

Macleay, W. (1887a) Description of a new species of *Hoplocephalus*. *Proceedings of the Linnaean Society of New South Wales*, Sydney (Series 2) (1887–1888) 1(4), 1111–1112.

Macleay, W. (1887b) On a new *Hoplocephalus* from the Gulf of Carpentaria. *Proceedings of the Linnaean Society of New South Wales*, Sydney (Series 2) (1887–1888) 2(2), 403–404.

Macleay, W. (1888a) Notes on Mr. Froggatt's collections made during the year 1887, in the vicinity of Derby, King's Sound, N.W., Australia. *Proceedings of the Linnaean Society of New South Wales*, Sydney (1887–1888) (Series 2) 2(4), 1017–1020.

Macleay, W. (1888b) Notes on some ophidians from King's Sound north-west Australia. *Proceedings of the Linnaean Society of New South Wales*, Sydney (1887–1888) (Series 2) 3(2), 416–418.

Madsen, T. & Osterkamp, M. (1982) Notes on the biology of the fish-eating snake *Lycodonomorphus bicolor* in Lake Tanganyika. *Journal of Herpetology* 16(2), 185–188, 2 figs.

Madsen, T. & Shine, R. (2001) Conflicting conclusions from long-term versus short-term studies on growth and reproduction of a tropical snake. *Herpetologica* 57(2), 147–156, 5 figs.

Maduwage, K., Silva, A., Manamendra-Arachchi & Pethiyagoda, R. (2009) A taxonomic revision of the South Asian hump-nosed vipers (Squamata: Viperidae: Hypnale). *Zootaxa* (2232), 1–28, 9 figs.

Maes, H.H. (1989) *Phylogeny of the snake subfamily Crotalinae using scale micro-ornamentation and hemipenial morphology.* Master's Thesis, California State University, Fullerton, 340 pp., 57 figs.

Mägdefrau, H., Mägdefrau, K. & Schlüter, A. (1991) Herpetologische Daten vom Guaiquinima-Tepui, Venezuela. *Herpetofauna*, Sydney 13(70), 13–26.

Magalhães, O. (1925) Contribuição para o estudo dos ophidios brasileiros. *Memorias do Instituto Oswaldo Cruz*, Rio de Janeiro 18(1), 151–161, pls. 7–12.

Maglio, V.J. (1970) West Indian xenodontine colubrid snakes: their probable origin, phylogeny, and zoogeography. *Bulletin of the Museum of Comparative Zoology* 141(1), 1–53, 37 figs.

Mahajan, K.K. & Agrawal, H.P. (1976) Report on a collection of reptile fauna from Simla Hills, Himachal Pradesh. *Records of the Zoological Survey of India*, Calcutta 71(1–4), 301–312.

Mahdi, N. & Georg, P.V. (1969) A systematic list of the vertebrates of Iraq. *Iraq Natural History Museum Publications*, Baghdad 26, 1–104.

Mahendra, B.C. (1936) On two collections of the ophidian genus, *Cylindrophis* Wagler. *Proceedings of the Indian Academy of Science*, Bangalore 4(3), 230–238, figs. a–f.

Mahendra, B.C. (1938a) The taxonomic description of *Rhinophis travancoricus* Boul. *The Journal of the Bombay Natural History Society* (1938–1939) 40(2), 388–390.

Mahendra, B.C. (1938b) The lepidosis of *Xenopeltis unicolor* Reinw. *Current Science*, Bangalore 6(11), 559–560.

Mahendra, B.C. (1984) Handbook of the snakes of India, Ceylon, Burma, Bangladesh, and Pakistan. *Annals of Zoology*, Agra 22B, 412 pp., 132 figs.

Mahlow, K., Tillack, F., Schmidtler, J.F. & Müller, J. (2013) An annotated checklist, description and key to the dwarf snakes of the genus *Eirenis* Jan, 1863 (Reptilia, Squamata, Colubridae), with special emphasis on dentition. *Vertebrate Zoology* 63(1), 41–85, 18 figs., 9 pls.

Mahnert, V. (1976) Catalogue des types de poissons, amphibiens et reptiles du Muséum d'Histoire naturelle de Genève. *Revue Suisse de Zoologie*, Genève 83(2), 471–496.

Mahony, S., Hasan, M.K., Kabir, M.M., Ahmed, M. & Hossain, M.K. (2009) A catalogue of amphibians and reptiles in the collection of Jahangirnagar University, Dhaka, Bangladesh. *Hamadryad* 34(1), 80–94.

Mahrdt, C.R. & Grismer, L.L. (1995) Geographic distribution: Serpentes: *Trimorphodon biscutatus* (Lyre snake). *Herpetological Review* 26(3), 157.

Mahrdt, C.R., Beaman, K.R., Rosen, P.C. & Holm, P.A. (2001a) *Chionactis. Catalogue of American amphibians and reptiles*, pp. 730.1–730.6, 4 figs., 3 col. photos.

Mahrdt, C.R., Beaman, K.R., Rosen, P.C. & Holm, P.A. (2001b) *Chionactis occipitalis. Catalogue of American amphibians and reptiles*, pp. 731.1–731.12, 4 figs.

Mahrdt, C.R., Beaman, K.R., Rosen, P.C. & Holm, P.A. (2001c) *Chionactis palarostris. Catalogue of American amphibians and reptiles*, pp. 732.1–732.5, 1 col. fig.

Maisano, J.A. & Rieppel, O. (2007) The skull of the Round Island boa, *Casarea dussumieri* Schlegel based on high-resolution x-ray computed tomography. *Journal of Morphology* 268, 371–384, 5 figs.

Majupuria, T.C. (1969) The functional minute anatomy of the hemipenes of the Indian rat-snake, *Ptyas mucosus* (Linn.). *Zoologischer Anzeiger*, Leipzig 181(5–6), 386–397, 15 figs.

Majupuria, T.C. (1981) Reptiles. *In:* Majupuria, T.C. (Ed.), *Wild is beautiful: introduction to the magnificent, rich and varied fauna and wildlife of Nepal.* S. Devi, Lashkar, pp. 147–177, 5 figs.

Maki, M. (1931) *A monograph of the snakes of Japan.* Dai-ichi Shobo, Tokyo, 240 pp., 158 figs., 86 pls.

Maki, M. (1933) *A monograph of the snakes of Japan.* Dai-ichi Shobo, Tokyo, 248 pp., 376 figs. (in Japanese)

Maki, M. (1931–1933) *A monograph of the snakes of Japan. Atlas.* Dai-ichi Shobo, Tokyo, 86 col. pls. [pls. 1–39 = 1931, pls. 40–86 = 1933]

Maki, M. (1935) A new poisonous snake (*Calliophis iwasakii*) from Loo-Choo. *Transactions of the Natural History Society of Formosa*, Taihoku 25, 216–219, figs.

Maki, M. (1937) A new subspecies, *Amblycephalus formosensis iwasakii*, belonging to Amblycephalidae from Ishigaki-jima. *Transactions of the Natural History Society of Formosa*, Taihoku 27(169), 217–218, 1 fig.

Makovicky, P.J., Apesteguía, S. & Agnolín, F.L. (2005) The earliest dromaeosaurid theropod from South America. *Nature* 437, 1007–1011.

Makowsky, R., Marshall, J.C., McVay, J., Chippindale, P. & Rissler, L.J. (2010) Phylogeographic analysis and environmental niche modeling of the plainbellied water snake (*Nerodia erythrogaster*) reveals low levels of genetic and ecological differentiation. *Molecular Phylogenetics and Evolution* 55(3), 985–995, 6 figs.

Maley, A.J., Savit, A.Z., Heinz, H.M., Powell, R. & Henderson, R.W. (2006) *Alsophis rufiventris. Catalogue of American Amphibians and Reptiles*, pp. 818.1–818.4, 1 fig.

Malhotra, A. & Davis, K. (1991) A report on a herpetological survey of the Srivilliputtur Reserve Forest, Tamil Nadu. *The Journal of the Bombay Natural History Society* 88(2), 157–166.

Malhotra, A. & Thorpe, R.S. (1997) New perspectives on the evolution of south-east Asian pitvipers (genus *Trimeresurus*) from molecular studies. *In:* Thorpe, R.S., Wüster, W. & Malhotra, A. (Eds.), *Venomous snakes: Ecology, evolution and snakebite.* Zoological Society of London, Clarendon Press, Oxford, pp. 115–128, 5 figs.

Malhotra, A. & Thorpe, R.S. (2000) A phylogeny of the *Trimeresurus* group of pit vipers: new evidence from a mitochondrial gene tree. *Molecular Phylogenetics and Evolution* 16(2), 199–211, 4 figs.

Malhotra, A. & Thorpe, R.S. (2004a) Reassessment of the validity and diagnosis of the pitvipers *Trimeresurus venustus* G. Vogel, 1991. *Herpetological Journal* 14, 21–33, 5 figs.

Malhotra, A. & Thorpe, R.S. (2004b) Maximizing information in systematic revisions: a combined molecular and morphological analysis of a cryptic green pitviper complex (*Trimeresurus stejnegeri*). *Biological Journal of the Linnean Society,* London 82(2), 219–235, 5 figs.

Malhotra, A. & Thorpe, R.S. (2004c) A phylogeny of four mitochondrial gene regions suggests a revised taxonomy for Asian pitvipers (*Trimeresurus* and *Ovophis*). *Molecular Phylogenetics and Evolution* 32(1), 83–100, 3 figs.

Malhotra, A., Thorpe, R.S. & Stuart, B.L. (2004) A morphometric analysis of *Trimeresurus vogeli* (David, Vidal and Pauwels, 2001), with new data on diagnostic characteristics, distribution and natural history. *Herpetological Journal* 14, 65–77, 3 figs.

Malhotra, A., Creer, S., Pook, C.E. & Thorpe, R.S. (2009) Inclusion of nuclear intron sequence data helps to identify the Asian sister group of New World pitvipers. *Molecular Phylogenetics and Evolution* 54(1), 172–178, 1 fig.

Malhotra, A., Dawson, K., Guo, P. & Thorpe, R.S. (2011a) Phylogenetic structure and species boundaries in the mountain pitviper *Ovophis monticola* (Serpentes: Viperidae: Crotalinae) in Asia. *Molecular Phylogenetics and Evolution* 59(2), 444–457, 5 figs.

Malhotra, A., Thorpe, R.S., Mrinalini, & Stuart, B.L. (2011b) Two new species of pitviper of the genus *Cryptelytrops* Cope, 1860 (Squamata: Viperidae: Crotalinae) from southeast Asia. *Zootaxa* (2757), 1–23, 8 figs.

Malkmus, R. (1979a) Zur Faunistik und Okologie der Amphibien und Reptilien in der Serra de Sintra Portugal. *Nachrichten des Naturwissenschafliches Musweum der Stadt Aschaffenburg* (88), 1–56, illust.

Malkmus, R. (1979b) Beitrag zur vertikalen Verbreitung der herpetofauna Portugals. *Boletim da Sociedade Portuguesa de Ciencias Naturais*, Lisboa (Series 2) 19, 125–145, 2 figs.

Malkmus, R. (1982) Beitrag zur Verbreitung der Amphibien und Reptilien in Portugal. *Salamandra*, Frankfurt-am-Main 18(3–4), 218–299.

Malkmus, R. (1983) Nachtrag zur Verbreitung der Amphibien und Reptilien Portugals. *Salamandra*, Frankfurt-am-Main 19(1–2), 71–83, illust.

Malkmus, R. (1987) Herpetologische beobachtungen am Mount Kinabalu, Borneo. *Mitteilungen aus dem Zoologischen Museum in Berlin* 63(2), 269–292, 4 figs.

Malkmus, R. (1989) Mount Kinabalu, Borneo #2. *Mitteilungen aus dem Zoologischen Museum in Berlin* 65(2), 179–200.

Malkmus, R. (1991) Zur herpetofauna der portugiesischen Litoralzone (Amphibia et Reptilia). *Faunistische Abhandlungen, Staatliches Museum für Tierkunde in Dresden* 18(1), 71–83, illust.

Malkmus, R. (1992) Herpetologische Beobachtungen am Mount Kinabalu, Nord Borneo. 3. *Mitteilungen aus dem Zoologischen Museum in Berlin* 68(1), 101–138, illust.

Malkmus, R. (1993a) Bemerkungen zu einer kleinen Sammlung von Amphibien und Reptilien aus Nordost-Sulawesi. *Mitteilungen aus dem Zoologischen Museum in Berlin* 69(1), 175–184, 6 figs.

Malkmus, R. (1993b) Zur herpetofauna der Serra do Reboredo. *Nachrichten des Naturwissenschafliches Musweum der Stadt Aschaffenburg* 100, 73–90, illust.

Malkmus, R. (1993c) Die Serra de Monchique—eine Klima Insel im Suden Portugals. *Bemerkungen zur Flora und Herpetofauna. Natur und Museum*, Frankfurt-am-Main 123(6), 171–182, illust.

Malkmus, R. (1994) Een zeldzame slang van Mt. Kinabalu, Schmidts Gronslang (*Calamaria schmidti*). *Lacerta*, Gravenhage 52(5), 111–113, illust.

Malkmus, R. (1995a) *Coronella austriaca acutirostris* subspec. nov. aus dem nordwesten der Iberischen Halbinsel (Reptilia, Serpentes, Colubridae). *Zoologische Abhandlung, Staatliches Museums für Tierkunde in Dresden* 48(15), 265–278, 8 figs.

Malkmus, R. (1995b) *Die Amphibien und Reptilien Portugals, Madeiras und der Azoren*. Westarp Wissenschaften, Magedeburg, 192 pp., 39 figs., 3 col. pls.

Malkmus, R. (2000) Herpetologische Beobachtungen auf Sulawesi. *Sauria*, Berlin 22(2), 11–17, 9 figs.

Malkmus, R. (2004) *Amphibians and reptiles of Portugal, Madeira and the Azores-Archipelago: distribution and natural history notes*. A.R.G. Gantner Verlag, Königstein, 447 pp., col. illust.

Malkmus, R., Manthey, U., Vogel, G., Hoffmann, P. & Kosuch, J. (2002) *Amphibians & reptiles of Mount Kinabalu (North Borneo)*. A.R.G. Gantner Verlag K.G., Königstein, 424 pp., 118 figs., 384 col. pls.

Mallow, D., Ludwig, D. & Nilson, G. (2003) *True vipers. Natural history and toxinology of Old World vipers*. Krieger Publishing Company, Malabar, 359 pp., 59 col. photos.

Malmberg, A.J. (1872) *Fauna Fennica. Suomen eläimistö, Nuorisolle. I–IV. Nisäkkäät, linnut, matelijat ja sammakot*. J.C. Frenckell'in ja Pojan kirjapainossa, Helsingissä, 160 pp.

Malnate, E.V. (1939) A study of the yellow-lipped snake, *Rhadinaea flavilata* (Cope). *Zoologica*, New York 24(3), 359–366, fig. 1, pl. 1.

Malnate, E.V. (1953) The taxonomic status of the Tibetan colubrid snake *Natrix baileyi*. *Copeia* 1953(2), 92–96, 2 figs.

Malnate, E.V. (1960a) Systematic division and evolution of the colubrid snake genus *Natrix*, with comments on the subfamily Natricinae. *Proceedings of the Academy of Natural Sciences of Philadelphia* 112(3), 41–71, 2 figs.

Malnate, E.V. (1960b) The identity of the natricine snake *Tropidontus baramensis* Smith. *Notulae Naturae*, Philadelphia (339), 1–2, 2 figs.

Malnate, E.V. (1961) *Amphiesmoides*, a new genus for *Tropidonotus ornaticeps* Werner (Serpentes, Colubridae). *Notulae Naturae*, Philadelphia (341), 1–7.

Malnate, E.V. (1962) The relationships of five species of the Asiatic natricine snake genus *Amphiesma*. *Proceedings of the Academy of Natural Sciences of Philadelphia* 114(8), 251–299, 8 figs., 1 pl.

Malnate, E.V. (1963) A new race of *Amphiesma pryeri* (Serpentes, Natricinae) from the southern Riukiu Island of Miyako-Shima. *Notulae Naturae*, Philadelphia (360), 1–6, 2 figs.

Malnate, E.V. (1966) *Amphiesma platyceps* (Blyth) and *Amphiesma sieboldii* (Guenther): sibling species (Reptilia, Serpentes). *The Journal of the Bombay Natural History Society* 63(1), 1–17, 3 figs.

Malnate, E.V. (1968) *Natrix dunni*, a new species of watersnake from Malaysia. *Notulae Naturae* (410), 1–5.

Malnate, E.V. (1971) A catalog of primary types in the herpetological collections of the Academy of Natural Sciences, Philadelphia (ANSP). *Proceedings of the Academy of Natural Sciences of Philadelphia* 123(9), 345–375.

Malnate, E.V. (1990) A review and comparison of hemipenial structure in the genus *Agkistrodon* (*sensu lato*). *In:* Gloyd, H.K. & Conant, R. (Eds.), *Snakes of the* Agkistrodon *complex*. Society for the Study of Amphibians and Reptiles, pp. 583–588, 2 figs.

Malnate, E.V. & Minton, S.A. (1965) A redescription of the Natricine snake *Xenochrophis cerasogaster*, with comments on its taxonomic status. *Proceedings of the Academy of Natural Sciences of Philadelphia* 117(2), 19–43, 3 figs., 1 pl.

Malnate, E.V. & Munsterman, H.E. (1960) Interpopulation variation in the colubrid snake *Natrix pryeri* from the Riukiu Islands, with description of a new subspecies. *Proceedings of the California Academy of Sciences* (Series 4) 31(2), 51–67, 2 figs.

Malnate, E.V. & Romer, J.D. (1969) A snake new to the fauna of Hong Kong and China, *Amphiesma atemporalis* (Bourret). *Notulae Naturae*, Philadelphia (424), 1–8, 3 figs.

Malnate, E.V. & Underwood, G. (1988) Australasian natricine snakes of the genus *Tropidonophis* Jan, 1863. *Proceedings of the Academy of Natural Sciences of Philadelphia* 140(1), 59–201, 32 figs.

Maluquer, J. (1917) Les serps de Catalunya. *Musei Barcinonensis Scientiarum Naturalium Opera, Série Zoológica* 7, 1–87, 8 pls.

Manaças, S. (1957) Sàurios e ofídios da Guiné Portuguesa. *Anais da Junta de Investigações do Ultramar*, Lisboa (1955) 10(4), 185–213.

Manaças, S. (1958) Anfíbios e répteis das Ilhas de São Tomé e do Príncipe e do Ilhéu das Rolas. *Conferemcia Internacional dos Africanistas Ocidentais* (*Zoologia e Biologia Animal*) [6th session, São Tome], Lisboa 4, 179–192.

Manaças, S. (1959) Ofídios de Moçambique. *Memórias do Junta da Investigações do Ultramar*, Lisboa (Series 2) 8, 135–160.

Manaças, S. (1961) Alguns sáurios e ofídios de Moçambique. *Memórias do Junta da Investigações do Ultramar*, Lisboa (Series 2) (23), 143–163.

Manaças, S. (1982) Ofideos venenosos da Guine, S. Tome, Angola e Mocambique. *Garcia de Orta* (Série de *Zoológica*) 10(1–2), 13–45, 2 figs.

Mancilla-Moreno, M. (1998) Geographic distribution: Serpentes: *Chersodromus liebmanni*. *Herpetological Review* 29(1), 52–53.

Mancilla-Moreno, M. & Ballardo, W.S. (1995) Geographic distribution: Serpentes: *Leptophis mexicanus septentrionalis* (Tamaulipan parrot snake). *Herpetological Review* 26(3), 156.

Mancilla-Moreno, M. & Ramirez-Bautista, A. (1998) Geographic distribution: Serpentes: *Ramphotyphlops braminus*. *Herpetological Review* 29(1), 54.

Mandaville, J.P. (1967) The hooded *Malpolon, M. moilensis* (Reuss), and notes on other snakes of north-eastern Arabia. *The Journal of the Bombay Natural History Society* 64(1), 115–118, 1 pl.

Mandaville, J.P. (1974) New snake records for the Eastern Province. *Journal of the Saudi Arabia Natural History Society* 1(11), 8.

Mane, Y. (1999) Une espèce nouvelle du genre *Elapsoidea* (Serpentes, Elapidae) du Sénégal. *Bulletin de la Société Hérpetologique de France* 91, 13–18.

Manier, M.K. (2004) Geographic variation in the long-nosed snake, *Rhinocheilus lecontei* (Colubridae): beyond the subspecies debate. *Biological Journal of the Linnean Society*, London 83(1), 65–85, 7 figs.

Manjunatha-Kini, R., Stefanssson, S. & Evans, H.J. (1987) Non-phospholipase anticoagulants from *Naja nigricollis* venom. *In:* Gopalakrishnakone, P. & Tan, C.K. (Eds.), *Progress in venom and toxin research*. National University of Singapore, Singapore, pp. 175–185, 4 figs.

Mansueti, R. (1941) A descriptive catalogue of the amphibians and reptiles found in and around Baltimore city, Maryland, within a radius of twenty miles. *Proceedings of the Natural History Society of Maryland* (7), 1–53, 2 pls.

Manthey, U. & Grossmann, W. (1997) *Amphibien & Reptilien Südostasiens*. Natur und Tier, Berlin, 512 pp., 158 figs., 355 col. pls.

Manzani, P.P., Moreno, R.A. & Mattar, M.N. (1997) Geographical distribution. *Bothrops fonsecai*. *Herpetological Review* 28(4), 210.

Manzanilla, J. & Natera, M. (2003) Geographic distribution: Serpentes: *Porthidium lansbergii*. *Herpetological Review* 34(4), 389.

Manzanilla, J. & Sánchez, D. (2005) Una nueva especie de *Thamnodynastes* (Serpentes: Colubridae) del macizo del Turimiquire, noreste de Venezuela. *Memoria de la Fundación La Salle de Ciencias Naturales* (2004) 161–162, 61–75, 3 figs.

Manzanilla, J., Mijares-Urrutia, A. & Rivero, R. (1998) Geographic distribution: Serpentes: *Rhadinaea fulviceps*. *Herpetological Review* 29(2), 115.

Mao, S.-H. (1961) Color variations of Taiwan (Formosa) many-banded krait. *Herpetologica* 17(4), 275–277, 2 figs.

Mao, S.-H. (1962) Sexual dimorphism of Taiwan bamboo vipers. *Bulletin of the Institute of Zoology, Academia Sinica* 1, 41–46, 5 figs.

Mao, S.-H. (1963) Variation in color pattern and hood marking of Taiwan cobras. *Bulletin of the Institute of Zoology, Academia Sinica* 2, 55–56, 28 figs.

Mao, S.-H. (1964) The taxonomic status of the green snake, *Opheodrys major*, on Taiwan. *Bulletin of the Institute of Zoology, Academia Sinica* 3, 25–30, 6 figs.

Mao, S.-H. (1965a) The taxonomic status of the white-stomach snake, *Natrix percarinata suriki* Maki, on Taiwan. *Bulletin of the Institute of Zoology, Academia Sinica* 4, 19–27, 5 figs.

Mao, S.-H. (1965b) The status of the Taiwan snakes of the colubrid genus *Zaocys*. *Bulletin of the Institute of Zoology, Academia Sinica* 4, 87–94, 5 figs.

Mao, S.-H. (1993) Common terrestrial venomous snakes of Taiwan. *National Museum of Natural Science, Taipei, Special Publication* (5), ix + 108 pp., 93 figs.

Mao, S.-H. & Chen, B.-Y. (1975) *Disteira stokesi* (Gray): an addition to the sea snake fauna of Taiwan. *Herpetologica* 31(4), 456–458, 4 figs.

Mao, S.-H. & Chen, B.-Y. (1980) Sea snakes of Taiwan. A natural history of sea snakes. *National Science Council Special Publication* (4), xi + 64 pp., 37 figs.

Maradjo, M. (1977) *Fauna Indonesia. 1. Snakes*. P.T. Karya Nusantara, Jakarta, 32 pp., col. illust. (in Indonesian)

Marais, J. (1985) *Snake versus man, a guide to dangerous and common harmless snakes of southern Africa*. Macmillan South Africa, Johannesburg, viii + 102 pp., illust., 62 col. pls.

Marais, J. (1997) *Snakes*. Barnes & Noble Books, New York, 144 pp., col. illust.

Marais, J. (2004) *A complete guide to the snakes of southern Africa*. Second edition. Struik Publishers, Cape Town, 312 pp., col. illust. [first edition, 1992, Blanford, xi + 208 pp., 301 figs., 15 illust., 208 col. pls.]

Marais, J. (2007) *What's that snake? A starter's guide to snakes of southern Africa*. Struik Publications, Cape Town, 95 pp., col. illust.

Marchi, G. (1901) *Serpenti del Trentino*. G.R. Monauni, Trento, 94 pp., 21 pls.

Marcus, H. (1937) Atmungssystem (Organe der Luftatmung). III. Schwimmblase und Lungen. B. Lungen. *In:* Bolk, L., Göppert, E., Kallius, E. & Lubosch, W. (Eds.), *Handbuch der vergleichenden Anatomie der Wirbeltiere. Dritter Band*. Urban & Schwarzenberg, Berlin, pp. 909–988, figs. 902–936.

Marcy, R.B. (1854) *Exploration of the Red River of Louisiana, in the year 1852*. Robert Armstrong, Washington, D.C. (1853), xv + 320 pp., 66 pls.

Marczak, S., Alloush, M., Jones, R.M., Martineau, K., Oliva, M.A. & Kaiser, K. (2010) *Sibon dimidiata*. *Herpetological Review* 41(1), 111.

Marelli, C.A. (1924) *Elenco sistemático de la fauna de la Provincia de Buenos Aires (Procordados y vertebrados)*. L. J. Rosso y Cia, Buenos Aires, pp. 584–596.

Marelli, C.A. (1931) List of the species of reptiles and amphibians. *In: Los vertebrados exhibidos en los Zoológicos del Plata*. La Plata, pp. 185–201.

Marián, M. (1988) *A Bakony Hegység Kétéktü és Hüllofaunája (Amphibia, Reptilia). Herpetológiai Alapuetés. (The amphibian and reptilian fauna of the Bakony Mountians [Amphibia, Reptilia]. A herpetological survey)*. Bakony Termeszettud Kut Eredmenyei, 102 pp, illust. (in Hungarian)

Mariani, G. (1988) *I serpenti Italiani, con i serpenti dell'areale Europea*. Lucchetti Editore, Bergamo, 63 pp., 7 figs., 27 col. figs.

Marineros, L. (2000) *Guía de las serpientes de Honduras*. Secretaría de Recursos Naturales y Ambiente, Dirección General de Biodiversidad, 252 pp., figs.

Maritz, B. & Alexander, G.J. (2010) Geographic distribution: Reptilia: Squamata: Elapidae: *Dendroaspis polylepis polylepis*. African Herp News (51), 27–28.

Maritz, B. & Alexander, G.J. (2012) Movement patterns in the smallest viper, *Bitis schneideri*. *Copeia* 2012(4), 732–737, 3 figs.

Markel, R.G. (1990) *Kingsnakes and milksnakes*. T.F.H. Publications, Neptune City, 144 pp., col. illust.

Markezich, A.L. (1976) *A reassessment of variation in, and a redescription of the South American snake,* Leimadophis poecilogyrus *(Wied)*. Master's Thesis, University of Illinois, Chicago, 129 pp., 35 figs.

Markezich, A.L. (2001) Geographic distribution: Serpentes: *Clelia clelia*. *Herpetological Review* 32(2), 122.

Markezich, A.L. (2002) New distribution records of reptiles from western Venezuela. *Herpetological Review* 33(1), 69–74.

Markezich, A.L. & Barrio-Amorós, C.L. (2004) A new species of *Atractus* (Serpentes: Colubridae) from northeastern Venezuela. *Bulletin of the Maryland Herpetological Society* 40(3), 111–121, 2 figs.

Markezich, A.L. & Dixon, J.R. (1979) A new South American species of snake and comments on the genus *Umbrivaga*. *Copeia* 1979(4), 698–701, 2 figs.

Markezich, A.L. & Dixon, J.R. (2001) Geographic distribution: Serpentes: *Hydrops triangularis*. *Herpetological Review* 32(2), 123.

Markezich, A.L. & Rossman, D.A. (1992) Geographic distribution: Serpentes: *Helicops pastazae*. *Herpetological Review* 23(4), 124.

Markezich, A.L. & Taphorn, D.C. (1993) A variational analysis of populations of *Bothrops* (Serpentes, Viperidae) from western Venezuela. *Journal of Herpetology* 27(3), 248–254, 2 figs.

Marques, L.B. & Lema, T. de. (1992) Estudo comparativo da osteologia craniana de *Clelia occipitolutea* (Duméril, Bibron et Duméril, 1854) e *C. rustica* (Cope, 1878) (Serpentes, Colubridae, Xenodontinae, Pseudoboini). *Acta Biológica Leopoldensia* 14(1), 27–54, 8 figs.

Marques, O.A.V. (1996a) Reproduction, seasonal activity and growth of the coral snake, *Micrurus corallinus* (Elapidae), in the southeastern Atlantic forest in Brazil. *Amphibia-Reptilia,* Leiden 17(3), 277–285, 6 figs.

Marques, O.A.V. (1996b) Biologia reproductiva da cobra-coral *Erythrolamprus aesculapii* Linnaeus (Colubridae), no sudeste do Brasil. *Revista Brasileira de Zoologia*, Rio de Janeiro 13(3), 747–753.

Marques, O.A.V. (1996c) Geographic distribution: Serpentes: *Imantodes cenchoa*. *Herpetological Review* 27(1), 34.

Marques, O.A.V. (2002) Natural history of the coral snake *Micrurus decoratus* (Elapidae) from the Atlantic forest in southeast Brazil, with comments on possible mimicry. *Amphibia-Reptilia,* Leiden 23(2), 228–232, 1 fig.

Marques, O.A.V. & Calleffo, M.E. (1997) Geographic distribution: Serpentes: *Pseustes sulphureus*. *Herpetological Review* 28(3), 160.

Marques, O.A.V. & Castro-Souza, V. (1993) Nota sobre a atividade alimentar de *Liophis miliaris* no ambiente marinho (Serpentes, Colubridae). *Revista Brasileira de Biologia*, Rio de Janeiro 53, 645–648.

Marques, O.A.V. & Franco, F.L. (1998) Geographic distribution: Serpentes: *Philodryas viridissimus*. *Herpetological Review* 29(1), 54.

Marques, O.A.V. & Puorto, G. (1996) Geographic distribution: Serpentes: *Chironius laevicollis*. *Herpetological Review* 27(4), 212.

Marques, O.A.V. & Puorto, G. (1998) Feeding, reproduction and growth in the crowned snake *Tantilla melanocephala* (Colubridae), from southeastern Brazil. *Amphibia-Reptilia,* Leiden 19(3), 311–318, 3 figs.

Marques, O.A.V., Esterovic, A. & Sazima, I. (2001) *Serpentes da Mata Atlântica. Guia illustrado para a Serra do Mar*. Holos, Editora, São Paulo 184 pp., 56 mostly col. figs., 106 unnumbered photos.

Marques, O.A.V., Martins, M. & Sazima, I. (2002) A new insular species of pitviper from Brazil, with comments on evolutionary biology and conservation of the *Bothrops jararaca* group (Serpentes, Viperidae). *Herpetologica* 58(3), 303–312, 3 figs.

Marques, O.A.V., Esterovic, A. & Sazima, I. (2004) *Snakes of the Brazilian Atlantic forest: an illustrated field guide for the Serra do Mar range*. Holos, Editora, São Paulo 204 pp., 56 mostly col. figs., 106 unnumbered photos.

Marques, O.A.V., Esterovic, A., Strüssmann, C. & Sazima, I. (2005) *Serpentes do Pantanal. Guia illustrado*. Holos, Editora, São Paulo, 179 pp., 65 mostly col. figs., 107 unnumbered photos.

Marques, O.A.V., Pizzatto, L. & Almeida-Santos, S.M. (2013) Reproductive strategies of New World coral snakes, genus *Micrurus*. *Herpetologica* 69(1), 58–66, 4 figs.

Marschall, A. de. (1873) *Nomenclator zoologicus, continens nomina systematica generum animalium tan viventium quam fossilium, secundum ordinem alphabeticum disposita*. Typus Carolini Ueberreuter (M. Salzer), Vindobonae, iv + 482 pp.

Marsh, O.C. (1869) Description of a new and gigantic fossil serpent (*Dinophis grandis*), from the Tertiary of New Jersey. *The American Journal of Science and Arts,* New Haven (Series 2) 48(144), 397–400.

Marsh, O.C. (1871) Description of some new fossil serpents, from the Tertiary deposits of Wyoming. *The American Journal of Science and Arts,* New Haven (Series 3) 1(5), 322–327.

Marsh, O.C. (1877) *Introduction and succession of vertebrate life in America, an address delivered before the American Association for the Advancement of Science, at Nashville, Tenn., Aug. 30, 1877*. Othinel C. Marsh, New Haven, 50 pp. [reprinted in 1878 in *Proceedings of the American Association for the Advancement of Science* (1877), 211–258]

Marsh, O.C. (1885) Names of extinct reptiles. *The American Journal of Science,* New Haven (Series 3) 29(170), 169.

Marsh, O.C. (1892) Notice of new reptiles from the Laramie formation. *The American Journal of Science,* New Haven (Series 3) 43(257), 449–453, 3 figs.

Martens, E. von. (1876) *Die Preußisch Expedition nach Ost-Asien. Nacht amtlichen quellen*. Zoologischer Thiel. Verlag der Königlichen Geheimen Ober-Hofbuchdruckerei, Berlin, xii + 412 pp., 15 pls.

Martens, H. (1993) Three species of snake new for Syria. *Zoology in the Middle East* 9, 49–58, 2 figs.

Martens, H. (1997) A review of "Zoogeography of amphibians and reptiles of Syria, with additional new records." *Herpetozoa* 10(3–4), 99–106.

Martin, P.S. (1955) Herpetological records from the Gómez Farías region of southwestern Tamaulipas, México. *Copeia* 1955(3), 173–180.

Martin, P.S. (1958) A biogeography of reptiles and amphibians in the Gomez Farias region, Tamaulipas, México. *Miscellaneous Publications, Museum of Zoology, University of Michigan* (101), 102 pp., 7 figs., 7 pls.

Martin, W.C.L. (1838) On some snakes collected during the Euphrates expedition. *Proceedings of the Zoological Society of London* 6(1), 81–84.

Martin, W.H. (2000) Distribution and habitat relationships of the eastern diamondback rattlesnake (*Crotalus adamanteus*). *Herpetological Natural History* 7(1), 9–34, 3 figs.

Martin del Campo, R. (1935) Nota acerca de la distribución geográfica de los reptiles ponzoñosos en México. *Anales del Instituto de Biologia, Universidad Nacional Autónoma de México* 6(3–4), 219–300.

Martinez, A.M. (2011) Geographic distribution: Serpentes: *Tantillita lintoni Herpetological Review* 42(1), 116.

Martinez, A.M. (2011) Geographic distribution: Serpentes: *Tantillita lintoni*. *Herpetological Review* 42(2), 245.

Martinez, J. & Behler, J.L. (1988) Serpentes. *Erpeton tentaculatum* (Tentacled snake). Reproduction. *Herpetological Review* 19(2), 35.

Martínez, L.R. & Avila, L.J. (2003) Geographic distribution: Serpentes: *Bothrops neuwiedi diporus*. *Herpetological Review* 34(4), 386.

Martínez, V.M. (1983) Panamá: nuevo ámbito de distribución para la serpiente venenosa *Bothrops picadoi* (Dunn). *Conciencia* 10(1), 26–27.

Martínez, V.M. & Bolaños, R. (1982) The bushmaster, *Lachesis muta muta* (Linnaeus) [Ophidia, Viperidae] in Panamá. *Revista de Biologia Tropical*, San José 30(1), 100–101, 1 fig.

Martínez-Coronel, M. & López-Hernández, I.D. (2008) Geographic distribution: Serpentes: *Leptotyphlops maximus*. *Herpetological Review* 39(3), 372.

Martínez-Rica, J.-P. (Ed.). (1989) El atlas provisional de los anfibios y reptiles de España y Portugal (aparep). Presentación y situación actual. *Monografias de Herpetologia,* Madrid (1), 1–73, 45 figs.

Martins, A.R., Silveira, A.L. & Freire-Bruno, S. (2010) New records of *Typhlops brongersmianus* (Serpentes, Typhlopidae) in southeastern Brazil. *Herpetology Notes* 3, 247–248, 2 figs.

Martins, M. & Oliveira, M.E. (1993) The snakes of the genus *Atractus* Wagler (Reptilia, Squamata, Colubridae) from the Manaus region, central Amazonia, Brazil. *Zoologische Mededelingen*, Leiden 67(2), 21–40, 8 figs.

Martins, M. & Oliveira, M.E. (1998) Natural history of snakes in forests of the Manaus region, central Amazonia, Brazil. *Herpetological Natural History* 6(2), 78–150, 7 figs., 113 col. pls.

Martins, M., Araujo, M.S., Sawaya, R.J. & Nuñes, R. (2001) Diversity and evolution of macrohabitat use, body size and morphology in a monophyletic group of Neotropical pitvipers (*Bothrops*). *Journal of Zoology*, London 254(4), 529–538, 6 figs.

Martof, B.S. (1956) *Amphibians and reptiles of Georgia: a guide*. University of Georgia Press, Athens, 94 pp.

Martof, B.S., Palmer, W.M., Bailey, J.R. & Harrison, J.R., III. (1980) *Amphibians and reptiles of the Carolinas and Virginia*. University of North Carolina Press, Chapel Hill, 264 pp., col. photos.

Marx, H. (1953a) The elapid genus of snakes *Walterinnesia*. *Fieldiana: Zoology* 34(16), 189–196, figs. 38–40.

Marx, H. (1953b) A new worm snake from Colombia. Genus *Anomalepis*. *Fieldiana: Zoology* 34(17), 197–198.

Marx, H. (1956) *Keys to the lizards and snakes of Egypt*. United States Naval Medical Research Unit No. 3, Cairo, 8 pp.

Marx, H. (1958a) Catalogue of type specimens of reptiles and amphibians in Chicago Natural History Museum. *Fieldiana: Zoology* 36(4), 407–496.

Marx, H. (1958b) Egyptian snakes of the genus *Psammophis*. *Fieldiana: Zoology* 39(18), 191–200, figs. 30–31.

Marx, H. (1959) Review of the colubrid snake genus *Spalerosophis*. *Fieldiana: Zoology* 39(30), 347–361, figs. 58–59.

Marx, H. (1960) A new colubrid snake of the genus *Atractus*. *Fieldiana: Zoology* 39(38), 411–413, fig. 71.

Marx, H. (1968) *Checklist of the reptiles and amphibians of Egypt*. Special Publication, United States Naval Medical Research Unit No. 3, Cairo, iii + 91 pp., 11 figs.

Marx, H. (1976) Supplementary catalogue of type specimens of reptiles and amphibians in Field Museum of Natural History. *Fieldiana: Zoology* 69(2), 33–94.

Marx, H. (1988) The colubrid snake, *Psammophis schokari*, from the Arabian Peninsula. *Fieldiana: Zoology* (Series 2) (40), i–iii + 1–16, 5 figs.

Marx, H. & Inger, R.F. (1955) Notes on snakes of the genus *Calamaria*. *Fieldiana: Zoology* 37, 167–209, figs. 21–27.

Marx, H. & Rabb, G.B. (1965) Relationships and zoogeography of the viperine snakes (family Viperidae). *Fieldiana: Zoology* 44(21), 161–206, figs. 32–46.

Marx, H. & Rabb, G.B. (1972) Phyletic analysis of fifty characters of advanced snakes. *Fieldiana: Zoology* (Series 2) (63), i–viii + 1–321 pp., 52 figs.

Marx, H., Rabb, G.B. & Arnold, S.J. (1982) *Pythonodipsas* and *Spalerosophis*: colubrid snake genera convergent to the vipers. *Copeia* 1982(3), 553–561, 1 fig.

Marx, H., Ashe, J.S. & Watrous, L.E. (1988) Phylogeny of the viperine snakes (Viperinae): Part I. Character analysis. *Fieldiana: Zoology* (Series 2) (51), i–iv + 1–16 pp., 1 fig.

Maryan, B. (1994) Natural history notes on the spotted mulga snake *(Pseudechis butleri). Monitor,* Ardeer 6(1), 4–8, 2 figs., 2 col. pls.

Masci, P. & Kendall, P. (1995) *The taipan: the world's most dangerous snake*. Kangaroo Press, Kenthurst, 90 pp., illust.

Maslin, T.P. (1942) Evidence for the separation of the crotalid genera *Trimeresurus* and *Bothrops*, with a key to the genus *Trimeresurus*. *Copeia* 1932(1), 18–24, 2 figs.

Maslin, T.P. (1945) A phylogenetic study of the Indian crotalid snake *Trimeresurus macrolepis* Beddome with a consideration of evolutionary trends within the genus. *Stanford University Bulletin* (7), 5–8.

Maslin, T.P. (1950) Snakes of the Kiukiang-Lushan area, Kiangsi, China. *Proceedings of the California Academy of Sciences* (Series 4) 26(12), 419–466, 10 figs.

Maslin, T.P. (1959) *An annotated check list of the amphibians and reptiles of Colorado*. University of Colorado Studies, Biology 6, vi + 98 pp.

Maslov, D.A. & Kotlobay, A.A. (1998) Geographic distribution: Serpentes: *Dinodon rufozonatum*. *Herpetological Review* 29(2), 111–112.

Mason, F. (1852) *Tenasserim, or notes on the fauna, flora, minerals and nations of British Burmah and Pegu, with systematic catalògues of the known minerals, plants, mammals, fishes, mollusks, sea-nettles, corals, sea-urchins, worms, insects, crabs, reptiles, and birds; with vernacular names*. Second edition. American Mission Press, Maulmein, viii + 712 pp. [first edition, 1850, viii + 712 pp.]

Mason, F. (1860) *Burmah, its people and natural productions, or notes on the nations, fauna, flora, and minerals of Tenasserim, Pegu and Burmah, with systematic catalogues of the known mammals, birds, fish, reptiles, insects, mollusks, crustaceans, annalids, radiates, plants and minerals, with vernacular names*. Third edition. Thos. Stowe Ranney, Rangoon, xvii + 913 pp.

Mason, G.E. (1888) Description of a new earth-snake of the genus *Silybura* from the Bombay Presidency with remarks on little known Uropeltidae. *Annals and Magazine of Natural History*, London (Series 6) 1(3), 184–186.

Masood, M.F. (2012) Ecological distribution of snake's fauna of Jazan region of Saudi Arabia, 15 pp.

Masroor, R. (2012) *A contribution to the herpetology of northern Pakistan*. Society for the Study of Amphibians and Reptiles, 217 pp., 107 col. illust.

Masroor, R. & Mebert, K. (2012) Geographic distribution: Serpentes: *Natrix tessellata*. *Herpetological Review* 43(3), 621.

Massalongo, A.B. (1854) *Saggio di un'erpetologia popolare Veronese*. Gius. Antonelli, Verona, 56 pp.

Massalongo, A.B. (1859) *Specimen photographicum animalium quorumdam plantarumque fossilium agri Veronensis*. Vicentini-Franchini, Veronae, 101 pp., 40 pls.

Massary, J.-C. & Ineich, I. (1994) *Sepedon rhombeata* Lichtenstein, 1823 (Serpentes, Viperidae) and its type locality. *Journal of African Zoology*, Tervuren 108(5), 503–504.

Massip-Gibert, J.M. (1993) *Els amfibis i els reptiles*. Fauna Comarcal, Banyoles, 287 pp., illust.

Mateo, J.A., Pleguezuelos, J.M., Fahd, S., Geniez, P. & Martínez-Media, F.J. (2003) *Los anfibios, los reptiles y el estrecho de Gibraltar: un ensayo sobre la herpetofauna de ceuta y su entorno*. Instituto de Estudios de Ceuties, Ceuta, 388 pp., 80 col. photos.

Mathew, R. (1983) On a collection of snakes from north-east India (Reptilia, Serpentes). *Records of the Zoological Survey of India*, Calcutta 80(3–4), 449–458.

Mathew, R. (1992) Additional records of snakes from northeast India, with four new records. *Records of the Zoological Survey of India*, Calcutta 91(3–4), 287–292.

Mathew, R. (1995) Reptilia. *In:* Gosh, A.K. (Ed.), *Fauna of Meghalaya, Part 1 (Vertebrates). State Fauna Series* 4. Zoological Survey of India, Kolkata, pp. 379–454, 3 figs., 38 col. pls.

Mathew, R. (2007) Reptilia. *In:* Alfred, J.R.B. (Ed.), *Fauna of Mizoram. State Fauna Series* 14. Zoological Survey of India, Kolkata, pp. 545–577, 28 col. pls.

Matias-Ferrer, N. & Murillo, S. (2004a) Geographic distribution: Serpentes: *Conophis vittatus*. *Herpetological Review* 35(2), 190.

Matias-Ferrer, N. & Murillo, S. (2004b) Geographic distribution: Serpentes: *Crotalus atrox*. *Herpetological Review* 35(2), 190.

Matias-Ferrer, N. & Murillo, S. (2004c) Geographic distribution: Serpentes: *Crotalus scutulatus*. *Herpetological Review* 35(2), 191.

Matlock, R.S. & Rehmeier, R.L. (2002) Status of the western diamondback rattlesnake (*Crotalus atrox*) in Kansas. *Southwestern Naturalist* 47(2), 212–214, 1 fig.

Matschie, P. (1890) Verzeichniss von Reptilien von Bismarckburg im Togolande. *Zoologischer Jahrbücher, Abteilung für Systematik, Ökologie und Geographie der Tiere*, Jena 5, 612–618.

Matschie, P. (1893a) Einige anscheinend neue Reptilien und Amphibien aus West-Afrika. *Sitzunsberichte der Gesellschaft Naturforschender Freunde zu Berlin* 1893, 170–175.

Matschie, P. (1893b) Beiträge zur fauna des Togolandes. Die Reptilien und Amphibien des Togogebietes. *Mitteilungen von Forschungsreisenden und Gelehrten aus den Deutschen Schutzgebieten* 6(3), 207–215.

Matsuda, B.M., Green, D.M. & Gregory, P.T. (2006) *Amphibians and reptiles of British Columbia*. Royal British Columbia Museum, Victoria, 266 pp., figs., 16 col. pls.

Matsui, M. & Ota, H. (1995) On Chinese herpetology. *Herpetologica* 51(2), 234–250.

Matsui, M., Hikida, T. & Ota, H. (1984) A small collection of amphibians and reptiles from Padang, Sumatra. *In:* Hotta, M. (Ed.), *Forest ecology and flora of G. Gadut, West Sumatra (1980–1984)*. Sumatra Nature Study, Kyoto, pp. 121–123.

Matsui, T. (1985) *Amphibians and reptiles in Japan*. Shogakukan, Tokyo, 158 pp., col. illust. (in Japanese)

Mattei, R. & Barrio-Amorós, C.L. (1999) Geographic distribution: Serpentes: *Oxyrhopus formosus*. *Herpetological Review* 30(1), 55.

Matthey, R. (1949) *Les chromosomes des vértébres*. F. Rouge, Lausaunne, 356 pp., illust.

Mattison, C. (1986) *Snakes of the world*. Facts on File Publications, New York, 190 pp., col. illust.

Mattison, C. (1995) *The encyclopedia of snakes*. Blanford, London, 256 pp., col. illust.

Mattison, C. (1999) *Snake: the essential visual guide to the world of snakes*. Dorling Kindersley Publishing, London, 192 pp., col. illust.

Matz, G. (1974) Les boidés ou serpents constricteurs. 4. *Eryx* Daudin 1803. *Aquarama*, Strasbourg 8(26), 53–55, 1 illust.

Matz, G. (1976a) *Spalerosophis diadema* (Schlegel). *Aquarama*, Strasbourg 10(35), 34–35, 2 illust.

Matz, G. (1976b) Les colubrides de France. *Aquarama*, Strasbourg 10(36), 34–38, 7 illust.

Matz, G. (1980) Les boides ou serpents constricteurs. 6. *Liasis* Gray et *Python* Daudin. *Aquarama*, Strasbourg 14(54), 69–70, 1 illust.

Matz, G. (1981) Les boïdés ou serpents constricteurs. 7. *Eunectes* Wagler. *Aquarama*, Strasbourg 15(61), 37–39, 3 illust.

Matz, G. (1983a) *Guide des amphibiens et reptiles d'Europe*. Delachaux & Niestlé, Paris, 292 pp., 175 illust., 44 col. pls.

Matz, G. (1983b) *Amphibien und Reptilien*. BLV Verlagsgesellschaft, München, 235 pp., illust. (translation of 1983a)

Matz, G. (1983c) *Python reticulatus* x *Python molurus*. *Aquarama*, Strasbourg 17(70), 30–45, 4 illust.

Matz, G. (2004) *Epicrates maurus* Gray, 1849 description des sous-espéces. *Situla* 10, 2–9.

Matz, G. & Matz, J. (1970) Les boïdés ou serpents constricteurs [1. *Boa* Linnaeus]. *Aquarama*, Strasbourg 4(11), 38–41, 7 illust.

Matz, G. & Matz, J. (1971) Les boïdés ou serpents constricteurs. 2. *Python* Daudin 1903. *Aquarama*, Strasbourg 5(15), 29–33, 8 illust.

Matz, G. & Matz, J. (1973) Les boidés ou serpents constricteurs. 3. *Corallus* Daudin et *Epicrates* Wagler. *Aquarama*, Strasbourg 7(22), 20–31, 8 pls.

Matz, G., Matz, J. & Vauderhaege, M. (1982) Les Boïdés ou Serpents constricteurs. 8. Reproduction de deux boïnés, *Epicrates c. maurus* Gray et *Acrantophis dumerili* Jan. *Aquarama*, Strasbourg 16(63), 49–51, 66–67, 10 illust.

Mauduyt, L. (1844) *Herpétologie de la Vienne, ou tableau méthodique, indicative et descriptif des reptiles tant vivants que fossils observe jusqu'à present dans ce département*. F.-A. Suarin, Poitiers, 62 pp.

Maung Maung Aye (1987) Clinico-pathological correlation in Russell's viper bite. *In:* Gopalakrishnakone, P. & Tan, C.K. (Eds.), *Progress in venom and toxin research*. National University of Singapore, Singapore, pp. 78–107, figs.

Maura-Gonçalves, J. (1956) Estudes sôbre venenos de serpentes Brasileiras. II. *Crotalus terrificas crotaminicus*, subespécie biológica. *Anais da Academia Brasileira de Ciências*, Rio de Janeiro 28(3), 365- 367, 1 fig.

Maxwell, G. (2005) *The more complete Chondro: a comprehensive guide to the care and breeding of green tree pythons*. Second edition. ECO Herpetologiocal Publishing, Lansing, 317 pp., 257 col. photos. [first edition, 2003, 247 pp., col. illust.]

Mayer-Goyenechea, I.G. (1995) *Revision taxonomica de los generos Conopsis Gunther y Toluca Kennicott (Reptilia, Colubridae)*. Doctoral Dissertation, Universidad Nacional Autonoma de Mexico, Mexico, D.F., 72 pp.

Mayet, V. (1903) *Exploration Scientifique de la Tunisie. Catalogue raisonné des reptiles et batraciens de la Tunisie*. Imprimirie Nationale, Paris, pp. 1–32.

Mayol-Serra, J. (2003) *Rèptils I amfibis de les Balears*. Editorial Moll, Mallorca, 249 pp., figs., 8 col. pls.

Mayorga, H. (1967) Informe sobre una colección de anfibios (Salientia) procedentes de Nicaragua. *Caribbean Journal of Science* 7(1–2), 69–77.

Mazzotti, S. & Stagni, G. (1993) *Gli anfibi e i rettili dell'Emilia-Romagna (Amphibia, Reptilia).* Museo Civico di Storia Naturale di Ferrara, 147 pp., 29 figs., 55 col. figs.

McAfee, A. & Gilardi, J. (2000) Geographical distribution. *Hypsiglena torquata. Herpetological Review* 31(1), 55.

McAllister, C.T. (1985) *Nerodia rhombifera. Catalogue of American amphibians and reptiles*, pp. 376.1–376.4.

McAllister, K.R. (1995) Distribution of amphibians and reptiles in Washington State. *Northwest Fauna* 3, 81–112.

McAlpine, D.F. (1981) Activity patterns of the keel-scaled boa *Casarea dussumieri* at the Jersey Wildlife Preservation Trust. *Dodo, Journal of the Jersey Wildlife Preservation Trust* 18,74–78, 8 figs.

McAtee, M.L. (1907) A list of the mammals, reptiles and batrachians of Monroe County, Indiana. *Proceedings of the Biological Society of Washington* 20, 1–16.

McCallum, S. (1986) Reminisces of venomous snakes. Part I. *Herptile* 11(1), 22–23.

McCann, C. (1938) The reptiles and amphibians of Cutch State. *The Journal of the Bombay Natural History Society* (1938–1939) 40(3), 425–427, 1 pl.

McCann, C. (1966) The marine turtles and snakes occuring in New Zealand. *Records of the Dominion Museum*, Wellington 5(21), 201–215, 2 figs.

McCarthy, C.J. (1985) Monophyly of elapid snakes (Serpentes, Elapidae). An assessment of the evidence. *Zoological Journal of the Linnean Society*, London 83(1), 79–93, 2 figs.

McCarthy, C.J. (1986) Relationships of the laticaudine sea snakes (Serpentes, Elapidae, Laticaudinae). *Bulletin of the British Museum, Natural History (Zoology)*, London 50(2), 127–161, 7 figs.

McCauley, R.H. (1945) *The reptiles of Maryland and the District of Columbia.* R.H. McCauley, Hagerstown, 194 pp., 46 figs., 48 pls.

McCleary, R.J.R. & McDiarmid, R.W. (1993) *Phyllorhynchus decurtatus* (Cope). Spotted leaf-nosed snake. *Catalogue of American amphibians and reptiles*, pp. 580.1–580.7, 2 figs.

McCoid, M.J. & Dixon, J.R. (1981) Notes on Mexican *Rhadinaea* (Serpentes, Colubridae). *Southwestern Naturalist* 26(4), 436–437, 1 fig.

McConnell, G.J. (2013) *A field guide to the snakes of Costa Rica.* Edition Chimaira, Frankfurt-am-Main, 222 pp., 300 col. photos.

McCoy, C.J. (1969) Snakes of the genus *Coniophanes* (Colubridae) from the Yucatan Peninsula, Mexico. *Copeia* 1969(4), 847–849.

McCoy, C.J. (1970) The snake fauna of Middlesex, British Honduras. *Journal of Herpetology* 4(3–4), 135–140, 1 fig.

McCoy, C.J. (1971) Comments on Bolivian *Atractus* (Serpentes, Colubridae). *Herpetologica* 27(3), 314–316.

McCoy, C.J. (1980) Identification guide to Pennsylvania snakes. *Carnegie Museum of Natural History, Education Bulletin* (1), 1–11, illust.

McCoy, C.J. (1982) Amphibians and reptiles in Pennsylvania. *Carnegie Museum of Natural History, Special Publication* (6), 1–91.

McCoy, C.J. (1986) Results of the Carnegie Museum of Natural History Expeditions to Belize. I. Systematic status and geographic distribution of *Sibon neilli* (Reptilia, Serpentes). *Annals of the Carnegie Museum* 55(6), 117–123, 1 fig.

McCoy, C.J. & Censky, E.J. (1982) Herpetological type-specimens in the Carnegie Museum of Natural History suppliment. *Annals of the Carnegie Museum* 51(15), 317–330.

McCoy, C.J. & Censky, E.J. (1992) Biology of the Yucatan hog-nosed pitviper, *Porthidium yucatanicum. In:* Campbell, J.A. & Brodie, E.D., Jr. (Eds.), *Biology of the pitvipers.* Selva, Tyler, pp. 217–222, 7 figs.

McCoy, C.J. & Flores-Villela, O.A. (1988) The identity of *Coluber nutkensis* (Reptilia, Serpentes). *Canadian Field-Naturalist* 102, 716–718, 1 fig.

McCoy, C.J. & Richmond, N. (1966) Herpetological type-specimens in Carnegie Museum. *Annals of the Carnegie Museum* (1965–1966) 38(10), 233–264.

McCoy, F. (1867) On the recent zoology and palaeontology of Victoria. *Annals and Magazine of Natural History*, London (Series 3) 20(117), 175–202.

McCoy, F. (1878a) *Pseudechis porphyriacus* (Shaw sp.). The black snake. *Natural history of Victoria. Prodromus of the zoology of Victoria; or figures and descripions of the living species of all classes of the Victorian indigenous animals*, Melbourne 1, 5–6, pl. 1.

McCoy, F. (1878b) *Hoplocephalus superbus* (Günth.). The copperhead snake. *Natural history of Victoria. Prodromus of the zoology of Victoria; or figures and descriptions of the living species of all classes of the Victorian indigenous animals*, Melbourne 1, 7–9, pl. 2.

McCoy, F. (1878c) *Hoplocephalus curtus* (Schlegel sp.). The tiger snake. *Natural history of Victoria. Prodromus of the zoology of Victoria; or figures and descriptions of the living species of all classes of the Victorian indigenous animals*, Melbourne 1, 11–14, pl. 3.

McCoy, F. (1878d) *Hoplocephalus flagellum* (McCoy). The little whip snake. *Natural history of Victoria. Prodromus of the zoology of Victoria; or figures and descriptions of the living species of all classes of the Victorian indigenous animals*, Melbourne 2, 7–8, pl. 11.

McCoy, F. (1878e) *Hoplocephalus coronoides* (Günth.). The white-lipped snake. *Natural history of Victoria. Prodromus of the zoology of Victoria; or figures and descriptions of the living species of all classes of the Victorian indigenous animals*, Melbourne 2, 8–9, pl. 11.

McCoy, F. (1878f) *Acanthophis antarctica* (Shaw sp.). The death adder. *Natural history of Victoria. Prodromus of the zoology of Victoria; or figures and descriptions of the living species of all classes of the Victorian indigenous animals*, Melbourne 2, 11–12, pl. 12.

McCoy, F. (1878g) *Morelia variegata* (Gray). The carpet snake. *Natural history of Victoria. Prodromus of the zoology of Victoria; or figures and descriptions of the living species of all classes of the Victorian indigenous animals*, Melbourne 2, 13–14, pl. 13.

McCoy, F. (1879a) *Diemenia superciliosa* (Fischer). The common black snake. *Natural history of Victoria. Prodromus of the zoology of Victoria; or figures and descriptions of the living species of all classes of the Victorian indigenous animals*, Melbourne 3, 11–12, pl. 23.

McCoy, F. (1879b) *Diemenia microlepidota* (McCoy). The small-scaled brown snake. *Natural history of Victoria. Prodromus of the zoology of Victoria; or figures and descriptions of the living species of all classes of the Victorian indigenous animals*, Melbourne 3, 12–13, pl. 23.

McCoy, F. (1879c) *Diemenia aspidorhyncha* (McCoy). The shield-fronted brown snake. *Natural history of Victoria. Prodromus of the zoology of Victoria; or figures and descriptions of the living species of all classes of the Victorian indigenous animals*, Melbourne 3, 13–14, pl. 23.

McCoy, F. (1879d) *Furina bicucullata* (McCoy). The two-hooded furina-snake. *Natural history of Victoria. Prodromus of the zoology of Victoria; or figures and descriptions of the living species of all classes of the Victorian indigenous animals*, Melbourne 4, 13–15, pl. 32.

McCoy, F. (1881) *Vermicella annulata* (Gray). The black and white ringed snake. *Natural history of Victoria. Prodromus of the zoology of Victoria; or figures and descriptions of the living species of all classes of the Victorian indigenous animals*, Melbourne 6, 11–12, pl. 52.

McCoy, F. (1885) *Typhlops nigrescens* (Gray sp.). The blackish Australian worm-snake. *Natural history of Victoria. Prodromus of the zoology of Victoria; or figures and descriptions of the living species of all classes of the Victorian indigenous animals*, Melbourne 11, 7–10, pl. 103.

McCoy, F. (1887) *Pseudechys australis* (Gray). The brown Pseudechys. *Natural history of Victoria. Prodromus of the zoology of Victoria; or figures and descriptions of the living species of all classes of the Victorian indigenous animals*, Melbourne 15, 159–160, pl. 142.

McCoy, M. (1980) *Reptiles of the Solomon Islands*. Wau Ecology Institute Handbook (7), vii + 80 pp., 8 figs., 20 pls.

McCoy, M. (2006) *Reptiles of the Solomon Islands*. Pensoft, Sofia, 147 pp., 128 col. photos.

McCranie, J.R. (1976) *Crotalus polystictus. Catalogue of American amphibians and reptiles*, pp. 180.1–180.2.

McCranie, J.R. (1980a) *Crotalus adamanteus. Catalogue of American amphibians and reptiles*, pp. 252.1–252.2.

McCranie, J.R. (1980b) *Crotalus pricei. Catalogue of American amphibians and reptiles*, pp. 266.1–266.2.

McCranie, J.R. (1980c) *Drymarchon. D. corais. Catalogue of American amphibians and reptiles*, pp. 267.1–267.4.

McCranie, J.R. (1981) *Crotalus basiliscus. Catalogue of American amphibians and reptiles*, pp. 283.1–283.2.

McCranie, J.R. (1983a) *Crotalus pusillust. Catalogue of American amphibians and reptiles*, pp. 313.1–313.2, 2 figs.

McCranie, J.R. (1983b) *Nerodia taxispilota. Catalogue of American amphibians and reptiles*, pp. 331.1–331.2.

McCranie, J.R. (1984) *Crotalus vergrandis. Catalogue of American amphibians and reptiles*, pp. 350.1–350.2.

McCranie, J.R. (1986) *Crotalus unicolor. Catalogue of American amphibians and reptiles*, pp. 389.1–389.2.

McCranie, J.R. (1988) Description of the hemipenis of *Sistrurus ravus* (Serpentes, Viperidae). *Herpetologica* 44(1), 123–126, illust.

McCranie, J.R. (1990) *Nerodia erythrogaster* (Forster). *Catalogue of American amphibians and reptiles*, pp. 500.1–500.8, 1 fig.

McCranie, J.R. (1991) *Crotalus intermedius. Catalogue of American amphibians and reptiles*, pp. 519.1–519.4, 3 figs.

McCranie, J.R. (1993a) *Crotalus durissus* Linnaeus. *Catalogue of American amphibians and reptiles*, pp. 577.1–577.11, 1 fig.

McCranie, J.R. (1993b) Additions to the herpetofauna of Honduras. *Caribbean Journal of Science*, Mayagüez 29(3–4), 254–255.

McCranie, J.R. (1996) Geographic distribution: Serpentes: *Ungaliophis continentalis. Herpetological Review* 27(1), 36.

McCranie, J.R. (1999) Notes on the type Series of *Oxybelis wilsoni* Villa and McCranie. *Herpetological Review* 30(1), 11–12.

McCranie, J.R. (2001) Benque Viejo and *Porthidium yucatanicum* (Squamata: Viperidae) in Belize. *Caribbean Journal of Science* 37, 284–285.

McCranie, J.R. (2004a) Geographic distribution: Serpentes: *Rhadinaea decorata. Herpetological Review* 35(3), 294.

McCranie, J.R. (2004b) *Enuliophis, E. sclateri. Catalogue of American amphibians and reptiles*, pp. 799.1–799.3.

McCranie, J.R. (2006a) New species of *Sibon* (Squamata, Colubridae) from northeastern Honduras. *Journal of Herpetology* 40(1), 16–21, 5 figs.

McCranie, J.R. (2006b) New species of the colubrid genus *Rhadinaea* (*godmani* group) from Parque Nacional El Casuco, Honduras. *Proceedings of the Biological Society of Washington* 119(4), 528–533, 5 figs.

McCranie, J.R. (2006c) Northeastern Honduras and the systematics of the colubrid snake *Scaphiodontophis. Caribbean Journal of Science* 42(1), 136–138, 2 figs.

McCranie, J.R. (2007a) A new species of *Sibon* (Squamata, Colubridae) from La Mosquitia, northeastern Honduras. *Herpetologica* 63(2), 213–218, 3 figs.

McCranie, J.R. (2007b) Notes on the poorly known colubrid snake *Rhadinaea anachoreta* Smith & Campbell. *Herpetological Bulletin* 101, 25–26.

McCranie, J.R. (2011a) *The snakes of Honduras. Systematics, distribution, and conservation*. Society for the Study of Amphibians and Reptiles, 714 pp., 65 figs., 20 col. pls.

McCranie, J.R. (2011b) A new species of *Tantilla* of the *taeniata* species group (Reptilia, Squamata, Colubridae, Colubrinae) from northeastern Honduras. *Zootaxa* (3037), 37–44, 3 figs.

McCranie, J.R. & Castañeda, F.E. (2004a) Notes on the second specimen of *Geophis damiani* Wilson, McCranie, and Williams and *Rhadinaea tolpanorum* Holm and Cruz D. (Colubridae). *Herpetological Review* 35(4), 341.

McCranie, J.R. & Castañeda, F.E. (2004b) A new species of snake of the genus *Omoadiphas* (Reptilia: Squamata: Colubridae) from the Cordillera Nombre de Dios in northern Honduras. *Proceedings of the Biological Society of Washington* 117(3), 311–316.

McCranie, J.R. & Castañeda, F.E. (2005) The herpetofauna of Parque Nacional Pico Bonito, Honduras. *Phyllomedusa* 4(1), 3–16, 1 fig.

McCranie, J.R. & Castañeda, F.E. (2007) The colubrid snake *Geophis rhodogaster* (Cope) in Honduras. *Herpetological Bulletin* (99), 19, 1 fig.

McCranie, J.R. & Cruz-Diaz, G.A. (2010) A third new species of the genus *Omoadiphas* (Reptilia, Squamata, Colubridae, Dipsadinae) from Honduras. *Zootaxa* (2690), 53–58, 2 figs.

McCranie, J.R. & Köhler, G. (1999) Two new species of colubrid snakes of the genus *Enulius* from Islas de la Bahia, Honduras. *Caribbean Journal of Science* 35(1–2), 14–22, 6 figs.

McCranie, J.R. & McAllister, C.T. (1988) *Nerodia valida. Catalogue of American amphibians and reptiles*, 431.1–431.3.

McCranie, J.R. & Orellana, L.V. (2011) Geographic distribution: Serpentes: *Boa constrictor. Herpetological Review* 42(3), 392–393.

McCranie, J.R. & Villa, J. (1993) A new genus for the snake *Enulius sclateri* (Colubridae, Xenodontinae). *Amphibia-Reptilia*, Leiden 14(3), 261–267, 3 figs.

McCranie, J.R. & Wilson, L.D. (1984) New herpetological records for the Mexican state of Aguascalientes. *Herpetological Review* 15, 22.

McCranie, J.R. & Wilson, L.D. (1987) The biogeography of the herpetofauna of the pine-oak woodlands of the Sierra Madre Occidental of México. *Milwaukee Public Museum, Contributions in Biology and Geology* (72), 1–30, 7 figs.

McCranie, J.R. & Wilson, L.D. (1991a) *Rhadinaea kinkelini. Catalogue of American amphibians and reptiles*, pp. 523.1–523.2, 2 figs.

McCranie, J.R. & Wilson, L.D. (1991b) *Rhadinaea montecristi. Catalogue of American amphibians and reptiles*, pp. 524.1–524.2, 1 fig.

McCranie, J.R. & Wilson, L.D. (1992) *Rhadinaea godmani. Catalogue of American amphibians and reptiles*, pp. 546.1–546.3, 1 fig.

McCranie, J.R. & Wilson, L.D. (1993) *Leptophis modestus* (Günther). *Catalogue of American amphibians and reptiles*, pp. 578.1–578.2, 1 fig.

McCranie, J.R. & Wilson, L.D. (1995) Two new species of colubrid snakes of the genus *Ninia* from Central America. *Journal of Herpetology* 29(2), 224–232, 2 figs.

McCranie, J.R. & Wilson, L.D. (2001a) Taxonomic status of *Typhlops stadelmani* Schmidt (Serpentes, Typhlopidae). *Copeia* 2001(3), 820–822.

McCranie, J.R. & Wilson, L.D. (2001b) The herpetofauna of the Mexican state of Aguascalientes. *Courier Forschungsinstitut Senckenberg*, Frankfurt-am-Main (230), 1–57, 2 figs., 5 pls.

McCranie, J.R. & Wilson, L.D. (2001c) A significant range extension for the montane snake *Rhadinaea lachrymans* (Colubridae). *Southwestern Naturalist* 46(1), 113–114.

McCranie, J.R., Wilson, L.D. & Gotte, S.W. (2001) Three new country records for Honduran snakes. *Herpetological Review* 32(1), 62–63.

McCranie, J.R., Townsend, J.H. & Wilson, L.D. (2003) Three snakes new to the herpetofauna of Honduras. *Herpetological Review* 34(4), 391–392.

McCranie, J.R., Wilson, L.D. & Köhler, G. (2005) *Amphibians & Reptiles of the Bay Islands and Cayos Cochinos, Honduras*. Bibliomania, Salt Lake City, x + 210 pp., color photos.

McCranie, J.R., Townsend, J.H. & Wilson, L.D. (2006) *The amphibians and reptiles of the Honduran Mosquitia*. Krieger Publishing Company, Melbourne, x + 291 pp., 171 col. pls.

McCranie, J.R., Marineros, L. & Orellana, L.V. (2012a) Geographic distribution: Serpentes: *Tantilla taeniata*. *Herpetological Review* 43(4), 623.

McCranie, J.R., Marineros, L. & Orellana, L.V. (2012b) Geographic distribution: Serpentes: *Tantilla taeniata*. *Herpetological Review* 43(4), 623.

McCrystal, H.K. & McCoid, M.J. (1986) *Crotalus mitchellii*. *Catalogue of American amphibians and reptiles*, pp. 388.1–388.4.

McCurley, K. (2005) *The complete ball python: a comprehensive guide to care, breeding, and genetic mutation*. ECO Herpetological Publishing, Lansing, 304 pp., 305 col. photos.

McDaniel, V.R. & Karges, J.P. (1983) *Farancia abacura*. *Catalogue of American amphibians and reptiles*, pp. 314.1–314.2.

McDiarmid, R.W. (1968) Variation, distribution and systematic status of the black-headed snake *Tantilla yaquia* Smith. *Bulletin of the Southern California Academy of Sciences* 67(3), 159–177.

McDiarmid, R.W. (1977) *Tantilla yaquia*. *Catalogue of American amphibians and reptiles*, pp. 198.1–198.2.

McDiarmid, R.W. (1992) Systematic status of the San Luis Potosi black-headed snake, *Tantilla deviatrix* Barbour (Colubridae). *Southwestern Naturalist* 37(3), 303–307.

McDiarmid, R.W. (2012) Reptile biodiversity and natural history: an overview. *In:* McDiarmid, R.W., Foster, M.S., Guyer, C., Gibbons, J.W. & Chernoff, N. (Eds.), *Reptile biodiversity: standard methods for inventory and monitoring*. University of California Press, Berkeley, pp. 7–23, 1 fig.

McDiarmid, R.W. & Bezy, R.L. (1971) The colubrid snake *Enulius oligostichus* in western México. *Copeia* 1971(2), 350–351, 1 fig.

McDiarmid, R.W. & Folke, S.H. (1991) *Tantilla bocourti*. *Catalogue of American amphibians and reptiles*, pp. 526.1–526.3, 2 figs.

McDiarmid, R.W. & McCleary, R.J.R. (1993) *Phyllorhynchus* Stejneger. Leaf-nosed snake. *Catalogue of American amphibians and reptiles*, pp. 579.1–579.5, 4 figs.

McDiarmid, R.W. & Scott, N.J., Jr. (1970) Geographic variation and systematic status of Mexican lyre snakes of the *Trimorphodon tau* group (Colubridae). *Contributions in Science, Los Angeles County Museum* (179), 1–43, 2 figs., 2 pls.

McDiarmid, R.W., Copp, J.F. & Breedlove, D.E. (1976) Notes on the herpetofauna of western Mexico: new records from Sinaloa and the Tres Marías Islands. *Contributions in Science, Los Angeles County Museum* (275), 1–17, 2 figs.

McDiarmid, R.W., Touré, T. & Savage, J.M. (1996) The proper name of the Neotropical tree boa referred to as *Corallus enhydris* (Serpentes, Boidae). *Journal of Herpetology* 30(3), 320–326.

McDiarmid, R.W., Campbell, J.A. & Touré, T. (1999) *Snake species of the world. A taxonomic and geographic reference*. The Herpetologist's League, xi + 511 pp.

McDermid, E.M., Board, P.G. & Agar, N.S. (1977) Studies on the blood of Australian elapid snakes—III. Electrophoretic analysis of serum proteins and red cell enzymes. *Comparative Biochemistry and Physiology* 56B, 361–365, 2 figs.

McDowell, S.B., Jr. (1961) Systematic division and evolution of the colubrid snake genus *Natrix*, with comments on the subfamily Natricinae, by E.V. Malnate. (Review). *Copeia* 1961(4), 502–506.

McDowell, S.B., Jr. (1967) *Aspidomorphus*: a genus of New Guinea snakes of the family Elapidae, with notes on related genera. *Journal of Zoology*, London 151(4), 497–543, 15 figs.

McDowell, S.B., Jr. (1968) Affinities of the snakes usually called *Elaps lacteus* and *E. dorsalis*. *Journal of the Linnaean Society (Zoology)*, London 47(313), 561–578, 4 figs.

McDowell, S.B., Jr. (1969a) *Toxicocalamus*: a New Guinea genus of snakes of the family Elapidae. *Journal of Zoology*, London 159(4), 443–511, 9 figs.

McDowell, S.B., Jr. (1969b) Notes on the Australian sea-snake *Ephalophis greyi* M. Smith (Serpentes, Elapidae, Hydrophinae) and the origin and classification of sea-snakes. *Journal of the Linnaean Society (Zoology)*, London 48(3), 333–349, 1 fig.

McDowell, S.B., Jr. (1970) On the status and relationships of the Solomon Island elapid snakes. *Journal of Zoology*, London 161(2), 145–190, 7 figs.

McDowell, S.B., Jr. (1972a) The genera of sea-snakes of the *Hydrophis* group (Serpentes, Elapidae). *Transactions of the Zoological Society of London* 32, 189– 247, 1 fig.

McDowell, S.B., Jr. (1972b) The species of *Stegonotus* (Serpentes, Colubridae) in Papua New Guinea. *Zoologische Mededelingen*, Leiden 47(2), 6–26, 3 figs.

McDowell, S.B., Jr. (1974a) A catalogue of the snakes of New Guinea and the Solomons, with special reference to those in the Bernice P. Bishop Museum. Part 1. Scolecophidia. *Journal of Herpetology* 8(1), 1–57, 20 figs.

McDowell, S.B., Jr. (1974b) Additional notes on the rare and primitive sea-snake, *Ephalophis greyi*. *Journal of Herpetology* 8(2), 123–128, 1 fig.

McDowell, S.B., Jr. (1975a) A catalogue of snakes of New Guinea and the Solomons, with special reference to those in the Bernice P. Bishop Museum. Part II. Anilioides and Pythoninae. *Journal of Herpetology* 9(1), 1–79, 23 figs.

McDowell, S.B., Jr. (1975b) *List of snake species for New Guinea (including Bismarck Archipelago, Bouganville, and eastern Moluccas).* Unpublished manuscript, 8 pp.

McDowell, S.B., Jr. (1979) A catalogue of the snakes of New Guinea and the Solomons with special reference to those in the Bernice P. Bishop Museum. Part III. Boinae and Acrochordoidea (Reptilia, Serpentes) *Journal of Herpetology* 13(1), 1–92, 29 figs.

McDowell, S.B., Jr. (1984) Results of the Archbold Expeditions. No. 112. The snakes of the Huon Peninsula, Papua New Guinea. *American Museum Novitates* (2775), 1–28.

McDowell, S.B., Jr. (1985) The terrestrial Australian elapids: general summary. *In:* Grigg, G., Shine, R. & Ehmann, H. (Eds.), *The biology of Australasian frogs and reptiles.* Royal Zoological Society of New South Wales, Chipping Norton, pp. 261–264.

McDowell, S.B., Jr. (1986) The architecture of the corner of the mouth of colubrid snakes. *Journal of Herpetology* 20(3), 353–407, 14 figs.

McDowell, S.B., Jr. (1987) Systematics. *In:* Seigel, R.A., Collins, J.T. & Novak, S.S. (Eds.), *Snakes: Ecology and evolutionary biology.* MacMilliam Publishing, New York, pp. 3–50.

McGuire, J.A. & Grismer, L.L. (1993) The taxonomy and biogeography of *Thamnophis hammondii* and *T. digueti* (Reptilia, Squamata, Colubridae) in Baja California, México. *Herpetologica* 49(3), 354–365, 6 figs.

McGuire, J.A. & Koch, U. (2005) *Bungarus flaviceps.* Vertical distribution. *Herpetological Review* 36(1), 68.

McGurty, B.M. (1988) Natural History of the California mountain kingsnake, *Lampropeltis zonata. Proceedings of the Conference on California Herpetology, Special Publication, Southwestern Herpetological Society* (4), 73–88, 7 figs.

McKay, J.L. (2006) *A field guide to the amphibians and reptiles of Bali.* Krieger Publishing Company, Malabar, vii + 138 pp.,163 col. pls.

McKay, J.L. & Lilley, R. (2012) New distributional records from the Lesser Sundas, Indonesia. *Herpetological Review* 43(1), 109–111, 2 figs.

McKeown, M.S. (1974) Check-list of amphibians and reptiles of Santa Barbara County, California. *Occasional Paper, Santa Barbara Museum of Natural History* (9), 1–11, figs.

McKeown, M.S. (1996) *A field guide to reptiles and amphibians of the Hawaiian Islands.* Diamond Head Publishing Co., Los Osos, 172 pp., color figs.

McKeown, M.S. (1998) The Madagascar ground boa *(Acrantophis madagascariensis):* a brief note on its natural history, captive management and breeding. *The Vivarium* 9(5), 10–13, 51, 7 col. photos.

McLachlan, G.R. (1978) A population of *Typhlops braminus* (Daudin) on the Cape Peninsula (Reptilia; Typhlopidae). *Zoologica Africana* 13(2), 353–354, 1 fig.

McLachlan, G.R. (1984) A bush snake *(Philothamnus semivariegatus)* from Springbok. *Journal of the Herpetological Association of Africa* 30: 11, 1 fig.

McLain, R.B. (1899) *Contribution to Neotropical herpetology.* Robert Baird McLain, Wheeling, 5 pp., 1 pl., 4 figs.

McLeod, D.S., Siler, C.D., Diesmos, A.C., Diesmos, M.L., Garcia, V.S., Arkonceo, A.O., Balaquit, K.L., Uy, C.C., Villaseran, M.M., Yarra, E.C. & Brown, R.M. (2011) Amphibians and reptiles of Luzon Island, V: the herpetofauna of Angat Dam watershed, Bulacan Province, Luzon Island, Philippines. *Asian Herpetological Research* (Series 2) 2(4), 177–198, 8 col. figs.

Meade, L. (1991) Kentucky snakes: their systematics, variation, and distribution. *Journal of the Tennessee Academy of Sciences* 66(4), 175–182, illust.

Meade, L. (2005) *Kentucky Snakes. Their identification, variation, and distribution.* Kentucky State Nature Preserves Commission and Center for North American Herpetology, 323 pp., 17 col. pls.

Means, D.B. (2004a) Geographic distribution: Serpentes: *Bothriopsis taeniata. Herpetological Review* 35(4), 410.

Means, D.B. (2004b) Gluttons of Chappell. *Fauna* 5(1), 14–22, 9 col. photos.

Means, D.B. (2004c) The loathed cottonmouth moccasin: is it aggressive? *Fauna* 5(2), 14–22, 9 col photos.

Means, D.B. (2007) Geographic distribution: Serpentes: *Atractus zidoki. Herpetological Review* 38(4), 484–485.

Means, D.B. & Kalamandeen, M. (2007) Geographic distribution: Serpentes: *Oxyrhopus melanogenys. Herpetological Review* 38(1), 105.

Mebert, K. (2008) Good species despite massive hybridization: genetic research on the contact zone between the water snakes *Nerodia sipedon* and *N. fasciata* in the Carolinas, USA. *Molecular Ecology* 17(8), 1918–1929, 4 figs.

Mecham, J.S. (1983) *Nerodia harteri. Catalogue of American amphibians and reptiles*, pp. 330.1–330.2.

Meckel, J.F. (1829) *System der vergleichenden Anatomie. Vierter Theil.* Bengerschen Buchhandlung, Halle, 758 pp.

Medeiros, T.T., Dias, I.R., Vila-Nova, M.F. & Suzart-Angólo, J. (2010) Geographic distribution: Serpentes: *Oxyrhopus clathratus. Herpetological Review* 41(4), 517.

Medem, F. (1965) Bibliografia comentada de reptiles de Colombia. *Revista de la Academia Colombiana de Ciencias*, Bogotá 12(47), 299–346.

Medem, F. (1969) El desarrollo de la herpetologia en Colombia. *Revista de la Academia Colombiana de Ciencias*, Bogotá (1968) 13, 149–199.

Medem, F. (1979) Los anfibios y reptiles de la Islas Gorgona y Gorgonilla. *In:* Prahl, H. von, Guhl, F. & Grogl, M. (Eds.), *Gorgona.* Universidad de Los Andes, Bogotá, pp. 189–218.

Medica, P.A. (1975) *Rhinocheilus. R. lecontei. Catalogue of American amphibians and reptiles*, pp. 175.1–175.4.

Meek, S.E. (1905) An annotated list of a collection of reptiles from southern California and northern Lower California. *Field Columbian Museum Publication 104* (Zoological Series) 7(1), 1–19, 3 pls.

Meek, S.E. & Elliot, D.G. (1899) Notes on a collection of cold-blooded vertebrates from the Olympic Mountains. *Field Columbian Museum Publication 31* (Zoological Series) 1(12), 225–236.

Meerman, J.C. (1992) Geographic distribution: Serpentes: *Typhlops microstomus. Herpetological Review* 23(3), 92.

Meerwarth, H. (1901) Die Westindischen Reptilien und Batracher des Naturhistorischen Museums im Hamburg. *Mitteilungen aus dem Naturhistorischen Museums in Hamburg* 18, 1–41.

Megson, S., Mitchell, P. & D'Cruze, N. (2009) Reptilia, Serpentes, Colubridae, *Heteroliodon fohy*: Distribution extension. *Check List* 5(3), 692–694, 2 figs.

Méhely, L. von. (1894) *Vipera ursinii* Bonap., eine verkannte Giftschlange Europas. *Zoologischer Anzeiger*, Leipzig 17, 57–61, 65–71.

Méhely, L. von. (1895) Beiträge zur herpetologie von Neu-Guinea. *Természetrajzi Füzetek*, Budapest 18, 73–79, 128–136.

Méhely, L. von. (1897) Zur herpetologie von Ceylon. *Természetrajzi Füzetek*, Budapest 20, 55–70.

Méhely, L. von. (1898) An account of the reptiles and batrachians collected by Mr. Lewis Biró in New Guinea. *Természetrajzi Füzetek*, Budapest 21, 165–178, pl. 12.

Méhely, L. von. (1911) Systematisch-phylogenetische Studien an Viperiden. *Annales Historico-Naturales Musei Nationalis Hungarici* 9, 186–243, 6 figs., pls. 3–5.

Méhely, L. von. (1991) *Herpetologia Hungarica*. Hungarian Natural History Museum, Budapest, 22 col. pls.

Mehrtens, J. (1987) *Living snakes of the world in color*. Sterling Publishing, New York, 256 pp., 542 col. photos.

Meier, M. & Schnieper, R. (1986) *Amphibien und Reptilien der Schweiz: Frösch und Kröten, Molche und Salamander, Eidechsen und Schlangen*. Mondo-Verlag, Lausanne, 152 pp., col. illust.

Meik, J.M., Lawling, A.M. & Pires-da Silva, A. (2010) Body size evolution in insular speckled rattlesnakes (Viperidae): *Crotalus mitchellii*). *PloS One* 5(3), 1–8, 3 figs.

Meirte, D. (1992) Clé de determination des serpents d'Afrique. *Annales du Musée Royal de l'Afrique Centrale* (267), 152 pp., 79 figs., 2 photos.

Meirte, D. (1993) New records of *Leioheterodon madagascariensis* (Reptilia, Colubridae) from the Comoros. *Journal of the Herpetological Association of Africa*, Stellenbosch (42), 21–23.

Meise, W. (1938) Eine neue Korallenschlange aus dem Amazonasgebiet. *Zoologischer Anzeiger*, Leipzig 123(1–2), 20–22.

Meise, W. & Hennig, W. (1932) Die Schlangengattung *Dendrophis*. *Zoologischer Anzeiger*, Leipzig 99(11–12), 273–297.

Meise, W. & Hennig, W. (1935) Zur Kenntnis von *Dendrophis* und *Chrysopelea*. Ein Beitrag zur systematischen Bervertung der opisthoglypha. *Zoologischer Anzeiger*, Leipzig 109(5–6), 138–150.

Meisner, C.F.A. (1820) Ueber die in der Schweiz einheimischen Schlangen überhaupt und die *Vipera* insbesondere. *Museum Naturgeschichte Bern* 1(11–12), 81–95, 1 pl.

Mejenes-López, S.M.A, Mendoza-Quijano, F., Madero-Vega, C. & Molleda-Contreras, A.K. (1999) Geographic distribution: Serpentes: *Coniophanes imperialis imperialis*. *Herpetological Review* 30(4), 235.

Melgarejo, A.R. (1980) Comportamiento depredador de *Sibynomorphus turgidus* (Cope) (Serpentes: Dipsadinae). *Resúmenes, Jornadas de Ciencias Naturales*, Montevideo 1, 127–128.

Melgarejo, A.R. & Meneghel, M.D. (1984) Presencia en el Uruguay de *Liotyphlops ternetzii* (Boulenger, 1896) (Serpentes, Anomalepididae). *Boletin de la Sociedad Zoologica del Uruguay*, Montevideo 2, 12–17, 2 figs.

Mell, R. (1922) Beiträge zur Fauna Sinica. I. Die vertebraten Südchinas, Feldlisten und Feldnoten der Säuger, Vögel, Reptilien, Batrachier. *Archiv für Naturgeschichte*, Berlin 88A(10), 1–146, pl. 4.

Mell, R. (1929) *Beiträge zur fauna Sinica. IV. Grundzuge einer Okologie der chinesischen Reptilien und einer herpetologischen Tiergeographie Chinas*. Walter de Gruyter, Berlin, 282 pp., 34 figs., 7 pls.

Mell, R. (1930) Schlangen, *In:* Beiträge zur Lurch- und Kriechtierfauna Kwangsi's. *Sitzungsberichte der Gesellschaft Naturforschender Freunde zu Berlin* 1929, 310–326.

Mell, R. (1931a) List of Chinese snakes. *Lingnan Science Journal*, Canton (1929) 8, 119–219.

Mell, R. (1931b) The distribution of Chinese reptiles in relation to zoogeographical boundaries. *Lingnan Science Journal*, Canton (1929) 8, 221–258.

Mellado, J. & Mateo, J.A. (1992) New records of Moroccan herpetofauna. *Herpetological Journal*, London 2, 58–61, 1 fig.

Mellink, E. (1996) Geographic distribution: Serpentes: *Eridiphas slevini*. *Herpetological Review* 27(2), 88.

Mellish, J. (1876) Notes on the serpents of Prince Edward Island. *Proceedings and Transactions of the Nova Scotian Institute of Natural Science* 4, 163–167.

Mello, O. (1926) Contribuição para o conhecimento dos ofidios brasileiros. *Memorias do Instituto Oswaldo Cruz*, Rio de Janeiro 19(1), 127–130, pls. 27–29.

Mendelson, J.R., III & Kizirian, D.A. (1995a) Geographic variation in *Rhadinaea hempsteadae* (Serpentes, Colubridae) with the description of a new species from Chiapas, Mexico. *Herpetologica* 51(3), 301–313, 4 figs.

Mendelson, J.R., III & Kizirian, D.A. (1995b) An anomalous specimen referable to *Rhadinaea kinkelini* (Serpentes, Colubridae), from Chiapas, Mexico. *Journal of Herpetology* 29(4), 624–625, 1 fig.

Mendelson, J.R., III, Williams, B.L., Sheil, C.A. & Mulcahy, D.G. (2005) Systematics of the *Bufo coccifer* complex (Anura: Bufonidae) in Mesoamerica. *Scientific Papers, Natural History Museum, University of Kansas* (38), 1–27.

Mendelssohn, H. (1963) On the biology of the venomous snakes of Israel. Part I. *Israel Journal of Zoology*, Jerusalem 12(1–4), 143–170, 9 figs.

Mendelssohn, H. (1965) On the biology of the venomous snakes of Israel. II. *Israel Journal of Zoology*, Jerusalem 14, 185–212, 24 figs.

Mendez de la Cruz, F.R., Camarillo, J.L., Villagrán-Santa Cruz, M. & Aguilar-Cortez, R. (1992) Observaciones sobre el status de los anfibios y reptiles de la Sierra de Guadalupe (Distrito Federal-Estado de México). *Anales Instituto de Biologia, Universidad Nacional Autónoma de México* 63(2), 249–256, 1 fig.

Mendoza, J.S. & Lucero, M.F. (2009) Geographic distribution: Serpentes: *Drymarchon caudomaculatus*. *Herpetological Review* 40(4), 455.

Mendoza, J.S., Berrio, V. & Gomez, D.A. (2012) Reptilia, Squamata, Viperidae, *Bothrops venezuelensis* Leybold, 1873: Distribution extension and first country record. *Check List* 8(2), 310–312, 5 figs.

Mendoza-Hernández, A.A., Pérez-Ramos, E. & García-Vázquez, U.O. (2011) *Drymobius chloroticus*. Geographic distribution: Serpentes: *Herpetological Review* 42(1), 114.

Mendoza-Quijano, F. (1990) *Estudio herpetofaunistico en el transecto Zacualtipán-Zoquizoquipan-San Juan Meztitlán, Hidalgo*. Master's Thesis, Universidad Nacional Autónoma de México, Tlalnepantla, 97 pp.

Mendoza-Quijano, F. & Hernández-Hernández, B. (2001) Geographic distribution: Serpentes: *Spilotes pullatus mexicanus*. *Herpetological Review* 33(4), 279–280.

Mendoza-Quijano, F. & Smith, H.M. (1993) A new species of hooknose snake, *Ficimia* (Reptilia, Serpentes). *Journal of Herpetology* 27(4), 406–410, 2 figs.

Mendoza-Quijano, F., Campos-Rodriguez, J.I., Lopez-Vidal, J.C., Smith, H.M. & Chiszar, D. (2003) *Adelphicos quadrivirgatum* (Serpentes: Colubridae) in Hidalgo, Mexico, with comments on its relationships to *A. visoninum*. *Bulletin of the Maryland Herpetological Society* 39(4), 77–84.

Mendoza-Quijano, F., Campos-Rodriguez, J.I., Canseco-Márques, L., Smith, H.M. & Chiszar, D. (2005) New variational data on *Adelphicos quadrivirgatum* (Serpentes, Colubridae) in Mexico. *Bulletin of the Maryland Herpetological Society* 41(1), 31–32.

Mendoza-Paz, R.F., Mendoza-Quijano, F., Bryson, R.W., Jr. & Smith, E.N. (2006) Geographic distribution: Serpentes: *Ophryacus undulatus*. *Herpetological Review* 37(1), 109.

Meneghel, M.D. & Achaval, F. (1983) Dispersion geografica de *Liophis flavifrenatus* (Cope, 1862) (Serpentes, Colubridae), y su presencia en la Republica Oriental del Uruguay. *Boletin de la Sociedad Zoologica del Uruguay*, Montevideo 1, 15–24, 1 fig.

Meneghel, M.D. & Achaval, F. (1997) Geographic distribution: Serpentes: *Leptophis ahaetulla marginatus*. *Herpetological Review* 28(2), 98.

Meneghel, M.D. & Melgarejo, A.R. (1984) Identification especifica de ejemplares Uraguayos del genero *Sibynomorphus* (Serpentes, Dipsadinae). *Boletin de la Sociedad Zoologica del Uruguay*, Montevideo 2, 12–17, 2 figs.

Menegon, M. (1996) *I serpenti e l'habitat.* Siste Edizioni, Trento, 32 pp., illust.

Menegon, M., Davenport, T.R.B. & Howell, K.M. (2011) Description of a new and critically endangered species of *Atheris* (Serpentes: Viperidae) from the southern highlands of Tanzania, with an overview of the country's tree viper fauna. *Zootaxa* (3120), 43–54, 7 figs.

Meneses G.O. (1974) Ofidios y ofidismo en el Peru. I. Las serpientes venenosas del Perú. *Revista de Instituto de Zoonosis y Investigaciones Pecuarias*, Lima 2(3–4), 69–77.

Ménétriés, E. (1832) *Catalogue raisonné des objets de zoologie recueillis dans un voyage au Caucase et jusqu'aux frontières actuelles de la Perse entrepris par ordre de S.M. l'Empereur*. Académie Impérial des Sciences, St. Pétersbourg, 271 + xxxiii + iv pp.

Ménez, A. (1985) Molecular immunology of snake toxins. *Pharmacological Therapeutics* 30, 91–113, 10 figs.

Mengden, G.A. (1983) The taxonomy of Australian elapid snakes: a review. *Records of the Australian Museum*, Sydney 35, 195–222.

Mengden, G.A. (1985a) Australian elapid phylogeny: a summary of the chromosomal and electrophoretic data. *In:* Grigg, G., Shine, R. & Ehmann, H. (Eds.), *The biology of Australasian frogs and reptiles*. Royal Zoological Society of New South Wales, Chipping Norton, pp. 185–192.

Mengden, G.A. (1985b) A chromosomal and electrophoretic analysis of the genus *Pseudonaja*. *In:* Grigg, G., Shine, R. & Ehmann, H. (Eds.), *The biology of Australasian frogs and reptiles*. Royal Zoological Society of New South Wales, Chipping Norton, pp. 193–208, 7 figs.

Mengden, G.A., Shine, R. & Moritz, C. (1986) Phylogenetic relationships within the Australian venomous snakes of the genus *Pseudechis*. *Herpetologica* 42(2), 215–229, 5 figs.

Mense, M. (2006) *Rautenpythons:* Morelia bredli, Morelia carinata *und der* Morelia-spilota-*Komplex*. Natur und Tier-Verlag, Münster, 207 pp., 205 col. photos.

Menzies, J.I. (1966) The snakes of Sierra Leone. *Copeia* 1966(2), 169–179, 6 figs.

Mercadante, O.A. (Ed.). (2005) *Serpentes ilhoas em Alcatrazes e Queimada Grande*. Editora Magma, São Paulo, 176 pp., col. photos.

Mercurio, V. & Andreone, F. (2005) Description of a new *Geodipsas* snake from northern Madagascar (Squamata, Colubridae). *Zootaxa* (1093), 61–68, 3 figs.

Merker, G. & Merker, W. (2005) Alterna, *The gray-banded kingsnake*. LM Digital, 80 pp., col. illust.

Merrem, B. (1790) *Beytraege zur Naturgeschichte*. Erstes Heft. Meyerschen Buchhandlung, Duisburg, vii + 47 pp., 12 pls.

Merrem, B. (1820) *Versuch eines Systems der Amphibien*. Johann Christian Krieger, Marburg, xv + 191 pp., 1 pl.

Mertens, R. (1921a) Zur Kenntnis der Reptilienfauna von Malta. *Zoologischer Anzeiger*, Leipzig 53, 236–240.

Mertens, R. (1921b) Reptilien, Amphibien und Fische aus Bialowies. *Senckenbergiana*, Frankfurt-am-Main 3(5), 146–148.

Mertens, R. (1922) Verzeichniss der Typen in der Herpetologischen Sammlung des Senckenbergischen Museums. *Senckenbergiana*, Frankfurt-am-Main (1920–1922) 4(6), 162–183.

Mertens, R. (1924) Über einige Reptilien aus Borneo. *Zoologischer Anzeiger*, Leipzig 40, 155–159.

Mertens, R. (1925) Amphibiens und Reptilien aus dem nördlichen und östlichen Spanien, gesammelt von Dr. F. Haas. *Abhandlungen der Senckenbergischen Naturforschenden Gesellschaft*, Frankfurt-am-Main 39(1), 27–129, pls. 2–4.

Mertens, R. (1926) Über die Rassen einiger indo-australischer Reptilien. *Senckenbergiana*, Frankfurt–am–Main 8, 272–279.

Mertens, R. (1927a) Herpetologische Mitteillungen XVIII. Zur Verbreitung der *Vipera russelii* Shaw. *Senckenbergiana*, Frankfurt-am-Main 9(5), 182–184.

Mertens, R. (1927b) Neue Amphibien und Reptilien aus dem Indo-Australischen Archipel gesammelt während der Sunda-Expedition Rensch. *Senckenbergiana*, Frankfurt-am-Main 9(6), 234–242.

Mertens, R. (1928) Herpetologische mitteilungen. XXII. Zur herpetofauna der Insel Sumba. *Senckenbergiana*, Frankfurt-am-Main 10, 229–231.

Mertens, R. (1929a) Bemerkungen über *Typhlops florensis*, seine rassen und seine verbreitung. *Treubia*, Buitenzorg 11(2), 293– 296.

Mertens, R. (1929b) Beiträge zur herpetologie Tunesiens. *Senckenbergiana*, Frankfurt-am-Main 11(5–6), 291–310.

Mertens, R. (1930) Die Amphibien und Reptilien der Inseln Bali, Lombok, Sumbawa und Flores (Beiträge zur Fauna der Kleinen Sunda-Inseln, I). *Abhandlungen der Senckenbergischen Naturforschenden Gesellschaft*, Frankfurt-am-Main 42(3), 115–344, 9 pls.

Mertens, R. (1933) Die Reptilien der Madagaskar-Expedition Prof. Dr. H. Bluntschli's. *Senckenbergiana*, Frankfurt-am-Main 15(3–4), 260–274.

Mertens, R. (1934a) Die Schlangengattung *Dendrelaphis* Boulenger in systematischer und zoogeographischer. Beziehung, I. *Archiv für Naturgeschichte*, Berlin (2)3(2), 187–204.

Mertens, R. (1934b) Die Insel-Reptilien: ihre Asubreitung, Variation und Artbildung. *Zoologica*, Stuttgart 32(84), 1–209, 6 pls.

Mertens, R. (1936) Eine neue Natter der Gattung *Helicops* aus Inner-Afrika. *Zoologischer Anzeiger*, Leipzig 114(9–10), 284–285.

Mertens, R. (1937) Ueber eine herpetologische Sammlung aus dem Gebiete des Njarasa-Grabens Ost-Afrika. *Veröffentlichungen Deutschen Kolonial- und Übersee-Museum in Bremen* 2(1), 1–9.

Mertens, R. (1938) Herpetologische Ergebnisse einer Reise nach Kamerun. *Abhandlungen der Senckenbergischen Naturforschenden Gesellschaft*, Frankfurt-am-Main (442), 1–52.

Mertens, R. (1939) Herpetologische Ergebnisse einer Reise nach der Insel Hispaniola, Westindien. *Abhandlungen der Senckenbergischen Naturforschenden Gesellschaft*, Frankfurt-am-Main (449), 1–84, 10 pls.

Mertens, R. (1940a) Bemerkungen über einige Schlangen aus Iran. *Senckenbergiana*, Frankfurt-am-Main 22(3–4), 244–259.

Mertens, R. (1940b) Zur herpetologie Kameruns und Deutsch-Ostafrikas. *Zoologischer Anzeiger*, Leipzig 131(9–10), 239–250.

Mertens, R. (1941a) Zur Kenntnis der Herpetofauna von Fernando-Poo. 3. Beitrag zu den wissenschaftlichen Ergebnissen der Forschungsreise H. Eidmann nach Spanisch-Guinea 1939/40. *Zoologischer Anzeiger*, Leipzig 135(11–12), 275–281.

Mertens, R. (1941b) Eine neue Korallennatter aus Mexiko. *Senckenbergiana*, Frankfurt-am-Main 23(46), 216–217, 1 fig.

Mertens, R. (1942) Amphibien und Reptilien I (Ausbeute der Hamburger Südperu-Expedition). *In:* Titschack, E. (Ed.), *Beiträge zur Fauna Perus* 2, 277–287. [second edition in 1952 with minor changes]

Mertens, R. (1943) Systematische und ökologische Bemerkungen über die Regenbogenschlange, *Xenopeltis unicolor* Reinwardt. *Zoologische Garten*, Frankfurt-am-Main 15(5–6), 213–220, 3 figs.

Mertens, R. (1947a) Die Lurche und Kriechtiere des Rhein-Main-Gebietes. *Senckenbergischen Naturforschenden Gesellschaft*, Frankfurt-am-Main 16, 1–144, 32 pls.

Mertens, R. (1947b) Studien zur Eldonomie und taxonomie der ringlenatter (*Natrix natrix*). *Abhandlungen der Senckenbergischen Naturforschenden Gesellschaft*, Frankfurt-am-Main (476), 1–38.

Mertens, R. (1951) Über de boiden-typus *Plastoseryx bronni*. *Senckenbergiana*, Frankfurt-am-Main 32(1–4), 205–206.

Mertens, R. (1952a) Neues über die Reptilien fauna von El Salvador. *Zoologischer Anzeiger*, Leipzig 148(3–4), 87–93.

Mertens, R. (1952b) Weitere neue Reptilien aus El Salvador. *Zoologischer Anzeiger*, Leipzig 149(5–6), 133–138.

Mertens, R. (1952c) Die amphibien und reptilien von El Salvador, auf Grund der Reisen von R. Mertens und A. Zilch. *Abhandlungen der Senckenbergischen Naturforschenden Gesellschaft*, Frankfurt-am-Main (487), 1–83, 16 pls.

Mertens, R. (1952d) Amphibien und Reptilien aus der Türkei. *Revue de la Faculté des Sciences de l'Université d'Istanbul* 17B(7), 41–75.

Mertens, R. (1952e) Nomenklatorisches über die Schlangengattung *Eteirodipsas*. *Senckenbergiana*, Frankfurt-am-Main 32(5–6), 307–308.

Mertens, R. (1954a) Eine neue Natter der Gattung *Elapomorphus*. *Senckenbergiana*, Frankfurt-am-Main 34(5), 183–185, 1 fig.

Mertens, R. (1954b) Neue Schlangenrassen aus Südwest- und Südafrika. *Zoologischer Anzeiger*, Leipzig 152(9–10), 213–219.

Mertens, R. (1955) Die Amphibien und Reptilien Südwestafrikas. Aus den Ergebnissen einer im Jahre 1952 ausgeführten Reise. *Abhandlungen der Senckenbergischen Naturforschenden Gesellschaft*, Frankfurt-am-Main (490), 1–172, 24 pls.

Mertens, R. (1956) Die viper von Montecristo. *Senckenbergiana Biologica*, Frankfurt-am-Main 37(3–4), 221–224, 2 figs.

Mertens, R. (1957a) Amphibien und reptilien aus dem Äussersten western Javas und von Benachbarten Eilanden. *Treubia*, Bogor 24(1), 83–105.

Mertens, R. (1957b) Die Amphibien und Reptilien Korsikas. *Senckenbergiana Biologica*, Frankfurt-am-Main 38(3–4), 175–192, 1 pl.

Mertens, R. (1957c) Die Würfelnatter (*Natrix tessellata*) der Schlangeninsel. *Senckenbergiana Biologica*, Frankfurt-am-Main 38(5–6), 271–275.

Mertens, R. (1958) *Bitis heraldica*, eine oft verkannte Otter aus Angola. *Senckenbergiana Biologica*, Frankfurt-am-Main 39(3–4), 145–148, 4 figs.

Mertens, R. (1959a) Amphibien und reptilien von Karimundjawa, Bawean und den Kangean-Inseln. *Treubia*, Bogor 25(1), 1–15.

Mertens, R. (1959b) Eine neue Wassernatter aus West-Pakistan. *Senckenbergiana Biologica*, Frankfurt-am-Main 40(3–4), 117–120, 1 pl.

Mertens, R. (1960) Über die Schlangen der Galapagos. *Senckenbergiana Biologica*, Frankfurt-am-Main 41(3–4), 133–141, 2 figs.

Mertens, R. (1961a) Die Amphibien und Reptilien der Insel Korfu. *Senckenbergiana Biologica*, Frankfurt-am-Main 42(1–2), 1–29.

Mertens, R. (1961b) Was ist "*Naja celebensis* Ahl" in Wirklichkeit? *Senckenbergiana Biologica*, Frankfurt-am-Main 42(5–6), 419–420.

Mertens, R. (1961c) Zur Synonymie, Taxionomie und Verbreitung der Vierstreifennatter, *Elaphe quaturolineata*. *Bonner Zoologische Beiträge*, Bonn 12, 324–331.

Mertens, R. (1963) Die Rassengliederung der afrikanischen Wassernatter *Limnophis bicolor*. *Senckenbergiana Biologica*, Frankfurt-am-Main 44(6), 437–439, 2 figs.

Mertens, R. (1964) *Welches Tier ist das? Kriechtiere und Lurche*. Kosmos, Stuttgart, 98 pp., 48 pls.

Mertens, R. (1965a) Eine neue Natter von einer Insel des Roten Meeres. *Senckenbergiana Biologica*, Frankfurt-am-Main 46(1), 5–9, 3 figs.

Mertens, R. (1965b) Zur Kenntnis der neotropischen Natterngattung *Pseudoeryx*. *Senckenbergiana Biologica*, Frankfurt-am-Main 46(4), 279–285, 3 figs.

Mertens, R. (1965c) Die Reptilien von Fernando Poo. *Bonner Zoologische Beiträge*, Bonn (1964) 15(3–4), 211–238.

Mertens, R. (1967a) Die herpetologische Sektion des Natur-Museums und Forschungs-Institutes Senckenberg in Frankfurt a. M. nebst einem Verzeichnis ihrer Typen. *Senckenbergiana Biologica*, Frankfurt-am-Main 48(1), 1–106.

Mertens, R. (1967b) Über *Lachesis libanotica* und den status von *Vipera bornmuelleri*. *Senckenbergiana Biologica*, Frankfurt-am-Main 48(3), 153–159, 4 figs.

Mertens, R. (1968a) Eine schwarze Zornnatter von den Cycladen, *Coluber gemonensis gyarosensis* n. subsp. *Senckenbergiana Biologica*, Frankfurt-am-Main 49(3–4), 181–189, 8 figs.

Mertens, R. (1968b) Die Arten und Unterarten der Schmuckbaumschlangen (*Chrysopelea*). *Senckenbergiana Biologica*, Frankfurt-am-Main 49(3–4), 191–217, 7 figs.

Mertens, R. (1968c) *Kriechtiere und Lurche*. Kosmos, Stuttgart, 48 pp.

Mertens, R. (1969a) Die Amphibien und Reptilien West-Pakistans. *Stuttgarter Beiträge zu Naturkunde* (197), 1–96.

Mertens, R. (1969b) Zur Synonymie und Variabilität der Würfelnatter (*Natrix tessellata*). *Senckenbergiana Biologica*, Frankfurt-am-Main 50(3–4), 125–131, 3 figs.

Mertens, R. (1969c) Über die Variabilität der Achtstreifennatter, *Oligodon octolineatus*. *Senckenbergiana Biologica*, Frankfurt-am-Main 50(5–6), 339–345, 2 figs.

Mertens, R. (1970) Die Amphibien und Reptilien West-Pakistans 1. Nachtrag. *Stuttgarter Beiträge zu Naturkunde* (216), 1–5.

Mertens, R. (1971) Die Herpetofauna Südwest-Afrikas. *Abhandlungen der Senckenbergischen Naturforschenden Gesellschaft*, Frankfurt-am-Main (529), 1–110.

Mertens, R. (1972a) Eine neue Schlanknatter der Gattung *Leptophis* aus México. *Senckenbergiana Biologica*, Frankfurt-am-Main 53(5–6), 341–342, 1 fig.

Mertens, R. (1972b) Herpetofauna tobagana. *Stuttgarter Beiträge zu Naturkunde* (252), 1–22.

Mertens, R. (1972c) Madagaskars Herpetofauna und die Kontinentadrift. Studien uber die Reptilienfauna Madagaskars 6. *Zoologische Mededelingen*, Leiden 46(7), 91–98.

Mertens, R. (1973) Bemerkenswerte Schlanknattern der neotropischen Gattung *Leptophis*. *Studies on Neotropical Fauna and Environment* 8, 141–154, 8 figs.

Mertens, R. & Müller, L. (1940) Die Amphibien und Reptilien Europas (Zweite Liste, nach dem Stand vom I. Januar 1940). Second edition. *Abhandlungen der Senckenbergischen Naturforschenden Gesellschaft*, Frankfurt-am-Main (451), 1–56. [first edition, 1928, 41(1), 1–62]

Mertens, R. & Rosenberg, H. (1943) *Elaphe flavirufa* (Cope): die mexikanische Nachtnatter. *Wochenschrift für Aquarien- und Terrarienkunde*, Braunschweig 37(3), 60–62, 3 figs.

Mertens, R. & Wermuth, H. (1960) *Die Amphibien und Reptilien Europas (Dritte Liste, nach dem Stand vom 1. Januar 1960)*. Waldemar Kramer, Frankfurt-am-Main, xi + 264 pp., 46 figs.

Mertens, R., Darevsky, I.S. & Klemmer, K. (1967) *Vipera latifii*, eine neue Giftschlange aus dem Iran. *Senckenbergiana Biologica*, Frankfurt-am-Main 48(3), 161–168, 5 figs.

Mertz, W.H., Reyes-Velasco, J. & Grünwald, C.I. (2011) Geographic distribution: Serpentes: *Leptodeira uribei*. *Herpetological Review* 42(1), 114–116.

Mertz, W.H., Shedo, J.D., Reyes-Velasco, J. & Grünwald, C.I. (2012) Geographic distribution: Serpentes: *Coniophanes michoacanensis*. *Herpetological Review* 43(3), 447–448.

Meshaka, W.E., Jr. & J.T. Collins (2009) *A pocket guide to Pennsylvania snakes*. Commonwealth of Pennsylvania, 56 pp., 48 col. figs.

Meshaka, W.E., Jr., Butterfield, B.P. & Hauge, J.B. (2004) *The exotic amphibians and reptiles of Florida*. Krieger Publishing Company, Malabar, 155 pp., col. illust.

Metaxa, L. (1823) *Monografia de' serpenti di Roma e suoi contorni*. Stamperia de Romanis, Roma, 48 pp., 1 pl.

Methuen, P.A. (1919) Descriptions of a new snake from the Transvaal, together with a new diagnosis and key of the genus *Xenocalamus*, and of some Batrachia from Madagascar. *Proceedings of the Zoological Society of London* 89(3–4), 349–355, 1 fig.

Methuen, P.A. & Hewitt, J. (1913) On a collection of reptiles from Madagascar made during the year 1911. *Annals of the Transvaal Museum*, Pretoria (1912) 3(4), 183–193, pls. 5–11.

Metzger, G.A., Kraus, F., Allison, A. & Parkinson, C.L. (2010) Uncovering cryptic diversity in *Aspidomorphus* (Serpentes: Elapidae): Evidence from mitochondrial and nuclear markers. *Molecular Phylogenetics and Evolution* 54(2), 405–416, 5 figs.

Mey, D. & Schmidt, K. (2002) Die Amphibien und Reptilien des Wartburgkreises und der Stadt Eisenach (Thüringen). *Naturschutz im Wartburgkreis* 10, 1–132, col. illust.

Meyer, A. (2001) Reptilienarten im Portrait. *In:* Amphibien und Reptilien in Ob- und Nidwalden. Mit Beiträgen aus der Naturwissenschaften, der Sagenwelt und der Volksmedizin. *Naturforschende Gesellschaft Ob- und Nidwalden,* Grafenort 2, 110–147, 20 col. photos.

Meyer, A.B. (1874) Übersicht der von mir auf Neu-Guinea und den Inseln Jobi, Mysore und Mafoor im Jahre 1873 gesammelten Amphibien. *Monatsberichte der Køniglich Akademie Wissenschaften zu Berlin* (1875) 1874(2), 128–140.

Meyer, A.B. (1886) Die Giftdrüsen bei der Gattung *Adeniophis*. *Sitzungsberichte der Königlich Preußischen Akademie der Wissenschaften zu Berlin* 30, 611–614.

Meyer, A.B. (1887) Verzeichniss der von mir in den Jahren 1870–1873 Ostindischen Archipel Gesammelten Reptilien und Batrachier. *Abhandlungen und Berichte des Königlichen Zoologischen Anthropologisch-Ethnographischen Museums zu Dresden* 1886–1887(2), 1–16.

Meyer, H. von. (1844a) No title (letter on fossil vertebrates). *Neues Jahrbuch für Mineralogie, Geognosie, Geologie und Petrefakten-Kunde* 12, 329–340.

Meyer, H. von. (1844b) No title (letter on fossil reptiles). *Neues Jahrbuch für Mineralogie, Geognosie, Geologie und Petrefakten-Kunde* 12, 564–567.

Meyer, H. von. (1845) *Zur Fauna der Vorwelt. Fossile Saeugethiere, Voegel und Reptilien aus dem Molasse-Mergel von Oeningen*. Frankfurt-am-Main, viii + 52 pp., 12 pls.

Meyer, H. von. (1855) No title (letter on fossil reptiles). *Neues Jahrbuch für Mineralogie, Geognosie, Geologie und Petrefakten-Kunde* 23, 326–337.

Meyer, H. von. (1860) *Coluber* (*Tropidonotus*?) *atavus* aus der Braunkohle des Siebengebirges. *Palaeontographica* 7(4), 232–240, 1 pl.

Meyer, J.R. (1969) *A biogeographic study of the amphibians and reptiles of Honduras*. Doctoral Dissertation, University of Southern California, Los Angeles, 589 pp., illust.

Meyer, J.R. & Wilson, L.D. (1971) Taxonomic studies and notes on some Honduran amphibians and reptiles. *Bulletin of the Southern California Academy of Sciences* 70(3), 106–114.

Meylan, P.A. (1982) The Squamate reptiles of the Inglis IA fauna (Irvintonian, Citris County, Florida). *Bulletin of the Florida State Museum, Biological Sciences* 27(3), 1–85, 15 figs., 5 pls.

Meylan, P.A. (1985) *Heterodon simus*. *Catalogue of American amphibians and reptiles*, pp. 375.1–375.2.

Meylan, P.A. (1987) Fossil snakes from Laetoli. *In:* Leakey, M.D. & Harris, J.M. (Eds.), *The Pliocene site of Laetoli, northern Tanzania*. Oxford University Press, Oxford, pp. 78–82, fig. 3.12.

Meylan, P.A. (1995) Pleistocene amphibians and reptiles from the Leisey Shell Pit, Hillsborough County, Florida. *Bulletin of the Florida Museum of Natural History* 37(9), 273–297, 3 figs.

Meza-Parral, A., Rojas, A.S. & Valverde, T.R. (2010) *Nerodia rhombifer werleri* (Hallowell, 1852) (Serpentes: Colubridae), Nuevo registro para la ofidiofauna de Puebla, México. *Acta Zoológica Mexicana* 26(1), 243–244, 1 fig.

Mezzena, R. & Dolce, S. (1977) *Anfibi e rettili del Carso: guida alla mostra erpetologica permanente del Museo Civico di Storia Naturale di Trieste*. Museo Civico di Storia Naturale, Trieste, 218 pp., 72 figs., 96 photos.

Michaud, E.J. & Dixon, J.R. (1987) Taxonomic revision of the *Liophis lineatus* complex (Reptilia, Colubridae) of Central and South America. *Contributions in Biology and Geology, Milwaukee Public Museum* (71), 1–26, 10 figs.

Mijangos, A., Huacuz, D. & Quijada-Mascareñas, A. (2006) Geographic distribution: Serpentes: *Pseudoleptodeira uribei*. *Herpetological Review* 37(4), 500.

Mijares-Urrutia, A. & Arends, A. (1993) New distributional records of amphibians and reptiles for the state of Falcón, Venezuela. *Herpetological Review* 24(4), 157–158.

Mijares-Urrutia, A. & Arends, A. (1999) Additional new regional and local records of amphibians and reptiles from the state of Falcón, Venezuela. *Herpetological Review* 30(2), 115.

Mijares-Urrutia, A. & Arends, A. (2000) Herpetofauna of Estado Falcón, northwestern Venezuela: a checklist with geographical and ecological data. *Smithsonian Herpetological Information Series* (123), 1–30, 1 fig.

Mijares-Urrutia, A., Markezich, A.L. & Arends, A. (1995) Hallazgo de *Leptodeira bakeri* Ruthven (Serpentes, Colubridae) en la Peninsula de Paraguaná, noroeste de Venezuela, con comentarios diagnósticos y biológicos. *Caribbean Journal of Science*, Mayagüez 31(1–2), 77–82, 1 fig.

Mijaria-Uriona, J. (2001) *Ofidios en la Amazonia Boliviana: culebras - víboras - serpientes.* Artes Gráficas "Sagitario," San Pedro, 153 pp., 93 figs., 151 col. photos.

Mikan, J.C. (1820–1825) *Delectus florae et faunae brasiliensis.* Typis Antonii Strauss, Vindobonae, 47 pp., 24 pls.

Mikátová, B., Vlasín, M. & Zavadil, V. (Eds.). (2001) *Atlas of the distribution of reptiles in the Czech Republic.* Praha, Brno, 257 pp., illust. (in Czech)

Mikhailowsky, M. (1904) Sur l'herpetologie de la province Transcaspei. *Annuaire du Musée Zoologique de l'Académie Impériale des Sciences de St. Pétersbourg* 9, 39–44. (in Russian)

Milanovich, J.R., Anfinson, J., Baker, T.J., Baxter, R.J., Brandebura, S.C., Deviney, S.A., Hiler, W.R., Learned, J., Matthews, M.W., McDowell, C.R., Neal, R.G., Stephens, N.T., Stewart, J.P., Wheeler, B.A. & Trauth, S.E. (2005) Geographic distribution: Serpentes: *Sonora semiannulata. Herpetological Review* 36(3), 340.

Miles, M.A., Thomson, A.G. & Walters, G.A. (1978) Amphibians and reptiles from the vicinity of Boughari, Casamance (Senegal), and the Gambia. *Bulletin de l'Institut Fondamental d'Afrique Noire*, Dakar 40A(2), 437–456, 5 figs.

Miller, B. (1982) The speckled brown snake *Pseudonaja guttata* Parker: an addition to the fauna of South Australia. *Transactions of the Royal Society of South Australia*, Adelaide 106(1–2), 79–80, 2 figs.

Miller, B.T. (1995) Geographic distribution: Serpentes: *Virginia valeriae* (smooth earth snake). *Herpetological Review* 26(3), 157.

Miller, D.J. (1979) A life history of the gray-banded kingsnake, *Lampropeltis mexicana alterna*, in Texas. *Contribution, Chihuahuan Desert Research Institute* (87), 1–48, 18 figs.

Miller, D.M., Young, R.A., Gatlin, T.W. & Richardson, J.A. (1982) Amphibians and reptiles of the Grand Canyon National Park. *Grand Canyon Natural History Association, Monograph* 4, 1–144, 88 figs., 43 pls.

Miller, H.W., Jr. (1955) Some Eocene reptiles from New Jersey. *Notulae Naturae*, Philadelphia (268), 1–5, 10 figs.

Miller, R.W. & Zyla, J.D. (1992) Additional rainbow snakes, *Farancia erytrogramma*, from Charles County, Maryland. *Bulletin of the Maryland Herpetological Society* 28(3), 99–101.

Miller, T.J. & Smith, H.M. (1979) The lesser African rock python. *Bulletin of the Maryland Herpetological Society* 15(3), 70–84, figs.

Millet de la Turtaudière, P.A. (1828) *Faune de Maine et Loire ou description méthodique des animaux qu'oon rencontre dans toute l'étendue du Département du Maine et Loire, tant sédentaires que de passage, avec des observations sur leur moeurs, habitudes etc. Tome II.* Rosier, Paris, pp. 381–773, 6 pls.

Mills, R.C. (1948) Check-list of the reptiles and amphibians of Canada. *Herpetologica* 4(2), 1–15.

Mills, T. (1995) Natural history of the scarlet kingsnake, *Lampropeltis triangulum elapsoides,* in South Carolina. *The Vivarium* 7(2), 24–29, 9 col. photos.

Milner, A.C., Milner, A.R. & Estes, R. (1982) Amphibians and squamates from the upper Eocene of Hordle Cliff, Hampshire—a preliminary report. *Tertiary Research* 4(4), 149–154.

Milstead, W.W. (1953) Geographic variation in the garter snake, *Thamnophis cyrtopsis. Texas Journal of Science* 5(3), 348–379, 6 figs., 1 pl.

Milstead, W.W. (1956) A small collection of amphibians and reptiles from Argentina. *Revista Brasileira de Biologia*, Rio de Janiero 16, 321–325.

Mincham, H. (1970) *Reptiles of Australia & New Zealand.* Rigby Instant Book, Kent Town, 64 pp., 32 col. photos.

Minoli, I., Alvares, D.J. & Avila, L.J. (2011) New records and geographic distribution map of *Bothropoides diporus* Cope, 1862 (Reptilia, Viperidae). *Check List* 7(5), 608–609, 2 figs.

Minter, C. (2005) *Venomous snakes of the Southeast: an introduction to the venomous snakes of the southeastern United States.* SeaRay Books, xiv + 59 pp., illust.

Minton, S.A., Jr. (1956) A new snake of the genus *Tantilla* from west Texas. *Fieldiana: Zoology* 34(39), 449–452, fig.

Minton, S.A., Jr. (1962) An annotated key to the amphibians and reptiles of Sind and Las Bela, West Pakistan. *American Museum Novitates* (2081), 1–60, 72 figs.

Minton, S.A., Jr. (1966) A contribution to the herpetology of West Pakistan. *Bulletin of the American Museum of Natural History* 134(2), 27–184, 12 figs., pls. 9–36.

Minton, S.A., Jr. (1968) Antigenic relationships of the venom of *Atractaspis microlepidota* to that of other snakes. *Toxicon* 6(1), 59–64, 2 figs.

Minton, S.A., Jr. (1975) Geographic distribution of sea snakes. *In:* Dunson, W.A. (Ed.), *The biology of sea snakes.* University Park Press, Baltimore, pp. 21–32.

Minton, S.A., Jr. (1980) *Thamnophis butleri. Catalogue of American amphibians and reptiles*, pp. 258.1–258.2.

Minton, S.A., Jr. (1983) *Sistrurus catenatus. Catalogue of American amphibians and reptiles*, pp. 332.1–332.2.

Minton, S.A., Jr. (1990) Immunologic relationships in *Agkistrodon* and related genera. *In:* Gloyd, H.K. and Conant, R. (Ed.), *Snakes of the* Agkistrodon *complex. A monographic review.* Society for the Study of Amphibians and Reptiles, pp. 589–599, 9 figs.

Minton, S.A., Jr. (1992) Serologic relationships among pitvipers: evidence from plasma albumins and immunodiffusion. *In:* Campbell, J.A. & Brodie, E.D., Jr. (Eds.), *Biology of the pitvipers.* Selva, Tyler, pp. 155–161, 5 figs.

Minton, S.A., Jr. (2001) *Amphibians and reptiles of Indiana.* Second edition. Indiana Academy of Science, Indianapolis, xiv + 404 pp., illust., col. photos. [first edition, 1972, v + 346 pp., 7 figs., 65 pls.]

Minton, S.A., Jr. & Dunson, W.A. (1985) Sea snakes collected at Chesterfield Reefs, Coral Sea. *Atoll Research Bulletin,* Washington, D.C. (292), 101–108, 2 figs.

Minton, S.A., Jr. & Smith, H.M. (1960) A new subspecies of *Coniophanes fissidens* and notes on Central American amphibians and reptiles. *Herpetologica* 16(2), 103–111, 1 fig.

Minton, S.A., Jr., Dowling, H.G. & Russell, F.E. (Eds.). (1968) *Poisonous snakes of the world. A manual for use by U.S. amphibious forces.* Second edition. Department of the Navy and Bureau of Medical Surgery, Washington, 212 pp., 120 figs., 8 col. pls.

Miranda, A. (1993) Manejo de fauna Silvestre. *Ciencias, Revista de Difusión,* México City (7), 103–110, 7 figs.

Miranda, F., Rochat, H. & Lissitzky, S. (1964) Sur les neurotoxins de deux espèces de scorpions nord-africains. III. Déterminations preliminaries aux etudes de structure sur les neurotoxins (scorpamins) d'*Androctonus australis* (L.) et de *Buthus occitanus* (Am.). *Toxicon* 2(2), 123–138, 2 figs.

Miralles, A. & David, P. (2010) First record of *Ahaetulla mycterizans* (Linnaeus, 1758) (Reptilia, Squamata, Colubridae) from Sumatra, Indonesia, with an expanded definition. *Zoosystema* 32(3), 449–456, 3 figs.

Miranda, B.H.V. & Fernandes, D.S. (2012) Morphological variation of *Sordellina punctata* (Peters, 1880) with notes on the phylogenetic position of the genus within the family Dipsadidae (Serpentes). *International Journal of Zoology* 2012 (827820), 1-10, 14 figs.

Miranda, D.B., Albuquerque, S. de & Venâncio, N.M. (2009) Reptilia, Colubridae, *Rhinobothryum lentiginosum* (Scopoli 1785): first record from state of Acre, Brazil. *Check List* 5(4), 917–918, 1 fig.

Miranda, L.C., Ribera-A., M.O., Sarniento-T., J., Salinas, E. & Navia-R., C. (Eds.). (1991) *Plan de manejo de la reserva de la biósfera Estación Biológica del Beni 1991*. Academia Nacional de Ciencias de Bolivia, La Paz, 556 pp.

Miranda, M.E. & Couturier, G.A. (1981) Una nueva subspecie de *Tachymenis peruviana* Wiegmann, 1835 (Ophidia, Boigidae) para la Argentina. *Comunicaciones del Museo Argentino de Ciencias Naturales Bernardino Rivadavia*, Buenos Aires 4(10), 79–83, fig.

Miranda, M.E. & Couturier, G.A. (1984) Consideraciones taxonomicas en torno a *Liophis typhlus* (Linnaeus), 1758 (Serpentes, Colubridae) en la Argentina. *Historia Natural*, Corrientes 4(23), 225–228, 3 figs.

Miranda, M.E. & Tio-Vallejo, M. (1985) El genero *Leptotyphlops* en la Argentina (Reptilia, Ophidia). I. Grupos *septemstriatus, dulcis* y *albifrons*. *Physis*, Buenos Aires 43(105), 71–72.

Miranda, M.E., Couturier, G.A. & Williams, J.D. (1983) *Guia de los ofidios Bonaerenses*. Second edition. Jardín Zoológica de La Plata, La Plata, 71 pp., illust.

Miranda-Ribeiro, A. de. (1915) *Lachesis lutzi*. Uma variedade de *L. pictus* Tschudi. *Archivos do Museu Nacional*, Rio de Janeiro 17(3), 1–4, 1 pl.

Miranda-Ribeiro, P. de. (1955) Tipos des espécies e subespecies do Prof. Alipio de Miranda Ribeiro depositados no Museu Nacional. *Arquivos do Museu Nacional*, Rio de Janeiro 42(1), 389–417.

Mirtschin, P.J. (1981) South Australian records of the inland taipan (*Oxyuranus microlepidotus* McCoy, 1879). *Herpetofauna*, Sydney 13(1), 20–23, 2 figs.

Mirtschin, P.J. (1982a) Seasonal colour changes in the inland taipan, *Oxyuranus microlepidotus* (McCoy). *Herpetofauna*, Sydney 14(2), 97–99, 2 figs.

Mirtschin, P.J. (1982b) Occurrence and distribution of the inland taipan, *Oxyuranus microlepidotus* (Reptilia, Elapidae) in South Australia. *Transactions of the Royal Society of South Australia*, Adelaide 106(3–4), 213–214, 1 fig.

Mirtschin, P.J. & Davis, R. (1992a) *Dangerous snakes of Australia; an illustrated guide to Australia's most venomous snakes*. Second edition. Rigby, Adelaide, 208 pp., 85 figs., 48 col. pls. (first edition, 1982)

Mirtschin, PJ. & Davis, R (1992b) *Snakes of Australia: dangerous & harmless*. Hill of Content, Melbourne, 216 pp., col. illust.

Mirtschin, P.J., Crowe, G.R. & Davis, R. (1990) Dangerous snakes of Australia. *In:* Gopalakrishnakone, P. & Chou, L.M. (Eds.), *Snakes of medical importance (Asia-Pacific region)*. National University of Singapore, Singapore, pp. 1–173, 119 figs.

Mishima, S. & Yamazato, S. (1970) Feeding of the snakes kept indoors at the Japan Snake Institute. *The Snake* 2(1), 59–68.

Misonne, X. (1980) Gaston de Witte (1897–1980). Notice biographique et liste bibliographique. *Bulletin de l'Institut Royal des Sciences Naturelles de Belgique*, Bruxelles 52(20), 1–8.

Mistry, V., Vogel, G. & Tillack, F. (2007) Rediscovery of *Dinodon gammiei* (Blanford, 1878) (Serpentes: Colubridae) with discussion of its validity. *Hamadryad* 31(2), 265–273, 6 figs.

Mitchell, F.J. (1951) The South Australian reptile fauna. Part 1. Ophidia. *Records of the South Australian Museum*, Adelaide 9, 545–557, figs.

Mitchell, F.J. (1955) Preliminary account of the Reptilia and Amphibia collected by the National Geographic Society-Commonwealth Government-Smithsonian Institution Expedition to Arnhem Land (April to November, 1948). *Records of the South Australian Museum*, Adelaide Mus. 11, 373–408, 7 figs., 1 pl.

Mitchell, F.J. (1964) Reptiles and amphibians of Arnhem Land. *In:* Specht, R.L. (Ed.), *Records of the American-Australian scientific expedition to Arnhem Land. Vol. 4. Zoology*. Melbourne University Press, pp. 309–343, 7 figs., 2 pls.

Mitchell, F.J. (1965) Australian geckos assigned to the genus *Gehyra* Gray (Reptilia, Gekkonidae). *Senckenbergiana Biologica* 46(4), 287–319, 11 figs.

Mitchell, J.C. (1974) The snakes of Virginia. *Virginia Wildlife* 35, 1–8.

Mitchell, J.C. (1977) Geographic variation of *Elaphe guttata* (Reptilia, Serpentes) in the Atlantic coastal plain. *Copeia* 1977(1), 33–41, 6 figs.

Mitchell, J.C. (1982a) *Farancia. Catalogue of American amphibians and reptiles*, pp. 292.1–292.2.

Mitchell, J.C. (1982b) *Farancia erytrogramma. Catalogue of American amphibians and reptiles*, pp. 293.1–293.2.

Mitchell, J.C. (1994) *The reptiles of Virginia*. Smithsonian Institution Press, Washington, D.C., xv + 352 pp., 8 figs., 63 col. pls.

Mitchell, J.C. & Anderson, J.M. (1994) Amphibians and reptiles of Assateague and Chincoteague islands. *Virginia Museum of Natural History, Special Publication* (2), vii + 120 pp., 6 figs., 27 col. pls.

Mitchell, J.C. & Reay, K.K. (1999) *Atlas of amphibians and reptiles in Virginia*. Virginia Department of Game and Inland Fish, Richmond, 122 pp., illust.

Mitchell, J.D. (1903) The poisonous snakes of Texas, with notes on their habits. *Transactions of the Texas Academy of Sciences* 5(2), 19–48, 9 pls.

Mittleman, M.B. (1947) Geographic variation in the sea snake, *Hydrophis ornatus* (Gray). *Proceedings of the Biological Society of Washington* 60, 1–6.

Mittleman, M.B. (1949) Geographic variation in Marcy's garter snake, *Thamnophis marcianus* (Baird and Girard). *Bulletin of the Chicago Academy of Sciences* 8(10), 235–249, 1 fig.

Miyata, K. (1980) *Patterns of diversity in tropical herpetofaunas*. Doctoral Dissertation, Harvard University, Cambridge, 787 pp.

Miyata, K. (1982) A check list of the amphibians and reptiles of Ecuador with a bibliography of Ecuadorian herpetology. *Smithsonian Herpetological Information Service* (54), 1–70.

Mizukami, N., Mihara, Y., Toriba, M. & Moriguchi, H. (1989) *Dinodon orientalis* collected at Sapporo, Hokkaido. *The Snake* 21(2), 114–115, 2 figs. (in Japanese)

Mlynarski, M. (1961) Serpents pliocènes et pléistocènes de la Pologne avec la revue critique des Colubridés fossiles. *Folia Quaternaria*, Kraków (1960) 4, 1–45, 6 pls. (in Polish)

Mlynarski, M. (1964) Die jungpliozäne Reptlilienfauna von Rebielice Królewskie, Polen. *Senckenbergiana Biologica*, Frankfurt-am-Main 43(4), 325–347, 43 figs.

Mlynarski, M. (1966) *Plazy I gady Polski. Atlas.* Wydawnictwa Szkolne I Pedagogiczne, Warszawa, 78 pp., 27 col. pls.

Mlynarski, M. (1971) *Nasze gady.* Panstwowe, Zaklady, Wydawnictw & Szkolnych, Warszawa, 179 pp., illust.

Mlynarski, M., Szyndlar, Z., Estes, R. & Sanchiz, B. (1982) Lower vertebrate fauna from the Miocene of Opole (Poland). *Estudios Geológicos*, Madrid 38, 103–119, 12 figs.

Mlynarski, M., Szyndlar, Z., Estes, R. & Sanchiz, B. (1984) Notes on the amphibians and reptilian fauna of the Polish Miocene. *Acta Zoologica Cracoviensia*, Krákow 27(8), 127–148, 2 figs.

Mlynarski, M., Szyndlar, Z., Estes, R. & Sanchiz, B. (1985) Amphibians and reptiles from the Pliocene locality of Weze II near Dzialoszyn (Poland). *Acta Palaeontologica Polonica*, Warszawa (1984) 29(3–4), 209–226, 7 figs.

Mo, D., Chen, H. & Li, F. (1984) *Opisthotropis premaxillaris* (Angel), a record new to China. *Acta Herpetologica Sinica*, Chengdu 3(1), 28.

Mocafico, G. (2007) *Serpens.* Steidl, Göttingen, 261 pp., 109 col. photos.

Mociño-Deloya, E., Setser, K., Reguera, S., Feriche, M. & Pleguezuelos, J.M. (2009) Natural history notes. *Thamnophis scaliger. Herpetological Review* 40(4), 443.

Mocquard, F. (1886) Sur une nouvelle espèce d'*Atractaspis* (*A. leucura*). *Bulletin de la Société Philomathique de Paris* (1885–1886) (Série 7) 10(1), 14–18, pl. 5.

Mocquard, F. (1887a) Du genre *Heterolepis* et des especes qui le composent dont trois nouvelles. *Bulletin de la Société Philomathique de Paris* (1886–1887) (Série 7) 11(1), 5–34, 2 pls.

Mocquard, F. (1887b) Sur une nouvelle espèce d'Elaps, *E. heterochilus. Bulletin de la Société Philomathique de Paris* (1886–1887) (Série 7) 11(1), 39–41.

Mocquard, F. (1887c) Sur les Ophidiens repportés du Congo par la Mission de Brazza. *Bulletin de la Société Philomathique de Paris* (1886–1887) (Série 7) 11(1), 62–92.

Mocquard, F. (1887d) Contribution à l'histoire du genre *Psammodynastes. Bulletin de la Société Philomathique de Paris* (1886–1887) (Série 7) 11(4), 172–180, pls. 3–4.

Mocquard, F. (1888) Sur une collection de reptiles et de batraciens rapportés des pays Comalis et de Zanzibar par M.G. Révoil. *Mémoires, Publiés par la Société Philomathique à l'occasion du Centenaire sa foundation 1788–1888,* 109–134, pls. 11–12.

Mocquard, F. (1889) Sur une collection de reptiles du Congo. *Bulletin de la Société Philomathique de Paris* (1888–1889) (Série 8) 1(4), 145–148, pl. 2.

Mocquard, F. (1890a) Diagnoses d'especes nouvelles de reptiles et de batraciens des iles Bornéo et Palawan. (Suite.) Ophidiens. *Le Naturaliste*, Paris (Série 2) 12(80), 154–155.

Mocquard, F. (1890b) Recherches sur la faune herpétologique des iles de Bornéo et de Palawan. *Nouvelles Archives du Muséum National d'Histoire Naturelle*, Paris (Série 3) 2, 115–168, pls. 7–11.

Mocquard, F. (1891) Séance du 22 Novembre 1980. *Bulletin de la Société Philomathique de Paris* (1890–1891) (8)3(1), 8–9.

Mocquard, F. (1892a) Description de deux ophidiens et d'un batracien d'espces nouvelles. *Le Naturaliste*, Paris (Série 2) 14(117), 35.

Mocquard, F. (1892b) Rectification. *Le Naturaliste*, Paris (Série 2) 14(120), 64.

Mocquard, F. (1892c) Voyage de M. Chaper à Borneo. Nouvelle contribution à la fauna herpétologique de Bornéo. *Mémoires de la Société Zoologique de France*, Paris 5, 190–206, pl. 7.

Mocquard, F. (1894a) Diagnosis de quelques reptiles nouveaux de Madagascar. *Compte-Rendu Sommaire des Séances de la Société Philomathique de Paris* 1894 (9), 3–5.

Mocquard, F. (1894b) Reptiles nouveaux ou insuffisamment connus de Madagascar. *Compte-Rendu Sommaire des Séances de la Société Philomathique de Paris* 1894 (17), 3–10.

Mocquard, F. (1895a) Sur les reptiles recueillis à Madagascar de 1867 à 1885 par M. Grandidier. *Bulletin de la Société Philomathique de Paris* (1894–1895) (Série 8) 7(3), 93–111.

Mocquard, F. (1895b) Sur une collection de reptiles recueillis à Madagascar par Mm. Alluaud et Belly. *Bulletin de la Société Philomathique de Paris* (1894–1895) (Série 8) 7(3), 112–135.

Mocquard, F. (1896a) Sur quelques reptiles et batraciens nouveaux du Haut Oubanghi. *Compte-Rendu de Séances de Troisième Congrès International de Zoologie*, 231–234.

Mocquard, F. (1896b) Sur une collection de reptiles du Haut–Oubanghi. *Compte-Rendu Sommaire des Séances de la Société Philomathique de Paris* 1896, 44–45.

Mocquard, F. (1897a) Note préliminaire sur une collection de reptiles recueillie par M. Haug à Lambaréné. *Bulletin du Muséum d'Histoire Naturelle*, Paris 3(1), 54–55.

Mocquard, F. (1897b) Note sur quelques reptiles de Tanga, don de M. Gierra. *Bulletin du Muséum d'Histoire Naturelle*, Paris 3(4), 122–123.

Mocquard, F. (1897c) Notes herpétologiques. *Bulletin du Muséum d'Histoire Naturelle*, Paris 3(6), 211–217.

Mocquard, F. (1897d) Sur une collection de reptiles recueillie par M. Haug, à Lambaréné. *Bulletin de la Société Philomathique de Paris* (1896–1897) (Série 8) 9(1), 5–20.

Mocquard, F. (1899) Contribution à la faune herpétologique de la Basse-Californie. *Nouvelles Archives du Muséum d'Histoire Naturelle*, Paris (Série 4) 1, 297–344, pls. 11–13.

Mocquard, F. (1901) Note préliminaire sur une collection de reptiles et de batraciens recueillis par M. Alluaud dans le sud de Madagascar. *Bulletin du Muséum d'Histoire Naturelle*, Paris 7(6), 251–255.

Mocquard, F. (1902a) Sur une collection de reptiles et de batraciens recueillis par M. Alluaud dans le sud de Madagascar. *Bulletin de la Société Philomathique de Paris* (1901–1902) (Série 9) 4(1), 5–25, 2 pls.

Mocquard, F. (1902b) Sur des reptiles et batraciens de l'Afrique orientale anglaise, du Gabon et de la Guinée française (région de Kouroussa). *Bulletin du Muséum d'Histoire Naturelle*, Paris 8(6), 404–417.

Mocquard, F. (1903) Description d'espèces nouvelles de la collection du Muséum. *Bulletin du Muséum d'Histoire Naturelle*, Paris 9(5), 209–221.

Mocquard, F. (1904a) Descriptions de quelques reptiles et d'un batracien nouveaux de la collection du Museum. *Bulletin du Muséum d'Histoire Naturelle*, Paris 10(6), 301–309.

Mocquard, F. (1904b) Serpents recueillis par M.A. Pavie en Indo-Chine. *In:* Pavie, A. (Ed.), *Mission Pavie Indochine* (Série 2) 3, 481–482.

Mocquard, F. (1905a) Diagnoses de quelques espèces nouvelles de reptiles. *Bulletin du Muséum d'Histoire Naturelle*, Paris 11(2), 76–79.

Mocquard, F. (1905b) Note préliminaire sur une collection de reptiles et de batraciens offerte au Muséum par M. Maurice de Rothschild. *Bulletin du Muséum d'Histoire Naturelle*, Paris 11(5), 285–288.

Mocquard, F. (1905c) Description de deux nouvelles espèces de reptiles. *Bulletin du Muséum d'Histoire Naturelle*, Paris 11(5), 288–289.

Mocquard, F. (1905d) Sur une collection de reptiles recueillie dans le Haut-Tonkin, par M. le Docteur Louis Vaillant. *Bulletin de la Société Philomathique de Paris* (Série 9) 7(5), 317–322.

Mocquard, F. (1906) Descriptions de quelques reptiles et d'un batracien d'espèces nouvelles. *Bulletin du Muséum d'Histoire Naturelle*, Paris 12(5), 247–253.

Mocquard, F. (1908) Description de quelques reptiles et d'un batracien nouveaux de la collection du Muséum. *Bulletin du Muséum d'Histoire Naturelle*, Paris 14, 259–262.

Mocquard, F. (1909) Synopsis des familles, genres et espèces des reptiles ecailleaux et des batraciens de Madagascar. *Nouvelles Archives du Muséum d'Histoire Naturelle*, Paris (Série 5) 1, 1–110.

Mocquard, F. (1910) Voyage de M. le D'Louis Vallant dans l'Asie centrale (Mission Pelliot). *Bulletin du Muséum d'Histoire Naturelle*, Paris 16(3), 145–152,

Moeller, D., Kossoy, A., Hamilton, W. & Miller, L.L. (2000) Geographical distribution. *Leptotyphlops dulcis. Herpetological Review* 31(1), 56.

Mohabey, D.M., Head, J.J. & Wilson, J.A. (2011) A new species of the snake *Madtsoia* from the Upper Cretaceous of India and its paleobiogeographical implications. *Journal of Vertebrate Paleontology* 31(3), 588–595, 3 figs.

Mohamed, A.H. & Kamel, A. (1963) Preparation of Egyptian black snake antivenin. *Toxicon* 1(2), 77–79.

Mohamed, A.H., El-Serougi, M.S. & Kamel, A. (1964) Effects of *Walterinessia aegypta* venom on blood sodium, potassium and catecholamines and urine 17-ketosteroids. *Toxicon* 2(2), 103–107, 4 figs.

Mohamed, A.H., Saleh, A.M. & Ahmed, F. (1971) Nerve growth factor from Egyptian snake venoms. *Toxicon* 9(3), 201–204, 5 figs.

Mohapatra, P.P., Das, A., Tillack, F. & Dutta, S.K. (2009) Taxonomy, natural history, and distribution of *Boiga forsteni* (Duméril, Bibron et Duméril, 1854) (Serpentes: Colubridae) from Orissa, India. *Russian Journal of Herpetology* 16(4), 243–252, 5 figs.

Mohapatra, P.P., Das, A. & Dutta, S.K. (2010a) Geographic distribution: Serpentes: *Psammodynastes pulverulentus. Herpetological Review* 41(1), 111.

Mohapatra, P.P., Dutta, S.K. & Parida, S.P. (2010b) Report of *Xenochrophis schnurrenbergeri* Kramer, 1977 (Serpentes: Natricidae) from Orissa, India. *Russian Journal of Herpetology* 17(2), 94–96, 1 fig.

Mojsisovics, F.G.H.A. von. (1889) Nachträgliche bemerkungen zu meines Arbeit "Über die geographisische verbreitung einiger westpalearktischen Schlangen." *Mitteilungen Naturwissenschaftlicher Vereines für Steiermark*, Graz 25, 270–273.

Mole, R.R. (1924) The Trinidad snakes. *Proceedings of the Zoological Society of London* 95(1), 235–278, 10 pls.

Mole, R.R. & Urich, F.W. (1894) A preliminary list of the reptiles and batrachians of the island of Trinidad. *Journal of the Field Naturalist's Club*, Port-of-Spain 2(3), 77–90.

Moler, P.E. (Ed.). (1992) *Rare and endangered biota of Florida.* Volume III. *Amphibians and reptiles.* University Press of Florida, Gainesville, xxviii + 291 pp., illust.

Molina, C.R., Pérez-Emán, J. & Sharpe, C.J. (2002) Geographical distribution. *Micrurus psyches. Herpetological Review* 33(3), 228.

Molina, G.I. (1782) *Saggio sulla Storia Naturale del Chili.* Stamperia di S. Tommaso d'Aquino, Bologna, 375 pp .

Molina R., Molina, C.R. & Rivas-Fuenemayer, G.A. (1996) Geographical distribution. *Rhinobothryum lentiginosum. Herpetological Review* 27(3), 155.

Mollier, P., Chwetzoff, S., Frachon, P. & Ménez, A. (1989) Immunological properties of notexin, a potent presynaptic and myotoxic component from the venomof the Australian tiger snake *Notechis scutatus scutatus. Federation of European Biochemical Societies Letters* 250(2), 479–482, 1 fig.

Monard, A. (1935) Mission Scientifique Suisse dans l'Angola. Résultats Scientifique. Reptiles. *Bulletin de la Société Neuchâteloise des Sciences Naturelles* 55, 89–111, 5 figs.

Monard, A. (1937) Contribution a l'herpetologie d'Angola. *Arquivos do Museu Bocage*, Lisboa (8), 19–153.

Monard, A. (1940) Résultats de la mission scientifique du Dr Monard en Guinée Portugaise 1937–1938. VIII. Reptiles. *Arquivos do Museu Bocage*, Lisboa 11, 147–180, 7 figs., 1 pl.

Monard, A. (1951) Resultats de la mission zoologique Suisse au Cameroun. Reptiles. *Mémoirs de l'Institut Français d'Afrique Noire (Sciences Naturelles)*, Dakar (1), 123–170.

Moncuit, T. & Daoues, K. (2002). *Reptiles.* Barnes & Noble Books, New York, 185 pp., col. illust.

Montanelli, S. & Alvaréz, B.B. (1998) Geographic distribution: *Tantilla melanocephala. Herpetological Review* 29(3), 179.

Montague, P.D. (1914) A report on the fauna of the Monte Bello Islands. *Proceedings of the Zoological Society of London* 84(3), 625–652, pl.

Montaquim, M.A., Sarker, A.H., Khan, M.A.R. & Husain, K.Z. (1980) List of the snakes of Bangladesh. *Bangladesh Journal of Zoology* 8(2), 127–129.

Montechiaro, L., Baptista de Oliveira, R., Funk-Pontes, G.M. & Di-Bernardo, M. (2006) Geographic distribution: Serpentes: *Dipsas indica petersi. Herpetological Review* 37(1), 108.

Monteil, V. (1951) Contribution a l'étude de la fauna du Sahara occidental du Sanglier au Phacochère. *Institut des Hautes Études Marocaines, Notes et etudes Documentaires* 9, 169 pp.

Monteiro, C., Montgomery, C.E., Spina, F., Sawaya, R.J. & Martins, M. (2006) Feeding, reproduction, and morphology of *Bothrops mattogrossensis* (Serpentes, Viperidae, Crotalinae) in the Brazilian Pantanal. *Journal of Herpetology* 40(3), 408–413, 2 figs.

Montero, R., Scrocchi, G., Montaño, M.E. & Fernández-S, I.M. (1995) Nuevas citas de saurios, anfisbenidos y ofidios para Bolivia. *Cuadernos de Herpetologia* 9(1), 7–13, 1 fig.

Montingelli, G.G. (2009) *Revisão taxonômica do gênero Mastigodryas Amaral, 1934 (Serpentes: Colubridae).* Doctoral Dissertation, Instituto de Biociências da Universidade de São Paulo, São Paulo, 339 pp.

Montingelli, G.G. & Zaher, H. (2011) New species of *Mastigodryas* Amaral, 1934 from Brazilian Amazonia and Guyana (Serpentes: Colubridae). *Journal of Herpetology* 45(1), 111–119, 6 figs.

Montrousier, P. (1860) [Note communicate in] Melanges et nouvelles. *Revue et Magasin de Zoologie Pure et Appliquée*, Paris (Série 2) 12, 95.

Monzel, M. (2009) *Genetische und morphologische Differenzierung der Gattung* Bothrops *in der südostlichen Neotropis.* Edition Chimaira, Frankfurt-am-Main, 195 pp., 26 col. photos.

Monzó-Giménez, J.C. (2002) *Anfibios y reptiles del entorno de Pinoso (Alicante).* Second edition. Concejalía de Medio Ambiente, Ayuntamiento de Pinoso, Pinoso, 151 pp., col. illust. [first edition, 2001]

Mooi, R.D, Wiens, J.P. & Casper, G.S. (2011) Extreme color variation within populations of the common gartersnake, *Thamnophis sirtalis* in central North America, with implications for subspecies status. *Copeia* 2011(2), 187–200, 6 figs.

Moonen, J., Eriks, W. & Deursen, K. van. (1979) *Surinaamse Slangeninkleur.* Kersten, Paramaribo, 119 pp., col. photos.

Moore, J.P. (1893) The eggs of *Pityophis melanoleucus*. *The American Naturalist* 27(322), 878–885, 2 figs., pls. 19–20.

Mora, J.M. (1987) Predation by *Loxocemus bicolor* on the eggs of *Ctenosaura similis* and *Iguana iguana*. *Journal of Herpetology* 21(4), 334–335.

Morafka, D.J. (1977) *A biogeographical analysis of the Chihuahuan Desert through its herpetofauna*. W. Junk B. V., The Hague, viii + 313 pp.

Morales, V.R. & McDiarmid, R.W. (1996) Annotated checklist of the amphibians and reptiles of Pakitza, Manu National Park Reserve Zone, with comments on the herpetofauna of Madre de Dios, Peru. *In:* Wilson, D.E. & Sandoval, A. (Eds.), *Manu*. Smithsonian Institution, Washington, pp. 503–522.

Morata, D.H., Avila, R.W., Kawashita-Ribeiro, R.A. & Carvalho, M.A. de. (2011) Squamata, Elapidae, *Micrurus surinamensis* (Cuvier, 1817): new records and distribution map in the state of Mato Grosso, Brazil, with notes on diet and activity period. *Check List* 7(3), 350–351, 3 figs.

Morato, S.A., Moura-Leite, J.C. de, Costa-Prudente, A.L. & Bérnils, R.S. (1995) A new species of *Pseudoboa* Schneider, 1801 from southeastern Brazil (Serpentes: Colubridae: Xenodontinae: Pseudoboini). *Biociências* 3(2), 253–264, 4 figs.

Morato, S.A., Franco, F.L. & Sanchez, E.J. (2003) Uma nova especie de *Clelia* (Serpentes, Colubridae) do sul do Brazil. *Phyllomedusa* 2(2), 93–100, 2 figs.

Morato, S.A., Lima, A.M.X. de, Staut, D.C.P., Faria, R.G., Souza-Alves, J.P. de, Gouveia, S.F., Cavilha-Scupino, M.R., Gomes, R. & Silva, M.J. da (2011) Amphibians and reptiles of the Refúgia de Vida Silvestre Mata do Junco, municipality of Capela, state of Sergipe, northeastern Brazil. *Check List* 7(6), 750–762, 5 figs.

Moreno, J.A. (1991) Appendix II. Status recomendacioanes de designation. *In:* Moreno, J.A. (Ed.), Status y distribucion de los Reptiles y Anfibios de la Region de Puerto Rico. *Publicación Cientifica Miscelanea,* San Juan (1), 65–67. [not seen]

Moreno-Arias, R.A. (2010) Geographic distribution: Serpentes: *Sibon annulatus*. *Herpetological Review* 41(3), 382.

Moreno-Arias, R.A., Medina-Rangel, G.F. & Rances-Caicedo, J. (2006) Geographic distribution: Serpentes: *Dipsas pratti*. *Herpetological Review* 37(1), 108.

Moreno-Arias, R.A., Medina-Rangel, G.F. & Castaño-Mora, O.V. (2008) Lowland reptiles of Yacopí (Cundinamarca, Colombia). *Revista de la Academia Colombiana de Ciencias* 32(122), 93–1033 figs.

Morgan, D.R. (1988) The lowland viper (*Atheris superciliaris*) in captivity: a preliminary perspective. *Journal of the Herpetological Association of Africa*, Stellenbosch 34, 38–40, illust.

Morgan, E.C. (1973) *Snakes of the subfamily Sibynophiinae*. Doctoral Dissertation, University of Southwestern Louisiana, Lafayette, 259 pp., 32 figs.

Morgan, G.S. & Lucas, S.G. (1999) Type specimens of fossil vertebrates in the New Mexico Museum of Natural History and Science. *In:* Lucas, S.G. (Ed.), *New Mexico's fossil record 2.* New Mexico Museum of Natural History and Science Bulletin 16, 253–259.

Mori, A. (1993) A note on the sidewinding locomotion in two colubrid snakes, *Opisthotropis typica* and *Pseudoxenodon macrops*. *The Snake* 25, 67–70.

Mori, A. (1994) Ecological and morphological characteristics of the Japanese rat snake, *Elaphe climacophora*, on Rammuri-jima Island, a possible case of insular gigantism. *The Snake* 26(1), 11–18.

Mori, A., Ota, H. & Kamezaki, N. (1999) Foraging on sea turtle nesting beaches: flexible foraging tactics by *Dinodon semicarinatum* (Serpentes: Colubridae). *In:* Ota, H. (Ed.), *Tropical island herpetofauna: origin, current diversity, and conservation.* Elsevier, Amsterdam, pp. 99–128, 13 figs.

Mori, A., Ikeuchi, I. & Hasegawa, M. (2006) Herpetofauna of Ampijoroa, Ankarafantsika Strict Nature Reserve, a dry forest in northwestern Madagascar. *Herpetological Natural History* 10(1), 31–60, 59 figs.

Mori, M. (1982) *Japans Schlangen. Vol. I.* Igaku-Shoin, Tokyo, 80 pp., 70 col. pls.

Mori, M. (1984) *Japans Schlangen. Vol. II.* Igaku-Shoin, Tokyo, 102 pp., 104 col. pls.

Mori, M. (1986) *Japans Schlangen. Vol. III.* Igaku-Shoin, Tokyo, 123 pp., 107 col. pls.

Mori, T. (1928) *A hand-list of the Manchurian and eastern Mongolian Vertebrata*. Tamezo Mori Publisher, Tokyo (1927), 186 pp.

Moriarty, J.J. (1998) Geographic distribution: Serpentes: *Pituophis catenifer sayi*. *Herpetological Review* 29(2), 115.

Moriarty, J.J. & Jones, D. (Eds.). (1997) *Minnesota's amphibians and reptiles: their conservation and status. Proceedings of a symposium.* Minnesota Herpetological Society Special Publication, Lanesboro, 75 pp., figs.

Morice, A. (1875a) *Coup d'oeil sur la faune de la Cochinchine française*. H. Georg, Lyon, 101 pp.

Morice, A. (1875b) Note sur l'*Herpeton tentaculatum*. *Annales des Sciences Naturelles*, Paris (Série 6) 2(5), 1–11, pl. 20.

Moriguchi, H. & Naito, S. (1979) A new locality record of *Achalinus spinalis werneri* from Tokuno-shima. *Japanese Journal of Herpetology*, Kyoto 8(1), 36.

Morisi, A. (1983) *Guida agli anfibi e rettili della provincia di Cuneo*. Museo Civico F. Eusebio, Alba, 64 pp., 24 figs.

Moritz. (1917) Aesherichnae zmeae *Coelopeltis monspessulana tucomena* iz Stavropoleskij gubernii. *Dubitedv Prirodv,* Petrograd, 1917(1), 18–23.

Moro, S.A. (1996) Osteología craneal y musculatura mandibular de tres especies de *Bothrops* (Serpentes, Crotalidae). *Acta Zoologica Lilloana* 43(2), 293–316, 56 figs.

Moroz, C., Grotto, L., Goldblum, N. & Vries, A. de. (1967) Enhancement of immunogenicity of snake venom neurotoxins. *In:* Russell, F.E. & Saunders, P.R. (Eds.), *Animal toxins.* Pergamon Press, Oxford, pp. 299–302.

Morris, M.A. (1982) Activity, reproduction, and growth of *Opheodrys aestivus* in Illinois (Serpentes, Colubridae). *Natural History Miscellanea*, Chicago (214), 1–11, 6 figs.

Morrison, C. (2003) *A field guide to the herpetofauna of Fiji.* University of the South Pacific, Suva, 121 pp., col. illust.

Morse, M. (1904) Batrachians and reptiles of Ohio. *Ohio State University, University Bulletin* (Series 8) (18), 92–144, 3 figs., 1 pl.

Mosauer, W. (1934) The reptiles and amphibians of Tunisia. *Publications of the University of California at Los Angeles in Biology* 1, 49–64, illust.

Mosauer, W. & Wallis, K. (1927) *Macroprotodon cucullatus* Geoffr. subspecies nova *melanocephala* and *Tropidonotus* (*Natrix*) *viperinus* Latr. aberratio nova *nigra*, zwei Schlangenfunde aus Tunesien. *Zoologischer Anzeiger*, Leipzig 72, 305–310, 2 figs.

Motley, J. & Dillwyn, L.L. (1855) *Contributions to the natural history of Labuan, and the adjacent coasts of Borneo. Part III. Reptilia.* John van Voorst, London, pp. 39–49, 5 pls.

Mount, R.H. (1975) *The reptiles & amphibians of Alabama.* Agricultural Experiment Station, Auburn University, Auburn, vii + 347 pp., 350 figs.

Mount, R.H. & Schwaner, T.D. (1970) Taxonomic and distributional relationships between the water snakes *Natrix taxispilota* (Holbrook) and *Natrix rhombifera* (Hallowell). *Herpetologica* 26(1), 76–82, 1 fig.

Moura-Gonçalves, J. (1956) Estudos sôbre venenos de serpente Brasileiras. II. *Crotalus terrificus crotaminicus*, subspécie biológica. *Anais da Academia Brasileira da Ciencias*, Rio de Janeiro 28(3), 365–367, 1 fig.

Moura-Leite, J.C. de & Bernarde, P.S. (1999) Geographic distribution: Serpentes: *Waglerophis merremii*. *Herpetological Review* 30(1), 56.

Moura-Leite, J.C. de, Morata, S.A.A. & Bérnils, R.S. (1996) New records of reptiles from the state of Paraná, Brazil. *Herpetological Review* 27(4), 216–217.

Mouton, P. le F.N. & van Wyk, J.H. (1993) Reptile fauna of the Katse Dam catchment area and a biogeographical assessment of species composition in the Lesotho highlands. *Koedoe*, Pretoria 36(1), 67–78, 1 fig.

Mrsic, N. (1980) Herpetofauna of the Nepal valley of Langtang and its vicinity. *Bioloski Vestnik*, Ljubljana 28(1), 67–82, 5 figs.

Mrsic, N. (1997) *Plazilci (Reptilia) Slovenije.* Zavod Republike Slovenije za Solstvo, Ljubljana, 167 pp., 328 figs.

Muelleman, P.J., White, L.A., Henderson, R.W. & Powell, R. (2009) Activity patterns of *Alsophis sibonius* and *Liophis juliae* (Dipsadidae) in Cabrits National Park, Dominica, West Indies. *South American Journal of Herpetology* 4(1), 55-60,2 figs.

Mueses-Cisnernos, J. & Cepeda-Quilindo, B. (2006) Geographic distribution: Serpentes: *Tantilla alticola*. *Herpetological Review* 37(4), 501.

Mukherjee, D. & Bhupathy, S. (2007) A new species of wolf snake (Serpentes, Colubridae, *Lycodon*) from Anaikatti Hills, Western Ghats, Tamil Nadu, India. *Russian Journal of Herpetology* 14(1), 21–26, 4 figs.

Mulcahy, D.G. (2008) Phylogeography and species boundaries of the western North American nightsnake (*Hypsiglena torquata*): revisiting the subspecies concept. *Molecular Phylogenetics and Evolution* 46(3), 1095–1115, 3 figs.

Mulcahy, D.G. & Archibald, M.A. (2003) Geographic variation in the Baja California nightsnake (*Eridiphas slevini*) with comments on taxonomy and diet. *Journal of Herpetology* 37(3), 566–571, 2 figs.

Mulcahy, D.G., Beckstead, T.H. & Sites, J.W., Jr. (2011) Molecular systematics of the Leptodeirini (Colubroidea: Dipsadidae) revisited: species trees analyses and multi-locus data. *Copeia* 2011(3), 407–417, 4 figs.

Mulder, J. (1994) Additional information on *Vipera albizona* (Reptilia, Serpentes, Viperidae). *Deinsea* 1, 77–83, 3 figs.

Mulks, M. (2004) *Zonata: the California mountain kingsnake*. LM Digital, 64 pp., col. photos.

Müller, A.H. (1968) *Lehrbuch der Paläozoologie. Band III. Vertebraten. Teil 2. Reptilien und Vögel.* Vebgustav Fischer Verlag, Jena, 657 pp., 728 figs.

Müller, F. (1877) Mittheilungen aus der herpetologischen Sammlung des Basler Museums. I. Ueber einige seltene und neue Reptilien aus Guatemala. *Verhandlungen der Naturforschenden Gesellschaft in Basel* (1878) 6(3), 389–411, 3 pls.

Müller, F. (1878a) Mittheilungen aus der herpetologischen Sammlung des Basler Museums. I. Ueber einige seltene und neue Reptilien aus Guatémala. *Verhandlungen der Naturforschenden Gesellschaft in Basel* (1877) 6(3), 389–411, 3 pls.

Müller, F. (1878b) Katalog der im Museum und Universitätskabinet zu Basel aufgestellten Amphibien und Reptilien nebst Anmerkungen. *Verhandlungen der Naturforschenden Gesellschaft in Basel* (1877) 6(4), 557–709, 3 pls.

Müller, F. (1880) Erster Nachtrag zum Katalog der herpetologischen Sammlung des Basler Museums. *Verhandlungen der Naturforschenden Gesellschaft in Basel* (1882–1885) 7(1), 120–165, 1 pl. [preprint in 1880]

Müller, F. (1884) Dritter Nachtrag zum Katalog der herpetologischen Sammlung des Basler Museums. *Verhandlungen der Naturforschenden Gesellschaft in Basel* (1882–1885) 7(2), 274–324, 3 pls.

Müller, F. (1885) Vierter Nachtrag zum Katalog der herpetologischen Sammlung des Basler Museums. *Verhandlungen der Naturforschenden Gesellschaft in Basel* (1882–1885) 7(3), 668–717, 2 pls.

Müller, F. (1887) Fünfter Nachtrag zum Katalog der herpetologischen Sammlung des Basler Museums. *Verhandlungen der Naturforschenden Gesellschaft in Basel* (1887–1890) 8(2), 249–296, 3 pls.

Müller, F. (1893) Siebenter Nachtrag zum Katalog der herpetologischen Sammlung des Basler Museums. *Verhandlungen der Naturforschenden Gesellschaft in Basel* (1892) 10(1), 195–215, pls. 3–4. [19 Sept. 1893]

Müller, F. (1895) Reptilien und Amphibien aus Celebes. *Verhandlungen der Naturforschenden Gesellschaft in Basel* (1892) 10(3), 825–843, 862–869, figs. [19 June 1895]

Müller, H.W. (1991) *Gonyosoma oxycephala* (Boie). *Sauria*, Berlin 13(1–4), 197–200, illust.

Müller, J. (1998) Untermiozäne Kieferfragmente von Schlangen (Reptilien: Serpentes: Erycinae) aus der französischen Lokalität Poncenat. *Neues Jahrbuch für Geologie und Paläontologie Abhandlungen,* Stuttgart 1998(2), 119-128.

Müller, J.P. (1832) Beiträge zur Anatomie und Naturgeschichte der Amphibien. *Zeitschrift für Physiologie,* Heidelberg 4(2), 190–275, pls. 18– 22.

Müller, J.W. (1865) Amphibia. *In: Reisen in den Vereinigten Staaten, Canada und Mexiko. III. Beiträge zur Geschichte, Statistik und Zoologie von Mexiko. Dritte Abtheilung. Die Wirbelthiere Mexikos.* Brockhaus, Leipzig, pp. 595–620

Müller, L. (1907) Über einen neuen gecko aus Kamerun und eine neue colubrine Schlange aus Centralchina. *Zoologischer Anzeiger*, Leipzig 31(25), 824–830.

Müller, L. (1910) Beiträge zur Herpetologie Kameruns. *Abhandlungen der Mathematisch-Physikalische Klassen der Königlich Bayerisch Akadnmie der Wissenschaften*, München 24(3), 545–625, 1 pl.

Müller, L. (1911) Zwei neue Schlangen aus dem Katangadistrikt, Kongostaat. *Zoologischer Anzeiger*, Leipzig 38(14–15), 357–360.

Müller, L. (1923) Neue oder seltene Reptilien und Batrachier der Zoologischen Sammlung des bayrischen Staats. *Zoologischer Anzeiger*, Leipzig 57(7–8), 145–156.

Müller, L. (1924) Ueber neue oder seltene Mittel- und südamerikanische Amphibien und Reptilien. *Mitteilungen aus dem Zoologischen Museum in Berlin* 11(1), 75–93.

Müller, L. (1925) Ueber eine neue opisthoglyphe Schlange aus Argentinien. *Mitteilungen aus dem Zoologischen Museum in Berlin* 12(1), 103–104.

Müller, L. (1926) Neue Reptilien und Batrachier der zoologischen Sammlung der bayerischen Staats. *Zoologischer Anzeiger*, Leipzig 65(7–8), 193–200.

Müller, L. (1927) Amphibien und Reptilien der Ausbeute Prof. Breslau's in Brasilien 1913–14. *Abhandlungen der Senckenbergischen Naturforschenden Gesellschaft*, Frankfurt-am-Main 40(3), 259–304.

Müller, L. (1928) Herpetologische Mitteilungen. *Zoologischer Anzeiger*, Leipzig 77(3–6), 61–84.

Müller, L. (1932) Beiträge zur Herpetologie des südosteuropäischen Halbinsel. 1. Herpetologisch Neues aus Bulgarien. *Zoologischer Anzeiger*, Leipzig 100(11–12), 299–309, illust.

Müller, L. (1939) Bemerkungen über einige von Herrn Fr. Fuss in der Umgebung von Akschchir, Anatolien, gesammelte Schlangen. *Zoologischer Anzeiger*, Leipzig 127(3–4), 83–95, photos.

Müller, L. (1941) Über die in der Sammlung des Zoologischen Institutes der Universität Halle a. Saale Aufbewahrten Amphibien- und Reptilientypen. *Zeitschrift für Naturwissenchaften*, Halle 94, 182–205.

Müller, L. & Wettstein, O. (1933) Amphibien und Reptilien vom Lebanon. *Sitzungsberichte der Akademie der Wissenschaften in Wien., Mathematische- Naturwissenschaftliche Klasse* 142, 135–144.

Müller, P. (1968a) *Die Herpetofauna der Insel von São Sebastião (Brasilien)*. Saarbrücker Zeitung, Gutenbergstrasse, 68 pp., 48 figs.

Müller, P. (1968b) Bemerkungen zur den beiden südamerikanischen Natterarten *Spilotes pullatus* und *Pseustes sulphureus*. *Aqua Terra*, Solothurn 5(12), 108–112, 6 figs.

Müller, P. (1968c) Beitrag zur Herpetofauna der Insel Campeche (27 42'S/48 28'W) (Ergebnisse einer Reise nach Brasilien im Jahre 1967). *Salamandra*, Frankfurt-am-Main 4(2–3), 47–55.

Müller, P. (1969a) Über die nördliche Arealgrenze von *Lystrophis semicinctus* (Duméril, Bibron und Duméril). *Aqua Terra*, Solothurn 6(8), 96–99, 4 figs.

Müller, P. (1969b) Zur Verbreitung der Gattung *Chironius* auf den Südbrasilianischen Inseln (Serpentes, Colubridae). *Senckenbergiana Biologica*, Frankfurt-am-Main 50(3–4), 133–141, 2 figs.

Müller, P. (1969c) Einige Bemerkungen zur Verbreitung von *Vipera aspis* (Serpentes, Viperidae) in Spanien. *Salamandra*, Frankfurt-am-Main 5(1–2), 57–62, 4 figs.

Müller, P. (1970) Über die *Eunectes*-Arten von Marajó. *Salamandra*, Frankfurt-am-Main 6(3–4), 140–141, 1 fig.

Müller, P. (1975) *Cyclagras gigas* von Marajó (Brasilien) (Serpentes, Colubridae). *Salamandra*, Frankfurt-am-Main 11(1), 60–61.

Müller, P. & Ritter, C. (1978) Erstnachweis von *Uromacerina ricardinii* (Peracca, 1897) für den Staat von Santa Catarina (Brasilien) (Reptilia, Serpentes, Colubridae). *Salamandra*, Frankfurt-am-Main 14(1), 44.

Müller, S. (1844) Bijdragen tot de kennis van Timor en eenige andere naburige Eilanden. *Verhandelingen Natuurlijke Geschichte Nederlandse Overz Bez, Land- en Volkenkunde* (7), 193–224, pls. 51–53, 56–57, 59, 64–66, 74.

Müller, S. (1857) *Reizen en Onderzoekingen in den Indischen Archipel, gedaan op last der Nederlandische Indische Regering, tusschen de Jaren 1828 en 1836*. Tweede Afdeeling. Frederik Müller, Amsterdam, viii + 350 pp.

Müller, S. & Schlegel, H. (1844a) Reptilia. Over de in den Indischen Archipel voorkomende soorten van het slangengeslacht *Trigonocephalus*. *In:* Temminck, C.J. (Ed.), *Verhandlingen over de natuurlijke geschiedenis der Nederlandsche overzeesche bezittingen, door de Leden der Natuurkundige Commissie in Indië en andere Schrijvers. Zoologie* (1842–1845). S. & J. Luchtmans & C.C. van der Hoek, Leiden, pp. 49–58, pl. 7. [text 1842 and pl. 1845 *fide* David & Vogel, 1996: 226]

Müller, S. & Schlegel, H. (1844b) Reptilia. Over de brilslangen van den Indischen Archipel. *In:* Temminck, C.J. (Ed.), *Verhandlingen over de natuurlijke geschiedenis der Nederlandsche overzeesche bezittingen, door de Leden der Natuurkundige Commissie in Indië en andere Schrijvers. Zoologie* (1842–1845). S. & J. Luchtmans & C.C. van der Hoek, Leiden, pp. 69–72, pl. 10.

Muñiz-Martínez, R. & Valdez-Lares, R. (2011) Geographic distribution: Serpentes: *Drymarchon melanurus*. *Herpetological Review* 42(3), 393.

Munjal, D. & Elliott, W.B. (1972) Further studies on the properties of phospholipase A from honeybee *(Apis mellifera)* venom. *Toxicon* 10(4), 367–375, 5 figs.

Muñoz-Alonso, L.A. (1988) Herpetofauna de la tierras altas de Chiapas. *Resúmen Simposio Internacional del Instituto Nacional de Investigaciónes de Reciente Bióta* 88, 13.

Muñoz-Saravia, A., Rivas-Torrico, L.R., Gonzales, L. & Quispe, J. (2009) Reptilia, Serpentes, Elapidae, *Micrurus serranus*: Distribution extension. *Check List* 5(3), 510–512, 2 figs.

Munyer, E.A. (1967) Behavior of an eastern hognose snake, *Heterodon platyrhinos*, in water. *Copeia* 1967(3), 668–670, 1 fig.

Murillo, S., Alvarez-Solorzano, T. & Matias-Ferrer, N. (2003) Geographic distribution: Serpentes: *Adelphicos quadrivirgatus*. *Herpetological Review* 34(2), 167.

Murphy, J.B. & Schlager, N. (2003) *Grzimek's animal life encyclopedia. Volume 7. Reptiles.* Second edition. Thomson-Gale Group, Farmington Hills, 593 pp., col. figs.

Murphy, J.C. (1984) Fea's viper, *Azemiops feae*: a clue to the evolution of advanced snakes. *Bulletin of the Chicago Herpetological Society* 16( ), 136–138.

Murphy, J.C. (1990) A model for regional herpetological society field studies, the snakes of southern California, a proposal and some thoughts on collecting. *Bulletin of the Chicago Herpetological Society* 25(3), 42–45.

Murphy, J.C. (1996) Crossing Bond's Line: The herpetofaunal exchange between the eastern Caribbean and mainland South America. *In:* Powell, R. & Henderson, R.W. (Eds.), *Contributions to West Indian herpetology: a tribute to Albert Schwartz*. Society for the Study of Amphibians and Reptiles, pp. 207–216.

Murphy, J.C. (1997) *Amphibians and reptiles of Trinidad and Tobago*. Krieger Publishing Company, Malabar, 245 pp., 172 col. photos.

Murphy, J.C. (2007a) A review of *Enhydris maculosa* (Blanford, 1879) and the description of a related species (Serpentes, Homalopsidae). *Hamadryad* 31(2), 281–287, 1 fig.

Murphy, J.C. (2007b) *Homalopsid snakes. Evolution in the mud.* Krieger Publishing Company, Malabar, viii + 249 pp., 47 figs., 38 pls.

Murphy, J.C. (2011) The nomenclature and systematics of some Australasian homalopsid snakes (Squamata: Serpentes: Homalosidae). *Raffles Bulletin of Zoology* 59(2), 229–236, 3 figs.

Murphy, J.C. (2012) Synonymised and forgotten, the bird's head stout-tailed snakes, *Calamophis* Meyer (Squamata: Serpentes: Homalopsidae). *Raffles Bulletin of Zoology* 60(2), 515–523, 5 figs.

Murphy, J.C. & Henderson, R.W. (1997) *Tales of giant snakes. A historical natural history of anacondas and pythons*. Krieger Publishing, Malabar, ix + 221 pp., illust.

Murphy, J.C. & Voris, H.K. (1994) A key to the homalopsine snakes. *The Snake* 26(2), 123–133, unnumb. illust.

Murphy, J.C. & Voris, H.K. (2005) A new Thai *Enhydris* (Serpentes, Colubridae, Homallopsinae). *Raffles Bulletin of Zoology* 53(1), 143–147, figs. 1a, 2.

Murphy, J.C., Voris, H.K. & Karns, D.R. (1994) A field guide and key to the snakes of the Danum Valley: a Bornean tropical ecosystem. *Bulletin of the Chicago Herpetological Society* 29(7), 133–151, 24 figs.

Murphy, J.C., Voris, H.K. & Auliya, M. (2005) A new species of *Enhydris* (Serpentes, Colubridae, Homalopsinae) from the Kapuas River system, West Kalimantan, Indonesia. *Raffles Bulletin of Zoology* 53(2), 271–275, 2 figs.

Murphy, J.C., Hill, J. & Karns, D.R. (2006) Geographic distribution: Serpentes: *Coelognathus flavolineatus*. *Herpetological Review* 37(2), 242.

Murphy, J.C., Chan-Ard, T., Mekchai, S., Cota, M. & Voris, H.K. (2008) The rediscovery of Angel's stream snake, *Paratapinophis praemaxillaris* Angel, 1929 (Reptilia: Serpentes: Natricidae). *The Natural History Journal of Chulalongkorn University* 8(2), 169–183, 4 figs.

Murphy, J.C., Voris, H.K. & Karns, D.R. (2012a) The dog-faced water snakes, a revision of the genus *Cerberus* Cuvier, (Squamata, Serpentes, Homalopsidae), with the description of a new species. *Zootaxa* (3484), 1–34, 10 figs.

Murphy, J.C., Mumpuni, Lang, R. de, Gower, D.J. & Sanders, K.L. (2012b) The Molluccan short-tailed snakes of the genus *Brachyorrhos* Kuhl (Squamata: Serpentes: Homalopsidae), and the status of *Calamophis* Meyer. *Raffles Bulletin of Zoology* 60(2), 501–514, 4 figs.

Murphy, R.W. (1975) Two new blind snakes (Serpentes, Leptotyphlopidae) from Baja California, México with a contribution to the biogeography of peninsular and insular herpetofauna. *Proceedings of the California Academy of Sciences* (Series 4) 40(5), 93–107, 4 figs.

Murphy, R.W. (1988) The problematic phylogenetic analysis of interlocus heteropolymer isozyme characters: a case study from sea snakes and cobras. *Canadian Journal of Zoology*, Ottawa 66(12), 2628–2633, illust.

Murphy, R.W. & Crabtree, B. (1985) Generic relationships of the Santa Catalina Island rattleless rattlesnake, *Crotalus catalinesis* (Serpentes, Viperidae). *Acta Zoologica Mexicana*, México (9), 1–16.

Murphy, R.W. & Méndez de la Cruz, F.R. (2010) The herpetofauna of Baja California and its associated islands: a conservation assessment and priorities. *In:* Wilson, L.D., Townsend, J.H. & Johnson, J.D. (Eds.), *Conservation of Mesoamerican amphibians and reptiles*. Eagle Mountain Publishing, Eagle Mountain, pp. 238–273, 2 figs.

Murphy, R.W. & Ottley, J.R., Jr. (1980) A genetic evaluation of the leafnose snake, *Phyllorhynchus arenicolus*. *Journal of Herpetology* 14(3), 263–268.

Murphy, R.W. & Ottley, J.R., Jr. (1984) Distribution of amphibians and reptiles on islands in the Gulf of California. *Annals of the Carnegie Museum* 53(8), 207–230, figs.

Murphy, R.W., Kovac, V., Haddrath, O., Allen, G.S., Fishbein, A. & Mandrak, N.E. (1995) MtDNA gene sequence, allozyme, and morphological uniformity among red diamond rattlesnakes, *Crotalus ruber* and *Crotalus exsul*. *Canadian Journal of Zoology*, Ottawa 73, 270–281, 3 figs.

Murphy, R.W., Fu, J., Lathrop, A., Feltham, J.V. & Kovac, V. (2002) Phylogeny of the rattlesnakes (*Crotalus* and *Sistrurus*) inferred from sequences of five mitochondrial DNA genes. *In:* Schuett, G.W., Höggren, M., Douglas, M.E. & Greene, H.W., (Eds.), *Biology of the vipers*. Eagle Mountain Press, Salt Lake City, pp. 69–92, 6 figs.

Murray, J.A. (1884) *The vertebrate zoology of the Sind*. London, xvi + 424 pp.

Murray, J.A. (1886) *The reptiles of Sind; a systematic account*. Bombay, vi + 92 pp., 8 figs., 7 pls.

Murray, J.A. (1887) Description of three new species of *Hydrophis* from the Bombay harbour and the Mekran coast. *The Journal of the Bombay Natural History Society* 2, 32–35.

Murray, J.A. (1892) *Zoology of Beloochistan and southern Afghanistan*. Bombay, 83 pp.

Murthy, T.S.N. (1972a) A note on some snakes from Anaimalais and Cardamom Hills, Western Ghats. *Journal of the University of Poona (Science and Technology)* (42), 95–102.

Murthy, T.S.N. (1972b) New generic names for some snakes of the families Colubridae and Elapidae (Reptilia: Serpentes). *Current Science,* Bangalore 41(13), 504.

Murthy, T.S.N. (1972c) A rare snake (*Sibynophis subpunctatus*) (Duméril & Bibron) from Anaimalais. *Current Science*, Bangalore 41(22), 826.

Murthy, T.S.N. (1973) A systematic index and a bibliography to the uropeltid snakes of India and Sri Lanka. *Indian Journal of Zootomy*, Bhopal 14(2), 103–109.

Murthy, T.S.N. (1974) Some earth-snakes (Serpentes, Uropeltidae) from Anaimalais and Palnis, Western Gnats, India. *Journal of the University of Poona (Science and Technology)* (46), 7–9.

Murthy, T.S.N. (1975) Occurrence of the uropeltid snake *Brachyophidium rhodogaster* Wall (Serpentes, Uropeltidae) from Tenmalai, Kerala, India. *British Journal of Herpetology*, London 5(4), 471.

Murthy, T.S.N. (1977a) On sea snakes occurring in Madras waters. *Journal of the Marine Biology Association of India*, Madras 19(1), 68–72, 2 pls.

Murthy, T.S.N. (1977b) Systematic index, distribution and bibliography of the sea-snakes of India. *Indian Journal of Zootomy*, Bhopal 18(2), 131–134.

Murthy, T.S.N. (1977c) Some recent collections of lizards and snakes from Western Ghats. *Newsletter, Zoological Society of India,* Calcutta 3(2), 70–72.

Murthy, T.S.N. (1978) The herpetofauna of Madras and its vicinity. *Hamadryad*, Madras 3(2), 7–12.

Murthy, T.S.N. (1980) Some ecological observations on the uropeltid snakes of South Indian hills. *The Snake* 12(1), 56–59, 4 figs.

Murthy, T.S.N. (1981a) Systematic index, distribution and bibliography of the sea-snakes of India. *Indian Journal of Zootomy*, Bhopal (1977) 18(2), 131–134.

Murthy, T.S.N. (1981b) Reptiles of the Silent Valley and New Amarambalam area, Kerala. *The Snake* 13(1), 42–52, 2 pls.

Murthy, T.S.N. (1981c) Checklist and key to the uropeltid snakes of India. *The Snake* 13(2), 142–150, 9 figs.

Murthy, T.S.N. (1982a) A contribution to the ophiology of Western Ghats, India. *The Snake* 14(1), 50–56, 7 pls.

Murthy, T.S.N. (1982b) An illustrated field guide to the rough tailed snakes of India. *The Snake* 14(2), 119–135, 20 figs.

Murthy, T.S.N. (1983a) A historical resume and bibliography of the snakes of India. *The Snake* 15(2), 113–135, 3 figs.

Murthy, T.S.N. (1983b) Some recent records of snakes from Western Gnats, India. *Indian Journal of Zootomy* 24(2), 77–80.

Murthy, T.S.N. (1985) Classification and distribution of the reptiles of India. *The Snake* 17(1), 48–71.

Murthy, T.S.N. (1986) Occurrence of the sea snake, *Enhydria schistosa,* in the Chilka Lake, Orissa. *Records of the Zoological Survey of India,* Calcutta 83(3–4), 189–190.

Murthy, T.S.N. (1987) Herpetofauna of the Chilka Lagoon, Orissa, India. *British Herpetological Society Bulletin* 21–22, 8–12.

Murthy, T.S.N. (1988) An overview of the reptiles in the estuaries of India, with brief notes on their habits and distribution. *British Herpetological Society Bulletin*, London (24), 17–19.

Murthy, T.S.N. (1989) A collection of reptiles from the Kalakad Wildlife Sanctuary, Tamil Nadu, India. *British Herpetological Society Bulletin*, London (28), 37–40.

Murthy, T.S.N. (1990a) Reptiles of Kalakad Sanctuary, India. *The Snake* 22(1), 44–59, 4 figs.

Murthy, T.S.N. (1990b) Venomous snakes of medical importance in India (Part A). *In*: Gopalakrishnakone, P. & Chou, L.M. (Eds.), *Snakes of medical importance (Asia-Pacific region).* National University of Singapore, Singapore, pp. 281–297, 10 figs.

Murthy, T.S.N. (1990c) A pocket book of the amphibians and reptiles of the Chilka Lagoon, Orissa. *Records of the Zoological Survey of India, Occasional Papers*, Calcutta (125), 1–35, 23 col. figs., 24 pls.

Murthy, T.S.N. (1990d) Illustrated guide to the snakes of the Western Ghats, India. *Records of the Zoological Survey of India, Occasional Papers*, Calcutta (114), 1–76, 10 figs., 57 pls., 11 col. pls.

Murthy, T.S.N. (1990e) *The snake book of India.* Second edition. International Book Distributors, Dehra Dun, xvi + 101 pp., 10 figs. [first edition, 1986, xvi + 101 pp., 10 figs., 26 pls.]

Murthy, T.S.N. (1991) A rare wolf snake, *Lycodon flavomaculatus*, from the Nilgiri Biosphere Reserve, South India. *The Snake* 23(1), 104–106, 2 figs.

Murthy, T.S.N. (1992) Marine reptiles of India: An overview. *Greater Cincinnati Herpetological Society, Contributions in Herpetology* 1992, 35–38.

Murthy, T.S.N. (1995a) Reptilia and, Amphibia. *In:* Alfred, J.R.B. (Ed.), *Fauna of Chilka Lake, Orissa. Wetland Ecosystem Series* 1. Zoological Survey of India, Calcutta, pp. 507–560, 3 pls.

Murthy, T.S.N. (1995b) *Illustrated encyclopedia of the reptiles of India.* B.R. Publishing Corporation, Delhi 113 pp., 140 photos.

Murthy, T.S.N. (2001) Reptilia. *In:* Alfred, J.R.B. (Ed.), *Fauna of Nilgiri Biosphere Reserve. Fauna of Conservation Areas* 11. Zoological Survey of India, Kolkata, pp. 239–243.

Murthy, T.S.N. (2007) *Pictorial handbook on marine reptiles of India.* Zoological Survey of India, Kolkata, 64 pp., col. illust.

Murthy, T.S.N. (2009) *A pocket book on Indian reptiles (crocodiles, testudines, lizards and snakes).* Nature Books India, Delhi, 88 pp., col. illust.

Murthy, T.S.N. & Acharjyo, L.N. (1987) Report on a collection of snakes in and around the Nandankanan Biological Park, Orissa, India. *The Snake* 19(2), 139–142, 3 figs.

Murthy, T.S.N. & Chakrapany, S. (1983) Rediscovery of the blind snake *Typhlops oatesii* in Andamans, India. *The Snake* 15(1), 48–49, 2 figs.

Murthy, T.S.N. & Chandrasekhar, S.V.A. (1989) Remarks on the colour pattern of the Indian trinket snake *Elaphe helena* (Daudin) (Serpentes, Colubridae). *The Snake* 21(1), 51–53, 3 figs.

Murthy, T.S.N. & Pillai, R.S. (1986) Snakes. *In:* Majupuria, T.C. (Ed.), *Wildlife wealth of India (resources & management).* Tecpress Service, Bangkok, pp. 224–244, illust.

Murthy, T.S.N. & Rao, K.V.R. (1988) Snakes of the Chilka Lake, Orissa, India. *The Snake* 20(1), 67–73, 7 figs.

Murthy, T.S.N. & Rao, K.V.R. (1989) An addition to the ophidian fauna of the Chilka Lake, India—*Typhlops acutus* (Dum. & Bibr.), with notes on its current distribution. *The Snake* 21(2), 101–102, 1 fig.

Murthy, T.S.N. & Sharma, B.D. (1976) A contribution to the herpetology of Jammu and Kashmir. *British Journal of Herpetology*, London 5(6), 533–538, 11 figs.

Murthy, T.S.N., Sharma, B.D. & Sharma, T. (1979) Second report on the herpetofauna of Jammu and Kashmir, India. *The Snake* 11(3), 234–241, 3 pls.

Murthy, T.S.N., Sanyal, D.P. & Duttagupta, B. (1993) Rare snakes of India. *The Snake* 25(2), 135–140.

Mushinsky, H.R. (1987) Foraging ecology. *In:* Seigel, R.A., Collins, J.T. & Novak, S.S. (Eds.), *Snakes: ecology and evolutionary biology.* Macmillan Publishing, New York, pp. 302–334.

Muskhelishvili, T.A. (1970a) *Lizardlike reptiles of eastern Georgia.* Mezhiereba, Tsibili 242 pp. (in Russian)

Muskhelishvili, T.A. (1970b) Possible dispersal routes of reptiles and some peculiarities of their spread in the territory of Georgia. *Soobshcheniya-Akademii Nauk Gruzinskoj S.S.R.,* Tbilisi 58, 209–212. (in Russian)

Mutschmann, F. (1995) *Die Strumpfbandnattern: Biologie, Verbreitung, Haltung.* Westarp Wissenschaften, Magdeburg, 172 pp., 101 figs., 2 col. pls.

Mutton, N. & Julander, J. (2011) *The complete carpet python. A comprehensive guide to the natural history, care, and breeding of the* Morelia spilota *complex.* ECO Herpetological Publishing, Rodeo, 339 pp., col. illust.

Myer, P.A. (1988) *The systematics of* Nerodia clarkii *and* Nerodia fasciata. Master's Thesis, Louisiana State University, Baton Rouge, 130 pp., 26 figs.

Myers, C.W. (1966a) A new species of colubrid snake, genus *Coniophanes*, from Darién, Panamá. *Copeia* 1966(4), 665–668, 2 figs.

Myers, C.W. (1966b) *Lygophis boursieri* (Jan), a snake new to the fauna of Colombia. *Copeia* 1966(4), 886–888.

Myers, C.W. (1967) The pine woods snake, *Rhadinaea flavilata* (Cope). *Bulletin of the Florida State Museum, Biological Sciences* 11(2), 47–97, 12 figs.

Myers, C.W. (1969a) Snakes of the genus *Coniophianes* in Panamá. *American Museum Novitates* (2372), 1–28, 10 figs.

Myers, C.W. (1969b) South American snakes related to *Lygophis boursieri*: A reappraisal of *Rhadinaea antioquiensis*, *Rhadinaea tristriata*, *Coronella whymperi*, and *Liophis atahuallpae*. *American Museum Novitates* (2385), 1–27, 8 figs.

Myers, C.W. (1973) A new genus for Andean snakes related to *Lygophis boursieri* and a new species (Colubridae). *American Museum Novitates* (2522), 1–37, 11 figs.

Myers, C.W. (1974) The systematics of *Rhadinaea* (Colubridae), a genus of New World snakes. *Bulletin of the American Museum of Natural History* 153(1), 1–262, 51 figs.

Myers, C.W. (1982) Blunt-headed vine snakes (*Imantodes*) in Panamá, including a new species and other revisionary notes. *American Museum Novitates* (2738), 1–50, 20 figs.

Myers, C.W. (1986) An enigmatic new snake from the Peruvian Andes, with notes on the Xenodontini (Colubridae, Xenodontinae). *American Museum Novitates* (2853), 1–12, 8 figs.

Myers, C.W. (2000) A history of herpetology at the American Museum of Natural History. *Bulletin of the American Museum of Natural History* (252), 1–232, 64 figs.

Myers, C.W. (2003) Rare snakes—Five new species from eastern Panama, reviews of northern *Atractus* and southern *Geophis* (Colubridae, Dipsadinae). *American Museum Novitates* (3391), 1–47, 20 figs.

Myers, C.W. (2011) A new genus and new tribe for *Enicognathus melanauchen* Jan, 1863, a neglected South American snake (Colubridae: Xenodontinae), with taxonomic notes on some Dipsadinae. *American Museum Novitates* (3715), 1–33, 8 figs.

Myers, C.W. & Böhme, W. (1996) On the type specimens of two Colombian poison frogs described by A.A. Berthold (1845), and their bearing on the locality "Provinz Popayan." *American Museum Novitates* (3185), 1–20.

Myers, C.W. & Cadle, J.E. (1994) A new genus for South American snakes related to *Rhadinaea obtusa* Cope (Colubridae) and resurrection of *Taeniophallus* Cope for the "*Rhadinaea*" *brevirostris* group. *American Museum Novitates* (3102), 1–33, 10 figs.

Myers, C.W. & Campbell, J.A. (1981) A new genus and species of colubrid snake from the Sierra Madre del Sur of Guerrero, México. *American Museum Novitates* (2708), 1–20, 12 figs.

Myers, C.W. & Donnelly, M.A. (1996) A new herpetofauna from Cerro Yaví, Venezuela: first results of the Robert G. Goelet American Museum—Terramar Expedition to the northwestern Tepuis. *American Museum Novitates* (3172), 1–56, 31 figs.

Myers, C.W. & Donnelly, M.A. (1997) A tepui herpetofauna on a granitic mountain (Tamacuari) in the borderland between Venezuela and Brazil: report from the Phipps Tapirapecó expedition. *American Museum Novitates* (3213), 1–71, 50 figs.

Myers, C.W. & Donnelly, M.A. (2001) Herpetofauna of the Yutajé-Corocoro Massif, Venezuela: second report from the Robert G. Goelet American Museum-Terramar Expedition to the northwestern Tepuis. *Bulletin of the American Museum of Natural History* (261), 1–85, 50 figs.

Myers, C.W. & Donnelly, M.A. (2008) The summit herpetofauna of Auyantepui, Venezuela: report from the Robert G. Goelet American Museum-Terramar Expedition. *Bulletin of the American Museum of Natural History* (308), 1–147, 80 figs.

Myers, C.W. & Hoogmoed, M.S. (1974) Zoogeographic and taxonomic status of the South American snake *Tachymenis surinamensis* (Colubridae). *Zoologische Mededelingen*, Leiden 48(17), 187–194, pl. 1, 2 figs.

Myers, C.W. & Ibañez, R. (2012) *Rhadinaea "sargenti,"* a misspelled patronym honoring Commander William M. Sergeant (1902–1983). *Herpetological Review* 43(4), 591.

Myers, C.W. & Rand, A.S. (1969) Checklist of amphibians and reptiles of Barro Colorado Island, Panamá, with comments on faunal change and sampling. *Smithsonian Contributions to Zoology* (10), 1–11, 2 figs.

Myers, C.W. & Schargel, W.E. (2006) Morphological extremes—Two new snakes of the genus *Atractus* from northwestern South America (Colubridae, Dipsadinae). *American Museum Novitates* (3532), 1–13, 7 figs.

Myers, C.W., Ibañez-D., R. & Cadle, J.E. (2007) On the uniquely fragmented distribution of a rare Panamanian snake, *Dipsas nicholsi* (Colubridae, Dipsadinae). *American Museum Novitates* (3554), 1–18, 4 figs.

Myers, G.S. (1947) Murray's *Reptiles of Sind,* with a note on three forgotten descriptions of Indian sea-snakes, published therein. *Herpetologica* 3(5), 167–168.

Myers, G.S. (1949) Status of the generic name *Lioheterodon* applied to Madagascan serpents. *Herpetologica* 5(6), 146.

Nabhitabhata, J., Chan-Ard, T. & Chuaynkern, Y. (2004) *Checklist of amphibians and reptiles in Thailand.* Office of Environmental Policy and Planning, Bangkok, Biodiversity Series (2000) 9, 152 pp., color illust.

Nagai, K. (1928) Tokara-Habu to Erabu-uangi (*Trimeresurus elegans* and *Laticauda semifasciata*). *In:* Kagoshimeken Hakubutsu-Chosa ("Survey on the natural history in Kagoshima Prefecture"). *Kagoshimeken Kyoiku Chosa Kai* 3, v + 64 pp., 14 pls. (in Japanese)

Nagy, Z.T., Joger, U., Guicking, D. & Wink, M. (2002) Phylogeography of the European whip snake *Coluber* (*Hierophis*) *viridiflavus* as inferred from nucleotide sequences of the mitochondrial cytochrome b gene and ISSR genomic fingerprinting. *Biota* 3(1–2), 109–118, 4 figs.

Nagy, Z.T., Joger, U., Wink, M., Glaw, F. & Vences, M. (2003a) Multiple colonization of Madagascar and Socotra by colubrid snakes: evidence for nuclear and mitochondrial gene frequencies. *Proceedings of Royal Society of London* 270, 2613–2621, 3 figs.

Nagy, Z.T., Schmidtler, J.F., Joger, U. & Wink, M. (2003b) Systematik der Zwergnattern (Reptilia: Colubridae: *Eirenis*) und verwandter Gruppen anhand von DNA-Sequenzen und morphologischen Daten. *Salamandra* 39(3–4), 149–168, 8 figs.

Nagy, Z.T., Lawson, R., Joger, U. & Wenk, M. (2004) Molecular systematics of racers, whipsnakes and relatives (Reptilia, Colubridae) using mitochondrial and nuclear markers. *Journal of Zoological Systematics and Evolutionary Research* 42, 223–233, 4 figs.

Nagy, Z.T., Glaw, F., Andreone, F., Wink, M. & Vences, M. (2007) Species boundaries in Malagasy snakes of the genus *Madagascarophis* (Serpentes: Colubridae: *sensu lato*) assessed by nuclear and mitochondrial markers. *Organisms, Diversity & Evolution* 7, 241–251, 6 figs.

Nagy, Z.T., Bellaagh, M., Wink, M., Paunovic, A. & Korsós, Z. (2010) Phylogeography of the Caspian whipsnake in Europe with emphasis on the westernmost population. *Amphibia-Reptilia,* Leiden 31(4), 455–461, 2 figs.

Nakada, K., Nakada, F., Ito, E. & Inoue, F. (1984) Quantification of myonecrosis and comparison of necrotic activity of snake venoms by determination of creatine phosphokinase activity in mice sera. *Toxicon* 22(6), 921–930, 6 figs.

Nakamura, K. (1938) On the blind snake *Typhlopus braminus. Zoological Magazine,* Tokyo 50(4), 192. (in Japanese)

Nakamura, K. & Ueno, S.-I. (1963) *Japanese reptiles and amphibians in color.* Hoikusha Publishing Company, Osaka, ix + 214 pp., 38 figs., 131 col. figs. (in Japanese)

Namiye, M. (1903) Reptilia and Amphibia of Shikoku. *Dobutsugaku Zasshi*, Tokyo 15(178), 284–286.

Nanayakkara, G.L.A. (1989) Variation of the rostral appendage of *Hypnale hypnale* of Sri Lanka. *The Snake*, Nittagun 21(1), 41–43, 1 fig., 3 photos.

Nanayakkara, G.L.A. (1994) The form and function of the procryptic adaptations of the green pit viper (*Trimeresurus trigonocephalus*) in Sri Lanka. *The Snake* 26(1), 75–78, 2 figs.

Nande, R. & Deshmukh, S. (2007) Snakes of Amravati District including Melghat, Maharashtra, with important records of the Indian egg-eater, montant trinket snake, and Indian smooth snake. *Zoos' Print Journal* 22(12), 2920–2924.

Nanhoe, L.M.R. & Ouboter, P.E. (1987) The distribution of reptiles and amphibians in the Annapurna-Dhaulagiri region (Nepal). *Zoologische Verhandelingen*, Leiden (240), 1–105.

Narayan, G. & Rosalind, L. (1990) King cobra *Ophiophagus hannah* in grassland: an unusual habitat. *The Journal of the Bombay Natural History Society* 87(2), 309–310.

Nascimento, F.P. do & Lima-Verde, A.S. (1989) Ocorrencia de ofídios de ambientes florestais em enclaves e matas úmidas do Ceará (Ophidia, Colubridae). *Boletim Museu Paraense Emílio Goeldi* (*Zoologia*), Belém 5(1), 95–100.

Nascimento, F.P. do, de Ávila-Pires, T.C.S. & da Cunha, O.R. (1987) Os répteis da área de Carajás, Pará, Brasil (Squamata). II. *Boletim Museu Paraense Emílio Goeldi* (*Zoologia*), Belém 3(1), 33–65, figs. 4–5.

Nascimento, F.P. do, de Ávila-Pires, T.C.S. & da Cunha, O.R. (1988) Répteis Squamata de Rondonia e Mato Grosso coletados através do programa Polonoroeste. *Boletim Museu Paraense Emílio Goeldi (Zoologia)*, Belém 4(1), 21–66.

Nascimento, F.P. do, de Ávila-Pires, T.C.S., Flock dos Santos, I.N.F. & Lima, A.C.M. (1991) Répteis de Marajó e Mexiana, Pará, Brasil. I. Revisão bibliográfica e novos registros. *Boletim Museu Paraense Emílio Goeldi (Zoologia)*, Belém 7(1), 25–41.

Natera-Mumaw, M. (2008) Nuevos registros geográficos y notas bio-ecológicas de *Dendrophidion dendrophis* (Schlegel, 1837) y *Dendrophidion nuchale* (Peters, 1863) (Serpentes: Colubridae) en Venezuela, con comentarios sobre la taxonomía de *Dendrophidion nuchale*. *Herpetotropicos* 4(1), 11–16, 3 figs.

Natera-Mumaw, M. & Battiston, P. (2008) Nuevos registros de distribución geográfica con notas bioecológicas sobre *Dipsas indica* Laurenti, 1768 (Serpentes: Colubridae) en Venezuela. *Herpetotropicos* 4(1), 3–5, 1 fig.

Natera-Mumaw, M., Acosta, J.C., Battiston, P. & Hidalgo, O.O. (2007a) Distribution extension and new state record for *Micrurus lemniscatus diutius* Burger, 1955 (Reptilia: Elapidae) in Venezuela. *Herpetotropicos* 3(1), 59, 1 fig.

Natera-Mumaw, M., Gutierrez, S., Hidalgo, O.O. & Battiston, P. (2007b) New local and regional records for *Liophis poecilogyrus schotti* (Schlegel, 1837) in Venezuela. *Herpetotropicos* 3(1), 60, 1 fig.

Naulleau, G. (1987a) Les serpents de France. Second edition. *Revue Française d'Aquariologie*, Nancy (1984) 11(3–4), 1–56, 103 col. figs.

Naulleau, G. (1987b) Repartition des viperes en France: morphologie, systematique, ecologie, ethologie et venin. *In: Serpents, venins, envenimations*. Édition Foundation Marcel Mérieux, Lyon, pp. 39–73, illust.

Naulleau, G. (1997) *La vipère aspic*. Eveil Nature, Saint Yrieix sur Charente, 72 pp., figs., col. photos.

Naumov, B. & Tomovic, L. (2007) A review of distribution and conservation status of *Zamensis situla* (Linnaeus, 1758) (Reptilia: Colubridae) in Bulgaria. *Acta Herpetologica* 2(1), 7–10, 1 fig.

Navarrete, L.F., López-Johnston, J.C. & Dávila, A.B. (2009) *Guía e las serpientes de Venezuela: biología, venenos, conservación y listado de especies*. Second edition. Zoocriadero Ecopets, Bioreptilia, y Grupo Río Verde, Caracas, 102 pp., col. photos. [first edition, 2006, 75 pp., col. photos]

Navarro-Cornejo, G. (2012) Revisión del género *Philodryas* (Ophidia, Dipsadidae) en Bolivia: taxonomía y distribución. Doctoral Dissertation, Santa Cruz de la Sierra, Bolivia. 90 pp., illus.

Neang, T., Grismer, L.L. & Daltry, J.C. (2012) A new species of kukri snake (Colubridae: *Oligodon* Fitzinger, 1826) from the Phnom Samkos Wildlife Sanctuary, Cardamom Mountains, southwest Cambodia. *Zootaxa* (3388), 41–55, 4 figs.

Neave, S.A. (Ed.). (1939) *Nomenclator zoologicus: a list of the names of genera and subgenera in zoolopgy from the tenth edition of Linnaeus 1758 to the end of 1935. Vol. II (D–L)*. Zoological Society of London, London, 1025 pp.

Neave, S.A. (Ed.). (1940a) *Nomenclator zoologicus: a list of the names of genera and subgenera in zoolopgy from the tenth edition of Linnaeus 1758 to the end of 1935. Vol. III (M–P)*. Zoological Society of London, London, 1065 pp.

Neave, S.A. (Ed.). (1940b) *Nomenclator zoologicus: a list of the names of genera and subgenera in zoolopgy from the tenth edition of Linnaeus 1758 to the end of 1935. Vol. IV (Q–Z)*. Zoological Society of London, London, 758 pp.

Necas, P., Modry, D. & Zavadil, V. (1997) *Czech recent and fossil amphibians and reptiles*. Edition Chimaira, Frankfurt-am-Main, 94 pp., illust., 34 col. pls.

Necker, W.L. (1939) Revised check list of reptiles and amphibians of the Chicago region. *Chicago Academy of Sciences Leaflet* (11), 1–4.

Negi, S.S. (1992) *Himalayan wildlife: habitat and conservation*. Indus Publishing Company, New Delhi, 207 pp.

Neill, W.T. (1949a) A new subspecies of rat snake (genus *Elaphe*), and notes on related forms. *Herpetologica* 5, 2nd supplement, 1–12.

Neill, W.T. (1949b) *A checklist of the amphibians and reptiles of Georgia*. Ross Allen's Reptile Institute, Silver Springs, 4 pp.

Neill, W.T. (1957) Objections to wholesale revision of type localities. *Copeia* 1957, 140–141.

Neill, W.T. (1963) A new subspecies of the queen snake *Natrix septemvittata*, from southern Alabama. *Herpetologica* 19(1), 1–9, 2 figs.

Neill, W.T. (1964) Taxonomy, natural history, and zoogeography of the rainbow snake, *Farancia erytrogramma*. *American Midland Naturalist* 71(2), 257–295, 11 figs.

Neill, W.T. (1965) New and noteworthy amphibians and reptiles from British Honduras. *Bulletin of the Florida State Museum, Biological Sciences* 9, 77–130, 8 figs.

Neill, W.T. & Allen, R. (1949) A new kingsnake (genus *Lampropeltis*) from Florida. *Herpetologica* 5(5), 101–105.

Neill, W.T. & Allen, R. (1959) Studies on the amphibians and reptiles of British Honduras. *Publications of the Research Division, Ross Allen's Reptile Institute* 2(1), 1–76, 32 figs.

Neill, W.T. & Allen, R. (1960) Noteworthy snakes from British Honduras. *Herpetologica* 16(3), 145–162, 4 figs.

Nellis, D.W., Norton, R.L. & MacLean, W.P. (1983) On the biogeography of the Virgin Islands boa, *Epicrates monensis granti*. *Journal of Herpetology* 17(4), 413–417, 1 fig.

Nelson, C.E. (1966) Systematics and distribution of snakes of the Central American genus *Hydromorphus* (Colubridae). *Texas Journal of Science* 18(4), 365–371, 9 figs.

Nelson, C.E. & Meyer, J.R. (1967) Variation and distribution of the Middle American snake genus, *Loxocemus* Cope (Boidae?). *Southwestern Naturalist* 12(4), 439–453, 5 figs.

Nenda, S.J. & Cacivio, P.M. (2007) Reptilia, Xenodontidae, *Lystrophis dorbignyi, Lystrophis pulcher,* and *Lystrophis semicinctus*: distribution extension, new provinces records in Argentina. *Check List* 3(2), 126–130, 6 figs.

Nenda, S.J. & Di Pietro, D.O. (2009) Serpentes, Echinantherini, *Taeniophallus poecilopon* (Cope, 1863): rediscovery in Argentina. *Check List* 5(3), 503–506, 2 figs.

Nessov, L.A. & Udovitschenko, N.I. (1984) Sea snakes and cartilagenous fishes of the Paleogene of South Kazakhstan. *Paleontologitchesky Sbornik* (21), 69–74, 20 figs. (in Russian)

Nessov, L.A., Zhegallo, V.I. & Averianov, A.O. (1998) A new locality of late Cretaceous snakes, mammals and other vertebrates in Africa (western Libya). *Annales de Paléontologie*, Paris 84, 265–274, 2 figs., 2 pls.

Netting, M.G. (1932) The poisonous snakes of Pennsylvania. *Vertebrate Zoology Pamphlet, Carnegie Museum* (1), 1–15, illust.

Netting, M.G. (1936a) Hand list of the amphibians and reptiles of Pennsylvania. *Herpetology Leaflet, Carnegie Museum* (1), 4 pp.

Netting, M.G. (1936b) Notes on a collection of reptiles from Barro Colorado Island, Panama Canal Zone. *Annals of the Carnegie Museum* (1935–1938) 25, 113–120.

Netting, M.G. & Goin, C.J. (1944) Another new boa of the genus *Epicrates* from the Bahamas. *Annals of the Carnegie Museum* (1943–1944) 30, 71–76, 1 fig.
Nevárez, M. (1999a) Geographic distribution: Serpentes: *Tantilla nigriceps fumiceps*. *Herpetological Review* 30(2), 114.
Nevárez, M. (1999b) Geographic distribution: Serpentes: *Ramphotyphlops braminus*. *Herpetological Review* 30(2), 114.
Nguyen, Q.T., Koch, A. & Ziegler, T. (2010a) A new species of reed snake, *Calamaria* Boie, 1827 (Squamata: Colubridae), from central Vietnam. *Hamadryad* (2009) 34(1), 1–8, 8 figs.
Nguyen, Q.T., David, P., Tran, T.T., Luu, Q.V., Le, K.Q. & Ziegler, T. (2010b) *Amphiesmoides ornaticeps* (Werner, 1924), an addition to the snake fauna of Vietnam, with a redescription and comments on the genus *Amphiesmoides* Malnate, 1961 (Squamata, Natricidae). *Revue Suisse de Zoologie* 117(1), 45–56, 6 figs.
Nguyen, V.S. (2007) *Fauna of Vietnam. 14. Serpentes.* Science and Technics Publishing House, Hanoi, 247 pp., 146 figs. (in Vietnamese)
Nguyen, V.S. & Ho, T.C. (1996) *A checklist of reptiles and amphibians in Vietnam.* Science and Technology Publishing House, Hanoi, 264 pp. (in Vietnamese)
Nguyen V.S. & Ho, T. C. (2005) *A checklist of amphibians and reptiles of Vietnam.* Nha xuat ban nong nghiep [Agriculture Publishing House], Hanoi, 180 pp. (in Vietnamese)
Nguyen V.S., Ho, T.C. & Nguyen, Q.T. (2009) *Herpetofauna of Vietnam.* Edition Chimaira, Frankfurt-am-Main, 768 pp., 637 col. pls.
Nicéforo-María, H. (1929) Observaciones acerca de algunos nombres científicos que emplea el Dr. Evaristo García en su obra titulada Los Ofidios Venenosos del Cauca. *Revista de la Sociedad Colombiana de Ciencias Naturales* 18(103), 189–191.
Nicéforo-María, H. (1930a) Los reptiles de Villevivencio en el Museo de La Salle. *Revista de la Sociedad Colombiana de Ciencias Naturales*, Bogotá 19(105), 40–54.
Nicéforo-María, H. (1930b) Los reptiles y batracios de Honda (Tolima) en el Museo de La Salle. *Revista de la Sociedad Colombiana de Ciencias Naturales*, Bogotá 19(105), 96–104.
Nicéforo-María, H. (1931) Dos nuevas especies de Ofidios colombianos. *Revista de la Sociedad Colombiana de Ciencias Naturales*, Bogotá 20(110), 48–52.
Nicéforo-María, H. (1933a) Serpientes de Villavicencio. *In: Libro Conmemorativo del secundo centenario de Don Jose Celestino Bruno Mutis y Bosio, 1732–1932.* Imprenta Nacional, Bogotá, pp. 199–237.
Nicéforo-María, H. (1933b) Algunos ofidios de Sasaima. *In: Libro Conmemorativo del secundo centenario de Don Jose Celestino Bruno Mutis y Bosio, 1732–1932.* Imprenta Nacional, Bogotá, pp. 239–249.
Nicéforo-María, H. (1938) Las serpientes colombianas de hocico proboscidiforme grupo *Bothrops lansbergii-nasuta-hyoprora*. *Revista Academia Colombiana de Ciencias Exactas, Fisicas y Naturales*, Bogotá 2(7), 417–421, 3 pls.
Nicéforo-María, H. (1939) Contribución al studio de la ofiolgía Colombiana. *Revista de la Academia Colombiana de Ciencias Exactas, Fisicas, y Naturales* 3(9–10), 91–94.
Nicéforo-María, H. (1942) Los ofidios de Colombia. *Revista Academia Colombiana de Ciencias Exactas, Fisicas y Naturales*, Bogotá 5(17), 84–101, 12 figs., 3 pls.
Nicéforo-María, H. (1950) Contribucion al conocimiento de los ofidios de Colombia. *Revista Academia Colombiana de Ciencias Exactas, Fisicas y Naturales*, Bogotá 7(28), 517–518, 3 figs.
Nicéforo-María, H. (1964) Herpetologia. *Boletín del Instituto de La Salle*, Bogotá (?), 5–11, 2 figs.
Nicéforo-María, H. (1970) Contribucion al conocimiento de los ofidios de Colombia. *Boletín del Instituto de La Salle*, Bogotá (210), 309–314, 6 figs.
Nicéforo-María, H. (1975) Contribución al estudio de las serpientes de Colombia II. *Boletín del Instituto de La Salle*, Bogotá (215), 1–4, 2 figs.
Nicholls, L. (1929a) The identification of the land snakes of Ceylon. *Ceylon Journal of Science*, Colombo 2, 91–142, 12 pls.
Nicholls, L. (1929b) A new species of earth snake of the genus *Silybura* (*Silybura phillipsi*). *Ceylon Journal of Science*, Colombo 15B, 153–155, 1 pl.
Nicholson, E. (1874) *Indian Snakes. An elementary treatise on ophiology with a descriptive catalogue of the snakes found in India and the adjoining countries.* Second edition. Higginbotham, Madras, vii + 188 pp., 20 pls. [first edition, 1870, ii + 118, 2 pls.]
Nickerson, M.A. & Heringhi, H.L. (1966) Three noteworthy colubrids from southern Sonora, Mexico. *Great Basin Naturalist* 26(3–4), 136–140.
Nicodemo P.R., Jr. & Bain, R.H. (2007) Geographic distribution: Serpentes: *Dendrelaphis ngansonensis*. *Herpetological Review* 38(3), 355.
Nicoli, R.M. (1963) Phylogenèse et systématique. Le phylum des Pentastomidae. *Annales de Parasitologie Humaine et Comparée*, Paris 38(3), 483–516.
Nieto-Montes de Oca, A. (2003) A new species of the *Geophis dubius* group (Squamata, Colubridae) from the Sierra de Juárez of Oaxaca, Mexico. *Herpetologica* 59(4), 572–585, 2 figs.
Nieto-Montes de Oca, A. & Mendelson, J.R., III (1997) Variation in *Rhadinaea marcellae* (Squamata, Colubridae), a poorly known species from the Sierra Madre Oriental of México. *Journal of Herpetology* 31(1), 124–127, 2 figs.
Nietzke, G. & Kornacker, P.M. (2002) *Die Terrarientiere. Band 3. Krokodile und Schlangen.* Fourth edition. Eugen Ulmer, Stuttgart, 375 pp., col. illust.
Nikolsky, A.M. (1896) A note on the two new reptiles from Turkestan. *Annuaire du Musée Zoologique de l'Académie Impériale des Sciences de St. Pétersbourg* 1, xiv–xv.
Nikolsky, A.M. (1899a) *Contia satunini* n. sp. et *Agama ruderata* Oliv. provenant du Transcaucase. *Annuaire du Musée Zoologique de l'Académie Impériale des Sciences de St. Pétersbourg* 4, 449–451. (in Russian)
Nikolsky, A.M. (1899b) *A travel to Turkistan by the founder of the Society A.P. Fedchenko, made under the auspices of the Imperial Society of Naturalists commissioned by Turkestan General-Governor K.P. Von Kaufmann. Zoological investigations. Part VII. Reptiles and amphibians of Turkestan (Herpetologia Turanica).* Moscow University, Moscow 2(23), 1–69, 9 pls. (in Russian)
Nikolsky, A.M. (1903a) *Contia transcaspica* n. sp. (Ophidia, Colubridae). *Annuaire du Musée Zoologique de l'Académie Impériale des Sciences de St. Pétersbourg* 8, 11–13. (in Russian)
Nikolsky, A.M. (1903b) *Alsophylax persicus* sp. n., *Contia bicolor* sp. n., *Bufo persicus* sp. n. *Annuaire du Musée Zoologique de l'Académie Impériale des Sciences de St. Pétersbourg* 8, 95–98. (in Russian).
Nikolsky, A.M. (1905) Herpetologia rossica. *Mémoires de l'Académie Impériale des Sciences de St. Pétersbourg* (Series 8) 17(1), 1–517, 2 pls. (in Russian)

Nikolsky, A.M. (1907) Reptiles et amphibiens, recueillis par Mr. N. A. Zaroudny en Perse en 1903–1904. *Annuaire du Musée Zoologique de l'Académie Impériale des Sciences de St. Pétersbourg* (1905) 10(3–4), 260–301, 10 figs., 1 pl. (in Russian)

Nikolsky, A.M. (1908) Materials connected with the herpetology of Russian Turkestan. *Annuaire du Musée Zoologique de l'Académie Impériale des Sciences de Petrograd* 13, 336–344. (in Russian)

Nikolsky, A.M. (1909a) Eine neue Vipern-Art aus dem Kaukasus, *Vipera kaznakovi* sp. nov. (Vorläufige Mitteilung). *Mitteilungen des Kaukasischen Museums*, Tiflis 4, 173–174. (in Russian)

Nikolsky, A.M. (1909b) Novae species reptilium e Caucaso. *Mitteilungen des Kaukasischen Museums*, Tiflis 4(3), 301–306. (in Russian)

Nikolsky, A.M. (1910) De nova g. Viperae specie e Caucaso, *Vipera kaznakovi. Mitteilungen des Kaukasischen Museums*, Tiflis 5(1), 81–84, 1 pl. (in Russian)

Nikolsky, A.M. (1913) Reptiles and amphibians of the Caucasus. (Herpetologia Caucasica). *Mitteilungen des Kaukasischen Museums*, Tiflis, 272 pp., 3 pls. (in Russian)

Nikolsky, A.M. (1914) Reptiles new to the Russian fauna from the southern part of the Maratime province. *Annuaire du Musée Zoologique de l'Académie Impériale des Sciences de St. Pétersbourg* 19, 88–91. (in Russian)

Nikolsky, A.M. (1916) *Faune de la Russie et des pays limitrophes fondée principalement sur les collections du Musée Zoologique de l'Académie Impériale des Science de Petrograd reptiles (Reptilia). Volume II. Ophidia.* Musée Zoologique de l'Académie Impériale des Sciences de Petrograd, Petrograd, 349 pp., 64 figs., 8 pls. (in Russian) [English translation by A. Mercado, 1964, Israel Program for Scientific Translations]

Nikolsky, A.M. (1925) Notice sur la faune herpétologique de la Sibérie Orientale. *Comptes Rendus de l'Académie des Sciences de l'Union des Républiques Soviétiques Socialistes,* Leningrad (A) 1925, 123–124. (in Russian).

Nikolsky, A.M. (1927) *Coluber* (*Vipera*) *sachalinensis* (Tzar.) et son histoire. *Annuaire du Musée Zoologique de l'Académie Impériale*, Leningrad 27, 247–260.

Nikolsky, A.M. (1931) A new snake subspecies from Turkestan, *Ancistrodon halys paramonovi*, subsp. nov. *Travaux du Musée Zoologique*, Kiev 10, 115–117. (in Russian)

Nilson, G. (1983) A new subspecies of the Asiatic pit viper *Agkistrodon halys* Pallas, 1776 (Serpentes, Viperidae) from Afghanistan. *Bonner Zoologische Beiträge*, Bonn 34(4), 469–476, 2 figs.

Nilson, G. & Andrén, C. (1981a) Morphology and taxonomic status of the grass snake, *Natrix natrix* (L.) (Reptilia, Squamata, Colubridae), on the island of Gotland, Sweden. *Zoological Journal of the Linnean Society*, London 72(4), 355–368, 8 figs.

Nilson, G. & Andrén, C. (1981b) Die herpetofauna des Kavir-Schutzgebietes, Kavir-Wüste, Iran. *Salamandra* 17(3–4), 130–146.

Nilson, G. & Andrén, C. (1984a) Systematics of the *Vipera xanthina* complex (Reptilia, Viperidae). II. An overlooked viper within the *xanthina* species-group in Iran. *Bonner Zoologische Beiträge*, Bonn 35 (1–3), 175–183, 3 figs.

Nilson, G. & Andrén, C. (1984b) A taxonomic account of the Iranian ratsnakes of the *Elaphe longissima* species-group. *Amphibia-Reptilia*, Leiden 5(2), 157–171, 9 figs.

Nilson, G. & Andrén, C. (1985a) Systematics of the *Vipera xanthina* complex (Reptilia, Viperidae). I. A new Iranian viper in the *raddei* species-group. *Amphibia-Reptilia*, Leiden 6(2), 207–214, 2 figs.

Nilson, G. & Andrén, C. (1985b) Systematics of the *Vipera xanthina* complex (Reptilia, Viperidae). III. Taxonomic status of the Bulgar Dagh viper in south Turkey. *Journal of Herpetology* 19(2), 276–283, 3 figs.

Nilson, G. & Andrén, C. (1986) The mountain vipers of the Middle East—the *Vipera xanthina* complex (Reptilia, Viperidae). *Bonner Zoologische Monographien*, Bonn (20), 1–90, 33 figs.

Nilson, G. & Andrén, C. (1988a) A new subspecies of the subalpine meadow viper, *Vipera ursinii* (Bonaparte) (Reptilia, Viperidae), from Greece. *Zoologica Scripta*, Oxford 17(3), 311–314, 5 figs.

Nilson, G. & Andrén, C. (1988b) *Vipera lebetina transmediterranea*: a new subspecies of viper from North Africa, with remarks on the taxonomy of *Vipera lebetina* and *Vipera mauritanica* (Reptilia, Viperidae). *Bonner Zoologische Beiträge*, Bonn 39(4), 371–379, 1 fig.

Nilson, G. & Andrén, C. (1992) The species concept in the *Vipera xanthina* complex, reflecting evolutionary history or hiding biological diversity? *Amphibia-Reptilia*, Leiden 13(4), 421–424.

Nilson, G. & Andrén, C. (1997) Evolution and biogeography of Palaearctic vipers. *In:* Thorpe, R.S., Wüster, W. & Malhotra, A. (Eds.), *Ecology, evolution and snakebite of venomous snakes.* Zoological Society of London, Clarendon Press, Oxford, pp. 31–42, 4 figs.

Nilson, G. & Andrén, C. (2001) The meadow and steppe vipers of Europe and Asia—the *Vipera* (*Acridophaga*) *ursinii* complex. *Acta Zoologica Academiae Scientiarum Hungaricae* 47, 87–267.

Nilson, G. & Flärdh, B. (1988) Die Vipern der Türkei. *Salamandra*, Frankfurt-am-Main 24(4), 215–247, 27 figs., 1 pl.

Nilson, G. & Flärdh, B. (1990) *Vipera albizona*, a new mountain viper from central Turkey, with comments on isolating effects of the Anatolian "Diagonal." *Amphibia-Reptilia*, Leiden 11(3), 285–294, 5 figs.

Nilson, G. & Rastegar-Pouyani, N. (2007) *Walterinnesia aegyptia* Lataste, 1887 (Ophidia, Elapidae) and the status of *Naja morgani* Mocquard, 1905. *Russian Journal of Herpetology* 14(1), 7–14, 7 col. figs.

Nilson, G. & Rastegar-Pouyani, N. (2011) On the occurrence of *Eirenis coronelloides* (Jan, 1862) in western Iran (Reptilia: Colubridae). *Zoology in the Middle East* 54, 133–135, 1 fig.

Nilson, G., Andrén, C. & Joger, U. (1993) A re-evaluation of the taxonomic status of the Moldavian steppe viper based on immunological investigations, with a discussion of the hypothesis of secondary intergradation between *Vipera ursinii rakosiensis* and *Vipera* (*ursinii*) *renardi*. *Amphibia-Reptilia*, Leiden 14(1), 45–57, 6 figs.

Nilson, G., Höggren, M., Tuniyev, B., Orlov, N. & Andrén, C. (1994) Phylogeny of the vipers of the Caucasus (Reptilia, Viperidae). *Zoologica Scripta*, Oxford 23(4), 353–360, 3 figs.

Nilson, G., Tuniyev, B., Orlov, N., Höggren, M. & Andrén, C. (1995) Systematics of the vipers of the Caucasus: polymorphism or sibling species? *Asiatic Herpetological Research*, Berkeley 6, 1–26, 22 figs., pl. 1 (col.).

Nilson, G., Tuniyev, B., Andrén, C. Orlov, N., Joger, U. & Herrmann, H.-W. (1999a) Taxonomic position of the *Vipera xanthina* complex. *In:* Joger, U. (Ed.), Phylogeny and systematics of the Viperidae. *Kaupia, Darmstädter Beiträge zur Naturgeschichte* 8, 99–102.

Nilson, G., Tuniyev, B., Andrén, C. & Orlov, N. (1999b) Vipers of Caucasus: taxonomic considerations. *In:* Joger, U. (Ed.), Phylogeny and systematics of the Viperidae. *Kaupia, Darmstädter Beiträge zur Naturgeschichte* 8, 103–106, 1 fig.

Nilson, G., Andrén, C., Ioannidis, Y. & Dimaki, M. (1999c) Ecology and conservation of the Milos viper, *Macrovipera schweizeri* (Werner, 1935). *Amphibia-Reptilia*, Leiden 20(4), 355–375, 2 figs.

Ninni, A.P. (1880) Sopra alcune varieta del *Tropidonotus natrix* (Linné) Schlegel osservato nel Veneto. Nota del Socio. *Atti della Società Italiana di Scienze Naturali*, Milano 23, 1–6

Nishimura, M. (1986) Fluctuations of body weight of habu, *Trimeresurus flavoviridis*, under rearing conditions. *The Snake* 18, 92–100, 8 figs.

Nishimura, M. & Kamura, T. (1992) Creamy white material ejected by male habu, *Trimeresurus flavoviridis* (Viperidae). *Japanese Journal of Herpetology*, Tokyo 14(4), 163–165.

Nishimura, M. & Kamura, T. (1993) Sex ratio and body size among hatchlings of habu, *Trimeresurus flavoviridis*, from the Okinawa Islands, Japan. *Amphibia-Reptilia*, Leiden 14(3), 275–283, 3 figs.

Nishimura, M., Araki, Y., Ueda, H. & Kawashima, Y. (1991) Frequencies of prey items of habu, *Trimeresurus flavoviridis* (Viperidae), in the Okinawa Islands. 1. *The Snake* 23( ), 81–83.

Niyomwan, P. (1999) *Species diversity, morphology and habitat types of blind snakes (family Typhlopidae) in Thailand*. Published Master's Thesis, Chulalongkorn University, Bangkok, 141 pp., 18 col. pls., 60 figs., 4 maps. (in Thai)

Niyomwan, P., Thirakhupt, K. & Nabhitabhata, J. (2001) A key to the blind snakes in Thailand. *The Natural History Journal of Chulalongkorn*, Bangkok 1(1), 47–52, 12 figs.

Noble, E.M. (1966) Occurrence of *Psammophis angolensis* in Ethiopia. *Copeia* 1966(1), 125–126, 1 fig.

Nobuo-Yuki, R. (1992) Sobre a ocorrência de *Philodryas mattogrossensis* Koslowsky, 1898 em Uberlândia, Minas Gerais, Brasil (Serpentes, Colubridae). *Acta Biológica Leopoldensia* 14(2), 123–126.

Nobusaka, R., Nakamoto, E. & Sawai, Y. (1994) Studies on the reproduction of habu (*Trimeresurus flavoviridis*) and coloration of the offsprings. *The Snake* 26(1), 1–9, 9 figs.

Noguchi, N. (1909) *Snake venoms: an investigation of venomous snakes with special reference to the phenomena of their venoms*. Carnegie Institution, Washington, D.C., 315 pp., 22 pls.

Nogueira, C. (2001) New records of squamate reptiles in central Brazilian cerrado. II. Brasília region. *Herpetological Review* 32(4), 285–287.

Nogueira, C. & Valdujo, P.H. (2001) Geographic distribution: Serpentes: *Bothrops alternatus*. *Herpetological Review* 32(1), 58.

Noonan, B.P. & Chippindale, P.T. (2006) Dispersal and vicariance: the complex evolutionary history of boid snakes. *Molecular Phylogenetics and Evolution* 40(2), 347–358, 5 figs.

Nopcsa, F.B. (1923) *Eidolosaurus* und *Pachyophis*. Zwei neue Neocom-Reptilien. *Palaeontographica*, Stuttgart 65, 97–154, pls. 7–8.

Nopcsa, F.B. (1925) Ergebnisse der Forschungsreisen Prof. E. Stromers in den Wüsten Ägyptens. II. Wirbeltier-Reste der Baharîje-Stufe (unterstes Cenoman). 5. Die *Symoliophis*-Reste. *Abhandlungen Bayerische Akademie der Wissenschaften, Mathematische-Physikalische Klasse* 30(4), 5–27, 1 pl.

Nordgren, M. & Villa, R. (2011) Geographic distribution: Serpentes: *Micrurus proximans*. *Herpetological Review* 42(3), 394–395.

Norman, B.R., Barron, A.D. & Feldman, C.R. (2012) Geographic distribution: Serpentes: *Contia longicaudae*. *Herpetological Review* 43(2), 307.

Norman, D.R. (1994) *Amphibians and reptiles of the Paraguayan Chaco. Volume I*. David R. Norman, Heredia, 281 pp., 32 col. pls.

Norsham, Y. & Lim, B.L. (2002) A new species of mountain reed snake, *Macrocalamus gentingensis*, from Genting highlands, Pahang, Peninsular Malaysia. *Hamadryad* 27(1), 83–89, 3 figs.

Notestein, F.N. (1905) The Ophidia of Michigan with analytical key. *Seventh Annual Report of the Michigan Academy of Science* 7, 111–125, 2 figs.

Novo-Rodriguez, J. & de Arazoza-Rodriguez, F. (1986) Cría en cautiverio de *Alsophis cantherigerus* (Ophidia, Colubridae). *Reporte de Investigaciónes del Instituto de Zoología, Academia de Ciencias de Cuba*, Habana (30), 1–15.

Nuñez, H. (1992) Geographical data of Chilien lizards and snakes in the Museo Nacional de Historia Natural Santiago, Chile. *Smithsonian Herpetological Information Service* (91), 1–29.

Nussbaum, R.A. (1980) The Brahminy blind snake (*Ramphotyphlops braminus*) in the Seychelles Archipelago; distribution, variation, and further evidence for parthenogenesis. *Herpetologica* 36(3), 215–221.

Nussbaum, R.A. (1984) Snakes of the Seychelles. *In:* Stoddart, D.R. (Ed.), *Biogeography and ecology of the Seychelles Islands*. W. Junk, The Hague, 361–377.

Nussbaum, R.A. & Hoyer, R.F. (1974) Geographic variation and the validity of subspecies in the rubber boa, *Charina bottae* (Blainville). *Northwest Science* 48(4), 219–229.

Nussbaum, R.A. & Raxworthy, C.J. (2000) Revision of the Madagascan snake genus *Heteroliodon* Boettger (Reptilia, Squamata, Colubridae). *Herpetologica* 36(4), 489–499, 4 figs.

Nussbaum, R.A., Brodie, E.D., Jr. & Storm, R.M. (1983) *Amphibians & Reptiles of the Pacific Northwest*. University Press of Idaho, Moscow, 332 pp., 60 pls.

Nussbaum, R.A., Andreone, F. & Raxworthy, C.J. (1998) New rain-forest species of *Pseudoxyrhopus* Günther (Squamata, Colubridae) from northern Madagascar. *Copeia* 1998(1), 128–132, 2 figs.

Nutaphand, W. (1971) *Poisonous snakes of Thailand*. Thai Zoological Center, Bangkok, 68 pp., illust. (in Thai)

Nutaphand, W. (1985) *Subfamily Boiginae*. Thai Zoological Center, Pata Zoo, Bangkok (year No. 1, Booklet No. 1), 1–10, figs. (apparently this was followed by a new issue in the same year with the same title, but with corrections). (*Boiga saengsomi*) (in Thai)

Nutaphand, W. (1986a) Wolf snakes: wolf snakes of Thailand. *Bulletin of the Thai Zoological Center*, Bangkok 1(9), 153–164, 7 figs. (in Thai)

Nutaphand, W. (1986b) Cobras: cobras of Thailand. *Bulletin of the Thai Zoological Center*, Bangkok 1(12), 201–216, 9 figs. (in Thai)

Nutaphand, W. (1991a) Ways of catching small snakes without hurting them. *Bulletin of the Thai Zoological Center*, Bangkok 6(69), 109–120, 7 figs. (in Thai)

Nutaphand, W. (1991b) Snakes in Thailand. *Bulletin of the Thai Zoological Center*, Bangkok 6(71), 133–144, 13 figs. (in Thai)

Nutaphand, W. (2001) *Snakes in Thailand*. Amarin Printing & Publishing Public Company, Bangkok, 320 pp., col. illust. (in Thai)

Nutaphand, W. & Tumvipart, V. (1982) *Poisonous snakes of Thailand and the cure from snakebites*. Bangkok, 162 pp., 82 figs., 8 illust., 62 col. pls. (in Thai)

Nutaphand, W., Cox, M.J., Trutnau, L. & Smith, H.M. (1991a) The status of the Thai palm viper, *Trimeresurus wiroti*. *Bulletin of the Maryland Herpetological Society* 27(3), 146–156, 1 fig.

Nutaphand, W., Cox, M.J., Smith, H.M. & Chiszar, D. (1991b) The original description, type specimen and status of the colubrid snake *Boiga saengsoni* Nutphand. *Bulletin of the Maryland Herpetological Society* 27(3), 157–165, 2 figs.

Obadi, N. (1997) *Fauna of Yemen. Part 2. Amphibians and Reptiles*. 151 pp., 59 figs. (in Arabic)

O'Brian, C. (2002) Geographic distribution: Serpentes: *Coronella austriaca*. *Herpetological Review* 33(3), 227.

Obst, F.J. (1963) Amphibien und Reptilien aus der Mongolei. *Mitteilungen aus dem Zoologischen Museum in Berlin* 39, 361–370, 4 figs.

Obst, F.J. (1976) Die Würfelnatter bei Meissen-ein erloschenes Vorkommen (Reptilia, Ophidia, Colubridae). *Zoologische Abhandlungen, Staatliches Museums für Tierkunde in Dresden* 34(4), 47–52, 2 figs.

Obst, F.J. (1977) Herpetologische taxa von Eduard Poeppig (1798–1868). *Zoologische Abhandlungen, Staatliches Museums für Tierkunde in Dresden* 34(12), 167–170.

Obst, F.J. (1981) Bemerkenswerte Schlangen aus Bulgarien in unserer Sammlung, *Coluber rubriceps* Venzmer und *Elaphe situla* (L.). *Faunistik Abhandlungen, Staatliches Museums für Tierkunde in Dresden* 8(17), 177–180, 1 fig.

Obst, F.J. (1983) Zur Kenntnis der Schlangengattung *Vipera* (Reptilia, Serpentes, Viperidae). *Zoologische Abhandlungen, Staatliches Museums für Tierkunde in Dresden* (1982) 38(13), 229–235.

Obst, F.J. & Szczerbak, N.N. (1993) *Elaphe dione* (Pallas, 1773)—steppennatter. *In:* Böhme, W. (Ed.), *Handbuch der Reptilien und Amphibien Europas. Band 3/I. Schlangen (Serpentes) I. (Typhlopidae, Boidae, Colubridae 1, Colubrinae)*. AULA-Verlag, Wiesbaden, pp. 295–315, figs. 57–59.

Obst, F.J., Richter, K. & Jacob, N. (1984) *Lexikon der Terraristik und Herpetologie*. Hannover, Leipzig, 466 pp., illust.

Obst, F.J., Richter, K. & Jacob, N. (1988) *The completely illustrated atlas of reptiles and amphibians for the terrarium*. T.F.H. Publications, Neptune City, 831 pp., figs., col. illust.

Obst, F.J., Szczerbak, N.N. & Böhme, W. (1993) *Elaphe situla* (Linnaeus, 1758)—Leopardnatter. *In:* Böhme, W. (Ed.), *Handbuch der Reptilien und Amphibien Europas. Band 3/I. Schlangen (Serpentes) I. (Typhlopidae, Boidae, Colubridae 1, Colubrinae)*. AULA-Verlag, Wiesbaden, pp. 431–453, figs. 82–85.

Ocaranza, F. (1930) Sistemática de los animals ponzoñosos de la América Latina y acción biológica de sus venenos. *Medicina*, México City, 10(121), 357–374.

Ochomogo, H.H. (1957) *Contribución al estudio de las serpientes venenosas de Nicaragua*. Master's Thesis, Universidad Nacional de Nicaragua, Leon, 81 pp.

Odhner, T. (1908) Reptilien und Batrachier, gesammelt von Dr. I. Trägardh in Natal und Zululand, 1904–05. *Arkiv för Zoologi*, Stockholm 4(18), 1–7.

Odum, R.A. (1979) The distribution and status of the New Jersey timber rattlesnake including an analysis of Pine Barrens population. *Herp* 15(1), 27–35, illust.

Ogilby, J.D. (1890a) Description of a new snake from the Herbert River District. *Proceedings of the Linnaean Society of New South Wales*, Sydney (Series 2) 5(1), 51–52.

Ogilby, J.D. (1890b) Description of a new snake belonging to the genus *Hoplocephalus*. *Proceedings of the Linnaean Society of New South Wales*, Sydney (Series 2) 4, 1027–1030.

Ogilby, J.D. (1891) Descriptions of three new Papuan snakes. *Records of the Australian Museum*, Sydney 1(9), 190–194.

Ogilby, J.D. (1892) On some undescribed reptiles and fishes from Australia. *Records of the Australian Museum*, Sydney 2, 23–26.

Ogilby, J.D. (1894) Description of a new Australian snake. *Proceedings of the Linnaean Society of New South Wales*, Sydney (Series 2) 9, 261–262.

Ogilby, J.D. (1898) Ophidia and Pisces. *In*: Contributions to a knowledge of the fauna of British New Guinea. (T. Steel, Ed.). *Proceedings of the Linnaean Society of New South Wales*, Sydney (Series 2) 13, 361–363, 1 fig.

Okada, Y. (1932) Notes on the Reptilia and Amphibia of Kôtôsho (Botel Tobago). *Bulletin of the Biogeographical Society of Japan*, Tokyo 3(1), 13–23.

Okada, Y. (1935) *Report of the first scientific expedition to Manchoukuo under the leadership of Shigeyasu Tokunaga, June–October 1933. Section V. Division II. Part II. Amphibia and Reptilia of Jehol*. Tokyo, 76 pp., 21 figs., 7 col. pls.

Okada, Y. (1938a) *A catalogue of vertebrates of Japan*. Maruzen Company, Tokyo, 412 pp.

Okada, Y. (1938b) Reptiles of the Tohoku-Districts, the northern part of Honsyu, Japan. *Saito Ho-on-kai Museum, Saito Gratitude Foundation Research Bulletin*, Sendai 15, 67–84.

Okada, Y. (1953) Reptiles. *In:* Uchida, K. (Ed.), *Illustrated encyclopedia of the fauna of Japan (exclusive of insects)*. Hokuryukan, Tokyo, pp. 223–271. (in Japanese)

Okada, Y. & Takara, T. (1958) A new species of *Liopeltis* (Ophidia, Colubridae) from Ryukyu Islands. (*L. kikuzatoi* sp. nov.). *Bulletin of the Biogeographical Society of Japan*, Tokyo 20(3), 9–11, 1 fig. (in Japanese)

Oken, L. (1816) *Lehrbuch der Naturgeschichte. Dritter Theil. Zoologie. Zweite Abtheilung. Fleischthiere*. August Schmid und Co., Jena, 1270 pp.

Oken, L. (1817) Cuviers und Okens Zoologien neben einander gestellt. *Isis von Oken*, Jena 8(8), 1145–1186.

Oken, L. (1836) *Allgemeine Naturgeschichte. Amphibien oder Lurche*. Hoffmann'sche, Stuttgart 6, 419–683.

Olazarri, T., Mones, A., Ximénez, A. & Philippi, M.E. (1970) Liste de los ejemplares-tipos depositados en el Museo Nacional de Historia Natural de Montevideo, Uruguay. *Comunicaciones Zoológicas del Museo de Historia Natural de Montevideo* 10(131), 1–12.

Oldfield, B. & Moriarty, J.J. (1994) *Amphibians and reptiles native to Minnesota*. University of Minnesota Press, Minneapolis, 240 pp., illust.

Oldham, J.C. & Smith, H.M. (1991) The generic status of the smooth green snake, *Opheodrys vernalis*. *Bulletin of the Maryland Herpetological Society* 27(4), 201–215, 2 figs.

Olgun, K., Avci, A., Ilgaz, Ç., Üzüm, N. & Yilmaz, C. (2007) A new species of *Rhynchocalamus* (Reptilia: Serpentes: Colubridae) from Turkey. *Zootaxa* (1399), 57–68, 9 figs.

Oliver, B. (1998) Captive reproduction of the amethystine python *(Morelia amethistina)*. *Monitor*, Ardeer 10(1), 18–21, 3 figs.

Oliver, J.A. (1937) Notes on a collection of amphibians and reptiles from the state of Colima, Mexico. *Occasional Papers of the Museum of Zoology, University of Michigan* (360), 1–28, 1 fig., 1 pl.

Oliver, J.A. (1942a) A check list of the snakes of the genus *Leptophis*, with descriptions of new forms. *Occasional Papers of the Museum of Zoology, University of Michigan* (462), 1–19.

Oliver, J.A. (1942b) A new snake of the genus *Imantodes* from Mexico. *Copeia* 1942(1), 1–2.

Oliver, J.A. (1948) The relationships and zoogeography of the genus *Thalerophis* Oliver. *Bulletin of the American Museum of Natural History* 92(4), 157–280, 13 figs., pls. 16–19.

Oliver, J.A. (1956) Reproduction in the king cobra, *Ophiophagus hannah* Cantor. *Zoologica* 41(4), 145–152, 5 pls.

Oliver, J.A. & Shaw, C.E. (1953) The amphibians and reptiles of the Hawaiian Islands. *Zoologica* 38, 65–95, 19 figs.

Oliver, W.L.R. (1982) The coney and the yellow snake: the distribution and status of the Jamaican hutia *Geocapromys browni* and the Jamaican boa *Epicrates subflavus*. *Dodo, Journal of the Jersey Wildlife Preservation Trust* 19, 6–33.

Olivier, E. (1894) Herpétologie Algérienne ou catalogue raisonné des reptiles et des batraciens observés jusqu'a ce jour en Algérie. *Mémoires de la Société Zoologique de France*, Paris 7, 98–131.

Olivier, E. (1896a) Matériaux pour la faune de la Tunisie. Catalogue des Reptiles. *Revue Scientifique du Bourbonnais et du Centre de la France,* Moulins 9, 117–128.

Olivier, E. (1896b) Les reptiles de Tunisie. *Compte Rendu de la Association Française pour l'Avancement des Sciences,* 25, 196.

Olivier, E. (1899) *Les serpents du nord de l'Afrique*. Augustin Challamel, Paris, 27 pp., 33 figs.

Olivier, G.A. (1801) *Voyage dans l'Empire Othoman, l'Egypte et de la Perse, fait par ordre de Gouvernement, pendant les six premières années de la République. Avec Atlas. Tome premier.* H. Agasse, Paris, xiii + 432 pp., Atlas (1801–1807), 50 pls.

Olori, J.C. & Bell, C.J. (2012) Comparative skull morphology of uropeltid snakes (Alethinophidia: Uropeltidae) with special reference to disarticulated elements and variation. PLoS ONE 7(3), 1–52, 30 figs.

Olson, R.E. (1977) Evidence for the species status of Baird's ratsnake. *Texas Journal of Science* 29(1–2), 79–84, 3 figs.

Ondrias, J.C. (1968) Liste des amphibiens et des reptiles de la Gréce. *Biologia Gallo-Hellenica* 1, 111–135.

Onn, C.K. (2009) The rediscovery of *Enhydris pahangensis* Tweedie, 1946. *Hamadryad* 34(1), 174–175, 1 fig.

Onn, C.K., Rooijen, J. von, Grismer, L.L., Belabut, D., Akil, M.A.M.M., Jamaludin, H., Gregory, R. & Ahmad, N. (2010) First report on the herpetofauna of Pulau Pangkor, Perak, Malaysia. *Russian Journal of Herpetology* 17(2), 139–146, 12 figs.

Oppel, M. (1811a) Suite du Ier. mémoire sur la classification des reptiles. *Annales du Museum d'Histoire Naturelle*, Paris (1810) 16, 376–393.

Oppel, M. (1811b) *Die Ordungen, Familien und Gattungen der Reptilien als Prodrom einer Naturgeschichte derselben.* Joseph Lindauer, München, xii + 87 pp.

Orcés, G. (1942) Los ofidios venenosus del Ecuador. *Flora*, Quito 2, 147–155.

Orcés, G. (1943) Los ofidios venenosus del Ecuador (conclusion). *Flora*, Quito 3, 165–170.

Orcés, G. (1948) Notas sobre las ofidios venenosas del Ecuador. *Publicaciónes de la Escuela Politécnica Nacional*, Quito 3, 231–250.

Orcés, G. & Almendáriz, A. (1987) Sistematica y distribución de las serpientes Dipsadinae del grupo *oreas*. *Politecnica (Biologia)*, Quito 12(4), 135–144, 1 fig.

Orcés, G. & Almendáriz, A. (1989a) Presencia en el Ecuador de los colubridos del genero *Sibynomorphus*. *Politecnica (Biologia 2)*, Quito 14(3), 57–67, 3 figs.

Orcés, G. & Almendáriz, A. (1989b) Acerca de la sistematica de *Spilotes megalolepis* Günther (Serpentes-Colubridae). *Politecnica (Biologia 3)*, Quito 14(3), 70–73, 1 fig.

Orcés, G. & Matheus, J.C. (1988) Hallazgo en el Ecuador del reptil apodo *Amphisbaena alba* (Amphisbaenidae) y de las serpientes del genero *Hydrophs* (Colubridae). *Publicaciones del Museo Ecuatoriano de Ciencias Naturales* (Série 2) 8(6), 81–82.

Orejas-Miranda, B.R. (1958) Dos especies de ofidios nuevos para el Uruguay. *Comunicaciones Zoológicas del Museo de Historia Natural de Montevideo* 4(79), 1–6, 2 pls.

Orejas-Miranda, B.R. (1959a) Uma nueva subspécie del genero *Philodryas* del Uruguay. *Comunicaciones Zoológicas del Museo de Historia Natural de Montevideo* 4(82), 1–3, 1 pl.

Orejas-Miranda, B.R. (1959b) El genero *Philodryas* en el Uruguay. *Actas y Trabalhos del Primer Congreso Sudamericano de Zoología,* 4, 103–109, 3 pls.

Orejas-Miranda, B.R. (1961) Una nueva especie de ofidio de la familia Leptotyphlopidae. *Acta Biologica Venezuelica*, Caracas 3(5), 83–97, 3 figs.

Orejas-Miranda, B.R. (1962) Description del hemipenis de *Leptotyphlops munoai* Orejas-Miranda, 1961. *Comunicaciones Zoológicas del Museo de Historia Natural de Montevideo* 7(97), 1–5, 2 pls.

Orejas-Miranda, B.R. (1964) Dos neuvos Leptotyphlopidae de Sur America. *Comunicaciones Zoológicas del Museo de Historia Natural de Montevideo* 8(103), 1–7, 3 pls.

Orejas-Miranda, B.R. (1966a) The snake genus *Lystrophis* in Uruguay. *Copeia* 1966(2), 193–205, 9 figs.

Orejas-Miranda, B.R. (1966b) Notas sobre la familia Leptotyphlopidae, I–II. I. Revalidación de *Leptotyphlops cupinensis* Bailey & Carvalho, 1946. *Comunicaciones Zoológicas del Museo de Historia Natural de Montevideo* 9(108), 1–3.

Orejas-Miranda, B.R. (1967) El género "*Leptotyphlops*" en la región Amazónica. *Atas do Simpósio sobre Biota Amazonica (Zoología),* 5, 421–422, 2 figs.

Orejas-Miranda, B.R. (1969) Tres nuevos *Leptotyphlops* (Reptilia, Serpentes). *Comunicaciones Zoológicas del Museo de Historia Natural de Montevideo* 10(124), 1–11, 2 pls.

Orejas-Miranda, B.R. & Achaval, F. (1969) Descripcion del hemipenis de *Leptotyphlops australis* Freiberg y Orejas-Miranda, 1968. *Neotropica*, La Plata 15(48), 132–134, 2 figs.

Orejas-Miranda, B.R. & Peters, G. (1970) Eine neue Schlangkblindschlange (Serpentes, Leptotyphlopidae) aus Ecuador. *Mitteilungen aus dem Zoologischen Museum in Berlin* (1969) 46(2), 439–441, 1 fig.

Orejas-Miranda, B.R. & Zug, G.R. (1974) A new tricolor *Leptotyphlops* (Reptilia, Serpentes) from Perú. *Proceedings of the Biological Society of Washington* 87(16), 167–173, 3 figs.

Orejas-Miranda, B.R., Roux-Estève, R. & Guibé, J. (1970) Un nouveau genre de leptotyphlopides (Ophidia) *Rhinoleptus koniagui* (Villiers). *Comunicaciones Zoológicas del Museo de Historia Natural de Montevideo* 10(127), 1–4, 3 figs.

Orellana, L.V. (2011) Geographic distribution: Serpentes: *Enulius flavitorques*. *Herpetological Review* 42(1), 114.

Orellana, L.V. & Gutsche, A. (2011) Geographic distribution: Serpentes: *Drymarchon melanurus*. *Herpetological Review* 42(1), 114.

Orlov, N.L. (1995) Rare snakes of the mountainous forests of northern Indonesia. *Russian Journal of Herpetology*, Moscow 2(2), 179–183, 4 col. figs.

Orlov, N.L. (1997) Viperid snakes (Viperidae Bonaparte, 1840) of Tam-Dao Mountain Ridge (Vinh-Phu and Bac-Thai provinces, Vietnam. *Russian Journal of Herpetology*, Moscow 4(1), 64–74, 3 figs., 7 col. figs.

Orlov, N.L. (2000) Distribution, biology and comparative morphology of the snakes of *Xenopeltis* genus (Serpentes, Macrostomata, Xenopeltidae) in Vietnam. *Russian Journal of Herpetology*, Moscow 7(2), 103–114, 20 figs.

Orlov, N.L. (2009) A new species of the genus *Calamaria* (Squamata: Ophidia: Colubridae) from the Central Highlands (Ngoc Linh Nature Reserve, Ngoc Linh Mountain, Kon Tum Province), Vietnam. *Russian Journal of Herpetology* 16(2), 146–154, 8 figs.

Orlov, N.L. & Barabanov, A.V. (1999) Analysis of nomenclature classification, and distribution of the *Agkistrodon halys-Agkistrodon intermedius* complexes: a critical review. *Russian Journal of Herpetology*, Moscow 6(3), 167–192, 24 figs.

Orlov, N.L. & Barabanov, A.V. (2000) About type localities for some species of the genus *Gloydius* Hoge et Romano-Hoge, 1981 (Crotalinae, Viperidae, Serpentes). *Russian Journal of Herpetology*, Moscow 7(2), 159–160, 1 fig.

Orlov, N.L. & Helfenberger, N. (1997) New mountain species of *Trimeresurus* (Serpentes, Viperidae, Crotalinae) of the "green" pit viper group from the Himalayas. *Russian Journal of Herpetology*, Moscow 4(2), 195–197, 8 figs.

Orlov, N.L. & Ryabov, S.A. (2002) A new species of the genus *Boiga* (Serpentes, Colubridae, Colubrinae) from Tanahjampea Island and description of "black form" of *Boiga cynodon* complex from Sumatra (Indonesia). *Russian Journal of Herpetology* 9(1), 33–56, 25 figs.

Orlov, N.L. & Ryabov, S.A. (2004) Revalidization and change of taxonomic status of *Dinodon rufozonatum meridionale* Bourret, 1935 (Serpentes, Colubridae, Colubrinae). *Russian Journal of Herpetology* 11(3), 181–197, 23 figs.

Orlov, N.L. & Tuniyev, B.S. (1986a) A new species of the grass snake, *Natrix megalocephala* sp. nov. from the Caucasus. *Proceedings of the Zoological Institute*, Leningrad 158, 116–139, 6 figs. (in Russian)

Orlov, N.L. & Tuniyev, B.S. (1986b) Present ranges, possible ways of their formation and phylogeny of the three species of vipers from the Eurosiberic group (*V. kaznakowi* complex) in the Caucasus. *Proceedings of the Zoological Institute*, Leningrad 157, 104–135. (in Russian)

Orlov, N.L. & Tuniyev, B.S. (1990) Three species in the *Vipera kaznakowi* complex (Eurosiberian group) in the Caucasus: their present distribution, possible genesis and phylogeny. *Asiatic Herpetological Research*, Berkeley 3, 1–36, 16 figs.

Orlov, N.L. & Tuniyev, B.S. (1992) A new species of grass snake, *Natrix megalocephala*, from the Caucasus (Ophidia, Colubridae). *Asiatic Herpetological Research*, Berkeley 4, 42–54, 10 figs.

Orlov, N.L., Darevsky, I.S. & Murphy, R.W. (1998) A new species of mountain stream snake, genus *Opisthotropis* Gunther, 1872 (Serpentes, Colubridae, Natricinae), from the tropical rain forests of southern Vietnam. *Russian Journal of Herpetology*, Moscow 5(1), 61–64, 4 figs.

Orlov, N.L., Murphy, R.W. & Papenfuss, T.J. (2000) List of snakes of Tam-Dao Mountain ridge (Tonkin, Vietnam). *Russian Journal of Herpetology*, Moscow 7(1), 69–80.

Orlov, N.L., Ryabov, S.A., Shiryaev, K.A. & Van Sang, N. (2001) On the biology of pit vipers of *Protobothrops* genus (Serpentes, Viperidae, Crotalinae). *Russian Journal of Herpetology*, Moscow 8(2), 159–164, 5 col. figs.

Orlov, N.L., Kudryavtzev, S.V., Ryabov, S.A. & Shumakov, O.V. (2003a) A new species of genus *Boiga* (Serpentes, Colubridae, Colubrinae) and color atlas of boigas from Bengkulu Province (Sumatra, Indonesia). *Russian Journal of Herpetology* 10(1), 33–52, 24 col. figs.

Orlov, N.L., Ryabov, S.A., Van Sang, N. & Truong, N.Q. (2003b) New records and data on the poorly known snakes of Vietnam. *Russian Journal of Herpetology* 10(3), 217–240, 35 col. figs.

Orlov, N.L., Ryabov, S.A., Thanh, B.N. & Cue, H.T. (2004) A new species of *Trimeresurus* (Ophidia, Viperidae, Crotalinae) from Karst region in central Vietnam. *Russian Journal of Herpetology* 11(2), 139–149, 14 col. figs.

Orlov, N.L., Ryabov, S.A. & Tao, N.T. (2009a) Two new species of genera *Protobothrops* Hoge et Romano-Hoge, 1983 and *Viridovipera* Malhotra et Thorpe, 2004 (Ophidia: Viperidae: Crotalinae) from Karst region in northeastern Vietnam. Part I. Description of a new species of *Protobothrops* genus. *Russian Journal of Herpetology* 16(1), 69–82, 24 col. figs.

Orlov, N.L., Kharin, V.E., Ananjeva, N.B., Tao, N,T. & Truong, N.Q. (2009b) A new genus and species of colubrid snake (Squamata, Ophidia, Colubridae) from south Vietnam (Lam Dong Province). *Russian Journal of Herpetology* 16(3), 228–240, 20 figs.

Orlov, N.L., Nguyen, T.Q., Nguyen, T.T., Ananjeva, N.B. & Hu, C.T. (2010a) A new species of the genus *Calamaria* (Squamata: Ophidia: Colubridae) from Thua Thien-Hue Province, Vietnam. *Russian Journal of Herpetology* 17(3), 236–242, 7 figs.

Orlov, N.L., Ryabov, S.A., Nguyen, T.T. & Nguyen, T.Q. (2010b) Rediscovery and redescription of two rare snake species: *Oligodon lacroixi* Angel et Bourret, 1933 and *Maculophis belllus chapaensis* (Bourret, 1934) (Squamata: Ophidia: Colubridae). *Russian Journal of Herpetology* 17(4), 310–322, 17 figs.

Orlov, N.L., Nguyen, T.T., Nazarov, R.A. & Melnikov, D.A. (2011) The northernmost record of colubrid snake *Dryocalamus davisonii* (Blanford, 1878) [Serpentes: Colubridae: Colubrinae]. *Russian Journal of Herpetology* 18(2), 154–156, 4 figs.

Orlova, V.F. & Semenov, D.V. (1999) *Nature in Russia: life of animals. Amphibians and reptiles.* Izdatel'stvo AST, Moscow, 478 pp., illust. (in Russia)

Orofino, R.P. de, Pizzatto, L. & Marques, O.A.V. (2010) Reproductive biology and food habits of *Pseudoboa nigra* (Serpentes: Dipsadidae) from the Brazilian cerrado. *Phyllomedusa* 9(1), 53–61, 3 figs.

Orrego-Aravena, R. (1971) *Reptiles de La Pampa.* Biblioteca Pampeana, Santa Rosa, 68 pp.

Ortega-Andrade, H.M., Bermingham, J., Aulestia, C. & Paucar, C. (2010) Herpetofauna of the Bilsa Biological Station, province of Esmeraldas, Ecuador. *Check List* 6(1), 119–154, illust.

Ortenburger, A.I. (1923) A note on the genera *Coluber* and *Masticophis*, and a description of a new species of *Masticophis*. *Occasional Papers of the Museum of Zoology, University of Michigan* (139), 1–14.

Ortenburger, A.I. (1927) A key to the snakes of Oklahoma. *Proceedings of the Oklahoma Academy of Sciences* 6, 197–218.

Ortenburger, A.I. (1928) The whip snakes and racers. Genera *Masticophis* and *Coluber*. *Memoirs, University of Michigan Museum of Zoology* (1), xviii + 247 pp., 64 figs., 35 pls.

Ortenburger, A.I. (1930) A key to the lizards and snakes of Oklahoma. *Publications of the University of Oklahoma, Biological Survey* 2(4), 209-239.

Ortiz, J.C. (1973) Étude sur le statut taxinomique de *Tachymenis peruviania* Wiegmann et *Tachymenis chilensis* (Schlegel) (Serpentes, Colubridae). *Bulletin du Muséum National d'Histoire Naturelle*, Paris (Série 3) Zool. 111(146), 1021–1039, 3 figs.

Orton, G.L. & Morrison, F.D. (1946) Some amphibians and reptiles from the Gold Coast. *Copeia* 1946(1), 15–17.

Osborn, H.F. (1930) Biogeographical memoir of Edward Drinker Cope, 1840–1897. *National Academy of Science Biographical Memoir* 13: 124–317, 1 pl.

O'Shea, M.T. (1989) The herpetofauna of Ilha de Maracá, of Roraima, northern Brazil. *In:* Coote, J. (Ed.), *Reptiles, proceedings of the 1988 U.K. Herpetological Societies Symposium on captive breeding*. British Herpetological Society, London, pp. 51–72, 1 fig., 3 col. pls.
O'Shea, M.T. (1990) The highly and potentially dangerous elapids of Papua New Guinea. *In:* Gopalakrishnakone, P. & Clou, L.M. (Eds.), *Snakes of medical importance (Asia–Pacific region).* National University of Singapore, Singapore, pp. 585–640, 8 figs.
O'Shea, M.T. (1996) *A guide to the snakes of Papua New Guinea.* Independent Publishing, Port Moresby, xii + 239 pp., figs., col. illust.
O'Shea, M.T. (1998) Herpetological results of two short field excursions to the Royal Bardia region of western Nepal, including range extensions for Assamese/Indo-Chinese snake taxa *In:* de Silva, A. (Ed.), *Biology and conservation of the amphibians, reptiles and their habitats in South Asia.* Amphibia and Reptile Research Organization of Sri Lanka (ARROS), Peradeniya, pp. 306–317, 3 figs.
O'Shea, M.T. (2003) Geographic distribution: Serpentes: *Dryocalamus gracilis. Herpetological Review* 34(3), 264.
O'Shea, M.T. (2005) *Boas and pythons of the world.* Princeton University Press, Princeton, 160 pp., col. illust.
O'Shea, M.T. (2007) *Venomous snakes of the world.* Princeton University Press, Princeton, 160 pp., col. illust.
O'Shea, M.T. & Lazell, J.D., Jr. (2008) Geographic distribution: Serpentes: *Python reticulatus. Herpetological Review* 39(4), 486.
O'Shea, M.T. & Stimson, A.F. (1993) An aberrant specimen of *Drymobius rhombifer* (Colubridae, Colubrinae), a new generic record for Brazil. *Herpetological Journal*, London 3(2), 70–71, 1 fig.
O'Shea, M.T., Sprackland, R.G. & Biglale, I.H. (2004) First record for the genus *Antaresia* (Squamata, Pythonidae) from Papua New Guinea. *Herpetological Review* 35(3), 225–227, 3 figs.
O'Shea, M.T., Sanchez, C., Heacox, S., Kathriner, A., Lopes-Carvalho, V., Varela-Ribeiro, A., Afranio-Soares, Z., Lemos de Araujo, L. & Kaiser, H. (2012) First update to herpetofaunal records from Timor-Leste. *Asian Herpetologoical Research* (Series 2) 3(2), 114–126, 8 figs.
Oshima, M. (1910) An annotated list of Formosan snakes, with descriptions of four new species and one new subspecies. *Annotationes Zoologicae Japonenses*, Tokyo 7(3), 185–207.
Oshima, M. (1914) Monograph of the Formosan sea-snakes. *Dobutsugaku Zasshi* (*Zoological Magazine*), Tokyo 26, 423–434. (in Japanese)
Oshima, M. (1916a) A correction of the scientific names of Formosan snakes. *Dobutsugaku Zasshi* (*Zoological Magazine*), Tokyo 28(328), 84–86. (in Japanese)
Oshima, M. (1916b) Key to the snakes of Formosa and the Riu-Kius. *Transactions of the Natural History Society of Formosa*, Taihoku 6(23), 1–12. (in Japanese)
Oshima, M. (1920) Notes on the venomous snakes from the islands of Formosa and Riu Kiu. *Annual Report of the Institute of Science, Government of Formosa*, Taihoku 8(2), 1–11, 1–99, 16 pls. (in Japanese)
Oshima, M. (1944) *Poisonous snakes in East and South Asia.* Hokuryukan, Tokyo, 330 + 19 pp., 131 figs., 18 pls. (in Japanese)
Osman, H.E.S.I. & El Sir, N.T. (1988) The snakes of the Sudan. 1. The snakes of Khartoum Province. *The Snake* 20(1), 74–79.
Ostrovskikh, S.V. (1997) Different forms of melanism and its development with age in the populations of steppe viper *Vipera renardi* (Christoph, 1861). *Russian Journal of Herpetology* 4(2), 186–191, 5 figs.
Ota, H. (1982) Notes on a specimen of Miyara's dwarf snake (*Calamaria pavimentata miyarai*) newly collected, and on some of the external characters of this subspecies. *The Snake*, Nittagun 14(1), 40–43, 1 fig. (in Japanese)
Ota, H. (1983) Herpetofauna of Kanmurijima Island. *In:* Kansai Environmental Assessment Center (Ed.), *Report on the fauna and flora of Kanmurijima Island.* Kansai Environmental Assessment Center Co., Osaka, pp. 33–34. (in Japanese)
Ota, H. (1985) Notes on the specimens of reptiles deposited in Shiretoko Museum, Shari, eastern Hokkaido, Japan. *Bulletin of the Shiretoko Museum*, Shari 7, 1–6, 2 figs. (in Japanese)
Ota, H. (1986) A review of reptiles and amphibians of the Amami Group, Ryukyu Archipelago. *Memoirs of the Faculty of Science, Kyoto University* (*Biology*), Kyoto 11(1–2), 57–71, 1 fig.
Ota, H. (1988) Taxonomic notes on *Lycodon ruhstrati* (Colubridae, Ophidia) from East Asia. *Journal of the Taiwan Museum* 41(1), 85–91, 2 figs.
Ota, H. (1991) Systematics and biogeography of terrestrial reptiles of Taiwan. *In:* Lin, Y.-S. & Chang, K.-H. (Eds.), *Proceedings of the First International Symposium of Wildlife Conservation.* Council of Agriculture, Taipei, pp. 47–112, 43 figs.
Ota, H. (1993) Kikuzato's stream snake *(Opisthotropis kikuzatoi):* its status and conservation. *Survey Reports on National Monuments of Okinawa,* Naha (33), xi + 96 pp., 62 col. figs.
Ota, H. (1997) An annotated catalogue of type specimens of amphibians and reptiles in the National Science Museum, Tokyo. *Bulletin of the National Science Museum*, Tokyo 23(4), 221–232.
Ota, H. (2000) A long overlooked holotype: taxonomic notes on *Lycodon tessellatus* Jan 1863 (Squamata Colubridae), with a revised key to the Philippine species of the genus. *Tropical Zoology,* Firenze 13(2), 299–304, 1 fig.
Ota, H. & Endo, H. (1999) *A catalogue of amphibian and reptile specimens from the east Asian islands deposited in the National Science Museum, Tokyo*. National Science Museum, Tokyo, 295 pp.
Ota, H. & Iwanaga, S. (1997) A systematic review of the snakes allied to *Amphiesma pryeri* (Boulenger) (Squamata, Colubridae) in the Ryukyu Archipelago, Japan. *Zoological Journal of the Linnean Society*, London 121(3), 339–360, illust.
Ota, H. & Mori, A. (1985a) Notes on the fourth collected specimen of *Rhabdophis tigrinus formosanus* and the morphological features characteristic of the subspecies. *Japanese Journal of Herpetology*, Kyoto 11(2), 41–45, 3 figs.
Ota, H. & Mori, A. (1985b) On the fourth specimen of *Opisthotropis kikuzatoi. The Snake*, Nittagun 17(2), 160–162, 2 figs.
Ota, H. & Ross, C.A. (1994) Four new species of *Lycodon* (Serpentes, Colubridae) from the northern Philippines. *Copeia* 1994(1), 159–174, 12 figs.
Ota, H. & Toyama, M. (1989a) Taxonomic re-definition of *Achalinus formosanus* Boulenger (Xenoderminae, Colubridae, Ophidia), with description of a new subspecies. *Copeia* 1989(3), 597–602, 3 figs.
Ota, H. & Toyama, M. (1989b) Two additional specimens of *Achalinus formosanus chigirai* (Colubridae, Ophidia) from the Yaeyama Group. Ryukyu Archipelago. *Japanese Journal of Herpetology*, Kyoto 13(2), 40–43, 1 fig.

Ota, H. & Weidenhöfer, T. (1990) A noteworthy collection of reptiles from the northern part of Vietnam. *Japanese Journal of Herpetology*, Kyoto 13(4), 152. (in Japanese)

Ota, H., Takahashi, H. & Kamezaki, N. (1985) On the specimens of yellow lipped sea krait *Laticauda colubrina* from the Yaeyama Group, Ryukyu Archipelago. *The Snake*, Nittagun 17(2), 156–159, 3 figs.

Ota, H., Toriba, M. & Takahashi, H. (1986a) The alteration pattern of dorsal scale rows in the yellow-lipped sea snake *Laticauda colubrina*, with special reference to sexual dimorphism. *Japanese Journal of Herpetology*, Kyoto 11(3), 145–151, 1 fig.

Ota, H., Sakihara, E. & Shimamura, H. (1986b) A case of abnormal colour pattern in the Sakishima Oriental king snake *Dinodon rufozonatus walli* from Haterumajima Island of the Yaeyama Group, Ryukyu Archipelago. *Biological Magazine*, Okinawa 24, 17–20, 1 fig.

Ota, H., Hikida, T. & Barcelo, J. (1987) On a small collection of lizards and snakes from Cameroon, West Africa. *African Study Monographs* 8(2), 111–123, 3 figs.

Ota, H., Hikida, T., Matsui, M., Mori, A. & Wynn, A.H. (1991a) Morphological variation, karyotype and reproduction of the parthenogenetic blind snake, *Ramphotyphlops braminus*, from the insular region of East Asia and Saipan. *Amphibia-Reptilia*, Leiden 12(2), 181–193, 5 figs.

Ota, H., Iwanaga, S., Itoman, K., Nishimura, M. & Mori, A. (1991b) Reproductive mode of a natricine snake, *Amphiesma pryeri* (Colubridae, Squamata), from the Ryukyu Archipelago, with special reference to the viviparity of *A. p. ishigakiensis*. *Biological Magazine*, Okinawa 29, 37–43, 3 figs.

Ota, H., Kamezaki, N., Mikida, T., Matsui, M., Mori, A., Hayashi, T. & Tanabe, S. (1991c) Karyotype of *Achalinus spinalis* from Japan, the first karyological description of a xenodermine snake (Colubridae). *Japanese Journal of Herpetology*, Kyoto 14(1), 12–14, 1 fig.

Ota, H., Shiroma, M. & Hikida, T. (1995) Geographic variation in the endemic Ryukyu green snake *Cyclophiops semicarinatus* (Serpentes, Colubridae). *Journal of Herpetology* 29(1), 44–50, 4 figs.

Ota, H., Lin, J.-T., Hirata, T. & Chen, S.-L. (1997) Systematic review of colubrid snakes of the genus *Pareas* in the East Asian islands. *Journal of Herpetology* 31(1), 79–87, 5 figs.

Ota, H., Ito, A. & Lin, J.-T. (1999) Systematic review of the elapid snakes allied to *Hemibungarus japonicus* (Günther, 1868) in the East Asian Islands, with description of a new subspecies from the Central Ryukyus. *Journal of Herpetology* 33(4), 675–687, 6 figs.

Ottley, J.R. (1978) A new subspecies of the snake *Lichanura trivirgata* from Cedros Island, Mexico. *Great Basin Naturalist* 38(4), 411–416, 2 figs.

Ottley, J.R. & Tanner, W.W. (1978) New range and a new subspecies for the snake *Eridiphas slevini*. *Great Basin Naturalist* 38(4), 406–410, 2 figs.

Ottley, J.R., Murphy, R.W. & Smith, G.V. (1980) The taxonomic status of the rosy boa *Lichanura roseofusca* (Serpentes, Boidae). *Great Basin Naturalist* 40(1), 59–62, 3 figs.

Ouyang, C. & Lin, S.S.Y. (1970) Relationship between pharmacological actions and enzymatic activities of the venom of *Trimeresurus gramineus*. *Toxicon* 8(2), 183–191, 1 fig.

Over, W.H. (1923) Amphibians and reptiles of South Dakota. *South Dakota Geological and Natural History Survey Bulletin* 12, 1–34, 18 pls.

Owadally, A.W. & Lambert, M. (1988) Herpetology in Mauritius. A history of extinction, future hope for conservation. *British Herpetological Society Bulletin*, London (23), 11–20.

Owen, R. (1840) The remains of an extinct species of serpent found at Sheppey. *Proceedings of the Geological Society of London* (1839–1840) 3(66), 165–166.

Owen, R. (1841a) Report on British fossil reptiles, Part II. *Report of the Eleventh Meeting of the British Association for the Advancement of Science*, John Murray, London (1842), pp. 60–204.

Owen, R. (1841b) Description of some ophidiolites (*Palaeophis toliapicus*) from the London Clay at Sheppey, indicative of an extinct species of serpent. *Transactions of the Geological Society of London* (1842) (Series 2) 6, 209–210, pl. 22.

Owen, R. (1850) Monograph on the fossil Reptilia of the London Clay, and of the Bracklesham and other Tertiary beds. Part III. Ophidia (*Palaeophis*, &c.). *Palaeontographical Society Monograph*, London (1849) 3(2), 51–68, pls. 12–16.

Owen, R. (1857) On the fossil vertebrae of a serpent (*Laophis crotaloides*, Ow.) discovered by Capt. Spratt, R.N., in a Tertiary formation at Salonica. *Quarterly Journal of the Geological Society of London* 13, 196–199, pl. 4.

Owen, R. (1859) *Descriptive catalogue of the specimens of natural history in spirit contained in the Museum of the Royal College of Surgeons. Vertebrata.* Royal College of Surgeons, London, 147 pp.

Owen, R. (1840–1845) *Odontography; or, a treatise on the comparative anatomy of the teeth; their physiological relations, mode of development, and microscopic structure, in the vertebrate animals.* 2 vols. Hippolyte Baillière, London, lxxiv + 655 pp., 168 pls.

Oxtoby, G.P. (1988) Enkele observaties betreffende de Maleise rattenslang (*Elaphe taeniura ridleyi*) en de Taiwanese rattenslang (*E. t. friesi*) in het terrarium. *Lacerta*, Gravenhage 46(10–11), 151–159, 8 figs.

Pack, H.J. (1930) *Snakes of Utah.* Utah Agricultural Experiment Station, Utah State Agricultural College, Logan (221), 3–30, 10 figs.

Padua-Almeida, A. de, Gasparini, J.L. & Germano, V.J. (2006) Geographic distribution: Serpentes: *Liophis meridionsalis*. *Herpetological Review* 37(4), 498.

Page, T.J. (1859) *La Plata, the Argentine Conferation, Paraguay. Being a narrative of the exploration of the tributaries of the river La Plata and adjacent countries during the years 1853, '54, '55, and '56, under the orders of the government.* Harper and Brothers, New York, 632 pp., 46 pls.

Pagenstecher, H.A. (1881) *Allgemeine Zoologie oder Grundesetze des thierischen Baus und Lebens. Vierter Theil.* Paul Parey, Berlin, 959 pp.

Pagini, E. & Lema, T. de. (1987) Reencontro de *Calamondontophis paucidens* (Amaral, 1936) e contribuição ao conhecimento do gênero e da espécie (Serpentes, Colubridae, Tachymeninae). *Comunicações do Museu de Ciências e Tecnologia da PUCRS (Zoologia),* Porto Alegre (47), 198–208, 10 figs.

Paik, N.-K. (1982) *Systematic studies on the suborder Serpentes (Reptilia) in Korea.* Kangreung National University, Kangreung, 86 pp.

Paik, N.-K. & Yang, S.-Y. (1986) Taxonomic study on two subspecies of European grass snake (*Rhabdophis tigrina*) in Korea. *Korean Journal of Systematic Zoology*, Seoul 2(1), 79–92, 5 figs.

Paik, N.-K. & Yang, S.-Y. (1989) Geographic variation and interspecific hybridization between two species of the genus *Agkistrodon* (Crotalidae) in Korea. *Korean Journal of Systematic Zoology*, Seoul 5(1), 77–88, 4 figs.

Paik, N.-K., Kim, Y.-J. & Yang, S.-Y. (1979) Biochemical variation and systematic status of the genus *Agkistrodon* (Crotalidae) in Korea. *Korean Journal of Zoology* 22(4), 153–164, 2 figs.

Paik, N.-K., Min, M.-S., Shim, J.-H., Yang, S.-Y. & Toriba, M. (1993) Isozyme analysis of the species of the genus *Agkistrodon* in Far Eastern Region. *The Snake* 25(2), 99–104, 1 fig.

Paik, N.-K., Park, C.-S., Jung, E.-K., Lee, H.-Y. & Toriba, M. (1998) MtDNA analysis of the genus *Agkistrodon* species in eastern Asia. *Russian Journal of Herpetology* 5(2), 97–102, 3 figs.

Pakenham, R.H.W. (1947) Modifications to the reptile and amphibian list of the Zanzibar Protectorate. *Annals and Magazine of Natural History* (Series 11) 14(110), 134–140.

Pakenham, R.H.W. (1983) The reptiles and amphibians of Zanzibar and Pemba Islands. *Journal of the East Africa Natural History Society and National Museum*, Nairobi (177), 1–40, 2 figs.

Palacky, J. (1898) La distribution des ophidiens sur le globe. *Mémoires de la Société Zoologique de France* 11, 88–125.

Palisot de Beauvois, A.M.F.J. (1799) Memoir on Amphibia. Serpents. *Transactions of the American Philosophical Society*, Philadelphia 4(42), 362–381, 1 pl.

Pallas, P.S. (1771–1776) *Reise durch verschiedene Provinzen des Russischen Reichs*. Three vols. Kayserlichen Academie der Wissenschaften, St. Petersburg. [vol. 1, 504 pp., 25 pls. (1771); vol. 2, 744 pp., 41 pls. (1773); vol. 3, 760 pp., 48 pls. (1776)]

Pallas, P.S. (1799) *Bemerkungen auf einer Reise in die südlichen Statthalterschaften der Russischen Reichs, in dem jahren 1793 und 1794*. Volume 1. G. Martini, Leipzig, xxxii + 516 pp.

Pallas, P.S. (1814) *Zoographia Rosso-Asiatica, sistens omnium animalium in extenso Imperio Rossico et adjacentibus maribus observatorum recensionem, domicilia, mores et descriptiones, anatomen atque icones plurimorum. Volumen tertium. Animalia monocardia seu frigidi sanguinis imperii Rosso-Asiatici.* Officina Caesarea Academiae Scientiarum, Petropoli, (1811) vii + 428 pp + cxxv pp.

Pallas, P.S. (1831) *Zoographia Rosso-Asiatica, sistens omnium animalium in extenso Imperio Rossico et adjacentibus maribus observatorum recensionem, domicilia, mores et descriptiones, anatomen atque icones plurimorum. Volumen tertium. Animalia monocardia seu frigidi sanguinis imperii Rosso-Asiatici. Atlas.* Officina Caesarea Academiae Scientiarum, Petropoli, vii + 428 + cxxv pp., 7 pls. [reprint of Pallas, 1814 with plates]

Palmer, W.M. (1974) *Poisonous snakes of North Carolina*. State Museum of North Carolina, Raleigh, 22 pp.

Palmer, W.M. (1978) *Sistrurus miliarius*. *Catalogue of American amphibians and reptiles*, pp. 220.1–220.2.

Palmer, W.M. & Braswell, A.L. (1995) *Reptiles of North Carolina*. University of North Carolina Press, Chapel Hill, 412 pp., 127 figs., 76 col. pls.

Pantuwatana, S., Imlarp, S. & Marshall, J.T., Jr. (1969) Vertebrate ecology of Bang Phra. *The Natural History Bulletin of the Siam Society* 23(1–2), 133–183, pls. 15–18.

Papavero, N. (1971) *Essays on the history of Neotropical dipterology, with special reference to collectors (1750 – 1905). Vol. I.* Museu de Zoologia, Universidade de São Paulo, São Paulo, vii + 216 pp., 10 pls.

Papavero, N. (1973) *Essays on the history of Neotropical dipterology, with special reference to collectors (1750 – 1905). Vol. II.* Museu de Zoologia, Universidade de São Paulo, São Paulo, iii + 217–446, 10 pls.

Papenfuss, T.J. (1969) Preliminary analysis of the reptiles of arid central west Africa. *Wasmann Journal of Biology* 27, 249–325, 7 figs.

Paphiles, P. & Valakos, E.D. (2012) *Amphibians and reptiles of Greece: an identification guide.* Patakes, Athens, 197 pp., 23 pls. (in Greek)

Paracuellos, M. (2001) Geographic distribution: Serpentes: *Natrix natrix*. *Herpetological Review* 32(1), 61.

Paraskev, K.P. (1956) *Reptiles of Kazakhstan*. Institute of Zoology, Kazakh Academy of Sciences, Alma Ata, 228 pp., 69 figs., 36 pls. (in Russian)

Parent, G.H. (1979) Atlas provisoire commenté de l'herpétofaune de la Belgique et du Grand-Duché de Luxembourg. *Les Naturalistes Belges* 60(9–10), 251–333, illust.

Parent, G.H. (1982) Bibliographie de l'herpetofaune francaise. *Inventaires de Faune et de Flore* 1982(17–18), 1–431.

Parent, G.H. (1983) *Protegeons nos batraciens et reptiles*. J. Duculot, Region Wallone, 172 pp.

Parent, G.H. (1984) Atlas des batraciens et reptiles de Belgique. *Cahiers d'Ethologie Appliquée* 4(3), 1–198, illust.

Parker, F. (1972) Snakes of the elapid genus *Suta* in New Guinea. *Papua New Guinea Scientific Society Proceedings* (1971) 23, 13–17, 1 fig.

Parker, F. (1982) The snakes of Western Province. (*Wildlife in Papua New Guinea* 82/1). Division of Wildlife, Konedobu, 78 pp.

Parker, H.W. (1925a) A collection of reptiles and batrachians from Tonkin. *Annals and Magazine of Natural History*, London (Series 9) 15(86), 300–306.

Parker, H.W. (1925b) New and rare reptiles and batrachians from Madagascar. *Annals and Magazine of Natural History*, London (Series 9) 16(94), 390–394.

Parker, H.W. (1926a) New reptiles and a new frog from Queenland. *Annals and Magazine of Natural History*, London (Series 9) 17(102), 665–670, 3 figs.

Parker, H.W. (1926b) The reptiles and batrachians of Gorgona Island, Colombia. *Annals and Magazine of Natural History*, London (Series 9) 17(101), 549–554.

Parker, H.W. (1926c) Description of a new snake from Trinidad. *Annals and Magazine of Natural History*, London (Series 9) 18(104), 205–207.

Parker, H.W. (1927a) Parallel evolution in some opisthoglyphous snakes, with the description of a new species. *Annals and Magazine of Natural History*, London (Series 9) 20(115), 81–86, 3 figs.

Parker, H.W. (1927b) A new blind snake from Madagascar. *Annals and Magazine of Natural History*, London (Series 9) 19(111), 379–380, 1 fig.

Parker, H.W. (1928) Notes on reptiles and batrachians from Mato Grosso and E. Bolivia. *Annals and Magazine of Natural History*, London (Series 10) 2(7), 96–99, 1 fig.

Parker, H.W. (1929) A new opisthoglyphous snake from West Africa. *Annals and Magazine of Natural History*, London (Series 10) 4(23), 449–451.

Parker, H.W. (1930a) A new colubrine snake from Ecuador. *Annals and Magazine of Natural History*, London (Series 10) 5(26), 207–209, 2 figs.

Parker, H.W. (1930b) Two new reptiles from southern Ecuador. *Annals and Magazine of Natural History*, London (Series 10) 5(29), 568–571.

Parker, H.W. (1930c) Three new reptiles from southern Arabia. *Annals and Magazine of Natural History*, London (Series 10) 6(35), 594–598, 1 fig.

Parker, H.W. (1930d) Three new reptiles from Somaliland. *Annals and Magazine of Natural History*, London (Series 10) 6(35), 603–606.

Parker, H.W. (1931a) Some reptiles and amphibians from S.E. Arabia. *Annals and Magazine of Natural History*, London (Series 10) 8(47), 514–522.

Parker, H.W. (1931b) A new species of blind snake from N.W. Australia. *Annals and Magazine of Natural History*, London (Series 10) 8(48), 604–605, 2 figs.

Parker, H.W. (1931c) Reports of an expedition to Brazil and Paraguay in 1926–27, supported by the Trustees of the Percy Sladen Memorial Fund and the Executive Committee of the Carnegie Trust for Scotland. Amphibia and Reptilia. *Journal of the Linnaean Society (Zoology)*, London 37(252), 285–289, pl. 16.

Parker, H.W. (1932a) Some new or rare reptiles and amphibians from southern Ecuador. *Annals and Magazine of Natural History*, London (Series 10) 9(49), 21–26.

Parker, H.W. (1932b) Two collections of reptiles and amphibians from British Somaliland. *Proceedings of the Zoological Society of London* 102(2), 335–367, 3 figs.

Parker, H.W. (1932c) Scientific results of the Cambridge Expedition to the East African lakes, 1930–1. 5. Reptiles and amphibians. *Journal of the Linnaean Society (Zoology)*, London 38(257), 213–229, 10 figs.

Parker, H.W. (1933a) Some amphibians and reptiles from the Lesser Antilles. *Annals and Magazine of Natural History*, London (Series 10) 11(61), 151–158.

Parker, H.W. (1933b) Some West African snakes. *Annals and Magazine of Natural History*, London (Series 10) 12(71), 544–548, 2 figs.

Parker, H.W. (1934) Reptiles and amphibians from southern Ecuador. *Annals and Magazine of Natural History*, London (Series 10) 14(80), 264–273.

Parker, H.W. (1935) The frogs, lizards, and snakes of British Guiana. *Proceedings of the Zoological Society of London* 105(3), 505–530.

Parker, H.W. (1936a) A collection of reptiles and amphibians from the mountains of British New Guinea. *Annals and Magazine of Natural History*, London (Series 10) 17(97), 66–93.

Parker, H.W. (1936b) Some extinct snakes of the West Indies. *Annals and Magazine of Natural History*, London (Series 10) 18(104), 227–233.

Parker, H.W. (1936c) Reptiles and amphibians collected by the Lake Rudolph Rift Valley Expedition, 1934. *Annals and Magazine of Natural History*, London (Series 10) 18(108), 594–609.

Parker, H.W. (1936d) Dr. Karl Jordan's expedition to South-West Africa and Angola, herpetological collections. *Novitates Zoologicae*, London (1936–1937) 40(1), 115–146, 2 figs.

Parker, H.W. (1936e) Revised list of reptiles (excluding chelonians) and amphibians collected in the Seychelles. *Transactions of the Linnaean Society of London* 19(4), 444–446.

Parker, H.W. (1938a) Reptiles and amphibians from the southern Hejaz. *Annals and Magazine of Natural History*, London (Series 11) 1(5), 481–492.

Parker, H.W. (1938b) The vertical distribution of some reptiles and amphibians in southern Ecuador. *Annals and Magazine of Natural History*, London (Series 11) 2(11), 438–450.

Parker, H.W. (1939) Reptiles and amphibians from Bougainville, Solomon Islands. *Bulletin du Musée Royal d'Histoire Naturelle de Belgique*, Bruxelles 15(60), 1–5, 2 figs.

Parker, H.W. (1940) Undescribed anatomical structures and new species of reptiles and amphibians. *Annals and Magazine of Natural History*, London (Series 11) 5(27), 257–274, 3 figs.

Parker, H.W. (1941) British Museum (Natural History) expedition to South-West Arabia, 1937–8. Reptiles and Amphibians. *Bulletin of the British Museum (Natural History)*, London 1(2), 3–6.

Parker, H.W. (1949) The snakes of Somaliland and the Sokotra Islands. *Zoologische Verhandelingen*, Leiden (6), 1–115, 11 figs.

Parker, H.W. & Grandison, A.G.C. (1977) *Snakes: a natural history.* Second edition. British Museum (Natural History), London, 108 pp., 29 figs., 16 col. pls. [first edition, 1965, Parker, H.W., 95 pp., 18 figs., 6 pls.]

Parker, H.W., Bellairs, A.A. & Saint-Girons, H. (1971) Les amphibiens et les reptiles. *In:* Carrington, R., Matthews, L.H. & Young, J.Z. (Eds.), *Le grande encyclopédie de la nature. Tome 10.* Editions Rencontre, Paris, pp. 391–767.

Parker, W.S. (1982) *Masticophis taeniatus. Catalogue of American amphibians and reptiles*, pp. 304.1–304.4.

Parker, W.S. & Brown, W.S. (1980) Comparative ecology of two colubrid snakes, *Masticophis t. taeniatus* and *Pituophis melanoleucus deserticola*, in northern Utah. *Milwaukee Public Museum, Publications in Biology and Geology* (7), 1–104, 35 figs.

Parkinson, C.L. (1999) Molecular systematics and biogeographical history of pitvipers as determined by mitochondrial ribosomal DNA sequences. *Copeia* 1999(3), 576–586, 4 figs.

Parkinson, C.L., Moody, S.M & Ahlquist, J.E. (1997) Phylogenetic relationships of the "*Agkistrodon* complex" based on mitochondrial DNA sequence data. *In:* Thorpe, R.S., Wüster, W. & Malhotra, A. (Eds.), *Ecology, evolution and snakebite venomous snakes*. Zoological Society of London, Clarendon Press, Oxford, pp. 63–78, 4 figs.

Parkinson, C.L., Zumudio, K.R. & Greene, H.W. (2000) Phylogeography of the pitviper clade *Agkistrodon*: historical ecology, species status, and conservation of cantils. *Molecular Ecology* 9(4), 411–420, 4 figs.

Parmalee, P.W. (1955) Reptiles of Illinois. *Illinois State Museum Popular Science Series* 5, 1–88, illust.

Parmley, D. (1984) Herpetofauna of the Coffee Ranch local fauna (Hemphillian Land Mammal Age) of Texas. *In:* Horner, N.V. (Ed.), *Festschrift for Watter W. Dalguest.* Midwestern State University Press, Wichita Falls, pp. 97–106.

Parmley, D. (1987) *Lampropeltis similis* from the Coffee Ranch Local fauna (Hemphillian Land Mammal age) of Texas. *Texas Journal of Science* 39(2), 123–128, 2 figs.

Parmley, D. (1988) Early Hemphillian (late Miocene) snakes from the Higgins Local fauna of Lipscomb County, Texas. *Journal of Vertebrate Paleontology* 8(3), 322–327, 4 figs.

Parmley, D. (1990) Late Pleistocene snakes from Fowlkes Cave, Culberson County, Texas. *Journal of Herpetology* 24(3), 266–274, 5 figs.

Parmley, D. (1994) Reevaluation of the extinct kingsnake *Lampropeltis intermedius* Brattstrom (Colubridae) with comments on the ancestry of *Lampropeltis triangulum. Herpetological Natural History* 2(2), 83–88, 7 figs.

Parmley, D. (2002) First record of a longnose snake (*Rhinecheilus lecontei*) from the Miocene of North America. *Herpetological Natural History* 9(1), 83–85, 1 fig.

Parmley, D. & Case, G.R. (1988) Palaeopheid snakes from the Gulf coastal region of North America. *Journal of Vertebrate Paleontology* 8(3), 334–339, 5 figs.

Parmley, D. & Holman, J.A. (1995) Hemphillian (late Miocene) snakes from Nebraska, with comments on Arikareean through Blancan snakes of midcontinental North America. *Journal of Vertebrate Paleontology* 15(1), 79–95, 10 figs.

Parmley, D. & Holman, D. (2003) *Nebraskophis* Holman from the late Eocene of Georgia (of USA), the oldest known North American colubrid snake. *Acta Zoologica Cravoviensia* 46(1), 1–8.

Parmley, D. & Holman, J.A. (2007) Earliest fossil record of a pigmy rattlesnake (Viperidae, *Sistrurus* Garman). *Journal of Herpetology* 41(1), 141–144, 1 fig.

Parmley, D. & Hunter, K.B. (2010) Fossil snakes of the Clarendonian (Late Miocene) Pratt Slide Local Fauna of Nebraska, wilth the description of a new Natricine colubrid. *Journal of Herpetology* 44(4), 526–543, 9 figs.

Parmley, D. & Peck, D. (2002) Amphibians and reptiles of the late Hemphillian White Cone Local Fauna, Navajo County, Arizona. *Journal of Vertebrate Paleontology* 22(1), 175–178.

Parmley, D. & Walker, D. (1998) *Elaphe vulpina* (Colubridae) from the Blancan of Washington, with a review of the species' paleodistribution. *Herpetological Natural History* 6(1), 41–46, 3 figs.

Parmley, D. & Walker, D. (2003) Snakes of the Pliocene Taunton local fauna of Adams County, Washington with the description of a new colubrid. *Journal of Herpetology* 37(2), 235–244, 3 figs.

Paolucci, L. (1880) Collezione ereptologica. *In: Catalogo sistematico delle collezione del gabinetto di Storia Naturale del R. Istituto Tecnico di Ancona.* Del Commercio, Anciona, 3 pp.

Parpinelli, L. & Marques, O.A.V. (2008) Seasonal and daily activity in the pale-headed blindsnake *Liotyphlops beui* (Serpentes: Anomalepidae) in southeastern Brazil. *South American Journal of Herpetology* 3(3), 207–212, 5 figs.

Parris, D.C. & Holman, J.A. (1978) An Oligocene snake from a coprolite. *Herpetologica* 34(3), 258–264.

Parsons, T.S. (1970) The nose and Jacobson's organ. *In:* Gans, C. & Parsons, T.S. (Eds.), *Biology of the Reptilia. Volume 2 (Morphology B).* Academic Press, London, pp. 99–191, 52 figs.

Pasqual, J.R.H. (1962) The biting mechanism of *Atractaspis*. *Nigerian Field*, Woodchester 27(3), 137–142, 2 figs.

Passos, P. (2001) Geographic distribution: Serpentes: *Drymobius rhombifer*. *Herpetological Review* 32(1), 9.

Passos, P. (2003) Geographic distribution: Serpentes: *Micrurus averyi*. *Herpetological Review* 34(4), 389.

Passos, P. (2008) *Revisão taxonómica do gênero* Atractus *Wagler, 1828 (Serpentes: Colubridae: Dipsadinae).* Doctoral Dissertation, Museo National, Universidade Federal do Rio de Janeiro, Rio de Janeiro, 671 pp.

Passos, P. & Arredondo, J.C. (2009) Rediscovery and redescription of the Andean earth-snake *Atractus wagleri* (Reptilia: Serpentes: Colubridae). *Zootaxa* (1969), 59–68, 16 figs.

Passos, P. & Brandão, F. (2002) Geographic distribution: Serpentes: *Drymoluber dichrous*. *Herpetological Review* 33(4), 324.

Passos, P. & Fernandes, D.S. (2001) Geographic distribution: Serpentes: *Liophis atraventer*. *Herpetological Review* 32(1), 60.

Passos, P. & Fernandes, D.S. (2002a) Geographic distribution: Serpentes: *Philodryas patagoniensis*. *Herpetological Review* 33(1), 69.

Passos, P. & Fernandes, D.S. (2002b) Geographic distribution: Serpentes: *Psomophis joberti*. *Herpetological Review* 33(1), 69.

Passos, P. & Fernandes, D.S. (2005) Variation and taxonomic status of the aquatic coral snake *Micrurus surinamensis* (Cuvier, 1817) (Serpentes, Elapidae). *Zootaxa* (953), 1–14, 10 figs.

Passos, P. & Fernandes, R. (2008a) A new species of the colubrid snake genus *Atractus* (Reptilia: Serpentes) from the central Amazon of Brazil. *Zootaxa* (1849), 59–66, 6 figs.

Passos, P. & Fernandes, R. (2008b) Revision of the *Epicrates cenchria* complex (Serpentes: Boidae). *Herpetological Monographs* (22), 1–30, 13 figs.

Passos, P. & Franco, F.L. (2002) Geographic distribution: Serpentes: *Epicrates cenchria maurus*. *Herpetological Review* 33(1), 67.

Passos, P. & Lynch, J.D. (2010) Revision of *Atractus* (Serpentes: Dipsadidae) from middle and upper Magdalena drainage of Colombia. *Herpetological Monographs* (24), 149–173, 16 figs.

Passos, P. & Prudente, A.L.C. (2012) Morphological variation, polymorphism, and taxonomy of the *Atractus torquatus* complex (Serpentes: Dipsadiae). *Zootaxa* (3407), 1–21, 13 figs.

Passos, P., Fernandes, D.S. & Caramaschi, U. (2004) The taxonomic status of *Leptognathus incertus* Jan, 1863, with revalidation of *Dipsas alternans* (Fischer, 1885) (Serpentes, Colubridae, Dipsadinae). *Amphibia-Reptilia,* Leiden 25(4), 381–393, 5 figs.

Passos, P., Caramaschi, U. & Pinto, R.R. (2005a) Rediscovery and redescription of *Leptotyphlops salgueiroi* Amaral, 1954 (Squamata, Serpentes, Leptotyphlopidae). *Boletim do Museu Nacional*, Rio de Janeiro (520), 1–10, 7 figs.

Passos, P., Fernandes, R. & Porto, M. (2005b) Geographical variation and taxonomy of the snail-eating snake *Dipsas albifrons* (Sauvage, 1884), with comments on the systematic status of *Dipsas albifrons cavalheiroi* Hoge, 1950 (Serpentes, Colubridae, Dipsadinae). *Zootaxa* (1013), 19–34, 6 figs.

Passos, P., Fernandes, R. & Zanella, N. (2005c) A new species of *Atractus* (Serpentes, Colubridae) from southern Brazil. *Herpetologica* 61(2), 209–218, 2 figs.

Passos, P., Caramaschi, U. & Pinto, R.R. (2006) Redescription of *Leptotyphlops koppesi* Amaral, 1954, and description of a new species of the *Leptotyphlops dulcis* group from central Brazil (Serpentes, Leptotyphlopidae). *Amphibia-Reptilia,* Leiden 27(3), 347–353, 4 figs.

Passos, P., Cisneros-Heredia, D.F. & Salazar-V., D. (2007a) Rediscovery and redescription of the rare Andean snake *Atractus modestus*. *Herpetological Journal* 17, 1–6, 4 figs.

Passos, P., Fernandes, D.S & Borges-Nojosa, D.M. (2007b) A new species of *Atractus* (Serpentes, Dipsadinae) from a relictual forest in northeastern Brazil. *Copeia* 2007(4), 788–797, 4 figs.

Passos, P., Lynch, J.D. & Fernandes, R. (2008) Taxonomic status of *Atractus sanctaemartae* and *Atractus nebularis*, and description of a new *Atractus* from the Atlantic coast of Colombia. *Herpetological Journal* 18, 175–186, 5 figs.

Passos, P., Aguayo, R. & Scrocchi, G. (2009a) Rediscovery of the rare *Atractus bocki*, with assessment of the taxonomic status of *Atractus canedii* (Serpentes: Colubridae: Dipsadinae). *Journal of Herpetology* 43(4), 710–715, 4 figs.

Passos, P., Fuenmayor, G.R. & Barrio-Amorgós, C.L. (2009b) Description of two new species from Venezuela in the highly diverse dipsadine genus *Atractus* (Serpentes: Colubridae). *Amphibia-Reptilia,* Leiden 30(2), 233–243, 4 figs.

Passos, P., Arredondo, J.C., Fernandes, R. & Lynch, J.D. (2009c) Three new *Atractus* (Serpentes: Dipsadidae) from the Andes of Colombia. *Copeia* 2009(3), 425–436, 8 figs.

Passos, P., Chiesse, A., Torres-Carvajal, O. & Savage, J.M. (2009d) Testing species boundaries within the *Atractus occipitoalbus* complex (Serpentes: Dipsadidae). *Herpetologica* 65(4), 384–403, 8 figs.

Passos, P., Mueses-Cisneros, J.J., Lynch, J.D. & Fernandes, R. (2009e) Pacific lowland snakes of the genus *Atractus* (Serpentes: Dipsadidae), with description of three new species. *Zootaxa* (2293), 1–34, 18 figs.

Passos, P., Dobiey, M. & Ventegas, P.J. (2010a) Variation and natural history notes on giant groundsnake, *Atractus gigas* (Serpentes: Dipsadidae). *South American Journal of Herpetology* 5(2), 73–82, 6 figs.

Passos, P., Fernandes, R., Bérnils, R.S. & Moura-Leite, J.C. de. (2010b) Taxonomic revision of the Brazilian Atlantic Forest *Atractus* (Reptilia: Serpentes: Dipsadidae). *Zootaxa* (2364), 1–63, 20 figs.

Passos, P., Ramos, L. & Pereira, D.N. (2012a) Distribution, natural history, and morphology of the rare snake, *Caaeteboia amarali* (Serpentes: Dipsadidae). *Salamandra* 48(1), 51–57, 7 figs.

Passos, P., Cisneros-Heredia, D., Rivera, D.E., Aguilar, C. & Schargel, W.E. (2012b) Rediscovery of *Atractus microrhynchus* and reappraisal of the taxonomic status of *A. emersoni* and *A. natans* (Serpentes: Dipsadidae). *Herpetologica* 63(3), 375–392, 9 figs..

Pasteur, G. (1967) Un serpent endémique du Maghreb: *Spalerosophis dolichospilus* (Werner), colubridé. *Bulletin du Muséum National d'Histoire Naturelle*, Paris (Série 2) 39(3), 444–451.

Pasteur, G. & Bons, J. (1960) Catalogue des reptiles actuels du Maroc. Revision de formes d'Afrique d'Europe et d'Asie. *Travaux de l'Institut Scientifique Chérifien, Series Zoologíque*, Rabat (21), 1–132, 3 figs., 5 pls.

Paterson, W. (1789) *A narrative of four journeys into the country of the Hottentots, and Caffraria. In the years one thousand seven hundred and seventy-seven, eight, and nine.* J. Johnson, London, xii + 171 pp., 17 pls.

Paterson, W. (1790) *A narrative of four journeys into the country of the Hottentots and Caffraria, in the years 1777, 1778, 1779. Second edition.* J. Johnson, London, xii + 175 pp., 19 pls.

Patterson, R.W. (1987) *Reptiles of southern Africa.* C. Struik, Cape Town, 128 pp., 232 col. photos.

Patterson, R.W. & Erasmus, H. (1978) Hatching of Anchieta's dwarf python *Python anchietae. International Zoo Yearbook* 18, 99–101, illust.

Paula, A. de, Zanella, N. & Guaragni, S.A. (2011) *Taeniophallus affinis* (Günther, 1858) (Squamata, Serpentes, Dipsadidae): distribution extension, Rio Grande do Sul, Brazil. *Check List* 7(3), 285–285, 2 figs.

Paulson, D.R. (1968) Variation in some snakes from the Florida Keys. *Quarterly Journal of the Florida Academy of Sciences* (1966) 29, 295–308, 1 fig.

Paulson, D.R. (1971) Key to Washington state reptiles and amphibians. *Pacific Northwest Herpetological Society* 5(1), 27–33.

Paulson, D.R. (1972) Reptiles of Washington. *Pacific Northwest Herpetological Society* 5(2), 13–15.

Pauwels, O.S.G. (1996) Geographic distribution: Serpentes: *Grayia tholloni. Herpetological Review* 27(2), 88.

Pauwels, O.S.G. & Lavoué, S. (2004) Geographic distribution: Serpentes: *Boulengerina annulata annulata. Herpetological Review* 35(3), 290.

Pauwels, O.S.G. & Meirte, D. (1996) Contribution to the knowledge of the Gambian herpetofauna. *British Herpetological Society Bulletin* 56, 27–34, illust.

Pauwels, O.S.G. & Meirte, D. (1997) The status of *Cryptophidion annamense*. *Cryptozoology* 12, 95–100.

Pauwels, O.S.G. & David, P. (2008a) Miscellanea herpetolica Gabonica I. *Hamadryad* 34(1), 22–27.

Pauwels, O.S.G. & David, P. (2008b) Miscellanea herpetolica Gabonica II. *Hamadryad* 34(1), 19–24.

Pauwels, O.S.G. & Salle, B. (2009) Miscellanea herpetolica Gabonica III. *Hamadryad* 32(1), 13–18.

Pauwels, O.S.G. & Sumontha, M. (2003) Geographic distribution: Serpentes: *Typhlops muelleri. Herpetological Review* 34(2), 171.

Pauwels, O.S.G. & Sumontha, M. (2007) Geographic distribution: Serpentes: *Ahaetulla fasciolata. Herpetological Review* 38(2), 220.

Pauwels, O.S.G. & Sumontha, M. (2008) Geographic distribution: Serpentes: *Fordonia leucobalia. Herpetological Review* 39(3), 371.

Pauwels, O.S.G. & Vande weghe, J.P. (2008) *Reptiles du Gabon*, Smithsonian Institution, Washington, D.C., 272 pp.

Pauwels, O.S.G., Grootaert, P. & Siyomwan, P. (2001a) Geographic distribution: Serpentes: *Ramphotyphlops braminus. Herpetological Review* 33(4), 279.

Pauwels, O.S.G., David, P. & Nutphand, W. (2001) First record of *Xenochrophis punctulatus* (Günther, 1858) (Serpentes: Colubridae: Natricinae) from Thailand. *Hamadryad* 26(2), 247–252.

Pauwels, O.S.G., Laohawat, O.-A., Naaktae, W., Puangjit, C., Wisutharom, T., Chimsunchart, C. & David, P. (2002a) Reptile and amphibian diversity in Phang-aga province, southern Thailand. *The Natural History Journal of Chalalongkorn University*, Bangkok 2(1), 25–30.

Pauwels, O.S.G., Wallach, V., David, P. & Chanhome, L. (2002b) A new species of *Oligodon* Fitzinger, 1826 (Serpentes, Colubridae) from southern Peninsular Thailand. *The Natural History Journal of Chulalongkorn University*, Bangkok 2(2), 7–18, 5 figs.

Pauwels, O.S.G., David, P. & Chanhome, L. (2004a) Geographic distribution: Serpentes: *Elaphe taeniura ridleyi. Herpetological Review* 35(3), 291.

Pauwels, O.S.G., Sumontha, M. & David, P. (2004b) Geographic distribution: Serpentes: *Xenochrophis punctulatus. Herpetological Review* 35(3), 294.

Pauwels, O.S.G., David, P., Chanhome, L., Vogel, G., Chan-ard, T. & Orlov, N.L. (2005) On the status of *Boiga ocellata* Kroon, 1973, with the designation of a neotype for *Boiga siamensis* Nootpand, 1971 (Serpentes, Colubridae). *Russian Journal of Herpetology* 12(2), 102–106, 2 figs.

Pauwels, O.S.G., Kunya, K., David, P. & Sumontha, M. (2009a) First record of the Yunnan keelback *Sinonatrix yunnanensis* Rao & Yang, 1998 (Serpentes: Natricidae) from Thailand. *Salamandra* 45(3), 165–169, 3 figs.

Pauwels, O.S.G., Kunya, K., David, P., Waengsothorn, S. & Sumontha, M. (2009b) On the occurrence of *Amphiesma khasiense* (Serpentes: Natricidae) in Thailand. *Hamadryad* 34(1), 75–79, 4 figs.

Pauwels, O.S.G., Tobi, E. & Sarrazin, P. (2011) Geographic distribution: Serpentes: *Ramphotyphlops braminus. Herpetological Review* 42(3), 395.

Pauwels, O.S.G., Sumontha, M. & Wallach, V. (2012) Geographic distribution: Serpentes: *Ramphotyphlops braminus. Herpetological Review* 43(3), 450.

Pavloff, P.A. (1926) The Manchurian Reptilia and Amphibia based upon collections stored in the M.R.S. Museum. *Miscellaneous Papers of the Manchuria Research Society, Natural History Section* 13A(1), 1–22, 2 pls. (in Russian)

Pavloff, P.A. (1932) Listes des sauriens et serpents des collections du Musée Hoang ho Pai ho de Tien Tsin. *Publications du Musée Hoang ho Pai ho de Tien Tsin* 12, 1–27, errata.

Pavón-Vázquez, C.J., García-Vázquez, U.O., Blancas-Hernández, J.C. & Nieto-Montes de Oca, A. (2011) A new species of the *Geophis sieboldi* group (Squamata: Colubridae) exhibiting color pattern polymorphism from Guerrero Mexico. *Herpetologica* 67(3), 332–343, 4 figs.

Pearson, D.J. (1993) Distribution, status and conservation of pythons in Western Australia. *In:* Lunney, D. & Ayers, D. (Eds.), *Herpetology in Australia: a diverse discipline*. Royal Zoological Society of New South Wales, Mosman, pp. 383–395, 7 figs.

Pefaur, J.E. (1992) Checklist and bibliography (1960-1985) of the Venezuelan herpetofauna. *Smithsonian Herpetological Information Service* (89), 1–54.

Pefaur, J.E. & Rivero, J.A. (2000) Distribution, species-richness, endemism, and conservation of Venezuelan amphibians and reptiles. *Amphibian & Reptile Conservation* 2(2), 42–70, 2 figs., 46 col. pls.

Pefaur, J.E., Davila, J., Lopez, E. & Nuñez, A. (1978) Distribucion y clasificacion de los reptiles del Departamento de Arequipa. *Bulletin de l'Institut Français d'Études Andines* 7(1–2), 129–139, illust.

Pel, H.S. (1851) Over de jagt aan de Goudkust, volgens eene tien jarige, eigene ondervinding. *Nederlandsch Tijdschrift voor Jagtkunde* 1(4–5), 149–173.

Pellegrin, J. (1909) Reptiles du Soudan récoltés par la mission Tilho-Gaillard. Description d'une espèce nouvelle. *Bulletin du Muséum d'Histoire Naturelle*, Paris 15(7), 413–415.

Pellegrin, J. (1910) Description d'une variété nouvelle de l'*Oligodon herberti* Boulenger, provenant du Tonkin. *Bulletin de la Société Zoologique de France*, Paris 35(2), 30–32.

Pellegrin, J. (1925) Liste des reptiles, batraciens et poissons d'eau douce des collections du Musée de L'Institut Scientifique Chérifien a Rabat. *Bulletin de la Société des Sciences Naturelles du Maroc*, Rabat 5(7–8), 315–321.

Pellegrin, J. (1927) Les reptiles et les batraciens de l'Afrique du nord française. *Compte Rendu de l'Association Française pour l'Avancement des Sciences*, Paris 51, 260–264.

Peñaranda-B., E.A., Calvo-C., A. & Peñaranda, J. (1994) Biología reproductiva de *Bothrops neuwiedi* (Serpentes, Viperidae) en cautiverio: resultados preliminares. *Ecología en Bolivia* 23, 57–70.

Peng, F.-T. & Chang, F.-W. (1968) A survey of suborder Serpentes in Kiangsi Province. *In: Communist Chinese Scientific Abstracts (30th Annual China Zoological Society Conference, December 1965). No. 2.* Joint Publications Research Service, Washington, pp. 276–277.

Peng, G. & Fuji, Z. (2001) Comparative studies on hemipenes of four species of *Trimeresurus* (*sensu stricto*) (Serpentes, Crotalinae). *Amphibia-Reptilia,* Leiden 22(1), 113–117, 1 fig.

Pennant, T. (1792) *Arctic zoology. Vol. II. Class II. Birds. Class III. Reptiles. Class IV. Fish. Class V. Insects.* Robert Faulder, London, 418 + 13 pp.

Pentecost, E.D. & Vogt, R.C. (1976) Amphibians and reptiles of the Lake Michigan drainage basin. *Environmental Status of the Lake Michigan Region* 16, 1–69, 58 figs.

Peracca, M.G. (1892) Descrizione di nuove specie di Rettili e Anfibi di Madagascar. *Bolletino dei Musei di Zoologia ed Anatomia Comparata della Università di Torino* 7(112), 1–5.

Peracca, M.G. (1894) Viaggio del Dr. E. Festa in Palestina, nel Libano e regioni vicine. VI. Rettili ed anfibi. *Bolletino dei Musei di Zoologia ed Anatomia Comparata della Università di Torino* 9(167), 1–20.

Peracca, M.G. (1896a) Sopra un nuovo genere ed una nuova specie di colubride aglifo dell'America meridionale. *Bolletino dei Musei di Zoologia ed Anatomia Comparata della Università di Torino* 11(231), 1–7.

Peracca, M.G. (1896b) Sopra alcuni Ofidii nuovi o poco noti dell'America meriodionale. *Bolletino dei Musei di Zoologia ed Anatomia Comparata della Università di Torino* 11(252), 1–4, 2 figs.

Peracca, M.G. (1896c) Rettili ed anfibi reccolti a Kazungula e sulla strada da Kazungula a Buluwaio dal Rev. Luigi Jalla, Missionario Valdese nell'alto Zambese. *Bolletino dei Musei di Zoologia ed Anatomia Comparata della Università di Torino* 11(255), 1–5, 2 figs.

Percca, M.G. (1896d) Nuovo genere di colubride aglifo dell'America meridionale. *Bolletino dei Musei di Zoologia ed Anatomia Comparata della Università di Torino* 11(266), 1–2, 1 fig.

Peracca, M.G. (1897a) Intorno ad alcuni ofidii raccolti a Maldi (Eritrea) dal Capitano A. Gasca. *Bolletino dei Musei di Zoologia ed Anatomia Comparata della Università di Torino* 12(273), 1–3, 1 fig.

Peracca, M.G. (1897b) Viaggio del Dott. Alfredo Borelli nel Chaco Boliviano e nella Repubblica Argentina. Rettili ed anfibi. *Bolletino dei Musei di Zoologia ed Anatomia Comparata della Università di Torino* 12(274), 1–19.

Peracca, M.G. (1897c) Sopra un nuovo genere di colubride opistoglifo della Repubblica Argentina. *Bolletino dei Musei di Zoologia ed Anatomia Comparata della Università di Torino* 12(278), 1–2.

Peracca, M.G. (1897d) Intorno ad una nuova specie di ofidio di S. Paulo (Brasile) riferibile al gen. *Uromacer* D. & B. *Bolletino dei Musei di Zoologia ed Anatomia Comparata della Università di Torino* 12(282), 1–2, 1 fig.

Peracca, M.G. (1897e) Intorno ad una piccola raccolta di rettili di Conocacco (Perù Orientale). *Bolletino dei Musei di Zoologia ed Anatomia Comparata della Università di Torino* 12(284), 1–7, 1 fig.

Peracca, M.G. (1897f) Viaggio del Dr. Enrico Festa nell'Ecuador e regioni vicine. IV. Rettili. *Bolletino dei Musei di Zoologia ed Anatomia Comparata della Università di Torino* 12(300), 1–20.

Peracca, M.G. (1899) Reptiles et batraciens de l'Archipel Malais. *Revue Suisse de Zoologie* 7(2), 321–330.

Peracca, M.G. (1904a) Viaggio del Dr. A. Borelli nel Matto Grosso brasiliano e nel Paraguay, 1899. IX Rettili ed amfibii. *Bolletino dei Musei di Zoologia ed Anatomia Comparata della Università di Torino* 19(460), 1–15.

Peracca, M.G. (1904b) Nouvelles espèces d'ophidiens d'Asie et d'Amerique, faisant partie de la collection du Museum d'Histoire naturelle de Genève. *Revue Suisse de Zoologie*, Genèva 12(3–4), 663–668.

Peracca, M.G. (1909) Rettili ed amfibii. *In:* Filippi, F. de & Amedeo di Savoia, L. (Eds.), *Risultati scientifici della Spedizione Condotta da S.A.R. il Duca degli Abruzzi al Ruwenzori. Volume I. Zoologia, botanica.* Ulrico Hoepli, Milano, 165–180.

Peracca, M.G. (1910) Descrizione di alcune nuove specie di ofidii del Museo Zoologico della R. Universita di Napoli. *Annuario del Museo Zoologico della Reale Università di Napoli* (Series 2) 3(12), 1–3.

Peracca, M.G. (1912a) Rettili ed anfibii raccolti durante I viaggi di S.A.R. la Duchessa Elena d'Aosta nella regione dei grandi laghi dell'Africa equatoriale. *Annuario del Museo Zoologico della Reale Università di Napoli* (Series 2) 3(25), 1–8.

Peracca, M.G. (1912b) Reptiles et batraciens de Colombie. *In:* Fuhrmann, O. & Mayor, E. (Eds.), Voyage d'exploration scientifique en Colombie. *Mémoires de la Société Neuchâteloise des Sciences Naturelles* (1914), 5, 96–111.

Pereira-Filho, G.A. & Montingelli, G.G. (2006) Geographic distribution: Serpentes: *Hydrodynastes gigas. Herpetological Review* 37(4), 497.

Pérez, C.H.F. (2012) *Philodryas agassizi* (Jan, 1863) (Squamata: Dipsadidae): distribution extension. *Check List* 8(1), 143–144, 2 figs.

Pérez, D.R. & Avila, L.J. (2000) Geographic distribution: Serpentes: *Bothrops neuwiedi diporus*. *Herpetological Review* 31(4), 254.

Pérez, D.R., Martinez, L.R. & Avila, L.J. (2003) Geographic distribution: Serpentes: *Bothrops neuwiedi diporus*. *Herpetological Review* 34(4), 386.

Pérez, F.P. (2001) *Reptiles de Bizkaia*. Diputación Foral de Bizkaia, Bilbao, 157 pp., col. illust.

Pérez-Bravo, G. (1978) Segundo hallazgo de *Helicops hogei* Lancini, 1964 (Serpentes, Colubridae). *Memórias do Instituto Butantan*, São Paulo (1976–1977) 40–41, 313–315, 2 figs.

Pérez-Higareda, G. (1981) The occurrence of *Pituophis melanoleucus sayi* in Veracruz (Serpentes, Colubridae). *Bulletin of the Maryland Herpetological Society* 17(2), 64–67, 1 fig.

Pérez-Higareda, G. (1985) A new subspecies of the genus *Tantillita* from southern Veracruz, Mexico (Serpentes, Colubridae). *Bulletin of the Maryland Herpetological Society* 21(1), 38–40, 1 fig.

Pérez-Higareda, G. & Smith, H.M. (1986) The status of the Los Tuxtlas (Mexico) false coral snakes (*Pliocercus*). *Bulletin of the Maryland Herpetological Society* 22(3), 125–130.

Pérez-Higareda, G. & Smith, H.M. (1988) Notes on two species of *Geophis* (Serpentes) of southern Mexico. *Southwestern Naturalist* 33(3), 388–390.

Pérez-Higareda, G. & Smith, H.M. (1989) The distribution of *Leptophis diplotropis* (Serpentes, Colubridae). *Bulletin of the Maryland Herpetological Society* 25(3), 73–76.

Pérez-Higareda, G. & Smith, H.M. (1990) The endemic coral snakes of the Los Tuxtlas region, southern Veracruz, Mexico. *Bulletin of the Maryland Herpetological Society* 26(1), 5–13, 3 figs.

Pérez-Higareda, G. & Smith, H.M. (1991) *Ofidiofauna de Veracruz: Análisis taxonómico y zoogeográfico. Ofidiofauna of Veracruz: taxonomical and zoogeographical analysis*. Universidad Nacional Autónoma de México, Mexico City, 122 pp., 1 fig., 46 col. photos.

Pérez-Higareda, G. & Smith, H.M. (1999) A major range extension of the Yucatecan [*sic*] blind snake, *Typhlops microstomus*, (Serpentes, Typhlopidae). *Bulletin of the Maryland Herpetological Society* 35(3), 79–80.

Pérez-Higareda, G., Smith, H.M. & Smith, R.B. (1985a) A new species of *Tantilla* from Veracruz, México. *Journal of Herpetology* 19(2), 290–292, 2 figs.

Pérez-Higareda, G., Smith, H.M. & Juliá-Zertuche, J. (1985b) A new jumping viper, *Porthidium olmec*, from southern Veracruz, México (Serpentes, Viperidae). *Bulletin of the Maryland Herpetological Society* 21(3), 97–106, 1 fig.

Pérez-Higareda, G., Vogt, R.C. & Flores-Villela, O.A. (1987) *Lista anotada de los anfibios y reptiles de la region de Los Tuxtlas, Veracruz. Estacion de Biologia Tropical "Los Tuxtlas," Veracruz*, 23 pp.

Pérez-Higareda, G., Smith, H.M. & López-Luna, M.A. (2001) A new *Geophis* (Reptilia, Serpentes) from southern Veracruz, Mexico. *Bulletin of the Maryland Herpetological Society* 37(2), 43–48, 1 fig.

Pérez-Higareda, G., López-Luna, M.A. & Smith, H.M. (2002a) A new species of *Conophis* (Reptilia, Serpentes) from Los Tuxtlas, an area of high endemism in southern Veracruz, Mexico. *Bulletin of the Maryland Herpetological Society* 38(1), 27–31, 4 figs.

Pérez-Higareda, G., López-Luna, M.A. & Smith, H.M. (2002b) A new snake related to *Sibon sanniola* (Serpentes: Dipsadidae) from Los Tuxtlas, Veracruz, Mexico. *Bulletin of the Maryland Herpetological Society* 38(2), 62–65.

Pérez-Higareda, G., López-Luna, M.A., Chiszar, D. & Smith, H.M. (2002c) Additions to and notes on the herpetofauna of Veracruz, Mexico. *Bulletin of the Chicago Herpetological Society* 37(4), 67–68, 2 figs.

Pérez-Higareda, G., López-Luna, M.A. & Smith, H.M. (2007) *Serpientes de la region de Los Tuxtlas, Veracruz, México: guía de identificación ilustrada*. Universidad Nacional Autónoma de México, Mexico City, 189 pp., 82 figs., 53 col. pls.

Perez-Ramos, E., Saldaña de la Riva, L. & Uribe-Peña, Z. (2000) A checklist of the reptiles and amphibians of Guerrero, Mexico. *Anales del Instituto Biologia, Universidad Nacional Autónoma de México, Série Zoología* 71(1), 21–40.

Pérez-Santos, C.E. (1983) *Aportacion al conocimiento de la zoogeografia de los ofidia en Colombia*. Tesina de Licenciatura. Universidad Complutense de Madrid.

Pérez-Santos, C.E. (1986) *Zoogeografia de los Ophidia en Colombia*. 2 vols. Doctoral Dissertation, Universidad Complutense de Madrid, Madrid, 971 pp.

Pérez-Santos, C.E. & Moreno, A.G. (1986) Distribución altitudinal de las serpientes del Colombia. *Revista Española de Herpetología* 1, 11–27.

Pérez-Santos, C.E. & Moreno, A.G. (1988) Ofidios de Colombia. *Museo Regionale di Scienze Naturali Torino, Monografias* 6, 6–517, 121 figs., 87 pls.

Pérez-Santos, C.E. & Moreno, A.G. (1989a) Addenda y corrigenda al libro "Ofidios de Colombia." *Bolletino dei Museo Regionale di Scienze Naturali Torino* 7(1), 1–17, 23 pls.

Pérez-Santos, C.E. & Moreno, A.G. (1989b) Una nueva especie de *Thamnodynastes* (Serpentes, Colubridae) en el norte de Colombia. *Bolletino dei Museo Regionale di Scienze Naturali Torino* 7(1), 1–9, 5 figs.

Pérez-Santos, C.E. & Moreno, A.G. (1991) Serpientes de Ecuador. *Museo Regionale di Scienze Naturali Torino, Monografias* 11, 538 pp., 114 figs., 150 col. pls.

Perez-Santos, C.E., Moreno, A.G. & Gearhart, G. (1993) Checklist of the snakes of Panama. *Revista Española de Herpetología* 7, 113–122.

Perkins, C.B. (1949a) The snakes of San Diego County with descriptions and key. Second edition. *Bulletin of the Zoological Society of San Diego* (23), 1–77, illust.

Perkins, C.B. (1949b) A key to the snakes of the United States. *Bulletin of the Zoological Society of San Diego* (24), 4–79, illust.

Pernetta, J.C. (1977) Observations on the habits and morphology of the sea snake *Laticauda colubrina* (Schneider) in Fiji. *Canadian Journal of Zoology*, Ottawa 55(10), 1612–1619, 2 figs.

Perret, J.-L. (1960) Une nouvelle et remarquable espece d'*Atractaspis* (Viperidae) et quelques autres serpents d'Afrique (*A. coalescens* sp. nov.). *Revue Suisse de Zoologie*, Genèva 67, 129–139.

Perret, J.-L. (1961) Études herpétologiques Africaines III. *Bulletin de la Société Neuchâteloise des Sciences Naturelle* (Series 3) 84, 133–138, 1 pl.

Perret, J.-L. & Mertens, R. (1957a) Étude d'une collection herpétologique faite au Cameroun de 1952 à 1955. *Bulletin de l'Institut Française d'Afrique Noire*, Dakar 19A(2), 548–601.

Perret, J.-L. & Mertens, R. (1957b) Revision du matériel herpétologique du Cameroun, étudié par A. Monard. *Revue Suisse de Zoologie*, Genève 64(3), 73–78.

Perrier, J.O.E. (1928) Développement embryogénique des vertébrés allantoïdiens: les reptiles. *In:* Grassé, P.P. (Ed.), *Traité de zoologie. Fascicule VIII (Développement embryogénique des vertébrés allantoïdiens). Les reptiles.* Masson et Cie., Éditeurs, Paris, pp. 2885–3118.

Perry, P.S. & Dexter, M.H. (1989) *Snakes and lizards of Minnesota.* Minnesota Department of Natural Resources, 25 pp., illust.

Persons, T.B. & Rosen, P.C. (2001) Geographic distribution: Serpentes: *Thamnophis cyrtopsis. Herpetological Review* 32(2), 125.

Pesantes-Segura, O. (2000) *Prevención y tratamiento de accidents por serpientes venenosas.* Centro Ecológico Recreacional Huastipa, Sedapal, Lima, 38 pp.

Pestov, M.V. (1999) *Amphibians and reptiles of the Nizhny Novgorod region.* Dront, Nizhny Novgorod, 44 pp. (in Russian)

Pestov, M.V., Mannapova, E.I., Ushakov, V.A., Katunov, L.P., Bakka, S.V., Deoelinskii, A.A. & Turutina, D.V. (2001) *Amphibians and reptiles of the Nizhny Novgorod region: materials for the kadaster.* Dront, Nizhny Novrogod, 178 pp., illust. (in Russian)

Peters, H.-J. (1997) Die Schlangen—Teil I. *Tarantulas of the World* 11, 3-11.

Peters, J.A. (1948) A new snake of the genus *Typhlops* from the Solomon Islands. *Occasional Papers of the Museum of Zoology, University of Michigan* (508), 1–3, 1 pl.

Peters, J.A. (1950) A new snake of the genus *Coniophanes* from Veracruz, Mexico. *Copeia* 1950(4), 279–280.

Peters, J.A. (1951) The proper citation for certain species described by Tschudi. *Copeia* 1951(1), 70.

Peters, J.A. (1952) Catalogue of type specimens in the herpetological collections of the University of Michigan Museum of Zoology. *Occasional Papers of the Museum of Zoology, University of Michigan* (539), 1–55.

Peters, J.A. (1953) A fossil snake of the genus *Heterodon* from the Pliocene of Kansas. *Journal of Paleontology* 27(3), 328–331, 6 figs.

Peters, J.A. (1954) The amphibians and reptiles of the coast and coastal sierra of Michoacán, México. *Occasional Papers of the Museum of Zoology, University of Michigan* (554), 1–37.

Peters, J.A. (1955) Herpetological type localities in Ecuador. *Revista Ecuatoriana de Entomología y Parasitología*, Quito (1954–1955) 2(3–4), 335–352.

Peters, J.A. (1956) An analysis of variation in a South American snake, Catesby's snail-sucker (*Dipsas catesbyi* Sentzen). *American Museum Novitates* (1783), 1–41, 18 figs.

Peters, J.A. (1957a) Taxonomic notes on Ecuadorian snakes in the American Museum of Natural History. *American Museum Novitates* (1851), 1–13.

Peters, J.A. (1957b) A new snake species of the snake genus *Sibon* from Ecuador. *Copeia* 1957(2), 109–111. (*Sibon dunni*)

Peters, J.A. (1960a) The snakes of Ecuador. A checklist and key. *Bulletin of the Museum of Comparative Zoology* 122(9), 491–541.

Peters, J.A. (1960b) The snakes of the subfamily Dipsadinae. *Miscellaneous Publications Museum of Zoology, University of Michigan* (114), 1–224, 11 figs., 8 pls.

Peters, J.A. (1963) Taxonomic notes on Ecuadorian snakes. *Beiträge zur Neotropischen Fauna*, Jena 3(1), 57–67.

Peters, J.A. (1964) Supplemental notes on snakes of the subfamily Dipsadinae (Reptilia, Colubridae). *Beiträge zur Neotropischen Fauna*, Jena 4(1), 45–50.

Peters, J.A. (1965) Liste der rezenten Amphibien and Reptilien, Colubridae (Dipsadinae). *Das Tierreich*, Berlin (81), 1–19.

Peters, J.A. (1968) A replacement name for *Bothrops lansbergii venezuelensis* Roze, 1959 (Viperidae, Serpentes). *Proceedings of the Biological Society of Washington* 81, 319–322.

Peters, J.A. & Broadley, D. (1967) The scientific name of the African puff adder. *Copeia* 1967(4), 864–865.

Peters, J.A. & Orejas-Miranda, B.R. (1970) Catalogue of the Neotropical Squamata. Part I. Snakes. *Bulletin of the United States National Museum* (297), 1–347, 41 figs.

Peters, J.A. & Orejas-Miranda, B.R. (1972) The taxonomic validity of *Apostolepis tenuis* Ruthven and *Apostolepis vittata* (Cope) (Serpentes, Colubridae). *Copeia* 1972(3), 588–590.

Peters, J.A. & Orcés-V, G. (1960) *Leptophis cupreus* Cope. A valid South American colubrid species. *Beiträge zur Neotropischen Fauna*, Jena 2(2), 139–141.

Peters, J.L. (1938) The name of the western coral king snake. *Copeia* 1938(2), 93.

Peters, W.C.H. (1854) Diagnosen neuer Batrachier, welche zusammen mit der früher (24. Juli und 17. August) gegebenen Übersicht der Schlangen und Eidechsen mitgetheilt werden. *Berichte über die Bekanntmachunggeeigneten Verhandlungen der Königlich-Preußischen Akademie der Wissenschaften zu Berlin* 1854(11), 614–628.

Peters, W.C.H. (1855) Uebersicht der auf seiner Reise gesammelten Amphibien. *Archiv für Naturgeschichte*, Berlin 21(1), 43–58. [reprint of Peters, 1854]

Peters, W.C.H. (1857a) Über *Amblyodipsas*, eine neue Schlangengattung aus Mossambique. *Monatsberichte der königlich Akademie der Wissenschaften zu Berlin* 1856(12), 592–595, 1 fig.

Peters, W.C.H. (1857b) Vier neue amerikanische Schlangen aus der Familie der Typhlopinen vor und machte dar über einige vorläusige Mittheilungen. *Monatsberichte der königlich Akademie der Wissenschaften zu Berlin* 1857(8), 402–403.

Peters, W.C.H. (1858a) Über die Typhlopinen oder blödsichtigen Schlangen und über neue dahin gehörige Arten. *Monatsberichte der königlich Akademie der Wissenschaften zu Berlin* 1857(11), 508–509.

Peters, W.C.H. (1858b) Über eine neue Schlanengattung, *Rhoptrura. Monatsberichte der königlich Akademie der Wissenschaften zu Berlin* 1858(6), 340–341.

Peters, W.C.H. (1859a) Nachrichten von Hrn. Fedor Jagor, der auf eigne Kosten nach Ostindien und den Philippinen gereist ist und von der Akademie mit Instructionen für das Sammeln naturwissenschaftlicher Gegenstände versehen wurde. *Monatsberichte der königlich Akademie der Wissenschaften zu Berlin* 1859(3), 269–270, 1 pl.

Peters, W.C.H. (1859b) Über die von Hrn. Dr. Hoffmann in Costa Rica gesammelten und an das Königl. zoologische Museum gesandten Schlangen. *Monatsberichte der königlich Akademie der Wissenschaften zu Berlin* 1859(3), 275–278, 1 pl.

Peters, W.C.H. (1859c) Eine neue Schlange aus der Familie der Uropeltacea, *Plectrurus ceylonicus* und einige Notizen über die hieher gehörigen Arten des K. zoologischen Museums. *Monatsberichte der königlich Akademie der Wissenschaften zu Berlin* 1859(5), 388–389.

Peters, W.C.H. (1860a) Über eine neue zu der Gattung *Onychocephalus* gehörige Wurmschlange, *Onychocephalus macrurus*, und verglich sie mit den bisher von dieser Gattung bekannt gewordenen Arten. *Monatsberichte der königlich Akademie der Wissenschaften zu Berlin* 1860(2), 77–83, 1 pl.

Peters, W.C.H. (1860b) Eine neue Gattung von Riesenschlangen, welche von einem geborenen Preußen, Hrn. Carl Reiss, in Guayaquil nebst mehreren anderen werthvollen Naturalien dem zoologischen Museum zugesandt worden ist. *Monatsberichte der königlich Akademie der Wissenschaften zu Berlin* 1860(4), 200–202, 1 pl.

Peters, W.C.H. (1860c) Eine neue Gattung von Schlangen, *Elapochrus*, aus Mexico. *Monatsberichte der königlich Akademie der Wissenschaften zu Berlin* 1860(6), 293–294, 1 pl.

Peters, W.C.H. (1860d) Drei neue Schlangen des k. zoologischen Museums aus America vor und fügtehieran Bemerkungen über die generelle Unterscheidung von anderen bereits bekannten Arten. *Monatsberichte der königlich Akademie der Wissenschaften zu Berlin* 1860(10), 517–521, 1 pl.

Peters, W.C.H. (1861a) Die Beschreibung von zwei neuen Schlangen, vor *Mizodon variegatus* aus Westafrika und *Bothriopsis quadriscutata. Monatsberichte der königlich Akademie der Wissenschaften zu Berlin* 1861(3), 358–360.

Peters, W.C.H. (1861b) Über eine Sammlung von Schlangen aus Huanusco in Mexico, welche das Königl. zoologische Museum kürzlich von Hrn. Dr. Hille erworben hat. *Monatsberichte der königlich Akademie der Wissenschaften zu Berlin* 1861, 460–462.

Peters, W.C.H. (1861c) Eine zweite Übersicht (vergl. Monatsberichte 1859 p. 269) der von Hrn. F. Jagor auf Malacca, Java, Borneo und den Philippinen gesammelten und dem Kgl. zoologischen Museum übersandten Schlangen. *Monatsberichte der königlich Akademie der Wissenschaften zu Berlin* 1861(4), 683–691.

Peters, W.C.H. (1861d) Über eine neue Art der Schlangengattung *Silybura, S. macrolepis. Monatsberichte der königlich Akademie der Wissenschaften zu Berlin* 1861(10), 901–905.

Peters, W.C.H. (1861e) Über neue Schlangen des Königl. zoologischen Museums, *Typhlops striolatus, Geophidium dubium, Streptophorus (Ninia) maculatus, Elaps hippocrepis. Monatsberichte der königlich Akademie der Wissenschaften zu Berlin* 1861(10), 922–925.

Peters, W.C.H. (1861f) Über zwei neue von Hrn. Dr. Gundlach auf Cuba entdeckte Schlangen, *Tropidonotus cubanus* und *Cryptodacus vittatus. Monatsberichte der königlich Akademie der Wissenschaften zu Berlin* 1861(11), 1001–1004.

Peters, W.C.H. (1861g) On the Asiatic snake called *Taphrometopon lineolatum* by Professor Brandt. *Proceedings of the Zoological Society of London* 29(1), 47–50, 5 figs.

Peters, W.C.H. (1861h) *De Serpentum familia Uropeltaceorum*. G. Reimer, Berlin, 22 pp., 2 pls.

Peters, W.C.H. (1862a) Über die von dem so früh in Afrika verstorbenen Freiherrn von Barnim und Dr. Hartmann auf ihrer Reise durch Aegypten, Nubien und dem Sennâr gesammelten Amphibien. *Monatsberichte der königlich Akademie der Wissenschaften zu Berlin* 1862(1), 271–279, 1 pl.

Peters, W.C.H. (1862b) Präparate von zur craniologischen Unterscheidung der Schlangengattung *Elaps* und machte eine mittheilung über eine neue Art der Gattung *Simotes, S. semicinctus. Monatsberichte der königlich Akademie der Wissenschaften zu Berlin* 1862(11), 635–638.

Peters, W.C.H. (1862c) Über die craniologischen Verschiedenheiten der Grubenottern (Trigonocephali) und über eine neue Art der Gattung *Bothriechis. Monatsberichte der königlich Akademie der Wissenschaften zu Berlin* 1862(12), 670–674.

Peters, W.C.H. (1863a) Eine übersicht der von Hrn. Richard Schomburgk an das zoologische Museum eingesandten Amphibien, aus Buchsfelde bei Adelaide in Südaustralien. *Monatsberichte der königlich Akademie der Wissenschaften zu Berlin* 1863(4), 228–236.

Peters, W.C.H. (1863b) Über die von Hrn. Dr. F. Jagor in Siam gesammelten Amphibien. *Monatsberichte der königlich Akademie der Wissenschaften zu Berlin* 1863(6), 245–247.

Peters, W.C.H. (1863c) Über einige neue oder weniger bekannte Schlangenarten des zoologischen Museums zu Berlin. *Monatsberichte der königlich Akademie der Wissenschaften zu Berlin* 1863(6), 272–289.

Peters, W.C.H. (1863d) Über eine neue Schlangengattung, *Styporhynchus*, und verschiedene andere Amphibien des zoologischen Museums. *Monatsberichte der königlich Akademie der Wissenschaften zu Berlin* 1863(10), 399–413.

Peters, W.C.H. (1864a) Über neue Amphibien (*Typhloscincus, Typhlops, Asthenodipsas, Ogmodon*). *Monatsberichte der königlich Akademie der Wissenschaften zu Berlin* 1864(4), 271–276, 1 pl.

Peters, W.C.H. (1864b) Über einige neue Säugethiere (*Mormops, Macrotus, Vesperus, Molossus, Capromys*), Amphibien (*Platydactylus, Otocryptis, Euprepes, Ungalia, Dromicus, Tropidonotus, Xenodon, Hylodes*) und Fische (*Sillago, Sebastes, Channa, Myctophum, Carassius, Barbus, Capoëta, Poecilia, Saurenchelys, Leptocephalus*). *Monatsberichte der königlich Akademie der Wissenschaften zu Berlin* 1864(6), 381–399.

Peters, W.C.H. (1864c) Über eine neue Art der Baumviper, *Atheris polylepis*, aus Liberia. *Monatsberichte der königlich Akademie der Wissenschaften zu Berlin* 1864(11), 642–645.

Peters, W.C.H. (1864d) [Der] Kieferapparat einer neuen Schlangengattung, *Ogmodon*, von den Fidgiinseln. *Sitzungsberichte der Gesellschaft Naturforschender Freunde zu Berlin* 1864(4), 8.

Peters, W.C.H. (1865) Einen ferneren Nachtrag zu seiner Abhandlung über Typhlopina. *Monatsberichte der königlich Akademie der Wissenschaften zu Berlin* 1865(6), 259–263, 1 pl.

Peters, W.C.H. (1866) Über neue Amphibien (*Amphibolurus, Lygosoma, Cyclodus, Masticophis, Crotaphopeltis*) und Fische (*Diagramma, Hapalogenys*) des Kgl. zoologischen Museum. *Monatsberichte der königlich Akademie der Wissenschaften zu Berlin* 1866(2), 86–96.

Peters, W.C.H. (1867a) Eine vorläufige Übersicht der aus dem Nachlass des Baron Carl von der Decken stammenden und auf seiner ostafrikanischen Reise gesammelten Säugethiere und Amphibien. *Monatsberichte der königlich Akademie der Wissenschaften zu Berlin* 1866, 884–892.

Peters, W.C.H. (1867b) Herpetologische Notizen. *Monatsberichte der königlich Akademie der Wissenschaften zu Berlin* 1867(1), 13–37.

Peters, W.C.H. (1867c) Über eine Sammlung von Flederthieren und Amphibien aus Otjimbinque in Südwestafrica, welch Hr. Missionär Hahn dem zoologischen Museum zugesandt hat. *Monatsberichte der königlich Akademie der Wissenschaften zu Berlin* 1867(4), 234–237, 1 pl.

Peters, W.C.H. (1867d) Über Flederthiere (*Pteropus Gouldii, Rhinolophus Deckenii, Vespertilio lobipes, Vesperugo Temminckii*) und Amphibien (*Hypsilurus Godeffroyi, Lygosoma scutatum, Stenostoma narirostre, Onychocephalus unguirostris, Ahaetulla polylepis, Pseudechis scutellatus, Hoplobatrachus Reinhardtii, Hyla coriacea*). *Monatsberichte der königlich Akademie der Wissenschaften zu Berlin* 1867(11), 703–712, 3 figs.

Peters, W.C.H. (1868a) Über eine neue Nagergattung, *Chiropodomys penicillatus*, so wie über einige neue oder weniger bekannte Amphibien und Fische. *Monatsberichte der königlich Akademie der Wissenschaften zu Berlin* 1868(7), 448–461, 2 pls.

Peters, W.C.H. (1868b) Über neue Säugethiere (*Colobus, Rhinolophus, Vesperus*) und neue oder weniger bekannte Amphibien (*Hemidactylus, Herpetodryas, Spilotes, Elaphis, Lamprophis, Erythrolamprus*). *Monatsberichte der königlich Akademie der Wissenschaften zu Berlin* 1868(12), 637–642.

Peters, W.C.H. (1869) Über neue Gattungen und neue oder weniger bekannte Arten von Amphibien (*Eremias, Dicrodon, Euprepes, Lygosoma, Typhlops, Eryx, Rhynchonyx, Elapomorphus, Achalinus, Coronella, Dromicus, Xenopholis, Anoplodipsas, Spilotes, Tropidonotus*). *Monatsberichte der königlich Akademie der Wissenschaften zu Berlin* 1869(5), 432–445, 1 pl.

Peters, W.C.H. (1870a) Über mexicanische Amphibien, welche Hr. Berkenbusch in Puebla auf Veranlassung des Hrn. Legationsraths von Schlözer dem zoologischen Museum zugesandt hat. *Monatsberichte der königlich Akademie der Wissenschaften zu Berlin* 1869(12), 874–881.

Peters, W.C.H. (1870b) Über neue Amphibien (*Hemidactylus, Urosaura, Tropidolepisma, Geophis, Uriechis, Scaphiophis, Hoplocephalus, Rana, Entomoglossus, Cystignathus, Hylodes, Arthroleptis, Phyllobates, Cophomantis*) des Königlich zoologischen Museums. *Monatsberichte der königlich Akademie der Wissenschaften zu Berlin* 1870(8), 641–652, 2 pls.

Peters, W.C.H. (1871a) Über eine von Hrn. Dr. Robert Abendroth in dem Hochlande von Peru gemachte Sammlung von Amphibien, welche derselbe dem Königl. zoologischen Museum geschenkt hat. *Monatsberichte der königlich Akademie der Wissenschaften zu Berlin* 1871(8), 397–404.

Peters, W.C.H. (1871b) Über neue Reptilien aus Ostafrika und Sarawak (Borneo), vorzüglich aus der Sammlung des Hrn. Marquis J. Doria zu Genua. *Monatsberichte der königlich Akademie der Wissenschaften zu Berlin* 1871(11), 566–581.

Peters, W.C.H. (1872a) Über einige von Hrn. Dr. A.B. Meyer bei Gorontalo und auf den Togian-Inseln gesammelte Amphibien. *Monatsberichte der königlich Akademie der Wissenschaften zu Berlin* 1872(7), 581–585.

Peters, W.C.H. (1872b) Über drei neue Schlangenarten (*Calamaria bitorques, Stenognathus brevirostris* und *Hemibungarus gemianulis*) von den Philippinen. *Monatsberichte der königlich Akademie der Wissenschaften zu Berlin* 1872(7), 585–587.

Peters, W.C.H. (1872c) Über eine neue von Hrn. Dr. A.B. Meyer auf Luzon entdeckte Art von Eidechsen (*Lygosoma* (*Hinulia*) *leucospilos*) und eine von demselben in Nordcelebes gefundene neue Schlangengattung (*Allophis nigricaudus*). *Monatsberichte der königlich Akademie der Wissenschaften zu Berlin* 1872(8), 684–687.

Peters, W.C.H. (1872d) Übersicht der von den Herren Mse. G. Doria und Dr. O. Beccari in Sarawack auf Borneo von 1865 bis 1868 gesammelten Amphibien. *Annali del Museo Civico di Storia Naturale di Genova* 3, 27–45, pls. 2–6.

Peters, W.C.H. (1873a) Über den *Hydrus fasciatus* Schneider und einige andere Seeschlangen. *Monatsberichte der königlich Akademie der Wissenschaften zu Berlin* 1872(12), 848–861, 2 pls.

Peters, W.C.H. (1873b) Über zwei Giftschlangen aus Afrika und über neue oder weniger bekannte Gattungen und Arten von Batrachiern. *Monatsberichte der königlich Akademie der Wissenschaften zu Berlin* 1873(5), 411–418, 4 pls.

Peters, W.C.H. (1873c) Über eine neue Schildrötenart, *Cinosternon Effeldtii* und einige andere neue oder weniger bekannte Amphibien. *Monatsberichte der königlich Akademie der Wissenschaften zu Berlin* 1873(10), 603–618, 1 pl.

Peters, W.C.H. (1874a) Über eine von Hrn. F. Pollen und van Dam auf Madagascar und anderen ostafrikanischen Inseln gemachte Sammlung von Amphibien. *Monatsberichte der königlich Akademie der Wissenschaften zu Berlin* 1873(12), 792–795.

Peters, W.C.H. (1874b) Über neue Reptilien (*Peropus, Agama, Euprepes, Lygosoma, Typhlops, Heterolepis*) der herpetologischen Sammlung des Berliner zoologischen Museum. *Monatsberichte der königlich Akademie der Wissenschaften zu Berlin* 1874(2), 159–164, 1 pl.

Peters, W.C.H. (1874c) Über einige neue Reptilien (*Lacerta, Eremias, Diploglossus, Euprepes, Lygosoma, Sepsina, Ablepharus, Simotes, Onychocephalus*). *Monatsberichte der königlich Akademie der Wissenschaften zu Berlin* 1874(6), 368–377, 1 pl.

Peters, W.C.H. (1875) Über die von Hrn. Professor Dr. R. Buchholz in Westafrika gesammelten Amphibien. *Monatsberichte der königlich Akademie der Wissenschaften zu Berlin* 1875(3), 196–212, 3 pls.

Peters, W.C.H. (1876a) Eine zweite Mittheilung über die von Hrn. Professor Dr. R. Buchholz in Westafrica gesammelten Amphibien. *Monatsberichte der königlich Akademie der Wissenschaften zu Berlin* 1876(2), 117–123, 1 pl.

Peters, W.C.H. (1876b) Über die von S.M.S. Gazelle mitgebrachten Amphibien. *Monatsberichte der königlich Akademie der Wissenschaften zu Berlin* 1876(8), 528–535, 1 pl.

Peters, W.C.H. (1877a) Herpetologische Notizen. II. Bemerkungen über neue oder weniger bekannte Amphibien. *Monatsberichte der königlich Akademie der Wissenschaften zu Berlin* 1877(6), 415–423, 1 pl.

Peters, W.C.H. (1877b) Über die von Hrn. Prof. Dr. K. Möbius 1874 auf den Maskarenen und Seychellen, sowie über die von Hrn. Dr. Sachs im vorigen Jahr in Venezuela gesammelten Amphibien. *Monatsberichte der königlich Akademie der Wissenschaften zu Berlin* 1877(7), 455–460, 1 pl.

Peters, W.C.H. (1877c) Übersicht der Amphibien aus Chinchoxo (Westafrika), welche von der Africanischen Gesellschaft dem Berliner zoologischen Museum übergeben sind. *Monatsberichte der königlich Akademie der Wissenschaften zu Berlin* 1877(10), 611–620, 1 pl.

Peters, W.C.H. (1877d) Nachtrag [Übersich der Amphibien aus Chinchoxo (Westafrika), welche von der Africanischen Gesellshaft dem Berliner zoologischen Museum übergeben sind]. *Monatsberichte der königlich Akademie der Wissenschaften zu Berlin* 1877(10), 620–621.

Peters, W.C.H. (1878) Über die von Hrn. J.M. Hildebrandt während seiner letzten ostafrikanischen Reise gesammelten Säugethiere und Amphibien. *Monatsberichte der königlich Akademie der Wissenschaften zu Berlin* 1878(3), 194–209, 2 pls.

Peters, W.C.H. (1879) Über neue Amphibien des Kgl. zoologischen Museums (*Euprepes, Acontias, Typhlops, Zamenis, Spilotes, Oedipus*). *Monatsberichte der königlich Akademie der Wissenschaften zu Berlin* 1879(8), 773–779, 1 pl.

Peters, W.C.H. (1880a) Über neue oder weniger bekannte Amphibien des Berliner zoologischen Museums (*Leptosoma dispar, Monopeltis* [*Phractogonus*] *jugularis, Typhlops depressus, Leptocalamus trilineatus, Xenodon punctatus, Elapomorphus erythronotus, Hylomantis fallax*). *Monatsberichte der königlich Akademie der Wissenschaften zu Berlin* 1880(2), 217–224, 1 pl.

Peters, W.C.H. (1880b) Über die von Hrn. Gerhard Rohlfs und Dr. A. Stecker auf der Reise nach der Oase Kurfa gesammelten Amphibien. *Monatsberichte der königlich Akademie der Wissenschaften zu Berlin* 1880(3), 305–309, 1 pl.

Peters, W.C.H. (1880c) Über die von Hrn. J.M. Hildebrandt auf Nossi-Bé und Madagascar gesammelten Säugethiere und Amphibien. *Monatsberichte der königlich Akademie der Wissenschaften zu Berlin* 1880(6), 508–511.

Peters, W.C.H. (1881a) Eine neue Gattung von Geckonen, *Scalabotes thomensis,* welche Hr. Professor Dr. Greeff in Marburg auf der westafrikanischen Insel St. Thomé entdeckt hat, und über die Stellung von *Elaps Sundevallii* Smith, eine von Wahlberg im Kafferlande gefundene Art von Schlangen. *Monatsberichte der königlich Akademie der Wissenschaften zu Berlin* 1880(10), 795–798, 1 pl.

Peters, W.C.H. (1881b) Über das Vorkommen schildförmiger Verbreiterungen der Dornfortsätze bei Schlangen und über neue oder weniger bekannte Arten dieser Abtheilung der Reptilien. *Sitzungsberichte der Gesellschaft Naturforschender Freunde zu Berlin* 1881(3), 49–52.

Peters, W.C.H. (1881c) Einige herpetologische Mittheilungen, 1. Uebersicht der zu den Familien der Typhlopes und Stenostomi gehorigen Gattungen oder Untergattungen. *Sitzungsberichte der Gesellschaft Naturforschender Freunde zu Berlin* 1881(4), 69–71.

Peters, W.C.H. (1881d) Herpetologische Mittheilungen. 2. Ueber zwei neue Arten der Schlangengattung *Psammophis* und über die Synonymie von zwei Arten der Lycodonten. *Sitzungsberichte der Gesellschaft Naturforschender Freunde zu Berlin* 1881(6), 88–90.

Peters, W.C.H. (1881e) Zwei neue von Herrn Major von Mechow während seiner letzten Expedition nach West-Afrika entdeckte Schlangen von und gab Uebersicht der von ihm mitgebrachten herpetologischen Sammlung. *Sitzungsberichte der Gesellschaft Naturforschender Freunde zu Berlin* 1881(9), 147–150.

Peters, W.C.H. (1882a) Über eine neue Gattung und Art der Vipernattern, *Dinodipsas angulifera*, aus Südamerika. *Monatsberichte der königlich Akademie der Wissenschaften zu Berlin* 1882(40), 893–896, pl. 15.

Peters, W.C.H. (1882b) Eine neue Art oder bemerkenswerthe Varietät der Schlangengattung *Macroprotodon* Guichenot, *M. maroccanus*, aus Marocco. *Sitzungsberichte der Gesellschaft Naturforschender Freunde zu Berlin* 1882(2), 27.

Peters, W.C.H. (1882c) Den Namen der Batrachiergattung *Hylonomus* in *Hyloscirtus* zu ändern und legte zwei neue Arten von Schlangen, *Microsoma notatum* und *Liophis Ygraecum* vor. *Sitzungsberichte der Gesellschaft Naturforschender Freunde zu Berlin* 1882(8), 127–129.

Peters, W.C.H. (1882d) *Naturwissenschaftliche Reise nach Mossambique auf Befehl seiner Majestät des Königs Friedrich Wilhelm IV. In den Jahren 1842 bis 1848 ausgeführt von Wilhelm C.H. Peters. Zoology. III. Amphibien*. Berlin, xv + 191 pp., 33 pls.

Peters, W.C.H. (1883) Über *Opisthoplus degener*, eine neue Gattung und Art der Schlangen mit ganz eigenthümlicher Bezahnung. *Sitzungsberichte der Gesellschaft Naturforschender Freunde zu Berlin* 1882(52), 1147–1150, 4 figs.

Peters, W.C.H. & Doria, G. (1878) Catalogo dei rettili e dei batraci raccolti da O. Beccari, L.M. d'Albertis e A.A. Bruijn nella sotto-regione Austro-Malese. *Annali del Museo Civico di Storia Naturale di Genova* 13, 323–450, 7 pls.

Petersen, R.C. & Fritsch, R.W., II. (1986) *Connecticut's venomous snakes: the timber rattlesnake and northern copperhead.* Second edition. State Geological and Natural History Survey of Connecticut Bulletin (111), vii + 48 pp., 18 figs., 8 col. pls. [first edition, 1970, Bulletin (103), 1–40]

Pethon, P. (1969) List of type specimens of fishes, amphibians and reptiles in the Zoological Museum, University of Oslo. *Rhizocrinus, Occasional Papers, Zoological Museum, University of Oslo* 1(1), 1–17.

Petkovski, S., Sidorovska, V. & Dzukie, G. (2000) The biodiversity richness of the Macedonian snake fauna (Reptilia: Serpentes). *Ekologija Zastita na Zivotnata Sredina* 7(1–2), 41–54. (in Slovakian)

Petrusev, S., Dzukic, G. & Petkovski, S. (1990) Further reports on the morphological characters and distribution of *Eryx jaculus turcicus* (Olivier, 1801) in Yugoslavia. *Fragmenta Balcanica Musei Macedonici Scientiarum Naturalium*, Skopje 14(15), 139–150, 4 figs.

Petzold, H.G. (1967) Some remarks on the breeding biology and the keeping of *Tretanorhinus variabilis*, a water snake of Cuba. *Herpetologica* 23(3), 242–246.

Petzold, H.G. (1983) *Die Anacondas: Gattung Eunectes*. A. Ziemsen [Die Neue Brehm-Bücherei Nr. 554], Wittenberg Lutherstadt, 142 pp., 40 figs.

Pfaff, J.R. (1950) Amphibians and reptiles. *In: List of Danish Vertebrates*. Dansk Videnskabs Forlag, Copenhagen, pp. 62–66.

Pfeffer, G.J. (1893) Ostafrikanische Reptilien und Amphibien, gesammelt von Herrn Dr. F. Stuhlmann im Jahre 1888 und 1889. *Jahrbuch der Hamburgischen Wissenschaftlichen Anstalten*, Hamburg (1892) 10(1), 69–105, 2 pls.

Pflaumer, C. (1943) Sobre las culebras chilenas. *Revista Universitaria*, Santiago 28(1), 137–141.

Philibosian, R. & Yntema, J.A. (1976) Record and status of some reptiles and amphibians in the Virgin Islands. I. 1968–1975. *Herpetologica* 32(1), 81–85.

Phelps, T. (1981) *Poisonous snakes.* Blanford Press, Poole, 237 pp., 70 figs., 33 col. pls.

Phelps, T. (2010) *Old World vipers: a natural history of the Azemiopinae and Viperinae.* Edition Chimaira, Frankfurt-am-Main, 558 pp., 556 col. photos.

Philippi, R.A. (1873) Über dia *Boa* der westlichen Provinzen der Argentinischen Republik. *Zeitschrift Gesammten Naturwissenschaften*, 41, 127–130, pl. 3.

Philippi, R.A. (1899) Sobre las serpientes de Chile. *Anales de la Universidad de Chile*, Santiago 104, 715–723.

Phillips, C.A. (2003) *Herpetological type-specimens in the University of Illinois Museum of Natural History: an update to the primary types.* Available online at http://www.inhs.illinois.edu/animals_plants/herps/UIPrimaryTypes.pdf

Phillips, C.A., Brandon, R.A. & Moll, E.O. (1999) *Field guide to amphibians and reptiles of Illinois.* Illinois Natural History Survey, Champaign, xiv + 282 pp., col. illust.

Phillips, S. (1993) Aspects of the distribution, ecology and morphology of Ingram's brown snake *Pseudonaja ingrami* (Elapidae). *In:* Lunney, D. & Ayers, D. (Eds.), *Herpetology in Australia; a diverse discipline*. Royal Zoological Society of New South Wales, Mosman, pp. 349–352, 2 figs.

Phillips, W.W.A. (1958) Some observations on the fauna of the Maldive Islands. Part IV—Amphibians and reptiles. *The Journal of the Bombay Natural History Society* 55(2), 217–220.

Phipson, H.M. (1887) The poisonous snakes of the Bombay Presidency. *The Journal of the Bombay Natural History Society* 2(4), 244–250.

Phipson, H.M. (1888) The poisonous snakes of the Bombay Presidency. *The Journal of the Bombay Natural History Society* 3(1), 49–53.

Phisalix, M. (1902) Relation des parentée entre le deux espécie indigenes vipere (*Vipera aspis* et *Vipera berus*). Utilité des charactères physiologique dans la classification. *Bulletin du Muséum d'Histoire Naturelle*, Paris 8(1), 102–106, 2 figs.

Phisalix, M. (1917a) La glande parotide venimeuse des colubridés aglyphes. *Bulletin du Muséum d'Histoire Naturelle*, Paris 23, 331–336.

Phisalix, M. (1917b) Sur la présense de glandes parotides dans des familles de serpents ou elle n'a pas encore été signalée jusqu'ici. *Bulletin du Muséum d'Histoire Naturelle*, Paris 23, 337–342.

Phisalix, M. (1917c) Propriétés venimeuses de la salive parotidienne chez des colubrides des genres *Tropidonotus, Zamenis, Helicops, Lycodon* et *Dendrophis. Bulletin du Muséum d'Histoire Naturelle*, Paris 23, 343–350.

Phisalix, M. (1922) *Animaux venimeux et venins: la fonction venimeuse chez tous les animaux; les appareils venimeux, les venins et leurs propriétés; les fonctions et usages des venins; l'envenimation et son traitement. Tome II.* Masson & Cie., Éditeurs, Paris, xii + 864 pp., 289 figs., 13 pls.

Picado, C. (1976) *Serpientes venenosas de Costa Rica: sus venenos seroterapica anti-ofídica.* Second edition. Editorial Universidad de Costa Rica, San José, x + 241 pp., 58 figs. [first edition, 1931, Secretaria de Salubridad y Protección Social, 219 pp., 58 figs.]

Pickwell, G. (1947) *Amphibians and reptiles of the Pacific states.* Stanford University Press, Stanford, xiv + 235 pp., 20 figs., 64 pls. [reprinted in 1972 by Dover Books, Dover]

Pickwell, G.V. & Culotta, W.A. (1980) *Pelamis. P. platurus. Catalogue of American amphibians and reptiles*, pp. 255.1–255.4.

Pickwell, G., Bezy, R.L. & Fitch, J.E. (1983) Northern occurrences of the sea snake, *Pelamis platurus*, in the eastern Pacific, with a record of predation on the species. *Caifornia Fish and Game* 69(3), 172–177.

Pienaar, U. de V., Haake, W.D. & Jacobsen, N.H.G. (1983) *The reptile fauna of the Kruger National Park.* Third edition. National Parks Board of South Africa, Pretoria, ii + 231 pp., 102 col. pls. [first edition, Pienaar & FitzSimons, 1966, Koedoe (1), 1–223, 63 figs., 100 pls.; second edition, 1978, vii + 222 pp., 99 pls.]

Pietsch, T.W. & Anderson, W.D., Jr. (Eds.). (1997) *Collection building in ichthyology and herpetology.* American Society of Ichthyologists and Herpetologists Special Publication (3), 1–593, illust.

Pifano, F. & Römer, M. (1949a) Ofidios ponzoñosos de Venezuela. I. Nueva comprobación en Venezuela del *Bothrops medusa* (Sternfeld, 1920) Amaral, 1929 y redescripción de la especie. *Archivos Venezolanos de Patología Tropical y Parisitología Medica*, Caracas 1(2), 277–300, 5 figs.

Pifano, F. & Römer, M. (1949b) Ofidios ponzoñosos de Venezuela. II. Sobre las sierpentes ponzoñosos Venezolanas del grupo *Bothrops lansbergii. Archivos Venezolanos de Patología Tropical y Parisitología Medica*, Caracas 1(2), 301–326, 6 figs.

Pifano, F., Römer, M. & Sandner-Montilla, F. (1950) Serpientes ponzoñosas de Venezuela. III. *Bothrops schlegelii* (Berthold, 1846) Jan, 1875, su existencia en Venezuela. *Archivos Venezolanos de Patología Tropical y Parisitología Medica*, Caracas 2(1), 257–263, 2 figs.

Piguet, P. (1955) Au sujet de *Coluber diadema* (Schlegel). *Bulletin de la Société d'Histoire Naturelle de l'Afrique du Nord*, Alger (1954) 45, 335–338. [not seen]

Pillet, J.-M. (1994) Nouvelles données sur la répartition et l'ecologie de la vipère naine du Haut Atlas *Vipera monticola* Saint Girons, 1954. *Revue Suisse de Zoologie*, Genève 101(3), 645–653, 4 figs.

Pinder, M.J. & Mitchell, J.C. (2002) *A guide to the snakes of Virginia.* Virginia Department of Game and Inland Fisherioes, Wildlife Diversity Special Publication (21), 1–32, figs., 64 col. photos.

Pinney, R. (1981) *The snake book.* Doubleday & Company, New York, 248 pp., illust.

Pinou, T. & Dowling, H.G. (1994) The phylogenetic relationships of the Central American snake *Tretanorhinus*, data from morphology and karyology. *Amphibia-Reptilia*, Leiden 15(3), 297–305, 4 figs.

Pinou, T., Hass, C.A. & Maxson, L.R. (1995) Geographic variation of serum albumin in the monotypic snake genus *Diadophis* (Colubridae, Xenodontinae). *Journal of Herpetology* 29(1), 105–110.

Pinou, T., Vicario, S., Marschner, M. & Coccone, A. (2004) Relict snakes of North America and their relationships within Caenophidia, using likelihood-based Bayesian methods on mitochondrial sequences. *Molecular Phylogenetics and Evolution* 32(2), 563–574, 1 fig.

Pinston, H., Craney, E., Pepin, D., Montadert, M. & Duquet, M. (2000) *Amphibiens et reptiles de Franche-Comté: atlas commenté de répartition.* Groupe Naturaliste de Franche-Comté, Bescancon, 116 pp., illust.

Pinto, R.R. & Curcio, F.F. (2011) On the generic identity of *Siagonodon brasiliensis,* with the description of a new leptotyphlopid from central Brazil (Serpentes: Leptotyphlopidae). *Copeia* 2011(1), 53–63, 6 figs.

Pinto, R.R., Passos, P. & Caramaschi, U. (2005) Geographic distribution: Serpentes: *Leptotyphlops brasiliensis. Herpetological Review* 36(2), 203.

Pinto, R.R., Marques, O.A.V. & Fernandes, R. (2010a) Reproductive biology of two sympatric colubrid snakes, *Chironius flavolineatus* and *Chironius quadricarinatus,* from the Brazialian Cerrado domain. *Amphibia-Reptilia,* Leiden 11(3), 463–473, 6 figs.

Pinto, R.R., Passos, P., Caiceido Portilla, J.R., Arredondo, J.C. & Fernandes, R. (2010b) Taxonomy of the threadsnakes of the tribe Epictini (Squamata: Serpentes: Leptotyphlopidae) in Colombia. *Zootaxa* (2724), 1–28, 14 figs.

Pinya, S. & Carretero, M.A. (2011) The Balearic herpetofauna: a species update and a review of the evidence. *Acta Herpetologica* 6(1), 59–80, 2 figs.

Pires-Lima, F.M., Queiroz, I.B. de & Acuña-Juncá, F. (2000) Geographic distribution: Serpentes: *Atractus potschii. Herpetological Review* 31(4), 254.

Pisani, G.R. (1972) Remarks on the holotype of *Virginia valeriae elegans* (Kennicott). *Herpetologica* 28(4), 348.

Pisani, G.R. & Collins, J.T. (1972) The smooth earth snake, *Virginia valeriae* (Baird and Girard), in Kentucky. *Transactions of the Kentucky Academy of Sciences* 32(1–2), 16–25, 4 figs.

Pisani, G.R., Collins, J.T. & Edwards, S.R. (1972) A re-evaluation of the subspecies of *Crotalus horridus. Transactions of the Kansas Academy of Sciences* 75(3), 255–263, 4 figs.

Pitman, C.R.S. (1935–1938) A guide to the snakes of Uganda. *The Uganda Journal* 3, 47–78, 130–148, pls. 1–2 (1935); 3, 211–229, 259–281, pls. 3–5 (1936); 4, 41–64, 126–150, pls. 6–10 (1936); 4, 220–246, 319–349, pls. 11–13 (1937); 5, 27–45, 93–129, pl. 14 (1937); 5, 160–244, pls. 15–17 (1938).

Pitman, C.R.S. (1938) *A guide to the snakes of Uganda.* Uganda Society, Kampala, xxi + 362 pp., col. pls. 1–13 & A–W. [reprint of Pitman, 1935–1938]

Pitman, C.R.S. (1965) Some additional field notes on *Vipera hindii* Boulenger received from C.J.P. Ionides on 15th November 1964. *Journal of the East Africa Natural History Society and National Museum*, Nairobi 25(2), 163.

Pitman, C.R.S. (1974) *A guide to the snakes of Uganda*. Second edition. Wheldon & Wesley, London, xvii + 290 pp., 13 figs., 3 b&w pls., 22 pls., 19 col. pls.

Pitts, R.M., Swataski, F.J. & Scharninghausen, J.S. (1992) Distributional records for several snakes from Honduras. *Texas Journal of Science* 44(3), 210–311.

Pizzatto, L., Marques, O.A. & Martins, M. (2007) Ecomorphology of boine snakes, with emphasis on South American forms. *In:* Henderson, R.W. & Powell, R. (Eds.), *Biology of the boas and pythons*. Eagle Mountain Publishing, Eagle Mountain, pp. 35–49, 14 figs., col. illust.

Placyk, J.S., Jr., Burghardt, G.M., Small, R.L., King, R.B., Casper, G.S. & Robinson, J.W. (2007) Post-glacial recolonization of the Great Lakes region by the common garter snake (*Thamnophis sirtalis*) inferred from mtDNA sequences. *Molecular Phylogenetics and Evolution* 43(2), 452–467, 5 figs.

Plateau, F. (1866) Sur la vision des poissons et des amphibies. *Académie Royale de Belgique, Mémoires Couronnés et Mémoires des Savants Étrangers* 33, 1–45, 1 pl.

Platt, D.R. (1969) Natural history of the hognose snakes, *Heterodon platyrhinos* and *Heterodon nasicus*. *University of Kansas Publications, Museum of Natural History* 18(4), 253–420, 26 figs., 7 pls.

Platt, D.R. (1983) *Heterodon*. *Catalogue of American amphibians and reptiles*, pp. 315.1– 315.2.

Platt, S.G., Meerman, J.C. & Rainwater, T.R. (1999) Diversity, observations, and conservation of the herpetofauna of Turneffe, Lighthouse, and Glovers Atolls, Belize. *British Herpetological Society Bulletin* (66), 1–13.

Pleguezuelos, J.M. (1997) Distribución y biogeografía de los anfibios y reptiles en España y Portugal. *Monografiás de Herpetologia* 3, 1–542, col. illust.

Pleguezuelos, J.M. & Feriche, M. (1999) Reproductive ecology of the horseshoe whip snake (*Coluber hippocrepis*) in the Iberian Peninsula. *Journal of Herpetology* 33(2), 202–207, 3 figs.

Pleguezuelos, J.M. & Feriche, M. (2003) *Anfibios y reptiles*. Diputación de Granada, Granada, 185 pp., col. illust.

Pleguezuelos, J.M. & Moreno, M. (1989) Folidosis, biometria y coloracion de ofidios en el se de la península Iberica, *Malpolon monspessulanus* (Hermann). *Revista Española de Herpetología* 3(2), 183–196, 4 figs.

Pleguezuelos, J.M., Honrubia, S. & Castillo, S. (1994) Diet of the false smooth snake, *Macroprotodon cucullatus* (Serpentes, Colubridae) in the western Mediterranean area. *Herpetological Journal*, London 4(3), 98–105.

Plummer, M.V. (1981) Habitat utilization, diet and movements of a temperate arboreal snake (*Opheodrys aestivus*). *Journal of Herpetology* 15(4), 425–432, 2 figs.

Plummer, M.V. (1984) Female reproduction in an Arkansas population of rough green snakes (*Opheodrys aestivus*). *In:* Seigel, R.A., Hunt, L.E., Knight, J.L., Malaret, L. & Zuschlag, N.L. (Eds.), *Vertebrate ecology and systematics: a tribute to Henry S. Fitch*. Museum of Natural History, University of Kansas, Lawrence, pp. 105–113, 5 figs.

Poche, F. (1903) Einige nothwendige Anderungen in der herpetologischen nomenclatur. *Zoologischer Anzeiger*, Leipzig 26(708), 698–703.

Poche, F. (1904) Richtigstellung eines Gattungsnamens unter den Colubriden. *Zoologischer Anzeiger*, Leipzig 28, 37–38.

Polcyn, M.J., Jacobs, L.L. & Habek, A. (2005) A morphological model and CT assessment of the skull of *Pachyrhachis problematicus* (Squamata, Serpentes), a 98 million year old snake with legs from the Middle East. *Palaeontologia Electronica* 8(1), 1–25, 10 figs.

Pollini, C. (1818) Ulivi e ad alcuni serpenti del Veronese. *Biblioteca Italiana o sia Giornale di Letteratura Scienze ed Arti*, Milano 9, 236–240.

Pomel, A. (1853) *Catalogue méthodique et descriptif des vertébrés fossiles découverts dans le bassin hydrographique supérieur de la Loire, et surtout dans la vallée de son affluent principal, l'Allier*. J.-B. Baillière, Paris, 193 pp.

Ponce, A., Solis, F.A., Ibáñez-D., R. & Jaramillo, C.A. (2008) Geographic distribution: Serpentes: *Ninia sebae*. *Herpetological Review* 39(2), 372.

Ponce-Campos, P. (2003) Geographic distribution: Serpentes: *Enulius oligostichus*. *Herpetological Review* 34(2), 169.

Ponce-Campos, P. & Huerta-Ortega, S. (1998) Geographic distribution: Serpentes: *Dryadophis cliftoni*. *Herpetological Review* 29(3), 176.

Ponce-Campos, P. & Huerta-Ortega, S. (1999) Geographic distribution: Serpentes: *Gyalopion canum*. *Herpetological Review* 30(2), 113.

Ponce-Campos, P. & Huerta-Ortega, S. (2000) Geographic distribution: Serpentes: *Tantilla bocourti bocourti*. *Herpetological Review* 31(2), 115.

Ponce-Campos. & Smith, H.M. (2001) A review of the stripeless snake (*Coniophanes lateritius*) complex of Mexico. *Bulletin of the Maryland Herpetological Society* 37(1), 10–17, 1 fig.

Ponce-Campos, P. & Ustach, P.C. (2004) Geographic distribution: Serpentes: *Pituophis lineaticollis lineaticollis*. *Herpetological Review* 35(2), 192–193.

Ponce-Campos, P., Huerta-Ortega, S.M. & de La Mora-Navarro, H. (2001) Geographic distribution: Serpentes: *Gyalopion canum*. *Herpetological Review* 32(1), 9.

Ponce-Campos, P., Smith, H.M., Harris, H.S. & Chiszar, D. (2004) A review of the taxonomic status of the members of the *Sonora michoacanensis* group (Serpentes: Colubridae). *Bulletin of the Maryland Herpetological Society* 40(3), 144–150, 1 fig.

Pons, A.R. (1965) *Rhinobothryum Bovallii*, Anderson. Género y especies de ofidio nuevos para Venezuela. *Kasmera* 2(1), 99–103.

Pook, C.E., Wüster, W. & Thorpe, R.S. (2000) Historical biogeography of the western rattlesnake (Serpentes, Viperidae, *Crotalus viridis*), inferred from mitochondrial DNA sequence information. *Molecular Phylogenetics and Evolution* 15(2), 269–282, 6 figs.

Pook, C.E., Joger, U., Stümpel, N. & Wüster, W. (2009) When continents collide: phylogeny, historical biogeography and systematics of the medically important viper genus *Echis* (Squamata: Serpentes: Viperidae). *Molecular Phylogenetics and Evolution* 53(3), 792–807, 4 figs.

Pope, C.H. (1928a) Seven new reptiles from Fukien Province, China. *American Museum Novitates* (320), 1–6.

Pope, C.H. (1928b) Four new snakes and a new lizard from south China. *American Museum Novitates* (325), 1–4.

Pope, C.H. (1929) Notes on reptiles from Fukien and other Chinese Provinces. *Bulletin of the American Museum of Natural History* 58(8), 335–487, 19 figs.

Pope, C.H. (1935) *The reptiles of China*. American Museum of Natural History, New York, lii + 604 pp., 78 figs., 27 pls.

Pope, C.H. (1944) *Amphibians and reptiles of the Chicago area*. Chicago Natural History Museum, Chicago, 275 pp., 50 figs., 12 pls.

Pope, C.H. (1961) *The giant snakes: the natural history of the boa constrictor, the anaconda, and the largest pythons, including comparative facts about other snakes and basic information on reptiles in general*. Alfred A. Knopf, New York, xv + 290 pp.

Pope, C.H. & Pope, S.H. (1933) A study of the green pit-vipers of southeastern Asia and Malaysia, commonly identified as *Trimeresurus gramineus* (Shaw), with description of a new species from peninsular India. *American Museum Novitates* (620), 1–12.

Pope, T.E.B. & Dickinson, W.E. (1928) The amphibians and reptiles of Wisconsin. *Bulletin of the Milwaukee Public Museum*, Milwaukee 8(1), 1–138, 27 figs., 21 pls.

Porej, D. (2001) Herpetofauna of the Muthurajawela Marsh, Sri Lanka, with notes on natural history. *Herpetological Natural History* 8(1), 27–35, 3 figs.

Porras, L.W. (1996) Exquisite!!! A kaleidoscope of colors: Honduran milk snakes, *Lampropeltis triangulum hondurensis*. *The Vivarium* 7(5), 28–37, 26 col. photos.

Porras, L.W., McCranie, J.R. & Wilson, L.D. (1979). New distributional records for *Tantilla oolitica* Telford (Reptilia, Serpentes, Colubridae) from the Florida Keys. *Journal of Herpetology* 13(2), 218–220, 1 fig..

Porras, L.W., McCranie, J.R. & Wilson, L.D. (1981) The systematics and distribution of the hognose viper *Bothrops nasuta* Bocourt (Serpentes, Viperidae). *Tulane Studies in Zoology* 22(2), 85–108, illust.

Porter, K.R. (1972) *Herpetology*. W.B. Saunders Company, Philadelphia, xi + 524 pp., illust.

Portevin, G. (1942) *Ce qu'il faut savoir des reptiles et batraciens de France. Savoir en Histoire Naturelle*, Volume XI. Paul Lechevalier, Paris, 133 pp., 40 figs., 8 pls.

Portis, A. (1890) *I rettili Pliocenici del Valdarno superiore e di alcune altre località Plioceniche di Toscana*. Tipi dei Successori le Monnier, Firenze, 32 pp., 2 pls.

Portis, A. (1901) Il *Palaeopython sardus* Port. Nuovo pitonide del Miocene medio della Sardegna. *Bollettino della Società Geologica Italiana,* Roma 20(2), 247–253, 1 fig.

Porto, M. & Caramaschi, U. (1988) Notes on the taxonomic status, and distribution of *Ptychophis flavovirgatus* Gomes, 1915 (Ophidia, Colubridae) (Note). *Anais da Academia Brasileira de Ciencias*, Rio de Janeiro 60(4), 471–475.

Porto, M. & Fernandes, R. (1996) Variation and natural history of the snail-eating snake *Dipsas neivai* (Colubridae, Xenodontinae). *Journal of Herpetology* 30(2), 269–271, 2 figs.

Posada-Arango, A. (1889a) Apuntamientos para la ofiologia colombiana. *Anales de la Academia Medicina de Medellin* 2(2), 45–49.

Posada-Arango, A. (1889b) Note sur quelques solénoglyphes de Colombie. *Bulletin de la Société Zoologique de France*, Paris 14, 343–345.

Posada-Arango, A. (1909) *Con algunos otros escritos suyos sobre diversos temas. Las serpientes*. Estudios Científicos. C.A. Molina, Medellin, 252–258.

Possani, L.D., Varela-Julía, M., Alagon, A., Julía-Zertuche, J. & Fletcher, P.L., Jr. (1978) Aislamento y caracaracterización de una fosfolipasa del venano de *Micrurus fulvius microgalbineos. Congresso Nacional de Zoología,* México City 1, 20–21.

Pottier, G. (2004) *Guide des reptiles & amphibiens de Midi-Pyrénées*. Toulouse, 126 pp., 148 col. photos.

Pottier, G. (2008) *Atlas de repartition des reptiles et amphibiens de Midi-Pyrénées*. 138 pp., col. photos.

Poulakakis, N., Kapli, P., Kardamaki, A., Skourtanioti, E., Göcmen, B., Ilgaz, Ç., Kumlutas, Y., Avci, A. & Lymberakis, P. (2013) Comparative phylogeography of six herpetofauna species in Cyprus: late Miocene to Pleistocene colonization routes. *Biological Journal of the Linnean Society,* London 108(3), 619–635, 4 figs.

Powell, R. (1990) *Elaphe vulpina. Catalogue of American amphibians and reptiles*, pp. 470.1–470.3.

Powell, R. & Henderson, R.W. (1994a) *Ialtris* Cope. *Catalogue of American amphibians and reptiles*, pp. 590.1–590.2.

Powell, R. & Henderson, R.W. (1994b) *Ialtris agyrtes* Schwartz and Rossman. *Catalogue of American amphibians and reptiles*, pp. 591.1–591.2, 2 figs.

Powell, R. & Henderson, R.W. (1994c) *Ialtris parishi* Cochran. *Catalogue of American amphibians and reptiles*, pp. 593.1–593.2, 2 figs.

Powell, R. & Henderson, R.W. (1998a) *Alsophis anomalus*. *Catalogue of American amphibians and reptiles*, pp. 659.1–659.2, 2 figs.

Powell, R. & Henderson, R.W. (1998b) *Alsophis melanichus*. *Catalogue of American amphibians and reptiles*, pp. 660.1–660.2, 2 figs.

Powell, R. & Henderson, R.W. (1999) Herpetology of Navassa Island, West Indies. *Caribbean Journal of Science*, Mayagüez 35(1–2), 1–13, 4 figs.

Powell, R. & Henderson, R.W. (2002) *Tropidophis bucculentus*. *Catalogue of American amphibians and reptiles*, pp. 760.1–760.3, 2 col. figs.

Powell, R. & Henderson, R.W. (Eds.). (2012) Island lists of West Indian amphibians and reptiles. *Bulletin of the Florida Museum of Natural History* 51(2), 85–166, 9 figs.

Powell, R. & Wittenberg, R.D. (1998) *Bothrops caribbaeus*. *Catalogue of American amphibians and reptiles*, pp. 676.1–676.4, 3 figs.

Powell, R., Collins, J.T. & Fish, L.D. (1992a) *Virginia valeriae* Baird and Girard. Smooth Earth Snake. *Catalogue of American amphibians and reptiles*, pp. 552.1–552.6, 5 figs.

Powell, R., Passaro, R.J. & Henderson, R.W. (1992b) Noteworthy herpetological records from Saint Maarten, Netherlands Antilles. *Caribbean Journal of Science*, Mayagüez 28(3–4), 234–235.

Powell, R., Collins, J.T. & Fish, L.D. (1994) *Virginia striatula* (Linnaeus). Rough earth snake. *Catalogue of American amphibians and reptiles*, pp. 599.1–599.6, 4 figs.

Powell, R., Henderson, R.W., Adler, K. & Dundee, H.A. (1996) An annotated checklist of West Indian amphibians and reptiles. *In:* Powell, R. & Henderson, R.W. (Eds.), *Contributions to West Indian herpetology: a tribute to Albert Schwartz*. Society for the Study of Amphibians and Reptiles, pp. 51–93, 8 col. pls.

Powell, R., Ottenwalder, J.A. & Incháustegui, S.J. (1999) The Hispaniolan herpetofauna, diversity, endemism, and historical perspective, with comments on Navassa Island. *In:* Crother, B.I. (Ed.), *Caribbean amphibians and reptiles*. Academic Press, San Diego, pp. 93–168, 4 figs.

Powell, R., Henderson, R.W. & Parmerlee, J.S., Jr. (2005) *The reptiles and amphibians of the Dutch Caribbean: St. Eustatius, Saba, and St. Maarten*. St Eustatius National Parks Foundation (STENAPA), Gallows Bay, 192 pp.

Powell, R., Collins, J.T. & Hooper, E.D., Jr. (2012) *A key to amphibians & reptiles of the continental United States and Canada*. Second edition. University Press of Kansas, Lawrence, 152 pp., 279 figs. [first edition, 1998, 131 pp., 257 figs.]

Powers, A.L. & Banta, B.H. (1974a) Description of a *Phyllorhynchus* from Cerralvo Island, Gulf of California, Mexico. *Great Basin Naturalist* 34(3), 241–244, 4 figs.

Powers, A.L. & Banta, B.H. (1974b) *Chilomeniscus stramineus* (Cope) recorded from Cerralvo Island, Gulf of California, Mexico. *Journal of Herpetology* 8(4), 386–387, 1 fig.

Pozuelo, M. (1974) Biogeografía en la evolución de un grupo de formas de *Coluber* en el Paleártico Occidental. *Doñana, Acta Zoología*, Sevilla 1, 29–49, 6 figs.

Pozzi, A. (1966) Geonemia e catalogo ragionato degli anfibi e dei rettili della Jugoslavia. *Natura*, Milano 57(1), 55 pp., 22 figs.

Pradhan, M.S. (2004) Reptilia. *In:* Alfred, J.R.B. (Ed.), *Fauna of Pench National Park (Maharashtra). Conservation Area Series* 20. Zoological Survey of India, Kolkata, pp. 93–122.

Pradhan, M.S. (2005) Reptilia. *In:* Alfred, J.R.B. (Ed.), *Fauna of Melghat Tiger Reserve. Conservation Area Series* 24. Zoological Survey of India, Kolkata, pp. 165–221.

Pradhan, M.S. (2006) Reptilia. *In:* Alfred, J.R.B. (Ed.), *Fauna of Tadoba-Andhari Tiger Reserve (Maharashtra). Conservation Area Series* 25. Zoological Survey of India, Kolkata, pp. 99–127.

Pradhan, M.S. (2008) Reptilia. *In:* Alfred, J.R.B. (Ed.), *Fauna of Goa. State Fauna Series* 16. Zoological Survey of India, Kolkata, pp. 281–364

Prado, A. (1939). Notas ofiologicas. I. Sôbre as serpents do grupo *Bothrops lansbergii*, com a descrição de uma nova espécie. *Memórias do Instituto Butantan* 12, 1–4.

Prado, A. (1940a) Notas ofiologicas. 2. Observações sobre os ofidios da fauna amazonica, com a descrição de um novo genero e espécie. *Memórias do Instituto Butantan*, São Paulo (1939) 13, 1–6, 1 pl.

Prado, A. (1940b) Notas ofiologicas. 4. Cinco espécies novas de serpentes colombianas do genero *Atractus* Wagler. *Memórias do Instituto Butantan*, São Paulo (1939) 13, 15–19, 5 figs.

Prado, A. (1941a) Notas ofiológicas. 5. Observações sobre serpentes de Colômbia. *Memórias do Instituto Butantan*, São Paulo (1940) 14, 1–11.

Prado, A. (1941b) Notas ofiológicas. 6. Uma nova espécie de colubrideo áglifo de Colômbia. *Memórias do Instituto Butantan*, São Paulo (1940) 14, 13–15.

Prado, A. (1941c) Notas ofiológicas. 7. Sobre a determinação de *Elapomorphus trilineatus* Boulenger e afins. *Memórias do Instituto Butantan*, São Paulo (1940) 14, 19–21, 2 figs.

Prado, A. (1941d) Notas ofiológicas. 8. Dois novos *Atractus* da Colombia. *Memórias do Instituto Butantan*, São Paulo (1940) 14, 25–27, 1 pl.

Prado, A. (1941e) Notas ofiológicas. 10. Outras serpentes da Colombia, com a descrição de uma nova espécie de Boideo. *Memórias do Instituto Butantan*, São Paulo (1940) 14, 35–41, 1 pl.

Prado, A. (1941f) Algumas serpentes colombianas com a descrição, de uma nova especie do gênero *Dipsas*. *Ciência*, México 2(10–12), 345–346.

Prado, A. (1942a) Notas ofiológicas. 12. Considerações em torno de dois *Atractus* da Colômbia, com a descrição de uma nova espécie. *Memórias do Instituto Butantan*, São Paulo (1941) 15, 377–380, 1 pl.

Prado, A. (1942b) Serpentes do genero *Dryophylax*, com a descrição de uma nova espécie. *Ciencia*, México 3(7), 204–205, 2 figs.

Prado, A. (1943) Notas ofiológicas. 14. Comentários acêrca de algumas serpentes opistóglifas do gênero *Apostolepis*, com a descrição de uma nova espécie. *Memórias do Instituto Butantan*, São Paulo (1942) 16, 7–9, 1 pl.

Prado, A. (1944a) Serpentes da Colombia, com a descrição de duas novas espécies de *Atractus*. *Ciencia*, México 5(4–5), 111, 2 figs.

Prado, A. (1944b) A posição do gênero *Rhadinaea* em sistemática, com a descrição de uma nova espécie. *Memórias do Instituto Butantan* 17, 11–19, pl.

Prado, A. (1945) Um novo *Atractus* da Colombia. *Ciencia*, México 6(2), 61, 1 fig.

Prado, A. (1946a) Notas ofiológicas. 18. A posição do gênero *Rhadinaea* em sistemática (continuação). *Memórias do Instituto Butantan*, São Paulo (1944–1945) 18, 105–107.

Prado, A. (1946b) Notas ofiológicas. 19. *Atractus* da Côlombia com a redescrição de tres novas espécies. *Memórias do Instituto Butantan*, São Paulo (1944) 18, 109–111, 1 pl.

Prado, A. & Hoge, A.R. (1947) Observações sobre serpentes do Perú, com a descrição de uma nova espécie. *Ciencia*, México 8(6–9), 180, 1 fig.

Prado, A. & Hoge, A.R. (1948) Notas ofiológicas. 21. Observações sobre serpentes do Peru. *Memórias do Instituto Butantan*, São Paulo (1947) 20, 283–296.

Prasad, J.N. (1992) Review of distribution of Condanarous sand-snake *Psammophis condanarus* (Merrem). *The Journal of the Bombay Natural History Society* 89(3), 382.

Prater, S.H. (1926) *The snakes of Bombay Island and Salsette*. Prince of Wales' Museum (Natural History), Bombay, 35 pp., 1 fig., 6 pls.

Pratt, A.E. (1892) *To the snows of Tibet through China*. London, xviii + 268 pp., 29 pls.

Pregill, G.K. & Berrian, J.E. (1984) Type specimens of amphibians and reptiles in the San Diego Natural History Museum. *Transactions of the San Diego Society of Natural History* 20(10), 151–164.

Pregill, G.K. & Crother, B.I. (1999) Ecological and historical biogeography of the Caribbean. *In:* Crother, B.I., (Ed), *Caribbean amphibians and reptiles*. Academic Press, San Diego, pp. 335–356, 2 figs.

Pregill, G.K., Steadman, S.L., Olson, S.L. & Grady, F.V. (1988) Late Holocene fossil vertebrates from Burma Quarry, Antigua, Lesser Antilles. *Smithsonian Contributions to Zoology* (463), 1–27.

Preston, W.B. (1982) *The amphibians and reptiles of Manitoba*. Manitoba Museum of Man and Nature, Winnipeg, 128 pp., illust.

Prestt, I. (1971) An ecological study of the viper *Vipera berus* in southern Britain. *Journal of Zoology*, London 164(3), 373–418, 15 figs.

Price, A.H. (1980) *Crotalus molossus*. *Catalogue of American amphibians and reptiles*, pp. 242.1–242.2.

Price, A.H. (1982) *Crotalus scutulatus*. *Catalogue of American amphibians and reptiles*, pp. 291.1–291.2.

Price, A.H. (2009) *Venomous snakes of Texas: a field guide*. Second edition. University of Texas Press, Austin, 116 pp., 36 figs., 39 col. figs. [first edition, 1998, Texas Parks and Wildlife Press, Austin, 112 pp., 29 figs., 16 col. figs.]

Price, M.S. (2009) *A guide to the rock rattlesnakes of the United States*. ECO Herpetological Publishing, Rodeo, 160 pp., col. illust.

Price, R.M. (1981) *Analysis of the ecological and taxonomic correspondence of dorsal snake scale microdermatoglyphics*. Doctoral Dissertation, New York University, New York, xx + 164 pp., 159 figs.

Price, R.M. (1983) Microdermatoglyphics, the *Liodytes-Regina* problem. *Journal of Herpetology* 17(3), 292–294, 6 figs.

Price, R.M. (1987a) Microdermatoglyphics, suggested taxonomic affinities of the viperid genera *Azemiops* and *Pseudocerastes*. *The Snake* 19(1), 47–50, 5 figs.

Price, R.M. (1987b) Disjunct occurrence of mole snakes in peninsular Florida, and the description of a new subspecies of *Lampropeltis calligaster*. *Bulletin of the Chicago Herpetological Society* 22(9), 148.

Price, R.M. (1990a) *Bogertophis*. *Catalogue of American amphibians and reptiles*, pp. 497.1–497.2, 2 figs.

Price, R.M. (1990b) *Bogertophis rosaliae* (Mocquard). *Catalogue of American amphibians and reptiles*, pp. 498.1–498.3, 4 figs.

Price, R.M. (1991) *Senticolis, S. triaspis. Catalogue of American amphibians and reptiles*, pp. 525.1–525.4, 4 figs.

Price, R.M. & Russo, P. (1991) Revisionary comments on the genus *Boa* with the description of a new subspecies of *Boa constrictor* from Peru. *The Snake*, Nittagun 25(1), 29–35.

Prigioni, C., Borteiro, C. & Kolenc, F. (2011) Amphibia and Reptilia, Quebrada de los Cuervos, Departamento de Trienta y Tres, Uruguay. *Check List* 7(6), 763–767, 3 figs.

Pringle, J.A. (1954) The cranial development of certain South African snakes and the relationship of these groups. *Proceedings of the Zoological Society of London* 123(4), 813–865, 42 figs.

Pringle, J.A. (1955) A new sub-species of the spitting cobra, *Naja nigricollis* Reinhardt, from the Cape Province. *Annals of the Natal Museum*, Pietermaritzburg 13, 253–254.

Pritzel, C. (2011) *Cornsnake morph guide: a collector's, breeder's, and buyers's guide to morphology, genetics, pricing, & identification.* Second edition. Charles Pritzel, Palmyra, 130 pp., col. illust. [first edition, 2006, 108 pp., col. illust.]

Procter, J.B. (1918) On the variation of the pit-viper, *Lachesis atrox*. *Proceedings of the Zoological Society of London* 88(1–2), 163–182, 5 figs.

Procter, J.B. (1921) Further lizards and snakes from Persia and Mesopotamia. *The Journal of the Bombay Natural History Society* (1921–1922) 28(1), 251–253.

Procter, J.B. (1922) Description of a new *Typhlops* from Tanganyika Territory. *Annals and Magazine of Natural History*, London (Series 9) 9(54), 685–686.

Procter, J.B. (1923a) On a new genus and species of colubrine snake from S.E. Brazil. *Annals and Magazine of Natural History*, London (Series 9) 11(62), 227–230.

Procter, J.B. (1923b) On new and rare reptiles from South America. *Proceedings of the Zoological Society of London* 93(4), 1061–1067, 2 figs.

Procter, J.B. (1924) Description of a new *Typhlops* from S. India and notes on *Brachyophidium* and *Platyplectrurus*. *Annals and Magazine of Natural History*, London (Series 9) 13(73), 139–142, 2 figs.

Prudente, A.L.C. & Feio, R.N. (2001) Geographic distribution: Serpentes: *Siphlophis longicaudatus*. *Herpetological Review* 2(2), 125.

Prudente, A.L.C. & Passos, P. (2008) New species of *Atractus* Wagler, 1828 (Serpentes: Dipsadinae) from Guyana Plateau in northern Brazil. *Journal of Herpetology* 42(4), 723–732, 3 figs.

Prudente, A.L.C. & Passos, P. (2010) New cryptic species of *Atractus* (Serpentes: Dipsadidae) from Brazilian Amazonia. *Copeia* 2010(3), 397–404, 4 figs.

Prudente, A.L.C. & Santos-Costa, M.C. (2006) A new species of *Atractus* Wagler, 1828 (Serpentes: Colubridae) from eastern Amazonia, Brazil. *Zootaxa* (1285), 21–29, 4 figs.

Prudente, A.L.C., Silva, M.A.A. da, Roche, W.A. da & Franco, F.L. (2008) Morphological variations in *Xenoxybelis boulengeri* (Procter, 1923) (Serpentes, Xenodontinae, Philodryadini). *Zootaxa* (1743), 53–61, 3 figs.

Prudente, A.L.C., Maschio, G.F., Santos-Costas, M.C. dos & Feitosa, D.T. (2010) Serpentes da Bacia Petrolifera de Urucu, Muncicipio de Coari, Amazonas, Brasil. *Acta Amazonica* 40(2), 381–386, 1 fig.

Puorto, G. & Ferrarezzi, H. (1994) Uma nova espécie de *Phalotris* Cope, 1862, com comentários sobre o grupo *bilineatus* (Serpentes, Colubridae, Xenodontinae). *Memórias do Instituto Butantan*, São Paulo (1993) 55(1), 39–46, 2 figs.

Puorto, G. & Henderson, R.W. (1994) Ecologically significant distribution records for the common tree boa (*Corallus enydris*) in Brasil. *Herpetological Natural History* 2(2), 89–91, 2 figs.

Puorto, G., da Graça Salomão, M., Theakston, R.D.G., Thorpe, R.S., Warrell, D.A. & Wüster, W. (2001) Combining mitochondral DNA sequences and morphological data to infer species boundaries, phylogeography of lanceheaded pitvipers in the Brazilian Atlantic forest, and the status of *Bothrops pradoi* (Squamata, Serpentes, Viperidae). *Journal of Evolutionary Biology* 14(4), 527–538, 3 figs.

Purananda, C. (1957) Poisonous snakes of Siam. *Natural History Bulletin of the Siam Society* 18, 1–14, 18 figs.

Pusch, G.G. (1837) *Polens paläontologie oder Abbildung und Beschreibung der vorzüglichsten und der unbeschriebenen Petrefakten ans den Gebirgsformationen in Polen, Volhynien und den Karpathen nebst einigen allgemeinen Beiträgen zur Petrefaktenkunde Europäischen Auer-Ochsen.* E. Schweizerbart's, Stuttgart, pp. 167–218, 16 pls.

Pusch, G.G. (1842) Fossile Batracher-und Ophidier-Reste aus Podolien. *Neues Jahrbuch für Mineralogie, Geognosie, Geologie und Petrefakten-Kunde*, Stuttgart 1842, 179–180.

Pyron, R.A. & Burbrink, F.T. (2009a) Neogene diversification and taxonomic stability in the snake tribe Lampropeltini (Serpentes: Colubridae). *Molecular Phylogenetics and Evolution* 52(3), 524–529, 2 figs.

Pyron, R.A. & Burbrink, F.T. (2009b) Lineage diversification in a widespread species: roles for niche divergence and conservation in the common kingsnake, *Lampropeltis getula*. *Molecular Ecology* 18(19), 3443–3457, 4 figs.

Pyron, R.A. & Burbrink, F.T. (2009c) Systematics of the common kingsnake (*Lampropeltis getula*: Serpentes: Colubridae) and the burden of heritage in taxonomy. *Zootaxa* (2241), 22–32, 3 figs.

Pyron, R.A., Burbrink, F.T., Colli, G.R., Nieto Montes de Oca, A., Vitt, L.J., Kuczynski, C.A. & Wiens, J.J. (2011) The phylogeny of advanced snakes (Colubroidea), with discovery of a new subfamily and comparison of support methods for likelihood trees. *Molecular Phylogenetics and Evolution* 58(3), 329–342, 3 figs.

Pyron, R.A., Dushantha-Kandambi, H.K., Hendry, C.R., Pushpamal, V., Burbrink, F.T. & Somaweera, R. (2013) Genus-level phylogeny of snakes reveals the origins of species richness in Sri Lanka. *Molecular Phylogenetics and Evolution* 66, 969–978, 1 fig.

Quaini, R.O. & Arzamendia, V. (1998) Geographic distribution: Serpentes: *Philodryas olfersii*. *Herpetological Review* 29(1), 54.

Queiroz, A. de & Lawson, R. (1994) Phylogenetic relationships of the garter snakes based on DNA sequence and allozyme variation. *Biological Journal of the Linnean Society,* London 53(3), 209–229, 4 figs.

Queiroz, A. de. & Smith, H.M. (1996) Geographic distribution: Serpentes: *Thamnophis eques*. *Herpetological Review* 27(3), 155.

Queiroz, A. de, Lawson, R. & Lemos-Espinal, J.A. (2002) Phylogenetic relationships of North American garter snakes (*Thamnophis*) based on four mitochondrial genes: how much DNA sequence is enough? *Molecular Phylogenetics and Evolution* 22(2), 315–329.

Quelch, J.J. (1898a) The poisonous snakes of British Guiana. *Timehri* (2)12, 26–36.

Quelch, J.J. (1898b) The boa constrictors of British Guiana. *Annals and Magazine of Natural History*, London (Series 7) 1(4), 296–308.

Quetz, P.-C. (2003) Die Amphibien und Reptilien in Stuttgart: Verbreitung, Gefährdung und Schutz. *Schriftenreihe des Amtes für Umweltschutz* (2002) 1, 1–296, illust., 4 col. pls.

Quijada, B.B. (1916) Catálogo sistemático de los reptiles Chilenos i estranjeros conservados en el Museo Nacional de Historia Naturale. *Boletín del Museo Nacional de Chile*, Santiago (9), 22–47.

Quijada-Mascareñas, A. & Canseco-Marques, L. (2007) Geographic distribution: Serpentes: *Ramphotyphlops braminus*. *Herpetological Review* 38(4), 490.

Quijada-Mascareñas, A. & Enderson, E.F. (2007) Geographic distribution: Serpentes: *Ralmphotyphlops braminus*. *Herpetological Review* 38(4), 490.

Quijada-Mascareñas, A., Ferguson, J.E., Pook, C.E., da Graça Salomão, M., Thorpe, R.S. & Wüster, W. (2007) Phylogeographic patterns of trans-Amazonian vicariants and Amazonian biogeography, the Neotropical rattlesnake (*Crotalus durissus* complex) as an example. *Journal of Biogeography* 2007, 1296–1312, 8 figs.

Quijano, F.M., Avila, R.R. & Narvaez, L.C. (1998) Geographic distribution: Serpentes: *Elaphe guttata meahllmorum*. *Herpetological Review* 29(2), 112.

Quinn, H.R. (1983) Two new subspecies of *Lampropeltis triangulum* from México. *Transactions of the Kansas Academy of Sciences* 86(4), 113–135, 5 figs.

Quinn, H.R. (1987) *Morphology, isozymes, and mitochondrial DNA as systematic indicators in* Crotalus. Doctoral Dissertation, University of Houston, University Park, xiii + 213 pp., 11 figs.

Quintero-Diaz, G., Vázquez-Diaz, J. & Smith, H.M. (1999) Geographic distribution: Serpentes: *Thamnophis scaliger*. *Herpetological Review* 30(4), 237.

Quintero-Diaz, G., Vázquez-Diaz, J. & Smith, H.M. (2000) Geographic distribution: Serpentes: *Micrurus distans zweifeli*. *Herpetological Review* 31(2), 114.

Quintero-Díaz, G., Vázquez-Díaz, J., Sigala-Rodríguez, J.J., Bryson, R.W., Jr. & Salmon, G.T. (2001) Geographic distribution: Serpentes: *Lampropeltis mexicana*. *Herpetological Review* 33(4), 278.

Quintero-Díaz, G., Sigala-Rodríguez, J.T., Vázquez-Díaz, J., Carrasco, J.L., Sigala-Rodríguez, J.I., Román, L.L. & de La Roza, R.G. (2007) Geographic distribution: Serpentes: *Masticophis flagellum*. *Herpetological Review* 38(4), 488.

Quyet, L.K. & Ziegler, T. (2004) Geographic distribution: Serpentes: *Euprepiophis mandarinus*. *Herpetological Review* 35(3), 291–292.

Quyet, L.K. & Ziegler, T. (2009) Geographic distribution: Serpentes: *Protobothrops cornutus*. *Herpetological Review* 40(1), 115.

Rabor, D.S. (1981) *Philippine reptiles & amphibians*. Pundasyon sa Pagpapaunlad ng Kaalaman sa Pagtuturo ng Agham, Manila, 79 pp., 33 figs.

Rabosky, D.L., Aplin, K.P., Donnellan, S.C. & Hedges, S.B. (2004) Molecular phylogeny of blindsnakes (*Ramphotyphlops*) from western Australia and resurrection of *Ramphotyphlops bicolor* (Peters, 1857). *Australian Journal of Zoology* 52(5), 531–548, 6 figs.

Radcliffe, C.W. & Maslin, T.P. (1975) A new subspecies of the red rattlesnake, *Crotalus ruber*, from San Lorenzo, Sur Island, Baja California Norte, Mexico. *Copeia* 1975(3), 490–495, 1 fig.

Raddi, G. (1820) Di alcune specie nuove di rettili e piante Brasiliane. *Memorie Matematica et di Fisica della Società Italiana della Scienze*, Modena 18(2), 313–342, 3 pls.

Radhakrishnan, C. (1997) Notes on a new distributional record and the ecology of *Rhabdops olivaceus* (Beddome) (Reptilia, Serpentes, Colubridae). *The Journal of the Bombay Natural History Society* 94(1), 167–168.

Radovanovic, M. (1941) Zur Kenntnis der Herpetofauna des Balkans. *Zoologischer Anzeiger*, Leipzig 136(7–8), 145–159.

Radovanovic, M. (1951) *Vodozemci i gmizavci nase zemlje*. Izdavacko Preduzec'e Narodne Republike Srbije, Beograd, 249 pp., 94 figs. (in Serbian)

Radovanovic, M. (1964) Die Verbreitung der Amphibien und Reptilien in Jugoslawien. *Senckenbergiana Biologica*, Frankfurt-am-Main 45(3–5), 553–561, 1 fig.

Radovanovic, M. & Martino, K. (1950) Snakes of the Balkan Peninsula. *Serbian Academy of Sciences, Institute of Ecology and Biogeography*, Belgrade 1, 1–45, 9 + 31 figs. (in Serbian)

Raffles, T.S.B. (1822) Descriptive catalogue of a zoological collection, made on account of the Honorable East India Company, in the island of Sumatra and its vicinity, under the direction of Sir Thomas Samford Raffles, Lieutenant-Governor of Fort Marlborough; with additional notices illustrative of the natural history of those countries. *Transactions of the Linnean Society of London* 13(2), 239–340.

Rafinesque-Schmaltz, C.S. (1810) *Caratteri di alcuni nuovi generi e nuove specie di animali e piante della Sicilia con varie osservazioni sopra i medesimi*. Sanfilippo, Palermo, 105 pp., 20 pls.

Rafinesque-Schmaltz, C.S. (1814) Prodrome di erpetologia siciliana *In: Specchio della Scienze o giornale enciclopedico di Sicilia* 2, 65–67, 102–104.

Rafinesque-Schmaltz, C.S. (1815) *Analyse de la nature ou tableau de l'univers et des corps organisés*. Constantine Samuel Rafinesque-Schmaltz, Palermo, 224 pp.

Rafinesque-Schmaltz, C.S. (1817) Dissertation on water snakes, sea snakes and sea serpents. *American Monthly Magazine and Critical Review*, New York 1, 431–435.

Rafinesque-Schmaltz, C.S. (1818a) Further discoveries in natural history, made during a journey through the western region of the United States. *American Monthly Magazine and Critical Review*, New York 3(6), 445-447.

Rafinesque-Schmaltz, C.S. (1818b) Further account of discoveries in natural history, in the western states. *American Monthly Magazine and Critical Review*, New York 4(1), 39–42.

Rafinesque-Schmaltz, C.S. (1819a) Prodrome de 70 nouveaux genres d'animaux découverts dans l'intérieur des États–Unis d'Amérique, durant l'année 1818. *Journal de Physique, de Chimie, d'Histoire Naturelle et des Arts*, Paris 88, 417–429.

Rafinesque-Schmaltz, C.S. (1819b) A journal of the progress of vegetation near Philadelphia, between the 20th of February and the 20th of May, 1816, with occasional zoological remarks. *The American Journal of Science and Arts*, New Haven 1, 77–82.

Rafinesque-Schmaltz, C.S. (1820) *Annals of nature or annual synopsis of new genera and species of animals, plants, &c. discovered in North America*. Constantine Samuel Rafinesque-Schmaltz, Lexington 1, 1–16.

Rage, J.-C. (1972) *Eryx* Daudin et *Gongylophis* Wagler (Serpentes, Boidae). Étude ostéologique. *Bulletin du Muséum d'Histoire Naturelle*, Paris (Série 3) 57(78), 893–898, 1 fig.

Rage, J.-C. (1973a) Présence de *Dunnophis* (Reptilia, Serpentes) dans l'Eocène et l'Oligocéne européens. *Compte Rendu Sommaire des Séances de la Société Géologique de France*, Paris 3, 77–78, figs. a–e.

Rage, J.-C. (1973b) Fossil snakes from Olduvai, Tanzania. *In:* Leakey, L.S.B., Savage, R.J.G. & Coryndon, S.C. (Eds.), *Fossil vertebrates of Africa*. Academic Press, London, 3, 1–6, 1 fig., 1 pl.

Rage, J.-C. (1974) Les serpents des phosphorites du Quercy. *Palaeovertebrata*, Montpellier 6(3–4), 274–303, 8 figs.

Rage, J.-C. (1975a) Un caenophidien primitif, (Reptilia, Serpentes) dans l'Eocène inférieur. *Compte Rendu Sommaire des Séances de la Société Géologique de France*, Paris 1975 (2), 46–47, 1 fig.

Rage, J.-C. (1975b) Un serpent du Paléocène du Niger. Étude préliminaire sur l'origine des caenophidiens (Reptilia, Serpentes). *Comptes Rendus de l'Académie des Sciences*, Paris 281D(9), 515–518, 1 fig.

Rage, J.-C. (1976) Les squamates du Miocène de Beñi Mellal, Maroc. *Géologie Méditerranéene* 3(2), 57–69, 7 figs., 1 pl.

Rage, J.-C. (1977a) La Position phylétique de *Dinilysia patagonica*, serpent du Crétacé supérieur. *Comptes Rendus de l'Académie des Sciences*, Paris 284D, 1765–1768, 1 fig.

Rage, J.-C. (1977b) An Erycine snake (Boidae) of the genus *Calamagras* from the French lower Eocene, with comments on the phylogeny of the Erycinae. *Herpetologica* 33(4), 459–463, 4 figs.

Rage, J.-C. (1978) La poche a phosphate de Ste-Neboule (Lot) et sa faune de vertebres du Ludien Supérieur. 5. Squamates. *Palaeovertebrata*, Montpellier 8(2–4), 201–215, 7 figs.

Rage, J.-C. (1980) Un serpent marin nouveau de l'Eocène de Belgique. Le probleme des serpents marins du Paleogene. *Comptes Rendus de l'Académie des Sciences*, Paris 291D(5), 469–471, figs.

Rage, J.-C. (1981) Les Colubridae (Reptilia, Serpentes) du neogene Francais. Le Probleme des especes du Miocene de Sansan (Gers.). *Geobios*, Lyons 14(4), 537–542, 1 fig.

Rage, J.-C. (1982) L'histoire des serpentes. *Pour la Science*, Paris (54), 16–27, 8 figs.

Rage, J.-C. (1983a) *Palaeophis colossaeus* nov. sp. (le plus grand Serpent connu?) de l'Eocène du Mali et le problème du genre chez les Palaeopheinae. *Comptes Rendus de l'Académie des Sciences*, Paris (Série 2) 296(22), 1742–1744, 1 fig.

Rage, J.-C. (1983b) Les serpents aquatiques de l'Éocène européen. Définition des espèces et aspects stratigraphiques. *Bulletin du Muséum National d'Histoire Naturelle*, Paris (Série 4) 5C(2), 213–241, 10 figs.

Rage, J.-C. (1984a) The fossil snake *Cheilophis huerfanoensis* Gilmore, 1938, from Eocene of Colorado: redescription and reappraisal of relationships. *Journal of Vertebrate Paleontology* 3(4), 219–222, 2 figs.

Rage, J.-C. (1984b) Serpentes. *In:* Wellnhofer, P. (Ed.), *Handbuch der Palaoherpetologie*. Gustav Fischer, Stuttgart (11), i–xi + 1–80, 37 figs.

Rage, J.-C. (1987) Fossil history. *In:* Seigel, R.A., Collins, J.T. & Novak, S.S. (Eds.), *Snakes, ecology and evolutionary biology*. Macmillan Publishing Company, New York, pp. 51–76, illust.

Rage, J.-C. (1988a) The oldest known colubrid snakes. The state of the art. *Acta Zoologica Cracoviensia*, Kraków 31(13), 457–474, 2 figs.

Rage, J.-C. (1988b) Un Serpent primitif (Reptilia, Squamata) dans le Cénomanien (base du Crétacé supérieur). *Comptes Rendus de l'Académie des Sciences*, Paris (Série 2) 307, 1027–1032, 1 fig.

Rage, J.-C. (1988c) Le gisement du Breton (Phosphorites du Quercy, Tarn-et-Garonne, France) et sa faune de Vertebres de l'Eocène supérieur. I. Amphibien et reptiles. *Palaeontographica*, Stuttgart 205A, 3–27, 21 figs.

Rage, J.-C. (1991) Squamate reptiles from the early Paleocene of the Tiupampa area (Santa Lucia formation), Bolivia. *In:* Suarez-Soruco, R. (Ed.), *Fosiles y facies de Bolivia. Vol. I. Vertebrados*. Revista Téchica de YPFB, Santa Cruz, pp. 503–508, 2 figs.

Rage, J.-C. (1994) Snake diversity. *In:* Bauchot, R. (Ed.), *Snakes: a natural history*. Sterling Publishing Co., New York, pp. 34-47.

Rage, J.-C. (1996) Les Madtsoiidae (Reptilia, Serpentes) du Crétacé supérieur d'Europe: rémoins gondwaniens d'une dispersion transtéthysienne. *Comptes Rendus de l'Académie des Sciences*, Paris 322(IIa), 603–608.

Rage, J.-C. (1998) Fossil snakes from the Palaeocene of São José de Itaboraí, Brazil. Part I. Madtsoiidae, Aniliidae. *Palaeovertebrata*, Montpellier 27(3–4), 109–144, 20 figs.

Rage, J.-C. (1999) Squamates (Reptilia) from the Upper Cretaceous of Laño (Basque Country, Spain). *Estudios del Museo de Ciencias Naturales de Alava* 14(1), 121–133, 12 figs.

Rage, J.-C. (2001) Fossil snakes from the Palaeocene of São José de Itaboraí, Brazil. Part II. Boidae. *Palaeovertebrata*, Montpellier 30(3–4), 111–150, 16 figs.

Rage, J.-C. (2011) Fossil snakes from the Palaeocene of São José de Itaboraí, Brazil. Part III. Ungaliophiinae, booids *increate sedis*, and Caenophidia. Summary, update, and discussion of the snake fauna from the locality. *Palaeovertebrata*, Montpellier (2008) 36(1–4), 37–73, 12 figs.

Rage, J.-C. & Albino, A.M. (1989) *Dinilysia patagonica* (Reptilia, Serpentes), matériel vertébral additionnel du Crétacé supérieur d'Argentine. Étude complémentaire des vertèbres, variations intraspécifiques et intracolumnaires. *Neues Jahrbuch für Geologie und Paläontologie, Monatsheft*, Stuttgart 1989(7), 433–447, 5 figs.

Rage, J.-C. & Augé, M. (2010) Squamate reptiles from the middle Eocene of Lissieu (France). A landmark in the middle Eocene of France. *Geobios*, Lyon 43, 253–268, 5 figs.

Rage, J.-C. & Bailon, S. (2005) Amphibians and squamate reptiles from the late early Miocene (MN 4) of Béon 1 (Montréal-du-Gers, southwestern France). *Geodiversitas* 27(3), 413–441, 6 figs.

Rage, J.-C. & Escuillié, F. (2000) Un nouveau serpent bipède du Cénomanien (Crétacé). Implications phylétiques. *Comptes Rendus de l'Académie des Sciences*, Paris (IIa), 330, 513–520, 3 figs.

Rage, J.-C. & Escuillié, F. (2002) *Eupodophis*, new name for the genus *Podophis* Rage and Escuillié, 2000, an extinct bipedal snake, preoccupied by *Podophis* Wiegmann, 1834 (Lacertilia, Scincidae). *Amphibia-Reptilia*, Leiden 23(2), 232–233.

Rage, J.-C. & Escuillié, F. (2003) Le Cénomanien: étage des serpents bìpedes. *Camets de Géologie* 2003, 1–11, 6 figs.

Rage, J.-C. & Ford, R.L. (1980) Amphibians and squamates from the upper Eocene of the Isle of Wight. *Tertiary Research*, Rotterdam 3(2), 47–60, 4 figs., 1 pl.

Rage, J.-C. & Holman, J.A. (1984) Des serpents (Reptilia, Squamata) de type Nord-Américain dans le Miocène Français. Évolution parallele ou dispersion? *Geobios*, Lyon 17(1), 89–104, 6 figs.

Rage, J.-C. & Prasad, G.V.R. (1992) New snakes from the late Cretaceous (Maastrichtian) of Naskal, India. *Neues Jahrbuch für Geologie und Paläontologie, Abhandlungen*, Stuttgart, 187(1), 83–97, 7 figs.

Rage, J.-C. & Rocek, Z. (1983) *Dolniceophis lehmani* (Serpentes, Colubridae), a new fossil snake from the lower Miocene of Czechoslovakia. *Casopis pro Mineralogii a Géologii*, Prague 28(1), 17–21, 2 pls.

Rage, J.-C. & Truc, G. (1970) *Elaphe scalaris* (la couleuvre a echelons) son extension en provence. *Bulletin Mensual de la Société Linnéenne de Lyon* 39(6), 206–208, 1 pl.

Rage, J.-C. & Werner, C. (1999) Mid-Cretaceous (Cenomanian) snakes from Wadi Abu Hashim, Sudan: the earliest snake assemblage. *Palaeontologia Africana* 35, 85–110, 22 figs.

Rage, J.-C. & Wouters, G. (1979) Decouverte du plus ancien Palaeopheide (Reptilia, Serpentes) dans le Maestrichtien du Maroc. *Geobios*, Lyon 12(2), 293–296, 1 fig.

Rage, J.-C., Buffetaut, E., Buffetaut-Tong, H., Chaimanee, Y., Ducrocq, S., Jaeger, J.-J. & Suteethorn, V. (1992) A colubrid snake in the late Eocene of Thailand: the oldest known Colubridae (Reptilia, Serpentes). *Comptes Rendus de l'Académie des Sciences*, Paris (Série 4) 314, 1085–1089, 3 figs.

Rage, J.-C., Gupta, S.S. & Prasad, G.V.R. (2001) Amphibians and squamates from the Neogene Siwalik beds of Jammu and Kashmir, India. *Paläontologische Zeitschrift*, Stuttgart 75(2), 197–205, 10 figs.

Rage, J.-C., Bajpai, S., Thewissen, J.G.M. & Tiwari, B.N. (2003) Early Eocene snakes from Kutch, western India, with a review of the Palaeophiidae. *Geodiversitas* 25(4), 695–716, 6 figs.

Rage, J.-C., Folie, A., Rana, R.S., Singh, H., Rose, K.D. & Smith, T. (2008) A diverse snake fauna from the early Eocene of Vastan Lignite Mine, Gujarat, India. *Acta Palaeontologica Polonica* 53(3), 391–403.

Rahman, S.C. (2012a) Geographic distribution: Serpentes: *Boiga ochracea*. *Herpetological Review* 43(1), 105.

Rahman, S.C.(2012b) Geographic distribution: Serpentes: *Boiga siamensis*. *Herpetological Review* 43(1), 105.

Rahman, S.C. (2012c) Geographic distribution: Serpentes: *Oligodon cyclurus*. *Herpetological Review* 43(1), 107.

Rahman, S.C. (2012d) Geographic distribution: Serpentes: *Sinomicrurus macclellandi*. *Herpetological Review* 43(1), 107.

Rahman, S.C. & Ahmed, M. (2012) Geographic distribution: Serpentes: *Rhabdophis himalayanus*. *Herpetological Review* 43(1), 107.

Raï, M.M. (1965) *Recherches sur les Colubridés d'Iran*. Doctoral Dissertation, Université Montpellier, Montpellier, 104 pp., 4 pls.

Raï, M.M. (1970) Sur *Naja morgani* Mocquard, Elapidae récolté en Iran. *Actes de la Société Linnéenne de Bordeaux* (1969) 106A(5), 2 pp., 1 pl.

Raï, M.M. (1978) Une nouvelle récolte de *Telescopus rhinopoma* Blanford 1874 (Serpentes, Colubridae), une espèce très rare. *Canadian Journal of Zoology*, Ottawa 56(1), 146–149, 1 fig.

Raj, B.S. (1926) Parturitions of electric rays and a sea snake in the Marine Aquarium, Madras. *The Journal of the Bombay Natural History Society* (1927–1928) 31(3), 828.

Rajabizadeh, M. & Rastegar-Pouyani, N. (2009) Two new records of reptiles (Reptilia: Squamata) from southeastern Iran. *Turkish Journal of Zoology* 33, 103–104, 3 figs.

Rajabizadeh, M., Nilson, G. & Kami, H.G. (2011) A new species of mountain viper (Ophidia: Viperidae) from the central Zagros Mountains, Iran. *Russian Journal of Herpetology* 18(3), 235–240, 5 figs.

Rajabizadeh, M., Schmidtler, J.F., Orlov, N. & Soleimani, G. (2012) Review of taxonomy and distribution of the *Eirenis medus* group (Chernov, 1940) (Ophidia: Colubridae) with description of a new species of the genus *Eirenis* from Kerman Province, southeastern Iran. *Russian Journal of Herpetology* 19(4), 307–313, 8 figs.

Rajendran, M.V. (1985) *Studies in uropeltid snakes*. Madurai Kamaraj University, 132 pp., 18 figs., 23 pls.

Rajendran, M.V. (1986) *Snakes of our land*. Jaya Publications, Palayamkottai, 135 pp. + 88 pp., illust.

Ramadhan, G., Iskandar, D.T. & Subasri, D.R. (2010) A new species of cat snake (Serpentes: Colubridae) morphologically similar to *Boiga cynodon* from the Nusa Tenggara Islands, Indonesia. *Asian Herpetological Research* (Series 2) 1(1), 22–30, 2 figs.

Ramírez-Bautista, A. (1994) Manual y claves illustradas de los anfibios y reptiles de la región de Chamela, Jalisco, México. *Instituto de Biología, Universidad Nacional Autonoma de México* (23), 127 pp.,13 figs., col. photos.

Ramírez-Bautista, A. & Smith, H.M. (1992) A new chromospecies of snake (*Pseudoleptodeira*) from Mexico. *Bulletin of the Maryland Herpetological Society* 28(3), 83–98, 9 figs.

Ramírez-Bautista, A., Mancilla-Moreno, M., Smith, H.M., Chiszar, D. & van Breukelen, F. (1998) Morphological variation and relationship of *Rhadinaea bogertorum* (Squamata, Colubridae), an endemic snake of the Sierra de Juarezz, Oaxaca, Mexico. *Bulletin of the Maryland Herpetological Society* 34(4), 99–106.

Ramirez-Bautista, A., Hernández-Ibarra, X., Torres-Cervantes, R. & Smith, H.M. (1999) External morphological variation in Hardy's hooknose snake *Ficimia hardyi* (Squamata, Colubridae). *Bulletin of the Maryland Herpetological Society* 35(3), 81–84.

Ramírez-Bautista, A., Hernández-Salinas, U., Mendoza-Quijano, F., Cruz-Elizalde, R., Vite-Silva, V.D. & Leyte-Manrique, A. (2010) *Lista anotada de los anfibios y reptiles del estado de Hidalgo, México*. Universidad Autónoma del Estado de Hidalgo, x + 104 pp., 124 figs.

Ramón de la Sagra, D. (1838–1843) *Historia física, política y natural de la isla de Cuba. Tomo II. Segunda parte—Reptiles y peces. Reptiles [J.T. Cocteau y G. Bibron]*. Arthus Bertrand, Paris, i + 143 pp., *Atlas* (1840), 36 col. pls.

Ramón de la Sagra, D. (1841–1843) *Histoire physique, politique, et naturelle de l'ile de Cuba. Reptiles. Par Mm. Cocteau et Bibron*. Arthus Betrand, Paris, xviii + 242 pp., *Atlas*, 31 col. pls. [French translation of Ramón de la Sagra, 1838–1843]

Ramsay, E.P. (1877) Description of a supposed new species of *Acanthophis*, from North Australia. *Proceedings of the Linnaean Society of New South Wales*, Sydney (1877–1878) 2(1), 72–73.

Ramsey, L.W. (1953) The lined snake, *Tropidoclonion lineatum* (Hallowell). *Herpetologica* 9(1), 7–24, 4 figs.

Rand, A.S. (1957) Notes on amphibians and reptiles from El Salvador. *Fieldiana: Zoology* 34(42), 505–534.

Rand, A.S. & Myers, C.W. (1990) The herpetofauna of Barro Colorado Island, Panamá: an ecological summary. *In:* Gentry, A.H. (Ed.), *Four Neotropical rainforests*. Yale University Press, New Haven, pp. 386–409.

Rand, A.S. & Ortleb, E.P. (1969) Defensive display in the colubrid snake *Pseustes poecilonotus shropshirei*. *Herpetologica* 25(1), 46–48, fig. 1.

Ranzani, C. (1836) De serpente Monspessulano generis *Coelopeltis wagleri*. *Novi Comment* [Emygdii ad Ulmo et J. Tiocchi], Bononiae 2: 229–240, 1 pl.

Ranzani, C. (1837) Descrizione di un serpente il quale appartiene ad una nuova specie del genere *Calamaria* di Boie. *Memorie di Matematica et di Fisica della Società Italiana della Scienze*, Modena (1836) 21, 101–113, pl. 3.

Rao, C.R.N. (1917) Notes on *Lachesis anamallensis* and allied forms. *Records of the Indian Museum*, Calcutta 13, 11–15, pl. 3.

Rao, D.-Q. & Yang, D.-T. (1992) Phylogenetic systematics of Pareinae (Serpents) of southeastern Asia and adjacent islands with relationships between it and the geology changes. *Acta Zoologica Sinica*, Beijing 38(2),139–150, 6 figs. (in Chinese)

Rao, D.-Q. & Yang, D.-T. (1998) A new species of *Sinonatrix* (Serpentes, Colubridae) of China with preliminary survey on *Sinonatrix*. *Russian Journal of Herpetology* 5(1), 70–73, 1 fig.

Rasmussen, A.R. (1987) Persian Gulf sea snake *Hydrophis lapemoides* (Gray): new records from Phuket Island, Andaman Sea, and the southern part of the Straits of Malacca. *Natural History Bulletin of the Siam Society*, Bangkok 35(1), 57–58.

Rasmussen, A.R. (1989a) Sea snakes *Thalassophina viperina* (Schmidt) and *Laticauda laticaudata* (Linne'), new records from Phuket Island, Andaman Sea, with remarks on subspecies of *L. laticaudata*. *Natural History Bulletin of the Siam Society*, Bangkok 37(1), 99–103, 2 figs.

Rasmussen, A.R. (1989b) An analysis of *Hydrophis ornatus* (Gray), *H. lamberti* Smith, and *H. inornatus* (Gray) (Hydrophiidae, Serpentes) based on samples from various localities, with remarks on feeding and breeding biology of *H. ornatus*. *Amphibia-Reptilia*, Leiden 10(4), 397–417, 12 figs.

Rasmussen, A.R. (1992) Rediscovery and redescription of *Hydrophis bituberculatus* Peters, 1872 (Serpentes, Hydrophidae). *Herpetologica* 48(1), 85–97, 10 figs.

Rasmussen, A.R. (1993) The status of the Persian Gulf sea snake *Hydrophis lapemoides* (Gray, 1849) (Serpentes, Hydrophiidae). *Bulletin of the Natural History Museum (Zoology)*, London 59(2), 97–105, 5 figs.

Rasmussen, A.R. (1994) A cladistic analysis of *Hydrophis* subgenus *Chitulia* (McDowell, 1972) (Serpentes, Hydrophiidae). *Zoological Journal of the Linnean Society*, London 111(2), 161–178, 9 figs.

Rasmussen, A.R. (1997) Systematics of sea snakes, a critical review. *In:* Thorpe, R.S., Wüster, W. & Malhotra, A. (Eds.), *Venomous snakes: Ecology, evolution and snakebite*. Zoological Society of London & Clarendon Press, Oxford, pp. 15–30, 4 figs.

Rasmussen, A.R. (2011) Sea snakes (Serpentes: subfamilies Hydrophiinae and Laticaudinae) in Vietnam: a comprehensive checklist and an updated identification key. *Zootaxa* (2894), 1–20, 12 figs.

Rasmussen, A.R. & Anderson, M. (1990) The sea snake *Kerilia jerdoni* Gray (1949), first records from Andaman Sea, Phuket Island, Thailand, with remarks on the two subspecies. *The Snake* 22(2), 131–133, 1 fig.

Rasmussen, A.R. & Ineich, I. (2000) Sea snakes of New Caledonia and surrounding waters (Serpentes: Elapidae): first report on the occurrence of *Lapemis curtus* and description of a new species of the genus *Hydrophis*. *Hamadryad* 25(2), 91–99.

Rasmussen, A.R. & Ineich, I. (2010) Species diversity in the genus *Emydocephalus* Krefft, 1869 (Serpentes, Elapidae, Hydrophiinae): insight from morphology and anatomy. *Herpetological Review* 41(3), 285–290, 1 fig.

Rasmussen, A.R. & Smith, L. (1997) The taxonomic status of the sea snake *Hydrophis czeblukovi* (Kharin, 1984) from north-west Australian waters. *Amphibia-Reptilia,* Leiden 18(4), 419–425, 1 fig.

Rasmussen, A.R., Auliya, M. & Böhme, W. (2001) A new species of the sea snake genus *Hydrophis* (Serpentes, Elapidae) from a river in west Kalimantan (Indonesia, Borneo). *Herpetologica* 57(1), 23–32, 5 figs.

Rasmussen, A.R., Gravlund, P., Nguyen, C.V. & Chanhome, L. (2007) A resurrection of *Hydrophis pachycercos* Fischer, 1855 (Serpentes, Elapidae) with a new neotype from the South China Sea. *Hamadryad* 31(2), 288–298, 4 figs.

Rasmussen, A.R., Elmberg, J., Sanders, K.L. & Gravlund, P. (2012) Rediscovery of the rare sea snake *Hydrophis parviceps* Smith, 1935: identification and conservation status. *Copeia* 2012(2), 276–282, 3 figs.

Rasmussen, J.B. (1975) Geographical variation, including an evolutionary trend, in *Psammodynastes pulverulentus* (Boie, 1827) (Boiginae, Homalopsidae, Serpentes). *Videnskabelige Meddelelser fra Dansk Naturhistorisk Forening i Kjøbenhavn* 138, 39–64, 10 figs.

Rasmussen, J.B. (1979) An intergeneric analysis of some boigine snakes—Bogert's (1940) group XIII and XIV (Boiginae, Serpentes). *Videnskabelige Meddelelser fra Dansk Naturhistorisk Forening i Kjøbenhavn* 141, 97–155, 22 figs.

Rasmussen, J.B. (1981) The snakes from the rainforest of the Usambara Mountains, Tanzania: a checklist and key. *Salamandra*, Frankfurt-am-Main 17(3–4), 173–188.

Rasmussen, J.B. (1982a) A new record of the rare *Opisthotropis premaxillaris* (Serpentes, Colubridae). *Amphibia-Reptilia*, Wiesbaden 3(2–3), 279–280.

Rasmussen, J.B. (1982b) The snake fauna of the Usambara Mountains. Appexdix 7. *In:* Rodgers, W.A. & Homewood, L.M. (Eds.), Species richness and endemism in the Usambara mountains forests, Tanzania. *Biological Journal of the Linnean Society,* London 18(3), 197–248.

Rasmussen, J.B. (1985a) A new species of *Crotaphopeltis* from east Africa, with remarks on the identity of *Dipsas hippocrepis* Reinhardt, 1843 (Serpentes, Boiginae). *Steenstrupia*, København 11(4), 113–129, 10 figs.

Rasmussen, J.B. (1985b) A re-evaluation of the systematic of the African rear-fanged snakes of Bogert's groups XIII–XVI, including a discussion of some evolutionary trends within Caenophidia. *In:* Schuchmann, K.-L. (Ed.), *Proceedings of the International Symposium on African Vertebrates*. Zoologisches Forschungsinstitut und Museum Alexander Koenig, Bonn, pp. 531–548, 6 figs.

Rasmussen, J.B. (1986) On the taxonomic status of *Dipsadoboa werneri* (Boulenger), *D. shrevei* (Loveridge), and *Crotaphopeltis hotamboeia kageleri* Uthmoller (Boiginae, Serpentes). *Amphibia-Reptilia*, Leiden 7(1), 51–73, 13 figs.

Rasmussen, J.B. (1989a) On the taxonomic status of *Dipsadoboa aulica aulica* Günther and *D. aulcia flavida* Broadley and Stevens, with the description of a new subspecies of *D. flavida* Broadley and Stevens (Boiginae, Serpentes). *Amphibia-Reptilia*, Leiden 10(1), 35–62, 18 figs.

Rasmussen, J.B. (1989b) A taxonomic review of the *Dipsadoboa duchesnei* complex. *Bonner Zoologische Beiträge*, Bonn 40(3–4), 249–264, 9 figs.

Rasmussen, J.B. (1990) The retina of *Psammodynastes pulverulentus* (Boie, 1827) and *Telescopus fallax* (Fleishmann, 1831) with a discussion of their phylogenetic significance (Colubroidea, Serpentes). *Zeitschrift für Zoologische Systematik und Evolutionsforschung*, Hamburg 28, 269–276, 5 figs.

Rasmussen, J.B. (1991) Snakes (Reptilia, Serpentes) from the Kouilou River basin, including a tentative key to the snakes of République du Congo. *Tauraco Research Report* 4, 175–188, 5 figs.

Rasmussen, J.B. (1993a) The current taxonomic status of Tornier's cat-snake (*Crotaphopeltis tornieri*). *Amphibia-Reptilia*, Leiden 14(4), 395–409, 7 figs.

Rasmussen, J.B. (1993b) A taxonomic review of the *Dipsadoboa unicolor* complex, including a phylogenetic analysis of the genus (Serpentes, Dipsadidae, Boiginae). *Steenstrupia*, Copenhagen 19(4), 129–196, 46 figs.

Rasmussen, J.B. (1996) Maxillary tooth number in the African tree-snakes genus *Dipsadoboa*. *Journal of Herpetology* 30(2), 297–300, 2 figs.

Rasmussen, J.B. (1997a) On two little known African water snakes (*Crotaphopeltis degeni* and *C. barotseensis*). *Amphibia-Reptilia* 18(2), 191–206, 5 figs.

Rasmussen, J.B. (1997b) Geographic distribution: Serpentes: *Rhinotyphlops praeocularis*. *Herpetological Review* 28(3), 160.

Rasmussen, J.B. (2002) A review of the African members of the genus *Micrelaps* Boettger 1880 (Serpentes Atractaspididae). *Tropical Zoology*, Firenze 15(1), 71–87, 6 figs.

Rasmussen, J.B. (2004) A review of Whyte's water-snake, *Lycodonomorphus whytii* (Serpentes: Colubridae: Lamprophiinae). *African Journal of Herpetology* 53(2), 155–162, 4 figs.

Rasmussen, J.B. (2005a) A review of the slender stiletto-snake, *Atractaspis aterrima* Günther 1863 (Serpentes Atractaspididae). *Tropical Zoology* 18(1), 137–148.

Rasmussen, J.B. (2005b) On the identification and distribution of the two-striped night adder (*Causus bilineatus*) and related forms. *African Journal of Herpetology* 54(1), 1–15, 8 figs.

Rasmussen, J.B. & Howell, K.M. (1982) The current status of the rare Usambara Mountain forest-viper, *Atheris ceratophorus* Werner, 1895, including a probable new record of *A. nitschei rungweensis* Bogert, 1940, and a discussion of its validity (Reptilia, Serpentes, Viperidae). *Amphibia-Reptilia*, Wiesbaden 3(2–3), 269–277, 3 figs.

Rasmussen, J.B. & Howell, K.M. (1998) A review of Barbour's short-headed viper *Adenorhinos barbouri* (Serpentes: Viperidae). *African Journal of Herpetology* 47(2), 69–75, illust.

Rasmussen, J.B. & Largen, M.J. (1992) A review of *Pseudoboodon* Peracca with the description of a new species from southwest Ethiopia (Serpentes, Dipsadidae, Lycodontinae, Boadontini). *Steenstrupia*, København 18(3), 65–80, 8 figs.

Rasmussen, J.B., Howell, K.M. & Andersen, M. (1995) A review of the Usambara forest snake, *Geodipsas vauerocegae* and the Uluguru forest snake *G. procterae*. *Amphibia-Reptilia*, Leiden 16(2), 123–136, 6 figs.

Rasmussen, J.B., Chirio, L. & Ineich, I. (2000) The Herald snakes (*Crotaphopeltis*) of the Central African Republic, including a systemic review of *C. hippocrepis*. *Zoosystema* 22(3), 585–600, 4 figs.

Rathke, H. (1837) Zur fauna der Krym—ein Beitrag. *Mémoires présentés a l'Académie Impériale des Sciences de Saint-Pétersbourg par Divers Savans, et Lus dans ses Assemblees* 3, 291–454, 10 pls.

Ratnikov, V.Y. (2004) Identification of some European species of *Elaphe* (Colubridae, Serpentes) on the basis of vertebrae. *Russian Journal of Herpetology* 11(2), 91–98, 8 figs.

Rato, C., Zuffi, M.A.L., Corti, C., Fornasiero, S., Gentilli, A., Razzetti, E., Scali, S., Carretero, M.A. & Harris, D.J. (2009) Phylogeography of the European whip snake, *Hierophis viridiflavus* (Colubridae), using mtDNA and nuclear DNA sequences. *Amphibia-Reptilia* 30(2), 283–289, 2 figs.

Raun, G.G. (1965) A guide to Texas snakes. *Texas Memorial Museum Notes* (9), 85 pp., 44 figs.

Raw, L.R.G. (1973) A review of the dusky-bellied water snake, *Lycodonomorphus laevissimus* (Günther), with descriptions of two new subspecies. *Annals of the Natal Museum*, Pietermaritzburg 21(3), 713–718, 2 figs.

Rawlings, L.H. & Donnellan, S.C. (2003) Phylogeographic analysis of the green python, *Morelia viridis*, reveals cryptic diversity. *Molecular Phylogenetics and Evolution* 27(1), 36-44, 2 figs.

Rawlings, L.H., Rabosky, D.L., Donnellan, S.C. & Hutchinson, M.N. (2008) *Python* phylogenetics: inference from morphology and mitochondrial DNA. *Biological Journal of the Linnean Society*, London 93(3), 603–619, 5 figs.

Rawlinson, P.A. (1966) Reptiles of the Victorian Mallee. *Proceedings of the Royal Society of Victoria* 79, 605-619.

Rawlinson, P.A. (1967) The vertebrate fauna of the Bass Strait Islands, 2. The Reptilia of Flinders and Kings Islands. *Proceedings of the Royal Society of Victoria*, Melbourne 80, 211–224.

Rawlinson, P.A. (1974) Biogeography and ecology of the reptiles of Tasmania and the Bass Strait area. *In:* Williams, W.D. (Ed.), *Biogeography and ecology in Tasmania*. Dr. W. Junk, The Hague, pp. 291–338.

Rawlinson, P.A. (1991) Taxonomy and distribution of the Australian tiger snakes (*Notechis*) and copperheads (*Austrelaps*) (Serpentes, Elapidae). *Proceedings of the Royal Society of Victoria*, Melbourne 103(2), 125–135, 2 figs.

Raxworthy, C.J. (1988) Reptiles, rainforest and conservation in Madagascar. *Biological Conservation* 43(3), 181–211, illust.

Raxworthy, C.J. (1990) Herpetofauna of the threatened rain forests of Madagascar — a unique habitat. *British Herpetological Society Bulletin*, London (33), 1–3, illust.

Raxworthy, C.J. & Attuquayefio, D.K. (2000) Herpetofaunal communities at Muni Lagoon in Ghana. *Biodiversity and Conservation* 9, 501–510, 1 fig.

Raxworthy, C.J. & Nussbaum, R.A. (1994) A review of the Madagascan snake genera *Pseudoxyrhopus*, *Pararhadinaea*, and *Heteroliodon* (Squamata, Colubridae). *Miscellaneous Publications, Museum of Zoology, University of Michigan* (182), 1–37, 25 figs.

Raxworthy, C.J. & Nussbaum, R.A. (1996) Montane amphibian and reptile communities in Madagascar. *Conservation Biology* 10(3), 750–756, 2 figs.

Ray, J.M. (2011) Geographic distribution: Serpentes: *Atropoides mexicanus*. *Herpetological Review* 42(3), 392.

Razoumowsky, G. de. (1789) *Histoire Naturelle du Jorat et de ses Environs; et celle de trois lacs de Neufchatel, Morat et Bienne; précédées d'un essai sur le climat, les productions, le commerce, les animaux de la partie du pays de Vaud ou de la Suisse Romande, qui entre dans le plan de cet ouvrage. Tome premier.* Jean Mourer, Lausanne, xvi + 322 pp.

Razzetti, E. & Msuya, C.A. (2002) *Field guide to the amphibians and reptiles of Arusha National Park (Tanzania).* Tanzania National Parks, Arusha, 84 pp., col. illust.

Razzetti, E., Sindaco, R., Grieco, C., Pella, F., Ziliani, U., Pupin, F., Riservato, E., Pellitteri-Rosa, D., Butikofer, L., Suleiman, A.S., Al-Aseily, B.A., Carugati, C., Boncompagni, E. & Fasola, M. (2011) Annotated checklist and distribution of the Socotran Archipelago herpetofauna (Reptilia). *Zootaxa* (2826), 1–44, 32 figs.

Reagan, D.P. (1984) Ecology of the Puerto Rican boa (*Epicrates inornatus*) in the Luquillo Mountains of Puerto Rico. *Caribbean Journal of Science* 20(3–4), 119–127.

Reati, G.J. (1996) Serpientes de la Provincia de Cordoba, Argentina. *In:* Tada, I.E. de & Bucher, E.H. (Eds.), *Biodiversidad de Córdoba. Fauna. Vol. I.* Universidad Nacional de Río Cuarto, Córdoba, pp. 239–254, 5 figs.

Recchio, I.M. (1997) Incubation of western blind snake eggs, *Leptotyphlops humilis. The Vivarium* 9(1), 26.

Recchio, I.M., Rodriguez, C.M. & Lazcano, D. (2007) Geographic distribution: Serpentes: *Geophis dugesii aquilonaris*. *Herpetological Review* 38(1), 103.

Recinos, A. (1954) *Monografia del departamento de Huehuetenango*. Second edition. Editorial del Ministerio de Educación Pública, Guatemala City, xvi + 518 pp., 66 illust. [first edition, 1913, 269 pp.]

Recoder, R.S., Teixeira Junior, M., Camacho, A., Sales-Nuñes, P.M., Mott, T., Valduio, P.H., Ghellere, J.M., Nogueira, C. & Rodrigues, M.T. (2011) Répteis da Estação Ecológica Serra Geral do Tocantins, Brasil Central. *Biota Neotropica* 11(1), 1–19, 8 figs.

Redfield, J.A., Holmes, J.C. & Holmes, R.D. (1978) Sea snakes of the eastern Gulf of Carpentaria. *Australian Journal of Marine & Freshwater Research*, Melbourne 29(3), 325–334, 1 fig.

Redkozubov, O. (2003) Pliocene and lower Pleistocene fauna reptiles of Republic Moldova. *In:* Ananjeva, N., Cherepanov, G., Iohansen, L., Khalikov, R. & Tsinenko, O. (Eds.), *Programme & abstracts, 12th Ordinary General Meeting, Societas Europaea Herpetologica*. Societas Europaea Herpetologica, St.-Petersburg, pp. 138–139.

Redmer, M. & Ballard, S.R. (1995) Recent distributional records for amphibians and reptiles in Illinois. *Herpetological Review* 16(1), 49–53.

Redmond, W.H., Echternacht, A.C. & Scott, A.F. (1990) Annotated checklist and bibliography of amphibians and reptiles of Tennessee (1835 through 1989). *Miscellaneous Publications, Center for Field Biology, Austin Peay State University*, Clarksville (4), 1–173.

Reed, C.A. & Marx, H. (1959) A herpetological collection from northeastern Iraq. *Transactions of the Kansas Academy of Sciences* 62(1), 91–122, 6 figs.

Reed, C.F. (1956a) The herpetofauna of Harford County, Maryland. *Journal of the Washington Academy of Sciences* 46(2), 58–60.

Reed, C.F. (1956b) *Contributions to the herpetology of Maryland and Delmarva No. 8 An annotated check list of the snakes of Maryland and Delmarva*. Reed Herpetorium, Baltimore, 20 pp.

Reed, C.F. (1957a) Contributions to the herpetofauna of Virgina, 2, The reptiles and amphibians of Northern Neck. *Journal of the Washington Academy of Sciences* 47, 21–23, 1 fig.

Reed, C.F. (1957b) Contributions to the herpetology of Maryland and Delmarva, 12, The herpetofauna of Anne Arundel County, Md. *Journal of the Washington Academy of Sciences* 47, 64–66.

Reed, C.F. (1957c) Contributions to the herpetology of Virginia, 3, The herpetofauna of Accomac and Northampton counties, Virginia. *Journal of the Washington Academy of Sciences* 47, 89–91.

Reed, C.F. (1957d) Contributions to the herpetology of Maryland and Delmarva, 15, The herpetofauna of Somerset County, Md. *Journal of the Washington Academy of Sciences* 47, 127–128.

Reed, C.F. (1958a) Contribution to the herpetology of Maryland and Delmarva. No. 17, Southeastern herptiles with northern limits on coastal Maryland, Delmarva, and New Jersey. *Journal of the Washington Academy of Sciences* 48, 28–32.

Reed, C.F. (1958b) Contributions to the herpetology of Maryland and Delmarva, 13, Piedmont herpetofauna on coastal Delmarva. *Journal of the Washington Academy of Sciences* 48, 95–99.

Reed, L.C. & Longnecker, O.M. (1932) The geology of Hemphill County, Texas. *University of Texas Bulletin* (3231), 1–98.

Reeves, C. (1933) *Manual of the vertebrate animals of northeastern and central China, exclusive of the birds*. Chung HWA Book Co., Shanghai, 806 pp., illust.

Regenass, U. & Kramer, E. (1981) Zur Systematik der grunen Grubenottern der Gattung *Trimeresurus* (Serpentes, Crotalidae). *Revue Suisse de Zoologie*, Genève 88(1), 163–205, 12 figs., 1 pl.

Regende de Carvalho, R., Jr. & Barreto-Nascimento, L. (2005) Geographic distribution: Serpentes: *Bothrops leucurus*. *Herpetological Review* 36(4), 469.

Rehák, I. (1985) *Coluber rubriceps thracius* ssp. n. from Bulgaria (Reptilia, Squamata, Colubridae). *Vestnik Ceské Zoologické Spolecnosti v Praze* 49(4), 276–280, illust.

Rehák, I. & Obst, F.J. (1993) *Coluber rubriceps* (Venzmer, 1919)—Rötliche Schlanknatter. *In:* Böhme, W. (Ed.), *Handbuch der Reptilien und Amphibien Europas. Band 3/I. Schlangen (Serpentes) I. (Typhlopidae, Boidae, Colubridae 1, Colubrinae)*. AULA-Verlag, Wiesbaden, pp. 156–166, figs. 36–38.

Reichenow, A. (1874) Eine Sammlung Lurche und Kriechthiere von Westafrika. *Archiv für Naturgeschichte,* Berlin 40(1), 287–298, 1 pl.

Reichle, S. & Embert, D. (2005) New species of *Clelia* (Colubridae) from the inter-Andean dry valleys of Bolivia. *Journal of Herpetology* 39(3), 379–383, 2 figs.

Reichling, S.B. (1995) The taxonomic status of the Louisiana pine snake (*Pituophis melanoleucus ruthveni*) and its relevance to the evolutionary species concept. *Journal of Herpetology* 29(2), 186–198, 6 figs.

Reid, H.A. (1966) Clinical notes on snakebite. *In:* Manson-Bahr, P.H. (Ed.), *Manson's tropical diseases: a manual of the diseases of warm climates.* Sixteenth edition. Williams & Wilkins, Baltimore, pp. 757–760.

Reinhardt, J.C.H. (1836) Den physiske Classe. *Oversigt over det Kongelige danske Videnskabernes Selskabs Forhandlinge og dets Medlemmers Arbeider*, Kjöbenhavn (1836–1837) 3, 6–7.

Reinhardt, J.T. (1843) Beskrivelse af nogle nye Slangearter. *Kongelige Danske Videnskabernes Selskabs, Naturvidenskabelige og Mathematiske Afhandlinger*, Kjöbenhavn 10, 233–279, 3 pls.

Reinhardt, J.T. (1844) Description of a new species of venomous snake, *Elaps Macclellandi. Calcutta Journal of Natural History, and Miscellany of the Arts and Sciences in India* 4, 532–534.

Reinhardt, J.T. (1861a) Herpetologiske meddelelser. I. Om nogle smaa Gruber, hvormed Skjaellene hos mange Slanger ere forsynede. *Videnskabelige Meddelelser fra den Naturhistoriske Forening i Kjöbenhavn* (1860) 22, 209–228.

Reinhardt, J.T. (1861b) Herpetologiske meddelelser. II. Beskrivelser af nogle nye til calamariernes Familie henhörende Slanger. *Videnskabelige Meddelelser fra den Naturhistoriske Forening i Kjöbenhavn* (1860) 22, 229–246, pls. 3–4.

Reinhardt, J.T. (1861c) Ueber einige kleine Gruben an den Schuppen mancher Schlangen. *Archiv für Naturgeschichte,* Berlin 27(1), 127–147, pl. 5.

Reinhardt, J.T. (1863) En ny Slaegt af Slangefamilien Rhachiodontidae. *Oversigt over det Kongelige danske Videnskabernes Selskabs Forhandlinger og dets Merdlemmrs Arbeider Kjöbenhavn* 1(2), 198–210, 1 pl. [English translation by Neilson and Gans, 1955, Carnegie Museum, 8 pp.]

Reinhardt, J.T. (1866) To nye Homalopsider. *Videnskabelige Meddelelser fra den Naturhistoriske Forening i Kjöbenhavn* (1867–1868) 28, 151–161, 6 figs.

Reinhardt, J.T. & Lütken, C.F. (1862) Bidrag til det vestindiske öriges og navnligen til de dansk–vestindiske öers Herpetologie. *Videnskabelige Meddelelser fra den Naturhistoriske Forening i Kjöbenhavn* (1862–1863) 24(10–18), 153–291.

Reitinger, F.F. & Lee, J.K.S. (1978) *Common snakes of south east Asia and Hong Kong*. Heinemann, Hong Kong, 114 pp., 26 figs., 31 col. pls.

Reitinger, F.F., Lee, J.K.S. & Lee, D. (1978) Recording amphibians and reptiles in the New Territories (1973–1977). *Memoirs of the Hong Kong Natural History Society* (13), 39–41.

Reitz, C.J. (1978) Poisonous South African snakes and snakebite. *South Africa Department of Health, Pretoria,* 64 + 64 pp., col. illust.

Rendahl, H. (1918) Zwei neue Arten der Gattung *Typhlops* aus dem Kaplande. *Arkiv für Zoologi*, Stockholm 11(7), 1–6.

Rendahl, H. (1933) Die Unterarten des *Ancistrodon halys* Pall. nebst einigen Bemerkungen zur Herpetologie Zentralasiens. *Arkiv für Zoologi*, Stockholm 25A(8), 1–33, 4 figs.

Rendahl, H. (1937) Beiträge zur Herpetologie von Birma. *Arkiv für Zoologi*, Stockholm 29A(10), 1–29, 5 figs.

Rendahl, H. (1939) Zur herpetologie der Seychellen. I. Reptilien. *Zoologische Jahrbuch, Abteilung für Systematik, Ökologie und Geographie der Tiere*, Jena 72, 255–328, illust.

Rendahl, H. & Vestergren, G. (1940) Notes on Colombian snakes. *Arkiv für Zoologi*, Stockholm 33A(1), 1–16, 4 figs.

Rendahl, H. & Vestergren, G. (1941) On a small collection of snakes from Ecuador. *Arkiv für Zoologi*, Stockholm 33A(5), 1–16, 10 figs.

Renjifo, J.M. & Lundberg, M. (1999) *Guía de campo: anfibios y reptiles de Urrá.* Skanska, Medellín. 96 pp., 97 col. pls.

Renjifo, J.M. & Lundberg, M. (2003) Una nueva especies de serpiente coral (Elapidae, *Micrurus*), de la Región de Urrá, Municipio de Tierra Alta, Córdoba, noroccidente de Colombia. *Revista de la Academia Colombiana de Ciencias* 27(102), 141–144.

Renner, M. & Vitzthum, S. (2007) *Amphibiens et reptiles de Lorraine: les observer, les identifier, les protéger.* Editions Serpenoise, Metz, 272 pp., 400 col. photos.

Renoult, J.P. & Raselimanana, A.P. (2009) A new species of Malagasy blind snake of the genus *Typhlops* Oppel (Serpentes: Typhlopidae). *Zootaxa* (2290), 65–68, 1 fig.

Resetar, A.R. & Marx, H. (1981) A redescription and generic reallocation of the African colubrid snake *Elapocalamus gracilis* Boulenger with a discussion of the union of the brille and postocular shielf. *Journal of Herpetology* 15(1), 83–89, 1 fig.

Restrepo, J.H. & Wright, J.W. (1987) A new species of the colubrid snake genus *Geophis* from Colombia. *Journal of Herpetology* 21 (3), 191–196, 3 figs.

Reuss, A. (1834) Zoologische Miscellen. Reptilien. Ophidier. *Museum Senckenbergianum*, Frankfurt-am-Main 1, 127–162, pls. 6–9.

Reuss, A.F.T. (1923) Systematische Beobachtungen an *Vipera* (*Pelias*) *berus* L. in Gefangenschaft. I. Paarung und Zucht; Verebungasymmetrischer Kopfzeichnung. II. Kämpfe der Ottern, Tänze und Liebeswerbungen. Das Signum der Weibchen. *Wochenschrift für Aquarien- und Terrarienkunde* (19), 295–296.

Reuss, A.F.T. (1924a) Systematische Beobachtungen an *Vipera* (*Pelias*) *berus* L. II. Eine neue Mutation der Kreuzotter aus der Mark Brandenburg. *Wochenschrift für Aquarien- und Terrarienkunde* (25), 545–547.

Reuss, A.F.T. (1924b) Systematische Beobachtungen an *Vipera* (*Pelias*) *berus* L. IIa. Eine neue gestreifte Aberration von *Vipera (Pelias) berus* L. III. Temperaturformen der Kreuzotter. *Wochenschrift für Aquarien- und Terrarienkunde* (31), 665–666.

Reuss, A.F.T. (1925a) Vergleichende Beobachtungen an einigen Giftschlangen. *Blätter für Aquarien- und Terrarienkunde* (8), 211–215.

Reuss, A.F.T. (1925b) No title (under Vereinsnachrichten-Berlin. "Lacerta"). *Wochenschrift für Aquarien- und Terrarienkunde* (28), 495.

Reuss, A.F.T. (1925c) Presentation-Lautäusserungen bei Schlangen (under Vereinsnachrichten-Berlin. "Lacerta") *Wochenschrift für Aquarien- und Terrarienkunde* (43), 720–721.

Reuss, A.F.T. (1926) No title (under Vereinsnachrichten-Berlin. "Lacerta"). *Wochenschrift für Aquarien- und Terrarienkunde* (23), 344–345.

Reuss, A.F.T. (1927a) Sechs europäische Giftschlangengattungen. *Zoologischer Anzeiger*, Leipzig 73(5–8), 124–129.

Reuss, A.F.T. (1927b) Beschreibung neuer Vipern aus Jugoslawien. *Zoologischer Anzeiger*, Leipzig 71(9–10), 215–223, 8 figs.

Reuss, A.F.T. (1927c) Über vermutliche Stammformen der europäischen Vipern in Südosteuropa. *Der Naturforscher*, Berlin 3, 639–642, pls. 89–90.

Reuss, A.F.T. (1927d) Vergleichende Bilder alter und neuer europäischer Giftschlangen. *Der Naturforscher*, Berlin 4, 127–129, pls. 20–21.

Reuss, A.F.T. (1929a) No title (under Vereinsnachrichten-Berlin. "Lacerta"). *Wochenschrift für Aquarien- und Terrarienkunde*, Braunschweig (5), 64.

Reuss, A.F.T. (1929b) *Canis familiaris* var. *catus*. *Wochenschrift für Aquarien- und Terrarienkunde*, Braunschweig (15), 217–220 & (16), 235–237.

Reuss, A.F.T. (1930a) No title (under Vereinsnachrichten-Berlin. "Lacerta"). *Wochenschrift für Aquarien- und Terrarienkunde,* Braunschweig (18), 296–297.

Reuss, A.F.T. (1930b) No title (under Vereinsnachrichten-Berlin. "Lacerta"). *Wochenschrift für Aquarien- und Terrarienkunde,* Braunschweig (24), 397–398.

Reuss, A.F.T. (1930c) Über eine neurotoxische Otterngruppe Europas, *Mesocoronis* 1927, und über ihre Stellung unter den Solenoglyphen der Welt. *Glasnik Zemaljskog Muzeja u Bosni I Hercegovini*, Sarajevo 42, 57–114, 6 pls.

Reuss, A.F.T. (1932) No title (under Vereinsnachrichten-Berlin. "Lacerta"). *Nachrichtenblatt für Aquarien- und Terrarien-Vereine*, Berlin (4), 4–5.

Reuss, A.F.T. (1933a) No title (under Vereinsnachrichten-Berlin. "Lacerta"). *Nachrichtenblatt für Aquarien- und Terrarien-Vereine*, Berlin (8), 98–99.

Reuss, A.F.T. (1933b) No title (under Vereinsnachrichten-Berlin. "Lacerta"). *Nachrichtenblatt für Aquarien- und Terrarien-Vereine*, Berlin (19), 283–284.

Reuss, A.F.T. (1933c) Fortsetzung und Schluss der Originalberichte (under Vereinsnachrichten-Berlin. "Lacerta"). *Nachrichtenblatt für Aquarien- und Terrarien-Vereine*, Berlin (26), 372–373.

Reuss, A.F.T. (1935a) Erläuterungen zur Typenfrage in der Zoologie an hand von Bildern europäischer Giftschlangen. *Nachrichtenblatt für Aquarien- und Terrarien-Vereine*, Berlin (16), 215–218, illust.

Reuss, A.F.T. (1935b) Europäisch Giftschlangen. Uebersichtskarte der Giftschlangenzuwanderung in Europa seit der Eiszeit (Quartärzeit). Mit einer Einführung in die entsprechende Aufteilung der europäischen Ottern unter Berücksichtigung von Gift- und Heilserumfragen. *Deutscher Almanach für Aquarien- und Terrarienfreunde,* Berlin 1935, 145–166.

Reuss, A.F.T. (1936). Der neue Rohstoff, das neue Rauschgift: Ophiotoxin. Eine Ergänzung zu "Europäische Giftschlangen" in diesem "Almanach," Ausgabe 1935. *Deutscher Almanach für Aquarien- und Terrarienfreunde,* Berlin 1936, 141–147.

Reuss, A.F.T. (1937) Observations on four species European Toxicophidia. *Comptes Rendus du XIIe Congrès International de Zoologie*, Lisbonne, 1935, 1787–1804.

Reuss, A.F.T. (1938a) Zoogeographische Toxicophiologie im Frühjahr 1938. Mit eine Beitrag zur Kunde der Schlangenbisse und zwei Neubeschreibungen, *Mesovip. morathi,* spec. nov., und *leopontica,* suspec. nova. *Das Aquarium. Die Fachzeitschrift des Naturfreundes*, Berlin 12(5), 78–80, 2 figs.

Reuss, A.F.T. (1938b) *Mesovipera stemmler morathi.* Eine neue Nasenotter aus Südfrankreich, Norditalien und der Südschweiz. *Zeitschrift für Aquarien- und Terrarien-Vereine*, Berlin (15), 166.

Reuss, A.F.T. (1938c) Ein Beitrag zur Kunde der Giftschlangen und der zugehörigen Unterscheidung von Giftschlangen (Toxophidia). *Das Aquarium. Die Fachzeitschrift des Naturfreundes,* Berlin 12(6), 95–99.

Reuss, A.F.T. (1939) Berichtigungen und Ergänzungen zu meinen Arbeiten über Toxicophidier, 1938. *Zeitschrift für Aquarien- und Terrarien-Vereine*, Berlin (1), 13–14.

Rey, L.R. & Lions, M.L. (1997a) Geographic distribution: Serpentes: *Atractus taeniatus. Herpetological Review* 28(1), 51.

Rey, L.R. & Lions, M.L. (1997b) Geographic distribution: Serpentes: *Oxyrhopus rhombifer inaequifasciatus. Herpetological Review* 28(3), 160.

Reyes-Velasco, J. & Mulcahy, D.G. (2010) Additional taxonomic remarks on the genus *Pseudoleptodeira* (Serpentes: Colubridae) and the phylogenetic placement of "*P. uribei.*" *Herpetologica* 66(1), 99–110, 5 figs.

Reyes-Velasco, J., Grünwald, C.J. & Jones, J.M. (2008a) Geographic distribution: Serpentes: *Dipsas gaigeae. Herpetological Review* 39(2), 241.

Reyes-Velasco, J., Grünwald, C.L. & Jones, J.M. (2008b) Geographic distribution: Serpentes: *Tropidodipsas annulifera. Herpetological Review* 39(4), 487.

Reyes-Velasco, J., Hermosillo-Lopez, I.A., Grünwald, C.L. & Avila-Lopez, O.A. (2009) New state records for amphibians and reptiles from Colima, Mexico. *Herpetological Review* 40(1), 117–120, 1 fig.

Reyes-Velasco, J., Grünwald, C.I., Jones, J.M. & Weatherman, G.N. (2010) Rediscovery of the rare Autlán long-tailed rattlesnake, *Crotalus lannomi. Herpetological Review* 41(1), 10–25, 4 figs.

Reyes-Velasco, J., Grünwald, C.I., Jones, J.M., Price, M.S. & Fisher, J.T. (2012) New distributional records for the herpetofauna of Mexico. *Herpetological Review* 43(3), 451–453.

Reynolds, R.G. (2012) *Epicrates chrysogaster. Catalogue of American amphibians and reptiles,* pp. 898.1–898.5, 4 figs.

Reynolds, R.G. & Riggs, B.M. (2011) Geographic distribution: Serpentes: *Pantherophis guttatus. Herpetological Review* 42(2), 243.

Reynolds, R.G., Gerber, G.P. & Burgess, J. (2010) Geographic distribution: Serpentes: *Tropidophis greenwayi greenwayi. Herpetological Review* 41(4), 520.

Reynolds, R.G., Batchasingh, J. & Riggs, B.M. (2011) Geographic distribution: Serpentes: *Ramphotyphlops braminus. Herpetological Review* 42(2), 244.

Reynolds, R.P. & Foster, M.S. (1992) Four new species of frog and one new species of snake from the Chapare region of Bolivia, with notes on other species. *Herpetological Monographs* (6), 83–104, 9 figs.

Reynolds, R.P. & MacCulloch, R.D. (2012) Preliminary checklist of amphibians and reptiles from Baramita, Guyana. *Check List* 8(2), 211–214, 3 figs.

Reynolds, R.P., Fritts, T., Gotte, S., Icochea, J. & Tello, G. (1997) Amphibians and reptiles I: biodiversity assessment in the lower Urubamba region. *In:* Dallmeier, F. & Alonso, A. (Eds.), *Biodiversity assessment and long-term monitoring, lower Urubamba region, San Martin-3 and Cashiriari-2 well sites.* Smithsonian Institution Monitoring & Assessment of Biodiversity Program (1), 213–221.

Reza, A.H.M. (2010a) Colubrid snake *Lycodon zawi* (Serpentes: Colubridae) from Lawachara National Park in Bangladesh. *Russian Journal of Herpetology* 17(1), 75–77, 1 fig.

Reza, A.H.M. (2010b) First record of *Amphiesma venningi* (Wall, 1910) (Serpentes, Colubridae, Natricinae) from Bangladesh, with notes on its taxonomy, natural history, biogeography and other sympatric species. *Hamadryad* 35(1), 64–72, 4 figs.

Rezaie-Atagholipour, M., Riyahi-Bakhtiari, A., Rajabizadeh, M. & Ghezellou, P. (2012) Status of the annulated sea snake, *Hydrophis cyanocinctus,* in the Hara Protected Area of the Persian Gulf. *Zoology in the Middle East* 57, 53–70, 5 figs.

Rheubert, J.L., McMahan, C.D., Sever, D.M., Bundy, M.R., Siegel, D.S. & Gribbins, K.M. (2010) Ultrastructure of the reproductive system of the black swamp snake *(Seminatrix pygaea).* VII. Spermatozoon morphology and evolutionary trends of sperm characters in snakes. *Journal of Zoological Systematics and Evolutionary Research* 48(4), 366–375, 6 figs.

Rhoads, D. (2008) *The complete Suboc: a comprehensive guide to the natural history, care, and breeding of the Trans-Pecos ratsnake.* ECO Herpetological Publishing, Lansing, 291 pp., col. illust.

Rhoads, D. & Salmon, G.T. (2012) A much belated obituary of an American zoo collector, with discussion of the type locality for *Bogertophis subocularis* and *Lampropeltis alterna. Herpetological Review* 43(2), 270–273, 6 figs.

Rhoads, S.N. (1895) Contributions to the zoology of Tennessee. No. 1, Reptiles and amphibians. *Proceedings of the Academy of Natural Sciences of Philadelphia* 47, 376–407.

Rhumbler, L. (1910) Über eine zweckmässige Weiterbildung der Linnéschen binären nomenklatur. *Zoologischer Anzeiger*, Leipzig 36(26), 453–471.

Ribeiro, L. (1939) *Brazilian medical contributions.* Livraria José Olympio, Rio de Janeiro, 181 pp., illust.

Ribeiro, S.C., Valença, N.S.M.S. & Guarnieri, M.C. (2011) Geographic distribution: *Xenopholis undulatus. Herpetological Review* 42(1), 116.

Ribeiro, S.C., Roberto, I.J., Lima-Sales, D., Ávila, R.W. & Oliveira-Almeida, W. de. (2012) Amphibians and reptiles from the Araripe bioregion, northeastern Brazil. *Salamandra* 48(3), 133–146, 5 figs.

Richmond, N.D. (1954) The ground snake, *Haldea valeriae*, in Pennsylvania and West Virginia with the description of a new subspecies. *Annals of the Carnegie Museum* (1952–1956) 33(15), 251–260.

Richmond, N.D. (1955) The blind snakes (*Typhlops*) of Bimini, Bahama Islands, British West Indies, with description of new species. *American Museum Novitates* (1734), 1–7, 1 fig.

Richmond, N.D. (1964) The blind snakes (*Typhlops*) of Haiti with descriptions of three new species. *Breviora* (202), 1–12, 4 figs.

Richmond, N.D. (1965) A new species of blind snake, *Typhlops*, from Trinidad. *Proceedings of the Biological Society of Washington* 78, 121–124, 1 fig.

Richmond, N.D. (1966) The blind snakes, *Typhlops*, of Guadeloupe and Dominica with the description of a new species. *Herpetologica* 22(2), 129–132, 1 fig.

Ridley, H.N. (1899) The habits of Malay reptiles. *Journal of the Straits Branch of the Royal Asiatic Society*, Singapore (32), 185–210.

Riemer, W.J. (1954) A new subspecies of the snake *Masticophis lateralis* from California. *Copeia* 1954(1), 45–48.

Rieppel, O. (1978) A functional and phylogenetic interpretation of the skull of Erycinae (Reptilia, Serpentes). *Journal of Zoology*, London 186(2), 185–208, 12 figs.

Rieppel, O. (1979) A cladistic classification of primitive snakes based on skull structure. *Zeitschrift für Zoologische Systematik und Evolutionsforschungs*, Hamburg 17, 140–150, 5 figs.

Rieppel, O. & Zaher, H. (2000) The intramandibular joint in squamates, and the phylogenetic relationships of the fossil snake *Pachyrhachis problematicus*. *Fieldiana: Geology* (43), 1–69, 17 figs.

Rieppel, O., Zaher, H., Tchernov, E. & Polcyn, M.J. (2003) The anatomy and relationships of *Haasiophis terrasanctus*: a fossil snake with well-developed hind limbs from the mid-Cretaceous of the Middle East. *Journal of Paleontology* 77(3), 536–558, 9 figs.

Ripa, D. (1994) Reproduction of the Central American bushmaster (*Lachesis muta stenophrys*) and the black-headed bushmaster (*Lachesis muta melanocephala*) for the first time in captivity. *Bulletin of the Chicago Herpetological Society* 29(8), 165–183, 12 figs.

Ripa, D. (1997) Range extension for *Bothrops leucurus*. *Bulletin of the Chicago Herpetological Society* 32(2), 25–26.

Ripa, D. (1999) Keys to understanding the bushmasters (genus *Lachesis* Daudin, 1803). *Bulletin of the Chicago Herpetological Society* 34(3), 45–92, 19 figs.

Ripa, D. (2001) *The bushmasters (genus* Lachesis *Daudin, 1803): morphology in evolution and behavior.* Ripa Ecologia, Wilmington. (publ. on CD–Rom only)

Ripa, D. (2004) Resurrecting Garcia's *Bothrops* [*Lachesis*] *acrochordus*. When are discovering a species and naming it different? Reflections on the taxonomic "Big Bang"; or how I learned to stop worrying and love the Linnaean bomb. *Bulletin of the Chicago Herpetological Society* 39(7), 122–134.

Risso, A. (1826) *Histoire naturelle des principales productions de l'Europe meridionale et particulièrement de celles des environs de Nice et des Alpes maritimes. Tome troisième.* F.-G. Levrault, Paris, xvi + 480, 16 pls.

Ritgen, F.A. (1828) Versuch einer Natürlichen Eintheilung der Amphibien. *Nova Acta Academiae Caesareae Leopoldino-Carolinae*, Bonn 14, 246–284.

Ritter, C. (1823) Vorläusiger Bericht einer naturhistorischen Reise nach St. Domingo (Hayti). *Archiv für Geschichte, Statistik, Literatur und Kunst,* Wien 14(84–85), 437–441.

Ritter, C. (1824a) Sur un voyage fait a Saint-Domingue de 1819 à 1821; par un botaniste austrichien. *Isis von Oken* 1, 61.

Ritter, C. (1824b) Nuovo viaggio nell'interno del Brasile. *Antologia* (Luglio, Agosto, Settembre, 1824), Firenze 15(44), 206.

Riva de la Vina, I., Castroviejo-Bolibar, J. & Cabot-Nieves, J. (1991) *Pseustes sulphureus* (Wagler, 1824) (Serpentes, Colubridae) nueva especie para Bolivia y datos sobre la herpetofauna boliviana. *Acta Zoologica Lilloana*, Tucumán 41, 215–218.

Rivas, J.A., Muñoz, M.C., Thorbjarnarson, J.B., Burghardt, G.M., Holmstrom, W. & Calle, P.P. (2007) Natural history of the green anaconda (*Eunectes murinus*) in the Venezuelan llanos. *In:* Henderson, R.W. & Powell, R. (Eds.), *Biology of the boas and pythons*. Eagle Mountain Publishing, Eagle Mountain, pp. 129–138, 13 figs.

Rivas-Fuenmayor, G. (2001) Geographic distribution: Serpentes: *Typhlops reticulatus*. *Herpetological Review* 32(4), 280.

Rivas-Fuenmayor, G.A. (2002) Geographic distribution: Serpentes: *Oxyrhopus doliatus*. *Herpetological Review* 33(2), 150.

Rivas-Fuenmayor, G.A. & Barrio-Amorás, C.L. (2005) New amphibian and reptile records from Cójedes State, Venezuela. *Herpetological Review* 36(2), 205–208.

Rivas-Fuenmayor, G.A. & Fuentes, O. (2000) Geographic distribution *Hydrops triangularis*. *Herpetological Review* 31(3), 186.

Rivas-Fuenmayor, G.A. & Fuentes, O. (2002) Geographic distribution: Serpentes: *Liophis epinephelus opisthotaenius*. *Herpetological Review* 33(1), 68.

Rivas-Fuenmayor, G.A. & La Marca, E. (2001a) Geographic distribution: Serpentes: *Chironius scurrulus*. *Herpetological Review* 32(1), 58–59.

Rivas-Fuenmayor, G. & La Marca, E. (2001b) Geographic distribution: Serpentes: *Clelia clelia*. *Herpetological Review* 32(1), 59.

Rivas-Fuenmayor, G.A. & La Marca, E. (2001c) Geographic distribution: Serpentes: *Pseudoboa coronata*. *Herpetological Review* 32(2), 124.

Rivas-Fuenmayor, G.A. & Molina-R, C.R. (1998) Geographic distribution, *Pseustes shopshirei*. *Herpetological Review* 29(3), 178.

Rivas-Fuenmayor, G.A. & Molina-R, C.R. (2003) New records of reptiles for the Orinoco Delta, Delta Amacura state, Venezuela. *Herpetological Review* 34(2), 171–173.

Rivas-Fuenmayor, G.A. & Oliveros, O. (1998) Herpetofauna del Estado Sucre, Venezuela: lista preliminaire de reptiles. *Memoria de la Sociedad de Ciencias Naturales La Salle* (1997) 57(147), 67–80, illust.

Rivas-Fuenmayor, G.A., Molina, C.R. & Fuentes, O. (2001) Geographic distribution: Serpentes: *Typhlops reticulatus*. *Herpetological Review* 32(2), 126–127.

Rivas-Fuenmayor, G.A., Fuentes, O. & Barrio, C.L. (2002) Geographic distribution: Serpentes: *Liophis poecilogyrus schotti*. *Herpetological Review* 33(1), 68.

Rivas-Fuenmayor, G.A., Ugueto, G., Rivero, R. & Miralles, A. (2005) The herpetofauna of Isla de Margarita, Venezuela: new records and comments. *Caribbean Journal of Science* 41(2), 346–351.

Rivas-Fuenmayor, G.A., Molina, C.R., Ugueto, G.N., Barros, T.R., Barrio-Amorós, C.L. & Kok, P.J.R. (2012) Reptiles of Venezuela: an updated and commented checklist. *Zootaxa* (3211), 1–64, 10 col. figs.

Rivera, P.C., Di Cola, V., Martinez, J.J., Gardenal, C.N. & Chiaraviglio, M. (2011) Species delimitation in the continental forms of the genus *Epicrates* (Serpentes, Boidae) integrating phylogenetics and environmental niche models. *PLoS ONE* 6(9), 1–13, 3 figs.

Rivera-Rodriguez, M.J., Bentz, E.J., Scantlebury, D.P., John, R.R., Quinn, D.P., Parmerlee, J.S., Jr., Henderson, R.W. & Powell, R. (2011) Rediscovery of the Grenada Bank endemic *Typhlops tasymicris* (Squamata: Typhlopidae). *Journal of Herpetology* 45(2), 167–168, 2 figs.

Rivero, J.A. (1998) *The amphibians and reptiles of Puerto Rico.* Second edition. University of Puerto Rico, San Juan, 510 pp., 66 pls. [first edition, 1978, x + 152 pp., 49 pls., vi + 144 pp.]

Rivero, J.A., Joglar, R. & Vazquez, I. (1982) Cinco nuevos ejemplares del culebron de la Mona, *Epicrates m. monensis* (Ophidia, Boidae). *Caribbean Journal of Science*, Mayagüez 17(1–4), 7–11, 2 figs.

Robb, J. (1960) The internal anatomy of *Typhlops* Schneider (Reptilia). *Australian Journal of Zoology*, Melbourne 8(2), 181–216, 24 figs., 3 pls.

Robb, J. (1967) The generic status of the Australian typhlopids (Reptilia, Squamata). *Annals and Magazine of Natural History*, London (Series 13) (1966) 9(106–108), 675–679, 6 figs. [9 March 1967]

Robb, J. (1972) A new species of the genus *Ramphotyphlops* (Serpentes, Typhlopidae) from western Australia. *Journal of the Royal Society of Western Australia*, Perth 55(2), 39–40, 1 fig.

Robb, J. (1980) *New Zealand amphibians and reptiles in colour.* Collins, Auckland, 128 pp., illust.

Robb, J. & Smith, H.M. (1966) The systematic position of the group of snake genera allied to *Anomalepis*. *Natural History Miscellanea*, Chicago (184), 1–8.

Roberto, I.J. & Veiga, S. (2009) Geographic distribution: Serpentes: *Leptotyphlops borapeliotes*. *Herpetological Review* 40(2), 238.

Roberto, I.J., Ribeiro, S.C. & Lima-Melo, J.C. (2009a) Geographic distribution: Serpentes: *Thamnodynastes almae*. *Herpetological Review* 40(2), 239.

Roberto, I.J., Pinto de Brito, P.T., Ribeiro, S.C. & Lima-Melo, J.C. (2009b) Geographic distribution: Serpentes: *Thamnodynastes sertanejo*. *Herpetological Review* 40(2), 239.

Roberto I.J., Ribeiro, S.C., Sousa-Delfino, M.M. de, & Oliveira-Almeida, W. de (2009c) Reptilia, Colubridae, *Helicops angulatus*: distributional extension and rediscovery in the state of Ceará. *Check List* 5(1), 118–121, 2 figs.

Robertson, W.B. & Tyson, E.L. (1950) Herpetological notes from eastern North Carolina. *Journal of the Elisha Mitchell Science Society* 66(2), 130–147.

Robinette, J.W. & Trauth, S.E. (1992) Reproduction in the western mud snake *Farancia abacura reinwardtii* (Serpentes, Colubridae) in Arkansas. *Proceedings of the Arkansas Academy of Sciences* 46, 61–64, 4 figs.

Robinson, H.C. & Kloss, C.B. (1916) The natural history of Kedah Peak. *Journal of the Federated Malay States Museum*, Kuala Lumpur 6(4), 219–244.

Robinson, H.C. & Kloss, C.B. (1920) Reptiles and batrachians collected in Korinchi, West Sumatra. *Journal of the Federated Malay States Museum*, Kuala Lumpur 8(2), 297–306.

Robinson, M.D. & Hughes, D.A. (1978) Observations on the natural history of Peringuey's adder, *Bitis peringueyi* (Boulenger) (Reptilia: Viperidae). *Annals of the Transvaal Museum* 31(16), 189–193, 3 pls.

Robosky, D.L., Aplin, K.P., Donnellan, S.C. & Hedges, S.B. (2004) Molecular phylogeny of blindsnakes (*Ramphotyphlops*) from western Australia and resurrection of *Ramphotyphlops bicolor* (Peters, 1857). *Australian Journal of Zoology* 52, 531–548, 6 figs.

Rocha, C.F.D. & Pontes, J.A.L. (2008) *Serpentes da Serra do Mendanha, Rio de Janeiro, RJ: ecología e conservaçao.* Editora Technica, 147 pp., 100 col. photos.

Rochebrune, A.T. de. (1880) Revision des ophidiens fossiles du Museum d'Histoire Naturelle. *Nouvelles Archives du Muséum d'Histoire Naturelle*, Paris (Série 2) 3, 271–296, pl. 12.

Rochebrune, A.T. de. (1884) Faune ophiologique des phosphorites du Quercy. *Mémoires de la Société des Sciences Naturelles de Saone-et-Loire* 5, 149–164, 2 pls.

Rochebrune, A.T. de. (1885) *Fauna de la Sénégambie, Reptiles.* Octave Douin, Paris, (1884) 221 pp., 20 pls.

Rochebrune, A.T. de. (1886) Vertebratorum novorum vel minus cognitorum orae Africae occidentalis incolarum. *Bulletin de la Société Philomathique de Paris* (7) 9, 86–99.

Rodda, G.H., Sawai, Y., Chiszar, D. & Tanaka, H. (1999) *Problem snake management: the habu and the brown treesnake.* Cornell University Press, Ithaca, 534 pp., illust.

Roddy, H.J. (1928) *Reptiles of Lancaster County and the state of Pennsylvania.* Science Press, Lancaster, v + 53 pp., illust.

Rödel, M.-O. (1994) Beiträge zur Kenntnis der Verbreitung, Habitatwahl und Biologie griechischer Amphibien und Reptilien: Daten aus 7 Exkursionen von 1987 bis 1991 (Amphibia et Reptilia). *Faunistiche Abhandlungen, Staatliche Museum für Tierkunde in Dresden* 19(29), 227–246, 1 fig.

Rödel, M.-O. & Bussmann, R. (1992) Die Amphibien und Reptilien der Julischen Alpen. *Sauria*, Berlin 12(4), 15–21, illust.

Rödel, M.-O., Grabow, K., Böckheler, C. & Mahsberg, D. (1995) Die Schlangen des Comóe-Nationalparks, Elfenbeinküste (Reptilia, Squamata, Serpentes). *Stuttgarter Beiträge zur Naturkunde, Stuttgart* (528)A, 1–18, 2 figs.

Rödel, M.-O., Kouadio, K. & Mahsberg, D. (1999) Die Schlangenfauna des Comoé-Nationalparks, Elfenbeinküste, Ergänzungen und Ausblick. *Salamandra* 35(3), 165–180, 5 figs.

Rodgers, W.A. & Homewood, K.M. (1982) Species richness and endemism in the Usambara mountains forests, Tanzania. *Biological Journal of the Linnean Society,* London 18(3), 197–242, 5 figs.

Rodrigues, L.B. & Cadle, J.E. (1990) A preliminary overview of the herpetofauna of Cocha Cashu, Manu National Park, Peru. *In:* Gentry, A.H. (Ed.), *Four neotropical rainforests.* Yale University Press, New Haven, pp. 410–425, illust.

Rodrigues, M.T. (1991) Herpetofauna das dunas interiores do Rio São Francisco. Bahia, Brasil. IV. Uma nova espècie de *Typhlops* (Ophidia, Typhlopidae). *Papeis Avulsos Zoologia*, São Paulo 37(22), 343–346, 2 figs.

Rodrigues, M.T. (1993a) Herpetofauna das dunas interiores do Rio São Francisco, Bahia, Brasil. V. Duas novas espécies de *Apostolepis* (Ophidia, Colubridae). *Memórias do Instituto Butantan*, São Paulo (1992) 54(2), 53–59, 3 figs.

Rodrigues, M.T. (1993b) Herpetofauna of palaeoquaternary sand dunes of the middle São Francisco River, Bahia, Brazil. VI. Two new species of *Phimophis* (Serpentes, Colubridae), with notes on the origin of psammophilic adaptations. *Papeis Avulos Zoologia*, Museu de Zoologia da Universidade de São Paulo 38(11), 187–198, 4 figs.

Rodrigues, M.T. & Junca, F.A. (2002) Herpetofauna of the Quaternary sand dunes of the middle Rio São Francisco: Bahia, Brazil. VII. *Typhlops amoipira* sp. nov., a possible relative of *Typhlops yonenagae* (Serpentes, Typhlopidae). *Papeis Avulos Zoologia*, Museu de Zoologia da Universidade de São Paulo 42(13), 325–333.

Rodrigues, M.T. & Puorto, G. (1994) On the second specimen of *Leptotyphlops brasiliensis* Laurent, 1949 (Serpentes, Leptotyphlopidae). *Journal of Herpetology* 28(3), 393–394, 1 fig.

Rodrigues-França, F.G., França, R.C. de, Souza-Germano, C.E. de & Oliveira-Filho, J.M. de (2012) *Bothropoides neuwiedi* (Wagler, 1824) (Squamata: Serpentes: Viperidae): distribution extension for the Atlantic forest, first vouchered record for Paraíba, and geographic distribution map. *Check List* 8(1), 170–171, 3 figs.

Rodrigues-Machado, S. (1993) A new genus of Amazonian vine snake (Xenodontinae, Alsophiini). *Acta Biologica Leopoldensia* 15(2), 99–108, 4 figs.

Rodríguez, J.P. & Rojas-Suárez, F. (1995. *Libro rojo de la fauna Venezolana.* Provita, Fundación Polar, Wildlife Conservation Society, Profauna–MARNR, UICN, Caracas 444 pp.

Rodríguez-Robles, J.A. & Greene, H.W. (1996) Ecological patterns in Greater Antillean macrostomatan snake assemblages, with comments on body-size evolution in *Epicrates* (Boidae). *In:* Powell, R. & Henderson, R. (Eds.), *Contributions to West Indian herpetology: a tribute to Albert Schwartz.* Society for the Study of Amphibians and Reptiles, pp. 339–357, 7 figs.

Rodríguez-Robles, J.A. & Greene, H.W. (1999) Food habits of the long-nosed snake (*Rhinocheilus lecontei*): a "specialist" predator? *Journal of Zoology*, London 248(4), 489–499, 3 figs.

Rodríguez-Robles, J.A. & de Jesús-Escobar, J.M. (1999) Molecular systematics of New World lampropeltinine snakes (Colubridae): implications for biogeography and evolution of food habits. *Biological Journal of the Linnean Society*, London 68(3), 355–385, 7 figs.

Rodríguez-Robles, J.A. & de Jesús-Escobar, J.M. (2000) Molecular systematics of New World gopher, bull, and pinesnakes (*Pituophis*, Colubridae): a transcontinental species complex. *Molecular Phylogenetics and Evolution* 14(1), 35–50, 6 figs.

Rodríguez-Robles, J.A., Denardo, D.F. & Staub, R.E. (1999) Phylogeography of the California mountain king snake, *Lampropeltis zonata* (Colubridae). *Molecular Ecology* 8(11), 1923–1934, 5 figs.

Rodríguez-Robles, J.A., Stewart, G.R. & Papenfuss, T.J. (2001) Mitochondrial DNA-based phylogeography of North American rubber boas, *Charina bottae* (Serpentes, Boidae). *Molecular Phylogenetics and Evolution* 18(2), 227–237, 6 figs.

Rodriguez-Robles, J.A., Good, D.A. & Wake, D.B. (2002) Brief history of herpetology in the Museum of Vertebrate Zoology, University of California, Berkeley, with a list of type specimens of recent amphibians and reptiles. *University of California Publications in Zoology* (131), xv + 119 pp., 1 fig. 1 pl.

Rodríguez-Schettino, L. (2000) Cuban reptiles: original citations, holotypes, and geographic range. *Smithsonian Herpetological Information Service* (125), 1–26.

Rodríguez-Schettino, L. (2003) *Anfibios y reptiles de Cuba*. Instituto de Ecología y Sistemática, La Habana, vi + 169 pp., col. illust.

Rodríguez-Torres, J. & Vázquez-Díaz, J. (1990) Nuevo registro de distribución de *Geophis dugesi aquilonaris* para el estado de Jalisco, México. *Boletin de la Sociedad Herpetológica Méxicana* 2(1), 9–12, 1 fig.

Roemer, F. (1870) Ueber *Python euboïcus*: eine fossile Riesenschlange aus tertïarem Kalkschiefer von Kumi auf der Insel Euboea. *Zeitschrift der Deutschen Geologischen Gesellschaft*, Berlin 22(13), 582–590, pl. 13.

Rogé, J.-P. & Sauvanet, J. (1987) *La documentation Guyanaise. Les serpents*. Saga, Cayenne, 32 pp., 19 figs., 36 col. illust.

Rogers, K.L. (1976) Herpetofauna of the Beck Ranch local fauna (upper Pliocene, Blancan) of Texas. *Publications of the Museum, Michigan State University, Paleontological Series* 1(5), 163–200, 20 figs.

Rogers, L. (1903) On the physiological action of the poison of the Hydrophidae. *Proceedings of the Royal Society of London* 71, 481–496.

Roitman, M.I. (1967) The blood coagulation activity of *Vipera libertina turanica* venom and its neutralization by specific antiserum. *Toxicon* 5(1), 70.

Rojas-Morales, J.A. (2012a) On the geographic distribution of the false coral snake, *Rhinobothryum bovallii* (Serpentes: Dipsadidae), in Colombia — a biogeographical perspective. *Salamandra* 48(4), 243–248.

Rojas-Morales, J.A. (2012b) Snakes of an urban-rural landscape in the central Andes of Colombia: species composition, distribution, and natural history. *Phyllomedusa* 11(2), 135–154, 3 figs.

Rojas-Morales, J.A. & González-Durán. (2011) Description of the colouration in life of *Tricheilostoma joshuai* (Serpentes, Leptotyphlopidae). A species tolerant of disturbed habitats? *Salamandra* 47(4), 237–240, 2 figs.

Rojas-Runjaic, F.J.M. (2006) Reptilia, Squamata, Colubridae, *Chironius exoletus*: distribution extension, new state record. *Check List* 2(3), 82–83, 2 figs.

Rojas-Runjaic, F.J.M. & Infante-Rivero, E.E. (2004) Geographic distribution: Serpentes: *Liotyphlops albirostris*. *Herpetological Review* 35(4), 411.

Rojas-Runjaic, F.J.M. & Natera-Mumaw. (2008) First record of the forest racer snake *Dendrophidion percarinatum* (Cope, 1893) (Serpentes: Colubridae) from Venezuela. *Caribbean Journal of Science* 44(1), 128–130, 1 fig.

Rojas-Runjaic, F.J.M., Infante-Rivero, E.E., Barrio, C.L. & Barros-Blanco, T.R. (2007) New distributional records of amphibians and reptiles from Estado Zulia in the Maracaibo Basin, Venezuela. *Herpetological Review* 38(2), 235–237.

Rojas-Runjaic, F.J.M., Natera-Mumaw, M. & Infante-Rivero, E.E. (2008) Reptilia, Squamata, Colubridae, *Urotheca fulviceps*: Distribution extension. *Check List* 4(4), 431–433, 2 figs.

Rollinat, R. (1934) *La vie des reptiles de la France Centrale*. Librairie Delagrave, Paris, 343 pp., 13 figs., 24 b & w pls., 11 col. pls.

Roman, B. (1968) Serpentes de Haute-Volta. Une espèce nouvelle de *Naja* (cracheur) en Haute-Volta. *Carrefour Africain* (299), 2.

Roman, B. (1969) Les *Naja* de Haute-Volta. *Revue de Zoologie et Botanique Africaine*, Bruxelles 79(1–2), 1–14, 2 figs.

Roman, B. (1972) Deux sous-espèces de la vipere *Echis carinatus* (Schneider) dans les territoires de Haute-Volta et du Niger, *Echis carinatus ocellatus* Stemmler — *Echis carinatus leucogaster* n. ssp. *Notes et Documents Voltaiques*, Ouagadougou 5(4), 3–13, 1 pl.

Roman, B. (1973a) Une collection de Serpents peu commune. *Notes et Documents Voltaiques*, Ouagadougou 6(2), 36–58, 3 figs.

Roman, B. (1973b) Viperidés et elapidés de Haute-Volta. *Notes et Documents Voltaiques*, Ouagadougou 6(4), 1–49.

Roman, B. (1974a) Influence du climat sur la dispersion des serpents en Haute-Volta et dans l'ouest du Niger (première étude). *Notes et Documents Voltaiques*, Ouagadougou 7(3), 44–53, 2 figs.

Roman, B. (1974b) Deux espèces du genre *Crotaphopeltis* (colubridés opistoglyphes) dans le territoire de Haute-Volta, *Crotaphopeltis hotamboeia* (Laurenti), et *Crotaphopeltis acarina* n. sp. *Notes et Documents Voltaiques*, Ouagadougou 8(1), 3–15, 4 pls.

Roman, B. (1975) La vipère *Echis carinatus leucogaster*, Roman, 1972 de Haute Volta et du Niger elevée au rang d'espèce, *Echis leucogaster*. *Notes et Documents Voltaiques*, Ouagadougou 8(4), 3–20.

Roman, B. (1976) *Études Scientifiques. Serpents mortels de l'Ouest Africain*. Editions et Publications des Pères Jesuites, Faggala—Le Caire, 56 pp., 3 figs., 18 illust., 6 col. figs.

Roman, B. (1977) Le genre *Telescopus* (colubridés opistoglyphes) Wagler, 1830 en Haute-Volta et dans l'ouest du Niger. *Notes et Documents Voltaiques*, Ouagadougou 10, 92–105.

Roman, B. (1980) *Serpents de Haute-Volta*. Centre National de la Recherche Scientifique et Technologique, Ouagadougou, 129 pp., 10 figs., 86 illust.

Roman, B. (1984) *Serpents des pays de l'entente*. Centre National de la Recherche Scientifique et Technologique, Ouagadougou, 45 pp., 33 figs.

Romano-Hoge, S.A.R.W.L. (1972) Notes on *Leptomicrurus* Schmidt (Serpentes, Elapidae). *Memórias do Instituto Butantan*, São Paulo (1971) 35, 111–115, 2 figs.

Romano-Hoge, S.A.R.W.L. & Hoge, A.R. (1973) Nota sobre *Xenodon* e *Ophis* Serpentes Colubridae. *Memórias do Instituto Butantan*, São Paulo (1972) 36, 209–214, 3 pls.

Romer, A.S. (1956) *Osteology of the reptiles: a comparative summary of the reptile skeleton, living and fossil, with a classification of the reptile family*. The University of Chicago Press, Chicago, 772 pp., 248 figs.

Romer, J.D. (1945) A new record of a rare snake (*Natrix xenura*) from Assam. *The Journal of the Bombay Natural History Society* (1944–1945) 45(3), 430–431.

Romer, J.D. (1952) A new record of a little-known snake, *Opisthotropis andersoni* (Boulenger), from Hong Kong. *Copeia* 1952(1), 46–47, fig. 1.
Romer, J.D. (1953a) Coloration of the snake *Elaphe moellendorffi*. *Copeia* 1953(2), 123–124.
Romer, J.D. (1953b) Reptiles and amphibians collected in the Port Harcourt area of Nigeria. *Copeia* 1953(2), 121–123.
Romer, J.D. (1954) Notes on sea snakes (Hydrophiidae) occurring in or near Hong Kong territorial waters. *Hong Kong University Fisheries Journal* (1), 35–37, 2 pls.
Romer, J.D. (1958) Occurrence of the sea snake *Hydrophis ornatus ornatus* (Gray) near Hong Kong. *Hong Kong University Fisheries Journal* (2), 129, 1 pl.
Romer, J.D. (1961) Annotated checklist with keys to the snakes of Hong Kong. *Memoirs of the Hong Kong Natural History Society* (5), 1–14.
Romer, J.D. (1963) Notes on the white-lipped pit viper, *Trimeresurus albolabris* Gray, in Hong Kong. *In:* Keegan, H.L. & Macfarlane, W.V. (Eds.), *Venomous and poisonous animals and noxious plants of the Pacific region*. Macmillan Company, New York, pp. 379–384.
Romer, J.D. (1966) Two colubrid snakes not previously recorded from Hong Kong. *Memoirs of the Hong Kong Natural History Society* (7), 13–14.
Romer, J.D. (1970) Revised checklist of snakes of Hong Kong. *Memoirs of the Hong Kong Natural History Society* (8), 1–22.
Romer, J.D. (1972) *Illustrated guide to the venomous snakes of Hong Kong with recommendations for first aid treatment of bites*. Third edition. J.R. Lee, Hong Kong, 24 pp., 5 figs., 8 col. pls. [first edition, 1959, 12 pp., 6 col. pls.; second edition, 1965, 26 pp., 5 figs., 8 col. pls.]
Romer, J.D. (1974a) Notes on a specimen of the snake *Natrix percarinata* (Boulenger) from Hong Kong. *Memoirs of the Hong Kong Natural History Society* (9), 20.
Romer, J.D. (1978) Reptiles new to Hong Kong. *Journal of the Hong Kong Branch of the Royal Asiatic Society* (1977) 17, 232–234.
Romer, J.D. (1979a) Second revised annotated checklist with keys to the snakes of Hong Kong. Third edition. *Memoirs of the Hong Kong Natural History Society* (14), 1–23, 2 figs. [first edition, 1965, Memoir (5), 1–14, 2 figs.; second edition, 1974, Memoir (8), 1–22, 2 figs.]
Romer, J.D. (1979b) *The red-necked keelback, a venomous snake of Hong Kong*. Urban Council, Hong Kong, 4 pp., 1 col. pl.
Romer, J.D. (1983) *Illustrated guide to the venomous snakes of Hong Kong*. Urban Council, Hong Kong, 20 pp.
Rooij, N. de. (1913) Praeda itineris a L.F. de Beaufort in Archipelago indico facti annis 1909–1910. III. Reptilien. *Bijdragen tot de Dierkunde*, Amsterdam 19, 15–29, 3 figs.
Rooij, N. de. (1917) *The reptiles of the Indo-Australian Archipelago. II. Ophidia*. E.J. Brill, Leiden, xiv + 334 pp., 117 figs. [reprinted in 1970 by E.J. Brill, Leiden]
Rooij, N. de. (1919) Reptilien aus Nord-Neuguinea gesammelt von Dr. P.N. Van Kampen und Dr. K. Gjellerup in den Jahren 1910 und 1911. *Bijdragen tot de Dierkunde*, Amsterdam 21, 81–95.
Rooij, N. de. (1922) Fauna Simalurensis. Reptilia. *Zoologische Mededelingen*, Leiden 6(4), 217–238, 8 figs.
Rooijen, J. van. (2007) The land snakes of the Santubong Peninsula, Sarawak, Borneo: a preliminary list of species with natural history notes. *Russian Journal of Herpetology* 14(1), 27–38, 15 figs.
Rooijen, J. van & Vogel, G. (2008a) Contributions to a review of the *Dendrelaphis pictus* complex (Squamata, Colubridae)—1. Description of a sympatric species. *Amphibia-Reptilia* 29(1), 101–115, 5 figs.
Rooijen, J. van & Vogel, G. (2008b) A new species of *Dendrelaphis* (Serpentes, Colubridae) from Java, Indonesia. *Raffles Bulletin of Zoology* 56(1), 189–197, 8 figs.
Rooijen, J. van & Vogel, G. (2008c) An investigation into the taxonomy of *Dendrelaphis tristis* (Daudin, 1803): revalidation of *Dipsas schokari* (Kuhl, 1820) (Serpentes, Colubridae). *Contributions to Zoology* 77(1), 33–43, 11 figs.
Rooijen, J. van & Vogel, G. (2009) A multivariate investigation into the population systematics of *Dendrelaphis tristis* (Daudin, 1803) and *Dendrelaphis schokari* (Kuhl, 1820), revalidation of *Dendrophis chairecacos* Boie, 1827 (Serpentes: Colubridae). *Herpetological Journal* 19, 193–200, 3 figs.
Rooijen, J. van & Vogel, G. (2010) On the discovery and origin of a Javan population of the Indochinese colubrid snake *Dendrelaphis subocularis* (Boulenger, 1888): a multivariate study. *Contributions to Zoology* 79(3), 85–92, 5 figs.
Rooijen, J. van & Vogel, G. (2012) A revision of the taxonomy of *Dendrelaphis caudolineatus* (Gray, 1834) (Serpentes: Colubridae). *Zootaxa* (3272), 1–25, 16 figs.
Rooijen, J. van, Wood, P.L., Jr., Grismer, J.L., Grismer, L.L. & Grossmann, W. (2011) Color pattern dimorphism in the colubrid snake *Oligodon purpurascens* (Schlegel, 1837) (Reptilia: Squamata). *Russian Journal of Herpetology* 18(3), 215–220, 3 figs.
Room, R.M. (1974) Lizards and snakes from the northern district of Papua, New Guinea. *British Journal of Herpetology* 5(3), 438–446, 1 fig.
Rose, F.L. & Selcer, K.W. (1989) Genetic divergence of the allopatric populations of *Nerodia harteri*. *Journal of Herpetology* 23(3), 261–267, 1 fig.
Rose, W. (1929) *Veld & vlei: an account of South African frogs, toads, lizards, snakes and tortoises*. The Specialty Press of South Africa, Cape Town, xxiii + 240 pp., 127 figs.
Rose, W. (1950) *The reptiles and amphibians of southern Africa*. Maskew Miller, Cape Town, xxv + 378 pp., 210 figs.
Rose, W. (1955) *Snakes—mainly South African*. Cape Town, xvi + 213 pp., 89 figs., 7 pls.
Rose, W. (1962) *The reptiles and amphibians of southern Africa*. Maskew Miller, Cape Town, xxix + 494 pp., 294 figs.
Rosén, N. (1905a) List of the snakes in the zoological museums of Lund and Malmo, with descriptions of new species and a new genus. *Annals and Magazine of Natural History*, London (Series 7) 15(86), 168–181, 2 pls.
Rosén, N. (1905b) Reply to Mr. G.A. Boulenger. *Annals and Magazine of Natural History*, London (Series 7) 16(92), 126–129.
Rosén, N. (1911) *Contributions to the fauna of the Bahamas. I. A general account of the fauna, with remarks on the physiography of the islands. II. The reptiles. III. The fishes.* H. & Kan Ohlsson, Lund, 45 pp., 13 figs., 1 pl.
Rosen, P.C. & Schwalbe, C.R. (1988) *Status of the Mexican and narrow-headed garter snakes (*Thamnophis eques megalops *and* Thamnophis rufipunctatus*) in Arizona*. Report to the United States Fish and Wildlife Service, 69 pp.
Rosenberg, H. (1955) Beobachtungen an einer schneckenfressenden Schlange, *Dipsas mikani* Schlegel. *Der Zoologische Garten*, Frankfurt-am-Main 22(1–3), 5–11, 5 figs.
Rosenfeld, G. (1963) Unfälle durch Giftschlangen. *In: Die Giftschlangen der Erde: Wirkungen und Antigenität der Gifte Therapie von Giftschlangenbissen*. N.G. Elwert Universtäts- und Verlags-Buchhandlung, Marburg, pp. 161–202, 10 figs.

Rosenfeld, G., Kelen, E.M.A. & Nudel, F. (1964) Hemolytic activity of animal venoms. 1. Classification in different types and activities. *Memórias do Instituto Butantan,* São Paulo (1960–1962) 30, 117–132, 2 figs.

Rösler, H. & Wranik, W. (2004) A key and annotated checklist to the reptiles of the Socotra Archipelago. *Fauna of Saudi Arabia* 20, 505–534.

Rösler, H. & Wranik, W. (2006) The reptiles of the Socotra Archipelago with special remarks on the slender blind snakes (Leptotyphlopidae: *Leptotyphlops*). *In:* Vences, M., Köhler, J., Ziegler, T. & Böhme, W. (Eds.), *Herpetologica Bonnensis II.* Proceedings of the 13th Congress of the Societas Europaea Herpetologica, Bonn, pp. 125–128.

Ross, A.M. (1878) *Catalogue of mammals, birds, reptiles, and fishes of the Dominion of Canada.* Canadian Naturalists' Union, Montreal, 13 pp.

Ross, C.A. & Gonzales, P.C. (1992) Amphibians and reptiles of Catanduanes Island, Philippines. *National Museum Papers,* Manila 2(2), 50–76, 3 figs.

Ross, C.A. & Lazell, J.D., Jr. (1991) Amphibians and reptiles of Dinagat and Siargao Islands, Philippines. *The Philippine Journal of Science*, Manila (1990) 119(3), 257–281, 1 fig.

Ross, C.A., Alcala, A.C. & Sison, R.V. (1987) Distribution of *Zaocys luzonensis* (Serpentes, Colubridae) in the Visayan Islands, Philippines. *Silliman Journal*, Dumaguete City 34(1–4), 29–31.

Ross, A.M. (1878) Catalogue of mammals, birds, reptiles, and fishes of the Dominion of Canada. Collected and classified by Alexander Milton Ross, Kt. *Canadian Naturalists' Union,* Montreal, 13 pp.

Rosselot, B. (1978) *Les serpents dangereux du Burundi.* République Française Ministère de la Coopération, Paris (1977), 63 pp., 38 figs.

Rossi, J.V. (1992) *Snakes of the United States and Canada. Keeping them healthy in captivity. Volume 1. Eastern area.* Krieger Publishing Company, Malabar, xiv + 209 pp., 100 col. pls.

Rossi, J.V. & Rossi, R. (1993) Notes on the captive maintenance and feeding behavior of a juvenile short-tailed snake (*Stilosoma extenuatum*). *Herpetological Review* 24(3), 100–101.

Rossi, J.V. & Rossi, R. (1995) *Snakes of the United States and Canada. Keeping them healthy in captivity. Vol. 2. Western area.* Krieger Publishing Company, Malabar, xvi + 325 pp., 273 col. photos.

Rossi, J.V. & Rossi, R. (2003) *Snakes of the United States and Canada. Keeping them healthy in captivity.* Second edition. Krieger Publishing Company, Malabar, xxxvi + 520 pp., 106 col. pls. [first edition, Rossi, 1992 & Rossi & Rossi, 1995]

Rossman, D.A. (1956) Notes on food of a captive black swamp snake, *Seminatrix pygaea pygaea* (Cope). *Herpetologica* 12, 154–155.

Rossman, D.A. (1961) Nomenclatural status of the Neotropical subspecies of the colubrid snake, *Thamnophis sauritus. Notulae Naturae*, Philadelphia (340), 1–2.

Rossman, D.A. (1962) *Thamnophis proximus* (Say): a valid species of garter snake. *Copeia* 1962(4), 741–748, 2 figs.

Rossman, D.A. (1963a) The colubrid snake genus *Thamnophis*: a revision of the *sauritus* group. *Bulletin of the Florida State Museum, Biological Sciences* 7(3), 99–178, 10 figs.

Rossman, D.A. (1963b) Relationships and taxonomic status of the North American Natricine snake genera *Liodytes*, *Regina*, and *Clonophis*. *Occasional Papers, Museum of Zoology, Louisiana State University* (29), 1–29, 6 figs.

Rossman, D.A. (1965a) A new subspecies of the common garter snake, *Thamnophis sirtalis*, from the Florida Gulf coast. *Proceedings of the Louisiana Academy of Sciences* 27, 67–73, 1 fig.

Rossman, D.A. (1965b) Two new colubrid snakes of the genus *Rhadinaea* from southern México. *Occasional Papers, Museum of Zoology, Louisiana State University* (32), 1–8, 2 figs.

Rossman, D.A. (1965c) Identity and relationships of the Mexican garter snake *Thamnophis sumichrasti* (Cope). *Copeia* 1965(2), 242–244, 2 figs.

Rossman, D.A. (1966) Evidence for conspecificity of the Mexican garter snake *Thamnophis phenax* (Cope) and *Thamnophis sumichrasti* (Cope). *Herpetologica* 22(4), 303–305, 1 fig.

Rossman, D.A. (1969) A new Natricine snake of the genus *Thamnophis* from northern México. *Occasional Papers, Museum of Zoology, Louisiana State University* (39), 1–4, 1 fig.

Rossman, D.A. (1970a) *Thamnophis proximus. Catalogue of American amphibians and reptiles*, pp. 98.1–98.3.

Rossman, D.A. (1970b) *Thamnophis sauritus. Catalogue of American amphibians and reptiles*, pp. 99.1–99.2.

Rossman, D.A. (1971) Systematics of the neotropical populations of *Thamnophis marcianus* (Serpentes, Colubridae). *Occasional Papers, Museum of Zoology, Louisiana State University* (41), 1–13, 5 figs.

Rossman, D.A. (1973) Evidence for the conspecificity of *Carphophis amoenus* (Say) and *Carphophis vermis* (Kennicott). *Journal of Herpetology* 7(2), 140–141.

Rossman, D.A. (1975) Redescription of the South American colubrid snake *Helicops hagmanni* Roux. *Herpetologica* 31(4), 414–418, 5 figs.

Rossman, D.A. (1976) Revision of the South American colubrid snakes of the *Helicops pastazae* complex. *Occasional Papers, Museum of Zoology, Louisiana State University* (50), 1–15, 7 figs.

Rossman, D.A. (1979) Morphological evidence for taxonomic partitioning of the *Thamnophis elegans* complex (Serpentes, Colubridae). *Occasional Papers, Museum of Zoology, Louisiana State University* (55), 1–12, 3 figs.

Rossman, D.A. (1985) *Liodytes* resurrected, reexamined, and reinterred. *Journal of Herpetology* 19(1), 169–171.

Rossman, D.A. (1991) Identity of the garter snake *Thamnophis sumichrasti cerebrosus* Smith. *Herpetological Review* 22(3), 80–81.

Rossman, D.A. (1992) Taxonomic status and relationships of the Tamaulipan montane garter snake, *Thamnophis mendax* Walker, 1955. *Proceedings of the Louisiana Academy of Sciences* 55, 1–14, 5 figs.

Rossman, D.A. (1996) Taxonomic status of the southern Durango spotted garter snake, *Thamnophis nigronuchalis* Thompson, 1957. *Proceedings of the Louisiana Academy of Sciences* 58, 1–10, 4 figs.

Rossman, D.A. (2000) Identity of the snake *Calopisma septemvittatum* Fischer, 1879, and redescription of the lectotype of *Helicops infrataeniatus* Jan, 1865. *Mitteilungen aus dem Hamburgisches Zoologisches Museum und Institut* 97, 123–129.

Rossman, D.A. (2002a) Variation in the xenodontid water snake *Helicops scalaris* Jan, and the status *H. hogei* Lancini. *Occasional Papers of the Museum of Natural Science, Louisiana State University* (78), 1–18.

Rossman, D.A. (2002b) Morphological variation in the endemic Colombian water snake, *Helicops danieli* Amaral, 1937 (Serpentes: Xenodontidae). *Revista de la Academia Colombiana de Ciencias* 26(101), 589–594.

Rossman, D.A. (2010) Morphology variation in the striped water snake *Helicops trivittatus* (Gray, 1849) (Reptilia: Serpentes: Xenodontidae) of eastern Amazonia. *Boletim Museu Paraense Emílio Goeldi (Ciencias Naturais)*, Belém 5(3), 271–278, 4 figs.

Rossman, D.A. & Abe, A.S. (1979) Comments on the taxonomic status of *Helicops yacu* (Serpentes, Colubridae). *Proceedings of the Louisiana Academy of Sciences* 42, 7–9.

Rossman, D.A. & Blaney, R.M. (1968) A new natricine snake of the genus *Adelophis* from western Mexico. *Occasional Papers, Museum of Zoology, Louisiana State University* (35), 1–12, 6 figs.

Rossman, D.A. & Burbrink, F.T. (2005) Species limits within the Mexican garter snakes of the *Thamnophis godmani* complex. *Occasional Papers of the Museum of Natural Science, Louisiana State University* (79), 1–43, 12 figs.

Rossman, D.A. & Dixon, J.R. (1975) A new colubrid snake of the genus *Helicops* from Peru. *Herpetologica* 31(4), 412–414, 1 fig. (*Helicops yacu*)

Rossman, D.A. & Eberle, W.G. (1977) Partition of the genus *Natrix*, with preliminary observations on evolutionary trends in natricine snakes. *Herpetologica* 33(1), 34–43, 4 figs.

Rossman, D.A. & Erwin, R.L. (1980) Geographic variation in the snake *Storeria occipitomaculata* (Storer) (Serpentes, Colubridae) in southeastern United States. *Brimleyana* (4), 95–102, 5 figs.

Rossman, D.A. & Good, D.A. (1993) Herpetological type specimens in the Museum of Natural Science, Louisiana State University. *Occasional Papers of the Museum of Natural Science, Louisiana State University* (66), 1–18.

Rossman, D.A. & Kizirian, D.A. (1993) Variation in the Peruvian dipsadine snakes *Sibynomorphus oneilli* and *S. vagus*. *Journal of Herpetology* 27(1), 87–90, 2 figs.

Rossman, D.A. & Lara-Gongora, G. (1991) Taxonomic status of the Mexican garter snake *Thamnophis scaliger*. Abstract, *Annual Meeting of the Herpetologist's League and the Society for the Study of Amphibians and Reptiles*, State College, p. 93.

Rossman, D.A. & Lara-Gongora, G. (1997) Variation in the Mexican garter snake *Thamnophis scalaris* Cope and the taxonomic status of *T. scaliger* (Jan). *Occasional Papers of the Museum of Natural Science, Louisiana State University* (74), 1–14.

Rossman, D.A. & Powell, R. (1985) *Clonophis. C. kirtlandii. Catalogue of American amphibians and reptiles*, pp. 364.1–364.2.

Rossman, D.A. & Schaefer, G.C. (1974) Generic status of *Opheodrys mayae*, a colubrid snake endemic to the Yucatan Peninsula. *Occasional Papers, Museum of Zoology, Louisiana State University* (45), 1–12, 4 figs.

Rossman, D.A. & Scott, N.J. (1968) Identity of *Helicops wettsteini* Amaral (Serpentes, Colubridae). *Herpetologica* 24(3), 262–263.

Rossman, D.A. & Stewart, G.R. (1987) Taxonomic reevaluation of *Thamnophis couchii* (Serpentes, Colubridae). *Occasional Papers, Museum of Zoology, Louisiana State University* (63), 1–25, 10 figs.

Rossman, D.A. & Thomas, R. (1979) A new dipsadine snake of the genus *Sibynomorphus* from Peru. *Occasional Papers, Museum of Zoology, Louisiana State University* (54), 1–6, 3 figs.

Rossman, D.A. & Wallach, V. (1987) *Adelophis. A. copei, A. foxi. Catalogue of American amphibians and reptiles*, pp. 408.1–408.2.

Rossman, D.A. & Wallach, V. (1991) *Virginia. Catalogue of American amphibians and reptiles*, pp. 529.1–529.4.

Rossman, D.A. & Williams, K.L. (1966) Defensive behavior of the South American colubrid snakes *Pseustes sulphureus* (Wagler) and *Spilotes pullatus* (Linnaeus). *Proceedings of the Louisiana Academy of Sciences* 29, 152–156, 2 figs.

Rossman, D.A., Liner, E.A., Treviño, C.H. & Chaney, A.H. (1989) Redescription of the garter snake *Thamnophis exsul* Rossman, 1969 (Serpentes, Colubridae). *Proceedings of the Biological Society of Washington* 102(2), 507–514, 5 figs.

Rossman, D.A., Ford, N.B. & Seigel, R.A. (1996) *The garter snakes, evolution and ecology*. University of Oklahoma Press, Norman, xx + 332 pp., 15 col. pls.

Rossman, N.J., Rossman, D.A. & Keith, N.K. (1982) Comparative visceral topography of the New World snake tribe Thamnophini (Colubridae, Natricinae). *Tulane Studies in Zoology and Botany* 23(2), 123–164, 20 figs.

Rostombekov, V. (1928) Ophidia de l'expedition de l'Ourmie en 1916. *Bulletin du Musée Géorgie*, Tiflis 4, 121–133, figs.

Roth, H.D. (1944) Some notes on our snakes. *Timehri* 26, 78–91.

Roth-Monzon, A.J., Mendoza-Hernandez, A.A. & Flores-Villela, O.A. (2011) *Lampropeltis ruthveni* (Serpentes: Colubridae) from the state of Hidalgo, Mexico. *Southwestern Naturalist* 56(3), 430–431.

Rotivo, S.M., Papenfuss, T.J. & Vásquez-Almazán, C.R. (2012) A new species of *Sibon* (Squamata: Colubridae) from the mountains of eastern Guatemala. *Zootaxa* (3266), 62–68, 3 figs.

Roussel, M.-R. & Villiers, A. (1965) Serpents du Mayo-Kebbi (Tchad). *Bulletin de l'Institut Française d'Afrique Noire A*, Dakar 27(4), 1522–1533, 1 fig.

Roux, J. (1907a) Sur quelques reptiles sud-africains. *Revue Suisse de Zoologie*, Genève 15(1), 75–86, 4 figs.

Roux, J. (1907b) Diagnosen neuer Reptilien aus Asien und Amerika. *Zoologischer Anzeiger*, Leipzig 31(24), 762–765.

Roux, J. (1907c) Beiträge zur Kenntnis der Fauna von Süd-Afrika. Ergebnisse einer Reise von Prof. Max Weber im Jahre 1894. VIII. Ophidia (Schlangen). *Zoologische Jahrbücher, Abteilung für Systematik, Ökologie und Geographie der Tiere*, Jena 25(5–6), 732–741, pl. 27.

Roux, J. (1910a) Eine neue *Helicops*-Art aus Brasilien. *Zoologischer Anzeiger*, Leipzig 36(25), 439–440.

Roux, J. (1910b) Reptilien und Amphibien der Aru- und Kei-Inseln. *In:* Merton, H. (Ed.), Ergebnisse einer Zoologischen Forschungsreise in den südöstlichen Molukken (Aru- und Kei-Inseln) im Auftrag der Senckenbergischen Naturforschenden Gesellschaft ausgeführt von Dr. Hugo Merton. Wissenschaftliche Ergebnisse. Band I. *Abhandlungen der Senckenbergischen Naturforschenden Gesellschaft*, Frankfurt-am-Main (1910–1911) 33(3), 209–247, pls. 13–14.

Roux, J. (1911) Elbert-Sunda-Expedition des Frankfurter Vereins für Geographie und Statistik. Reptilien und Amphibien. *Zoologische Jahrbücher, Abteilung für Systematik, Ökologie und Geographie der Tiere*, Jena 30(5), 495–508.

Roux, J. (1913) Les reptiles de la Nouvelle-Calédonie et des îles Loyalty. *In:* Sarasin, F. & Roux, J. (Eds.), *Nova Caledonia. A. Zoologie*. C.W. Kreidels, Wiesbaden 1(2), 79–160, pls. 4–5.

Roux, J. (1914) Note sur une espèce nouvelle d'*Oligodon* provenant de Sumatra. *Revue Suisse de Zoologie*, Genève 22(2), 27–29.

Roux, J. (1919a) Sur un nouveau serpent (*Simotes musyi*) provenant de la Chine. *Revue Suisse de Zoologie*, Genève 27(3), 61–63, 2 figs.

Roux, J. (1919b) Note sur quelques reptiles provenant de la Nouvelle-Guinée. *Revue Suisse de Zoologie*, Genève 27(10), 347–351.

Roux, J. (1926) Notes d'erpétologie sud-américaine. *Revue Suisse de Zoologie*, Genève 33(4), 291–299. (*Typhlops lehneri*)

Roux, J. (1927) Contribution à l'erpétologie du Vénézuéla. *Verhandlungen der Naturforschenden Gesellschaft in Basel* 38, 252–261.

Roux, J. (1928) Reptiles et amphibiens de l'Inde meridionale. *Revue Suisse de Zoologie*, Genève 35(21), 439–471.

Roux, J. (1931) Reptilien und Amphibien in zoogeographischer Hinsicht. *In:* Elbert, J. (Ed.), *"Die Sunda-Expedition" des vereins für Geographie und Statistik zu Frankfurt a. M.* Band II. Hermann Minjon, Frankfurt-am-Main, pp. 305–308.

Roux, J. (1934) Contribution à la connaissance de la faune erpétologique des îles Salomon. 1. Note sur *Parapistocalamus*, nouveau genre de serpent protéroglyphe. *Verhandlungen der Naturforschenden Gesellschaft in Basel* 45, 77–80, 2 figs.

Roux, J. (1936) Mission scientifique de l'Omo. Reptilia et Amphibia. *Mémoires du Muséum National d'Histoire Naturelle*, Paris (1935) 3(25), 157–190.

Roux, J. (1939) Récoltes de R. Paulian et A. Villiers dans le Haut Atlas Marocain, 1938 (Troisième note, reptiles). *Bulletin de la Société des Sciences Naturelles et Physiques du Maroc*, Rabat 19(1), 11–22.

Roux-Estève, R. (1965) Les Serpentes de la region de la Maboke-Boukoko. *Cahiers de la Maboké (Zoologie)*, Paris 3(1), 51–92, 39 figs.

Roux-Estève, R. (1969a) Les serpents de la région de Lamto (Cote-d'Ivoire). *Annales de l' Université d'Abidjan* (Écologie) 2(1), 81–140, 32 figs.

Roux-Estève, R. (1969b) Étude comparée de *Meizodon coronatus* (Schlegel) et de *Meizodon regularis* Fischer (Colubrides–Serpentes). *Bulletin du Muséum National d'Histoire Naturelle*, Paris 41(2), 395–409.

Roux-Estève, R. (1974a) Revision systematique des Typhlopidae d'Afrique. Reptilia-Serpentes. *Mémoires du Muséum National d'Histoire Naturelle*, Paris (Série 2) (Zool.) 87A, 1–313, 185 figs.

Roux-Estève, R. (1974b) Recherches sur la morphologie, la biogeographie et la phylogenie des Typhlopidae d'Afrique. *Bulletin de l'Institut Fondamental d'Afrique Noire*, Dakar 36A(2), 428–508, 45 figs.

Roux-Estève, R. (1975) Typhlopidae et Leptotyphlopidae. *In:* Saeger, H. de, Roux-Estève, R. & Witte, G.-F. de (Eds.), Exploration du Parc National des Virunga. *Fondacion pour Favoriser les Recherches Scientifiques en Afrique*, Bruxelles (Série 2) (24), 62–108.

Roux-Estève, R. (1979a) Une nouvelle espèce de *Leptotyphlops* (Serpentes) du Cameroun: *Leptotyphlops perreti*. *Revue Suisse de Zoologie*, Genève 86(2), 463–466, fig.

Roux-Estève, R. (1979b) Liste des amphibiens et reptiles des collections du Museum national d'Histoire naturelle de Paris, recoltes par Lesueur (1778–1846). *Bulletin Trimestrial de la Société Géologique de Normandie et des Amis du Muséum de Havre* 66(3), 25–29.

Roux-Estève, R. (1980) Une nouvelle espèce de Typhlopidae (Serpentes) du Centre-Est de Madagascar, *Typhlops domerguei*. *Bulletin du Muséum National d'Histoire Naturelle*, Paris (Série 4) 2(1), 321–323, 1 fig.

Roux-Estève, R. (1983) Les specimens-types du genre *Micrurus* (Elapidae) conservés au Muséum national d'Histoire naturelle de Paris. *Memórias do Instituto Butantan*, São Paulo (1982) 46, 79–94.

Roux-Estève, R. & Guibé, J. (1965a) Contribution a l'étude du genre *Boaedon* (Serpentes, Colubridae). *Bulletin du Muséum National d'Histoire Naturelle*, Paris (1964) (Série 2) 36(6), 761–774, 5 figs.

Roux-Estève, R. & Guibé, J. (1965b) Étude comparée de *Boaedon fuliginosus* (Boié) et *B. lineatus* D. et B. (Ophidiens). *Bulletin de l'Institut Fondamental d'Afrique Noire*, Dakar 27A(1), 397–409, 4 figs.

Royans, D. & Knight, S. (2009) An overview of the Indian trinket snake. *The Herptile* 34(2), 35–44, illust.

Royle, J. (1989) *An introduction to the common snakes of Malawi*. Wildlife Society of Malawi, 47 pp., illust.

Roze, J.A. (1952a) Colección de reptiles del Professor Scorza, de Venezuela. *Acta Biologica Venezuelica*, Caracas 1(5), 93–114, 3 figs.

Roze, J.A. (1952b) Contribución al conocimiento de los ofidios de las familias Typhlopidae y Leptotyphlopidae en Venezuela. *Memorias de la Sociedad de Ciencias Naturales La Salle*, Caracas 12(32), 143–158, 7 figs.

Roze, J.A. (1953a) Ofidios de Camuri Chico, Macuto, D.F. Venezuela, colectados por el Rvdo. Padre Cornelius Vogl. *Boletin de la Sociedad Venezolana de Ciencias Naturales*, Caracas 14(79), 200–211, 2 figs.

Roze, J.A. (1953b) The *Rassenkreis Coluber (Masticophis) mentovarius* (Dumeril, Bibron et Dumeril), 1854. *Herpetologica* 9(3), 113–120.

Roze, J.A. (1954) Nota preliminar sobre los ofidios de la Expedición Franco-Venezolana al Alto Orinoco. *Archivos Venezolanos de Patología Tropical y Parisitología Medica*, Caracas 2(2), 227–234, 4 figs.

Roze, J.A. (1955) Revision de la corales (Serpentes, Elapidae) de Venezuela. *Acta Biologica Venezuelica*, Caracas 1(17), 453–500, 4 figs.

Roze, J.A. (1956) La herpetofauna de las islas Los Roques y La Orchila. *In: El Archipiélago de Los Roques y La Orchila*. Sociedad de Ciencias Naturales La Salle, Caracas, pp. 79–86.

Roze, J.A. (1957a) Ofidios coleccionados por la Expedicion Franco-Venezolana al alto Orinoco; 1951 a 1952. *Boletin del Museo de Ciencias Naturales*, Caracas (1955) 1(3–4), 179–195, 2 pls.

Roze, J.A. (1957b) Notas sobre *Hydrops lehmanni* Dunn, 1944, y los generos neotropicales, *Pseudoeryx*, *Hydrops*, y *Helicops* (Colubridae). *Acta Biologica Venezuelica*, Caracas 2(3), 17–26, 5 figs.

Roze, J.A. (1957c) Resumen de una revision del genero *Hydrops* (Wagler), 1830 (Serpentes, Colubridae). *Acta Biologica Venezuelica*, Caracas 2(8), 51–95, 14 figs.

Roze, J.A. (1958a) A new species of the genus *Urotheca* (Serpentes, Colubridae) from Venezuela. *Breviora* (88), 1–5.

Roze, J.A. (1958b) Resultados zoologicos de la expedición de la Universidad Central de Venezuela a la region del Auyantepui en la Guayana Venezolana, Abril de 1956. 5. Los reptiles del Auyantepui, Venezuela, basándose en las colecciónes de las expediciones de Phelps-Tate, del American Museum of Natural History, 1937–1938, y de la Universidad Central de Venezuela, 1956. *Acta Biologica Venezuelica*, Caracas 2(22), 243–270, 11 figs.

Roze, J.A. (1958c) Los reptiles del Chimanta Tepui (Estado Bolivar, Venezuela) colectados por la Expedicion Botanica del Chicago Natural History Museum. *Acta Biologica Venezuelica*, Caracas 2(25), 299–314, 4 figs.

Roze, J.A. (1958d) On Hallowell's type specimens of reptiles from Venezuela in the collection of the Academy of Natural Sciences of Philadelphia. *Notulae Naturae*, Philadelphia (309), 1–4.

Roze, J.A. (1958e) Un nuevo genero de corales venenosas (*Leptomicrurus*) para Venezuela. *Acta Cientifica Venezolana*, Caracas 9(6–7), 128–130, 2 figs.

Roze, J.A. (1959a) El genero *Erythrolamprus* Wagler (Serpentes, Colubridae) en Venezuela. *Acta Biologica Venezuelica*, Caracas 2(35), 523–534, 1 fig.

Roze, J.A. (1959b) Una nueva especie del genero *Drymarchon* (Serpentes, Colubridae) de la Isla de Margarita, Venezuela. *Novedades Cientificas, Contribuciones Ocasionales del Museo de Historia Natural La Salle* (*Zoológica*), Caracas (25), 1–4, 2 figs.

Roze, J.A. (1959c) Taxonomic notes on a collection of Venezuelan reptiles in the American Museum of Natural History. *American Museum Novitates* (1934), 1–14, 1 fig.

Roze, J.A. (1961) El genero *Atractus* (Serpentes, Colubridae) en Venezuela. *Acta Biologica Venezuelica*, Caracas 3(7), 103–119.

Roze, J.A. (1962a) Contribución al conocimiento de los ofidios del cordon litoral. Los ofidios de Curupao, Estado Miranda (Venezuela). *Acta Biologica Venezuelica*, Caracas 3(11), 161–172, 3 figs.

Roze, J.A. (1962b) Notas taxonomicas sobre una pequeña colección de serpientes del Río Palmar, Estado Zulia (Venezuela). *Acta Biologica Venezuelica*, Caracas 3(13), 195–200, 2 figs.

Roze, J.A. (1964) The snakes of the *Leimadophis-Urotheca-Liophis* complex from Parque Nacional Henri Pittier (Rancho Grande), Venezuela, with a description of a new genus and species (Reptilia, Colubridae). *Senckenbergiana Biologica*, Frankfurt-am-Main 45(3–5), 533–542, 5 figs.

Roze, J.A. (1965) La herpetologica de la Isla de Margaritas, Venezuela. *Memorias de la Sociedad de Ciencias Naturales La Salle*, Caracas (1964) 24(69), 209–241.

Roze, J.A. (1966a) *La taxonomia y zoogeografia de los ofidios de Venezuela*. Universidad Central de Venezuela, Caracas, 362 pp., 79 figs.

Roze, J.A. (1966b) On the synonymy and type specimens of the coral snakes *Micrurus corallinus* and *M. ibiboboca* (Marcgravii). *Copeia* 1966(2), 369–371.

Roze, J.A. (1967) A check list of the New World venomous coral snakes (Elapidae), with descriptions of new forms. *American Museum Novitates* (2287), 1–60, 17 figs.

Roze, J.A. (1969) Una nueva coral falsa del género *Scaphiodontophis* (Serpentes, Colubridae) de Colombia. *Caldasia*, Bogotá 10(48), 355–363, 3 figs.

Roze, J.A. (1974) *Micruroides*. *M. euryxanthus*. *Catalogue of American amphibians and reptiles*, pp. 163.1–163.4.

Roze, J.A. (1983) New World coral snakes (Elapidae): a taxonomic and biological summary. *Memórias do Instituto Butantan*, São Paulo (1982) 46, 305–338.

Roze, J.A. (1987) Summary of coral snakes (Elapidae) from Cerro de la Neblina, Venezuela, with description of a new subspecies. *Revue Française d'Aquariologie*, Nancy 14(3), 109–112, 3 figs.

Roze, J.A. (1989) New species and subspecies of coral snakes, genus *Micrurus* (Elapidae), with notes on type specimens of several species. *American Museum Novitates* (2932), 1–15, 7 figs.

Roze, J.A. (1994) Notes on taxonomy of venomous coral snakes (Elapidae) of South America. *Bulletin of the Maryland Herpetological Society* 30(4), 177–185.

Roze, J.A. (1996) *Coral snakes of the Americas. Biology, identification, and venoms*. Krieger Publishing, Malabar, xii + 328 pp., 61 figs., 60 col. pls., 49 col. drawings.

Roze, J.A. & Bernal-Carlo, A. (1988) Las serpientes corales venenonas del género *Leptomicrurus* (Serpentes, Elapidae) de Suramérica con descripción de una nueva subespecie. *Bollettino del Museo Regionale di Scienze Naturali*, Torino (1987) 5(2), 573–608, 14 figs.

Roze, J.A. & Silva, NJ. da. (1990) Coral snakes (Serpentes, Elapidae) from hydroelectric power plant of Samuel, Rondonia, Brazil, with description of a new species. *Bulletin of the Maryland Herpetological Society* 26(4), 169–176, 1 fig.

Roze, J.A. & Tilger, G.M. (1983) *Micrurus fulvius*. *Catalogue of American amphibians and reptiles*, pp. 316.1–316.4.

Rubio, M. (1998) *Rattlesnake: portrait of a predator*. Smithsonian Institution Press, Washington, D.C., 240 pp., col. illust.

Rubio, M. (2010) *Rattlesnakes of the United States & Canada*. ECO Herpetological Publishing, Rodeo, 307 pp., col. illust.

Rudman, S.M. & Powell, R. (2010) *Typhlops dominicanus*. *Catalogue of American amphibians and reptiles*, pp. 879.1–879.3, 1 fig.

Rueda-Almonacid, J.V. & Rueda-Martínez, J.N. (2004) Primer registro de una serpiente del género *Amastridium* Cope, (Serpentes: Colubridae), para Sudamérica. *Revista de la Academia Colombiana de Ciencias Exactas, Físicas y Naturales* 28(107), 291293.

Rufino, N. (1998a) Geographic distribution, *Liophis poecilogyrus schotti*. *Herpetological Review* 29(3), 177.

Rufino, N. (1998b) Geographic distribution, *Micrurus collaris*. *Herpetological Review* 29(3), 178.

Rufino, N. (1999) Geographic distribution: Serpentes: *Liophis almadensis*. *Herpetological Review* 30(2), 113–114.

Ruiter, L.C. de. (1958) Zeeslangen II. *Lacerta*, Leiden 17(1–2), 1–2.

Ruiz-Pérez, G.A. (1996) *Claves preliminares para reconocer a los reptiles de Nicaragua*. Cedaprode, Managua, 88 pp., 18 figs.

Rüppell, E. (1845) Verzeichniss der in dem Museum der Senckenbergischen naturforschenden Gesellschaft aufgestellten Sammlungen. Dritte Abtheilung, Amphibien. *Museum Senckenbergianum*, Frankfurt-am-Main 3, 293–316.

Russell, A.P. & Bauer, A.H. (2000) *The amphibians and reptiles of Alberta: a field guide and primer of boreal herpetology*. Second edition. University of Calgary Press, Calgary, xii + 279 pp., figs., 16 col. pls. [first edition, 1993, x + 279 pp., figs., 16 col. pls.]

Russell, P. (1796) *An account of Indian serpents, collected on the coast of Coromandel; containing descriptions and drawings of each species; together with experiments and remarks on their several poisons*. George Nicol, London, vii + 91 pp., 46 pls.

Russell, P. (1801–1809/1810). *A continuation of an account of Indian serpents; containing descriptions and figures, from specimens and drawings, transmitted from various parts of India, to the Hon. the Court of Directors of the East India Company*. G. and W. Nicol, London, v + 57 pp., 42 pls. [issued in five sections: pp. i–v + 1–12, pls. 1–10 (1801), pp. 13–20, pls. 11–18 (1802), pp. 21–28, pls. 19–24 (1804), pp. 29–38, pls. 25–32 (1807), and pp. 39–57, pls. 33–42 (1809/1810) *fide* Zhao and Adler (1993, 396–397)]

Russo, V.P. (2007) *The complete boa constrictor. A comprehensive guide to the care, breeding, and geographical races*. ECO Herpetological Publishing, Lansing, Rodeo, 281 pp., 450 col. photos.

Rustanov, A.K. & Szczerbak, N.N. (1986) Geographic zonation of the herpetofauna of central Asia. *Izvestiya Akademiya Nauk Turkmenskoy SSR* (*Biologiya*) 1986(3), 13–20, illust.

Ruthven, A.G. (1907) A collection of reptiles and amphibians from southern New Mexico and Arizona. *Bulletin of the American Museum of Natural History* 23(23), 483–603, 19 figs.

Ruthven, A.G. (1908) Variations and genetic relationships of the garter-snakes. *Bulletin of the United States National Museum* (61), i–xii, 1–201 pp., 82 figs.

Ruthven, A.G. (1912) The reptiles of Michigan. *In:* Ruthven, A.G., Thompson, C. & Thompson, H. (Eds.), The herpetology of Michigan. *Michigan Geological and Biological Survey, Biological Series* 3 (10), 63–161, figs. 21–55, pls. 7–12.

Ruthven, A.G. (1922) The amphibians and reptiles of the Sierra Nevada de Santa Marta, Colombia. *Miscellaneous Publications, Museum of Zoology, University of Michigan* (8), 1–69, 12 pls.

Ruthven, A.G. (1923) The reptiles of the Dutch Leeward islands. *Occasional Papers of the Museum of Zoology, University of Michigan* (143), 1–10.

Ruthven, A.G. (1927) Description of an apparently new species of *Apostolepis* from Bolivia. *Occasional Papers of the Museum of Zoology, University of Michigan* (188), 1–2.

Ruthven, A.G. (1936) *Leptodeira bakeri*, new species. *Occasional Papers of the Museum of Zoology, University of Michigan* (330), 1–2.

Ruthven, A.G. & Gaige, H.T. (1935) Observations on *Typhlops* from Puerto Rico and some of the adjacent islands. *Occasional Papers of the Museum of Zoology, University of Michigan* (307), 1–12, 2 figs.

Ruthven, A.G., Thompson, C. & Gaige, H.T. (1928) The herpetology of Michigan. *Michigan Handbook Series* (3), University of Michigan, Ann Arbor, ix + 228 pp., 2 figs., 20 pls.

Ryabov, S.A., Orlov, N.L., Panteleev, D.J & Shiryaev, K.A. (2002) *Trimeresurus hageni*, *Trimeresurus puniceus*, and *Trimeresurus sumatranus* (Ophidia, Viperidae, Crotalinae): the data on reproductive biology and methods of captive breeding in laboratory conditions. *Russian Journal of Herpetology* 9(3), 243–254, 10 col. figs.

Ryabov, S.A., Orlov, N.L., Korshunov, I.S. & Nguyen, T.T. (2011) The first captive breeding of rare colubrid snake *Maculophis bellus chapaensis* (Bourret, 1934) [Ophidia: Colubridae: Colubrinae] in terrarium. *Russian Journal of Herpetology* 18(1), 80–82, 6 col. figs.

Ryan, M. (2002) Geographic distribution: Serpentes: *Micrurus clarki*. *Herpetological Review* 33(4), 324.

Rystedt, F.M. (1823) *Fauna Americae Meridionalis quam Venia. Exp. Facult. Med. Upsal. praeside C.P. Thunberg. 10 Junii MDCCCXXIII.* Excudebant Palmblad et C., Upsaliae, iii + 7 pp.

Sabath, M.D. & Sabath, L.E. (1969) Morphological integradation in Gulf coastal brown snakes, *Storeria dekayi* and *Storeria tropica*. *American Midland Naturalist* 81(1), 148–155, 1 fig.

Sackett, J.T. (1940a) Zoological results of the George Vanderbilt Sumatran Expedition, 1936–1939. Part IV. The reptiles. *Notulae Naturae*, Philadelphia (41), 1–3.

Sackett, J.T. (1940b) Preliminary report on results of the West Indies-Guatemala Expedition of 1940 for the Academy of Natural Sciences of Philadelphia. Part I. A new blind snake of the genus *Typhlops*. *Notulae Naturae*, Philadelphia (48), 1–2.

Sadlier, R.A. (1990) The terrestrial and semiaquatic reptiles (Lacertilia, Serpentes) of the Magela Creek region, Northern Territory. *Australian Museum, Technical Memorandum, Syndey* (32), i–vi, 1–77, 69 pls.

Sahi, D.N. & Duda, P.L. (1986) A checklist and keys to the amphibians and reptiles of Jammu and Kashmir State, India. *Bulletin of the Chicago Herpetological Society* (1985) 20(3–4), 86–97, 1 fig.

Said-Aliyev, S.A. (1979) *Amphibians and reptiles of Tadzhikistan.* Donnish, Dushanbe, 145 pp., 58 figs.

Saint-Aubain, M.L. de. (1984) A comparative study of the functional anatomy of pulmonary vaso-constriction in lung-breathing vertebrates. *Zeitschrift für Zoologische Systematik und Evolutionsforschung,* Berlin 22, 118–136, 17 figs.

Saint-Girons, H. (1951) Les serpents de l'Afrique palaerctique. *Comptes Rendus des Séances de la Société de Biogeographie*, Paris (243), 99–102, 2 figs.

Saint-Girons, H. (1952a) Étude biogeographique des serpents palearctiques. *Comptes Rendus des Séances de la Société de Biogeographie*, Paris (250), 50–67, 7 figs.

Saint-Girons, H. (1952b) Écologie et ethologie des vipères de France. *Annales des Sciences Naturelles (Zoologie)*, Paris (Série 11) 14, 263–343, 5 figs.

Saint-Girons, H. (1953) Une Vipère naine, *Vipera latastei montana. Bulletin de la Société Zoologique de France*, Paris 78(1), 24–28, 2 figs.

Saint-Girons, H. (1954) Note rectificative au sujet de *Vipera latastei montana*, n. subsp. *Bulletin de la Société Zoologique de France*, Paris (1953) 78(5–6), 475.

Saint-Girons, H. (1956) Les serpents du Maroc. *Variétés Scientifiques recueillies par la Société des Sciences Naturelles et Physiques dur Maroc*, Chérifien (8), 1–29, 3 pls.

Saint-Girons, H. (1964) Notes sur l'ecologie et la structure des populations des Laticaudinae (Serpents, Hydrophidae) en Nouvelle Caledonie. *La Terre et la Vie* 1964(2), 185–214, 2 figs., 4 pls.

Saint-Girons, H. (1967) Morphologie comparee de l'hypophyse chez les Squamata: données complémentaires et apport à la phylogénie des reptiles. *Annales de Sciences Naturelles (Zoologie)* (Série 12) 9, 229–308, 51 figs..

Saint-Girons, H. (1971a) Notes sur les *Enhydris* du groupe *innominata*, *smithi* et *longicauda* (Serpentes, Homalopsinae). *Bulletin de la Société des Sciences Naturelles et Physiques du Maroc*, Rabat 51, 221–234, 3 figs., 2 pls.

Saint-Girons, H. (1971b) Le cycle sexuel des serpents du Cambodge. *Annales des Sciences Naturelles (Zoologie)*, Paris (Série 12) 13, 543–572, 22 figs.

Saint-Girons, H. (1972a) Les serpents du Cambodge. *Mémoires du Muséum National d'Histoire Naturelle*, Paris (Série 2) 74A, 1–170, 42 pls.

Saint-Girons, H. (1972b) Étude de la collection de serpents de Cochinchine du Docteur Morice, déposée en 1873 et 1874 au Muséum d'Histoire Naturelle de Lyon. *Nouvelles Archives du Muséum National d'Histoire Naturelle*, Paris 9, 3–12.

Saint-Girons, H. (1977) Systématique de *Vipera latastei latastei* Bosca, 1878 et description de *Vipera latastei gaditana*, subsp. n. (Reptilia, Viperidae). *Revue Suisse de Zoologie*, Genève 84(3), 599–607, 2 figs.

Saint-Girons, H. (1978) Morphologie externe comparée et systématique de vipères d'Europe (Reptilia, Viperidae). *Revue Suisse de Zoologie*, Genève 85(3), 565–595, 2 figs., 2 pls.

Saint-Girons, H. (1980) Biogéographie et évolution des vipères Européennes. *Comptes Rendus des Séances de la Société de Biogeographie*, Paris (496), 146–172, 5 figs.

Saint-Girons, H. (1990) Notes on ecology and population structure of the Laticaudinae (Serpents, Hydrophidae) in New Caledonia. *Bulletin of the Chicago Herpetological Society* 25(11), 197–209, 2 figs., 4 illust. [translation of Saint-Girons, 1964, by Fanci, I.M., Gritis, P.A. & Voris, H.K.]

Saint-Girons, H. & Détrait, J. (1980) Communautés antigéniques des venins et systématique des Elapidae. *Bijdragen tot de Dierkunde,* 50(1), 96–104, 3 figs.

Saint-Girons, H. & Duguy, R. (1976) Écologie et position systématique de *Vipera seoanei* Lataste, 1879. *Bulletin de la Société Zoologique de France*, Paris 101(2), 325–339, 2 figs.

Saint-Girons, H. & Pfeffer, P. (1972) Notes sur l'ecologie des serpents du Cambodge. *Zoologische Mededelingen*, Leiden 47(6), 65–87, 2 figs.

Sajdak, R.A. (2010) *Hunters in the trees: a natural history of arboreal snakes.* Krieger Publishing Company, Malabar, 171 pp., col. illust.

Sajdak, R.A. & Henderson, R.W. (1982) Notes on the eggs and young of *Antillophis parvifrons stygius* (Reptilia, Serpentes, Colubridae). *Florida Scientist* 45(3), 200–204.

Sajdak, R.A. & Henderson, R.W. (1990) Status of West Indian racers in the Lesser Antilles. *Oryx*, London 25(1), 33–38, illust.

Salazar, M. & Barquero, M.D. (2012) First country record for *Urotheca decipiens* and *Urotheca pachyura* and range extensions of *Urotheca guentheri* in Nicaragua. *Herpetological Bulletin* (121), 33–35, 2 figs.

Salazar-Bravo, J., Vargas, J., Jimenez-Ruiz, A. & Savage, J.M. (2010) A new record of *Atractus boettgeri* (Serpentes: Colubridae), with notes on taxonomy and natural history. *Revista Mexicana de Biodiversidad* 81, 925–929, 2 figs.

Salazar-Valenzuela, D., Carrillo, E.D. & Adás, A.S. (2010) *Tricheilostoma anthracinum.* Geographic distribution: Serpentes: *Herpetological Review* 41(1), 111–112.

Saleh, M.A. (1997) *Amphibians and reptiles of Egypt.* Publication of National Biodiversity Unit (6), viii + 234 pp, 6 figs., 113 col. pls.

Salinas, U.H. & Ramírez-Bautista, A. (2010) Geographic distribution: Serpentes: *Ramphotyphlops braminus. Herpetological Review* 41(4), 519.

Salinas, U.H., Ramírez, J.P., Goyenechea, I., Ramírez-Bautista, A. & Vite-Silva, V. (2010) Geographic distribution: Serpentes: *Epicta goudotii. Herpetological Review* 41(4), 516.

Salles, R. de O.L., Weber, L.N. & Silva-Soares, T. (2010) Reptiles, Squamata, Parque Natural Municipal da Taquara, municipality of Duque de Caxias, state of Rio de Janeiro, southeastern Brazil. *Check List* 6(2), 280–286, 5 figs.

Salmon, G.T., Bryson, R.W., Jr. & Lazcano, D. (2001) Geographic distribution: Serpentes: *Lampropeltis mexicana. Herpetological Review* 32(2), 123.

Salmon, G.T., Liner, E.A., Forks, J.E. & Lazcano, D. (2004) Geographic distribution: Serpentes: *Lampropeltis alterna. Herpetological Review* 35(3), 292.

Salomão, M.G., Wüster, W., Thorpe, R.S. & BBBSP. (1997) DNA evolution of South American pitvipers of the genus *Bothrops* (Reptilia, Serpentes, Viperidae). *In:* Thorpe, R.S., Wüster, W. & Malhotra, A. (Eds.), *Venomous Snakes. Ecology, evolution and snakebite.* Zoological Society of London, Clarendon Press, Oxford, pp. 89–98, 2 figs.

Salomão, M.G., Wüster, W., Thorpe, R.S., Touzet, J.-M. & BBBSP. (1999) MtDNA phylogeny of Neotropical pitvipers of the genus *Bothrops* (Squamata, Serpentes, Viperidae). *In:* Joger, U. (Ed.), Phylogeny and systematics of the Viperidae. *Kaupia, Darmstädter Beiträge zur Naturgeschichte* 8, 127–134, 4 figs.

Salvado, R. (1851) *Memorie storiche dell'Australia particolarmente della Missione Benedettina di Nuova Norcia e degli usi e costumi degli Australiani.* Sacra Congregatio de Propaganda Fide, Roma, xii + 388 pp.

Salvador, A. (1974) *Guia de los anfibios y reptiles Españoles.* Instituto Nacional para la Conservación de la Naturaleza de Madrid, 282 pp., 34 figs., 66 col. illust.

Salvador, A. (1985) *Guia de Campo de los anfibios y reptiles de la Peninsula Iberica, Islas Baleares y Canarias.* Santiago Garcia, Leon, Spain, 255 pp., 104 col. pls.

Salvador, A. (Ed.). (1998) *Fauna Iberica. Vol. 10. Reptiles.* Museo Nacional de Ciencias Naturales, Madrid, 706 pp., 83 figs., 47 maps.

Salvador, A. & Peris, S. (1975) Contribucion al estudio de la fauna herpetologica de Rio de Oro. *Boletín de la Estación Central de Ecología*, Madrid 4(8), 49–60, illust.

Salvador, A. & Pleguezuelos, J.M. (2002) *Reptiles Españoles: identidficación, historia natural y distribución.* Canseco Editores, Talavera de la Reina, 493 pp., figs., col. illust.

Salvin, O. (1860) On the reptiles of Guatemala. *Proceedings of the Zoological Society of London* 28(1), 451–461, pl. 32.

Salvin, O. (1861) On a collection of reptiles from Guatemala. *Proceedings of the Zoological Society of London* 29(1), 227–229.

Samarawickrama, V.A.M.P.K., Samarawickrama, V.A.P., Wijesena, N.M. & Orlov, N.L. (2005) A new species of *Boiga* (Serpentes, Colubridae, Colubrinae) from Sri Lanka. *Russian Journal of Herpetology* 12(3), 213–222, 14 figs.

Sampaio, P.R.M. & Lima-Maciel, J.M. (2012) Geographic distribution: Serpentes: *Imantodes lentiferus. Herpetological Review* 43(2), 307–308.

San, S.-H. & Lee, Y.-U. (2007) *Field guide to reptiles and amphibians of Korea*, Slow and Steady, 175 pp., 372 col. photos. (in Korean)

Sanchez, C., Lopes-Carvalho, V., Kathriner, A., O'Shea, M. & Kaiser, H. (2012) First report on the herpetofauna of the Oecusse District, an enclave of Timor-Leste. *Herpetology Notes* 5, 137–149, 18 figs.

Sanchez, D., de Sousa, L., Esqueda, L.F. & Manzanilla, J. (2004) Especie nueva de *Atractus* (Serpentes, Colubridae) del Macizo del Turimiquire, Tramo Oriental de La Cordillera de La Costa, Venezuela. *Universidad Oriente, Saber* 16(2), 89–95, 4 figs.

Sanchez, H., Castaño, O. & Cardenas, G. (1995) Diversidad de los reptiles en Colombia. *In:* Rangel, J.O. (Ed.), *Colombia diversidad biotica I, Clima, centros de concentracion de especies, Fauna, reptiles, aracnidos, himenopteros.* Universidad Nacional de Colombia, Bogotá, pp. 277–325.

Sánchez, J.D., Daza, J.M. & Alzate, A.A. (2009) Geographic distribution: Serpentes: *Corallus caninus. Herpetological Review* 40(1), 113.

Sanchez de la Vega, G., Mata-Silva, V., Molina-Garcia, E., Garcia-Mendez, A. & Buenrostro-Silva, A. (2012) Geographic distribution: Serpentes: *Crotalus culminatus. Herpetological Review* 43(1), 105.

Sánchez-Luna, M., González-Hernández, A.J., Cabrera-Guzmán, E. & Reynoso, V.H. (2009) Geographic distribution: Serpentes: *Leptophis diplotropis. Herpetological Review* 40(4), 456.

Sanchíz, F.B. (1980) Nota preliminar sobre los reptiles Pleistocenicos de la Cueva de la Paloma (Soto de las Regueras, Asturias). *In:* Hotos-Gómez, M., Martínez-Navarrete, M.I., Chapa-Brunet, T., Casiaños, P. & Sanchíz, F.B. (Eds.), *La Cueva de la Paloma: Soto de las Regueras (Asturias).* Excavaciones Arqueológicas en España, Ministerio de Cultura, Madrid, 116, 109–114, 7 figs.

Sanders, K.L., Malhotra, A. & Thorpe, R.S. (2002) A contribution to the systematics of two commonly confused pitvipers from the Sunda region: *Trimeresurus hageni* and *T. sumatranus. Bulletin of the Natural History Museum of London (Zoology)* 68(2), 107–111, 1 fig.

Sanders, K.L., Malhotra, A., Gumprecht, A., Thorpe, R.S. & Kuch, U. (2004) *Popeia inornata*, a new species of pitviper from West Malaysia (Squamata, Viperidae, Crotalinae). *Russian Journal of Herpetology* 11(3), 163–176, 8 figs.

Sanders, K.L., Malhotra, A. & Thorpe, R.S. (2006) Combining molecular, morphological and ecological data to infer species boundaries in a cryptic tropical pitviper. *Biological Journal of the Linnean Society,* London 87(3), 343–364, 6 figs.

Sanders, K.L., Lee, M.S.Y., Leys, R., Foster, R. & Keogh, J. (2008) Molecular phylogeny and divergence dates for Australasian elapids and sea snakes (Hydrophiinae): evidence from seven genes for rapid evolutionary radiations. *Journal of Evolutionary Biology* 21, 682–695, 4 figs.

Sanders, K.L., Lee, M.S.Y., Mumpuni, Bertozzi, T. & Rasmussen, A.R. (2013) Multilocus phylogeny and recent rapid radiation of the viviparous sea snakes (Elapidae: Hydrophiinae). *Molecular Phylogenetics and Evolution* 66(3), 575–591.

Sanderson, W.E. (1983) *Systematics of the water snakes of the* Nerodia cyclopion *complex.* Master's Thesis, Louisiana State University, Baton Rouge, viii + 102 pp., 22 figs.

Sanderson, W.E. (1993) Additional evidence for the specific status of *Nerodia cyclopion* and *Nerodia floridana* (Reptilia, Colubridae). *Brimleyana* 19, 83–94, 4 figs.

Sandner-Montilla, F. (1952) Serpientes *Bothrops* de Venezuela. *Monografias Cientificas del Instituto de Terapeutica Experimental de los Laboratorios "Veros" Ltda*, Bogotá 9(21), 8 pp., 4 figs.

Sandner-Montilla, F. (1961) Ofidios ponzoñosos de Venezuela. *Bothrops venezuelae* Sp. nov. *Novedades Cientificas, Contribuciones Ocasionales del Museo de Historia Natural La Salle*, Caracas (*Zoología*) (30), 1–36, figs. a–q, 1 pl.

Sandner-Montilla, F. (1965) *Manual de las serpentes ponzõnosas de Venezuela.* García, Caracas, 108 pp., 69 figs., 8 pls. [reprinted in 1975]

Sandner-Montilla, F. (1976) Una nueva especie del genero *Bothrops* (Serpientes, Crotalidae, Lachesinae) de la Gran Sabana, Edo. Bolivar, Venezuela. *Memorias Cientificas de Ofidiologia, Contribuciones Periodicas del Instituto Venezolano de Ofidiologia*, Caracas (1), 1–4, 1 fig.

Sandner-Montilla, F. (1978) *Bothrops castelnaudi* Duméril, Bribron et Duméril, 1854. Su existencia en Venezuela y el paso a sinonimia de *Botrops lichenosus* Roze, 1958. *Memorias Cientificas de Ofidiologia, Contribuciones Periodicas del Instituto Venezolano de Ofidiologia*, Caracas (2), 1–8, 4 figs.

Sandner-Montilla, F. (1979) Una nueva especie del genero *Bothrops* (Serpentes, Crotalidae, Bothropinae) de la region de Guanare, Estado Portuguesa, Venezuela. *Memorias Cientificas de Ofidiologia, Contribuciones Periodicas del Instituto Venezolano de Ofidiologia*, Caracas (4), 19 pp., 17 figs.

Sandner-Montilla, F. (1980) Una nueva especie del genero *Crotalus* (Serpentes, Crotalidae, Crotalinae) del sur del Estado Guarico, Venezuela. *Memorias Cientificas de Ofidiologia, Contribuciones Periodicas del Instituto Venezolano de Ofidiologia*, Caracas (5), 1–12, 9 figs.

Sandner-Montilla, F. (1981) Una nueva subespecie de *Bothrops lanceolatus* (Lacépède, 1789) Fam. Crotalidae. *Memorias Cientificas de Ofidiologia, Contribuciones Periodicas del Instituto Venezolano de Ofidiologia*, Caracas (6), 1–15, 10 figs.

Sandner-Montilla, F. (1983) Serpientes Crotalinae de Venezuela. *Memórias do Instituto Butantan*, São Paulo (1982) 46, 193–194.

Sandner-Montilla, F. (1985) La creacion de la familia Micruridae fam. nov. para las corales de America de la superfamilia Elapoidea y la proposicion para la creacion de la familia Oxyuranidae de la misma superfamilia, para proteroglifas Australianas. *Memorias Cientificas de Ofidiologia, Contribuciones Periodicas del Instituto Venezolano de Ofidiologia*, Caracas (8), 1–21, 22 figs.

Sandner-Montilla, F. (1989) Una nueva subespecie de *Bothrops lansbergi* (Schlegel, 1841) de la familia Crotalidae: *Bothrops lansbergi hutmanni* n. ssp. *Memorias Cientificas de Ofidiologia, Contribuciones Periodicas del Instituto Venezolano de Ofidiologia*, Caracas (9), 1–16.

Sandner-Montilla, F. (1990) Una nueva subespecie de *Bothrops lanceolatus* (Lacépède, 1789), género *Bothrops* Wagler, familia Crotalidae Gray, del estado Carabobo y regiones edya center, *Bothrops lanceolatus nacaritae*, ssp. nov. *Memorias Cientificas de Ofidiologia, Contribuciones Periodicas del Instituto Venezolano de Ofidiologia*, Caracas 10, 33–37.

Sandner-Montilla, F. & Römer, M. (1961) Ofidios ponzoñosos de Venezuela. *Bothrops pifanoi* Sp. nov. *Novedades Cientificas, Contribuciones Ocasionales del Museo de Historia Natural La Salle*, Caracas (*Zoología*) (29), 1–15, 5 figs.

Sandoval, P.E. (1991) Panorama de la herpetología en Bolivia. *Acta Zoologica Lilloana*, 41, 5–12, 1 fig.

Santos, A.P. dos, Jr. (2003) Geographic distribution: Serpentes: *Siphlophis worontzowi. Herpetological Review* 34(4), 390.

Santos, A.P. dos, Jr. & Lema, T. de. (2004) Geographic distribution: Serpentes: *Echinanthera occipitalis. Herpetological Review* 35(1), 83.

Santos, A.P. dos, Jr., Di-Bernardo, M. & Lema, T. de. (2008) New species of the *Taeniophallus occipitalis* group (Serpentes, Colubridae) from eastern Amazonia, Brazil. *Journal of Herpetology* 42(3), 419–426, 4 figs.

Santos, E. (1943) *As cobras venenosas como conhece-las e evitá-las tratamento do ofidismo.* Edição da Chácaras e Quintais, São Paulo, 107 pp.

Santos-Jordão, R. dos & Fernandes-Bizerra, A. (1996) Reproduçao, dimorfismo sexual e atividade de *Simophis rhinostoma* (Serpentes, Colubridae). *Revista Brasileira de Biologia* 56(3), 507–512.

Sanyal, D.P. (1993) Reptilia. *In:* Gosh, A.K. (Ed.), *Fauna of Orissa. Part 4. Reptilia, Amphibia, freshwater fishes, birds, mammals. State Fauna Series* 1. Zoological Survey of India, Calcutta, pp. 51–74.

Sanyal, D.P. & Gayen, N.C. (1985) Fauna of Namdapha, Arunachal Pradesh: Reptilia. *Records of the Zoological Survey of India*, Calcutta 82(1–4), 297–302.

Sanyal, D.P. & Gayen, N.C. (2006) Reptilia. *In:* Alfred, J.R.B. (Ed.), *Fauna of Arunachal Pradesh (Part-1). State Fauna Series* 13. Zoological Survey of India, Kolkata, pp. 247–284.

Sanyal, D.P., Dattagupta, B. & Gayen, N.C. (1993) Reptilia. *In:* Gosh, A.K. (Ed.), *Fauna of Andhra Pradesh. Part I (Reptilia, Amphibia, Fishes). State Fauna Series* 5. Zoological Survey of India, Calcutta, pp. 1–63.

Sanyal, D.P., Dattagupta, B. & Gayen, N.C. (2002) Reptilia. *In:* Alfred, J.R.B. (Ed.), *Fauna of Tripura (Part-1). Vertebrates. State Fauna Series* 7. Zoological Survey of India, Kolkata, pp. 159–177.

Sanyal, D.P., Sur, S. & Gayen, N.C. (2006) Reptilia. *In:* Alfred, J.R.B. (Ed.), *Fauna of Sikkim (Part-1) Vertebrates. State Fauna Series* 9. Zoological Survey of India, Kolkata, pp. 157–171.

Sanz, J.L. & Sanchíz, F.B. (1980) Nota preliminary sobre los Squamata (Reptilia) del Pleistoceno medio de Aridos – 1 (Arganda, Madrid). *Ocupaciones Achelenses en el Valle del Jarama, Publicaciones de la Excelentisima Diputación Provincial de Madrid,* pp. 127–138, 4 figs.

Sarasin, F. 1910. Über die Geschichte der Tierwelt von Ceylon. *Zoologische Jahrbücher*, (supplement) Jena 12(1), 125–160, 3 pls.

Sarauw, G.F.L. (1903) Om Danmarks sjaeldne Slanger. *Videnskabelige Meddelelser fra Dansk Naturhistorisk e Forening i Kjobenhavn* 1903, 5–42, 2 figs.

Sarkar, N.K. & Devi, A. (1968) Enzymes in snake venoms. *In:* Bücherl, W., Buckley, E.E. & Deulofeu, V. (Eds.), *Venomous animals and their venoms. Volume I. Venomous vertebrates.* Academic Press, New York, pp. 167–216, 4 figs.

Sasa, M. (1993) Distribution and reproduction of the gray earth snake *Geophis brachycephalus* (Serpentes, Colubridae) in Costa Rica. *Revista de Biologia Tropical*, San José 41(2), 295–297, 1 fig.

Sasa, M. (1997) *Cerrophidion godmani* in Costa Rica: a case of extremely low allozyme variation? *Journal of Herpetology* 31(4), 569–572, 1 fig.

Sasa, M. & Barrantes, R. (1998) Allozyme variation in populations of *Bothrops asper* (Serpentes, Viperidac) in Costa Rica. *Herpetologica* 54(4), 462–469.

Sasa, M. & Solórzano, A. (1995) The reptiles and amphibians of Santa Rosa National Park, Costa Rica, with comments about the herpetofauna of xerophytic areas. *Herpetological Natural History* 3(2), 113–126, 11 figs.

Sasa, M., Chaves, G. & Porras, L. (2010) The Costa Rican herpetofauna: conservation status and future perspectives. *In:* Wilson, L.D., Townsend, J.H. & Johnson, J.A. (Eds.), *Conservation of Mesoamerican amphibians and reptiles.* Eagle Mountain Publishing, Eagle Mountain, pp. 510–603, 23 figs.

Satishkumar, V.M. (2007) Reptiles. *In:* Alfred, J.R.B. (Ed.), *Fauna of Chennai coast. Marine Ecosystem Series* 1. Zoological Survey of India, Kolkata, pp. 289–294.

Sauer, F. (1989) *Kriechtiere und Lurche im Mittelmeerraum nach Farbfotos erkannt.* Fauna-Verlag, Eichenweg, 162 pp., 72 col. pls.

Sauvage, H.E. (1877) Sur quelques ophidiens d'espèces nouvelles ou peu connues de la collection du Muséum. *Bulletin de la Société Philomathique de Paris* (1876–1877) (Série 7) 1, 107–115.

Sauvage, H.E. (1878) Essai sur la faune herpétologique de la Nouvelle-Guinée, suivi de la description de quelques espèces nouvelles ou peu connues. *Bulletin de la Société Philomathique de Paris* (Série 7) 2, 25–44.

Sauvage, H.E. (1879a) Notice sur quelques reptiles nouveaux ou peu connus de la Nouvelle-Guinée. *Bulletin de la Société Philomathique de Paris* (1878–1879) (Série 7) 3, 47–61.

Sauvage, H.E. (1879b) Sur un Rhabdosomien de genre nouveau provenant de Ternate. *Bulletin de la Société Philomathique de Paris* (1878–1879) (Série 7) 3, 61–63.

Sauvage, H.E. (1880a) Sur l'existence d'un reptile du type ophidien dans les couches à *Ostrea columba* des Charentes. *Comptes Rendus de l'Académie des Sciences*, Paris 91, 671–672.

Sauvage, H.E. (1880b) "...sur l'existence d'un reptile du type ophidien dans les couches à *Ostrea columba* des Charentes." (Actually an editorial comment on publication by Sauvage earlier in the year). *Le Naturaliste*, Paris (40), 313.

Sauvage, H.E. (1881) On the existence of a reptile of the ophidian type in the beds with *Ostrea columba*, of the Charente. *Annals and Magazine of Natural History*, London (Series 5) 7(38), 205–206.

Sauvage, H.E. (1884a) Sur quelques reptiles de la collection du Muséum d'histoire naturelle. *Bulletin de la Société Philomathique de Paris* (1883–1884) (Série 7) 8, 142–147.

Sauvage, H.E. (1884b) Notice sur une collection de reptiles et de poisons recueillie a Majumba, Congo. *Bulletin de la Société Zoologique de France* 9, 199–204.

Sauvage, H.E. (1898) *Vertébrés fossils du Portugal—Contributions à l'Étude des poisons et des reptiles du Jurassique et du Crétacique.* Direction des Travaux Géologiques du Portugal 1897–1898. Lisbonne, 47 pp., 10 pls.

Savage, J.M. (1950) Two new blind snakes (genus *Typhlops*) from the Philippines. *Proceedings of the California Zoological Club* 1(10), 49–54, 2 figs.

Savage, J.M. (1955) Description of new colubrid snakes, genus *Atractus*, from Ecuador. *Proceedings of the Biological Society of Washington* 68, 11–20, 1 fig.

Savage, J.M. (1960) A revision of the Ecuadorian snakes of the colubrid genus *Atractus. Miscellaneous Publications, Museum of Zoology, University of Michigan* (112), 1–86, 9 figs.

Savage, J.M. (1966) The origins and history of the Central American herpetofauna. *Copeia* 1966(4), 719–766, 26 figs.

Savage, J.M. (1970) On the trail of the golden frog: with Warszewicz and Gabb in Central America. *Proceedings of the California Academy of Sciences* (Series 4) 38, 273–287.

Savage, J.M. (1974) Type localities for species of amphibians and reptiles described from Costa Rica. *Revista de Biologia Tropical*, San José 22(1), 71–122.

Savage, J.M. (1980a) *A handlist with preliminary keys to the herpetofauna of Costa Rica.* Allan Hancock Foundation, Los Angeles, 111 pp.

Savage, J.M. (1980b) *A preliminary handlist of the herpetofauna of Costa Rica.* Allan Hancock Foundation, Los Angeles, 21 pp.

Savage, J.M. (1981) A new species of the secretive colubrid snake genus *Geophis* from Costa Rica. *Copeia* 1981(3), 549–553, 2 figs.

Savage, J.M. (1982) The enigma of the Central American herpetofauna: dispersals or vicariance? *Annals of the Missouri Botanical Garden* 69, 464–547, 25 figs.

Savage, J.M. (2002) *The amphibians and reptiles of Costa Rica. A herpetofauna between two continents, between two seas.* Chicago University Press, xx + 934 pp., figs., 516 col. pls.

Savage, J.M. (2011) The correct species-group name for an *Oxyrhopus* (Squamata: Dipsadidae) variously called *Coluber petalarius*, *C. pethola*, *C. petola*, or *C. petolarius* by early authors. *Proceedings of the Biological Society of Washington* 124(3), 223–225.

Savage, J.M. & Bolaños, F. (2009) A checklist of the amphibians and reptiles of Costa Rica: additions and nomenclatural revisions. *Zootaxa* (2005), 1–23.

Savage, J.M. & Boundy, J. (2012) On the type species of the snake generic name *Anilios* Gray, 1845 (Serpentes: Typhlopidae). *Herpetological Review* 43(4), 537–538.

Savage, J.M. & Cliff, F.S. (1953) A new subspecies of sidewinder, *Crotalus cerastes*, from Arizona. *Natural History Miscellanea*, Chicago (119), 1–7, 1 fig.

Savage, J.M. & Cliff, F.S. (1954) A new snake *Phyllorhynchus arenicola*, from the Gulf of California, México. *Proceedings of the Biological Society of Washington* 67, 69–76, 2 figs.

Savage, J.M. & Crother, B.I. (1989) The status of *Pliocercus* and *Urotheca* (Serpentes, Colubridae), with a review of included species of coral snake mimics. *Zoological Journal of the Linnean Society*, London 95(4), 335–362, 4 figs.

Savage, J.M. & Donnelly, M.A. (1988) Variation and systematics in the colubrid snakes of the genus *Hydromorphus. Amphibia-Reptilia*, Leiden 9(3), 289–300, 3 figs.

Savage, J.M. & Lahanas, P.N. (1989) A new species of colubrid snake (genus *Urotheca*) from the Cordillera de Talamanca of Costa Rica. *Copeia* 1989(4), 892–896, 2 figs.

Savage, J.M. & Lahanas, P.N. (1991) On the species of the colubrid snake genus *Ninia* in Costa Rica and western Panamá. *Herpetologica* 47(1), 37–53, 8 figs.

Savage, J.M. & McDiarmid, R.W. (1992) Rediscovery of the Central American colubrid snake, *Sibon argus*, with comments on related species from the region. *Copeia* 1992(2), 421–432, 5 figs.

Savage, J.M. & Oliver, J.A. (1956) Proposed addition to the "Official List of Generic Names in Zoology" of "*Ahaetulla*" Link, 1807, with "*Ahaetulla mycterizana*" Link, 1807, as type species (class Reptilia). *Bulletin of Zoological Nomenclature* 12(5), 147–152.

Savage, J.M. & Scott, N.J., Jr. (1985) The *Imantodes* (Serpentes, Colubridae) of Costa Rica: two or three species? *Revista de Biologia Tropical*, San José 33(2), 107–132, 5 figs.

Savage, J.M. & Slowinski, J.B. (1992) The colouration of the venomous coral snakes (family Elapidae) and their mimics (families Aniliidae and Colubridae). *Biological Journal of the Linnean Society*, London 45(3), 235–254, 6 figs.

Savage, J.M. & Slowinski, J.B. (1996) Evolution of coloration, urotomy and coral snake mimicry in the snake genus *Scaphiodontophis* (Serpentes, Colubridae). *Biological Journal of the Linnean Society*, London 57(2), 129–194, 20 figs.

Savage, J.M. & Vial, J.L. (1973) The venomous coral snakes (genus *Micrurus*) of Costa Rica. *Revista de Biologia Tropical*, San José 21(2), 295–349, 7 figs.

Savage, J.M. & Villa, J. (1986) *Introduction to the herpetofauna of Costa Rica*. Society for the Study of Amphibians and Reptiles, 207 pp.

Savage, J.M. & Watling, J.I. (2008) Not so rare snakes: a revision of the *Geophis sieboldi* group (Colubridae: Dipsadinae) in lower Central America and Colombia. *Zoological Journal of the Linnean Society*, London 153(3), 561–599, 18 figs.

Savage, J.M., Campbell, J.A. & Lamar, W.M. (2005) On names for Neotropical rattlesnakes (Reptilia, Serpentes, Viperidae, *Crotalus*). *Herpetological Review* 36(4), 369–371.

Savage, J.M., David, P. & Vogel, G. (2012) *Tropidolaemus* Wagler, 1830 and *Cophias wagleri* F. Boie, 1827 (currently *Tropidolaemus wagleri*) (Reptilia, Squamata, Viperidae): proposed conservation. *Bulletin of Zoological Nomenclature* 69(2), 116–121.

Savitzky, A.H. & Smith, H.M. (1971) A new snake from México of the *taeniata* group of *Tantilla*. *Journal of Herpetology* 5(3–4), 167–171, 1 fig.

Savitzky, B.A., Savitzky, A.H., Belcher, R.T. & Ewers, S. (2002) Geographic distribution: Serpentes: *Ramphotyphlops braminus*. *Herpetological Review* 33(2), 150–151.

Sawai, Y. (1976) Medical treatment of snakebites. 2. Southeast Asia. *The Snake* 8(1), 1–30, 38 figs.

Sawai, Y. (1979) Studies on snakebites in the Asian areas. *In:* Eaker, D. & Wadström,T. (Eds.), *Natural toxins*. Pergamon Press, Oxford, pp. 25–32, 2 figs.

Sawaya, R.J. & Sazima, I. (2003) A new species of *Tantilla* (Serpentes, Colubridae) from southeastern Brazil. *Herpetologica* 59(1), 119–126, 4 figs.

Say, T. (1819) Notes on herpetology. *American Journal of Science*, New York (1818) 1, 256–265.

Say, T. (1825) Descriptions of three new species of *Coluber*, inhabiting the United States. *Journal of the Academy of Natural Sciences of Philadelphia* 4(2), 237–241.

Sayed, S.M. (1972) Checklist of marine fauna of Pakistan. *Records of the Zoological Survey of Pakistan Marine Fauna* (Suppl.) 4(1–2), 1–68.

Sazima, I. (1989) Feeding behavior of the snail-eating snake, *Dipsas indica*. *Journal of Herpetology* 23(4), 464–468, 2 figs.

Sazima, I. (1992) Natural history of the jararaca pitviper, *Bothrops jararaca*, in southeastern Brazil. *In:* Campbell, J.A. & Brodie, E.D., Jr. (Eds.), *Biology of the pitvipers*. Selva, Tyler, pp. 199–216, 18 figs.

Scanferla, C.A. (2006) The oldest record of *Clelia* (Serpentes-Colubridae) in South America. *Comptes Rendus Palevol* 5, 721–724, 2 figs.

Scanferla, C.A. & Nenda, S.J. (2005) El registro más antiguo del género *Bothrops* (Serpentes, Viperidae), proveniente del Pleistoceno inferior a medio de Argentina. *Revista del Museo Argentina de Ciencias Naturales* 7(2), 177–181, 2 figs.

Scanferla, C.A., Martin de los Reyes, L. & Cenizo, M.M. (2005) Sobre el primer registro fósil del género *Lystrophis* Cope, 1885 (Serpentes-Colubridae-Xenodontinae). *Studia Geologica Salmanticensia* 41, 93–101, 3 figs.

Scanferla, C.A., Agnolin, F. & Voglino, D. (2009) *Boiruna* cf. *B. maculata* (Ophidia: Colubroidea) from the early to middle Pleistocene of Argentina, and the effects of Pleistocene extinctions on South American reptiles. *South American Journal of Herpetology* 4(3), 259–267, 4 figs.

Scanlon, J.D. (1992) A new large madtsoiid snake from the Miocene of the Northern Territory. *The Beagle, Records of the Northern Terrritory Museum of Arts and Sciences* 9(1), 49–60, 1 fig.

Scanlon, J.D. (1993) Madtsoiid snakes from the Eocene Tingamarra Fauna of eastern Queensland. *Kaupia, Darmstädter Beiträge zur Naturgeschichte* 3, 3–8, 3 figs.

Scanlon, J.D. (1995) First records from Wellington Caves, New South Wales, of the extinct Madtsoiid snake *Wonambi naracoortensis* Smith, 1976. *Proceedings of the Linnaean Society of New South Wales* 115, 233–238, 1 fig.

Scanlon, J.D. (1997) *Nanowana* gen. nov., small madtsoiid snakes from the Miocene of Riversleigh, sympatric species with divergently specialised dentition. *Memoirs of the Queensland Museum*, Brisbane 41(2), 393–412, 12 figs.

Scanlon, J.D. (2003a) The Australian elapid genus *Cacophis*: morphology and phylogeny of rainforest crowned snakes. *Herpetological Journal* 13(1), 1–20, 6 figs.

Scanlon, J.D. (2003b) The basicranial morphology of madtsoiid snakes (Squamata, Ophidia) and the earliest Alethinophidia (Serpentes). *Journal of Vertebrate Paleontology* 23(4), 971–976, 1 fig.

Scanlon, J.D. (2005) Australia's oldest known snakes, *Patagoniophis*, *Alamitophis*, and cf. *Madtsoia* (Squamata, Madtsoiidae) from the Eocene of Queensland. *Memoirs of the Queensland Museum* 51(1), 215–235, 8 figs.

Scanlon, J.D. (2006) Skull of the large non-macrostomatan snake *Yurlunggur* from the Australian Oligo-Miocene. *Nature* 439, 839–842, 2 figs.

Scanlon, J.D. & Lee, M.S.Y. (2000) The Pleistocene serpent *Wonambi* and the early evolution of snakes. *Nature* 403(6768), 416–420.

Scanlon, J.D. & Lee, M.S.Y. (2002) Varanoid-like dentition in primitive snakes (Madtsoiidae). *Journal of Herpetology* 36(1), 100–106, 4 figs.

Scanlon, J.D. & Lee, M.S.Y. (2004) Phylogeny of Australasian venomous snakes (Colubridae, Elapidae, Hydrophiinae) based on phenotypic and molecular evidence. *Zoologica Scripta* 33, 335–366, 10 figs.

Scanlon, J.D. & Mackness, B.S. (2002) A new giant python from the Pliocene Bluff Downs Local Fauna of northeastern Australia. *Alcheringa* 25, 425–437, 2 figs.

Scanlon, J.D. & Shine, R. (1988) Dentition and diet in snakes: adaptations to oophagy in the Australian elapid genus *Simoselaps*. *Journal of Zoology*, London 216(3), 519–528, 2 figs.

Scanlon, J.D., Lee, M.S.Y., Caldwell, M.W. & Shine, R. (1999) The palaeoecology of the primitive snake *Pachyrhachis*. *Historical Biology* 13, 127–152, 7 figs.

Scartozzoni, R.R., Trevine, V.C. & Germano, V.J. (2010) Reptilia, Squamata, Serpentes, Dipsadidae, *Pseudoeryx plicatilis* (Linnaeus, 1758): new records and geographic distribution map. *Check List* 6(4), 534–537, 1 fig.

Schaal, S. & Baszio, S. (2004) *Messelophis ermannorum* n. sp., eine neue Zwergboa (Serpentes: Boidae: Tropidopheinae) aus dem Mittel-Eozän von Messel. *Courier Forschungsinstitut Senckenberg*, Frankfurt-am-Main (252), 67–77.

Schaefer, E.F. (1998) Geographic distribution: Serpentes: *Phalotris lemniscatus*. *Herpetological Review* 29(3), 178.

Schaefer, E.F. & Alvarez, B.B. (2000) Geographic distribution: Serpentes: *Tantilla melanocephala*. *Herpetological Review* 31(2), 115.

Schaefer, N. (1970) A new species of house snake from Swaziland, with notes on the status of the two genera *Lamprophis* and *Boaedon*. *Annals of the Cape Province Museum (Natural History)*, Grahamstown 8(14), 205–208, 1 fig.

Schargel, W.E. & Castoe, T.A. (2003) The hemipenes of some snakes of the semifossorial genus *Atractus*, with comments on variation in the genus. *Journal of Herpetology* 37(4), 718–721, 1 fig.

Schargel, W.E. & García-Pérez, J.E. (2002) A new species and a new record of *Atractus* (Serpentes, Colubridae) from the Andes of Venezuela. *Journal of Herpetology* 36(3), 398–402, 2 figs.

Schargel, W.E., Fuenmayor, G.R. & Myers, C.W. (2005) An enigmatic new snake from cloud forest of the Peninsula de Paria, Venezuela (Colubridae, genus *Taeniophallus*?). *American Museum Novitates* (3484), 1–22, 10 figs.

Schargel, W.E., Fuenmayor, G.R., Barros-Blanco, T.R., Péfaur, J.E. & Navarrete, L.F. (2007) A new aquatic snake (Colubridae, *Pseudoeryx*) from the Lake Maracaibo Basin, northwestern Venezuela: a relic of the past course of the Orinoco River. *Herpetologica* 63(2), 236–244, 5 figs.

Schätti, B. (1982) Bemerkungen zur Okologie, Verbreitung und intraspezifischen Variation der Vipernatter, *Natrix maura* (Linne, 1758) (Reptilia, Serpentes). *Revue Suisse de Zoologie*, Genève 89(2), 521–542, illust.

Schätti, B. (1985a) Systematics of east African species of *Meizodon* Fischer, 1856 (Reptilia, Serpentes, Colubridae). *Monitore Zoologico Italiano, Supplemento*, Firenze (Series 2) 20(8), 149–175, 6 figs.

Schätti, B. (1985b) Eine neue Zornnatter aus Zypern, *Coluber cypriensis* n. sp. (Reptilia, Serpentes, Colubridae). *Revue Suisse de Zoologie*, Genève 92(2), 471–477, 2 figs.

Schätti, B. (1986a) Catalogue des types et des examplaires figures du Musée d'histoire naturelle de Neuchatel. II. Ophidiens. *Bibliothèques et Musées de la Ville de Neuchâtel*, Neuchâtel 1985, 98–108.

Schätti, B. (1986b) Morphological evidence for a partition of the genus *Coluber* (Reptilia, Serpentes). *In:* Rocek, Z. (Ed.), *Studies in herpetology*. Charles University, Prague, pp. 235–238, 2 figs.

Schätti, B. (1986c) Herpetologische Ausbeute einer Sammelreise nach Mali (Amphibia, Reptilia). *Revue Suisse de Zoologie*, Genève 93(3), 765–778, 3 figs.

Schätti, B. (1986d) Morphologie und Systematik von *Coluber algirus* und *C. hippocrepis* (Reptilia, Colubridae). *Bonner Zoologische Beiträge*, Bonn 37(4), 281–293, illust.

Schätti, B. (1987) The phylogenetic significance of morphological characters in the Holarctic racers of the genus *Coluber* Linnaeus, 1758 (Reptilia, Serpentes). *Amphibia-Reptilia*, Leiden 8(4), 401–418, 3 figs.

Schätti, B. (1988a) *Systematik und evolution der Schlangengattung Hierophis Fitzinger, 1843 (Reptilia, Serpentes)*. Doctoral Dissertation, Universität Zurich, Zurich, 50 pp., 8 figs.

Schätti, B. (1988b) Systematics and phylogenetic relationships of *Coluber florulentus* Geoffroy, 1827 (Reptilia Serpentes). *Tropical Zoology*, Firenze 1(1), 95–116, illust.

Schätti, B. (1989) Amphibien und Reptilien aus der Arabischen Republik Jemen und Djibouti. *Revue Suisse de Zoologie*, Genève 94(4), 905–937, illust.

Schätti, B. (1993a) *Coluber* Linnaeus, 1758 – Zornnatter. *In:* Böhme, W. (Ed.), *Handbuch der Reptilien und Amphibien Europas. Band 3/I. Schlangen (Serpentes) I. (Typhlopidae, Boidae, Colubridae 1, Colubrinae)*. AULA-Verlag, Wiesbaden, pp. 69–73.

Schätti, B. (1993b) *Coluber hippocrepis* (Linnaeus, 1758) – Hufeisennatter. *In:* Böhme, W (Ed.), *Handbuch der Reptilien und Amphibien Europas. Band 3/I. Schlangen (Serpentes) I. (Typhlopidae, Boidae, Colubridae, 1, Colubrinae)*. AULA-Verlag, Wiesbaden, pp. 115–130, figs. 23–26.

Schätti, B. (2001a) A new species of *Coluber* (*sensu lato*) from the Dahlak Islands, Eritrea, with a review of the herpetofauna of the archipelago. *Russian Journal of Herpetology* 8(2), 139–148, 3 figs.

Schätti, B. (2001b) Morphologie und Verbreitung von *Coluber* (*s.l.*) *andreanus* (Werner, 1917) (Reptilia: Serpentes: Colubridae) *Revue Suisse de Zoologie*, Genève 108(3), 487–493, 1 fig.

Schätti, B. (2004) A preliminary conribution to *Platyceps najadum* (Eichwald, 1831) and systematic status of *Coluber atayevi* Tunijev & Shammakov, 1993 (Reptilia: Squamata: Colubrinae). *Revue Suisse de Zoologie*, Genève 111(3), 525–538, 3 figs.

Schätti, B. (2006a) Northeast African racers of the *Platyceps rhodorachis* complex (Reptilia: Squamata: Colubrinae). *Revue Suisse de Zoologie,* Genève 113(1), 77–86, 1 fig.

Schätti, B. (2006b) Racer populations from the Levant to the Caspian region referred to *Platyceps ventromaculatus* (Gray, 1834) (Reptilia: Squamata: Colubrinae). *Revue Suisse de Zoologie*, Genève 113(3), 675–691, 1 fig., 1 pl.

Schätti, B. (2007) Geographic distribution: Serpentes: *Platyceps taylori*. *Herpetological Review* 38(3), 355.

Schätti, B. & Agarian, A. (1985) Ein neues Konzept für den *Coluber ravergieri-C. nummifer*-Komplex (Reptilia, Serpentes, Colubridae). *Zoologische Abhandlungen, Staatliches Museum für Tierkunde*, Dresden 40(9), 109–123, 5 figs.

Schätti, B. & Baran, I. (1988) Bemerkungen zur Verbreitung von *Elaphe hohenackeri* (Strauch, 1873) und *Vipera xanthina* (Gray, 1849) in Sud-Anatolien (Serpentes, Colubridae, Viperidae). *Salamandra*, Frankfurt-am-Main 24(4), 306–309, illust.

Schätti, B. & Charvet, C. (2003) Systematics of *Platyceps brevis* (Boulenger 1893) and related East African racers (Serpentes Colubridae). *Tropical Zoology,* Firenze 16(1), 93–111, 4 figs.

Schätti, B. & Desvoignes, A. (1999) The herpetofauna of southern Yemen and the Sokotra Archipelago. *Museum d'Histoire Naturelle, Genève, Instruments Biodiversitatis* IV, 1–178.

Schätti, B. & Gasperetti, J. (1994) A contribution to the herpetofauna of southwest Arabia. *Fauna of Saudi Arabia*, Riyadh 14, 348–423, 1 fig., 7 pls.

Schätti, B. & Ineich, B. (2004) A new racer of the genus *Platyceps* Blyth from Djibouti. *Revue Suisse de Zoologie*, Genève 111(4), 685–690.

Schätti, B. & Kramer, E. (1991) A new pitviper from Ecuador, *Bothriechis mahnerti* n. sp. *Revue Suisse de Zoologie*, Genève 98(1), 9–14, 1 fig., 1 pl.

Schätti, B. & Kramer, E. (1993) Ecuadorianische Grubenottern der Gattungen *Bothriechis*, *Bothrops* und *Porthidium* (Serpentes, Viperidae). *Revue Suisse de Zoologie*, Genève 100(2), 235–278, 4 figs.

Schätti, B. & Lanza, B. (1989) *Coluber messanai*, a new species of snake from northern Somalia (Reptilia, Serpentes). *Bollettino del Museo Regionale di Scienze Naturali*, Torino 7(2), 413–421, 6 figs.

Schätti, B. & Loumont, C. (1992) Ein Beitrag zur Herpetofauna von São Tomé (Golf von Guinea) (Amphibia et Reptilia). *Zoologische Abhandlungen, Staatliches Museum für Tierkunde*, Dresden 47(4), 23–36, 5 figs.

Schätti, B. & McCarthy, C. (1987) Relationships of the snake genera *Pythonodipsas* Günther and *Spalerosophis* Jan (Reptilia, Colubridae). *Bonner Zoologische Beiträge*, Bonn 38(3), 265–268, 1 fig.

Schätti, B. & McCarthy, C. (2001) *Coluber (sensu lato) schmidtleri* n. sp. from the southern Zagros Mountains in Iran. *Herpetozoa* 14(1–2), 81–89, 5 figs.

Schätti, B. & McCarthy, C. (2004) Saharo-Arabian racers of the *Platyceps rhodorhachis* complex—description of a new species (Reptilia: Squamata: Colubrinae). *Revue Suisse de Zoologie*, Genève 111(4), 691–705, 2 figs.

Schätti, B. & Perret, J.-L. (1997) Catalogue révisé des types d'amphibiens et de reptiles du Muséum d'histoire de Genéve. *Revue Suisse de Zoologie*, Genève 104(2), 357–370.

Schätti, B. & Schmitz, A. (2006) Re-assessing *Platyceps ventromaculatus* (Gray, 1834) (Reptilia: Squamata: Colubrinae). *Revue Suisse de Zoologie*, Genève 113(4), 747–768, 3 figs.

Schätti, B. & Sigg, H. (1989) Die herpetofauna der Insel Zypern. Teil 2. Schildkroten, Echsen und Schlangen. *Herpetofauna*, Sydney 11(62), 17–26, illust.

Schätti, B. & Stutz, A. (2005) Morphology and systematic status of *Coluber karelini mintonorum* Mertens, 1969 (Reptilia: Squamata: Colubrinae). *Revue Suisse de Zoologie*, Genève 112(2), 409–420, 1 fig., 1 pl.

Schätti, B. & Trape, J.-F. (2008) *Bamanophis*, a new genus for the West African colubrid *Periops dorri* Lataste, 1888 (Reptilia: Squamata: Colubrinae). *Revue Suisse de Zoologie*, Genève 115(4), 595–615, 6 figs.

Schätti, B. & Utiger, U. (2001) *Hemerophis*, a new genus for *Zamenis socotrae* Günther, and a contribution to the phylogeny of Old World racers, whip snakes, and related genera (Reptilia, Squamata, Colubridae). *Revue Suisse de Zoologie*, Genève 108(4), 919–948, 10 figs.

Schätti. B. & Vanni, S. (1986) Intraspecific variation in *Coluber viridiflavus* Lacépède, 1789, and validity of its subspecies (Reptilia, Serpentes, Colubridae). *Revue Suisse de Zoologie*, Genève 93(1), 219–232, illust.

Schätti, B. & Wilson, L.D. (1986) *Coluber. Catalogue of American amphibians and reptiles*, pp. 399.1–399.4.

Schätti, B., Kramer, E. & Touzet, J.-M. (1990) Systematics remarks on a rare crotalid snake from Ecuador, *Bothriechis albocarinata* (Shreve), with some comments on the generic arrangement of arboreal Neotropical pitvipers. *Revue Suisse de Zoologie*, Genève 97(4), 877–885, 3 figs.

Schätti, B., Baran, I. & Sigg, H. (1991) Rediscovery of the Bolkar viper: morphological variation and systematic implications on the "*Vipera xanthina* complex." *Amphibia-Reptilia*, Leiden 12(3), 305–327, 6 figs.

Schätti, B., Baran, I. & Sigg, H. (1992) The "*Vipera xanthina* complex"—a reply to Nilson and Andrén. *Amphibia-Reptilia*, Leiden 13(4), 425.

Schätti, B., Baran, I. & Maunoir, P. (2001) Taxonomie, morphologie und verbreitung der Masken-Schlanknatter *Coluber* (s.l.) *collaris* (Müller, 1878). *Revue Suisse de Zoologie*, Genève 108(1), 11–30, 4 figs.

Schätti, B., Stutz, A. & Charvet, C. (2005) Morphologie,Verbreitung und Systematik der Schlanknatter *Platyceps najudum* (Eichwald, 1831) (Reptilia: Squamata: Colubrinae). *Revue Suisse de Zoologie*, Genève 112(3), 573–625, 8 figs., 1 pl.

Schätti, B., Tillack, F. & Kucharzewski, C. (2009a) Identät und Status aus Ägypten beschriebener Schlangenarten: *Coluber tyria* Linnaeus, 1758, *Coluber cahirinus* Gmelin, 1789 und *Coluber geoffroyii* Gray, 1831 (Squamata: Serpentes: Colubridae). *Herpetozoa* 22(1–2), 43–54.

Schätti, B., Tillack, F. & Helfenberger, N. (2009b) A contribution to *Spalerosophis microlepis* Jan, 1865, with a short review of the genus and a key to the species. *Herpetozoa* 22(3–4), 115–135, 4 figs.

Schätti, B., Tillack, F. & Schulz, K.-D. (2010a) History, status, and validity of the racer *Zamenis persicus* Jan, 1863, and the scientific name of the ratsnake *Z. persicus* (Werner, 1913) (Squamata, Colubrinae). *Zoosystematics and Evolution* 86(2), 275–281.

Schätti, B., Ineich, I. & Kucharzewski, C. (2010b) Nominal taxa of *Spalerosophis diadema* (Schlegel, 1837) from Iraq to Pakistan — two centuries of confusion (Reptilia: Squamata: Colubrinae). *Revue Suisse de Zoologie,* Genève 117(4), 637–664, 2 figs.

Schätti, B., Kucharzewski, C., Masroor, R. & Rastegar-Pouyani, E. (2012) *Platyceps karelini* (Brandt, 1838) from Iran to Pakistan and revalidation of *Coluber chesneii* Martin, 1838 (Reptilia: Squamata: Colubrinae). *Revue Suisse de Zoologie*, Genève 119(4), 441–483, 10 figs.

Scheidt, V. (1996) An introduction to the snakes of the genus *Lamprophis:* the house snakes. *The Vivarium* 8(3), 2631, 48, 12 col. photos.

Schenkel, E. (1901) Achter Nachtrag zum Katalog der herpetologischen Sammlung des Basler Museums. *Verhandlungen der Naturforschenden Gesellschaft in Basel* (1901–1902) 13(1), 142–199, 7 figs.

Scherlinski, H. (1939) Nachtragzum Katalog der Wolterstoff-Sammlung im Museum für Naturkunde und Vorgeschichte zu Madgeburg. *Abhandlungen und Berichte aus dem Museum für Naturkunde und Vorgeschichte und dem Naturwissenschaftlichen Verein*, Magdeburg 7(1), 31–57.

Scheurer, T. (2007) *Die Afrikanische Eierschlange,* Dasypeltis scabra. Natur und Tier-Verlag, Münster, 61 pp., col. illust.

Schiemenz, H. (1995) *Die Kreuzotter.* Die Neue Brehm-Bücherei, Heidelberg, 108 pp., 64 figs., 2 col. pls.

Schiemenz, H. & Günther, R. (1994) *Verbreitungsatlas der amphibien und reptilien Ostdeutschlands (Gebiet der ehemaligen DDR).* Verlag Natur und Text, Rangsdorf, 143 pp., 54 figs.

Schinz, H.R. (1822) *Das Thierreich eingetheilt nach dem Bau der Thiere als Grundlage ihrer Naturgeschichte und der vergleichenden Anatomie von dem Herrn Ritter von Cuvier... Vol. 2.* J.G. Cotta, Stuttgart und Tübingen, 189 pp.

Schinz, H.R. (1833) *Naturgeschichte und Abbildungen der Reptilien.* Brodtmann's lithographischer Anstalt, Schasshausen, iv + 240 pp., 102 pls.

Schinz, H.R. (1840) *Europäischev Fauna oder Berzeichniss der Wirbelthiere Europa's. Zweiter Band. Reptilien und Fische.* E. Schweizerbarts Verlagshandlung, Stuttgart, 535 pp.

Schlegel, H. (1826a) Notice sur l'Erpétologie de l'ile de Java; par M. Böié (Ouvrage manuscrit). *Bulletin des Sciences Naturelles et de Géologie*, Paris 9(2), 233–240.

Schlegel, H. (1826b) Erpetologische Nachrichten. *Isis von Oken*, Jena (1827) 20(3), 281–294.

Schlegel, H. (1828) Auszuge aus Riesen von Heinr. Boie zu Java, an Hn. Schlegel, conservator anim. Vertebr. am Konigl. niederl. Museum. *Isis von Oken*, Jena 21(10), 1025–1035.

Schlegel, H. (1837) *Essai sur la physionomie des serpents. I. Partie générale. II. Partie descriptive. Atlas, 21 planches et 3 cartes.* Arnz Arnz & Comp., Leiden, xxviii + 251 + xvi pp., 606 pp., 21 pls.

Schlegel, H. (1837–1844) *Abbildungen neuer oder unvollstandig bekannter Amphibien, nach der Natur oder dem Leben entworfen herausgegeben und mit einem erläuternden Texte begleitet.* Arnz & Comp., Düsseldorf, xiv + 141 pp., 50 pls.

Schlegel, H. (1838) Reptilia. *In:* Siebold, P.F. von (Ed.), *Fauna Japonica, auctore Ph. Fr. de Siebold. Reptilia elaborantibus C.J. Temminck et H. Schlegel. Cum mappa geographico-zoologica et tabulis lithogr. XXVIII.* J.G. Lalau, Lugduni Batavorum, xxii + 140 pp., 27 pls.

Schlegel, H. (1841) Description d'une nouvelle espece du genre Trigonocephale (*Trigonocephalus Lansbergii*). *Magasin de Zoologie , d'Anatomie Comparée et de Palaeontologie*, Paris 1841, 1–3, pl. 1, 4 figs.

Schlegel, H. (1843) *Essay on the physiognomy of serpents* (translated by T.S. Traill). Maclachian, Stewart, Edinburgh, vii + 254 pp., 21 pls.

Schlegel, H. (1848a) Description d'une nouvelle espece du genre *Eryx: Eryx reinhardtii. Bijdragen tot de Dierkunde*, Amsterdam 1, 1–3, 1 pl.

Schlegel, H. (1848b) Over *Elaps jamesonii* Traillust. *Bijdragen tot de Dierkunde*, Amsterdam 1, 5.

Schlegel, H. (1855) Over eenige nieuwe soorten van vergiftige Slangen van de Goudkust. *Verslagen en Mededeelingen der Koninklijke Akademie van Weteschappen*, Amsterdam 3, 312–317.

Schlegel, H. (1872) *De Dierentuin van het Koninklijk Zoologisch Genootschap Natura Artis Magistra te Amsterdam. Reptilia.* Gebr. Van Es, Amsterdam, (4), 64 + iii pp., illust.

Schlegel, H. & Müller, S. (1844a) Reptilia. Over de slangensoorten van het geslacht *Homalopsis*, uit den Indischen Archipel. *In:* Temminck, C.J. (Ed.), *Verhandlingen over de naturlijke geschiedenis der Nederlandsche overzeesche bezittingen, door de Leden der Natuurkundige Commissie in Indië en andere Schrijvers. Zoologie* (1839–1845). S. & J. Luchtmans & C.C. van der Hoek, Leiden, 59–62, pl. 8. [dated 1845 *fide* Koch, 2012: 368]

Schlegel, H. & Müller, S. (1844b) Reptilia. Over de slangen van het geslacht *Elaps*, welke den Indischen Archipel bewonen. *In:* Temminck, C.J. (Ed.), *Verhandlingen over de naturlijke geschiedenis der Nederlandsche overzeesche bezittingen, door de Leden der Natuurkundige Commissie in Indië en andere Schrijvers. Zoologie* (1839–1845). S. & J. Luchtmans & C.C. van der Hoek, Leiden, pp. 63–68, pl. 9. [dated 1845 *fide* Koch, 2012: 368]

Schleich, H.H. (1977) Distributional maps of reptiles of Iran. *Herpetological Review* 8(4), 126–129.

Schleich, H.H. (1984) Reptilien und amphibien des Kouf National Parks in NE-Libyen. *LVAR Mitteilungen*, München 4(1), 9–10.

Schleich, H.H. & Kästle, W. (2002) *Amphibians and Reptiles of Nepal*. A.R.G. Gantner Verlag, x + 1200 pp., figs., 376 col. pls.

Schleich, H.H. & Maskey, T.M. (1992) Der Chitwan-Nationalpark in süden Nepals. *Natur & Museum* 122(8), 248–257.

Schleich, H.H., Kästle, W. & Kabisch, K. (1996) *Amphibians and Reptiles of North Africa*. Koeltz Scientific Books, Koenigstein, 627 pp., figs., 63 col. pls.

Schleip, W.D. (2008) Revision of the genus *Leiopython* Hubrecht 1879 (Serpentes: Pythonidae) with the redescription of taxa recently described by Hoser (2000) and the description of new species. *Journal of Herpetology* 42(4), 645–667, 6 figs.

Schleip, W.D. & O'Shea, M. (2010) Annotated checklist of the recent and extinct pythons (Serpentes, Pythonidae), with notes on nomenclature, taxonomy, and distribution. *ZooKeys* (66), 29–79.

Schlosser, M. (1916) Neue Funde fossiler Säugetiere in der Eichstätter Geg. *Abhandlungen der Bayerische Akademie der Wissenschaften, Mathematische-Physikalische Klasse*, München 28(6), 1–78, 5 pls.

Schmida, G. (1985) *The cold-blooded Australians. A unique photographic study of Australia's reptiles, amphibians & freshwater fish.* Doubleday, Sydney, 205 pp., col. illust.

Schmidt, D. (1996a) *Riesenschlangen.* Bede-Verlag GmbH, Ruhmannsfelden, 95 pp., col. illust.

Schmidt, D. (1996b) *Wassernattern: die Unterfamilie Natricinae.* Bede-Verlag GmbH, Ruhmannsfelden, 86 pp., col. illust.

Schmidt, D. (1998) *Trugnattern: Boiginae.* Bede-Verlag GmbH, Ruhmannsfelden, 88 pp., col. illust.

Schmidt, D. (2000) *Kornnattern und Erdnattern:* Elaphe guttata *und* Elaphe obsoleta. Natur und Tier-Verlag, Münster, 199 pp., col. illust.

Schmidt, D. (2001) *Atlas der Schlangen: Biologie, Arten, Terraristik.* Bede-Verlag GmbH, Ruhmannsfelden, 350 pp., col. illust.

Schmidt, D. (2004a) *Die Erdnatter,* Pantherophis obsoletus (Elaphe obsoleta). Natur und Tier-Verlag, Münster, 62 pp., col. illust.

Schmidt, D. (2004b) *Die Gebänderte Wassernatter,* Nerodia fasciata. Natur und Tier-Verlag, Münster, 61 pp., col. illust.

Schmidt, D. (2004c) *Die Kettennatter,* Lampropeltis getula. Natur und Tier-Verlag, Münster, 62 pp., col. illust.

Schmidt, D. (2006a) *Die Dreiecksnatter,* Lampropeltis triangulum. Natur und Tier-Verlag, Münster, 62 pp., col. illust.

Schmidt, D. (2006b) *Die Kupferkopf,* Agkistrodon contortrix. Natur und Tier-Verlag, Münster, 61 pp., col. illust.

Schmidt, D. (2007) *Die Faux-Corail,* Lampropeltis triangulum. Natur und Tier-Verlag, Münster, 62 pp., col. illust.

Schmidt, D. (2008) *Die Texas-Klapperschlange,* Crotalus atrox. Natur und Tier-Verlag, Münster, 63 pp., col. illust.

Schmidt, K.P. (1920) Contributions to the herpetology of Porto Rico. *Annals of the New York Academy of Sciences* 28, 167–200.

Schmidt, K.P. (1921a) Notes on the herpetology of Santo Domingo. *Bulletin of the American Museum of Natural History* 44(2), 7–20, 12 figs.

Schmidt, K.P. (1921b) The herpetology of Navassa Island. *Bulletin of the American Museum of Natural History* 44(18), 555–559, 5 figs., pls. 25–26.

Schmidt, K.P. (1922) The amphibians and reptiles of Lower California and the neighboring islands. *Bulletin of the American Museum of Natural History* 46(11), 607–707, 1 fig.

Schmidt, K.P. (1923) Contributions to the herpetology of the Belgian Congo based on the collection of the American Museum Congo Expedition, 1909–1915. Part II. Snakes. *Bulletin of the American Museum of Natural History* 49(1), 1–146, 15 figs., 22 pls.

Schmidt, K.P. (1925a) New reptiles and a new salamander from China. *American Museum Novitates* (157), 1–5.

Schmidt, K.P. (1925b) New Chinese amphibians and reptiles. *American Museum Novitates* (175), 1–3.

Schmidt, K.P. (1926a) The amphibians and reptiles of Mona Island, West Indies. *Publication, Field Museum of Natural History (Zoological Series)* (1917–1929) 12(12), 149–163, 6 figs.

Schmidt, K.P. (1926b) Amphibians and reptiles of the James Simpson-Roosevelt Asiatic Expedition. *Publication, Field Museum of Natural History (Zoological Series)* (1917–1929) 12(13), 167–173, 2 figs.

Schmidt, K.P. (1927a) The reptiles of Hainan. *Bulletin of the American Museum of Natural History* 54(3), 395–465, 17 figs., pl. 27.

Schmidt, K.P. (1927b) Notes on Chinese reptiles. *Bulletin of the American Museum of Natural History* 54(4), 467–551, 22 figs., pls. 28–30.

Schmidt, K.P. (1927c) New reptilian generic names. *Copeia* 1927(163), 58–59.

Schmidt, K.P. (1928a) Amphibians and land reptiles of Porto Rico, with a list of those reported from the Virgin Islands. *Scientific Survey of Porto Rico and the Virgin Islands*, New York 10(1), 1–160, 52 figs., 4 pls.

Schmidt, K.P. (1928b) Reptiles collected in Salvador for the California Institute of Technology. *Publication, Field Museum of Natural History (Zoological Series)* (1917–1929) 12(16), 193–201, 1 fig.

Schmidt, K.P. (1928c) Notes on American coral snakes. *Bulletin of the Antivenin Institute of America* 2(3), 63–64.

Schmidt, K.P. (1930) Reptiles of Marshall Field North Arabian Desert Expeditions, 1927–1928. *Publication, Field Museum of Natural History (Zoological Series)* 17(6), 223–230, 2 figs., pl. 2.

Schmidt, K.P. (1932a) A new subspecies of coral snake from Guatemala. *Proceedings of the California Academy of Sciences* (Series 4) 20(7), 265–267.

Schmidt, K.P. (1932b) Stomach contents of some American coral snakes, with the description of a new species of *Geophis*. *Copeia* 1932(1), 6–9.

Schmidt, K.P. (1932c) Reptiles and amphibians from the Solomon Islands. *Publication, Field Museum of Natural History (Zoological Series)* 18(9), 175–190, pl. 8.

Schmidt, K.P. (1933a) Amphibians and reptiles collected by the Smithsonian Biological Survey of the Panama Canal Zone. *Smithsonian Miscellaneous Collections* 89(1), 1–20.

Schmidt, K.P. (1933b) A new snake from Arabia. *Zoological Series of Field Museum of Natural History* 20 (320), 9–10.

Schmidt, K.P. (1933c) Preliminary account of the coral snakes of Central America and Mexico. *Zoological Series of Field Museum of Natural History* 20(324), 29–40.

Schmidt, K.P. (1933d) The reptiles of the Pulitzer Angola Expedition. *Annals of the Carnegie Museum* (1933–1934) 22, 1–15, 2 pls.

Schmidt, K.P. (1936a) Notes on snakes from Yucatan. *Zoological Series of Field Museum of Natural History* 20(18), 167–187, figs. 20–23.

Schmidt, K.P. (1936b) Preliminary account of coral snakes of South America. *Zoological Series of Field Museum of Natural History* 20(19), 189– 203.

Schmidt, K.P. (1936c) Notes on Central American and Mexican coral snakes. *Zoological Series of Field Museum of Natural History* 20(20), 205–216, figs. 24–27.

Schmidt, K.P. (1936d) New amphibians and reptiles from Honduras in the Museum of Comparative Zoology. *Proceedings of the Biological Society of Washington* 49, 43–50.

Schmidt, K.P. (1937) The history of *Elaps collaris* Schlegel, 1837–1937. *Zoological Series of Field Museum of Natural History* 20(26), 361–364.

Schmidt, K.P. (1939a) A new coral snake from British Guiana. *Zoological Series of Field Museum of Natural History* 24(6), 45–47, fig. 5.

Schmidt, K.P. (1939b) Reptiles and amphibians from southwestern Asia. *Zoological Series of Field Museum of Natural History* 24(7), 49–92, fig. 6.

Schmidt, K.P. (1940) Notes on Texas snakes of the genus *Salvadora*. *Zoological Series of Field Museum of Natural History* 24(12), 143–150, figs. 13–15.

Schmidt, K.P. (1941) The amphibians and reptiles of British Honduras. *Zoological Series of Field Museum of Natural History* 22(8), 475–510, fig. 38.

Schmidt, K.P. (1943) Corollary and commentary for "climate and evolution." *American Midland Naturalist* 30, 241–253.

Schmidt, K.P. (1952) The Surinam coral snake, *Micrurus surinamensis*. *Fieldiana: Zoology* 34(4), 25–34, figs. 4–6.

Schmidt, K.P. (1953a) *A check list of North American amphibians and reptiles*. Sixth edition. University of Chicago Press, Chicago, viii + 280 pp.

Schmidt, K.P. (1953b) Hemprich's coral snake, *Micrurus hemprichi*. *Fieldiana: Zoology* 34(13), 165–170, figs. 31–32.

Schmidt, K.P. (1953c) The Amazonian coral snake, *Micrurus spixi*. *Fieldiana: Zoology* 34(14), 171–180, figs. 33–35.

Schmidt, K.P. (1953d) Amphibians and reptiles of Yemen. *Fieldiana: Zoology* 34(24), 253–261, fig. 43.

Schmidt, K.P. (1954) The annellated coral snake, *Micrurus annellatus* Peters. *Fieldiana: Zoology* 34(30), 319–325, figs. 62–63.

Schmidt, K.P. (1955a) Coral snakes of the genus *Micrurus* in Colombia. *Fieldiana: Zoology* 34(34), 337–359, figs. 65–69.

Schmidt, K.P. (1955b) Amphibians and reptiles from Iran. *Videnskabelige Meddelelser fra Dansk Naturhistorisk Forening i København* 117, 193–207, 1 fig.

Schmidt, K.P. (1957) The venomous coral snakes of Trinidad. *Fieldiana: Zoology* 39(8), 55–63, figs. 7–9.

Schmidt, K.P. (1958) Some rare or little-known Mexican coral snakes. *Fieldiana: Zoology* 39(19), 201–212, figs. 32–34. [published posthumously]

Schmidt, K.P. & Andrews, E.W. (1936) Notes on snakes from Yucatán. *Zoological Series of Field Museum of Natural History* 20(18), 167–187, figs. 20– 23.

Schmidt, K.P. & Conant, R. (1956) Appendix to Opinion 385. Request that the International Commission on Zoological Nomenclature should direct that the nominal species "*Coluber sirtalis*" Linnaeus 1758, and "*Coluber saurita*" Linnaeus 1766, should be interpreted by the neotypes here designated by the present authors instead of, as hitherto proposed, by reference to specified previously published figures and descriptions. *Opinions and Declarations rendered by the International Commission on Zoological Nomenclature* 12(6), 224–230.

Schmidt, K.P. & Davis, D.D. (1941) *Field book of snakes of the United States and Canada*. G.P. Putnam's Sons, New York, xiii + 365 pp., 103 figs., 34 pls.

Schmidt, K.P. & Inger, R.F. (1951) Amphibians and reptiles of the Hopkins-Branner Expedition to Brazil. *Fieldiana: Zoology* 31(42), 439–465, fig. 85.

Schmidt, K.P. & Inger, R.F. (1957) *Living reptiles of the world*. Doubleday & Company, New York, 287 pp., 266121 figs., 145 col. photos.

Schmidt, K.P. & Marx, H. (1956) The herpetology of Sinai. *Fieldiana: Zoology* 39(4), 21–40, figs. 3–6.

Schmidt, K.P. & Necker, W.L. (1936) The scientific name of the American smooth green snake. *Herpetologica* 1(1), 63–64.

Schmidt, K.P. & Owens, D.W. (1944) Amphibians and reptiles of northern Coahuila, Mexico. *Zoological Series of Field Museum of Natural History* 29(16), 97–115.
Schmidt, K.P. & Rand, A.S. (1957) Geographic variation in the Central American colubrine snake, *Ninia sebae*. *Fieldiana: Zoology* 39(10), 73–84.
Schmidt, K.P. & Schmidt, F.J.W. (1925) New coral snakes from Peru. Reports on results of the Captain Marshall Field Expeditions. *Publication, Zoological Series of Field Museum of Natural History* 12(10), 129–134.
Schmidt, K.P. & Shannon, F.A. (1947) Notes on amphibians and reptiles of Michoacán, Mexico. *Fieldiana: Zoology* 31(9), 63–85, fig. 8.
Schmidt, K.P. & Smith, H.M. (1943) Notes on coral snakes from Mexico. *Zoological Series of Field Museum of Natural History* 29(2), 25–31.
Schmidt, K.P. & Smith, T.F. (1944) Amphibians and reptiles of the Big Bend region of Texas. *Zoological Series of Field Museum of Natural History* 29(5), 75–96.
Schmidt, K.P. & Stuart, L.C. (1941) The herpetological fauna of the Salamá Basin, Baja Verapaz, Guatemala. *Zoological Series of Field Museum of Natural History* 24(21), 233–247, figs. 21–22.
Schmidt, K.P. & Walker, W.F. (1943a) Peruvian snakes from the University of Arequipa. *Zoological Series of Field Museum of Natural History* 24(26), 279–296.
Schmidt, K.P. & Walker, W.F. (1943b) Snakes of the Peruvian coastal region. *Zoological Series of Field Museum of Natural History* 24(27), 297–324.
Schmidt, K.P. & Walker, W.F. (1943c) Three new snakes from the Peruvian Andes. *Zoological Series of Field Museum of Natural History* 24(28), 325–329.
Schmidt, K.P. & Walker, W.F. (1945) Ofidios de la región costera del Perú. *Boletín del Museo de Historia Naturale "Javier Prado," Universidad Mayor de San Marcos* 14(34–35), 293–329.
Schmidt, P. (1846) Beschreibung zweier neuen Reptilien, aus dem Naturhistorischen Museum zu Hamburg. *Abhandlungen aus dem Gebiete der Naturwissenschaften, Naturwissenschaftlicher Verein*, Hamburg 1, 164–172, 7 figs.
Schmidt, P. (1852) Beiträge zur ferneren Kenntniss der Meerschlangen. *Abhandlungen aus dem Gebiete der Naturwissenschaften Vereins in Hamburg* 2(2), 69–86, 7 pls.
Schmidt, T. (2002) *Grasnattern*. Natur und Tier-Verlag, Münster, 78 pp., col. illust.
Schmidt, T. (2009) *Die Westliche Hakennasennatter,* Heterodon nasicus. Natur und Tier-Verlag, Münster, 78 pp., col. illust.
Schmidt, W. (1999) *Kornnattern*. Natur und Tier-Verlag, Münster, 87 pp., col. illust.
Schmidtler, J.F. (1988) *Eirenis barani* n. sp. aus dem mediterranen Süden der Türkei (Serpentes, Colubridae). *Salamandra*, Frankfurt-am-Main 24(4), 203–214, 8 figs.
Schmidtler, J.F. (1993) Zur Systematik und Phylogenie des *Eirenis-modestus*-Komplexes in Süd-Anatohèn. *Spixiana*, München 16(1), 79–96, 14 figs.
Schmidtler, J.F. (1997) Die Zwergnattern (*Eirenis modestus*-Komplex) des Antitaurus in Süd-Anatolien und ihre geographischen Beziehungen zur begleitenden Herpetofauna. *Salamandra*, Frankfurt-am-Main 33(1), 33–60, 13 figs.
Schmidtler, J.F. & Baran, I. (1993a) *Eirenis collaris* (Ménétriés, 1832)—Halsband-Zwergnatter. *In:* Böhme, W. (Ed.), *Handbuch der Reptilien und Amphibien Europas. Band 3/I. Schlangen (Serpentes) I. (Typhlopidae, Boidae, Colubridae 1, Colubrinae)*. AULA-Verlag, Wiesbaden, pp. 267–278, figs. 52–53.
Schmidtler, J.F. & Baran, I. (1993b) *Eirenis modestus* (Martin, 1838)—Kopfbinden-Zwergnatter. *In:* Böhme, W. (Ed.), *Handbuch der Reptilien und Amphibien Europas. Band 3/I. Schlangen (Serpentes) I. (Typhlopidae, Boidae, Colubridae 1, Colubrinae)*. AULA-Verlag, Wiesbaden, pp. 279–292, figs. 54–56.
Schmidtler, J.F. & Eiselt, J. (1991) Zur Systematik und Verbreitung ostanatolischer Zwergnattern; mit Beschreibung von *Eirenis hakkariensis* n. sp. *Salamandra*, Frankfurt-am-Main 27(4), 225–237, illust.
Schmidtler, J.F., Göçmen, B., Yildiz, M.Z., Akman, B., Yalçinkaya, D. & Nagy, Z.T. (2009) Rediscovery of the snake genus *Eirenis* on Cyprus (Reptilia: Colubridae). *Salamandra*, Frankfurt-am-Main 45(1), 15–23, 5 figs.
Schmidtler, J.J. & Lanza, B. (1990) A new dwarf-snake (*Eirenis*) from Lake Van in eastern Turkey. *Amphibia-Reptilia*, Leiden 11(4), 363–371, 6 figs.
Schmidtler, J.J. & Schmidtler, J.F. (1978) Eine neue Zwergnatter aus der Türkei; mit einer Übersicht über die Gattung *Eirenis* (Colubridae, Reptilia). *Annalen des Naturhistorischen Museums in Wien* 81, 383–400, 1 fig., 3 pls.
Schneider, B. (1975) Eine mittelpleistozäne Herpetofauna von der Insel Chios, Ägäis. *Senckenbergiana Biologica*, Frankfurt-am-Main 56(4–6), 191–198, 7 figs.
Schneider, B. (1980) Eine melanistische Schlanknatter, *Coluber najadum kalymnensis* n. subsp. (Colubridae, Serpentes), von der Insel Kalymnos (Dodekanes, Agais). *Bonner Zoologische Beiträge*, Bonn (1979) 30(3–4), 380–384, 2 figs.
Schneider, B. (1983) Zur herpetofauna der Inseln Kaylmnos und Telentos (Dodekanes, Ägäis). *Salamandra*, Frankfurt-am-Main 19(1–2), 61–70, 3 figs.
Schneider, B. (1986) Zur herpetofauna der Inseln Limnos (Hellespontische Inseln, Nordägäis, Griechenland). *Salamandra*, Frankfurt-am-Main 22(4), 276–280, 1 fig.
Schneider, J.G. (1799) *Historiae amphibiorum naturalis et literariae. Fasciculus primus, continens Ranas, Calamitas, Bufones, Salamandras et Hydros in genera et species descriptos notisque suis distinctos*. Friederici Frommanni, Jenae, 264 pp., 2 pls.
Schneider, J.G. (1801) *Historiae amphibiorum naturalis et literariae. Fasciculus secundus, continens Crocodilos, Scincos, Chamaesauras, Boas, Pseudoboas, Elapes, Angues, Amphisbaenas et Caecilias*. Friederici Frommann, Jenae, 364 pp., 2 pls.
Schneider, J.G. (1821) Beytrag zur Klassifikation und kritische Uebersicht der Arten aus der Gattung der Riesenschlangen *(Boa). Abhandlung Denkschriften der Königlichen Akademie der Wissenschaften zu München* (1818–1820) 7, 89–134, pls. 6–8.
Schnurrenberger, H. (1959) Observations on behavior in two Libyan species of viperine snakes. *Herpetologica* 15(2), 70–72, 2 figs.
Schoenherr, A.A. (1976) *The herpetofauna of the San Gabriel Mountains, Los Angeles County, California, including distribution and biogeography.* Special Publication, Southwestern Herpetologists Society, Fullerton, 88 pp., 25 figs.
Schöpf, J.D. (1788) *Reise durch der mittlern und südlichen vereinigten nordamerikanischen Staaten nach Ost-florida und den Bahama-inseln unternommen in den Jahren 1783 und 1784. Erste Theil.* Johann Jacob Palm, Erlangen, xxxii + 551 pp.
Schoppe, S. & Cervancia, M. (2009) Herpetological surveys along Pagdanan Range and Dumaran Island, northern Palawan, Philippines. *Hamadryad* 34(1), 95–106.
Schouten, G.B. (1937) Fauna herpetológica del Paraguay. *Novena Reunión de la Sociedad Argentina de Patología Regional* 2, 1218–1232.

Schouten, J.R. & Thevenot, M. (1988) Amphibians and reptiles of the Khnifiss–La'youne region. *In:* Dakki, M. & de Ligny, W. (Eds.), *The Khnifiss lagoon and its surrounding environment (Province of La'youne, Morocco).* Travaux de l'Institut Scientifique, Rabat, pp. 105–113, 4 figs.

Schreiber, E. (1912) *Herpetologia Europaea. Eine systematische Bearbeitung der Amphibien und Reptilien welche bisher in Europa aufgefunden sind.* Second edition. Gustav Fischer, Jena, x + 960 pp., 176 figs. [first edition, 1875, Friedrich Vieweg Braunschweig, xvii + 639 pp.]

Schreiber, G. (1950) Análise do polimorfismo da "boipeva" (*Xenodon merremii* Wagl. – Oph. Colubr.). *Revista da Universidade de Minas Gerais,* Belo Horizonte (8), 26–46, 7 figs.

Schuett, G.W., Höggren, M., Douglas, M.E. & Greene, H.W. (Eds.). (2002) *Biology of the vipers.* Eagle Mountain Press, Salt Lake City, xii + 578 pp., 16 col. pls.

Schultze, H.-P., Hunt, L., Chorn, J. & Neuner, A.M. (1985) Type and figured specimens of fossil vertebrates in the collection of the University of Kansas Museum of Natural History. Part 2. Fossil amphibians and reptiles. *Miscellaneous Publications, Museum of Natural History, University of Kansas* (77), 1–66.

Schulz, K.-D. (1985a) Die hinterasiatischen Kletternattern der Gattung *Elaphe.* Teil I. *Elaphe rufodorsata. Sauria,* Berlin 7(2), 21–25, 3 figs., 2 photos.

Schulz, K.-D. (1985b) Die hinterasiatischen Kletternattern der Gattung *Elaphe.* Teil II. *Elaphe climacophora* (Boie, 1826). *Sauria,* Berlin 7(3), 7–8, 2 photos.

Schulz, K.-D. (1885c) Die hinterasiatischen Kletternattern der Gattung *Elaphe.* Teil III. *Elaphe schrencki* (Strauch, 1873). *Sauria,* Berlin 7(4), 3–6, 4 figs.

Schulz, K.-D. (1986a) Die hinterasiatischen Kletternattern der Gattung *Elaphe.* Teil IV. *Elaphe dione* (Pallas, 1773). *Sauria,* Berlin 8(1), 27–30, 1 fig., 6 photos.

Schulz, K.-D. (1986b) Die hinterasiatischen Kletternattern der Gattung *Elaphe.* Teil V. *Elaphe bimaculata* (Schmidt, 1925). *Sauria,* Berlin 8(2), 23–26, 1 fig., 4 photos.

Schulz, K.-D. (1986c) Die hinterasiatischen Kletternattern der Gattung *Elaphe.* Teil VI. *Elaphe hodgsoni* (Günther, 1860). *Sauria,* Berlin 8(3), 29–30, 1 fig., 1 photo.

Schulz, K.-D. (1986d) Die hinterasiatischen Kletternattern der Gattung *Elaphe.* Teil VII. *Elaphe radiata* (Schlegel, 1837). *Sauria,* Berlin 8(4), 3–6, 2 figs., 4 photos.

Schulz, K.-D. (1987a) Die hinterasiatischen Kletternattern der Gattung *Elaphe.* Teil VIII. *Elaphe taeniura* (Cope, 1861). *Sauria,* Berlin 9(1), 21–30, 3 figs., 7 col. photos.

Schulz, K.-D. (1987b) Die hinterasiatischen Kletternattern der Gattung *Elaphe.* Teil IX. *Elaphe quadrivirgata* (Boie, 1826). *Sauria,* Berlin 9(2), 3–6, 4 col. photos.

Schulz, K.-D. (1987c) Die hinterasiatischen Kletternattern der Gattung *Elaphe.* Teil X. *Elaphe flavolineata* (Schlegel, 1837). *Sauria,* Berlin 9(3), 21–23, 2 figs., 2 col. photos.

Schulz, K.-D. (1988a) Die hinterasiatischen Kletternattern der Gattung *Elaphe.* Teil XII. *Elaphe conspiciliata* (Boie, 1826). *Sauria,* Berlin 10(1), 21–23, illust.

Schulz, K.-D. (1988b) Die hinterasiatischen Kletternattern der Gattung *Elaphe.* Teil XIII. *Elaphe moellendorffi* (Boettger, 1886). *Sauria,* Berlin 10(2), 27–29, 1 fig., 3 col. photos.

Schulz, K.-D. (1988c) Die hinterasiatischen Kletternattern der Gattung *Elaphe.* Teil XIV. *Elaphe subradiata* (Schlegel, 1837) und *Elaphe erythrura* (Duméril & Bibron, 1854). *Sauria,* Berlin 10(3), 17–20, 13 figs., 2 col. photos.

Schulz, K.-D. (1988d) Contribution to the knowledge of *Elaphe schrencki* (Strauch, 1873). *Litteratura Serpentium* (English edition) 8(5), 213–224, 6 photos.

Schulz, K.-D. (1988e) Asian rat snakes of the *Elaphe* genus. Part I. *Elaphe rufodorsata. Snake Keep* 2(2), 4–7, 2 figs.

Schulz, K.-D. (1988f) Asian rat snakes of the *Elaphe* genus. Part II. *Elaphe climacophora* (Boie, 1826). *Snake Keep* 2(3), 6–7.

Schulz, K.-D. (1988g) Asian rat snakes of the *Elaphe* genus. Part III. *Elaphe schrencki. Snake Keep* 2(4), 4–6, 1 fig.

Schulz, K.-D. (1988h) Asian rat snakes of the *Elaphe* genus. Part IV. *Elaphe dione* (Pallas, 1773). *Snake Keep* 2(5), 4–6, 1 fig.

Schulz, K.-D. (1988i) Asian rat snakes of the *Elaphe* genus. Part V. *Elaphe bimaculata* (Schmidt, 1925). *Snake Keep* 2(6), 11–14, 1 fig., 3 photos.

Schulz, K.-D. (1988j) Asian rat snakes of the *Elaphe* genus. Part VI. *Elaphe hodgsoni* (Gunther, 1860). *Snake Keep* 2(7), 10–11, 1 fig., 1 photo.

Schulz, K.-D. (1988k) Asian rat snakes of the *Elaphe* genus. Part VII. *Elaphe radiata* (Schlegel, 1837). *Snake Keep* 2(8), 12–15, 2 figs., 4 photos.

Schulz, K.-D. (1988L) Asian rat snakes of the *Elaphe* genus. Part VIII(a). *Elaphe taeniura* (Cope, 1861). *Snake Keep* 2(9), 13–15, 1 fig., 1 photo.

Schulz, K.-D. (1988m) Asian rat snakes of the *Elaphe* genus. Part VII(b). *Elaphe taeniura* (Cope, 1861). *Snake Keep* 2(10), 4–9, 2 figs., 5 photos.

Schulz, K.-D. (1988n) Asian rat snakes of the *Elaphe* genus. Part IX. *Elaphe quadrivirgata* (Boie, 1826). *Snake Keep* 2(11), 4–6, 2 photos.

Schulz, K.-D. (1988o) Asian rat snakes of the *Elaphe* genus. Part X. *Elaphe flavolineata* (Schlegel, 1837). *Snake Keep* 2(12), 17–18, 2 figs., 1 photo.

Schulz, K.-D. (1989a) Die hinderasiatischen Kletternattern der Gattung *Elaphe.* Teil XV. *Elaphe davidi* (Sauvage, 1884). *Sauria,* Berlin 11(1), 21–22, 4 figs., 1 col. photo.

Schulz, K.-D. (1989b) Die hinderasiatischen Kletternattern der Gattung *Elaphe.* Teil XVI. *Elaphe perlacea* Stejneger, 1929. *Sauria,* Berlin 11(2), 15–16, 4 figs., 1 col. photo.

Schulz, K.-D. (1989c) Die hinderasiatischen Kletternattern der Gattung *Elaphe.* Teil XVII. *Elaphe porphyracea* (Cantor, 1839). *Sauria,* Berlin 11(4), 21–24, 5 figs., 5 col. photos.

Schulz, K.-D. (1989d) Asian rat snakes of the *Elaphe* genus. Part 12. *Elape conspicillata* (Boie, 1826). *Snake Keep* 3(2), 11–13, illust.

Schulz, K.-D. (1989e) Asian rat snakes of the *Elaphe* genus. Part 13. *Elaphe moellendorffi* (Boettger, 1886). *Snake Keep* 3(3), 18–21, illust.

Schulz, K.-D. (1989f) Asian rat snakes of the *Elaphe* genus. Part XIV. *Elaphe subradiata* (Schlegel, 1837) and *Elaphe erythrura* (Dumeril & Bibron, 1854). *Snake Keep* 3(4), 14–17, 11 figs., 2 photos.

Schulz, K.-D. (1990a) *Acanthophis pyrrhus* Boulenger. *Sauria,* Berlin 12(2), 1–2, 1 fig.

Schulz, K.-D. (1990b) Die hinderasiatischen Kletternattern der Gattung *Elaphe.* Teil XIX. *Elaphe cantoris* (Cantor, 1839). *Sauria,* Berlin 12(3), 19–20, 1 fig., 1 photo, 2 col. photos.

Schulz, K.-D. (1990c) Die hinderasiatischen Kletternattern der Gattung *Elaphe.* Teil XVIII. *Elaphe mandarina* (Cantor, 1842). *Sauria,* Berlin 12(2), 25–29, 1 fig., 3 photos, 7 col. photos.

Schulz, K.-D. (1991) Die hinderasiatischen Kletternattern der Gattung *Elaphe.* Teil XX. *Elaphe prasina* (Blyth, 1854) und *Elaphe frenata* (Gray, 1853). *Sauria,* Berlin 13(2), 15–17, 2 figs., 1 photo, 2 col. photos.

Schulz, K.-D. (1992a) The far-east Asian ratsnakes of the genus *Elaphe.* Part XXI. *Elaphe leonardi* (Wall, 1921). *Sauria* [E], Kempton Park 14(1), 11–13, 1 fig., 3 photos, 1 col. photo.

Schulz, K.-D. (1992b) Die geographischen Unterarten der Indischen Schmucknatter *Elaphe helena* (Daudin, 1803). *Salamandra*, Frankfurt-am-Main 28(1), 14–24, 2 figs.

Schulz, K.-D. (1992c) Die hinterasiatischen Kletternattern der Gattung *Elaphe*. Teil XXII. *Elaphe carinata* (Günther, 1864). *Sauria*, Berlin 14(4), 3–13, 7 figs., 6 col. photos.

Schulz, K.-D. (1992d) Variation, distribution and biology of *Elaphe triaspis* (Cope, 1866), Part I. *Litteratura Serpentium* (English edition) 12(2), 32–41, photos 1–8.

Schulz, K.-D. (1992e) Variation, distribution and biology of *Elaphe triaspis* (Cope, 1866). Part 2, with remarks about care and breeding of the southern subspecies *Elaphe triaspis mutabilis* (Cope, 1885). *Litteratura Serpentium* (English edition) 12(3), 54–68, 7 figs., photos 9–16.

Schulz, K.-D. (1992f) Die hinterasiatischen Kletternattern der Gattung *Elaphe*. Teil 22. *Elaphe carinata* (Gunther, 1864). *Sauria*, Berlin 14(4), 3–13, illust.

Schulz, K.-D. (1993) Remarkes on the taxonomic status of the various subspecies of the Mexican ratsnake *Elaphe flavirufa* (Cope, 1867). *Litteratura Serpentium* (English edition) 13(1), 14–29, 11 photos.

Schulz, K.-D. (1996) *A monograph of the colubrid snakes of the genus* Elaphe *Fitzinger*. Koeltz Scientific Books, Koenigstein, iii + 439 pp., 121 figs., 58 col. pls.

Schulz, K.-D. & Helfenberger, N. (1998) Eine Revision des Unterarten-Komplexes der Roten Bambusnatter *Elaphe porphyracea* (Cantor, 1839). *Sauria*, Berlin 20(1), 25–45, 14 col. figs.

Schulz, K.-D. & Münzenmaier, J. (1990) Die hinderasiatischen Kletternattern der Gattung *Elaphe*. Teil XVIII. *Elaphe mandarina* (Cantor, 1842). *Sauria*, Berlin 12(2), 25–29, illust.

Schulz, K.-D. & Philippes, D. (1991) The systematic-taxonomic position of *Elaphe bairdi* (Yarrow, 1880). *Litteratura Serpentium* (English edition)11(6), 138–142, 5 photos.

Schulz, K.-D., Smith, H.M., Auth, D.L., Chiszar, D. & van Breukelen, F. (1997) A second specimen of *Elaphe flavirufa matudai* (Reptilia, Serpentes). *Bulletin of the Chicago Herpetological Society* 32(11), 225–226, 1 fig.

Schulz, K.-D., Helfenberger, N., Rao, D.-Q. & Cen, J. (2000) Eine verkannte Colubriden-art, *Elaphe bella* (Stanley, 1917). *Sauria*, Berlin 22(1), 11–18, 8 col. figs.

Schulze, F.E., Kükenthal, W. & Hediger, K. (1927–1929). *Nomenclator animalium generum et subgenerum. Zweiter Band (C–E).* Verlag Preußischen Akademie der Wissenschaften, Berlin, pp. 477–1298.

Schulze, F.E., Kükenthal, W., Hediger, K. & Hesse, R. (1932–1935). *Nomenclator animalium generum et subgenerum. Vierter Band (N–P).* Verlag Preußischen Akademie der Wissenschaften, Berlin, pp. 2185–2988.

Schulze, F.E., Kükenthal, W., Hediger, K. & Hesse, R. (1936–1954). *Nomenclator animalium generum et subgenerum. Fünfte Band (Q–Z).* Verlag Preußischen Akademie der Wissenschaften, Berlin, pp. 2989–3716.

Schüz, E. (1929) Verzeichnis der Typen des Staatlichen Museums für Tierkunde in Dresden. Part 1. Fische, Amphibien und Reptilien mit einem Anhang, Die Schlangen der papuanischen Ausbeute Dr. Schlaginhaufens 1909. *Abhandlungen und Berichte aus dem Staatlichen Museum für Tierkunde und Völkerkunde in Dresden*, Leipzig 17(2), 1–16, 1 fig.

Schwab, D. (1984) Notes on the amphibians and reptiles of the Great Dismal Swamp of Virginia and North Carolina. *Bulletin of the Chicago Herpetological Society* 19(3), 85–93, 1 fig.

Schwaner, T.D. (1985a) *Snakes in South Australia: a species list and overview.* South Australian Museum, Adelaide, 8 pp., 38 col. photos.

Schwaner, T.D. (1985b) Population structure of black tiger snakes, *Notechis ater niger*, on offshore islands of South Australia. *In:* Grigg, G., Shine, R. & Ehmann, H. (Eds.), *Biology of Australasian frogs and reptiles.* Royal Zoological Society of New South Wales, Chipping Norton, pp. 35–46, 6 figs.

Schwaner, T.D. (1989) A field study of thermoregulation in black tiger snakes (*Notechis ater niger*, Elapidae) on the Franklin Islands, South Australia. *Herpetologica* 45(4), 393–401, 2 figs.

Schwaner, T.D. & Mount, R.H. (1976) Systematics and ecological relationships of the water snakes *Natrix sipedon* and *N. fasciata* in Alabama and the Florida Panhandle. *Occasional Papers of the Museum of Natural History, University of Kansas* (45), 1–44, 36 figs.

Schwaner, T.D. & Sarre, S.D. (1990) Body size and sexual dimorphism in mainland and island tiger snakes. *Journal of Herpetology* 24(3), 320–322, 1 fig.

Schwaner, T.D., Baverstock, P.R., Dessauer, H.C. & Mengden, G.A. (1985) Immunological evidence for the phylogenetic relationships of Australian elapid snakes. *In:* Grigg, G., Shine, R. & Ehmann, H. (Eds.), *Biology of Australasian frogs and reptiles.* Royal Zoological Society of New South Wales, Chipping Norton, pp. 177–184.

Schwardt, H.H. (1938) *Reptiles of Arkansas.* Agricultural Experiment Station, University of Arkansas, Fayetteville, 1–47, 53 figs.

Schwartz, A. (1953) A new subspecies of crowned snake (*Tantilla coronata*) from the southern Appalachian Mountains. *Herpetologica* 9(3), 153–157.

Schwartz, A. (1957) A new species of boa (genus *Tropidophis*) from western Cuba. *American Museum Novitates* (1839), 1–8, 1 fig.

Schwartz, A. (1963) A new subspecies of *Tropidophis greenwayi* from the Caicos Bank. *Breviora* (194), 1–6.

Schwartz, A. (1965) A review of the colubrid snake genus *Arrhyton* with a description of a new subspecies from southern Oriente Province, Cuba. *Proceedings of the Biological Society of Washington* 78, 99–113, 16 figs.

Schwartz, A. (1966) Snakes of the genus *Alsophis* in Puerto Rico and the Virgin Islands. *Studies on the Fauna of Curaçao and other Caribbean Islands*, Utrecht 23(90), 177–227, figs. 44–57.

Schwartz, A. (1967) A review of the genus *Dromicus* in Puerto Rico and the Virgin Islands. *Stahlia*, University of Puerto Rico, Rio Piedras (9), 1–14, 4 figs.

Schwartz, A. (1970a) A systematic review of *Uromacer catesbyi* Schlegel (Serpentes, Colubridae). *Tulane Studies in Zoology* 16(4), 131–149, 1 fig.

Schwartz, A. (1970b) *Darlingtonia haetiana* (Serpentes, Colubridae): a new subspecies. *Herpetologica* 26(3), 324–331, 1 fig.

Schwartz, A. (1971) A systematic review of the Hispaniolan snake genus *Hypsirhynchus. Studies on the Fauna of Curaçao and other Caribbean Islands*, Utrecht 35(128), 63–94, 3 figs.

Schwartz, A. (1975) Variation in the Antillean boid snake *Tropidophis haetianus* Cope. *Journal of Herpetology* 9(3), 303–311.

Schwartz, A. (1976) Variation in the Hispaniolan colubrid snake *Uromacer frenatus* Günther (Reptilia, Serpentes, Colubridae). *Journal of Herpetology* 10(4), 319–327.

Schwartz, A. (1979a) The status of *Uromacer frenatus* and *Uromacer dorsalis* (Reptilia, Serpentes, Colubridae). *Herpetologica* 35(3), 207–216, 1 fig.

Schwartz, A. (1979b) The herpetofauna of Ile A Cabrit, Haiti, with the description of two new subspecies. *Herpetologica* 35(3), 248–255, 1 fig.

Schwartz, A. (1986) *Darlingtonia. D. haetiana. Catalogue of American amphibians and reptiles*, pp. 390.1–390.2.
Schwartz, A. & Garrido, O.H. (1975) A reconsideration of some Cuban *Tropidophis* (Serpentes, Boidae). *Proceedings of the Biological Society of Washington* 88(9), 77–89, 1 fig.
Schwartz, A. & Garrido, O.H. (1981) A review of the Cuban members of the genus *Arrhyton* (Reptilia, Serpentes, Colubridae). *Annals of the Carnegie Museum* 50(7), 207–230, 3 figs.
Schwartz, A. & Henderson, R.W. (1984a) *Uromacer. Catalogue of American amphibians and reptiles*, p. 355.1.
Schwartz, A. & Henderson, R.W. (1984b) *Uromacer catesbyi. Catalogue of American amphibians and reptiles*, pp. 356.1–356.2.
Schwartz, A. & Henderson, R.W. (1984c) *Uromacer oxyrhynchus. Catalogue of American amphibians and reptiles*, pp. 358.1–358.2.
Schwartz, A. & Henderson, R.W. (1985) *A guide to the identification of the amphibians and reptiles of the West Indies exclusive of Hispaniola*. Milwaukee Public Museum, Milwaukee, 165 pp., 87 figs., 19 pls.
Schwartz, A. & Henderson, R.W. (1988) West Indian amphibians and reptiles: a checklist. *Contributions in Biology and Geology, Milwaukee Public Museum*, Milwaukee (74), 1–264, 7 figs.
Schwartz, A. & Henderson, R.W. (1991) *Amphibians and reptiles of the West Indies. Descriptions, distributions, and natural history*. University Press of Florida, Gainesville, xvi + 720 pp.
Schwartz, A. & Marsh, R.J. (1960) A review of the *pardalis-maculatus* complex of the boid genus *Tropidophis* of the West Indies. *Bulletin of the Museum of Comparative Zoology* 123(2), 49–85, 10 figs.
Schwartz, A. & Ogren, L.H. (1956) A collection of reptiles and amphibians from Cuba, with the descriptions of two new forms. *Herpetologica* 12(2), 91–110.
Schwartz, A. & Rossman, D.A. (1976) A review of the Hispaniolan colubrid snake genus *Ialtris*. *Studies on the Fauna of Curaçao and other Caribbean Islands*, Utrecht 50(165), 76–102, figs. 22–25.
Schwartz, A. & Thomas, R. (1960) Four new snakes (*Tropidophis*, *Dromicus*, *Alsophis*) from the Isla de Pinos and Cuba. *Herpetologica* 16(2), 73–90, 8 figs.
Schwartz, A. & Thomas, R. (1966) The genus *Darlingtonia* (Serpentes) in Hispaniola, including a new species from the Dominican Republic. *Breviora* (229), 1–10, 4 figs.
Schwartz, A. & Thomas, R. (1975) A check-list of West Indian amphibians and reptiles. *Special Publication, Carnegie Museum of Natural History* (1), 216 pp.
Schwartz, A., Thomas, R. & Ober, L.D. (1978) First supplement to a check-list of West Indian amphibians and reptiles. *Special Publication, Carnegie Museum of Natural History* (5), 1–35.
Schwartz, V. & Golden, D.M. (2002) *Field guide to reptiles and amphibians of New Jersey.* New Jersey Division of Fish and Wildlife, Trenton, 89 pp., col. illust.
Schwarz, E. (1936) Untersuchungen über Systematik und Verbreitung der europäischen und mediterranen Ottern. *In:* Bieling, R., Demintz, A., Schaumann, O., Schlossberger, H., Schuckmann, W. & Schwartz, E. (Eds.), *Die europäischen und mediterranen Ottern und ihre Gifte Grundlagen zur Darstellung eines wirksamen Schlangenserums*. Behringwerke, Marburg-Lahn, pp. 159–262, 35 pls.
Schweiger, M. (1992) Herpetologische Beobachtungen im Gebiet von Quarzazate (Marokko). *Herpetozoa*, Wien 5(1–2), 13–31, illust.
Schweiger, M. (1994) Erstnachweis von *Elaphe longissima* (Laurenti, 1768) für die zentrale Osttürkei (Squamata, Serpentes, Colubridae). *Herpetozoa*, Wien 7(3–4), 149–151.
Schwetz, J. (1934) Note sur les serpents de Stanleyville (Congo Belge). *Revue de Zoologie et Botanique Africaine*, Tervuren 25, 379–384.
Schwinford, G. (1999) Phenotypical color and pattern variations in the banded rock rattlesnake, *Crotalus lepidus klauberi. The Vivarium* 10(4), 41–43, 5 col. photos.
Sclater, W.L. (1891a) Notes on the collection of snakes in the Indian Museum with descriptions of several new species. *Journal of the Asiatic Society of Bengal (Natural History),* Calcutta 60(3), 230–250, pl. 6.
Sclater, W.L. (1891b) *List of snakes in the Indian Museum*. Indian Museum, Calcutta, x + 70 pp.
Sclater, W.L. (1898) List of the reptiles and batrachians of South Africa, with descriptions of new species. *Annals of the South African Museum* 1, 95–111, pl. 5.
Scolaro, A. (2005) *Reptiles Patagónicos sur: una guía de campo.* Universidad Nacioanl de la Patagonia, San Juan Bosco, 79 pp., col. illust.
Scolaro, A. (2006) *Reptiles Patagónicos norte: una guía de campo.* Universidad Nacioanl de la Patagonia, San Juan Bosco, 112 pp., col. illust.
Scolaro, A. & Cei, J.M. (1979) The southernmost population of *Elapomorphus bilineatus* in Argentine Patagonia. *Copeia* 1979(4), 745–747.
Scopoli, J.A. (1788) *Deliciae Florae et Faunae Insubricae. Vol. III.* Monast. S. Salvatoris, Ticino, ii + 87 pp., 25 pls.
Scortecci, G. (1928) Rettili dell'Eritrea esistenti nelle collezioni del Museo Civico di Milano. *Atti della Società Italiana di Scienze Naturali e del Museo Civico di Storia Naturale di Milano* 67(3–4), 290–338, 8 figs., 3 pls.
Scortecci, G. (1929) Primo contributo alla conoscenza dei rettili edegli anfibi della Somalia italiana. *Atti della Società Italiana di Scienze Naturali e del Museo Civico di Storia Naturale di Milano* 68(3–4), 245–278, figs., pl.
Scortecci, G. (1930a) Rettili ed anfibi raccolti dal Prof. E. Zavattari in Eritrea. *Atti della Società Italiana di Scienze Naturali e del Museo Civico di Storia Naturale di Milano* 69(3–4), 193– 217, 3 figs.
Scortecci, G. (1930b) Contributo alla conoscenza dei rettili e degli anfibi della Somalia, dell'Eritrea e dell'Abissinia. *Bolletino dei Musei di Zoologia ed Anatomia Comparata della Università di Torino* (Series 3) 41(10), 1–26, 9 figs.
Scortecci, G. (1931) Terzo contributo alla conoscenza dei rettili della Somalia Italiana (Ofidi). *Atti della Società Italiana di Scienze Naturali e del Museo Civico di Storia Naturale di Milano* 70, 203–215.
Scortecci, G. (1932a) Rettili dello Yemen. *Atti della Società Italiana di Scienze Naturali e del Museo Civico di Storia Naturale di Milano* 71, 39–49, 1 fig.
Scortecci, G. (1932b) Descrizione preliminare di un nuovo ofidio ed un anfibio della Somalia Italiana. *Atti della Società Italiana di Scienze Naturali e del Museo Civico di Storia Naturale di Milano* 71, 58–60.
Scortecci, G. (1932c) Nuove specie di anfibi e rettili della Somalia Italiana (Descrizione preliminare). *Atti della Società Italiana di Scienze Naturali e del Museo Civico di Storia Naturale di Milano* 71, 264–269.
Scortecci, G. (1933) *Leptotyphlops yemenicus* sp. n. *Atti della Società Italiana di Scienze Naturali e del Museo Civico di Storia Naturale di Milano* 72, 165–166.

Scortecci, G. (1934) Ofidi velenosi della Somalia Italiana. *Rassegna Economica delle Colonie* 22(1–4), 63–96, 22(5–6), 325–367, 36 figs. [reprinted in 1934 by Instituto Poligrafico della Stato, Roma, 83 pp., 36 figs.]

Scortecci, G. (1935) Un nuovo genere e una nuova specie di colubridi opistoglifi della Peninsola dei Somali. *Annali dell Museo Civico di Storia Naturali di "Giacoma Doria,"* Genova (1935–1937) 59, 1–5, 3 figs. [reprinted in 1937]

Scortecci, G. (1939a) *Gli ofidi velenosi dell'Africa italiana*. Instituto Sieroterapico Milanese, Milano, xv + 292 pp., 151 figs.

Scortecci, G. (1939b) Rettili, Ofidi. Spedizione zoologica del Marcheso Saverio Patrizi nel Basso Giuba e nell'regiuba. *Annali dell Museo Civico di Storia Naturali di "Giacoma Doria,"* Genova (1935–1939) 58, 263–291.

Scortecci, G. (1943) Reptilia. *In: Missione biologica Sayan–Omo. Zoologia* 7(1), 267–305.

Scott, A.F. & Snyder, D.H. (1968) The amphibians and reptiles of Montgomery County, Tennessee. *Journal of the Tennessee Academy of Science* 43(3), 79–84, 1 fig.

Scott, C. (1996) *Snake lovers' lifelist & journal*. University of Texas Press, Austin, 246 pp., col. pls.

Scott, F., Parker, F. & Menzies, J.J. (1977) A checklist of the amphibians and reptiles of Papua New Guinea. *Wildlife in Papua New Guinea*, Konedobu (77–3), 1–18.

Scott, N.J., Jr. (1962) *The reptiles of southern Sinaloa, Mexico: a taxonomic and ecological study*. Master's Thesis, Humboldt State College, Arcata, xii + 92 pp., 11 figs.

Scott, N.J., Jr. (1967) The colubrid snake, *Tropidodipsas annulifera*, with reference to the status of *Geatractas*, *Exelencophis*, *Chersodromus annulatus*, and *Tropidodipsas malacodryas*. *Copeia* 1967(2), 280–287, 3 figs.

Scott, N.J., Jr. (1969) *A zoogeographic analysis of the snakes of Costa Rica*. Doctoral Dissertation, University of Southern California, Los Angeles, xiv + 390 pp., 13 figs.

Scott, N.J., Jr. & McDiarmid, R.W. (1984a) *Trimorphodon*. *Catalogue of American amphibians and reptiles*, pp. 352.1–352.2.

Scott, N.J., Jr. & McDiarmid, R.W. (1984b) *Trimorphodon biscutatus*. *Catalogue of American amphibians and reptiles*, pp. 353.1–353.4.

Scott, N.J., Jr. & McDiarmid, R.W. (1984c) *Trimorphodon tau*. *Catalogue of American amphibians and reptiles*, pp. 354.1–354.2.

Scott, N.J., Jr., Lovett, J.W. & Auburn University Cooperative Wildlife Research Unit. (1975) A collection of reptiles and amphibians from the Chaco of Paraguay. *Occasional Papers, University of Connecticut (Biological Science Series)* 2(16), 257–266.

Scott, N.J., Jr., Savage, J.M. & Robinson, D.C. (1983) Checklist of reptiles and amphibians. *In:* Jansen, D.H. (Ed.), *Costa Rican natural history*. University of Chicago Press, Chicago, pp. 367–374.

Scott, N.J., Jr., Maxwell, T.C., Thornton, O.W., Jr., Fitzgerald, L.A. & Flury, J.W. (1989) Distribution, habitat and future of Harter's water snake, *Nerodia harteri*, in Texas. *Journal of Herpetology* 23(4), 373–389, illust.

Scott, N.J., Giraudo, A.R., Scrocchi, G., Aquino, A.L., Cacciali, P. & Motte, M. (2006) The genera *Boiruna* and *Clelia* (Serpentes: Pseudoboini) in Paraguay and Argentina. *Papéis Avulsos de Zoologia* 46(9), 77–105, 14 figs.

Scrocchi, G.J. (1990a) Contribucion al conociemiento de Leptotyphlopidae de Argentina. 2. Nuevo datos sobre *Leptotyphlops australis* Freiberg y Orejas Miranda, 1968. *Acta Zoologica Lilloana*, Tucumán 39(2), 113–114, illust.

Scrocchi, G.J. (1990b) El genero *Micrurus* (Serpentes, Elapidae) en la Republica Argentina. *Bollettino del Museo Regionale di Scienze Naturali*, Torino 8(2), 343–368, 1 fig.

Scrocchi, G.J. (1992a) Analisis preliminar de la osteologia craneal del genero *Micrurus* Wagler (Ophidia, Elapidae). *Acta Zoologica Lilloana*, Tucumán 41, 311–327, illust.

Scrocchi, G. (1992b) *Rachidelus brazili* (Serpentes: Colubridae) en la region central de la Argentina. *Boletin de la Asociación Herpetológica Argentina* 8(1), 5–6.

Scrocchi, G.J. & Cei, J.M. (1991) A new species of the genus *Atractus* from the northwestern Argentina (Serpentes, Colubridae). *Bollettino del Museo Regionale di Scienze Naturali*, Torino 9(1), 205–208, pls. a–b.

Scrocchi, G.J. & Cruz, F.B. (1993) Description of a new species of the genus *Lystrophis* Cope and a revalidation of *Lystrophis pulcher* (Jan, 1863) (Serpentes, Colubridae). *Papéis Avulsos Zoologia*, São Paulo 38(10), 171–186, illust.

Scrocchi, G.J. & Viñas, M. (1990) El genero *Clelia* (Serpentes, Colubridae) en la Republic Argentina, revision y comentarios. *Bollettino del Museo Regionale di Scienze Naturali*, Torino 8(2), 487–499, 4 figs.

Scrocchi, G.J., Porto, M. & Rey, L. (1993) Descripción de una especie nueva y situación del genero *Sibynomorphus* (Serpentes, Colubridae) en la Argentina. *Revista Brasileira de Biologia*, Rio de Janeiro 53(2), 197–208, 5 figs.

Scrocchi, G.J., Lucia-Ferreira, U., Giraudo, A.R., Avila, R.W. & Motte, M. (2005) A new species of *Hydrops* (Serpentes, Colubridae, Hydropsini) from Argentina, Brazil, and Paraguay. *Herpetologica* 61(4), 468–477, 4 figs.

Scudder, S.H. (1882) Nomenclator zoologicus. An alphabetical list of all generic names that have been employed by naturalists for Recent and fossil animals from the earliest times to the close of the year 1879. *Bulletin of the United States National Museum* (19), 1–376 + 1–340.

Seba, A. (1734) *Locupletissimi rerum naturalium thesauri accurata descriptio, et iconibus artificiosissimis expressio, per universam physices historiam. Opus, cui, in hoc rerum henere, nullum par exstitit. Ex toto terrarum orbe collegit, digessit, descripsit, et depingendum curarit Albertus Seba, etzela oostfrisius, Academiae Caesareae Leopoldino Carolinae Naturae Curiosorum Collega Xenocrates dictus; Societatis Regiae Anglicanae, et Instituti Bononiensis, sodalis*. Tomus I. Janssonio-Waesbergias, J. Wetstenium, and Gul. Smith, Amstelaedami, 178 pp., 111 pls. [reprinted in 2001 by Taschen, Köln, 582 pp., 448 pls.]

Seba, A. (1735) *Locupletissimi rerum naturalium thesauri accurata descriptio, et iconibus artificiosissimis expressio, per universam phtsices historiam. Opus, cui, in hoc rerum henere, nullum par exstitit. Ex toto terrarum orbe collegit, digessit, descripsit, et depingendum curarit Albertus Seba, etzela oostfrisius, Academiae Caesareae Leopoldino Carolinae Naturae Curiosorum Colleha Xenocrates dectus; Societatis Regiae Anglicanae, et Instituti Bononiensis, sodalis*. Tomus II. Janssonio-Waesbergias, J. Wetstenium, and Gul. Smith, Amstelaedami, 154 pp., 114 pls. [reprinted in 2001 by Taschen, Köln, 582 pp., 448 pls.]

Secor, S.M. (1994) Natural history of the sidewinder, *Crotalus cerastes*. *In:* Brown, P.R. & Wright, J.W. (Eds.), Herpetology of the North American Deserts. Proceedings of a symposium. *Southwestern Herpetological Society, Special Publication* (5), 281–301, 9 figs.

Seetzen, U.J. (1795) Naturalien so zu verkaufen. *Intelligenzblatt der Allgemeinen Literatur-Zeitung*, Jena 40, 320.

Seetzen, U.J. (1796) Ophiologische Fragmente. *Meyer's Zoologische Archiv* 2, 49–74.

Seetzen, U.J. (1855) *Ulrich Jasper Seetzen's Reisen durch Syrien, Palästina, Phönicien, die Transjordan-Länder, Arabia Petraea und Unter-Aegypten. Herausgegeben und commentirt von Professor Dr. Fr. Kruse in Verbindung mit Prof. Dr. Hunrichs, Dr. G. Fr. Hermann Müller und mehreren andern Gelehrten.* Dritter Band. G. Reimer, Berlin, ii + 502 pp.

Segniagbeto, G.H., Trape, J.-F., David, P., Ohler, A., Dubois, A. & Glitho, I.A. (2011) The snake fauna of Togo: systematics, distribution and biogeography, with remarks on selected taxonomic problems. *Zoosystema* 33(3), 325–360, 12 figs.

Seidel, M.E. & Franz, R. (1994) Amphibians and reptiles (exclusive of marine turtles) of the Cayman Islands. *In:* Brunt, M.A. & Davies, J.A. (Eds.), *The Cayman Islands: natural history and biogeography.* Kluwer Academic Publishers, Norwell, pp. 407–433, 2 figs., 1 pl.

Seiss, C.F. (1903) The snakes of Pennsylvania and New Jersey. *The Atlantic Slope Naturalist,* Narberth 1(5), 59–61.

Señaris, J.C. (1998) A new species of *Typhlophis* (Serpentes, Anomalepididae) from Bolivar State, Venezuela. *Amphibia-Reptilia* 19(3), 303–310, 2 figs.

Sengoku, S. (1977) *Amphibians and reptiles: color guide.* Ie no Hikari Kyokai, Tokyo, 206 pp., col. pls. (in Japanese)

Sengoku, S. (1993a) *Animals of the earth. Vol. 5. Amphibia and Reptilia. No. 9. Primitive and common snakes.* Asahi-Shinbun, Tokyo, pp. 257–288, col. illust. (in Japanese)

Sengoku, S. (1993b) *Animals of the earth. Vol. 5. Amphibia and Reptilia. No. 10. Venomous snakes.* Asahi-Shinbun, Tokyo, pp. 289–320, col. illust. (in Japanese)

Senna, A. (1886) Sulla distribuzione geografica generale degli ofidi. Nota del socio. *Atti della Società Italiana di Scienze Naturali,* Milano 29, 477–516.

Seoane, V.L. (1881) Neue Boidengattung und Art von den Philippinen. *Abhandlungen der Senckenbergischen Naturforschenden Gesellschaft,* Frankfurt-am-Main (1880) 12(1–2), 217–223, 1 pl.

Seoane, V.L. (1884) *Identidad de* Lacerta schreiberi *(Bedriaga),* Lacerta viridis, *var.* gadovii *(Boulenger) é investigaciones herpetológicas de Galicia.* La Coruña, San Nicolás, 1–19.

Sergent, E. (1919) Sur les serpents venimeux d'Algerie. *Bulletin de la Société d'Histoire Naturelle de Afrique du Nord,* Algiers 10(2), 38–45, 7 figs.

Serié, P. (1915a) Description d'une variété de *Philodryas baroni* Berg. *Anales del Museo Nacional de Historia Natural de Buenos Aires* 26, 227–230, 1 pl.

Serié, P. (1915b) Suplemento a la fauna erpetologica argentina. *Anales del Museo Nacional de Historia Natural de Buenos Aires* 27, 93–109.

Serié, P. (1915c) Notas sobre erpetologia del Paraguay. I. Coleccion de ofidios del Dr. C. Fiebrig. *Boletim de la Sociedad Physis,* Buenos Aires 1, 573–582.

Serié, P. (1916) Sobre tres supuestos nuevos trignocephales del Paraguay. *Physis,* Buenos Aires 2(10), 171–174.

Serié, P. (1919) Las serpientes de la Argentina. *Sociedad "Luz,"* Buenos Aires (Série 2) 4(66), 1–30, 7 figs.

Serié, P. (1921) Catalogo de los ofidios argentinos. *Anales de la Sociedad Científica Argentina,* Buenos Aires 92, 145–172.

Serié, P. (1936) Nueva enumeracion sistematica y distribucion de los ofidios argentinos. *Obra del Cincuentenario del Museo de La Plata* 2, 33–61.

Setser, K., Mociño-Deloya, E. & Lazcano, D. (2009) Natural history notes. *Crotalus polystictus. Herpetological Review* 40(4), 440–441.

Sette, V. (1821) Notizia sopra una nuova vipera, etc. *Bibliotheque Universelle des Sciences Belles-Lettres, et Arts,* Genève 16, 50–61.

Sexton, O. & Heatwole, H. (1965) Life history notes of some Panamanian snakes. *Caribbean Journal of Science* 5, 39–43.

Shaffer, L.L. (1992) *Pennsylvania amphibians and reptiles.* Pennsylvania Fish Commission, Harrisburg, 161 pp., col. illust.

Shah, K.B. (1999) New records and distribution of some herpetofauna of Nepal. *Journal of the Natural History Museum,* Kathmandu 18, 99–111.

Shah, K.B. & Tiwari, S. (2004) *Herpetofauna of Nepal: a conservation companion.* IUCN Nepal, Kathmandu, viii + 237 pp., col. illust.

Shammakov, S. (1981) *The reptiles of the Turkmenian lowlands.* Academy of Sciences of the Turkmenistan Soviet Socialist Republic, Ashabad, 311 pp., 71 figs. (in Russian)

Shammakov, S. and C. Ataev. (1971) *The snakes of Turkmenistan.* Ylym, Ashkhabad 1971, 1–28, illust. (in Russian)

Shammakov, S., Ataev, C. & Rustamov, E.A. (1993) Herpetogeographical map of Turkmenistan. *Asiatic Herpetological Research,* Berkeley 5, 127–136, 2 figs.

Shannon, F.A. (1956) The reptiles and amphibians of Korea. *Herpetologica* 12(1), 22–49, 1 fig., 1 pl.

Shannon, F.A. & Humphrey, F.L. (1959a) A new subspecies of *Phyllorhynchus browni* from Sinaloa. *Herpetologica* 15(3), 145–148.

Shannon, F.A. & Humphrey, F.L. (1959b) Two new species of *Tropidodipsas* from the west coast of Mexico. *Herpetologica* 15(4), 217–222, 2 figs.

Shannon, F.A. & Humphrey, F.L. (1964) Remarks on *Leptodeira* and *Pseudoleptodeira* from the west coast of Mexico. *Herpetologica* 19(4), 262–269.

Shannon, F.A. & Smith, H.M. (1950) Herpetological results of the University of Illinois Field Expedition, Spring 1949. I. Introduction, Testudines, Serpentes. *Transactions of the Kansas Academy of Sciences* 52(4), 494–509.

Shaposhnikov, A.V. (2001) About new record of *Eryx miliaris* on the Mt. Bol'Shoi Bogdo in Astrakhan region, Russia. *Russian Journal of Herpetology* 8(1), 79.

Sharma, B.D. (1998a) Fauna of Indian snakes. *In:* Sharma, B.D. & Kumari, T. (Eds.), *Snakes in India: a source book.* Asiatic Publishing House, Delhi, pp. 87–108, 33 figs.

Sharma, B.D. (1998b) Scientific and vernacular names of some common snakes of India. *In:* Sharma, B.D. & Kumari, T. (Eds.), *Snakes in India: a source book.* Asiatic Publishing House, Delhi, pp. 109–114.

Sharma, B.D. & Sharma, R.C. (1975) Some new records of snakes (Reptilia, Serpentes) from Jammu and Kashmir State. *Current Science,* Bangalore 44(17), 646–647.

Sharma, B.D. & Sharma, T. (1976) Snakes of Jammu and Kashmir State, India. *Newsletter of the Madras Snake Park Trust and Conservation Centre* 1(2), 8–10.

Sharma, G.P. & Nakhasi, U. (1980) Karyological studies on six species of Indian snakes (Colubridae, Reptilia). *Cytobios* 27, 177–192, 17 figs.

Sharma, R.C. (1971) The reptile fauna of the Nagarjunasagar Dam area (Andhra Pradesh, India). *Records of the Zoological Survey of India,* Calcutta 63(1–4), 77–93, 1 fig.

Sharma, R.C. (1974) A report on the snakes of Rajasthan. *Indian Journal of Zootomy,* Bhopal 2(1), 19–27, 1 pl.

Sharma, R.C. (1976a) Records of the reptiles of Goa. *Records of the Zoological Survey of India,* Calcutta (1975) 71(1–4), 149–167, 1 fig., 2 pls.

Sharma, R.C. (1976b) Some observations on ecology and systematics of *Coluber bholanathi*, a new species of snake (Reptilia, Squamata, Colubridae) from India. *Comparative Physiology and Ecology*, Jodhpur 1(3), 105–107, 1 fig.

Sharma, R.C. (1982) Taxonomic and ecological studies on the reptiles of Gujarat. *Records of the Zoological Survey of India*, Calcutta 80(1–2), 85–106, illust.

Sharma, R.C. (2000) Reptilia. *In:* Alfred, J.R.B. (Ed.), *Fauna of Gujarat, (Part-1) Vertebrates. State Fauna Series* 8. Zoological Survey of India, Kolkata, pp. 243–297.

Sharma, R.C. (2003) *Handbook — Indian snakes.* Zoological Survey of India, Kolkata, xx + 292 pp., 69 col. pls.

Sharma, R.C. (2007) *The fauna of India and adjacent countries. Reptilia. Volume III (Serpentes).* Zoological Survey of India, Kolkata, xxvi + 410 pp., 187 figs.

Sharma, R.C. & Vazirani, T.G. (1977) Food and feeding habits of some reptiles of Rajasthan. *Records of the Zoological Survey of India*, Calcutta 73(1–4), 77–93, illust.

Sharma, S.K., Pandey, D.P., Shah, K.B., Tillack, F., Chappuis, F., Thapa, C.L., Airol, E. & Kuch, U. (2013) *Venomous snakes of Nepal: a photographic guide.* B.P. Kolrala Institute of Health Sciences, Dharan, 85 pp., 113 col. photos.

Shaw, C.E. & Campbell, S. (1974) *Snakes of the American West.* Alfred A. Knopf, New York, xii + 328 pp., 70 pls.

Shaw, G. (1794) *Zoology of New Holland. Vol. I.* James Sowerby, London, 33 pp., 12 pls.

Shaw, G. (1802) *General zoology, or systematic natural history. Vol. III. Part II. Amphibia.* G. Kearsley, London, viii + pp. 313–615, pls. 87–140.

Shaw, G. & Nodder, F.P. (1789–1813) *The naturalist's miscellany; or colored figures of natural objects; drawn and described immediately from nature.* Georgius Shaw and Fredericus P. Nodder, London, 24 vols., 1068 pls. [1: pls. 1–37, 1789–1790; 2: pls. 53–74, 1790–1791; 3: pls. 75–110, 1791–1792; 4: pls. 111–145, 1791–1792; 5: pls. 146–182, 1792–1793; 6: pls. 183–218, 1793–1794; 7: pls. 219–254, 1794–1795; 8: pls. 255–300, 1795–1796; 9: 301–348, 1796–1797; 10: pls. 349–396, 1798–1799; 11: pls. 397–444, 1799–1800; 12: pls. 445–492, 1800–1801; 13: pls. 493–540, 1801–1802; 14: pls. 541–588, 1802–1803; 15: pls. 589–635, 1803–1804; 16: pls. 636–684, 1804–1805; 17: pls. 685–732, 1805–1806; 18: pls. 733–780, 1806–1807; 19: pls. 781–828, 1807–1808; 20: pls. 829–878, 1808–1809; 21: pls. 879–924, 1809–1810; 22: pls. 925–972, 1810–1811; 23: pls. 973–1020, 1811–1812; 24: pls. 1021–1064, 1812–1813; pls. 1065–1068 unpublished]

Shaw, G.E. & Shebbeare, E.O. (1928–1931) The snakes of northern Bengal and Sikkim. Preface. *Journal of the Darjeeling Natural History Society* 2(4), 64–74, 3 pls. (1928a); 3(1), 88–94 (1928b); 3(2), 111–116 (1928c); 3(3), 51–57 (1929a); 3(4), 75–82 (1929b); 4(1), 3–7 (1929c); 4(1), 28–32 (1929d); 4(3), 53–58 (1930a); 4(4), 78–83 (1930b); 5(1), 3–8 (1930c); 5(2), 23–29 (1930d); 5(3), 68–74, 1 pl. (1931).

Shaw, G.E., Shebbeare, E.O. & Barker, P.E. (1938–1947) The snakes of northern Bengal and Sikkim. *Journal of the Darjeeling Natural History Society* (Series 2) 12(3), 105–112, 3 pls. (1938a); 12(4), 166–173, pl. 4 (1938b); 13(2), 64–73, 12 figs. (1938c); 13(3), 114–123 (1939a); 13(4), 150–159 (1939b); 14(2), 67–79 (1939c); 14(3), 106–112 (1940a); 14(4), 137–145, figs. a–e (1940b); 15(2), 60–68 (1940c); 16(2), 57–67 (1941); 16(4), 113–121, 1 fig., 2 pls. (1942); 21(1), 13–18 (1946a); 21(2), 58–65 (1946b); 22(1), 17–25, 2 figs. (1947a); 22(2), 61–70, 1 fig. (1947b).

Shaw, G.E., Shebbeare, E.O. & Barker, P.E. (1999) *The snakes of Sikkim and Bengal.* Asiatic Publishing House, Delhi, 125 pp., 16 pls. [reprint of G.E. Shaw et al., 1938–1947]

Shea, G.M. (1990) On the status of *Katophis plumbea* Macleay (Serpentes, Colubridae). *Journal of Herpetology* 24(3), 313–314, 1 fig.

Shea, G.M. (1995) The effect of an inadequate key: *Ramphotyphlops broomi* (Squamata: Typhlopidae) does not occur in New South Wales. *Herpetofauna*, Sydney 25(1), 15–18.

Shea, G.M. (1996) Correction of the incorrect original spelling of the species name of a hydrophiid snake. *The Snake* 27(2), 157.

Shea, G.M. (1998) Geographic variation in scalation and size of the black whip snakes (Squamata: Elapidae: *Demansia vestigiata* complex): evidence for two broadley sympatric species. *The Beagle, Records of the Museums and Art Galleries of the Northern Territory* 14, 41–61.

Shea, G.M. (1999) Waite's blind snakes (Squamata: Scolecophidia: Typhlopidae): identification of sources and correction of errors. *Records of the Australian Museum* 51(1), 43–56.

Shea, G.M. & Horner, P. (1996) A new species of *Ramphotyphlops* (Squamata, Typhlopidae) from the Darwin area, with notes on two similar species from Northern Australia. *The Beagle, Records of the Museums and Art Galleries of the Northern Territory* 13, 53–60, 4 figs.

Shea, G.M. & Sadlier, R.A. (1999) A catalogue of the non-fossil amphibian and reptile type specimens in the collection of the Australian Museum: types currently, previously and purportedly present. *Technichal Reports of the Australian Museum* (15), 91 pp.

Shea, G.M. & Scanlon, J.D. (2007) Revision of the small tropical whipsnakes previously referred to *Demansia olivacea* (Gray, 1842) and *Demansia torquata* (Günther, 1852) (Squamata, Elapidae). *Records of the Australian Museum* 59(2), 117–142, 21 figs.

Sheil, C.A. (1998) *Emmochliophis miops*, redescription of *Synophis miops* (Boulenger, 1898). *Journal of Herpetology* 32(4), 604–607, 1 fig.

Sheil, C.A. & Grant, T. (2001) A new species of colubrid snake (*Synophis*) from western Colombia. *Journal of Herpetology* 35(2), 204–209, 2 figs.

Sheldon, A.B. (2006) *Amphibians & reptiles of the North Woods: a field guide to all 46 northern herps.* Kollath-Stensaas Publishing, Duluth, 148 pp., col. illust.

Shelford, R. (1901a) On two new snakes from Borneo. *Annals and Magazine of Natural History*, London (Series 7) 8(48), 516–517.

Shelford, R. (1901b) A list of the reptiles of Borneo. *Journal of the Straits Branch of the Royal Asiatic Society* (35), 43–68.

Sheplan, B.R. & Schwartz, A. (1974) Hispaniolan boas of the genus *Epicrates* (Serpentes, Boidae) and their Antillean relationships. *Annals of the Carnegie Museum* (1974–1975) 45(5), 57–143.

Sheppard, R. (1804) Description of the British lizards; and of a new British species of viper. *Transactions of the Linnean Society of London* 7, 49–56.

Sherborn, C.D. (1895) On the dates of Shaw and Nodder's "Naturalist's Miscellany." *Annals and Magazine of Natural History*, London (Series 6) 15(88), 375–376.

Shetty, S. & Prasad, K.V.D. (1996) Geographic variation in the number of bands in *Laticauda colubrina*. *Hamadryad* 21, 44–45.

Shi, H.-T. (2011) *The Amphibia and Reptilia fauna of Hainan.* Huayu Nature Book Trade Co., Beijing, 285 pp., 29 col. pls. (in Chinese)

Shibata, Y. (1960) Amphibia and Reptilia collected from Tokara Islands. *Bulletin, Osaka Museum of Natural History* (12), 58–62.

Shibata, Y. (1981) Amphibians and reptiles of the Koshiki Islands, Kyushu, Japan. *Bulletin, Osaka Museum of Natural History* (34), 17–22, 1 fig.

Shibata, Y. (1988) First record of *Dinodon orientalis* from Danjo Islands, Japan (Reptilia, Colubridae). *Bulletin, Osaka Museum of Natural History* (42), 25–32, 2 figs.

Shine, R. (1977a) Reproduction in Australian elapid snakes. 2. Female reproductive cycles. *Australian Journal of Zoology*, Melbourne 25(4), 655–666, illust.

Shine, R. (1977b) Habitats, diets, and sympatry in snakes, a study from Australia. *Canadian Journal of Zoology*, Ottawa 55(7), 1118–1128, illust.

Shine, R. (1980a) Reproduction, feeding and growth in the Australian burrowing snake *Vermicella annulata*. *Journal of Herpetology* 14(1), 71–77, 5 figs.

Shine, R. (1980b) Comparative ecology of three Australian snake species of the genus *Cacophis* (Serpentes, Elapidae). *Copeia* 1980(4), 831–838, 6 figs.

Shine, R. (1980c) Ecology of eastern Australian whipsnakes of the genus *Demansia*. *Journal of Herpetology* 14(4), 381–389, 3 figs.

Shine, R. (1980d) Ecology of the Australian death adder *Acanthophis antarcticus* (Elapidae): evidence for convergence with the Viperidae. *Herpetologica* 36(4), 281–289, 5 figs.

Shine, R. (1981a) Venomous snakes in cold climates: ecology of the Australian genus *Drysdalia* (Serpentes, Elapidae). *Copeia* 1981(1), 14–25.

Shine, R. (1981b) Ecology of Australian elapid snakes of the genera *Furina* and *Glyphodon*. *Journal of Herpetology* 15(2), 219–224, 4 figs.

Shine, R. (1982) Ecology of an Australian elapid snake *Echiopsis curta*. *Journal of Herpetology* 16(4), 388–393, 3 figs.

Shine, R. (1983a) Arboreality in snakes: ecology of the Australian elapid genus *Hoplocephalus*. *Copeia* 1983(1), 198–205, 3 figs.

Shine, R. (1983b) Food habits and reproductive biology of Australian elapid snakes of the genus *Denisonia*. *Journal of Herpetology* 17(2), 171–175, 1 fig.

Shine, R. (1984a) Ecology of small fossorial Australian snakes of the genera *Neelaps* and *Simoselaps* (Serpentes, Elapidae). *In:* Seigel, R.S., Hunt, L.E., Knight, J.L, Malaret, L. & Zuschlag, N.L. (Eds.), *Vertebrate Ecology and Systematics: a tribute to Henry S. Fitch.* University of Kansas Museum of Natural History, pp. 173–183.

Shine, R. (1984b) Reproductive biology and food habits of the Australian elapid snakes of the genus *Cryptophis*. *Journal of Herpetology* 18(1), 33–39, 1 fig.

Shine, R. (1985a) Ecological evidence on the phylogeny of Australasian elapid snakes. *In:* Grigg, G., Shine, R. & Ehmann, G.H. (Eds.), The ecology of Australasian frogs and reptiles. *Royal Zoological Society of New South Wales*, Chipping Norton, pp. 255–260, pl. 9.

Shine, R. (1985b) A radiotelemetric study of movements, thermoregulation and habitat utilization of Arafura filesnakes (Serpentes, Acrochordidae). *Herpetologica* 41(3), 351–361, 5 figs.

Shine, R. (1986a) Sexual differences in morphology and niche utilization in an aquatic snake, *Acrochordus arafurae*. *Oecologia*, Berlin 69, 260–267, 3 figs.

Shine, R. (1986b) Natural history of two monotypic snake genera of southwestern Australia, *Elapognathus* and *Rhinoplocephalus* (Elapidae). *Journal of Herpetology* 20(3), 436–439, 1 fig.

Shine, R. (1987a) Food habits and reproductive biology of Australian snakes of the genus *Hemiaspis* (Elapidae). *Journal of Herpetology* 21(1), 71–74, 1 fig.

Shine, R. (1987b) Ecological ramifications of prey size, food habits and reproductive biology of Australian copperhead snakes (*Austrelaps*, Elapidae). *Journal of Herpetology* 21(1), 21–28, 2 figs.

Shine, R. (1987c) Ecological comparisons of island and mainland populations of Australian tigersnakes (*Notechis*, Elapidae). *Herpetologica* 43(2), 233–240, 1 fig.

Shine, R. (1987d) Intraspecific variation in thermoregulation, movements and habitat use by Australian blacksnakes, *Pseudechis porphyriacus* (Elapidae). *Journal of Herpetology* 21(3), 165–177, 3 figs.

Shine, R. (1987e) The evolution of viviparity: ecological correlates of reproductive mode within a genus of Australian snakes (*Pseudechis*, Elapidae). *Copeia* 1987(3), 551–563, 3 figs.

Shine, R. (1987f) Reproductive mode may determine geographic distributions in Australian venomous snakes (*Pseudechis*, Elapidae). *Oecologia*, Berlin 71, 608–612.

Shine, R. (1988) Food habits and reproductive biology of small Australian snakes of the genera *Unechis* and *Suta* (Elapidae). *Journal of Herpetology* 22(3), 307–315, 2 figs.

Shine, R. (1989) Constraints, allometry and adaptation, food habits and reproductive biology of Australian brownsnakes (*Pseudonaja*, Elapidae). *Herpetologica* 45(2), 195–207, 2 figs.

Shine, R. (1991) *Australian snakes. A natural history*. A.H. & A.W. Reed, Sydney, 214 pp., col. photos.

Shine, R. & Allen, S. (1980) Ritual combat in the Australian copperhead, *Austrelaps superbus* (Serpentes, Elapidae). *Victorian Naturalist* 97(5), 188–190, 1 fig.

Shine, R. & Charles, N. (1982) Ecology of the Australian elapid snake *Tropidechis carinatus*. *Journal of Herpetology* 16(4), 383–387, 4 figs.

Shine, R. & Covacevich, J. (1983) Ecology of highly venomous snakes: the Australian genus *Oxyuranus* (Elapidae). *Journal of Herpetology* 17(1), 60–69, 6 figs.

Shine, R. & Webb, J.K. (1990) Natural History of Australian typhlopid snakes. *Journal of Herpetology* 24(4), 357–363, 2 figs.

Shine, R., Grigg, G.C., Shine, T.G. & Harlow, P. (1981) Mating and male combat in Australian blacksnakes, *Pseudechis porphyriacus*. *Journal of Herpetology* 15(1), 101–107, 1 fig.

Shine, R., Branch, W.R., Harlow, P.S. & Webb, J.K. (1996a) Sexual dimorphism, reproductive biology, and food habits of two species of African filesnakes (*Mehelya*, Colubridae). *Journal of Zoology*, London 240(2), 327–340, 2 figs.

Shine, R., Haagner, G.V., Branch, W.R., Harlow, P.S. & Webb, J.K. (1996b) Natural history of the African shieldnose snake *Aspidelaps scutatus* (Serpentes, Elapidae). *Journal of Herpetology* 30(3), 361–366, 1 fig.

Shine, R., Sun, L.-A., Zhao, E. & Bonnet, X. (2002a) A review of 30 years of ecological research on the Shedao pitviper *Gloydius shedaoensis*. *Herpetological Natural History* 9(1), 1–14, 5 figs.

Shine, R., Reed, R.N., Shetty, S., Lemaster, M. & Mason, R.T. (2002b) Reproductive isolating mechanisms between two sympatric sibling species of sea snakes. *Evolution* 56(8), 1655–1662, 2 figs.

Shiroma, H. (1993) Follicular growth and fat bodies cycle of female habus, *Trimeresurus flavoviridis*, in the Okinawa Islands. *Japanese Journal of Herpetology* 15(2), 53–58, 3 figs.

Shiroma, H. & Kamura, T. (1987) Mating season, proportion of reproductive females, and sexual of the habu, *Trimeresurus flavoviridis*. *Japanese Journal of Herpetology*, Kyoto 12(1), 16–21, illust.

Shiryaev, K.A., Orlov, N.L., Ryabov, S.A. & Truong, N.-Q. (2007) Captive breeding and reproductive biology of the *Protobothrops cornutus* and *Triceratolepidophis sieversorum*. *Russian Journal of Herpetology* 14(1), 57–64, 11 col. pls.

Shofner, R.M. (2007) A modern checklist of the amphibians, reptiles, and turtles of Utah. *Journal of Kansas Herpetology* (21), 16–18.

Shoup, C.S. (1974) *A bibliography of the zoology of Tennessee and the Tennessee Valley region*. United States Atomic Energy Commission, Technical Information Center, Springfield, 255 pp.

Showler, D. (1994) Reptiles on the island of Socotra, Republic of Yemen. *British Herpetological Society Bulletin* 48, 7–12.

Shrestha, R. & Shah, K. (1984) *Snakes of Nepal*. Tribhuvan University, Kathmandu, 82 pp., 57 figs. (in Nepalese)

Shrestha, T.K. (1981) Eco-geographic distribution of reptiles and need for their conservation. *In:* Majupuria, T.C. (Ed.), *Wild is beautiful: introduction to the magnificent, rich and varied fauna and wildlife of Nepal*. S. Devi, Lashkar, pp. 178–183.

Shrestha, T.K. (2001) *Herpetology of Nepal: a study of amphibians and reptiles of Trans-Himalayan region of Nepal, India, Pakistan and Bhutan*. Mrs. Bimala Shrestha, Kathmandu, 280 pp., illust.

Shrestha, T.K. & Majupuria, T.C. (1977) Redescription of coral snake *Calliophis maccellandi* from Kathmandu. *Journal of the Natural History Museum*, Kathmandu 1(2–4), 103–112, 2 figs.

Shreve, B. (1934) Notes on Ecuadorian snakes. *Occasional Papers of the Boston Society of Natural History* (8), 125–131.

Shreve, B. (1938) A new *Agkistrodon* from Mexico. *Copeia* 1938(1), 9.

Shreve, B. (1947) On the Venezuelan reptiles and amphibians collected by Dr. H.G. Kugler. *Bulletin of the Museum of Comparative Zoology* 99(5), 517–537.

Shreve, B. (1951) A new snake of the genus *Tropidodipsas* from Honduras. *Copeia* 1951(1), 52.

Shreve, B. (1953) Notes on the races of *Micrurus frontalis* (Duméril, Bibron and Duméril). *Breviora* (16), 1–6.

Shreve, B. (1964) A new species of the snake *Leptotyphlops* from Colombia. *Breviora* (211), 1–4.

Shreve, B. & Gans, C. (1958) *Thamnophis bovalli* Dunn rediscovered (Reptilia, Serpentes). *Breviora* (83), 1–7, 3 figs.

Shuntov, V.P. (1962) Sea snakes (Hydrophiidae) of the Gulf of Tonkin (Northern Viet-Nam). *Zoologischeskii Zhurnal*, Moscow 41(8), 1203–1209, 4 figs. (in Russian)

Shuntov, V.P. (1963) Sea snakes. *Priroda*, Moskow 3, 103–104, 1 fig. (in Russian)

Shuntov, V.P. (1966) Distribution of sea snakes in the South China Sea and East Indian Ocean. *Zoologischeskii Zhurnal*, Moscow 45(12), 1882–1886, 3 figs. (in Russian)

Shuntov, V.P. (1971) Sea snakes of the North Australian shelf. *Ékologiya* 2(4), 65–71, 4 figs. (in Russian)

Shupe, S. (2004) *U.S. guide to venomous asnakes and their mimics*. Stoeger Publishing, Accokeek, 160 pp., 96 col. photos.

Shupe, S. & Zimmerer, E. (1998) Geographic distribution: Serpentes: *Clonophis kirtlandii*. *Herpetological Review* 29(1), 53.

Shwayat, S.N., Disi, A.M. & Amr, Z.S. (2009) Snakes of the genus *Eirenis* in Jordan (Reptilia: Squamata: Colubridae). *Vertebrate Zoology* 59(1), 91–101.

Siddiqui, M.S.U. (1969) *Fauna of Pakistan*. Agricultural Research Council, Government of Pakistan, Karachi, 160 pp.

Siebold, P.F. von. (1827) De historiae naturalis in Japonica statu. *Isis von Oken*, Jena 20, 135–143.

Sievert, G. & Sievert, L. (1988) *A field guide to reptiles of Oklahoma*. Oklahoma Department of Wildlife Conservation, Oklahoma City, 96 pp., col. illust.

Sievert, J. (2008) *Die Seitenwinder-Klapperschlange,* Crotalus cerastes. Natur und Tier-Verlag, Münster, 63 pp., col. illust.

Sifford, D.H. & Johnson, B.D. (1978) Fractionation of *Agkistrodon bilineatus* venom by ion exchange chromatography. *In:* Rosenberg, P. (Ed.), *Toxins: animal, plant and microbial*. Pergamon Press, Oxford, pp. 231–242, 3 figs.

Sigala-Rodríguez, J.J., Vázquez, J., Quintero-Díaz, E., Sigala-Rodríguez, J.I. & Ardit, R.R. (2008) Geographic distribution: Serpentes: *Coluber* (=*Masticophis*) *taeniatus*. *Herpetological Review* 39(2), 240.

Sights, W.P. (1949) Annotated list of reptiles taken in western Bengal. *Herpetologica* 5, 81–83.

Siler, C.D., Welton, L.J., Siler, J.M., Brown, J., Bucol, A., Diesmos, A.C. & Brown, R.M. (2011) Amphibians and reptiles, Luzon Island, Aurora Province and Aurora Memorial National Park, northern Philippines: new island distribution records. *Check List* 7(2), 182–195, 37 figs.

Silva, A. de. (1976) Venomous snakes of Sri Lanka. *The Snake* 8(1), 31-42, 15 figs.

Silva, A. de. (1982) Rare snakes of Sri Lanka. *Loris*, Colombo 16(1), 19–22.

Silva, A. de. (1983) *Haplocercus ceylonensis* Gunther, a relict snake of Sri Lanka. *Loris*, Colombo 16(3), 105–107, 3 figs.

Silva, A. de. (1990a) Venomous snakes, their bites and treatment in Sri Lanka. *In:* Gopalakrishnakone, P. & Chou, L.M. (Eds.), *Snakes of medical importance (Asia-Pacific region)*. National University of Singapore, Singapore, pp. 479–556, 38 figs.

Silva, A. de. (1990b) *Colour guide to the snakes of Sri Lanka*. R. & A. Publishing Ltd., Portishead, vi + 130 pp., 12 pls.

Silva, A. de. (1994) An introduction to the herpetofauna of Sri Lanka. *Lyriocephalus* 1, 3–19.

Silva, A. de. (1996) *The herpetofauna of Sri Lanka: a brief review.* Anslem de Silva, Peradeniya, 99 pp., 15 pls.

Silva, A. de. (Ed.). (1998) *Biology and conservation of the amphibians, reptiles and their habitats in South Asia*. Amphibian and Reptile Research Organization of Sri Lanka, Peradeniya, 364 pp., illust., 1 col. pl.

Silva, A. de. (2001) *The herpetofauna of Sri Lanka: an historical overview, current status with checklists*. Amphibian and Reptile Research Organization of Sri Lanka, Colombo, 91 pp., 10 figs., 15 col. pls.

Silva, A. de. (Ed.). (2004) The herpetology of Sri Lanka: current research. *Lyriocephalus* 5(1–2), 1–222, illust.

Silva, A. de. (2009) *Snakes of Sri Lanka: a coloured atlas (with accounts on snakes in archaeology, history, folklore, venomous snakes and snakebite)*. Vijitha Yapa Publications, Colombo, 129 pp., 25 figs., 52 col. pls.

Silva, A. de. & Pradeep, D.M.N. (2005) First report on some reproductive biology of *Boiga barnesii* (Reptilia: Serpentes: Colubridae). *Lyriocephalus* 6(1–2), 303–304, pl. 15, figs. 1–4.

Silva, A. de. & Velarathne, S. (2005) *Dryocalamus gracilis* (Scarce bridal snake): additional observations. *Lyriocephalus* 6(1–2), 329–330, pl. 15, figs. 7–8.

Silva, A. de, Goonewardene, S., Bauer, A., Austin, C., Drake, J. & Silva, P. de. (2005) Notes on the snakes inhabiting The Knuckles Massif with special reference to *Uropeltis melanogaster* (Gray, 1858) and *Uropeltis phillipsi* (Nicholls, 1929). *Lyriocephalus* 6(1–2), 149–159, 1 fig., pl. 10.

Silva, A. de., Freed, P., Rudge, J., Rasmussen, A.R. & Sanders, K. (2010) Some observations on the wart snake *Acrochordus granulatus* (Schneider, 1799) (Squamata: Serpentes) in Sri Lanka. *Lyriocephalus* 7(1–2), 203–206, pl. 10, figs. 5–6.

Silva, A. de, Ukuwela, K.D.B., Sivaruban, A. & Sanders, K.L. (2011) Preliminary observations on the reproductive biology of six species of Sri Lankan sea snakes (Elapidae: Hydrophiinae). *Salamandra* 47(4), 193–198, 2 figs.

Silva, F.M. da, Menks, A.C., Prudente, A.L.C., Carlos, J., Costa, L., Travassos, A.E.M. & Galatti, U. (2011) Squamate reptiles from Municipality of Barcarena and surroundings, state of Pará, north of Brazil. *Check List* 7(3), 220–226, 3 figs.

Silva, M., Jr. (1956) *O ofidismo no Brasil*. Serviço Nacional de Educaçao Sanitária, Ministério de Saúde, Rio de Janeiro, 346 + 6 pp., 199 figs.

Silva, N.J. da, Jr. (1993) The snakes from Samuel Hydroelectric Power Plant and vicinity, Rondônia, Brazil. *Herpetological Natural History* 1(1), 37–86, 17 figs.

Silva, N.J. da, Jr. (1996) Geographic distribution: Serpentes: *Micrurus lemniscatus* (coral snake). *Herpetological Review* 27(1), 34.

Silva, N.J. da, Jr. (2001) Geographic distribution: Serpentes: *Anilius scytale*. *Herpetological Review* 33(4), 227.

Silva, N.J. da, Jr. & Sites, J.W., Jr. (1999) Revision of the *Micrurus frontalis* complex (Serpentes, Elapidae). *Herpetological Monographs* (13), 142–194, 33 figs.

Silva, N.J. da, Jr. & Sites, J.W., Jr. (200l) Phylogeny of South American triad coral snakes (Elapidae, *Micrurus*) based on molecular characters. *Herpetologica* 57(1), 1–22, 3 figs.

Silva, N.J. da, Jr. & Souza, I.F.E. (2002) Geographic distribution: Serpentes: *Hydrodynastes bicinctus*. *Herpetological Review* 33(3), 227.

Silva, N.J. da, Jr., Rodrigues-Silva, H.L., Ribeiro, R.S., Souza, I. & do Amaral-Souza, C. (2005) Uma nova espécie do género *Atractus* Wagler, 1928 (Colubridae, Dipsadinae) do Cerrado do Brasil Central. *Papeis Avulsos do Departamento de Zoologia, Universidade de São Paulo* 45(3), 33–39, 2 figs.

Silva, N.J. da, Jr., Domingos-Cintra, C.E., Rodrigues-Silva, H.L., Costa, M.C., Amaral-Souza, C. do, Pachéco, A.A., Jr. & Gonçalves, F.A. (2009) Herpetofauna, Ponte de Pedra Hydroelectric Power Plant, states of Mato Grosso and Mato Grosso do Sul, Brazil. *Check List* 5(3), 518–525, 2 figs.

Silva, N.S. de. (2000) *Snakes of Sri Lanka: information and importance*. National Zoological Gardens, Colombo, 31 pp., 24 col. figs. (in Sinhalese)

Silva, P.H.D.H. de. (1969) Taxonomic studies on Ceylon snakes of the family Colubridae. *Spolia Zeylanica*, Colombo 31(2), 431–546, 53 figs.

Silva, P.H.D.H. de. (1971) Taxonomic study of the snakes of the family Colubridae Part II. Genus *Aspidura*. *Spolia Zeylanica*, Colombo 32(1), 107–114, 5 figs.

Silva, P.H.D.H. de. (1980a) *Snake fauna of Sri Lanka with special reference to skull, dentition and venom in snakes*. Sri Lanka National Museum, Colombo, ix + 472 pp., 129 figs., 57 pls.

Silva, P.H.D.H. de. (1980b) An annotated bibliography of snakes of Sri Lanka. *The Snake* 12(1–2), 61–108.

Silva, V.X. da & Rodrigues, M.T. (2008) Taxonomic revision of the *Bothrops neuwiedi* complex (Serpentes, Viperidae) with description of a new species. *Phyllomedusa* 7(1), 45–90, 17 figs.

Silva-Haad, J.J. (1994) Los *Micrurus* de la Amazonia Colombiana. Biología y toxicología experimental de sus venenos. *Colombia Amazonica* 7(1-2), 41-138, 41 figs. (23 col.)

Silva-Haad, J.J. (2004) Las serpientes del género *Atractus* Wagler, 1828 (Colubridae, Xenodontinae) en la Amazonia Colombiana. *Revista de la Academia Colombiana de Ciencias* 28(108), 409–446, 52 figs.

Silva Haad, J.J., Franco, F.L. & Maldonado, J. (2008) Una nueva especie de *Liotyphlops* Peters, 1881 (Serpentes, Scolecophidia, Anomalepidae) del sur de la Amazonia Colombiana. *Biota Colombiana* 9(2), 295–300, 7 figs.

Silva-Leite, R.R., Loebmann, D. & Costa-Silva, P. da. (2009) Geographic distribution: Serpentes: *Hydrodynastes gigas*. *Herpetological Review* 40(3), 365.

Silva-Leite, R.R. da, Roberto, J.J., Loebmann, D., Nascimento, J. da S., and Costa-Silva, P. da. (2010) Geographic distribution: Serpentes: *Eunectes murinus*. *Herpetological Review* 41(2), 244–245.

Silva-Rubio, E. da. (1995) Contribución al atlas herpetológico de la provincia de Badajoz. II. Reptiles. *Revista Española de Herpetologia* 9, 49–56, 22 figs.

Silveira, A.L. (2010) Reptilia, Squamata, Serpentes, Leptotyphlopidae, *Siagonodon brasiliensis* (Laurent, 1949): distribution extension and geographic distribution map. *Check List* 6(4), 532–533, 4 figs.

Silveira, A.L. & Cotta, G.A. (2006) Geographic distribution: Serpentes: *Thamnodynastes sertanejo*. *Herpetological Review* 37(3), 364.

Silveira, A.L. & Evers, P.R., Jr. (2007) Geographic distribution: Serpentes: *Bothrops alternatus*. *Herpetological Review* 38(3), 354.

Silveira, A.L., Silvério-Pires, M.R. & Cotta, G.A. (2004a) Geographic distribution: Serpentes: *Echinanthera melanostigma*. *Herpetological Review* 35(4), 410.

Silveira, A.L., Silvério-Pires, M.R. & Cotta, G.A. (2004b) Geographic distribution: Serpentes: *Leptotyphlops dimidiatus*. *Herpetological Review* 35(4), 411.

Silveira, A.L., Costa, E.R. & Salles, L.M. (2004c) Geographic distribution: Serpentes: *Lystrophis nattereri*. *Herpetological Review* 35(4), 412.

Silveira-Vasconcelos, T. da & Gomes dos Santos, T. (2009) Reptilia, Serpentes, Dipsadidae, *Phalotris lativittatus*: Distribution extension and geographic distribution map. *Check List* 5(1), 5–7, 2 figs.

Simbotwe, M. & Mubamba, R. (1993) *A guide to reptiles, amphibians and fishes of Zambia*. Wildlife Conservation Society of Zambia, Zambia, 89 pp., col. illust.

Simon, H. (1979) *Easy identifcation guide to North American snakes*. Dodd & Mead, New York., 128 pp., col. illust.

Simpson, G.G. (1933. A new fossil snake from the Notostylops beds of Patagonia. *Bulletin of the American Museum of Natural History* 67(1), 1–22, 6 figs.

Simpson, J.W. (1975) Distribution of collagenolytic enzyme activity among snake venoms. *Comparative Biochemistry and Physiology* 51B, 425–428, 1 fig.

Simpson, K.N.G. (1973) Amphibians, reptiles and mammals of the Murray River region between Mildura and Renmark, Australia. *Memoirs of the National Museum of Victoria*, Melbourne (34), 275–279, pl. 36.

Sinclair, R., Hon, W. & Ferguson, B. (1970) *Amphibians and reptiles of Tennessee*. Second edition. Tennessee Game and Fish Commission, Nashville, 28 pp., illust. [first edition, 1965, 28 pp., illust.]

Sinclair, W.F. (1889) How a snake climbs. *Journal of the Bombay Natural History Society* 4, 311.

Sindaco, R., Fedrighi, N. & Venchi, A. (1995) Contribution to the herpetology of Jordan. *Bollettino del Museo Regionale di Scienze Naturali*, Torino 23(1), 389–405.

Sindaco, R., Venchi, A., Carpaneto, G.M. & Bologno, M. (2000) The reptiles of Anatolia: a checklist and zoogeographical analysis. *Biogeographia*, Bologna 21, 441–554, 86 figs. + 30 col. figs.

Sindaco, R., Doria, G., Razzetti, E. & Bernini, F. (2006) *Atlas of Italian amphibians and reptiles*. Edizioni Polistampa, Firenze, 789 pp., col. illust.

Singh, L. (1972) Evolution of karyotypes in snakes. *Chromosoma*, Berlin 38(2), 185–236, 20 figs.

Sison, R. & Gonzales, P.C. (1995) New island records from Panay, Philippines. *Herpetological Review* 26(1), 48.

Sivan, N. & Werner, Y.L. (1992) Survey of the reptiles of the Golan Plateau and Mt. Hermon, Israel. *Israel Journal of Zoology*, Jerusalem 37(4), 193–211, illust.

Sivan, N. & Werner, Y.L. (2003) Revision of the Middle-eastern dwarf-snakes commonly assigned to *Eirenis coronella* (Colubridae). *Zoology in the Middle East* 28, 39–59.

Sjöstedt, Y. (1896) *Atractaspis reticulata*, eine neue Schlange aus Kamerum. *Zoologischer Anzeiger*, Leipzig 19(520), 516–517.

Sjöstedt, Y. (1897) Reptilien aus Kamerun West-Afrika. *Bihang till Kongliga Svenska Vetenskaps-Akademiens, Handlingar*, Stockholm 23(4), 36 pp., 3 pls.

Skelton-Bourgeois, M. (1961) Reptiles et batraciens d'Afrique oriental. *Revue de Zoologie et Botanique Africaines* 63, 309–338.

Skinner, A. (2009) A multivariate morphometric analysis and systematic review of *Pseudonaja* (Serpentes, Elapidae, Hydrophiinae). *Zoological Journal of the Linnean Society* 155(1), 171–197, 16 figs.

Skinner, A., Donnellan, S.C., Hutchinson, M.N. & Hutchinson, R.G. (2005) A phylogenetic analysis of *Pseudonaja* (Hydrophiinae, Elapidae, Serpentes) based on mitochondria DNA sequences. *Molecular Phylogenetics and Evolution* 37(2), 558–571.

Skinner, H.A. (1973) *Snakes and us: an introduction to East African herpetology*. East African Literature Bureau, Nairobi, 146 pp., 29 figs., 64 pls.

Sklenár, J. & Rocek, Z. (1979) *Zoogeografie obojzivelniku a plazu vychodnich Cech*. KMVC, Hradec Kralove, 65 pp. (in Czech)

Skoczylas, R. (1978) Physiology of the digestive tract. *In:* Gans, C. & Gans, K.A. (Eds.), *Biology of the Reptilia. Volume 8 (Physiology B)*. Academic Press, London, pp. 589–717.

Slack, J.H. (Ed.). (1862) *Handbook to the Museum of the Academy of Natural Sciences of Philadelphia*. Academy of Natural Sciences, Philadelphia, viii + 116 pp.

Slater, K.R. (1956) On the New Guinea taipan. *Memoirs of the National Museum of Victoria*, Melbourne 20, 201–205, 2 pls.

Slater, K.R. (1968) *A guide to the dangerous snakes of Papua*. V.P. Bloink, Port Moresby, 22 pp., 4 figs., 7 illust.

Slater, P. (1997) *A first field guide to Australian frogs & reptiles*. Steve Parish Publishing, Fortitude Valley, 56 pp., col. illust.

Sleen, W.G.N. van der. (1932) Zeeslangen. *De Topische Natuur*, Batavia 17(6), 8597.

Slevin, J.R. (1925) Contributions to Oriental herpetology. II. Korea or Chosen. *Proceedings of the California Academy of Sciences* (Series 4) 14(5), 89–100.

Slevin, J.R. (1926) Expedition to the Revillagigedo Islands, Mexico, in 1925, III. Notes on a collection of reptiles and amphibians from the Tres Marias and Revillagigedo Islands, and west coast of Mexico, with description of a new species of Tantilla. *Proceedings of the California Academy of Sciences* (Series 4) 15(3), 195–207, pl. 22.

Slevin, J.R. (1934a) The Templeton Crocker Expedition to western Polynesian and Melanesian Islands, 1933. No. 15. Notes on the reptiles and amphibians, with the description of a new species of sea-snake. *Proceedings of the California Academy of Sciences* (Series 4) 21(15), 183–188.

Slevin, J.R. (1934b) *A handbook of reptiles and amphibians of the Pacific States including certain eastern species*. California Academy of Sciences, San Francisco, 173 pp., 11 pls.

Slevin, J.R. (1935) An account of the reptiles inhabiting the Galápagos Islands. *Bulletin of the New York Zoological Society* 38, 3–25.

Slevin, J.R. (1936) A new Central American snake. *Proceedings of the California Academy of Sciences* (Series 4) 23(4), 79–81.

Slevin, J.R. (1939) Notes on a collection of reptiles and amphibians from Guatemala. I. Snakes. *Proceedings of the California Academy of Sciences* (Series 4) 23(26), 393–414, pls. 37–38.

Slevin, J.R. (1942) Notes on a collection of reptiles from Boquete, Panama, with the description of a new species of *Hydromorphus*. *Proceedings of the California Academy of Sciences* (Series 4) 23(32), 463–480, pls. 39–42.

Slevin, J.R. & Leviton, A.E. (1956) Holotype specimens of reptiles and amphibians in the collection of the California Academy of Sciences. *Proceedings of the California Academy of Sciences* (Series 4) 28(14), 529–560.

Slip, D.J. & Shine, R. (1988a) Feeding habits of the diamond python, *Morelia s. spilota*: ambush predation by a boid snake. *Journal of Herpetology* 22(3), 323–330, 1 fig.

Slip, D.J. & Shine, R. (1988b) Thermoregulation of free-ranging diamond pythons, *Morelia spilota* (Serpentes, Boidae). *Copeia* 1988(4), 984–995, 6 figs.

Slip, D.J. & Shine, R. (1988c) Reptilian endothermy, a field study of thermoregulation by brooding diamond pythons. *Journal of Zoology*, London 216(2), 367–378.

Slip, D.J. & Shine, R. (1988d) The reproductive biology and mating system of diamond pythons, *Morelia spilota* (Serpentes, Boidae). *Herpetologica* 44(4), 396–404, 1 fig.

Slip, D.J. & Shine, R. (1988e) Habitat use, movements and activity patterns of free-ranging diamond pythons, *Morelia s. spilota* (Serpentes, Boidae): a radiotelemetric study. *Australian Wildlife Research*, East Melbourne 15, 515–531.

Sloan, A.J. (1965) Holotype specimens of reptiles in the collection of the San Diego Society of Natural History. *Transactions of the San Diego Society of Natural History* 14(1), 1–8.

Slowinski, J.B. (1989) The interrelationships of laticaudine sea snakes based on the amino acid sequences of short-chained neurotoxins. *Copeia* 1989(3), 783–788, 2 figs.

Slowinski, J.B. (1991) *The phylogenetic relationships of the New World coral snakes (Elapidae,* Leptomicrurus, Micruroides *and* Micrurus*) based on biochemical and morphological data.* Doctoral Dissertation, University of Miami, Miami, 146 pp.

Slowinski, J.B. (1994) A phylogenetic analysis of *Bungarus* (Elapidae) based on morphological characters. *Journal of Herpetology* 28(4), 440– 446, figs. 5.

Slowinski, J.B. (1995) A phylogenetic analysis of the New World coral snakes (Elapidae, *Leptomicrurus*, *Micruroides*, and *Micrurus*) based on allozymic and morphological characters. *Journal of Herpetology* 29(3), 325–338, 6 figs.

Slowinski, J.B. & Keogh, J.S. (2000) Phylogenetic relationships of elapid snakes based on cytochrome b mtDNA sequences. *Molecular Phylogenetics and Evolution* 15(1), 157–164, 2 figs.

Slowinski, J.B. & Lawson, R. (2005) Elapid relationships. *In:* Donnelly, M.A., Crother, B.I., Guyer, C., Wake, M.H. & White, M.E. (Eds.), *Ecology & Evolution in the tropics: a herpetological perspective*. University of Chicago Press, Chicago, pp. 174–189.

Slowinski, J.B. & Wüster, W. (2000) A new cobra (Elapidae, *Naja*) from Myanmar (Burma). *Herpetologica* 56(2), 257–270, 1 fig.

Slowinski, J.B., Boundy, J. & Lawson, R. (2001a) The phylogenetic relationships of Asian coral snakes (Elapidae, *Calliophis* and *Maticora*) based on morphological and molecular characteristics. *Herpetologica* 57(2), 233–245, 5 figs.

Slowinski, J.B., Pawar, S.S., Win, H., Thin, T., Gyi, S.W., Oo, S.W. & Tun, H. (2001b) A new *Lycodon* (Serpentes: Colubridae) from northeast India and Myanmar (Burma). *Proceedings of the California Academy of Sciences* (Series 4) 52(20), 397–405.

Sly, G.R. (1976) New locality records for some peninsular Malaysia Amphibia, Reptilia and Mammalia. *Malayan Nature Journal* 29(3), 155–157.

Smedley, N. (1928) Some reptiles and amphibians from the Anamba Islands. *Journal of the Malayan Branch of the Asiatic Society* 6(3), 76–77.

Smedley, N. (1932a) Amphibians and reptiles from the Cameron Highlands, Malay Peninsula. *Bulletin of the Raffles Museum*, Singapore (1931) (6), 105–123, 5 figs., pl. 2.

Smedley, N. (1932b) Notes on the herpetological collections in the Selangor Museum. *Bulletin of the Raffles Museum*, Singapore (7), 9–17, 2 figs.

Smedt, J. de. (2006) *The vipers of Europe*. JDS Verlag, Frankfurt-am-Main, 339 pp., 386 col. pls.

Smith, A. (1826) On the snakes of southern Africa. *The Edinburgh New Philosophical Journal* 1, 248–254.

Smith, A. (1828) Descriptions of new or imperfectly known objects of the animal kingdom, found in the south of Africa. *South African Commercial Advertiser*, Cape Town [1 Nov.] 3(144), 2.

Smith, A. (1829) Contributions to the natural history of South Africa. *The Zoological Journal*, London 4(16), 433–444.

Smith, A. (1831) Contributions to the natural history of South Africa, No. 1. *South African Quarterly Journal*, Cape Town 1(5), 9–24.

Smith, A. (1833) Secretary's report. *South African Quartlerly Journal,* Cape Town (2) 1(3): 80.

Smith, A. (1837) The characters of two new genera of South African reptiles with descriptions of species belonging to each. *Magazine of Zoology and Botany* 1(2), 141–145.

Smith, A. (1838) Contributions to the natural history of southern Africa. *Magazine of Natural History*, London 2, 92–94.

Smith, A. (1838–1849) *Illustrations of the zoology of South Africa; consisting chiefly of figures and descriptions of the objects of natural history collected during an expedition into the interior of South Africa, in the years 1834, 1835, and 1836; fitted out by "The Cape of Good Hope Association for Exploring Central Africa," together with a summary of African zoology, and an inquiry into the geographical ranges of species in that quarter of the globe. Reptilia.* Smith, Elder & Company, London, unnumbered text, 78 pls., 28 pp. appendix.

Smith, A.G. (1949) The subspecies of the plains garter snake, *Thamnophis radix. Bulletin of the Chicago Academy of Sciences* 8(14), 285–300.

Smith, B.E. (1993) Notes on a collection of squamate reptiles from eastern Mindanao, Philippine Islands 2, Serpentes. *Asiatic Herpetological Research*, Berkeley 5, 96–102.

Smith, E.N. (1995) *Geophis rhodogaster* (Colubridae), an addition to the snake fauna of Mexico. *Southwestern Naturalist* 40(1), 123–124.

Smith, E.N. & Acevedo, M.E. (1997) The northernmost distribution of *Corallus annulatus* (Boidae), with comments on its natural history. *Southwestern Naturalist* 42(3), 347–349.

Smith, E.N. & Campbell, J.A. (1994) A new species of *Rhadinaea* (Colubridae) from the Caribbean versant of Guatemala. *Occasional Papers of the Museum of Natural History Kansas* (167), 1–9, 3 figs.

Smith, E.N. & Campbell, J.A. (1996) The systematic status of Guatemalan populations of snakes allied with *Ninia maculata* (Reptilia, Colubridae). *Proceedings of the Biological Society of Washington* 109(4), 749–754, 1 fig.

Smith, E.N. & Ferrari-Castro, J.A. (2008) A new species of jumping pitviper of the genus *Atropoides* (Serpentes: Viperidae: Crotalinae) from the Sierra de Botaderos and the Sierra La Muralla, Honduras. *Zootaxa* (1948), 57–68, 7 figs.

Smith, E.N., Manamendra-Arachchi, K. & Samaweera, R. (2008) A new species of coralsnake of the genus *Calliophis* (Squamata: Elapidae) from the Central Province of Sri Lanka. *Zootaxa* (1847), 19–33, 8 figs.

Smith, E.N., Ogale, H., Deepak, V.H. & Giri, V.B. (2012) A new species of coralsnake of the genus *Calliophis* (Squamata: Elapidae) from the west coast of peninsular India. *Zootaxa* (3437), 51–68, 11 figs.

Smith, H.M. (1936) A preoccupied generic name of a South African ophidian. *Copeia* 1936(1), 70.

Smith, H.M. (1938) A review of the snake genus *Farancia. Copeia* 1938(3), 110–117, 2 figs., 1 pl.

Smith, H.M. (1939a) Notes on the snakes of the genus *Salvadora. University of Kansas Science Bulletin* (1938) 25(12), 229–237, pl. 22.

Smith, H.M. (1939b) Notes on Mexican reptiles and amphibians. *Zoological Series of Field Museum of Natural History* 24(3), 15–35, fig. 4.

Smith, H.M. (1940) Descriptions of new lizards and snakes from Mexico and Guatemala. *Proceedings of the Biological Society of Washington* 53, 55–64, 2 figs.

Smith, H.M. (1941a) A new genus of Mexican snakes related to *Rhadinaea. Copeia* 1941(1), 7–10, 1 fig.

Smith, H.M. (1941b) On the Mexican snakes of the genus *Pliocercus. Proceedings of the Biological Society of Washington* 54, 119–124.

Smith, H.M. (1941c) Notes on Mexican snakes of the genus *Geophis. Smithsonian Miscellaneous Collections* 99(19), 1–6.

Smith, H.M. (1941d) Further notes on Mexican snakes of the genus *Salvadora. Smithsonian Miscellaneous Collections* 99(20), 1–12, 7 figs.

Smith, H.M. (1941e) Notes on snakes of the genus *Conophis. Journal of the Washington Academy of Sciences* 31(3), 117–124.

Smith, H.M. (1941f) A new genus of Central American snakes related to *Tantilla. Journal of the Washington Academy of Sciences* 31(3), 115–117.

Smith, H.M. (1941g) A new *Geophis* from Mexico. *Proceedings of the New England Zoological Club* 28, 49–55.

Smith, H.M. (1941h) A new name for the Mexican snakes of the genus *Dendrophidion. Proceedings of the Biological Society of Washington* 54, 73–76.

Smith, H.M. (1941i) Notes on Mexican snakes of the genus *Trimeresurus. Zoologica* 26(1), 61–64.

Smith, H.M. (1941j) Notes on the Mexican snakes of the genus *Masticophis. Journal of the Washington Academy of Sciences* 31(9), 388–398, 1 fig.

Smith, H.M. (1941k) A new *Leptodeira* from Mexico. *Proceedings of the Biological Society of Washington* 54, 115–118.

Smith, H.M. (1941l) Notes on Mexican snakes of the genus *Elaphe*. *Copeia* 1941(3), 132–136, 2 figs.
Smith, H.M. (1941m) A review of the subspecies of the indigo snake (*Drymarchon corais*). *Journal of the Washington Academy of Sciences* 31(11), 466– 481, 2 figs.
Smith, H.M. (1941n) The Mexican subspecies of the snake *Coniophanes fissidens*. *Proceedings of the United States National Museum* (1941–1942) 91(3127), 103–111, fig. 33.
Smith, H.M. (1941o) Notes on the snake genus *Trimorphodon*. *Proceedings of the United States National Museum* (1941–1942) 91(3130), 149–168, figs. 37–38.
Smith, H.M. (1942a) A new name for a Chinese snake. *Copeia* 1942(1), 52.
Smith, H.M. (1942b) A résumé of Mexican snakes of the genus *Tantilla*. *Zoologica* 27(7), 33–42.
Smith, H.M. (1942c) Notes on *Masticophis mentovarius*. *Copeia* 1942(2), 85–88, 1 fig.
Smith, H.M. (1942d) A review of the snake genus *Adelphicos*. *Proceedings of the Rochester Academy of Sciences* 8, 175–195, 6 figs.
Smith, H.M. (1942e) Remarks on the Mexican king snakes of the *triangulum* group. *Proceedings of the Rochester Academy of Sciences* 8, 196–207, 1 fig., 1 pl.
Smith, H.M. (1942f) A new race of *Ninia* from Mexico. *Copeia* 1942(3), 152–154.
Smith, H.M. (1942g) The synonymy of the garter snakes (*Thamnophis*) with notes on Mexican and Central American species. *Zoologica* 27(3–4), 97–123.
Smith, H.M. (1942h) Additional notes on Mexican snakes of the genus *Pliocercus*. *Proceedings of the Biological Society of Washington* 55, 159–164.
Smith, H.M. (1942i) Descriptions of new species and subspecies of Mexican snakes of the genus *Rhadinaea*. *Proceedings of the Biological Society of Washington* 55, 185–192, pl. 3.
Smith, H.M. (1942j) Mexican herpetological miscellany. 6. The Mexican subspecies of *Drymobius margaritiferus*. *Proceedings of the United States National Museum* (1942–1943) 92(3153), 380–384.
Smith, H.M. (1942k) Mexican herpetological miscellany. 7. Notes on Mexican *Imantodes*. *Proceedings of the United States National Museum* (1942–1943) 92(3153), 384–391, pl. 37.
Smith, H.M. (1942l) Mexican herpetological miscellany. 8. Two new snakes of the genus *Clelia*. *Proceedings of the United States National Museum* (1942–1943) 92(3153), 391– 394.
Smith, H.M. (1942m) Mexican herpetological miscellany. 9. Additional notes on *Conophis*. *Proceedings of the United States National Museum* (1942–1943) 92(3153), 394–395.
Smith, H.M. (1943a) Summary of the collections of snakes and crocodilians made in Mexico under the Walter Rathbone Bacon traveling scholarship. *Proceedings of the United States National Museum* 93(3169), 393–504, pl. 32.
Smith, H.M. (1943b) Another Mexican snake of the genus *Pliocercus*. *Journal of the Washington Academy of Sciences* 33(11), 344–345
Smith, H.M. (1943c) A new snake of the genus *Tropidodipsas* from Mexico. *Journal of the Washington Academy of Sciences* 33(12), 371–373.
Smith, H.M. (1944) Snakes of the Hoogstraal Expedition to northern Mexico. *Zoological Series of Field Museum of Natural History* 29(8), 135–152, figs. 18–19.
Smith, H.M. (1946) Preliminary notes and speculations on the *triseriatus* group of rattlesnakes in Mexico. *University of Kansas Science Bulletin* 31(3), 75–101, 23 figs., pl. 3.
Smith, H.M. (1947) Notes on Mexican amphibians and reptiles. *Journal of the Washington Academy of Sciences* 37(11), 408–412.
Smith, H.M. (1951) The identity of the ophidian name *Coluber eques* Reuss. *Copeia* 1951, 138–140, 1 pl.
Smith, H.M. (1956) *Handbook of amphibians and reptiles of Kansas*. Second edition. University of Kansas Museum of Natural History, Lawrence, 356 pp., 253 figs. [first edition, 1950, 336 pp., 232 figs.]
Smith, H.M. (1958) Handlist of the snakes of Panama. *Herpetologica* 14(4), 222–224.
Smith, H.M. (1960) New and noteworthy reptiles from Oaxaca, Mexico. *Transactions of the Kansas Academy of Sciences* 62(4), 265–272, 2 pls.
Smith, H.M. (1962) The subspecies of *Tantilla schistosa* of Middle America (Reptilia, Serpentes). *Herpetologica* 18(1), 13–18.
Smith, H.M. (1963a) The identity of the Black Hills population of *Storeria occipitomaculata*, the red-bellied snake. *Herpetologica* 19(1), 17–21, 1 fig.
Smith, H.M. (1963b) *Dryadophis* Stuart, 1939 (Reptilia, Serpentes): proposed validation under the Plenary Powers. *Bulletin of Zoological Nomenclature*, London 20(3), 230.
Smith, H.M. (1963c) *Coluber subocularis* Brown, 1901 (Reptilia, Serpentes): proposed validation under the Plenary Powers. *Bulletin of Zoological Nomenclature*, London 20(3), 231.
Smith, H.M. (1963d) *Spilotes melanurus* Duméril, Bibron & Duméril, 1854 (Reptilia, Serpentes): proposed validation under the Plenary Powers. *Bulletin of Zoological Nomenclature*, London 20(3), 232–233.
Smith, H.M. (1965) Two new colubrid snakes from the United States and Mexico. *Journal of the Ohio Herpetological Society* 5(1), 1–4.
Smith, H.M. (1971) The snake genus *Amastridium* in Oaxaca, Mexico. *Great Basin Naturalist* 31, 254–255.
Smith, H.M. & Braestrup, F.W. (1963) The identity of *Chersodromus nigricans* Reinhardt (Reptilia,Serpentes) and the travels in Mexico of F.W. Liebmann. *Herpetologica* (1962) 18(4), 233–340.
Smith, H.M. & Brodie, E.D., Jr. (1982) *A guide to field identification reptiles of North America*. Golden Press, New York, 240 pp., illust.
Smith, H.M. & Burger, W.L. (1950) A new snake (*Tantilla*) from México. *Herpetologica* 6(5), 117–119. (*Tantilla flavilineata*)
Smith, H.M. & Chiszar, D. (1988) Occupation of *Ovophis* Burger (Serpentes, Viperidae). *Bulletin of the Chicago Herpetological Society* 24(1), 23–26, 1 fig.
Smith, H.M. & Chiszar, D. (1992a) A second locality for *Geophis sallei* (Reptilia, Serpentes). *Bulletin of the Maryland Herpetological Society* 28(1), 16–18.
Smith, H.M. & Chiszar, D. (1992b) A Mexican genus of *Tropidopheine* snakes. *Bulletin of the Maryland Herpetological Society* 28(1), 19–28, 5 figs.
Smith, H.M. & Chiszar, D. (1993) Apparent intergradation in Texas between the subspecies of the Texas blind snake (*Leptotyphlops dulcis*). *Bulletin of the Maryland Herpetological Society* 29(4), 143–155, 2 figs.
Smith, H.M. & Chiszar, D. (1994) Variation in the lined snake (*Tropidoclonion lineatum*) in northern Texas. *Bulletin of the Maryland Herpetological Society* 30(1), 6–14, 2 figs.
Smith, H.M. & Chiszar, D. (1996) *Species-group taxa of the false coral snake genus* Pliocercus. Ramus Publishing, Pottsville, 112 pp., 25 figs.

Smith, H.M. & Chiszar, D. (2000b) The identity of the isolated Los Tuxtlas population of the false coral snake, *Pliocercus elapoides*, of southern Veracruz, Mexico. *Journal of the Colorado-Wyoming Academy of Sciences* 32(1), 18.

Smith, H.M. & Chiszar, D. (2001a) Reassessment of the nominal species of false coral snakes, *Pliocercus psychoides* (Reptilia, Serpentes). *Southwestern Naturalist* 46(1), 114–116.

Smith, H.M. & Chiszar, D. (2001b) *Pliocercus*. *Catalogue of American amphibians and reptiles*, pp. 735.1–735.9, 2 figs.

Smith, H.M. & Chiszar, D. (2001c) *Pliocercus andrewsi*. *Catalogue of American amphibians and reptiles*, pp. 736.1–736.3, 3 color figs.

Smith, H.M. & Chiszar, D. (2001d) *Pliocercus bicolor*. *Catalogue of American amphibians and reptiles*, pp. 737.1–737.4, 6 figs.

Smith, H.M. & Chiszar, D. (2001e) *Pliocercus dimidiatus*. *Catalogue of American amphibians and reptiles*, pp. 738.1–738.3, 2 color figs.

Smith, H.M. & Chiszar, D. (2001f) *Pliocercus elapoides*. *Catalogue of American amphibians and reptiles*, pp. 739.1–739.11, 18 figs.

Smith, H.M. & Chiszar, D. (2001g) *Pliocercus euryzonus*. *Catalogue of American amphibians and reptiles*, pp. 740.1–740.4, 6 figs.

Smith, H.M. & Chiszar, D. (2001h) A new subspecies of cantil (*Agkistrodon bilineatus*) from central Veracruz, Mexico (Reptilia, Serpentes). *Bulletin of the Chicago Herpetological Society* 37(4), 130–136, 2 figs.

Smith, H.M. & Chiszar, D. (2001i) The *nomen protectum* Rule of the 1999 Code and the name for the garter snake *Thamnophis melanogaster* (Peters). *Herpetological Review* 32(3), 160–161.

Smith, H.M. & Chiszar, D. (2003) The checkered nomenclatural history of the narrow-headed garter snake, *Thamnophis rufipunctatus* (Cope). *Bulletin of the Chicago Herpetological Society* 38(3), 46–48.

Smith, H.M. & Chrapliwy, P.S. (1957) A new subspecies of the Central American false coral snakes (*Pliocercus*). *Herpetologica* 13(3), 233–135, 1 pl.

Smith, H.M. & Chrapliwy, P.S. (1958) New and noteworthy Mexican herptiles from the Lidicker collection. *Herpetologica* 13(4), 267–271.

Smith, H.M. & Flores-Villela, O. (1993) Variation in two species (*G. bicolor*, *G. duellmani*) of Mexican earth snakes (*Geophis*). *Bulletin of the Maryland Herpetological Society* 29(1), 20–23.

Smith, H.M. & Flores-Villela, O. (1994) Deletion of the coral snake *Micrurus fulvius* from the herpetofaunal concept of Michoacán, Mexico. *Bulletin of the Maryland Herpetological Society* 30(2), 76–77.

Smith, H.M. & Grant, C. (1958a) Noteworthy herptiles from Jalisco, México. *Herpetologica* 14(1), 18–23, 2 figs.

Smith, H.M. & Grant, C. (1958b) New and noteworthy snakes from Panamá. *Herpetologica* 14(4), 207–215.

Smith, H.M. & Grant, C. (1958c) The proper names for some Cuban snakes: an analysis of dates of publication of Ramon de la Sagra's Historia Natural de Cuba, and of Fitzinger's Systema Reptilium. *Herpetologica* 14(4), 215–222.

Smith, H.M. & Holland, R.L. (1969) Two new snakes of the genus *Geophis* from México. *Transactions of the Kansas Academy of Sciences* 72(1), 47–53, 3 figs.

Smith, H.M. & Huheey, J.E. (1960) The watersnake genus *Regina*. *Transactions of the Kansas Academy of Sciences* 63, 156–164.

Smith, H.M. & Kennedy, J.P. (1951) *Pituophis melanoleucus ruthveni* in eastern Texas and its bearing on the status of *P. catenifer*. *Herpetologica* 7(3), 93–96, 1 fig.

Smith, H.M. & Kohler, A.J. (1978) A survey of herpetological introductions in the United States and Canada. *Transactions of the Kansas Academy of Science* 80(1–2), 1–24.

Smith, H.M. & Landy, M.J. (1965) New and unusual snakes of the genus *Pliocercus* from Oaxaca, Mexico. *Natural History Miscellanea*, Chicago (183), 1–4, 1 fig.

Smith, H.M. & Langebartel, D.A. (1950) Notes on a collection of reptiles and amphibians from the Isthmus of Tehuantepec, Oaxaca. *Journal of the Washington Academy of Sciences* (1949) 39(12), 409–416, 2 figs.

Smith, H.M. & Langebartel, D.A. (1951) A new geographical race of leaf-nosed snake from Mexico. *Herpetologica* 7(3), 181–183.

Smith, H.M. & Larsen, K.R. (1975) Some overlooked rattlesnake *nomina nuda*. *Southwestern Naturalist* 20(2), 285–286.

Smith, H.M. & Laufe, L.E. (1945a) A new South American *Leptotyphlops*. *Proceedings of the Biological Society of Washington* 58, 29–32, figs. a–b, pl. 5.

Smith, H.M. & Laufe, L.E. (1945b) Notes on a herpetological collection from Oaxaca. *Herpetologica* 3(1), 1–13, 1 fig.

Smith, H.M. & Lemos-Espinal, J.A. (2006a) Geographic distribution: Serpentes: *Imantodes gemmistratus*. *Herpetological Review* 37(1), 109.

Smith, H.M. & Lemos-Espinal, J.A. (2006b) Geographic distribution: Serpentes: *Gyalopion quadrangulare*. *Herpetological Review* 37(1), 109.

Smith, H.M. & Lemos-Espinal, J.A. (2006c) A second specimen and first record for Chihuahua, Mexico, of *Tropidodipsas repleta* (Serpentes, Colubridae). *Bulletin of the Maryland Herpetological Society* 42(4), 171–172, 1 fig.

Smith, H.M. & List, J.C. (1958) The type of the blind snake *Stenostoma albifrons*. *Herpetologica* 13(4), 271.

Smith, H.M. & Moll, E.O. (1969) A taxonomic rearrangement of the pit vipers of the *Bothrops nigroviridis* complex of southern Mexico. *Journal of Herpetology* 3(3–4), 151–155, 1 fig.

Smith, H.M. & Necker, W.L. (1943) Alfredo Dugès' types of Mexican reptiles and amphibians. *Anales de la Escuela Nacional de Ciencias Biologicas*, México 3, 180–219, 7 pls.

Smith, H.M. & Shannon, F.A. (1947) Notes on amphibians and reptiles of Michoacán, Mexico. *Fieldiana: Zoology* 31(9), 63–85, fig. 8.

Smith, H.M. & Smith, P.W. (1951) A new snake (*Tantilla*) from the Isthmus of Tehuantepec, Mexico. *Proceedings of the Biological Society of Washington* 64, 97–100, 2 figs.

Smith, H.M. & Smith, P.W. (1964) Range extension of a snake (*Geophis*) in Central America. *Herpetologica* 20(1), 72.

Smith, H.M. & Smith, R.B. (1973) *Synopsis of the herpetofauna of Mexico. Volume II. Analysis of the literature exclusive of the Mexican axolotl.* Eric Lundberg, Augusta, xxxiii + 367 pp.

Smith, H.M. & Smith, R.B. (1976) *Synopsis of the herpetofauna of Mexico. Volume III. Source analysis and index for Mexican reptiles.* John Johnson, North Bennington, 996 pp.

Smith, H.M. & Smith, R.B. (1993) *Synopsis of the herpetofauna of Mexico. Volume VII. Bibliographic addendum IV and index, bibliographic addenda II–IV 1979–1991.* University Press of Colorado, Boulder, ix + 1082 pp.

Smith, H.M. & Tanner, W.W. (1944) Description of a new snake from Mexico. *Copeia* 1944(3), 131–136, 1 fig.

Smith, H.M. & Taylor, E.H. (1941) A review of the snakes of the genus *Ficimia*. *Journal of the Washington Academy of Sciences* 31(8), 356–368, 17 figs.

Smith, H.M. & Taylor, E.H. (1945) An annotated checklist and key to the snakes of México. *Bulletin of the United States National Museum* (187), iv + 239 pp.

Smith, H.M. & Taylor, E.H. (1950a) Type localities of Mexican reptiles and amphibians. *University of Kansas Science Bulletin* 33(8), 313–380.

Smith, H.M. & Taylor, E.H. (1950b) An annotated checklist and key to the reptiles of Mexico exclusive of the snakes. *Bulletin of the United States National Museum* (199), v + 253 pp.

Smith, H.M. & Wallach, V. (1993) The date of publication of *Jaltris* Cope (Reptilia, Serpentes). *Bulletin of the Maryland Herpetological Society* 29(3), 80–84.

Smith, H.M. & Wallach, V. (1994) *Lycognathophis* Boulenger, 1893 (Reptilia, Serpentes): proposed conservation. *Bulletin of Zoological Nomenclature*, London 51(4), 330–331.

Smith, H.M. & Wallach, V. (1997) *Diemenia atra* Macleay, 1884 (currently *Demansia atra*, Reptilia, Serpentes): proposed conservation of the specific name. *Bulletin of Zoological Nomenclature*, London 54(1), 31–14.

Smith, H.M. & Werler, J.E. (1970) The status of the northern red black-headed snake, *Tantilla diabola* Fouquette and Potter. *Journal of Herpetology* 3(3), 172–173, 1 fig.

Smith, H.M. & White, F.N. (1955) Adrenal enlargement and its significance in the hognose snakes *(Heterodon)*. *Herpetologica* 11(2), 137–144.

Smith, H.M. & Williams, K.L. (1963) New and noteworthy amphibians and reptiles from southern México. *Herpetologica* 19(1), 22–27, 1 fig.

Smith, H.M. & Williams, K.L. (1966a) A new snake (*Geophis*) from México. *Journal of the Ohio Herpetological Society* 5(3), 90–92, 1 fig.

Smith, H.M. & Williams, K.L. (1966b) The ratsnake of the Bay Islands, Honduras. *Natural History Miscellanea*, Chicago (185), 1–2.

Smith, H.M. & Williams, K.L. (1966c) A new snake (*Tantilla*) from Las Islas de La Bahia, Honduras. *Southwestern Naturalist* 11(4), 483–487, 1 fig.

Smith, H.M., Nixon, C.W. & Smith, P.W. (1950) Mexican and Central American garter snakes (*Thamnophis*) in the British Museum (Natural History). *Zoological Journal of the Linnean Society,* London 41(282), 571–584.

Smith, H.M., Langebartel, D.A. & Williams, K.L. (1964) Herpetological type-specimens in the University of Illinois Museum of Natural History. *Illinois Biological Monographs*, Urbana (32), 1–80.

Smith, H.M., Lynch, J.D. & Altig, R. (1965a) New and noteworthy herpetozoa from southern Mexico. *Natural History Miscellanea*, Chicago (180), 1–4, 2 figs.

Smith, H.M., Maslin, T.P. & Brown, R.L. (1965b) *Summary of the distribution of the herpetofauna of Colorado. A supplement to an annotated check list of the amphibians and reptiles of Colorado*. University of Colorado Press, Boulder (15), 1–52.

Smith, H.M., Arndt, R.G. & Sherbrooke, W.C. (1967) A new snake of the genus *Enulius* from Mexico. *Natural History Miscellanea*, Chicago (186), 1–4.

Smith, H.M., Williams, K.L. & Pérez-Higareda, G. (1986a) The specific name for the Linnaean *Oxyrhopus*, or the calico false coral snake. *Bulletin of the Maryland Herpetological Society* 22(1), 10–13.

Smith, H.M., Fitzgerald, K., Pérez-Higareda, G. & Chiszar, D. (1986b) A taxonomic rearrangement of the snakes of the genus *Scaphiodontophis*. *Bulletin of the Maryland Herpetological Society* 22(4), 159–166, 2 figs.

Smith, H.M., Chiszar, D. & Pérez-Higareda, G. (1989) Variation in the northern subspecies of *Pliocercus elapoides*. *Bulletin of the Chicago Herpetological Society* 24(6), 112–116, 2 figs.

Smith, H.M., Hammerson, G.A., Roth, J.J. & Chiszar, D. (1991) Distributional addenda for the smooth green snake (*Opheodrys vernalis*) in western Colorado, and the status of its subspecies. *Bulletin of the Maryland Herpetological Society* 27(2), 99–106, 1 fig.

Smith, H.M., Byers, G.W., Williams, K.L. & Wallach, V. (1992) The nomenclatural status of *Limnophis* Günther, 1865 (Reptilia, Serpentes). *Bulletin of the Maryland Herpetological Society* 28(1), 4–7.

Smith, H.M., Flores-Villela, O. & Chiszar, D. (1993a) New variational extremes for *Tantilla calamarina* and a locality record correction for *Conophis vittatus viduus* (Reptilia, Serpentes). *Bulletin of the Maryland Herpetological Society* 29(1), 1–3.

Smith, H.M., Dixon, J.R. & Wallach, V. (1993b) *Dromicus giganteus* is a *Nomen Nudum*. *Bulletin of the Maryland Herpetological Society* 29(3), 77–79.

Smith, H.M., Flores-Villela, O. & Chiszar, D. (1993c) The generic allocation of *Tantilla canula* (Repatilia, Serpentes). *Bulletin of the Maryland Herpetological Society* 29(3), 126–129.

Smith, H.M., Dixon, J.R. & Wallach, V. (1994a) *Coluber poecilogyrus* Wied-Neuwied [1824] (currently *Liophis poecilogyrus*) (Reptilia, Serpentes): proposed conservation of the specific name. *Bulletin of Zoological Nomenclature*, London 51(3), 250–252.

Smith, H.M., Chiszar, D., Staley, J.R., II & Tepedelen, K. (1994b) Populational relationships in the corn snake *Elaphe guttata* (Reptilia, Serpentes). *Texas Journal of Science* 46(3), 259–292, 6 figs.

Smith, H.M., Ota, H. & Wallach, V. (1995a) *Phyllophis carinata* Günther, 1864 (currently *Elaphe carinata*, Reptilia, Serpentes): proposed conservation of the specific name. *Bulletin of Zoological Nomenclature* 52(2), 166–169.

Smith, H.M., Ramírez-Bautista, A. & Chiszar, D. (1995b) The nomenclatural status of Müller's two varietal names of *Dipsas cenchoa* (Reptilia, Serpentes). *Bulletin of the Maryland Herpetological Society* 31(4), 198–203.

Smith, H.M., Wallach, V. & Chiszar, D. (1995c) Observations of the snake genus *Pliocercus*, I. *Bulletin of the Maryland Herpetological Society* 31(4), 204–214.

Smith, H.M., Pérez-Higareda, G. & Chiszar, D. (1996) Observations on the snake genus *Pliocercus*, II. The endemic species of Los Tuxtlas, Veracruz, Mexico. *Bulletin of the Maryland Herpetological Society* 32(2), 75–79.

Smith, H.M., Chiszar, D. & Auth, D.L. (1997) Geographic distribution: Serpentes: *Tantilla calamarina*. *Herpetological Review* 28(3), 160.

Smith, H.M., Chiszar, D. & Breukelen, F.V. (1998a) Resurrection of *Tantilla triseriata* (Reptilia, Serpentes) of Mexico. *Southwestern Naturalist* 43(3), 374–375.

Smith, H.M., van Breukelen, F., Auth, D.L. & Chiszar, D. (1998b) A subspecies of the Texas blind snake (*Leptotyphlops dulcis*) without supraoculars. *Southwestern Naturalist* 43(4), 437–440, 2 figs.

Smith, H.M., Brown, L.E., Chiszar, D., Grismer, L.L., Allen, G.S., Fishbein, A., Hollingsworth, B.D., McGuire, J.A., Wallach, V., Strimple, P. & Liner, E.A. (1998c) *Crotalus ruber* Cope, 1892 (Reptilia, Serpentes): proposed precedence of the specific name over that of *Crotalus exsul* Garman, 1884. *Bulletin of Zoological Nomenclature* 55(4), 229–232.

Smith, H.M., Chiszar, D., Clary, J. & Martelli, J.-L. (1999) Geographic distribution: Serpentes: *Lampropeltis triangulum dixoni*. *Herpetological Review* 30(2), 113.

Smith, H.M., Chiszar, D. & Humphrey, R. (2000) *Pseudoleptodeira* (Serpentes) in Chiapas, Mexico. *Bulletin of the Maryland Herpetological Society* 36(4), 119–110.
Smith, H.M., Moreno, M.M. & Chiszar, D. (2001a) *Rhadinaea hesperia* (Serpentes) in Oaxaca, Mexico. *Bulletin of the Maryland Herpetological Society* 37(1), 1–2.
Smith, H.M., Chiszar, D. & Moreno, M.M. (2001b) Nomenclature of the earth snakes (*Adelphicos*) of the *A. quadrivirgatus* complex. *Bulletin of the Maryland Herpetological Society* 37(2), 39–41.
Smith, H.M., Chiszar, D., Tepedelen, K. & Breukelen, F. van (2001c) A revision of the bevelnosed boas (*Candoia carinata* complex) (Reptilia, Serpentes). *Hamadryad* 26(2), 283–315, 9 figs.
Smith, H.M., Chiszar, D., Eckerman, C.M. & Walley, H.D. (2003) The taxonomic status of the Mexican hognosed snake *Heterodon kennerlyi* Kennicott (1860). *Journal of Kansas Herpetology* (5), 17–20, 1 fig.
Smith, H.M., Lemos-Espinal, J.A., Hartman, D. & Chiszar, D. (2005a) A new species of *Tropidodipsas* (Serpentes, Colubridae) from Sonora, Mexico. *Bulletin of the Maryland Herpetological Society* 41(1), 39–41.
Smith, H.M., Chiszar, D. & Lemos-Espinal, J.D. (2005b) Geographic distribution: Serpentes: *Pituophis deppei deppei*. *Herpetological Review* 36(1), 83.
Smith, H.M., Lemos-Espinal, J.A. & Heimes, P. (2005c) Amphibians and reptiles from northwestern Mexico. *Bulletin of the Chicago Herpetological Society* 40(11), 206–212.
Smith, H.M., Lemos-Espinal, J.A. & Chiszar, D. (2006) *Procinura. P. aemula. Catalogue of American amphibians and reptiles*, pp. 820.1–820.4, 1 fig.
Smith, L.A. (1974) The sea snakes of western Australia (Serpentes, Elapidae, Hydrophiinae) with a description of a new subspecies. *Records of the Western Australian Museum*, Perth 3(2), 93–110.
Smith, L.A. (1976) The reptiles of Barrow Island. *Western Australian Naturalist*, Perth 13(6), 125–136.
Smith, L.A. (1978) The elapid snakes *Denisonia pallidiceps* and *Denisonia suta* in the Kimberleys of Western Australia. *Western Australian Naturalist*, Perth 14(3), 75.
Smith, L.A. (1980) Taxonomy of *Denisonia punctata* and *Denisonia fasciata* (Serpentes, Elapidae). *Records of the Western Australian Museum*, Perth 8(2), 327–333, 1 fig., 2 pls.
Smith, L.A. (1981a) A revision of the python genera *Aspidites* and *Python* (Serpentes, Boidae) in Western Australia. *Records of the Western Australian Museum*, Perth 9(2), 211–226, 7 figs.
Smith, L.A. (1981b) A revision of the *Liasis olivaceus* species-group (Serpentes, Boidae) in Western Australia. *Records of the Western Australian Museum*, Perth 9(2), 227–233, 2 figs.
Smith, L.A. (1982) Variation in *Pseudechis australis* (Serpentes, Elapidae) in Western Australia and description of a new species of *Pseudechis*. *Records of the Western Australian Museum*, Perth 10(1), 35–45, 2 figs.
Smith, L.A. (1985) A revision of the *Liasis childreni* species-group (Serpentes, Boidae). *Records of the Western Australian Museum*, Perth 12(3), 257–276, 7 figs.
Smith, L.A. (1986) A new species of *Hydrophis* (Serpentes, Hydrophiidae) from north-west Australian waters. *Records of the Western Australian Museum*, Perth 13(1), 151–153, 2 figs.
Smith, L.A. (1988) *Lycodon aulicus capucinus*, a colubrid snake introduced to Christmas Island, Indian Ocean. *Records of the Western Australian Museum*, Perth 14, 251–252.
Smith, L.A. & Sidik, I. (1998) Description of a new species of *Cylindrophis* (Serpentes, Cylindrophiidae) from Yamdena Island, Tanimbar Archipelago, Indonesia. *Raffles Bulletin of Zoology* 46(2), 419–424, 4 figs.
Smith, M.A. (1914a) Occurrence of the krait (*Bungarus candidus*) and the small-spotted coral snake (*Calliophis maculiceps*) in Siam. A new color variety of the latter. *Journal of the Natural History Society of Siam*, Bangkok 1(2), 123–125.
Smith, M.A. (1914b) The snakes of Bangkok. *Journal of the Natural History Society of Siam*, Bangkok 1(1), 5–18, 3 pls.
Smith, M.A. (1914c) The snakes of Bangkok. *Journal of the Natural History Society of Siam*, Bangkok 1(2), 93–104, 7 figs.
Smith, M.A. (1914d) On the breeding habits of *Hypsirhina enhydris* and *Herpeton tentaculatum* (the tentacle snake). *Journal of the Natural History Society of Siam*, Bangkok 1(2), 126–127.
Smith, M.A. (1915a) Reptilies and batrachians. *In:* Gardiner, K.G. Notes on the fauna and flora of Ratburi and Petchaburi Districts. (continued). *Journal of the Natural History Society of Siam,* Bangkok 1(3), 131–156.
Smith, M.A. (1915b) The snakes of Bangkok (conclusion). *Journal of the Natural History Society of Siam*, Bangkok 1(3), 173–187, 4 pls.
Smith, M.A. (1915c) List of the snakes at present known to inhabit Siam. *Journal of the Natural History Society of Siam*, Bangkok 1(3), 211–215.
Smith, M.A. (1915d) A new snake from Bangkok. *Journal of the Natural History Society of Siam*, Bangkok 1(3), 255–256.
Smith, M.A. (1915e) Notes on some snakes from Siam. *The Journal of the Bombay Natural History Society* (1914–1915) 23(4), 784–789.
Smith, M.A. (1916) Descriptions of three new lizards and a new snake from Siam. *Journal of the Natural History Society of Siam*, Bangkok 2(1), 44–47, 1 pl.
Smith, M.A. (1917a) Description of new reptiles and a new batrachian from Siam. *Journal of the Natural History Society of Siam*, Bangkok 2(3), 221–225, 2 pls.
Smith, M.A. (1917b) Descriptions of a new snake and a new frog from Siam. *Journal of the Natural History Society of Siam*, Bangkok 2(4), 276–278.
Smith, M.A. (1917c) Preliminary diagnoses of four new sea snakes. *Journal of the Natural History Society of Siam*, Bangkok 2(4), 340–342.
Smith, M.A. (1918) Description of a new snake (*Opisthotropis spenceri*) from Siam. *Journal of the Natural History Society of Siam*, Bangkok 3(1), 13, 1 pl.
Smith, M.A. (1920) On sea-snakes from the coasts of the Malay Peninsula, Siam and Cochin China. *Journal of the Federated Malay States Museums*, Kuala Lumpur 10, 1–63, 1 pl.
Smith, M.A. (1921a) New or little-known reptiles and batrachians from southern Annam (Indo-China). *Proceedings of the Zoological Society of London* 91(2), 423–440, 2 pls.
Smith, M.A. (1921b) Two new batrachians and a new snake from Borneo and the Malay Peninsula. *Journal of the Federated Malay States Museums*, Kuala Lumpur 10(3), 197–199, pl. 2.
Smith, M.A. (1922a) Notes on reptiles and batrachians from Siam and Indo-China (No. 1). *Journal of the Natural History Society of Siam*, Bangkok 4(4), 203–214, pl. 2
Smith, M.A. (1922b) On a collection of reptiles and batrachians from the mountains of Pahang, Malay Peninsula. *Journal of the Federated Malay States Museums*, Kuala Lumpur 10(4), 263–282.
Smith, M.A. (1923a) On a collection of reptiles and batrachians from the island of Hainan. *Journal of the Natural History Society of Siam*, Bangkok 6, 195–212, 2 figs.

Smith, M.A. (1923b) The poisonous land snakes of Siam. *Journal of the Natural History Society of Siam*, Bangkok 6, 55–64, 1 pl.

Smith, M.A. (1925a) Contributions to the herpetology of Borneo. *Sarawak Museum Journal*, Kuching 3(1), 15–34, 1 fig.

Smith, M.A. (1925b) On a collection of reptiles and amphibians from Mt. Murud, Borneo. *Sarawak Museum Journal*, Kuching 3(8), 5–14, 1 pl.

Smith, M.A. (1926a) *Monograph of the sea-snakes (Hydrophiidae)*. British Museum (Natural History), London, xvii + 130 pp., 35 figs., 2 pls.

Smith, M.A. (1926b) "Spolia Mentawia": Reptiles and amphibians. *Annals and Magazine of Natural History*, London (Series 9) 18(103), 76–80.

Smith, M.A. (1927) Contributions to the herpetology of the Indo-Australian region. *Proceedings of the Zoological Society of London* 97(1), 199–225, 4 figs., 2 pls.

Smith, M.A. (1928a) Two vipers new to Siam. *Journal of the Natural History Society of Siam*, Bangkok, Supplement (7), 194.

Smith, M.A. (1928b) The status of some recently described genera and species of snakes. *Annals and Magazine of Natural History*, London (Series 10) 1(4), 494–497.

Smith, M.A. (1929) Remarks on three rare reptiles from the Indo-Chinese region. *Journal of the Natural History Society of Siam*, Bangkok Supplement 8(1), 49–50.

Smith, M.A. (1930a) The Reptilia and Amphibia of the Malay Peninsula. *Bulletin of the Raffles Museum*, Singapore (3), xviii + 149 pp., 13 figs.

Smith, M.A. (1930b) Two new snakes from Tonkin, Indo-China. *Annals and Magazine of Natural History*, London (Series 10) 6(36), 681–683, 1 fig.

Smith, M.A. (1930c) Amphibia and Reptilia. *In: Zoological Record* (1929) 66(14), 1–40.

Smith, M.A. (1931a) Description of a new genus of sea-snake from the coast of Australia, with a note on the structures providing for complete closure of the mouth in aquatic snakes. *Proceedings of the Zoological Society of London* 101(2), 397–398, 1 pl.

Smith, M.A. (1931b) The herpetology of Mt. Kinabalu, North Borneo, 13,455 ft. *Bulletin of the Raffles Museum*, Singapore (5), 3–33, 2 pls.

Smith, M.A. (1935) The sea snakes (Hydrophiidae). *Dana-Report Number* 8. Carlsberg Foundation, Copenhagen, 6 pp., 2 figs.

Smith, M.A. (1937) The names of two Indian vipers. *The Journal of the Bombay Natural History Society* (1936–1937) 39(4), 730–731.

Smith, M.A. (1938) The nucho-dorsal glands of snakes. *Proceedings of the Zoological Society of London* 107B(4), 575–583, 1 pl.

Smith, M.A. (1939) A revision of the Acrochordinae (snakes). *Annals and Magazine of Natural History*, London (Series 11) 3(15), 393–395.

Smith, M.A. (1940a) Contributions to the herpetology of Afghanistan. *Annals and Magazine of Natural History*, London (Series 11) 5(28), 382–384.

Smith, M.A. (1940b) A new snake of the genus *Anomochilus* from the Malay Peninsula. *Annals and Magazine of Natural History*, London (Series 11) 6(35), 447–449, figs. a–c.

Smith, M.A. (1940c) The amphibians and reptiles obtained by Mr. Ronald Kaulback in upper Burma. *Records of the Indian Museum*, Calcutta 42(3), 465–486, 1 pl.

Smith, M.A. (1940d) A facsimile of R.H. Beddome's articles on Indian reptiles 1862–1870. *Journal of the Society for the Bibliography of Natural History*, London 1(10), 273–334, 10 pls.

Smith, M.A. (1941) The herpetology of the Andaman and Nicobar Islands. *Proceedings of the Linnaean Society of London* 153, 150–158.

Smith, M.A. (1943) *The fauna of British India, Ceylon and Burma, including the whole of the Indo-Chinese sub-region. Reptilia and Amphibia. Volume III. Serpentes*. Taylor and Francis, London, xii + 583 pp., 166 figs.

Smith, M.A. (1949) A new species of pit viper from South India, *Trimeresurus huttoni* sp. nov. *The Journal of the Bombay Natural History Society* (1948–1949) 48(3), 596.

Smith, M.A. (1951) On a collection of amphibians and reptiles from Nepal. *Annals and Magazine of Natural History*, London (Series 12) 4(43), 726–728.

Smith, M.A. (1954) *The British amphibians and reptiles*. Second edition, Collins, London, xiv + 322 pp., 91 figs., 32 pls. [first edition, 1951, 315 pp., 88 figs., 33 pls., 18 col. pls.]

Smith, M.A. & Battersby, J.C. (1953) On a collection of amphibians and reptiles from Nepal. *Annals and Magazine of Natural History*, London (Series 12) 6(69), 702–704.

Smith, M.A. & Kloss, C.B. (1915) On reptiles and batrachians from the coast and islands of south-east Siam. *Journal of the Natural History Society of Siam*, Bangkok 1(4), 237–249.

Smith, M.A. & Procter, J.B. (1921) On a collection of reptiles and batrachians from the island of Ceram, Indo-Australian Archipelago. *Annals and Magazine of Natural History*, London (9) 7(40), 352–355.

Smith, M.J. (1975) The vertebrae of four Australian elapid snakes (Squamata, Elapidae). *Transactions of the Royal Society of South Australia*, Adelaide 99(2), 71–84, 9 figs.

Smith, M.J. (1976) Small fossil vertebrates from Victoria Cave, Naracoorte, South Australia. IV. Reptiles. *Transactions of the Royal Society of South Australia*, Adelaide 100(1), 39–51, 6 figs.

Smith, M.J. (1985) *Wonambi naracoortensis* Smith, 1976, the giant Australian python. *In:* Rich, P.V. & van Tets, G.P. (Eds.), *Kadimakara. Extinct vertebrates of Australia*. Pioneer Design Studio, Lilydale, Victoria, pp. 156–159.

Smith, M.J. & Plane, M. (1985) Pythonine snakes (Boidae) from the Miocene of Australia. *BMR Journal of Australian Geology and Geophysics* 9, 191–195, 2 figs.

Smith, P.W. (1961) The amphibians and reptiles of Illinois. *Illinois Natural History Survey Bulletin*, Urbana 28(1), 1–298, 252 figs.

Smith, P.W. & Smith, H.M. (1963) The systematic status of the lined snake of Iowa. *Proceedings of the Biological Society of Washington* 76, 297–304, 1 fig.

Smith, T. (1989) *Elaphe mandarina* (Cantor, 1842): a progress report on a problematic species in captivity. *British Herpetological Society Bulletin*, London (29), 42–48, 1 fig., 2 pls.

Smith, T. (1998) The blue beauty snake: breeding an unusual color morph of the stripe-tailed ratsnake, *(Elaphe taeniura). The Vivarium* 9(1), 38–42, 70, 7 col. photos.

Smith, W.H. (1879) *Catalogue of the Reptilia and Amphibia of Michigan*. Science News (Supplement) 1879, iii–viii.

Smith, W.H. (1882) Report on the amphibians and reptiles of Ohio. *Report of the Geological Survey of Ohio* 4, 629–734, 8 figs.

Smout, A.-M. & Pritchard, S. (1995) *The amphibians & reptiles of Fife: a provisional atlas*. Fife Nature, Glenrothes, 20 pp., illust.

Snellings, E.H.K., Jr. (2012) *The diminutive giant: the dusky pygmy rattlesnake,* Sistrurus miliarius barbouri. ECO Herpetological Publishing, Rodeo, 60 pp., col. illust.

Snyder, D.H. (1972) *Amphibians and reptiles of the Land Between the Lakes*. Tennessee Valley Authority, Golden Pond, 90 pp., 20 figs., 90 col. photos.

Sobolevsky, N.I. (1929) The herpetofauna of the Talysyh and of the Lenkoran lowland. *Mémoires de la Société Impériale des Amis des Sciences Naturelles de Moscou* 1929, 7–141. (in Russian)

Sochurek, E. (1956) Einiges über den Seefrosch und über die Würfelnatter. *Bürgenländische Heimatblätter* 18(2), 88–91, 1 pl.

Sochurek, E. (1974) *Vipera ammodytes gregorwallneri* n. ssp. *Cerastes cerastes karlhartli* n. ssp. *Herpetologische Blätter*, Wien 1(1), 1–3, 2 figs.

Sochurek, E. (1976) Zur systematischen Stellung der alpinen Hornotter. *Carinthia II*, Klagenfurt 166/86, 447–452, 4 figs.

Sochurek, E. (1979a) Kritische Liste der Giftschlangen Europas mit Beschreibung einer neuen Unterart. *Mitteilungen aus dem Zoologischen Gesellschaft Braunau*, Braunau-am-Inn 3(8–9), 213–218.

Sochurek, E. (1979b) Die Schlangen Nordafrikas. *Mitteilungen aus dem Zoologischen Gesellschaft Braunau*, Braunau-am-Inn 3(8–9), 219–226.

Soderberg, P. (1965) The seven major poisonous snakes of Thailand. Part I, the cobra (*Naja naja kaouthia*). *Conservation News, South East Asia*, Bangkok (6), 19–21, 1 fig., 5 illust.

Soderberg, P. (1966a) Two little known snakes from Khao Yai. *Natural History Bulletin of the Siam Society*, Bangkok 21(1–2), 35–43, 2 figs.

Soderberg, P. (1966b) Confirmation of two colour phases of the tentacled snake (*Erpeton tentaculatum*). *Natural History Bulletin of the Siam Society*, Bangkok 21, 342.

Soderberg, P. (1966c) The seven major poisonous snakes of Thailand. Part II, king cobra and the banded krait. *Conservation News, South East Asia*, Bangkok (7), 60–61, 3 illust.

Soderberg, P. (1967a) Notes on a collection of herpetological specimens recently donated to the centre for Thai National Reference Collections. *Natural History Bulletin of the Siam Society*, Bangkok 22(1–2), 151–166.

Soderberg, P. (1967b) An annotated list of the snakes of Chon Buri Province, Thailand. *The Bulletin of the Student Science Society, International School*, Bangkok 2(1), 24–29, 6 pls.

Soderberg, P. (1973) On eleven Asian elapid snakes with special reference to their occurrence in Thailand. *Natural History Bulletin of the Siam Society*, Bangkok 24(3–4), 203–298, pls. 28–36.

Soini, P. (1974) Polychromatism in a population of *Micrurus langsdorffi*. *Journal of Herpetology* 8(3), 267–269.

Sokolov, V.E. (Ed.). (1988) *Dictionary of animal names in five languages: amphibians and reptiles.* Russky Yazyk Publishers, Moscow, 555 pp. (in Latin, Russian, English, German, French)

Solano-Zavaleta, I., García-Vázquez, U. & Canseco-Márquez, L. (2006) Geographic distribution: Serpentes: *Ramphotyphlops braminus*. *Herpetological Review* 37(4), 500.

Solís, F.A. (1991) Comentarios sobre dos especies de serpientes en Panama, *Tantilla alticola* (Boulenger) y *Bothrops picadoi* (Dunn). *Scientia*, San José 6(2), 107–110, 1 fig.

Solís, F.A., Ibáñez-D., R. & Jaramillo, C.A. (1996) Establishing a new uniformity on the checklists of Panamanian snakes. Estableciendo una uniformidad sobre las especies de serpientes presentes en Panama. *Revista de Biologia Tropical*, San José 44(1), 19–22.

Sollas, I.B.J. (1906) Reptilia and Batrachia. *In: Zoological Record* (1905) 42(4), 1–40.

Solórzano, A. (1995) Una nueva especie de serpiente venenosa terrestre del género *Porthidium* (Serpentes, Viperidae), del Suroeste de Costa Rica. *Revista de Biologia Tropical*, San José 42(3), 695–701, 3 figs.

Solórzano, A. (2001a) Geographic distribution: Serpentes: *Coniophanes bipunctatus*. *Herpetological Review* 33(4), 278.

Solórzano, A. (2001b) Una nueve especie de serpiente del genero *Sibon* (Serpentes: Colubridae) de la vertiente de Caribe de Costa Rica. *Revista de Biologia Tropical*, San José 49(3–4), 1111–1120.

Solórzano, A. (2004) *Serpientes de Costa Rica: distribución, taxonomía e historía natural. Snakes of Costa Rica: distribution, taxonomy, and natural history.* Instituto Nacional de Biodiversidad (NBio), Santo Domingo de Heredía, 792 pp., 300 figs.

Solórzano, A. & Cerdas, L. (1985) Confirmación de la presencia de *Micrurus clarki* Schmidt (Elapidae) en Costa Rica. *Revista de Biologia Tropical*, San José (1984) 32(2), 317–318, 1 fig.

Solórzano, A. & Cerdas, L. (1986) A new subspecies of the bushmaster, *Lachesis muta*, from southeastern Costa Rica. *Journal of Herpetology* 20(3), 463–466, 3 figs.

Solorzano, A., Santana, M. & Greene, H.W. (1987) Reproductive biology of *Lachesis muta stenophrys* (Serpientes: Viperidae) in Costa Rica. *In: Program and Abstracts. 1987 Joint Annual Meeting, Society for the Study of Amphibians and Reptiles, Herpetologists' League, and Comite Herpetologico Nacional, Veracruz, Mexico.* Instituto de Biologia, Universidad Autonoma de Mexico. Non-paginated.

Solórzano, A., Gutierrez, J.M. & Cerdas, L. (1989) Notes on the natural history and karyotype of the colubrid snake, *Hydromorphus concolor* Peters, from Costa Rica. *Journal of Herpetology* 23(3), 314–315, fig. 1.

Solórzano, A., Gómez, L.D., Monge-Nájera, J. & Crother, B.I. (1998). Redescription and validation of *Bothriechis supraciliaris* (Serpentes: Viperidae). *Revista de Biologia Tropical* 46(2), 453–462, 4 figs.

Soman, P.W. (1962) Extension of range of olivaceus smooth snake [*Rhabdops olivaceus* (Beddome)]. *The Journal of the Bombay Natural History Society* 59(3), 966.

Somaweera, R. (2006) *The snakes of Sri Lanka.* Wildlife Heritage Trust of Sri Lanka, Colombo, 8 + 297 pp., col. illust. (in Tamil)

Somaweera, R. & Somaweera, N. (2009) An overview of Sri Lankan sea snakes with an annotated checklist and a field key. *Taprobanica,* Homagama 1(1), 43–54, 9 pls.

Song, B. & Shi, Y. (2002) *Amphibia & Reptilia animals of the world in colored illustrations.* China Forestry Publishing House, Beijing, 536 pp., 1307 col. photos. (in Chinese)

Song, G.S. (1961) A new subspecies of a Korean snake. *Saeng-mul*, Pyongyang (2), 58–60. (in Korean)

Song, M.-T. (1987a) Survey of the reptiles of southern Shaanxi. *Acta Herpetologica Sinica*, Chengdu (Series 2) 6(1), 59–64, 6 figs.

Song, M.-T. (1987b) The herpetofauna of Shaanxi Province. *Acta Herpetologica Sinica*, Chengdu (Series 2) 6(4), 63–73, 1 fig.

Song, M.-T. & Chen, F.-G. (1985) A new subspecies of pit-viper from Mt. Qin Ling. *Animal World*, Xian 2(2), 99–103. (in Chinese)

Song, M.-T. & Fang, R.-S. (1982) On collections of amphibians and reptiles from Shangnan-Luonan area, Shaanxi. *Acta Herpetologica Sinica*, Chengdu (Series 2) 1(1), 88–89.

Song, Z.M. (1990) A new snake record to Gansu Province, China. *In:* Zhao, E.-M. (Ed.), *From water onto land.* Society for the Study of Amphibians and Reptiles, p. 221 (in Chinese)

Sonnemann, N. (2000) A breeding history of a captive children's python *(Antaresia childreni). Monitor,* Somerville 11(1), 6–8, 3 figs.

Sonnini, C.S. & Latreille, P.A. (1801a) *Histoire naturelle des reptiles, avec figures dessinees d'apres nature. Tome III. Seconde partie. Serpens.* Deterville, Paris, 335 pp. [reprinted in 1826]

Sonnini, C.S. & Latreille, P.A. (1801b) *Histoire naturelle des reptiles, avec figures dessinees d'apres nature. Tome IV. Seconde partie. Serpens.* Deterville, Paris, 410 pp. [reprinted in 1826]

Soorae, P.S., Al Quarqaz, M. & Gardner, A.S. (2009) An overview and checklist of the native and alien herpetofauna of the United Arab Emirates. *Herpetological Conservation and Biology* 5(3), 529–536, 2 figs.

Sordelli, F. (1886) Rettili di Orta-Keuei (Adrianopoli) raccolti e donati al Civico Museo di Milano dal sig. cav. Luigi De Magistris. *Rendiconti del Reale Istituto Lombardo di Scienze e Lettere,* (Série 2) 19(7), 295-304.

Sorensen, D. (1988) The genus *Eryx. Bulletin of the Chicago Herpetological Society* 23(2), 21–25, 4 figs.

Sosa, R.A., Henderson, R.W. & Powell, R. (1995) *Hypsirhynchus. H. ferox. Catalogue of American amphibians and reptiles*, pp. 617.1–617.4, 1 fig.

Soto, J.G., Perez, J.C. & Minton, S.A. (1988) Proteolytic, hemorrhagic and hemolytic activities of snake venoms. *Toxicon* 26(9), 875–882.

Soubeiran, J.L. (1855) *De la vipière, de son venin et de sa morsure.* Victor Masson, Paris, 156 pp., 9 figs.

Soulé, M. & Sloan, A.J. (1986) Biogeography and distribution of the reptiles and amphibians on islands in the Gulf of California, Mexico. *Transactions of the San Diego Society of Natural History* 14, 137–156, 4 figs.

Sourav, S.H. & Purkayastha, J. (2011) *Xenochrophis cerasogaster* (Cantor, 1839) (Serpentes: Colubridae): distribution extension in Bangladesh. *Check List* 7(3), 352–353, 1 fig.

Sousa, P.A.G. & Freire, E.M.X. (2008) Geographic distribution: Serpentes: *Philodryas patagoniensis. Herpetological Review* 39(2), 243.

Souza, I.F.E. & Silva, N.J. da, Jr. (2002a) Geographic distribution: Serpentes: *Leptophis ahaetulla. Herpetological Review* 33(3), 227–228.

Souza, I.F.E. & Silva, N.J. da, Jr. (2002b) Geographic distribution: Serpentes: *Leptotyphlops koppesi. Herpetological Review* 33(3), 228.

Sowerby, A de C. (1930) *The naturalist in Manchuria. Volume IV. The cold-blooded vertebrates and tunicates of the Manchurian region.* Tientsin Press, Tientsin, 321 pp.

Sowerby, A de C. (1943) Amphibians and reptiles recorded from or known to occur in the Shanghai area. *Notes d'Herpétologie, Musée Heude*, Chang-Hai (1), 1–16, 4 pls.

Sowig, P. (1985) Beiträge zur Kenntnis der Verbreitung und Ökologie der Amphibien und Reptilien Kretas. *Salamandra*, Frankfurt-am-Main 21(4), 252–262, 6 figs.

Spangler, J.A. & Mount, R.H. (1969) The taxonomic status of the natricine snake *Regina septemvittata mabila* (Neill). *Herpetologica* 25, 113–119.

Sparreboom, M. (1981) *De amfibieën en reptielen van Nederland, België en Luxemburg.* A.A. Balkema, Rotterdam, 284 pp., 78 figs.

Sparrman, A. (1795) *Coluber ferruginosus*: en aldeles ny Hugg-Orm, funnen i Södermanland och beskrifven. *Kongliga Vetenskaps-Akademiens nya Handlingar*, Stockholm 16(7–9), 180–183, pl. vii.

Spawls, S. (1978) A checklist of the snakes of Kenya. *Journal of the East Africa Natural History Society and National Museum*, Nairobi 31(167), 1–18.

Spawls, S. (1983) A new *Psammophis* from northern Ghana. *British Journal of Herpetology*, London 6, 311–312, 2 figs.

Spawls, S. (1992) Activity patterns in nocturnal West African savanna snakes. *Journal of the Herpetological Association of Africa*, Stellenbosch (40), 61–66.

Spawls, S. (2004) A new species of *Pseudoboodon* (Reptilia, Serpentes) from the Central Highlands of Ethiopia; with notes on some other members of the genus. *African Journal of Herpetology* 53(1), 13–19, 6 figs.

Spawls, S. & Branch, B. (1995) *The dangerous snakes of Africa.* Ralph Curtis, Sanibel Island, 192 pp., col. illust.

Spawls, S., Howell, K.M., Drewes, R.C. & Ashe, J. (2002) *A field guide to the reptiles of East Africa.* Natural World, Academic Press, San Diego, 543 pp., col. illust.

Spawls, S., Howell, K.M. & Drewes, R.C. (2006) *Reptiles and amphibians of East Africa.* Princeton University Press, Princeton, 240 pp., col. illust.

Speake, D.W., McGlincy, J.A. & Colvin, T.R. (1978) Ecology and management of the eastern indigo snake in Georgia: a progress report. *In:* Odum, R.R. & Landers, L. (Eds.), Proceedings of the Rare and Endangered Wildlife Symposium. *Georgia Department of Natural Resources, Game and Fish Division, Technical Bulletin* WL4, pp. 64–73, 2 figs.

Speck, F.G. (1923) Reptile lore of the northern Indians. *Journal of American Folklore,* Boston 36(141), 273–280.

Spellerberg, I.F. (2002) *Amphibians and reptiles of north-west Europe: their natural history, ecology and conservation.* Science Publishers, Enfield, xii + 203 pp., 6 figs., 28 col. pls.

Spellerberg, I.F. & Phelps, T. (1977) Biology, general ecology and behavior of the snake, *Coronella austriaca* Laurenti. *Biological Journal of the Linnean Society,* London 9(2), 133–164.

Speybroeck, J., Beukema, W. & Crochet, P.-A. (2010) A tentative species list of the European herpetofauna (Amphibia and Reptilia)—an update. *Zootaxa* (2492), 1–27.

Spiteri, D.E. (1988) The geographic variability of the species *Lichanura trivirgata* and a description of a new subspecies. *In:* De Lisle, H.F., Brown, P.R., Kaufman, B. & McGurty, B.M. (Eds.), Proceedings of the conference on California herpetology. *Southwest Herpetological Society Special Publication* (4), 113–130, 14 figs.

Spiteri, D.E. (1991) The subspecies of *Lichanura trivirgata*: why the confusion? *Bulletin of the Chicago Herpetological Society* 26(7), 153–156, 1 fig., 7 photos.

Spiteri, D.E. (1993) The current taxonomy and captive breeding of the rosy boa (*Lichanura trivirgata*). *The Vivarium* 5(3), 18–19, 27, 34, 1 fig., 9 col. photos.

Spiteri, D.E. (1994) A further analysis of the Cedros Island boa, *Lichanura trivirgata bostici. Bulletin of the Chicago Herpetological Society* 29(11), 248–250, 2 figs.

Spix, J.B. von. (1824) *Serpentum brasiliensium species novae ou histoire naturelle des espèces nouvelles de serpens, recueillies et observées pendant le voyage dans l'intérieur du Brésil dans les années 1817, 1818, 1819, 1820, exécuté par ordre de sa majesté le Roi de Bavière.* Typis F. S. Hübschmanni, Monachii, viii + 75 pp., 26 pls.

Spix, J.B. von & von Martius, C.F.P. (1823) *Reise in Brasilien auf Befehl Sr. Majestat Maximilian Joseph I. Konigs von Baiern in den Jahren 1817 bis 1820 gemacht und beschrieben.* Volume 1. Lindauer, Munchen, xiv + 412 pp., 1 atlas.

Srinivasulu, C. & Das, I. (2008) The herpetofauna of Nallamala Hills, Eastern Ghats, India: an annotated checklist, with remarks on nomenclature, taxonomy, habitat use, adaptive types and biogeography. *Asiatic Herpetological Research* 11, 110–131, 1 fig.

St. John, A. (2002) *Reptiles of the Northwest: California to Alaska & Rockies to the Coast.* Lone Pine Publishing, Renton, 272 pp., col. illust.

Stafford, P.J. (1986) *Pythons and boas*. T.F.H. Publications, Neptune City, 192 pp., illust.

Stafford, P.J. (1996) Geographic distribution: Serpentes: *Stenorrhina degenhardtii*. *Herpetological Review* 27(4), 214–215.
Stafford, P.J. (2000a) On the status of the coral snake *Micrurus nigrocinctus* (Serpentes, Elapidae) in Belize, and its northermost distribution in Atlantic Middle America. *Herpetological Review* 31(2), 114.
Stafford, P.J. (2000b) *Snakes*. Smithsonian Institution Press, Washington, 112 pp., col. illust.
Stafford, P.J. (2002) Record of the colubrid snake *Dendrophidion nuchale* (Peters) from Nicaragua. *Southwestern Naturalist* 47(4), 615–616.
Stafford, P.J. (2004) A new species of *Tantilla* (Serpentes, Colubridae) of the Taeniata group from southern Belize. *Journal of Herpetology* 38(1), 43–52, 4 figs.
Stafford, P.J. & Henderson, R.W. (1996) *Kaleidoscopic tree boas: the genus* Corallus *of tropical America*. Krieger Publishing Co., Malabar, 86 pp., 53 col. pls.
Stafford, P.J. & Meyer, J.R. (2000) *A guide to the reptiles of Belize*. Academic Press, San Diego, xix + 356 pp., 31 figs., 181 col. pls.
Stafford, P.J., Walker, P., Edgar, P. & Penn, M.G. (2010) Distribution and conservation of the herpetofauna of Belize. *In:* Wilson, L.D., Townsend, J.H. & Johnson, J.D. (Eds.), *Conservation of Mesoamerican amphibians and reptiles*. Eagle Mountain Publications, Eagle Mountain, pp. 370–404, 3 figs.
Standhardt, B.R. (1986) *Vertebrate paleontology of the Cretaceous/Tertiary transition of Big Bend National Park, Texas*. Doctoral Dissertation, Louisiana State University, Baton Rouge, 298 pp.
Stanek, V.J. (1962) *Introducing non-poisonous snakes*. Golden Pleasure Books, London, 80 pp., 96 photos.
Stanger, M., Clayton, M., Schodde, R., Wombey, J. & Mason, I. (1998) *CSIRO list of Australian vertebrates. A reference with conservation status*. CSIRO Publishing, Collingwood, 123 pp.
Stanley, A.S. (1915) The collections of Chinese reptiles in the Shanghai Museum. *Journal of the North China Branch of the Royal Asiatic Society*, Shanghai (1914) 45, 21–31.
Stanley, A.S. (1917) Two new species of Chinese snakes. *Journal of the North China Branch of the Royal Asiatic Society*, Shanghai (1916) 47, 83–84, 7 unnumbered figs., 1 pl.
Stansbury, H. (1852) *Exploration and survey of the Valley of the Great Salt Lake of Utah, including a reconnaissance of a new route through the Rocky Mountains*. Lippincott & Grambo, Philadelphia, 487 pp., pls. [reprinted in 1853 by Robert Armstrong, Washington]
Stansbury, H. (1854) *Die Mormonen-Ansiedlungen, die Felsenbirge und der Grosse Salzsee, nebst einer Beschreibung der Auswanderer-Strasse und der intersuchungs. Expedition von Howard Stansbury, Captain im Corps der geographischen Ingenieure (Vereinigte-Staaten-Armee)*. Deutsch bearbeitet von Dr. Knottenkamp. Mit einer Karte. Franck'sche Verlagshandlung, Stuttgart, viii + 293 pp.
Starace, F. (1995) Contribution to the study of the ophidians of French Guyana. Presence of a new colubrid in French Guyana, *Xenopholis scalaris* (Wuchener, 1861). *Litteratura Serpentium* 15(6), 153–157, 2 photos.
Starace, F. (1998) *Guide des serpents et amphisbènes de Guyane*. Ibis Rouge Editions, Guadeloupe, 449 pp., col. illust.
Staszko, R. & Walls, J.G. (1994) *Rat snakes, a hobbyist's guide to* Elaphe *and kin*. T.F.H. Publications, Neptune City, 208 pp., col. pls.
Steadman, D.W. (1986) Holocene vertebrate fossils from Isla Floreana, Galapagoes. *Smithsonian Contributions to Zoology* (413), 1–103, 25 figs., 4 pls.
Steadman, D.W., Pregill, G.K. & Olson, S.L. (1984) Fossil vertebrates from Antigua, Lesser Antilles. Evidence for late Holocene human-caused extinctions in the West Indies. *Proceedings of the National Academys of Sciences* 81, 4448–4451, 2 figs.
Stebbing, E.P. (1908) *A manual of elementary forest zoology for India*. Superintendent Government Printing, Calcutta, 229 pp., 329 figs.
Stebbins, R.C. (1954) *Amphibians and reptiles of western North America*. McGraw-Hill, New York, xxii + 528 pp., 104 pls.
Stebbins, R.C. (1959) *Reptiles and amphibians of the San Francisco Bay region*. University of California Press, Berkeley, 72 pp., figs., 8 col. pls.
Stebbins, R.C. (2003) *Western reptiles and amphibians*. Third edition. Houghton Mifflin Company, Boston, xiii + 533 pp., 204 figs., 56 pls. [first edition,1966, 279 pp., 190 figs., 39 pls.; second edition, 1985, xiv + 336 pp., 40 figs., 48 pls.]
Stebbins, R.C. & McGinnis, S.M. (2012) *Field guide to amphibians and reptiles of California*. Second edition. University of California Press, Berkeley, 538 pp., 39 figs., 125 col. pls. [first edition, 1972, 152 pp., figs., 8 col. pls.]
Steehouder, T. (1992a) Successful breeding of an Asian ratsnake (*Elaphe rufodorsata*). *Litteratura Serpentium* 12(2), 42–45.
Steehouder, T. (1992b) A successful cross-breeding with *Psammophis sibilans* and *Psammophis subtaeniatus* in the terrarium. *Litteratura Serpentium* 12(6), 134–144, 3 figs.
Steenis-Kruseman, M.J. van. (1950) *Malaysian plant collectors and collections: being a cyclopaedia of botanical exploration in Malaysia and a guide to the concerned literature up to the year 1950*. Noordhoff-Kolff, Djakarta, 639 pp., 3 pls.
Stein, K.J. & Helmy, I.M. (1994) Some new distribution records for the snakes of Egypt (Squamata, Serpentes). *Bulletin of the Maryland Herpetological Society* 30(1), 15–26, 1 fig.
Steindachner, F. (1864) Über eine neue *Epicrates*-Art aus Columbien. *Denkschriften der Kaiserlichen Akademie der Wissenschaften in Wien* 22(2), 89–93, 1 pl.
Steindachner, F. (1867a) Reptilien. *In: Reise der Österreichischen Fregatte Novara um die Erde in den Jahren 1857, 1858, 1859 unter dem Befehlen des Commodore B. von Wüllerstorf-Urbair. Zoologischer Theil. Erster Band*. Kaiserlich-königlichen Hof- und Staatsdruckerel, Wien, 98 pp, 3 pls.
Steindachner, F. (1867b) Herpetologische Notizen. *Sitzungsberichte der Akademie der Wissenschaften in Wien, Mathematisch-Naturwissenschaftliche Klasse*, Wien 55(1), 265–273, 3 pls.
Steindachner, F. (1867c) Ueber drei neue Schlangenarten. *Verhandlungen der Zoologisch-Botanischen Gesellschaft in Wien* 17, 513–516, pl.
Steindachner, F. (1870a) Herpetologischen Notizen (II). I. Reptilien gesammelt während einer Reise in Senegambien (October bis December 1868). *Sitzungsberichte der Akademie der Wissenschaften in Wien, Mathematisch-Naturwissenschaftliche Klasse*, Wien 62(3), 326–335.
Steindachner, F. (1870b) Herpetologischen Notizen (II). II. Über einige neue oder seltene Reptilien des Wiener Museums. *Sitzungsberichte der Akademie der Wissenschaften in Wien, Mathematisch-Naturwissenschaftliche Klasse*, Wien 62, 336–350, 8 pls.
Steindachner, F. (1876) Die Schlangen und Eidechsen der Galapagos-Inseln. *In: Festschrift zur Feier des Fünfundzwanzigjährigen Bestehens K.K. Zoologisch-Botanischen Gesellschaft in Wien*, pp. 303–330, 7 pls.
Steindachner, F. (1880a) Über eine peruanische Ungalia-Art, *Ungalia Taczanowskyi*. *Sitzungsberichte der Akademie der Wissenschaften in Wien, Mathematisch-Naturwissenschaftliche Klasse*, Wien (1879) 80(1), 522–525, 2 figs.

Steindachner, F. (1880b) Über eine neue Pythonart (*Python breitensteini*) aus Borneo. Vorläufige Mittheilung. *Sitzungsberichte der Akademie der Wissenschaften in Wien, Mathematisch-Naturwissenschaftliche Klasse*, Wien 82(2), 267–268.

Steindachner, F. (1891) Über einige neue und seltene Reptilien- und Amphibien-Arten. *Sitzungsberichte der Akademie der Wissenschaften in Wien, Mathematisch-Naturwissenschaftliche Klasse*, Wien 100(1), 289–315, 2 pls.

Steindachner, F. (1901) Herpetologische und ichthyologische Ergebnisse einer Reise nach Südamerika mit einer Einleitung von Therese Prinzessin von Bayern. *Anzeiger der Akademie der Wissenschaften in Wien, Mathematisch-Naturwissenschaftliche Klasse*, Wien 38(18), 194–196.

Steindachner, F. (1902) Herpetologische und ichthyologische Ergebnisse einer Reise nach Südamerika, mit einer Einleitlung von Therese Prinzessin von Bayern. *Denkschriften der Kaiserlichen Akademie der Wissenschaften in Wien, Mathematisch-Naturwissenschaftliche Klasse*, Wien 38, 89–148, 5 pls.

Steindachner, F. (1903a) Batrachier und Reptilien aus Südarabien und Sokotra. *Sitzungsberichte der Mathematisch-Naturwissenschaftliche Klasse der Kaiserlichen Akademie der Wissenschaften*, Wien, 112(1), 7–14.

Steindachner, F. (1903b) Über einige neue Reptilien– und Fischarten des Hofmuseums in Wien. *Sitzungsberichte der Mathematisch-Naturwissenschaftliche Klasse der Kaiserlichen Akademie der Wissenschaften*, Wien, 112(1), 15–22.

Steindachner, F. (1906) Bemerkungen zu *Liparophis Bedoti*, Peracca und *Lachesis monticola* (Gthr.). *Sitzungsberichte der Akademie der Wissenschaften in Wien, Mathematisch-Naturwissenschaftliche Klasse*, Wien 115(1), 905–909, 1 pl.

Steindachner, F. (1913a) Über zwei neue Schlangenarten von Formosa. *Anzeiger der Akademie der Wissenschaften in Wien, Mathematisch-Naturwissenschaftliche Klasse*, Wien 50(12), 218–220.

Steindachner, F. (1913b) Bericht über die von Hans Sauter auf Formosa gesammelten Schlangenarten. *Denkschriften der Kaiserlichen Akademie der Wissenschaften in Wien* 90, 319–361, 21 figs., 4 pls.

Steinheil, F. (1913a) *Die Europaeischen Schlangen. Kupferdrucktafeln nach Photographien der lebenden Tiere. Erstes Heft.* Gustav Fischer, Jena, 18 pp., 5 pls.

Steinheil, F. (1913b) *Die Europaeischen Schlangen. Kupferdrucktafeln nach Photographien der lebenden Tiere. Zweites Heft.* Gustav Fischer, Jena, 10 pp., pls. 6–10.

Steinheil, F. (1913c) *Die Europaeischen Schlangen. Kupferdrucktafeln nach Photographien der lebenden Tiere. Drittes Heft.* Gustav Fischer, Jena, 4 pp., pls. 11–15.

Steinheil, F. (1914a) *Die Europaeischen Schlangen. Kupferdrucktafeln nach Photographien der lebenden Tiere. Viertes Heft.* Gustav Fischer, Jena, 9 pp., pls. 16–20.

Steinheil, F. (1914b) *Die Europaeischen Schlangen. Kupferdrucktafeln nach Photographien der lebenden Tiere. Fünftes Heft.* Gustav Fischer, Jena, 3 pp., pls. 21–25.

Steinheil, F. (1914c) *Die Europaeischen Schlangen. Kupferdrucktafeln nach Photographien der lebenden Tiere. Sechstes Heft.* Gustav Fischer, Jena, 3 pp., pls. 26–30.

Steinheil, F. & Müller, L. (1927) *Die Europaeischen Schlangen. Kupferdrucktafeln nach Photographien der lebenden Tiere. Siebentes Heft.* Gustav Fischer, Jena, 18 pp., pls. 31–35.

Steinheil, F. & Müller, L. (1928) *Die Europaeischen Schlangen. Kupferdrucktafeln nach Photographien der lebenden Tiere. Achtes Heft.* Gustav Fischer, Jena, 7 pp., pls. 36–40.

Steinheil, F. & Müller, L. (1931) *Die Europaeischen Schlangen. Kupferdrucktafeln nach Photographien der lebenden Tiere. Neuntes Heft.* Gustav Fischer, Jena, 8 pp., pls. 41–45.

Stejneger, L.H. (1889) Diagnosis of a new species of snake (*Lichanura orcutti*), from San Diego County, California. *The Western American Scientist,* San Diego 6(46), 83.

Stejneger, L.H. (1890a) Description of two new species of snakes from California. *Proceedings of the United States National Museum* (1890) 12(766), 95–99, 3 figs.

Stejneger, L.H. (1890b) On a new genus and species of colubrine snakes from North America. *Proceedings of the United States National Museum Museum* (1890–1891) 13(802), 151–155, 1 fig.

Stejneger, L.H. (1891) On the snakes of the Californian genus *Lichanura. Proceedings of the United States National Museum Museum* (1891–1892) 14(878), 511–515.

Stejneger, L.H. (1893) Annotated list of the reptiles and batrachians collected by the Death Valley Expedition in 1891, with description of new species. *North American Fauna* (7), 159–228, 4 pls.

Stejneger, L.H. (1894a) Description of a new species of blind-snake (Typhlopidae) from the Congo Free State. *Proceedings of the United States National Museum* (1893–1894) 16(969), 709–710.

Stejneger, L.H. (1894b) On some collections of reptiles and batrachians from East Africa and the adjacent islands, recently received from Dr. W.L. Abbott and Mr. William Astor Chanler, with descriptions of new species. *Proceedings of the United States National Museum* (1893–1894) 16(970), 711–741.

Stejneger, L.H. (1895) The poisonous snakes of North America. Annual Report of the Smithsonian Institution (1893), 337–488, 19 pls.

Stejneger, L.H. (1896) Description of a new species of snake (*Tantilla eiseni*) from California. *Proceedings of the United States National Museum* (1895–1896) 18(1044), 117–118.

Stejneger, L.H. (1898) On a collection of batrachians and reptiles from Formosa & adjacent islands. *Journal of the Science College, Imperial University, Tokyo*, 12(3), 215–225.

Stejneger, L.H. (1901a) A new systematic name for the yellow boa of Jamaica. *Proceedings of the United States National Museum* (1900–1901) 23(1218), 469–470.

Stejneger, L.H. (1901b) Description of a new species of snake from Clarion Island, west coast of México. *Proceedings of the United States National Museum* (1900–1901) 23(1231), 715–717.

Stejneger, L.H. (1901c) An annotated list of batrachians and reptiles collected in vicinity of La Guaira, Venezuela, with descriptions of two new species of snakes. *Proceedings of the United States National Museum Museum* (1901–1902) 24(1248), 179–192, 33 figs.

Stejneger, L.H. (1901d) Diagnoses of eight new batrachians and reptiles from the Riu Kiu Archipelago, Japan. *Proceedings of the Biological Society of Washington* 14, 189–191.

Stejneger, L.H. (1902a) A new opisthoglyph snake from Formosa. *Proceedings of the Biological Society of Washington* 15, 15–17.

Stejneger, L.H. (1902b) The reptiles of the Huachuca Mountains, Arizona. *Proceedings of the United States National Museum* (1902–1903) 25(1282), 149–158.

Stejneger, L.H. (1903) A new hognose snake from Florida. *Proceedings of the Biological Society of Washington* 16, 123–124.

Stejneger, L.H. (1904) The herpetology of Porto Rico. *Annual Report of the United States National Museum* 1902, 549–724, 196 figs., 1 pl.

Stejneger, L.H. (1907a) Herpetology of Japan and adjacent territory. *Bulletin of the United States National Museum* (58), i–xx, 1–577, 409 figs., 35 pls.

Stejneger, L.H. (1907b) A new calamarine snake from the Philippine Islands. *Smithsonian Miscellaneous Collections* 50, 30–31.

Stejneger, L.H. (1909) Description of a new snake from Panamá. *Proceedings of the United States National Museum* (1909) 36(1681), 457–458.

Stejneger, L.H. (1910) The batrachians and reptiles of Formosa. *Proceedings of the United States National Museum* (1910) 38(1731), 91–114.

Stejneger, L.H. (1925) Chinese amphibians and reptiles in the United States National Museum. *Proceedings of the United States National Museum* (1924–1926) 66(2562), 1–115, 6 figs.

Stejneger, L.H. (1926) Identity of Hallowell's snake genera *Megalops* and *Aepidea*. *Proceedings of the United States National Museum* (1926–1927) 69(2643), 1–3.

Stejneger, L.H. (1927) The green pit viper, *Trimeresurus gramineus*, in China. *Proceedings of the United States National Museum* (1927–1928) 72(2715), 1–10, 2 figs.

Stejneger, L.H. (1929) A new snake from China. *Proceedings of the Biological Society of Washington* 42, 129–130.

Stejneger, L.H. & Barbour, T. (1943) A check list of North American amphibians and reptiles. Fifth edition. *Bulletin of the Museum of Comparative Zoology* 93(1), xix + 260 pp. [first edition, Stejneger, 1917, Harvard University Press, Cambridge, iv + 125 pp.; second edition, 1923, x + 171 pp.; third edition, 1933, xiv + 185 pp; fourth edition, 1939, xvi + 207 pp.]

Stemmler, O. (1959) Schlangen auf Sardinien. *Zeitschrift für Vivaristik*, Manheim 5(3–4), 51–61.

Stemmler, O. (1965) Zur Biologie der Rassen von *Echis carinatus* (Schneider) 1801. *Salamandra*, Frankfurt-am-Main 1(1–2), 29–46, 8 figs.

Stemmler, O. (1967) Die Reptilien der Schweiz mit besonderer Berücksichtigung der Basler Region. *Veröffentlichungen aus dem Naturhistorischen Museum Basel* (5), 1–32, illust.

Stemmler, O. (1969a) Über das Paarungverhalten von *Eryx johnii johnii* (Serpentes, Boidae). *Salamandra*, Frankfurt-am-Main 5, 23–31, 3 figs.

Stemmler, O. (1969b) Die Sandrasselotter aus Pakistan, *Echis carinatus sochureki* subsp. nov. *Aqua Terra*, Solothurn 6(10), 118–124, 4 figs.

Stemmler, O. (1970) Die Sandrasselotter aus Westafrika, *Echis carinatus ocellatus* subsp. nov. (Serpentes, Viperidae). *Revue Suisse de Zoologie*, Genève 77(2), 273–282, 2 figs.

Stemmler, O. (1971a) Zur Haltung von *Echis carinatus leakeyi* Stemmler and Sochurek 1969, der Kenya-Sandrasselotter (Reptilia, Serpentes). *Zoologische Garten*, Leipzig (Series 2) 40(4–5), 200–210, 5 figs.

Stemmler, O. (1971b) Die Eierschlange, *Dasypeltis scabra* (Linnaeus, 1758), eine weitere aethiopische Form in der marokkanischen Herpetofauna (Reptilia, Colubridae). *Zoologische Abhandlungen, Staatliches Museum für Tierkunde*, Dresden 32(6), 69–73, 2 figs.

Stemmler, O. (1972a) Herpetologische Beobachtungen in Marokko. Im Hochtal des Oued Rhirhaia. *Aqua Terra* 9(1), 8–12, 4 pls.

Stemmler, O. (1972b) Bericht über eine zweite Herpetologische sammelreise nach Marokko im Juli und August 1970. *Monitore Zoologico Italiano, Supplemento*, Firenze 4(6), 123–158, 12 pls.

Stemmler, O. (1972c) Herpetologische Beobachtungen in Marokko 13. In den Arganiers-Wäldern des Sous-Tales. *Aqua Terra* (4), 44–47, 3 pls.

Stemmler, O. (1972e) Herpetologische Beobachtungen in Marokko 14. Die Gegend von Imi-n-Ifri (Haut-Atlas). *Aqua Terra* 9(5), 56–60, 3 pls.

Stemmler, O. & Hotz, H. (1973) Bericht über eine herpetologische Sammelreise nach Marokko im Juli 1969. *Verhandlungen der Naturforschenden Gesellschaft in Basel* 83(1), 125–160, 3 figs., 28 pls.

Stemmler, O. & Sochurek, E. (1969) Die Sandrasselotter von Kenya, *Echis carinatus leakeyi* subsp. nov. *Aqua Terra*, Solothurn 6(9), 89–94, 2 figs.

Stemmler, O., Haefelfinger, H.-R. & Kramer, E. (1971) *Der Reptilien der Schweiz*. Second edition. Veröffentlichungen aus dem Naturhistorischen Museum Basel (5), 55 pp., 32 figs. [first edition, 1967, 32 pp., 56 figs.]

Stepánek, O. (1934) Poznámky k herpetologii krety. Sur l'herpétologie de l'ile de Créte. *Sbornik Zoologicke Oddeleni Národniho Musea v Praze* 1(2), 7–10.

Stepánek, O. (1944) Zur herpetologie Griechenlands. *Vestnik Ceské Zoologicke Spolecnosti v Praze* 9, 123–147.

Stepánek, O. (1949) *Obojzivelníci a plazi zemí ceskych se zretelem k faune strední Evropy*. Tiskem B. Styblové V Narodni, Praha, 122 pp., illust.

Sterer, Y. (1992) A mixed litter of horned and hornless horned vipers *Cerastes cerastes cerastes* (Ophidia: Viperidae). *Israel Journal of Zoology* 37(4), 247–249.

Sternfeld, R. (1908a) Neue und ungenügend bekannte afrikanische Schlangen. *Sitzungsberichte der Gesellschaft Naturforschender Freunde zu Berlin* 1908(4), 92–95.

Sternfeld, R. (1908b) Die Schlangenfauna von Kamerun. Mit einer Bestimmungstabelle. *Mitteilungen aus dem Zoologischen Museum in Berlin* 3(4), 397–432, 6 figs.

Sternfeld, R. (1908c) Die Schlangenfauna Togos. Mit einer Bestimmungs tabelle. Bearbetit nach dem Material des Berliner Zoologischen Museums. *Mitteilungen aus dem Zoologischen Museum in Berlin* 4(1), 207–236, 3 figs.

Sternfeld, R. (1908d) Zur Schlangenfauna Ostafrikas. I. Schlangen aus Süd-Abessinien. *Mitteilungen aus dem Zoologischen Museum in Berlin* 4(1), 239–241, 2 figs.

Sternfeld, R. (1908e) Zur Schlangenfauna Ostafrikas. II. Schlangen aus Britisch-Ostafrika. *Mitteilungen aus dem Zoologischen Museum in Berlin* 4(1), 242–245, 7 figs.

Sternfeld, R. (1908f) Zur Schlangenfauna Ostafrikas. III. Schlangen aus Portugiesisch-Ostafrika. *Mitteilungen aus dem Zoologischen Museum in Berlin* 4(1), 246–247.

Sternfeld, R. (1909a) *Die Fauna der deutschen Kolonien. Herausgegeben mit Unterstützung des Reichs-Kolonialamtes vom Zoologischen Museum in Berlin. Reihe I, Kamerun. Heft. 1. Die Schlangen Kameruns*. R. Friedländer & Sohn, Berlin, iv + 28 pp., 34 figs.

Sternfeld, R. (1909b) *Die Fauna der deutschen Kolonien. Herausgegeben mit Unterstützung des Reichs-Koloniatlamtes vom Zoologischen Museum in Berlin. Reihe II. Togo. Heft 1. Die Schlangen Togos*. R. Friedländer & Sohn, Berlin, iv + 29 pp., 42 figs.

Sternfeld, R. (1910a) Zur Schlangenfauna Deutsch-Südwestafrika. *Mitteilungen aus dem Zoologischen Museum in Berlin* 5(1), 51–60, 6 figs.

Sternfeld, R. (1910b) Neue Beiträge zur Schlangenfauna Kameruns. *Mitteilungen aus dem Zoologischen Museum in Berlin* 5(1), 61–65.

Sternfeld, R. (1910c) Neue Schlangen aus Kamerun, Abessinien und Deutsch-Ostafrika. *Mitteilungen aus dem Zoologischen Museum in Berlin* 5(1), 67–70.

Sternfeld, R. (1910d) *Die Fauna der deutschen Kolonien. Herausgeben mit Unterstützung des Reichs-Kolonialamtes vom Zoologischen Museum in Berlin. Reihe III. Deutsch-Ostafrika. Heft 2. Die Schlangen Deutsch-Ostafrikas.* R. Friedländer & Sohn, Berlin, iv + 47 pp., 54 figs.

Sternfeld, R. (1910e) *Die Fauna der deutschen Kolonien. Herausgeben mit Unterstützung des Reichs-Kolonialamtes vom Zoologischen Museum in Berlin. Reihe IV, Deutsch-Südwestafrika. Heft 1. Die Schlangen Deutsch-Südwestafrika.* R. Friedländer & Sohn, Berlin, iv + 45 pp., 50 figs.

Sternfeld, R. (1911) Zur Reptilienfauna Deutsch-Ostafrikas. *Sitzungsberichte der Gesellschaft Naturforschender Freunde zu Berlin* 1911(4), 245–251.

Sternfeld, R. (1913a) Reptilia. *In: Wissenschaftliche Ergebnisse der deutschen Zentral-Afrika-Expedition 1907–1908 unter Führung Adolf Friedrichs Herzogs zu Mecklenburg. Band IV, Zoologie II.* (1912) Klinkhardt & Biermann, Leipzig, pp. 197–280, 4 figs., pls. 6–8.

Sternfeld, R. (1913b) Beiträge zur Schlangenfauna Neuguineas und der benachbarten Inselgruppen. *Sitzungsberichte der Gesellschaft Naturforschender Freunde zu Berlin* 1913 (8), 384–389, 2 figs.

Sternfeld, R. (1913c) *Die Reptilien und Amphibien Mitteleuropas.* (1912) Quelle & Meyer, Leipzig, 81 pp., 19 figs., 30 pls.

Sternfeld, R. (1914) *Die Fauna der deutschen Kolinien. Herausgeben mit Unterstützung des Reichs-Kolonialamtes vom Zoologischen Museum in Berlin. Reihe VI, Deutsche Schutzgebiete in der Südsee. Heft 1. Die Schlangen der deutschen Schutzgebiete in der Südsee.* R. Friedländer & Sohn, Berlin, iv + 47 pp., 36 figs.

Sternfeld, R. (1917) Reptilia und Amphibia. *In: Ergebnisse der zweiten Deutschen Zentral-Afrika-Expedition 1910–1911 unter Führung Adolf Friedrichs, Herzogs zu Mecklenburg. Band I. Zoologie.* Lieferung 11. Klinkhardt & Biermann, Leipzig, pp. 407–509, pls. 22–24.

Sternfeld, R. (1919) Neue Schlangen und Echsen aus Zentralaustralien. *Senckenbergiana*, Frankfurt-am-Main (1918) 1(3), 76–83.

Sternfeld, R. (1920a) Eine neue Schlange der Gattung *Lachesis* aus Südamerika. *Senckenbergiana*, Frankfurt-am-Main 2(6), 179–181, 2 figs.

Sternfeld, R. (1920b) Zur systematik der Schlangen-Gattung *Spilotes*. *Senckenbergiana*, Frankfurt-am-Main 2(6), 181–186, 4 figs.

Sternfeld, R. (1920c) Zur Tiergeographie Papuasiens und der pazifischen Inselwelt. *Abhandlungen der Senckenbergischen Naturforschenden Gesellschaft*, Frankfurt-am-Main 36(4), 373–436, pl. 31.

Sternfeld, R. (1925) Beiträge zur Herpetologie Inner-Australiens. *Abhandlungen der Senckenbergischen Naturforschenden Gesellschaft*, Frankfurt-am-Main 38(3), 221–251.

Sternfeld, R. & Steiner, G. (1952) *Die Reptilien und Amphibien Mitteleuropas.* Quelle & Meyer, Heidelberg, 94 pp., 30 pls.

Steveni (1835) *Coluber cruentatus. Bulletin de la Société Impériale des Naturalistes de Moscou* 8, 317–318, pl.

Stevens, R.A. (1973) A report on the lowland viper, *Atheris superciliaris* (Peters), from the Lake Chilwa floodplain, Malawi. *Arnoldia (Rhodesia)*, Salisbury 6(22), 1–22, 7 figs., 2 pls.

Stevens, R.A. (1974) An annotated check list of the amphibians and reptiles known to occur in south-central Malawi. *Arnoldia (Rhodesia)* 6(30), 1–22.

Steward, J.W. (1971) *The snakes of Europe.* Davis & Charles, Newton Abbot, Devon, 238 pp., 6 figs., illust.

Stewart, G.D. (1954) A small collection of reptiles from central Korea. *Copeia* 1954(1), 65–67.

Stewart, G.D. (1970) *Preliminary diagnosis of two new subspecies of snakes from Korea.* G.D. Stewart, Perry Point, Maryland, 1 p.

Stewart, G.D. (1973) The East African puff adder. *Baltimore University Communication* (1), 1–4, 1 pl.

Stewart, G.R. (1977) *Charina. C. bottae. Catalogue of American amphibians and reptiles*, pp. 205.1–205.2.

Stewart, G.R. (1988) The rubber boa (*Charina bottae*) in California, with particular reference to the southern subspecies, *C. b. umbratica. In:* De Lisle, H.F., Brown, P.R., Kaufman, B. & McGurty, B.M. (Eds.), Proceedings of the conference on California herpetology. *Southwest Herpetological Society Special Publication* (4), 131–138, 5 figs.

Stewart, M.M. (1968) A collection of reptiles from northern Malawi, Africa. *Journal of the Herpetological Association of Africa*, Stellenbosch (2), 22–29.

Stewart, M.M. (1969) Observations on reptiles in northern Malawi. *Society of Malawi Journal* 22(2), 1–11, 1 fig. [reprint of Stewart, 1968]

Stewart, M.M. & Wilson, V.J. (1966) Herpetofauna of the Nyika Plateau (Malawi and Zambia). *Annals of the Natal Museum*, Pietermaritzburg 18, 287–313, 5 figs.

Stickel, W.H. (1938) The snakes of the genus *Sonora* in the United States and Lower California. *Copeia* 1938(4), 182–190.

Stickel, W.H. (1941) The subspecies of the spade-nosed snake, *Sonora occipitalis. Bulletin of the Chicago Academy of Sciences* 6(7), 135–140, 2 figs.

Stickel, W.H. (1943) The Mexican snakes of the genera *Sonora* and *Chionactis* with notes on the status of other colubrid genera. *Proceedings of the Biological Society of Washingtion* 56, 109–127.

Stille, B. (1987) Dorsal scale microdermatoglyphics and rattlesnake (*Crotalus* and *Sistrurus*) phylogeny (Reptilia, Viperidae, Crotalinae). *Herpetologica* 43(1), 98–104, 4 figs.

Stimson, A.F. (1969) Liste der rezenten Amphibien und Reptilien: Boidae (Boinae + Bolyeriinae + Loxoceminae + Pythoninae). *Das Tierreich*, Berlin (89), i–xi + 1–49.

Stirling, E.C. & Zietz, A. (1893) Vertebrata, Reptilia (collected by the Elder Exploring Expedition). *Transactions of the Royal Society of South Australia*, Adelaide 16, 159–176, pls.

Stojanov, A.J., Tzankov, N. & Naumov, B. (2011) *Die Amphibien und Reptilien Bulgariens.* Edition Chimaira, Frankfurt-am-Main, 582 pp., 300 col. photos.

Stoliczka, F. (1870a) Observations on some Indian and Malayan Amphibia and Reptilia. (Abstract.). *Proceedings of the Asiatic Society of Bengal (Natural History),* Calcutta 1870(4), 103–109. [April 1870]

Stoliczka, F. (1870b) Observations on some Indian and Malayan Amphibia and Reptilia. (Abstract.). *Annals and Magazine of Natural History*, London (Series 4) 6(31), 105–109. [July 1870]

Stoliczka, F. (1870c) Observations on some Indian and Malayan Amphibia and Reptilia. [Pt. 2] *Journal of the Asiatic Society of Bengal (Natural History),* Calcutta 39(2), 134–157, pl. 9.

Stoliczka, F. (1870d) Observations on some Indian and Malayan Amphibia and Reptilia. [Pt. 2] *Journal of the Asiatic Society of Bengal (Natural History),* Calcutta 39(3), 159–228, pls. 10–12. [September 1870]

Stoliczka, F. (1871) Notes on some Indian and Burmese ophidians. *Journal of the Asiatic Society of Bengal (Natural History),* Calcutta 40(2), 421–445, pls. 25–26.

Stoliczka, F. (1872a) Notes on the reptilian and amphibian fauna of Kachh [Cutch]. *Proceedings of the Asiatic Society of Bengal (Natural History),* Calcutta 1872(5), 71–85.

Stoliczka, F. (1872b) Notes on reptiles, collected by Surgeon F. Day in Sind. *Proceedings of the Asiatic Society of Bengal (Natural History),* Calcutta 1872(1), 85–92.

Stoliczka, F. (1872c) Notes on a few Barmese species of Sauria, Ophidia and Batrachia. *Proceedings of the Asiatic Society of Bengal (Natural History),* Calcutta 1872(5), 143–147.

Stoliczka, F. (1873) Notes on some species of Malayan Reptilia and Amphibia. *Journal of the Asiatic Society of Bengal (Natural History),* Calcutta 42(2), 111–126, pl. 11.

Stoops, E.D. & Wright, A. (1993) *Snakes and other reptiles of the Southwest.* Second edition. Golden West Publishers, Phoenix, 102 pp., B&W illust., 63 col. photos.

Storer, D.H. (1839) Reptiles of Massachusetts. *In:* Storer, D.H. & Peabody, W. (Eds.), *Reports on the fishes, reptiles and birds of Massachusetts.* Dutton and Wentworth, Boston, pp. 203–256, 410–416, pl. 4. [reprinted in 1840 by Journal of Natural History, Boston 3(1–2), 1–64, 1 pl.]

Storer, D.H. (1840) A report on the reptiles of Massachusetts. *Boston Journal of Natural History* 3(1–2), 1–64, pl. 1.

Storm, R.M., Leonard, W.P., Brown, H.A., Bury, R.B., Darda, D.M., Diller, L.V. & Peterson, C.R. (1995) *Reptiles of Washington and Oregon.* Seattle Audubon Society, Seattle, viii + 176 pp., col. photos.

Storr, G.M. (1964) *Denisonia monachus*, a new elapid snake from Western Australia. *Western Australian Naturalist*, Perth 9(4), 89–90.

Storr, G.M. (1965) The physiography, vegetation and vertebrate fauna of the Wallabi group, Houtman Abrolhos. *Journal of the Royal Society of Western Australia*, Perth 48, 1–14, 3 figs.

Storr, G.M. (1968) The genus *Vermicella* (Serpentes, Elapidae) in Western Australia and the Northern Territory. *Journal of the Royal Society of Western Australia*, Perth (1967) 50(3), 80–92.

Storr, G.M. (1978) Whip snakes (*Demansia*, Elapidae) of western Australia. *Records of the Western Australian Museum*, Perth 6(3), 287–301, 4 figs.

Storr, G.M. (1979) Revisionary notes on the genus *Vermicella* (Serpentes, Elapidae). *Records of the Western Australian Museum*, Perth 8(1), 75–79.

Storr, G.M. (1980) A new *Brachyaspis* (Serpentes, Elapidae) from Western Australia. *Records of the Western Australian Museum*, Perth 8(3), 397–399, 1 fig.

Storr, G.M. (1981a) The genus *Furina* (Serpentes, Elapidae) in Western Australia. *Records of the Western Australian Museum*, Perth 9(2), 119–123, 2 figs.

Storr, G.M. (1981b) The genus *Acanthophis* (Serpentes, Elapidae) in Western Australia. *Records of the Western Australian Museum*, Perth 9(2), 203–210, 4 figs.

Storr, G.M. (1981c) The *Denisonia gouldii* species-group (Serpentes, Elapidae) in Western Australia. *Records of the Western Australian Museum*, Perth 8(4), 501–515, 6 figs.

Storr, G.M. (1981d) The genus *Ramphotyphlops* (Serpentes, Typhlopidae) in Western Australia. *Records of the Western Australian Museum*, Perth 9(3), 235–271, 34 figs.

Storr, G.M. (1982) The genus *Notechis* (Serpentes, Elapidae) in Western Australia. *Records of the Western Australian Museum*, Perth 9(4), 325–340, 6 figs.

Storr, G.M. (1983) A new *Ramphotyphlops* (Serpentes, Typhlopidae) from Western Australia. *Records of the Western Australian Museum*, Perth 10(4), 315–317, 2 figs.

Storr, G.M. (1984a) Revision of *Denisonia suta* (Serpentes, Elapidae) and the description of a new species closely related to it. *Records of the Western Australian Museum*, Perth 11(3), 249–257, 1 fig.

Storr, G.M. (1984b) A new *Ramphotyphlops* (Serpentes, Typhlopidae) from central Australia. *Records of the Western Australian Museum*, Perth 11(3), 313–314, 1 fig.

Storr, G.M. (1985) Phylogenetic relationships of Australian elapid snakes, external morphology with an emphasis on species in Western Australia. *In:* Grigg, G., Shine, R. & Ehmann, H. (Eds), *Biology of Australasian frogs and reptiles.* Royal Zoological Society of New South Wales, Chipping Norton, pp. 221–222.

Storr, G.M. (1988) A new *Rhinoplocephalus* (Serpentes, Elapidae) from Western Australia. *Records of the Western Australian Museum*, Perth 14(1), 137–138.

Storr, G.M. (1989) A new *Pseudonaja* (Serpentes, Elapidae) from Western Australia. *Records of the Western Australian Museum*, Perth 14(3), 421–423, figs.

Storr, G.M. & Hanlon, T.M.S. (1980) Herpetofauna of the Exmouth region, Western Australia. *Records of the Western Australian Museum*, Perth 8(3), 423–439.

Storr, G.M. & Harold, G. (1978) Additions to the herpetofauna of the Shark Bay region, Western Australia. *Western Australian Naturalist*, Perth 14(8), 449–467, illust.

Storr, G.M. & Harold, G. (1980) Herpetofauna of the Zuytdorp coast and hinterland, Western Australia. *Records of the Western Australian Museum*, Perth 8(3), 359–374, illust.

Storr, G.M. & Harold, G. (1985) Herpetofauna of the Onslow region, Western Australia. *Records of the Western Australian Museum*, Perth 12(3), 277–291, illust.

Storr, G.M., Hanlon, T.M.S. & Harold, G. (1981) Herpetofauna of the shores and hinterland of the Great Australian Bight, Western Australia. *Records of the Western Australian Museum*, Perth 9(1), 23–39, illust.

Storr, G.M., Hanlon, T.M.S. & Dunlop, J.N. (1983) Herpetofauna of the Geraldton region, Western Australia. *Records of the Western Australian Museum*, Perth 10(3), 215–234, illust.

Storr, G.M., Smith, L.A. & Johnstone, R.E. (2002) *Snakes of Western Australia.* Second edition. Western Australian Museum, Perth, 309 pp., figs., 54 col. pls. [first edition, 187 pp., 112 figs., 24 col. pls.]

Strand, E. (1928) Miscellanea nomenclatorica zoologica et palaeontologica. I–II. *Archiv für Naturgeschichte*, Berlin (1926) 92(8A), 30–75.

Strauch, A. (1862) Essai d'une erpétologie de Algérie. *Mémoires de l'Académie Impériale des Sciences de St.-Pétersbourg* (Series 7) 4(7), 1–86.

Strauch, A. (1868) Concerning poisonous snakes found in Russia. Trudy Russkikh Estestvoispytatelei, St.-Pétersbourg 1, 271–297. (in Russian)

Strauch, A. (1869) Synopsis der Viperiden, nebst Bemerkungen über die Geographische Verbreitung dieser Giftschlangen-Familie. *Memoires de l'Académie des Sciences de St.-Pétersbourg* (Série 7) 14(6), 1–144, 2 pls..

Strauch, A. (1873) Die Schlangen des Russischen Reichs, in Systematischer und Zoogeographischer Beziehung. *Mémoires de l'Académie Impériale des Sciences de St.-Pétersbourg* (Series 7) 21(4), 1–287, 6 pls.

Strauch, A. (1876) Reptiles and Amphibians. *In*: Przevalski, N.M. (Ed.), *Mongol, i Strana Tangut* 2(3), 1–55, 8 pls., (in Russian)

Strauch, A. (1884) Bemerkungen über die Schlangengattung *Elapomorphus* aus der Familie der Calamariden. *Bulletin de l'Académie Impériale des Sciences de St.-Pétersbourg* 29, 541–590.

Strauch, A. (1885) Bemerkungen über die Schlangengattung *Elapomorphus* aus der Familie der Calamariden. *Mélanges Biologiques tires du Bulletin de l'Académie Impériale des Sciences de St.-Pétersbourg* (1884–1888) 12, 141–211.

Strecker, J.K. (1915) Reptiles and amphibians of Texas. *Baylor University Bulletin*, Waco 18(4), 1–82.

Street, D. (1979) *Reptiles of northern and central Europe*. B.T. Batsford, London, xi + 268 pp., 44 pls.

Strijbosch, H. & van Gelder, J.J. (1993) Okologie und Biologie der Schlingnatter, *Coronella austriaca* Laurenti, 1768 in den Niederlanden. *Mertensiella* 3, 39–57, illust.

Strimple, P.D. (1992) *Crotalus mitchelli*, the speckled rattlesnake. *Litteratura Serpentium* (English version) 12(2), 26–31.

Strimple, P.D. (1993) Overview of the natural history of the green anaconda (*Eunectes murinus*). *Herpetological Natural History* 1(1), 25–35.

Strimple, P.D. (1994) Husbandry of juvenile green anacondas *(Eunectes murinus)* in captivity. *The Vivarium* 6(1), 52–54, 3 col. photos.

Strimple, P.D. (1998) A little-known Mexican pit-viper. *The Vivarium* 10(1), 41, 53–54, 1 col. photo.

Stromer, E. (1910) Reptilien- und Fischreste aus dem marinen Alttertiär von Südtogo (Westafrika). *Monatsberichten der Deutschen Geologischen Gesellschaft* 62(7), 475–507, 4 figs., 1 pl.

Strüssmann, C. & Andre de Carvalho, M.A. (1998) New herpetological records for the state of Mato Grosso, western Brazil. *Herpetological Review* 29(3), 183–185.

Strüssmann, C. & Sazima, I. (1991) Esquadrinhar com a cauda, uma tática de caça de serpente *Hydrodynastes gigas* no Pantanal, Mato Grosso. *Memórias do Instituto Butantan*, São Paulo (1990) 52(2), 57–61, 2 figs.

Stuart, B.L. (1999) Amphibians and reptiles. *In:* Duckworth, J.W., Salter, R.E. & Khounboline, K. (Eds.), *Wildlife in Lao PDR: 1999 status report.* IUCN Wildlife Conservation Society, Centre for Protected Areas and Watershed Management, Vientiane, pp. 43–67.

Stuart, B.L. & Chuaynkern, Y. (2007) A new *Opisthotropis* (Serpentes: Colubridae: Natricinae) from northeastern Thailand. *Current Herpetology* 26(1), 35–40.

Stuart, B.L. & Emmett, D.A. (2006) A collection of amphibians and reptiles from the Cardamom Mountains, southwestern Cambodia. *Fieldiana: Zoology* (Series 2) (109), 1–27, 8 figs.

Stuart, B.L., Sok, K. & Neang, T. (2006) A collection of amphibians and reptiles from hilly eastern Cambodia. *Raffles Bulletin of Zoology* 54(1), 129–155, 16 figs.

Stuart, C.T. (1976) A note on *Pythonodipsas carinata* Günther (western keeled snake) in South West Africa, with new distributional records. *Madoqua*, Windhoek 9(4), 63–64.

Stuart, C.T. (1980) An annotated preliminary list of amphibians and reptiles known to occur in the Namib Desert Park, Namibia. *Journal of the Herpetological Association of Africa*, Stellenbosch (24), 4–13.

Stuart, L.C. (1933) Studies on neotropical Colubridae. II. Some new species and subspecies of *Eudryas* Fitzinger, with an annotated list of the forms of *Eudryas boddaertii* (Sentzen). *Occasional Papers of the Museum of Zoology, University of Michigan* (254), 1–10.

Stuart, L.C. (1934a) Studies on neotropical Colubrinae. IV. An apparently new species of *Coluber* from Guatemala. *Occasional Papers of the Museum of Zoology, University of Michigan* (284), 1–4.

Stuart, L.C. (1934b) A contribution to a knowledge of the herpetological fauna of El Peten, Guatemala. *Occasional Papers of the Museum of Zoology, University of Michigan* (292), 1–18, 1 fig.

Stuart, L.C. (1935a) The king snake of the Tres Marías Islands. *Occasional Papers of the Museum of Zoology, University of Michigan* (323), 1–3.

Stuart, L.C. (1935b) A contribution to a knowledge of the herpetology of a portion of the savanna region of central Peten, Guatemala. *Miscellaneous Publications, Museum of Zoology, University of Michigan* (29), 7–56, 1 fig., 4 pls.

Stuart, L.C. (1935c) Studies of neotropical Colubridae. V. A new snake of the genus *Lampropeltis* from Yucatán. *Occasional Papers of the Museum of Zoology, University of Michigan* (309), 1–6.

Stuart, L.C. (1938) Studies on neotropical Colubrinae. VI. A new species of *Eudryas* from South America. *Copeia* 1938(1), 7–8.

Stuart, L.C. (1939) A new name for the genus *Eudryas* Fitzinger, 1843. *Copeia* 1939(1), 55.

Stuart, L.C. (1941a) Studies of Neotropical Colubrinae. VIII. A revision of the genus *Dryadophis* Stuart, 1939. *Miscellaneous Publications, Museum of Zoology, University of Michigan* (49), 9–106, 12 figs., 4 pls.

Stuart, L.C. (1941b) Some new snakes from Guatemala. *Occasional Papers of the Museum of Zoology, University of Michigan* (452), 1–7.

Stuart, L.C. (1942) A new *Tropidodipsas* (Ophidia) from Alta Verapaz, Guatemala. *Proceedings of the Biological Society of Washington* 55, 177–180.

Stuart, L.C. (1943) Comments on the herpetofauna of the Sierra de los Cuchumatanes of Guatemala. *Occasional Papers of the Museum of Zoology, University of Michigan* (471), 1–28, 7 figs.

Stuart, L.C. (1948) The amphibians and reptiles of Alta Verapaz, Guatemala. *Miscellaneous Publications, Museum of Zoology, University of Michigan* (69), 1–109.

Stuart, L.C. (1949) A new *Trimetopon* (Ophidia) from Guatemala. *Proceedings of the Biological Society of Washington* 62, 165–168.

Stuart, L.C. (1950) A geographical study of the herpetofauna of Alta Verapaz, Guatemala. *Contributions from the Laboratory of Vertebrate Biology, University of Michigan* (45), 1–77, 9 pls.

Stuart, L.C. (1951) The herpetofauna of the Guatemalan Plateau, with special reference to its distribution on the southwestern highlands. *Contributions from the Laboratory of Vertebrate Biology, University of Michigan* (49), 1–71, 7 pls.

Stuart, L.C. (1954a) Descriptions of some new amphibians and reptiles from Guatemala. *Proceedings of the Biological Society of Washington* 67, 159–177.

Stuart, L.C. (1954b) Further notes on the status of the generic names *Peropodum* and *Ungaliophis*. *Herpetologica* 10(2), 79–82.

Stuart, L.C. (1954c) Herpetofauna of the southeastern Highlands of Guatemala. *Contributions from the Laboratory of Vertebrate Biology, University of Michigan* (68), 1–65, 4 pls.

Stuart, L.C. (1958) A study of the herpetofauna of the Uaxactun-Tikal area of northern El Petén, Guatemala. *Contributions from the Laboratory of Vertebrate Biology, University of Michigan* (75), 1–30.

Stuart, L.C. (1963) A checklist of the herpetofauna of Guatemala. *Miscellaneous Publications, Museum of Zoology, University of Michigan* (122), 1–150, 1 pl.

Stuart, L.C. & Bailey, J.R. (1941) Three new species of the genus *Rhadinaea* from Guatemala. *Occasional Papers of the Museum of Zoology, University of Michigan* (442), 1–11, 1 fig.

Stucki-Stirn, M.C. (1979) *Snake report 721, a comparative study of the herpetological fauna of the former West Cameroon/Africa, with a classification and synopsis of 95 different snakes and description of some new subspecies.* Herpeto-Verlag, Teuffenthal, 650 pp., illust.

Stuebing, R.B. (1991) A checklist of the snakes of Borneo. *Raffles Bulletin of Zoology*, Singapore 39(2), 323–362.

Stuebing, R.B. (1994a) A checklist of the snakes of Borneo, addenda and corrigenda. *Raffles Bulletin of Zoology*, Singapore 42(4), 931–936.

Stuebing, R.B. (1994b) A new species of *Cylindrophis* (Serpentes, Cylindrophidae) from Sarawak, western Borneo. *Raffles Bulletin of Zoology*, Singapore 42(4), 967–973, 2 figs.

Stuebing, R.B. & Goh, R. (1993) A new record of Leonard's pipe snake, *Anomochilus leonardi* Smith (Serpentes, Uropeltidae, Cylindrophinae), from Sabah, northwestern Borneo. *Raffles Bulletin of Zoology,* Singapore 41(2), 311–314, 1 fig.

Stuebing, R.B. & Inger, R.F. (1998) Additional records on two rare snakes from Borneo, with the confirmation of *Trimeresurus malcolmi* Loveridge as a distinct species. *Raffles Bulletin of Zoology*, Singapore 46(2), 325–328.

Steubing, R.B. & Lian, T.F. (2002) Notes on the fire-lipped keelback *Rhabdophis murudensis* (Smith, 1925) (Ophidia, Colubridae, Natricinae) from northern Borneo. *Raffles Bulletin of Zoology*, Singapore 50(1), 227–230, 1 fig.

Stugren, B. (1955) La vipère de steppe, *Vipera ursinii* (Bonap.) de la Finatele Clujului. *Studii si Cercetari Stiintifice, Filiala Cluj* (Series 2) 6(1–2), 59–77, 11 figs.

Stull, O.G. (1928) A revision of the genus *Tropidophis. Occasional Papers of the Museum of Zoology, University of Michigan* (195), 1–49, 9 figs.

Stull, O.G. (1929) The description of a new subspecies of *Pituophis melanoleucus* from Louisiana. *Occasional Papers of the Museum of Zoology, University of Michigan* (205), 1–3.

Stull, O.G. (1932a) An annotated list of the forms of the genus *Pituophis. Occasional Papers of the Museum of Zoology, University of Michigan* (250), 1–5.

Stull, O.G. (1932b) Five new subspecies of the family Boidae. *Occasional Papers of the Boston Society of Natural History* (8), 25–30, 2 pls.

Stull, O.G. (1933) Two new subspecies of the family Boidae. *Occasional Papers of the Museum of Zoology, University of Michigan* (267), 1–4.

Stull, O.G. (1935) A check list of the family Boidae. *Proceedings of the Boston Society of Natural History* 40(8), 387–408.

Stull, O.G. (1938) Three new subspecies of the family Boidae. *Occasional Papers of the Boston Society of Natural History* (8), 297–300.

Stull, O.G. (1940) Variations and relationships in the snakes of the genus *Pituophis. Bulletin of the United States National Museum* (175), i–vi + 1–225, 84 figs.

Stull, O.G. (1956) Description of a new subspecies of the boid snake, *Enygrus carinatus. Copeia* 1956(3), 185–186.

Stumpel, A.H.P. (1995) *Masticophis taeniatus taeniatus* (desert striped whipsnake). Elevation Record. *Herpetological Review* 26(2), 102.

Sturm, J. (1799) *Deutschlands Fauna in Abbildungen nach der Natur mit Beschreibungen. III. Abtheilung. Die Amphibien.* 2. Heft. Kosten des Berfassers, Nürnberg, 38 pp, 12 color pls.

Suárez-Velázquez, A. & Luna-Reyes, R. (2009) *Mordeduras por serpientes en Chiapas.* Instituto de Salud Pública en el Estado de Chiapas, Tuxtla Gutiérrez. 121 pp., col. illust.

Subba-Rao, M.V. (2012) *Conservation of an endangered king cobra in Salur Wildlife Sanctuary, Andhra Pradesh, India.* B.R. Publishing Corporation, Delhi, 125 pp., figs., 32 col. pls.

Suckow, G.A. (1797) *Anfangsgründe der theoretischen und angewandten Naturgeschichte der Thiere. Dritter Theil, von den Amphibien.* Weissmann, Leipzig, 298 pp.

Sullivan, R.M. & Holman, A. (1996) Squamata. *In:* Prothero, D.R. & Emry, R.J. (Eds.), *The terrestrial Eocene-Oligocene transition in North America.* Cambridge University Press, Cambridge, pp. 354–372.

Sullivan, R.M. & Lucas, S.G. (1988) Fossil Squamata from the San José Formation, early Eocene, San Juan Basin, New Mexico. *Journal of Paleontology* 62(4), 631–639, 12 figs.

Sülter, M.M. (1962) A contribution to the cranial morphology of *Causus rhombeatus* (Lichtenstein) with special reference to cranial kinesis. *Annale Universiteit van Stellenbosch* 37A(1), 1–40, 18 figs.

Sumichrast, F. (1880) Contribution à l'histoire naturelle de Mexique. I. Notes sur une collection de reptiles et e batrachiens de la partie occidentale de l'Isthmus de Tehuantepec. *Bulletin de la Société Zoologique de France,* Paris 15, 162–190.

Sumichrast, F. (1882) Enumeracion de las especies de reptiles observados en la parte meridional de la Republica Mexicana. *La Naturaleza*, México City 6, 31–45.

Sumontha, M., Kunya, K., Pauwels, O.S.G., Nitikul, A. & Punnadee, S. (2011) *Trimeresurus (Popeia) phuketensis*, a new pitviper (Squamata: Viperidae) from Phuket Island, southwestern Thailand. *Russian Journal of Herpetology* 18(2), 185–194, 9 figs.

Sun, A.-L. (1961) Notes on fossil snakes from Shanwang, Shantung. *Vertebrata Palasiatica*, Peking 1961(4), 306–312, 2 figs., 2 pls.

Sun, X., Liu, S.-Y. & Huang, S. (2011) Tibetan Plateau relict snakes of the genus *Thermophis* and their relationship to New World relict snakes. *Asian Herpetological Research* (Series 2) 2(3), 161–168, 1 fig.

Sundersingh, P.P.H. (1960) Reptilian fauna of Pilani with a note on the food contents. Part 1. Ophidia. *Proceedings of the Rajasthan Academy of Science* 7, 52–59, 1 fig.

Sunyer, J. & Köhler, G. (2010) Conservation status of the herpetofauna of Nicaragua. *In:* Wilson, L.D., Townsend, J.H. & Johnson, J.D. (Eds.), *Conservation of Mesoamerican amphibians and reptiles.* Eagle Mountain Publications, Eagle Mountain, pp. 489–509, 2 figs.

Sunyer, J., Townsend, J.H., Wilson, L.D., Travers, S.L., Obando, L.A., Páiz, G., Griffith, D.M. & Köhler, G. (2009) Three new country records of reptiles from Nicaragua. *Salamandra* 45(3), 186–190, 3 figs.

Supriatna, J. (1981) *Ular berbisa Indonesia.* Penerbit Bhratara Karya Aksara, Jakarta, 75 pp., 40 figs.

Sur, S., Gayen, N.C. & Raha, S. (2007) Reptilia. *In:* Alfred, J.R.B. (Ed.), *Fauna of Madhya Pradesh (including Chhattisgarh), (Part-1). State Fauna Series* 15. Zoological Survey of India, Kolkata, pp. 79–111.

Sura, P. (1981a) Captive breeding of *Elaphe rufodorsata* and *Rhabdophis tigrinus* from the Korean People's Democratic Republic. *British Herpetological Society Bulletin*, London (3), 20–24, 4 pls.

Sura, P. (1981b) Notes on the reptiles of Bulgaria. *British Herpetological Society Bulletin*, London (3), 25–28, 5 pls.

Sure, S.A. (1926) Neue Giftschlangen. *Scherl's Magazin,* Berlin 2(5), 447-451.

Surface, H.A. (1906) The serpents of Pennsylvania. *Pennsylvania State Department of Agriculture, Monthly Bulletin of the Division of Zoology* 4(4–5), 113–202, figs. 4–23, pls. 15–42.

Sutherland, S.K. (1981) *Venomous creatures of Australia. A field guide with notes on first aid.* Oxford University Press, Melbourne, xi + 131 pp.

Suvatti, C. (1950) *Fauna of Thailand.* Department of Fisheries, Bangkok, ii + 1100 pp.

Suzart-Argôlo, A.J. (2004a) *As serpents dos Cacauais do sudeste da Bahia.* Editora de UESC, Ilhéus, 258 pp., 61 col. figs.

Suzart-Argôlo, A.J. (2004b) Geographic distribution: Serpentes: *Drymoluber brazili. Herpetological Review* 35(2), 191.

Suzuki, T. & Iwanaga, S. (1970) Snake venoms: bradykinin, kallidine and kallikrein. *In:* Erdös, E.G. (Ed.), *Handbook of Experimental Pharmacology.* Springer-Verlag, Berlin, 25, 193–212.

Svetovidov, A.N. (1976) On the dates of publication of P.S. Pallas' "Zoographia Rosso-Asiatica." *Zoologicheskii Zhurnal* 55(4), 596599, 3 figs. (in Russian)

Swainson, W. (1839) On the natural history and classification of fishes, amphibians and reptiles, or monocardian animals. *In: The Cabinet Cyclopedia, Natural History. Volume II.* Longman, Orme, Brown, Green & Longmans, London, 452 pp., 135 figs.

Swan, G. (1990) *A field guide to the snakes and lizards of New South Wales.* Three Sisters Productions, Winmalee, 224 pp., 35 figs., 188 col. pls.

Swan, G. (1995) *A photographic guide to snakes & other reptiles of Australia.* Ralph Curtis Books, Sanibel Island, 144 pp., col. pls.

Swan, G., Shea, G.M. & Sadlier, R. (2004) *A field guide to reptiles of New South Wales.* Second edition. Reed New Holland, Sydney, 302 pp., col. illust.

Swan, L.W. (1947) The distribution of amphibians and reptiles in the mountains of southeastern Asia. Master's Thesis, Stanford University, Palo Alto, 91 pp., 15 figs.

Swan, L.W. & Leviton, A.E. (1962) The herpetology of Nepal: A history, check list, and zoogeographical analysis of the herpetofauna. *Proceedings of the California Academy of Sciences* (Series 4) 32(6), 103–147, 4 figs.

Swan, M. (2007) *Keeping & breeding Australian pythons.* Mike Swan Herp Books, Lilydale, 336 pp., col. illust.

Swanson, S. (2008) *A wild Australia guide: snakes (identification, habitat, behaviour, diet, conservation status).* Steve Parish Publishing, Archerfield, 96 pp., col. illust.

Swanson, S. (2012) *Field guide to Australian reptiles.* Second edition. Steve Parish Publishing, Brisbane, 336 pp., 660 col. photos. [first edition, 2007, Steve Parish Publishing, Archerfield, 272 pp., col. illust.]

Swarth, H.S. (1921) The type locality of *Crotalus willardi* Meek. *Copeia* 1921(100), 83.

Sweeney, R.C.H. (1961) *Snakes of Nyasaland.* Nyasaland Society, Zomba, xi + 200 pp., 43 figs.

Sweet, S.S. (1984) Species groups within *Pituophis melanoleucus*: evidence from morphology. *Program and abstracts, 1984 combined meetings (ASIH, HL and SSAR), 28 July–3 Aug.*, Norman, 200 (abstract).

Sweet, S.S. & Parker, W.S. (1990) *Pituophis melanoleucus. Catalogue of American amphibians and reptiles*, pp. 474.1–474.8.

Swinhoe, R. (1863) A list of the Formosan reptiles; with notes on a few of the species, and some remarks on a fish (*Orthagoriscus*, sp.). *Annals and Magazine of Natural History*, London (Series 3) 12(69), 219–226.

Swinton, W.E. (1926) *Daunophis langi*, gen. et. sp. n. (Pliocene, Burma). *Annals and Magazine of Natural History*, London (Series 9) 17(99), 342–348, 1 pl.

Switak, K.H. (1998) Extraordinary feeding behavior for the yellow-bellied sea snake, *Pelamis platurus* (Linnaeus, 1766). *Monitor,* Ardeer 9(2), 49–50, 1 fig.

Sworder, G.H. (1923) A list of the snakes of Singapore Island. *Singapore Naturalist* (1922) 2, 55–73.

Szczerbak, N.N. (1966) *Herpetologia Taurica: amphibians and reptiles of the Crimea.* Naukova Dumka, Kiev, 240 pp., 72 figs., 1 pl. (in Russian)

Szczerbak, N.N. (1994) Zoogeography analysis of the reptiles of Turkmenistan. *In:* Fet, V. & Atamuradov, K.I. (Eds.), *Biogeography and ecology of Turkmenistan.* Kluwer Academy Publishers, Dordrecht, pp. 307–328.

Szczerbak, N.N. (2003) *Guide to the reptiles of the eastern Palearctic.* Krieger Publishing Company, Malabar, 260 pp., 194 col. pls.

Szczerbak, N.N. & Böhme, W. (1993a) *Coluber caspius* Gmelin, 1789—Kapische Pfeilnatter oder Springnatter. *In:* Böhme, W. (Ed.), *Handbuch der Reptilien und Amphibien Europas. Band 3/I. Schlangen (Serpentes) I. (Typhlopidae, Boidae, Colubridae 1, Colubrinae).* AULA-Verlag, Wiesbaden, pp. 83–96, figs. 15–18.

Szczerbak, N.N. & Böhme, W. (1993b) *Coluber schmidti* Nikolskij, 1908—Schmidts pfeilnatter. *In:* Böhme, W. (Ed.), *Handbuch der Reptilien und Amphibien Europas. Band 3/I. Schlangen (Serpentes) I. (Typhlopidae, Boidae, Colubridae 1, Colubrinae).* AULA-Verlag, Wiesbaden, pp. 167–175, fig. 39.

Szunyoghy, J. von. (1932) Beiträge zur vergleichenden Formenlehre des Colubridenschädels, nebst einer kraniologischen Synopsis der fossilen Schlangen Ungarns mit Nomencklatorischen, Systematischen und Phyletischen Bemerkungen. *Acta Zoologica*, Stockholm 13, 1–56, 116 figs., 7 pls.

Szyndlar, Z. (1981) Early Pleistocene reptile fauna from Kozi Grzbiet in the Holy Cross Mts. *Acta Géologica Polonica*, Warszawa 31(1–2), 81–101.

Szyndlar, Z. (1984a) Fossil snakes from Poland. *Acta Zoologica Cracoviensia*, Kraków 28(1), 1–156, 62 figs.

Szyndlar, Z. (1984b) A description of a small collection of amphibians and reptiles from the People's Democratic Republic of Korea with notes on the distribution of the herpetofauna in that country. *Acta Zoologica Cracoviensia.*, Kraków 27(1), 3–18, illust.

Szyndlar, Z. (1984c) Past and present distribution of Aesculapian snake *Elaphe longissima* (Laurenti) at the territory of Poland. *Przeglad Zoologiczny* 28(4), 513–523, 2 figs.

Szyndlar, Z. (1985a) Ophidian fauna (Reptilia, Serpentes) from the uppermost Miocene of Algora (Spain). *Estudios Géologicos*, Madrid 41(5–6), 447–465, 11 figs.

Szyndlar, Z. (1985b) Occurrence of *Elaphe davidi* (Sauvage, 1884) (Serpentes, Colubridae) in the Democratic People's Republic of Korea. *The Snake* 17(2), 163–167, 5 figs.

Szyndlar, Z. (1986) Neogene and Quaternany snakes of eastern Europe, fossil record. *In:* Rocek, Z. (Ed.), *Studies in Herpetology.* Charles University, Prague, pp. 695–698, 2 figs.

Szyndlar, Z. (1987) Snakes from the lower Miocene locality of Dolnice (Czechoslovakia). *Journal of Vertebrate Paleontology* 7(1), 55–71, 11 figs.

Szyndlar, Z. (1988) Two new extinct species of the genera *Malpolon* and *Vipera* (Reptilia, Serpentes) from the Pliocene of Layna (Spain). *Acta Zoologica Cracoviensia*, Kraków 31(27), 687–706, 8 figs.

Szyndlar, Z. (1991a) A review of Neogene and Quaternary snakes of central and eastern Europe. Part I. Scolecophidia, Boidae, Colubrinae. *Estudios Geologicos*, Madrid 47(1–2), 103–126, 22 figs.

Szyndlar, Z. (1991b) A review of Neogene and Quaternary snakes of central and eastern Europe. Part II. Natricinae, Elapidae, Viperidae. *Estudios Geologicos*, Madrid 47(3–4), 237–266, 19 figs.

Szyndlar, Z. (1991c) Ancestry of the grass snake (*Natrix natrix*), Paleontological evidence. *Journal of Herpetology* 25(4), 412–418, 4 figs.

Szyndlar, Z. (1992) "*Coluber*" *kargii* von Meyer, 1845 (Serpentes, Viperidae), from the Miocene of Öhningen (Germany). *Neues Jahrbuch für Geologie und Paläontologie, Monatsheft*, Stuttgart 11, 692–700, 3 figs.

Szyndlar, Z. (1994) Oligocene snakes of southern Germany. *Journal of Vertebrate Paleontology* 14(1), 24–37, 6 figs.

Szyndlar, Z. (1995) The vertebrate locality Maramena (Macedonia, Greece) at the Turolian-Ruscinian boundary (Neogene). 4. Serpentes (Squamata, Reptilia). *Münchner Geowissenschaftliche Abhandlungen* 28A, 39–42, 1 fig.

Szyndlar, Z. (1997) A review of the fossil snakes (Reptilia, Serpentes) described by Hermann von Meyer. *Neues Jahrbuch für Geologie und Paläontologie, Abhandlungen* 203(2), 211–219.

Szyndlar, Z. (1998) Vertebrates from the early Miocene lignite deposits of the opencast mine Oberdorf (western Styrian Basin, Austria) 3. Reptilia 2: Serpentes. *Annalen des Naturhistorischen Museums in Wien* 99, 31–38, 4 figs.

Szyndlar, Z. (1999) Serpientes. *In:* Montoya, P., Alberdi, M.T., Blázquez, A.M., Barbadillo, L.J., Fumanal, M.P., Made, J. van der, Marín, J.M., Molina, A., Morales, J., Murelaga, X., Peñalver, E., Robles, F., Ruiz-Bustos, A., Sánchez,A., Sanchiz, B., Soria, D. & Szyndlar, Z. (Eds.), La fauna del Pleistoceno inferior de la Sierra de Quibas (Abanilla, Murcia). La fauna del Plesitoceno inferior de la Sierra de Quibas (Abanilla, Murcia). *Estudios Geologicos* 55(3–4), 137–139, figs. 7–9.

Szyndlar, Z. (2005) Snake fauna from the late Miocene of Rudabanya. *Palaeontographia Italica* (2004) 90, 31–52, 12 figs.

Szyndlar, Z. (2009) Snake fauna (Reptilia: Serpentes) from the early/middle Miocene of Sandelzhausen and Rothenstein 13 (Germany). *Paläontologische Zeitschrift* 83, 55–66, 4 figs.

Szyndlar, Z. (2012) Early Oligocene to Pliocene Colubridae of Europe: a review. *Bulletin de la Société Géologique de France*, Paris 183(6), 661–681.

Szyndlar, Z. & Alférez, F. (2005) Iberian snake fauna of the early/middle Miocene transition. *Revista Española de Herpetologia* 19, 57–70, 2 figs.

Szyndlar, Z. & Böhme, W. (1993) Die fossilen Schlangen Deutschlands Geschichte der faunen und ihrer Erforschung. *Mertensiella* 3, 381–432, 7 figs.

Szyndlar, Z. & Böhme, W. (1996) Redescription of *Tropidonotus atavus* von Meyer, 1855 from the upper Oligocene of Rott (Germany) and its allocation to *Rottophis* gen. nov. (Serpentes, Boidae). *Palaeontographica* 240A, 145–161, 4 figs.

Szyndlar, Z. & Hosgör, I. (2012) Boine snake *Bavarioboa* from the Oligocene/Miocene of eastern Turkey with comments on connections between European and Asiatic faunas. *Acta Palaeontologica Polonica* 57(3), 667–671, 3 figs.

Szyndlar, Z. & Hung, D.O. (1987) Reptiles of the Democratic People's Republic of Korea. Part I. Serpentes. *Chinese Herpetological Research,* Chongging 1987, 22–59, 6 figs.

Szyndlar, Z. & Nguyen, V.S. (1996) Terrestrial snake fauna of Vietnam: distributional records. *The Snake* 27(2), 91–98.

Szyndlar, Z. & Rage, J.-C. (1990) West Palearctic cobras of the genus *Naja* (Serpentes, Elapidae): interrelationships among extinct and extant species. *Amphibia-Reptilia*, Leiden 11(4), 385–400, 4 figs.

Szyndlar, Z. & Rage, J.-C. (1999) Oldest fossil vipers (Serpentes, Viperidae) from the Old World. *In:* Joger, U. (Ed.), Phylogeny and systematics of the Viperidae. *Kaupia, Darmstädter Beiträge zur Naturgeschichte* 8, 9–20, 9 figs.

Szyndlar, Z & Rage, J.-C. (2002) Fossil record of the true vipers. *In:* Schuett, G., Höggren, M., Douglas, M.E. & Greene, H.W. (Eds.), *Biology of the vipers*. Eagle Mountain Publishing, Eagle Mountain, pp. 419–444, 8 figs..

Szyndlar, Z. & Rage, J.-C. (2003) *Non-erycine Booidea from the Oligocene and Miocene of Europe.* Institute of Systematics and Evolution of Animals, Polish Academy of Sciences, Kraków, 109 pp., 50 figs.

Szyndlar, Z. & Schleich, H.H. (1993) Description of Miocene snakes from Petersbuch 2 with comments on the lower and middle Miocene ophidian faunas of southern Germany. *Stuttgarter Beiträge zur Naturkunde*, Stuttgart (Series B) (192), 1–47, 10 figs.

Szyndlar, Z. & Schleich, H.H. (1994) Two species of the genus *Eryx* (Serpentes; Boidae; Erycinae) from the Spanish Neogene with comments on the past distribution of the genus in Europe. *Amphibia-Reptilia*, Leiden 15(3), 233–248, 3 figs.

Szyndlar, Z. & Zerova, G.A. (1990) Neogene cobras of the genus *Naja* (Serpentes, Elapidae) of east Europe. *Annalen des Naturhistorischen Museums in Wien*, (1987) 91A, 53–61, illust.

Szyndlar, Z. & Zerova, G.A. (1992) Miocene snake fauna from Cherevichnoie (Ukraine, USSR), with description of a new species of *Vipera*. *Neues Jahrbuch für Geologie und Paläontologie, Monatsheft*, Stuttgart 184(1), 87–99, 24 figs.

Szyndlar, Z., Smith, R. & Rage, J.-C. (2008) A new dwarf boa (Serpentes, Booidea, "Tropidophiidae") from the early Oligocene of Belgium: a case of the isolation of Western European snake faunas. *Zoological Journal of the Linnean Society,* London 152(2), 393–406, 4 figs.

Taggart, T.W., Crother, B.I. & White, M.E. (2001) Palm-pitviper (*Bothriechis*) phylogeny: mtDNA, and consilience. *Cladistics* 17(4), 355–370, 8 figs.

Takahashi, S. (1922) *Japanese venomous snakes*. Tabuchi Sekihan Insatsuzyo, Taihoku, 32 pp. (in Japanese)

Takahashi, S. (1930) *Terrestrial snakes in Japan*. Shunyô–do, Tokyo, 172 unnumbered pp., 202 figs., 65 col. pls. (in Japanese)

Takahashi, S. (1935) Notes on the Hydridae of Kotosho (Botel Tobago Island). *Transactions of the Natural History Society of Taiwan* 25, 142–144. (in Japanese)

Takara, T. (1962) Studies on the terrestrial snakes in the Ryukyu Archipelago. *Science Bulletin of the Agriculture and Home Economics Division, University of the Ryukyus* (9), 1–202, 96 figs., 22 pls. (in Japanese)

Talukdar, S.K. & Dattagupta, B. (1980) Notes on the occurrence of the sea snakes, *Hydrophis mamillaris* (Daudin) and *Microcephalophis gracilis* (Shaw) from West Bengal. *Journal of the Marine Biology Association of India*, Mandapam Camp, Madras 18(2), 389–391.

Talukdar, S.K., Sanyal, D.P. & Duttagupta, B. (1980) Redescription of and lectotype designation for *Calamaria fusca* Blyth, 1854 (Serpentes, Colubridae). *Bulletin of the Zoological Survey of India*, Calcutta 3(1–2), 93–95, pl. 2, figs. a–c.

Tamayo, J.L. (1962) *Geografía general de México. Tomo III. Geografía biologica y humana.* Second edition. Instituto Mexicano de Investigaciónes Económicas, México City, 633 pp., illust.

Tamiya, N. (1987) Non-divergence theory of evolution: amino acid sequence comparison of proteroglyphous snake venom components. *In:* Gopalakrishnakone, P. & Tan, C.K. (Eds.), *Progress in venom and toxin research.* National University of Singapore, Singapore, pp. 137–145, 1 fig.

Tamiya, N. & Puffer, H. (1974) Lethality of sea snake venoms. *Toxicon* 12(1), 85–87.

Tamiya, N., Maeda, N. & Cogger, H.G. (1983a) Neurotoxins from the venoms of the sea snakes *Hydrophis ornatus* and *Hydrophis lapemoides*. *Biochemical Journal* 213, 31–38, 6 figs.

Tamiya, N., Sato, A., Kim, H.S., Teruuchi, T., Takasaki, C., Ishikawa, Y., Guinea, M.L., McCoy, M., Heatwole, H. & Cogger, H.G. (1983b) Neurotoxins of sea snakes genus *Laticauda*. *Toxicon*, London (Supplement 3), 445–447, 2 figs.

Tanner, V.M. (1929) A distributional list of the amphibians and reptiles of Utah. *Copeia* 1929(171), 46–52.

Tanner, V.M. (1938) A new subspecies of worm snake from Utah. *Proceedings of the Utah Academy of Sciences* 15, 149–150.

Tanner, V.M. (1948) Pacific islands herpetology. No. I. Mariana Islands. A new species of *Typhlops*. *Great Basin Naturalist* 9(1–2), 1–20.

Tanner, V.M. (1949) Pacific islands herpetology. No. II. Philippine Islands. *Great Basin Naturalist* 9(3–4), 25–39.

Tanner, V.M. (1950) Pacific islands herpetology. No. III. Morotai Island. *Great Basin Naturalist* 10(1–4), 1–30, 13 figs.

Tanner, V.M. (1951) Pacific islands herpetology. No. V. Guadalcanal, Solomon Islands. A check list of species, illustration. *Great Basin Naturalist* 11(3–4), 53–86.

Tanner, V.M. (1953) Pacific islands herpetology. No. VIII. Korea. *Great Basin Naturalist* 13, 67–73.

Tanner, V.M. & Tanner, W.W. (1939) Notes on *Charina bottae* in Utah. Reproduction. *Great Basin Naturalist* 1, 27–30.

Tanner, W.W. (1941) The reptiles and amphibians of Idaho. No. I. *Great Basin Naturalist* 2(2), 87–97.

Tanner, W.W. (1943) Two new species of *Hypsiglena* from western North America. *Great Basin Naturalist* 4(1–2), 49–54.

Tanner, W.W. (1946) A taxonomic study of the genus *Hypsiglena*. *Great Basin Naturalist* (1944) 5(3–4), 25–92, 4 pls.

Tanner, W.W. (1953) A study of taxonomy and phylogeny of *Lampropeltis pyromelana* Cope. *Great Basin Naturalist* 13(1–2), 47–66, 2 figs.

Tanner, W.W. (1954) Additional notes on the genus *Hypsiglena* with a description of a new subspecies. *Herpetologica* 10(1), 54–56, 1 fig.

Tanner, W.W. (1959) A new *Thamnophis* from western Chihuahua with notes on four other species. *Herpetologica* 15(4), 165–172, 2 figs.

Tanner, W.W. (1961) A new subspecies of *Conopsis nasus* from Chihuahua, México. *Herpetologica* 17(1), 13–18, 8 figs,

Tanner, W.W. (1966a) The night snakes of Baja California. *Transactions of the San Diego Society of Natural History* 14, 189–196, 2 figs.

Tanner, W.W. (1966b) A new rattlesnake from western México. *Herpetologica* 22(4), 298–302, fig. 1.

Tanner, W.W. (1967) *Contia tenuis* Baird and Girard in continental British Columbia. *Herpetologica* 23, 323.

Tanner, W.W. (1970) A catalogue of the fish, amphibian, and reptile types in the Brigham Young University Museum of Natural History. *Great Basin Naturalist* 30(4), 219–226.

Tanner, W.W. (1981) A new *Hypsiglena* from Tiburon Island, Sonora, Mexico. *Great Basin Naturalist* 41(1), 139–142, 1 fig.

Tanner, W.W. (1983) *Lampropeltis pyromelana* (Cope). *Catalogue of American amphibians and reptiles*, pp. 342.1–342.2.

Tanner, W.W. (1985) Snakes of western Chihuahua. *Great Basin Naturalist* 45(4), 615–676, 9 figs.

Tanner, W.W. (1986) *Crotalus lannomi*. *Catalogue of American amphibians and reptiles*, pp. 387.1.

Tanner, W.W. (1988) Status of *Thamnophis sirtalis* in Chihuahua, México (Reptilia, Colubridae). *Great Basin Naturalist* 48(4), 499–507, 3 figs.

Tanner, W.W. (1990) *Thamnophis rufipunctatus*. *Catalogue of American amphibians and reptiles*, pp. 505.1–505.2, 1 fig.

Tanner, W.W. & Avery, D.F. (1982) Buccal floor of reptiles: a summary. *The Great Basin Naturalist,* Provo 42(3), 273–349, 29 figs.

Tanner, W.W. & Banta, B.H. (1962a) Description of a new *Hypsiglena* from San Martin Island, México, with a resumé of the reptile fauna of the island. *Herpetologica* 18(1), 21–25, 1 fig.

Tanner, W.W. & Banta, B.H. (1962b) The distribution of *Tantilla utahensis* Blanchard. *Great Basin Naturalist* 22(4), 116–118, 1 fig.

Tanner, W.W. & Banta, B.H. (1966) A systematic review of the Great Basin reptiles in the collections of Brigham Young University and the University of Utah. *Great Basin Naturalist* 26(3–4), 87–135, 1 fig.

Tanner, W.W. & Loomis, R.B. (1957) A taxonomic and distributional study of the western subspecies of the milk snake, *Lampropeltis doliata*. *Transactions of the Kansas Academy of Sciences* 60(1), 12–42, 4 pls.

Tanner, W.W. & Lowe, C.H. (1989) Variations in *Thamnophis elegans* with descriptions of new subspecies. *Great Basin Naturalist* 49(4), 511–516, 7 figs.

Tanner, W.W. & Robison, G.W., Jr. (1960a) Herpetological notes for northwestern Jalisco, Mexico. *Herpetologica* 16(1), 59–62, 1 pl.

Tanner, W.W. & Robison, G.W., Jr. (1960b) New and unusual serpents from Chihuahua, Mexico. *Herpetologica* 16(1), 67–70.

Tanner, W.W., Dixon, J.R. & Harris, H.S., Jr. (1972) A new subspecies of *Crotalus lepidus* from western Mexico. *Great Basin Naturalist* 32(1), 16–24, 3 figs.

Tanzer, E.C. (1970) Polymorphism in the *Mexicana* complex of kingsnakes, with notes on their natural history. *Herpetologica* 26(4), 419–428, 2 figs.

Tarashchuk, V.I. (1959) *Fauna of Ukraine. Volume 7. Amphibians and reptiles*. Publishing House of the Academy of Sciences of the Ukrainian Soviet Socialistic Republic, Kiev, 246 pp., 82 figs. (in Russian)

Tarawneh, K.A., Al-Oran, R., Lhony, S., Bedour, A. & Al-Bassal, M.A. (1999) *Vertebrates of Jordan.* Biology Department, Mu'tah University, Karak, 115 pp., 29 col. pls. (in Arabic)

Tatarinov, L.P. (1963) The first occurrence in the USSR of the ancient sea snakes. *Paleontologicheskii Zhurnal*, Moscou 1963(2), 109–115. (in Russian)

Tatarinov, L.P. (1988) Structure of the skull in the sea snake, "*Archaeophis*" *turkmenicus* from the lower Eocene of Turkmeniya. *Paleontologicheskii Zhurnal*, Moscou 1988(1), 75–82, 1 fig. (in Russian; an English translation in *Paleontological Journal* 22(1), 73–79, 1 fig.)

Taub, A.M. (1965) *Comparative histological studies on the parotid gland of colubrid snakes.* Doctoral Dissertation, State University of New York, Buffalo, 239 pp., 3 figs., 8 pls.

Taub, A.M. (1966) Ophidian cephalic glands. *Journal of Morphology* 118(4), 529–542, 3 pls.

Taylor, E.H. (1917) Snakes and lizards known from Negros, with descriptions of new species and new subspecies. *The Philippine Journal of Science*, Manila 12D(6), 353–381, 2 figs., 2 pls.

Taylor, E.H. (1918a) Reptiles of the Sulu Archipelago. *The Philippine Journal of Science*, Manila 13D(5), 233–267, 11 figs., 3 pls.

Taylor, E.H. (1918b) Two new snakes of the genus *Holarchus*, with descriptions of other Philippine species. *The Philippine Journal of Science*, Manila 13D(6), 359–367, 2 pls.

Taylor, E.H. (1919) New or rare Philippine reptiles. *The Philippine Journal of Science*, Manila (1918) 14D(1), 105–123, 4 figs., 2 pls.

Taylor, E.H. (1922a) The snakes of the Philippine Islands. *Philippine Bureau of Science Publications*, Manila, 312 pp., 32 figs., 37 pls.

Taylor, E.H. (1922b) Additions to the herpetological fauna of the Philippine Islands, I. *The Philippine Journal of Science*, Manila 21D(2), 161–206, 7 pls.

Taylor, E.H. (1922c) Additions to the herpetological fauna of the Philippine Islands, II. *The Philippine Journal of Science*, Manila 21D(3), 257–303, 4 pls.

Taylor, E.H. (1922d) Herpetological fauna of Mount Makiling. *Philippine Agriculturist*, Manila 11, 127–139. [not seen]

Taylor, E.H. (1923) Additions to the herpetological fauna of the Philippine Islands, III. *The Philippine Journal of Science*, Manila (1922) 22D(5), 515–557, 3 pls.

Taylor, E.H. (1925) Additions to the herpetological fauna of the Philippines, IV. *The Philippine Journal of Science*, Manila 26D(1), 97–111.

Taylor, E.H. (1929) A revised checklist of the snakes of Kansas. *University of Kansas Science Bulletin* 19(5), 53–62.

Taylor, E.H. (1931) Notes on two specimens of the rare snake *Ficimia cana* and the description of a new species of *Ficimia* from Texas. *Copeia* 1931(1), 4–7.

Taylor, E.H. (1933) Notes on type specimens of reptiles and amphibians in the "Alfredo Dugès" Museum, Guanajuato, Mexico. *Copeia* 1933(2), 97.

Taylor, E.H. (1934) Zoological results of the third De Schauensee Siamese Expedition, Part III. Amphibians and reptiles. *Proceedings of the Academy of Natural Sciences Philadelphia* 86, 281–310, 4 figs., pl. 17.

Taylor, E.H. (1936a) Description of a new Sonoran snake of the genus *Ficimia*, with notes on other Mexican species. *Proceedings of the Biological Society of Washington* 49, 51–54.

Taylor, E.H. (1936b) Notes on the herpetological fauna of the Mexican state of Sinaloa. *University of Kansas Science Bulletin* 24(20), 505–530, pls. 44–46.

Taylor, E.H. (1937a) Notes and comments on certain American and Mexican snakes of the genus *Tantilla*, with descriptions of new species. *Transactions of the Kansas Academy of Sciences* (1936) 39, 335–348, 6 figs.

Taylor, E.H. (1937b) A new snake of the genus *Sonora* from Mexico, with comments on *S. michoacanensis*. *Herpetologica* 1(3), 69–72, 2 figs.

Taylor, E.H. (1938) Notes on the Mexican snakes of the genus *Leptodeira*, with a proposal of a new snake genus, *Pseudoleptodeira*. *University of Kansas Science Bulletin* 25(15), 315–344, 6 figs., pls. 30–34.

Taylor, E.H. (1939a) On Mexican snakes of the genera *Trimorphodon* and *Hypsiglena*. *University of Kansas Science Bulletin* (1938) 25(16), 357–383, 1 fig., pls. 35–38.

Taylor, E.H. (1939b) Two new species of the genus *Anomalepis* Jan, with a proposal of a new family of snakes. *Proceedings of the New England Zoological Club* 17, 87–96, pl. 5.

Taylor, E.H. (1939c) On North American snakes of the genus *Leptotyphlops*. *Copeia* 1939(1), 1–7, 1 pl.

Taylor, E.H. (1940a) Two new snakes of the genus *Thamnophis* from Mexico. *Herpetologica* 1(7), 183–189, 2 figs., pl. 19.

Taylor, E.H. (1940b) Mexican snakes of the genus *Typhlops*. *University of Kansas Science Bulletin* (1939) 26(13), 441–444, 2 figs.

Taylor, E.H. (1940c) Some Mexican serpents. *University of Kansas Science Bulletin* 26(14), 445–487, 9 figs., pls. 49–52.

Taylor, E.H. (1940d) Herpetological miscellany, No. I. *University of Kansas Science Bulletin* (1939) 26(15), 489–571, 7 figs., pls. 53–63.

Taylor, E.H. (1940e) A new *Lampropeltis* from western Mexico. *Copeia* 1940(4), 253–255, 2 figs.

Taylor, E.H. (1942a) Herpetological miscellany, No. II. *University of Kansas Science Bulletin* (1941) 27(7), 105–139, 7 figs., pls. 3–6.

Taylor, E.H. (1942b) Mexican snakes of the genera *Adelophis* and *Storeria*. *Herpetologica* 2(4), 75–79.

Taylor, E.H. (1944a) Present location of certain herpetological and other type specimens. *University of Kansas Science Bulletin* 30(11), 117–187.

Taylor, E.H. (1944b) Two new species of crotalid snakes from Mexico. *University of Kansas Science Bulletin* 30(4), 47–56, 2 figs., pl. 6–7.

Taylor, E.H. (1947a) Publication dates of Kelaart's Prodromus Fauna Zeylanicae. *Herpetologica* 4(1), 26.

Taylor, E.H. (1947b) Comments on Ceylonese snakes of the genus *Typhlops* with descriptions of new species. *University of Kansas Science Bulletin* 31(13), 283–298, 3 figs.

Taylor, E.H. (1949) A preliminary account of the herpetology of the state of San Luis Potosí, México. *University of Kansas Science Bulletin* 33(2), 169–215.

Taylor, E.H. (1950a) Second contribution to the herpetology of San Luis Potosí. *University of Kansas Science Bulletin* 33(11), 441–457, pls. 4–9.

Taylor, E.H. (1950b) A brief review of Ceylonese snakes. *University of Kansas Science Bulletin* 33(14), 519–603, 3 figs., pls. 12–25.

Taylor, E.H. (1951) A brief review of the snakes of Costa Rica. *University of Kansas Science Bulletin* 34(1), 1–188, 7 figs., 23 pls.

Taylor, E.H. (1952) Third contribution to the herpetology of the Mexican state of San Luis Potosí. *University of Kansas Science Bulletin* 34(13), 793–815, 1 fig.

Taylor, E.H. (1953a) Fourth contribution to the herpetology of San Luis Potosí. *University of Kansas Science Bulletin* 35(13), 1587–1614, 1 fig., pls. 118–119.

Taylor, E.H. (1953b) Report on a collection of Ceylonese serpents. *University of Kansas Science Bulletin* 35(14), 1615–1624.

Taylor, E.H. (1954) Further studies on the serpents of Costa Rica. *University of Kansas Science Bulletin* 36(11), 673–801, 40 figs.

Taylor, E.H. (1955) Additions to the known herpetological fauna of Costa Rica with comments on other species. No. II. *University of Kansas Science Bulletin* 37(13), 499–575, 19 figs.

Taylor, E.H. (1962) New oriental reptiles. *University of Kansas Science Bulletin* 43(7), 209–263, 18 figs.

Taylor, E.H. (1963) New and rare oriental serpents. *Copeia* 1963(2), 429–433.

Taylor, E.H. (1965) The serpents of Thailand and adjacent waters. *University of Kansas Science Bulletin* 45(9), 609–1096, 125 figs.

Taylor, E.H. & Elbel, R.E. (1958) Contribution to the herpetology of Thailand. *University of Kansas Science Bulletin* 38(13), 1033–1189, 36 figs.

Taylor, E.H. & Smith, H.M. (1939) Miscellaneous notes on Mexican snakes. *University of Kansas Science Bulletin* (1938) 25(13), 239–258, 4 figs., pl. 23.

Taylor, E.H. & Smith, H.M. (1942a) Concerning the snake genus *Pseudoficimia*. *University of Kansas Science Bulletin* 28(12), 241–251, 4 figs., pl. 21.

Taylor, E.H. & Smith, H.M. (1942b) The snake genera *Conopsis* and *Toluca*. *University of Kansas Science Bulletin* 28(15), 325–363, 12 figs., pls. 30–35.

Taylor, E.H. & Smith, H.M. (1943) A review of American Sibynophine snakes, with a proposal of a new genus. *University of Kansas Science Bulletin* 29(6), 301–337, 9 figs., pls. 21–25.

Taylor, E.H. & Weyer, D. (1958) Report on a collection of amphibians and reptiles from Harbel, Republic of Liberia. *University of Kansas Science Bulletin* 38(14), 1191–1229.

Taylor, J. (1993) *The amphibians and reptiles of New Hampshire*. New Hampshire Fish and Game Department, 71 pp.

Taylor, R.H.R. (1948) The distribution of reptiles and Amphibia in the British Isles, with notes on species recently introduced. *British Journal of Herpetology*, London (1), 1–38, figs.

Taylor, R.H.R. (1963) The distribution of amphibians and reptiles in England and Wales, Scotland and Ireland and the Channel Isles: a revised survey. *British Journal of Herpetology* 3, 95–115.

Taylor, R.T., Flores, A., Flores, G. & Bolaños, R. (1974) Geographical distribution of Viperidae, Elapidae and Hydrophidae in Costa Rica. *Revista de Biologia Tropical*, San José 21(2), 383–397, 6 figs.

Taylor, W.E. (1892) Ophidia of Nebraska. *Annual Report Nebraska State Board of Agriculture, for the year 1891*, 310–357.

Tchang, T.-L. (1932) Notes on some Chinese snakes. *Bulletin of the Fan Memorial Institute of Biology*, Peiping 3(1), 1–23.

Tchernov, E., Rieppel, O., Zaher, H., Polcyn, M.J. & Jacobs, L.L. (2000) A fossil snake with limbs. *Science* 287(5460), 2010–2012, 3 figs.

Tedaldi, G. (1998) *Guida agli anfibi e ai rettili della Romagna*. Maggioli Editore, Rimini, 94 pp., illust.

Tekiela, S. (2003) *Reptiles & amphibians of Minnesota field guide*. Adventure Publications, Cambridge, MN, 140 pp., col. illust.

Tekiela, S. (2004a) *Reptiles & amphibians of Wisconsin field guide*. Adventure Publications, Cambridge, MN, 147 pp., col. illust.

Tekiela, S. (2004b) *Reptiles & amphibians of Michigan field guide*. Adventure Publications, Cambridge, MN, 145 pp., col. illust.

Telford, R.H. (1967) Fossil mammals from the Tertiary Carl Creek Limestone, northwestern Queensland. *Bulletin of the Australian Bureau of Mineral Resources*, Canberra 92, 217–236.

Telford, S.R., Jr. (1964) Prevalence of microfilariae in Mexican snakes. *Journal of Parasitology* 50(3), 40 (abstract).

Telford, S.R., Jr. (1966) Variation among the southeastern crowned snakes, genus *Tantilla*. *Bulletin of the Florida State Museum, Biological Sciences* 10(7), 261–304, 10 figs.

Telford, S.R., Jr. (1980a) *Tantilla oolitica*. *Catalogue of American amphibians and reptiles*, p. 256.1.

Telford, S.R., Jr. (1980b) *Tantilla relicta*. *Catalogue of American amphibians and reptiles*, pp. 257.1–257.2.

Telford, S.R., Jr. (1982) *Tantilla coronata*. *Catalogue of American amphibians and reptiles*, pp. 308.1–308.2.

Temminck, C.J. & Schlegel, H. (1838) Reptilia. *In:* Siebold, P.F. de (Ed.), *Fauna Japonica sive descriptio animalium, quae in itinere par Japoniam, jussu et auspiciis superiorum, qui summum in India Batava imperium tenent, fuscepto, annis 1823–1830 collegit, notis, observationibus et adumbrationibus illustravit. Regis auspiciis edita*. Lugduni Batavorum, Amstelodami, xxii + 140 pp., 27 pls.

Teng, C.-M., Kuo, Y.-P., Lee, L.-G. & Ouyang, C. (1987) Characterization of the anticoagulants from Taiwan cobra (*Naja naja atra*) snake venom. *Toxicon* 25(2), 201–210, 7 figs.

Tennant, A. (1984) *The snakes of Texas*. Texas Monthly Press, Austin, 561 pp., 128 col. photos.

Tennant, A. (2006) *Texas snakes*. Third edition. Taylor Trade Publishing, Lanham, xx + 263 pp., illust., 169 col. photos. [first edition, 1985, Texas Monthly Press, Austin, 260 pp., illust., 110 col. photos; second edition, 1998, Gulf Publishing Company, Houston, xix + 291 pp., 134 col. photos]

Tennant, A. (2003a) Snakes of Florida. Second edition. Taylor Trade Publishing, Lanham, xiii + 271 pp., illust., 128 col. photos. [first edition, 1997, Gulf Publishing Company, Houston, xiii + 255 pp., illust., 128 col. photos]

Tennant, A. (2003b) *Snakes of North America: eastern and central regions*. Second edition. Lone Star Books, Lanham, xxx + 604 pp., 243 col. photos. [first edition, Tennant & Bartlett, 2000, Gulf Publishing Company, Houston, xxv + 588 pp., 196 col. photos.]

Terent'ev, P.V. (1926) A list of reptiles from province of Terek (Caucasus). *Bulletin Scientific de l'Institut de l'Exploration Regional du Caucase du Nord,* Vladicaucase 1, 307–309. (in Russian)

Terent'ev, P.V. (1961) *Herpetology: a manual on amphibians and reptiles*. State Education-Teaching Publishing House of the Russian Federation, Moscow, 336 pp., 68 figs. (in Russian)

Terent'ev, P.V. (1965) Herpetology: a manual on amphibians and reptiles. Israel Program for Scientific Translation, Jerusalem, v + 313 pp., 68 figs. [English translation of Terent'ev, 1961]

Terent'ev, P.V. & Chernov, S.A. (1949) *Synopsis of the reptiles and amphibians of the USSR*. Third edition. State Education-Teaching Publishing House of the Russian Federation, Moscow, 288 pp., 122 figs. (in Russian) [first edition, 1936/1937, 96 pp., 32+ figs., 5 pls.; second edition, 1940, 184 pp., illust.]

Terent'ev, P.V. & Chernov, S.A. (1965) *Key to amphibians and reptiles*. Israel Program for Scientific Translation, Jerusalem, v + 313 pp., 122 figs. [English translation of Terent'ev & Chernov, 1949]

Terhivuo, J. (1993) Provisional atlas and status of populations for the herpetofauna of Finland in 1980–1992. *Annales Zoologici Fennici* 30(1), 55–69.

Terron, C.C. (1930) *Conopsis nasus heliae*, sub sp. nov. *Anales Instituto de Biologia* 1(2), 175–176, 2 figs.

Test, F.H., Sexton, O.J. & Heatwole, H. (1966) Reptiles of Rancho Grande and vicinity, Estado Aragua, Venezuela. *Miscellaneous Publications, Museum of Zoology, University of Michigan* (128), 1–63, 2 figs., 2 pls.

Teynié, A. & David, P. (2007) Additions to the snake fauna of southern Laos, with the second Laotian specimen of *Naja siamensis* (Laurenti, 1768) and the first country record of *Oligodon taeniatus* (Günther, 1861) (Squamata, Serpentes). *Russian Journal of Herpetology* 14(1), 39–44, 4 col. pls.

Thakur, S. & Watve, A. (2009) Occurrence of *Enhydris sieboldii* (Schlegel, 1837) in Bhopal, Madhya Pradesh state of India. *Russian Journal of Herpetology* 16(2), 159–160, 1 fig.

Thanh, V.N. & Ziegler, T. (2009) Geographic distribution: Serpentes: *Rhadinophis prasinus*. *Herpetological Review* 40(1), 116.

Theobald, W., Jr. (1868a) Catalogue of the reptiles of British Birma, embracing the provinces of Pegu, Martaban, and Tenasserim; with descriptions of new or little-known species. *Journal of the Linnaean Society (Zoology),* London 10(41), 4–67.

Theobald, W., Jr. (1868b) Catalogue of reptiles in the Museum of the Asiatic Society of Bengal, 1866. *Journal of the Asiatic Society of Bengal (Extra number)*, 37, vi + 88 + iii pp., 4 pls.

Theobald, W., Jr. (1876) *Descriptive catalogue of the reptiles of British India*. Thacker, Spink & Co., Calcutta, xiii + 238 pp. + xxxviii + xii.

Theobald, W., Jr. (1880) Reptilian fauna. *In:* Spearman, H.R. (Ed.), *Gazetteer of Burma.* Government Press, Rangoon, pp. 605–640.

Theodor, O. (1955) *On poisonous snakes and snake bite in Israel.* Israel Scientific Press, Jerusalem, 48 pp., 13 figs., 10 pls.

Theophilus, E., Captain, A., Tillack, F. & Kuch, U. (2008) Reptilia, Elapidae, *Bungarus niger*: distribution extension and first record for the state of Uttarakhand, India, with notes on snakebite in the Gori River Valley. *Check List* 4(4), 404–409, 3 figs.

Thermido, A.A. (1941) Répteis e batráquios das Colónias Portuguesas (Catálogo das colecções do Museu Zoológico de Coimbra). *Memórias y Estudos do Museu Zoológico da Universidade de Coimbra* (119), 1–31.

Thermido, A.A. (1942) Anfibios et Répteis de Portugal (Catálogo das colecções do Museu Zoológico de Coimbra). *Memórias y Estudos do Museu Zoológico da Universidade de Coimbra* (133), 1–49.

Thermido, A.A. (1945) Répteis do Brasil. (Catálogo das coleçoes do Museu Zoológico de Coimbra). *Memórias y estudos do Museu Zoologico da Universidade de Coimbra* (168), 1–15.

Thireau, M. (1991) Types and historically important specimens of rattlesnakes in the Muséum National d'Histoire Naturelle (Paris). *Smithsonian Herpetological Information Service* (87), 1–10.

Thiriet, J. & Vacher, J.-P. (2010) *Atlas de repartition des amphibiens et reptiles d'Alsace.* Colmar, 273 pp., 184 col. photos.

Thirion, J.-M., Grillet, P. & Geniez, P. (2002) *Les amphibiens et les reptiles du centre-ouest de la France: region Poitou-Charentes et départements limitrophes.* Collection Parthénope, Mèze, 144 pp., col. illust.

Thistleton, G.F. (1961) *Nature study for Africa: reptiles.* Evans Brothers, London, 112 pp., 82 figs.

Thomas, D.L., Eubanks, J. & Pecor, K.W. (2000) Geographic distribution: Serpentes: *Coluber constrictor latrunculus. Herpetological Review* 31(1), 54.

Thomas, L.A. (1997) Geographic distribution: Serpentes: *Ramphotyphlops braminus. Herpetological Review* 28(2), 98.

Thomas, R.A. (1975) *Philodryas olfersi* (Lichtenstein) new to Colombia and Venezuela. *Herpetological Review* 6(4), 108.

Thomas, R.A. (1976a) *A revision of the South American colubrid snake genus* Philodryas *Wagler, 1830.* Doctoral Dissertation, Texas A & M University, College Station, xiv + 324 pp., 31 figs.

Thomas, R.A. (1976b) *A checklist of Texas amphibians and reptiles.* Second edition. Texas Parks and Wildlife Department, Technichal Series, Austin (17), 1–16. [first edition, 1974, 15 pp.]

Thomas, R.A. (1977a) A new generic arrangement for *Incaspis* and mainland South American *Alsophis* and the status of two additional Peruvian species. *Copeia* 1977(4), 648–652, 6 figs.

Thomas, R.A. (1977b) The status of *Philodryas subcarinatus* Boulenger (Reptilia, Serpentes, Colubridae). *Journal of Herpetology* 11(2), 230–231.

Thomas, R.A. (1977c) *Philodryas nattereri* Steindachner, 1870 (Reptilia, Serpentes): proposed conservation. Z.N.(S.)2166. *Bulletin of Zoological Nomenclature* 33(3–4), 216–217.

Thomas, R.A. (1997) Galápagos terrestrial snakes: biogeography and systematics. *Herpetological Natural History* 5(1), 19–40, 17 figs.

Thomas, R.A. & Dixon, J.R. (1977) A new systematic arrangement for *Philodryas serra* (Schlegel) and *Philodryas pseudoserra* Amaral (Serpentes, Colubridae). *Pearce-Sellards Series* (27), 1–20, 8 figs.

Thomas, R.A. & Fernandes, R. (1996) The systematic status of *Platyinion lividum* Amaral, 1923 (Serpentes, Colubridae, Xenodontinae). *Herpetologica* 52(2), 271–275, 3 figs.

Thomas, R.A. & Ineich, I. (1999) The identity of the syntypes of *Dryophylax freminvillei* with comments on their locality data. *Journal of Herpetology* 33(1), 152–155, 2 figs.

Thomas, R.A. & Johnson, J.D. (1984) *Philodryas varius* (Jan, 1863), a senior synonym of *Philodryas borellii* Peracca (Serpentes, Colubridae). *Journal of Herpetology* 18(1), 80, 1 fig.

Thomas, R.A., Laurent, R.F. & Barrio, A. (1977) *Philodryas borellii* Peracca (Serpentes, Colubridae), a distinct species. *Herpetologica* 33(1), 82–86, 2 figs.

Thomas, R.A., Di-Bernardo, M. & Grillitsch, H. (2001) *Philodryas pallidus* Werner, 1926, a synonym of the colubrid snake *Liopholidophis varius* (Fischer, 1884) from Madagascar. *Journal of Herpetology* 35(1), 120–122, 2 figs.

Thomas, R.A., Bernils, R.S., de Moura-Leite, J.C. & Morato, S.A.A. (2006) Redescription of *Ditaxodon taeniatus* (Hensel, 1868) (Serpentes, Colubridae, Xenodontinae): variation, relationships, and distribution. *South American Journal of Herpetology* 1(2), 94–101, 5 figs.

Thomas, R. (1963) Cayman Islands *Tropidophis* (Reptilia, Serpentes). *Breviora* (195), 1–8, 3 figs.

Thomas, R. (1965a) The genus *Leptotyphlops* in the West Indies with description of a new species from Hispaniola (Serpentes, Leptotyphlopidae). *Breviora* (222), 1–12, 3 figs.

Thomas, R. (1965b) A new species of *Typhlops* from the Barahona Peninsula of Hispaniola. *Copeia* 1965(4), 436–439, 3 figs.

Thomas, R. (1966a) Leeward island *Typhlops* (Reptilia, Serpentes). *Proceedings of the Biological Society of Washington* 79, 255–266, 5 figs.

Thomas, R. (1966b) A reassessment of the Virgin Islands *Typhlops* with the description of two new subspecies. *Revista de Biologia Tropical*, San José (1966) 13(2), 187–201, 6 figs.

Thomas, R. (1966c) A reassessment of the herpetofauna of Navassa Island. *Journal of the Ohio Herpetological Society* 5(3), 73–89, 5 figs.

Thomas, R. (1968) The *Typhlops biminiensis* group of Antillean blind snakes. *Copeia* 1968(4), 713–722, 6 figs.

Thomas, R. (1974a) A new species of *Typhlops* (Serpentes, Typhlopidae) from Hispaniola. *Proceedings of the Biological Society of Washington* 87(2), 11–18, 3 figs.

Thomas, R. (1974b) A new species of Lesser Antillean *Typhlops* (Serpentes, Typhlopidae). *Occasional Papers, Museum of Zoology, Louisiana State University* (46), 1–5, 1 fig.

Thomas, R. (1975) The hemipenis of *Leptotyphlops tenella* Klauber (Serpentes, Leptotyphlopidae) and a new distributional record. *Journal of Herpetology* 9(2), 250–252, 1 fig.

Thomas, R. (1976) *Systematics of the Antillean blind snakes of the genus* Typhlops *(Serpentes, Typhlopidae).* Doctoral Dissertation, Louisiana State University, Baton Rouge, 288 pp., 77 figs.

Thomas, R. (1989) The relationships of Antillean *Typhlops* (Serpentes, Typhlopidae) and the description of three new Hispaniolan species. *In:* Wood, C.A. (Ed.), *Biogeography of the West Indies: past, present, and future.* Sandhill Crane Press, Gainesville, pp. 409–432, 7 figs.

Thomas, R. & Garrido, O.H. (1967) A new subspecies of *Dromicus andreae* (Serpentes, Colubridae). *Annals of the Carnegie Museum* (1964–1968) 39, 219–226, 3 figs.

Thomas, R. & Hedges, S.B. (2007) Eleven new species of snakes of the genus *Typhlops* (Serpentes, Typhlopidae) from Hispaniola and Cuba. *Zootaxa* (1400), 1–26, 11 figs.

Thomas, R. & Joglar, R. (1996) The herpetology of Puerto Rico: Past, present, and future. *Annals of the New York Academy of Sciences* 776, 181–200, 2 figs.
Thomas, R. & Powell, R. (1992) *Typhlops hectus* Thomas. *Catalogue of American amphibians and reptiles*, pp. 550.1–550.2, 1 fig.
Thomas, R. & Powell, R. (1994a) *Typhlops pusillus* Barbour. *Catalogue of American amphibians and reptiles*, pp. 595.1–595.2, 1 fig.
Thomas, R. & Powell, R. (1994b) *Typhlops schwartzi* Thomas. *Catalogue of American amphibians and reptiles*, pp. 596.1–596.2, 1 fig.
Thomas, R. & Powell, R. (1994c) *Typhlops sulcatus* Cope. *Catalogue of American amphibians and reptiles*, pp. 597.1–597.2, 1 fig.
Thomas, R. & Powell, R. (1994d) *Typhlops tetrathyreus* Thomas. *Catalogue of American amphibians and reptiles*, pp. 598.1–598.2, 1 fig.
Thomas, R. & Powell, R. (1995a) *Typhlops capitulatus*. *Catalogue of American amphibians and reptiles*, pp. 618.1–618.2, 1 fig.
Thomas, R. & Powell, R. (1995b) *Typhlops gonavensis*. *Catalogue of American amphibians and reptiles*, pp. 619.1–619.2, 1 fig.
Thomas, R. & Powell, R. (1995c) *Typhlops titanops*. *Catalogue of American amphibians and reptiles*, pp. 620.1–620.2, 1 fig.
Thomas, R. & Powell, R. (1999) The Puerto Rico area. *In:* Crother, B.I. (Ed.), *Caribbean amphibians and reptiles*. Academic Press, San Diego, pp. 169–179, 1 fig.
Thomas, R. & Schwartz, A. (1965) Hispaniolan snakes of the genus *Dromicus* (Colubridae). *Revista de Biologia Tropical*, San José 13(1), 59–83, 10 figs.
Thomas, R., McDiarmid, R.W. & Thompson, F.G. (1985) Three new species of thread snakes (Serpentes, Leptotyphlopidae) from Hispaniola. *Proceedings of the Biological Society of Washington* 98(1), 204–220, 9 figs.
Thompson, F.G. (1957) A new Mexican gartersnake (genus *Thamnophis*) with notes on related forms. *Occasional Papers of the Museum of Zoology, University of Michigan* (584), 1–10, 2 pls.
Thompson, J.C. (1912) *Herpetological Notices. On reptiles new to the island arcs of Asia*. Joseph C. Thompson, San Francisco (3), 1–5.
Thompson, J.C. (1913a) Contribution to the anatomy of the Ophidia. *Proceedings of the Zoological Society of London* 83(3), 414–425, figs. 77–78.
Thompson, J.C. (1913b) *Oxyrhopus trigeminus* Duméril and Bibron the type of *Erythroxyrhopus* gen. nov. *Proceedings of the Academy of Natural Sciences of Philadelphia* 65(1), 78–80.
Thompson, J.C. (1913c) Contributions to the synonymy of serpents of the family Elapidae. *Proceedings of the Academy of Natural Sciences of Philadelphia* 65(11), 508–515.
Thompson, J.C. (1913d) Notes on serpents of the family Colubridae. *Proceedings of the Academy of Natural Sciences of Philadelphia* 65(5), 213–218.
Thompson, J.S. & Crother, B.I. (1998) Allozyme variation among disjunct populations of the Florida green watersnake (*Nerodia floridana*). *Copeia* 1998(3), 715–719, 1 fig.
Thompson, S.W. (1982) Snakes of South Dakota. *South Dakota Conservation Digest* 49(1), 12–18, illust.
Thomson, D.F. (1933) Notes on Australian snakes of the genera *Pseudechis* and *Oxyuranus*. *Proceedings of the Zoological Society of London* 103(4), 855–860, 3 pls.
Thomson, D.F. (1934) A new snake from north Queensland. *Proceedings of the Zoological Society of London* 104(3), 529–531, 1 pl.
Thomson, D.F. (1950) The taipan—Australia's largest and most aggressive snake: a description of the capture of the first live specimen. *Illustrated London News* (Dec. 23), 127, 1048–1049, 9 illust.
Thorpe, R.S. (1979) Multivariate analysis of the population systematics of the ringed snake, *Natrix natrix* (L.). *Proceedings of the Royal Society of Edinburg* 78B, 1–62, 28 figs.
Thorpe, R.S. (1984a) Geographic variation in the western grass snake (*Natrix natrix helvetica*) in relation to hypothesized phylogeny and conventional subspecies. *Journal of Zoology*, London 203(3), 345–355, 6 figs.
Thorpe, R.S. (1984b) Multivariate patterns of geographic variation between the island and mainland populations of eastern grass snake (*Natrix natrix natrix*). *Journal of Zoology*, London 204(4), 551–561, figs.
Thorpe, R.S. (1989a) Pattern and function of sexual dimorphism. A biometric study of character variation in the grass snake (*Natrix natrix*, Colubridae) due to sex and its interaction with geography. *Copeia* 1989(1), 53–63, 2 figs.
Thorpe, R.S. (1989b) Geographic variation, multivariate analysis of sex character systems in snakes in relation to character number. *Copeia* 1989(1), 63–70, 7 figs.
Thorpe, R.S. & McCarthy, C.J. (1978) A preliminary study using multivariate analysis, of a species complex of African house snakes (*Boaedon fuliginosis*). *Journal of Zoology*, London 184(4), 489–506, figs.
Thorpe, R.S., Wüster, W. & Malhotra, A. (1997) *Venomous snakes. Ecology, evolution and snakebite*. Zoological Society of London, Clarendon Press, Oxford, xix + 276 pp.
Thorpe, R.S., Pook, C.E. & Malhotra, A. (2007) Phylogeography of the Russell's viper (*Daboia russelii*) complex in relation to variation in the colour pattern and symptoms of envenoming. *Herpetological Journal* 17, 209–218, 4 figs.
Thunberg, C.P. (1787) *Museum Naturalium Academiae Upsaliensis. Cujus partem primam. Typus Edmannianus, Fridericus Wilhelm*. Radloff, Upsala, 32 pp.
Thunberg, C.P. (1807) Beskrifning och teckning pa tvanne artforandringar af et nytt species ibland ocmar, *Boa variegata* kalladt. *Kongliga Svenska Vetenskaps-Akademiens, Handlingar*, Stockholm 28, 1–4.
Thys van den Audenaerde, D.F.E. (1965) Les serpents des environs de Leopoldville. *Revue de Zoologie et Botanique Africaine*, Tervuren 72(3–4), 366–388.
Thys van den Audenaerde, D.F.E. (1966) Les serpents des environs de Dundo (Angola). *Publicações Culturais da Companhia de Diamantes de Angola*, Dundo (76), 29–38.
Tian, W.-S., Jiang, Y.-M., Wu, G.-F., Hu, Q.-X., Zhao, E.-M. & Huang, Q.-Y. (1986) *Handbook of Chinese Amphibia and Reptilia*. Kolsuo–Choppan, Beijing, v + 164 pp., 20 pls. (in Chinese)
Tiedemann, F. & Grillitsch, H. (1999) Additions to the catalogues of type specimens of the herpetological collection at the Natutral History Museum Vienna (Austria) (Amphibia, Reptilia). *Herpetozoa* 12(3–4), 147–156.
Tiedemann, F. & Häupl, M. (1980) Typenkatalog der Herpetologischen Sammlung. Teil II: Reptilia. *Kataloge der Wissenschaftlichen Sammlungen des Naturhistorischen Museums in Wien: Vertebrata,* Wien 4(2), 1–80.
Tiedemann, F. & Häupl, M. (1994) Katalog der Typen der Herpetologischen Sammlung nach den Stand vom 1. Jänner 1994. Teil II: Reptilia. *Kataloge der wissenschaftlichen Sammlungen des Naturhistorischen Museums in Wien: Vertebrata,* Wien 10(4), 1–102.

Tiedemann, F., Cabela, A., Grillitsch, H. & Kollar, R. (1990) *Lurch und Kriechtiere Wiens*. J & V Edition, Wien, 200 pp., illust., 16 col. pls.

Tien, D. v. (1957) Rapport sur les recherches zoologiques dans la region de Vinh-Linh (Province de Quang-tri, centre Vietnam). *Zoologischeskii Zhurnal*, Moscow 36(8), 1209–1216. (in Russian)

Tien, D. v. (1960) Recherches zoologiques dans la région de Vinh-Linh (Province de Quáng-tri, Centre Vietnam). *Zoologischer Anzeiger*, Leipzig 164(5–6), 221–239, 4 figs. [French translation of Tien, 1957]

Tikader, B.K. & Das, A.K. (1985) Glimpses of animal life of Andaman and Nicobar Islands. *Zoological Survey of India*, Calcutta, 170 pp., illust.

Tillack, F. (2008) *Oligodon rhombifer* Werner, 1924, a junior synonym of *Oligodon ancorus* (Girard, 1857) (Reptilia: Squamata: Colubridae). *Russian Journal of Herpetology* 15(2), 122–128, 2 figs.

Tillack, F. & Grossmann, W. (2001) Ein neuer Nachweis zur Schlangenfauna Nepals, *Bungarus niger* Wall, 1908 (Reptilia, Serpentes, Elapidae). *Sauria* 23(1), 3–9.

Tillack, F. & Günther, R. (2009) Revision of the species of *Oligodon* from Sumatra and adjacent islands, with comments on the taxonomic status of *Oligodon subcarinatus* (Günther, 1872) and *Oligodon annulifer* (Boulenger, 1893) from Borneo (Reptilia, Squamata, Colubridae). *Russian Journal of Herpetology* 16(4), 265–294, 8 figs.

Tillack, F. & Shah, K.B. (2002) Zur Verbreitung von *Trachischium laeve* Peracca, 1904 mit ersten Nachweisen für den zentralen Nepal-Himalaya (Reptilia: Serpentes: Colubridae). *Sauria* 24(1), 39–44, 6 col. figs.

Tillack, F., Lorenz, M., Orlov, N.L., Helfenberger, N., Shah, K.B. & Eckert, W. (2003) Shah's grubenotter *Trimeresurus karanshahi* Orlov and Helfenberger, 1997—ein juniorsynonym von *Trimeresurus tibetanus* Huang, 1982 (Serpentes: Viperidae: Crotalinae), mit angaben zur verbreitung, biologie und der vorstellung neuer farbvariation aus Zentral-Nepal. *Sauria* 25(2), 3–15.

Tillack, F., Ziegler, T. & Quyet, L.K. (2004) Eine neue Art der Gattung *Boiga* Fitzinger, 1826 (Serpentes, Colubridae, Colubrinae) aus dem zentralen Vietnam. *Sauria* 26(4), 3–12, 11 figs.

Tillack, F., Scheidt, U., Ihle, E. & Ihle, T. (2006) First record of Blakeway's mountain snake, *Plagiopholis blakewayi* Boulenger, 1893, from Thailand, with remarks on the distribution of *Plagiopholis nuchalis* (Boulenger, 1893) (Reptilia: Squamata: Colubridae, Pseudoxenodontinae). *Veröffenlichungen Naturkundemuseum Erfurt* 25, 181–186, 3 col. figs.

Tilley, S.G. & Huheey, J.E. (2001) *Reptiles & amphibians of the Smokies*. Great Smoky Mountains Natural History Association, Gatlinburg, 143 pp., col. illust.

Tinkle, D.W. & Conant, R. (1961) The rediscovery of the water snake, *Natrix harteri*, in western Texas, with the description of a new subspecies. *Southwestern Naturalist* 6(1), 33–44, 4 figs.

Tinoco V., R.A. (1976) *Las serpientes de Colombia: ciencia, mitos y leyendas*. Libreria Nacional Limitada, Bogotá, xv + 105 pp., 61 pls.

Tipton, B.L. (2005) *Snakes of the Americas. Checklist and lexicon*. Krieger Publishing Co., Malabar, xiv + 477 pp. (plus compact disc).

Tirant, G. (1885a) *Notes sur les reptiles et les batraciens de la Cochinchine et du Cambodge*. Saigon, 104 pp.

Tirant, G. (1885b) Notes sur les reptiles de Cochinchine et du Cambodge. III. *Cochinchine française, Excursions et reconnaissances*, Saigon 8(20), 387–428.

Tirant, G. (1904) Autres serpents cités pour Indo-Chine. *In:* Pavie, A. (Ed.), *Mission Pavie Indo-Chine 1879–1895—Études diverses. III, Recherches sur l'Histoire Naturelle de l'Indo-chine orientale*. Ernest Leroux, Éditour, Paris, pp. 482–484

Tiranti, S.I. & Avila, L.J. (1997) Reptiles of La Pampa Province, Argentina: an annotated checklist. *Bulletin of the Maryland Herpetological Society* 33(3), 97–117.

Tirvengadum, D.D. & Bour, R. (1985) Checklist of the herpetofauna of the Mascarene Islands. *Atoll Research Bulletin* (292), 49–60.

Tiwari, K.K. & Biswas, S. (1973) Two new reptiles from the Great Nicobar Island. *Journal of the Zoology Society of India*, Calcutta 25(1–2), 57–63, 2 figs.

Tiwari, K.K. & Sharma, R.C. (1971) Reptiles of western Maharashtra. *Journal of the Zoology Society of India*, Calcutta (1970) 22(1–2), 101–115.

Tiwari, M. (1992) First record of the sunbeam snake *Xenopeltis unicolor* Reinwardt, 1827 (Serpentes, Xenopeltidae) from Great Nicobar Island. *The Journal of the Bombay Natural History Society* 89(3), 383.

Tiwari, S.K. (1985) *Readings in Indian zoogeography. Vol. 1*. Today & Tomorrow's Printers and Publishers, New Delhi, 604 pp.

Tobey, F.J. (1985) *Virginia's amphibians and reptiles: a distributional survey*. Virginia Herpetological Survey, 114 pp.

Toda, M. & Okada, S. (1999) New records of the colubrid snake *Dinodon orientale* (Hilgendorf, 1880) (Reptilia: Colubridae) from Shimo-koshikijima and Takeshima Islands. *Bulletin, Osaka Museum of Natural History* 53, 53–56.

Toda, M., Nishida, M., Tu, M.-C., Hikida, T. & Ota, H. (1999) Genetic variation, phylogeny and biogeography of the pit vipers of the genus *Trimeresurus* sensu lato (Reptilia: Viperidae) in the subtropical East Asian islands. *In:* Ota, H. (Ed.), *Tropical island herpetofauna: origin, current diversity, and conservation*. Elsevier, Amsterdam, pp. 249–270, 5 figs.

Tok, C.V. & Kumlutas, Y. (1996) On *Vipera ammodytes transcaucasiana* (Viperidae) from Persembe, Black Sea region of Turkey. *Zoology in the Middle East* 13, 47–50.

Tokar, A.A. (1986) Studien zur taxonomischen der zentralasiatischen Schlangen der Gattung *Eryx* (Daud, 1803). *In: Herpetological investigations in the Mongolian People's Republic*. Academia Nauk SSSR, Moscow, pp. 174–178, 242, illust. (in Russian)

Tokar, A.A. (1989a) A revision of the genus *Eryx* (Serpentes, Boidae) based upon osteological data. *Vestnik Zoologii*, Kiev (4), 46–55, 6 figs. (in Russian)

Tokar, A.A. (1989b) On the discovery of *Eryx elegans* (Gray) outside its normal area of distribution. *Vestnik Zoologii*, Kiev 1989(6), 25. (in Russian)

Tokar, A.A. (1991) A revision of the subspecies structure of javelin sand boa, *Eryx jaculus* (Linnaeus, 1758) (Reptilia, Boidae). *Herpetological Research*, Leningrad (1), 18–41, 9 figs.

Tokar, A.A. (1995a) Taxonomic revision of the genus *Gongylophis* Wagler 1830 (Serpentes, Boidae). *In: Abstracts, Second Asian Herpetological Meeting*. Ashgabat, Turkmenistan, p. 55.

Tokar, A.A. (1995b) Taxonomic revision of the genus *Gongylophis* Wagler, 1830, *G. conicus* (Schneider 1801) and *G. muelleri* Boulenger, 1892 (Serpentes Boidae). *Tropical Zoology*, Firenze 8(2), 347–360, 3 figs.

Tokar, A.A. & Obst, F.J. (1993) *Eryx jaculus* (Linnaeus, 1758)—Westliche Sandboa. *In:* Böhme, W. (Ed.), *Handbuch der Reptilien und Amphibien Europas. Band 3/I. Schlangen (Serpentes) I. (Typhlopidae, Boidae, Colubridae 1, Colubrinae).* AULA-Verlag, Wiesbaden, pp. 35–53, figs. 4–9.

Tokar, A.A., Tokar, A.N. & Golubev, M. (1995) Comments on the validity of *Coluber atayevi* Tuniyev et Shammakov, 1993 (Ophidia, Colubridae). *In: Abstracts, Second Asian Herpetological Meeting, Ashgabat, Turkmenistan*, 55.

Toktosunov, A.T., Mazik, E.Y., Kassymalieva, E.B. & Imanalieva, N.D. (1992) Specific characters of karyotypes of some snake species (Colubridae, Viperidae, Crotalidae) from the northern Tien-shan. *Zoologischeskii Zhurnal*, Moscow 71(5), 118–127, 3 figs. (in Russian)

Tolson, P.J. (1987) Phylogenetics of the boid snake genus *Epicrates* and Caribbean vicariance theory. *Occasional Papers of the Museum of Zoology, University of Michigan* (715), 1–68, 26 figs.

Tolson, P.J. (1988) Critical habitat, predator pressures, and the management of *Epicrates monensis* (Serpentes, Boidae) on the Puerto Rico Bank: a multivariate analysis. *In:* Szaro, R.C., Severson, K.E. & Patton, D.R. (Eds.), *Management of amphibians, reptiles, and small mammals in North America.* USDA Forest Service General Technical Report RM-166, pp. 228–238, 7 figs.

Tolson, P.J. & Henderson, R.W. (1993) *The natural history of West Indian boas.* R & A Publishing, Somerset, 125 pp, 108 pls.

Tolson, P.J., Garcia, M.A. & Ellsworth, C.L. (2007) Habitat use by the Mona boa (*Epicrates monensis monensis*) on Isla Mona, West Indies *In:* Henderson, R.W. & Powell, R. (Eds.), *Biology of the boas and pythons.* Eagle Mountain Publishing, Eagle Mountain, pp. 119–127, col. illust.

Tomas-Rodas, J. (1938) *Contribucion al estudio de las serpentes venenosas de Guatemala.* Tipographia Nacional, Guatemala, 152 pp., illust.

Tome, S. (2002). *Kace zakaj se jih bojimo?!?* Prirodoslovni Muzej Slovenije, Ljubljana, 69 pp., col. illust.

Tomovic, L. (2006) Systematics of the nose-horned viper (*Vipera ammodytes*, Linnaeus, 1758). *Herpetological Journal* 16, 191–201.

Tomovic, L. & Dzukic, G. (2003) Geographic variation and taxonomy of the nose-horned viper, *Vipera ammodytes* (L., 1758), in the central and eastern parts of the Balkans; a multivariate study. *Amphibia-Reptilia,* Leiden 24(3), 359–377, 6 figs.

Tomovic, L, Radojicic, J., Dzukic, G. & Kalezic, L.M. (2002) Sexual dimorphism of the sand viper (*Vipera ammodytes* L.) from the central part of Balkan Peninsula. *Russian Journal of Herpetology* 9, 69–76.

Tonge, S. (1986) The herpetofauna of Jersey. *British Herpetological Society Bulletin*, London (17), 18–21, 1 fig.

Tonge, S. (1990) The past, present and future of the herpetofauna of Mauritius. *Bulletin of the Chicago Herpetological Society* 25(12), 220–226, 7 figs.

Tonge, S. & Morgan, D. (1984) Notes on the herpetofauna of southern Malawi. *British Herpetological Society Bulletin*, London (9), 35–42, illust.

Toral-C., E., Morales-M., E.M. & Ortiz-P, A. (2004) Geographic distribution: Serpentes: *Siphlophis compressus. Herpetological Review* 35(1), 84.

Toriba, M. (1986) Preliminary study on the systematic status of a Danjo Islands snake. *Japanese Journal of Herpetology*, Kyoto 11(3), 124–136, 5 figs.

Toriba, M. (1987) Karyotypes of some species of the genus *Bungarus. Chinese Herpetological Research*, Chongging 1, 17–21, illust.

Toriba, M. (1988) Taxonomic status of *Agkistrodon caliginosus* Gloyd and *A. blomhoffii ussuriensis* (Emelianov). *The Snake*, Nittagun 20(1), 30–39, 11 figs.

Toriba, M. (1989a) Karyotypes of Japanese species of the genus *Trimeresurus. Japanese Journal of Herpetology*, Kyoto 13(1), 10–14, illust.

Toriba, M. (1989b) Karyotypes of Japanese species of the genus *Trimeresurus. In:* Matsui, M., Hikida, T. & Goris, R.C. (Eds.), *Current herpetology in East Asia.* Herpetological Society of Japan, Kyoto, pp. 122–127, 3 figs.

Toriba, M. (1989c) Karyotypes of two species of elapid snakes. *The Snake*, Nittagun 21(2), 87–89, 2 figs.

Toriba, M. (1990a) Karyotypes of some somes of Japanese colubrid snakes. *In:* Olmo, E. (Ed.), *Cytogenetics of amphibians and reptiles.* Advances in Life Sciences, Basel, pp. 255–259, 6 figs.

Toriba, M. (1990b) Venomous snakes of medical importance in Korea. *In:* Gopalakrishnakone, P. & Chou, L.M. (Eds.), *Snakes of medical importance (Asia-Pacific region).* National University of Singapore, Singapore, pp. 377–386, 4 figs.

Toriba, M. (1990c) Venomous snakes of medical importance in the Philippines. *In:* Gopalakrishnakone, P. & Chou, L.M. (Eds.), *Snakes of medical importance (Asia-Pacific region).* National University of Singapore, Singapore, pp. 463–469, 3 figs.

Toriba, M. (1991) Karyotypes of two colubrid snakes, *Entechinus semicarinatus* and *Dinodon orientalis. The Snake*, Nittagun 23(1), 1–5, 3 figs.

Toriba, M. (1992a) Reconsideration of color patterns of *Rhabdophis tigrinus* in Japan. *The Snake*, Nittagun 24(1), 16–22, 8 figs.

Toriba, M. (1992b) A comment on the status of the Thai palm viper, *Trimeresurus wiroti. The Snake*, Nittagun 24(1), 71–73.

Toriba, M. (1992c) Karyotype of red-tailed pipe snake, *Cylindrophis rufus* (Uropeltidae, Cylindrophiinae). *The Snake*, Nittagun 24(2), 143–146, 1 fig.

Toriba, M. (1993) Present location of the type specimens described by Maki. *The Snake*, Nittagun 25(1), 71–72.

Toriba, M. (1994a) Karyotype of *Trimeresurus erythrurus. The Snake*, Nittagun 26(2), 141–143, 1 fig.

Toriba, M. (1994b) Gender of the genus *Amphiesma* Dumeril, Bibron and Dumeril. *The Snake*, Nittagun 26(2), 145.

Toriba, M. & Nakamoto, E. (1987) Reproductive biology of Erabu sea snake, *Laticauda semifasciata* (Reinwardt). *The Snake*, Nittagun 19, 101–106, 1 fig.

Toriba, M. & Nakamoto, E. (1991) Occurrence of *Hydrophis ornatus* (Gray) at Amami-oshima Island, Japan. *The Snake*, Nittagun 23(2), 117–118.

Toriba, M. & Sawai, Y. (1981) New record of Persian Gulf seasnake, *Hydrophis lapemoides* (Gray), from Penang, Malaysia. *The Snake*, Nittagun 13(2), 134–136, 2 figs.

Toriba, M. & Sawai, Y. (1990) Venomous snakes of medical importance in Japan. *In:* Gopalakrishnakone, P. & Chou, L.M. (Eds.), *Snakes of medical importance (Asia-Pacific region).* National University of Singapore, Singapore, pp. 323–347, 12 figs.

Toriba, M. & Yosida, T.H. (1986) Karyotypes of the laticaudine sea snakes from Ryukyu Archipelago. *The Snake*, Nittagun 18(2), 70–76, 4 figs.

Toriba, M., Moriguchi, H., Sakai, A., Sugano, H. & Miyata, H. (1983) Reptiles of the western part of Tochigi Prefecture, Japan. *The Snake*, Nittagun 15(1), 43–47, 2 figs. (in Japanese)

Toriba, M., Hirabayashi, K. & Cho, K.A. (1990) A case of reproduction of Burmese green pit viper, *Trimeresurus erythrurus*, with notes on morphological features. *The Snake*, Nittagun 22(1), 8–10, 1 fig.

Tornier, G. (1896) *Reptilien und Amphibien Ost-Afrikas.* D. Reimer, Berlin, 164 pp., figs. a–e.

Tornier, G. (1901) Die Reptilien und Amphibien der deutschen Tiefseeexpedition 1898/99. *Zoologischer Anzeiger*, Leipzig 24, 61–66.

Tornier, G. (1902a) Herpetologisch neues aus Deutsch-Ost-Afrika. *Zoologische Jahrbücher, Abteilung für Systematik, Ökologie und Geographie der Tiere*, Jena 15(2), 578–590, 2 figs.

Tornier, G. (1902b) Herpetologisch neues aus Ost-Afrikas. *Zoologischer Anzeiger*, Leipzig 25(683–684), 700–704.

Tornier, G. (1904) Entstehen und Bedeutung der Farbkleimuster der Eidechsen und Schlangen. *Sitzungsberichte der Königlich Preußischen Akademie der Wissenschaften (Physikalisch-Mathematischen Classe)*, Berlin 40, 1203–1214.

Torr, G. (2000) *Pythons of Australia: a natural history*. University of New South Wales Press, Syndey, viii + 103 pp., 34 col. pls.

Torre-Loranca, M.A. de., López-Luna, M.A. & Vogt, R.C. (2000) Geographic distribution: Serpentes: *Geophis chalybeus*. *Herpetological Review* 31(3), 185–186.

Torres-Carvajal, O. (2004) Herpetofauna of Isla de La Plata, Ecuador. *Herpetological Review* 35(1), 85.

Torres-Carvajal, O., Yánez-Muñoz, M.H., Quirola, D., Smith, E.N. & Almendáriz, A. (2012) A new species of blunt-headed vine snake (Colubridae, *Imantodes*) from the Chocó region of Ecuador. *ZooKeys* (244), 91–110, 7 figs.

Tortonese, E. (1942) Gli anfibi e i Rettili del R. Museo Zoologico di Torino. *Bolletino dei Musei di Zoologia ed Anatomia Comparata della Università di Torino* (Series 4) 49(127), 203–222.

Tortonese, E. (1953) Spigolature di erpetologia pedemontana. *Natura*, Milano 44(1–2), 24–34.

Tortonese, E. & Lanza, B. (1968) *Pesci, Anfibi e Rettili*. Aldo Martello Ed., Milano, 176 pp.

Tóth, T., Grillitsch, H., Farkas, B., Gál, J. & Susic, G. (2006) Herpetofaunal data from Cres Island, Croatia. *Herpetozoa*, Wien 19(1–2), 27–58, 30 figs., 13 col. photos.

Touzet, J.-M. (1986) Mordeduras de ofidios venenosos en la communidad de los indigenas Siona-secoya de San Pablo de Kantesyia y datos sobre la fauna de reptiles y anfibios locales. *Publicaciones del Museo Ecuatoriano de Ciencias Naturales, Série Revistas* 7(5), 163–190, illust.

Tovar-Tovar, H. & Mendoza-Quijano, F. (2001) Geographic distribution: Serpentes: *Agkistrodon taylori*. *Herpetological Review* 33(4), 276–277.

Townsend, C.H.T. (1897) On the biogeography of Mexico, Texas, New Mexico, and Arizona. II. *Transactions of the Texas Academy of Science* 2(1), 33–86.

Townsend, J.H. (2006a) *Inventory and conservation assessment of the herptofauna of the Sierra de Omoa, Honduras, and a review of the* Geophis *(Squamata, Colubridae) of eastern Nuclear Central America*. Master's Thesis, University of Florida, Gainesville, xiv + 124 pp., 20 figs.

Townsend, J.H. (2006b) *Geophis dunni*. *Catalogue of American amphibians and reptiles*, pp. 838.1–838.3, 1 fig.

Townsend, J.H. (2009) Morphological variation in *Geophis nephodrymus* (Squamata: Colubridae), with comments on conservation of *Geophis* in eastern Nuclear Central America. *Herpetologica* 65(3), 292–302, 4 figs.

Townsend, J.H. & Wilson, L.D. (2006a) A new species of snake of the *Geophis dubius* group (Reptilia, Squamata, Colubridae) from the Sierra de Omoa of northwestern Honduras. *Proceedings of the Biological Society of Washington* 119, 150–159.

Townsend, J.H. & Wilson, L.D. (2006b) *Geophis fulvoguttatus*. *Catalogue of American amphibians and reptiles*, pp. 839.1–839.4, 1 fig.

Townsend, J.H. & Wilson, L.D. (2008) *The amphibians & reptiles of Cusuco National Park, Honduras. Los anfibios y reptiles del Parque Nacional Cusuco Honduras*. Bibliomania, Salt Lake City, xiv + 312 pp., col. illust.

Townsend, J.H. & Wilson, L.D. (2010) Conservation of the Honduran herpetofauna issues and imperatives. *In:* Wilson, L.D., Townsend, J.H. & Johnson, J.A. (Eds.), *Conservation of Mesoamerican amphibians and reptiles*. Eagle Mountain Publishing, Eagle Mountain, pp. 461–487, 2 figs.

Townsend, J.H., Wilson, L.D., Plenderleith, T.L., Talley, B.L. & Nifong, J.C. (2005a) *Ninia pavimentata* (Squamata, Colubridae), an addition to the snake fauna of Honduras. *Caribbean Journal of Science* 41, 869–870.

Townsend, J.H., Nifong, J.C. & Wilson, L.D. (2005b) First record of the colubrid snake *Rhadinaea anachoreta* Smith and Campbell from Honduras. *Herpetological Bulletin* 94, 2–3.

Townsend, J.H., Wilson, L.D., Talley, B.L., Fraser, D.C., Plenderleith, T.L. & Hughes, S.M. (2006) Additions to the herpetofauna of Parque Nacional El Cusuco, Honduras. *Herpetological Bulletin* (96), 29–39, 6 figs.

Townsend, J.H., Butler, J.M., Wilson, L.D., Ketzler, L.P., Slapcinsky, J. & Stewart, N.M. (2008a) Significant range extentions for the Central American colubrid snake *Ninia pavimentata* (Bocourt 1883). *Herpetological Bulletin* (106), 15–17, 1 fig.

Townsend, J.H., Wilson, L.D., Ketzler, L.P. & Luque-Montes, I.R. (2008b) The largest blindsnake in Mesoamerica: a new species of *Typhlops* (Squamata: Typhlopidae) from an isolated karstic mountain in Honduras. *Zootaxa* (1932), 18–26, 4 figs.

Toyama, M. (1983) Taxonomic reassignment of the colubrid snake *Opheodrys kikuzatoi*, from Kumejima Island, Ryukyu Archipelago. *Japanese Journal of Herpetology* 10, 33–38.

Traill, T.S. (1843) Description of the *Elaps jamesoni*, a new species of serpent from Demerama. *The Edinburgh New Philosophical Journal* 34(67), 53–55, 2 figs.

Tran, K. & Nguyen, Q.T. (1995) *Poisonous snakes of Vietnam*. Science and Technology Publishing House, Hanoi, 1–204. (in Vietnamese)

Trape, J.-F. (1981) *Liste des serpents de la Republique Populaire du Congo*. Office de la Recherche Scientifique et Technique Outre-Mer, Brazzaville, 11 pp.

Trape, J.-F. (1985) Les serpents de la région de Dimonika (Mayombe, République Populaire du Congo). *Revue de Zoologie Africaine*, Tervuren 99(2), 135–140, 1 fig.

Trape, J.-F. (1990) Présence de *Ramphotyphlops braminus* (Ophidia, Typhlopidae) au Sénégal. *Bulletin de la Société Herpétologique de France* (55), 40–41.

Trape, J.-F. (1997) Geographic distribution: Serpentes: *Haemorrhois* (=*Coluber*) *dorri*. *Herpetological Review* 28(2), 97–98.

Trape, J.-F. (2005) Note sur quelques serpents méconnus du Burkina Faso de la collection de Benigno Roman. *Bulletin de la Société Herpétologique de France* (116), 39–49.

Trape, J.-F. & Ba, H. (2006) Geographic distribution: Serpentes: *Ramphotyphlops braminus*. *Herpetological Review* 37(3), 363.

Trape, J.-F. & Balde. (2006) Geographic distribution: Serpentes: *Haemorrhois dorri*. *Herpetological Review* 37(3), 362.

Trape, J.-F. & Carme, B. (1982) Les serpents venimeux de la République Populaire du Congo. *Revue Médicale du Congo*, Brazzaville 2(2), 53–70, 22 figs.

Trape, J.-F. & Ineich, I. (2006) Geographic distribution: Serpentes: *Elapsoidea trapei*. *Herpetological Review* 37(3), 362.

Trape, J.-F. & Mané, Y. (1995a) Geographic distribution: Serpentes: *Afronatrix anoscopus* (African brown water snake). *Herpetological Review* 26(3), 156.

Trape, J.-F. & Mané, Y. (1995b) Geographic distribution: Serpentes: *Grayia tholloni* (Thollon's water snake). *Herpetological Review* 26(3), 156.

Trape, J.-F. & Mané, Y. (2000) Les serpents des environs de Dielmo (Sine-Saloum, Sénégal). *Bulletin de la Société Herpétologique de France* (95), 19–35, 2 figs.

Trape, J.-F. & Mané, Y. (2004) Les serpents des environs de Bandafassi (Sénégal oriental). *Bulletin de la Société Herpétologique de France* (109), 5–34, 3 figs.

Trape, J.-F. & Mané, Y. (2005) Une nouvelle espèce du genre *Mehelya* (Serpentes: Colubridae) de Haute-Casamance (Sénégal). *Bulletin de la Société Herpétologique de France* (115), 23–30, 2 figs.

Trape, J.-F. & Mané, Y. (2006a) Le genre *Dasypeltis* Wagler (Serpentes, Colubridae) en Afrique de l'Ouest: description de trois espèce nouvelles. *Bulletin de la Société Herpétologique de France* (119), 27–56, 6 figs.

Trape, J.-F. & Mané, Y. (2006b) *Guide des serpents d'Afrique Occidentale.* Institut de Recherche pour de Développement, Paris, 226 pp.

Trape, J.-F. & Roux-Estève, R. (1990) Note sur une collection de serpents du Congo avec description d'une espèce nouvelle. *Journal of African* Zoology, Tervuren 104(5), 375–383, 3 figs.

Trape, J.-F. & Roux-Estève, R. (1995) Les serpents du Congo: Liste commentée et clé détermination. *Journal of African Zoology,* Teruven 109(1), 31–50, 1 fig.

Trape, J.-F., Mané, Y. & Balde, C. (2005) Le mamba noir *Dendroaspis polylepis* (Serpentes: Elapidae) en Afrique de l'Ouest. *Bulletin de la Société Herpétologique de France* (115), 31–36.

Trape, J.-F., Mané, Y. & Ineich, I. (2006) *Atractaspis microlepidota, A. micropholis* et *A. watsoni* en Afrique occidentale et centrale. *Bulletin de la Société Herpétologique de France* (119), 15–26, 6 figs.

Trape, J.-F., Chirio, L., Broadley, D.G. & Wüster, W. (2009) Phylogeography and systematic revision of the Egyptian cobra (Serpentes: Elapidae: *Naja haje*) species complex, with the description of a new species from West Africa. *Zootaxa* (2236), 1–25, 9 figs.

Trape, S., Mediannikov, O. & Trape, J.F. (2012) When colour patterns reflect phylogeography: new species of *Dasypeltis* (Serpentes: Colubridae: Boigini) from West Africa. *Comptes Rendus Biologies,* Paris 335(7), 488–501, 8 figs.

Trapido, H. (1937) *The snakes of New Jersey. A guide.* Newark Museum, Newark, 2–60, 1 fig.

Trapido, H. (1941) A new species of *Natrix* from Texas. *American Midland Naturalist* 25(3), 673–680, 5 figs.

Trapido, H. (1944) The snakes of the genus *Storeria. American Midland Naturalist* 31(1), 1–83, 61 figs.

Trapp, B. (2007) *Amphibien und Reptilien des griechischen Festlandes.* Natur und Tier-Verlag, Münster, 279 pp., col. illust.

Trauth, S.E. (1991a) Distribution, scutellation, and reproduction in the queen snake, *Regina septemvittata* (Serpentes, Colubridae), from Arkansas. *Proceedings of the Arkansas Academy of Sciences* 45, 103–106, illust.

Trauth, S.E. (1991b) Posterior maxillary fangs of the flathead snake, *Tantilla gracilis* (Serpentes, Colubridae), using scanning electron microscopy. *Proceedings of the Arkansas Academy of Sciences* 45, 133–136, illust.

Trauth, S.E. (1993) Enlarged posterior maxillary teeth in the scarlet snake, *Cemophora coccinea* (Serpentes, Colubridae), using scanning electron microscopy. *Proceedings of the Arkansas Academy of Sciences* 47, 157–160, 2 figs.

Trauth, S.E. (1998) First records for the blackmask racer (*Coluber constrictor latrunculus*) in eastern Arkansas. *Journal of the Arkansas Academy of Science* 51, 174–177.

Trauth, S.E., Robison, H.W. & Plummer, M.V. (2004) *The amphibians and reptiles of Arkansas.* University of Arkansas Press, Fayetteville, 421 pp., 534 figs.

Travers, S.L., Townsend, J.H., Sunyer, J., Obando, L.A., Wilson, L.D. & Nickerson, M.A. (2011) New and noteworthy records of amphibians and reptiles from Reserva de la Biósfera Bosawas, Nicaragua. *Herpetological Review* 42(3), 399–401, 1 fig.

Trembath, V.L. & Lloyd, R. (2005) Geographic distribution: Serpentes: *Stegonotus cucullatus. Herpetological Review* 36(2), 204.

Trethewie, E.R. (1956) Pharmacological effects of Australian snake venoms. *In:* Buckley, E.E. & Porges, N. (Eds.), *Venoms.* American Association for the Advancement of Science, Washington, D.C., pp. 243–251, 9 figs.

Trethewie, E.R. & Khaled, L. (1973) Changes in antigenic character of fresh Australian snake venom upon aging. *Toxicon* 11(1), 55–58, 10 figs.

Treu, B. (2008) *Amurnattern,* Elaphe schrenckii & Elaphe anomala. Natur und Tier-Verlag, Münster, 61 pp., col. illust.

Treu, B. (2012) *Die Steppennatter,* Elaphe Dione. Natur und Tier-Verlag, Münster, 62 pp., col. illust.

Tristram, H.B. (1860) *The great Sahara: wanderings south of the Atlas Mountains.* John Murray, London, 453 pp.

Tristram, H.B. (1884) *The survey of western Palestine. The fauna and flora of Palestine.* Committee of the Palestine Exploration Fund, London, xxii + 455 pp., 20 pls.

Troncoso, J.F. & Ortiz, J.C. (1987) Catalogo herpetologico del Museo Regional de Concepción. *Comunicaciones del Museo Regional de Concepción* 1, 9–19.

Troost, G. (1836) On a new genus of serpents, and two new species of the genus *Heterodon,* inhabiting Tennessee. *Annals of the Lyceum of Natural History of New York* (1828–1836) 3, 174–190, 1 pl.

Troost, G. (1844) Neue Schlangensippe und zwo neue Gattungen in Tennessee. *Isis von Oken,* Leipzig 37, 113–116.

Troschel, F.H. (1848) Amphibien. *In:* Schomburgk, R.H. (Ed.), *Reisen in Britisch-Guiana in den Jahren 1840–1844. Versuch einer Fauna und Flora von Britisch-Guiana.* Dritter Theil. J.J. Weber, Leipzig, pp. 645–661.

Troschel, F.H. (1854) Einige fossile Wirbelthiere aus der Braunkohle von Rott. *Verhandlungen des Naturhistorischen Vereins der Prueßischen Rheinlande und Westfalens,* Bonn 11(1), xix.

Troschel, F.H. (1855) [No title]. *Sitzungsberichte der Niederrheinischen Gesellschaft für Natur- und Heilkunde,* Bonn 1854, 126–127.

Troschel, F.H. (1859) Eine fossile Schlange aus der Braunkohle von Rott (Siebengebirge). *Neues Jahrbuch für Mineralogie, Geognosie, Geologie und Petrefakten-Kunde* 1859, 237.

Troschel, F.H. (1877) Bericht über die Leistungen in der Herpetologie während des Jahres 1876. *Archiv für Naturgeschichte,* Berlin 43(2), 97–117.

Troschel, F.H. & Ruthe, J.F. (1848) *Handbuch der Zoologie. Dritte Auslage.* C.G. Luderip, Berlin, 651 pp.

Truitt, J.O. (1962) *A guide to the snakes of south Florida (Lake Okeechobee to Florida Keys).* Museum of Science and Natural History, Miami, v + 46 pp., illust.

Truong, N.Q., van Kien, D., Coc, H.T. & van Sang, N. (2007a) Geographic distribution: Serpentes: *Boiga kraepelini. Herpetological Review* 38(1), 102.

Truong, N.Q., van Kien, D., Coc, H.T. & van Sang, N. (2007b) Geographic distribution: Serpentes: *Sinonatrix aequifasciata*. *Herpetological Review* 38(1), 106.

Trutnau, L. (1975) *Europäische Amphibien und Reptilien*. Belser, Stuttgart, 212 pp., 135 col. photos.

Trutnau, L. (1998) *Schlangen im Terrarium. Band 2. Giftschlangen*. Second edition. Verlag Eugen Ulmer, Stuttgart, 367 pp., 125 col. photos. [first edition, 1981, 200 pp., 59 col. photos.]

Trutnau, L. (2002) *Schlangen im Terrarium. Band 1. Ungiftige Schlangen*. Second edition. Verlag Eugen Ulmer, Stuttgart, 632 pp., 251 col. photos. [first edition, 1979, 200 pp., 63 col. photos.]

Trutnau, L. (2004) *Venomous snakes*. Third edition. Krieger Publishing Company, Malabar, 340 pp., 129 col. photos. [English translation of Trutnau, 1998]

Tsai, T.-S. & Tu, M.-C. (2001) Reproductive cycle of female Chinese green tree vipers, *Trimeresurus stejnegeri stejnegeri* in northern Taiwan. *Herpetologica* 57(2), 157–168, 6 figs.

Tschudi, J.J. von. (1837) Neues Genus von Wasserschlangen. *Archiv für Naturgeschichte*, Berlin 3(1), 331–335, pl. 8.

Tschudi, J.J. von. (1845) Reptilium conspectum quae in Republica Peruana reperiuntur et pleraque observata vel collecta sunt in itinere. *Archiv für Naturgeschichte*, Berlin 11(1), 150–170. [reprinted in 1968 by Society for the Study of Amphibians and Reptiles]

Tschudi, J.J. von. (1846) *Untersuchungen uber die Fauna Peruana. Herpetologie*. Scheitlin und Zollikofer, St. Gallen (1845), 80 pp., 12 col. pls.

Tu, A.T. (1974) Sea snakes investigation in the Gulf of Thailand. *Journal of Herpetology* 8(3), 201–210.

Tu, A.T. (1976) Investigation of the sea snake, *Pelamis platurus* (Reptilia, Serpentes, Hydrophiidae), on the Pacific coast of Costa Rica, Central America. *Journal of Herpetology* 10, 13–18.

Tu, A.T. & Tu, T. (1970) Addenda—reptiles. Sea snakes from southeast Asia and far east and their venoms. *In:* Halstead, B.W. (Ed.), *Poisonous and venomous marine animals of the world. Vol. 3*. Government Printing Office, Washington, D.C., pp. 885–903, 10 pls.

Tuberville, T., Bodie, J.R., Jensen, J.B., LaClaire, L. & Gibbons, J.W. (2000) Apparent decline of the southern hognose snake, *Heterodon simus*. *Journal of the Elisha Mitchell Scientific Society* 116, 19–40.

Tuck, R.G., Jr. (1971) Amphibians and reptiles from Iran in the United States National Collection. *Bulletin of the Maryland Herpetological Society* 7(3), 48–86, 9 figs.

Tuck, R.G., Jr. (1974a) Some amphibians and reptiles from Iran. *Bulletin of the Maryland Herpetological Society* 10(2), 59–65, 11 figs.

Tuck, R.G., Jr. (1974b) Poisonous and dangerous snakes of Iran and a key to their identification. *Hunting and Nature*, Tehran (August-September), 44–52, illust. (in Persian)

Tuijl, L. van. (1995) Revised catalogue of the type specimens of recent amphibians and reptiles in the "Zoölogisch Museum" University of Amsterdam the Netherlands. *Bulletin, Zoölogisch Museum*, Amsterdam 14(8), 125–141.

Tuniyev, B.S. (1995) On the Mediterranean influence on the formation of herpetofauna of the Caucasian Isthmus and its main xerophylous refugia. *Russian Journal of Herpetology*, Moscow 2(2), 95–119, 10 figs.

Tuniyev, B.S. & Ostrovskikh, S.V. (2001) Two new species of vipers of "kaznakovi" complex (Ophidia, Viperinae) from the western Caucasus. *Russian Journal of Herpetology* 8(2), 117–126, 12 col. figs.

Tuniyev, B.S. & Shammakov, S.M. (1993) *Coluber atayevi* sp. nov. (Ophidia, Colubridae) from the Kopet-Dag Mountains of Turkmenistan. *Asiatic Herpetological Research*, Berkeley 5, 1–10, 3 figs.

Tuniyev, B.S., Atayev, C. & Shammakov, S. (1998) Report on fauna of amphibians and reptiles from eastern Kopet Dag in Turkmenistan. *Russian Journal of Herpetology* 5(1), 74–81, 7 figs.

Tuniyev, B.S, Dusej, G. & Flärdh, B. (1999) Zoogeographic analysis of the herpetofauna of south-western Turkmenistan. *Russian Journal of Herpetology* 6(2), 125–142, 4 figs.

Tuniyev, B.S., Orlov, N.L., Ananjeva, N.B. & Agasyan, A.L. (2009) *Snakes of the Caucasus: taxonomic diversity, distribution, conservation*. KMK Scientific Press, St. Petersburg, 222 pp., 80 col. photos. (in Russian)

Tuniyev, B.S., Nilson, G. & Andrén, C. (2010) A new species of viper (Reptilia, Viperidae) from the Altay and Saur Mountains, Kazakhstan. *Russian Journal of Herpetology* 17(2), 110–120, 12 figs.

Tuniyev, B.S., Avci, A., Tuniyev, B.S., Agasian, A.L. & Agasian, L.A. (2012) Description of a new species of shield-head vipers—*Pelias olguni* sp. nov., from basin of upper flow of the Kura River in Turkey. *Russian Journal of Herpetology* 19(4), 314–332, 22 col. figs.

Turell, M.J. & Middlebrook, J.L. (1988) Mosquito inoculation: an alternative bioassay for toxins. *Toxicon* 26(11), 1089–1094.

Turner, F.B. (1955) *Reptiles and amphibians of Yellowstone National Park*. Yellowstone Library and Museum Association, Yellowstone National Park, iv + 40 pp., 20 figs.

Tweedie, M.W.F. (1940a) Notes on Malayan reptiles. *Bulletin of the Raffles Museum*, Singapore (16), 83–87.

Tweedie, M.W.F. (1940b) The poisonous snakes of Malaya. I. *Malayan Nature Journal*, Kuala Lumpur 1(2), 45–53, illust.

Tweedie, M.W.F. (1941a) The poisonous snakes of Malaya. II. *Malayan Nature Journal*, Kuala Lumpur 1(3), 78–85, illust.

Tweedie, M.W.F. (1941b) The poisonous snakes of Malaya. III. *Malayan Nature Journal*, Kuala Lumpur 1(4), 128–134, illust.

Tweedie, M.W.F. (1941c) *Poisonous animals of Malaya*. Malaya Publishing House, Singapore, 90 pp., 7 figs.

Tweedie, M.W.F. (1946) A new snake from the Malay Peninsula. *Annals and Magazine of Natural History* (Series 11) 13(98), 142–144, 1 fig.

Tweedie, M.W.F. (1949) Reptiles from the Kelabit Plateau. *Sarawak Museum Journal*, Kuching 5, 154–155.

Tweedie, M.W.F. (1950) Notes on Malayan reptiles, No. 2. *Bulletin of the Raffles Museum*, Singapore (23), 191–199, 1 fig.

Tweedie, M.W.F. (1954) Notes on Malayan reptiles, No. 3. *Bulletin of the Raffles Museum*, Singapore (25), 107–117, 2 figs., 1 pl.

Tweedie, M.W.F. (1983) *The snakes of Malaya*. Third edition. Singapore National Printers, Singapore, 167 pp., 28 figs., 12 pls. [first edition, 1953, 139 pp., 26 figs., 12 pls.; second edition, 1957, 143 pp., 28 figs., 16 pls.]

Tzarevsky, S.F. (1915) Formae *Viperae ammodytes* L. in Europa et in Transcaucasia lectae. *Annuaire du Musée Zoologique de l'Académie Impériale des Sciences de Petrograd* 20(1), 130–136. (in Russian) [surname also translated as Carevskij, Czarevsky, Tsarewsky and Zarevskij *fide* Adler, 2012: 216]

Tzarevsky, S.F. (1916) Aperçu des représentants du genre *Eryx*, principalement de l'Empire Russe et des pays limitrophes. *Annuaire du Musée Zoologique de l'Académie Impériale des Sciences de Petrograd* (1915) 20(1), 340–388, 11 figs. (in Russian)

Tzarevsky, S.F. (1917) Formes nouvelles du genre *Vipera*, trouvées dans l'Empire Russe: *Vipera tigrina* n. sp. and *Vipera berus sacjalinensis* var. nova? *Annuaire du Musée Zoologique de l'Académie des Sciences de Petrograd* (1916) 21(1), 34–39. (in Russian)

Tzarevsky, S.F. (1926) The reptiles and amphibians of P.K. Kozlov's expedition in Mongolia and Sze-chuen in 1907–1909. *Annuaire du Musée Zoologique de l'Académie des Sciences de l'URSS*, 26(1–2), 79–86. (in Russian)

Uchiyama, R., Maeda, N., Numata, K. & Seki, S. (2002) *A photographic guide: amphibians and reptiles in Japan*. Heibonsha, Tokyo, 337 pp., col. illust. (in Japanese)

Ugueto, G.N. & Rivas-Fuenemayer, G.A. (2010) *Amphibians and reptiles of Margarita, Coche, and Cubagua*. Edition Chimaira, Frankfurt-am-Main, 350 pp., 145 col. figs.

Ugurtas, I.H., Papenfuss, T.J. & Orlov, N.L. (2001) New record of *Walterinnesia aegyptia* Lataste, 1887 (Ophidia, Elapidae, Bungarinae) in Turkey. *Russian Journal of Herpetology* 8(2), 239–245, 15 col. figs.

Újvári, B., Korsós, Z. & Péchy, T. (2000) Life history, population characteristics and conservation of the Hungarian meadow viper (*Vipera ursinii rakosiensis*). *Amphibia-Reptilia,* Leiden 21(3), 267–278, 4 figs.

Ukuwela, K.D.B. & Dawundasekara, P.K. (2012) Color variations, sexual dimorphism, and population structure of *Aspidura trachyprocta* (Serpentes: Colubridae) in an anthropogenic habitat in Sri Lanka. *Russian Journal of Herpetology* 19(1), 70–74, 4 figs.

Ulber, T.M. (1993) *Catalogue of valid species and synonyms. Volume 1 (Crotalidae)*. Herprint International, Pretoria, xxv + 1.1–172. pp.

Ulber, T.M. (1994) *Catalogue of valid species and synonyms. Volume 2 (Viperidae/Atractaspididae)*. Herprint International, Pretoria, xxv + 1.1–181.1 pp.

Ulber, T.M. (1995) *Catalogue of valid species and synonyms. Volume 3 (Elapidae)*. Herprint International, Pretoria, xxxviii + 1.1–276.2 pp.

Ulber, T.M. (1999) Sir Andrew Smith's "Illustrations of the zoology of South Africa" an annotated table of contents for the "Reptilia" volume. *Smithsonian Herpetological Information Service* (120), 1–17.

Ullenbruch, K., Grell, O. & Böhme, W. (2010) Reptiles from southern Benin, West Africa, with the description of a new *Hemidactylus* (Gekkonidae), and a country-wide checklist. *Bonn Zoologiocal Bulletin* 57(1), 31–54, 17 figs.

Ulloa, J.C. (2006) Geographic distribution: Serpentes: *Leptotyphlops borapeliotes*. *Herpetological Review* 37(4), 498.

Underwood, G. (1962) *Reptiles of the eastern Caribbean*. Caribbean Affairs, Department of Extra-Mural Studies, University of the West Indies, 184 pp.

Underwood, G. (1967) *A contribution to the classification of snakes*. British Museum (Natural History), London, x + 179 pp., 14 figs.

Underwood, G. (1968) On the status of some South African vipers. *Annals of the Cape Provincial Museum*, Grahamstown 6, 81–85.

Underwood, G. (1976) A systematic analysis of boid snakes. *In:* Bellairs, A.A. & Cox, C.B. (Eds.), Morphology and biology of reptiles. *Linnaean Society Symposium Series* 3, Academic Press, London, pp. 151– 175.

Underwood, G. (1993) A new snake from St. Lucia, West Indies. *Bulletin of the Natural History Museum (Zoology)*, London 59(1), 1–9, 5 figs.

Underwood, G. (1995) *A tale of old serpents*. St. Lucia National Trust, Castries, 29 pp., 6 figs.

Underwood, G. (1999) Morphological evidence on the affinities of Vipers. *In:* Joger, U. (Ed.), Phylogeny and systematics of the Viperidae. *Kaupia, Darmstädter Beiträge zur Naturgeschichte* 8, 3–8, 2 figs.

Underwood, G. & Kochva, E. (1993) On the affinities of the burrowing asps *Atractaspis* (Serpentes, Atractaspididae). *Zoological Journal of the Linnean Society*, London 107(1), 3–64, 21 figs.

Underwood, G. & Stimson, A.F. (1990) A classification of pythons (Serpentes, Pythoninae). *Journal of Zoology*, London 221(4), 565–603, 9 figs.

Ursenbacher, S., Conelli, A., Golay, P., Monney, J.-C., Zuffi, M.A.L., Thierry, G., Durand, T. & Fumagalli, L. (2006) Phylogeography of the asp viper (*Viper aspis*) inferred from mitochondrial DNA: evidence for multiple Mediterranean refugial areas. *Molecular Phylogenetics and Evolution* 38(2), 546–552, 2 figs.

Ursenbacher, S., Schweiger, S., Tomovic, L., Crnobrnja-Isailovic, J., Fumagalli, L. & Mayer, W. (2008) Molecular phylogeography of the nose-horned viper (*Vipera ammodytes,*Linnaeus (1758)): evidence for high generic diversity and multiple refugia in the Balkan peninsula. *Molecular Phylogenetics and Evolution* 46(3), 1116–1128, 4 figs.

Uthmöller, W. (1937) Beitrag zur kenntnis der Schlangenfauna des Kilimandscharo (Tanganyika-Territory, Ehemaliges Deutsch-Ostafrika). *Temminckia*, Leiden 2, 97–134.

Uthmöller, W. (1938) Über eine kleine Sammlung von Schlangen vom Kilimandjaro. Ergebnisse der Ostafrikareise Uthmöller-Bohmanne. II. *Zoologischer Anzeiger*, Leipzig 124(1–2), 41–48.

Uthmöller, W. (1939) Über eine neue Rasse von *Crotaphopeltis hotamboeia*. *Zoologischer Anzeiger*, Leipzig 125(5–6), 108–112.

Uthmöller, W. (1941a) Die Schlangen Ostafrikas in ihrem Lebensraum unter besonderer Berucksichtigung der vulkane Kilimandjaro, Meru, Hanang, der Grossen Ostafrikanischen Bruchstufe und des Usambara-Gebirges. *Archiv für Naturgeschichte*, Berlin (Series 2) 10, 1–70, 20 figs., 2 pls.

Uthmöller, W. (1941b) Beitrag zur Kenntnis der Schlangenfauna Nordost-Ostafrikas (Tanganyika–Territory). Ergebnisse der Ostafrika-Expedition Uthmöller-Bohmann. V. A. Schlangen. I. *Zoologischer Anzeiger*, Leipzig 135(11–12), 225–242.

Utiger, U. & Schätti, B. (2004) Morphology and phylogenetic relationships of the Cyprus racer, *Hierophis cypriensis,* and the systematic status of *Coluber gemonensis gyarosensis* Mertens Reptilia: Squamata: Colubrinae). *Revue Suisse de Zoologie*, Genève 111(1), 225–238, 4 figs.

Utiger, U., Helfenberger, N., Schätti, B., Schmidt, C., Ruf, M. & Ziswiler, V. (2002) Molecular systematics and phylogeny of Old World ratsnakes, *Elaphe* AUCT., and related genera (Reptilia, Squamata. Colubridae). *Russian Journal of Herpetology* 9(2), 105–124, 3 figs.

Utiger, U., Schätti, B. & Helfenberger, N. (2005) The Oriental colubrine genus *Coelognathus* Fitzinger, 1843 and classification of Old and New World racers and ratsnakes (Reptilia, Squamata, Colubridae, Colubrinae). *Russian Journal of Herpetology* 12(1), 39–60, 5 figs.

Utsunomiya, T., Utsunomiya, Y., Okawa, H., Okada, S. & Naitou, J. (1996) *A field guide to amphibians and reptiles of Hiroshima Prefecture.* Hiba Society of Natural History, Hiroshima, 168 pp., 355 col. photos. (in Japanese)

Uzzell, T.M. & Starrett, P. (1958) Snakes from El Salvador. *Copeia* 1958(4), 339–342.

Vacher, J.-P. & Geniez, M. (Eds.) (2010) *Les reptiles de France, Belgique, Luxembourg et Suisse.* Paris, 543 pp., 676 col. photos.

Valakos, E.D. & Mylonas, M. (1992) Distribution and ecological aspects of the herpetofauna of Strofadhes Islands (Ionian Archipelago, Greece). *Herpetozoa*, Wien 5(1–2), 33–39, 2 figs.

Valakos, E.D. & Papapanagiotou, D. (1985) The herpetofauna of Kastellorizo (Megisti) Island (SE Aegean Archipelago). *British Herpetological Society Bulletin*, London (13), 9–12, 1 fig.

Valakos, E.D., Dimaki, M. & Pafilis, P. (2004) *Natural history of Lesvos: reptiles and amphibians.* Natural History Collection of Vrissas-Lesvos, Mytilene, 128 pp., col. illust.

Valakos, E.D., Pafilis, P., Lymberakis, P., Maragou, P., Sottiropoulos, K. & Foufopoulos, J. (2008) *The amphibians and reptiles of Greece.* Edition Chimaira, Frankfurt-am-Main, 480 pp., 411 col. figs.

Valdez-Lares, R. & Muñiz-Martinez, R. (2011a) Geographic distribution: Serpentes: *Leptodeira splendida. Herpetological Review* 42(3), 394.

Valdez-Lares, R. & Muñiz-Martinez, R. (2011b) Geographic distribution: Serpentes: *Leptophis diplotropis. Herpetological Review* 42(3), 394.

Valdivieso, D. & Tamsitt, J.R. (1963a) Records and observations on Columbian reptiles. *Herpetologica* 19(1), 28–39.

Valdivieso, D. & Tamsitt, J.R. (1963b) A check list and key to the amphibian and reptiles of Providencia and San Andres. *Caribbean Journal of Science*, Mayagüez 3(2–3), 77–79.

Valdujo, P.H. & Nogueira, C. (1999) Geographical distribution. *Philodryas livida. Herpetological Review* 30(1), 55.

Valdujo, P.H. & Nogueira, C. (2001) New records of squamate reptiles in the central Brazilian Cerrado, Emas National Park region. *Herpetological Review* 32(2), 128–130.

Valdujo, P.H., Nogueira, C., Baumgarten, L., Guinmarães Rodrigues, F.H., Brandão, R.A., Eterovic, A., Ramos-Neto, M.B. & Marques, O.A.V. (2009) Squamate reptiles from Parque Nacional das Emas and surroundings, Cerrado of central Brazil. *Check List* 5(3), 405–417, 5 figs.

Valencia-Hernández, A.A. (2006) *Taxonomía y distribución del género* Crotalus *(Linneo, 1758) en el estado de Hidalgo.* Doctoral Dissertation, Universidad Autónoma del Estado de Hidalgo, Pachuca de Soto, viii + 104 pp., 39 figs.

Valenzuela-Dellarossa, G., Núñez, H., Heibi, C. & Ortiz, J.C. (2010) Reptilia, Serpentes, Colubridae, *Tachymenis* Wiegmann, 1836: latitudinal and altitudinal distribution extension in Chile. *Check List* 6(2), 5–6, 4 figs.

Valverde, J.A. (1989a) Notas sobre vertebrados. VI. *Psammophis sibilans teknae* (Colubridae, Serpentes) una nueva forma del Marruecos occidental. *Actas IX Bienal de la Real Sociedad Española de Historia Natural*, 207–213.

Valverde, J.A. (1989b) Notas sobre vertebrados. VII. Una nueva cobra del NW de Africa, *Naja haje legionis*, ssp. nov. (Elapidae, Serpentes). *Actas IX Bienal de la Real Sociedad Española de Historia Natural*, 216–230.

Vance, T.L. (1982) A field key to the reptiles of Arkansas. HERP, *Bulletin of the New York Herpetological Society* 17(2), 32–48.

Vance, T.L. (1985) Annotated checklist and bibliography of Arkansas reptiles. *Smithsonian Herpetological Information Service* (63), 1–45.

Van Denburgh, J. (1895a) A review of the herpetology of Lower California. Part I. Reptiles. *Proceedings of the California Academy of Sciences* (1895–1896) (Series 2) 5, 556–561, pls. 4–14.

Van Denburgh, J. (1895b) Description of a new rattlesnake (*Crotalus pricei*) from Arizona. *Proceedings of the California Academy of Sciences* (1895–1896) (Series 2) 5, 856–857, pl. 50.

Van Denburgh, J. (1897) The reptiles of the Pacific coast and Great Basin. An account of the species known to inhabit California, and Oregon, Washington, Idaho and Nevada. *Occasional Papers of the California Academy of Sciences* (5), 1–236, figs.

Van Denburgh, J. (1909) New and previously unrecorded species of reptiles and amphibians from the island of Formosa. *Proceedings of the California Academy of Sciences* (1908–1913) (Series 4) 3(3), 49–56.

Van Denburgh, J. (1912a) *Advance diagnoses of new reptiles and amphibians from the Loo Choo Islands and Formosa.* John Van Denburgh, San Francisco, 8 pp.

Van Denburgh, J. (1912b) Expedition of the California Academy of Sciences to the Galapagos Islands, 1905–1906. IV. The snakes of the Galapagos Islands. *Proceedings of the California Academy of Sciences* (1908–1913) (Series 4) 1(4), 323–374, pls. 23–30.

Van Denburgh, J. (1912c) Concerning certain species of reptiles and amphibians from China, Japan, the Loo Choo Islands, and Formosa. *Proceedings of the California Academy of Sciences* (1908–1913) (Series 4) 3(10), 187–258.

Van Denburgh, J. (1920a) A further study of variation in the gopher-snakes of western North America. *Proceedings of the California Academy of Sciences* (Series 4) 10(1), 1–27, 2 pls.

Van Denburgh, J. (1920b) Description of a new species of rattle-snake (*Crotalus lucasensis*) from Lower California. *Proceedings of the California Academy of Sciences* (Series 4) 10(2), 29–30, pl. 3.

Van Denburgh, J. (1920c) Description of a new subspecies of boa (*Charina bottae utahensis*) from Utah. *Proceedings of the California Academy of Sciences* (Series 4) 10(3), 31–32.

Van Denburgh, J. (1922) The reptiles of western North America. An account of the species known to inhabit California and Oregon, Washington, Idaho, Utah, Nevada, Arizona, British Columbia, Sonora and Lower California. Vol. II. Snakes and turtles. *Occasional Papers of the California Academy of Sciences* (10), 617–1028, pls. 58–128.

Van Denburgh, J. (1923) A new subspecies of watersnake (*Natrix vibakari ruthveni*) from eastern Asia. *Proceedings of the California Academy of Sciences* (Series 4) 13(2), 3–4.

Van Denburgh, J. (1924) Notes on the herpetology of New Mexico, with a list of species known from that state. *Proceedings of the California Academy of Sciences* (Series 4) 13(12), 189–230.

Van Denburgh, J. & Slevin, J.R. (1913) A list of the amphibians and reptiles of Arizona, with notes on the species in the collection of the Academy. *Proceedings of the California Academy of Sciences* (1908–1913) (Series 4) 3, 391–454, pls. 17–28.

Van Denburgh, J. & Slevin, J.R. (1915) A list of the amphibians and reptiles of Utah, with notes on the species in the collection of the Academy. *Proceedings of the California Academy of Sciences* (Series 4) 5(4), 99–110, pls. 12–14.

Van Denburgh, J. & Slevin, J.R. (1918) The garter-snakes of western North America. *Proceedings of the California Academy of Sciences* (Series 4) 8(6), 181–270, pls. 7–17.

Van Denburgh, J. & Slevin, J.R. (1919) The gopher snakes of western North America. *Proceedings of the California Academy of Sciences* (Series 4) 9(6), 197–220, pls. 11–13.

Van Denburgh, J. & Slevin, J.R. (1921a) Preliminary diagnoses of new species of reptiles from islands in the Gulf of California, México. *Proceedings of the California Academy of Sciences* (Series 4) 11(6), 95–98.

Van Denburgh, J. & Slevin, J.R. (1921b) Preliminary diagnoses of more new species of reptiles from islands in the Gulf of California, México. *Proceedings of the California Academy of Sciences* (Series 4) 11(17), 395–398.
Van Denburgh, J. & Slevin, J.R. (1923) Preliminary diagnosis of four new snakes from lower California, Mexico. *Proceedings of the California Academy of Sciences* (Series 4) 13(1), 1–2.
Van Denburgh, J. & Thompson, J.L. (1908) Description of a new species of sea snake from the Philippine Islands, with a note on the palatine teeth in the Proteroglypha. *Proceedings of the California Academy of Sciences* (1908–1913) (Series 4) 3, 41–48, 1 pl.
Van der Winden, J., Bogaerts, S. & Strijbosch, H. (1997) Herpetofauna des Göksu Deltas und des umliegenden Gebirges, Türkei. *Salamandra* 33(1), 9–24.
Van Devender, R.W. (1980) Preliminary checklist of the herpetofauna of Monteverde, Puntarenas Province, Costa Rica and vicinity. *Brenesia* 17, 319–326.
Van Devender, R.W. & Cole, C.J. (1977) Notes on a colubrid snake, *Tantilla vermiformis*, from Central America. *American Museum Novitates* (2625), 1–12, 8 figs.
Van Devender, T.R. & Bradley, G.L. (1994) Late Quaternary amphibians and reptiles from Maravillas Canyon Cave, Texas, with discussion of the biogeography and evolution of the Chihuahuan desert herpetofauna. *In:* Brown, P.R. & Wright, J.W. (Eds.), Herpetology of the North American deserts. Proceedings of a symposium. *Southwestern Herpetological Society, Special Publication* (5), pp. 23–53, 5 figs.
Van Devender, T.R & Lowe, C.H., Jr. (1977) Amphibians and reptiles of Yepómera, Chihuahua, Mexico. *Journal of Herpetology* 11(1), 41–50.
Van Devender, T.R. & Worthington, R.D. (1977) The herpetofauna of Howell's Ridge Cave and the paleoecology of the northwestern Chihuahuan Desert. *In:* Wauer, R.H. & Riskin, D.H. (Eds.), *Transactions of the Symposium on the Biological Resources of the Chihuahuan Desert Region, United States and Mexico.* National Park Service Transactions and Proceedings 3, 85–106.
Van Devender, T.R., Lowe, C.H. & Lawler, H.E. (1994) Factors influencing the distribution of the Neotropical vine snake (*Oxybelis aeneus*) in Arizona and Sonora, México. *Herpetological Natural History* 2(1), 25–42, 1 fig.
Vandoni, C. (1914) *I rettili d'Italia. Con appendice pei collezionisti di rettili ed anfibi.* Ulrico Hoepli, Milano, xii + 274 pp., 55 figs.
Van Kampen, P.N. & Helmans, J. (1927) Amphibians and reptiles. *In:* Boschma, J. (Ed.), *Fauna van Nederland.* A.J. Sijthoff, Leiden, 63 pp., illust. (in Dutch)
Van Vleet, A.H. (1902) Snakes of Oklahoma. *Territory of Oklahoma Department of Geology and Natural History, Second Biennial Report* (1901–1902), 167–173.
Vanzolini, P.E. (1947) Sobre um novo gênero e espécie de colubrídeo opistóglifo. *Papeis Avulsos do Departamento de Zoologia*, São Paulo 8(14), 181–187, 8 figs.
Vanzolini, P.E. (1948) Notas sobre os ofídios e lagartos da Cachoeira de Emas, no municipio de Pirassununga, Estado de São Paulo. *Revista Brasileira de Biologia*, Rio de Janeiro 8(3), 377–400, 6 figs.
Vanzolini, P.E. (1952) Fossil snakes and lizards from the lower Miocene of Florida. *Journal of Paleontology* 26(3), 452–457, pls. 55–56.
Vanzolini, P.E. (1961) A biometrical note on *Bothrops moojeni* Hoge, 1966 (Serpentes, Viperidae). *Anais da Academia Brasileira de Ciencias*, Rio de Janeiro 63(4), 389–401, illust.
Vanzolini, P.E. (1969) Taxonomic index to Andersson and Lönnberg papers on present location of Linnaean reptile and amphibian types. *Smithsonian Herpetological Information Service* 1969, 1–10.
Vanzolini, P.E. (1972) *Typhlops brongersmai* spec. nov. from the coast of Bahia, Brasil (Serpentes, Typhlopidae). *Zoologische Mededelingen*, Leiden 47(3), 27–29, 1 fig.
Vanzolini, P.E. (1976) *Typhlops brongersmianus*, a new name for *Typhlops brongersmai* Vanzolini, 1972, preoccupied (Serpentes, Typhlopidae). *Papeis Avulsos do Departamento de Zoologia*, São Paulo 29(24), 247.
Vanzolini, P.E. (1977) *An annotated bibliography of the land and fresh-water reptiles of South America (1758–1975). Vol. 1.* Museu de Zoologia, Universidad de São Paulo, São Paulo, iv + 186 pp.
Vanzolini, P.E. (1978) *An annotated bibliography of the land and fresh-water reptiles of South America (1758–1975). Vol. 2.* Museu de Zoologia, Universidad de São Paulo, São Paulo, 136 pp.
Vanzolini, P.E. (1980) Reptilia. *In:* Hurlbert, S.H., Rodriguez, S.H. & Santos, N.D. dos (Eds.), *Aquatic biota of tropical South America (being a compilation of taxonomic bibliographies for the fauna and flora of inland waters of the tropical portion of South America). Part 2. Anarthropoda.* San Diego State University, San Diego, pp. 246–261.
Vanzolini, P.E. (1981) The scientific and political contexts of the Bavarian Expedition to Brasil. *In:* Spix, J.B. von & Wagler, J.G., *Herpetology of Brazil.* Society for the Study of Amphibians and Reptiles, pp. ix–xxix, 2 figs. [reprint of Spix, 1824]
Vanzolini, P.E. (1985) *Micrurus averyi* Schmidt, 1939, in central Amazonia (Serpentes, Elapidae). *Papeis Avulsos Zoologia*, São Paulo 36(8), 77–85, illust.
Vanzolini, P.E. (1986a) Addenda and corrigenda to the Catalogue of Neotropical Squamata. *Smithsonian Herpetological Information Service* (70), 1–25.
Vanzolini, P.E. (1986b) *Levantamento herpetológico da área do estado de Rondônia sab a influência da rodovia BR 364.* Assessoria Editorial, Brasilia, 50 pp., 6 figs.
Vanzolini, P.E. (1991) A biometrical note on *Bothrops moojeni* Hoge, 1966 (Serpentes, Viperidae). *Anais da Academia Brasileira de Ciencias*, Rio de Janeiro 63(4), 389–401.
Vanzolini, P.E. (1996) A new (and very old) species of *Leptotyphlops* from northeastern Brasil (Serpentes, Leptotyphlopidae). *Papeis Avulsos Zoologia*, São Paulo 39(15), 281–291, 1 fig.
Vanzolini, P.E., Romos-Costa, A.M.M. & Vitt, L.J. (1980) *Repteis das Caatingas.* Academia Brasileira de Ciencias, Rio de Janeiro, 161 pp., 102 figs., 40 pls.
Vas-Silva, W., Guedes, A.G., Azevedo-Silva, P.L. de, Barbosa, R.S., Aloisio, G.R. & Gomes de Oliveira, F.C. (2007) Herpetofauna, Espora Hydroelectric Power Plant, state of Goiás, Brazil. *Check List* 3(4), 338–345, 2 figs.
Vasiliu, G.D. & Sova, C. (1968) *Fauna vertebratica Romaniae (index).* Muzeul Judetean Bacau, Bacau, 296 pp.
Vaughan, R.K., Dixon, J.R. & Thomas, R.A. (1996) A reevaluation of populations of the corn snake *Elaphe guttata* (Reptilia, Serpentes, Colubridae) in Texas. *Texas Journal of Science* 48(3), 175–190, 5 figs.
Vaz-Ferreira, R. & Sierra de Soriano, B. (1960) Notas sobre reptiles del Uruguay. *Revista de la Facultad Humanidades y Ciencias (Série Zoologica Vertebrata)*, Montevideo (18), 133–206, 10 pls.

Vaz-Ferreira, R., Covelo de Zollesi, L. & Achaval, F. (1976) Incubation de huevos de reptiles en la naturaleza y en cautividad. *Primer Congres de Parques Zoológicos del Cono Sur,* Durazno, 4 pp.

Vázquez-Díaz, J. & Quintero-Díaz, G. (1997b) Anfibios y reptiles de Aguascalientes. *Cuadernos de Aguascalientes* 81, 1–32.

Vázquez-Díaz, J. & Quintero-Díaz, G. (1998) Geographic distribution: Serpentes: *Heterodon nasicus. Herpetological Review* 29(2), 113.

Vázquez-Díaz, J. & Quintero-Díaz, G. (1999) Geographic distribution: Serpentes: *Geophis dugesi. Herpetological Review* 30(4), 235.

Vázquez-Díaz, J. & Quintero-Díaz, G. (2001) Geographic distribution: Serpentes: *Ramphotyphlops braminus. Herpetological Review* 33(4), 279.

Vázquez-Díaz, J. & Quintero-Díaz, G. (2005) *Anfibios y reptiles de Aguascalientes.* Second edition. Centro de Invesigaciones y Estudios Multidisciplinarios de Aguascalientes, Aguascalientes, 318 pp., figs. 72 col. photos [first edition, 1997a, 145 pp., figs., col. illust.]

Vázquez-Díaz, J., Quintero-Díaz, G. & Ramirez-Bautista, A. (1998) Geographic distribution: Serpentes: *Oxybelis aeneus* (Mexican vine snake). *Herpetological Review* 29(2), 115.

Vázquez-Díaz, J., Quintero-Díaz, G. & Smith, H.M. (1999) Geographic distribution: Serpentes: *Rhadinaea hesperia. Herpetological Review* 30(4), 236.

Vedmederja, V.I. (1984) Distribution, variation and ecological features of *Vipera kaznakowi. In: The species and its productivity within its distribution range. Part 5. Questions of herpetology. Materials from the 4th All-Union Symposium, 3–7 April 1984.* Urals Science Centre, Academy of Science, USSR, Sverdlovsk, pp. 8–9. (in Russian)

Vedmederja, V.I., Grubant, V.N. & Rudayeva, A.B. (1986a) On the naming of the black viper of the steppe-forest region of the European USSR. *Vestnik Kharkov University* (288), 83–85. (in Russian)

Vedmederja, V.I., Orlov, N.L. & Tuniyev, B.S. (1986b) On taxonomy of three viper species of the *Vipera kaznakowi* complex. *Proceedings of the Zoological Institute, USSR, Academy of Science,* Leningrad 157, 55–61, 6 pls. (in Russian)

Vedmederja, V.I, Zinenko, O. & Barabanov, A. (2009) An annotated type catalogue of amphibians and reptiles in the Museum of Nature at V. N. Karazin Kharkiv National University (Kharkiv, Ukraine). *Russian Journal of Herpetology* 16(3), 203–212, 10 figs.

Vega-López, A.A. & Alvarez-S, T. (1992) La herpetofauna de los volcanes Popocatepetl e Iztacchihuatl. *Acta Zoologica Mexicana* 51, 128 pp., 46 figs.

Veith, G. (1991a) Die Reptilien Bosniens und der Herzegowina. Teil I. *Herpetozoa,* Vienna 3(3–4), 97–196.

Veith, G. (1991b) Die Reptilien Bosniens und der Herzegowina Teil 2. *Herpetozoa* 4(1–2), 1–96, illust.

Velasco, A.L. (1890a) Geografía y estadística del estado de Veracruz-Llave. *In:* Velasco, A.L. (Ed.), *Geografía y estadística de la República Mexicana. Volume 3.* Secretario Fomento, México City, 240 pp.

Velasco, A.L. (1890b) Geografía y estadística del estado de Guanajuato. *In:* Velasco, A.L. (Ed.), *Geografía y estadística de la República Mexicana. Volume 5.* Secretario Fomento, México City, 300 pp.

Velasco, A.L. (1890c) Geografía y estadística del estado de Michoacán de Ocampo. *In:* Velasco, A.L. (Ed.), *Geografía y estadística de la República Mexicana. Volume 6.* Secretario Fomento, México City, 233 pp.

Velasco, A.L. (1890d) Geografía y estadística del estado de Morelos. *In:* Velasco, A.L. (Ed.), *Geografía y estadística de la República Mexicana. Volume 7.* Secretario Fomento, México City, 140 pp.

Velasco, A.L. (1891) Geografía y estadística del estado de Oaxaca de Juarez. *In:* Velasco, A.L. (Ed.), *Geografía y estadística de la República Mexicana. Volume 9.* Secretario Fomento, México City, 404 pp.

Velasco, A.L. (1892a) Geografía y estadística del estado de Guerrero. *In:* Velasco, A.L. (Ed.), *Geografía y estadística de la República Mexicana. Volume 10.* Secretario Fomento, México City, 248 pp.

Velasco, A.L. (1892b) Geografía y estadística del estado de Tamaulipas. *In:* Velasco, A.L. (Ed.), *Geografía y estadística de la República Mexicana. Volume 12.* Secretario Fomento, México City, 204 pp.

Velasco, A.L. (1893) Geografía y estadística del estado de Durango. *In:* Velasco, A.L. (Ed.), *Geografía y estadística de la República Mexicana. Volume 13.* Secretario Fomento, México City, 196 pp.

Velasco, A.L. (1895) Geografía y estadística del estado de Campeche. *In:* Velasco, A.L. (Ed.), *Geografía y estadística de la República Mexicana. Volume 16.* Secretario Fomento, México City, 140 pp.

Velasco, A.L. (1896) Geografía y estadística del estado de Colima. *In:* Velasco, A.L. (Ed.), *Geografía y estadística de la República Mexicana. Volume 18.* Secretario Fomento, México City, 142 pp.

Velasco, A.L. (1897) Geografía y estadística del estado de Coahuila de Zaragoza. *In:* Velasco, A.L. (Ed.), *Geografía y estadística de la República Mexicana. Volume 19.* Secretario Fomento, México City, 202 pp.

Velasco, A.L. (1898) Geografía y estadística del estado de Chiapas. *In:* Velasco, A.L. (Ed.), *Geografía y estadística de la República Mexicana. Volume 20.* Secretario Fomento, México City, 164 pp.

Vellard, J. (1941) Serpents venimeaux du Venezuela. *Annales des Sciences Naturelles (Zoologie),* Paris (Série 11) 3, 193–225.

Vellard, J. (1943) Una nuevo forma de *Oxyrhopus, O. rhombifer septentrionalis. Acta Zoologica Lilloana,* Tucumán 1, 89–91, 1 pl.

Velloso-Calleffo, M.E. (1997) Geographic distribution: Serpentes: *Micrurus averyi. Herpetological Review* 28(4), 210.

Veloso, A. & Navarro, J. (1988) Lista sistematica y distribucion geographica de anfibios y reptiles de Chile. *Bollettino del Museo Regionale di Scienze Naturali,* Torino 6(2), 481–539, 3 figs., 6 pls.

Vences, M. & Glaw, F. (2003) Phylogeography, systematics and conservation status of boid snakes from Madagascar (*Sanzinia* and *Acrantophis*). *Salamandra* 39, 181–206.

Vences, M., Franzen, M., Fläschendräger, A., Schmitt, R. & Regös, J. (1998) Beobachtungen zur herpetofauna von Nicaragua: kommentierte artenliste der reptilien. *Salamandra* 34(1), 17–42.

Vences, M., Glaw, F., Kosuch, J., Böhme, W. & Veith, M. (2001) Phylogeny of South American and Malagasy Boine snakes: molecular evidence for the validity of *Sanzinia* and *Acrantophis* and biogeographic implications. *Copeia* 2001(4), 1151–1154.

Venchi, A. & Sindaco, R. (2006) Annotated checklist of the reptiles of the Mediterranean countries, with keys to species identification. Part 2 – Snakes. *Annali del Museo Civico di Storia Naturale "Giacomo Doria,"* Genova 98, 259–363, 90 figs.

Venczel, M. (1994) Late Miocene snakes from Polgárdi (Hungary). *Acta Zoologica Cracoviensia,* Kraków 37(1), 1–29, 13 figs.

Venczel, M. (1998) Late Miocene snakes (Reptilia, Serpentes) from Polgárdi (Hungary): a second contribution. *Acta Zoologica Cracoviensia* 41(1), 1–22, 13 figs.

Venczel, M. (2001) Anurans and squamates from the Lower Pliocene (MN 14) Osztramos 1 locality (Northern Hungary). *Fragmenta Palaeontologica Hungarica* 19, 79–90.

Venczel, M. & Sen, S. (1994) Pleistocene amphibians and reptiles from Emirkaya-2, Turkey. *Herpetological Journal*, London 4, 159–165, 7 figs.

Venczel, M. & Stiucă, E. (2008) Late middle Miocene amphibians and squamate reptiles from Taut, Romania. *Geodiversitas* 30, 731–763, 16 figs.

Venegas, P.J. (2005) Geographic distribution: Serpentes: *Leptophis depressirostris*. *Herpetological Review* 36(3), 339.

Venning, F.E.W. (1910) A collection of the Ophidia from the Chin Hills, (with notes on the same by Major F. Wall, I.M.S., C.M.Z.S.). *The Journal of the Bombay Natural History Society* (1910–1911) 20(2), 331–344.

Venning, F.E.W. (1911) Further notes on snakes from the Chin Hills. (With notes by Major F. Wall, I.M.S., C.M.Z.S.). *The Journal of the Bombay Natural History Society* (1910–1911) 20(3), 770–775.

Venning, F.E.W. (1914) *Simotes splendidus* Gunth, in Burma. *The Journal of the Bombay Natural History Society* (1914–1915) 23(1), 164–166.

Venzmer, G. (1918) Beiträge zur Kenntnis der Reptilien- und Amphibienfauna des cilicischen Taurus. Teil I. *Sitzungsberichte der Gesellschaft Naturforschender Freunde zu Berlin* 1918(7), 209–251.

Venzmer, G. (1919) Zur Schlangenfauna Süd-Kleinasiens, speziell des cilicischen Taurus. *Archiv für Naturgeschichte*, Berlin (1917) 83A(11), 95–122.

Verhoeven, K. (1995) Slangen van Rara Avis, Costa Rica. *Lacerta* 54(2), 45–53, 2 figs, 4 col. photos.

Vermersch, T. & Kuntz, R. (1986) *Snakes of south-central Texas*. Eakin Press, Austin, xiii + 137 pp., illust.

Vernet, R. & Grenot, C. (1972) Place des reptiles dans l'écosystème de l'erg au Sahara nord-occidental. *Naturalistes Orléanais* (Série 3) (5), 49–63, figs., 9 pls.

Vesely, M. & Köhler, G. (2009) Geographic distribution: Serpentes: *Ramphotyphlops braminus*. *Herpetological Review* 40(1), 116.

Vesey-FitzGerald, L.D.E.F. (1948) Reptiles and amphibians from the Seychelles Archipelago. *Annals and Magazine of Natural History*, London (1947) (Series 11) 14(116), 577–583.

Vesey-FitzGerald, L.D.E.F. (1958) The snakes of northern Rhodesia and the Tanganyika borderlands. *Proceedings and Transactions, Rhodesia Scientific Association*, Salisbury 46, 17–102, 28 photos.

Vesey-FitzGerald, L.D.E.F. (1975) A guide to the snakes of the Tanzania and Kenya border-lands. *Journal of the East Africa Natural History Society and National Museum*, Nairobi (149), 1–26, 20 figs.

Vetencourt, F.H. (1960) Ofidios venenosos de Venezuela. *Revista de Medicina Veterinaria y Parasitológia*, Maracay (1959–1960) 18, 161–192.

Vial, J.L. & Jiménez-Porras, J.M. (1967) The ecogeography of the bushmaster, *Lachesis muta*, in Central America. *American Midland Naturalist* 78(1), 182–187, 1 fig.

Vial, J.L., Berger, T.J. & McWilliams, W.T., Jr. (1977) Quantitative demography of copperheads, *Agkistrodon contortrix* (Serpentes, Viperidae). *Researches on Population Ecology, Society of Population Ecology* 18(2), 223–234, 2 figs.

Viaud-Grand-Marais, A. (1880a) *Note sur l'envenimation ophidienne étudiée dans les differents groupes de serpents.* Mme. Vve. Camille Mellinet, Nantes, 28 pp.

Viaud-Grand-Marais, A. (1880b) L'envenimation ophidienne étudiée dans les différents groupes de serpents. *Gazette des Hopitaux Civils et Militaires* 53(113), 901–902.

Vidal, N., Lecointre, G., Vie, J.C. & Gasc, J.P. (1997) Molecular systematics of pitvipers: paraphyly of the *Bothrops* complex. *Comptes Rendus de l'Académie des Sciences*, Paris 320, 95–101, 2 figs.

Vidal, N., Lecointre, G., Vie, J.C. & Gasc, J.P. (1999) What can mitochondrial gene sequences tell us about intergeneric relationships of pitvipers? *In:* Joger, U. (Ed.), Phylogeny and systematics of the Viperidae. *Kaupia, Darmstädter Beiträge zur Naturgeschichte* 8, 107–112, 1 fig.

Vidal, N., Kindl, S.G., Wong, A. & Hedges, S.B. (2000) Phylogenetic relationships of xenodontine snakes inferred from 12S and 16S ribosomal RNA sequences. *Molecular Phylogenetics and Evolution* 14(3), 389–402, 2 figs.

Vidal, N., Henderson, R.W., Delmas, A.-S. & Hedges, S.B. (2005) A phylogenetic study of the emerald treeboa (*Corallus caninus*). *Journal of Herpetology* 39(3), 500–503, 2 figs.

Vidal, N., Branch, W.R., Pauwels, O.S.G., Hedges, S.B., Broadley, D.G., Wink, M., Cruaud, C., Joger, U. & Nagy, Z.T. (2008) Dissecting the major African snake radiation: a molecular phylogeny of the Lamprophiidae Fitzinger (Serpentes, Caenophidia). *Zootaxa* (1945), 51–66, 2 figs.

Vidal, N., Dewynter, M. & Gower, D.J. (2010a) Dissecting the major American snake radiation: a molecular phylogeny of the Dipsadidae Bonaparte (Serpentes, Caenophidia). *Comptes Rendus Biologies* 333, 48–55, 1 fig.

Vidal, N., Marin, J., Morini, M., Donnellan, S., Branch, W.R., Thomas, R., Vences, M., Wynn, A., Cruaud, C. & Hedges, S.B. (2010b) Blindsnake evolutionary tree reveals long history on Gondwana. *Biology Letters* 2010, 1–4, 1 fig.

Vieites, D.R., Ratsoavina, F.M., Randrianianina, R.D., Nagy, Z.T., Glaw, F. & Vences, M. (2010) A rhapsody of colours from Madagascar: discovery of a remarkable new snake of the genus *Liophidium* and its phylogenetic relationships. *Salamandra* 46(1), 1–10, 4 figs.

Vigle, G.O. (2008) The amphibians and reptiles of the Estación Biológica Jatun Sacha in the lowland rainforest of Amazonian Ecuador: a 20 year record. *Breviora* (514), 1–30, 4 figs.

Vijayakumar, S.P. & David, P. (2006) Taxonomy, natural history, and distribution of the snakes of the Nicobar islands (India), based on new materials and with an emphasis on endemic species. *Russian Journal of Herpetology* 13(1), 11–40, 23 figs.

Villa, J. (1962) *Las Serpientes venenosas de Nicaragua*. Editorial Novedades, Managua, 94 pp., 4 figs., 19 pls.

Villa, J. (1968) A new colubrid snake from the Corn Islands, Nicaragua. *Revista de Biologia Tropical*, San José 15(1), 117–121, 4 figs.

Villa, J. (1969) Two new insular subspecies of the natricid snake *Tretanorhinus nigroluteus* Cope from Honduras and Nicaragua. *Journal of Herpetology* 3(3–4), 145–150, 7 figs.

Villa, J. (1970a) Notas sobre la historia natural de la serpiente de los pantanos, *Tretanorhinus nigroluteus*. *Revista de Biologia Tropical*, San José 17(1), 97–104, 1 fig.

Villa, J. (1970b) The snake *Hydromorphus* in Nicaragua, with a description of its hemipenis. *Caribbean Journal of Science*, Mayagüez 10(3–4), 119–121.

Villa, J. (1971a) Notes on some Nicaraguan reptiles. *Journal of Herpetology* 5(1–2), 45–48, 1 fig.

Villa, J. (1971b) *Crisantophis*, a new genus for *Conophis nevermanni* Dunn. *Journal of Herpetology* 5(3–4), 173–177, 1 fig.

Villa, J. (1972) Snakes of the Corn Islands, Caribbean, Nicaragua. *Brenesia* 1, 14–18.

Villa, J. (1983) *Nicaraguan fishes, amphibians & reptiles: a checklist and bibliography.* Escuela de Ecología y Recursos Naturales, Universidad Centroamericana, Managua, 53 pp.

Villa, J. (1984) The venomous snakes of Nicaragua, a synopsis. *Milwaukee Publ. Museum, Contributions in Biology and Geology* (55), 1–41, 10 figs., 1 pl.

Villa, J. (1988a) *Crisantophis, C. nevermanni. Catalogue of American amphibians and reptiles*, pp. 429.1–429.2.

Villa, J. (1988b) *Typhlops costaricensis. Catalogue of American amphibians and reptiles*, pp. 435.1–435.2.

Villa, J. (1990a) *Hydromorphus, H. concolor, H. dunni. Catalogue of American amphibians and reptiles*, pp. 472.1–472.2, 1 fig.

Villa, J. (1990b) *Leptotyphlops nasalis. Catalogue of American amphibians and reptiles*, p. 473.1, 1 fig.

Villa, J. (1993) Geographic distribution: Serpentes: *Masticophis mentovarius mentovarius* (Sabanera). *Herpetological Review* 24(3), 110.

Villa, J. & McCranie, J.R. (1995) *Oxybelis wilsoni*, a new species of vine snake from Isla de Roatán, Honduras (Serpentes, Colubridae). *Revista de Biologia Tropical*, San José 43(1–3), 297–305, 2 figs.

Villa, J. & Wilson, L.D. (1990) *Ungaliophis, U. continentalis, U. panamensis. Catalogue of American amphibians and reptiles*, pp. 480.1–480.4, 4 figs.

Villa, J.D., Wilson, L.D. & Johnson, J.D. (1988) *Middle American herpetology, a bibliographic checklist.* University of Missouri Press, Columbia, xxxii + 131 pp., 91 col. photos.

Villa, R.A., Bryson, Jr., R.W. & Ramirez-Chaparro, R. (2012a) Geographic distribution: Serpentes: *Rhadinaea laureata. Herpetological Review* 43(1), 107.

Villa, R.A., Van Devender, T.R. & Bygott, D. (2012b) Geographic distribution: Serpentes: *Geophis dugesi. Herpetological Review* 43(2), 307.

Villa, R.A., Bryson, Jr., R.W. & Ramírez-Chaparro, R. (2012c) Geographic distribution: Serpentes: *Rhadinaea laureata. Herpetological Review* 43(2), 308–309.

Villada, M.M. (1875) El *Diadophis punctatus*, var. *Dougesii. La Naturaleza,* México City 3, 226–230, pl.

Villiers, A. (1948) Note sur quelques ophidiens de la Guinée portugaise. *Boletim Cultural da Guiné Portugaise*, Janeiro (9), 201–204.

Villiers, A. (1950a) *Catalogue VI. La collection de serpents de l'I.F.A.N.* Institut Française d'Afrique Noire, Dakar, 155 pp., 12 figs.

Villiers, A. (1950b) Contribution à l'étude du peuplement de la Mauritanie. Ophidiens. *Bulletin de l'Institut Française d'Afrique Noire*, Dakar 12A(4), 984–998, 2 figs.

Villiers, A. (1950c) Reptiles ophidiens et chéloniens. *In:* Chopard, L. & Villiers, A. (Eds.), Contribution a l'étude de l'Aïr. *Mémoires de l'Institut Française d'Afrique Noire (Sciences Naturelles)*, Dakar (10), pp. 337–344.

Villiers, A. (1950d) Les serpents de l'Ouest Africain: les serpents minute. *Notes Africaines* (45), 23–26, 10 figs.

Villiers, A. (1951a) Mission A. Villiers au Togo et au Dahomey (1950). II. Ophidiens. *Études Damoméenes* 5, 17–43, 2 figs.

Villiers, A. (1951b) La collection de serpents de l'IFAN. (acquisitions 1950). *Bulletin de l'Institut Française d'Afrique Noire*, Dakar 13A(3), 813–836, 3 figs.

Villiers, A. (1952a) A propos de *Leptotyphlops macrorhynchus* Jan. *Bulletin de l'Institut Française d'Afrique Noire*, Dakar 14A(1), 243–246.

Villiers, A. (1952b) La collection de serpents de l'IFAN (acquisitions 1951). *Bulletin de l'Institut Française d'Afrique Noire*, Dakar 14A(4), 881–898.

Villiers, A. (1953) La collection de serpents de L'IFAN (acquisitions 1952). *Bulletin de l'Institut Française d'Afrique Noire*, Dakar 15A(3), 1103–1127, 5 figs.

Villiers, A. (1954a) La collection de serpents de l'IFAN (acquisitions 1953). *Bulletin de l'Institut Française d'Afrique Noire*, Dakar 16A(4), 1234–1247, 2 figs.

Villiers, A. (1954b) Un mamba noir au Senegal. *Notes Africaines*, Dakar April, 59–61, 3 figs.

Villiers, A. (1956a) La collection de serpents de l'IFAN. (acquisitions 1954–1955). *Bulletin de l'Institut Française d'Afrique Noire*, Dakar 18A(3), 877–883.

Villiers, A. (1956b) Le Parc National du Niokolo-Koba. Fascicule I. V. Reptiles. *Mémoires de l'Institut Française d'Afrique Noire (Sciences Naturelles)*, Dakar (48), 143–162, 3 figs.

Villiers, A. (1963) Serpents africains des collections du Muséum de Paris. *Bulletin de l'Institut Française d'Afrique Noire*, Dakar 25A(4), 1367–1373, 3 figs.

Villiers, A. (1965) Serpents récoltes au Mali et en Haute-Volta par le Dr. Lamontellerie. *Bulletin de l'Institut Française d'Afrique Noire*, Dakar 27A(3), 1192–1195.

Villiers, A. (1966) Contribution à la faune du Congo (Brazzaville). Mission A. Villiers et A. Descarpentries. XLII. Reptiles ophidiens. *Bulletin de l'Institut Fondamental d'Afrique Noire*, Dakar 28A(4), 1720–1760, 3 figs.

Villiers, A. & Condamin, M. (2005) *Les serpentes de l'ouest Africain.* Fourth edition. Les Nouvelles Éditions Africains du Sénégal, Dakar, 205 pp., 307 figs., 8 col. figs. [first edition, A. Villiers, 1950 by I.F.A.N., 148 pp., 190 figs.; second edition, A. Villiers, 1963, 190 pp., 258 figs., third edition, A. Villiers, 1975, 195 pp., 282 figs.]

Viñas, M. & de Olmedo, E.V. (1988) Sobre *Lystrophis histricus* (Jan) en la Argentina (Serpentes Colubridae). *Revista del Museo Argentino de Ciencias Naturales Bernardino Rivadavia (Zoologia)*, Buenos Aires 15(1), 1–6, 1 fig.

Viñas, M., Dameri, G. & Gnida, G. (1989) Presencia de *Pseudablabes agassizii* en la Sierra de la Ventana (Bs. As.) y confirmación para la pcia. De La Pampa. *Boletin de la Asociación Herpetologica de Argentina* 5(1–2), 13–14.

Vinciguerra, D. (1892) Rettili e batraci di Engano. *Annali del Museo Civico di Storia Naturali di Genova* (Serie 2) 12, 517–526.

Vinciguerra, D. (1931) Spedizione del Barone Raimondo Franchetti in Dancalia. Rettili, Batraci e Pesci. *Annali del Museo Civico di Storia Naturali di Genova* 55, 96–108, 1 col. pl.

Vinson, J.-M. (1949) L'ile Ronde et l'ile aux serpents. *Proceedings of the Royal Society of Arts and Sciences of Mauritius*, Port Louis 1(1), 32–52, 2 figs., 4 pls.

Vinson, J.-M. (1975) Notes on the reptiles of Round Island. *Mauritius Institute Bulletin* 8(1), 49–67, pl. 9.

Viosca, P., Jr. (1944) Distribution of certain cold-blooded animals in Louisiana in relationship to the geology and physiography of the state. *Proceedings of the Louisiana Academy of Sciences* 8, 47–62.

Viosca, P., Jr. (1949) Amphibians and reptiles of Louisiana. *Popular Science Bulletin, Louisiana Academy of Science,* Baton Rouge (1), 1–12.

Visinoni, A. (1994) Contributi alla conoscenza dell'erpetofauna Boliviana, *Lachesis muta muta* (Linnaeus, 1766) (Serpentes, Viperidae) in Bolivia. *Bollettino dei Museo Civico di Storia Naturale di Venezia* 43(1992), 209–219, 2 figs.

Visinoni, A. (1996) Contributi per la conoscenza dell'erpetofauna boliviana. II. *Leptomicrurus narduccii narduccii* (Jan, 1863) endemico dell'oriente boliviano (Serpentes, Elapidae). *Atti della Società Italiana di Scienze Naturali e del Museo Civico di Storia Naturali di Milano* 136(1), 86–93, 2 figs.

Visser, J. (1967) *Poisonous snakes of southern Africa and the treatment of snakebite*. Howard Timmins, Cape Town, 60 pp., 5 figs., 60 pls.

Visser, J. (1972) *What snake is that?* Purnell Pocketbook, Purnell & Sons, Cape Town, 84 pp.

Visser, J. (1978) Notes on two rare house snakes—1. *Lamprophis fiskii* Boulenger (1887) and *L. swazicus* Schaefer (1970). *Journal of the Herpetological Association of Africa*, Stellenbosch (19), 10–13, 1 photo.

Visser, J. (1979a) Notes on two rare house snakes—Part 2. The generic status of *Lamprophis fiskii* Boulenger (1887) and *Lamprophis swazicus* Schaefer (1970). *Journal of the Herpetological Association of Africa*, Stellenbosch (21), 31–37, 8 figs.

Visser, J. (1979b) *Common snakes of South Africa*. Purnell & Sons, Cape Town, xix + 74 pp. + xxvi, figs., 37 col. pls.

Visser, J. (2010) Variation in hemipenes of different colour phases of *Pseudaspis cana* (Linneaus) warrents re-evaluation of the species. *African Herp News* (51), 2–3, 1 fig.

Visser, J. & Chapman, D.S. (1978) *Snakes and snakebite. Venomous snakes and management of snakebite in southern Africa*. Purnell & Sons, Cape Town, 1–152, 24 pls.

Vitt, L.J. (1983) Ecology of an anuran-eating guild of terrestrial tropical snakes. *Herpetologica* 39(1), 52–66, 5 figs.

Vitt, L.J. (1987) Communities. *In:* Seigel, R.A., Collins, J.T. & Novak, S.S. (Eds.), *Snakes: ecology and evolutionary biology*. Macmillan Company, New York, pp. 335–365, 10 figs.

Vitt, L.J. (1996) Ecological observations on the tropical colubrid snake *Leptodeira annulata*. *Herpetological Natural History* 4(1), 69–76, 4 figs.

Vitt, L.J. & Laerm, J. (1982) Checklist of amphibians and reptiles of Georgia. *Georgia Journal of Science* 40(3–4), 75–83.

Vives-Balmaña, M.V. (1984) *Els amfibis i els reptiles de Catalunya*. Ketres Editora, Collecció Ventall, Barcelona, 229 pp., illust.

Vo, V.C. & Nguyen, D.M. (1993) *Venomous snakes of Vietnam and treatment of snakebites*. Science and Technics Publishing House, Hanoi, 225 pp., 82 figs. (in Vietnamese)

Vogel, G. (1990a) Über eine verkannte *Trimeresurus* Art aus Thailand. *Sauria*, Berlin 12(2), 11–13, 3 figs.

Vogel, G. (1990b) Die Bronzenattern Thailands. *Herpetofauna* 12(65), 10–16, 6 figs.

Vogel, G. (1991) Eine neue *Trimeresurus*-Art aus Thailand, *Trimeresurus venustus* sp. nov. (Reptilia, Serpentes, Crotalidae). *Sauria*, Berlin 13(1), 23–28, 8 figs.

Vogel, G. (1994) *Boiga multitemporalis* Bourret, 1935—ein junior-synonym von *Boiga kraepelini* Stejneger, 1902 (Serpentes, Colubridae). *Sauria*, Berlin 16(1), 15–20, 5 figs.

Vogel, G. (1995) *Dendrelaphis striatus* (Cohn), neu für die Fauna Borneos (Serpentes, Colubridae). *Mitteilungen aus dem Zoologische Museum in Berlin* 71(1), 147–149, 2 figs.

Vogel, G. (2000) Zur Bestimmung der Unterarten von *Boiga dendrophila* (Boie, 1827) nebst einigen Neuigkeiten zu *Boiga dendrophila gemmicincta* (Duméril, Bibron & Duméril, 1854). *Sauria*, Berlin 22(1), 27–43, 15 col. figs.

Vogel, G. (2006) *Venomous snakes of Asia*. Edition Chimaira, Frankfurt-am-Main, 148 pp., 520 col. photos.

Vogel, G. (2010) On the distribution of *Pareas hamptoni* (Boulenger, 1905) in Thailand (Serpentes: Pareatinae). *Russian Journal of Herpetology* 17(3), 219–222, 1 fig.

Vogel, G. & Brachtel, N. (2008) Contribution to the knowledge of *Lycodon ruhstrati* (Fischer, 1886) in Vietnam—taxonomy and biology of a little known species. *Salamandra* 44(4), 207–224, 7 figs.

Vogel, G. & David, P. (1999) A revision of the genus *Macrocalamus* (Serpentes, Colubridae), with description of a new species and a key to the genus. *The Raffles Bulletin of Zoology* 47(2), 309–332, 4 figs., 17 pls.

Vogel, G. & David, P. (2006) On the taxonomy of the *Xenochrophis piscator* complex (Serpentes, Natricidae). *Proceedings of the 13th Congress, Societas Europaea Herpetologica*, pp. 241–246, 1 fig.

Vogel, G. & David, P. (2010) A new species of the genus *Lycodon* (Boie, 1826) from Yunnan Province, China (Serpentes: Colubridae). *Bonn Zoological Bulletin* 57(2), 289–296, 6 figs.

Vogel, G. & David, P. (2012) A revision of the species group of *Xenochrophis piscator* (Schneider, 1799) (Squamata: Natricidae). *Zootaxa* (3473), 1–60, 47 figs.

Vogel, G. & Köhler, G. (1996) Notes on the occurrence of *Elapoidis fusca* Boie, 1827 on Borneo (Reptilia, Serpentes, Colubridae). *Senckenbergiana Biologica* 76(1–2), 61–63, 1 fig.

Vogel, G. & Luo, H. (2011) A new species of the genus *Lycodon* (Boie, 1826) from the southwestern mountains of China (Squamata: Colubridae). *Zootaxa* (2807), 29–40, 5 figs.

Vogel, G. & Pauwels, O.S.G. (2004) Geographic distribution: Serpentes: *Elaphe prasina*. *Herpetological Review* 35(4), 410.

Vogel, G. & Rooijen, J. van. (2007) A new species of *Dendrelaphis* (Serpentes, Colubridae) from southeast Asia. *Zootaxa* (1394), 25–45, 16 figs.

Vogel, G. & Rooijen, J. van. (2008) Contributions to a review of the *Dendrelaphis pictus* (Gmelin, 1789) complex—2. The eastern forms (Serpentes: Colubridae). *Herpetozoa* 21(1–2), 3–29, 17 figs.

Vogel, G. & Rooijen, J. van. (2011a) Contribution to a review of the *Dendrelaphis pictus* (Gmelin, 1789) complex (Serpentes: Colubridae)—3. The Indian forms, with the description of a new species from the Western Gnats. *Journal of Herpetology* 45(1), 100–110, 9 figs.

Vogel, G. & Rooijen, J. van. (2011b) Description of a new species of the genus *Dendrelaphis* Boulenger, 1890 from Myanmar (Squamata: Serpentes: Colubridae). *Bonn Zoological Bulletin* 60(1), 17–24, 3 figs.

Vogel, G. & Rooijen, J. van. (2011c) A new species of *Dendrelaphis* (Serpentes: Colubridae) from the Western Ghats – India. *Taprobanica,* Homagama 3(2), 77–85.

Vogel, G., David, P. & Pauwels, O.S.G. (2004) A review of morphological variation in *Trimeresurus popeiorum* (Serpentes, Viperidae, Crotalinae), with the description of two new species. *Zootaxa* (727), 1–63, 20 col. figs.

Vogel, G., David, P., Lutz, M., Roolijn, J. van. & Vidal, N. (2007) Revision of the *Tropidolaemus wagleri*-complex (Serpentes, Viperidae, Crotalinae). I. Definition of included taxa and redescription of *Tropidolaemus wagleri* (Boie, 1827). *Zootaxa* (1644), 1–40, 30 col. figs.

Vogel, G., David, P., Pauwels, O.S.G., Sumontha, M., Norval, G., Hendrix, R., Vu, N.T. & Ziegler, T. (2009) A revision of *Lycodon ruhstrati* (Fischer, 1886) auctorum (Squamata Colubridae), with the description of a new species from Thailand and a new subspecies from the Asian mainland. *Tropical Zoology,* Firenze 22(2), 131–182, 21 figs.

Vogel, G., Rooijen, J. van & Hauser, S. (2012a) A new species of *Dendrelaphis* Boulenger, 1890 (Squamata: Colubridae) from Thailand and Myanmar. *Zootaxa* (3392), 35–46, 9 figs.

Vogel, G., Nguyen, Q.T., Kingsada, P. & Ziegler, T. (2012b) A new species of the genus *Lycodon* Boie, 1826 from Laos (Squamata: Colubridae). *North-Western Journal of Zoology,* Oradea 8(2), 344–352, 5 figs.

Vogel, Z. (1966) *Reptiles and amphibians: their care and behavior.* Paul Hamlyn, London, 228 pp., 253 figs.

Vogrin, N. (1997) On the herpetofauna of Lake Vrana and its vicinity (Dalmacia, Croatia) (Amphibia, Reptilia). *Herpetozoa* 10(1–2), 85–88, 1 fig.

Vogt, C.C. (1851) *Zoologische Briefe. Naturgeschichte der lebenden und untergegangenen Thiere für Lehrer, höhere Schlüler und Gebildete aller Stande. Zweiter Band.* Literarische Anstalt, Frankfurt-am-Main, 640 pp.

Vogt, R.C. (1981) *Natural history of amphibians and reptiles of Wisconsin.* Milwaukee Public Museum, Milwaukee, 205 pp., illust.

Vogt, T. (1922) Zur Reptilien- und Amphibienfauna Südchinas. *In:* Mell, R. (Ed.), Beiträge zur Fauna Sinica. *Archiv für Naturgeschichte*, Berlin 88A(10), 135–146.

Vogt, T. (1932a) *In:* Weigelt's, Das Lebensbild zur Bildungszeit der mitteleozänen Braunkohlen des Geiseltals. *Paläontologie*, Berlin 14, 12–14.

Vogt, T. (1932b) Beitrag zur Reptilien fauna der echemaligen Kolonie Deutsch-Neuguinea. *Sitzunsberichte der Gesellschaft Naturforschender Freunde zu Berlin* 1832, 281–294.

Voigtländer, F. (1860) Einige anatomische Notizen über den Scelettbau von *Scotophis alleghaniensis* B. und G., *Elaps fulvius* Cuv. und *Heterodon platyrrhinos* Catr. *Denkschriften der naturwissenschaftlichen Gesellschaft Isis zu Dresden* 1860, 63–67, 1 pl.

Völkl, W. & Käsewieter, D. (2003) *Die Schlingnatter.* Laurenti Verlag, Bielefeld, 151 pp., figs., 7 col. pls.

Völkl, W. & Thiesmeier, B. (2002) *Die Kreuzotter.* Laurenti Verlag, Bielefeld, 159 pp., figs., 8 col. pls.

Vølsoe, H. (1939) The sea snakes of the Iranian Gulf and the Gulf of Oman. With a summary of the biology of the sea snakes. *In:* Jessen, K. & Spank, R (Eds.), *Danish Scientific Investigations in Iran, Part I.* Copenhagen, pp. 9–45.

Vølsoe, H. (1958a) Herpetology of Rennell Island. *In:* Wolff, T. (Ed.), *The natural history of Rennell Island, British Solomon Islands, Vol. 1 (Vertebrates). Scientific results of the Danish Rennell Expedition, 1951, and the British Museum (Natural History) Expedition, 1953.* Danish Science Press, Copenhagen, pp. 121–134, 1 fig.

Vølsoe, H. (1958b) Additional specimens of snakes from Rennell Island. *In:* Wolff, T. (Ed.), *The natural history of Rennell Island, British Solomon Islands, Vol. 1 (Vertebrates). Scientific results of the Danish Rennell Expedition, 1951, and the British Museum (Natural History) Expedition, 1953.* Danish Science Press, Copenhagen, pp. 227–228.

Volz, W. (1904) Schlangen von Palembang (Sumatra), (Reise von Dr. Walter Volz.). *Zoologische Jahrbücher (Abtheilung für Systematik, Geologie und Biologie der Thiere)*, Jena 20(5), 491–508.

Voorhies, M.R., Holman, J.A. & Xue, X.-X. (1987) The Hottell Ranch rhino quarries (basal Ogallala, medial Barstovian), Banner County, Nebraska. Part I, Geologic setting, faunal lists, lower vertebrates. *Contributions to Geology, University of Wyoming* 25(1), 55–69, 9 figs.

Voris, H.K. (1966) Fish eggs as the apparent sole food item for a genus of sea snake, *Emydocephalus* (Krefft). *Ecology* 47(1), 152–154.

Voris, H.K. (1969) *The evolution of the Hydrophiidae with a critique on methods of approach.* Doctoral Dissertation, University of Chicago, Chicago, 429 pp.

Voris, H.K. (1972) The role of sea snakes (Hydrophiidae) in the trophic structure of coastal ocean communities. *Journal of the Marine Biology Association of India, Mandapam Camp*, Madras 14(2), 429–442, 2 pls.

Voris, H.K. (1977) A phylogeny of the sea snakes (Hydrophiidae). *Fieldiana: Zoology* 70(4), 79–169.

Voris, H.K. (1985) Population size estimates for a marine snake (*Enhydrina schistosa*) in Malaysia. *Copeia* 1985(4), 955–961.

Voris, H.K. & Glodek, G.S. (1980) Habitat, diet, and reproduction of the file snake, *Acrochordus granulatus*, in the Straits of Malacca. *Journal of Herpetology* 14(1), 108–111, 2 figs.

Voris, H.K. & Jayne, B.C. (1979) Growth, reproduction and population structure of a marine snake, *Enhydrina schistosa* (Hydrophiidae). *Copeia* 1979(2), 307–318, 4 figs.

Voris, H.K. & Moffett, M.W. (1981) Size and proportion relationship between the beaked sea snake and its prey. *Biotropica* 13(1), 15–19, illust.

Voris, H.K. & Voris, H.H. (1983) Feeding strategies in marine snakes, an analysis of evolutionary, morphological, behavioural and ecological relationships. *American Zoologist* 23(2), 411–425, illust.

Voris, H.K., Voris, H.H. & Lim, B.L. (1978) The food and feeding behavior of a marine snake, *Enhydrina schistosa* (Hydrophiidae). *Copeia* 1978(1), 134–146, 1 fig.

Voris, H.K., Alfaro, M.E., Karns, D.R., Starnes, G.L., Thompson, E. & Murphy, J.C. (2002) Phylogenetic relationships of the Oriental-Australian rear-fanged water snakes (Colubridae, Homalopsinae) based on mitochondrial DNA sequences. *Copeia* 2002(4), 906–915, 1 fig.

Vosjoli, P. de. (1997) A new look at *Boa constrictors. The Vivarium* 9(1), 37–43, 6 col. pls.

Vosjoli, P. de., Klingenberg, R. & Ronne, J. (1998) The Boa constrictor *manual.* Advanced Vivarium Systems, Santee, 84 pp., col. illust.

Vosmaer, A. (1774) *Beschryving van twee verschillende en voor als nog zeer weinig bekende platstaart slangen, zynde de bruinrug uit Mexico, en de gerinde uit de Indische Zeen.* 8 pp., 1 col. pl.

Vozenilek, P. (1966) Herpetofauno en Cehoslovakio. *Scienca Revuo*, Praha 17(3–4), 127–128.

Vrcibradic, D., Rocha, C.F.D., Kiefer, M.C., Hatano, F.H., Fontes, A.F., Almeida-Gomes, M., Siqueira, C.C., Pontes, J.A.L., Borges-Junior, V.N.T., Gil, L.O., Klaion, T., Rubiao, E.C.N. & Van Sluys, M. (2011) Herpetofauna, Estacao Ecologica Estadual do Paraiso, state of Rio de Janeiro, southeastern Brazil. *Check List* 7(6), 745–749, 2 figs.

Vullo, R., Rage, J.-C. & Neraudeau, D. (2011) Anuran and Squamate remains from the Cenomanian (Late Cretaceous) of Charentes, western France. *Journal of Vertebrate Paleontology* 31(2), 279–291, 8 figs.

Vuoto, J.A. (1996) Ampliación del area de distribución de *Waglerophis merremii* (Wagler, 1824) (Serpentes: Colubridae) sobre los provincias de Entre Rios, Santa Fe y Buenos Aires, Argentina. *Cuadernos de Herpetologia* 10(1–2), 59–70.

Vyas, R. (1989) Melanistic form of the royal snake (*Spalerosophis diadema* Schlegel). *The Journal of the Bombay Natural History Society* 86(1), 112, 1 fig.

Vyas, R. (1993) *Studies on some snakes (Ophidia: Reptilia) of Gujarat State, India.* Doctoral Dissertaion, Bhavnagar University, Bhavnagar, xxxiv + 187 pp.

Vyas, R. (1996) An identification key and checklist of snakes of Gujarat, India. *Journal of Animal Morphology and Physiology* 43(2), 183–190.

Vyas, R. (1998) The reptiles of Gujarat State: updated distribution. *Tigerpaper* 25(1), 8–14.

Vyas, R. (2008) Discussion on the snake fauna of Gujarat State, with some notable records. *The Journal of the Bombay Natural History Society* 105(3), 344–347.

Vyas, R. (2011) Preliminary survey on reptiles of Jassore Wildlife Sanctuary, Gujarat State, India. *Russian Journal of Herpetology* 18(3), 211–214, 4 figs.

Vyas, R. & Patel, S.S. (2007) New distributional records of the endemic snake, *Coronella brachyura* (Günther, 1866) (Serpentes, Colubridae, Colubrinae), from Gujarat State, India. *Sauria* 29(3), 47–50, 4 figs.

Wacha, R.S. & Christiansen, J.L. (1974) Systematics of the eimerian parasites from North American snakes of the family Colubridae, and their prevalence in the colubrids of Iowa. *Journal of Protozoology* 21(4), 483–489, 7 figs.

Wade, E. (1988) Intraspecific variation in the colubrid snake genus *Macroprotodon*. *Herpetological Journal* 1(6), 237–245, 6 figs.

Wade, E. (2001) Review of the false smooth snake genus *Macroprotodon* (Serpentes, Colubridae) in Algeria with a description of a new species. *Bulletin of the Natural History Museum of London* (*Zoology*) 67(1), 85–107.

Wagler, J.G. (1824) *Serpentum brasiliensium species novae ou Histoire naturelle des espèces nouvelles de serpens, recueillies et observées pendant le voyage dans l'intérieur du Brésil dans les années 1817, 1818, 1819, 1820, exécuté par ordre de Sa Majesté le Roi de Bavière, publiée par Jean de Spix,...décrite d'après les notes du voyageur par Jean Wagler.* Monachii, Franc. Seraph. Hübschmann, viii + 75 pp.

Wagler, J.G. (1828–1833) *Descriptiones et icones amphibiorum.* Fasc. I. J.G. Cotta, München, unnumbered text, 36 pls.

Wagler, J.G. (1828) Auszüge aus seinem Systema Amphibiorum. *Isis von Oken*, Leipzig 21(7), 740–744, 4 figs., pl. 10.

Wagler, J.G. (1830) *Natürliches System der Amphibien, mit vorangehender Classification der Saugthiere und Vogel. Ein Beitrag zur vergleichenden Zoologie.* J.G. Cotta schen Buchhandlung. Munchen, Stuttgart und Tubingen, vi + 354 pp.

Wagler, J.G. (1833) Deutung der in Seba's Thesauro Rerum Naturalium T. 1 et 2. enthaltenen Abbildungen von Lurchen, mit critischen Bemerkungen. *Isis von Oken*, Leipzig 27(9), 885–905.

Wagner, F.W., Jr. (1975) *A revision of the Asian colubrid snakes* Oligodon cinereus *(Günther),* Oligodon joynsoni *(Smith), and* Oligodon cyclurus *(Cantor).* Master's Thesis, Louisiana State University, Baton Rouge, 97 pp., 22 figs.

Wagner, F.W., Jr. (1976) A new species of *Oligodon* from Poulo Nias. Unpublished manuscript, Baton Rouge, 7 pp., 3 figs.

Wagner, P. & Wilms, T.M. (2010) A crowned devil: new species of *Cerastes* Laurenti, 1768 (Ophidia, Viperidae) from Tunisia, with two nomenclatural comments. *Bonn Zoological Bulletin* 57(2), 297–306, 5 figs.

Wagner, P., Bauer, A.M. & Böhme, W. (2012) Amphibians and reptiles collected by Moritz Wagner, with a focus on the ZFMK collection. *Bonn Zoological Bulletin* 61(2), 216–240, 4 figs.

Waite, E.R. (1893) Notes on Australian Typhlopidae. *Records of the Australian Museum*, Sydney 2(5), 57–62, pl. 15.

Waite, E.R. (1894a) Notes on Australian Typhlopidae. *Proceedings of the Linnaean Society of New South Wales*, Sydney (Series 2) 9(1), 9–14, 1 pl.

Waite, E.R. (1894b) Re-description of *Aspidites ramsayi*, Macleay. *Proceedings of the Linnaean Society of New South Wales*, Sydney (Series 2) 9, 715–717, 1 pl.

Waite, E.R. (1897a) Notes on Australian Typhlopidae. *Transactions of the Royal Society of South Australia*, Adelaide 21, 25–27, pl. 3.

Waite, E.R. (1897b) A new blind snakes from the Duke of York Island. *Records of the Australian Museum*, Sydney 3(3), 69–70, 3 figs.

Waite, E.R. (1898) *A popular account of Australian Snakes*. Thomas Shine, Sydney, 71 pp.

Waite, E.R. (1899) Notes on snakes. *Records of the Australian Museum* 3, 104–105.

Waite, E.R. (1918a) Review of the Australian blind snakes. *Records of the South Australian Museum*, Adelaide 1(1), 1–34, 24 figs., 1 pl.

Waite, E.R. (1918b) Description of a new blind snake from the Solomon Islands, with a list of species from the Austro-Malayan and Polynesian sub-regions. *Records of the South Australian Museum*, Adelaide 1(1), 35–38, fig. 25.

Waite, E.R. (1925) Field notes on some Australian reptiles and amphibians. *Records of the South Australian Museum*, Adelaide 3(1), 17–32, 15 figs.

Waite, E.R. (1928) *The reptiles and amphibians of South Australia.* Harrison Weir, Adelaide, 270 pp., 192 figs.

Waite, E.R. & Longman, H.A. (1920) Descriptions of little-known Australian snakes. *Records of the South Australian Museum*, Adelaide 1(3), 173–180, figs. 32–38, pl. 27.

Wake, D.B. & Kluge, A.G. (1961) The Machris expedition to Tchad Africa. Amphibians and Reptiles. *Contributions in Science, Los Angeles County Museum* (40), 1–12.

Walker, C.F. (1955) A new gartersnake (*Thamnophis*) from Tamaulipas. *Copeia* 1955(2), 110–113, 1 fig.

Walker, W.F. (1945) A study of the snake, *Tachymenis peruviana* Wiegmann and its allies. *Bulletin of the Museum of Comparative Zoology* 96(1), 1–55, 5 pls.

Wall, F. (1898) Notes on two specimens of *Hypsirhina sieboldii*. *The Journal of the Bombay Natural History Society* (1897–1898) 11(4), 732–734.

Wall, F. (1903) A prodromus of the snakes hitherto recorded from China, Japan, and the Loo Choo Islands; with some notes. *Proceedings of the Zoological Society of London* 73(1), 84–102.

Wall, F. (1905a) Notes on snakes collected at Hakgala, Ceylon. *Spolia Zeylanica*, Colombo 3(10), 144–147.

Wall, F. (1905b) Description of a new snake from Burma. *Oligodon McDougalli*. *The Journal of the Bombay Natural History Society* (1904–1905) 16(2), 251–252.

Wall, F. (1905c) Notes on snakes collected in Cannanore from 5th November 1903 to 5th August 1904. *The Journal of the Bombay Natural History Society* (1904–1905) 16(2), 292–317.

Wall, F. (1905d) The breeding of Russell's viper (*Vipera russellii*). *The Journal of the Bombay Natural History Society* (1904–1905) 16(2), 374–375.

Wall, F. (1905e) Notes on some Bangalore snakes. *The Journal of the Bombay Natural History Society* (1904–1905) 16(2), 389–394.

Wall, F. (1905f) Notes on a collection of snakes from Japan and the Loo Choo Islands. *Proceedings of the Zoological Society of London* 75(4), 511–517.

Wall, F. (1905g) A popular treatise on the common Indian snakes. Illustrated by coloured plates and diagrams. Part I. *The Journal of the Bombay Natural History Society* (1904–1905) 16(4), 533–554, 12 figs., pl. 1.

Wall, F. (1906a) A popular treatise on the common Indian snakes. Illustrated by coloured plates and diagrams. Part II. *The Journal of the Bombay Natural History Society* (1906–1907) 17(1), 1–9, 13 figs., pl. 2.

Wall, F. (1906b) A new snake (*Melanelaps mcphersoni*) from the Aden Hinterland. *The Journal of the Bombay Natural History Society* (1906–1907) 17(1), 27–28, 3 figs.

Wall, F. (1906c) A new Himalayan snake *Lycodon mackinnoni. The Journal of the Bombay Natural History Society* (1906–1907) 17(1), 29–30.

Wall, F. (1906d) The poisonous snakes of India and how to recognize them. *The Journal of the Bombay Natural History Society* (1906–1907) 17(1), 51–71, 15 figs., 17(2), 299–333, 40 figs.

Wall, F. (1906e) A popular treatise on the common Indian snakes. Illustrated by coloured plates and diagrams. Part III. *The Journal of the Bombay Natural History Society* (1906–1907) 17(2), 259–273, 3 figs., pl. 3.

Wall, F. (1906f) A descriptive list of the sea-snakes in the Indian Museum. *Memoirs of the Asiatic Society of Bengal*, Calcutta 1, 277–299, illust.

Wall, F. (1907a) A new krait from Oudh (*Bungarus walli*). *The Journal of the Bombay Natural History Society* (1906) (1906–1907) 17(3), 608–611, 1 pl.

Wall, F. (1907b) Some new Asian snakes. *The Journal of the Bombay Natural History Society* (1906–1907) 17(3), 612–619, 2 figs., 2 pls.

Wall, F. (1907c) Note on the breeding of Russell's viper (*Vipera russelli*) in captivity. *The Journal of the Bombay Natural History Society* (1906–1907) 17(3), 808.

Wall, F. (1907d) A popular treatise on the common Indian snakes. Illustrated by coloured plates and diagrams. Part IV. *The Journal of the Bombay Natural History Society* (1906–1907) 17(4), 857–869, pl. 4.

Wall, F. (1907e) A popular treatise on the common Indian snakes. Part V. *The Journal of the Bombay Natural History Society* (1907–1908) 18(1), 1–17, 3 figs., pl. 5.

Wall, F. (1907f) Notes on snakes collected in Fyzabad. *The Journal of the Bombay Natural History Society* (1907–1908) 18(1), 101–129, 6 figs.

Wall, F. (1907g) Reports on a collection of Batrachia, reptiles and fish from Nepal and the western Himalayas. Ophidia. *Records of the Indian Museum*, Calcutta 1(2), 155–157.

Wall, F. (1907h) On the Hydrophiidae in the Colombo Museum. *Spolia Zeylanica*, Colombo 4(16), 166–172.

Wall, F. (1908a) A popular treatise on the common Indian snakes. Illustrated by coloured plates and diagrams. Part VI. *The Journal of the Bombay Natural History Society* (1907–1908) 18(2), 227–243, 2 figs., pl. 6.

Wall, F. (1908b) Two new snakes from Assam. *The Journal of the Bombay Natural History Society* (1907–1908) 18(2), 272–274, 1 pl.

Wall, F. (1908c) Notes on a collection of snakes from the Khasi Hills, Assam. *The Journal of the Bombay Natural History Society* (1907–1908) 18(2), 312–337, 7 figs.

Wall, F. (1908d) Remarks upon the snake *Contia angusticeps. The Journal of the Bombay Natural History Society* (1907–1908) 18(3), 501–503, 3 figs.

Wall, F. (1908e) A popular treatise on the common Indian snakes. Illustrated by coloured plates and diagrams. Part VII. *The Journal of the Bombay Natural History Society* (1907–1908) 18(3), 525–554, 6 figs., pl. 7.

Wall, F. (1908f) A popular treatise on the common Indian snakes. Illustrated by coloured plates and diagrams. Part VIII. *The Journal of the Bombay Natural History Society* (1907–1908) 18(4), 711–735, 7 figs., pl. 8.

Wall, F. (1908g) Remarks on some recently acquired snakes. *The Journal of the Bombay Natural History Society* (1907–1908) 18(4), 778–784.

Wall, F. (1908h) A new pit viper of the genus *Ancistrodon. The Journal of the Bombay Natural History Society* (1907–1908) 18(4), 792–794.

Wall, F. (1908i) Notes on a collection of snakes from Persia. *The Journal of the Bombay Natural History Society* (1907–1908) 18(4), 795–805, 3 figs.

Wall, F. (1908j) A new colour variety of the common green whip snake (*Dryophis mycterizans*). *The Journal of the Bombay Natural History Society* (1907–1908) 18(4), 919.

Wall, F. (1908k) Notes on a gravid female of Siebold's watersnake (*Hypsirhina sieboldii*). *The Journal of the Bombay Natural History Society* (1907–1908) 18(4), 920.

Wall, F. (1908l) Miscellanea. Reptiles. Remarks on *Simotes splendidus. Records of the Indian Museum*, Calcutta 2(1), 105–106, 2 figs.

Wall, F. (1909a) A monograph of the sea snakes. *Memoirs of the Asiatic Society of Bengal*, Calcutta 2(8), 169–251, 65 figs., pls. 7–10.

Wall, F. (1909b) A popular treatise on the common Indian snakes. Illustrated by coloured plates and diagrams. Part IX. *The Journal of the Bombay Natural History Society* (1909–1910) 19(1), 87–106, 9 figs., pl. 9.

Wall, F. (1909c) A popular treatise on the common Indian snakes. Illustrated by coloured plates and diagrams. Part X. *The Journal of the Bombay Natural History Society* (1909–1910) 19(2), 287–298, 9 figs., pl. 10.

Wall, F. (1909d) Notes on snakes from the neighbourhood of Darjeeling. *The Journal of the Bombay Natural History Society* (1909–1910) 19(2), 337–357, 10 figs.

Wall, F. (1909e) A popular treatise on the common Indian snakes. Illustrated by coloured plates and diagrams. Part XI. *The Journal of the Bombay Natural History Society* (1909–1910) 19(3), 555–563, 9 figs., pl. 11.

Wall, F. (1909f) Notes on snakes collected in upper Assam. *The Journal of the Bombay Natural History Society* (1909–1910) 19(3), 608–623.

Wall, F. (1909g) Notes on a collection of snakes from Matheran. *The Journal of the Bombay Natural History Society* (1909–1910) 19(3), 756–757.

Wall, F. (1909h) Notes on a collection of snakes made in Baxa Dooars. *The Journal of the Bombay Natural History Society* (1909–1910) 19(3), 757.

Wall, F. (1909i) Remarks on some little known Indian Ophidia. *Records of the Indian Museum*, Calcutta 3(2), 145–150, 1 fig.

Wall, F. (1909j) Remarks on some forms of *Dipsadomorphus. Records of the Indian Museum*, Calcutta 3(2), 151–155.

Wall, F. (1909k) A new colour variety of Macclelland's coral snake (*Callophis macclellandi*) and extension of the habitat of this species. *The Journal of the Bombay Natural History Society* (1909–1910) 19(1), 266.

Wall, F. (1910a) A popular treatise on the common Indian snakes. Illustrated by coloured plates and diagrams. Part XII. *The Journal of the Bombay Natural History Society* (1909–1910) 19(4), 775–792, 9 figs., pl. 12.

Wall, F. (1910b) A new blind snake from the Western Himalayas. *The Journal of the Bombay Natural History Society* (1909–1910) 19(4), 805–806.

Wall, F. (1910c) Notes on snakes collected in Upper Assam. Part II. *The Journal of the Bombay Natural History Society* (1909–1910) 19(4), 825–845, 8 figs., 1 pl.

Wall, F. (1910d) Notes on snakes collected in the Jalpaiguri district. *The Journal of the Bombay Natural History Society* (1909–1910) 19(4), 897–900, 2 figs.

Wall, F. (1910e) A new snake from Assam. *Oligodon erythrorhachis. The Journal of the Bombay Natural History Society* (1909–1910) 19(4), 923–924, 1 pl.

Wall, F. (1910f) A second specimen of the snake *Oligodon erythrogaster* from the eastern Himalayas. *The Journal of the Bombay Natural History Society* (1909–1910) 19(4), 1000–1001, 4 figs.

Wall, F. (1910g) A popular treatise on the common Indian snakes. Illustrated by coloured plates and diagrams. Part XIII. *The Journal of the Bombay Natural History Society* (1910–1911) 20(1), 65–79, 8 figs., pl. 13.

Wall, F. (1910h) Remarks on the varieties and distribution of the common green whipsnake (*Dryophis mycterizans*). *The Journal of the Bombay Natural History Society* (1910–1911) 20(1), 229–230.

Wall, F. (1910i) Notes on the viviparous habit of Jerdon's pit viper (*Lachesis jerdoni*) and observations on the foetal tooth in the unborn embryo. *The Journal of the Bombay Natural History Society* (1910–1911) 20(1), 231–234, 5 figs.

Wall, F. (1910j) A new *Tropidonotus* from the Chin Hills. *The Journal of the Bombay Natural History Society* (1910–1911) 20(2), 345–346.

Wall, F. (1910k) Remarks on some recently acquired Ceylon snakes. *Spolia Zeylanica*, Colombo 7(25), 35–38.

Wall, F. (1911a) A popular treatise on the common Indian snakes. Illustrated by coloured plates and diagrams. Part XIV. *The Journal of the Bombay Natural History Society* (1910–1911) 20(3), 603–633, 15 figs., pl. 14.

Wall, F. (1911b) Notes on a brood of young sea-snakes (*Distira spiralis*, Shaw). *The Journal of the Bombay Natural History Society* (1910–1911) 20(3), 858–863, 8 figs.

Wall, F. (1911c) A popular treatise on the common Indian snakes. Illustrated by coloured plates and diagrams. Part XV. *The Journal of the Bombay Natural History Society* (1910–1911) 20(4), 933–953, 8 figs., pl. 15.

Wall, F. (1911d) Remarks on the snake collection in the Quetta Museum. *The Journal of the Bombay Natural History Society* (1910–1911) 20(4), 1033–1042.

Wall, F. (1911e) A new snake, *Simotes juglandifer*, hitherto unrecognized as a distinct species. *The Journal of the Bombay Natural History Society* (1910–1911) 20(4), 1162–1164.

Wall, F. (1911f) A popular treatise on the common Indian snakes. Illustrated by coloured plates and diagrams. Part XVI. *The Journal of the Bombay Natural History Society* (1911–1912) 21(1), 1–19, 1 fig., pl. 16.

Wall, F. (1911g) Reptiles collected in Chitral. *The Journal of the Bombay Natural History Society* (1911–1912) 21(1), 132–145.

Wall, F. (1911h) A new snake from the Western Himalayas. *The Journal of the Bombay Natural History Society* (1911–1912) 21(1), 201–202.

Wall, F. (1911i) Remarks on two rare blind snakes. *The Journal of the Bombay Natural History Society* (1911–1912) 21(1), 278–279, 2 figs.

Wall, F. (1911j) Remarks on the greater, and lesser black kraits (*Bungarus niger*, and *B. lividus*). *The Journal of the Bombay Natural History Society* (1911–1912) 21(1), 281–283.

Wall, F. (1911k) The development of embryo in the eggs of the oviparous viper, *Lachesis monticola*, prior to oviposition. *The Journal of the Bombay Natural History Society* (1911–1912) 21(1), 284–285.

Wall, F. (1912a) A popular treatise on the common Indian snakes. Illustrated by coloured plates and diagrams. Part XVII. *The Journal of the Bombay Natural History Society* (1911–1912) 21(2), 447–475, pl. 17.

Wall, F. (1912b) A new snake *Psammophis triticeus* from Baluchistan. *The Journal of the Bombay Natural History Society* (1911–1912) 21(2), 634–636, 6 figs.

Wall, F. (1912c) The breeding of the false Himalayan viper (*Psammodynastes pulverulentus*). *The Journal of the Bombay Natural History Society* (1911–1912) 21(2), 686.

Wall, F. (1912d) A popular treatise on the common Indian snakes. Illustrated by coloured plates and diagrams. Part XVIII. *The Journal of the Bombay Natural History Society* (1911–1912) 21(3), 1009–1021, 6 figs., pl. 18.

Wall, F. (1913a) A popular treatise on the common Indian snakes. Illustrated by coloured plates and diagrams. Part XIX. *The Journal of the Bombay Natural History Society* (1913–1914) 22(1), 22–28, 3 figs., pl. 19.

Wall, F. (1913b) A popular treatise on the common Indian snakes. Illustrated by coloured plates and diagrams. Part XX. *The Journal of the Bombay Natural History Society* (1913–1914) 22(2), 243–259, pl. 21.

Wall, F. (1913c) A rare snake *Elachistodon westermanni* from the Jalpaiguri District. *The Journal of the Bombay Natural History Society* (1913–1914) 22(2), 400–401, 3 figs.

Wall, F. (1913d) On the common (*Bungarus caeruleus*) and Sind kraits (*Bungarus sindanus*). *The Journal of the Bombay Natural History Society* (1913–1914) 22(2), 401–403, pls. C–D.

Wall, F. (1913e) Some new snakes from the Oriental region. *The Journal of the Bombay Natural History Society* (1913–1914) 22(3), 514–516, 6 figs.

Wall, F. (1913f) A popular treatise on the common Indian snakes. Illustrated by coloured plates and diagrams. Part XX. *The Journal of the Bombay Natural History Society* (1913–1914) 22(3), 550–568, 4 figs., 3 photos.

Wall, F. (1913g) Varieties of *Hemibungarus nigrescens* and *Hydrophis torquatus*. *The Journal of the Bombay Natural History Society* (1913–1914) 22(3), 638–639.

Wall, F. (1913h) Notes on some interesting snakes recently presented to the Society. *The Journal of the Bombay Natural History Society* (1913–1914) 22(3),639.

Wall, F. (1913i) *The poisonous terrestrial snakes of our British Indian Dominions (including Ceylon) and how to recognize them, with symptoms of snake poisoning and treatment.* Third edition. Bombay Natural History Society, Bombay, iv + 149 pp., 42 figs.

Wall, F. (1914a) A popular treatise on the common Indian snakes. Illustrated by coloured plates and diagrams. Part XX. *The Journal of the Bombay Natural History Society* (1913–1914) 22(4), 749–760, 4 figs., pl. 20.

Wall, F. (1914b) A popular treatise on the common Indian snakes. Illustrated by coloured plates and diagrams. Part XXII. *The Journal of the Bombay Natural History Society* (1914–1915) 23(1), 34–43, 6 figs., pl. 22.

Wall, F. (1914c) A new snake of the genus *Tropidonotus* from the Eastern Himalayas. *The Journal of the Bombay Natural History Society* (1914–1915) 23(1), 166–167.

Wall, F. (1914d) A new snake from Baluchistan. *The Journal of the Bombay Natural History Society* (1914–1915) 23(1), 167–168.

Wall, F. (1914e) Remarks to show that the snake hitherto known as *Zamenis mucosus* has been misplaced, and should be included in the genus *Zaocys*. *The Journal of the Bombay Natural History Society* (1914–1915) 23(1), 168–169.

Wall, F. (1914f) A popular treatise on the common Indian snakes. Illustrated by coloured plates and diagrams. Part XXIII. *The Journal of the Bombay Natural History Society* (1914–1915) 23(2), 206–215, 6 figs., pl. 23.

Wall, F. (1914g) Remarks on the sea-snakes in our Society's collection. *The Journal of the Bombay Natural History Society* (1914–1915) 23(2), 374–377, pls. A–B.

Wall, F. (1917) Notes on the Russell's viper. *The Journal of the Bombay Natural History Society* (1917–1918) 25(2), 307–308, 1 pl.

Wall, F. (1918a) A popular treatise on the common Indian snakes. Illustrated by coloured plates and diagrams. Part XXIV. *The Journal of the Bombay Natural History Society* (1917–1918) 25(3), 375–382, 10 figs., pl. 24.

Wall, F. (1918b) Notes on a collection of sea snakes from Madras. *The Journal of the Bombay Natural History Society* (1917–1918) 25(4), 599–607.

Wall, F. (1918c) A popular treatise on the common Indian snakes. Illustrated by coloured plates and diagrams. Part XXV. *The Journal of the Bombay Natural History Society* (1917–1918) 25(4), 628–632, 8 figs., pl. 25.

Wall, F. (1918d) Notes on a gravid *Hydrophis cyanocinctus* and her brood. *The Journal of the Bombay Natural History Society* (1917–1918) 25(4), 754–756, 13 figs.

Wall, F. (1918e) A popular treatise on the common Indian snakes. Illustrated by coloured plates and diagrams. Part XXVI. *The Journal of the Bombay Natural History Society* (1918–1920) 26(1), 89–97, 6 figs., pl. 26.

Wall, F. (1919a) A popular treatise on the common Indian snakes. Illustrated by coloured plates and diagrams. Part XXVII. *The Journal of the Bombay Natural History Society* (1918–1920) 26(2), 430–437, 10 figs., pl. 27.

Wall, F. (1919b) Notes on a collection of snakes made in the Nilgiri Hills and the adjacent Wynaad. *The Journal of the Bombay Natural History Society* (1918–1920) 26(2), 552–584, 20 figs.

Wall, F. (1919c) A popular treatise on the common Indian snakes. Illustrated by coloured plates and diagrams. Part XXVIII. *The Journal of the Bombay Natural History Society* (1918–1920) 26(3), 803–810, 7 figs., pl. 28.

Wall, F. (1919d) Notes on the snake *Trirhinopholis nuchalis* (Boulenger). *The Journal of the Bombay Natural History Society* (1918–1920) 26(3), 863.

Wall, F. (1919e) Notes on some recent additions to our Society's snake collection. *The Journal of the Bombay Natural History Society* (1918–1920) 26(3), 865–867.

Wall, F. (1921a) A new snake of the family Uropeltidae. *The Journal of the Bombay Natural History Society* (1921–1922) 28(1), 41–42, figs. a–d.

Wall, F. (1921b) Notes on some notable additions to the Bombay Natural History Society's snakes collection. *The Journal of the Bombay Natural History Society* (1921–1922) 28(1), 43–44, 1 pl.

Wall, F. (1921c) *Phayrea isabellina* Theobald, re-described. *Records of the Indian Museum*, Calcutta 22(2), 109–110.

Wall, F. (1921d) Remarks on a specimen of *Calamaria javanica*. *Records of the Indian Museum*, Calcutta 22(5), 729.

Wall, F. (1921e) Notes on some Ceylon snakes. *Spolia Zeylanica*, Colombo 11(43–44), 396–406.

Wall, F. (1921f) Remarks on the Indian species of *Dendrophis* and *Dendrelaphis*. *Records of the Indian Museum*, Calcutta 22(2), 151–162.

Wall, F. (1921g) *Ophidia taprobanica or the snakes of Ceylon*. H. R. Cottle, Colombo, xxii + 582 pp., 98 figs.

Wall, F. (1922a) Hatching of cobras (*Naia tripudians*) with remarks on the oodont, genitalia, etc. *The Journal of the Bombay Natural History Society* (1921–1922) 28(2), 553–556, 2 figs.

Wall, F. (1922b) Acquisition of four more specimens of the snake (*Brachyophidium rhodogastor*, Wall). *The Journal of the Bombay Natural History Society* (1921–1922) 28(2), 556–557.

Wall, F. (1922c) A review of the Indian species of *Amblycephalus*. *Records of the Indian Museum*, Calcutta 24(1), 19–27, figs.

Wall, F. (1922d) A new snake from the northern frontier of Assam. *Records of the Indian Museum*, Calcutta 24(1), 29–30.

Wall, F. (1923a) Notes on a collection of snakes from Shembaganur, Palnai Hills (Circa 7,000 feet). *The Journal of the Bombay Natural History Society* (1923–1924) 29(2), 388–398.

Wall, F. (1923b) Notes on a collection of snakes from Sinlum Kaba. *The Journal of the Bombay Natural History Society* (1923–1924) 29(2), 466–468.

Wall, F. (1923c) A hand-list of snakes of the Indian empire. Part I. *The Journal of the Bombay Natural History Society* (1923–1924) 29(2), 345–361.

Wall, F. (1923d) A hand-list of snakes of the Indian empire. Part II. *The Journal of the Bombay Natural History Society* (1923–1924) 29(3), 598–632.

Wall, F. (1923e) A review of the Asiatic species of *Contia* in Indian Museums. *The Journal of the Bombay Natural History Society* (1923–1924) 29(3), 769–779.

Wall, F. (1923f) The affinities of *Eryx jaculus* and *Eryx johni*. *The Journal of the Bombay Natural History Society* (1923–1924) 29(3), 842–843.

Wall, F. (1923g) A review of the Indian species of the genus *Oligodon* suppressing the genus *Simotes* (Ophidia). *Records of the Indian Museum*, Calcutta 25(3), 305–334.

Wall, F. (1923h) Notes on snakes collected on Annasigalla Estate from August 1920, to December, 1921. *Spolia Zeylanica*, Colombo 12(46), 252–270.

Wall, F. (1923i) *How to identify the snakes of India (including Burma and Ceylon)*. Frank Wall, Karachi, 56 pp., 70 figs.

Wall, F. (1924a) A hand-list of the snakes of the Indian empire. Part III. *The Journal of the Bombay Natural History Society* (1923–1924) 29(4), 864–878.

Wall, F. (1924b) A hand-list of the snakes of the Indian empire. Part IV. *The Journal of the Bombay Natural History Society* (1924–1925) 30(1), 12–24.

Wall, F. (1924c) The hamadryad or king cobra, *Naia hannah* (Cantor). *The Journal of the Bombay Natural History Society* (1924–1925) 30(1), 189–195, 1 pl.

Wall, F. (1924d) Notes on Ceylon snakes collected by Mr. W.W.A. Phillips. *Spolia Zeylanica*, Colombo 13(1), 71–88.

Wall, F. (1924e) Notes on two specimens of the sea snake *Lioselasma lapemidoides* (Gray). *Spolia Zeylanica*, Colombo 13(1), 89–92, 1 fig.

Wall, F. (1924f) Reply to Miss Procter's remarks on the snake *Brachyophidium rhodogaster*, Wall. *Annals and Magazine of Natural History*, London (Series 9) 14(80), 200.

Wall, F. (1925a) A hand-list of the snakes of the Indian empire. Part V. *The Journal of the Bombay Natural History Society* (1924–1925) 30(2), 242–252.

Wall, F. (1925b) Two new Burmese snakes. *The Journal of the Bombay Natural History Society* (1924–1925) 30(3), 587–588, 1 pl.

Wall, F. (1925c) A new snake from Burma. *The Journal of the Bombay Natural History Society* (1924–1925) 30(4), 734.

Wall, F. (1925d) Notes on snakes collected in Burma in 1924. *The Journal of the Bombay Natural History Society* (1924–1925) 30(4), 805–821.

Wall, F. (1926a) The reticulate python *Python reticulatus* (Schneider). *The Journal of the Bombay Natural History Society* (1926–1927) 31(1), 84–90.

Wall, F. (1926b) Snakes collected in Burma in 1925. *The Journal of the Bombay Natural History Society* (1926–1927) 31(3–4), 558–566, 1 fig.

Wall, F. (1928) *The poisonous terrestrial snakes of our British Indian Dominions (including Ceylon) and how to recognize them, with symptoms of snake poisoning and treatment.* Fourth edition. Bombay Natural History Society, x + 173 pp., 40 figs. [first edition, 1906, 56 pp., 55 figs.; second edition, 1907, i + 59 pp., 40 figs.; third edition, 1913, iv + 149 pp., 42 figs.]

Wall, F. & Evans, G.H. (1900) Notes on Ophidia collected in Burma from May to December, 1899. *The Journal of the Bombay Natural History Society* (1900–1901) 13(2), 343–354, 1 fig.

Wall, F. & Evans, G.H. (1901) Burmese snakes. Notes on specimens including 45 species of ophidian fauna collected in Burma from 1st January to 30th June, 1900. *The Journal of the Bombay Natural History Society* (1900–1901) 13(4), 611–620.

Wallace, A.R. (1876) *The geographical distribution of animals with a study of the relations of living and extinct faunas as elucidating the past changes of the earth's surface.* Two volumes. Harper & Brothers, New York, 503 + 607 pp.

Wallach, V. (1980) Report on bites by the West African nigh adder, *Causus maculates. Journal of the Herpetological Association of Africa* (22), 3–8.

Wallach, V. (1984) A new name for *Ophiomorphus colberti* Haas 1980. *Journal of Herpetology* 18(3), 329.

Wallach, V. (1985) A cladistic analysis of the terrestrial Australian Elapidae. *In:* Grigg, G., Shine, R. & Ehmann, H. (Eds.), *The biology of Australasian frogs and reptiles*. Royal Zoological Society of New South Wales, Chipping Norton, pp. 223–253, 5 figs.

Wallach, V. (1986) Status of the fossil snake, *Scytalophis* de Rochebrune. *Journal of Herpetology* 20(3), 448–449.

Wallach, V. (1988) Status and redescription of the genus *Padangia* Werner, with comparative visceral data on *Collorhabdium* Smedley and other genera (Serpentes, Colubridae). *Amphibia-Reptilia*, Leiden 9(1), 61–76, 3 figs.

Wallach, V. (1991) *Comparative visceral topography of African colubrid snakes of the subfamilies Aparallactinae and Atractaspidinae*. Master's Thesis, Louisiana State University, Baton Rouge, xix + 490 pp., 144 figs.

Wallach, V. (1993a) The supralabial imbrication pattern of the Typhlopoidea (Reptilia, Serpentes). *Journal of Herpetology* 27(2), 214–218, 1 fig.

Wallach, V. (1993b) A new species of blind snake, *Typhlops marxi*, from the Philippines (Serpentes, Typhlopidae). *Raffles Bulletin of Zoology*, Singapore 41(2), 263–278, 2 figs.

Wallach, V. (1993c) Presence of a left lung in the Typhlopidae (Reptilia, Serpentes). *Journal of the Herpetological Association of Africa*, Stellenbosch 42, 2 pp.

Wallach, V. (1994a) *Aparallactus lineatus* (Peters) and *Aparallactus niger* Boulenger, two valid species from West Africa. *Journal of Herpetology* 28(1), 95–99, 2 figs.

Wallach, V. (1994b) The status of the Indian endemic *Typhlops acutus* (Duméril & Bibron) and the identity of *Typhlops psittacus* Werner (Reptilia, Serpentes, Typhlopidae). *Bulletin de l'Institut Royal des Sciences Naturelles de Belgique*, Bruxelles 64, 209–229, 3 figs.

Wallach, V. (1995a) A new genus for the *Ramphotyphlops subocularis* species group (Serpentes, Typhlopidae), with description of a new species. *Asiatic Herpetological Research*, Berkeley 6, 132–150, 5 figs.

Wallach, V. (1995b) Revalidation of the genus *Tropidodipsas* Günther, with notes on the Dipsadini and Nothopsini (Serpentes, Colubridae). *Journal of Herpetology* 29(3), 476–481.

Wallach, V. (1996a) Notes and corrections on two scolecophidians, *Ramphotyphlops albiceps* and *Leptotyphlops brasiliensis*. *Herpetological Review* 27(1), 10.

Wallach, V. (1996b) The systematic status of the *Ramphotyphlops flaviventer* (Peters) complex (Serpentes, Typhlopidae). *Amphibia-Reptilia*, Leiden 17(4), 341–359, 5 figs.

Wallach, V. (1996c) *Leptotyphlops drewesi* n. sp., a worm snake from central Kenya (Serpentes, Leptotyphlopidae). *Journal of African Zoology* 110(6), 425–431, 1 fig.

Wallach, V. (1996d) Two new blind snakes of the *Typhlops ater* species group from Papua New Guinea (Serpentes, Typhlopidae). *Russian Journal of Herpetology*, Moscow 3(2), 107–118, 2 figs.

Wallach, V. (1998a) The lungs of snakes. *In:* Gans, C. & Gaunt, A.S. (Eds.), *Biology of the Reptilia. Vol. 19 (Morphology G).* Society for the Study of Amphibians and Reptiles, pp. 93–295, 27 figs.

Wallach, V. (1998b) *The visceral anatomy of blindsnakes and wormsnakes and its systematic implications (Serpentes: Anomalepididae, Typhlopidae, Leptotyphlopidae).* Doctoral Dissertation, Northeastern University, Boston, xxvi + 611 pp., 10 figs.

Wallach, V. (1999a) *Typhlops meszoelyi*, a new species of blind snake from northeastern India (Serpentes, Typhlopidae). *Herpetologica* 55(2), 185–191, 4 figs.

Wallach, V. (1999b) Geographic distribution: Serpentes: *Ramphotyphlops braminus. Herpetological Review* 30(4), 236.

Wallach, V. (2000a) Natural history notes. *Typhlops tasymicris.* Maximum length and second known specimen. *Herpetological Review* 31(3), 180.

Wallach, V. (2000b) Critical review of some recent descriptions of Pakistani *Typhlops* by M.S. Khan, 1999 (Serpentes, Typhlopidae). *Hamadryad* 25(2), 129–143.

Wallach, V. (2001) *Typhlops roxaneae*, a new species of Thai blindsnake from the *T. diardii* species group, with a synopsis of the Typhlopidae of Thailand (Serpentes, Scolecophidia). *Raffles Bulletin of Zoology,* Singapore 49(1), 39–49, 1 fig.

Wallach, V. (2002) *Typhlops etheridgei*, a new species of African blindsnake in the *Typhlops vermicularis* species group from Mauritana (Serpentes, Typhlopidae). *Hamadryad* 27(1), 108–127, 1 fig.

Wallach, V. (2003) Scolecophidia miscellanea. *Hamadryad* 27(2), 227–245, 1 fig.

Wallach, V. (2005a) *Letheobia pauwelsi*, a new species of blindsnake from Gabon (Serpentes, Typhlopidae). *African Journal of Herpetology* 54(1), 85–91, 1 fig.

Wallach, V. (2005b) The nomenclatorural status of Australian *Ramphotyphlops* (Serpentes, Typhlopidae). *Bulletin of the Maryland Herpetological Society* 42(1), 8–24.

Wallach, V. (2009) *Ramphotyphlops braminus* (Daudin): a synopsis of morphology, taxonomy, nomenclature and distribution (Serpentes: Typhlopidae). *Hamadryad* 34(1), 34–62, 3 figs.

Wallach, V. (2011) Synonymy and preliminary identifications of the snake illustrations of Albertus Seba's "Thesaurus" (1734–1735). *Hamadryad* 35(2), 1–190.

Wallach, V. & Bauer, A.M. (1996) On the identity and status of *Simotes semicinctus* Peters, 1862 (Serpentes, Colubridae). *Hamadryad* 21(1), 13–18, 2 figs.

Wallach, V. & Boundy, J. (2005) *Leptotyphlops greenwelli*, a new wormsnake of the *L. bicolor* species group from Nigeria (Serpentes, Leptotyphlopidae). *Annals of the Carnegie Museum* 74(1), 39–44, 1 fig.

Wallach, V. & Glaw, F. (2009) A new mid-altitude rainforest species of *Typhlops* (Serpentes: Typhlopidae) from Madagascar with notes on the taxonomic status of *T. boettgeri* Boulenger, *T. microcephalus* Werner, and *T. capensis* Rendahl. *Zootaxa* (2294), 23–38, 4 figs.

Wallach, V. & Günther, R. (1997) Typhlopidae vs. Anomalepididae, The identity of *Typhlops mutilatus* Werner (Reptilia, Serpentes). *Mitteilungen aus dem Zoologischen Museum in Berlin* 73(2), 333–342.

Wallach, V. & Günther, R. (1998) Visceral anatomy of the Malaysian snake genus *Xenophidion*, including a cladistic analysis and allocation to a new family (Serpentes, Xenophidiidae). *Amphibia-Reptilia,* Leiden 19(4), 385–405, 2 figs.

Wallach, V. & Hahn, D.E. (1997) *Leptotyphlops broadleyi*, a new species of worm snake from Côte d'Ivoire (Serpentes, Leptotyphlopidae). *African Journal of Herpetology* 46(2), 103–109, 1 fig.

Wallach, V. & Ineich, I. (1996) Redescription of a rare Malagasy blind snake, *Typhlops grandidieri* Mocquard, with placement in a new genus (Serpentes, Typhlopidae). *Journal of Herpetology* 30(3), 367–376, 2 figs.

Wallach, V. & Jones, G.S. (1994) *Cryptophidion annamense*, a new genus and species of cryptozoic snake from Vietnam (Reptilia, Serpentes). *Cryptozoology* (1992) 11, 1–37, 3 figs.

Wallach, V. & Jones, G.S. (1997) *Cryptophidion* is a valid taxon. *Cryptozoology* 12, 102–113, 2 figs.

Wallach, V. & Lanza, B. (2004) Taxonomic reassessment of Giuseppe Scortecci's scolecophidian type and non-type specimens from Eritrea, Ethiopia, Somalia, and Yemen, in the Museo Civico di Storia Naturale in Milano (Reptilia; Serpentes). *Atti della Società Italiana di Scienze Naturali e del Museo Civico di Storia Naturali di Milano* 145(1), 79–91.

Wallach, V. & Pauwels, O.S.G. (2004) *Typhlops lazelli*, a new species of Chinese blindsnake from Hong Kong (Serpentes, Typhlopidae). *Breviora* (512), 1–21, 1 fig.

Wallach, V. & Pauwels, O.S.G. (2008) The systematic status of *Cathetorhinus melanocephalus* Duméril & Bibron, 1844 (Serpentes, Typhlopidae). *Hamadryad* 32(2), 116–124, 1 fig.

Wallach, V. & Smith, H.M. (1992) *Boella tenella* is *Epicrates inornatus* (Reptilia, Serpentes). *Bulletin of the Maryland Herpetological Society* 28(4), 162–170.

Wallach, V., Brown, R.M., Diesmos, A.C. & Gee, C.V.A. (2007a) An enigmatic new species of blind snake from Luzon Island, northern Phillipines, with a synopsis of the genus *Acutotyphlops* (Serpentes, Typhlopidae). *Journal of Herpetology* 41(4), 690–702, 4 figs.

Wallach, V., Mercurio, V. & Andreone, F. (2007b) Rediscovery of the enigmatic blind snake genus *Xenotyphlops* in northern Madagascar, with description of a new species (Serpentes, Typhlopidae). *Zootaxa* (1402), 59–68, 4 figs.

Wallach, V., Wüster, W. & Broadley, D.G. (2009) In praise of subgenera: taxonomic status of cobras of the genus *Naja* Laurenti (Serpentes: Elapidae). *Zootaxa* (2236), 26–36, 1 fig.

Wallach, V., Lanza, B. & Nistri, A. (2010) *Aprosdoketophis andreonei*, a new genus and species of snake from Somalia (Serpentes: Colubridae: Boiginae). *African Journal of Herpetology* 59(2), 95–110, 3 figs.

Waller, T. & Buongermini, E. (1998) Geographic distribution: Serpentes: *Epicrates cenchria alvarezi*. *Herpetological Review* 29(4), 249.

Walley, H.D. (1999) *Rhadinaea flavilata. Catalogue of American amphibians and reptiles*, pp. 699.1–699.5.

Walley, H.D. (2003) Geographic distribution: Serpentes: *Liochlorophis. L. vernalis. Catalogue of American amphibians and reptiles*, pp. 776.1–776.13.

Walley, H.D. & Eckerman, C.M. (1999) *Heterodon nasicus*. *Catalogue of American amphibians and reptiles*, pp. 698.1–698.10.

Walley, H.D. & Janis, L. (2002) Remarks on the type designation of *Leptophis majalis* Baird and Girard 1853 (=*Opheodrys aestivus majalis* Grobman 1984). *Bulletin of the Chicago Herpetological Society* 37(7), 128.

Walley, H.D. & Plummer, M.V. (2000) *Opheodrys aestivus* (Linnaeus). *Catalogue of American amphibians and reptiles*, pp. 718.1–718.14.

Walley, H.D., Wusterbarth, T.L & Stanford, K.M. (2003) *Thamnophis radix*. *Catalogue of American amphibians and reptiles*, pp. 779.1–779.13, 3 figs.

Walley, H.G., King, R.B., Ray, J.M. & Wusterbarth, T.L. (2012) *Nerodia sipedon*. *Catalogue of American amphibians and reptiles*, pp. 899.1–889.58, 4 figs.

Wallin, L. (1991) *Catalogue of type specimens. 2. General zoology.* Uppsala University Zoology Museum, 80 pp.

Wallin, L. (1992) *Catalogue of type specimens. 4. Linnaean specimens.* Uppsala University Zoology Museum, 233 pp.

Wallin, L. (1992) Linnaean specimens in the Zoological Museum of Uppsala University. *Archives of Natural History* 19(2), 219–230, 4 figs.

Wallin, L. (1996) *Catalogue of type specimens. 2. General zoology.* Uppsala University Zoology Museum, Uppsala, 75 pp.

Walls, J.G. (1998a) *The living boas. A complete guide to the boas of the world.* T.F.H. Publications, Neptune City, 288 pp., figs & col. illust.

Walls, J.G. (1998b) *The living pythons. A complete guide to the pythons of the world.* T.F.H. Publications, Neptune City, 256 pp., figs. & col. illust.

Wandolleck, B. (1911) Die Amphibien und Reptilien der papuanischen Ausbeute Dr. Schlaginhaufens. *Abhandlungen und Berichte des königlichen Zoologischen und Anthropologische-Ethnographischen Museums zu Dresden* 13, 1–15, 1 pl.

Wang, C.-S. (1962) The reptiles of Botel-Tobago. *Quarterly Journal of the Taiwan Museum*, Taipei 15(3–4), 141–191, 7 figs., 3 pls.

Wang, C.-S. & Wang, Y-M. (1956) The reptile of Taiwan. *Quarterly Journal of the Taiwan Museum*, Taipei 9(1), 1–86.

Wang, P.-X. (1986) A new record of snakes of Gansu—*Elaphe schrenckii. Acta Herpetologica Sinica*, Chengdu (Series 2) 5(3), 238.

Wang, R.-F., He, W.-S., Shi, L.-M. & Shi, Q.-H. (1991) Observations on mitotic chromosomes and synaptonemal complex of *Elaphe taeniura* (Reptilia). *Acta Zoologica Sinica*, Chengdu 37(4), 438–442, 2 figs. (in Chinese)

Wang, X.-J. & Zhao, E.-M. (2007) Revised taxonomic status of *Gloydius brevicaudus siniticus. Sichuan Journal of Zoology* 26(2), 250–254, 1 fig.

Wangyal, J.T. (2011) Snakes and lizards from the Bumdeling Wildlife Sanctuary region of Bhutan: review of herpetofaunal information and new country records. *Herpetological Review* 42(1), 117–123, 1 fig.

Wangyal, J.T. (2012) New records of snakes and lizards from Bhutan. *Hamadryad* 36(1), 25–31, 13 figs.

Warburg, M.R. (1964) Observations on microclimate in habitats of some desert vipers in the Negev, Arava and Dead Sea regions. *Vie et Milieu*, Paris 15, 1017–1041, 21 figs.

Wareham, D.C. (2008) *The reptiles and amphibians of Dorset.* British Herpetological Society, London, 120 pp., illust.

Warner, J.K. & Combrink, K. (2010a) *Prosymna bivittata.* Geographic distribution: Serpentes: *Herpetological Review* 41(1), 111.

Warner, J.K. & Combrink, K. (2010b) *Rhinotyphlops lalandi.* Geographic distribution: Serpentes: *Herpetological Review* 41(1), 111.

Warrell, D.A., Ormerod, L.D. & Davidson, N.M. (1976) Bites by the night adder *(Causus maculates)* and burrowing vipers (genus *Atractaspis)* in Nigeria. *American Journal of Tropical Medicine and Hygiene* 25(3), 517–524, 3 figs.

Warrell, D.A., Looareesuwan, S., Stimson, A.F. & Hutton, R.A. (1992) Rediscovery and redefinition of Malcolm Smith's *Trimeresurus kanburiensis* in Thailand, with a report of envenoming. *Transactions of the Royal Society of Tropical Medicine and Hygiene* 86, 95–99.

Wasco, D.K. (2009) *Spatial and feeding ecology of the fer-de-lance* (Bothrops asper) *in Costa Rica.* Doctoral Dissertation, University of Miami, Miami, x + 116 pp., illust.

Watermolen, D.J. (1992) Wisconsin herpetology: A bibliographical update with taxonomic, subject, and geographical indices. *Wisconsin Department of Natural Resources, Wisconsin Endangered Research Report Series*, Madison (87), 1–18.

Watharow, S. (2000) Egg incubation methods and juvenile dispersal of eastern brown snakes *(Pseudonaja textilis). Monitor,* Somerville 11(1), 16–17, 1 fig., 1 col. photo.

Watkins-Colwell, G.J. & Leenders, T. (2002) Geographic distribution: Serpentes: *Gloydius blomhofii brevicaudus. Herpetological Review* 33(3), 227.

Watkins-Colwell, G.J. & de Leon, J. (2003) Geographic distribution: Serpentes: *Typhlops angolensis. Herpetological Review* 34(2), 171.

Wauer, R.H. (1964) *Reptiles and amphibians of Zion National Park.* Zion Natural History Association, 54 pp., illust.

Webb, J.K. & Shine, R. (1992) To find an ant: trail-following in Australian blindsnakes (Typhlopidae). *Animal Behavior* 43, 941–948.

Webb, J.K. & Shine, R. (1993a) Dietary habits of Australian blindsnakes (Typhlopidae). *Copeia* 1993(3), 762–770.

Webb, J.K. & Shine, R. (1993b) Prey-size selection, gape limitation and predator vulnerability in Australian blindsnakes (Typhlopidae). *Animal Behavior* 45, 1117–1126.

Webb, R.G. (1961) A new kingsnake from México, with remarks on the *mexicana* group of the genus *Lampropeltis. Copeia* 1961(3), 326– 333, 3 figs.

Webb, R.G. (1966) Resurrected names for Mexican populations of black-necked garter snakes *Thamnophis cyrtopsis* (Kennicott). *Tulane Studies in Zoology* 13(2), 55–70, 6 figs.

Webb, R.G. (1970) *Reptiles of Oklahoma.* University of Oklahoma Press, Norman, xi + 370 pp., 120 figs.

Webb, R.G. (1976) A review of the garter snake *Thamnophis elegans* in México. *Contributions in Science, Los Angeles County Museum* (284), 1–13, 3 figs.

Webb, R.G. (1977) Comments on snakes of the genus *Geophis* (Colubridae) from the Mexican states of Durango and Sinaloa. *Southwestern Naturalist* 21(4), 548–551, 1 fig.

Webb, R.G. (1978) A review of the Mexican garter snake *Thamnophis cyrtopsis postremus* Smith with comments on *Thamnophis vicinus* Smith. *Contributions in Biology and Geology, Milwaukee Public Museum* (19), 1–13, 3 figs.

Webb, R.G. (1980) *Thamnophis cyrtopsis. Catalogue of American amphibians and reptiles*, pp. 245.1–245.4.

Webb, R.G. (1982) Taxonomic status of some Neotropical garter snakes (genus *Thamnophis*). *Bulletin of the Southern California Academy of Sciences* 81(1), 26–40, 8 figs.

Webb, R.G. (1984) Herpetogeography in the Mazatlán-Durango region of the Sierra Madre Occidental, Mexico. *In:* Seigel, R.S., Hunt, L.E., Knight, J.L, Malaret, L. & Zuschlag, N.L. (Eds.), *Vertebrate ecology and systematics: a tribute to Henry S. Fitch.* University of Kansas Museum of Natural History, pp. 217–241, 6 figs.

Webb, R.G. (1990) Description of a new subspecies of *Bogertophis subocularis* (Brown) from northern México (Serpentes, Colubridae). *Texas Journal of Science* 42(3), 227–243, 5 figs.

Webb, R.G. (2001) Proposed deletion of two colubrid snakes from the Mexican state of Durango herpetofauna. *Herpetological Review* 32(3), 159–160.

Webb, R.G. (2002) Geographic distribution: Serpentes: *Thamnophis melanogaster chihuahuaensis. Herpetological Review* 33(1), 69.

Webb, R.G. & Eckerman, C.M. (1998) Neotype and type locality of the western hognose snake, *Heterodon nasicus* (Serpentes: Colubridae). *Texas Journal of Science* 50, 99–106.

Webb, R.G. & Ferguson, G.M. (1986) Morphological variation in the Trans-Pecos rat snake (*Elaphe subocularis*). *Southwestern Naturalist* 31, 118–121.

Webb, R.G., Jones, J.K., Jr. & Byers, G.W. (1962) Some reptiles and amphibians from Korea. *University of Kansas Publications, Museum of Natural History* 15(2), 149–173.

Weems, R.G. (1999) Reptile remains from the Fisher/Sullivan Site. *In:* Weems, R.G. & Grimsley, G.J. (Eds.), Early Eocene vertebrates and plants from the Fisher/Sullivan Site (Nanjemoy Formation), Stafford County, Virginia. *Virginia Division of Mineral Resources Publication* 152, 101–121.

Wegener, J.E., Swoboda, S., Hawlitschek, O. Franzen, M., Wallach, V., Vences, M., Nagy, Z.T., Hedges, S.B., Köhler, J. & Glaw, F. (in press) Morphological variation and taxonomic reassessment of the endemic Malagasy blind snake family Xenotyphlopidae (Serpentes, Scolecophidia). *Spixiana.* [24 pp., 6 figs.]

Wegner, A.M.R. (1954) The snakes of the genus *Maticora* Gray with special reference to *Maticora bivirgata bivirgata* (Boie) and *M. intestinalis intestinalis* (Laurenti). *Penggemar Alam*, Bogor 34(1–2), 55–58, 2 figs.

Wehekind, L. (1960) Trinidad snakes. *Royal Victoria Institute Museum Miscellaneous Papers*, Trinidad 1, 8 pp.

Wei, G., Xu, N., Wang, J. & Li, D. (1992) Karyotype, C-band and Ag-NORs of *Elaphe mandarina. The Snake* 24, 151–155.

Weidensaul, S. (1991) *Snakes of the world.* Chartwell Books, Seacaucus, 128 pp., col. illust.

Weier, M. & Vitt, R. (1999) *Der Grüne Baumpython: Lebensweise, Pflege, Zucht, Erkrankungen.* Herpeton, Verlag Elke Köhler, Offenbach, 110 pp., 19 figs., 51 col. photos.

Weigel, C.E. (1782) Beschreibung einer Schlange. *Schriften der Gesellschaft Naturforschender Freunde zu Berlin* 3, 190–193.

Weigel, J. (1990) *Australian reptile park's guide to snakes of South-East Australia.* John Weigel Postscript, 103 pp., col. photos.

Weigel, J. & Russell, T. (1993) A record of a third specimen of the rough-scaled python *Morelia carinata*. *Herpetofauna*, Sydney 23(2), 1–5, 2 figs.

Weinstein, S.A., Minton, S.A. & Wilde, C.E. (1985) The distribution among phidian venoms of a toxin isolated from the venom of the Mojave rattlesnake *(Crotalus scutulatus scutulatus)*. *Toxicon* 23(5), 825–844.

Weissenberg, S., Ovadia, M., Fleminger, G. & Kochva, E. (1991) Antihemorrhagic factors from the blood serum of the western diamondback rattlesnake *Crotalus atrox*. *Toxicon* 29(7), 807–818, 5 figs.

Weissmair, W. & Moser, J. (2008) *Atlas der Amphibien und Reptilien Oberösterreichs*. Biologiezentrum der Oberösterreichischen Landesmuseum, Linz, 132 pp. col. illust.

Welch, K.R.G. (1980) A comment on the European colubrid genus *Haemorrhois*. *Bulletin of the South West Herpetological Society* 1980, 14–15.

Welch, K.R.G. (1982) *Herpetology of Africa: a checklist and bibliography of the orders Amphisbaenia, Sauria and Serpentes*. Robert E. Krieger, Malabar, 293 pp.

Welch, K.R.G. (1983a) A comment on the generic allocation of the Old World species of the genus *Coluber*. *Litteratura Serpentium*, Utrecht 3, 104–110.

Welch, K.R.G. (1983b) *Herpetology of Europe and southwest Asia: A checklist and bibliography of the orders Amphisbaenia, Sauria and Serpentes*. Robert E. Krieger, Malabar, viii + 135 pp.

Welch, K.R.G. (1988) *Snakes of the Orient: a checklist*. Robert E. Krieger, Malabar, vii + 183 pp.

Welch, K.R.G. (1994a) *Snakes of the world: a checklist. 1. Venomous snakes*. R & A Research and Information Limited, KCM Books, Bristol, 135 pp.

Welch, K.R.G. (1994b) *Snakes of the world: a checklist. 2. Boas, pythons, shield-tails and worm snakes*. R & A Research and Information Limited, KCM Books, Bristol, 89 pp.

Weller, W.H. (1930) Description of a new species of colubrine snake of the genus *Leptophis*. *Proceedings of the Junior Society of Natural Science*, Cincinnati 1(2), 7.

Wellman, J. (1959) Notes on the variation in and distribution of the Mexican colubrid snake *Coniophanes lateritius*. *Herpetologica* 15(1), 127–128, 1 fig.

Wellman, J. (1963) A revision of snakes of the genus *Conophis* (family Colubridae), from Middle America. *University of Kansas Publications, Museum of Natural History* 15(6), 251–295, 9 figs.

Wells, R.W. (2002) Taxonomy of the genus *Pseudonaja* (Reptilia: Elapidae) in Australia. *Australian Biodiversity Record* (7), 1–48.

Wells, R.W. (2007) Some taxonomic and nomenclatural considerations on the class Reptilia in Australia. The sea snakes of Australia. An introduction to the members of the families Hydrophiidae and Laticaudidae in Australia, with a new familial and generic arrangement. *Australian Biodiversity Record* (8), 1–124.

Wells, R.W. & Wellington, C.R. (1984) A synopsis of the class Reptilia in Australia. *Australian Journal of Herpetology*, Marrickville (1983) 1(3–4), 73–129.

Wells, R.W. & Wellington, C.R. (1985) A classification of the Amphibia and Reptilia of Australia. *Australian Journal of Herpetology*, Marrickville (1984) Supplement Series (1), 1–61.

Wells, R.W. & Wellington, C.R. (1987) A new species of proteroglyphous snake (Serpentes, Oxyuranidae) from Australia. *The Australian Herpetologist* (503), 1–8, 1 fig.

Welsh, H.H., Jr. (1976) *Ecogeographic distribution of the herpetofauna of the Sierra San Pedro Martir region, Baja California de Norte, Mexico*. Master's Thesis, Humboldt State University, Arcata, California, 169 pp.

Welsh, J.H. (1966) Acetylcholine in snake venoms. *In:* Russell, F.E. & Saunders, P.R. (Eds.), *Animal toxins*. Pergamon Press, Oxford, pp. 363–368.

Wen, Y.-T. (1983) Several new records of reptiles from Guangxi. *Acta Herpetologica Sinica*, Chengdu (Series 2) 2(3), 72–74.

Wen, Y.-T. (1998) Description of a new species of *Boiga* (Serpentes: Colubridae) from China. *Sichuan Journal of Zoology* 17(2), 51–52. (in Chinese)

Werler, J.E. (1978) *Poisonous snakes of Texas*. Fifth edition. Bulletin, Texas Parks and Wildlife Department (31), 1–53. 17 col. pls. [first edition, 1950, 40 pp., illust.; second edition, 1952, 40 pp., illust.; third edition, 1963, 62 pp., col. pls.; fourth edition, 1964, 62 pp., 16 col. pls.]

Werler, J.E. & Dixon, J.R. (2000) *Texas snakes. Identification, distribution, and natural history*. University of Texas Press, Austin, xvi + 437 pp., 70 figs., 200 col. photos.

Werler, J.E. & Keegan, H.L. (1963) Venomous snakes of the Pacific area. *In:* Keegan, H.L. & MacFarlane, W.V. (Eds.), *Venomous and poisonous animals and noxious plants of the Pacific region*. Macmillan Company, New York, pp. 219–325.

Werler, J.E. & Smith, H.M. (1952) Notes on a collection of reptiles and amphibians from Mexico, 1951–1952. *Texas Journal of Science* 4(4), 551–573.

Werman, S.D. (1984) Taxonomic comments on the Costa Rican pit viper, *Bothrops picadoi*. *Journal of Herpetology* 18(2), 207–210, 1 fig.

Werman, S.D. (1986a) Phylogenetic relationships of the true viper *Eristocophis mcmahoni*, based on parsimony analysis of allozyme characters. *Copeia* 1986(4), 1014–1020.

Werman, S.D. (1986b) *Phylogenetic systematics and biogeography of Neotropical pit vipers: A cladistic analysis of biochemical and anatomical characteristics*. Doctoral Dissertation, University of Miami, Miami, 182 pp.

Werman, S.D. (1992) Phylogenetic relationships of Central and South American pitvipers of the genus *Bothrops* (*sensu lato*): cladistic analysis of biochemical and anatomical characters. *In:* Campbell, J.A. & Brodie, E.D., Jr. (Eds.), *Biology of the pitvipers*. Selva, Tyler, pp. 21–40, 3 figs.

Werman, S.D. (1997) Systematic implications of lactate dehydrogenase isozyme phenotypes in Neotropical pitvipers (Viperidae, Crotalinae). *In:* Thorpe, R.S., Wüster, W. & Malhotra, A (Eds.), *Venomous Snakes. Ecology, evolution and snakebite*. Zoological Society of London, Clarendon Press, Oxford, pp. 79–88, 5 figs.

Werman, S.D. (1999) Molecular phylogenetics and morphological evolution in Neotropical pitvipers: an evaluation of mitochondrial DNA sequence information and the comparative morphology of the cranium and palatomaxillary arch. *In:* Joger, U. (Ed.), Phylogeny and systematics of the Viperidae. *Kaupia, Darmstädter Beiträge zur Naturgeschichte* 8, 113–120, 4 figs.

Werner, C. & Rage, J.-C. (1994) Mid-Cretaceous snakes from Sudan: A preliminary report on an unexpected diverse snake fauna. *Comptes Rendus de l'Académie des Sciences, La Terre et Planetes* 319(2), 247–252, illust.

Werner, F. (1890) Untersuchungen ueber die Zeichung der Schlangen. Verlag von K. Krawani, Wien, 122 pp., 8 pls.

Werner, F. (1891) Beiträge zur Kenntniss der Reptilien und Amphibien von Istrien und Dalmatien. *Verhandlungen der Zoologisch-Botanischen Gesellschaft in Wien* 41, 751–768.

Werner, F. (1893) Bemerkungen über Reptilien und Batrachier aus dem tropischen Asien und von der Sinai-Halbinsel. *Verhandlungen der Zoologisch-Botanischen Gesellschaft in Wien* 43, 349–359.

Werner, F. (1895) Ueber einige interessante Reptilien aus Afrika. *Verhandlungen der Zoologisch-Botanischen Gesellschaft in Wien* 45, 92–94.

Werner, F. (1896a) Ueber einige Reptilien aus Usambara (Deutsch-Ostafrica). *Verhandlungen der Zoologisch-Botanischen Gesellschaft in Wien* (1895) 45, 190–195, pl. 5.

Werner, F. (1896b) Zweiter Beitrag zur Herpetologie der indo-orientalischen Region. *Verhandlungen der Zoologisch-Botanischen Gesellschaft in Wien* 46, 6–24, 1 pl.

Werner, F. (1896c) Beiträge zur Kenntniss der Reptilien und Batrachier von Central America und Chile, sowie einiger seltenerer Schlangenarten. *Verhandlungen der Zoologisch-Botanischen Gesellschaft in Wien* 46, 344–365, pl. 6.

Werner, F. (1897a) Ueber einige neue oder seltene Reptilien und Frösche der zoologischen Sammlung des Staates in München. *Sitzungsberichte der Akademie der Wissenschaften in Wien, Mathematisch-Naturwissenschaftliche Klasse* 27(2), 203–220, 2 figs.

Werner, F. (1897b) Ueber Reptilien und Batrachier aus Togoland, Kamerun und Tunis aus dem Kgl. Museum für Naturkunde in Berlin. *Verhandlungen der Zoologisch-Botanischen Gesellschaft in Wien* 47, 395–408, pl. 2. (*Buhoma depressiceps*)

Werner, F. (1897c) Über einige noch unbeschriebene Reptilian und Batrachier. *Zoologischer Anzeiger*, Leipzig 20(537), 261–267, 2 figs.

Werner, F. (1897d) *Die Amphibien und Reptilien Osterreichs-Ungarns und der Occupationsländer.* Pichlers Wwe. und Sohn, Wien, 162 pp.

Werner, F. (1898a) Über einige neue Reptilien und einen neuen Frosch aus dem cilicischen Taurus. *Zoologischer Anzeiger*, Leipzig 21(555), 217–223.

Werner, F. (1898b) Verläufige Mittheilung über die von Herrn Prof. F. Dahl im Bismarck-archipel gesammelten Reptilien und Batrachier. *Zoologischer Anzeiger*, Leipzig 21(571), 552–556.

Werner, F. (1898c) Die reptilien und batrachier der Sammlung Plate. *Zoologische Jahrbücher, Abteilung für Systematik, Ökologie und Geographie der Tiere*, Jena, Supplement 4, 244–278.

Werner, F. (1898d) Ueber reptilien aus Syrien und Südafrika. *Jahresbericht und Abhandlungen des Naturwissenschaftlichen Vereins in Magdeburg* 1898, 127–148.

Werner, F. (1899a) Beschreibung einiger neuer Schlangen und Batrachier. *Zoologischer Anzeiger*, Leipzig 22(581), 114–117.

Werner, F. (1899b) Beiträge zur Herpetologie der pacifischen Inselwelt und von Kleinasien. *Zoologischer Anzeiger*, Leipzig 22(596), 371–378.

Werner, F. (1899c) Beschreibung neuer Reptilien und Batrachier. *Zoologischer Anzeiger*, Leipzig 22(602), 479–484.

Werner, F. (1899d) Allerlei aus dem Kreichthierleben in Käfig II. *Zoologische Garten*, Frankfurt-am-Main 40, 12–24.

Werner, F. (1900a) Beschreibung einiger noch unbekannter neotropischer und indischer Reptilian. *Zoologischer Anzeiger*, Leipzig 23(612), 196–198, 5 figs.

Werner, F. (1900b) Reptilien und Batrachier aus Sumatra, gesammelt von Herrn Gustav Schneider jr. im Jahre 1897–98. *Zoologische Jahrbücher, Abteilung für Systematik, Ökologie und Geographie der Tiere*, Jena 13, 479–508, pls. 31–35.

Werner, F. (1900c) *Die Reptilien- und Batrachier fauna des Bismarck-Archipels. Auf Grundlage des von Prof. Dr. F. Dahl gesammelten Materials bearbeitat.* R. Friedländer & Sohn, Berlin, 132 pp., 41 figs.

Werner, F. (1901a) Ueber Reptilien und Batrachier aus Ecuador und Neu-Guinea. *Verhandlungen der Zoologisch-Botanischen Gesellschaft in Wien* 51, 593–614.

Werner, F. (1901b) Bemerkungen über einige seltenere Schlangen-Arten. *Verhandlungen der Zoologisch-Botanischen Gesellschaft in Wien* 51, 634–639.

Werner, F. (1901c) Reptilien und Batrachier aus Peru und Bolivien. *Abhandlungen und Berichte des königlichen Zoologischen und Anthropologische-Ethnographischen Museums zu Dresden* (1900–1901) 9(2), 1–13.

Werner, F. (1901d) Neue reptilien des Königsberger Zoologischen Museums. *Zoologischer Anzeiger*, Leipzig 24(643), 297–301.

Werner, F. (1902) Ueber westafrikanische Reptilien. *Verhandlungen der Zoologisch-Botanischen Gesellschaft in Wien* 52(1), 332–348.

Werner, F. (1903a) Ueber Reptilien und Batrachier aus Guatemala und China in der zoologischen Staats-Sammlung in München nebst einem Anhang über seltene Formen aus anderen Gegenden. *Abhandlungen der Bayerische Akademie der Wissenschaften, II Klasse*, Wien 22(2), 341–384, 1 pl., 4 figs.

Werner, F. (1903b) Neue Reptilien und Batrachier aus dem naturhistorischen Museum in Brüssel. Nebst Bemerkungen über einige andere Arten. *Zoologischer Anzeiger*, Leipzig (1902) 26(693), 246–253.

Werner, F. (1903c) Ueber Reptilien und Batrachien aus West-Asien (Anatolien und Persien). *Zoologische Jahrbücher, Abteilung für Systematik, Ökologie und Geographie der Tiere*, Jena 19, 329–346, pls. 23–24.

Werner, F. (1904) Beschreibung neuer reptilien aus den gattungen *Acanthosaura, Calotes, Gastriopholis* und *Typhlops*. *Zoologischer Anzeiger*, Leipzig 27(15), 461–464.

Werner, F. (1905) Einige für Kleinasien neue Reptilien. *Zoologischer Anzeiger*, Leipzig 29(13), 411–413.

Werner, F. (1906) Neue Reptilien aus Deutsch-Ostafrika. *Zoologischer Anzeiger*, Leipzig 30(1–2), 53–55, 3 figs.

Werner, F. (1907a) Ergebnisse der mit Subvention aus der Erbschaft Treitl unter-nommenen zoologischen Forschungsreise Dr. Franz Werner's nach dem ägyptischen Sudan und Nord-Uganda. XII. Die Reptilien und Amphibien. *Anzeiger der Kaiserlichen Akademie der Wissenschaften, Mathematisch-Naturwissenschaftliche Klasse,* Wien 26, 477–483.

Werner, F. (1907b) *In:* Krefft, P., *Das Terrarium: ein Handbuch der häuslichen Reptilien- und Amphibienpflege*. F. Pfenninstorff, Berlin, 631 pp.

Werner, F. (1907c) Ergebnisse der mit Subvention aus der Erbschaft Treitl unternommenen zoologischen Forschungsreise Dr. Franz Werner's nach dem ägyptischen Sudan und Nord-Uganda. XII. Die Reptilien und Amphibien. *Sitzungsberichte der Akademie der Wissenschaften in Wien, Mathematisch-Naturwissenschaftliche Klasse* 116(1), 1823–1926, 4 pls.

Werner, F. (1908) Ergebnisse der mit Subvention aus der Erbschaft Treitl unternommenen zoologischen Forschungsreise Dr. Franz Werner's nach dem ägyptischen Sudan und Nord-Uganda. XII. Die Reptilien und Amphibien. *Sitzungsberichte der Akademie der Wissenschaften in Wien, Mathematisch-Naturwissenschaftliche Klasse* (1907) 116(1), 1823–1926, 4 pls.

Werner, F. (1909a) Reptilien, Batrachier und Fische von Tripolis und Barka. *Zoologische Jahrbücher, Abteilung für Systematik, Ökologie und Geographie der Tiere*, Jena 27(6), 595–646, pl. 30.

Werner, F. (1909b) Über neue oder seltene Reptilien des Naturhistorischen Museums in Hamburg. I. Schlangen. *Mitteilungen aus dem Naturhistorischen Museum in Hamburg* 26, 205–247, 14 figs.

Werner, F. (1909c) Neue oder seltenere Reptilien des Musée Royal d'Histoire Naturelle de Belgique in Brüssel. *Zoologische Jahrbücher, Abteilung für Systematik, Ökologie und Geographie der Tiere*, Jena 28(3), 263–285, 2 figs.

Werner, F. (1909d) Beschreibung neuer Reptilien aus dem Kgl. Naturalienkabinet in Stuttgart. *Jahreshefte des Vereins für Vaterländlische Naturkunde in Württemburg* 65, 55–63.

Werner, F. (1909e) Reptilia exkl. Geckonidae und Scincidae. *In:* Michaelsen, W. & Hartmeyer, R. (Eds.), *Die Fauna Südwest-Australiens*. Gustav, Jena 2, 251–278, pls. 14–15.

Werner, F. (1909f) Die eidechsennattern. *Natur und Haus* 16, 34.

Werner, F. (1910) Reptilia und Amphibia. *In:* Schultz, L. (Ed.), Zoologische und anthropologische Ergebnisse einer Forschungsreise in Westlich und Zentralen Südafrika ausgeführt in den Jahren 1903–1905. *Denkschriften der Medicinisch-Naturwissenschaftlichen Gesellschaft zu Jena* 16, 279–370.

Werner, F. (1913) Neue oder seltene Reptilien und Frösche des Naturhistorischen Museums in Hamburg. *Mitteilungen aus dem Naturhistorischen Museum in Hamburg* (1912) 30, 1–51.

Werner, F. (1915) Reptilia und Amphibia. *In:* Michaelsen, W. (Ed.), Beiträge zur Kenntnis der Land- und Süßwasserfauna Deutsch-Südwestafrikas. *Ergebnisse der Hamburger Deutsch-Südwestafrikanischen Studienreise* 1911, 3, 323–376, pl. 7.

Werner, F. (1916) Bemerkungen über einige niedere Wirbeltiere der Anden von Kolumbien mit Beschreibungen neuer Arten. *Zoologischer Anzeiger*, Leipzig 47(10), 301–304 & 47(11), 305–311.

Werner, F. (1917a) Über einige neue Reptilien und einen neuen Frosch des Zoologischen Museums in Hamburg. *Mitteilungen aus dem Zoologischen Museum in Hamburg* (1916) 34, 31–36.

Werner, F. (1917b) Versuch einer Synopsis der Schlangenfamilie der Glauconiiden. *Mitteilungen aus dem Zoologischen Museum in Hamburg* 34, 191–208.

Werner, F. (1917c) Reptilien aus Persien (Provinz Fars). *Verhandlungen der Zoologisch-Botanischen Gesellschaft in Wien* 67(5–6), 191–220.

Werner, F. (1919) Wissenschaftliche ergebnisse der mit unterstützung der Kaiserlichen Akademie der Wissenschaften in Wien aus der erbschaft treitl von F. Werner unternommenen. Zoologischen Expedition nach dem Anglo-Äegyptischen Sudan (Kordofan) 1914. IV. Bearbeitung der Fische, Amphibien und Reptilien. *Denkschriften der Kaiserlichen Akademie der Wissenschaften in Wien, Mathematische-Naturwissenschaftliche Klasse* 96, 437–509, 2 pls.

Werner, F. (1921a) Synopsis der Schlangenfamilie der Typhlopiden auf Grund des Boulenger'schen Schlangenkatalogs (1893–1896). *Archiv für Naturgeschichte*, Berlin 87A(7), 266–338, 21 figs.

Werner, F. (1921b) Synopsis der Schlangenfamilie der Boiden auf Grund des Boulenger'schen Schlangenkatalogs (1893/96). *Archiv für Naturgeschichte*, Berlin 87A(8), 230–265, 3 figs.

Werner, F. (1922a) Neue Reptilien aus Sud-China, gesammelt von Dr. H. Handel–Mazzetti. *Anzeiger der Akademie der Wissenschaften in Wien* 59(24–25), 220–222.

Werner, F. (1922b) Synopsis der Schlangenfamilien der Amblycephaliden und Viperiden nebst Übersicht über kleineren Familien und die Colubriden der Acrochordinengruppe. Auf Grund des Boulengerschen Schlangenkatalogs (1893–1896). *Archiv für Naturgeschichte*, Berlin 88A(8), 185–244, 6 figs.

Werner, F. (1922c) *Das Tierreich. III. Reptilien und Amphibien. Erster Band: Reptilien.* Walter de Gruyter & Co., Berlin, 140 pp., 32 figs.

Werner, F. (1923a) Neue Schlangere des Naturhistorischen Museums in Wien. *Annalen des Naturhistorischen Museums in Wien* 36, 160–166.

Werner, F. (1923b) Übersicht der Gattungen und Arten der Schlangen der Familie Colubridae. I. Teil. *Archiv für Naturgeschichte*, Berlin 89A(8), 138–199.

Werner, F. (1924a) Neue oder wenig bekannte Schlangen aus dem Naturhistorischen Staatsmuseum in Wien. *Anzeiger der Akademie der Wissenschaften in Wien* (4), 1–3.

Werner, F. (1924b) Neue oder wenig bekannte Schlangen aus dem Naturhistorischen Staatsmuseum in Wien. *Sitzungsberichte der Akademie der Wissenschaften in Wien, Mathematisch-Naturwissenschaftliche Klasse*, 133(1–3), 29–56, 9 figs.

Werner, F. (1924c) Übersicht der Gattungen und Arten der Schlangen der Familie Colubridae. II. Teil (Dipsadomorphinae und Hydrophiinae). *Archiv für Naturgeschichte*, Berlin 90A(12), 108–166, 10 figs.

Werner, F. (1924d) Ergebnisse der Expedition Dr. Handel-Mazzetti's nach China 1914 bis 1918 auf Kosten der Akademie der Weissenschaften in Wien. Über Reptilien und Amphibien aus Südchina. *Denkschriften der Kaiserlichen Akademie der Wissenschaften in Wien, Mathematisch-Naturwissenschaftliche Klasse* 99, 39–58.

Werner, F. (1925) Neue oder wenig bekannte Schlangen, aus dem Wiener naturhistorischen Staatmuseum. 2. Teil. *Sitzungsberichte der Akademie der Wissenschaften in Wien, Mathematisch-Naturwissenschaftliche Klasse* 134(1–2), 45–66, 4 figs.

Werner, F. (1926) Synonymisches über Schlangen von Hainan und von chinesischen Festland. *Zoologischer Anzeiger*, Leipzig 67(5–6), 141– 144.

Werner, F. (1927) Neue oder wenig bekannte Schlangen aus dem Wiener naturhistorischen Staatsmuseum (3 teil). *Sitzungsberichte der Akademie der Wissenschaften in Wien, Mathematisch-Naturwissenschaftliche Klasse*, (1926) 135(7–8), 243–257.

Werner, F. (1929a) Übersicht der Gattungen und Arten der Schlangen der Familie Colubridae. III. Teil (Colubrinae). Mit einem Nachtrag zu den übrigen Familien. *Zoologische Jahrbücher, Abteilung für Systematik, Ökologie und Geographie der Tiere*, Jena 57(1–2), 1–196, 48 figs.

Werner, F. (1929b) Wissenschaftliche Ergebnisse einer zoologischen Forschungsreise nach Westalgerien und Marokko. II. Reptilien und Amphibien. *Sitzungsberichte der Akademie der Wissenschaften in Wien, Mathematisch-Naturwissenschaftliche Klasse* 138(1–2), 1–34, 4 pls.

Werner, F. (1929c) A harmless viper *(Vipera ursinii* Bonaparte). *Bulletin of the Antivenin Institute of America* 3(3), 77–79, 2 figs.

Werner, F. (1930a) Contribution to the knowledge of the reptiles and amphibians of Greece, especially the Aegean Islands. *Occasional Papers of the Museum of Zoology, University of Michigan* (211), 1–47, 6 pls.

Werner, F. (1930b) Boidenstudien im Wiener Naturhistorischen Museum. *Zoologischer Anzeiger*, Leipzig 87(7–8), 198–207, figs.

Werner, F. (1930c) Über *Vipera (Mesocoronis) bosnensis* Wern. u. T. Reuß. *Zoologischer Anzeiger*, Leipzig 88(1–4), 89–93.

Werner, F. (1931) Ergebnisse einer zoologischen Forschangsreise nach Marokko. *Sitzungsberichte der Akademie der Wissenschaften in Wien, Mathematisch-Naturwissenschaftliche Klasse* 140(3–4), 271–318, pls.

Werner, F. (1932) Eine neue Schlange aus dem Cycladenarchipel. *Anzeiger der Akademie der Wissenschaften in Wien, Mathematisch-Naturwissenschaftliche Klasse* 69(20), 232.

Werner, F. (1933) Ergebnisse einer zoologischen Studien- und Sammelreise nach den Inseln des Agäischen Meeres. I. Reptilien und Amphibien. *Sitzungsberichte der Akademie der Wissenschaften in Wien, Mathematisch-Naturwissenschaftliche Klasse* 142(3–4), 103–133, 12 figs.

Werner, F. (1935) Reptilien der Ägäichen inseln. *Sitzungsberichte der Akademie der Wissenschaften in Wien, Mathematisch-Naturwissenschaftliche Klasse* 144(1–2), 81–117, figs.

Werner, F. (1936) Nova species Boidarum. *Annales Historico-Naturales Musei Nationalis Hungarici*, Budapest 30, 105, 1 pl.

Werner, F. (1937) Über reptilien aus dem Hoggar-Gebirge (West-Sahara). *Zoologischer Anzeiger*, Leipzig 118(1–2), 31–35.

Werner, F. (1938a) Reptilien aus Iran und Belutschistan. *Zoologischer Anzeiger*, Leipzig 121(9–10), 265–271, 3 figs.

Werner, F. (1938b) Eine verkannte viper (*Vipera palaestinae* n. sp.) *Zoologischer Anzeiger*, Leipzig 122(11–12), 313–318, 4 figs.

Werner, F. (1939) Die Amphibien und Reptilien von Syrien. *Abhandlungen und Berichte für Naturkunde und Vorgeschichte und dem Naturwissenschaftliche Verein in Magdeburg* 7(1), 211–223, 14 figs., 15 pls.

Werner, Y.L. (1965) Dictionary of the reptiles and amphibians of Israel. *Memoirs of the Academy of the Hebrew Language* (1963–1964) 10–11, 1–6. (in Hebrew)

Werner, Y.L. (1971) Lizards and snakes from Transjordan, recently acquired by the British Museum (Natural History). *Bulletin of the British Museum (Natural History)*, London 21(6), 215–256, 8 figs., 6 pls.

Werner, Y.L. (1973) *The reptiles of the Sinai Peninsula.* Department of Zoology, Hebrew University of Jerusalem, Jerusalem iv + 46 pp., 1 pl. (in Hebrew)

Werner, Y.L. (1982) Herpetofaunal survey of the Sinai Peninsula (1967–77), with emphasis on the Saharan sand community. *In:* Scott, N.J. (Ed.), *Herpetological communities.* United States Department of the Interior, Fish and Wildlife Service Report 13, 153–161, 1 fig.

Werner, Y.L. (1983) Lizards and snakes from eastern lower Egypt in the Hebrew University of Jerusalem and Tel Aviv University, with range extensions. *Herpetological Review* 14(1), 29–31, 2 figs.

Werner, Y.L. (1987) Ecological zoogeography of the Saharo-Arabian, Saharan and Arabian reptiles in the sand deserts of southern Israel. *In:* Krupp, F., Schneider, W. & Kinzelbach, R. (Eds.), *Proceedings of the symposium on the fauna and zoogeography of the Middle East.* Dr. Ludwig Reichert Verlag, Wiesbaden, Beihefte zum Tübinger Atlas des vorderen Orients, Reihe A (Naturwissenschaften) (28), 272–295.

Werner, Y.L. (1988) The zoogeography of Israel. 13. Herpetofaunal survey of Israel (1950–85), with comments on Sinai and Jordan and on zoogeographical heterogeneity. *Monographiae Biologicae* 62, 355–388.

Werner, Y.L. (1994) Head size variation in *Cerastes* (Ophidia, Viperidae) parallels body size variation in potential prey (Rodentia, Gerbillinae). *The Snake* 26(1), 57–60.

Werner, Y.L. (1995) *A guide to the reptiles and amphibians of Israel.* Nature Reserves Authority, Yefe-Nof Library, Jerusalem, 86 pp., col. photos.

Werner, Y.L. & Sivan, N. (1991) Addition of *Coluber sinai* to the herpetofaunal list of Israel with comments on *C. elegantissimus. British Herpetological Society Bulletin* (36), 27–35, illust.

Werner, Y.L. & Sivan, N. (1992) Systematics and zoogeography of *Cerastes* (Ophidia: Viperidae) in the Levant: 2. Taxonomy, ecology, and zoogeography. *The Snake* 24(1), 34–49, 7 figs.

Werner, Y.L., Le Verdier, A., Rosenman, D. & Sivan, N. (1991) Systematics and zoogeography of *Cerastes* (Ophidia: Viperidae) in the Levant: 1. Distinguishing Arabian from African "*Cerastes cerastes*." *The Snake* 23(2), 90–100, 4 figs.

Werner, Y.L., Sivan, N., Kushnir, V. & Motro, U. (1999) A statistical approach to variation in *Cerastes* (Ophidia, Viperidae), with the description of two new endemic subspecies. *In:* Joger, U. (Ed.), Phylogeny and systematics of the Viperidae. *Kaupia, Darmstädter Beiträge zur Naturgeschichte* 8, 83–97, 8 figs.

West, J.A. (1979) The occurrence of some exotic reptiles and amphibians in New Zealand. *Herpetofauna*, Sydney 10(2), 4–9, illust.

West, R.M., Hutchison, J.H. & Munthe, J. (1991) Miocene vertebrates from the Si Walik Group, western Nepal. *Journal of Vertebrate Paleontology* 11(1), 108–129, illust.

Westermann, J.H. (1942) Snakes from Bangka and Billiton. *Treubia*, Buitenzorg 18(3), 611–619, 2 figs.

Westgate, J.W. (1984) Lower vertebrates from the late Eocene Crow Creek local fauna, St. Francis County, Arkansas. *Journal of Vertebrate Paleontology* 4(4), 536–546, 4 figs.

Westgate, J.W. & Ward, J.F. (1981) The giant aquatic snake *Pterosphenus schucherti* (Palaeonphidae) in Arkansas and Mississippi. *Journal of Vertebrate Paleontology* 1(2), 161–164, 1 fig.

Westphal-Castelnau, A. (1870) *Catalogue de la collection de reptiles, complété par son fils Alfred Westphal-Castelnau. 1869.* J. Martel Ainé, Monpellier, 58 pp.

Wetherbee, D.K. (1986a) *Zoological exploration of Dominican Republic for endemic species.* David Kenneth Wetherbee, Shelbourne, 332 pp.

Wetherbee, D.K. (1986b) *Ruiz de Velva as herpetologist in Peru, 1778–1788.* David K. Wetherbee, Shelburne, 24 pp.

Wetherbee, D.K. (1987) *Further contributions to the history of zoology in Hispaniola.* David Kenneth Wetherbee, Shelburne, 114 pp.

Wettstein, O. (1928) Beiträge zur Wirbeltierfauna der kroatischen Gebirge. *Annalen des Naturhistorischen Museums in Wien* 42, 1–45.

Wettstein, O. (1930) Eine neue colubride Schlange aus Brasilien. *Zoologischer Anzeiger*, Leipzig 88(1–4), 83–94, 1 fig.

Wettstein, O. (1934) Ergebnisse der österreichischen biologischen Costa Rica Expedition 1930. Die Amphibien und Reptilien. *Sitzungsberichte der Akademie der Wissenschaften in Wien, Mathematisch-Naturwissenschaftliche Klasse* 143(1–2), 1–39, figs.

Wettstein, O. (1937) Vierzehn neue Reptilienrassen von den südlichen Ägäischen inseln. *Zoologischer Anzeiger*, Leipzig 118(3–4), 79–90, 2 figs.

Wettstein, O. (1938) Über einige Reptilien aus Ost.-Tibet. *Zoologischer Anzeiger,* Leipzig 122(9–10), 255–257.

Wettstein, O. (1941) Franz Werner Werner als Mensch und Forscher. *Annalen des Naturhistorischen Museums in Wien* 51, 8–53.

Wettstein, O. (1953a) Dreizehn neue Reptilienrassen von den Ägäichsen Inseln. *Anzeiger der Osterreichischen Akademie der Wissenschaften*, Wien (1952) 89, 251–256.

Wettstein, O. (1953b) Herpetologia Aegaea. *Sitzunsberichte der Osterreichischen Akademie der Wissenschaften, Mathematisch-Naturwissenschaftliche Klasse*, Abteilung I, 162(9–10), 651–833, 8 pls.

Weyenbergh, H. (1876a) *Coronella bachmanni* n. sp. *Periódico Zoológico*, Cordobá 2, 193–195.

Weyenbergh, H. (1876b) Die Thierwelt Argentiniens. *In:* Napp, R. *Die Argentinische Republik. Im auftrag des Argentin, Central Comité's für die Philadelphia-Ausstellung und mit dem beistand mehrerer mitarbeiter.* Sociedad Anónima, Buenos Aires, pp. 150–190.

Weyenbergh, H. (1876c) The Argentine fauna. *In:* Napp, R. *The Argentine Republic, written in German by Richard Napp assisted by several fellow-writers for the Central Argentine Commission on the Centenary Exhibition at Philadelphia.* Sociedad Anónima, Buenos Aires, pp. 137–174. [English translation of Weyenbergh, 1876b]

Weyer, D. (1990) *Snakes of Belize.* Belize Audobon Society, Belize City, 54 pp., 29 figs.

Wharton, C.H. (1960) Birth and behavior of a brood of cottonmouths, *Agkistrodon piscivorus piscivorus*, with notes on tail-luring. *Herpetologica* 16(2), 125–129.

Wharton, C.H. (1963) *Georgia snake tribe.* Georgia State Museum of Science and Industry, Atlanta, 20 pp., figs.

Wharton, C.H. (1966) Reproduction and growth in the cottonmouth, *Agkistrodon piscivorus* Lacépède, of Cedar Keys, Florida. *Copeia* 1966(2), 149–161.

Wharton, C.H. (1969) The cottonmouth moccasin on Sea Horse Key, Florida. *Bulletin of the Florida State Museum, Biological Sciences* 14, 227–272.

Wheeler, G.C. (1947) The amphibians and reptiles of North Dakota. *American Midland Naturalist* 38, 162–190.

Wheeler, G.C. (1954) *The amphibians and reptiles of the North Dakota Badlands.* Theodore Roosevelt Natural History Association, Medora, 56 pp.

Wheeler, G.C. & Wheeler, J. (1966) *The amphibians and reptiles of North Dakota.* University of North Dakota Press, Grand Forks, vii + 104 pp., 24 figs.

Whistler, D.P. & Wright, J.W. (1989) A late Miocene rear-fanged colubrid snake from California with comments on the phylogeny of North American snakes. *Herpetologica* 45(3), 350–367, 2 figs.

Whitaker, R. (1969) The dog-faced watersnake (*Cerberus rhynchops*) in the Bombay area and notes on its habits. *The Journal of the Bombay Natural History Society* 66(2), 386.

Whitaker, R. (1975) The olive keelback (*Atretium schistosum* Russell) feeding on mosquito larvae. *The Journal of the Bombay Natural History Society* 72(1), 209.

Whitaker, R. (1978a) *Common Indian snakes. A field guide.* Macmillan, Delhi, xiv + 154 pp., illust.

Whitaker, R. (1978b) Herpetological survey in the Andamans. *Hamadryad*, Madras 3(1), 9–16.

Whitaker, R. (1978c) Birth record of the Andaman pit viper (*Trimeresurus purpureomaculatus*). *The Journal of the Bombay Natural History Society* 75(1), 233, 1 pl.

Whitaker, R. (1982) Reptiles and amphibians seen in Kodaikanal (Palni Hills, Tamil Nadu) 16 to 22 September 81. *Hamadryad*, Madras 7(1), 7.

Whitaker, R. (1999) *Snakes of Kodagu.* Coorg Wildlife Society, Madikeri, 58 pp., 23 col. photos.

Whitaker, R. & Captain, A. (2004) *Snakes of India: the field guide.* Draco Books, Chennai, xiv + 481 pp., 495 col. photos.

Whitaker, R. & Dattatri, S. (1982) A new species of *Oligodon* from the Palni Hills, South India (Serpentes, Colubridae). *The Journal of the Bombay Natural History Society* 79(3), 630–631, 1 pl.

Whitaker, R. & Whitaker, Z. (1977) Collection of a rare snake in the Nilgiris. *The Journal of the Bombay Natural History Society* 74(3), 539.

Whitaker, R., Whitaker, Z. & Mills, D. (1982) *Reptiles of Papua New Guinea.* Wildlife of Papua New Guinea, Konedobu (82/2), 1–53, illust.

Whitaker, Z. & Whitaker, R.. (1986) *The snakes around us.* National Book Trust, New Delhi, 62 pp., 51 illust.

White, J. (1790) Journal of a voyage to new South Wales with sixty-five plates on non descript animals, birds, lizards, serpents, curious cones of trees and other natural productions. J. Debrett, London, 261+ pp. 65 pls. [snake plates include: *Pseudonaja textilis,* pl. 32, fig. 1, pl. 43; *Dendrelaphis punctulata,* pl. 44; *Morelia spilota,* pls. 45, 46, fig. 1; *Vermicella annulata,* pl. 46, fig. 2 *fide* Bauer (1999: 16); reprinted in 1962 by Angus & Robertson, and 1971 by Arno Press]

White, J. (1981) Ophidian envenomation, a South Australian perspective. *Records of the Adelaide Children's Hospital,* Adelaide 2(3), 311–421, 55 figs.

White, J.E., Jr. & White, A.W. (2002) *Amphibians and reptiles of Delmarva.* Tidewater Publishing, Centreville, xvi + 248 pp., 96 col. pls.

White, L.R., Parmerlee, J.S., Jr. & Powell, R. (1992) *Typhlops syntherus* Thomas. *Catalogue of American amphibians and reptiles*, pp. 551.1–551.2, 1 fig.

Whiteman, H.H., Mills, T.M., Scott, D.E. & Gibbons, J.W. (1995) Confirmation of a range extension for the pine woods snake (*Rhadinaea flavilata*). *Herpetological Review* 26(3), 158.

Whitten, A.J. & McCarthy, C.J. (1993) List of amphibians and reptiles of Jawa and Bali. *Tropical Biodiversity* 1, 169–177.

Wickramaratna, J.C., Fry, B.G. & Hodgson, W.C. (2003a) Species-dependent variations in the in vitro myotoxicity of death adder (*Acanthophis*) venoms. *Toxicological Science* 74, 352–360, 5 figs.

Wickramaratna, J.C., Fry, B.G., Aguilar, M.-I., Manjunatha, R. & Hodgson, W.C. (2003b) Isolation and pharmacological characterization of a phospholipase A myotoxin from the venom of the Irian Jayan death adder (*Acanthophis rugosus*). *British Journal of Pharmacology* 138, 333–342, 7 figs.

Wickramasinghe, L.J.M., Vidanapathirana, D.R., Wickramasinghe, N. & Ranwella, P.N. (2009) A new species of *Rhinophis* Hemprich, 1820 (Reptilia: Serpentes: Uropeltidae) from Rakwana Massif, Sri Lanka. *Zootaxa* (2044), 1–22, 14 figs.

Wied-Neuwied, A.P.M. zu. (1820a) *Reise nach Brasilien in den Jahren 1815 bis 1817.* Erster Band. Heinrich Ludwig Brönner, Frankfurt-am-Main, xxxiv + 380 pp., 11 pls.

Wied-Neuwied, A.P.M. zu. (1820b) Ueber die Cobra Coral oder Cobra Coraes der brasilianer. *Nova Acta Physico-Medica Academiae Caesareae Leopoldino-Carolinae Naturae Curiosorum*, Halle 10(1), 105–110, 1 pl.

Wied-Neuwied, A.P.M. zu. (1821) *Reise nach Brasilien in den Jahren 1815 bis 1817.* Zweyter Band. Heinrich Ludwig Brönner, Frankfurt-am-Main, xviii + 345 pp., 8 pls.

Wied-Neuwied, A.P.M. zu. (1822–1831) *Abbildungen zur Naturgeschichte Brasilien,* Weimar, 90 col. pls. [issued in 15 parts with 6 pls. each: lief. 1 in 1822, lief. 2–4 in 1823; lief. 5–8 in 1824; lief. 9 in 1825; life. 10–11 in 1827, lief. 12 in 1828/29, 13 in 1829, lief. 14 in 1829/30, and lief. 15 in 1831]

Wied-Neuwied, A.P.M. zu. (1824) Verzeichniss der Amphibien, welche im zwerten Bande der Naturgeschichte Brasiliens vom Prinz Mac von Neuwied werden beschrieben werden. (Nach Merrem's Versuch eines Systems der Amphibien.) *Isis von Oken*, Jena 14(6), 661–674.

Wied-Neuwied, A.P.M. zu. (1825a) Über *Coluber lichtensteinii* eine neue brasilianische Natter-Art. *Nova Acta Physico-Medica Academiae Caesareae Leopoldino-Carolinae Naturae Curiosorum*, Halle 12(2), 491–502.

Wied-Neuwied, A.P.M. zu. (1825b) Beiträge zur Nat. Gesch. p. Brasilien. *Isis von Oken*, Jena 17(11), 1335–1340.

Wied-Neuwied, A.P.M. zu. (1825c) *Beiträge zur Naturgeschichte von Brasilien. I. Verzeichniss der auf einer Reise zwischen dem 13ten und 23sten Grade südlicher Briete im östlichen Brasilien beobachteten Amphibien, Säugthiere und Vögel Gr. H.S. priv.* Laudes-Industrie-Comptoirs, Weimar, xxii + 614 pp., 3 pls.

Wied-Neuwied, A.P.M. zu. (1828) Abbildungen zur Naturgeschichte Brasilien von Maximilian, Prinzen von Wied. *Isis von Oken*, Leipzig 21(1), 86.

Wied-Neuwied, A.P.M. zu. (1838–1841) *Reise in das innere Nord-America in den Jahren 1832 bis 1834. Erster & zweiter Band. Atlas.* J. Hölscher, Coblenz. xiii + 653 p., xviii + 687 pp., 84 col. pls. [issued in 20 parts: vol. 1(1-6): 1–392 in 1838, vol. 1(7–14): 393–653 in 1839, vol. 2(15–17): 169–504 in 1840, and vol. 2(18–20): 505–687 in 1841]

Wied-Neuwied, A.P.M. zu. (1865) *Verzeichniss der Reptilien, welche auf einer Reise im nördlichen America beobachtet wurden.* E. Blochmann & Sohn, Dresden, 141 pp., 7 pls.

Wiedl, H.-J. (2003) *Snakes of Cyprus.* United Nations Development Programme, 40 pp., col. illust.

Wiegmann, A.F.A. (1834) Amphibien. *In:* Meyen, F.J.F. (Ed.), Beiträge zur Zoologie, gesammelt auf einer Reise um die Erde. *Nova Acta Academiae Caesareae Leopoldino-Carolinae Germinicae Naturae Curiosorum*, Halle 17(1), 183–268, 268a–d, pls. 19–20.

Wiest, J.A., Jr. (1978) *Revision of the Neotropical snake genus Chironius Fitzinger (Serpentes, Colubridae).* Doctoral Dissertation, Texas A & M University, College Station, 370 pp., 64 figs.

Wilcox, T.P., Zwickl, D.J., Heath, T.A. & Hillis, D.M. (2002) Phylogenetic relationships of the dwarf boas and a comparison of Bayesian and bootstrap measures of phylogenetic support. *Molecular Phylogenetics and Evolution* 25(3), 361–371, 6 figs.

Wilkey, R. (2002a) A checklist of African snakes. Vol. I. The family Viperidae (cobras, mambas and allied species). Rupert Wilkey, London, 77 pp.

Wilkey, R. (2002b) A checklist of African snakes. Vol. I. The family Elapidae (adders and vipers). Rupert Wilkey, London, 72 pp.

Willard, B. (1972) *An analysis of variation within some Mexican taxa of the colubrid snake genus* Leptodeira. Master's Thesis, University of Kansas, Lawrence, 128 pp., 18 figs.

Willard, D.E. (1977) Constricting methods of snakes. *Copeia* 1977(2), 379–382.

Willey, A. (1903a) Contribution to the fauna of Ceylon. *Spolia Zeylanica*, Colombo 1(1), 1–13, illust.

Willey, A. (1903b) Some rare snakes of Ceylon. *Spolia Zeylanica*, Colombo 1(3), 81–89, 14 figs.

Willey, A. (1905) Guide to the antiquities, minerals, and natural history collections in the Colombo Museum. *Spolia Zeylanica*, Colombo 3(9), 1–49, illust.

Willey, A. (1906) Terrestrial Colubridae of Ceylon. *Spolia Zeylanica*, Colombo 3(12), 227–234, 2 figs.

Williams, D.J., O'Shea, M., Daguerre, R.L., Pook, C.E., Wüster, W., Hayden, C.J., Mcvay, J.D., Paiva, O., Matainaho, T., Winkel, K.D. & Austin, C.C. (2008) Origin of the eastern brownsnake, *Pseudonaja textiles* (Duméril, Bibron and Duméril) (Serpentes, Elapidae, Hydrophiinae) in New Guinea: evidence of multiple dispersals from Australia, and comments on the status of *Pseudonaja textiles pughi* Hoser, 2003. *Zootaxa* (1703), 47–61, 3 figs.

Williams, E.E. & Parker, F. (1964) The snake genus *Parapistocalamus* on Bougainville, Solomon Islands. *Senckenbergiana Biologica*, Frankfurt-am-Main 45(3–5), 543–552, 5 figs.

Williams, J.D. (1983) Los corales. *In: Fauna Argentina.* Centro Editor de America Latina S.A., Buenos Aires, 56, 1–32, col. illust.

Williams, J.D. & Couturier, G.A. (1984) Primera cita del genero *Hydrops* Wagler, 1830 para la republica Argentina (Serpentes, Colubridae). *Historia Natural*, Corrientes 4(7), 61–66, 3 figs.

Williams, J.D. & Francini, F. (1991) A checklist of the Argentine snakes. *Bollettino del Museo Regionale di Scienze Naturali*, Torino 9(1), 55–90.

Williams, J.D. & Scrocchi, G. (1994) Ofidios de agua dulce de la Republic Argentina. *In:* Castellanos, Z.A. de (Ed.), *Fauna de agua dulce de la República Argentina* 42, pp 4–55.

Williams, K.L. (1985) *Cemophora. C. coccinea. Catalogue of American amphibians and reptiles*, pp. 374.1–374.4.

Williams, K.L. (1987) Comments on the snake generic names *Antaresia* and *Australiasis. Bulletin of the Maryland Herpetological Society* 23(2), 69–71.

Williams, K.L. (1988) Systematics and natural history of the American milk snake, *Lampropeltis triangulum.* Second edition. *Milwaukee Public Museum,* Milwaukee, 176 pp., 66 figs., 25 col. pls. [first edition, 1978, Milwaukee Pub. Mus. Publ. Biol. Geol. (2), 1–258, 70 figs., 8 col. pls.]

Williams, K.L. (1994) *Lampropeltis triangulum* (Lacepède). Milk snake. *Catalogue of American amphibians and reptiles*, pp. 594.1–594.10, 2 col. figs.

Williams, K.L. & Fugler, C.M. (1968) An additional record of *Pseudoeryx plicatilis* from Ecuador. *Journal of Herpetology* 1(1–4), 104.

Williams, K.L. & Smith, H.M. (1966) A record of the snake *Leptophis modestus* from Mexico. *Journal of the Ohio Herpetological Society* 5(3), 103–104.

Williams, K.L. & Wallach, V. (1989) *Snakes of the world. Volume I. Synopsis of snake generic names.* Krieger Publishing Company, Malabar, 234 pp.

Williams, K.L. & Wilson, L.D. (1967) A review of the colubrid snake genus *Cemophora* Cope. *Tulane Studies in Zoology* 13(4), 103–124, 6 figs.

Williams, K.L., Chrapliwy, P.S. & Smith, H.M. (1961) Snakes from northern Mexico. *Natural History Miscellanea*, Chicago (177), 1–8, 1 fig.

Williams, K.L., Brown, B.C. & Wilson, L.D. (1966) A new subspecies of the colubrid snake *Cemophora coccinea* (Blumenbach) from southern Texas. *Texas Journal of Science* 18, 85–88.

Williamson, G.K. & Moulis, R.A. (1979) Distribution of Georgia amphibians and reptiles in the Savannah Science Museum Collection. *Savannah Science Museum, Special Publication* (1), 154 + 7 pp.

Williamson, G.K. & Moulis, R.A. (1994) Distribution of amphibians and reptiles in Georgia. Vol. 1. Maps. Vol. 2. Locality data. *Savannah Science Museum, Special Publication* (3), 712 pp.

Williamson, M.A., Hyder, P.W. & Applegarth, J.S. (1994) *Snakes, lizards, turtles, frogs, toads & salamanders of New Mexico. A field guide.* Sunstone Press, Santa Fe, iv + 176 pp., col. illust.

Williston, S.W. (1914) *Water reptiles of the past and present.* University of Chicago Press, Chicago, vii + 300 pp., 131 figs.

Williston, S.W. & Gregory, W.K. (1925) *The osteology of the reptiles.* Harvard University Press, Cambridge, xiii + 300 pp., 191 figs.

Wilms, T. & Hulbert, F. (2000) On the herpetofauna of the Sultanate of Oman, with comments on the relationship between the Afrotropiocal and Saharo-Sindian faunas. *Bonner Zoologische Monographien,* Bonn (46), 367–380, 4 figs.

Wilson, J.A., Mohabey, D.M., Peters, S.E. & Head, J.J. (2010) Predation upon hatchling dinosaurs by a new snake from the late Cretaceous of India. *PLoS Biology* 8(3), 1–10, 6 figs.

Wilson, L.D. (1967) Generic reallocation and review of *Coluber fasciolatus* Shaw (Serpentes, Colubridae). *Herpetologica* 23(4), 260–275, 8 figs.

Wilson, L.D. (1970a) The racer *Coluber constrictor* (Serpentes, Colubridae) in Louisiana and eastern Texas. *Texas Journal of Science* 22(1), 67–85, 3 figs.

Wilson, L.D. (1970b) The coachwhip snake, *Masticophis flagellum* (Shaw): taxonomy and distribution. *Tulane Studies in Zoology and Botany* 16(2), 31–99, 31 figs.

Wilson, L.D. (1970c) A review of the *chloroticus* group of the colubrid snake genus *Drymobius*, with notes on a twin-striped form of *D. chloroticus* (Cope) from southern México. *Journal of Herpetology* 4(3–4), 155–163, 4 figs.

Wilson, L.D. (1973a) *Masticophis. Catalogue of American amphibians and reptiles*, pp. 144.1– 144.2.

Wilson, L.D. (1973b) *Masticophis flagellum. Catalogue of American amphibians and reptiles*, pp. 145.1–145.4.

Wilson, L.D. (1974) *Drymobius margaritiferus. Catalogue of American amphibians and reptiles*, pp. 172.1–172.2.

Wilson, L.D. (1975a) *Drymobius. Catalogue of American amphibians and reptiles*, pp. 170.1–170.2.

Wilson, L.D. (1975b) *Drymobius chloroticus. Catalogue of American amphibians and reptiles*, p. 171.1.

Wilson, L.D. (1975c) *Drymobius melanotropis. Catalogue of American amphibians and reptiles*, p. 173.1.

Wilson, L.D. (1976) Variation in the South American colubrid snake *Tantilla semicincta* (Duméril, Bibron, and Duméril) with comments on pattern dimorphism. *Bulletin of the Southern California Academy of Sciences* 75(1), 42–48, 4 figs.

Wilson, L.D. (1978) *Coluber constrictor. Catalogue of American amphibians and reptiles*, pp. 218.1–218.4.

Wilson, L.D. (1979) A new snake of the genus *Tantilla* from Ecuador. *Herpetologica* 35(3), 274–276, 2 figs.

Wilson, L.D. (1982a) A review of the colubrid snakes of the genus *Tantilla* of Central America. *Contributions in Biology and Geology, Milwaukee Public Museum* (52), 1–77, 23 figs.

Wilson, L.D. (1982b) *Tantilla. Catalogue of American amphibians and reptiles*, pp. 307.1– 307.4.

Wilson, L.D. (1983a) A new species of *Tantilla* of the *taeniata* group from Chiapas, Mexico. *Journal of Herpetology* 17(1), 54–59, 2 figs.

Wilson, L.D. (1983b) *Tantilla taeniata. Catalogue of American amphibians and reptiles*, pp. 344.1–344.2.

Wilson, L.D. (1984) The status of *Micrurus ruatanus* (Günther), a coral snake endemic to the Bay Islands of Honduras. *Herpetological Review* 15(3), 67.

Wilson, L.D. (1985a) Rediscovery of *Tantilla bairdi* Stuart and a definite Guatemalan locality for *Tantilla taeniata* (Bocourt). *Herpetological Review* 16(2), 105.

Wilson, L.D. (1985b) *Tantilla briggsi. Catalogue of American amphibians and reptiles*, p. 365.1.

Wilson, L.D. (1985c) *Tantilla cuesta. Catalogue of American amphibians and reptiles*, p. 366.1.

Wilson, L.D. (1985d) *Tantilla cuniculator. Catalogue of American amphibians and reptiles*, p. 367.1.

Wilson, L.D. (1985e) *Tantilla flavilineata. Catalogue of American amphibians and reptiles*, p. 368.1.

Wilson, L.D. (1985f) *Tantilla jani. Catalogue of American amphibians and reptiles*, p. 369.1.

Wilson, L.D. (1985g) *Tantilla reticulata. Catalogue of American amphibians and reptiles*, p. 370.1.

Wilson, L.D. (1985h) *Tantilla albiceps. Catalogue of American amphibians and reptiles*, p. 377.1.

Wilson, L.D. (1985i) *Tantilla andinista. Catalogue of American amphibians and reptiles*, p. 378.1.

Wilson, L.D. (1985j) *Tantilla annulata. Catalogue of American amphibians and reptiles*, p. 379.1.

Wilson, L.D. (1985k) *Tantilla bairdi. Catalogue of American amphibians and reptiles*, p. 380.1.

Wilson, L.D. (1986a) The status of the name *Homalocranium breve* Günther, 1895 (Colubridae). *Southwestern Naturalist* 31(2), 243–244.

Wilson, L.D. (1986b) *Tantilla alticola. Catalogue of American amphibians and reptiles*, p. 400.1.

Wilson, L.D. (1987a) *Tantilla schistosa. Catalogue of American amphibians and reptiles*, pp. 409.1–409.2.

Wilson, L.D. (1987b) *Tantilla vermiformis. Catalogue of American amphibians and reptiles*, p. 410.1.

Wilson, L.D. (1987c) A résumé of the colubrid snakes of the genus *Tantilla* of South America. *Contributions in Biology and Geology, Milwaukee Public Museum* (68), 1–35, 12 figs.

Wilson, L.D. (1987d) *Geagras, G. redimitus. Catalogue of American amphibians and reptiles*, pp. 430.1–430.2.

Wilson, L.D. (1988a) *Tantilla brevicauda. Catalogue of American amphibians and reptiles*, p. 432.1.

Wilson, L.D. (1988b) *Tantilla calamarina. Catalogue of American amphibians and reptiles*, pp. 433.1–433.2.

Wilson, L.D. (1988c) *Tantilla canula. Catalogue of American amphibians and reptiles*, p. 434.1.

Wilson, L.D. (1988d) *Amastridium, A. veliferum. Catalogue of American amphibians and reptiles*, pp. 449.1–449.3.

Wilson, L.D. (1988e) *Tantilla cascadae. Catalogue of American amphibians and reptiles*, p. 451.1.

Wilson, L.D. (1988f) *Tantilla deppei. Catalogue of American amphibians and reptiles*, p. 452.1.

Wilson, L.D. (1988g) *Tantilla equatoriana. Catalogue of American amphibians and reptiles*, p. 453.1.

Wilson, L.D. (1988h) *Tantilla moesta. Catalogue of American amphibians and reptiles*, p. 454.1.

Wilson, L.D. (1988i) *Tantillita, T. brevissima, T. lintoni. Catalogue of American amphibians and reptiles*, pp. 455.1–455.2.

Wilson, L.D. (1988j) The status of *Tantilla excubitor* Wilson. *Journal of Herpetology* 22(4), 469–470.

Wilson, L.D. (1990a) *Tantilla capistrata. Catalogue of American amphibians and reptiles*, p. 475.1, 1 fig.

Wilson, L.D. (1990b) *Tantilla lempira. Catalogue of American amphibians and reptiles*, p. 476.1, 1 fig.

Wilson, L.D. (1990c) *Tantilla miyatai. Catalogue of American amphibians and reptiles*, p. 477.1, 1 fig.

Wilson, L.D. (1990d) *Tantilla semicincta. Catalogue of American amphibians and reptiles*, p. 478.1, 1 fig.

Wilson, L.D. (1990e) *Tantilla tayrae. Catalogue of American amphibians and reptiles*, p. 479.1, 1 fig.

Wilson, L.D. (1990f) *Tantilla coronadoi* Hartweg. *Catalogue of American amphibians and reptiles*, p. 501.1, 1 fig.

Wilson, L.D. (1990g) *Tantilla insulamontana* Wilson and Mena. *Catalogue of American amphibians and reptiles*, p. 502.1, 1 fig.

Wilson, L.D. (1990h) *Tantilla oaxacae* Wilson and Meyer. *Catalogue of American amphibians and reptiles*, p. 503.1, 2 figs.

Wilson, L.D. (1990i) *Tantilla striata* Dunn. *Catalogue of American amphibians and reptiles*, p. 504.1, 1 fig.
Wilson, L.D. (1991a) *Tantilla petersi*. *Catalogue of American amphibians and reptiles*, p. 527.1, 2 figs.
Wilson, L.D. (1991b) *Tantilla shawi*. *Catalogue of American amphibians and reptiles*, p. 528.1.
Wilson, L.D. (1992a) *Tantilla melanocephala* (Linnaeus). *Catalogue of American amphibians and reptiles*, pp. 547.1–547.3, 1 fig.
Wilson, L.D. (1992b) *Tantilla nigra* (Boulenger). *Catalogue of American amphibians and reptiles*, p. 548.1, 1 fig.
Wilson, L.D. (1999) Checklist and key to the species of the genus *Tantilla* (Serpentes, Colubridae), with some commentary on distribution. *Smithsonian Herpetological Information Service* (122), 1–34.
Wilson, L.D. & Campbell, J.A. (2000) A new species of the *calamarina* group of the colubrid snake genus *Tantilla* (Reptilia, Squamata) from Guerrero, Mexico, with a review of and key to members of the group. *Proceedings of the Biological Society of Washington* 113(3), 820–827, 1 fig.
Wilson, L.D. & Cruz-Diaz, G.A. (1993) The herpetofauna of the Cayos Cochinos, Honduras. *Herpetological Natural History* 1(1), 13–24.
Wilson, L.D. & Hahn, D.E. (1973) The herpetofauna of the Islas de la Bahía, Honduras. *Bulletin of the Florida State Museum, Biological Sciences* 17(2), 93–150, 3 figs.
Wilson, L.D. & Johnson, J.D. (2010) Herpetofaunal diversity and endemicity. *In:* Wilson, L.D., Townsend, J.H. & Johnson, J.D. (Eds.), *Conservation of Mesoamerican amphibians and reptiles*. Eagle Mountain Publishing, Eagle Mountain, pp. 30–236, 9 figs.
Wilson, L.D. & McCranie, J.R. (1984) *Bothrops nasuta*. *Catalogue of American amphibians and reptiles*, pp. 349.1–349.2.
Wilson, L.D. & McCranie, J.R. (1992) *Bothriechis marchi* (Barbour and Loveridge). March's Palm-pitviper. *Catalogue of American amphibians and reptiles*, pp. 544.1–544.2, 3 figs.
Wilson, L.D. & McCranie, J.R. (1994) Second update on the list of amphibians and reptiles known from Honduras. *Herpetological Review* 25(4), 146–150.
Wilson, L.D. & McCranie, J.R. (1999) The systematic status of Honduran populations of the *Tantilla taeniata* group (Serpentes, Colubridae), with notes on other populations. *Amphibia-Reptilia,* Leiden 20(3), 326–329.
Wilson, L.D. & McCranie, J.R. (2004a) The conservation status of the herpetofauna of Honduras. *Amphibian and Reptilian Conservation* 3(1), 6–33, 50 col. pls.
Wilson, L.D. & McCranie, J.R. (2004b) The herpetofauna of the cloud forest of Honduras. *Amphibian and Reptilian Conservation* 3(1), 34–48, 3 figs.
Wilson, L.D. & Mena, C.E. (1980) Systematics of the *melanocephala* group of the colubrid snake genus *Tantilla*. *Memoirs of the San Diego Society of Natural History* (11), 5–58, 15 figs.
Wilson, L.D. & Meyer, J.R. (1969) A review of the colubrid snake genus *Amastridium*. *Bulletin of the Southern California Academy of Sciences* 68(3), 145–159, 3 figs.
Wilson, L.D. & Meyer, J.R. (1971) A revision of the *taeniatus* group of the colubrid snake genus *Tantilla*. *Herpetologica* 27(1), 11–40, 22 figs.
Wilson, L.D. & Meyer, J.R. (1981) Systematics of the *calamarina* group of the colubrid snake genus *Tantilla*. *Contributions in Biology and Geology, Milwaukee Public Museum* (42), 1–25, 9 figs.
Wilson, L.D. & Meyer, J.R. (1985) *Snakes of Honduras*. Milwaukee Public Museum, x + 150 pp., 120 figs. [first edition, 1982, 159 pp., 96 figs.]
Wilson, L.D. & Porras, L. (1983) The ecological impact of man on the south Florida herpetofauna. *University of Kansas Museum of Natural History Special Publication* 9, vi + 89 pp., 35 figs.
Wilson, L.D. & Townsend, J.H. (2007) A checklist and key to the snakes of the genus *Geophis* (Squamata, Colubridae, Dipsadinae), with commentary on distribution and conservation. *Zootaxa* (1395), 1–31.
Wilson, L.D. & Townsend, J.H. (2010) The herpetofauna of Mesoamerica: biodiversity significance, conservation status, and future challenges. *In:* Wilson, L.D., Townsend, J.H. & Johnson, J.A. (Eds.), *Conservation of Mesoamerican amphibians and reptiles*. Eagle Mountain Publishing, Eagle Mountain, pp. 760–812.
Wilson, L.D. & Williams, K.L. (2002) *Scolecophis*. *Catalogue of American amphibians and reptiles*, pp. 758.1–758.3, 2 figs.
Wilson, L.D., McCranie, J.R. & Porras, L. (1978) Two snakes, *Leptophis modestus* and *Pelamis platurus*, new to the herpetofauna of Honduras. *Herpetological Review* 9(2), 63–64.
Wilson, L.D., McCranie, J.R. & Porras, L. (1979) New departmental records for reptiles and amphibians from Honduras. *Herpetological Review* 10(1), 25.
Wilson, L.D., Porras, L. & McCranie, J.R. (1986) Distribution and taxonomic comments on some members of the Honduran herpetofauna. *Contributions in Biology and Geology, Milwaukee Public Museum* (66), 1–18, illust.
Wilson, L.D., McCranie, J.R. & Slowinski, J.B. (1992) *Micrurus ruatanus* (Günther). Babaspul (local dialect), Coral (Spanish), Roatan Coral Snake (English). *Catalogue of American amphibians and reptiles*, pp. 545.1–545.2, 1 fig.
Wilson, L.D., McCranie, J.R. & Espinal, M.R. (1996) Coral snake mimics of the genus *Pliocercus* (family Colubridae) in Honduras and their mimetic relationships with *Micrurus* (family *Elapidae*). *Herpetological Natural History* 4(1), 57–63, 1 fig.
Wilson, L.D., McCranie, J.R. & Williams, K.L. (1998) A new species of *Geophis* of the *sieboldi* group (Reptilia, Squamata, Colubridae) from northern Honduras. *Proceedings of the Biological Society of Washington* 111(2), 410–417, 2 figs.
Wilson, L.D., Vaughan, R.K. & Dixon, J.R. (1999) Another new species of *Tantilla* of the *taeniata* group from Chiapas, Mexico. *Journal of Herpetology* 33(1), 1–5, 2 figs.
Wilson, L.D., McCranie, J.R. & Espinal, M.R. (2001) The ecogeography of the Honduran herpetofauna and the design of biotic reserves. *In:* Johnson, J.D., Webb, R.G. & Flores-Villela, O.A. (Eds.), *Mesoamerican herpetology: systematics, zoogeography, and conservation*. University of Texas at El Paso, El Paso, pp. 109–158, 3 figs.
Wilson, L.D., McCranie, J.R., Gotte, S. & Townsend, J.H. (2003) Distributional comments on some members of the herpetofauna of the Mosquitia, Honduras. *Herpetological Bulletin* 84, 15–19.
Wilson, L.D., Townsend, J.H. & Johnson, J.D. (2010) *Conservation of Mesoamerican amphibians and reptiles*. Eagle Mountain Publishing, Eagle Mountain, xvii + 812 pp., illust.
Wilson, R.L. (1968) Systematics and faunal analysis of a lower Pliocene vertebrate assemblage from Trego County, Kansas. *Contributions from the Museum of Paleontology, University of Michigan* 22(7), 75–126, 17 figs.
Wilson, S. (2005) *A field guide to reptiles of Queensland*. Reed New Holland, Sydney, 256 pp., col. illust.
Wilson, S. & Knowles, D.G. (1988) *Australian reptiles. A photographic reference to the terrestrial reptiles of Australia*. William Collins Pty. Ltd., Sydney, 447 pp., 846 photos.

Wilson, S. & Swan, G. (2003) *Reptiles of Australia*. Princeton University, Princeton, 480 pp., col. illust.

Wilson, V.J. (1965) The snakes of the eastern Province of Zambia. *Puku, Occasional Papers of the Department of Game and Fisheries*, Zambia (3), 149–170.

Wingerde, J. van. (1986) The distribution of *Vipera xanthina* (Gray, 1849) on the east Aegean islands and in Thrace. *Litteratura Serpentium* (English edition) 6(4), 131–139, illust.

Winokur, R.M. (1977) The integumentary tentacles of the snake *Erpeton tentaculatum*: structure, function, evolution. *Herpetologica* 33(2), 247–253, 6 figs.

Winters, R. (2011) *A treasury of endemic fauna of Mauritius and Rodrigues*. Mauritania, Grand Baie, 95 pp., col. illust.

Wise, C.J., Biegel, P.A. & Linnell, D. (2008) Natural history notes. *Thamnophis elegans*. *Herpetological Review* 39(3), 360.

Witte, G.-F. de. (1922) Description d'un ophidien nouveau récolté au Congo par le Dr. Schouteden. *Revue Zoologique Africaine*, Tervuren 10, 318–319.

Witte, G.-F. de. (1930a) Un Serpent nouveau du Congo Belge (*Rhinocalamus rodhaini* sp. n.). *Revue de Zoologie et Botanique Africaine*, Tervuren 19(1), 1–3, 5 figs.

Witte, G.-F. de. (1930b) Description d'un vipéride nouveau du Kivu (*Atractaspis Schoutendeni* sp. n.). *Revue de Zoologie et Botanique Africaine*, Tervuren 19, 224–225, figs.

Witte, G.-F. de. (1930c) Mission saharienne Augiéras-Draper, 1927. Reptiles et batraciens. *Bulletin du Muséum National d'Histoire Naturelle*, Paris (Série 2) 2, 614–618.

Witte, G.-F. de. (1933a) Reptiles recoltes au Congo Belge par le Dr. H. Schouteden et par M.G.–F. de Witte. *Annales du Musée Royal du Congo Belge* (Série in Octavao, *Science Zoologique*), Tervuren (Series 1) 3(2), 57–99, 4 pls.

Witte, G.-F. de. (1933b) Descriptions de reptiles nouveaux provenant du Katanja (1930–31). *Revue de Zoologie et Botanique Africaine*, Tervuren 23(2), 185–191.

Witte, G.-F. de. (1933c) Descriptions d'un *Typhlops* nouveau provenant du Ruanda. *Revue de Zoologie et Botanique Africaine*, Tervuren 23(2), 206–207, 3 figs.

Witte, G.-F. de. (1933d) Descriptions d'un *Typhlops* nouveau du Congo Belge. *Revue de Zoologie et Botanique Africaine*, Tervuren 24(1), 104–105, 3 figs.

Witte, G.-F. de. (1941) Exploration du Parc National Albert. Mission G.F. de Witte (1933–1935). Batracians et reptiles. *Institut des Parcs Nationaux du Congo Belge*, Bruxelles (33), xvii + 261 pp., 51 figs., 76 pls.

Witte, G.-F. de. (1948) *Faune de Belgique: amphibiens et reptiles*. Second edition. Musée Royal d'Histoire Naturelle du Belgique, Bruxelles, 321 pp., 207 figs. [first edition, 1942, 123 pp., 138 figs.]

Witte, G.-F. de. (1951) Description d'un serpent nouveau du Katanga *Calamelaps ventrimaculatus katangensis* subsp. nov. *Revue de Zoologie et Botanique Africaine*, Tervuren 45(1–2), 17–18.

Witte, G.-F. de. (1952) Amphibiens et reptiles. *In: Exploration hydrobiologique du Lac Tanganika (1946–1947). Resultats scientifiques*, 3(3), 3–22, 1 pl.

Witte, G.-F. de. (1953) Reptiles. Exploration des Parc National de l'Upemba. Mission de G.F. de Witte. *Institut des Parcs Nationaux du Congo Belge*, Bruxelles (6), 1–322, 41 pls.

Witte, G.-F. de. (1955) La population de reptiles de la région de Mutsora (Kivu Nord). *Bulletin des Séances Académie Royale des Science Coloniales,* Bruxelles (Série 2) 1(2), 203–225.

Witte, G.-F. de. (1959) Contribution à la faune herpétologique du Congo Belge: description de trois serpents nouveaux. *Revue de Zoologie et Botanique Africaine*, Tervuren 60(3–4), 348–351.

Witte, G.-F. de. (1961) Contribution à la fauna herpétologique du Congo. Description d'une sous-espèce nouvelle du genre *Typhlops*. *Revue de Zoologie et Botanique Africaine*, Tervuren 64, 183–184.

Witte, G.-F. de. (1962) Genera des serpents du Congo et du Ruanda-Urundi. *Annales du Musée Royal de l'Afrique Centrale* (*Science Zoologique*), Tervuren (Série in Octavo) (104), 1–203, 94 figs., 15 pls.

Witte, G.-F. de. (1963) The colubrid snake genera *Chamaelycus* Boulenger and *Oophilositum* Parker. *Copeia* 1963(4), 634–636.

Witte, G.-F. de. (1966) Exploration du Parc National de la Garamba. Reptiles. *Institut du Parcs Nationaux du Congo*, Bruxelles (48), 108 pp., 5 pls.

Witte, G.-F. de. (1975) Serpents, Boidae, Colubridae, Elapidae et Viperidae. *In:* Saeger, H. de, Roux-Estève, R. & Witte, G.-F. de (Eds.), Exploration du Parc National des Virunga. *Fondacion pour Favoriser les Recherches Scientifiques en Afrique*, Bruxelles (Série 2) (24), 56–61.

Witte, G.-F. de & Laurent, R.F. (1942) Contribution a la faune herpetologique du Congo Belge. *Revue de Zoologie et Botanique Africaine*, Tervuren 36(2), 101–115.

Witte, G.-F. de & Laurent, R.F. (1943) Contribution à la systématique des Boiginae du Congo belge (Rept.). *Revue de Zoologie et Botanique Africaine*, Tervuren 37(1–2), 157–189, 319 (errata slip).

Witte, G.-F. de & Laurent, R.F. (1947) Revision d'un groupe de Colubridae africains, genres *Calamelaps*, *Miodon*, *Aparallactus* et formes affines. *Mémoires du Musée Royal d'Histoire Naturelle de Belgique*, Bruxelles (Série 2) 29, 1–134, 130 figs.

Wittmann, B. (1954) *Europas giftschlangen*. Hippolyt-Verlag, Wien, 185 pp., 89 figs., 16 col. pls.

Woerkom, A.B. van. (1982) De slangen van het genus *Malpolon*. *Litteratura Serpentium* 2(4), 162–179, 4 figs., 2 photos.

Wogan, G.O.U., Vindum, J.V., Wilkinson, J.A., Koo, M.S., Slowinski, J.B., Win, H., Thin, W.K., San, L.O., Kyi, S.L. & Awan, K.S. (2008) New country records and range extensions for Myanmar amphibians and reptiles. *Hamadryad* 33, 83–96.

Wöllner, K. (2011) *Der Netzpython,* Python reticulatus, Broghammerus reticulatus. Natur und Tier-Verlag, Münster, 62 pp., col. illust.

Wong, H. (1997) Geographical distribution, *Chilomeniscus cinctus*. *Herpetological Review* 28(1), 51.

Wood, D.A., Fisher, R.N. & Reeder, T.W. (2008a) Novel patterns of historical isolation, dispersal, and secondary contact across Baja California in the rosy boa (*Lichanura trivirgata*). *Molecular Phylogenetics and Evolution* 46(2), 484–502, 5 figs.

Wood, D.A., Meik, J.M., Holycross, A.T., Fisher, R.N. & Vandergast, A.G. (2008b) Molecular and phenotypic diversity in *Chionactis occipitalis* (western shovel-nosed snake), with emphasis on the status of *C. o. klauberi* (Tucson shovel-nosed snake). *Conservation Genetics* 9, 1489–1507, 5 figs.

Wood, D.A., Vandergast, A.G., Lemos Espinal, J.A., Fisher, R.N. & A.T. Holycross (2011) Refugial isolation and divergence in the narrowheaded gartersnake species complex (*Thamnophis rufipunctatus*) as revealed by multilocus DNA sequence data. *Molecular Ecology* 20(17), 3856–3878, 4 figs.

Wood, J.T. (1949) Observations on *Natrix septemvittata* (Say) in southwestern Ohio. *American Midland Naturalist* 42, 744–750.

Wood, K.R. & Law, W. (2010) Geographic distribution: Serpentes: *Ramphotyphlops braminus*. *Herpetological Review* 41(4), 518.

Wood, P.L. (2007) Geographic distribution: Serpentes: *Dendrelaphis caudolineatus*. *Herpetological Review* 38(1), 103.

Wood, P.L., Elias, N. & Morgan, D. (2003a) Geographic distribution: Serpentes: *Dendrelaphis striatus*. *Herpetological Review* 34(3), 264.

Wood, P.L., Youmans, T.M. & Szutz, T. (2003b) Geographic distribution: Serpentes: *Elaphe flavolineata*. *Herpetological Review* 34(4), 388.

Wood, P.L., Grismer, L.L., Youmans, T.M., Nasir, N., Ahmad, N. & Senawi, J. (2008) Additions to the herpetofauna of Endau-Rompin, Johor, West Malaysia. *Herpetological Review* 39(1), 113–121, 16 figs.

Woodburne, M.O. (1987) (Editor), *Cenozoic mammals of North America: geochronology and biostratigraphy.* University of California Press, Berkeley, 336 pp.

Woodbury, A.M. (1929) A new rattlesnake from Utah. *Bulletin of the University of Utah* 20(6), 3 pp., 2 figs.

Woodbury, A.M. (1931) *A descriptive catalog of the reptiles of Utah.* University of Utah, Salt Lake City, x + 1–129, 58 figs.

Woodbury, A.M. & Woodbury, D.M. (1942) Studies of the rat snake, *Elaphe laeta*, with description of a new subspecies. *Proceedings of the Biological Society of Washington* 55, 133–142, 2 figs.

Woodhouse, S.W. (1853) Report on the natural history of the country passed over by the exploring expedition under the command of Brevet Captain L. Sitgreaves, U.S. Topographical Engineers, during the year 1851. *In:* Sitgreaves, L. (Ed.), *Report of an expedition down the Zuni and Colorado Rivers.* Robert Armstrong, Washington, pp. 33–40.

Woodward, A.S. (1901) On some extinct reptiles from Patagonia, of the genera *Miolania*, *Dinilysia*, and *Genyodectes*. *Proceedings of the Zoological Society of London* 70(2), 169–184, pls. 15–20.

Woolrich-Piña, G.A., López, L.O. & Lemos-Espinal, J.A. (2005) *Anfibios y reptiles del Valle de Zapotitlán, Salinas, Puebla.* Universidad Nacional Autónoma de México, Mexico City, 54 pp., 35 figs., 42 col. photos.

Worrell, E. (1945) The orange-naped whipsnake. *Proceedings of the Royal Zoological Society of New South Wales*, Mosman (1944–1945), 32–33.

Worrell, E. (1955) A new elapine snake from Queensland. *Proceedings of the Royal Zoological Society of New South Wales*, Mosman (1953–54), 41–43, 3 figs., 2 pls.

Worrell, E. (1956) A new snake from Queensland. *Australian Zoologist*, Mosman 12(3), 202–205, 3 figs.

Worrell, E. (1958) A new Papuan python. *Proceedings of the Royal Zoological Society of New South Wales*, Sidney (1956–1957), 26–27, 4 figs.

Worrell, E. (1960) A new elapine snake from Western Australia. *Western Australian Naturalist*, Perth 7(5), 132–134, 2 figs.

Worrell, E. (1961a) A new generic name for a nominal species of *Denisonia*. *Proceedings of the Royal Zoological Society of New South Wales*, Sydney (1958–1959), 54–55, 2 figs.

Worrell, E. (1961b) A new insular brown snake. *Proceedings of the Royal Zoological Society of New South Wales*, Sidney (1958–1959), 56–58.

Worrell, E. (1961c) Herpetological name changes. *Western Australian Naturalist*, Perth 8(11), 18–27, 4 figs.

Worrell, E. (1963a) A new elapine generic name (with skull diagrams of type species recently separated from the Australian genus *Denisonia*). *Australian Reptile Park Records*, Gosford (1), 1–7, 27 figs.

Worrell, E. (1963b) Two new subspecies of the elapine genus *Notechis* from Bass Strait. *Australian Reptile Park Records*, Gosford (2), 2–11, 6 figs.

Worrell, E. (1967) *Australian snakes, crocodiles, tortoises, lizards.* Angus and Robertson, Sydney, 64 pp., illust.

Worrell, E. (1969) *Dangerous snakes of Australia and New Guinea.* Sixth edition. Angus and Robertson, Sydney, 65 pp., 3 figs., 34 photos. [first edition, 1952, 64 pp., 3 figs., 32 photos; second edition, 1953, 64 pp., 3 figs., 32 photos; third edition, 1957, 68 pp., 3 figs.; fourth edition, 1961, 68 pp., 3 figs., 36 photos; fifth edition, 1963, 66 pp.]

Worrell, E. (1970) *Reptiles of Australia: crocodiles – turtles – tortoises – lizards – snakes. Describing all Australian species, their appearance, their haunts, their habits, with over 330 illustrations, many in full colour.* Angus and Robertson, Sydney, xv + 169 pp., 70 pls. [first edition, 1963, xv + 207 pp., 70 pls.]

Worthington, R.D. (1980) *Elaphe subocularis*. *Catalogue of American amphibians and reptiles*, pp. 268.1–268.2.

Worthington, R.D. & Arvizo, R.B. (1974) Western records of the Davis Mountains kingsnake, *Lampropeltis mexicana alterna*, in Texas. *Southwestern Naturalist* 19(3), 330–331.

Wright, A.H. & Bishop, S.C. (1916) A biological reconnaissance of the Okefinokee Swamp in Georgia. II. Snakes. *Proceedings of the Academy of Natural Sciences of Philadelphia* 67, 139–192.

Wright, A.H. & Wright, A.A. (1952) List of the snakes of the United States and Canada by states and provinces. *American Midland Naturalist* 48(3), 574–603.

Wright, A.H. & Wright, A.A. (1957) *Handbook of snakes of the United States and Canada.* Comstock Publishing Associates, Ithaca, 2 volumes, xviii + 1105 pp., 304 figs.

Wright, B.A. (1941) Habit and habitat studies of the massasauga rattlesnake (*Sistrurus catenatus catenatus* Raf.) in northeastern Illinois. *American Midland Naturalist* 25, 659–672.

Wu, L., Li, D.-J., Xu, R.-H., Dong, Q. & Liu, J.-S. (1979) New species and new records of reptiles from Guizhou Province, China. *Journal of the Zunyi Medical College* (1978) 1979(2), 104–110, 3 figs. (in Chinese, with English summary)

Wu, L., Li, D.-J. & Liu, J.-C. (1985) *Reptiles of Guizhou.* Guizhou People's Publishing House, Guiyang City, iv + 436 pp., 111 figs., 149 pls., 16 col. pls. (in Chinese)

Wu, Y.-F., Wu, M.-L. & Cao, Y.-P. (2009) *The fauna of Hebei, China: Amphibia, Reptilia and Mammalia.* Hebei Science and Technology Publishing House, Shijiazhuang, 409 pp., 12 col. pls. (in Chinese)

Wucherer, O.E.H. (1861a) On the ophidians of the Province of Bahia, Brazil (Part I). *Proceedings of the Zoological Society of London* 29(1), 113–116.

Wucherer, O.E.H. (1861b) On the ophidians of the Province of Bahia, Brazil (Part II). *Proceedings of the Zoological Society of London* 29(1), 322–325, pl. 19.

Wucherer, O.E.H. (1861c) Description of a new species of *Elapomorphus* from Brazil. *Proceedings of the Zoological Society of London* 29(1), 325–326.

Wucherer, O.E.H. (1862) Description of a new species of *Elapomorphus* from Brazil. *Annals and Magazine of Natural History* (Series 3) 9(52), 318–319. [reprint of Wucherer, 1861c]

Wüster, W. (1988) (Abstract) Population systematics of cobras in southeast Asia. *1988 SSAR-ASIH meetings*, Ann Arbor, 193.

Wüster, W. (1996) Taxonomic changes and toxinology: systematic revisions of the Asiatic cobras (*Naja naja* species complex). *Toxicon* 34(4), 399–406, 1 fig.

Wüster, W. (1998a) The cobras of the genus *Naja* in India. *Hamadryad* 23(1), 15–32, 1 fig.

Wüster, W. (1998b) The genus *Daboia* (Serpentes, Viperidae), Russell's viper. *Hamadryad* 23(1), 33–40, 1 fig.

Wüster, W. & Broadley, D.G. (2003) A new species of spitting cobra (*Naja*) from north-eastern Africa (Serpentes, Elapidae). *Journal of Zoology*, London 259(4), 345–359, 8 figs.

Wüster, W. & Broadley, D.G. (2007) Get an eyeful of this: a new species of giant spitting cobra from eastern and north-eastern Africa (Squamata, Serpentes, Elapidae, *Naja*). *Zootaxa* (1532), 51–68, 7 figs.

Wüster, W. & Thorpe, R.S. (1987) Geographic variation in the cobras of the genus *Naja* in south-east Asia: a multivariate analysis. *Proceedings of the 4th Ordinary General Meeting of the Societas Europaea Herpetologica*, Nijmegen, 449–452, 2 figs.

Wüster, W. & Thorpe, R.S. (1989) Population affinities of the Asiatic cobra (*Naja naja*) species complex in south-east Asia: reliability and random resampling. *Biological Journal of the Linnean Society*, London 36(4), 391–409, 6 figs.

Wüster, W. & Thorpe, R.S. (1990) Systematics and biogeography of the Asiatic cobra (*Naja naja*) species complex in the Philippine islands. *In:* Peters, G. & Hutterer, R. (Eds.), *Vertebrates in the tropics*. Museum Alexander Koenig, Bonn, pp. 333–344, 3 figs.

Wüster, W. & Thorpe, R.S. (1991) Asiatic cobras: systematics and snakebite. *Experientia*, Basel 47, 205–209, 3 figs.

Wüster, W. & Thorpe, R.S. (1992a) Asiatic cobras: population systematics of the *Naja naja* species complex (Serpentes, Elapidae) in India and central Asia. *Herpetologica* 48(1), 69–85, 7 figs.

Wüster, W. & Thorpe, R.S. (1992b) Dentitional phenomena in cobras revisited: spitting and fang structure in the Asiatic species of *Naja* (Serpentes, Elapidae). *Herpetologica* 48(4), 424–434, 5 figs.

Wüster, W. & Thorpe, R.S. (1994) *Naja siamensis*, a cryptic species of venomous snake revealed by mtDNA sequencing. *Experientia*, Basel 50, 75–79, 2 figs.

Wüster, W., Otsuka, S., Malhotra, A. & Thorpe, R.S. (1992a) Population systematics of Russell's viper: a multivariate study. *Biological Journal of the Linnean Society*, London 47(1), 97–113, 7 figs.

Wüster, W., Otsuka, S., Thorpe, R.S. & Malhotra, A. (1992b) Morphological variation in Russell's viper in Burma and Thailand. *Herpetological Journal* 2, 99–101, 1 fig.

Wüster, W., Thorpe, R.S., Cox, M.J., Jintakune, P. & Nabhitabhata, J. (1995) Population systematics of the snake genus *Naja* (Reptilia, Serpentes, Elapidae) in Indonesia: multivariate morphometrics and comparative mitochondrial DNA sequencing (cytochrome oxidase I). *Journal of Evolutionary Biology*, Basel 8, 493–510, 3 figs.

Wüster, W., Thorpe, R.S., Puorto, G. & BBBSP. (1996) Systematics of the *Bothrops atrox* complex (Reptilia, Serpentes, Viperidae) in Brazil: a multivariate analysis. *Herpetologica* 52(2), 263–271, 2 figs.

Wüster, W., Graça-Salomão, M. de, Thorpe, R.S., Puorto, G., Furtado, M.F.D., Hoge, S.A., Theakston, R.D.G. & Warrell, D.A. (1997a) Systematics of the *Bothrops atrox* complex: new insights from multivariate analysis and mitochondrial DNA sequence information. *In:* Thorpe, R.S., Wüster, W. & Malhotra, A. (Eds.), *Venomous Snakes. Ecology, evolution and snakebite*. Zoological Society of London, Clarendon Press, Oxford, pp. 99–113, 5 figs.

Wüster, W., Warrell, D.A., Cox, M.J., Jintakune, P. & Narhitabhata, J. (1997b) Redescription of *Naja siamensis* (Serpentes, Elapidae), a widely overlooked spitting cobra from S. E. Asia: geographic variation, medical importance and designation of a neotype. *Journal of Zoology*, London 243(4), 771–788, 2 figs., 4 pls.

Wüster, W., Golay, P. & Warrell, D.A. (1997c) Synopsis of recent developments in venomous snake systematics. *Toxicon* 35(3), 319–340.

Wüster, W., Golay, P. & Warrell, D.A. (1998) Synopsis of recent developments in venomous snake systematics, no. 2. *Toxicon* 36(2), 299–307.

Wüster, W., Graça-Salomão, M. da, Duckett, G.J., Thorpe, R.S. & BBBSP. (1999) Mitochondrial DNA phylogeny of the *Bothrops atrox* species complex (Squamata, Serpentes, Viperidae). *In:* Joger, U. (Ed.), Phylogeny and systematics of the Viperidae. *Kaupia, Darmstädter Beiträge zur Naturgeschichte* 8, 135–144, 7 figs.

Wüster, W., Yrausquin, J.L. & Mijares-Urrutia, A. (2001a) A new species of indigo snake from north-western Venezuela (Serpentes, Colubridae, *Drymarchon*). *Herpetological Journal* 11, 157–165, 4 figs.

Wüster, W., Bush, B., Keogh, J.S., O'Shea, M. & Shine, R. (2001b) Taxonomic contributions in the "amateur" literature: comments on recent descriptions of new genera and species by Raymond Hoser. *Litturatura Serpentium* 21(3), 67–91, 3 figs.

Wüster, W., Thorpe, R.S., Graça-Salomäo, M. da, Thomas, L., Puorto, G., Theakston, R.D.G. & Warrell, D.A. (2002) Origin and phylogenetic position of the Lesser Antillean species of *Bothrops* (Serpentes, Viperidae): biogeographical and medical implications. *Bulletin of the Natural History Museum*, London (*Zoology*) 68(2), 101–106, 1 fig., 1 pl.

Wüster, W., Duarte, M.R. & Graça-Salomäo, M. da (2005a) Morphological correlates of incipient arboreality and ornithophagy in island pitvipers, and the phylogenetic position of *Bothrops insularis*. *Journal of Zoology*, London 266(1), 1–10, 5 figs.

Wüster, W., Dumbrell, A.J., Hay, C., Pook, C.E., Williams, D.J. & Fry, B.G. (2005b) Snakes across the Strait: Trans-Torresian phylogeographic relationships in three genera of Australasian snakes (Serpentes: Elapidae: *Acanthophis*, *Oxyuranus* and *Pseudechis*). *Molecular Phylogenetics and Evolution* 34(1), 1–14.

Wüster, W., Ferguson, J.E., Quijada-Marcareñas, J.A., Pook, C.E., da Graça-Salomão, M. & Thorpe, R.S. (2005c) Tracing an invasion: landbridges, refugia, and the phylogeography of the Neotropical rattlesnake (Serpentes, Viperidae, *Crotalus durissus*). *Molecular Ecology* 14(4), 1095–1108, 4 figs.

Wüster, W., Crookes, S., Ineich, I., Mané, Y., Pook, C.E., Trape, J.-F. & Broadley, D.G. (2007) The phylogeny of cobras from mitochondrial DNA sequences, evolution of venom spitting and the phylogeography of the African spitting cobras (Serpentes, Elapidae, *Naja nigricollis* complex). *Molecular Phylogenetics and Evolution* 45(2), 437–453, 4 figs.

Wüster, W., Peppin, L., Pook, C.E. & Walker, D.E. (2008) A nesting of vipers: phylogeny and historical biogeography of the Viperidae (Squamata: Serpentes). *Molecular Phylogenetics and Evolution* 49(2), 445–459, 4 figs.

Wütschert, R. (1984) Neues über die Reptilienfauna der Insel Korfu. *Salamandra*, Frankfurt-am-Main 20(4), 221–228, 5 figs.

Wycherley, J. & Anstis, R. (2001) *Amphibians and reptiles of Surrey*. Surrey Wildlife Trust, Woking, 112 pp., figs., 32 col. pls.

Wyder, J.F. (1817) Bemerkungen über die schweizer Schlangen. *Isis von Oken*, Jena 1(132), 1050–1054.

Wynn, A.H. & Leviton, A.E. (1993) Two new species of blind snake, genus *Typhlops* (Reptilia, Squamata, Typhlopidae), from the Philippine Archipelago. *Proceedings of the Biological Society of Washington* 106(1), 34–45, 3 figs.

Wynn, D.E. & Moody, S.M. (2006) Ohio turtle, lizard, and snake atlas. *Ohio Biological Survey, Miscellaneous Contribution* (10), iv + 81 pp., 4 figs.

Xiang, G.-S. (2007) *Natural portraits of herpetofauna of Shei-Pa National Park*. Miaoli Xian, Minguo, 192 pp., illust. (in Chinese)

Xiang, G.-S. & Li, P.-X. (2009) *Colored illustrations of amphibians and reptiles of Taiwan*. Owl Press, Taipei, 336 pp., col. illust. (in Chinese)

Xu, R.-G. (2001) *Fauna Inner Mongolia. Volume 2. Amphibia and Reptilia*. Inner Mongolia University Press, Hohhot, 247 pp., 47 figs., 21 col. photos. (in Chinese)

Xie, R.-G., Liu, Z.-H., Qiu, S.-M., Zheng, W.R., Nie, J.-X. & Ma, W.H. (1998) Surveys of snakes in Deqing County, Guangdong Province, China. *Russian Journal of Herpetology* 5(1), 82.

Xu, Y., Liu, Q., Myers, E.A., Wang, L., Huang, S., He, Y., Peng, P.-H. & Guo, P. (2012) Molecular phylogeny of the genus *Gloydius* (Serpentes: Crotalinae). *Asian Herpetological Research* (Series 2) 3(2), 127–132, 2 figs.

Yakovleva, I.D. (1961) *Identification guide for reptiles of Kirgizia*. Institute of Zoology and Parisitology, Kazakh Academy of Sciences, Frunze, 112 pp., 21 figs. (in Russian)

Yakovleva, I.D. (1964) *Reptiles of Kirghizia*. Institute of Zoology and Parisitology, Kazakh Academy of Sciences, Frunze 272 pp., 81+ figs. (in Russian)

Yamamoto, E. (1981) On *Achalinus* (*Achalinus*) *spinalis* at Oda-cho, Ehime Pref. *The Snake* 13, 137–141, 6 figs. (in Japanese)

Yamamoto, E. (1984) *Achalinus spinalis* Peters at Oda-cho, Ehime Pref., Japan. II. *The Snake* 16(1), 49–55, 1 fig. (in Japanese)

Yang, D.-T. (2008) (Ed.). *Amphibia and Reptilia of Yunnan*. Yunnan Science and Technology Press, Kunming, 411 pp., 245 col. pls. (in Chinese)

Yang, D.-T. & Inger, R.F. (1986) (translators). Key to snakes and lizards of China. *Smithsonian Herpetological Information Service* (71), 1–21.

Yang D.-T & Su, C.-G. (1984) A preliminary study on the subspecies differentiation of *Elaphe carinata* (Serpentiformes, Colubridae). *Zoological Research* 5(2), 159–163. (in Chinese)

Yang, D.-T., Su, C.-G. & Li, S.-M. (1983) A study on amphibians and reptiles from Hengduanshan Mountains of Yunnan. *Acta Herpetologica Sinica*, Chengdu (Series 2) 2(3), 37–49, 1 fig.

Yang, J.-H., Wang, Y.-Y., Zhang, B., Lau, M.W.-N. & Chou, W.-H. (2011a) Revisions of the diagnostic characters of *Opisthotropis maculosa* Stuart and Chuayukern, 2007 with notes on its distribution and variation, and a key to the genus *Opisthotropis* (Squamata: Natricidae). *Zootaxa* (2785), 61–68, 2 figs.

Yang, J.-H., Orlov, N.L. & Wang, Y.-Y. (2011b) A new species of pitviper of the genus *Protobothrops* from China (Squamata: Viperidae). *Zootaxa* (2936), 59–68, 3 figs.

Yang, Y.J., Huang, M.-H. & Zhong, X.-X. (1986) A new record of the snakes of Zhejiang Province—*Elaphe schrenckii* (Strauch). *Acta Herpetologica Sinica*, Chengdu (Series 2) 5(4), 303.

Yanosky, A.A., Dixon, J.R. & Mercolli, C. (1992) First specific record for *Liophis dilepis* (Cope, 1862) in Argentina. *Revista Española Herpetología* 7, 33–35.

Yanosky, A.A., Dixon, J.R. & Mercolli, C. (1993) The herpetofauna of El Bagual Ecological Reserve (Formosa, Argentina) with comments on its herpetological collection. *Bulletin of the Maryland Herpetological Society* 29(4), 160–171.

Yanosky, A.A., Dixon, J.R. & Mercolli, C. (1997) Ecology of the snake community at El Bagual Ecological Reserve, northeastern Argentina. *Herpetological Natural History* 4(2), 97–110, 5 figs.

Yao, C.-Y. (2012) *Amphibians and reptiles of Gansu*. Gansu Science and Technology Press, Lanzhou, iv + 164 pp., 15 figs., 125 col. photos. (in Chinese)

Yao, C.-Y. & Cheng, S.-Z. (1981) Three new records of reptiles from Gansu. *Acta Herpetologica Sinica*, Chengdu 5(4), 65–66. (in Chinese)

Yarlequé-Chocas, A. (2000) *Las serpientes Peruanas y sus venenos*. Universidad Nacional Mayor de San Marcos, Fordo Ed., Lima, 78 pp., 16 col. pls.

Yaron, Z. (1985) Reptilian placentation and gestation: structure, function, and endocrine control. *In:* Gans, C. & Billett, F. (Eds.), *Biology of the Reptilia. Volume 15 (Development B)*. John Wiley & Sons, New York, pp. 527–605, 32 figs.

Yarrow, H.C. (1875) Report upon the collections of batrachians and reptiles made in portions of Nevada, Utah, California, Colorado, New Mexico, and Arizona, during the years 1871, 1872, 1873, and 1874. *In:* Wheeler, G.M. (Ed.), *Report upon the geographical and geological explorations and surveys west of the one hundredth meridian, Vol. V. Zoology*. Government Printing Office, Washington, D.C., 5, 509–584, pls. 16–25.

Yarrow, H.C. (1882) Descriptions of new species of reptiles and amphibians in the United States National Museum. *Proceedings of the United States National Museum* (1882–1883) 5(299), 438–443.

Yarrow, H.C. (1883a) Description of new species of reptiles in the United States National Museum. *Proceedings of the United States National Museum* (1883) 6(10), 152–154.

Yarrow, H.C. (1883b) Check list of North American Reptilia and Batrachia, with catalogue of specimens in U.S. National Museum. *Bulletin of the United States National Museum* (1882) (24), 1–249.

Yildiz, M.Z., Göçmen, B. & Akman, B. (2009) Taxonomical comments on Sudan beaked worm snake, *Leptotyphlops macrorhynchus* (Jan & Sordelli, 1860) (Serpentes: Leptotyphlopidae) from Anatolia, Turkey. *Biharean Biologist* 3(2), 151–156, 3 figs.

Yingling, R.P. (1972) *A review of the colubrid snake* Imantodes gemmistratus. M.S. Thesis, San Diego State College, 87 pp.

Yingling, R.P. (1982) *Lichanura. L. trivirgata. Catalogue of American amphibians and reptiles*, pp. 294.1–294.2.

Yorks, D.T., Henderson, R.W. & Powell, R. (2003) *Typhlops tasymicris. Catalogue of American amphibians and reptiles*, pp. 780.1–780.2, 1 fig.

Youmans, T.M., Escobar, R., III & Grismer, J.L. (2002) Geographic distribution: Serpentes: *Eridiphas sleveni. Herpetological Review* 33(4), 324.

Youmans, T.M., Wood, P.L. & Szutz, T.R. (2003) Geographic distribution: Serpentes: *Pareas vertebralis. Herpetological Review* 34(4), 389.

Young, B.A. & Kardong, K.V. (1996) Dentitional surface features in snakes (Reptilia, Serpentes). *Amphibia-Reptilia,* Leiden 17(3), 261–276, 4 figs.

Young, B.A. & Wallach, V. (1998) Description of a papillate tactile organ in the Typhlopidae. *South African Journal of Zoology* 33(4), 249–253.

Young, K.R. & León, B. (1999) Peru's humid eastern montane forests: an over view of their physical settings, biological diversity, human use and settlement, and conservation needs. *DIVA Technical Report no. 5. Center for Research on the Cultural and Biological Diversity of Andean Rainforests*, Ronde, Denmark, 97 pp., 73 figs.

Young, M.T. (2011) *The guide to Colorado reptiles and amphibians*. Fulcrum Publishing, Golden, vi + 170 pp., col. photos.

Yu, M.-J. (1996) Checklist of vertebrates of Taiwan. *Biological Bulletin*, Tunghai University, Taichung, 211 pp.

Yuan, H. (1983a) A new record of snake of Shaanxi Province—*Azemiops feae* Boulenger. *Acta Herpetologica Sinica*, Chengdu (Series 2) 1(2), 77–78. (in Chinese)

Yuan, H. (1983b) A new species of the genus *Oligodon* from Shaanxi, China. *Acta Herpetologica Sinica*, Chengdu (Series 2) 2(2), 65–67, 1 fig. (in Chinese)

Yuan, H. (1983c) A herpetological survey in the Huoditang Forest of the Qinling Mountains. *Acta Herpetologica Sinica*, Chengdu (Series 2) 2(3), 70–71.

Yuan, H. (1985) Reptilian fauna of Shaanxi Province. *Acta Herpetologica Sinica*, Chengdu (Series 2) 4(2), 133–139, 1 fig.

Yuki, R.N. (1997a) Geographical distribution. *Hydrops triangularis*. *Herpetological Review* 28(1), 52.

Yuki, R.N. (1997b) Geographical distribution. *Bothrops neuwiedi*. *Herpetological Review* 28(3), 158.

Yuki, R.N. (1999) Geographical distribution. *Taeniophallus brevirostris*. *Herpetological Review* 30(1), 55.

Yuki, R.N. & Lema, T. de. (2006) Análise comparativa entre as cobras d'água meridionalis *(Helicops carinicaudus* (Wied, 1825) e *Helicops infrataeniatus* Jan, 1865), com a descrição do crânio e hemipênis (Serpentes: Colubridae: Xenodontinae). *Comunicações do Museu de Ciências e Tecnologia da PUCRS (Zoologia),* Porte Alegre (2005) 18(2), 85–128, 13 figs.

Yuki, R.N. & de Santos, R.M. (1996) Snakes from Marajó and Mexiana islands, Pará State, Brazil. *Boletim do Museu Paraense Emílio Goeldi* (*Zoologia*), Belém 12(1), 41–53.

Zaher, H. (1994) Les Tropidopheoidea (Serpentes; Alethinophidia) sont-ils réellement monophylétiques? Arguments en faveur de leur polyphylétisme. *Comptes Rendus de l'Académie des Sciences*, Paris 317, 471–478, 3 figs.

Zaher, H. (1996) A new genus and species of Pseudoboine snake, with a revision of the genus *Clelia* (Serpentes, Xenodontinae). *Bollettino del Museo Regionale di Scienze Naturali*, Torino 14(2), 289–337, 9 figs., 8 pls.

Zaher, H. (1998) The phylogenetic position of *Pachyrhachis* within snakes (Squamata, Lepidosauria). *Journal of Vertebrate Paleontology* 18(1), 1–3.

Zaher, H. (1999) Hemipenial morphology of the South American Xenodontinae snakes, with a proposal for a monophyletic Xenodontinae and a reapaisal of colubrid hemipenis. *Bulletin of the American Museum of Natural History* (240), 1–168, figs.

Zaher, H. & Caramaschi, U. (1993) Sur le statut taxonomique d'*Oxyrhopus trigeminus* et *O. guibei* (Serpentes, Xenodontinae). *Bulletin du Muséum National d'Histoire Naturelle*, Paris (4) 14A(3–4), 805–827, 8 figs.

Zaher, H. & Prudente, A.L.C. (1999) Intraspecific variation of the hemipenis in *Siphlophis* and *Tripanurgos*. *Journal of Herpetology* 33(4), 698–702, 2 figs.

Zaher, H., & Rieppel, O. (1999) The phylogenetic relationships of *Pachyrhachis problematicus*, and the evolution of limblessness in snakes (Lepidosauria, Squamata). *Comptes Rendus des Séances de l'Académie des Sciences*, Paris 329, 831–837.

Zaher, H. & Scanferla, C.A. (2012) The skull of the Upper Cretaceous snake *Dinilysia patagonica* Smith-Woodward, 1901, and its phylogenetic position revisited. *Zoological Journal of the Linnean Society,* London 164(1), 194–238.

Zaher, H., Souza, I., Gower, D.J., Hinget-Zaher, E. & Silva, N.J. da, Jr. (2005) Redescription of *Atractus albuquerquei* (Serpentes, Colubridae, Dipsadinae), with comments on geographical distribution and infraspecific variation. *Papeis Avulsos Zoologia, Universidade de São Paulo* 45(2), 19–32, 6 figs.

Zaher, H., Oliveira, N.E. & Franco, F.L. (2008a) A new brightly colored species of *Pseudoboa* Schneider, 1801 from the Amazon Basin (Serpentes, Xenodontinae). *Zootaxa* (1674), 27–37, 6 figs.

Zaher, H., Scrocchi, G. & Masiero, R. (2008b) Rediscovery and redescription of the type of *Philodryas laticeps* Werner, 1900 and the taxonomic status of *P. oligolepis* Gomes, 1921 (Serpentes, Colubridae). *Zootaxa* (1940), 25–40, 7 figs.

Zaher, H., Grazziotin, F.G., Cadle, J.E., Murphy, R.W., Moura-Leite, J.C. de & Bonatto, S.L. (2009), Molecular phylogeny of advanced snakes (Serpentes, Caenophidia) with an emphasis on South American xenodontines: a revised classification and descriptions of new taxa. *Papeis Avulsos Zoologia, Universidade de São Paulo* 49(11), 115–153, 1 fig.

Zaher, H., Grazziotin, F.G., Graboski, R., Fuentes, R.G., Sánchez-Martinez, P., Montingelli, G.G., Zhang, Y.-P. & Murphy, R.W. (2012) Phylogenetic relationships of the genus *Sibynophis* (Serpentes: Colubroidea). *Papeis Avulsos Zoologia, Universidade de São Paulo* 52(12), 141–149, 4 figs.

Zaldivar-Riverón, A. & Pérez-Ramos, E. (2001) Geographic distribution: Serpentes: *Rhadinaea taeniata aemula*. *Herpetological Review* 32(3), 196.

Zamprogno, C. (1997) Geographic distribution: Serpentes: *Uromacerina ricardinii*. *Herpetological Review* 28(4), 211.

Zamudio, K.R. & Greene, H.W. (1997) Phylogeography of the bushmaster (*Lachesis muta*, Viperidae): implications for neotropical biogeography, systematics, and conservation. *Biological Journal of the Linnean Society,* London 62(3), 421–442, 5 figs.

Zanini-Cechin, S.T. (1989) *Tomodon dorsatus* Duméril, Bibron & Duméril, 1854 um sinõnimo senior de *Opisthoplus degener* Peters, 1882 (Serpentes, Colubridae, Tachymeninae). *Comunicações do Museu de Ciências e Tecnologia da PUCRS (Zoologia),* Porte Alegre 2(11), 203–211, 2 figs.

Zappalorti, R.T., Johnson, E.W. & Leszczynski, Z. (1983) The ecology of the northern pine snake, *Pituophis melanoleucus melanoleucus* (Daudin) (Reptilia, Serpentes, Colubridae), in southern New Jersey, with special notes on habitat and nesting behavior. *Bulletin of the Chicago Herpetological Society* 18(3–4), 57–72, 3 figs.

Zavattari, E. (1922) Vertebrati di Cirenaica raccolti dal Gen. Med. Prof. Francesco Testi. *Atti della Societé dei Naturalisti e Matematici di Modena* (Series 5) 6, 13–22.

Zavattari, E. (1929) Anfibi e rettili della Cirenaica raccolti. Not. Econ. Cirenaica. Governo della Cirenaica. *Camera di Commercio* 5–6, 86–88.

Zavattari, E. (1930a) *Gli animali velenosi della Colonia Eritrea*. Premiate Tip. Rilinite Donati, Chiara, 37 pp., 15 figs.

Zavattari, E. (1930b) Erpetologia della Cirenaica. *Archivio Zoologico Italiano*, Napoli 14(2–4), 253–289.

Zavattari, E. (1933) Sui serpenti velenosi della colonia Eritrea. *Istituto di Anatomia I Fisiologia Comparata della Università di Pavia* 1933(1), 9–15.

Zavattari, E. (1934) *Prodromo della fauna libica*. Cooperativa, Pavia, viii + 1234 pp. [not seen]

Zavattari, E. (1936) I Vertebrati della Libia. *In:* Strand, E. (Ed.), *Enthaltend, dem Jubilar gewidmete Arbeiten ausländischer Zoologen und Palaeontologen*. Izdevniceiba "Latvija," Riga, pp. 526–560.

Zeiner, D.C., Laudenslayer, W.F., Jr. & Mayer, K.E. (1988) *California's wildlife. Volume I. Amphibians and reptiles.* State of California, Department of Fish and Game, Sacramento, 272 pp., illust.

Zenneck, J. (1898) Die Zeichung der Boiden. *Zeitschrift für Wissenschaftliche Zoologie*, Leipzig 64(1–2), 1–384, 28 figs., 8 pls.

Zerova, G.A. (1987) Lézards et serpents des gisements du Miocène supérieur au Pléistocène inférieur d'Ukraine. *In:* Skomorovskaia, E.I. (Ed.), *Sur quelques groupes de vertébrés du Cénozoïque supérieur d'Ukraine*. Institut Zoologi Akademii Nauk Ukrainskoy S.S.R., pp. 12–18. (in Russian)

Zerova, G.A. (1989) The first find of a fossil sand boa of the genus *Albaneryx* (Serpentes, Boidae) in the USSR. *Vestnik Zoologii*, Kiev 1989(5), 30–35, 12 figs. (in Russian)

Zerova, G.A. (1992) *Vipera (Daboia) ukrainica*—a new viper (Serpentes; Viperidae) from the middle Sarmatian (upper Miocene) of the Ukraine. *Neues Jahrbuch für Geologie und Palaeontologie, Monatsheft.*, Stuttgart 184(2), 235–249, 10 figs.

Zerova, G.A. (1993) Late Cainozoic localities of snakes and lizards of Ukraine. *Revue de Paléobiologie* 7, 273–280.

Zerova, G.A., Lungu, A.N. & Chkhikvadze, V.M. (1987) Large fossil vipers from northern Black seaside and Transcaucasas. *Proceedings of the Zoological Institute, USSR Academy of Sciences*, Leningrad (1986) 158, 89–99, 3 figs., pl. 6. (in Russian)

Zhang, F.-J. (1993) Division of the genus *Trimeresurus* (*sensu lato*) based on the morphology of their skulls. *Proceedings of the First Asian Herpetological Meetings*, China Forestry Press, Beijing, pp. 48–57, 373, 5 figs. (in Chinese)

Zhang, F.-J. (1998) Description of the distinct pit viper of genus *Ermia* (Serpentes, Viperidae) of China. *Russian Journal of Herpetology* 5(1), 83–84, 1 fig.

Zhang, F.-J. & Zhao, E.-M. (1990) Morphological characteristics of the skulls of six species of genus *Trimeresurus* (Serpentes, Viperidae), with reference to interspecific relationships. *In:* Zhao, E.-M. (Ed.), *From water onto land*. China Publishing House, Beijing, pp. 79–86. (in Chinese)

Zhang, J., Jiang, K., Vogel, G. & Rao, D. (2011) A new species of the genus *Lycodon* (Squamata, Colubridae) from Sichuan Province, China. *Zootaxa* (2982), 59–68, 5 figs.

Zhang, Y.-X. (2009) *Herpetology in Guangxi*. Guangxi Normal University Press, Guilin, 170 pp., 173 figs. (in Chinese)

Zhang, Y.-X. & Tang, Z.-J. (1990) Five new snake records to Guangxi. *In:* Zhao, E.-M. (Ed.), *From water onto land*. China Publishing House, Beijing, pp. 225–226. (in Chinese)

Zhao, E.-M. (1972) Key to Chinese snakes. *Materials for Herpetological Research*, Chengdu 1(1), 1–60, 6 pls. (in Chinese)

Zhao, E.-M. (1978) *Trimeresurus xiangchengensis*—a new species of the genus *Trimeresurus* from China. *Proceedings of Symposium, China Society of Zoology–China Society of Entomology*, Chengdu unpaginated, unpubl. abstract. (in Chinese)

Zhao, E.-M. (1979a) A new snake of the genus *Trimeresurus* from Sichuan, China. *Acta Zootaxonomica Sinica*, Beijing 4(4), 422–424, figs. (in Chinese)

Zhao, E.-M. (1979b) A new *Agkistrodon* from Shedao (Snake Island), Liaoning. *Acta Herpetologica Sinica*, Chengdu 1(1), 4–6, 2 figs. (in Chinese)

Zhao, E.-M. (1980) Taxonomic study of pit viper of Shedao (Snake Island), by morphological and experimental methods and a preliminary discussion on the origin of Snake-Island pit-viper on Shedao. *Acta Herpetologica Sinica*, Chengdu 1(4), 1–16, 4 pls. (in Chinese)

Zhao, E.-M. (1982) (Editor) *Dienagkistrodon acutus* its morphology, ecology, toxicology and applications. *Acta Herpetologica Sinica*, Chengdu 6, i–x + 1–124, 75 figs. (in Chinese)

Zhao, E.-M. (1983) Some comments on Chinese snake taxonomy and nomenclature by Japanese herpetologist. *Acta Herpetologica Sinica*, Chengdu (Series 2) 2(4), 79–80. (in Chinese)

Zhao, E.-M. (1987a) A taxonomic study on Chinese species of the genus *Sibynophis*. *Chinese Herpetological Research*, Chongging 1, 1–6, 3 figs. (in Chinese)

Zhao, E.-M. (1987b) The distribution of *Python molurus bivittatus* Schlegel over Sichuan Province being reconfirmed. *Acta Herpetologica Sinica*, Chengdu (Series 2) 6(3), 78.

Zhao, E.-M. (1990a) The validity of *Elaphe perlacea*, a rare endemic snake from Sichuan Province, China. *Asiatic Herpetological Research*, Berkeley 3, 101–103, 2 figs.

Zhao, E.-M. (1990b) Venomous snakes of China. *In:* Gopalakrishnakone, P. & Chou, L.M. (Eds.), *Snakes of medical importance (Asia-Pacific region)*. National Univsity of Singapore, Singapore, pp. 243–268, 12 figs.

Zhao, E.-M. (1995) Infraspecific classification of some Chinese snakes. *Sichuan Journal of Zoology* 14(3), 107–112.

Zhao, E.-M. (1997a) A new species of *Rhabdophis* (Serpentes, Colubridae) from Hainan Island, China. *Asiatic Herpetological Research* 7, 166–169, 1 fig.

Zhao, E.-M. (1997b) Infraspecific classification of some Chinese snakes. *Asiatic Herpetological Research* 7, 170–172.

Zhao, E.-M. (Ed.). (1998) *Amphibia and Reptilia. China red data book of endangered animals. Volume II.* Science Press, Beijing, xiv + 330, 4 pls.

Zhao, E.-M. (1999) Diagnosis of a new frog and a new snake from China. *Sichuan Journal of Zoology* 18(3), 1–2.

Zhao, E.-M. (2002) *Colored atlas of Sichuan reptiles*. Chengdu Institute of Biology, Chengdu, 292 pp., col. illust. (in Chinese)

Zhao, E.-M. (2004) A re-description of *Opisthotropis cheni* Zhao, 1999 (Serpentes: Colubridae). *Sichuan Journal of Zoology* 24(3), 342–343.

Zhao, E.-M. (2006) *Snakes of China*. Vols. I–II. Anhui Science and Technology Publishing House, Hefei, 390 + 287 pp., 194 col. pls. (in Chinese)

Zhao, E.-M. & Adler, K. (1989) Asian pipe snake—a snake family new to China. *Sichuan Journal of Zoology*, Chengdu 8(2), 26. (in Chinese)

Zhao, E.-M. & Adler, K. (1993) *Herpetology of China*. Society for the Study of Amphibians and Reptiles, Contributions to Herpetology (10), 1–522, 42 figs., 48 col. pls.

Zhao, E.M. & Chen, Y.-H. (1990) Description of a new species of the genus *Trimeresurus*. *Sichuan Journal of Zoology*, Chengdu 9(1), 11–12, photo. (in Chinese)

Zhao, E.-M. & Huang, K.C. (1982) A survey of amphibians and reptiles in Liaoning Province. *Acta Herpetologica Sinica*, Chengdu (Series 2) 1(1), 1–23, 1 pl.

Zhao, E.-M. & Jiang, Y.-M. (1977) A survey of reptiles in Xizang Autonomous Region, with faunal analysis and descriptions of new forms. *Acta Zoologica Sinica*, Chengdu 23(1), 64–71, 2 pls. (in Chinese)

Zhao, E.-M. & Jiang, Y.-M. (1979) Preliminary studies on snakes of northern Xinjiang. *Acta Herpetologica Sinica*, Chengdu 2(1), 1–24, 5 figs.

Zhao, E.-M. & Jiang, Y.-M. (1981) A new species and a new subspecies of snakes from Sichuan. *Acta Herpetologica Sinica*, Chengdu 5(7), 53–58, 1 pl. (in Chinese)

Zhao, E.-M. & Jiang, Y.-M. (1986) Partition of Chinese *Natrix* names and a suggestion of their Chinese names. *Acta Herpetologica Sinica*, Chengdu (Series 2) 5(3), 239–240. (in Chinese)

Zhao, E.-M. & Li, S.-Q. (1985) Herpetological survey of Mount Namjagbarwa, Tibet. *Acta Herpetologica Sinica*, Chengdu (Series 2) 4(2), 103–108, 1 fig.

Zhao, E.-M. & Li, S.-Q. (1987) The herpetological composition and characteristics of Mount Namjagbarwa, Tibet. *Acta Herpetologica Sinica*, Chengdu (Series 2) 6(2), 36–42.

Zhao, E.-M. & Yang, D.-T. (1997) Amphibians and reptiles of the Huengduan Mountains region. *In: The Series of scientific expedition to the Huengduan Mountains Qinhai-Xizang Plateau.* Science Press, Beijing, xiv + 303 pp, 8 pls. (in Chinese)

Zhao, E.-M. & Zhao, G. (1981) Notes on Fea's viper (*Azemiops feae* Boulenger) from China. *Acta Herpetologica Sinica*, Chengdu 5(11), 71–75.

Zhao, E.-M. & Zhao, H. (1994) *Chinese herpetological literature: catalogue and indices.* Herpetologiocal Series (5), Chengdu University of Science and Technology, Chengdu, 398 pp.

Zhao, E.-M., Jiang, Y.-M. & Huang, Q.-Y. (1978) Three new snake species in China. *Materials for Herpetological Research*, Chengdu 4, 21. (in Chinese)

Zhao, E.-M., Jiang, Y.-M. & Huang, Q.-Y. (1979) Two new colubrid snakes from China. *Nature or Ziran Zazhi*, Shanghai 2(5), 313, 2 figs. (in Chinese)

Zhao, E.-M., Liu, X.-B. & Kang, S.-H. (1986a) Five snake species new to Sichuan. *Acta Herpetologica Sinica*, Chengdu (Series 2) 5(2), 157.

Zhao, E.-M., Jiang, Y.-M. & Li, S.-Q. (1986b) Reptilian faunal analysis and zoogeographical division of Xizang Autonomous Region. *Acta Herpetologica Sinica*, Chengdu (Series 2) 5(3), 199–203.

Zhao, E.-M., Huang, M.-H. & Zong, Y. (1998) *Fauna Sinica. Reptilia. Vol. 3. Squamata: Serpentes.* Science Press, Beijing, 522 pp., 100 figs., 8 pls., 30 col. photos. (in Chinese)

Zhao, E.-M., Chang, H.-W. & Zhao, H. (Eds.). (2000) *Taxonomic bibliography of Chinese Amphibia and Reptilia, including karyological literature.* Herpetological Series (11), Kaohsiung Fu-Wen Publishing Co., Taipei, 260 pp.

Zhao, E.-M., Zhao, H., Shi, H.-T. & Wang, L.-J. (2004) Surveys of terrestrial snakes on Hainan. *Sichuan Journal of Zoology*, 24(3), 315–322, pls. 9-10. (in Chinese)

Zhao, W.-G. (2008) *The Amphibia and Reptilia fauna of Heilongjiang.* Science Press, Beijing, 251 pp., 6 col. pls. (in Chinese)

Zhong, C.-F. (1981) Two new records of reptiles from Jiangxi Province. *Acta Herpetologica Sinica*, Chengdu 5(15), 95–98, 3 figs.

Zhong, C.-F. (1983) The venomous snakes of Jiangxi Province. *Acta Herpetologica Sinica*, Chengdu (Series 2) 2(4), 75–76.

Zhong, C.-F. (1984) A new record of the snakes of Jiangxi Province—*Opisthotropis kuatunensis* Pope. *Acta Herpetologica Sinica*, Chengdu (Series 2) 3(2), 20.

Zhong, C.-F. (1985) Preliminary survey of the reptiles of Jingdezhen. *Acta Herpetologica Sinica*, Chengdu (Series 2) 4(2), 144–146.

Zhong, C.-F. (1990) Survey of reptiles in Ruijin County, Jiangxi Province. *In:* Zhao, E.-M. (Ed.), *From water onto land.* China Publishing House, Beijing, pp. 236–238. (in Chinese)

Zhong, C.-F. (1993) First records for *Ophisaurus harti* and *Python molurus bivittatus* from Jiangxi Province, China. *Asiatic Herpetological Research*, Berkeley 5, 103–104, 2 figs.

Zhong, C.-F. & Wu, G.-F. (1981) A preliminary list and its geographical distribution of Reptilia of Jiangxi Province. *Acta Herpetologica Sinica*, Chengdu 5(16), 99–110. (in Chinese)

Zhong, C.-F. & Zhou, W.-X. (1988) *Ophiophagus hannah* (Cantor) found in Ruijin County, Ziangxi. *Acta Herpetologica Sinica*, Chengdu (Series 2) 7, 153, 1 fig.

Zhou, Y.-F. (1985) Survey of the sea snakes of Liaoning Province. *Acta Herpetologica Sinica*, Chengdu (Series 2) 4(2), 158–159.

Ziegler, T. (2002) *Die Amphibien und Reptilien eines Tieflandfeuchtwald-Schutzgebietes in Vietnam.* Natur und Tier-Verlag, Münster, 342 pp., 382 col. photos.

Ziegler, T. & Le, K.Q. (2005) A new species of reed snake, *Calamaria* (Squamata, Colubridae), from the central Truong Son (Annamite mountain range), Vietnam. *Zootaxa* (1042), 27–38.

Ziegler, T. & Le, K.Q. (2006) A new natricine snake of the genus *Amphiesma* (Squamata, Colubridae, Natricinae) from the central Truong Son, Vietnam. *Zootaxa* (1225), 39–56, 10 figs.

Ziegler, T. & Nguyen, T.Q. (2010) New discoveries of amphibians and reptiles from Vietnam. *Bonn Zoological Bulletin* 57(2), 137–147, 6 figs.

Ziegler, T. & Phong, V.B. (2005) Geographic distribution: Serpentes: *Cylindrophis ruffus*. *Herpetological Review* 36(1), 81.

Ziegler, T. & Quang, H.X. (2005) Geographic distribution: Serpentes: *Sinonatrix aequifasciata*. *Herpetological Review* 36(1), 84.

Ziegler, T. & Vogel, G. (1999) On the knowledge and specific status of *Dendrelaphis ngansonensis* (Bourret, 1935) (Reptilia, Serpentes, Colubridae). *Russian Journal of Herpetology* 6(3), 199–208, 8 col. figs.

Ziegler, T., Vences, M., Glaw, F. & Böhme, W. (1996) Remarks on the genital morphology of the Malagasy snake genus *Liophidium* (Reptilia, Serpentes, Colubridae). *Acta Biologica Benrodis* 8, 157–159, 1 fig.

Ziegler, T., Vences, M., Glaw, F. & Böhme, W. (1997) Genital morphology and systematics of *Geodipsas* Boulenger, 1896 (Reptilia, Serpentes, Colubridae), with description of a new genus. *Revue Suisse de Zoologie*, Genève 104(1), 95–114, 5 figs.

Ziegler, T., Herrmann, H.-W., David, P., Orlov, N.L. & Pauwels, O.S.G. (2001) *Triceratolepidophis sieversorum*, a new genus and species of pitviper (Reptilia, Serpentes, Viperidae, Crotalinae) from Vietnam. *Russian Journal of Herpetology* (2000) 7(3), 199–214, 10 figs.

Ziegler, T., Lehr, E. & Unger, J. (2003) Zur Kenntnis von *Psomophis genimaculatus* (Boettger, 1885) in Paraguay unter besonderer Berücksichtigung des Defensiv-Verhaltens. *Herpetofauna* 23, 19–23, 6 figs.

Ziegler, T., Thanh, V.N., Le, K.Q., Truong, N.Q., Hallermann, J., Khoi, L.V. & Hoang, T.M. (2006a) Neue Verbreitungsnachweise einiger wenig bekannter vietnamesischer Amphibien und Reptilien. *Sauria* 28(2), 29–40, 10 figs.

Ziegler, T., Ohler, A., Vu, N.T., Le, K.Q., Nguyen, X.T., Dinh, H.T. & Bui, N.T. (2006b) Review of the amphibian and reptile diversity of Phong Nha—Ke Bang National Park and adjacent areas, central Truong Son, Vietnam. *Proceedings of the 13th Congress of the Societas Europaea Herpetologica*, pp. 247–262, 18 figs.

Ziegler, T., Hendrix, R., Vu, N.T., Vogt, M., Forster, B. & Dang, N.K. (2007) The diversity of a snake community in a karst forest ecosystem in the central Truong Son, Vietnam, with an identification key. *Zootaxa* (1493), 1–40, 20 figs.

Ziegler, T., David, P. & Vu, N.T. (2008a) A new natricine snake of the genus *Opisthotropis* from Tam Dao, Vinh Phuc Province, northern Vietnam (Squamata, Colubridae). *Zoosystematics and Evolution* 84(2), 197–203, 4 figs.

Ziegler, T., Nguyen, V.S. and Nguyen, Q.T. (2008b) A new reed snake of the genus *Calamaria* Boie (Squamata: Colubridae) from Vietnam. *Current Herpetology* 27(2), 71–80, 3 figs.

Ziegler, T., David, P., Miralles, A., Doan, V.K. & Nguyen, Q.T. (2008c) A new species of the snake genus *Fimbrios* from Phong Nha—Ke Bang National Park, Truong Son, central Vietnam (Squamata, Xenodermatidae). *Zootaxa* (1729), 37–48, 9 figs.

Ziegler, T., Orlov, N.L., Giang, T.T., Nguyen, Q.T., Nguyen, T.T., Le, K.Q., Nguyen, V.K. & Thanh, V.N. (2010) New records of cat snakes, *Boiga* Fitzinger, 1826 (Squamata, Serpentes, Colubridae), from Vietnam, inclusive of an extended diagnosis of *Boiga bourreti* Tillack, Le & Ziegler, 2004. *Zoosystematics and Evolution* 86(2), 263–274, 9 figs.

Zigno, A. de. (1882) Nuove aggiunte alla fauna eocena del Veneto. *Memorie dell'Reale Istituto Veneto di Scienze, Lettore ed Arti*, Venezia 21, 775–790.

Zigno, A. de. (1890) Ofidani trovati allo stato fossile e descrizione di due colubri scoperti nei terreni Terkziani del Veneto. *Atti e Memorie della Reale Accademia di Scienze, Lettre ed Arti in Padova* (1889–1890) (2) 6, 109–114, 9 figs.

Zillig, L.D. (1958) The status of *Haldea* Baird and Girard and *Virginia* Baird and Girard. *Copeia* 1958(2), 152.

Zimmerman, B.L. & Rodrigues, M.T. (1990) Frogs, snakes and lizards of the INPA-WWF Reserves near Manaus, Brazil. *In:* Gentry, A.H. (Ed.), *Four neotropical rainforests*. Yale University, New Haven, pp. 426–454, illust.

Zinenko, O., Turcanu, V. & Strugariu, A. (2010) Distributional and morphological variation in *Vipera berus nikolskii* Vedmederja, Grubant et Rudaeva, 1986 in western Ukraine, the Republic of Moldovia and Romania. *Amphibia-Reptilia*, Leiden 31(1), 51–67, 3 figs.

Zinner, H. (1967) Herpetological collection trips to the Lebanon 1965–1966. *Israel Journal of Zoology*, Jerusalem 16, 49–58, 5 figs.

Zinner, H. (1971) On ecology and the significance of semantic coloration in the nocturnal desert-elapid *Walterinnesia aegyptia* Lataste (reptiles Ophidia). *Oecologia*, Berlin 7, 267–275, 4 figs.

Zinner, H. (1972) *Systematics and evolution of the species group* Coluber jugularis *Linnaeus, 1758*–Coluber caspius *Gmelin, 1789 (Reptilia, Serpentes)*. Doctoral Dissertation, Hebrew University, Jerusalem, 41 pp.

Zinner, H. (1977) The status of *Telescopus hoogstraali* Schmidt and Marx 1956 and the *Telescopus fallax* Fleischmann 1831 complex (Reptilia, Serpentes, Colubridae). *Journal of Herpetology* 11(2), 207–212, 9 figs.

Zippel, K.C., Parmerlee, J.S., Jr. & Powell, R. (1994) *Ialtris dorsalis* (Günther). *Catalogue of American amphibians and reptiles*, pp. 592.1–592.3, 3 figs.

Zlotkin, E., Menashé, M., Miranda, F. & Lissitzky, S. (1975) Toxic proteins in cobra venom specifically active on arthropods. *Toxicon* 13(2), 135–136.

Zong, Y. & Ma, J. (1983) A new species of the genus *Achalinopsis* from Jiangxi and the restoration of this genus. *Acta Herpetologica Sinica*, Chengdu (Series 2) 2(2), 61–63, 3 figs. (in Chinese)

Zou, S.-C. & Chen, C.-F. (1998) Studies of snakes of the Jiangsu Province. *Russian Journal of Herpetology* 5(1), 85–88, 1 fig.

Zuffi, M.A.L. (2002) A critique of the systematic position of the asp viper subspecies *Vipera aspis aspis* (Linnaeus, 1758), *Vipera aspis atra* Meisner, 1820, *Vipera aspis francisciredi* Laurenti, 1768, *Vipera aspis hugyi* Schinz, 1833 and *Vipera aspis zinnikeri* Kramer, 1958. *Amphibia-Reptilia*, Leiden 23(2), 191–213, 8 figs.

Zuffi, M.A.L. (2003) On the subspecies of *Vipera aspis*: a reply to Filippi. *Amphibia-Reptilia*, Leiden 24(2), 238–241.

Zuffi, M.A.L. & Bonnet, X. (1999) Italian subspecies of the asp viper, *Vipera aspis*: pattern of variability and distribution. *Italian Journal of Zoology* 66, 87–95, 2 figs.

Zuffi, M.A.L., Gentilli, A., Razzetti, E. & Scali, S. (2000) Transition-hybridization areas in parapatric species of *Vipera aspis* group from northern Italy. *Biota*, 3(1–2), 191–196, 2 figs.

Zug, G.R. (1987) Amphibians and reptiles of the Outamba-Kilimi region, Sierra Leone. *Journal of the Herpetological Association of Africa*, Stellenbosch (33), 1–4, illust.

Zug, G.R. & Ineich, I. (1993) Review of the biology and morphology of the Fijian bola, *Ogmodon vitianus* (Elapidae). *The Snake* 25(1), 9–20, 5 figs.

Zug, G.R. & Mitchell, J. (1995) Amphibians and reptiles of the Royal Chitwan National Park, Nepal. *Asiatic Herpetological Research*, Berkeley 6, 172–180, 3 figs.

Zug, G.R., Win, H., Thin, T., Min, T.Z., Lhon, W.Z. & Kyaw, K. (1998) Herpetofauna of the Chatthin Wildlife Sanctuary, north-central Myanmar with preliminary observations of their natural history. *Hamadryad* 23(2), 111–120, 2 figs.

Zug, G.R., Gotte, S.W. & Jacobs, J.F. (2011) Pythons in Burma: short-tailed python (Reptilia: Squamata). *Proceedings of the Biological Society of Washington* 124(2), 112–136, 3 figs.

Zulueta, A. de. (1908) Nota sobre batracios y reptiles de Mogador, con descripción de la forma joven de *Saurodactylus mauritanicus* (Dum. et Bibr.). *Boletín de la Real Sociedad Española de Historia Natural* 8, 451–457.

Zulueta, A. de. (1909) Nota sobre reptiles de Melilla (Marruecos). *Boletín de la Real Sociedad Española de Historia Natural* 9, 351–354.

Zweifel, R.G. (1952) Pattern variation and evolution of the mountain kingsnake, *Lampropeltis zonata*. *Copeia* 1952, 152–168, 6 figs., 1 pl.

Zweifel, R.G. (1954a) A new species of *Chersodromus* from Mexico. *Herpetologica* 10(1), 17–19, 1 fig.

Zweifel, R.G. (1954b) Notes on the distribution of some reptiles in western Mexico. *Herpetologica* 10(3), 145–149, 1 fig.

Zweifel, R.G. (1958) Results of the Puritan-American Museum of natural history Expedition to western Mexico. 2. Notes on reptiles and amphibians from the Pacific coastal islands of Baja California. *American Museum Novitates* (1895), 1–17, 1 fig.

Zweifel, R.G. (1959a) The provenance of reptiles and amphibians collected in western México by J.J. Major. *American Museum Novitates* (1949), 1–9.

Zweifel, R.G. (1959b) Additions to the herpetofauna of Nayarit, Mexico. *American Museum Novitates* (1953), 1–13.

Zweifel, R.G. (1959c) Snakes of the genus *Imantodes* in western México. *American Museum Novitates* (1961), 1–18, 3 figs.

Zweifel, R.G. (1960) Results of the Puritan-American Museum of Natural History expedition to western México. 9. Herpetology of the Tres Marías Islands. *Bulletin of the American Museum of Natural History* 119(2), 77–128, 5 figs., pls. 41–44.

Zweifel, R.G. (1974) *Lampropeltis zonata*. *Catalogue of American amphibians and reptiles*, pp. 174.1–174.4.

Zweifel, R.G. (1980) Aspects of the biology of a laboratory population of kingsnakes. *Contributions to Herpetology*, Society for the Study of Amphibians and Reptiles 1, 141–152, illust.

Zweifel, R.G. (1981a) Color pattern morphs of the king snake (*Lampropeltis getulus*) in southern California: distribution and evolutionary status. *Bulletin of the Southern California Academy of Sciences* 80(2), 70–81, 2 figs.

Zweifel, R.G. (1981b) Genetics of color pattern polymorphism in the California king snake. *Journal of Heredity* 72, 238–244, 2 figs.

Zweifel, R.G. & Dessauer, H.C. (1983) Multiple insemination demonstated experimentally in the kingsnake (*Lampropeltis getulus*). *Experientia*, Basel 39(3), 317–319, illust.

Zweifel, R.G. & Norris, K.S. (1955) Contributions to the herpetology of Sonora, Mexico: descriptions of new subspecies of snakes (*Micruroides euryxanthus* and *Lampropeltis getulus* and miscellaneous collecting notes). *American Midland Naturalist* 54(1), 230–249, 3 figs., 1 pl.

Index

B

D

E

F

G

H

J

K

M

P

V

W

For Product Safety Concerns and Information,
please contact our EU representative GPSR@taylorandfrancis.com
Taylor & Francis Verlag GmbH, Kaufingerstraße 24,
80331 München, Germany

Printed by Integrated Books International,
United States of America